The Electrical Engineering Handbook Series

Series Editor
Richard C. Dorf
University of California, Davis

Titles Included in the Series

T H E

MEASUREMENT,
INSTRUMENTATION,
AND
SENSORS
H A N D B O O K

Editor-in-Chief

John G. Webster

 CRC PRESS  IEEE PRESS

A CRC Handbook Published in Cooperation with IEEE Press

Library of Congress Cataloging-in-Publication Data

The measurement, instrumentation, and sensors handbook ; John G. Webster, editor-in-chief.
    p.   cm. — (Electrical engineering handbook series)
    Includes bibliographical references and index.
    ISBN 0-8493-8347-1
    1. Physical measurements—Handbooks, manuals, etc. 2. Mensuration—Handbooks, manuals, etc.
  3. Scientific apparatus and instruments—Handbooks, manuals, etc.   I. Webster, John G., 1932- .
  II. Series.
  QC39.M393 1999
  530.8—dc21
                                                                         98-31681
                                                                         CIP

© 1999 by CRC Press LLC

No claim to original U.S. Government works
International Standard Book Number 0-8493-8347-1
Library of Congress Card Number 98-31681
Printed in the United States of America 1 2 3 4 5 6 7 8 9 0
Printed on acid-free paper

Preface

Introduction

The purpose of *The Measurement, Instrumentation, and Sensors Handbook* is to provide a reference that is both concise and useful for engineers in industry, scientists, designers, managers, research personnel and students, as well as many others who have measurement problems. The *Handbook* covers an extensive range of topics that comprise the subject of measurement, instrumentation, and sensors.

The *Handbook* describes the use of instruments and techniques for practical measurements required in engineering, physics, chemistry, and the life sciences. It includes sensors, techniques, hardware, and software. It also includes information processing systems, automatic data acquisition, reduction and analysis and their incorporation for control purposes.

Articles include descriptive information for professionals , students ,and workers interested in measurement. Articles include equations to assist engineers and scientists who seek to discover applications and solve problems that arise in fields not in their specialty. They include specialized information needed by informed specialists who seek to learn advanced applications of the subject, evaluative opinions, and possible areas for future study. Thus, the *Handbook* serves the reference needs of the broadest group of users — from the advanced high school science student to industrial and university professionals.

Organization

The *Handbook* is organized according to the *measurement problem*. Section I includes general instrumentation topics, such as accuracy and standards. Section II covers spatial variables, such as displacement and position. Section III includes time and frequency. Section IV covers solid mechanical variables such as mass and strain. Section V comprises fluid mechanical variables such as pressure, flow, and velocity. Section VI covers thermal mechanical variables such as temperature and heat flux. Section VII includes electromagnetic variables such as voltage and capacitance. Section VIII covers optical variables such as photometry and image sensors. Section IX includes radiation such as x rays and dosimetry. Section X covers chemical variables in composition and environmental measurements. Section XI includes biomedical variables such as blood flow and medical imaging. Section XII comprises signal processing such as amplifiers and computers. Section XIII covers display such as cathode ray tube and recorder. Section XIV includes control such as optimal control and motion control. The Appendix contains conversion factors to SI units.

Locating Your Topic

To find out how to measure a given variable, skim the Table of Contents, turn to that section and find the chapters that describe different methods of making the measurement. Consider the alternative methods of making the measurement and each of their advantages and disadvantages. Select a method, sensor, and signal processing method. Many articles list a number of vendors to contact for more information. You can also visit the http://www.sensorsmag.com site under Buyer's Guide to obtain a list of vendors.

For more detailed information, consult the index, since certain principles of measurement may appear in more than one chapter.

Acknowledgments

I appreciate the help of the many people who worked on this handbook. David Beams assisted me by searching books, journals, and the Web for all types of measurements, then helped me to organize the outline. The Advisory Board made suggestions for revision and suggested many of the authors. Searching the INSPEC database yielded other authors who had published on a measurement method. At CRC Press, Felicia Shapiro, Associate Production Manager; Kristen Maus, Developmental Editor; Suzanne Lassandro, Book Group Production Director; and Susan Fox, Project Editor, produced the book.

John G. Webster
Editor-in-Chief

Editor-in-Chief

John G. Webster received the B.E.E. degree from Cornell University, Ithaca, NY, in 1953, and the M.S.E.E. and Ph.D. degrees from the University of Rochester, Rochester, NY, in 1965 and 1967, respectively.

He is Professor of Electrical and Computer Engineering at the University of Wisconsin-Madison. In the field of medical instrumentation he teaches undergraduate and graduate courses, and does research on RF cardiac ablation and measurement of vigilance.

He is author of *Transducers and Sensors*, An IEEE/EAB Individual Learning Program (Piscataway, NJ: IEEE, 1989). He is co-author, with B. Jacobson, of *Medicine and Clinical Engineering* (Englewood Cliffs, NJ: Prentice-Hall, 1977), with R. Pallás-Areny, of *Sensors and Signal Conditioning* (New York: Wiley, 1991), and with R. Pallas-Areny, of *Analog Signal Conditioning* (New York: Wiley, 1999). He is editor of *Encyclopedia of Medical Devices and Instrumentation* (New York: Wiley, 1988), *Tactile Sensors for Robotics and Medicine* (New York: Wiley, 1988), *Electrical Impedance Tomography* (Bristol, UK: Adam Hilger, 1990), *Teaching Design in Electrical Engineering* (Piscataway, NJ: Educational Activities Board, IEEE, 1990), *Prevention of Pressure Sores: Engineering and Clinical Aspects* (Bristol, UK: Adam Hilger, 1991), *Design of Cardiac Pacemakers* (Piscataway, NJ: IEEE Press, 1995), *Design of Pulse Oximeters* (Bristol, UK: IOP Publishing, 1997), *Medical Instrumentation: Application and Design, Third Edition* (New York: Wiley, 1998), and *Encyclopedia of Electrical and Electronics Engineering* (New York, Wiley, 1999). He is co-editor, with A. M. Cook, of *Clinical Engineering: Principles and Practices* (Englewood Cliffs, NJ: Prentice-Hall, 1979) and *Therapeutic Medical Devices: Application and Design* (Englewood Cliffs, NJ: Prentice-Hall, 1982), with W. J. Tompkins, of *Design of Microcomputer-Based Medical Instrumentation* (Englewood Cliffs, NJ: Prentice-Hall, 1981) and *Interfacing Sensors to the IBM PC* (Englewood Cliffs, NJ: Prentice Hall, 1988), and with A. M. Cook, W. J. Tompkins, and G. C. Vanderheiden, *Electronic Devices for Rehabilitation* (London: Chapman & Hall, 1985).

Dr. Webster has been a member of the IEEE-EMBS Administrative Committee and the NIH Surgery and Bioengineering Study Section. He is a fellow of the Institute of Electrical and Electronics Engineers, the Instrument Society of America, and the American Institute of Medical and Biological Engineering. He is the recipient of the AAMI Foundation Laufman-Greatbatch Prize and the ASEE/Biomedical Engineering Division, Theo C. Pilkington Outstanding Educator Award.

Advisory Board

Contributors

Rene G. Aarnink
University Hospital Nijmegen
Nijmegen, The Netherlands

Mushtaq Ali
The National Grid Company
Leatherhead, Surrey, England

Joseph H. Altman
Pittsford, New York

A. Ambrosini
Institute of Radioastronomy
National Research Council
Via Fiorentina
Villa Fontano, Italy

Jeff P. Anderson
LTV Steel Corporation
Independence, Ohio

Keith Antonelli
Kinetic Sciences, Inc.
Vancouver, B.C., Canada

John C. Armitage
Ottawa–Carleton Institute for
    Physics
Carleton University
Ottawa, Ontario, Canada

Pasquale Arpaia
Università di Napoli Federico II
Naples, Italy

Per Ask
Department of Biomedical
    Engineering
Linkoping University
Linkoping, Sweden

Marc J. Assael
Faculty of Chemical Engineering
Aristotle University of Thessaloniki
Thessalonika, Greece

Viktor P. Astakhov
Mechanical Engineering
    Department
Concordia University
Montreal, Quebec, Canada

Francesco Avallone
Università di Napoli Federico II
Naples, Italy

Aldo Baccigalupi
Università di Napoli Federico II
Naples, Italy

William E. Baker
Mechanical Engineering
    Department
University of New Mexico
Albuquerque, New Mexico

W. John Ballantyne
Systems Engineering Department
Spar Aerospace Ltd.
Brampton, Ontario, Canada

Amit Bandyopadhyay
Department of Ceramic Science and
    Engineering
Rutgers University
Piscataway, New Jersey

Partha P. Banerjee
Electrical and Computer
    Engineering Department
University of Alabama – Huntsville
Huntsville, Alabama

William A. Barrow
Planar Systems
Beaverton, Oregon

Cipriano Bartoletti
University of Rome "La Sapieriza"
Rome, Italy

L. Basano
Dipartimento de Fisica
Università di Genova
Genova, Italy

M.W. Bautz
Pennsylvania State University
University Park, Pennsylvania

William H. Bayles
The Fredericks Company
Huntington Valley, Pennsylvania

David M. Beams
Department of Electrical
    Engineering
University of Texas at Tyler
Tyler, Texas

K. Beilenhoff
Institut für Hochfrequenztechnik,
    Technische Universität Darmstadt
Muenchen, Germany

B. Benhabib
Department of Mechanical and
    Industrial Engineering
University of Toronto
Toronto, Ontario, Canada

Michael Bennett
Willison Associates
Manchester, England

Vikram Bhatia
Virginia Tech
Blacksburg, Virginia

Richard J. Blotzer
LTV Steel Corporation
Independence, Ohio

A. Bonen
University of Toronto
Toronto, Ontario, Canada

C. Bortolotti
Institute of Radioastronomy
National Research Council
Via Fiorentina
Villa Fontano, Italy

Howard M. Brady
The Fredericks Company
Huntington Valley, Pennsylvania

Arnaldo Brandolini
Dipartimento di Elettrotecnica
Politecnico di Milano
Milano, Italy

John C. Brasunas
NASA/Goddard Space Flight Center
Greenbelt, Maryland

Detlef Brumbi
Krohue Messtechnik GmbH
Diusburg, Germany

Christophe Bruttin
Rittmeyer Ltd.
Zug, Switzerland

Saps Buchman
Stanford University
Stanford, California

Wolfgang P. Buerner
Los Angeles Scientific
    Instrumentation
Los Angeles, California

B.E. Burke
Pennsylvania State University
University Park, Pennsylvania

Jim Burns
Burns Engineering Inc.
Minnetonka, MN

Barrett Caldwell
University of Wisconsin–Madison
Madison, Wisconsin

Robert B. Campbell
Sandia National Laboratories
Livermore, California

Claudio de Capua
Università di Napoli Federico II
Napoli, Italy

Kevin H.L. Chau
Micromachined Products Division
Analog Devices
Cambridge, Massachusetts

Adam Chrzanowski
University of New Brunswick
Fredericton, N.B., Canada

Richard O. Claus
Bradley Department of Electrical
    Engineering
Virginia Tech
Blacksburg, Virginia

Charles B. Coulbourn
Los Angeles Scientific
    Instrumentation Co.
Los Angeles, California

Bert M. Coursey
Ionizing Radiaiton Division, Physics
    Laboratory
National Institute of Standards and
    Technology
Gaithersburg, Maryland

Robert F. Cromp
NASA/Goddard Space Flight Center
Greenbelt, Maryland

Robert M. Crovella
NVIDIA Corporation
Plano, Texas

Brian Culshaw
Department of Electronic and
    Electrical Engineering
University of Strathclyde
Royal College Building
Glasgow, England

G. Mark Cushman
NASA/Goddard Space Flight Center
Greenbelt, Maryland

N. D'Amico
Institute of Radioastronomy
National Research Council
Via Fiorentina
Villa Fontano, Italy

Larry S. Darken
Oxford Instruments, Inc.
Oak Ridge, Tennessee

David Dayton
ILC Data Device Corporation
Bohemia, New York

Timothy R. DeGrado
Duke University Medical Center
Durham, North Carolina

Alfons Dehé
Institut für Hochfrequenztechnik,
    Technische Universität Darmstadt
Muenchen, Germany

Ronald H. Dieck
Pratt & Whitney
Palm Beach Gardens, Florida

Thomas E. Diller
Virginia Polytechnic Institute
Blacksburg, Virginia

Madhu S. Dixit
Centre for Research in Particle
    Physics
Carleton University
Ottawa, Ontario, Canada

James T. Dobbins III
Duke University Medical Center
Durham, North Carolina

Achim Dreher
German Aerospace Center
Wessling, Germany

Emil Drubetsky
The Fredericks Company
Huntington Valley, Pennsylvania

Jacques Dubeau
Centre for Research in Particle
  Physics
Carleton University
Ottawa, Ontario, Canada

Maria Eklund
Nynas Naphthenics AB
Nynashamn, Sweden

M.A. Elbestawi
Mechanical Engineering
McMaster University
Hamilton, Ontario, Canada

Halit Eren
Curtin Unversity of Technology
Perth, WA, Australia

Alessandro Ferrero
Dipartimento di Elettrotecnica
Politecnico di Milan
Milano, Italy

Richard S. Figliola
Department of Mechanical
  Engineering
Clemson University
Clemson, South Carolina

Michael Fisch
Department of Physics
John Carroll University
University Heights, Ohio

Jacob Fraden
Advanced Monitors Corporation
San Diego, California

Randy Frank
Semiconductor Products Sector
Transporation Systems Group
Motorola, Inc.
Phoenix, Arizona

Larry A. Franks
Sandia National Laboratories
Livermore, California

Richard Frayne
University of Wisconsin
Madison, Wisconsin

K. Fricke
Institut für Hochfrequenztechnik,
  Technische Universität Darmstadt
Muenchen, Germany

Mark Fritz
Denver Instrument Company
Arvada, Colorado

Chun Che Fung
Curtin University of Technology
Perth, WA, Australia

Alessandro Gandelli
Dipartimento di Elettrotecnica
Politecnico di Milano
Milano, Italy

John D. Garrison
San Diego State University
San Diego, California

Ivan J. Garshelis
Magnova, Inc.
Pittsfield, Massachusetts

Daryl Gerke
Kimmel Gerke Associates, Ltd.
Mesa, Arizona

W.A. Gillespie
University of Abertay Dundee
Dundee, Scotland

Paolo Giordano
Rittmeyer Ltd.
Zug, Switzerland

Olaf Glük
Institut für Schicht-und
  Ionentechnik
Forschungszentrum Julich GmbH
Germany

Ron Goehner
The Fredericks Company
Huntington Valley, Pennsylvania

James Goh
Curtin University of Technology
Perth, WA, Australia

J.A. Gregory
Pennsylvania State University
University Park, Pennsylvania

R.E. Griffiths
Pennsylvania State University
University Park, Pennsylvania

Steven M. Grimes
Department of Physics and
  Astronomy
Ohio University
Athens, Ohio

G. Grueff
Institute of Radioastronomy
National Research Council
Via Fiorentina
Villa Fontano, Italy

J.Y. Gui
General Electric Research and
  Development Center
General Electric Company
Schenectady, New York

Anthony Guiseppi–Elie
Abtech Scientific, Inc.
Yardley, Pennsylvania

Reinhard Haak
Universitaet Erlangen–Nuernberg
Erlangen, Germany

Sean M. Hames
Duke University Medical Center
Durham, North Carolina

R. John Hansman, Jr.
Department of Aeronautics and
  Astronautics
Massachusetts Institute of
  Technology
Cambridge, Massachusetts

Daniel Harrison
Department of Physics
John Carroll University
University Heights, Ohio

H.L. Hartnagel
Institut für Hochfrequenztechnik,
  Technische Universität Darmstadt
Muenchen, Germany

Bruce H. Hasegawa
University of California
San Francisco, California

Emil Hazarian
Denver Institute Company
Arvada, Colorado

Michael B. Heaney
Huladyne Research
Palo Alto, California

Albert D. Helfrick
Embry–Riddle Aeronautical
    University
Dayton Beach, Florida

David A. Hill
National Institute of Standards and
    Technology
U.S. Department of Commerce
Boulder, Colorado

Thomas Hossle
Rittmeyer Ltd.
Zug, Switzerland

C.H. Houpis
Air Force Institute of Technology
Wright–Patterson AFB, Ohio

Zaki D. Husain
Daniel Flow Products, Inc.
Bellaire, Texas

Alan L. Huston
Naval Research Laboratory
Washington, D.C.

Robert M. Hyatt, Jr.
Howell Electric Motors
Plainfield, New Jersey

Stanley S. Ipson
Department of Electronic and
    Electrical Engineering
University of Bradford
Bradford, W. Yorkshire, England

Rahman Jamal
National Instruments Germany
Applications Engineering
Munchen, Germany

Ralph B. James
Sandia National Laboratories
Livermore, California

Victor F. Janas
Center for Ceramic Research
Rutgers University
Piscataway, New Jersey

Roger Jones
Primary Children's Medical Center
Salt Lake City, Utah

Brian L. Justus
Optical Science Department
Naval Research Laboratory
Washington, D.C.

Motohisa Kanda
National Institute of Standards and
    Technology
Boulder, Colorado

Mohammad A. Karim
University of Tennessee, Knoxville
Knoxville, Tennessee

Nils Karlsson
National Defense Research
    Establishment
Stockholm, Sweden

Sam S. Khalilieh
Electrical Engineering
Earth Tech
Grand Rapids, Michigan

Andre Kholkine
Rutgers University
Piscataway, New Jersey

William Kimmel
Kimmel Gerke Associates, Ltd.
Mesa, Arizona

John A. Kleppe
Electrical Engineering Department
University of Nevada
Reno, Nevada

H. Klingbeil
Institut für Hochfrequenztechnik,
    Technische Universität Darmstadt
Muenchen, Germany

James Ko
Kinetic Sciences Inc.
Vancouver, B.C., Canada

Hebert Köchner
Universitaet Erlangen–Nuernberg
Erlangen, Germany

Wei Ling Kong
Curtin Institute of Technology
Perth, WA, Australia

M. Kostic
Northern Illinois University
DeKalb, Illinois

R.L. Kraft
Penn State University
University Park, Pennsylvania

V. Krozer
Institut für Hochfrequenztechnik,
    Technische Universität Darmstadt
Muenchen, Germany

Shyan Ku
Kinetic Sciences Inc.
Vancouver, B.C., Canada

H.L. Kwok
Penn State University
University Park, Pennsylvania

C.K. Laird
School of Applied Chemistry
Kingston University
Kingston Upon Thames, England

Brook Lakew
NASA/Goddard Space Flight Center
Greenbelt, Maryland

Carmine Landi
Università de L'Aquila
L'Aquila, Italy

Jacqueline Le Moigne
NASA/Goddard Space Flight Center
Greenbelt, Maryland

G.E. LeBlanc
School of Geography and Geology
McMaster University
Hamilton, Ontario, Canada

W. Marshall Leach, Jr.
School of Electrical and Computer
    Engineering
Georgia Institute of Technology
Atlanta, Georgia

Kathleen M. Leonard
Department of Civil and
  Environmental Engineering
The University of Alabama in
  Huntsville
Huntsville, Alabama

Yufeng Li
Samsung Information Systems
  America
HDD R & D Center
San Jose, California

E. B. Loewenstein
National Instruments
Austin, Texas

Robert Lofthus
Xerox Corporation
Rochester, New York

Michael A. Lombardi
Time and Frequency Division
National Institute of Standards and
  Technology
Boulder, Colorado

Michael Z. Lowenstein
Harmonics Limited
Mequou, Wisconsin

Albert Lozano–Nieto
Commonwealth College
Wilkes Barre Campus
Penn State University
Lehman, Pennsylvania

D.H. Lumb
Penn State University
University Park, Pennsylvania

Christopher S. Lynch
Mechanical Engineering
  Department
The Georgia Institute of Technology
Atlanta, Georgia

A.M. MacLeod
School of Engineering
University of Abertay Dundee
Dundee, Scotland

Steven A. Macintyre
Macintyre Electronic Design
Herndon, Virginia

Tolestyn Madaj
Technical University of Gdansk
Gdansk, Poland

Kin F. Man
Jet Propulsion Lab
California Institute of Technology
Pasadena, California

Dimitris E. Manolakis
Department of Automation
Technological Education Institute
Thessaloniki, Greece

Robert T. Marcus
Datacolor International
Middletown, New Jersey

S. Mariotti
Institute of Radioastronomy
National Research Council
Via Fiorentina
Villa Fontano, Italy

Wade M. Mattar
The Foxboro Company
Foxboro, Massachusetts

J.R. René Mayer
Mechanical Engineering
Ecole Polytechnique de Montreal
Montreal, Quebec, Canada

Edward McConnell
Data Acquisition
National Instruments
Austin, Texas

P.F. Martin
University of Abertay Dundee
Dundee, Scotland

Robert T. McGrath
Department of Engineering Science
  and Mechanics
Pennsylvania State University
University Park, Pennsylvania

John McInroy
Department of Electrical
  Engineering
University of Wyoming
Laramie, Wyoming

Douglas P. McNutt
The MacNauchtan Laboratory
Colorado Springs, Colorado

G.H. Meeten
Department of Fluid Chemistry and
  Physics
Schlumberger Cambridge Research
Cambridge, England

Adrian Melling
Universitaet Erlangen–Nuernberg
Erlangen, Germany

Rajan K. Menon
Laser Velocimetry Products
TSI Inc.
St. Paul, Minnesota

Hans Mes
Centre for Research in Particle
  Physics
Carleton University
Ottawa, Ontario, Canada

John Mester
W.W. Hansen Experimental Physics
  Laboratory
Stanford University
Stanford, California

Jaroslaw Mikielewicz
Institute of Fluid Flow Machinery
Gdansk, Poland

Harold M. Miller
Data Industrial Corporation
Mattapoisett, Massachusetts

Mark A. Miller
Naval Research Laboratory
Washington, D.C.

Jeffrey P. Mills
Illinois Institute of Technology
Chicago, Illinois

Devendra Misra
Electrical Engineering and
    Computer Science Department
University of Wisconsin–Milwaukee
Milwaukee, Wisconsin

William C. Moffatt
Sandia National Laboratories
Livermore, California

Stelio Montebugnoli
Institute of Radioastronomy
National Research Council
Villa Fontano, Italy

Roger Morgan
School of Engineering
Liverpool John Moores University
Liverpool, England

Armelle M. Moulin
University of Cambridge
Cambridge, England

Jerry Murphy
Electronic Measurements Division
Hewlett–Packard
Colorado Springs, Colorado

Steven A. Murray
SPAWAR Systems Center
San Diego, California

Soe-Mie F. Nee
Research and Technology Division
U.S. Naval Air Warfare Center
China Lake, California

Nam-Trung Nguyen
Berkeley Sensor and Actuator Center
University of California at Berkeley
Berkeley, California

J.V. Nicholas
The New Zealand Institute for
    Industrial Research and
    Development
Lower Hutt, New Zealand

Seiji Nishifuji
Electrical and Electronic
    Engineering
Yamaguchi University
Ube, Japan

John A. Nousek
Department of Astronomy and
    Astrophysics
Pennsylvania State University
University Park, Pennsylvania

David S. Nyce
MTS Systems Corp.
Cary, North Carolina

Peter O'Shea
Department of Computer and
    Electrical Engineering
Royal Melbourne Institute of
    Technology
Melbourne, Victoria, Australia

F. Gerald Oakham
Centre for Research in Particle
    Physics
Carleton University
Ottawa, Ontario, Canada

P. Åke Öberg
Department of Biomedical
    Engineering
Linkoping University Hospital
Linkoping, Sweden

Chima Okereke
Independent Consultant
Formerly of Department of
    Electronic and Electrical
    Engineering
University of Bradford
Bradford, W. Yorkshire, U.K

John G. Olin
Sierra Instruments, Inc.
Monterey, California

A. Orfei
Institute of Radioastronomy
National Research Council
Via Fiorentina
Villa Fontano, Italy

P. Ottonello
Dipartimento di Fisica
Universita di Genova
Genova, Italy

M. Pachter
Air Force Institute of Technology
Wright–Patterson AFB, Ohio

Behrooz Pahlavanpour
The National Grid Company
Leatherhead, Surrey, England

Ramón Pallás-Areny
Universitat Politecnica de Catalunya
Barcelona, Spain

Ronney B. Panerai
University of Leicester
Leicester Royal Infirmary
Leicester, U.K.

Franco Pavese
CNR
Instituto di Metrologia "G.
    Colonnetti"
Torino, Italy

Peder C. Pedersen
Electrical and Computer
    Engineering
Worcester Polytechnic Institute
Worcester, Massachusetts

Teklic Ole Pedersen
Linkoping Universitet
Linkoping, Sweden

B.W. Petley
National Physical Laboratory
Middlesex, U.K.

Rekha Philip-Chandy
School of Engineering
Liverpool John Moores University
Liverpool, England

Thad Pickenpaugh
AFRL/SNHI
Wright–Patterson AFB, Ohio

Charles P. Pinney
Pinney Technologies, Inc.
Albuquerque, New Mexico

Luca Podestà
University of Rome "La Sapieriza"
Rome, Italy

Rodney Pratt
University of South Australia
Adelaide, S. Australia

Per Rasmussen
GRAS Sound and Vibration
Vedback, Denmark

R.P. Reed
Proteun Services
Albuquerque, New Mexico

Shyam Rithalia
Department of Rehabilitation
University of Salford
Salford, U.K.

Gordon W. Roberts
Department of Electrical
   Engineering
McGill University
Montreal, Quebec, Canada

Stephen B.W. Roeder
Department of Chemistry
San Diego State University
San Diego, California

Herbert M. Runciman
Pilkington Optronics
Glasgow, Scotland, U.K.

Terry L. Rusch
Marshfield Medical Research
   Foundation
Tampa, Florida

Ricardo Saad
University of Toronto
Toronto, Ontario, Canada

Giancarlo Sacerdoti
University of Rome "La Sapieriza"
Rome, Italy

Ahmad Safari
Department of Ceramic Science and
   Engineering
Rutgers University
Piscataway, New Jersey

Robert J. Sandberg
Mechanical Engineering
   Department
University of Wisconsin
Madison, Wisconsin

Ravi Sankar
Department of Electrical
   Engineering
University of South Florida
Tampa, Florida

Meyer Sapoff
MS Consultants
Princeton, New Jersey

Kalluri R. Sarma
Advanced Displays Department
Honeywell, Inc.
Phoenix, Arizona

Michael J. Schöning
Institut für Schicht-und
   Ionentechnik
Forschungszentrum Julich GmbH
Germany

Fritz Schuermeyer
USAF Wright Laboratory
Wright–Patterson AFB, Ohio

Stuart Schweid
Xerox Corporation
Rochester, New York

Patricia J. Scully
School of Engineering
Liverpool John Moores University
Liverpool, England

R.A. Secco
The University of Western Ontario
Ontario, Canada

James M. Secord
Geodesy and Geomatics Engineering
University of New Brunswick
Fredericton, N.B., Canada

DeWayne B. Sharp
Shape of Things
San Luis Obispo, California

Kanai S. Sheh
Radiation Monitoring Devices, Inc.
Watertown, Massachusetts

Norman F. Sheppard, Jr.
Gamera Bioscience Corporation
Medford, Massachusetts

Christopher J. Sherman
Merrimack, New Hampshire

F. Greg Shinskey
Process Control Consultant
N. Sandwich, New Hampshire

K.C. Smith
University of Toronto
Toronto, Ontario, Canada

Michael R. Squillante
Radiation Monitoring Devices, Inc.
Watertown, Massachusetts

Jan Stasiek
Mechanical Engineering
   Department
Technical University of Gdansk
Gdansk, Poland

Mark A. Stedham
Electrical and Computer
   Engineering
University of Alabama at Huntsville
Huntsville, Alabama

Robert Steer
Frequency Devices
Haverhill, Massachusetts

Robert J. Stephenson
University of Cambridge
Cambridge, England

T.J. Sumner
Imperial College
London, England

Haiyin Sun
Coherent Auburn Group
Auburn, California

Mark Sun
NeoPath, Inc.
Redmond, Washington

Peter H. Sydenham
University of South Australia
Mawsons Lakes, South Australia
and
University College, London
London, UK

Micha Szyper
University of Mining and Metallurgy
Cracow, Poland

Shogo Tanaka
Electrical and Electronic
  Engineering
Yamaguchi University
Ube, Japan

Nitish V. Thakor
Biomedical Engineering
  Department
Johns Hopkins University Medical
  School
Baltimore, Maryland

David B. Thiessen
California Institute of Technology
Pasadena, California

Richard Thorn
School of Engineering
University of Derby
Derby, U.K.

Marion Thust
Institut für Schicht-und
  Ionentechnik
Forschungszentrum Julich GmbH
Germany

G. Tomassetti
Institute of Radioastronomy
National Research Council
Via Fiorentina
Villa Fontano, Italy

Michael F. Toner
Nortel Networks
Nepean, Ontario, Canada

E.E. Uzgiris
General Electric Research and
  Development Center
General Electric Company
Schenectady, New York

Sander van Herwaarden
Xensor Integration
Delft, The Netherlands

Hans-Peter Vaterlaus
Instrument Department
Rittmeyer Ltd.
Zug, Switzerland

Ramanapathy Veerasingam
Penn State University
University Park, Pennsylvania

Herman Vermariën
Laboratory of Physiology
Vrije Universiteit
Brussels, Belgium

James H. Vignos
The Foxboro Company
Foxboro, Massachusetts

Gert J.W. Visscher
Institute of Agricultural and
  Environmental Engineering
Wageningen, The Netherlands

David Wadlow
Sensors Research Consulting, Inc.
Basking Ridge, New Jersey

William A. Wakeham
Imperial College
London, England

Anbo Wang
Bradley Department of Electrical
  Engineering
Virgina Tech
Blacksburg, Virginia

Donald J. Wass
Daniel Flow Products, Inc.
Houston, Texas

Mark E. Welland
University of Cambridge
Cambridge, England

Grover C. Wetsel
Erik Jonsson School of Engineering
  and Computer Science
University of Texas at Dallas
Richardson, Texas

Hessel Wijkstra
University Hospital Nijmegen
Nijmegen, The Netherlands

Jesse Yoder
Automation Research Corporation
Dedham, Massachusetts

Bernhard Günther Zagar
Electrical Engineering
Technical University Graz
Graz, Austria

James A. Zagzebski
Department of Medical Physics
University of Wisconsin
Madison, Wisconsin

Contents

VI Mechanical Variables Measurement — Thermal

I

Measurement
Characteristics

1

Characteristics of Instrumentation

R. John Hansman, Jr.

Massachusetts Institute of Technology

In addressing measurement problems, it is often useful to have a conceptual model of the measurement process. This chapter presents some of the fundamental concepts of measurement in the context of a simple generalized instrument model.

In abstract terms, an *instrument* is a device that transforms a *physical variable* of interest (the *measurand*) into a form that is suitable for recording (the *measurement*). In order for the measurement to have broad and consistent meaning, it is common to employ a standard system of *units* by which the measurement from one instrument can be compared with the measurement of another.

An example of a basic instrument is a ruler. In this case the measurand is the length of some object and the measurement is the number of units (meters, inches, etc.) that represent the length.

1.1   Simple Instrument Model

Figure 1.1 presents a generalized model of a simple instrument. The physical process to be measured is in the left of the figure and the measurand is represented by an observable physical variable X. Note that the observable variable X need not necessarily be the measurand but simply related to the measurand in some known way. For example, the mass of an object is often measured by the process of *weighing*, where the measurand is the mass but the physical measurement variable is the downward force the mass exerts in the Earth's gravitational field. There are many possible physical measurement variables. A few are shown in Table 1.1.

The key functional element of the instrument model shown in Figure 1.1 is the *sensor*, which has the function of converting the *physical variable input* into a *signal variable output*. Signal variables have the property that they can be manipulated in a transmission system, such as an electrical or mechanical circuit. Because of this property, the signal variable can be transmitted to an output or recording device that can be remote from the sensor. In electrical circuits, voltage is a common signal variable. In mechanical systems, displacement or force are commonly used as signal variables. Other examples of signal variable are shown in Table 1.1. The signal output from the sensor can be displayed, recorded, or used as an input signal to some secondary device or system. In a basic instrument, the signal is transmitted to a *display* or recording device where the measurement can be read by a human observer. The observed output is the measurement M. There are many types of display devices, ranging from simple scales and dial gages to sophisticated computer display systems. The signal can also be used directly by some larger

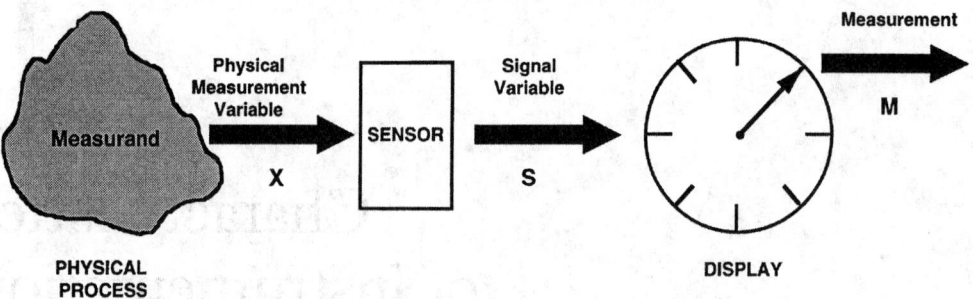

FIGURE 1.1 Simple instrument model.

TABLE 1.1

Common physical variables	Typical signal variables
• Force	• Voltage
• Length	• Displacement
• Temperature	• Current
• Acceleration	• Force
• Velocity	• Pressure
• Pressure	• Light
• Frequency	• Frequency
• Capacity	
• Resistance	
• Time	
• ...	

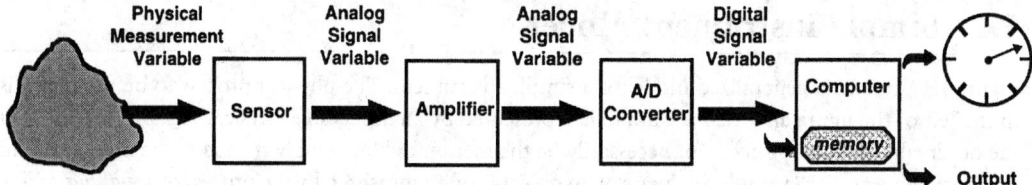

FIGURE 1.2 Instrument model with amplifier, analog to digital converter, and computer output.

system of which the instrument is a part. For example, the output signal of the sensor may be used as the input signal of a closed loop control system.

If the signal output from the sensor is small, it is sometimes necessary to amplify the output shown in Figure 1.2. The amplified output can then be transmitted to the display device or recorded, depending on the particular measurement application. In many cases it is necessary for the instrument to provide a digital signal output so that it can interface with a computer-based data acquisition or communications system. If the sensor does not inherently provide a digital output, then the analog output of the sensor is converted by an analog to digital converter (ADC) as shown in Figure 1.2. The digital signal is typically sent to a computer processor that can display, store, or transmit the data as output to some other system, which will use the measurement.

Passive and Active Sensors

As discussed above, sensors convert physical variables to signal variables. Sensors are often transducers in that they are devices that convert input energy of one form into output energy of another form. Sensors

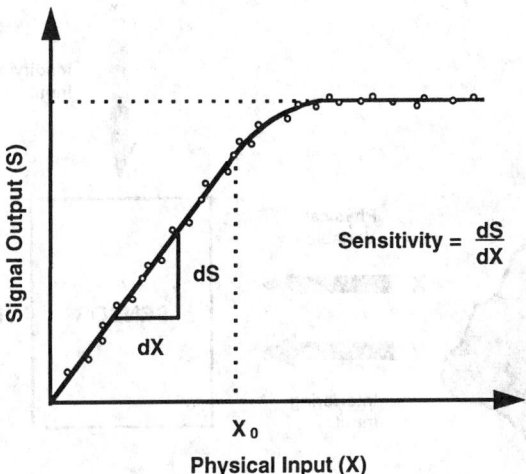

FIGURE 1.3 Calibration curve example.

can be categorized into two broad classes depending on how they interact with the environment they are measuring. *Passive sensors* do not add energy as part of the measurement process but may remove energy in their operation. One example of a passive sensor is a thermocouple, which converts a physical temperature into a voltage signal. In this case, the temperature gradient in the environment generates a thermoelectric voltage that becomes the signal variable. Another passive transducer is a pressure gage where the pressure being measured exerts a force on a mechanical system (diaphragm, aneroid or Borden pressure gage) that converts the pressure force into a displacement, which can be used as a signal variable. For example, the displacement of the diaphragm can be transmitted through a mechanical gearing system to the displacement of an indicating needle on the display of the gage.

Active sensors add energy to the measurement environment as part of the measurement process. An example of an active sensor is a radar or sonar system, where the distance to some object is measured by actively sending out a radio (radar) or acoustic (sonar) wave to reflect off of some object and measure its range from the sensor.

Calibration

The relationship between the physical measurement variable input and the signal variable (output) for a specific sensor is known as the *calibration* of the sensor. Typically, a sensor (or an entire instrument system) is calibrated by providing a known physical input to the system and recording the output. The data are plotted on a calibration curve such as the example shown in Figure 1.3. In this example, the sensor has a linear response for values of the physical input less than X_0. The *sensitivity* of the device is determined by the slope of the calibration curve. In this example, for values of the physical input greater than X_0, the calibration curve becomes less sensitive until it reaches a limiting value of the output signal. This behavior is referred to as *saturation,* and the sensor cannot be used for measurements greater than its saturation value. In some cases, the sensor will not respond to very small values of the physical input variable. The difference between the smallest and largest physical inputs that can reliably be measured by an instrument determines the *dynamic range* of the device.

Modifying and Interfering Inputs

In some cases, the sensor output will be influenced by physical variables other than the intended measurand. In Figure 1.4, X is the intended measurand, Y is an *interfering input,* and Z is a *modifying input.* The interfering input Y causes the sensor to respond in the same manner as the linear superposition of Y and the intended measurand X. The measured signal output is therefore a combination of X and Y,

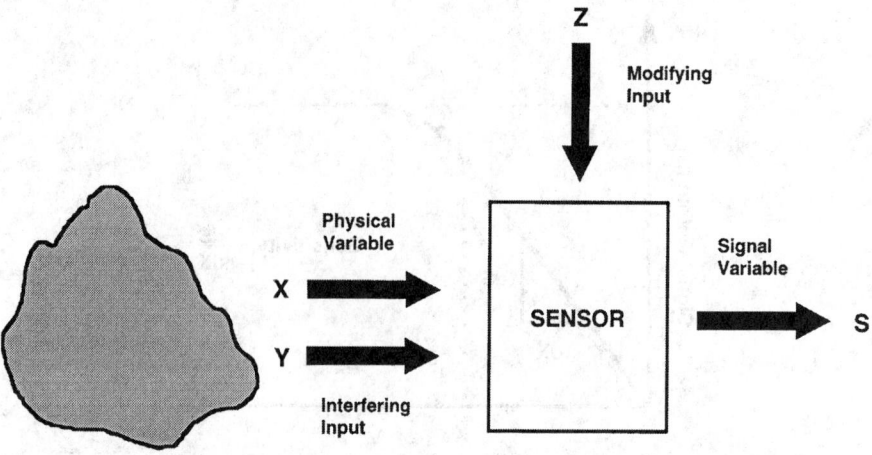

FIGURE 1.4   Interfering inputs.

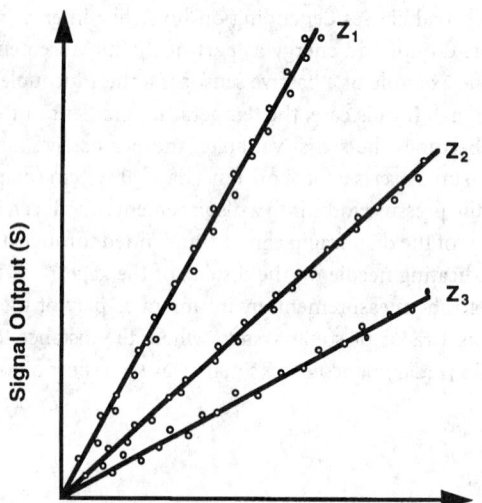

FIGURE 1.5   Illustration of the effect of a modifying input on a calibration curve.

with Y interfering with the intended measurand X. An example of an interfering input would be a structural vibration within a force measurement system.

Modifying inputs changes the behavior of the sensor or measurement system, thereby modifying the input/output relationship and calibration of the device. This is shown schematically in Figure 1.5. For various values of Z in Figure 1.5, the slope of the calibration curve changes. Consequently, changing Z will result in a change of the apparent measurement even if the physical input variable X remains constant. A common example of a modifying input is temperature; it is for this reason that many devices are calibrated at specified temperatures.

Accuracy and Error

The *accuracy* of an instrument is defined as the difference between the *true value* of the measurand and the *measured value* indicated by the instrument. Typically, the true value is defined in reference to some absolute or agreed upon standard. For any particular measurement there will be some error due to

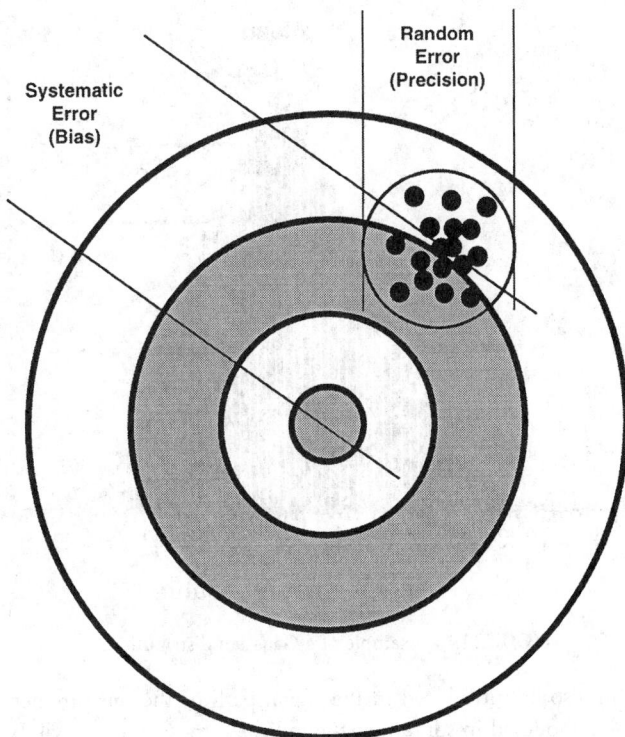

FIGURE 1.6   Target analogy of measurement accuracy.

systematic (*bias*) and *random* (*noise*) error sources. The combination of systematic and random error can be visualized by considering the analogy of the target shown in Figure 1.6. The total error in each shot results from both systematic and random errors. The systematic (bias) error results in the grouping of shots being offset from the bulls eye (presumably a misalignment of the gunsight or wind). The size of the grouping is determined by random error sources and is a measure of the *precision* of the shooting.

Systematic Error Sources (Bias)

There are a variety of factors that can result in systematic measurement errors. One class of cause factors are those that change the input–output response of a sensor resulting in miscalibration. The modifying inputs and interfering inputs discussed above can result in sensor miscalibration. For example, if temperature is a modifying input, using the sensor at a temperature other than the calibrated temperature will result in a systematic error. In many cases, if the systematic error source is known, it can be corrected for by the use of *compensation methods.*

There are other factors that can also cause a change in sensor calibration resulting in systematic errors. In some sensors, aging of the components will change the sensor response and hence the calibration. Damage or abuse of the sensor can also change the calibration. In order to prevent these systematic errors, sensors should be periodically recalibrated.

Systematic errors can also be introduced if the measurement process itself changes the intended measurand. This issue, defined as *invasiveness,* is a key concern in many measurement problems. Interaction between measurement and measurement device is always present; however, in many cases, it can be reduced to an insignificant level. For example, in electronic systems, the energy drain of a measuring device can be made negligible by making the input impedance very high. An extreme example of invasiveness would be to use a large warm thermometer to measure the temperature of a small volume of cold fluid. Heat would be transferred from the thermometer and would warm the fluid, resulting in an inaccurate measurement.

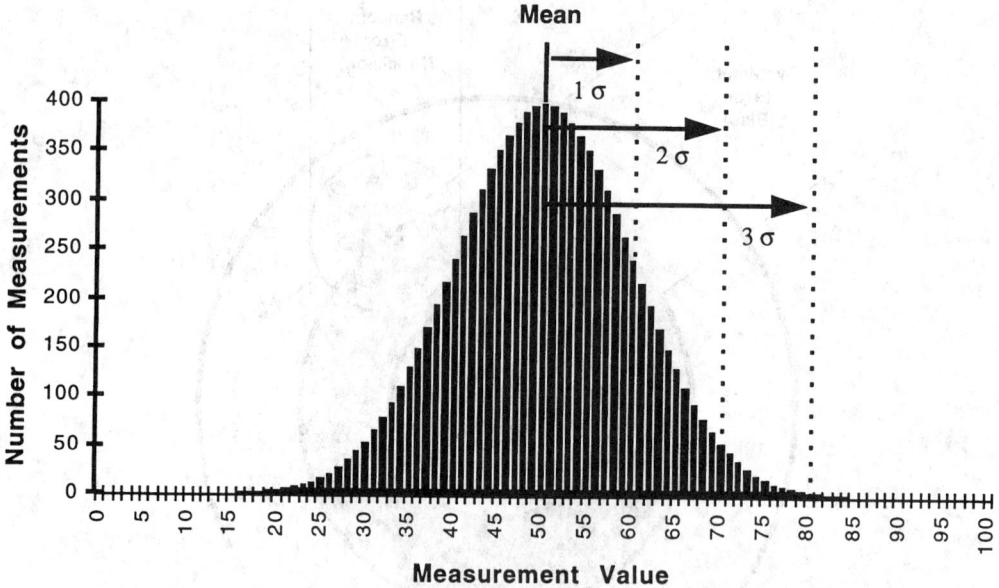

FIGURE 1.7   Example of a Gaussian distribution.

Systematic errors can also be introduced in the signal path of the measurement process shown in Figure 1.3. If the signal is modified in some way, the indicated measurement will be different from the sensed value. In physical signal paths such as mechanical systems that transmit force or displacement, friction can modify the propagation of the signal. In electrical circuits, resistance or attenuation can also modify the signal, resulting in a systematic error.

Finally, systematic errors or bias can be introduced by human observers when reading the measurement. A common example of observer bias error is *parallax error*. This is the error that results when an observer reads a dial from a non-normal angle. Because the indicating needle is above the dial face, the apparent reading will be shifted from the correct value.

Random Error Sources (Noise)

If systematic errors can be removed from a measurement, some error will remain due to the random error sources that define the precision of the measurement. Random error is sometimes referred to as *noise*, which is defined as a signal that carries no useful information. If a measurement with true random error is repeated a large number of times, it will exhibit a *Gaussian distribution*, as demonstrated in the example in Figure 1.7 by plotting the number of times values within specific ranges are measured. The Gaussian distribution is centered on the true value (presuming no systematic errors), so the mean or average of all the measurements will yield a good estimate of the true value.

The precision of the measurement is normally quantified by the standard deviation (σ) that indicates the width of the Gaussian distribution. Given a large number of measurements, a total of 68% of the measurements will fall within $\pm 1\sigma$ of the mean; 95% will fall within $\pm 2\sigma$; and 99.7% will fall within $\pm 3\sigma$. The smaller the standard deviation, the more precise the measurement. For many applications, it is common to refer to the 2σ value when reporting the precision of a measurement. However, for some applications such as navigation, it is common to report the 3σ value, which defines the limit of likely uncertainty in the measurement.

There are a variety of sources of randomness that can degrade the precision of the measurement — starting with the repeatability of the measurand itself. For example, if the height of a rough surface is to be measured, the measured value will depend on the exact location at which the measurement is taken. Repeated measurements will reflect the randomness of the surface roughness.

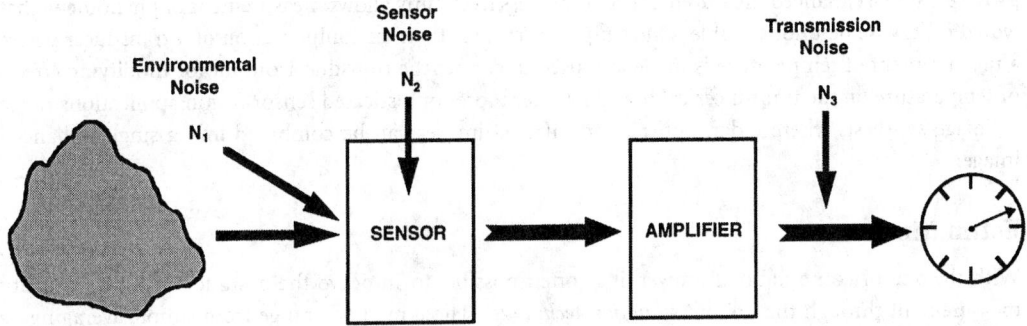

FIGURE 1.8   Instrument model with noise sources.

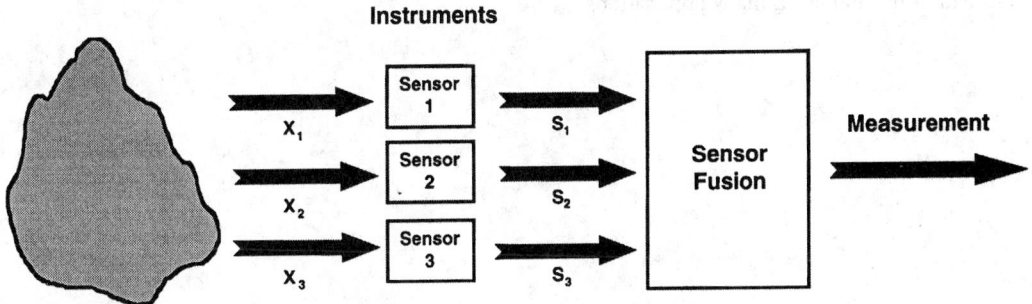

FIGURE 1.9   Example of sensor fusion.

Random error generating noise can also be introduced at each stage in the measurement process, as shown schematically in Figure 1.8. Random interfering inputs will result in noise from the measurement environment N_1 that are introduced before the sensor, as shown in the figure. An example would be background noise received by a microphone. Sensor noise N_2 can also be introduced within the sensor. An example of this would be thermal noise within a sensitive transducer, such as an infrared sensor. Random motion of electrons, due to temperature, appear as voltage signals, which are apparently due to the high sensitivity of the device. For very sensitive measurements with transducers of this type (e.g., infrared detectors), it is common to cool the detector to minimize this noise source.

Noise N_3 can also be introduced in the transmission path between the transducer and the amplifier. A common example of transmission noise in the U.S. is 60 Hz interference from the electric power grid that is introduced if the transmission path is not well grounded, or if an inadvertent electric grand loop causes the wiring to act as an antenna.

It is important to note that the noise will be amplified along with the signal as it passes through the amplifier in Figure 1.8. As a consequence, the figure of merit when analyzing noise is not the level of the combined noise sources, but the *signal to noise ratio* (*SNR*), defined as the ratio of the signal power to the power in the combined noise sources. It is common to report SNR in decibel units.

The SNR is ideally much greater than 1 (0 dB). However, it is sometimes possible to interpret a signal that is lower than the noise level if some identifying characteristics of that signal are known and sufficient signal processing power is available. The human ability to hear a voice in a loud noise environment is an example of this signal processing capability.

Sensor Fusion

The process of *sensor fusion* is modeled in Figure 1.9. In this case, two or more sensors are used to observe the environment and their output signals are combined in some manner (typically in a processor) to provide a single enhanced measurement. This process frequently allows measurement of phenomena that would otherwise be unobservable. One simple example is thermal compensation of a transducer where a measurement of temperature is made and used to correct the transducer output for modifying effects of temperature on the transducer calibration. Other more sophisticated sensor fusion applications range to image synthesis where radar, optical, and infrared images can be combined into a single enhanced image.

Estimation

With the use of computational power, it is often possible to improve the accuracy of a poor quality measurement through the use of *estimation techniques*. These methods range from simple averaging or low-pass filtering to cancel out random fluctuating errors to more sophisticated techniques such as Wiener or Kalman filtering and model-based estimation techniques. The increasing capability and lowering cost of computation makes it increasingly attractive to use lower performance sensors with more sophisticated estimation techniques in many applications.

2

Operational Modes
of Instrumentation

Richard S. Figliola
Clemson University

2.1 Null Instrument

The null method is one possible mode of operation for a measuring instrument. A **null instrument** uses the null method for measurement. In this method, the instrument exerts an influence on the measured system so as to oppose the effect of the measurand. The influence and the measurand are balanced until they are equal but opposite in value, yielding a null measurement. Typically, this is accomplished by some type of feedback operation that allows the comparison of the measurand against a known standard value. Key features of a null instrument include: an iterative balancing operation using some type of comparator, either a manual or automatic feedback used to achieve balance, and a null deflection at parity.

A null instrument offers certain intrinsic advantages over other modes of operation (e.g., see deflection instruments). By balancing the unknown input against a known standard input, the null method minimizes interaction between the measuring system and the measurand. As each input comes from a separate source, the significance of any measuring influence on the measurand by the measurement process is reduced. In effect, the measured system sees a very high input impedance, thereby minimizing loading errors. This is particularly effective when the measurand is a very small value. Hence, the null operation can achieve a high accuracy for small input values and a low loading error. In practice, the null instrument will not achieve perfect parity due to the usable resolution of the balance and detection methods, but this is limited only by the state of the art of the circuit or scheme being employed.

A disadvantage of null instruments is that an iterative balancing operation requires more time to execute than simply measuring sensor input. Thus, this method might not offer the fastest measurement possible when high-speed measurements are required. However, the user should weigh achievable accuracy against needed speed of measurement when considering operational modes. Further, the design of the comparator and balance loop can become involved such that highly accurate devices are generally not the lowest cost measuring alternative.

An equal arm balance scale is a good mechanical example of a manual balance-feedback null instrument, as shown in Figure 2.1. This scale compares the unknown weight of an object on one side against a set of standard or known weights. Known values of weight are iteratively added to one side to exert an influence to oppose the effect of the unknown weight on the opposite side. Until parity, a high or low value is noted by the indicator providing the feedback logic to the operator for adding or removing

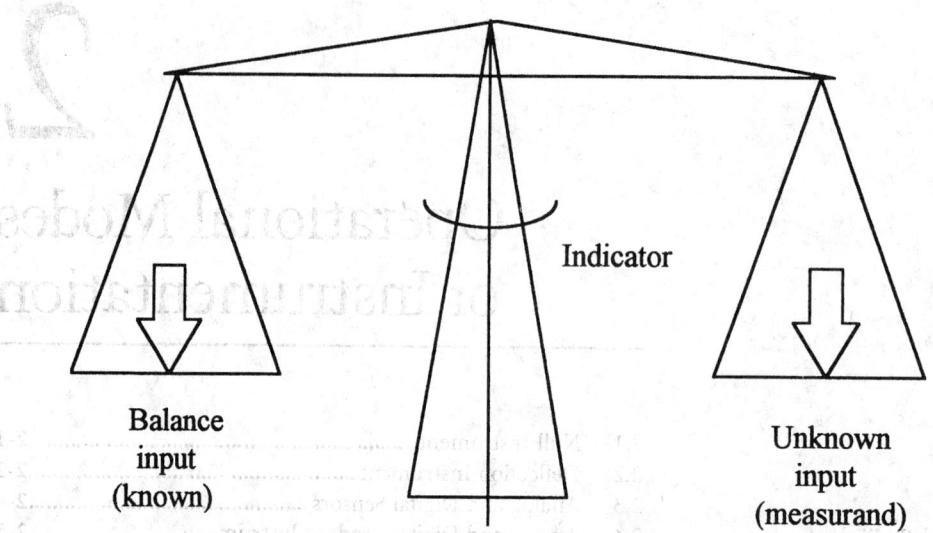

FIGURE 2.1   The measurand and the known quantities balance one another in a null instrument.

weights in a balancing iteration. At true parity, the scale indicator is null; that is, it indicates a zero deflection. Then, the unknown input or measurand is deduced to have a value equal to the balance input, the amount of known weights used to balance the scale. Factors influencing the overall measurement accuracy include the accuracy of the standard weights used and resolution of the output indicator, and the friction at the fulcrum. Null instruments exist for measurement of most variables. Other common examples include bridge circuits, often employed for highly accurate resistance measurements and found in load cells, temperature-compensated transducers, and voltage balancing potentiometers used for highly accurate low-voltage measurements.

Within the null instrument, the iteration and feedback mechanism is a loop that can be controlled either manually or automatically. Essential to the null instrument are two inputs: the measurand and the balance input. The null instrument includes a differential comparator, which compares and computes the difference between these two inputs. This is illustrated in Figure 2.2. A nonzero output from the comparator provides the error signal and drives the logic for the feedback correction. Repeated corrections provide for an iteration toward eventual parity between the inputs and results in the null condition where the measurand is exactly opposed by the balance input. At parity, the error signal is driven to zero by the opposed influence of the balance input and the indicated deflection is at null, thus lending the name to the method. It is the magnitude of the balance input that drives the output reading in terms of the measurand.

2.2   Deflection Instrument

The deflection method is one possible mode of operation for a measuring instrument. A **deflection instrument** uses the deflection method for measurement. A deflection instrument is influenced by the measurand so as to bring about a proportional response within the instrument. This response is an output reading that is a deflection or a deviation from the initial condition of the instrument. In a typical form, the measurand acts directly on a prime element or primary circuit so as to convert its information into a detectable form. The name is derived from a common form of instrument where there is a physical deflection of a prime element that is linked to an output scale, such as a pointer or other type of readout, which deflects to indicate the measured value. The magnitude of the deflection of the prime element brings about a deflection in the output scale that is designed to be proportional in magnitude to the value of the measurand.

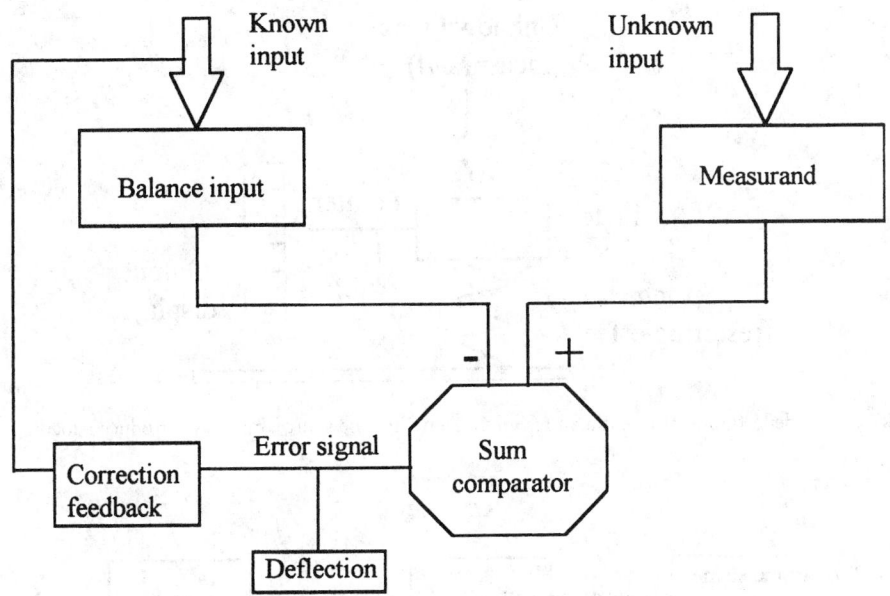

FIGURE 2.2   A null instrument requires input from two sources for comparison.

Deflection instruments are the most common of measuring instruments. The relationship between the measurand and the prime element or measuring circuit can be a direct one, with no balancing mechanism or comparator circuits used. The proportional response can be manipulated through signal conditioning methods between the prime element and the output scale so that the output reading is a direct indication of the measurand. Effective designs can achieve a high accuracy, yet sufficient accuracy for less demanding uses can be achieved at moderate costs.

An attractive feature of the deflection instrument is that it can be designed for either static or dynamic measurements or both. An advantage to deflection design for dynamic measurements is in the high dynamic response that can be achieved. A disadvantage of deflection instruments is that by deriving its energy from the measurand, the act of measurement will influence the measurand and change the value of the variable being measured. This change is called a loading error. Hence, the user must ensure that the resulting error is acceptable. This usually involves a careful look at the instrument input impedance for the intended measurement.

A spring scale is a good, simple example of a deflection instrument. As shown in Figure 2.3, the input weight or measurand acts on a plate-spring. The plate-spring serves as a prime element. The original position of the spring is influenced by the applied weight and responds with a translational displacement, a deflection x. The final value of this deflection is a position that is at equilibrium between the downward force of the weight, W, and the upward restoring force of the spring, kx. That is, the input force is balanced against the restoring force. A mechanical coupler is connected directly or by linkage to a pointer. The pointer position is mapped out on a corresponding scale that serves as the readout scale. For example, at equilibrium $W = kx$ or by measuring the deflection of the pointer the weight is deduced by $x = W/k$.

The flow diagram logic for a deflection instrument is rather linear, as shown if Figure 2.4. The input signal is sensed by the prime element or primary circuit and thereby deflected from its initial setting. The deflection signal is transmitted to signal conditioners that act to condition the signal into a desired form. Examples of signal conditioning are to multiply the deflection signal by some scaler magnitude, such as in amplification or filtering, or to transform the signal by some arithmetic function. The conditioned signal is then transferred to the output scale, which provides the indicated value corresponding to the measurand value.

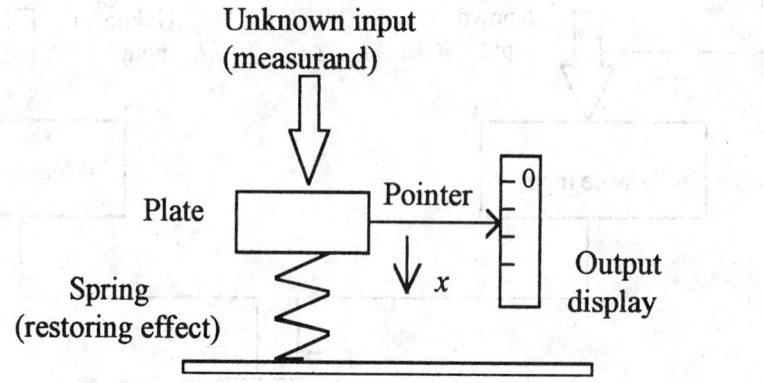

FIGURE 2.3   A deflection instrument requires input from only one source, but may introduce a loading error.

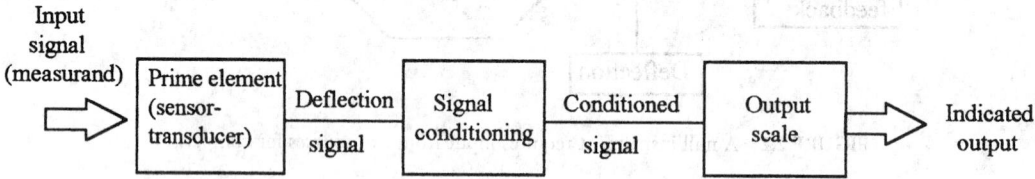

FIGURE 2.4   The logic flow chart for a deflection instrument is straightforward.

2.3   Analog and Digital Sensors

Analog sensors provide a signal that is continuous in both its magnitude and its temporal (time) or spatial (space) content. The defining word for analog is "continuous." If a sensor provides a continuous output signal that is directly proportional to the input signal, then it is analog.

Most physical variables, such as current, temperature, displacement, acceleration, speed, pressure, light intensity, and strain, tend to be continuous in nature and are readily measured by an analog sensor and represented by an analog signal. For example, the temperature within a room can take on any value within its range, will vary in a continuous manner in between any two points in the room, and may vary continuously with time at any position within the room. An analog sensor, such as a bulb thermometer or a thermocouple, will continuously respond to such temperature changes. Such a continuous signal is shown in Figure 2.5, where the signal magnitude is analogous to the measured variable (temperature) and the signal is continuous in both magnitude and time.

Digital sensors provide a signal that is a direct digital representation of the measurand. Digital sensors are basically binary ("on" or "off") devices. Essentially, a digital signal exists at only discrete values of

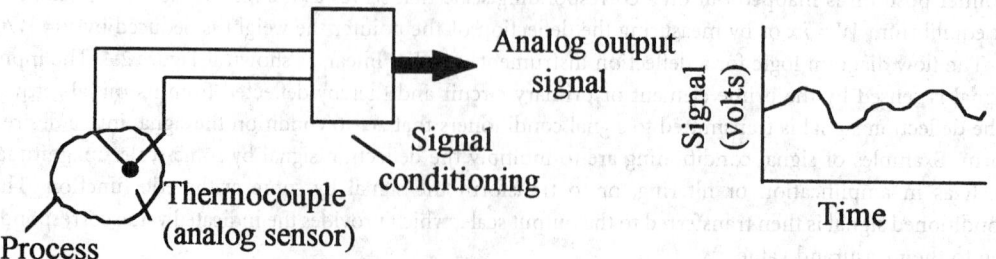

FIGURE 2.5   A thermocouple provides an analog signal for processing.

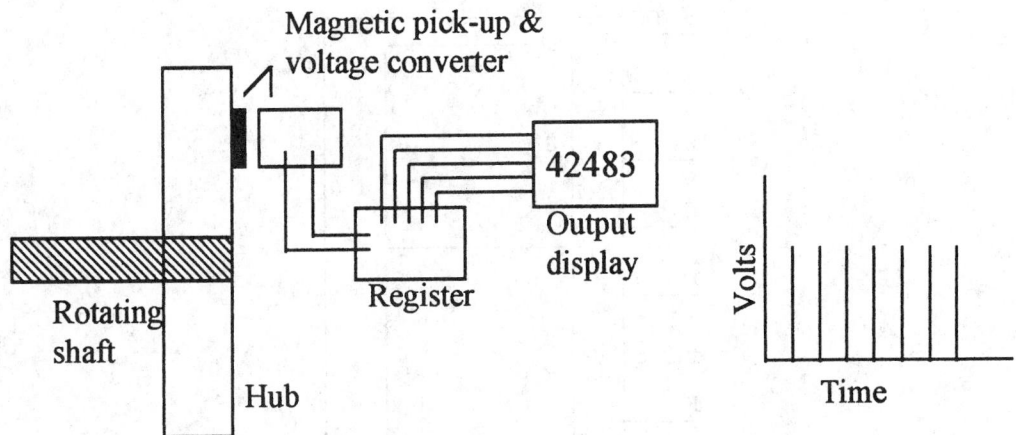

FIGURE 2.6 A rotating shaft with a revolution counter produces a digital signal.

time (or space). And within that discrete period, the signal can represent only a discrete number of magnitude values. A common variation is the **discrete sampled signal** representation, which represents a sensor output in a form that is discrete both in time or space and in magnitude.

Digital sensors use some variation of a binary numbering system to represent and transmit the signal information in digital form. A binary numbering system is a number system using the base 2. The simplest binary signal is a single bit that has only one of two possible values, a 1 or a 0. Bits are like electrical "on-off" switches and are used to convey logical and numerical information. With appropriate input, the value of the bit transmitted is reset corresponding to the behavior of the measured variable. A digital sensor that transmits information one bit at a time uses serial transmission. By combining bits or transmitting bits in groups, it is also possible to define logical commands or integer numbers beyond a 0 or 1. A digital sensor that transmits bits in groups uses parallel transmission. With any digital device, an M-bit signal can express 2^M different numbers. This also provides the limit for the different values that a digital device can discern. For example, a 2-bit device can express 2^2 or 4 different numbers, 00, 01, 10, and 11, corresponding to the values of 0, 1, 2, and 3, respectively. Thus, the resolution in a magnitude discerned by a digital sensor is inherently limited to 1 part in 2^M.

The concept of a digital sensor is illustrated by the revolution counter in Figure 2.6. Such devices are widely used to sense the revolutions per minute of a rotating shaft. In this example, the sensor is a magnetic pick-up/voltage converter that outputs a pulse with each pass of a magnetic stud mounted to a hub on the rotating shaft. The output from the pick-up normally is "off" but is momentarily turned "on" by the passing stud. This pulse is a voltage spike sent to a digital register whose value is increased by a single count with each spike. The register can send the information to an output device, such as the digital display shown. The output from the sensor can be viewed in terms of voltage spikes with time. The count rate is related to the rotational speed of the shaft. As seen, the signal is discrete in time. A single stud with pick-up will increase the count by one for each full rotation of the shaft. Fractions of a rotation can be resolved by increasing the number of studs on the hub. In this example, the continuous rotation of the shaft is analog but the revolution count is digital. The amplitude of the voltage spike is set to activate the counter and is not related to the shaft rotational speed.

2.4 Analog and Digital Readout Instruments

An **analog readout instrument** provides an output indication that is continuous and directly analogous to the behavior of the measurand. Typically, this might be the deflection of a pointer or an ink trace on a graduated scale, or the intensity of a light beam or a sound wave. This indicated deflection may be driven by changes in voltage or current, or by mechanical, magnetic, or optical means, or combinations

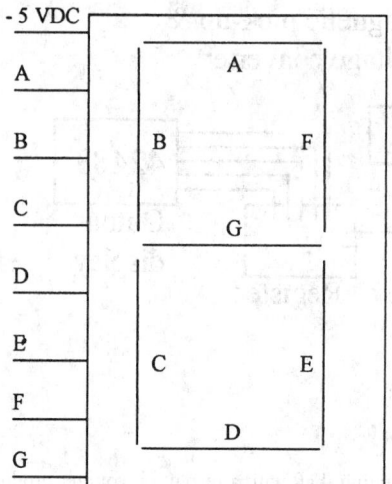

FIGURE 2.7 A seven-segment display chip can display any digit from 0 to 9.

of these. The resolution of an analog readout is defined by the smallest usable increment on its readout scale. The span of the readout is defined by the difference between the minimum and maximum values that it can indicate. Its range specifies the minimum and maximum values that it can indicate.

A **digital readout instrument** provides an output indication that is discrete. The value of the digital output is directly related to the value of the measurand. The digital readout is typically in the form of a numerical value that is either a fixed number or a number that is updated periodically. One means of displaying a digital number is the seven-segment digital display chip, shown in Figure 2.7, whose output can be updated by altering the grounding inputs A through G. The resolution of a digital readout is given by its least count, the equivalent amount of the smallest change resolved by the least significant digit in the readout. The span and range are defined as for analog instruments.

Many digital devices combine features of an analog sensor with a digital readout or, in general, convert an analog signal to a discrete signal, which is indicated through a digital output. In such situations, an analog to digital converter (ADC) is required. This hybrid device has its analog side specified in terms of its full-scale analog range, E_{FSR}, which defines the analog voltage span over which the device will operate. The digital side is specified in terms of the bit size of its register. An M-bit device will output an M-bit binary number. The resolution of such a device is given by $E_{FSR}/2^M$.

2.5 Input Impedance

In the ideal sense, the very act of measurement should not alter the value of the measured signal. Any such alteration is a **loading error**. Loading errors can occur at any junction along the signal chain but can be minimized by impedance matching of the source with the measuring instrument. The measuring instrument input impedance controls the energy that is drawn from the source, or measured system, by a measuring instrument. The power loss through the measuring instrument is estimated by $P = E^2/Z_2$ where Z_2 is the input impedance of the measuring instrument, and E is the source voltage potential being measured. Thus, to minimize the power loss, the input impedance should be large.

This same logic holds for the two instruments in a signal chain as the subsequent instrument draws energy from the previous instrument in the chain. As a general example, consider the situation in Figure 2.8 in which the output signal from one instrument provides the input signal to a subsequent device in a signal chain. The open circuit potential, E_1, is present at the output terminal of source device 1 having output impedance, Z_1. Device 2 has an **input impedance** Z_2 at its input terminals. Connecting

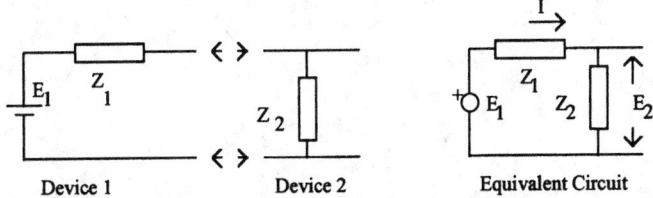

FIGURE 2.8 An equivalent circuit is formed by applying a measuring instrument to the output terminals of an instrument.

the output terminals of device 1 to the input terminals of device 2 creates the equivalent circuit also shown in Figure 2.7. The potential actually sensed by device 2 will be

$$E_2 = E_1 \frac{1}{1 + Z_1/Z_2}$$

The difference between the actual potential E_1 at the output terminals of device 1 and the measured potential E_2 is a **loading error** brought on by the input impedance of measuring device 2. It is clear that a high input impedance Z_2 relative to Z_1 minimizes this error. A general rule is for the input impedance to be at least 100 times the source impedance to reduce the loading error to 1%.

In general, null instruments and null methods will minimize loading errors. They provide the equivalent of a very high input impedance to the measurement, minimizing energy drain from the measured system. Deflection instruments and deflection measuring techniques will derive energy from the process being measured and therefore require attention to proper selection of input impedance.

Defining Terms

Analog sensor: Sensors that output a signal that is continuous in both magnitude and time (or space).
Deflection instrument: A measuring device whose output deflects proportional to the magnitude of the measurand.
Digital sensor: Sensors that output a signal that is discrete (noncontinuous) in time and/or magnitude.
Input impedance: The impedance measured across the input terminals of a device.
Loading error: That difference between the measurand and the measuring system output attributed to the act of measuring the measurand.
Measurand: A physical quantity, property, or condition being measured. Often, it is referred to as a measured value.
Null instrument: A measuring device that balances the measurand against a known value, thus achieving a null condition. A null instrument minimizes measurement loading errors.
Readout: This is the display of a measuring system.
Resolution: This is the least count or smallest detectable change in measurand capable.
Sensor: The portion of a measurement system that responds directly to the physical variable being measured.

Further Information

E. O. Doebelin, *Measurement Systems, 4th ed.*, New York: McGraw-Hill, 1990.
R. S. Figliola and D. E. Beasley, *Theory and Design for Mechanical Measurements, 2nd ed.*, New York: Wiley, 1995.
D. Wobschall, *Circuit Design for Electronic Instrumentation: Analog and Digital Devices from Sensor to Display, 2nd ed.*, New York: McGraw-Hill, 1987.

$$E = E_x \frac{1}{1 + \frac{Z_1}{Z_2}}$$

3

Static and Dynamic Characteristics of Instrumentation

Peter H. Sydenham
University of South Australia

Before we can begin to develop an understanding of the static and time changing characteristics of measurements, it is necessary to build a framework for understanding the process involved, setting down the main words used to describe concepts as we progress.

Measurement is the process by which relevant information about a system of interest is interpreted using the human thinking ability to define what is believed to be the new knowledge gained. This information may be obtained for purposes of controlling the behavior of the system (as in engineering applications) or for learning more about it (as in scientific investigations).

The basic entity needed to develop the knowledge is called *data*, and it is obtained with physical assemblies known as sensors that are used to observe or sense system variables. The terms *information* and *knowledge* tend to be used interchangeably to describe the entity resulting after data from one or more sensors have been processed to give more meaningful understanding. The individual variables being sensed are called *measurands*.

The most obvious way to make observations is to use the human senses of seeing, feeling, and hearing. This is often quite adequate or may be the only means possible. In many cases, however, sensors are used that have been devised by man to enhance or replace our natural sensors. The number and variety of sensors is very large indeed. Examples of man-made sensors are those used to measure temperature, pressure, or length. The process of sensing is often called *transduction*, being made with transducers. These man-made sensor assemblies, when coupled with the means to process the data into knowledge, are generally known as (measuring) instrumentation.

The degree of perfection of a measurement can only be determined if the goal of the measurement can be defined without error. Furthermore, instrumentation cannot be made to operate perfectly. Because of these two reasons alone, measuring instrumentation cannot give ideal sensing performance and it must be selected to suit the allowable error in a given situation.

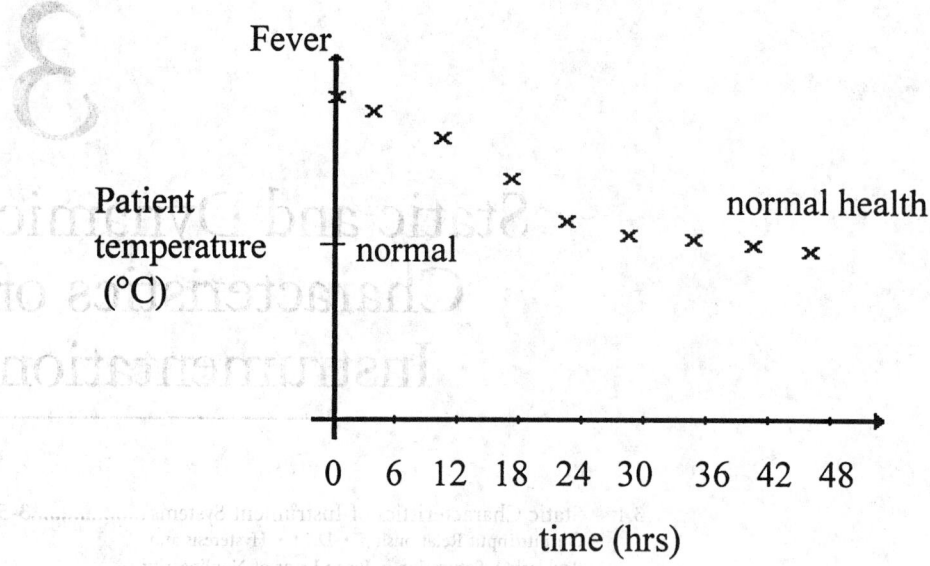

FIGURE 3.1   A patient's temperature chart shows changes taking place over time.

Measurement is a process of mapping actually occurring variables into equivalent values. Deviations from perfect measurement mappings are called *errors*: what we get as the result of measurement is not exactly what is being measured. A certain amount of error is allowable provided it is below the level of uncertainty we can accept in a given situation. As an example, consider two different needs to measure the measurand, time. The uncertainty to which we must measure it for daily purposes of attending a meeting is around a 1 min in 24 h. In orbiting satellite control, the time uncertainty needed must be as small as milliseconds in years. Instrumentation used for the former case costs a few dollars and is the watch we wear; the latter instrumentation costs thousands of dollars and is the size of a suitcase.

We often record measurand values as though they are constant entities, but they usually change in value as time passes. These "dynamic" variations will occur either as changes in the measurand itself or where the measuring instrumentation takes time to follow the changes in the measurand — in which case it may introduce unacceptable error.

For example, when a fever thermometer is used to measure a person's body temperature, we are looking to see if the person is at the normally expected value and, if it is not, to then look for changes over time as an indicator of his or her health. Figure 3.1 shows a chart of a patient's temperature. Obviously, if the thermometer gives errors in its use, wrong conclusions could be drawn. It could be in error due to incorrect calibration of the thermometer or because no allowance for the dynamic response of the thermometer itself was made.

Instrumentation, therefore, will only give adequately correct information if we understand the static and dynamic characteristics of both the measurand and the instrumentation. This, in turn, allows us to then decide if the error arising is small enough to accept.

As an example, consider the electronic signal amplifier in a sound system. It will be commonly quoted as having an amplification constant after feedback if applied to the basic amplifier of, say, 10. The actual amplification value is dependent on the frequency of the input signal, usually falling off as the frequency increases. The frequency response of the basic amplifier, before it is configured with feedback that markedly alters the response and lowers the amplification to get a stable operation, is shown as a graph of amplification gain versus input frequency. An example of the open loop gain of the basic amplifier is given in Figure 3.2. This lack of uniform gain over the frequency range results in error — the sound output is not a true enough representation of the input.

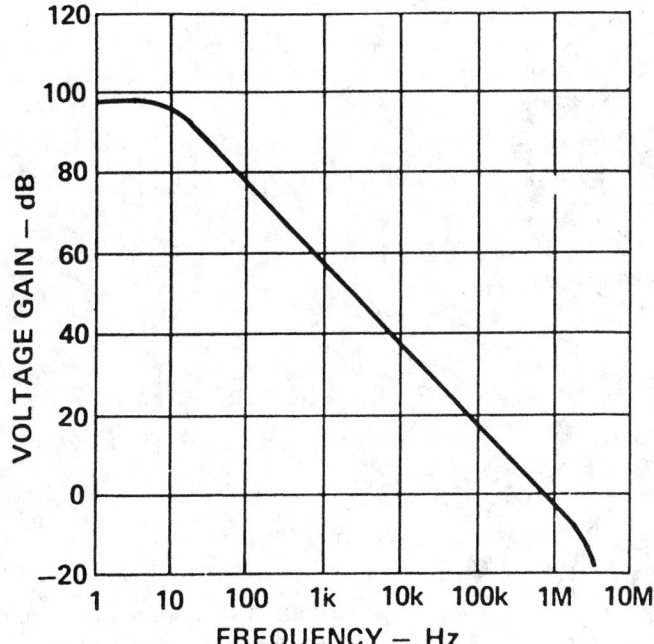

FIGURE 3.2 This graph shows how the amplification of an amplifier changes with input frequency.

Before we can delve more deeply into the static and dynamic characteristics of instrumentation, it is necessary to understand the difference in meaning between several basic terms used to describe the results of a measurement activity.

The correct terms to use are set down in documents called *standards*. Several standardized metrology terminologies exist but they are not consistent. It will be found that books on instrumentation and statements of instrument performance often use terms in different ways. Users of measurement information need to be constantly diligent in making sure that the statements made are interpreted correctly.

The three companion concepts about a measurement that need to be well understood are its *discrimination*, its *precision*, and its *accuracy*. These are too often used interchangeably — which is quite wrong to do because they cover quite different concepts, as will now be explained.

When making a measurement, the smallest increment that can be discerned is called the *discrimination*. (Although now officially declared as wrong to use, the term *resolution* still finds its way into books and reports as meaning discrimination.) The discrimination of a measurement is important to know because it tells if the sensing process is able to sense fine enough changes of the measurand.

Even if the discrimination is satisfactory, the value obtained from a repeated measurement will rarely give exactly the same value each time the same measurement is made under conditions of constant value of measurand. This is because errors arise in real systems. The spread of values obtained indicates the precision of the set of the measurements. The word *precision* is not a word describing a quality of the measurement and is incorrectly used as such. Two terms that should be used here are: *repeatability*, which describes the variation for a set of measurements made in a very short period; and the *reproducibility*, which is the same concept but now used for measurements made over a long period. As these terms describe the outcome of a set of values, there is need to be able to quote a single value to describe the overall result of the set. This is done using statistical methods that provide for calculation of the "mean value" of the set and the associated spread of values, called its *variance*.

The *accuracy* of a measurement is covered in more depth elsewhere so only an introduction to it is required here. Accuracy is the closeness of a measurement to the value defined to be the true value. This

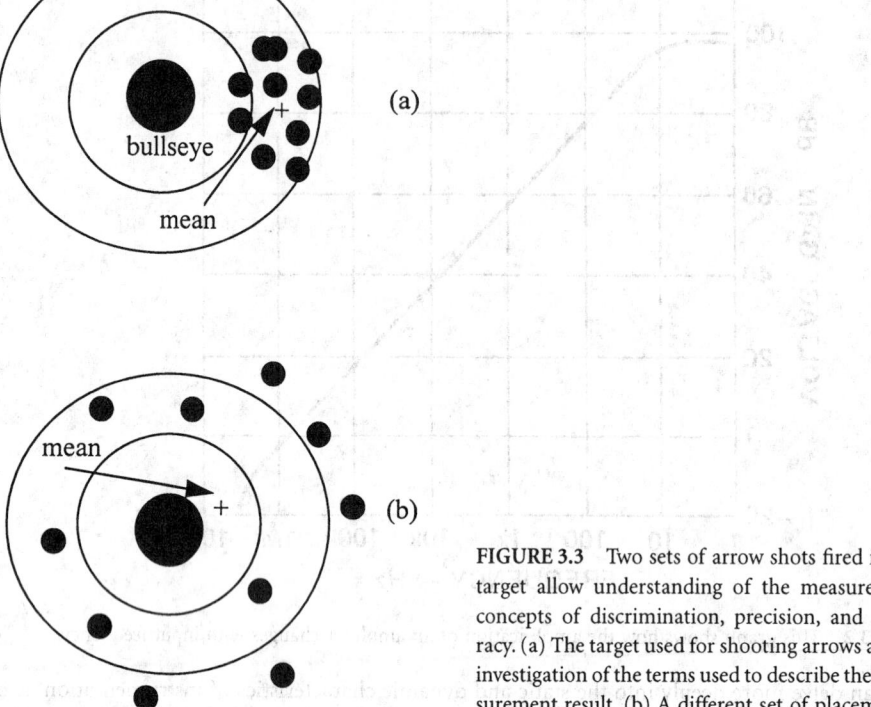

FIGURE 3.3 Two sets of arrow shots fired into a target allow understanding of the measurement concepts of discrimination, precision, and accuracy. (a) The target used for shooting arrows allows investigation of the terms used to describe the measurement result. (b) A different set of placements.

concept will become clearer when the following illustrative example is studied for it brings together the three terms into a single perspective of a typical measurement.

Consider then the situation of scoring an archer shooting arrows into a target as shown in Figure 3.3(a). The target has a central point — the bulls-eye. The objective for a perfect result is to get all arrows into the bulls-eye. The rings around the bulls-eye allow us to set up numeric measures of less-perfect shooting performance.

Discrimination is the distance at which we can just distinguish (i.e., discriminate) the placement of one arrow from another when they are very close. For an arrow, it is the thickness of the hole that decides the discrimination. Two close-by positions of the two arrows in Figure 3.3(a) cannot be separated easily. Use of thinner arrows would allow finer detail to be decided.

Repeatability is determined by measuring the spread of values of a set of arrows fired into the target over a short period. The smaller the spread, the more precise is the shooter. The shooter in Figure 3.3(a) is more precise than the shooter in Figure 3.3(b).

If the shooter returned to shoot each day over a long period, the results may not be the same each time for a shoot made over a short period. The mean and variance of the values are now called the *reproducibility* of the archer's performance.

Accuracy remains to be explained. This number describes how well the mean (the average) value of the shots sits with respect to the bulls-eye position. The set in Figure 3.3(b) is more accurate than the set in Figure 3.3(a) because the mean is nearer the bulls-eye (but less precise!).

At first sight, it might seem that the three concepts of discrimination, precision, and accuracy have a strict relationship in that a better measurement is always that with all three aspects made as high as is affordable. This is not so. They need to be set up to suit the needs of the application.

We are now in a position to explore the commonly met terms used to describe aspects of the static and the dynamic performance of measuring instrumentation.

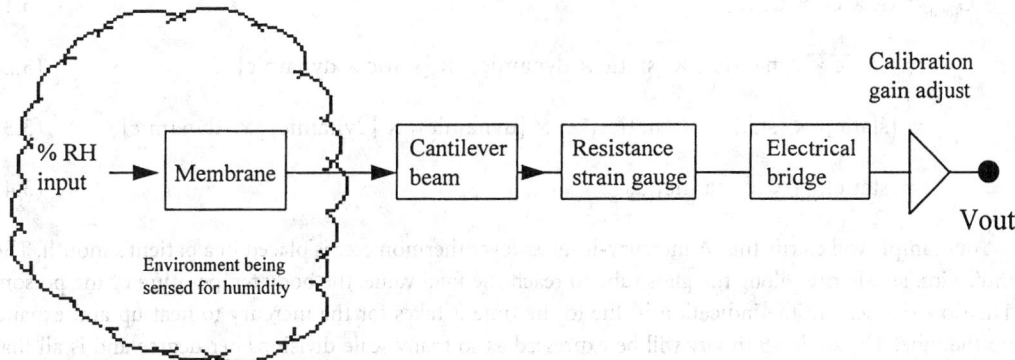

FIGURE 3.4 Instruments are formed from a connection of blocks. Each block can be represented by a conceptual and mathematical model. This example is of one type of humidity sensor.

3.1 Static Characteristics of Instrument Systems

Output/Input Relationship

Instrument systems are usually built up from a serial linkage of distinguishable building blocks. The actual physical assembly may not appear to be so but it can be broken down into a representative diagram of connected blocks. Figure 3.4 shows the block diagram representation of a humidity sensor. The sensor is activated by an input physical parameter and provides an output signal to the next block that processes the signal into a more appropriate state.

A key generic entity is, therefore, the relationship between the input and output of the block. As was pointed out earlier, all signals have a time characteristic, so we must consider the behavior of a block in terms of both the static and dynamic states.

The behavior of the static regime alone and the combined static and dynamic regime can be found through use of an appropriate mathematical model of each block. The mathematical description of system responses is easy to set up and use if the elements all act as linear systems and where addition of signals can be carried out in a linear additive manner. If nonlinearity exists in elements, then it becomes considerably more difficult — perhaps even quite impractical — to provide an easy to follow mathematical explanation. Fortunately, general description of instrument systems responses can be usually be adequately covered using the linear treatment.

The output/input ratio of the whole cascaded chain of blocks 1, 2, 3, etc. is given as:

$$[\text{output/input}]_{\text{total}} = [\text{output/input}]_1 \times [\text{output/input}]_2 \times [\text{output/input}]_3 \ldots$$

The output/input ratio of a block that includes both the static and dynamic characteristics is called the *transfer function* and is given the symbol G.

The equation for G can be written as two parts multiplied together. One expresses the static behavior of the block, that is, the value it has after all transient (time varying) effects have settled to their final state. The other part tells us how that value responds when the block is in its dynamic state. The static part is known as the *transfer characteristic* and is often all that is needed to be known for block description.

The static and dynamic response of the cascade of blocks is simply the multiplication of all individual blocks. As each block has its own part for the static and dynamic behavior, the cascade equations can be rearranged to separate the static from the dynamic parts and then by multiplying the static set and the dynamic set we get the overall response in the static and dynamic states. This is shown by the sequence of Equations 3.1 to 3.4.

$$G_{total} = G_1 \times G_2 \times G_3 \dots \tag{3.1}$$

$$= [\text{static} \times \text{dynamic}]_1 \times [\text{static} \times \text{dynamic}]_2 \times [\text{static} \times \text{dynamic}]_3 \dots \tag{3.2}$$

$$= [\text{static}]_1 \times [\text{static}]_2 \times [\text{static}]_3 \dots \times [\text{dynamic}]_1 \times [\text{dynamic}]_2 \times [\text{dynamic}]_3 \dots \tag{3.3}$$

$$= [\text{static}]_{total} \times [\text{dynamic}]_{total} \tag{3.4}$$

An example will clarify this. A mercury-in-glass fever thermometer is placed in a patient's mouth. The indication slowly rises along the glass tube to reach the final value, the body temperature of the person. The slow rise seen in the indication is due to the time it takes for the mercury to heat up and expand up the tube. The static *sensitivity* will be expressed as so many scale divisions per degree and is all that is of interest in this application. The dynamic characteristic will be a time varying function that settles to unity after the transient effects have settled. This is merely an annoyance in this application but has to be allowed by waiting long enough before taking a reading. The wrong value will be viewed if taken before the transient has settled.

At this stage, we will now consider only the nature of the static characteristics of a chain; dynamic response is examined later.

If a sensor is the first stage of the chain, the static value of the gain for that stage is called the *sensitivity*. Where a sensor is not at the input, it is called the *amplification factor* or *gain*. It can take a value less than unity where it is then called the *attenuation*.

Sometimes, the instantaneous value of the signal is rapidly changing, yet the measurement aspect part is static. This arises when using ac signals in some forms of instrumentation where the amplitude of the waveform, not its frequency, is of interest. Here, the static value is referred to as its *steady state* transfer characteristic.

Sensitivity may be found from a plot of the input and output signals, wherein it is the slope of the graph. Such a graph, see Figure 3.5, tells much about the static behavior of the block.

The intercept value on the *y*-axis is the *offset* value being the output when the input is set to zero. Offset is not usually a desired situation and is seen as an error quantity. Where it is deliberately set up, it is called the *bias*.

The range on the *x*-axis, from zero to a safe maximum for use, is called the *range* or *span* and is often expressed as the zone between the 0% and 100% points. The ratio of the span that the output will cover

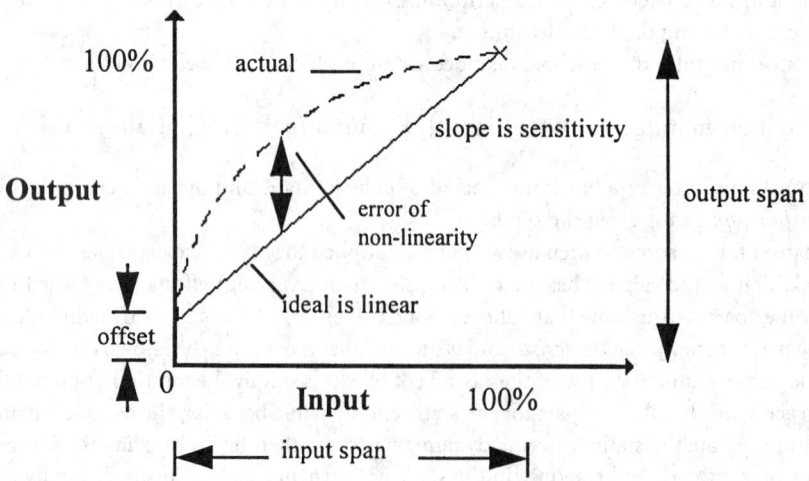

FIGURE 3.5 The graph relating input to output variables for an instrument block shows several distinctive static performance characteristics.

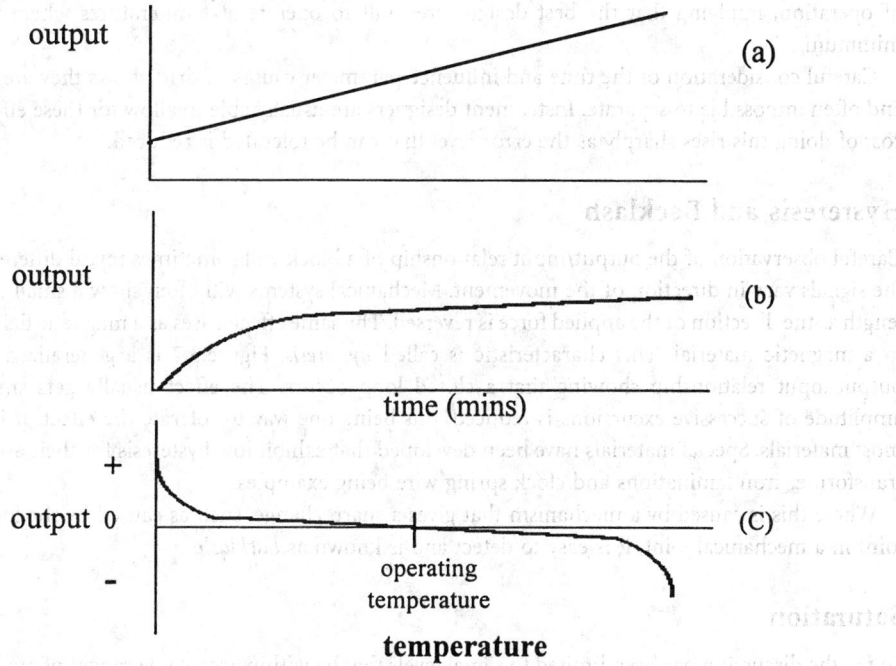

(a)

(b)

time (mins)

+

output 0

(C)

−

operating
temperature

temperature

FIGURE 3.6 Drift in the performance of an instrument takes many forms: (*a*) drift over time for a spring balance; (*b*) how an electronic amplifier might settle over time to a final value after power is supplied; (*c*) drift, due to temperature, of an electronic amplifier varies with the actual temperature of operation.

for the related input range is known as the *dynamic range*. This can be a confusing term because it does not describe dynamic time behavior. It is particularly useful when describing the capability of such instruments as flow rate sensors — a simple orifice plate type may only be able to handle dynamic ranges of 3 to 4, whereas the laser Doppler method covers as much as 10^7 variation.

Drift

It is now necessary to consider a major problem of instrument performance called *instrument drift*. This is caused by variations taking place in the parts of the instrumentation over time. Prime sources occur as chemical structural changes and changing mechanical stresses. Drift is a complex phenomenon for which the observed effects are that the sensitivity and offset values vary. It also can alter the accuracy of the instrument differently at the various amplitudes of the signal present.

Detailed description of drift is not at all easy but it is possible to work satisfactorily with simplified values that give the average of a set of observations, this usually being quoted in a conservative manner. The first graph (*a*) in Figure 3.6 shows typical steady drift of a measuring spring component of a weighing balance. Figure 3.6(*b*) shows how an electronic amplifier might settle down after being turned on.

Drift is also caused by variations in environmental parameters such as temperature, pressure, and humidity that operate on the components. These are known as *influence parameters*. An example is the change of the resistance of an electrical resistor, this resistor forming the critical part of an electronic amplifier that sets its gain as its operating temperature changes.

Unfortunately, the observed effects of influence parameter induced drift often are the same as for time varying drift. Appropriate testing of blocks such as electronic amplifiers does allow the two to be separated to some extent. For example, altering only the temperature of the amplifier over a short period will quickly show its temperature dependence.

Drift due to influence parameters is graphed in much the same way as for time drift. Figure 3.6(*c*) shows the drift of an amplifier as temperature varies. Note that it depends significantly on the temperature of operation, implying that the best designs are built to operate at temperatures where the effect is minimum.

Careful consideration of the time and influence parameter causes of drift shows they are interrelated and often impossible to separate. Instrument designers are usually able to allow for these effects, but the cost of doing this rises sharply as the error level that can be tolerated is reduced.

Hysteresis and Backlash

Careful observation of the output/input relationship of a block will sometimes reveal different results as the signals vary in direction of the movement. Mechanical systems will often show a small difference in length as the direction of the applied force is reversed. The same effect arises as a magnetic field is reversed in a magnetic material. This characteristic is called *hysteresis*. Figure 3.7 is a generalized plot of the output/input relationship showing that a closed loop occurs. The effect usually gets smaller as the amplitude of successive excursions is reduced, this being one way to tolerate the effect. It is present in most materials. Special materials have been developed that exhibit low hysteresis for their application — transformer iron laminations and clock spring wire being examples.

Where this is caused by a mechanism that gives a sharp change, such as caused by the looseness of a joint in a mechanical joint, it is easy to detect and is known as *backlash*.

Saturation

So far, the discussion has been limited to signal levels that lie within acceptable ranges of amplitude. Real system blocks will sometimes have input signal levels that are larger than allowed. Here, the dominant errors that arise — *saturation* and *crossover distortion* — are investigated.

As mentioned above, the information bearing property of the signal can be carried as the instantaneous value of the signal or be carried as some characteristic of a rapidly varying ac signal. If the signal form is not amplified faithfully, the output will not have the same linearity and characteristics.

The gain of a block will usually fall off with increasing size of signal amplitude. A varying amplitude input signal, such as the steadily rising linear signal shown in Figure 3.8, will be amplified differently according to the gain/amplitude curve of the block. In uncompensated electronic amplifiers, the larger amplitudes are usually less amplified than at the median points.

At very low levels of input signal, two unwanted effects may arise. The first is that small signals are often amplified more than at the median levels. The second error characteristic arises in electronic amplifiers because the semiconductor elements possess a dead-zone in which no output occurs until a small threshold is exceeded. This effect causes crossover distortion in amplifiers.

If the signal is an ac waveform, see Figure 3.9, then the different levels of a cycle of the signal may not all be amplified equally. Figure 3.9(*a*) shows what occurs because the basic electronic amplifying elements are only able to amplify one polarity of signal. The signal is said to be *rectified*. Figure 3.9(*b*) shows the effect when the signal is too large and the top is not amplified. This is called *saturation* or *clipping*. (As with many physical effects, this effect is sometimes deliberately invoked in circuitry, an example being where it is used as a simple means to convert sine-waveform signals into a square waveform.) Crossover distortion is evident in Figure 3.9(*c*) as the signal passes from negative to positive polarity.

Where input signals are small, such as in sensitive sensor use, the form of analysis called *small signal* behavior is needed to reveal distortions. If the signals are comparatively large, as for digital signal considerations, a *large signal* analysis is used. Design difficulties arise when signals cover a wide dynamic range because it is not easy to allow for all of the various effects in a single design.

Bias

Sometimes, the electronic signal processing situation calls for the input signal to be processed at a higher average voltage or current than arises normally. Here a dc value is added to the input signal to raise the level to a higher state as shown in Figure 3.10. A need for this is met where only one polarity of signal

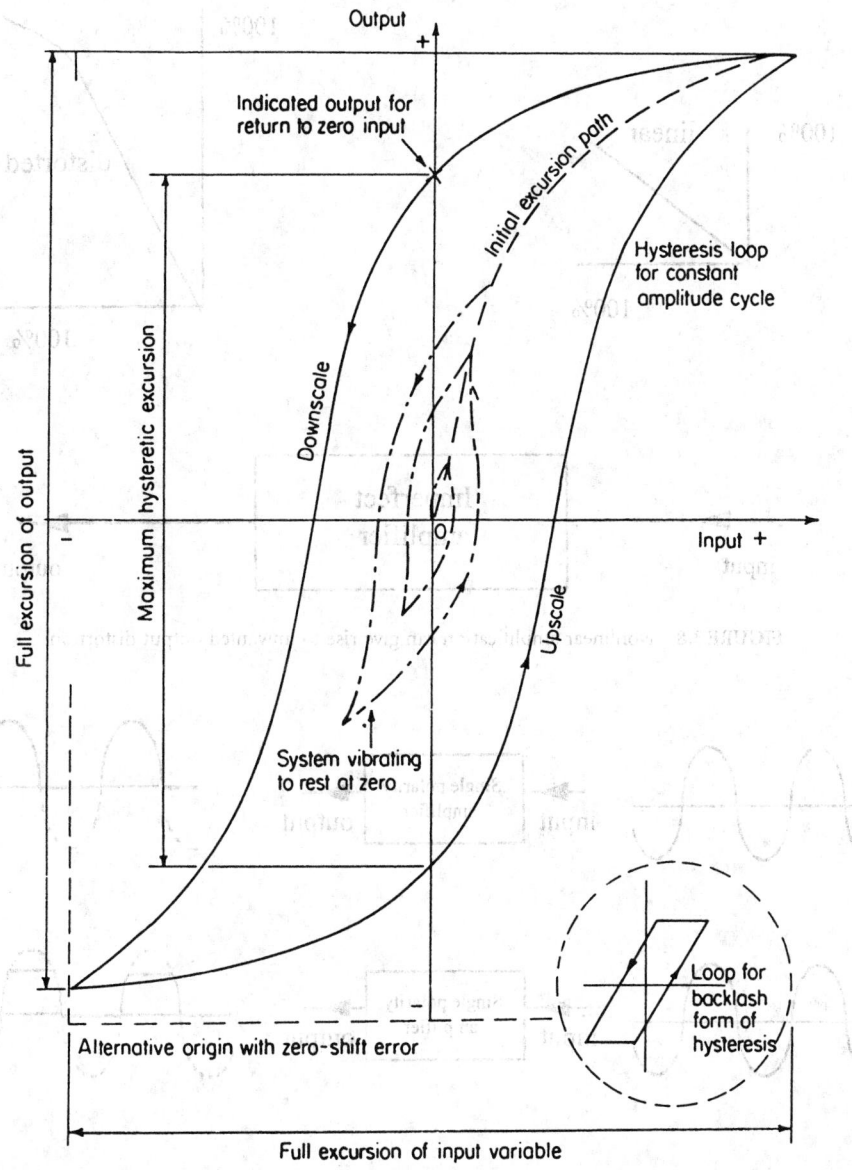

FIGURE 3.7 Generalized graph of output/input relationship where hysteresis is present. (From P. H. Sydenham, *Handbook of Measurement Science*, Vol. 2, Chichester, U.K., John Wiley & Sons, 1983. With permission.)

can be amplified by a single semiconductor element. Raising the level of all of the waveform equally takes all parts into the reasonably linear zone of an amplifier, allowing more faithful replication. If bias were not used here, then the lower half cycle would not be amplified, resulting in only the top half appearing in the output.

Error of Nonlinearity

Ideally, it is often desired that a strictly linear relationship exists between input and output signals in amplifiers. Practical units, however, will always have some degree of nonconformity, which is called the *nonlinearity*. If an instrument block has constant gain for all input signal levels, then the relationship graphing the input against the output will be a straight line; the relationship is then said to be linear.

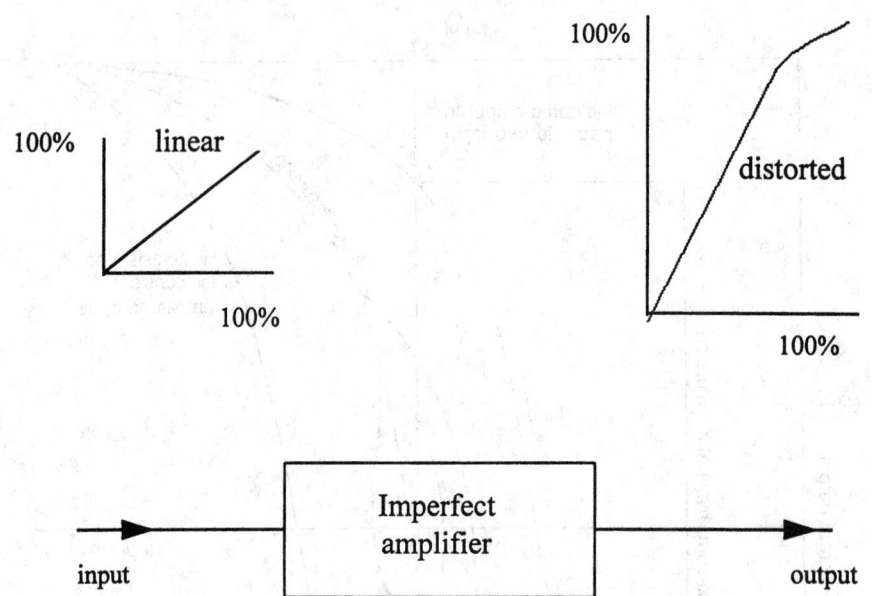

FIGURE 3.8   Nonlinear amplification can give rise to unwanted output distortion.

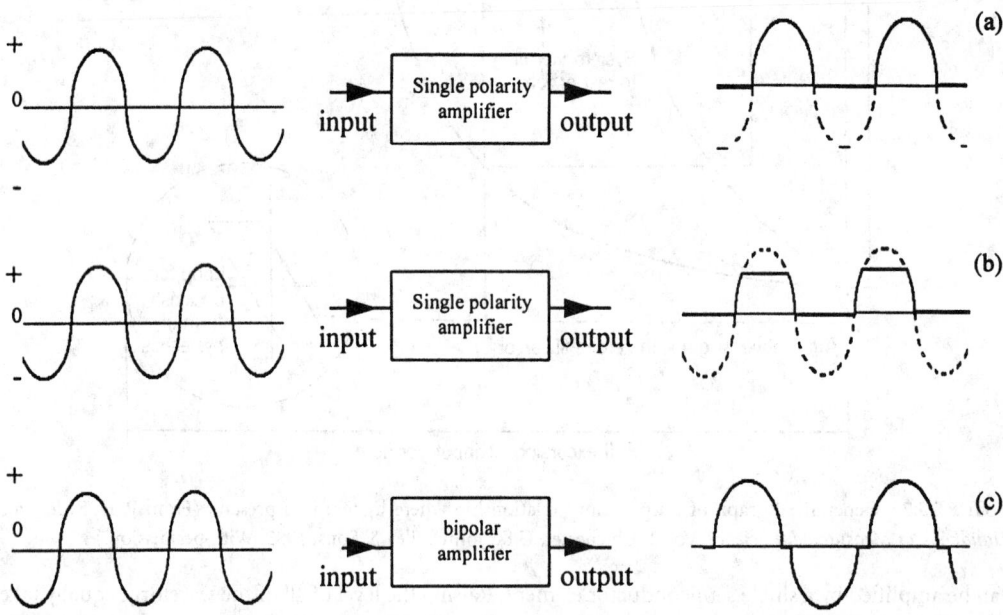

FIGURE 3.9   Blocks can incorrectly alter the shape of waveforms if saturation and crossover effects are not controlled: (*a*) rectification; (*b*) saturation; and (*c*) crossover distortion.

Linearity is the general term used to describe how close the actual response is compared with that ideal line. The correct way to describe the error here is as the *error of nonlinearity*. Note, however, that not all responses are required to be linear; another common one follows a logarithmic relationship.

    Detailed description of this error is not easy for that would need a statement of the error values at all points of the plot. Practice has found that a shorthand statement can be made by quoting the maximum departure from the ideal as a ratio formed with the 100% value.

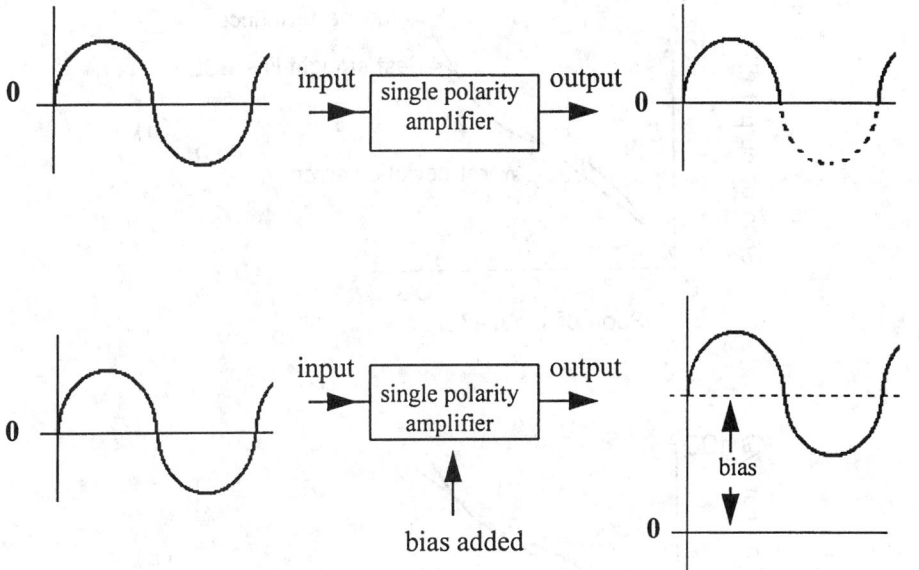

FIGURE 3.10 Bias is where a signal has all of its value raised by an equal amount. Shown here is an ac input waveform biased to be all of positive polarity.

Difficulties arise in expressing error of nonlinearity for there exist many ways to express this error. Figure 3.11 shows the four cases that usually arise. The difference arises in the way in which the ideal (called the "best fit") straight line can be set up. Figure 3.11(*a*) shows the line positioned by the usually calculated statistical averaging method of least squares fit; other forms of line fitting calculation are also used. This will yield the smallest magnitude of error calculation for the various kinds of line fitting but may not be appropriate for how the stage under assessment is used. Other, possibly more reasonable, options exist. Figure 3.11(*b*) constrains the best fit line to pass through the zero point. Figure 3.11(*c*) places the line between the expected 0% and the 100% points. There is still one more option, that where the theoretical line is not necessarily one of the above, yet is the ideal placement, Figure 3.11(*d*).

In practice then, instrument systems linearity can be expressed in several ways. Good certification practice requires that the method used to ascertain the error is stated along with the numerical result, but this is often not done. Note also that the error is the worst case and that part of the response may be much more linear.

The description of instrument performance is not a simple task. To accomplish this fully would require very detailed statements recording the performance at each and every point. That is often too cumbersome, so the instrument industry has developed many short-form statements that provide an adequate guide to the performance. This guide will be seen to be generally a conservative statement.

Many other descriptors exist for the static regime of an instrument. The reader is referred to the many standards documents that exist on instrument terminology; for example, see Reference [3].

3.2 Dynamic Characteristics of Instrument Systems

Dealing with Dynamic States

Measurement outcomes are rarely static over time. They will possess a dynamic component that must be understood for correct interpretation of the results. For example, a trace made on an ink pen chart recorder will be subject to the speed at which the pen can follow the input signal changes.

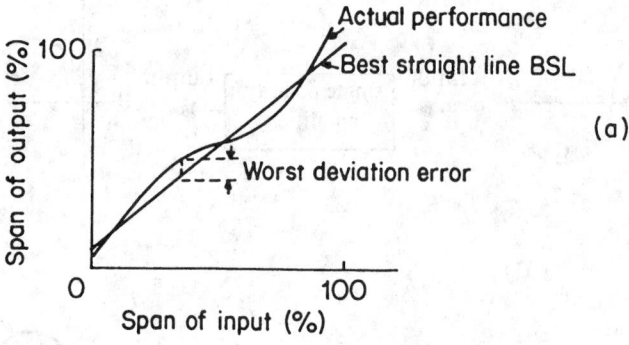

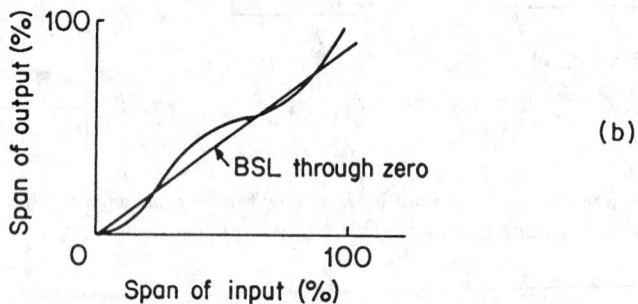

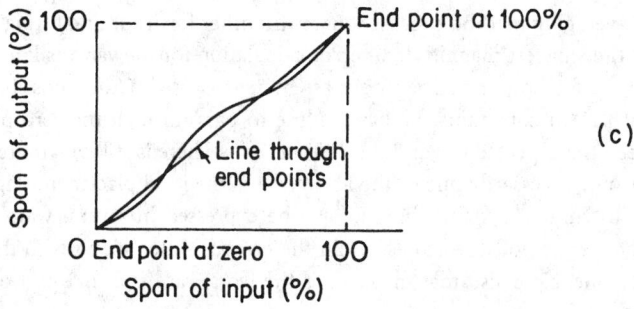

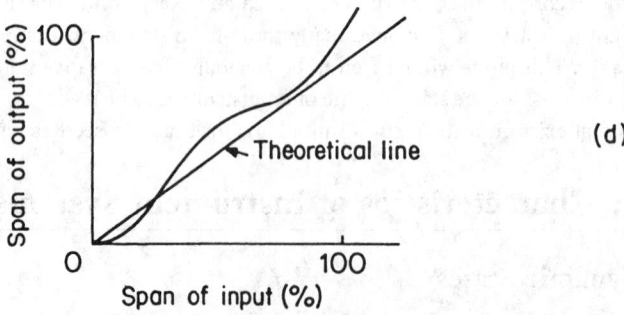

FIGURE 3.11   Error of nonlinearity can be expressed in four different ways: (*a*) best fit line (based on selected method used to decide this); (*b*) best fit line through zero; (*c*) line joining 0% and 100% points; and (*d*) theoretical line. (From P. H. Sydenham, *Handbook of Measurement Science,* Vol. 2, Chichester, U.K., John Wiley & Sons, 1983. With permission.)

To properly appreciate instrumentation design and its use, it is now necessary to develop insight into the most commonly encountered types of dynamic response and to develop the mathematical modeling basis that allows us to make concise statements about responses.

If the transfer relationship for a block follows linear laws of performance, then a generic mathematical method of dynamic description can be used. Unfortunately, simple mathematical methods have not been found that can describe all types of instrument responses in a simplistic and uniform manner. If the behavior is nonlinear, then description with mathematical models becomes very difficult and might be impracticable. The behavior of nonlinear systems can, however, be studied as segments of linear behavior joined end to end. Here, digital computers are effectively used to model systems of any kind provided the user is prepared to spend time setting up an adequate model.

Now the mathematics used to describe linear dynamic systems can be introduced. This gives valuable insight into the expected behavior of instrumentation, and it is usually found that the response can be approximated as linear.

The modeled response at the output of a block G_{result} is obtained by multiplying the mathematical expression for the input signal G_{input} by the transfer function of the block under investigation $G_{response}$, as shown in Equation 3.5.

$$G_{result} = G_{input} \times G_{response} \tag{3.5}$$

To proceed, one needs to understand commonly encountered input functions and the various types of block characteristics. We begin with the former set: the so-called *forcing functions*.

Forcing Functions

Let us first develop an understanding of the various types of input signal used to perform tests. The most commonly used signals are shown in Figure 3.12. These each possess different valuable test features. For example, the sine-wave is the basis of analysis of all complex wave-shapes because they can be formed as a combination of various sine-waves, each having individual responses that add to give all other wave-shapes. The step function has intuitively obvious uses because input transients of this kind are commonly encountered. The ramp test function is used to present a more realistic input for those systems where it is not possible to obtain instantaneous step input changes, such as attempting to move a large mass by a limited size of force. Forcing functions are also chosen because they can be easily described by a simple mathematical expression, thus making mathematical analysis relatively straightforward.

Characteristic Equation Development

The behavior of a block that exhibits linear behavior is mathematically represented in the general form of expression given as Equation 3.6.

$$........ a_2 d^2 y / dt^2 + a_1 dy / dt + a_0 y = x(t) \tag{3.6}$$

Here, the coefficients a_2, a_1, and a_0 are constants dependent on the particular block of interest. The left-hand side of the equation is known as the *characteristic equation*. It is specific to the internal properties of the block and is not altered by the way the block is used.

The specific combination of forcing function input and block characteristic equation collectively decides the combined output response. Connections around the block, such as feedback from the output to the input, can alter the overall behavior significantly: such systems, however, are not dealt with in this section being in the domain of feedback control systems.

Solution of the combined behavior is obtained using Laplace transform methods to obtain the output responses in the time or the complex frequency domain. These mathematical methods might not be familiar to the reader, but this is not a serious difficulty for the cases most encountered in practice are

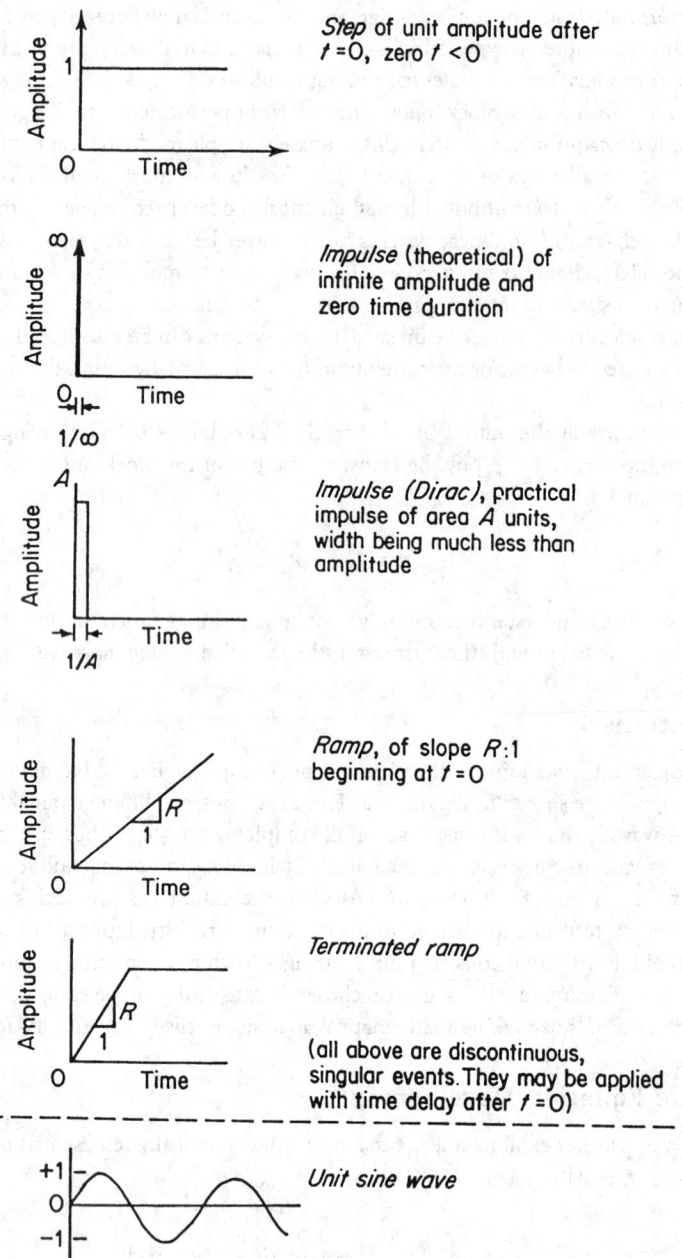

FIGURE 3.12 The dynamic response of a block can be investigated using a range of simple input forcing functions. (From P. H. Sydenham, *Handbook of Measurement Science*, Vol. 2, Chichester, U.K., John Wiley & Sons, 1983. With permission.)

well documented in terms that are easily comprehended, the mathematical process having been performed to yield results that can be used without the same level of mathematical ability. More depth of explanation can be obtained from [1] or any one of the many texts on energy systems analysis. Space here only allows an introduction; this account is linked to [1], Chapter 17, to allow the reader to access a fuller description where needed.

The next step in understanding block behavior is to investigate the nature of Equation 3.6 as the number of derivative terms in the expression increases, Equations 3.7 to 3.10.

Zero order $\qquad a_0 y = x(t)$ $\hfill$ (3.7)

First order $\qquad a_1\, dy/dt + a_0 y = x(t)$ $\hfill$ (3.8)

Second order $\qquad a_2\, d^2 y/dt^2 + a_1\, dy/dt + a_0 y = x(t)$ $\hfill$ (3.9)

nth order $\qquad a_n\, d^n y/dt^n + a_{n-1} d^{n-1} y/dt^{n-1} + \ldots + a_0 y = x(t)$ $\hfill$ (3.10)

Note that specific names have been given to each order. The zero-order situation is not usually dealt with in texts because it has no time-dependent term and is thus seen to be trivial. It is an amplifier (or attenuator) of the forcing function with gain of a_0. It has infinite bandwidth without change in the amplification constant.

The highest order usually necessary to consider in first-cut instrument analysis is the second-order class. Higher-order systems do occur in practice and need analysis that is not easily summarized here. They also need deep expertise in their study. Computer-aided tools for systems analysis can be used to study the responses of systems.

Another step is now to rewrite the equations after Laplace transformation into the frequency domain. We then get the set of output/input Equations 3.11 to 3.14.

Zero order $\qquad Y(s)/X(s) = 1$ $\hfill$ (3.11)

First order $\qquad Y(s)/X(s) = 1/(\tau s + 1)$ $\hfill$ (3.12)

Second order $\qquad Y(s)/X(s) = 1/(\tau_1 s + 1)(\tau_2 s + 1)$ $\hfill$ (3.13)

nth order $\qquad Y(s)/X(s) = 1/(\tau_1 s + 1)(\tau_2 s + 1) \ldots (\tau_n s + 1)$ $\hfill$ (3.14)

The terms $\tau_1, \ldots, \tau_n$ are called *time constants*. They are key system performance parameters.

Response of the Different Linear Systems Types

Space restrictions do not allow a detailed study of all of the various options. A selection is presented to show how they are analyzed and reported, that leading to how the instrumentation person can use certain standard charts in the study of the characteristics of blocks.

Zero-Order Blocks

To investigate the response of a block, multiply its frequency domain forms of equation for the characteristic equation with that of the chosen forcing function equation.

This is an interesting case because Equation 3.7 shows that the zero-order block has no frequency-dependent term (it has no time derivative term), so the output for all given inputs can only be of the same time form as the input. What can be changed is the amplitude given as the coefficient a_0. A shift in time (phase shift) of the output waveform with the input also does not occur as it can for the higher-order blocks.

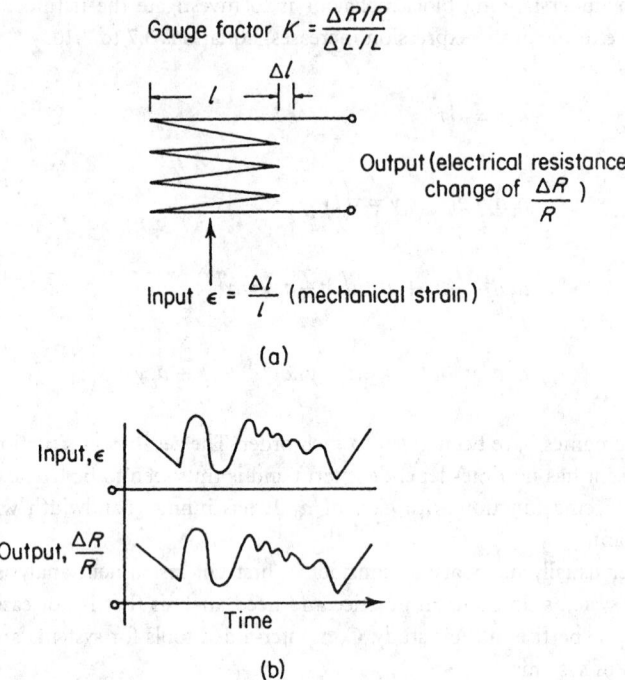

FIGURE 3.13 Input and output responses for a zero-order block: (*a*) strain gage physical and mathematical model; and (*b*) responses. (From P. H. Sydenham, *Handbook of Measurement Science*, Vol. 2, Chichester, U.K., John Wiley & Sons, 1983. With permission.)

This is the response often desired in instruments because it means that the block does not alter the time response. However, this is not always so because, in systems, design blocks are often chosen for their ability to change the time shape of signals in a known manner.

Although somewhat obvious, Figure 3.13, a resistive strain gage, is given to illustrate zero-order behavior.

First-Order Blocks

Here, Equation 3.8 is the relevant characteristic equation. There is a time-dependent term, so analysis is needed to see how this type of block behaves under dynamic conditions. The output response is different for each type of forcing function applied. Space limitations only allow the most commonly encountered cases — the step and the sine-wave input — to be introduced here. It is also only possible here to outline the method of analysis and to give the standardized charts that plot generalized behavior.

The step response of the first-order system is obtained by multiplying Equation 3.12 by the frequency domain equation for a step of amplitude A. The result is then transformed back into the time domain using Laplace transforms to yield the expression for the output, $y(t)$

$$y(t) = AK\left(1 - e^{t/\tau}\right) \tag{3.15}$$

where A is the amplitude of the step, K the static gain of the first-order block, t the time in consistent units, and τ the time constant associated with the block itself.

This is a tidy outcome because Equation 3.15 covers the step response for all first-order blocks, thus allowing it to be graphed in normalized manner, as given in Figure 3.14. The shape of the response is always of the same form. This means that the step response of a first-order system can be described as having "a step of AK amplitude with the time constant τ."

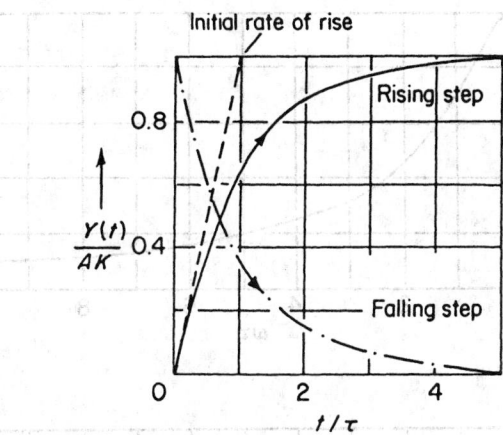

FIGURE 3.14 The step response for all first-order systems is covered by these two normalized graphs. (From P. H. Sydenham, *Handbook of Measurement Science,* Vol. 2, Chichester, U.K., John Wiley & Sons, 1983. With permission.)

If the input is a sine-wave, the output response is quite different; but again, it will be found that there is a general solution for all situations of this kind. As before, the input forcing equation is multiplied by the characteristic equation for the first-order block and Laplace transformation is used to get back to the time domain response. After rearrangement into two parts, this yields:

$$y(t) = \left[AK\tau\omega e^{-t/\tau} / (\tau^2\omega^2 + 1) \right] + \left[AK / (\tau^2\omega^2 + 1)^{1/2} \cdot \sin(\omega t + \phi) \right] \tag{3.16}$$

where ω is the signal frequency in angular radians, $\phi = \tan^{-1}(-\omega t)$, A the amplitude of the sine-wave input, K the gain of the first-order block, t the time in consistent units, and τ the time constant associated with the block.

The left side of the right-hand bracketed part is a short-lived, normally ignored, time transient that rapidly decays to zero, leaving a steady-state output that is the parameter of usual interest. Study of the steady-state part is best done by plotting it in a normalized way, as has been done in Figure 3.15.

These plots show that the amplitude of the output is always reduced as the frequency of the input signal rises and that there is always a phase lag action between the input and the output that can range from 0 to 90° but never be more than 90°. The extent of these effects depends on the particular coefficients of the block and input signal. These effects must be well understood when interpreting measurement results because substantial errors can arise with using first-order systems in an instrument chain.

Second-Order Blocks

If the second-order differential term is present, the response of a block is quite different, again responding in quite a spectacular manner with features that can either be wanted or unwanted.

As before, to obtain the output response, the block's characteristic function is multiplied by the chosen forcing function. However, to make the results more meaningful, we first carry out some simple substitution transformations.

The steps begin by transforming the second-order differential Equation 3.6 into its Laplace form to obtain:

$$X(s) = a_2 s^2 Y(s) + a_1 s Y(s) + a_0 Y(s) \tag{3.17}$$

This is then rearranged to yield:

$$G(s) = Y(s)/X(s) = 1/a_0 \cdot 1/\left\{ (a_2/a_0)s^2 + (a_1/a_0)s + 1 \right\} \tag{3.18}$$

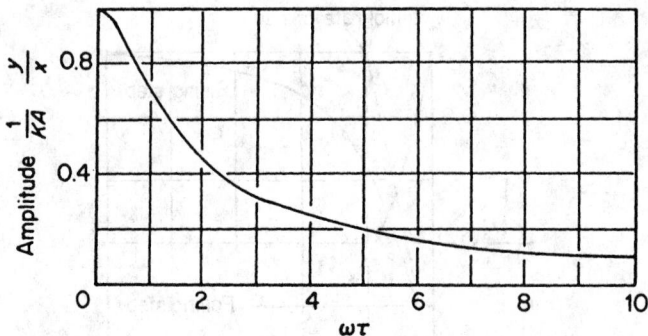

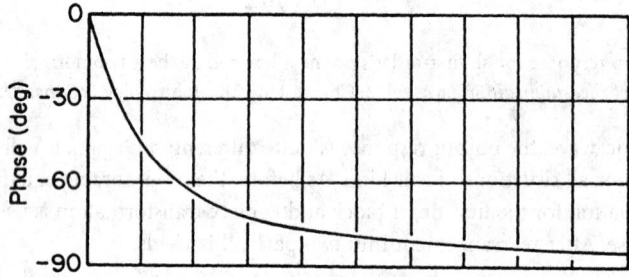

FIGURE 3.15   The amplitude and phase shift of the output of all first-order systems to a sine-wave input is shown by these two normalized curves; (*a*) amplitude and (*b*) phase. (From P. H. Sydenham, *Handbook of Measurement Science*, Vol. 2, Chichester, U.K., John Wiley & Sons, 1983. With permission.)

The coefficients can then be expressed in system performance terms as follows.

Angular natural frequency

$$\omega_n = \left(a_0/a_2\right)^{1/2} \tag{3.19}$$

Damping ratio

$$\zeta = a_1 \Big/ 2\left(a_0 \cdot a_2\right)^{1/2} \tag{3.20}$$

Static gain

$$K = 1/a_0 \tag{3.21}$$

These three variables have practical relevance, as will be seen when the various responses are plotted.

Using these transformed variables, the characteristic equation can be rewritten in two forms ready for investigation of output behavior to step and sine-wave inputs, as:

$$G(s) = K \Big/ \left\{\left(1/\omega_n^2\right)s^2 + \left(2\zeta/\omega_n\right)s + 1\right\} \tag{3.22}$$

and then as:

$$G(s) = K \Big/ \left(\tau^2 s^2 + 2\zeta\tau s + 1\right) \tag{3.23}$$

We are now ready to consider the behavior of the second-order system to the various forcing inputs.

First consider the step input. After forming the product of the forcing and characteristic functions, the time domain form can be plotted as shown in Figure 3.16.

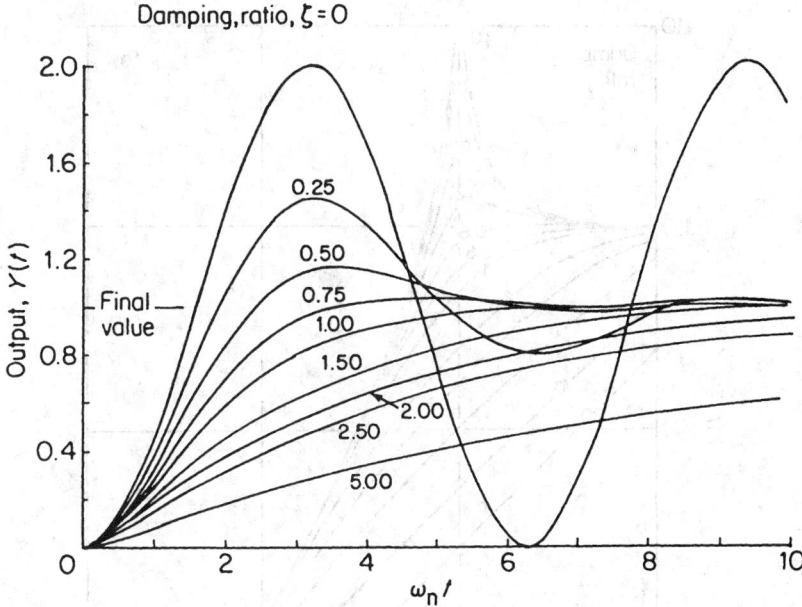

FIGURE 3.16 The response of second-order systems to a step input is seen from this normalized plot. (From P. H. Sydenham, *Handbook of Measurement Science,* Vol. 2, Chichester, U.K., John Wiley & Sons, 1983. With permission.)

This clearly shows that the response is strongly dependent on the damping ratio ζ value. If it is less than unity, it exhibits an oscillatory movement settling down to the final value. If the damping value is greater than unity, the response moves to the final value without oscillation. The often preferred state is to use a damping factor of unity, *critical damping*. The choice of response depends strongly on the applications, for all levels of damping ratio have use in practice, ranging from needing an oscillation that never ceases (zero damping) to the other extreme where a very gradual rate of change is desired.

A similar analysis is used to see how the second-order system responds to the sine-wave input. The two response plots obtained are shown in Figure 3.17: one for the amplitude response, and the other showing how the phase shifts as the frequency changes.

The most unexpected result is seen at the point where the gain rises to infinity for the zero damping state. This is called *resonance* and it occurs at the block's *natural frequency* for the zero damping state. Resonance can be a desirable feature, as in detecting a particular frequency in a radio frequency detection circuit, or it may be most undesirable, as when a mechanical system resonates, possibly to destruction. It can be seen that it is mostly controlled by the damping ratio. Note also that the phase shift for the second-order system ranges from 0 to 180°. This has important implications if the block is part of a feedback loop because as the frequency rises, the phase shift from the block will pass from stable negative feedback (less than 90°) to positive feedback (greater than 90°), causing unwanted oscillation.

More detail of the various other situations, including how to deal with higher orders, cascaded blocks of similar kind, and ramp inputs are covered elsewhere [1].

3.3 Calibration of Measurements

We have already introduced the concept of accuracy in making a measurement and how the uncertainty inherent in all measurements must be kept sufficiently small. The process and apparatus used to find out if a measurement is accurate enough is called *calibration*. It is achieved by comparing the result of a measurement with a method possessing a measurement performance that is generally agreed to have less uncertainty than that in the result obtained. The error arising within the calibration apparatus and process

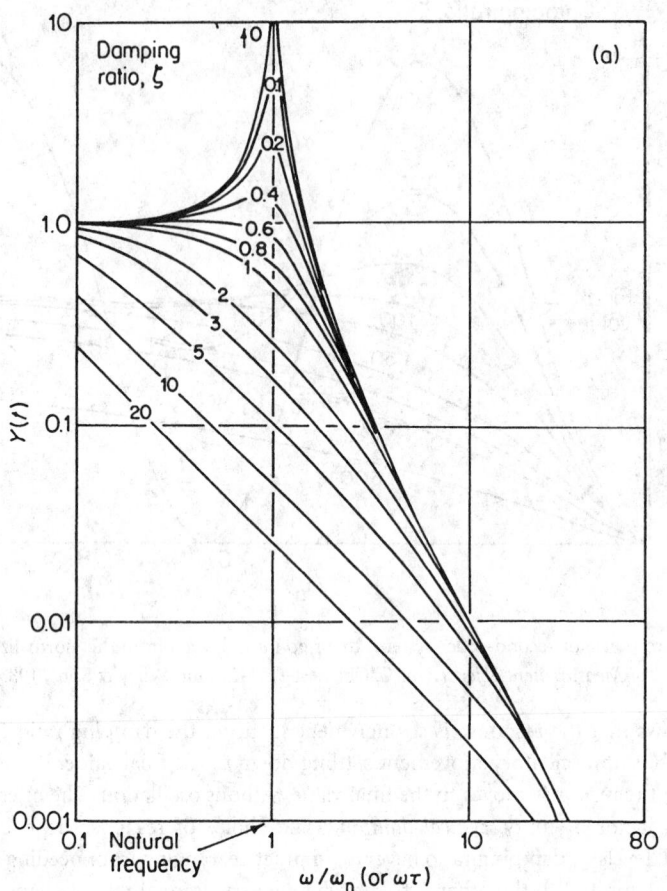

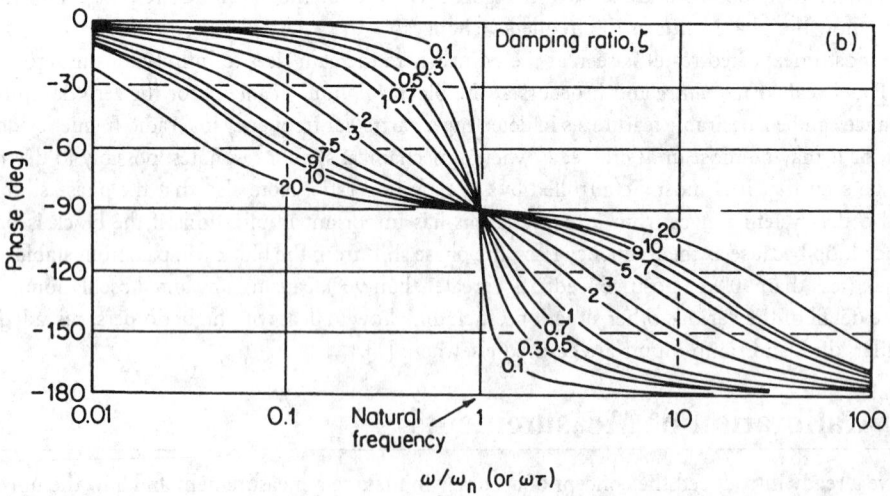

FIGURE 3.17   These two plots allow the behavior of second-order blocks with sine-wave inputs to be ascertained: (*a*) amplitude and (*b*) phase. (From P. H. Sydenham, *Handbook of Measurement Science,* Vol. 2, Chichester, U.K., John Wiley & Sons, 1983. With permission.)

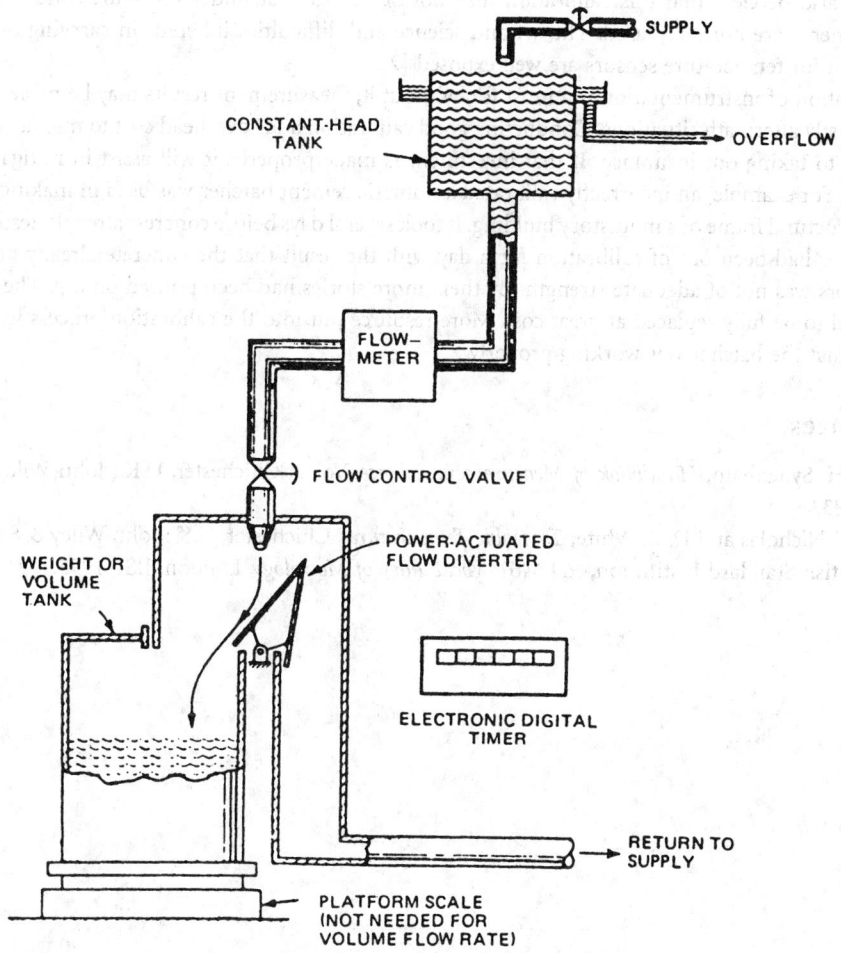

FIGURE 3.18   This practical example illustrates how flow meters are calibrated by passing a known quantity of fluid through the meter over a given time. (Originally published in P. H. Sydenham, *Transducers in Measurement and Control*, Adam Hilger, Bristol, IOP Publishing, Bristol, 1984. Copyright P. H. Sydenham.)

of comparison must necessarily be less than that required. This means that calibration is often an expensive process. Conducting a good calibration requires specialist expertise.

The method and apparatus for performing measurement instrumentation calibrations vary widely. An illustrative example of the comparison concept underlying them all is given in the calibration of flow meters, shown diagrammatically in Figure 3.18.

By the use of an overflowing vessel, the top tank provides a flow of water that remains constant because it comes from a constant height. The meter to be calibrated is placed in the downstream pipe.

The downstream is either deflected into the weigh tank or back to the supply. To make a measurement, the water is first set to flow to the supply. At the start of a test period, the water is rapidly and precisely deflected into the tank. After a given period, the water is again sent back to the supply. This then has filled the tank with a given amount of water for a given time period of flow. Calculations are then undertaken to work out the quantity of water flowing per unit time period, which is the *flow rate*. The meter was already registering a flow rate as a constant value. This is then compared with the weighed method to yield the error. Some thought will soon reveal many sources of error in the test apparatus, such as that the temperature of the water decides the volume that flows through and thus this must be allowed for in the calculations.

It will also be clear that this calibration may not be carried out under the same conditions as the measurements are normally used. The art and science and difficulties inherent in carrying out quality calibration for temperature sensors are well exposed [2].

Calibration of instrumentation is a must for, without it, measurement results may be misleading and lead to costly aftermath situations. Conducting good calibration adds overhead cost to measurement but it is akin to taking out insurance. If that investment is made properly, it will assist in mitigating later penalties. For example, an incorrectly calibrated automatic cement batcher was used in making concrete for the structural frame of a multistory building. It took several days before concrete strength tests revealed the batcher had been out of calibration for a day with the result that the concrete already poured for three floors was not of adequate strength. By then, more stories had been poured on top. The defective floors had to be fully replaced at great cost. More resource put into the calibration process would have ensured that the batcher was working properly.

References

1. P. H. Sydenham, *Handbook of Measurement Science,* Vol. 2, Chichester, U.K.: John Wiley & Sons, 1983.
2. J. V. Nicholas and D. R. White, *Traceable Temperatures,* Chichester, U.K.: John Wiley & Sons, 1994.
3. British Standard Institution, *PD 6461: Vocabulary of Metrology,* London: BSI, 1995.

4

Measurement Accuracy

Ronald H. Dieck

Pratt & Whitney

All test measurements are taken so that data may be acquired that are useful in decision making. No tests are run and no measurements made when the "answer" is already known. For data to be useful, it is necessary that their measurement errors be small in comparison to the changes or effect under evaluation. Measurement error is unknown and unknowable. This chapter addresses the techniques used to estimate, with some confidence, the expected limits of the measurement errors.

4.1 Error: The Normal Distribution and the Uniform Distribution

Error is defined as the difference between the measured value and the true value of the meaurand [1]. That is,

$$E = \left(measured\right) - \left(true\right) \tag{4.1}$$

where E = the measurement error
(measured) = the value obtained by a measurement
(true) = the true value of the measurand

It is only possible to estimate, with some confidence, the expected limits of error. The most common method for estimating those limits is to use the *normal distribution* [2]. It is

$$Y = \frac{1}{\sigma\sqrt{2\pi}} \, e^{-\frac{1}{2}(X-\mu)^2 / \sigma^2} \tag{4.2}$$

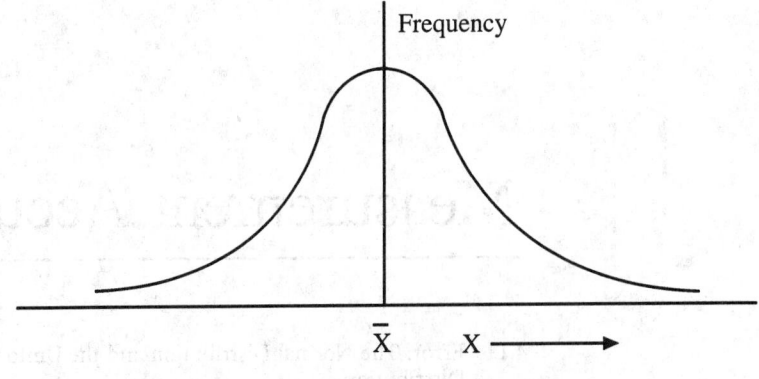

FIGURE 4.1

where X = the input variable, here the value obtained by a measurement
   μ = the average of the population of the X variable
   σ = the standard deviation of the population, expressed as:

$$\sigma = \sqrt{\frac{\sum_{i=1}^{n}\left(X_i - \mu\right)^2}{n}} \qquad (4.3)$$

where X_i = the ith X measurement
   n = the number of data points measured from the population

Typically, neither n, μ, nor σ are known.

Figure 4.1 illustrates this distribution. Here, for an infinite population ($N = \infty$), the standard deviation, σ, would be used to estimate the expected limits of a particular error with some confidence. That is, the average, plus or minus 2σ divided by the square root of the number of data points, would contain the true average, μ, 95% of the time.

However, in test measurements, one typically cannot sample the entire population and must make do with a sample of N data points. The sample standard deviation, S_X, is then used to estimate σ_X, the expected limits of a particular error. (That sample standard deviation divided by the square root of the number of data points is the starting point for the confidence interval estimate on μ.) For a large dataset (defined as having 30 or more degrees of freedom), plus or minus $2S_X$ divided by the square root of the number of data points in the reported average, M, would contain the true average, μ, 95% of the time. That S_X divided by the square root of the number of data points in the reported average is called the *standard deviation of the average* and is written as:

$$S_{\bar{X}} = \sqrt{\frac{\sum_{i=1}^{N}\left(X_i - \bar{X}\right)^2}{N-1}} \Bigg/ \sqrt{M} = S_X \big/ \sqrt{M} \qquad (4.4)$$

where $S_{\bar{X}}$ = the standard deviation of the average; the sample standard deviation of the data divided by the square root of M
   S_X = the sample standard deviation
   $\bar{X}$ = the sample average, that is,

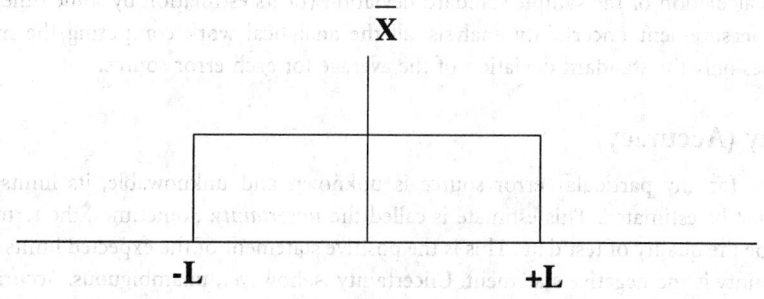

FIGURE 4.2

$$\bar{X} = \sum_{i=1}^{M}\left(X_i/N\right) \qquad (4.5)$$

X_i = the ith data point used to calculate the sample standard deviation and the average, $\bar{X}$, from the data

N = the number of data points used to calculate the standard deviation

$(N-1)$ = the degrees of freedom of S_X and $S_{\bar{X}}$

M = the number of data points in the reported average test result

Note in Equation 4.4 that N does not necessarily equal M. It is possible to obtain S_X from historical data with many degrees of freedom ($[N-1]$ greater than 30) and to run the test only M times. The test result, or average, would therefore be based on M measurements, and the standard deviation of the average would still be calculated with Equation 4.4. In that case, there would be two averages, $\bar{X}$. One $\bar{X}$ would be from the historical data used to calculate the sample standard deviation, and the other $\bar{X}$, the average test result for M measurements.

Note that the sample standard deviation, S_X, is simply:

$$S_X = \sqrt{\frac{\sum_{i=1}^{N}\left(X_i - \bar{X}\right)^2}{N-1}} \qquad (4.6)$$

In some cases, a particular error distribution may be assumed or known to be *a uniform or rectangular distribution*, Figure 4.2, instead of a normal distribution. For those cases, the sample standard deviation of the data is calculated as:

$$S_X = L/\sqrt{3} \qquad (4.7)$$

where L = the plus/minus limits of the uniform distribution for a particular error [3].

For those cases, the standard deviation of the average is written as:

$$S_{\bar{X}} = \frac{L/\sqrt{3}}{\sqrt{M}} \qquad (4.8)$$

Although the calculation of the sample standard deviation (or its estimation by some other process) is required for measurement uncertainty analysis, all the analytical work computing the measurement uncertainty uses only the standard deviation of the average for each error source.

Uncertainty (Accuracy)

Since the error for any particular error source is unknown and unknowable, its limits, at a given confidence, must be estimated. This estimate is called the *uncertainty*. Sometimes, the term *accuracy* is used to describe the quality of test data. This is the positive statement of the expected limits of the data's errors. Uncertainty is the negative statement. Uncertainty is, however, unambiguous. Accuracy is sometimes ambiguous. (For example,: what is twice the accuracy of ±2%? ±1% or ±4%?) For this reason, this chapter will use the term *uncertainty* throughout to describe the quality of test data.

4.2 Measurement Uncertainty Model

Purpose

One needs an estimate of the uncertainty of test results to make informed decisions. Ideally, the uncertainty of a well-run experiment will be much less than the change or test result expected. In this way, it will be known, with high confidence, that the change or result observed is real or acceptable and not a result of the errors of the test or measurement process. The limits of those errors are estimated with uncertainty, and those error sources and their limit estimators, the uncertainties, may be grouped into classifications to ease their understanding.

Classifying Error and Uncertainty Sources

There are two classification systems in use. The final total uncertainty calculated at a confidence is identical no matter what classification system is used. The two classifications utilized are the *ISO classifications* and the *engineering classifications*. The former groups errors and their uncertainties by type, depending on whether or not there is data available to calculate the sample standard deviation for a particular error and its uncertainty. The latter classification groups errors and their uncertainties by their effect on the experiment or test. That is, the engineering classification groups errors and uncertainties by *random* and *systematic* types, with subscripts used to denote whether there are data to calculate a standard deviation or not for a particular error or uncertainty source. For this reason, engineering classification groups usually are more useful and recommended.

ISO Classifications

This error and uncertainty classification system is not recommended in this chapter, but will yield a total uncertainty in complete agreement with the recommended classification system — the engineering classification system. In this ISO system, errors and uncertainties are classified as Type A if there are data to calculate a sample standard deviation and Type B if there is not [4]. In the latter case, the sample standard deviation might be obtained from experience or manufacturer's specifications, to name two examples.

The impact of multiple sources of error is estimated by root-sum-squaring their corresponding multiple uncertainties. The operating equations are

Type A, data for the calculation of the standard deviation:

$$U_A = \left[\sum_{i=1}^{N_A} \left(\theta_i U_{A_i} \right)^2 \right]^{1/2} \tag{4.9}$$

where U_{A_i} = the standard deviation (based on data) of the average for uncertainty source i of Type A each with its own degrees of freedom. U_A is in units of the test or measurement result. It is an $S_{\bar{x}}$.

N_A = the number of parameters with a Type A uncertainty

θ_i = the sensitivity of the test or measurement result, R, to the ith Type A uncertainty. θ_i is the partial derivative of the result with respect to each ith independent measurement.

The uncertainty of each error source in units of that source, when multiplied by the sensitivity for that source, converts that uncertainty to result units. Then the effect of several error sources may be estimated by root-sum-squaring their uncertainties as they are now all in the same units. The sensitivities, θ_i, are obtained for a measurement result, R, which is a function of several parameters, P_i. The basic equations are

$$R = \text{the measurement result}$$

where $R = f(P_1, P_2, P_3 \ldots P_N)$

P = a measurement parameter used to calculate the result, R

$\theta_i = \partial R / \partial P_i$

Obtaining the θ_i is often called error propagation or uncertainty propagation.

Type B (no data for standard deviation) calculation

$$U_B = \left[\sum_{i=1}^{N_B} \left(\theta_i U_{B_i} \right)^2 \right]^{1/2} \tag{4.10}$$

where U_{B_i} = the standard deviation (based on an estimate, not data) of the average for uncertainty source i of Type B; U_B is in units of the test or measurement result, R. It is an $S_{\bar{x}}$

N_B = the number of parameters with a Type B uncertainty

θ_i = the sensitivity of the test or measurement result to the ith Type B uncertainty R

For these uncertainties, it is assumed that the U_{B_i} represent one standard deviation of the average for one uncertainty source with an assumed normal distribution. (They also represent one standard deviation as the square root of the "M" by which they are divided is one, that is, there is only one Type B error sampled from each of these distributions.) The degrees of freedom associated with this standard deviation (also standard deviation of the average) is infinity.

Note that θ_i, the sensitivity of the test or measurement result to the ith Type B uncertainty, is actually the change in the result, R, that would result from a change, of the size of the Type B uncertainty, in the ith input parameter used to calculate that result.

The degrees of freedom of the U_A and the U_{B_i} are needed to compute the degrees of freedom of the combined total uncertainty. It is calculated with the Welch–Satterthwaite approximation. The general formula for degrees of freedom [5] is

$$df_R = \nu_R = \frac{\left[\sum_{i=1}^{N} \left(S_{\bar{X}_i} \right)^2 \right]^2}{\left[\sum_{i=1}^{N} \frac{\left(S_{\bar{X}_i} \right)^4}{\left(\nu_i \right)} \right]} \tag{4.11}$$

where $df_R = \nu_R$ = degrees of freedom for the result

ν_i = the degrees of freedom of the ith standard deviation of the average

For the ISO model, Equation 4.11 becomes:

$$df_{R,ISO} = \nu_{R,ISO} = \frac{\left[\displaystyle\sum_{i=1}^{N_A}\left(\theta_i U_{A_i}\right)^2 + \sum_{i=1}^{N_B}\left(\theta_i U_{B_i}\right)^2\right]^2}{\left[\displaystyle\sum_{i=1}^{N_A}\frac{\left(\theta_i U_{A_i}\right)^4}{\left(\nu_i\right)} + \sum_{i=1}^{N_B}\frac{\left(\theta_i U_{B_i}\right)^4}{\left(\nu_i\right)}\right]} \qquad (4.12)$$

The degrees of freedom calculated with Equation 4.12 is often a fraction. This should be truncated to the next lower whole number to be conservative.

Note that in Equations 4.9, 4.10, and 4.12, N_A and N_B need not be equal. They are only the total number of parameters with uncertainty sources of Type A and B, respectively.

In computing a total uncertainty, the uncertainties noted by Equations 4.10 and 4.11 are combined. For the ISO model [3], this is calculated as:

$$U_{R,ISO} = \pm t_{95}\left[\left(U_A\right)^2 + \left(U_B\right)^2\right]^{1/2} \qquad (4.13)$$

where t_{95} = Student's t for ν_R degrees of freedom

Student's t is obtained from Table 4.1.

Note that alternative confidences are permissible. 95% is recommended by the ASME [6], but 99% or 99.7% or any other confidence is obtained by choosing the appropriate Student's t. 95% confidence is, however, recommended for uncertainty analysis.

In all the above, the errors were assumed to be independent. Independent sources of error are those that have no relationship to each other. That is, an error in a measurement from one source cannot be used to predict the magnitude or direction of an error from the other, independent, error source. Nonindependent error sources are related. That is, if it were possible to know the error in a measurement from one source, one could calculate or predict an error magnitude and direction from the other,

TABLE 4.1 Student's t Statistic for 95% Confidence, t_{95}, Degrees of Freedom, ν. This is Frequently Written as: $t_{95,\nu}$

ν	t_{95}	ν	t_{95}	ν	t_{95}
1	12.706	11	2.201	21	2.080
2	4.303	12	2.179	22	2.074
3	3.182	13	2.160	23	2.069
4	2.776	14	2.145	24	2.064
5	2.571	15	2.131	25	2.060
6	2.447	16	2.120	26	2.056
7	2.365	17	2.110	27	2.052
8	2.306	18	2.101	28	2.048
9	2.262	19	2.093	29	2.045
10	2.228	20	2.086	≥ 30	2.000

nonindependent error source. These are sometimes called *dependent error sources.* Their degree of dependence may be estimated with the linear correlation coefficient. If they are nonindependent, whether Type A or Type B, Equation 4.13 becomes [7]:

$$U_{R,\text{ISO}} = t_{95} \left\{ \sum_{T=A}^{B} \sum_{i=1}^{N_{i,T}} \left[\left(\theta_i U_{i,T} \right)^2 + \sum_{j=1}^{N_{j,T}} \theta_i \theta_j U_{(i,T),(j,T)} \left(1 - \delta_{i,j} \right) \right] \right\}^{1/2}$$ (4.14)

where: $U_{i,T}$ = the ith elemental uncertainty of Type T (can be Type A or B)
$U_{R,\text{ISO}}$ = the total uncertainty of the measurement or test result
θ_i = the sensitivity of the test or measurement result to the ith Type T uncertainty
θ_j = the sensitivity of the test or measurement result to the jth Type T uncertainty
$U_{(i,T),(j,T)}$ = the covariance of U_{i,T_i} on $U_{j,T}$

$$= \sum_{l=1}^{K} U_{i,T}(l) U_{j,T}(l)$$ (4.15)

= the sum of the products of the elemental systematic uncertainties that arise from a common source (l)
l = an index or counter for common uncertainty sources
K = the number of common source pairs of uncertainties
$\delta_{i,j}$ = the Kronecker delta. $\delta_{i,j} = 1$ if $i = j$, and $\delta_{i,j} = 0$ if not [7]
T = an index or counter for the ISO uncertainty type, A or B

This ISO classification equation will yield the same total uncertainty as the engineering classification, but the ISO classification does not provide insight into how to improve an experiment's or test's uncertainty. That is, whether to possibly take more data because the random uncertainties are too high or calibrate better because the systematic uncertainties are too large. The engineering classification now presented is therefore the preferred approach.

Engineering Classification

The engineering classification recognizes that experiments and tests have two major types of errors whose limits are estimated with uncertainties at some chosen confidence. These error types may be grouped as *random* and *systematic.* Their corresponding limit estimators are the random uncertainty and systematic uncertainties, respectively.

Random

The general expression for random uncertainty is the ($1 S_{\bar{X}}$) standard deviation of the average [6]:

$$S_{\bar{X},R} = \left[\sum_{T=A}^{B} \sum_{i=1}^{N_{i,T}} \left(\theta_i S_{\bar{X}_{i,T}} \right)^2 \right]^{1/2} = \left[\sum_{T=A}^{B} \sum_{i=1}^{N_{i,T}} \left(\theta_i S_{X_{i,T}} / \sqrt{M_{i,T}} \right)^2 \right]^{1/2}$$ (4.16)

where: $S_{X_{i,T}}$ = the sample standard deviation of the ith random error source of Type T
$S_{\bar{X}_{i,T}}$ = the random uncertainty (standard deviation of the average) of the ith parameter random error source of Type T
$S_{\bar{X},R}$ = the random uncertainty of the measurement or test result
$N_{i,T}$ = the total number of random uncertainties, Types A and B, combined
$M_{i,T}$ = the number of data points averaged for the ith error source, Type A or B
θ_i = the sensitivity of the test or measurement result to the ith random uncertainty

Note that $S_{\bar{X},R}$ is in units of the test or measurement result because of the use of the sensitivities, θ_i. Here, the elemental random uncertainties have been root-sum-squared with due consideration for their sensitivities, or influence coefficients. Since these are all random uncertainties, there is, by definition, no correlation in their corresponding error data so these can always be treated as independent uncertainty sources.

Systematic

The systematic uncertainty of the result, B_R, is the root-sum-square of the elemental systematic uncertainties with due consideration for those that are correlated [7]. The general equation is

$$B_R = \left\{ \sum_{T=A}^{B} \sum_{i=1}^{N_T} \left[\left(\theta_i B_{i,T} \right)^2 + \sum_{j=1}^{N_T} \theta_i \theta_j B_{(i,T),(j,T)} \left(1 - \delta_{i,j} \right) \right] \right\}^{1/2} \tag{4.17}$$

where: $B_{i,T}$ = the ith parameter elemental systematic uncertainty of Type T
      B_R = the systematic uncertainty of the measurement or test result
      N = the total number of systematic uncertainties
      θ_i = the sensitivity of the test or measurement result to the ith systematic uncertainty
      θ_j = the sensitivity of the test or measurement result to the jth systematic uncertainty
      $B_{(i,T),(j,T)}$ = the covariance of B_i on B_j

$$= \sum_{i=1}^{M} B_{i,T}(l) B_{j,T}(l) \tag{4.18}$$

               = the sum of the products of the elemental systematic uncertainties that arise from a common source (l)
      l = an index or counter for common uncertainty sources
      $\delta_{i,j}$ = the Kronecker delta. $\delta_{i,j} = 1$ if $i = j$, and $\delta_{i,j} = 0$ if not [7]
      T = an index or counter for the ISO uncertainty type, A or B

Here, each $B_{i,T}$ and $B_{j,T}$ are estimated as $2S_X$ for an assumed normal distribution of errors at 95% confidence with infinite degrees of freedom [6].

The random uncertainty, Equation 4.16, and the systematic uncertainty, Equation 4.17, must be combined to obtain a total uncertainty:

$$U_{R,ENG} = t_{95} \left[\left(B_R / 2 \right)^2 + \left(S_{\bar{X},R} \right)^2 \right]^{1/2} \tag{4.19}$$

Note that B_R is in units of the test or measurement result as was $S_{\bar{X},R_i}$.

The degrees of freedom will be needed for the engineering system total uncertainty. It is accomplished with the Welch–Satterthwaite approximation, the general form of which is Equation 4.10, and the specific formulation here is

$$df_R = \nu_R = \frac{\left\{ \sum_{T=A}^{B} \left[\sum_{i=1}^{N_{S_{\bar{X}_{i,T}}}} \left(\theta_i S_{\bar{X}_{i,T}} \right)^2 + \sum_{i=1}^{N_{B_{i,T}}} \left(\theta_i B_{i,T/t} \right)^2 \right] \right\}^2}{\left\{ \sum_{T=A}^{B} \left[\sum_{i=1}^{N_{S_{\bar{X}_{i,T}}}} \frac{\left(\theta_i S_{\bar{X}_{i,T}} \right)^4}{\left(\nu_{i,T} \right)} + \sum_{i=1}^{N_{B_{i,T}}} \frac{\left(\theta_i B_{i,T/t} \right)^4}{\left(\nu_{i,T} \right)} \right] \right\}} \tag{4.20}$$

where $N_{S_{\bar{X}_{i,T}}}$ = the number of random uncertainties of Type T
$N_{B_{i,T}}$ = the number of systematic uncertainties of Type T
$\nu_{i,T}$ = the degrees of freedom for the ith uncertainty of Type T
$\nu_{i,T}$ = infinity for all systematic uncertainties
t = Student's t associated with the d.f. for each B_i

Symmetrical Systematic Uncertainties

Most times, all elemental uncertainties will be symmetrical. That is, their ± limits about the measured average will be the same. That is, they will be ±3°C or ±2.05 kPa and the like and not +2.0°C, –1.0°C or, +1.5 kPa, –0.55 kPa. The symmetrical measurement uncertainty may therefore be calculated as follows. (For an elegant treatment of nonsymmetrical uncertainties, see that section in Reference [6].)

Note that throughout these uncertainty calculations, all the uncertainties are expressed in engineering units. All the equations will work with relative units as well. That approach may be seen in Reference [6] also. However, it is often easier to express all the uncertainties and the uncertainty estimation calculations in engineering units and then, at the end, with the total uncertainty, convert the result into relative terms. That is what this section recommends.

4.3   Calculation of Total Uncertainty

ISO Total (Expanded) Uncertainty

The ISO total uncertainty for independent uncertainty sources (the most common) is Equation 4.13:

$$U_{R,\text{ISO}} = \pm t_{95}\left[\left(U_A\right)^2 + \left(U_B\right)^2\right]^{1/2} \tag{4.21}$$

where: $U_{R,\text{ISO}}$ = the measurement uncertainty of the result
U_A = the Type A uncertainty for the result
U_B = the Type B uncertainty for the result
t_{95} = Student's t_{95} is the recommended multiplier to assure 95% confidence

The ISO uncertainty with some nonindependent uncertainty sources is Equation 4.14:

$$U_{R,\text{ISO}} = \left\{\sum_{T=A}^{B}\sum_{i=1}^{N_{i,T}}\left[\left(\theta_i U_{i,T}\right)^2 + \sum_{j=1}^{N_{j,T}}\theta_i\theta_j U_{(i,T),(j,T)}\left(1-\delta_{i,j}\right)\right]\right\}^{1/2} \tag{4.22}$$

Engineering System Total Uncertainty

The engineering system equation for total uncertainty for independent uncertainty sources (the most common) is

$$U_{R,\text{ENG}} = \pm t_{95}\left[\left(B_R/2\right)^2 + \left(S_{\bar{X},R}\right)^2\right]^{1/2} \tag{4.23}$$

Here, just the first term of Equation 4.23 is needed as all the systematic uncertainty sources are independent.

The engineering system equation for uncertainty for nonindependent uncertainty sources (those with correlated systematic uncertainties) is also Equation 4.23; but remember to use the full expression for B_R, Equation 4.17:

TABLE 4.2 Temperature Measurement Uncertainties, F

Defined measurement process	Systematic uncertainty, B_i	d.f. for B_i	Standard deviation, $S_{\bar{X},i}$	Number of data points, N_i	Random uncertainty, $S_{\bar{X},i}$	Degrees of freedom, d.f., v_i
Calibration of tc	0.06_B	∞	0.3_A	10	0.095_A	9
Reference junction	0.07_A	12	0.1_A	5	0.045_A	4
Data acquisition	0.10_B	∞	0.6_A	12	0.173_A	11
RSS						

$$B_R = \left\{ \sum_{T=A}^{B} \sum_{i=1}^{N_T} \left[\left(\theta_i B_{i,T} \right)^2 + \sum_{j=1}^{N_T} \theta_i \theta_j B_{(i,T),(j,T)} \left(1 - \delta_{i,j} \right) \right] \right\}^{1/2} \tag{4.24}$$

The degrees of freedom for Equations 4.21 through 4.24 is calculated with the Welch–Satterthwaite approximation, Equation 4.12 for the ISO system and Equation 4.20 for the engineering system.

High Degrees of Freedom Approximation

It is often the case that it is assumed that the degrees of freedom are 30 or higher. In these cases, the equations for uncertainty simplify further by setting t_{95} equal to 2.000. This approach is recommended for a first-time user of uncertainty analysis procedures as it is a fast way to get to an approximation of the measurement uncertainty.

Calculation Example

The following calculation example is taken where all the uncertainties are independent and are in the units of the test result — temperature. It is a simple example that illustrates the combination of measurement uncertainties in their most basic case. More detailed examples are given in many of the references cited. Their review may be needed to assure a more comprehensive understanding of uncertainty analysis.

It has been shown [8] that there is often little difference in the uncertainties calculated with the different models. The data from Table 4.2 [9] will be used to calculate measurement uncertainty with these two models. These data are all in temperature units and thus the influence coefficients, or sensitivities, are all unity.

Note the use of subscripts "A" and "B" to denote where data exist to calculate a standard deviation. Note too that in this example, all errors (and therefore uncertainties) are independent and that all degrees of freedom for the systematic uncertainties are infinity except for the reference junction whose degrees of freedom are 12. Also note that B_R is calculated as:

$$B_R = 2\left[\left(\frac{0.06}{2} \right)^2 + \left(\frac{0.07}{2.18} \right)^2 + \left(\frac{0.1}{2} \right)^2 \right]^{1/2} = 0.13 \tag{4.25}$$

Each uncertainty model will now be used to derive a measurement uncertainty.

For the U_{ISO} model one obtains, via Equation 4.13, the expression:

$$U_A = \left[\left(0.095 \right)^2 + \left(0.045 \right)^2 + \left(0.173 \right)^2 + \left(\frac{0.07}{2.18} \right)^2 \right]^{1/2} = 0.21 \tag{4.27}$$

$$U_B = \left[\left(\frac{0.06}{2}\right)^2 + \left(\frac{0.10}{2}\right)^2\right]^{1/2} = 0.058 \tag{4.28}$$

$$U_{R,ISO} = \pm K\left[\left(U_A\right)^2 + \left(U_B\right)^2\right]^{1/2} = \pm K\left[\left(0.21\right)^2 + \left(0.058\right)^2\right]^{1/2} \tag{4.29}$$

Here, remember that the 0.21 is the root sum square of the $1S_{\bar{x}}$ Type A uncertainties in Table 4.2, and 0.058 that for the $1S_{\bar{x}}$ Type B uncertainties. Also note that in most cases, the Type B uncertainties have infinite degrees of freedom and represent an equivalent $2S_X$. That is why they are divided by 2 — to get an equivalent $1S_X$. Where there are less than 30 degrees of freedom, one needs to divide by the appropriate Student's *t* that gave the 95% confidence interval. For the reference junction systematic uncertainty above, that was 2.18.

If "*K*" is taken as Student's t_{95}, the degrees of freedom must first be calculated. Remember that all the systematic components of Type "B" have infinite degrees of freedom except for the 0.07, which has 12 degrees of freedom. Also, all the B_i in Table 4.1 represent an equivalent $2S_X$ except for 0.07, which represents $2.18S_X$, as its degrees of freedom are 12 and not infinity. To use their data here, divide them all but the 0.07 by 2 and the 0.07 by 2.18 so they all now represent $1S_X$, as do the random components. All Type A uncertainties, whether systematic or random in Table 4.1, have degrees of freedom as noted in the table. The degrees of freedom for U_{ISO} is then:

$$df_R = \nu_R = \frac{\left[\left(0.095\right)^2 + \left(0.045\right)^2 + \left(0.173\right)^2 + \left(0.06/2\right)^2 + \left(0.07/2.18\right)^2 + \left(0.10/2\right)^2\right]^2}{\left[\frac{\left(0.095\right)^4}{9} + \frac{\left(0.045\right)^4}{4} + \frac{\left(0.173\right)^4}{11} + \frac{\left(0.06\right)^4}{\infty} + \frac{\left(0.07/2.18\right)^4}{12} + \frac{\left(0.10/2\right)^4}{\infty}\right]} = 22.51 \approx 22 \tag{4.30}$$

t_{95} is therefore 2.07. $U_{R,ISO}$ is then:

$$U_{R,ISO} = \pm 2.07\left[\left(0.21\right)^2 + \left(0.058\right)^2\right]^{1/2} = 0.45 \text{ for 95\% confidence} \tag{4.31}$$

For a detailed comparison to the engineering system, here denoted as the $U_{R,ENG}$ model, three significant figures are carried so as not to be affected by round-off errors. Then:

$$U_{R,ISO} = \pm 2.074\left[\left(0.205\right)^2 + \left(0.0583\right)^2\right]^{1/2} = 0.442 \text{ for 95\% confidence} \tag{4.32}$$

For the engineering system, $U_{R,ENG}$, model, Equation 4.23, one obtains the expression:

$$U_{R,ENG} = \pm t_{95}\left[\left(0.13/2\right)^2 + \left(0.20\right)^2\right]^{1/2} \tag{4.33}$$

Here, the (0.13/2) is the $B_R/2$ and the 0.20 is as before the random component. To obtain the proper t_{95}, the degrees of freedom need to be calculated just as in Equation 4.30. There, the degrees of freedom were 22 and t_{95} equals 2.07. $U_{R,ENG}$ is then:

$$U_{R,\text{ENG}} = \pm 2.07\left[\left(0.13/2\right)^2 + \left(0.20\right)^2\right]^{1/2} = 0.44 \text{ for } 95\% \text{ confidence} \tag{4.34}$$

Carrying four significant figures for a comparison to $U_{R,ISO}$ not affected by round-off errors, one obtains:

$$U_{R,\text{ENG}} = \pm 2.074\left[\left(0.133/2\right)^2 + \left(0.202\right)^2\right]^{1/2} = 0.442 \text{ for } 95\% \text{ confidence} \tag{4.35}$$

This is identical to $U_{R,ISO}$, Equation 4.32, as predicted.

4.4 Summary

Although these formulae for uncertainty calculations will not handle every conceivable situation, they will provide, for most experimenters, a useful estimate of test or measurement uncertainty. For more detailed treatment or specific applications of these principles, consult the references and the recommended "Further Information" section at the end of this chapter.

Defining Terms

Accuracy: The antithesis of uncertainty. An expression of the maximum possible limit of error at a defined confidence.

Confidence: A statistical expression of percent likelihood.

Correlation: The relationship between two datasets. It is not necessarily evidence of cause and effect.

Degrees of freedom: The amount of room left for error. It may also be expressed as the number of independent opportunities for error contributions to the composite error.

Error: [Error] = [Measured] − [True]. It is the difference between the measured value and the true value.

Influence coefficient: *See* sensitivity.

Measurement uncertainty: The maximum possible error, at a specified confidence, that may reasonably occur. Errors larger than the measurement uncertainty should rarely occur.

Non-symmetrical uncertainty: An uncertainty for which there is an uneven likelihood that the true value lies on one side of the average or the other.

Propagation of uncertainty: An analytical technique for evaluating the impact of an error source (and its uncertainty) on the test result. It employs the use of influence coefficients.

Random error: An error that causes scatter in the test result.

Random uncertainty: An estimate of the limits of random error, usually one standard deviation of the average.

Sensitivity: An expression of the influence an error source has on a test or measured result. It is the ratio of the change in the result to an incremental change in an input variable or parameter measured.

Standard deviation of the average or mean: The standard deviation of the data divided by the number of measurements in the average.

Systematic error: An error that is constant for the duration of a test or measurement.

Systematic uncertainty: An estimate of the limits of systematic error, usually taken as 95% confidence for an assumed normal error distribution.

True value: The desired result of an experimental measurement.

Welch–Satterthwaite: The approximation method for determining the number of degrees of freedom in the random uncertainty of a result.

References

1. American National Standards Institute/American Society of Mechanical Engineers (ANSI/ASME) PTC 19.1-1985, *Instruments and Apparatus, Part 1, Measurement Uncertainty*, 1985, 64.
2. E. O. Doebelin, *Measurement Systems, Application and Design*, 4th ed., New York: McGraw-Hill, 1990, 38 ff.
3. International Standards Organization, *Guide to the Expression of Uncertainty Measurement*, 1993, 23.
4. International Standards Organization, *Guide to the Expression of Uncertainty in Measurement*, 1993, 10 and 11.
5. R. H. Dieck, *Measurement Uncertainty, Methods and Applications*, ISA, Research Triangle Park, NC, 1992, 45.
6. American National Standards Institute/American Society of Mechanical Engineers (ANSI/ASME) PTC 19.1-1998 *Instruments and Apparatus, Part 1, Measurement Uncertainty.*
7. K. K. Brown, H. W. Coleman, W. G. Steele, and R. P. Taylor, Evaluation of Correlated Bias Approximations in Experimental Uncertainty analysis, *Proc. 32nd Aerospace Sciences Meeting & Exhibit*, Reno, NV, AIAA paper no. 94-0772, Jan 10-13, 1996.
8. W. T. Strike, III and R. H. Dieck, Rocket Impulse Uncertainty; An Uncertainty Model Comparison, *Proc. 41st Int. Instrumentation Symposium*, Denver, CO, May 1995.
9. R. H. Dieck, Measurement Uncertainty Models, *Proc. 42nd Int. Instrumentation Symposium*, San Diego, CA, May 1996.

Further Information

ICRPG Handbook for Estimating the Uncertainty in Measurements Made with Liquid Propellant Rocket Engine Systems, Chemical Propulsion Information Agency, No. 180, 30 April 1969.

R. B. Abernethy, et al., *Handbook-Gas Turbine Measurement Uncertainty*, AEDC, 1973.

R. B. Abernethy and B. Ringhiser, The History and Statistical Development of the New ASME-SAE-AIAA-ISO Measurement Uncertainty Methodology, *Proc. AIAA/SAE/ASME/ASME 21st Joint Propulsion Conf.*, Monterey, CA, July 8-10, 1985.

W. G. Steele, R. A. Ferguson, and R. P. Taylor, Comparison of ANSI/ASME and ISO Models for Calculation of Uncertainty, *Proc. 40th Int. Instrumentation Symp.*, Paper Number 94-1014, pp. 410-438, 1994.

W. T. Strike, III and R. H. Dieck, Rocket Impulse Uncertainty; An Uncertainty Model Comparison, *Proceedings of the 41st International Instrumentation Symposium*, Denver, CO, May 1995.

5
Measurement Standards

DeWayne B. Sharp
Shape of Things

Measurement standards are those devices, artifacts, procedures, instruments, systems, protocols, or processes that are used to define (or to realize) measurement units and on which all lower echelon (less accurate) measurements depend. A measurement standard may also be said to store, embody, or otherwise provide a physical quantity that serves as the basis for the measurement of the quantity. Another definition of a standard is the physical embodiment of a measurement unit, by which its assigned value is defined, and to which it can be compared for calibration purposes. In general, it is not independent of physical environmental conditions, and it is a true embodiment of the unit only under specified conditions. Another definition of a standard is a unit of known quantity or dimension to which other measurement units can be compared.

5.1 A Historical Perspective

Many early standards were based on the human body: the length of man's hand, the width of his thumb, the distance between outstretched fingertips, the length of one's foot, a certain number of paces, etc. In the beginning, while groups were small, such standards were convenient and uniform enough to serve as the basis for measurements.

The logical person to impose a single standard was the ruler of the country — hence, our own 12-inch or other short measuring stick is still called a *ruler*. The establishment of measurement standards thus became the prerogative of the king or emperor, and this right has since been assumed by all governments.

History is replete with examples that show the importance of measurements and standards. In a report to the U.S. Congress in 1821, John Quincy Adams said, "Weights and measures may be ranked among the necessaries to every individual of human society." Our founding fathers thought them so important that the United States Constitution expressly gives the Congress the power to fix uniform standards of weights and measures. The need for weights and measures (standards) dates back to earliest recorded history and are even mentioned in the Old Testament of the Bible. Originally, they were locally decreed to serve the parochial needs of commerce, trade, land division, and taxation. Because the standards were defined by local or regional authorities, differences arose that often caused problems in commerce and early scientific investigation. The rapid growth of science in the late 17th century highlighted a number of serious deficiencies in the system of units then in use and, in 1790, led the French National Assembly to direct the French Academy of Sciences to "deduce an invariable standard for all measures and all the weights." The Academy proposed a system of units, the metric system, to define the unit of length in terms of the earth's circumference, with the units of volume and mass being derived from the unit of length. Additionally, they proposed that all multiples of each unit be a multiple of 10.

In 1875, the U.S. and 16 other countries signed the "Treaty of the Meter," establishing a common set of units of measure. It also established an International Bureau of Weights and Measures (called the BIPM). That bureau is located in the Parisian suburb of Sèvres. It serves as the worldwide repository of all the units that maintain our complex international system of weights and measures. It is through this system that compatibility between measurements made thousands of miles apart is currently maintained.

The system of units set up by the BIPM is based on the meter and kilogram instead of the yard and the pound. It is called the Système International d'Unités (SI) or the International System of Units. It is used in almost all scientific work in the U.S. and is the only system of measurement units in most countries of the world today.

Even a common system of units does not guarantee measurement agreement, however. Therein lies the crux of the problem. We must make measurements, and we must know how accurately (or, to be more correct, with what uncertainty) we made those measurements. In order to know that, there must be standards. Even more important, everyone must agree on the values of those standards and use the same standards.

As the level of scientific sophistication improved, the basis for the measurement system changed dramatically. The earliest standards were based on the human body, and then attempts were made to base them on "natural" phenomena. At one time, the basis for length was supposed to be a fraction of the circumference of the earth but it was "maintained" by the use of a platinum/iridium bar. Time was maintained by a pendulum clock but was defined as a fraction of the day and so on. Today, the meter is no longer defined by an artifact. Now, the meter is the distance that light travels in an exactly defined fraction of a second. Since the speed of light in a vacuum is now defined as a constant of nature with a specified numerical value ($299, 792, 458$ m/s), the definition of the unit of length is no longer independent of the definition of the unit of time.

Prior to 1960, the second was defined as 1/86,400th of a mean solar day. Between 1960 and 1967, the second was defined in terms of the unit of time implicit in the calculation of the ephemerides: "The second is the fraction 1/31, 556, 925.9747 of the tropical year for January 0 at 12 hours of ephemeris time." With the advent of crystal oscillators and, later, atomic clocks, better ways were found of defining the second. This, in turn, allowed a better understanding of things about natural phenomena that would not have been possible before. For example, it is now known that the earth does not rotate on its axis in a uniform manner. In fact, it is erratically slowing down. Since the second is maintained by atomic clocks it is necessary to add "leap seconds" periodically so that the solar day does not gradually change with respect to the time used every day. It was decided that a constant frequency standard was preferred over a constant length of the day.

5.2   What Are Standards?

One problem with standards is that there are several kinds. In addition to "measurement standards," there are "standards of practice or protocol standards" that are produced by the various standards bodies such as the International Organization for Standardization (ISO), the International Electrotechnical Commission (IEC), the American National Standards Institute (ANSI), and the Standards Council of Canada (SCC). See Figure 5.1.

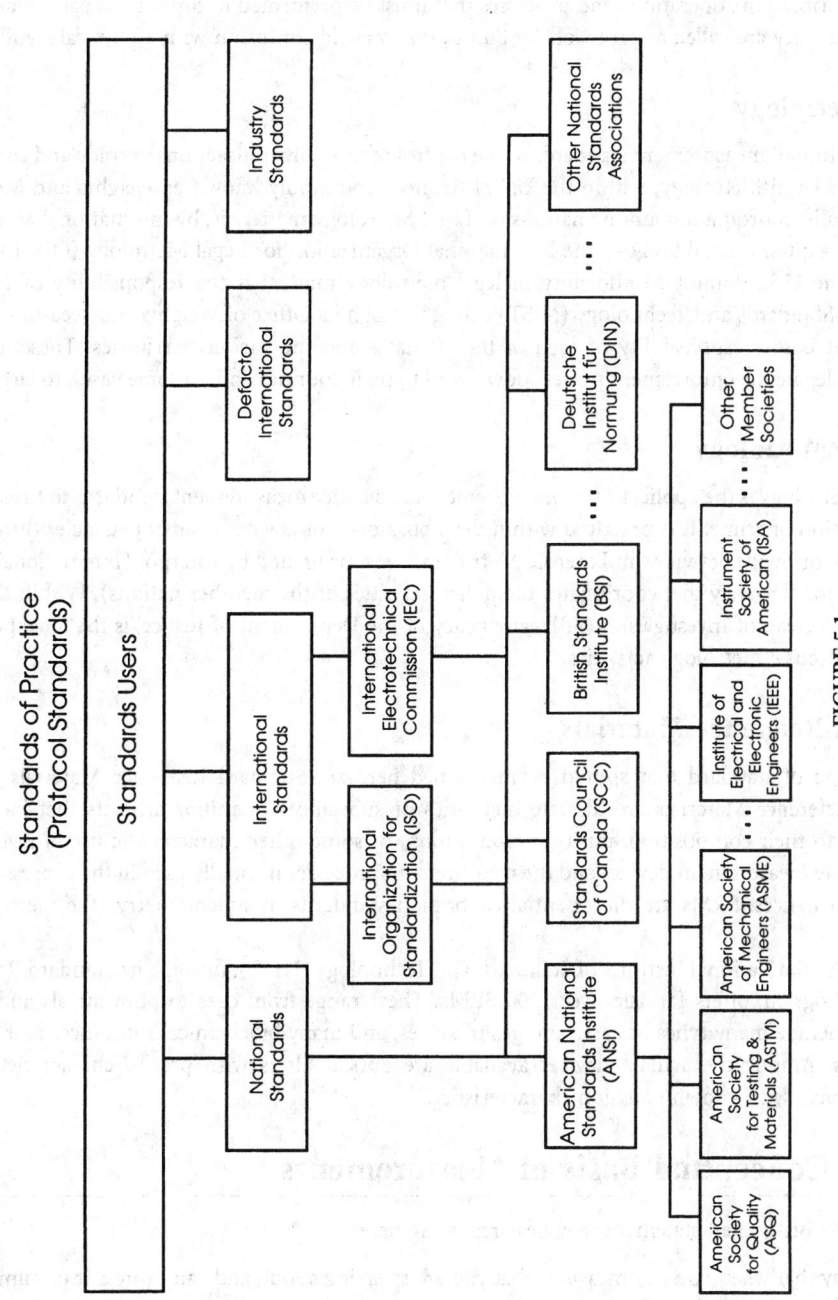

FIGURE 5.1

Standards of Practice (Protocol Standards)

These standards define everything from the dimensions and electrical characteristics of a flashlight battery to the shape of the threads on a machine screw and from the size and shape of an IBM punched card to the Quality Assurance Requirements for Measuring Equipment. Such standards can be defined as documents describing the operations and processes that must be performed in order for a particular end to be achieved. They are called a "protocol" by Europeans to avoid confusion with a physical standard.

Legal Metrology

The application of measurement standards to the control of the daily transactions of trade and commerce is known as Legal Metrology; within the U.S., it is more commonly known as Weights and Measures. Internationally, coordination among nations on Legal Metrology matters is, by international agreement, handled by a quasi-official body — the International Organization for Legal Metrology (OIML).

Within the U.S., domestic uniformity in legal metrology matters is the responsibility of National Institute of Standards and Technology (NIST) acting through its Office of Weights and Measures. Actual enforcement is the responsibility of each of the 50 states and the various territories. These, in turn, generally delegate the enforcement powers downward to their counties and, in some cases, to large cities.

Forensic Metrology

Forensic Metrology is the application of measurements and hence measurement standards to the solution and prevention of crime. It is practiced within the laboratories of law enforcement agencies throughout the world. Worldwide activities in Forensic Metrology are coordinated by Interpol (*International Police*; the international agency that coordinates the police activities of the member nations). Within the U.S., the Federal Bureau of Investigation (FBI), an agency of the Department of Justice, is the focal point for most U.S. forensic metrology activities.

Standard Reference Materials

Another type of standard that should be mentioned here are Standard Reference Materials (SRM). Standard Reference Materials are discrete quantities of substances or minor artifacts that have been certified as to their composition, purity, concentration, or some other characteristic useful in the calibration of the measurement devices and the measurement processes normally used in the process control of those substances. SRMs are the essential calibration standards in stoichiometry (the metrology of chemistry).

In the U.S., the National Institute of Standards and Technology (NIST), through its Standard Reference Materials Program, offers for sale over 1300 SRMs. These range from ores to pure metals and alloys. They also include many types of gases and gas mixtures; and many biochemical substances and organic compounds. Among the artifact devices available are optical filters with precise characteristics and standard lamps with known emission characteristics.

5.3   A Conceptual Basis of Measurements

Lord Kelvin's oft-quoted statement may bear repeating here:

> I often say that when you can measure what you are speaking about, and can express it in numbers, you know something about it; but when you cannot measure it, cannot express it in numbers, your knowledge is of a meager and unsatisfactory kind; it may be the beginnings of knowledge, but you have scarcely, in your thoughts, advanced to the stage of science, whatever the matter may be. So therefore, if science is measurement, then without metrology there can be no science.

William Thomson (Lord Kelvin), May 6, 1886

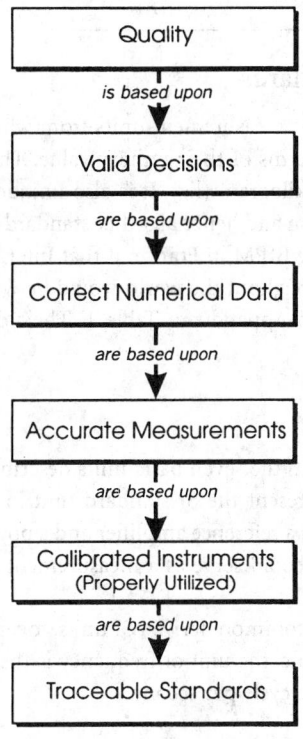

FIGURE 5.2

Lord Kelvin's statement has been quoted so many times that it has almost become trite, but looking at Figure 5.2 will show an interesting hierarchy. In order to achieve quality or "to do things right," it is necessary to make some decisions. The correct decisions cannot be made unless there are good numerical data on which to base those decisions. Those numerical data, in turn, must come from measurements and if "correct" decisions are really needed, they must be based on the "right" numbers. The only way to get "good" numerical data is to make accurate measurements using calibrated instruments that have been properly utilized. Finally, if it is important to compare those measurements to other measurements made at other places and other times, the instruments must be calibrated using traceable standards.

5.4 The Need for Standards

Standards define the units and scales in use, and allow comparison of measurements made in different times and places. For example, buyers of fuel oil are charged by a unit of liquid volume. In the U.S., this would be the gallon; but in most other parts of the world, it would be the liter. It is important for the buyer that the quantity ordered is actually received and the refiner expects to be paid for the quantity shipped. Both parties are interested in accurate measurements of the volume and, therefore, need to agree on the units, conditions, and method(s) of measurement to be used.

Persons needing to measure a mass cannot borrow the primary standard maintained in France or even the national standard from the National Institute of Standards and Technology (NIST) in the U.S. They must use lower-level standards that can be checked against those national or international standards. Everyday measuring devices, such as scales and balances, can be checked (calibrated) against working level mass standards from time to time to verify their accuracy. These working-level standards are, in turn, calibrated against higher-level mass standards. This chain of calibrations or checking is called "traceability." A proper chain of traceability must include a statement of uncertainty at every step.

5.5 Types of Standards

Basic or Fundamental Standards

In the SI system, there are seven basic measurement units from which all other units are derived. All of the units except one are defined in terms of their unitary value. The one exception is the unit of mass. It is defined as 1000 grams (g) or 1 kilogram (kg). It is also unique in that it is the only unit currently based on an artifact. The U.S. kilogram and hence all other standards of mass are based on one particular platinum/iridium cylinder kept at the BIPM in France. If that International Prototype Kilogram were to change, all other mass standards throughout the world would be wrong.

The seven basic units are listed in Appendix 1, Table 1. Their definitions are listed in Appendix 1, Table 2.

Derived Standards

All of the other units are derived from the seven basic units described in Appendix 1, Table 1. Measurement standards are devices that represent the SI standard unit in a measurement. (For example, one might use a zener diode together with a reference amplifier and a power source to supply a known voltage to calibrate a digital voltmeter. This could serve as a measurement standard for voltage and be used as a reference in a measurement.)

Appendix 1, Table 3 lists the most common derived SI units, together with the base units that are used to define the derived unit. For example, the unit of frequency is the hertz; it is defined as the reciprocal of time. That is, 1 hertz (1 Hz) is one cycle per second.

The Measurement Assurance System

Figure 5.3 illustrates the interrelationship of the various categories of standards throughout the world. While it gives more detail to U.S. structure, similar structures exist in other nations. Indeed, a variety of regional organizations exist that help relate measurements made in different parts of the world to each other.

5.6 Numbers, Dimensions, and Units

A measurement is always expressed as a multiple (or submultiple) of some unit quantity. That is, both a numeric value and a unit are required. If electric current were the measured quantity, it might be expressed as some number of milliamperes or even microamperes. It is easy to take for granted the existence of the units used, because their names form an indispensable part of the vocabulary.

5.7 Multiplication Factors

Since it is inconvenient to use whole units in many cases, a set of multiplication factors has been defined that can be used in conjunction with the units to bring a value being measured to a more reasonable size. It would be difficult to have to refer to large distances in terms of the meter; thus, one defines longer distances in terms of kilometers. Short distances are stated in terms of millimeters, micrometers, nanometers, etc. See Appendix 1, Table 4.

Defining Terms

Most of the definitions in this listing were taken from the *International Vocabulary of Basic and General Terms in Metrology*, published by the ISO, 1993 (VIM) [7]. They are indicated by the inclusion (in brackets) of their number designation in the VIM. The remainder of the definitions are not intended to

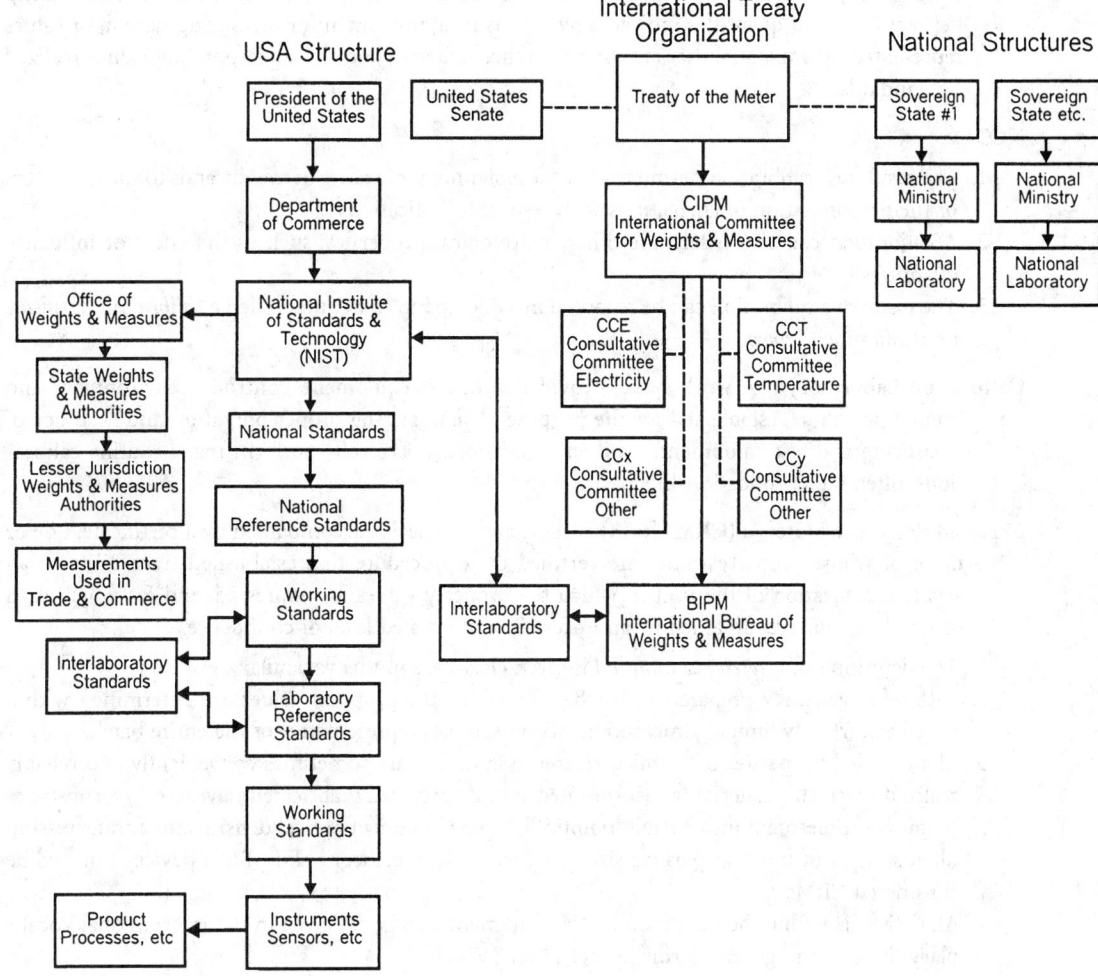

FIGURE 5.3

represent any official agency but are ones widely accepted and are included to help in the understanding of this material. More detailed and rigorous definitions can be found in other works available from ANSI, IEC, ISO, and NIST. Words enclosed in parentheses "(…)" may be omitted from the term if it is unlikely that such omission will cause confusion.

Accuracy of measurement [3.5]:* The closeness of the agreement between the result of a measurement and a true value of the measurand.

NOTES:

1. *Accuracy* is a qualitative concept.
2. The term *precision* should not be used for *accuracy*. (Precision only implies repeatability.)

Note, that to say an instrument is accurate to 5% (a common way of stating it) is wrong. One would not find such an instrument very useful if it, in fact, were only accurate 5% of the time. What is meant when such a statement is made is that the instrument's inaccuracy is less than 5% and it is accurate to better than 95%. Unfortunately, this statement is almost as imprecise as "accurate to 5%." An instrument would not be useful if it were accurate only 95% of the time; but this is not what is implied by "5% accuracy." What is meant is that, (almost) all of the time, its indication is within 5% of the "true" value.

Calibration [6.11]: A set of operations that establish, under specified conditions, the relationship between values of quantities indicated by a measuring instrument or measuring system, or values represented by a material measure or a reference material, and the corresponding values realized by standards.

NOTES:

1. The result of a calibration permits either the assignment of values of measurands to the indicators or the determination of corrections with respect to indications.
2. A calibration can also determine other metrological properties, such as the effect of influence quantities.
3. The result of a calibration can be recorded in a document, sometimes called a *calibration certificate* or a *calibration report*.

Calibration Laboratory: A work space, provided with test equipment, controlled environment and trained personnel, established for the purpose of maintaining proper operation and accuracy of measuring and test equipment. *Calibration laboratories* typically perform many routine calibrations, often on a production-line basis.

Certified Reference Material (CRM) [6.14]: A *reference material*, accompanied by a certificate, one or more of whose property values are certified by a procedure that established traceability to an accurate realization of the unit in which the property values are expressed, and for which each certified value is accompanied by an uncertainty at a stated level of confidence.

1. The definition of a *reference material* is given elsewhere in this vocabulary.
2. CRMs are generally prepared in batches for which the property values are determined within stated uncertainty limits by measurements on samples representative of the entire batch.
3. The certified properties of certified reference materials are sometimes conveniently and reliably realized when the material is incorporated into a specifically fabricated device, e.g., a substance of known triple-point into a triple-point cell, a glass of known optical density into a transmission filter, spheres of uniform particle size mounted on a microscope slide. Such devices can also be considered CRMs.
4. All CRMs lie within the definition of "measurement standards" given in the International Vocabulary of basic and general terms in metrology (VIM).
5. Some RMs and CRMs have properties that, because they cannot be correlated with an established chemical structure or for other reasons, cannot be determined by exactly defined physical and chemical measurement methods. Such materials include certain biological materials such as vaccines to which an International unit has been assigned by the World Health Organization.

This definition, including the Notes, is taken from ISO Guide 30:1992.

Coherent (derived) unit (of measurement) [1.10]: A derived unit of measurement that may be expressed as a product of powers of base units with the proportionality factor one (1).

NOTE: Coherency can be determined only with respect to the base units of a particular system. A unit can be coherent with respect to one system but not to another.

Coherent system of units (of measurement) [1.11]: A system of units of measurement in which all of the derived units are coherent.

Conservation of a (measurement) standard [6.12]: A set of operations, necessary to preserve the metrological characteristics of a measurement standard within appropriate limits.

NOTE: The operations commonly include periodic calibration, storage under suitable conditions, and care in use.

Interlaboratory Standard: A device that travels between laboratories for the sole purpose of relating the magnitude of the physical unit represented by the standards maintained in the respective laboratories.

International (measurement) standard [6.2]:   A standard recognized by an international agreement to serve internationally as the basis for assigning values to other standards of the quantity concerned.

International System of Units (SI) [1.12]:   The coherent system of units adopted and recommended by the General Conference on Weights and Measures (CGPM).

NOTE: The SI is based at present on the following seven base units: meter, kilogram, second, ampere, kelvin, mole, and candela.

Measurand [2.6]:   A particular quantity subject to measurement.

EXAMPLE: Vapor pressure of a given sample of water at 20°C.

NOTE: The specification of a measurand may require statements about quantities such as time, temperature, and pressure.

Measurement [2.1]:   A set of operations having the object of determining a value of a quantity.

NOTE: The operations may be performed automatically.

Method of Measurement [2.4]:   A logical sequence of operations, described generically, used in the performance of measurements.

NOTE: Methods of measurement may be qualified in various ways, such as:
  • Substitution method
  • Differential method
  • Null method

Metrology [2.2]:   The science of measurement.

NOTE: Metrology includes all aspects, both theoretical and practical, with reference to measurements, whatever their uncertainty, and in whatever fields of science or technology they occur.

National (measurement) Standard [6.3]:   A standard recognized by a national decision to serve, in a country, as the basis for assigning values to other standards of the quantity concerned.

National Reference Standard: A standard maintained by national laboratories such as the National Institute of Standards and Technology (NIST) in Gaithersburg, MD; the National Research Council (NRC) located in Ottawa, Canada; the National Physical Laboratory (NPL) in Teddington, U.K.; the Physikalisch-Technische Bundesanstalt (PTB) at Braunschweig, Germany; and which are the legal standards of their respective countries.

National Institute of Standards and Technology (NIST):   The U.S. national standards laboratory, responsible for maintaining the physical standards upon which measurements in the U.S. are based.

Primary Standard [6.4]:   A standard that is designated or widely acknowledged as having the highest metrological qualities and whose value is accepted without reference to other standards of the same quantity.

NOTE: The concept of primary standard is equally valid for base quantities and derived quantities.

Principle of Measurement [2.3]:   The scientific base of a measurement.

EXAMPLES:
  • The thermoelectric effect applied to the measurement of temperature
  • The Josephson effect applied to the measurement of electric potential difference
  • The Doppler effect applied to the measurement of velocity
  • The Raman effect applied to the measurement of the wave number of molecular vibrations

Reference Standard [6.6]:   A standard, generally having the highest metrological quality available at a given location or in a given organization, from which measurements made there are derived.

Reference Material [6.13]:   A material or substance, one or more of whose property values are sufficiently homogeneous and well established to be used for the calibration of an apparatus, the assessment of a measurement method, or for assigning values to materials.

NOTE: A reference material can be in the form of a pure or mixed gas, liquid or solid. Examples are water for the calibration of viscometers, sapphire as a heat-capacity calibrant in calorimetry, and solutions used for calibration in chemical analysis.

This definition, including the Note, is taken from ISO Guide 30:1992.

Repeatability (of results of measurements) [3.6]:   The closeness of the agreement between the results of successive measurements of the same measurand carried out under the same conditions of measurement.

NOTES:

1. These conditions are called *repeatability conditions*.
2. Repeatability conditions include:

    a. The same measurement process
    b. The same observer
    c. The same measuring instrument, used under the same conditions
    d. The same location
    e. Repetition over a short period of time

3. Repeatability can be expressed quantitatively in terms of the dispersion of characteristics of the results.

Reproducibility (of results of measurements) [3.7]:   The closeness of the agreement between the results of measurements of the same measurand carried out under changed conditions of measurement.

NOTES:

1. A valid statement of reproducibility requires specification of the conditions changed.
2. The changed conditions include:

    a. Principle of measurement
    b. Method of measurement
    c. Observer
    d. Measuring instrument
    e. Reference standard
    f. Location
    g. Condition of use
    h. Time

3. Reproducibility can be expressed quantitatively in terms of the dispersion characteristics of the results.
4. Results here are usually understood to be corrected results.

Secondary Standard [6.5]:   A standard whose value is assigned by comparison with a primary standard of the same quantity.

Standards Laboratory: A work space, provided with equipment and standards, a properly controlled environment, and trained personnel, established for the purpose of maintaining traceability of standards and measuring equipment used by the organization it supports. Standards laboratories typically perform fewer, more specialized and higher accuracy measurements than Calibration Laboratories.

Tolerance: In metrology, the limits of the range of values (the uncertainty) that apply to a properly functioning measuring instrument.

Traceability [6.10]: The property of the result of a measurement or the value of a standard whereby it can be related to stated references, usually national or international standards, through an unbroken chain of comparisons all having stated uncertainties.

NOTE:
1. The concept is often expressed by the adjective *traceable.*
2. The unbroken chain of comparisons is called a *traceability chain.*

Even though the ISO has published (and accepted) the definition listed above, many practitioners endeavor to make this term more meaningful. They feel that the definition should introduce the aspect of evidence being presented on a continuing basis, to overcome the idea that if valid traceability is achieved, it could last forever. A definition similar to the following one would meet that requirement.

Traceability is a characteristic of a calibration or a measurement. A traceable measurement or calibration is achieved only when each instrument and standard, in a hierarchy stretching back to the national (or international) standard was itself properly calibrated and the results properly documented including statements of uncertainty on a continuing basis. The documentation must provide the information needed to show that all the calibrations in the chain of calibrations were appropriately performed.

Transfer Standard [6.8]: A standard used as an intermediary to compare standards.

NOTE: The term *transfer device* should be used when the intermediary is not a standard.

Traveling Standard [6.9]: A standard, sometimes of special construction, intended for transport between locations.

EXAMPLE: A portable battery-operated cesium frequency standard.

Uncertainty of Measurement [3.9]: A parameter, associated with the result of a measurement, that characterizes the dispersion of the values that could reasonably be attributed to the measurand.

NOTES:
1. The parameter can be, for example, a standard deviation (or a given multiple of it), or the half-width of an interval having a stated level of confidence.
2. Uncertainty of measurement comprises, in general, many components. Some of these components can be evaluated from the statistical distribution of the results of series of measurements and can be characterized by experimental standard deviations. The other components, which can also be characterized by standard deviations, are evaluated from assumed probability distributions based on experience or other information.
3. It is understood that the result of the measurement is the best estimate of the value of the measurand, and that all components of uncertainty (including those arising from systematic effects) such as components associated with corrections and reference standards, contribute to the dispersion.

This definition is that of the *Guide to the Expression of Uncertainty in Measurement,* in which its rationale is detailed (see, in particular, 2.2.4 and annex D).[4]

Value (of a quantity) [1.18]: The magnitude of a particular quantity generally expressed as a unit of measurement multiplied by a number.

EXAMPLES:
- Length of a rod: 5.34 m or 534 cm
- Mass of a body: 0.152 kg or 152 g
- Amount of substance of a sample of water (H_2O): 0.012 mol or 12 mmol

NOTES:
1. The value of a quantity can be positive, negative, or zero.
2. The value of a quantity can be expressed in more than one way.

3. The values of quantities of dimension one are generally expressed as pure numbers.
4. A quantity that cannot be expressed as a unit of measurement multiplied by a number can be expressed by reference to a conventional reference scale or to a measurement procedure or both.

Working Standard [6.7]:   A standard that is used routinely to calibrated or check material measures, measuring instruments or reference materials.

NOTES:

1. A working standard is usually calibrated against a *reference standard*.
2. A working standard used routinely to ensure that a measurement is being carried out correctly is called a *check standard*.

References

1. NIST Special Publication 250 Appendix, Fee Schedule, U.S. Dept of Commerce, Technology Administration, National Institute of Standards and Technology, Calibration Program, Bldg. 820, Room 232, Gaithersburg, MD, 20899-0001.
2. B. N. Taylor, NIST Special Publication 811, 1995 edition, *Guide for the Use of the International System of Units (SI)*, U.S. Department of Commerce, National Institute of Standards and Technology, Gaithersburg, MD, 20899-0001, 1995.
3. H. A. Klein, *The Science of Measurement: A Historical Survey*, New York: Dover Publications, Inc., 1974.
4. B. N. Taylor and C. E. Kuyatt, Guidelines for Evaluating and Expressing the Uncertainty of NIST Measurement Results, NIST Technical Note 1297. (1994 ed.).
5. R. C. Cochrane, Measures for Progress — History of the National Bureau of Standards, published by the United States Department of Commerce. (1966) Library of Congress Catalog Card Number: 65-62472.
6. NIST Standard Reference Material Catalog, NIST Special Publication 260, NIST CODEN:XNBSAV, Available from the Superintendent of Documents, Washington, D.C. 20402.
7. *International Vocabulary of Basic and General Terms in Metrology*, ISO, 1993.

II

Spatial Variables Measurement

6

Displacement Measurement, Linear and Angular

Keith Antonelli
Kinetic Sciences Inc.

Viktor P. Astakhov
Concordia University

Amit Bandyopadhyay
Rutgers University

Vikram Bhatia
Virginia Tech

Richard O. Claus
Virginia Tech

David Dayton
ILC Data Device Corp.

Halit Eren
Curtin University of Technology

Robert M. Hyatt, Jr.
Howell Electric Motors

Victor F. Janas
Rutgers University

Nils Karlsson
National Defense Research Establishment

Andrei Kholkine
Rutgers University

James Ko
Kinetic Sciences, Inc.

Wei Ling Kong

Shyan Ku
Kinetic Sciences Inc.

J.R. René Mayer
Ecole Polytechnique de Montreal

David S. Nyce
MTS Systems Corp.

Teklic Ole Pedersen
Linkopings Universitet

Ahmad Safari
Rutgers University

Anbo Wang
Virginia Tech

Grover C. Wetsel
University of Texas at Dallas

Bernhard Günther Zagar
Technical University Graz

6.1 Resistive Displacement Sensors

Keith Antonelli, James Ko, and Shyan Ku

Resistive displacement sensors are commonly termed potentiometers or "pots." A pot is an electromechanical device containing an electrically conductive *wiper* that slides against a fixed *resistive element* according to the position or angle of an external shaft. See Figure 6.1. Electrically, the resistive element is "divided" at the point of wiper contact. To measure displacement, a pot is typically wired in a "voltage divider" configuration, as shown in Figure 6.2. The circuit's output, a function of the wiper's position, is an analog voltage available for direct use or digitization. Calibration maps the output voltage to units of displacement.

Table 6.1 lists some attributes inherent to pots. This chapter describes the different types of pots available, their electrical and mechanical characteristics, and practical approaches to using them for precision measurement. Sources and typical prices are also discussed. Versatile, inexpensive, and easy-to-use, pots are a popular choice for precision measurement.

Precision Potentiometers

Pots are available in great variety, with specific kinds optimized for specific applications. Position measurement requires a high-quality pot designed for extended operation. Avoid pots classified as trimmers, rheostats, attenuators, volume controls, panel controls, etc. Instead, look for *precision potentiometers*.

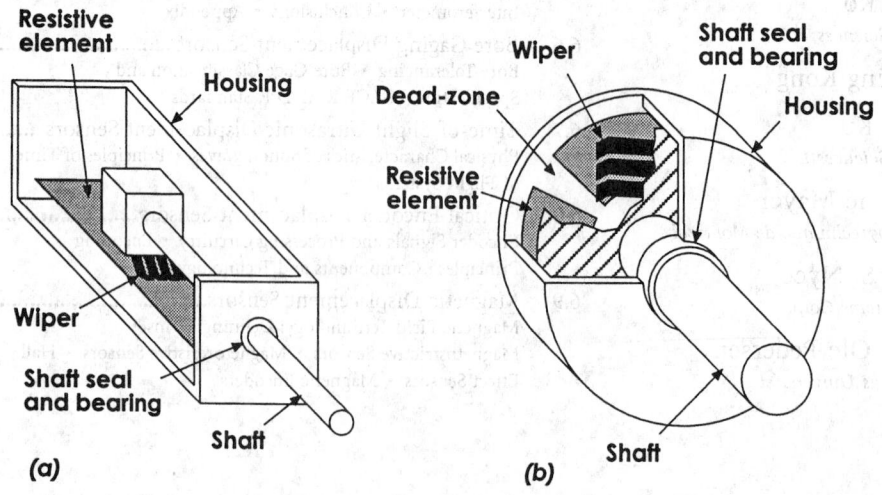

FIGURE 6.1 Representative cutaways of linear-motion (*a*) and rotary (*b*) potentiometers.

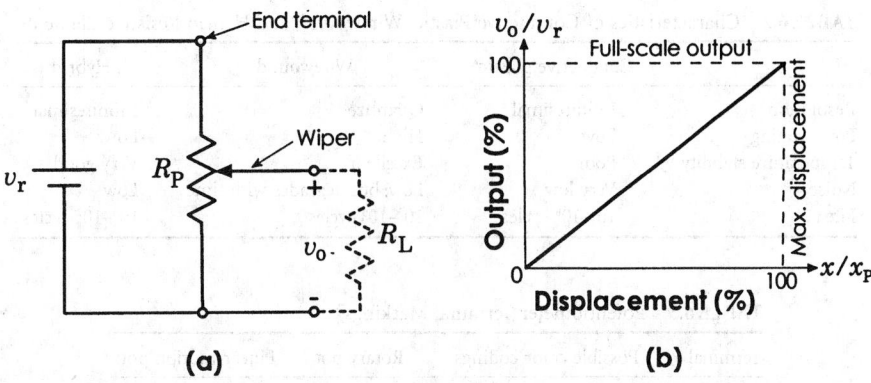

FIGURE 6.2 (*a*) Schematic diagrams depict a potentiometer as a resistor with an arrow representing the wiper. This schematic shows a pot used as a variable voltage divider — the preferred configuration for precision measurement. R_P is the total resistance of the pot, R_L is the load resistance, v_r is the reference or supply voltage, and v_o is the output voltage. (*b*) shows an ideal linear output function where x represents the wiper position, and x_P is its maximum position.

TABLE 6.1 Fundamental Potentiometer Characteristics

Advantages	Disadvantages
Easy to use	Limited bandwidth
Low cost	Frictional loading
Nonelectronic	Inertial loading
High-amplitude output signal	Wear
Proven technology	

Types of Precision Potentiometers

Precision pots are available in *rotary*, *linear-motion*, and *string pot* forms. String pots — also called *cable pots, yo-yo pots, cable extension transducers*, and *draw wire transducers* — measure the extended length of a spring-loaded cable. Rotary pots are available with single- or multiturn abilities: commonly 3, 5, or 10 turns. Linear-motion pots are available with maximum strokes ranging from roughly 5 mm to over 4 m [1, 2]. String pots are available with maximum extensions exceeding 50 m [3]. Pot manufacturers usually specify a pot's type, dimensions, resistive element composition, electrical and mechanical parameters, and mounting method.

Resistive Element

Broadly, a pot's resistive element can be classified as either *wirewound*, or *nonwirewound*. Wirewound elements contain tight coils of resistive wire that quantize measurement in step-like increments. In contrast, nonwirewound elements present a continuous sheet of resistive material capable of essentially unlimited measurement resolution.

Wirewound elements offer excellent temperature stability and high power dissipation abilities. The coils quantize measurement according to wire size and spacing. Providing the resolution limits are acceptable, wirewound elements can be a satisfactory choice for precision measurement; however, conductive plastic or hybrid elements will usually perform better and for considerably more cycles. These and other popular nonwirewound elements are described in more detail below.

Conductive plastic elements feature a smooth film with unlimited resolution, low friction, low noise, and long operational life. They are sensitive to temperature and other environmental factors and their power dissipation abilities are low; however, they are an excellent choice for most precision measurement applications.

TABLE 6.2   Characteristics of Conductive Plastic, Wirewound, and Hybrid Resistive Elements

	Conductive plastic	Wirewound	Hybrid
Resolution	Infinitesimal	Quantized	Infinitesimal
Power rating	Low	High	Low
Temperature stability	Poor	Excellent	Very good
Noise	Very low	Low, but degrades with time	Low
Life	10^6–10^8 cycles	10^5–10^6 cycles	10^6–10^7 cycles

TABLE 6.3   Potentiometer Terminal Markings

Terminal	Possible color codings			Rotary pot	Linear-motion pot
1	Yellow	Red	Black	CCW limit	Fully retracted limit
2	Red	Green	White	Wiper	Wiper
3	Green	Black	Red	CW limit	Fully extended limit

Hybrid elements feature a wirewound core with a conductive plastic coating, combining wirewound and conductive plastic technologies to realize some of the more desirable attributes of both. The plastic limits power dissipation abilities in exchange for low noise, long life, and unlimited resolution. Like wirewounds, hybrids offer excellent temperature stability. They make an excellent choice for precision measurement.

Cermet elements, made from a ceramic-metal alloy, offer unlimited resolution and reasonable noise levels. Their advantages include high power dissipation abilities and excellent stability in adverse conditions. Cermet elements are rarely applied to precision measurement because conductive plastic elements offer lower noise, lower friction, and longer life.

Carbon composition elements, molded under pressure from a carbon–plastic mixture, are inexpensive and very popular for general use, but not for precision measurement. They offer unlimited resolution and low noise, but are sensitive to environmental stresses (e.g., temperature, humidity) and are subject to wear.

Table 6.2 summarizes the distinguishing characteristics of the preferred resistive elements for precision measurement.

Electrical Characteristics

Before selecting a pot and integrating it into a measurement system, the following electrical characteristics should be considered.

Terminals and Taps

Table 6.3 shows the conventional markings found on the pot housing [4, 5]; CW and CCW indicate clockwise and counter-clockwise rotation as seen from the front end. Soldering studs and eyelets, integral connectors, and flying leads are common means for electrical connection. In addition to the wiper and end terminals, a pot may possess one or more terminals for *taps*. A tap enables an electrical connection to be made with a particular point along the resistive element. Sometimes, a *shunt resistor* is connected to a tap in order to modify the output function. End terminations and taps can exhibit different electrical characteristics depending on how they are manufactured. See [2] for more details.

Taper

Pots are available in a variety of different tapers that determine the shape of the output function. With a linear-taper pot, the output varies linearly with wiper motion, as shown in Figure 6.2. (Note that a pot with a linear taper should not be confused with a linear-motion pot, which is sometimes called a "linear pot.") Linear-taper pots are the most commonly available, and are widely used in sensing and control

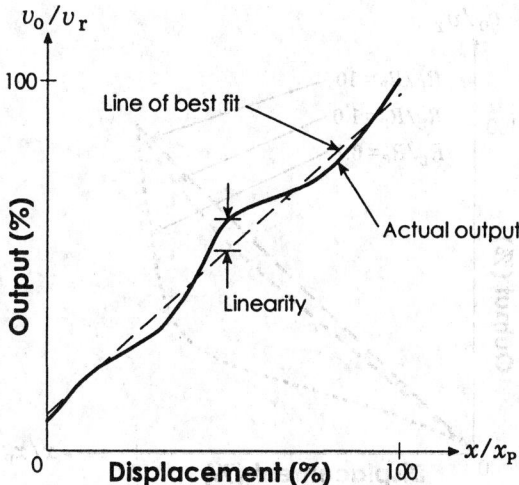

FIGURE 6.3 Independent linearity is the maximum amount by which the actual output function deviates from a line of best fit.

applications. Pots with nonlinear tapers (e.g., logarithmic, sine, cosine, tangent, square, cube) can also be useful, especially where computer control is not involved. Nonstandard tapers can be custom-manufactured or alternatively, certain types of output functions can be produced using shunt resistors, by combining outputs from ganged pots or by other means. (Refer to [6, 7] for more details.) Of course, if a computer is involved, the output function can always be altered through a software lookup table or mapping function.

Electrical Travel

Figure 6.2 shows how the ideal output of a pot changes with wiper position. In practice, there is a small region at both ends where output remains constant until the wiper hits a mechanical stop. *Mechanical travel* is the total motion range of the wiper, and *electrical travel* is the slightly smaller motion range over which the electrical output is "valid." Thus, when using a pot as a sensor, it is important to ensure that the wiper motion falls within the electrical travel limits.

Linearity

Linearity is the maximum deviation of the output function from an ideal straight line. *Independent linearity* is commonly specified, where the straight line is defined as the line that minimizes the linearity error over a series of sampled points, not necessarily measured over the full range of the pot. See Figure 6.3. Other linearity metrics, such as *terminal-based linearity, absolute linearity,* and *zero-based linearity,* are also sometimes used. Refer to [8] for more details. Pots are commonly available with independent linearities ranging from under 0.1% to 1%. When dealing with nonlinear output functions, *conformity* is specified since it is the more general term used to describe deviation from any ideal function. Conformity and linearity are usually expressed as a percentage of full-scale output (FSO).

Electrical Loading

Loading can significantly affect the linearity of measurements, regardless of a pot's quality and construction. Consider an ideal linear pot connected to an infinite load impedance (i.e., as in Figure 6.2). Since no current flows through the load, the output changes perfectly linearly as the wiper travels along the length of the pot. However, if the load impedance is finite, the load draws some current, thereby affecting the output as illustrated in Figure 6.4. Circuit analysis shows that:

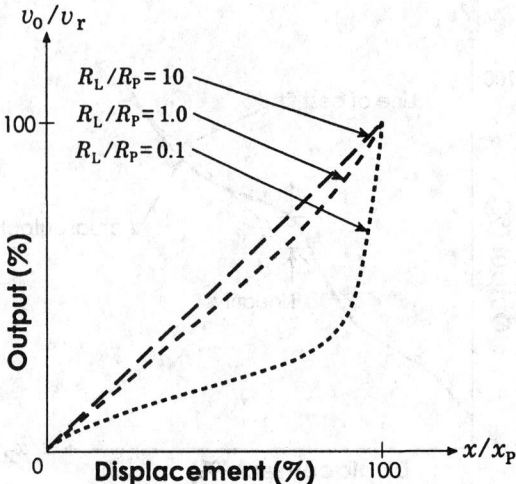

FIGURE 6.4 Linearity can be greatly influenced by the ratio of load resistance, R_L, to potentiometer resistance, R_P.

$$\frac{v_o}{v_r} = \frac{\left(x/x_P\right)\left(R_L/R_P\right)}{\left(R_L/R_P\right)+\left(x/x_P\right)-\left(x/x_P\right)^2} \tag{6.1}$$

Therefore, R_L/R_P should be maximized to reduce loading effects (this also involves other trade-offs, to be discussed). A minimum R_L/R_P value of 10 is sometimes used as a guideline since loading error is then limited to under 1% of full-scale output. Also, some manufacturers recommend a minimum load impedance or maximum wiper current in order to minimize loading effects and prevent damage to the wiper contacts. The following are some additional strategies that can be taken:

- Use a regulated voltage source whose output is stable with load variations
- Use high input-impedance signal conditioning or data acquisition circuitry
- Use only a portion of the pot's full travel

Resolution

Resolution defines the smallest possible change in output that can be produced and detected. In wirewound pots, the motion of the wiper over the coil generates a quantized response. Therefore, the best attainable resolution is $r = (1/N) \times 100\%$, where N is the number of turns in the coil. Nonwirewound pots produce a smooth response with essentially unlimited resolution. Hybrid pots also fall into this category. In practice, resolution is always limited by factors such as:

- Electrical noise, usually specified as *noise* for wirewound pots and *smoothness* for nonwirewound pots, both expressed as a percentage of full-scale output [10]
- Stability of the voltage supply, which can introduce additional noise into the measurement signal
- Analog-to-digital converter (ADC) resolution, usually expressed in "bits" (e.g., 10 mm travel digitized using a 12-bit ADC results in 10 mm/4096 = 0.0024 mm resolution at best)
- Mechanical effects such as stiction

Power Rating

The power dissipated by a pot is $P = v_r^2/R_P$. Therefore, power rating determines the maximum voltage that can be applied to the pot at a given temperature. With greater voltage supplied to the pot, greater output (and noise) is produced but more power is dissipated, leading to greater thermal effects. In general, wirewound and cermet pots are better able to dissipate heat, and thus have the highest power ratings.

Temperature Coefficient

As temperature increases, pot resistance also increases. However, a pot connected as shown in Figure 6.2 will divide the voltage equally well, regardless of its total resistance. Thus, temperature effects are not usually a major concern as long as the changes in resistance are uniform and the pot operates within its ratings. However, an increase in pot resistance also increases loading nonlinearities. Therefore, temperature coefficients can become an important consideration. The temperature coefficient, typically specified in ppm $°C^{-1}$, can be expressed as $\alpha = (\Delta R_p/R_p)/\Delta t$, where Δt is the change in temperature and ΔR_p is the corresponding change in total resistance. In general, wirewound pots possess the lowest temperature coefficients. Temperature-compensating signal-conditioning circuitry can also be used.

Resistance

Since a pot divides voltage equally well regardless of its total resistance, resistance tolerance is not usually a major concern. However, total resistance can have a great impact on loading effects. If resistance is large, less current flows through the pot, thus reducing temperature effects, but also increasing loading.

AC Excitation

Pots can operate using either a dc or an ac voltage source. However, wirewound pots are susceptible to capacitive and inductive effects that can be substantial at moderate to high frequencies.

Mechanical Characteristics

The following mechanical characteristics influence measurement quality and system reliability, and thus should be considered when selecting a pot.

Mechanical Loading

A pot adds inertia and friction to the moving parts of the system that it is measuring. As a result, it increases the force required to move these parts. This effect is referred to as *mechanical loading*. To quantify mechanical loading, rotary pot manufacturers commonly list three values: the equivalent *mass moment of inertia* of the pot's rotating parts, the *dynamic* (or *running*) *torque* required to maintain rotation in a pot shaft, and the *starting torque* required to initiate shaft rotation. For linear-motion pots, the three analogous loading terms are *mass, starting force*, and *dynamic* (or *running*) *force*.

In extreme cases, mechanical loading can adversely affect the operating characteristics of a system. When including a pot in a design, ensure that the inertia added to the system is insignificant or that the inertia is considered when analyzing the data from the pot. The starting and running force or torque values might also be considered, although they are generally small due to the use of bearings and low-friction resistive elements.

Mechanical Travel

Distinguished from electrical travel, *mechanical travel* is the wiper's total motion range. A mechanical stop delimits mechanical travel at each end of the wiper's range of motion. Stops can withstand small loads only and therefore should not be used as mechanical limits for the system. Manufacturers list maximum loads as the *static stopping strength* (for static loads) and the *dynamic stopping strength* (for moving loads).

Rotary pots are also available without mechanical stops. The shaft of such an "unlimited travel" pot can be rotated continuously in either direction; however, electrical travel is always less than 360° due to the discontinuity or "dead-zone" where the resistive element begins and ends. (See Figure 6.1.) Multiple revolutions can be measured with an unlimited travel pot in conjunction with a counter: the counter maintains the number of full revolutions while the pot measures subrevolution angular displacement.

Operating Temperature

When operated within its specified temperature range, a pot maintains good electrical linearity and mechanical integrity. Depending on construction, pots can operate at temperatures from as low as $-65°C$

to as high as 150°C. Operating outside specified limits can cause material failure, either directly from temperature or from thermally induced misalignment.

Vibration, Shock, and Acceleration

Vibration, shock, and acceleration are all potential sources of contact discontinuities between the wiper and the resistive element. In general, a contact failure is considered to be a discontinuity equal to or greater than 0.1 ms [2]. The values quoted in specification sheets are in *g*s and depend greatly on the particular laboratory test. Some characterization tests use sinusoidal vibration, random vibration, sinusoidal shock, sawtooth shock, or acceleration to excite the pot. Manufacturers use mechanical design strategies to eliminate weaknesses in a pot's dynamic response. For example, one technique minimizes vibration-induced contact discontinuities using multiple wipers of differing resonant frequencies.

Speed

Exceeding a pot's specified maximum speed can cause premature wear or discontinuous values through effects such as wiper bounce. As a general rule, the slower the shaft motion, the longer the unit will last (in total number of cycles). Speed limitations depend on the materials involved. For rotary pots, wirewound models have preferred maximum speeds on the order of 100 rpm, while conductive plastic models have allowable speeds as high as 2000 rpm. Linear-motion pots have preferred maximum velocities up to 10 m s^{-1}.

Life

Despite constant mechanical wear, a pot's expected lifetime is on the order of a million cycles when used under proper conditions. A quality film pot can last into the hundreds of millions of cycles. Of wirewound, hybrid, and conductive plastic pots, the uneven surface of a wirewound resistive element inherently experiences the most wear and thus has the shortest expected operating life. Hybrids improve on this by using a wirewound construction in combination with a smooth conductive film coating. Conductive plastic pots generally have the longest life expectancy due to the smooth surface of their resistive element.

Contamination and Seals

Foreign material contaminating pots can promote wear and increase friction between the wiper and the resistive element. Consequences range from increased mechanical loading to outright failure (e.g., seizing, contact discontinuity). Fortunately, sealed pots are available from most manufacturers for industrial applications where dirt and liquids are often unavoidable. To aid selection, specifications often include the type of *case sealing* (i.e., mechanisms and materials) and the *seal resistance* to cleaning solvents and other commonly encountered fluids.

Misalignment

Shaft misalignment in a pot can prematurely wear its bearing surfaces and increase its mechanical loading effects. A good design minimizes misalignment. (See *Implementation*, below.) Manufacturers list a number of alignment tolerances. In linear-motion pots, *shaft misalignment* is the maximum amount a shaft can deviate from its axis. The degree to which a shaft can rotate around its axis is listed under *shaft rotation*. In rotary pots, *shaft end play* and *shaft radial play* both describe the amount of shaft deflection due to a radial load. *Shaft runout* denotes the shaft diameter eccentricity when a shaft is rotated under a radial load.

Mechanical Mounting Methods

Hardware features on a pot's housing determine the mounting method. Options vary with manufacturer, and among rotary, linear-motion, and string pots. Offerings include custom bases, holes, tabs, flanges, and brackets — all of which secure with machine screws — and threaded studs, which secure with nuts. Linear-motion pots are available with rod or slider actuation, some with internal or external return springs. Mounting is typically accomplished by movable clamps, often supplied by the pot manufacturer. Other linear-motion pots mount via a threaded housing. For rotary pots, the two most popular mounting methods are the *bushing mount* and the *servo mount*. See Figure 6.5.

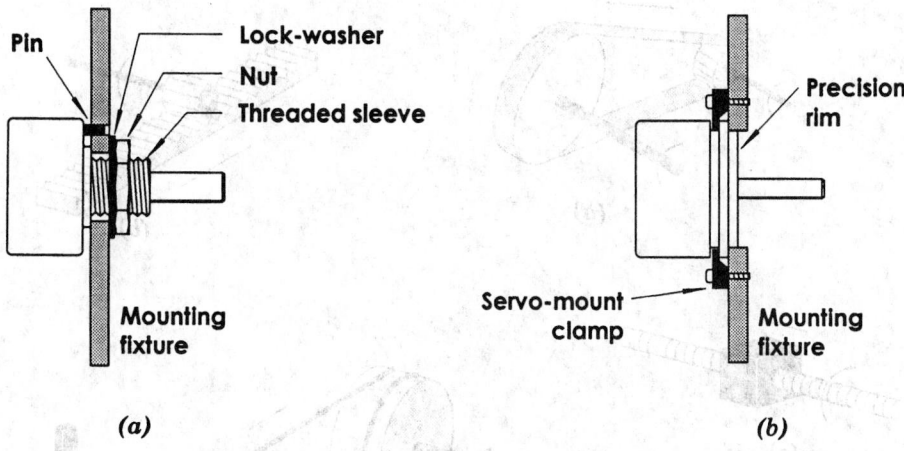

FIGURE 6.5   The two most common rotary pot mounts are the bushing mount (*a*), and the servo mount (*b*).

TABLE 6.4   Sources of Small Mechanical Components

PIC Design
   86 Benson Road, P.O. Box 1004
   Middlebury, CT 06762-1004
   Tel: (800) 243-6125, (203) 758-8272; Fax: (203) 758-8271
   www.penton.com/md/mfg/pic/
Stock Drive Products/Sterling Instrument
   2101 Jericho Turnpike, Box 5416
   New Hyde Park, NY 11042-5416
   Tel: (516) 328-3300; Fax: (800) 737-7436, (516) 326-8827
   www.sdp-si.com
W.M. Berg, Inc.
   499 Ocean Ave.
   East Rockaway, NY 11518
   Tel: (800) 232-2374, (516) 599-5010; Fax: (800) 455-2374, (516) 599-3274
   www.wmberg.com

Bushing mount
The pot provides a shaft-concentric, threaded sleeve that invades a hole in a mounting fixture and secures with a nut and lock-washer. An off-axis tab or pin prevents housing rotation. Implementing a bushing mount requires little more than drilling a hole; however, limited rotational freedom and considerable play before tightening complicate precise setup.

Servo mount
The pot provides a flanged, shaft-concentric, precision-machined rim that slips into a precision-bored hole in a mounting fixture. The flange secures with symmetrically arranged, quick-releasing *servo mount clamps*, available from Timber-Top, Inc. [9] and also from the sources listed in Table 6.4. (These clamps are also called *synchro mount clamps* and *motor mount cleats*, since servo-mounting synchros and stepper motors are also available.) Servo mounts are precise and easy to adjust, but entail the expense of precision machining.

Measurement Techniques

To measure displacement, a pot must attach to mechanical fixtures and components. The housing typically mounts to a stationary reference frame, while the shaft couples to a moving element. The *input motion* (i.e., the motion of interest) can couple directly or indirectly to the pot's shaft. A direct connection, although straightforward, carries certain limitations:

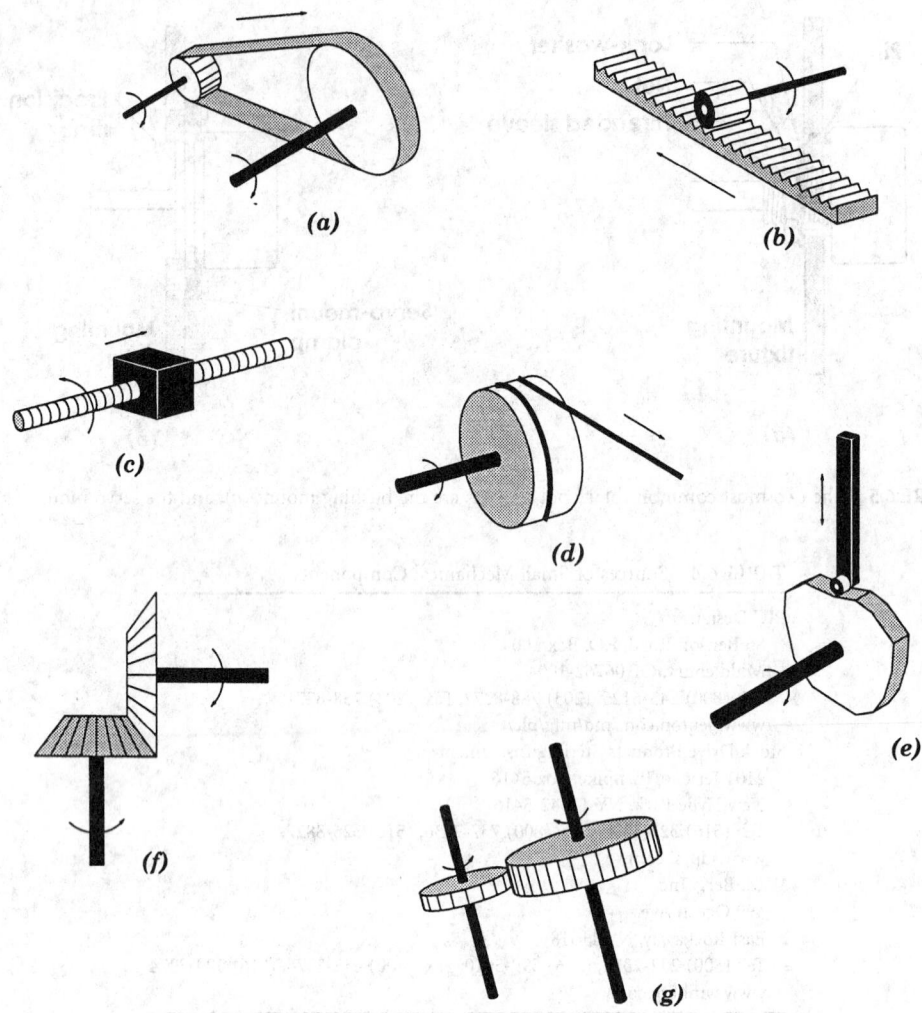

FIGURE 6.6 Mechanisms that extend a precision potentiometer's capabilities include belts and pulleys (*a*), rack-and-pinions (*b*), lead-screws (*c*), cabled drums (*d*), cams (*e*), bevel gears (*f*), and spur gears (*g*).

- The input motion maps 1:1 to the shaft motion
- The input motion cannot exceed the pot's mechanical travel limits
- Angle measurement requires a rotary pot; position measurement requires a linear-motion pot
- The pot must mount close to the motion source
- The input motion must be near-perfectly collinear or coaxial with the shaft axis

Figure 6.6 shows ways to overcome these limitations. Mechanisms with a mechanical advantage scale motion and adjust travel limits. Mechanisms that convert between linear and rotary motion enable any type of pot to measure any kind of motion. Transmission mechanisms distance a pot from the measured motion. Compliant mechanisms compensate for misalignment. Examples and more details follow. Most of the described mechanisms can be realized with components available from the sources in Table 6.4.

Gears scale the mapping between input and pot shaft motions according to gear ratio. They also displace rotation axes to a parallel or perpendicular plane according to type of gear (e.g., spur vs. bevel). Gears introduce backlash. Friction rollers are a variation on the gear theme, immune to backlash but prone to slippage. The ratio of roller diameters scales the mapping between input and pot shaft motions.

Rack-and-pinion mechanisms convert between linear and rotary motion. Mapping is determined by the rack's *linear pitch* (i.e., tooth-to-tooth spacing) compared to the number of teeth on the pinion. Backlash is inevitable.

Lead-screws convert rotary motion to linear motion via the screw principle. Certain low-friction types (e.g., ball-screws) are also capable of the reverse transformation (i.e., linear to rotary). Either way, mapping is controlled by the screw's *lead* — the distance the nut travels in one revolution. Lead-screws are subject to backlash.

Cabled drums convert between linear and rotary motion according to the drum circumference, since one turn of diameter D wraps or unwraps a length πD of cable. An external force (e.g., supplied by a spring or a weight) might be necessary to maintain cable tension.

Pulleys can direct a string pot's cable over a complex path to a motion source. Mapping is 1:1 unless the routing provides a mechanical advantage.

Pulleys and *belts* transmit rotary motion scaled according to relative pulley diameters. The belt converts between linear and rotary motion. (See Figure 6.6(a)) The empty area between pulleys provides a convenient passageway for other components. Sprocket wheels and chain have similar characteristics. Matched pulley-belt systems are available that operate with negligible slip, backlash, and stretch.

Cams map rotary motion into linear motion according to the function "programmed" into the cam profile. See [10] for more information.

Linkages can be designed to convert, scale, and transmit motion. Design and analysis can be quite complex and mapping characteristics tend to be highly nonlinear. See [11] for details.

Flexible shafts transmit rotary motion between two non-parallel axes with a 1:1 mapping, subject to torsional windup and hysteresis if the motion reverses.

Conduit, like a bicycle brake cable or *Bowden cable*, can route a cable over an arbitrary path to connect a pot to a remote motion source. The conduit should be incompressible and fixed at both ends. Mapping is 1:1 with some mechanical slop. Lubrication helps mitigate friction.

A mechanism's mapping characteristics impact measurement resolution and accuracy. Consider a stepper motor turning a lead-screw to translate a nut. A linear-motion pot could measure the nut's position directly to some resolution. Alternatively, a rotary pot linked to the lead-screw could measure the position with increased resolution if the mechanism mapped the same amount of nut travel to considerably more wiper travel. Weighing the resolution increase against the uncertainty due to backlash would determine which approach was more accurate.

Implementation

Integrating a pot into a measurement system requires consideration of various design issues, including the impact of the pot's physical characteristics, error sources, space restrictions, and wire-routing. The pot's shaft type and bearings must be taken into consideration and protected against excessive loading. A good design will:

- Give the pot mount the ability to accommodate minor misalignment
- Protect the shaft from thrust, side, and bending loads (i.e., not use the pot as a bearing)
- Provide hard limit stops within the pot's travel range (i.e., not use the pot's limit stops)
- Protect the pot from contaminants
- Strain-relieve the pot's electrical connections

A thorough treatment of precision design issues appears in [12].

Coupling to the Pot

Successful implementation also requires practical techniques for mechanical attachment. A string pot's cable terminator usually fastens to other components with a screw. For other types of pots, coupling technique is partly influenced by the nature of the shaft. Rotary shafts come with various endings, including plain, single-flatted, double-flatted, slotted, and knurled. Linear-motion shafts usually terminate in threads, but are also available with roller ends (to follow surfaces) or with spherical bearing ends (to accommodate misalignment). With care, a shaft can be cut, drilled, filed, threaded, etc.

In a typical measurement application, the pot shaft couples to a mechanical component (e.g., a gear, a pulley), or to another shaft of the same or different diameter. Successful couplings provide a positive link to the shaft without stressing the pot's mechanics. Satisfying these objectives with rotary and linear-motion pots requires a balance between careful alignment and compliant couplings. Alignment is not as critical with a string pot. Useful coupling methods include the following.

Compliant couplings. It is generally wise to put a compliant coupling between a pot's shaft and any other shafting. A compliant coupling joins two misaligned shafts of the same or different diameter. Offerings from the companies in Table 6.4 include bellows couplings, flex couplings, spring couplings, spider couplings, Oldham couplings, wafer spring couplings, flexible shafts, and universal joints. Each type has idiosyncrasies that impact measurement error; manufacturer catalogs provide details.

Sleeve couplings. Less expensive than a compliant coupling, a rigid sleeve coupling joins two shafts of the same or different diameter with the requirement that the shafts be perfectly aligned. Perfect alignment is difficult to achieve initially, and impossible to maintain as the system ages. Imperfect alignment accelerates wear and risks damaging the pot. Sleeve couplings are available from the companies listed in Table 6.4.

Press fits. A press fit is particularly convenient when the bore of a small plastic part is nominally the same as the shaft diameter. Carefully force the part onto the shaft. Friction holds the part in place, but repeated reassembly will compromise the fit.

Shrink fits. Components with a bore slightly under the shaft diameter can be heated to expand sufficiently to slip over the shaft. A firm grip results as the part cools and the bore contracts.

Pinning. Small hubbed components can be pinned to a shaft. The pin should extend through the hub partway into the shaft, and the component should fit on the shaft without play. Use roll pins or spiral pins combined with a thread-locking compound (e.g., Loctite 242).

Set-screws. Small components are available with hubs that secure with set-screws. The component should fit on the shaft without play. For best results, use two set-screws against a shaft with perpendicular flats. Dimple a plain shaft using the component's screw hole(s) as a drill guide. Apply a thread-locking compound (e.g., Loctite 242) to prevent the set-screws from working loose.

Clamping. Small components are also available with split hubs that grip a shaft when squeezed by a matching hub clamp. Clamping results in a secure fit without marring the shaft.

Adhesives. Retaining compounds (e.g., Loctite 609) can secure small components to a shaft. Follow manufacturer's instructions for best results.

Spring-loaded contact. A spring-loaded shaft will maintain positive contact against a surface that moves at reasonable speeds and without sudden acceleration.

Costs and Sources

Precision pots are inexpensive compared to other displacement measurement technologies. Table 6.5 lists approximate costs for off-the-shelf units in single quantity. Higher quality generally commands a higher price; however, excellent pots are often available at bargain prices due to volume production or surplus conditions. Electronic supply houses offer low-cost pots (i.e., under $20) that can suffice for short-term projects. Regardless of price, always check the manufacturer's specifications to confirm a pot's suitability for a given application.

Table 6.6 lists several sources of precision pots. Most manufacturers publish catalogs, and many have Web sites. In addition to a standard product line, most manufacturers will custom-build pots for high-volume applications.

TABLE 6.5 Typical Single-quantity Prices ($US) for Commercially Available Pots

Potentiometer type	Approximate price range
Rotary	$10–$350
Linear-motion	$20–$2000
String	$250–$1000

TABLE 6.6 Sources of Precision Pots

Company	Potentiometer types
Betatronix, Inc. 110 Nicon Court Hauppauge, NY 11788 Tel: (516) 582-6740; Fax (516) 582-6038 www.betatronix.com	Exotic linear-motion, rotary
BI Technologies Corp. 4200 Bonita Place Fullerton, CA 92635 Tel: (714) 447-2345; Fax: (714) 447-2500	Rotary
Bourns, Inc. Sensors & Controls Division 2533 N. 1500 W. Ogden, UT 84404 Tel: (801) 786-6200; Fax: (801) 786-6203 www.bourns.com	Mostly rotary, some linear-motion
Celesco Transducer Products, Inc. 7800 Deering Avenue Canoga Park, CA 91309 Tel: (800) 423-5483, (818) 884-6860 Fax: (818) 340-1175 www.celesco.com	String
Data Instruments 100 Discovery Way Acton, MA 01720-3648 Tel: (800) 333-3282, (978) 264-9550 Fax: (978) 263-0630 www.datainstruments.com	Linear-motion, rotary
Duncan Electronics Division BEI Sensors & Systems Company 15771 Red Hill Avenue Tustin, CA 92680 Tel: (714) 258-7500; Fax: (714) 258-8120 www.beisensors.com	Linear-motion, rotary
Dynamation Transducers Corp. 348 Marshall Street Holliston, MA 01746-1441 Tel: (508) 429-8440; Fax: (508) 429-1317	Linear-motion, rotary
JDK Controls, Inc. 424 Crown Pt. Circle Grass Valley, CA 95945 Tel: (530) 273-4608; Fax: (530) 273-0769	Rotary, "do-it-yourself" rotor/wiper assemblies
Midori America Corp. 2555 E. Chapman Ave, Suite 400 Fullerton, CA 92631 Tel: (714) 449-0997; Fax: (714) 449-0139 www.midori.com	Linear-motion, rotary, string; also magneto-resistive
New England Instrument 245 Railroad Street Woonsocket, RI 02895-1129 Tel: (401) 769-0703; Fax: (401) 769-0037	Linear-motion, rotary, resistive elements
Novotechnik U.S., Inc. 237 Cedar Hill Street Marlborough, MA 01752 Tel: (800) 667-7492, (508) 485-2244 Fax: (508) 485-2430 www.novitechnik.com	Linear-motion, rotary

TABLE 6.6 (continued)   Sources of Precision Pots

Company	Potentiometer types
Servo Systems, Co. 115 Main Road, PO Box 97 Montville, NJ 07045-0097 Tel: (800) 922-1103, (973) 335-1007 Fax: (973) 335-1661 www.servosystems.com	Linear-motion, rotary (surplus)
SpaceAge Control, Inc. 38850 20th Street East Palmdale, CA 93550 Tel: (805) 273-3000; Fax: (805) 273-4240 www.spaceagecontrol.com	String
Spectrol Electronics Corp. 4051 Greystone Drive Ontario, CA 91761 Tel: (909) 923-3313; Fax: (909) 923-6765	Rotary
UniMeasure, Inc. 501 S.W. 2nd Street Corvallis, OR 97333 Tel: (541) 757-3158 Fax: (541) 757-0858 www.unimeasure.com	String
Axsys Technologies, Inc. Vernitron Sensor Systems Division Precision Potentiometer Division 2800 Anvil Street North St. Petersburg, FL 33710 Tel: (813) 347-2181; Fax: (813) 347-7520 www.axsys.com	Linear-motion, rotary

Evaluation

Precision pots are a mature technology, effectively static except for occasional developments in materials, packaging, manufacturing, etc. Recent potentiometric innovations — including momentary-contact membrane pots [13] and solid-state digital pots [14] — are unavailing to precision measurement.

The variable voltage divider is the traditional configuration for precision measurement. The circuit's output, a high-amplitude dc voltage, is independent of variations in the pot's total resistance, and is highly compatible with other circuits and systems. Other forms of output are possible with a precision pot configured as a variable resistor. For example, paired with a capacitor, a pot could supply a position-dependent RC time constant to modulate an oscillator's output frequency or duty cycle. In this setup, the pot's stability and ac characteristics would be important.

An alternative resistive displacement sensing technology is the *magneto-resistive potentiometer*, available in rotary and linear-motion forms. Magneto-resistive pots incorporate a noncontacting, permanent magnet "wiper" that rides above a pair of magneto-resistive elements. The elements, configured as a voltage divider, change their resistances according to the strength of the applied magnetic field, and thus divide the voltage as a function of the magnet's position. The output is approximately linear over a limited range of motion (e.g., 90° in rotary models). Magneto-resistive pots offer unlimited resolution and exceptionally long life, but may require temperature compensation circuitry. See [15] for more information.

References

1. Bourns, Inc., *Electronic Components RC4 Solutions Guide*, 1995, 304.
2. Vernitron Motion Control Group, (New York, NY) *Precision Potentiometers*, Catalog #752, 1993.

3. UniMeasure, Inc., *Position & Velocity Transducers*, UniMeasure document No. 400113-27A, Corvallis, OR.

4. Instrument Society of America (Research Triangle Park, NC), *ISA-S37.12-1977 (R1982) Specifications and Tests for Potentiometric Displacement Transducers*, 1982.

5. E. C. Jordan (Ed.), *Reference Data for Engineers: Radio, Electronics, Computer, and Communications*, 7th ed., Indianapolis: H.W. Sams, 1985, 5–16.

6. D. C. Greenwood, *Manual of Electromechanical Devices*, New York: McGraw-Hill, 1965, 297–299.

7. E. S. Charkey, *Electromechanical System Components*, New York: Wiley-Interscience, 1972, 302–303.

8. Variable Resistive Components Institute (Vista, CA), *VRCI-P-100A Standard for Wirewound and Nonwirewound Precision Potentiometers*, 1988.

9. Timber-Top, Inc., P.O. Box 517, Watertown, CT 06795, Tel: (860)-274-6706; Fax (860)-274-8041.

10. J. Angeles and C. S. López-Cajún, *Optimization of Cam Mechanisms*, Dordrecht, The Netherlands: Kluwer Academic, 1991.

11. P. W. Jensen, *Classical and Modern Mechanisms for Engineers and Inventors*, New York: Marcel Dekker, 1991.

12. A. H. Slocum, *Precision Machine Design*, Englewood Cliffs, NJ: Prentice Hall, 1992.

13. Spectra Symbol Inc., data sheet: *SoftPot® (Membrane Potentiometer)*, Salt Lake City, UT, 1996.

14. Dallas Semiconductor Corp., *Digital Potentiometer Overview*, web page: www.dalsemi.com/Prod_info/Dig_Pots/, December 1997.

15. Midori America Corp., (Fullerton, CA) *Midori Position Sensors 1995 Catalog*.

6.2 Inductive Displacement Sensors

Halit Eren

Inductive sensors are widely used in industry in many diverse applications. They are robust and compact, and are less affected by environmental factors (e.g., humidity, dust) in comparison to their capacitive counterparts.

Inductive sensors are primarily based on the principles of magnetic circuits. They can be classified as self-generating or passive. The self-generating types utilize an electrical generator principle; that is, when there is a relative motion between a conductor and a magnetic field, a voltage is induced in the conductor. Or, a varying magnetic field linking a stationary conductor produces voltage in the conductor. In instrumentation applications, the magnetic field may be varying with some frequency and the conductor may also be moving at the same time. In inductive sensors, the relative motion between field and conductor is supplied by changes in the measurand, usually by means of some mechanical motion. On the other hand, the passive transducer requires an external source of power. In this case, the action of the transducer is simply the modulation of the excitation signal.

For the explanation of the basic principles of inductive sensors, a simple magnetic circuit is shown in Figure 6.7. The magnetic circuit consists of a core, made from a ferromagnetic materia,l with a coil of n number of turns wound on it. The coil acts as a source of magnetomotive force (mmf) which drives the flux Φ through the magnetic circuit. If one assumes that the air gap is zero, the equation for the magnetic circuit can be expressed as:

$$\text{mmf} = \text{Flux} \times \text{Reluctance} = \Phi \times \mathfrak{R} \quad \text{A - turns} \tag{6.2}$$

such that the reluctance $\mathfrak{R}$ limits the flux in a magnetic circuit just as resistance limits the current in an electric circuit. By writing the mmf in terms of current, the magnetic flux may be expressed as:

$$\Phi = ni/\mathfrak{R} \quad \text{weber} \tag{6.3}$$

In Figure 6.7, the flux linking a single turn is by Equation 6.3; but the total flux linking by the entire n number of the turns of the coil is

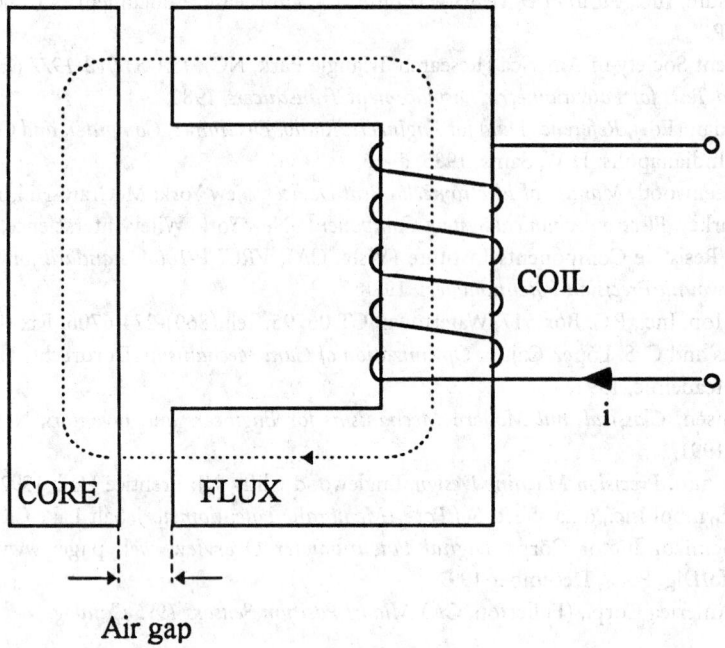

FIGURE 6.7 A basic inductive sensor consists of a magnetic circuit made from a ferromagnetic core with a coil wound on it. The coil acts as a source of magnetomotive force (mmf) that drives the flux through the magnetic circuit and the air gap. The presence of the air gap causes a large increase in circuit reluctance and a corresponding decrease in the flux. Hence, a small variation in the air gap results in a measurable change in inductance.

$$\Psi = n\Phi = n^2 \, i / \Re \qquad \text{weber} \tag{6.4}$$

Equation 6.4 leads to self inductance L of the coil, which is described as the total flux (Ψ weber) per unit current for that particular coil; that is

$$L = \Psi / I = n^2 / \Re \tag{6.5}$$

This indicates that the self inductance of an inductive element can be calculated by magnetic circuit properties. Expressing $\Re$ in terms of dimensions as:

$$\Re = l / \mu \mu_0 \, A \tag{6.6}$$

where l = the total length of the flux path
  μ = the relative permeability of the magnetic circuit material
  μ_0 = the permeability of free space (= $4\pi \times 10^{-7}$ H/m)
  A = the cross-sectional area of the flux path

The arrangement illustrated in Figure 6.7 becomes a basic inductive sensor if the air gap is allowed to vary. In this case, the ferromagnetic core is separated into two parts by the air gap. The total reluctance of the circuit now is the addition of the reluctance of core and the reluctance of air gap. The relative permeability of air is close to unity, and the relative permeability of the ferromagnetic material is of the order of a few thousand, indicating that the presence of the air gap causes a large increase in circuit reluctance and a corresponding decrease in the flux. Hence, a small variation in the air gap causes a measurable change in inductance. Most of the inductive transducers are based on these principles and are discussed below in greater detail.

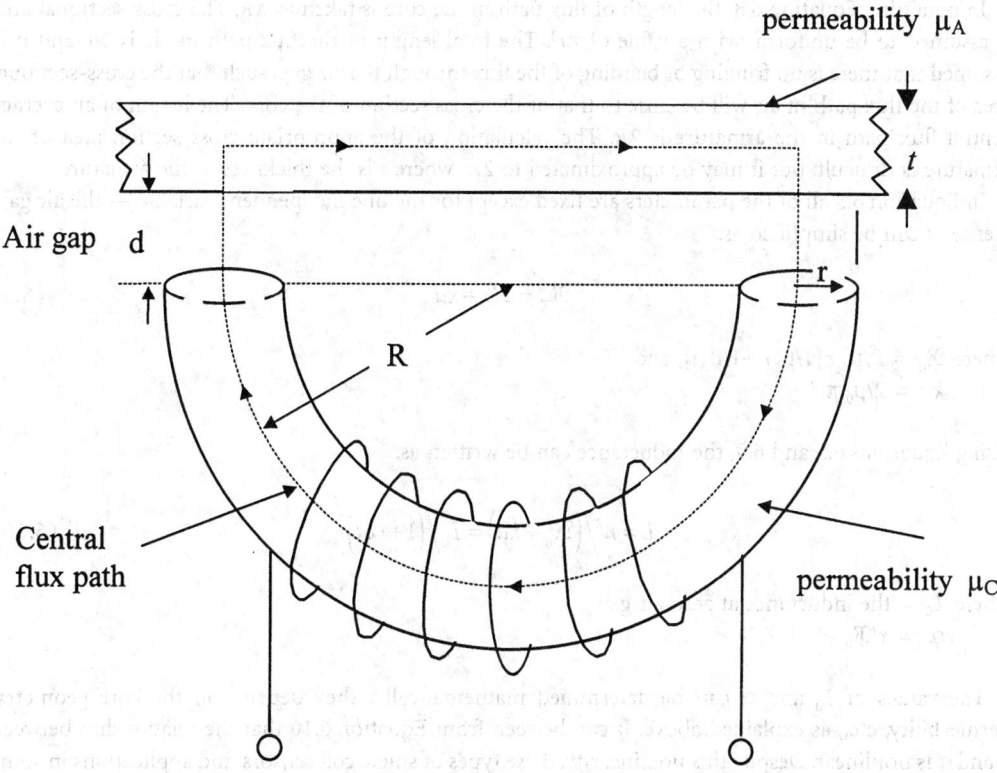

permeability μ_A

Air gap d

R

Central
flux path

permeability μ_C

FIGURE 6.8 A typical single-coil, variable-reluctance displacement sensor. The sensor consists of three elements: a ferromagnetic core in the shape of a semicircular ring, a variable air gap, and a ferromagnetic plate. The reluctance of the coil is dependent on the single variable. The reluctance increases nonlinearly with increasing gap.

Linear and Rotary Variable-Reluctance Transducer

The variable-reluctance transducers are based on change in the reluctance of a magnetic flux path. This type of transducer finds application particularly in acceleration measurements. However, they can be constructed to be suitable for sensing displacements as well as velocities. They come in many different forms, as described below.

The Single-Coil Linear Variable-Reluctance Sensor

A typical single-coil variable-reluctance displacement sensor is illustrated in Figure 6.8. The sensor consists of three elements: a ferromagnetic core in the shape of a semicircular ring, a variable air gap, and a ferromagnetic plate. The total reluctance of the magnetic circuit is the sum of the individual reluctances:

$$\Re_T = \Re_C + \Re_G + \Re_A \qquad (6.7)$$

where $\Re_C$, $\Re_G$, and $\Re_A$ are the reluctances of the core, air gap, and armature, respectively.

Each one of these reluctances can be determined by using the properties of materials involved, as in Equation 6.6. In this particular case, the reluctance $\Re_T$ can be approximated as:

$$\Re_T = R/\mu_C \mu_0 \, r^2 + 2d/\mu_0 \, \pi r^2 + R/\mu_A \mu_0 \, rt \qquad (6.8)$$

In obtaining Equation 6.8, the length of flux path in the core is taken as πR. The cross-sectional area is assumed to be uniform, with a value of πr^2. The total length of the flux path in air is $2d$, and it is assumed that there is no fringing or bending of the flux through the air gap, such that the cross-sectional area of the flux path in air will be close to that of the cross section of the core. The length of an average central flux path in the armature is $2R$. The calculation of the appropriate cross section area of the armature is difficult, but it may be approximated to $2rt$, where t is the thickness of the armature.

In Equation 6.8 all of the parameters are fixed except for the one independent variable — the air gap. Hence, it can be simplified as:

$$\Re_T = \Re_0 + kd \tag{6.9}$$

where $\Re_0 = R/\mu_0\, r\,[1/\mu_C r + 1/\mu_A t]$, and
$\quad\ k = 2/\mu_0\, \pi r^2$

Using Equations 6.5 and 6.9, the inductance can be written as:

$$L = n^2 \big/ \left(\Re_0 + kd \right) = L_0 \big/ \left(1 + \alpha d \right) \tag{6.10}$$

where L_0 = the inductance at zero air gap
$\quad\ \alpha = k/\Re_0$

The values of L_0 and α can be determined mathematically: they depend on the core geometry, permeability, etc., as explained above. It can be seen from Equation 6.10 that the relationship between L and α is nonlinear. Despite this nonlinearity, these types of single coil sensors find applications in some areas, such as force measurements and telemetry. In force measurements, the resultant change in inductance can be made to be a measure of the magnitude of the applied force. The coil usually forms one of the components of an *LC* oscillator, for which the output frequency varies with the applied force. Hence, the coil modulates the frequency of the local oscillator.

The Variable-Differential Reluctance Sensor

The problem of the nonlinearity can be overcome by modifying the single coil system into a variable-differential reluctance sensor (also known as push-pull sensor), as shown in Figure 6.9. This sensor consists of an armature moving between two identical cores, and separated by a fixed distance of $2d$. Now, Equation 6.10 can be written for both coils as:

$$L_1 = L_{01} \big/ \left[1 + \alpha \left(d - x \right) \right],$$
$$L_2 = L_{02} \big/ \left[1 + \alpha \left(d + x \right) \right] \tag{6.11}$$

Although the relationship between L_1 and L_2 is still nonlinear, the sensor can be incorporated into an ac deflection bridge to give a linear output for small movements. The hysteresis errors of these transducers are almost entirely limited to the mechanical components. These sensors respond to both static and dynamic measurements. They have continuous resolution and high outputs, but they may give erratic performance in response to external magnetic fields. A typical sensor of this type has an input span of 1 cm, a coil inductance of 25 mH, and a coil resistance of 75 Ω. The resistance of the coil must be carefully considered when designing oscillator circuits. The maximum nonlinearity is 0.5%.

A typical commercially available variable differential sensor is shown in Figure 6.10. The iron core is located halfway between the two E-shaped frames. The flux generated by primary coils depends on the reluctance of the magnetic path, the main reluctance being the air gap. Any motion of the core increases the air gap on one side and decreases it on the other side, thus causing reluctance to change, in accordance

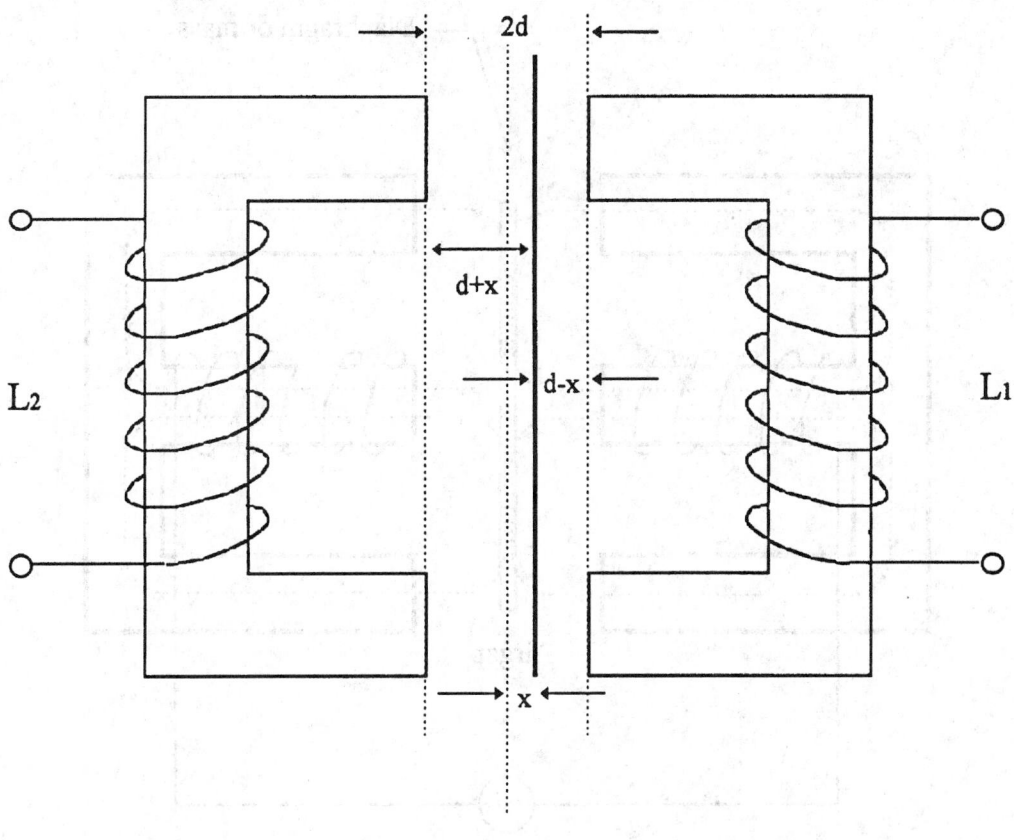

Reference line

FIGURE 6.9 A variable-differential reluctance sensor consists of an armature moving between two identical cores separated by a fixed distance. The armature moves in the air gap in response to a mechanical input. This movement alters the reluctance of coils 1 and 2, thus altering their inductive properties. This arrangement overcomes the problem of nonlinearity inherent in single coil sensors.

with the principles explained above, and thereby inducing more voltage on one of the coils than on the other. Motion in the other direction reverses the action with a 180° phase shift occurring at null. The output voltage can be modified, depending on the requirements in signal processing, by means of rectification, demodulation, or filtering. In these instruments, full-scale motion may be extremely small — on the order of few thousandths of a centimeter.

In general, variable reluctance transducers have small ranges and are used in specialized applications such as pressure transducers. Magnetic forces imposed on the armature are quite large and this severely limits their application. However, the armature can be constructed as a diaphragm; hence, suitable for pressure measurements.

Variable-Reluctance Tachogenerators

Another example of a variable reluctance sensor is shown in Figure 6.11. These sensors are based on Faraday's law of electromagnetic induction; therefore, they may also be referred to as electromagnetic sensors. Basically, the induced emf in the sensor depends on the linear or angular velocity of the motion.

The variable-reluctance tachogenerator consists of a ferromagnetic, toothed wheel attached to a rotating shaft, and a coil wound onto a permanent magnet, extended by a soft iron pole piece. The wheel moves in close proximity to the pole piece, causing the flux linked by the coil to change, thus inducing an emf in the coil. The reluctance of the circuit depends on the width of the air gap between the rotating wheel and the pole piece. When the tooth is close to the pole piece, the reluctance is minimum and it

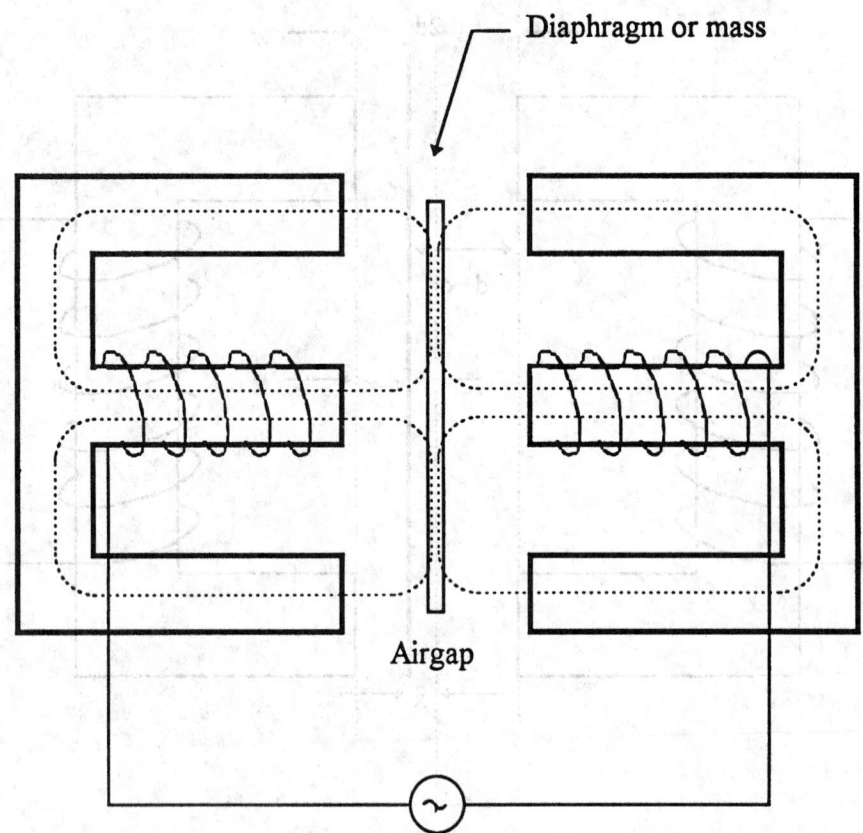

FIGURE 6.10 A typical commercial variable differential sensor. The iron core is located half-way between the two E frames. Motion of the core increases the air gap for one of the E frames while decreasing the other side. This causes reluctances to change, thus inducing more voltage on one side than the other. Motion in the other direction reverses the action, with a 180° phase shift occurring at null. The output voltage can be processed, depending on the requirements, by means of rectification, demodulation, or filtering. The full-scale motion may be extremely small, on the order of few thousandths of a centimeter.

increases as the tooth moves away from the pole. When the wheel rotates with a velocity ω, the flux may mathematically be expressed as:

$$\Psi(\theta) = A + B \cos m\theta \tag{6.12}$$

where A = the mean flux
         B = the amplitude of the flux variation
         m = the number of teeth

The induced emf is given by:

$$E = -d\Psi(\theta)/dt = -\left(d\Psi(\theta)/d\theta\right) \times \left(d\theta/dt\right) \tag{6.13}$$

or

$$E = bm\omega \sin m\omega t \tag{6.14}$$

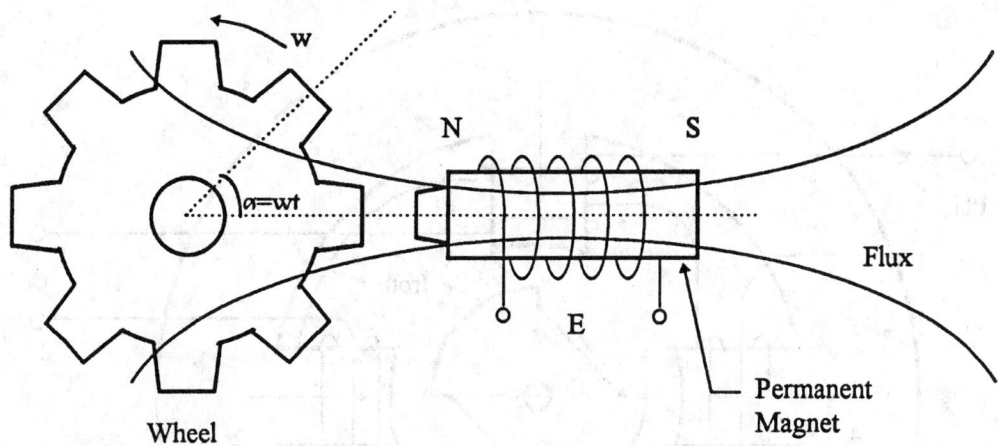

FIGURE 6.11 A variable-reluctance tachogenerator is a sensor which is based on Faraday's law of electromagnetic induction. It consists of a ferromagnetic toothed wheel attached to the rotating shaft and a coil wound onto a permanent magnet extended by a soft iron pole piece. The wheel rotates in close proximity to the pole piece, thus causing the flux linked by the coil to change. The change in flux causes an output in the coil similar to a square waveform whose frequency depends on the speed of the rotation of the wheel and the number of teeth.

Both amplitude and frequency of the generated voltage at the coil are proportional to the angular velocity of the wheel. In principle, the angular velocity ω can be found from either the amplitude or the frequency of the signal. In practice, the amplitude measured may be influenced by loading effects and electrical interference. In signal processing, the frequency is the preferred option because it can be converted into digital signals easily.

The variable-reluctance tachogenerators are most suitable for measuring angular velocities. They are also used in the volume flow rate measurements and the total volume flow determination of fluids.

Microsyn

Another commonly used example of variable-reluctance transducer is the Microsyn, as illustrated in Figure 6.12. In this arrangement, the coils are connected in such a manner that at the null position of the rotary element, the voltages induced in coils 1 and 3 are balanced by voltages induced in coils 2 and 4. The motion of the rotor in the clockwise direction increases the reluctance of coils 1 and 3 while decreasing the reluctance of coils 2 and 4, thus giving a net output voltage e_0. The movement in the counterclockwise direction causes a similar effect in coils 2 and 4 with a 180° phase shift. A direction-sensitive output can be obtained by using phase-sensitive demodulators, as explained in LVDT section of this chapter.

Microsyn transducers are used extensively in applications involving gyroscopes. By the use of microsyns, very small motions can be detected, giving output signals as low as 0.01° of changes in angles. The sensitivity of the device can be made as high as 5 V per degree of rotation. The nonlinearity may vary from 0.5% to 1.0% full scale. The main advantage of these transducers is that the rotor does not have windings and slip-rings. The magnetic reaction torque is also negligible.

Synchros

The term *synchro* is associated with a family of electromechanical devices that can be discussed under different headings. They are used primarily in angle measurements and are commonly applied in control engineering as parts of servomechanisms, machine tools, antennas, etc.

The construction of synchros is similar to that of wound-rotor induction motors, as shown in Figure 6.13. The rotation of the motor changes the mutual inductance between the rotor coil and the three stator coils. The three voltage signals from these coils define the angular position of the rotor.

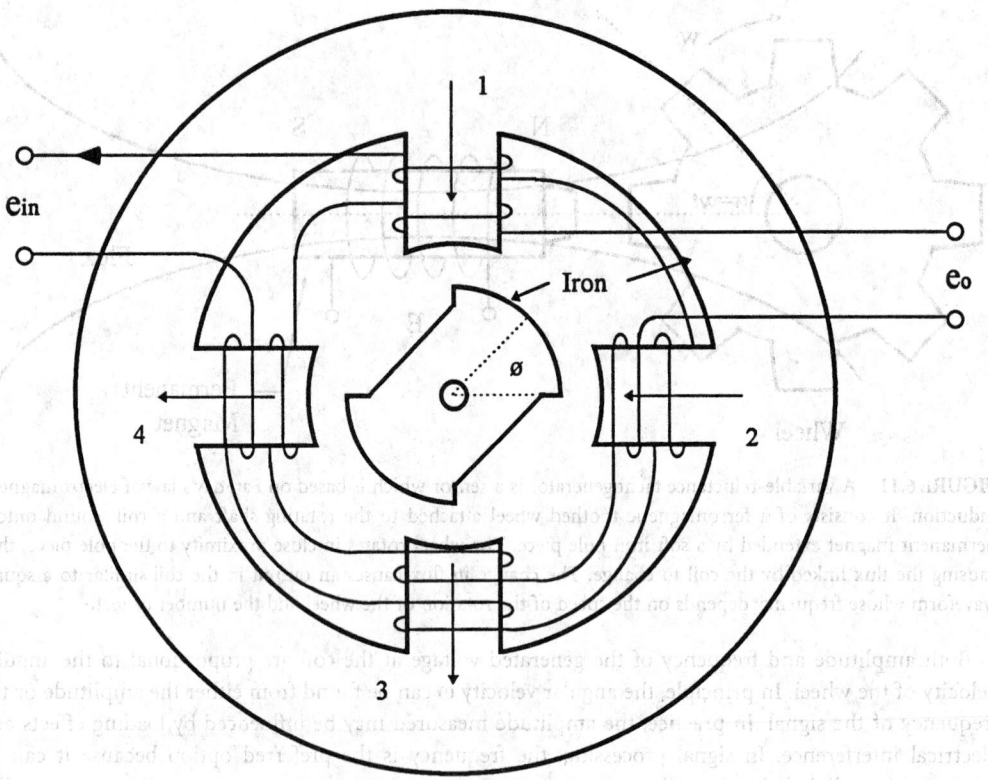

FIGURE 6.12 A microsyn is a variable reluctance transducer that consists of a ferromagnetic rotor and a stator carrying four coils. The stator coils are connected such that at the null position, the voltages induced in coils 1 and 3 are balanced by voltages induced in coils 2 and 4. The motion of the rotor in one direction increases the reluctance of two opposite coils while decreasing the reluctance in others, resulting in a net output voltage e_o. The movement in the opposite direction reverses this effect with a 180° phase shift.

Synchros are used in connection with variety of devices, including: control transformers, Scott T transformers, resolvers, phase-sensitive demodulators, analog to digital converters, etc.

In some cases, a control transformer is attached to the outputs of the stator coils such that the output of the transformer produces a resultant mmf aligned in the same direction as that of the rotor of the synchro. In other words, the synchro rotor acts as a search coil in detecting the direction of the stator field of the control transformer. When the axis of this coil is aligned with the field, the maximum voltage is supplied to the transformer.

In other cases, ac signals from the synchros are first applied to a Scott T transformer, which produces ac voltages with amplitudes proportional to the sine and cosine of the synchro shaft angle. It is also possible to use phase-sensitive demodulations to convert the output signals to make them suitable for digital signal processing.

Linear-Variable Inductor

There is a little distinction between variable-reluctance and variable-inductance transducers. Mathematically, the principles of linear-variable inductors are very similar to the variable-reluctance type of transducer. The distinction is mainly in the pickups rather than principles of operations. A typical linear variable inductor consists of a movable iron core that provides the mechanical input and two coils forming two legs of a bridge network. A typical example of such a transducer is the variable coupling transducer, which is discussed next.

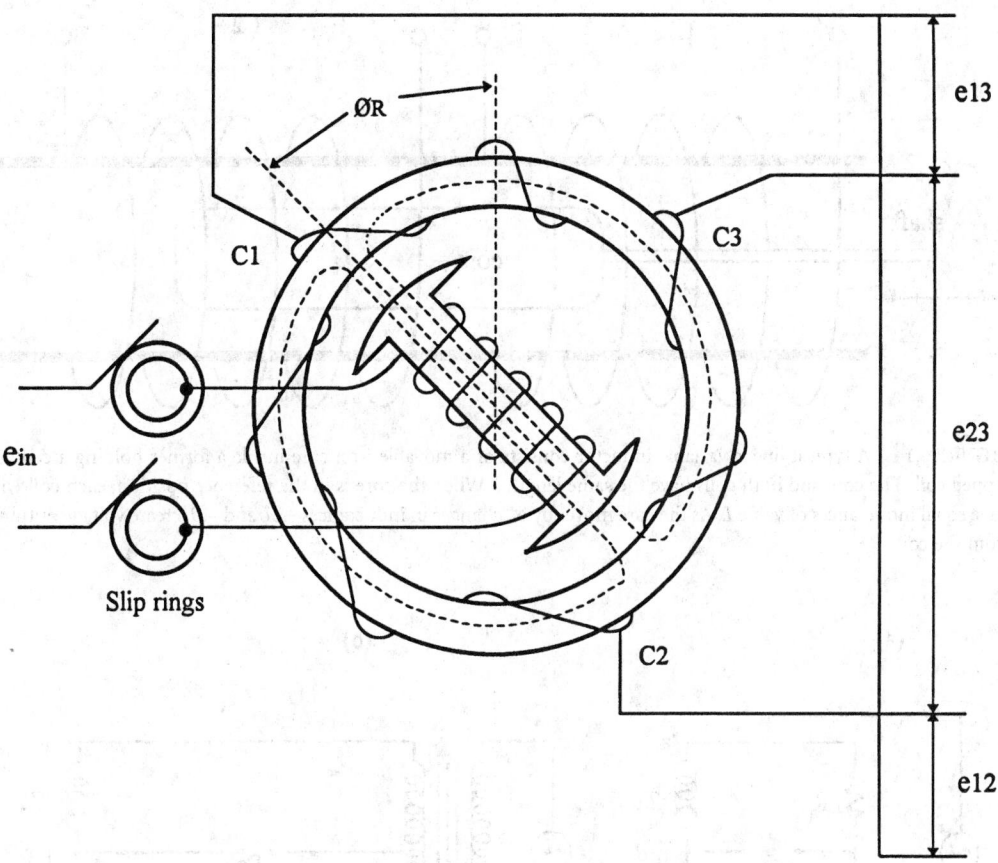

FIGURE 6.13 A synchro is similar to a wound-rotor induction motor. The rotation of the rotor changes the mutual inductance between the rotor coil and the stator coils. The voltages from these coils define the angular position of the rotor. They are primarily used in angle measurements and are commonly applied in control engineering as parts of servomechanisms, machine tools, antennas, etc.

Variable-Coupling Transducers

These transducers consist of a former holding a center tapped coil and a ferromagnetic plunger, as shown in Figure 6.14.

The plunger and the two coils have the same length l. As the plunger moves, the inductances of the coils change. The two inductances are usually placed to form two arms of a bridge circuit with two equal balancing resistors, as shown in Figure 6.15. The bridge is excited with ac of 5 V to 25 V with a frequency of 50 Hz to 5 kHz. At the selected excitation frequency, the total transducer impedance at null conditions is set in the 100 Ω to 1000 Ω range. The resistors are set to have about the same value as transducer impedances. The load for the bridge output must be at least 10 times the resistance, R, value. When the plunger is in the reference position, each coil will have equal inductances of value L. As the plunger moves by δL, changes in inductances $+\delta L$ and $-\delta L$ creates a voltage output from the bridge. By constructing the bridge carefully, the output voltage can be made as a linear function displacement of the moving plunger within a rated range.

In some transducers, in order to reduce power losses due to heating of resistors, center-tapped transformers can be used as a part of the bridge network, as shown in Figure 6.15(b). In this case, the circuit becomes more inductive and extra care must be taken to avoid the mutual coupling between the transformer and the transducer.

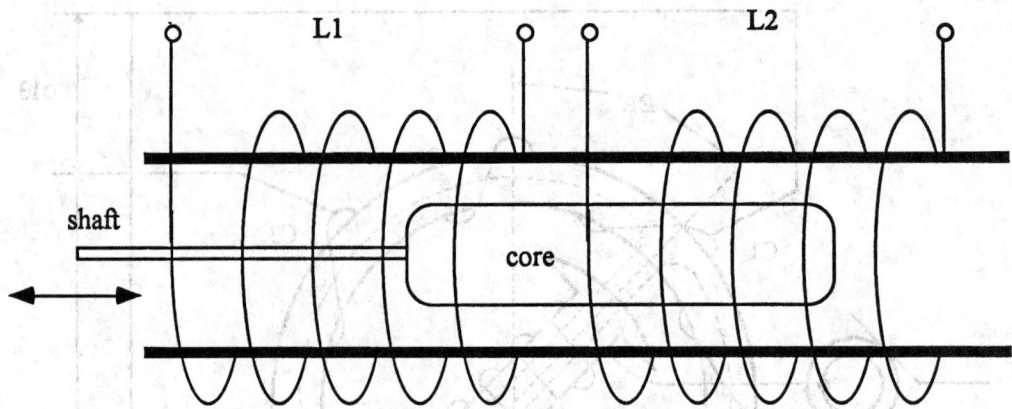

FIGURE 6.14 A typical linear-variable inductor consists of a movable iron core inside a former holding a center-tapped coil. The core and both coils have the same length l. When the core is in the reference position, each coil will have equal inductances of value L. As the core moves by δl, changes in inductances $+\delta L$ and $-\delta L$ create voltage outputs from the coils.

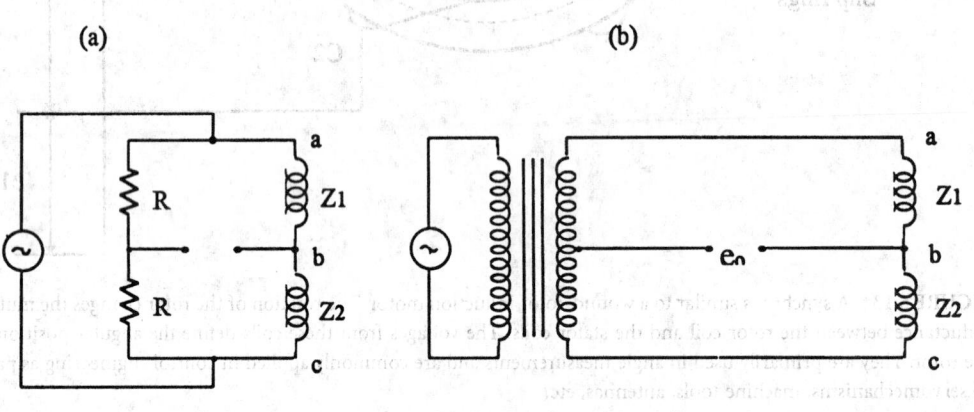

FIGURE 6.15 The two coils of a linear-variable inductor are usually placed to form two arms of a bridge circuit, also having two equal balancing resistors as in circuit (a). The bridge is excited with ac of 5 V to 25 V with a frequency of 50 Hz to 5 kHz. At a selected excitation frequency, the total transducer impedance at null conditions is set in the 100 Ω to 1000 Ω range. By careful construction of the bridge, the output voltage can be made a linear function displacement of the core within a limited range. In some cases, in order to reduce power losses due to heating of resistors, center-tapped transformers may be used as a part of the bridge network (b).

It is particularly easy to construct transducers of this type, by simply winding a center-tapped coil on a suitable former. The variable-inductance transducers are commercially available in strokes from about 2 mm to 500 cm. The sensitivity ranges between 1% full scale to 0.02% in long stroke special constructions. These devices are also known as linear displacement transducers or LDTs, and they are available in various shape and sizes.

Apart from linear-variable inductors, there are rotary types available too. Their cores are specially shaped for rotational applications. Their nonlinearity can vary between 0.5% to 1% full scale over a range of 90° rotation. Their sensitivity can be up to 100 mV per degree of rotation.

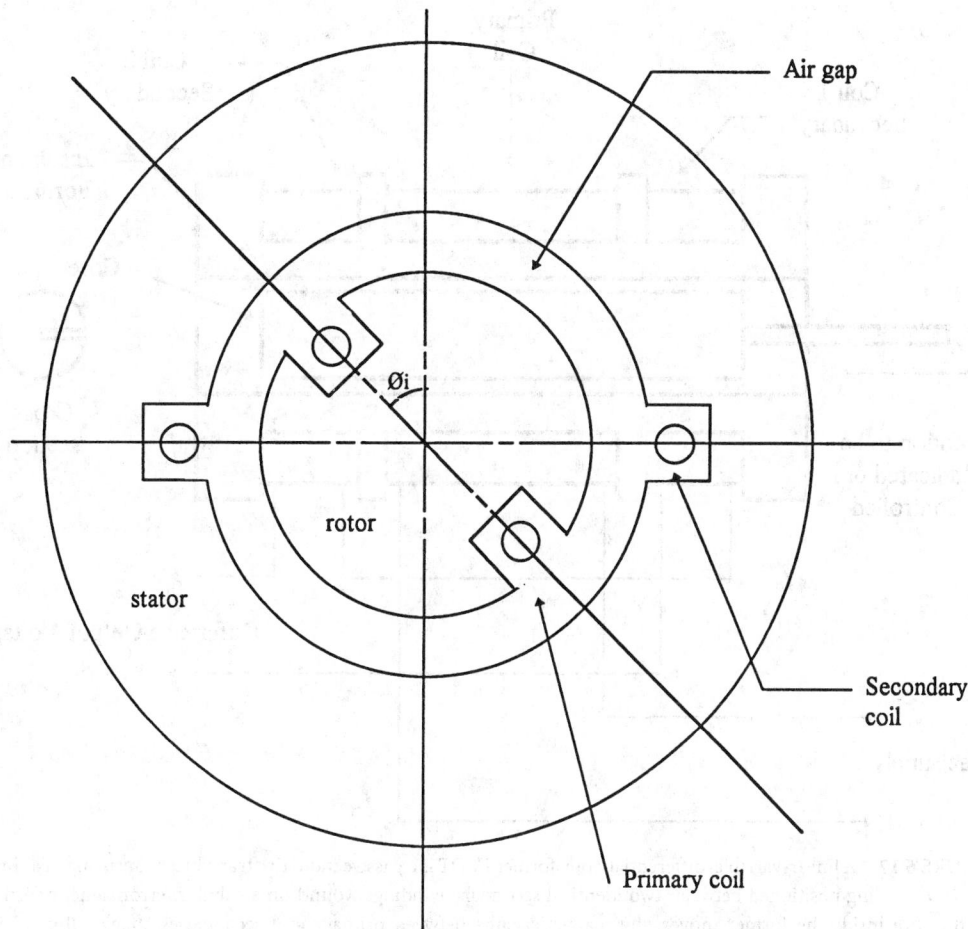

FIGURE 6.16 An induction potentiometer is a linear-variable inductor with two concentrated windings wound on the stator and on the rotor. The rotor winding is excited with ac, inducing voltage in the stator windings. The amplitude of the output voltage is dependent on the relative positions of the coils, as determined by the angle of rotation. For concentrated coils, the variation of the amplitude is sinusoidal, but linearity is restricted in the region of the null position. Different types of induction potentiometers are available with distributed coils that give linear voltages over an angle of 180° of rotation.

Induction Potentiometer

One version of a rotary type linear inductor is the induction potentiometer shown in Figure 6.16. Two concentrated windings are wound on the stator and rotor. The rotor winding is excited with an ac, thus inducing voltage in the stator windings. The amplitude of the output voltage is dependent on the mutual inductance between the two coils, where mutual inductance itself is dependent on the angle of rotation. For concentrated coil type induction potentiometers, the variation of the amplitude is sinusoidal, but linearity is restricted in the region of the null position. A linear distribution over an angle of 180° can be obtained by carefully designed distributed coils.

Standard commercial induction pots operate in a 50 to 400 Hz frequency range. They are small in size, from 1 cm to 6 cm, and their sensitivity can be on the order of 1 V/deg rotation. Although the ranges of induction pots are limited to less than 60° of rotation, it is possible to measure displacements in angles from 0° to full rotation by suitable arrangement of a number of induction pots. As in the case of most

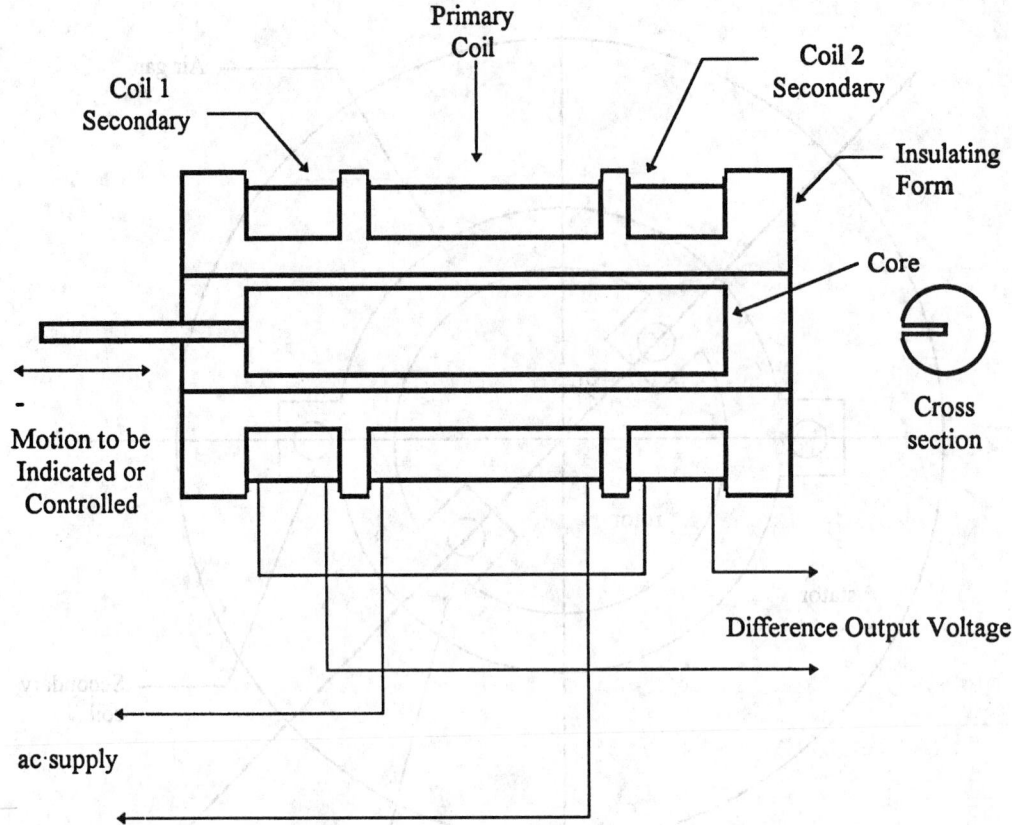

FIGURE 6.17 A linear-variable-differential-transformer LVDT is a passive inductive transducer consisting of a single primary winding positioned between two identical secondary windings wound on a tubular ferromagnetic former. As the core inside the former moves, the magnetic paths between primary and secondaries change, thus giving secondary outputs proportional to the movement. The two secondaries are made as identical as possible by having equal sizes, shapes, and number of turns.

inductive sensors, the output of an induction pot may need phase-sensitive demodulators and suitable filters. In many cases, additional dummy coils are used to improve linearity and accuracy.

Linear Variable-Differential Transformer (LVDT)

The linear variable-differential transformer, LVDT, is a passive inductive transducer finding many applications. It consists of a single primary winding positioned between two identical secondary windings wound on a tubular ferromagnetic former, as shown in Figure 6.17. The primary winding is energized by a high-frequency 50 Hz to 20 kHz ac voltage. The two secondary windings are made identical by having an equal number of turns and similar geometry. They are connected in series opposition so that the induced output voltages oppose each other.

In many applications, the outputs are connected in opposing form, as shown in Figure 6.18(a). The output voltages of individual secondaries v_1 and v_2 at null position are illustrated in Figure 6.18(b). However, in opposing connection, any displacement in the core position x from the null point causes amplitude of the voltage output v_o and the phase difference α to change. The output waveform v_o in relation to core position is shown in Figure 6.18(c). When the core is positioned in the middle, there is

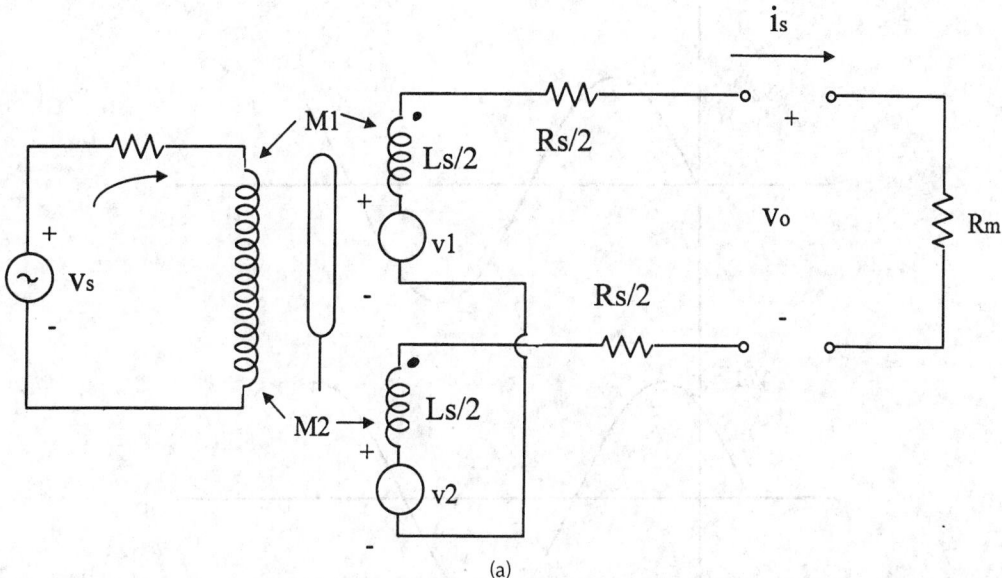

(a)

FIGURE 6.18 The voltages induced in the secondaries of a linear-variable differential-transformer (a) may be processed in a number of ways. The output voltages of individual secondaries v_1 and v_2 at null position are illustrated in (b). In this case, the voltages of individual coils are equal and in phase with each other. Sometimes, the outputs are connected opposing each other, and the output waveform v_o becomes a function of core position x and phase angle α, as in (c). Note the phase shift of 180° as the core position changes above and below the null position.

an equal coupling between the primary and secondary windings, thus giving a null point or reference point of the sensor. As long as the core remains near the center of the coil arrangement, output is very linear. The linear ranges of commercial differential transformers are clearly specified, and the devices are seldom used outside this linear range.

The ferromagnetic core or plunger moves freely inside the former, thus altering the mutual inductance between the primary and secondaries. With the core in the center, or at the reference position, the induced emfs in the secondaries are equal; and since they oppose each other, the output voltage is zero. When the core moves, say to the left, from the center, more magnetic flux links with the left-hand coil than with the right-hand coil. The voltage induced in the left-hand coil is therefore larger than the induced voltage on the right-hand coil. The magnitude of the output voltage is then larger than at the null position and is equal to the difference between the two secondary voltages. The net output voltage is in phase with the voltage of the left-hand coil. The output of the device is then an indication of the displacement of the core. Similarly, movement in the opposite direction to the right from the center reverses this effect, and the output voltage is now in phase with the emf of the right-hand coil.

For mathematical analysis of the operation of LVDTs, Figure 6.18(a) can be used. The voltages induced in the secondary coils are dependent on the mutual inductance between the primary and individual secondary coils. Assuming that there is no cross-coupling between the secondaries, the induced voltages may be written as:

$$v_1 = M_1 s i_p \quad \text{and} \quad v_2 = M_2 s i_p \tag{6.15}$$

where M_1 and M_2 are the mutual inductances between primary and secondary coils for a fixed core position; s is the Laplace operator; and i_p is the primary current

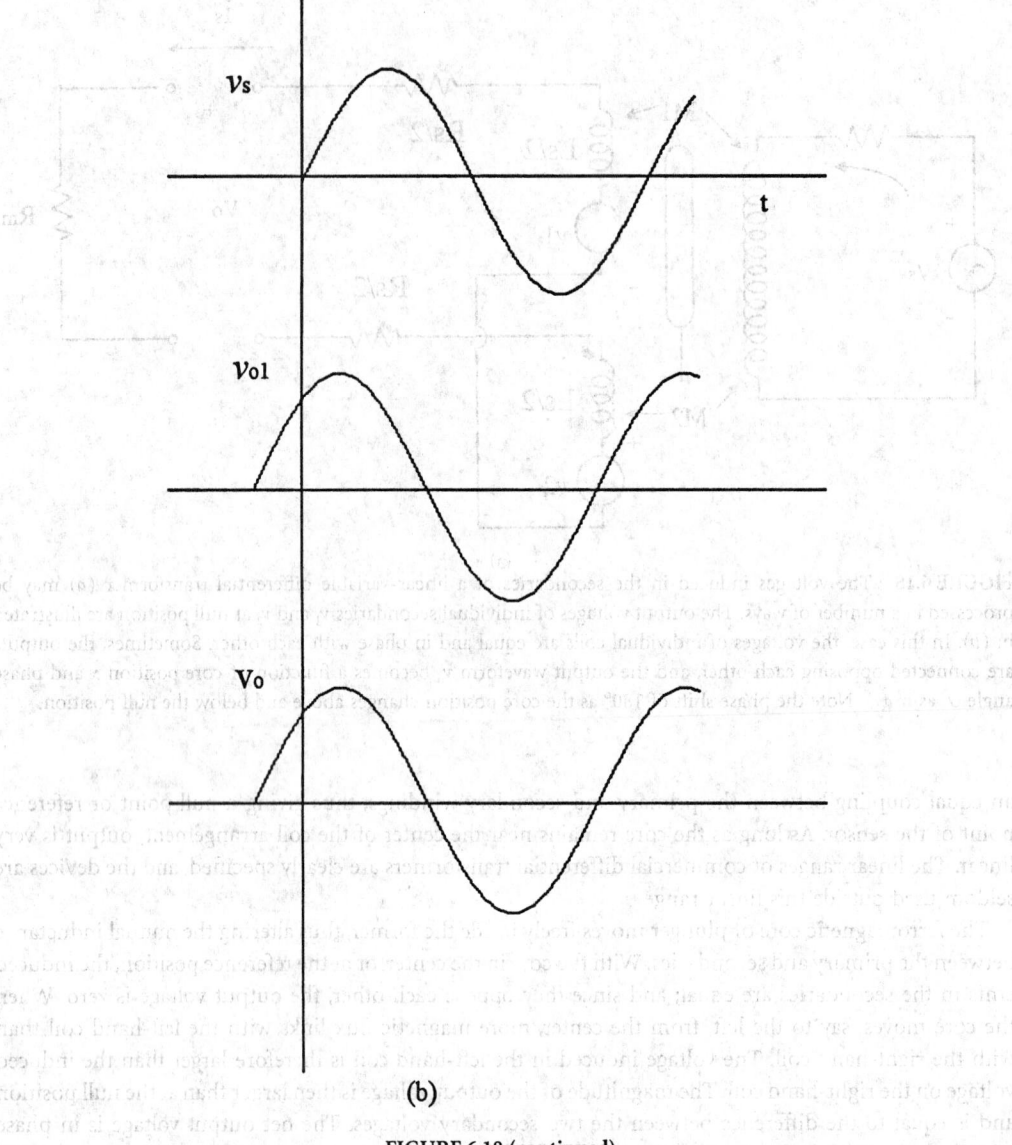

(b)

FIGURE 6.18 (continued)

In the case of opposing connection, no load output voltage v_o without any secondary current may be written as:

$$v_o = v_1 - v_2 = \left(M_1 - M_2\right) si_p \qquad (6.16)$$

writing

$$v_s = i_p \left(R + sL_p\right) \qquad (6.17)$$

Substituting i_p in Equation 6.15 gives the transfer function of the transducer as:

$$v_o/v_s = \left(M_1 - M_2\right) s / \left(R + sL_p\right) \qquad (6.18)$$

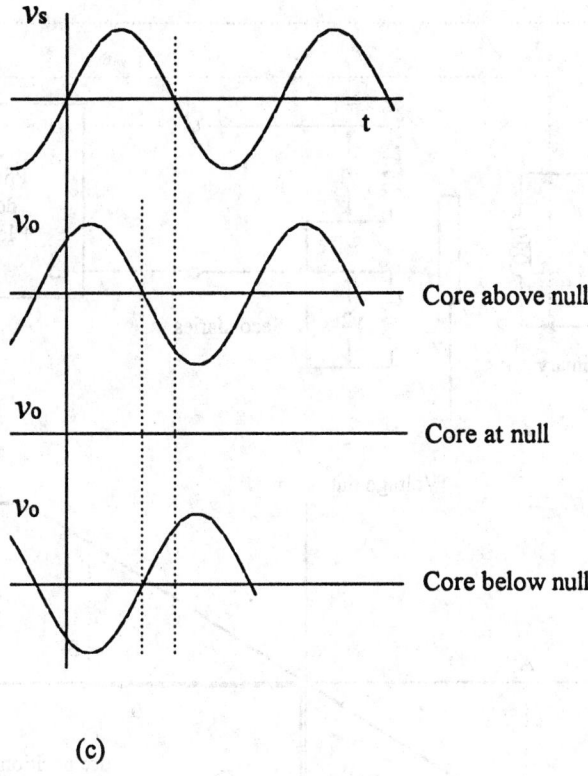

(c)

FIGURE 6.18 (continued)

However, if there is a current due to output signal processing, then describing equations may be modified as:

$$v_o = R_m\, i_s \tag{6.19}$$

where $i_s = (M_1 - M_2)\, si_p/(R_s + R_m + sL_s)$
and

$$v_s = i_p\left(R + sL_p\right) - \left(M_1 - M_2\right) si_s \tag{6.20}$$

Eliminating i_p and i_s from Equations 4.19 and 4.20 results in a transfer function as:

$$v_o/v_s = R_m\left(M_1 - M_2\right)s \Big/ \left\{ \left[\left(M_1 - M_2\right)^2 + L_s L_p\right]s^2 + \left[L_p\left(R + R_m\right) + RL_s\right]s + \left(R_s + R_m\right) + R\right\} \tag{6.21}$$

This is a second-order system, which indicates that due to the effect of the numerator of Eq. 6.21, the phase angle of the system changes from +90° at low frequencies to –90° at high frequencies. In practical applications, the supply frequency is selected such that at the null position of the core, the phase angle of the system is 0°.

The amplitudes of the output voltages of secondary coils are dependent on the position of the core. These outputs may directly be processed from each individual secondary coil for slow movements of the core, and when the direction of the movement of the core does not bear any importance. However, for fast movements of the core, the signals can be converted to dc and the direction of the movement from the null position can be detected. There are many options to do this; however, a *phase-sensitive demodulator*

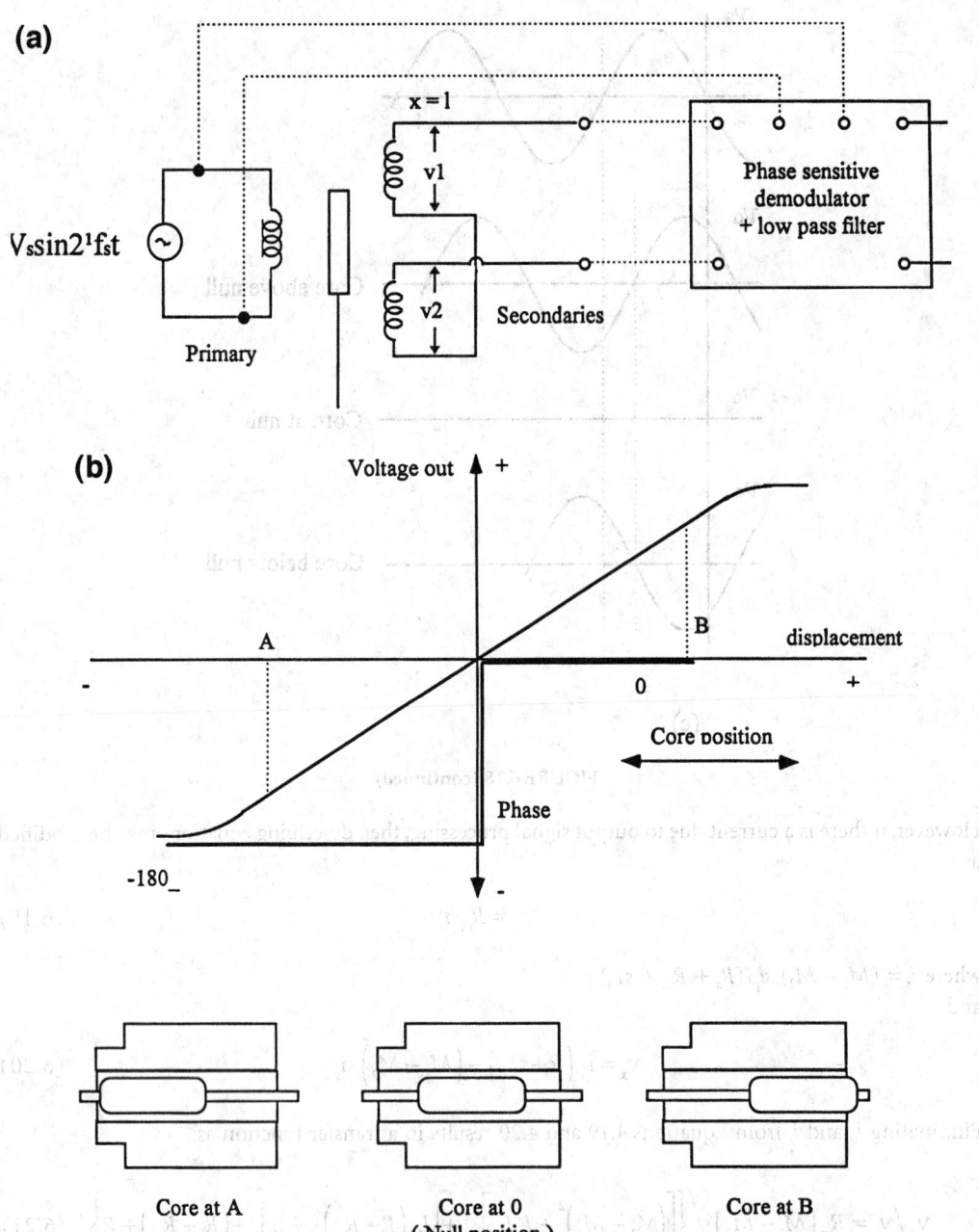

FIGURE 6.19 Phase-sensitive demodulator and (*a*) are commonly used to obtain displacement proportional signals from LVDTs and other differential type inductive sensors. They convert the ac outputs from the sensors into dc values and also indicate the direction of movement of the core from the null position. A typical output of the phase-sensitive demodulator is shown in (*b*). The relationship between output voltage v_o and phase angle α. is also shown against core position *x*.

and filter arrangement is commonly used, as shown in Figure 6.19(*a*). A typical output of the phase-sensitive demodulator is illustrated in Figure 6.19(*b*), in relation to output voltage v_o, displacement *x*, and phase angle α.

The phase-sensitive demodulators are used extensively in differential type inductive sensors. They basically convert the ac outputs to dc values and also indicate the direction of movement of the core

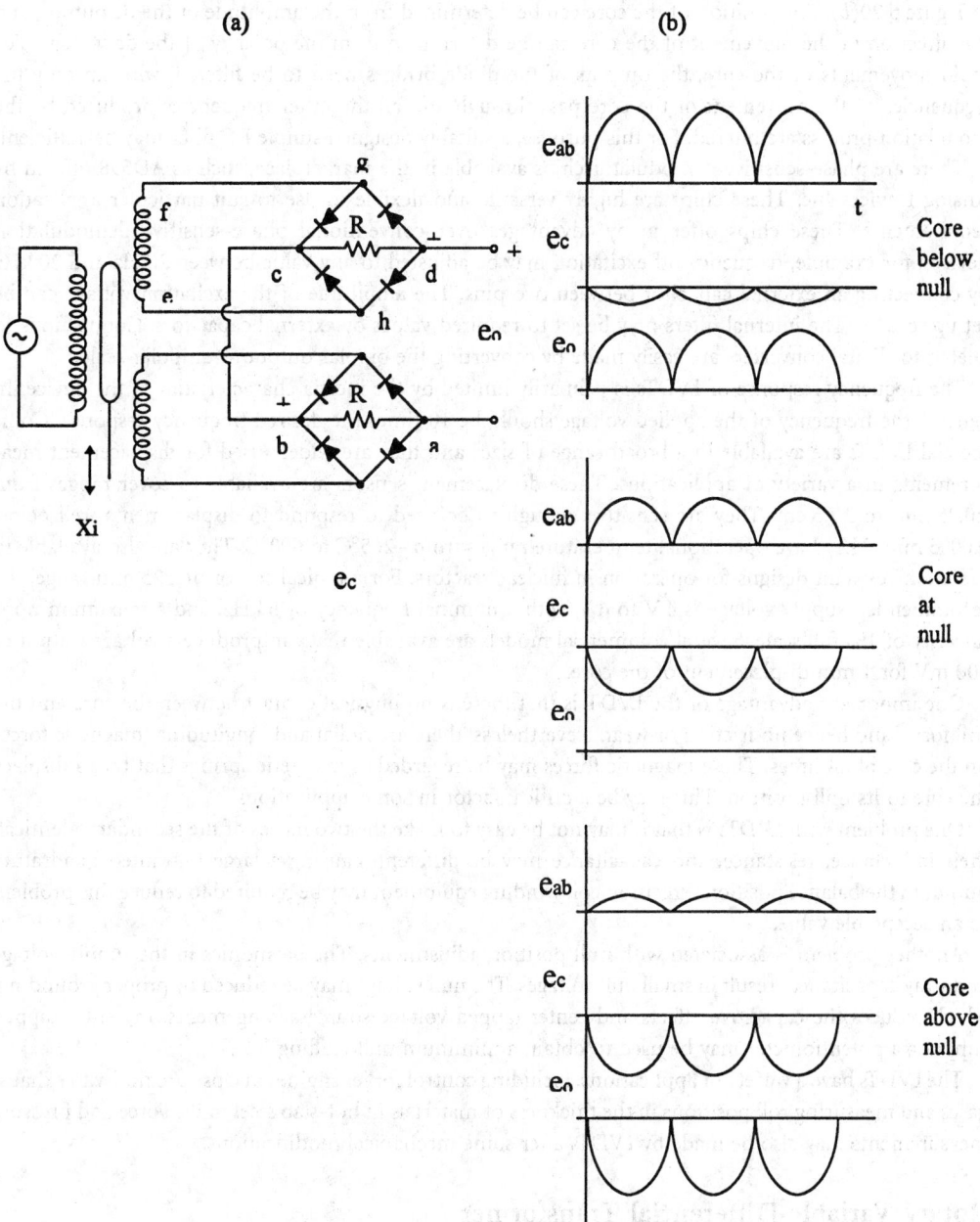

FIGURE 6.20 A typical phase-sensitive demodulation circuit based on diode bridges as in (*a*). Bridge 1 acts as a rectification circuit for secondary 1, and bridge 2 acts as a rectifier for secondary 2 where the net output voltage is the difference between the two bridges, as in (*b*). The position of the core can be determined from the amplitude of the dc output, and the direction of the movement of the core can be determined from the polarity of the voltage. For rapid movements of the core, the output of the diode bridges must be filtered, for this, a suitably designed simple *RC* filter may be sufficient.

from the null position. A typical phase-sensitive demodulation circuit may be constructed, based on diodes shown in Figure 6.20(*a*). This arrangement is useful for very slow displacements, usually less than 1 or 2 Hz. In this figure, bridge 1 acts as a rectification circuit for secondary 1, and bridge 2 acts as a rectifier for secondary 2. The net output voltage is the difference between the outputs of two bridges, as in Figure 6.20(*b*). The position of the core can be determined from the amplitude of the dc output, and the direction of the movement of the core can be determined from the polarity of the dc voltage. For rapid movements of the core, the outputs of the diode bridges need to be filtered, wherein only the frequencies of the movement of the core pass through and all the other frequencies produced by the modulation process are filtered. For this purpose, a suitably designed simple RC filter may be sufficient.

There are phase-sensitive demodulator chips available in the marketplace, such as AD598 offered by Analog Devices Inc. These chips are highly versatile and flexible to use to suit particular application requirements. These chips offer many advantages over conventional phase-sensitive demodulation devices; for example, frequency off excitation may be adjusted to any value between 20 Hz and 20 kHz by connecting an external capacitor between two pins. The amplitude of the excitation voltage can be set up to 24 V. The internal filters may be set to required values by external capacitors. Connections to analog-to-digital converters are easily made by converting the bipolar output to unipolar scale.

The frequency response of LVDTs is primarily limited by the inertia characteristics of the device. In general, the frequency of the applied voltage should be 10 times the desired frequency response. Commercial LVDTs are available in a broad range of sizes and they are widely used for displacement measurements in a variety of applications. These displacement sensors are available to cover ranges from ±0.25 mm to ±7.5 cm. They are sensitive enough to be used to respond to displacements well below 0.0005 mm. They have operational temperature ranges from −265°C to 600°C. They are also available in radiation-resistant designs for operation in nuclear reactors. For a typical sensor of ±25 mm range, the recommended supply voltage is 4 V to 6 V, with a nominal frequency of 5 kHz, and a maximum non-linearity of 1% full scale. Several commercial models are available that can produce a voltage output of 300 mV for 1 mm displacement of the core.

One important advantage of the LVDT is that there is no physical contact between the core and the coil form, and hence no friction or wear. Nevertheless, there are radial and longitudinal magnetic forces on the core at all times. These magnetic forces may be regarded as magnetic springs that try to displace the core to its null position. This may be a critical factor in some applications.

One problem with LVDTs is that it may not be easy to make the two halves of the secondary identical; their inductance, resistance, and capacitance may be different, causing a large unwanted quadrature output in the balance position. Precision coil winding equipment may be required to reduce this problem to an acceptable value.

Another problem is associated with null position adjustments. The harmonics in the supply voltage and stray capacitances result in small null voltages. The null voltage may be reduced by proper grounding, which reduces the capacitive effects and center-tapped voltage source arrangements. In center-tapped supplies, a potentiometer may be used to obtain a minimum null reading.

The LVDTs have a variety of applications, including control for jet engines in close proximity to exhaust gases and measuring roll positions in the thickness of materials in hot-slab steel mills. Force and pressure measurements may also be made by LVDTs after some mechanical modifications.

Rotary Variable-Differential Transformer

A variation from the linear-variable differential transformer is the rotary core differential transformer shown in Figures 6.21(*a*) and 6.21(*b*). Here, the primary winding is wound on the center leg of an E core, and the secondary windings are wound on the outer legs of the core. The armature is rotated by an externally applied force about a pivot point above the center leg of the core. When the armature is displaced from its reference or balance position, the reluctance of the magnetic circuit through one secondary coil decreases, simultaneously increasing the reluctance through the other coil. The induced emfs in the secondary windings, which are equal in the reference position of the armature, are now different in magnitude and phase as a result of the applied displacement. The induced emfs in the secondary coils are made to oppose each other and the transformer operates in the same manner as an LVDT. The rotating variable transformers may be sensitive to vibrations. If a dc output is required, a demodulator network can be used, as in the case of LVDTs.

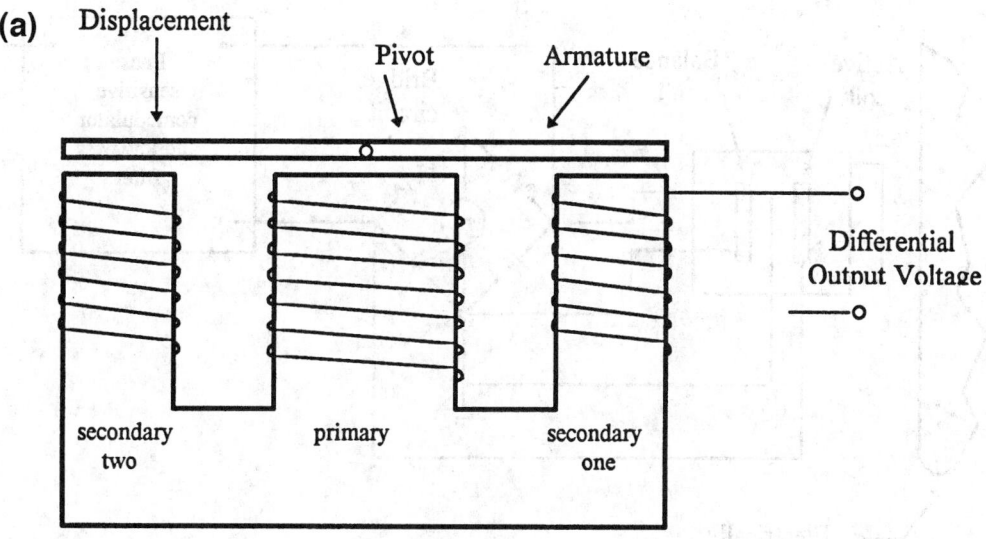

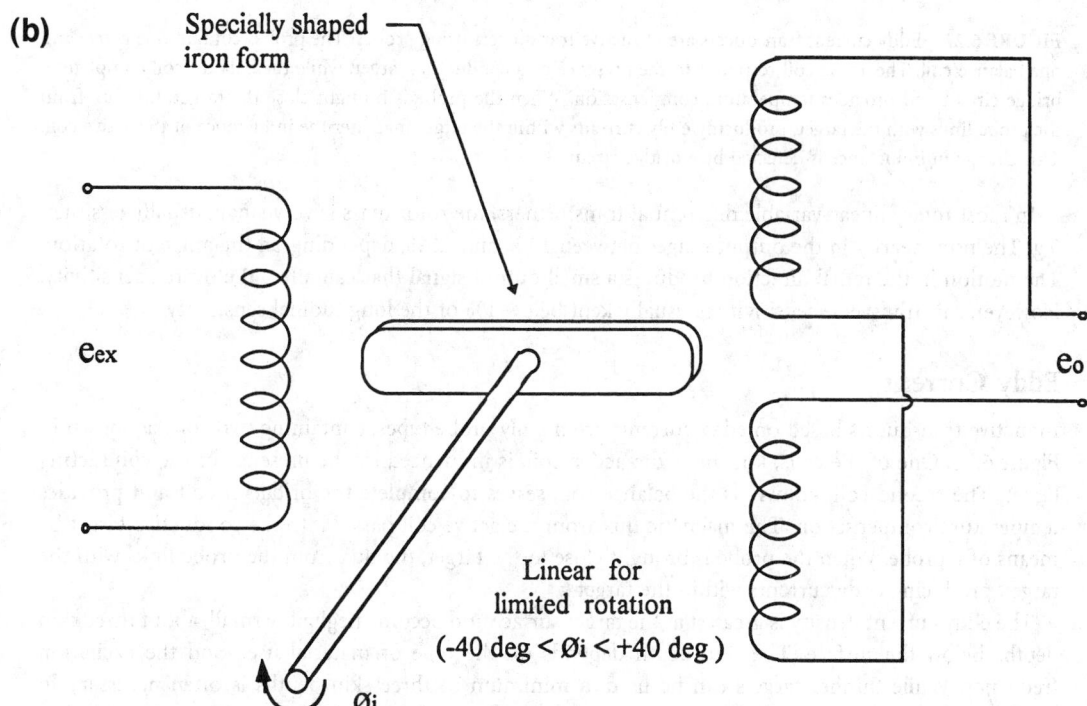

FIGURE 6.21 A rotary core differential transformer has an E-shaped core, carrying the primary winding on the center leg and the two secondaries on the outer legs, as in (*a*). The armature is rotated by an externally applied force about a pivot point above the center leg of the core (*b*). When the armature is displaced from its reference or balance position, the reluctance of the magnetic circuit through one secondary coil is decreased, increasing the reluctance through the other coil. The induced emfs in the secondary windings are different in magnitude and phase as a result of the applied displacement.

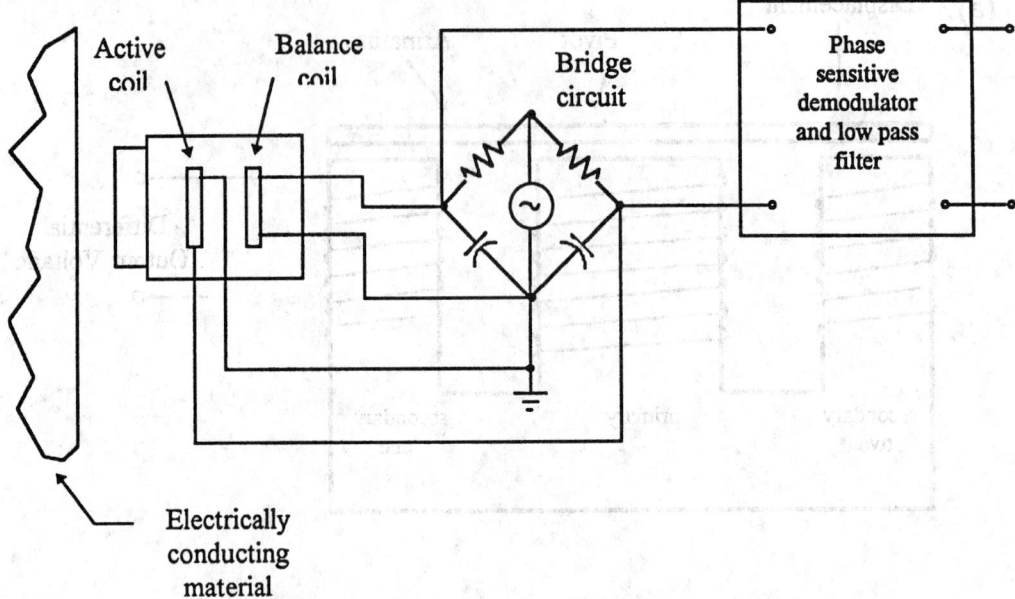

FIGURE 6.22 Eddy current transducers are inductive transducers using probes. The probes contain one active and one balance coil. The active coil responds to the presence of a conducting target, while the balance coil completes a bridge circuit and provides temperature compensation. When the probe is brought close the target, the flux from the probe links with the target, producing eddy currents within the target that alter the inductance of the active coil. This change in inductance is detected by a bridge circuit.

In most rotary linear-variable differential transformers, the rotor mass is very small, usually less than 5 g. The nonlinearity in the output ranges between ±1% and ±3%, depending on the angle of rotation. The motion in the radial direction produces a small output signal that can affect the overall sensitivity. However, this transverse sensitivity is usually kept below 1% of the longitudinal sensitivity.

Eddy Current

Inductive transducers based on eddy currents are mainly probe types, containing two coils as shown in Figure 6.22. One of the coils, known as the active coil, is influenced by the presence of the conducting target. The second coil, known as the balance coil, serves to complete the bridge circuit and provides temperature compensation. The magnetic flux from the active coil passes into the conductive target by means of a probe. When the probe is brought close to the target, the flux from the probe links with the target, producing eddy currents within the target.

The eddy current density is greatest at the target surface and become negligibly small, about three skin depths below the surface. The skin depth depends on the type of material used and the excitation frequency. While thinner targets can be used, a minimum of three skin depths is often necessary to minimize the temperature effects. As the target comes closer to the probe, the eddy currents become stronger, causing the impedance of the active coil to change and altering the balance of the bridge in relation to the target position. This unbalance voltage of the bridge may be demodulated, filtered, and linearized to produce a dc output proportional to target displacement. The bridge oscillation may be as high as 1 MHz. High frequencies allow the use of thin targets and provide good system frequency response.

Probes are commercially available with full-scale diameter ranging from 0.25 to 30 mm with a non-linearity of 0.5% and a maximum resolution of 0.0001 mm. Targets are usually supplied by the clients, involving noncontact measurements of machine parts. For nonconductive targets, conductive materials of sufficient thickness must be attached to the surface by means of commercially available adhesives. Since the target material, shape, etc. influence the output, it is necessary to calibrate the system statistically for a specific target. The recommended measuring range of a given probe begins at a standoff distance equal to about 20% of the stated range of the probe. In some cases, a standoff distance of 10% is recommended for which the system is calibrated as standard. A distance greater than 10% of the measuring range can be used as long as the calibrated measuring range is reduced by the same amount.

Flat targets must be of the same diameter as the probe or larger. If the target diameter is smaller than the probe diameter, the output drops considerably, thus becoming unreliable. Curved-surface targets can behave similar to flat surfaces if the diameter exceeds about three or four times the diameter of the probe. In this case, the target essentially becomes an infinite plane. This also allows some cross-axis movement without affecting the system output. Target diameters comparable to the sensor could result in detrimental effects in measurements due to cross-axis movements.

For curved or irregular shaped targets, the system must be calibrated using the exact target that is seen in the operation. This tends to eliminate any errors caused by the curved surfaces during application. However, special multiprobe systems are available for orbital motions of rotating shafts. If the curved (shaft) target is about 10 times larger than the sensor diameter, it acts as an infinite plane and does not need special calibrations. Care must be exercised to deal with electrical runout due to factors such as inhomegeneities in hardness, etc., particularly valid for ferrous targets. However, nonferrous targets are free from electrical runout concerns.

Shielding and Sensitivity of Inductive Sensors to Electromagnetic Interference

Magnetic fields are produced by currents in wires and more strongly by the coils. The fields due to coils are important due to magnetic coupling, particularly when there are two or more coils in the circuit. The magnetic coupling between coils may be controlled by large spacing between coils, orientation of coils, the shape of the coils, and by shielding.

Inductive sensors come in different shape and sizes. While some sensors have closed cores such as toroidal shapes, others have open cores and air gaps between cores and coils. Closed cores can have practically zero external fields, except for small leakage fluxes. Even if the sensors do not have closed cores, most variable inductor sensors have rather limited external fields, due to two neighboring sets of coils connected in opposite directions that minimize the external fields.

Inductive sensors are made from closed conductors. This implies that if the conductor moves in a magnetic field, a current will flow. Alternatively, a magnetic change produces current in a stationary closed conductor. Unless adequate measures are taken, there may be external magnetic fields linking (interference) with the sensor coils, thus producing currents and unwanted responses.

Due to inherent operations, inductive sensors are designed to have high sensitivity to magnetic flux changes. External electromagnetic interference and external fields can seriously affect the performance of the sensors. It is known that moderate magnetic fields are found near power transformers, electrical motors, and power lines. These small fields produce current in the inductive sensors elements. One way of eliminating external effects is accomplished by magnetic shielding of the sensors and by grounding appropriately. In magnetic shielding, one or more shells of high-permeability magnetic materials surround the part to be shielded. Multiple shells may be used to obtain very complete shielding. The ends of each individual shell are separated by insulation so that the shell does not act as a single shorted turn, thus accommodating high current flows. Similarly, in the case of multiple shielding, shells are isolated from each other by proper insulation.

Alternating magnetic fields are also screened by interposing highly conductive metal sheets such as copper or aluminum on the path of the magnetic flux. The eddy currents induced in the shield give a counter mmf that tends to cancel the interfering magnetic field. This type of shielding is particularly effective at high frequencies. Nevertheless, appropriate grounding must still be observed.

In many inductive sensors, stray capacitances can be a problem, especially at the null position of the moving core. If the capacitive effect is greater than a certain value, say 1% of the full-scale output, this effect may be reduced by the use of center-tapped supply and appropriate grounding.

References

1. J. P. Bentley, *Principles of Measurement Systems*, 2nd ed., United Kingdom: Longman Scientific and Technical, 1988.
2. E. O. Doebelin, *Measurement Systems: Application and Design*, 4th ed., New York: McGraw-Hill, 1990.
3. J. P. Holman, *Experimental Methods for Engineers*, 5th ed., New York: McGraw-Hill, 1989.
4. W. J. Tompkins and J. G. Webster, *Interfacing Sensors to the IBM PC*, Englewood Cliffs, NJ: Prentice-Hall, 1988.

Appendix to Section 6.2

LIST OF MANUFACTURERS

Adsen Tech. Inc.
18310 Bedford Circle
La Puente, CA 91744
Fax: (818) 854-2776

Dynalco Controls
3690 N.W. 53rd Street
Ft. Lauderdale, FL 33309
Tel: (954) 739-4300 & (800) 368-6666
Fax: (954) 484-3376

Electro Corporation
1845 57th Street
Sarasato, FL 34243
Tel: (813) 355-8411 & (800) 446-5762
Fax: (813) 355-3120

Honeywell
Dept 722
11 W. Spring Street
Freeport, IL 61032
Tel: (800) 537-6945
Fax: (815) 235-5988

Kaman Inst. Co.
1500 Garden of the Gods Rd.
Colorado Springs, CO 80907
Tel: (719) 599-1132 & (800) 552-6267
Fax: (719) 599-1823

Kavlico Corporation
14501 Los Angeles Avenue
Moorpark, CA 93021
Tel: (805) 523-2000
Fax: (805) 523-7125

Lucas
1000 Lucas Way
Hampton, VA 23666
Tel: (800) 745-8008
Fax: (800) 745-8004

Motion Sensors Inc.
786 Pitts Chapel Rd.
Alizabeth City, NC 27909
Tel: (919) 331-2080
Fax: (919) 331-1666

Rechner Electronics Ind. Inc.
8651 Buffalo Ave.
Niagara Falls, NY 14304
Tel: (800) 544-4106
Fax: (716) 283-2127

Reed Switch Developments Co. Inc.
P.O. Drawer 085297
Racine, WI 53408
Tel: (414) 637-8848
Fax: (414) 637-8861

Smith Research & Technology Inc.
205 Sutton Lane, Dept. TR-95
Colorado Springs, CO 80907
Tel: (719) 634-2259
Fax: (719) 634-2601

Smith Systems Inc.
6 Mill Creek Dr.
Box 667
Brevard, NC 28712
Tel: (704) 884-3490
Fax: (704) 877-3100

Standex Electronics
4538 Camberwell Rd.
Dept. 301L
Cincinnati, OH 45209
Tel: (513) 871-3777
Fax: (513) 871-3779

Turck Inc.
3000 Campus Drive
Minneapolis, MN 55441
Tel: (612) 553-7300 & (800) 544-7769
Fax: (612) 553-0708

Xolox Sensor Products
6932 Gettysburg Pike
Ft. Wayne, IN 46804
Tel: (800) 348-0744
Fax: (219) 432-0828

6.3   Capacitive Sensors—Displacement

Halit Eren and Wei Ling Kong

Capacitive sensors are extensively used in industrial and scientific applications. They are based on changes in capacitance in response to physical variations. These sensors find many diverse applications — from humidity and moisture measurements to displacement sensing. In some cases, the basic operational and sensing principles are common in dissimilar applications; and in other cases, different principles can be used for the same applications. For example, capacitive microphones are based on variations of spacing between plates in response to acoustical pressure, thus turning audio signals to variations in capacitance. On the other hand, a capacitive level indicator makes use of the changes in the relative permittivity between the plates. However, capacitive sensors are best known to be associated with displacement measurements for rotational or translational motions, as will be described next. Other applications of capacitance sensors such as humidity and moisture will be discussed.

Capacitive Displacement Sensors

The measurement of distances or displacements is an important aspect of many industrial, scientific, and engineering systems. The displacement is basically the vector representing a change in position of a body or point with respect to a reference point. Capacitive displacement sensors satisfy the requirements of applications where high linearity and wide ranges (from a few centimeters to a couple of nanometers) are needed.

The basic sensing element of a typical displacement sensor consists of two simple electrodes with capacitance C. The capacitance is a function of the distance d (cm) between the electrodes of a structure, the surface area A (cm^2) of the electrodes, and the permittivity ε (8.85×10^{-12} F m^{-1} for air) of the dielectric between the electrodes; therefore:

$$C = f\left(d, A, \varepsilon\right) \tag{6.22}$$

There are three basic methods for realizing a capacitive displacement sensor: by varying d, A, or ε, as discussed below.

Variable Distance Displacement Sensors

A capacitor displacement sensor, made from two flat coplanar plates with a variable distance x apart, is illustrated in Figure 6.23. Ignoring fringe effects, the capacitance of this arrangement can be expressed by:

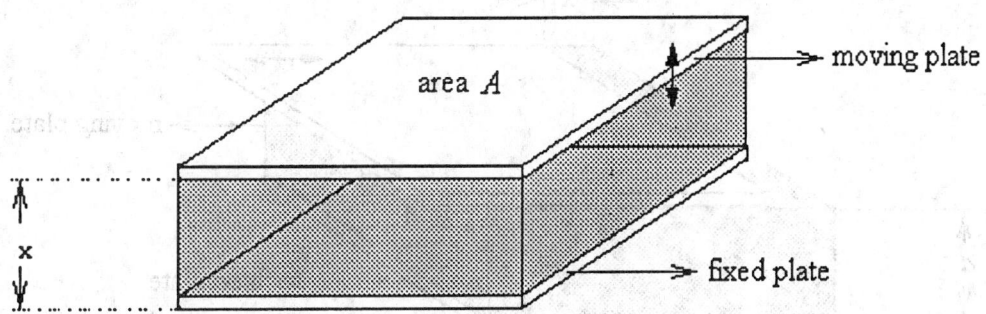

FIGURE 6.23   A variable distance capacitive displacement sensor. One of the plates of the capacitor moves to vary the distance between plates in response to changes in a physical variable. The outputs of these transducers are nonlinear with respect to distance x having a hyperbolic transfer function characteristic. Appropriate signal processing must be employed for linearization.

$$C(x) = \varepsilon A / x = \varepsilon_r \varepsilon_0 A / x \qquad (6.23)$$

where ε = the dielectric constant or permittivity
   ε_r = the relative dielectric constant (in air and vacuum $\varepsilon_r \approx 1$)
   ε_0 = 8.854188 × 10⁻¹² F/m⁻¹, the dielectric constant of vacuum
   x = the distance of the plates in m
   A = the effective area of the plates in m²

The capacitance of this transducer is nonlinear with respect to distance x, having a hyperbolic transfer function characteristic. The sensitivity of capacitance to changes in plate separation is

$$dC/dx = -\varepsilon_r \varepsilon_0 A / x^2 \qquad (6.24)$$

Equation 6.24 indicates that the sensitivity increases as x decreases. Nevertheless, from Equations 6.23 and 6.24, it follows that the percent change in C is proportional to the percent change in x. This can be expressed as:

$$dC/C = -dx/x \qquad (6.25)$$

This type of sensor is often used for measuring small incremental displacements without making contact with the object.

Variable Area Displacement Sensors

Alternatively, the displacements may be sensed by varying the surface area of the electrodes of a flat plate capacitor, as illustrated in Figure 6.24. In this case, the capacitance would be:

$$C = \varepsilon_r \varepsilon_0 (A - wx) / d \qquad (6.26)$$

where w = the width
   wx = the reduction in the area due to movement of the plate

Then, the transducer output is linear with displacement x. This type of sensor is normally implemented as a rotating capacitor for measuring angular displacement. The rotating capacitor structures are also used as an output transducer for measuring electric voltages as capacitive voltmeters.

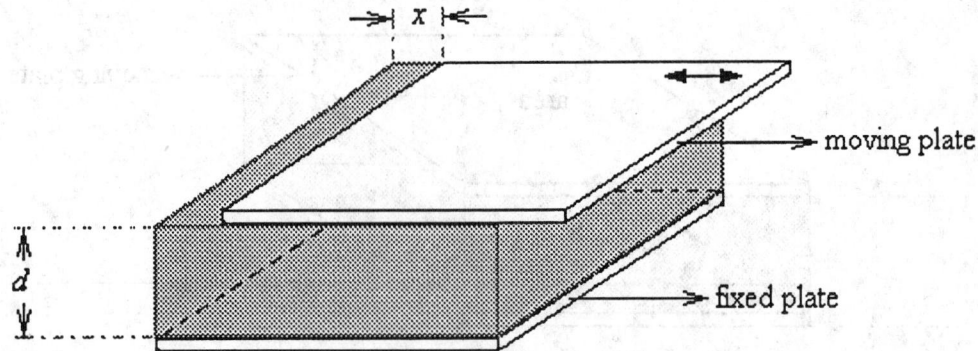

FIGURE 6.24 A variable area capacitive displacement sensor. The sensor operates on the variation in the effective area between plates of a flat-plate capacitor. The transducer output is linear with respect to displacement x. This type of sensor is normally implemented as a rotating capacitor for measuring angular displacement.

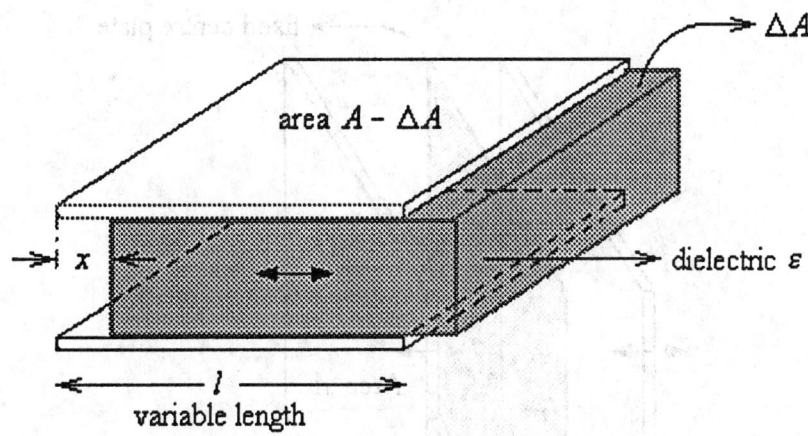

FIGURE 6.25 A variable dielectric capacitive displacement sensor. The dielectric material between the two parallel plate capacitors moves, varying the effective dielectric constant. The output of the sensor is linear.

Variable Dielectric Displacement Sensors

In some cases, the displacement may be sensed by the relative movement of the dielectric material between the plates, as shown in Figure 6.25. The corresponding equations would be:

$$C = \varepsilon_0 \, w \left[\varepsilon_2 l - \left(\varepsilon_2 - \varepsilon_1 \right) x \right] \tag{6.27}$$

where ε_1 = the relative permittivity of the dielectric material
ε_2 = the permittivity of the displacing material (e.g., liquid)

In this case, the output of the transducer is also linear. This type of transducer is predominantly used in the form of two concentric cylinders for measuring the level of fluids in tanks. A nonconducting fluid forms the dielectric material. Further discussion will be included in the level measurements section.

Differential Capacitive Sensors

Some of the nonlinearity in capacitive sensors can be eliminated using differential capacitive arrangements. These sensors are basically three-terminal capacitors, as shown in Figure 6.26. Slight variations in the construction of these sensors find many different applications, including differential pressure measurements. In some versions, the central plate moves in response to physical variables with respect to the fixed plates. In others, the central plate is fixed and outer plates are allowed to move. The output from the center plate is zero at the central position and increases as it moves left or right. The range is equal to twice the separation d. For a displacement d, one obtains:

$$2\delta C = C_1 - C_2 = \varepsilon_r \varepsilon_0 lw / \left(d - \delta d \right) - \varepsilon_r \varepsilon_0 lw / \left(d + \delta d \right) = 2\varepsilon_r \varepsilon_0 lw\delta d / \left(d^2 + \delta d^2 \right) \tag{6.28}$$

and

$$C_1 + C_2 = 2C = \varepsilon_r \varepsilon_0 lw / \left(d - \delta d \right) + \varepsilon_r \varepsilon_0 lw / \left(d + \delta d \right) = 2\varepsilon_r \varepsilon_0 lwd / \left(d^2 + \delta d^2 \right) \tag{6.29}$$

Giving approximately:

$$\delta C / C = \delta d / d \tag{6.30}$$

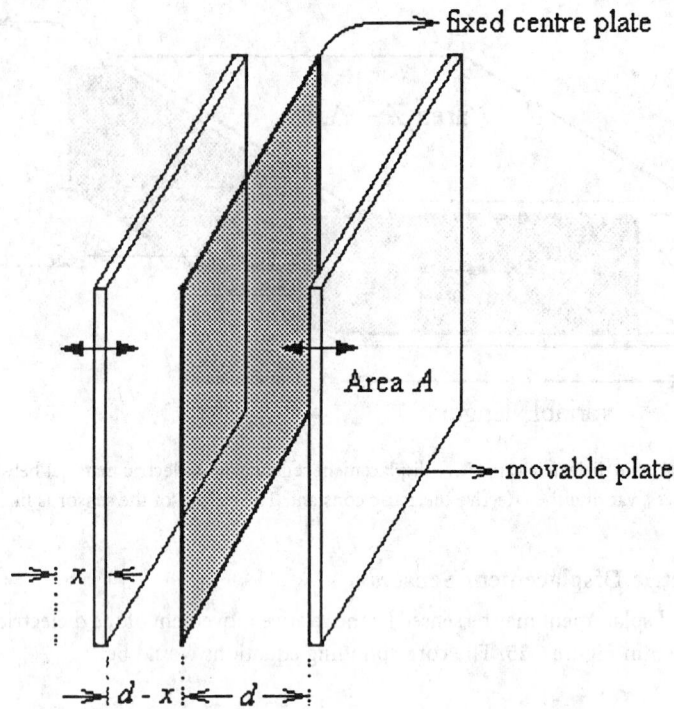

FIGURE 6.26 A differential capacitive sensor. They are essentially three terminal capacitors with one fixed center plate and two outer plates. The response to physical variables is linear. In some versions, the central plate moves in response to physical variable with respect to two outer plates, and in the others, the central plate is fixed and outer plates are allowed to move.

This indicates that the response of the device is more linear than the response of the two plate types. However, in practice some nonlinearity is still observed due to defects in the structure. Therefore, the outputs of these type of sensors still need to be processed carefully, as explained in the signal processing section.

In some differential capacitive sensors, the two spherical depressions are ground into glass disks; then, these are gold-plated to form the fixed plates of a differential capacitor. A thin, stainless-steel diaphragm is clamped between the disks and serves as a movable plate. With equal pressure applied to both ports, the diaphragm is then in neutral position and the output is balanced at a corresponding bridge. If one pressure is greater than the other, the diaphragm deflects proportionally, giving an output due to the differential pressure. For the opposite pressure difference, there is a phase change of 180°. A direction-sensitive dc output can be obtained by conventional phase-sensitive demodulation and appropriate filtering. Details of signal processing are given at the end of this chapter. In general, the differential capacitors exhibit better linearity than single-capacitor types.

Integrated Circuit Smart Capacitive Position Sensors

Figure 6.27 shows a typical microstructure capacitive displacement sensor. The sensor consists of two electrodes with capacitance, C_x. Since the system is simple, the determination of the capacitance between the two electrodes is straightforward. The smaller electrode is surrounded by a guard electrode to make C_x independent of lateral and rotational movements of the system parallel to the electrode surface. However, the use of a guard electrode introduces relative deviations in the capacitance C_x between the two electrodes. This is partially true if the size of the guard electrode is smaller than:

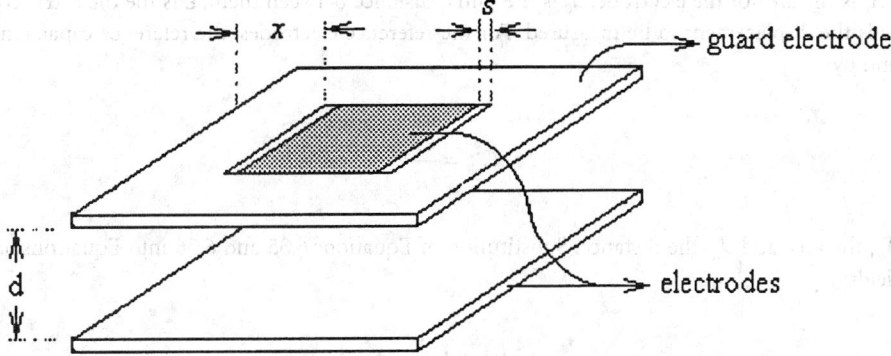

FIGURE 6.27 A typical smart capacitive position sensor. This type of microstructure position sensor contains three electrodes, two of which are fixed and the third electrode moves infinitesimally relative to the others. Although the response is highly nonlinear the integrated chip contains linearization circuits. They feature a 0 mm to 1 mm measuring range with 1 μm accuracy.

$$\delta < \exp\left(-\pi x/d\right) \tag{6.31}$$

where x is the width of the guard and d is the distance between the electrodes. Since this deviation introduces nonlinearity, δ is required to be less than 100 ppm. Another form of deviation also exists between the small electrode and the surrounding guard, particularly for gaps

$$\delta < \exp\left(-\pi d/s\right) \tag{6.32}$$

where s is the width of the gap. When the gap width, s, is less than 1/3 of the distance between electrodes, this deviation is negligible.

For signal processing, the system uses the three-signal concept. The capacitor C_x is connected to an inverting operational amplifier and oscillator. If the external movements are linear, by taking into account the parasitic capacitors and offsetting effects, the following equation can be written:

$$M_x = mC_x + M_{off} \tag{6.33}$$

where m is the unknown gain and M_{off} is the unknown offset. By performing the measurement of a reference C_{ref}, by measuring the offset, M_{off}, and by making $m = 0$, the parameters m and M_{off} can be eliminated. The final measurement result for the position, P_{os}, can be defined as:

$$P_{os} = \frac{M_{ref} - M_{off}}{M_x - M_{off}} \tag{6.34}$$

In this case, the sensor capacitance C_x can be simplified to:

$$C_x = \frac{\varepsilon A_x}{d_0 + \Delta d} \tag{6.35}$$

where A_x is the area of the electrode, d_0 is the initial distance between them, ε is the dielectric constant, and Δd is the displacement to be measured. For the reference electrodes, the reference capacitance may be found by:

$$C_{ref} = \frac{\varepsilon A_{ref}}{d_{ref}} \tag{6.36}$$

with A_{ref} the area and d_{ref} the distance. Substitution of Equations 6.35 and 6.36 into Equations 6.33 and 6.34 yields:

$$P_{os} = \frac{A_{ref}(d_0 + \Delta d)}{A_x d_{ref}} = a_1 \frac{\Delta d}{d_{ref}} + a_0 \tag{6.37}$$

P_{os} is a value representing the position if the stable constants a_1 and a_0 are unknown. The constant $a_1 = A_{ref}/A_x$ becomes a stable constant so long as there is good mechanical matching between the electrode areas. The constant $a_0 = (A_{ref}\, d_0)/(A_x\, d_{ref})$ is also a stable constant for fixed d_0 and d_{ref}. These constants are usually determined by calibration repeated over a certain time span. In many applications, these calibrations are omitted if the displacement sensor is part of a larger system where an overall calibration is necessary. This overall calibration usually eliminates the requirement for a separate determination of a_1 and a_0.

The accuracy of this type of system could be as small as 1 μm over a 1 mm range. The total measuring time is better than 0.1 s. The capacitance range is from 1 pF to 50 fF. Interested readers should refer to [4] at the end of this chapter.

Capacitive Pressure Sensors

A commonly used two-plate capacitive pressure sensor is made from one fixed metal plate and one flexible diaphragm, as shown in Figure 6.28. The flat circular diaphragm is clamped around its circumference and bent into a curve by an applied pressure P. The vertical displacement y of this system at any radius r is given by:

$$y = 3\left(1 - v^2\right)\left(a^2 - r^2\right) P/16\ Et^3 \tag{6.38}$$

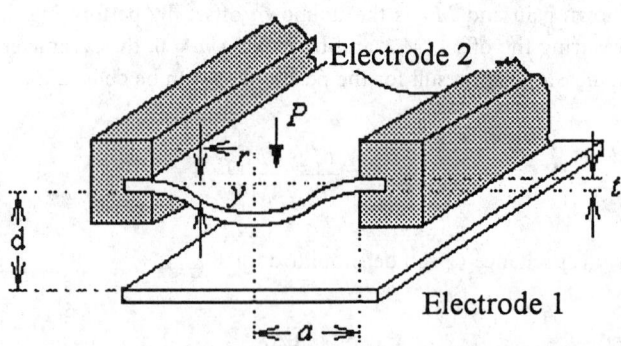

FIGURE 6.28 A capacitive pressure sensor. These pressure sensors are made from a fixed metal plate and a flexible diaphragm. The flat flexible diaphragm is clamped around its circumference. The bending of the flexible plate is proportional to the applied pressure P. The deformation of the diaphragm results in changes in capacitance.

where *a* = the radius of diaphragm
    t = the thickness of diaphragm
    E = Young's modulus
    v = Poisson's ratio

Deformation of the diaphragm means that the average separation of the plates is reduced. Hence, the resulting increase in the capacitance ΔC can be calculated by:

$$\Delta C/C = \left(1 - v^2\right) a^4 P / 16\ Et^3 \tag{6.39}$$

where *d* is the initial separation of the plates and *C* is the capacitance at zero pressure.

Another type of sensor is the differential capacitance pressure sensor shown in Figure 6.29. The capacitances C_1 and C_2 of the sensor change with respect to the fixed central plate in response to the applied pressures P_1 and P_2. Hence, the output of the sensor is proportional to $(P_1 - P_2)$. The signals are processed using one of the techniques described in the "Signal Processing" section of this chapter.

Capacitive Accelerometers and Force Transducers

In recent years, capacitive-type micromachined accelerometers, as illustrated in Figure 6.30, are gaining popularity. These accelerometers use the proof mass as one plate of the capacitor and use the other plate as the base. When the sensor is accelerated, the proof mass tends to move; thus, the voltage across the capacitor changes. This change in voltage corresponds to the applied acceleration.

In Figure 6.30, let $F(x)$ be the positive force in the direction in which *x* increases. Neglecting all losses (due to friction, resistance, etc.), the energy balance of the system can be written for an infinitesimally small displacement d*x*, electrical energy dE_e, and field energy dE_f of the electrical field between the electrodes as:

$$dE_m + dE_e = dE_f \tag{6.40}$$

in which:

$$dE_m = F\left(x\right) dx \tag{6.41}$$

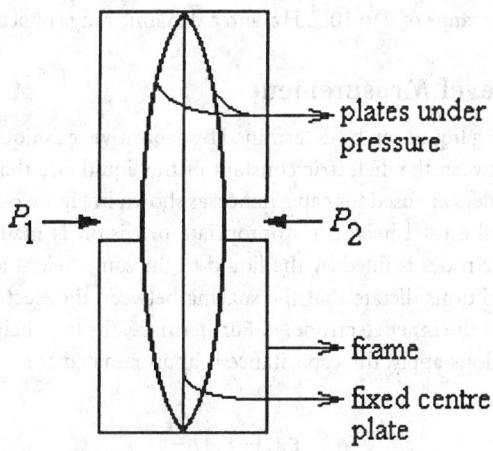

$P_1 \longrightarrow$

$\longleftarrow P_2$

plates under pressure

frame

fixed centre plate

FIGURE 6.29 A differential capacitive pressure sensor. The capacitances C_1 and C_2 of the sensor changes due to deformation in the outer plates, with respect to the fixed central plate in response to the applied pressures P_1 and P_2.

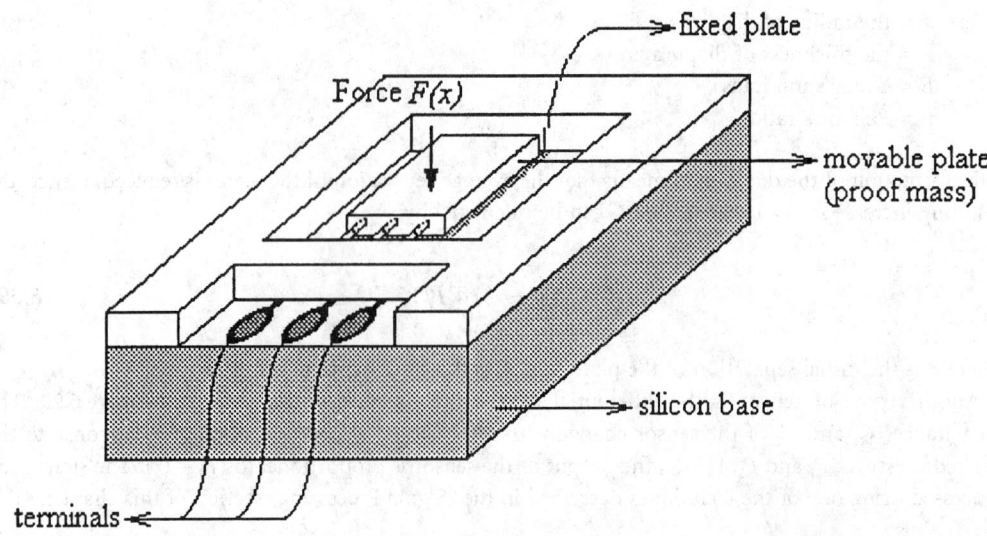

FIGURE 6.30 A capacitive force transducer. A typical capacitive micromachined accelerometer has one of the plates as the proof mass. The other plate is fixed, thus forming the base. When the sensor is accelerated, the proof mass tends to move, thus varying the distance between the plates and altering the voltage across the capacitor. This change in voltage is made to be directly proportional to the applied acceleration.

Also,

$$dE_m = d(QV) = Q\,dV + V\,dQ \tag{6.42}$$

If the supply voltage V across the capacitor is kept constant, it follows that $dV = 0$. Since $Q = VC(x)$, the Coulomb force is given by:

$$F(x) = -V^2 \frac{dC(x)}{dx} \tag{6.43}$$

Thus, if the movable electrode has complete freedom of motion, it will have assumed a position in which the capacitance is maximal; also, if C is a linear function of x, the force $F(x)$ becomes independent of x.

Capacitive silicon accelerometers are available in a wide range of specifications. A typical lightweight sensor will have a frequency range of 0 to 1000 Hz, and a dynamic range of acceleration of ± 2 g to ± 500 g.

Capacitive Liquid Level Measurement

The level of a nonconducting liquid can be determined by capacitive techniques. The method is generally based on the difference between the dielectric constant of the liquid and that of the gas or air above it. Two concentric metal cylinders are used for capacitance, as shown in Figure 6.31. The height of the liquid, h, is measured relative to the total height, l. Appropriate provision is made to ensure that the space between the cylindrical electrodes is filled by the liquid to the same height as the rest of the container. The usual operational conditions dictate that the spacing between the electrodes, $s = r_2 - r_1$, should be much less than the radius of the inner electrode, r_1. Furthermore, the tank height should be much greater than r_2. When these conditions apply, the capacitance is approximated by:

$$C = \frac{\varepsilon_1(l) + \varepsilon_g(h-1)}{4.6 \log[1 - (s/r)]} \tag{6.44}$$

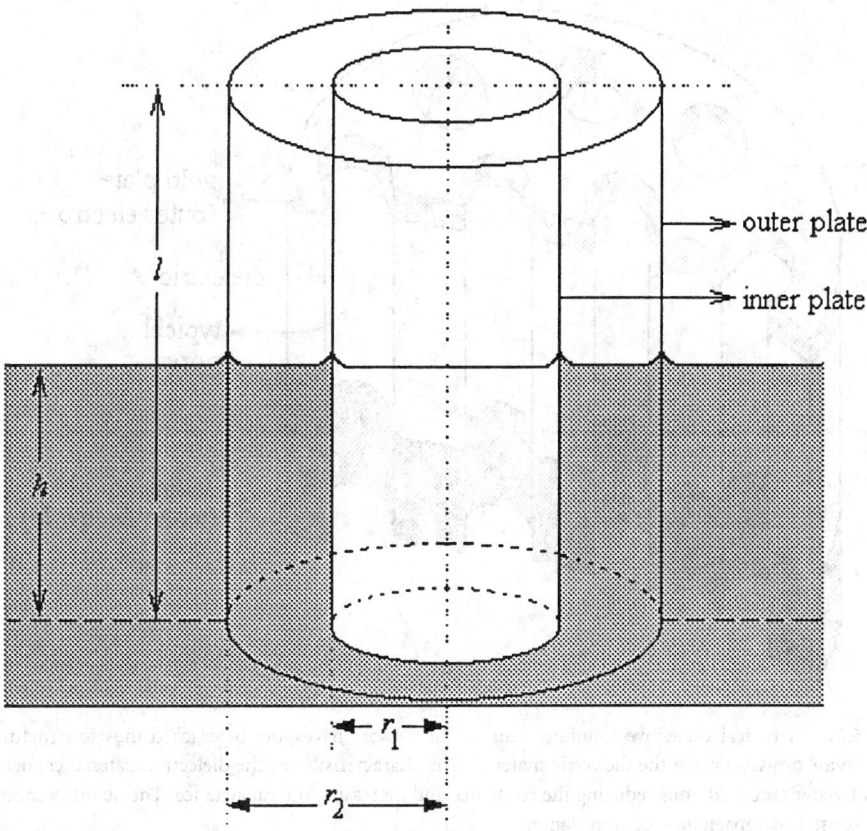

FIGURE 6.31 A capacitive liquid level sensor. Two concentric metal cylinders are used as electrodes of a capacitor. The value of the capacitance depends on the permittivity of the liquid and that of the gas or air above it. The total permittivity changes depending on the liquid level. These devices are usually applied in nonconducting liquid applications.

where ε_l and ε_g are the dielectric constants of the liquid and gas (or air), respectively. The denominator of the above equation contains only terms that relate to the fixed system. Therefore, they become a single constant. A typical application is the measurement of the amount of gasoline in a tank in airplanes. The dielectric constant for most compounds commonly found in gasoline is approximately equal to 2, while that of air is approximately unity. A linear change in capacitance with gasoline level is expected for this situation. Quite high accuracy can be achieved if the denominator is kept quite small, thus accentuating the level differences. These sensors often incorporate an ac deflection bridge.

Capacitive Humidity and Moisture Sensors

The permittivities of atmospheric air, of some gases, and of many solid materials are functions of moisture content and temperature. Capacitive humidity devices are based on the changes in the permittivity of the dielectric material between plates of capacitors. The main disadvantage of this type sensor is that a relatively small change in humidity results in a capacitance large enough for a sensitive detection.

Capacitive humidity sensors enjoy wide dynamic ranges, from 0.1 ppm to saturation points. They can function in saturated environments for long periods of time, a characteristic that would adversely affect many other humidity sensors. Their ability to function accurately and reliably extends over a wide range of temperatures and pressures. Capacitive humidity sensors also exhibit low hysteresis and high stability with minimal maintenance requirements. These features make capacitive humidity sensors viable for many specific operating conditions and ideally suitable for a system where uncertainty of unaccounted conditions exists during operations.

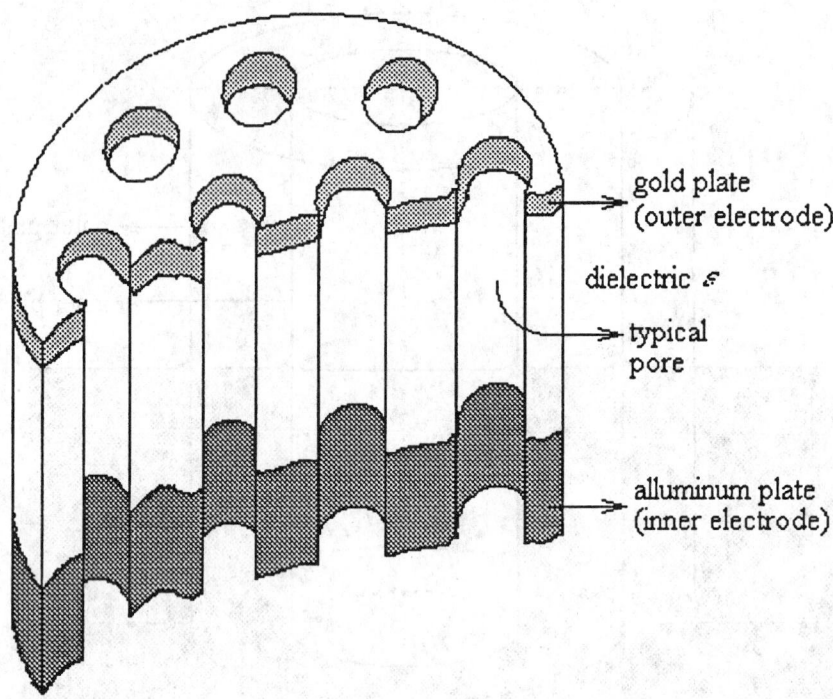

FIGURE 6.32 A typical capacitive humidity sensor. The sensors have pore or cracked mosaic structure for the moisture in air or gas to reach the dielectric material. The characteristics of the dielectric material change with the amount of water absorbed, thus reducing the resistance and increasing the capacitance. The quantity measured can be either resistance, capacitance, or impedance.

There are many types of capacitive humidity sensors. Aluminum, tantalum, silicon, and polymer types are introduced here.

Aluminum Type Capacitive Humidity Sensors

The majority of capacitive humidity sensors are aluminum oxide type sensors. In these type of sensors, high-purity aluminum is chemically oxidized to produce a prefilled insulating layer of partially hydrated aluminum oxide, which acts as the dielectric. A water-permeable but conductive gold film is deposited onto the oxide layer, usually by vacuum deposition, which forms the second electrode of the capacitor.

In another type, the aluminum-aluminum oxide sensor has a pore structure as illustrated in Figure 6.32. The oxide, with its pore structure, forms the active sensing material. Moisture in the air reaching the pores reduces the resistance and increases the capacitance. The decreased resistance can be thought of as being due to an increase in the conduction through the oxide. An increase in capacitance can be viewed as due to an increase in the dielectric constant. The quantity measured can be either resistance, capacitance, or impedance. High humidities are best measured by capacitance because resistance changes are vanishingly small in this region.

In addition to the kind of transducer design illustrated here, there are many others available with a number of substantial modifications for particular properties, such as increased sensitivity or faster response. Although most of these modifications result in a change in physical dimensions or appearance, the sensing material of the transducer — the aluminum oxide — remains the same.

In some versions, the oxide layer is formed by parallel tubular pores that are hexagonally packed and perpendicular to the plane of the base layer. These pores stop just before the aluminum layer, forming a very thin pore base. Water absorbed in these tubules is directly related to the moisture content of the gas in contact with it. The porous nature of the oxide layer produces a large area for the absorption of water vapor. At low humidities, the capacitance is due entirely to the mixed dielectric formed between the oxide, water vapor, and air. However, at higher humidities, parallel conductance paths through the absorbed water are formed down the pore surfaces. Near saturation, this pore surface resistance becomes negligible, implying that the measured capacitance is virtually that between the very thin pore base and the aluminum core.

Tantalum Type Capacitive Humidity Sensors

In some versions of capacitive humidity sensors, one of the capacitor plates consists of a layer of tantalum deposited on a glass substrate. A layer of polymer dielectric is then added, followed by a second plate made from a thin layer of chromium. The chromium layer is under high tensile stress such that it cracks into a fine mosaic structure that allows water molecules to pass into the dielectric. The stress in the chromium also causes the polymer to crack into a mosaic structure. A sensor of this type has an input range of 0% to 100% relative humidity, RH. The capacitance is 375 pF at 0% RH and a linear sensitivity of 1.7 pF per % RH. The error is usually less than 2% due to nonlinearity and 1% due to hysteresis.

Silicon Type Capacitive Humidity Sensors

In other capacitive humidity sensors, silicon is used as the dielectric. The structure and operation of silicon humidity sensors are very similar to the aluminum oxide types. Some silicon-type humidity sensors also use the aluminum base and a thin-film gold layer as the two electrodes. The silicon dielectric has a very large surface area, which means that the sensitivity is still relatively large even if the sensing area is very small. This is an important feature with the increasing trend of miniaturization. Both sensor types are now typically found as extremely small wafer-shaped elements, placed on a mechanical mount with connecting lead wires. The formation of porous silicon is a very simple anodization process and, since no elaborate equipment is needed, devices can be made at relatively low cost. Also, by controlling the formation conditions, the structure of the porous silicon can easily be modified so devices can be tailored to suit particular applications.

In both the silicon and the aluminum oxide capacitive humidity sensors, the radii of the pores in the dielectric are such that they are specifically suited for water molecules. Most possible contaminants are too large in size to pollute the dielectric. However, contaminants can block the flow of water vapor into the sensor material, thus affecting the accuracy of the instrument. For example, in dust-contaminated streams, it may be possible to provide a simple physical barrier such as a sintered metal or plastic hoods for the sensor heads. Many sensors come with some form of casing to provide protection.

Polymer Type Capacitive Humidity Sensors

In some sensors, the dielectric consists of a polymer material that has the ability to absorb water molecules. The absorption of water vapor of the material results in changes in the dielectric constant of the capacitor. By careful design, the capacitance can be made directly proportional to percentage relative humidity of the surrounding gas or atmosphere.

In general, an important key feature of capacitive humidity sensors is the chemical stability. Often, humidity sensing is required in an air sample that contains vapor contaminants (e.g., carbon monoxide) or the measurements are performed on a gas sample other than air (e.g., vaporized benzene). The performance of these sensors, and in particular the silicon types, is not affected by many of these gases. Hydrocarbons, carbon dioxide, carbon monoxide, and CFCs do not cause interference. However, the ionic nature of the aluminum oxide dielectric makes it susceptible to certain highly polar, corrosive gases such as ammonia, sulphur trioxide, and chlorine. Silicon is inert; its stable nature means that these polar gases affect the sensor element to a far lesser degree.

Capacitive Moisture Sensors

Capacitive moisture measurements are based on the changes in the permittivity of granular or powder type dielectric materials such as wheat and other grains containing water. Usually, the sensor consists of a large cylindrical chamber, (e.g., 150 mm deep and 100 mm in diameter), as shown in Figure 6.33. The chamber is filled with samples under test. The variations in capacitance with respect to water content are processed. The capacitor is incorporated into an oscillatory circuit operating at a suitable frequency.

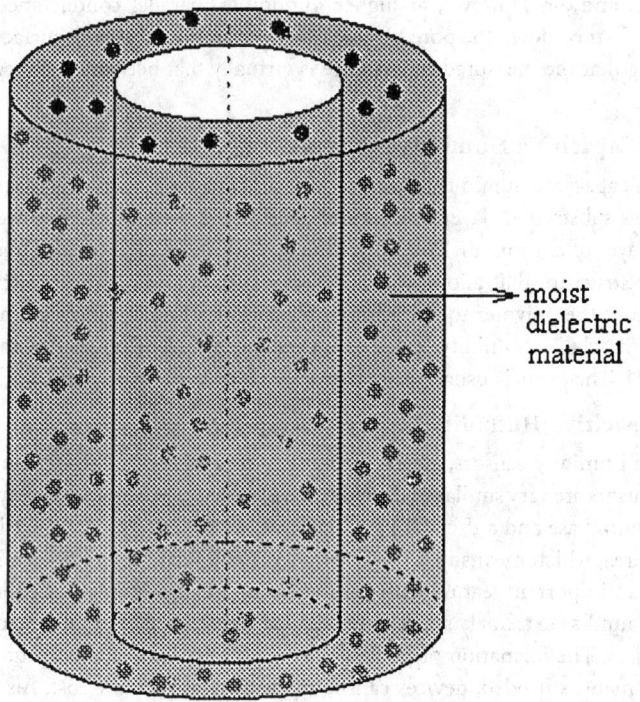

→ moist
  dielectric
  material

FIGURE 6.33 A capacitive moisture sensor. The permittivity of material between two cylindrical or parallel plates with fixed dimensions changes, depending on the moisture level of the materials in the chamber. The variations in capacitance values with respect to water content is processed. The capacitor is incorporated as a part of an oscillatory circuit operating at a suitable frequency, usually at radio frequencies.

Capacitive moisture sensors must be calibrated for samples made from different materials, as the materials themselves demonstrate different permittivities. Accurate temperature is necessary as the dielectric constant may be highly dependent on temperature. Most of these devices are built to operate at temperature ranges of 0°C to 50°C, supported by tight temperature compensation circuits. Once calibrated for a specific application, they are suitable for measuring moisture in the range of 0% to 40%.

Signal Processing

Generally, capacitive type pickups require relatively complex circuitry in comparison to many other sensor types, but they have the advantage of mechanical simplicity. They are also sensitive, having minimum mechanical loading effects. For signal processing, these sensors are usually incorporated either in ac deflection bridge circuits or oscillator circuits. In practice, capacitive sensors are not pure capacitances but have associated resistances representing losses in the dielectric. This can have an important influence in the design of circuits, particularly in oscillator circuits. Some of the signal processing circuits are discussed below.

Operational Amplifiers and Charge Amplifiers

One method of eliminating the nonlinearity of the relationship between the physical variable, (e.g., two-plate displacement sensors) and capacitance C is through the use of operational amplifiers, as illustrated in Figure 6.34. In this circuit, if the input impedance of the operational amplifier is high, the output is not saturated, and the input voltage is small, it is possible to write:

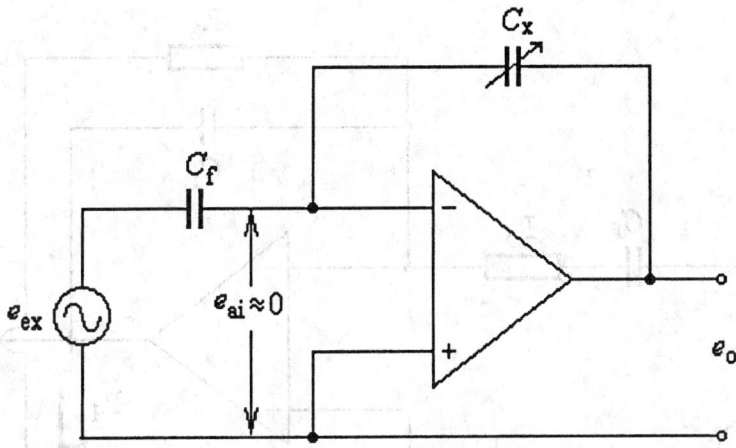

FIGURE 6.34 An operational amplifier signal processor. This method is useful to eliminate the nonlinearity in the signals generated by capacitive sensors. By this type of arrangement, the output voltage can be made directly proportional to variations in the signal representing the nonlinear operation of the device.

$$1/C_f = \int i_f \, dt = e_{ex} - e_{ai} = e_{ex} \tag{6.45}$$

$$1/C_x = \int i_x \, dt = e_0 - e_{ai} = e_0 \tag{6.46}$$

$$i_f + i_x - i_{ai} = 0 = i_f + i_x \tag{6.47}$$

Manipulation of these equations yields:

$$e_0 = -C_f \, e_{ex}/C_x \tag{6.48}$$

Substituting the value of C_x yields:

$$e_0 = -C_f x \, e_{ex}/\varepsilon A \tag{6.49}$$

Equation 6.49 shows that the output voltage is directly proportional to the plate separation x, thus giving linearity for all variations in motion.

However, a practical circuit requires a resistance across C_f to limit output drift. The value of this resistance must be greater than the impedance of C_f at the lowest frequency of interest. Also, because the transducer impedance is assumed to be purely capacitive, the effective gain is independent of frequency.

A practical charge amplifier circuit is depicted in Figure 6.35. In this case, the effective feedback resistance R_{ef} is given by:

$$R_{ef} = R_3 \left(R_1 + R_2 \right)/R_2 \tag{6.50}$$

It is possible to reduce the output drift substantially by selecting the resistors suitably. The accuracy of this circuit can be improved further by cascading two or more amplifiers. In this way, a substantial improvement in the signal-to-noise ratio can also be achieved. In the inverting input, the use of resistor R_4 is necessary because of bias currents.

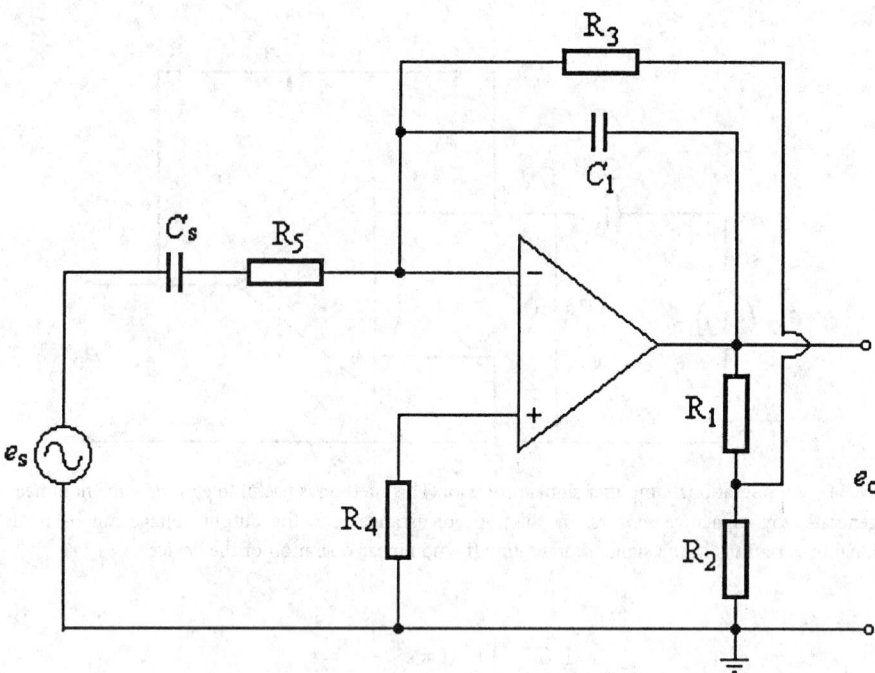

FIGURE 6.35   A practical charge amplifier. The effective feedback resistance is a function of other resistances. It is possible to reduce the output drift substantially by selecting the resistors suitably. The accuracy of this circuit can be improved further by cascading two or more amplifiers, thereby substantially improving the signal-to-noise ratio.

Pulse Width Modulation

As in the case of some capacitive vibrational displacement sensors, the output of the sensor may be an amplitude-modulated wave as shown in Figure 6.36. When rectified, the average value of this wave gives the mean separation of the plates. The vibration amplitude around this mean position may be extracted by a demodulator and a low-pass filter circuit. The output of the low-pass filter is a direct indication of vibrations, and the waveform can be viewed on an oscilloscope.

Square Wave Linearization

Another linearization technique applied in capacitive pressure transducers and accelerometers is pulse width modulation. The transducer consists of two differential capacitors as shown in Figure 6.37. The voltages of these capacitors, e_1 and e_2, switch back and forth with a high excitation frequency (e.g., 400 kHz) between excitation voltage and ground. The system is arranged in such a way that the output voltage is the average voltage difference between e_1 and e_2. At null position, $e_1 = e_2$, the output is a symmetrical square wave with zero average value. As the relative positions of the plates change, due to vibration, the average value of the output voltage shifts from the zero average value and becomes positive or negative depending on the direction of the displacement. Hence, the output voltage can be expressed by:

$$e_o = e_{ex} \left(C_1 - C_2 \right) / \left(C_1 + C_2 \right) \tag{6.51}$$

Substituting:

$$C_1 = C_0 \, x_0 / \left(x_0 - x_i \right) \quad \text{and} \quad C_2 = C_0 \, x_0 / \left(x_0 + x_i \right)$$

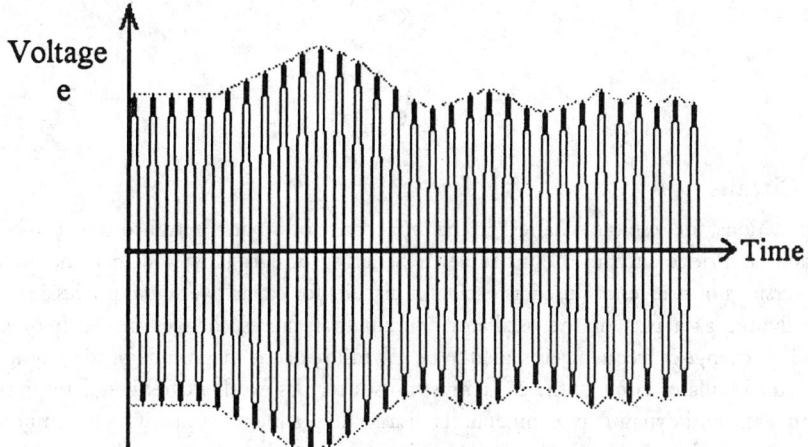

FIGURE 6.36 A amplitude modulated signal. It is possible to configure some sensors to give a amplitude-modulated signals, as in the case of capacitive vibrational displacement sensors. When rectified, the average value of this wave gives the mean separation of the plates. The vibration amplitude around this mean position can be extracted by a demodulator and low-pass filter circuit. The output of the low-pass filter is a direct indication of vibrations.

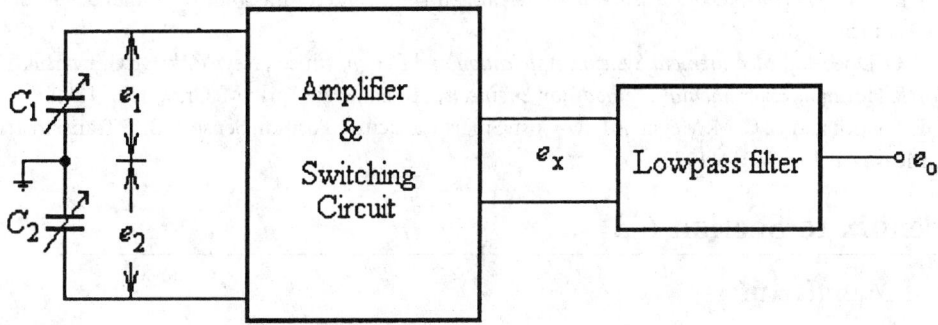

FIGURE 6.37 Block diagram of a square-wave linearization circuit. This is particularly useful for differential capacitance type sensors. The voltages of these two capacitors are made to switch back and forth with a high excitation frequency between excitation voltage and ground. As the relative positions of the plates change due to vibration, the average value of the output voltage becomes positive or negative, depending on the direction of the displacement.

yields

$$e_o = e_{ex} \, x_i / x_0 \tag{6.52}$$

Thus, the output is directly proportional to the variable x_i.

Feedback Linearization

Linearization of a capacitance transducer can also be obtained using a feedback system that adjusts capacitor current amplitude so that it stays constant at a reference value for all displacements. This is accomplished by obtaining a dc signal proportional to capacitor current from a demodulator, comparing this current with the reference current, and adjusting the voltage amplitude of the system excitation oscillator until the two currents agree. If the capacitor current is kept constant irrespective of capacitor motion, then the voltage amplitude is linearly related to x as:

$$e = K\, x_i \tag{6.53}$$

where

$$K = \left| i_c \right| \big/ \omega\, C_0\, x_0 \tag{6.54}$$

Oscillator Circuits

In many applications, the resultant changes in the capacitance of capacitive transducers can be measured with a suitable ac bridge such as Wein bridge or Schering bridge. However, in a majority of cases, improvised versions of bridges are used as oscillator circuits for capacitive signal processing. The transducer is configured as a part of the oscillatory circuit that causes changes in the frequency of the oscillations. This change in frequency is scaled to be a measure of the magnitude of the physical variable.

As part of the oscillator circuits, the capacitive transducer has excellent frequency response and can measure both static and dynamic phenomena. Its disadvantages include sensitivity to temperature variations and the possibility of erratic or distorted signals due to long lead lengths. Also, the receiving instrumentation may be large and complex, and it often includes a second fixed-frequency oscillator for heterodyning purposes. The difference frequency thus produced can be read by an appropriate output device such as an electronic counter.

References

1. J. P. Bentley, *Principles of Measurement Systems,* 2nd ed., United Kingdom: Longman Scientific and Technical, 1988.
2. E. O. Doebelin, *Measurement Systems: Application and Design,* 4th ed., New York: McGraw-Hill, 1990.
3. J. P. Holman, *Experimental Methods for Engineers,* 5th ed., New York: McGraw-Hill, 1989.
4. F. T. Noth and G. C. M. Meijer, A Low-Cost, Smart Capacitive Position Sensor, IEEE Trans. Instrum. Meas., 41, 1041-1044, 1992.

Appendix to Section 6.3

List of Manufacturers

ANALITE Inc.
  24-T Newtown Plaza
  Plainview, NY 11803
  Tel: (800) 229-3357

FSI/FORK Standards Inc.
  668 Western Avenue
  Lombard, IL 60148-2097
  Tel: (708) 932-9380

Gordon Engineering Corp.
  67 Del Mar Drive
  Brookfiled, CT 06804
  Tel: (203) 775-4501

Hecon Corp.
  15-T Meridian Rd.
  Eatontown, NJ 07724
  Tel: (800) 524-1669

Kistler Instrumentation Corp.
  Amherst, NY 14228-2171
  Tel: (716) 691-5100
  Fax: (716) 691-5226

Locon Sensor Systems, Inc.
  1750 S. Eber Road
  P.O. Box 789
  Holland, OH 43526
  Tel: (419) 865-7651
  Fax: (419) 865-7756

Rechner Electronic Industries Inc.
  8651 Buffalo Avenue, Box 7
  Niagara Falls, NY 14304
  Tel: (800) 544-4106

RDP Electrosense, Inc.
  2216-Dept. B
  Pottstown, PA
  Tel: (800) 334-5838

6.4   Piezoelectric Transducers and Sensors

Ahmad Safari, Victor F. Janas, Amit Bandyopadhyay,
and Andrei Kholkine

Piezoelectricity, discovered in Rochelle salt in 1880 by Jacques and Pierre Curie, is the term used to describe the ability of certain materials to develop an electric charge that is proportional to a direct applied mechanical stress. These materials also show the converse effect; that is, they will deform (strain) proportionally to an applied electric field. Some crystalline materials show piezoelectric behavior due to their unique crystal structure. The lattice structure of a crystal is described by the Bravais unit cell [1]. There are 230 microscopic symmetry types (space groups) in nature, based on the several symmetry elements such as translation, inversion center, mirror plane, or rotation axes. Combinations of these symmetry elements yield the macroscopic symmetry known as point groups. All natural crystals can be grouped into 32 different classes (point groups) based on their symmetry elements. The 32 point groups can be further classified into two subgroups: (1) crystals with a center of symmetry, and (2) crystals with no center of symmetry. The 11 centrosymmetric subgroups do not show piezoelectricity. Of the 21 non-centrosymmetric groups, 20 show the piezoelectric effect along unique directional axes. An important class of piezoelectric materials includes ferroelectrics, in which the piezoelectric effect is closely related to the ferroelectric polarization that can be reversed by the application of sufficiently high electric field [2, 3]. To induce piezoelectric properties in ferroelectric materials, a poling procedure is often required, which consists of the temporary application of a strong electric field. Poling is analogous to the magnetizing of a permanent magnet.

Governing Equations and Coefficients

The phenomenological master equation describing the deformations of an insulating crystal subject to both elastic and electric stress is given by:

$$x_{ij} = s_{ijkl} X_{kl} + d_{mij} E_m + M_{mnij} E_m E_n \,, \tag{6.55}$$

where x_{ij} are components of elastic strain, s_{ijkl} is the elastic compliance tensor, X_{kl} are the stress components, d_{mij} are the piezoelectric tensor components, M_{mnij} is the electrostrictive tensor, and E_m and E_n are components of the electric field.

Neglecting the second-order effects (electrostriction) and assuming that the material is under no stress ($X_{kl} = 0$), the elastic strain is given by:

$$x_{ij} = d_{mij} E_m \tag{6.56}$$

Equation 6.56 is the mathematical definition of the converse piezoelectric effect, where induced strain is directly proportional to the first power of the field. The thermodynamically equivalent direct piezoelectric effect is given by:

$$P_m = d_{mij} X_{ij} \,, \tag{6.57}$$

where P_m is the component of electrical polarization. The difference between the direct and converse piezoelectric effect is shown schematically in Figure 6.38. The converse effect describes the actuating function of a piezoelectric, where a controlled electric field accurately changes the shape of a piezoelectric material. The sensing function of a piezoelectric is described by the direct effect, where a controlled stress on a piezoelectric material yields a charge proportional to the stress.

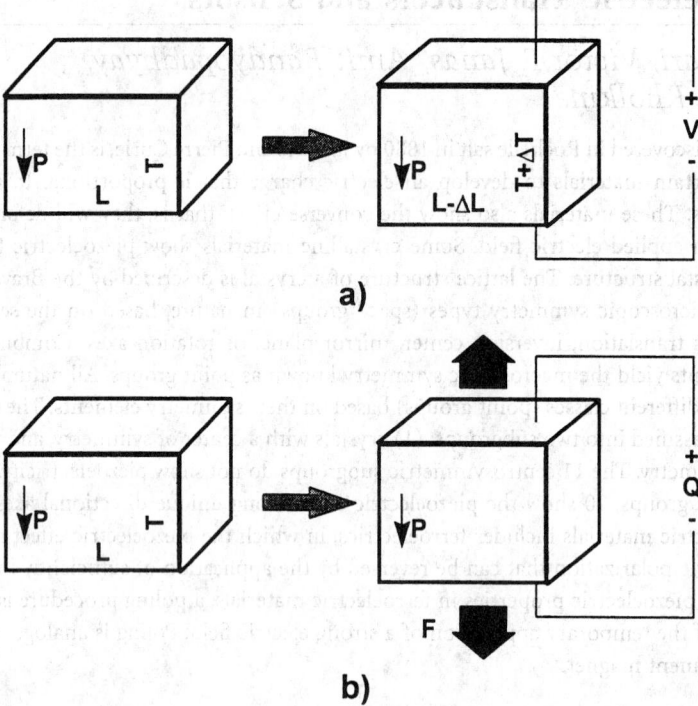

FIGURE 6.38 Schematic representations of the direct and converse piezoelectric effect: (*a*) an electric field applied to the material changes its shape; (*b*) a stress on the material yields a surface charge.

The third rank tensor d_{mij} may be simplified using matrix notation [1] to d_{ij}, a second rank tensor. In this form, d_{ij} is simply known as the piezoelectric charge coefficient, with units of coulombs per newton (CN^{-1}) or meters per volt ($m\,V^{-1}$). Another set of moduli that may be used to characterize piezoelectric materials are the piezoelectric voltage coefficients, g_{ij}, defined in matrix notation as:

$$E_i = g_{ij} X_j, \tag{6.58}$$

where E_i is the component of electric field arising due to the stress X_j. The d_{ij} and g_{ij} coefficients are related by:

$$g_{ij} = d_{ij} / (\varepsilon_0 \varepsilon_{ii}), \tag{6.59}$$

where ε_{ii} is the dielectric constant (relative permittivity), and ε_0 is the permittivity of free space ($8.854 \times 10^{-12}\ F\,m^{-1}$). Another key property of piezoelectric materials is their electromechanical coupling coefficient k, defined as:

$$k^2 = \text{resulting mechanical energy/input electrical energy} \tag{6.60}$$

or

$$k^2 = \text{resulting electrical energy/input mechanical energy} \tag{6.61}$$

The value of k represents the efficiency of the piezoelectric in converting one form of energy to another. Since energy conversion is never complete, the value of k^2 is always less than unity, so k is always less than 1.

Two final important parameters for piezoelectric (ferroelectric) materials are their Curie point (T_0) and Curie temperature (T_c). The Curie point is the temperature above which the material loses its ferroelectric and piezoelectric behavior. The Curie temperature is defined by the Curie-Weiss law:

$$\varepsilon = \varepsilon_0 + \frac{C}{T - T_c} \quad \text{for } T > T_c, \tag{6.62}$$

where ε is the dielectric constant of the material, C is the Curie–Weiss constant, and T is the temperature. It represents the temperature where the material tends to have its highest dielectric constant. The Curie temperature is always lower (often within 10°C) than the Curie point.

Piezoelectric Materials

Single Crystals

A number of single-crystal materials have demonstrated piezoelectricity. These materials dominate certain applications, such as frequency-stabilized oscillators in watches and radars, and surface acoustic wave devices in television filters and analog signal correlators. A list of single-crystal piezoelectric materials includes quartz, lithium niobate and lithium tantalate, ammonium dihydrogen sulfate, lithium sulfate monohydrate, and Rochelle salt. Recently, it was discovered that some relaxor-based ferroelectric single crystals of lead zinc niobate and lead magnesium niobate, and their solid solutions with lead titanate possess superior piezoelectric properties when compared to other piezoelectric materials.

Quartz, which is crystalline SiO_2, has a low value of d_{11} ($|2.3 \times 10^{-12}|$ C N^{-1}) [1]. Right-handed quartz develops a positive charge when put under compression, and a negative charge when put under tension. The coupling coefficient k for quartz is also very low, typically around 0.1 [4]. In addition, the dielectric constant ε for quartz is small (~4) [5]. The Curie point, however, is relatively high ($T_0 \sim 573$°C) [5], so quartz is stable for high-temperature applications. Despite the low piezoelectric properties, quartz is very abundant and inexpensive, and has found a strong position in low-cost or high-temperature applications.

Piezoelectric behavior in lithium niobate ($LiNbO_3$) and lithium tantalate ($LiTaO_3$) was first studied in the mid-1960s [2]. Under shear, the d_{15} of $LiNbO_3$ and $LiTaO_3$ are 73 and 26 ($\times 10^{-12}$ C N^{-1}), respectively. Both have ε values of approximately 40. If cut correctly, they have coupling coefficient (k) values of 0.65 and 0.4, respectively. In addition, the Curie points for both are extremely high ($T_0 \sim 1210$°C for $LiNbO_3$, and 620°C for $LiTaO_3$). Both $LiNbO_3$ and $LiTaO_3$ are commonly used in infrared detectors.

Rochelle salt ($KNaC_4H_4O_6 \cdot H_2O$) was first found to be piezoelectric in 1880 [2]. The d_{31} and k_{31} are 275×10^{-12} C N^{-1} and 0.65, respectively. The relative dielectric constant is approximately 350. Rochelle salt has two Curie points (lower $T_0 \sim 18$°C, and upper $T_0 \sim 24$°C). It is highly soluble in water, and is still extensively used in electroacoustic transducers.

Lead zinc niobate, $Pb(Zn_{1/3}Nb_{2/3})O_3$, and lead magnesium niobate, $Pb(Mg_{1/3}Nb_{2/3})O_3$, are typical relaxor materials characterized by the broad frequency-dependent maximum of dielectric constant vs. temperature. The solid solutions of these materials with lead titanate, $PbTiO_3$, were shown to posses excellent piezoelectric properties when oriented along the [001] direction [6]. The piezoelectric charge coefficient d_{33} of 25×10^{-10} C N^{-1}, coupling coefficient k of more than 0.9, and ultrahigh strain of 1.7% were achieved in $Pb(Zn_{1/3}Nb_{2/3})O_3$-$PbTiO_3$ solid solution. These single-crystal relaxor materials are now being intensively investigated and show great promise for future generations of piezoelectric transducers and sensors.

Other types of piezoelectric materials dominate the market for transducers. These materials include piezoelectric ceramics, piezoelectric polymers, and composites of piezoelectric ceramic with inactive polymers. The focus of the remainder of this chapter will be on non-single-crystal piezoelectric materials.

Piezoelectric Ceramics

In polycrystalline ceramics with polar grains, the randomness of the grains, as shown schematically in Figure 6.39(a), yields a nonpiezoelectric material. Piezoelectric behavior is induced by "poling" the

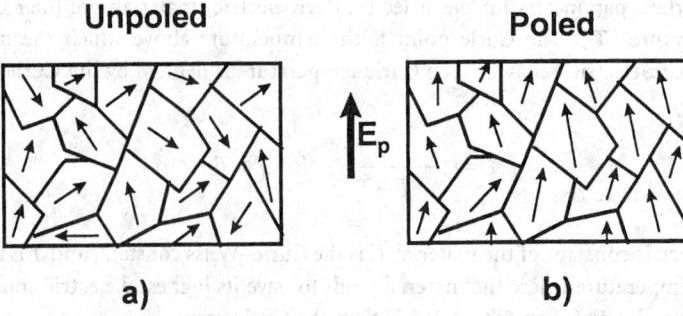

FIGURE 6.39 Schematic of the poling process in piezoelectric ceramics: (*a*) in the absence of an electric field, the domains have random orientation of polarization; (*b*) the polarization within the domains are aligned in the direction of the electric field.

ceramic. By applying a strong dc electric field at a temperature just below the Curie temperature, the spontaneous polarization in each grain gets oriented toward the direction of the applied field. This is schematically shown in Figure 6.39(*b*). Although all of the domains in a ceramic can never be fully aligned along the poling axis due to symmetry limitations, the ceramic ends up with a net polarization along the poling axis.

The largest class of piezoelectric ceramics is made up of mixed oxides containing corner-sharing octahedra of O^{2-} ions. The largest structure type, built with corner-shared oxygen octahedra, is the perovskite family, which is discussed in the following section.

Perovskites

Perovskite is the name given to a group of materials with general formula ABO_3 having the same structure as the mineral calcium titanate ($CaTiO_3$). Piezoelectric ceramics having this structure include barium titanate ($BaTiO_3$), lead titanate ($PbTiO_3$), lead zirconate titanate ($PbZr_xTi_{1-x}O_3$, or PZT), lead lanthanum zirconate titanate [$Pb_{1-x}La_x(Zr_yT_{1-y})_{1-x/4}O_3$, or PLZT], and lead magnesium niobate [$PbMg_{1/3}Nb_{2/3}O_3$, or PMN]. Several of these ceramics are discussed below.

The piezoelectric effect in $BaTiO_3$ was discovered in the 1940s [4], and it became the first piezoelectric ceramic developed. It replaced Rochelle salt because it is more stable, has a wider temperature range of operation, and is easily manufacturable. The Curie point, T_0, is about 130°C. Above 130°C, a nonpiezo-electric cubic phase is stable, where the center of positive charge (Ba^{2+} and Ti^{4+}) coincides with the center of the negative charge (O^{2-}) (Figure 6.40(*a*)). When cooled below the Curie point, a tetragonal structure (shown in Figure 6.40(*b*)) develops where the center of positive charge is displaced relative to the O^{2-} ions, leading to the formation of electric dipoles. Barium titanate has a relative dielectric constant ε_{33} of 1400 when unpoled, and 1900 when poled [2, 4]. The d_{15} and d_{33} coefficients of $BaTiO_3$ are 270 and 191×10^{-12} C N^{-1}, respectively. The k for $BaTiO_3$ is approximately 0.5. The large room temperature dielectric constant in barium titanate has led to its wide use in multilayer capacitor applications.

Lead titanate, $PbTiO_3$, first reported to be ferroelectric in 1950 [4], has a similar structure to $BaTiO_3$, but with a significantly higher Curie point ($T_0 = 490°C$). Pure lead titanate is difficult to fabricate in bulk form. When cooled through the Curie point, the grains go through a cubic to tetragonal phase change, leading to large strain and ceramic fracturing. This spontaneous strain can be decreased by the addition of dopants such as Ca, Sr, Ba, Sn, and W. Calcium-doped $PbTiO_3$ [7] has a relative dielectric constant ε_{33} of 200, a d_{33} of 65×10^{-12} C/N, and a k of approximately 0.5. The addition of calcium results in a lowering of the Curie point to 225°C. The main applications of lead titanate are hydrophones and sonobuoys.

Lead zirconate titanate (PZT) is a binary solid solution of $PbZrO_3$ (an antiferroelectric orthorhombic structure) and $PbTiO_3$ (a ferroelectric tetragonal perovskite structure) [2–4]. It has a perovskite structure, with the Zr^{4+} and Ti^{4+} ions occupying the B site of the ABO_3 structure at random. At the morphotropic

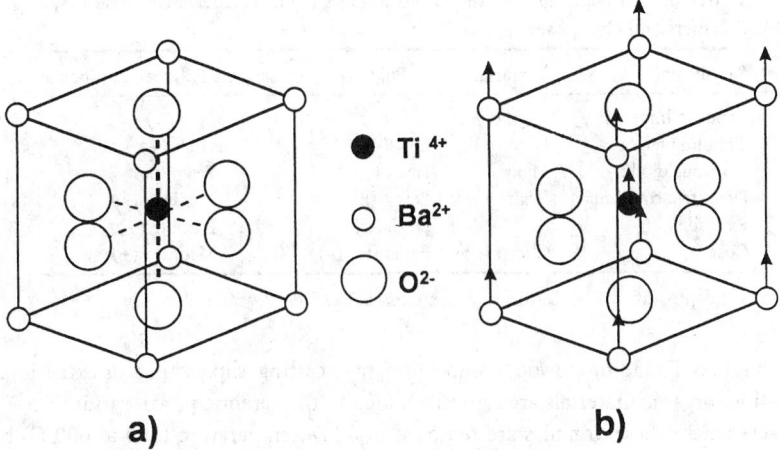

FIGURE 6.40 The crystal structure of $BaTiO_3$: (*a*) above the Curie point, the cell is cubic; (*b*) below the Curie point, the cell is tetragonal with Ba^{2+} and Ti^{4+} ions displaced relative to O^{2-} ions.

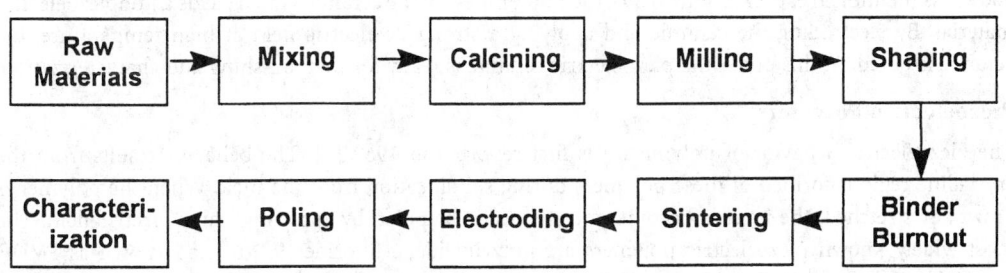

FIGURE 6.41 Flow chart for the processing of piezoelectric ceramics.

phase boundary (MPB) separating the tetragonal and orthorhombic phases, PZT shows excellent piezoelectric properties. At room temperature, the MPB is at a Zr/Ti ratio of 52/48, resulting in a piezoelectric ceramic which is extremely easy to pole. Piezoelectric PZT at the MPB is usually doped by a variety of ions to form what are known as "hard" and "soft" PZTs. Hard PZT is doped with acceptor ions, such as K^+ or Na^+ at the A site, or Fe^{3+}, Al^{3+}, or Mn^{3+} at the B site. This doping lowers the piezoelectric properties, and makes the PZT more difficult to pole or depole. Typical piezoelectric properties of hard PZT include [5, 7]: Curie point, T_0, of 365°C, ε_{33} of 1700–1750 (poled), a piezoelectric charge coefficient d_{33} of 360 to 370×10^{-12} C N^{-1}, and a coupling coefficient of about 0.7. Soft PZT is doped with donor ions such as La^{3+} at the A site, or Nb^{5+} or Sb^{5+} at the B site. It has very high piezoelectric properties, and is easy to pole or depole. Typical piezoelectric properties of soft PZT include [5, 7]: Curie point, T_0, of 210°C, relative dielectric constant ε_{33} of 3200–3400 (poled), a d_{33} of 580 to 600×10^{-12} C N^{-1}, and a coupling coefficient k_{33} of 0.7.

Processing of Piezoelectric Ceramics

The electromechanical properties of piezoelectric ceramics are largely influenced by their processing conditions. Each step of the process must be carefully controlled to yield the best product. Figure 6.41 is a flow chart of a typical oxide manufacturing process for piezoelectric ceramics. The high-purity raw materials are accurately weighed according to their desired ratio, and mechanically or chemically mixed. During the calcination step, the solid phases react to yield the piezoelectric phase. After calcining, the solid mixture is ground into fine particles by milling. Shaping is accomplished by a variety of ceramic

TABLE 6.7 Advantages (+) and Disadvantages (−) of Piezoelectric Ceramics, Polymers and Composites

Parameter	Ceramic	Polymer	Ceramic/Polymer Composite
Acoustic impedance	High (−)	Low (+)	Low (+)
Coupling factor	High (+)	Low (−)	High (+)
Spurious modes	Many (−)	Few (+)	Few (+)
Dielectric constant	High (+)	Low (−)	Medium (+)
Flexibility	Stiff (−)	Flexible (+)	Flexible (+)
Cost	Cheap (+)	Expensive (−)	Medium (+)

Adapted from T. R. Gururaja, *Amer. Ceram. Soc. Bull.*, 73, 50, 1994.

processing techniques, including powder compaction, tape casting, slip casting, or extrusion. During the shaping operation, organic materials are typically added to the ceramic powder to improve its flow and binding characteristics. These organics are removed in a low temperature (500 to 600°C) binder burn-off step.

After burnout, the ceramic structure is sintered to an optimum density at an elevated temperature. For the lead-containing piezoelectric ceramics ($PbTiO_3$, PZT, PLZT), sintering is performed in sealed crucibles with an optimized PbO atmosphere. This is because lead loss occurs in these ceramics above 800°C. As mentioned earlier (Figure 6.39), the randomness of the ceramic grains yields a nonpiezoelectric material. By electroding the ceramic and applying a strong dc electric field at high temperature, the ceramic is poled. At this point, the piezoelectric ceramic is ready for final finishing and characterization.

Piezoelectric Polymers

The piezoelectric behavior of polymers was first reported in 1969 [8]. The behavior results from the crystalline regions formed in these polymers during solidification from the melt. When the polymer is drawn, or stretched, the regions become polar, and can be poled by applying a high electric field. The most widely known piezoelectric polymers are polyvinylidene fluoride [9, 10], also known as PVDF, polyvinylidene fluoride — trifluoroethylene copolymer, or P(VDF-TrFE) [9, 10], and odd-number nylons, such as Nylon-11 [11].

The electromechanical properties of piezoelectric polymers are significantly lower than those of piezoelectric ceramics. The d_{33} values for PVDF and P(VDF-TrFE) are approximately 33 ($\times 10^{-12}$ C N^{-1}), and the dielectric constant ε is in the range 6 to 12 [12, 13]. They both have a coupling coefficient (k) of 0.20, and a Curie point (T_0) of approximately 100°C. For Nylon-11, ε is around 2 [11], while k is approximately 0.11.

Piezoelectric Ceramic/Polymer Composites

As mentioned above, a number of single-crystal, ceramic, and polymer materials exhibit piezoelectric behavior. In addition to the monolithic materials, composites of piezoelectric ceramics with polymers have also been formed. Table 6.7 [14] summarizes the advantages and disadvantages of each type of material. Ceramics are less expensive and easier to fabricate than polymers or composites. They also have relatively high dielectric constants and good electromechanical coupling. However, they have high acoustic impedance, and are therefore a poor acoustic match to water, the media through which it is typically transmitting or receiving a signal. Also, since they are stiff and brittle, monolithic ceramics cannot be formed onto curved surfaces, limiting design flexibility in the transducer. Finally, they have a high degree of noise associated with their resonant modes. Piezoelectric polymers are acoustically well matched to water, are very flexible, and have few spurious modes. However, applications for these polymers are limited by their low electromechanical coupling, low dielectric constant, and high cost of fabrication. Piezoelectric ceramic/polymer composites have shown superior properties when compared to single-phase materials. As shown in Table 6.7, they combine high coupling, low impedance, few spurious modes, and an intermediate dielectric constant. In addition, they are flexible and moderately priced.

TABLE 6.8 Suppliers of Piezoelectric Materials and Sensors

Name	Address	Ceramic	Polymer	Composite
AMP Sensors	950 Forge Ave. Morristown, PA 19403 Phone: (610) 650-1500 Fax: (610) 650-1509		X	
Krautkramer Branson	50 Industrial Park Rd. Lewistown, PA 17044 Phone: (717) 242-0327 Fax: (717) 242-2606			X
Materials Systems, Inc.	531 Great Road Littleton, MA 01460 Phone: (508) 486-0404 Fax: (508) 486-0706			X
Morgan Matroc, Inc.	232 Forbes Rd. Bedford, OH 44146 Phone: (216) 232-8600 Fax: (216) 232-8731	X		
Sensor Technology Ltd.	20 Stewart Rd. P.O. Box 97 Collingwood, Ontario, Canada Phone: +1 (705) 444-1440 Fax: +1 (705) 444-6787	X		
Staveley Sensors, Inc.	91 Prestige Park Circle East Hartford, CT 06108 Phone: (860) 289-5428 Fax: (860) 289-3189			X
Valpey-Fisher Corporation	75 South Street Hopkinton, MA 01748 Phone: (508) 435-6831 Fax: (508) 435-5289		X	
Vermon U.S.A.	6288 SR 103 North Bldg. 37 Lewistown, PA 17044 Phone: (717) 248-6838 Fax: (717) 248-7066	X		
TRS Ceramics, Inc.	2820 E. College Ave. State College, PA 16801 Phone: (814) 238-7485 Fax: (814) 238-7539	X		

Suppliers of Piezoelectric Materials

Table 6.8 lists a number of the suppliers of piezoelectric materials, their addresses, and whether they supply piezoelectric ceramic, polymers, or composites. Most of them tailor the material to specific applications.

Measurements of Piezoelectric Effect

Different means have been proposed to characterize the piezoelectric properties of materials. The resonance technique involves the measurement of the characteristic frequencies when the suitably shaped specimen (usually ceramic) is driven by a sinusoidally varying electric field. To a first approximation, the behavior of a piezoelectric sample close to its fundamental resonance frequency can be represented by an equivalent circuit as shown in Figure 6.42(a). The schematic behavior of the reactance of the sample as a function of frequency is represented in Figure 6.42(b). By measuring the characteristic frequencies

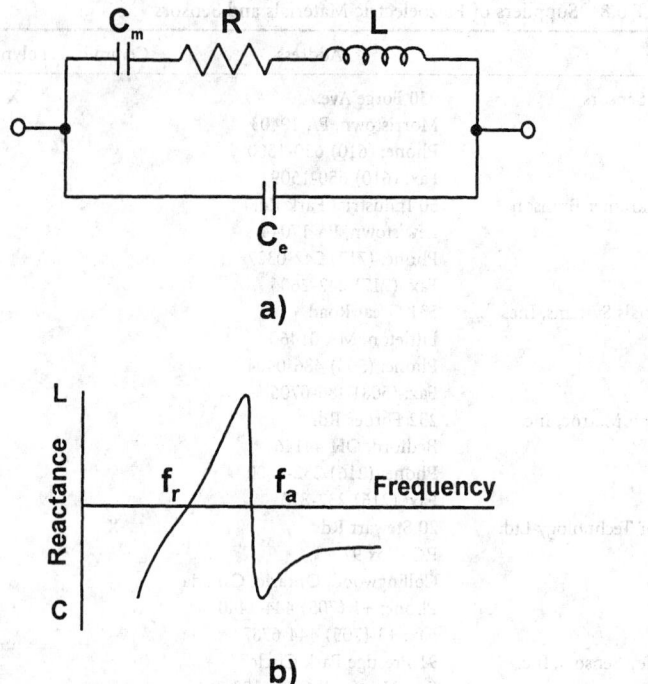

FIGURE 6.42 (*a*) Equivalent circuit of the piezoelectric sample near its fundamental electromechanical resonance (top branch represents the mechanical part and bottom branch represents the electrical part of the circuit); (*b*) electrical reactance of the sample as a function of frequency.

of the sample, the material constants including piezoelectric coefficients can be calculated. The equations used for the calculations of the electromechanical properties are described in the IEEE Standard on piezoelectricity [15]. The simplest example of piezoelectric measurements by resonance technique relates to a piezoelectric ceramic rod (typically 6 mm in diameter and 15 mm long) poled along its length. It can be shown that the coupling coefficient k_{33} is expressed as a function of the series and parallel resonance frequencies, f_s and f_p, respectively:

$$k_{33}^2 = \frac{\pi}{2}\frac{f_s}{f_p}\tan\left(\frac{\pi}{2}\frac{f_p - f_s}{f_p}\right) \tag{6.63}$$

The longitudinal piezoelectric coefficient d_{33} is calculated using k_{33}, elastic compliance s_{33}^E and low-frequency dielectric constant ε_{33}^X:

$$d_{33} = k_{33}\left(\varepsilon_{33}^X s_{33}^E\right)^{1/2} \tag{6.64}$$

Similarly, other electromechanical coupling coefficients and piezoelectric moduli can be derived using different vibration modes of the sample. The disadvantage of the resonance technique is that measurements are limited to the specific frequencies determined by the electromechanical resonance. It is used mostly for the rapid evaluation of the piezoelectric properties of ceramic samples whose dimensions can be easily adjusted for specific resonance conditions.

Subresonance techniques are frequently used to evaluate piezoelectric properties of materials at frequencies much lower than the fundamental resonance frequency of the sample. They include both the measurement of piezoelectric charge under the action of external mechanical force (direct effect) and the

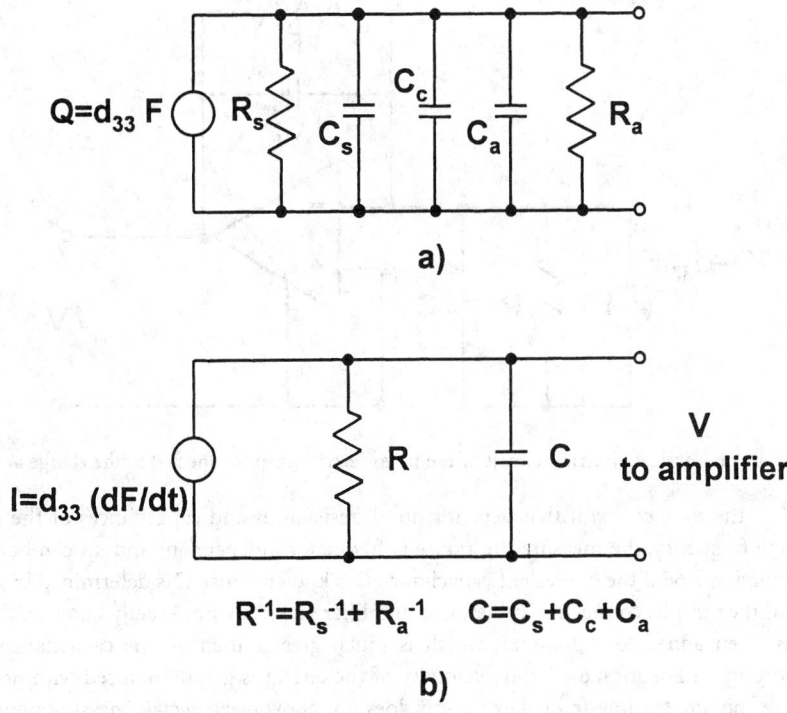

FIGURE 6.43 Full (*a*) and simplified (*b*) equivalent electrical circuits of the piezoelectric sensor connected to the voltage amplifier.

measurement of electric field-induced displacements (converse effect). In the latter case, the displacements are much smaller than in resonance; however, they still can be measured by using strain gages, capacitive sensors, LVDT (linear variable differential transformer) sensors or by optical interferometry [16, 17].

A direct method is widely used to evaluate the sensor capabilities of piezoelectric materials at sufficiently low frequency. The mechanical deformations can be applied in different modes such as thickness expansion, transverse expansion, thickness shear, and face shear to obtain different components of the piezoelectric tensor. In the simplest case, the metal electrodes are placed onto the major surfaces of the piezoelectric transducer normal to its poling direction (direction of ferroelectric polarization) and the mechanical force is applied along this direction (Figure 6.38(*b*)). Thus, the charge is produced on the electrode plates under mechanical loading, which is proportional to the longitudinal piezoelectric coefficient d_{33} of the material. To relate the output voltage of the transducer to the piezoelectric charge, it is necessary to consider the equivalent circuit (Figure 6.43(*a*)). A circuit includes the charge generator, $Q = d_{33}F$, leakage resistor of the transducer, R_s, transducer capacitance, C_s, capacitance of the connecting cables, C_c, and input resistance and capacitance of the amplifier, R_a and C_a, respectively. Here, F denotes the force applied to the transducer (tensile or compressive). All the resistances and capacitances shown in Figure 6.43(*a*) can be combined, as shown in Figure 6.43(*b*). A charge generator can be converted to a current generator, I, according to:

$$I = \frac{dQ}{dt} = d_{33}\frac{dF}{dt} \tag{6.65}$$

Assuming that the amplifier does not draw any current, the output voltage V at a given frequency ω can be calculated:

$$V = \frac{d_{33}F}{C}\frac{j\omega\tau}{1+j\omega\tau}, \tag{6.66}$$

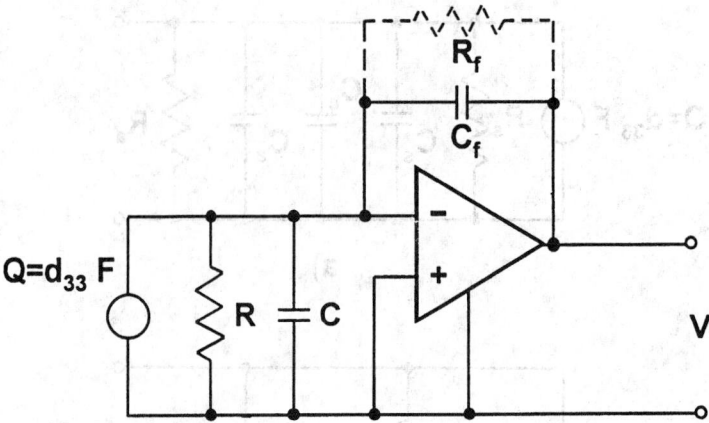

FIGURE 6.44   Equivalent electrical circuit of the piezoelectric sensor connected to the charge amplifier.

where $\tau = RC$ is the time constant that depends on all resistances and capacitances of the circuit. For sufficiently high frequency, the measured response is frequency independent and d_{33} can be easily evaluated from Equation 6.66 if the equivalent capacitance C is known. Since C is determined by the parallel capacitances of the sample, connecting cables, and amplifier (typically not exactly known), the standard capacitance is often added to the circuit, which is much greater than all the capacitances involved. However, according to Equation 6.66, the sensitivity of the circuit is greatly reduced with decreasing C. If τ is not large enough, the low-frequency cut-off does not allow piezoelectric measurements in quasi-static or low-frequency conditions.

To overcome the difficulties of using voltage amplifiers for piezoelectric measurements, a so-called charge amplifier was proposed. The idealized circuit of a charge amplifier connected with the piezoelectric transducer is shown in Figure 6.44. Note that a FET-input operational amplifier is used with a capacitor C_f in the feedback loop. Assuming that the input current and voltage of the operational amplifier are negligible, one can relate the charge on the transducer with the output voltage:

$$V = -Q/C_f = -d_{33}F/C_f \tag{6.67}$$

Equation 6.67 gives frequency-independent response where the output voltage is determined only by the piezoelectric coefficient d_{33} and the known capacitance C_f. Unfortunately, this advantage is difficult to realize since even a small input current of the amplifier will charge the feedback capacitor, leading to saturation of the amplifier. Therefore, a shunt resistor R_f is added to the circuit (dotted line in Figure 6.44), which prevents such charging. If one takes into account the RC circuit of the feedback loop, the output voltage will have the form of Equation 6.66, i.e., becomes frequency dependent. In this case, the time constant τ is determined by the parameters of the feedback loop and does not depend on the capacitance of the transducer, connecting cables, or the input capacitance of the amplifier. This gives an important advantage to the charge amplifier when it is compared to the ordinary voltage amplifier.

Applications

The direct and converse piezoelectric effects in a number of materials have led to their use in electromechanical transducers. Electromechanical transducers convert electrical energy to mechanical energy, and vice versa. These transducers have found applications where they are used in either passive or active modes. In the passive (sensor) mode, the transducer only receives signals. Here, the direct piezoelectric properties of the material are being exploited to obtain a voltage from an external stress. Applications in the passive mode include hydrophones, or underwater listening devices, microphones, phonograph pickups, gas igniters, dynamic strain gages, and vibrational sensors. In the active (actuator) mode, the

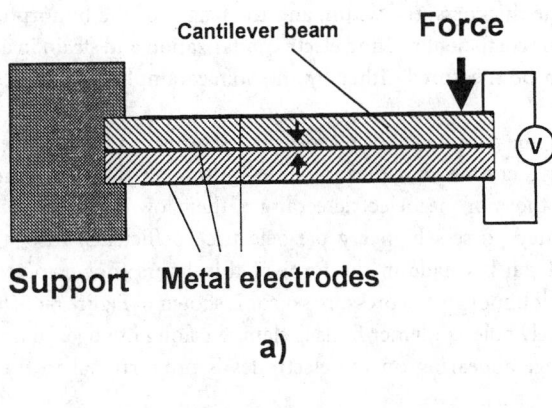

a)

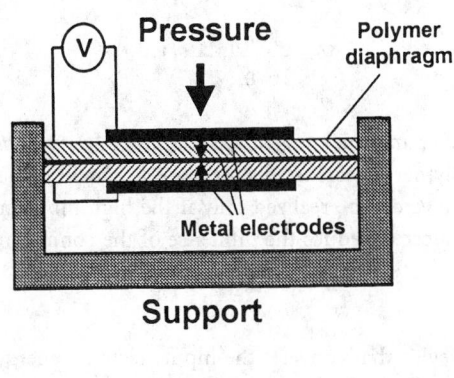

b)

FIGURE 6.45 Schematic designs of the displacement sensor based on piezoelectric ceramic (*a*) and of the pressure sensor based on piezoelectric polymer film (*b*). Arrows indicate the directions of ferroelectric polarization in the piezoelectric material.

transducer, using the converse piezoelectric properties of the material, changes its dimensions and sends an acoustic signal into a medium. Active mode applications include nondestructive evaluation, fish/depth finders, ink jet printers, micropositioners, micropumps, and medical ultrasonic imaging. Often, the same transducer is used for both sensor and actuator functions.

Two examples of piezoelectric sensors are given below. The first example is the ceramic transducer, which relates the deformation of the piezoelectric sensor to the output voltage via direct piezoelectric effect. Piezoceramics have high Young's moduli; therefore, large forces are required to generate strains in the transducer to produce measurable electric response. Compliance of the piezoelectric sensor can be greatly enhanced by making long strips or thin plates of the material and mounting them as cantilevers or diaphragms. Displacement of the cantilever end will result in a beam bending, leading to the mechanical stress in the piezoelectric material and the electric charge on the electrodes. A common configuration of the piezoelectric bender is shown in Figure 6.45(*a*). Two beams poled in opposite directions are cemented together with one common electrode in the middle and two electrodes on the outer surfaces. Bending of such a bimorph will cause the upper beam to stretch and the lower beam to compress, resulting in a piezoelectric charge of the same polarity for two beams connected in series. To the first approximation, the charge Q appearing on the electrodes is proportional to the displacement Δl of the end of the bimorph via Equation 6.68 [18]:

$$Q = \frac{3}{8}\frac{Hw}{L}e_{31}\Delta l, \tag{6.68}$$

where H, w, and L are the thickness, the width, and the length of the bimorph, respectively, and e_{31} is the transverse piezoelectric coefficient relating electric polarization and strain in a deformed piezoelectric material. The charge can be measured either by the voltage amplifier (Figure 6.43) or by the charge amplifier (Figure 6.44).

In certain applications, the parameters of piezoelectric sensors can be improved by using ferroelectric polymers instead of single crystals and piezoceramics. Although the electromechanical properties of polymers are inferior to those of piezoelectric ceramics, their low dielectric constant offers the higher voltage response since they possess higher g piezoelectric coefficients. Also, the polymers are more mechanically robust and can be made in the form of thin layers (down to several micrometers). An example using the polymer bimorph as a pressure sensor is shown in Figure 6.45(b). A circular diaphragm composed of two oppositely poled polymer films is clamped along its edges to a rigid surround, forming a microphone. The voltage appearing on the electrodes is proportional to the applied pressure p by Equation 6.69 [19]:

$$V = \frac{3}{16} \frac{d_{31}}{\varepsilon_{33}} \frac{D^2}{h} \left(1 - \nu\right) p \tag{6.69}$$

where D and h are the diameter and thickness of the diaphragm, respectively, and ν is the Poisson ratio. The high d_{31}/ε_{33} value for polymer sensors is advantageous to obtain higher voltage response. According to Equation 6.66, this advantage can be realized only if the high input impedance amplifier is used in close proximity to the transducer to reduce the influence of the connecting cables.

Defining Terms

Piezoelectric transducer: Device that converts the input electrical energy into mechanical energy and vice versa via piezoelectric effect.

Coupling coefficients: Materials constants that describe an ability of piezoelectric materials to convert electrical energy into mechanical energy and vice versa.

Piezoelectric coefficients: Materials constants that are used to describe the linear coupling between electrical and mechanical parameters of the piezoelectric.

Ferroelectrics: Subgroup of piezoelectric materials possessing a net dipole moment (ferroelectric polarization) that can be reversed by the application of sufficiently high electric field.

Poling: Process of aligning the ferroelectric polarization along a unique (poling) direction.

Piezoelectric composites: Materials containing two or more components with different piezoelectric properties.

Charge amplifier: An operational amplifier used to convert the input charge into output voltage by means of the capacitor in the feedback loop.

References

1. J. F. Nye, *Physical Properties of Crystals*, Oxford: Oxford University Press, 1985.
2. Y. Xu, *Ferroelectric Materials and Their Applications*, Amsterdam: North-Holland, 1991.
3. L. E. Cross, Ferroelectric ceramics: tailoring properties for specific applications, In N. Setter and E. L. Colla (ed.), *Ferroelectric Ceramics: Tutorial Reviews, Theory, Processing, and Applications*, Basel: Birkhauser, 1993.
4. B. Jaffe, W. R. Cook, Jr., and H. Jaffe, *Piezoelectric Ceramics*, Marietta, OH: R. A. N., 1971.
5. *The User's Guide to Ultrasound & Optical Products*, Hopkinton, MA: Valpey-Fisher Corporation, 1996.
6. S.-E. Park and T. R. Shrout, Relaxor based ferroelectric single crystals with high piezoelectric performance, *Proc. of the 8th US-Japan Seminar on Dielectric and Piezoelectric Ceramics*: October 15-18, Plymouth, MA, 1997, 235.

7. *Piezoelectric Products*, Sensor Technology Limited, Collingwood, Ontario, Canada, 1991.
8. H. Kawai, The piezoelectricity of poly(vinylidene fluoride), *Japan. J. Appl. Phys.*, 8, 975, 1969.
9. L. F. Brown, Ferroelectric polymers: Current and future ultrasonic applications, *Proc. 1992 IEEE Ultrasonics Symposium*: IEEE, New York, 1992, 539.
10. T. Furukawa, Recent advances in ferroelectric polymers, *Ferroelectrics*, 104, 229, 1990.
11. L. F. Brown, J. I. Scheinbeim, and B. A. Newman, High frequency dielectric and electromechanical properties of ferroelectric nylons, *Proc. 1994 IEEE Ultrasonics Symposium*: IEEE, New York, 1995, 337.
12. *Properties of Raytheon Polyvinylidene Fluoride (PVDF)*, Raytheon Research Division, Lexington, MA, 1990.
13. *Standard and Custom Piezo Film Components*, Atochem Sensors Inc., Valley Forge, PA, 1991.
14. T. R. Gururaja, Piezoelectric transducers for medical ultrasonic imaging, *Amer. Ceram. Soc. Bull.*, 73, 50, 1994.
15. IEEE Standards on Piezoelectricity, *IEEE Std.* 176, 1978.
16. W. Y. Pan and L. E. Cross, A sensitive double beam laser interferometer for studying high-frequency piezoelectric and electrostrictive strains, *Rev. Sci. Instrum.*, 60, 2701, 1989.
17. A. L. Kholkin, Ch. Wuethrich, D. V. Taylor, and N. Setter, Interferometric measurements of electric field-induced displacements in piezoelectric thin films, *Rev. Sci. Instrum.*, 67, 1935, 1996.
18. A. J. Moulson and J. M. Herbert, *Electroceramics: Materials, Properties, Applications*, London: Chapman and Hall, 1990.
19. J. M. Herbert, *Ferroelectric Transducers and Sensors*, New York: Gordon and Breach, 1982.

6.5 Laser Interferometer Displacement Sensors

Bernhard Günther Zagar

In the past few years, very high precision, numerically controlled machine tools have been developed. To achieve the potential precision of these tools, length and displacement measurements whose resolution exceeds the least significant digit of the tool must be made. The measurement equipment typically would not rely on mechanical scales.

Laser interferometers compare the changes in optical path length to the wavelength of light, which can be chosen from atomic constants that can be determined with very little uncertainty.

In 1983, there was a redefinition of the meter [1] that was previously defined in 1960. The old definition was based on the wavelength of a certain radiation (the krypton-86 standard) that could not be realized to better than 4 parts in 10^9. The new definition, being based on frequency but not related to any particular radiation, opened the way to significant improvements in the precision with which the meter can be realized. As recommended in resolution 2 for the practical realization of the meter, the wavelength in vacuum λ_v of a plane electromagnetic wave of frequency f is $\lambda_v = c/f$, where c is the speed of light in vacuum, $c = 299, 792, 458$ m s^{-1} exactly. This way, the wavelength is related to *frequency* and *time*, which can be measured with the highest precision of all units within the Système International (SI).

In order to be independent of any environmental parameters, the meter is defined using the speed of light in a vacuum. However, interferometers usually must operate in ambient air. Thus, environmental parameters that influence the speed of light in a particular medium (air) will affect and degrade the precision of the measurement.

Three major factors limit the absolute accuracy attainable with laser interferometers operating in ambient air: (1) the uncertainties of the vacuum wavelength, λ_v, of the laser source; (2) the uncertainty of the refractive index of the ambient air; and (3) the least count resolution of the interferometer.

This chapter section is organized as follows. First, some basic laser principles are detailed, including ways to stabilize the vacuum wavelength of the laser. The effect most often used to stabilize lasers in commercial interferometers is the Zeeman effect, which yields relative uncertainties of 10^{-8}.

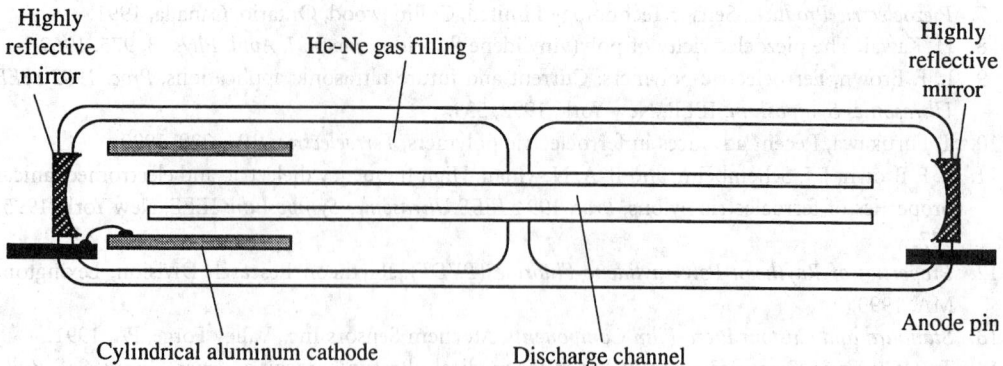

FIGURE 6.46   Schematics of the helium–neon laser (Reprinted with permission of University Science Books [3]).

Second, the refractive index of air as another major factor limiting the attainable accuracy of laser interferometers operated in air is addressed. It is shown that it cannot be determined currently with uncertainty better than 5×10^{-8}.

And finally, the chapter section describes the most widely used Michelson interferometer and two of its variants for long-travel length measurement and gives their resolution.

Helium–Neon Laser

In order to attain the best possible accuracy, great care must be taken to ensure the highest wavelength stability of the light source. Almost all interferometric dimensional gages utilize a helium–neon laser because it has proven reliable, its emitted wavelength is in the visible range at about 633 nm, and it can be stabilized sufficiently well utilizing the Zeeman effect and to an even higher degree with the use of a very well-defined iodine absorption line also at ≈633 nm [1, 2].

The helium–neon laser consists of a discharge tube as shown in Figure 6.46 [3] filled with the single-isotope gases helium (He^3) at a partial pressure of ≈105 Pa and neon (Ne^{20}) with a partial pressure of ≈13 Pa. It is pumped electrically using a voltage on the order of kilovolts with a current of a few milliamperes to excite both helium and neon atoms. Since the helium gas is the majority component, it dominates the discharge properties of the laser tube. Neutral helium atoms collide with free electrons that are accelerated by the axial voltage and become excited and remain in two rather long-lived metastable states. These are close enough to the energy levels of certain excited states of neon atoms so that collisional energy transfer can take place between these two groups of atoms. Excited helium atoms may drop down to the ground state, while simultaneously neon atoms take up almost exactly the same amount of energy. Therefore helium only serves to excite neon atoms, they do not contribute to the emission of light. The excited neon atoms remain in the excited state for a rather long period of time (on the order of 10^{-3} s). They return to lower energetic levels by stimulated emission of highly coherent light. This stimulated emission comes into effect when light emitted by some neon atoms also prompts other atoms to emit. The mirrors of the laser cavity, by reflecting most of the incident light cause the light to traverse multiple paths through the active laser volume, thereby greatly amplifying the light if the cavity length L is an integer multiple m of half the wavelength λ.

$$L = m\frac{\lambda}{2} \tag{6.70}$$

The emitted light is fairly monochromatic, but still has some finite spectral linewidth determined by the random emissions of Ne^{20}.

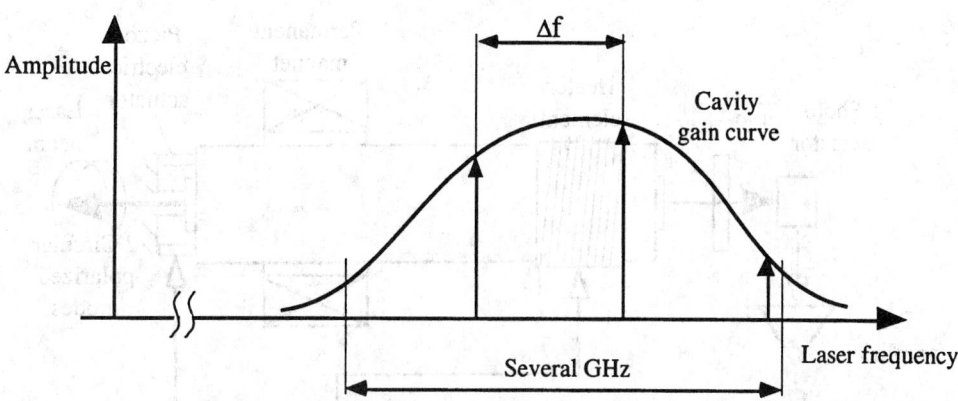

FIGURE 6.47 A He–Ne laser can have multiple resonating modes (shown for ≈20 cm cavity length) (Reprinted with permission of University Science Books [3]).

Brewster angle [4] end windows of the discharge tube transmit light of the proper linear polarization if desired. The end mirrors have to be carefully polished. Their curvature radii have to satisfy the condition for stability. They have wavelength–selective dielectric coatings of very high reflectivity sometimes exceeding 99%.

Unless special precautions are taken, a He–Ne laser will emit several axial modes as shown schematically in Figure 6.47, resulting in a beat frequency that limits the temporal coherence and renders the laser unsuitable for interferometric purposes. Also, due to thermal expansion of the laser tube, the end mirrors will change their relative distance, thereby effectively tuning the wavelength within the linewidth of the gain curve.

Another important property of a high-quality He–Ne laser is its Gaussian cross-sectional profile, which is maintained along a propagating wave, i.e., fundamental lateral mode. It is also a necessary condition for the wavefronts to remain quasiplanar.

Frequency Stabilization of He–Ne Lasers

The resonant frequency (and the wavelength λ) of the laser is determined in part by the distance between the two end mirrors and also by the refractive index n_M of the active medium (the He–Ne mixture). Since the linewidth of the gain profile of the active medium is usually in the gigahertz range, multiple axial modes can resonate in the cavity as is shown in Figure 6.47. The frequency difference Δf between two adjacent longitudinal modes is the *free spectral range* (FSR), which is given by Equation 6.71 and depends only on the cavity length, L, the refractive index of the active medium, n_M, and the speed of light in vacuum, c.

$$\Delta f = \frac{c}{2 n_M L} \tag{6.71}$$

Due to thermal expansion of the laser cavity and/or thermally induced change in refractive index of the active medium, all resonating laser modes will move within the envelope of the gain profile. The effort undertaken in stabilizing the laser wavelength or equivalently stabilizing its frequency is aimed at locking the modes with respect to the gain profile and reducing the number of modes resonating simultaneously.

Longitudinal Zeeman Effect

One of the most often used effects in stabilizing the frequency of a He–Ne laser for distance measurements is the Zeeman effect [3, 5].

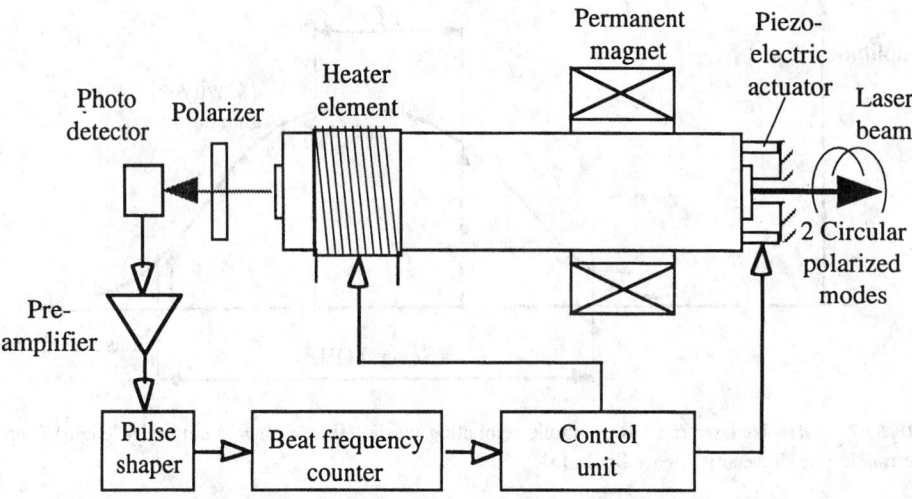

FIGURE 6.48 Laser wavelengths can be stabilized by utilizing the Zeeman-effect (Courtesy of Spindler & Hoyer Inc.).

The tube of the laser whose frequency is to be stabilized is mounted inside an axial permanent magnet as shown in Figure 6.48. Because of a short cavity length chosen, the free spectral range Δf of the laser is large. Therefore, the laser emission appears as a single longitudinal mode when the cavity length is properly controlled. An externally applied magnetic field in longitudinal direction causes the transition frequencies of the neon atoms to split symmetrically to the nominal frequency, the separation in frequency being proportional to the magnetic field strength. This phenomenon is called the *longitudinal Zeeman effect*.

With the applied field strength, the single longitudinal laser mode of the He–Ne laser is split into right and left circularly polarized modes with a frequency difference of typically 1.5 MHz. The frequency difference depends on the laser frequency and exhibits an extreme value when the laser frequency matches exactly the atomic transition frequency. Frequency stabilization is accomplished with control of the cavity length so that the frequency is locked to the extreme value corresponding to the nominal frequency of the laser. In order to determine the actual laser frequency within the gain profile, the cavity length is periodically modulated with a piezoelectric actuator shown in Figure 6.48. The frequency difference is constantly monitored at the minimum and maximum elongation of the modulation cycle. In the case when the extremal values are identical, the difference frequency will assume a maximum and thus the nominal value of the laser frequency is attained. This is achieved by controlling the length of the laser cavity. The cavity length is controlled both by using the thermal expansion of the laser tube, which is in close proximity to an electric heater whose operating current can be varied on a less dynamic scale, and on a very short time basis with a piezoelectric actuator attached to one of the end faces of the cavity [5].

This fast actuator allows for only a limited amplitude, typically on the order of less than a wavelength. Therefore, the thermal expansion must be used to attain greater modulation amplitude. To obtain the highest possible stability of the laser frequency, the frequency differences are measured within a very short time interval, and the determined deviations from the nominal value are fed back in a closed control loop to the electromechanical and thermal positioning devices.

The short- and long-term frequency stability $\Delta f/f$ of such a stabilized laser can be as high as 2×10^{-8}, depending also on the parameters of the feedback loop.

Another phenomenon that exceeds the stabilization attainable with the Zeeman effect by far uses a particular absorption line of an iodine isotope ($^{127}I_2$, transition 11-5, R(127), component i) and locks the He–Ne laser [2, 6] in that very well-defined frequency. Its wavelength $\lambda_1 = 632.9913981$ nm is very close to one of the nominal wavelengths of the He–Ne laser, so this can easily be accomplished; however, the necessary equipment including the sophisticated electronics is rather involved [2]. For this reason, iodine-stabilized He–Ne lasers are only used where very high precision is required.

TABLE 6.9 Refractive Indices n of Various Gaseous Compounds Their Maximum Workplace Concentration (MWC) and the Change in Refractive Index Caused by that Particular Concentration (for $T = 20°C$, $P = 101.315$ hPa, $\lambda = 633$ nm)

Gas	Refractive index $(n-1)10^4$	Concentration for $\Delta n/n = 10^{-7}$ in air (ppm)	MWC mg m^{-3}	$\Delta n/n \times 10^7$ due to MWC
Air	2.72			
Propane	10.3	130	1800	8
Butane	12.9	98	2350	10
Ethanol	8.1	190	1900	5
Ethyl acetate	13.0	97	1400	4
Dimethylketone	10.2	130	2400	9
Octane	23	50	2350	10
Chlorofluorocarbons: e.g., R12	10.3	130	5000	7

Reprinted with permission of VDI Verein Deutscher Ingenieure, G. Wilkening, Kompensation der Luftbrechzahl.

The overall estimated relative uncertainty of such a stabilized laser is $\pm 10^{-9}$ (which results from an estimated relative standard deviation of 3.4×10^{-10} [1]), which makes it suitable for the practical realization of the definition of primary and secondary standards.

Refractive Index of Air

The length scale of a displacement interferometer that is operated in ambient air is given by $\lambda_A = \lambda_V/n_A$, where λ_A is the wavelength of the source in air, λ_V is the wavelength in vacuum and n_A is the refractive index of air. The vacuum wavelength λ_V is related to the speed of light by $\lambda_V = c/f_0$. The constant c, the speed of light in vacuum, was redefined in 1983 [1]. It is now defined to be exactly $c = 299,792,458$ m s^{-1}.

Thus, in order to measure distances interferometrically, it is essential to know the refractive index of air with an accuracy that is not less than the stability and the degree of certainty of the vacuum wavelength of the light source.

According to Edlén [7, 8] and Owens[9], the refractive index is a function of the atmospheric pressure P, the temperature T, the relative humidity H, or alternatively the partial pressure of the water vapor in air e_s, and the carbon dioxide concentration by volume D. The standard composition by volume of dry air is given by 78.03% N_2, 20.99% O_2, 0.933% Ar, 0.035% CO_2, and some other trace components, mainly noble gases [10]. The only component of the above list that might have a variability is CO_2, which follows a long-term increasing behavior presumably associated with the combustion of fossil fuel [11]. The current value is ≈ 350 ppm by volume and is increasing by ≈ 1.4 ppm per year. The CO_2 concentration in air can also change in an industrial environment due to CO_2 emitters and can therefore show significant local variations.

More recently, dependencies of the refractive index on nonnatural gaseous compounds like hydrocarbons in the air have been published [12] as well as corrections to Edlén's formulations [13–15]. Corrections to the refractive index due to those compounds may be necessary for very high-precision measurements in an industrial environment where chemical solvents or oils are in use. In Table 6.9, some nonnatural compounds, their maximum workplace concentrations, and their effect on the refractive index if present at the location of the interferometer are listed.

Jones [16] combined a precise determination of the density of moist air with Edlén's formulation to yield a somewhat simpler representation. For a typical iodine-stabilized He–Ne laser that has a vacuum wavelength of $\lambda_V = 632.9913$ nm, the Jones formulation is given by [17]:

$$n\left(P,T,H,D\right) = 1 + A - B \qquad (6.72)$$

where

$$A = 78.603\left[1 + 0.540\left(D - 0.0003\right)\right]\frac{P}{TZ}\times 10^{-8}$$

$$B = \left(0.00042066\ f_E\ e_s\ H\right)\times 10^{-8} \tag{6.73}$$

In Equation 6.73, P is the atmospheric pressure in pascals, T is the absolute temperature in kelvin, H is the relative humidity in %, and D is the concentration of CO_2 in percent by volume. There are three additional factors in Jones' formulation that take the nonideal behavior of moist air as compared to an ideal gas into account. They are Z, a compressibility factor that reflects the nonideality of the air–water vapor mixture and which, for air containing reasonable amounts of CO_2 at a temperature between 15°C and 28°C and pressure of between 7×10^4 Pa and 11×10^4 Pa, lies in the range between 0.99949 and 0.99979. f_E is an enhancement factor that expresses the fact that the effective saturation vapor pressure of water in air is greater than the saturation vapor pressure e_s. For the pressure and temperature ranges given above, f_E is bounded between 1.0030 and 1.0046 [16]. e_s is the saturation vapor pressure over a plane surface of pure liquid water and according to Jones is about 1705 Pa at a temperature of 15.0°C and about 3779 Pa for 28.0°C. Tables of Z, f_E and e_s are included in the Appendix of Jones' paper [16].

Table 6.10 gives an overview of the changes in environmental parameters that would cause a relative index change of 10^{-7}.

Edlén [8] and Jones [16] estimate that their empirical expressions for the dependency of the refractive index of air on the listed parameters has an absolute uncertainty of 5×10^{-8}.

Besides this fundamental limitation, there are some practical considerations that must be taken into account regarding the precision with which the environmental parameters can be measured. Estler [17] states that atmospheric pressure P can currently be determined with an uncertainty of ≈ 2.7 Pa, which can be assumed to be constant for the entire optical path of the interferometer if it is oriented horizontally. Please note that at sea level, the pressure gradient is ≈ -13 Pa m^{-1}, resulting in a pressure-induced change in n_A of 3.4×10^{-8} m^{-1} if the measuring equipment is not kept level.

In an exceptionally well-controlled laboratory environment where special care is devoted to keep temperature gradients from affecting the refractive index along the optical path as much as possible, uncertainties of the temperature measurement can be as low as 0.01°C according to [17]. Humidity measured with high accuracy dew-point hygrometers can have uncertainties down to 0.5%. Changes in carbon dioxide concentrations have to be very significant (20% of the natural concentration) to cause a $\Delta n/n$ of 10^{-8}.

Michelson Interferometer

The basis for most interferometers used in interferometric dimensional gages is the classical Michelson interferometer [4] which is shown in Figure 6.49. The coherent monochromatic light of a wavelength-stabilized He–Ne laser is incident onto a beam splitter which splits the light into two equally intense beams (**1**) and (**2**).

TABLE 6.10   Parameters of Standard Air and Their Deviation to Cause a $\Delta n/n$ of 10^{-7}

Parameter	Standard value	Variation for $\Delta n/n = +1 \times 10^{-7}$
Pressure P	101.3 kPa	+37.3 Pa
Temperature T	20.0°C	−0.1°C
Humidity H	40%	−10.0%
CO_2 concentration	350 ppm	+670 ppm

Reprinted with permission of *J. Applied Optics* [17].

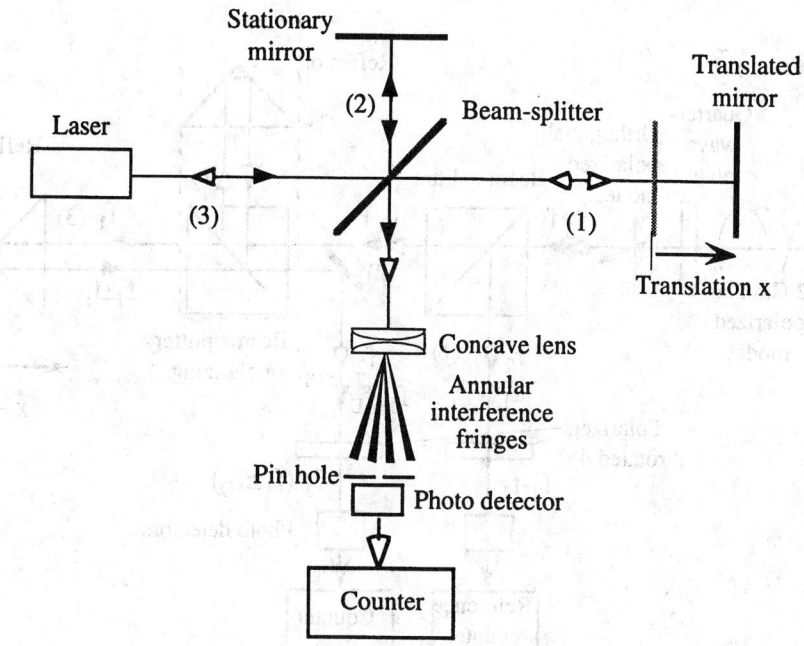

FIGURE 6.49 Schematics of the basic Michelson interferometer.

They are reflected off of both the stationary and the translatable mirror whose displacement x is to be measured, and recombined at the splitter, where they are redirected toward a concave lens. Due to the coherence of the laser light, the wavefronts have a well-defined phase relation with respect to each other. This phase is determined by the difference between the optical path lengths of the two beams in arms 1 and 2. If this path difference is continuously changed by translating one of the mirrors, a sinusoidal intensity variation can be observed at a fixed location in space behind the lens used to introduce a beam divergence, resulting in an annular fringe pattern. The pinhole is used to define an exact location of observation and the photodetector picks up the varying intensity for further processing. In the most basic signal processing setup, the number of bright and dark cycles are fed into a counter, which then counts changes in optical path length in integer multiples of $\lambda_A/2$. More sophisticated signal processing not only counts cycles but also determines relative phase changes in the sinusoidal varying intensity so that resolutions of $\lambda_A/512$ can ultimately be achieved.

When moving the mirror, one must guarantee a smooth motion without backward jitter of the mirror carriage to avoid double counts of interference fringes. Very high-quality linear bearings (such as air bearings) are necessary to accomplish just that.

As can be seen in Figure 6.49, the light reflected off both mirrors essentially retraces its own path and is at least partially incident onto the active volume of the laser source (3), thereby forming an external laser resonator which is able to detune the laser, effectively modulating its output power as well as its wavelength. To avoid this effect, commercial versions of Michelson interferometers employ corner-cube reflectors instead of plane mirrors as well as optical isolators, as shown in Figures 6.50 and 6.51. Some authors, however, report using optical arrangements that utilize this effect in conjunction with laser diodes to realize low-cost, short-travel displacement sensors [18, 19]. These setups will not be discussed here, however.

Two-Frequency Heterodyne Interferometer

Figure 6.50 shows the commercially available two-frequency Michelson interferometer operating with a Zeeman-stabilized He–Ne laser source. This laser emits two longitudinal modes with frequencies f_1 and f_2 that are both circularly polarized in opposite directions. By passing the modes through a quarter-wave

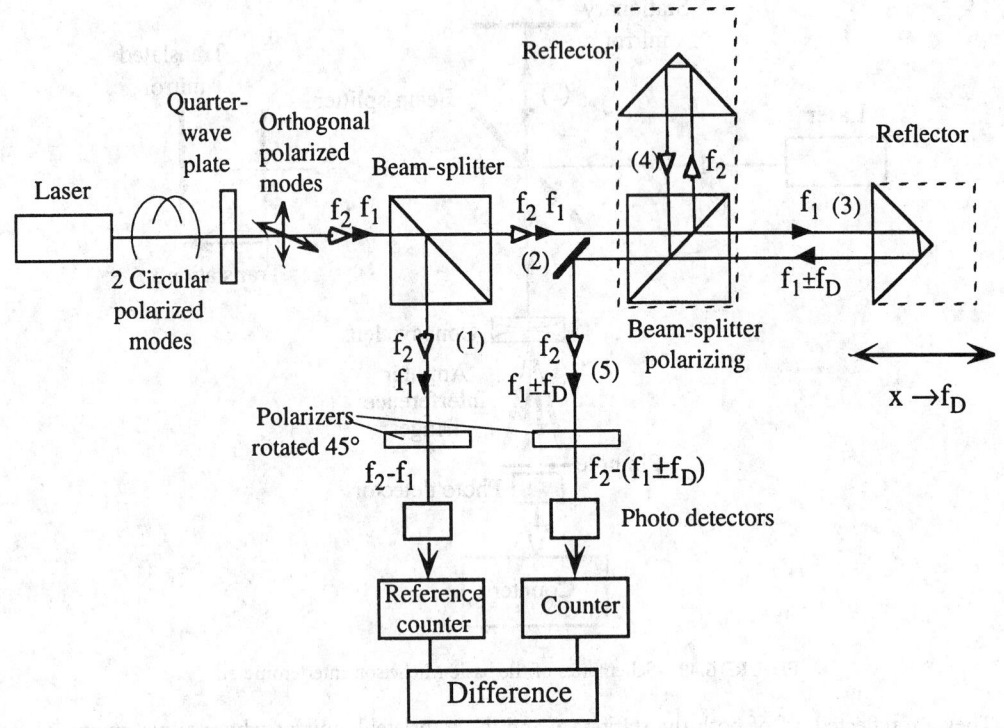

FIGURE 6.50   Two-frequency heterodyne interferometer (Courtesy of Spindler & Hoyer Inc.).

plate, two orthogonal linearly polarized waves are generated. Both are split by a nonpolarizing beam splitter. There is a polarizer located in arm **1** of that splitter, which is rotated by 45° with respect to both polarized waves impingng on it, thus effectively allowing them to interfere behind it yielding a difference frequency of $f_2 - f_1$ that is picked up by a photodetector and counted by a reference counter (frequency difference typically 1.5 MHz).

The orthogonal polarized waves in **2** are further split by a polarizing splitter. Spectral component $f_1 < f_2$ is transmitted into measuring arm **3** and frequency component f_2 is reflected into reference arm **4** of the interferometer. Due to the velocity v of the reflector in arm **3** resulting in a displacement x, the frequency f_1 is Doppler-shifted by f_D (Equation 6.74). Movement of the reflector toward the interferometer results in a positive Doppler frequency $f_D > 0$. After recombining both waves from **3** and **4** in the beam splitter again, they are sent through a polarizer in arm **5** that also is rotated by 45° with respect to the direction of polarization of both recombined waves, thereby allowing them to interfere, yielding a difference frequency of $f_2 - f_1 - f_D$ at the location of the photodetector, which is counted by a second counter. By continuously forming the difference of both counts, the measurand (the displacement of x in multiples of $\lambda_A/2$) is calculated.

With this type of interferometer, the direction of motion is given by the sign of the resulting count. One disadvantage of the two-mode heterodyne interferometer is its limited dynamic range for the velocity v of the reflector moving toward the interferometer, since the Doppler frequency f_D, given by:

$$f_D = \frac{2}{\lambda_A} v \qquad\qquad (6.74)$$

is bound to be less than the initial frequency difference between f_2 and f_1 for stationary reflectors. Given a typical Zeeman effect-induced frequency difference f_z of 1.5 MHz, the velocity v is therefore bound to be less than:

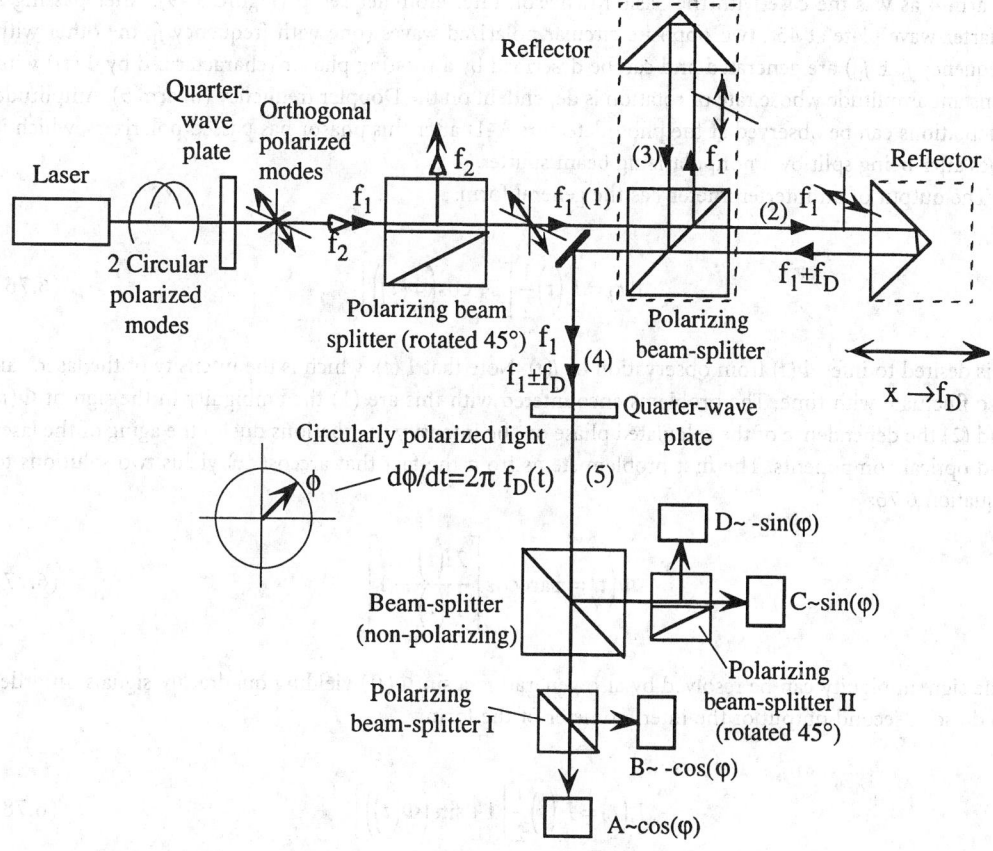

FIGURE 6.51 Single-mode homodyne interferometer (Courtesy of Spindler & Hoyer Inc.).

$$v < \frac{f_z}{2} \lambda_A = 0.474 \text{ m s}^{-1} \tag{6.75}$$

There is no such bound if the reflector is traveling away from the interferometer. By electronically interpolating the output signal of the photodetector, subwavelength resolution can be obtained [22].

Single-Mode Homodyne Interferometer

An interferometer setup that has no limitation on the maximum velocity in the above sense is the single-mode homodyne interferometer shown in Figure 6.51.

As in the two-frequency heterodyne interferometer, a Zeeman effect-stabilized laser source that emits two frequency-displaced circularly polarized axial modes is usually used. After passing through a quarter-wave plate, two orthogonal polarized waves are generated, only one of which (f_1) is further used. The other (f_2) is reflected out of the optical path by an appropriately oriented polarizing beam splitter. The plane of polarization in arm **1** of the interferometer is tilted by 45° with respect to the plane defined by the two arms **2** and **3**. The second polarizing beam splitter will transmit one horizontally oriented component into the measuring arm **2** and reflect a vertically oriented component into the reference arm **3** of the interferometer. The two arms of the interferometer maintain their orthogonal polarizations. The frequency of the wavefront in arm **2** is shifted by the Doppler effect (Equation 6.74) due to the motion of the reflector. The light reflected by the two triple mirrors is recombined in the polarizing beam splitter and redirected by a mirror. Since in this particular optical setup, the polarization states of the two beams in the two arms of the interferometer are orthogonal, there is no interference after the redirecting mirror in arm **4** as was the case with the basic Michelson interferometer setup (Figure 6.49). After passing a quarter-wave plate at 45°, two opposite circular polarized waves (one with frequency f_1, the other with frequency $f_1 \pm f_D$) are generated and can be described by a rotating phasor (characterized by $\Phi(t)$) with constant amplitude whose rate of rotation is dependent on the Doppler frequency (in arm 5). Amplitude fluctuations can be observed at the photodetectors A–D after this phasor has passed polarizers, which it does after being split by a nonpolarizing beam splitter.

The output of an interferometer has the general form:

$$I(t) = I_o(t) \frac{1}{2} \left[1 + \cos\left(\Phi(t) \right) \right] \tag{6.76}$$

It is desired to infer $\Phi(t)$ from observation of $I(t)$. Note that $I_o(t)$, which is the intensity of the laser, can also fluctuate with time. The problems encountered with this are (1) the ambiguity in the sign of $\Phi(t)$ and (2) the dependence of the calculated phase on the intensity fluctuations due to the aging of the laser and optical components. The first problem stems from the fact that arccos(…) yields two solutions to Equation 6.76:

$$\Phi(t) = \pm \arccos \left[\frac{2I(t)}{I_o(t)} - 1 \right] \tag{6.77}$$

The sign ambiguity can be resolved by also generating a $\sin(\Phi(t))$ yielding quadrature signals. In order to do so, a second output of the interferometer of the form:

$$I_2(t) = I_o(t) \frac{1}{2} \left[1 + \sin\left(\Phi(t) \right) \right] \tag{6.78}$$

is sought. Equations 6.76 and 6.78 will determine $\Phi(t)$ unambiguously only in the region $[0, 2\pi)$, but there is still an ambiguity modulo 2π. The second problem can be dealt with by adding two more outputs of the form:

$$I_3(t) = I_o(t) \frac{1}{2} \left[1 - \cos\left(\Phi(t) \right) \right] \tag{6.79}$$

$$I_4(t) = I_o(t) \frac{1}{2} \left[1 - \sin\left(\Phi(t) \right) \right] \tag{6.80}$$

Taking Equations 6.76 to 6.79 and 6.78 to 6.80, it is possible to obtain a zero crossing at the linear most sensitive point of inflection of the fringe where the effect of intensity fluctuations on the phase measurement is minimal. The setup of Figure 6.51 attempts to obtain these four outputs. Signal A represents the intensity variations as given in Equation 6.76 and signal B due to the nature of the splitting action is shifted with respect to A by 180° (Equation 6.79). There is another arm to the right of the nonpolarizing beam splitter incorporating the polarizing beam splitter II, which is rotated by 45° with respect to beam splitter I so that the attached detectors C and D are generating the signals defined by Equations 6.78 and 6.80.

Since the Doppler frequency is time dependent according to the velocity v of the measuring reflector, the distance, x, traveled by the reflector up to time T is given by Equation 6.81.

$$x = \int_0^T v(t)\, dt = \int_0^T f_D(t) \frac{\lambda_A}{2}\, dt = \int_0^T \frac{\partial \Phi(t)}{\partial t} \frac{\lambda_A}{4\pi}\, dt \tag{6.81}$$

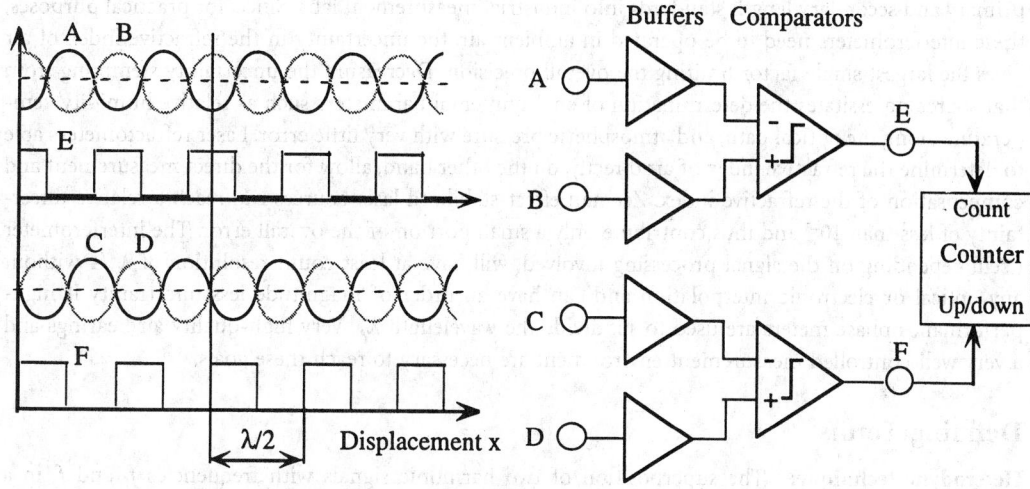

FIGURE 6.52 A simple signal conditioning circuit yielding quadrature signals E and F.

It should be noted that the resolution of this interferometer is limited to $\lambda_A/4$ if no special hardware is used to interpolate between interference fringes.

This particular interferometer is not limited with respect to a maximum unambiguous Doppler frequency.

Interferometer Signal Processing

At the photodetectors ends the optical path of the interferometers. Since the photodetectors are sensitive to light intensity values only the sinusoidal signal caused by motion of the reflector will be superimposed onto a pedestal signal proportional to the mean intensity over time or displacement, respectively. The electronics attached to the photodetectors is aimed to reliably detect zero-crossings of the sinusoidal component of the signals (A through D) even under low contrast conditions by subtracting out the pedestal signal component in the comparators. Low fringe contrast can be the result of small reflector tilt and/or vibration during periods of reflector movement, nonideal interferometer alignment, and imperfections in optical components like unequal splitting of the beam in beam splitters. Figure 6.52 shows a simple conditioning unit.

The main purpose of that set up is to produce digital quadrature signals (E and F in Figure 6.52) that can most easily be used to perform signal interpolation digitally [20, 21] and also allow for the determination of the direction of movement by using, for example, signal E as count signal and F as an up/down indicator.

Both comparators compare sinusoidal components having a possibly slowly time-varying dc pedestal amplified by buffer amplifiers. All photodetector signals A through D will be affected by this dc pedestal so that forming their difference in the comparator will yield zero-crossing signals independent of the pedestal. The two comparator units are necessary to generate quadrature signals for direction detection. Furthermore, this electronics allows easily for a fringe interpolation by a factor of 4, bringing the least-count resolution of the interferometer down to ≈ 160 nm. This interpolation can be done if instead of feeding the signal E into the count input, the exored signal E$\oplus$F is used. In this case, the up/down terminal of the counter circuit needs to be connected to the signal E&F. If an even higher resolution is sought, digital interpolation of these quadrature signals can be performed [20–22].

Conclusions

Using laser interferometers operating with highly stabilized laser sources, relative uncertainties in length measurements as low as 5×10^{-8} can be realized, which makes these kinds of equipment suitable to convey primary and secondary length standards into industrial measurement labs. Since, for practical purposes, these interferometers need to be operated in ambient air, the uncertainty in the refractive index of air gives the largest single factor limiting the overall precision. Decreasing the uncertainty stemming from that source necessitates the determination of environmental parameters such as relative humidity, temperature along the optical path, and atmospheric pressure with very little error. Laser refractometers able to determine the refractive index of air directly, on the other hand, allow for the direct measurement and compensation of the refractive index. Zeeman effect-stabilized laser sources can reach a relative uncertainty of less than 10^{-8} and thus contribute only a small portion of the overall error. The interferometer itself, depending on the signal processing involved, will have at least count resolution of $\lambda_A/4$ without any optical or electronic interpolation and can have an order of magnitude less uncertainty if high-performance phase meters are used to subdivide the wavelength λ_A. Very high-quality air bearings and a very well-controlled measurement environment are necessary to reach these goals.

Defining Terms

Heterodyne technique: The superposition of two harmonic signals with frequencies f_1 and f_2 in a nonlinear device results in a signal containing both sum and differences frequencies. For interferometric purposes where each individual frequency is in the 10^{14} Hz range, the frequency difference might be low enough to be registered by photodetectors. They also serve as nonlinear devices because their response is to light intensity only, which is proportional to the light amplitude squared.

Interference: A phenomenon that strikingly illustrates the wave nature of light. It occurs when radiation follows more than one path from its source to the point of detection. It may be described as the local departures of the resultant intensity from the law of addition as the point of detection is moved, for the intensity oscillates about the sum of the separate intensities.

Laser: The acronym of *Light Amplification by Stimulated Emission of Radiation*. It was originally used to describe light amplification, but has more recently come to mean an optical oscillator.

Refractive index: A number specifying the ratio of the propagation velocities of light in vacuum *c* to that in a medium. The refractive index for any medium is always somewhat larger than 1. Its value is dependent on the composition of the medium and the wavelength of the incident radiation.

Zeeman effect: An effect that is observed if excited atoms emit their radiation in the presence of a magnetic field. The longitudinal Zeeman effect causes a single emission line to split symmetrically into right and left circularly polarized lines.

References

1. Documents concerning the new definition of the metre, *Metrologia*, 19, 163-177, 1984.
2. E. Jaatinen and N. Brown, A simple external iodine stabilizer applied to 633 nm 612 nm and 543 nm He–Ne lasers, *Metrologia*, 32, 95-101, 1995.
3. A. E. Siegman, *Lasers*, Mill Valley, CA: University Science Books, 1986.
4. W. R. Steel, *Interferometry*, 2nd ed., (Cambridge studies in modern optics), Cambridge, U.K.: Cambridge University Press, 1985.
5. W. R. C. Rowley, The performance of a longitudinal Zeeman-stabilized He–Ne laser (633 nm) with thermal modulation and control, *Meas. Sci. Technol.*, 1, 348-351, 1990.
6. Wolfgang Demtröder, *Laser Spectroscopy: Basic Concepts and Instrumentation*, 2nd ed., Berlin: Springer, 1996.
7. B. Edlén, The dispersion of standard air, *J. Opt. Soc. Amer.*, 43(5), 339-344, 1953.
8. B. Edlén, The refractive index of air, *Metrologia*, 2(2), 71-80, 1966.
9. J. C. Owens, Optical refractive index of air: dependence on pressure, temperature and composition, *Appl. Optics*, 6(1), 51-59, 1967.
10. H. D. Baehr, *Thermodynamik*, 6th ed., Berlin: Springer, 1988.

11. R. Revelle, Carbon dioxide and world climate, *Sci. Amer.*, 247(2), 35, 1982.

12. K. P. Birch, F. Reinboth, R. W. Ward, and G. Wilkening, Evaluation of the effect of variations in the refractive index of air upon the uncertainty of industrial length measurement, *Metrologia*, 30(1), 7-14, 1993.

13. K. P. Birch and M. J. Downs, An updated Edlén equation for the refractive index of air, *Metrologia*, 30, 155-162, 1993.

14. K. P. Birch and M. J. Downs, Corrections to the updated Edlén equation for the refractive index of air, *Metrologia*, 31, 315-316, 1994.

15. P. E. Ciddor, Refractive index of air: new equations for the visible and near infrared, *Appl. Optics*, 35(9), 1566-1573, 1996.

16. F. E. Jones, The refractivity of air, *J. National Bureau of Standards*, 86(1), 27-32, 1981.

17. W. T. Estler, High-accuracy displacement interferometry in air, *J. Appl. Optics*, 24(6), 808-815, 1985.

18. J. A. Smith, U. W. Rathe, and C. P. Burger, Lasers with optical feedback as displacement sensors, *Opt. Eng.*, 34(9), 2802-2810, 1995.

19. N. Takahashi, S. Kakuma, and R. Ohaba, Active heterodyne interferometric displacement measurement using optical feedback effects of laser diodes, *Opt. Eng.*, 35, 802-907, 1996.

20. K. Oka, M. Tsukada, and Y. Ohtsuka, Real-time phase demodulator for optical heterodyne detection processes, *Meas. Sci. Technol.*, 2, 106-110, 1991.

21. J. Waller, X. H. Shi, N. C. Altoveros, J. Howard, B. D. Blackwell, and G. B. Warr, Digital interface for quadrature demodulation of interferometer signals, *Rev. Sci. Instrum.*, 66, 1171-1174, 1995.

22. J. A. Smith and C. P. Burger, Digital phase demodulation in heterodyne sensors, *Opt. Eng.*, 34, 2793-2801, 1995.

Appendix to Section 6.5

In the appendix below some companies are listed that manufacture either complete interferometer systems or major components thereof, such as beam splitters, retroreflectors, refractometers, wavemeters, etc. This list is by no means exhaustive. Furthermore, no price information is included because the system cost is too much dependent on the particular choice of system components.

Companies that produce interferometers or significant components.

Manufacturer	Sub-systems	Complete systems	Manufacturer	Sub-systems	Complete systems
Aerotech Inc. 101 Zeta Drive Pittsburgh, PA 15238 Tel: (412) 963-7470	*	*	Oriel Instruments Inc. 250 Long Beach Blvd. Stratford, CT 06497-0872 Tel: (203) 380-4364	—	*
Burleigh Inc. Burleigh Park Fishers, NY 14453-0755 Tel: (716) 924-9355	—	*	Polytec PI Inc. Auburn, MA 01501 Tel: (508) 832-3456	*	*
Hewlett-Packard Inc. Test & Measurement Customer Business Center, P.O. Box 4026 Englewood, CO 80155-4026 Tel: (800) 829-4444	*	—	Spindler & Hoyer Inc. 459 Fortune Blvd. Milford, MA 01757-1745 Tel: (508) 478-6200	*	*
Melles Griot Inc. 4665 Nautilus Court South Boulder, CO 80301 Tel: (303) 581-0337	*	*	Zygo Corporation Middlefield, CT 06455-0448 Tel: (860) 347-8506	*	*

6.6   Bore Gaging Displacement Sensors

Viktor P. Astakhov

Dimensions are a part of the total specification assigned to parts designed by engineering. However, the engineer in industry is constantly faced with the fact that no two objects in the material world can ever be made exactly the same. The small variations that occur in repetitive production must be considered in the design. To inform the workman how much variation from exact size is permissible, the designer uses a tolerance or limit dimension technique. A *tolerance* is defined as the total permissible variation of size, or the difference between the limits of size. *Limit dimensions* are the maximum and minimum permissible dimensions. Proper tolerancing practice ensures that the finished product functions in its intended manner and operates for its expected life.

Bore Tolerancing

All bore dimensions applied to the drawing, except those specifically labeled as basic, gage, reference, maximum, or minimum, will have an exact tolerance, either applied directly to the dimension or indicated by means of general tolerance notes. For any directly tolerated decimal dimension, the tolerance has the same number of decimal places as the decimal portion of the dimension.

Engineering tolerances may broadly be divided into three groups: (1) *size tolerances* assigned to dimensions such as length, diameter, and angle; (2) *geometric tolerances* used to control a hole shape in the longitudinal and transverse directions; and (3) *positional tolerances* used to control the relative position of mating features. Interested readers may refer to [1, 2].

The ISO system of limits and fits (ISO Recommendation R 286) covers standard tolerances and deviations for sizes up to 3150 mm. The system is based on a series of tolerances graded to suit all classes of work from the finest to the most coarse, along with different types of fits that range from coarse clearance to heavy interference. Here, *fit* is the general term used to signify the range of tightness that may result from the application of a specific combination of tolerances in the design of mating parts.

There are 18 tolerance grades intended to meet the requirements of different classes of parts. These tolerance grades are referred to as ITs and range from IT 01, IT 02 (reserved for the future), and IT 1, to IT 16 (for today's use). In each grade, the tolerance values increase with size according to a formula that relates the value of a given constant to the mean diameter of a particular size range. The system provides 27 different fundamental deviations for sizes up to and including 500 mm, and 14 for larger sizes to give different type of fits ranging from coarse clearance to heavy interference. Interested readers may refer to [3].

Bore Gage Classification and Specification

To measure the above-listed tolerances, modern manufacturing requires the use of gages. A *gage* is defined as a device for investigating the dimensional fitness of a part for specific function. *Gaging* is defined by ANSI as a process of measuring manufactured materials to assure the specified uniformity of size and contour required by industries. Gaging thereby assures the proper functioning and interchangeability of parts; that is, one part will fit in the same place as any similar part and perform the same function, whether the part is for the original assembly or replacement in service.

Bore gages may by classified as follows:

1. Master gages
2. Inspection gages
3. Manufacturer's gages
4. Gages that control dimensions
5. Gages that control various parameters of bore geometry
6. Fixed limit working gages

7. Variable indicating gages
8. Post-process gages
9. In-process gages

Master gages are made to their basic dimensions as accurately as possible and are used for reference, such as for checking or setting inspection of manufacturer's gages. *Inspection gages* are used by inspectors to check the manufactured products. *Manufacturer's gages* are used for inspection of parts during production.

Post-process gages are used for inspecting parts after being manufactured. Basically, this kind of gage accomplishes two things: (1) it controls the dimensions of a product within the prescribed limitations, and (2) it segregates or rejects products that are outside these limits. Post-process gaging with feedback is a technique to improve part accuracy by using the results of part inspection to compensate for repeatable errors in the machine tool path. The process is normally applied to CNC (computer numerically controlled) machines using inspection data to modify the part program, and on tracer machines using the same data to modify the part template.

In-process gages are used for inspecting parts during the machining cycle. In today's manufacturing strategy, in-process gages and data-collection software provide faster feedback on quality. Indeed, the data-collection and distribution aspect of 100% inspection has become as important as the gaging technology itself. Software specifically designed to capture information from multiple gages, measure dozens of products types and sizes, and make it available to both roving inspectors and supervising quality personnel as needed, is quickly becoming part of quality control strategies. In conjunction with computer numerically controlled (CNC) units, in-process gaging can automatically compensate for workpiece misalignment, tool length variations, and errors due to tool wear.

Gages That Control Dimensions

Gages that control dimensions are used to control bore diameter. These gages can be either post-process or in-process gages. Further, these gages can be either *fixed limit gages* or *variable indicating gages*.

A *plug gage* is a fixed limit working bore gage. These inexpensive gages do not actually measure dimensions or geometry. They simply tell the operator whether the bore is oversized or undersized. The actual design of most plug gages is standard, being covered by American Gage Design (AGD) standards. However, there are many cases where a special plug gage must be designed.

A plug gage is usually made up of two members. One member is called the go end, and the other the no-go or not-go end. The gage commonly has two parts: the gaging member, and a handle with the sign, go or no-go, and the gagemaker's tolerance marked on it. There are generally three types of AGD standard plug gages. First is the single-end plug gage (Figure 6.53(*a*)); the second is the double-end (Figure 6.53(*b*)); and the third is the progressive gage (Figure 6.53(*c*)). Interested readers may refer to [4].

Fixed-limit gage tolerance is generally determined from the amount of workpiece tolerance. A 10% rule is generally used for determining the amount of gage tolerance for fixed, limit-type gages. Four classes of gagemakers' tolerances have been established by the American Gage Design Committee and are in general use [4]. These four classes establish maximum variation for any designed gage size. The degree of accuracy needed determines the class of gage to be used. Table 6.11 shows these four classes of gagemakers' tolerances. Referring to Table 6.11, class XX gages are used primarily as master gages and for final close tolerance inspection. Class X gages are used for some types of master gage work and as close tolerance inspection and working gages. Class Y gages are used as inspection and working gages. Class Z are used as working gages where part tolerances are large. Table 6.12 shows the diameter ranges and prices of the plug gages manufactured by the Flexbar Machine Corp.

Variable indicating gages allow the user to inspect some bore parameters and get numbers for charting and statistical process control (commonly abbreviated as SPC). These gages have one primary advantage over fixed gages: they show how much a hole is oversized or undersized. When using a variable indicating gage, a master ring gage to the nominal dimension to be checked must be used to preset the gage to zero. Then, in applying the gage, the variation from zero is read from the dial scale. Figure 6.54 shows industry's

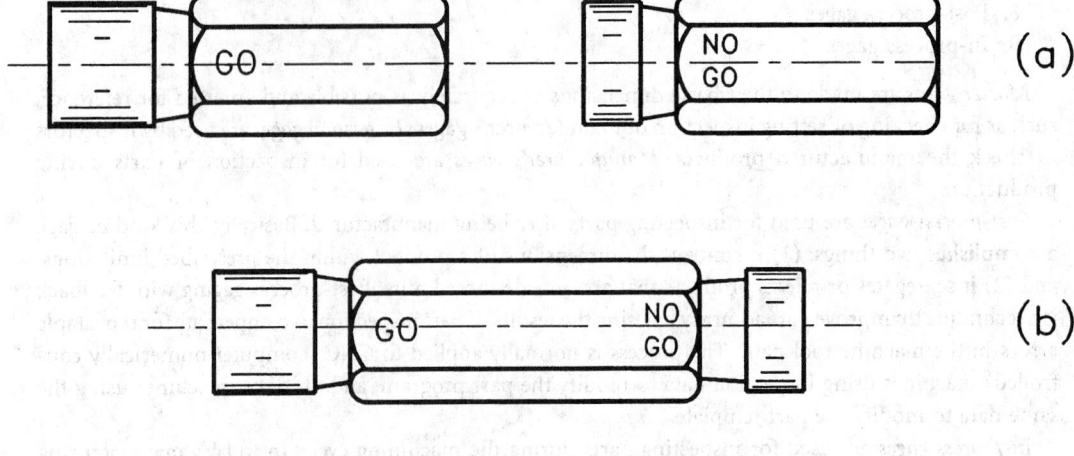

FIGURE 6.53 AGD cylindrical plug gages to inspect the diameter of holes: (*a*) two separate gage members; (*b*) two gage members mounted on single handle with one gage member on each end; (*c*) progressive gage.

TABLE 6.11   Standard Gagemakers' Tolerances

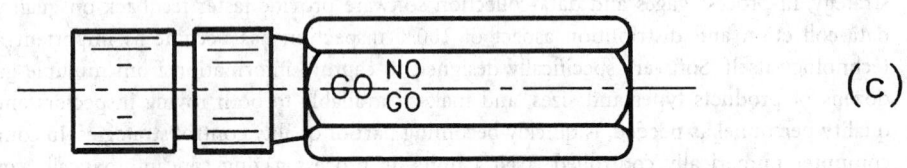

Above	To and including	Class			
		XX	X	Y	Z
0.010 in.	0.825 in.	0.00002 in.	0.00004 in.	0.00007 in.	0.00010 in.
0.254 mm	20.95 mm	0.00051 mm	0.00102 mm	0.00178 mm	0.00254 mm
0.825 in.	1.510 in.	0.00003 in.	0.00006 in.	0.00009 in.	0.00012 in.
20.95 mm	38.35 mm	0.00076 mm	0.00152 mm	0.00229 mm	0.00305 mm
1.510 in.	2.510 in.	0.00004 in.	0.00008 in.	0.00012 in.	0.00016 in.
38.35 mm	63.75 mm	0.00102 mm	0.00203 mm	0.00305 mm	0.00406 mm
2.510 in.	4.510 in.	0.00005 in.	0.00010 in.	0.00015 in.	0.00020 in.
63.75 mm	114.55 mm	0.00127 mm	0.00254 mm	0.00381 mm	0.00508 mm

most popular dial and electronic bore gages. Figure 6.55 shows a set of dial bore gages for the range of 35 mm to 150 mm. Figure 6.56 shows the Intrimic® plus (Brown & Sharpe) internal micrometer. Intrimic® plus provides simple, accurate inside measurement capability. The micrometer features automatic shut-off, electronic memory mode, instantaneous inch/metric conversion, and a standard direct output for SPC applications. Tolerance classification at a glance allows quick, efficient sorting while inspecting. The display shows the inspector, in color, if any dimension is within tolerance (green), out-of-tolerance (red), or reworkable (yellow). Table 6.13 presents the basic ranges and prices for these micrometers.

Gages That Control Geometry

Gages that control various parameters of bore geometry are used for complex comparisons of part shape to an ideal shape. All these gages are post-process gages. Two major categories of geometry gages are in use: gages with manual probe head systems and form measuring machines.

TABLE 6.12 Premium Quality Hardened Steel GO/NO GO Plug Gages by the Flexbar Machine Corp.

| Size range | Class | Price ($) | | | Handle | |
		1	2–4	5–10	No	Price ($)
0.01 in. to 0.030 in.	XX	35.45	28.65	22.26	1W	8.00
0.25 mm to 0.762 mm	X	31.15	23.35	17.00		
	Y	28.35	22.15	15.05		
	Z	25.65	20.70	12.55		
0.03 in. to 0.075 in.	XX	19.25	15.30	13.63	1W	8.00
0.762 mm to 1.91 mm	X	15.05	11.70	10.60		
	Y	13.90	10.80	9.40		
	Z	11.40	9.35	8.15		
0.075 in. to 1.80 in.	XX	21.15	17.00	14.80	2W	8.30
1.91 mm to 4.57 mm	X	18.15	14.45	12.85		
	Y	17.00	13.90	11.25		
	Z	14.00	11.10	9.65		
0.180 in. to 0.281 in.	XX	22.00	17.80	15.30	3W	9.00
4.57 mm to 7.14 mm	X	18.90	15.30	13.10		
	Y	17.85	13.55	11.95		
	Z	14.30	11.40	9.90		
0.281 in. to 0.406 in.	XX	24.20	19.50	17.80	4W	9.35
7.17 mm to 10.31 mm	X	21.15	17.00	14.80		
	Y	19.30	15.65	13.80		
	Z	14.90	12.00	10.50		
0.406 in. to 0.510 in.	XX	25.35	20.35	17.80	5W	9.70
10.31 mm to 12.95 mm	X	22.25	17.80	15.80		
	Y	20.45	16.50	14.55		
	Z	16.30	13.15	11.30		
0.510 in. to 0.635 in.	XX	26.70	21.45	18.50	6W	11.40
12.95 mm to 16.13 mm	X	23.70	19.25	16.30		
	Y	21.85	17.60	15.50		
	Z	17.50	14.30	12.55		
0.635 in. to 0.760 in.	XX	28.15	22.60	19.85	7W	13.50
16.13 mm to 19.30 mm	X	25.35	20.05	17.50		
	Y	23.25	18.45	16.40		
	Z	18.65	15.15	13.45		
0.780 in. to 1.010 in.	XX	44.25	34.25	32.80	8W	20.00
19.30 mm to 25.65 mm	X	39.56	29.80	27.30		
	Y	36.30	27.25	24.90		
	Z	32.40	25.65	24.20		

Gages with Manual Probe Head Systems

Geometry gages with manual probe head systems are rapidly becoming common in many high-precision metalworking applications. The simplest form of manual probe head systems in common use is the air plug gage (spindle) (Figure 6.57). Compressed air from the air gage indicating unit is pressed to the plug gage and allowed to escape from two or more jets in the periphery. When the air plug gage is inserted into a hole, the air escaping from the jets is limited by the clearance between the jet faces and the hole. The small changes in clearance, arising when the air plug gage is inserted in successive holes of different sizes, produce changes in the flow rate or back pressure in the circuit. The magnification and datum setting of systems with variable control orifices and zero bleeds is carried out with master holes. Some errors of form that can be detected with air plug gages are: (1) taper, (2) bell mouthing, (3) barreling, (4) ovality, and (5) lobing. Dearborn (Dearborne Gage Company, MI) open-orifice air spindles are available as standard to use in measuring thru, blind, and counterbored holes ranging in diameter from 0.070 in. (2 mm) to 6.000 in. (154 mm).

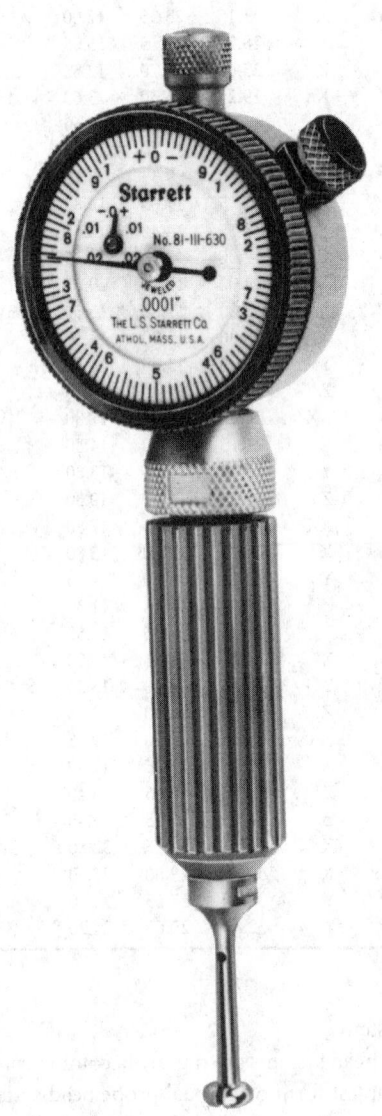

FIGURE 6.54 Industry's most popular dial and electronic bore gages (Courtesy of The L.S. Starrett Co.).

Another type of geometry gage with manual probe head system is the electronic bore gage. These gages measure bores at various depths to determine conditions such as bellmouth, taper, convexity, or concavity. They are also able to determine out-of-roundness conditions when equipped with a 3-point measuring system. Figure 6.58 shows a TRIOMATIC® electronic bore gage (Brown & Sharpe), and

FIGURE 6.54 (continued)

Table 6.14 presents the basic ranges and prices for these gages. TRIOMATIC® electronic bore gages feature automatic shut-off, electronic memory mode, instantaneous inch/metric conversion, and a standard direct output to a data handling system for SPC and statistical quality control. Mechanically, TRIOM-ATIC® electronic bore gages use the time-tested, three contact points interchangeable heads. The contact points are spaced 120° apart, ensuring proper centering and alignment that are especially essential for deep holes. The tips of the points are made of tungsten-carbide to resist wear, and extend to the surface of the heard for measuring at the bottom of blind holes or the surface of steps within a hole. Since the tips are connected to a cone-actuated electronic measuring system, these gages are referred to as *electronic gages.*

Form Measuring Instruments

Most modern form measuring instruments stage the workpiece on a turntable and provide a means to position a gage head against the part (Figure 6.59). As the turntable rotates, the gage head measures deviation from the true circle. Those gages where the gage head is supported by a simple, rigid, manual, or motorized stand that does not provide precise control over positioning are capable of performing the following measurements: roundness, concentricity, circular runout, circular flatness, perpendicularity, plane runout, top and bottom face runout, circular parallelism, and coaxiality. Modern fully automatic machines are the most sophisticated measuring instruments. Normally, they are equipped with a Win-dows™-based, PC-compatible graphical user interface to perform real-time data acquisition and processing. Mitutoyo/MTI Corp. produces a wide range of these machines (Table 6.15). The RA-600 series (Figure 6.60) features an innovative, fully automatic method that enables the machine to perform centering and leveling automatically if any deviation is detected during preliminary measurement. In addition, these machines can measure thickness, squareness, cylindricity, spiral cylindricity, straightness, total

FIGURE 6.54 (continued)

runout, vertical straightness, and vertical parallelism. The machines are supplied with MeasurLink® data acquisition software for Windows™, which allows immediate measurement data analysis and feedback for variable, attribute, and short inspection runs. The software gives the quality control/production manager the ability to create traceability lists of unlimited size. Information such as machine center, operator, materials used, assignable causes, and other relevant data can be stored and attached to measurement values. See Table 6.16 for a lsit of companies that make bore gages.

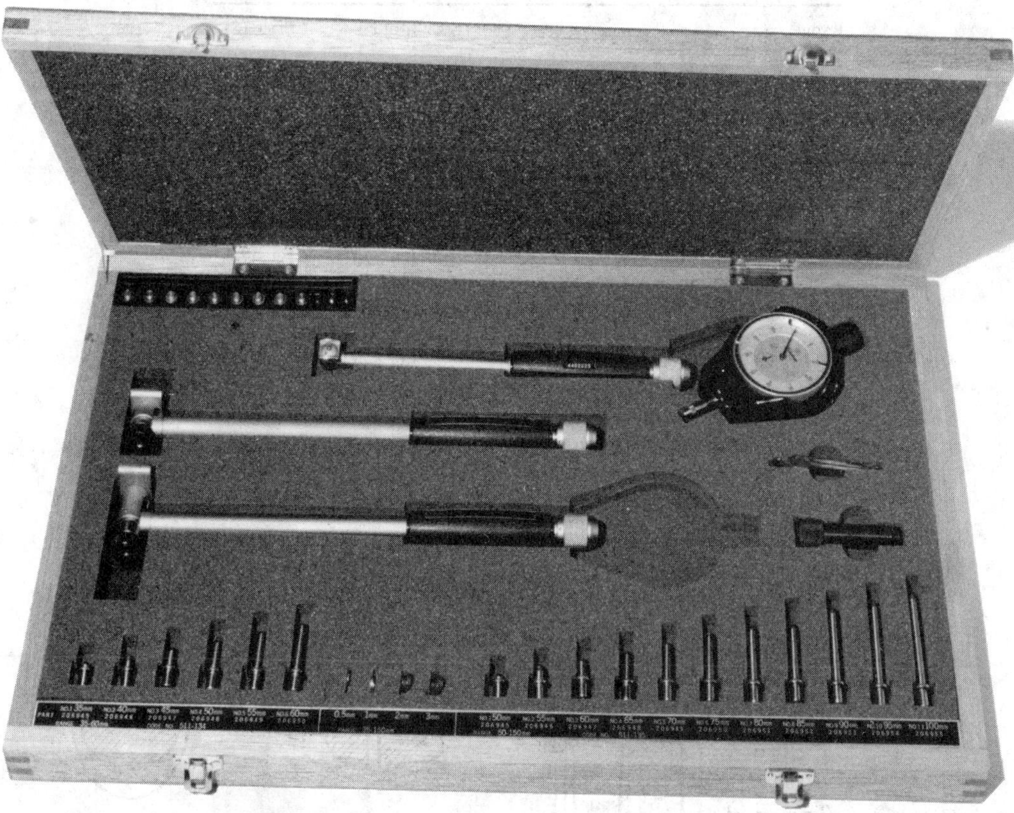

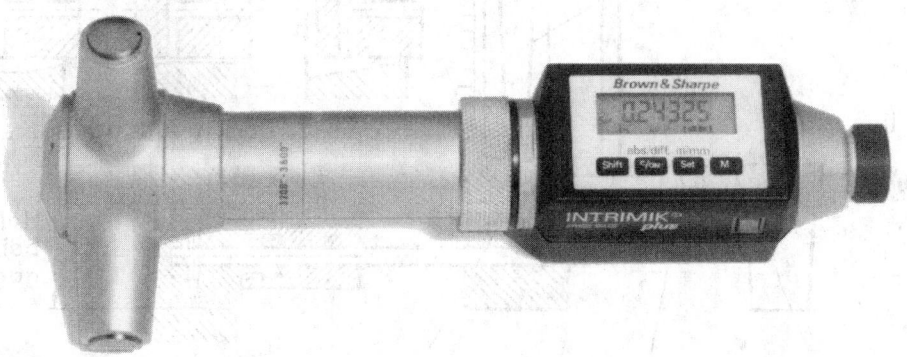

FIGURE 6.55 Set of dial bore gages (Courtesy of MITUTOYO/MTI Corporation).

FIGURE 6.56 Intrimic® plus internal micrometer (Courtesy of Brown & Sharpe Manufacturing Company).

GAGE R AND R Standards

Gage Repeatability and Reproducibility (GAGE R AND R) capability standards have direct implications for parts makers and for gage manufacturers. Repeatability is the ability of an operator using a single gage to obtain the same measurements during a series of tests. Reproducibility is the ability of different

TABLE 6.13   Intrimik® Plus Internal Micrometers by Brown & Sharpe

Range	B&S Tool No.	Price ($)
0.275 in. to 0.350 in. (6–8 mm)	599-290-35	745.40
0.350 in. to 0.425 in. (8–10 mm)	599-290-42	745.40
0.425 in. to 0.500 in. (10–12 mm)	599-290-50	745.40
0.500 in. to 0.600 in. (12–14 mm)	599-290-60	826.70
0.600 in. to 0.700 in. (14–17 mm)	599-290-70	826.70
0.700 in. to 0.800 in. (17–20 mm)	599-290-80	826.70
0.800 in. to 1.0 in. (20–25 mm)	599-290-100	843.60
1.0 in. to 1.2 in. (25–30 mm)	599-290-120	843.60
1.2 in. to 1.4 in. (30–35 mm)	599-290-140	854.30
1.4 in. to 1.6 in. (35–40 mm)	599-290-160	854.30
1.6 in. to 2.0 in. (40–50 mm)	599-290-200	933.10
2.0 in. to 2.4 in. (50–60 mm)	599-290-240	933.10
2.4 in. to 2.8 in. (60–70 mm)	599-290-280	933.10
2.8 in. to 3.2 in. (70–80 mm)	599-290-320	950.20
3.2 in. to 3.6 in. (80–90 mm)	599-290-360	950.20
3.6 in. to 4.0 in. (90–100 mm)	599-290-400	950.20
Intrimik Plus Complete Set #5	599-290-5	3374.60

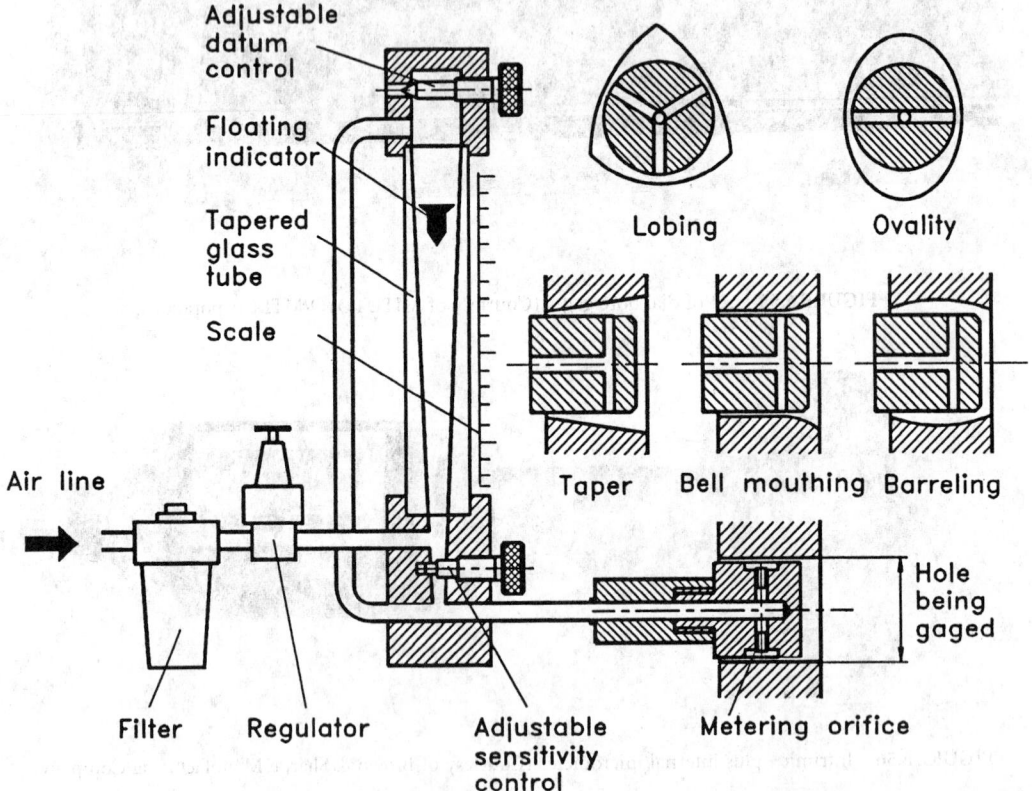

FIGURE 6.57   Air plug gage.

operators to obtain similar results with the same gage. GAGE R AND R blends these two factors together to determine a measuring system's reliability and its suitability for a particular measuring application. For example, a gage design that meets the GAGE R AND R standards for 50 mm (2 in.) bores may be unsatisfactory on 250 mm (10 in.) bores. A gage that meets a tolerance of 2 μm, may not be satisfactory

FIGURE 6.58 TRIOMATIC® electronic bore gage (Courtesy of Brown & Sharpe Manufacturing Company).

TABLE 6.14 TRIOMATIC® Electronic Bore Gages by Brown & Sharpe

Range	B&S Tool No.	Price ($)
0.600 in. to 0.800 in. (12–15 mm)	62-32005	1492.60
0.800 in. to 1.000 in. (20–25 mm)	62-32006	1501.20
1.000 in. to 1.200 in. (25–30 mm)	62-32007	1515.40
1.200 in. to 1.600 in. (30–40 mm)	62-32008	1578.10
1.600 in. to 2.000 in. (40–50 mm)	62-32009	1612.40
2.000 in. to 2.400 in. (50–60 mm)	62-32010	1639.70
2.400 in. to 2.800 in. (60–70 mm)	62-32011	1697.10
2.800 in. to 3.200 in. (70–80 mm)	62-32012	1695.90
3.200 in. to 3.600 in. (80–90 mm)	62-32013	1702.10
3.600 in. to 4.000 in. (90–100 mm)	62-32014	1759.60
4.000 in. to 4.400 in. (100–110 mm)	62-32015	1759.60
4.400 in. to 4.800 in. (110–120 mm)	62-32016	1759.60
TRIOMATIC II Means Sets		
0.600 in. to 1.200 in. (12–30 mm)	62-32001	2161.00
1.200 in. to 2.400 in. (30–60 mm)	62-32002	2663.80
2.400 in. to 3.600 in. (60–90 mm)	62-32003	2509.30
3.600 in. to 4.800 in. (90–120 mm)	62-32004	2752.70

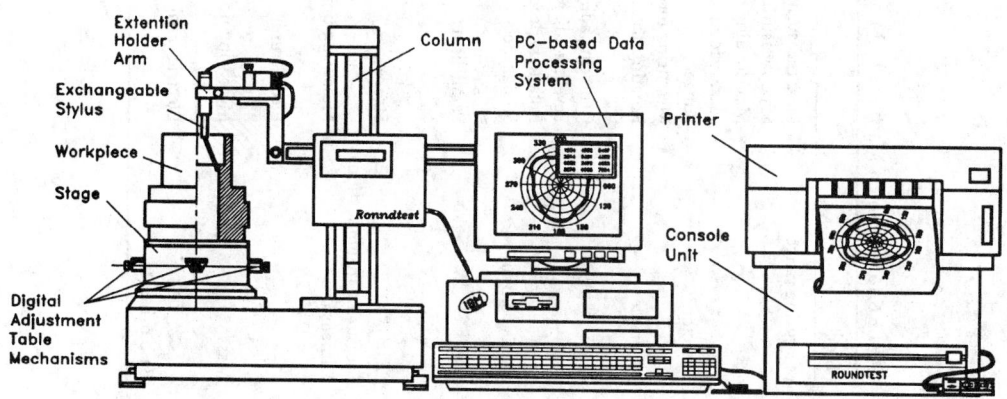

FIGURE 6.59 Form measuring machine.

TABLE 6.15 Rountest Machines by Mitutoyo/MIT Corp.

Model		RA-112	RA-334	RA-434	RA-661
Measuring range	Max. measuring dia.	11 in. (280 mm)	11.8 in. (300 mm)	11.8 in.(300 mm)	
	Max. measuring height	8.6 in. (220 mm)	27.6 in. (700 mm)	13.8 in. (350 mm)	
	Max. loading dia.	—	21.7 in. (550 mm)	20.4 in. (520 mm)	
	Max. loading capacity	22 lb (10 kg)	66.1 lb (30 kg)	132 lb (60 kg)	
Detector	Range	±0.01 in. (±250 μm)	±0.012 in. (± 300 μm)	±0.012 in. (± 300 μm)	
	Measuring force		7–10 gf	7–10 gf	
Turntable	Rotating accuracy	(1.6 + 0.3H) μinch	(1.6 + 0.6H) μinch	(1.6 + 0.6H) μinch	
		(0.04 + 0.3H) μm	(0.04 + 0.6H) μm	(0.04 + 0.6H) μm	
	Centering adj. range	±0.08 in. (±2 mm)	±0.2 in. (±5 mm)	±0.2 in. (±5 mm)	
	Leveling adj. range		±1°	±1°	
	Rotating speed		6 rpm	2, 4, 6 rpm	
Z-axis column	Straightness	—	—	40 μinch/7.9 in.	8 μinch/8 in.
				1 μm/200 mm	0.2 μm/200 mm
	Parallelism	—	—	120 μinch/7.9 in.	
				80 μinch/13.8 in.	
				3 μm/200 mm	2 μm/200 mm
	Stroke	10 in. (25 mm)		18.9 in. (480 mm)	13.8 in.
Measuring magnifications		100—20,000×		100–50,000×	100–100,000×
Dimensions W × D × H	Measuring unit	9.9 × 15.9 × 21.5 in.		24.4 × 19.7 × 36.2 in.	28.7 × 23.2 × 62.2 in.
		251 × 404 × 576 mm		620 × 500 × 920 mm	730 × 590 × 1580 mm
	Electric unit	11.4 × 11.8 × 3.6 in.		9.8 × 16.1 × 13 in.	30.7 × 23.3 × 28.8 in.
		290 × 300 × 92 mm		250 × 410 × 330 mm	780 × 592 × 732 mm
Mass	Measuring unit	61.7 lb (28 kg)	276 lb (125 kg)		298 lb (135 kg)
					770 lb (350 kg)
	Electric (analyzer) unit	11 lb (5 kg)		24.2 lb (11 kg)	110 lb (50 kg)
Base price ($)		17,000	45,000	60,000	60,000

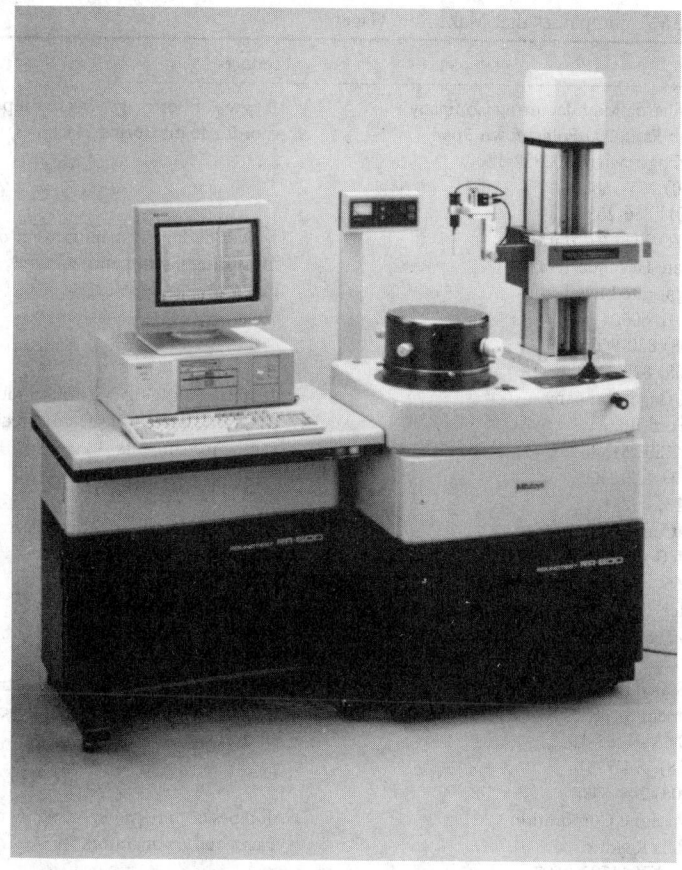

FIGURE 6.60 Fully automatic form measuring machine RA-600 (Courtesy of MITUTOYO/MTI Corporation).

at a tolerance of 1 μm. The GAGE R AND R standards will help parts makers identify the best gage for each application and at each tolerance.

The growing acceptance of ISO 9004 as an international quality philosophy is creating significant changes for manufacturers. The most important change that ISO 9004 will create is the need for international GAGE R AND R capability standards and the guarantee of a level playing field. Previously, gage manufacturers and parts makers had many different sets of measuring criteria. Now, in establishing GAGE R AND R capability standards along with ISO 9004 standards, there will be an international set of standards that applies to everyone. A set of measurements from Europe will mean the same as a set from North America or a set from Asia.

The factors that influence GAGE R AND R capability when gaging a bore include:

1. Variation resulting from the bore gage: This variation includes linearity, repeatability, stability, and calibration.
2. Variation resulting from the operation using a bore gage: This variation includes repeatability by an individual operator using the gage and reproducibility by different operators using the same gage.
3. Variation resulting from the production line: Part surface finish, application setup errors, and temperature changes cause this variation.

Dyer company (Dyer, Lancaster, PA 17604) reported the following ranges of these variations for its 830 and 230 series bore gages: linearity ±0.001 mm (±0.000040 in.); repeatability <±0.00025 mm (±0.000010 in.); stability — bore gages are made from a special alloy steel that resists temperature changes

TABLE 6.16 Companies that Make Bore Gages

Company	Products
Brown & Sharpe Manufacturing Company Precision Park, 200 Frenchtown Road North Kingstown, RI 02852-1700 Tel: (800) 283-3600 Fax: (401) 886-2553	All types of bore gages (from calipers to coordinate measuring machines)
MITUTOYO/MTI Corporation Corporate Headquarters 965 Corporate Blvd. Aurora, IL 60504 Tel: (708) 820-9666 Fax: (708) 820-1393	The world's largest manufacturer of precise measuring instruments; all types of bore gages
Deadborn Gage Company 32300 Ford Road Garden City, MI 48135 Tel: (313) 422-8300 Fax: (313) 422-4445	Air spindles, gaging systems, modular electronic and air-electronic precision gaging instruments
The Dyer Company 1500 McGovernville Road Box 4966 Lancaster, PA 17604 Tel: (800) 631-333 Fax: (717) 569-6721	Dimensional mechanical and electronic bore gages for use in the shop floor
The L.S. Starrett Co. 121 Cresent Street Athol, MA 01331-1915 Tel: (508) 294-3551 Fax: (508) 249-8495	Dimensional mechanical and electronic bore gages, coordinate measuring machines, optical comparators
Flexbar Machine Corporation 250 Gibbs Road Islandia, NY 11722-2697 Tel: (800) 883-5554 Fax: (516) 582-8487	All types of bore gages, including deep-hole gages and comparators
Federal Products Co. 1144 Eddy Street P.O. Box 9400 Providence, RI 02940 Tel: (800) FED-GAGE, (401) 784-3100 Fax: (401) 784-3246 Internet: www.gardnerweb.com/federal/index.html	All types of bore gages: indicator gages, air gages, electronic gaging products, dimensional standards, geometry measurement, laboratory gages
Comtorgage Corporation 58 N.S. Industrial Drive Slatersville, RI 02876 Tel: (401) 765-0900 Fax: (401) 765-2846	Dial indicating — expansion plug system dedicated to one specific size. Customized bore gages
Marposs Corporation Auburn Hills, MI 48326-2954 Tel: (800) 811-0403 Fax: (810) 370-0990	In-process and post-process gaging for grinders and lathes, automatic gaging and special gaging machines
Rank Taylor Hobson Inc. 2100 Golf Road, Suite 350 Rolling Meadows, IL 60008 Tel: (800) 464-7265 Fax: (847) 290-1430	Metrology systems for precision metalworking and high tolerance engineering products
Sterling Mfg. & Engineering Inc. 7539 19th Mile Road Sterling Heights, MI 48314 Tel: (800) 373-0098 Fax: (810) 254-3601	Custom-designed dimensional gages including air, manual, electronic and computerized systems

TABLE 6.16 (continued)   Companies that Make Bore Gages

Company	Products

TRAVERS™ TOOL Co. Inc.
  128-15 26th Ave. P.O. Box 541 550
  Flushing, NY 11354-0108
  Tel: (800) 221-0270
  Fax: (800) 722-0703
	Global Sales Distributer
	  1 or 2 day delivery at ground rates

or "hand heat." The sealed construction ensures performance reliability even under damp or oily conditions; repeatability by an individual operator — the gages are non-tipping and self-centering in both the axial and radial planes of the bore. The operator can keep his/her hands on or off the gage. The operator has no influence when measuring the bore. The automatic alignment of the bore gage results in a highly reproducible reading.

Defining Terms

Dimension: A numerical value expressed in appropriate units of measure and indicated on drawings with lines, symbols, and notes to define the geometrical characteristics of an object.

Tolerance: The total permissible variation of size, form, or location.

Fit: The general term used to signify the range of tightness which may result from the application of a specific combination of allowances and tolerances in the design of mating parts.

ISO System of Limits and Fits: A standardized system of limits and fits, a group of tolerances considered as corresponding to the same level of accuracy for all basic sizes.

Gage: A device for investigating the dimensional fitness of a part for specified function.

Gaging: A process of measuring manufacturing materials to assure the specified uniformity of size and contour required by industries.

References

1. J. W. Greve (ed.), *Handbook of Industrial Metrology*, ASTM Publications Committee, Englewood Cliffs, NJ: Prentice-Hall, 1967.
2. M. F. Spotts, *Dimensioning and Tolerancing for Quantity Production*, Englewood Cliffs, NJ: Prentice-Hall, 1983.
3. E. R. Friesth, *Metrication for Manufacturing*, New York: Industrial Press Inc., 1978.
4. J. G. Nee, *Fundamentals of Tool Design*, 4th ed., Dearborn, MI: Society of Manufacturing Engineers, 1998.

Further Information

F. T. Farago, *Handbook of Dimensional Measurement*, 3rd ed., New York: Industrial Press Inc., 1994, provides extensive definitions of terms, methods, and measuring setups.

J. Dally, *Instrumentation for Engineering Measurements*, 2nd ed., New York: Wiley, 1993, provides detailed characteristics of measuring tools and machines.

R. S. Sirohi, *Mechanical Measurements*, 3rd ed., New York: Wiley, 1991, provides a textbook presentation of the theory of mechanical measurements.

J. D. Meadows, *Geometric Dimensioning and Tolerancing: Applications and Techniques for Use in Design, Manufacturing, and Inspection*, New York: M. Dekker, 1995, presents basic rules and procedures for good dimensioning practice, providing the background that aids in the solution of dimensioning problems as they arise in day-to-day work.

6.7   Time-of-Flight Ultrasonic Displacement Sensors

Teklic Ole Pedersen and Nils Karlsson

Ultrasound is an acoustic wave with a frequency higher than the audible range of the human ear, which is 20 kHz. Ultrasound can be within the audible range for some animals, like dogs, bats, or dolphins. In the years around 1883, Sir Francis Galton performed the first known experiments with whistles generating ultrasound. Many decades later, people started to find ultrasound applications in engineering, medicine, and daily life. The basic principle for the use of ultrasound as a measurement tool is the *time-of-flight technique*. The pulse-echo method is one example. In the pulse-echo method, a pulse of ultrasound is transmitted in a medium. When the pulse reaches an another medium, it is totally or partially reflected, and the elapsed time from emission to detection of the reflected pulse is measured. This time depends on the distance and the velocity of the sound. When sound travels with a known velocity *c*, the time *t* elapsed between the outgoing signal and its incoming echo is a measure of the distance *d* to the object causing the echo.

$$d = \frac{ct}{2} \tag{6.82}$$

Figure 6.61 shows a simple pulse-echo system. The transmitter and the receiver could be the same device, but they are separated for clarity in this figure.

The oscillator generates an electric signal with a typical frequency of 40 kHz. This electric signal is transformed into mechanical vibrations of the same frequency in the transmitter. These vibrations generate sound waves that are reflected by the object. The reflected sound echo causes an electric signal in the receiver. For precise measurements, the speed of sound is a crucial parameter. A typical value in air at 1 atm pressure and room temperature is 343 m s^{-1}, but the speed of sound is influenced by air pressure, air temperature, and the chemical composition of air (water, CO_2, etc.). For example, the speed of sound is proportional to the square root of absolute temperature. Measuring distances in an environment with large temperature gradients can result in erroneously calculated distances. As an advantage, ultrasound waves are robust against other disturbances such as light, smoke, and electromagnetic interference [1–4].

Physical Characteristics of Sound Waves

Sound is a vibration in matter. It propagates as a longitudinal wave, i.e., the displacement in the material is in the direction of the sound wave propagation. A plane wave that propagates in the *x* direction can be described by

$$\Delta x = A \sin \omega \left(t - \frac{x}{c} \right) \tag{6.83}$$

where *A* is the amplitude, $\omega = 2\pi f$, *f* being the frequency of the wave and Δx is the displacement of a particle at time *t* at the position *x*.

The velocity of sound depends on the medium in which it propagates. In a homogeneous and isotropic solid, the velocity depends on the density ρ and the modulus of elasticity *E* according to Equation 6.84.

$$c = \sqrt{\frac{E}{\rho}} \tag{6.84}$$

In a liquid, the velocity depends on the density and the adiabatic compressibility *K*, Equation 6.85.

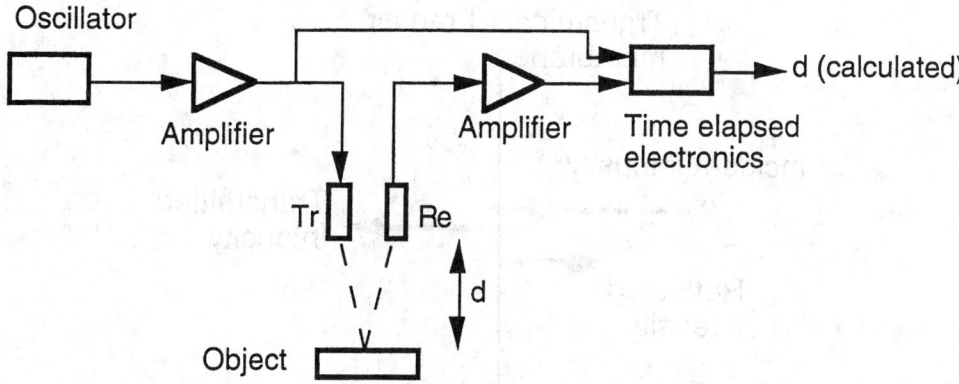

FIGURE 6.61 Principle of a pulse-echo ultrasound system for distance measurements (Tr = transmitter, Re = receiver).

$$c = \sqrt{\frac{1}{K\rho}} \tag{6.85}$$

In gases, the velocity of sound is described by Equation 6.86. Here g represents the ratio of the specific heat at constant pressure (c_p) to the specific heat at constant volume (c_v), p is pressure, R is the universal gas constant, T is the absolute temperature, and M is the molecular weight.

$$c = \sqrt{\frac{gRT}{M}} = \sqrt{\frac{c_p}{c_v} \frac{p}{\rho}} \tag{6.86}$$

An important quantity is the *specific acoustic impedance*. It is, in general, a complex quantity but in the far field (Figure 6.63), the imaginary component disappears, leaving a real quantity. This real quantity is the product of the density ρ and the sound speed c in the medium. This product is called the characteristic impedance R_a (Equation 6.87).

$$R_a = \rho c \tag{6.87}$$

The characteristic impedance is thus independent of the sound frequency.

An acoustic wave has an intensity I (rate of flow of energy per unit area), which can be expressed in watts per square meter (W m^{-2}). A usually unwanted phenomenon arises when the sound wave has to pass from one medium with characteristic impedance R_1 to another medium with characteristic impedance R_2. If R_1 and R_2 have different values, a part of the wave intensity will reflect at the boundary between the two media (see Figure 6.61 and 6.62). The two media are said to be mismatched, or poorly coupled, if a major part of the wave intensity is reflected and a minor part is transmitted. The relative amounts of reflected and transmitted wave intensities can be defined by:

$$\text{Reflection coefficient} = \frac{I_{\text{refl}}}{I_{\text{incident}}} \tag{6.88a}$$

$$\text{Transmission coefficient} = \frac{I_{\text{trans}}}{I_{\text{incident}}} \tag{6.88b}$$

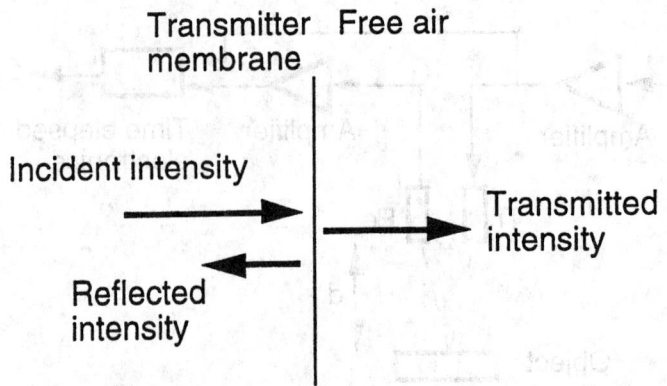

FIGURE 6.62   Reflection and transmission of a sound wave at the interface between media of different characteristic impedances.

It can be shown [1] that these coefficients have simple relations to the previously mentioned characteristic impedances.

$$\text{Reflection coefficient} = \frac{(R_1 - R_2)^2}{(R_1 + R_2)^2} \tag{6.89a}$$

$$\text{Transmission coefficient} = \frac{4R_1R_2}{(R_1 + R_2)^2} \tag{6.89b}$$

The practical importance of the acoustic impedance is realized when the ultrasonic pulse-echo system shown in Figure 6.61 is considered. First, the electric energy is converted into mechanical vibrations of a membrane in the transmitter. Second, the vibrations (the sound wave) have to pass through the boundary between the membrane (usually a solid material) and free air. Because the transmitter membrane and the free air have different characteristic impedances, much of the acoustic intensity is reflected (Figure 6.62).

The transmitted ultrasound in free air will first propagate in a parallel beam (near field of the transducer); but after a distance L, the beam diverges (the far field of the transducer). See Equation 6.90 and Figure 6.63.

$$L \approx \frac{D^2}{4\lambda} \tag{6.90}$$

D is the diameter of the circular transmitter and λ is the wavelength of the ultrasound.

The sound intensity in the near field is complicated due to interference effects of sound originating from different parts of the transducer membrane. In the far field, the intensity is approximately uniform and the beam spread follows:

$$\sin\beta = 1.22\frac{\lambda}{D} \tag{6.91}$$

where β is the half lobe angle.

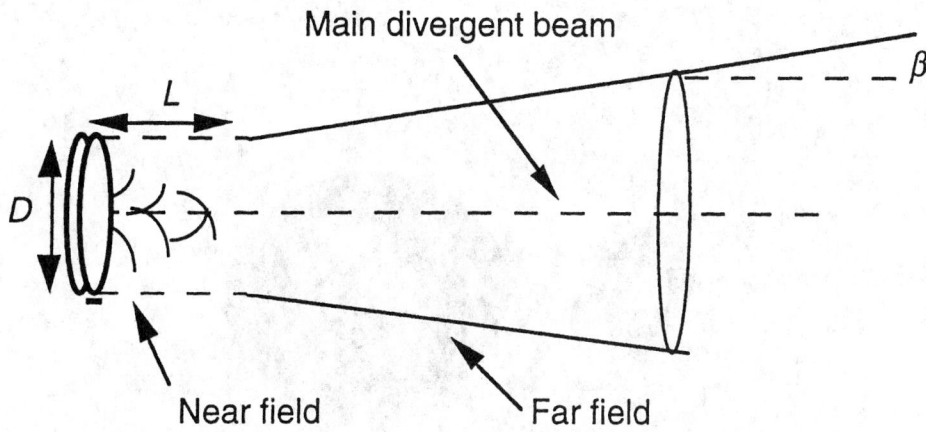

FIGURE 6.63   Illustration of the ultrasound beam in the near field and the far field of the transducer.

To get a narrow beam, the transmitter membrane diameter must be large with respect to the wavelength. High-frequency ultrasound cannot be the general solution as ultrasound of a high frequency is absorbed faster than ultrasound of a low frequency [1–4].

Ultrasound Transducers

Most ultrasound transducers convert electric energy to mechanical energy and vice versa. The most common types of in-air transducers are [5–7]:

1. Mechanical
2. Electromagnetic
3. Piezoelectric
4. Electrostatic
5. Magnetostrictive

The simplest type, mechanical transducers such as whistles and sirens, are used up to approximately 50 kHz. This type works only as a transmitter.

Electromagnetic transducers such as loudspeakers and microphones can be used for ultrasonic wave generation, but they are mainly suited for lower frequencies.

The piezoelectric transducer (Figure 6.64) is more suitable for use in ultrasonics and is quite common. It uses a property of piezoelectric crystals: they change dimensions when they are exposed to an electric field. When an alternating voltage is applied over the piezoelectric material, it changes its dimensions with the frequency of the voltage. The transducer is mainly suited for use at frequencies near the mechanical resonance frequency of the crystal. The piezoelectric transducer can be both a transmitter and a receiver: when a piezolectric material is forced to vibrate by a sound pulse, it generates a voltage. Some natural crystals, such as quartz, are piezoelectric. Ceramics can be polarized to become piezoelectric; so can some polymers like PVDF (polyvinylidene fluoride). Polymers are suitable as transducers in air since their acoustic impedance is low [8–10] compared with other standard piezoelectric materials.

The electrostatic transducer (Figure 6.64) is a plate capacitor with one plate fixed and the other free to vibrate as a membrane. When a voltage is applied between the plates, the electrostatic forces tend to attract or repel the plates relative to each other depending on the polarity of the voltage. This transducer can be used both as a transmitter and a receiver [11].

The magnetostrictive transducer is based on the phenomenon of magnetostriction, which means that the dimensions of a ferromagnetic rod change due to the changes of an externally applied magnetic field. This transducer can also act as both a receiver and a transmitter.

FIGURE 6.64   Ultrasonic transducers piezoelectric (left) and electrostatic (right).

Principles of Time-of-Flight Systems

There are several techniques for ultrasonic range measurements [12–15].

The previously described *pulse echo method* is the simplest one. Usually, this method has a low signal-to-noise ratio (SNR) because of the low transmitted energy due to the short duration of the pulse. Multireflections are detectable.

In the *phase angle method*, the phase angle is measured between the continuous transmitted signal and the continuous received signal and is used as a measure of the distance. The method is relatively insensitive to disturbances. Multireflections are not detectable in a meaningful way. When the distance is longer than one wavelength, another method must be used to monitor the distance.

The *frequency modulation method* uses transmitted signals that are linearly frequency modulated. Thus, detected signals are a delayed replica of the transmitted signal at an earlier frequency. The frequency shift is proportional to the time-of-flight. The method is robust against disturbing signals, and multireflections are detectable.

The *correlation method* (Figure 6.65) determines the cross-correlation function between transmitted and received signals. When the transmitted signal is a random sequence, i.e., white Gaussian noise, the cross-correlation function estimates the impulse response of the system, which, in turn, is a good indicator of all possible time delays. The method is robust against disturbances, and multireflections are detectable.

Industrial acoustic noise can affect the received signals in an ultrasound time-of-flight system. The noise can be generated from leaking compressed air pipes, noisy machines, or other ultrasonic systems. This noise is not correlated with the relevant echo signals of the transmitted noise and can therefore be eliminated by the use of correlation methods. Disturbances correlated with the relevant echo signal (e.g., unwanted reflections) will not be eliminated by the use of correlation methods.

The impulse response $h(t, t_0)$ is used as a sensitive indicator of time delay between transmitted signal at time t_0 and received signal at time t. The impulse response is given by [14]:

$$h(t, t_0) = F^{-1}\left[\frac{S_{xy}}{S_{xx}}\right] \tag{6.92}$$

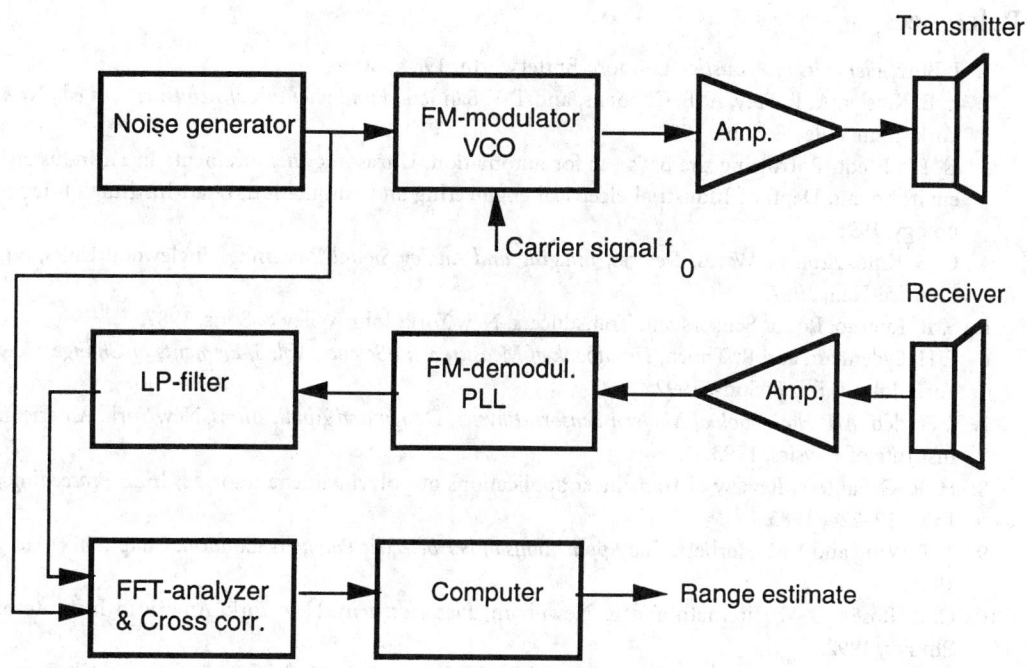

FIGURE 6.65 Diagram of a correlation-based time-of-flight system.

where F^{-1} is the inverse Fourier transform, x is the transmitted signal, y is the received signal, $S_{xy}(f)$ is the cross-spectral density function [the Fourier transform of the cross-correlation function of the transmitted signal $x(t)$ and the received signal $y(t)$] and S_{xx} is the power density function (the Fourier transform of the auto-correlation function of the transmitted signal).

To analyze the transfer channel by data acquisition requires a high sampling rate, theoretically at least two times the highest frequency component in the received signal (and in practice as high as 10 times the highest frequency). One way to reduce the sampling rate is to first convert the signal from its bandpass characteristics around a center frequency f_0 (approx. 50 kHz) to lowpass characteristics from dc to $B/2$, where B is the appropriate bandwidth. The accuracy of the range estimate, and hence the time interval $t - t_0$, can be improved by processing the estimate of the impulse response $h(t - t_0)$ with a curve-fitting (least square) method and digital filtering in a computer. A block diagram of a correlation-based time-of-flight system is shown in Figure 6.65. Further details and complete design examples can be found in the literature [12–15].

Table 6.17 lists some advantages and drawbacks of the described time-of-flight methods.

TABLE 6.17 Advantages and Disadvantages of Time-of-Flight Methods

Method	Main advantage	Main disadvantage
Pulse echo method	Simple	Low signal-to-noise ratio
Phase angle method	Rather insensitive to disturbances	Cannot be used directly at distances longer than the wavelength of the ultrasound
Frequency modulation method	Robust against disturbances; multireflections detectable	Can give ambigous results measurements on long and short distances can give the same result (compare with phase angle method)
Correlation method	Very robust against disturbances	Make relatively high demands on hardware and/or computations

References

1. J. Blitz, *Elements of Acoustics*, London: Butterworth, 1964.
2. L. E. Kinsler, A. R. Frey, A. B. Coppens, and T. V. Sanders, *Fundamentals of Acoustics, 3rd ed.*, New York: John Wiley & Sons, 1982.
3. G. Lindstedt, Borrowing the bat's ear for automation. Ultrasonic measurements in an industrial environment, Dept. of Industrial electrical engineering and automation, Lund Institute of Technology, 1996.
4. G. S. Kino, *Acoustic Waves: Devices, Imaging and Analog Signal Processing*, Englewood Cliffs, NJ: Prentice-Hall, 1987.
5. S. R. Ruocco, Robot Sensors and Transducers, New York: John Wiley & Sons, 1987.
6. P. H. Sydenham and R. Thorn, *Handbook of Measurement Science, Vol. 3, Elements of Change*, New York: John Wiley & Sons, 1992.
7. J. Fraden, *AIP Handbook of Modern Sensors, Physics, Design and Applications*, New York: American Institute of Physics, 1993.
8. H. R. Gallantree, Review of transducer applications of polyvinylidene fluoride, *IEEE Proceedings*, 130, 219–224, 1983.
9. T. T. Wang and J. M. Herbert, *The Applications of Ferroelectric Polymers*, London: Chapman & Hall, 1988.
10. C. Z. Rosen, B. V. Hiremath and R. Newnham, *Piezoelectricity*, New York: American Institute of Physics, 1992.
11. P. Mattila, F. Tsuzuki, H. Väätäjä, and K. Sasaki, Electroacoustic Model for Electrostatic Ultrasonic Transducers with V-Grooved Backplates, in *IEEE Trans. Ultrasonics, Ferroelectrics and Frequency Control*, Vol. 42, No. 1, January, 1995.
12. P. Holmberg, Instrumentation, Measurements and Applied Signal Processing for Industrial Robot Applications, Ph.D. dissertation No. 334, Dept. of Physics and Measurement Technology, Linköping University, 1994.
13. J. A. Kleppe, *Engineering Applications of Acoustics*, Boston, 1989.
14. J. S. Bendat and A. G. Piersol, *Random Data Analysis and Measurement Procedures, 2nd ed.*, New York: John Wiley & Sons, 1986.
15. P. Holmberg, Robust ultrasonic range finder — an FFT analysis, *Meas. Sci. Technol.*, 3, 1025–1037, 1992.

6.8   Optical Encoder Displacement Sensors

J. R. René Mayer

The detection of angular and linear motion is a key function in a multitude of systems such as machine tools, industrial robots, a variety of instruments, computer mice, etc. Although they are one of many techniques capable of such measurements, the ease with which they are interfaced to digital systems has made them very popular.

Optical encoders are used to measure either angular or linear positions. Those used for angular detection are commonly called rotary or shaft encoders, since they usually detect the rotation of a shaft. Optical encoders encompass a variety of devices, all of which use light as the means to transform movement into electrical signals. All devices have two basic building blocks: a main grating and a detection system. It is the position of one with respect to the other that is detected. The main grating represents the measurement standard. For linear measurements, the main grating, commonly called the scale, is one or more sets of parallel lines of constant or specially coded pitch supported by a substrate. Similarly, a rotary encoder has a grating with radial lines on a disk.

Both linear and rotary encoders can, in principle, be absolute or incremental, although in practice, linear absolute encoders employing optical principles are quite uncommon and have drastically limited

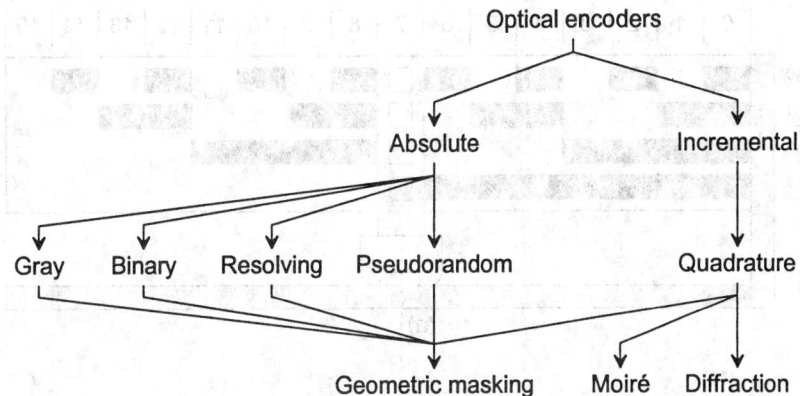

FIGURE 6.66 Classifications of optical encoders based on (1) the nature of the final information provided, (2) the type of signals generated, and (3) the technology used to generate the signals.

performance characteristics (accuracy, resolution, and/or maximum operating speed). Figure 6.66 shows a simplified classification of optical encoders. This classification refers to the nature of the information generated. The incremental encoder detects movement relative to a reference point. As a result, some form of reference signal is usually supplied by the encoder at a fixed position in order to define a reference position. The current position is then incremented (or decremented) as appropriate. Multiple reference marks can also be used, where the distance between successive marks is unique so that as soon as two successive marks have been detected, it becomes possible to establish absolute position from then on. The reference point can also be mechanical. Should power be lost or a signal transmission error occur, then the absolute position is lost and the encoder must return to one or more reference points in order to reset its counters. Unfortunately, a loss of count may not be detected until a reference point is reaccessed. Furthermore, reading errors may accumulate. On the other hand, absolute encoders produce a set of binary signals from which the absolute position can be deduced without the knowledge of the previous motion history. The current position is known right from powering-on. In the case of absolute rotary encoders, single and multiturn devices are available. Multiturn devices use an internal mechanical transmission system to drive a second grating that serves as turn counter.

Most incremental encoders use quadrature signals as output to carry the motion information. Some encoders use one square-wave signal, which is used for position in one direction only. Also, this single square wave can be fed into either a PLC (programmable logic controller) or another electronic interface that converts this signal to a rate or RPM (revolution per minute) for speed indication. However, whenever bidirectional operation is required, quadrature signals are necessary. Quadrature signals come in analog or digital form. The analog form consists simply of a sine and a cosine signal. The number of sinusoidal cycles per unit change of the measured variable (a revolution or 360° for a rotary encoder) determines the basic resolution of the encoder prior to interpolation. The digital form consists of two square-wave trains, 90° (often called electrical degree) out of phase. The 90° phase lag is indispensable in order to detect the motion direction and hence increment or decrement the position counter accordingly. The main optical techniques to generate the quadrature signals are geometric masking, Moiré fringes, and diffraction based. For linear encoders, the basic resolution is related to the distance traveled by the grating in order for the encoder to produce one full quadrature cycle. For rotary encoders, the basic resolution is usually described as the number of quadrature cycles per turn. The resolution of an encoder system can be increased by electronic means. With analog quadrature signals, it is possible to interpolate within each quadrature cycle. The limit of the interpolation factor depends on the quality (mark space, quadrature separation, and jitter) of the basic signals. With square-wave signals, multiplication by a factor of two or four is easily achieved. Increasing the resolution in this manner does not, however, improve the trueness, often called accuracy (or linearity) of the measurement.

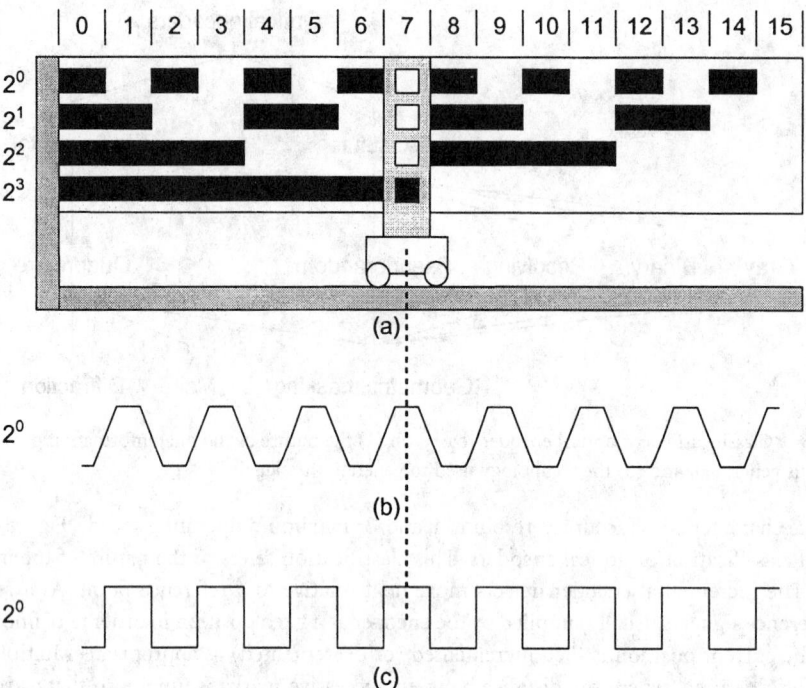

FIGURE 6.67 (*a*) Absolute encoders using a natural binary code of four digits. Four tracks are required. The moving read head has four apertures and is shown in position 7 along the scale. (*b*) The output of the read head aperture corresponding to the least significant track. It represents the proportion of light area covering the aperture. (*c*) The binary digit obtained after squaring the raw output signal.

Absolute encoders are classified according to the type of code used. The main four codes are Gray, binary (usually read by vee-scan detection), optical resolving, and pseudorandom. All absolute encoders use geometric masking to generate the code.

Encoder Signals and Processing Circuitry

Absolute Encoders

Direct Binary

Figure 6.67(*a*) illustrates the concept of an absolute linear optical encoder using a direct binary encoded scale. The fixed scale has n tracks (here n is 4), each providing one bit of a direct binary number. The lowest track (first track from the center of the disk for a rotary encoder) is the most significant digit and has a weight, 2^{n-1} (here 2^3), while the upper track is the least significant digit with a weight 2^0. The track providing the least significant digit has 2^{n-1} cycles of light and dark records, while the most significant track has 2^0 or 1 such cycle. For each track, the moving read head has a readout unit consisting of a light source, a mask, and a photodetector. Figure 6.67(*b*) shows the output from the photodetector, which represents the total intensity of light reaching its surface. As the mask passes over a clear region of the grating, the photodetector output increases, and then decreases. In theory, a truncated triangular wave is obtained, which can easily be converted to a square wave (Figure 6.67(*c*)) by a suitably chosen thresholding level. The result is a high or 1 for a light record and a low or 0 for a dark one. The position, in base 10, corresponding to the reading head position in Figure 6.67 is

$$1 \times 2^0 + 1 \times 2^1 + 1 \times 2^2 + 0 \times 2^3 = 7 \tag{6.93}$$

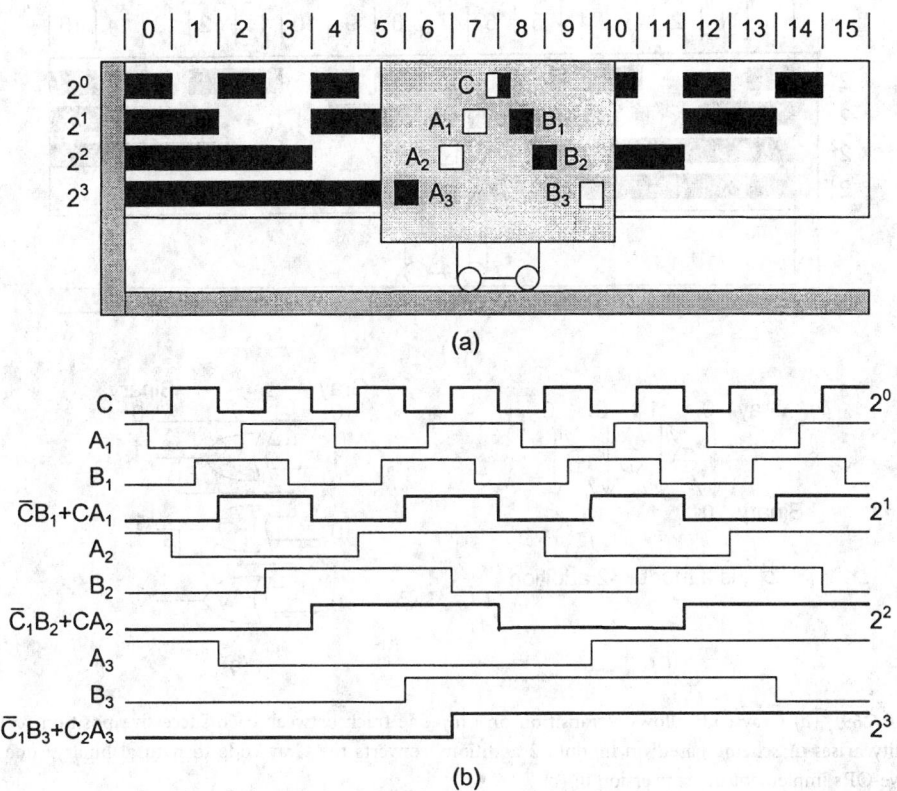

FIGURE 6.68 The vee-scan configuration of reading units in (*a*) removes the ambiguity associated with a natural binary scale. Simple combinational logic is then used to generate the natural binary readout in (*b*).

The code configuration just described is not suitable for practical use because some transitions require that two or more bit values change simultaneously. For example, from position 7 to position 8, all bits change values. Unless the change is simultaneous, an incorrect position readout results at some position. This would require that the scale is geometrically perfect, that the read head be perfectly aligned with the scale, and that the electronics are perfectly adjusted and stable over time. This problem is solved either by the use of a vee-scan detection method or the use of a unit-distance code such as the Gray code.

Vee-scan

The vee-scan method uses a V-shape pattern of readout units that removes the potential reading ambiguity of direct binary scales. Stephens et al. [1] indicate the read points at which transitions are detected in Figure 6.68(*a*). They also describe the conversion of the thresholded output signals to binary code using combinational logic. The primary advantage is that the location tolerance of the transition point of each reading unit need only be ±1/8 of the cycle length for that particular track. For example, ±45° for the most significant track of a rotary encoder disk. Figure 6.68(*b*) shows a direct binary word obtained through logic combinations of the vee-scan readings.

Gray Code

The use of vee-scan requires additional reading heads as well as processing electronics. The Gray code is a unit-distance code and so only one bit of data changes between representations of two consecutive numbers or successive positions. This removes the possibility of ambiguous readout. It has the following advantages: (1) it is easily converted to direct binary code, and (2) the finest tracks are twice the width of equivalent direct binary code tracks. Figure 6.69(*a*) shows a Gray code linear scale. Figure 6.69(*b*) shows a scheme for the conversion from Gray code to binary code and proceeds as follows: (1) the most significant bit (msb) of the binary code equals the msb of the Gray-coded number; (2) add (modulo-2)

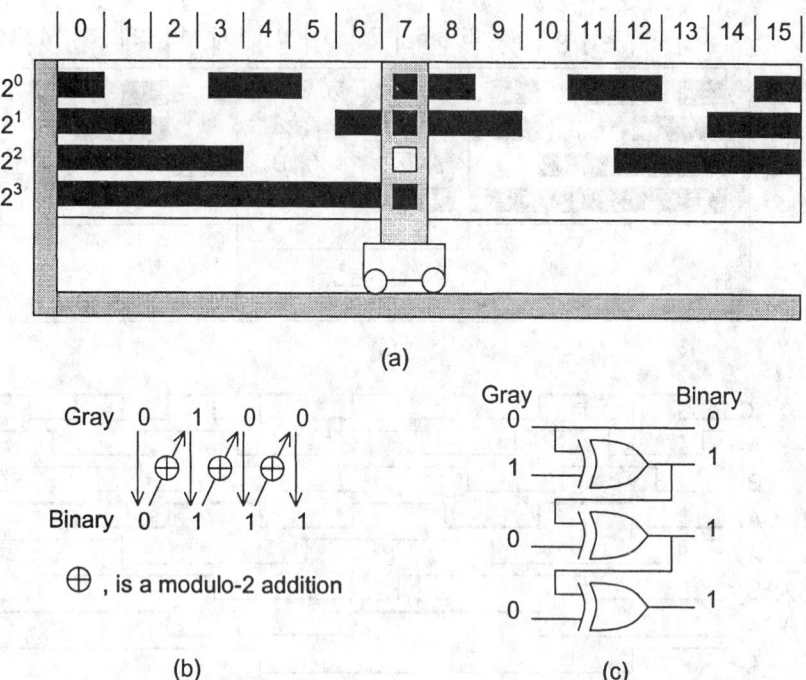

FIGURE 6.69   (*a*) Gray code allows a transition on only one track between each successive position so that no ambiguity arises. A scheme based on modulo-2 additions converts the Gray code to natural binary code in (*b*). Exclusive-ORs implement the conversion in (*c*).

the msb of the binary number to the next significant bit of the Gray-coded number to obtain the next binary bit; and (3) repeat step (2) until all bits of the Gray-coded number have been added modulo-2. The resultant number is the binary equivalent of the Gray-coded number. The modulo-2 addition is equivalent to the action of an exclusive-OR. Figure 6.69(*c*) shows a simple circuit using combinational logic to perform Gray to binary conversion. Sente et al. [2] suggest the use of an external ROM to convert the code where the input coded word is the address and the data is the output coded word. Using a 16-bit ROM for a 12-bit code, Sente et al. [2] suggest using the remaining 4 bits to implement a direction signal. Stephens et al. [1] describe the use of vee-scan with a gray code scale for even better robustness.

Pseudorandom Code

Pseudorandom encoding allows the use of only two tracks to produce an absolute encoder. One track contains the pattern used to identify the current position, while the other is used to synchronize the reading of the encoded track and remove ambiguity problems. A pseudorandom binary sequence (PRBS) is a series of binary records or numbers, generated in such a way that any consecutive series of n digits is unique. Such code is called chain code, and it has the property that the first $n-1$ digits of an n-bit word are identical to the last $n-1$ digits of the previous code word. This allows their partial overlapping on a single track. A PRBS of length 2^n-1 is defined by:

$$\mathrm{XN}(j)\big|\,j=0,1,\ldots,2^{n-1} \tag{6.94}$$

The code can be generated by reading the nth stage of a feedback shift register after j shifts. The register must be initialized so that at least one of the registers is nonzero and the feedback connection implements the formula

$$X(0)=X(n)\oplus c(n-1)\,X(n-1)\oplus\ldots\oplus c(1)\,X(1) \tag{6.95}$$

TABLE 6.18 Shift-Register Feedback Connections
for Generating Pseudorandom Binary Sequences

n	Length	Direct sequence	Reverse sequence
4	15	1, 4	3, 4
5	31	2, 5	3, 5
6	63	1, 6	5, 6
7	127	3, 7	4, 7
8	255	2, 3, 4, 8	4, 5, 6, 8
9	511	4, 9	5, 9
10	1023	3, 10	7, 10
11	2047	2, 11	9, 11
12	4095	1, 4, 6, 12	6, 8, 11, 12
13	8191	1, 3, 4, 13	9, 10, 12, 13
14	16,383	1, 6, 10, 14	4, 8, 13, 14

where the c coefficients are 0 or 1. The feedback registers for which c is 1 are listed in Table 6.18 for values of n from 4 to 14. The following is a PRBS for $n = 4$ with the pseudocode obtained using all registers set to 1 initially, 111101011001000. For a rotary encoder disk, the 15 sectors would have a 24° width.

Petriu [3, 4] describes a possible configuration of a PRBS disk that uses a PRBS track and a synchronization track (Figure 6.70), together with the processing method to reconstitute the position in natural binary. Table 6.18 gives the reverse feedback configuration. The shift register is initially loaded with the current n-tuple. Then the reverse logic is applied recurrently until the initial sequence of the PRBS is reached. At this point, the n-bit counter represents the value of j. For a rotary encoder $(j * 360)/(2^n - 1)$ is the current angular position; whereas for a linear encoder, the position is $j * P$ where P is the scale record length or pitch. Petriu [4] suggests that in order to allow nonambiguous bidirectional reading, $n + 1$ heads are used on the PRBS track. The synchronization track has a series of 0s and 1s in records of the same width as the PRBS track and in phase. There is a $P/2$ shift, where P is the record's length between the A head on the synchronization track and the $n + 1$ read heads on the PRBS track. The $n + 1$ records are updated on a trigger from the A signal. This ensures that the $n + 1$ heads are closely aligned with the PRBS record mid-position. A second read head called B on the synchronization track is shifted by $P/2$ relative to A. A and B are in quadrature, which allows their simultaneous use to generate a motion direction signal. The correct n-tuple, i.e., the lower or upper subset, is selected on the basis of the moving direction and is then converted to natural binary by reverse feedback. Petriu [4] also suggests a simple means of increasing the resolution by a factor of 2 using some additional electronics. He also proposes a scheme to use an arbitrary (not $2^n - 1$) number of sectors, but this requires a third track and some additional correction electronics to handle the last $n - 1$ records of a disk, since these are no longer PRBS patterns. Tomlinson [5] proposes another method for truncation of the PRBS sequence that does not require a third track. Instead, particular codes were removed by applying additional logic in the direct and reverse feedback logic.

Ross and Taylor [6] and Arsic and Denic [7] suggest ways of reducing the number of reading heads by accumulating readings into a shift register so that a minimum of two heads are sufficient to read the PRBS track. However, on start-up, the correct position is not known until the encoder has moved so that all registers have been updated. This type of encoder is therefore not completely absolute because it does not indicate its correct position on start-up. Finally, Arazi [8] mentions the use of a ROM that stores the translation table, as an alternative to a logic circuit.

Optical Resolving
This method has similarities with its electromagnetic counterpart and depends on the generation of a sine and a cosine signal pair per encoder shaft revolution. The resolution and accuracy of this encoder depend on its ability to generate signals that conform to their ideal waveforms and the resolving power of the electronic circuit responsible for performing the rectangular to polar conversion to produce angular

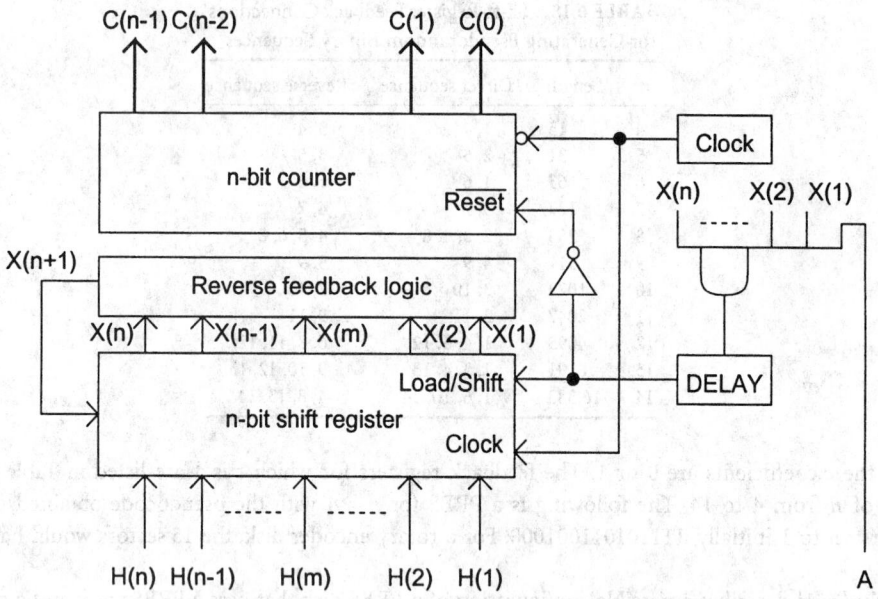

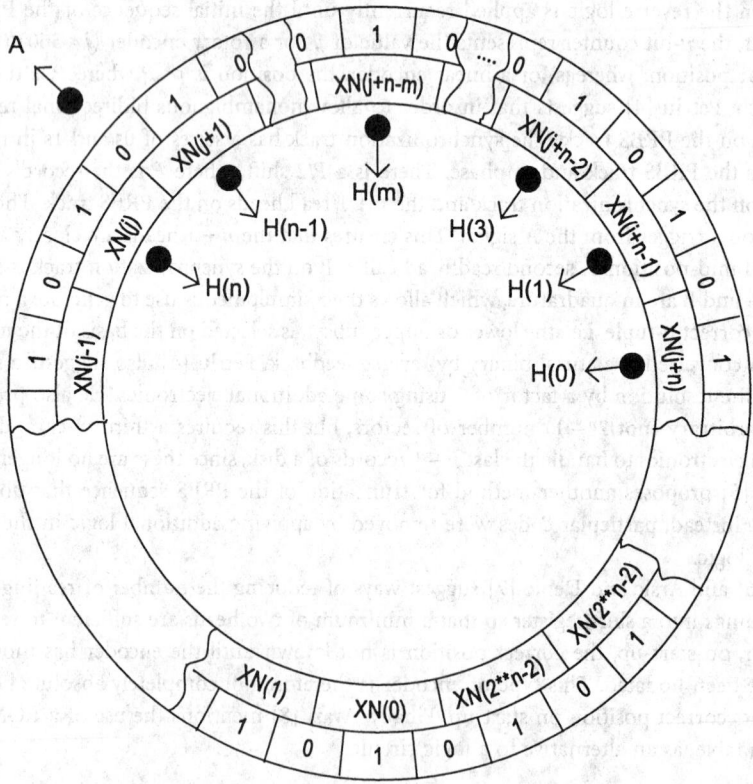

FIGURE 6.70 Pseudorandom shaft encoder with a simple synchronization track to validate the readout from the n-tuple read by the reading heads $H(i)$, $i = n$, ..., 1. The circuit is a simplified code conversion to binary based on the reverse feedback logic. The counter counts the number of steps required to return to the initial sequence.

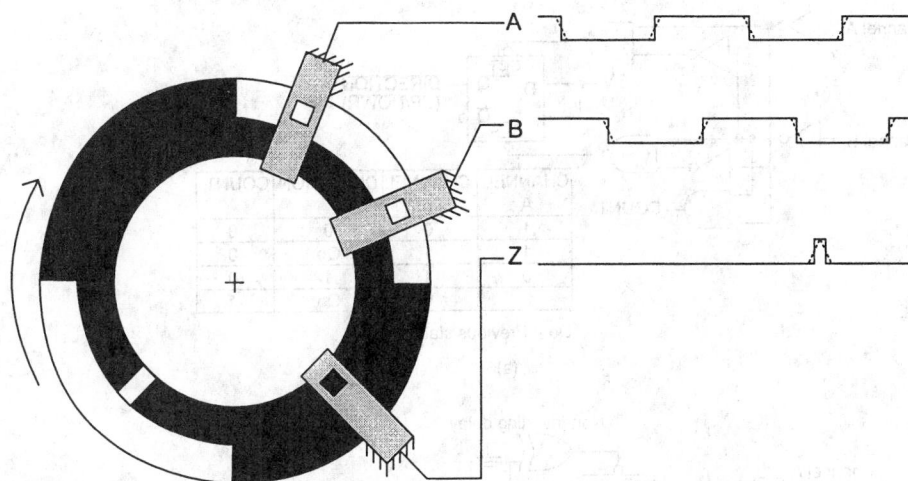

FIGURE 6.71 The lowest possible resolution for a rotary incremental encoder. The outer track provides $n = 2$ cycles of the alternating black and clear records. Two heads A and B displaced by 90° give, after squaring, two square waves in quadrature.

output data. The methods for accomplishing this conversion are similar to those described in the sections on analog quadrature signal interpolation. The code disk of such a device does not incorporate a series of parallel lines of constant pitch, but rather depends on special code tracks having analog profile signal outputs. These can be effected by changing the code tracks' cross-sections or by graduating their optical density around the revolution. In the latter case, a masking reticle need not be employed.

Incremental Encoders Quadrature Signals

Digital Quadrature Signals

Figure 6.71 illustrates the concept of an incremental rotary optical encoder. The shaft-mounted disk has a series of alternating dark and light sectors of equal length and in equal numbers. The dark and light code is detected by a stationary mask with two apertures, A and B, displaced one quarter of a cycle from each other. When a light sector covers a window, a 1 signal is produced, and a 0 results from a dark sector. At a transition, a rising or falling signal occurs. These signals require some pretreatment to square the signals and avoid problems associated with slow movement at the transition positions, resulting in slow rise time and fall time, in the presence of low noise. The resulting cleaned A and B signals are two square waves 90° out of phase and are called quadrature signals. Since the A and B signals have only four possible states, they clearly do not provide a means of distinguishing more than four different locations. As a result, absolute position discrimination is only possible within one quadrature cycle. Instead, the quadrature signals are used to increment or decrement a counter that gives the actual position. The counter is initialized (usually reset) on a z-phase signal produced on a separate track; the innermost track in Figure 6.71. It is also possible to have a number of z-phase signals over the encoder range with a distance coded variation between the z-phase markers. The code can be used to initialize the counter at nonzero positions.

Figure 6.72(*a*) shows a simple circuit from Conner [9] to provide one count per cycle of the quadrature signal. The direction information necessary to decide whether to increment or decrement the counters is also produced. Note that Schmitt triggers provide some amount of hysteresis and redress the falling and rising edges from the encoder. Further circuitry is required to increase the amount of hysteresis in very noisy environments. Additional treatment may be required before feeding these signals to counters. The reason is that if, following a count, there is a small movement in the opposite direction without a reverse count signal being issued followed by a forward motion, then the forward motion produces a second count at the same position. Kuzdrall [10] proposes a more reliable circuit in Figure 6.72(*b*). The

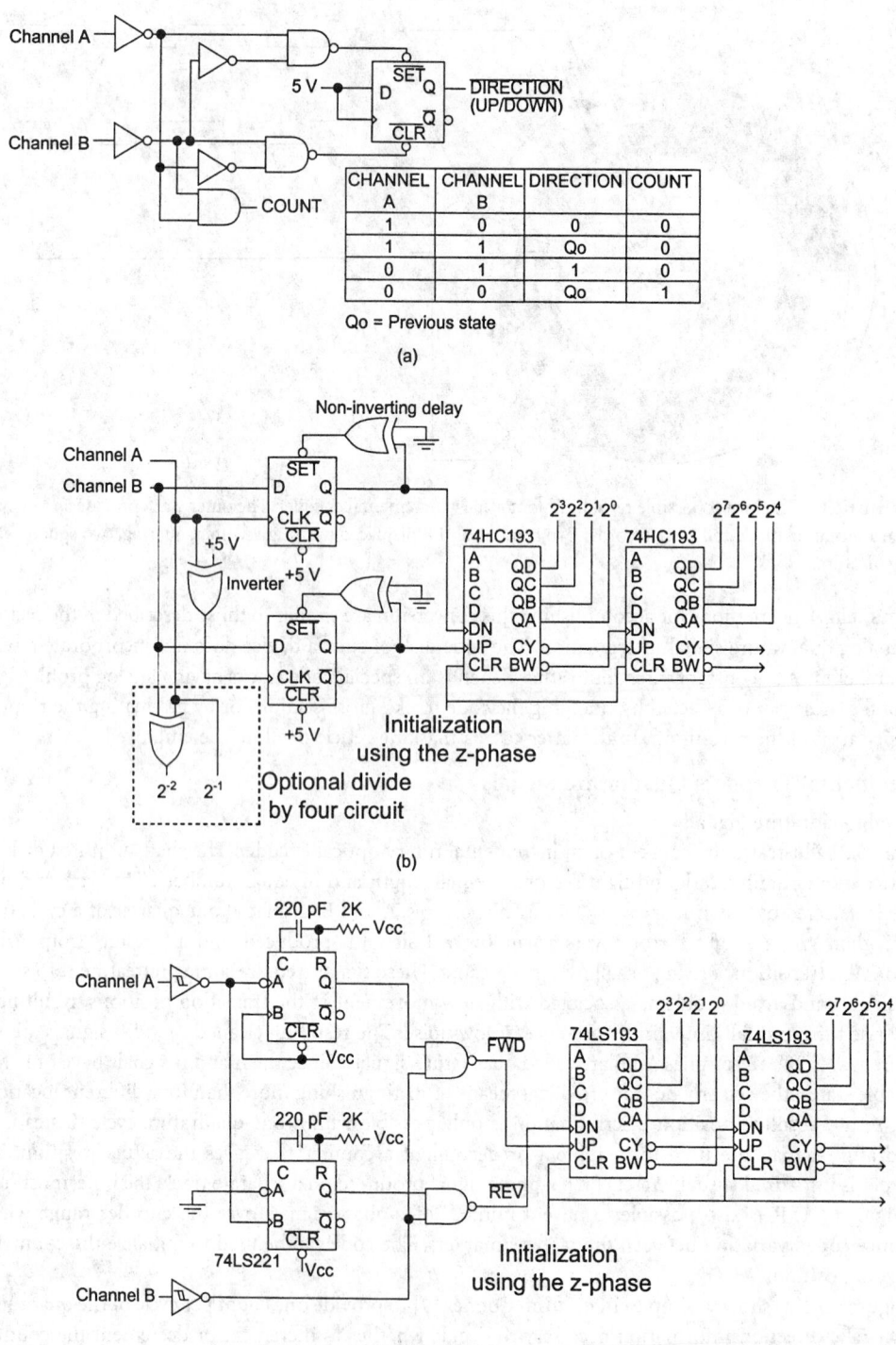

CHANNEL A	CHANNEL B	DIRECTION	COUNT
1	0	0	0
1	1	Qo	0
0	1	1	0
0	0	Qo	1

Qo = Previous state

(a)

(b)

(c)

FIGURE 6.72 Circuit producing a direction and a count signal as described by Conner [9] in (*a*). Caution must be exercised when using this circuit for counting purposes since multiple counts are possible when the encoder oscillates around the A = 0, B = 0 and the A = 0, B = 1 states. Kuzdrall's circuit [10] in (*b*), counts up and down at the same states transition but depending on the direction of movement.

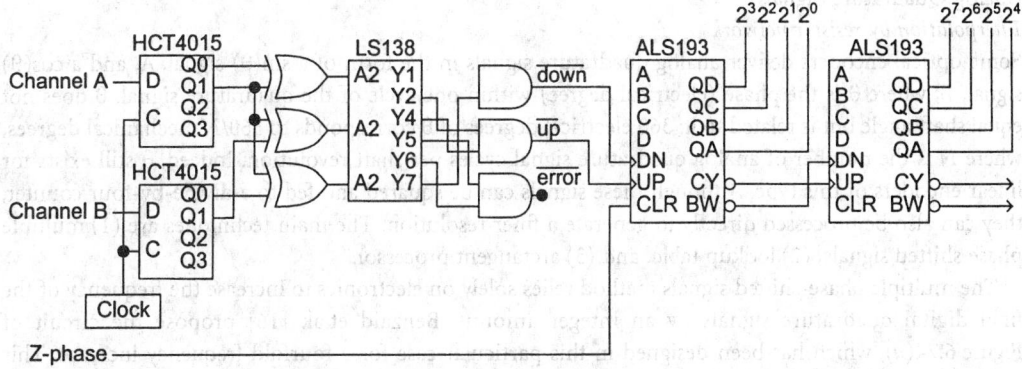

FIGURE 6.73 Divide-by-four circuit producing four counts per cycle of the quadrature signals. Based on Butler's design [14], except for the LS138 demultiplexer that replaces the suggested 4051 multiplexer because of the latter's unequal fall time and rise time. The clocked shift registers generate a 4-bit code to the demultiplexer for each clock cycle. The clock frequency should be 8 times the maximum frequency of the quadrature signals.

circuit drives a 74HC193 counter with a down clock and an up clock. Both flip-flops are normally in the high state. Whenever one of them switches to a low state, it is soon reset by the use of that output as a Set signal. That low output is only possible when phase B is low and there is a transition on A. Depending on the direction of that transition, one of the flip-flops produces the brief low state, causing an appropriate up or down count. The problem associated with small oscillations around a position is reduced since the up and down counts occur at the same encoder position. Venugopal [11] proposes a circuit, in Figure 6.72(*c*), that produces a similar effect. A count can only occur when B is low and there is a transition on A. Depending on the transition direction, one of two monostables triggers and effectively allows the count at the appropriate input of the 74LS193 counter.

In cases where a noisy environment is present, Holle [12] describes a digital filter to clean the A and B signals further, a filter that uses a small number of logic gates to form a 4-bit delay filter. Wigmore [13] proposes other circuits for count and direction detection.

Tables 6.23 and 6.24 list commercial chips and their suppliers' details. Some chips produce count and direction signals or up and down clocking signals. Others also include counters.

The above circuits do not fully exploit the information contained in the quadrature signals since only one of the four edges (or states) within one quadrature cycle is used to count. Figure 6.73 shows a slightly modified version of a divide-by-four counter circuit proposed by Butler [14]. Two 4-bit shift registers, three exclusive-OR gates, and an eight channel demultiplexer derive the up and down count signals. The clocked shift registers generate a 4-bit code to the demultiplexer for each clock cycle. To ensure that no encoder transitions are missed, the clock frequency should be at least 8 NS, where N is the number of cycles produced by the encoder for each shaft revolutions and S is the maximum speed in revolutions per second. The up and down signals can be fed to cascaded 74ALS193 counters. This circuit analyzes the current and previous states of the A and B channels to either count up, count down, or issue an error flag when an improper sequence occurs.

Kuzdrall [10] proposes to view a quadrature signal cycle as a Gray code so that the two least significant bits of the count are obtained by a Gray code to binary code conversion as in Figure 6.72(*b*). Phase A generates bit 1 of the natural binary position, and phase B is exclusive-ORed with phase A to produce the least significant bit. The counter provides the remaining bits of the natural binary position. Marty [15] proposes a state machine that stores both the actual A and B values and the previous values using D-type flip-flops (in a similar way to Butler) to form a hexadecimal number. In all, eight different hex-digits are generated, four for each direction of motion. A 4-line to 16-line decoder then feeds into two 4-input NAND gates to produce an up or down count signal. As for Butler, this last circuit is not dependent on propagation delays as with Kuzdrall's circuit.

Analog Quadrature Signals

Interpolation by resistor network

Some optical encoders deliver analog quadrature signals in the form of a $\sin(\theta)$ signal, A, and a $\cos(\theta)$ signal, B, where θ is the phase (electrical degree) within one cycle of the quadrature signal. θ does not equal shaft angle but is related to it: 360 electrical degrees of θ corresponds to $360/N$ mechanical degrees, where N is the number of analog quadrature signal cycles per shaft revolution. Indeed, θ still exists for linear encoders of this type. Although these signals can be squared and fed to a divide-by-four counter, they can also be processed directly to generate a finer resolution. The main techniques are (1) multiple phase shifted signals, (2) lookup table, and (3) arctangent processor.

The multiple phase-shifted signals method relies solely on electronics to increase the frequency of the final digital quadrature signals by an integer amount. Benzaid et al. [16] propose the circuit of Figure 6.74(a), which has been designed in this particular case for a fourfold frequency increase. This can then be followed by a digital divide-by-four circuit (not shown). The A and B signals are combined to produce an additional six phase-shifted signals, three of which are shifted by $\pi/8$, $\pi/4$, and $3\pi/8$, respectively, from the A signal, and three others that are shifted similarly from the B signal. The result is a total of 8 available sinusoidal signals, phase-shifted by $\alpha = i\,\pi/8$ with $i = 0, 1, ..., 7$. This can be generalized, saying that for an m-fold increase in resolution, $2m$ signals that are phase-shifted by $\alpha i\,\pi/2m$ with $i = 0, 1, ..., 2m - 1$ are required. The phase-shifted sinusoidal signals are then squared using TTL converters. Figure 6.74(b) shows the resulting square waves. The addition, modulo-2, of the signals for even values of i gives A′, and similarly for the signals with odd values of i gives B′. The modulo-2 sum is performed via exclusive-ORs.

The vector additions of the initial A and B signals result in phase-shifted signals. The weights of A and B are calculated from the trigonometric relations:

$$\sin(\theta + \alpha) = \cos(\alpha)\,\sin(\theta) + \sin(\alpha)\,\cos(\theta) \qquad (6.96)$$

and

$$\cos(\theta + \alpha) = -\sin(\alpha)\,\sin(\theta) + \cos(\alpha)\,\cos(\theta) \qquad (6.97)$$

where $\sin(\theta) = A$ and $\cos(\theta) = B$.

Thus,

$$\sin(\theta + \alpha) = \cos(\alpha)\,A + \sin(\alpha)\,B \qquad (6.98)$$

and

$$\cos(\theta + \alpha) = -\sin(\alpha)\,A + \cos(\alpha)\,B \qquad (6.99)$$

Note that the amplitudes of the phase-shifted signals are not critical, since it is their zero-crossing points that produce the square-wave transition. Also, in the circuit of Figure 6.74(a), the weights were slightly modified to simplify the implementation, which results in a small variation of the duty cycle of A′ and B′.

The phase-shifted signals can alternatively be produced using voltage dividers and Schmitt triggers, where the divider resistors are in the ratio $\tan(\alpha)$ [17]. As m increases, the precision of the weights become more stringent and the speed of the electronics processing the high-frequency square signals might limit the upper value of m possible with this method.

Interpolation by Sampling and Numerical Processing

A number of methods digitize the analog quadrature signals in order to perform a digital interpolation within one cycle of the quadrature signals. These techniques permit an even higher interpolation. The signals are periodically sampled in sample-and-hold circuitry and digitized by an analog to digital converter (ADC). One technique uses an interpolation table, while the other performs arctangent calculations.

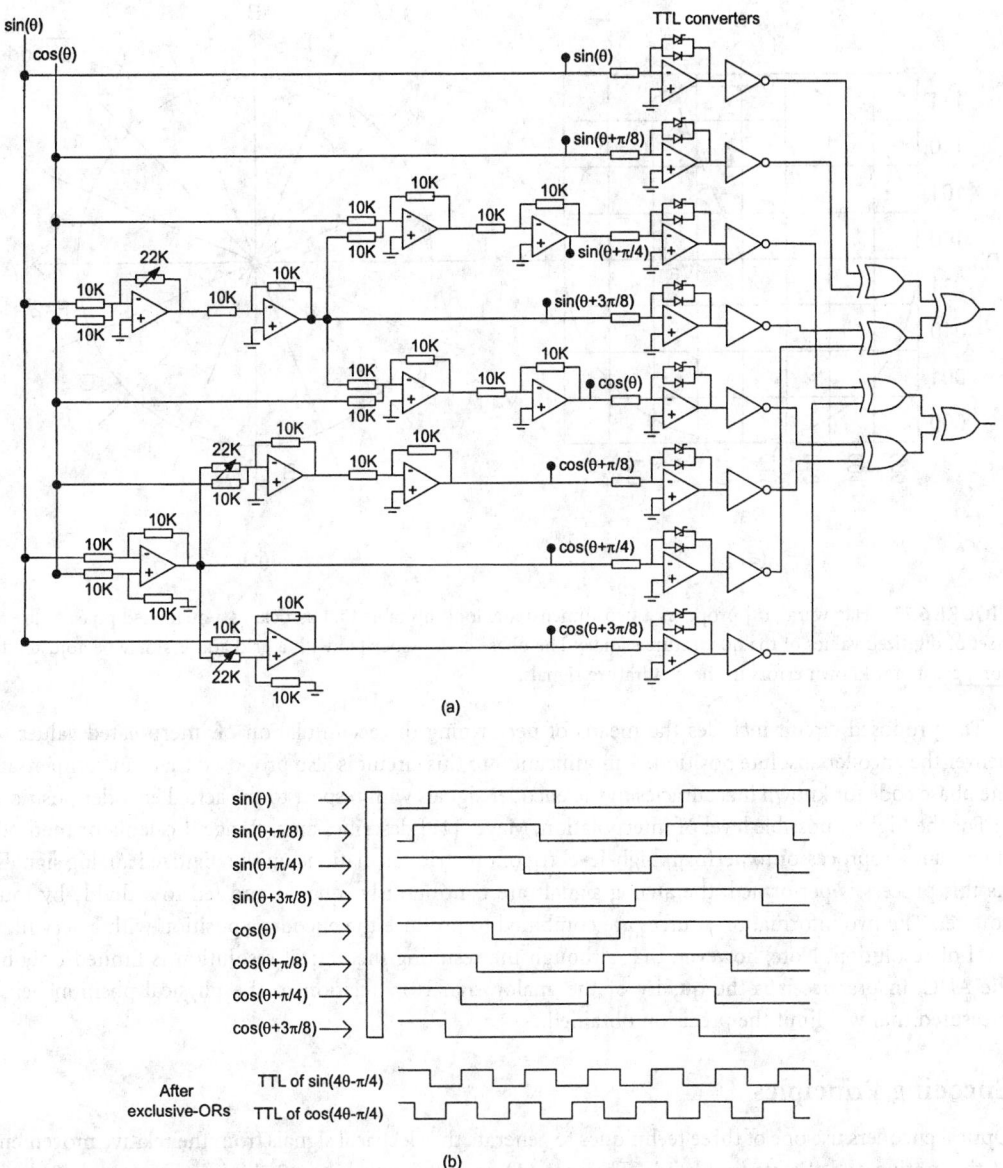

FIGURE 6.74 Interpolation circuit by Benzaid et al. [16]. The sine and cosine signals from the encoder are vector-combined in (*a*) to produce various phase-shifted signals. Following TTL conversion, they are exclusive-ORed. The squared phase-shifted signals and the final result are in (*b*).

Hagiwara [18] uses an *m*-bit ADC to digitize the analog A and B signals into the binary numbers *Da* and *Db*. Figure 6.75(*a*) shows how these binary numbers are used as the addresses of a grid, here built for *m* = 3 as a 2^m by 2^m grid. Hagiwara then associates a phase angle with the center of each cell of the matrix using simple arctangent calculations. The angle is calculated with respect to the center of the grid. This phase angle is then associated with one of $2^n - 1$ phase codes using:

$$\text{Phase code} = \text{integer part of}\left(2^n \theta / 2\pi\right) \tag{6.100}$$

as shown in Figure 6.75(*b*).

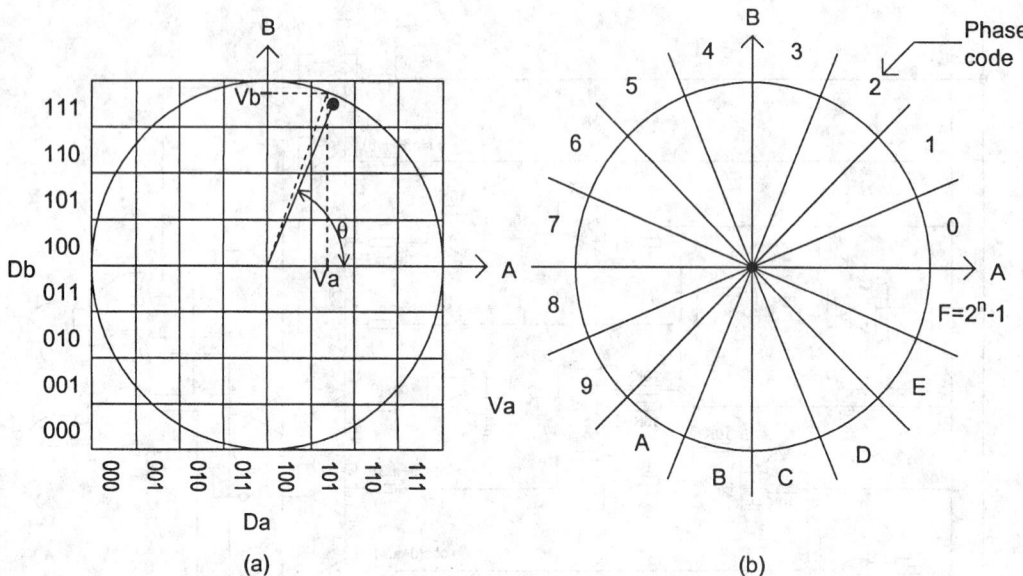

FIGURE 6.75 Hagiwara [18] proposes a two-dimensional look-up table that associates a quantized phase value to a set of digitized values of the quadrature signals. The phase code associated with a grid address may be adjusted to compensate for known errors in the quadrature signals.

The proposed circuit includes the means of performing the accumulation of interpolated values to deliver the encoder absolute position. A modification to this circuit is also proposed that can compensate the phase code for known inaccuracies in the encoder signals with respect to the actual encoder position.

For the highest possible level of interpolation, Mayer [19] describes an arctangent calculator method. It uses a microprocessor to perform high-level trigonometric calculations on the digitized analog signals. As this process is performed, the analog signals are continuously squared and fed to a divide-by-four counter. The two information sources are combined to produce the encoder's position with a very high level of resolution. Note, however, that although the resulting theoretical resolution is limited only by the ADC, in practice it is the quality of the analog signals in relation to the physical position being measured that will limit the precision obtained.

Encoding Principles

Optical encoders use one of three techniques to generate the electrical signals from the relative movement of the grating and the reading heads. They are (1) geometric masking, (2) Moiré effects, and (3) laser interference. Note that absolute encoders mainly use geometric masking. Geometric masking relies on geometric optics theory and considers light as traveling in a straight line. However, as the grating period reduces and the resolution of the encoder increases, Moiré fringe effects are observed and used to produce the signals. Although diffraction effects then become nonnegligible, their influence can be controlled by careful design. Finally, for very high resolution, diffraction effects are directly exploited to perform the measurements.

Geometric Masking

Geometric masking is applied to absolute and incremental encoders. The electromagnetic field associated with the propagation of visible light is characterized by very rapid oscillations (frequencies of the order of 10^{14} s^{-1}). It may therefore be expected that a good first-order approximation to the propagation laws of light is obtained by neglecting the wavelength of light. In this case, diffraction phenomena may be ignored and light may be thought to propagate in a straight line. Geometry can then be used to analyze

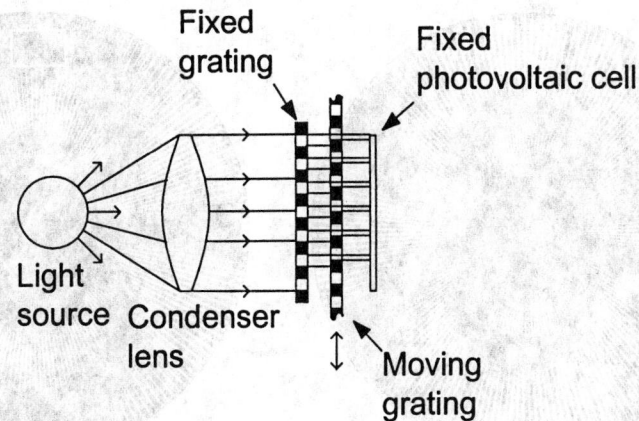

FIGURE 6.76 The reading head contains a light source that can be an incandescent bulb or a light emitting diode. The light may require a condenser in order to collimate it. This redirects the light rays from the source so that they travel perpendicular to the gratings. The stationary index grating structures the light beam into bands that then cross the moving grating. Depending on the relative position of the fixed and the moving gratings, more or less light reaches the photodetector. The detector produces a signal proportional to the total amount of light on its sensitive surface.

the behavior of light. The approximation is valid whenever light rays propagate through an encoder grating with fairly coarse pitch (say, more than 10 μm). Figure 6.76 shows a portion of an encoder scale being scanned by a reading head with multiple slits in order to send more light onto the photosensitive area. The light source is collimated (rays are made parallel to each other after having been emitted in multiple directions from a small source), passes through the stationary grating or index grating, and propagates through the moving grating. In the case of an incremental encoder, a second head would read from the same moving grating but be displaced by $n + 1/4$ pitch in order to produce a second signal with a phase of 90°. When the slits of the two gratings are aligned, a maximum amount of light reaches the detector. Similarly, a minimum amount of light is transmitted when the gratings are out of phase by 180°. Depending on the design of the grating, such as the duty cycle of the transmitting and opaque sectors, various cyclic signal patterns are obtained. Generally, it is noteworthy that optical encoders can also be used as a reflective as opposed to transmissive design. Also, in some absolute optical encoders, the encoding principle does not rely on a primary grating, per se, or the detection system does not necessarily incorporate any masking element.

Moiré Fringes

Moiré fringe methods are primarily associated with incremental encoders. Moiré fringes are observed when light passes through two similar periodic patterns brought close to each other and with their line pattern nearly parallel. Figure 6.77 shows linear and radial gratings producing Moiré fringes. Take a linear grating with sinusoidal amplitude transmittance as proposed by Gasvik [20]:

$$f(x, y) = a + a \cos(2\pi x/P) \tag{6.101}$$

where P is the grating period, a is the amplitude, and x is measured perpendicularly to the grating lines. It is also possible to represent a square wave type grating using a Fourier series in the case of a radial grating and a Fourier integral for a linear grating. When two linear gratings, a and b, are laid in contact, the resulting transmittance, f_c, is the product of their individual transmittances f_a and f_b:

$$f_c = f_a \times f_b \tag{6.102}$$

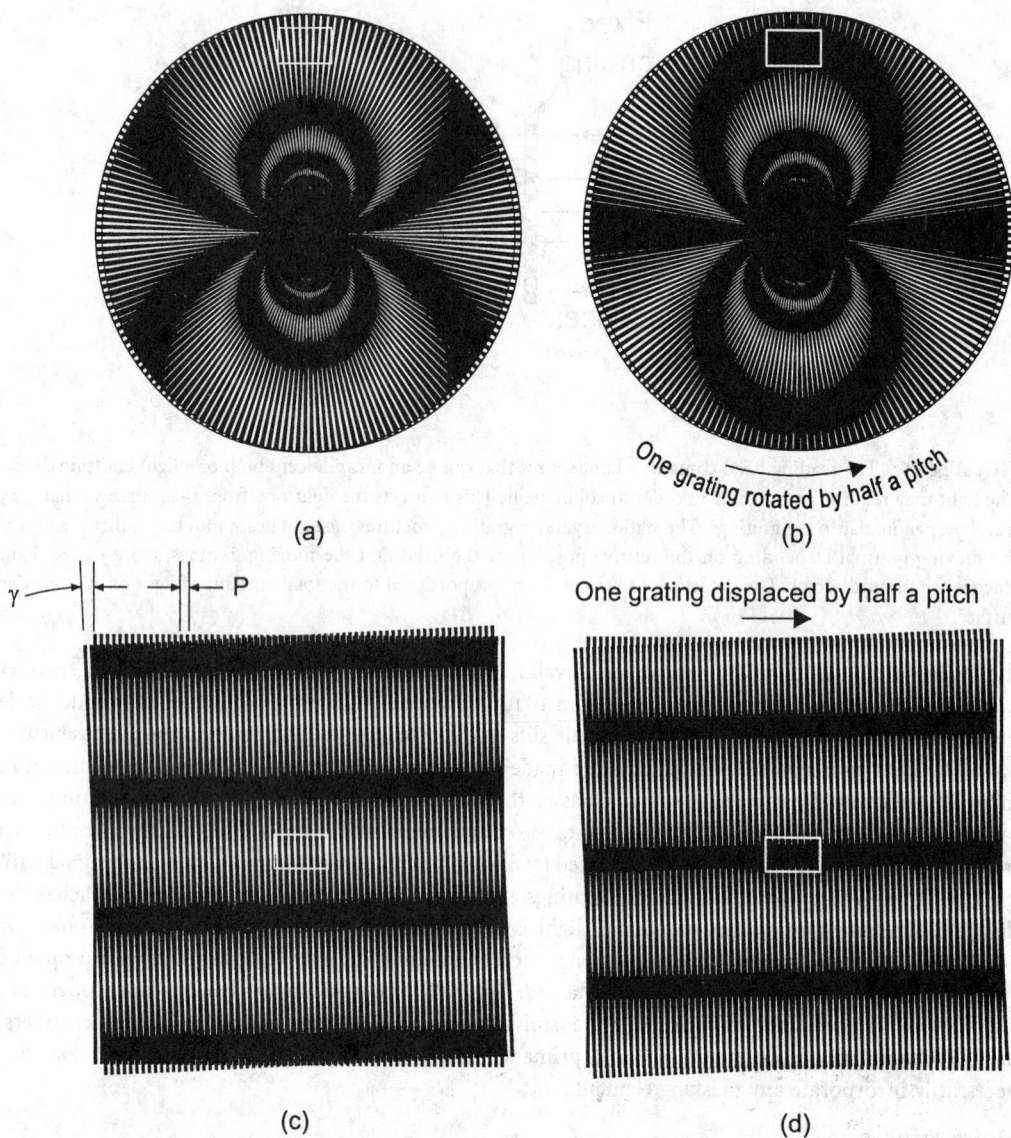

FIGURE 6.77 Moiré fringes produced when two gratings of equal pitch are overimposed. In (*a*), two radial gratings with 180 cycles have their centers displaced horizontally by approximately half of the grating pitch (as measured on the perimeter). This could be due to encoder assembly errors. Ditchburn [21] explains that the fringes are always circular and pass by the two centers. In (*b*), the same displacement was followed by a rotation of half a pitch as a result of the normal movement of the encoder disk. This causes a phase change of 180° of the fringe pattern at any fixed observation point, as represented by the rectangular observation region. The width of the fringes depends on the offset displacement between the two grating centers. If they coincide, the whole pattern changes from a dark to a relatively bright fringe. In (*c*), two linear gratings are rotated by one tenth (in radian) of their grating pitch. This could be due to encoder assembly or mounting errors. As a result, Moiré fringes occur at 90° to the grating and with a period 10 times larger. The normal displacement of one grating by half of the grating pitch causes a phase change of 180° of the Moiré fringe pattern at any fixed observation point, as represented by the rectangular observation region.

In the case when a misalignment γ exists between the two gratings, the separation of the Moiré fringe is d with:

$$d = P/\left[2\sin\left(\gamma/2\right)\right] \tag{6.103}$$

and for small γ,

$$d = P/\gamma \tag{6.104}$$

It follows that when the two gratings are moved relative to each other by a distance x' in the direction perpendicular to their lines, then the phase $\varphi(x')$ of the Moiré fringes at a stationary location changes by:

$$\varphi\left(x'\right) = 2\pi x'/P \tag{6.105}$$

In effect, every time x' equals P, the entire Moiré fringe pattern undergoes one complete cycle of fluctuation. If the two gratings are adjusted to near parallelism, then the fringes are "fluffed out" (see Burch [22]) to an infinite width. Under such conditions, there is no fringe motion to be observed and the entire region simply goes from brightness to darkness. These effects are mathematically predicted using a convolution integral.

$$f_c\left(x'\right) = \int_{x_1}^{x_2} f_a\left(x\right) f_b\left(x-x'\right)/dx \tag{6.106}$$

with x_1 to x_2 representing the region covered by the photodetector. If $g_a(k)$ and $g_b(k)$ are the exponential Fourier transforms of f_a and f_b, then the photodetector response f_c will have the Fourier transform $g_c(k)$,

$$g_c\left(k\right) = g_a\left(k\right) g_b\left(-k\right) \tag{6.107}$$

Moiré effects can be obtained by transmission or by reflection at the grating.

It must be kept in mind, however, that as the grating pitch reduces, diffraction effects become significant; careful design and adjustment of the gap between the stationary and moving grating and also lighting and detection arrangements become critical. Since a large number of lines of the gratings are used to generate the photodetector response, small local imperfections of the grating pitch are averaged out, resulting in an improved measurement accuracy.

Diffraction-Based Encoders

Diffraction-based encoders are used for the highest levels of precision. They successfully exploit diffraction effects and are referred to as physical optics encoders as opposed to geometrical optics encoders. Figure 6.78 shows diffraction effects observed when coherent light encounters a pattern of slits [20]. The slit is a few wavelengths wide, so strong diffraction effects are observed. Each slit may be thought of as acting as a coherent light source producing circular wavefronts. Along certain directions, portions of these wavefronts are in phase with each other. These directions are given by:

$$\sin\theta_m = m\lambda/P \tag{6.108}$$

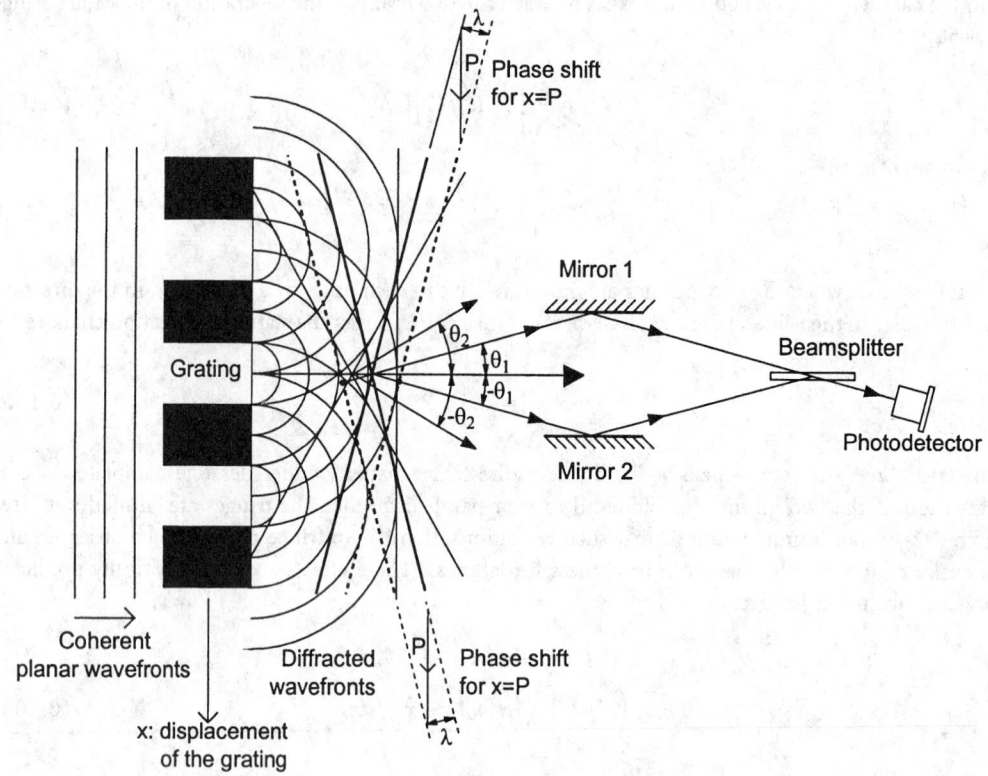

FIGURE 6.78 The grating diffracts the coherent planar wavefront coming from the left into a series of circular wavefronts. The coherent circular wavefronts are in phase along certain directions resulting in diffracted planar wavefronts. These planar wavefronts correspond to the common tangents of the circles. For example, the third innermost circle of the uppermost slit is in phase with the fourth innermost circle of the middle slit and also with the fifth innermost circle of the lowest slit. This diffracted wavefront corresponds to the first order diffraction ($m = 1$) and makes an angle θ_1 with the initial planar wavefront direction. A similar order wavefront ($m = -1$) has a direction $-\theta_1$. These two wavefronts are initially in phase. A displacement of the grating downward by x causes the $m = 1$ wavefront to move by a distance $x\lambda/P$ and the $m = -1$ wavefront by $-x\lambda/P$, thus causing a relative phase shift between the two wavefronts of $2x\lambda/P$. This relative phase shift produces a movement of interference fringes at the photodetector.

where m is a positive or negative integer and is the order of diffraction, and P is the diffraction grating period. Suppose now that the slits are moving by a distance x as shown in Figure 6.78. Then the phase of the wavefront at a stationary location will increase for m positive and decrease for m negative by $2\pi m x/P$. When the slit pattern has moved by one pattern cycle, then the two wavefronts will have developed a relative phase change of $2m$ cycles. The two phase-shifted wavefronts may be recombined to produce interference fringes.

Rotary Encoders

The Canon laser rotary encoder uses an optical configuration that generates four cycles of interference fringes per cycle of the diffraction grating. This is achieved by splitting the original wavefront in two and interrogating the grating at two diametrically opposed locations. Reflecting both diffracted beams back through the grating results in a further doubling of the resolution. The two beams are finally recombined for interference fringe counting. Furthermore, as explained by Nishimura and Ishizuka [23], using diametrically opposed portions of the disk attenuates the effect of eccentricity. An encoder with an external diameter of 36 mm produces 81,000 analog quadrature cycles per revolution. The grating has a pitch of approximately 5 μm and the laser light has a wavelength of 780 nm.

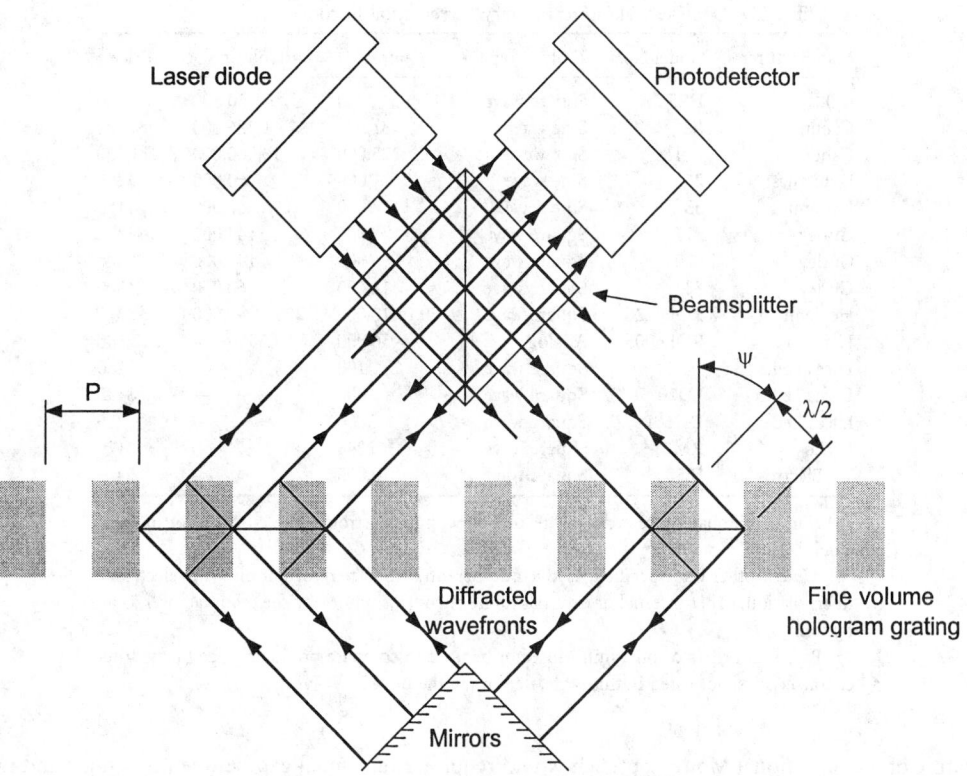

FIGURE 6.79   The coherent planar wavefront of the laser diode is split into two at the beam splitter. Both beams can be regarded as being partially reflected at the grating by minute mirrors separated by a pitch, P. Maximum reflection is achieved when $2P \sin \psi = \lambda$. Each reflected portion of the incident beam is phase shifted by one λ with respect to that reflected by an adjacent mirror so that the reflected wavefront remains coherent. The reflected wavefronts are truly reflected at the orthogonal mirrors for a second pass through the grating, after which they are recombined by the splitter to interfere at the photodetector. A displacement of the grating by x along its axis causes one beam path to be shortened by $2x\lambda/P$, while the other is lengthened by the same amount. A grating motion by P causes four fringe cycles.

Linear Encoders

Sony markets a laser linear encoder with a volume hologram grating of 0.55 μm pitch [24]. Figure 6.79 gives some insight into the principle of operation of this type of encoder. In theory, for maximum intensity of the reflected light beam through the hologram grating, the path length difference from successive plane mirrors must be equal to λ [20]. The Sony Laserscale encoder uses a semiconductor laser as the coherent light source. The beam splitter produces two beams to interrogate the grating at two separate locations. The beams are then diffracted by the hologram grating, followed by a reflection at the mirrors and pass a second time through the grating. They are finally recombined at the photodetector to produce interference fringes. Because the two beams are diffracted by the grating in opposite directions, and because they pass twice through the grating, four signal cycles are obtained when the grating moves by one pitch.

The Heidenhain company produces both linear and rotary encoders of very high resolution, also using the principles of diffraction. Their approach, unlike those previously mentioned, uses an index grating and a reflecting scale grating [17, 25].

The Renishaw company uses a fine reflective grating of 20 μm pitch that diffuses the light from an infrared light-emitting diode. In order to avoid the problems caused by diffraction effects, the index

TABLE 6.19   Commercial Optical Rotary Incremental Encoders

Manufacturer	Model No.	Output type	Counts[a]	Resolution[b]	Price[c]
BEI	H25	Square wave	2540	50,800	$340
Canon	K1	Sine wave	81,000	1,296,000	$2700
Canon	X-1M	Sine wave	225,000	18,000,000	$14,000
Dynamics	25	Sine wave	3000	60,000	$290
Dynamics	35	Sine wave	9000	360,000	$1150
Gurley	911	Square wave	1800	144,000	$1300
Gurley	920	Square wave	4500	144,000	$500
Gurley	835	Square wave	11,250	360,000	$1500
Heidenhain	ROD 426	Square wave	50 to 10,000	200 to 40,000	$370
Heidehain	ROD 905	Analog	36,000	0.035 arcsec	$12,600
Lucas Ledex	LD20	Square wave	100	—	$180
Lucas Ledex	LD20	Square wave	1000	—	$195
Lucas Ledex	DG60L	Square wave	5000	—	$245
Renco	RM21	Square wave	2048	—	$176
TR Electronic	IE58	Square wave	10,000	—	$428

[a] Number of quadrature cycles per revolution without electronic divide-by-four or interpolation.

[b] Unless otherwise specified, is the number of counts per revolution with electronic interpolation, either internal or external to the encoder, supplied by the manufacturer as an option.

[c] Based on orders of one unit. Must not be used to compare products since many other characteristics, not listed in this table, determine the price.

grating of a conventional Moiré approach would require a very small gap between the index and main gratings. Instead, the index grating is located at a distance of 2.5 mm. The index grating is then able to diffract the diffused light from the main grating and to image it 2.5 mm further. There, a fringe pattern of the Moiré type is produced and photoelectrically analyzed [26].

Components and Technology

The choice of an encoder demands careful consideration of a number of factors, such as: (1) the required resolution, repeatability, and accuracy (linearity); (2) the maximum and minimum operating speeds; (3) the environmental conditions: temperature range, relative humidity (condensing or not condensing), contaminants such as dust, water, oil, etc.; (4) minimum friction torque (or force) acceptable; (5) maximum inertia acceptable; (6) available space; (7) position to be known immediately after a power loss; (8) range in degrees (or mm); (9) mounting and shaft loading; and (10) price.

Rotary encoders also sometimes employ reflective tapes attached to, or markings directly etched into, the surface of a drum or spindle to be read from the side. In special cases, tapes or engravings of this type can be made to conform to the surface of an elliptical cross-section cam or other irregular contour. Cylindrical primary gratings employing transmissive readout have also been produced.

The main gratings (scale) come in a wide variety of materials, both for the substrate and for the marking. Flexible scales based on a metal tape substrate are also available from Renishaw, allowing very long (tens of meters) continuous reading of linear motion. Tables 6.19 through 6.21 list a number of encoders currently available on the market. The list is not exhaustive in terms of suppliers and does not cover all the models of the suppliers listed. Table 6.22 gives the address and telephone numbers of the suppliers. Tables 6.23 and 6.24 list some suppliers of quadrature decoding circuits.

TABLE 6.20 Commercial Optical Rotary Absolute Encoders

Manufacturer	Model Number	Steps per turn	No. of turn	Price[a]
BEI	M25	65,536	1	$2130
BEI	MT40	512	16	$1240
BEI	MT40	65,530	512	$5000
Gurley	25/04S	131,072	1	$1900
Heidenhain	ROC 424	4096	4096	
Lucas Ledex	AG60E	360 or 512	1	$486
Lucas Ledex	AG661	4096	4096	$1260
TR Electronic	CE65[b]	8192	4096	$1408

[a] Based on orders of one unit. Must not be used to compare products since many other characteristics, not listed in this table, determine the price.

[b] Programmable output.

TABLE 6.21 Commercial Optical Linear Incremental Encoders

Manufacturer	Model No.	Output type	Pitch[a]	Resolution[b]	Length (mm)	Price[c](length)
Canon	ML-16+	Sine wave	1.6 μm	0.4 μm	To 300	$1525 (50 mm)
Canon	ML-08+	Sine wave	0.8 μm	0.2 μm	To 150	$3100
Gurley	LE18	Square wave	20 μm	0.1 μm	To 1500	$750 (1000 mm)
Gurley	LE25	Square wave	20 μm	0.1 μm	To 3000	$800 (1000 mm)
Heidenhain	LS603	Sine wave	20 μm	5 μm	To 3040	$932 (1020 mm)
Heidenhain	LIP401	Sine wave	2 μm	0.005 μm	To 420	$4000 (100 mm)
Renishaw	RG2	RS422A	20 μm	0.5 μm	To 60,000	$640 + $360/1000 mm
Sony	BS75A-30NS	Square wave	0.14 μm	0.05 μm	30	$2628

[a] Period of the quadrature cycle without electronic divide-by-four or interpolation.

[b] With electronic interpolation supplied by the manufacturer.

[c] Based on orders of one unit. Must not be used to compare products since many other characteristics, not listed in this table, determine the price.

TABLE 6.22 Companies that Make Optical Encoders

BEI Sensors and Motion Systems Company
  Encoder Systems Division
  13100 Telfair Avenue
  Sylmar, CA
  Tel: (848) 341-6161
Canon USA Inc.
  Components Division
  New York Headquarters :
  One Canon Plaza
  Lake Success, NY 11042
  Tel: (516) 488-6700
DR. JOHANNES HEIDDENHAIN GmbH
  DR.-Johannes-Heidenhain-Strasse 5
  D83301 Traunreut, Deutschland
  Tel: (08669)31-0
Gurley Precision Instruments Inc.
  514 Fulton Street
  Troy, NY 12181-0088
  Tel: (518) 272-6300
Ledex Products
  Lucas Control Systems Products
  801 Scholz Drive
  P.O. Box 427
  Vandalia, OH 45377-0427
  Tel: (513) 454-2345

Renco Encoders Inc.
  26 Coromar Drive
  Goleta, CA 93117
  Tel: (805) 968-1525
Renishaw plc,
  Transducer Systems Division
  Old Town, Wotton-under-Edge
  Gloucestershiire GL12 7DH
  United Kingdom
  Tel: +44 1453 844302
TR Electronic GmbH
  Eglishalde 6
  Postfach 1552
  D-7218 Trossingen
  Germany
  Tel: 0 74 25/228-0
Sony Magnescale Inc.
  Toyo Building, 9-17
  Nishigotanda 3-chome
  Shinagawa-ku, Tokyo
  141 Japan
  Tel: (03)-3490-9481

TABLE 6.23 Commercial Digital Quadrature Signal Decoder Circuits

Manufacturer	Model No.	Output	Decoding factor	Counter	Price
Hewlett Packard	HCTL-2000	Count	12-bit	×4	$12.75
Hewlett Packard	HCTL-2016	Count	16-bit	×4	$12.75
Hewlett Packard	HCTL-2020	Count	16-bit & cascade o/p	×4	$14.55
U.S. Digital Corp.	LS7083	Up and Down clock		×1 or ×4	$3.05
U.S. Digital Corp.	LS7084	Count and direction		×1 or ×4	$3.60

TABLE 6.24 Companies that Make Divide-by-Four Decoders

U.S. Digital Corporation
3800 N.E. 68th Street, Suite A3
Vancouver, WA 98661-1353
Tel: (360) 696-2468
Hewlett-Packard Company
Direct Marketing Organization
5301 Stevens Creek Boulevard
P.O. Box 58059, MS 51LSJ
Santa Clara, CA 95052-8059
Tel: (408) 246-4300

References

1. P. E. Stephens and G. G. Davies, New developments in optical shaft-angle encoder design, *Marconi Rev.*, 46 (228), 26-42, 1983.
2. P. Sente and H. Buyse, From smart sensors to smart actuators: application of digital encoders for position and speed measurements in numerical control systems, *Measurement*, 15(1), 25-32, 1995.
3. E. M. Petriu, Absolute-type position transducers using a pseudorandom encoding, *IEEE Trans. Instrum. Meas.*, IM-36, 950-955, 1987.
4. E. M. Petriu, Scanning method for absolute pseudorandom position encoders, *Electron. Lett.*, 24, 1236-1237, 1988.
5. G. H. Tomlinson, Absolute-type shaft encoder using shift register sequences, *Electron. Lett.*, 23, 398-400, 1987.
6. J. N. Ross and P. A. Taylor, Incremental digital position encoder with error detection and correction, *Electron. Lett.*, 25, 1436-1437, 1989.
7. M. Arsic and D. Denic, New pseudorandom code reading method applied to position encoders, *Electron. Lett.*, 29, 893-894, 1993.
8. B. Arazi, Position recovery using binary sequences, *Electron. Lett.*, 20, 61-62, 1984.
9. D. Conner, Long-lived devices offer high resolution, *EDN*, 35 (9), 57-64, 1990.
10. J. A. Kuzdrall, Build an error-free encoder interface, *Electron. Design*, September 17, 81-86, 1992.
11. P. Venugopal, Reflective optical SMT module reduces encoder size, *Power Conversion and Intelligent Motion*, 21(5), 60-62, 1995.
12. S. Holle, Incremental encoder basics, *Sensors*, 7(4), 22-30, 1990.
13. T. Wigmore, Optical shaft encoder from sharp, *Elektor Electron.*, 15(169), 60-62, 1989.
14. M. M. Butler, Simplified multiplier improves standard shaft encoder, *Electronics*, November 20, 128-129, 1980.
15. B. Marty, Design a robust quadrature encoder, *Electron. Design*, June 24, 71-72, 74, 76, 1993.
16. O. Benzaid and B. M. Bird, Interpolation techniques for incremental encoders, *Proc. 23rd Int. Intelligent Motion Conf.*, Jun 22-24, 165-172, 1993.
17. Heidenhain General Catalog, Dr. Johannes Heidenhain GmbH, DR.-Johannes-Heidenhain-Strasse 5, D83301 Traunreut, Deutschland, November 1993, 8.

18. N. Hagiwara, Y. Suzuki, and H. Murase, A method of improving the resolution and accuracy of rotary encoders using a code compensation technique, *IEEE Trans. Instrum. Meas.*, 41(1), 98-101, 1992.

19. J. R. R. Mayer, High-resolution of rotary encoder analog quadrature signals, *IEEE Trans. Instrum. Meas.*, 43(3), 494-498.

20. K. J. Gasvik, *Optical Metrology*, New York: John Wiley & Sons, 1987.

21. R. W. Ditchburn, *Light Volume 1*, New York: Academic Press, 1976.

22. J. M. Burch, The metrological applications of diffraction gratings, in E. Wolf (Eds.) *Progress in Optics, Volume II*, Amsterdam: North-Holland Publishing, 1963.

23. T. Nishimura and K. Ishizuka (From Canon, Inc., Tokyo), Laser Rotary Encoders, Motion: Official Journal of the Electronic Motion Control Association, September/October 1986, Reprint obtained from Canon.

24. Anonymous from Sony Magnescale America Inc., Hologram technology goes to work, *Machine Design,* January 12 1995.

25. Anonymous from Heidenhain, Encoding systems Vorsprung durch Heidenhain, *Engineering Materials and Design,* September 1989 :53-54.

26. Jim Henshaw of Renishaw, Linear encoder offers superior flexibility, *Design Engineering,* September 1995.

6.9 Magnetic Displacement Sensors

David S. Nyce

Several types of linear and angular displacement measuring devices rely on electromagnetic fields, and the magnetic properties of materials, in the operation of their basic sensing elements. Some may not commonly be referred to as magnetic sensors, but are instead named according to their specific sensing technique. Magnetic sensors presented here use a permanent magnet, or an ac or dc powered electromagnet. Together with various materials used to sense the magnetic field, the combination is arranged to obtain a response indicating angular or linear displacement. The sensor is either caused to operate by a magnetic field, or the properties of the sensor are derived from the use of a magnetic field. Types of magnetic sensors presented in this section include magnetostrictive, magnetoresistive, Hall effect, and magnetic encoders. Some versions of synchro/resolvers and related sensors meet those requirements, but are included in Section 6.10 and thus will not be included in this section. Inductive proximity sensors measure displacement over a very limited range, and are covered in Section 6.2. LVDTs meet these requirements, but are also in Section 6.2.

An important aspect of magnetic sensors is that they utilize a noncontact sensing element. There is no mechanical connection or linkage between the stationary members and the movable members of the sensor. In some devices that sense a position magnet or core, the sensor can even be designed to allow removal of the magnet or core from the sensitive element, when readings are not required. Noncontact implies that the lifetime of the sensing element is not limited to a finite number of cycles by friction-induced wear. This is important in some industrial machinery. Sensors presented here utilize noncontact sensing techniques.

Displacement refers to a change in position, rather than an absolute position. In common industrial practice, however, displacement sensors are typically labeled as either incremental or absolute. An incremental sensor indicates the amount of change between the present location and a previous location. If the information that describes the current location is lost, due to power loss or other disturbance, the system must be reset. During the reset, the sensor must be in a reference position. Magnetic encoders can be designed as either incremental or absolute reading. Optical encoders, inductosyns, and synchro/resolvers are types of displacement sensors that can be designed as either incremental or absolute reading, but are covered in other chapters.

Most displacement position sensors described in this section are absolute reading. They supply a reading of distance or angle from a fixed datum, rather than from a previous position. Consecutive readings can be subtracted to give an incremental indication. An absolute sensor indicates the current position without the need for knowledge of the previous position. It never needs to be reset to a reference location in order derive the measured location. Absolute reading displacement sensors are also commonly called *position sensors*.

Magnetic sensor types will be described here based on the technology employed, rather than the application. Relative usefulness for making linear or angular measurements will be indicated for each type of sensor.

Noncontact magnetic sensor technology for displacement measurement includes magnetostrictive, magnetoresistive, Hall effect, and magnetic encoders.

Magnetic Field Terminology: Defining Terms

Magnetic field intensity (H), or magnetizing force: The force that drives the generation of magnetic flux in a material. H is measured in A m^{-1}.

Magnetic flux density (B): The amount of magnetic flux resulting from the applied magnetizing force. B is measured in N/(A-m).

Magnetic permeability (μ): The ability of a material to support magnetic lines of flux. The μ of a material is the product of the relative permeability of that material and the permeability of free space. The relative permeability of most nonferrous materials is near unity. In free space, magnetic flux density is related to magnetic field intensity by the formula:

$$B = \mu_0\,H$$

where μ_0 is the permeability of free space, having the value $4\pi \times 10^{-7}$ H m^{-1}. In other materials, the magnetic flux density at a point is related to the magnetic intensity at the same point by:

$$B = \mu H$$

where

$$\mu = \mu_0\mu_r$$

and μ_r is the relative permeability [1].

Hysteresis: A phenomenon in which the state of a system does not reversibly follow changes in an external parameter [2]. In a displacement sensor, it is the difference in output readings obtained at a given point when approaching that point from upscale and downscale readings. Figure 6.80 is a typical output vs. input graph.

Magnetic hysteresis: Depicted in the hysteresis loop, Figure 6.81. When a ferromagnetic material is placed in an alternating magnetic field, the flux density (B) lags behind the magnetizing force (H) that causes it. The area under the hysteresis loop is the hysteresis loss per cycle, and is high for permanent magnets and low for high permeability, low-loss magnetic materials [3].

Magnetic saturation: The upper limit to the ability of ferromagnetic materials to carry flux.

Magnetization curve: Shows the amount of magnetizing force needed for a ferromagnetic material to become saturated. It is a graph with B as the ordinate and H as the abscissa (also known as the B–H curve). A magnetization curve for a specific material would look the same as Figure 6.81, with the addition of calibration marks and the curve adjusted to describe the characteristic of that material.

Magnetostrictive Sensors

A magnetostrictive displacement sensor uses a ferromagnetic element to detect the location of a position magnet that is displaced along its length. The position magnet is attached to a member whose position

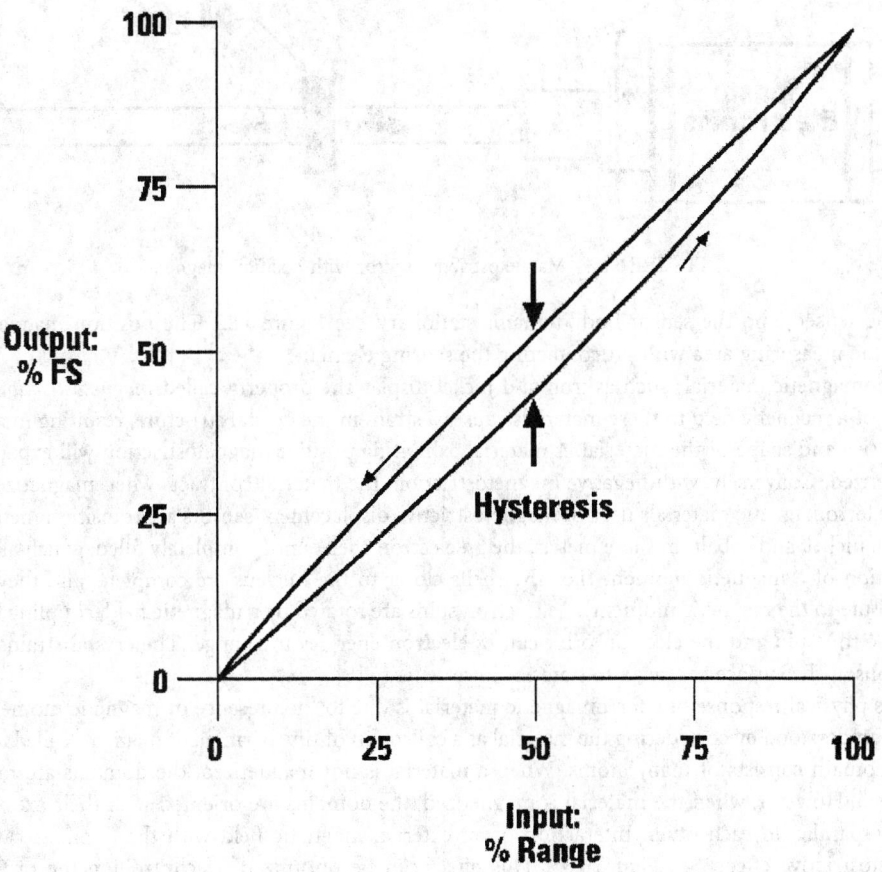

FIGURE 6.80 Hysteresis: output vs. input.

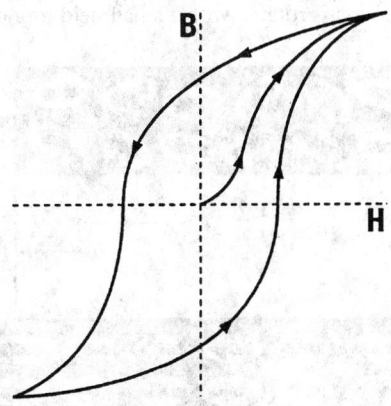

FIGURE 6.81 Magnetic hysteresis.

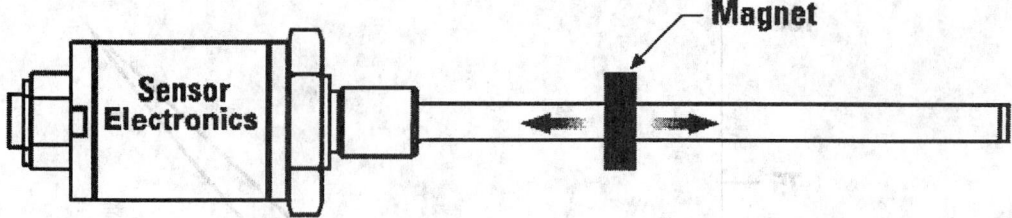

FIGURE 6.82   Magnetostrictive sensor with position magnet.

is to be sensed, and the sensor body remains stationary, see Figure 6.82. The position magnet moves along the measuring area without contacting the sensing element.

Ferromagnetic materials such as iron and nickel display the property called *magnetostriction*. Application of a magnetic field to these materials causes a strain in the crystal structure, resulting in a change in the size and shape of the material. A material exhibiting positive magnetostriction will expand when magnetized. Conversely, with negative magnetostriction, the material contracts when magnetized [4].

The ferromagnetic materials used in magnetostrictive displacement sensors are transition metals, such as iron, nickel, and cobalt. In these metals, the 3*d* electron shell is not completely filled, which allows the formation of a magnetic moment (i.e., the shells closer to the nucleus are complete, and they do not contribute to the magnetic moment). As electron spins are rotated by a magnetic field, coupling between the electron spin and the electron orbit causes electron energies to change. The crystal strains so that electrons at the surface can relax to states of lower energy [5].

This physical response of a ferromagnetic material is due to the presence of magnetic moments, and can be understood by considering the material as a collection of tiny permanent magnets, called *domains*. Each domain consists of many atoms. When a material is not magnetized, the domains are randomly arranged. However, when the material is magnetized, the domains are oriented with their axes approximately parallel to each other. Interaction of an external magnetic field with the domains causes the magnetostrictive effect. See Figure 6.83. This effect can be optimized by controlling the ordering of domains through alloy selection, thermal annealing, cold working, and magnetic field strength.

While application of a magnetic field causes the physical strain, as described above, the reverse is also true: exerting stress causes the magnetic properties (permeability, susceptibility) to change. This is called the *Villari effect*.

In magnetostrictive sensors, uniform distortions of length, as shown in Figure 6.83, offer limited usefulness. Usually, the magnetization is rotated with a small field to induce a local distortion, using the

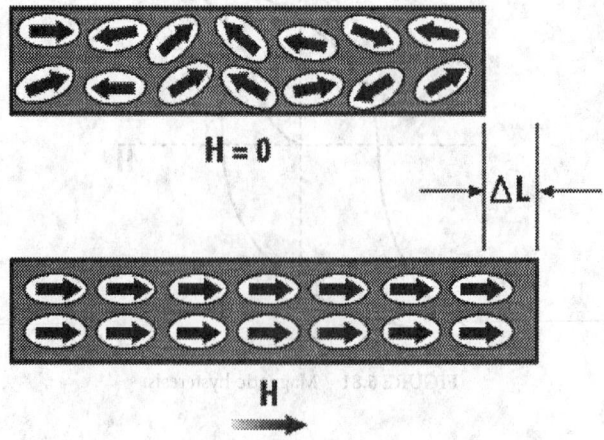

FIGURE 6.83   Magnetic domains: alignment with magnetic field, "*H*", causes dimensional changes.

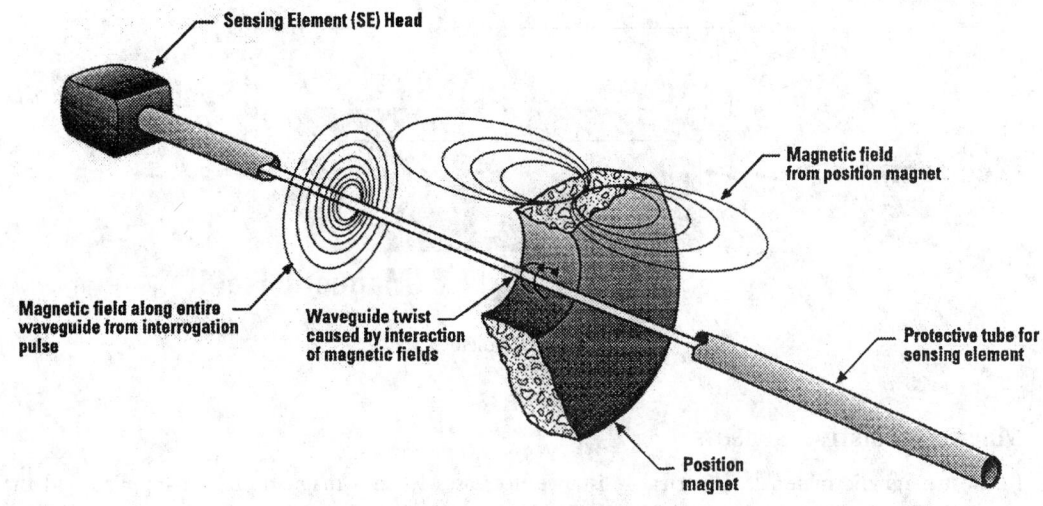

FIGURE 6.84 Operation of magnetostrictive position sensor.

Wiedemann effect. This is a mechanical torsion that occurs at a point along a magnetostrictive wire when an electric current is passed through the wire while it is subjected to an axial magnetic field. The torsion occurs at the location of the axial magnetic field, which is usually provided by a small permanent magnet called the position magnet.

In a displacement sensor, a ferromagnetic wire or tube called the waveguide is used as the sensing element, see Figure 6.84. The sensor measures the distance between the position magnet and the pickup. To start a measurement, a current pulse I (called the interrogation pulse), is applied to the waveguide. This causes a magnetic field to instantly surround it along its full length.

In a magnetostrictive position sensor, the current is a pulse of approximately 1 to 2 µs duration. A torsional mechanical wave is launched at the location of the position magnet due to the Wiedemann effect. Portions of this wave travel both toward and away from the pickup. The wave traveling along the waveguide toward the pickup is detected when it arrives at the pickup. The time measurement between application of the current pulse (launching of the torsion wave at the position magnet) until its detection by the pickup represents the location of the position magnet. The speed of the wave is typically about 3000 m s^{-1}. The portion of the wave traveling away from the pickup could act as an interfering signal after it is reflected from the waveguide tip. So instead, it is damped by a damping element when it reaches the end of the waveguide opposite the pickup. Damping is usually accomplished by attaching one of various configurations of elastomeric materials to the end of the waveguide. The end of the waveguide within the damping element is unusable for position determination, and therefore called the "dead zone."

The time measurement can be buffered and used directly as the sensor output, or it can be conditioned inside the sensor to provide various output types, including analog voltage or current, pulse width modulation, CANbus, SSI, HART, Profibus, etc. Magnetostrictive position sensors can be made as short as 1 cm long or up to more than 30 m long. Resolution of those produced by MTS Systems Corp. is as fine as 1 µm. Temperature coefficients of 2 to 5 ppm °C^{-1} can be achieved. The sensors are inherently stable, since the measurement relies on the physical properties of the waveguide material. Longer sensors become very cost effective because the same electronics package can drive sensors of varying length; only the waveguide and its packaging are increased in length to make the sensor longer.

The magnetostrictive wire can be straight for a linear sensor, or shaped to provide curved or rotary measurements. Curved sensors are often used to measure angular or nonlinear motion in industrial applications, although rotary magnetostrictive sensors are not yet very popular.

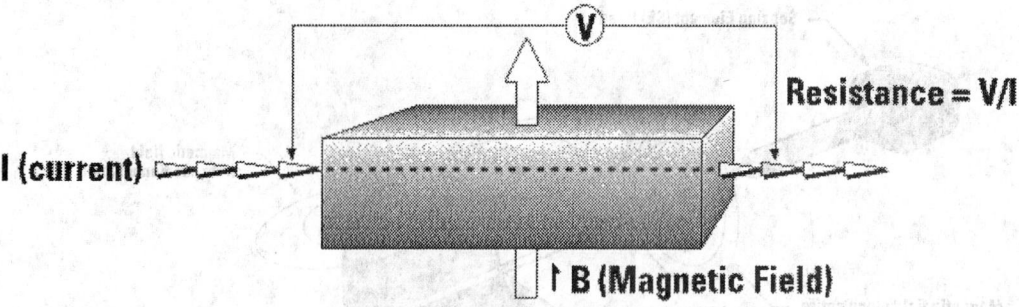

FIGURE 6.85   Magnetoresistance.

Magnetoresistive Sensors

In most magnetic materials, electrical resistance decreases when a magnetic field is applied and the magnetization is *perpendicular* to the current flow (a current will be flowing any time electrical resistance is measured) (see Figure 6.85). The resistance decreases as the magnetic flux density increases, until the material reaches magnetic saturation. The rate of resistance decrease is less as the material nears saturation. The amount of resistance change is on the order of about 1% at room temperature (0.3% in iron, 2% in nickel). When the magnetic field is *parallel* to the current, the resistance increases with increasing magnetic field strength. Sensitivity is greatest when the magnetic field is perpendicular to the current flow. These are properties of the phenomenon called *magnetoresistance* (MR). The MR effect is due to the combination of two component parts. These are: a reduction in forward carrier velocity as a result of the carriers being forced to move sideways as well as forward, and a reduction in the effective cross-sectional area of the conductor as a result of the carriers being crowded to one side [6].

When a position magnet is brought close to a single MR sensing element, the resistance change is maximum as the magnet passes over the approximate center of the element and then reduces until the magnet is past the element. The resistance changes according to:

$$\text{Resistivity} = \text{Voltage}\big/\big(\text{carrier density} \times \text{carrier velocity}\big) \qquad (6.111)$$

By using multiple MR elements arranged along a line, a longer displacement measuring device can be fashioned. The signals from the string of sensors are decoded to find which elements are being affected by the magnet. Then the individual readings are used to determine the magnet position more precisely. Relatively high-performance sensors can be manufactured. Temperature sensitivity of the MR elements needs to be compensated, and longer sensors contain many individual sensing elements. Because of this, longer sensors become more difficult to manufacture, and are expensive.

Anisotropic MR materials are capable of resistance changes in the range of 1% or 2%. The MR of a conductor body can be increased by making it a composite of two or more layers of materials having different levels of magnetoresistance. Multilayered structures of exotic materials (sometimes more than 10 layers) have enabled development of materials that exhibit much greater magnetoresistive effect, and saturate at larger applied fields. This has been named Giant MagnetoResistance (GMR). Some commercial sensors based on GMR are currently available. The GMR elements can be arranged in a four-element bridge connection for greater sensitivity. In this arrangement, two of the elements are shielded from the applied magnetic field. The other two elements are sensitive to the applied field. Sensitivity can also be increased by incorporating flux concentrators on the sensitive elements. In a bridge connection, the output voltage can vary by more than 5% of the supply voltage [7]. Rotary sensors can be constructed by attaching a pole piece to a rotating shaft. One or more permanent magnets and the pole piece are arranged to cause the magnetic field around the MR element to change with angular displacement.

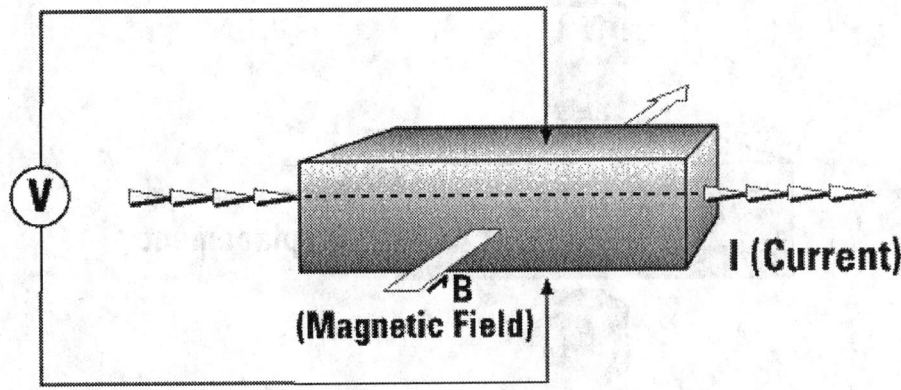

FIGURE 6.86 Hall effect.

Further research is being conducted on MR materials to improve the sensitivity by lowering the strength of magnetic field needed, and increasing the amount of resistance change. The next higher level of MR performance is being called Colossal MagnetoResistance (CMR). CMR is not yet practical for industrial sensors because of severe limitations on the operating temperature range.

Although MR, GMR, and CMR are limited for use in displacement sensors at this time by cost, temperature, and fabrication constraints, much research is in progress. Maybe Humongous Magneto-Resistance (HMR) is next?

Hall Effect Sensors

The Hall effect is a property exhibited in a conductor affected by a magnetic field. A voltage potential V_H, called the Hall voltage, appears across the conductor when a magnetic field is applied at right angles to the current flow. Its direction is perpendicular to both the magnetic field and current. The magnitude of the Hall voltage is proportional to both the magnetic flux density and the current. The magnetic field causes a gradient of carrier concentration across the conductor. The larger number of carriers on one side of the conductor, compared to the other side, causes the voltage potential V_H. A pictorial representation is shown in Figure 6.86. The amplitude of the voltage varies with the current and magnetic field according to: [8]

$$V_H = K_H \beta I / z \tag{6.112}$$

where V_H = Hall voltage
K_H = Hall constant
β = magnetic flux density
I = current flowing through the conductor
z = thickness of the conductor

Sensors utilizing the Hall effect typically are constructed of semiconductor material, giving the advantage of allowing conditioning electronics to be deposited right on the same material. Either *p*- or *n*-type semiconductor material can be used, with the associated polarity of current flow. The greatest output is achieved with a large Hall constant, which requires high carrier mobility. Low resistivity will limit thermal noise voltage, for a more useful signal-to-noise ratio (SNR). These conditions are optimized using an *n*-type semiconductor [6].

A displacement sensor can be made with a Hall sensing element and a movable magnet, with an output proportional to the distance between the two. Two magnets can be arranged with one Hall sensor as in Figure 6.87 to yield a near-zero field intensity when the sensor is equidistant between the magnets. These

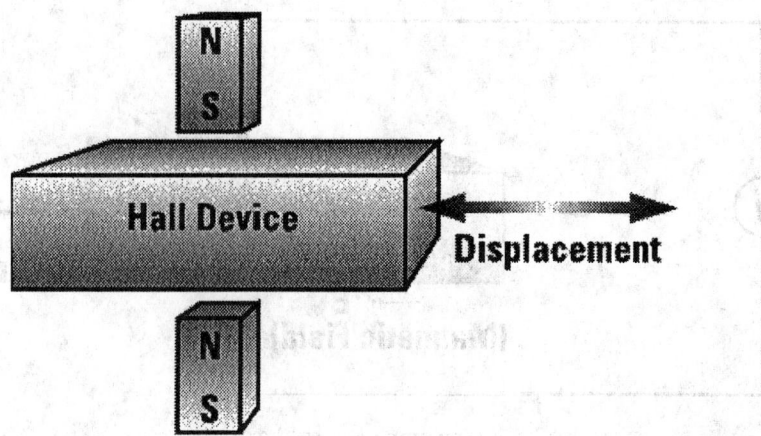

FIGURE 6.87   Two magnet hall sensor.

single Hall effect device configurations have a very limited linear range. Longer range displacement sensors can be built using multiple Hall sensors spaced along a carrier substrate. A magnet is moved along in close proximity to the carrier. As the magnet approaches and then moves away from each Hall element, the respective sensors will have increasing or decreasing outputs. The output from the battery of sensors is derived by reading the individual outputs of the sensors closest to the magnet, and also decoding those particular sensors being read. This method can produce relatively high-performance displacement sensors of up to several meters long. Longer sensors become increasingly more difficult to produce and are expensive because of the large number of sensors being multiplexed. Rotary, as well as linear displacement, sensors can be produced by mechanical arrangement of the sensing elements to cause magnetic field variation with the desired angular or linear input.

Magnetic Encoders

Magnetic encoders use a strip or disk of magnetic media onto which digital information is stored. This information is recorded at the location it describes, and is in the form of a collection of magnetized and nonmagnetized areas. A magnetic encoder includes this sensing element, as well as one or more read heads, electronics, and a mechanical enclosure with input shaft and bushings. The input shaft moves in and out for a linear sensor. It has wipes to prevent ingestion of foreign material, and bushings designed to accept side-loading. An angular sensor has a shaft that rotates, and includes bushings to withstand thrust and side-loading. The encoded media is implemented as either a strip in a linear sensor, or as a disk in an angular sensor.

As a read head passes above the encoded area, it picks up the magnetic variations and reads the position information. The information, digital ones and zeroes, will usually be encoded in several parallel tracks to represent the binary digits of the position information. A standard binary code presents a problem for encoders in that some numbers require the changing of several of the bits at one time to indicate a single increment of the number represented. If all the changing bits are not perfectly aligned with each other, instantaneous erroneous readings will result. To avoid this problem, a special adaptation of the binary code called "Gray code" is used. See Table 6.25. A single increment of the number represented causes a change of only 1 bit in the Gray code.

The read head incorporates a ferromagnetic core wound with input and output windings. A read pulse is applied to the input winding, and information is read on the output winding. If the core is above a magnetized area of the magnetic media, the core becomes saturated, no output pulse is generated, and a logic 0 results [9]. If the core is above a nonmagnetized area when the read pulse is applied, an output pulse occurs and produces a logic 1. Another arrangement that is practical for angular, but not linear, encoders, uses a ring-shaped multipole permanent magnet. The magnet is rotated past a pair of sensors

TABLE 6.25 Gray Code

Base$_{10}$ number	"Natural" binary	Gray code	Binary Coded Decimal (BCD)	
			tens	units
0	0000	0000	0000	0000
1	0001	0001	0000	0001
2	0010	0011	0000	0010
3	0011	0010	0000	0011
4	0100	0110	0000	0100
5	0101	0111	0000	0101
6	0110	0101	0000	0110
7	0111	0100	0000	0111
8	1000	1100	0000	1000
9	1001	1101	0000	1001
10	1010	1111	0001	0000

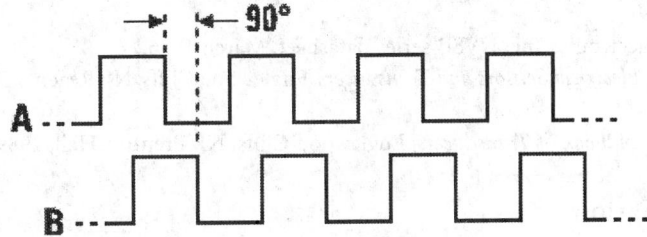

FIGURE 6.88 Quadrature output.

to yield an incremental reading with sine and cosine outputs (called "quadrature" output. See Figure 6.88). The waveforms can be square, sinusoidal, or triangular. A and B outputs are used to indicate the displacement and the direction of the displacement. The number of transitions or "counts" is proportional to the displacement magnitude. The direction of displacement (i.e., + or −) can be found by comparing the two phases. For example, in Figure 6.88, at the time the A phase changes from a logic 0 to a 1, the status of the B phase will indicate the direction of travel. A logic 0 on the B phase could equal the positive direction; a logic 1 could equal the negative direction.

Magnetic encoders can be incremental or absolute. In an incremental configuration, equally spaced pulses encoded on the magnetic media are read from one or more tracks. The pulses are collected by an up/down counter, and the counter output represents the position. Quadrature outputs can be coded to tell the direction of displacement, as described above. The zero position is set by resetting the counter.

Absolute magnetic encoders have the digital code representing the position encoded directly at that position. No counter is needed. The Gray code can be interpreted to yield the position in engineering units. Nonlinear coding, such as sine or cosine, is sometimes used. Table 6.26 provides a list of sensors and manufacturers.

References

1. O. Esbach, *Handbook of Engineering Fundamentals,* New York: John Wiley & Sons, 1975, 957.
2. R. Lerner and G. Trigg, *Encyclopedia of Physics,* New York: VCH Publishers, 1990, 529.
3. P. Neelakanta, *Handbook of Electromagnetic Materials,* Boca Raton, FL: CRC Press, 1995, 333.
4. D. S. Nyce, Magnetostriction-based linear position sensors, *Sensors,* 11(4), 22, 1994.
5. R. Philippe, *Electrical and Magnetic Properties of Materials,* Norwood, MA: Artech House, 1988.
6. H. Burke, *Handbook of Magnetic Phenomena,* New York: Van Nostrand Reinhold, 1986.

TABLE 6.26 Sensors and Manufacturers

Technology	Manufacturers	Description	Price
Magnetostrictive	MTS Systems Corp. Cary, NC & Germany	Lengths to 20 m; 2 μm resolution; CAN, SSI, Profibus, HART	$150–$3000
	Balluff Germany	Lengths to 3.5 m, 20 μm resolution no standard interfaces in head	$400–$2300
Magnetoresistive	Nonvolatile Electronics Eden Prairie, MN	GMR sensors with flux concentrator and shield	$2.50–$6.00
	Midori America	Rotary MR sensors	$64–$500
	Fullerton, CA	Linear MR up to 30 mm	$67–$200
Hall Effect	Optec Technology, Inc. Carrollton, TX	Linear position	$5–$50
	Spectec Emigrant, MT	Standard and custom sensors	Approx. $90
Magnetic encoder	Heidenhain Schaumburg, IL	Rotary and linear encoders	$300–$2000
	Sony Precision Technology America Orange, CA	Rotary and linear encoders	$100–$2000

7. Nonvolatile Electronics Inc. NVSB series datasheet. March 1996.
8. J. R. Carstens, *Electrical Sensors and Transducers,* Englewood Cliffs, NJ: Regents/Prentice-Hall, 1992, p. 125.
9. H. Norton, *Handbook of Transducers,* Englewood Cliffs, NJ: Prentice-Hall, 1989, 106-112.

Further Information

B. D. Cullity, *Introduction to Magnetic Materials,* Reading, MA: Addison-Wesley, 1972.
D. Craik, *Magnetism Principles and Applications,* New York: John Wiley & Sons, 1995.
P. Lorrain and D. Corson, *Electromagnetic Fields and Waves,* San Francisco: W.H. Freeman, 1962.
R. Boll, *Soft Magnetic Materials,* London: Heyden & Son, 1977.
H. Olson, *Dynamical Analogies,* New York: D. Van Nostrand, 1943.
D. Askeland, *The Science and Engineering of Materials,* Boston: PWS-Kent Publishing, 1989.
R. Rose, L. Shepard, and J. Wulff, *The Structure and Properties of Materials,* New York: John Wiley & Sons, 1966.
J. Shackelford, *Introduction to Materials Science for Engineers,* New York: Macmillan, 1985.
D. Jiles, *Introduction to Magnetism and Magnetic Materials,* London: Chapman and Hall, 1991.
F. Mazda, *Electronics Engineer's Reference Book, 6th ed.,* London: Butterworth, 1989.
E. Herceg, *Handbook of Measurement and Control,* New Jersey: Schaevitz Engineering, 1976.

6.10   Synchro/Resolver Displacement Sensors

Robert M. Hyatt, Jr. and David Dayton

Most electromagnetic position transducers are based on transformer technology. Transformers work by exciting the primary winding with a continuously changing voltage and inducing a voltage in the secondary winding by subjecting it to the changing magnetic field set up by the primary. They are ac-only devices, which make all electromagnetically coupled position sensors ac transformer coupled. They are inductive by nature, consisting of wound coils. By varying the amount of coupling from the primary (excited) winding of a transformer to the secondary (coupled) winding with respect to either linear or rotary displacement, an analog signal can be generated that represents the displacement. This coupling variation is accomplished by moving either one of the windings or a core element that provides a flux path between the two windings.

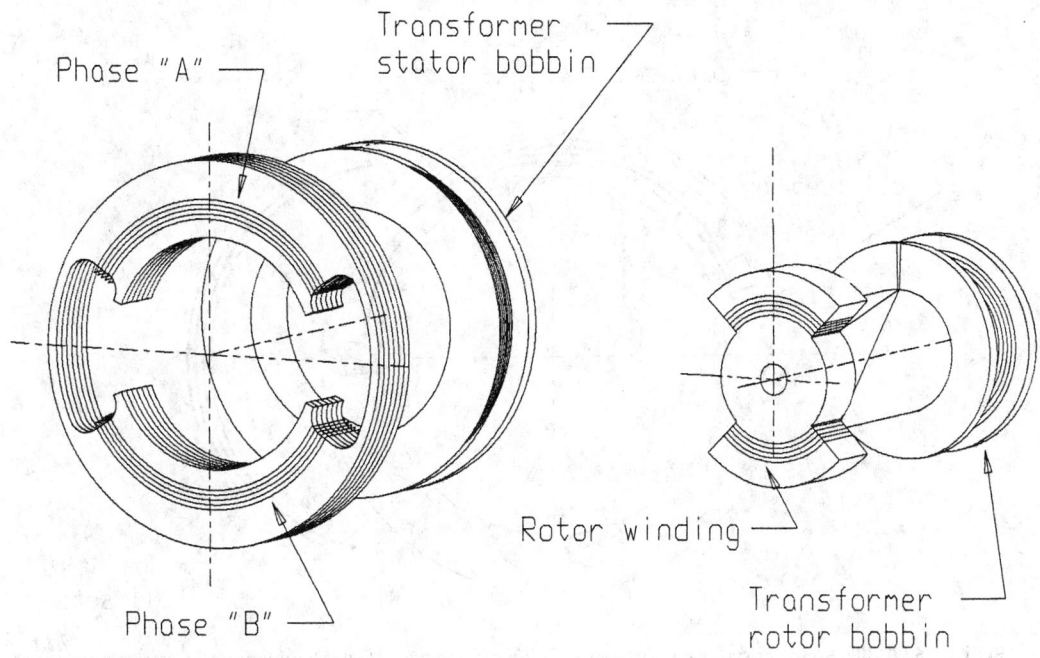

FIGURE 6.89   The induction potentiometer has windings on the rotor and the stator.

One of the simplest forms of electromagnetic position transducers is the LVDT, which is described in Section 6.2 on inductive sensors. If the displacement of the core in an LVDT-type unit is changed from linear to rotary, the device becomes an RVDT (rotary variable differential transformer).

Induction Potentiometers

The component designer can "boost" the output, increase accuracy, and achieve a slightly greater angular range if windings are placed on the rotor as shown in Figure 6.89. The disadvantages to this method are (1) additional windings, (2) more physical space required, (3) greater variation over temperature, and (4) greater phase shift due to the additional windings.

The advantage of the induction pot design is greater sensitivity (more volts per degree), resulting in better signal to noise and higher accuracy in most cases.

Resolvers

If the two-slot lamination in the stator stack shown in Figure 6.89 is replaced by a multislot lamination (see Figure 6.90), and two sets of windings are designed in concentric coil sets and distributed in each quadrant of the laminated stack; a close approximation to a sine wave can be generated on one of the secondary windings and a close approximation to a cosine wave can be generated on the other set of windings. Rotary transformers of this design are called *resolvers*. Using a multislot rotor lamination and distributing the windings in the rotor, the sine-cosine waveforms can be improved even further.

A resolver effectively amplitude modulates the ac excitation signal placed on the rotor windings in proportion to the sine and the cosine of the angle of mechanical rotation. This sine-cosine electrical output information measured across the stator windings may be used for position and velocity data. In this manner, the resolver is an analog trigonometric function generator. Most resolvers have two primary windings that are located at right angles to each other in the stator, and two secondary windings also at right angles to each other, located on the rotor (see Figure 6.91).

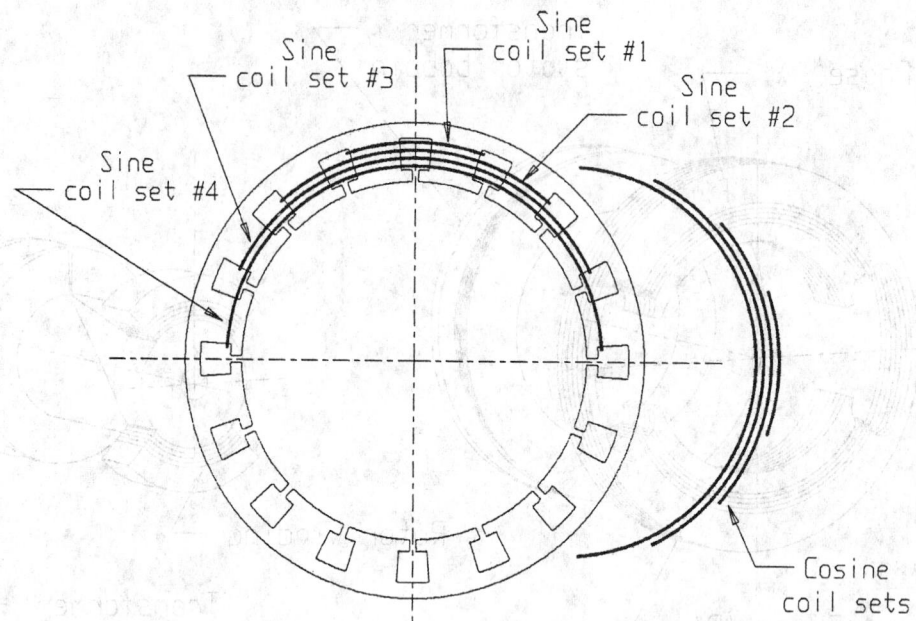

FIGURE 6.90   The resolver stator has distributed coil windings on a 16-slot lamination to generate a sine wave.

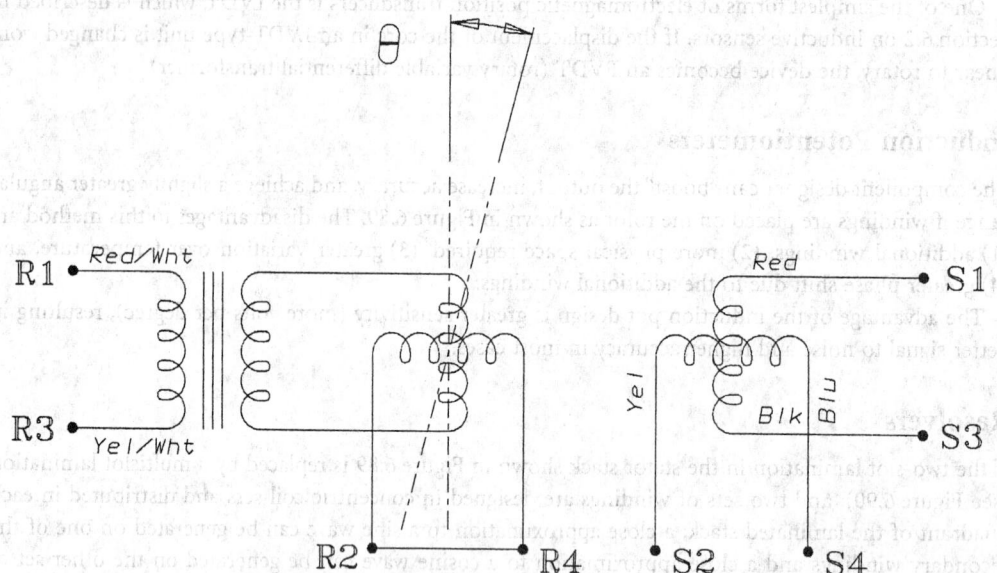

FIGURE 6.91   A brushless resolver modulates the ac excitation on the rotor by the rotation angle.

If the rotor winding (R1-R3) is excited with the rated input voltage (see Figure 6.92), the amplitude of the output winding of the stator (S2-S4) will be proportional to the sine of the rotor angle θ, and the amplitude of the output of the second stator winding (S1-S3) will be proportional to cosine θ. (See Figure 6.93.) This is commonly called the "control transmitter" mode and is used with most "state-of-the-art" resolver to digital converters.

In the control transmitter mode, electrical zero may be defined as the position of the rotor with respect to the stator at which there is minimum voltage across S2-S4 when the rotor winding R1-R3 is excited

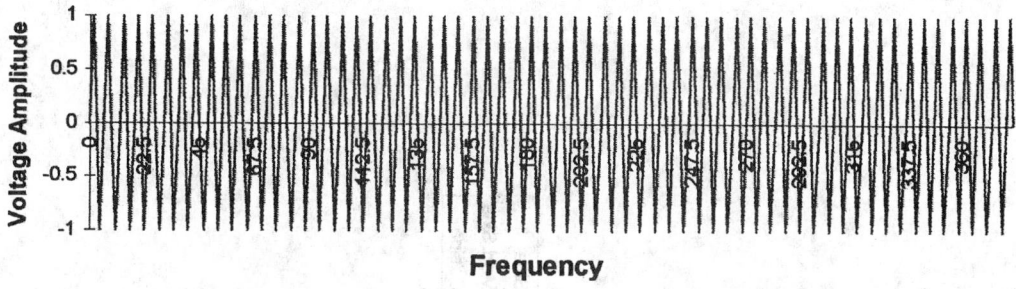

FIGURE 6.92   The resolver rotor winding is excited with the rated input voltage.

with rated voltage. Nulls will occur across S2-S4 at the 0° and 180° positions, and will occur across S1-S3 at the 90° and 270° positions.

If the stator winding S1-S3 is excited with the rated input voltage and stator winding S2-S4 is excited with the rated input voltage electrically shifted by exactly 90°, then the output sensed on the rotor winding R1-R3 does not vary with rotor rotation in amplitude or frequency from the input reference signal. It is the sum of both inputs. It does, however, vary in time phase from the rated input by the angle of the shaft from a referenced "zero" point (see Figure 6.94). This is a "phase analog" output and the device is termed a "control transformer." By measuring the time difference between the zero crossing of the reference voltage waveform and the output voltage waveform, the phase angle (which is the physical angular displacement of the output shaft) can be calculated.

Because the resolver is an analog device and the outputs are continuous through 360°, the theoretical resolution of a resolver is infinite. There are, however, ambiguities in output voltages caused by inherent variations in the transformation of the voltage from primary to secondary through 360° of rotation. These ambiguities result in inaccuracy when determining the true angular position. The types of error signals that are found in resolvers are shown in Figure 6.95.

As a rule, the larger the diameter of the stator laminations, the better the accuracy and the higher the absolute resolution of the device. This is a function of the number of magnetic poles that can be fit into the device, which is a direct function of the number of slots in the stator and rotor laminations. With multispeed units (see the section on multispeeds below), the resolution increases as a multiple of the speeds. For most angular excursions, the multispeed resolver can exceed the positioning accuracy capability of any other component in its size, weight, and price range.

Operating Parameters and Specifications for Resolvers

There are seven functional parameters that define the operation of a resolver in the analog mode. These are (1) accuracy, (2) operating voltage amplitude, (3) operating frequency, (4) phase shift of the output voltage from the referenced input voltage, (5) maximum allowable current draw, (6) the transformation ratio of output voltage over input voltage, and (7) the null voltage. Although impedance controls the functional parameters, it is transparent to the user. The lamination and coil design are usually developed to minimize null voltage and input current, and the impedance is a direct fallout of the inherent design of the resolver. The following procedure can be used to measure the seven values for most resolvers.

Equipment Needed for Testing Resolvers

A mechanical index stand that can position the shaft of the resolver to an angular accuracy that is an order of magnitude greater than the specified accuracy of the resolver.

An ac signal generator capable of up to 24 Vrms at 10 kHz.

A phase angle voltmeter (PAV) capable of measuring "in phase" and "quadrature" voltage components for determining the transformation ratio and the null voltage as well as the phase angle between the output voltage and the reference input voltage.

A 1 Ω resistor used to measure input current with the PAV.

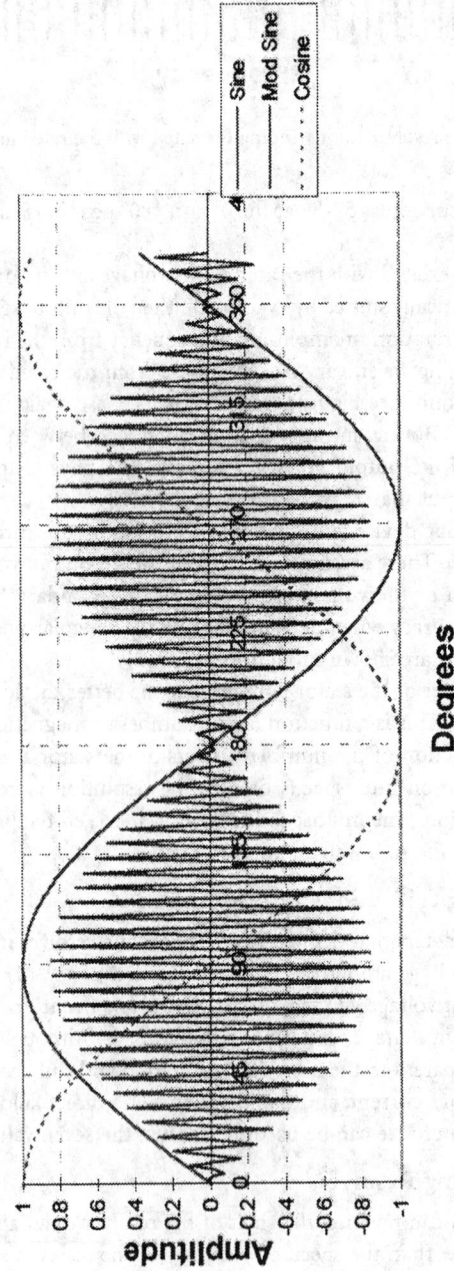

FIGURE 6.93 The single-speed resolver stator output is the sine or cosine of the angle.

Input Voltage
Phase Analog Output
Amplitude
1
0.8
0.6
0.4
0.2
0
-0.2
-0.4
-0.6
-0.8
-1
0
45
90
135
180
225
270
315
360
405
Degrees

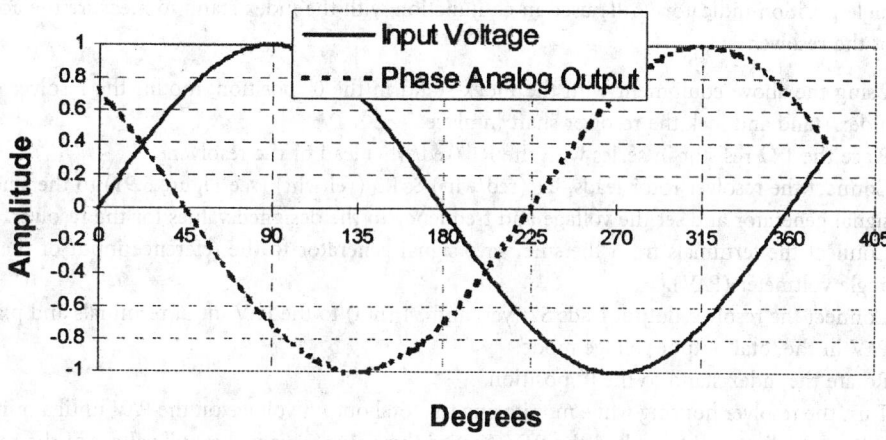

FIGURE 6.94 Resolver stator windings excited at 0° and 90° yield a phase shift with rotor rotation. Phase analog output at 225° vs. input voltage.

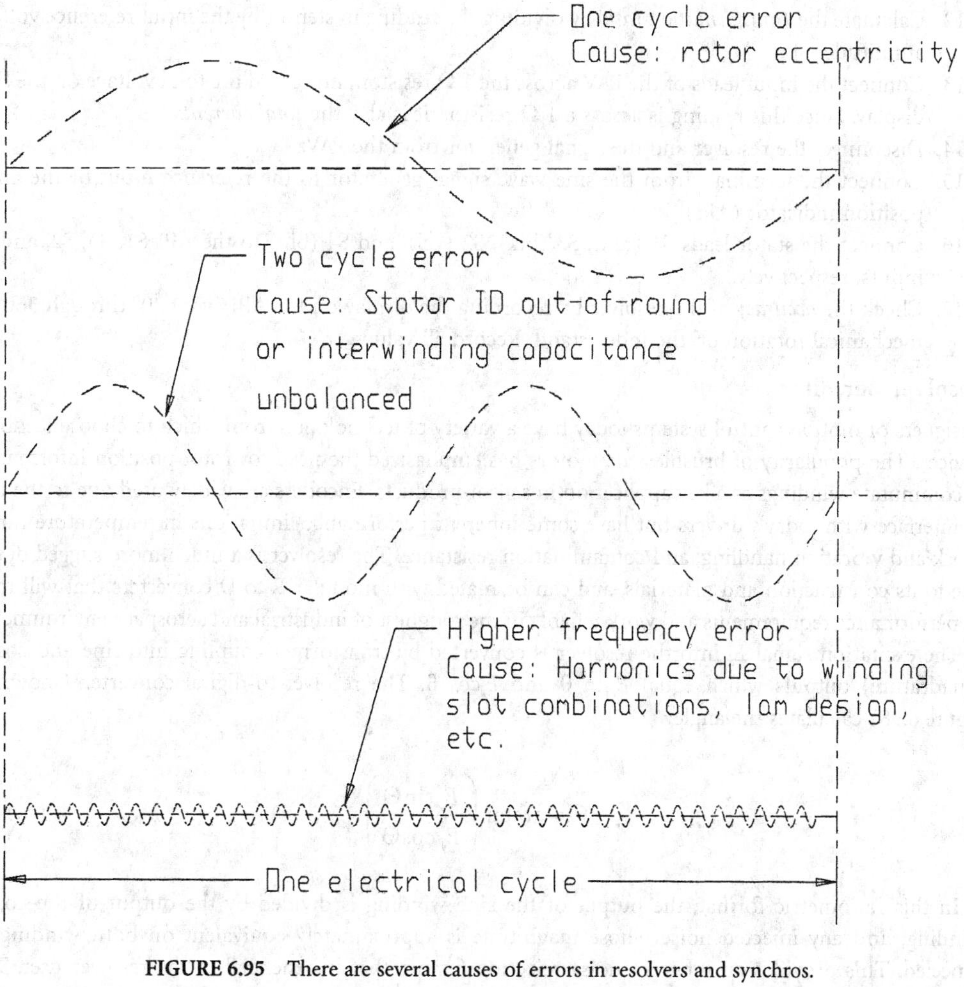

One cycle error
Cause: rotor eccentricity
Two cycle error
Cause: Stator ID out-of-round
or interwinding capacitance
unbalanced
Higher frequency error
Cause: Harmonics due to winding
slot combinations, lam design,
etc.
One electrical cycle

FIGURE 6.95 There are several causes of errors in resolvers and synchros.

An angle position indicator (API) used in conjunction with the index stand to measure the accuracy of the resolver.

1. Using the above equipment with the index stand in the 0° position, mount the resolver on the index stand and lock the resolver shaft in place.
2. Place the 1 Ω resistor in series with the R1 (red/wht) lead of the resolver.
3. Connect the resolver rotor leads, R1 (red/wht) & R3 (yel/wht) (see Figure 6.91) to the sine wave signal generator and set the voltage and frequency to the designed values for the resolver.
4. Connect the terminals from the sine wave signal generator to the reference input of the phase angle voltmeter (PAV).
5. Connect the resolver output leads S2 (yel) and S4 (blu) to the PAV input terminals and place the PAV in the total output voltage mode.
6. Rotate the index stand to the 0° position.
7. Turn the resolver housing while monitoring the total output voltage on the PAV until a minimum voltage reading is obtained on the PAV. Record this value. This is the *null voltage* of the resolver.
8. Turn the index stand to the 90° position.
9. Change the output display on the PAV to show the *phase angle* between the reference voltage and the voltage on leads S2-S4. Record this phase angle reading.
10. Change the output display on the PAV to show the *total output voltage* of the resolver.
11. Record this voltage.
12. Calculate the *transformation ratio* by dividing the reading in step 11 by the input reference voltage amplitude.
13. Connect the input leads of the PAV across the 1 Ω resistor and record the total voltage on the PAV display. Since this reading is across a 1 Ω resistor, it is also the *total current*.
14. Disconnect the resolver and the signal generator from the PAV.
15. Connect the terminals from the sine wave signal generator to the reference input of the angle position indicator (API).
16. Connect the stator leads S1 (red), S3 (blk), S2 (yel), and S4 (blu) to the API S1, S3, S2, and S4 inputs, respectively.
17. Check the *accuracy* of the resolver by recording the display on the API every 20° through 360° of mechanical rotation on the index stand. Record all values.

Resolver Benefits

Designers of motion control systems today have a variety of technologies from which to choose feedback devices. The popularity of brushless dc motors has emphasized the need for rotor position information to commutate windings as a key application in sensor products. Encoders are widely used due to the ease of interface with today's drivers but have some inherent performance limitations in temperature range, shock and vibration handling, and contamination resistance. The resolver is a much more rugged device due to its construction and materials, and can be mated with modular R to D converters that will meet all performance requirements and work reliably in the roughest of industrial and aerospace environments.

The excitation signal E_x into the resolver is converted by transformer coupling into sine and cosine (quadrature) outputs which equal $E_x \sin \theta$ and $E_x \cos \theta$. The resolver-to-digital converter (shown in Figure 6.98) calculates the angle:

$$\Theta = \arctan\left(\frac{E_x \sin\Theta}{E_x \cos\Theta}\right) \qquad (6.113)$$

In this ratiometric format, the output of the sine winding is divided by the output of the cosine winding, and any injected noise whose magnitude is approximately equivalent on both windings is canceled. This provides an inherent noise rejection feature that is beneficial to the resolver user. This feature also results in a large degree of temperature compensation.

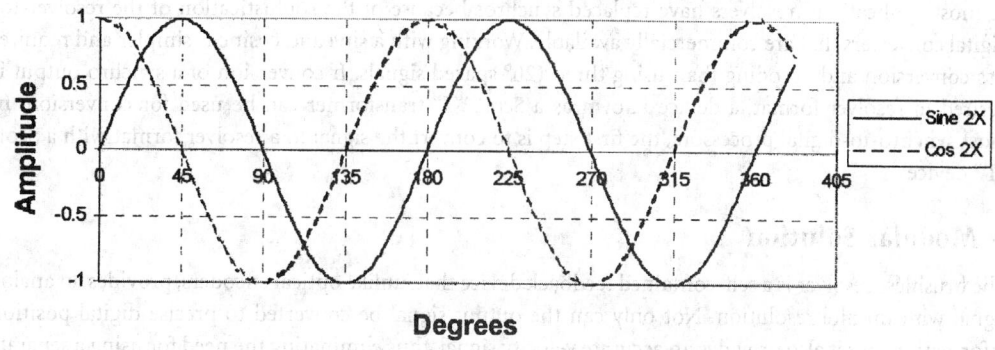

Degrees

FIGURE 6.96   A two-speed resolver yields two electrical cycles for one rotation.

Multispeed Units

The relationship for multispeeds is that the speed (2×, 3×, etc.) designates how many full sinusoidal cycles the resolver output electrically completes in 360° of mechanical rotation. The 2× electrical output is such that the full sinusoidal cycle for a 2× resolver occurs in 180° instead of 360°. A 2× resolver output is shown in Figure 6.96. A full 3× cycle is completed in 120°. The number of speeds selected for use is a function of the system requirements. Increasing the number of magnetic poles in the rotor and stator creates multispeed units. Each speed has several winding and slot combinations. The optimum combination is selected by the resolver designer based on system demands.

Applications

Resolvers are often used in conjunction with motors, and because of their inherent similarity of design (copper windings on iron lamination stacks), their environmental resistance is quite similar. They are ideal to design into industrial applications where dust and airborne liquids can obscure optical encoder signals. NC machines, coil winders, presses, and positioning tables are uses where resolvers excel. The resolver's inherent resistance to shock and vibration makes it uniquely suited to moving platforms, and their reliability under these conditions lends a welcome hand to the designers of robots, gantries, and automotive transfer lines.

Heat sensitivity is always a problem for motion control systems designers. Resolvers used for sensing the position of valves in high-temperature applications such as aircraft engines, petrochemical refining, and chemical processing have continually proven their reliability.

Moving devices to precise positions with smooth and accurate control can be a real challenge in the electromagnetic noise environment of the modern industrial facility. Emitted and conducted EMI from adjacent equipment, and input voltage variations with unwanted current spikes on input power lines can rob digital systems of their signal integrity. The analog resolver continues to function without information loss or signal interruption. Digitizing the signal can be done at a remote interface under more controlled conditions than on the factory floor. Only robust materials can perform well in harsh environments.

Synchros

As long ago as World War II, synchros were used in analog positioning systems to provide data and to control the physical position of mechanical devices such as radar antennae, indicator needles on instrumentation, and fire control mechanisms in military equipment. The term "synchro" defines an electromagnetic position transducer that has a set of three phase output windings that are electrically and mechanically spaced by 120° instead of the 90° spacing found in a resolver. In the rotor primary mode, the synchro is excited by a single-phase ac signal on the rotor. As the rotor moves 360°, the three amplitude modulated sine waves on the three phases of the output have a discrete set of amplitudes for each angular position. By interpreting these amplitudes, a table can be established to decode the exact rotary position.

In most applications, resolvers have replaced synchros because of the sophistication of the resolver-to-digital converters that are commercially available. Working with a sine and cosine is simpler and requires less conversion and decoding than using three 120° spaced signals. If conversion of a synchro output is desired in resolver format, a device known as a Scott "T" transformer can be used for conversion. In most synchro-to-digital processors, the first step is to convert the signal to a resolver format with a Scott "T" device.

A Modular Solution

The brushless resolver is a self-contained feedback device that, unlike optical encoders, provides an analog signal with infinite resolution. Not only can the output signal be converted to precise digital position information, but it also provides an accurate velocity signal, thus eliminating the need for using a separate tachometer. Reliability is enhanced using the same resolver for speed feedback and commutation. Piece part count can be reduced and the complexity of using Hall-effect devices for timing signals for commutation can be eliminated.

A modular approach allows the designer to easily select a single or multispeed resolver and appropriate electronics that will meet almost any desired level of resolution and accuracy. The resolvers are designed in the most commonly used frame sizes: 8, 11, 15, and 21. Housed models feature high-quality, motor-grade ball bearings. Heavy-duty industrial grade units are enclosed in rugged black painted aluminum housings with either flange, face, or servo-type mounting, and utilize MS-style connectors.

The Sensible Design Alternative for Shaft Angle Encoding

The requirement for velocity and position feedback plays an important role in today's motion control systems. With the development of low-cost monolithic resolver-to-digital converters, a resolver-based system provides design engineers with the building blocks to handle a wide variety of applications. A resolver's small size, rugged design, and the ability to provide a very high degree of accuracy under severe conditions, make this the ideal transducer for absolute position sensing. These devices are also well suited for use in extremely hostile environments such as continuous mechanical shock and vibration, humidity, oil mist, coolants, and solvents. Absolute position sensing vs. incremental position sensing is a necessity when working in an environment where there is the possibility of power loss. Whenever power is supplied to an absolute system, it is capable of reading its position immediately; this eliminates the need for a "go home" or reference starting point.

Resolver-to-Digital Converters

A monolithic resolver-to-digital converter requires only six external passive components to set the bandwidth and maximum tracking rate. The bandwidth controls how quickly the converter will react to a large change in position on the resolver output. The converter can also be programmed to provide either 10, 12, 14, or 16 bits of parallel data. A resolver-based system can provide high dynamic capability and high resolution for today's motion control systems where precision feedback for both position and velocity is required.

Closed Loop Feedback

In a typical closed loop servo model as in Figure 6.97, the position sensor plays an important role by constantly updating the position and velocity information. Selection of a machine control strategy will often be based on performance, total application cost, and technology comfort. The accuracy of the system is determined by the smallest resolution of the position-sensing device. A resolver-to-digital converter in the 16-bit mode has 2^{16} (65,536) counts per revolution, which is equivalent to a resolution of 20 arc seconds. The overall accuracy of the resolver-to-digital converter is ±2.3 arc minutes. An accuracy specification defines the maximum error in achieving a desired position. System accuracy must be smaller than the tolerance on the desired measurement. An important feature of the resolver-to-digital converter

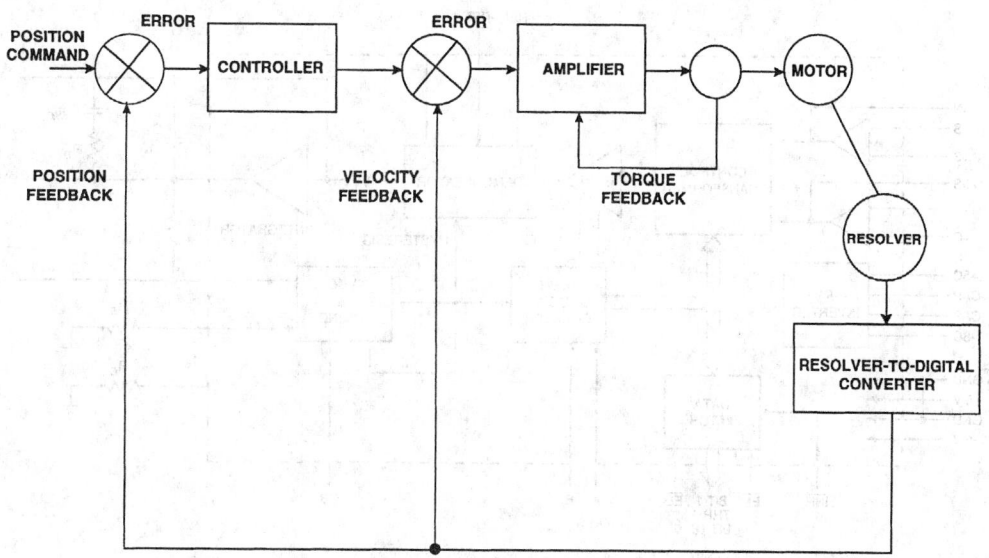

FIGURE 6.97 A closed-loop servo model uses a resolver-to-digital converter.

is repeatability. With a repeatability specification of ±1 LSB (least significant bit) in the 16-bit mode, this provides an accurate measurement when determining position from point to point. For example, moving from point A to point B and back to point A, the converter in the 16-bit mode will be accurate within 20 arc seconds of the original position. The error curve of a resolver-to-digital converter is repeatable within ±1 LSB. The combination of high precision resolvers (±20 arc seconds) with a resolver-to-digital converter provides accurate absolute position information for precision feedback for motion control.

Type II Servo Loop

The motor speed is monitored using the velocity output signal generated by the resolver-to-digital converter. This signal is a dc voltage proportional to the rate of speed, positive for increasing angles and negative for decreasing angles, with a typical linearity specification of 0.25% and a typical reversal error of 0.75%. The error processing is performed using the industry standard technique for type II tracking, resolver-to-digital converters (see Figure 6.98).

The dc error is integrated, yielding a velocity voltage that drives a voltage-controlled oscillator (VCO). This VCO is an incremental integrator (constant voltage input to position rate output) that together with the velocity integrator, forms a type II critically damped, servo feedback loop. This information allows the motor to maintain constant speeds under varying loads when it is interfaced with a programmable logic controller (PLC). The PLC-based architecture is used for I/O intensive control applications. The PLC provides a low-cost option for those developers familiar with its ladder logic programming language. Integration of the motion, I/O, operator's interface, and communication are usually supported through additional cards that are plugged into the backplane.

Applications

Specific applications require unique profiles to control the speed and acceleration of the motor to perform the task at hand. By reducing the accelerations and decelerations that occur during each operation, it is possible to lower the cost and use more efficient motors. Industrial applications include the following:

Ballscrew positioning
Motor commutation
Robotics positioning

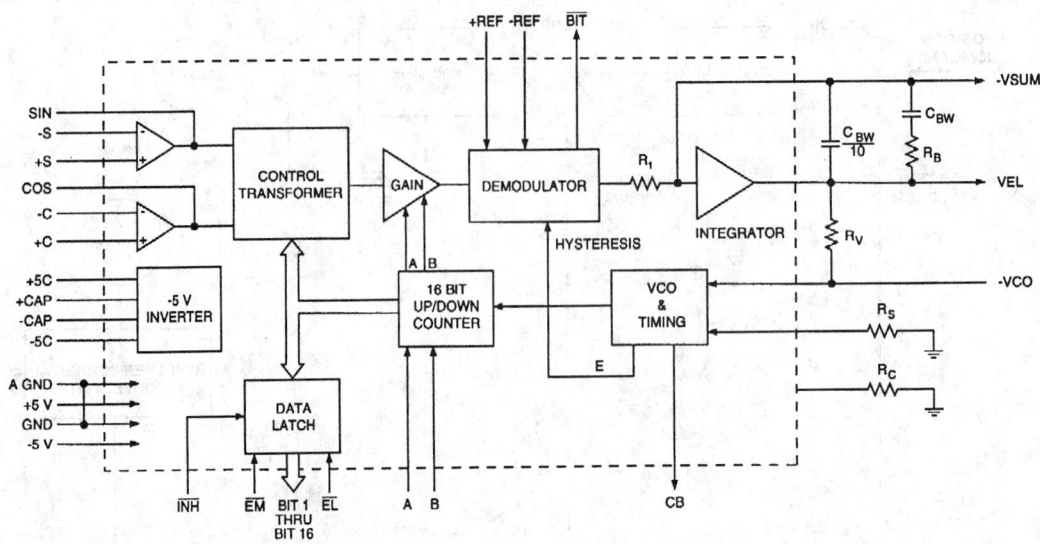

FIGURE 6.98   Error processing uses type II tracking resolver-to-digital converters.

Machine vision systems
X–Y tables
Component insertion
Remote video controls
Web guides
Pick and place machines

Resolver-to-Digital Conversion

For a resolver-to-digital converter, the resolver information is presented to a solid-state resolver conditioner that reduces the signal amplitude to 2 V rms sine and cosine; the amplitude of one being proportional to the sine of θ (the angle to be digitized), and the amplitude of the other being proportional to
the cosine of θ. (The amplitudes referred to are, of course, the carrier amplitudes at the reference
frequency, i.e., the cosine wave is actually $\cos \theta \cos \omega t$; but the carrier term, $\cos \omega t$, will be ignored in
this discussion because it will be removed in the demodulator, and at any rate contains no data). A
quadrant selector circuit in the control transformer enables selection of the quadrant in which θ lies,
and automatically sets the polarities of the sine θ and cos θ appropriately, for computational significance.
The sin θ, cos θ outputs of the quadrant selector are then fed to the sine and cosine multipliers, also
contained in the control transformer. These multipliers are digitally programmed resistive networks. The
transfer function of each of these networks is determined by a digital input (which switches in propor-
tioned resistors), so that the instantaneous value of the output is the product of the instantaneous value
of the analog input and the sine (or cosine) of the digitally encoded angle. If the instantaneous value of
the analog input of the sine multiplier is cos θ, and the digitally encoded "word" presented to the sine
multiplier is ϕ, then the output code is $\cos \theta \sin \phi$. Thus, the two outputs of the multipliers are

From the sine multiplier: $\cos \theta \sin \phi$

From the cosine multiplier: $\sin \theta \cos \phi$

These outputs are fed to an operational subtractor, at the differencing junction shown, so that the input
fed to the demodulator is

$$\sin \theta \cos \phi - \cos \theta \sin \phi = \sin (\theta - \phi) \qquad (6.114)$$

The right-hand side of this trigonometric identity indicates that the differencing-junction output represents a carrier-frequency sine wave with an amplitude proportional to the sine of the difference between θ (the angle to be digitized) and ϕ (the angle stored in digital form in the up/down counter). This point is the ac error signal brought out as (e). The demodulator is also presented with the reference voltage, which has been isolated from the reference source, and appropriately scaled, by the reference conditioner. The output of the demodulator is then, an analog dc level, proportional to $\sin(\theta - \phi)$, in other words, to the sine of the "error" between the actual angular position of the resolver and the digitally encoded angle, ϕ, which is the output of the counter. This point dc error is sometimes brought out as (D) while an addition of a threshold detector will give a built-in-test (BIT) flag. When the ac error signal exceeds 100 LSBs, the BIT flag will indicate a tracking error. This angular error signal is then fed into the error processor and VCO. This circuit consists essentially of an analog integrator whose output (the time integral to the error) controls the frequency of a voltage-controlled oscillator (VCO). The VCO produces clock pulses that are counted by the up/down counter. The "sense" of the error (ϕ too high or ϕ too low) is determined by the polarity of (ϕ), and is used to generate a control counter signal (U), which determines whether the counter increments upward or downward. Finally, note that the up/down counter, like any counter, is functionally an incremental integrator; therefore, the tracking converter constitutes in itself a closed-loop servomechanism (continuously attempting to null the error to zero) with two integrators in series. This called a "Type II" servo loop, which has decided advantages over Type 1 or Type 0 loops. In order to appreciate the value of a Type II servo behavior of this tracking converter, consider first the shaft is not moving. Ignoring inaccuracies, drifts, and the inevitable quantizing error, the error should be zero ($\theta = \phi$), and the digital output represents the true shaft angle of the resolver. Now, start the resolver shaft moving, and allow it to accelerate uniformly, from $d\theta/dt = 0$ to $d\theta/dt = V$. During the acceleration, an error will develop because the converter cannot instantaneously respond to the change of angular velocity. However, since the VCO is controlled by an integrator, the output of which is the integral of the error, the greater the lag (between θ and ϕ), the faster the counter will be called on to catch up. When the velocity becomes constant at V, the VCO will have settled to a rate of counting that exactly corresponds to the rate of change in θ per unit time and instantaneously $\theta = \phi$. Therefore, $d\phi/dt$ will always track $d\theta/dt$ without a velocity or position error. the only error will be momentary (transient) error, during acceleration or deceleration. Furthermore, the information produced by the tracking converter is always "fresh," being continually updated, and always available at the output of the counter. Since $d\theta/dt$ tracks the input velocity it can be brought out as velocity, a dc voltage proportional to the rate of rotation, which is of sufficient linearity in modern converters to eliminate the need for a tachometer in many systems.

Bandwidth Optimization

When using a low-cost monolithic converter for position and velocity feedback, it is important to understand the dynamic response for a changing input. When considering what bandwidth to set the converter, several parameters must be taken into consideration. The ability to track step responses and accelerations will determine what bandwidth to select. The lower the bandwidth of the resolver-to-digital converter, the greater the noise immunity; high frequency noise will be rejected. The relationship between the maximum tracking rate and bandwidth determines the settling time for small and large steps. For a small step input, the bandwidth determines the converter settling time. When one has a large step, the maximum velocity slew rate and bandwidth together, determine the settling time.

Encoder Emulation

Today's resolver-to-digital converters also have the ability to emulate the output of an optical incremental encoder. By providing the outputs A, B, and Zero Index, the encoder can be replaced with a resolver and resolver-to-digital converter without changing the existing interface hardware.

Determining Position Lag Error Due to Acceleration

As the bandwidth and the maximum tracking rate are varied, one can determine the acceleration constant (K_a) and large step settling time.

EXAMPLE:

Resolution:	16 bit
Bandwidth:	100 Hz
Reference:	1000 Hz
Max tracking:	10 rps

$$BW = \frac{\sqrt{2}A}{\pi} \qquad (6.115)$$

If the bandwidth = 100 Hz

$$A = 222$$
$$K_a = A^2$$
$$K_a = 49{,}284° \text{ s}^{-2}$$

The lag in degrees during an acceleration is:

$$\frac{Acceleration}{K_a} \qquad (6.116)$$

EXAMPLE:
Acceleration = $19{,}000° \text{ s}^{-2}$
$K_a = 49{,}284° \text{ s}^{-2}$
$LAG = 19{,}000/49{,}284 = 0.38°$
In 16 bit 1 LSB = $0.0055°$
Acceleration (1 LSB lag) = $K_a \times 0.0055$ (16 bit)
$49{,}284 \times 0.0055 = 270° \text{ s}^{-2}$

Large Step Settling Time

To determine the settling time for a large step response (179°), one must take into account the maximum tracking rate and bandwidth.

EXAMPLE:
A 179° step with 100 Hz BW and 10 rps max tracking
Max tracking at 10 rps = $3600° \text{ s}^{-1}$
$179/3600 = 49$ ms

Then one must add the settling time due to bandwidth limitations; this is approximately 11 time constants (16-bit mode).

Time Constants

Resolution	# of counts/rotation	# of time constants
10	1024	7
12	4096	8
14	16384	10
16	65536	11

Time constant	= $1/A$
A	= 222
$1/A$	= 4.5 ms
11 time constants	= 45 ms

TABLE 6.27　List of Resolver and Synchro Suppliers

Company	Location	Phone number	Types of resolvers
API Harowe	West Chester, PA	(800) 566-5274	Brushless frameless, housed, & heavy-duty units
Admotec, Inc.	Norwich, VT	(802) 649-5800	Rotasyn solid rotor resolvers
Neotech, Inc.	Hatfield, PA	(215) 822-5520	Housed brush & brushless units
Vernitron Corp.	San Diego, CA	(800) 777-3393	Brushless, segment, brushed, & housed units
Servo Systems	Montville, NJ	(973) 335-1007	Brushed & brushless units
American Electronics	Fullerton, CA	(714) 871-3020	Housed units
Computer Conversions	East Northport, NY	(516) 261-3300	Explosion proof & specialty units
Poltron Corp.	Gaithersburg, MD	(301) 208-6597	Resolvers
Tamagawa Trading Co.	Tokyo, Japan	011-81-37-383-175	Brushless frameless & housed units
MPC Products	Skokie, IL	(800) 323-4302	Housed resolvers
Transicoil, Inc.	Trooper, PA	(800) 323-7115	Housed brushless & brushed resolvers
Litton Poly-Scientific	Blacksburg, VA	(800) 336-2112	Brushed and brushless resolvers
Kearfott Guidance & Navigation Corp.	Wayne, NJ	(973) 785-6000	Brushed & brushless resolvers
Novatronics, Inc.	Stratford, Ontario, Canada	(519) 271-3880	Resolvers
Muirhead Vactric	Lake Zurich, IL	(847) 726-0270	Resolvers

Therefore, the approximate settling time for a large step would be 94 ms. This is an approximation. Synchros and resolvers are used in a wide variety of dynamic conditions. Understanding how the converter reacts to these input changes will allow one to optimize the bandwidth and maximum tracking rate for each application. Table 6.27 provides a list of resolver and synchro suppliers.

Further Information

Synchro/Resolver Conversion Handbook, 4th ed., Bohemia NY: ILC Data Device Corp., 1994.

Analog Devices, Inc., *Analog-Digital Conversion Handbook, 3rd ed.*, Englewood Cliffs, NJ: Prentice-Hall, 1986.

Synchro & Resolver Conversion, East Molesey, UK: Memory Devices Ltd., 1980.

6.11　Optical Fiber Displacement Sensors

Richard O. Claus, Vikram Bhatia, and Anbo Wang

The objective of this section is to present a rigorous theoretical and experimental analysis of short gage length optical fiber sensors for measurement of cyclical strain on or in materials. Four different types of sensors are evaluated systematically on the basis of various performance criteria such as strain resolution, dynamic range, cross-sensitivity to other ambient perturbations, simplicity of fabrication, and complexity of demodulation process. The sensing methods that would be investigated include well-established technologies, (e.g., fiber Bragg gratings), and rapidly evolving measurement techniques such as long-period gratings. Other than the grating-based sensors, two popular versions of Fabry–Perot interferometric sensors (intrinsic and extrinsic) will be evaluated for their suitability. A theoretical study of the cross-sensitivities of these sensors to an arbitrary combination of strain vectors and temperature, similar to that proposed by Sirkis in his SPIE paper in 1991 [1], will be developed.

The outline of this section is as follows. The principle of operation and fabrication process of each of the four sensors are discussed separately. Sensitivity to strain and other simultaneous perturbations such as temperature are analyzed. The overall cost and performance of a sensing technique depend heavily on the signal demodulation process. The detection schemes for all four sensors are discussed and compared on the basis of their complexity. Finally, a theoretical analysis of the cross-sensitivities of the four sensing schemes is presented and their performances are compared.

Strain measurements using optical fiber sensors in both embedded and surface-mounted configurations have been reported by researchers in the past [2]. Fiber optic sensors are small in size, immune to electromagnetic interference, and can be easily integrated with existing optical fiber communication links. Such sensors can typically be easily multiplexed, resulting in distributed networks that can be used for health monitoring of integrated, high-performance materials and structures. Optical fiber sensors for strain measurements should possess certain important characteristics. These sensors should either be insensitive to ambient fluctuations in temperature and pressure, or should have demodulation techniques that compensate for changes in the output signal due to the undesired perturbations. In the embedded configuration, the sensors for axial strain measurements should have minimum cross-sensitivity to other strain states. The sensor signal should itself be simple and easy to demodulate. Nonlinearities in the output demand expensive decoding procedures or require precalibrating the sensor. The sensor should ideally provide an absolute and real-time strain measurement in a form that can be easily processed. For environments where large strain magnitudes are expected, the sensor should have a large dynamic range while at the same time maintaining the desired sensitivity. A discussion of each of the four sensing schemes individually, along with their relative merits and demerits, follows.

Extrinsic Fabry–Perot Interferometric Sensor

The extrinsic Fabry–Perot interferometric (EFPI) sensor, proposed by Murphy et al., is one of the most popular fiber optic sensors used for applications in health monitoring of smart materials and structures [3]. As the name suggests, the EFPI is an interferometric sensor in which the detected intensity is modulated by the parameter under measurement. The simplest configuration of an EFPI is shown in Figure 6.99.

The EFPI system consists of a single-mode laser diode that illuminates a Fabry–Perot cavity through a fused biconical tapered coupler. The cavity is formed between an input single-mode fiber and a reflecting single-mode or multimode fiber. Since the cavity is external to the lead-in/lead-out fiber, the EFPI sensor is independent of transverse strain and small ambient temperature fluctuations. The input fiber and the reflecting fiber are aligned using a hollow-core silica fiber. For uncoated fiber ends, a 4% Fresnel reflection results at both ends. The first reflection, R_1, called the reference reflection, is independent of the applied perturbation. The second reflection, R_2, termed the sensing reflection, is dependent on the length of the cavity, d, which in turn is modulated by the applied perturbation. These two reflections interfere (provided $2d < L_c$, the laser diode's coherence length), and the intensity I at the detector varies as a function of the cavity length:

$$I = I_0 \cos\left(\frac{4\pi}{\lambda} d\right) \tag{6.117}$$

where, I_0 is the maximum value of the output intensity and λ is the laser diode center wavelength.

The typical EFPI transfer function curve is shown in Figure 6.100. Small perturbations that result in operation around the quiescent-point or Q-point of the sensor lead to a linear variation in output intensity. A fringe in the output signal is defined as the change in intensity from a maximum to a maximum or from a minimum to a minimum. Each fringe corresponds to a change in the cavity length by one half of the operating wavelength, λ. The change in the cavity length, Δd, is then employed to calculate the strain ε using the expression:

$$\varepsilon = \frac{\Delta d}{L} \tag{6.118}$$

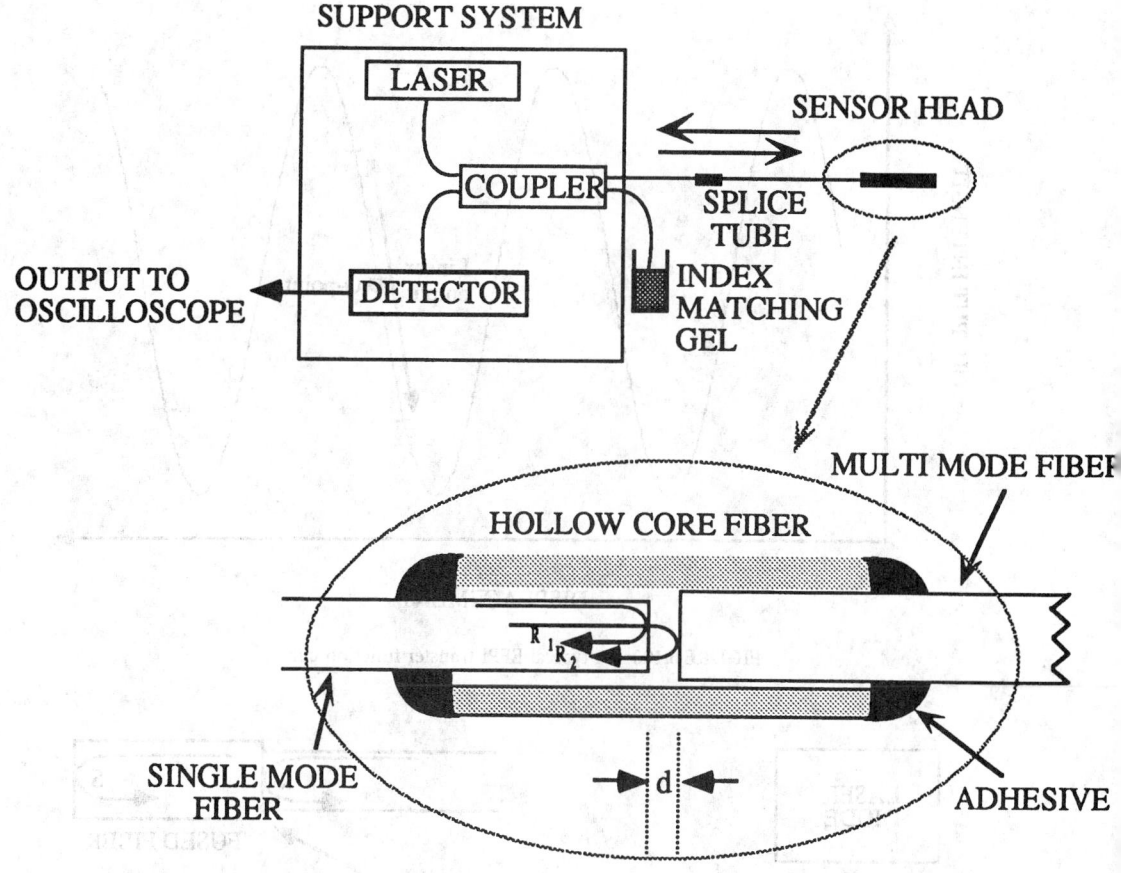

FIGURE 6.99 A simple configuration of an extrinsic Fabry–Perot interferometric (EFPI) sensing system.

where, L is defined as the gage length of the sensor and is typically the distance between two points where the input and reflecting fibers are bonded to the hollow-core fiber. Matching of the two reflection signal amplitudes allows good fringe visibility in the output signal.

The EFPI sensor has been extensively used for measuring fatigue loading on F-15 aircraft wings, detection of crack formation and propagation in civil structures, and cure and lifetime monitoring in concrete and composite specimens [2, 4]. The temperature insensitivity of this sensor makes it attractive for a large number of applications. The EFPI sensor is capable of measuring sub-Angstrom displacements with strain resolution better than 1 microstrain and a dynamic range greater than 10,000 $\mu\varepsilon$. Although the change in output intensity of the EFPI is nonlinear corresponding to the magnitude of the parameter being measured, for small perturbations its operation can be limited to that around the Q-point of the transfer function curve. Moreover, the large bandwidth available with this sensor simplifies the measurement of highly cyclical strain. The EFPI sensor is capable of providing single-ended operation and is hence suitable for applications where access to the test area is limited. The sensor requires simple and inexpensive fabrication equipment and an assembly time of less than 10 min. Additionally, since the cavity is external to the fibers, transverse strain components that tend to influence intrinsic sensors through the Poisson's effect have negligible effect on the EFPI sensor output. The sensitivity to only axial strain and insensitivity to input polarization state have made the EFPI sensor the most preferred fiber optic sensor for embedded applications [1]. Thus, overall, the EFPI sensing system is very well suited to measurement of small magnitudes of cyclical strain.

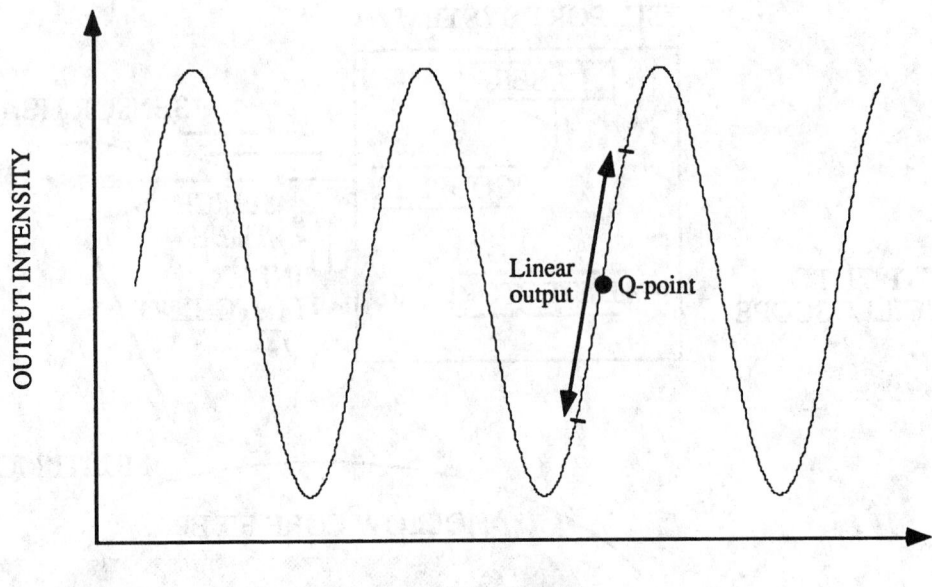

FIGURE 6.100   A typical EFPI transfer function curve.

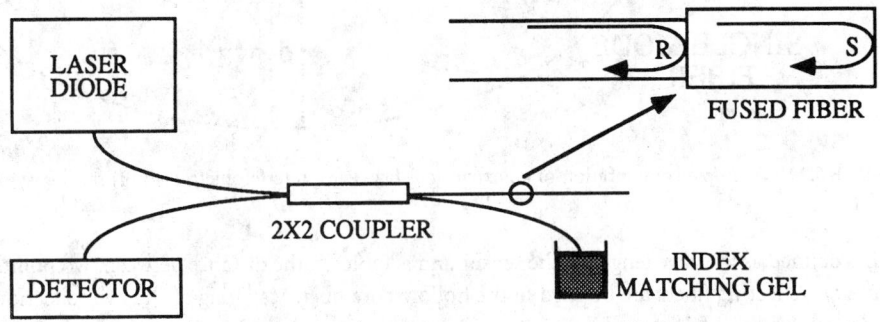

FIGURE 6.101   An intrinsic Fabry-Perot interferometric sensor (IFPI).

Although a version of the EFPI sensor that provides absolute output has been demonstrated, it lacks the bandwidth typically desired during the measurement of cyclical strain [5]. We have also recently proposed a small cavity length/high finesse EFPI sensor for measurement of small perturbations [6]. This configuration has a simple output that can be demodulated using an optical filter/photodetector combination.

Intrinsic Fabry–Perot Interferometric Sensor

The intrinsic Fabry–Perot interferometric (IFPI) sensor is similar in operation to its extrinsic counterpart but significant differences exist in the configurations of the two sensors [7]. The basic IFPI sensor is shown in Figure 6.101. An optically isolated laser diode is used as the optical source to one of the input arms of a bidirectional 2 × 2 coupler. The Fabry–Perot cavity is formed by fusing a small length of a single-mode fiber to one of the output legs of the coupler. As shown in Figure 6.101, the reference (R) and sensing (S) reflections interfere at the detector face to provide a sinusoidal intensity variation. The cavity can also be obtained by introducing two Fresnel reflectors — discontinuities in refractive index — along the length of a single fiber. Photosensitivity in germanosilicate fibers has been used in the past to fabricate broadband reflectors that enclose an IFPI cavity [8]. Since the cavity is formed within an optical fiber, changes in the refractive index of the fiber due to the applied perturbation can significantly alter the phase of the sensing signal, S. Thus, the intrinsic cavity results in the sensor being sensitive to ambient temperature fluctuations and all states of strain.

The IFPI sensor, like all other interferometric signals, has a nonlinear output that complicates the measurement of large magnitude strain. This can again be overcome by operating the sensor in the linear regime around the Q-point of the sinusoidal transfer function curve. The main limitation of the IFPI strain sensor is that the photoelastic effect-induced change in index of refraction results in a nonlinear relationship between the applied perturbation and the change in cavity length. In fact, for most IFPI sensors, the change in propagation constant of the fundamental mode dominates the change in cavity length. Thus, IFPIs are highly susceptible to temperature changes and transverse strain components [1]. In embedded applications, the sensitivity to all the strain components can result in erroneous outputs. The fabrication process of an IFPI strain sensor is more complicated than that of the EFPI sensor since the sensing cavity must be formed within the optical fiber by some special procedure. The strain resolution of the IFPIs is also expected to be around 1 µε with an operating range greater than 10,000 µε. IFPI sensors also suffer from drift in the output signal due to variations in the polarization state of the input light.

Thus, the preliminary analysis shows that the extrinsic version of the Fabry–Perot optical fiber sensor seems to have an overall advantage over its intrinsic version. The extrinsic sensor has negligible cross-sensitivity to temperature and transverse strain. Although the strain sensitivity, dynamic range, and bandwidth of the two sensors are comparable, the IFPIs can be expensive and cumbersome to fabricate due to the intrinsic nature of the sensing cavity.

The extrinsic and intrinsic Fabry–Perot interferometric sensors possess nonlinear sinusoidal outputs that complicate the signal processing at the detection end. Although intensity-based sensors have a simple output variation, they suffer from limited sensitivity to strain or other perturbations of interest. Grating-based sensors have recently become popular as transducers that provide wavelength-encoded output signals that can typically be easily demodulated to derive information about the perturbation under investigation. The advantages and drawbacks of Bragg grating sensing technology are discussed first. The basic operating mechanism of the Bragg grating-based strain sensor is elucidated and the expressions for strain resolution is obtained. These sensors are then compared to the recently developed long-period gratings in terms of fabrication process, cross-sensitivity to other parameters, and simplicity of signal demodulation.

Fiber Bragg Grating Sensor

The phenomenon of photosensitivity in optical fibers was discovered by Hill and co-workers in 1978 [9]. It was found that permanent refractive index changes could be induced in fibers by exposing the germanium-doped core to intense light at 488 or 514 nm. The sinusoidal modulation of index of refraction in the core due to the spatial variation in the writing beam gives rise to a refractive index grating that can be used to couple the energy in the fundamental guided mode to various guided and lossy modes. Later Meltz et al. proposed that photosensitivity is more efficient if the fiber is side-exposed to fringe pattern at wavelengths close to the absorption wavelength (242 nm) of the germanium defects in the fiber [10]. The side-writing process simplified the fabrication of Bragg gratings, and these devices have recently emerged as highly versatile components for communication and sensing systems. Recently, loading the fibers with hydrogen has been reported to result in two orders of magnitude higher index change in germanosilicate fibers [11].

Principle of Operation

Bragg gratings are based on the phase-matching condition between spatial modes propagating in optical fibers. This phase-matching condition is given by:

$$k_g + k_c = k_B \tag{6.119}$$

where, k_g, k_c, and k_B are, respectively, the wave-vectors of the coupled guided mode, the resulting coupling mode, and the grating. For a first-order interaction, $k_B = 2\pi/\Lambda$, where Λ is the grating periodicity. Since it is customary to use propagation constants while dealing with optical fiber modes, this condition reduces to the widely used equation for mode coupling due to a periodic perturbation:

$$\Delta\beta = \frac{2\pi}{\Lambda} \tag{6.120}$$

where, $\Delta\beta$ is the difference in the propagation constants of the two modes involved in mode coupling (both assumed to travel in the same direction).

Fiber Bragg gratings (FBGs) involve the coupling of the forward-propagating fundamental LP_{01} optical fiber waveguide propagation mode to the reverse-propagating LP_{01} mode [12]. Consider a single mode fiber with β_{01} and $-\beta_{01}$ as the propagation constant of the forward- and reverse-propagating fundamental LP_{01} modes. To satisfy the phase-matching condition,

$$\Delta\beta = \beta_{01} - \left(-\beta_{01}\right) = \frac{2\pi}{\Lambda} \tag{6.121}$$

where, $\beta_{01} = 2\pi n_{eff}/\lambda$ (n_{eff} is the effective index of the fundamental mode and λ is the free-space wavelength). Equation 6.121 reduces to [12]:

$$\lambda_B = 2\Lambda n_{eff} \tag{6.122}$$

where λ_B is termed the Bragg wavelength. The Bragg wavelength is the wavelength at which the forward-propagating LP_{01} mode couples to the reverse-propagating LP_{01} mode. This coupling is wavelength dependent since the propagation constants of the two modes are a function of the wavelength. Hence, if an FBG is interrogated by a broadband optical source, the wavelength at which phase-matching occurs is found to be reflected back. This wavelength is a function of the grating periodicity (Λ) and the effective index (n_{eff}) of the fundamental mode (Equation 6.122). Since strain and temperature effects can modulate both these parameters, the Bragg wavelength shifts with these external perturbations. This spectral shift is utilized to fabricate FBGs for sensing applications.

Figure 6.102 shows the mode coupling mechanism in fiber Bragg gratings using the β-plot. Since the difference in propagation constants ($\Delta\beta$) between the modes involved in coupling is large, Equation 6.120 reveals that only a small value of periodicity, Λ, is needed to induce this mode coupling. Typically for telecommunication applications, the value of λ_B is around 1.5 μm. From Equation 6.122, Λ is determined to be 0.5 μm (for n_{eff} = 1.5). Due to the small periodicities (of the order of 1 μm), FBGs are classified as short-period gratings.

Fabrication Techniques

Fiber Bragg gratings have commonly been manufactured using two side-exposure techniques: the interferometric method and the phase mask method. The interferometric method, depicted in Figure 6.103

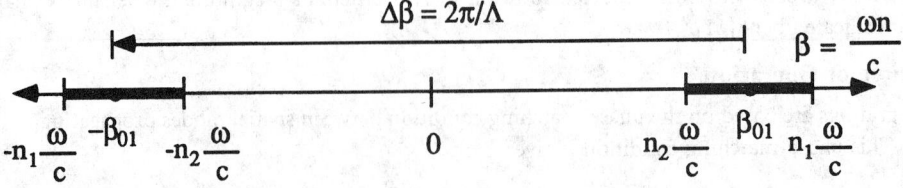

FIGURE 6.102   Mode coupling mechanism in a fiber Bragg grating.

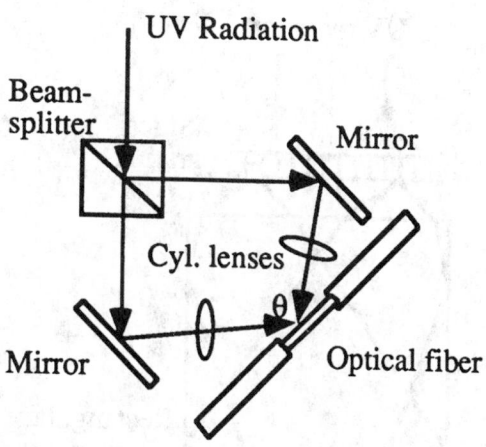

FIGURE 6.103 The interferometric fiber Bragg grating.

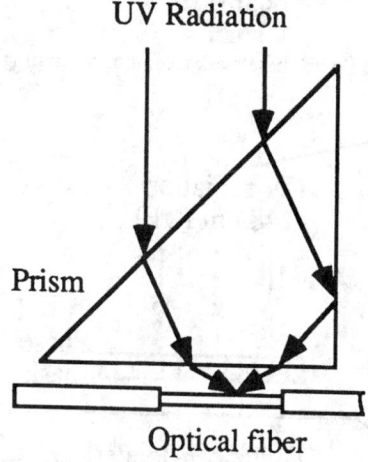

FIGURE 6.104 The novel interferometer technique.

comprises a UV beam at 244 or 248 nm spilt in two equal parts by a beam splitter [10]. The two beams are then focused on a portion of Ge-doped fiber (whose protective coating has been removed) using cylindrical lenses, and the periodicity of the resulting interference pattern and, hence, the Bragg wavelength are varied by altering the mutual angle, θ. The limitation of this method is that any relative vibration of the pairs of mirrors and lenses can lead to the degradation of the quality of the final grating and, hence, the entire system has a stringent stability requirement. To overcome this drawback, Kashyap et al. have proposed the novel interferometer technique where the path difference between the interfering UV beams is produced by the propagation through a right-angled prism (Figure 6.104) [12]. This technique is inherently stable because both beams are perturbed similarly by any prism vibration.

The phase mask technique has recently gained popularity as an efficient holographic side-writing procedure for grating fabrication [13]. In this method, as shown in Figure 6.105, an incident UV beam is diffracted into −1, 0, and +1 orders by a relief grating generated on a silica plate by e-beam exposure and plasma etching. The two first diffraction orders undergo total internal reflection at the glass/air interface of a rectangular prism and interfere on the bare fiber surface placed directly behind the mask. This technique is wavelength specific since the periodicity of the resulting two-beam interference pattern is uniquely determined by the diffraction angle of −1 and +1 orders and, thus, the properties of the phase

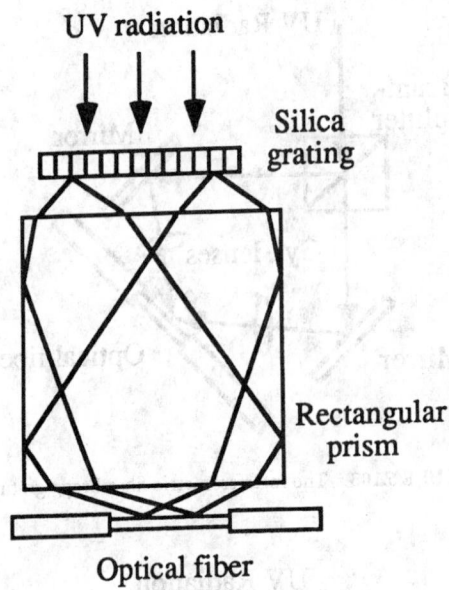

FIGURE 6.105   The phase mask technique for grating fabrication.

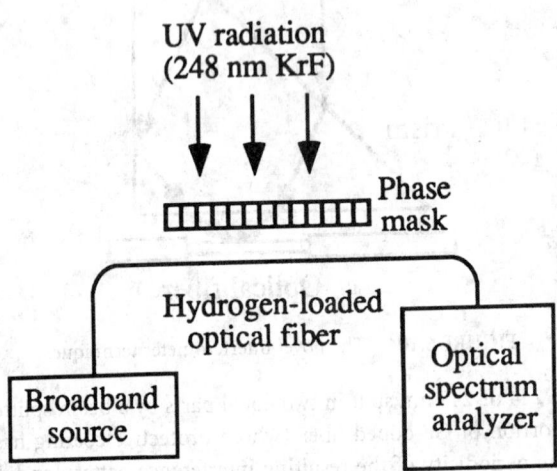

FIGURE 6.106   Diagram of the setup for monitoring the growth of the grating in transmission during fabrication.

mask. Obviously, different phase masks are required for fabrication of gratings at different Bragg wavelengths. The setup for actively monitoring the growth of the grating in transmission during fabrication is shown in Figure 6.106.

Bragg Grating Sensors

From Equation 6.122 we see that a change in the value of n_{eff} and/or Λ can cause the Bragg wavelength, λ, to shift. This fractional change in the resonance wavelength, $\Delta\lambda/\lambda$, is given by the expression:

$$\frac{\Delta\lambda}{\lambda} = \frac{\Delta\Lambda}{\Lambda} + \frac{\Delta n_{eff}}{n_{eff}} \qquad (6.123)$$

where, $\Delta\Lambda/\Lambda$ and $\Delta n_{eff}/n_{eff}$ are the fractional changes in the periodicity and the effective index, respectively. The relative magnitudes of the two changes depend on the type of perturbation the grating is subjected to; for most applications, the effect due to change in effective index is the dominating mechanism.

Any axial strain, ε, applied to the grating changes the periodicity and the effective index and results in a shift in the Bragg wavelength, given by:

$$\frac{1}{\lambda}\frac{\Delta\lambda}{\varepsilon} = \frac{1}{\Lambda}\frac{\Delta\Lambda}{\varepsilon} + \frac{1}{n_{eff}}\frac{\Delta n_{eff}}{\varepsilon} \tag{6.124}$$

The first term on the right-hand side is unity, while the second term has its origin in the photoelastic effect. An axial strain on the fiber serves to change the refractive index of both the core and the cladding. This results in the variation in the value of the effective index of glass. The photoelastic or strain-optic coefficient that relates the change in index of refraction due to mechanical displacement is about −0.27. Thus, the variation in n_{eff} and Λ due to strain have contrasting effects on the Bragg peak. The fractional change in the Bragg wavelength due to axial strain is $0.73\,\varepsilon$ or 73% of the applied strain. At 1550 and 1300 nm, the shifts in the resonance wavelength are 11 nm/%ε and 9 nm/%ε, respectively. With temperature, a FBG at 1500 nm shifts by 1.6 nm for every 100°C rise in temperature [9].

Limitations of Bragg Grating Strain Sensors

The major limitation of Bragg grating sensors is the complex and expensive fabrication technique. Although side-writing is commonly used to manufacture these gratings, the requirement of expensive phase masks increases the cost of the sensing system. In the interferometric technique, stability of the setup is a critical factor in obtaining high-quality gratings. Since index changes of the order of 10^{-3} are required to fabricate these gratings, laser pulses of high energy levels are necessary. This might reduce laser operating lifetime and lead to increased maintenance expense. Additionally, introducing hydrogen or deuterium into the fiber allows increased index modulation as a result of the irridiation process.

The second major limitation of Bragg gratings is their limited bandwidth. The typical value of the full-width at half maximum (FWHM) is between 0.1 and 1 nm. Although higher bandwidths potentially can be obtained by chirping the index or periodicity along the grating length, this adds to the cost of the grating fabrication. The limited bandwidth requires high-resolution spectrum analyzers to monitor the grating spectrum. Kersey et al. have proposed an unbalanced Mach–Zender interferometer to detect the perturbation-induced wavelength shift [14]. Two unequal arms of the Mach–Zender interferometer are excited by the backreflection from a Bragg grating sensor element. Any change in the input optical wavelength modulates the phase difference between the two arms and results in a time-varying sinusoidal intensity at the output. This interference signal can be related to the shift in the Bragg peak and, hence, the magnitude of the perturbation can be obtained. Recently, modal interferometers have also been proposed to demodulate the output of a Bragg grating sensor [15]. The unbalanced interferometers are also susceptible to external perturbations and hence need to be isolated from the parameter under investigation. Moreover, the nonlinear output might require fringe counting equipment, which can be complex and expensive. Additionally, a change in the perturbation polarity at the maxima or minima of the transfer function curve will not be detected by this demodulation scheme. To overcome this limitation, two unbalanced interferometers can be employed for dynamic measurements.

The cross-sensitivity to temperature fluctuations leads to erroneous strain measurements in applications where the ambient temperature has a temporal variation. Thus, a reference grating that measures the temperature change must be utilized to compensate for the output of the strain sensor. Recently, temperature-independent sensing has been demonstrated using chirped gratings written in tapered optical fibers [16].

Last, the sensitivity of fiber Bragg grating strain sensors might not be adequate for certain applications. This sensitivity of the sensor depends on the minimum detectable wavelength shift at the detection end. Although excellent wavelength resolution can be obtained with unbalanced interferometric detection techniques, standard spectrum analyzers typically provide a resolution of 0.1 nm. At 1300 nm, this minimum detectable change in wavelength corresponds to a strain resolution of 111 με. Hence, in applications where strain smaller than 100 με is anticipated, Bragg grating sensors might not be practical. The dynamic range of strain measurement can be as much as 15,000 με.

Long-Period Grating Sensor

This section discusses the use of novel long-period gratings as strain sensing devices. The principle of operation of these gratings, their fabrication process, preliminary strain tests, demodulation process, and cross-sensitivity to ambient temperature are analyzed.

Principle of Operation

Long-period gratings that couple the fundamental guided mode to different guided modes have been demonstrated in the past [17, 18]. Gratings with longer periodicities that involve coupling of a guided mode to forward-propagating cladding modes were recently proposed by Vengsarkar et al. [19, 20]. As stated previously, fiber gratings satisfy the Bragg phase-matching condition between the guided and cladding or radiation modes or, another guided mode. This wavelength-dependent phase-matching condition is given by:

$$\beta_{01} - \beta = \Delta\beta = \frac{2\pi}{\Lambda} \tag{6.125}$$

where Λ is the periodicity of the grating, β_{01} and β are the propagation constants of the fundamental guided mode and the mode to which coupling occurs, respectively.

For conventional fiber Bragg gratings, the coupling of the forward propagating LP_{01} mode occurs to the reverse propagating LP_{01} mode ($\beta = -\beta_{01}$). Since $\Delta\beta$ is large in this case (Figure 6.107(a)), the grating periodicity is small, typically of the order of 1 μm. Unblazed long-period gratings having index variations parallel to the long axis of the fiber couple the fundamental mode to the discrete and circularly-symmetric, forward-propagating cladding modes ($\beta = \beta^n$), resulting in smaller values of $\Delta\beta$ (Figure 6.107(b)) and

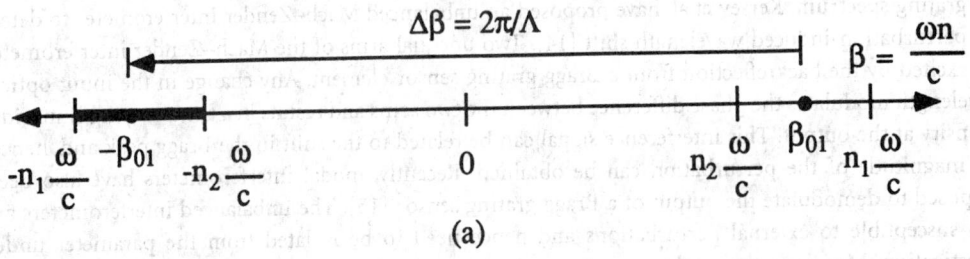

(a)

(b)

FIGURE 6.107

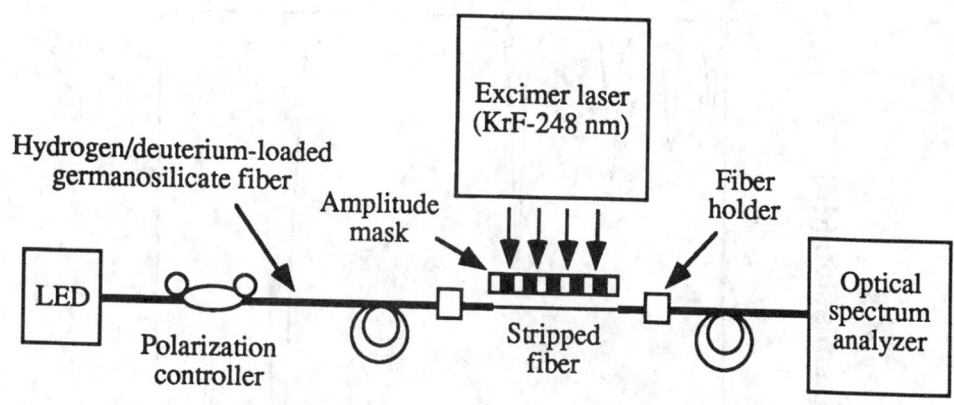

FIGURE 6.108 Setup used to fabricate long-period gratings.

hence periodicities ranging in hundreds of micrometers [19]. The cladding modes attenuate rapidly as they propagate along the length of the fiber due to the lossy cladding-coating interface and bends in the fiber. Since $\Delta\beta$ is discrete and a function of the wavelength, this coupling to the cladding modes is highly selective, leading to a wavelength-dependent loss. As a result, any modulation of the core and cladding guiding properties modifies the spectral response of long-period gratings, and this phenomenon can be utilized for sensing purposes. Moreover, since the cladding modes interact with the fiber jacket or any other material surrounding the cladding, changes in the properties of these ambient materials can also be detected.

Fabrication Procedure

To fabricate long-period gratings, hydrogen-loaded (3.4 mol%) germanosilicate fibers are exposed to 248 nm UV radiation from a KrF excimer laser, through a chrome-plated amplitude mask possessing a periodic rectangular transmittance function. Figure 6.108 shows the setup used to fabricate the gratings. The laser was pulsed at 20 Hz with a 8 ns pulse duration. The typical writing times for an energy of 100 mJ cm^{-2} pulse^{-1} and a 2.5 cm exposed length vary between 6 to 15 min for different fibers. The coupling wavelength, λ_p, shifts to higher values during exposure, due to the photoinduced enhancement of the refractive index of the fiber core and the resulting increase in β_{01}. After writing, the gratings are annealed at 150°C for 10 h to remove the unreacted hydrogen. This high-temperature annealing causes λ_p to move to shorter wavelengths due to the decay of UV-induced defects and diffusion of molecular hydrogen from the fiber. Figure 6.109 depicts the typical transmittance of a grating. Various attenuation bands correspond to coupling to discrete cladding modes of different orders. A number of gratings can be fabricated at the same time by placing more than one fiber behind the amplitude mask. Moreover, the stability requirements during the writing process are not as severe as those for short-period Bragg gratings.

For coupling to the highest-order cladding-mode, the maximum isolation (loss in transmission intensity) is typically in the 5 to 20 dB range on wavelengths, depending on fiber parameters, duration of UV exposure, and mask periodicity. The desired fundamental coupling wavelength can easily be varied using inexpensive amplitude masks of different periodicities. The insertion loss, polarization-mode dispersion, backreflection, and polarization-dependent loss of a typical grating are 0.2 dB, 0.01 ps, −80 dB, and 0.02 dB, respectively. The negligible polarization sensitivity and backreflection of these devices eliminate the need for expensive polarizers and isolators.

Preliminary experiments were performed to examine the strain sensitivity of long-period gratings written in different fibers [21, 22]. Gratings were fabricated in four different types of fibers: standard dispersion-shifted fiber (DSF), standard 1550 nm fiber, and conventional 980 and 1050 nm single-mode fibers, which for the sake of brevity are referred to as fibers A, B, C, and D, respectively. The strain sensitivity of gratings written in different fibers was determined by axially straining the gratings between

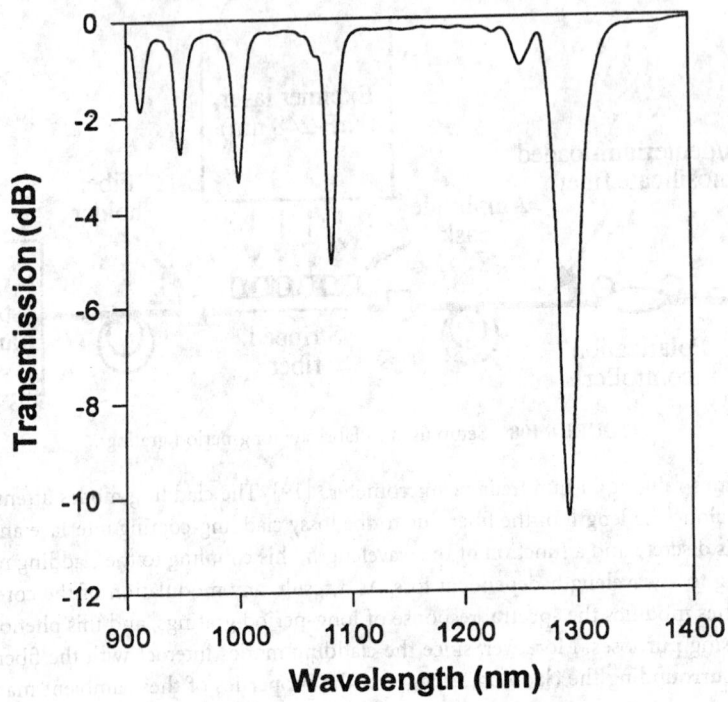

FIGURE 6.109   Typical transmission of a grating.

two longitudinally separated translation stages. The shift in the peak loss wavelength of the grating in fiber D as a function of the applied strain is depicted in Figure 6.110, along with that for a Bragg grating (about 9 nm %ε^{-1}, at 1300 nm) [9]. The strain coefficients of wavelength shift (β) for fibers A, B, C, and D are shown in Table 6.28. Fiber D has a coefficient 15.2 nm %ε^{-1}, which gives it a strain-induced shift that is 50% larger than that for a conventional Bragg grating. The strain resolution of this fiber for a 0.1 nm detectable wavelength shift is 65.75 $\mu\varepsilon$.

The demodulation scheme of a sensor determines the overall simplicity and sensitivity of the sensing system. Short-period Bragg grating sensors were shown to possess signal processing techniques that are complex and expensive to implement. A simple demodulation method to extract information from long-period gratings is possible. The wide bandwidth of the resonance bands enables the wavelength shift due to the external perturbation to be converted into an intensity variation that can be easily detected.

Figure 6.111 shows the shift induced by strain in a grating written in fiber C. The increase in the loss at 1317 nm is about 1.6 dB. A laser diode centered at 1317 nm was used as the optical source, and the change in transmitted intensity was monitored as a function of applied strain. The transmitted intensity is plotted in Figure 6.112 for three different trials. The repeatability of the experiment demonstrates the feasibility of using this simple scheme to utilize the high sensitivity of long-period gratings. The transmission of a laser diode centered on the slope of the grating spectrum on either side of the resonance wavelength can be used as a measure of the applied perturbation. A simple detector and amplifier combination at the output can be used to determine the transmission through the detector. On the other hand, a broadband source can also be used to interrogate the grating. At the output, an optical bandpass filter can be used to transmit only a fixed bandwidth of the signal to the detector. The bandpass filter should again be centered on either side of the peak loss band of the resonance band. These schemes are easy to implement, and unlike conventional Bragg gratings, the requirement of complex and expensive interferometric demodulation schemes is not necessary [22].

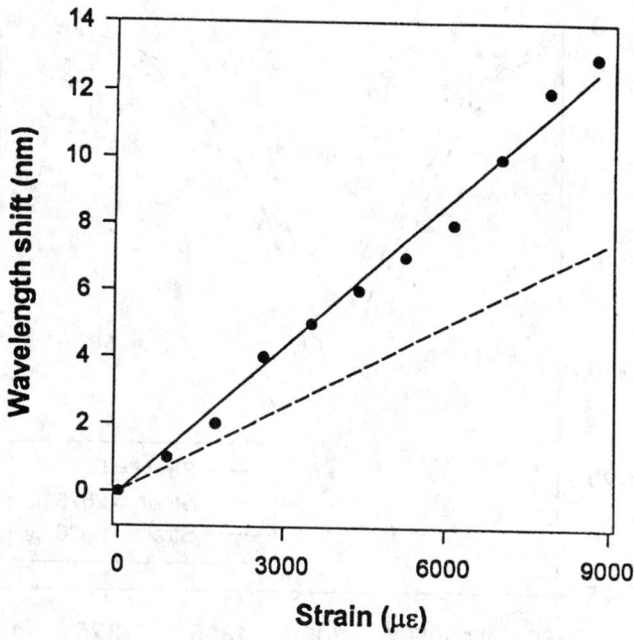

FIGURE 6.110 Shift in peak loss wavelength as a function of the applied strain.

TABLE 6.28 Strain Sensitivity of Long-Period Gratings Written in Four Different Types of Fibers

Type of fiber	Strain sensitivity (nm %ε^{-1})
A — Standard dispersion-shifted fiber (DSF)	−7.27
B — Standard 1550 nm communication fiber	4.73
C — Converntional 980 nm single-mode fiber	4.29
D — Conventional 1060 nm single-mode fiber	15.21

Note: The values correspond to the shift in the highest order resonance wavelength.

Temperature Sensitivity of Long-Period Gratings

Gratings written in different fibers were also tested for their cross-sensitivity to temperature [22]. The temperature coefficients of wavelength shift for different fibers are shown in Table 6.29. The temperature sensitivity of a fiber Bragg grating is 0.014 nm °C^{-1}. Hence, the temperature sensitivity of a long-period grating is typically an order of magnitude higher than that of a Bragg grating. This large cross-sensitivity to ambient temperature can degrade the strain sensing performance of the system unless the output signal is adequately compensated. Multiparameter sensing using long-period gratings has been proposed to obtain precise strain measurements in environments with temperature fluctuations [21].

In summary, long-period grating sensors are highly versatile. These sensors can easily be used in conjunction with simple and inexpensive detection techniques. Experimental results prove that these methods can be used effectively without sacrificing the enhanced resolution of the sensors. Long-period grating sensors are insensitive to the input polarization and do not require coherent optical sources. The cross-sensitivity to temperature is a major concern while using these gratings for strain measurements.

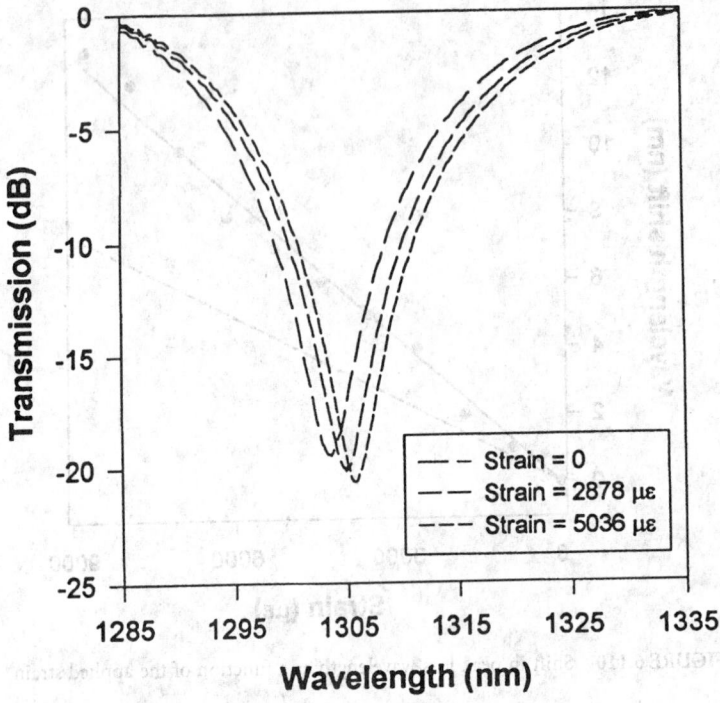

FIGURE 6.111 The shift induced by strain in a grating written in fiber C.

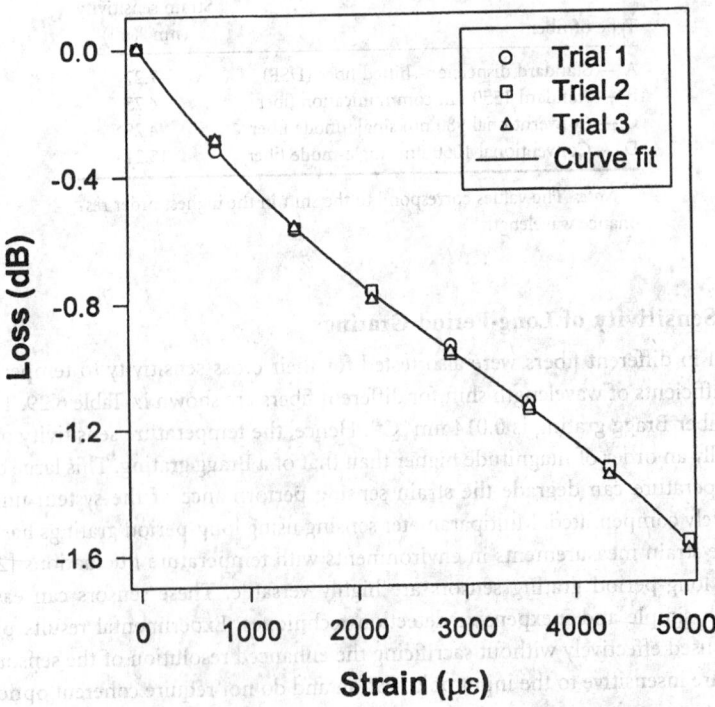

FIGURE 6.112 Plot of the change in transmitted intensity as a function of strain, for three different trials.

Table 6.29 Temperature Sensitivity of Long-Period Gratings Written in Four Different Types of Fibers

Type of fiber	Temperature sensitivity (nm °C^{-1})
A — Standard dispersion-shifted fiber (DSF)	0.062
B — Standard 1550 nm communication fiber	0.058
C — Converntional 980 nm single-mode fiber	0.154
D — Conventional 1060 nm single-mode fiber	0.111

Note: The values correspond to the shift in the highest order resonance wavelength.

Comparison of Sensing Schemes

Based on the above results, the interferometric sensors have a high sensitivity and bandwidth, but are limited by the nonlinearity in their output signals. Conversely, intrinsic sensors are susceptible to ambient temperature changes while the grating-based sensors are simpler to multiplex. Each may be used in specific applications.

Conclusion

We have investigated the performance of four different interferometric and grating-based sensors. This analysis was based on the sensor head fabrication and cost, signal processing, cross-sensitivity to temperature, resolution, and operating range. The relative merits and demerits of the various sensing schemes were also discussed.

References

1. J. Sirkis, Phase-strain-temperature model for structurally embedded interferometric optical fiber strain sensors with applications, *Fiber Optic Smart Structures and Skins IV,* SPIE, Vol. 1588, 1991.
2. R. O. Claus, M. F. Gunther, A. Wang, and K. A. Murphy, Extrinsic Fabry-Perot sensor for strain and crack opening displacement measurements from –200 to 900°C, *J. Smart Materials and Structures,* 1, 237-242, 1992.
3. K. A. Murphy, M. F. Gunther, A. M. Vengsarkar, and R. O. Claus, Fabry-Perot fiber optic sensors in full-scale fatigue testing on an F-15 aircraft, *Appl. Optics,* 31, 431-433, 1991.
4. V. Bhatia, C. A. Schmid, K. A. Murphy, R. O. Claus, T. A. Tran, J. A. Greene, and M. S. Miller, Optical fiber sensing technique for edge-induced and internal delamination detection in composites, *J. Smart Materials Structures,* 4, 164-169, 1995.
5. V. Bhatia, M. J. de Vries, K. A. Murphy, R. O. Claus, T. A. Tran, and J. A. Greene, Extrinsic Fabry-Perot interferometers for absolute measurements, *Fiberoptic Product News,* 9(Dec.), 12-13, 1994.
6. V. Bhatia, M. B. Sen, K. A. Murphy, and R. O. Claus, Wavelength-tracked white light interferometry for highly sensitive strain and temperature measurements, *Electron. Lett.,* 32, 247-249, 1996.
7. C. E. Lee and H. F. Taylor, Fiber-optic Fabry-Perot temperature sensor using a low-coherence light source, *J. Lightwave Technol.,* 9, 129-134, 1991.
8. J. A. Greene, T. A. Tran, K. A. Murphy, A. J. Plante, V. Bhatia, M. B. Sen, and R. O. Claus, Photoinduced Fresnel reflectors for point-wise and distributed sensing applications, *Proc. Conf. Smart Structures and Materials,* SPIE'95, paper 2444-05, February 1995.
9. K. O. Hill, Y. Fuijii, D. C. Johnson, and B. S. Kawasaki, Photosensitivity in optical fiber waveguides: applications to reflection filter fabrication, *Appl. Phys. Lett.,* 32, 647, 1978.
10. G. Meltz, W. W. Morey, and W. H. Glenn, Formation of Bragg gratings in optical fibers by transverse holographic method, *Optics Lett.,* 14, 823, 1989.

11. P. J. Lemaire, A. M. Vengsarkar, W. A. Reed, V. Mizrahi, and K. S. Kranz, Refractive index changes in optical fibers sensitized with molecular hydrogen, in *Proc. Conf. Optical Fiber Communications, OFC'94*, Technical Digest, paper TuL1, 47, 1994.

12. R. Kashyap, Photosensitive optical fibers: devices and applications, *Optical Fiber Technol.*, 1, 17-34, 1994.

13. D. Z. Anderson, V. Mizrahi, T. Ergodan, and A. E. White, Phase-mask method for volume manufacturing of fiber phase gratings, in *Proc. Conf. Optical Fiber Communication*, post-deadline paper PD16, 1993, p. 68.

14. A. D. Kersey and T. A. Berkoff, Fiber-optic Bragg-grating differential-temperature sensor, *IEEE Photonics Technol. Lett.*, 4, 1183-1185, 1992.

15. V. Bhatia, M. B. Sen, K. A. Murphy, A. Wang, R. O. Claus, M. E. Jones, J. L. Grace, and J. A. Greene, Demodulation of wavelength-encoded optical fiber sensor signals using fiber modal interferometers, *SPIE Photonics East*, Philadelphia, PA, paper 2594-09, October 1995.

16. M. G. Xu, L. Dong, L. Reekie, J. A. Tucknott, and J. L. Cruz, Chirped fiber gratings for temperature-independent strain sensing, in *Proc. First OSA Topical Meet. Photosensitivity and Quadratic Non-linearity in Glass Waveguides: Fundamentals and Applications*, paper PMB2, 1995.

17. K. O. Hill, B. Malo, K. Vineberg, F. Bilodeau, D. Johnson, and I. Skinner, Efficient mode-conversion in telecommunication fiber using externally written gratings, *Electron. Lett.*, 26, 1270-1272, 1990.

18. F. Bilodeau, K. O. Hill, B. Malo, D. Johnson, and I. Skinner, Efficient narrowband $LP_{01} \leftrightarrow LP_{02}$ mode convertors fabricated in photosensitive fiber: spectral response, *Electron. Lett.*, 27, 682-684, 1991.

19. A. M. Vengsarkar, P. J. Lemaire, J. B. Judkins, V. Bhatia, J. E. Sipe, and T. E. Ergodan, Long-period fiber gratings as band-rejection filters, *Proc. Conf. Optical Fiber Communications*, OFC '95, post-deadline paper, PD4-2, 1995.

20. A. M. Vengsarkar, P. J. Lemaire, J. B. Judkins, V. Bhatia, J. E. Sipe, and T. E. Ergodan, Long-period fiber gratings as band-rejection filters, *J. Lightwave Technol.*, 14, 58-65, 1996.

21. V. Bhatia, M. B. Burford, K. A. Murphy, and A. M. Vengsarkar, Long-period fiber grating sensors, *Proc. Conf. Optical Fiber Communication*, paper ThP1, February 1996.

22. V. Bhatia and A. M. Vengsarkar, Optical fiber long-period grating sensors, *Optics Lett.*, 21, 692-694, 1996.

23. C. D. Butter and G. B. Hocker, Fiber optics strain gage, *Appl. Optics*, 17, 2867-2869, 1978.

24. J. S. Sirkis and H. W. Haslach, Interferometric strain measurement by arbitrarily configured, surface mounted, optical fiber, *J. Lightwave Technol.*, 8, 1497-1503, 1990.

6.12 Optical Beam Deflection Sensing

Grover C. Wetsel

Measurements of the intensity of the light reflected and transmitted by a sample have been sources of information concerning the structure of matter for over a century. In recent decades, it has been found that measurement of the position of an optical beam that has scattered from a sample is an important and versatile means of characterizing materials and the motion of devices. Surely, the availability of a well-collimated beam from a laser has been crucial in the development of techniques and applications of *optical beam deflection* (OBD) sensing; however, the development and ready availability of various types of *position sensing detectors* (PSDs) have also been important factors. Optical beam deflection may be caused, for example, by propagation of a laser beam through a refractive-index gradient or by reflection from a displaced surface. A PSD provides an electronic signal that is a function of the laser beam position on the detector.

In this section, applications of optical beam deflection sensing are reviewed, the theories of operation of the three most common types of OBD sensors are developed, and typical operational characteristics of the devices are presented. The advantages and disadvantages of the various PSDs are also discussed.

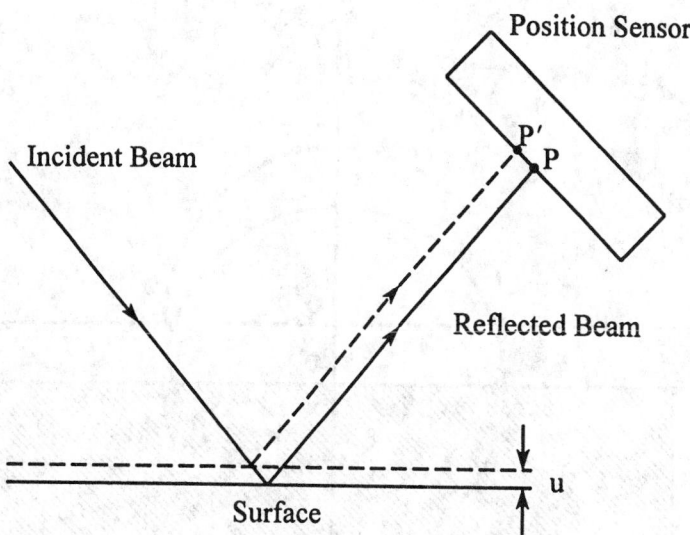

FIGURE 6.113 A schematic diagram of the basic optical-beam-deflection (OBD) sensing configuration.

A schematic diagram of the basic OBD sensing configuration is illustrated in Figure 6.113. In this case, the displacement, *u*, of the surface causes the position of the reflected beam on the PSD to move from point P to point P′; the positional change produces a change in the output voltage of the PSD. The output voltage, *V*, of the PSD electronics can be calibrated in terms of the actual displacement, *u*, by measuring *V* versus *u* for known displacements.

OBD sensing has been used in a variety of applications, including photothermal optical beam deflection (PTOBD) spectroscopy [1], absolute measurement of optical attenuation [2], PTOBD imaging of surface and subsurface structure [3], photothermal displacement spectroscopy [4], atomic-force microscopy [5], and materials characterization [6]. It has also been used as an uncomplicated, sensitive, and accurate method of measurement of surface motion for scanning tunneling microscope scanner transducers [7] and ultrasonic transducer imaging [8].

Theory

The three basic types of devices for OBD sensing are (1) a photodetector behind a sharply edged screen (a knife edge); (2) a small array of photodetectors separated by relatively small, insensitive areas (bicell, quadcell); and (3) a continuous solid-state position sensor (one or two dimensional). Sensing characteristics of a device are determined by the effect of optical beam displacement on the photodetector power distribution. Since laser beams are commonly used in OBD sensing, the analysis involves the assumption that the spatial distribution of the intensity (*I*) in the plane perpendicular to the direction of wave propagation is axially symmetric with a Gaussian radial variation.

Knife-Edge Photodetector

The essential features of a PSD are represented by a photodetector shadowed by a semi-infinite knife edge, $y < 0$, as illustrated in Figure 6.114. As can be anticipated from the symmetry of the arrangement and proved mathematically, the maximal deflection sensitivity occurs when the undeflected beam is centered on the knife edge. The intensity of the light reaching the photodetector due to the displacement (*u*) of the center of the beam is given in the reference frame of the displaced beam by:

$$I(r') = \frac{aP}{\pi} e^{-ar'^2}$$

(6.126)

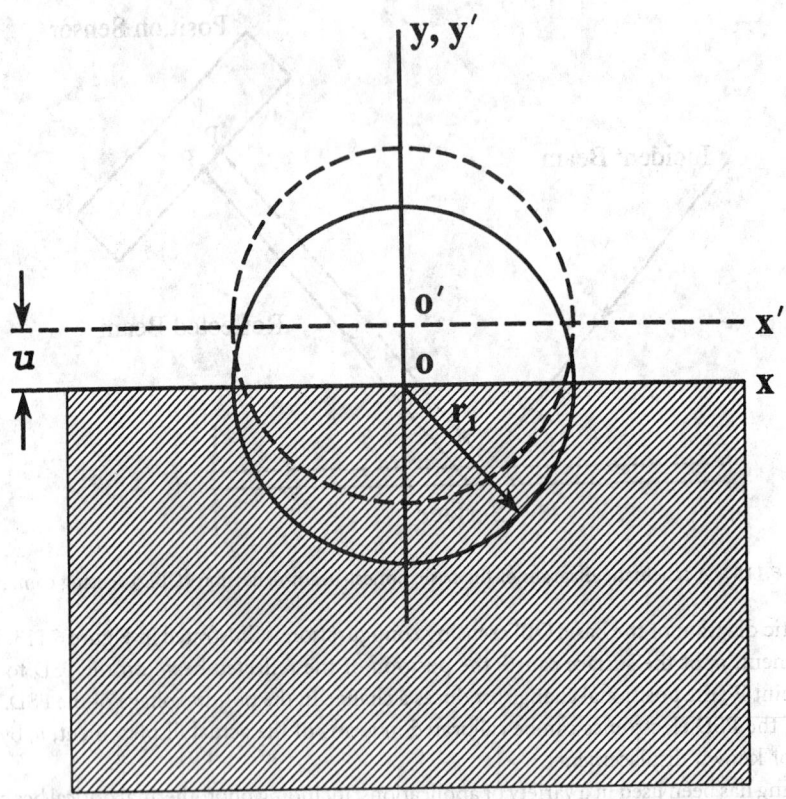

FIGURE 6.114 Essential features of a position-sending detector (PSD), as represented by a photodetector shadowed by a semiinfinite knife edge.

where P is the total incident beam power, $a = 2/r_1^2$, r_1 is the Gaussian beam radius, and $r'^2 = x'^2 + y'^2$. In terms of the coordinates (x,y) of the undeflected beam, the rectangular coordinates of the deflected beam are $x' = x$ and $y' = y - u$. The power (P_d) on the detector is thus given by:

$$P_d = \frac{aP}{\pi} \int_0^\infty e^{-a(y-u)^2} \, dy \int_{-\infty}^\infty e^{-ax^2} \, dx = \frac{P}{2}\left[1 + erf\left(\sqrt{2}\,\frac{u}{r_1} \right) \right] \qquad (6.127)$$

where *erf* is the error function. One can see by inspection of Equation 6.127 that the essential characteristics of this position sensor are determined by u/r_1. The normalized response, P_d/P, is shown in Figure 6.115 as a function of u/r_1. When $u = r_1$, then $P_d = 97.7\% \, P$.

The deflection sensitivity is given by the slope of Equation 6.127,

$$\frac{dP_d}{du} = \sqrt{\frac{2}{\pi}}\,\frac{P}{r_1}\,e^{-2\left(\frac{u}{r_1}\right)^2}, \quad with \quad \left(\frac{dP_d}{du} \right)_{max} = \left(\frac{dP_d}{du} \right)_{u=0} = \sqrt{\frac{2}{\pi}}\,\frac{P}{r_1} \qquad (6.128)$$

Define the small-signal position sensor sensitivity (units of $\mathrm{m^{-1}}$):

$$\alpha_{KE} \equiv \frac{1}{P}\left(\frac{dP_d}{du} \right)_{u=0} = \frac{1}{r_1}\sqrt{\frac{2}{\pi}} \qquad (6.129)$$

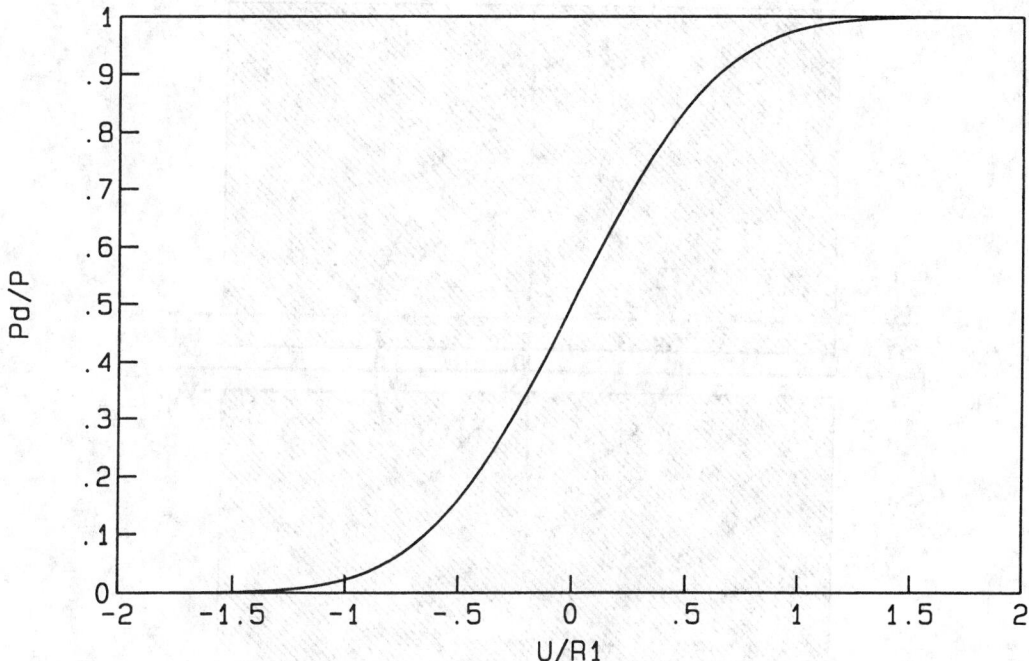

FIGURE 6.115   The normalized response P_d/P as a function of u/r_1 (u = displacement of center of beam; r_1 = Gaussian beam radius; P_d = power on the detector; P = total incident beam power).

The optical power reaching the photodetector for small signals is then given by:

$$P_d \cong \frac{P}{2} + \left(\frac{dP_d}{du}\right)_{u=0} u = \frac{P}{2} + \alpha u P = \frac{P}{2}\left[1 + 2\alpha u\right] \tag{6.130}$$

The photodetector signal will be linear in displacement if $u \le 0.387 r_1$.

The photocurrent is:

$$I = KP_d \cong \frac{KP}{2}\left[1 + 2\alpha u\right] \tag{6.131}$$

where K is the photodetector responsivity in A/W. The position sensor voltage is obtained using a transimpedance amplifier with gain Z:

$$V = KZP_d \cong \frac{KZP}{2}\left[1 + 2\alpha u\right] \tag{6.132}$$

Bicell Detector

The deflection of a Gaussian beam initially centered in the insensitive gap of a bicell detector is illustrated in Figure 6.116. The power incident on the upper half of the bicell is given by:

$$P_2 = \frac{aP}{\pi} \int_{y_2}^{\infty} e^{-a(y-u)^2} dy \int_{-\infty}^{\infty} e^{-ax^2} dx = \frac{P}{2}\left[1 - erf\left(\frac{\sqrt{2}}{r_1}(y_2 - u)\right)\right] \tag{6.133}$$

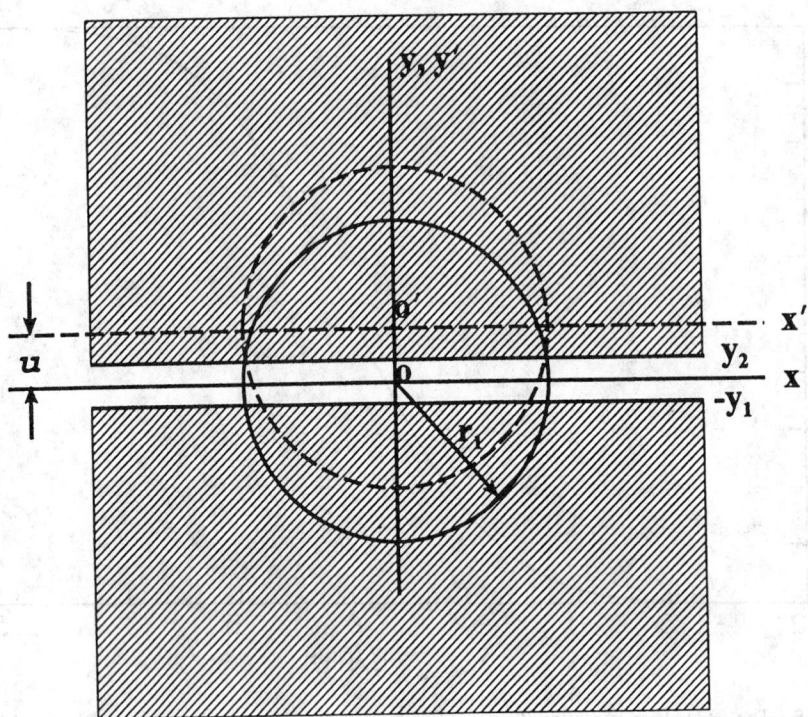

FIGURE 6.116 Deflection of a Gaussian beam initially centered in the insensitive gap of a bicell detector.

The power incident on the lower half of the bicell is given by:

$$P_1 = \frac{aP}{\pi} \int_{-\infty}^{-y_1} e^{-a(y-u)^2} dy \int_{-\infty}^{\infty} e^{-ax^2} dx = \frac{P}{2}\left[1 - erf\left(\frac{\sqrt{2}}{r_1}(y_1 + u)\right)\right] \tag{6.134}$$

The photocurrent from each detector of the bicell is converted to voltage by identical transimpedance amplifiers: $V_2 = KZP_2$ and $V_1 = KZP_1$; a difference amplifier is then used to obtain the bicell signal voltage:

$$V = V_2 - V_1 = KZ(P_2 - P_1) = \frac{KZP}{2}\left[erf\left(\frac{\sqrt{2}}{r_1}(y_1 + u)\right) - erf\left(\frac{\sqrt{2}}{r_1}(y_2 - u)\right)\right] \tag{6.135}$$

The normalized response, $2V/(KZP)$, is shown in Figure 6.117 as a function of u/r_1 for $y_1 = y_2 = r_1/10$. Suppose that the beam is centered in the gap, $y_1 = y_2$; then, for small displacements, one obtains:

$$V \cong 2\sqrt{\frac{2}{\pi}}\frac{KZPu}{r_1} e^{-2(y_1/r_1)^2} \tag{6.136}$$

The small-signal sensitivity is:

$$\alpha_{BC} \equiv \frac{1}{KZP}\left(\frac{dV}{du}\right)_{u=0} = 2\sqrt{\frac{2}{\pi}}\frac{e^{-2(y_1/r_1)^2}}{r_1} \tag{6.137}$$

This quantity is optimized when $r_1 = 2y_1$, and the optimal sensitivity is $0.484/y_1$.

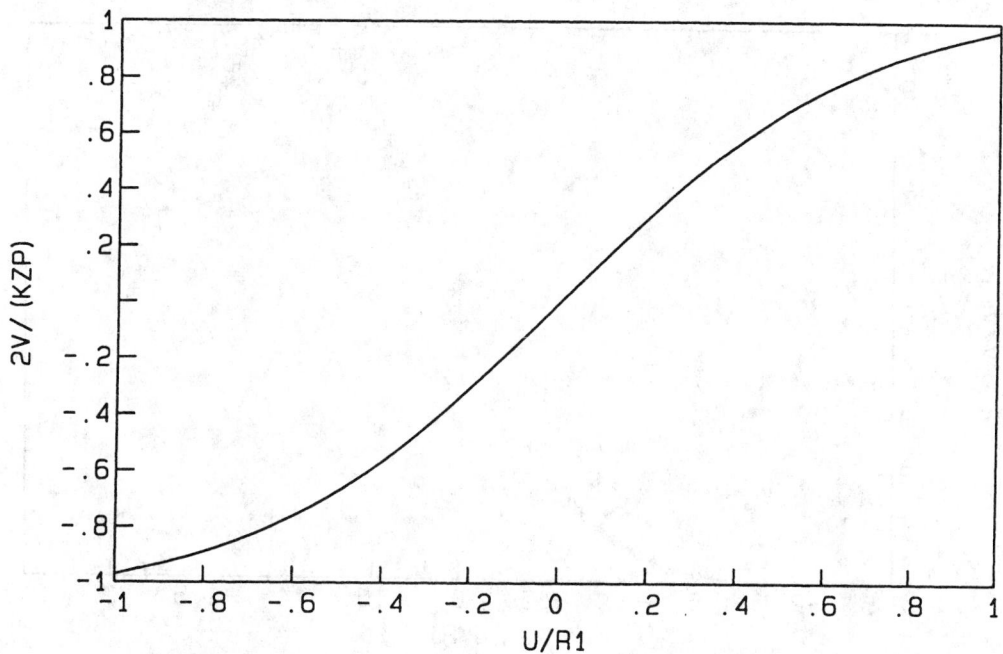

FIGURE 6.117 The normalized response $2V/KZP$ in a bicell detector as a function of u/r_1 for $y_1 = y_2 = r_1/10$.

Continuous Position Sensor

The position information in a continuous position sensor (also known as a lateral-effect photodiode) is derived from the divided path taken by photon-generated electrons to two back electrodes on the device. For a homogeneous device, the current to an electrode depends only on the distance of the centroid of the light beam from that electrode; the currents would be equal in an ideal device when the beam is located at its electrical center.

Consider the analysis of a one-dimensional continuous PSD. The current signal from each electrode is converted to a voltage signal by a transimpedance amplifier with gain, Z. Operational amplifiers are then used to provide the sum signal, $V_s = KPZA_s$, and the difference signal, $V_d = KPZA_d\alpha u$, where A_s and A_d are the sum and difference amplifier gains, respectively. The position sensor sensitivity, α, is then given by:

$$\alpha \equiv \frac{V_d A_s}{u V_s A_d} \tag{6.138}$$

which is determined by measuring V_d and V_s as a function of u.

Characterization of PSDs

The PSD characteristics presented here were measured by mounting the device to a translation stage with an optical encoder driven by a piezoelectric motor [9]; the position accuracy was ±0.1 μm. An appropriately attenuated He–Ne laser beam was directed at normal incidence to the PSD as it was translated past the beam. The PSD sum and difference voltages were measured with GP–IB digital voltmeters as a function of displacement; the PSD displacement and voltage measurements were computer controlled.

The operation of a knife-edge PSD can be evaluated using the signal from one side of a bicell as well as from a photodetector behind a knife edge. The transimpedance amplifier output voltage corresponding

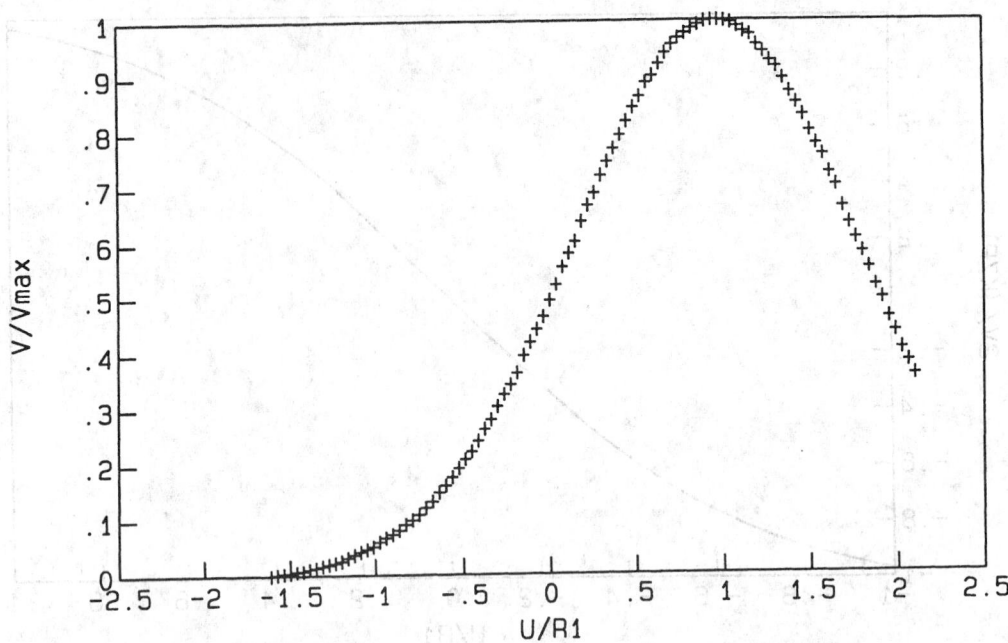

FIGURE 6.118 Transimpedance amplifier output voltage corresponding to the current signal from one cell of a United Detector Technology (UDT) SPOT2D bicell as a function of displacement. Data are plotted as normalized cell voltage vs. u/r_1.

to the current signal from one cell of a United Detector Technology [10] (UDT) SPOT2D bicell is shown as a function of displacement in Figure 6.118. The data are plotted as normalized cell voltage as a function of u/r_1. Since the signal from one cell of a bicell is equivalent to a knife-edge photodetector, the data can be compared with Figure 6.115. The data are in reasonable agreement for $u/r_1 < 1$ and a best fit allows the inference of $r_1 = 0.4$ mm; the disagreement for $u/r_1 \geq 1$ corresponds to the laser beam partially moving off the outside edge of the cell.

The operation of a bicell PSD is evaluated by the data of Figure 6.119 which shows the difference voltage as a function of displacement for the UDT SPOT2D. The deviations from the theoretical curve of figure 6.117 for $u \leq 0.4$ mm and $u \geq 1.1$ mm are due to the beam moving off the outside edges of the bicell. The linear region of the PSD is centered at the optimal quiescent point $(V_1 - V_2 = 0)$; a least-squares fit of a straight line gives a slope of $A_{12} = 4.88$ mV μm^{-1} for this device. For the data of Figures 6.118 and 6.119, $P = 0.37$ mW and $Z = 5$ kΩ; thus, the experimental value of the responsivity is determined to be $K_{exp} = 0.65$ A W^{-1}. The nominal gap size of the SPOT2D bicell is such that $y_1 = 63.5$ μm; thus, using this value of y_1 and $r_1 = 0.4$ mm in Equation 6.137, the calculated bicell sensitivity is 3.8×10^3 m^{-1}. The experimental value of sensitivity is calculated from A_{12}, K_{exp}, P, and Z to be 4.1×10^3 m^{-1}.

Evaluation of the operation of the x-axis of a UDT SC10 two-axis continuous PSD is illustrated in Figures 6.120 through 6.122. The PSD was oriented such that the y-axis was vertical; thus, ideally, V_{dy} is constant as the PSD is translated, except near the edges of the 10 mm × 10 mm device. As shown in Figure 6.120, V_{sx} is virtually constant except near the edges of the device; this means that the optical responsivity (K) does not greatly vary with the laser beam position on this detector. Variation of V_{dx} with displacement is shown in Figure 6.121; this PSD has a broad linear range. The linearity of continuous PSDs can often be improved by using the ratio, V_{dx}/V_{sx}, as shown in Figure 6.122; the sensitivity of this PSD is determined from a least-squares fit of the linear part of this characteristic to be 183 m^{-1}.

A newer class of PSDs is represented by the Sitek [11] 1L10 single-axis continuous PSD, which is typically more linear. An example of the V_d/V_s versus u characteristic is shown in Figure 6.123. The sensitivity is determined from a least-squares fit of the linear part of the characteristic to be 307 m^{-1}.

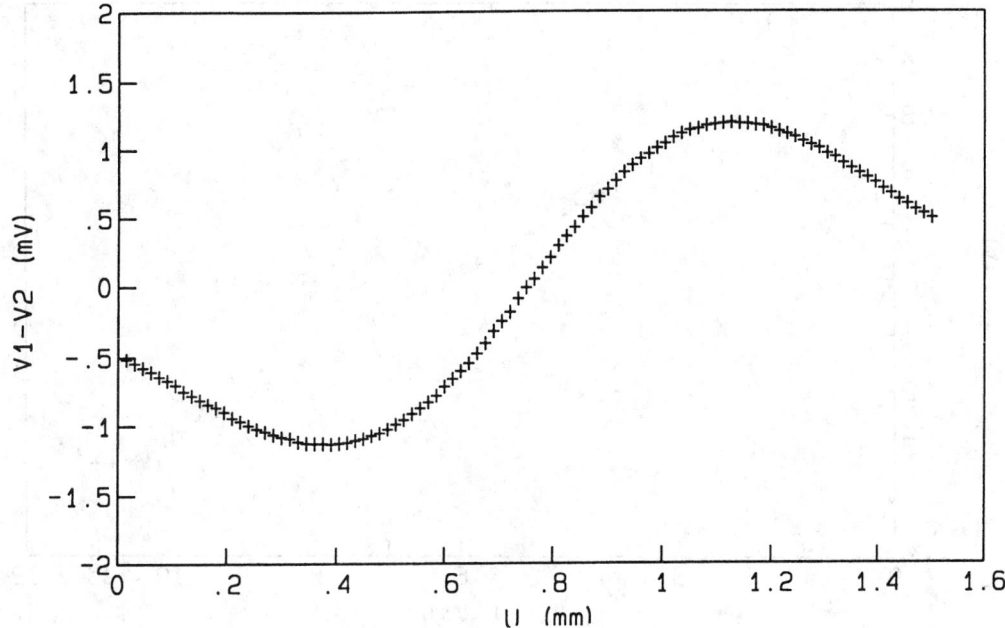

FIGURE 6.119 Difference voltage as a function of displacement for the United Detector Technology (UDT) SPOT2D bicell.

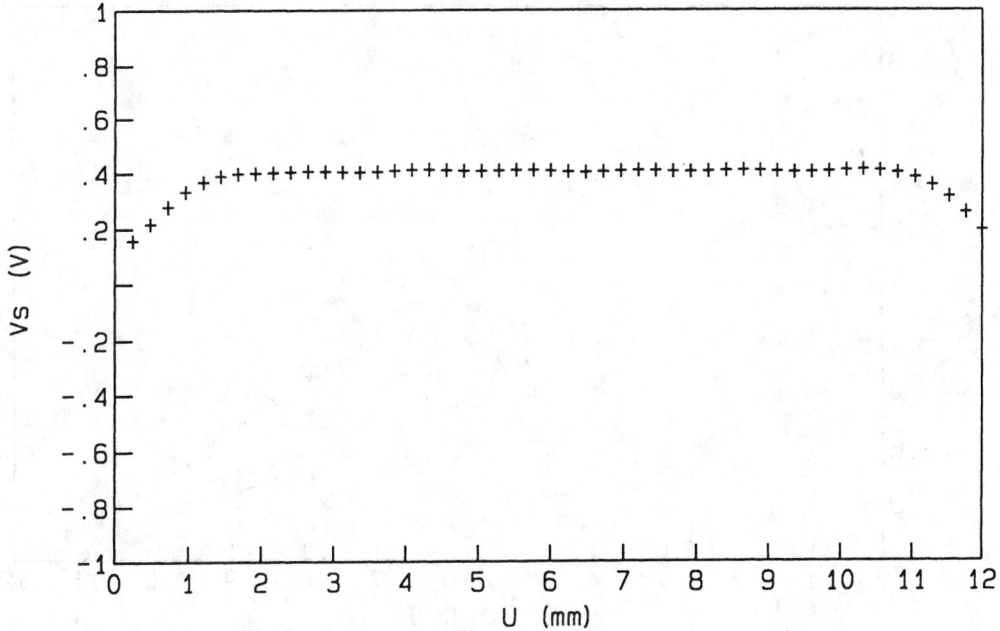

FIGURE 6.120 Sum voltage vs. displacement for x axis of UDT SCIO continuous PSD.

Summary

A PSD composed of a knife-edge photodetector is simple, has a sensitivity that depends only on the radius of the laser beam, and has a response time determined by the photodetector. Thus, for fast rise-time detection of small displacements, this type of PSD is to be preferred.

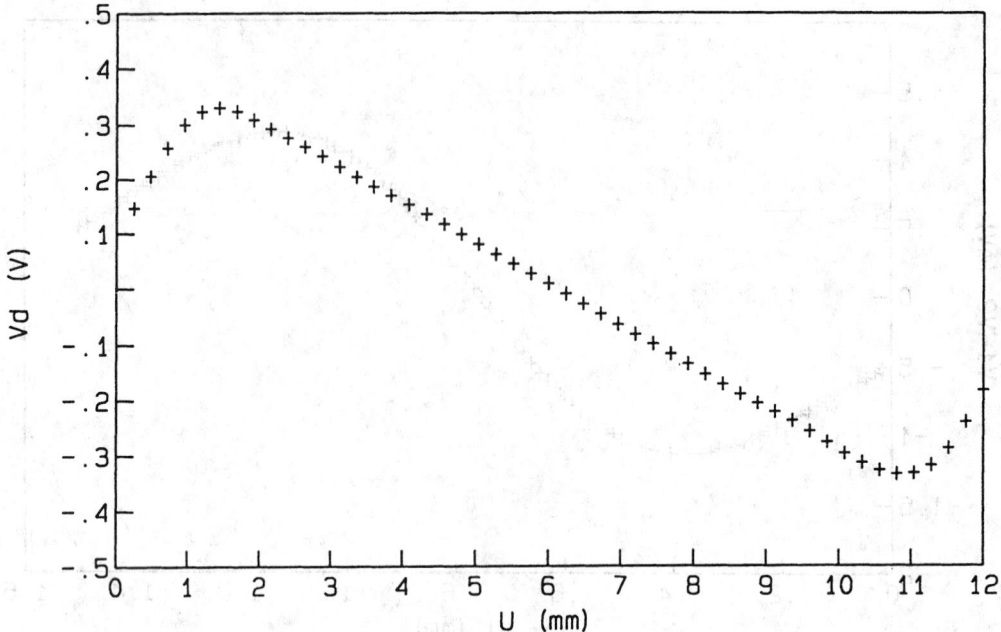

FIGURE 6.121   Difference voltage vs. displacement for x axis of UDT SCIO continuous PSD.

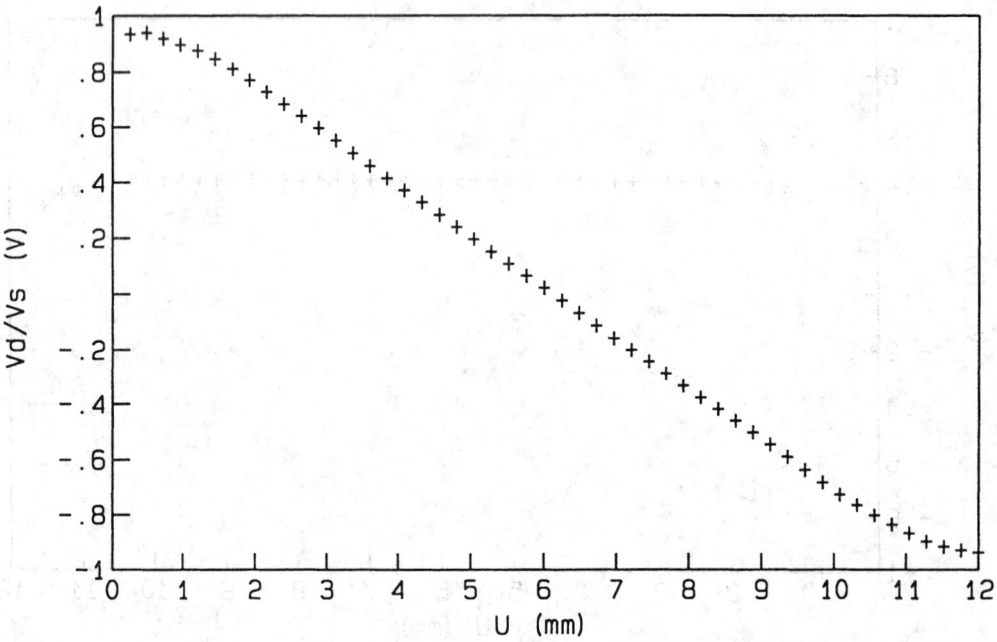

FIGURE 6.122   Ratio of difference voltage to sum voltage vs. displacement for x axis of UDT SCIO continuous PSD.

    The maximal sensitivity of a bicell is about 20% greater than that of a knife-edge photodetector. It has the advantages of small size and economy. The nominal risetime of the UDT SPOT2D is of the order of 10 ns. Furthermore, two-axis detection can be readily obtained with a quadcell, with an increase in risetime due to the increased capacitance of the detector.

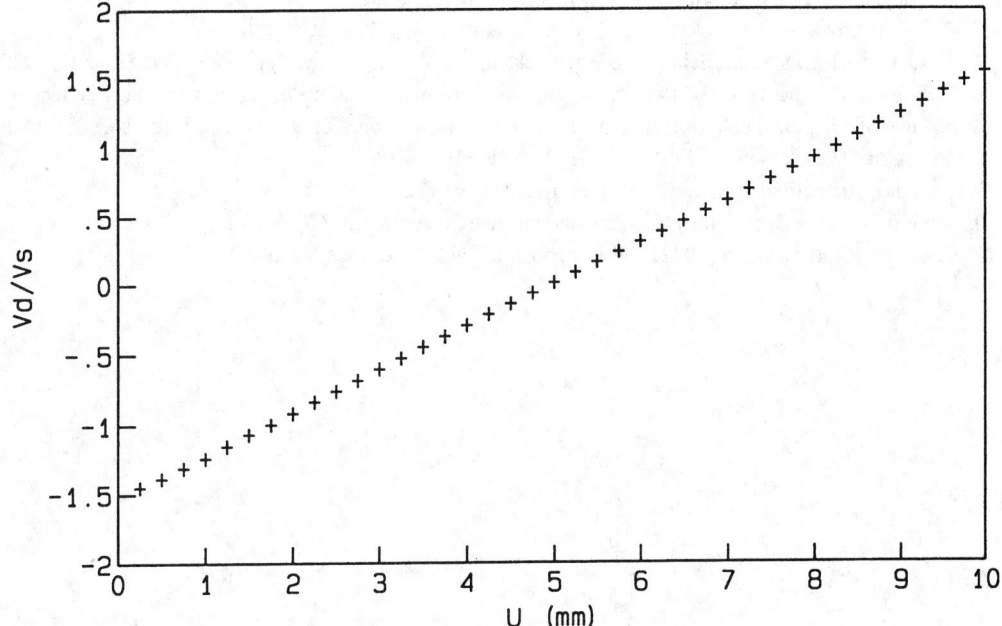

FIGURE 6.123   Ratio of difference voltage to sum voltage vs. displacement for Sitek 1L10 continuous PSD.

The disadvantages due to small laser beam sizes and small displacements are overcome by the continuous PSDs. Since these devices are typically much larger (available in several-inch diameters), they typically have longer risetimes than the other PSDs. However, the Sitek 1L10, with a 10 mm linear active range has a measured upper half-power frequency of 3 MHz.

In applications where the output signal from a PSD must be linearly proportional to the displacement of the beam, analog-divider operational amplifiers to obtain V_d/V_s in real time are used to extend the range of linearity of the device. Unfortunately, the frequency response of these amplifiers are often the frequency-response-limiting factors of the PSD system. In cases where high-frequency response is important, V_d alone can often be used if care is taken to operate in the linear range of the device. For large static displacements that are of the order of the size of the detector, V_d and V_s can be recorded with computer-controlled data acquisition, the calibration characteristic numerically fitted to a polynomial, and then any voltage from the detector can be related to beam position.

The noise limitations in OBD sensing are due to the laser, the nature of the reflecting surface, and the PSD. Lasers with good amplitude stability are to be preferred, but this is not an important contribution to noise when V_d/V_s is used to infer displacement. Laser beam-pointing stability, on the other hand, is important. If the reflecting surface is that of a typical solid, then negligible noise is introduced on reflection; this may not be true for a reflector such as a pellicle, where Brownian motion of the surface may be significant. The noise limitations of the PSD are the usual ones associated with the photodetector and the amplifiers.

References

1. A. C. Boccara, D. Fournier, and J. Badoz, *Appl. Phys. Lett.*, 30, 933, 1983.
2. G. C. Wetsel, Jr. and S. A. Stotts, *Appl. Phys. Lett.*, 42, 931, 1983.
3. e.g.: D. Fournier and A. C. Boccara, *Scanned Image Microscopy*, E. A. Ash, Ed., London: Academic Press, 1980, 347-351; J. C. Murphy and L. C. Aaamodt, *Appl. Phys. Lett.*, 39, 519, 1981; G. C. Wetsel, Jr. and F. A. McDonald, *Appl. Phys. Lett.*, 41, 926, 1982.
4. M. A. Olmstead, S. Kohn, N. M. Amer, D. Fournier, and A. C. Boccara, *Appl. Phys. A*, 132, 68, 1983.

5. G. Meyer and N. M. Amer, *Appl. Phys. Lett.*, 53, 1045, 1988.

6. J. C. Murphy and G. C. Wetsel, Jr., *Mater. Evaluation*, 44, 1224, 1986.

7. G. C. Wetsel, Jr., S. E. McBride, R. J. Warmack, and B. Van de Sande, *Appl. Phys. Lett.*, 55, 528, 1989.

8. S. E. McBride and G. C. Wetsel, Jr., Surface-displacement imaging using optical beam deflection, *Review of Progress in Quantitative Nondestructive Evaluation*, Vol. 9A, D. O. Thompson and D. E. Chimenti, (Eds.), New York: Plenum, 1990, 909-916.

9. Burleigh Instruments, Inc., Fishers, NY 14453.

10. United Detector Technology, 12525 Chadron Ave., Hawthorne, CA 90250.

11. On-Trak Photonics Inc., 20321 Lake Forest Dr., Lake Forest, CA 92630.

7

Thickness Measurement

John C. Brasunas
NASA/Goddard

G. Mark Cushman
NASA/Goddard

Brook Lakew
NASA/Goddard

One can measure thickness on many scales. The galaxy is a spiral disk about 100 Em (10^{20} m) thick. The solar system is pancake-like, about 1 Tm (10^{12} m) thick. The rings of Saturn are about 10 km thick. Closer to home, Earth's atmosphere is a spherical shell about 40 km thick; the weather occurs in the troposphere, about 12 km thick. The outermost shell of the solid Earth is the crust, about 35 km thick. The ocean has a mean depth of 3.9 km. In the Antarctic, the recently discovered objects believed to be microfossils indicative of ancient Martian life are less than 100 nm thick. In terms of the man-made environment, industry must contend with thickness varying from meters, for construction projects, to millimeters on assembly lines, to micrometers and nanometers for the solid-state, optical, and coatings industries. Perhaps the most familiar way of measuring thickness is by mechanical means, such as by ruler or caliper. Other means are sometimes called for, either because both sides of an object are not accessible, the dimension is either too big or too small for calipers, the object is too fragile, too hot, or too cold for direct contact, or the object is in motion on an assembly line — it may not even be a solid. Thickness may also be a function of position, as either the object may have originally been made with nonuniform thickness, deliberately or not, or the thickness may have become nonuniform with time due either to corrosion, cracking, or some other deterioration. The thickness may also be changing with time due to deliberate growth or etching, as example for thin films. Thus it follows that, in more general terms, measuring thickness might require measuring the topography or height profile of two surfaces and taking the difference. Alternatively, the measurement technique may produce a reading directly related to the difference. Table 7.1 lists some of the many techniques suited to determining thickness, together with the range of usefulness and some comments on accuracy and/or precision.

0-8493-8347-1/99/$0.00+$.50
© 1999 by CRC Press LLC

7-1

TABLE 7.1 Thickness Measuring Techniques

Technique	Range	Comments
Mechanical		
Caliper gage, micrometer	1 µm–100 mm	±3 µm accuracy
Electronic gages: LVDT	0–1 m	Precision depends on noise level
Pneumatic gaging	50 nm minimum	
Optical/focusing, shadowing, comparing		
Microscope	5 µm minimum	About 1% accuracy
Comparators/projectors	25–250 nm	
Laser caliper	100 µm–100 mm	Precision of 6 µm or better
Weighing	Range depends on area	
Capacitive gaging	From <1 µm to about 1 cm	
Inductive gaging (eddy current sensing)	0–1.5 mm	Precision of 2.5 µm
Magnetic induction	0–4 mm	10% accuracy
Hall effect gage	0–10 mm	1–3% accuracy
Far-field/time-of-flight		
Sonar/ultrasound	0.5–250 mm	25 µm accuracy
Radar	0.1 to few hundred km	
Lidar	10 m–5 km	
THz technology		
Far-field/resonance		
Resonant ultrasound		
Interferometry: spectral and spatial	1 nm–100 µm	Accuracy about λ/50
Ellipsometry	0.3 nm–10 µm	0.1 nm accuracy
Far-field/absorption, scattering, emission		
Gamma-ray backscatter	Range to 25 mm	0.5% precision
Beta-transmission	2 µm–1 mm	0.2% precision
Beta-backscatter	100 nm–50 µm	3 to 20% precision
X-ray fluorescence	0–30 µm	
Infrared absorption	Depends on material	
Scanning techniques: scanning probe microscopy		Precision better than 0.1 nm
Destructive techniques: electrolytic	15 nm–50 µm	

7.1 Descriptions of the Relatively Mature Measuring Techniques

The following descriptions will also refer to some of the relevant vendors, whose addresses are found in Table 7.2. Additional vendor information, with specific price or model number identification, is found in Table 7.3. The words "gage" and "gauge" are used interchangeably.

Mechanical

The fundamental tool for measuring thickness is the line-graduated instrument [1, 2]. It is the only mechanical means to make direct measurements. Graduated spacings that represent known distances are used as direct comparisons to the unknown distance. Instruments include bars, rules, and tapes generically called rulers; caliper gages, which employ a positive contact device for improved alignment of the distance boundaries; and micrometers, which typically have greater precision due to a combination of linear and circumferential scales. Caliper precision can be improved with vernier scales or linear transducers. Fixed gages are often used to measure objects on a pass/fail basis. An object of fixed geometry (length, tapered bore, thread, etc.) is compared to a test piece typically for part inspection. Variations include the master gage, an object used to represent the nominal dimension of the part; the limit gage, an object used to represent the limit condition for tolerance dimensioning; and gage blocks or Johansson blocks, an object of fixed length used as a dimensional reference standard. Dial indicators are used to sense displacement

TABLE 7.2 Vendor addresses.

Vendor	Address
Bomem	Quebec, Canada
Brown & Sharpe	North Kingstown, RI
CMI International	Elk Grove Village, IL
Conductus	Sunnyvale, CA
deFelsko	Ogdensburg, NY
Digilab	Cambridge, MA
Digital Instruments	Santa Barbara, CA
Electromatic	Cedarhurst, NY
Fischer	Windsor, CT
Hewlett Packard	Englewood, CO
Kta-Tator	Pittsburgh, PA
Magnetic Analysis Corp.	Mount Vernon, NY
Mattson	Madison, WI
Measurex	Cupertino, CA
Micro Photonics	Allentown, PA
Midac	Costa Mesa, CA
Mitutoyo	Plymouth, MI
Moore Products Co	Spring House, PA
NDC Systems	Irwindale, CA
Nicolet	Madison, WI
Ono Sokki	Addison, IL
Oxford Instruments	Concord, MA
Panametrics	Waltham, MA
Park Scientific Instruments	Sunnyvale, CA
Penny + Giles	Attleboro, MA
Perkin Elmer	Norwalk, CT
Phase-Shift Technology	Tucson, AZ
Rudolf Instruments	Fairfield, NJ
Scantron	Dist. by Micro Photonics
Schaevitz	Pennsauken, NJ
Sentech	Dist. by Micro Photonics
SolveTech	Claymont, DE
Starrett	Athol, MA
Stresstel	Scotts Valley, CA
Transicoil Inc.	Valley Forge, PA
Trans-Tek	Ellington, CT
Willrich Precision Instrument Co.	Cresskill, NJ
J.A. Woolam Co., Inc.	Lincoln, NE
Wyko	Tucson, AZ
Zygo	Middlefield, CT

from a reference plane and display the deviation thereof. The display can be electronically coupled for amplification and/or display purposes. The range of a measuring instrument may be extended if multiple copies of the object to be measured are available. For example, the thickness of a sheet of paper may be measured by a simple ruler if 500 sheets of paper are stacked. (Vendors: Brown & Sharpe, Starrett, Mitutoyo. Also see [3].)

Electronic Gages

A Linear Variable Differential Transformer (LVDT), utilizes multiple toroidal transformers to sense axial displacement of an iron core that is attached to a measuring contact, either directly or by another joint (such as a lever). The displacement has a direct correlation to the distance that other electronics display. Thus, the LVDT serves as a replacement for a lined ruler or micrometer, incorporating an electrical readout. (Vendors: Penny + Giles; Schaevitz; Transicoil Inc.; Trans-Tek.)

TABLE 7.3 Instruments for measuring thickness.

Manufacturer	Model Number	Price	Description
KTa-Tator	TI-12	$1595.	General-purpose ultrasonic gage, 0.75 mm to 75 mm range
NDC Systems	6100TC	$49,300.	Backscatter gamma gage for 60 in. web, 25 mm range
NDC Systems		$66,600.	Transmission beta gage, for continuous web products
Panametrics	25DL	$2200. to $3800.	Single-element ultrasonic gage, 50 mm range
Panametrics	26DL Plus	$1400. to $2500.	Dual-element ultrasonic gage, 250 mm range
Panametrics	8000	$6500.	Hall effect magnetic gage, for nonferrous materials, 6 mm range
DeFelsko	Positest 1000-N	$1995.	Eddy current sensor, Apple Newton read-out, measure out to 1.5 mm nonferrous, nonconducting coating on conducting substrate
Magnetic Analysis	Various	$1500. to $100,000.	Ultrasonic, time-of-flight gages
Fischer	Deltascope MP2C	$1200.	Magnetic induction gage, measure nonmagnetic coating on ferromagnetic substrate
Fischer	IsoScope MP1C	$1200.	Eddy current gage, measure nonconducting coating on nonferrous conducting substrate
Fischer	Fischerscope MMS	$6500.	Beta-backscatter system to measure coating thickness
Fischer	Fischerscope X-Ray 1020 video	$34,000.	X-ray fluorescence system to measure coating thickness
Fischer	Couloscope Sx	$2500.+ accessories	Electrolytic, destructive system to measure coating thickness
J.A. Woollam Co. Inc.	M-44	Application specific	Variable angle, multiwavelength spectroscopic ellipsometer
Rudolf Instruments	431A31WL633	$10,100.	Manual, HeNe wavelength ellipsometer
Rudolf Instruments	444A12	$34,000.	Automatic, HeNe wavelength ellipsometer
Hewlett Packard	HP8712C	$13,500.	RF vector network analyzer, measure transmission/reflection frequency response to 1.3 GHz, optional to 3 GHz
Stresstel	T-Mike Programmable	$995.	Dual-element ultrasonic system
Stresstel	TM1D	$1795.	Single-element ultrasonic system
Measurex	DMC480	Application specific	High-speed X-ray thickness gage
Bomem	MB series	$20,000. and up	1 cm^{-1} resolution Fourier transform spectrometer
Park Scientific Instruments	Autoprobe CP	$65,000.	Ambient scanning probe microscope
Park Scientific Instruments	Autoprobe VP2	$130,000.	UHV scanning probe microscope
Digital Instruments	Nanoscope IIIa/D3000	$90,000.	Small sample scanning probe microscope

Pneumatic Gaging

Pneumatic gages have pressurized air exiting gage orifices. The air velocity differential or backpressure is a function of the separation of the gage and the part. In the direct or open jet method, the pressurized air experiences backpressure due to the impedances posed by the measured part. The typical scenario is that the gage head and the measured part have similar geometry (i.e., a cylindrical gage in a bored hole). By placing two gages on either side of a flat plate, the thickness may be inferred. In the indirect or contact method, the pressurized air pushes on a contact piece that directly contacts the part. Tolerances as small as 50 nm can be measured. (Vendors: Willrich Precision Instrument Co.; Moore Products Co.)

Optical: Focusing, Shadowing, Comparing

This includes microscopes, which can determine thickness either by comparison with a known reference, or by focusing on the front and rear surfaces of a sample, noting the difference in focus position. Comparators project onto a screen what might be noted through a microscope. Laser calipers retrieve dimensions by measuring the shadowing of a laser beam. (Vendors: NDC Systems for laser caliper; Scantron for laser profilometer.)

Weighing

Given a plate of material with known density, first measure the area with some type of calibrated video system. Then, a measurement of weight can be simply converted to an estimate of the thickness. As is common with this technique and most of the following techniques, estimating the thickness requires knowledge of some other property of the material to be measured — in this case, the density.

Capacitive Gaging

Capacitive gaging is realized by inserting a nonmetallic material into a known electric field. Knowing the gage sensor area and the material's dielectric constant, the thickness can be determined. Submicron thickness levels can be achieved. (Vendors: Ono Sokki; SolveTech.)

Inductive Gaging (Eddy Current Sensing)

The principle here is that ac currents in a coil induce eddy currents in a nearby conducting plate [4, 5]. These eddy currents can be sensed by a pickup coil, which may be the exciting coil or a second coil. The presence of the eddy currents manifests itself as a modification of the apparent inductance and/or the loss of the pickup coil. This technique is appropriate for nonferrous metals, and is especially sensitive to thickness variations due to flaws such as cracks or corrosion. There is one particular instance in which it is common to measure thickness rather than variations. That would be the thickness of a nonconducting coating on a nonferrous conducting substrate. The coating thickness creates a gap (lift-off) between the exciting coil and the eddy currents, thereby affecting the eddy current signal. The range of this technique would be about 1 mm. Fischer has an instrument designed for measuring the thickness of a newly laid road surface coating to a depth of 40 cm, by burying a conductive plate below the road. (Vendors: Fischer; deFelsko; CMI International.)

Magnetic Induction

This technique is also used to measure coating thickness, in this case a nonmagnetic coating on a ferromagnetic substrate. The nonmagnetic coating creates a gap (lift-off) between the ferromagnetic substrate and a probe. One way to measure the gap and thereby the thickness is by measuring the force required to pull away a magnetic probe. Another technique would be to magnetically couple the ferromagnetic substrate to a transformer core, with a gap between the substrate and the core. This technique would have a range of about 4 mm. CMI International has an informative brochure describing the relative merits of measuring coating thickness via eddy current, magnetic induction, beta-backscatter, microresistance, and X-ray fluorescence; the choice of technique depends, among other things, on the material to be tested. (Vendors: Fischer; CMI International; Electromatic; deFelsko.)

Hall Effect Gage

This sensor measures the thickness of nonferrous materials with 1% accuracy by sandwiching the material being measured between a magnetic probe on one side and a small target steel ball on the other side [6].

It measures up to 10 mm. The Hall effect sensor is used to measure the magnetic field, as a dc measurement; ac Hall effect measurements can be made more precisely because they eliminate bias and are done with less noise. (Vendor: Panametrics.)

Far-Field/Time-of-Flight: Ultrasound, Radar, Lidar

Using 1940s sonar principles and today's microprocessor technology, high-frequency (1–20 MHz) ultrasound waves can be used to measure thickness by sending pulsed sound waves through a material and measuring the transit time of the reflected signal [5, 7]. Knowing the sound velocity of the material, materials from 0.5 mm to 250 mm can be measured, often as fine as 25 μm. Media include metal, glass, ceramic, liquid, rubber, fiberglass, plastic, and concrete. Ultrasound can also be used to measure living tissues, as is often done in the agricultural and medical fields. Fat layers of cattle and pigs can influence marketability. Skin burn depths can direct treatment procedures. The depth of foreign objects in the body is useful for microsurgery. Ultrasonic thickness determination has expanded to include mulitdimensional echolocation applications, such as imagery and acoustic microscopes that can resolve in the submicron level. The principles behind ultrasound also apply to electromagnetic waves. In the gigahertz range, this is called radar. Radar can be used to estimate the thickness of atmospheric layers such as cloud layers. The light-wave version of radar, called lidar, can be used to measure the thickness of water vapor layers in the lower atmosphere. (Ultrasonic vendors: KTa-Tator; deFelsko; Stresstel; Magnetic Analysis Corp.; Panametrics; Electromatic.)

Far-Field/Resonance: Ultrasound, Interferometry, Ellipsometry

The idea here is that when waves such as ultrasound impinge on a plane-parallel slab of material, there will be reflected power from both the front and rear surfaces; depending on whether the slab thickness is an odd or even number of quarter-wavelengths, the reflected beams will be in constructive or destructive interference. If the frequency is swept, the distance in frequency between successive maxima and minima may be related to the slab thickness, if the index of refraction is known. Since the natural, or resonant, modes of an object depend on the properties and dimensions of an object, knowledge of the properties enables estimation of dimensions from the resonant frequencies. Compared with time-of-flight ultrasound, resonant ultrasound is much less common. It has been used to characterize concrete, and is quite sensitive to flaws, as anyone who has heard a cracked bell would know. Resonant techniques are much more common with visible [8], infrared, or microwave [4] radiation. Spectral interferometry would be appropriate to characterize the thickness of transparent substrates with reasonably flat surfaces, sufficiently parallel to one another. A common way to do this would be to measure a transmission spectrum with a spectrometer such as a Fourier transform spectrometer (FTS). The successive maxima and minima are here called the Fabry–Perot effect, and their appearance in a spectrum is called channeling. Thickness can also be measured with spatial interferometry, which is essentially a way of measuring surface topography. An example would be the phenomenon of Newton's rings, which occur when the surface to be tested is in contact with an optical flat. Using a transparent optical flat, transmit monochromatic light such as a mercury lamp through the flat and onto the interface between the flat and the test surface. If there are variations in the height of the test surface, then the two return beams from the optical flat and the test surface will alternate between constructive and destructive interference, producing fringes or rings. The sensitivity is not limited to the scale of the wavelength λ: with sufficient stability and signal-to-noise, dimensions down to 1/1000 of a fringe can be measured. With sources of longer coherence length, such as lasers, the test surface and the optical flat need not be in direct contact.

Another optical way to measure thickness is with ellipsometry [9], typically used to measure properties of thin, transparent films from a few tenths of nanometers to several hundreds of nanometers thick. This includes metals, as long as the metal is sufficiently thin to be partially transparent. By measuring the change in polarization state for nonnormal incidence light, both the thickness and refractive index of a thin layer may be inferred. Additional information (e.g., the properties of multiple layers) can be obtained by varying the angle of incidence and by observing at multiple wavelengths. The ability to estimate both thickness and refractive index is an important advantage of this technique, as often the refractive index of a material in thin film form is not the same as the bulk value, and indeed may be a property of the deposition conditions. (FTS vendors: Bomem; Digilab; Mattson; Midac; Nicolet; Perkin-Elmer. FTS system pricing may range from about $15,000 to over $100,000, depending on the application. Spatial interferometer vendors: Zygo; Wyko, Phase-Shift Technology. Ellipsometer vendors: J.A. Woollam Co., Inc.; Rudolf Instruments; Sentech. The cost of an ellipsometer may range from $10,000 for a manual, single-wavelength system to $200,000 for an automatic, multiwavelength system. Microwave resonance vendor: Hewlett-Packard.)

Far-Field/Absorption, Scattering, Emission: Beta, Gamma, X-Ray, Infrared

These techniques depend on the extinction (scattering or absorption) or emission of photons or massive particles (electrons, protons, neutrons) when transiting the material to be measured. Typically, the extinction or emission shows an exponential dependence on thickness; the dependence becomes linear if the absorption is sufficiently low. These techniques, in particular gamma-ray backscatter and beta-ray transmission, are used to measure continuously moving web materials (paper, metals, fabrics) on assembly lines. Infrared absorption is also suitable if the moisture content is controlled. Beta-backscatter and X-ray fluorescence [10] are used for measuring coatings. In X-ray fluorescence, upon exposure to X-rays, certain elements fluoresce (emit) X-rays at characteristic wavelengths. The strength of this emission is related to thickness. These absorption/emission techniques may sometimes be better suited than time-of-flight ultrasound to the dimensional measurement of objects with complex shapes. (Gamma gage vendor; NDC Systems. X-ray absorption vendor: Measurex. X-ray fluorescence vendors: Fischer; NDC Systems; CMI International. Beta-backscatter vendor: Fischer; Electromatic; CMI International; Measurex. Infrared absorption vendor: NDC Systems. The prices for these systems will depend on the application; a typical system could cost $500,000.)

Destructive Techniques

Fischer markets a system that removes a coating into an electrolyte and then electrolytically deposits the removed coating. The electrical charge required for deposition is related to the coating thickness.

7.2 Future Directions in Thickness Measurement

Concerning Techniques Mentioned Above

Concerning capacitive sensors, the NASA Langley Research Center is developing sensors based on patterns of conductors sandwiched between insulating layers. The presence of ice over the conductors changes the capacitance, providing a way of sensing ice build-up on aircraft wings. With respect to eddy current sensing, one limitation is that a nonsuperconducting sense coil responds best to high-frequency excitations, and not at all to dc magnetic fields. This limits the technique to fairly high frequencies and thus low penetration depths, since the skin depth becomes shallower with increasing frequency. One possibility is to use a SQUID (superconducting quantum interference detector) as the sensor, since the SQUID is probably the most sensitive sensor of dc and low-frequency magnetic fields. One disadvantage of the SQUID has been the need for liquid helium for cooling for low-temperature superconductors; with the recent availability of high-temperature superconductors (HTS, above 90 K) and now HTS SQUIDS, cooling can be done with liquid nitrogen or single-stage mechanical coolers. In the area of spatial interferometry, work at Lawrence Livermore National Laboratory replaces the reference surface with a single-mode fiber in a process called phase-shifting diffraction interferometry. A measurement accuracy of 1.44 nm rms is quoted, with a goal of 0.1 nm rms. (HTS SQUID vendor: Conductus.)

THz Technology

With the availability of femtosecond pulsed lasers, Bell Labs has been investigating a technique using 100 fs pulses to pulse an antenna in the range of 0.1 THz to 3.0 THz. The terahertz pulses are sent through the material to be tested, detected, and the received pulse shape is analyzed to extract constituent information. This technique may also provide information on thickness.

Nanoscale-Scanning Probe Microscopy

Scanning probe microscopes (SPMs) are used in a wide variety of disciplines, including fundamental surface science, routine surface roughness analysis, and spectacular three-dimensional imaging — from atoms of silicon to micron-sized protrusions on the surface of a living cell [11]. The scanning probe microscope is an imaging tool with a vast dynamic range, spanning the realms of optical and electron microscopes. It is also a profiler with unprecedented 3-D resolution. In some cases, scanning probe microscopes can measure physical properties such as surface conductivity, static charge distribution, localized friction, magnetic fields, and elastic moduli. As a result, applications of SPMs are very diverse. The scanning tunneling microscope (STM), the progenitor of SPMs, utilizes a sharp conductive tip with a bias voltage applied between the tip and the sample. When the tip is within 1 nm of the sample, electrons from the sample begin to tunnel through the 1 nm gap into the tip. If the bias voltage is reversed, the tunneling occurs into the sample. The tunneling current is a function of the separation. Both the tip and the sample must be conductors or semiconductors.

The atomic force microscope (AFM) utilizes a small tip at the end of a cantilever. Forces between the tip and sample cause a deflection in the cantilever, which is translated into a signal. The tip or sample can be scanned covering a large area, producing a topographical map. AFMs can be used on insulators or conductors. AFMs are used in two modes: contact and noncontact. In contact mode, the tip is brought within about 200 pm — about the length of a chemical bond. The electron clouds of the tip and sample atoms interact, netting a repulsive force. For this reason, the contact mode is also called repulsive. Vertical resolution of about 50 pm can be achieved. In noncontact mode, a vibrating cantilever is used in the attractive regime of the van der Waals interactions. The cantilever is typically 2 nm to 20 nm away from the sample surface and has low total force. Noncontact AFM is subsequently less sensitive; thus, sensitive ac detection systems must be employed. The low force does have the advantage of not contaminating the sample surface and is preferred for applications involving silicon wafers and soft or elastic tissues. In noncontact mode, the cantilever is resonated with a small amplitude. As the tip comes near the sample surface, the resultant force changes the spring constant, translating into a deviation of the resonance frequency. This change in resonance (or vibrational amplitude) reflects changes in the sample topology.

Intermittent-contact mode is a combination of noncontact and contact modes and best suited for soft, adhesive, or fragile samples. Contact mode can damage the tip and the sample due to frictional or shear forces and/or create data artifacts from tip/surface adhesion. Noncontact mode produces lower amplitudes and hence lower resolution. Furthermore, surface monolayers of adsorbed gases such as water vapor can produce erroneous results. Intermittent-contact mode avoids these pitfalls by placing the tip in contact with the surface, providing high resolution and then removing the tip to prevent dragging and/or lateral forces. The cantilever is resonated via a piezoelectric crystal (50 kHz to 500 kHz in ambient, 5 kHz to 40 kHz in fluids) overcoming the tip/sample adhesion forces.

In magnetic force microscopy (MFM), the noncontact mode is employed using a tip coated with a ferromagnetic film. Both magnetic and van der Walls interactions are present, but at larger tip/sample separations, the magnetic forces dominate. Multiple scans as a function of tip/sample distance allow differentiation of magnetic forces and topographic information. Magnetic domain structures are resolved to 50 nm via this technique. Current applications of MFM include data storage devices, imaging of micromagnetic structures, IC analysis, imaging of magnetotactic bacteria, and magnetic geophysics. Lateral force microscopy (LFM) is used to generate profiles of changes in surface friction and/or height variations. The probe tip is deflected laterally, indicating some sort of twist. Electronics measure the cantilever deflection. To differentiate between the two effects, LFM and AFM images should be obtained simultaneously. Phase detection microscopy or phase imaging is an extension of intermittent-contact AFM. It utilizes the phase lag between the driving frequency (cantilever) and the output signal frequency, generating a map of specific mechanical properties such as adhesion, elasticity, and friction. Identification of contaminants, composite materials, and regions of hardness and low surface adhesion can be obtained at the nanometer scale. Additional techniques include force modulation microscopy, where a periodic signal is applied to the cantilever, generating a map of the sample's elastic modulus and/or contaminants; electrostatic force microscopy, where a charged tip is scanned over the sample, revealing the locally charged domains generating a map of the charge carrier density; scanning capacitance microscopy, where a charged tip, kept at a constant tip/sample distance, generates a map of capacitance correlated information such as dielectric material thickness and subsurface charge carrier distributions (i.e., dopant profiles of ion implanted semiconductors); thermal scanning microscopy, where the tip in noncontact mode and a bimetal cantilever are used to map the thermal conductivity of the sample. (Vendors: Park Scientific Instruments; Digital Instruments; Oxford Instruments.)

References

1. R. E. Green (ed.), *Machinery's Handbook, 24th ed.*, New York: Industrial Press, 1992.
2. F. T. Farago and M. A. Curtis, *Handbook of Dimensional Measurement, 3rd ed.*, New York: Industrial Press, 1994.
3. T. Busch, *Fundamentals of Dimensional Metrology, 2nd ed.*, Albany, NY: Delmar Publishers, 1989.
4. R. C. McMaster, P. McIntire, and M. L. Mester (eds.), *Nondestructive Testing Handbook, Vol. 4, 2nd ed.*, American Society for Nondestructive Testing, 1986.
5. D. E. Bray and D. McBride, *Nondestructive Testing Techniques*, New York: John Wiley & Sons, 1992.
6. M. Giannini and A. deChiara, Wall Thickness Gaging in the Blow Mold Industry, distributed by Panametrics.
7. A. S. Birks, R. E. Green, and P. McIntire (eds.), *Nondestructive Testing Handbook, Vol. 7, 2nd ed.*, American Society for Nondestructive Testing, 1991.
8. D. Malacara, *Optical Shop Testing*, New York: John Wiley & Sons, 1978.
9. J. A. Woollam and P. G. Snyder, Variable Angle Spectroscopic Ellipsometry, VASE, distributed by J.A. Woollam Co.
10. H. H. Behncke, Coating thickness measurement by the X-ray fluorescence method, *Metal Finishing*, May, 33-39, 1984.
11. R. Howland and L.Benatar, A Practical Guide to Scanning Probe Microscopy, Park Scientific Instruments, 1993.

8

Proximity Sensing for Robotics

R. E. Saad
University of Toronto

A. Bonen
University of Toronto

K. C. Smith
University of Toronto

B. Benhabib
University of Toronto

The objective of this chapter is to review the state-of-the-art in proximity-sensing technologies for robotics. Special attention is paid to the sensing needs of robotic manipulators for grasping applications, in contrast to the needs of mobile robots for navigation purposes. For a review of the application of proximity sensing to mobile robots, the reader is referred to [1].

Robotic sensors can be categorized into three groups: medium-range (object recognition and gross position/orientation estimation) sensors, short-range (proximity) sensors, and contact sensors. Recent literature [2–6] suggests that robotic end effectors should be equipped with both short-range proximity and contact sensors.

Proximity sensors should be able to measure the position and orientation (pose) of an object's surface. The range must be sufficiently large to compensate for uncertainties in the medium-range pose-estimation process, while maintaining sufficient accuracy to permit effective grasping of the object.

Transducers used by current proximity sensors vary in sophistication. Despite their great variety, however, these transducers and their accompanying electronic interface circuits (together comprising the proximity sensor) cannot presently meet the stringent robustness requirements of most industrial robotic applications. Novel sensing algorithms and techniques still must be developed in order to improve on their current characteristics, and, furthermore, to control both the sensing and grasping processes.

8.1 Proximity Definition

The term "proximity," quantified by "pose" in this chapter, refers to three geometrical parameters x, u, and v as shown in Figure 8.1, where:

x = the translation from the origin of the sensor's reference coordinate frame, F_p, to a target point on the surface of the object measured along X_p. This target point defines the origin of the surface-frame, F_r

u = the *vertical* orientation of the object's surface, defined as a rotation around Y_p (of the translated frame), thereby specifying the new Z_r

v = the *horizontal* orientation of the object's surface, defined as a rotation around Z_r, thereby specifying Y_r.

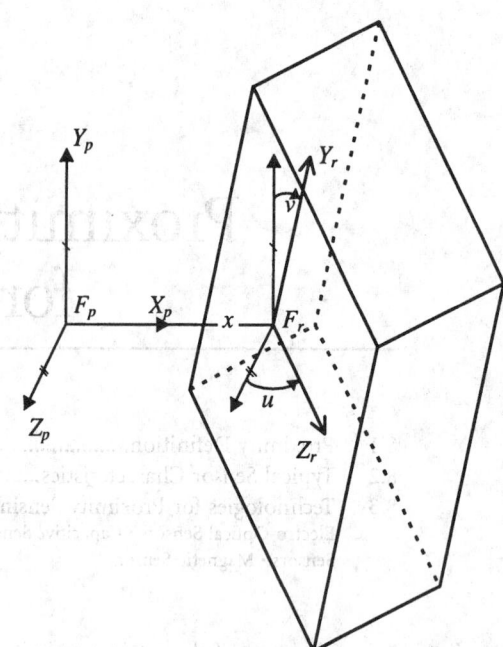

FIGURE 8.1   Proximity parameters.

8.2   Typical Sensor Characteristics

Conventionally, proximity sensors should be capable of measuring distances of up to 50 mm, and two degree-of-freedom orientations equivalent to an overall inclination of up to ±30°. The intended principal application of the sensor is to act as a guide for the robot. Thus, it would be desirable to have higher sensitivity and accuracy as the gripper approaches the object, namely when both the relative orientation and the distance approach near-zero values.

The signals received by the electronic interface circuit should be processed without limiting the required operating range of the sensor. The interface circuit should also minimize the effect of interference from the surroundings. It should therefore employ solutions to reduce background-noise interference and dynamic-range limitations.

The operation of the robot should not be slowed down by the sensor. Namely, a pose of the object should normally be estimated in 1 ms to 10 ms.

8.3   Technologies for Proximity Sensing

Proximity sensors have employed various transduction media, including sound waves, magnetic fields, electric fields, and light. Presently, electro-optical techniques seem to be the most appropriate for robotic-grasping applications. Such sensors are relatively small in size, have a large range of operation, and impose almost no restrictions on the object's material. However, recently, some new ultrasonic and capacitive proximity sensors have been fabricated directly as ICs, also showing the possibility of very-small-size proximity sensors based on these technologies [7, 8].

Brief descriptions of the principles of the primary technologies used by proximity sensors are given below, with the main emphasis being on optical transducers. A survey of commercial proximity sensors capable of measuring distances can be found in [6].

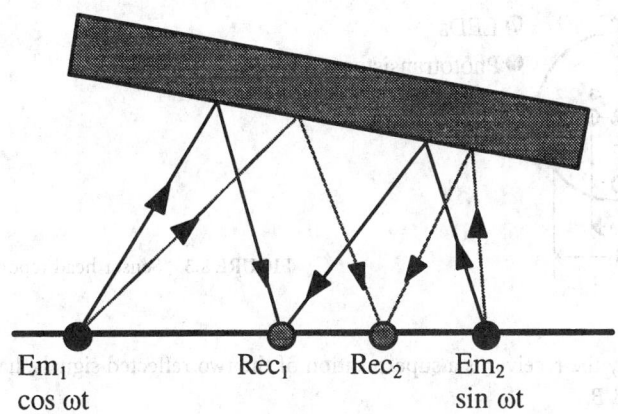

FIGURE 8.2 The basic phase-modulated proximity-sensor configuration.

Electro-Optical Sensors

Many proximity sensors use light, directly scattered from a target surface, to determine the distance and orientation of the target object from the gripper. The mechanism by which light is reflected can be explained by a model that specifies four different reflection phenomena. According to this model, light reflects from the surface primarily as a result of one or more of the following interactions:

1. *Single surface reflection*: Light waves that reflect specularly a single time off a planar microfacet, whose dimensions are significantly larger than the wavelength.
2. *Multiple surface reflection*: Light waves that reflect specularly at least twice between multiple microfacets.
3. *Reflection after penetration*: Light waves that penetrate into the material, refract, and then reflect back out as diffused light.
4. *Corner reflection*: Light waves that diffract from interfaces with surface details about the same size or smaller than the wavelength (such as from corners of microfacets).

The primary phenomenon (1) usually exists in both dielectrics and metal. However, due to the high conductivity of metal surfaces, most of the light reflects specularly off the interface between the metal and the air, while the portion that penetrates into the metal surface is absorbed. Accordingly, the reflection intensity originating from internal refraction in metals is practically zero. In dielectrics, however, a large portion of the light penetrates into the surface, and then reflects back out as diffused light (3). The secondary phenomena (2) and (4) exist both in metals and dielectrics and add to the diffused reflectance.

Common measurement techniques used in optical proximity sensing utilize one or more of the reflected components to determine the pose of the object in relation to the transducer.

Phase Modulation

A phase-modulated (PM) proximity sensor usually consists of two light sources and one or more photodetectors. The light sources are driven by modulated sinusoidal signals having a 90° phase relationship (Figure 8.2).

The emitter control voltages of the emitters, V_{em1} and V_{em2}, have amplitudes of a and b, respectively:

$$V_{em1}(t) = a \cdot \cos\omega t \tag{8.1}$$

$$V_{em2}(t) = b \cdot \sin\omega t \tag{8.2}$$

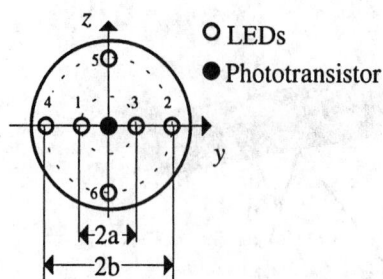

FIGURE 8.3   Sensor head reported in [9].

The signal detected by the receiver is a superposition of the two reflected signals, having corresponding attenuations of A and B.

$$V_{rec}(t) = A \cdot V_{em1} + B \cdot V_{em2} \qquad (8.3)$$

The signal attenuation is a function of the geometrical and electrical parameters of the sensor, the reflectivity characteristics of the object's surface, and the surface's distance and orientation with respect to the sensor. The combined signal at the receiver is therefore:

$$V_{rec}(t) = M \cdot \sin(\omega t + \phi) \qquad (8.4)$$

where M = the combined attenuation-function, and
ϕ = the combined phase-shift.

Usually, only the phase information ϕ is used, and the amplitude is completely neglected or used only for verifying the likelihood of error and its potential magnitude.

A proximity sensor that uses this technique has been reported in [9]. Figure 8.3 shows a *sensor head*; it comprises six light sources (LEDs) and a photodetector (a phototransistor). This sensor can measure the distance from the sensor's coordinate frame to the target point on the surface of the object (x), as well as the horizontal and vertical orientation of the object surface (u, v).

A simple model for the sensor was developed assuming that the light sources (LEDs) have low directivity, the photodetector (phototransistor) has high directivity, and the surface has diffused reflectivity. Figure 8.4 shows the basic configuration for the measurement of distance (x) and orientation (u and v). Table 8.1 shows the combinations of the driving signals in each LED needed for the measurement of distance and orientation.

For the measurement of the distance x, LED1 and LED3 are modulated by $K_1 \sin \omega t$, and LED2 and LED4 by $K_2 \cos \omega t$, respectively. The brightness detected by the photodetector can be calculated (using Lambert's law) to be:

$$L_p = C\left(G_1 \frac{\cos(\alpha - v)}{a^2 + x^2} + G_2 \frac{\cos(\beta - v)}{b^2 + x^2} + G_3 \frac{\cos(\alpha + v)}{a^2 + x^2} + G_4 \frac{\cos(\beta + v)}{b^2 + x^2} \right) \qquad (8.5)$$

where G_i (i = 1, 2, 3, 4) are the intensities of the light sources, and C is the reflection factor of the surface at point P. Considering that for this case: $G_1 = G_3 = K_1 \sin \omega t$, $G_2 = G_4 = K_2 \cos \omega t$, $\cos \alpha = x/\sqrt{a^2 + x^2}$ and $\cos \beta = x/\sqrt{b^2 + x^2}$, Equation 8.5 can be rewritten as:

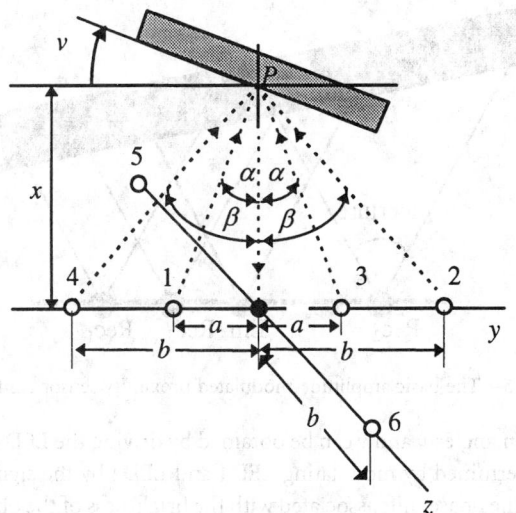

FIGURE 8.4 Measurement parameters for the sensor reported in [9].

TABLE 8.1 Combinations of the LED Driving Signals

Mode	LED1	LED2	LED3	LED4	LED5	LED6
Distance (x)	$\sin \omega t$	$\cos \omega t$	$\sin \omega t$	$\cos \omega t$	—	—
Orientation (u)	—	$\cos \omega t$	—	$\sin \omega t$	—	—
Orientation (v)	—	—	—	—	$\cos \omega t$	$\sin \omega t$

$$L_p = 2Cx \left\{ \frac{K_1}{\left(a^2 + x^2\right)^{3/2}} \sin \omega t + \frac{K_2}{\left(b^2 + x^2\right)^{3/2}} \cos \omega t \right\} \cos v = M \sin\left(\omega t + \phi_x\right) \qquad (8.6)$$

The amplitude M and the phase shift ϕ_x are given by:

$$M = 2Cx \left\{ \frac{K_1^2}{\left(a^2 + x^2\right)^3} + \frac{K_2^2}{\left(b^2 + x^2\right)^3} \right\}^{1/2} \cos v \qquad (8.7)$$

and

$$\phi_x = \tan^{-1} \left[\frac{K_2\left(a^2 + x^2\right)^{3/2}}{K_1\left(b^2 + x^2\right)^{3/2}} \right] \qquad (8.8)$$

From Equation 8.8, it can be observed that the distance (x) can be obtained from ϕ_x. It is important to note that, in theory, the calculation is not affected by the reflection factor (C) of the surface.

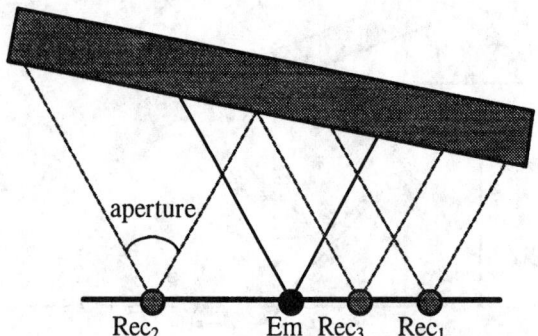

FIGURE 8.5   The basic amplitude-modulated proximity sensor configuration.

Similarly, the orientation angles u and v can be obtained by driving the LEDs as indicated in Table 8.1. For example, v can be determined by modulating LED4 and LED2 by the signals $K_1\sin\omega t$ and $K_2\cos\omega t$, respectively. For this case, the phase shift associated with the brightness of the object at point P is given by:

$$\phi_v = \tan^{-1}\left[\frac{K_2(x - b\tan v)}{K_1(x + b\tan v)}\right] \tag{8.9}$$

Note that, in order to recover v from Equation 8.9, x must be known. Accordingly, the distance x must be determined first. Correspondingly, the orientation angle (v) can be calculated from the new phase shift ϕ_v. The angle u can be calculated by modulating LED6 and LED5 with $K_1\sin\omega t$ and $K_2\cos\omega t$, respectively, and then determining the corresponding phase shift of the associate brightness at point P.

The pose-estimation results using this sensor were quite satisfactory and showed a good agreement between the theory and experiment.

In [10], an experimental setup of a PM distance sensor, similar to the one in [9], was reported for investigating the effect of the geometric and electronic parameters on the performance of the sensor. Optimal parameters were obtained for some targeted sensor-operation characteristics.

Amplitude Modulation

In amplitude-modulated (AM) sensors, the magnitude of the light reflected from a surface is utilized to determine the pose of the object.

AM transducers usually consist of one light source and several photodetectors (Figure 8.5). They were redesigned and optimized several times over the past decade to yield better measurement accuracy [11–14].

Many AM proximity sensors utilize optical fibers to illuminate and collect light from the surfaces of objects. The use of optical fibers, in a Y-guide configuration (Figure 8.6), facilitates the operation of sensitive low-noise circuitry in a shielded environment appropriately remote from the robot's electromagnetic interference sources.

AM transducers primarily use variations of the basic Y-guide transducer. Two important parameters can be varied in the design of Y-guides: the distance, d, between the emitting and receiving fibers (referred to hereafter as the emitter and the receiver, respectively), and the inclination angle, ϑ, of the receiver fiber with respect to the transducer's surface. The emitter is usually placed perpendicular to the transducer's surface, due to symmetry requirements, as will be explained later in this section.

The collection of a sufficient amount of reflected light requires the use of relatively wide-diameter fibers, typically having a 0.3 mm to 2 mm core size. This requirement demands the use of relatively low-grade plastic fibers. Although attenuations of up to 1 dB m⁻¹ are common in such plastic fibers, this loss rate is relatively insignificant for Y-guide applications because of the short length of the cables normally used. The numerical aperture (NA) of the plastic fibers, on the other hand, is an important parameter

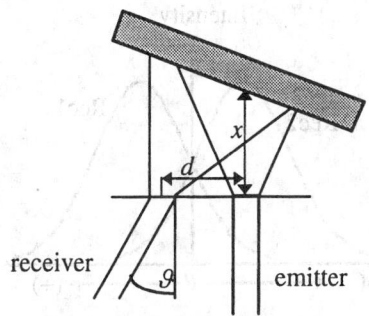

FIGURE 8.6 Y-guide transducer.

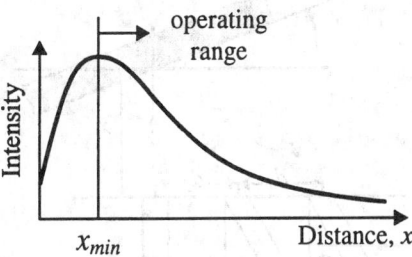

FIGURE 8.7 Y-guide response for distance measurement.

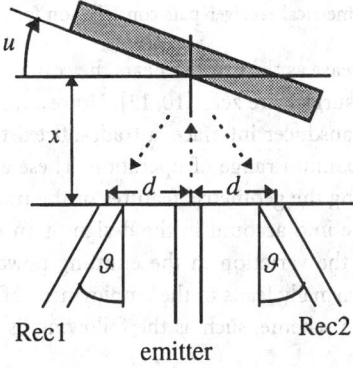

FIGURE 8.8 Typical receiver-pair constellation for orientation measurements.

in the transducer design, since lenses are rarely used in conjunction with AM-type transducers. In particular, the acceptance angle of the fiber is fixed and given by $\alpha = 2\sin^{-1}$ NA.

For a Y-guide, the intensity of the light reflected from the surface is not a monotonic function of the distance. Thus, the minimum operating distance of the transducer (x_{min}) is usually limited to a value that will guarantee a monotonic response (Figure 8.7).

For the measurement of surface orientation, a symmetrical three-fiber constellation (Figure 8.8) can be used. In this Y-guide configuration, the emitter is at the center and the two receivers are positioned symmetrically on either side [12]. The light intensities detected by the receivers, for the transducer shown in Figure 8.8, are illustrated in Figure 8.9 as a function of the surface orientation.

In the usual operating range of an AM transducer, the intensity of the light at the receiver is inversely related to the distance squared. As a result, it is conceptually possible to configure a transducer such that

Intensity

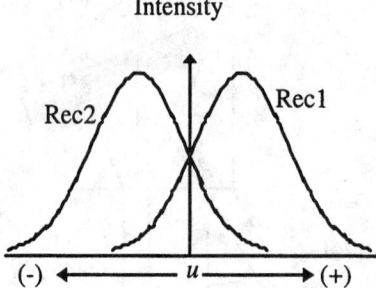

FIGURE 8.9   The light intensity detected by each receiver as a function of the surface orientation (u).

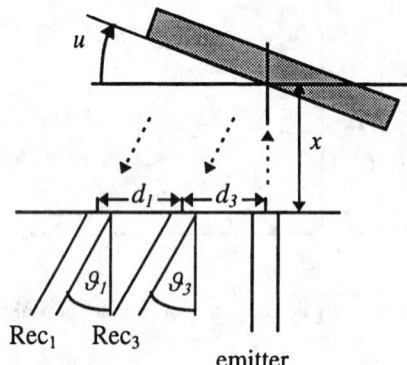

FIGURE 8.10   An asymmetrical receiver-pair constellation for distance measurements.

its sensitivity and accuracy will increase as the gripper nears the contact point, at which both the distance and the orientation of the object's surface are zero [10, 12]. However, in practice, because of the limited dynamic range of the electronic transducer interface, a trade-off exists between the desired maximum accuracy near contact, and the maximum range of operation. These and other considerations must be taken into account when establishing the geometric features of the transducer.

Another important factor to take into account in the design of an AM sensor is the need to reduce, as much as possible, the effect of the variation in the emitting power of the light source, P_o, on the transducer's measurements. This normally leads to the employment of a pair of receivers. A normalized differential voltage (DV) estimation scheme, such as the following, is then applied to the pair of measurements:

$$\text{DV} = \frac{V_{\text{rec}_1} - V_{\text{rec}_2}}{V_{\text{rec}_1} + V_{\text{rec}_2}} \tag{8.10}$$

where V_{rec_1}, V_{rec_2} are the voltages measured by receivers 1 and 2. However, in order to eliminate the effect of P_o on DV, each receiver must linearly convert the light intensity to a corresponding voltage measurement.

In order to use a DV scheme for the measurement of distance, an asymmetrical transducer configuration can be used (Figure 8.10). However, one must note that, although orientation measurements are not affected by variations in distance, distance measurements are significantly affected by the orientation of the surface, e.g., [15].

Accordingly, in using an AM proximity sensor with a DV scheme, the orientation is first approximated, and subsequently the distance is determined. The accuracies of the measured distance and orientation angle can be further improved by an iterative process.

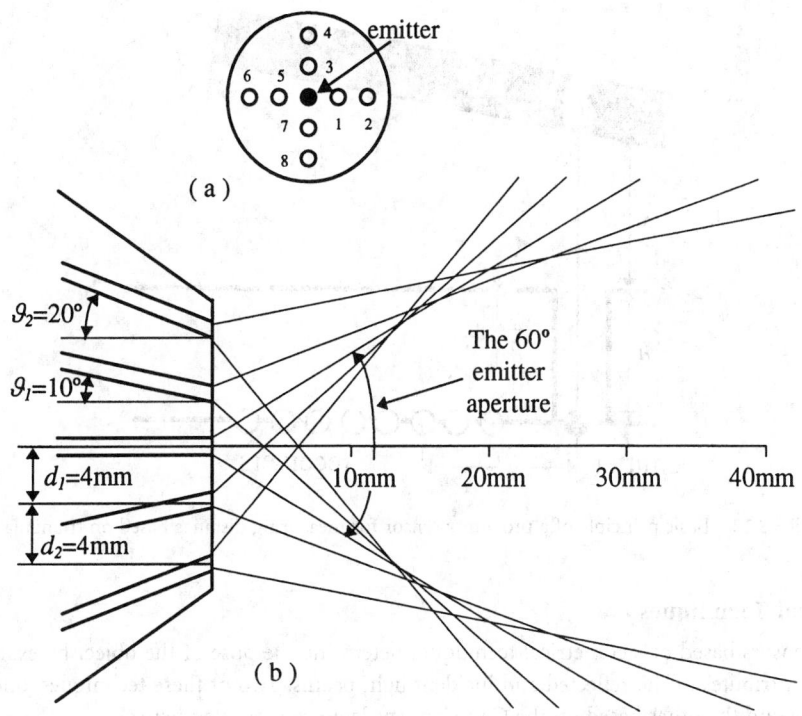

FIGURE 8.11 AM transducer design for the sensor reported in [14]: (*a*) top view; (*b*) front view.

Based on the above issues and observations, the outputs of the three receivers of the basic AM proximity sensor (Figure 8.5) can be paired for measuring both distance and orientation: the pair rec_1-rec_2 can be used for orientation measurement, while the pair rec_1-rec_3 can be used for distance measurement.

An experimental AM proximity sensor, capable of estimating the pose of an object with high accuracy, was reported in [14, 16, 17]. The transducer consists of one emitter, placed perpendicularly to the sensor head, and eight inclined receiver elements (Figure 8.11). The receivers of this transducer were paired for the specific measurements of distance, as well as of the vertical or horizontal orientation. However, the pose of the surface was determined with higher accuracy by using a polynomial fit technique (as opposed to the DV scheme described above), that provided relationships between the individual estimated parameters (x, u, and v) and all eight signals received.

The sensor presented in [14], and shown in Figure 8.11, operates in the range of 0 mm to 50 mm and ±20°. It can achieve an accuracy of 6.25 μm in distance and an accuracy of 0.02° in angular measurements in the near-contact region (0 mm to 6 mm range), using a general calibration-per-group strategy for different material groupings. This implies that the measured object's material belongs to a calibration group, which includes similar object surface characteristics; for example, machined metals. Better accuracies can be achieved using a calibration-per-surface strategy.

A similar configuration to the one shown in Figure 8.11 was reported earlier in [18] for the measurement of distances, where the orientations of each receiver pair relative to the emitter ϑ_1 and ϑ_3 were set at 10°. However, in this case, the apertures of the emitter and receiver were severely restricted by a collimating graded index (GRIN) lens. The emitter diameter was larger than that of the receivers in order to transmit more light. The measurements of the transducer were then processed in two phases: (1) the DVs of all the receiver pairs were processed independently to provide four distance estimations; and, (2) the four distance estimations were then averaged to provide a more accurate estimate, eliminating adverse effects due to variations in surface orientation.

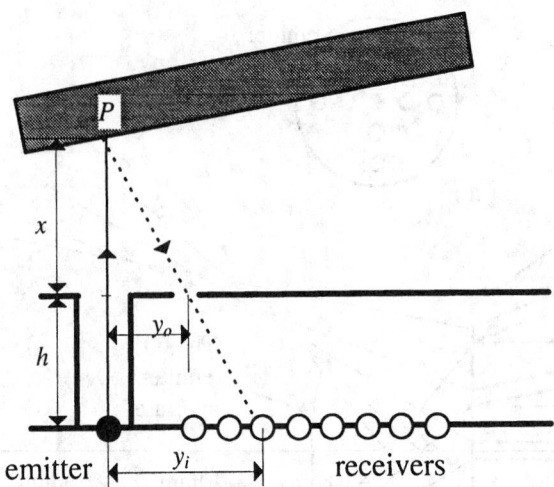

FIGURE 8.12 Basic principle of a proximity sensor for measuring distance based on triangulation.

Geometrical Techniques

Proximity sensors based on geometrical techniques determine the pose of the object by examining the geometrical attributes of the reflected and incident light beams. Two of these techniques, one based on triangulation and the other based on the Gaussian lens law, are presented here.

Figure 8.12 shows the basic configuration of a proximity transducer for measuring distance (x) based on the triangulation technique [19, 20]. The sensor head consists of a laser light source and a linear array of photodetectors (R_i, with i = 1, 2, ..., n). A narrow light beam illuminates point P, and the receivers detect the reflected light from the illuminated point through a transmitting slit. The geometry of the ray trajectory provides the basic information for the estimation of the distance (x). While the light source illuminates the surface of the object, the photodetector array is scanned to detect the light path used for making the output signal maximum. The light path obtained by this scanning is called the effective light path [19]. This light path is the one indicated in Figure 8.12. The distance (x) can be determined by accurately detecting the position (y_i) and precisely measuring the dimensions (h) and (y_o),

$$\frac{x}{x+h} = \frac{y_o}{y_i}$$ (8.11)

or

$$x = \frac{y_o h}{y_i - y_o}$$ (8.12)

In [26], it is claimed that such a sensor has the following properties: (1) the influence of irregularities, reflectivity, and orientation of the object is negligible; (2) the distance measurement is not affected by illumination from the environment and luminance of the object (their influence is eliminated by comparison of two sensor signals obtained in successive on-and-off states of the light source); and (3) the sensor head is sufficiently small to be used in a robot hand.

An experimental proximity sensor configuration, based on triangulation and capable of measuring both distance and orientation, is shown in Figure 8.13 [21]. The sensor uses six infrared LEDs as light sources, an objective lens, and an area-array detector chip for detecting spot positions. The directions of

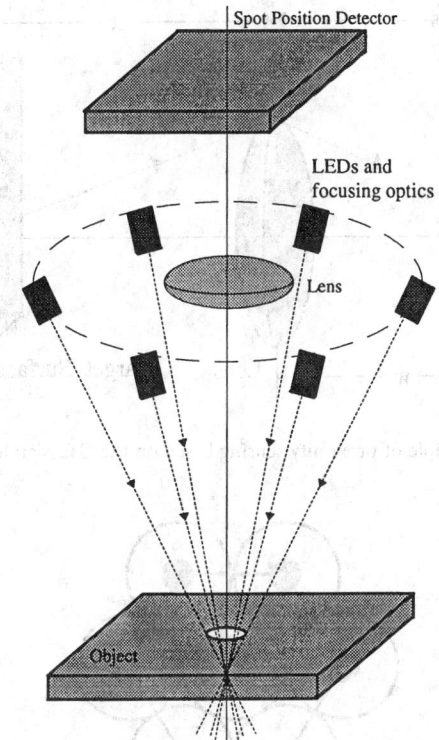

FIGURE 8.13 Multilight source proximity sensor reported in [21].

the beams are aligned to form a fixed cone of light. This sensor is of a type called scanning-emitter since each LED is sequentially pulsed to perform the measurements. As each LED is sequentially pulsed, the sensor IC detects the position of the spot projected by the reflected light beam from the object's surface. Knowledge of the spot's position, together with the camera's optics and the trajectory of each light beam, can be used to perform a triangulation to determine the three-dimensional coordinates of each light spot on the target surface. A set of six 3-D points are obtained for the six LEDs. Then, by fitting a plane to those points, the distance and orientation of the object's surface are approximated.

Another scanning-emitter-type proximity sensor was reported in [22]. In this case, a mechanical scanning system was utilized. One notes that inherent problems with sensors that use mechanical scanning devices include lower reliability and increased overall size.

Some recently reported triangulation sensors are sufficiently small in size to be mounted on a gripper [5]. However, they are still susceptible to errors due to distortion and separation of the light beam's reflection caused by surface irregularity, and can also have blind spots as a result of discontinuities associated with the shape of the sensed object.

Another group of geometrical electro-optical proximity sensors are those based on the Gaussian lens law [23, 24]. The basic configuration of such a transducer is shown in Figure 8.14. A light beam, collinear to the optical axis, forms a spot on the target's surface. The light scattered from the spot is collected by the lens. In Figure 8.14, PN represents the limiting ray that can be collected by the lens. The target distance (x) can be calculated in terms of the focal length of the lens (f) and the image position (w). Applying the Gaussian lens law, the distance x can be calculated as:

$$x = \frac{fw}{w-f} \qquad (8.13)$$

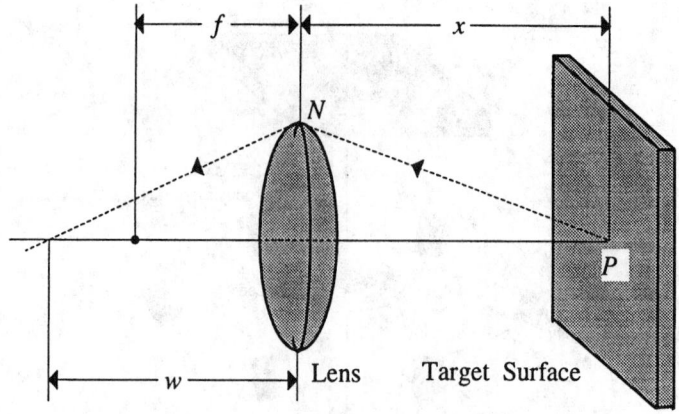

FIGURE 8.14   The principle of proximity sensing based on the Gaussian lens law reported in [7].

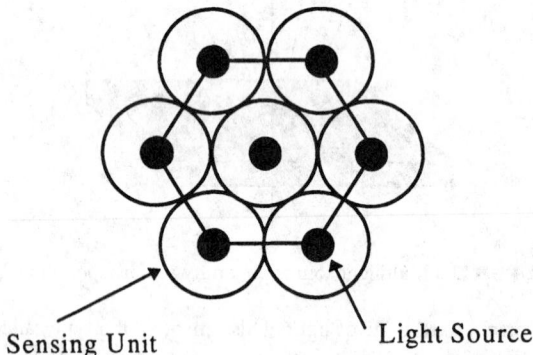

Sensing Unit   Light Source

FIGURE 8.15   Configuration of HexEye (top view).

A proximity sensor (the HexEye) based on this lens principle is shown in Figure 8.15 [23, 24]. The sensor consists of seven identical "sensing units" arranged in a hexagonal pattern. Each sensing unit in turn comprises four main parts: an objective lens, a conical mirror, six linear receiver arrays, and a laser diode light source.

The light beam generated by a light source forms a spot on the target surface, and the light flux scattered from the spot is collected by the objective lens and projected onto the receiver arrays in each sensing unit. The target distance is determined by the position, size, and shape of an image formed on individual receiver arrays of the unit.

The sensor operates in two modes, either distance or orientation. In the distance measurement mode, the seven light sources are activated one at a time to generate a light spot on the target surface, while all the sensing units receive the light flux scattered from the same light spot. The active unit determines the distance using the principle of the Gaussian lens law, while the nonactive units determine the distance based on triangulation. For the orientation measurement mode, the seven sensing units are grouped into orientation measurement units (OMUs). Each OMU comprises three neighboring sensing units around the center, resulting in a total of six possible OMUs. The local orientation of the surface is estimated by integrating the six orientation measurements.

The mapping between the light distribution on the arrays of light detectors and the target distance is obtained through calibration. Using the light distribution (instead of light intensity as in the case of AM sensors), it is intended that the mapping be independent of surface properties such as color, material, diffusion, and reflectance factors. However, it has been shown that the mapping may be corrupted by "noise" from several sources, including reflection patterns and ambient light.

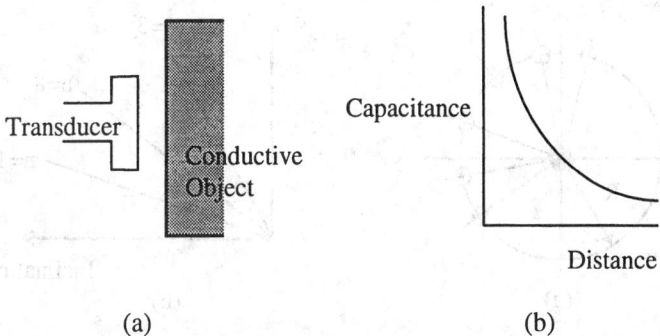

FIGURE 8.16 Capacitive proximity sensor based on the principle of parallel plates, (*a*) structure and (*b*) sensor response.

Using the Gaussian lens law has the following advantages over the triangulation principle: (1) the light source can be located at the center of the objective lens, allowing not only the sensor to be compact, but also the amount of light flux input to the lens to be maximized, and (2) the sensitivity can be optimized for a certain range of measurement distance by controlling *f*.

Time-of-Flight

Time-of-flight-measuring electro-optical sensors are radar-type systems. However, unlike regular radar, which transmit a pulse of radio-frequency energy, these sensors normally use a modulated light beam. The distance to the target is extracted from the measured phase shift of the reflected light. Two problems associated with such sensors are: difficulty in measuring short distances (which requires a very high modulation frequency), and the need for a mechanical scanning/switching system to get additional information (such as orientation) [5, 25].

Photothermal Effect

The photothermal effect transducer uses a strong light beam directed toward the object's surface. The distance to the object is extracted from measurements of the thermal wave generated by the light absorbed by the object. The detection scheme and signal processing are similar to those used in an AM sensor. Since the shape of the thermal wave generated at the surface is surface-texture independent, the photothermal sensor does not suffer from the surface robustness problem associated with AM sensors. However, the photothermal sensor is rather slow, and limited to highly absorbing surfaces [26].

Capacitive Sensors

Capacitive sensors generate and measure changes in an electric field caused by either a dielectric or conducting object in their proximity.

There are basically two types of capacitive proximity sensor. One type uses the principle of a parallel plate capacitor, the other uses the principle of fringing capacitances [8, 27, 28]. For the parallel plate type proximity sensor, the transducer forms one plate and the object measured forms the other plate. The structure of a parallel plate type proximity sensor and its typical response are shown in Figure 8.16 [8].

The parallel plate type proximity sensor is widely applied in industry. However, this type of a sensor has three major limitations: (1) the object being measured must be conductive; (2) the inverse gap-capacitance relationship is highly nonlinear and (3) the sensitivity drops significantly in the case of large gaps.

The second type of capacitive proximity sensor uses the principle of fringing capacitance [8]. The sensor has two "live" electrodes and the object being measured does not need to be part of the sensor system. The target object could be either conductive or nonconductive. However, the measurement of distances is affected by the type of object material. Therefore, separate calibrations must be carried out for different materials.

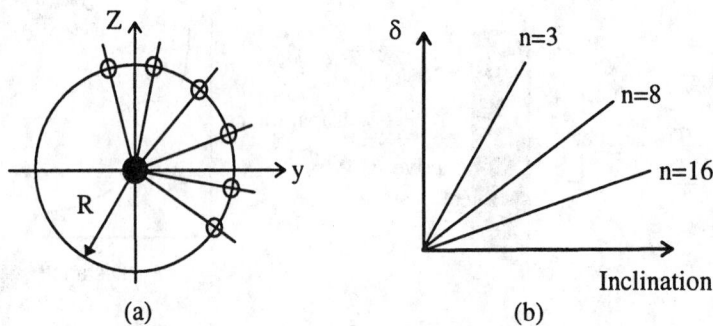

FIGURE 8.17   (*a*) Configuration of the ultrasonic sensor and (*b*) surface inclination versus δ.

In [30], an innovative capacitive microsensor was presented. Using micromachining technology, the electrode thickness can be significantly reduced and the fringing effect increased when compared with other capacitive sensors. Consequently, this sensor yields a better sensitivity. An array of such transducers can be implemented to measure the distance and orientation of an object.

Proximity capacitive sensors have the following general advantages: (1) low energy consumption and (2) simple structure. The major disadvantages, however, are that they are influenced by external signals and a calibration-per-surface technique must be carried out, since their operation directly depends on the object's material.

Ultrasonic Sensors

The basic principle underlying ultrasonic ranging sensors is the measurement of the time required for a sound wave to travel from the emitter to the object's surface and return to the detector. By using several such emitters and detectors, one can obtain information about the distance and orientation of the surface.

In [19], a novel method is proposed to measure the orientation angles of an object's surface using the phase differences of reflected echoes. Figure 8.17(*a*) shows the configuration of a planar sensor head with *n* receivers, which are equally spaced and located on a circle of radius *R* around the transmitter T. A linear relationship exists between the difference in lengths, δ, of two reflecting paths for an adjacent pair of receivers and the object inclinations.

In Figure 8.17 (*b*), the relationship between the inclination of the target surface and δ is shown. It can be observed that the measuring range of the sensor can be enlarged with an increase in the number of receivers. In [29], it is also shown that the measuring range can be enlarged by reducing *R*. However, it was noted that measurements carried out with a small sensor are potentially less accurate.

Experimental results using a transducer with six (*R* = 30 mm) and eight (*R* = 20 mm) receivers were reported in [29]. With the six-receiver transducer, the measuring range of the orientation angles was ±15°; while for the eight-receiver transducer, the maximum measuring range was ±30°. With the six-receiver transducer, the orientation angle could be determined with an accuracy of 0.5°, in the measuring range of ±15°, and 0.2° when the range was restricted to ±5°.

One of the major disadvantages of ultrasonic proximity sensors is that they are relatively large in size. However, implementing these sensors using micromachining could solve this problem. In [7], the generation and detection of ultrasound, for proximity sensing, was investigated using micromachined resonant membrane structures.

Magnetic Sensors

A magnetic-type sensor creates an alternating magnetic field, whose variation provides information about the object's position.

The simplest magnetic sensors are reed microswitches or Hall effect switches. However, the most commonly used sensors in robotics are based on the electromagnetic inductive principle, emphasizing eddy current generation. The basic principle consists of creating a magnetic field using appropriate coils around a core with high permeability and an oscillator with a frequency excitation high enough to minimize the penetration of the field inside a conductive material. The main problems with magnetic sensors are their high size/range ratio and difficulty in providing reliable distance measurements in varying magnetic environments.

Acknowledgments

The authors would like to thank Martin Bonert for careful review and critique of this chapter. We also acknowledge the financial support of Natural Sciences and Engineering Research Council of Canada.

References

1. H. R. Everett, *Sensors for Mobile Robots: Theory and Application*, Natick, MA: A. K. Peters, Ltd., 1995.
2. B. Espiau, An overview of local environment sensing in robotics applications, *Sensors and Sensory Systems for Advanced Robots*, NATO ASI Series, F43, 125-151, 1988.
3. W. D. Koenigsberg, Noncontact distance sensor technology, *SPIE, Intelligent Robots*, 449, 519-531, 1988.
4. Å. Wernersson, B. Boberg, B. Nilsson, J. Nygårds, and T. Rydberg, On sensor feedback for gripping an object within prescribed posture tolerances, *IEEE, Int. Conf. on Robotics and Automation*, Nice, France, 1992, 1654-1660.
5. A. Bradshaw, Sensors for mobile robots, *Measurement and Control*, 23(2), 48-52, 1990.
6. R. Volpe and R. Ivlev, A survey and experimental evaluation of proximity sensors for space robotics, *IEEE Int. Conf. on Robotics and Automation*, 4, 3466-3473, 1994.
7. O. Brand, H. Baltes, and U. Baldenweg, Ultrasound-transducer using membrane resonators realized with bipolar IC technology, *IEEE Conf. on Micro Electro Mechanical Systems*, Oiso, Japan, 1994, 33-38.
8. R. C. Luo and Z. Chen, Modeling and implementation of an innovative micro proximity sensor using micromachining technology, Proc. *IEEE/RSJ Int. Conf. Intelligent Robots and Systems*, Yokohama, Japan, 1993, 1709-1716.
9. R. Masuda, Multifunctional optical proximity sensor using phase modulation, *J. Robotic Systems*, 3(2), 137-147, 1986.
10. O. Partaatmadja, B. Benhabib, and A.A. Goldenberg, Analysis and design of a robotic distance sensor, *J. Robotic Systems*, 10, 427-445, 1993.
11. O. Partaatmadja, B. Benhabib, A. Sun, and A. A. Goldenberg, An electrooptical orientation sensor for robotics, *IEEE Trans. on Robotics and Automation*, 8, 111-119, 1992.
12. O. Partaatmadja, B. Benhabib, E. Kaizerman, and M.Q. Dai, A two-dimensional orientation sensor, *J. Robotic Systems*, 9, 365-383, 1992.
13. P. P. L. Regtien, Accurate optical proximity detector, *IEEE Conf. on Instrumentation and Measurement Technology*, San Jose, CA, 1990, 141-143.
14. A. Bonen, R. E. Saad, K. C. Smith, and B. Benhabib, Active-sensing via a novel robotic proximity sensor, *Int. Conf. on Recent Advances in Mechatronics (ICRAM'95)*, Istanbul, 1995, 1053-1058.
15. Y. F. Li, Characteristics and signal processing of a proximity sensor, *Robotica*, 12, 335-341, 1994.
16. A. Bonen, R. E. Saad, K. C. Smith, and B. Benhabib, A novel calibration technique for electro-optical proximity sensors, *Int. Conf. on Industrial Electronics, Control and Instrumentation (IECON'95)*, Orlando, FL, 1995, 1190-1195.
17. A. Bonen, R. E. Saad, K. C. Smith, and B. Benhabib, A novel optoelectronic interface-circuit design for sensing applications, *IEEE Trans. Instrum. Meas.*, 45, 580-584, 1996.
18. H. Bukow, Fiber optic distance sensor for robotic applications, *SME Conf., Sensors*, MS86-938, Detroit, MI, 1986.
19. T. Okada, Development of an Optical Distance Sensor for Robots, *Int. J. Robotics Res.*, 1, 3-14, 1982.

20. M. A. Kujoory, Real-Time Range and Elevation Finder, *Proc. IEEE*, 72(12), 1821-1822, 1984.
21. M. Fuhrman and T. Kanade, Optical proximity sensor using multiple cones of light for measuring surface shape, *Optical Eng.*, 23, 546-553, 1984.
22. T. Okada and U. Rembold, Proximity sensor using a spiral-shaped light-emitting mechanism, *IEEE Trans. on Robotics and Automation*, 7, 798-805, 1991.
23. S. Lee, Distributed optical proximity sensor system: HexEYE, *IEEE Int. Conf. Robotics and Automation*, 2, 1567-1572, Nice, France, 1992.
24. S. Lee and J. Desai, Implementation and evaluation of HexEye: a distributed optical proximity sensor system, *Proc. IEEE Int. Conf. Robotics and Automation*, 3, 2353-2360, Nagoya, Aichi, Japan, 1995.
25. S. Shinohara et al., Compact and high precision range finder with wide dynamic range using one sensor head, *IEEE Conf. Instrumentation and Measurement Technology*, Atlanta, GA, 1991, 126-130.
26. M. Ito, K. Hane, F. Matsuda, and T. Goto, Proximity sensing technique using the photothermal effect, *J. Japan Soc. Precision Eng.*, 58, 139-144, 1992.
27. B. E. Noltingk, A novel proximity gauge, *J. Scientific Instruments, Series 2*, 2, 356-360, 1969.
28. B. E. Noltingk, A. E. T. Nye, and H. J. Turner, Theory and application of a proximity gauge using fringing capacitance, *Proc. ACTA IMEKO*, 1976, 537-549.
29. S. Nakajima and Y. Takahashi, An ultrasonic orientation sensor with distributed receivers, *Advanced Robotics*, 4, 151-168, 1990.
30. A. Moldoveanu, Inductive proximity sensors, fundamentals and standards, *Sensors*, 10(6), 11-14, 1993.

9

Distance

W. John Ballantyne
Spar Aerospace Ltd.

The tools and techniques of distance measurement are possibly one of humankind's longest-running inventive pursuits. The scale shown in Figure 9.1 illustrates the enormous range of distances that science and engineering have an interest in measuring [1]. This chapter concerns itself with methods to measure a relatively small segment of this range — from centimeters to kilometers. Even within this limited segment, it would hardly be possible to list, much less describe, all of the distance measurement approaches that have been devised. Nevertheless, the small sampling of technologies that are covered here should be of help to a broad range of readers.

Distance measurement, at its most basic, is concerned with determining the length of a unidimensional line joining two points in three-dimensional space. Oftentimes, a collection of distance measurements is called for, so that the shape, the orientation, or the changes in position of an object can be resolved. Therefore, one must consider not only the measurement of distances, but also their spatial and temporal distributions. The terminology "ranging" will be used in reference to systems that perform single sensor-to-target measurements, "range-imaging" for systems that collect a dense map or grid of spatially distributed range measurements, and "position tracking" for systems that record the time history of distance measurement to one or several targets.

9.1 Basic Distinctions Between Range Measurement Techniques

Range measurement devices may be classified according to some basic distinctions. Generalizations can be made based on these broad classes, thereby facilitating the process of comparison and selection. The following subsections identify the fundamental bases for classification.

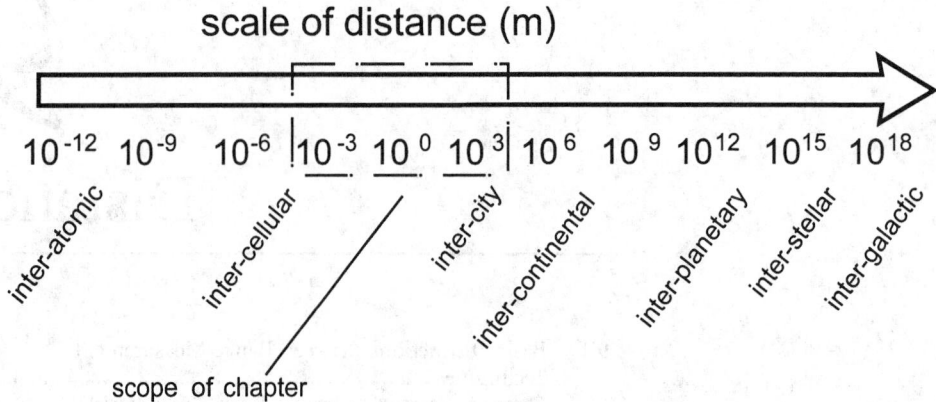

FIGURE 9.1   From the interatomic to the intergalactic, the range of measurable distances spans at least 30 orders of magnitude. The box outline indicates the relatively small segment that concerns this chapter.

Contact or Noncontact

A common approach to measuring the distance to a point on an object is through a calibrated mechanical device that simultaneously connects the selected point to a reference position. Any tape measure, feeler gage, or dial gage may be considered an example of a simple contacting measurement device. Mechanical/electronic devices are available that allow a user to "digitize" discrete point positions on a three-dimensional surface. A gimbaled probe on the end of an *X-Y-Z* positioner or articulated arm is used to touch a specific point, and sensory information of the linear positions or joint articulations provide an accurate position estimate. Mechanical, contact-based methods are widely used in industry and can be extremely accurate. Some coordinate measuring machines (CMMs), for example, can achieve 1 μm repeatability.

The chief disadvantage of mechanical approaches is that they are usually restricted to distances and work volumes up to a few meters at maximum. This is due to fundamental scaling laws for mechanical structures. As the requirement to span larger distances increases, the mass and mechanical tolerancing requirements on the machine make designs impractical. Also, mechanical approaches are too slow to make multiple measurements in rapid succession, as is typically required in range imaging or position tracking, when the measurement involves large sets of spatially or temporally distributed data.

Noncontact techniques for performing ranging, range imaging, and position tracking are many and varied. Besl [2] reviews and compares several of these. In the centimeters to meters range, most do not approach the accuracy of CMMs; but at larger scales and for large quantities of data, they become a practical necessity. The rest of this chapter will review noncontact approaches only.

Active or Passive

Noncontact distance measurement may be divided into *active* or *passive* techniques. Active techniques involve some form of controlled energy (field or wave) linking a known reference location to the unknown target location. The source of energy is typically associated with the reference location, but in some cases the target, or both target and reference, may be active. Passive techniques rely on an externally occurring source of energy (e.g., sunlight or target/background temperature contrast) to make the target detectable.

An active approach can often simplify the distance measurement problem because it allows a greater degree of control over the many factors that can influence a measurement. For example, the choice of the form of energy and the power level of the active source can minimize the effect of uncontrolled variables like ambient illumination, weather, and atmospheric conditions. Furthermore, an active approach provides an opportunity to selectively localize the measurement spatially and temporally, eliminating possible ambiguity about which target point was measured at a given time. In contrast, passive systems (e.g., stereo ranging) sometimes suffer from the so-called "correspondence problem," which is concerned with how to determine whether a given target point, detected from two or more viewpoints, or over two or more instants, is in fact the same physical point.

A common use of active approaches is to make range measurements "through" materials that are mechanically or optically impenetrable. Examples include medical imaging, where various forms of directed energy (ultrasound, X-rays) are used to build surface or volumetric maps of organs and bones; sonar, which penetrates water better than light does; and ground-penetrating radar, which can detect objects and their depth beneath ground surface.

Passive approaches, while not offering the same range of control and flexibility of active approaches, offer certain advantages. First, because they emit no energy, their existence cannot be detected by another remote detection system. This feature is very important in military applications. Second, passive systems can often collect multiple point range measurements more quickly because they are not limited by the rate at which they can direct an energy source toward a target point, as is the case with most active systems. For example, a stereo ranging system effectively collects all resolvable target points in its field of view simultaneously, while a scanning laser, radar, or sonar ranging system collects each measured point sequentially. Finally, the absence of a directed energy source is a simplification that can significantly reduce the size, cost, and hardware complexity of a device (although at the expense of increased signal processing complexity).

Time-of-Flight, Triangulation, or Field Based

There are many different classes and instances of noncontact ranging devices, but with very few exceptions they are based on one of the following three basic principles:

1. Energy propagates at a known, finite, speed (e.g., the speed of light, the speed of sound in air)
2. Energy propagates in straight lines through a homogeneous medium
3. Energy fields change in a continuous, monotonically decreasing, and predictable manner with distance from their source

The techniques associated with these basic phenomena are referred to as time-of-flight, triangulation, and field based, respectively.

Time-of-Flight

Time-of-flight (TOF) systems may be of the "round-trip" (i.e., echo, reflection) type or the "one-way" (i.e., cooperative target, active target) type. Round-trip systems effectively measure the time taken for an emitted energy pattern to travel from a reference source to a partially reflective target and back again. Depending on whether radio frequencies, light frequencies, or sound energy is used, these devices go by names such as radar, lidar, and sonar. One-way systems transmit a signal at the reference end and receive it at the target end or vice versa. Some form of synchronizing reference must be available to both ends in order to establish the time of flight.

A characteristic of many TOF systems is that their range resolution capability is based solely on the shortest time interval they can resolve, and not the absolute range being measured. That is, whether an object is near or far, the error on the measurement is basically constant.

Triangulation

Triangulation techniques were known and practiced by the Ancients. Triangulation is based on the idea that if one knows the length of one side of a triangle and two of its angles, the length of the other sides can be calculated. The known side is the "baseline." Lines of detection extend from either end of the baseline to the target point as shown in Figure 9.2. If the angles formed between these lines and the baseline can be determined, the distance is calculated as:

$$R = b \sin\alpha_{\text{left}} \sin\alpha_{\text{right}} / \sin\left(\alpha_{\text{right}} - \alpha_{\text{left}}\right) \qquad (9.1)$$

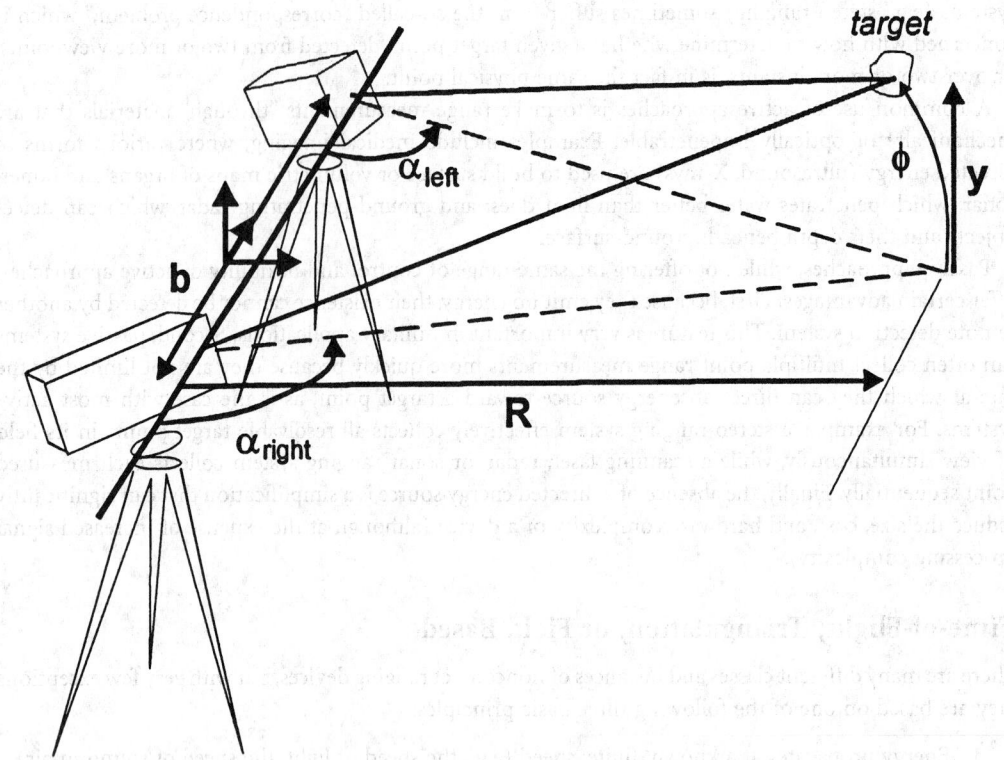

FIGURE 9.2 The basic triangulation geometry as used in classical surveying determines the distance to a remote point by sighting it from two locations separated by a known baseline. The pointing angles α_{left} and α_{right} are measured locally.

Classical surveying is a passive range-finding technique based on the above formula. A surveyor uses a precision pointing instrument to sight a target from two positions separated by a known baseline. Reference [3] notes that the distance to a nearby star may be calculated by observing it through a pointing instrument at 6-month intervals and using the diameter of Earth's solar orbit as the baseline. Stereo ranging, which compares the disparity (parallax) between common features within images from two cameras, is another form of passive triangulation. It is of interest to note that human vision estimates distance using a variety of cues, but two of the most important — stereopsis and motion parallax — are fundamentally triangulation based [4].

Active triangulation techniques use a projected light source, often laser, to create one side of the triangle, and the viewing axis of an optical detection means to create the second side. The separation between the projector and detector is the baseline.

A fundamental issue for all triangulation-based approaches is that their ability to estimate range diminishes with the square of the range being measured. This may be contrasted with TOF approaches, which have essentially constant error over their operating range. Figure 9.3 illustrates how, conceptually, there is a "crossover" distance where TOF techniques become preferable to triangulation techniques.

Field-Based Approaches

Whereas TOF and active triangulation techniques employ the wave propagation phenomena of a particular energy form, *field-based* approaches make use of the spatially distributed nature of an energy form. The intensity of any energy field changes as a function of distance from its source. Moreover, fields often exhibit vector characteristics (i.e., directionality). Therefore, if the location of a field generator is known and the spatial characteristics of the field that it produces are predictable, remote field measurements contain information that may be used to infer distance from the source.

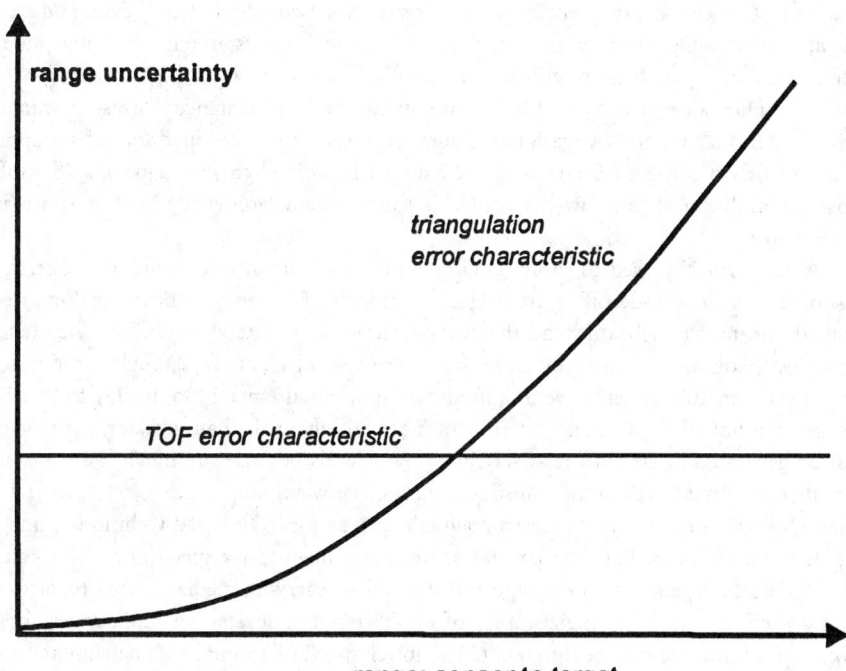

FIGURE 9.3 Time-of-flight (TOF) and active triangulation techniques tend to exhibit error characteristics related to their fundamental principles of operation. The dominant error source in TOF systems is usually the shortest measurable time interval, but this is a detection issue and is essentially independant of distance. Active triangulation systems are typically more accurate at close distances, but geometry considerations dictate that the effects of their error sources will increase with the square of distance.

An interesting distinction between field-based approaches and wave-based approaches is that the former, although they employ energy fields, do not rely on the propagation and conversion (and concomitant losses) of energy. That is, they may employ stationary fields, like those generated by a magnet or static charge. Such fields encode position information by their very shape. Sound and light, although having a wave nature, can be exploited in the same manner as stationary fields because of their distance-dependent intensity.

Field-based techniques must confront some basic issues that limit their range of application. First, the characteristics of most practically exploitable fields are typically influenced by objects or materials in the vicinity, and it is not always possible to ensure that these influences will remain constant. Second, the variation of fields through space is highly nonlinear (typically inverse square or inverse cube), implying that the sensitivity of a measurement is strongly affected by proximity to the source. Notwithstanding these concerns, devices have been developed and are available that perform very well in the situations for which they are intended [7].

Form of Energy

As discussed above, all noncontact, active ranging devices employ some form of energy. This is true whether time-of-flight, triangulation, or field-based principles apply. The following subsections describe the various forms of energy employed and some generalizations about the effectiveness of each in various situations.

Sound

Ranging systems based on sound energy are usually of the pulsed-echo TOF type and employ carrier frequencies in the so-called "ultrasonic" (beyond audible) range of frequencies. Besides being inaudible (an obvious benefit), ultrasonic frequencies are more readily focused into directed beams and are practical to generate and detect using piezoelectric transducers. Ultrasonic signals propagate through air, but long-distance transmission is much more effective in liquids, like water, where higher density-to-viscosity ratios result in higher wave velocity and lower attenuation per unit distance. Ultrasonic ranging techniques (or SONAR, for SOund NAvigation and Ranging) were first developed for subsea applications, where sound is vastly superior to electromagnetic energy (including light) in terms of achievable underwater transmission distances [5]. Low-cost, portable sonar systems are widely used by sport fisherman as "fish finders" [6].

The frequencies typically used in sonic ranging applications are at a few tens of kilohertz to a few hundred kilohertz. A basic trade-off in the choice of ultrasonic frequency is that while high frequencies can be shaped into narrower beams, and therefore achieve higher lateral resolution, they tend to fade more quickly with distance. It may be noted that beam widths narrow enough for range imaging applications (less than 10°) are effective in a fluid medium, but attenuate too quickly to be practical in air. Interestingly, although sound energy attenuates more rapidly in air than in water, useful short-range signals can be generated in air with relatively low power levels because the much lower density of air requires smaller dynamic forces in the transducer for a given wave amplitude.

When comparing sound energy to electromagnetic energy for TOF-based techniques, one needs to remember that sound, unlike light, propagates at not only much lower speeds, but with considerably more speed variation, depending on the type and state of the carrying media. Therefore, factors like air humidity and pressure will affect the accuracy of a TOF ranging device. For underwater applications, salinity and depth influence the measurement. The lower speed of sound has a detrimental impact on the rate at which range samples can be collected. For example, a target 10 m away takes at least 60 ms to measure through an air medium. This may not seem like a long time to wait for a single sample, but it becomes an issue if the application involves multiple sampling, as in motion tracking or collision avoidance sensing.

Stationary Magnetic Fields

Stationary or pseudostationary (i.e., low frequency) magnetic fields are only used in field-based approaches. An advantage of such fields is that they are easily and cheaply produced by either a permanent magnet or electrical coil. Since stationary fields do not transmit energy, the targets cannot be passive — they must actively sense the properties of the field at their particular location. A variety of sensing technologies may be used to make measurements of the direction and intensity of a magnetic field, including flux gate, Hall effect, and magnetostrictive type magnetometers. A comprehensive list of such technologies is given in [7].

Radio Frequencies

Echo-type TOF ranging systems based on the band of the electromagnetic spectrum between approximately 1 m and 1 mm wavelength are known as RADAR (RAdio Detection And Ranging). Radio waves can be used for long-distance detection in a variety of atmospheric conditions. As in the case of sound waves, there are trade-offs to be addressed in the choice of frequency. Long waves tend to propagate better over long distances, but short waves can be focused into narrow beams capable of better lateral discrimination. An interesting application of short-range radar is ground-penetrating radar, which can be used to locate and image subsurface objects [8]. Here, the frequency vs. range trade-off is particularly acute because of the need to balance reasonable imaging capability (narrow beam) with good depth penetration (long wave).

An example of a TOF one-way (active receiver) system that uses radio frequencies is the global positioning system (GPS). The distance between a receiver on land is determined by each of several orbiting satellites equipped with a transmitter and a very precise Cesium clock for synchronization. A good description of GPS and its use in vehicle navigation is available in [9].

Light Frequencies

Beyond the radio portion of the electromagnetic spectrum are the infrared, visible, and ultraviolet frequencies. These frequencies can be produced by lasers and detected by solid-state photosensitive devices and are useful for both TOF and active triangulation ranging. Echo-type TOF techniques are known as LIDAR (LIght Detection And Ranging), in keeping with the terminology introduced earlier.

While light frequencies attenuate more than radio frequencies through cloud and fog, they can have very narrow beam widths, allowing superior lateral resolution and target selectivity.

Coherent or Noncoherent Detection

Echo-type TOF devices, whether sonar, radar, or lidar, can be further classified according to whether the detection approach measures time-of-flight directly (noncoherent) or exploits an inherent periodicity in the emitted energy to ascertain the flight distance (coherent).

Noncoherent techniques face the problem of timing short intervals. This is not a serious challenge in the case of sound waves, where a meter round trip corresponds to 6 ms, but is somewhat more problematic for light and radio waves, where that distance equates to only 6 ns. Accuracy of noncoherent detection typically relies on the averaging of repeated measurements.

Coherent detection is achieved by combining a portion of the emitted signal with the reflected signal to produce a third signal indicating the amount of phase delay. The signals are continuous wave (CW) as opposed to pulsed. Coherent detection techniques are classified as amplitude modulated (AMCW) or frequency modulated (FMCW).

A basic issue with coherent detection techniques is the inability to distinguish between integral multiples of the basic modulation wavelength. Any coherent detection system must employ techniques to resolve the so-called "ambiguity interval." Noncoherent techniques do not face this problem.

Ranging, Range Imaging, or Position Tracking

Ranging devices are typically pointed toward a target to produce a single range reading. A common example of simple ranging is the feedback sensor used in auto-focus cameras. There are many active ranging devices currently available based on TOF (i.e., radar, sonar, lidar) and active triangulation principles.

Range imaging devices use the same principles as ranging devices, except that they include some form of scanning that is employed to generate an array of spatially distributed range samples. Sometimes, the scanning action is accomplished by means intrinsic to the sensor (e.g., spinning and nodding mirrors, or phased-array antenna) so that the reference location remains fixed. In this case, the data are recorded in the polar form (range, elevation, azimuth) as shown in Figure 9.4. In other cases, the sensor might scan on only one axis internally while the second scan dimension is realized by moving the sensor location through some set pattern. It is not uncommon to record the "intensity" or return energy associated with a range sample as well. The intensity map may be presented as a "gray scale" image and, like a black and white photograph, often contains additional information useful in interpreting a scene. Range images can be used to produce three-dimensional graphic representations of scenes and objects. A common use of range imaging is aerial terrain mapping.

Position tracking devices are used to measure the change in an object's position and orientation over time. Basic issues in position tracking are the acquiring of, and locking on to, specific target points. These issues can be avoided by employing active targets, and most systems available today are of this type.

9.2 Performance Limits of Ranging Systems

The performance characteristics of available ranging systems vary widely, as do the requirements of the applications for which they are designed. The following subsections review the most basic performance categories and the technical issues of performance limits.

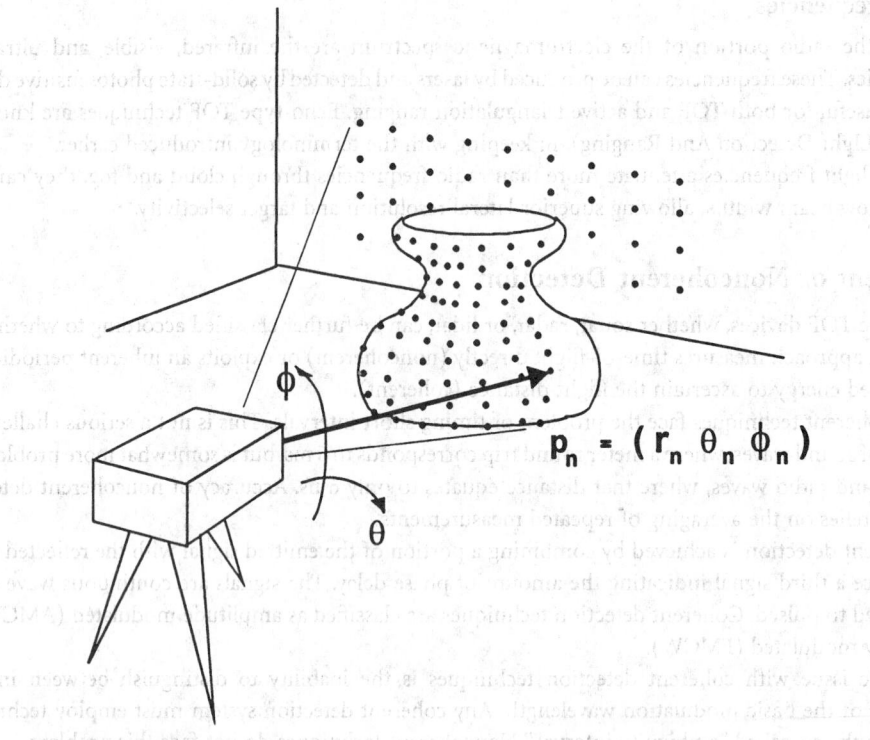

$$p_n = (r_n \ \theta_n \ \phi_n)$$

FIGURE 9.4 Range images are typically an array of individual range values sampled while changing the pointing direction (e.g., azimuth and elevation angles) of a ranging device. A digital range image of the polar form shown can be readily transformed into rectangular coordinates if required.

Range Accuracy

As illustrated in Figure 9.3, TOF and active triangulation techniques differ fundamentally in their error vs. distance characteristics. Currently available systems based on active triangulation achieve better repeatability and accuracy in the less than 1 m range than do TOF systems, but are seldom used at distances of several meters. Hymarc Ltd. and Perceptron Inc. each offer laser triangulation systems with 3σ accuracy of 25 mm and 50 mm, respectively [10, 11].

In principle, TOF systems could achieve accuracy rivaling active triangulation, but the most promising detection technique — a variation of laser interferometry, which solves the ambiguity interval problem [12] — has yet to make its commercial debut.

Depth of Field

Depth of field refers to the interval of distance through which a stationary reference ranging system can measure without resorting to a change in configuration. Large depth of field is often an important characteristic in practical applications. For example, if the distance to the target is poorly known a priori, then a large depth of field is desirable.

Passive optical triangulation approaches like stereography and photogrammetry tend to have restricted depth of field because they rely on camera-type imaging, which is inherently limited by depth of focus. Timed-interval TOF systems have excellent depth of field because they do not rely strongly on optical imaging except to concentrate the collected return energy on the detector. Some active triangulation systems do rely on optical imaging of the projected laser spot, but the design employed by Hymarc Ltd. regains a large depth of field by tilting the detector array with respect to the lens plane [13].

Maximum Range

Any active ranging, range imaging, or position tracking system has a practical maximum distance that it can measure. This is because the controlled energy, whether propagated as a wave or established as a field, must spread before reaching the detector. The spreading inevitably increases with distance and all detectors, no matter what form of energy they measure, require a certain minimum amount to exceed their inherent "noise floor."

The "classical radar range equation" is introduced in many texts on radar (e.g., [14]). Jelalian [15] points out that the equation is equally applicable to lidar, which, after all, just employs a higher frequency version of electromagnetic wave. In fact, the same idea applies to sonar and to active triangulation systems as well. The equation computes the power of the received signal as:

$$P_R = P_T G_T \big/ 4\pi R^2 \times \rho A \big/ 4\pi R^2 \times \pi D^2 \big/ 4 \times \eta_{atm} \eta_{sys} \qquad (9.2)$$

where P_R = power at the receiver
  P_T = power transmitted
  G_T = transmitter gain
  R = range to target
  ρ = reflectivity of target
  A = effective area of target
  D = diameter of collecting aperture
  η_{atm} = atmospheric transmission coefficient
  η_{sys} = system transmission coefficient

Equation 9.2 applies when the target area is smaller than the footprint of the incident beam, which is often the case for radar and sonar ranging. However, in the case of laser-based systems, the relatively narrow beam usually means that the laser spot is small compared to the target. For a transmitted beam that spreads with a solid angle θ_T, the illuminated patch area is:

$$\sigma_{spot} = \pi R^2 \theta_T^2 \qquad (9.3)$$

The definition of transmitter gain is based on the notion of the solid angle beam width as compared to an omnidirectional transmitter

$$G_T = 4\pi \big/ \theta_T^2 \qquad (9.4)$$

One can substitute for Equation 9.4 for G_T and Equation 9.3 for the variable s in Equation 9.2 to produce the range equation for a small spot size.

$$P_R = P_T \big/ R^2 \times \rho \big/ 4 \times \pi D^2 \big/ 4 \times \eta_{atm} \eta_{sys} \qquad (9.5)$$

The importance of this equation is primarily in the $1/R^2$ dependence. Any ranging system that works by bouncing energy off a diffuse reflective target encounters severe signal attenuation with increasing distance. Given a detector with a fixed noise floor, the only ways to improve maximum range are to increase the transmitted power or the collecting area. In practice, there are design constraints that limit both of these measures. For example, laser power must sometimes be limited for eye-safety considerations, and increased collecting area can imply a proportional increase in sensor packaging volume.

Lateral Resolution

In range imaging applications, it is generally desirable to use the narrowest possible beam width to provide good lateral discrimination of target surface features. Lasers, because of their short wavelength, can be optically collimated to produce much narrower beam widths than are possible with radio sources. However, even lasers cannot produce arbitrarily narrow beams. The interested reader is referred to [13] for a discussion of Gaussian beam propagation and optimal focusing. There are basically two ways to project laser light. The beam can be "focused down" to produce the smallest possible spot at a particular point inside the measurement range, in which case the beam will diverge as the distance from that point increases; or the beam can be focused at infinity or some very distant point so as to minimize the divergence through the entire measurement range. The former approach provides higher lateral resolution at the focus distance, but by implication restricts the practical depth of field. The latter compromises spot size for increased depth of field.

Rate of Acquisition

The rate at which a ranging sensor can acquire range samples is important when the target object is changing shape or position, or when the required sample density of a range image is very high. There are several potential factors that can limit sample acquisition rate: the amount of time required by the detector to integrate the weak return signal to a sufficient level (integration time); the time constant of any filtering or averaging that must be performed to realize an acceptably "clean" signal (smoothing time); the rate at which samples can be transferred through the signal processing stages (transfer time); and the velocity limits of mechanical scanning apparatus (scanning bandwidth). Acquisition rates vary widely: from tens of hertz for acoustic ranging devices to tens of kilohertz for some laser-based systems. It is worth noting that, in general, there is a trade-off between rate of acquisition, accuracy, and maximum range. Some systems permit control over basic parameters so that this trade-off may be optimized for a particular application. The reader should be aware that data sheets may not be clear as to whether stated performance figures for these three specifications are valid in combination.

9.3 Selected Examples of Ranging, Range Imaging, and Motion Tracking Systems

The following sections review selected examples of some specific ranging, range imaging, and position tracking sensor systems. The list is by no means exhaustive, but offers a reasonable sampling of available technologies.

Laser-Based Active Triangulation Ranging and Range Imaging Sensors

Active Triangulation Basics

Figure 9.5 illustrates the basic active triangulation geometry. In this so-called "pinhole camera" model, practical aspects like lenses for projection and detection and mirrors for scanning are eliminated for clarity. It can be shown by means of similar triangles that the range is inversely proportional to the deflection of the imaged spot.

$$R = bf/u \qquad (9.6)$$

where R = distance to object
$\quad\ b$ = baseline distance
$\quad\ f$ = lens to detector distance
$\quad\ u$ = detected spot position in the image plane

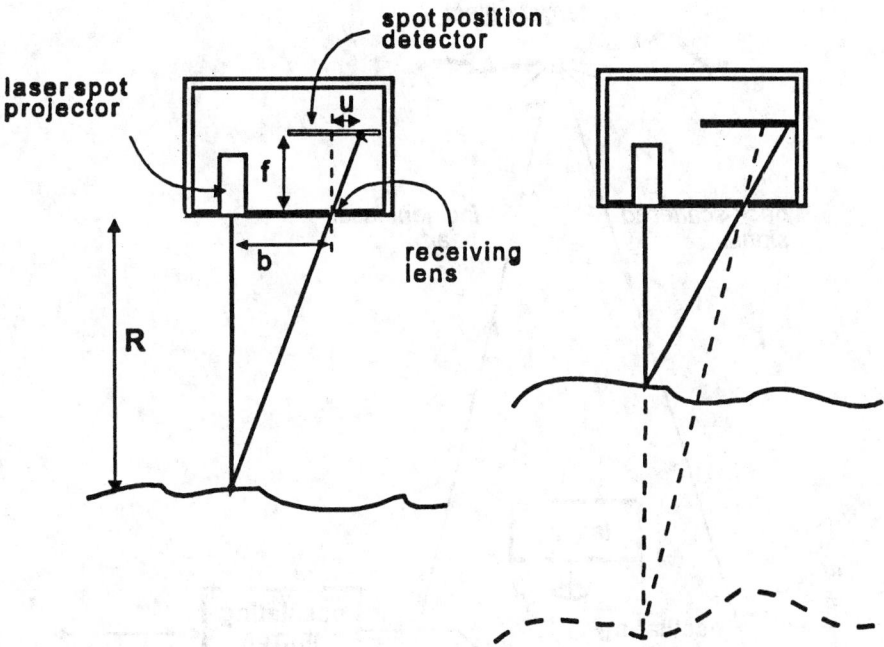

FIGURE 9.5 A simple pinhole camera model illustrates the basic active triangulation principle. As the distance R to the target surface changes, the spot position u on the detector changes, maintaining similarity between the large triangle outside the camera and the small triangle inside. There is an inverse relationship between R and u.

The sensitivity of the range measurement, or the incremental change in u with R, is

$$\left| du/dR \right| = bf/R^2 \tag{9.7}$$

The significance of Equation 9.7 is that range estimating performance is expected to fall as distance increases. Improvements in accuracy are realized by increasing the baseline or the lens to detector distance (i.e., the focal length).

Synchronized Scanning Principle

Lateral scanning of an active triangulation ranging sensor is accomplished by an elegant and effective technique developed at Canada's National Research Council and now marketed by Hymarc Ltd. under the name "HYSCAN" [10]. A two-sided oscillating mirror simultaneously steers the outgoing beam on one face and directs the collected light to the spot-imaging optics on the opposite face. By synchronously scanning both the beam and the axis of the detection system, rather than the beam only, as conventionally practiced, significant performance improvements are made. Figure 9.6 is a schematic illustration of the approach. Note also that the detector plane is tilted with respect to the lens plane. This feature increases the depth of focus so that the ranging performance is maximized over the measuring volume. The Hyscan sensor produces a single-axis sweep, or so-called "line scan." Full-dimension range images are acquired by translating the sensor over a target surface with a controlled motion pattern.

Light Plane Principle

Perceptron Inc. offers a similar line-scan system under the name "TriCam" [11]. In this case, the laser is not swept. Instead, the beam is transformed to a focused plane by means of cylindrical lenses. A two-dimensional detector is used to generate range profiles through the analysis of a deformation of the laser line as the sensor is translated over the object surface.

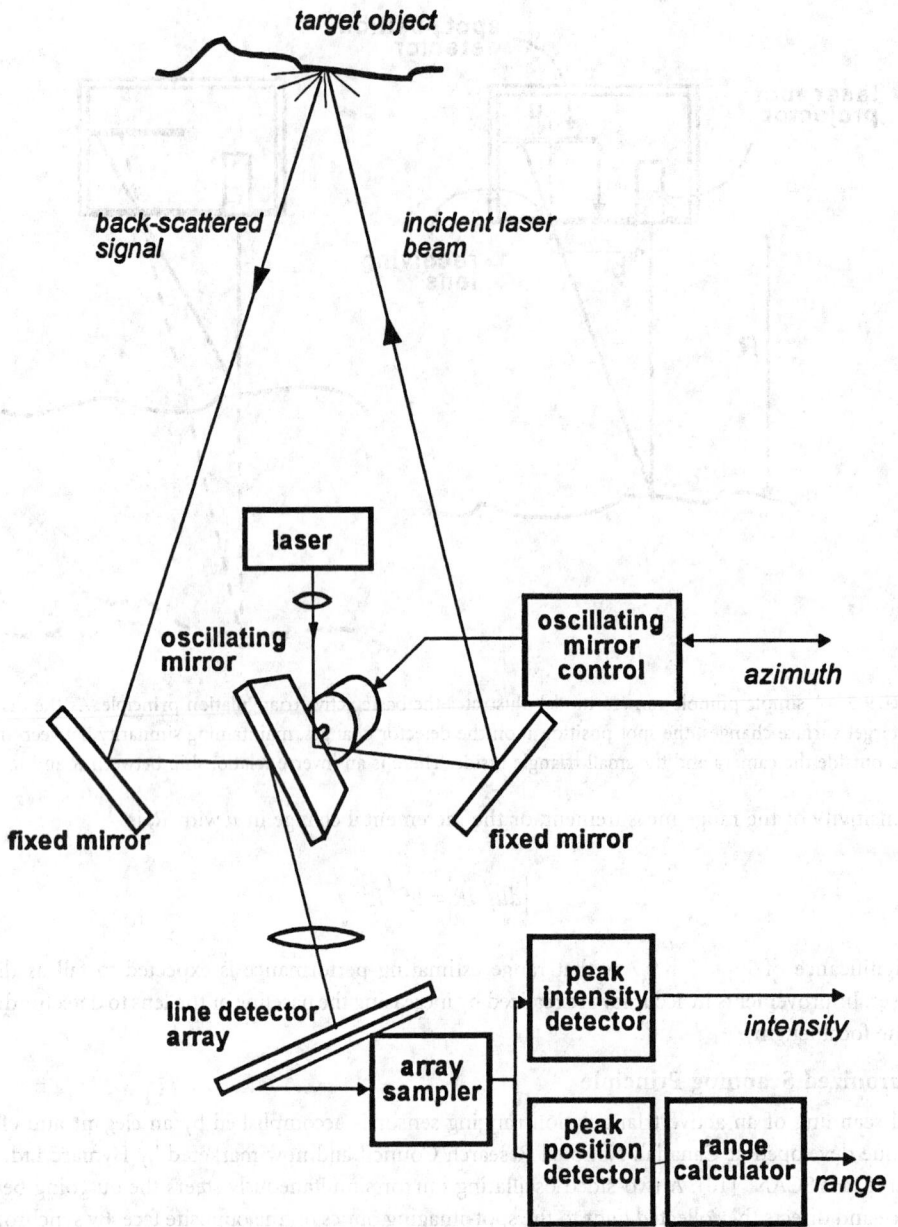

FIGURE 9.6   The Hymarc laser triangulaton line scanner uses the synchronized scanning principle. Both sides of an oscillating mirror are used to sweep both the projected beam and the axis of detection over the target. The detector array is tilted to the lens plane to maximize the depth of focus.

Laser-Based Lidar Range Imaging Sensors

AM Lidar (Phase-Based Detection)

Perceptron Inc. also offers a scanning lidar under the name "LASAR" that can produce high-resolution range images through a large measurement volume. The device uses a near-infrared laser that is projected through a collimating telescope to form a spot on the first surface encountered. The spot is swept over

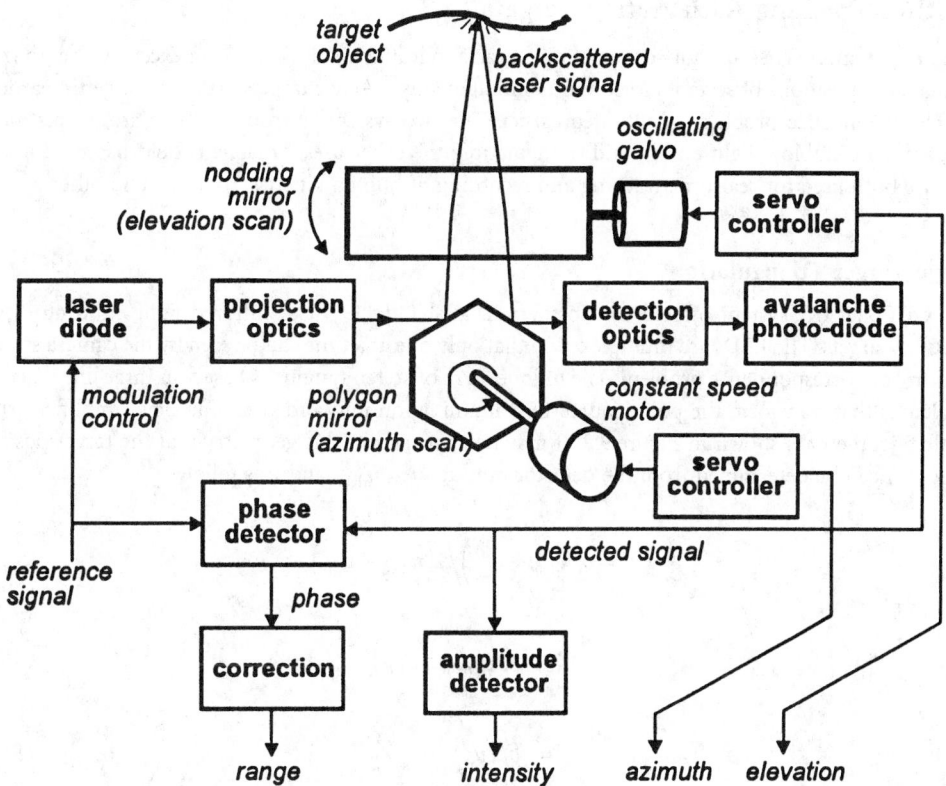

FIGURE 9.7 The Perceptron AM Lidar system described in U.S. patent 5,006,721 uses a rotating polygon mirror for synchronized scanning. A "nodding mirror" is also added to sweep at a slower rate in the orthogonal direction, producing a raster scan pattern. Range measurement is determined by comparing the phase of the outgoing and returning AM laser signal.

a programmable field of view in a raster pattern by means of a spinning polygon mirror and an oscillating "nodding mirror." Some of the backscattered light is collected and directed by means of an adjacent facet of the polygon mirror. The projected laser light is amplitude modulated at a reference frequency by controlling the power to the laser diode source. The return signal, although orders of magnitude weaker than the outgoing signal, is phase-compared to determine the range for a particular azimuth and elevation. The intensity of the return energy is also recorded. The Perceptron sales literature claims a maximum measurement volume of $60° \times 72° \times 40$ m, a range image grid resolution of 1024×2048 pixels and a maximum acquisition rate of 360,000 pixels/s. A schematic diagram of the LASAR™ system is shown in Figure 9.7. Details of the Perceptron technical approach may be found in [16].

Resonating Lidar (Frequency-Based Detection)

Acuity Research Inc. has developed a laser-based TOF ranging sensor based on a simple but effective idea. The detector controls the laser output such that the absence of a signal drives the laser on and the presence of a signal turns it off. The finite transit time of the light bounce turns this arrangement into a two-state resonator, with the period being proportional to the target distance. Rather than measuring the period, which is extremely short and difficult to time, the frequency is measured using conventional counting techniques for as many cycles as necessary to yield the required accuracy. The AccuRange 4000, as it is named, is also available in a $360°$ line-scanning arrangement suitable for robotic vehicle navigation applications [17]. Details of the technical approach may be found in [18].

Position Tracking with Active Targets

Active target approaches are not convenient in some applications, but they are an excellent way to track the changing positions of several target points simultaneously. Active targets are a way of getting around the "correspondence problem" mentioned earlier. The two systems introduced here are interesting to compare. One employs light energy and triangulation; the other uses a magnetic field-based approach. They are both used for real-time tracking and recording of human kinetics, robotics, and other moving objects.

Active Target Triangulation

The "OPTOTRAK" system offered by Northern Digital Ltd. [19] uses infrared light emitting diodes (LEDs) as targets. The LEDs are multiplexed so that only one at a time can be seen by the camera system, avoiding the correspondence problem. The unique form of stereo ranging is based on three line detectors with lenses that transform the point source LED illumination into a focused line. The simplified triangulating geometry is shown in Figure 9.8. It may be shown from this geometry that the target position (x_p, y_p, z_p) can be determined from the detector outputs u_{left}, u_{right}, and v as follows:

$$x_p = b\left(u_{\text{right}} + u_{\text{left}}\right) \Big/ 2\left(u_{\text{right}} - u_{\text{left}}\right) \tag{9.8}$$

$$y_p = bv \Big/ \left(u_{\text{right}} - u_{\text{left}}\right) \tag{9.9}$$

$$z_p = fb \Big/ \left(u_{\text{right}} - u_{\text{left}}\right) \tag{9.10}$$

where f and b are the lens-to-detector distance and the baseline separation respectively. In practice, the image space to object space mapping is much more complicated than Equations 9.8 to 9.10, and involves a camera model with more than 60 parameters that are determined through a calibration process.

OPTOTRACK offers high sampling rate, large measurement volume, and high accuracy compared to many other position tracking systems.

Magnetic Position Tracking

A position/orientation tracking sensor based on a three-axis magnetic dipole transmitter and a three-axis magnetic loop detector has been developed by Polhemus Inc. [20]. The transmitted fields are alternating current for ease of detection (i.e., transformer coupled) and time-multiplexed so that the field due to each axis can be distinguished from the others. Distance between transmitter and detector is determined by exploiting the $1/R^3$ relationship between field strength and distance from the source. Orientation of the detector is determined by exploiting the directionality of magnetic fields and the direction sensitivity of loop detectors.

An issue with respect to the use of ac fields is the distortions in field shape that occur if metal objects are present, and the consequent effect on sensor accuracy. These distortions result from eddy currents in the conducting metal. Ascension Technology Corp. has developed a variation on the Polhemus sensor based on dc magnetic fields. The switching transient due to time-multiplexing does produce an eddy current effect, but it is allowed to die out before measurement is made. Details of the dc technique are available in [21].

An important difference between optical and magnetic tracking technologies is that the former require an unbroken line of sight to the targets while the latter do not. This gives magnetic trackers an advantage in some applications. On the other hand, the $1/R^3$ field distribution characteristic of magnetic tracking

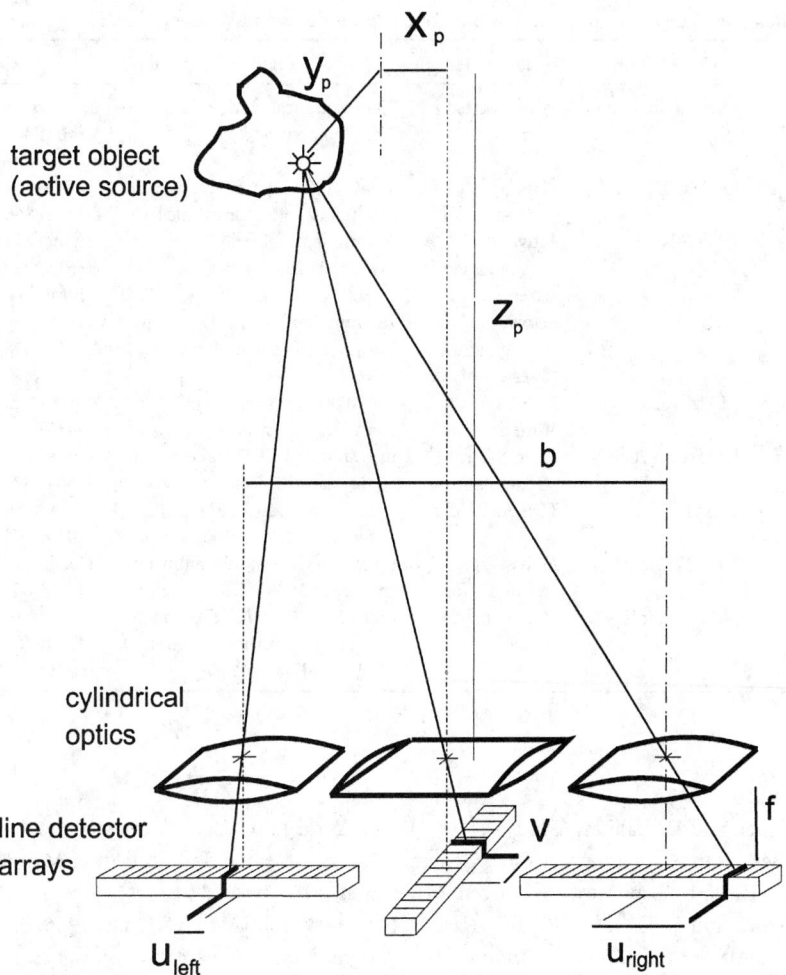

FIGURE 9.8 The OPTOTRAK position tracking system employs a novel arrangement of cylindrical optics and one-dimensional detectors to triangulate the 3-D position of an infrared LED target. Up to 255 individual multiplexed targets can be tracked by the system.

implies an extreme sensitivity loss with distance, whereas optical triangulation has a more benign $1/R$ characteristic. This, to some extent, explains why the volume of measurement and accuracy of optical triangulation systems is generally much better than for magnetic systems.

9.4 A Sampling of Commercial Ranging, Range Imaging, and Motion Tracking Products

Table 9.1 contains information collected from vendor literature. Be advised when comparing specifications that test conditions, standards, and interpretations can vary significantly. The specifications, therefore, should serve only as a rough guide.

TABLE 9.1 Ranging, Range Imaging, and Position Tracking Products and Vendors

Class	Trade Name	Principle	Features	Contact
Ranging (contact)	MicroScribe-3DX	Instrumented arm	50 in. spherical work volume, 0.3 mm accuracy	Immersion Corp. (408) 467-1900, info@immerse.com
Ranging (noncontact)	LASERVISION	TOF, laser	50 m range, 4.9 mm accuracy @ 15 m, integrated electronic level	ZIRCON Corp., (408) 866-8600
Range-Imaging (line scan)	HYSCAN	Active triangulation laser	40 mm depth of field, 70 mm swath, 0.025 mm accuracy, 10,000 points/s	Hymarc Ltd., (613) 727-1584, info@hymarc.com
Range-Imaging (line scan)	TriCam	Active triangulation laser	120 mm depth of field, 60 mm swath, 0.05 mm accuracy	Perceptron Inc., (810) 478-7710, inquiry@perceptron.com
Range-Imaging (line scan)	ALTM 1020	TOF laser time-interval	330-1000 m range, 15 cm accuracy, 20° swath	Optech Inc., (416) 661-5904
Range-Imaging (area scan)	Rangecam 7000	Laser or strobe triangulation	uses standard CCD camera and light plane projector	Range Vision Inc. (604) 473-9411
Range-Imaging (area scan)	LASAR	TOF, AM Lidar	2–40 m range, 60 × 70° max field of view, 360,000 samples/s	Perceptron Inc., (810) 478-7710
Position Tracking	OPTOTRAK	Active target triangulation	up to 255 targets, submillimeter accuracy, 5000 3 DoF samples/s	Northern Digital Inc., (519) 884-5142
Position Tracking	Flock of Birds	Magnetic field based	up to 30 position/orientation targets, approx. 10 mm acuracy, 144 6-DoF samples/s	Ascension Technology Corp. (802) 860-6440

References

1. R. Resnick and D. Halliday, *Physics (Part 1)*. New York: John Wiley & Sons, 1966. 4.
2. P. J. Besl, Range imaging sensors. General Motors Research Publication, GMR-6090, General Motors Research Laboratories, Warren, MI, March, 1988.
3. R. Resnick and D. Halliday, *Physics (Part 1)*. New York: John Wiley & Sons, 1966. 3.
4. D. F. McAllister (ed.), *Stereo Computer Graphics and Other True 3D Technologies*, Princeton, NJ: Princeton University Press, 1993. Ch. 4.
5. L. E. Kinsler and A. R. Frey, *Fundamentals of Acoustics, 2nd. ed.*, New York: John Wiley & Sons, 1962, Chs. 9, 15.
6. W. Diedrich, Foundations of reading sonar, *The In-Fisherman*, April-May, 42-56, 1996.
7. E. B. Blood, Device for quantitatively measuring the relative position and orientation of two bodies in the presence of metals utilizing direct current magnetic fields, U.S. Patent 4,945,305, Jul. 31, 1990.
8. W. J. Steinway and C. R. Barrett, Development status of a stepped-frequency ground penetrating radar, in *Underground and Obscured Object Imaging and Detection, SPIE Proceedings*, Vol. 1942, Orlando, FL, April 1993, 34-43.
9. J. Borenstein, H. R. Everett, and L. Feng, Where am I? Sensors and Methods for Autonomous Mobile Robot Positioning, 1995 Edition. University of Michigan report for the United States Dept. of Energy Robotics Technology Development Program, Ann Arbor, MI, 1995. Ch. 3.
10. Hymarc Ltd., 1995. Product Information, Hyscan 3D Laser Digitizing Systems. Ottawa, Ontario, Canada.
11. Perceptron Inc., 1995. Product Information, TriCam Non-Contact Measurement Solutions. Farmington Hills, MI.
12. F. E. Goodwin, Frequency Modulated Laser Radar, U.S. Patent 4,830,486, May 16, 1989.
13. F. Blais, M. Rioux, and J.-A. Beraldin, Practical considerations for a design of a high precision 3D laser scanner system, *SPIE Vol. 959, Optomechanical and Electro-Optical Design of Industrial Systems*, 1988.

14. D. K. Barton, *Radar System Analysis,* Englewood Cliffs, NJ: Prentice-Hall, 1964. Ch. 4.
15. A. V. Jelalian, *Laser Radar Systems,* Artech House, 1992. Ch. 1.
16. E. S. Cameron, R. P. Srumski, and J. K. West, Lidar Scanning System, U.S. Patent 5,006,721, Apr. 9, 1991.
17. Acuity Research Inc., 1995. Product Information, Accurange 4000. Menlo Park, CA.
18. R. R. Clark, Scanning rangefinder with range to frequency conversion, U.S. Patent 5,309,212, May 3, 1994.
19. Northern Digital Inc., 1990. Product Literature, OPTOTRACK 3D Motion Measurement System, Waterloo, Ontario, Canada.
20. F. H. Raab, E. B. Blood, T. O. Steiner, and H. R. Jones, Magnetic position and orientation tracking system, *IEEE Trans. Aerospace Electronic Systems,* Vol. AES-15, No. 5, September 1979.
21. E. B. Blood, Device for quantitatively measuring the relative position and orientation of two bodies in the presence of metals utilizing direct current magnetic fields, U.S. Patent 4,945,305, July 31, 1990.

10
Position, Location, Altitude Measurement

Dimitris E. Manolakis
Technological Education Institute

Mark Stedham
University of Alabama in Huntsville

Partha P. Banerjee
University of Alabama in Huntsville

Seiji Nishifuji
Yamaguchi 'University

Shogo Tanaka
Yamaguchi University

Halit Eren
Curtin University of Technology

C.C. Fung
Curtin University of Technology

Jacob Fraden
Advanced Monitors Corporation

10.1 Altitude Measurement

Dimitris E. Manolakis

Accurate monitoring of aircraft cruising height is required in order to reduce vertical separation to a minimum standard. Interest here focuses on the measurement of the distance between aircraft level and the sea surface level. This distance can be estimated onboard via barometric altimeters or it can be measured — either onboard or in ground stations — via electronic radio wave systems. The indication of the first equipment is referred to as pressure altitude, or simply altitude, whereas that of the second category is referred to as geometric height or simply height.

The altitude information at air traffic control (ATC) centers is based on pressure altitude measurement that the aircraft transponder system sends after it receives an appropriate interrogation — known as mode C interrogation — transmitted by a secondary surveillance radar. Actually, the altitude information is an atmospheric pressure measurement transformed to altitude indication through a formula expressing the pressure/altitude relationship. When a flight level is cleared for an aircraft, it actually means that the pilot must keep flying on an isobaric surface. However, the altimetry system may present systematic errors (biases) that are different for each airplane, and that significantly affect safety. Thus, the altimetry system performance as well as the aircraft height keeping performance must be monitored by an independent radar or satellite system.

Radar or satellite systems determine the position of an object through algorithms that are fed with range, or range difference, or range sum, or range and bearing measurements, and they estimate the object position vector employing appropriate techniques such as triangulation or trilateration. The primary radar measurements are contaminated by two kinds of errors: random and systematic errors. The effect of random errors can be reduced by the use of appropriate noise rejection filters, such as a Kalman filter. The second kind of error is usually removed by calibrating the instrument if this is possible; otherwise, suitable algorithms must be invented to anticipate it.

The estimation may be derived in ground stations or onboard the aircraft according to where the data acquisition and processing is performed. In the latter case, the vertical position estimation has to be downlinked to the appropriate ATC center. Also, the estimation may be performed off-line or on-line. Ground-based methods or systems are the Navigation Accuracy Measurement System (NAMS), the height estimation method with a single air traffic control radar, the method with one Secondary Surveillance Radar (SSR) and one omnidirectional radar, the Dual synchronized Autonomous Height Monitoring System (DAMS) and methods that use multiple SSRs and estimate the height by quadrilateration, or by the use of pseudorange measurements, or by the use of range difference measurements. On-board height measurement methods derive their estimates by trilateration using the Distance Measuring Equipment (DME) or the Global Positioning System (GPS) signals.

Ground-Based Height Estimation

The radars used to derive the original measurements are either primary or secondary surveillance radars. A primary radar sends a signal and scans for the arrival of its reflection. The range to the object reflecting the signal is derived from the time elapsed between transmission and reception. With secondary radar, the radar sends an interrogation to aircraft — to all aircraft or to a selected one — and the appropriate aircraft sends a reply via its transponder. The range to the aircraft is computed from the time elapsed between the signal transmission and the signal arrival, taking into account the nominal delay time of the transponder. Most of the methods estimating the aircraft height make use of the SSR equipment because it is cost effective, the transponder reply signal is stronger than that reflected to a primary radar, and the system can operate more reliably in dense traffic areas.

Any systematic errors in the primary radar and in the ground equipment of the SSR can be corrected by calibration. However, the problem encountered with SSR is that it involves the transponder delay time in the range measurement process. Thus, any systematic error in the transponder delay time causes range bias errors that are different for each aircraft and thus suitable methods must be used to anticipate for it in the subsequent measurement data processing.

Navigation Accuracy Measurement System

Nagaoka has proposed an off-line height estimation system [1–3]. It is composed of a primary marine radar located under an airlane and measures range R and depression angle β. Figure 10.1 shows the geometry of the system. The antenna rotates about a vertical axis and scans the area above it with rate equal to 1 rotation per 3 s. The principle exploited to derive the height estimate is that the range varies as the aircraft passes through the data acquisition area. The rate of change of range is mainly a function of the flight height z and secondarily of depression angle. It is easily derived from Figure 10.1, that the relation between the above quantities and the position x along the x axis at time t is:

$$R(t) = \sqrt{x^2(t) + R'(t)^2} = \sqrt{x^2(t) + \frac{z^2}{\cos^2 \beta(t)}} \tag{10.1}$$

Let x_0 denote the position of the aircraft at time t_0. Assuming the aircraft flies in straight and level flight, this means that the velocity V_x, the depression angle, and the height h remain constant during the data

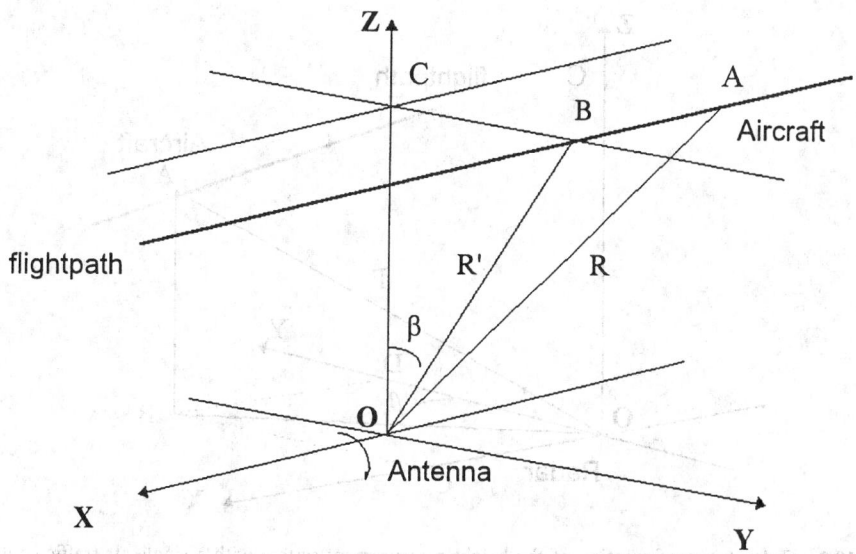

FIGURE 10.1 Geometry of the Navigation Accuracy Measurement System (NAMS). BA = x, CB = y, OC = z, R' = $z/\cos(\beta)$.

collection period. The above quantities and the range measurement at time t_i are related through Equation 10.2.

$$R_i = \sqrt{\left[x_0 + V_x\left(t_i - t_0\right)\right]^2 + \frac{z^2}{\cos^2\beta}} = f\left(t_i, \beta, \mathbf{q}\right) = f_i\left(\mathbf{q}\right) \tag{10.2}$$

where $\mathbf{q}$ is the unknown quantities vector, $\mathbf{q} = [x_0, V_x, z]^T$.

Measurements of R_i and β are collected at times t_0, t_1, ..., t_n, and their set is briefly expressed as a vector function of the unknown quantities with the following matrix equation:

$$\mathbf{R} = \mathbf{f}\left(\mathbf{q}\right) \tag{10.3}$$

where

$$\mathbf{R} = \begin{bmatrix} R_0 \\ R_1 \\ \cdot \\ \cdot \\ R_n \end{bmatrix} \qquad \mathbf{f} = \begin{bmatrix} f_0 \\ f_1 \\ \cdot \\ \cdot \\ f_n \end{bmatrix}$$

Equation 10.3 is nonlinear. Thus, a nonlinear least square method, such as the Gauss-Newton iterative method, must be used to estimate the unknown vector. Let $\mathbf{q}_k$ be the estimate at the kth iteration. Then, the next estimate is:

$$\hat{\mathbf{q}}_{k+1} = \hat{\mathbf{q}}_k + \left(\mathbf{F}^T\mathbf{F}\right)^{-1}\mathbf{F}^T\left(\mathbf{R} - \mathbf{f}\left(\hat{\mathbf{q}}_k\right)\right) \tag{10.4}$$

where $\mathbf{F}$ is the partial derivatives (Jacobian) matrix; that is:

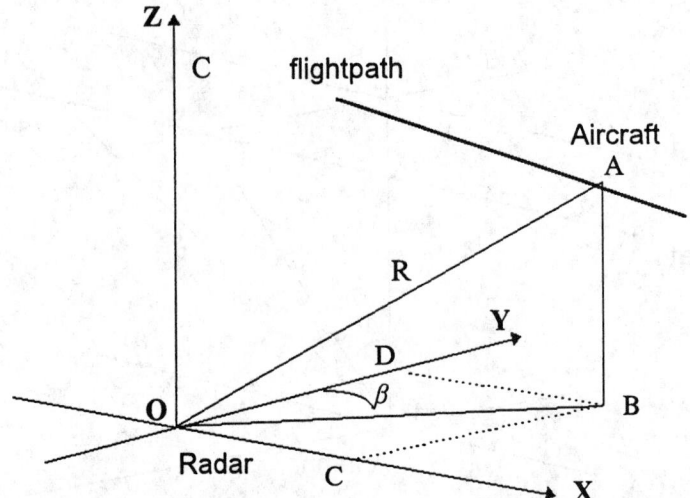

FIGURE 10.2 Geometric configuration of the height measurement system with a single air-traffic control radar. $OC = x$, $OD = y$, $AB = z$, $\beta = \angle\, BOD$.

$$F = \frac{\partial f}{\partial q} = \begin{bmatrix} \dfrac{\partial f_0}{\partial x_0} & \dfrac{\partial f_0}{\partial V_x} & \dfrac{\partial f_0}{\partial z} \\ \cdot & \cdot & \cdot \\ \cdot & \cdot & \cdot \\ \dfrac{\partial f_n}{\partial x_0} & \dfrac{\partial f_n}{\partial V_x} & \dfrac{\partial f_n}{\partial z} \end{bmatrix} \tag{10.5}$$

Assuming typical aircraft velocities and flight levels, and taking into account the antenna fan beam and rotation rate, it is determined that the observation time is approximately 1 min and the number of measurements is approximately 20. The standard deviation (SD) of the height estimation errors is $\sigma_z = 10$ m when the SD of the measurement errors of range and depression angle are $\sigma_R = 10$ m and $\sigma_B = 0.08°$, respectively.

The system has been developed and operates at the Electronic Navigation Institute of Tokyo, Japan. To estimate and anticipate for systematic errors in the depression angle and range measurements, it has been proposed to perform 40 experimental flights above the radar range. The basic assumption of the NAMS system (i.e., that the aircraft flies at constant height) cannot be validated by the NAMS itself. Thus, an SSR is required to confirm that the flight is indeed performed at the same level via the mode C interrogations [3].

Height Estimation with a Single Air-Traffic Control Radar

Nagaoka at the Electronic Navigation Institute of Tokyo investigated the possibility to derive a height estimation with a very simple and inexpensive method [4]. The idea is to use a single radar normally used for air traffic control. The radar antenna rotates around a horizontal axis once every 4 s and measures range R and bearing β, as it is shown in Figure 10.2.

Let (x, y, z) be the 3-D position of the aircraft. The bearing angle is defined as:

$$\beta = \cos^{-1}\left(\frac{y}{\sqrt{x^2 + y^2}} \right) \tag{10.6}$$

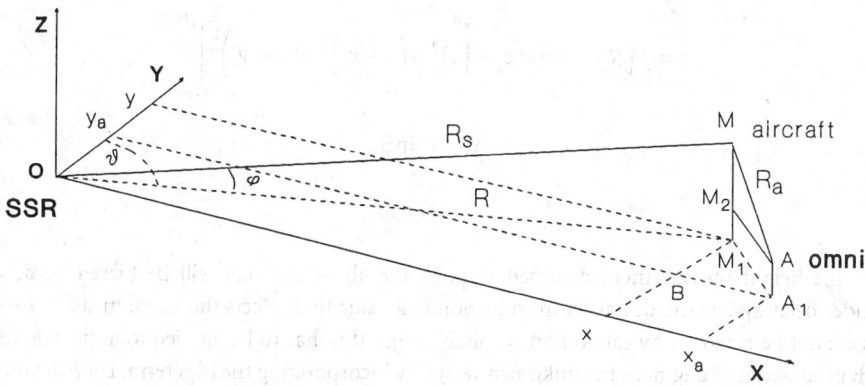

FIGURE 10.3 Geometry of the height estimation system with one SSR and one omnidirectional radar (SSROR). The distance x_a of the radars typically ranges from 80 km up to 160 km. $M_1M = z$.

Thus, the relation between the measurements and the unknown quantities is:

$$R = \left(z^2 + x^2 + y^2\right)^{1/2} = \left(z^2 + \frac{y^2}{\cos^2 \beta}\right)^{1/2} \tag{10.7}$$

The measurement principle is the same as that of the NAMS system, i.e., that during the time the aircraft passes through the data acquisition area, the range varies as a function of both the horizontal position and the height. When straight and level flight is assumed, the unknown quantities are the height z, the y-axis initial position y_0, and the velocity V_y. The expression relating the measurements R_i, and β_i and the unknown quantities vector $\mathbf{q} = [y_0, V_y, z]^T$ at time t_i is:

$$R_i = \left(z^2 + \frac{\left(y_o + V_y\left(t_i - t_o\right)\right)^2}{\cos^2 \beta_i}\right)^{1/2} = f\left(t_i, \beta_i, \mathbf{q}\right) = f_i\left(\mathbf{q}\right) \tag{10.8}$$

The measurements are collected as the aircraft flies close to the radar at a distance of 30 km. Approximately, the data acquisition area is 15 km, the observation time is 1 min and the total number of scans is 15. The total set of measurements is expressed as a nonlinear matrix function in the form of Equation 10.3; that is, $\mathbf{R} = \mathbf{f}(\mathbf{q})$ where $\mathbf{R}$ denotes the measurement vector. The unknown vector is estimated using a nonlinear least squares method like that of the NAMS system expressed by Equation 10.4.

The system performance is evaluated with Monte Carlo simulations. The best accuracy achieved is $\sigma_z = 100$ m under the assumption that the SD errors of range and bearing measurements are $\sigma_R = 100$ m and $\sigma_B = 0.06°$. The method can use either primary or secondary radar. However, in the latter case, the method has to include systematic delays by the transponder.

Height Estimation with one SSR and One Omni Radar (SSROR)

Manolakis et al. investigated the possibility to estimate the aircraft flight height with a system that consists of one standard SSR plus a secondary radar equipped with an omnidirectional antenna located far away from the SSR; hence, this system could be referred to as SSROR [5]. Figure 10.3 shows a typical geometric configuration of the system. The SSR measures bearing θ and slant range R_s, which is transformed to horizontal range R with correcting look-up tables. The omni radar measures range R_a. Let (x,y,z), (x_a, y_a, z_a) and $(0,0,0)$ be the coordinates of aircraft, omni radar, and SSR, respectively. Then, from Figure 10.3 it follows that the theoretical height function is:

$$z = f(R_a, x, y) = z_a + \left[R_a^2 - (x - x_a)^2 - (y - y_a)^2 \right]^{1/2} \tag{10.9}$$

$$x = R \sin\theta \tag{10.10}$$

$$y = R \cos\theta \tag{10.11}$$

However, the height measurement obtained through the above formula will be biased because of the transponder delay systematic deviation from its nominal value that affects the range measurements. Since this bias cannot be removed by calibration, a suitable algorithm has to be applied to anticipate for it. The approach proposed is to augment the unknown vector by incorporating the bias term. Let b denote the bias in range measurements. The biased measurements R_b and R_{ab} of R and R_a, respectively, are expressed as:

$$R_b = R + b \cos\varphi \tag{10.12}$$

$$R_{ab} = R_a + b \tag{10.13}$$

where φ is the elevation angle. Let s denote the squared height:

$$s = (z - z_a)^2 \tag{10.14}$$

and s_b denote the corresponding quantity derived from (biased) range measurements. Then, after some manipulations of the above equations, the following relation is obtained:

$$s_b = s + b\,a \tag{10.15}$$

$$a = 2 \left[R_a - \cos\varphi \left(R - x_a \sin\theta - y_a \cos\theta \right) \right] \tag{10.16}$$

The term a is the bias multiplying factor determined by the relative geometry of the system. Equation 10.15 is a linear relation between measurement s_b and the unknown quantities s and b. Figure 10.4 shows that the effect of the bias varies as the aircraft passes through the surveillance area. Its form is mainly determined by the flight height. Consequently, it is possible to estimate both the height and the bias by collecting data during the period the aircraft remains in the surveillance area. The measurement equation at time t_i is:

$$s_{bi} = s_i + b\,a_i + e_i \tag{10.17}$$

where e_i represents the effect of the random measurement errors; hence, it could be referred to as the equation error. Assuming level flight, $s_i = s$, and the set of collected data is expressed with the following linear matrix equation:

$$s_b = A\,q + e \tag{10.18}$$

$$s_b = \begin{bmatrix} s_{b0} \\ s_{b1} \\ \cdot \\ \cdot \\ s_{bn} \end{bmatrix} \quad A = \begin{bmatrix} 1 & a_0 \\ 1 & a_1 \\ \cdot & \cdot \\ 1 & a_n \end{bmatrix} \quad e = \begin{bmatrix} e_0 \\ e_1 \\ \cdot \\ e_n \end{bmatrix} \quad q = \begin{bmatrix} s \\ b \end{bmatrix}$$

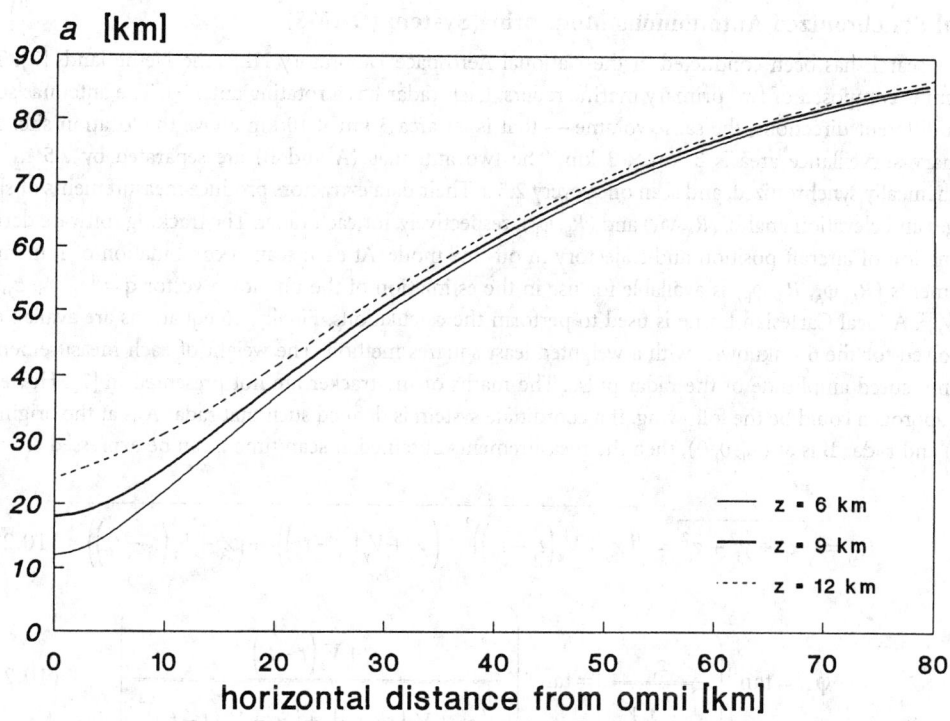

FIGURE 10.4 Range bias multiplier in squared height measurements as a function of the aircraft distance from the omni radar. Three different flight heights are examined. The omni radar is at position (150 km, 0, 0).

The best estimate $\hat{\mathbf{q}}$ of $\mathbf{q}$, minimizing the weighted sum of the squared errors $\mathbf{e}^{\mathrm{T}}\,\mathbf{W}\,\mathbf{e}$, is:

$$\hat{\mathbf{q}} = \left(\mathbf{A}^{\mathrm{T}}\mathbf{W}\mathbf{A}\right)^{-1}\mathbf{A}^{\mathrm{T}}\mathbf{W}\mathbf{s}_{\mathrm{b}} \qquad (10.19)$$

where $\mathbf{W}$ is the weighting matrix defined as the inverse of the equation error covariance matrix:

$$\mathbf{W}^{-1} = \mathrm{E}\left\{\mathbf{e}\,\mathbf{e}^{\mathrm{T}}\right\} \qquad (10.20)$$

where the notation $\mathrm{E}\{\ \}$ stands for the expected value operation. Thus, the estimate of the geometric height is:

$$\hat{z} = \sqrt{\hat{q}_1} = \sqrt{\hat{s}} \qquad (10.21)$$

In the case of nonlevel flights, an augmented unknown vector is used that comprises bias b, initial height z_0, and vertical velocity V_z [6]. In this case, however, the measurement equation is nonlinear in terms of the unknown quantities; hence, a nonlinear least squares iterative algorithm must be employed.

The system performance is a function of: (1) the range and bearing measurement standard deviation errors, σ_R and σ_θ, respectively; (2) the aircraft velocity, which affects the number of scans; (3) the omni radar scan rate; and (4) geometric terms such as the flight level and the distance between the radars. By assuming that $\sigma_R = 70$ m, $\sigma_\theta = 0.08°$, the SD of the height estimation error will be between 50 m and 100 m when the values of the other factors lie in reasonable ranges.

Dual Synchronized Autonomous Monitoring System (DAMS)

The research has been conducted in the National Aerospace Laboratory NLR, the Netherlands [7]. The system is composed of two primary marine radars. Each radar has a rotating antenna. The antennas scan, with different directions, the same volume — that is an area 3 km × 10 km above the location site. The primary surveillance area is 3 km × 3 km. The two antennas (A and B) are separated by 2.5 m, are mechanically synchronized, and scan once every 2.5 s. Their data extractors produce measurements of slant ranges and elevation angles, (R_A, φ_A) and (R_B, φ_B), respectively, for each radar. The tracking software derives estimation of aircraft position and trajectory in off-line mode. At each scan, a combination of four measurements $(R_A, \varphi_A, R_B, \varphi_B)$ is available for use in the estimation of the unknown vector $\mathbf{q} = [x_0, y_0, z_0, V_x, V_y, V_z]^T$. A local Cartesian frame is used to perform the calculations. Finally, 16 equations are available to be solved for the 6 unknowns with a weighted least squares method. The weight of each measurement is the measured amplitude of the radar pulse. The maths of the tracker are not presented in [7]. However, one approach could be the following. If a coordinate system is defined such that radar A is at the origin (0, 0, 0) and radar B is at $(x_B, 0, 0)$, then the measurements obtained at scan time t_i can be expressed as:

$$R_{Ai} = \sqrt{x_i^2 + y_i^2 + z_i^2} = \sqrt{\left(x_0 + V_x(t_i - t_0)\right)^2 + \left(y_0 + V_y(t_i - t_0)\right)^2 + \left(z_0 + V_z(t_i - t_0)\right)^2} \quad (10.22\text{a})$$

$$\varphi_{Ai} = \tan^{-1}\left(\frac{z_i}{\sqrt{x_i^2 + y_i^2}}\right) = \tan^{-1}\left(\frac{z_0 + V_z(t - t_i)}{\sqrt{\left(x_0 + V_x(t - t_i)\right)^2 + \left(y_0 + V_y(t - t_i)\right)^2}}\right) \quad (10.22\text{b})$$

$$R_{Bi} = \sqrt{\left(x_i - x_B\right)^2 + y_i^2 + z_i^2}$$
$$= \sqrt{\left(x_0 + V_x(t_i - t_0) - x_B\right)^2 + \left(y_0 + V_y(t_i - t_0)\right)^2 + \left(z_0 + V_z(t_i - t_0)\right)^2} \quad (10.22\text{c})$$

$$\varphi_{Bi} = \tan^{-1}\left(\frac{z_i}{\sqrt{\left(x_i - x_B\right)^2 + y_i^2}}\right) = \tan^{-1}\left(\frac{z_0 + V_z(t - t_i)}{\sqrt{\left(x_0 + V_x(t - t_i) - x_B\right)^2 + \left(y_0 + V_y(t - t_i)\right)^2}}\right) \quad (10.22\text{d})$$

Notice that x_B has a small value (2.5 m) compared to the magnitude of the other quantities; consequently, it can be neglected. The above set of equations at four different times yields 16 equations that can be solved with a nonlinear least squares method such as the Gauss–Newton iterative method presented in Equation 10.4.

The SD of the height estimation error will be less than 15 m in the primary surveillance area, and 30 m at the edges of the area when the SD of range and elevation angle measurements are smaller than 10 m and 0.1°, respectively.

Height Measurement by Quadrilateration

Rice proposed a system consisting of four synchronized receiving SSR stations S_i, $i = 0, 1, 2, 3$ that use SSR transmissions from the aircraft transponder and estimate the height by quadrilateration [8]. Figure 10.5 shows a typical configuration of systems composed of N SSRs. One of them is an active station, which means that this station has both an interrogator and a receiver. Let (x, y, z) and (x_i, y_i, z_i) denote the Cartesian coordinates of aircraft and station S_i, respectively. Also let R_i denote the range from station S_i to aircraft, c denote the velocity of the light, and T_s denote the time of signal transmission from aircraft transponder. The stations measure the time of arrival (TOA) T_i, $i = 0, 1, 2, 3$, of the aircraft transponder signal at each site. The following relations hold:

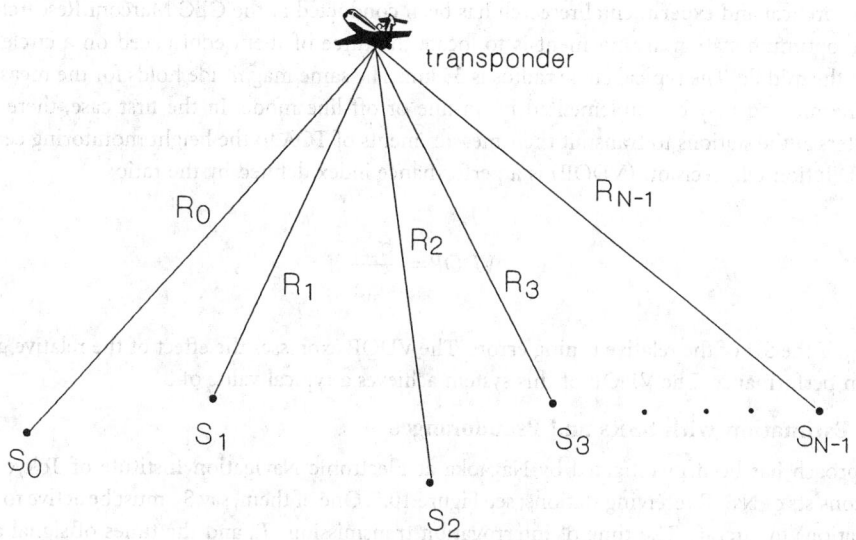

FIGURE 10.5 Typical configuration of the height estimation systems that are based on N SSR stations. One of the stations is active, i.e., it both transmits the interrogations and receives the replies, whereas the other stations are receivers only.

$$T_i = T_s + \frac{R_i}{c} \quad i = 0, 1, 2, 3 \tag{10.23}$$

$$R_i^2 = (x - x_i)^2 + (y - y_i)^2 + (z - z_i)^2, \quad i = 0, 1, 2, 3 \tag{10.24}$$

The above system of eight equations can be solved for the unknown quantities. The unknown quantities used by Rice are $(R_0, R_1, R_2, R_3, x, y, z, T_s)^T$. However, an equivalent approach is to substitute for R_i in Equation 10.23, which becomes:

$$T_i = T_s + \frac{1}{c} \sqrt{(x - x_i)^2 + (y - y_i)^2 + (z - z_i)^2} = f_i(\mathbf{q}) \quad i = 0, 1, 2, 3 \tag{10.25}$$

where $\mathbf{q} = [x, y, z, T_s]^T$ is the unknown vector. Thus, there are four nonlinear equations to be solved for $\mathbf{q}$. One suitable method, for example, is the Newton–Raphson method, which iteratively approximates the solution via the following formula:

$$\mathbf{q}_{k+1} = \mathbf{q}_k + \mathbf{F}(\mathbf{q}_k)^{-1}(\mathbf{T} - \mathbf{f}(\mathbf{q}_k)) \tag{10.26}$$

where $\mathbf{T} = [T_0, T_1, T_2, T_3]^T$ is the measurement vector and $\mathbf{F}$ is the Jacobian matrix:

$$\mathbf{F} = \frac{\partial \mathbf{f}}{\partial \mathbf{q}} = \begin{bmatrix} \dfrac{\partial f_0}{\partial x} & \dfrac{\partial f_0}{\partial y} & \dfrac{\partial f_0}{\partial z} & \dfrac{\partial f_0}{\partial T_s} \\ \vdots & & & \vdots \\ \dfrac{\partial f_3}{\partial x} & \dfrac{\partial f_3}{\partial y} & \dfrac{\partial f_3}{\partial z} & \dfrac{\partial f_3}{\partial T_s} \end{bmatrix} \tag{10.27}$$

Notice that the time of interrogation transmission, as well as the transponder nominal delay time, are not involved in the measurements. The measured quantities are only the TOAs at the station sites. Thus, the height estimate is not affected by any transponder bias.

The theoretical and experimental research has been conducted at the GEC Marconi Research Center, U.K. The optimum station arrangement is to locate the three of them equispaced on a circle and the fourth in the middle. The typical circle radius is 35 km. The same magnitude holds for the measurement range. The method may be implemented in on-line or off-line mode. In the first case, there must be transmitters at the stations to transmit their measurements of TOA to the height monitoring center. The Vertical Dilution Of Precision (VDOP) is a performance index defined by the ratio:

$$VDOP = \frac{\sigma_z}{\sigma_{rte}} \qquad (10.28)$$

where σ_{rte} is the SD of the relative timing errors. The VDOP expresses the effect of the relative geometry to system performance. The VDOP of this system achieves a typical value of 3.

Height Estimation with SSRs and Pseudoranges

This approach has been investigated by Nagaoka at Electronic Navigation Institute of Tokyo [9]. The system consists of N SSR receiving stations; see Figure 10.5. One of them, say S_0, must be active to transmit interrogations to aircraft. The time of interrogation transmission, T_t, and the times of signal arrival at the receiving stations T_i, i = 0, 1, ..., N − 1, are measured. Thus, N pseudorange measurements r_i are obtained where $r_i = c \, (T_i - T_t)$. Let T_D denote the transponder delay and D denote the distance corresponding to this delay, $D = c \, T_D$. Then, for each pseudorange measurement r_i, the following relation holds:

$$r_i = D + R_i + R_0 = D + \sqrt{\left(x-x_i\right)^2 + \left(y-y_i\right)^2 + \left(z-z_i\right)^2} + \sqrt{\left(x-x_0\right)^2 + \left(y-y_0\right)^2 + \left(z-z_0\right)^2}$$
$$= f_i\!\left(\mathbf{q}\right) \qquad i = 0, 1, ..., N-1 \qquad (10.29)$$

where $\mathbf{q} = [x, y, z, D]^T$ is the unknown vector. The set of N measurements $\rho = [\rho_0, \rho_1, \dots \rho_{N-1}]^T$, $N \ge 4$, and the unknown vector are related through Equation 10.30.

$$\mathbf{r} = \mathbf{f}\!\left(\mathbf{q}\right) \qquad (10.30)$$

The unknown vector $\mathbf{q}$ can be obtained from the solution of Equation 10.30 with a nonlinear weighted least squares method. Thus, the best estimate of $\mathbf{q}$ is iteratively calculated as:

$$\hat{\mathbf{q}}_{k+1} = \hat{\mathbf{q}}_k + \left(\mathbf{F}^T \mathbf{F}\right)^{-1} \mathbf{F}^T \left(\mathbf{r} - \mathbf{f}\!\left(\hat{\mathbf{q}}_k\right)\right) \qquad (10.31)$$

where $\mathbf{F}$ is the Jacobian matrix

$$\mathbf{F} = \frac{\partial \mathbf{f}}{\partial \mathbf{q}} = \begin{bmatrix} \dfrac{\partial f_0}{\partial x} & \dfrac{\partial f_0}{\partial y} & \dfrac{\partial f_0}{\partial z} & \dfrac{\partial f_0}{\partial D} \\ \cdot & \cdot & \cdot & \cdot \\ \dfrac{\partial f_{N-1}}{\partial x} & \dfrac{\partial f_{N-1}}{\partial y} & \dfrac{\partial f_{N-1}}{\partial z} & \dfrac{\partial f_{N-1}}{\partial D} \end{bmatrix} \qquad (10.32)$$

The estimate of $\mathbf{q}$ is free of the transponder delay systematic error because the estimation is based not on the nominal delay, but on the actual delay time, which is one of the parameters to be estimated, whereas the rest of the parameters are the aircraft 3-D position coordinates x, y, z.

The station arrangement proposed by Nagaoka, when there are four stations, is an equilateral triangle formed by the three stations, whereas the fourth station is located in the center. The VDOP, defined as the ratio σ_z/σ_R (where σ_R is the observation error) has a typical value of 4 when the aircraft is above the center at a height equal to the baseline radius. The VDOP increases as the aircraft flies higher and longer and as the baseline radius becomes smaller.

Height Measurement with SSRs and Range Differences

This approach has been proposed by Manolakis and Lefas [10, 11]. The system consists of $N-1$ receiving SSR stations S_i, $i = 1, N-1$, and one station, say S_0, which is both receiver and interrogator, see Figure 10.5. The stations receive the reply and the time difference of arrival (TDOA) between a reference station, say S_0, and station S_i is measured. A set of $N-1$ TDOA or equivalently range difference (RD) measurements is collected at each time the transponder sends a reply signal. The height estimation derived from this set of measurements is not affected by any transponder delay systematic error since this error is inherently subtracted from the measurements used. This system could be referred to as RD height monitoring unit (RDHMU). The systems that derive the position fix based on this kind of measurement are known as TDOA or RD or hyperbolic systems.

Let τ_i denote the TDOA between stations S_i and S_0, and d_i denote the corresponding RD measurement, $d_i = c\,\tau_i$. The following relation holds:

$$d_i = R_i - R_0 = \sqrt{\left(x - x_i\right)^2 + \left(y - y_i\right)^2 + \left(z - z_i\right)^2} - \sqrt{\left(x - x_0\right)^2 + \left(y - y_0\right)^2 + \left(z - z_0\right)^2}$$

$$= f_i\left(\mathbf{q}\right) \qquad i = 1, 2, \ldots, N-1 \tag{10.33}$$

where $\mathbf{q} = [x, y, z]^T$ is the unknown aircraft position vector. The vector of RD measurements $\mathbf{d} = [d_1, d_2, \ldots, d_{N-1}]$ is expressed as:

$$\mathbf{d} = \mathbf{f}\left(\mathbf{q}\right) \tag{10.34}$$

A commonly employed method to solve for $\mathbf{q}$ in this nonlinear equation is the Taylor series method or equivalently the Gauss-Newton iterative method. The best estimate of $\mathbf{q}$ is iteratively approximated as:

$$\hat{\mathbf{q}}_{k+1} = \hat{\mathbf{q}}_k + \left(\mathbf{F}^T\mathbf{F}\right)^{-1}\mathbf{F}^T\left(\mathbf{d} - \mathbf{f}\left(\hat{\mathbf{q}}_k\right)\right) \tag{10.35}$$

where $\mathbf{F}$ is the Jacobian matrix:

$$\mathbf{F} = \frac{\partial \mathbf{f}}{\partial \mathbf{q}} = \begin{bmatrix} \dfrac{\partial f_1}{\partial x} & \dfrac{\partial f_1}{\partial y} & \dfrac{\partial f_1}{\partial z} \\ \dfrac{\partial f_{N-1}}{\partial x} & \dfrac{\partial f_{N-1}}{\partial y} & \dfrac{\partial f_{N-1}}{\partial z} \end{bmatrix} \tag{10.36}$$

In the case of four stations the best arrangement is an equilateral triangle with the fourth station in the center. The SD of height estimation error σ_z will be 15 m when the baseline radius is 6 km, the flying height is 9 km, and σ_{TDOA} is 10 ns.

Work on proof of principles and system development of a HMU based on the concept of TDOA measurement of SSR signals has been conducted by Roke Manor Research Ltd., U.K. [12].

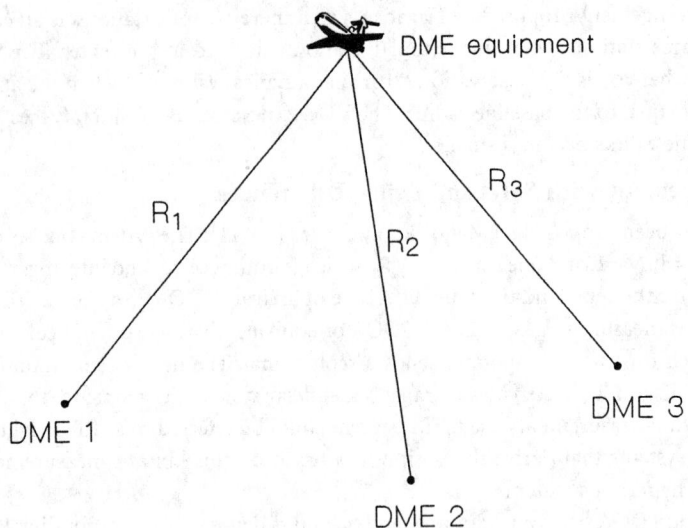

FIGURE 10.6 Configuration of the on-board height estimation system that utilizes the distance measurements derived from the DME equipment.

Onboard Derived Height Estimation

Height Measurement with Distance Measuring Equipment (DME)

This approach for deriving the geometric height onboard the aircraft using DME equipment was first reported by Rekkas et al. [13], whereas more efficient and general techniques have been proposed by Manolakis [14, 15]. Using the DME interrogation equipment, the distance from three DME ground stations is measured onboard (see Figure 10.6). The three stations are located under an airway. The height is then computed from the range measurement vector $\mathbf{R} = [R_1, R_2, R_3]^T$ by trilateration. An exact and efficient solution of the nonlinear measurement equation was derived in [15]. Specifically, the height is computed from the closed form:

$$z = g(\mathbf{R}) = \frac{-b(\mathbf{R}) + \sqrt{d(\mathbf{R})}}{2a} \tag{10.37}$$

where $b(\mathbf{R})$ and $d(\mathbf{R})$ are the following simple polynomial-type functions

$$b(\mathbf{R}) = b_0 + b_1 R_1^2 + b_2 R_2^2 + b_3 R_3^2 \tag{10.38}$$

$$d(\mathbf{R}) = d_{00} + d_{01} R_1^2 + d_{02} R_2^2 + d_{03} R_3^2 + d_{11} R_1^4 + d_{22} R_2^4 + d_{33} R_3^4 + d_{12} R_1^2 R_2^2 + d_{13} R_1^2 R_3^2 + d_{23} R_2^2 R_3^2 \tag{10.39}$$

The coefficients a, b_i, d_{ij} are analytically defined in the Appendix of [15]. An important aspect of these coefficients is that they are completely defined by the ground stations' coordinates (x_i, y_i, z_i), which are fixed. Thus, the coefficients are calculated only once at the moment the aircraft enters the data acquisition area. Then, every time a new set of range measurements is available, the height is computed from the above equations using the range measurements and the stored coefficients. Define the ratio σ_z/σ_R as the VDOP of this technique, where σ_R is the SD of the ranging error. The VDOP is 1 in the case where the stations form an equilateral triangle inscribed in a circle with 10 km radius and the aircraft is above the triangle center at a height of 8 km.

Estimation of Vertical Position with the Global Positioning System (GPS)

The research and development of the GPS has been coordinated by the U.S. Department of Defense. Another similar system is the Global Navigation Satellite System (GLONASS) developed by the former Soviet Union. The GPS is a satellite system providing users with accurate timing and ranging information. The system is available with reduced accuracy to civilian users. Many companies, mainly from the U.S., produce GPS receivers. Let (x, y, z) and (x_i, y_i, z_i) be the coordinates of the user and satellite s_i. The GPS receiver of the user derives the pseudorange measurement D_i, and the corresponding measurement equation is:

$$D_i = R_i + cT_b = \sqrt{\left(x - x_i\right)^2 + \left(y - y_i\right)^2 + \left(z - z_i\right)^2} + b = f_i\left(\mathbf{q}\right) \qquad (10.40)$$

where T_b is the user clock bias, and $\mathbf{q} = [x, y, z, b]^T$ is the unknown vector that incorporates the bias term b. Thus, in order to estimate the 3-D position of the aircraft, four pseudorange measurements are required at least; consequently, four satellites must be visible from the receiver. The set of N pseudorange measurements $\mathbf{D} = [D_1, D_2, ..., D_N]^T$ defines the following matrix measurement equation:

$$\mathbf{D} = \mathbf{f}\left(\mathbf{q}\right) \qquad (10.41)$$

which is solved for $\mathbf{q}$ with the Gauss–Newton least squares iterative method, that is:

$$\hat{\mathbf{q}}_{k+1} = \hat{\mathbf{q}}_k + \left(\mathbf{A}^T\mathbf{W}\mathbf{A}\right)^{-1}\mathbf{A}^T\mathbf{W}\left(\mathbf{D} - \mathbf{f}\left(\hat{\mathbf{q}}_k\right)\right) \qquad (10.42)$$

where $\mathbf{A}$ is the partial derivatives matrix:

$$\mathbf{A} = \frac{\partial \mathbf{f}}{\partial \mathbf{q}} = \begin{bmatrix} \dfrac{\partial f_1}{\partial x} & \dfrac{\partial f_1}{\partial y} & \dfrac{\partial f_1}{\partial z} & \dfrac{\partial f_1}{\partial b} \\ \cdot & \cdot & \cdot & \cdot \\ \dfrac{\partial f_N}{\partial x} & \dfrac{\partial f_N}{\partial y} & \dfrac{\partial f_N}{\partial z} & \dfrac{\partial f_N}{\partial b} \end{bmatrix} = \begin{bmatrix} a_{x1} & a_{y1} & a_{z1} & 1 \\ \cdot & & \cdot & \cdot \\ a_{xN} & a_{yN} & a_{zN} & 1 \end{bmatrix} \qquad (10.43)$$

The elements a_{xi}, a_{yi}, a_{zi}, of the partial derivatives matrix $\mathbf{A}$ are the direction cosines from the receiver to the satellite s_i. The weighting matrix is the inverse of the covariance matrix of the pseudorange measurement errors, $\mathbf{W}^{-1} = E(\delta\mathbf{D}\,\delta\mathbf{D}^T)$. The weighting is generally used to take into account the possible different performances of each satellite, although usually the same performance is assumed for all satellites; that is, $\mathbf{W} = \mathbf{I}$. The VDOP, defined as σ_z/σ_D, depends on the geometry which varies continuously, even in the case of a fixed receiver, because the satellites are not geostationary but move in such orbits as to complete a rotation in 12 h. The world mean value of VDOP is about 2 [16]. Typical VDOP values range from 1.5 to 7, depending on the area of the receiver and on the time of day. The ranging error for the precision positioning service (available only to U.S. military users) has been specified to be less than 6 m (SD). For the standard positioning service, normally available to civilian users, the specified ranging error is double (12 m, SD), whereas it will be about 40 m when selective availability is activated by the Department of Defense. The corresponding measured ranging errors found to be smaller than the specified ones. Namely, for the three operating conditions mentioned, the corresponding values for ranging errors were found to be 2.3 m, 6 m, and 20 m, respectively [17]. The multiplication of the ranging SD error by the VDOP yields the standard deviation of the height estimation error.

To anticipate for the error intentionally induced by the Department of Defense, the Differential GPS (DGPS) method has been developed. A station located at a precisely known position receives the satellite signals, computes its own position on the basis of pseudorange measurements, compares this position with the known position in order to estimate the included error in satellite signals, and finally transmits the appropriate corrections to the receivers in its neighborhood. The achieved accuracy is in the order of a few meters.

Special Topics

The performance analysis of the various parameter estimation systems is usually restricted to the variance analysis, and the estimation error is usually assumed to have zero mean value. However, it is proven that in all of the above systems, the estimation error does not have zero mean value due to the nonlinearity of the measurement equation. Another important aspect is that there are cases where it is not possible to obtain a solution due to the relative geometry of aircraft vs. stations that leads to large errors. In these cases, the successive iterations applied to solve the system of nonlinear equations may not converge. Also, even after convergence, the solutions need not necessarily be "the best" or the "correct ones."

Inherent Bias

The nonlinearity of the systems, joined with the measurement random errors, causes inherently biased estimations although the measured quantities are unbiased. For example, take the case of height estimation with DME measurements. The function $g(\mathbf{R})$ in Equation 10.37, which determines that the height z is nonlinear. In addition, the range measurements will be contaminated by additive zero mean value random errors. Let $\mathbf{R}_m$ denote the noisy measurement vector and z_m denote the height measurement derived from $g(\mathbf{R}_m)$, i.e., from the measurement function when it is fed with noisy measurements. Extending $g(\mathbf{R}_m)$ in a Taylor series around the actual values of ranges R_1, R_2, R_3 up to second-order terms and taking the expected values, it is derived that the expected value of z_m will not be the actual value z, but it will differ by an amount b_z, which is called inherent bias. Specifically, for the DME case, the inherent bias is evaluated as:

$$b_z = E\left[z_m\right] - z = \frac{1}{2}\left(\frac{\partial^2 g}{\partial R_1^2}\sigma_{R_1}^2 + \frac{\partial^2 g}{\partial R_2^2}\sigma_{R_2}^2 + \frac{\partial^2 g}{\partial R_3^2}\sigma_{R_3}^2\right) \tag{10.44}$$

Figure 10.7 shows the inherent bias generated in the DME system. The inherent bias becomes larger as the magnitude of the measurement errors and the system nonlinearity becomes stronger. This bias error is inherently generated in all position estimation algorithms and must be taken into account when precise position estimation is required. Biased height estimates have also been reported in [18] for the SSROR system, in [10] for the RDHMU, and in [3] for the NAMS system.

Existence and Uniqueness of Position Fix

In some systems, there are singular cases for which it is not possible to achieve a position fix. This fact has been reported by Abel and Chafee for the GPS system in [19], where it is shown that for some satellites/aircraft relative geometries, it is not possible to solve the relevant equation or there is more than one solution. Also, for the RDHMU, it has been shown by Manolakis and Lefas that there are some station arrangements for which it is not possible to derive height estimation when the aircraft is at specific areas [11]. For example, in the case of four stations, when the quadrilateral defined by the stations is inscribed in a circle, it is not possible to estimate the height when the aircraft is above the center. Also, when the quadrilateral is a rectangle or symmetric trapezoid, it is not possible to derive a position fix when the aircraft is above the line that passes from the middle of the parallel sides. From a mathematical point of view, this singularity is expressed by the singularity of the Jacobian matrix; consequently, this matrix cannot be inverted as is required by the relevant position estimation algorithm. The algorithm in this case diverges from the actual height and finally collapses. Notice that height estimation is achieved everywhere except at this singular point. However, when the aircraft is close to the singular region,

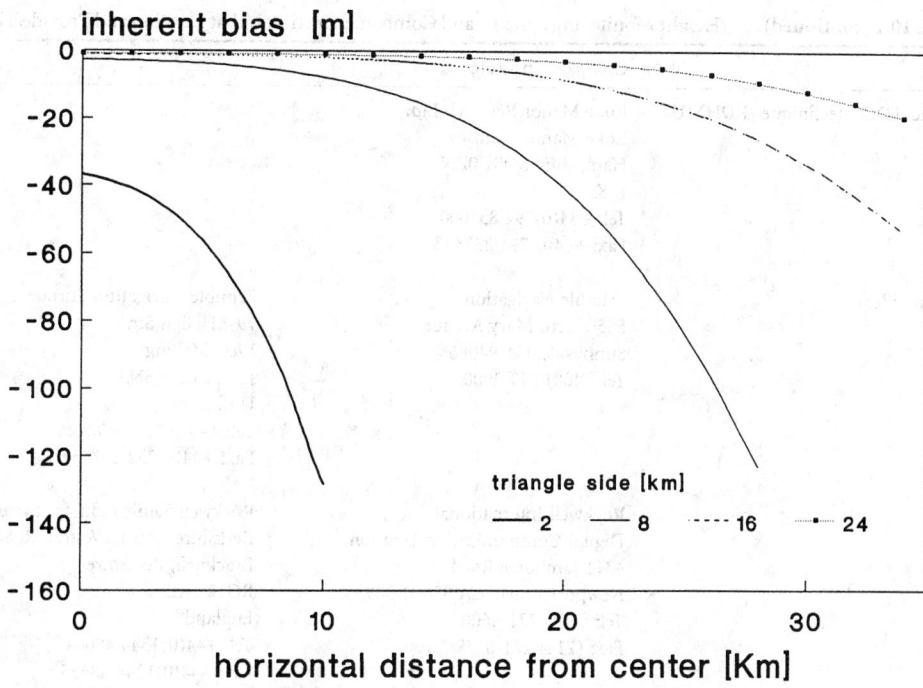

FIGURE 10.7 Bias inherently generated by the height estimation algorithm of the system based on DME measurements. The stations' sites form an equilateral triangle. The inherent bias is shown as function of the horizontal distance from the center for various magnitudes of the triangle side. The flight height is $z = 9$ km, and the SD of the distance measurement error is $\sigma_R = 90$ m.

although a position fix is achieved, it is not actually reliable since it is affected by large errors; for example, the VDOP could be larger than 600 in regions close to the singular region.

Table 10.1 presents the institutes and companies that either investigate and develop prototype height monitoring units or provide relevant systems in the market.

TABLE 10.1 Height Monitoring Systems and Companies/Institutes that Develop and Provide Them

System	Company/Institute
Navigation Accuracy Measurement System (NAMS)	Electronic Navigation Research Institute
	Ministry of Transport
	6-38-1 Shinkawa, Mitaka
	Tokyo, 181, Japan
	Tel: +81 422 413171
	Fax: 81-422-413176
DAMS height monitoring unit	National Aerospace Laboratory NLR
	Anthony Fokkerweg 2
	1059 CM Amsterdam
	The Netherlands
	Tel: +31 (0)20 511 3113
	Fax: +31 (0)20 511 3210
SSR and quadrilateration technique	GEC-Marconi Electronics Ltd.
	Marconi Research Laboratories
	West Hanningfield Road
	Great Baddow, Chelmsford
	Essex, England

TABLE 10.1 (continued)   Height Monitoring Systems and Companies/Institutes that Develop and Provide Them

System	Company/Institute	
SSR and TDOA technique (RDHMU)	Roke Manor Research Ltd. Roke Manor, Romsey Hampshire SO51 0ZN U.K. Tel: +44(0)794 833000 Fax: +44(0)794 833433	
GPS providers	Trimble Navigation 585 North Mary Avenue Sunnyvale, CA 94086 Tel: (408) 730-2900	Trimble Navigation Europe Ltd. 79-81 High Str. West Malling Kent ME19 6NA U.K. Tel: +44(0)732 849242 Fax: +44(0)732 847437
	Rockwell International Digital Communication Division 4311 Jamboree Road Newport Beach, CA 92660-3095 Tel: (714) 221-4600 Fax: (714) 221-6375	Rockwell Semiconductor Systems Berkshire Court, Western Road Bracknell, Berkshire RG12 1RE England Tel: +44(0)1344 48644 Fax: +44(0)1344 48655

References

1. S. Nagaoka, E. Yoshioka, and P. T. Muto, Radar estimation of the height of a cruising aircraft, *J. Navigation*, 32(3), 352-356, 1979.
2. S. Nagaoka, E. Yoshioka, and P. T. Muto, A simple radar for navigation accuracy measurements, *J. Navigation*, 34(3), 462-469, 1981.
3. S. Nagaoka, Possibility of detecting a non-level-flight aircraft by the navigation accuracy measurement system (NAMS), *ICAO, Review of the General Concept of Separation Panel, 7th Meeting*, Montreal, RGCSP-WP/180, 30/10/90.
4. S. Nagaoka, Height estimation of a cruising aircraft via a radar for air traffic control, *Electronics and Communications in Japan, Part 1 (Communications)*, USA, 71(11), 95-105, 1988.
5. D. E. Manolakis, C. C. Lefas, G. S. Stavrakakis, and C. M. Rekkas, Computation of aircraft geometric height under radar surveillance, *IEEE Trans. Aerosp. & Electr. Systems*, AES-28(1), 241-248, 1992.
6. D. E. Manolakis, Computation of aircraft geometric height under radar surveillance for non level flights, *Int. J. Systems Sci.*, 25(4), 619-627, 1994.
7. J. Brugman, J. Verpoorte, and A. J. L. Willekens, DAMS Height Monitoring Unit-Phase One, Report CR 92328 C, NLR, The Netherlands, 1992.
8. D. E. Rice, Height measurement by quadrilateration, *The Marconi Review*, XLVI, (228), 1-17, 1983.
9. S. Nagaoka, Possibility of geometric height measurement by using secondary surveillance radars, *ICAO, Review of the General Concept of Separation Panel, 7th Meeting*, Montreal, RGCSP-WP/181, 30/10/90.
10. D. E. Manolakis and C. C. Lefas, Aircraft geometric height computation using secondary surveillance radar range differences, *IEEE Proc.-Radar, Sonar and Navigation*, 141(2), 119-124, 1994.
11. D. E. Manolakis and C. C. Lefas, Station arrangement effects on ground referenced height computation by using time differences, *Navigation, J. Inst. Navigation*, 42(2), 409-420, 1995.
12. L. G. Hopkins, D. Sherry, and D. C. Rickard, Geometric Height Monitor Unit (HMU) Programme — Final Report on Phase 1, Proof of Principles, Roke Manor Research Ltd., Report No. 72/91/R161IU, Roke Manor, U.K.,1991.

13. C. M. Rekkas, C. C. Lefas, and N. J. Krikelis, Improving the accuracy of aircraft absolute aircraft altitude estimation using DME measurements, *Int. J. Systems Sci.*, 21(7), 1381-1392, 1990.

14. D. E. Manolakis, Efficient solution and performance analysis of 3-D position estimation by trilateration, *IEEE Trans. Aerosp. & Electr. Systems*, AES-32(4), 1239-1248, 1996.

15. D. E. Manolakis and A. I. Dounis, Advances in aircraft height computation using distance measuring equipment, *IEEE Proc.-Radar, Sonar and Navigation*, 143(1), 47-52, 1996.

16. J. L. Leva, Relationship between navigation vertical error, VDOP, and pseudorange error in GPS, *IEEE Trans. Aerosp. & Electr. Systems*, AES-30(4), 1138-1142, 1994.

17. B. W. Parkinson, History and operation of NAVSTAR, the Global Positioning System, *IEEE Trans. Aerosp. & Electr. Systems*, AES-30(4), 1145-1161, 1994.

18. D. E. Manolakis, C. C. Lefas, and A. I. Dounis, Inherent bias in height computation employing mixed type radar data, *IEEE Trans. Aerosp. & Electr. Systems*, AES-30(4), 1045-1049, 1994.

19. J. S. Abel and J. W. Chaffee, Existence and uniqueness of GPS solutions, *IEEE Trans. Aerosp. & Electr. Systems*, AES-27(6), 952-956, 1991.

10.2 Attitude Measurement

Mark A. Stedham, Partha P. Banerjee, Seiji Nishfuji, and Shogo Tanaka

In many practical situations, it is important to determine and measure the attitude of a particular vehicle, such as a ship, an airplane, a piece of mechanical equipment such as a crane lifter, or a spacecraft. For this reason, many attitude sensors have been developed with advanced computer and semiconductor technologies. This section first introduces the various attitude sensors with an explanation of their operating principles and then presents several methodologies for attitude measurement and determination, including ships and crane lifters, aircraft, and spacecraft applications.

Attitude Sensors for Ships, Aircraft, and Crane Lifters

There are many types of gyroscopes that, corresponding to the physical measurement mechanisms used, may be classified as two-axes *freedom gyro* and single-axis freedom gyro using precession, *vibratory gyro* using Coriolis' force, and *optic gyro* using Sagnac's effect. Among them, the two-axes freedom gyro has the longest history. It consists of a high-speed rotating rotor around a spin axis supported by two orthogonal axes. This type of gyro is generally classified as either a *free gyro*, a *vertical gyro* (VG), or a *directional gyro* (DG).

The single-axis freedom gyro has only one output axis in addition to the spin axis. Depending on the specifications (in which) the gyro is designed, there are two types of gyros, the *rate gyro and* the *rate integrating gyro.* Related to these rotating-type gyros is another type of gyro known as the *electrostatic gyro*, which makes use of a high-speed rotating sphere in a vacuum cavity. Because of its resistance-free property, the electrostatic gyro has the highest accuracy among existing gyros. There are also rotorless gyros. The first one is a vibratory gyro that uses Coriolis' force as the measurement principle. The second type is an optical one. Among optical gyros, there are two types: the *ring laser gyro* and the *fiber optic gyro*. Both rely on the Sagnac effect in their measurement mechanisms. The performance of gyros is evaluated by their drift rates, and the performance of various gyros is shown in Table 10.2, for reference, with their primary usages.

Recently, with the development of computer technology, many types of three-axes gyros have been developed that can measure not only the tilt angles but also the angular velocities and the accelerations along the three axes by combining several gyros and accelerometers. Accelerometers are often coupled with gyros to provide flight and ship navigation systems as well as attitude sensors for dynamic objects such as crane lifters. Examples include the *attitude and heading reference system* (AHRS), *inertial navigation system* (INS), *inertial measurement unit* (IMU), and *gyro compass* (GC), as well as the VGs and DGs discussed above [1].

TABLE 10.2   Performance of Different Types of Gyros

Type of the gyro	Degrees of freedom	Quantities to be detected	Accuracy (° h⁻¹)
Free gyro	2	Angle	1
Vertical gyro	2	Declination from horizontal plane	1
Directional gyro	2	Shift from reference direction	1
Rate gyro	1	Angular velocity	10
Rate integrating gyro	1	Angle	0.001–1
Ring laser gyro	1	Angular velocity	0.003
Fiber optic gyro	1	Angular velocity	0.01
Electrostatic gyro	2	Angle	0.00001–0.01

The principle of a servo-type accelerometer is explained below (see Figure 10.8). As soon as the shift of the beam caused by the acceleration α is detected by the deflection pickup, the current i is generated by the servo-amplifier, which produces a torque to keep the beam at the principle axis of the sensor. Since the torque and the current that generates the torque are proportional to α, the acceleration can be measured using the current. The measurement process forms a closed-loop system, so that the sensor is not only robust to disturbances, but also achieves a high measurement accuracy (see Table 10.3).

Similarly, an inclinometer is another inertial sensor that measures tilt angle to provide attitude information (see Figure 10.9). The principle of servo-type inclinometers is the same as that of the servo-type accelerometer, except that the beam in the accelerometer is replaced by a pendulum suspended from the supporting point in the sensor. When the sensor is placed on the inclined static surface of tilt angle β, the pendulum takes the angle β against the principle axis of the sensor, assuming the sensor has no force other than gravity acting on it. The sensor can, however, generate a torque $T_c = mg_l\sin\beta \cong mg_l\beta$ to keep the pendulum at the principle axis, then the tilt angle β can be accurately measured using the torque (and consequently the current producing the torque), where m and l are the pendulum mass and length of the pendulum to its mass center, respectively. One must note, however, that such a sensor is essentially designed to measure the tilt angles of static inclined surfaces. Thus, when applied to dynamic inclined surfaces, the accelerations will affect the torque, making the sensor unreliable. An intelligent attitude

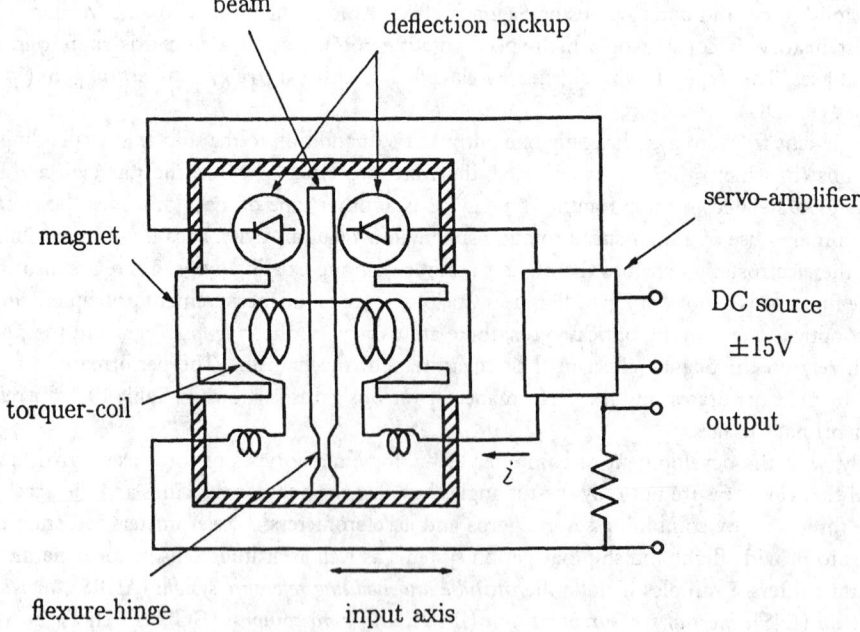

FIGURE 10.8   Servo-type accelerometer.

TABLE 10.3   Specification of a Servo-Type Accelerometer

Measurement range	±5 g
Resolution	Less than 5 μg (dc)
Sensitivity	2 V g^{-1}
Output resistance	560 Ω
Torquer current	3.5 mA g^{-1}
Case alignment	Less than ±1°
Frequency response	450 Hz (±3 dB)
Temperature range	−25 to +70°C
Power source	±15 V (dc)
Consumption current	Less than 15 mA
Size	28.4 mm × 24.5 mm
Mass	46 g (including the cable 10 g)

Note: g: gravitational acceleration (according to the type TA-25D-05 by TOKIMEC).

sensing system that overcomes such difficulty will be introduced later. Although application is limited to static inclined surfaces with minute tilt angles, a dielectric-type inclinometer employing electrodes and a bubble kept in an electrolyte can achieve high accuracies on the order of $10^{-4°}$.

Attitude Sensors for Spacecraft Applications

Attitude measurement for spacecraft usually requires two or more sensors for detecting the reference sources needed to satisfy attitude requirements. The choice of which sensors to employ is primarily influenced by the direction the spacecraft is usually pointing as well as the accuracy requirements for attitude determination [2]. Table 10.4 summarizes some performance parameters for these sensors as well as typical manufacturers.

Inertial measurement units generally consist of gyroscopes coupled with accelerometers, which together measure both rotational and translational motion. These IMUs may be either gimbal mounted (movement about a gimbal point, independent of the spacecraft) or a strapdown system (rigidly mounted to the spacecraft body), where expansive software is used to convert sensor outputs into reference frame measurements. IMUs tend to suffer gyro drift and other bias errors and, when used for spacecraft attitude measurements, are often used with one or more of the sensors discussed below.

Sun sensors detect the visible light from the sun, measuring the angle between the sun's radiation and the detector's photocell. The sun is a commonly chosen attitude reference source since it is by far the

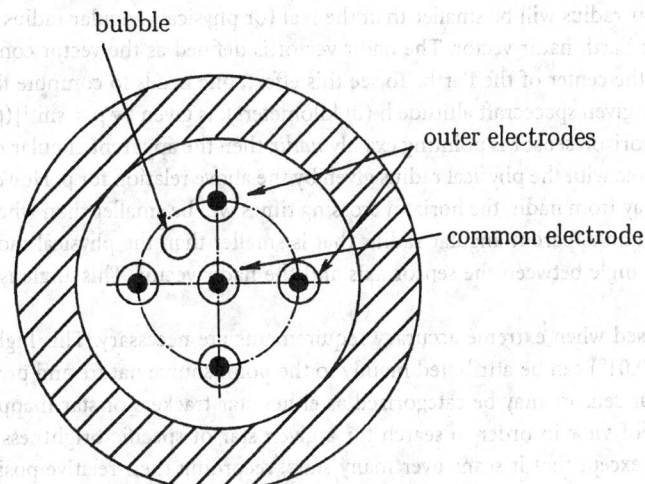

FIGURE 10.9   Dielectric-type inclinometer (front view).

TABLE 10.4   Spacecraft Attitude Determination Sensors

Sensor	Accuracy	Mass (kg)	Typical vendors
IMU	1 to 5×10^{-6} g	3 to 25	Northrop Grumman, Bendix, Kearfott, Honeywell, Hamilton, Standard, Litton, Teledyne
Sun Sensor	10^{-2} to 3°	0.5 to 2	Adcole, TRW, Ball Aerospace
Horizon Sensor	10^{-1} to 1°	2 to 5	Barnes, Ithaco, Lockheed Martin, Lockheed Barnes
Star Sensor	10^{-3} to 10^{-2}°	3 to 7	Ball Aerospace, Bendix, Honeywell, Hughes
Magnetometer	0.5 to 3°	~1	Schonstedt, Develco

Adapted from Larsen, W. J. and Wertz, J. R., Eds., *Space Mission Analysis and Design,* Torrance, CA: Microcosm Inc. and Dordrecht, The Netherlands: Kluwer Academic Publishers, 1992, p.360.

visually brightest object in the sky, having a total radiation per unit area of 1353 W m^{-2} at Earth distances [3]. Also, it is generally accepted as a valid point source for most attitude applications, having an angular radius of 0.25° at Earth distances. Increased measurement accuracy can be obtained by determining its centroid. Even though sun sensors are quite accurate (0.01° to 3.0°), they do require clear fields of view, and sometimes suffer periods of eclipse from both the Earth and the moon [4]. Also, sensitive equipment (such as imaging devices) must be protected from the powerful radiation of direct sunlight. When the sun is available, the angle between it and the sensor's primary axis is referred to as the *sun angle.*

For spacecraft in near-Earth orbits, the Earth is the second brightest object in the sky and covers as much as 40% of the sky. Earth *horizon sensors* detect the interface between the Earth's edge (or limb) and the space background. Horizon sensors can detect either of the Earth's visible limb (albedo sensor), infrared limb, or air glow. The infrared limb is the edge between the warm Earth and the cold space background. The air glow is a region of the atmosphere around the Earth that is visible to the spacecraft when it is on the night side of the Earth. Accuracies for horizon sensors are in the 0.1° to 1.0° range. Increased accuracy requires Earth oblate spheroid modeling [4]. Some problems associated with albedo detection include the distortion effects of the Earth's atmosphere, falsely identifying the day/night terminator crossing as the true Earth limb, and the considerable variability of the Earth's albedo in the visible spectrum (varies from land, sea, ice).

Most sensors used to detect the Earth's horizon are scanning sensors with narrow fields of view that measure the time between horizon crossings. In general, two horizon crossings occur per sensor scan period: one crossing when the sensor scans from the space background onto the Earth, followed by a second crossing when the sensor scans from the Earth back to space. The combination of horizon crossing times, scan rate, and spacecraft altitude allows for the computation of the Earth's apparent *angular radius.* The apparent angular radius will be smaller than the real (or physical) angular radius if the spacecraft is tilted away from the Earth nadir vector. The nadir vector is defined as the vector connecting the center of the spacecraft to the center of the Earth. To see this effect, one needs to compute the Earth's physical radius ρ, which for a given spacecraft altitude h (in kilometers), is given by $\rho = \sin^{-1}[(6371)/(6371 + h)]$.

If the spacecraft horizon sensor is pointing exactly *nadir,* then the apparent angular radius as measured by the sensor will agree with the physical radius given by the above relation for ρ. However, if the horizon sensor is pointed away from nadir, the horizon crossing times will be smaller than when pointing exactly nadir. This results in an apparent angular radius that is smaller than the physical radius by an amount proportional to the angle between the sensor axis and the nadir vector. This angle is referred to as the *nadir angle.*

Star sensors are used when extreme accuracy requirements are necessary. This high degree of sensor accuracy (0.003° to 0.01°) can be attributed mainly to the point source nature and precise fixed location of stars in space. Star sensors may be categorized as either star trackers or star mappers. A star tracker utilizes a wide field of view in order to search for a given star of specific brightness. A star mapper is similar to a tracker, except that it scans over many stars, recording their relative positions and angular separations. By comparing the recorded data with that from a *star catalog* (database), exact spacecraft orientation can be obtained. The angle between the star line-of-sight and the sensor's primary axis is referred to as the *star angle*.

The accuracy of star sensors is obtained with higher costs, however. Star sensors are generally heavier and consume more power than other types of attitude sensors. In addition, star sensors are quite sensitive to stray light sources such as sunlight reflected from the spacecraft or the Earth and sunlight scattered from dust particles and jet exhausts [4]. Most rely on optical shielding to reduce the effects of stray light.

Magnetic sensors (called *magnetometers*) measure both the magnitude and direction of the Earth's magnetic field. The difference in orientation between the measured field and the true field translates into attitude determination. Magnetometer accuracies (0.5° to 3.0°) are usually less than the other sensor types because of the uncertainty in the Earth's true field, which tends to change or shift over time. In addition, the Earth's magnetic field decreases with increasing altitude, and magnetometers are generally limited to altitudes of about 6000 km. For this reason, magnetometers are often used with one of the other sensor types already discussed for improved measurement accuracy [2].

Automatic On-Line Attitude Measurement for Ships and Crane Lifters

For on-line attitude measurement for ships and crane lifters, the first thing that comes to mind is to use gyros. However, because they often suffer from drifts, accurate attitude measurements might not be achieved using the gyros. Accordingly, one uses attitude on-line measurement systems that do not utilize gyros but servo-type accelerometers and inclinometers. The philosophy of the measurement systems introduced here is to make the best use of the system dynamics of the object and the sensors and to apply Kalman filters or adaptive filters to achieve high measurement accuracy.

Attitude Measurement for Ships

On-line accurate measurement of a ship's attitude is extremely important in exact search of the seabed patterns with sonars [5, 6]. It is also required by high-performance ships like hovercrafts from the viewpoint of suppressing swings by the waves. The measurement of a ship's attitude can usually be reduced to that of the heaving, rolling, and pitching of the ship. For such a measurement, a heave sensor has been used, whose output is given by double integration of the output of an accelerometer vertically directed with a gyroscope. However, since the initial values of heaving displacement and its velocity are unknown, the output will contain a bias that increases with time, and the accuracy of the sensor deteriorates considerably. From this viewpoint, one introduces a strapdown-type on-line measurement system that adequately processes the outputs of the two servo-type inclinometers and one accelerometer mounted on the ship [7].

Location of Sensors and Outputs
The two servo-type inclinometers and one servo-type accelerometer are located on the deck (at the point A) of vertical distance L from O, the intersection of rolling and pitching axes (see Figure 10.10).The two inclinometers are set in such a way that the rolling and the pitching angles are measured respectively. The accelerometer is set upward to the deck to obtain the information on the heaving. Because inclinometers were originally developed for the measurement of the tilt angles of static inclined surfaces, the rigid pendulum inside the sensor is considerably affected by the ship's acceleration other than the gravitational one. Applying Lagrange's equations of motion [8, 9] to rigid pendulums and calculating the torques to keep their deflections from the principal axes almost zero yields the sensor outputs [7]:

$$z_1(t) = \theta(t) - \frac{L}{g}\ddot{\theta}(t) + v_1(t)$$ (10.45)

$$z_2(t) = p(t) - \frac{L}{g}\ddot{p}(t) + v_2(t)$$ (10.46)

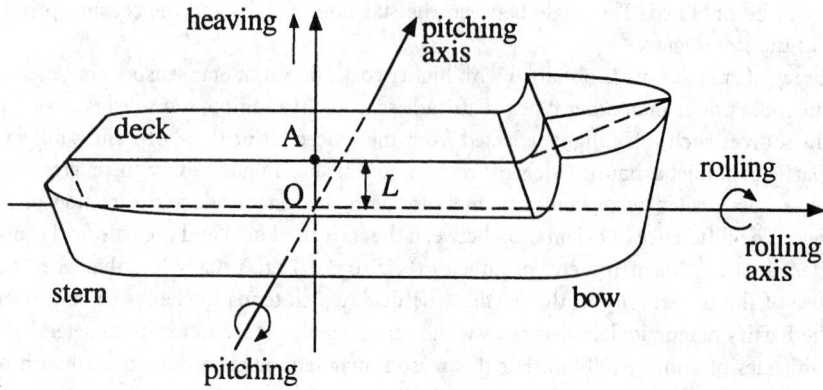

FIGURE 10.10 Location of sensors.

where $z_1(t)$, $z_2(t)$, $\theta(t)$, $p(t)$, and g denote, respectively, the outputs of the two inclinometers, the rolling and the pitching angles, and the gravitational acceleration ($v_1(t)$ and $v_2(t)$: noises of the outputs, including the approximation errors in deriving the outputs).

On the other hand, the accelerometer output is expressed as:

$$z_3(t) = \left(g + \alpha(t)\right)\cos\theta(t)\cos p(t) + v_3(t) \tag{10.47}$$

where $\alpha(t)$ and $v_3(t)$ represent, respectively, the heaving acceleration and the accelerometer noise.

Dynamics of Attitude Signals

It is well known that each of the heaving, rolling, and pitching in inshore seas has two dominant waves in a short interval. That is, a sinusoidal wave of long periodic length (in the range of 6 s to 10 s) and a sinusoidal wave of short periodic length (in the range of 2 s to 3 s) [10–12]. Thus, one model each of the signals in a short interval by a composite wave of the two dominant sinusoidal waves. For the heaving (in a short interval), the displacement is modeled by:

$$x(t) = a_1\sin(\omega_1 t + \varphi_1) + a_2\sin(\omega_2 t + \varphi_2) \tag{10.48}$$

with the parameters $\{a_i\}$, $\{\varphi_i\}$, and $\{\omega_i\}$ unknown. From the 4th-order differential equation satisfied by the $x(t)$, we obtain the linear dynamic equation [7]:

$$\dot{\mathbf{x}}(t) = A\mathbf{x}(t), \quad A \equiv \begin{bmatrix} 0 & 1 & 0 & 0 \\ 0 & 0 & 1 & 0 \\ 0 & 0 & 0 & 1 \\ -\omega_1^2\omega_2^2 & 0 & -(\omega_1^2 + \omega_2^2) & 0 \end{bmatrix} \tag{10.49}$$

where $\mathbf{x}(t) \equiv (x_1, x_2, x_3, x_4)^T$ ($x_n = d^{n-1}x/dt^{n-1}$ ($n = 1, \ldots, 4$)). On the other hand, the rolling and pitching angles can be modeled by:

$$x(t) = a_1\sin(\omega_1 t + \varphi_1) + a_2\sin(\omega_2 t + \varphi_2) + b \tag{10.50}$$

because there are usually some biases associated with them. From the 5th-order differential equation which Equation 10.50 satisfies, we get the similar state variable representation of the model as

Equation 10.49. In practice, the heaving, rolling, and pitching signals have many nondominant sinusoidal waves in addition to the dominant ones. Therefore, Equation 10.49 is modified by introducing a white Gaussian noise $w(t)$ with zero mean and adequate variance σ^2 as follows:

$$\dot{\mathbf{x}}(t) = A\mathbf{x}(t) + \Gamma w(t) \tag{10.51}$$

where $\Gamma = (0,1,0,0)^T$ for the heaving and $\Gamma = (0,1,0,0,0)^T$ for the rolling and pitching. The higher the order of the models, the better the measurement accuracy will be. If we consider the on-line measurement of the signals, Equation 10.51 will be sufficient.

On-Line Attitude Measurement
The observation Equations 10.45 and 10.46 are expressed using their own state vector $\mathbf{x}(t)$. The observation equations in a discretized form are:

$$y_k = H\mathbf{x}_k + v_k \tag{10.52}$$

where $H = [1,0,-L/g,0,0]$ and y_k, $\mathbf{x}_k$, and v_k, respectively, denote $y(t)$, $\mathbf{x}(t)$, and $v(t)$ of the corresponding signals at the k-th sampling instant [7, 9]. The discretized form of the dynamic Equation 10.51 is:

$$\mathbf{x}_{k+1} = F\mathbf{x}_k + \mathbf{w}_k \tag{10.53}$$

where

$$F \equiv \Phi(t)\Big|_{t=\Delta T}, \quad \Phi(t) \equiv L^{-1}\left\{(sI - A)^{-1}\right\}. \tag{10.54}$$

Here, L^{-1} and ΔT, respectively, denote the inverse Laplace transformation and the sampling period. The discretized transition noise $\mathbf{w}_k$ becomes a white Gaussian noise with zero mean and covariance:

$$W = \sigma^2 \int_0^{\Delta T} \Phi(\Delta T - \tau)\Gamma\Gamma^T\Phi^T(\Delta T - \tau)d\tau \tag{10.55}$$

The measurement of the rolling and pitching can thus be reduced to the state estimation of the linear discrete dynamic systems (Equations 10.52 and 10.53), if the angular frequencies ω_1 and ω_2 are given and v_k is assumed to have a white Gaussian property. The state estimation is achieved by a Kalman filter [7, 13]. However, difficulties in implementing the filter are that the exact values of the two angular frequencies are a priori unknown and also time variant. To overcome the difficulty, adequate candidates $\{(\omega_1^i, \omega_2^i); 1 \le i \le M\}$ for the parameters $\{\omega_1, \omega_2\}$ are set and a bank of Kalman filters is used. Then, the final estimate is obtained as the conditional expectation of the state estimate as follows:

$$\hat{\mathbf{x}}_{k/k}^0 \equiv \sum_{i=1}^{M} p_k^i \hat{\mathbf{x}}_{k/k}^i \tag{10.56}$$

where $\hat{\mathbf{x}}_{k/k}^i$ represents the state estimate $\hat{\mathbf{x}}_{k/k}$ for the i-th candidate $\Omega_i = (\omega_1^i, \omega_2^i)$, and p_k^i denotes the conditional posteriori probability of the i-th candidate calculated based on the Bayesian theorem:

$$p_k^i = \frac{p(y_k/\Omega_i, Y^{k-1})p_{k-1}^i}{\sum_{j=1}^{M} p_{k-1}^j p(y_k/\Omega_j, Y^{k-1})} \tag{10.57}$$

Here, $p(y_k/\Omega_i, Y^{k-1})$ represents the conditional Gaussian probability density function of y_k under Ω_i and $Y^{k-1} \equiv \{y_j; j \le k-1\}$, whose mean and variance are calculated recursively [7].

The proposed measurement system can adaptively and automatically select the most appropriate candidate versus time. It thus enables an accurate on-line measurement of the rolling and pitching whose dominant angular frequencies vary with time. The first, second, and third components of the final estimate $\hat{x}^o_{k/k}$ represent, respectively, the estimates of the displacement, velocity, and acceleration. The proposed system thus has an advantage in that it can measure not only the displacements, but also the velocities and the accelerations of the three signals. In order to improve the measurement accuracy of the rolling and pitching, one should place the inclinometers near the intersection O of the rolling and the pitching axes.

Finally, the dynamics of the heaving is given by Equation 10.51 similar to that of the rolling and pitching. Substituting the estimates $\hat{\theta}(t)$ and $\hat{p}(t)$ obtained above into Equation 10.47 and subtracting the effect of the gravitational acceleration, one can derive a linear observation equation for $\alpha(t)$:

$$y_k = \left[z_3(t) - g\cos\hat{\theta}(t)\cos\hat{p}(t) \right]\Big|_{t=k\Delta T} \tag{10.58}$$

$$= H_k \mathbf{x}_k + v_k$$

where $H_k = [0,0,\cos\hat{\theta}(t)\cos\hat{p}(t),0]\big|_{t=k\Delta T}$

$$v_k = v_3(k\Delta T)$$

Thus, the on-line measurement of the heaving is also realized by executing the same procedure as described before. The location of the rolling and pitching axes were assumed to be known; however, even when they are unknown, the attitude measurement system described above is effective, if we introduce the candidates on the location of the axes adding to the angular frequencies.

Attitude Measurement for Crane Lifters

Dynamics of Attitude Signals

An illustrative diagram of a crane lifter system is shown in Figure 10.11. One of the easiest ways to measure the attitude of the lifter is to set up a high-resolution camera on the bottom of the trolley and to track a mark on the top of the lifter. However, it increases the cost and also the difficulty in maintenance. Furthermore, sometimes the scheme does not work because of shadows and light reflection. As previously mentioned, for gyros not offering sufficiently accurate measurement, a high-sensitivity servo-type accelerometer is used to extract the attitude signals. When setting up the sensor on the lifter, however, there is a secondary swing signal adding to the primary one, due to the free suspension of the lifter and the structure of the lifter. Despite its small amplitude, the secondary one has a higher frequency and for this reason has a large magnitude on the sensor output. The important signal for practical applications, such as the attitude control of the lifter, is the primary one, which has a larger amplitude with a lower angular frequency of $\omega = \sqrt{g/\ell}$ (g: the gravitational acceleration; ℓ: the wire length from the primary supporting point to the center of gravity of the pulley). If we try to attenuate the secondary swing signal by passing the output through a low-pass filter, the phase lag is also introduced into the primary swing signal and the signal can no longer be used for the accurate attitude control of the lifter.

For the above reasons, we introduce an autonomous measurement system that measures both the primary and the secondary swings by modeling the lifter system with a double pendulum and applying a Kalman filter to it [14]. The dynamics of the trolley-lifter system is derived using Lagrange's equations of motion [8, 9].

$$\frac{d}{dt}\left(\frac{\partial T}{\partial \dot{x}}\right) - \frac{\partial T}{\partial x} + \frac{\partial V}{\partial x} = u - z\dot{x} \tag{10.59}$$

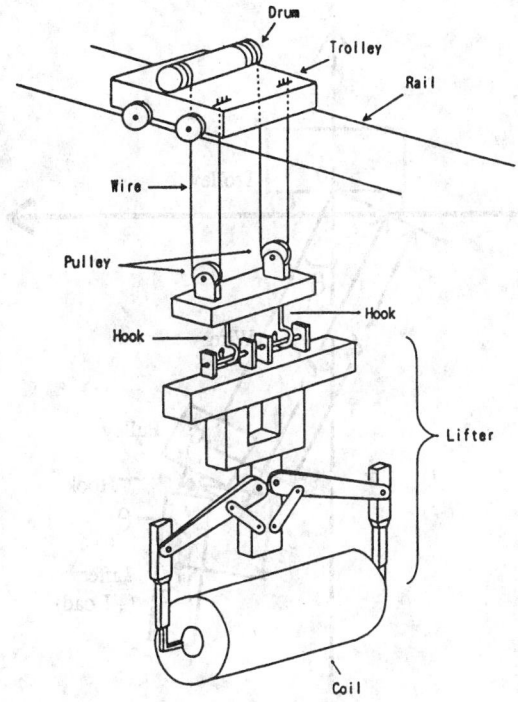

FIGURE 10.11 A crane lifter system.

$$\frac{d}{dt}\left(\frac{\partial T}{\partial \dot{\theta}_i}\right) - \frac{\partial T}{\partial \theta_i} + \frac{\partial V}{\partial \theta_i} = 0 \quad (i=1,2) \tag{10.60}$$

where T and V represent, respectively, the kinetic and the potential energies of the trolley-lifter system, and θ_1, θ_2 denote, respectively, the angles that the primary and the secondary pendulums take against the vertical line. The other variables x, u, and a represent, respectively, the location of the trolley, the driving force, and the coefficient of friction between the trolley and the rail. Considering that θ_i, $\dot{\theta}_i$ ($1 \le i \le 2$) are small, the dynamic equation of the trolley-lifter system can be expressed as [14]:

$$\dot{\mathbf{x}}(t) = A\mathbf{x}(t) + \mathbf{b}u(t) \tag{10.61}$$

where $\mathbf{x}(t)$ is the state vector $\mathbf{x}(t) = x, \dot{x}, \theta_1, \dot{\theta}_1, \theta_2, \dot{\theta}_2)^T$. Taking into account the approximation errors in deriving Equation 10.61, air resistance, friction in the wires, and microscopic swings at the other connection points, it is reasonable to introduce white Gaussian noises $w(t)$ ($1 \le i \le 3$) with zero mean and appropriate variances to the dynamic Equation 10.61 as in Equation 10.49 as follows [14]:

$$\dot{\mathbf{x}}(t) = A\mathbf{x}(t) + \mathbf{b}u(t) + \Gamma\mathbf{w}(t) \tag{10.62}$$

where

$$\Gamma = \begin{bmatrix} 0 & 1 & 0 & 0 & 0 & 0 \\ 0 & 0 & 0 & 1 & 0 & 0 \\ 0 & 0 & 0 & 0 & 0 & 1 \end{bmatrix}^T, \quad \omega(t) = \begin{bmatrix} \omega_1(t), & \omega_2(t), & \omega_3(t) \end{bmatrix}^T \tag{10.62a}$$

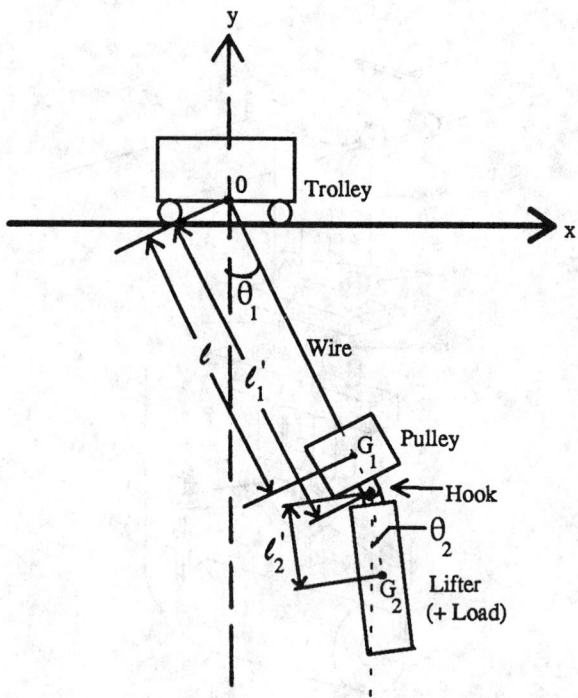

FIGURE 10.12 Dynamics of a trolley lifter system.

Sensor Outputs and On-Line Attitude Measurement
When a servo-type accelerometer is set up on the lifter (in the direction of the swing) at the place of the distance ℓ_2' from the secondary supporting point, the output of the sensor becomes [14]:

$$y(t) \cong -\ddot{x} - \ell_1'\ddot{\theta}_1 - \ell_2'\ddot{\theta}_2 - g\theta_2 \qquad (10.63)$$

where ℓ_1' is the distance between the primary and the secondary supporting points (see Figure 10.12). Substitution of Equation 10.61 into Equation 10.63 yields an output expressed in terms of the state vector $\mathbf{x}(t)$, as in Equation 10.52. Using a rotary-encoder to measure the location and the velocity of the trolley, and then combining these three sensor outputs with the dynamic Equation 10.62 and applying a Kalman filter enables the state vector to be estimated on-line. Using this approach, both angular displacement and velocity of the deflections θ_1, θ_2 of the two pendulums can be measured exactly.

Aircraft Attitude Determination

The determination of aircraft attitude requires the measurement of angles about three independent body axes. These angles are the roll, pitch, and yaw angles. There are two primary means employed today for measuring these angles; the first method uses VGs to measure the roll and pitch angles, and a DG to measure the yaw angle. The second method, more commonly used today, employs an IMU for full three-axis attitude determination coupled with a baro-altimeter to correct for vertical drift errors in the IMU. Both methods are described below.

Vertical and Directional Gyro Analysis

A VG is a two degree-of-freedom gyro with its spin axis mounted nominally vertical. It employs two specific force sensors mounted nominally horizontal on the inner gimbal. The two angles measured by the VG — roll and pitch — require nearly identical analyses [1]. Consider the situation shown in

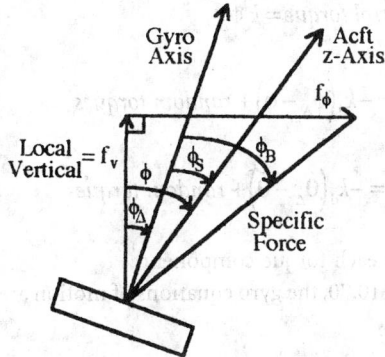

FIGURE 10.13   Vertical gyro analysis.

Figure 10.13, depicting an aircraft with a roll angle of ϕ with respect to the local vertical. The sensed roll angle ϕ_s is given by the difference in the actual roll angle and the gyro roll drift error ϕ_Δ:

$$\phi_s = \phi - \phi_\Delta \tag{10.64}$$

In order to compensate for this drift error, gyros employ a specific force sensor such as an electrolytic bubble device, which senses drift error. This correction device senses the angular difference between the specific force vector **f** acting on the aircraft roll axis and the gyro axis, as shown in Figure 10.13. Thus,

$$\phi_B = \tan^{-1}\left[\left(f_\phi / f_v\right) - \phi_\Delta\right] \cong \left(f_\phi / f_v\right) - \phi_\Delta \tag{10.65}$$

where f_ϕ is the side horizontal component of **f** and f_v = force of gravity is the vertical component. A similar analysis for the pitch angle θ yields:

$$\theta_s = \theta - \theta_\Delta \tag{10.66}$$

$$\theta_B = \tan^{-1}\left[f_\theta / f_v - \theta_\Delta\right] \cong f_\theta / f_v - \theta_\Delta \tag{10.67}$$

where f_θ is the back horizontal component of **f**. Next, define the gyro angular momentum vector by:

$$\mathbf{H}_{VG} = \left[J_x \dot{\phi}_\Delta, J_y \dot{\theta}_\Delta, -h\right] \tag{10.68}$$

where J_x and J_y are the sensor moments of inertia and h is the gyro spin angular momentum. In addition, define the inner gimbal axes angular velocity vector as:

$$\omega_{VG} = \left[\dot{\phi}_\Delta, \dot{\theta}_\Delta, 0\right] \tag{10.69}$$

Finally, define the gimbal torque vector by:

$$Q_{VG} = \left[Q_{cx} + Q_{dx}, Q_{cy} + Q_{dy}, 0\right] \tag{10.70}$$

where

$$Q_{cx} = \textit{gimbal roll control torque} = -k_c \theta_B \tag{10.71a}$$

$$Q_{cy} = \text{gimbal pitch control torque} = k_c \phi_B \qquad (10.71\text{b})$$

$$Q_{dx} = \text{gimbal roll disturbance torque} = -k_d\left(\dot{\phi}_\Delta - \dot{\phi}\right) + \text{random torques} \qquad (10.71\text{c})$$

$$Q_{dy} = \text{gimbal pitch disturbance torque} = -k_d\left(\dot{\theta}_\Delta - \dot{\theta}\right) + \text{random torques} \qquad (10.71\text{d})$$

and the k_c and k_d are constant scaling factors related to each torque component.

Using the vectors defined in Equations 10.68 through 10.70, the gyro equations of motion are given by:

$$\frac{\partial}{\partial t}\left(H_{VG}\right) + \left(\omega_{VG} \times H_{VG}\right) = Q_{VG} \qquad (10.72)$$

Taking the Laplace transform of the expansion of Equation 10.72, with the assumption that $J_x \cong J_y = J$, yields the following gyro equations of motion in the Laplace domain:

$$\begin{bmatrix} J_x s^2 + k_d s & -\left(hs + k_c\right) \\ hs + k_c & J_y s^2 + k_d s \end{bmatrix}\begin{bmatrix} \phi_\Delta(s) \\ \theta_\Delta(s) \end{bmatrix} \cong \begin{bmatrix} -k_c\theta_B + k_d s\phi(s) + \text{random torques} \\ \left(k_c/g\right)f_\phi(s) + k_d s\theta(s) + \text{random torques} \end{bmatrix} \qquad (10.73)$$

For normal gyro operation, $J_x \cong J_y \cong 0$ and $k_d/h \ll 1$; so these factors may be ignored in Equation 10.73. Thus, solving for the desired roll and pitch angles under these assumptions gives [1]:

$$\phi_s = \begin{cases} \phi & \omega \gg k_c/h \\ \phi - f_\phi/g & \omega \ll k_c/h \end{cases} \qquad (10.74)$$

$$\theta_s = \begin{cases} \theta & \omega \gg k_c/h \\ \theta - f_\theta/g & \omega \ll k_c/h \end{cases} \qquad (10.75)$$

A DG is a two degree-of-freedom gyro with its spin axis mounted nominally horizontal and pointing in the direction of magnetic north. It employs a single specific force sensor mounted on the inner gimbal [1]. The DG measures the third required aircraft angle, yaw, generally denoted by ψ. The sensed yaw angle ψ_s is given by the difference in the actual yaw angle ψ (angle between the aircraft z-axis and true north) and the gyro heading angle drift error ψ_Δ (angle between the gyro axis and true north):

$$\psi_s = \psi - \psi_\Delta \qquad (10.76)$$

Define the gyro angular momentum vector by:

$$H_{DG} = \left[J_y\dot{\theta}_\Delta, \; J_z\dot{\psi}_\Delta, \; -h\right] \qquad (10.77)$$

and the inner gimbal axes angular velocity vector as:

$$\omega_{DG} = \left[\dot{\theta}_\Delta, \; \dot{\psi}_\Delta, \; 0\right] \qquad (10.78)$$

and the gimbal torque vector as

$$Q_{DG} = \left[Q_{cy} + Q_{dy}, \, Q_{cz} + Q_{dz}, \, 0 \right] \tag{10.79}$$

Here, the torque vector components are given by:

$$Q_{cy} = k_c \left(M_\Delta - \psi_\Delta \right) \tag{10.80a}$$

$$Q_{cz} = -k_c \theta_B \tag{10.80b}$$

$$Q_{dy} = -k_d \left(\dot{\theta}_\Delta - \dot{\theta} \right) + random\ torques \tag{10.80c}$$

$$Q_{dz} = -k_d \left(\dot{\psi}_\Delta - \dot{\psi} \right) + random\ torques \tag{10.80d}$$

where M_Δ = magnetic compass heading error (from true north). Therefore, the DG equations of motion are given in Laplace domain as:

$$\begin{bmatrix} J_y s^2 + k_d s & -\left(hs + k_c \right) \\ hs + k_c & J_z s^2 + k_d s \end{bmatrix} \begin{bmatrix} \theta_\Delta(s) \\ \psi_\Delta(s) \end{bmatrix} \cong \begin{bmatrix} k_c M_\Delta(s) + k_d s \theta(s) + random\ torques \\ -k_c \theta_B + k_d s \psi(s) + random\ torques \end{bmatrix} \tag{10.81}$$

The desired yaw angle measurement for the DG is thus given as [1]:

$$\psi_s = \begin{cases} \psi & \omega \gg k_c / h \\ \psi - M_\Delta & \omega \ll k_c / h \end{cases} \tag{10.82}$$

As indicated in Table 10.2, the accuracies of both VGs and DGs are approximately 1°. An improvement of over 2 orders of magnitude can be obtained through the use of inertial measurement units, which are described next.

Inertial Measurement Units (IMUs)

Inertial measurement units consist of gyroscopes and accelerometers that together provide full three-axis attitude measurements. Most are mounted on stable gimbaled platforms that remain locally horizontal via torquing devices. An IMU aboard an aircraft cannot measure exactly the local vertical due to the fact that the specific force acting on the aircraft has a horizontal component due to vehicle motion. In addition, since the vehicle is moving with respect to the inertial reference frame, the Earth's magnetic pole cannot be determined precisely [1].

These problems (errors) are minimized by aligning the IMU to be exactly horizontal and north pointing while the aircraft is stationary. Once platform motion begins, the IMU may be constantly realigned by sensing changes in the direction of vertical and north, and then applying appropriate torques to the platform to keep it properly aligned. This realignment is accomplished by integrating the two orthogonal accelerometer outputs to determine the components of horizontal velocity. This data, combined with the Earth's rotation rate, yields the desired rates of change in local vertical and true north at the vehicle's current latitude and longitude. Performing a second integration of the sensor outputs yields an estimate of relative position.

Analysis in Bryson et al. [1], has shown that the pitch angle (variation in platform horizontal position) is given by the IMU sensor output as:

$$\theta(t) = \frac{-\varepsilon}{\omega_s}\sin(\omega_s t) - \frac{b}{g} \qquad (10.83)$$

where ε = gyro drift rate error, b = specific force sensor error, and $\omega_s \equiv$ Schuler frequency = $\sqrt{g/R}$, [g = force of gravity, R = Earth's radius]. Thus, the platform root-mean-square pitch angle becomes:

$$\theta_{rms} = \left[\frac{1}{2}\left(\frac{\varepsilon}{\omega_s}\right)^2 + \left(\frac{b}{g}\right)^2\right]^{\frac{1}{2}} \qquad (10.84)$$

Using typical values for ε ($\cong 0.015° \, h^{-1}$), ω_s ($\cong 0.71° \, h^{-1}$), and b ($\cong 0.01$) yields an rms pitch angle error of $\theta_{rms} = 0.01°$. Thus, it is apparent that under normal operating conditions the IMU provides a two orders-of-magnitude improvement in sensor accuracy when compared to the VG and DG.

Spacecraft Attitude Determination

Most spacecraft attitude determination techniques rely upon finding the orientation of a single axis in space (e.g., the spacecraft z-axis) plus the spacecraft rotation about this axis. This provides a full three-axis attitude solution. In order to achieve this, reference sources that are external to the spacecraft must be used. Specifically, full three-axis spacecraft attitude determination requires at least two external vector measurements. Commonly used reference sources for these external vector measurements include the sun, Earth, moon, stars, planets, and the Earth's magnetic field. In addition, IMUs are also used to provide the necessary attitude measurements.

Attitude Determination Methodology

The first step in attitude determination is to determine the angles between the spacecraft's primary axis and the two (or more) attitude reference sources. For example, suppose a particular spacecraft is using the sun and the Earth for attitude reference. The two angles in this case are referred to as the sun angle β_S and the nadir angle Γ_N. Since the orientation of even a single spacecraft axis is unknown at this point, these angles establish two *cones* along which the attitude vector $\mathbf{A}$ must lie. Since the attitude vector must lie on both cones, it must lie along the intersection between the two cones [4] (See Figure 10.14). The two vectors, notably $\mathbf{A}_1$ and $\mathbf{A}_2$, resulting from the intersection of these two cones may be determined by the following method derived by Grubin [15]. Let $\mathbf{S}$ represent the sun vector, $\mathbf{E}$ the spacecraft nadir vector, and $\mathbf{A}$ the desired attitude vector, each defined in Cartesian space as follows:

$$\mathbf{S} = \left(S_x, S_y, S_z\right) \qquad (10.85)$$

$$\mathbf{E} = \left(E_x, E_y, E_z\right) \qquad (10.86)$$

$$\mathbf{A} = \left(A_x, A_y, A_z\right) \qquad (10.87)$$

Let the vectors $\mathbf{S}$, $\mathbf{E}$, and $\mathbf{N}$ define a set of base unit vectors with:

$$\mathbf{N} = \frac{\mathbf{S} \times \mathbf{E}}{|\mathbf{S} \times \mathbf{E}|} = \left(N_x, N_y, N_z\right) \qquad (10.88)$$

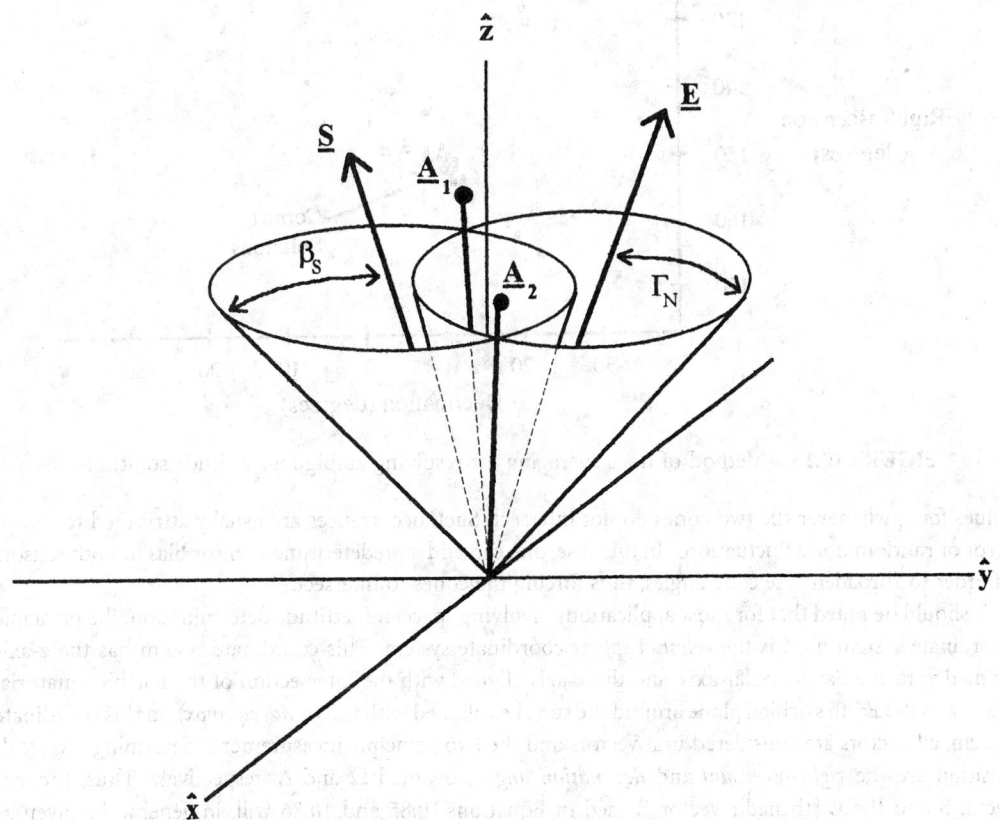

FIGURE 10.14 Relationship between reference vectors and single-axis attitude cones.

If we introduce a proper set of scaling factors as follows:

$$I_x = \frac{\left[\cos\beta_S - \left(\mathbf{S}\cdot\mathbf{E}\right)\cos\Gamma_N\right]}{1-\left(\mathbf{S}\cdot\mathbf{E}\right)^2} \tag{10.89a}$$

$$I_y = \frac{\left[\cos\Gamma_N - \left(\mathbf{S}\cdot\mathbf{E}\right)\cos\beta_S\right]}{1-\left(\mathbf{S}\cdot\mathbf{E}\right)^2} \tag{10.89b}$$

$$I_z = \sqrt{1 - I_x\cos\beta_S - I_y\cos\Gamma_N} \tag{10.89c}$$

then the two possible attitude vectors $\mathbf{A}_1$ and $\mathbf{A}_2$ are found to be:

$$\mathbf{A}_{1,2} = \left[\left(I_x S_x + I_y E_y \pm I_z N_x\right), \left(I_x S_y + I_y E_y \pm I_z N_y\right), \left(I_x S_z + I_y E_z \pm I_z N_z\right)\right] \tag{10.90}$$

In Equations 10.88 through 10.90, $\mathbf{S} \times \mathbf{E}$ represents the Cartesian vector product, and $\mathbf{S} \cdot \mathbf{E}$ represents the Cartesian scalar product. The radicand in Equation 10.89c may be negative, thus producing imaginary

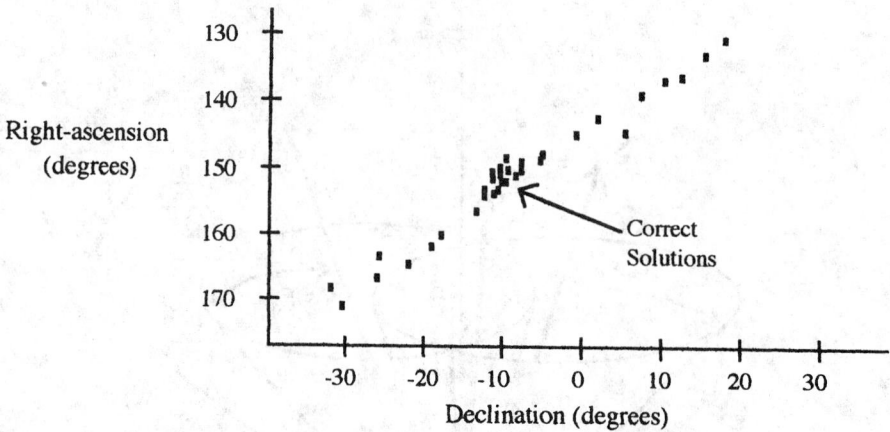

FIGURE 10.15   Method of trace averaging for resolving ambiguous attitude solutions.

values for I_z whenever the two cones do not intersect. Such occurrences are usually attributed to sensor error or random noise fluctuations. In this case, one can add a predetermined sensor bias to both sensors in order to "broaden" the cone angles, thus forcing the cones to intersect.

It should be noted that for most applications involving spacecraft attitude determination, the principle coordinate system used is the *celestial sphere* coordinate system. This coordinate system has the z-axis aligned with the Earth's polar axis, and the x-axis aligned with the intersection of the Earth's equatorial plane and the Earth's orbital plane around the sun (i.e., aligned with the *vernal equinox*). In this coordinate system, all vectors are considered unit vectors and the two principle measurements describing a vector's position are the *right-ascension* and *declination angles*, denoted Ω and Δ, respectively. Thus, the sun vector **S** and the Earth nadir vector **E** used in Equations 10.85 and 10.86 will, in general, be given as right-ascension and declination angles that can be converted to Cartesian coordinates via the following set of transformations:

$$x = \cos(\Omega)\cos(\Delta); \quad y = \sin(\Omega)\cos(\Delta); \quad z = \sin(\Delta) \tag{10.91a}$$

$$\Omega = \tan^{-1}(y/x); \quad \Delta = \sin^{-1}(z) \tag{10.91b}$$

The final step in measuring three-axis attitude is to determine which attitude solution is correct, $\mathbf{A}_1$ or $\mathbf{A}_2$, and then measure the rotation about this axis. The two ambiguous attitude solutions may be resolved by comparison with a priori attitude information, if available, or through the use of *trace averaging* [4]. Trace averaging is a method of plotting each attitude solution on a right-ascension versus declination plot and choosing the area of greatest concentration as the correct solution, as demonstrated in Figure 10.15. Since the attitude is assumed to change more slowly than the attitude sensor's sample rate, over short time intervals the data for the correct solution usually form a "cluster" near the correct attitude; the data for the incorrect solution are usually much more scattered.

Once the correct attitude vector has been obtained, the orientation of the remaining two orthogonal axes may be found by measuring the rotation, or phase angle, of the spacecraft about the preferred axis. Any sensor measurement that provides this phase angle may be used. An example of this technique is provided by the panoramic annular lens attitude determination system (PALADS), described in the next section. This imaging system uses a unique "three-dimensional" lens that provides simultaneous detection of two (or more) reference sources [16]. This information, combined with the orientation of the single axis, uniquely determines three-axis attitude.

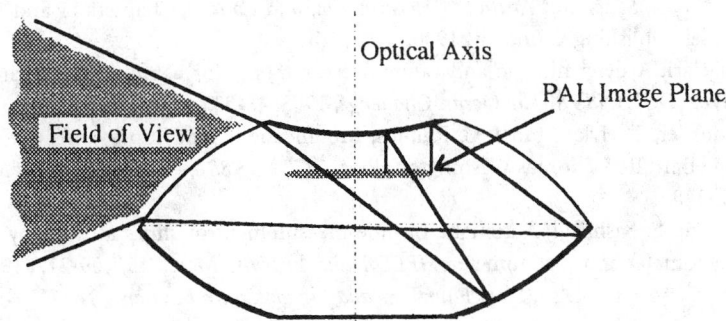

FIGURE 10.16 Panoramic annular lens ray diagram.

The three angles derived above, which are commonly referred to as *Euler angles*, define the orientation of the three spacecraft axes with respect to the chosen reference frame. A more formal treatment of the attitude solution usually requires specifying the components of the 3×3 attitude matrix **A**. Each component of the attitude matrix defines the angular relationship between a given spacecraft axis and a reference frame axis. Various methods exist for computing the attitude matrix **A** (see [4]); the preferred method depends on the particular application at hand.

PALADS

The primary component of PALADS is the panoramic annular lens (PAL), a single-element lens made from a high index of refraction glass with certain portions of the lens coated with a mirrored surface. Hence, it relies on both refraction and reflection in forming an image (Figure 10.16). The lens is unique in that it images a three-dimensional object field onto a two-dimensional image plane, whereas a "normal" lens is capable of only imaging two-dimensional object space onto an image plane. The combination of high index of refraction glass and mirrored surfaces provides the PAL with a field of view extending from approximately 65° to 110° from the optical axis. This 45° field of view covers the entire 360° surrounding the optical axis [17]. Any ray originating from outside the 45° field of view will not form a part of the image. The PAL may be attached to any high-quality imaging system using an appropriate *transfer lens*. As currently configured, the PALADS imaging system utilizes a Sony XC-73 charged-couple device (CCD), a black and white video camera coupled to the PAL via a Nikon *f*/1.4 transfer lens.

The hemispherical view provided by PALADS allows for single-sensor detection of multiple attitude reference sources, such as the Earth and the sun or moon. The position of each reference source in the image plane translates into a unique azimuth elevation angle between the PAL's optical axis and the reference source. Since the PAL has a 360° field of view surrounding the optical axis, it may detect several reference sources simultaneously. The data points associated with each source are extracted from the image plane using digital image processing techniques. Thus, it is easy to see how a single image from PALADS (containing two or more reference sources) provides the necessary angle data to determine three-axis spacecraft attitude.

References

1. A. E. Bryson, *Control of Spacecraft and Aircraft*, Chapter 10, Princeton, NJ: Princeton University Press, 1994.
2. W. J. Larson and J. R. Wertz (eds.), *Space Mission Analysis and Design*, Chapter 11, Torrance, CA: Microcosm Inc. and Dordrecht, The Netherlands: Kluwer Academic Publishers, 1992.
3. NASA Technical Memorandum NASA TM X-64757, *Terrestrial Environment (Climatic) Criteria Guidelines for Use in Aerospace Vehicle Development (1973 Revision)*, Marshall Space Flight Center, AL, 1973.

4. J. R. Wertz (ed.), *Spacecraft Attitude Determination and Control*, Chapters 11 and 12, The Netherlands: Reidel Publishing Company, 1980.
5. R. D. Angelari, A deterministic and random error model for a multibeam hydrographic sonar system, *Proc. OCEANS'78. The Ocean Challenge*, 1978, 48-53.
6. C. de Moustier, T. Hylas, and J. C. Phillips, Modifications and improvements to the Sea Beam system on board R/V Thomas Washington, *Proc. OCEANS'88 — A Partnership of Marine Interests*, 1988, 372-378.
7. S. Tanaka and S. Nishifuji, Automatic on-line measurement of ship's attitude by use of a servo-type accelerometer and inclinometers, *IEEE Trans. Instrum. Meas.*, 45, 209-217, 1996.
8. D. G. Shultz and J. L. Melsa, *State Functions and Linear Control Systems*, New York: McGraw-Hill, 1967.
9. Y. Takahashi, M. J. Rabins, and D. M. Auslander, *Control and Dynamic Systems*, Reading, MA: Addison-Wesley, 1971.
10. S. Tanaka and S. Nishifuji, On-line sensing system of dynamic ship's attitude by use of servo-type accelerometers, *IEEE J. Oceanic Eng.*, 20, 339-346, 1995.
11. S. Tanaka, On automatic attitude measurement system for ships using servo-type accelerometers (in Japanese), *Trans. SICE*, 27, 861-869, 1991.
12. D. E. Cartwright and M. S. Longuet-Higgins, The statistical distribution of the maxima of a random function, *Proc. Roy. Soc. London*, Ser. A, 237, 212-232, 1956.
13. R. E. Kalman, A new approach to linear filtering and prediction problems, *Trans. ASME, J. Basic Eng.*, 82, 35-45, 1960.
14. S. Tanaka, S. Kouno, and H. Hayashi, Automatic measurement and control of attitude for crane lifters (in Japanese), *Trans. SICE*, 32(1), 97-105, 1996.
15. C. Grubin, Simple algorithm for intersecting two conical surfaces, *J. Spacecraft Rockets*, 14(4), 251-252, 1977.
16. M. A. Stedham and P. P. Banerjee, The panoramic annular lens attitude determination system, *SPIE Proceedings, Space Guidance, Control, and Tracking II*, Orlando, FL, 17-18 April, 1995.
17. J. A. Gilbert, D. R. Matthys, and P. Greguss, Optical measurements through panoramic imaging systems, *Proc. Int. Conf. Hologram Interferometry and Speckle Metrology*, Baltimore, MD, November 4-7, 1990.

10.3 Inertial Navigation

Halit Eren and C. C. Fung

The Principles

The original meaning of the word *navigation* is "ship driving." In ancient times when sailing boats were used, navigation was a process of steering the ship in accordance with some means of directional information, and adjusting the sails to control the speed of the boat. The objective was to bring the vessel from location A to location B safely. At present, navigation is a combination of science and technology. No longer is the term limited to the control of a ship on the sea surface; it is applied to land, air, sea surface, underwater, and space.

The concept of inertial-navigator mechanization was first suggested by Schuler in Germany in 1923. His suggested navigation system was based on an Earth-radius pendulum. However, the first inertial guidance system based on acceleration was suggested by Boykow in 1938. The German A-4 rocket, toward the end of World War II, used an inertial guidance system based on flight-instrument type gyroscopes for attitude control and stabilization. In this system, body-mounted gyro-pendulum-integrating accelerometers were used to determine the velocity along the trajectory. The first fully operational inertial auto-navigator system in the U.S. was the XN-1 developed in 1950 to guide C-47 rocket. Presently, inertial navigation systems are well developed theoretically and technologically. They find diverse applications,

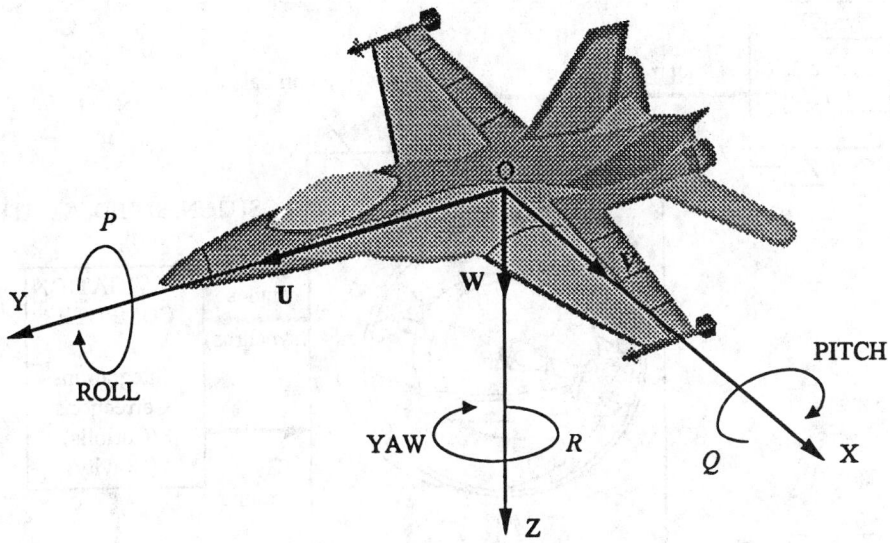

FIGURE 10.17 In inertial navigation, the movement of a vehicle, rocket, ship, aircraft, robot, etc. with respect to a reference axis is monitored. On the Earth's surface, the conventional reference is the Earth's fixed axes — North, East, and Down. A vehicle such as an aircraft or a marine vessel will have its own local axes, known as roll, pitch, and yaw.

allowing the choice of appropriate navigation devices, depending on cost, accuracy, human interface, global coverage, time delay, autonomy, etc.

Inertial navigation is a technique using a self-contained system to measure a vehicle's movement and determine how far it has moved from its starting point. Acceleration is a vector quantity involving magnitude and direction. A single accelerometer measures magnitude but not direction. Typically, it measures the component of accleration along a predetermined line or direction. The direction information is usually supplied by gyroscopes that provide a reference frame for the accelerometers. Unlike other positional methods that rely on external references, an *inertial navigation system* (INS) is compact and self-contained, as it is not required to communicate to any other stations or other references. This property enables the craft to navigate in an unknown territory.

Inertial navigation can be described as a process of directing the movement of a vehicle, rocket, ship, aircraft, robot, etc., from one point to another with respect to a reference axis. The vehicle's current position can be determined from "dead reckoning" with respect to a known initial starting reference position. On the Earth's surface, the conventional reference will be North, East, and Down. This is referred to as the *Earth's fixed axes*. A vehicle such as an aircraft or a marine vessel will have its own *local axes*: roll, pitch, and yaw, as shown in Figure 10.17.

The inertial sensors of the INS can be mounted in such a way that they stay leveled and pointing in a fixed direction. This system relies on a set of gimbals and sensors attached on three axes to monitor the angles at all times. This type of INS is based on a *navigational platform*. A sketch of a three-axis platform is shown in Figure 10.18. Another type of INS is the *strapdown system* that eliminates the use of gimbals. In this case, the gyros and accelerometers are mounted to the structure of the vehicle. The measurements received are made in reference to the local axes of roll, pitch, and yaw. The gyros measure the movement of angles in the three axes in a short time interval (e.g., 100 samples per second). The computer then uses this information to resolve the accelerometer outputs into the navigation axes. A schematic block diagram of the strapdown system is shown in Figure 10.19.

The controlling action is based on the sensing components of acceleration of the vehicle in known spatial directions, by instruments which mechanize Newtonian laws of motion. The first and second integration of the sensed acceleration determine velocity and position, respectively. A typical INS includes

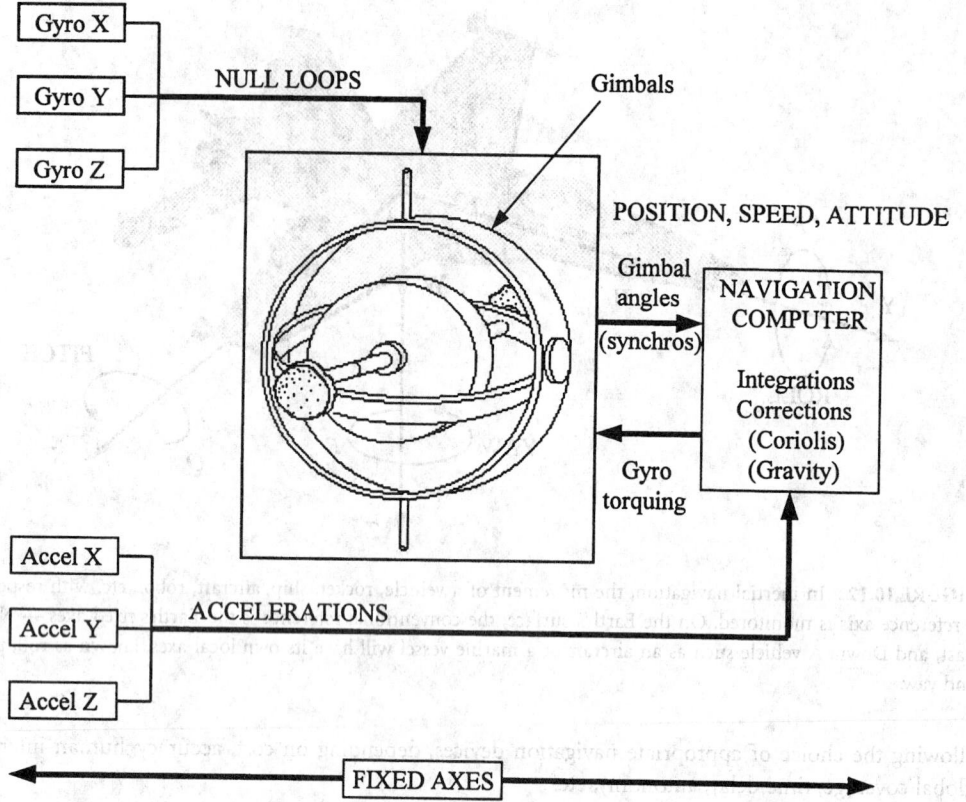

FIGURE 10.18 Some Inertial Navigation Systems, INS, are based on a navigational platform. The inertial sensors are mounted in such a way they can stay leveled at all times, pointing in a fixed direction. This system uses a set of gimbals and sensors attached on three axis in the *x*, *y*, and *z* directions to monitor the angles and accelerations constantly. The navigation computer makes corrections for coriolis, gravity, and other effects.

a set of gyros, a set of accelerometers, and appropriate signal processing units. Although the principle of the systems may be simple, the fabrication of a practical system demands a sophisticated technological base. The system accuracy is independent of altitude, terrain, and other physical variables, but is limited almost purely by the accuracy of its own components. Traditional INSs mainly relied on mechanical gyros and accelerometers, but today there are many different types available, such as optical gyroscopes, piezoelectric vibrating gyroscopes, active and passive resonating gyroscopes, etc. Also, micromachined gyroscopes and accelerometers are making an important impact on modern inertia navigation systems. A brief description and operational principles of gyroscopes and accelerometers suitable for inertial navigation are given below.

Major advances in INS over the years include the development of the *electrostatic gyro* (ESG) and the laser gyro. In ESG, the rotor spins at a speed above 200×10^3 rpm in a near-vacuum environment. The rotor is suspended by an electrostatic field; thus, it is free from bearing friction and other random torques due to mechanical supports. Hence, its operation results in a superior performance compared to others, closely resembling the peformance of a theoretical gyro. Although no system can claim to reach perfection, an ESG requires less frequent updates as compared to other mechanical gyros.

Gyroscopes

There are two broad categories: (1) mechanical gyroscopes and (2) optical gyroscopes. Within both of these categories, there are many different types available. Only the few basic types will be described to

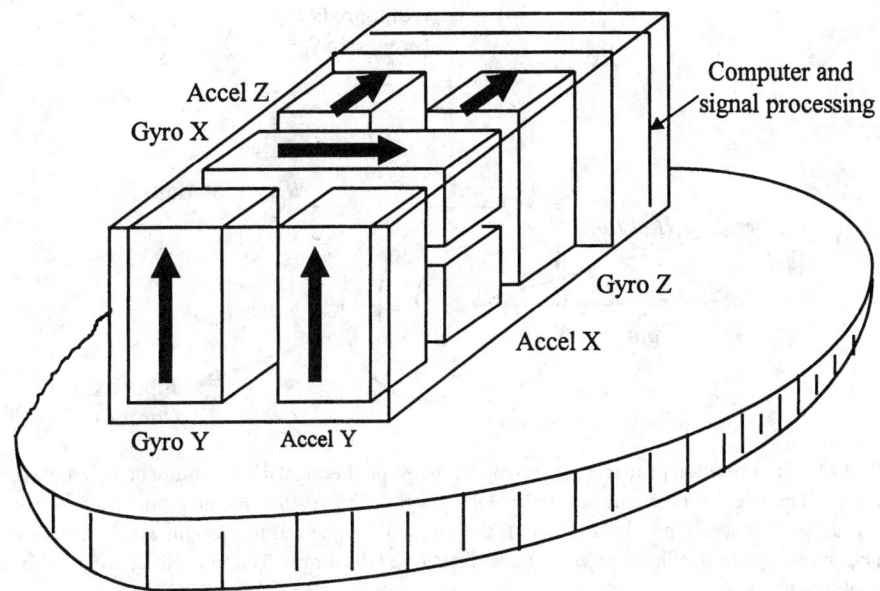

FIGURE 10.19 The use of a strapdown system eliminates the need for gimbals. The gyros and accelerometers are mounted rigidly on the structure of the vehicle, and the measurements are referenced to the local axes of roll, pitch, and yaw. The gyros measure the movement of angles in the three axes in short time intervals to be processed by the computer. This information is used, together with the accelerometer outputs, for predicting navigation axes.

illustrate the operating principles; detailed information may be found in the references listed at the end of this chapter.

Mechanical gyroscopes: The first mechanical gyroscope was built by Foucault in 1852, as a gimbaled wheel that stayed fixed in space due to angular momentum while the platform rotated around it. They operate on the basis of *conservation of angular momentum* by sensing the change in direction of an angular momentum. There are many different types, which are:

1. *Single degree of freedom gyroscopes*: include the rate, rate integrating, spinning rotor flywheel, electron, and particle gyros.
2. *Two degree of freedom gyroscopes*: incorporate the external gimbal types, two-axis floated, spherical free-rotor, electrically suspended, gas-bearing free-rotor gyros.
3. *Vibrating gyroscopes*: include the tuning fork, vibrating string, vibrating shell, hemispherical resonating, and vibrating cylinder gyros.
4. *Continuous linear momentum gyroscopes*: incorporate a steady stream of fluid, plasma, or electrons, which tends to maintain its established velocity vector as the platform turns. One typical example is based on a differential pair of hot-wire anemometers to detect the apparent lateral displacement of the flowing air column.

The operating principle of all mechanical gyroscopes is based on the conservation of angular momentum, as shown in Figure 10.20. The angular momentum is important since it provides an axis of reference. From Newton's second law, the angular momentum of a body will remain unchanged unless it is acted upon by a torque. The rate of change of angular momentum is equal to the magnitude of the torque, in vectorial form as:

$$T = dH/dt \qquad (10.92)$$

where H = angular momentum (= inertia × angular velocity, $I\omega$).

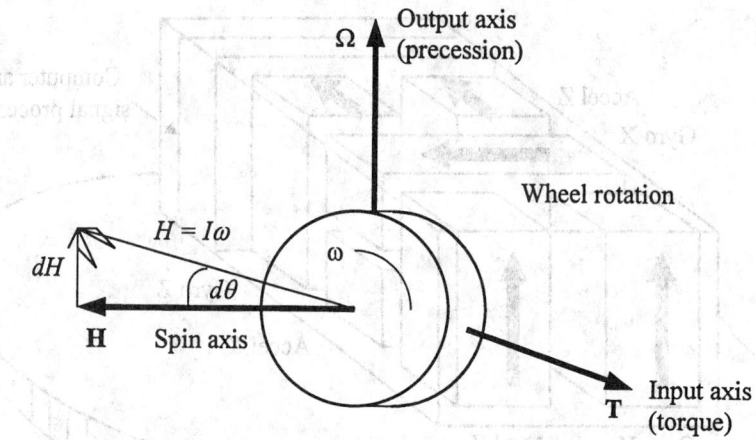

FIGURE 10.20   The operation principle of gyroscopes is based on the angular momentum of a carefully constructed rotating body. The angular momentum stabilizes the system. The angular momentum of a body will remain unchanged unless it is acted upon by a torque. If the torque is orthogonal to the spin axis, it cannot change the velocity, but it can change the direction in the same direction as the torque. The spin axis always tries to align with the external torque.

If a torque acts about the axis of rotation, it changes the angular velocity by:

$$T = I\, d\omega/dt = I\alpha \tag{10.93}$$

where I = inertia about the spin axis
      α = angular acceleration

If the torque is orthogonal to the spinning axis, it cannot change the magnitude of the angular velocity vector, but it can change direction in the same direction as torque T; then:

$$dH = H\, d\theta \tag{10.94}$$

where θ = angle of rotation.
   Therefore, from Equations 10.94 and 10.92:

$$T = dH/dt = H\, d\theta/dt = H\,\Omega \tag{10.95}$$

where Ω is the precession rate or the angular velocity of the spinning wheel about the axis normal to the plane of the spin and the input torque. Generally, the spin axis tries to align with the external input torque.
   These equations can be elaborated to describe the operating principles of mechanical gyros by taking into account the angular momentum in x, y, and z directions, nutation, coriolis accelerations, directions of other influencing torques and linear forces, etc. Here, the operation of the well-known *flywheel gyroscope* will be described as the basis for further discussions on inertial navigation systems.
   An example of a double-axis flywheel gyro is shown in Figure 10.21. In this type of gyroscope, an electrically driven rotor is suspended in a pair of precision low-friction bearings at both ends of the rotor axle. The rotor bearings are supported by a circular ring known as an *inner gimbal ring*, which in turn pivots on a second set of bearings that is attached to the *outer gimbal ring*. The pivoting action of the inner gimbal defines the horizontal axis of the gyro, which is perpendicular to the spin axis of the rotor. The outer gimbal ring is attached to the instrument frame by a third set of bearings that defines the vertical axis of the gyro that is perpendicular to both the horizontal axis and the spin axis. This type of

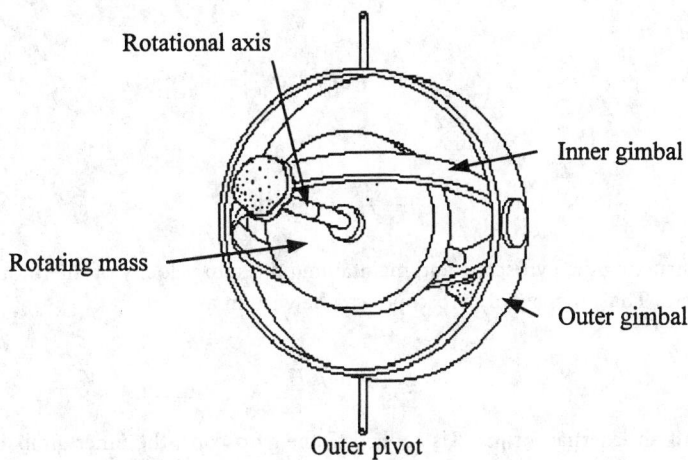

FIGURE 10.21 In a double-axis flywheel gyro, an electrically driven rotor is suspended by a pair of precision low-friction bearings at the rotor axle. The rotor bearings are supported by a circular inner gimbal ring. The inner gimbal ring in turn pivots on a second set of bearings attached to an outer gimbal ring. The pivoting action of the inner gimbal defines the horizontal axis of the gyro, which is perpendicular to the spin axis of the rotor. The outer gimbal ring is attached to the instrument frame by a third set of bearings. This arrangement always preserves the predetermined spin-axis direction in inertial space.

suspension has the property of always preserving the predetermined spin-axis direction in inertial space. Equations governing the two degrees of freedom gyroscope can be written using Equations 10.92 to 10.95. The torque with respect to an inertial reference frame can be expressed as:

$$T = \dot{H}_I \tag{10.96}$$

If the Earth is taken as a moving reference frame, then:

$$\dot{H}_I = \dot{H}_E + \omega_{IE}\, H \tag{10.97}$$

If the gyroscope itself is mounted on a vehicle (e.g., aircraft) that is moving with respect to the Earth, then:

$$\dot{H}_E = \dot{H}_B + \omega_{EB}\, H \tag{10.98}$$

The case of the gyroscope can be mounted on a platform so that it can rotate relative to the platform; then:

$$\dot{H}_B = \dot{H}_C + \omega_{BC}\, H \tag{10.99}$$

Finally, the inner gimbal can rotate relative to the case, hence:

$$\dot{H}_C = \dot{H}_G + \omega_{GC}\, H \tag{10.100}$$

Substituting Equations 10.97 to 10.100 into Equation 10.96 yields:

$$T = \dot{H}_G \left(\omega_{GC} + \omega_{BC} + \omega_{EB} + \omega_{IE} \right) H \tag{10.101}$$

But:

$$\left(\omega_{GC} + \omega_{BC} + \omega_{EB} + \omega_{IE}\right) = \omega_{IG} \tag{10.102}$$

Therefore,

$$T = \dot{H}_G + \omega_{IG} H \tag{10.103}$$

By carefully constructing the gyroscope and maintaining the spin velocity of the rotor constant, H_G can be made to be zero. Thus, the law of gyroscopes can be written as:

$$T = \omega_{IG} H \tag{10.104}$$

This means that if an external torque T is applied to the gyroscope, the inner gimbal will precess with respect to the inertial frame with a velocity ω such that Equation 10.104 is satisfied.

In most designs (e.g., rate gyros), the gimbal is hermetically sealed in a liquid and liquid is floated in the case, to unload the gimbal bearings and to provide viscous damping. A pick-off senses gimbal deflection by means of position transducers and it controls a servo system, with a servomotor driving the case to maintain pick-off null.

Optical gyroscopes are based on the inertial properties of light instead of Newton's law of motion. They operate on the Sagnac effect, which produces interferometer fringe shift against the rotation rate. In this case, two light waves circulate in opposite directions around a path of radius R, beginning at source S. A typical arrangement for the illustration of operation principles is shown in Figure 10.22. When the gyro is stationary, the two beams arrive at the detector at the same time and no phase difference will be recorded. Assume that the source is rotating with a velocity ω so that light traveling in the opposite direction to rotation returns to the source sooner than that traveling in the same direction. Thus, any rotation of the system about the spin axis causes the distance covered by the beam traveling in the direction of rotation to lengthen, and the distance traveled by the beam in the opposite direction to shorten. The two beams interfere to form a fringe pattern and the fringe position may be recorded, or the phase differences of the two beams may be sensed. This phase difference is directional and proportional to the angular velocity. Usually, photodetectors are used to measure the phase shift.

Two different types of optical gyros can be categorized: either passive or active, and resonant or nonresonant. In passive gyrosensors, the Sagnac phase is measured by some external means; whereas in active gyros, the Sagnac phase causes a frequency change internal to the gyro that is directly proportional to the rotation rate.

The Sagnac interferometer is the basis of the *interferometric fiber-optic gyro* (IFOG). A typical fiber-optic gyroscope is shown in Figure 10.22. However, the most widely used gyro is the active resonant *ring laser gyro* (RLG), which is applied extensively in aircraft navigation. Two different types of resonant passive gyros, the *resonant fiber-optic gyro* (RFOG) and the *micro-optic gyro* (MOG), are lower cost devices commonly used and comparable to RLGs.

Accelerometers

In inertial navigation, the absolute acceleration is measured in terms of three mutually perpendicular components of the total acceleration vector. Integrating these acceleration signals twice gives the displacement from an initial known starting location. Details of the acceleration and accelerometers are given elsewhere in this book (see Acceleration, Vibration, and Shock). Accelerometers are made from three basic elements: proof mass, suspension mechanism, and pick-off mechanism. Some accelerometers require electric or magnetic force generators and appropriate servo loops. Accelerometers measure not only real vehicular acceleration, but also respond to gravitational reaction forces. Acceleration due to gravity is a function of position — in particular, latitude and altitude — and is compensated by computers.

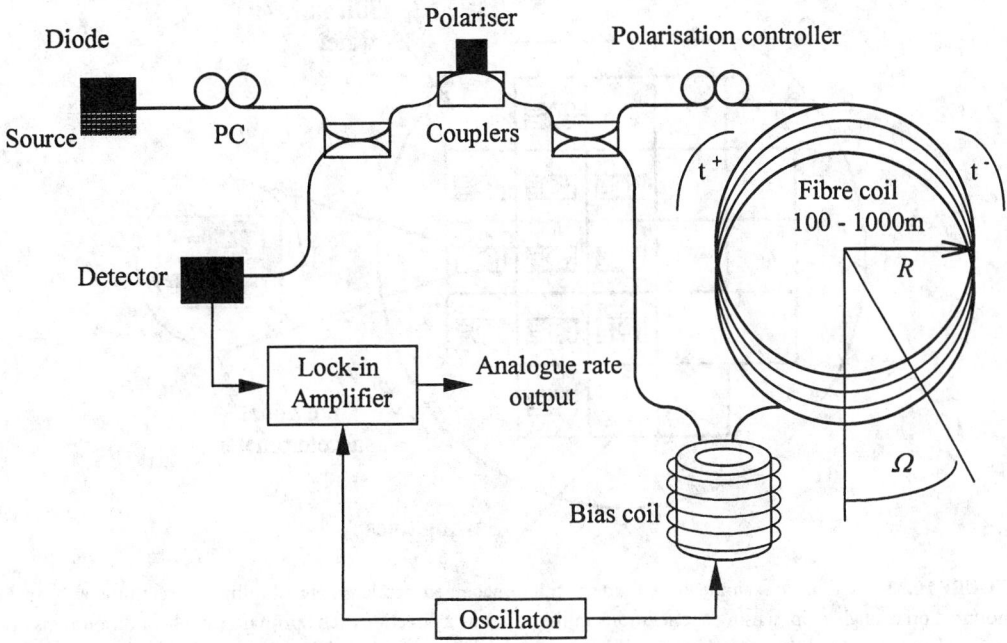

FIGURE 10.22 A typical fiber-optic gyroscope. This gyroscope is based on the inertial properties of light, making use of the Sagnac effect. The Sagnac effect describes interferometer fringe shift against rotation rate. Two light waves circulate in opposite directions around a path of radius R, beginning at source S. When the gyro is stationary, the two beams arrive at the detector at the same time and no phase difference is recorded. If the optical path is rotating with a velocity, the light traveling in the opposite direction to rotation returns to the source sooner than that traveling in the same direction. The two beams interfere to form a fringe pattern and the fringe position may be recorded, or the phase differences of the two beams may be sensed by photodetectors.

The most commonly used accelerometer in navigation systems is based on pendulous types. These accelerometers can be classified as:

1. Generic pendulous accelerometer
2. Q-flex type accelerometers
3. Micromachined accelerometers (A typical example of a modern micromachined accelerometer is given in Figure 10.23.)

Accelerations in the three axes are measured by suitably positioned accelerometers. Since accelerometers contain errors, the readings must be compensated by removing fixed biases or by applying scaling factors. The errors may be functions of operating temperature, vibration, or shock. Measurement of time must be precise as it is squared within the integration process for position determination. The Earth's rotation must also be considered and gravitational effects must be compensated appropriately.

Errors and Stabilization

Errors

In general, inertial navigation is an initial value process in which the location of the navigating object is determined by adding distances moved in known directions. Any errors in the system cause misrepresentation of the desired location by being off-target. The major disadvantage of an inertial guidance system is that its errors tend to grow with time. These errors in the deduced location are due to a number of reasons, including: imperfect knowledge of the starting conditions, errors in computation, and mainly errors generated by gyros and accelerometers.

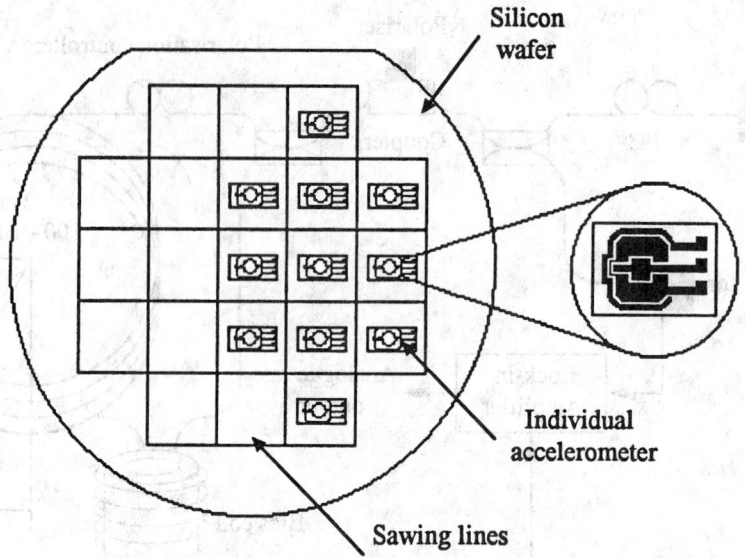

FIGURE 10.23 A typical example of a modern micromachined accelerometer. Multiple accelerometers can be mounted on a single chip, sensing accelerations in the *x*, *y*, and *z* directions. The primary signal conditioning is also provided in the same chip. The output from the chip is usually read in digital form.

If the error build-up with time becomes too large, external aids (e.g., LORAN, OMEGA) may be used to reset or update the system. Optimal use of the data from external aids must account for the geometry of the update and also for the accuracy of the update relative to the accuracy of the inertial system. The Kalman filter, for example, is one of the computational procedures frequently applied for optimally combining data from different sources.

Errors can broadly be classified as:

1. *System heading error:* A misalignment angle in the heading of an object traveling with a velocity can cause serious errors. For example, a vehicle traveling with velocity of 500 km h^{-1} in the same direction with 0.1° initial heading error will be off the target by approximately 873 m at the end of 1 h travel.

2. *Scale error:* Error in scaling can accumulate. In order to minimize scale errors, a scale factor is used. The scale factor is the ratio between changes in the input and output signals. It simply translates the gyro output (counts per second in the case of RLG) into a corresponding angle rotation. Some instruments may have different scale factors for positive and negative inputs, known as *scale factor asymmetry.* (Scale factors are measured in ° h^{-1} mA^{-1}, ° h^{-1} Hz^{-1}, or g Hz^{-1}.)

3. *Nonlinearity and composite errors:* In most cases, scale factors are not constant, but they can have second- or higher-order terms relating the output signals to the input. Statistical techniques can be employed to minimize these errors.

4. *Bias errors:* Zero offset or bias error is due to existence of some level of output signal for a zero input. Bias errors exist in accelerometers, gyros, tilt misalignments, etc.

5. *Random drift and random walk errors:* In some cases, the outputs of the devices can change due to disturbances inside the sensors, such as ball bearing noise in mechanical gyros. These disturbances may be related to temperature changes, aging, etc. White noise in optical gyros can cause a long-term accumulation in angle error known as the *random walk.*

6. *Dead band, threshold, resolution, and hysteresis errors:* These errors can be related to inherent operation of accelerometers and gyros. They can be due to stiction, minimum input required for an output, minimum measurable outputs, and nonrepeatability of variations in the output versus variations in the input.

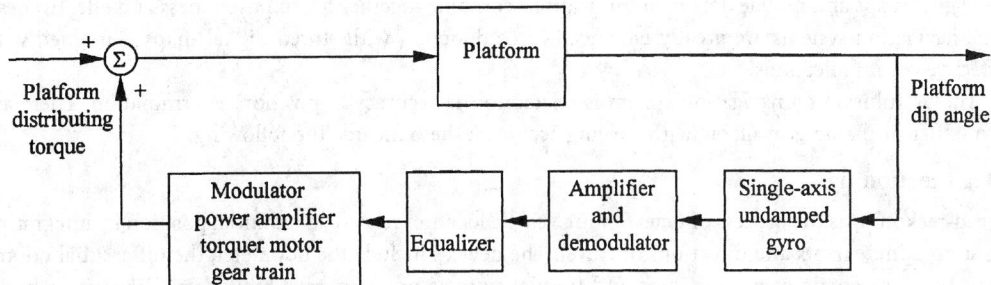

FIGURE 10.24 Stabilization is obtained using platforms designed to accurately maintain accelerometers and gyros leveled and oriented in the azimuth direction. In some cases, the platform is driven around its axis by servo amplifiers and electric motors. Sensitive pick-offs on the gyroscopes fed error signals are used to maintain a desired stability of the platform in the presence of disturbing torques.

It should be pointed out that this list is by no means exhaustive. Detailed error analysis can be found in the references cited.

Stabilization

The inertial navigation sensors must maintain angles within specified limits in spite of the disturbances imposed by the moving object. Accuracy requirements demand that the system must provide reliable and stable information in spite vibrations and other disturbing factors. One way of achieving stabilization is by using a stabilized platform. These platforms are designed to maintain accelerometers and gyros accurately leveled and oriented in the azimuth direction. In some cases, the platform is driven around its axis by servo amplifiers and electric motors. Usually, outputs of doubly integrating accelerometers are used directly to control the level-axis gyroscope precession rates. Sensitive pick-offs on the gyroscopes fed error signals are used to maintain a desired stable platform in the face of disturbing torques. The operation of a typical system, in block diagram form, is shown in Figure 10.24.

Unlike platform models, in a strapped-down system, gyroscopes and accelerometers are rigidly mounted to the vehicle structure so that they move with the vehicle. The accelerometers and gyroscopes are manufactured to measure accelerations and angles up to the maximum expected values. As the vehicle travels, the measured values are frequently transmitted to a computer. The computer uses these values to resolve the readings into the navigation axis sets and make deductions on the body axis sets.

Vehicular Inertial Navigation

In modern vehicular navigation, computerized maps and mobile communication equipment are integrated together with inertial and/or other electronic navigation systems. In recent years, in the wake of low-cost GPS systems, the vehicular navigation system has attracted much attention due to its large potential markets for consumer as well as business vehicles.

Automobile navigation systems are based on dead-reckoning, map matching, satellite positioning, and other navigational technologies. Map intelligent systems achieve high relative accuracy by matching dead-reckoned paths with road geometry encoded in a computerized map. This is also used to perform other functions such as vehicle routing and geocoding. Satellite-based navigation systems achieve high absolute accuracy with the support of dead-reckoning augmentation.

The capabilities and functions of automobile navigation systems depend on:

Choosing the necessary technology
Integrating the overall system
Resolving driver interface
Providing map data basis
Coordinating mobile communications

Digital maps and mobile data communications combine together for full usefulness and effectiveness. The navigation systems are greatly enhanced in conjunction with stored digital maps combined with effective communications.

The usefulness of a navigation system is related to the accuracy in position determination. There are a number of methods available with varying accuracy; these include the following:

Dead-reckoning

Dead-reckoning is the process of determining vehicle location relative to an initial position by integrating measured increments and directions of travel. The devices include the odometer, the differential odometer, and a magnetic compass. Gyros and inertial systems prove to have limited applications in harsh automotive environments. Although, dead-reckoning systems suffer from error accumulation, they are widely used inertial navigation systems, particularly in robotics and vehicular applications. Even the most precise navigation system requires periodic reinitialization and continuous calibrations by computers.

Radiolocation

In radiolocation, the global positioning system (GPS) is used extensively. Nevertheless, LORAN is gaining popularity as means of tracking land vehicle location from a central location. But its modest accuracy limits its global application in automotive navigation.

Map Matching

Artificial intelligence concepts are applied to match dead-reckoned vehicle paths, which are stored in computers. In map matching, sensed mathematical features of the vehicle paths are continuously associated with those encoded in a map database. Thus, a vehicle's dead-reckoned location can be initialized automatically at every turn to prevent accumulation of dead-reckoning errors.

The first application of map matching technology was in the Automatic Route Control System (ARCS), which used a differential odometer for dead-reckoning. In another system, the Etak map matching system, a solid-state flux gate compass is used as well as a differential odometer to dead-reckon paths for matching with digitized maps and aerial photographs. Further details on these technologies can be found in the references given at the end of this chapter.

In a map matching system, as long as the streets and road connectivities are accurately defined, the process identifies position relative to the road network as visually perceived by the vehicle driver.

Most of the dead-reckoning equipment commercially available is sufficiently robust to support map matching when operating in a defined road network. However, a good dead-reckoning accuracy is required to achieve reinitialization through map matching upon returning to the road network after off-road operations.

Proximity Beacon

This approach uses strategically located short-range transmitters, and reception of their location coded signal infers the receiving vehicle's instantaneous location. There are several variations of the proximity approach; some versions involve two-way communications with the equipped vehicle. Typically, the driver enters the destination code on the vehicle panel, for automatic transmission to the roadside unit, as the vehicle approaches an instrumented intersection. The roadside unit, which can be networked with a traffic management system, analyzes the destination code and transmits route instructions to the display on the vehicle panel. Proximity beacon systems are being tested in Germany and Japan. One of the most popular system is the ALI-SCOUT (see references) proximity beacon system, which uses dead-reckoning and map matching techniques between beacons to download updated map and traffic data in Berlin.

The approach to the interface between an on-board navigation system and a vehicle operator must take into account ergonomics and safety considerations as well as functional requirements. As a result of intensive research, especially in the aerospace industry, display of information for the operator is a well-developed area. In a well-known European system, Philips' CARIN, a color CRT map display is used to show vehicle location relative to the surroundings. Many other systems use short visual messages, symbolic graphics, and voice.

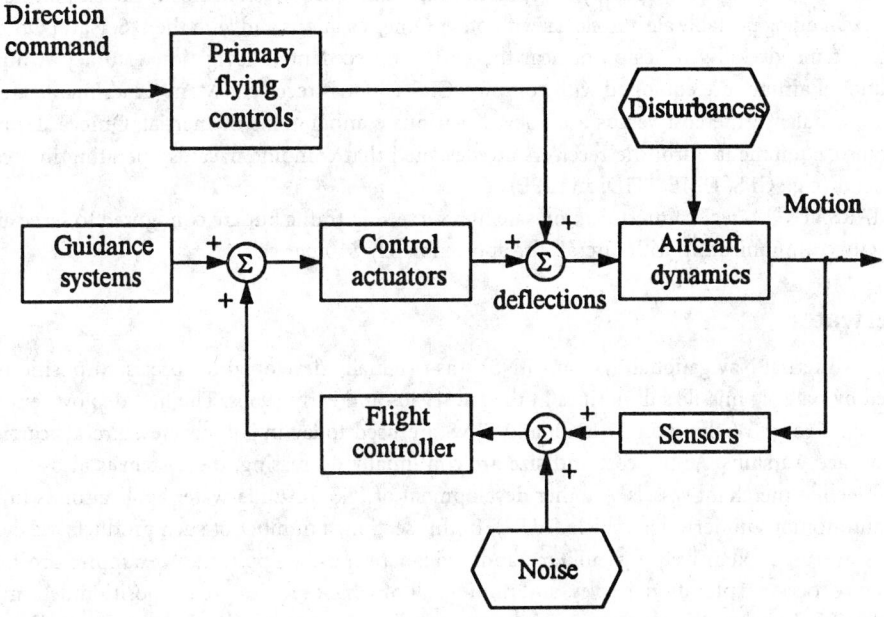

FIGURE 10.25   In many commercial aircraft, suitably located gyroscopes and accelerometers give signals for controlling the stability of the aircraft. Due to various instrumental errors and in-flight random disturbances such as gyro drift, scale factor errors, and accelerometer bias errors, the errors in the desired orientation increase with time. These accumulated errors need to be compensated periodically by external information such as Omega navigation systems.

Major potential roles for data communications in future automobile navigation need to provide current updates (road additions, closures, detours, etc.) for on-board map databases, and also provide real-time information on traffic conditions for on-board route generation.

Aircraft

The primary navigation aid for civil aircraft flying in the airspace of most of the developed countries is the VOR/DME system. The VOR (Very high frequency Omni Range) and the DME (Distance Measuring Equipment) enable on-board determination of an aircraft's bearing relative to North at the fixed ground station and slant range from the station, respectively. Further details can be found in references given at the end of this chapter.

Many commercial aircraft are equipped with precision Inertial Navigation Systems (INS) not only for navigation purposes, but also for stabilization of the aircraft at all times. Suitably located gyroscopes and accelerometers give signals to control the stability of the aircraft, as shown in Figure 10.25. Many aircraft INS utilize a gyro-stabilized platform on which the accelerometers are mounted. The platform is aligned before take-off to the desired orientation. Due to alignment errors and in-flight random disturbances such as gyro drift, scale factor errors, and accelerometer bias errors, the errors in the desired orientation increase with time. The errors need to be compensated periodically by external information such as Omega navigation systems.

Omega is a 10 kHz to 14 kHz navigation system that was primarily intended for updating submarine inertial navigators because of its very low frequencies penetrating beneath the ocean surface. It is also used by many general-aviation and military aircraft because of its worldwide coverage. Some airlines (e.g., American Airlines) have equipped their entire fleet with Omega receivers. The U.S. Coast Guard maintains two Omega stations and other countries maintain six more. Each station transmits eight consecutive sinusoidal tones with a 10 s repetition rate. Four are navigation tones common to all stations;

the other four tones uniquely identify the station. Each station has a cesium clock that is calibrated within 1 μs by exchanging portable atomic clocks with other Omega stations and with the U.S. Naval Observatory.

GPS systems give all vehicles on or near the Earth unprecedented navigation accuracy. A number of international airlines are equipped with confined GPS-Glonoss receivers. Many experiments are now in progress to balance the cost versus accuracy of various combinations of inertial, Omega, Loran, GPS, and Transit equipment. Airborne receivers are designed that combine navaids operating in a common radio band (e.g., GPS, DME, JTID, and IFF).

INMARSAT's "Marec" communication satellites serve ship traffic but are configured to serve air traffic by directly communicating with aircraft for approximately $10 per call.

Underwater

The Ship's Inertial Navigational System (SINS) was originally developed for precision position-finding required by ballistic missile submarines in the late 1950s and early 1960s. The first deployment was on-board U.S. George Washington in 1960, and SINS are used today in submarines, aircraft carriers, and other surface warships. As the cost and size are continually decreasing, the system is also deployed in naval as well as merchant vessels. Another development of INS for underwater application is in the area of the autonomous underwater vehicle (AUV). In this section, a number of such products are described.

AUVs are used extensively for military and civilian purposes. Application examples are mapping, surveillance, ocean exploration, survey, and mining, all of which require precise position determination. The desired features of such systems are: low power, high accuracy, small volume, light weight, and low cost. Two typical examples are the LN family produced by Litton Guidance and Control Systems, and the system developed by the Harbor Branch Oceanographic Institution Inc. (HBOI). The specifications of some of these systems are briefly described below to give examples of underwater INS.

The Litton LN-100 System

Litton's LN-100 is an example of the strapdown INS. The LN-100 system consists of three Litton Zero-Lock Gyros (ZLG), a Litton A4 accelerometer triad, power supply, supporting electronics, and a JIWAG standard 80960 computer. The single-board computer performs all the control, navigation, and interface functions.

The HBOI System

The HOBI system was developed with Kearfott Guidance and Navigation (KGN) and utilizes a Monolithic Ring Laser Gyroscope (MRLG), motion sensors, GPS input, and control unit. The inertial measurement unit is based on the Kearfott's T16-B three-axis ring laser gyro and three accelerometers.

Robotics

Closely related to the autonomous underwater vehicles, autonomous mobile robots also use INS extensively as a self-contained, independent navigation system. Typical applications are mining, unknown terrain exploration, and off-line path planning. There are many commercially available inertial navigation systems suitable for cost-effective utilization in the navigation of robots. Some of these are: gyrocompasses, rate gyros, gyrochip, piezoelectric vibrating gyros, ring laser gyros, interferometric, and other types of fiber-optic gyros. Three popular systems will be explained here.

The Honeywell Modular Azimuth Position System (MAPS)

Honeywell's H-726 Modular Azimuth Position System (MAPS) is a typical example of an inertial navigation system for land-based vehicles. It consists of a Dynamic Reference Unit (DRU) that provides processed information from the inertial sensor assembly, a Control Display Unit (CDU) that is used for human-machine interface, and a Vehicle Motion Sensor (VMS) that monitors the vehicle's directional and distance information. The inertial sensor assembly comprises three Honeywell GG1342 ring-laser gyros and three Sundstrand QA2000 accelerometers mounted to measure movements in three local axes. The inertial processor translates the information to the navigation processor that resolves the vehicle movement information from the VMS. The CDU provides mode selection, data display, waypoint information, and general operator interface.

Hitachi Fiber-Optic Gyroscopes

The Hitachi products are significant as they are relatively inexpensive and were designed for automotive applications. The open-loop Interferometric Fiber-Optic Gyros (IFOG) HOFG-4FT received the "Most Technologically Significant New Products of the Year" award in 1993, and is now installed in one of the production models from Toyota. The subsequent models of IFOG are HOFG-X, HOFG-1, and HGA-D. The HOFG-1 has been employed extensively in industrial mobile robots. The output of the system can be in serial form in RS-232 standard or as analog signals. Specifications of HOFG-1 include a range of $\pm 60°\,s^{-1}$, an update rate of 15 ms, and linearity of $\pm 0.5\%$. The power requirement is 10 to 16 V dc and 0.5 A.

References

K. R. Britting, *Inertial Navigation Systems Analysis,* New York: Wiley-Interscience, 1971.

M. Kayton, *Navigation-Land, Sea, Air and Space,* New York: IEEE Press, 1990.

A. Lawrance, *Modern Inertial Technology — Navigation, Guidance, and Control,* New York: Springer-Verlag, 1993.

Appendix — List of Manufacturers/Suppliers

Aerodata Co.
5550 Merric Rd.
Suit 205
Massapequa, NY
Tel: (616) 798-1873

AGV Products, Inc.
9307-E Monroe Rd.
Charlotte, NC 28270
Tel: (704) 825-1110
Fax: (704) 825-1111

American GNC Corporation
9131 Mason Avenue
Chatsworth, CA 91311
Tel: (818) 407-0092
Fax: (818) 407-0093
e-mail: agnc@kincyb.com

Astronautics Co.
P.O. Box 523
518 W Cherry St.
Milwaukee, WI

Cybermotion, Inc.
115 Sheraton Drive
Salem, VA 24153
Tel: (703) 562-7626
Fax: (703) 562-7632

First State Map Co.
12 Marry Ella Dr.
Wilmington, CT 19805
Tel: (800) 327-7992

Hitachi Cable America, Inc.
50 Main Street,
White Plains, NY 10606-1920
Tel: (914) 993-0990
Fax: (914) 993-0997

Honeywell, Inc.
Technology Center
3660 Technology Drive
Minnepolis, MN 55418
Tel: (612) 951-7715
Fax: (612) 951-7438

Ketema Inc.
790-T Greenfield Dr.
P.O. Box 666
El Cajon, CA

NASA Goddard Space Flight
  Center
Robotics Branch, Code 714.1
Greenbelt, MD 20771
Tel: (301) 286-4031
Fax: (301) 286-1613

Naval Command Control Center
RDT&E Division 5303
San Diego, CA 92152-7383
Tel: (619) 553-3672
Fax: (619) 553-6188

Navigation Science Co.
31127 Via Clinas
Suite 807
Westlake Village, CA 91326
Tel: (818) 991-9794
Fax: (818) 991-9896

Navigations Technologies Corp.
740 Arques Avenue
Sunnyvale, CA 94086
Tel: (408) 737-3200
Fax: (408) 737-3280

NSC
P. O. Box 4453
Thousand Oaks, CA 91359
Tel: (818) 991-9794
Fax: (818) 991-9896

Romarc Co.
512 Scott Rd.
Plumsteadville, PA 18949
Tel: (800) 755-2572

Schwartz Electro-Optics, Inc.
3404 N. Orange Blossom Trail
Orlando, FL 32804
Tel: (407) 298-1802
Fax: (407) 297-1794

Siemens Co.
1301 Avenue of the Americas
New York, NY 10019

Southern Avionics Co.
5000-T Belmont
Beaumont, TX 77707
Tel: (800) 280-0322
Fax: (409) 842-2987

Sperry Marine Inc.
1070 T Seminole Tr.
Charlottesville, VA 22901
Tel: (804) 974-2000
Fax: (804) 974-2259

Systron Donner Inertial
  Division
BEI Electronics
2700 Systron Drive
Concord, CA 94518-1399
Tel: (510) 682-6161
Fax: (510) 671-6590

Trackor Inc.
6500-T Trackor Lane
Austin, TX

Warren-Knight Inst. Co.
2045 Bennet Dr.
Philadelphia, PA 19116
Tel: (215) 484-9300

10.4   Satellite Navigation and Radiolocation

Halit Eren and C. C. Fung

Modern electronic navigation systems can be classified by range, scope, error, and cost. The range classifications are short, medium, and long ranges, within which exact limits are rather indefinite. The scope classifications can be either self-contained or externally supported, and active (transmitting) or passive (not transmitting) mode of operation.

Short-range systems include radiobeacons, radar, and Decca. Medium-range systems include Decca and certain types of extended-range radars. The long-range systems include Loran-C, Consol, and Omega. All these systems depend on active radio frequency (RF) transmissions, and all are externally supported with respect to the object being navigated, with the exception of the radar. In addition to these, there is another category of systems which are called *advanced navigation systems*; the transit satellite navigation systems, Glonass, and the Global Positioning Systems (GPS) are typical examples.

Utilization of electromagnetic radio waves is common to all navigation systems discussed here. Understanding of their behavior in the Earth's atmosphere is very important in the design, construction, and use of all kinds of navigation equipment — from satellites to simple hand-held receivers.

When an FM radio wave is generated within the Earth's atmosphere, the wave travels outward. The waves may be absorbed or reflected from surfaces of materials they encounter. The absorption and scattering of electromagnetic waves take place for many reasons, one of which is caused by excitation of electrons within the molecules in the propagation media. The behavior of an electromagnetic wave is dependent on its frequency and corresponding wavelength. Figure 10.26 shows the frequency spectrum of electromagnetic waves. They are classified as *audible waves* at the lower end of the spectrum, *radio waves* from 5 kHz to 300 GHz, and *visible light* and various other types of rays at the upper end of the spectrum.

For practical purposes, the radio wave spectrum is broken into eight bands of frequencies; these are: *very low frequency* (VFL) less than 30 kHz, *low frequency* (LF) 30 kHz to 300 kHz, *medium frequency* (MF) 300 kHz to 3 MHz, *high frequency* (HF) 3 MHz to 30 MHz, *very high frequency* (VHF) 30 MHz to 300 MHz, *ultra high frequency* (UHF) 300 MHz to 3 GHz, *super high frequency* (SHF) 3 GHz to 30 GHz, and *extremely high frequency* (EHF) 30 GHz to 300 GHz.

For easy identification, the frequencies above 1 GHz are further broken down by letter designators, as: L-band (1–2 GHz), S-band (2–4 GHz), C-band (4–8 GHz), X-band (8–12.5 GHz), and K-band (12.5–40 GHz). Since severe absorption of radar waves occurs near the resonant frequency of water vapor at 22.2 GHz, the K-band is subdivided into lower K-band (12.5–18 GHz) and upper K-band (26.5–40 GHz). Most navigation radars operate in the X- and S-bands, and many weapons fire control radars operate in the K-band range.

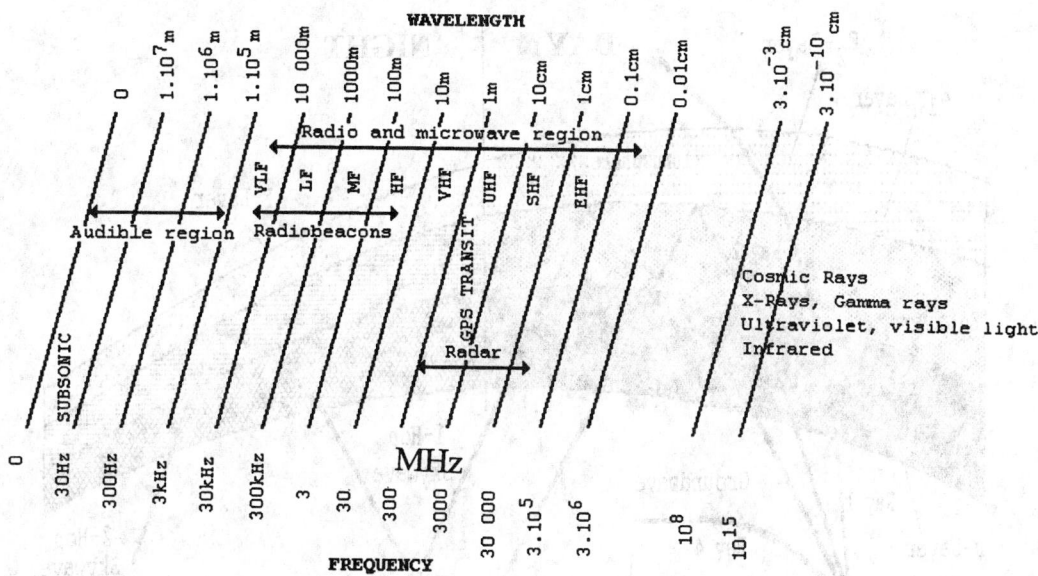

FIGURE 10.26 Electromagnetic wave frequency spectrum. Audible range can be heard if converted to sound waves. Radiobeacons operate in the VLF, LF, and MF ranges. Omega operating at VLF covers the entire world with only eight transmission stations. GPS, Transit, and Glonass use UHF frequencies. Wavelengths less than 10 cm are not suitable for satellite systems, but they are used in radars.

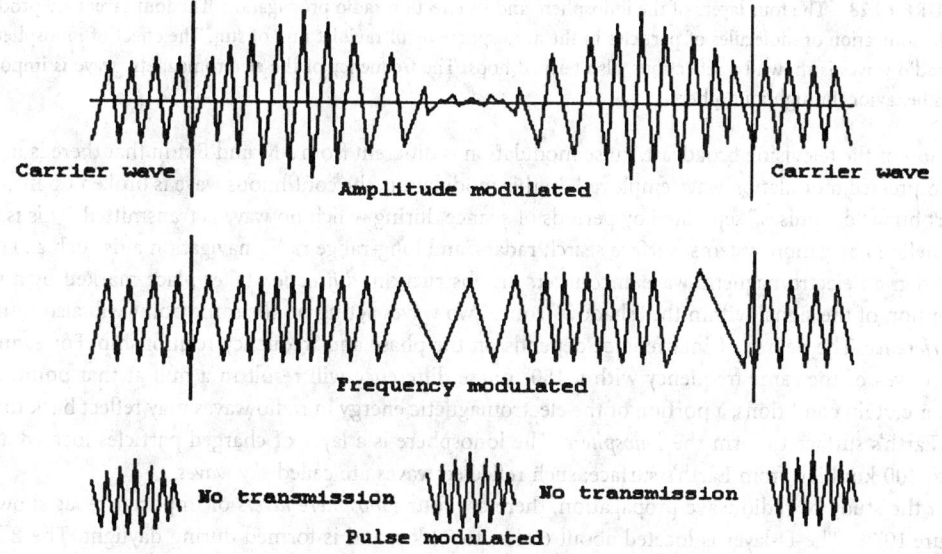

FIGURE 10.27 Amplitude, frequency, and pulse modulation of RF carrier waves. Amplitude modulation is suitable for broadcasting radio stations. Frequency modulation is used in commercial radio broadcasts. The pulse modulation is used in satellite systems, radars, and long-range navigation aids.

The radio waves are transmitted as continuous or modulated waves. A carrier wave (CW) is *modulated* to convey information in three basic forms: amplitude, frequency, and pulse modulation, as shown in Figure 10.27. The amplitude modulation (AM) modifies the amplitude of the carrier wave with a modulating signal. In frequency modulation (FM), the frequency of the carrier wave is altered in accordance with the frequency of the modulating wave. FM is used in commercial radio broadcasts and the sound

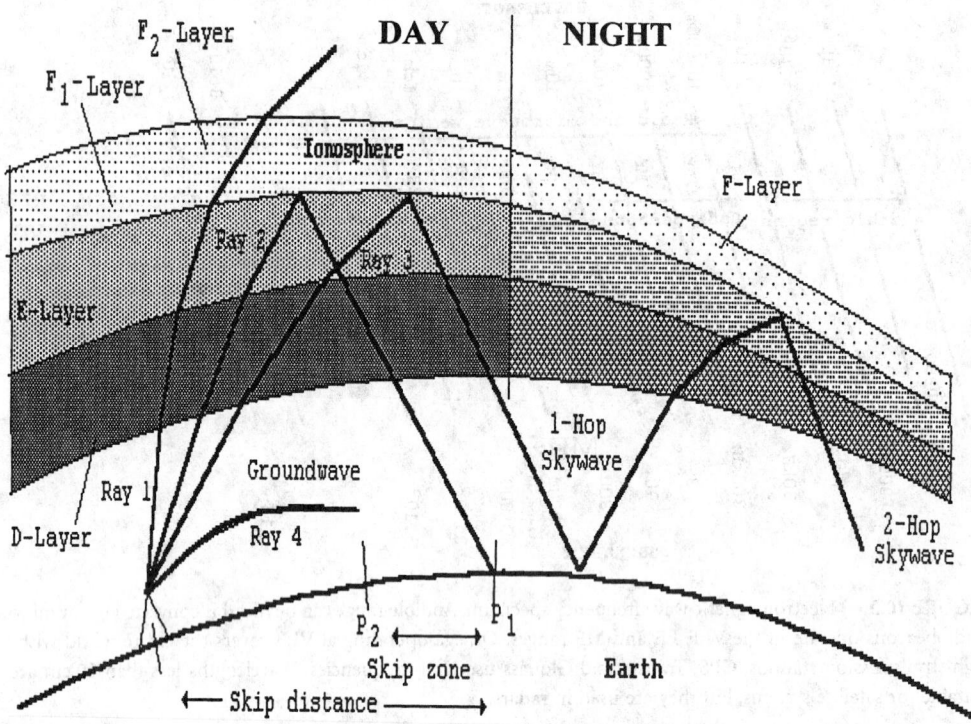

FIGURE 10.28   The four layers of the ionosphere and its effect on radio propagation. The four layers are produced by the ionization of molecules of particles in the atmosphere by ultraviolet rays of sun. The effect of ionosphere on the radio waves is shown by reflections, also termed hops. The frequency of the electromagnetic wave is important in its behavior through ionosphere.

portion of the television broadcast. Pulse modulation is different from AM and FM in that there is usually no impressed modulation wave employed. In this modulation, the continuous wave is broken up into very short bursts or "pulses," separated by periods of silence during which no wave is transmitted. This is used in satellite navigation systems, surface search radars, and long-range radio navigation aids such as Loran.

When an electromagnetic wave encounters an obstruction, *diffraction* takes place marked by a weak reception of the signal within the "shadow" zone. Two waves acting on the same point will also result in *interference*. The degree of interference depends on the phase and frequency relationship. For example, two waves of the same frequency with a 180° phase difference will result in a null at that point. Also, under certain conditions, a portion of the electromagnetic energy in radio waves may reflect back toward the Earth's surface to form the *ionosphere*. The ionosphere is a layer of charged particles located about 90 to 400 km high from Earth's surface; such reflected waves are called *sky waves*.

In the study of radio wave propagation, there are four *ionosphere layers* of importance, as shown in Figure 10.28. The D-layer is located about 60 km to 90 km and is formed during daylight. The E-layer is about 110 km. It persists through the night with decreased intensity. The F_1-layer is between 175 km and 200 km; it occurs only during daylight. The F_2-layer is between 250 km and 400 km; its strength is greatest in the day but it combines with the F_1-layer later to form a weak F-layer after dark. The layers in the ionosphere are variable, with the pattern seeming to have diurnal, seasonal, and sun spot periods. The layers may be highly conductive or may entirely hinder transmissions, depending on the frequency of the wave, its angle of incidence, height, and intensity on various layers at the time of transmission. In general, frequencies in the MF and HF bands are most suitable for ionosphere reflections during both day and night.

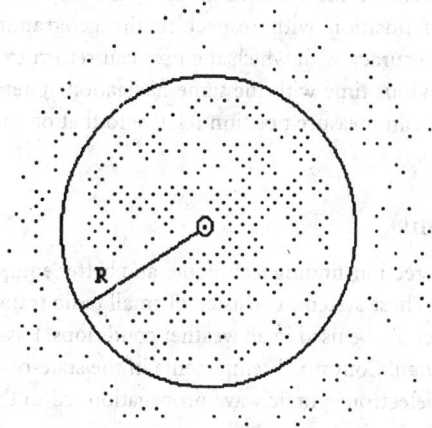

FIGURE 10.29 The rms radius circle that encompasses 68% of all measured positions. The variations in the measurements are due to a number of factors, including: ionosphere conditions, precise location of satellites, and inefficiencies in electronic circuits. (2 rms encompasses 95% of all indicated positions.)

Because of the higher resistance of the Earth's crust as compared to the atmosphere, the lower portions of radio waves parallel to the Earth's surface are slowed down, causing the waves to bend toward Earth. A wave of this type is termed a *ground wave*. The ground waves exist because they use the Earth's surface as a conductor. They occur at low frequency since LF causes more bending in conformity to Earth's shape. The ultimate range of such ground waves depends on the absorption effects. Sometimes, in the lower atmosphere, *surface ducting* occurs by multiple hopping, thus extending the range of a ground wave well beyond its normal limits. It is associated with higher radio and radar frequencies. This phenomenon is common in tropical latitudes. Behavior patterns of waves transmitted at various angles are illustrated in Figure 10.28.

Accuracy of Electronic Fix

There are a number of random effects that influence the accuracy of an electronic position determination; atmospheric disturbances along the transmission path, errors in transmitters and receivers, clocks, inaccuracy in electronic circuitry, gyro errors, etc. As a result, a series of positions determined at a given time and location usually results in a cluster of points near the true position. There are two measures commonly used to describe the accuracy: the first is the *circular error probable* (CEP) — a circle drawn on the true position whose circumference encompasses 50% of all indicated positions, and the second technique, more common, is the *root mean square* (rms), where:

$$\text{rms} = \sqrt{\sum_{n=1}^{N}\left(E_n\right)^2 \Big/ N} \qquad (10.105)$$

where E = the distance between actual and predicted positions
N = the number of predicted positions

A circle, shown in Figure 10.29, with one rms value is expected to contain 68% of all the indicated positions. Another circle of radius equal to 2 rms should contain 95% of all the indicated positions, for isotropic scattering, or errors.

In electronic navigation systems, three types of accuracy are important: (1) *predictable* or *absolute accuracy* — the accuracy of a position with respect to the geographic coordinates of the Earth; (2) *repeatable accuracy* — the accuracy with which the user can return to a position whose coordinates have been determined at a previous time with the same navigation system; and (3) *relative accuracy* — the accuracy with which a user can measure position relative to that of another user of the same system at the same time.

Radionavigation Systems

In the 1930s, improved radio direction-finding techniques and better equipment led to the establishment of systems called *radiobeacons*. These systems consisted of small radio transmitters located in fixed places to provide radio bearings that could be used in all weather conditions. Position findings by these beacons became known as *radionavigation*. Continued refinements in the state-of-the-art electronics technology and a better understanding of electromagnetic wave propagation led to the subsequent development of radar and longer-range radionavigation systems.

Essentially, radiobeacons are single transmitters, transmitting a continuous wave at low power, usually modulated by audible Morse code characters for identification. The transmitted signal is received by an on-board receiver incorporating a radio direction finder (RDF) to be processed further.

Short- and Medium-Range Radiolocation Systems

Most short- to medium-range navigation systems are designed to provide either a bearing to a fixed transmitter site, as in the case of radiobeacons, or a range and bearing from the transmitter to a natural or manufactured navigation aid, as in the case of radar.

Long-Range Hyperbolic Navigation Systems

Long-range electronic navigation systems are based on *hyperbolic systems*. In these systems, the lines of positions yield in segments of hyperbolas rather than as radial lines. The line connecting the master and secondary stations transmitting the same signal simultaneously is called the *baseline*. Hyperbola lines represent the locus of all points of the arrival of master and secondary pulses at specified time differences. This is illustrated in Figure 10.30. Any change in the position of the receiver near the baseline corresponds to a relatively large change in time difference of reception of pulses.

In practice, the secondary stations transmit pulses at fixed intervals, called *coding delays*, after having received a pulse from the master station. Currently, hyperbolic systems employ atomic time standards to regulate both master and secondary station transmissions to increase the system accuracy by eliminating random time errors due to atmospheric conditions.

In some systems, as an alternative to short pulses, continuous waves are propagated with the leading edges of each cycle representing a given time and distance interval. The corresponding hyperbolas then represent loci of constant phase differences, called *isophase lines*. The space between the hyperbolas are referred to as *lanes*. The position of the receiver within a lane is determined by the phase difference between the master and secondary signals. A disadvantage of this system is that it is not possible to distinguish one lane from another by the use of phase comparison alone. Hence, the lanes must be either counted as they are traversed from some fixed position, or they must be identified by some other means. For good accuracy, the user's distance from the stations can seldom exceed about six times the length of the baseline.

Radiobeacons

Most radiobeacon signals are limited to less than 320 km (200 miles), with a majority not receivable beyond about 32 km (20 miles). Often, radiobeacons located in a given area are grouped on a common frequency, such that each transmitter transmits only during a segment of a time-sharing plan. Radio bearings to the site of transmitter are determined by the use of a radio receiver equipped with a Radio Direction Finder (RDF). There are moderately priced, manually operated RDF receivers, and several more expensive fully automated models are also available. As a general rule, RDF bearings are normally

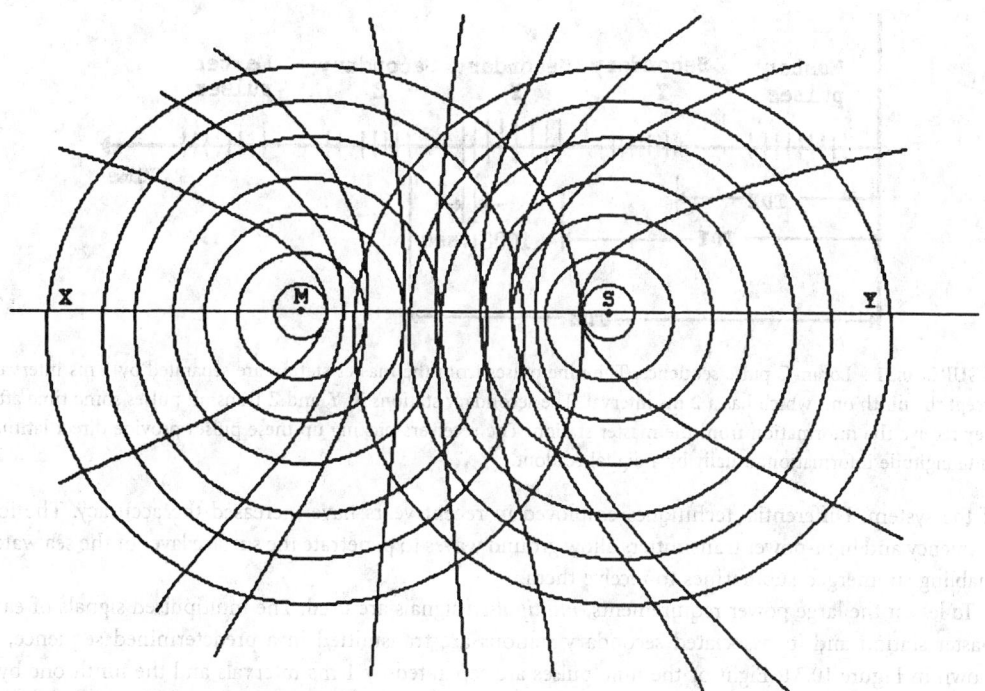

FIGURE 10.30 Hyperbolic patterns of two interacting radio waves propagated in opposite directions. These lines represent the locus of all points of a specified time difference between master and secondary pulses. The straight line MS is called the baseline. The maximum distance of the target object should not exceed 6 times the length of the baseline.

considered accurate only to within ±2°; for example, under 200 km (120 miles) to the transmitter in favorable conditions and ±5° to 10° when conditions are unfavorable.

Information on the locations, ranges, and using procedures of radio beacons are given in a number of publications such as DMAHTC Publication No. 117 *The Radio Navigation Aids.* Correct radiobeacon bearings are plotted on Mercator charts for position estimation. Because of possible inaccuracies, radiobeacons are not used universally. Navigators such as small boats and merchant ships not equipped with other systems use radiobeacons.

Loran-C

Loran was developed in the 1940s to be one of the first systems implementing a long-range hyperbolic system for both ships and aircraft. The system used master and slave stations transmitting sequential radio waves in the upper MF band with frequencies between 1850 kHz and 1950 kHz. Loran-A featured ground wave coverage out to between 700 km and 1250 km from the baseline by day, and up to 2200 km by night. It was the most widely used electronic navigation system until 1970. Later, a system employing synchronized pulses for both time-difference and phase comparison measurements was developed, known as Loran-C. Loran-C was configured to operate in a *chain* form consisting of more than one slave station usually located in triangles.

All stations in the system transmit a signal on a common carrier frequency in mid-LF band of 100 kHz ± 10 kHz. The ground wave range is about 1900 km. One-hop sky waves have a range of about 3600 km, and two-hop signals were noted to have been received about 6250 km from the ground station. One-hop sky waves are produced both by day and by night, while two-hop sky waves are formed only at night. Present Loran-C chains have baseline distances between 1500 km and 2300 km. The accuracy of the system varies from about ±200 m rms near the baseline to ±600 m rms near the extreme ranges

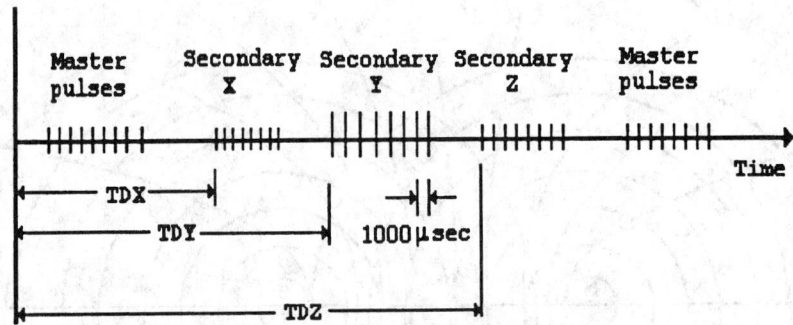

FIGURE 10.31 Loran-C pulse sequence. The nine pulses from the master station are separated by 1 ms intervals, except the ninth one, which has a 2 ms interval. The secondary stations X, Y, and Z transmit pulses some time after they receive the information from the master station. The receivers picking up these pulses provide direct latitude and longitude information, usually by a digital readout.

of the system. Differential techniques employed in recent years have increased the accuracy. The low frequency and high-power transmitters allow ground waves to penetrate the surface layer of the sea water, enabling submerged submarines to receive them.

To lessen the large power requirements, *multipulsed* signals are used. The multipulsed signals of each master station and its associated secondary stations are transmitted in a predetermined sequence, as shown in Figure 10.31. Eight of the nine pulses are separated by 1 ms intervals and the ninth one by a 2-ms interval. Integrated master and secondary pulses are compared at a sampling point at exactly 30 μs from their leading edges.

Loran-C receivers are fully automatic, suitable to be employed for marine, land vehicle, and aircraft applications. Most receivers provide direct lat/long digital readout, precise to a tenth of a minute of arc. They also provide auxiliary features such as destination selection, course and speed overground, etc. Once initialized, they automatically select the best chain and the most suitable master/secondary pulses.

There are 13 Loran-C chains worldwide. Each chain uses a different basic pulse repetition rate (PRR) and pulse repetition interval (PRI). The specific PRI used in a given chain is referred to as group repetition interval (GRI), often called *rate*. This system has enjoyed great expansion since 1970s, attracting many users. It has found applications in ships, aircraft, as well as land vehicles. Land vehicles are equipped with automatic vehicle location systems (AVLS). In one application in the U.S., the AVLS system is integrated with the city's emergency telephone system, a computer-aided dispatch system to control the position of emergency vehicles such as ambulances and fire trucks.

Nevertheless, the progress made in satellite-based global positioning systems (GPS) calls for termination of Loran-C in the very near future. However, complete system shut-down may not occur immediately, due to the vast number of current users of the system.

Decca

Decca is similar to Loran-C in that each chain is composed of one master and three slave stations arranged in *star* pattern, at an angle of about 120° between the baselines. It uses unmodulated continuous waves rather than the pulsed waves of Loran. The characteristic hyperbolic grid pattern is formed by phase comparisons of master and slave signals. All stations transmit in the LF-band between 70 kHz and 130 kHz. The nominal range is about 400 km both by day and by night. The system is extremely accurate within the operating range. The signals transmitted by each of the stations are all harmonics of a single fundamental frequency.

There are a wide variety of receivers available for Decca system, including automatic flight logs for aircraft. In general, the systems consist of four separate receivers, each of which can be set to receive one of the four signals transmitted by a given chain. The lane identification is accomplished by a signal transmitted by each master and slave station, transmitted once every 20 s for a duration of 0.6 s. By some

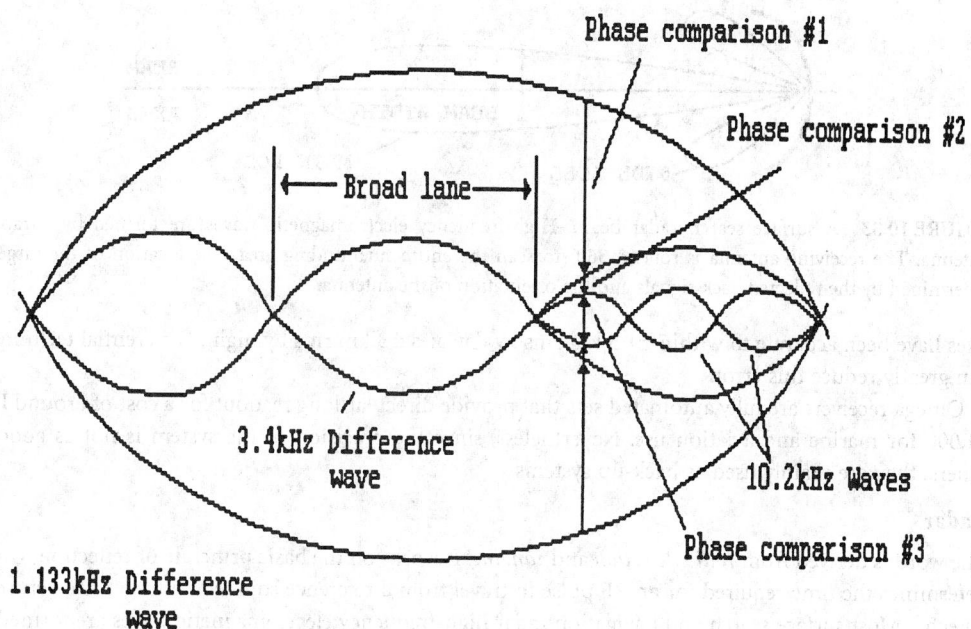

FIGURE 10.32   Three successive phase comparisons for lane resolution in Omega systems. Phase differences are compared in three stages with respect to three different signals transmitted by each station for accurate position finding. One wavelength is 25 km, representing two lanes. Accuracy of the Omega system is limited.

short signal interruptions and transmissions of harmonics simultaneously, zones and lanes are identified in a precise manner.

Consol

Consol is limited to the Eastern and Northern Atlantics. It is a hyperbolic system with extremely short baseline lengths, such that a collapsed hyperbolic pattern is formed. It employs three towers located three wavelengths apart. The operational frequencies are in the MF range between 250 kHz and 370 kHz. The system range is about 1900 km by day and 2400 km by night. The minimum range is about 40 km near the stations. One tower transmits a continuous wave, while other towers transmit waves with 180° phase shift by a *keying cycle*. The signals are modulated by dots and dashes such that receivers determine the position by counting them and printing on Consol grid patterns.

Omega

Omega is a hyperbolic navigation system that covers the entire world with only eight transmission stations located 7500 km to 9500 km apart. It transmits on frequencies in the VLF band from 10 kHz to 14 kHz at a power of 10 kW. The signals of at least three and usually four stations can be received at any position on Earth.

The 10 kHz to 14 kHz frequency band was chosen specifically to take advantage of several favorable propagation characteristics, such as: (1) to use the Earth's surface and ionosphere as a waveguide; (2) to enable submerged submarines to receive the signals; and (3) to form long baselines at 7500 km to 9500 km.

The basic frequency at which all eight stations transmit is 10.2 kHz. Each station transmits four navigation signals as well as a timing signal with atomic frequency standards ensuring that all stations are kept exactly in phase. Two continuous waves are in phase but traveling in opposite directions to produce a series of Omega lanes. Within each lane, a phase difference measurement would progress from 0° to 360° as the receiver moves across, as shown in Figure 10.32. Two Omega lanes complete one cycle, giving a wavelength of 25 km and lane of 12 km expanding as the distance from the baseline increases. Lanes are identified by three other signals transmitted by each station on a multiplexed basis. Omega

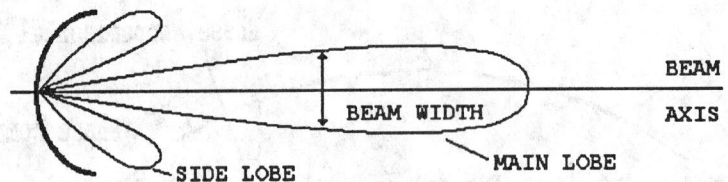

FIGURE 10.33 A surface search radar beam. High-frequency electromagnetic waves are formed by parabolic antenna. The receiving antenna is rotated 360° to scan the entire surrounding area. The location of the target is determined by the reflected back signals and the orientation of the antenna.

fixes have been accurate to within ±1.5 km rms by day and ±3 km rms by night. Differential techniques can greatly reduce this error.

Omega receivers are fully automated sets that provide direct lat/long readout for a cost of around U.S. $1,000 for marine and aviation use. Nevertheless, since the precision of the system is not as good as others, they are mainly used as back-up systems.

Radar

The word is derived from *radio detection* and *ranging*. It works on the basic principle of reflection, which determines the time required for an RF pulse to travel from a reference source to a target and return as an echo. Most surface search and navigation radar high-frequency electromagnetic waves are formed by a parabolic antenna into a beam form, as shown in Figure 10.33. The receiving antenna is rotated to scan the entire surrounding area, and the bearings to the target are determined by the orientation of the antenna at the moment the echo returns. A standard radar is made up of five components: transmitter, modulator, antenna, receiver, and indicator. They operate on pulse modulation.

Radars are extremely important devices for air control applications. Nowadays, airborne beacon radar systems are well developed in traffic alert and collision avoidance systems (TCAS). In this system, each plane constantly emits an interrogation signal, which is received by all nearby aircraft that are equipped appropriately. The signal triggers a transponder in the target aircraft, which then transmits some information concerning 3-D location and identification.

Satellite Relay Systems

The use of satellites is a highly developed technology utilized extensively throughout the world. In the past 2 decades, it has progressed from quasi-experimental in nature to one with routine provisions of new services. They take advantage of the unique characteristics of *geostationary satellite orbits* (GSO). The design of satellite systems is well understood, but the technology is still dynamic. The satellites are useful for long-distance communication services, for services across oceans or difficult terrain, and point-to-multipoint services such as television distribution.

Frequency allocation for satellites is controlled by the International Telecommunication Union (ITU). In the U.S., the Federal Communications Commission (FCC) makes the frequency allocations and assignments for nongovernment satellite usage. The FCC imposes a number of conditions regarding construction and maintenance of in-orbit satellites.

There are many satellite systems operated by different organizations and different countries mainly developed for communications and data transmissions; these include: Iridium of Motorola, Globalstar of Loral Corporation, Intelsat, CS-series of Japan, Turksat of Turkey, Aussat of Australia, Galaxy and Satcom of the U.S., Anik of Canada, TDF of France, etc. Some of the communication satellite systems are suitable for navigation purposes. However, satellite systems specifically designed for navigation are limited in number. The most established and readily accessible by civilian and commercial users are the GPS system of the U.S. and the Glonass of Russia.

The first generation of the satellite system was the *Navy satellite system* (Navsat), which became operational in January 1964, following the successful launch of the first transit satellite into polar orbit.

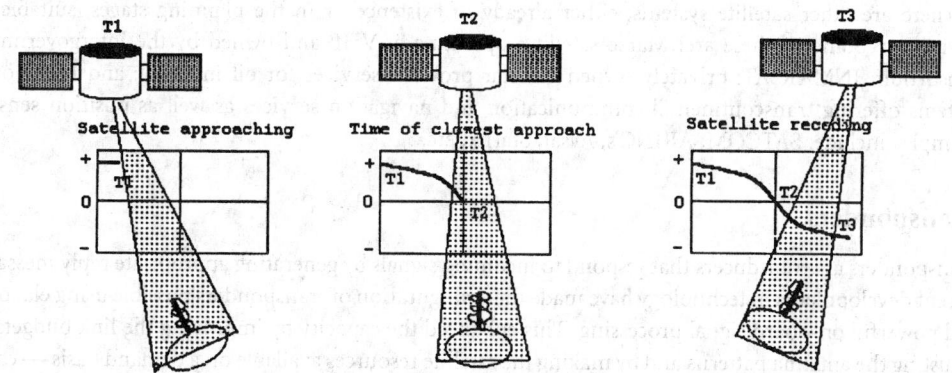

FIGURE 10.34 Transit satellite Doppler curve. As the satellite approaches the receiver, the frequency of the received signal increases due to Doppler shift. At the time of closest approach, the transmitted and received frequencies are the same. The frequencies received from a receding satellite result in lower values. This is also applicable in other position sensing satellites such as GPS, Glonass, Starfix, etc.

The system was declared open for private and commercial use in 1967. Civil designation of the name of the system is *Transit Navigation Satellite System*, or simply *Transit*. Later, this system evolved to become the modern Navsat GPS system, which will be discussed in detail. Most of the operational principles discussed here are inherited by the GPS system.

The Transit system consists of operational satellites, plus several orbiting spares, a network of ground tracking stations, a computing center, an injection station, naval observatory time signals, and receiver-computer combinations. The transit satellites are in circular polar orbits about 1075 km above ground with periods of revolution of about 107 min. Because of the rotation of the Earth beneath the satellites, every position on Earth comes within range of each satellite at least twice a day, at 12 h intervals. As originally intended, if at least five satellites are operational at any given time, the average time between fix opportunities would vary from about 95 min near the equator to about 35 min or less above 70° North and South.

The Transit system is based on the Doppler shift of two frequencies, 150 MHz and 400 MHz, transmitted simultaneously by each satellite moving its orbit at a tangential velocity of about 7.5 km s^{-1}. Two frequencies are used so that the effects of the ionosphere and atmospheric refraction on the incoming satellite transmission can be compensated for by the receivers. Each frequency is modulated by a repeating data signal lasting 2 min, conveying the current satellite time and its orbital parameters and other information. Within the receiver, a single Doppler signal is created by processing the two signals transmitted by the satellite. By plotting the frequency of this signal versus time, a characteristic curve of the type shown in Figure 10.34 is obtained. Since the frequency of the transmitted signal is compressed as the satellite approaches, according to what proportion of the velocity vector is seen by the user receiver, the curve begins at time T_1 at a frequency several cycles higher than the transmitted frequency.

Tracking stations record Doppler observations and memory readout received during each satellite pass to relay them to a computer center. Updated orbital position and time data communications are relayed to an "injection" station from the computer center for transmission to satellite in a burst once each 12 h. Enough data is supplied in this 15 s injection message to last for 16 h of consecutive 2 min broadcasts describing the current orbital positions of the satellite.

The system accuracy depends on the accuracy of the satellite orbit computation, the effect of ionosphere refraction, the precision of the receiver speed, and heading determination. Under optimal conditions, the system is capable of producing fixes with a maximum rms error of about 35 m for the stationary receivers anywhere on Earth. Nevertheless, if a site is occupied for several weeks, an accuracy better than 1 m can be achieved. The time signal transmitted as a "beep" at the end of each 2 min transmission cycle coincides with even minutes of Coordinated Universal Time, which can be used as a chronometer check.

There are other satellite systems, either already in existence or in the planning stages, suitable for navigation. Some of these are: Marec satellites operating at VHF and owned by the intergovernment consortium INMARSAT; privately owned Geostar provides services for oil industry; and many other systems offering transcontinental communication and navigation services as well as position sensing; examples include: SATCOM, ARINC's, Avsat, Starfix, etc.

Transponders

Transponders are transducers that respond to incoming signals by generating appropriate reply messages. Recent developments in technology have made the configuration of transponders possible using elaborate and powerful on-board signal processing. This enhanced the capacity by improving the link budgets, by adjusting the antenna patterns and by making the satellite resources available on a demand basis — called the "switch board in the sky concept."

Increased interest in deep sea exploration has brought acoustic transponders to the forefront as an important navigation tool. They provide three-dimensional position information for subservience vehicles and devices.

Some transponders are based on radar signals that respond to radar illumination. Transponders are programmed to identify friend or foe or, in some cases, simply inform ground stations about the position of aircraft.

Transponders are used for emergency warning. The U.S. and Canadian satellites carry Sarsat transponders, and Russian satellites carry Cospas transponders. They are also used as warning devices in collision avoidance systems in aircraft and land vehicles.

Global Satellite Navigation Systems

The GPS System

The Global Satellite Navigation Systems are second-generation satellites evolved primarily from the Naval Global Positioning System. They provide a continuous three-dimensional position-finding capability (i.e., latitude, longitude, and altitude), in contrast to the periodic two-dimensional information of the Transit system. Twenty-four operational satellites, as shown in Figure 10.35, constitute the system. Each satellite orbit is circular, about 2200 km high, and inclined at angles of 55° with respect to Earth's axis.

The position determination using the GPS system is based on the ability of the receivers to accurately determine the distance to the GPS satellites above the user's horizon at the time of fix. If accurate distances of two such satellites and the heights are known, then the position can be determined. In order to do this, the receiver would need to know the exact time at which the signal was broadcast and the exact time that it was received. If the propagation speeds through the atmosphere are known, the resulting range can be calculated. The measured ranges are called *pseudoranges*. Nowadays, normally, information is received from at least four satellites, leading to accurate calculations of the fix. The time errors plus propagation speed errors result in range errors, common to all GPS receivers. Time is the fourth parameter evaluated by the receiver if at least four satellites can be received at a given time. If a fifth satellite is received, an error matrix can be evaluated additionally.

Each GPS satellite broadcasts simultaneously on two frequencies for the determination and elimination of ionosphere and other atmospheric effects. The Navstar frequencies are at 1575.42 MHz and 1227.6 MHz, designated as L1 and L2 in the L-band of the UHF range. Both signals are modulated by 30 s navigation messages transmitted at 50 bits s^{-1}. The first 18 s of each 30 s frame contain *ephemeris* data for that particular satellite, which defines the position of the satellite as a function of time. The remaining 12 s is the *almanac* data, which define orbits and operational status of all satellites in the system. The GPS receivers store and use the ephemeris data to determine the pseudorange, and the almanac data to help determine the four best satellites to use for positional data at any given time. However, the "best four" philosophy has been overtaken slowly by an all-in-view philosophy.

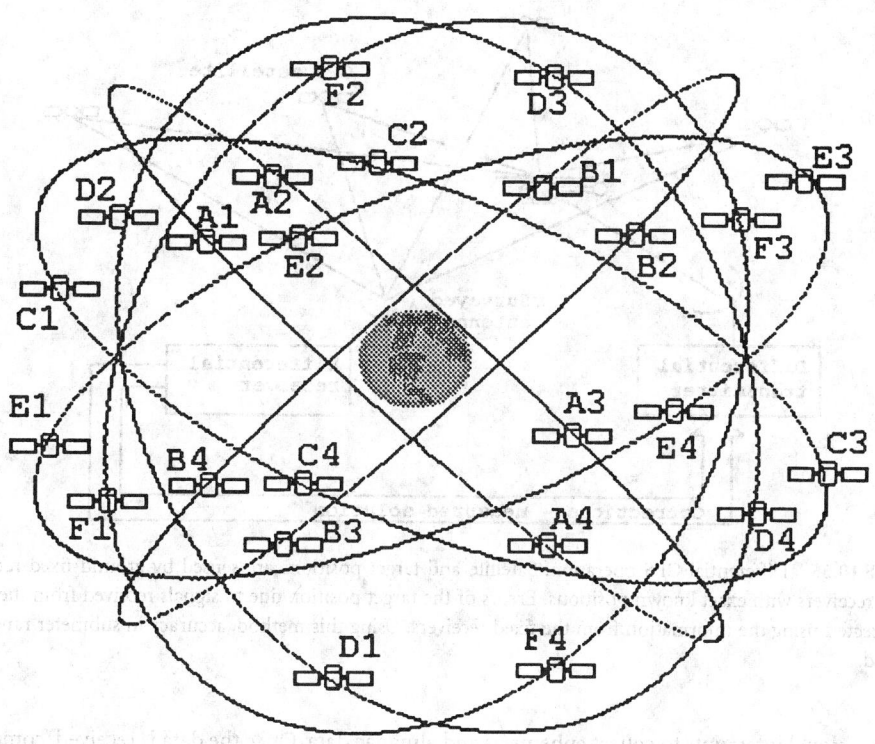

FIGURE 10.35 Operational GPS satellite coverage for navigation. Four satellites orbit in circular form. There are six such orbits inclined at angles of 60° from each other. In this arrangement, any point on Earth can see at least four satellites at any given time. This yields great accuracy in the position determination of the target, even only with C/A codes received.

The L1 and L2 satellite navigation signals are also modulated by two additional binary sequences called *C/A code* for acquisition of coarse navigation and the other *P-code* for precision ranging. The L1 signal is modulated both by the C/A and P-codes, and the L2 only by the P-code. Positional accuracies of about 20 m rms are usual in using C/A codes alone. The P-code, however, is not available for civilian users. The P-code is redesignated to be a Y-code, decipherable only by high-precision receivers having access to encrypted information in the satellite message. Nevertheless, it is fair to comment that civilians have figured out how to benefit the P/Y signals without actually knowing the codes, but at lower SNR. Further, C/A codes are degraded by insertion of random errors such that positional accuracy is limited to 50 m rms for horizontal values and 70 m for vertical values. These errors are intended to be lifted by the year 2006. Civilian users have access to the so-called *Standard Positioning Services* (SPS) accurate to 50 m rms, while U.S. and NATO military users will use *Precise Positioning Service* (PPS).

In enhancing SPS accuracy, differential techniques may be applied, as shown in Figure 10.36, to the encrypted GPS signals. Since the reference receiver is at a known location, it can calculate the correct ranges of pseudoranges at any time. The differences in the measured and calculated pseudoranges give the correction factors. Accuracy less than 1 m can be obtained in the stationary and moving measurements. Recently, differential receivers became commonly available, giving higher accuracy in sub-centimeter ranges. They are getting cheaper day by day and finding applications in many areas such as airplanes, common transport vehicles, cars, geological surveying, orienteering, farming, etc.

There are currently three basic types of GPS receivers designed and built to address various user communities. These are called *slow sequencing*, *fast sequencing*, and *continuous tracking* receivers. The least complicated and lowest cost receiver for most applications is the slow sequencing type, wherein only one measurement channel is used to receive sequential L1 C/A code from each satellite every 1.2 s,

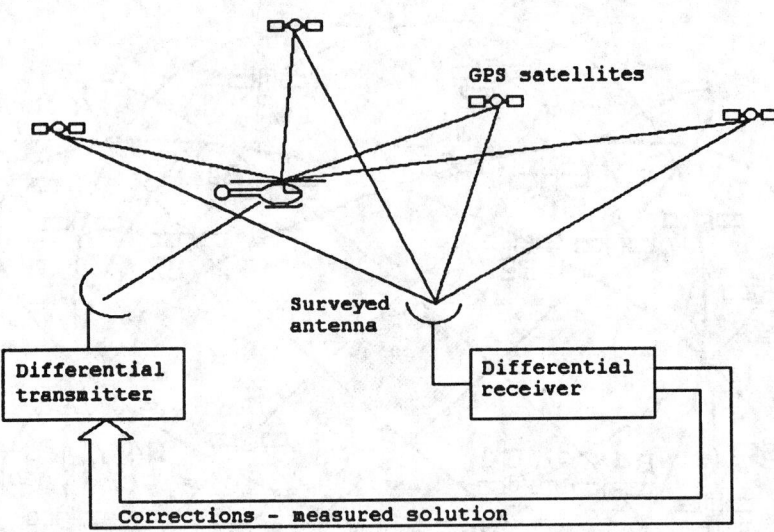

FIGURE 10.36 Differential GPS operation. Satellite and target positions are sensed by ground-fixed receivers or mobile receivers with exact known positions. Errors of the target position due to signals received from the satellites are corrected using the information from the fixed receivers. Using this method, accuracy in submeter range can be obtained.

with occasional interrupts to collect ephemeris and almanac data. Once the data is received, computation is carried out within 5 s, making this system suitable for stationary or near-stationary fixes.

Fast sequencing receivers have two channels: one for making continuous pseudorange measurements, and the other collection of the ephemeris and almanac data. This type is used in medium dynamic applications such as ground vehicles.

Continuous tracking receivers employ multiple channels (at least five) to track, compute, and process the pseudoranges to the various satellites being utilized simultaneously, thus obtaining the highest possible degree of accuracy and making it suitable for high dynamic applications such as aircraft and missiles. The parallel channel receivers are so cost effective nowadays that other types are likely to disappear.

A number of companies produce highly sophisticated GPS receivers. EURONAV GPS receivers operate on two and five channels for military applications. They provide features such as precise time, interfacing with digital flight instruments, RS-422 interface, altimeter input, self initialization, etc.

Software implementation satellite management functions, having different features, are offered by many manufacturers. In the case of DoD NAVSTAR GPS receivers, for example, three functional requirements are implemented: (1) database management of satellite almanac, ephemeris, and deterministic correction data; (2) computation of precise satellite position and velocity for use by navigation software; and (3) using satellite and receiver position data to periodically calculate the constellation of four satellites with optimum geometry for navigation. The DoD receivers are divided for three functions as Satellite Manager (SM), Satellite-Data-Base-Manager (SDBM) SV-Position Velocity Acceleration (SVPVA), and Select-Satellites (SS).

Differential navigation is also applied where one user set is navigating relative to another user set via a data link. In some cases, one user has been at a destination at some prior time and is navigating relative to coordinates measured at that point. The true values of this receiver's navigation fix are compared against the measured values, and the differences become the differential corrections. These corrections are transmitted to area user sets in real-time, or they may be recorded for post-mission use so that position fixes are free of GPS-related biases.

Differential navigation and GPS systems find applications in enroute navigations for commercial and civil aviation, military application, navigation of ships especially in shallow waters, in station keeping of aircraft, seismic geophysical explorations, land surveying, transport vehicles and traffic controls, etc.

The Glonass

There are a number of other satellite navigation systems similar to GPS of the U.S., such as Russian *Glonass*. The Glonass consists of 24 satellites orbiting in circular form 1500 km above the ground. The accuracy of the system is about 10 m rms. Glonass satellites transmit details of their own position and a time reference. The carrier frequencies are in L-band, around 1250 MHz (L2) and 1600 MHz (L1). Only the L1 frequency carries the Civil C/A code. The radio frequency carriers used by Glonass are channelized within bands 1240–1260 MHz and 1597–1617 MHz, the channel spacing being 0.4375 MHz at the lower frequencies and 0.5625 MHz at the higher frequencies. The number of channels is 24. Glonass data message is formatted in frames of 3000 bits, with a duration of 30 s. The ephemeris data are transmitted as a set of position, velocity, and acceleration coordinates in a Cartesian Earth-centered, Earth-fixed (ECEF) coordinate system. The new ephemeris data are available every half hour, valid for the following quarter-hour. The data are sent at a 50 baud rate and superimposed on a pseudorandom noise (PRN) code. The low-precision code has length 511 bits as compared to 1023 bits for Navstar. Glonass accuracy is as good as that for the GPS system. Glonass and GPS have different coordinate frames and different time frames that are being coordinated together.

The Starfix

Another interesting satellite-based system — privately funded, developed, launched, and maintained — is the Starfix positioning system. This system is designed primarily for oil surveying. The system consists of a master site, which generates satellite ephemeris data, and four satellites in geosynchronous orbits. The system said to have a precision of 2.5 m rms.

References

M. Kayton, *Navigation-Land, Sea, Air and Space,* New York: IEEE Press, 1990.

A. F. Inglis, *Electronic Communication Handbook,* New York: McGraw-Hill, 1988.

J. Everet, *VSATs — Very Small Aperture Terminals,* IEEE Telecommunication Series 28, London: Peter Peregrinus Ltd., 1992.

A. Leick, *GPS Satellite Surveying,* New York: John Wiley & Sons, 1990.

B. R. Elbert, *The Satellite Communication Applications Handbook,* Boston: Artech House, 1997.

Appendix — List of Manufacturers/Suppliers

Advanced Videotech Co.
1840 County Line Rd., Dept. G
Huntington Valley, PA 19006
Tel: (800) 221-8930
Fax: (800) 221-8932

AlliedSignal
101 Colombia Rd.
Dept. CAC
Morristown, NJ 07962
Tel: (800) 707-4555
Fax: (602) 496-1001

American GNC Corp.
9131 Mason Avenue
Chatsworth, CA 91311
Tel: (818) 407-0092
Fax: (818) 407-0093

Astroguide
Lasalle, IL 61301
Tel: (815) 224-2700
Fax: (815) 224-2701

Colombia Elect. Int. Inc.
P.O. Box 960-T
Somis, CA 93066
Tel: (805) 386-2312 or
    (800) 737-9662
Fax: (805) 386-2314

Comstream
10180 Barnes Canyon Rd.
San Diego, CA 92121
Tel: (800) 959-0811
Fax: (619) 458-9199

GE Co.
3135 Easton Tpke.
Fairfield, CT 06431
Tel: (800) 626-2004
Fax: (518) 869-2828

Orbitron
351-TR-S Peterson St.
Spring Green, WI 53588
Tel: (608) 588-2923
Fax: (608) 588-2257

STI
31069 Genstar Rd.
Hayward, CA 94544-7831
Tel: (800) 991-4947
Fax: (510) 471-9757

10.5 Occupancy Detection

Jacob Fraden

Occupancy sensors detect the presence of people in a monitored area. Motion detectors respond only to moving objects. A distinction between the two is that the occupancy sensors produce signals whenever an object is stationary or not, while the motion detectors are selectively sensitive to moving objects. The applications of these sensors include security, surveillance, energy management (electric lights control), personal safety, friendly home appliances, interactive toys, novelty products, etc. Depending on the applications, the presence of humans may be detected through any means that is associated with some kind of a human body's property or actions [1]. For example, a detector may be sensitive to body weight, heat, sounds, dielectric constant, etc. The following types of detectors are presently used for the occupancy and motion sensing of people:

1. *Air pressure sensors:* detect changes in air pressure resulting from opening doors and windows
2. *Capacitive:* detectors of human body capacitance
3. *Acoustic:* detectors of sound produced by people
4. *Photoelectric:* interruption of light beams by moving objects
5. *Optoelectric:* detection of variations in illumination or optical contrast in the protected area
6. *Pressure mat switches:* pressure-sensitive long strips used on floors beneath the carpets to detect the weight of an intruder
7. *Stress detectors:* strain gages embedded into floor beams, staircases, and other structural components
8. *Switch sensors:* electrical contacts connected to doors and windows
9. *Magnetic switches:* a noncontact version of switch sensors
10. *Vibration detectors:* react to the vibration of walls or other building structures; may also be attached to doors or windows to detect movements
11. *Glass breakage detectors:* sensors reacting to specific vibrations produced by shattered glass
12. *Infrared motion detectors:* devices sensitive to heat waves emanating from warm or cold moving objects
13. *Microwave detectors:* active sensors responsive to microwave electromagnetic signals reflected from objects
14. *Ultrasonic detectors:* similar to microwaves, except that instead of electromagnetic radiation, ultrasonic waves are used
15. *Video motion detectors:* video equipment that compares a stationary image stored in memory with the current image from the protected area
16. *Laser system detectors:* similar to photoelectric detectors, except that they use narrow light beams and combinations of reflectors
17. *Triboelectric detectors:* sensors capable of detecting static electric charges carried by moving objects

One of the major aggravations in detecting occupancy or intrusion is a false positive detection. The term "false positive" means that the system indicates an intrusion when there is none. In some noncritical applications where false positive detections occur once in a while, for example, in a toy or a motion switch controlling electric lights in a room, this may be not a serious problem: the lights will be erroneously turned on for a short time, which will unlikely do any harm. In other systems, especially those used for security purposes, the false positive detections, while generally not as dangerous as false negative ones (missing an intrusion), may become a serious problem. While selecting a sensor for critical applications, consideration should be given to its reliability, selectivity, and noise immunity. It is often good practice to form a multiple sensor arrangement with symmetrical interface circuits; this can dramatically improve the reliability of a system, especially in the presence of external transmitted noise. Another efficient way to reduce erroneous detections is to use sensors operating on different physical principles [2]; for example, combining capacitive and infrared detectors is an efficient combination as they are receptive to different kinds of transmitted noise.

Ultrasonic Sensors

Ultrasonic detectors are based on transmission to the object and receiving reflected acoustic waves. Ultrasonic waves are mechanical — they cover frequency range well beyond the capabilities of human ears, i.e., over 20 kHz. However, these frequencies may be quite perceptive by smaller animals, like dogs, cats, rodents, and insects. Indeed, the ultrasonic detectors are the biological ranging devices for bats and dolphins.

When the waves are incident on an object, part of their energy is reflected. In many practical cases, the ultrasonic energy is reflected in a diffuse manner. That is, regardless of the direction where the energy comes from, it is reflected almost uniformly within a wide solid angle, which may approach 180°. If an object moves, the frequency of the reflected waves will differ from the transmitted waves. This is called the Doppler effect (see below). To generate any mechanical waves, including ultrasonic, the movement of a surface is required. This movement creates compression and expansion of the medium, which can be a gas (air), a liquid, or a solid. The most common type of the excitation device that can generate surface movement in the ultrasonic range is a piezoelectric transducer operating in the so-called *motor* mode [3]. The name implies that the piezoelectric device directly converts electrical energy into mechanical energy.

Microwave Motion Detectors

Microwave detectors offer an attractive alternative to other detectors, when it is required to cover large areas and to operate over an extended temperature range under the influence of strong interferences (e.g., wind, acoustic noise, fog, dust, moisture, etc.). The operating principle of the microwave detector is based on radiation of electromagnetic radio frequency (RF) waves toward a protected area. The most common frequencies are 10.525 GHz (X-band) and 24.125 GHz (K-band). These wavelengths are long enough ($\lambda = 3$ cm at X-band) to pass freely through most contaminants, such as airborne dust, and short enough to be reflected by larger objects.

The microwave part of the detector consists of a Gunn oscillator, an antenna, and a mixer diode. The Gunn oscillator is a diode mounted in a small precision cavity that, on application of power, oscillates at microwave frequencies. The oscillator produces electromagnetic waves, part of which is directed through an iris into a waveguide and focusing antenna that directs the radiation toward the object. Focusing characteristics of the antenna are determined by the application. As a general rule, the narrower the directional diagram of the antenna, the more sensitive it is (the antenna has a higher gain). Another general rule is that a narrow beam antenna is much larger, while a wide-angle antenna can be quite small. A typical radiated power of the transmitter is 10 mW to 20 mW.

An antenna transmits the frequency f_0, which is defined by the wavelength λ_0 as:

$$f_0 = \frac{c_0}{\lambda_0} \tag{10.106}$$

where c_0 is the speed of light. When the target moves toward or away from the transmitting antenna, the frequency of the reflected radiation will change. Thus, if the target is moving away with velocity v, the reflected frequency will decrease, and it will increase for the approaching targets. This is called the *Doppler effect,* after the Austrian scientist Christian Johann Doppler (1803–1853). While the effect was first discovered for sound, it is applicable to electromagnetic radiation as well. However, in contrast to sound waves that may propagate with velocities dependent on movement of the source of the sound, electromagnetic waves propagate with speed of light, which is an absolute constant. The frequency of reflected electromagnetic waves can be predicted by the theory of relativity as:

$$f_r = f_0 \frac{\sqrt{1 - \left(v/c_0\right)^2}}{1 + v/c_0} \tag{10.107}$$

For practical purposes, however, the quantity $(v/c_0)^2$ is very small compared with unity; hence, it can be ignored. Then, the equation for the frequency of the reflected waves becomes identical to that for the acoustic waves:

$$f_r = f_0 \frac{1}{1 + v/c_0} \tag{10.108}$$

Due to a Doppler effect, the reflected waves have a different frequency f_r. A mixing diode combines the radiated (reference) and reflected frequencies and, being a nonlinear device, produces a signal that contains multiple harmonics of both frequencies.

The Doppler frequency in the mixer can be found from:

$$\Delta f = f_0 - f_r = f_0 \frac{1}{c_0/v + 1} \tag{10.109}$$

and since $c_0/v \gg 1$, the following holds after substituting Equation 10.106:

$$\Delta f \approx \frac{v}{\lambda_0} \tag{10.110}$$

Therefore, the signal frequency at the output of the mixer is linearly proportional to the velocity of a moving target. For example, if a person walks toward the detectors with a velocity of 0.6 m s^{-1}, a Doppler frequency for the X-band detector is $\Delta f = 0.6/0.03 = 20$ Hz.

Equation 10.110 holds true only for movements in the normal direction. When the target moves at angles Θ with respect to the detector, the Doppler frequency is:

$$\Delta f \approx \frac{v}{\lambda_0} \cos\Theta \tag{10.111}$$

Micropower Impulse Radar

In 1993, Lawrence Livermore National Laboratory developed a *micropower impulse radar* (MIR), which is a low-cost, noncontact ranging sensor [3]. The operating principle of the MIR is fundamentally the same as a conventional pulse radar system, but with several significant differences. The MIR consists of a noise generator whose output signal triggers a pulse generator. Each pulse has a fixed short duration, while the repetition of these pulses is random, according to triggering by the noise generator. The pulses are spaced randomly with respect to one another in a Gaussian noise-like pattern. It can be said that the pulses have the pulse frequency modulation (PFM) by white noise with maximum index of 20%. In turn, the square-wave pulses cause amplitude modulation (AM) of a radio transmitter. The radio transmitter produces short bursts of high-frequency radio signal that propagate from the transmitting antenna to the surrounding space. The electromagnetic waves reflect from the objects and propagate back to the radar. The same pulse generator that modulates the transmitter, gates (with a predetermined delay) the radio receiver to enable the output of the MIR only during a specific time window. Another reason for gating the receiver is to reduce its power consumption. The reflected pulses are received, demodulated (the square-wave shape is restored from the radio signal), and the time delay with respect to the transmitted pulses is measured. Since the pulses are spaced randomly, practically any number of identical MIR systems can operate in the same space without a frequency division (i.e., they work at the same carrier frequency within the same bandwidth). There is little chance that bursts from the interfering transmitters overlap and, if they do, the interference level is significantly reduced by the averaging circuit.

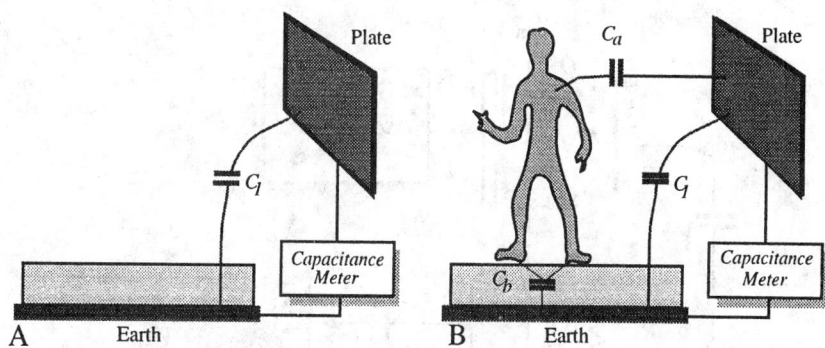

FIGURE 10.37 An intruder brings in additional capacitance to a detection circuit.

Capacitive Occupancy Detectors

Being a conductive medium with a high dielectric constant, a human body develops a coupling capacitance to its surroundings. This capacitance greatly depends on such factors as body size, clothing, materials, type of surrounding objects, weather, etc. However wide the coupling range is, the capacitance can vary from a few picofarads to several nanofarads. When a person moves, the coupling capacitance changes, thus making it possible to discriminate static objects from the moving ones. In effect, all objects form some degree of capacitive coupling with respect to one another. If a human (or for that purpose, anything) moves into the vicinity of the objects whose coupling capacitance with each other has been previously established, a new capacitive value arises between the objects as a result of the presence of an intruding body [3]. Figure 10.37 shows that the capacitance between a test plate and Earth is equal to C_1. When a person moves into the vicinity of the plate, it forms two additional capacitors: one between the plate and its own body C_a, and the other between the body and the Earth, C_b. Then, the resulting capacitance C between the plate and the Earth becomes larger by ΔC.

$$C = C_1 + \Delta C = C_1 + \frac{C_a C_b}{C_a + C_b} \qquad (10.112)$$

With the appropriate apparatus, this phenomenon can be used for occupancy detection [3]. What is required is to measure a capacitance between a test plate (the probe) and a reference plate (the Earth).

Figure 10.38 illustrates a circuit diagram for detecting variations in the probe capacitance C_p [4]. That capacitance is charged from a reference voltage source V_{ref} through a gate formed by transistor Q_1 when the output voltage of a control oscillator goes low. When it goes high, transistor Q_1 closes while Q_2 opens. The probe capacitance C_p discharges through a constant-current sink constructed with a transistor Q_3. A capacitor C_1 filters the voltage spikes across the transistor. The average voltage, e_p, represents a value of the capacitor C_p. When an intruder approaches the probe, the latter's capacitance increases, which results in a voltage rise across C_1. The voltage change passes through the capacitor C_2 to the input of a comparator with a fixed threshold V_T. The comparator produces the output signal V_{out} when the input voltage exceeds the threshold value.

When a capacitive occupancy (proximity) sensor is used near or on a metal device, its sensitivity can be severely reduced due to capacitive coupling between the electrode and the device's metallic parts. An effective way to reduce that stray capacitance is to use driven shields [3].

Triboelectric Detectors

Any object can accumulate, on its surface, static electricity. These naturally occurring charges arise from the triboelectric effect; that is, a process of charge separation due to object movements, friction of clothing

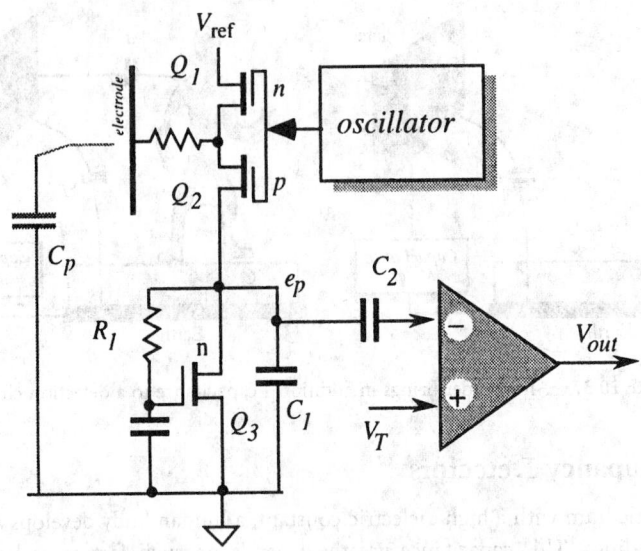

FIGURE 10.38   Circuit diagram for a capacitive intrusion detector.

fibers, air turbulence, atmosphere electricity, etc. Usually, air contains either positive or negative ions that can be attracted to the human body, thus changing its charge. Under idealized static conditions, an object is not charged — its bulk charge is equal to zero. In reality, any object that, at least temporarily, is isolated from the ground can exhibit some degree of its bulk charge imbalance. In other words, it becomes a carrier of electric charges that become a source of electric field. The field, or rather its changes, can be detected by an electronic circuit having a special pick-up electrode at its input [3, 5]. The electrode increases the capacitive coupling of the circuit's input with the environment, very much like in the capacitive detectors described above. The electrode can be fabricated in the form of a conductive surface that is well isolated from the ground.

If a charge carrier (a human or an animal) changes its position — moves away or a new charge carrying an object enters into the vicinity of the electrode — the static electric field is disturbed. The strength of the field depends on the atmospheric conditions and the nature of the objects. For example, a person in dry man-made clothes walking along a carpet carries a million times stronger charge than a wet intruder who has come from the rain.

It should be noted that contrary to a capacitive motion detector, which is an active sensor, a triboelectric detector is passive; that is, it does not generate or transmit any signal. There are several possible sources of interference that can cause spurious detections by the triboelectric detectors. That is, the detector may be subjected to transmitted noise resulting in false positive detection. Among the noise sources are 60 Hz or 50 Hz power line signals, electromagnetic fields generated by radio stations, power electric equipment, lightnings, etc. Most of these interferences generate electric fields that are distributed around the detector quite uniformly and can be compensated for by employing a differential input circuit with a significant common mode rejection ratio.

Optoelectronic Motion Detectors

Optoelectronic motion detectors rely on electromagnetic radiation in the optical range, specifically having wavelengths from 0.4 µm to 20 µm. This covers visible, near- and part of far-infrared spectral ranges. The detectors are primarily used for the indication of movement of people and animals. They operate over distance ranges up to several hundred meters and, depending on the particular need, can have either a narrow or wide field of view.

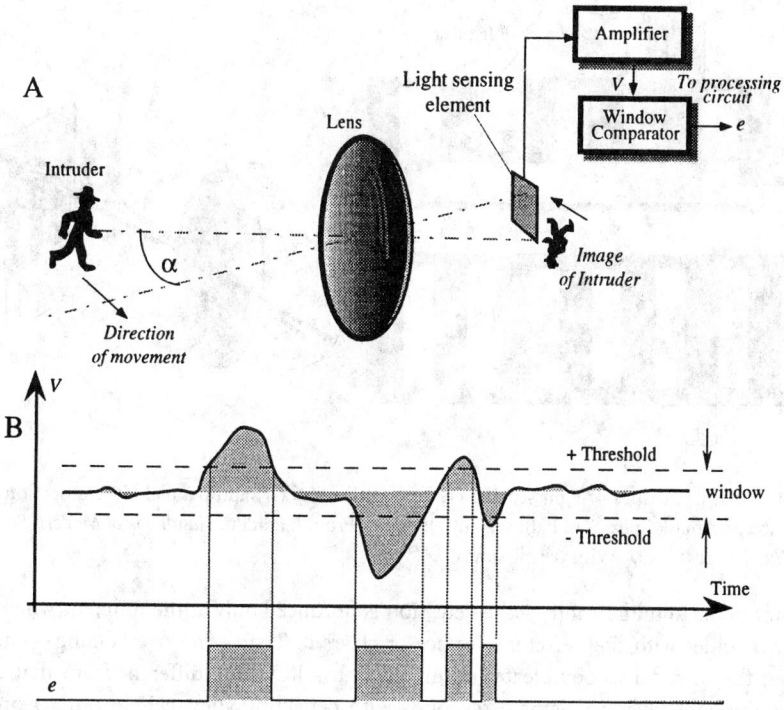

FIGURE 10.39 General arrangement of an optoelectronic motion detector. A lens forms an image of a moving object (intruder). When the image crosses the optical axis of the sensor, it superimposes with the sensitive element (*A*). The element responds with the signal that is amplified and compared with two thresholds in the window comparator (*B*). (From J. Fraden, *Handbook of Modern Sensors*, 2nd ed., Woodburg, NY: AIP Press, 1997. With permission.)

Most of the objects (apart from very hot) radiate electromagnetic waves only in the mid- and far-infrared spectral ranges. Hence, visible and near-infrared light motion detectors must rely on an additional source of light to illuminate the object. The light is reflected by the object's body toward the focusing device for subsequent detection. Such illumination can be sunlight or the invisible infrared light from an additional near-infrared light source (a projector).

The major application areas for the optoelectronic motion detectors are in security systems (to detect intruders), in energy management (to turn lights on and off), and in the so-called "smart" homes where they can control various appliances such as air conditioners, cooling fans, stereo players, etc. They can also be used in robots, toys, and novelty products. The most important advantage of an optoelectronic motion detector is simplicity and low cost.

Sensor Structures

A general structure of an optoelectronic motion detector is shown in Figure 10.39(*A*). Regardless what kind of sensing element is employed, the following components are essential: a focusing device (a lens or a focusing mirror), a light detecting element, and a threshold comparator. An optoelectronic motion detector resembles a photographic camera. Its focusing components create an image of its field of view on a focal plane. While there is no mechanical shutter like in a camera, in place of the film, a light sensitive element is used. The element converts the focused light into an electric signal. A focusing lens creates an image of the surroundings on a focal plane where the light sensitive element is positioned. If the area is unoccupied, the image is static and the output signal from the element is steady stable. When an "intruder" enters the room and keeps moving, his/her image on the focal plane also moves. In a certain moment, the intruder's body is displaced for an angle α and the image overlaps with the element. This

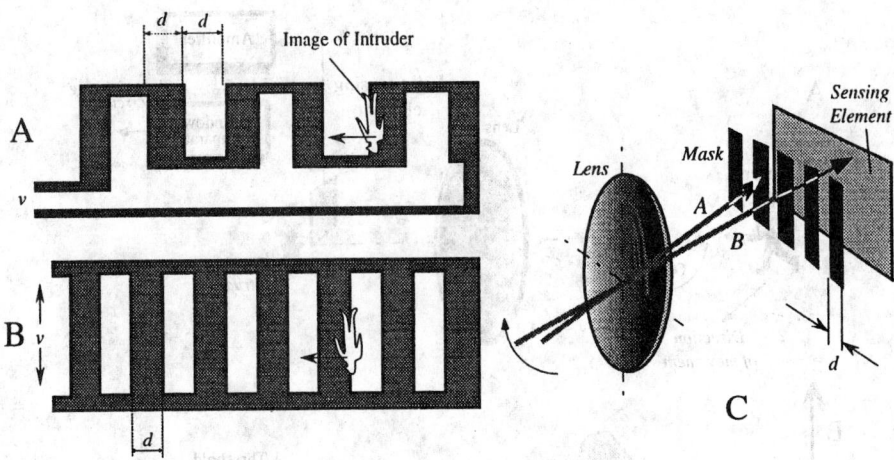

FIGURE 10.40 Complex shapes of light sensing element with series (A) and parallel (B) connection of segments. (C) shows use of a grid mask in front of light sensing element. (From J. Fraden, *Handbook of Modern Sensors*, 2nd ed., Woodburg, NY: AIP Press, 1997. With permission.)

is an important point to understand — the detection is produced only at the moment when the object's image either coincides with the detector's surface or clears it. That is, no overlapping — no detection. Assuming that the intruder's body creates an image with a light flux different from that of the static surroundings, the light-sensitive element responds with deflecting voltage V. In other words, to cause detection, a moving image will have a certain degree of optical contrast with its surroundings.

Figure 10.39(B) shows that the output signal is compared with two thresholds in the window comparator. The purpose of the comparator is to convert the analog signal V into two logic levels: ø, no motion detected and 1, motion is detected.

To increase area of coverage, an array of detectors can be placed in the focal plane of a focusing mirror or lens. Each individual detector covers a narrow field of view, while in combination they protect larger areas. All detectors in the array should either be multiplexed or otherwise interconnected to produce a combined detection signal; that is, they can be made into a complex sensor shape. An alternative solution is the use of a multiple-element focusing system.

Complex Sensor Shape

If the detector's surface area is sufficiently large to cover an entire angle of view, it may be optically broken into smaller elements, thus creating an equivalent of a multiple detector array. To break the surface area into several parts, one can shape the sensing element in an odd pattern, like the interdigitized shape shown if Figure 10.40(A) or parallel grid as in Figure 10.40(B). Each part of the sensing element acts as a separate light detector.

The parallel or serially connected detectors generate a combined output signal, for example, voltage v, when the image of the object moves along the element surface crossing alternatively sensitive and nonsensitive areas. This results in an alternate signal v at the detector terminals. Each sensitive and nonsensitive area must be sufficiently large to overlap with most of the object's image. An alternative solution to the complex shape of the sensor is use of the image distortion mask as shown in Figure 10.40(C); however, this solution requires a larger sensor surface area.

Facet Focusing Element

A cost-effective way of broadening the field of view while employing a small-area detector is to use multiple focusing devices. A focusing mirror or a lens may be divided into an array of smaller mirrors or lenses called facets, just like in the eye of a fly. Each facet creates its own image resulting in multiple images, as shown in Figure 10.41. When the object moves, the images also move across the element,

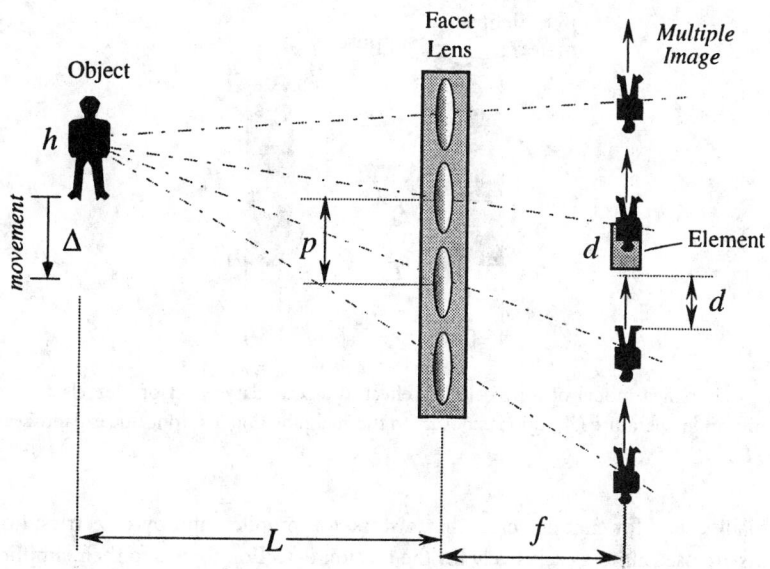

FIGURE 10.41 Facet lens creates multiple images near the sensing element.

resulting in an alternate signal. By combining multiple facets, it is possible to create any desirable detecting pattern in the field of view, in both horizontal and vertical planes. Positioning of the facet lens, focal distances, number, and a pitch of the facets (a distance between the optical axes of two adjacent facets) can by calculated in every case by applying rules of geometrical optics [3]. In the far-infrared spectral range (thermal radiation sensors), the polyethylene facet Fresnel lenses are used almost exclusively, thanks to their low cost and relatively high efficiency.

For the visible portion of the optical spectrum, a simple, very inexpensive, yet efficient motion detector can be developed for nondemanding applications, like light control or interactive toys, using simple photoresistors and pinhole lenses [3, 6, 7].

Far-Infrared Motion Detectors

A motion detector that perceives electromagnetic radiation that is naturally emitted by any object operates in the optical range of thermal radiation, also called far-infrared (FIR). Such detectors are responsive to radiative heat exchange between the sensing element and the moving object. The principle of thermal motion detection is based on the physical theory of emission of electromagnetic radiation from any object whose temperature is above absolute zero (see Chapter 32, Section 6, on *Infrared Thermometers*).

For IR motion detection, it is essential that a surface temperature of an object be different from that of the surrounding objects, so a thermal contrast would exist. All objects emanate thermal radiation from their surfaces and the intensity of that radiation is governed by the Stefan–Boltzmann law. If the object is warmer than the surroundings, its thermal radiation is shifted toward shorter wavelengths and its intensity becomes stronger. Many objects whose movement is to be detected are nonmetals, hence they radiate thermal energy quite uniformly within a hemisphere. Moreover, the dielectric objects generally have a high emissivity. Human skin is one of the best emitters, with emissivity over 90%, while most fabrics also have high emissivities, between 0.74 and 0.95 [3]. Below, two types of far-infrared motion detectors are described. The first utilizes a passive infrared (PIR) sensor, while the second has active far-infrared (AFIR) elements.

PIR Motion Detectors

These detectors became very popular for security and energy management systems. The PIR sensing element must be responsive to far-infrared radiation within a spectral range from 4 μm to 20 μm where most of the thermal power emanated by humans is concentrated. There are three types of sensing elements

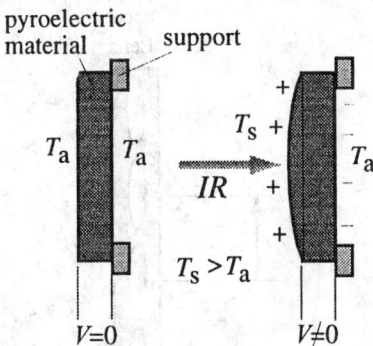

FIGURE 10.42 A simplified model of a pyroelectric effect as a secondary effect of piezoelectricity. Initially, the element has a uniform temperature (*A*); upon exposure to thermal radiation, its front side expands, causing a stress-induced charge (*B*).

that are potentially useful for that detector: thermistors, thermopiles, and pyroelectrics; however, pyro-electric elements are used almost exclusively for the motion detection thanks to their simplicity, low cost, high responsivity, and broad dynamic range. A pyroelectric effect is described in Chapter 32, Section 7 on *Pyroelectric Thermometers*. How this effect can be used in practical sensor design is discussed here.

A pyroelectric material generates an electric charge in response to thermal energy flow through its body. In a simplified way it may be described as a secondary effect of thermal expansion (Figure 10.42). Since all pyroelectrics are also piezoelectrics, the absorbed heat causes the front side of the sensing element to expand. The resulting thermally induced stress leads to the development of a piezoelectric charge on the element electrodes. This charge is manifested as voltage across the electrodes deposited on the opposite sides of the material. Unfortunately, the piezoelectric properties of the element also have a negative effect. If the sensor is subjected to a minute mechanical stress due to any external force, it also generates a charge that in most cases is indistinguishable from that caused by the infrared heat waves. Sources of such mechanical noise are wind, building vibrations, loud sound, etc.

To separate thermally induced charges from the piezoelectrically induced charges, a pyroelectric sensor is usually fabricated in symmetrical form (Figure 10.43(*A*)). Two identical elements are positioned inside the sensor's housing. The elements are connected to the electronic circuit in such a manner as to produce the out-of-phase signals when subjected to the same in-phase inputs. The idea is that interferences produced by, for example, the piezoelectric effect or spurious heat signals are applied to both electrodes simultaneously (in phase) and thus will be canceled at the input of the circuit, while the variable thermal radiation to be detected will be absorbed by only one element at a time, thus avoiding a cancellation.

One way to fabricate a differential sensor is to deposit two pairs of electrodes on both sides of a pyroelectric element. Each pair forms a capacitor that can be charged either by heat or by mechanical stress. The electrodes on the upper side of the sensor are connected together forming one continuous electrode, while the two bottom electrodes are separated, thus creating the opposite-serially connected capacitors. Depending on the side where the electrodes are positioned, the output signal will have either a positive or negative polarity for the thermal influx. In some applications, a more complex pattern of the sensing electrodes is required (for example, to form predetermined detection zones), so that more than one pair of electrodes is needed. In such a case, for better rejection of the in-phase signals (common mode rejection), the sensor should still have an even number of pairs where positions of the pairs alternate for better geometrical symmetry. Sometimes, such an alternating connection is called an interdigitized electrode.

A differential sensing element should be mounted in such a way as to ensure that both parts of the sensor generate the same signal if subjected to the same external factors. At any moment, the optical component must focus a thermal image of an object on the surface of one part of the sensor only, which is occupied by a single pair of electrodes. The element generates a charge only across the electrode pair that is subjected to a heat flux. When the thermal image moves from one electrode to another, the current

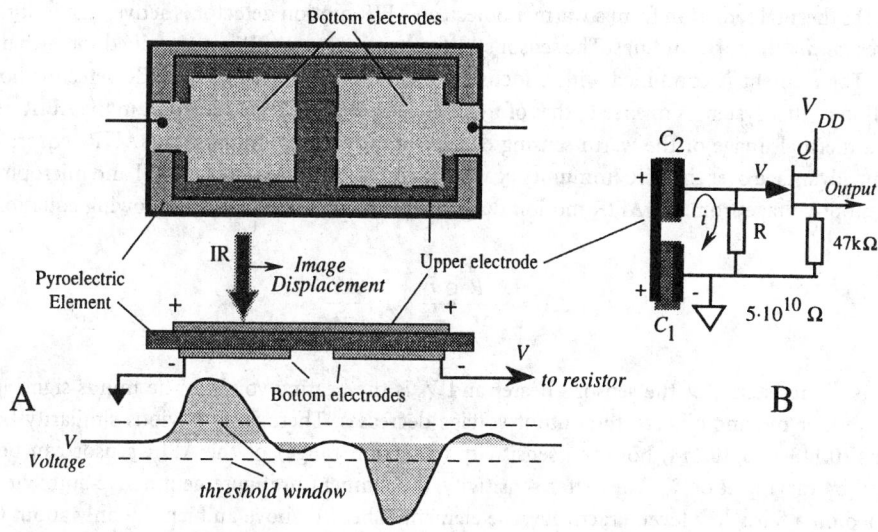

FIGURE 10.43 Dual pyroelectric sensor. (*A*) A sensing element with a front (upper) electrode and two bottom electrodes deposited on a common crystalline substrate. (*B*) A moving thermal image travels from the left part of the sensor to the right, generating an alternating voltage across bias resistor, *R*.

i flowing from the sensing element to the bias resistor *R* (Figure 10.43(*B*)) changes from zero, to positive, then to zero, to negative, and again to zero (Figure 10.43(*A*) lower portion). A JFET transistor *Q* is used as an impedance converter. The resistor *R* value must be very high. For example, a typical alternate current generated by the element in response to a moving person is on the order of 1 pA (10^{-12} A). If a desirable output voltage for a specific distance is $v = 50$ mV, then according to Ohm's law, the resistor value is $R = v/i = 50$ GΩ (5×10^{10} Ω). Such a resistor cannot be directly connected to a regular electronic circuit; hence, transistor *Q* serves as a voltage follower (the gain is close to unity). Its typical output impedance is on the order of several kilohms.

The output current *i* from the PIR sensor can be calculated on the basis of the Stefan–Boltzmann law as [3]:

$$i \approx \frac{2P\sigma a\gamma}{\pi h c} bT_s^3 \frac{\Delta T}{L^2} \tag{10.113}$$

where $\Delta T = (T_b - T_a)$ is the temperature gradient between the object and its surroundings, *P* is the pyroelectric coefficient, σ is the Stefan-Boltzmann constant, *a* is the lens area, γ is the lens transmission coefficient, *h* is the thickness, and *c* is the specific heat of the pyroelectric element, respectively, and *L* is the distance to the object.

There are several conclusions that can be drawn from Equation 10.113. The first part of the equation (the first ratio) characterizes a detector, while the rest relates to an object. The pyroelectric current *i* is directly proportional to the temperature difference (thermal contrast) between the object and its surroundings. It is also proportional to the surface area of the object that faces the detector. A contribution of the ambient temperature T_a is not as strong as it might appear from its third power. The ambient temperature must be entered in kelvin, hence its variations become relatively small with respect to the scale. The thinner the sensing element, the more sensitive the detector. The lens area also directly affects signal magnitude. On the other hand, pyroelectric current does not depend on the sensor's area as long as the lens focuses an entire image on a sensing element.

AFIR Motion Detectors

The AFIR motion detector is a new class of thermal sensors whose operating principle is based on balancing thermal power supplied to the sensing element [8, 9]. Contrary to a passive motion detector that absorbs thermal radiation from a warmer object, an AFIR motion detector is active; that is, it radiates heat waves *toward* the surroundings. The sensor's surface temperature (T_s) is maintained somewhat above ambient. The element is combined with a focusing system, very much like the PIR detector; however, the function of that system is inverse to that of the passive detectors. A focusing part in the AFIR detector projects a thermal image of the warm sensing element into its surroundings. The AFIR sensors have a significant advantage over the PIR: immunity against many interferences (such as RFI and microphonics).

The output voltage from the AFIR motion detector can be described by the following equation [3]:

$$\Delta V \approx -\frac{R}{V_0}\frac{\sigma a \gamma}{\pi}bT_s^3\frac{\Delta T}{L^2} \qquad (10.114)$$

where R is the resistance of the sensor's heater and V_0 is the heating voltage. The minus sign indicates that for warmer moving objects, the output voltage decreases. There is an obvious similarity between Equations 10.113 and 10.114; however, sensitivity (detection range) of the AFIR sensor can be easily controlled by varying R or V_0. For better sensitivity, the temperature increment above ambient can be maintained on a fairly low level. Practically, the element is heated above ambient by only about 0.2° C.

References

1. S. Blumenkrantz, *Personal and Organizational Security Handbook,* Government Data Publications, Washington, D.C.: 1989.
2. P. Ryser and G. Pfister, Optical fire and security technology: sensor principles and detection intelligence, *Transducers'91. Int. Conf. Solid-State Sensors Actuators,* 1991, 579-583.
3. J. Fraden, *Handbook of Modern Sensors,* 2nd ed., Woodburg, NY: AIP Press, 1997.
4. N. M. Calvin, *Capacitance proximity sensor.* U.S. Patent No. 4,345,167, 1982.
5. J. Fraden, *Apparatus and method for detecting movement of an object,* U.S. Patent No. 5,019,804, 1991.
6. J. Fraden, *Motion discontinuance detection system and method.* U.S. Patent No. 4,450,351, 1984.
7. J. Fraden, *Toy including motion-detecting means for activating same.* U.S. Patent No. 4,479,329, 1984.
8. J. Fraden, *Active infrared motion detector and method for detecting movement.* U.S. Patent No. 4,896,039, 1990.
9. J. Fraden, Active far infrared detectors, in *Temperature. Its Measurement and Control in Science and Industry,* Vol. 6, Woodburg, NY: American Institute of Physics, 1992, Part 2, 831-836.

11

Level Measurement

Level is defined as the filling height of a liquid or bulk material, for example, in a tank or reservoir. Generally, the position of the surface is measured relative to a reference plane, usually the tank bottom. If the product's surface is not flat (e.g., with foam, waves, turbulences, or with coarse-grained bulk material) level usually is defined as the average height of a bounded area.

Various classic and modern methods exist to measure product level in process and storage tanks in the chemical, petrochemical, pharmaceutical, water, and food industries, in mobile tanks on vehicles and ships, but also in natural reservoirs like seas, dams, lakes, and oceans. Typical tank heights are approximately between 0.5 m and 40 m.

Two different tasks can be distinguished: (1) continuous level measurements (level indication, LI), and (2) level switches (LS) (e.g., to detect an alarm limit to prevent overfilling). Figure 11.1 shows the principal operational modes of level measurement. Every continuous system can also be used as a programmable switch. Many level devices are mounted on top of the tank and measure primarily the distance d between their mounting position and the product's surface. The level L is then calculated, defining the tank height h as constant, as shown in Figure 11.1 and expressed as:

$$L = h - d \qquad (11.1)$$

The following examples describe primarily the measurement of liquids, but most of the methods can also be applied to solids (bulk material). The emphasis of this chapter will be general information about the measurement principles. The focus of the descriptions is on the methods most commonly practiced; other principles are mentioned less comprehensively. Readers interested in more detailed discussions may refer to [1–5].

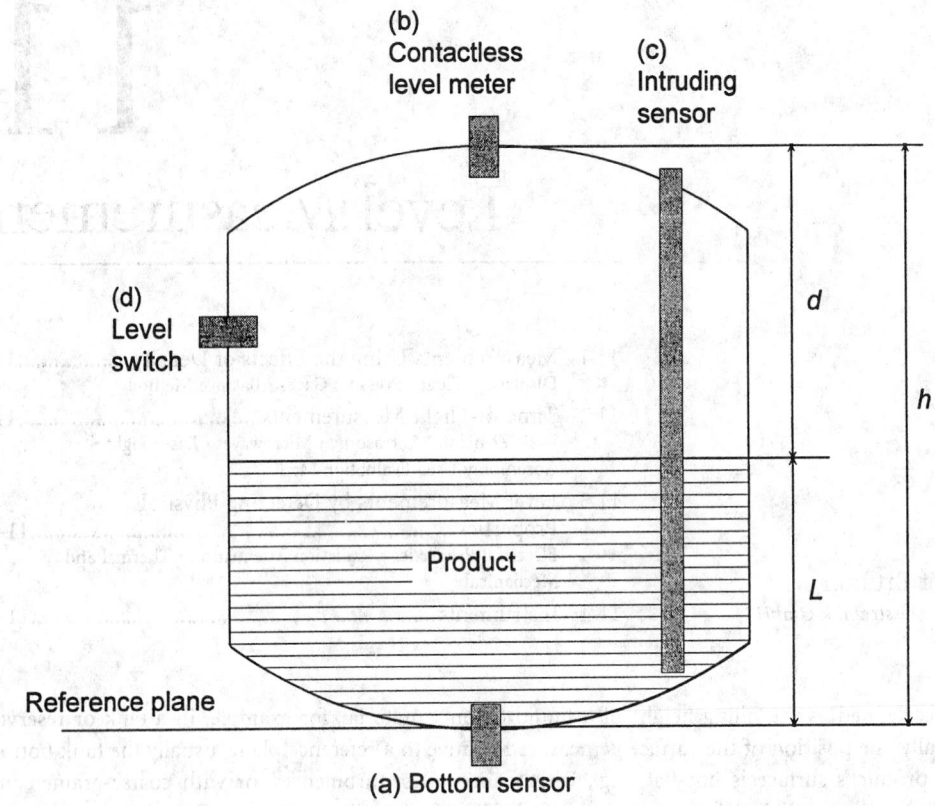

FIGURE 11.1 Representation of a tank with a liquid or solid material (hatched area), the product to be measured. The level sensor can be mounted (*a*) contacting product at the bottom, (*b*) as a contactless instrument on top, (*c*) as an intrusive sensor, or (*d*) at the sides as a level switch.

11.1 Measurements Using the Effects of Density

All methods described in this chapter have in common that the product in the tank has an effect due to its density ρ, (1) producing buoyancy to a solid submerged into the liquid, or (2) executing a force due to its weight.

Displacer

Displacers measure the buoyancy of a solid body that is partially submerged in the liquid. The change in weight is measured. Figure 11.2 illustrates the parameters used for these calculations. The cross section A of the body is assumed to be constant over its length b. The weight of force F_G due to gravity g and mass m is:

$$F_G = g\, m = g\, Ab\rho_D \qquad (11.2)$$

The buoyant force F_B accounts for the partial length L_d that is submerged with the remainder of the body in the atmosphere:

$$F_B = g\, A L_d\, \rho_L + g\, A\left(b - L_d\right)\rho_A \qquad (11.3)$$

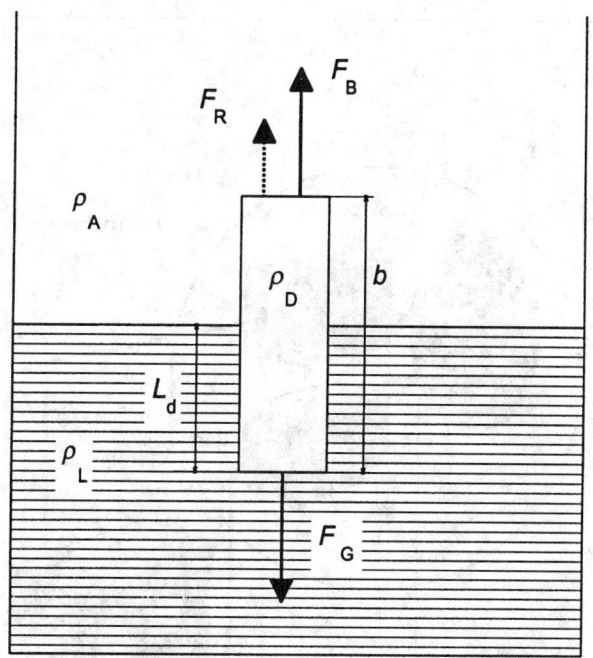

FIGURE 11.2   Quantities of a solid body immersed into a liquid. The forces F can be calculated from Equations 2, 3, and 4. ρ = density; b = length of the body; L_d = dipped length.

Combining Equations 11.2 and 11.3 gives the resulting force to be measured by an appropriate method (see Chapter 23 of this handbook):

$$F_R = F_G - F_B \tag{11.4}$$

The result for level L_d, related to the lower edge of the displacer is:

$$L_d = \frac{b(\rho_D - \rho_A) - \dfrac{F_R}{gA}}{\rho_L - \rho_A} \tag{11.5}$$

The density of the body should be higher than the density of the liquid; otherwise, the measurement operating range is limited (until the displacer floats on the liquid). In another version, a servo-gage moves the displacer up and down to detect the interface between the atmosphere and a liquid, or between two different liquids, by measuring the change in buoyancy. Figure 11.3 shows a special configuration, in which a small ball with volume V is mounted to a thin wire driven by a stepping motor and put into resonant vibration. The resulting force F can be measured from the resonating frequency f of the wire between points A and B:

$$F = \rho_W \, A_W \, 4 f^2 l^2 \tag{11.6}$$

where l = length of the wire between the points A and B
   ρ_w = density of the wire
   A_w = cross-sectional area of the wire

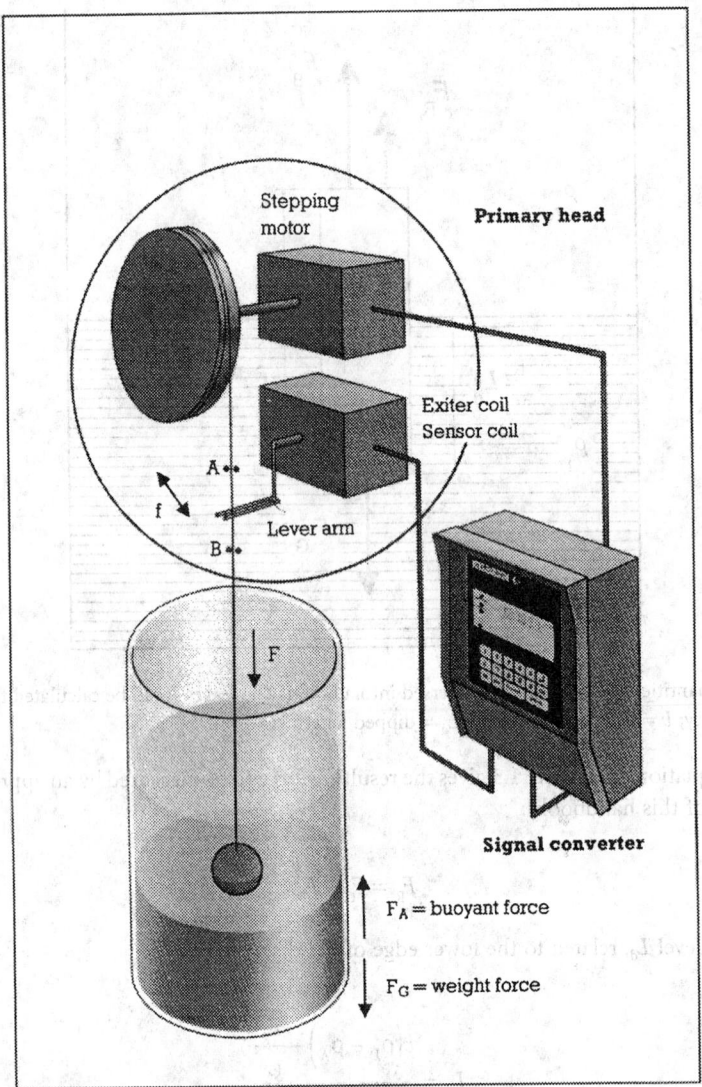

FIGURE 11.3 Level, interface and density sensor using the effects of buoyancy. A stepping motor drives the small ball attached to the thin wire to different heights in the liquid or to the interfaces. The resulting force F as a difference between weight force and buoyant force is measured from the resonant frequency of the wire-ball system. The lever arm excites the wire into oscillation and a sensor coil coupled to the lever arm measures its frequency. The signal converter controls the stepping motor and calculates the measured values [6].

And the surrounding density ρ_L can be calculated:

$$F = gV\left(\rho_D - \rho_L\right) \Leftrightarrow \rho_L = \rho_D - \frac{F}{gV} \tag{11.7}$$

Float

Floats are similar to displacers, but are swimming on the liquid's surface due to the buoyancy. Hence, the density of the float must be lower than the density of the liquid. Figure 11.4(*a*) shows the principle

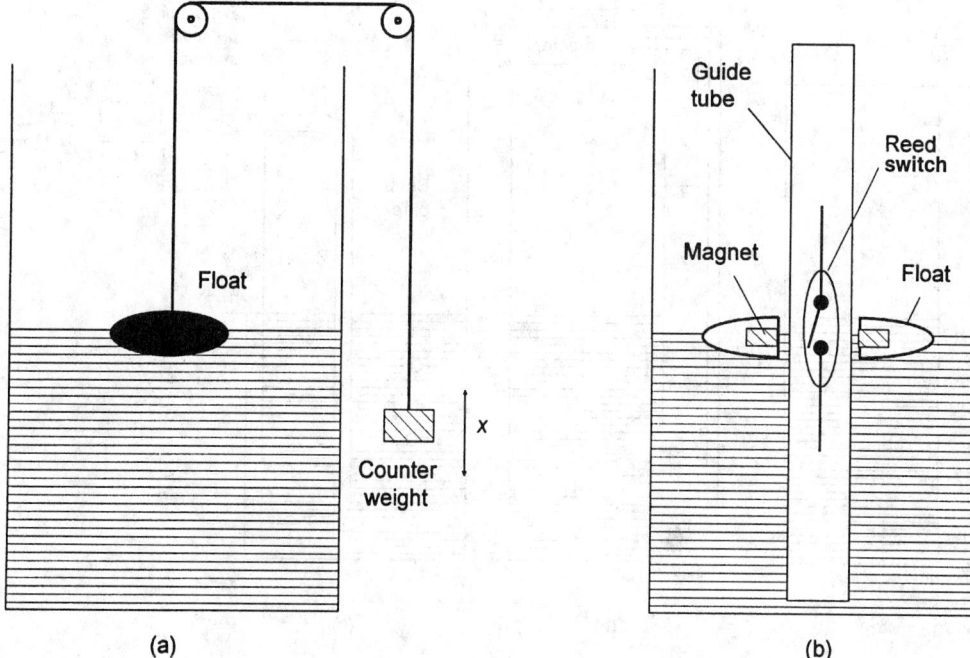

FIGURE 11.4 Principle of operation for float level meters. (*a*) A counter weight balances the float that swims on the liquid's surface. Its position represents the level. (*b*) The float contains a magnet that contacts a reed switch inside a guide tube. Using a bistable relay, this system is used as a level switch. One can also insert multiple relays into the tube to achieve different switching points for quasicontinuous operation.

of operation. The position of the float is (1) observed visually, or (2) transferred to an external display or to an angle transmitter. In general, the float is coupled to the transmitter magnetically. Figure 11.4(*b*) shows a level switch, using a reed relay magnetically coupled with the float. Also, a magnetostrictive linear sensor may determine the position of the float. For more information about this, refer to Chapter 6, Section 9 of this handbook.

If the float is very flat, it is called a "sensing plate". This plate is mechanically guided, e.g., by a servo control, on the surface until uplift is detected. For solids, specially shaped perpendicular floats are helpful.

Pressure Gages

A hydrostatic pressure p, caused by the weight of the product, is present at the bottom of a tank, in addition to the atmospheric pressure p_0:

$$p = p_0 + g\,\rho_L\,L \Leftrightarrow L = \frac{p - p_0}{g\,\rho_L} \tag{11.8}$$

Pressure gages at the bottom of the tank measure this pressure. In process tanks with varying atmospheric pressure, a differential pressure measurement is achieved by measuring the difference between the pressure at the bottom and that at the top of the tank, above the liquid. Figure 11.5(*a*) shows such a configuration with a differential pressure sensor. For techniques of pressure measurement, refer to Chapter 26, Section 1 of this handbook. Because measurement by hydrostatic pressure is proportional to the density, level errors result if density changes; see Equation 11.8. Primary pressure gaging is a mass measurement. Figure 11.5(*b*) shows a vertical arrangement with three sensors; the measurements of p_1 and p_2 are used to compensate for the influence of density ρ_L, and to calculate the level:

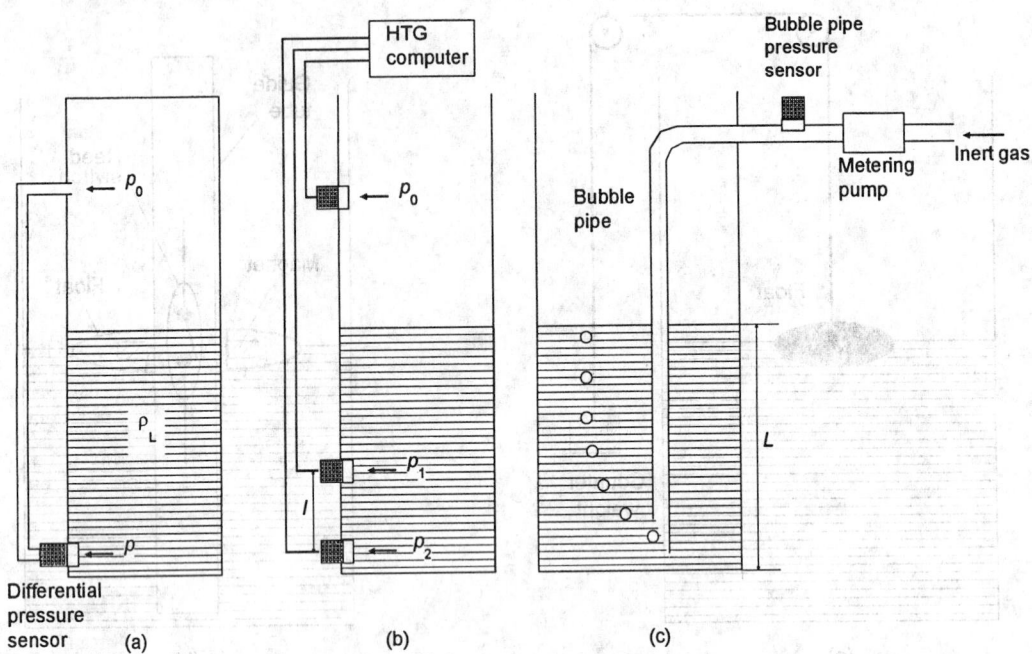

FIGURE 11.5 Level gaging by hydrostatic pressure measurement. The bottom pressure p is proportional to level. (a) The atmospheric pressure p_0 can be taken into consideration by a differential measurement. The low side of the differential pressure sensor is connected via a thin pipe to the top of the tank. (b) A differential measurement within the liquid is called "hydrostatic tank gaging, HTG" and can be used to for compensate errors due to density variations of the liquid. The signals from all three sensors are evaluated by a computer. (c) With a so-called "bubble tube," the sensor can be mounted on the top of the tank: an inert gas is injected into the tube such that bubbles of gas escape from the end of the tube. The flow rate of the gas is constant so the head pressure in the system can be measured at the inlet of the pipe.

$$\rho_L = \frac{p_2 - p_1}{g\,l} \Rightarrow L = \frac{p_2 - p_0}{p_2 - p_1}\,l \tag{11.9}$$

A system of this configuration is often called "hydrostatic tank gaging" (HTG). Figure 11.5(c) shows a further arrangement, called "bubble tube," in which the bottom pressure is transmitted to the top of the tank. This is often used for level gaging if the sensor cannot be mounted at the bottom of the tank. It requires a tank with pressure equalization due to the steady insertion of inert gas.

Balance Method

Here simply the weight F of the complete tank is measured, dependent on the level L:

$$F = F_0 + g\,AL\rho_L \tag{11.10}$$

where F_0 is the weight of the empty tank and A the cross-sectional area, which is assumed to be constant throughout the tank height. In order to measure the weight force correctly, it is necessary to isolate the complete tank mechanically. For precise measurements, the buoyancy in air must be taken into consideration:

$$F = F_0 + g\,AL\left(\rho_L - \rho_A\right) \Leftrightarrow L = \frac{F - F_0}{g\,A\left(\rho_L - \rho_A\right)} \tag{11.11}$$

For techniques of weight measurement, refer to Chapter 20 of this handbook.

This method has severe disadvantages when the tank is not inside a building. Outside, wind forces and the weight of snow and rain can cause errors.

11.2 Time-of-Flight Measurements

An indirect measurement of level is evaluating the time-of-flight of a wave propagating through the atmosphere above the liquid or solid. This is primarily a distance measurement; the level can then be calculated accordingly. The increasing demand of industry for nonintrusive continuous level gaging systems has been instrumental in accelerating the development of technologies using time-of-flight measurements [7].

Basic Principle

Although different types of physical waves (acoustic or electromagnetic) are applied, the principle of all these methods is the same: a modulated signal is emitted as a wave toward the product, reflected at its surface and received by a sensor, which in many cases is the same, (e.g., the ultrasonic piezoelectric transducer or the radar antenna). Figure 11.6 demonstrates the principle of operation. The measuring system evaluates the time-of-flight t of the signal:

$$t = \frac{2d}{v} \qquad (11.12)$$

where v is the propagation velocity of the waves.

One can generate an unmodulated pulse, a modulated burst as in Figure 11.6(b), or special forms. Table 11.1 lists the main properties of the three preferred types of waves, used for time-of-flight level gaging.

The very short time spans of only a few nanoseconds for radar and laser measurement techniques require the use of time expansion by sampling (see Chapter 85 of this handbook) or special evaluation methods (see below).

Ultrasonic

Ultrasonic waves are longitudinal acoustic waves with frequencies above 20 kHz. Ultrasonic waves need a propagation medium, which for level measurements is the atmosphere above the product being measured. Sound propagates with a velocity of about 340 m s^{-1} in air; but this value is highly dependent on temperature and composition of the gas, and also on its pressure (see Chapter 6, Section 7 of this handbook). In vacuum, ultrasonic waves cannot propagate. In practice, the reflection ratio is nearly 100% at the product's surface (e.g., at transitions gas/liquid or gas/solid). Piezoelectric transducers (see Chapter 26, Section 3 of this handbook) are utilized as emitter and detector for ultrasonic waves, a membrane coupling it to the atmosphere. The sensor is installed as in Figure 11.1(b), the signal form is as in Figure 11.6(b). Level gaging is, in principle, also possible with audible sound 16 Hz to 20 kHz or infrasonic waves less than 16 Hz.

Another procedure is to propagate the waves within the liquid by a sensor mounted at the bottom of the tank. The velocity of sound in the liquid must be known, considering the dependence on temperature and type of liquid. This method is similar to an echo sounder on ships for measuring the water depth. For more information about time-of-flight ultrasound evaluation techniques, refer to Chapter 6, Section 7 of this handbook.

Microwaves

Microwaves are generally understood to be electromagnetic waves with frequencies above 2 GHz and wavelengths of less than 15 cm. For technical purposes, microwave frequencies are used up to max. 120 GHz; in practice, the range around 10 GHz (X-band) is preferred.

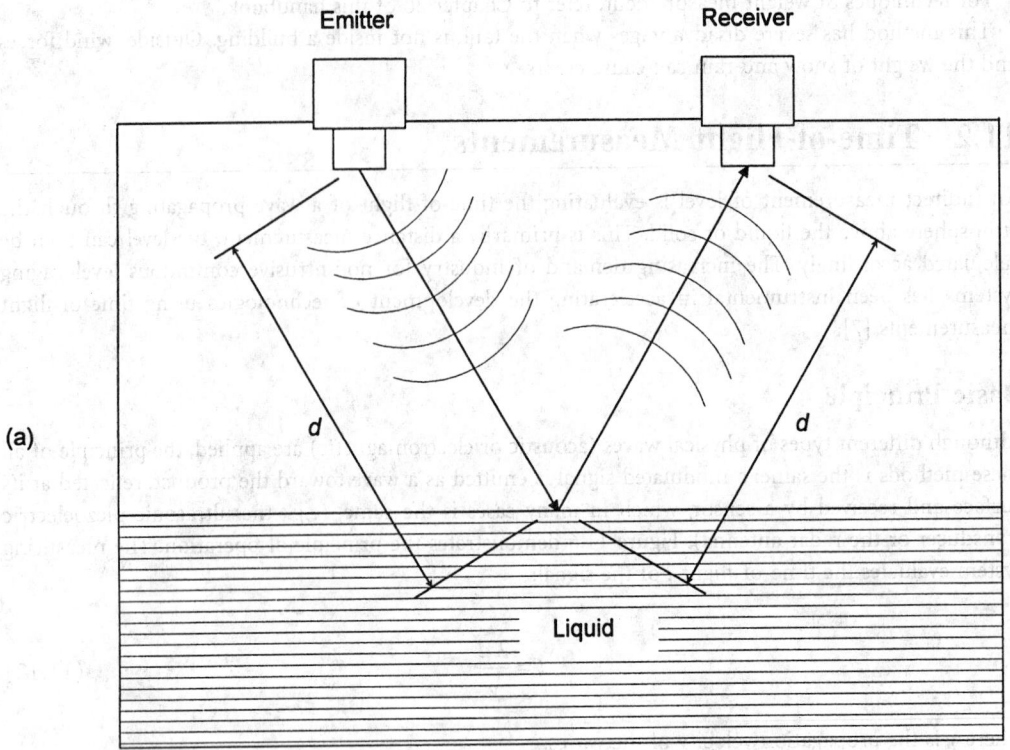

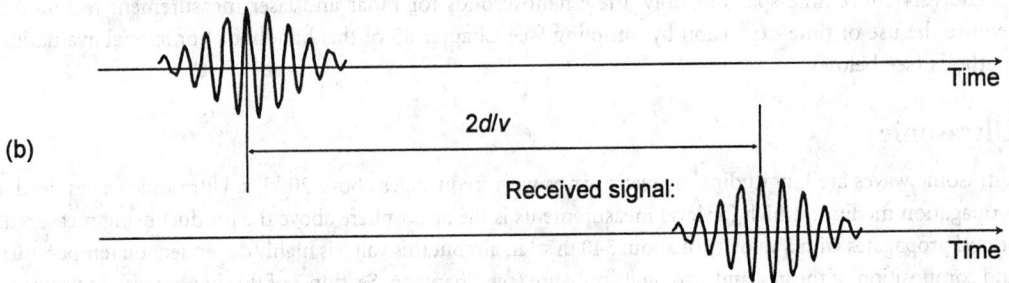

FIGURE 11.6 (*a*)Representation of time-of-flight measurements. The emitter couples a wave (ultrasonic or electromagnetic) into the atmosphere that propagates the wave toward the liquid. Its surface reflects the wave and a sensor receives it. (*b*) Due to the propagation velocity *v*, a time delay is measured between emission and receipt of the signal. This example is characterized by a modulated burst. The time scale is arbitrary.

TABLE 11.1 Properties of the wave types for time-of-flight measuring.

Principle	Wave Velocity	Avg. Carrier Frequency	Wavelength	Avg. Burst Time
Ultrasonic	340 m s^{-1}	50 kHz	7 mm	1 ms
Radar	300,000 km s^{-1}	10 GHz	3 cm	1 ns
Laser	300,000 km s^{-1}	300 THz	1 μm	1 ns

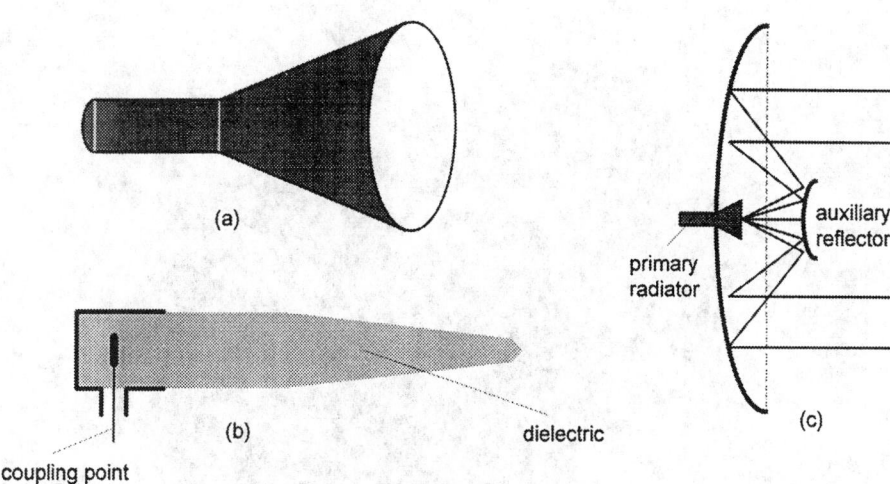

FIGURE 11.7 Practical antenna forms used for radar level instruments: (*a*) conical horn antenna, (*b*) dielectric rod antenna, and (*c*) parabolic mirror with a small antenna as primary radiator and an auxiliary reflector giving a very small beam angle (so-called Cassegrain model).

The usually applied time-of-flight measurements with microwaves are RADAR-based [8, 9]. The term "RADAR" is generally understood to mean a method by means of which short electromagnetic waves are used to detect distant objects and determine their location and movement. It is an acronym from RAdio Detection And Ranging. Figure 11.7 shows preferred antenna forms. They are usually combined with a compact sensor, as in Figure 11.8. For level measuring systems, a small radiation angle is desirable in order to avoid interfering reflections from the tank wall or tank internals as much as possible. The larger the aperture area, the smaller the radiation angle and the higher the antenna gain. The power balance is given by the general radar equation:

$$P_R = \frac{P_T\, G_T\, R G_R}{D^2} \tag{11.13}$$

where P_R = received power
P_T = transmitted power
G_T = transmitting antenna gain
R = reflection factor of target
G_R = receiving antenna gain
D^2 = propagation loss to and from the surface, due to power density decrease and atmospheric influences

The reflection factor R of the product's surface is dependent on the dielectric permittivity ε_r of the liquid or bulk material:

$$R = \frac{\left(\sqrt{\varepsilon_r} - 1\right)^2}{\left(\sqrt{\varepsilon_r} + 1\right)^2} \tag{11.14}$$

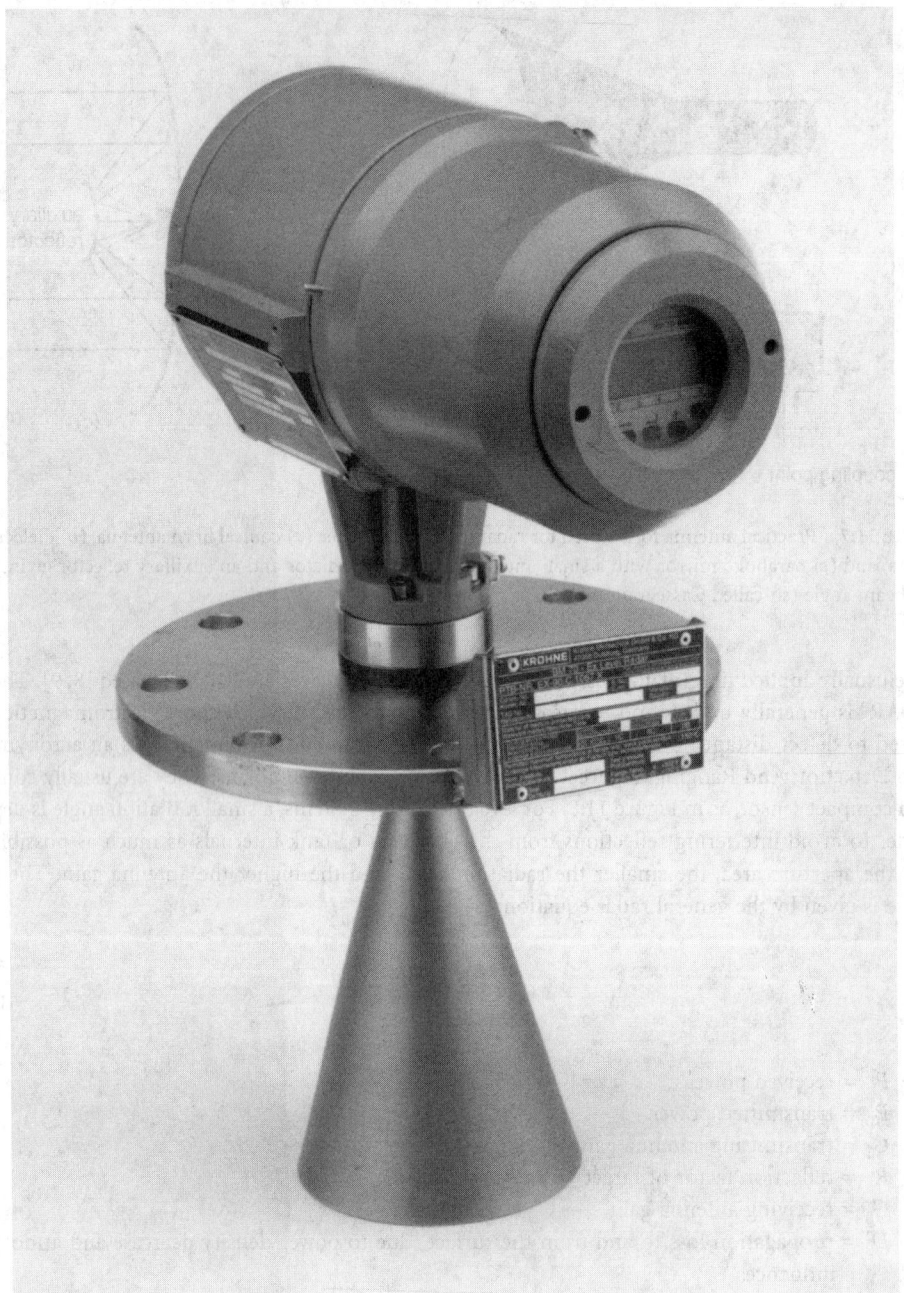

FIGURE 11.8 Design of a compact industrial level radar system. The converter above the flange includes the complete microwave circuitry, signal processing stages, microprocessor control, display, power supply, and output signal [6].

In level measurement situations, the reflecting area is so large that it intersects the beam cross section completely; therefore, D^2 is approximately proportional with distance d^2. Thus also, the received power decreases proportionately with d^2, as derived in [8]:

$$P_R \propto \frac{1}{d^2}$$

(11.15)

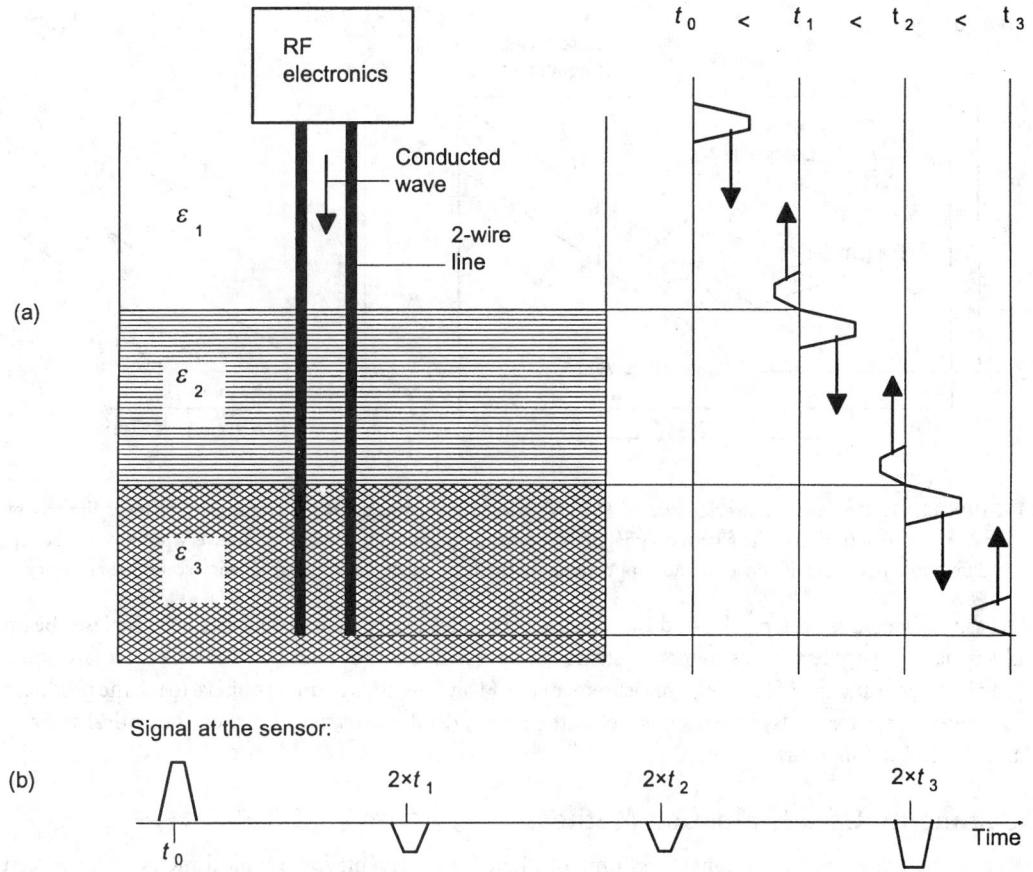

FIGURE 11.9 Principle of operation of a wire-conducting high-frequency level measurement system. (*a*) An electrical pulse is generated (time t_0) and a two-wire line guides the electromagnetic wave. At every position where the surrounding permittivity ε changes, a part of the wave is sharply reflected (time t_1) back to the sensor. The wave propagates along the entire line and is reflected a second time (t_2) at the interface between the two liquids, and a third time at the end of the line. (*b*) The signal delay times ($2t_1$, $2t_2$, and $2t_3$) represent the positions of the interfaces with respect to the end of the line, which can be used as a reference. The signal polarity is inverted due to the reflection from lower to higher permittivity. The time scale is arbitrary.

This is not the case if the waves propagate within an electromagnetic waveguide, like a vertical tube dipping into the liquid, called a stilling well. Here, the propagation is nearly without losses.

An alternative method using electromagnetic waves is to propagate them in a cable. Figure 11.9(*a*) illustrates the operation with a cable dipped into the liquid or bulk material. Where the dielectric permittivity of the surrounding medium changes, part of the wave is reflected. This method can be applied to interface measurements too. Figure 11.9 shows the signals in an application with a two-phase product. This method is called "time domain reflectometry" (TDR).

Laser/Light

Lasers and light-emitting diodes produce electromagnetic waves of very short wavelength (less than 2 μm), which can also be used for time-of-flight measurements, similar to the described microwave methods. Preferred laser signals are (1) short pulses of less than 1 ns duration, or (2) lasers with amplitude-modulated intensity with frequencies of some megahertz. For more details about laser operation and interferometry methods, refer to Chapter 6, Section 5 of this handbook.

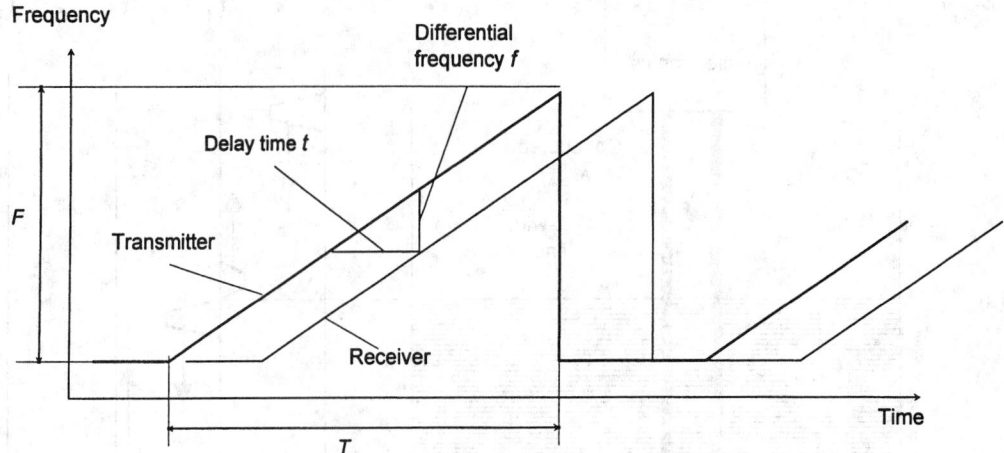

FIGURE 11.10 Operation characteristics of FMCW radar. The frequency of the transmitter changes linearly by time in an interval (sweep). The received signal has the same form, but is time-delayed. At every point of the sweep, the differential frequency is constant and proportional to the time delay. Time and frequency scales are arbitrary.

Laser systems are very precise and can achieve accuracies better than 1 mm. Because the laser beam is very narrow, such level measurement systems can be installed without influence of tank internals. Some practical disadvantages of laser level measurement are: (1) it functions as does your eye to see the product's surface and therefore fails if dust, smoke, etc. are present; (2) it is sensitive to dirt on the optical sensors; and (3) the equipment is expensive.

Commonly Used Evaluation Methods

Due to the great benefits of contactless time-of-flight measurement, some typical methods have been evaluated for level gaging within the last few years, mainly in radar techniques [8].

Frequency-Modulated Continuous Wave Radar

Because the flight times in typical level applications are very short (a resolution of 1 mm requires a 7 ps time resolution), it is difficult to evaluate information directly in the time domain. By modulation of the microwave signals, the time delay is transformed into the frequency domain, obtaining low-frequency signals. For general information about modulation, see Chapter 81 of this handbook.

Therefore, Frequency Modulated Continuous Wave (FMCW) radar has been established as the dominant technique. FMCW radar utilizes a linearly frequency-modulated microwave signal; the transmission frequency rises linearly in a time interval T. The frequency difference in this interval is called the frequency sweep F.

Figure 11.10 shows the principle of FMCW radar. Due to the time delay during signal propagation, the transmitted frequency changes such that the difference between the momentary transmitted frequency and the received frequency, a low-frequency signal is obtained. The frequency f of that signal (typically up to a few kilohertz) is proportional to the reflector distance d (see Figure 11.6); in this method, therefore, the delay t is transformed into a frequency f:

$$f = \frac{F}{T}t = \frac{F}{T}\frac{2d}{c} \Leftrightarrow d = \frac{fcT}{2F} \tag{11.16}$$

In Equation 11.16, c is the speed of light and F/T is the sweep velocity; see Figure 11.10.

Figure 11.11 shows a basic circuit block diagram of an FMCW radar system. Because the resultant signal frequencies are low, further signal processing is technically simple and very accurate. Normally,

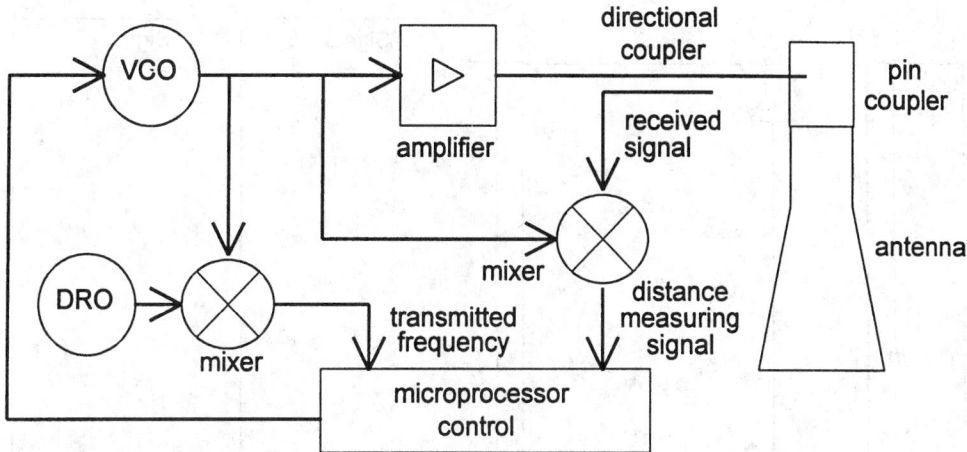

FIGURE 11.11 Basic circuit block diagram of an FMCW radar system: a microprocessor controls a voltage-controlled oscillator (VCO), such that the desired frequency sweep is obtained. This signal is amplified and fed into the transmitting antenna. The instantaneous frequency must be measured in order to ensure good sweep linearity. This is accomplished by counting the frequency after it has been mixed with the known frequency of a dielectric resonance oscillator (DRO). A directional coupler decouples the received signal, which is mixed with the transmission signal and processed by the microprocessor.

evaluation is by means of digital signal processing. For more information about signal processing techniques using spectrum analysis, refer to Chapter 83 of this handbook.

Time-of-Flight Through Product

Alternatively, the propagation time of the waves through a weakly absorbing liquid or bulk material of low permittivity ε_r can be measured, as well as the propagation through the air. In cases where the reflection from the interface between air and the upper surface of the product is poor, part of the signal travels through the liquid and is reflected a second time at the tank bottom or at an interface between two liquids (e.g., oil on water).

Figure 11.12 demonstrates this technique. The evaluation is done in the following four steps:

1. Where microwaves in the tank atmosphere of height d are propagated at the speed of light c, microwaves in the medium (relative permittivity $= \varepsilon_r$, height L) are propagated at a slower velocity v.
2. Hence, the reflection r_2 from the tank bottom appears to be shifted downward, and the apparent tank height h_v is greater than the true height h.
3. The transit time in the medium is $t_1 = L/v$, where for the same distance in an empty tank would be $t_0 = L/c$. The ratio of apparent "thickness layer" $(h_v - d)$ to true filling height $(h - d)$ therefore corresponds to the ratio of the wave propagation rates:

$$\frac{h_v - d}{h - d} = \frac{c}{v} = \sqrt{\varepsilon_r} \tag{11.17}$$

4. When ε_r, h, and h_v are known, distance d and, from that, filling height L can be calculated exactly:

$$L = h - d = \frac{h_v - h}{\sqrt{\varepsilon_r} - 1} \tag{11.18}$$

By this method, a direct level measurement — not a distance measurement — is attained. It can even be applied when signal r_1 from the surface of the medium is no longer measurable. The evaluation of

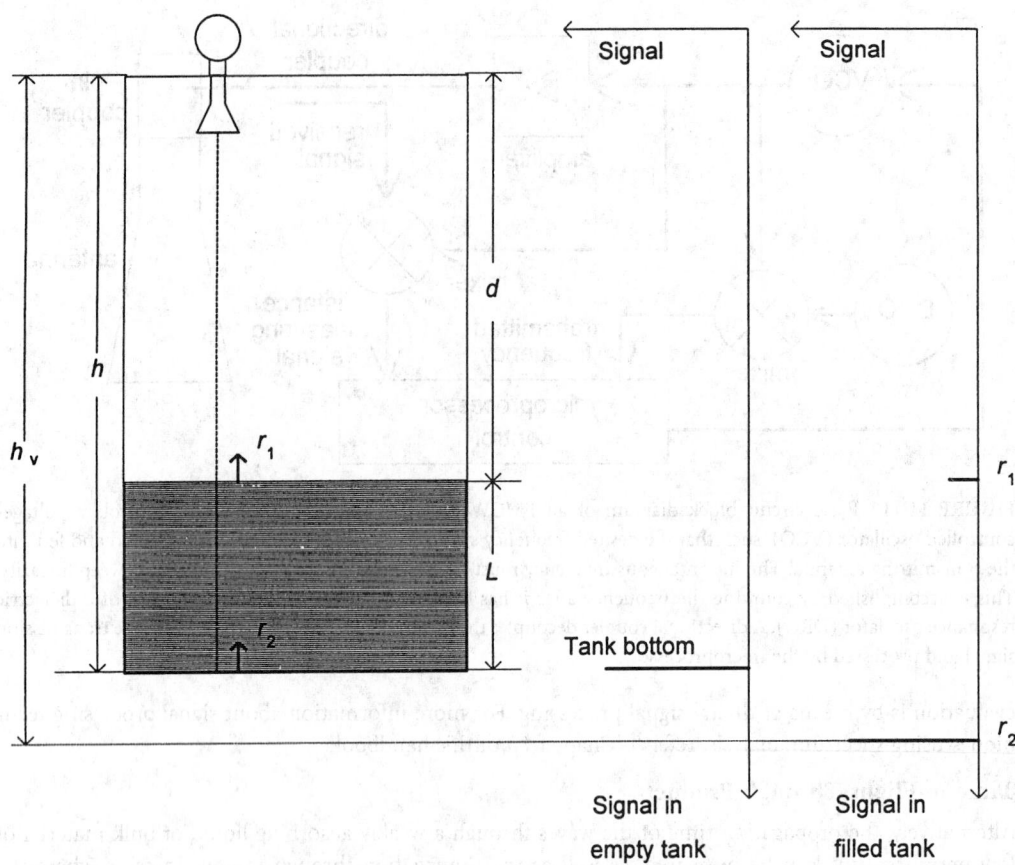

FIGURE 11.12 Representation of time-of-flight measurement through liquid: the wave is reflected once (r_1) at the product's surface and a second time (r_2) at the tank bottom. Due to the reduced wave velocity within the liquid, the reflection r_2 appears below the geometric position of the bottom. From that shift, the level can be calculated; see Equations 11.17 and 11.18.

the tank bottom reflection signal is known as "tank bottom tracing," and is used in the radar level system offered by Krohne in Figure 11.8.

11.3   Level Measurements by Detecting Physical Properties

To measure level, one can detect physical parameters that are significantly different between the atmosphere and the product; for example, conductivity, viscosity, or attenuation of any type of radiation. Two types are possible: (1) continuous measurement with an integral sensor, or (2) switching by a point measurement when the sensor comes in contact with the product.

Electrical Properties

The sensor must be in direct or indirect contact with the product to detect its electrical properties. For continuous measurement, only part of the intrusive sensor must be in contact with the product to detect the difference in dielectric permittivity or conductivity.

Capacitive

In most applications, a rod electrode is arranged vertically in the tank. The electrode can be (1) noninsulated if the liquid is nonconductive, or (2) insulated. The metallic vessel acts as a reference electrode.

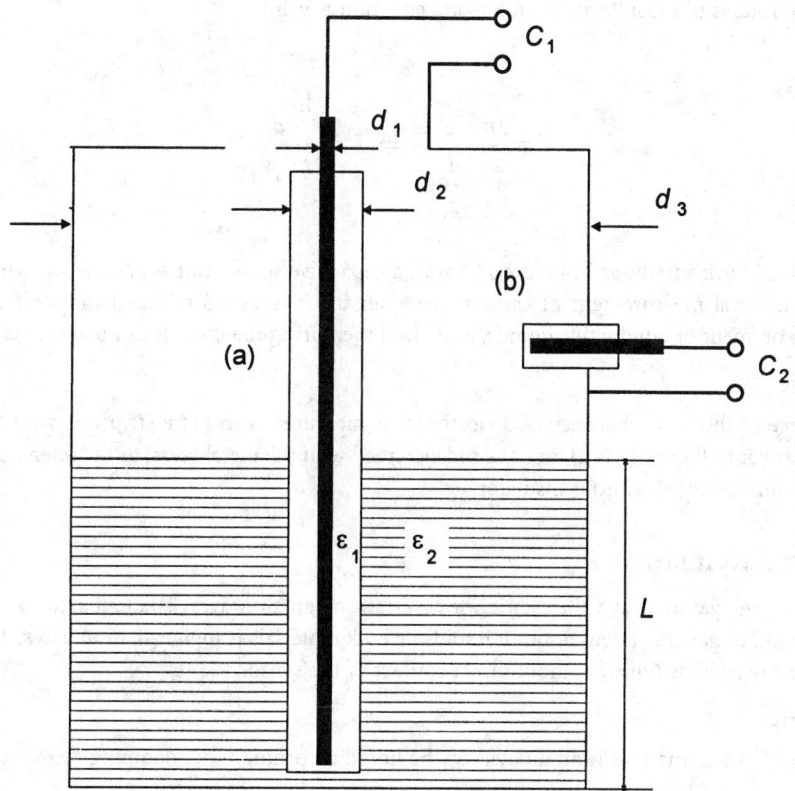

FIGURE 11.13 Principle of operation for a capacitance-type level device. (*a*) An insulated electrode protrudes into the liquid. The capacitance between the inner conductor and the tank walls is measured. (*b*) As a capacitance level switch, the electrode can be mounted at the appropriate position directly into the tank wall.

The result depends on the permittivity ε_2 of the product. Figure 11.13(*a*) shows an electrode concentrically mounted on a cylindrical tank. For such a rotationally symmetrical configuration, the capacitance C of an insulated electrode changes with level L according to:

$$C = \frac{2\pi\varepsilon_0 L}{\dfrac{1}{\varepsilon_1}\ln\dfrac{d_2}{d_1} + \dfrac{1}{\varepsilon_2}\ln\dfrac{d_3}{d_2}} \iff L = \frac{C\left(\dfrac{1}{\varepsilon_1}\ln\dfrac{d_2}{d_1} + \dfrac{1}{\varepsilon_2}\ln\dfrac{d_3}{d_2}\right)}{2\pi\varepsilon_0} \tag{11.19}$$

ε_0 is the dielectric constant of vacuum (8.85×10^{-12} As V^{-1}m^{-1}); ε_1 and ε_2 are the relative permittivities of the insulation material and the liquid, respectively.

If the liquid itself is highly conductive, Equation 11.19 simplifies to:

$$C = \frac{2\pi\varepsilon_0\varepsilon_1 L}{\ln\dfrac{d_2}{d_1}} \iff L = \frac{\ln\dfrac{d_2}{d_1}}{2\pi\varepsilon_0\varepsilon_1} \tag{11.20}$$

If the electrode is not insulated, the following equation is valid:

$$C = \frac{2\pi\varepsilon_0\varepsilon_2 L}{\ln\dfrac{d_3}{d_1}} \Leftrightarrow L = \frac{\ln\dfrac{d_3}{d_1}}{2\pi\varepsilon_0\varepsilon_2} \tag{11.21}$$

When arranged horizontally, as in Figure 11.13(b), a capacitive sensor can act as a level switch.

For the electrical measurement of capacitance, refer to Chapter 6.3 of this handbook. For a more precise measurement of conductive liquids, a method measuring the complex impedance is helpful.

Conductive

The resistance of the liquid between two electrodes is measured with (1) a strip line with two parallel electrodes similar to Figure 11.9(a), or (2) a rod electrode with the metal vessel as the reference electrode, similar to Figure 11.13(a) without insulator.

Radiation Attenuation

All radiation (e.g., gamma rays, ultrasonic waves, electromagnetic waves) is attenuated to some degree in any medium. In general, attenuation in liquids or bulk materials is higher than in gases. This effect is used to measure level or limits, without direct contact of the sensor.

Radiometric

The intensity I of gamma rays is attenuated by the liquid according to its damping factor α:

$$I = I_0 \exp(-\alpha d) \tag{11.22}$$

The source can be a radioactive material Co-60 or Cs-137, having half-lives of 5.23 years and 29.9 years, respectively. Emitter and sensor may take the form of (1) a point emitting the rays radially in all directions, (2) a rod emitting radially from a line, or (3) an array consisting of several point emitters in a row. Any combination of point/rod/array emitter with point/rod/array detector is possible. Figure 11.14 shows two different configurations. Radiation protection regulations must be considered. A real-time clock in the system must compensate for the decrease of intensity (dose rate) I by time t according to the half-life T_H of the applied material:

$$I = I_0\, 2^{-t/T_H} \tag{11.23}$$

For more information about radiation detection methods, refer to Chapter 66 of this handbook. Plastic scintillators and counting tubes are preferred for radiometric level gaging. The level–intensity characteristic is nonlinear, so the equipment should be calibrated in place. Mengelkamp [10] describes the radiometric techniques in more detail.

Ultrasonic Switch

A short ultrasonic transmission path can be used to detect products that dampen sonar waves. For instance, this method is applicable for the detection of slurries or to determine the interface between two different liquids. When combined with a servo system, the vertical profile of ultrasonic attenuation can be measured. Another application uses a noncontact sensor mounted on the outside of the vessel. It measures the acoustic impedance through the vessel wall that changes if liquid or gas is present behind the wall.

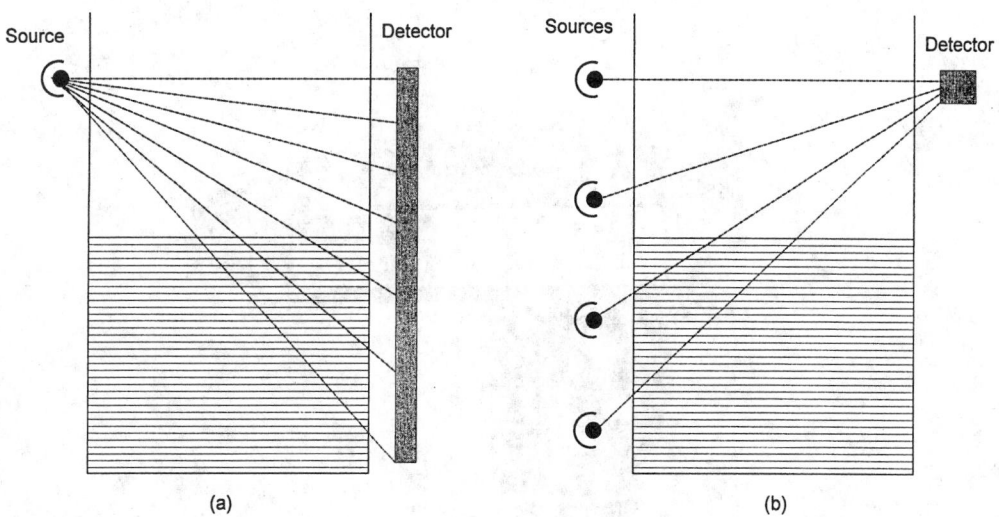

FIGURE 11.14 Representation of a radiometric continuous level gage. The rays are emitted by a radioactive source, propagate through the tank walls, and are damped by the liquid. In (*a*), a point source is combined with a rod detector (e.g., scintillator rod); (*b*), a source array is combined with a point detector.

Microwave Switch

Liquids and solids dampen microwaves in many cases, sometimes absorbing them completely. A simple unmodulated microwave source and an accompanying receiver are sufficient for level switching.

Photoelectric Barrier

A photoelectric barrier can act as a level switch for liquids that are not transparent, as well as most solids. But in closed nontransparent tanks, the coupling of the photoelectric components to the tank will not be possible in most cases.

Thermal and Mechanical

For some special applications, level sensors utilize the different heat dissipation properties and viscosities of the media.

Thermal

A self-heated resistor with a high temperature coefficient is immersed into the liquid. Heat dissipation causes the temperature to drop somewhat in the region where the liquid covers the sensor. Therefore, the resistance change is nearly linear with the level. This method is often used in automotive applications. In some applications with heated liquids (e.g., chemical reaction vessels), a simple temperature sensor can be used as a level switch by emitting a signal when the liquid contacts the sensor and heats it.

Viscosity

The effect of viscosity, which is significantly higher for liquids than for gases, dampens the movement of a body. These level sensors measure the degree of damping of a vibrating fork when dipped in a liquid. Normally, it is only used as a point level switch. Figure 11.15 shows such a "tuning fork," named according to the typical form with two or three vibrating paddles. The integrated electronics evaluate the power loss or the frequency shift of the mechanical resonance system. For solids, a sensor with a rotating paddle that stops when contacting the product is useful.

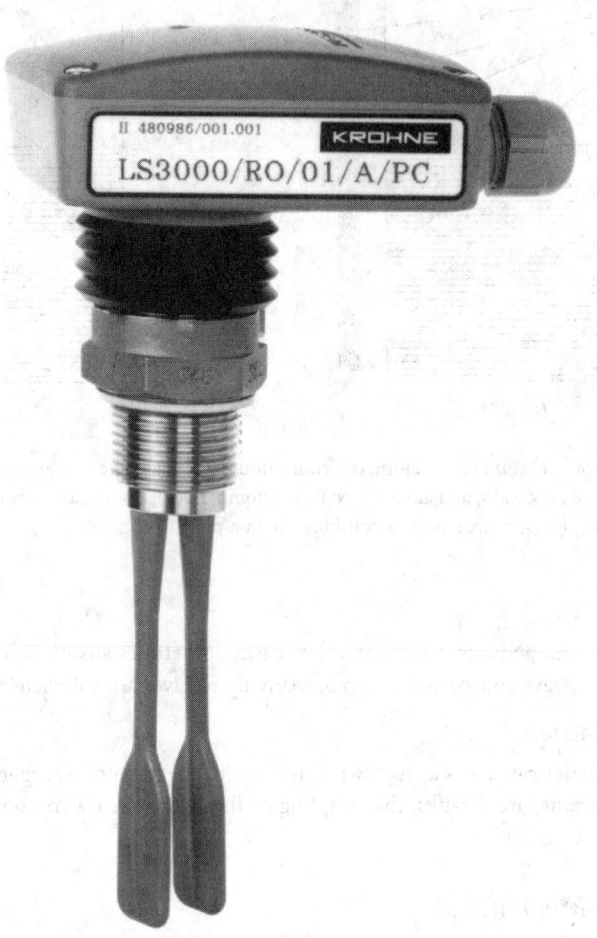

FIGURE 11.15 Design of a vibrating level switch. The switch reacts to product viscosity changes, which dampens the vibration of the paddles [6].

11.4 Instruments

Table 11.2 lists the most common techniques for level gaging with corresponding application range, accuracy, and average prices. Pricing as shown is for typical industrial applications and can vary widely, depending on materials of construction, measuring range, accuracy, approval requirements, etc. Table 11.3 provides contact information for manufacturers of level instruments. Most of these companies manufacture various instruments, using the different technologies shown in Table 11.2. This list is not comprehensive; additionally, some smaller companies provide equipment for specialized level measurements.

TABLE 11.2 Priority Level Measurement Techniques and Prices for Industrial Applications

Technique	Application Range	Attainable Accuracy	Avg. Price
Displacer/float	Continuous, liquids	1 mm	$2,000
Float	Switch, liquids	10 mm	$500
Pressure	Continuous, liquids	10 mm	$2,000
Ultrasonic	Continuous, liquids, solids	5 mm	$2,500
Radar	Continuous, liquids, solids	1 mm	$3,000
TDR	Continuous, liquids, solids	3 mm	$2,500
Laser	Continuous, liquids, solids	0.1 mm	$10,000
Radiometric	Continuous, liquids, solids	10 mm	$5,000
Capacitive	Switch, liquids	10 mm	$500
Viscosity	Switch, liquids	10 mm	$500

TABLE 11.3 Companies that Make Level Measuring Instruments

Berthold Systems, Inc.
101 Corporation Drive
Aliquippa, PA 15001
Tel: (508) 378-1900

Bindicator
1915 Dove Street
Port Huron, MI 48061
Tel: (810) 987-2700

Danfoss Instrumark
  International
485 Sinclair Frontage Road
Milpitas, CA 95035
Tel: (408) 262-0717

Drexelbrook Engineering
  Company
205 Keith Valley Road
Horsham, PA 19044
Tel: (215) 674-1234

Eckardt
10707 Addington Street
Houston, TX 77043
Tel: (713) 722-286

Endress & Hauser Instruments
2350 Endress Place
Greenwood, IN 46143
Tel: (317) 535-7138

Enraf, Inc.
500 Century Plaza Drive
Suite 120
Houston, TX 77073
Tel: (713) 443-4291

The Foxboro Company
38 Neponset Avenue
Foxboro, MA 02035
Tel: (508) 543-8750

Krohne America Inc.
7 Dearborn Road
Peabody, MA 01960
Tel: (508) 535-1720

MTS Systems Corporation
Sensors Division
3001 Sheldon Drive
Cary, NC 27513
Tel: (919) 677-0100

Milltronics
709 Stadium Drive
Arlington, TX 76011
Tel: (800) 569-2130

Rosemount, Inc.
Measurement Division
8200 Market Boulevard
Chanhassen, MN 55317
Tel: (612) 949-7000

Saab Tank Control
10235 West Little York
Suite 258
Houston, TX 7740
Tel: (713) 849-2092

TN Technologies, Inc.
2555 North Interstate 35
P.O. Box 800
Round Rock, TX 78680
Tel: (512) 388-9100

Vega HiTech Technologies, Inc.
P.O. Box 535
Newtown, PA 18940
Tel: (215) 968-1795

Whessoe Varec, Inc.
10800 Valley View Street
Cypress, CA 90630
Tel: (714) 761-1300

Yokogawa Corporation of
  America
2 Dart Road
Newnan, GA 30265
Tel: (404) 253-7000

References

1. D.M. Considine, *Process Instruments and Control Handbook,* 2nd ed., New York: McGraw-Hill, 1974.
2. E.B. Jones, *Instrument Technology, Vol. 1, Measurement of Pressure, Level and Temperature,* London: Butterworth & Co., 1974.

3. Verein Deutscher Ingenieure, Verband Deutscher Elektrotechniker (VDI/VDE), *Füllstandmessung von Flüssigkeiten und Feststoffen (Level Measurement of Liquids and Solids)*, VDI/VDE 3519, Part 1, Berlin: Beuth, 1984.

4. Verein Deutscher Ingenieure, Verband Deutscher Elektrotechniker (VDI/VDE), *Füllstandmessung von Flüssigkeiten und Feststoffen (Level Measurement of Liquids and Solids)*, VDI/VDE 3519, Part 2, Berlin: Beuth, 1984.

5. K.W. Bonfig (ed.), *Technische Füllstandsmessung und Grenzstandskontrolle*, Ehningen: Expert, 1990.

6. Krohne Messtechnik, *Technical Data Sheets of Level Measurement Products*, Duisburg: Krohne, 1996.

7. K. Blundy, Radar systems — setting a practical approach, *Control & Instrum.*, July 1996.

8. D. Brumbi, *Fundamentals of Radar Techniques for Level Gauging*, Duisburg: Krohne, 1995.

9. D. Brumbi, Measuring process and storage tank level with radar technology, *Int. Radar Conf. IEEE*, 256-260, 1995.

10. B. Mengelkamp, *Radiometrie, Füllstand- und Dichtemessung*, Berlin: Elitera, 1972.

12

Area Measurement

Charles B. Coulbourn
*Los Angeles Scientific
Instrumentation Co.*

Wolfgang P. Buerner
*Los Angeles Scientific
Instrumentation Co.*

One must often measure the area of enclosed regions on plan-size drawings. These areas might be either regular or irregular in shape and describe one of the following:

- Areas enclosed by map contours
- Cross section of the diastolic and systolic volumes of heart cavities
- Farm or forest land shown in aerial photographs
- Cross sections of proposed and existing roads
- Quantities of materials used in clothing manufacture
- Scientific measurements
- Swimming pools
- Quantities of ground cover

Tools for this type of measurement include planimeters, digitizer-computer setups, digitizers with built-in area measuring capability, and grid overlay transparencies.

12.1 Theory

Planimeter

A planimeter is a mechanical integrator that consists of a bar (tracer arm), a measuring wheel with its axis parallel to the bar, and a mechanism that constrains the movement of one end of the bar to a fixed track, Figure 12.1. The opposite end of the bar is equipped with a pointer for tracing the outline of an area. The measuring wheel, Figure 12.2, is calibrated with 1000 or more equal divisions per revolution. Each division equals one count. It accumulates counts, P, according to:

$$P = \frac{K}{\pi D} \int \sin \phi \, ds \qquad (12.1)$$

where K = number of counts per revolution of the measuring wheel
       D = diameter of the measuring wheel
       ϕ = angle between the measuring wheel axis and the direction of travel
       s = traced path

0-8493-8347-1/99/$0.00+$.50

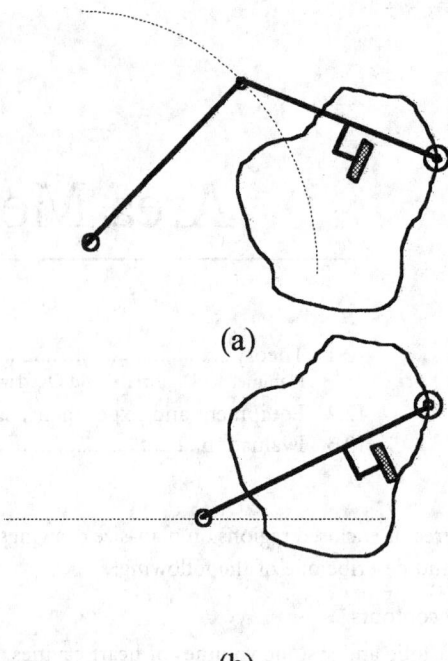

(a)

(b)

FIGURE 12.1   The constrained end of a polar planimeter (*a*) follows a circular path; the constrained end of a linear planimeter (*b*) follows a straight line path.

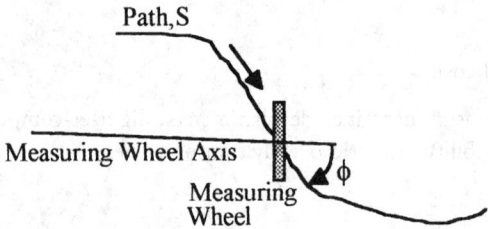

FIGURE 12.2   The rotation of a measuring wheel is proportional to the product of distance moved and the sine of the angle between the wheel axis and direction of travel.

The size of an area, *A*, traced is:

$$A = \frac{P}{K} \times \pi D \times L \qquad (12.2)$$

where *L* = length of bar
    P = accumulated counts (Equation 12.1)

Figure 12.3 shows how a basic wheel and bar mechanism determines the area of a parallelogram. The traced path is along the sloped line; however, the wheel registers an amount that is a function of the product of the distance traveled and the sine of the angle between the direction of travel and the axis of the measuring wheel (Equation 12.1). This is the altitude of the parallelogram. The product of the altitude (wheel reading converted to distance) and base (bar length) is the area.

Figure 12.4(*a*) illustrates the operation of a planimeter when the area of a four-sided figure is measured. Figures 12.4(*b*), (*c*), (*d*), and (*e*) show the initial and final positions of the bar as each side of the figure

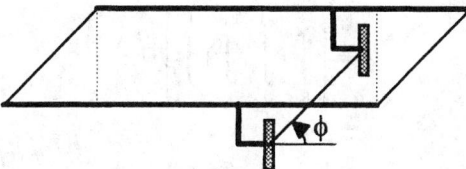

FIGURE 12.3 The area of this parallelogram is proportional to the product of tracer arm length and measuring wheel revolutions.

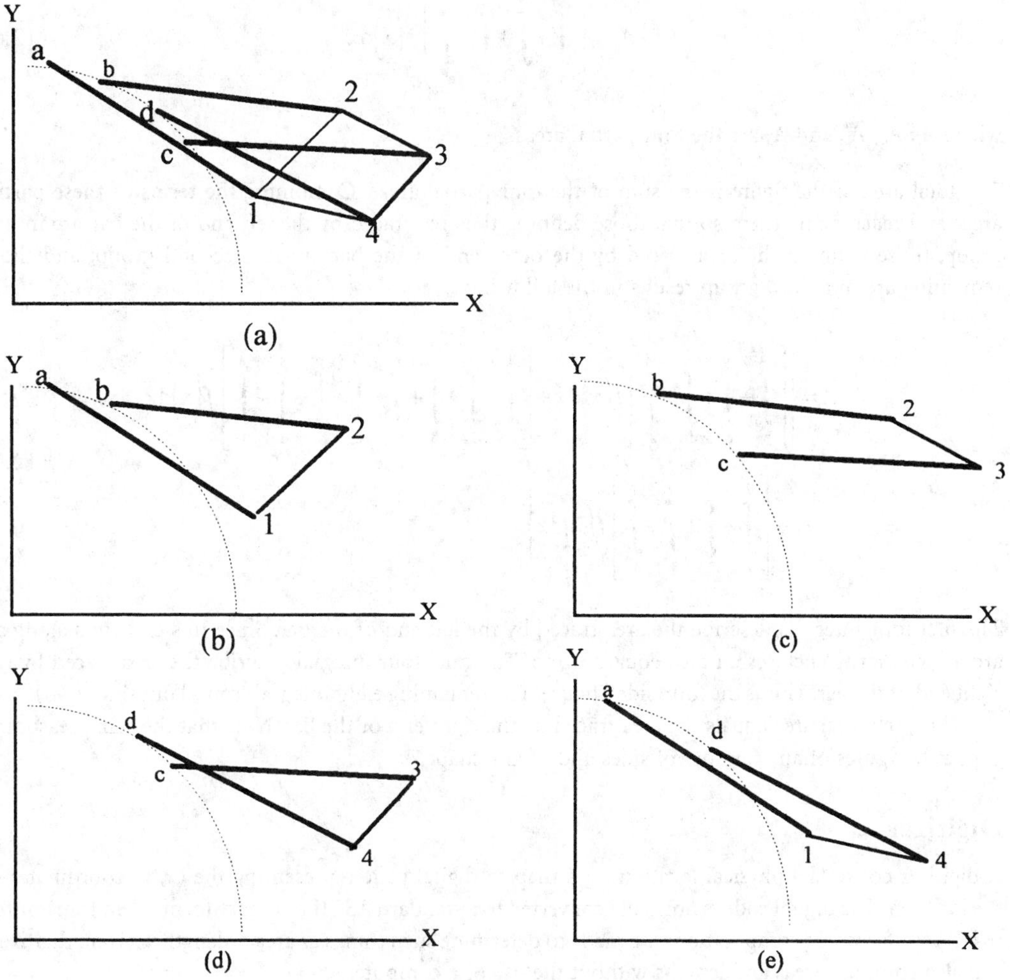

(a)

(b)

(c)

(d)

(e)

FIGURE 12.4 This schematic shows a planimeter pointing to each junction of a four-sided figure being traced in (*a*) and at the ends of each of the segments in (*b*) through (*e*). The constrained end of the tracer arm follows a circle.

is traced. Applying the general expression for the area under a curve, $A = \int f(x)dx$ for each of these partial areas gives:

$$A_a = \left(\int_a^1 + \int_1^2 + \int_2^b + \int_b^a \right) f(x)dx \qquad (12.3)$$

$$A_b = \left(\int_b^2 + \int_2^3 + \int_3^c + \int_c^b \right) f(x)dx \qquad (12.4)$$

$$A_c = \left(\int_c^3 + \int_3^4 + \int_4^d + \int_d^c \right) f(x)dx \qquad (12.5)$$

$$A_d = \left(\int_d^4 + \int_4^1 + \int_1^a + \int_a^d \right) f(x)dx \qquad (12.6)$$

where A_a, A_b, A_c, and A_d are the four partial areas.

The total area of the figure is the sum of the four partial areas. Combining the terms of these partial areas and rearranging them so that those defining the area traced by the left end of the bar are in one group, those defining the area traced by the other end of the bar are in a second group, and those remaining are in a third group results in the following:

$$A = \left\{ \left[\int_b^a + \int_c^b + \int_d^c + \int_a^d \right] f(x)dx + \left[\int_a^1 + \int_2^b + \int_b^2 + \int_3^c + \int_c^3 + \int_4^d + \int_d^4 + \int_1^a \right] f(x)dx \right.$$

$$\left. + \left[\int_1^2 + \int_2^3 + \int_3^4 + \int_4^1 \right] f(x)dx \right\} \qquad (12.7)$$

The first four integrals describe the area traced by the left end of the bar. Since this end runs along an arc, it necessarily encloses an area equal to zero. The final four integral describe the area traced by the right end of the bar. This is the four-sided figure. The remaining eight integrals cancel out since $\int_a^1 + \int_1^a = 0$, etc. Thus, the total area equals the area traced by the right end of the bar. Note that the same reasoning applies to figures of any number of sides and of any shape.

Digitizer

A digitizer converts a physical location on a map to digital code representing the (x, y) coordinates of the location. The digital code is normally converted to a standard ASCII or binary format and transmitted to a computer where computations are made to determine such things as area or length. Certain digitizers can also compute areas and lengths without the use of a computer.

Area, A, can be computed using the coordinate pairs that define the area boundary.

$$A = \frac{1}{2}(y_1 + y_2)(x_2 - x_1) + \frac{1}{2}(y_2 + y_3)(x_3 - x_2) + \ldots + \frac{1}{2}(y_{n-1} + y_n)(x_n - x_{n-1})$$

$$+ \frac{1}{2}(y_n + y_1)(x_1 - x_n) \qquad (12.8)$$

where x_1, x_2, x_3, etc. = sequentially measured x coordinates along the boundary.

y_1, y_2, y_3, etc. = are corresponding y coordinates

All digitized coordinate pairs must be used in the computations since the Δx intervals of this data are typically unequal.

In Equation 12.8, the final term deserves special consideration because it contains both the first and last coordinate pair of the series. When computations are performed from data stored in a computer, information is available for computing the final term so that no special actions are necessary. However, when the computations are performed by an embedded microprocessor with limited memory, normally each term is computed as the coordinates are read. In this case, one of two actions must be taken. First, the first coordinate pair is saved and then, once the measurement has been completed, a key is pressed to initiate computation of the final term. This is called "closing the area." Second, if the first coordinate pair is not saved, to prevent an error the final point digitized must coincide with the first point digitized so that $x_n = x_1$, $y_n = y_1$ and the final term is zero.

Popular types of digitizers in use today are tablets, sonic digitizers, and arm digitizers. Probably the most popular type is the tablet.

Tablet Digitizer

Tablet digitizers consist of a pointer and a work surface containing embedded wires configured as a grid. The horizontal wires are parallel and spaced by about 12 mm. The vertical wires are also parallel and spaced the same. Different sensing techniques are used to locate the pointer position relative to the grid wires.

One sensing technique employs grid wires made from magnetostrictive material that has the property of changing shape very rapidly when subjected to a magnetic field. Each set of grid wires, the horizontal and vertical, is independently energized by a send wire that lies perpendicular to that set of wires [1]. A pulse transmitted over the send wire has a magnetic field that changes the shape of each magnetostrictive wire, causing a strain wave to propagate down the wire. Coincidentally, the pulse starts a counter. A coil in the pointer senses the strain wave and sends a signal to stop the counter. The counter reading is thus proportional to the propagation time of the strain wave. The product of the propagation time and velocity of the strain wave equals the physical distance between send line and pointer detector. The velocity of the strain waves is slow enough so that any errors in time measurement can be made acceptable.

A second sensing technique uses a grid made of conductive wires and a pointer that emits a signal on the order of 57.6 kHz [2]. The vertical grid wires are sensed to determine the amplitude distribution of signal induced in each wire. The point of maximum signal strength determines the location of the pointer along the horizontal axis. The horizontal grid wires are likewise sensed to find the point of maximum signal strength that determines the pointer location along the vertical axis. Coupling between the pointer and grid can be by either electromagnetic or by electrostatic means [3].

Signals can also be applied to the grid wires and the pointer used as a receiver. In this case, the signal to each grid wire must be coded or else applied sequentially, first to one set of grid wires and then to the other. Amplitude profiles of the signals received by the pointer are determined to locate the pointer along each axis.

Sonic Digitizers

Sonic digitizers consist of a pointer and two or more microphones. The pointer, used to identify the point to be digitized, typically contains a spark gap that periodically emits a pulse of sonic energy [4]. The microphones in certain cases are mounted in a bar that is located along the top of the drawing area. In other cases, they are mounted in an L frame and located along the top and one side of the drawing area.

The microphones receive the sonic pulse emitted by the pointer. The time taken for the pulse to travel from the transmitter to each receiver is measured and the slant distance computed from the product of this elapsed time and the sonic velocity. The x and y coordinates of the pointer are computed from the slant ranges and the locations of the microphones.

Ambiguities exist since a single set of slant ranges describe two points; however, the ambiguities can be resolved either by using additional microphones or by ensuring that the ambiguous points are outside the work area. With the microphones aligned along the top of the work area, the ambiguous points are

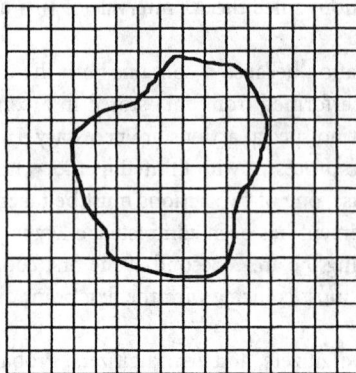

FIGURE 12.5   Grid overlays offer a simple and readily available method for measuring areas.

outside the work area; with them aligned along top and side, the ambiguities have to be resolved with additional microphones.

The sound wave velocity is slow enough so that errors in time measurements can be made acceptable.

Arm Digitizer

An arm digitizer consists of a base, two arms, two rotary encoders, and a pointer. The *base* must be anchored at a point removed from the work area, normally at the top of the user's desk. One end of the arm, called the *base arm*, swivels about a vertical axis at the base. The other end of the base arm hinges to one end of a second arm, called the *tracer arm*, about a vertical axis. On the other end of the tracer arm is the *pointer*. One of the encoders, the *base encoder*, detects the angle between the base and the base arm. The other encoder, the *arm encoder*, detects the angle between the base arm and the tracer arm.

To simplify implementation, each arm is made the same length and, for sufficient accuracy, each encoder provides in excess of 36,000 counts per revolution. The encoder output signals consist of dual square waves with a 90° phase relationship. The leading and trailing edges of each square wave are counted and the 90° phase relationship provides count direction.

Since each encoder is a relative rather than absolute counting device, the count registers of each encoder must be initialized at a known angle. The count register of the arm encoder is set to the angle that results when the pointer is moved to a precisely known location. This location is called "home." The count register for the base encoder is initialized when the rotation of the x–y coordinate system is set. Thereafter, the encoders add to or subtract from their count registers as the pointer is moved.

With the arm lengths and the angles precisely known, the *x*, *y* position of the pointer can be computed. However, *x* and *y* are relative to a coordinate system whose origin is at the axis of the base and whose rotation is unknown. The user must therefore select an origin, usually near the lower left corner of the drawing area, and the direction of the x-axis, usually parallel to the bottom edge of the drawing area. This information is sufficient for the processor to compute the correct translation and rotation or the coordinate system.

Grid Overlay

A grid overlay is simply a transparent sheet onto which a grid has been drawn. To use it, place it atop the drawing or photo of the area to be measured (Figure 12.5). Then count the number of squares that lie within the boundary of the area. Squares that are at least half enclosed should be counted. The unknown area equals the product of the number of blocks counted and the area of each block. The accuracy achieved is dependent on grid size, precision of the grid dimensions, and counting accuracy.

12.2   Equipment and Experiment

Many different types of planimeters and digitizers are manufactured in the U.S., Europe, and Japan. A representative sample of these instruments are described in this section.

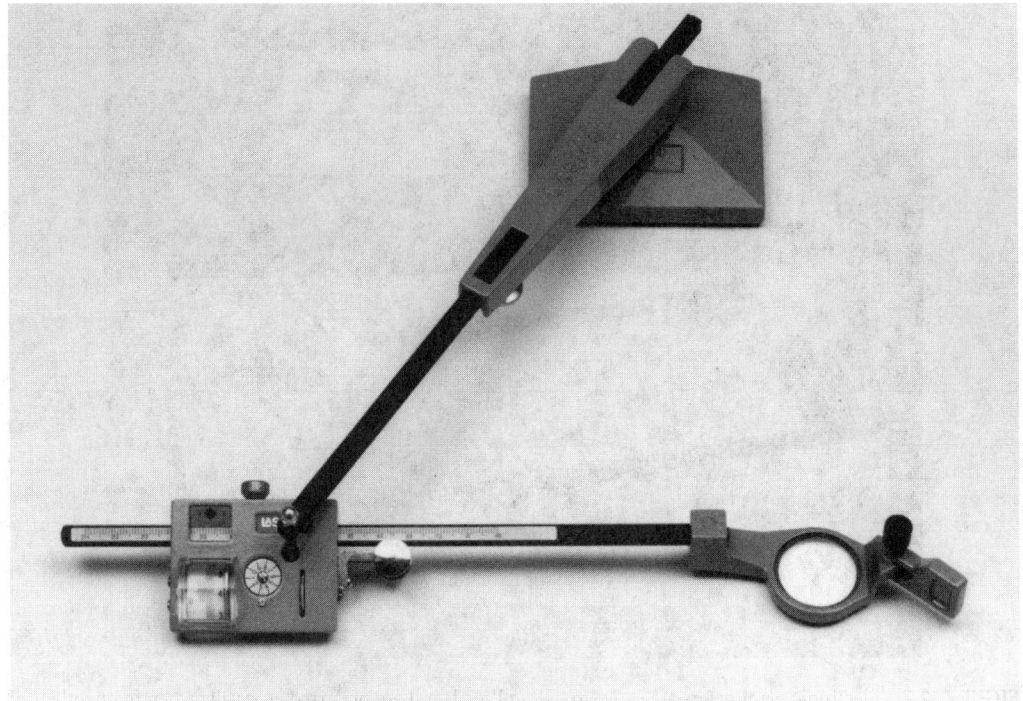

FIGURE 12.6 Mechanical planimeters are normally preferred when occasional use is required.

One of the simplest and least costly area measuring devices is the manual polar planimeter (Figure 12.6). It consists of a weight, two arms, a measuring wheel, and a pointer. The *weight* secures one end of a *pole arm*, allowing the other end to rotate along a fixed arc. The rotating end of the pole arm attaches to one end of a *tracer arm* and constrains its movement to the arc. At the other end of the tracer arm is a *pointer* used for tracing the periphery of an unknown area. The *measuring wheel* is located in the box at one end of the tracer arm. The location of the measuring wheel is not critical; however, its axle must be parallel to the tracer arm.

The length of both arms of the planimeter shown in Figure 12.6 can be adjusted. The length of the pole arm has absolutely no effect on measurement accuracy and is adjustable only for convenience. The effective length of the tracer arm directly affects the reading; a shorter arm results in a larger reading. By adjusting the tracer arm length, one can achieve a very limited range of scaling; however, this is usually not done. Rather, the arm length is adjusted according to the general size of areas to be measured: a shorter arm for smaller areas and a longer arm for larger areas. A shorter arm results in a greater number of counts per unit area, which is needed for smaller areas. Scaling is usually done by multiplying the result by an appropriate value.

One zeros the measuring wheel of the planimeter shown in Figure 12.6 by turning a small knurled wheel attached to the measuring wheel axle. On other models, one pushes a plunger to zero the wheel. The latter method is easier but is sensitive to misalignment.

A planimeter pointer is usually a lens with a small circle engraved in the center, although some planimeter models use a needle as the pointer. The best type of pointer is a matter of personal preference although lens pointers are much more popular.

Figure 12.7 shows two electronic planimeters with digital readouts. With these planimeters, one can measure length as well as area. To measure length, snap out the measuring wheel housing, attach an auxiliary handle, and roll the wheel along the line to be measured. Extremely high accuracy can be achieved.

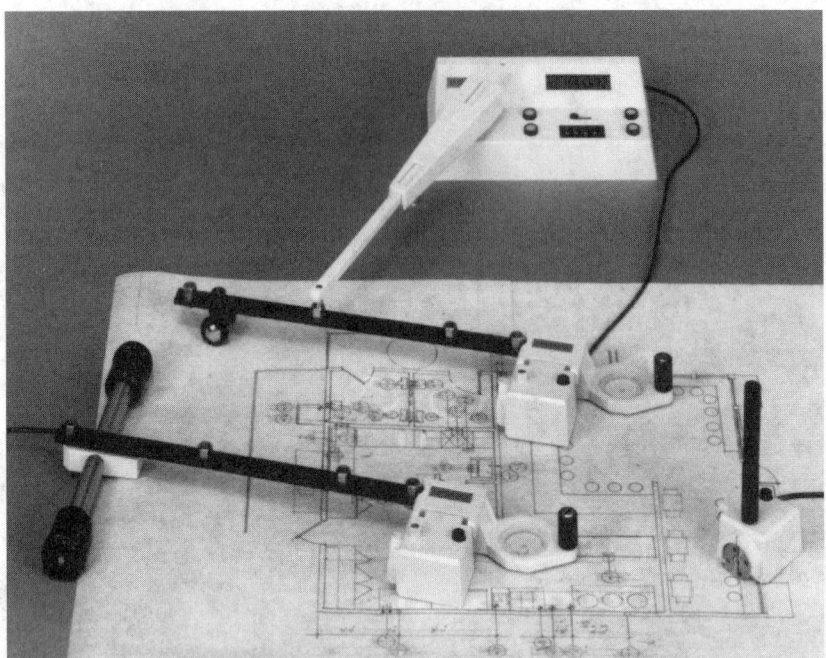

FIGURE 12.7   Electronic planimeters are easier to use and read, and are preferred especially when frequently used.

The upper planimeter in Figure 12.7 is a polar type. It consists of the same parts as the model in Figure 12.6, except that the measuring wheel is attached to a small optical encoder and the weight is packaged in the digital readout. The encoder provides two square-wave outputs that are in phase quadrature, that is, one is 90° out of phase with the other. Both outputs are fed to a processor that counts the pulses and uses the phase difference to determine count direction.

The processor has an electronic scale feature that translates the planimeter reading to real measurement units such as square feet, square meters, acres, or hectares. One can transmit processor data to a computer using an auxiliary interface unit (not shown).

The planimeter at the bottom of Figure 12.7 is a linear model since the path traveled by the constrained end of the tracer arm is a straight line. The straight line path is maintained by a rigid connection between the two carriage wheels and their axle. Other linear planimeters use an actual rail to guide the constrained end of the tracer arm along a straight line path.

The polar planimeter shown at the top of Figure 12.8 and the linear planimeter shown at the bottom of Figure 12.8 are both compact battery-operated models. The measuring wheel is built into the processor, which is attached to the pole arm of the planimeter. The effective length of the tracer arm for both instruments extends from the axis of the constrained end to the pointer, which for these instruments is a small circle engraved in the center of a lens. These planimeters provide electronic scaling and averaging of multiple readings. They cannot be used to measure length.

Figure 12.9 shows an arm digitizer that can be used either as a stand-alone area and length measuring device or to digitize a map or drawing. When operating as a digitizer, the arm digitizer displays the (x, y) coordinates as well as transmits them to a computer. The digitizer has a built-in interface and can transmit using any of over 24 different ASCII and binary codes, each with a choice of parameters. It can be set to measure x and y coordinates in either English or metric units.

Three other modes, in addition to the digitizer mode, are available for computing and displaying either area and length, area and item count, or item count and length. Measurements can also be made in either English or metric units. Any displayed item can be transmitted to a computer through the built-in interface.

FIGURE 12.8   Battery-operated planimeters have the advantages of electronics planimeters and are portable.

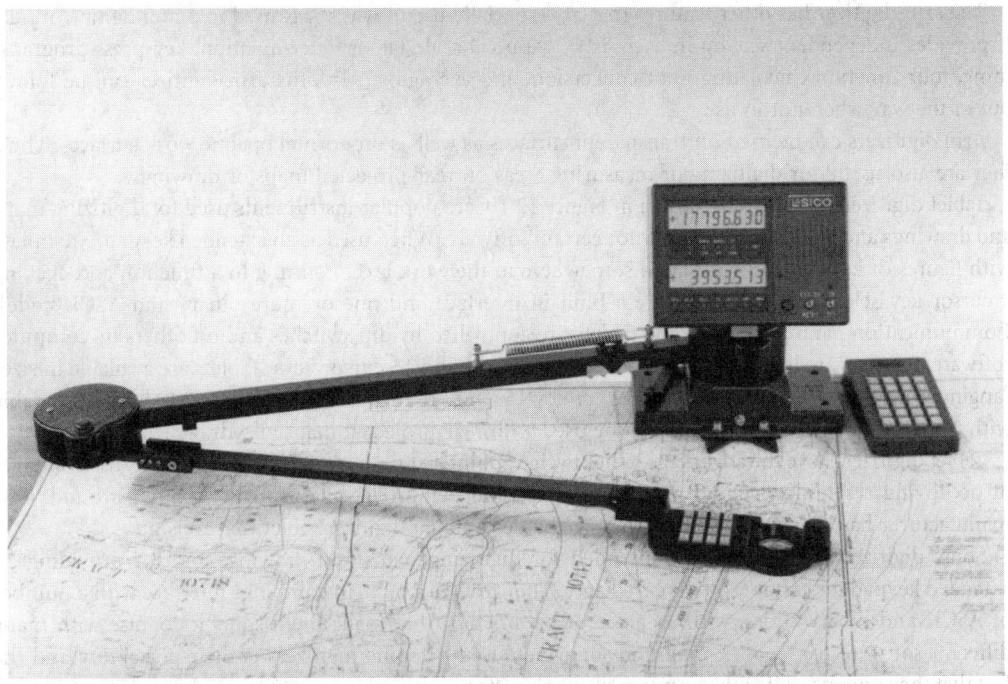

FIGURE 12.9   An arm digitizer lets one measure area, length, and coordinates, and transmit the displayed data to a computer; it requires minimum desk space.

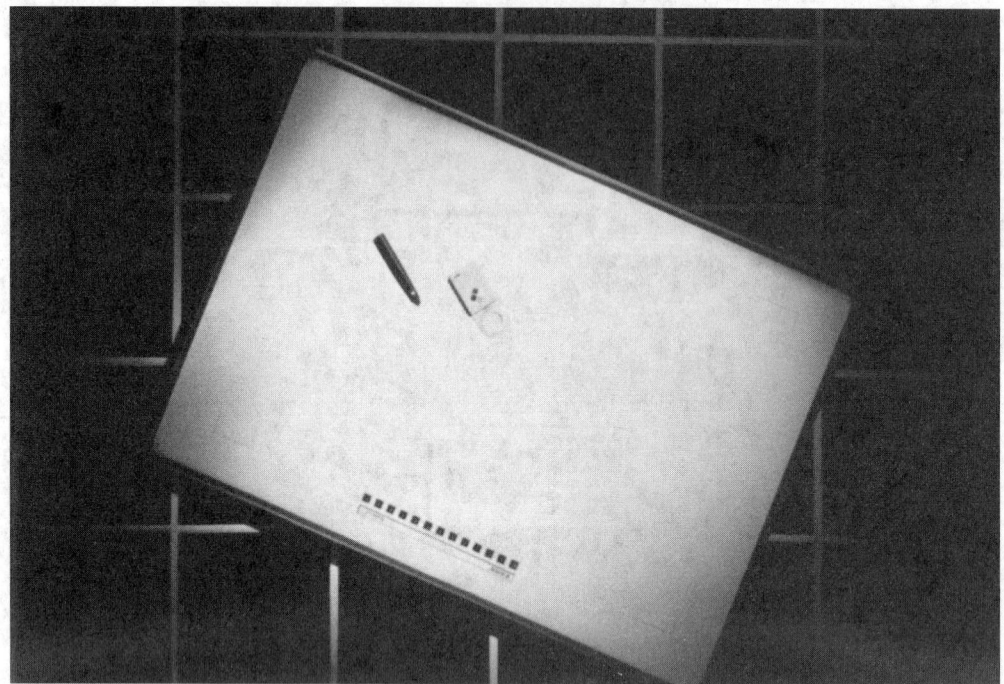

FIGURE 12.10 The popular pad digitizer includes a lightweight puck or pen-type pointer for selecting points whose coordinates are transmitted to a computer for processing.

The arm digitizer has other features that are especially useful when it is used in a nondigitizer mode. It provides independent scaling in each axis, empirical scale factor determination, key-press programming, four-function calculating, unit conversion, and averaging. Also, the arm digitizer can be folded out of the way when not in use.

Arm digitizers can be used on transparent surfaces as well as on normal opaque work surfaces. Thus, they are also useful for digitizing or measuring areas on rear-projected maps or drawings.

Tablet digitizers, like the one shown in Figure 12.10, are popular instruments used for digitizing maps and drawings and as extended keypads for certain software. When used as an extended keypad, a template with figures of key functions is placed somewhere in the work area. Pointing to a function and clicking a cursor key selects it. Tablets feature a built-in interface and one or more binary and ASCII codes. Communication parameters on some tablets are controlled by dip switches and on others by computer software. Pointers are lightweight and include between 1 and 16 cursor keys. Tablets are available in sizes ranging from that of a notebook up to 122 cm by 168 cm (48 in. by 66 in.) or more. Models are available with transparent work surfaces and can be used with rear-projected maps and drawings.

Tablet digitizers have the advantage of lightweight pointers but the larger tablets have the disadvantages of occupying a significant amount of floor space and being relatively difficult to move around. One manufacturer has overcome this disadvantage by designing a digitizer tablet that rolls up.

Sonic digitizers, like tablets, are intended for digitizing maps and drawings and for providing an extended keypad for certain software packages. They feature a built-in computer interface with a number of ASCII and binary codes, with a choice of parameters, that make them suitable for use with many different software packages. A major advantage enjoyed by sonic digitizers is their portability and the fact that they operate well with a transparent work surface required for rear-projected drawings and maps.

Table 12.1 provides a list of manufacturers of planimeters and Table 12.2 lists many of the manufacturers of graphic digitizers.

TABLE 12.1 Companies that Make Planimeters

Gebruder Haff Gmbh	Lasico Inc.
Tiroler Strasse 5	2451 Riverside Drive
D-87459 Pfronten	Los Angeles, CA 90039
Germany	Tel: (213) 662-2128
Tel: 49-8363-9122-0	
Koizumi Sokki Mfg. Co., Ltd.	Sokkia Corp.
1-132, Midori-Cho, Nagaoka-Shi	9111 Barton St.
Niigata 940-21	P.O. Box 2934
Japan	Overland Park, KS
Tel: (0) 258-27-1102	Tel: (800) 476-5542

TABLE 12.2 Companies that Make Digitizers

Altek Corporation	Lasico Inc.
12210 Plum Orchard St.	2451 Riverside Drive
Silver Spring, MD 20904	Los Angeles, CA 90039
Tel: (301) 572-2555	Tel: (213) 662-2128
Calcomp Technology Inc.	Numonics Corporation
2411 West La Palma Ave.	101 Commerce Drive
Anaheim, CA 92801-2589	Box 1005
Tel: (800) 445-6515	Montgomeryville, PA 18936
	Tel: (215) 362-2766
GTCO Corporation	
7125 Riverwood Drive	Wacom Technology Corp.
Columbia, MD 21046	501 S.E. Columbia Shores Blvd., Suite 300
Tel: (800) 344-4723	Vancouver, WA 98661
	Tel: (360) 750-8882

12.3 Evaluation

Each of the area measuring devices described in this section are excellent and have been thoroughly proven by use. However, some of the devices are more suited to certain tasks and operating environments than others.

To measure an area that is smaller than an equivalent circle of about 2 cm in diameter, a digitizer is probably the best choice. For areas of this magnitude, the resolution element of planimeters starts to become a significant portion of the total area. The resolution element of most digitizers is significantly less than that of a planimeter, and any measurement is always plus- or -minus a resolution element.

For measuring areas that are equivalent to a circle between 2 cm and 55 cm in diameter, either a planimeter or a digitizer will provide excellent results.

When measuring areas larger than an equivalent circle of about 55 cm in diameter, one can still use a planimeter; however, one must subdivide the large area into smaller areas and then individually measure each of the smaller areas. A digitizer can measure significantly larger areas, but at some point it too will reach a limit. Then, one can use the same technique of subdividing the large area into smaller areas that the digitizer can handle.

Area measuring instruments of the future will undoubtedly make even greater use of microprocessors to provide more features such as incorporation of slope and tilt correction, statistical operations, and determination of centroids, moment, etc. The instruments should become mechanically simpler and more reliable. Features such as conservation of office space and portability will be emphasized. The ultimate area measuring device will consist of a detached cursor for pointing and a small calculator-like device for operating on the results, displaying them, storing them, and sending them to a computer.

Scanners and associated software will also impact the field of area measurement, particularly as their coverage increases and their price decreases.

Defining Terms

Planimeter: A mechanical integrator used for measuring the area of enclosed regions on maps, plans, etc.

Pole arm: One of the two bars comprising a polar planimeter. One end of the pole arm is fixed and the other end is free to rotate. The length of the pole arm has no effect on the planimeter reading.

Tracer arm: The bar of a planimeter to which is attached the measuring wheel. One end of the tracer arm is constrained to a fixed path, while the other end traces the perimeter of an enclosed region whose area is being measured. The length of the pole arm is indirectly proportional to the planimeter reading.

Polar planimeter: A planimeter with a tracer arm whose constrained end follows a circle.

Linear planimeter: A planimeter with a tracer arm whose constrained end follows a straight line.

Measuring wheel: The planimeter wheel whose degree of rotation is directly proportional to area.

Digitizer: A device to convert data or an image to digital form. The digitizers discussed here convert images to digital form and are categorized as graphic digitizers.

Pointer: The part of a planimeter or digitizer that is used to follow the line being traced.

Resolution element: The smallest elemental area that can be discerned. When referred to in connection with area measurement, it is an area with a value of 1.

References

1. P. A. Santi, J. Fryhofer, and G. Hansen, Electronic planimetry, *Byte*, March 1980, 113–122.
2. K. Mandelberg, Anonymous, *Internet*, 4–96.
3. E. Jones, '89 Planning guide-digitizers, *Architectural & Engineering*, July 1980, 37–40.
4. P. E. Maybaum, Digitizing and computer graphics, *Keyboard*, Sept. 1978, 1–3.

Further Information

F. A. Willers, *Mathematische Instrumente*, Munchen und Berlin: Verlag von R. Oldenbourg, 1943.

13

Volume Measurement

René G. Aarnink
University Hospital Nijmegen

Hessel Wijkstra

For simple geometric shapes, volume measurements can be performed analytically by measuring the dimensions of the object in question and using the appropriate formula for that shape. Volume formulae are found in geometry and calculus textbooks as well as in reference books such as CRC Press's *Handbook of Mathematical Science* or other references.

Volume can also be measured by fluid displacement. The object whose volume is to be measured is placed in a container filled with fluid and the initial and final volumes are measured. The object's volume is equal to the final volume minus the initial volume. This technique is especially useful for irregularly shaped objects.

Fluids can also be used to measure volume of cavities within objects by filling the cavity entirely with a fluid and then measuring the volume of the fluid after it is removed from the cavity.

The remainder of this chapter is dedicated to the more specific problem of volume measurements in medical applications.

Quantitative volume information can be of importance to the clinician in the diagnosis of a variety of diseases or abnormalities. It can also improve the understanding of the physiology of the patient. Information on volume may be used in various applications such as cardiac monitoring, diagnosis of prostate diseases, follow-up during organ transplantation, surgery for tumor resection, blood flow measurements, plastic surgery, follow-up of preterm infants, sports performance analysis, etc. Because of this wide spectrum of applications, various techniques for volume measurements have been developed, some of which are useful in determining the amount of blood flow to the organ (dynamic), while others are used to obtain the size of the object (static). The techniques can be invasive or noninvasive, and are based on either direct or indirect measurements. Each technique has its own advantages and disadvantages, and the application determines the selection of volume measurement method.

One of the earliest techniques to measure (changes of) body volume was *plethysmography*, originally developed by Glisson (1622) and Swammerdam (1737) to demonstrate isovolumetric contraction of isolated muscle. The measuring technique consists of surrounding the organ or tissue with a rigid box filled with water or air. The displacement of the fluid or the change in air pressure indicates the volume changes of the tissue due to arterial in-flow. Two major types of plethysmography exist, and these can be distinguished by the technique used to measure the volume change. These are *volume plethysmography* (direct-measurement displacement plethysmography including water and air types), and *electrical plethysmography* (strain-gages, inductive and impedance plethysmographs). The physical condition in which the measurements should be performed determines the plethysmographic method chosen.

Advances in medical imaging have provided new possibilities for noninvasively extracting quantitatively useful diagnostic information. Images are constructed on a grid of small picture elements (pixels) that reflect the intensity of the image in the array occupied by the pixel. Most medical images represent a two-dimensional projection of a three-dimensional object.

Currently, the most commonly used medical imaging modalities are ultrasound, nuclear magnetic resonance imaging, X-rays and X-ray computer tomography. *Ultrasound imaging* is based on the transmission and reflection of high-frequency acoustic waves. Waves whose frequencies lie well above the hearing range can be transmitted through biological tissue and will be reflected if they cross a boundary between media of different acoustic properties. These reflected signals can be reconverted to electrical signals and displayed to obtain a two-dimensional section. *Magnetic resonance imaging* is based on the involvement of the interaction between magnetic moment (or spin) of nuclei and a magnetic field. The proton spins are excited by an external radio frequency signal and the return to an equilibrium distribution is used to construct cross-sectional images. *Computer tomography* or *CT-scanning* uses a rotating source of X-rays. The X-ray source and detector are scanned across a section of the object of interest, and the transmission measurements are used to reconstruct an image of the object.

These imaging techniques display cross-sections of views that can be used to estimate the size of specific components. One way to estimate the volume of internal objects is make certain assumptions concerning the shape and to apply formulae to estimate the volume with dimensions of the object such as length, height, and width. A more accurate technique is *step-section planimetry*, a clinical application of numerical integration. During this procedure, cross-sections of the object are recorded with a certain (fixed) interval. The area of the object is determined in every section, and the total volume is calculated by multiplying the contribution of each section with the interval and summarizing all contributions.

The volume of a fluid-filled region can be calculated if a known quantity of indicator is added to the fluid and the concentration measured after it has been dispersed uniformly throughout the fluid. The selection of the indicator used to measure the volume depends on the application and can be based on temperature, color, or radioactivity.

Finally, volume (changes) also be performed directly using *water-displacement volumetry*, a sensitive but time-consuming method to measure the volume of an extremity. This method is not suitable for patients in the immediate postoperative period.

13.1 Plethysmography Theory

Fluid in-flow can be measured by blocking the out-flow and then measuring the change in volume due to the in-flow over a defined time interval. Plethysmography enables the volume change to be determined in a reasonably accurate manner. A direct-volume displacement plethysmograph uses a rigid chamber usually filled with water, into which the limb or limb segment is placed. This type of plethysmograph, sometimes called *chamber plethysmography*, can be used in two ways. It can be used to measure the sequence of pulsations proportional to the individual volume changes with each heart beat (arterial plethysmography or rheography). Also, the total amount of blood flowing into the limb or digit can be measured by venous occlusion: by inflating the occluding cuff placed upstream of the limb or digit just above the venous pressure 5 kPa to 8 kPa (40 mm Hg to 60 mm Hg), arterial blood can enter the region but venous blood is unable to leave. The result is that the limb or digit increases its volume with each heart beat by the volume entering during that beat.

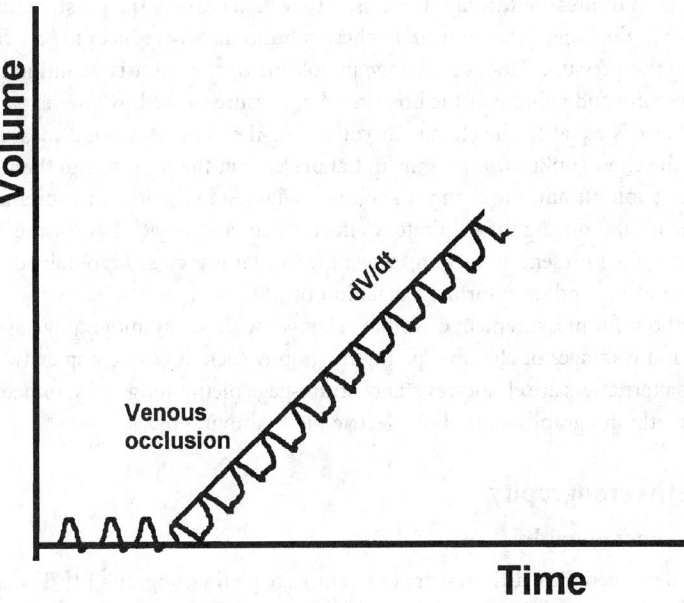

FIGURE 13.1 Typical recording from air plethysmograph during venous occlusion, revealing the changes in volume over time caused by arterial in-flow.

Air/Water Plethysmography or Chamber Plethysmography

Air plethysmography uses the relation between the volume change of a gas in a closed system and the corresponding pressure and temperature [1]. The relation between pressure and volume is described by Boyle's law, which can be written as:

$$P_i V_i = P_f V_f \tag{13.1}$$

with P_i and V_i the initial pressure and volume, respectively, and P_f and V_f the final pressure and volume, measured at constant temperature: the displacement of the fluid or the compressing of the air is a direct measure of the blood flow or original volume. The air plethysmograph uses the change in pressure that occurs in a cuff wrapped around the segment of interest due to the change in volume. By inflating the cuff to about 60 mm Hg, the arterial in-flow causes small increases in pressure. These small changes in pressure over the cardiac cycle can be monitored (see Figure 13.1). The measurement of blood flow is achieved by comparing the pressure changes to changes caused by removing known amounts of air from the system. The second measurement uses volume changes at various pressures between systolic and diastolic pressures, and the peak deflection is compared with the deflection caused by removal of known amounts of air from the system. In segmental plethysmography, two cuffs are used to measure the volume changes in a segment of a limb. Venous occlusion is established by the first cuff, while the second is inflated to a pressure that will exclude blood flow from those parts that should not be included in the measurement.

The technique can also be used for whole-body plethysmography [2], a common technique used to measure residual volume, functional residual capacity (FRC), and total lung volume, the parameters usually determined during pulmonary function testing [3]. In body plethysmography, the patient sits inside an airtight box, and inhales or exhales to a particular volume (usually the functional residual capacity, the volume that remains after a normal exhalation). Then, a shutter closes the breathing tube, while the subject tries to breathe through the closed tube. This causes the chest volume to expand and decompress the air in the lungs. The increase in chest volume slightly reduces the air volume of the box, thereby increasing the pressure. First, the change in volume of the chest is quantified using Boyle's law with the initial pressure and volume of the box, and the pressure in the box after expansion. The change in volume of the box is equal to the change in volume of the chest. A second measurement using the initial volume of the chest (unknown) and the initial pressure at the mouth, and the inspiratory volume (the unknown chest volume and the change in volume obtained in the first measurement) together with the pressure at the mouth during the inspiratory effort. By solving Boyle's law for the unknown volume, the original volume of gas present in the lungs when the shutter was closed is obtained, which is normally the volume present at the end of a normal exhalation or FRC.

Alternative methods for measurement of volume changes with plethysmography have been introduced to overcome the disadvantages of chamber plethysmography such as cost, complexity, and awkwardness of use. Important alternatives are elastic-resistance strain-gage plethysmography, impedance plethysmography, inductive plethysmography, and photoelectric plethysmography.

Electrical Plethysmography

Strain-Gage Plethysmography

In 1953, Whitney described the elastic-resistance strain-gage plethysmograph [4]. The strain-gage instrument quantifies the change in resistance of the gage as it is stretched due to the enlargement of the object (e.g., limb segment). The gage is typically a small elastic tube filled with mercury or an electrolyte or conductive paste. The ends of the tube are sealed with copper plugs or electrodes that make contact with the conductive column. These plugs are connected to a Wheatstone bridge circuit.

To illustrate the principle of the strain-gage method, a strain-gage of length l_0 is placed around the limb segment, a circular cross section with radius r_0. The length of the strain-gage can thus also be expressed as $l_0 = 2\pi r_0$. Expansion of the limb gives a new length $l_1 = 2\pi r_1$, with r_1 the new radius of the limb. The increase in length of the strain-gage is thus $\delta l = 2\pi(r_1 - r_0)$, while the change in cross-sectional area of the limb δA is $\pi(r_1^2 - r_0^2)$. This can also be expressed as:

$$\delta A = \pi \left[2r_0 \left(r_1 - r_0 \right) + \left(r_1 - r_0 \right)^2 \right] \tag{13.2}$$

Since $r_1 - r_0$ is usually small, $(r_1 - r_0)^2$ can be neglected. Consequently, δA can be written as $2\pi r_0(r_1 - r_0)$ which, on dividing by $A = \pi(r_0)^2$, gives:

$$\frac{\delta A}{A} = 2\frac{r_1 - r_0}{r_0} \tag{13.3}$$

But since $\delta V/V = \delta A/A$ and $\delta l/l = \delta r/r$:

$$\frac{\delta V}{V} = 2\frac{l_1 - l_0}{l_0} \tag{13.4}$$

Thus, the percentage increase in volume can be obtained by measuring the initial gage length and the change in length. In practice, this change in length of the strain-gage is recorded over a relatively short period to overcome the problem of back pressure that builds up in the limb or digit because of the venous occlusion, and the measurement is usually expressed as milliliters per minute per 100 g of tissue, or: volume flow rate = $2(\delta l/l_0) \times (100/t)$, where δl is the increase in gage length during time t. To measure $\delta l/t$, the gage must be calibrated by removing it from the limb and stretching it on a measuring jig until the pen recorder returns to the zero level; this is the length l_0. The gage can then be stretched accurately by typical amounts, to calibrate the output and calculate the volume flow rate.

Impedance Plethysmography

The physical principle of impedance plethysmography is the variation in electrical impedance of a tissue segment over the cardiac cycle due to a changing conductance. The change is caused by arterial in-flow to the segment while the venous drainage is occluded.

If a segment of limb is assumed to be cylindrical and of length l and volume V, the impedance Z of that part can be written as:

$$Z = \frac{\rho l^2}{V} \tag{13.5}$$

with ρ the specific resistivity of the tissue forming the cylinder.

Blood entering the tissue segment during the cardiac cycle introduces a parallel impedance Z_b described as $Z_b = Z_0 Z_1/(Z_0 - Z_1)$, with Z_0 the initial impedance and Z_1 the new value. A small change in volume δV is related to a small change in resistance δZ, such that:

$$\delta Z = Z_0 - Z_1 = \rho l^2 \left(\frac{1}{V_0} - \frac{1}{V_1} \right) = \rho l^2 \frac{(V_1 - V_0)}{V_1 V_0} \approx -\rho l^2 \frac{\delta V}{V^2} = -Z \frac{\delta V}{V} \tag{13.6}$$

when very small changes in volume are assumed [5]. Thus, δV can be written as:

$$\delta V = \rho \left(\frac{1}{Z_0^2} \right) \delta Z \tag{13.7}$$

and these parameters Z_0 and δZ can be obtained from the impedance plethysmogram. If it is assumed that $\rho = 150 \ \Omega$ cm, being the specific resistivity of soft tissue, the change in volume can be determined.

Two major techniques have been used to measure tissue impedance. One method uses two electrodes to measure both the voltage drop across the electrodes and the current flow. The main disadvantage of this technique is that the current distribution in the vicinity of the electrodes is not known, meaning the exact volume of the interrogated tissue is also unknown. In 1974, Kubicek et al. [6] described the use of four electrodes to measure tissue impedance: the outer electrodes supply a small-amplitude, high-frequency ac signal, while the inner two are used to measure the potential difference between two points on the same surface (see Figures 13.2 and 13.3).

Inductive Plethysmography

For measurement of volume changes during respiration, a system based on inductive plethysmography has been developed; the respiratory inductive plethysmograph was introduced in 1978. During inductive plethysmography, two coils are attached — one placed around the rib cage and the other around the abdomen. During respiration, the cross-sectional diameter of the coils changes, leading to a change in the inductances of the coils. By converting these inductances into proportional voltages, a measure for the changes in volume during respiration can be obtained. The conversion is achieved by using oscillators whose frequencies depend on a fixed internal capacitor and the inductance of each coil. These oscillator frequencies are converted to proportional voltages, voltages that are then recorded. After calibration on the patient, the system can be used for respiratory volume measurement [9].

Two general methods have been introduced for calibration of the respiratory inductive plethysmograph. These calibration methods are based on the following equation:

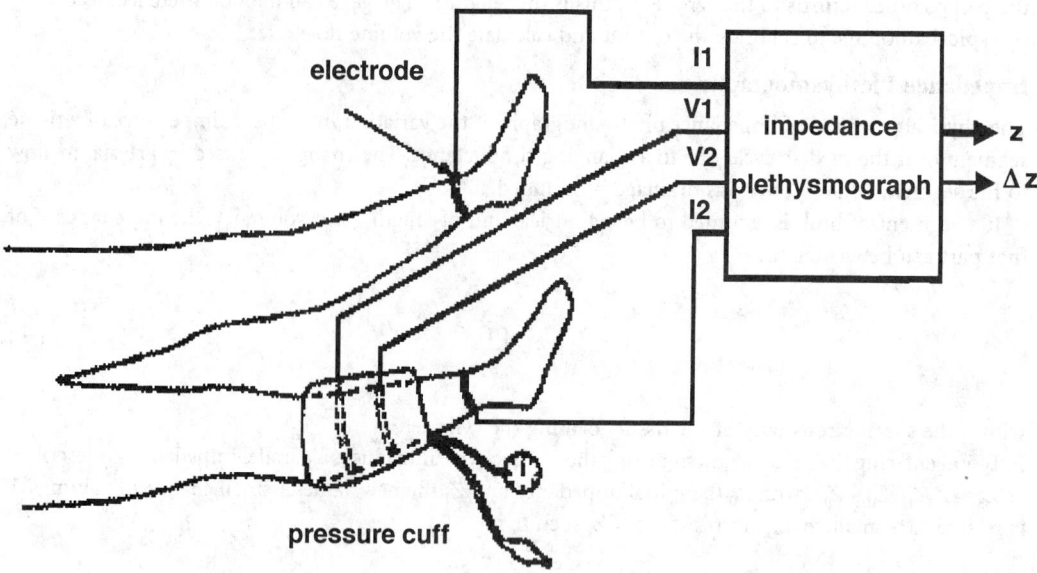

FIGURE 13.2   Electrode positioning for impedance plethysmography used for early detection of peripheral athero-sclerosis: the outer two supply current while the inner two measure voltage. The combination of these two is used to obtain the changes in impedance, and these changes are then used to obtain the changes in volume of the leg. (From R. Shankar and J. G. Webster, *IEEE Trans. Biomed. Eng.*, 38, 62-67, 1993. With permission.)

$$V_T = V_{RC} + V_{AB} \qquad (13.8)$$

with V_{RC} and V_{AB} the contributions of the rib cage (RC) and abdominal compartments (AB) to the tidal volume V_T. Calibration can be performed by determining the calibration constant in an isovolume measurement or by regression methods of differences in inductance changes due to a different body position or a different state of sleep [10]. Once it is calibrated, the respiratory inductive plethysmograph can be used for noninvasive monitoring of changes of thoracic volume and respiratory patterns.

13.2   Numerical Integration with Imaging

For diagnostic purposes in health care, a number of imaging modalities are currently in use (see also Chapter 79 on medical imaging). Webb nicely reviewed the scientific basis and physical principles of medical imaging [11].

With medical imaging modalities, diameters of objects can be determined in a noninvasive manner. By applying these diameters in an appropriate formula for volume calculation, the size of the object can be estimated. For example, in a clinical setting, the prostate volume V_p can be approximated with an ellipsoid volume calculation:

$$V_p = \frac{\pi}{6} H W L \qquad (13.9)$$

with H the height, W the width, and L the length of the prostate as illustrated in Figure 13.4. Variations have been introduced to find the best formulae to estimate the prostate volume [12, 13]. Also, for the determination of ventricular volumes, formulae have been derived, based on the following variation of

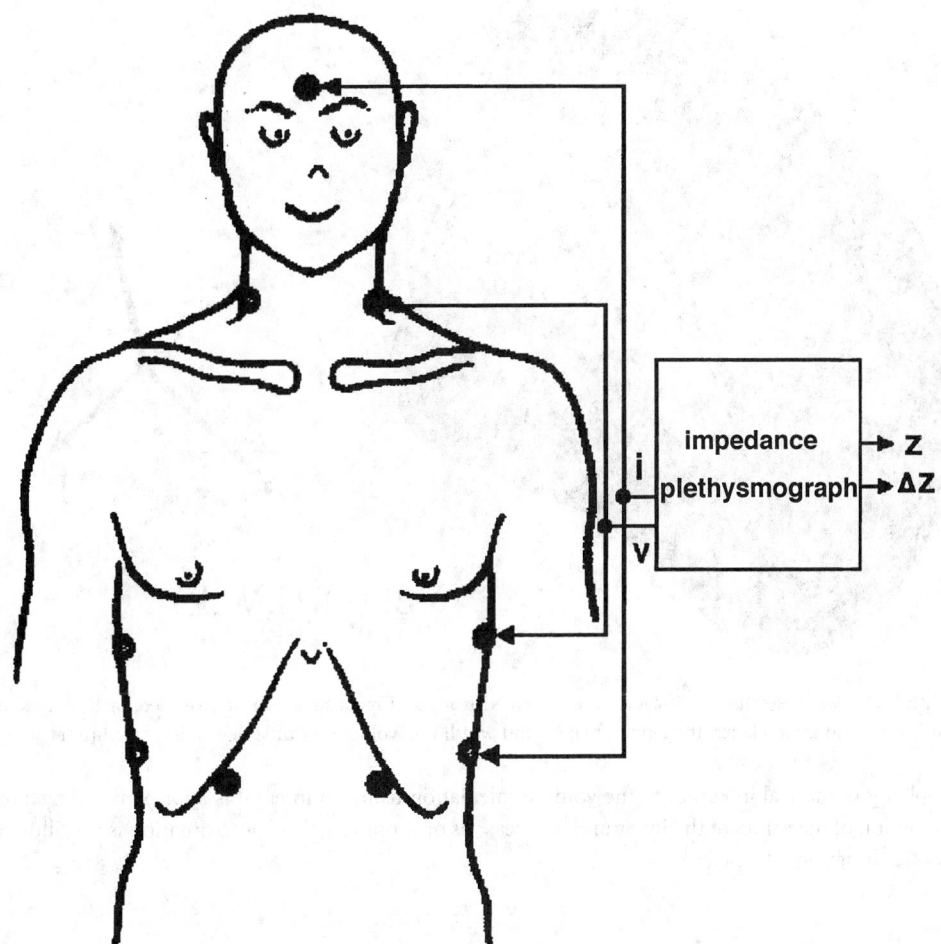

FIGURE 13.3 A second example of electrode positioning for impedance plethysmography, here used to estimate stroke volume: the outer two supply current while the inner two measure voltage. (From N. Verschoor et al., *Physiol. Meas.*, 17, 29-35, 1996. With permission.)

the formula for prostate volume measurements: (1) the volume is obtained with the same formula as for prostate volume measurements, (2) the two short axes are equal, (3) the long axis is twice the short axis, and (4) the internal diameter is used for all axes.

By displaying cross-sectional images of a 3-D object (e.g., an organ), volume calculations can also be performed by integrating areas of the organ over a sequence of medical images as illustrated in Figure 13.5. For ultrasound images, applications have been reported in the literature on prostate, heart, bladder, and kidney volume. The application of integration techniques involves the application of segmentation algorithms on sequences of images, either in transverse direction with fixed intersection distance or in the radial with images at a fixed intersection angle.

For a volume defined by the function $f(x,y,z)$, the volume can be calculated from:

$$V = \int_{x=x_1}^{x_2} \int_{y=y_1}^{y_2} \int_{z=z_1}^{z_2} f(x,y,z)\,\mathrm{d}z\mathrm{d}y\mathrm{d}x \qquad (13.10)$$

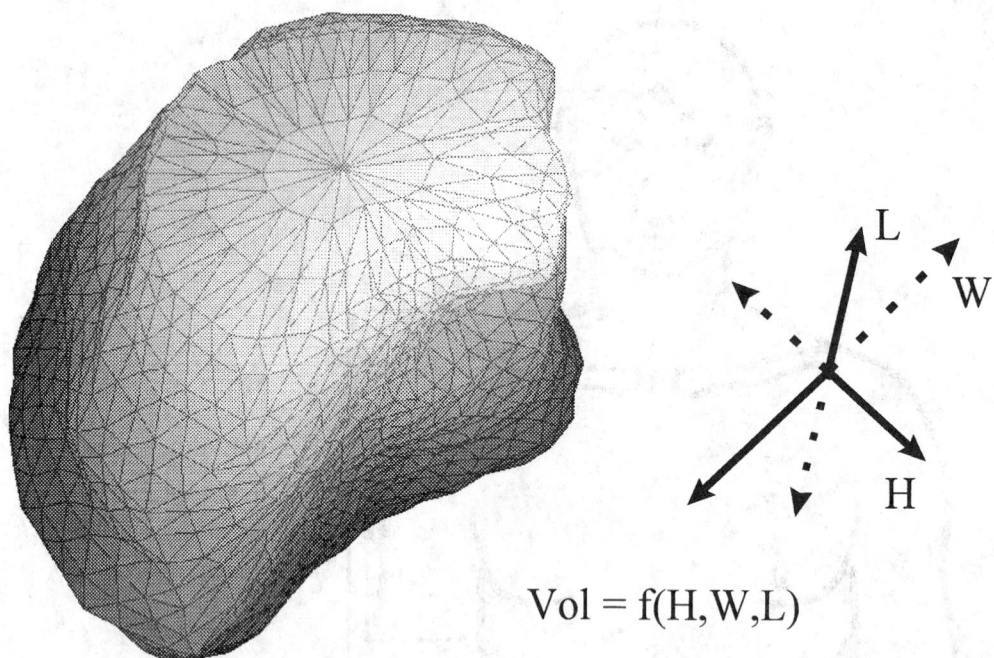

$$\text{Vol} = f(H, W, L)$$

FIGURE 13.4   Volume estimate of a prostate, obtained from 3-D reconstruction of cross-sectional images, using the dimensions in three planes: the length, height, and width in a volume formula describing an ellipsoid shape.

By applying numerical integration, the volume calculation using an integral is approximated by a linear combination of the values of the integrand. For reasons of simplicity, this approximation is first illustrated for a one-dimensional case:

$$\int_{x=a}^{b} f(x)\,dx \approx w_1 f(x_1) + w_2 f(x_2) + \ldots + w_{nf} f(x_n) \qquad (13.11)$$

In this equation, $x_1, x_2, \ldots, x_n$ are n points chosen in the interval of integration $[a,b]$, and the numbers $w_1, w_2, \ldots, w_n$ are n weights corresponding to these points. This approximation can also be expressed in the so-called Riemann sum:

$$\int_{x=a}^{b} f(x)\,dx \approx w_1 f(x_1) + w_2 f(x_2) + \ldots + w_n f(x_n) = \sum_{i=1}^{n} f(\xi_i)(x_i - x_{i-1}) \qquad (13.12)$$

For volume calculation, the Riemann sum needs to be extended to a three-dimensional case:

$$V = \int_{x=x_1}^{x_2} \int_{y=y_1}^{y_2} \int_{z=z_1}^{z_2} f(x,y,z)\,dz\,dy\,dx \approx \sum f(p_{ijk})(x_{i+1} - x_i)(y_{j+1} - y_j)(z_{k+1} - z_k) \qquad (13.13)$$

This three-dimensional case can be reduced by assuming that $(x_{i+1} - x_i)(y_{i+1} - y_i)f(p_{ijk})$ represents the surface S_k of a two-dimensional section at position z_k, while $(z_{k+1} - z_k)$ is constant for every k and equals h, the intersection distance:

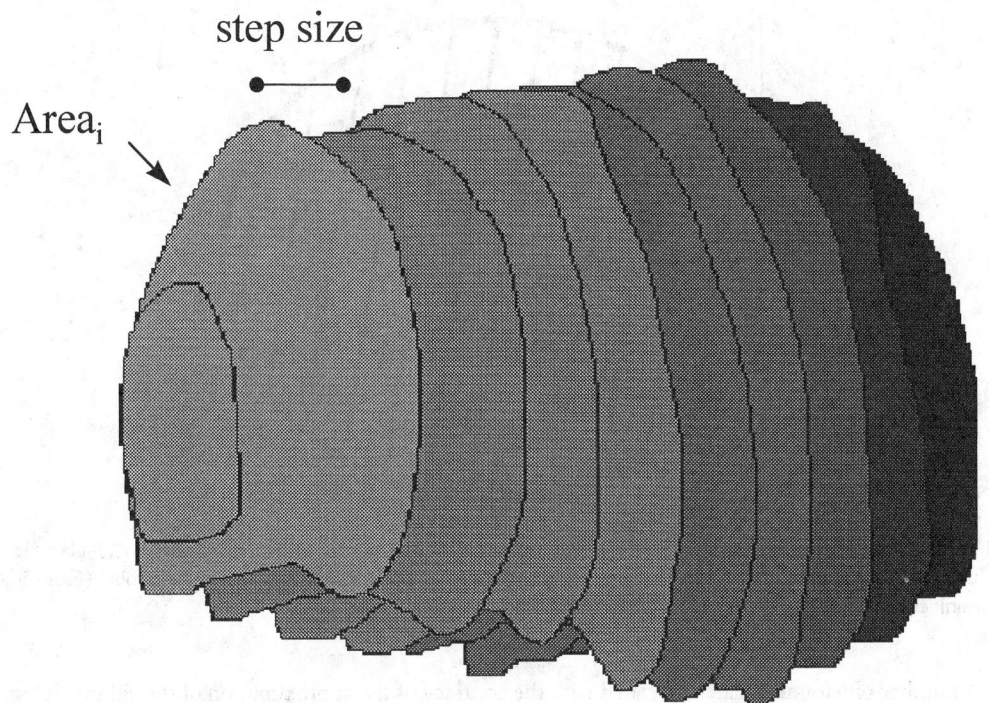

FIGURE 13.5 Outlines of cross-sectional images of the prostate used for planimetric volumetry of the prostate: transverse cross-sections obtained with a step size of 4 mm have been outlined by an expert observer, and the contribution of each section to the total volume is obtained by determining the area enclosed by the prostate contour. The volume is then determined by summarizing all contributions after multiplying them with the step size.

$$V \approx h \sum S(z_k) = h \sum_{k=0}^{n-1} S(a + kh) \qquad (13.14)$$

where a represents the position of the first section, n the number of sections, and h the step size, given by $h = (b - a)/n$, where b the position of the last section.

In general, three different forms of the approximation can be described, depending on the position of the section. The first form is known as the rectangular rule, and is presented in Equation 13.14. The midpoint rule is described by:

$$V \approx h \sum S(z_k) = h \sum_{k=0}^{n-1} S\left[a + \left(k + \frac{1}{2} h \right) \right] \qquad (13.15)$$

The trapezoidal rule is similar to the rectangular rule, except that it uses the average of the right-hand and left-hand Riemann sum:

$$T_n(f) = h\left[\frac{f(a)}{2} + f(a + h) + f(a + 2h) + \dots + f(a + (n-1)h) + \frac{f(b)}{2} \right] \qquad (13.16)$$

The trapezoidal and midpoint rules are exact for a linear function and converge at least as fast as n^{-2}, if the integrand has a continuous second derivative [14].

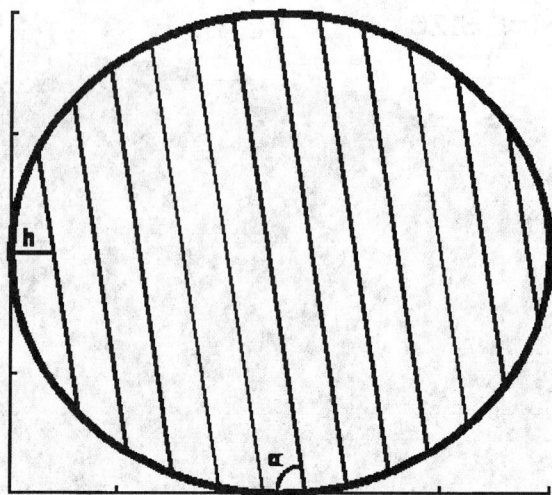

FIGURE 13.6   An ellipsoid-shaped model of the prostate that was used for theoretical analysis of planimetric volumetry. Indicated are the intersection distance *h* and the scan angle α with the axis of the probe. (From R. G. Aarnink et al., *Physiol. Meas.*, 16, 141-150, 1995. With permission.)

A number of parameters are associated with the accuracy of the approximation of the volume integral with a Riemann sum; first of all, the number of cross sections *n* is important as the sum converges with n^{-2}: the more sections, the more accurate the results. Furthermore, the selection of the position of the first section is important. While the interval of integration is mostly well defined in theoretical analyses, for *in vivo* measurements, the first section is often less well defined. This means not only that the selection of this first section is random with the first possible section and this section increased with the step size, but also that sections can be missed. The last effect that may influence the numerical results is the effect of taking oblique sections, cross sections that are not taken perpendicular to the axis along which the intersection distance is measured. All the above assumes that every section S_k can be exactly determined.

The influence of the parameters associated with these kinds of measurements can be tested in a theoretical model. In a mathematical model with known exact volume or surface, the parameters — including step size, selection of the first section, and the scan angle — can be tested. Figure 13.6 shows a model used to model the prostate in two dimensions. It consists of a spheroid function with a length of the long axis of 50 mm and a length of the short axis of 40 mm. In mathematical form, this function can be described as:

$$y = f\left(x\right) = y_m \pm y_o \sqrt{1 - \left(\frac{x - x_m}{x_o}\right)^2}$$ (13.17)

with x_0 half the diameter of the ellipse in *x*-direction (25 mm) and y_0 half the diameter of the ellipse in *y*-direction (20 mm) and (x_m, y_m) the center point of the ellipse. The analytical solution of the integral to obtain the surface of this model is $\pi x_0 y_0$.

Figure 13.6 also reveals the possible influence of the parameters *h* and α. Figure 13.7 shows the results for varying the step size *h* between 4 mm and 16 mm corresponding to between 13 and 3 sections used for numerical integration as a function of the position of the first section. The first section is selected (at random) between the first coordinate of the function, and this coordinate plus the step size *h*. The sections were taken perpendicular to the long axis. From this study, which is fully described by Aarnink et al. [15], it was concluded that if the ratio of step size to the longest diameter is 6 or more, then the

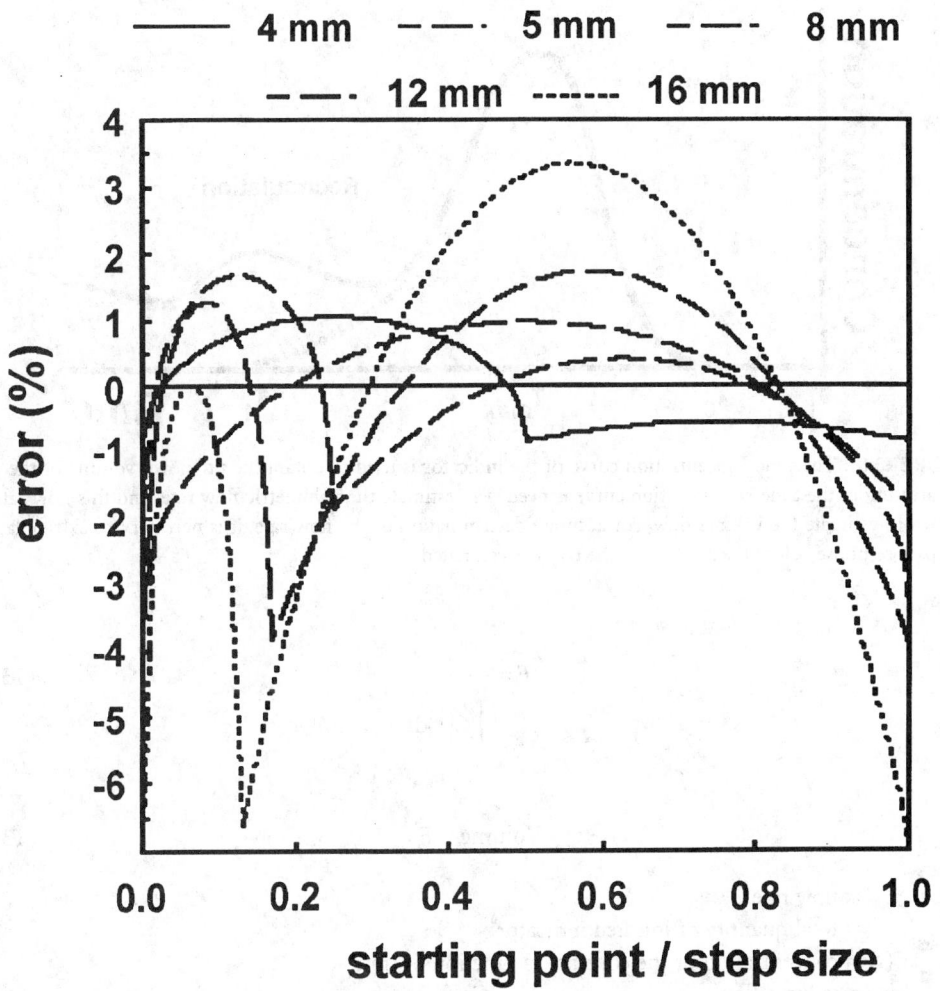

FIGURE 13.7 The errors in surface estimates by numerical integration for different step sizes h presented as percentage of the exact solution, as a function of the position of the first section, located between 0 and h. (From R. G. Aarnink et al., *Physiol. Meas.,* 16, 141-150, 1995. With permission.)

error in volume estimation is less than 5%. Although clinical application might introduce additional errors such as movement artifacts, this rule provides a good indication of the intersection distance that should be used for good approximations with numerical integration.

13.3 Indicator Dilution Methods

The principle of the indicator dilution theory to determine gas or fluid volume was originally developed by Stewart and Hamilton. It is based on the following concept: if the concentration of an indicator that is uniformly dispersed in an unknown volume is determined, and the volume of the indicator is known, the unknown volume can be determined. Assuming a single in-flow and single out-flow model, all input will eventually emerge through the output channel, and the volumetric flow rate can be used to identify the volume flow. It can be described with two equations, of which the second is used to determine the volume using the result of the first equation [16]:

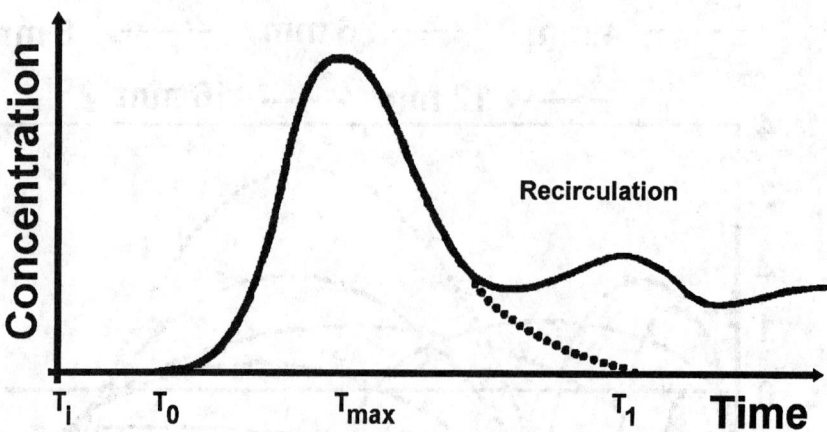

FIGURE 13.8 The time-concentration curve of the indicator is used to obtain the unknown volume of the fluid. The area under the time-concentration curve is needed to estimate the volumetric flow rate, and this flow rate can be used to estimate the volume flow. For accurate determination of the flow rate, it is necessary to extrapolate the first pass response before the area under the curve is estimated.

$$F = \frac{I_i}{\int_0^\infty C(t)\,dt} \tag{13.18}$$

$$\text{Volume} = F\, t_m \tag{13.19}$$

where
F = flow rate
I_i = total quantity of injected indicator
$C(t)$ = concentration of the contrast as function of time
Vol = volume flow
t_m = mean transit time

The integrated value of the concentration over time can be obtained from the concentration-time curve (Figure 13.8) using planimetry and an approximation described by the Riemann sum expression (Equation 13.13). Care must be taken to remove the influence of the effect of recirculation of the indicator. An indicator is suitable for use in the dilution method if it can be detected and measured, and remains within the circulation during at least its first circuit, without doing harm to the object. Subsequent removal from the circulation is an advantage. Dyes such as Coomassie blue and indocyanine green were initially used, using the peak spectral absorption at certain wavelengths to detect the indicator. Also, radioisotopes such as I-labeled human serum albumin have been used. Currently, the use of cooled saline has become routine in measurement of cardiac output in the intensive care unit.

Thermodilution

To perform cardiac output measurements, a cold bolus (with a temperature gradient of at least 10E) is injected into the right atrium and the resulting change in temperature of the blood in the pulmonary artery is detected using a rapidly responding thermistor. After recording the temperature-time curve, the blood flow is calculated using a modified Stewart-Hamilton equation. This equation can be described as follows [17]:

$$V_i \; SW_i \; SH_i \left(T_b - T_i \right) = F \int T_b \left(dt \right) SW_b \; SH_b \tag{13.20}$$

The terms on the left represent the cooling of the blood (with temperature T_b), as caused by the injection of a cold bolus with volume V_i, with temperature T_i, specific weight SW_i and specific heat SH_i. The same amount of "indicator" must appear in the blood (with specific weight and heat SW_b and SH_b, respectively) downstream to the point of injection, where it is detected in terms of the time-course of the temperature T_b in the flow F by means of a thermistor. The flow can be described as:

$$F = \frac{T_b - T_i}{\int T_b \left(dt \right)} K \tag{13.21}$$

where K (the so-called calculation constant) is described as:

$$K = V_i \frac{SW_i - SH_i}{SW_b - SH_b} C \tag{13.22}$$

This constant is introduced to the computer by the user. T_i is usually measured at the proximal end of the injection line; therefore, a "correction factor" C must be introduced to correct for the estimated losses of cold saline in the catheter, due to the dead space volume. Furthermore, the warming effect of the injected fluid during passing through the blood must be corrected; this warming effect depends on the speed of injection, the length of immersion of the catheter, and the temperature gradient. This warming effect reduces the actual amount of injected fluid, so the correction factor C is less than 1 [17].

Radionuclide Techniques

Radionuclide imaging is a common noninvasive technique used in the evaluation of cardiac function and disease. It can be compared to X-ray radiology, but now the radiation emanating from inside the human body is used to construct the image. Radionuclide imaging has proven useful to obtain ejection fraction and left ventricular volume. As in thermodilution, a small volume of indicator is injected peripherally; in this case, radioisotopes (normally indicated as radiopharmaceuticals) are used. The radioactive decay of radioisotopes leads to the emission of alpha, beta, gamma, and X radiation, depending on the radionuclide used. For *in vivo* imaging, only gamma- or X-radiation can be detected with detectors external to the body, while the minimal amount of energy of the emitted photons should be greater than 50 keV. Using a photon counter such as a Gamma camera, equipped with a collimator, the radioactivity can be recorded. Two techniques are commonly used: the first pass method and the dynamic recording method [11].

First Pass Method

To inject a radionuclide bolus into the blood system, a catheter is inserted in a vein, for example, an external jugular or an antecubital vein. The camera used to detect the radioactive bolus is usually positioned in the left anterior oblique position to obtain optimal right and left ventricular separation, and is tilted slightly in an attempt to separate left atrial activity from that of the left ventricle. First, the background radiation is obtained by counting the background emissions. Then, a region of interest is determined over the left ventricle and a time-activity curve is generated by counting the photon emissions over time. The radioactivity count over time usually reveals peaks at different moments; the first peak occurs when the radioactivity in the right ventricle is counted, the second peak is attributed to left ventricular activity. More peaks can occur during recirculation of the radioactivity. After correction for background emissions, the time-intensity curve is evaluated to determine the ejection fraction (*EF*) of the heart, which is given by:

$$EF = \frac{c_d - c_s}{c_d - c_b} \qquad (13.23)$$

where c_d is the end diastolic count, c_s is the end systolic count, and c_b is the background count [18]. Using a dynamic cardiac phantom, the accuracy of the *EF* measurement and *LV* volume estimation by radionuclide imaging has been determined [18]. The count-based method was found to give accurate results for evaluating cardiac *EF* and volume measurements.

Dynamic Recording Method

During dynamic recording, the data acquisition is triggered by another signal, usually the ECG, to obtain the gated activity during ventricular systole and diastole. The contraction of the left ventricle of the heart is used to align the acquisition data from different cardiac cycles. During dynamic recording, information about the cardiac function is averaged over many heart beats during a period when the radiopharmaceuticals are uniformly distributed in the blood pool. Estimations of end systolic and end diastolic volumes are made and the ejection fraction is calculated (see above). Also, other parameters can be obtained; for example, the amplitude and phase of the ejection from the left ventricle. These parameters may show malfunction in the cardiac cycle. These dynamic recordings may also be applied to the brain, lungs, liver, kidney, and vascular systems and they can be important to judge the function of these organs (e.g., after kidney transplantation).

Gas Dilution

Gas dilution is a method to determine lung volumes because standard techniques for lung measurements measure only the inhaled or exhaled air as a function of time (spirometry) and cannot be used to assess the absolute lung volume. The subject is connected to a spirometer that contains a known concentration of gas (e.g., helium). The subject is then asked to breathe for several minutes, to equalize the concentration of helium in the lung and the spirometer. Using the law of conservation of matter, the volume of the lung can be calculated. Since the total amount of helium is the same before and after measurement, the fractional concentration times the volume before equals the fractional concentration times the volume after: $C_1 \times V_1 = C_2 \times (V_1 + V_2)$. The volume after the measurement can be extracted from this equation and, by subtracting the volume of the spirometer, the lung volume is calculated.

13.4   Water Displacement Volumetry

The measurement of the volume of an extremity such as arm or leg can be measured using a water tank, which is illustrated schematically in Figure 13.9. The advantage of water displacement volumetry is the possibility for direct measurement of objects with an irregular form. In the clinical situation, volume determination of the leg can be valuable for monitoring the severity of edema or hematoma after surgery or trauma. A setup for water displacement can be developed by the user, and an example of such a system is described by Kaulesar Sukul et al. [19]. A special tank for water displacement volumetry of the leg was constructed consisting of two overflow tubes. The tank was filled to the lower overflow tube, and the overflow tube was subsequently closed. The patient was then asked to lower his leg into the tank, and the amount of overflow of the upper tube was measured. The total volume was then calculated by measuring the fluid volume delivered to the upper tube and the volume difference between the lower and upper overflow tube, the so-called "reserve volume." The volume of ankle and foot was then measured by filling the tank to the upper overflow tube and measuring the amount of water in the cylinder when the foot and ankle were immersed in the water.

FIGURE 13.9 A schematic overview of a volume measuring method using water displacement: the desired volume of the object is represented by the overflow volume in the small tank after lowering the object into the large tank. This technique is especially useful for irregular-shaped objects.

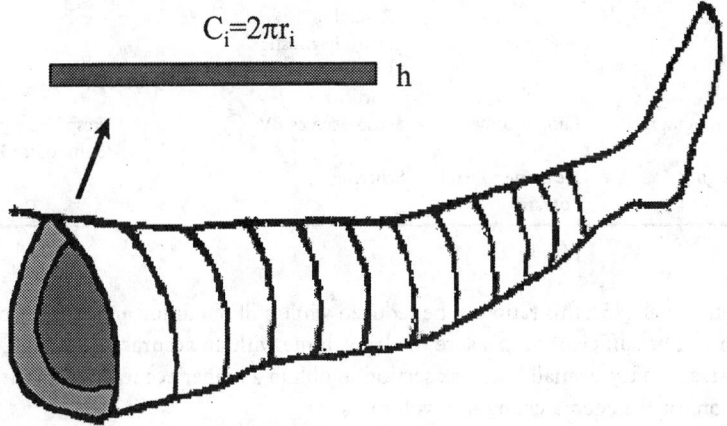

FIGURE 13.10 Illustration of the disk model to measure leg volume. The leg is divided in different sections with a fixed intersection distance the circumference of each section is obtained. This circumference is then used to describe the leg as perfect circles with an assumed radius calculated from the circumference. This radius is then used to calculate the contribution of that section to the volume and the total volume is obtained by summation.

Disadvantages of water displacement volumetry are hygiene problems, it is time-consuming, and not suitable for measurements of the volume of extremities of patients in the immediate postoperative period. Therefore, alternatives have been sought, with the disk model method as a promising one. The calculation of the volume of the leg is performed by dividing the leg into disks of thickness h (e.g., 3 cm) as illustrated in Figure 13.10. The total volume is equal to the sum of the individual disk volumes:

$$V = \sum_{i=1}^{n} \frac{C_i^2}{4\pi} h = \sum_{i=1}^{n} \pi r_i^2 h = h \sum_{i=1}^{n} \pi r_i^2 \qquad (13.24)$$

where C_i is the circumference of the disk at position i with assumed radius of r_i.

A study to compare water displacement volumetry with the results of the disk model method indicated that both methods give similar results. Consequently, because of the ease of application, the disk model method is the method of choice to measure volumes of extremities. Assuming the length of a leg to be 75 cm, a ratio between length and step size (3 cm was proposed in [19]) of 25 is obtained. According to

TABLE 13.1 Volume Measuring Techniques, Applications, and Equipment for Different Applications

Technique	Application	Companies	Products	Price (U.S.$)
Spirometry	Lung volume	Nellcor Puritan Bennett	Renaissance	
		Morgan Science		
		Spirometrics		
		CDX Corporation	Spiro 110S	
Whole-body plethysmography	Lung volume	Morgan Science		
		ACI Medical Inc.		25,000.00
Gas-dilution	Lung volume	Equilibrated Biosystems Inc.		
		Melville		
Thermodilution	Heart	Abbott Critical Care System		
		Baxter		
		American Edwards Laboratories		
Strain-gage plethysmography	Cardiac output	Parks Medical Electronics		
Impedance plethysmography	Perfusion studies	Ambulatory monitoring systems		
		Vitalog		
		RJL systems Detroit		
		Codman and Shurtleff Inc.		
		Randolph		
		Electrodiagnostic		
		Instrument Inc.		
		Burbank		
Inductive plethysmograph	Lung volume	SensorMedics BV	RespiTrace plus	
			SomnoStar PT	15,000.00
Radionuclide imaging	Heart, peripheral organs	Schering		

a study by Aarnink et al. [15], this ratio can be reduced while still obtaining an accuracy of >95%. A disk height of 10 cm will be sufficient to measure the leg volume with an accuracy of at least 95%. However, it might be necessary to use a small interdisk section to obtain a higher accuracy, which might be needed to accurately monitor the edema changes in volume.

13.5 Equipment and Experiments

Table 13.1 summarizes the different methods for volume measurement and their applications. The table serves as a starting point to evaluate the available equipment for volume measurement.

13.6 Evaluation

Plethysmography is an easy and noninvasive method to obtain knowledge for assessing vascular diseases, cardiac output disorders, or pulmonary disfunctions. Systems for direct measurement of volume changes have been developed, but physical properties have also been introduced as an indirect measure of the change in volume. Each system has its own advantages and disadvantages.

For *chamber plethysmography*, the water-filled type is more stable with respect to temperature change but thermal problems may occur if the water temperature is significantly different from that of the limb segment. Furthermore, the hydrostatic effect of the water may alter the blood flow. Also, the rubber sleeve between the limb segment and the water may influence the release of sweat. It is a cumbersome measurement that is not very useful during or after exercise. The *air displacement* type introduces problems of drift because of its high coefficient of expansion, although self-compensating systems have been proposed. Also, the thermal behavior of the body plethysmograph may produce variations in pressure inside the airtight box. Two sources of temperature variation can be mentioned: the chest volume variation induced by breathing produces heat and a temperature gradient may exist between the patient and the internal temperature of the box [20].

In *impedance plethysmography*, the changes in impedance in tissue are used, which are primarily due to changes in the conductivity of the current path with each pulsation of blood. Several theories attempt to explain the actual cause of these changes in tissue impedance. One explanation is that blood-filling of a segment of the body lowers the impedance of that segment. A second theory is that the increase in diameter due to additional blood in a segment of the body increases the cross-sectional area of the segment's conductivity path and thereby lowers the resistance of the path. A third explanation is based on the principle of pressure changes on the electrodes that occur with each blood pulsation and uses the changes in the impedance of the skin–electrode interface. The main difficulty with the procedure is the problem of relating the output resistance to any absolute volume measurement. Detection of the presence of arterial pulsations, measurement of pulse rate, and determination of time of arrival of a pulse at any given point in the peripheral circulation can all be satisfactorily handled by impedance plethysmography. Also, the impedance plethysmograph can measure time-variant changes in blood volume. A problem with impedance plethysmography may be the sensitivity to movement of the object. Research is being conducted to reduce the influences of these movement artifacts, either by different electrode configuration, electrode location, or using multiple sensors or different frequencies [21].

A low-cost *inductive plethysmograph* was designed by Cohen et al. to obtain a noninvasive measure of lung ventilation [22]. This plethysmograph indirectly monitors ventilation by measuring the cross-sectional area of the chest and abdomen. They attached commercially available elastic bands containing wire around the chest and abdomen and determined their inductances by measuring the frequency of an inductive-controlled oscillator.

New devices for plethysmographic measurements are under development, using different properties to obtain the quantitative knowledge required. For example, *acoustic plethysmography* measures body volume by determining changes in the resonant frequency of a Helmholtz resonator. A Helmholtz resonator consists of an enclosed volume of gas connected to its surroundings through a single opening. The gas can be forced to resonate acoustically by imposing periodic pressure fluctuations of the opening. The resonator frequency is inversely proportional to the square root of the volume of air in the resonator. An object placed in the resonator reduces the volume of air remaining in the resonator by its own volume, causing an increase in the resonator frequency. From a study to obtain density values of preterm infants, it was concluded that the acoustic plethysmograph can be used to measure total body volume of preterm infants [23].

In addition, volume measurements with diagnostic imaging modalities are well established in the clinical environment; for example, prostate ultrasonography or echocardiography. Besides important parameters such as step size and first step selection, accurate determination of the surface in different sections is important. While predominantly performed manually now, several attempts for automated detection of the surface have been reported [24–27]. Automatic detection of the surface in each section should be possible and would enable a system for automated measurement of the prostate, heart, liver, etc. Further research will indicate the usefulness of such a system in a clinical setting.

References

1. A. J. Comerota, R. N. Harada, A. R. Eze, and M. L. Katz, Air plethysmography: a clinical review, *Int. Angiology,* 14, 45-52, 1995.
2. A. B. Dubois, S. J. Botello, G. N. Beddell, R. Marshall, and J. H. Comroe, A rapid plethysmographic method for measuring thoracic gas volume: a comparison with a nitrogen washout method for measuring functional residual capacity in normal subjects, *J. Clin. Invest.,* 35, 322-326, 1956.
3. J. B. West, *Respiratory Physiology — The Essentials,* Baltimore, MD: Williams and Wilkins, 1987.
4. J. R. Whitney, The measurement of volume changes in human limbs, *J. Physiol.,* 121, 1-27, 1953.
5. J. Nyboer, S. Bagno, and L. F. Nims, The Impedance Plethysmograph: An Electrical Volume Recorder, National Research Council, Committee on Aviation Medicine, Rep. No. 149, 1943.

6. W. G. Kubicek, F. J. Kottke, M. V. Ramos, R. P. Patterson, D. A. Witsoe, J. W. Labree, W. Remole, T. E. Layman, H. Schoening, and J. T. Garamala, The Minnesota impedance cardiograph — Theory and applications, *Biomed. Eng.*, 9, 410-416, 1974.

7. R. Shankar and J. G. Webster, Noninvasive measurement of compliance of human leg arteries, *IEEE Trans. Biomed. Eng.*, 38, 62-67, 1993.

8. N. Verschoor, H. H. Woltjer, B. J. M. van der Meer, and P. M. J. M. de Vries, The lowering of stroke volume measured by means of impedance cardiography during endexpiratory breath holding, *Physiol. Meas.*, 17, 29-35, 1996.

9. M. A. Cohn, H. Watson, R. Weisshaut, F. Stott, and M. A. Sackner, A transducer for non-invasive monitoring of respiration, in *ISAM 1977, Proc. Sec. Int. Symp. Ambulatory Monitoring*, London: Academic Press, 1978, 119-128.

10. J. A. Adams, Respiratory inductive plethysmography, in J. Stocks, P. D. Sly, R. S. Tepper, and W. J. Morgan (eds.), *Infant Respiratory Function Testing*, New York: Wiley-Liss, 1996, 139-164.

11. S. Webb, *The Physics of Medical Imaging*, Bristol, U.K.: IOP Publishing, 1988, 204-221.

12. M. K. Terris and T. A. Stamey, Determination of prostate volume by transrectal ultrasound, *J. Urol.*, 145, 984-987, 1991.

13. R. G. Aarnink, J. J. M. C. H. de la Rosette, F. M. J. Debruyne, and H. Wijkstra, Formula-derived prostate volume determination, *Eur. Urol.*, 29, 399-402, 1996.

14. P. J. Davis and P. Rabinowitz, *Methods of Numerical Integration*, San Diego: Academic Press, 1975, 40-43.

15. R. G. Aarnink, R. J. B. Giesen, J. J. M. C. H. de la Rosette, A. L. Huynen, F. M. J. Debruyne, and H. Wijkstra, Planimetric volumetry of the prostate: how accurate is it?, *Physiol. Meas.*, 16, 141-150, 1995.

16. E. D. Trautman and R. S. Newbower. The development of indicator-dilution techniques, *IEEE Trans. Biomed. Eng.*, 31, 800-807, 1984.

17. A. Rubini, D. Del Monte, V. Catena, I. Ittar, M. Cesaro, D. Soranzo, G. Rattazzi, and G. L. Alatti, Cardiac output measurement by the thermodilution method: an *in vitro* test of accuracy of three commercially available automatic cardiac output computers, *Intensive Care Med.*, 21, 154-158, 1995.

18. S. Jang, R. J. Jaszczak, F. Li, J. F. Debatin, S. N. Nadel, A. J. Evans, K. L. Greer, and R. E. Coleman, Cardiac ejection fraction and volume measurements using dynamic cardiac phantoms and radio-nuclide imaging, *IEEE Trans. Nucl. Sci.*, 41, 2845-2849, 1994.

19. D. M. K. S. Kaulesar Sukul, P. T. den Hoed, E. J. Johannes, R. van Dolder, and E. Benda, Direct and indirect methods for the quantification of leg volume: comparison between water displacement volumetry, the disk model method and the frustum sign model method, using the correlation coefficient and the limits of agreement, *J. Biomed. Eng.*, 15, 477-480, 1993.

20. P. Saucez, M. Remy, C. Renotte, and M. Mauroy, Thermal behavior of the constant volume body plethysmograph, *IEEE Trans. Biomed. Eng.*, 42, 269-277, 1995.

21. J. Rosell, K. P. Cohen, and J. G. Webster, Reduction of motion artifacts using a two-frequency impedance plethysmograph and adaptive filtering, *IEEE Trans Biomed. Eng.*, 42, 1044-1048, 1995.

22. K. P. Cohen, D. Panescu, J. H. Booske, J. G. Webster, and W. L. Tompkins, Design of an inductive plethysmograph for ventilation measurement, *Physiol. Meas.*, 15, 217-229, 1994.

23. O. S. Valerio Jimenez, J. K. Moon, C. L. Jensen, F. A. Vohra, and H. P. Sheng, Pre-term infant volume measurements by acoustic plethysmography, *J. Biomed. Eng.*, 15, 91-98, 1993.

24. R. G. Aarnink, R. J. B. Giesen, A. L. Huynen, J. J. M. C. H. de la Rosette, F. M. J. Debruyne, and H. Wijkstra, A practical clinical method for contour determination in ultrasonographic prostate images, *Ultrasound Med. Biol.*, 20, 705-717, 1994.

25. C. H. Chu, E. J. Delp, and A. J. Buda, Detecting left ventricular endocardial and epicardial boundaries by digital two-dimensional echocardiography, *IEEE Trans. Med. Im.*, 7, 81-90 1988.

26. J. Feng, W. C. Lin, and C. T. Chen, Epicardial boundary detection using fuzzy reasoning, *IEEE Trans. Med. Im.*, 10, 187-199, 1991.

27. S. Lobregt and M. A. Viergever, A discrete contour model, *IEEE Trans. Med. Im.*, 14, 12-24, 1995.

Further Information

Anonymous, AARC Clinical Practice Guideline; Static lung volume, *Respir. Care,* 39, 830-836, 1994.

Anonymous, AARC Clinical Practice Guideline; Body plethysmography, *Respir. Care,* 39, 1184-1190, 1994.

E. F. Bernstein (ed.), *Noninvasive Diagnostic Techniques in Vascular Disease,* 3rd ed., St. Louis: Mosby, 1985.

P. J. Davis and P. Rabinowitz, *Methods of Numerical Integration,* London: Academic Press, 1975.

H. Feigenbaum, *Echocardiography,* 5th ed., Philadelphia: Lee & Febiger, 1993.

W. N. McDicken, *Diagnostic Ultrasonics: Principles and Use of Instruments,* 3rd ed., London: Crosby Lockwood Staples, 1991.

J. Nyboer, *Electrical Impedance Plethysmography,* Springfield, IL: Charles Thomas, 1959.

S. Webb, *The Physics of Medical Imaging,* Bristol: IOP Publishing, 1988.

J. B. West, *Respiratory Physiology — The Essentials,* Baltimore, MD: Williams and Wilkins, 1987.

14

Angle Measurement

Robert J. Sandberg
University of Wisconsin

An *angle* is defined as the figure formed by two lines or planes that intersect one another. Such lines or planes may be real, such as the edges and surfaces of a object, or they may be defined, such as lines from an observer to two distant stars.

The units of measurement of angle are degrees (°) (1° is defined as 1/360 of a circle), and radians (rad) (1 rad is defined as 1/(2 π) of a circle). One radian is equal to 57.29578°, and small angles may be expressed in the unit of milliradians (1 × 10^{-3} rad). A degree of angle is further divided into 60′ (minutes), and 1′ of angle is divided into 60″ (seconds). One second of angle, being 1/1,296,000 of a circle, is a very small unit of measurement when related to manufactured parts, but is a significant unit when related to much larger dimensions such as the Earth (1″ of angle equals approximately 30 m of a great circle), or in space travel (an included angle of 1″ represents about 9 km on the surface of the moon when it is observed from the Earth during its closest approach to Earth.)

Many terms are used to describe angles in many different fields of expertise. Table 14.1 lists some of these terms, along with very basic definitions.

Many devices and instruments are used to measure or set angles. The following paragraphs describe the variety of equipment involved. See Table 14.2 for a partial list of manufacturers and suppliers of this equipment. See Table 14.3 for a partial list of specific models and approximate prices of angle measurement devices and systems.

14.1 Angle Gage Blocks

Angle gage blocks are right triangle-shaped, hardened and ground steel, flat, about 7 mm thick and 60 mm long. They are supplied in sets that include blocks with one of the acute angles equal to 1, 2, 3, 4, 5, 6, 7, 8, 9, 10, 15, 20, or 30°. These blocks can be used in combination to set work pieces or measure objects in 1° increments. Other sets include blocks with as small as 1″ steps. Special angle gage blocks can be made to any acute angle with the aid of a sine bar and thickness gage blocks.

Angle gage blocks provide a durable, simple, and inexpensive method for measuring and setting angles; for example, positioning work pieces in a vice or fixture prior to a machining operation such as milling, drilling, or grinding.

Because they are not adjustable and made of hardened steel, their accuracy can be assumed by simple observation of their physical condition. Look for indications of wear, nicks, or dents before using.

TABLE 14.1   Defining Terms Relating to Angles

Term	Definition
Angle	A figure formed by two lines or planes that intersect one another.
Acute angle	An angle less than 90°.
Azimuth	The horizontal angle measured along the Earth's horizon, between a fixed reference (usually due south) and an object.
Bank	A lateral inclination.
Circle	A closed plane curve where all of its points are the same distance from its center point.
Declination = declivity	A negative slope.
Degree	Equal to 1/360 of a circle.
Goniometer	An instrument for measuring angles (from the Greek word *gonio*).
Incline = Slope = Bias = Slant = Gradient = Grade	The deviation, plus or minus, from horizontal as defined by gravity.
Latitude	An angle measured north or south from the equator on a meridian to a point on the earth.
Lean = List = Tilt	The deviation from vertical as defined by gravity.
Longitude	The angle between the prime meridian (through Greenwich, England) and the meridian of a given place on Earth. This angle is defined as positive moving west.
Milliradian	An angle equal to 1/1000 rad.
Minute	An angle equal to 1/60°.
Oblique angle	An obtuse or acute angle.
Obtuse angle	An angle greater than 90°.
Quadrant	One quarter of a circle (90°).
Radian	The angle subtended by an arc of a circle equal to the radius of that circle. One radian is equal to 57.29578°.
Rake	Equals the deviation in degrees from being perpendicular (90°) to a line or a plane.
Right angle	An angle of 90°.
Rise	A positive incline.
Second	An angle equal to 1/60′ (1/3600°).
Straight	An angle equal to 180°.
Taper	The change in diameter or thickness per unit length of axis.
Twist	The angle of turn per unit length of axis, as in a gun barrel or a screw thread.

14.2   Clinometers

A *clinometer* is an electronic device that measures vertical angle with respect to gravitational level. It is rectangular, with each side being a 90° to its adjacent sides. With a range of readings of at least ±45°, this shape allows measurements up to a full 360°. Floating zero can be set anywhere and resolutions of ±0.01° are obtainable. Some models will convert readings to inches per foot, % of grade, and millimeters per meter (mm m^{-1}).

A clinometer can be used anywhere the angle of a surface with respect to gravity or another surface needs to be measured. High accuracy and resolution are obtainable, but calibration should be checked periodically with respect to a known level surface and a known angle. Surfaces that are remote to one another or have an intervening obstruction pose no problem for a clinometer.

14.3   Optical Comparator

An *optical comparator* measures angles, along with other dimensions or profiles, by referencing a visual image (usually magnified) of an object to a reticule that is calibrated in the measurement units desired. A hand-held optical comparator is placed directly over the object to be measured and the operator's eye is moved to the proper distance above the comparator for good focus of the magnified image. Some models contain a battery- or line-powered light source. Reticules for these hand-held devices are generally graduated in 1° increments.

TABLE 14.2 A Partial List of Manufacturers and Supplies of Angle Measurement Equipment

Company	Address
Flexbar Machine Corporation (Representative for Erich Preissr & Co., West Germany)	250 Gibbs Road Islandia, NY 11722-2697 Tel: (800) 879-7575
Fred V. Fowler Co., Inc.	66 Rowe Street P.O. Box 299 Newton, MA 02166 Tel: (617) 332-7004
L. S. Starrett Company	121 Crescent Street Athol, MA 01331 Tel: See local distributor
Brown & Sharpe Mfg. Co.	931 Oakton Street Elk Grove Village, IL 60007 Tel: (312) 593-5950
Swiss Precision Instruments, Inc. SPI	2206 Lively Blvd. Elk Grove Village, IL 60007 Tel: (708) 981-1300
Edmund Scientific Co.	101 East Gloucester Pike Barrington, NJ 08007-1380 Tel: (609) 573-6250

Projection-type optical comparators are available as bench or floor models and are made for either horizontal or vertical beam viewing. They use a high-intensity light source and magnifying optics to display an image of an object onto a rear-projection, frosted glass screen that is inscribed with angular as well as linear markings. The image displayed is the result of light being projected past the object, referred to as a shadow graph, or of light being reflected off the surface of the object. The method used is determined by the shape of the viewed object and its surface quality.

Magnification of the optical system in these devices can range from 10× by 100×, with screen diameters ranging from 0.3 m to 1 m.

These instruments are useful for measuring profiles of parts after final matching for the purpose of quality control or duplication.

As the name implies, the image that is projected can be superimposed on a mask or outline drawing placed directly on the view screen so that any deviations from the required shape can easily be determined. Optical comparators are heavy, nonportable devices that require a fairly high amount of maintenance and are best used in a fairly dark room.

14.4 Protractor

A *protractor* is an instrument used for measuring and constructing angles. A direct-reading protractor usually is graduated in 1° increments and can be semicircular or circular in shape. The simplest models are of one-piece construction and made from metal or plastic. Other models include a blade or pointer pivoted in the center of the graduated circle.

More precise protractors are equipped with a vernier scale that allows an angle to be indicated to 5′ of arc. See Figure 14.1 for an explanation of how to read such a vernier scale.

TABLE 14.3   Instruments and Devices Used to Measure or Indicate Angles

Type	Manufacturer	Model	Description	Approx. price
Sine bar	Flexbar	16292	5 in. × 15/16 in. wide	$130.00
		16293	10 in. × 1 in. wide	
		16294	5 in. × 2 in. wide	
		12202	5 in. × 1 in. wide, economy	$30.00
	Fowler	52-455-010	5 in. center to center, 15/16 in. wide	
		52-455-015	10 in. C. to C., 1 in. wide	
		52-455-030	2.5 in. C. to C., 1 in. wide	
	SPI	30-712-4	10 in. C. to C., universal bench center	$3048.00
		98-379-1	5 in. C. to C., 1 in. wide, accuracy between rolls = 0.0003 in.	$31.00
		30-091-3	10 in. C. to C., 1 in. wide, accuracy between rolls = 0.0001 in.	$203.00
	Brown & Sharpe	598-291-121-1	5 in. C. to C., 1 in. wide	
		598-293-121-1	10 in. C. to C., 1 1/8 in. wide	
Sine plate	Flexbar	14612	5 in. C. to C., 6 in. × 3 in. × 2 in.	$320.00
		14615	10 in. C. to C., 12 in. × 6 in. × 2 5/8 in.	$1000.00
	Fowler	57-374-001	5 in. C. to C., 6 in. × 3 in. × 2 in.	
		57-374-004	10 in. C. to C., 12 in. × 6 in. × 2 5/8 in.	
	SPI	77-026-3	10 in. C. to C., 12 in. × 6 in. × 2 5/8 in.	$872.00
	Brown & Sharpe	599-925-10	10 in. C. to C., 12 in. × 6 in. × 2 3/8 in.	
Compound sine plate	Flexbar	14616	5 in. C. to C., 6 in. × 6 in. × 3 1/8 in.	$1100.00
	Fowler	57-375-001	5 in. C. to C., 6 in. × 6 in. × 3 1/8 in.	
	SPI	7-072-7	5 in. C. to C., 6 in. × 6 in. × 3 1/8 in.	$926.00
	Brown & Sharpe	599-926-5	5 in. C. to C., 6 in. × 6 in. × 3 1/2 in.	
Angle Computer	Flexbar	19860	3-axis with vernier protractors	$3750.00
Protractor-Direct	Flexbar	16337	Rectangular Head, 0–180°	$25.00
	Starrett	RP1224W	Head only, To fit 12 in., 18 in. & 24 in. blades	
		C183	Rectangular head, 0–180° 6 in. Blade	
	SPI	30-393-3	Rectangular head, 0–180°	$23.00
		31-804-8	Head only. To fit 12 in., 18 in. & 24 in. Blades	$39.00
Protractor-Vernier	Flexbar	16339	360° range, 1′ reading with magnifier, 12 in. & 6 in. blades incl.	$400.00
		16338	360° range, 5′ reading	$75.00
	Starrett	C364DZ	12 in. Blade, 0–90° range thru 360°, 5′ graduations	
	SPI	30-395-8	6 in. Blade, 0–90° range thru 360°, 5′ graduations	$65.00
		30-390-9	6 in. & 12 in. Blades, 0–90° range thru 360°, 1′ graduations with magnifier	$540.00
	Brown & Sharpe	599-490-8	8 in. Blade, 0–90° range thru 360°, Magnifier optional	
Protractor, Digital (Inclinometer)	Flexbar	17557	±45° range, ±0.1° resolution	$260.00
		17556	±60°, ±0.01° resolution, SPC output	$450.00
	Fowler	54-635-600	±45° range, ±0.01° resolution, RS232 output available	
	SPI	31-040-9	±45° range, resolution: ±0.01° (0 to ±10°), 0.1° (10° to 90°)	$329.00
Protractor, Dial Bevel		30-150-7	8 in. Blade, 1 3/8 in. diameter dial, geared to direct read to 5′	$527.30
	Brown & Sharpe	599-4977-8	8 in. Blade, dial read degrees and 5′	
Square-Reference	SPI	30-392-5	90° fixed angle	$55.00
Optical Comparator (Projector)	Fowler	53-912-000	12 in. screen diameter, 10×, 20×, 25× lens available, horizontal beam, with separate light source for surface illumination	
	Starrett	HB350	14 in. screen diameter, 10×, 20×, 25×, 31.25×, 50×, 100× lens available, horizontal beam	

TABLE 14.3 (continued)   Instruments and Devices Used to Measure or Indicate Angles

Type	Manufacturer	Model	Description	Approx. price
		VB300	12 in. screen diameter, 10× through 100× lens available, Vertical beam	
		HS1000	40 in. screen diameter, 10× thru 100× lens available, Horizontal beam	
	SPI	40-350-1	14 in. screen diameter, 10×, 20×, 50× lens available, Horizontal beam	$2995.00
Optical	SPI	40-145-3	10× Magnification, Pocket style	$57.50
Comparator			Additional Reticles	$11.00
(hand-held)		40-140-6	7× Magnification, pocket style with illuminator	$62.50
			Additional Reticles	$10.50
	Edmund Scientific	A2046	6× Magnification, pocket style, 360° protractor reticle, 1° increments	$58.75
Angle Plate	Fowler	52-456-000	Set of 2, 9/32 in. thick, steel, 30 × 60 × 90°, 45 × 45 × 90°	
	SPI	98-463	Set of 2, 5/16 in. thick, steel, 30 × 60 × 90°, 45 × 45 × 90°	$32.00
Angle Positioning Block	SPI	70-997-2	0 to 60°, 10′ vernier (for setting workpiece in a vice)	$122.00
Angle Gage	Fowler	52-470-180	18 leaves, spring steel, 1 thru 10, 14, 14.5, 15, 20, 25, 30, 35, and 45°	
	SPI	31-375-9	18 gage set, 5° thru 90° in 5° steps, 5′ accuracy	$49.60
Angle Gage Blocks	Starrett	Ag18.TR	18 block set, use in combination for steps of 1″, 1″ accuracy	
	Starrett AG16.LM		16 block set, use in combination for steps of 1″, 1/4″ accuracy	
	SPI	30-140-8	10 block set, 1, 2, 3, 4, 5, 10, 15, 20, 25, 30°, Accuracy = ±0.0001″ per inch.	$170.00
			1/4° and 1/2° blocks optional (each)	$18.00

14.5   Sine Bar

A *sine bar* is a device used to accurately measure angles or to position work pieces prior to grinding and other machining procedures. It is constructed from a precisely hardened and ground steel rectangular bar to which are attached two hardened and ground steel cylindrical rods of the same diameter. The axis of each rod is very accurately positioned parallel to the other and to the top surface of the bar.

A sine bar is used in conjunction with precision gage blocks that are placed under one of the two cylindrical rods to raise that rod above the other rod a distance H (see Figure 14.2) equal to the sine of the angle desired, times the distance D between the two rods. The standard distance between the rods is 250 mm (5 in.) or 500 mm (10 in.). The governing equation in using a sine bar is $\sin A = H/D$.

A work piece positioned using a sine bar is usually secured with the use of a precision vice. The vice may clamp directly to the work piece or, when using a sine bar that has tapped holes on its top surface, to the sine bar sides with the work piece bolted to the top of the sine bar.

14.6   Sine Plate

A variation of the sine bar is the *sine plate*. A sine plate consists of the three elements of a sine bar plus a bottom plate and side straps used to lock the plate in the desired position. In addition, one of the ground steel rods is arranged to form a hinge between the top and bottom plates. When using a sine plate, a work piece is secured to the top plate using bolts or clamps and the bottom plate is secured to a machine tool table using clamps or a magnetic chuck.

Compound sine plates are bidirectional, allowing angles to be measured and set in each of two orthogonal planes (true compound angles).

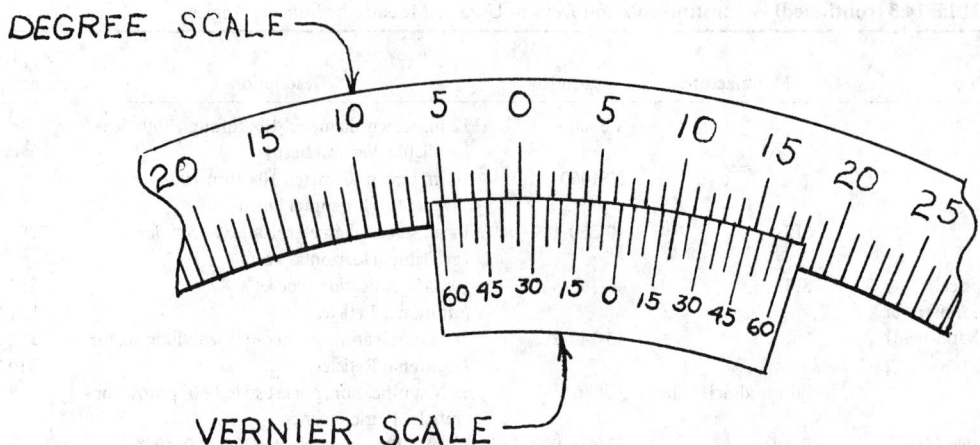

FIGURE 14.1 Vernier protractor. If the zero mark on the vernier scale is to the right of the zero mark on the main scale, as shown in this drawing, then the right side of the vernier scale must be used. Look for the mark on the vernier scale that best aligns with one of the marks of the protractor degree scale. Count the number of marks on the vernier scale from the zero mark to this mark. Each mark thus counted is, in this example, equal to 5′ of arc and, therefore, the number of minutes to be added to the number of degrees indicated is 5 times the vernier marks counted. (In this example, the fourth mark aligns the best with the main scale indicating 20′). The number of degree indicated is the degree mark just to the left of the zero mark on the vernier scale. The left side of the vernier is similarly used when the indicated angle is to the left of the zero mark on the degree scale.

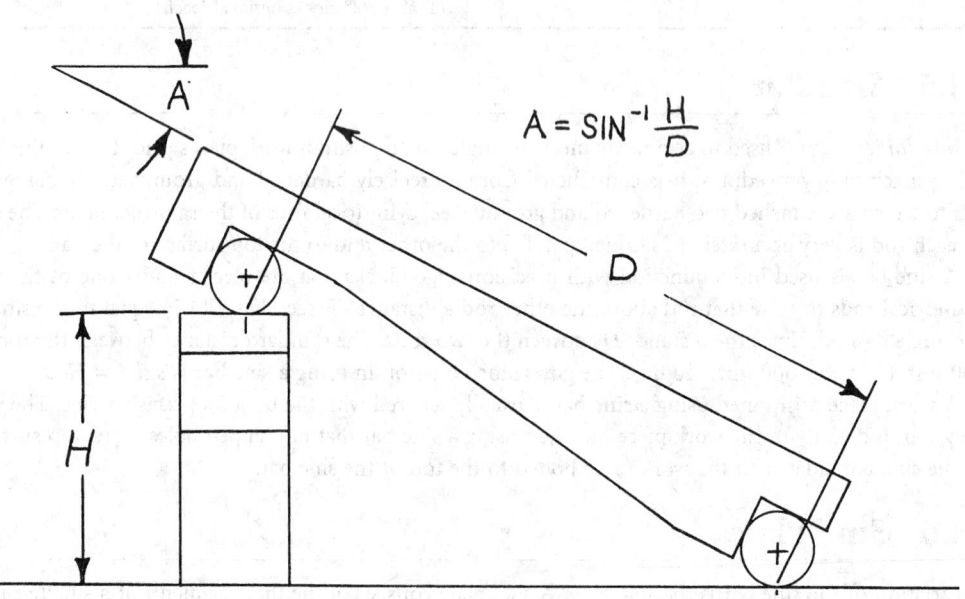

FIGURE 14.2 Sine bar. A sine bar is used in conjunction with precision gage blocks that are placed under one of the cylindrical rods, raising that end a distance, *H*, equal to the sine of the desired angle times the distance, *D*, between the two rods.

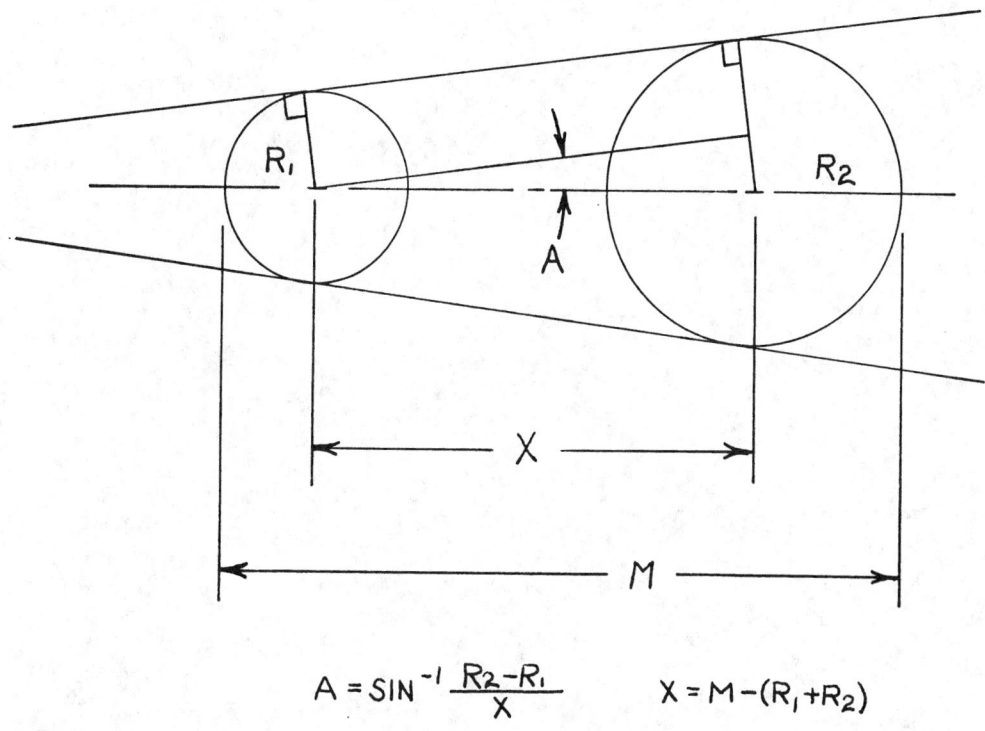

$$A = SIN^{-1} \frac{R_2 - R_1}{X} \qquad X = M - (R_1 + R_2)$$

FIGURE 14.3   Measuring tapers. Two balls (for holes) or gage pins (for slots) should be selected to fit the hole size to be measured. The distance between the balls or pins should be as large as possible to allow for the best accuracy of measurement. The position of the balls can be determined and the equations shown in this figure applied to determined the angle of the taper.

14.7 Taper

An accurate method that can be used to measure a tapered hole is described in Figure 14.3. In this method, one uses a pair of steel balls of proper diameters to match the hole size and angle of taper. This method can also be used to measure the angle between two adjacent planes, using either balls or gage pins.

Further Information

Further information on this subject can be found in company catalogs in both print and Internet formats. An additional reference is *Machinery's Handbook*, 25th ed., Industrial Press Inc., 200 Madison Avenue, New York, NY 10016-4078, 1996.

15

Tilt Measurement

Adam Chrzanowski
University of New Brunswick

James M. Secord
University of New Brunswick

If the relative position of two points is described by the three-dimensional orientation of a line joining them, then, in general, *tilt* is the angular amount that the orientation has changed, in a vertical plane, from a previous direction or from a reference direction. If the original or reference orientation is nearly horizontal, then the term "tilt" is usually used. If it is nearly vertical, then the change is often regarded as "inclination." Here, "tilt" will refer to either. The two points can be separated by a discernable amount, the base, or the tilt can be measured at a point with the reference orientation being defined by the direction of the force of gravity at that point. Thus, the same instrument that measures tilt at a point can be called either a *tiltmeter* or an *inclinometer* (or clinometer), depending on the interpretation of the results. The instrument used to measure a series of tilts along any vertical orientation is often called an inclinometer (e.g., Dunnicliff [1]).

Angular tilt is directly related to the linear amount of change subtending the length of the base. Consequently, angular tilt does not have to be measured directly but can be derived from the mechanical or other measurement of this linear change if the length of the base is known.

Therefore, the following discussion has been subdivided into:

1. Tiltmeters or inclinometers (angular tilt at a point or over a limited, relatively small base length)
2. Geodetic leveling (tilt derived from a height difference over an extended base of virtually limitless length)
3. Hydrostatic leveling (tilt derived from a height difference over an extended base of limited length)
4. Suspended and inverted pendula, or plumb lines (inclination from a difference in horizontal relative position over a vertical base or height difference)

15.1 Tiltmeters or Inclinometers

Considering the basic principle of operation, tiltmeters may be divided into: liquid (including spirit bubble type), vertical pendulum, or horizontal pendulum. Dunnicliff [1] provides a comprehensive review of tiltmeters and inclinometers according to the method in which the tilt is sensed (i.e., mechanical, with accelerometer transducer, with vibrating wire transducer, or with electrolytic transducer).

The sensitivity of tilt falls into two distinct groups: geodetic or geophysical special tiltmeters with a resolution of 10^{-8} rad (0.002") or even 10^{-9} rad; and engineering tiltmeters with resolutions from 0.1" to several seconds of arc, depending on the required range of tilt to be measured.

The first group includes instruments that are used mainly for geophysical studies of Earth tide phenomena and tectonic movements; for example, the Verbaander-Melchior [2] and the Zöllner [3, 4] horizontal pendulum tiltmeters, and the Rockwell Model TM-1 [5] liquid-bubble type. This category of instrument requires extremely stable mounting and a controlled environment. There are very few engineering projects where such sensitive instruments are required. However, deformation measurements of underground excavations for the storage of nuclear waste may be one of the few possible applications. An example is a mercury tiltmeter (Model 300) developed for that purpose by the Auckland Nuclear Accessory Co. Ltd. in New Zealand. In this instrument, the change in capacitance between electrodes and a mercury surface is proportional to the tilt. This tiltmeter, with a total range of 15″, is claimed to give a resolution of 10^{-9} rad (0.0002″), which corresponds to a relative vertical displacement of only 6×10^{-7} mm over its base length of 587 mm.

In the second group, there are many models of liquid or pendulum tiltmeters of reasonable price ($2000 to $5000) that satisfy most needs in engineering studies. Apart from a spirit level or level vial by itself, the simplest form of tiltmeter is a base that is tens of centimeters long and leveled by centering the bubble in the vial by direct viewing or by an optical coincidence viewing of the two ends of the bubble. Direct viewing gives a resolution of 1/5 of a vial division (usually 2 mm), which typically has a sensitivity of 10″ to 30″ per division. Coincidence viewing increases the setting accuracy to 0.03 of the sensitivity of the vial. The discrepancy from horizontal between the two measurement points can be determined by a dial gage or micrometer that has a resolution of 0.0005 in. or 0.02 mm. Huggenberger AG claim a sensitivity of 0.3″ (1×10^{-4} gon) over a range of ±21′ for their clinometer with a 100 mm base and coincidence centering of the bubble in the level vial. The clinometer can be attached to 1 m bases for either horizontal or vertical measurements.

If the vial is filled with an electrolytic liquid, the centering of the bubble can be done electrically. An example is the Electrolevel (by the British Aircraft Corp.), which uses the spirit bubble principle [6] and in which the movement of the bubble is sensed by three electrodes. A tilt produces a change in differential resistivity between the electrodes that is measured by means of a Wheatstone bridge. A resolution of 0.25″ is obtained over a total range of a few minutes of arc. Many other liquid types of tiltmeters with various ranges (up to 30°) are available from various companies. Holzhausen [7] and Egan and Holzhausen [8] discuss the application of electrolytic tiltmeters (resolution of 2″ over a range of ±1°, manufactured by Applied Geomechanics) in the monitoring of various hydroelectric power dams in the U.S.

The Rank Organization in the United Kingdom [9] makes a liquid-dampened pendulum-type electronic level, the Talyvel, which gives an accuracy of ±0.5″ over a total range of ±8′. A similar transducer of the pendulum type is used in the Niveltronic tiltmeter (range of ±150″ with an accuracy of ±0.2″) produced by Tesa S.A. in Switzerland. Of particular popularity are servo-accelerometer tiltmeters with horizontal pendula. They offer ruggedness, durability, and can operate in low temperatures. The output voltage is proportional to the sine of the angle of tilt. Schaevitz Engineering produces such a servo-accelerometer that employs a small-mass horizontal paddle (pendulum) which tries to move in the direction of tilt, due to the force of gravity. Any resultant motion is converted by position sensors to a signal input to the electronic amplifier whose current output is applied to the torque motor. This develops a torque that is equal and opposite to the original. The torque motor current produces a voltage output that is proportional to the sine of the angle of tilt.

The typical output voltage range for tiltmeters is ±5 V, which corresponds to the maximum range of tilt and readily allows for serial interfacing. The angular resolution of a tiltmeter depends on its range of tilt since a larger range would result in more angular tilt per unit voltage so a higher resolution tiltmeter would have a smaller range of measurable tilt. Typically, the resolution is 0.02% of the range (volts) [10].

There are many factors affecting the accuracy of tilt sensing, not just the resolution of the readout. A temperature change produces dimensional changes in the mechanical components and changes in the viscosity of the liquid in electrolytic tiltmeters and of the dampening oil in pendulum-type tiltmeters. Also, electric characteristics can alter with temperature changes. Drifts in tilt indications and fluctuations of the readout may also occur. Compensation for the effects of temperature changes can be incorporated

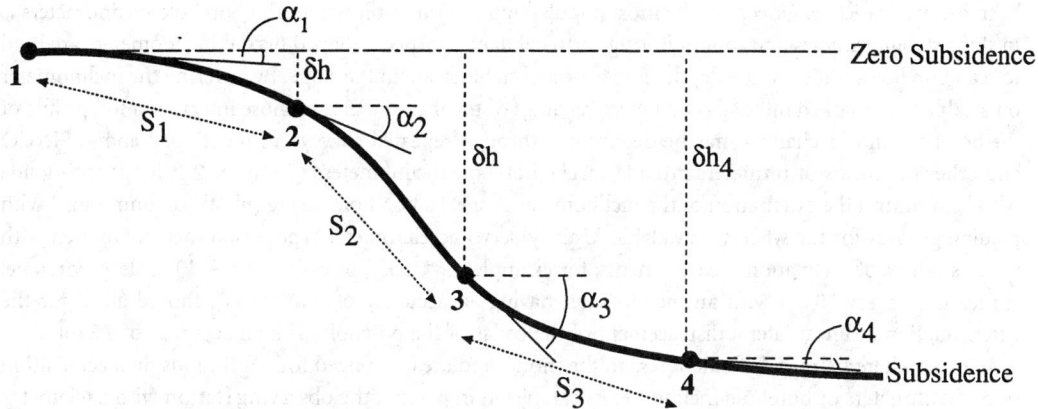

FIGURE 15.1 Ground subsidence derived from tilt measurements.

in the construction of an instrument, but at an increased cost. An alternative is to design a linear reaction by the instrument to the effects of temperature and to apply simple calibration corrections.

In less expensive models, compensation for the aforementioned sources of error is not very sophisticated, and such tiltmeters may show nonlinear output in reaction to changes in temperature and erratic drifts or other behavior that would be difficult to predict without testing. Consequently, very thorough testing and calibration are required even when the accuracy requirement is not very high [11]. Testing should investigate, at least, the linearity of output in reaction to induced tilts over the instrument's full range (±) and to changes in temperature. Some suggestions for testing and calibrating inclinometers, among other instruments, are given in Dunnicliff [1]. It is further emphasized that regular and up-to-date calibration is important in order to ensure continuity in the fidelity of the data being gathered. In most cases, the behavior being investigated changes with time and incorrect data cannot be recaptured.

Compensators for vertical circle readings in precision theodolites work on the same principle as some engineering tiltmeters. The liquid compensator of the Kern E2 electronic theodolite [12] gave a repeatability of better than 0.3" over a range of ±150" and was incorporated in their tiltmeter, NIVEL 20, in 1989. The same compensation system has been used in the currently available Leica TC2002 precision electronic total station [13]. Consequently, the theodolite may also be used as a tiltmeter, in some applications, giving the same accuracy as the Electrolevel, for example.

Tiltmeters have a wide range of applications. A series of tiltmeters, if arranged along a terrain profile in a mining area, may replace geodetic leveling in the determination of ground subsidence [11] as shown in Figure 15.1. For example, the subsidence (i.e., the variation from the previous or original profile) of point 4 (δh_4) with respect to point 1 may be estimated from the observed changes in tilt, from a base or original position, (α_i in radians) and known distances between the points as:

$$\delta h_4 = s_1 \left(\alpha_1 + \alpha_2 \right) / 2 + s_2 \left(\alpha_2 + \alpha_3 \right) / 2 + s_3 \left(\alpha_3 + \alpha_4 \right) / 2 \tag{15.1}$$

The fidelity of this method depends on the density of tilt measurements along the profile and the continuity of the profile (a constant slope of the terrain between measurement points is assumed). Similarly, deformation profiles of tall buildings may be determined by placing a series of tiltmeters at different levels of the structure [14]. Also, changes in borehole profiles can be created in a similar manner. The absolute profile of a borehole can be generated by considering the horizontal displacement in the direction of the orientation of the inclinometer (usually controlled by guide grooves in the borehole casing) for the *i*-th position, as it traverses a borehole with observation of α_i at a depth s_i. However, this would require calibration of the inclinometer to correct its output to show zero in its vertical position since the α_i are tilts from the vertical rather than angular changes from an original inclination.

In geomechanical engineering, the most popular application of tiltmeters and borehole inclinometers is in slope stability studies and in monitoring earth-fill dams. Torpedo-shaped biaxial inclinometers are used to scan boreholes drilled to the depth of an expected stable strata in the slope. By lowering the inclinometer on a cable with marked intervals and taking readings of the inclinometer at those intervals, a full profile of the borehole and its changes may be determined through repeated surveys, as mentioned above. SINCO and other producers of tiltmeters provide special borehole inclinometers (50 cm or 2 ft long) with guide wheels to control the orientation of the inclinometer. A special borehole casing (plastic or aluminum) with guiding grooves for the wheels is available. Usually, servo-accelerometer type inclinometers are used with various ranges of inclination measurements; for example, $\pm6°$, $\pm53°$, or even $\pm90°$. A 40 m deep borehole, if measured every 50 cm with an inclinometer having an accuracy of only $\pm100''$, should allow for the determination of linear lateral displacement of the collar of the borehole with an accuracy of ±2 mm.

In cases where there is difficult access to the monitored area or a need for continuous data acquisition or both, tiltmeters or borehole inclinometers can be left in place at the observing station with a telemetry monitoring system allowing for communication to the processing location. One example of a station setup of a telemetric monitoring of ground subsidence in a mining area near Sparwood, B.C. used a telemetry system developed for the Canadian Centre for Mining and Energy Technology (CANMET) by the University of New Brunswick [11, 15]. Terra Technology biaxial servo-accelerometer tiltmeters of $\pm1°$ range were used in the study. The telemetry system could work with up to 256 field stations. Each station accepted up to six sensors (not only tiltmeters but any type of instrument with electric output, e.g., temperature, power level or voltage). Another example is a fully automated borehole scanning system with a SINCO inclinometer and telemetric data acquisition that was also developed at the University of New Brunswick [16]. It has been used successfully in monitoring highwall stability at the Syncrude Canada Limited tarsands mining operation in northern Alberta.

15.2   Geodetic Leveling

Geodetic or differential leveling measures the height difference between two points using precision tilting levels, or precision automatic levels with compensators, with parallel plate micrometers and calibrated invar leveling staves or rods. Recent technology has provided digital automatic levels that use a CCD sensor in the instrument and bar codes, rather than linear graduations, on the staves [17]. Their ease of use and comparable precision have quickly made them rivals to the traditional optical instruments. In a single setup of the level, the height difference is the backsight rod reading minus the foresight rod reading. Any number of setup height differences can be combined to determine the height difference between two points of interest; however, the errors involved accumulate with the number of setups. With sight lengths limited to no more than 20 m, geodetic leveling can produce height differences with a precision of ±0.1 mm per setup, which is equivalent to a precision of $\pm0.5''$ in tilt. Although geodetic leveling is traditionally used to determine elevations, it is often used to monitor not only the settlement of sensitive structures but also to describe the tilt of components of a structure by determining the tilt between appropriate pairs of benchmarks (monumented in or on the structure) [18]. Since the level reference is created by an optical line of sight through a telescope (magnification up to 40×), a major source of systematic error is the effect of vertical atmospheric refraction. A vertical temperature gradient of even $1°C$ m^{-1} across the line of sight would bend the line of sight to be in error by 0.4 mm at 30 m. Gradients of this magnitude are commonly encountered in industrial settings and are usually even more evident outdoors. Less effect is realized if the sight lengths are kept short, but this must be weighted against the accumulation of random error with each additional setup (shorter sight lengths would require more setups to traverse the same height difference). The errors that seem to be insignificant in a single setup (or in a few setups) become magnified in height differences involving a large number of setups (e.g., rod scale error and settlement of the instrument or rods). Such errors have become quite noticeable in the monitoring of tectonic plate movement and undetected systematic effects can be misleading. Further discussion on precision leveling and sources of error is available in Vanicek et al. [19] and in Schomacker and Berry [20].

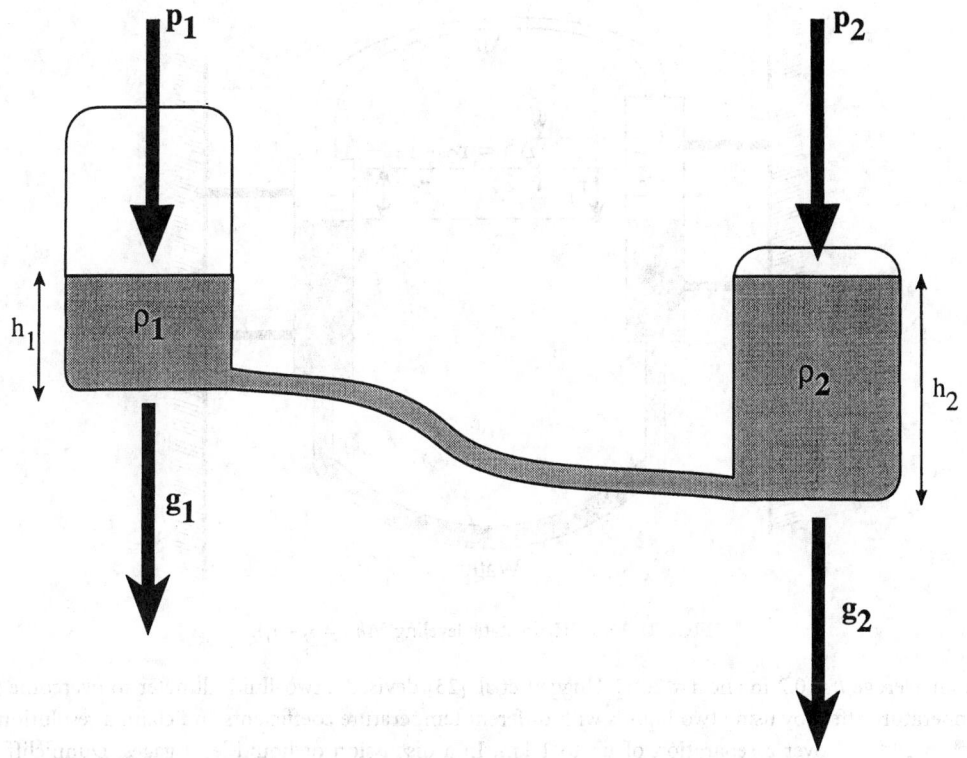

FIGURE 15.2 Hydrostatic equilibrium in connected vessels.

Having the elevations or heights, h_1 and h_2, of two points or having measured, or determined, the height difference between them, $\Delta h_{12^t} = h_2 - h_1$, at a time t, means that, if $\delta \Delta h = \Delta h_{12^{t2}} - \Delta h_{12^{t1}}$, the tilt, T_{12}, can be calculated if the horizontal separation s_{12} is known, since $T_{12} = \delta \Delta h / s_{12}$. The separation does not have to be known as precisely as the height difference since the total random error is $\sigma_{T^2} = \sigma_{\delta \Delta h^2}/s^2 + \sigma_{s^2}(\delta \Delta h^2/s^4)$. As an example, for two points that are 60 m apart with a height difference of 0.5 m (extreme in most structural cases) with the height difference known to ± 50 μm ($\sigma_{\delta \Delta h}$) and the distance known to ± 0.01 m (σ_s), the tilt would have a precision (σ_T) of $\pm 0.3''$. Further, neither the measurement of the height difference nor the determination of the separation have to be done directly between the two points. The leveling can be done along whatever route is convenient and the separation can be obtained in a variety of ways, for example, inversing from coordinated values for the points [21].

15.3 Hydrostatic Leveling

If two connected containers (Figure 15.2) are partially filled with a liquid, then the heights h_1 and h_2 are related through the hydrostatic equation (Bernoulli's equation, as given in [22]):

$$h_1 + P_1/\left(g_1\rho_1\right) = h_2 + P_2/\left(g_2\rho_2\right) = c \tag{15.2}$$

where P is the barometric pressure, g is the force of gravity, ρ is the density of the liquid which is a function of temperature, and c is a constant.

The above relationship has been employed in hydrostatic leveling, as shown schematically in Figure 15.3. The air tube connecting the two containers eliminates possible error due to different air pressures at two stations. The temperature of the liquid should also be maintained constant because, for example, a difference of 1.2°C between two containers may cause an error of 0.05 mm in a Δh determination

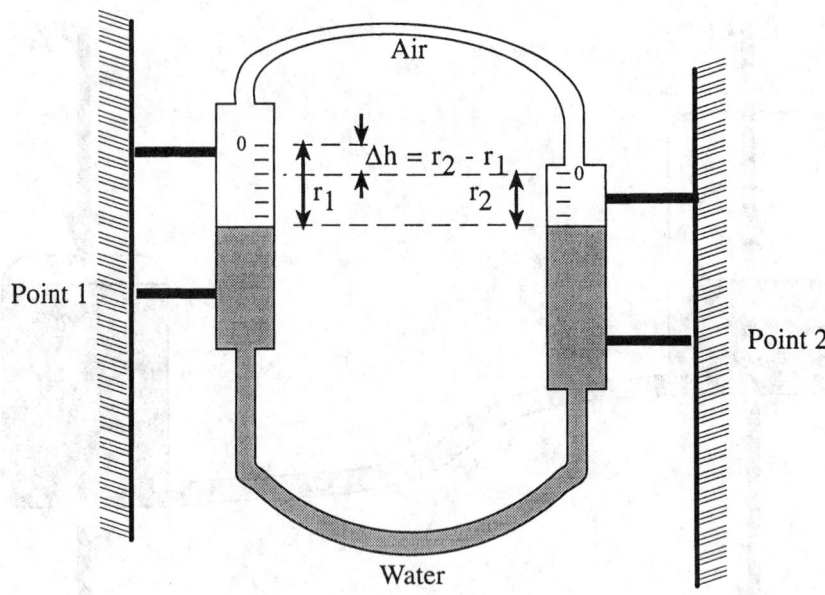

FIGURE 15.3   Hydrostatic leveling ($\Delta h_{12} = r_2 - r_1$).

for an average $h = 0.2$ m and $t = 20°C$. Huggett et al. [23] devised a two-fluid tiltmeter to overcome the temperature effect by using two liquids with different temperature coefficients and claim a resolution of 10^{-8} to 10^{-9} rad over a separation of up to 1 km. In a discussion of liquid level gages, Dunnicliff [1] emphasizes that care should be exercised to ensure that there is no discontinuity in the liquid since any gas (usually air, often entering when filling the tubing) in the liquid line will introduce an error in the level difference, especially in a vertical, more than in a horizontal, portion of tubing. He also mentions that the behavior of the liquid is influenced by the inside diameter and capillary effects of the tubing, while the outside diameter is likely what is quoted by manufacturers. Dunnicliff [1] also provides a comprehensive summary of the variety of liquid level gages.

Two examples of typical hydrostatic instruments used in precision leveling will be mentioned here. The ELWAAG 001, developed in Germany [24], is a fully automatic instrument with a traveling (by means of an electric stepping motor) sensor pin that closes the electric circuit on touching the surface of the liquid. A standard deviation of ±0.03 mm is claimed over distances of 40 m between the instruments [22]. Another automatic system, the Nivomatic Telenivelling System, is available from Telemac or Roctest Ltd. The Nivomatic uses inductance transducers that translate the up and down movements of its floats into electric signals (frequency changes in a resonant circuit). An accuracy of ±0.1 mm is claimed over a 24 m length. P & S Enterprises, Ltd. produces a Pellissier model H5 portable hydrostatic level/tiltmeter, for which they claim an accuracy of ±5 μm over a tube length of 14 m, for engineering and Earth tide measurements.

Hydrostatic levels may be used in a network formation of permanently installed instruments to monitor tilts in large structures. Robotti and Rossini [25] report on a DAG (Automatic Measuring Device of Levels and Inclinations) network monitoring system available from SIS Geotecnica (Italy) that offers an accuracy of about ±0.01 mm using inductive transducers in the measurement of liquid levels. Various systems of double liquid (e.g., water and mercury) settlement gages based on the principle of hydrostatic leveling are used for monitoring power dams [26] with extended networks of connecting tubing.

Instruments with direct measurement of liquid levels are limited in their vertical range by the height of their containers. This problem may be overcome if liquid pressures are measured instead of the changes in elevation of the liquid levels. Pneumatic pressure cells or pressure transducer cells may be used. Both Dunnicliff [1] and Hanna [26] give numerous examples of various settlement gages based on that principle. Meier [27] mentions the application of a differential pressure tiltmeter in the monitoring of a concrete dam.

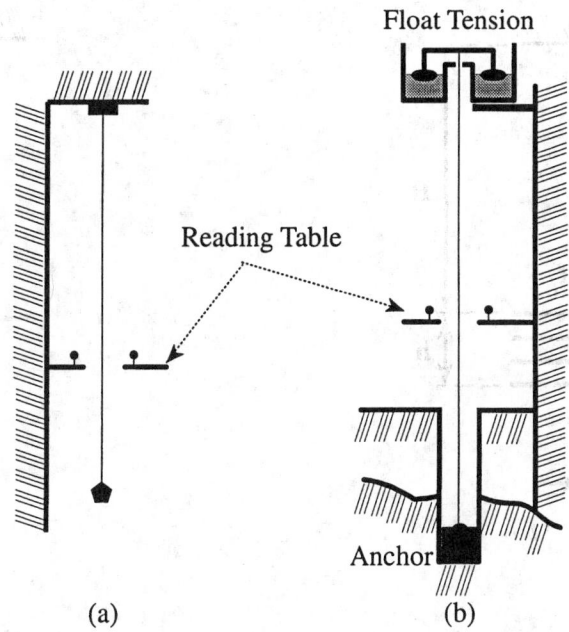

FIGURE 15.4 Inclination measurements with plumblines. (*a*) suspended pendulum; (*b*) inverted pendulum.

15.4 Suspended and Inverted Plumb Lines

Two types of mechanical plumbing are used in monitoring the stability of vertical structures:

1. Suspended pendula or plumb lines (Figure 15.4(*a*))
2. Floating pendula or inverted, or reversed, plumb lines (Figure 15.4(*b*))

Typical applications are in the monitoring of power dams and of the stability of reference survey pillars. Suspended pendula are also commonly used in mine orientation surveys and in monitoring the stability of mine shafts. Tilt, or inclination, is derived from differences in horizontal relative position combined with vertical separation in the same way as tilt is derived from geodetic leveling. So, similarly, the vertical separation between two reading tables or between a reading table and anchor point does not have to be known as precisely as the change in relative position. Two table readings, each ±0.02 mm, with a relative position difference of 100 mm and a vertical separation of 10 m, known to ±0.01 m, would result in a tilt precision of ±2″.

Inverted plumb lines have become standard instrumentation in large dams (e.g., Hydro Quebec uses them routinely). Their advantage over suspended plumb lines is in the possibility of monitoring the absolute displacements of structures with respect to deeply anchored points in the foundation rock which may be considered as stable. In power dams, the depth of anchors must be 30 m or even more below the foundation in order to obtain absolute displacements of the surface points. The main problem with inverted plumb lines is the drilling of vertical boreholes so that the vertical wire of the plumb line would have freedom to remain straight and vertical. A special technique for drilling vertical holes has been developed at Hydro Quebec [28].

Several types of recording devices that measure displacements of structural points with respect to vertical plumb lines are produced by different companies. The simplest are mechanical or electromechanical micrometers with which the plumb wire can be positioned with respect to reference lines of a recording (coordinating) table with an accuracy of ±0.2 mm or better. Traveling microscopes may give the same accuracy. Automatic sensing and recording is possible with a Telecoordinator from Huggenberger

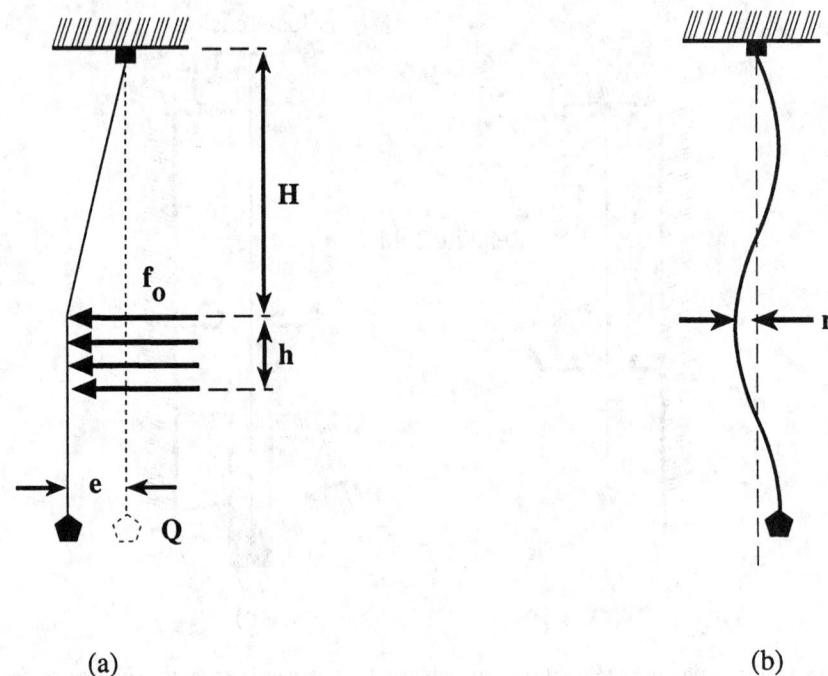

(a)   (b)

FIGURE 15.5   (*a*) Influence of air currents on a suspended plumbline. (*b*) Horizontal error due to the spiral shape of the wire.

AG in Switzerland. Telemac Co. (France) developed a system, Telependulum (marketed by Roctest), for continuous sensing of the position of the wire with remote reading and recording. A rigidly mounted reading table supports two pairs of induction type proximity sensors arranged on two mutually perpendicular axes. A hollow cylinder is fixed on the pendulum wire at the appropriate level, passing through the center of the table and between the sensors. Changes in the width of the gap between the target cylinder and the sensors are detected by the corresponding changes in the induction effect. The system has a resolution of ±0.01 mm.

An interesting Automated Vision System has been developed by Spectron Engineering (Denver, Colorado). The system uses solid state electronic cameras to image the plumb line with a resolution of about 3 µm over a range of about 75 mm. Several plumb lines at Glen Canyon dam and at Monticello dam, near Sacramento, California, have been using the system since 1982 [29].

Two sources of error, which may be often underestimated by the user, may strongly affect plumb line measurements:

1. The influence of air currents
2. The spiral shape of the wire

If the wire of a plumb line (Figure 15.5(*a*)), with pendulum mass Q, is exposed along a length h to an air current of speed v at a distance H from the anchor point, then the plumb line is deflected by an amount [30]:

$$e = f_o h H / Q \qquad (15.3)$$

where f_o is the acting force of air current per unit length of the wire. The value of f_o may be calculated approximately from [30]

$$f_o = 0.08 d v^2 / Q \qquad (15.4)$$

where *d* is the diameter of the wire in millimeters, *v* is in meters per second, and *Q* is in kilograms. As an example, if *H* = 50 m, *h* = 5 m, *d* = 1 mm, *Q* = 20 kg, and *v* = 1 m s^{-1} (only 3.6 km h^{-1}) then *e* = 1 mm.

The second source of error, which is usually underestimated in practice, is that the spiral shape (annealing) of the wire (Figure 15.5(*b*)) affects all wires unless they are specially straightened or suspended for a prolonged time (on the order of several months). If the wire changes its position (rotates) between two campaigns of measurements, then the recorded displacements could have a maximum error of 2*r*. The value of *r* can be calculated from [30]:

$$r = \left(\pi d^4 E\right)\Big/\left(64 RQ\right) \tag{15.5}$$

where *E* is Young's modulus of elasticity (about 2 × 10^{11} Pa for steel); *R* is the radius of the free spirals of the unloaded wire that, typically, is about 15 cm for wires up to 1.5 mm diameter; and *d* and *Q* are as above. For a plumb wire with *d* = 1 mm, *R* = 15 cm, and *Q* = 196 N (i.e., 20 kg), *r* = 0.3 mm.

If one plumb line cannot be established through all levels of a monitored structure, then a combination of suspended and inverted plumb lines may be used as long as they overlap at least at one level of the structure. At Hydro Quebec, the drill holes of the plumb lines are also used for monitoring height changes (vertical extension) by installing tensioned invar wires [31].

15.5 Integration of Observations

The above discussion has considered using individual instruments. Because many investigations, using other instrumentation as well as the measurement of tilt, involve the repetition of measurements, often over a long term, the fidelity of the measurements and their being referenced to the original or initial observation is vital to the investigation. It is risky to expect consistent behavior of instrumentation, particularly in environments with dramatic variations (especially in temperature), and over a long period of time. Any conclusion relating to the behavior of a structure is only as good as the data used in the analysis of the behavior. Two ways to ensure reliability are possible. One is to make regular testing and calibration a component of the observation regimen. The other is to analyze the observations as they are accumulated, either each observable as a temporal series of repeated measurements or observations of different locations or types together in an integrated analysis. The analytical tools for integrated analyses, as well as for calibration testing and temporal series analysis, have been developed [21, 32] and successfully implemented in several projects [15, 18, 33]. Proper calibration testing and correction, rigorous statistical analysis of trend in temporal series, and integrated analysis have proven to be valuable tools in the analysis of deformations and serve to enhance the monitoring of sensitive structures.

References

1. J. Dunnicliff, *Geotechnical Instrumentation for Monitoring Field Performance*, New York: John Wiley & Sons, 1988.
2. P. Melchior, *The Tides of the Planet Earth*, 2nd ed., Oxford, U.K.: Pergammon Press, 1983.
3. W. Torge, *Geodesy*, New York: Walter de Gruyter, 1991.
4. M. van Ruymbeke, Sur un pendule horizontal équipé d'un capteur de déplacement a capacité variable, *Bull. Géodesique*, 50, 281-290, 1976.
5. G. L. Cooper and W. T. Schmars, Selected applications of a biaxial tiltmeter in the ground motion environment, *J. Spacecraft Rockets*, 11, 530-535, 1974.
6. M. A. R. Cooper, *Modern Theodolites and Levels*, London: Crosby Lockwood, 1971.
7. G. R. Holzhausen, Low-cost, automated detection of precursors to dam falure: Coolidge Dam, Arizona, *Association of State Dam Safety Officials, 8th Annu. Conf.*, San Diego, CA, 1991.
8. N. H. Egan and G. R. Holzhausen, Evaluating structures using tilt (rotation) measurements, *Sensors Expo West Proc.*, 1991.

9. W. Caspary and A. Geiger, Untersuchungen zur Leistungsfähigkeit elektronischer Neigungsmesser, *Schriftenreihe, Vermessungswesen HSBW*, 3, München, Germany, 1978.
10. A. Chrzanowski, Geotechnical and other non-geodetic methods in deformation measurements, in Y. Bock (ed.), *Proc. Deformation Measurements Workshop*, Massachusetts Institute of Technology, Boston, 1986, 112-153.
11. A. Chrzanowski, W. Faig, B. Kurz, and A. Makosinski, Development, installation and operation of a subsidence monitoring and telemetry system, Contract report submitted to CANMET, Calgary, Canada, November 1980.
12. Kern & Co. Ltd., E2 Instruction Manual, Kern & Co. Ltd., Aarau, Germany, 1984.
13. Leica AG, Wild TC2002 User Manual, Leica AG, Heerbrugg, Germany, 1993.
14. H. Kahmen, *Elektronische Messverfahren in der Geodäsie*, Karlsruhe: Herbert Wichmann, 1978.
15. A. Chrzanowski and M. Y. Fisekci, Application of tiltmeters with a remote data acquisition in monitoring mining subsidence, *Proc. 4th Canadian Symp. Mine Surveying*, Banff, Alberta, Canada, June 1982.
16. A. Chrzanowski, A. Makosinski, A. Zielinski, and W. Faig, Highwall Monitoring System, Contract report to Syncrude Canada Limited, April 1988.
17. H. Ingensand, Wild NA2002, NA3000, Technical paper digital levels, Leica AG, Heerbrugg, Germany, 1993.
18. A. Chrzanowski and J. M. Secord, The 1989 Integrated Analysis of Deformation Measurements at the Mactaquac Generating Station, Contract Report to N.B. Power, May 1990.
19. P. Vanicek, R. O. Castle, and E. I. Balazs, Geodetic leveling and its applications, *Rev. Geophys. Space Phys.*, 18(2), 505-524, 1980.
20. M. C. Schomaker and R. M. Berry, *Geodetic Levelling*, NOAA Manual NOS NGS 3, U.S. Department of Commerce, National Oceanic and Atmospheric Administration, National Ocean Survey, Rockville, MD, 1981.
21. J. M. Secord, Development of the automatic data management and the analysis of integrated deformation measurements, Ph.D. dissertation, Department of Geodesy and Geomatics Engineering Technical Report 176, University of New Brunswick, Fredericton, 1995.
22. K. Schnädelbach, Neuere Verfahren zur präzisen Längen-und Höhenmessung, *Algemeine Vermessungs-Nachrichten*, 1/1980.
23. G. R. Huggett, L. E. Slater, and G. Pavlis, Precision leveling with a two-fluid tiltmeter, *Geophys. Res. Lett.*, 3(12), 754-756, 1976.
24. H. Thierbach and W. Barth, Eine neue automatische Präzisionsschlauchwaage, *Z. Vermessungswesen*, 100, 470-478, 1976.
25. F. Robotti and T. Rossini, Analysis of differential settlements on monumental structures by means of the DAG automatic measuring device of levels and inclinations, in *Land Subsidence*, IAHA Publication No. 151, 1984.
26. T. H. Hanna, *Field Instrumentation in Geotechnical Engineering*, Clausthal-Zellerfeld, Germany, Trans Tech Publications, 1985.
27. E. Meier, A differential pressure tiltmeter for large-scale ground monitoring, *Water Power & Dam Construction*, 43(1), 38-40, 1991.
28. L. Dubreuil and R. Hamelin, Le forage vertical des trous de pendules inversés, *2nd Canadian Symp. Mining Surveying Rock Deformation Measurements*, Kingston, Ontario, Canada, 1974.
29. G. Kanegis, Automated vision system installed at Glen Canyon Dam, Spectron Engineering, n.d. [circa 1983].
30. A. Chrzanowski, E. Derenyi, and P. Wilson, Underground survey measurements — Research for progress, *The Canadian Mining and Metallurgical Bulletin*, June 1967.
31. B. Boyer and R. Hamelin, Monitoring Survey: Recent Developments in the Use of Inverted Pendula, Report No. 4, Question 56, *XV Congress Int. Commission Large Dams, Lausanne*, 1985.
32. A. Chrzanowski, Y. Q. Chen, P. Romero, and J. Secord, Integration of geodetic and geotechnical deformation surveys in the geosciences, *Tectonophysics*, 130, 1986.
33. A. Chrzanowski, Y. Q. Chen, J. Secord, and A. Szostak-Chrzanowski, Problems and solutions in the integrated monitoring and analysis of dam deformations, *CISM J. ACSGC*, 45(4), 547-560, 1991.

Appendix

A Sampling of Possible Suppliers of Tilt Measuring Instrumentation

Applied Geomechanics Inc.
1336 Brommer Street
Santa Cruz, CA 95062

Auckland Nuclear Accessory Company Ltd.
P.O. Box 16066
Auckland, 3. New Zealand

Eastman Whipstock GmbH
Gutenbergstrasse 3
3005 Hannover-Westerfeld
West Germany

Geotechnical Instruments Ltd.
Station House, Old Warwich Road
Leamington Spa, Warwickshire CV31 3HR
England

Huggenberger AG
Holstrasse 176
CH-8040 Zürich
Switzerland

IRAD GAGE
Etna Road
Lebanon, NH 03766

Leica AG
CH-9435 Heerbrugg
Switzerland

Measurement Devices Limited
11211 Richmond Avenue, Suite 106, Building B
Houston, TX 77082

Maihak AG
Semperstrasse 38
D-2000 Hamburg 60
West Germany

Roctest Ltée Ltd.
665 Pine
Montreal, P.Q.
Canada J4P 2P4

RST Instruments Ltd.
1780 McLean Avenue
Port Coquitlam, B.C.
Canada V3C 4K9

Schaevitz Engineering
P.O. Box 505
Camden, NJ 08101

Serata Geomechanics, Inc.
4124 Lakeside Drive
Richmond, CA 94806

SINCO (Slope Indicator Co.)
3668 Albion Place N.
Seattle, WA 98103

Soil Instruments Ltd.
Bell Lane, Uckfield
East Sussex TN22 1Ql
England

Solexperts AG
Postfach 274
CH-8034 Zürich
Switzerland

Solinst Canada Ltd.
2440 Industrial St.
Burlington, Ontario
Canada L7P 1A5

SIS Geotecnica
Via Roma, 15
20090 Segrate (Mi)
Italy

Spectron Engineering
800 West 9th Avenue
Denver, CO 80204

Spectron Glass and Electronics Inc.
595 Old Willets Path
Hauppauge, NY 11788

Telemac
2 Rue Auguste Thomas
92600 Asnieres
France

Terrametrics
16027 West 5th Avenue
Golden, CO 80401

P & S Enterprises, Ltd.
240 South Monaco Pkwy, # 302
Denver, CO 80224

Edi Meier & Partner
8408 Winterthur
Switzerland

16

Velocity Measurement

Charles P. Pinney
Pinney Technologies, Inc.

William E. Baker
University of New Mexico

16.1 Introduction

The *linear velocity* of an object, or more correctly a particle, is defined as the time rate of change of position of the object. It is a vector quantity, meaning it has a direction as well as a magnitude, and the direction is associated with the direction of the change in position. The magnitude of velocity is called the speed (or pace), and it quantifies how fast an object is moving. This is what the speedometer in a car tells you; thus, the speedometer is well named. Linear velocity is always measured in terms of, or from, some reference object. Thus, the speedometer of a car tells how fast one is moving relative to the earth. Usually, linear velocity is identified using only the term "velocity." Common units for velocity include meters per second and miles per hour, but any similar combination of units of length per unit of time is correct.

The *rotational velocity* (or angular velocity) of an object is defined as the time rate of change of angular position, and it is a measure of how fast an object is turning. It is completely analogous to linear velocity, but for angular motion. Common units are revolutions per minute, but any angular unit of measurement per unit of time can be used. Rotational velocity is a vector quantity also, with the direction of the vector being the same as the direction of the axis about which object is turning. For example, with a car stopped at a stop light with the motor running, the rotational velocity of the crankshaft of the motor is given by a magnitude (rotational speed), say 800 rpm (rev/min), and a direction associated with the direction in which the crankshaft is pointing. The axis of rotation of the object may be moving, rather than fixed as when the car is turning a corner. The roll, yaw, or pitch velocity of an airplane would be given in terms of rotational speeds about each of the turning axes in the same manner as for a crankshaft.

Usually, the reference from which linear or rotational velocity is given is understood from the context of the problem. It is often not stated explicitly. The measurement method used defines the reference.

Applications for velocity measurement include:

1. Controlling the speed at which metal stock is fed into a machine tool. If the metal is fed too quickly the result could be premature tool wear or it could even lead to machine failure. Feeding the material too slowly will reduce the yield of the machine tool.
2. Measuring the approach speed of a robotic tool onto its target.
3. Monitoring the speed of a generator in an electric power station.
4. An airport radar system measuring the speed of an approaching aircraft using the Doppler effect.
5. Measuring an automobile's wheel speed in order to provide feedback to an antilock brake system.

0-8493-8347-1/99/$0.00+$.50
© 1999 by CRC Press LLC

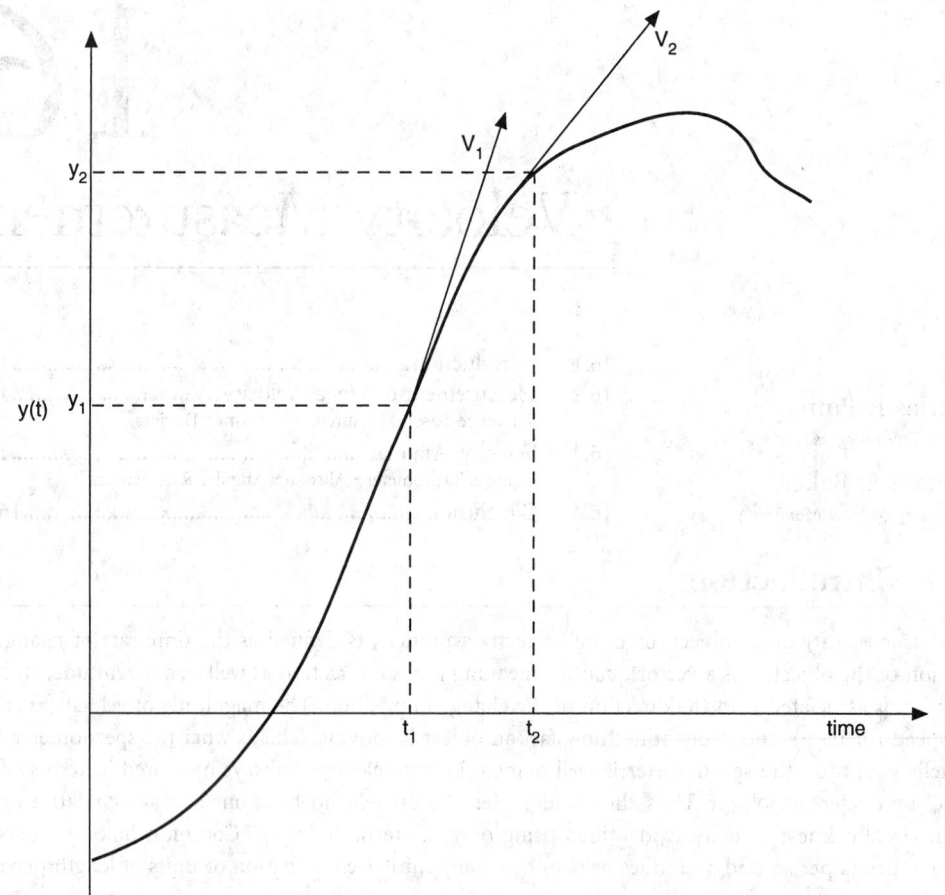

FIGURE 16.1   Position-time function for an object moving on a straight path.

16.2   Measurement of Linear Velocity

The problem of velocity measurement is somewhat different from that of measurement of other quantities in that there is not a large number of transducer types and transducer manufacturers from which to choose for a given problem. Frequently, the problem is such that the person must use his/her knowledge of measurement of other quantities and ingenuity to develop a velocity measurement method suitable for the problem at hand. Velocity is often obtained by differentiation of displacement or integration of acceleration. As background information for this, the necessary equations are given below.

Figure 16.1 shows a graph that represents the position of an object as a function of time as it moves along a straight, vertical path (y direction). The quantity to be measured could be an average velocity, and its magnitude would then be defined as follows:

$$\text{Average speed} = V_{avg} = \frac{y_2 - y_1}{t_2 - t_1} = \frac{\Delta y}{\Delta t} \tag{16.1}$$

for the time interval t_1 to t_2. As the time interval becomes small, the average speed becomes the instantaneous speed V_y, and the definition becomes:

$$V_y = \lim_{\Delta t \to 0} \frac{\Delta y}{\Delta t} = \frac{dy}{dt} \qquad (16.2)$$

which is the slope of the position–time curve. The subscript indicates the y component. This speed, when associated with the known direction, becomes the velocity.

Since acceleration is defined as the time rate of change of velocity, the speed of an object may also be given by:

$$V_y(t) = V_i - \int_{t_i}^{t} a_y(t)\,dt \qquad (16.3)$$

where $a_y(t)$ is the acceleration in the y direction (for Figure 16.1) and V_i is the speed at time t_i. Each of the above equations can be used as a basis for making a velocity measurement. Note that for motion in more than one dimension, there would be more than one component, and there would be corresponding equations for the other dimensions (x and y). However, velocity measurements are always done by individual component.

It is convenient in the discussion of techniques of measuring velocity to divide the methods into two categories: one will be called "referenced-based methods" and the other "seismic or inertial referenced transducers." Referenced-based methods refer to measurements made for which the instrumentation has component(s) on both the moving object and the reference frame for the measurement. The seismic transducers do not require contact with the reference frame. However, they give a speed relative to the transducer speed at the start of the test. The initial motion must be determined from other considerations in the test setup and added to the relative speed.

Reference-Based Measurement

Using Equation 16.1, one value of the average speed in a given direction of an object can be determined from the distance traveled in that direction and the time required. Determining the muzzle speed of a projectile is an example. Having two pickups spaced a known distance apart, and recording the time for the projectile edge to pass between them is a common way of doing this. Typical pickups would include proximity transducers (see displacement measurement), laser or collimated light beams with diode sensors, and electric contacts closed (or opened) by the moving object. Measuring the time interval can be done with an electronic counter or displaying the output of the pickups on an oscilloscope. In designing such a system, care must be exercised to minimize or eliminate the effect to the object passing through the window on the positions of the sensors and their response. For example, pressure from a muzzle blast can move the sensors on their supports. This could distort the distance between them during the measurement; but afterwards, the appearance of the setup could be unchanged, so the experimenter would be unaware of the error.

Using a series of equally spaced pickups can determine the average speed for a sequence of positions. For some applications, illumination of the path of motion of the object with a stroboscope flashing at a known rate and use of time exposure photography can give a single picture of the object at a sequence of positions. With knowledge of the length scale and the flash rate, the average speed at a sequence of positions can be calculated. If the plane of motion is the same as the plane of the photograph, then two components of the velocity can be determined. A variation of this method is to use video recording of the motion and the time base of the video for measurement of time increments. Availability of high-speed video cameras, to 12,000 frames per second, extends the range of applicable velocities, and digital recording can enhance the ease and accuracy of making the distance measurements.

Another variation of this method is to use some type of position transducer to record the position–time function of the moving object and then differentiate this function to get speed–time. Displacement

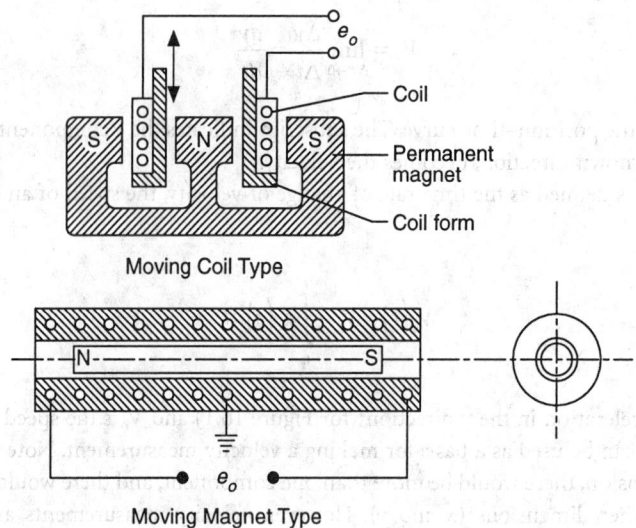

FIGURE 16.2   Velocity transducers (LVT).

transducers were discussed in an earlier chapter, and the selection of an acceptable transducer is important. Range, resolution, and mass loading are important parameters. Because differentiation of experimental data is a noise-generating process, particular care must be exercised to reduce the electric noise in the displacement data to a minimum. Also, the calculated speed–time function might require some smoothing to reduce the numerically introduced noise.

One type of velocity transducer is based on a linear generator. When a coil cuts the magnetic field lines around a magnet, a voltage is induced in the coil, and this voltage is dependent on the following relation:

$$e_i \propto BLV \qquad (16.4)$$

where e_i = induced voltage
 B = magnetic field strength
 L = length of wire in the coil
 V = speed of the coil relative to the magnet.

This relation is used as the basis for linear velocity transducers, called LVTs, and a schematic is shown in Figure 16.2. Manufacturers of these transducers include Trans-Tek Inc. of Ellington, CN; Robinson-Halpern Products of Valley Forge, PA; and the Macro Sensors, Div. of Howard A. Schaevitz Technologies, Inc. of Pennsauken, NJ. The working displacement ranges are from 0.5 in. to 24 in., and typical sensitivities are from 40 mV/ips (inches per second) to 600 mV/ips.

Conversion of Linear to Rotational Velocity

A rotational dc generator (discussed in the next section) can also be used to measure linear velocities by placing a rack on the moving object and having the rack drive the generator through a pinion gear. This is the same principle by which a speedometer converts the linear velocity of an automobile to an angular velocity gage on the dashboard of a car.

Doppler Shift

The Doppler shift is an apparent change in the frequency of waves occurring when the source and observer are in motion relative to each other. This phenomenon is applicable to waves in general; for example, sound, light, microwaves, etc. It was first observed for sound waves, and it is named after the Austrian mathematician and physicist Christian Doppler (1803–1853) who first published a paper on it for light waves in 1842. The frequency will increase when the source and observer approach each other (red shift) and decrease when they move apart (blue shift). This phenomenon was illustrated by having people listen to the pitch of an oncoming train. The high-pitched whistle would transition to a lower pitch as the train passed the observer.

Radar, which is named for radio detection and ranging, is another technique for detecting the position, motion, and nature of a remote object by means of radio waves reflected from its surface. Pulse radar systems use a single directional antenna to transmit and receive the waves. They transmit pulses of electromagnetic waves (usually microwaves), some of which are reflected by objects in the path of the beam. Reflections are received by the radar unit, processed electronically, and converted into images on a cathode-ray tube. The antenna must be connected only to the transmitter when sending and only to the receiver while receiving; this is accomplished by switching from one to the other and back again in the fraction of a microsecond between pulses. The distance of the object from the radar source is determined by measuring the time required for the radar signal to reach the target and return. The direction of the object with respect to the radar unit is determined from the direction in which the pulses were transmitted. In most units, the beam of pulses is continuously rotated at a constant speed, or it is scanned (swung back and forth) over a sector at a constant rate. Pulse radar is used primarily for aircraft and naval navigation and for military applications. In Doppler radar, or continuous-wave radar, two antennas are used — one to transmit and the other to receive. Because the time a continuous-wave signal takes to reach the target and return cannot be measured, Doppler radar cannot determine distance. The velocity of the object is determined using the Doppler effect. If the object is approaching the radar unit, the frequency of the returned signal is greater than the frequency of the transmitted signal. If the object is receding, the returned frequency is less; and if the object is not moving relative to the radar unit, the frequency of the returned signal is the same as the frequency of the transmitted signal.

One value of this Doppler technology is shown on the evening weather broadcast. Radar can measure wind rotation inside a thunderstorm and identify possible tornadoes. The VORAD system by Eaton is an on-board system for vehicle safety. It detects when a dangerous approach to another vehicle is taking place. It will automatically apply the brakes in an emergency situation.

Light Interference Methods

Velocity measurements can be made using light interference principles. Figure 16.3 shows the setup used by Michelson in the 1890s to demonstrate light interference. A beam of monochromatic light is split into two beams. One beam is directed onto a stationary mirror. The other beam is directed onto a moving target. The observer sees the superposition of the two beams. As the mirror moves in one direction, summation of the waves of the two beams will alternately reinforce and cancel each other. The amount of motion for one cycle of light intensity variation is the wavelength of the light being used. The frequency of these light-to-dark transitions is proportional to the velocity of the moving target. Highly accurate measurements are available with interferometer techniques. For example 1 m is 1,650,763.73 fringe counts for the orange light emitted by krypton-86.

Refinements of this principle are needed for convenience of use. Lasers are used as the light source, for example. One commercial supplier of this type of device, commonly called a Laser Doppler Vibrometer, is Polytec PI, Inc. of Costa Mesa, CA. The basic principle gives velocity parallel to the laser beam, but Polytec PI also has a unit that utilizes scattered laser light which permits measurement of the in-plane velocity. It is called a laser surface velocimeter.

VISAR System

Another application of interferometry to the measurement of velocity–time profiles is a device called VISAR, for "velocity interference system for any reflector." Earlier interferometer systems had required that the target have a highly polished reflecting surface and that there be very little surface tilt during a test. The VISAR system functions with either specularly or diffusely reflecting surfaces, and is quite insensitive to tilting of the target. It was developed for shock wave research work, and is useful for measurement of very high speeds. Reference [3] gives a detailed description of the principles of operation.

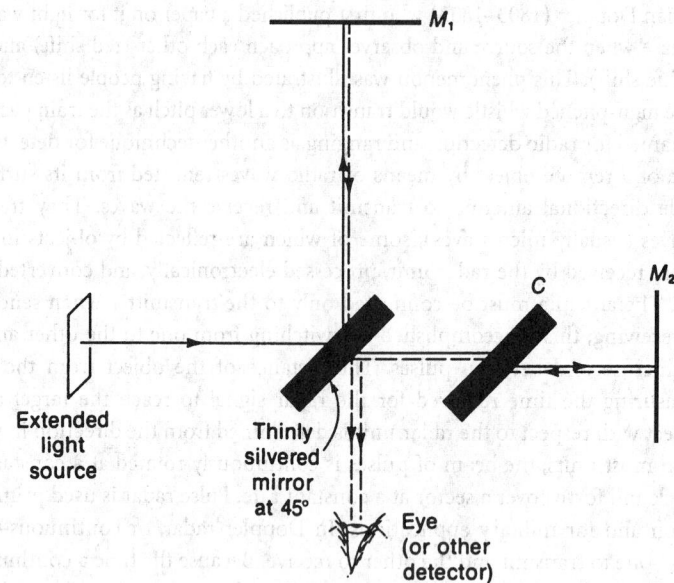

FIGURE 16.3 The basic components of a Michelson interferometer. The clear glass slab C is called a compensating plate. It has the same dimensions and orientation as the 45° mirror in order to make the light paths in glass equal along the two arms, a condition necessary when a white-light source is used. (From A. Hudson and R. Nelson, *University Physics*, New York, Harcourt Brace Jovanovich, 1982. With permission.)

The signal from the VISAR is generated with a photodiode or other light-sensitive device, and is basically a measure of the rate of fringe variation. Additional data reduction is required to obtain speeds. The sensitivities of the devices are specified in "fringes per meter/second." Typical sensitivities are in the range of 100 m/s to 4000 m/s per fringe.

The first VISARs were laboratory devices, individually assembled from the needed optical components. Commercial units are now available. Valyn International of Albuquerque, NM, makes VISARs and components. Figure 16.4 shows a schematic of a test setup. This unit can measure speeds from 100 m s^{-1} to 10 km s^{-1} or more. The standard measurement range, given as depth of field, of the VISAR is 12 mm, but systems measuring over 10 m have been used. Applications for the VISAR include:

In-bore projectile velocity
Flyer plate velocity
Flying foil velocity
Hugoniot equation of state
Structural response to shock loading

Seismic Devices

The devices discussed in the previous section required a link of some type between the reference and the moving object. Seismic devices do not have this requirement. A seismic device refers to a transducer, which is based on a mass attached to the transducer base, usually with a linear spring. The base is attached to the surface whose motion is desired, and the motion of the seismic mass relative to the base is recorded with a motion transducer. Figure 16.5 shows the principal components of this transducer type. Use of the governing equation of motion for the seismic mass permits the determination of the motion of the base from the relative motion function.

If the motion transducer in the seismic instrument measures the displacement of the mass relative to the base, the output of the transducer is proportional to the acceleration of the transducer base for a specific frequency range and the device is called an *accelerometer*. Acceleration measurement using this

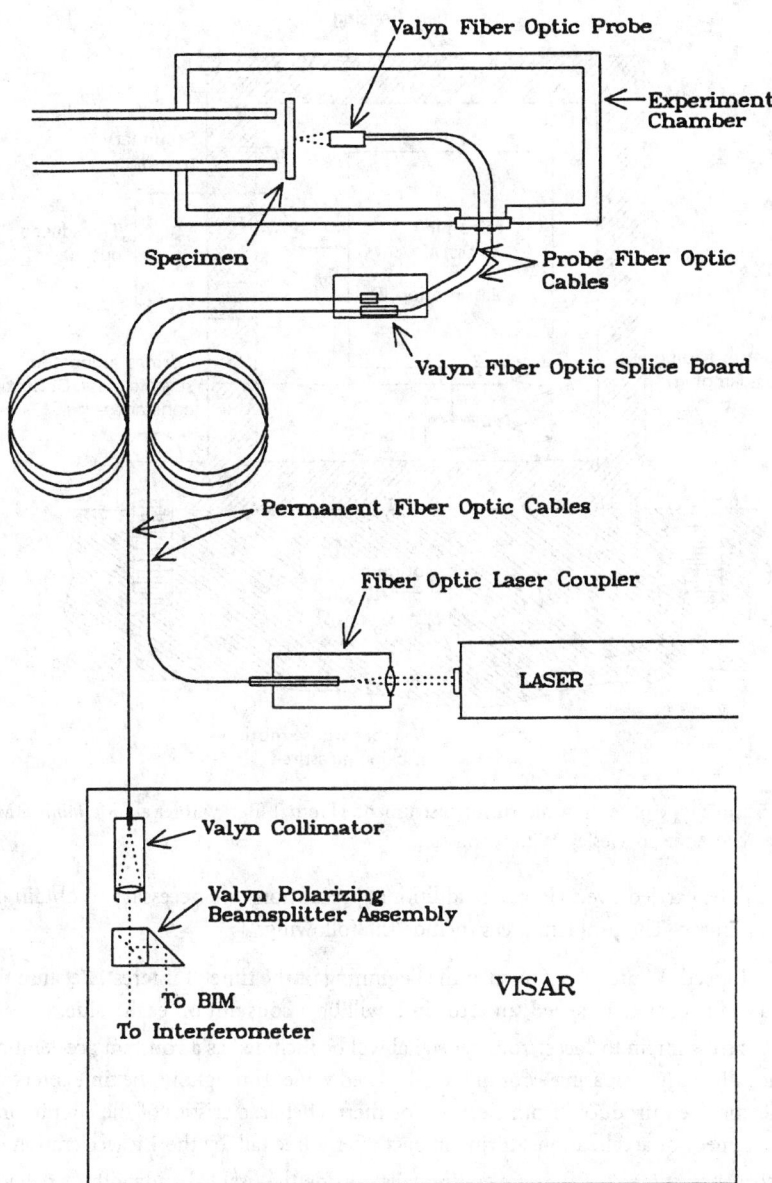

VISAR TEST CONFIGURATION

FIGURE 16.4 Schematic diagram showing how fiber optic components, available from Valyn, can transport laser light to and from a shock experiment, minimizing any laser light beam hazards. (Courtesy: Valyn International, Albuquerque, NM.)

type of device (or with other types of accelerometers) permits the determination of the velocity–time function by integration through the application of Equation 16.3. In this equation, $a_y(t)$ would be determined from the output of the accelerometer.

The simplicity of this concept is evident and, because integration is a smoothing process, the numerically introduced noise encountered with a "differentiation of displacement" method of speed measurement does not occur. However, other errors can be introduced. First, any error in the acceleration

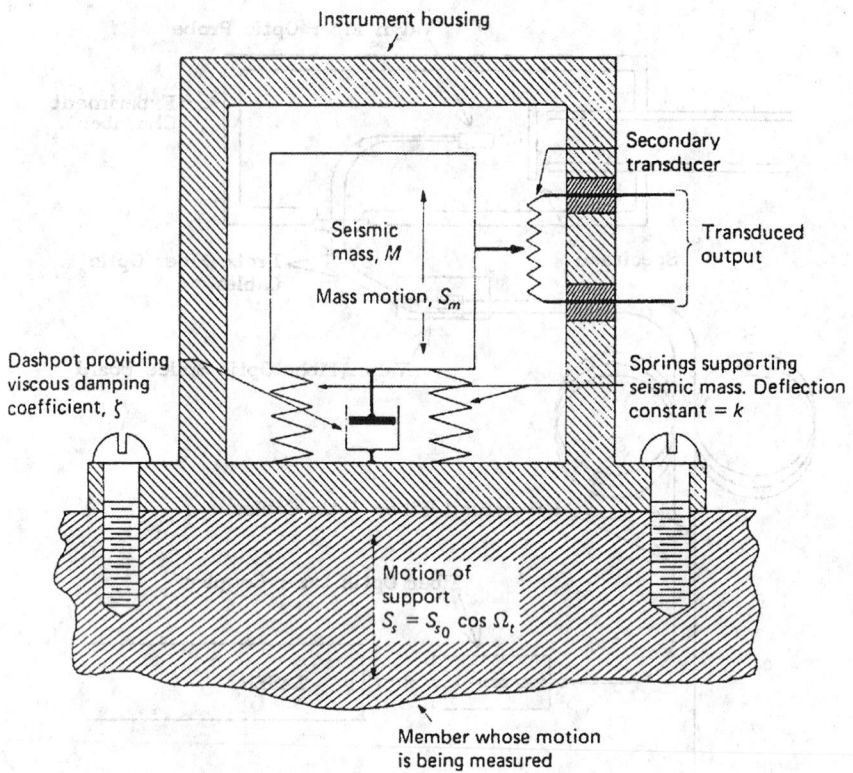

FIGURE 16.5 Seismic type of motion-measuring instrument. (From T. Beckwith et al., *Mechanical Measurements,* 4th ed., Reading, MA: Addison-Wesley. With permission.)

measurement will be carried over. However, additional precautions are necessary to obtain good results for speed measurement. The problem areas include the following:

- The initial speed, V_i, must be known at the beginning of the time of interest. Because this quantity is added to the change in speed, an error in it will be a constant on each value.
- A bias, or zero shift, in the accelerometer signal will be included as a constant acceleration, and thus introduce a linearly increasing error in the calculated values throughout the time interval of interest. This bias may be introduced from electrical or thermal characteristics of the circuit, or, in the case of measurement of accelerations during impact after a free fall, by the 1 g acceleration of gravity.
- If the frequency content of the acceleration falls outside the usable bandwidth of the accelerometer and recording circuit, errors in acceleration occur. The low-frequency cutoff depends on the recording equipment and circuit, and the high frequency cutoff depends on the natural frequency and damping of the accelerometer, as well as the bandwidth of each piece of equipment in the recording circuit.
- Accelerometer theory is based on harmonic excitation of the system. For many velocity measurement applications, the input is a transient. Combination of these two factors can result in inaccurate accelerometer data; for example, ringing may occur, and cause errors in the calculated speeds. This problem is accentuated for lightly damped accelerometers.

When this method of speed measurement must be used, a series of check tests should be conducted to evaluate the accuracy of the method for that particular system.

A variation of the above method is to put an integrating circuit in the accelerometer and perform the integration with an analog circuit. Then, the output of the "velocity" transducer is proportional to the change in speed. This type of device is subject to all of the potential error sources discussed above. A manufacturer of this type of transducer is Wilcoxon Research of Gaithersburg, MD.

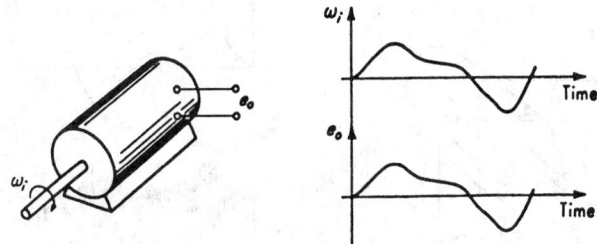

FIGURE 16.6 Permanent-magnet dc tach-generator. (With permission.)

It can be shown that if the electromechanical transducer in a seismic instrument gives an output which is proportional to the speed on one end relative to the other end, then the output of the seismic transducer is proportional to the speed of the transducer in an inertial reference frame, i.e., relative to the earth, for input motion frequencies well above the natural period of the seismic mass. Thus, use of a linear velocity transducer as the motion measurement transducer in a seismic instrument makes it a "seismic velocity transducer." This type of device is called several different names, including seismometer, geophone, and vibrometer, as well as velocity transducer.

The natural frequency and damping in these instruments are selected to match the application. As with an accelerometer, the usable bandwidth depends on these two characteristics. The low-frequency limit for this type of transducer is dependent on the accuracy required in the measurement. The governing equation is given in Doebelin [3]. As an example, it can be used to show that if an accuracy of 5% is required, the lowest data frequency must be 4.6 times the natural frequency of the transducer, and that the upper data frequency is not limited. In fact, the higher the upper frequency, the more accurate the results.

Seismometers are used for recording and studying motion from earthquakes, and these devices can be quite large. Natural periods can be in the range of 10 s to 50 s, and damping is normally selected as 0.7 of critical to extend the frequency range as much as possible. Geophones are commonly used for oil well logging and related work. Their natural periods are in the vicinity of 10 s. Manufacturers of these devices include Teledyne Brown Engineering and GeoSpace Corporation of Houston, TX.

16.3 Velocity: Angular

Measurement of angular velocity is often applied to rotating machinery such as pumps, engines, and generators. The most familiar unit of measurement in rotating machinery applications is revolutions per minute (rpm). In most cases, the measurement of rpm involves the generation of a pulse train or sine wave whose frequency is proportional to angular velocity. The measuring technologies using pulse trains and waves include ac and dc generator tachometers, optical sensors, variable reluctance sensors, rotating magnet sensors, Wiegand effect sensors, stroboscopy, etc.

These types of measurements are taken with respect to the base of the item being measured. They are relative measurements because one is measuring the motion between two bodies.

Another class of measurement problem is that of moving or inertial bodies. In this case, a measurement of absolute motion is performed. Some fixed reference must be stated or implied. This reference is often the Earth. A universal reference is sometimes required for celestial measurements. These inertial measurements are typically taken with gyroscope-type devices.

Relative: Tachometer

Electrical (dc and ac) Tachometer Generator

A rotating generator produces a voltage signal proportional to the rotational velocity of the input shaft. A dc generator produces a voltage level proportional to speed, as in Figure 16.6. The ac generator produces

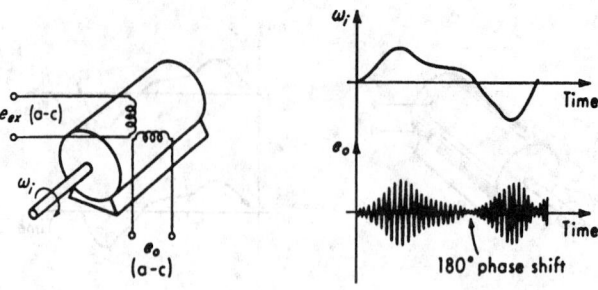

FIGURE 16.7   Ac tach-generator. (With permission.)

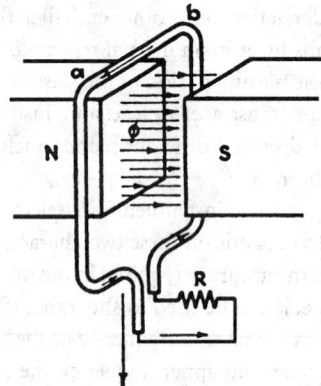

FIGURE 16.8   Generated electromotive force. Moving conductor. (From E. A. Loew, *Direct and Alternating Currents: Theory and Machinery,* New York: McGraw-Hill, 1946. With permission.)

an ac voltage output with a frequency proportional to rotational speed, as shown in Figure 16.7. In a simple two-phase motor, the ac voltage is applied to one phase of the motor and the measurement is taken off the other. Typical operating frequencies are 60 Hz and 400 Hz. This carrier frequency should be 5 to 10 times the required frequency response of the ac generator tachometer. The direction of travel is determined by the phase of the signal with opposite directions being 180° out of phase. The basic dc generator is shown in Figure 16.8.

Sources of tachometer generators include the GE Company of Fairfield, CT; Kollmorgen Motion Technologies Group of Radford, VA; Sierracin/Magnedyne of Vista, CA; and Micro Mo Electronics of Clearwater, FL.

Counter Types

An entire class of angular velocity measuring techniques exists that uses pulses generated by electromechanical interaction. The common thread is a pulse-to-voltage converter giving a voltage output proportional to velocity.

Rotating Magnet Sensors: Passive speed sensors convert mechanical motion to ac voltage without an external power source. These self-contained magnetic sensors produce a magnetic field that, when in the proximity of ferrous objects in motion, generates a voltage.

When a magnetic sensor is mounted in the proximity of a ferrous target, such as gear teeth on a rotating shaft, the voltage output frequency is directly proportional to the rotational speed of the target. A frequency-to-voltage converter can then convert the signal to a voltage. An engineering unit conversion from voltage to velocity then provides an actual velocity measurement.

Typical applications for these types of sensors include:

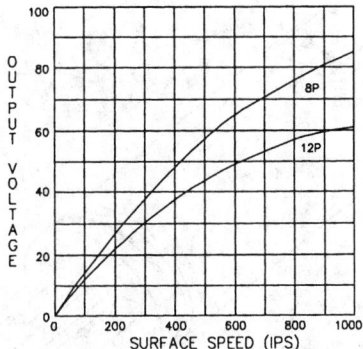

FIGURE 16.9   Magnetic speed sensor output voltage vs. speed. (Courtesy: Smith Systems, Inc., Brevard, NC.)

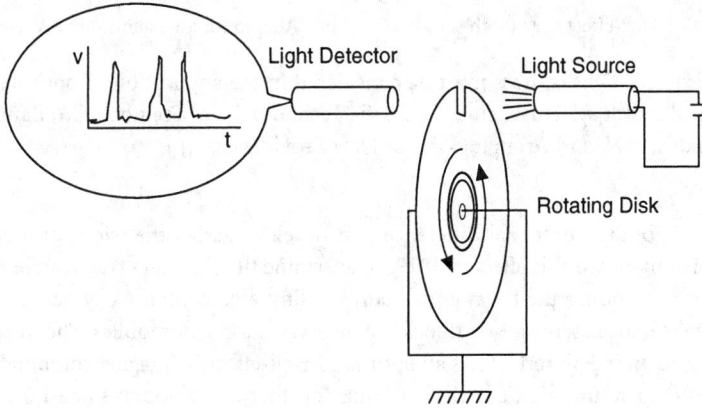

FIGURE 16.10   A slotted disk provides one pulse output for each rotation.

Transmission speed
Engine rpm
Over/under speed
Wheel speed
Pump shaft speed
Multiple engine synchronization
Feedback for speed control
Crankshaft position/engine timing
Computer peripheral speeds

The typical specifications for magnetic speed sensors are given by a graph of output voltage versus surface speed in inches per second, as in Figure 16.9.

Sources for magnetic sensors include Smith Systems of Brevard, NC; Optek Technology of Carrolton, TX; Allied-Signal of Morristown, NJ; and Baluff of Florence, KY.

Optical Sensors

Optical methods of angular velocity detection employ a light emitter and a light detector. A light-emitting diode (LED) paired with a light-sensitive diode is the most common arrangement.

A slotted disk is placed in the axis of a rotating shaft. Each slot or slit will allow the light to pass through the disk. Figure 16.10 shows a typical arrangement. The detector will generate a pulse train with a rate proportional to the angular velocity.

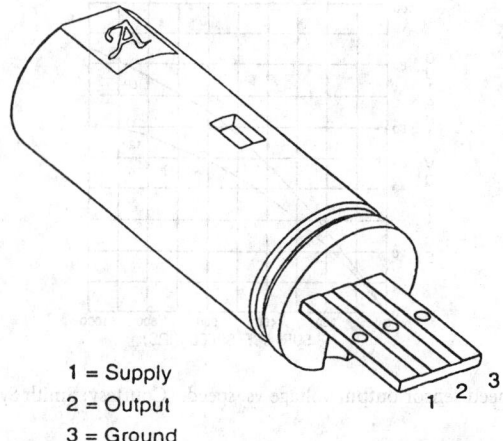

1 = Supply
2 = Output
3 = Ground

FIGURE 16.11   Hall-effect gear tooth sensor. (Courtesy: Allegro Microsystems, Inc., Worcester, MA.)

The effects of external light sources must be considered in the application of optical sensors.

Sources of optical sensor systems include Scientific Technologies of Fremont, CA; Banner Engineering Corp. of Minneapolis, MN; and Aromat Corp. of New Providence, NJ.

Hall Effect

The Hall effect describes the potential difference that develops across the width of a current-carrying conductor. E.H. Hall first used this effect in 1879 to determine the sign of current carriers in conductors. Hall effect devices are finding their way into many sensing applications. A typical Hall effect sensor application is the wheel speed sensor for antilock braking systems in automobiles. The Allegro ATS632LSC gear-tooth sensor, shown in Figure 16.11, is an optimized Hall-effect IC/magnet combination. The sensor consists of a high-temperature plastic shell that holds together a compound samarium–cobalt magnet, a single-element self-calibrating Hall effect IC, and a voltage regulator. The operation of this circuit is shown in Figure 16.12.

Wiegand Effect

The Wiegand effect is useful for proximity sensing, tachometry, rotary shaft encoding, and speed sensing in applications such as:

Electronic indexing for water, gas, and electric meters and remote metering systems
Measuring shaft speed in engines and other machinery
Tachometers, speedometers, and other rotational counting devices

Wiegand effect technology employs unique magnetic properties of specially processed, small-diameter ferromagnetic wire. By causing the magnetic field of this wire to suddenly reverse, a sharp, uniform voltage pulse is generated. This pulse is referred to as a Wiegand pulse. Sensors utilizing this effect require only a few simple components to produce sharply defined voltage pulses in response to changes in the applied magnetic field. These sensors consist of a short length of Wiegand wire, a sensing coil, and alternating magnetic fields that generally are derived from small permanent magnets.

The major advantages of the Wiegand effect based sensors are:

No external power requirement
Two-wire operation
Noncontact with no wear
20 kHz pulse rate
High-level voltage output pulse
Wide operating temperature range (e.g., −40°C to +125°C)

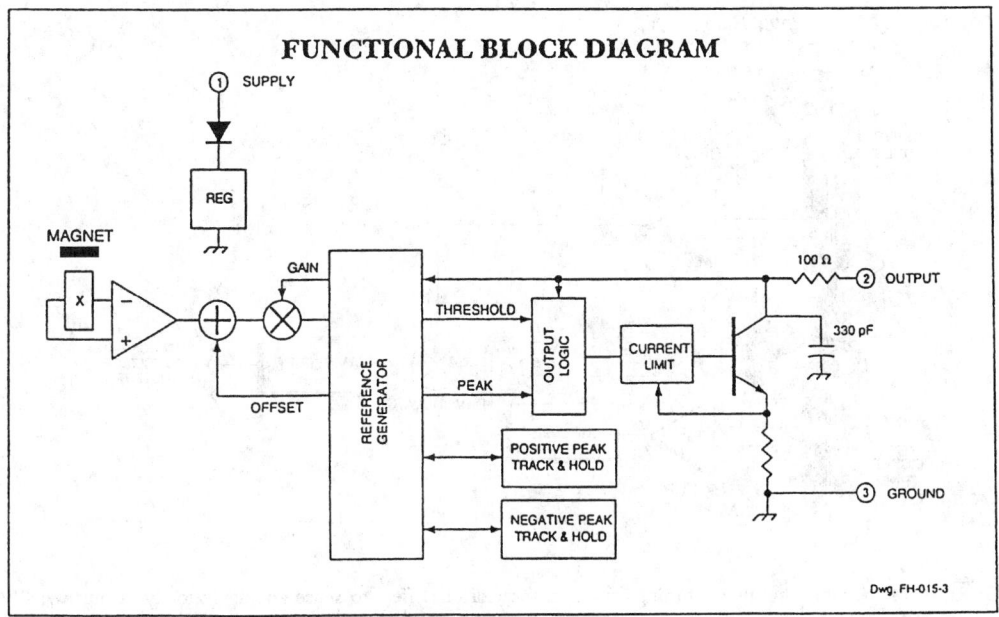

FIGURE 16.12   Hall-effect gear tooth sensor circuit. (Courtesy: Allegro Microsystems, Inc., Worcester, MA.)

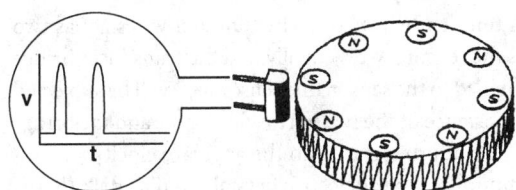

FIGURE 16.13   Small magnets cause sudden reversal in the ferromagnetic wire in a Wiegand sensor. (Courtesy: HID Corporation, North Haven, CT.)

When an alternating magnetic field of proper strength is applied to the Wiegand wire, the magnetic field of the core switches polarity and then reverses, causing the Wiegand pulse to be generated, as shown in Figure 16.13. The magnetic switching action of the Wiegand wire induces a voltage across the pickup coil of approximately 10 μs duration. These alternating magnetic fields are typically produced by magnets that are affixed to the rotating or moving equipment, by a stationary read head and moving Wiegand wires, or by an alternating current generated field.

Absolute: Angular Rate Sensors

Gyroscopes

Many absolute angular rate-measuring devices fall under the designation of gyroscope. A mechanical gyroscope is a device consisting of a spinning mass, typically a disk or wheel, mounted on a base so that its axis can turn freely in one or more directions and thereby maintain its orientation regardless of any movement of the base. It is important to make an initial distinction between angular velocity gyros and rate-integrating gyros. Angular velocity gyros are used to measure motion and as signal inputs to stabilization systems. Rate-integrating gyros are used as the basis for highly accurate inertial navigation systems. They allow a stable platform to maintain a fixed attitude with reference. These devices can be very complex. Three gyros are often teamed with three double-integrated accelerometers to provide an accurate measurement of absolute vehicle motion.

Ricardo Dao of Humphrey Inc. provided an excellent comparison of angular rate sensors in an article in *Measurements & Control* [14]. The five different technologies are summarized below.

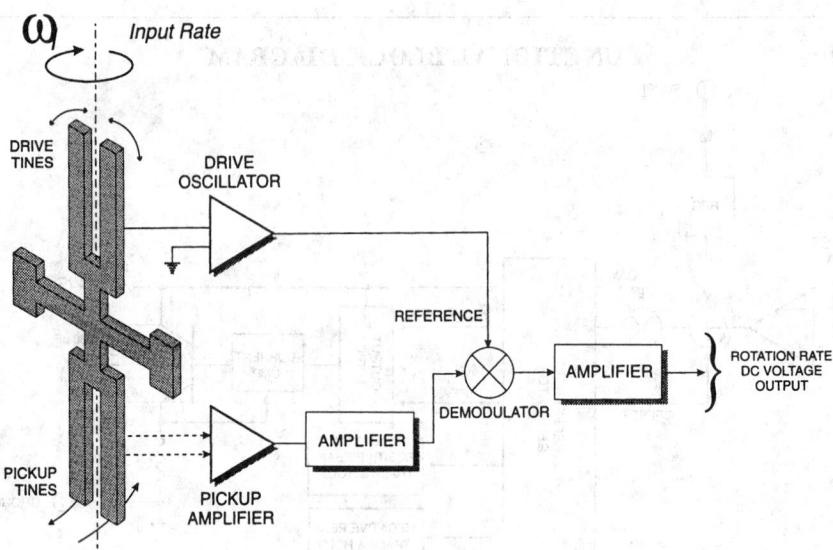

FIGURE 16.14 A vibrating quartz tuning fork uses the Coriolis effect to sense angular velocity. (Courtesy: BEI Sensors and Systems Co., Concord, CA.)

Spinning mass: The traditional gyro consists of a spinning wheel in a gimbaled frame. The principle of conservation of angular momentum provides the measurement tool.

Fluidic: A stream of helium gas flows past two thin tungsten wires [14]. The tungsten wires act as two arms of a Wheatstone bridge. At rest, the gas flow cools the sensing wires equally and the transducer bridge is balanced with zero output. When angular motion is applied to the sensor, one sensor wire will be subjected to increased flow while the other will see less flow. The resistance of the two wires will change and the bridge will be unbalanced. The sensor will produce a voltage output proportional to the angular velocity.

A pump is used to circulate the helium gas. This pump is a piezoelectric crystal circular disk that is excited with an external circuit. The pump produces a laminar flow of relatively high-velocity gas across the two parallel sensing wires.

Piezoelectric vibration: A number of angular velocity sensors have been developed that use micromachined quartz elements. A number of shapes are used, but the operating principle is similar for each. The quartz element vibrates at its natural frequency. Angular motion causes a secondary vibration that, when demodulated, is proportional to angular vibration. A description of one design follows.

The QRS and GyroChip™ family of products uses a vibrating quartz tuning fork to sense angular velocity [15, 16]. Using the Coriolis effect, a rotational motion about the sensor's longitudinal axis produces a dc voltage proportional to the rate of rotation. Figure 16.14 shows that the sensor consists of a microminiature double-ended quartz tuning fork and supporting structure, all fabricated chemically from a single wafer of monocrystalline piezoelectric quartz (similar to quartz watch crystals).

Use of piezoelectric quartz material simplifies the active element, resulting in exceptional stability over temperature and time. The drive tines, being the active portion of the sensor, are driven by an oscillator circuit at a precise amplitude, causing the tines to move toward and away from each another at a high frequency.

Each tine will have a Coriolis force acting on it of: $\{F = 2m\ W_i \times V_r\}$ where the tine mass is m, the instantaneous radial velocity is V_r, and the input rate is W_i. This force is perpendicular to both the input rate and the instantaneous radial velocity.

The two drive tines move in opposite directions, and the resultant forces are perpendicular to the plane of the fork assembly and in opposite directions. This produces a torque that is proportional to the

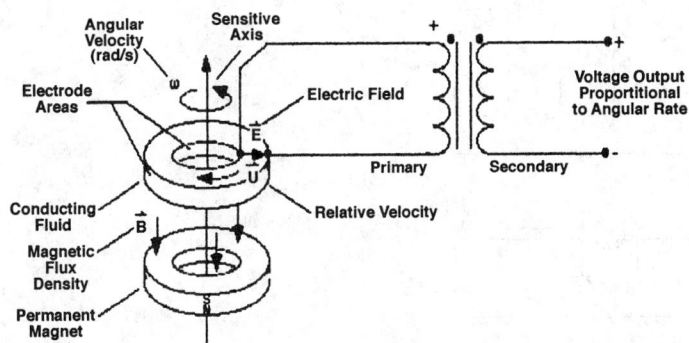

FIGURE 16.15 Magnetohydrodynamic angular rate sensor. (Courtesy: ATA Sensors, Albuquerque, NM.)

input rotational rate. Since the radial velocity is sinusoidal, the torque produced is also sinusoidal at the same frequency of the drive tines, and in-phase with the radial velocity of the tine.

The pickup tines, being the sensing portion of the sensor, respond to the oscillating torque by moving in and out of plane, producing a signal at the pickup amplifier. After amplification, those signals are demodulated into a dc signal that is proportional to the rotation of the sensor.

The output signal of the GyroChip™ reverses sign with the reversal of the input rate since the oscillating torque produced by the Coriolis effect reverses phase when the direction of rotation reverses. The GyroChip™ will generate a signal only with rotation about the axis of symmetry of the fork; that is, the only motion that will, by Coriolis sensing, produce an oscillating torque at the frequency of the drive tines. This also means that the GyroChip™ can truly sense a zero rate input.

MHD effect: The magnetohydrodynamic angular rate sensor is used to measure angular vibrations in the frequency range of 1 Hz to 1000 Hz. It is used where there is a high shock environment and a high rate of angular motion such as 10 to 250 rad s⁻¹. It does not measure a constant or dc velocity. It is used to measure impacts shorter than 1 s duration and vibrations between 1 Hz and 1000 Hz.

The principle of operation is illustrated in Figure 16.15 [17, 18]. A permanent magnet is attached to the outer case of the sensor. When the case turns, a moving magnetic field is produced (*B*). There is also a conductive fluid inside the sensor. When the sensor case turns, the fluid tends to stay in one place, according to Newton's first law. This produces a relative motion (*U*) between a magnetic field and conductor. This motion will produce a voltage (*E*) across the conductor proportional to relative velocity according to Faraday's law.

Since the fluid is constrained to move in an angular path, the voltage signal will be proportional to angular velocity about the center axis of the sensor. Due to this constraint, the sensor is insensitive to linear motion. The voltage signal is amplified through a transformer or an amplifier for output to a measuring device.

Fiber optic/laser: A beam of light is directed around the axis of rotation. A phase shift of the optical or laser beam is detected to measure angular velocity. The principle of operation is similar to the Doppler shift.

Differenced and integrated accelerometers: An array of accelerometers can be used to measure angular motion. The output of the accelerometers is differenced when they are aligned, or summed when they are mounted in opposite directions. This differencing will eliminate the linear component of motion. As shown in Figure 16.16, the magnitude of the differenced signals, a_1 and a_2, is divided by the distance between the two sensors, *l*. This gives a measure of angular acceleration. The angular acceleration is integrated over time to give angular velocity. It is important to address the same concerns in this process as when integration was discussed in the linear section. It is assumed that there is a rigid mounting structure between the two accelerometers.

This technique is commonly applied to crash testing of anthropomorphic test devices (ATDs). The ATDs are used in automotive crash testing and aerospace egress system testing.

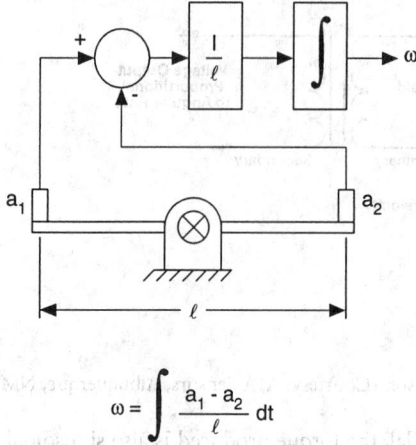

$$\omega = \int \frac{a_1 - a_2}{\ell} \, dt$$

FIGURE 16.16   Angular acceleration by differencing accelerometers and integration.

16.4   Conclusion

But alas, as Poincaré [20] stated, there is no way to determine absolute velocity.

References

1. J. W. Dally, W. F. Riley, and K. G. McConnell, *Instrumentation for Engineering Measurements*, New York: John Wiley & Sons, 1984.
2. A. Hudson and R. Nelson, *University Physics*, New York: Harcourt Brace Jovanovitch, 1982.
3. E. O. Doebelin, *Measurement Systems, 4th ed.*, New York: McGraw-Hill, 1990.
4. Trans-Tek, Inc., [Online], 1998. Available *http://transtekinc.com/lvt.htm*
5. L. M. Barker and R. E. Hollenbach, Laser interferometer for measuring high velocities of any reflecting surface, *J. Appl. Phys.*, 43(11), 1972.
6. Valyn International, [Online], 1998. Available *www.valynvisar.com*
7. L. C. Rodman, J. H. Bell, and R. D. Mehta, A 3-component laser-doppler velocimeter data acquisition and reduction system, NASA contractor report; NASA CR-177390.
8. E. A. Loew, *Direct and Alternating Currents, 3rd edition*, New York: McGraw-Hill, 1946.
9. P. Emerald, Low duty cycle operation of hall effect sensors for circuit power conservation, *Sensors*, 15(3), 1998.
10. Allegro MicroSystems, Inc., [Online], 1998. Available *http://www.allegromicro.com/*
11. The Hall effect sensor: basic principles of operation and application, *Sensors*, May 1988.
12. J. B. Scarborough, *The Gyroscope: Theory and Applications*, New York: Interscience, 1958.
13. R. F. Deimel, *Mechanics of the Gyroscope: The Dynamics of Rotation*, New York: Dover Publications, 1950.
14. E. Dao, Fluidic angular rate sensors, *Measurements & Control*, February 1994, 126-131.
15. Systron Donner Inertial Division, [Online], 1998. Available *http://www.systron.com*
16. S. Orloski and B. Zimmerman, Dead reckoning in an emergency vehicle location system using a quartz rotation sensor and GPS, *Proc. Sensor Expo*, Chicago, IL, September 1995, Peterborough, NH: Helmers Publishing, 1995.
17. G. Kulikovskiy and G. A. Lyubimov, *Magnetohydrodynamics*, Reading, MA: Addison-Wesley, 1965.
18. Applied Technology Associates, [Online], 1998. Available *http://www.aptec.com/bus-opp.html*
19. G. Unger et al., NASA's first in-space optical gyroscope: a technology experiment on the X-ray timing Explorer spacecraft, NASA technical memorandum; 109242.
20. R.P. Feynman, R.B. Leighton, and M. Sands, *The Feynman Lectures on Physics*, Reading, MA: Addison-Wesley, 1963, 15-5.

17

Acceleration, Vibration, and Shock Measurement

Halit Eren
Curtin University of Technology

Acceleration is measured by accelerometers as an important parameter for general-purpose absolute motion measurements, and vibration and shock sensing. Accelerometers are commercially available in a wide variety of ranges and types to meet diverse application requirements. They are manufactured small in size, light in weight, rugged, and robust to operate in harsh environment. They can be configured as active or passive sensors. An active accelerometer (e.g., piezoelectric) gives an output without the need for an external power supply, while a passive accelerometer only changes its electric properties (e.g., capacitance) and requires an external electrical power. In applications, the choice of active or passive type accelerometer is important, since active sensors cannot measure static or dc mode operations. For

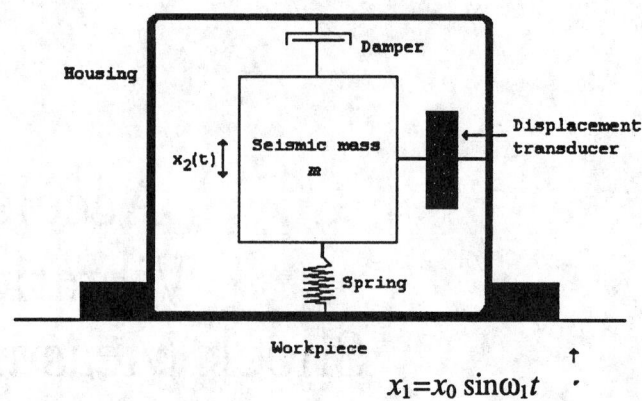

$$x_1 = x_0 \sin\omega_1 t$$

FIGURE 17.1 A typical deflection-type seismic accelerometer. In this basic accelerometer, the seismic mass is suspended by a spring or cantilever inside a rigid frame. The frame is connected to the vibrating structure; as vibrations take place, the mass tends to remain fixed so that relative displacements can be picked up. They are manufactured in many different types and sizes and they exhibit diverse characteristics.

true static measurements, passive sensors must be used. In general, accelerometers are preferred over displacement and velocity sensors for the following reasons:

1. They have a wide frequency range from zero to very high values. Steady accelerations can easily be measured.
2. Acceleration is more frequently needed since destructive forces are often related to acceleration rather than to velocity or displacement.
3. Measurement of transients and shocks can readily be made, more easily than displacement or velocity sensing.
4. Displacement and velocity can be obtained by simple integration of acceleration by electronic circuitry. Integration is preferred over differentiation.

Accelerometers can be classified in a number of ways, such as *deflection* or *null-balance* types, mechanical or electrical types, dynamic or kinematic types. The majority of industrial accelerometers can be classified as either deflection type or null-balance type. Those used in vibration and shock measurements are usually the deflection types, whereas those used for measurements of motions of vehicles, aircraft, etc. for navigation purposes may be either type. In general, null-balance types are used when extreme accuracy is needed.

A large number of practical accelerometers are of the deflection type; the general configuration is shown in Figure 17.1. There are many different deflection-type accelerometers. Although their principles of operation are similar, they only differ in minor details, such as the spring elements used, types of damping provided, and types of relative motion transducers employed. These types of accelerometers behave as second-order systems; the detailed mathematical analysis will be given in later sections.

Accelerometers can be classified as *dynamic*, meaning that the operation is based on measuring the force required to constrain a seismic mass to track the motion of the accelerated base, such as spring-constrained-slug types. Another type is the *kinematic* accelerometer, which is based on timing the passage of an unconstrained proof mass from spaced points marked on the accelerated base; this type is found in highly specific applications such as interspace spacecraft and in gravimetry type measurements.

For practical purposes, accelerometers can also be classified as *mechanical* or *electrical*, depending on whether the restoring force or other measuring mechanism is based on mechanical properties, (e.g., the law of motion, distortion of a spring or fluid dynamics, etc.) or on electrical or magnetic forces.

Calibrations of accelerometers are necessary in acceleration, vibration, and shock sensing. The calibration methods can broadly be classified to be *static* or *dynamic*. Static calibration is conducted at one or several levels of constant acceleration. For example, if a tilting table calibration method is selected, the vertical component of the free fall is used without a choice of magnitude. On the other hand, if a centrifuge is selected, it produces a constant acceleration as a function of the speed of rotation, and the magnitudes can be chosen in a wide range from 0 to well over 50,000 g. The dynamic calibration is usually done using an electrodynamic shaker. The electrodynamic shaker is designed to oscillate in a sinusoidal motion with variable frequencies and amplitudes. They are stabilized at selected levels of calibration. This is an absolute method that consists of measuring the displacement with a laser inter-ferometer and a precise frequency meter for accurate frequency measurements. The shaker must be driven by a power amplifier, thus giving a sinusoidal output with minimal distortion. The National Bureau of Standards uses this method as a reference standard. Precision accelerometers, mostly of the piezoelectric type, are calibrated by the absolute method and then used as the working standard. A preferred method is back-to-back calibration, where the test specimen is directly mounted on the working standard that, in turn, is mounted on an electrodynamic shaker.

Before providing details of different type of accelerometers, the common features such as accelerometer dynamics, velocity, distance, shock frequency responses, etc. will be introduced in the next section.

17.1 Accelerometer Dynamics: Frequency Response, Damping, Damping Ratio, and Linearity

This section concerns the physical properties of acceleration, vibration, and shock measurements in which accelerometers are commonly used. A full understanding of accelerometer dynamics is necessary in relation to characteristics of acceleration, vibration, and shock. The vibrations can be periodic, stationary random, nonstationary random, or transient.

Periodic Vibrations

In periodic vibrations, the motion of an object repeats itself in an oscillatory manner. This can be represented by a sinusoidal waveform:

$$x(t) = X_{peak} \sin(\omega t) \tag{17.1}$$

where $x(t)$ = time-dependent displacement
ω = $2\pi ft$ = angular frequency
X_{peak} = maximum displacement from a reference point

The velocity of the object is the time rate of change of displacement:

$$u(t) = dx/dt = \omega X_{peak} \cos(\omega t) = U_{peak} \sin(\omega t + \pi/2) \tag{17.2}$$

where $u(t)$ = time-dependent velocity
$U_{peak} = \omega X_{peak}$ = maximum velocity

The acceleration of the object is the time rate change of velocity:

$$a(t) = du/dt = d^2 u/dt^2 = -\omega^2 X_{peak} \sin(\omega t) = A_{peak} \sin(\omega t + \pi) \tag{17.3}$$

where $a(t)$ = time-dependent acceleration
$A_{peak} = \omega^2 X_{peak} = \omega U_{peak}$ = maximum acceleration

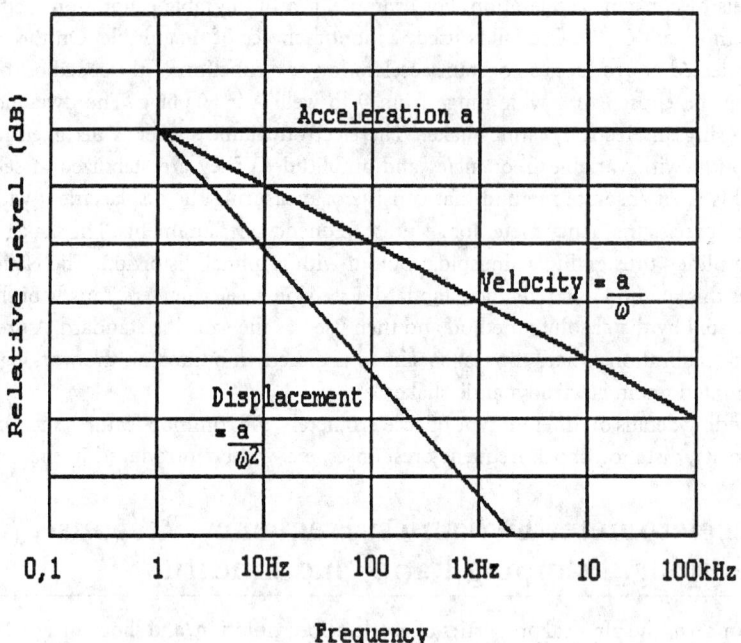

FIGURE 17.2 Logarithmic relationship between acceleration, velocity, and displacement. Velocity at a particular frequency can be obtained by dividing acceleration by a factor proportional to frequency. For displacement, acceleration must be divided by a factor proportional to the square of the frequency. Phase angles need to be determined separately, but they can be neglected in time-averaged measurements.

From the above equations, it can be seen that the basic form and the period of vibration remains the same in acceleration, velocity, and displacement. But velocity leads displacement by a phase angle of 90° and acceleration leads velocity by another 90°. The amplitudes of the three quantities are related as a function of frequency, as shown in Figure 17.2.

In nature, vibrations can be periodic but not necessarily sinusoidal. If they are periodic but nonsinusoidal, they can be expressed as a combination of a number of pure sinusoidal curves, described by Fourier analysis as:

$$x(t) = X_0 + X_1 \sin(\omega_1 t + \phi_1) + X_2 \sin(\omega_2 t + \phi_2) + \dots X_n \sin(\omega_n t + \phi_n) \qquad (17.4)$$

where $\omega_1, \omega_2, \dots, \omega_n$ = frequencies (rad s^{-1})

$X_0, X_1, \dots, X_n$ = maximum amplitudes of respective frequencies

$\phi_1, \phi_2, \dots, \phi_n$ = phase angles

The number of terms may be infinite: the higher the number of elements, the better the approximation. These elements constitute the *frequency spectrum*. The vibrations can be represented in time domain or frequency domain, both of which are extremely useful in the analysis. As an example, in Figure 17.3, the time response of the seismic mass of an accelerometer is given against a rectangular pattern of excitation of the base.

Stationary Random Vibrations

Random vibrations are often met in nature where they constitute irregular cycles of motion that never repeat themselves exactly. Theoretically, an infinitely long time record is necessary to obtain a complete description of these vibrations. However, statistical methods and probability theory can be used for the

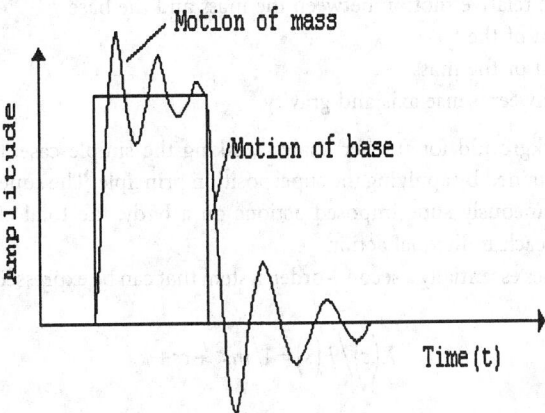

FIGURE 17.3 Time response of a shock excitation of a single degree-of-freedom system. As the duration of the shock pulse increases, sustained oscillations get shorter in time but larger in amplitude. The maximum system response may be as high as twice the magnitude of the shock pulse.

analysis by taking representative samples. Mathematical tools such as probability distributions, probability densities, frequency spectra, cross- and auto-correlations, Digital Fourier Transforms (DFT), Fast Fourier Transforms (FFT), auto spectral analysis, RMS values, and digital filter analysis are some of the techniques that can be employed. Interested readers should refer to references for further information.

Transients and Shocks

Often, short-duration and sudden-occurrence vibrations need to be measured. Shock and transient vibrations may be described in terms of force, acceleration, velocity, or displacement. As in the case of random transients and shocks, statistical methods and Fourier Transforms are used in the analysis.

Nonstationary Random Vibrations

In this case, the statistical properties of vibrations vary in time. Methods such as time averaging and other statistical techniques can be employed. A majority of accelerometers described here can be viewed and analyzed as seismic instruments consisting of a mass, a spring, and a damper arrangement, as shown in Figure 17.1. Taking only the mass-spring system, if the system behaves linearly in a time invariant manner, the basic second-order differential equation for the motion of the mass alone under the influence of a force can be written as:

$$f(t) = m\,\mathrm{d}^2x/\mathrm{d}t^2 + c\,\mathrm{d}x/\mathrm{d}t + kx \tag{17.5}$$

where $f(t)$ = force
  m = mass
  c = velocity constant
  k = spring constant

Nevertheless, in seismic accelerometers, the base of the arrangement is also in motion. Therefore, Equation 17.5 can be generalized by taking the effect motion of the base into account. Then, Equation 17.5 can be modified as:

$$m\,\mathrm{d}^2z/\mathrm{d}t^2 + c\,\mathrm{d}z/\mathrm{d}t + kz = mg\,\cos(\theta) - m\,\mathrm{d}^2x_1/\mathrm{d}t^2 \tag{17.6}$$

where $z = x_2 - x_1$ = the relative motion between the mass and the base
     x_1 = displacement of the base
     x_2 = displacement of the mass
     θ = the angle between sense axis and gravity

In order to lay a background for further analysis, taking the simple case, the complete solution to Equation 17.5 can be obtained by applying the superposition principle. The superposition principle states that if there are simultaneously superimposed actions on a body, the total effect can be obtained by summing the effects of each individual action.

Equation 17.5 describes essentially a second-order system that can be expressed in Laplace transform as:

$$X(s)/F(s) = 1/ms^2 + cs + k \qquad (17.7)$$

or

$$X(s)/F(s) = K/\left[s^2/\omega_n^2 + 2\zeta s/\omega_n + 1\right] \qquad (17.8)$$

where s = the Laplace operator
    $K = 1/k$ = static sensitivity
    $\omega_n = \sqrt{k/m}$ = undamped critical frequency, rad/s
    $\zeta = c/2\sqrt{km}$ = damping ratio

As can be seen, in the performance of accelerometers, important parameters are the static sensitivity, the natural frequency, and the damping ratio, which are functions of mass, velocity, and spring constants. Accelerometers are designed to have different characteristics by suitable selection of these parameters.

Once the response is expressed in the form of Equations 17.7 and 17.8, analysis can be taken further, either in the time domain or in the frequency domain. The time response of a typical second-order system for a unit-step input is given in Figure 17.4. The Bode plot gain phase responses are depicted in Figure 17.5. Detailed discussions about frequency response, damping, damping ratio, and linearity are made in relevant sections, and further information can be obtained in the references.

Systems in which a single structure moves in more than one direction are termed *multi-degree-of-freedom systems*. In this case, the accelerations become functions of dimensions as d^2x/dt^2, d^2y/dt^2, and d^2z/dt^2. Hence, in multichannel vibration tests, multiple transducers must be used to create uniaxial, biaxial, or triaxial sensing points for measurements. Mathematically, a linear multidegree-of-freedom system can be described by a set of coupled second-order linear differential equations; and when the frequency response is plotted, it normally shows one resonance peak per degree of freedom.

Frequently, acceleration and vibration measurements of thin plates or small masses are required. Attaching an accelerometer with a comparable mass onto a thin plate or a small test piece can cause "mass loading." Since acceleration is dependent on the mass, the vibration characteristics of the loaded test piece could be altered, thus yielding wrong measurements. In such cases, a correct interpretation of the results of the measuring instruments must be made. Some experimental techniques are also available for the correction of the test results in the form of performing repetitive tests conducted by sequentially adding small known masses and by observing the differences.

The following sections discuss different types of accelerometers.

17.2 Electromechanical Force-Balance (Servo) Accelerometers

Electromechanical accelerometers, essentially servo or null-balance types, rely on the principle of feedback. In these instruments, acceleration-sensitive mass is kept very close to a neutral position or zero displacement point by sensing the displacement and feeding back this displacement. A proportional

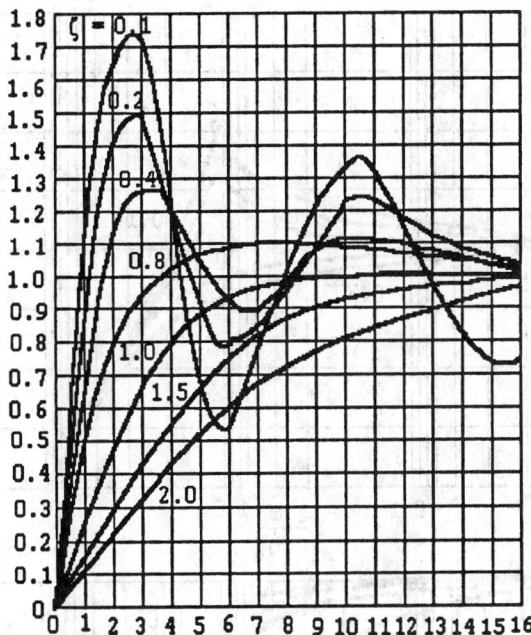

FIGURE 17.4 Unit step time responses of a second-order system with various damping ratios. The maximum overshoot, delay, rise, settling times, and frequency of oscillations depend on the damping ratio. Smaller damping ratios give faster response but larger overshoot. In many applications, a damping ratio of 0.707 is preferred.

magnetic force is generated to oppose the motion of the mass displaced from the neutral, thus restoring neutral position — just as a mechanical spring in a conventional accelerometer would do. The advantages of this approach are the better linearity and elimination of hysteresis effects as compared to mechanical springs. Also, in some cases, electric damping can be provided, which is much less sensitive to temperature variations.

One very important feature of null-balance type instruments is the capability of testing the static and dynamic performances of the devices by introducing electrically excited test forces into the system. This remote self-checking feature can be quite convenient in complex and expensive tests where it is extremely critical that the system operates correctly before the test commences. They are also useful in acceleration control systems, since the reference value of acceleration can be introduced by means of a proportional current from an external source. They are usually used for general-purpose motion measurements and monitoring low-frequency vibrations. They are specifically applied in measurements requiring better accuracy than achieved by those accelerometers based on mechanical springs as the force-to-displacement transducer.

There are a number of different types of electromechanical accelerometers: coil-and-magnetic types, induction types, etc.

Coil-and-Magnetic Type Accelerometers

These accelerometers are based on Ampere's law; that is: "a current carrying conductor disposed within a magnetic field experiences a force proportional to the current, the length of the conductor within the field, the magnetic field density, and the sine of the angle between the conductor and the field."

Figure 17.6 illustrates one form of accelerometer making use of the above principle. The coil is located within the cylindrical gap defined by a permanent magnet and a cylindrical soft iron flux return path. It is mounted by means of an arm situated on a minimum friction bearing so as to constitute an acceleration-sensitive seismic mass. A pick-off mechanism senses the displacement of the coil under

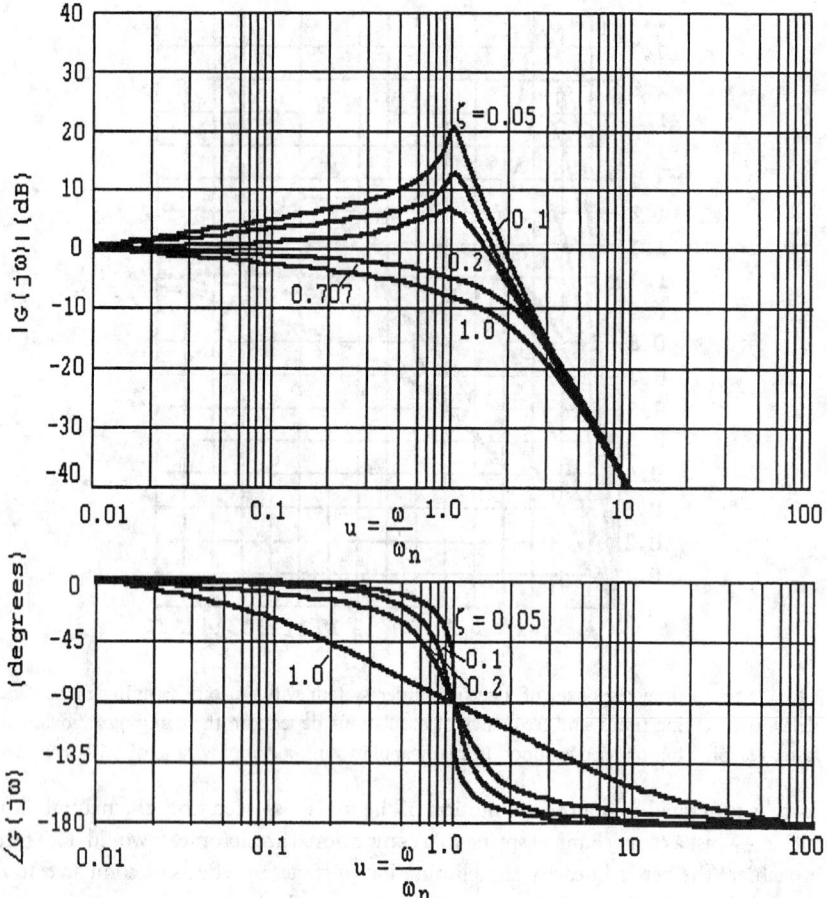

FIGURE 17.5 Bode plots of gains and phase angles against frequency of a second-order system. Curves are functions of frequencies as well as damping ratios. These plots can be obtained theoretically or by practical tests conducted in the frequency range.

acceleration and causes the coil to be supplied with a direct current via a suitable servo-controller to restore or maintain a null condition.

Assuming a downward acceleration with the field being radial (90°) and using Ampere's law, the force experienced by the coil may be written as:

$$F = ma = ilB \tag{17.9}$$

or the current

$$i = ma/lB \tag{17.10}$$

where B = the effective flux density
l = the total effective length of the conductor in the magnetic field

Current in the restoring circuit is linearly proportional to acceleration, provided: (1) armature reaction effects are negligible and fully neutralized by the compensating coil in opposition to the moving coil, and (2) the gain of the servo system is large enough to prevent displacement of the coil from the region in which the magnetic field is constant.

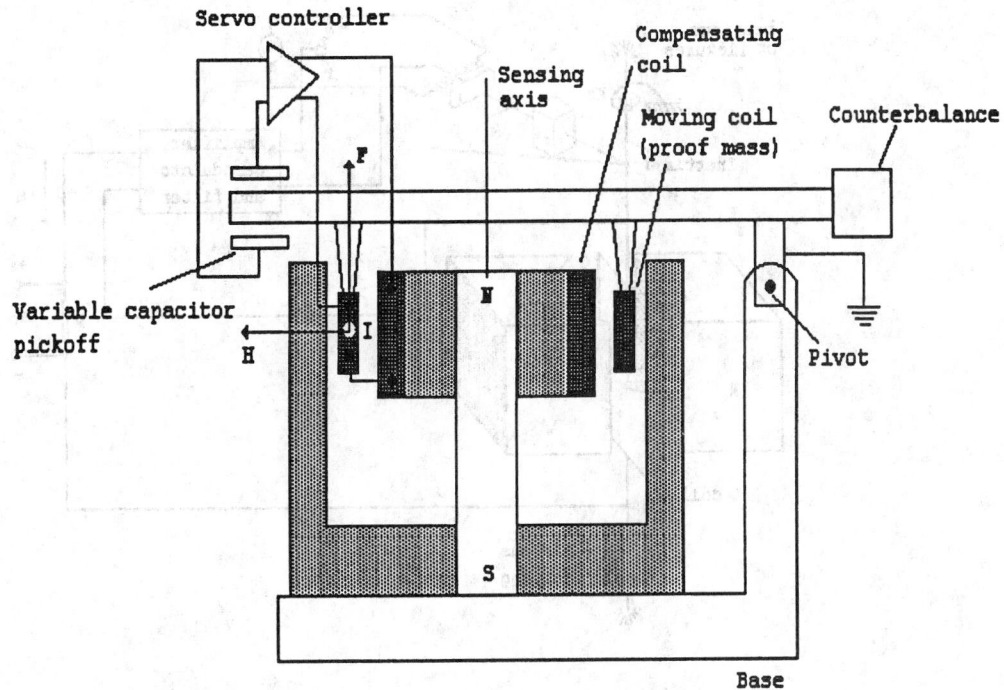

FIGURE 17.6   A basic coil and permanent magnet accelerometer. The coil is supported by an arm with minimum friction bearings to form a proof mass in a magnetic field. Displacement of the coil due to acceleration induces an electric potential in the coil to be sensed and processed. A servo system maintains the coil in a null position.

In these accelerometers, the magnetic structure must be shielded adequately to make the system insensitive to external disturbances or Earth's magnetic field. Also, in the presence of acceleration, there will be a temperature rise due to i^2R losses. The effect of these i^2R losses on the performance is determined by the thermal design and heat transfer properties of the accelerometer. In many applications, special care must be exercised in choosing the appropriate accelerometer such that the temperature rises caused by unexpected accelerations cannot affect excessively the scale factors or the bias conditions.

A simplified version of another type of servo-accelerometer is given in Figure 17.7. The acceleration a of the instrument case causes an inertial force F on the sensitive mass m, tending to make it pivot in its bearings or flexure mount. The rotation θ from neutral is sensed by an inductive pickup and amplified, demodulated, and filtered to produce a current i_a directly proportional to the motion from the null. This current is passed through a precision stable resistor R to produce the output voltage signal and is applied to a coil suspended in a magnetic field. The current through the coil produces magnetic torque on the coil, which takes action to return the mass to neutral. The current required to produce magnetic torque that just balances the inertial torque due to acceleration is directly proportional to acceleration a. Therefore, the output voltage e_0 becomes a measure of acceleration a. Since a nonzero displacement θ is necessary to produce the current i_a, the mass is not exactly returned to null, but becomes very close to zero because of the high gain amplifier. Analysis of the block diagram reveals that:

$$e_0 / R = \left(mra - e_0 K_c / R \right) \times \left(K_p K_a / K_s \right) \Big/ \left(s^2 / \omega^2_{nl} + 2\zeta_1 s / \omega_{nl} + 1 \right) \qquad (17.11)$$

Rearranging this expression gives:

$$mrRK_p K_a / K_s = \left(s^2 / \omega^2_{nl} + 2\zeta_1 s / \omega_{nl} + 1 + K_c K_p K_a / K_s \right) e_0 \qquad (17.12)$$

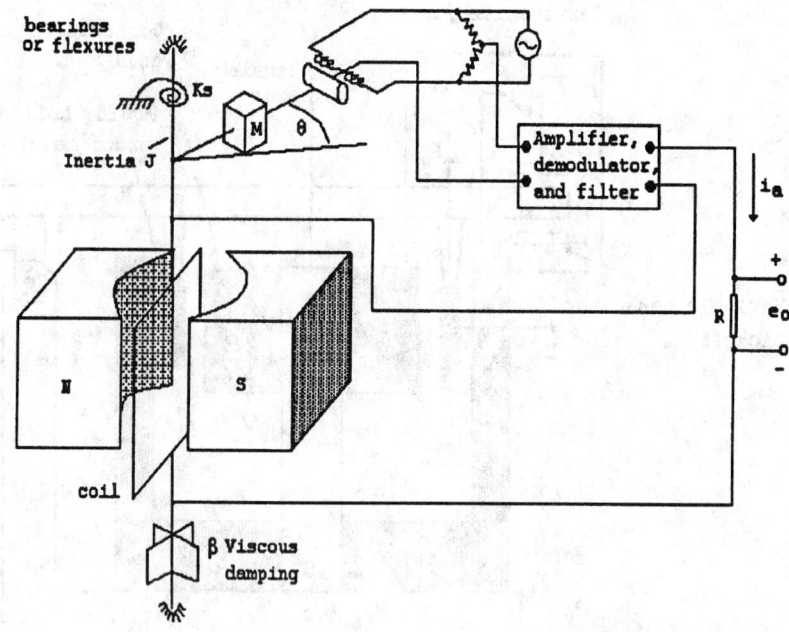

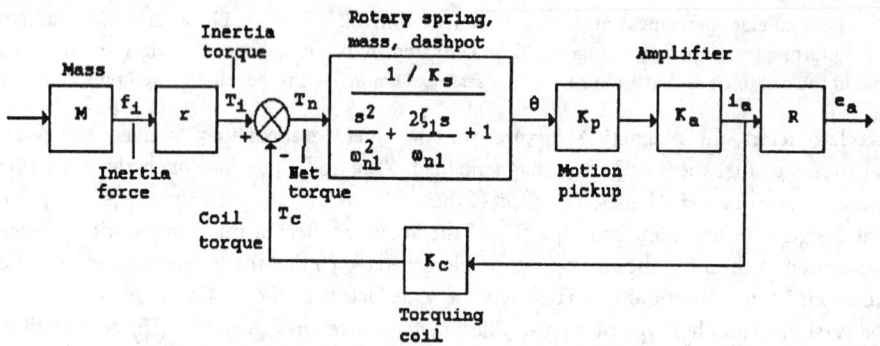

FIGURE 17.7 A simplified version of a rotational type servo-accelerometer. Acceleration of the instrument case causes an inertial force on the sensitive mass, tending to make it pivot in its bearings or flexure mount. The rotation from neutral is sensed by inductive sensing and amplified, demodulated, and filtered to produce a current directly proportional to the motion from the null. The block diagram representation is useful in analysis.

By designing the amplifier gain, K_a is made large enough so that $K_c K_p K_a a / K_s \gg 1.0$; then:

$$e_0 / a\,(s) = K \big/ \big(s^2 / \omega^2_{nl} + 2\zeta_1 s / \omega_{nl} + 1 + K_c K_p K_a a / K_s\big)\, e_0 \qquad (17.13)$$

where

$$K \cong MrR / K_c, \qquad \big(\mathrm{V\ m^{-1}\ s^{-2}}\big) \qquad (17.14)$$

$$\omega_n \cong \omega_{nl}\sqrt{K_c K_p K_a / K_s} \qquad \mathrm{rad/s} \qquad (17.15)$$

$$\zeta \cong \zeta_1 / \sqrt{K_c K_p K_a / K_s} \qquad (17.16)$$

Equation 17.14 shows that the sensitivity depends on the values of m, r, R, and K_c, all of which can be made constant. In this case, a high-gain feedback is useful in shifting the requirements for accuracy and stability from mechanical components to a selected few parameters where the requirements can easily be met. As in all feedback systems, the gain cannot be made arbitrarily high because of dynamic instability; however, a sufficiently high gain can be achieved to obtain good performance. An excellent comprehensive treatment of this topic is given by Doebelin, 1990; interested readers should refer to [3].

Induction Type Accelerometers

The cross-product relationship of current, magnetic field, and force gives the basis for induction type electromagnetic accelerometers, which are essentially generators rather than motors. One type of instrument, cup-and-magnet, includes a pendulous element with a pick-off and a servo-controller driving a tachometer coupling and a permanent magnet and a flux return ring, closely spaced with respect to an electrically conductive cylinder attached to the pendulus element. A rate proportional drag-force is obtained by electromagnetic induction effects between magnet and conductor. The pick-off senses pendulum deflection under acceleration and causes the servo-controller to turn the rotor in a sense to drag the pendulus element toward null. Under steady-state conditions, motor speed is a measure of the acceleration acting on the instrument. Stable servo operation is achieved employing a time-lead network to compensate the inertial time lag of the motor and magnet combination. The accuracy of servo type accelerometers is ultimately limited by consistency and stability of the scale factors of coupling devices and magnet-and-cup as a function of time and temperature.

Another accelerometer based on induction types is the *eddy current induction torque generation*. It was noted that the force-generating mechanism of an induction accelerometer consists of a stable magnetic field, usually supplied by a permanent magnet, which penetrates orthogonally through a uniform conduction sheet. The movement of the conducting sheet relative to the magnetic field in response to an acceleration results in a generated electromotive potential in each circuit in the conductor. This action is in accordance with the law of Faraday's principle. In induction-type accelerometers, the induced eddy currents are confined to the conductor sheet, making the system essentially a drag coupling.

A typical commercial instrument based on the servo-accelerometer principle might have a micromachined quartz flexure suspension, differential capacitance angle pick-off, air squeeze film, plus servo lead compensation for system damping. Of the various available models, a 30 g range unit has threshold and resolution of 1 µg, frequency response flat within 0.05% at 10 Hz and 2% at 100 Hz, natural frequency 500 Hz, damping ratio 0.3 to 0.8, and transverse or cross-axis sensitivity 0.1%. If, for example, the output current is about 1.3 mA g^{-1} and a 250 Ω readout resistor would give about ± 10 V full scale at 30 g. These accelerometers are good with respect to precision, and are used in many applications, such as aircraft and missile control systems, the measurement of tilt angles, axle angular bending in weight and balance systems, etc.

17.3 Piezoelectric Accelerometers

Piezoelectric accelerometers are used widely for general-purpose acceleration, shock, and vibration measurements. They basically are motion transducers with large output signals and comparatively small sizes. They are available with very high natural frequencies and are therefore suitable for high-frequency applications and shock measurements.

These devices utilize a mass in direct contact with the piezoelectric component, or crystal, as shown in Figure 17.8. When a varying motion is applied to the accelerometer, the crystal experiences a varying force excitation ($F = ma$), causing a proportional electric charge q to be developed across it.

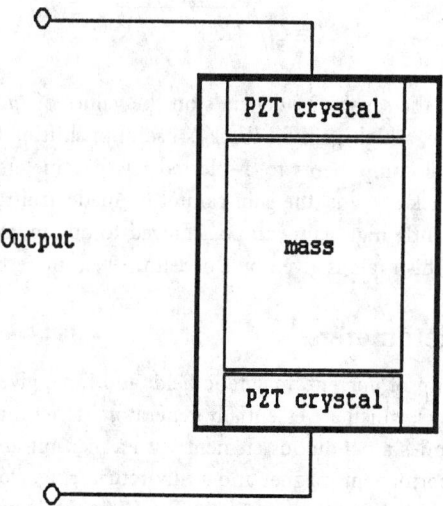

FIGURE 17.8 A compression-type piezoelectric accelerometer. The crystals are under compression at all times, either by a mass or mass and spring arrangement. Acceleration causes a deformation of the crystal, thus producing a proportional electric signal. They are small in size and widely used. They demonstrate poor performance at low frequencies.

$$q = d_{ij} F = d_{ij} ma \qquad (17.17)$$

where q = the charge developed

d_{ij} = the material's piezoelectric coefficient

As Equation 17.17 shows, the output from the piezoelectric material is dependent on its mechanical properties, d_{ij}. Two commonly used piezoelectric crystals are lead-zirconate titanate ceramic (PZT) and quartz. They are both self-generating materials and produce a large electric charge for their size. The piezoelectric strain constant of PZT is about 150 times that of quartz. As a result, PZTs are much more sensitive and smaller in size than their quartz counterparts. In the accelerometers, the mechanical spring constants for the piezoelectric components are high, and the inertial masses attached to them are small. Therefore, these accelerometers are useful for high frequency applications. Figure 17.9 illustrates a typical frequency response for a PZT device. Since piezoelectric accelerometers have comparatively low mechanical impedances, their effects on the motion of most structures is negligible. They are also manufactured to be rugged and they have outputs that are stable with time and environment.

Mathematically, their transfer function approximates to a third-order system as:

$$e_0(s)/a(s) = \left(K_q/C\omega_n^2\right)\tau s/\left[(\tau s+1)\left(s^2/\omega_n^2 + 2\zeta s/\omega_n + 1\right)\right] \qquad (17.18)$$

where K_q = the piezoelectric constant related to charge (C cm)

τ = the time constant of the crystal

It is worth noting that the crystal itself does not have a time constant τ, but the time constant is observed when the accelerometer is connected into an electric circuit (e.g., an *RC* circuit).

The low-frequency response is limited by the piezoelectric characteristic $\tau s/(\tau s + 1)$, while the high-frequency response is related to mechanical response. The damping factor ζ is very small, usually less than 0.01 or near zero. Accurate low-frequency response requires large τ, which is usually achieved by

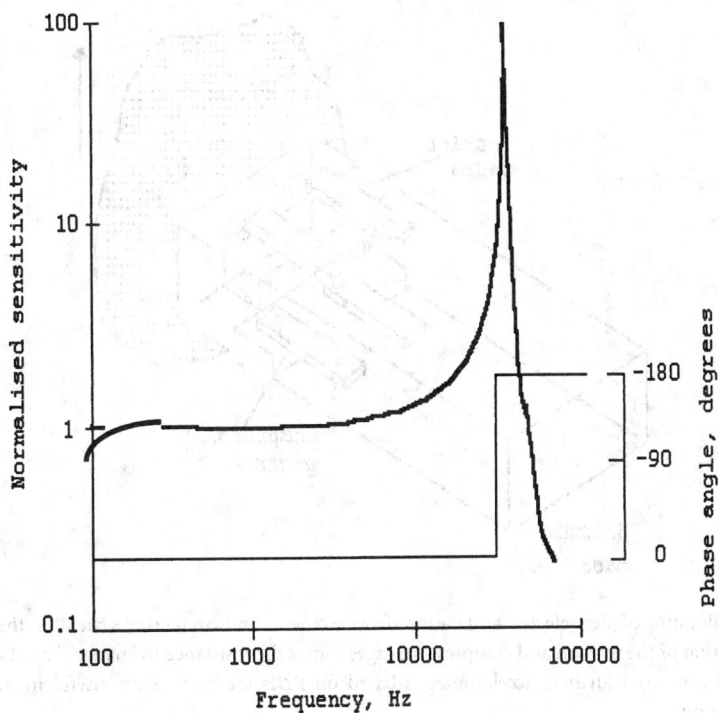

FIGURE 17.9 Frequency response of a typical piezoelectric accelerometer. Measurements are normally confined to the linear portion of the response curve. The upper frequency of the accelerometer is limited by the resonance of the PZT crystal. The phase angle is constant up to the resonance frequency.

the use of high-impedance voltage amplifiers. At very low frequencies, thermal effects can have severe influences on the operation characteristics.

In piezoelectric accelerometers, two basic design configurations are used: compression types and shear stress types. In compression-type accelerometers, the crystal is held in compression by a preload element; therefore, the vibration varies the stress in compressed mode. In the shear accelerometer, vibration simply deforms the crystal in shear mode. The compression type has a relatively good mass/sensitivity ratio and hence exhibits better performance. But, since the housing acts as an integral part of the spring mass system, it may produce spurious interfaces in the accelerometer output, if excited in its proper natural frequency.

Microelectronic circuits have allowed the design of piezoelectric accelerometers with charge amplifiers and other signal conditioning circuits built into the instrument housing. This arrangement allows greater sensitivity, high-frequency response, and smaller size accelerometers, thus lowering the initial and implementation costs.

Piezoelectric accelerometers are available in a wide range of specifications and are offered by a large number of manufacturers. For example, the specifications of a shock accelerometer may have 0.004 pC g^{-1} in sensitivity and a natural frequency of up to 250,000 Hz, while a unit designed for low-level seismic measurements might have 1000 pC g^{-1} in sensitivity and only 7000 Hz natural frequency. They are manufactured as small as 3 × 3 mm in dimensions with about 0.5 g in mass, including cables. They have excellent temperature ranges, and some of them are designed to survive intensive radiation environment of nuclear reactors. However, piezoelectric accelerometers tend to have larger cross-axis sensitivity than other types; about 2% to 4%. In some cases, large cross-axis sensitivity can be used during installation for the correct orientation of the device. These accelerometers can be mounted with threaded studs, with cement or wax adhesives, or with magnetic holders.

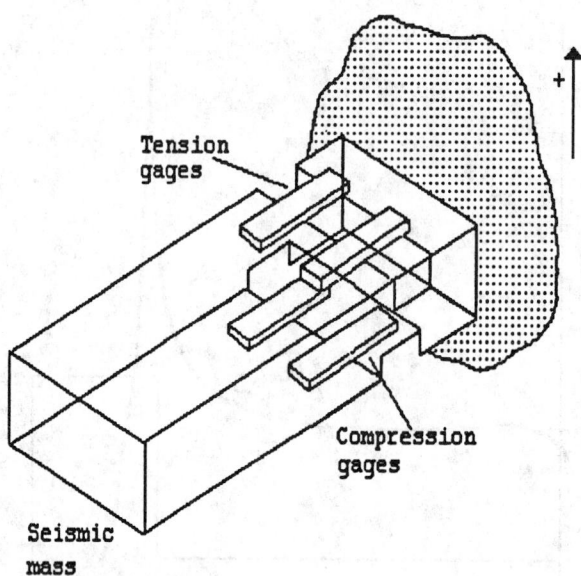

FIGURE 17.10 Bonding of piezoelectric and piezoresistive elements onto an inertial system. As the inertial member vibrates, deformation of the tension and compression gages causes the resistance to change. The change in resistance is picked up and processed further. Accelerometers based on PZTs are particularly useful in medium- to high-frequency applications.

17.4 Piezoresistive Accelerometers

Piezoresistive accelerometers are essentially semiconductor strain gages with large gage factors. High gage factors are obtained because the material resistivity is dependent primarily on the stress, not only on dimensions. The increased sensitivity is critical in vibration measurement because it allows the miniaturization of the accelerometer. Most piezoresistive accelerometers use two or four active gages arranged in a Wheatstone bridge. Extra-precision resistors are used, as part of the circuit, in series with the input to control the sensitivity, balancing, and offsetting temperature effects. The mechanical construction of a piezoresistive accelerometer is shown in Figure 17.10.

In some applications, overload stops are necessary to protect the gages from high-amplitude inputs. These instruments are useful for acquiring vibration information at low frequencies (e.g., below 1 Hz). In fact, the piezoresistive sensors are inherently true static acceleration measurement devices. Typical characteristics of piezoresistive accelerometers may be 100 mV g^{-1} in sensitivity, 0 to 750 Hz in frequency range, 2500 Hz in resonance frequency, 25 g in amplitude range, 2000 g in shock rating, 0 to 95°C in temperature range, with a total mass of about 25 g.

17.5 Differential-Capacitance Accelerometers

Differential-capacitance accelerometers are based on the principle of change of capacitance in proportion to applied acceleration. They come in different shapes and sizes. In one type, the seismic mass of the accelerometer is made as the movable element of an electrical oscillator as shown in Figure 17.11. The seismic mass is supported by a resilient parallel-motion beam arrangement from the base. The system is characterized to have a certain defined nominal frequency when undisturbed. If the instrument is accelerated, the frequency varies above and below the nominal value, depending on the direction of acceleration.

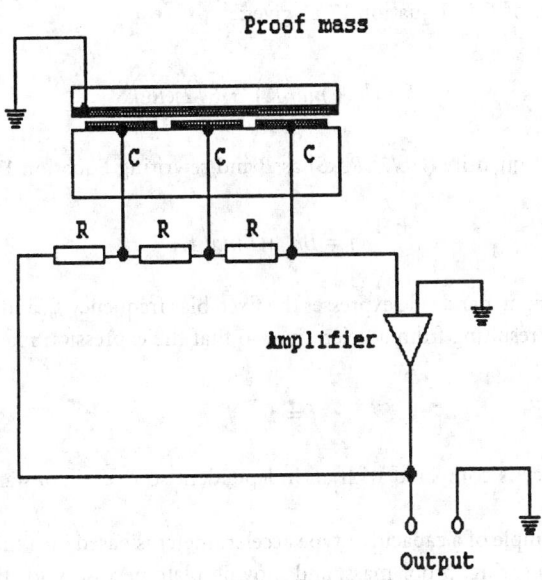

FIGURE 17.11 A typical differential capacitive accelerometer. The proof mass is constrained in its null position by a spring. Under acceleration, variable frequencies are obtained in the electric circuit. In a slightly different version, the proof mass may be constrained by an electrostatic-feedback-force, thus resulting in convenient mechanical simplicity.

The seismic mass carries an electrode located in opposition to a number of base-fixed electrodes that define variable capacitors. The base-fixed electrodes are resistance coupled in the feedback path of a wide-band, phase-inverting amplifier. The gain of the amplifier is made of such a value to ensure maintenance of oscillations over the range of variation of capacitance determined by the applied acceleration. The value of the capacitance C for each of the variable capacitor is given by:

$$C = \varepsilon k S / h \tag{17.19}$$

where k = dielectric constant
ε = capacitivity of free space
S = area of electrode
h = variable gap

Denoting the magnitude of the gap for zero acceleration as h_0, the value of h in the presence of acceleration a may be written as:

$$h = h_0 + ma/K \tag{17.20}$$

where m = the value of the proof mass and K is the spring constant. Thus,

$$C = \varepsilon k S / \left(h_0 + ma/K \right) \tag{17.21}$$

If, for example, the frequency of oscillation of the resistance-capacitance type circuit is given by the expression:

$$f = \sqrt{6} / 2\pi RC \tag{17.22}$$

Substituting this value of C in Equation 17.21 gives:

$$f = \left(h_0 + ma/K \right) \; \sqrt{6}/2\pi R\epsilon kS \qquad (17.23)$$

Denoting the constant quantity $(\sqrt{6}/2\pi R\epsilon kS)$ as B and rewriting Equation 17.23 gives:

$$f = Bh_0 + Bma/K \qquad (17.24)$$

The first term on the right-hand side expresses the fixed bias frequency f_0, and the second term denotes the change in frequency resulting from acceleration, so that the expression may be written as:

$$f = f_0 + f_a \qquad (17.25)$$

If the output frequency is compared with an independent source of constant frequency f_0, f_a can be determined.

A commonly used example of a capacitive-type accelerometer is based on a thin diaphragm with spiral flexures that provide the spring, proof mass, and moving plate necessary for the differential capacitor, as shown in Figure 17.12. Plate motion between the electrodes pumps air parallel to the plate surface

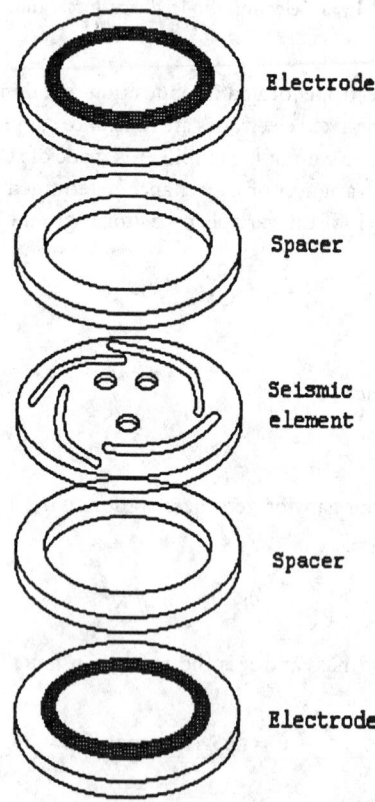

Electrode

Spacer

Seismic element

Spacer

Electrode

FIGURE 17.12 Diaphragm-type capacitive accelerometer. The seismic element is cushioned between the electrodes. Motion of the mass between the electrodes causes air movement passing through the holes, which provides a squeeze film damping. In some cases, oil can be used as the damping element.

and through holes in the plate to provide squeeze film damping. Since air viscosity is less temperature sensitive than oil, the desired damping ratio of 0.7 hardly changes more than 15%. A family of such instruments are readily available, having full-scale ranges from ±0.2 g (4 Hz flat response) to ±1000 g (3000 Hz), cross-axis sensitivity less than 1%, and full-scale output of ±1.5 V. The size of a typical device is about 25 mm³ with a mass of 50 g.

17.6 Strain-Gage Accelerometers

Strain gage accelerometers are based on resistance properties of electrical conductors. If a conductor is stretched or compressed, its resistance alters due to two reasons: dimensional changes and the changes in the fundamental property of material called *piezoresistance*. This indicates that the resistivity ρ of the conductor depends on the mechanical strain applied onto it. The dependence is expressed as the gage factor

$$\left(dR/R\right)/\left(dL/L\right)=1+2v+\left(d\rho/\rho\right)/\left(dL/L\right) \tag{17.26}$$

where
$\qquad 1 =$ resistance change due to length
$\qquad 2v =$ resistance change due to area
$\qquad (d\rho/\rho)/(dL/L) =$ resistance change due to piezoresistivity

In acceleration measurements, the resistance strain gages can be selected from different types, including unbonded metal-wire gages, bonded metal-wire gages, bonded metal-foil gages, vacuum-deposited thin-metal-film gages, bonded semiconductor gages, diffused semiconductor gages, etc. But, usually, bonded and unbonded metal-wire gages find wider applications in accelerometers. Occasionally, bonded semiconductor gages, known as piezoresistive transducers, are used but suffer from high-temperature sensitivities, nonlinearity, and some mounting difficulties. Nevertheless, in recent years, they have found new application areas with the development of micromachine transducer technology, which is discussed in detail in the micro-accelerometer section.

Unbonded-strain-gage accelerometers use the strain wires as the spring element and as the motion transducer, using similar arrangements as in Figure 17.10. They are useful for general-purpose motion and vibration measurements from low to medium frequencies. They are available in wide ranges and characteristics, typically ±5 g to ±200 g full scale, natural frequency 17 Hz to 800 Hz, 10 V excitation voltage ac or dc, full-scale output ±20 mV to ±50 mV, resolution less than 0.1%, inaccuracy less than 1% full scale, and cross-axis sensitivity less than 2%. Their damping ratio (using silicone oil damping) is 0.6 to 0.8 at room temperature. These instruments are small and lightweight, usually with a mass of less than 25 g.

Bonded-strain-gage accelerometers generally use a mass supported by a thin flexure beam. The strain gages are cemented onto the beam to achieve maximum sensitivity, temperature compensation, and sensitivity to both cross-axis and angular accelerations. Their characteristics are similar to unbonded-strain gage accelerometers, but have larger sizes and weights. Often, silicone oil is used for damping. Semiconductor strain gages are widely used as strain sensors in cantilever-beam/mass types of accelerometers. They allow high outputs (0.2 V to 0.5 V full scale). Typically, a ±25 g acceleration unit has a flat response from 0 Hz to 750 Hz, a damping ratio of 0.7, a mass of 28 g, and an operational temperature of −18°C to +93°C. A triaxial ±20,000 g model has flat response from 0 kHz to 15 kHz, a damping ratio 0.01, a compensation temperature range of 0°C to 45°C, 13 × 10 × 13 mm in size, and 10 g in mass.

17.7 Seismic Accelerometers

These accelerometers make use of a seismic mass that is suspended by a spring or a lever inside a rigid frame. The schematic diagram of a typical instrument is shown in Figure 17.1. The frame carrying the seismic mass is connected firmly to the vibrating source whose characteristics are to be measured. As the system vibrates, the mass tends to remain fixed in its position so that the motion can be registered as a relative displacement between the mass and the frame. This displacement is sensed by an appropriate transducer and the output signal is processed further. Nevertheless, the seismic mass does not remain absolutely steady; but for selected frequencies, it can satisfactorily act as a reference position.

By proper selection of mass, spring, and damper combinations, the seismic instruments may be used for either acceleration or displacement measurements. In general, a large mass and soft spring are suitable for vibration and displacement measurements, while a relatively small mass and stiff spring are used in accelerometers.

The following equation may be written by using Newton's second law of motion to describe the response of seismic arrangements similar to shown in Figure 17.1.

$$md^2x_2/dt^2 + cdx_2/dt + kx_2 = cdx_1/dt + kx_1 + mg\,\cos\!\left(\theta\right) \tag{17.27}$$

where x_1 = the displacement of the vibration frame
$\qquad x_2$ = the displacement of the seismic mass
$\qquad c$ = velocity constant
$\qquad k$ = spring constant

Taking md^2x_1/dt^2 from both sides of the equation and rearranging gives:

$$md^2z/dt^2 + c\,dz/dt + kz = mg\,\cos\!\left(\theta\right) - md^2x_1/dt^2 \tag{17.28}$$

where $z = x_2 - x_1$ is the relative motion between the mass and the base
$\qquad \theta$ = the angle between sense axis and gravity

In Equation 17.27, it is assumed that the damping force on the seismic mass is proportional to velocity only. If a harmonic vibratory motion is impressed on the instrument such that:

$$x_1 = x_0\,\sin\omega_1 t \tag{17.29}$$

where ω_1 is the frequency of vibration, in rad s^{-1}.

Writing

$$md^2x_1/dt^2 = m\,x_0\,\omega_1^2\,\sin\omega_1 t$$

modifies Equation 17.28 as:

$$md^2z/dt^2 + c\,dz/dt + kz = mg\,\cos\!\left(\theta\right) + m\,a_1\,\sin\omega_1 t \tag{17.30}$$

where $a_1 = m\,x_0\,\omega_1^2$

Equation 17.30 will have transient and steady-state solutions. The steady-state solution of the differential Equation 17.30 can be determined as:

$$z = \left[mg\,\cos\!\left(\theta\right)/k\right] + \left[m\,a_1\,\sin\omega_1\,t/\!\left(k - m\,\omega_1^2 + j c\omega_1\right)\right] \tag{17.31}$$

Rearranging Equation 17.31 results in:

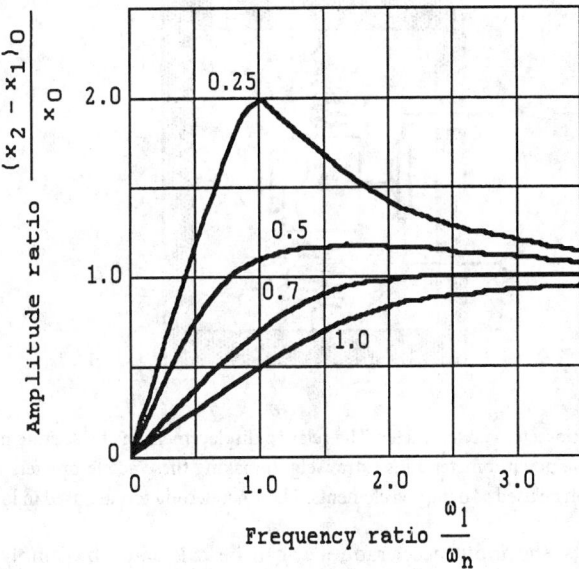

FIGURE 17.13 A typical displacement of a seismic instrument. Amplitude becomes large at low damping ratios. The instrument constants should be selected such that, in measurements, the frequency of vibration is much higher than the natural frequency (e.g., greater than 2). Optimum results are obtained when the value of instrument constant c/c_c is about 0.7.

$$z = \left[mg\cos(\theta)/\omega_m \right] + \left\{ a_1\sin(\omega_1 - \phi) \Big/ \left[\omega_m^2\left(1 - r^2\right)^2 + \left(2\zeta r\right)^2 \right]^{1/2} \right\} \qquad (17.32)$$

where $\omega_n = \sqrt{k/m}$ = the natural frequency of the seismic mass

ζ = $c/2\sqrt{km}$ = the damping ratio, also can be written in terms of critical damping ratio as $\zeta = c/c_c$, where $(c_c = 2\sqrt{km})$

ϕ = $\tan^{-1}(c\omega_1/(k - m\omega_1^2))$ = the phase angle

r = ω_1/ω_m = the frequency ratio

A plot of Equation 17.32, $(x_1 - x_2)_0/x_0$ against frequency ratio ω_1/ω_n, is illustrated in Figure 17.13. This figure shows that the output amplitude is equal to the input amplitude when $c/c_c = 0.7$ and $\omega_1/\omega_n > 2$. The output becomes essentially a linear function of the input at high frequency ratios. For satisfactory system performance, the instrument constant c/c_c and ω_n should carefully be calculated or obtained from calibrations. In this way, the anticipated accuracy of measurement can be predicted for frequencies of interest. A comprehensive treatment of the analysis is given by McConnell [1].

If the seismic instrument has a low natural frequency and a displacement sensor is used to measure the relative motion z, then the output is proportional to the displacement of the transducer case. If the velocity sensor is used to measure the relative motion, the signal is proportional to the velocity of the transducer. This is valid for frequencies significantly above the natural frequency of the transducer. However, if the instrument has a high natural frequency and the displacement sensor is used, the measured output is proportional to the acceleration:

$$kz = m\,d^2x_1/dt^2 \qquad (17.33)$$

This equation is true since displacement x_2 becomes negligible in comparison to x_1.

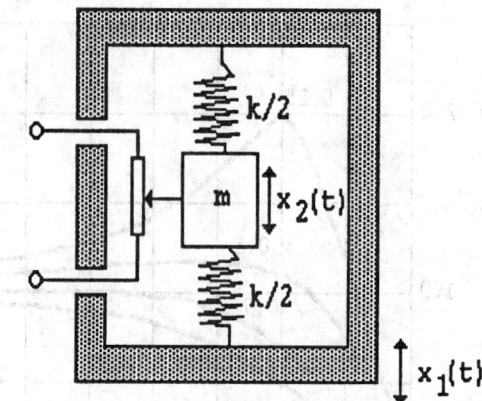

FIGURE 17.14 A potentiometer accelerometer. The relative displacement of the seismic mass is sensed by a poten-tiometer arrangement. The potentiometer adds extra weight, making these accelerometers relatively heavier. Suitable liquids filling the frame can be used as damping elements. These accelerometers are used in low-frequency applications.

In these instruments, the input acceleration a_0 can be calculated by simply measuring $(x_1 - x_2)_0$. Generally, in acceleration measurements, unsatisfactory performance is observed at frequency ratios above 0.4. Thus, in such applications, the frequency of acceleration must be kept well below the natural frequency of the instrument. This can be accomplished by constructing the instrument to have a low natural frequency by selecting soft springs and large masses.

Seismic instruments are constructed in a variety of ways. Figure 17.14 illustrates the use of a voltage divider potentiometer for sensing the relative displacement between the frame and the seismic mass. In the majority of potentiometric instruments, the device is filled with a viscous liquid that interacts continuously with the frame and the seismic mass to provide damping. These accelerometers have low frequency of operation (less than 100 Hz) and are mainly intended for slow varying acceleration and low-frequency vibrations. A typical family of such instrument offers many different models, covering the range of ±1 g to ±50 g full scale. The natural frequency ranges from 12 Hz to 89 Hz, and the damping ratio ζ can be kept between 0.5 to 0.8 using a temperature-compensated liquid damping arrangement. Potentiometer resistance can be selected in the range of 1000 Ω to 10,000 Ω, with corresponding resolu-tion of 0.45% to 0.25% of full scale. The cross-axis sensitivity is less than ±1%. The overall accuracy is ±1% of full scale or less at room temperatures. The size is about 50 mm³; with a mass of about 1/2 kg.

Linear variable differential transformers (LVDT) offer another convenient means to measure the relative displacement between the seismic mass and the accelerometer housing. These devices have higher natural frequencies than potentiometer devices, up to 300 Hz. Since the LVDT has lower resistance to the motion, it offers much better resolution. A typical family of liquid-damped differential-transformer accelerometers exhibits the following characteristics: full scale range from ±2 g to ±700 g, natural fre-quency from 35 Hz to 620 Hz, nonlinearity 1% of full scale, the full scale output is about 1 V with an LVDT excitation of 10 V at 2000 Hz, damping ratio 0.6 to 0.7, residual voltage at null is less than 1%, and hysteresis less than 1% full scale; the size is 50 mm³, with a mass of about 120 g.

Electric resistance strain gages are also used for displacement sensing of the seismic mass as shown in Figure 17.15. In this case, the seismic mass is mounted on a cantilever beam rather than on springs. Resistance strain gages are bonded on each side of the beam to sense the strain in the beam resulting from the vibrational displacement of the mass. Damping for the system is provided by a viscous liquid that entirely fills the housing. The output of the strain gages is connected to an appropriate bridge circuit. The natural frequency of such a system is about 300 Hz. The low natural frequency is due to the need for a sufficiently large cantilever beam to accommodate the mounting of the strain gages. Other types of seismic instruments using piezoelectric transducers and seismic masses are discussed in detail in the section dedicated to piezoelectric-type accelerometers.

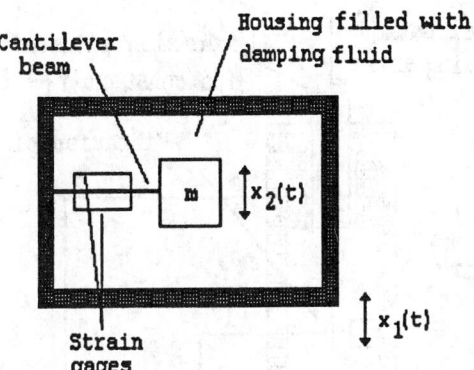

FIGURE 17.15 A strain gage seismic instrument. The displacement of the proof mass is sensed by piezoresistive strain gages. The natural frequency of the system is low, due to the need of a long lever beam to accommodate strain gages. The signal is processed by bridge circuits.

Seismic vibration instruments are affected seriously by the temperature changes. Devices employing variable resistance displacement sensors will require correction factors to account for resistance change due to temperature. The damping of the instrument may also be affected by changes in the viscosity of the fluid due to temperature. For example, the viscosity of silicone oil, often used in these instruments, is strongly dependent on temperature. One way of eliminating the temperature effect is by using an electrical resistance heater in the fluid to maintain the temperature at a constant value regardless of surrounding temperatures.

17.8 Inertial Types, Cantilever, and Suspended-Mass Configuration

There are a number of different inertial accelerometers, including gyropendulum, reaction-rotor, vibrating string, and centrifugal-force-balance types. In many of them, the force required to constrain the mass in the presence of the acceleration is supplied by an inertial system.

A vibrating string type instrument, Figure 17.16, makes use of proof mass supported longitudinally by a pair of tensioned, transversely vibrating strings with uniform cross-section, and equal lengths and masses. The frequency of vibration of the strings is set to several thousand cycles per second. The proof mass is supported radially in such a way that the acceleration normal to strings does not affect the string tension. In the presence of acceleration along the sensing axis, a differential tension exists on the two strings, thus altering the frequency of vibration. From the second law of motion, the frequencies can be written as:

$$f_1{}^2 = T_1 / (4m_s l) \quad \text{and} \quad f_2{}^2 = T_2 / (4m_s l) \tag{17.34}$$

where T is the tension, and m_s and l are the masses and lengths of strings, respectively.

Quantity $(T_1 - T_2)$ is proportional to ma, where a is the acceleration along the axis of the strings. An expression for the difference of the frequency-squared terms may be written as:

$$f_1{}^2 - f_2{}^2 = (T_1 - T_2) / (4m_s l) = ma / (4m_s l) \tag{17.35}$$

Hence,

$$f_1 - f_2 = ma / \left[(f_1 + f_2) 4m_s l \right] \tag{17.36}$$

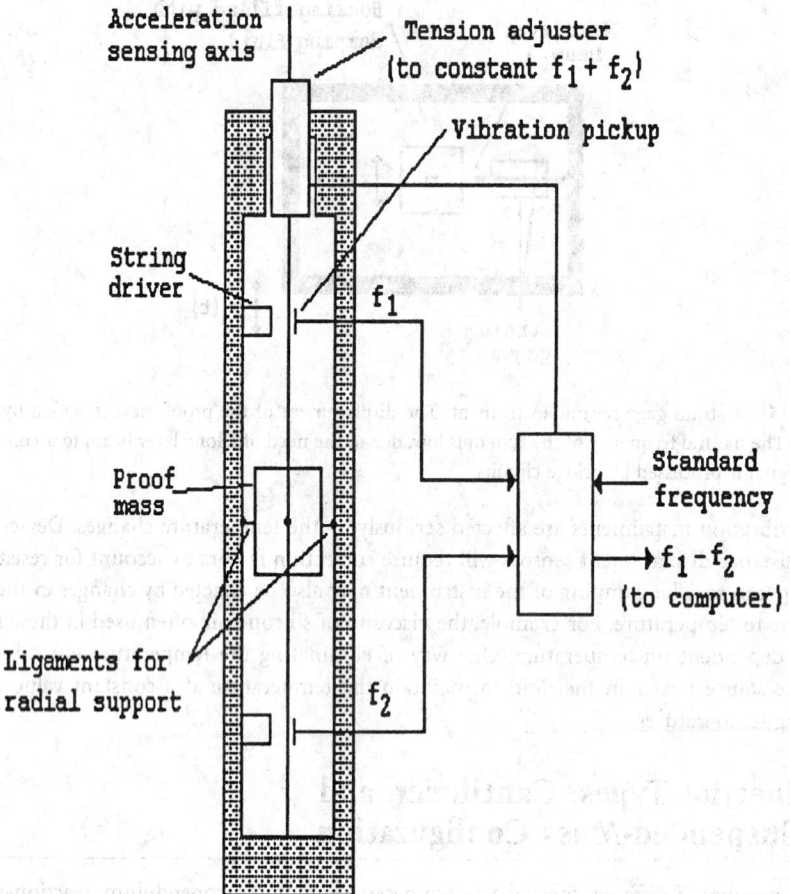

FIGURE 17.16 A vibrating string accelerometer. A proof mass is attached to two strings of equal mass and length and supported radially by suitable bearings. The vibration frequencies of strings are dependent on the tension imposed by the acceleration of the system in the direction of the sensing axis.

The sum of frequencies ($f_1 + f_2$) can be held constant by servoing the tension in the strings with reference to the frequency of a standard oscillator. Then, the difference between the frequencies becomes linearly proportional to acceleration. In some versions, the beam-like property of the vibratory elements is used by gripping them at nodal points corresponding to the fundamental mode of vibration of the beam. Improved versions of these devices lead to cantilever-type accelerometers, as discussed next.

In cantilever-type accelerometers, a small cantilever beam mounted on the block is placed against the vibrating surface, and an appropriate mechanism is provided for varying the beam length. The beam length is adjusted such that its natural frequency is equal to the frequency of the vibrating surface — hence the resonance condition obtained. Slight variations of cantilever beam-type arrangements are finding new applications in microaccelerometers.

In a different type suspended mass configuration, a pendulum is used that is pivoted to a shaft rotating about a vertical axis. Pick-offs are provided for the pendulum and the shaft speed. The system is servo-controlled to maintain it at null position. Gravitational acceleration is balanced by the centrifugal acceleration. Shaft speed is proportional to the square root of the local value of the acceleration.

All inertial force accelerometers described above have the property of absolute instruments. That is, their scale factors can be predetermined solely by establishing mass, length, and time quantities, as distinguished from voltage, spring stiffness, etc.

17.9 Electrostatic Force Feedback Accelerometers

Electrostatic accelerometers are based on Coulomb's law between two charged electrodes. They measure the voltage in terms of force required to sustain a movable electrode of known area, mass, and separation from an affixed electrode. The field between the electrodes is given by:

$$E = Q/\varepsilon kS \qquad (17.37)$$

where E is the intensity or potential gradient (dV/dx), Q is charge, S area of the conductor, and k is the dielectric constant of the space outside the conductor.

Using this expression, it can be shown that the force per unit area of the charged conductor (in N m^{-2}) is given by:

$$F/S = Q^2 / \left(2\varepsilon kS^2\right) = \varepsilon kE^2/2 \qquad (17.38)$$

In an electrostatic-force-feedback-type accelerometer (similar in structure to that in Figure 17.10), an electrode of mass m and area S is mounted on a light pivoted arm for moving relative to the fixed electrodes. The nominal gap, h, between the pivoted and fixed electrodes is maintained by means of a force balancing servo system capable of varying the electrode potential in response to signals from a pick-off that senses relative changes in the gaps.

Considering one movable electrode and one stationary electrode, and assuming that the movable electrode is maintained at a bias potential V_1 and the stationary one at a potential V_2, the electrical intensity E in the gap can be expressed as:

$$E_1 = \left(V_1 - V_2\right)/h \qquad (17.39)$$

so that the force of attraction may be found as:

$$F_1 = \varepsilon kE^2 S / \left(2h^2\right) = \varepsilon k \left(V_1 - V_2\right)^2 S / \left(2h^2\right) \qquad (17.40)$$

In the presence of acceleration, if V_2 is adjusted to restrain the movable electrode to null position; the expression relating acceleration and electrical potential may be given by:

$$a = F_1/m = \varepsilon k \left(V_1 - V_2\right)^2 S / \left(2h^2 m\right) \qquad (17.41)$$

The device thus far described can measure acceleration in one direction only, and the output is of quadratic character; that is:

$$\left(V_1 - V_2\right) = D\sqrt{a} \qquad (17.42)$$

where D = the constant of proportionality.

The output can be linearized in a number of ways; for example, by quarter-square method. If the servo controller applies a potential $-V_2$ to other fixed electrode, the force of attraction between this electrode and the movable electrode becomes:

$$a = F_1/m = \varepsilon k \left(V_1 + V_2\right)^2 S / \left(2h^2 m\right) \qquad (17.43)$$

and the force balance equation of the movable electrode when the instrument experiences a downward acceleration a is:

$$ma = F_1 - F_2 = \left[\left(V_1 + V_2 \right)^2 - \left(V_1 - V_2 \right)^2 \right] \varepsilon k S \big/ \left(2h^2 m \right)$$

or

$$= \varepsilon k S \left(4 V_1 V_2 \right) \big/ \left(2h^2 m \right) \qquad (17.44)$$

Hence, if the bias potential V_1 is held constant and the gain of the control loop is high so that variations in the gap are negligible, the acceleration becomes a linear function of the controller output voltage V_2 as:

$$a = V_2 \left[\varepsilon k S \left(2 V_1 \right) \big/ \left(h^2 m \right) \right] \qquad (17.45)$$

The principal difficulty in mechanizing the electrostatic force accelerometer is the relatively high electric field intensity required to obtain adequate force. Also, extremely good bearings are necessary. Damping can be provided electrically, or by viscosity of gaseous atmosphere in the interelectrode space if the gap h is sufficiently small.

The main advantages of the electrostatic accelerometers include extreme mechanical simplicity, low power requirements, absence of inherent sources of hysteresis errors, zero temperature coefficients, and ease of shielding from stray fields.

17.10 Microaccelerometers

By the end of the 1970s, it became apparent that the essentially planar processing IC (integrated circuit) technology could be modified to fabricate three-dimensional electromechanical structures, called micromachining. Accelerometers and pressure sensors were among the first IC sensors. The first accelerometer was developed in 1979. Since then, technology has been progressing steadily to the point where an extremely diverse range of accelerometers is readily available. Most sensors use bulk micromachining rather than surface micromachining techniques. In bulk micromachining, the flexures, resonant beams, and all other critical components of the accelerometer are made from bulk silicon in order to exploit the full mechanical properties of single-crystal silicon. With proper design and film process, bulk micromachining yields an extremely stable and robust accelerometer.

The selective etching of multiple layers of deposited thin films, or surface micromachining, allows movable microstructures to be fabricated on silicon wafers. With surface micromachining, layers of structure material are disposed and patterned, as shown in Figure 17.17. These structures are formed by polysilicon and a sacrificial material such as silicon dioxide. The sacrificial material acts as an intermediate spacer layer and is etched away to produce a free-standing structure. Surface machining technology also allows smaller and more complex structures to be built in multiple layers on a single substrate.

The operational principles of microaccelerometers are very similar to capacitive force-balance-type accelerometers or vibrating beam types, as discussed earlier. Nevertheless, manufacturing techniques may change from one manufacturer to another. In general, vibrating beam accelerometers are preferred because of better air gap properties and improved bias performance characteristics.

The vibrating beam accelerometers, also called resonant beam force transducers, are made in such a way that an acceleration along a positive input axis places the vibrating beam in tension. Thus, the resonant frequency of the vibrating beam increases or decreases with the applied acceleration. A mechanically coupled beam structure, also known as a double-ended tuning fork (DETF), is shown in Figure 17.18.

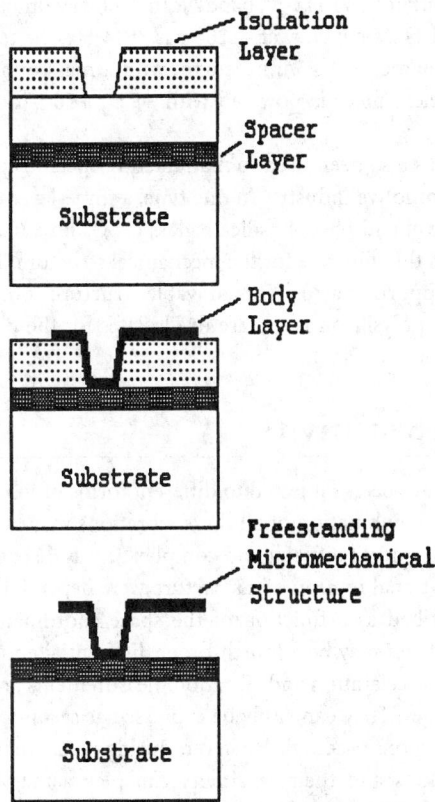

FIGURE 17.17 Steps of surface micromachining. The acceleration-sensitive, three-dimensional structure is formed on a substrate and a sacrificial element. The sacrificial element is etched to leave a free-standing structure. The spacing between the structure and substrate is about 2 μm.

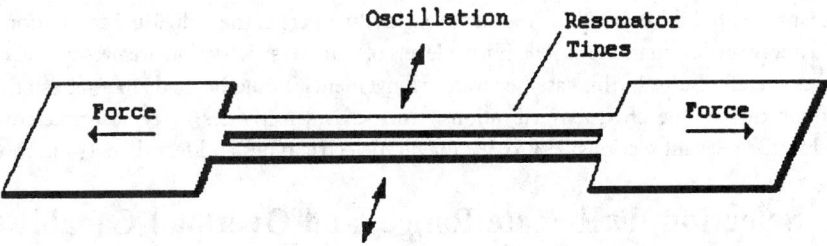

FIGURE 17.18 A double-ended tuning fork (DETF) acceleration transducer. Two beams are vibrated 180° out of phase to eliminate reaction forces at the beam ends. The resonant frequency of the beam is altered by acceleration. The signal processing circuits are also integrated in the same chip.

In DETF, an electronic oscillator capacitively couples energy into two vibrating beams to keep them oscillating at their resonant frequency. The beams vibrate 180° out of phase to cancel reaction forces at the ends. The dynamic cancellation effect of the DETF design prevents energy from being lost through the ends of the beam. Hence, the dynamically balanced DETF resonator has a high Q factor, which leads to a stable oscillator circuit. The acceleration signal is an output from the oscillator as a frequency-modulated square wave that can be used for digital interface.

The frequency of resonance of the system must be much higher than any input acceleration and this limits the measurable range. In a typical military micromachine accelerometer, the following characteristics are given: range ± 1200 g, sensitivity 1.11 Hz g^{-1}, bandwidth 2500 Hz, unloaded DETF frequency 9952 Hz, frequency at $+1200$ g is 11221 Hz, frequency at -1200 g is 8544 Hz, the temperature sensitivity 5 mg °C. Accelerometer size is 6 mm diameter × 4.3 mm length, with a mass of about 9 g, and it has a turn on time less then 60 s. The accelerometer is powered with +9 to +16 V dc and the nominal output is a 9000 Hz square wave.

Surface micromachining has also been used to manufacture specific application accelerometers, such as air-bag applications in automotive industry. In one type, a three-layer differential capacitor is created by alternate layers of polysilicon and phosphosilicate glass (PSG) on a 0.38 mm thick and 100 mm long wafer. A silicon wafer serves as the substrate for the mechanical structure. The trampoline-shaped middle layer is suspended by four supporting arms. This movable structure is the seismic mass for the accelerometer. The upper and lower polysilicon layers are fixed plates for the differential capacitors. The glass is sacrificially etched by hydrofluoric acid.

17.11   Cross-Axis Sensitivity

A vibrational structure may have been subjected to different forms of vibrations, such as compressional, torsional, transverse, etc.; or a combination of all these vibrations may take place simultaneously, which makes the analysis and measurements difficult and complex. It was discussed earlier that the differential equations governing the vibrational motion of a structure were dependent on the number of degrees of freedom, which can be described as a function of the space coordinates $f(x,y,z,t)$. For example, the transverse vibrations of structures may be a fourth-order differential equation.

Fortunately, most common acceleration and vibration measurements are simple in nature, being either compressional or torsional types. They can easily be expressed as second-order differential equations, as explained in the frequency response section. However, during measurements, most accelerometers are affected by transverse vibrations and their sensitivity can play a major role in the accuracy of the measurements.

The transverse, also known as cross-axis sensitivity, of an accelerometer is its response to acceleration in a plane perpendicular to the main accelerometer axis, as shown in Figure 17.19. The cross-axis sensitivity is normally expressed in percent of the main axis sensitivity and should be as low as possible. There is not a single value of cross-axis sensitivity, but it varies depending on the direction. The direction of minimum sensitivity is usually supplied by the manufacturer.

The measurement of the maximum cross-axis sensitivity is part of the individual calibration procedure and should always be less than 3% to 4%. If high levels of transverse vibration are present, this may result in erroneous overall results. In this case, separate arrangements should be made to establish the level and frequency contents of the cross-axis vibrations. Cross-axis sensitivities of typical accelerometers are mentioned in the relevant sections: 2% to 3% for piezoelectric types and less than 1% in most others.

17.12   Selection, Full-Scale Range, and Overload Capability

Ultimate care must be exercised for the selection of the correct accelerometer to meet the requirements of a particular application. At first glance, there may seem to be a confusingly large repertoire of accelerometers available; however, they can be classified into two main groups. The first group are the general-purpose accelerometers offered in various sensitivities, frequencies, full scale, and overload ranges, with different mechanical and electrical connection options. The second group of accelerometers are the special types that have characteristics targeted toward a particular application.

In deciding the application type (e.g., general purpose or special) and the accelerometer to be employed, the following characteristics need to be considered: transient response or cross-axis sensitivity; frequency range; sensitivity, mass and dynamic range; cross-axis response; and environmental conditions such as temperature, cable noise, etc. Some useful hints about these characteristics are given below.

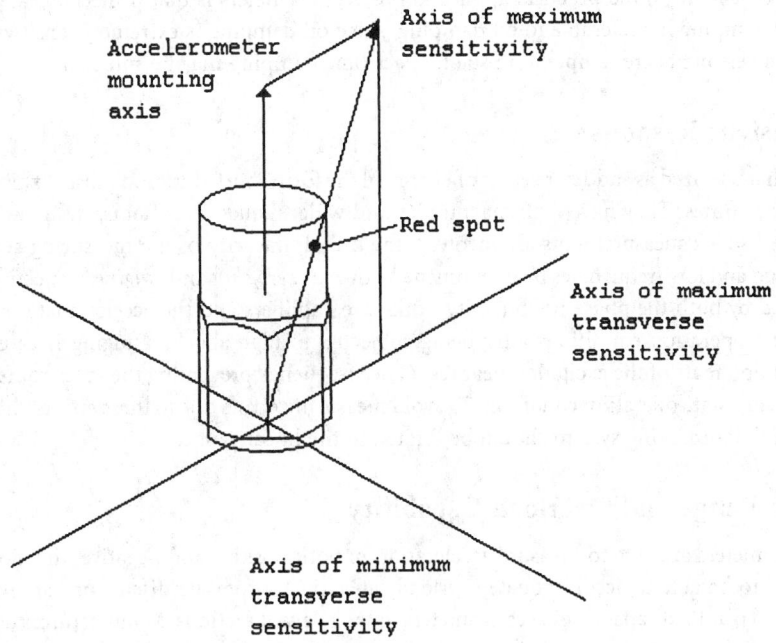

FIGURE 17.19 Vectorial illustration of cross-axis sensitivity. Accelerometers may sense vibrations not only in the direction of main axis, but also perpendicular to the main axis. These cross-axis responses are minimized in many accelerometers to a value less than 1%. Sometimes, this sensitivity is used to determine the correct orientation of the device.

The Frequency Range

Acceleration measurements are normally confined to using the linear portion of the response curve. The response is limited at the low frequencies as well as at the high frequencies by the natural resonances. As a rule of thumb, the upper frequency limit for the measurement can be set to one third of the accelerometer's resonance frequency such that the vibrations measured will be less than 1 dB in linearity. It should be noted that an accelerometer's useful frequency range is significantly higher, that is, to 1/2 or 2/3 of its resonant frequency. The measurement frequencies may be set to higher values in applications where lower linearity (e.g., 3 dB) may be acceptable, as in the case of monitoring internal conditions of machines since the reputability is more important than the linearity. The lower measuring frequency limit is determined by two factors. The first is the low-frequency cut-off of the associated preamplifiers. The second is the effect of ambient temperature fluctuations to which the accelerometer could be sensitive.

The Sensitivity, Mass, and Dynamic Range

Ideally, the higher the transducer sensitivity, the better; but compromises might have to be made for sensitivity versus frequency, range, overload capacity, and size.

Accelerometer mass becomes important when using it on small and light test objects. The accelerometer should not load the structural member, since additional mass can significantly change the levels and frequency presence at measuring points and invalidate the results. As a general rule, the accelerometer mass should not be greater than one tenth the effective mass of the part or the structure that is mounted onto for measurements.

The dynamic range of the accelerometer should match the high or low acceleration levels of the measured objects. General-purpose accelerometers can be linear up to 5000 g to 10,000 g, which is well into the range of most mechanical shocks. Special accelerometers can measure up to 100,000 g.

An important point in the practical application of accelerometers is that if mechanical damping is a problem, air damping is preferable to oil damping, since oil damping is extremely sensitive to viscosity changes. If the elements are temperature stable, electronic damping may be sufficient.

The Transient Response

Shocks are characterized as sudden releases of energy in the form short-duration pulses exhibiting various shapes and rise times. They have high magnitudes and wide frequency contents. In applications where transients and shock measurements are involved, the overall linearity of the measuring system may be limited to high and low frequencies by phenomena known as *zero shift* and *ringing*, respectively. The zero shift is caused by both the phase nonlinearity in the preamplifiers and the accelerometer not returning to steady-state operation conditions after being subjected to high shocks. Ringing is caused by high-frequency components of the excitation near resonance frequency preventing the accelerometer to return back to its steady-state operation condition. To avoid measuring errors due to these effects, the operational frequency of the measuring system should be limited to the linear range.

Full-Scale Range and Overload Capability

Most accelerometers are able to measure acceleration in both positive and negative directions. They are also designed to be able to accommodate overload capacity. Appropriate discussions are made on full-scale range and overload capacity of accelerometers in the relevant sections. Manufacturers usually supply information on these two characteristics.

Environmental Conditions

In the selection and implementation of accelerometers, environmental conditions such as temperature ranges, temperature transients, cable noise, magnetic field effects, humidity, acoustic noise, etc. need to be considered. Manufacturers also supply information on environmental conditions.

17.13 Signal Conditioning

Common signal conditioners are appropriate for interfacing accelerometers to computers or other instruments for further signal processing. Caution needs to be exercised to provide appropriate electric load to self-generating accelerometers. Generally, the generated raw signals are amplified and filtered suitably by the circuits within the accelerometer casing supplied by manufacturers. Nevertheless, piezoelectric and piezoresistive transducers require special signal conditioners with certain characteristics, as discussed next. Examples of signal conditioning circuits are also given for microaccelerometers.

Signal Conditioning Piezoelectric Accelerometers

The piezoelectric accelerometer supplies a very small energy to the signal conditioner. It has a high capacitive source impedance. The equivalent circuit of a piezoelectric accelerometer can be regarded as an active capacitor that charges itself when loaded mechanically. The configuration of external signal conditioning elements is dependent on the equivalent circuit selected. The charge amplifier design of the conditioning circuit is the most common approach since the system gain and low-frequency responses are well defined. The performance of the circuit is independent of cable length and capacitance of the accelerometer.

The charge amplifier consists of a charge converter output voltage that occurs as a result of the charge input signal returning through the feedback capacitor to maintain the input voltage at the input level close to zero, as shown in Figure 17.20. An important point about charge amplifiers is that their sensitivities can be standardized. They basically convert the input charge to voltage first and then amplify this voltage. With the help of basic operational-type feedback, the amplifier input is maintained at essentially

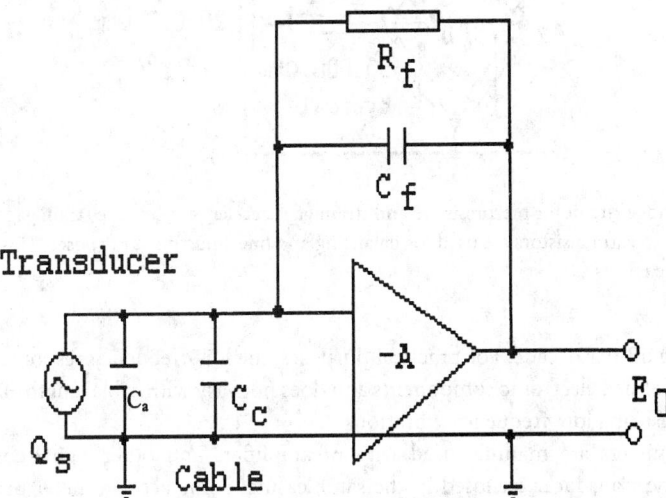

FIGURE 17.20 A typical charge amplifier. The transducer charge, which is proportional to acceleration, is first converted to voltage form to be amplified. The output voltage is a function of the input charge. The response of the amplifier can be approximated by a first-order system. In PZT transducers, the preamplifier is integrated within the same casing.

zero volts; therefore, it looks like a short circuit to the input. The charge converter output voltage that occurs as a result of a charge input signal is returned through the feedback capacitor to maintain the voltage at the input level near zero. Thus, the charge input is stored in the feedback capacitor, producing a voltage across it, that is equal to the value of the charge input divided by the capacitance of the feedback capacitor. The complete transfer function of the circuit describing the relationship between the output voltage and the input acceleration magnitude can be determined by:

$$E_o/a_0 = S_a j R_f C_f \omega \Big/ \Big\{ 1 + j R_f C_f \Big[1 + \big(C_a + C_c \big) \big/ \big(1 + G \big) \times C_f \Big] \omega \Big\} \tag{17.46}$$

where E_o = charge converter output (V)
a_0 = magnitude of acceleration (m s^{-2})
S_a = accelerometer sensitivity (mV g^{-1})
C_a = accelerometer capacitance (F)
C_c = cable capacitance (F)
C_f = feedback capacitance (F)
R_f = feedback loop resistance
G = amplifier open loop gain

In most applications, since C_f is selected to be large compared to $(C_a + C_c)/(1 + G)$, the system gain becomes independent of the cable length. In this case, the denominator of the equation can be simplified to give a first-order system with roll-off at:

$$f_{-3 \text{ dB}} = 1 \Big/ \big(2 \pi R_f C_f \big) \tag{17.47}$$

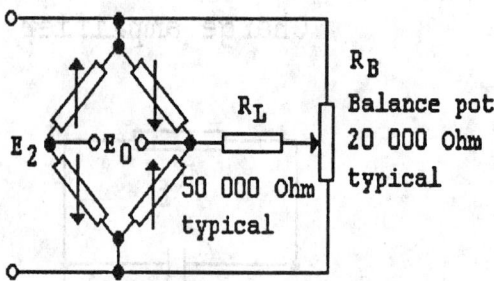

FIGURE 17.21   Bridge circuit for piezoresistive and strain gage accelerometers. The strain gages form the four arms of the bridge. The two extra resistors are used for balancing and fine adjustment purposes. This type of arrangement reduces temperature effects.

with a slope of 10 dB per decade. For practical purposes, the low-frequency response of this system is a function of well-defined electronic components and does not vary with cable length. This is an important feature when measuring low-frequency vibrations.

Many accelerometers are manufactured with preamplifiers and other signal-conditioning circuits integrated with the transducer enclosed in the same casing. Some accelerometer preamplifiers include integrators to convert the acceleration proportional outputs to either velocity or displacement proportional signals. To attenuate noise and vibration signals that lie outside the frequency range of interest, most preamplifiers are equipped with a range of high-pass and low-pass filters. This avoids interference from electric noise or signals inside the linear portion of the accelerometer frequency range. Nevertheless, it is worth mentioning that these devices usually have two time constants, external and internal. The mixture of these two time constants can lead to problems, particularly at low frequencies. The internal time constant is usually fixed by the manufacturer in design and construction. Special care must be observed to take care of the effect of external time constants in many applications by mainly observing impedance matching.

Signal Conditioning of Piezoresistive Transducers

Piezoresistive transducers generally have high amplitude outputs, low output impedances, and low intrinsic noise. Most of these transducers are designed for constant voltage excitations. They are usually calibrated for constant current excitations to make them independent of external influences. Many piezoresistive transducers are configured as full-bridge devices. Some have four active piezoresistive arms and, together with two fixed precision resistors permit shunt calibration in the signal conditioner, as shown in Figure 17.21.

Microaccelerometers

In microaccelerometers, signal conditioning circuitry is integrated within the same chip with the sensor as shown in Figure 17.22. A typical example of the signal conditioning circuitry is given in Figure 17.23 in block diagram form. In this type of accelerometer, the electronic system is essentially a crystal-controlled oscillator circuit and the output signal of the oscillator is a frequency modulated acceleration signal. Some circuits provide a buffered square-wave output that can directly be interfaced digitally. In this case, the need for analog to digital conversion is eliminated, thus removing one of the major sources of error. In other types of accelerometers, signal conditioning circuits such as analog to digital converters (ADC) are retained within the chip.

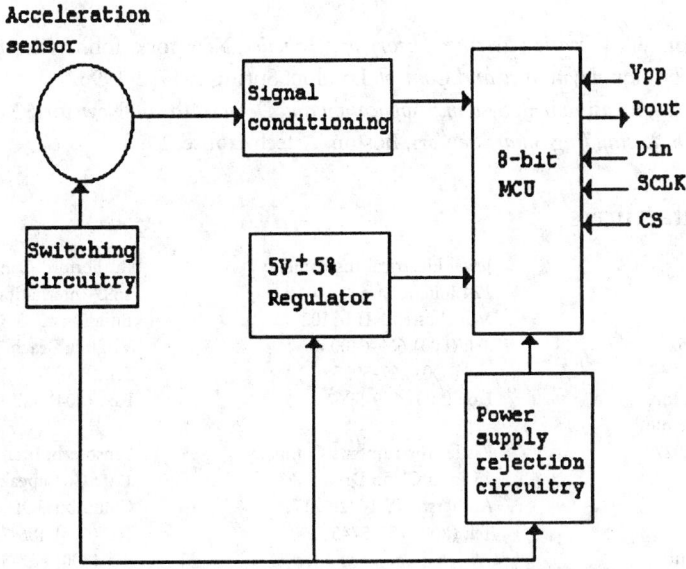

FIGURE 17.22   A block diagram of an accelerometer combined with MCU. The signal conditioning, switching, and power supply circuits are integrated to form a microaccelerometer. The device can directly be interfaced with a digital signal processor or a computer. In some cases, ADCs and memory are also integrated.

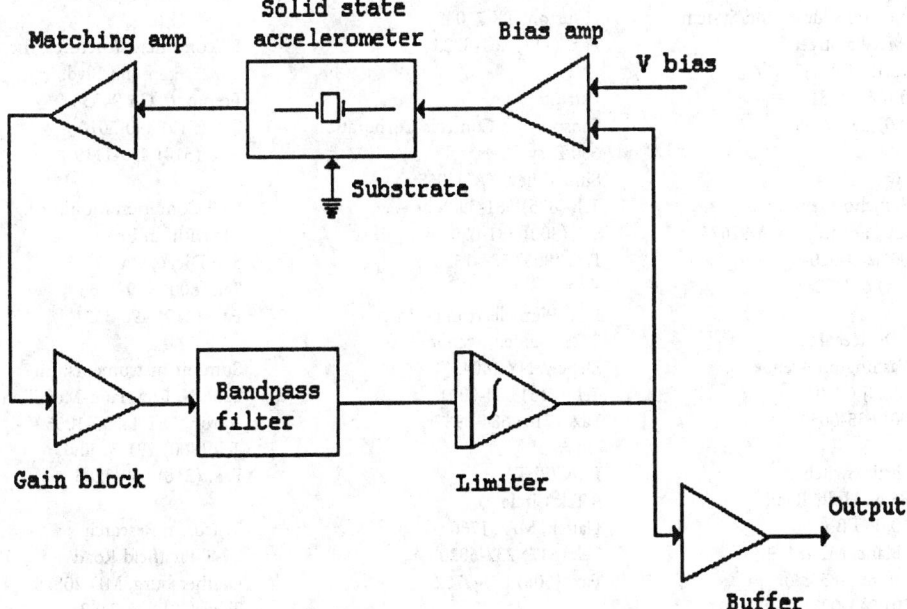

FIGURE 17.23   Block diagram of a signal-conditioning circuit of a microaccelerometer. The output signal of the oscillator is a frequency-modulated acceleration signal. The circuit provides a buffered square-wave frequency output that can be read directly into a digital device.

References

1. K. G. McConnell, *Vibration Testing: Theory and Practice*, New York: John Wiley & Sons, 1995.
2. *Machine Vibration: Dynamics and Control*, London: Springer, 1992-1996.
3. E. O. Doebelin, *Measurement Systems: Application and Design*, 4th ed., New York: McGraw-Hill, 1990.
4. R. Frank, *Understanding Smart Sensors*, Boston: Artech House, 1996.

List of Manufacturers

Allied Signal, Inc.
101 Colombia Road
Dept. CAC
Morristown, NJ 07962

Bokam Engineering, Inc.
9552 Smoke Tree Avenue
Fountain Valley, CA 92708
Tel: (714) 962-3121
Fax: (714) 962-5002

CEC Vibration Products
Division of Sensortronics
196 University Parkway
Pomona, CA 91768
Tel: (909) 468-1345 or
     (800) 468-1345
Fax: (909) 468-1346

Dytran Instrument, Inc.
Dynamic Transducers and Systems
21592 Marilla Street
Chatsworth, CA 91311
Tel: (800) 899-7818
Fax: (800) 899-7088

Endevco
30700 Rancho Viejo Road
San Juan Capistrona, CA 92675
Tel: (800) 289-8204
Fax: (714) 661-7231

Entran Devices, Inc.
10-T Washington Avenue
Fairfield, NJ 07004
Tel: (800) 635-0650

First Inertia Switch
G-10386 N. Holly Road
Dept. 10, PO Box 704
Grand Blanc, MI 48439
Tel: (810) 695-8333 or
     (800) 543-0081
Fax: (810) 695-0589

Instrumented Sensor Technology
4701 A Moor Street
Okemos, MI 48864
Tel: (517) 349-8487
Fax: (517) 349-8469

Jewel Electrical Instruments
124 Joliette Street
Manchester, NH 03102
Tel: (603) 669-6400 or
     (800) 227-5955
Fax: (603) 669-5962

Kistler Instruments Company
75 John Glenn Drive
Amherst, NY 14228-2171
Tel: (800) 755-5745

Lucas Control Production, Inc.
1000 Lucas Way
Hampton, VA 23666
Tel: (800) 745-8008
Fax: (800) 745-8004

Metrix Instrument Company
1711 Townhurst
Houston, TX 77043
Fax: (713) 461-8223

Patriot
Sensors and Controls Corporation
650 Easy Street
Simi Valley, CA 93065
Tel: (805) 581-3985 or
     (800) 581-0701
Fax: (805) 583-1526

PCB Piezoelectronics, Inc.
3425 Walden Avenue
Depew, NY 14043
Tel: (716) 684-0001
Fax: (716) 684-0987

PMC/BETA
9 Tek Circle
Natick, MA 91760
Tel: (617) 237-6020
Fax: (508) 651-9762

Rutherford Controls
2697 International Parkway
Building #3, Suite 122
Virginia Beach, VA 23452
Tel: (800) 899-5625
Fax: (804) 427-9549

Sensotech, Inc.
1202 Chesapeak Ave.
Columbus, OH 43212
Tel: (614) 486-7723 or
     (800) 867-3890
Fax: (614) 486-0506

Setra
45 Nagog Park
Acton, MA 01720
Tel: (508) 263-1400 or
     (800) 257-3872
Fax: (508) 264-0292

Silicon Microstructures, Inc.
46725 Fremond Blvd.
Fremond, CA 94358
Tel: (510) 490-5010
Fax: (510) 490-1119

SKF Condition Monitoring
4141 Ruffin Road
San Diego, CA 92123
Tel: (800) 959-1366
Fax: (619) 496-3531

Summit Instruments, Inc.
2236 N. Cleveland-Massillon Road
Akron, OH 44333-1255
Tel: (800) 291-3730
Fax: (216) 659-3286

Wilcoxon Research
21-T Firstfield Road
Gaithersburg, MD 20878
Tel: (800) 842-7367
Fax: (301) 330-8873

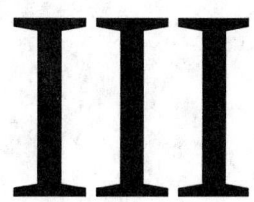

Time and Frequency Measurement

18

Time Measurement

Michael A. Lombardi

National Institute of Standards and Technology

Time measurements can be divided into two general categories. The first category is *time-of-day* measurements. Time-of-day is labeled with a unique expression containing the year, month, day, hour, minute, second, etc., down to the smallest unit of measurement that we choose. When we ask the everyday question, "What time is it?", we are asking for a time-of-day measurement.

The second type of time measurement (and the one more commonly referred to by metrologists) is a *time interval* measurement. A time interval measurement requires measuring the interval that elapses between two events. Time interval is one of the four basic standards of measurement (the others are length, mass, and temperature). Of these four basic standards, time interval can be measured with the most resolution and the least amount of uncertainty.

Timekeeping involves both types of measurements. First, we must find a *periodic event* that repeats at a constant rate. For example, the pendulum in a clock may swing back and forth at a rate of once per second. Once we know that the pendulum swings back and forth every second, we can establish the second as our basic unit of time interval. We can then develop a system of timekeeping, or a *time scale*. A time scale is an unambiguous way to order events. It is created by measuring a small time unit (like the second) and then counting the number of elapsed seconds to establish longer time intervals, like minutes, hours, and days. The device that does the counting is called a *clock*.

There are many types of periodic events that can form the basis for many types of clocks. Let's continue our discussion by looking at the evolution of clocks and timekeeping.

TABLE 18.1   Relationship of Frequency Uncertainty to Timing Uncertainty

Frequency Uncertainty	Measurement Period	Timing Uncertainty
$\pm 1.00 \times 10^{-3}$	1 s	± 1 ms
$\pm 1.00 \times 10^{-6}$	1 s	± 1 μs
$\pm 1.00 \times 10^{-9}$	1 s	± 1 ns
$\pm 2.78 \times 10^{-7}$	1 h	± 1 ms
$\pm 2.78 \times 10^{-10}$	1 h	± 1 μs
$\pm 2.78 \times 10^{-13}$	1 h	± 1 ns
$\pm 1.16 \times 10^{-8}$	1 day	± 1 ms
$\pm 1.16 \times 10^{-11}$	1 day	± 1 μs
$\pm 1.16 \times 10^{-14}$	1 day	± 1 ns

18.1   The Evolution of Clocks and Timekeeping

All clocks share several common features. Each clock has a device that produces the periodic event mentioned previously. This device is called the *resonator*. In the case of the pendulum clock, the pendulum is the resonator. Of course, the resonator needs an energy source, a mainspring or motor, for example, before it can move back and forth. Taken together, the energy source and the resonator form an *oscillator*. Another part of the clock counts the "swings" of the oscillator and converts them to time units like hours, minutes, and seconds, or smaller units like milliseconds (ms), microseconds (μs), and nanoseconds (ns). And finally, part of the clock must display or record the results.

The frequency uncertainty of a clock's resonator relates directly to the timing uncertainty of the clock. This relationship is shown in Table 18.1.

Throughout history, clock designers have searched for more stable frequency sources to use as a resonator. As early as 3500 B.C., time was kept by observing the movement of an object's shadow between sunrise and sunset. This simple clock is called a *sundial*, and the resonance frequency is the apparent motion of the sun. Later, waterclocks, hourglasses, and calibrated candles allowed dividing the day into smaller units of time. In the early 14th century, mechanical clocks began to appear. Early models used a *verge and foliet mechanism* for a resonator and had an uncertainty of about 15 min/day ($\cong 1 \times 10^{-2}$). A major breakthrough occurred in 1656, when the Dutch scientist Christiaan Huygens built the first *pendulum clock*. The pendulum, a mechanism with a "natural" period of oscillation, had been studied by Galileo Galilei as early as 1582, but Huygens was the first to invent an escapement that kept the pendulum swinging. The uncertainty of Huygens's clock was less than 1 min/day ($\cong 7 \times 10^{-4}$), and later reduced to about 10 s/day ($\cong 1 \times 10^{-4}$). Huygens later developed the spring and balance wheel assembly still found in some of today's wristwatches.

Pendulum technology continued to improve until the 20th century. By 1761, John Harrison had built a marine chronometer using a spring and balance wheel escapement. One of Harrison's clocks gained only 54 s during a 5-month voyage to Jamaica, or about 1/3 s/day ($\cong 4 \times 10^{-6}$). The practical performance limit of mechanical clocks was reached in 1921, when W.H. Shortt demonstrated a clock with two pendulums — one a slave and the other a master. The slave pendulum moved the clock's hands, and freed the master pendulum of mechanical tasks that would disturb its regularity. The Shortt clock had an uncertainty of just a few seconds per year ($\cong 1 \times 10^{-7}$) and became the reference used by laboratories [1,2].

In 1927, Joseph W. Horton and Warren A. Marrison of Bell Laboratories built the first clock based on a quartz crystal oscillator. By the 1940s, quartz clocks had replaced the Shortt pendulum as primary laboratory standards. Quartz clocks work because of the piezoelectric property of quartz crystals. When an electric current is applied to a quartz crystal, it resonates at a constant frequency. With no gears or escapements to disturb their resonance frequency, quartz clocks can easily outperform pendulum clocks. Uncertainties of ± 100 μs/day ($\cong 1 \times 10^{-9}$) are possible, and quartz oscillators are used extensively in wristwatches, wall and desk clocks, and electronic circuits [2–4].

The resonance frequency of quartz relies upon a mechanical vibration that is a function of the size and shape of the quartz crystal. No two crystals can be precisely alike or produce exactly the same frequency. Quartz oscillators are also sensitive to environmental parameters like temperature, humidity, pressure, and vibration [5]. These shortcomings make quartz clocks inadequate for many applications, and led to the development of atomic oscillators.

18.2 Atomic Oscillators

Atomic oscillators use the quantized energy levels in atoms and molecules as the source of their resonance frequency. The laws of quantum mechanics dictate that the energies of a bound system, such as an atom, have certain discrete values. An electromagnetic field can boost an atom from one energy level to a higher one. Or an atom at a high energy level can drop to a lower level by emitting electromagnetic energy. The resonance frequency (f) of an atomic oscillator is the difference between the two energy levels divided by Planck's constant (h):

$$f = \frac{E_2 - E_1}{h}$$
(18.1)

Time is kept by observing and counting the frequencies at which electromagnetic energy is emitted or absorbed by the atoms. In essence, the atom serves as a pendulum whose oscillations are counted to mark the passage of time [3].

There are three major types of atomic oscillators. The least expensive and most common type is the *rubidium oscillator*, based on the 6.835 GHz resonance of $^{Rb}87$. Rubidium oscillators range in price from about $2000 to $8000. They are well-suited for applications that require a small, high-performance oscillator with an extremely fast warm-up time. The frequency uncertainty of a rubidium oscillator ranges from about $\pm 5 \times 10^{-10}$ to $\pm 5 \times 10^{-12}$.

The second type of atomic oscillator, the *cesium beam*, serves as the primary reference for most precision timing services. As will be seen in Section 18.3, the resonance frequency of cesium (9.1926 GHz) is used to define the SI second. The primary frequency standard in the United States is a cesium oscillator called NIST-7 with a frequency uncertainty of about $\pm 5 \times 10^{-15}$. Commercially available cesium oscillators differ in quality, but their frequency uncertainty should be at least $\pm 5 \times 10^{-12}$. The price of a cesium oscillator is high, ranging from about $30,000 to $80,000.

A third type of atomic oscillator, the *hydrogen maser*, is based on the 1.42 GHz resonance frequency of the hydrogen atom. Although the performance of hydrogen masers is superior to cesium in some ways, they are not widely used due to their high cost. Few are built, and most are owned by national standards laboratories. The price of a hydrogen maser often exceeds $200,000 [2,3,6].

Table 18.2 summarizes the evolution of clock design and performance.

18.3 Time Scales and the SI Definition of the Second

As observed, the uncertainty of all clocks depends upon the irregularity of some type of periodic motion. By quantifying this periodic motion, one can define the second, which is the basic unit of time interval in the International System of Units (SI). Since atomic time standards are so clearly superior to their predecessors, they are used to define the SI second. Since 1971, the cesium atom has been used to define the second:

The duration of 9,192,631,770 periods of the radiation corresponding to the transition between two hyperfine levels of the ground state of the cesium-133 atom.

International Atomic Time (TAI) is an atomic time scale that conforms as closely as possible to the SI definition of the second. TAI is maintained by the Bureau International des Poids et Measures (BIPM) in Sevres, France. As of 1996, it is created by averaging data obtained from about 250 laboratory and commercial atomic standards located at more than 40 different laboratories. Most of these standards are

TABLE 18.2   The Evolution of Clock Design and Performance

Type of Clock	Resonator	Date [Ref.]	Typical Timing Uncertainty (24 h)	Typical Frequency Uncertainty (24 h)
Sundial	Apparent motion of sun	3500 B.C.	NA	NA
Verge escapement	Verge and foliet mechanism	14th century	±15 min	$\pm1 \times 10^{-2}$
Pendulum	Pendulum	1656	±10 s	$\pm7 \times 10^{-4}$
Harrison chronometer	Pendulum	1761	±400 ms	$\pm4 \times 10^{-6}$
Shortt pendulum	Two pendulums, slave and master	1921	±10 ms	$\pm1 \times 10^{-7}$
Quartz crystal	Quartz crystal	1927 [4]	±100 μs	$\pm1 \times 10^{-9}$
Rubidium gas cell	Rubidium atomic resonance (6834.682608 MHz)	1958 [7]	±1 μs	$\pm1 \times 10^{-11}$
Cesium beam	Cesium atomic resonance (9192.63177 MHz)	1952 [8]	±10 ns	$\pm1 \times 10^{-13}$
Hydrogen maser	Hydrogen atomic resonance (1420.405752 MHz)	1960 [9]	±10 ns	$\pm1 \times 10^{-13}$

based on cesium, although the number of contributing hydrogen masers is increasing. The National Institute of Standards and Technology (NIST) and the United States Naval Observatory (USNO) are among the many laboratories that contribute to TAI.

Before the acceptance of atomic time scales, astronomical time scales were used. These time scales are based on the *mean solar day*, or one revolution of the Earth on its axis. Until 1956, the *mean solar second* served as the SI second. The mean solar second is defined as 1/86,400 of the mean solar day, where 86,400 is the number of seconds in the day. This mean solar second provides the basis for Universal Time (UT). Several variations of UT have been defined:

UT0: The original mean solar time scale, based on the rotation of the Earth on its axis. UT0 was first kept by pendulum clocks. As better clocks based on quartz oscillators became available, astronomers noticed errors in UT0 due to polar motion, which led to the UT1 time scale.

UT1: The most widely used astronomical time scale, UT1 is an improved version of UT0 that corrects for the shift in longitude of the observing station due to polar motion. Since the Earth's rate of rotation is not uniform, UT1 is not completely predictable, and has an uncertainty of ±3 ms per day.

UT2: Mostly of historical interest, UT2 is a smoothed version of UT1 that corrects for known deviations in the Earth's rotation caused by angular momenta of the Earth's core, mantle, oceans, and atmosphere.

The *ephemeris second* served as the SI second from 1956 to 1971. The ephemeris second was a fraction of the tropical year, or the interval between the annual vernal equinoxes on or about March 21. The tropical year was defined as 31,556,925.9747 ephemeris seconds. Determining the precise instant of the equinox is difficult, and this limited the uncertainty of Ephemeris Time (ET) to $\pm$ 50 ms over a 9-year interval. ET was primarily used by astronomers as the time-independent variable for planetary ephemerides. In 1984, ET was replaced by *Terrestrial Time* (TT), which is equal to TAI + 32.184 s. The uncertainty of TT is ±10 μs [10, 11].

18.4   Coordinated Universal Time (UTC)

Since January 1, 1972, all national laboratories and broadcast time services distribute Coordinated Universal Time (UTC), which differs from TAI by an integer number of seconds. The difference between UTC and TAI increases when *leap seconds* are inserted in UTC. When necessary, leap seconds are added to the UTC time scale on either June 30 or December 31. Their purpose is to keep atomic time (UTC) within ±0.9 s of astronomical time (UT1). Many time services broadcast a *DUT1 correction*, or the current value of UT1 minus UTC. By applying this correction to UTC, one can obtain UT1.

Leap seconds occur slightly less than once per year because UT1 is currently changing by about 800 ms per year with respect to UTC. The first leap second was introduced on June 30, 1972. So far, all leap seconds have been positive, which indicates that the mean solar second (defined by the Earth's rotation) is longer than the atomic second (defined by cesium). UTC is running faster than UT1 for two reasons. The first, and most important reason, is that the definition of the atomic second caused it to be slightly shorter than the mean solar second. The second reason is that the speed of the Earth's rotation is generally decreasing. When a positive leap second is added to UTC, the sequence of events is:

$$23 \text{ h } 59 \text{ m } 59 \text{ s}$$

$$23 \text{ h } 59 \text{ m } 60 \text{ s}$$

$$0 \text{ h } 0 \text{ m } 0 \text{ s}$$

The insertion of the leap second creates a minute that is 61 s long. This effectively "stops" the UTC time scale for 1 s, so that UT1 has a chance to catch up. Unless a dramatic, unforeseen change in the Earth's rotation rate takes place, future leap seconds will continue to be positive [10, 11].

18.5 Introduction to Time Transfer

Many applications require different clocks at different locations to be set to the same time (*synchronization*), or to run at the same rate (*syntonization*). A common application is to transfer time from one location and synchronize a clock at another location. This requires a 1 pulse per second (pps) output referenced to UTC. Once we have an on-time pulse, we know the arrival time of each second and can syntonize a local clock by making it run at the same rate. However, we still must know the *time-of-day* before we can synchronize the clock. For example, is it 12:31:38 or 12:31:48? To get the time-of-day, we need a *time code* referenced to UTC. A time code provides the UTC hour, minute, and second, and often provides date information like month, day, and year.

To summarize, synchronization requires two things: an on-time pulse and a time code. Many *time transfer signals* meet both requirements. These signals originate from a UTC clock referenced to one or more cesium oscillators. The time signal from this clock is then distributed (or transferred) to users.

Time can be transferred through many different mediums, including coaxial cables, optical fibers, radio signals (at numerous places in the spectrum), telephone lines, and computer networks. Before discussing the available time transfer signals, the methods used to transfer time are examined.

Time Transfer Methods

The single largest contributor to time transfer uncertainty is *path delay*, or the delay introduced as the signal travels from the transmitter (source) to the receiver (destination). To illustrate the path delay problem, consider a time signal broadcast from a radio station. Assume that the time signal is nearly perfect at its source, with an uncertainty of ±100 ns of UTC. If a receiving site is set up 1000 km away, we need to calculate how long it takes for the signal to get to the site. Radio signals travel at the speed of light ($\cong 3.3$ μs km^{-1}). Therefore, by the time the signal gets to the site, it is already 3.3 ms late. We can compensate for this path delay by making a 3.3 ms adjustment to the clock. This is called *calibrating the path*.

There is always a limit to how well we can calibrate a path. For example, to find the path length, we need coordinates for both the receiver and transmitter, and software to calculate the delay. Even then, we are assuming that the signal took the shortest path between the transmitter and receiver. Of course, this is not true. Radio signals can bounce between the Earth and ionosphere, and travel much further than the distance between antennae. A good path delay estimate requires knowledge of the variables that influence radio propagation: the time of year, the time of day, the position of the sun, the solar index, etc. Even then, path delay estimates are so inexact that it might be difficult to recover time with ±1 ms uncertainty (10,000 times worse than the transmitted time) [12, 13].

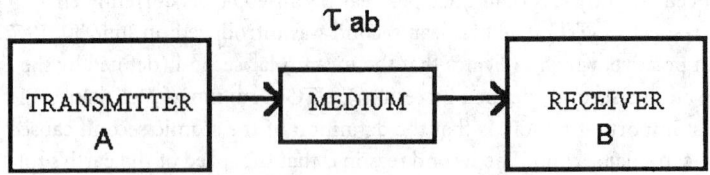

FIGURE 18.1   One-way time transfer.

Designers of time transfer systems have developed many innovative ways to deal with the problem of path delay. The more sophisticated methods have a *self-calibrating path* that automatically compensates for path delay. The various time transfer systems can be divided into five general categories:

1. *One-way method (user calibrates path)*: This is the simplest and most common kind of time transfer system, a one-way system where the user is responsible for calibrating the path (if required). As illustrated in Figure 18.1, the signal from the transmitter to the receiver is delayed τ_{ab} by the medium. To obtain the best results, the user must estimate τ_{ab} and calibrate the path by compensating for the delay.

    Often, the user of a one-way system only requires timing uncertainty of ± 1 s, so no effort is made to calibrate the path. For example, the user of a high-frequency (HF) broadcast can synchronize a clock at the 1 s level without worrying about the effect of path delay.

2. *One-way method (self-calibrating path)*: This method is a variation of the simple one-way method shown in Figure 18.1. However, the time transfer system (and not the user) is responsible for estimating and removing the τ_{ab} delay.

    One of two techniques is commonly used to reduce the size of τ_{ab}. The first technique is to make a rough estimate of τ_{ab} and to send the time out early by this amount. For example, if it is known that τ_{ab} will be at least 20 ms for all users, we can send out the time 20 ms early. This advancement of the timing signal will reduce the uncertainty for all users. For users where τ_{ab} is $\cong 20$ ms, it will remove nearly all of the uncertainty caused by path delay.

    A more sophisticated technique is to compute τ_{ab} in software. A correction for τ_{ab} can be computed and applied if the coordinates of both the transmitter and receiver are known. If the transmitter is stationary, a constant can be used for the transmitter position. If the transmitter is moving (a satellite, for example) it must broadcast its coordinates in addition to broadcasting a time signal. The receiver's coordinates must be computed by the receiver (in the case of radionavigation systems), or input by the user. Then, a software-controlled receiver can compute the distance between the transmitter and receiver and compensate for the path delay by correcting for τ_{ab}. Even if this method is used, uncertainty is still introduced by position errors for either the transmitter or receiver and by variations in the transmission speed along the path.

    Both techniques can be illustrated using the GOES satellite time service (discussed later) as an example. Because the GOES satellites are in geostationary orbit, it takes about 245 ms to 275 ms for a signal to travel from Earth to the satellite, and back to a random user on Earth. To compensate for this delay, the time kept by the station clock on Earth is advanced by 260 ms. This removes most of τ_{ab}. The timing uncertainty for a given user on Earth should now be ± 15 ms. The satellite also sends its coordinates along with the timing information. If a microprocessor-controlled receiver is used, and if the coordinates of the receiver are known, the receiver can make an even better estimate of τ_{ab}, typically within ± 100 μs.

3. *Common-view method*: The common-view method involves a single reference transmitter (R) and two receivers (A and B). The transmitter is in "common view" to both receivers. Both receivers compare the simultaneously received signal to their local clock and record the data (Figure 18.2). Receiver A receives the signal over the path τ_{ra} and compares the reference to its local clock (R – Clock A). Receiver B receives the signal over the path τ_{rb} and records (R – Clock B). The two receivers then exchange and difference the data. Errors from the two paths (τ_{ra} and τ_{rb}) that are

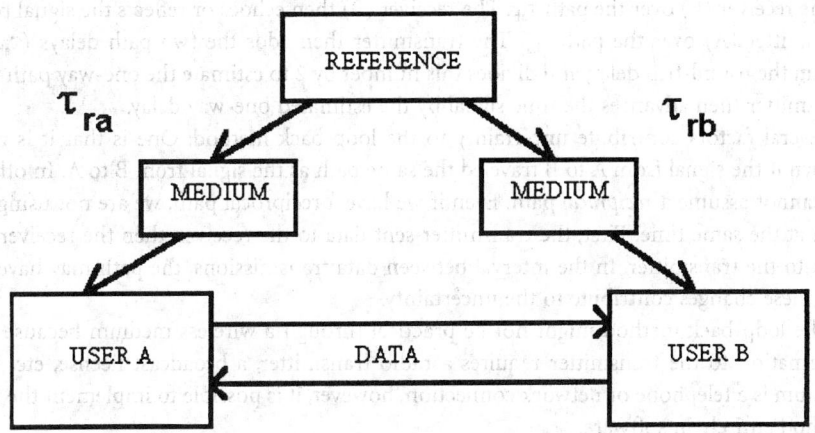

FIGURE 18.2 Common-view time transfer.

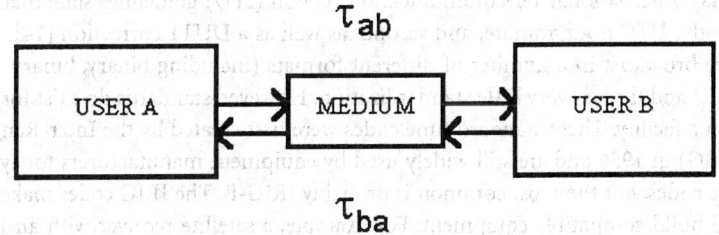

FIGURE 18.3 Two-way time transfer.

common to the reference cancel out, and the uncertainty caused by path delay is nearly eliminated. The result of the measurement is (Clock A – Clock B) – $(\tau_{ra} - \tau_{rb})$.

Keep in mind that the common-view technique does not synchronize clocks in *real time*. The data must be exchanged with another user, and the results might not be known until long after the measurements are completed.

4. *Two-way method*: The two-way method requires two users to both transmit and receive through the same medium at the same time (Figure 18.3). Sites A and B simultaneously exchange time signals through the same medium and compare the received signals with their own clocks. Site A records A – (B + τ_{ba}) and site B records B – (A + τ_{ab}), where τ_{ba} is the path delay from A to B, and τ_{ab} is the path delay from A to B. The difference between these two sets of readings produces 2(A – B) – $(\tau_{ba} - \tau_{ab})$. If the path is reciprocal ($\tau_{ab} = \tau_{ba}$), then the difference, A – B, is known perfectly because the path between A and B has been measured. When properly implemented using a satellite or fiber optic links, the two-way method outperforms all other time transfer methods and is capable of ±1 ns uncertainty.

Two-way time transfer has many potential applications in telecommunications networks. However, when a wireless medium is used, there are some restrictions that limit its usefulness. It might require expensive equipment and government licensing so that users can transmit. And like the common-view method, the two-way method requires users to exchange data. However, because users can transmit, it is possible to include the data with the timing information and to compute the results in real time.

5. *Loop-back method*: Like the two-way method, the loop-back method requires the receiver to send information back to the transmitter. For example, a time signal is sent from the transmitter (A)

to the receiver (B) over the path τ_{ab}. The receiver (B) then echoes or reflects the signal back to the transmitter (A) over the path τ_{ba}. The transmitter then adds the two path delays ($\tau_{ab} + \tau_{ba}$) to obtain the round-trip delay, and divides this number by 2 to estimate the one-way path delay. The transmitter then advances the time signal by the estimated one-way delay.

Several factors contribute uncertainty to the loop-back method. One is that it is not always known if the signal from A to B traveled the same path as the signal from B to A. In other words, we cannot assume a *reciprocal* path. Even if we have a reciprocal path, we are not using the same path at the same time. First, the transmitter sent data to the receiver, then the receiver sent data back to the transmitter. In the interval between data transmissions, the path may have changed, and these changes contribute to the uncertainty.

The loop-back method might not be practical through a wireless medium because returning information to the transmitter requires a radio transmitter, a broadcast license, etc. When the medium is a telephone or network connection, however, it is possible to implement the loop-back method entirely in software.

Time Codes

A *time code* is a message containing time-of-day information, that allows the user to set a clock to the correct time-of-day. International Telecommunications Union (ITU) guidelines state that all time codes should distribute the UTC hour, minute, and second, as well as a DUT1 correction [14].

Time codes are broadcast in a number of different formats (including binary, binary coded decimal [BCD], and ASCII) and there is very little standardization. However, standards do exist for redistributing time codes within a facility. These standard time codes were first created by the Inter-Range Instrumentation Group (IRIG) in 1956 and are still widely used by equipment manufacturers today. IRIG defined a number of time codes, but the most common is probably IRIG-B. The IRIG codes make it possible for manufacturers to build compatible equipment. For example, a satellite receiver with an IRIG-B output can drive a large time-of-day display that accepts an IRIG-B input. Or, it can provide a timing reference to a network server that can read IRIG-B.

The IRIG time code formats are serial, width-modulated codes that can be used in either dc level shift or amplitude-modulated (AM) form. For example, IRIG-B has a 1 s frame period and can be transmitted as either a dc level shift modulation envelope or as a modulated 1000 Hz carrier. BCD and straight binary time data (days, hours, minutes, seconds) is included within the 1 s frame. Simple IRIG-B decoders retrieve just the encoded data and provide 1 s resolution. Other decoders count carrier cycles and provide timing resolution equal to the period of the 1000 Hz cycle (1 ms). More advanced decoders phase lock an oscillator to the time code and provide resolution limited only by the time code signal-to-noise ratio (typically ± 2 µs).

18.6 Radio Time Transfer Signals

Many types of receivers receive time codes transmitted by radio. The costs vary widely, from less than $500 to $15,000 or more. Radio clocks come in several different forms. Some are standalone (or rack mount) devices with a digital time display. These often have a computer interface like RS-232 or IEEE-488. Others are available as cards that plug directly into a computer's bus. When selecting a radio clock, make sure that the signal is usable in the area and that the appropriate type of antenna can be mounted.

When reviewing radio time signals, please remember that the stated uncertainty values refer to the raw signal. Additional delays are introduced before the signal is processed by the receiver and used to synchronize a clock. For example, there is cable delay between the antenna and receiver. There are equipment delays introduced by hardware, and processing delays introduced by software. If unknown, these delays can cause *synchronization errors*. Depending on the application, synchronization errors may or may not be important. However, they must be measured and accounted for when performing an uncertainty analysis of a timing system.

HF Radio Signals (including WWV and WWVH)

High-frequency (HF) or *shortwave* radio broadcasts are commonly used for time transfer at moderate performance levels. These stations are popular for several reasons: they provide worldwide coverage, they work with low-cost receivers, and they provide an audio announcement that lets you "listen" to the time.

To use an HF time signal, you need a shortwave radio. Many types of shortwave radios are available, ranging from simple portables that cost less than $100 to communication-type receivers costing many thousands of dollars. A few companies manufacture dedicated HF timing receivers that automatically find the best signal to use by scanning several different frequencies. Some of them have a built-in computer interface (usually RS-232) so you can use them to set a computer clock.

There are many HF time and frequency stations located around the world, including the NIST-operated stations, WWV and WWVH. WWV is near Fort Collins, CO, and WWVH is on the island of Kauai, HI. Both stations broadcast continuous time and frequency signals on 2.5, 5, 10, and 15 MHz. WWV also broadcasts on 20 MHz. All frequencies carry the same program, and at least one frequency should be usable at all times. The stations can also be heard by telephone: dial (303) 499-7111 for WWV and (808) 335-4363 for WWVH.

The audio portion of the WWV/WWVH broadcast includes seconds pulses or ticks produced by a double-sideband, 100% modulated signal on each RF carrier. The first pulse of every hour is an 800 ms pulse of 1500 Hz. The first pulse of every minute is an 800 ms pulse of 1000 Hz at WWV and 1200 Hz at WWVH. The remaining seconds pulses are brief audio bursts (5 ms pulses of 1000 Hz at WWV and 1200 Hz at WWVH) that sound like the ticking of a clock. All pulses occur at the beginning of each second. The 29th and 59th seconds pulses are omitted. Each tick is preceded by 10 ms of silence and followed by 25 ms of silence to avoid interference from other time stations and to make it easier to hear the tick. At the start of each minute, a voice announces the current UTC hour and minute. WWV uses a male voice to announce the time, and WWVH uses a female voice.

In addition to audio, a time code is also sent on a 100 Hz subcarrier. The time code is a modified version of IRIG-H and is sent once per minute in BCD format, at a 1 bit per second (bps) rate. Within 1 min, enough bits are sent to express the minute, hour, and day of year, the DUT1 correction, and a Daylight Saving Time (DST) indicator. The coded time information refers to the time at the start of the 1-min frame.

WWV and WWVH are best suited for synchronization at the 1 s (or fraction of a second) level. The actual uncertainty depends on the user's distance from the transmitter, but should be less than 30 ms. Although ± 1 ms uncertainty is possible with a well-calibrated path, there are other signals available that are easier to use and more reliable at the 1 ms level [15].

LF Radio Signals (including WWVB and LORAN-C)

Before the development of satellite signals, low-frequency (LF) signals were the method of choice for time transfer. While the use of LF signals has diminished, they still have one major advantage — they can often be received indoors without an external antenna. This makes them ideal for many consumer electronic applications.

Many countries have time services in the LF band from 30 kHz to 300 kHz, as well as in the VLF (very low frequency) band from 3 kHz to 30 kHz. These signals lack the bandwidth needed to provide voice announcements, but often provide an on-time pulse and/or a time code. As with HF signals, the user must calibrate the path to get the best results. However, because part of the LF signal is *groundwave* and follows the curvature of the Earth, a good path delay estimate is much easier to make. Two examples of LF signals used for time transfer are WWVB and LORAN-C. WWVB transmits a binary time code on a 60 kHz carrier. LORAN-C transmits on-time pulses at 100 kHz but has no time code.

WWVB

WWVB is an LF radio station (60 kHz) operated by NIST from the same site as WWV near Ft. Collins, CO. The signal currently covers most of North America, and a power increase (6 dB and scheduled for 1998) would increase the coverage area and improve the signal-to-noise ratio within the United States.

Although far more stable than an HF path, the WWVB path is influenced by the path length, and by daily and seasonal changes. Path length is important because part of the signal travels along the ground (*groundwave*), and another part is reflected from the ionosphere (*skywave*). The groundwave path is more stable and considerably easier to estimate than the skywave path. If the path is relatively short (less than 1000 km), then it is often possible for a receiver to continuously track the groundwave signal, because it always arrives first. If the path length increases, a mixture of groundwave and skywave will be received. And over a very long path, the groundwave could become so weak that it will only be possible to receive the skywave. In this instance, the path becomes much less stable.

Time of day is also important. During the day, the receiver might be able to distinguish between groundwave and skywave and path stability might vary by only a few microseconds. However, if some skywave is being received, *diurnal* phase shifts will occur at sunrise and sunset. For example, as the path changes from all darkness to all daylight, the ionosphere lowers. This shortens the path between the transmitter and receiver, and the path delay decreases until the entire path is in sunlight. The path delay then stabilizes until either the transmitter or receiver enters darkness. Then the ionosphere rises, increasing the path delay.

The WWVB time code is synchronized with the 60 kHz carrier and is broadcast once per minute. The time code is sent in BCD format. Bits are sent by shifting the power of the carrier. The carrier power is reduced 10 dB at the start of each second. If full power is restored 200 ms later, it represents a 0 bit. If full power is restored 500 ms later, it represents a 1 bit. Reference markers and position identifiers are sent by restoring full power 800 ms later. The time code provides year, day, hour, minute, and second information, a DUT1 correction, and information about Daylight Saving Time, leap years, and leap seconds [15].

LORAN-C

LORAN-C is a ground-based radionavigation system. Most of the system is operated by the U.S. Department of Transportation (DOT), but some stations are operated by foreign governments. The system consists of groups of stations (called chains). Each chain has one master station, and from two to five secondary stations. The stations are high power, typically 275 to 1800 kW, and broadcast on a carrier frequency of 100 kHz using a bandwidth from 90 kHz to 110 kHz.

Because there are many LORAN-C chains using the same carrier frequency, the chains transmit pulses so that individual stations can be identified. Each chain transmits a pulse group consisting of pulses from all of the individual stations. The pulse group is sent at a unique Group Repetition Interval (GRI). For example, the 7980 chain transmits pulses every 79.8 ms. By looking for pulse groups spaced at this interval, the receiver can identify the 7980 chain.

Once a specific station within the chain is identified, the pulse shape allows the receiver to locate and track a specific groundwave cycle of the carrier. Generally, a receiver within 1500 km of the transmitter can track the same groundwave cycle indefinitely, and avoid reception of the skywave. Since the receiver can distinguish between groundwave and skywave, the diurnal phase shifts are typically quite small (<500 ns). However, if the path length exceeds 1500 km, the receiver could lose lock, and "jump" to another cycle of the carrier. Each cycle jump introduces a 10 μs timing error, equal to the period of 100 kHz.

LORAN-C does not deliver a time code, but can deliver an on-time pulse referenced to UTC. This is possible because the arrival time of a pulse group coincides with the UTC second at a regular interval. This *time of coincidence* (TOC) occurs once every 4 to 16 min, depending on the chain being tracked. To get a synchronized 1 pps output, one needs a timing receiver with TOC capability and a good path delay estimate. One also needs a TOC table for the chain being tracked (available from the United States Naval Observatory). Once a LORAN-C clock is set on time, it can produce a 1 pps output with an uncertainty of ±500 ns [12, 16].

Geostationary Operational Environmental Satellite (GOES)

NIST provides a continuous time code through the GOES (Geostationary Operational Environmental Satellite) satellites. These satellites are operated by the National Oceanic and Atmospheric Administration (NOAA). The service provides coverage to the entire Western Hemisphere.

Two satellites are used to broadcast time. GOES/East at 75° West longitude broadcasts on a carrier frequency of 468.8375 MHz. GOES/West at 135° West longitude broadcasts on a carrier frequency of 468.825 MHz. The satellites are in geostationary orbit 36,000 km above the equator. The GOES master clock is synchronized to UTC(NIST) and located at NOAA's facility in Wallops Island, VA. The satellites serve as transponders that relay signals from the master clock.

The GOES time code includes the year, day-of-year, hour, minute, and second, the DUT1 correction, satellite position information, and Daylight Saving Time and leap second indicators. The time code is interlaced with messages used by NOAA to communicate with systems gathering weather data. A 50 bit message is sent every 0.5 s, but only the first 4 bits (one BCD word) contains timing information. A complete time code frame consists of 60 BCD words and takes 30 s to receive.

By using the satellite position information, GOES receiving equipment can measure and compensate for path delay if the receiver's coordinates are known. The timing uncertainty of the GOES service is ±100 μs [15].

Global Positioning System (GPS)

The Global Positioning System (GPS) is a radionavigation system developed and operated by the U.S. Department of Defense (DOD). It consists of a constellation of 24 Earth-orbiting satellites (21 primary satellites and 3 in-orbit spares). The 24 satellites orbit the Earth in six fixed planes inclined 55° from the equator. Each satellite is 20,200 km above the Earth and has an 11-h, 58-min orbital period, which means a satellite will pass over the same place on Earth 4 min earlier each day. Since the satellites continually orbit the Earth, GPS should be usable anywhere on the Earth's surface.

Each GPS satellite broadcasts on two carrier frequencies: L1 at 1575.42 MHz and L2 at 1227.6 MHz. Each satellite broadcasts a spread spectrum waveform, called a *pseudo random noise* (PRN) code on L1 and L2, and each satellite is identified by the PRN code it transmits. There are two types of PRN codes. The first type is a *coarse acquisition* code (called the C/A code) with a chip rate of 1023 chips per millisecond. The second is a *precision* code (called the P code) with a chip rate of 10230 chips per millisecond. The C/A code repeats every millisecond. The P code only repeats every 267 days, but for practical reasons is reset every week. The C/A code is broadcast on L1, and the P code is broadcast on both L1 and L2

For national security reasons, the DOD started the *Selective Availability* (SA) program in 1990. SA intentionally increases the positioning and timing uncertainty of GPS by adding about 300 ns of noise to both the C/A code and the P code. The resulting signal is distributed through the *Standard Positioning Service* (SPS). The SPS is intended for worldwide use, and can be used free of charge by anyone with a GPS receiver. The *Precise Positioning Service* (PPS) is only available to users authorized by the United States military. PPS users require a special receiver that employs cryptographic logic to remove the effects of SA. Since PPS use is restricted, nearly all civilian GPS users use the SPS.

Using GPS in One-Way Mode

GPS has the best price-performance ratio of any current time transfer system. Receivers range in price from less than $500 for an OEM timing board, to $20,000 or more for the most sophisticated models. The price often depends on the quality of the receiver's timebase oscillator. Lower priced models have a low-quality timebase that must be constantly steered to follow the GPS signal. Higher priced receivers have better timebases (some have internal rubidium oscillators), and can ignore many of the GPS path variations because their oscillator allows them to coast for longer intervals.

TABLE 18.3   Timing Uncertainty of GPS in One-Way Mode

Service	Uncertainty (ns) 50th percentile	Uncertainty (ns) 1σ	Uncertainty (ns) 2σ
SPS	±115	±170	±340
PPS	±68	±100	±200

To use most receivers, you simply mount the antenna, connect the antenna to the receiver, and turn the receiver on. The antenna is often a small cone or disk (normally about 15 cm in diameter) and must be mounted outdoors where it has a clear, unobstructed view of the sky. Once the receiver is turned on, it performs a sky search to find out which satellites are currently above the horizon and visible from the antenna site. The receiver then collects two blocks of data (called the *almanac* and *ephemeris*) from the satellites it finds. Once this is done, it can compute a 3-dimensional coordinate (latitude, longitude, and altitude) as long as four satellites are in view. The receiver can then compensate for path delay, and synchronize its on-time pulse.

If the antenna has a clear view of the sky, at least four satellites should be in view at all times, and the receiver should always be able to compute its position. The simplest GPS receivers have just one channel and look at multiple satellites using a sequencing scheme that rapidly switches between satellites. More sophisticated models have parallel tracking capability and can assign a separate channel to each satellite in view. These receivers typically track from 5 to 12 satellites at once (although more than 8 will only be available in rare instances). By averaging data from multiple satellites, a receiver can remove some of the effects of SA and reduce the timing uncertainty.

GPS Performance in One-Way Mode

Most GPS timing receivers provide a 1 pps on-time pulse. GPS also broadcasts three pieces of time code information: the number of weeks since GPS time began (January 5, 1980); the current second in the current week; and the number of leap seconds since GPS time began. By using the first two pieces of information, a GPS receiver can recover GPS time. By adding the leap second information, the receiver can recover UTC. GPS time differs from UTC by the number of leap seconds that have occurred since January 5, 1980.

Most GPS receivers output UTC in the traditional time-of-day format: month, day, year, hour, minute, and second. Table 18.3 lists the UTC uncertainty specifications for both the SPS and PPS.

Since nearly all GPS receivers are limited to using the SPS, the top row in the table is of most interest. It shows there is a 50% probability that a given on-time pulse from GPS will be within ±115 ns of UTC. The 1σ uncertainty of GPS (~68% probability) is ±170 ns, and the 2σ uncertainty (95%) is ±340 ns.

To achieve the uncertainties shown in Table 18.3, one must calibrate receiver and antenna delays, and estimate synchronization errors. For this reason, some manufacturers of GPS equipment quote a timing uncertainty of ±1 µs. This specification should be easy to support, even if receiver and antenna delays are roughly estimated or ignored. Other manufacturers use averaging techniques or algorithms that attempt to "remove" SA. These manufacturers might quote an uncertainty specification of ±100 ns or less [17, 18].

Using GPS in Common-View Mode

The *common-view* mode is used to synchronize or compare time standards or time scales at two or more locations. Common-view GPS is the method used by the BIPM to collect data from laboratories who contribute to TAI.

Common-view time transfer requires a specially designed GPS receiver that can read a tracking schedule. This schedule tells the receiver when to start making measurements and which satellite to track. Another user at another location uses the same schedule and makes simultaneous measurements from

TABLE 18.4 Performance of Radio Time Transfer Signals

Radio Time Transfer Signal	Performance without Path Calibration	Typical Limit of System	Items Needed to Reach Performance Limit
HF (WWV/WWVH)	–30 ms (signal delay depends on distance from transmitter)	±1 ms	Path delay estimate, radio propagation model
GOES Satellite	±16 ms	±100 μs	Receiver that corrects for path delay
WWVB	–10 ms (signal delay depends upon distance from transmitter)	±100 μs	Path delay estimate, knowledge of equipment and antenna delays
LORAN-C	±10 μs	±500 ns	Path delay estimate, receiver that is TOC capable, TOC table, knowledge of equipment and antenna delays
GPS (one-way mode)	±340 ns	±100 ns	Commercially available GPS receiver that averages data from multiple satellites. Knowledge of equipment and antenna delays.
GPS (common-view mode)	±20 ns	±5 ns	Common-view receiver, tracking schedule, and another user or laboratory that will exchange data; knowledge of equipment and antenna delays

the same satellite. The tracking schedule must be designed so that it chooses satellites visible to both users at reasonable elevation angles.

Each site measures the difference between its local clock and the satellite. Typically, measurements are made during a satellite pass lasting for less than 15 min. The individual measurements at each site are estimates of (Clock A – GPS) and (Clock B – GPS). If these results are subtracted, the GPS clock drops out and an estimate of (Clock A – Clock B) remains. Since GPS clock errors (including SA) do not influence the results, the uncertainty of GPS common view can be as small as ±5 ns [19].

Although common view is important for international timing comparisons, it is impractical for the average user for several reasons. First, it requires exchanging data with another user. Second, common view does not provide continuous data. It only provides data during comparisons lasting for less than 15 min. And finally, common view does not work in real time; one will not know the timing uncertainty until after the data exchange.

Table 18.4 summarizes the various radio time transfer signals.

Table 18.5 lists some manufacturers of radio clocks. It includes radio clocks that use high-frequency (HF) radio signals, WWVB or other low-frequency (LF) radio signals, LORAN-C (LORAN), the Global Positioning System (GPS), and GPS common-view (GPS-CV). A wide variety of equipment is available for a wide variety of applications. Contact the manufacturer for specific details.

18.7 Computer Time Transfer Signals

One of the most common time transfer problems involves synchronizing computer clocks. Radio clocks like those described in the last section are often used for computer timekeeping. However, using a dial-up or Internet time service is often more convenient and less expensive than purchasing a radio clock [20].

Dial-Up Time Setting Services

Dial-up time services allow computers to synchronize their clocks using an ordinary telephone line. To illustrate how these services work, take a look at NIST's Automated Computer Time Service (ACTS), which went online in 1988.

TABLE 18.5 Sources of Radio Clocks

Radio Clock Manufacturer	Location	Type of Receivers Sold
Absolute Time	San Jose, CA	GPS
Arbiter	Paso Robles, CA	GPS
Allen Osborne	Westlake Village, CA	GPS, GPS-CV
Ashtech	Sunnyvale, CA	GPS
Austron	Austin, TX	LORAN, GPS, GPS-CV
Bancomm	San Jose, CA	GPS
Chrono-Log	Havertown, PA	HF, GPS
Datum	Anaheim, CA	GPS
Efratom	Irvine, CA	GPS
ESE	El Segundo, CA	GPS
Franklin Instrument	Warminster, PA	LF
Frequency and Time Systems	Beverly, MA	GPS
Garmin	Lenexa, KS	GPS
Heath	Benton Harbor, MI	HF
Hewlett-Packard	Santa Clara, CA	GPS
Macrodyne	Clifton Park, NY	GPS
Magellan	San Dimas, CA	GPS
Motorola	Northbrook, IL	GPS
Odetics	Anaheim, CA	GPS
Rockwell	Newport Beach, CA	GPS
Spectracom	East Rochester, NY	LF, GPS
Spectrum Geophysical	West Covina, CA	GPS
Stanford Research	Sunnyvale, CA	LORAN
Stanford Telecomm	Santa Clara, CA	GPS
Telecom Solutions	San Jose, CA	GPS
Tracor	Austin, TX	LF
Trak Systems	Tampa, FL	GPS
Trimble	Sunnyvale, CA	GPS
True Time	Santa Rosa, CA	HF, LF, GPS

ACTS requires a computer, a modem, and some simple software. When a computer connects to ACTS by telephone, it receives an ASCII time code. The information in this time code is used to synchronize the computer clock. ACTS is usable at modem speeds up to 9600 baud with 8 data bits, 1 stop bit, and no parity. To receive the full time code, one must connect at 1200 baud or higher. The full time code is transmitted once per second and contains more information than the 300 baud time code, which is transmitted every 2 s. Table 18.6 describes the full ACTS time code.

The last character in the time code is the on-time marker (OTM). The values enclosed in the time code refer to the arrival time of the OTM. In other words, if the time code states it is 12:45:45, it means it is 12:45:45 when the OTM arrives. To compensate for the path delay between NIST and the user, ACTS sends the OTM out 45 ms early. The 45 ms advance was chosen based on experiments conducted using 1200 baud modems. It allows 8 ms for sending the OTM at 1200 baud, 7 ms for transmission time between NIST and a typical user, and 30 ms for modem processing delay.

Advancing the OTM by 45 ms always reduces the amount of path delay. However, ACTS can calibrate the actual path by using the *loop-back* method. The loop-back method is implemented if the user's software returns the OTM to ACTS after it is received. Each time the OTM is returned, ACTS measures the round-trip path delay, and divides this quantity by 2 to get the one-way path delay. Then, after three consistent measurements, ACTS advances the OTM by the amount of the one-way path delay. For example, if the one-way path delay is 50.4 ms, ACTS sends the OTM out 50.4 ms (instead of 45 ms) early. At this point, the path is calibrated, and the OTM changes from an asterisk to a pound sign (#). If the loop-back option is used, the uncertainty of ACTS is ±5 ms [21].

TABLE 18.6 The Automated Computer Time Service (ACTS) Time Code

Code	Meaning
JJJJJ	The Modified Julian Date (MJD). The MJD is the last five digits of the Julian Date, which is a count of the number of days since January 1, 4713 B.C.
YR-MO-DA	The year, month, and day.
HH:MM:SS	Hour, minute, and second. The time is always transmitted as UTC, and an offset must be applied to obtain local time.
TT	A two-digit code that indicates whether the United States is on Standard Time (ST) or Daylight Saving Time (DST). It also indicates when ST or DST is approaching. The code is set to 00 when ST is in effect, or to 50 when DST is in effect. During the month in which a time change occurs, this number will begin deincrementing by 1 each day until the change occurs. For example, if TT is set to 5, it means the change from DST to ST will take place in 5 days.
L	A one-digit code that indicates whether a leap second will be added or subtracted at midnight on the last day of the current month. If the code is 0, no leap second will occur this month. If the code is 1, a positive leap second will be added at the end of the month. If the code is 2, a second will be deleted at the end of the month.
DUT1	The current difference between the UT1 time scale and UTC. This number can be added to UTC to obtain UT1. The correction ranges from –0.8 to 0.8 s.
msADV	A five-digit code that displays the number of milliseconds that NIST advances the time code. It defaults to 45.0 ms, but will change to reflect the actual one-way line delay if the on-time marker (OTM) is echoed back to NIST.
UTC(NIST)	A label that indicates one is receiving UTC from the National Institute of Standards and Technology (NIST).
<OTM>	A single character sent at the end of each time code. The OTM is originally an asterisk (*) and changes to a pound sign (#) if ACTS has successfully calibrated the path.

TABLE 18.7 Dial-Up Time Setting Services

Organization	Location	Telephone Number
Federal Institute of Physics and Metrology (PTB)	Germany	011-49-53-1-512038
National Research Council (NRC)	Canada	(613) 745-3900, Ottawa
		(416) 445-9408, Toronto
National Center for Metrology (CENAM)	Mexico	011-52-42-110505
National Institute of Standards and Technology (NIST)	Boulder, CO	(303) 494-4774
National Observatory at Rio de Janeiro (ONRJ)	Brazil	011-55-21-580-0677
Technical University of Graz (TUG)	Austria	011-433-16472366
Telecommunications Laboratory (TL)	Taiwan	011-886-3-424-5490
United States Naval Observatory (USNO)	Washington, D.C.	(202) 762-1594

Some dial-up services similar to ACTS are listed in Table 18.7. These services all transmit time codes in ASCII, but several different formats are used. Due to this lack of standardization, software that accesses multiple services must be able to interpret several different time code formats.

Network Time Setting Services

Computers connected to the Internet can be synchronized without the expense of using a dial-up service. The Internet time servers provide a higher level of standardization than the dial-up services. Several standard timing protocols are defined in a series of RFC (Request for Comments) documents. One can obtain these documents from a number of Internet sites. The four major timing protocols are the Time Protocol, the Daytime Protocol, the Network Time Protocol (NTP), and the Simple Network Time Protocol (SNTP). Table 18.8 summarizes the various protocols and their port assignments, or the port on which the time server "listens" for a request from the client [22–26].

TABLE 18.8   Internet Time Protocols

Protocol Name	Document	Format	Port Assignments
Time Protocol	RFC-868	Unformatted 32-bit binary number contains time in UTC seconds since January 1, 1900.	Port 37 tcp/ip, udp/ip
Daytime Protocol	RFC-867	Exact format not specified in standard. Only requirement is that time code is sent as ASCII characters. Often is similar to time codes sent by dial-up services like ACTS.	Port 13 tcp/ip, udp/ip
Network Time Protocol (NTP)	RFC-1305	Server responds to each query with a data packet in NTP format. The data packet includes a 64-bit timestamp containing the time in UTC seconds since January 1, 1900, with a resolution of 200 ps, and an uncertainty of 1 to 50 ms. NTP software runs continuously on the client machine as a background task that periodically gets updates from the server.	Port 123 udp/ip
Simple Network Time Protocol (SNTP)	RFC-1769	A version of NTP that does not change the specification, but simplifies some design features. It is intended for machines where the full performance of NTP is "not needed or justified."	Port 123 udp/ip

NIST operates a Network Time Service that distributes time using the Time, Daytime, and NTP formats from multiple servers [27]. Small computers (PCs) normally use the Daytime Protocol. Large computers and workstations normally use NTP, and NTP software is often included with the operating system. The Daytime Protocol time code is very similar to ACTS. Like ACTS, the Daytime Protocol time code is sent early (by 50 ms), but the server does not calibrate the path. However, the timing uncertainty should be ±50 ms at most locations.

Computer software to access the various dial-up and network time services is available for all major operating systems. One can often obtain evaluation copies (shareware) from the Internet or another on-line service.

18.8   Future Developments

Both the realization of the SI second and the performance of time transfer techniques will continue to improve. One promising development is the increased use of *cesium-fountain* standards. These devices work by laser cooling the atoms and then lofting them vertically. The resonance frequency is detected as the atoms rise and fall under the influence of gravity. Many laboratories are working on this concept, which should lead to substantial improvement over existing atomic-beam cesium standards [28]. In the longer term, a *trapped-ion* standard could lead to improvements of several orders of magnitude. This standard derives its resonance frequency from the systematic energy shifts in transitions in certain ions. The frequency uncertainty of such a device could eventually reach $\pm 1 \times 10^{-18}$ [29].

The future of time transfer should involve more and more reliance on satellite-based systems. Ground-based systems such as LORAN-C are expected to be phased out [30]. The timing uncertainty of GPS will improve if the Selective Availability (SA) program is discontinued (as expected) in the early part of the next century [31]. GLONASS, the Russian counterpart to GPS, might become more widely used [32]. And, in the near future, a time transfer service from the geostationary INMARSAT satellites could be available. This service uses technology similar to GPS, but should provide better performance.

References

1. J. Jesperson and J. Fitz-Randolph, From sundials to atomic clocks: understanding time and frequency, *Nat. Bur. of Stan. Monograph 155*, 1977.
2. C. Hackman and D. B. Sullivan, Resource letter: TFM-1: time and frequency measurement, *Am. J. Phys.*, 63(4), 306-317, 1995.
3. W. M. Itano and N. F. Ramsey, Accurate measurement of time, *Sci. Am.*, 269(1), 56-65, 1993.

4. W. A. Marrison, The evolution of the quartz crystal clock, *Bell Systems Tech. J.*, 27(3), 510-588, 1948.

5. F. L. Walls and J. Gagnepain, Environmental sensitivities of quartz oscillators, *IEEE Trans. Ultrason., Ferroelectr., Freq. Control*, 39, 241-249, 1992.

6. L. Lewis, An introduction to frequency standards, *Proc. IEEE*, 79(7), 927-935, 1991.

7. P. L. Bender, E. C. Beaty, and A. R. Chi, Optical detection of narrow Rb^{87} hyperfine absorption lines, *Phys. Rev. Lett.*, 1(9), 311-313, 1958.

8. J. Sherwood, H. Lyons, R. McCracken, and P. Kusch, High frequency lines in the hfs spectrum of cesium, *Bull. Am. Phys. Soc.*, 27, 43, 1952.

9. H. Goldenberg, D. Kleppner, and N. Ramsey, Atomic hydrogen maser, *Phys. Rev. Lett.*, 5, 361-362, 1960.

10. P. K. Seidelmann, ed., *Explanatory Supplement to the Astronomical Almanac*, Mill Valley, CA: University Science Books, 1992.

11. T. J. Quinn, The BIPM and the accurate measurement of time, *Proc. IEEE*, 79(7), 894-905, 1991.

12. G. Kamas and M. A. Lombardi, Time and frequency users manual, *Natl. Inst. of Stan. Special Publ. 559*, 1991.

13. D. B. Sullivan and J. Levine, Time generation and distribution, *Proc. IEEE*, 79(7), 906-914, 1991.

14. Radiocommunication Study Group 7, Time signals and frequency standard emissions, *Int. Telecom. Union-Radiocommunications (ITU-R)*, TF Series, 1994.

15. R. Beehler and M. A. Lombardi, NIST time and frequency services, *Natl. Inst. of Stan. Special Publication 432*, 1991.

16. G. Hefley, The development of LORAN-C navigation and timing, *Natl. Bur. of Stan. Monograph 129*, 1972.

17. B. Hoffmann-Wellenhof, H. Lichtenegger, and J. Collins, *GPS: Theory and Practice*, 3rd ed., New York: Springer-Verlag, 1994.

18. ARINC Researc Corporation, NAVSTAR Global Positioning System: User's Overview, NAVSTAR GPS Joint Program Office, Los Angeles, YEE-82-009D, March 1991.

19. W. Lewandowski, G. Petit, and C. Thomas, Precision and accuracy of GPS time transfer, *IEEE Trans. Instrum. Meas.*, 42(2), 474-479, 1993.

20. M. A. Lombardi, Keeping time on your PC, *BYTE Magazine*, 18(11), 57-62, 1993.

21. J. Levine, M. Weiss, D. D. Davis, D. W. Allan, and D. B. Sullivan, The NIST automated computer time service, *Natl. Inst. of Stan. J. Res.*, 94, 311-321, 1989.

22. D. L. Mills, Internet time synchronization: the network time protocol, *IEEE Trans. Comm.*, 39, 1482-1493, 1991.

23. D. L. Mills, Network Time Protocol (Version 3) Specification, Implementation, and Analysis, RFC 1305, University of Delaware, March 1992.

24. D. L. Mills, Simple Network Time Protocol (SNTP), RFC 1769, University of Delaware, March 1995.

25. J. Postel, Daytime Protocol, RFC 867, USC/Information Sciences Institute, May 1983.

26. J. Postel and K. Harrenstien, Time Protocol, RFC 868, USC/Information Sciences Institute, May 1983.

27. J. Levine, The NIST Internet Time Service, *Proc. 25th Annu. Precise Time and Time Interval (PTTI) Meeting*, 1993, 505-511.

28. A. Clairon, P. Laurent, G. Santarelli, S. Ghezali, S. N. Lea, and M. Bahoura, A cesium fountain frequency standard: preliminary results, *IEEE Trans. Instrum. Meas.*, 44, 128-131, 1995.

29. W. M. Itano, Atomic ion frequency standards, *Proc. IEEE*, 79, 936-942, 1991.

30. U.S. Dept. of Transportation, Federal Radionavigation Plan, DOT-VNTSC-RSPA-95-1, DOD-4650.5, 1994 (new version published every 2 years).

31. G. Gibbons, A national GPS policy, *GPS World*, 7(5), 48-50, 1996.

32. P. Daly, N. B. Koshelyaevsky, W. Lewandowski, G. Petit, and C. Thomas, Comparison of GLONASS and GPS time transfers, *Metrologia*, 30(2), 89-94, 1993.

19

Frequency Measurement

Michael A. Lombardi
National Institute of Standards and Technology

Frequency is the rate of occurrence of a repetitive event. If T is the period of a repetitive event, then the frequency $f = 1/T$. The International System of Units (SI) states that the period should always be expressed in units of seconds (s), and the frequency should always be expressed in hertz (Hz). The frequency of electric signals often is measured in units of kilohertz (kHz) or megahertz (MHz), where 1 kHz equals 1000 (10^3) cycles per second, and 1 MHz equals 1 million (10^6) cycles per second.

Average frequency over a time interval can be measured very precisely. Time interval is one of the four basic standards of measurement (the others are length, mass, and temperature). Of these four basic standards, time interval (and frequency) can be measured with the most resolution and the least uncertainty. In some fields of metrology, 1 part per million (1×10^{-6}) is considered quite an accomplishment. In frequency metrology, measurements of 1 part per billion (1×10^{-9}) are routine, and even 1 part per trillion (1×10^{-12}) is commonplace.

Devices that produce a known frequency are called *frequency standards*. These devices must be calibrated so that they remain within the tolerance required by the user's application. The discussion begins with an overview of frequency calibrations.

19.1 Overview of Frequency Measurements and Calibration

Frequency calibrations measure the performance of frequency standards. The frequency standard being calibrated is called the *device under test* (DUT). In most cases, the DUT is a *quartz, rubidium,* or *cesium* oscillator. In order to perform the calibration, the DUT must be compared to a *standard* or *reference*. The standard should outperform the DUT by a specified ratio in order for the calibration to be valid. This ratio is called the *test uncertainty ratio* (TUR). A TUR of 10:1 is preferred, but not always possible. If a smaller TUR is used (5:1, for example), then the calibration will take longer to perform.

Once the calibration is completed, the metrologist should be able to state how close the DUT's output is to its *nameplate frequency*. The nameplate frequency is labeled on the oscillator output. For example, a DUT with an output labeled "5 MHz" is supposed to produce a 5 MHz frequency. The calibration measures the difference between the actual frequency and the nameplate frequency. This difference is called the *frequency offset*. There is a high probability that the frequency offset will stay within a certain range of values, called the *frequency uncertainty*. The user specifies a frequency offset and an associated uncertainty requirement that the DUT must meet or exceed. In many cases, users base their requirements on the specifications published by the manufacturer. In other cases, they may "relax" the requirements and use a less demanding specification. Once the DUT meets specifications, it has been successfully calibrated. If the DUT cannot meet specifications, it fails calibration and is repaired or removed from service.

The reference used for the calibration must be *traceable*. The International Organization for Standardization (ISO) definition for *traceability* is:

> The property of the result of a measurement or the value of a standard whereby it can be related to stated references, usually national or international standards, through an unbroken chain of comparisons all having stated uncertainties [1].

In the United States, the "unbroken chain of comparisons" should trace back to the National Institute of Standards and Technology (NIST). In some fields of calibration, traceability is established by sending the standard to NIST (or to a NIST-traceable laboratory) for calibration, or by sending a set of reference materials (such as a set of artifact standards used for mass calibrations) to the user. Neither method is practical when making frequency calibrations. Oscillators are sensitive to changing environmental conditions and especially to being turned on and off. If an oscillator is calibrated and then turned off, the calibration could be invalid when the oscillator is turned back on. In addition, the vibrations and temperature changes encountered during shipment can also change the results. For these reasons, laboratories should make their calibrations on-site.

Fortunately, one can use *transfer standards* to deliver a frequency reference from the national standard to the calibration laboratory. Transfer standards are devices that receive and process radio signals that provide frequency traceable to NIST. The radio signal is a link back to the national standard. Several different types of signals are available, including NIST radio stations WWV and WWVB, and radionavigation signals from LORAN-C and GPS. Each signal delivers NIST traceability at a known level of uncertainty. The ability to use transfer standards is a tremendous advantage. It allows traceable calibrations to be made simultaneously at a number of sites as long as each site is equipped with a radio receiver. It also eliminates the difficult and undesirable practice of moving frequency standards from one place to another.

Once a traceable transfer standard is in place, the next step is developing the technical procedure used to make the calibration. This procedure is called the *calibration method*. The method should be defined and documented by the laboratory, and ideally a measurement system that automates the procedure should be built. ISO/IEC Guide 25, *General Requirements for the Competence of Calibration and Testing Laboratories*, states:

> The laboratory shall use appropriate methods and procedures for all calibrations and tests and related activities within its responsibility (including sampling, handling, transport and storage, preparation of items, estimation of uncertainty of measurement, and analysis of calibration and/or test data). They shall be consistent with the accuracy required, and with any standard specifications relevant to the calibrations or test concerned.

In addition, Guide 25 states:

> The laboratory shall, wherever possible, select methods that have been published in international or national standards, those published by reputable technical organizations or in relevant scientific texts or journals [2,3].

Calibration laboratories, therefore, should automate the frequency calibration process using a well-documented and established method. This helps guarantee that each calibration will be of consistently high quality, and is essential if the laboratory is seeking ISO registration or laboratory accreditation.

Having provided an overview of frequency calibrations, we can take a more detailed look at the topics introduced. This chapter begins by looking at the specifications used to describe a frequency calibration, followed by a discussion of the various types of frequency standards and transfer standards. A discussion of some established calibration methods concludes this chapter.

19.2 The Specifications: Frequency Uncertainty and Stability

This section looks at the two main specifications of a frequency calibration: uncertainty and stability.

Frequency Uncertainty

As noted earlier, a frequency calibration measures whether a DUT meets or exceeds its uncertainty requirement. According to ISO, uncertainty is defined as:

Parameter, associated with the result of a measurement, that characterizes the dispersion of values that could reasonably be attributed to the measurand [1].

When we make a frequency calibration, the measurand is a DUT that is supposed to produce a specific frequency. For example, a DUT with an output labeled 5 MHz is supposed to produce a 5 MHz frequency. Of course, the DUT will actually produce a frequency that is not exactly 5 MHz. After calibration of the DUT, we can state its frequency offset and the associated uncertainty.

Measuring the frequency offset requires comparing the DUT to a reference. This is normally done by making a *phase comparison* between the frequency produced by the DUT and the frequency produced by the reference. There are several calibration methods (described later) that allow this determination. Once the amount of phase deviation and the *measurement period* are known, we can estimate the frequency offset of the DUT. The measurement period is the length of time over which phase comparisons are made. Frequency offset is estimated as follows, where Δt is the amount of phase deviation and T is the measurement period:

$$f\left(offset\right) = \frac{-\Delta t}{T} \tag{19.1}$$

If we measure $+1\ \mu s$ of phase deviation over a measurement period of 24 h, the equation becomes:

$$f\left(offset\right) = \frac{-\Delta t}{T} = \frac{1\ \mu s}{86,400,000,000\ \mu s} = -1.16 \times 10^{-11} \tag{19.2}$$

The smaller the frequency offset, the closer the DUT is to producing the same frequency as the reference. An oscillator that accumulates $+1\ \mu s$ of phase deviation per day has a frequency offset of about -1×10^{-11} with respect to the reference. Table 19.1 lists the approximate offset values for some standard units of phase deviation and some standard measurement periods.

The frequency offset values in Table 19.1 can be converted to units of frequency (Hz) if the nameplate frequency is known. To illustrate this, consider an oscillator with a nameplate frequency of 5 MHz that is high in frequency by 1.16×10^{-11}. To find the frequency offset in hertz, multiply the nameplate frequency by the dimensionless offset value:

$$\left(5 \times 10^{6}\right)\left(+1.16 \times 10^{-11}\right) = 5.80 \times 10^{-5} = +0.0000580\ Hz \tag{19.3}$$

TABLE 19. 1 Frequency Offset Values for Given
Amounts of Phase Deviation

Measurement Period	Phase Deviation	Frequency Offset
1 s	±1 ms	±1.00 × 10⁻³
1 s	±1 µs	±1.00 × 10⁻⁶
1 s	±1 ns	±1.00 × 10⁻⁹
1 h	±1 ms	±2.78 × 10⁻⁷
1 h	±1 µs	±2.78 × 10⁻¹⁰
1 h	±1 ns	±2.78 × 10⁻¹³
1 day	±1 ms	±1.16 × 10⁻⁸
1 day	±1 µs	±1.16 × 10⁻¹¹
1 day	±1 ns	±1.16 × 10⁻¹⁴

The nameplate frequency is 5 MHz, or 5,000,000 Hz. Therefore, the actual frequency being produced by the frequency standard is:

$$5,000,000 \text{ Hz} + 0.0000580 \text{ Hz} = 5,000,000.0000580 \text{ Hz} \qquad (19.4)$$

To do a complete uncertainty analysis, the measurement period must be long enough to ensure that one is measuring the frequency offset of the DUT, and that other sources are not contributing a significant amount of uncertainty to the measurement. In other words, must be sure that Δt is really a measure of the DUT's phase deviation from the reference and is not being contaminated by noise from the reference or the measurement system. This is why a 10:1 TUR is desirable. If a 10:1 TUR is maintained, many frequency calibration systems are capable of measuring a 1×10^{-10} frequency offset in 1 s [4].

Of course, a 10:1 TUR is not always possible, and the simple equation given for frequency offset (Equation 19.1) is often too simple. When using transfer standards such as LORAN-C or GPS receivers (discussed later), radio path noise contributes to the phase deviation. For this reason, a measurement period of at least 24 h is normally used when calibrating frequency standards using a transfer standard. This period is selected because changes in path delay between the source and receiver often have a cyclical variation that averages out over 24 h. In addition to averaging, curve-fitting algorithms and other statistical methods are often used to improve the uncertainty estimate and to show the *confidence level* of the measurement [5]. Figure 19.1 shows two simple graphs of phase comparisons between a DUT and a reference frequency. Frequency offset is a linear function, and the top graph shows no discernible variation in the linearity of the phase. This indicates that a TUR of 10:1 or better is being maintained. The bottom graph shows a small amount of variation in the linearity, which could mean that a TUR of less than 10:1 exists and that some uncertainty is being contributed by the reference.

To summarize, frequency offset indicates how accurately the DUT produces its nameplate frequency. Notice that the term *accuracy* (or *frequency accuracy*) often appears on oscillator specification sheets instead of the term frequency offset, since frequency accuracy and frequency offset are nearly equivalent terms. Frequency uncertainty indicates the limits (upper and lower) of the measured frequency offset. Typically, a 2σ uncertainty test is used. This indicates that there is a 95.4% probability that the frequency offset will remain within the stated limits during the measurement period. Think of frequency offset as the result of a measurement made at a given point in time, and frequency uncertainty as the possible dispersion of values over a given measurement period.

Stability

Before beginning a discussion of *stability*, it is important to mention a distinction between frequency offset and stability. Frequency offset is a measure of how well an oscillator produces its nameplate frequency, or how well an oscillator is adjusted. It does not tell us about the inherent quality of an oscillator. For example, a high-quality oscillator that needs adjustment could produce a frequency with

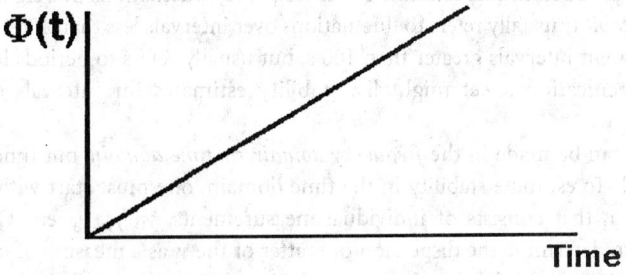

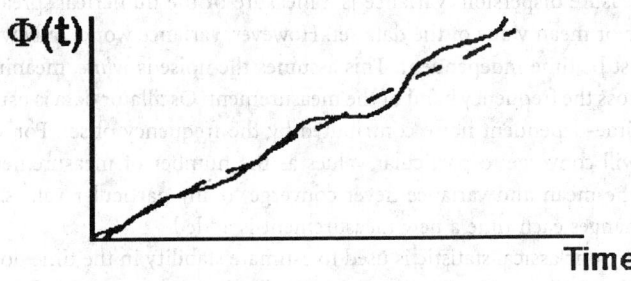

FIGURE 19.1 Simple phase comparison graphs.

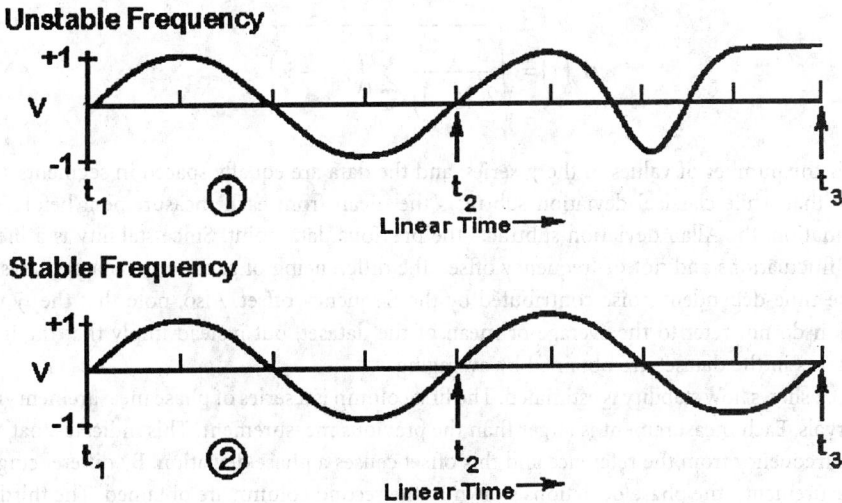

FIGURE 19.2 Comparison of unstable and stable frequencies.

a large offset. A low-quality oscillator may be well adjusted and produce (temporarily at least) a frequency very close to its nameplate value.

Stability, on the other hand, indicates how well an oscillator can produce the same frequency over a given period of time. It does not indicate whether the frequency is "right" or "wrong," only whether it stays the same. An oscillator with a large frequency offset could still be very stable. Even if one adjusts the oscillator and moves it closer to the correct frequency, the stability usually does not change. Figure 19.2 illustrates this by displaying two oscillating signals that are of the same frequency between t_1 and t_2. However, it is clear that signal 1 is unstable and is fluctuating in frequency between t_2 and t_3.

Stability is defined as the statistical estimate of the frequency fluctuations of a signal over a given time interval. *Short-term stability* usually refers to fluctuations over intervals less than 100 s. *Long-term stability* can refer to measurement intervals greater than 100 s, but usually refers to periods longer than 1 day. A typical oscillator specification sheet might list stability estimates for intervals of 1, 10, 100, and 1000 s [6, 7].

Stability estimates can be made in the *frequency domain* or *time domain*, but time domain estimates are more widely used. To estimate stability in the time domain, one must start with a set of frequency offset measurements y_i that consists of individual measurements, y_1, y_2, y_3, etc. Once this dataset is obtained, one needs to determine the dispersion or scatter of the y_i as a measure of oscillator noise. The larger the dispersion, or scatter, of the y_i, the greater the instability of the output signal of the oscillator.

Normally, classical statistics such as *standard deviation* (or *variance*, the square of the standard deviation) are used to measure dispersion. Variance is a measure of the numerical spread of a dataset with respect to the average or mean value of the data set. However, variance works only with stationary data, where the results must be time independent. This assumes the noise is *white*, meaning that its power is evenly distributed across the frequency band of the measurement. Oscillator data is usually nonstationary, because it contains time-dependent noise contributed by the frequency offset. For stationary data, the mean and variance will converge to particular values as the number of measurements increases. With nonstationary data, the mean and variance never converge to any particular values. Instead, there is a moving mean that changes each time a new measurement is added.

For these reasons, a nonclassical statistic is used to estimate stability in the time domain. This statistic is often called the *Allan variance*, but because it is actually the square root of the variance, its proper name is the *Allan deviation*. By recommendation of the Institute of Electrical and Electronics Engineers (IEEE), the Allan deviation is used by manufacturers of frequency standards as a standard specification for stability. The equation for the Allan deviation is:

$$\sigma_y(\tau) = \sqrt{\frac{1}{2(M-1)} \sum_{i=1}^{M-1} \left(\bar{y}_{i+1} - \bar{y}_i \right)^2} \tag{19.5}$$

where M is the number of values in the y_i series, and the data are equally spaced in segments τ seconds long. Note that while classical deviation subtracts the mean from each measurement before squaring their summation, the Allan deviation subtracts the previous data point. Since stability is a measure of frequency fluctuations and not of frequency offset, the differencing of successive data points is done to remove the time-dependent noise contributed by the frequency offset. Also, note that the $\bar{y}$ values in the equation do not refer to the average or mean of the dataset, but instead imply that the individual measurements in the dataset are obtained by averaging.

Table 19.2 shows how stability is estimated. The first column is a series of phase measurements recorded at 1 s intervals. Each measurement is larger than the previous measurement. This indicates that the DUT is offset in frequency from the reference and this offset causes a phase deviation. By differencing the raw phase measurements, the phase deviations shown in the second column are obtained. The third column divides the phase deviation (Δt) by the 1 s measurement period to get the frequency offset. Since phase deviation is about 4 ns s^{-1}, it indicates that the DUT has a frequency offset of about 4×10^{-9}. The frequency offset values in the third column form the y_i data series. The last two columns show the first differences of the y_i and the squares of the first differences. Since the sum of the squares equals 2.2×10^{-21}, the equation (where $\tau = 1$ s) becomes:

$$\sigma_y(\tau) = \sqrt{\frac{2.2 \times 10^{-21}}{2(9-1)}} = 1.17 \times 10^{-11} \tag{19.6}$$

TABLE 19. 2 Using Phase Measurements to Estimate Stability (data recorded at 1 s intervals)

Phase Measurements (ns)	Phase Deviation (ns), Δt	Frequency Offset $\Delta t/1$ s (y_i)	First Differences $(y_{i+1} - y_i)$	First Difference Squared $(y_{i+1} - y_i)^2$
3321.44	—	—	—	—
3325.51	4.07	4.07×10^{-9}	—	—
3329.55	4.04	4.04×10^{-9}	-3×10^{-11}	9×10^{-22}
3333.60	4.05	4.05×10^{-9}	$+1 \times 10^{-11}$	1×10^{-22}
3337.65	4.05	4.06×10^{-9}	$+2 \times 10^{-11}$	4×10^{-22}
3341.69	4.04	4.04×10^{-9}	-2×10^{-11}	4×10^{-22}
3345.74	4.05	4.05×10^{-9}	$+1 \times 10^{-11}$	1×10^{-22}
3349.80	4.06	4.06×10^{-9}	$+1 \times 10^{-11}$	1×10^{-22}
3353.85	4.05	4.05×10^{-9}	-1×10^{-11}	1×10^{-22}
3357.89	4.04	4.04×10^{-9}	-1×10^{-11}	1×10^{-22}

Using the same data, the Allan deviation for $\tau = 2$ s can be computed by averaging pairs of adjacent values and using these new averages as data values. For $\tau = 4$ s, take the average of each set of four adjacent values and use these new averages as data values. More data must be acquired for longer averaging times. Keep in mind that the confidence level of a stability estimate improves as the averaging time increases. In the above example, we have eight samples for our $\tau = 1$ s estimate. However, one would have only two samples for an estimate of $\tau = 4$ s. The confidence level of the estimate (1σ) can be roughly estimated as:

$$\frac{1}{\sqrt{M}} \times 100\% \tag{19.7}$$

In the example, where $M = 9$, the error margin is 33%. With just two samples, the estimate could be in error by as much as 70%. With 10^4 samples, the error margin is reduced to 1%.

A sample Allan deviation graph is shown in Figure 19.3. It shows the stability improving as the measurement period gets longer. Part of this improvement is because measurement system noise becomes less of a factor as the measurement period gets longer. At some point, however, the oscillator will reach its *flicker floor*, and from a practical point of view, no further gains will be made by averaging. The flicker floor is the point where the white noise processes begin to be dominated by nonstationary processes such as frequency drift. Most quartz and rubidium oscillators reach their flicker floor at a measurement period of 10^3 s or less, but cesium oscillators might not reach their flicker floor for 10^5 s (more than 1 day). Figure 19.3 shows a sample Allan deviation graph of a quartz oscillator that is stable to about 5×10^{-12} at 100 s and is approaching its flicker floor [8-10].

Do not confuse stability with frequency offset when reading a specifications sheet. For example, a DUT with a frequency offset of 1×10^{-8} might still reach a stability of 1×10^{-12} in 1000 s. This means that the output frequency of the DUT is stable, even though it is not particularly close to its nameplate frequency. To help clarify this point, Figure 19.4 is a graphical representation of the relationship between frequency offset (accuracy) and stability.

19.3 Frequency Standards

Frequency standards all have an internal device that produces a periodic, repetitive event. This device is called the *resonator*. Of course, the resonator must be driven by an energy source. Taken together, the energy source and the resonator form an *oscillator*. There are two main types of oscillators used as frequency standards: quartz oscillators and atomic oscillators.

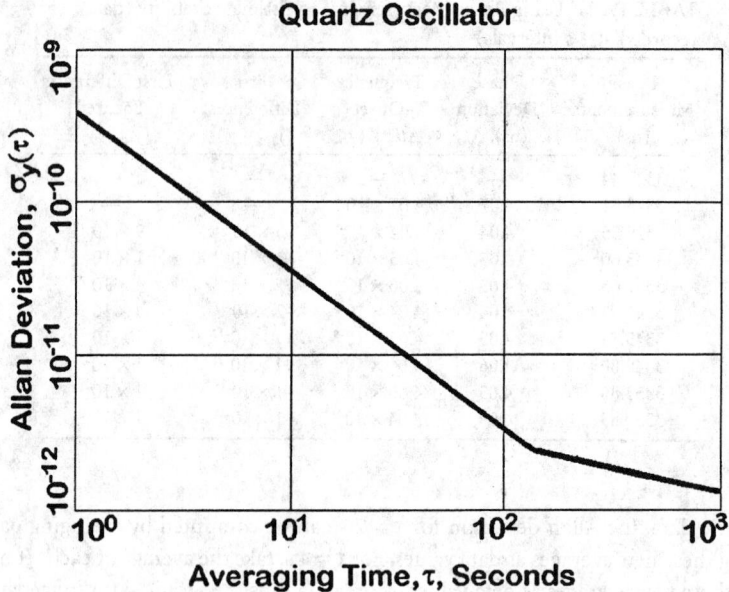

FIGURE 19.3 A sample Allan deviation graph.

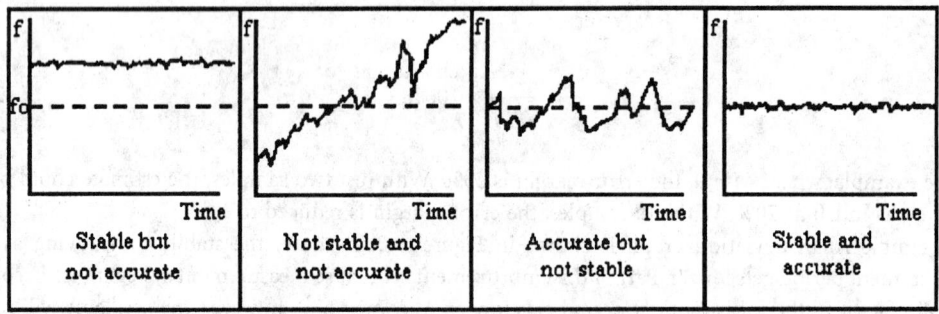

FIGURE 19.4 The relationship between frequency uncertainty (accuracy) and stability.

Quartz Oscillators

Quartz crystal oscillators first appeared during the 1920s and quickly replaced pendulum devices as laboratory standards for time and frequency [11]. Today, more than 10^9 quartz oscillators are manufactured annually for applications ranging from inexpensive wristwatches and clocks to communications networks and space tracking systems [12]. However, calibration and standards laboratories usually calibrate only the more expensive varieties of quartz oscillators, such as those found inside electronic instruments (like frequency counters) or those designed as standalone units. The cost of a high-quality quartz oscillator ranges from a few hundred to a few thousand dollars.

The quartz crystal inside the oscillator can be made of natural or synthetic quartz, but all modern devices are made of synthetic material. The crystal serves as a mechanical resonator that creates an oscillating voltage due to the *piezoelectric effect*. This effect causes the crystal to expand or contract as voltages are applied. The crystal has a *resonance frequency* that is determined by its physical dimensions and the type of crystal used. No two crystals can be exactly alike or produce exactly the same frequency. The output frequency of a quartz oscillator is either the fundamental resonance frequency or a multiple

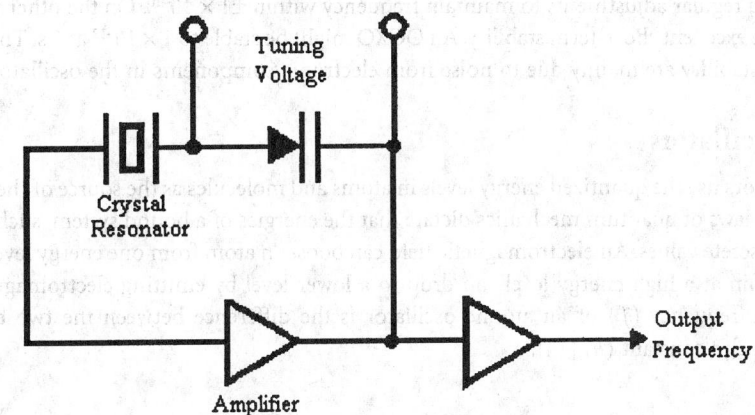

FIGURE 19.5 Block diagram of quartz oscillator.

of that frequency. Figure 19.5 is a simplified circuit diagram that shows the basic elements of a quartz oscillator. The amplifier provides the energy needed to sustain oscillation.

Quartz oscillators are sensitive to environmental parameters such as temperature, humidity, pressure, and vibration [12, 13]. When environmental parameters change, the fundamental resonance frequency also changes. There are several types of quartz oscillator designs that attempt to reduce the environmental problems. The *oven-controlled crystal oscillator* (OCXO) encloses the crystal in a temperature-controlled chamber called an oven. When an OCXO is first turned on, it goes through a "warm-up" period while the temperatures of the crystal resonator and its oven stabilize. During this time, the performance of the oscillator continuously changes until it reaches its normal operating temperature. The temperature within the oven then remains constant, even when the outside temperature varies. An alternative solution to the temperature problem is the *temperature-compensated crystal oscillator* (TCXO). In a TXCO, the output signal from a special temperature sensor (called a thermistor) generates a correction voltage that is applied to a voltage-variable reactance (called a varactor). The varactor then produces a frequency change equal and opposite to the frequency change produced by temperature. This technique does not work as well as oven control, but is much less expensive. Therefore, TCXOs are normally used in small, usually portable units when high performance over a wide temperature range is not required. A third type of quartz oscillator is the *microcomputer-compensated crystal oscillator* (MCXO). The MCXO uses a microprocessor and compensates for temperature using digital techniques. The MCXO falls between a TCXO and an OCXO in both price and performance.

All quartz oscillators are subject to *aging*, which is defined as "a systematic change in frequency with time due to internal changes in the oscillator." Aging is usually observed as a nearly linear change in the resonance frequency. Aging can be positive or negative and, occasionally, a reversal in aging direction is observed. Often, the resonance frequency decreases, which may indicate that the crystal is getting larger. Aging has many possible causes, including contamination of the crystal due to deposits of foreign material, changes in the oscillator circuitry, or changes in the quartz material or crystal structure. The vibrating motion of the crystal can also contribute to aging. High-quality quartz oscillators age at a rate of 5×10^{-9} per year or less.

In spite of the temperature and aging problems, the best OCXOs can achieve frequency offsets as small as 1×10^{-10}. Less expensive oscillators produce less impressive results. Small ovenized oscillators (such as those used as timebases in frequency counters) typically are offset in frequency by $\pm 1 \times 10^{-9}$ and cost just a few hundred dollars. The lowest priced quartz oscillators, such as those found in wristwatches and electronic circuits, cost less than \$1. However, because they lack temperature control, these oscillators have a frequency offset of about $\pm 1 \times 10^{-6}$ in the best case and may be offset as much as $\pm 1 \times 10^{-4}$.

Since the frequency offset of a quartz oscillator changes substantially over long periods of time, regular adjustments might be needed to keep the frequency within tolerance. For example, even the best quartz oscillators need regular adjustments to maintain frequency within $\pm 1 \times 10^{-10}$. On the other hand, quartz oscillators have excellent short-term stability. An OCXO might be stable to 1×10^{-12} at 1 s. The limitations in short-term stability are mainly due to noise from electronic components in the oscillator circuits.

Atomic Oscillators

Atomic oscillators use the quantized energy levels in atoms and molecules as the source of their resonance frequency. The laws of quantum mechanics dictate that the energies of a bound system, such as an atom, have certain discrete values. An electromagnetic field can boost an atom from one energy level to a higher one. Or, an atom at a high energy level can drop to a lower level by emitting electromagnetic energy. The resonance frequency (f) of an atomic oscillator is the difference between the two energy levels divided by Planck's constant (h) [14]:

$$f = \frac{E_2 - E_1}{h} \qquad (19.8)$$

All atomic oscillators are *intrinsic standards*, since their frequency is inherently derived from a fundamental natural phenomenon. There are three main types of atomic oscillators: rubidium standards, cesium standards, and hydrogen masers (discussed individually in the following sections). All three types contain an internal quartz oscillator that is locked to a resonance frequency generated by the atom of interest. Locking the quartz oscillator to the atomic frequency is advantageous. Most of the factors that degrade the long-term performance of a quartz oscillator disappear, because the atomic resonance frequency is much less sensitive to environmental conditions than the quartz resonance frequency. As a result, the long-term stability and uncertainty of an atomic oscillator are much better than those of a quartz oscillator, but the short-term stability is unchanged.

Rubidium Oscillators

Rubidium oscillators are the lowest priced members of the *atomic* oscillator group. They offer perhaps the best price/performance ratio of any oscillator. They perform much better than a quartz oscillator and cost much less than a cesium oscillator.

A rubidium oscillator operates at the resonance frequency of the rubidium atom (^{87}Rb), which is 6,834,682,608 Hz. This frequency is synthesized from a lower quartz frequency (typically 5 MHz) and the quartz frequency is steered by the rubidium resonance. The result is a very stable frequency source with the short-term stability of quartz, but much better long-term stability. Since rubidium oscillators are more stable than quartz oscillators, they can be kept within tolerance with fewer adjustments. The initial price of a rubidium oscillator (typically from \$3000 to \$8000) is higher than that of a quartz oscillator, but because fewer adjustments are needed, labor costs are reduced. As a result, a rubidium oscillator might be less expensive to own than a quartz oscillator in the long run.

The frequency offset of a rubidium oscillator ranges from 5×10^{-10} to 5×10^{-12} after its warm-up period. Maintaining frequency within $\pm 1 \times 10^{-11}$ can be done routinely with a rubidium oscillator but is impractical with even the best quartz oscillators. The performance of a well-maintained rubidium oscillator can approach the performance of a cesium oscillator, and a rubidium oscillator is much smaller, more reliable, and less expensive.

Cesium Oscillators

Cesium oscillators are *primary frequency standards* because the SI second is based on the resonance frequency of the cesium atom (^{133}Cs), which is 9,192,631,770 Hz. This means that a cesium oscillator that is working properly should be very close to its nameplate frequency without any adjustment, and there should be no change in frequency due to aging. The time scale followed by all major countries, Coordinated Universal Time (UTC), is derived primarily from averaging the performance of a large ensemble of cesium oscillators.

Cesium oscillators are the workhorses in most modern time and frequency distribution systems. The primary frequency standard for the United States is a cesium oscillator named NIST-7 with a frequency uncertainty of about $\pm 5 \times 10^{-15}$. Commercially available cesium oscillators differ in quality, but their frequency uncertainty should still be $\pm 5 \times 10^{-12}$ or less.

The two major drawbacks of cesium oscillators involve reliability and cost. Reliability is a major issue. The major component of a cesium oscillator, called the *beam tube*, has a life expectancy of about 3 years to 10 years. The beam tube is needed to produce the resonance frequency of the cesium atom, and this frequency is then used to discipline a quartz oscillator. When the beam tube fails, the cesium oscillator performs like an undisciplined quartz oscillator. For this reason, a cesium oscillator needs to be constantly monitored to make sure that it is still delivering a cesium-derived frequency. Cost is also a major issue. The initial purchase price of a cesium oscillator ranges from \$30,000 to \$80,000 and maintenance costs are high. The cost of a replacement beam tube is a substantial fraction of the cost of the entire oscillator. Laboratories that use cesium oscillators need to budget not only for their initial purchase, but for the cost of maintaining them afterward.

Hydrogen Masers

The *hydrogen maser* is the most elaborate and most expensive commercially available frequency standard. Few masers are built and most are owned by observatories or national standards laboratories. The word "maser" is an acronym that stands for Microwave Amplification by Stimulated Emission of Radiation. Masers derive their frequency from the resonance frequency of the hydrogen atom, which is 1,420,405,752 Hz.

There are two types of hydrogen masers. The first type, called an *active maser*, oscillates spontaneously and a quartz oscillator is phase-locked to this active oscillation. The second type, called a *passive maser*, frequency-locks a quartz oscillator to the atomic reference. The "passive" technique is also used by rubidium and cesium oscillators. Since active masers derive the output frequency more directly from the atomic resonance, they have better short-term stability than passive masers. However, both types of maser have better short-term stability than a cesium oscillator. Even so, the frequency uncertainty of a maser is still greater than that of a cesium oscillator because its operation is more critically dependent on a complex set of environmental conditions. Although the performance of a hydrogen maser is excellent, its cost is still very high, typically \$200,000 or more [15, 16].

Table 19.3 summarizes the characteristics of the different types of oscillators.

19.4 Transfer Standards

To briefly review, a frequency calibration compares the *device under test* (DUT) to a reference. The DUT is usually a quartz, rubidium, or cesium oscillator. The reference is an oscillator of higher performance than the DUT or a *transfer standard* that receives a radio signal. All transfer standards receive a signal that has a cesium oscillator at its source, and this signal delivers a cesium-derived frequency to the user. This benefits many users, because cesium oscillators are expensive both to buy and maintain, and not all calibration laboratories can afford them. Even if a laboratory already has a cesium oscillator, it still needs to check its performance. The only practical way to do this is by comparing the cesium to a transfer standard.

Transfer standards also provide *traceability*. Most transfer standards receive signals traceable to the national frequency standard maintained by NIST. Some signals, such as those transmitted by HF (high frequency) radio stations WWV and WWVH and the LF (low frequency) station WWVB, are traceable because they are directly controlled by NIST. Other signals, like the LORAN-C and Global Positioning System (GPS) satellite broadcasts, are traceable because their reference is regularly compared to NIST. Some signals broadcast from outside the United States are also considered to be traceable. This is because NIST compares its frequency standard to the standards maintained in other countries.

Some compromises are made when using a transfer standard. Even if the radio signal is referenced to a nearly perfect frequency, its performance is degraded as it travels along the radio path between the transmitter and receiver. To illustrate, consider a laboratory that has a rack-mounted frequency standard

TABLE 19.3 Summary of Oscillator Types

Oscillator type	Quartz (TCXO)	Quartz (MCXO)	Quartz (OCXO)	Rubidium	Cesium	Hydrogen maser
Primary standard	No	No	No	No	Yes	No
Intrinsic standard	No	No	No	Yes	Yes	Yes
Resonance frequency	Mechanical (varies)	Mechanical (varies)	Mechanical (varies)	6.834682608 GHz	9.1926177 GHz	1.420405752 GHz
Leading cause of failure	None	None	None	Rubidium lamp (15 years)	Cesium beam tube (3 to 10 years)	Hydrogen depletion (7 years +)
Power (W)	0.05	0.04	0.6	20	30	>100
Stability, $\sigma_y(\tau)$, $\tau = 1$ s	1×10^{-9}	3×10^{-10}	1×10^{-12}	5×10^{-11}–5×10^{-12}	5×10^{-11}–5×10^{-12}	$\equiv 1 \times 10^{-12}$
Aging/year	5×10^{-7}	3×10^{-8}	5×10^{-9}	2×10^{-10}	None	$\equiv 1 \times 10^{-13}$
Frequency Offset after Warm-up	1×10^{-6}	1×10^{-8}	1×10^{-8}–1×10^{-10}	5×10^{-10}–5×10^{-12}	5×10^{-12}–1×10^{-14}	1×10^{-12}–1×10^{-13}
Warm-up time	<10 s–1×10^{-6}	<10 s–1×10^{-8}	<5 min–1×10^{-8}	<5 min–5×10^{-10}	30 min–5×10^{-12}	24 h–1×10^{-12}
Cost	$100	$1000	$2000	$3000–$8000	$30,000–$80,000	$200,000–$300,000

TABLE 19.4 Traceability Levels Provided by Various Transfer Standards

Transfer standard	Frequency uncertainty over 24 h measurement period (with respect to NIST)
HF receiver (WWV and WWVH)	$\pm 1 \times 10^{-7}$
LF receiver (LORAN-C and WWVB)	$\pm 1 \times 10^{-12}$
Global Positioning System (GPS) Receiver	$\pm 5 \times 10^{-13}$

that produces a 5 MHz signal. Metrologists need to use this signal on their work bench, so they run a length of coaxial cable from the frequency standard to their bench. The signal is delayed as it travels from the standard to the bench, but because the cable is of fixed length, the delay is constant. Constant delays do not change the frequency. The frequency that goes into one end of the cable is essentially the same frequency that comes out the other end. However, what if the cable length were constantly changing? This would generally cause the frequency to fluctuate. Over long periods of time, these fluctuations will average out, but the short-term frequency stability would still be very poor. This is exactly what happens when one uses a transfer standard. The "cable" is actually a radio path that might be thousands of kilometers in length. The length of the radio path is constantly changing and appears to introduce frequency fluctuations, even though the source of the frequency (a cesium oscillator) is not changing. This makes transfer standards unsuitable as a reference when making short-term stability measurements. However, transfer standards are well suited for measuring frequency uncertainty, because one can minimize and even eliminate the effects of these frequency fluctuations if one averages for a long enough measurement interval. Eventually, the performance of a cesium oscillator will be recovered.

Some radio signals have path variations that are so pronounced that they are not well suited for high-level frequency calibrations. To illustrate this, consider the signal broadcast from WWV, located in Fort Collins, CO. WWV is an HF radio station (often called a *shortwave* station) that transmits on 2.5, 5, 10, 15, and 20 MHz. WWV is referenced to the national frequency standard at NIST, but by the time the signal gets to the receiver, much of its potential performance has been lost. Most shortwave users receive the *skywave*, or the part of the signal that travels up to the ionosphere and is reflected back to Earth. Since the height of the ionosphere constantly changes, the path delay constantly changes, often by as much as 500 μs to 1000 μs. Because there is so much variability in the path, averaging leads to only limited improvement. Therefore, even though WWV is traceable to NIST, its frequency uncertainty is limited to $\pm 1 \times 10^{-7}$ when averaged for 1 day.

Other radio signals have more stable paths and much lower uncertainty values. Low-frequency (LF) radio stations (like NIST radio station WWVB and LORAN-C) can provide traceability to NIST with a frequency uncertainty of $\pm 1 \times 10^{-12}$ per day. An LF path is much more stable than an HF path, but still experiences a path delay change when the height of the ionosphere changes at sunrise and sunset. Currently, the most widely used signals originate from the Global Positioning System (GPS) satellites. GPS signals have the advantage of an unobstructed path between the transmitter and receiver. The frequency uncertainty of GPS is about $\pm 5 \times 10^{-13}$/day. WWVB, LORAN-C, and GPS are described in the next three sections.

Table 19.4 shows some of the transfer standards available and the level of NIST traceability they provide when averaged for a measurement period of at least 24 h [17].

WWVB

Many countries broadcast time and frequency signals in the LF band from 30 kHz to 300 kHz, as well as in the VLF (very low frequency) band from 3 kHz to 30 kHz. Because part of the LF signal is *groundwave* and follows the curvature of the Earth, the path stability of these signals is acceptable for many applications. One such station is WWVB, which is operated by NIST. WWVB transmits on 60 kHz from the same site as WWV in Fort Collins, CO. The signal currently covers most of North America, and a power increase (6 dB, scheduled for 1998) will increase the coverage area and improve the signal-to-noise ratio within the United States.

Although far more stable than an HF path, the WWVB path length is influenced by environmental conditions along the path and by daily and seasonal changes. Path length is important because part of the signal travels along the ground (*groundwave*) and another part is reflected from the ionosphere (*skywave*). The groundwave path is far more stable than the skywave path. If the path is relatively short (less than 1000 km), the receiver might continuously track the groundwave signal because it always arrives first. For longer paths, a mixture of groundwave and skywave is received. And over a very long path, the groundwave might become so weak that it will only be possible to receive the skywave. In this instance, the path becomes much less stable.

The characteristics of an LF path are different at different times of day. During the daylight and nighttime hours, the receiver might be able to distinguish between groundwave and skywave, and path stability might vary by only a few hundred nanoseconds. However, if some skywave is being received, *diurnal* phase shifts will occur at sunrise and sunset. For example, as the path changes from all darkness to all daylight, the ionosphere lowers and the path gets shorter. The path length then stabilizes until either the transmitter or receiver enters darkness. At this point, the ionosphere rises and the path gets longer.

WWVB receivers have several advantages when used as a transfer standard. They are low cost and easy to use, and the received signals are directly traceable to NIST. With a good receiver and antenna system, one can achieve a frequency uncertainty of $\pm 1 \times 10^{-12}$ by averaging for 1 day [18].

LORAN-C

LORAN-C is a radionavigation system that operates in the LF band. Most of the system is operated by the U.S. Department of Transportation (DOT), but some stations are operated by other governments. The system consists of groups of stations called *chains*. Each chain has one master station and from two to five secondary stations. The stations operate at high power, typically 275 kW to 800 kW, and broadcast on a carrier frequency of 100 kHz using the 90 kHz to 110 kHz band.

Since all LORAN-C chains use the same carrier frequency, the chains transmit pulses so that individual stations can be identified. Each chain transmits a pulse group that includes pulses from all of the individual stations. The pulse group is sent at a unique Group Repetition Interval (GRI). For example, the 7980 chain transmits pulses every 79.8 ms. When the pulses leave the transmitter, they radiate in all directions. The *groundwave* travels parallel to the surface of the Earth. The *skywave* travels upward and is reflected off the ionosphere. The pulse shape was designed so that the receiver can distinguish between groundwave and skywave and lock onto the more stable groundwave signal. Most receivers stay locked to the ground-wave by tracking the third cycle of the pulse. The third cycle was chosen for two reasons. First, it arrives early in the pulse so we know that it is groundwave. Second, it has more amplitude than the first and second cycles in the pulse, which makes it easier for the receiver to stay locked. Generally, a receiver within 1500 km of the transmitter can track the same groundwave cycle indefinitely and avoid skywave reception. The variations in groundwave path delay are typically quite small (<500 ns per day). However, if the path length exceeds 1500 km, the receiver might lose lock, and jump to another cycle of the carrier. Each cycle jump introduces a 10 µs phase step, equal to the period of 100 kHz.

Commercially available LORAN-C receivers designed as transfer standards range in price from about $3000 to $10,000. These receivers typically provide several different frequency outputs (usually 10 MHz, 1 Hz, and the GRI pulse). LORAN-C navigation receivers are mass produced and inexpensive but do not provide a frequency output. However, it is sometimes possible to find and amplify the GRI pulse on the circuit board and use it as a reference frequency.

LORAN-C Performance

The frequency uncertainty of LORAN-C is degraded by variations in the radio path. The size of these variations depends on the signal strength, the distance between the receiver and the transmitter, the weather and atmospheric conditions, and the quality of the receiver and antenna. Figure 19.6 shows the type of path variations one can typically expect. It shows a 100 s phase comparison between the LORAN-C

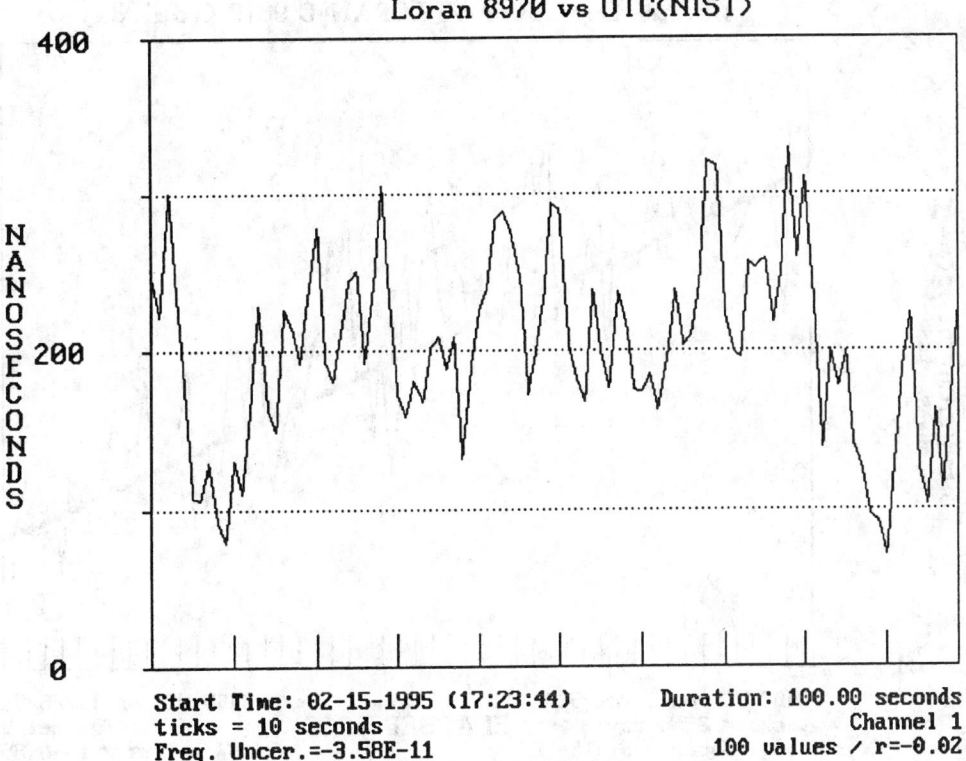

FIGURE 19.6   LORAN-C path variations during 100 s measurement period.

8970 chain as received in Boulder, CO, and the NIST national frequency standard. The 8970 signal is broadcast from the master station in Dana, IN, a site about 1512 km from Boulder. The frequency offset between the NIST frequency standard and the cesium standard at the LORAN-C transmitter would produce a phase shift of <1 ns per 100 s. Therefore, the graph simply shows the LORAN-C path noise or the difference in the arrival times of the GRI pulses due to path variations. The range of this path noise is about 200 ns.

The path variations cause the short-term stability of LORAN-C to be poor. However, because the path variations average out over time, the long-term stability is much better. This means that we can use LORAN-C to calibrate nearly any frequency standard if we average for a long enough interval. The better the frequency standard, the longer it takes to obtain a TUR approaching 10:1. For example, a quartz oscillator with a frequency uncertainty of 1×10^{-7} could be calibrated in 100 s. The quartz would accumulate 10 μs of phase shift in 100 s, and this would completely hide the path noise shown in Figure 19.6. However, a rubidium oscillator with a frequency uncertainty of 1×10^{-11} would accumulate only 1 ns of phase shift in 100 s, or about 1 μs per day. For this reason, a measurement period of at least 24 h is recommended when using LORAN-C to calibrate atomic oscillators. Figure 19.7 shows the results of a 96 h calibration of a cesium oscillator using LORAN-C. The thick line is a least squares estimate of the frequency offset. Although the path noise is clearly visible, the slope of the line indicates that the cesium oscillator is low in frequency by $\cong 3.4 \times 10^{-12}$.

Global Positioning System (GPS)

The Global Positioning System (GPS) is a radionavigation system developed and operated by the U.S. Department of Defense (DOD). The system consists of a constellation of 24 Earth orbiting satellites (21 primary satellites and 3 in-orbit spares). Each satellite carries four atomic frequency standards (two

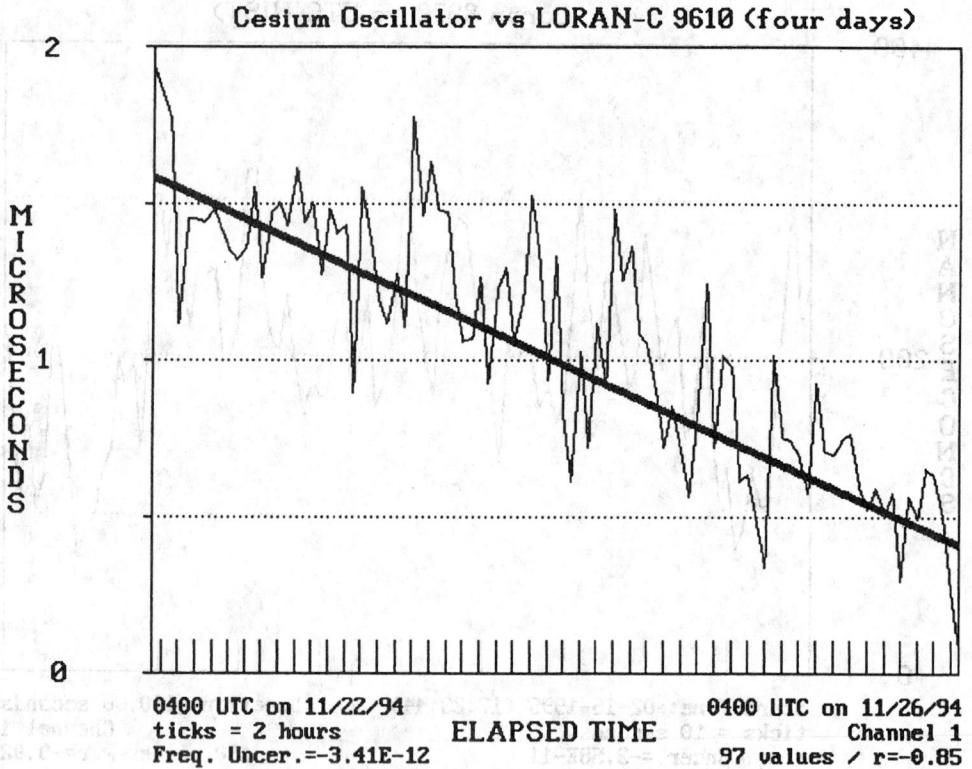

FIGURE 19.7   LORAN-C compared to cesium oscillator over 96 h interval.

rubidiums and two cesiums) that are referenced to the United States Naval Observatory (USNO) and traceable to NIST. The 24 satellites orbit the Earth in six fixed planes inclined 55° from the equator. Each satellite is 20,200 km above the Earth and has an 11 h, 58 min orbital period, which means a satellite will pass over the same place on Earth 4 min earlier each day. Since the satellites continually orbit the Earth, GPS should be usable anywhere on the Earth's surface.

Each GPS satellite broadcasts on two carrier frequencies: L1 at 1575.42 MHz and L2 at 1227.6 MHz. Each satellite broadcasts a spread spectrum waveform called a *pseudo-random noise* (PRN) code on L1 and L2, and each satellite is identified by the PRN code it transmits. There are two types of PRN codes. The first type is a *coarse acquisition* code (called the C/A code) with a chip rate of 1023 chips per millisecond. The second is a *precision* code (called the P code) with a chip rate of 10230 chips per millisecond. The C/A code repeats every millisecond. The P code repeats only every 267 days, but for practical reasons is reset every week. The C/A code is broadcast on L1, and the P code is broadcast on both L1 and L2 [19, 20].

For national security reasons, the DOD started the *Selective Availability* (SA) program in 1990. SA intentionally increases the positioning and timing uncertainty of GPS by adding about 300 ns of noise to both the C/A code and the P code. The resulting signal is distributed through the *Standard Positioning Service* (SPS). The SPS is intended for worldwide use, and can be used free of charge by anyone with a GPS receiver. The *Precise Positioning Service* (PPS) is available only to users authorized by the United States military. PPS users require a special receiver that employs cryptographic logic to remove the effects of SA. Because PPS use is restricted, nearly all civilian GPS users use the SPS.

GPS Receiving Equipment

At this writing (1996), GPS receivers have become common-place in the consumer electronics market, and some models cost $200 or less. However, receivers suitable for use as a transfer standard are much less common, and more expensive. Receivers range in price from less than $100 for an OEM timing board, to $20,000 or more for the most sophisticated models. The price often depends on the quality of the receiver's timebase oscillator. Lower priced models have a low-quality timebase that must be constantly steered to follow the GPS signal. Higher priced receivers have better timebases (some have internal rubidium oscillators) and can ignore many of the GPS path variations because their oscillator allows them to coast for longer intervals [21].

Most GPS receivers provide a 1 *pulse per second* (pps) output. Some receivers also provide a 1 kHz output (derived from the C/A code) and at least one standard frequency output (1, 5, or 10 MHz). To use these receivers, one simply mounts the antenna, connects the antenna to the receiver, and turns the receiver on. The antenna is often a small cone or disk (normally about 15 cm in diameter) that must be mounted outdoors where it has a clear, unobstructed view of the sky. Once the receiver is turned on, it performs a sky search to find out which satellites are currently above the horizon and visible from the antenna site. It then computes its three-dimensional coordinate (latitude, longitude, and altitude as long as four satellites are in view) and begins producing a frequency signal. The simplest GPS receivers have just one channel and look at multiple satellites using a sequencing scheme that rapidly switches between satellites. More sophisticated models have parallel tracking capability and can assign a separate channel to each satellite in view. These receivers typically track from 5 to 12 satellites at once (although more than 8 will only be available in rare instances). By averaging data from multiple satellites, a receiver can remove some of the effects of SA and reduce the frequency uncertainty.

GPS Performance

GPS has many technical advantages over LORAN-C. The signals are usually easier to receive, the equipment is often less expensive, and the coverage area is much larger. In addition, the performance of GPS is superior to that of LORAN-C. However, like all transfer standards, the short-term stability of a GPS receiver is poor (made worse by SA), and it lengthens the time required to make a calibration. As with LORAN-C, a measurement period of at least 24 h is recommended when calibrating atomic frequency standards using GPS.

To illustrate this, Figure 19.8 shows a 100 s comparison between GPS and a cesium oscillator. The cesium oscillator has a frequency offset of $\cong 1 \times 10^{-13}$, and its accumulated phase shift during the 100 s measurement period is <1 ns. Therefore, the noise on the graph can be attributed to GPS path variations and SA.

Figure 19.9 shows a 1-week comparison between a GPS receiver and the same cesium oscillator used in Figure 19.8. The range of the data is 550 ns. The thick line is a least squares estimate of the frequency offset. Although the GPS path noise is still clearly visible, one can see the linear trend contributed by the frequency offset of the cesium; this trend implies that the cesium oscillator is low in frequency by $\cong 5 \times 10^{-13}$.

19.5 Calibration Methods

To review, frequency standards are normally calibrated by comparing them to a traceable reference frequency. In this section, a discussion of how this comparison is made, is presented. To begin, look at the electric signal produced by a frequency standard. This signal can take several forms, as illustrated in Figure 19.10. The dashed lines represent the supply voltage inputs (ranging from 5 V to 15 V for CMOS), and the bold solid lines represent the output voltage.

If the output frequency is an oscillating sine wave, it might look like the one shown in Figure 19.11. This signal produces one cycle (2π radians of phase) in one period. Frequency calibration systems compare a signal like the one shown in Figure 19.11 to a higher quality reference signal. The system then measures and records the change in phase between the two signals. If the two frequencies were exactly the same, their phase relationship would not change. Because the two frequencies are not exactly the same, their phase relationship will change; and by measuring the rate of change, one can determine the frequency offset of the DUT. Under normal circumstances, the phase changes in an orderly, predictable fashion. However, external factors like power outages, component failures, or human errors can cause a sudden

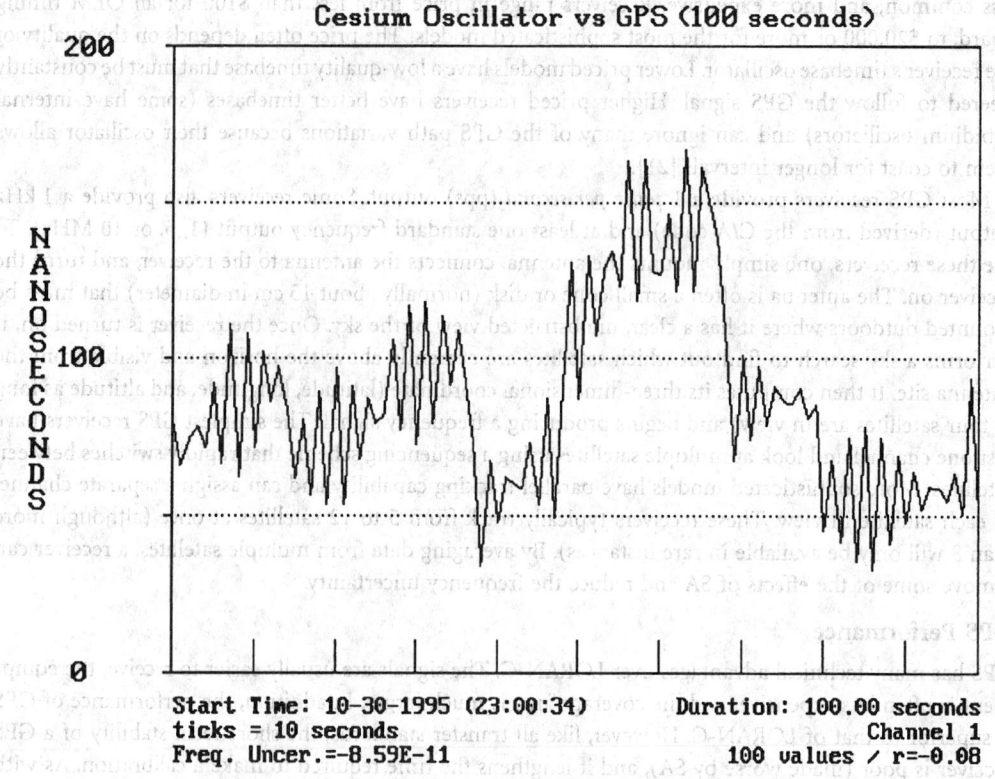

FIGURE 19.8   GPS compared to cesium oscillator over 100 s interval.

phase change, or *phase step*. A calibration system measures the total amount of phase shift (caused either by the frequency offset of the DUT or a phase step) over a given measurement period.

Figure 19.12 shows a phase comparison between two sinusoidal frequencies. The top sine wave represents a signal from the DUT, and the bottom sine wave represents a signal from the reference. Vertical lines have been drawn through the points where each sine passes through zero. The bottom of the figure shows "bars" that indicate the size of the phase difference between the two signals. If the phase relationship between the signals is changing, the phase difference will either increase or decrease to indicate that the DUT has a frequency offset (high or low) with respect to the reference. Equation 19.1 estimates the frequency offset. In Figure 19.12, both Δt and T are increasing at the same rate and the phase difference "bars" are getting wider. This indicates that the DUT has a constant frequency offset.

There are several types of calibration systems that can be used to make a phase comparison. The simplest type of system involves directly counting and displaying the frequency output of the DUT with a device called a *frequency counter*. This method has many applications but is unsuitable for measuring high-performance devices. The DUT is compared to the counter's timebase (typically a TCXO), and the uncertainty of the system is limited by the performance of the timebase, typically $\cong 1 \times 10^{-8}$. Some counters allow use of an external timebase, which can improve the results. The biggest limitation is that frequency counters display frequency in hertz and have a limited amount of resolution. Detecting small changes in frequency could take many days or weeks, which makes it difficult or impossible to use this method to adjust a precision oscillator or to measure stability [22]. For this reason, most high-performance calibration systems collect time series data that can be used to estimate both frequency uncertainty and stability. A discussion follows on how phase comparisons are made using two different methods, the *time interval method* and the *dual mixer time difference method* [9, 10].

The time interval method uses a device called a *time interval counter* (TIC) to measure the time interval between two signals. A TIC has inputs for two electrical signals. One signal starts the counter and the

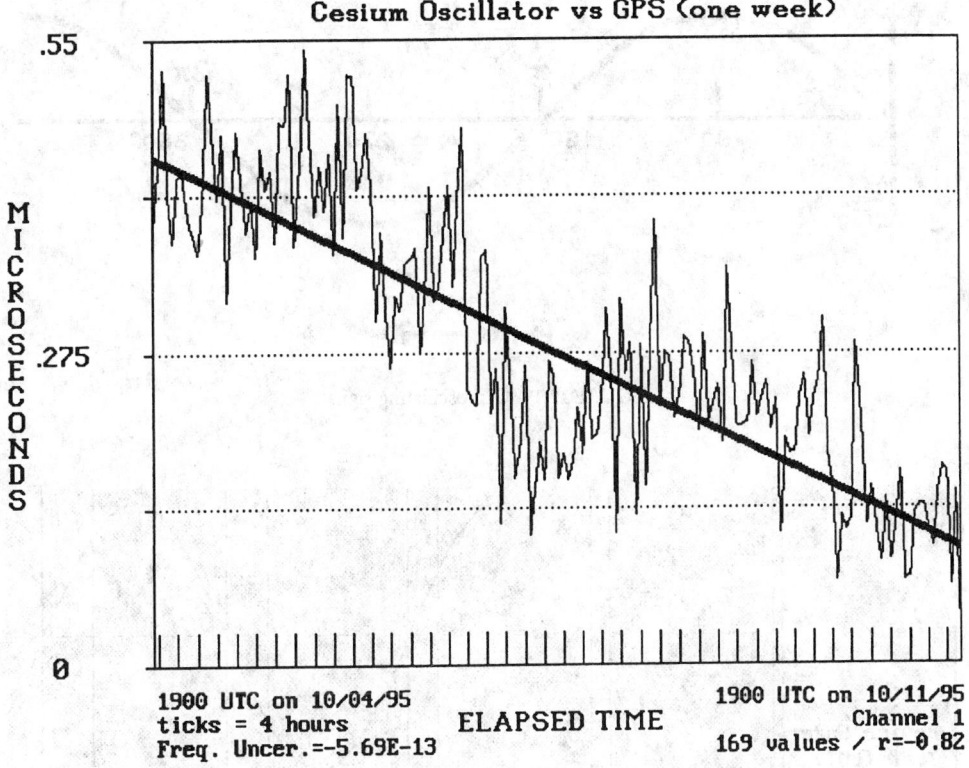

FIGURE 19.9   GPS compared to cesium oscillator over 1 week interval.

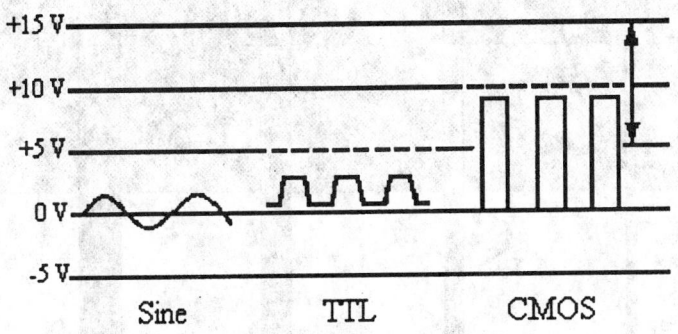

FIGURE 19.10   Oscillator outputs.

other signal stops it. If the two signals have the same frequency, the time interval will not change. If the two signals have different frequencies, the time interval will change, although usually very slowly. By looking at the rate of change, one can calibrate the device. It is exactly as if there were two clocks. By reading them each day, one can determine the amount of time one clock gained or lost relative to the other clock. It takes two time interval measurements to produce any useful information. By subtracting the first measurement from the second, one can tell whether time was gained or lost.

TICs differ in specification and design details, but they all contain several basic parts known as the *timebase*, the *main gate*, and the *counting assembly*. The timebase provides evenly spaced pulses used to measure time interval. The timebase is usually an internal quartz oscillator that can often be phase locked to an external reference. It must be stable because timebase errors will directly affect the measurements.

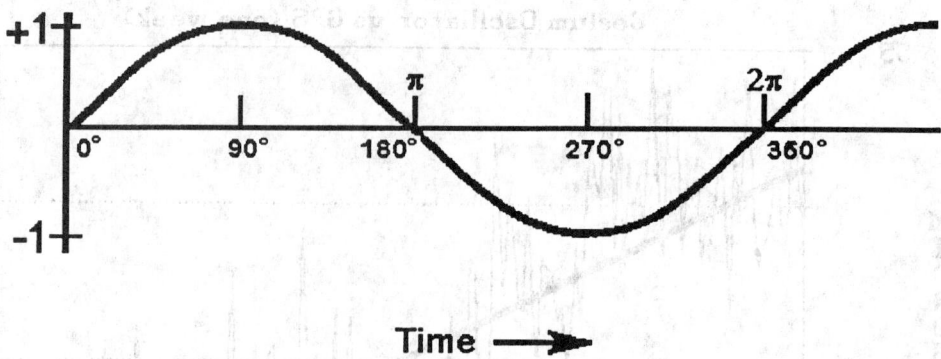

FIGURE 19.11   An oscillating signal.

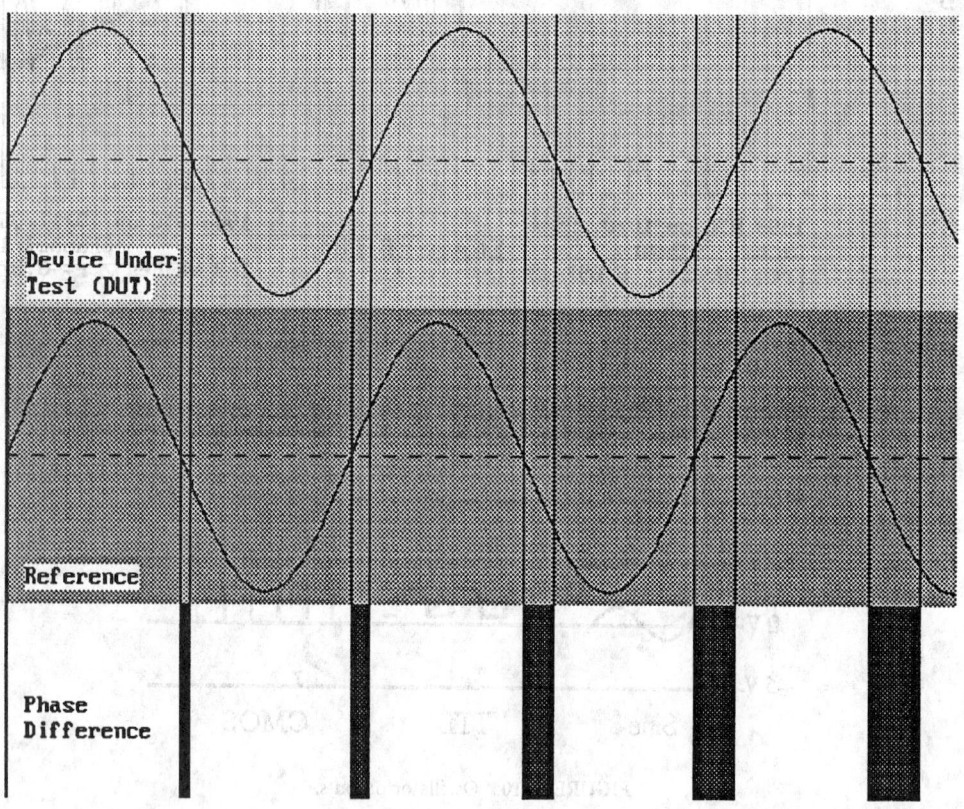

FIGURE 19.12   Two signals with a changing phase relationship.

The main gate controls the time at which the count begins and ends. Pulses that pass through the gate are routed to the counting assembly, where they are displayed on the TIC's front panel or read by computer. The counter can then be reset (or *armed*) to begin another measurement. The stop and start inputs are usually provided with level controls that set the amplitude limit (or *trigger level*) at which the counter responds to input signals. If the trigger levels are set improperly, a TIC might stop or start when it detects noise or other unwanted signals and produce invalid measurements.

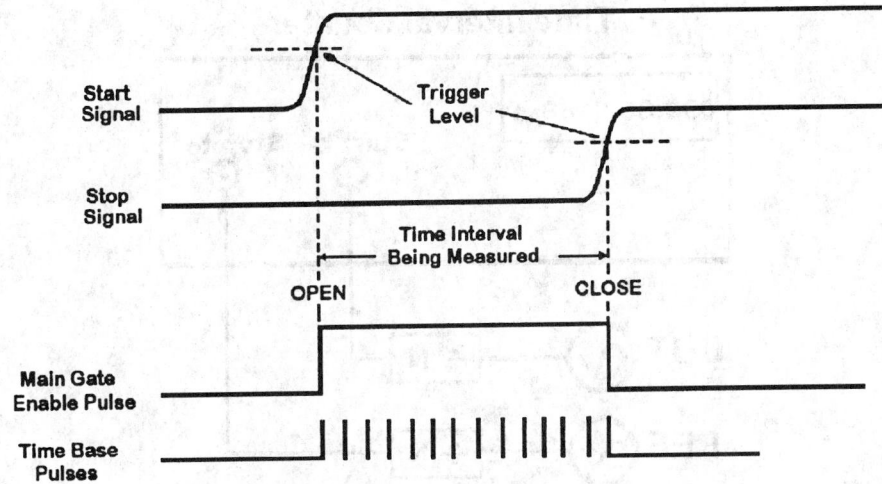

FIGURE 19.13   Measuring time interval.

Figure 19.13 illustrates how a TIC measures the interval between two signals. Input A is the start pulse and Input B is the stop pulse. The TIC begins measuring a time interval when the start signal reaches its trigger level and stops measuring when the stop signal reaches its trigger level. The time interval between the start and stop signals is measured by counting cycles from the timebase. The measurements produced by a TIC are in time units: milliseconds, microseconds, nanoseconds, etc. These measurements assign a value to the phase difference between the reference and the DUT.

The most important specification of a TIC is *resolution*; that is, the degree to which a measurement can be determined. For example, if a TIC has a resolution of 10 ns, it can produce a reading of 3340 ns or 3350 ns, but not a reading of 3345 ns. Any finer measurement would require more resolution. In simple TIC designs, the resolution is limited to the period of the TIC's timebase frequency. For example, a TIC with a 10 MHz timebase is limited to a resolution of 100 ns if the unit cannot resolve time intervals smaller than the period of one cycle. To improve this situation, some TIC designers have multiplied the timebase frequency to get more cycles and thus more resolution. For example, multiplying the timebase frequency to 100 MHz makes 10 ns resolution possible, and 1 ns counters have even been built using a 1 GHz timebase. However, a more common way to increase resolution is to detect parts of a timebase cycle through interpolation and not be limited by the number of whole cycles. Interpolation has made 1 ns TICs commonplace, and 10 ps TICs are available [23].

A time interval system is shown in Figure 19.14. This system uses a TIC to measure and record the difference between two signals. Low-frequency start and stop signals must be used (typically 1 Hz). Since oscillators typically produce frequencies like 1, 5, and 10 MHz, the solution is to use a *frequency divider* to convert them to a lower frequency. A frequency divider could be a stand-alone instrument, a small circuit, or just a single chip. Most divider chips divide by multiples of 10, so it is common to find circuits that divide by a thousand, a million, etc. For example, dividing a 1 MHz signal by 10^6 produces a 1 Hz signal. Using low-frequency signals reduces the problem of counter overflows and underflows and helps prevent errors that could be made if the start and stop signals are too close together. For example, a TIC might make errors when attempting to measure a time interval of <100 ns.

The time interval method is probably the most common method in use today. It has many advantages, including low cost, simple design, and excellent performance when measuring long-term frequency uncertainty or stability. There are, however, two problems that can limit its performance when measuring short-term stability. The first problem is that after a TIC completes a measurement, the data must be processed before the counter can reset and prepare for the next measurement. During this delay, called *dead time*, some information is lost. Dead time has become less of a problem with modern TICs where

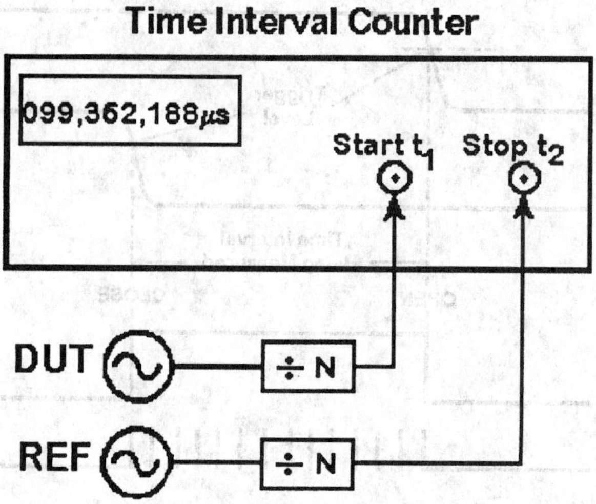

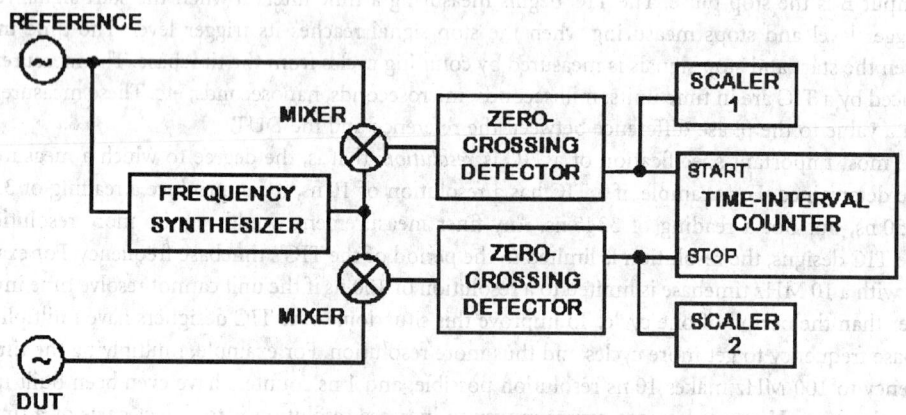

FIGURE 19.14   Time interval system.

FIGURE 19.15   Dual mixer time difference system.

faster hardware and software has reduced the processing time. A more significant problem when measuring short-term stability is that the commonly available frequency dividers are sensitive to temperature changes and are often unstable to ±1 ns. Using more stable dividers adds to the system cost [10, 22].

A more complex system, better suited for measuring short-term stability, uses the Dual Mixer Time Difference (DMTD) method, shown in Figure 19.15. This method uses a frequency synthesizer and two mixers in parallel. The frequency synthesizer is locked to the reference oscillator and produces a frequency lower than the frequency output of the DUT. The synthesized frequency is then heterodyned (or mixed) both with the reference frequency and the output of the DUT to produce two beat frequencies. The beat frequencies are out of phase by an amount proportional to the time difference between the DUT and reference. A *zero crossing detector* is used to determine the zero crossing of each beat frequency cycle. An event counter (or *scaler*) accumulates the coarse phase difference between the oscillators by counting the number of whole cycles. A TIC provides the fine-grain resolution needed to count fractional cycles. The resulting system is much better suited for measuring short-term stability than a time interval system. The oscillator, synthesizer, and mixer combination provides the same function as a divider, but the noise is reduced, often to ±5 ps or less.

Other variations of these calibration methods are used and might be equally valid for a particular application [9, 10, 21, 22]. Keep in mind that a well-defined and documented calibration method is mandatory for a laboratory seeking ISO registration or compliance with a laboratory accreditation program.

19.6 Future Developments

In future years, the performance of frequency standards will continue to improve. One promising development is the *cesium-fountain* standard. This device works by laser cooling cesium atoms and then lofting them vertically through a microwave cavity. The resonance frequency is detected as the atoms rise and fall under the influence of gravity. Many laboratories are working on this concept, which should reduce the frequency uncertainty realized with existing atomic-beam cesium standards [24]. Eventually, a *trapped-ion* standard could lead to improvements of several orders of magnitude. This standard derives its resonance frequency from the systematic energy shifts in transitions in certain ions that are held motionless in an electromagnetic trap. The frequency uncertainty of such a device could be as small as $\pm 1 \times 10^{-18}$ [25]. Also, new statistical tools could improve the ability to estimate oscillator stability, particularly at long term [26].

The future of transfer standards should involve more and more reliance on satellite receivers, and their performance should continue to improve. Ground-based systems such as LORAN-C are expected to be phased out [27]. The frequency uncertainty of GPS receivers will improve when the Selective Availability (SA) program is discontinued in the early part of the next century [28].

References

1. International Organization for Standardization (ISO), *International Vocabulary of Basic and General Terms in Metrology (VIM)*, Geneve, Switzerland, 1993.
2. ISO/IEC Guide 25, *General Requirements for the Competence of Calibration and Testing Laboratories*, International Organization for Standardization (ISO), 1990.
3. ANSI/NCSL Z540-1-1994, *Calibration Laboratories and Measuring and Test Equipment — General Requirements*, American National Standards Institute, 1994.
4. M. A. Lombardi, An introduction to frequency calibration. Part I, *Cal Lab Int. J. Metrol.*, January-February, 17-28, 1996.
5. B. N. Taylor and C. E. Kuyatt, Guidelines for evaluating and expressing the uncertainty of NIST measurement results, *Natl. Inst. of Stan. and Technol. Tech. Note 1297*, 1994.
6. IEEE, IEEE Standard Definitions of Physical Quantities for Fundamental Frequency and Time Metrology, *IEEE 1139*, Piscataway, NJ, 1988.
7. D. W. Allan, H. Hellwig, P. Kartaschoff, J. Vanier, J. Vig, G. M. R. Winkler, and N. F. Yannoni, Standard Terminology for Fundamental Frequency and Time Metrology, *Characterization of Clocks and Oscillators — Natl. Inst. of Stan. and Technol. Tech. Note 1337*, 1990, 139-145.
8. J. Jesperson, Introduction to the time domain characterization of frequency standards, *Proc. 25th Annu. Precise Time and Time Interval (PTTI) Meeting*, Pasadena, CA, December 1991, 83-102.
9. S. R. Stein, Frequency and time — their measurement and characterization, *Precision Frequency Control*, Vol. 2, E. A. Gerber and A. Ballato, Eds., Academic Press, New York, 1985, 191-232.
10. D. A. Howe, D. W. Allan, and J. A. Barnes, Properties of signal sources and measurement methods, *Characterization of Clocks and Oscillators*, D. B. Sullivan, D. W. Allan, D. A. Howe, and F. L. Walls, Eds., *Natl. Inst. of Stan. Technol. Tech. Note 1337*, 1990, 14-60.
11. W. A. Marrison, The evolution of the quartz crystal clock, *Bell Systems Tech.*, 27(3), 510-588, 1948.
12. J. R. Vig, Introduction to Quartz Frequency Standards, *Army Research and Development Technical Report*, SLCET-TR-92-1, October 1992.
13. F. L. Walls and J. Gagnepain, Environmental sensitivities of quartz oscillators, *IEEE Trans. Ultrason., Ferroelectr., Freq. Control*, 39, 241-249, March 1992.

14. W. M. Itano and N. F. Ramsey, Accurate measurement of time, *Sci. Am.*, 269(1), 56-65, 1993.

15. L. Lewis, An introduction to frequency standards, *Proc. IEEE*, 79(7), 927-935, 1991.

16. J. Vanier and C. Audoin, *The Quantum Physics of Atomic Frequency Standards*, Adam Hilger, Bristol, England, 2 Vols., 1989.

17. M. A. Lombardi, An introduction to frequency calibration. Part II, *Cal Lab Int. J. Metrology*, March-April, 28-34, 1996.

18. R. Beehler and M. A. Lombardi, NIST time and frequency services, *Natl. Inst. of Stan. and Technol. Special Publ. 432*, 1991.

19. B. Hoffmann-Wellenhof, H. Lichtenegger, and J. Collins, *GPS: Theory and Practice*, 3rd ed., Springer-Verlag, New York, 1994.

20. ARINC Researc Corporation, NAVSTAR Global Positioning System: User's Overview, NAVSTAR GPS Joint Program Office, Los Angeles, YEE-82-009D, March 1991.

21. T. N. Osterdock and J. A. Kusters, Using a new GPS frequency reference in frequency calibration operations, *IEEE Int. Freq. Control Symp.*, 1993, 33-39.

22. R. J. Hesselberth, Precise frequency measurement and calibration, *Proc. Natl. Conf. Stan. Lab.*, 1988, 47-1–47-7.

23. V. S. Zhang, D. D. Davis, and M. A. Lombardi, High resolution time interval counter, *Precise Time and Time Interval Conf. (PTTI)*, 1994, 191-200.

24. A. Clairon, P. Laurent, G. Santarelli, S. Ghezali, S. N. Lea, and M. Bahoura, A cesium fountain frequency standard: preliminary results, *IEEE Trans. Instrum. Meas.*, 44(2), 128-131, 1995.

25. W. M. Itano, Atomic ion frequency standards, *Proc. IEEE*, 79(7), 936-942, 1991.

26. D. A. Howe, An extension of the Allan variance with increased confidence at long term, *IEEE Int. Freq. Control Symp.*, 1995, 321-330.

27. U.S. Dept. of Transportation, Federal Radionavigation Plan, *DOT-VNTSC-RSPA-95-1*, DOD-4650.5, 1994 (new version published every 2 years).

28. G. Gibbons, A national GPS policy, *GPS World*, 7(5), 48-50, 1996.

IV

Mechanical Variables Measurement — Solid

IV

Mechanical Variables Measurement — Solid

20

Mass and Weight Measurement

Mark Fritz
Denver Instrument Company

Emil Hazarian
Denver Instrument Company

Mass and weight are often used interchangeably; however, they are different. Mass is a quantitative measure of inertia of a body at rest. As a physical quantity, mass is the product of density and volume.

Weight or weight force is the force with which a body is attracted toward the Earth. Weight force is determined by the product of the mass and the acceleration of gravity.

$$M = V \times D \qquad (20.1)$$

where M = mass
V = volume
D = density

Note: In most books, the symbol for density is the Greek letter ρ.

$$W = M \times G \qquad (20.2)$$

where W = weight
G = gravity

The embodiment of units of mass are called weights; this increases the confusion between mass and weight. In the International System of Units (SI), the modernized metric measurement system, the unit for mass is called the kilogram and the unit for force is called the newton. In the United States, the customary system the unit for mass is called the slug and the unit for force is called the pound. When using the U.S. customary units of measure, people are using the unit pound to designate the mass of an object because, in the United States, the pound has been defined in terms of the kilogram since 1893.

20.1   Weighing Instruments

Weighing is one of the oldest known measurements, dating back to before written history. The equal arm balance was probably the first instrument used for weighing. It is a simple device in which two pans are suspended from a beam equal distance from a central pivot point. The standard is placed in one pan and the object to be measured is placed in the second pan; when the beam is level, the unknown is equal to the standard. This method of measurement is still in wide use throughout the world today. Figure 20.1 shows an Ainsworth equal arm balance.

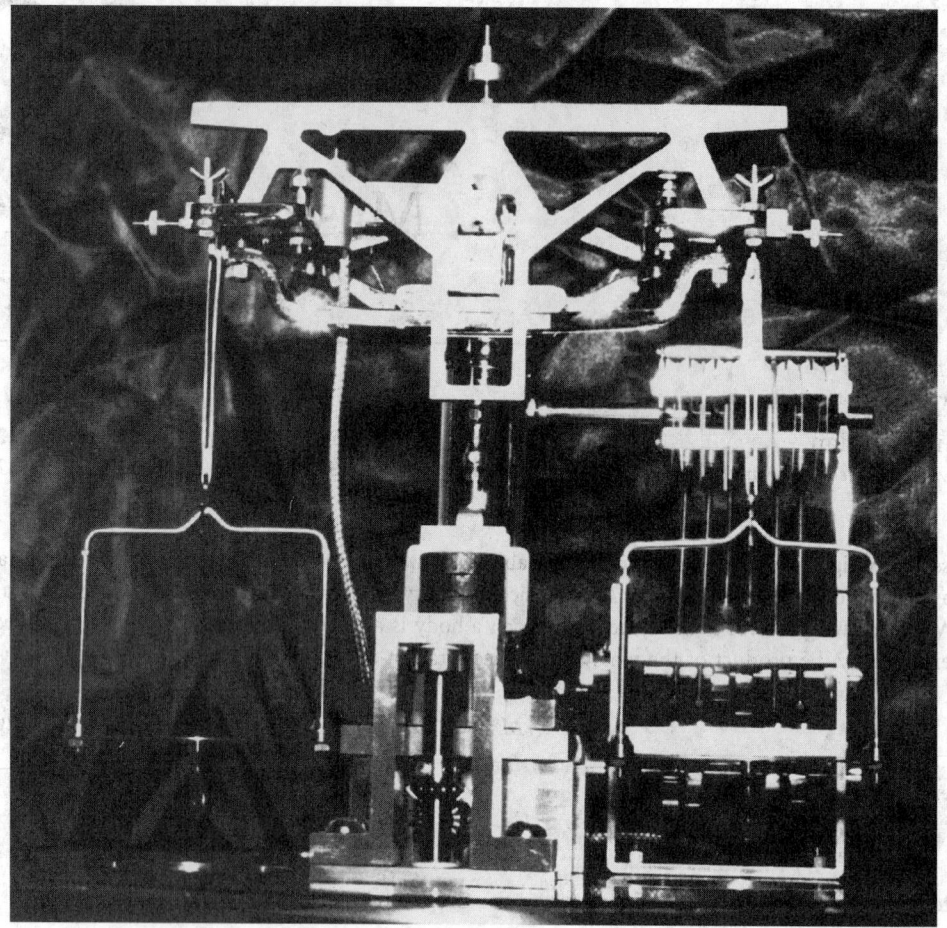

FIGURE 20.1   Ainsworth FV series equal arm balance. (Courtesy Denver Instrument Company.)

A balance is a measuring instrument used to determine the mass of an object by measuring the force exerted by the object on its support within the gravitational field of the Earth. One places a standard of known value on one pan of the balance. One then adds the unknown material to the second pan, until the gravitational force on the unknown material equals the gravitational force on the standard. This can be expressed mathematically as:

$$S \times G = U \times G \qquad (20.3)$$

where S = mass of the standard
G = gravity
U = mass of the unknown

Given the short distance between pans, one assumes that the gravitational forces acting on them are equal. Another assumption is that the two arms of the balance are equal.

Since the gravitational force is equal, it can be removed from the equation and the standard and the unknown are said to be equal. This leads to one of the characteristics of the equal arm balance as well as of other weighing devices, the requirement to have a set of standards that allows for every possible measured value. The balance can only indicate if objects are equal and has a limited capability to determine how much difference there is between two objects.

Probably, the first attempt to produce direct reading balances was the introduction of the single pan substitution balance. A substitution balance is, in principle, similar to an equal arm balance. The object to be measured is placed in the weighing pan; inside the balance are a series of calibrated weights that can be added to the standard side of the balance, through the use of dials and levers. The standard weights can be added and subtracted through the use of the balance's mechanical system, to equal a large variety of weighing loads. Very small differences between the standard weights and the load are read out on an optical scale.

The spring scale is probably the least expensive device for making mass measurements. The force of gravity is once again used as the reference. The scale is placed so that the unknown object is suspended by the spring and the force of gravity can freely act on the object. The elasticity of the spring is assumed to be linear and the force required to stretch it is marked on the scale. When the force of gravity and the elastic force of the spring balance, the force is read from the scale, which has been calibrated in units of mass. Capacity can be increased by increasing the strength of the spring. However, as the strength of the spring increases, the resolution of the scale decreases. This decrease in resolution limits spring scales to relatively small loads of no more than a few kilograms. There are two kinds of springs used: spiral and cantilevered springs.

The torsion balance is a precise adaptation of the spring concept used to determine the mass indirectly. The vertical force produced by the load produces a torque on a wire or beam. This torque produces an angular deflection. As long as the balance is operated in the linear range, the angular deflection is proportional to the torque. Therefore, the angular deflection is also proportional to the applied load. When the torsion spring constant is calibrated, the angular deflection can be read as a mass unit. Unlike the crude spring scales, it is possible to make torsion balances capable of measuring in the microgram region. The torsion element could be a band, a wire, or a string.

The beam balance is probably the next step in accuracy and cost. The beam balance uses the same principle of operation as the equal arm balance. However, it normally has only one pan and the beam is offset. A set of sliding weights are mounted on the beam. As the weights slide out the beam, they gain a mechanical advantage due to the inequality of the distance from the pivot point of the balance. The weights move out along the beam until the balance is in equilibrium. Along the beam, there are notched positions that are marked to correspond to the force applied by the sliding weights. By adding up the forces indicated by the position of each weight, the mass of the unknown material is determined. Beam balances and scales are available in a wide range of accuracy's load capacities. Beam balances are available to measure in the milligram range and large beam scales are made with capacities to weigh trucks and trains. Once again, the disadvantage is that as load increases the resolution decreases. Figure 20.2 shows an example of a two pan beam balance.

The next progression in cost and accuracy is the strain gage load cell. A strain gage is an electrically resistive wire element that changes resistance when the length of the wire element changes. The gage is bonded to a steel cylinder that will shorten when compressed or lengthen when stretched. Because the gage is bonded to the cylinder, the length of the wire will lengthen or contract with the cylinder. The electrical resistance is proportional to the length of the wire element of the gage. By measuring the resistance of the strain gage, it is possible to determine the load on the load cell. The electric resistance is converted into a mass unit readout by the electric circuitry in the readout device.

The force restorative load cell is the heart of an electronic balance, shown in Figure 20.3. The force restorative load cell uses the principle of the equal arm balance. However, in most cases, the fulcrum is offset so it is no longer an equal arm balance, as one side is designed to have a mechanical advantage. This side of the balance is attached to an electric coil. The coil is suspended in a magnetic field. The other side is still connected to a weighing pan. Attached to the beam is a null indicating device, consisting of a photodiode and a light-emitting diode (LED) that are used to determine when the balance is in equilibrium. When a load is placed on the weighing pan, the balance goes out of equilibrium. The LED photodiode circuit detects that the balance is no longer in equilibrium, and the electric current in the coil is increased to bring the balance back to equilibrium. The electric current is then measured across a precision sense resistor and converted into a mass unit reading and displayed on the digital readout.

FIGURE 20.2 Beam balance. (Courtesy Denver Instrument Company.)

FIGURE 20.3 Force restorative load cell. (Courtesy Denver Instrument Company.)

A variation of the latter is the new generation of industrial scales, laboratory balances, and mass comparators. Mass comparators are no longer called balances because they always perform a comparison between known masses (standards) and unknown masses. These weighing devices are employing the electromagnetic force compensation principle in conjunction with joint flexures elements replacing the traditional knife-edge joints. Some of the advantages include a computer interfacing capability and a maintenance-free feature because there are no moving parts.

Another measuring method used in the weighing technology is the vibrating cord. A wire or cord of known length, which vibrates transversely, is tensioned by the force F. The vibration frequency changes in direct proportion to the load F. The piezoelectric effect is also used in weighing technology. Such weighing devices consist of the presence of an electric voltage at the surface of a crystal when the crystal is under load. Balances employing the gyroscopic effect are also used. This measuring device uses the output signal of a gyrodynamic cell similar to the frequency. Balances wherein the weight force of the load changes the reference distance of the capacitive or inductive converters are also known. As well, balances using the radioactivity changes of a body as a function of its mass under certain conditions exist.

20.2 Weighing Techniques

When relatively low orders of accuracy are required, reading mass or weight values directly from the weighing instrument are adequate. Except for the equal arm balance and some torsion balances, most modern weighing instruments have direct readout capability. For most commercial transactions and for simple scientific experiments, this direct readout will provide acceptable results.

In the case of equal arm balances, the balance will have a pointer and a scale. When relatively low accuracy is needed, the pointer and scale are used to indicate when the balance is close to equilibrium. The same is true when using a torsion balance. However, the equal arm balances of smaller (e.g., 30 g) or larger (e.g., 900 kg) capacity are also used for high-accuracy applications. Only the new generation of electronic balances are equal or better in terms of accuracy and benefit from other features.

Weighing is a deceptively simple process. Most people have been making and using weighing measurements for most of their lives. We have all gone to the market and purchased food that is priced by weight. We have weighed ourselves many times, and most of us have made weight or mass measurements in school. What could be simpler? One places an object on the weighing pan or platform and reads the result.

Unfortunately, the weighing process is very susceptible to error. There are errors caused by imperfections in the weighing instrument; errors caused by biases in the standards used; errors caused by the weighing method; errors caused by the operator; and errors caused by environmental factors. In the case of the equal arm balance, any difference between the lengths of the arms will result in a bias in the measurement result. Nearly all weighing devices will have some degree of error caused by small amounts of nonlinearity in the device. All standards have some amount of bias and uncertainty. Mass is the only base quantity in the International System of Units (SI) defined in relation with a physical artifact. The international prototype of the kilogram is kept at Sevres in France, under the custody of the International Bureau of Weights and Measures. All weighing measurements originate from this international standard. The international prototype of the kilogram is, by international agreement, exact; however, over the last century, it has changed in value. What one does not know is the exact magnitude or direction of the change. Finally, environmental factors such as temperature, barometric pressure, and humidity can affect the weighing process.

There are many weighing techniques used to reduce the errors in the weighing process. The simplest technique is *substitution weighing*. The substitution technique is used to eliminate some of the errors introduced by the weighing device. The single-substitution technique is one where a known standard and an unknown object are both weighed on the same device. The weighing device is only used to determine the difference between the standard and the unknown. First, the standard is weighed and the weighing device's indication is noted. (In the case of an equal arm balance, tare weights are added to the second pan to bring the balance to equilibrium.) The standard is then removed from the weighing device and the unknown is placed in the same position. Again, the weighing device's indication is noted. The first noted indication is subtracted from the second indication. This gives the difference between the standard and the unknown. The difference is then added to the known value of the standard to calculate the value of the unknown object. A variation of this technique is to use a small weight of known value to offset the weighing device by a small amount. The amount of offset is then divided by the known value of the small weight to calibrate the readout of the weighing device. The weighing results of this measurement is calculated as follows:

$$U = S + \left(O_2 - O_1\right)\left(SW\right) / \left(O_3 - O_2\right)$$ (20.4)

where U = value of the unknown
      S = known value of the standard
      SW = small sensitivity weight used to calibrate the scale divisions
      O_1 = first observation (standard)
      O_2 = second observation (unknown)
      O_3 = third observation (unknown + SW)

These techniques remove most of the errors introduced by the weighing device, and are adequate when results to a few tenths of a gram are considered acceptable.

If results better than a few tenths of a gram are required, environmental factors begin to cause significant errors in the weighing process. Differences in density between the standard and the unknown object and air density combine together to cause significant errors in the weighing process.

It is the buoyant force that generates the confusion in weighing. What is called the "true mass" of an object is the mass determined in vacuum. The terms "true mass" and "mass in vacuum" are referring to the same notion of inertial mass or mass in the Newtonian sense. In practical life, the measurements are performed in the surrounding air environment. Therefore, the objects participating in the measurement process adhere to the Archimedean principle being lifted with a force equal to the weight of the displaced volume of air. Applying the buoyancy correction to the measurement requires the introduction of the term "apparent mass." The "apparent mass" of an object is defined in terms of "normal temperature" and "normal air density," conventionally chosen as 20°C and 1.2 mg cm^{-3}, respectively. Because of these conventional values, the "apparent mass" is also called the "conventional mass." The reference material is either brass (8.4 g cm^{-3}) or stainless steel (8.0 g cm^{-3}), for which one obtains an "apparent mass versus brass" and an "apparent mass versus stainless steel," respectively. The latter is preferred for reporting the "apparent mass" of an object.

Calibration reports from the National Institute of Standards and Technology will report mass in three ways: True Mass, Apparent Mass versus Brass, and Apparent Mass versus Stainless Steel. Conventional mass is defined as the mass of an object with a density of 8.0 g cm^{-3}, at 20°C, in air with a density of 1.2 mg cm^{-3}. However, most scientific weighings are of materials with densities that are different from 8.0 g cm^{-3}. This results in significant measurement errors.

As an example, use the case of a chemist weighing 1 liter of water. The chemist will first weigh a mass standard, a 1 kg weight made of stainless steel; then the chemist will weigh the water. The 1 kg mass standard made of 8.0 g cm^{-3} stainless steel will have a volume of 125 cm^3. The same mass of water will have a volume approximately equal to 1000 cm^3 (Volume = Mass/Density). The mass standard will displace 125 cm^3 of air, which will exert a buoyant force of 150 mg (125 cm^3 × 1.2 mg cm^{-3}). However, the water will displace 1000 cm^3 air, which will exert a buoyant force of 1200 mg (1000 cm^3 × 1.2 mg cm^{-3}). Thus, the chemist has introduced a significant error into the measurement by not taking the differing densities and air buoyancy into consideration.

Using 1.2 mg cm^{-3} for the density of air is adequate for measurements made close to sea level; it must be noted that air density decreases with altitude. For example, the air density in Denver, CO, is approximately 0.98 mg cm^{-3}. Therefore, to make accurate mass measurements, one must measure the air density at the time of the measurement if environmental errors in the measurement are to be reduced.

Air density can be calculated to an acceptable value using the following equations:

$$\rho_A \cong 0.0034848 / \left(t + 273.15\right)\left(P - 0.0037960 \times U \times e_s\right)$$ (20.5)

where ρ_A = air density in mg cm^{-3}
t = temperature in °C
P = barometric pressure in pascals
U = relative humidity in percent
e_s = saturation vapor pressure

$$e_s \cong \left(1.7526 \times 10^{11}\right) \times e^{\left(-5315.56 / \left(273.15 + t\right)\right)} \tag{20.6}$$

where $e \cong 2.7182818$
t = temperature in °C

To apply an air buoyancy correction to the single substitution technique, use the following formulae:

$$M_u = \left(M_s\left(1 - \rho A/\rho_s\right) + \left(O_2 - O_1\right)\left(M_{SW}\left(1 - \rho_A/\rho_{SW}\right)/\left(O_3 - O_2\right)\right)\right) / \left(1 - \rho_A/\rho_u\right) \tag{20.7}$$

where M_u = mass of the unknown (in a vacuum)
M_s = mass of the standard (in a vacuum)
M_{sw} = mass of the sensitivity weight
ρ_A = air density
ρ_s = density of the standard
ρ_u = density of the unknown
ρ_{sw} = density of the sensitivity weight
O_1 = first observation (standard)
O_2 = second observation (unknown)
O_3 = third observation (unknown + SW)

$$CM = M_u\left(1 - 0.0012/\rho_u\right)/0.99985 \tag{20.8}$$

where CM = conventional mass
M_u = mass of the unknown in a vacuum
ρ_u = density of the unknown

When very precise measurements are needed, the double-substitution technique coupled with an air buoyancy correction will provide acceptable results for nearly all scientific applications. The double-substitution technique is similar to the single-substitution technique using the sensitivity weight. In the double-substitution technique, the sensitivity weight is weighed with both the mass standard and the unknown. The main advantage of this technique over single substitution is that any drift in the weighing device is accounted for in the technique. Because of the precision of this weighing technique, it is only appropriate to use it on precision balances or mass comparators. As in the case of single substitution, one places the standard on the balance pan and takes a reading. The standard is then removed and the unknown object is placed on the balance pan and a second reading is taken. The third step is to add the small sensitivity weight to the pan with the unknown object and take a third reading. Then remove the unknown object and return the standard to the pan with the sensitivity weight and take a fourth reading. The mass is calculated using the following formulae:

$$M_u = \frac{\left(M_s\left(1 - \rho_A/\rho_s\right) + \left(O_2 - O_1 + O_3 - O_4\right)\right)/2\left(M_{SW}\left(1 - \rho_A/\rho_{SW}\right)/\left(O_3 - O_2\right)\right)}{\left(1 - \rho_A/\rho_u\right)} \tag{20.9}$$

where M_u = mass of the unknown (in a vacuum)
  M_s = mass of the standard (in a vacuum)
  M_{sw} = mass of the sensitivity weight
  ρ_A = air density
  ρ_s = density of the standard
  ρ_u = density of the unknown
  ρ_{sw} = density of the sensitivity weight
  O_1 = first observation (standard)
  O_2 = second observation (unknown)
  O_3 = third observation (unknown + sensitivity weight)
  O_4 = fourth observation (standard + sensitivity weight)

$$CM = M_u\left(1 - 0.0012/\rho_u\right)/0.99985 \tag{20.10}$$

where CM = conventional mass
  M_u = mass of the unknown in a vacuum
  ρ_u = density of the unknown

To achieve the highest levels of accuracy, advanced weighing designs have been developed. These advanced weighing designs incorporate redundant weighing, drift compensation, statistical checks, and multiple standards. The simplest of these designs is the three-in-one design. It uses two standards to calibrate one unknown weight. In its simplest form, one would perform three double substitutions. The first compares the first standard and the unknown weight; the second double substitution compares the first standard against the second standard, which is called the check standard; and the third and final comparison compares the second (or check standard) against the unknown weight. These comparisons would then result in the following:

  O_1 = reading with standard on the balance
  O_2 = reading with unknown on the balance
  O_3 = reading with unknown and sensitivity weight on the balance
  O_4 = reading with standard and sensitivity weight on the balance
  O_5 = reading with standard on the balance
  O_6 = reading with check standard on the balance
  O_7 = reading with check standard and sensitivity weight on the balance
  O_8 = reading with standard and sensitivity weight on the balance
  O_9 = reading with check standard on the balance
  O_{10} = reading with unknown on the balance
  O_{11} = reading with unknown and sensitivity weight on the balance
  O_{12} = reading with check standard and sensitivity weight on the balance

The measured differences are calculated using the following formulae:

$$a = \left[\left(O_1 - O_2 + O_4 - O_3\right)/2\right] \times \left[M_{sw}\left(1 - \rho_A/\rho_{sw}\right)/O_3 - O_2\right] \tag{20.11}$$

$$b = \left[\left(O_5 - O_6 + O_8 - O_7\right)/2\right] \times \left[M_{sw}\left(1 - \rho_A/\rho_{sw}\right)/O_7 - O_6\right] \tag{20.12}$$

$$c = \left[\left(O_9 - O_{10} + O_{12} - O_{11}\right)/2\right] \times \left[M_{sw}\left(1 - \rho_A/\rho_{sw}\right)/O_{11} - O_{10}\right] \tag{20.13}$$

where a = difference between standard and unknown
 b = difference between standard and check standard
 c = difference between check standard and unknown
 M_{sw} = mass of sensitivity weight
 ρ_A = air density calculated using Equations 20.5 and 20.6
 ρ_{sw} = density of sensitivity weight

The least-squares measured difference is computed for the unknown from:

$$d_u = \left(-2a - b - c\right)/3 \tag{20.14}$$

Using the least-squares measured difference, the mass of the unknown is computed as:

$$U = \left(S\left(1 - \rho_A/\rho_S\right) + d_u\right)/\left(1 - \rho_A/\rho_U\right) \tag{20.15}$$

where U = mass of unknown
 S = mass of the standard
 d_u = least-squares measured difference of the unknown
 ρ_A = air density calculated using Equations 20.5 and 20.6
 ρ_S = density of the standard
 ρ_U = density of the unknown

The conventional mass of the unknown is now calculated as:

$$CU = U\left(1 - 0.0012/\rho_U\right)/0.99985 \tag{20.16}$$

where CU = conventional mass
 U = mass of unknown
 ρ_U = density of unknown

The least-squares measured difference is now computed for the check standard as:

$$d_{CS} = \left(-a - 2b - c\right)/3 \tag{20.17}$$

Using the least-squares measured difference, the mass of the check standard is computed from:

$$CS = \left(S\left(1 - \rho_A/\rho_S\right) + d_{CS}\right)/\left(1 - \rho_A/\rho CS\right) \tag{20.18}$$

where CS = mass of check standard
 S = mass of the standard
 d_{CS} = least-squares measured difference of the check standard
 ρ_A = air density calculated using Equations 20.5 and 20.6
 ρ_S = density of the standard
 ρCS = density of unknown

The mass of the check standard must lie within the control limits for the check standard. If it is out of the control limits, the measurement must be repeated.

The short-term standard deviation of the process is now computed:

$$\text{Short-term standard deviation} = 0.577\left(a - b + c\right) \tag{20.19}$$

The short-term standard deviation is divided by the historical pooled short-time standard deviation to calculate the *F*-statistic:

$$F\text{-statistic} = \text{short-term standard deviation/pooled short-time standard deviation}$$

The *F*-statistic must be less than the value obtained from the student *t*-variant at the 99% confidence level for the number of degrees of freedom of the historical pooled standard deviation. If this test fails, the measurement is considered to be out of statistical control and must be repeated.

By measuring a check standard and by computing the short-term standard deviation of the process and comparing them to historical results, one obtains a high level of confidence in the computed value of the unknown.

There are many different weighing designs that are valid; the three-in-one (three equal weights) and four equal weights are the ones that can be easily calculated without the use of a computer. Primary calibration laboratories — private and government — are using these multiple intercomparisons, state-of-the-art mass calibration methods under the Mass Measurement Assurance Program using the Mass Code computer program provided by the National Institute of Standards and Technology. A full discussion of these designs can be found in the *National Bureau of Standards Technical Note 952*.

References

1. J. K. Taylor and H. V. Oppermann, *Handbook for the Quality Assurance of Metrological Measurements*, NBS Handbook 145, Washington, D.C.: U.S. Department of Commerce, National Bureau of Standards, 1986.
2. L. V. Judson, *Weights and Measures Standards of the United States, A Brief History*, NBS Special Publication 447, Washington, D.C.: Department of Commerce, National Bureau of Standards, 1976.
3. P. E. Pontius, *Mass and Mass Values*, NBS Monograph 133, Washington, D.C.: Department of Commerce, National Bureau of Standards, 1974.
4. K.B. Jaeger and R. S. Davis, *A Primer for Mass Metrology*, NBS Special Publication 700-1, Washington, D.C.: Department of Commerce, National Bureau of Standards, 1984.
5. J. M. Cameron, M. C. Croarkin, and R. C. Raybold, *Designs for the Calibration of Standards of Mass*, NBS Technical Note 952, Washington, D.C.: Department of Commerce, National Bureau of Standards, 1977.
6. G. L. Harris (Ed.), *Selective Publications for the Advanced Mass Measurements Workshop*, NISTIR 4941, Washington, D.C.: Department of Commerce, National Institute of Standards and Technology, 1992.
7. Metron Corporation, *Physical Measurements*, NAVAIR 17-35QAL-2, California: U.S. Navy, 1976.
8. R. S. Cohen, *Physical Science*, New York: Holt, Rinehart and Winston, 1976.
9. D. B. Prowse, The Calibration of Balances, CSIRO Division of Applied Physics, Australia, 1995.
10. E. Hazarian, Techniques of mass measurement, *Southern California Edison Mass Seminar Notebook*, Los Angeles, CA, 1994.
11. E. Hazarian, Analysis of mechanical convertors of electronic balances, *Measurement Sci. Conf.*, Anaheim, CA, 1993.
12. B. N. Taylor, *Guide for the use of the International System of Units (SI)*, NIST SP811, 1995.

21

Density Measurement

Halit Eren
Curtin University of Technology

Density is a significant part of measurement and instrumentation. Density measurements are made for at least two important reasons: (1) for the determination of mass and volume of products, and (2) the quality of the product. In many industrial applications, density measurement ascertains the value of the product.

Density is defined as the mass of a given volume of a substance under fixed conditions. However, ultimate care must be exercised in measurements because density varies with pressure and temperature. The variation is much greater in gases.

In many modern applications, the densities of products are obtained by sampling techniques. In measurements, there are two basic concepts: *static density measurements* and *dynamic (on-line) density measurements*. Within each concept, there are many different methods employed. These methods are based on different physical principles. In many cases, the application itself and the characteristics of the process determine the best suitable method to be used. Generally, static methods are well developed, lower in cost, and more accurate. Dynamic samplers are expensive, highly automated, and use microprocessor-based signal processing devices. Nevertheless, nowadays, many static methods are also computerized, offering easy to use, flexible, and self-calibrating features.

There is no single universally applicable density measurement technique. Different methods must be employed for different products and materials. In many cases, density is normalized under reference conditions.

The density of a substance is determined by dividing the density of that substance by the density of a standard substance obtained under the same conditions. This dimensionless ratio is called the *specific gravity* (SG), also termed the *relative density*. The specific gravities of liquid and gases under reference conditions are given by:

$$\text{Liquid SG} = \text{density of liquid} / \text{density of water} \qquad (21.1)$$

$$\text{Gas SG} = \text{density of gas} / \text{density of air} \qquad (21.2)$$

Commonly accepted sets of conditions are *normal temperature and pressure* (NTP) and *standard temperature and pressure* (STP). NTP is usually taken as the temperature of 0.00°C and a pressure of 760 mm Hg. The NTP is accepted as 15.00 or 15.56°C and 101.325 kPa.

21.1 Solid Density

If the uniformity is maintained, the determination of density of solids is a simple task. Once the volume of the solid and its mass are known, the density can be found using the basic ratio: density = mass/volume ($kg\ m^{-3}$).

However, in many applications, solids have different constituents and are made up from different materials having different ratios. Their volumes can also change often. In these cases, dynamic methods are employed, such as radioactive absorption types, ultrasonic, and other techniques. Some of these methods are described below.

21.2 Fluid Density

The measurement of densities of fluids is much more complex than for solids. For fluid densities, many different techniques are available. This is mainly due to complexities in processes, variations of fluid densities during the processes, and diverse characteristics of the process and the fluids themselves. Some of these methods are custom designed and applicable to special cases only. Others are very similar in principles and technology, and applicable to many different type of fluids. At present, apart from conventional methods, there are many novel and unusual techniques undergoing extensive development and research. For example, densitometers based on electromagnetic principles [1] can be given as part of an intelligent instrumentation system.

Fluids can be divided to liquids and gases. Extra care and further considerations are necessary in gas density measurements. For example, perfect gases contain an equal number of molecules under the same conditions and volumes. Therefore, molecular weights can be used in density measurements.

Depending on the application, fluid densities can be measured both in *static* and *dynamic* forms. In general, static density measurements of fluids are well developed, precise, and have greater resolution than most dynamic techniques. Pycnometers and buoyancy are examples of static techniques that can be adapted to cover small density ranges with a high resolution and precision. Nowadays, many manufacturers offer dynamic instruments previously known to be static. Also, many static density measurement devices are computerized and come with appropriate hardware and software. In general, static-type measurements are employed in laboratory conditions, and dynamic methods are employed for real-time measurements where the properties of a fluid vary from time to time.

In this chapter, the discussion will concentrate on the commonly applied, modern density measuring devices. These devices include:

1. Pycnometric densitometers
2. Buoyancy-type densitometers
3. Hydrometers
4. Hydrostatic weighing densitometers
5. Balance-type densitometers
6. Column-type densitometers
7. Vibrating element densitometers
8. Radioactive densitometers
9. Refractometer and index of reflection densitometers
10. Coriolis densitometers
11. Absorption-type densitometers

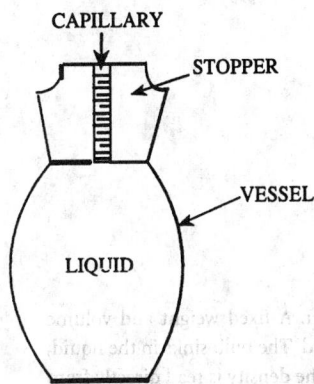

FIGURE 21.1 A pycnometer. A fixed volume container is filled with liquid and weighed accurately; capillary is used to determine the exact volume of the liquid.

Pycnometric Densitometers

Pycnometers are static devices. They are manufactured as fixed volume vessels that can be filled with the sample liquid. The density of the fluid is measured by weighing the sample. The simplest version consists of a vessel in the shape of a bottle with a long stopper containing a capillary hole, as shown in Figure 21.1. The capillary is used to determine the exact volume of the liquid, thus giving high resolution when filling the pycnometer. The bottle is first weighed empty, and then with distilled-aerated water to determine the volume of the bottle. The bottle is then filled with the process fluid and weighed again. The density is determined by dividing the mass by the volume. The specific gravity of the liquid is found by the ratio of the fluid mass to water mass. When pycnometers are used, for good precision, ultimate care must be exercised during the measurements; that is, the bottle must be cleaned after each measurement, the temperature must be kept constant, and precision balances must be used. In some cases, to ensure filling of the pycnometer, twin capillary tubes are used. The two capillaries, made of glass, are positioned such that the fluid can be driven into the vessel under vacuum conditions. Accurate filling to graduation marks on the capillary is then made.

The pycnometers have to be lightweight, strong enough to contain samples, and they need to be nonmagnetic for accurate weighing to eliminate possible ambient magnetic effects. Very high-resolution balances must be used to detect small differences in weights of gases and liquids. Although many pycnometers are made of glass, they are also made of metals to give enough strength for the density measurements of gases and liquids at extremely high pressures. In many cases, metal pycnometers are necessary for taking samples from the line of some rugged processes.

Pycnometers have advantages and disadvantages. Advantages are that if used correctly, they are accurate; and they can be used for both density and specific gravity measurements. The disadvantages include:

1. Great care must be exercised for accurate results.
2. The sample must be taken off-line, with consequent time lag in results. This creates problems of relating samples to the materials that exist in the actual process.
3. High-precision pycnometers are expensive. They require precision weighing scales and controlled laboratory conditions. Specialized techniques must be employed to take samples in high-pressure processes and hostile conditions, such as offshore installations.
4. Their good performances might depend on the skill of operator.

Buoyancy-Type Densitometers

The buoyancy method basically uses Archimedes principle. A suspended sinker, with a known mass and volume attached to a fine wire, is totally immersed in the sample liquid. A precision force balance is used to measure the force to support the sinker. Once the mass, volume, and supporting weight of the sinker

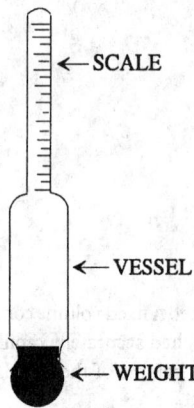

←SCALE

←VESSEL

←WEIGHT

FIGURE 21.2   Hydrometer. A fixed weight and volume bulb is placed into the liquid. The bulb sinks in the liquid, depending on its density. The density is read directly from the scale. Temperature correction is necessary.

are known, the density of the liquid can be calculated. However, some corrections need to be made for surface tension on the suspension wire, the cubicle expansion coefficient of the sinker, and the temperature of process. Buoyancy-type densitometers give accurate results and are used for the calibration of the other liquid density transducers.

One advanced version of the buoyancy technique is the magnetic suspension system. The sinker is fully enclosed in a pressure vessel, thus eliminating surface tension errors. Their uses can also be extended to applications such as the specific gravity measurements under low vapor pressures and density measurements of hazardous fluids.

Hydrometers

Hydrometers are the most commonly used devices for measurement of the density of liquids. They are so commonly used that their specifications and procedure of use are described by national and international standards, such as ISO 387. The buoyancy principle is used as the main technique of operation. The volume of fixed mass is converted to a linear distance by a sealed bulb-shaped glass tube containing a long stem measurement scale, shown in Figure 21.2. The bulb is ballasted with a lead shot and pitch, the mass of which depends on the density range of the liquid to be measured. The bulb is simply placed into the liquid and the density is read from the scale. The scale is graduated in density units such as kg m^{-3}. However, many alternative scales are offered by manufacturers, such as specific gravity, API gravity, Brix, Brine, etc. Hydrometers can be calibrated for different ranges for surface tensions and temperatures. Temperature corrections can be made for set temperature such as 15°C, 20°C, or 25°C. ISO 387 covers a density range of 600 kg m^{-3} to 2000 kg m^{-3}. Hydrometers have a number of advantages and disadvantages. The advantages include:

1. Relatively low cost and easy to use
2. Good resolution for small range
3. Traceable to national and international standards

The disadvantages include:

1. They have small span; therefore, a number of meters are required to cover a significant range.
2. They are made from glass and fragile. Metal and plastic versions are not as accurate.
3. The fluid needs to be an off-line sample, not representing the exact conditions of the process. There are pressure hydrometers for low vapor pressure hydrocarbons, but this adds a need for accurately determining the pressure too.
4. If good precision is required, they are difficult to use, needing surface tension and temperature corrections. Further corrections could be required for opaque fluids.

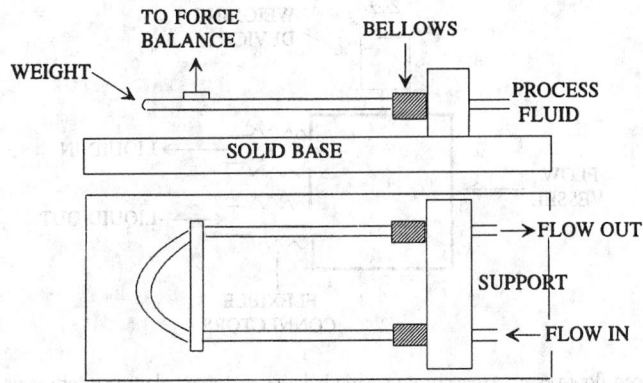

FIGURE 21.3 Hydrostatic weighing. The total weight of a fixed-volume tube containing liquid is determined accurately. The density is calculated using mass: volume ratio.

Hydrostatic Weighing Densitometers

The most common device using a hydrostatic weighing method consists of a U-tube that is pivoted on flexible end couplings. A typical example is shown in Figure 21.3. The total weight of the tube changes, depending on the density of fluid flowing through it. The change in the weight needs to be measured accurately, and there are a number of methods employed to do this. The most common commercial meters use a force balance system. The connectors are stainless steel bellows. In some cases, rubber or PTFE are used, depending on the process fluid characteristics and the accuracy required. There are temperature and pressure limitations due to bellows, and the structure of the system may lead to a reading offset. The meter must be securely mounted on a horizontal plane for optimal accuracy.

The advantages of this method include:

1. They give continuous reading and can be calibrated accurately.
2. They are rugged and can be used for two-phase liquids such as slurries, sugar solutions, powders, etc.

The disadvantages of these meters include:

1. They must be installed horizontally on a solid base. These meters are not flexible enough to adapt to any process; thus, the process must be designed around it.
2. They are bulky and cumbersome to use.
3. They are unsuitable for gas density measurements.

Balance-Type Densitometers

Balance-type densitometers are suitable for liquid and gas density measurements. Manufacturers offer many different types; four of the most commonly used ones are discussed below.

Balanced-Flow Vessel

A fixed volume vessel as shown in Figure 21.4 is employed for the measurements. While the liquid is flowing continuously through the vessel, it is weighed automatically by a sensitive scale — a spring balance system or a pneumatic force balance transmitter. Because the volume and the weight of the liquid are known, the density or specific gravity can easily be calculated and scaled in respective units. In the design process, extra care must be exercised for the flexible end connections.

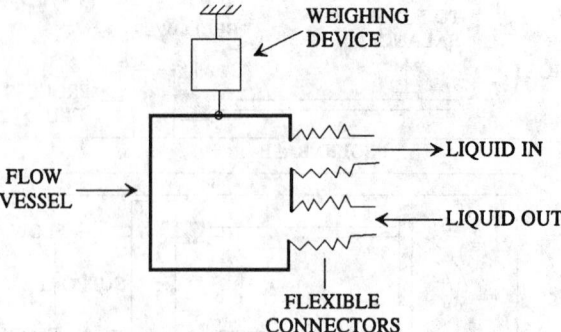

FIGURE 21.4 Balanced flow vessel. An accurate spring balance or force balance system is used to weigh the vessel as the liquid flows through it.

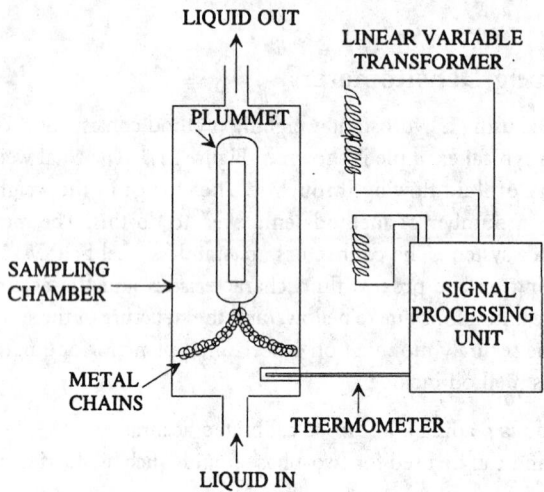

FIGURE 21.5 Chain balance float. The fixed volume and weight plummet totally suspended in the liquid assumes equilibrium position, depending on the density. The force exerted by the chains on the plummet is a function of plummet position; hence, the measured force is proportional to the density of the liquid.

Chain Balanced Float

In this system, a self-centering, fixed-volume, submerged plummet is used for density measurements, as illustrated in Figure 21.5. The plummet is located entirely under the liquid surface. At balance, the plummet operates without friction and is not affected by surface contamination. Under steady-state conditions, the plummet assumes a stable position. The effective weight of the chain on the plummet varies, depending on the position of the plummet, which in turn is a function of the density of the liquid. The plummet contains a metallic transformer core that transmits changes in the position to be measured by a pickup coil. The voltage differential, a function of plummet displacement, is calibrated as a measure of variations in specific gravity. A resistance thermometer bridge is used for the compensation of temperature effects on density.

Gas Specific Gravity Balance

A tall column of gas is weighed by the floating bottom of the vessel. This weight is translated into the motion of an indicating pointer, which moves over a scale graduated in units of density or specific gravity. This method can be employed for any gas density measurement.

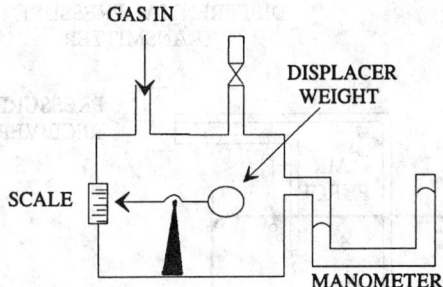

FIGURE 21.6 Buoyancy gas balance. The position of the balance beam is adjusted by a set pressure air, air is then displaced by gas of the same pressure. The difference in the reading of the balance beam gives the SG of the gas. The pressures are read on the manometer.

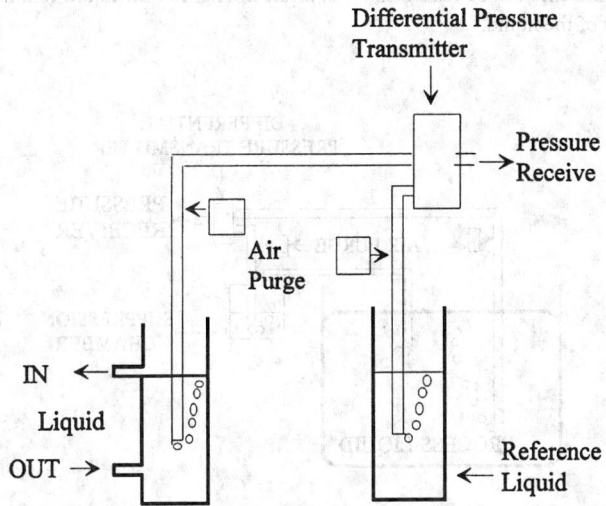

FIGURE 21.7 Reference column densitometer. Two identical tubes, having the same distance from the surface, are placed in water and liquid. Water with known density characteristics is used as the reference. The pressures necessary to displace the fluids inside the tubes are proportional to the densities of the fluids. The pressure difference at the differential pressure transmitter is translated into relative densities.

Buoyancy Gas Balance

In this instrument, a displacer is mounted on a balance beam in a vessel, as shown in Figure 21.6. The displacer is balanced for air, and the manometer reading is noted at the exact balance pressure. The air is then displaced by gas, and the pressure is adjusted until the same balance is restored. The ratio of the pressure of air to the pressure of gas is then the density of the gas relative to air. This method is commonly applied under laboratory conditions and is not suitable for continuous measurements.

Column-Type Densitometers

There are number of different versions of column methods. As a typical example, a reference column method is illustrated in Figure 21.7. A known head of sample liquid and water from the respective bubbler pipes are used. A differential pressure measuring device compares the pressure differences, proportional to relative densities of the liquid and the water. By varying the depth of immersion of the pipes, a wide

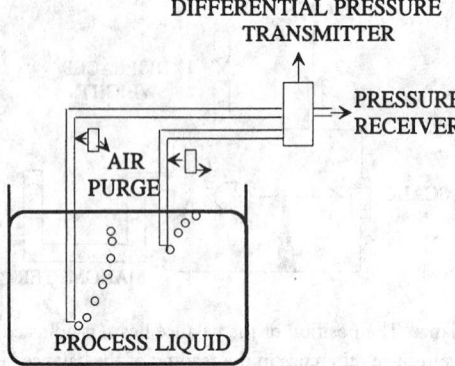

FIGURE 21.8   Two-tube column densitometer. The pressure difference at the differential pressure transmitter depends on the relative positions of the openings of the pipes and the density of liquid. Once the relative positions are fixed, the pressure difference can be related to the equivalent weight of the liquid column at the openings of the pipes, hence the density of the liquid.

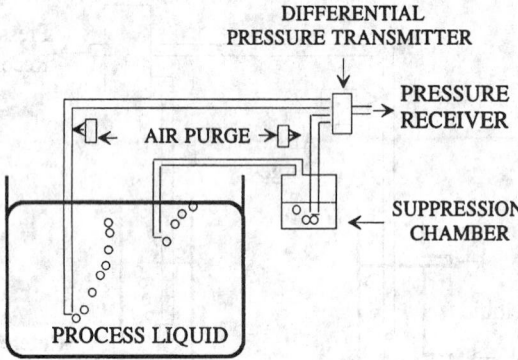

FIGURE 21.9   Suppression-type, two-tube column densitometer. Operation principle is the same as in Figures 21.7 and 21.8. In this case, the suppression chamber affords greater accuracy in readings.

range of measurements can be obtained. Both columns must be maintained at the same temperature to avoid the necessity for corrections of temperature effects.

A simpler and more widely used method of density measurement is achieved by the installation of two bubbler tubes as illustrated in Figure 21.8. The tubes are located in the sample fluid such that the end of one tube is higher than that of the other. The pressure required to bubble air into the fluid from both tubes is equal to the pressure of the fluid at the end of the bubbler tubes. The outlet of one tube higher than the other and the distances of the openings of the tubes are fixed; hence, the difference in the pressure is the same as the weight of a column of liquid between the ends. Therefore, the differential pressure measurement is equivalent to the weight of the constant volume of the liquid, and calibrations can be made that have a direct relationship to the density of the liquid. This method is accurate to within 0.1% to 1% specific gravity. It must be used with liquids that do not crystallize or settle in the measuring chamber during measurements.

Another version is the range suppression type, which has an additional constant pressure drop chamber as shown in Figure 21.9. This chamber is in series with the low-pressure side to give advantages in scaling and accurate readings of densities.

Vibrating Element Densitometers

If a body containing or surrounded by a fluid is set to resonance at its natural frequency, then the frequency of oscillation of the body will vary as the fluid properties and conditions change. The natural frequency is directly proportional to the stiffness of the body and inversely proportional to the combined mass of the body and the fluid. It is also dependent on the shape, size, and elasticity of the material, induced stress, mass, and mass distribution of the body. Basically, the vibration of the body can be equated to motion of a mass attached to a mechanical spring. Hence, an expression for the frequency can be written as:

$$\text{Resonant frequency} = \text{SQRT}\left(K/\left(M + k\rho\right)\right) \tag{21.3}$$

where K is the system stiffness, M is the transducer mass, k is the system constant, and ρ is the fluid density.

A factor common to all types of vibrating element densitometers is the problem of setting the element in vibration and maintaining its natural resonance. There are two drives for the purpose.

Magnetic Drives

Magnetic drives of the vibrating element and the pickup sensors of vibrations are usually achieved using small coil assemblies. Signals picked up by the sensors are amplified and fed back as a drive to maintain the disturbing forces on the vibrating body of the meter.

In order to achieve steady drives, the vibrating element sensor can be made from nonmagnetic materials. In this case, small magnetic armatures are attached.

The main advantage of magnetic drive and pickup systems is they are noncontact methods. They use conventional copper windings and they are reliable within the temperature range of −200 to +200°C.

Piezoelectric Drives

A wide range of piezoelectric materials are available to meet the requirements of driving vibrating elements. These materials demonstrate good temperature characteristics as do magnetic drive types. They also have the advantage of being low in cost. They have high impedance, making the signal conditioning circuitry relatively easy. They do not require magnetic sensors.

The piezoelectric drives are mechanically fixed on the vibrating element by adhesives. Therefore, attention must be paid to the careful placement of the mount in order to reduce the strain experienced by the piezo element due to thermal and pressure stresses while the instrument is in service.

A number of different types of densitometers have been developed that utilize this phenomenon. The three main commercial types are introduced here.

Vibrating Tube Densitometers

These devices are suitable for highly viscous liquids or slurry applications. The mode of operation of vibration tube meters is based on the transverse vibration of tubes as shown in Figure 21.10. The tube and the driving mechanisms are constrained to vibrate on a single plane. As the liquid moves inside the tube, the density of the entire mass of the liquid is measured. The tube length is approximately 20 times greater than the tube diameter.

A major design problem with the vibrating tube method is the conflict to limit the vibrating element to a finite length and accurately fix the nodes. Special attention must be paid to avoid any exchange of vibrational energy outside the sensory tube. The single tube has the disadvantage of presenting obstruction to the flow, thus experiencing some pressure losses. The twin tube, on the other hand, offers very small blockage (Figure 21.11) and can easily be inspected and cleaned. Its compact size is another distinct advantage. In some densitometers, the twin tube is designed to achieve a good dynamic balance, with the two tubes vibrating in antiphase. Their nodes are fixed at the ends, demonstrating maximum sensitivity to installation defects, clamping, and mass loading.

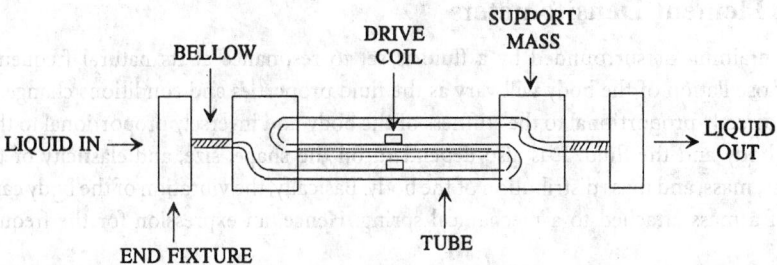

FIGURE 21.10 Vibrating tube densitometer. Tube containing fluid is vibrated at resonant frequency by electromagnetic vibrators. The resonant frequency, which is a function of the density of the fluid, is measured accurately. The tube is isolated from the fixtures by carefully designed bellows.

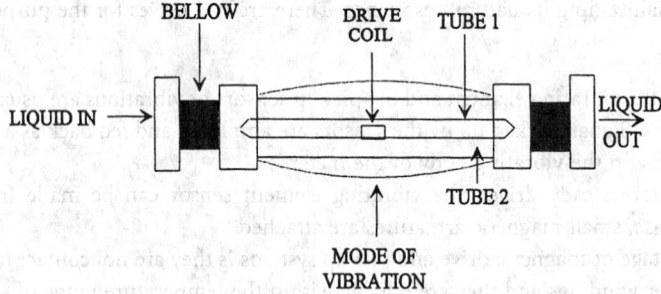

FIGURE 21.11 Two-tube vibrating densitometer. Two tubes are vibrated in antiphase for greater accuracy. Twin-tube densitometers are compact in size and easy to use.

The main design problems of the vibrating tube sensors are in minimizing the influence of end padding and overcoming the effects of pressure and temperature. Bellows are used at both ends of the sensor tubes to isolate the sensors from external vibrations. Bellows also minimize the end loadings due to differential expansions and installation stresses.

The fluid runs through the tubes; therefore, no pressure balance is required. Nevertheless, in some applications, the pressure stresses the tubes, resulting in stiffness changes. Some manufacturers modify the tubes to minimize the pressure effects. In these cases, corrections are necessary only when high accuracy is mandatory. The changes in the Young's modulus with temperature can be reduced to near-zero using Ni-span-C materials whenever corrosive properties of fluids permit. Usually, manufacturers provide pressure and temperature correction coefficients for their products.

It is customary to calibrate each vibration element densitometer against others as a transfer of standards. Often, the buoyancy method is used for calibration purposes. The temperature and pressure coefficients are normally found by exercising the transducer over a range of temperatures and pressures on a liquid with well-known properties. Prior to calibration, the vibration tube densitometers are subjected to a programmed burn-in cycle to stabilize them against temperatures and pressures.

Vibrating Cylinder Densitometers

A thin-walled cylinder, with a 3:1 length:diameter ratio, is fixed with stiff ends. The thickness of the cylinder wall varies from 25 μm to 300 μm, depending on the density range and type of fluid used. The cylinder can be excited to vibrate in a hoop mode by magnetic drives mounted either in or outside the cylinder.

For good magnetic properties, the cylinder is made of corrosion-resistant magnetic materials. Steel such as FV520 is often used for this purpose. Such materials have good corrosion-resistance characteristics; unfortunately, due to their poor thermoelastic properties, they need extensive temperature corrections.

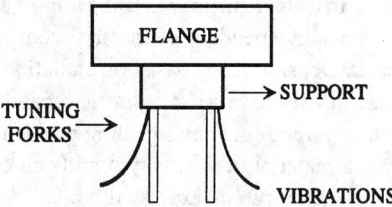

FIGURE 21.12 Tuning fork densitometer. Twin forks are inserted into the fluid and the natural frequencies are measured accurately. The natural frequency of the forks is a function of the density of the fluid.

Nickel-iron alloys such as Ni-span-C are often used to avoid temperature effects. Once correctly treated, the Ni-span-C alloy has near-zero Young's modulus properties. Because the cylinder is completely immersed in the fluid, there are no pressure coefficients.

The change in the resonant frequency is determined by the local mass loading of the fluid in contact with the cylinder. The curve of frequency against density is nonlinear and has a parabolic shape, thus requiring linearization to obtain practical outputs. The resonant frequency range varies from 2 kHz to 5 kHz, depending on the density range of the instrument. The cylinders need precision manufacturing and thus are very expensive to construct. Each meter needs to be calibrated individually for different temperatures and densities to suit specific applications. In the case of gas density applications, gases with well-known properties (e.g., pure argon or nitrogen) are used for calibrations. In this case, the meters are subjected to a gas environment with controlled temperature and pressure. The calibration curves are achieved by repetitions to suit the requirements of individual customers for particular applications. In the case of liquids, the meters are calibrated with liquids of known density, or they are calibrated against another standard (e.g., pycnometer or buoyancy type densitometers).

Vibration cylinder-type densitometers have zero pressure coefficients and they are ideal for liquefied gas products or refined liquids. Due to relatively small clearances between cylinder and housing, they require regular cleaning. They are not suitable for liquids or slurries with high viscous properties.

Tuning Fork Densitometers

These densitometers make use of the natural frequency of low-mass tuning forks, shown in Figure 21.12. In some cases, the fluid is taken into a small chamber in which the electromechanically driven forks are situated. In other cases, the fork is inserted directly into the liquid. Calibration is necessary in each application.

The advantages of vibrating element meters include:

1. They are suitable for both liquids and gases with reasonable accuracy.
2. They can be designed for real-time measurements.
3. They can easily be interfaced because they operate on frequencies and are inherently digital.
4. They are relatively robust and easy to install.
5. Programmable and computerized versions are available. Programmed versions make all the corrections automatically. They provide the output of live density, density at reference conditions, relative density, specific gravity, concentration, solid contents, etc.

The disadvantages include:

1. They do not relate directly to primary measurements; therefore, they must be calibrated.
2. They all have problems in measuring multiphase liquids.

Radioactive Densitometers

As radioactive isotopes decay, they emit radiation in the form of particles or waves. This physical phenomenon can be used for the purposes of density measurement. For example, γ rays are passed through the samples and their rate of arrivals are measured using ion- or scintillation-based detection [2]. Generally, γ-ray mass absorption rate is independent of material composition; hence they can be programmed for a wide range materials. Densitometers based on radiation methods can provide accuracy up to +0.0001 g mL⁻¹. Many of these devices have self-diagnostic capabilities and are able to compensate for drift caused by source decay, thus pinpointing any signaling problems.

If γ rays of intensity J_0 penetrate a material of a density ρ and thickness d then the intensity of the radiation after passing through the material can be expressed by:

$$J = J_0 \exp\left(n\rho d\right) \qquad (21.4)$$

where n is the mass absorption coefficient.

The accuracy of the density measurement depends on the accuracy of the measurement of the intensity of the radiation and the path length d. A longer path length through the material gives a stronger detection signal.

For accurate operations, there are many arrangements for relative locations of transmitters and detectors, some of which are illustrated in Figures 21.13 and 21.14. Generally, the source is mounted in a lead container clamped onto the pipe or the container wall. In many applications, the detector is also clamped onto the wall.

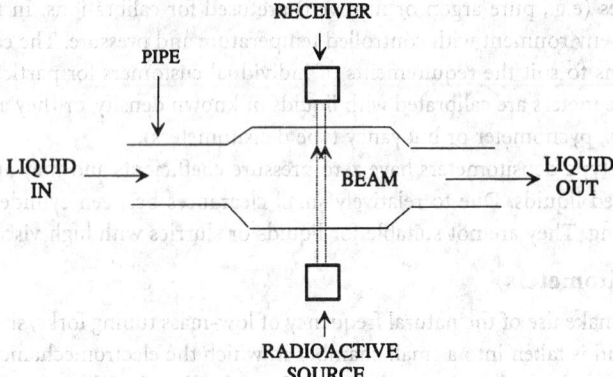

FIGURE 21.13 Fixing radioactive densitometer on an enlarged pipe. The pipe is enlarged to give longer beam length through the liquid, and hence better attenuation of the radioactive energy.

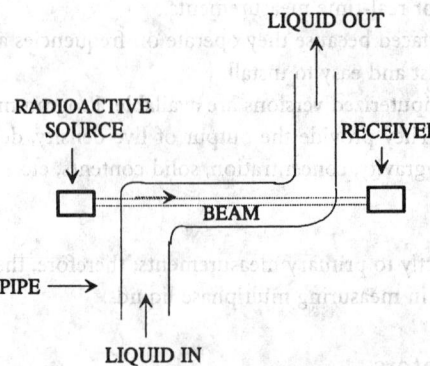

FIGURE 21.14 Fixing radioactive densitometer on an elongated pipe. Elongated path yields a longer path length of the radioactive energy through the liquid; hence, a stronger attenuation.

The advantages of using radioactive methods include:

1. The sensor does not touch the sample; hence, there is no blockage to the path of the liquid.
2. Multiphase liquids can be measured.
3. They come in programmable forms and are easy to interface.
4. They are most suitable in difficult applications, such as mining and heavy process industries.

The disadvantages include:

1. A radioactive source is needed; hence, there is difficulty in handling.
2. For reasonable accuracy, a minimum path length is required.
3. There could be long time constants, making them unsuitable in some applications.
4. They are suitable only for solid and liquid density measurements.

Refractometer and Index of Refraction Densitometers

Refractometers are essentially optical instruments operating on the principles of refraction of light traveling in liquid media. Depending on the characteristics of the samples, measurement of refractive index can be made in a variety of ways (e.g., critical angle, collimation, and displacement techniques). Usually, an in-line sensing head is employed, whereby a sensing window (commonly known as a prism) is wetted by the product to be measured. In some versions, the sensing probes must be installed inside the pipelines or in tanks and vessels. They are most effective in reaction-type process applications where blending and mixing of liquids take place. For example, refractometers can measure dissolved soluble solids accurately.

Infrared diodes, lasers, and other lights may be used as sources. However, this measurement technique is not recommended in applications in processes containing suspended solids, high turbidity, entrained air, heavy colors, poor transparency and opacity, or extremely high flow rates. The readings are automatically corrected for variations in process temperature. The processing circuitry can include signal outputs adjustable in both frequency and duration.

Another version of a refractometer is the index of refraction type densitometer. For example, in the case of position-sensitive detectors, the index of refraction of liquid under test is determined by measuring the lateral displacement of a laser beam. When the laser beam impinges on the cell at an angle of incidence, as in Figure 21.15, the axis of the emerging beam is displaced by the cell wall and by the inner liquid. The lateral displacement can accurately be determined by position-sensitive detectors. For maximum sensitivity, the devices need to be calibrated with the help of interferometers.

Refractometers are often used for the control of adulteration of liquids of common use (e.g., edible oils, wines, and gasoline). They also find application in pulp and paper, food and beverage, sugar, dairy, and other chemical industries.

Coriolis Densitometers

The Coriolis density metering systems are similar to vibrating tube methods, but with slight variations in the design. They are comprised of a sensor and a signal-processing transmitter. Each sensor consists of one or two flow tubes enclosed in a sensor housing. They are manufactured in various sizes and shapes [3]. The sensor tubes are securely anchored at the fluid inlet and outlet points and force is vibrated at the free end, as shown in Figure 21.16. The sensor operates by applying Newton's second law of motion ($F = ma$).

Inside the housing, the tubes are vibrated in their natural frequencies using drive coils and a feedback circuit. This resonant frequency of the assembly is a function of the geometry of the element, material of construction, and mass of the tube assembly. The tube mass comprises two parts: the mass of the tube itself and the mass of the fluid inside the tube. The mass of the tube is fixed for a given sensor. The mass of fluid in the tube is equal to the fluid density multiplied by volume. Because the tube volume is constant, the frequency of oscillation can be related directly to the fluid density. Therefore, for a given geometry

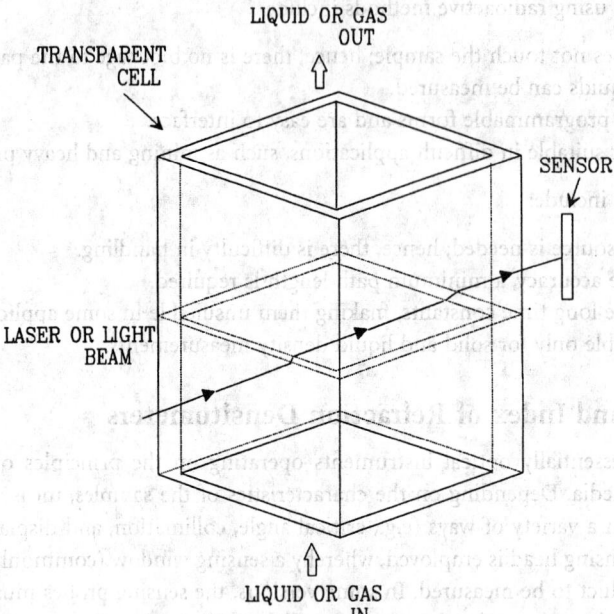

FIGURE 21.15 Index of refraction-type densitometer. The angle of refraction of the beam depends on the shape, size, and thickness of the container, and the density of fluid in the container. Because the container has the fixed characteristics, the position of the beam can be related to density of the fluid. Accurate measurement of the position of the beam is necessary.

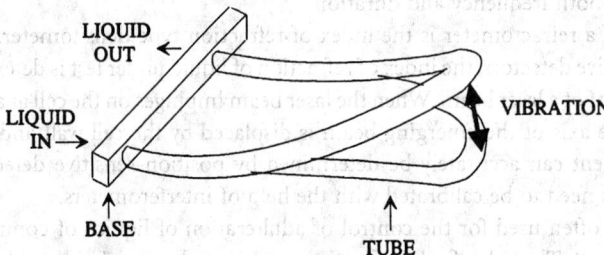

FIGURE 21.16 Coriolis densitometer. Vibration of the tube is detected and related to the mass and flow rate of the fluid. Further calibrations and calculations must be made to determine the densities.

of tube and the material of the construction, the density of the fluid can be determined by measuring the resonant frequency of vibration. Temperature sensors are used for overcoming the effects of changes in modulus of elasticity of the tube. The fluid density is calculated using a linear relationship between the density and the tube period and calibration constants.

Special peripherals, based on microprocessors, are offered by various manufacturers for a variety of measurements. However, all density peripherals employ the natural frequency of the sensor coupled with the sensor temperature to calculate on-line density of process fluid. Optional communication, interfacing facilities, and appropriate software are also offered.

Absorption-Type Densitometers

Absorption techniques are also used for density measurements in specific applications. X-rays, visible light, UV light, and sonic absorptions are typical examples of this method. Essentially, attenuation and phase shift of a generated beam going through the sample is sensed and related to the density of the sample. Most absorption-type densitometers are custom designed for applications having particular characteristics. Two typical examples are: (1) UV absorption or X-ray absorptions are used for determining the local densities of mercury deposits in arc discharge lamps, and (2) ultrasonic density sensors are used in connection with difficult density measurements (e.g., density measurement of slurries). The lime slurry, for example, is a very difficult material to handle. It has a strong tendency to settle out and coat all equipment with which it comes in contact. An ultrasonic density control sensor can fully be emerged into an agitated slurry, thus avoiding the problems of coating and clogging. Inasmuch as the attenuation of the ultrasonic beam is proportional to the suspended solids, the resultant electronic signal is proportional to the specific gravity of the slurry. Such devices can give accuracy up to 0.01%. The ultrasonic device measures the percentage of the suspended solids in the slurry by providing a close approximation of the specific gravity.

References

1. H. Eren, Particle concentration characteristics and density measurements of slurries using electro-magnetic flowmeters, *IEEE Trans. Instr. Meas.*, 44, 783-786, 1995.
2. Micro Motion Product Catalogue, Mount Prospect, IL: Fisher-Rosemount, 1995.
3. Kay-Ray, Solution for Process Measurement, Mount Prospect, IL: Fisher-Rosemount, 1995.

Appendix

List of Manufacturers

ABB K-Flow Inc.
Drawer M Box 849
Millville, NJ 08332
Tel: (800) 825-3569

American Density Materials Inc.
Rd. 2, Box 38E
Belvidere, J 07823
Tel: (908) 475-2373

Anton Paar U.S.A.
13, Maple Leaf Ct.
Ashland, VA 23005
Tel: (800) 221-0174

Arco Instrument Company, Inc.
1745 Production Circle
Riverside, CA 92509
Tel: (909) 788-2823
Fax: (909) 788-2409

Cambridge Applied Systems, Inc.
196 Boston Avenue
Medford, MA 02155
Tel: (617) 393-6500

Dynatron
Automation Products, Inc.
3032 Max Roy Street
Houston, TX 77008
Tel: (800) 231-2062
Fax: (713) 869-7332

Kay-Ray/Sensall, Fisher-Rosemount
1400 Business Center Dr.
Mount Prospect, IL 60056
Tel: (708) 803-5100
Fax: (708) 803-5466

McGee Engineering Co., Inc.
Tujunga Canyon Blvd.
Tujunga, CA 91042
Tel: (800) 353-6675

Porous Materials, Inc.
Cornell Business & Technology Park
Ithaca, NY 14850
Tel: (800) 825-5764

Princo Instruments Inc
1020 Industrial Hwy., Dept L
Southampton, PA 18966-4095
Tel: (800) 496-5343

Quantachrome Corp.
1900-T Corporate Drive
Boynton Beach, FL 33426
Tel: (800) 966-1238

Tricor Systems, Inc.
400-T River Ridge Rd.
Elgin, IL 60123
Tel: (800) 575-0161

X-rite, Inc.
3100-T 44th St. S.W
Grandville, MI 49418
Tel: (800) 545-0694

22

Strain Measurement

Christopher S. Lynch
The Georgia Institute of Technology

This chapter begins with a review of the fundamental definitions of strain and ways it can be measured. This is followed by a review of the many types of strain sensors and their application, and sources for strain sensors and signal conditioners. Next, a more detailed look is taken at operating principles of various strain measurement techniques and the associated signal conditioning.

22.1   Fundamental Definitions of Strain

Stress and strain are defined in many elementary textbooks about the mechanics of deformable bodies [1, 2]. The terms *stress* and *strain* are used to describe loads on and deformations of solid materials. The simplest types of solids to describe are homogeneous and isotropic. *Homogeneous* means the material properties are the same at different locations and *isotropic* means the material properties are independent of direction in the material. An annealed steel bar is homogeneous and isotropic, whereas a human femur is not homogeneous because the marrow has very different properties from the bone, and it is not isotropic because its properties are different along the length and along the cross-section.

The concepts of stress and strain are introduced in the context of a long homogeneous isotropic bar subjected to a tensile load (Figure 22.1). The stress σ, is the applied force F, divided by the cross-sectional area A. The resulting strain ε, is the length change ΔL, divided by the initial length L. The bar elongates in the direction the force is pulling (longitudinal strain ε_L) and contracts in the direction perpendicular to the force (transverse strain ε_t).

When the strain is not too large, many solid materials behave like linear springs; that is, the displacement is proportional to the applied force. If the same force is applied to a thicker piece of material, the spring is stiffer and the displacement is smaller. This leads to a relation between force and displacement that depends on the dimensions of the material. Material properties, such as the density and specific heat, must be defined in a manner that is independent of the shape and size of the specimen. Elastic material properties are defined in terms of stress and strain. In the linear range of material response, the stress is proportional to the strain (Figure 22.2). The ratio of stress to strain for the bar under tension is an elastic constant called the Young's modulus E. The negative ratio of the transverse strain to longitudinal strain is the Poisson's ratio v.

Forces can be applied to a material in a manner that will cause distortion rather than elongation (Figure 22.3). A force applied tangent to a surface divided by the cross-sectional area is described as a shear stress τ. The distortion can be measured by the angle change produced. This is the shear strain γ

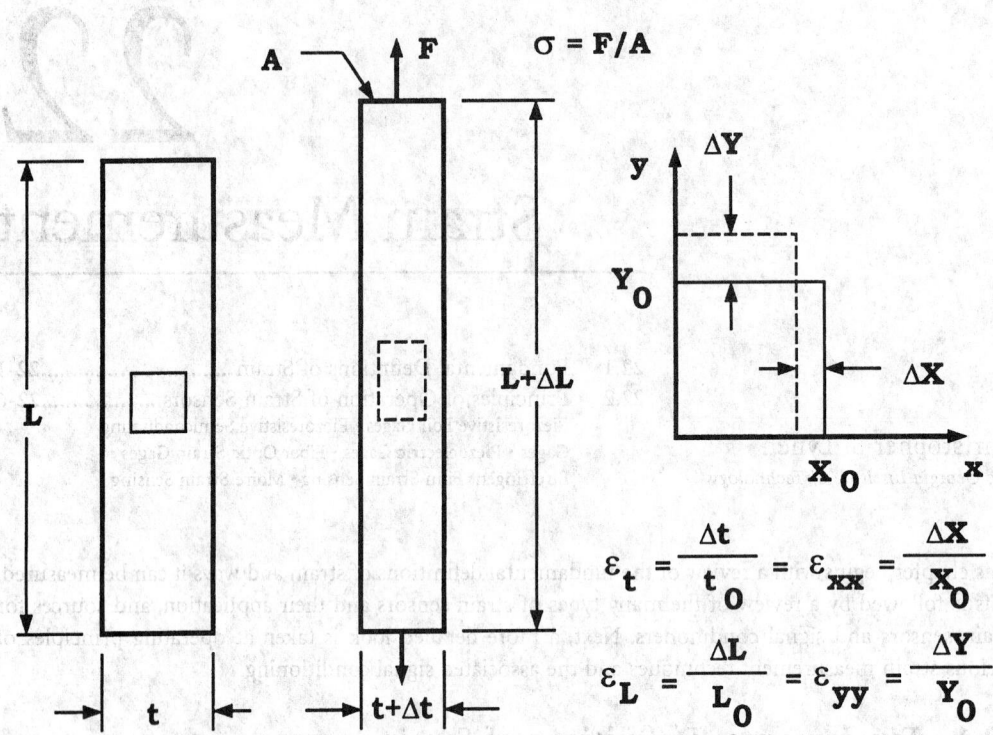

FIGURE 22.1 When a homogeneous isotropic bar is stretched by a uniaxial force, it elongates in the direction of the force and contracts perpendicular to the force. The relative elongation and contraction are defined as the longitudinal and transverse strains, respectively.

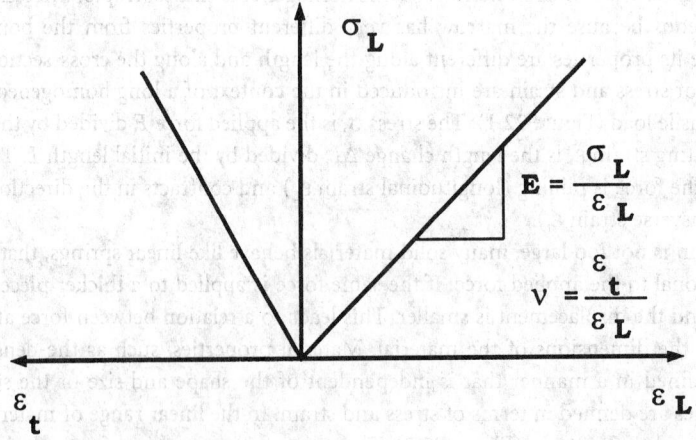

FIGURE 22.2 The uniaxial force shown in Figure 22.1 produces uniaxial stress in the bar. When the material response is linear, the slope of the stress vs. strain curve is the Young's modulus. The negative ratio of the transverse to longitudinal strain is the Poisson's ratio.

when the angle change is small. When the relation between shear stress and shear strain is linear, the ratio of the shear stress to shear strain is the shear modulus G.

Temperature change also induces strain. This is thermal expansion. In most materials, thermal strain increases with temperature. Over a limited temperature range, the relationship between thermal strain

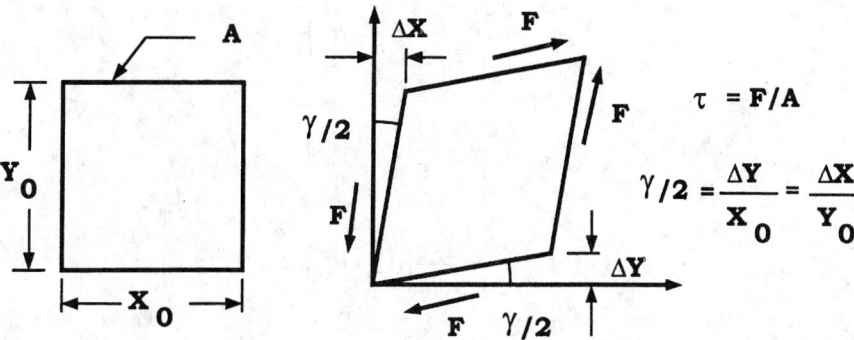

FIGURE 22.3 When a block of material is subjected to forces parallel to the sides as shown, it distorts. The force per unit area is the shear stress τ, and the angle change is the shear strain γ.

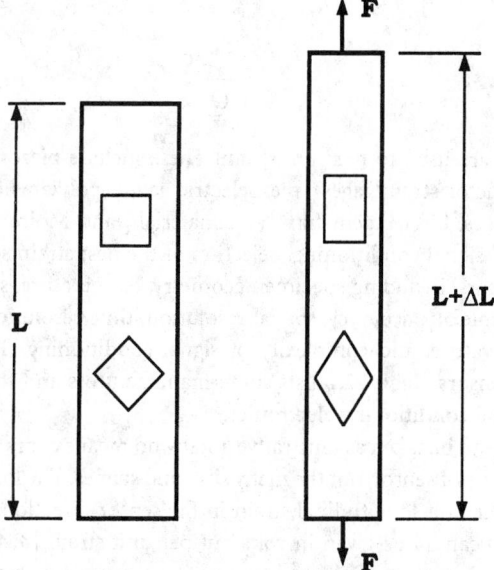

FIGURE 22.4 Some elements of a bar under uniaxial tension undergo elongation and contraction. These elements lie in principal directions. Other elements undergo distortion as well.

and temperature is linear. In this case, the strain divided by the temperature change is the thermal expansion coefficient α. In isotropic materials, thermal expansion only produces elongation strain, no shear strain.

Principal directions in a material are directions that undergo elongation but no shear. On any particular surface of a solid, there are always at least two principal directions in which the strain is purely elongation. This is seen if two squares are drawn on the bar under uniform tension (Figure 22.4). When the bar is stretched, the square aligned with the load is elongated, whereas the square at 45° is distorted (angles have changed) and elongated. If the principal directions are known, as with the bar under tension, then strain gages can be applied in these directions. If the principal directions are not known, such as near a hole or notch, in an anisotropic specimen, or in a structure with complicated geometry, then additional strain gages are needed to fully characterize the strain state.

The elastic and thermal properties can be combined to give Hooke's law, Equations 22.1 to 22.6.

$$\varepsilon_{xx} = \frac{\sigma_{xx}}{E} - \frac{\nu}{E}\left(\sigma_{yy} + \sigma_{zz}\right) \tag{22.1}$$

$$\varepsilon_{yy} = \frac{\sigma_{yy}}{E} - \frac{\nu}{E}\left(\sigma_{xx} + \sigma_{zz}\right) \tag{22.2}$$

$$\varepsilon_{xx} = \frac{\sigma_{zz}}{E} - \frac{\nu}{E}\left(\sigma_{xx} + \sigma_{yy}\right) \tag{22.3}$$

$$\gamma_{xy} = \frac{\tau_{xy}}{G} \tag{22.4}$$

$$\gamma_{xz} = \frac{\tau_{xz}}{G} \tag{22.5}$$

$$\gamma_{yz} = \frac{\tau_{yz}}{G} \tag{22.6}$$

Several types of sensors are used to measure strain. These include piezoresistive gages (foil or wire strain gages and semiconductor strain gages), piezoelectric gages (polyvinylidene fluoride (PVDF) film and quartz), fiber optic gages, birefringent films and materials, and Moiré grids. Each type of sensor requires its own specialized signal conditioning. Selection of the best strain sensor for a given measurement is based on many factors, including specimen geometry, temperature, strain rate, frequency, magnitude, as well as cost, complexity, accuracy, spatial resolution, time resolution, sensitivity to transverse strain, sensitivity to temperature, and complexity of signal conditioning. Table 22.1 describes typical characteristics of several sensors. Table 22.2 lists some manufacturers and the approximate cost of the sensors and associated signal conditioning electronics.

The data in Table 22.1 are to be taken as illustrative and by no means complete. The sensor description section describes only the type of sensor, not the many sizes and shapes. The longitudinal strain sensitivity is given as sensor output per unit longitudinal strain in the sensor direction. If the signal conditioning is included, the sensitivities can all be given in volts out per unit strain [3, 4], but this is a function of amplification and the quality of the signal conditioner. The temperature sensitivity is given as output change due to a temperature change. In many cases, higher strain resolution can be achieved, but resolving smaller strain is more difficult and may require vibration and thermal isolation. For the Moiré technique, the strain resolution is a function of the length of the viewing area. This technique can resolve a displacement of 100 nm (1/4 fringe order). This is divided by the viewing length to obtain the strain resolution. The spatial resolution corresponds to the gage length for most of the sensor types. The measurable strain range listed is the upper limit for the various sensors. Accuracy and reliability are usually reduced when sensors are used at the upper limit of their capability.

Manufacturers of the various sensors provide technical information that includes details of using the sensors, complete calibration or characterization data, and details of signal conditioning. The extensive technical notes and technical tips provided by Measurements Group, Inc. address such issues as thermal effects [5], transverse sensitivity corrections [6], soldering techniques [7], Rosettes [8], and gage fatigue [9]. Strain gage catalogs include information about gage materials, sizes, and selection. Manufacturers of other sensors provide similar information.

TABLE 22.1 Comparison of Strain Sensors

Description	Longitudinal strain sensitivity	Transverse strain sensitivity	Temperature sensitivity	Strain resolution	Spatial resolution	Time resolution	Measurable strain range
Piezoresistive constantan foil	$\Delta R/R/\Delta\epsilon_L = 2.1$	$\Delta R/R/\Delta\epsilon_t = <0.02$	$\Delta R/R/\Delta T = 2\times10^{-6}/°C$	<1 μstrain[a]	5–100 mm[b]	<1 μs[c]	0–3%
Annealed constantan foil[d]	$\Delta R/R/\Delta\epsilon_L = 2.1$	$\Delta R/R/\Delta\epsilon_t = <0.02$	$\Delta R/R/\Delta T = 2\times10^{-6}/°C$	<11 μstrain	5–100 mm	<1 μs	0–10%
Piezoresistive semiconductor	$\Delta R/R/\Delta\epsilon_L = 150$	$\Delta R/R/\Delta\epsilon_t = ???$	$\Delta R/R/\Delta T = 1.7\times10^{-3}/°C$	<0.1 μstrain	1–15 mm	<1 μs	0–0.1%
Piezoelectric PVDF	$\Delta Q/A/\Delta\epsilon_L = 120$ nC/m²/με	$\Delta Q/A/\Delta\epsilon_t = 60$ nC/m²/με	$\Delta Q/A/\Delta T = -27$ μC/m²/°C	1–10 μstrain	Gage size	<1 μs	0–30%
Piezoelectric quartz	$\Delta Q/A/\Delta\epsilon_L = 150$ nC/m²/με bonded to steel	Near zero	$\Delta Q/A/\Delta T = 0$	<0.01 μstrain 20 mm gage	Gage size	<10 μs	0–0.1%
Fiber optic Fabry-Perot	2 to 1000 μstrain/V			<1 μstrain	2–10 mm	<20 μs	
Birefringent Film	$K^e = 0.15–0.002$				0.5 mm[f]	<5 μs	0.05–5%
Moiré	1 fringe order/417 nm displ.	1 fringe order/417 nm displ.	Not defined	41.7 με over 10 mm	full field[g]	Limited by signal conditioning	0.005–5%

a With good signal conditioning.

b Equal to grid area.

c Gage response is within 100 ns. Most signal conditioning limits response time to far less than this.

d Annealed foil has a low yield stress and a large strain to failure. It also has hysteresis in the unload and a zero shift under cyclic load.

e This technique measures a difference in principal strains. $\epsilon_2 - \epsilon_1 = N\lambda/2tK$

f Approximately the film thickness.

g The spatial strain resolution depends on the strain level. This is a displacement measurement technique.

TABLE 22.2 Sources and Prices of Strain Sensors

Supplier	Address	Sensor Types	Sensor Cost	Signal Conditioning	Cost
Micro Measurements	P.O. Box 27777 Raleigh, NC 27611	Piezoresistive foil Birefringent film	From $5.00 From $10.00	Wheatstone bridge Polariscope	From $500 $5000 to 10,000
Texas Measurements	P.O. Box 2618 College Station, TX 77841	Piezoresistive foil and wire Load cells	From $5.00		
Omega Engineering	P.O. Box 4047 Stamford, CT 06907-0047	Piezoresistive foil	From $5.00	Strain meter Wheatstone bridge	From $550 From $2700
Dynasen, Inc.	20 Arnold Pl. Goleta, CA 93117	Piezoresistive foil Specialty gages for shock wave measurements Piezoelectric PVDF (calibrated)	From $55.00 From $55.00	2-Channel pulsed Wheatstone bridge Passive charge integrator	$5000 $250.00
Entran Sensors and Electronics	Entran Devices, Inc. 10 Washington Ave. Fairfield, NJ 07004-3877	Piezoresistive semiconductor	From $15.00		
Amp Inc.	Piezo Film Sensors P.O. Box 799 Valley Forge, PA 19482	Piezoelectric PVDF (not calibrated)	From $5.00		
Kistler Instrument Corp.	Amherst, NY 14228-2171	Piezoelectric quartz		Electrometer Charge amplifier	
F&S Inc.	Fiber and Sensor Technologies P.O. Box 11704 Blacksburg, VA 24062-1704	Fabry–Perot strain sensors	From $75	Electronics	From $3500.00
Photomechanics, Inc.	512 Princeton Dr. Vestal, NY 13850-2912	Moiré interferometer			$60,000

22.2   Principles of Operation of Strain Sensors

Piezoresistive Foil Gages

Piezoresistive foil and wire gages comprise a thin insulating substrate (usually polyimide film), a foil or wire grid (usually constantan) bonded to the substrate, lead wires to connect the grid to a resistance measuring circuit, and often an insulating encapsulation (another sheet of polyimide film) (Figure 22.5). The grid is laid out in a single direction so that strain will stretch the legs of the grid in the length direction. The gages are designed so that strain in the width or transverse direction separates the legs of the grid without straining them. This makes the gage sensitive to strain only along its length. There is always some sensitivity to transverse strain, and almost no sensitivity to shear strain. In most cases, the transverse sensitivity can be neglected.

When piezoresistive foil or wire strain gages are bonded to a specimen and the specimen is strained, the gage strains as well. The resistance change is related to the strain by a gage factor, Equation 22.7.

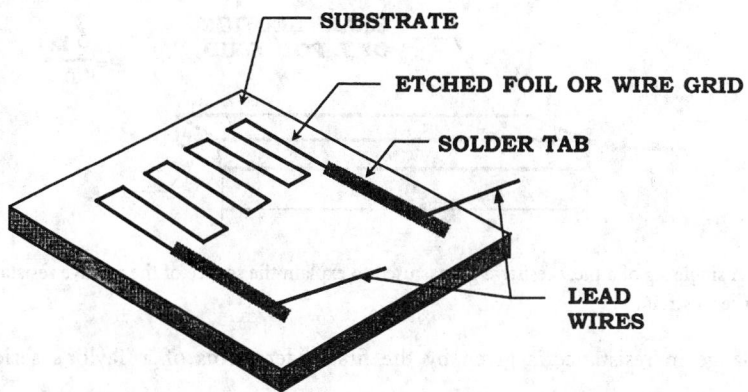

FIGURE 22.5 Gage construction of a foil or wire piezoresistive gage.

$$\frac{\Delta R}{R} = G_L \varepsilon_L \qquad (22.7)$$

where $\Delta R/R$ = Relative resistance change
  G = Gage factor
  ε = Strain

These gages respond to the average strain over the area covered by the grid [10]. The resistance change is also sensitive to temperature. If the temperature changes during the measurement period, a correction must be made to distinguish the strain response from the thermal response. The gage response to longitudinal strain, transverse strain, and temperature change is given by Equation 22.8.

$$\frac{\Delta R}{R} = G_L \varepsilon_L + G_t \varepsilon_t + G_T \Delta T \qquad (22.8)$$

where G_L, G_t, and G_T are the longitudinal, transverse, and temperature sensitivity, respectively. Micromeasurements, Inc. uses a different notation. Their gage data is provided as $G_L = F_G$, $G_t = K_t F_G$, $G_T = \beta_g$. When a strain gage is bonded to a specimen and the temperature changes, the strain used in Equation 22.8 is the total strain, thermal plus stress induced, as given by Equation 22.7.

The temperature contribution to gage output must be removed if the gages are used in tests where the temperature changes. A scheme referred to as self-temperature compensation (STC) can be used. This is accomplished by selecting a piezoresistive material whose thermal output can be canceled by the strain induced by thermal expansion of the test specimen. Gage manufacturers specify STC numbers that match the thermal expansion coefficients of common specimen materials.

Strain of piezoresistive materials produces a relative resistance change. The resistance change is the result of changes in resistivity and dimensional changes. Consider a single leg of the grid of a strain gage with a rectangular cross-section (Figure 22.6). The resistance is given by Equation 22.9.

$$R = \rho \frac{L}{A} \qquad (22.9)$$

where R = Resistance
  ρ = Resistivity
  L = Length
  A = Area of the cross-section

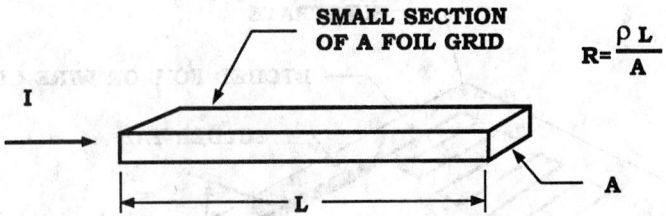

FIGURE 22.6 A single leg of a piezoresistive gage is used to explain the source of the relative resistance change that occurs in response to strain.

A small change in resistance is given by the first-order terms of a Taylor's series expansion, Equation 22.10.

$$\Delta R = \frac{\partial R}{\partial \rho} \Delta \rho + \frac{\partial R}{\partial L} \Delta L + \frac{\partial R}{\partial A} \Delta A \tag{22.10}$$

Differentiating Equation 22.9 to obtain each term of Equation 22.10 and then dividing by the initial resistance leads to Equation 22.11.

$$\frac{\Delta R}{R_0} = \frac{\Delta \rho}{\rho_0} + \frac{\Delta L}{L_0} - \frac{\Delta A}{A_0} \tag{22.11}$$

The relative resistance change is due to a change in resistivity, a change in length strain, and a change in area strain.

The strain gage is a composite material. The metal in a strain gage is like a metal fiber in a polymer matrix. When the polymer matrix is deformed, the metal is dragged along in the length direction; but in the width and thickness directions, the strain is not passed to the metal. This results in a stress state called uniaxial stress. This state was discussed in the examples above. The mathematical details involve an inclusion problem [11, 12]. Accepting that the stress state is uniaxial, the relationship between the area change and the length change in Equation 22.11 is found from the Poisson's ratio. The area strain is the sum of the width and thickness strain, Equation 22.12.

$$\frac{\Delta A}{A_0} = \frac{\Delta w}{w_0} + \frac{\Delta t}{t_0} \tag{22.12}$$

The definition of the Poisson's ratio gives Equation 22.13.

$$\frac{\Delta A}{A_0} = -2v \frac{\Delta L}{L_0} \tag{22.13}$$

where $\Delta w/w$ = width strain and $\Delta t/t$ = thickness strain. Substitution of Equation 22.13 into Equation 22.11 gives Equation 22.14 for the relative resistance change.

$$\frac{\Delta R}{R_0} = \frac{\Delta \rho}{\rho_0} + \frac{\Delta L}{L_0}(1 + 2v) \tag{22.14}$$

The relative resistivity changes in response to stress. The resistivity is a second-order tensor [13], and the contribution to the overall resistance change can be found in terms of strain using the elastic constitutive law [14]. The results lead to an elastic gage factor just over 2 for constantan gages. If the strain is large, the foil or wire in the gage will experience plastic deformation. When the deformation is plastic, the resistivity change is negligible and the dimensional change dominates. In this case, Poisson's ratio is 0.5 and the gage factor is 2. This effect is utilized in manufacturing gages for measuring strains larger than 1.5%. In this case, annealed foil is used. The annealed foil undergoes plastic deformation without failure. These gages are capable of measuring strain in excess of 10%. When metals undergo plastic deformation, they do not unload to the initial strain. This shows up as hysteresis in the gage response, that is, on unload, the resistance does not return to its initial value.

Foil and wire strain gages can be obtained in several configurations. They can be constructed with different backing materials, and left open faced or fully encapsulated. Backing materials include polyimide and glass fiber-reinforced phenolic resin. Gages can be obtained with solder tabs for attaching lead wires, or with lead wires attached. They come in many sizes, and in multiple gage configurations called rosettes.

Strain gages are mounted to test specimens with adhesives using a procedure that is suitable for bonding most types of strain sensors. This is accomplished in a step-by-step procedure [15] that starts with surface preparation. An overview of the procedure is briefly described. To successfully mount strain gages, the surface is first degreased. The surface is abraded with a fine emery cloth or 400 grit paper to remove any loose paint, rust, or deposits. Gage layout lines are drawn (not scribed) on the surface in a cross pattern with pen or pencil, one line in the grid direction and one in the transverse direction. The surface is then cleaned with isopropyl alcohol. This can be done with an ultrasonic cleaner or with wipes. If wiped, the paper or gauze wipe should be folded and a small amount of alcohol applied. The surface should be wiped with one pass and the wipe discarded. This should be repeated, wiping in the other direction. The final step is to neutralize the surface, bringing the alkalinity to a pH of 7 to 7.5. A surface neutralizer is available from most adhesive suppliers. The final step is to apply the gage.

Gage application is accomplished with cellophane tape, quick-set glue, and a catalyst. The gage is placed on a clean glass or plastic surface with bonding side down, using tweezers. (Never touch the gage. Oils from skin prevent proper adhesion.) The gage is then taped down with a 100 mm piece of cellophane tape. The tape is then peeled up with the gage attached. The gage can now be taped onto its desired location on the test specimen. Once the gage has been properly aligned, the tape is peeled back from one side, lifting the gage from the surface. The tape should remain adhered to the surface about 1 cm from the gage. Note that one side of the tape is still attached to the specimen so that the gage can be easily returned to its desired position. A thin coating of catalyst is applied to the exposed gage surface. A drop of glue is placed at the joint of the tape and the specimen. Holding the tape at about a 30° angle from the surface, the tape can be slowly wiped down onto the surface. This moves the glue line forward. After the glue line has passed the gage, the gage should be pressed in place and held for approximately 1 min. The tape can now be peeled back to expose the gage, and lead wires can be attached.

The relative resistance change of piezoresistive gages is usually measured using a Wheatstone bridge [16]. This allows a small change of resistance to be measured relative to an initial zero value, rather than relative to a large resistance value, with a corresponding increase in sensitivity and resolution. The Wheatstone bridge is a combination of four resistors and a voltage source (Figure 22.7). One to four of the resistors in the bridge can be strain gages. The output of the bridge is the difference between the voltage at points B and D. Paths ABC and ADC are voltage dividers so that V_B and V_D are given by Equations 22.15a and b.

$$V_B = V_{in} \frac{R_2}{R_1 + R_2}$$

(22.15a)

$$V_D = V_{in} \frac{R_3}{R_3 + R_4}$$

(22.15b)

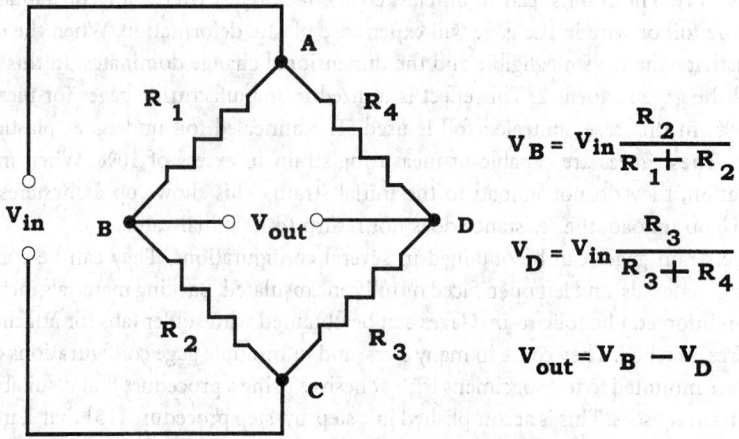

$$V_B = V_{in} \frac{R_2}{R_1 + R_2}$$

$$V_D = V_{in} \frac{R_3}{R_3 + R_4}$$

$$V_{out} = V_B - V_D$$

FIGURE 22.7 The Wheatstone bridge is used to measure the relative resistance change of piezoresistive strain gages.

The bridge output, Equation 22.16, is zero when the balance condition, Equation 22.17, is met.

$$V_0 = V_B - V_D \qquad (22.16)$$

$$R_1 R_3 = R_2 R_4 \qquad (22.17)$$

Wheatstone bridge signal conditioners are constructed with a way to "balance" the bridge by adjusting the ratio of the resistances so that the bridge output is initially zero.

The balance condition is no longer met if the resistance values undergo small changes ΔR_1, ΔR_2, ΔR_3, ΔR_4. If the values $R_1 + \Delta R_1$, etc. are substituted into Equation 22.15, the results substituted into Equation 22.16, condition (22.17) used, and the higher order terms neglected, the result is Equation 22.18 for the bridge output.

$$V_{out} = V_{in} \frac{R_1 R_3}{(R_1 + R_2)(R_3 + R_4)} \left(-\frac{\Delta R_1}{R_1} + \frac{\Delta R_2}{R_2} - \frac{\Delta R_3}{R_3} + \frac{\Delta R_4}{R_4} \right) \qquad (22.18)$$

The Wheatstone bridge can be used to directly cancel the effect of thermal drift. If R_1 is a strain gage bonded to a specimen and R_2 is a strain gage held onto a specimen with heat sink compound (a thermally conductive grease available at any electronics store), then R_1 will respond to strain plus temperature, and R_2 will only respond to temperature. Since the bridge subtracts the output of R_1 from that of R_2, the temperature effect cancels.

The sensitivity of a measuring system is the output per unit change in the quantity to be measured. If the resistance change is from a strain gage, the sensitivity of the Wheatstone bridge system is proportional to the input voltage. Increasing the voltage increases the sensitivity. There is a practical limitation to increasing the voltage to large values. The power dissipated (heat) in the gage is $P = I^2 R$, where I, the current through the gage, can be found from the input voltage and the bridge resistances. This heat must go somewhere or the temperature of the gage will continuously rise and the resistance will change due to heating. If the gage is mounted on a good thermal conductor, more power can be conducted away than if the gage is mounted on an insulator. The specimen must act as a heat sink.

Heat sinking ability is proportional to the thermal conductivity of the specimen material. A reasonable temperature gradient to allow the gage to induce in a material is 40°C per meter (about 1°C per 25 mm).

For thick specimens (thickness several times the largest gage dimension), this can be conducted away to the grips or convected to the surrounding atmosphere. If the four bridge resistances are approximately equal, the power to the gage in terms of the bridge voltage is given by Equation 22.19.

$$P_g = \frac{V_{in}^2}{4R} \qquad (22.19)$$

The power per unit grid area, A_g, or power density to the gage can be equated to the thermal conductivity of the specimen and the allowable temperature gradient in the specimen by Equation 22.20.

$$\frac{P_g}{A_g} = \frac{V_{in}^2}{4RA_g} = K\nabla T \qquad (22.20)$$

Thermal conductivities of most materials can be found in tables or on the Web. Some typical values are Al: $K = 204$ W m^{-1} °C^{-1}, steel: $K = 20$ to 70 W m^{-1} °C^{-1}, glass: $K = 0.78$ W m^{-1} °C^{-1} [17].

The acceptable bridge voltage can be calculated from Equation 22.21.

$$V_{in} = \sqrt{K\nabla T 4RA_g} \qquad (22.21)$$

A sample calculation shows that for a 0.010 m × 0.010 m 120 Ω grid bonded to a thick piece of aluminum with a thermal conductivity of 204 W m^{-1} °C^{-1} and an acceptable temperature gradient of 40°C per meter, the maximum bridge voltage is 19 V. If thin specimens are used, the allowable temperature gradient will be smaller. If smaller gages are used for better spatial resolution, the bridge excitation voltage must be reduced with a corresponding reduction in sensitivity.

A considerably higher bridge voltage can be used if the bridge voltage is pulsed for a short duration. This dissipates substantially less energy in the gage and thus increases the sensitivity by a factor of 10 to 100. Wheatstone bridge pulse power supplies with variable pulse width from 10 μs and excitation of 350 V are commercially available [18].

The strain measurement required is often in a complex loading situation where the directions of principal strain are not known. In this case, three strain gages must be bonded to the test specimen at three angles. This is called a strain rosette. The angle between the rosette and the principal directions, as well as the magnitude of the principal strains, can be determined from the output of the rosette gages. This is most easily accomplished using the construct of a Mohr's circle (Figure 22.8).

A common rosette is the 0–45–90° pattern. The rosette is bonded to the specimen with relative rotations of 0°, 45°, and 90°. These will be referred to as the x, x', and y directions. The principal directions are labeled the 1 and 2 directions. The unknown angle between the x direction and the 1 direction is labeled θ. The Mohr's circle is drawn with the elongational strain on the horizontal axis and the shear strain on the vertical axis. The center of the circle is labeled C and the radius R. The principal directions correspond to zero shear strain. The principal values are given by Equations 22.22 and 22.23.

$$\varepsilon_1 = C + R \qquad (22.22)$$

$$\varepsilon_2 = C - R \qquad (22.23)$$

A rotation through an angle 2θ on the Mohr's circle corresponds to a rotation of the rosette of θ relative to the principal directions. The center of the circle is given by Equation 22.24 and the output of the strain gages is given by Equations 22.25 to 22.27.

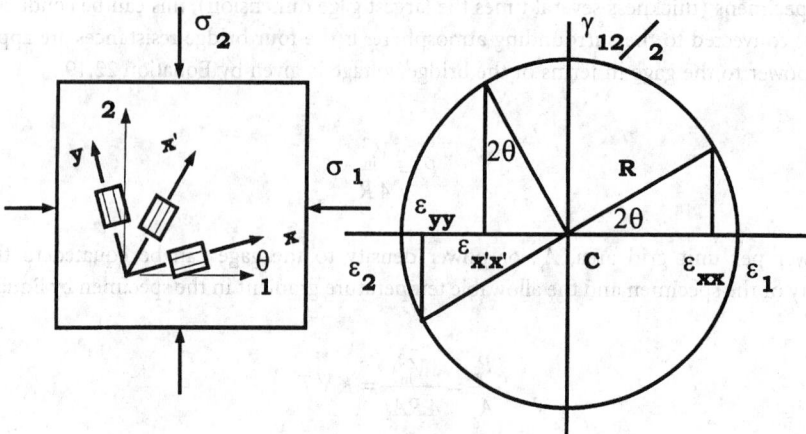

FIGURE 22.8 A three-element rosette is used to measure strain when the principal directions are not known. The Mohr's circle is used to find the principal directions and the principal strain values.

$$C=\frac{\varepsilon_{xx}+\varepsilon_{yy}}{2} \qquad (22.24)$$

$$\varepsilon_{xx}-C=R\cos2\theta \qquad (22.25)$$

$$\varepsilon_{x'x'}-C=-R\sin2\theta \qquad (22.26)$$

$$\varepsilon_{yy}-C=-R\cos2\theta \qquad (22.27)$$

Dividing Equation 22.25 by Equation 22.26 leads to θ and then to R, Equations 22.28 and 22.29.

$$R^2=\left(\varepsilon_{xx}-C\right)^2+\left(\varepsilon_{x'x'}-C\right)^2 \qquad (22.28)$$

$$\tan2\theta=\frac{C-\varepsilon_{x'x'}}{\varepsilon_{xx}-C} \qquad (22.29)$$

The principal directions and principal strain values have been found from the output of the three rosette gages.

Piezoresistive Semiconducting Gages

Piezoresistive semiconductor strain gages, like piezoresistive foil and wire gages, undergo a resistance change in response to strain, but with nearly an order of magnitude larger gage factor [19]. The coupling between resistance change and temperature is very large, so these gages have to be temperature compensated. The change of resistance with a small temperature change can be an order of magnitude larger than that induced by strain. Semiconductor strain gages are typically used to manufacture transducers such as load cells. They are fragile and require great care in their application.

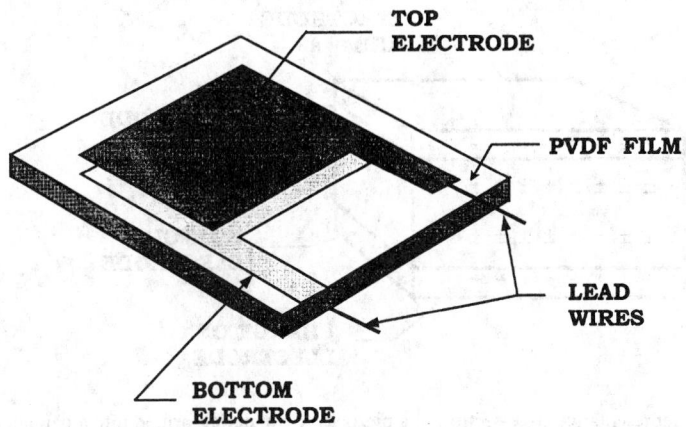

FIGURE 22.9   Typical gage construction of a piezoelectric gage.

Piezoelectric Gages

Piezoelectric strain gages are, effectively, parallel plate capacitors whose dielectric changes polarization in response to strain [14]. When the polarization changes, a charge proportional to the strain is produced on the electrodes. PVDF film strain gages are inexpensive, but not very accurate and subject to depoling by moderate temperature. They make good sensors for dynamic measurements such as frequency and logarithm decrement, but not for quantitative measurements of strain. When used for quasistatic measurements, the charge tends to drain through the measuring instrument. This causes the signal to decay with a time constant dependent on the input impedance of the measuring instrument. Quartz gages are very accurate, but also lose charge through the measuring instrument. Time constants can be relatively long (seconds to hours) with electrometers or charge amplifiers.

The PVDF gage consists of a thin piezoelectric film with metal electrodes (Figure 22.9). Lead wires connect the electrodes to a charge measuring circuit. Gages can be obtained with the electrodes encapsulated between insulating layers of polyimide.

The gage output can be described in terms of a net dipole moment per unit volume. If the net dipole moment is the total charge, Q, on the electrodes multiplied by spacing, d, between the electrodes, then the polarization is given by Equation 22.30.

$$P = \frac{Qd}{V} \qquad (22.30)$$

From Equation 22.30, it is seen that the polarization P (approximately equal to the electric displacement D) is the charge per unit electrode area (Figure 22.10).

A Taylor's series expansion of Equation 22.30 gives Equation 22.31.

$$\Delta P = \frac{\partial P}{\partial V} \Delta V + \frac{\partial P}{\partial (Qd)} \Delta Qd \qquad (22.31)$$

Which, after differentiating Equation 22.30 and substituting becomes Equation 22.31.

$$\Delta P = \frac{\Delta Qd}{V_0} - \frac{\Delta V}{V_0} P_0 \qquad (22.32)$$

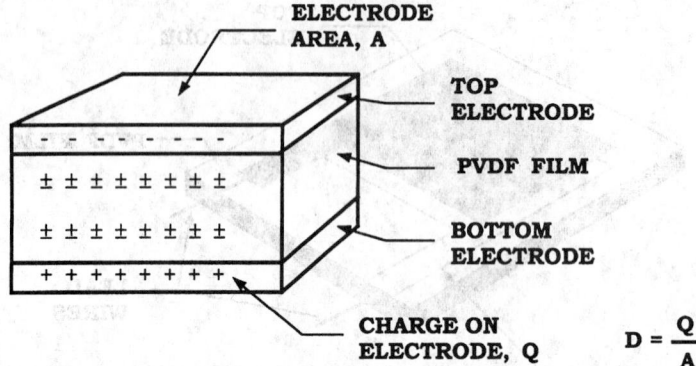

$$D = \frac{Q}{A}$$

FIGURE 22.10 A representative cross-section of a piezoelectric material formed into a parallel plate capacitor. The piezoelectric material is polarized. This results in charge on the electrodes. When the material is strained, the polarization changes and charge flows.

For PVDF film, the second term in Equation 22.32 dominates. The output is proportional to the remanent polarization P_0. The remanent polarization slowly decays with time, has a strong dependence on temperature, and decays rapidly at temperatures around 50°C. This makes accuracy a problem. If the sensors are kept at low temperature, accuracy can be maintained within ±3%.

Strain sensors can also be constructed from piezoelectric ceramics like lead zirconate titanate (PZT) or barium titanate. Ceramics are brittle and can be depoled by strain so should only be used at strains less than 200 microstrain. PZT loses some of its polarization with time and thus has accuracy problems, but remains polar to temperatures of 150°C or higher. "Hard" PZT (usually iron doped) is the best composition for polarization stability and low hysteresis. Quartz has the best accuracy. It is not polar, but polarization is induced by strain. Quartz has excellent resolution and accuracy over a broad temperature range but is limited to low strain levels. It is also brittle, so is limited to small strain.

Two circuits are commonly used for piezoelectric signal conditioning: the electrometer and the charge amplifier (Figure 22.11). In the electrometer circuit, the piezoelectric sensor is connected to a capacitor with a capacitance value C, at least 1000 times that of the sensor C_g. There is always some resistance in

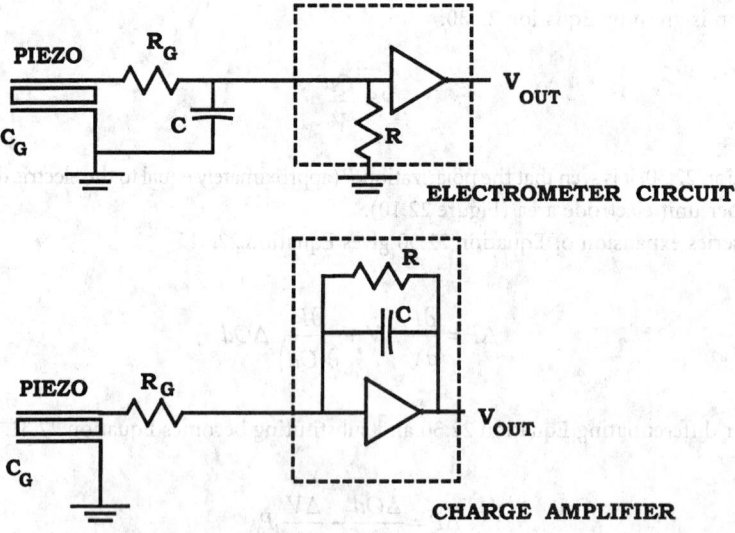

FIGURE 22.11 The electrometer and charge amplifier are the most common circuits used to measure charge from piezoelectric transducers.

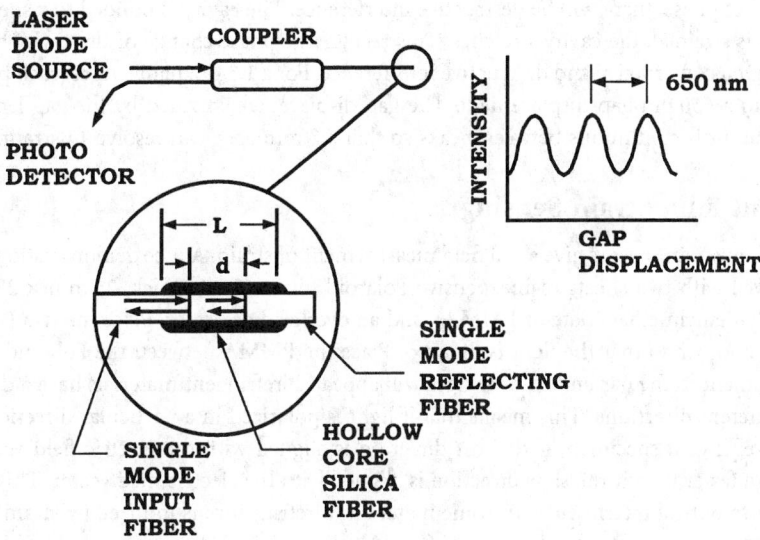

FIGURE 22.12 A schematic of the Fabry–Perot fiber optic strain gage. When the cavity elongates, alternating constructive and destructive interference occur.

the cable that connects the sensor to the capacitor. The circuit is simply two capacitors in parallel connected by a resistance. The charge equilibrates with a time constant given by $R_g C_g$. This time constant limits the fastest risetime that can be resolved to about 50 ns, effectively instantaneous for most applications. The charge is measured by measuring the voltage on the capacitor, then using Equation 22.33.

$$Q = CV \qquad (22.33)$$

The difficulty is that measuring devices drain the charge, causing a time decay with a time constant *RC*. This causes the signal to be lost rapidly if conventional op amps are used. FET input op amps have a very high input impedance and can extend this time constant to many hours. The charge amplifier is another circuit used to measure charge. This is usually an FET input op amp with a capacitor feedback. This does not really amplify charge, but produces a voltage proportional to the input charge. Again, the time constant can be many hours, allowing use of piezoelectric sensors for near static measurements.

High input impedance electrometer and charge amplifier signal conditioners for near static measurements are commercially available [20] as well as low-cost capacitive terminators for high-frequency (high kilohertz to megahertz) measurements [18]. An advantage of piezoelectric sensors is that they are active sensors that do not require any external energy source.

Fiber Optic Strain Gages

Fiber optic strain gages are miniature interferometers [21, 22]. Many commercially available sensors are based on the Fabry–Perot interferometer. The Fabry–Perot interferometer measures the change in the size of a very small cavity.

Fabry–Perot strain sensors (Figure 22.12) comprise a laser light source, single-mode optical fibers, a coupler (the fiber optic equivalent of a beam splitter), a cavity that senses strain, and a photodetector. Light leaves the laser diode. It passes down the fiber, through the coupler, and to the cavity. The end of the fiber is the equivalent of a partially silvered mirror. Some of the light is reflected back up the fiber and some is transmitted. The transmitted light crosses the cavity and then is reflected from the opposite end back into the fiber where it recombines with the first reflected beam. The two beams have a phase difference related to twice the cavity length. The recombined beam passes through the coupler to the photodetector. If the two reflected beams are in phase, there will be constructive interference. If the two beams are out of phase, there will be destructive interference. The cavity is bonded to a specimen. When the specimen is strained, the cavity stretches. This results in a phase change of the cavity beam, causing a cycling between constructive and destructive interference. For a 1.3 μm light source, each peak in output corresponds to a 650 nm gap displacement. The gap displacement divided by the gap length gives the strain. The output is continuous between peaks so that a 3 mm gage can resolve 1 μstrain.

Birefringent Film Strain Sensing

Birefringent film strain sensors give a full field measurement of strain. A nice demonstration of this effect can be achieved with two sheets of inexpensive Polaroid film, a 6 mm thick, 25 mm × 200 mm bar of Plexiglas (polymethylmethacrylate or PMMA), and an overhead projector. Place the two Polaroid sheets at 90° to one another so that the light is blocked. Place the PMMA between the Polaroid sheets. Apply a bending moment to the bar and color fringes will appear. Birefringent materials have a different speed of light in different directions. This means that if light is polarized in a particular direction and passed through a birefringent specimen, if the fast direction is aligned with the electric field vector, the light passes through faster than if the slow direction is aligned with the electric field vector. This effect can be used to produce optical interference. In some materials, birefringence is induced by strain. The fast and slow directions correspond to the directions of principal strain, and the amount of birefringence corresponds to the magnitude of the strain. One component of the electric field vector travels through the specimen faster than the other. They emerge with a phase difference. This changes the relative amplitude and thus rotates the polarization of the light. If there is no birefringence, no light passes through the second polarizer. As the birefringence increases with strain, light passes through. As it further increases, the polarization rotation will be a full 180° and again no light will pass through. This produces a fringe that corresponds to a constant difference in principal strains. The difference in principal strains is given by Equation 22.34.

$$\varepsilon_2 - \varepsilon_1 = \frac{N\lambda}{tK} \tag{22.34}$$

where ε_1, ε_2 = Principal strains
  N = Fringe order
  λ = Wavelength
  t = Specimen thickness
  K = Strain-optical coefficient of the photoelastic material

A similar technique can be used with a birefringent plastic film with a silvered backing laminated to the surface of a specimen. Polarized light is passed through the film; it reflects from the backing, passes back through the film, and through the second polarizer. In this case, because light passes twice through the film, the equation governing the difference in principal strains is Equation 22.35.

$$\varepsilon_2 - \varepsilon_1 = \frac{N\lambda}{2tK} \tag{22.35}$$

If the polarizers align with principal strain directions, no birefringence is observed. Rotation of both polarizers allows the principal directions to be found at various locations on the test specimen. If a full view of the fringes is desired, quarter wave plates are used (Figure 22.13). In this arrangement, light is passed through the first polarizer, resulting in plane polarization; through the quarter wave plate, resulting in circular polarization; through the test specimen, resulting in phase changes; through the second quarter wave plate to return to plane polarization; and then through the final polarizer.

The optical systems for viewing birefringence are commercially available as "Polariscopes" [23]. Optical components to construct custom systems are available from many optical components suppliers.

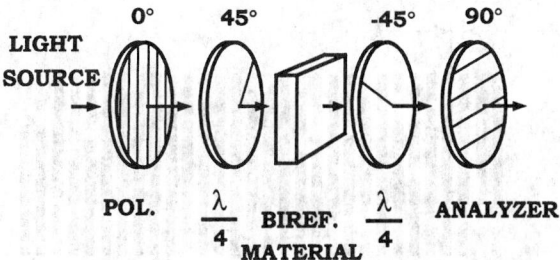

FIGURE 22.13   A schematic of the polariscope, a system for measuring birefringence. This technique gives a full field measure of the difference in principal strains.

Moiré Strain Sensing

Moiré interference is another technique that gives a full field measurement, but it measures displacement rather than strain. The strain field must be computed from the displacement field. This technique is based on the interference obtained when two transparent plates are covered with equally spaced stripes. If the plates are held over one another, they can be aligned so that no light will pass through or so that all light will pass through. If one of the plates is stretched, the spacing of the lines is wider on the stretched plate. Now, if one plate is placed over the other, in some regions light will pass through and in some regions it will not (Figure 22.14). The dark and light bands produced give information about the displacement field.

Moiré is defined as a series of broad dark and light patterns formed by the superposition of two regular gratings [24]. The dark or light regions are called fringes. Examples of pure extension and pure rotation are shown. In both cases, some of the light that would emerge from the first grating is obstructed by the superimposed grating. At the centers of the dark fringes, the bar of one grating covers the space of the other and no light comes through. The emergent intensity, I, is zero. Proceeding from there toward the next dark fringe, the amount of obstruction diminishes linearly and the amount of light increases linearly until the bar of one grating falls above the bar of the other. There, the maximum amount of light passes through the gratings.

Both geometric interference and optical interference are used. This discussion is restricted to geometric interference. Geometric moiré takes advantage of the interference of two gratings to determine displacements and rotations in the plane of view. In-plane moiré is typically conducted with two gratings, one applied to the specimen (specimen grating) and the other put in contact with the specimen grating (reference grating). When the specimen is strained, interference patterns or fringes occur. N is the moiré fringe order. Each fringe corresponds to an increase or decrease of specimen displacement by one grating pitch. The relationship between displacement and fringes is $\delta = gN$, where δ is component of the displacement perpendicular to the reference grating lines, g is reference grating pitch, and N is the fringe order.

For convenience, a zero-order fringe is designated assuming the displacement there is zero. With the reference grating at $0°$ and $90°$, the fringe orders N_x and N_y are obtained. The displacements in x, y directions are then obtained from Equations 22.36 and 22.37.

$$u_x\left(x,y\right) = gN_x\left(x,y\right) \tag{22.36}$$

$$u_y\left(x,y\right) = gN_y\left(x,y\right) \tag{22.37}$$

Differentiation of Equations 22.36 and 22.37 gives the strains, Equations 22.38 through 22.40.

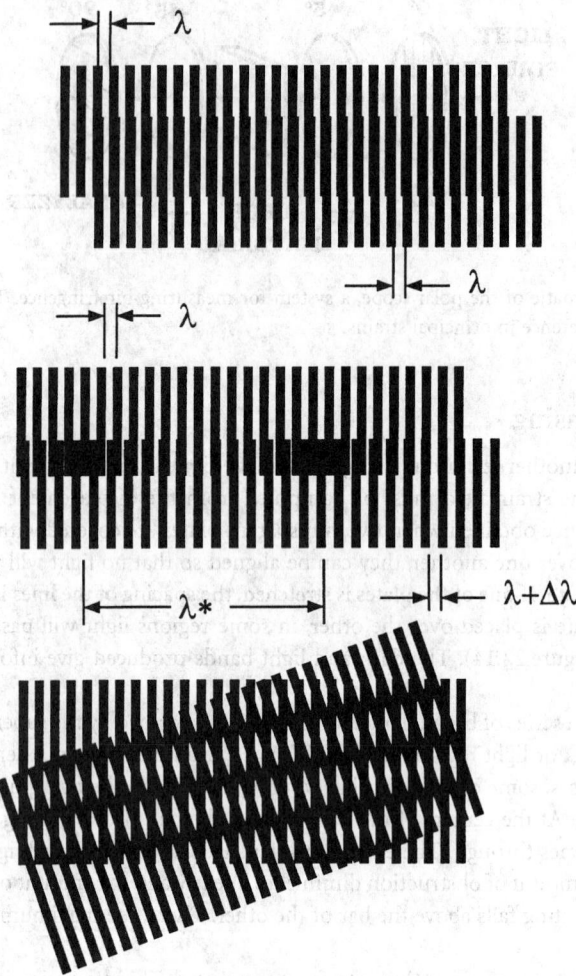

FIGURE 22.14 A demonstration of moiré fringes formed by overlapping gratings. The fringes are the result of stretching and relative rotation of the gratings. The fringe patterns are used to determine displacement fields.

$$\varepsilon_x = \frac{\partial u_x}{\partial x} = g\frac{\partial N_x}{\partial x} \tag{22.38}$$

$$\varepsilon_{xy} = \frac{1}{2}\left(\frac{\partial u_y}{\partial x} + \frac{\partial u_x}{\partial y}\right) = \frac{1}{2}\left(g\frac{\partial N_y}{\partial x} + g\frac{\partial N_x}{\partial y}\right) \tag{22.39}$$

$$\varepsilon_y = \frac{\partial u_y}{\partial y} = g\frac{\partial N_y}{\partial y} \tag{22.40}$$

In most cases, the sensitivity of geometric moiré is not adequate for determination of strain distributions. Strain analysis should be conducted with high-sensitivity measurement of displacement using moiré interferometry [24, 25]. Moiré interferometers are commercially available [26]. Out-of-plane measurement can be conducted with one grating (the reference grating). The reference grating is made to interfere with either its reflection or its shadow [27, 28].

References

1. N. E. Dowling, *Mechanical Behavior of Materials,* Englewood Cliffs, NJ: Prentice-Hall, 1993, 99-108.
2. R. C. Craig, *Mechanics of Materials,* New York: John Wiley & Sons, 1996.
3. A. Vengsarkar, Fiber optic sensors: a comparative evaluation, *The Photonics Design and Applications Handbook,* 1991, 114-116.
4. H. U. Eisenhut, Force measurement on presses with piezoelectric strain transducers and their static calibration up to 5 MN, *New Industrial Applications of the Piezoelectric Measurement Principle,* July 1992, 1-16.
5. TN-501-4, Strain Gauge Temperature Effects, Measurements Group, Inc., Raleigh, NC 27611.
6. TN-509, Transverse Sensitivity Errors, Measurements Group, Inc., Raleigh, NC 27611.
7. TT-609, Soldering Techniques, Measurements Group, Inc., Raleigh, NC 27611.
8. TN-515, Strain Gage Rosettes, Measurements Group, Inc., Raleigh, NC 27611.
9. TN-08-1, Fatigue of Strain Gages, Measurements Group, Inc., Raleigh, NC 27611.
10. C. C. Perry and H. R. Lissner, *The Strain Gage Primer,* New York: McGraw-Hill, 1962.
11. J. B. Aidun and Y. M. Gupta, Analysis of Lugrangian gauge measurements of simple and nonsimple plain waves, *J. Appl. Phys.,* 69, 6998-7014, 1991.
12. Y. M. Gupta, Stress measurement using piezoresistance gauges: modeling the gauge as an elastic-plastic inclusion, *J. Appl. Phys.,* 54, 6256-6266, 1983.
13. D. Y. Chen, Y. M. Gupta, and M. H. Miles, Quasistatic experiments to determine material constants for the piezoresistance foils used in shock wave experiments, *J. Appl. Phys.,* 55, 3984, 1984.
14. C. S. Lynch, Strain compensated thin film stress gauges for stress wave measurements in the presence of lateral strain, *Rev. Sci. Instrum.,* 66(11), 1-8, 1995.
15. B-129-7 M-Line Accessories Instruction Bulletin, Measurements Group, Inc., Raleigh, NC 27611.
16. J. W. Dally, W. F. Riley, and K. G. McConnell, *Instrumentation for Engineering Measurements,* 2nd ed., New York: John Wiley & Sons, 1993.
17. J. P. Holman, *Heat Transfer,* 7th ed., New York: McGraw Hill, 1990.
18. Dynasen, Inc. 20 Arnold Pl., Goleta, CA 93117.
19. M. Dean (ed.) and R. D. Douglas (assoc. ed.), *Semiconductor and Conventional Strain Gages,* New York: Academic Press, 1962.
20. Kistler, Instruments Corp., Amhurst, NY, 14228-2171.
21. J. S. Sirkis, Unified approach to phase strain temperature models for smart structure interferometric optical fiber sensors. 1. Development, *Opt. Eng.,* 32(4), 752-761, 1993.
22. J. S. Sirkis, Unified approach to phase strain temperature models for smart structure interferometric optical fiber sensors. 2. Applications, *Optical Engineering,* 32(4), 762-773, 1993.
23. Photoelastic Division, Measurements Group, Inc., P.O. Box 27777, Raleigh, NC 27611.
24. T. Valis, D. Hogg, and R. M. Measures, Composite material embedded fiber-optic Fabry-Perot strain rosette, *SPIE,* 1370, 154-161, 1990.
25. D. Post, B. Han, and P. Lfju, *High Sensitivity Moiré,* New York: Springer-Verlag, 1994.
26. V. J. Parks, Geometric Moiré, *Handbook on Experimental Mechanics,* A. S. Kobayashi, Ed., VCH Publisher, Inc., 1993.
27. Photomechanics, Inc. 512 Princeton Dr. Vestal, NY, 13850-2912.
28. T. Y. Kao and F. P. Chiang, Family of grating techniques of slope and curvature measurements for static and dynamic flexure of plates, *Opt. Eng.,* 21, 721-742, 1982.
29. D. R. Andrews, Shadow moiré contouring of impact craters, *Opt. Eng.,* 21, 650-654, 1982.

23

Force Measurement

M. A. Elbestawi
McMaster University

Force, which is a vector quantity, can be defined as an action that will cause an acceleration or a certain reaction of a body. This chapter will outline the methods that can be employed to determine the magnitude of these forces.

23.1 General Considerations

The determination or measurement of forces must yield to the following considerations: if the forces acting on a body do not produce any acceleration, they must form a *system of forces in equilibrium*. The system is then considered to be in static equilibrium. The forces experienced by a body can be classified into two categories: internal, where the individual particles of a body act on each other, and external otherwise. If a body is supported by other bodies while subject to the action of forces, deformations and/or displacements will be produced at the points of support or contact. The internal forces will be distributed throughout the body until equilibrium is established, and then the body is said to be in a state of tension, compression, or shear. In considering a body at a definite section, it is evident that all the internal forces act in pairs, the two forces being equal and opposite, whereas the external forces act singly.

23.2 Hooke's Law

The basis for force measurement results from the physical behavior of a body under external forces. Therefore, it is useful to review briefly the mechanical behavior of materials. When a metal is loaded in uniaxial tension, uniaxial compression, or simple shear (Figure 23.1), it will behave elastically until a critical value of normal stress (S) or shear stress (τ) is reached, and then it will deform plastically [1]. In the elastic region, the atoms are temporarily displaced but return to their equilibrium positions when the load is removed. Stress (S or τ) and strain (e or γ) in the elastic region are defined as indicated in Figure 23.2.

$$v = -\frac{e_2}{e_1} \qquad (23.1)$$

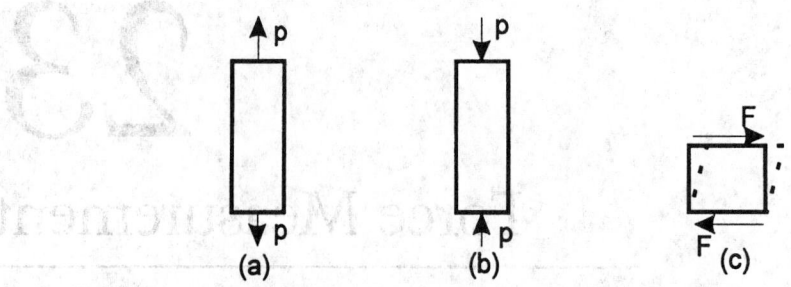

FIGURE 23.1 When a metal is loaded in uniaxial tension (*a*) uniaxial compression (*b*), or simple shear (*c*), it will behave elastically until a critical value of normal stress or shear stress is reached.

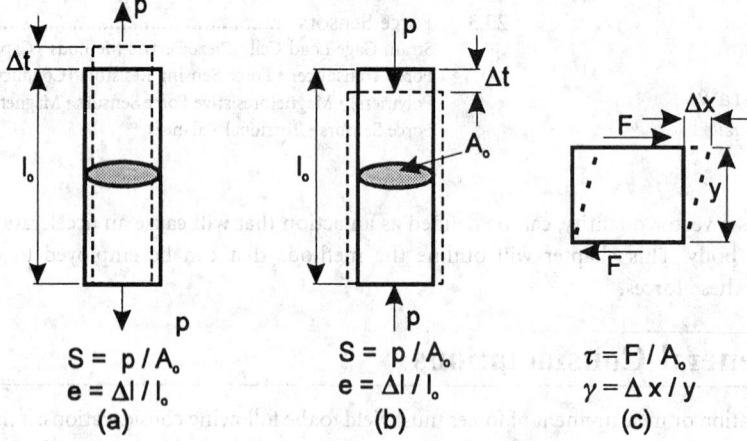

FIGURE 23.2 Elastic stress and strain for: (*a*) uniaxial tension; (*b*) uniaxial compression; (*c*) simple shear [1].

Poisson's ratio (v) is the ratio of transverse (e_2) to direct (e_1) strain in tension or compression. In the elastic region, v is between 1/4 and 1/3 for metals. The relation between stress and strain in the elastic region is given by Hooke's law:

$$S = E\,e\left(\text{tension or compression}\right) \qquad (23.2)$$

$$\tau = G\gamma\left(\text{simple shear}\right) \qquad (23.3)$$

where E and G are the Young's and shear modulus of elasticity, respectively. A small change in specific volume ($\Delta Vol/Vol$) can be related to the elastic deformation, which is shown to be as follows for an isotropic material (same properties in all directions).

$$\frac{\Delta Vol}{Vol} = e_1\left(1 - 2v\right) \qquad (23.4)$$

The bulk modulus (K = reciprocal of compressibility) is defined as follows:

$$K = \Delta p \Big/ \left(\frac{\Delta Vol}{Vol}\right) \qquad (23.5)$$

where Δp is the pressure acting at a particular point. For an elastic solid loaded in uniaxial compression (S):

$$K = S \bigg/ \left(\frac{\Delta Vol}{Vol} \right) = \frac{S}{e_1 (1 - 2v)} = \frac{E}{1 - 2v} \qquad (23.6)$$

Thus, an elastic solid is compressible as long as v is less than 1/2, which is normally the case for metals. Hooke's law (Equation 23.2) for uniaxial tension can be generalized for a three-dimensional elastic condition.

The theory of elasticity is well established and is used as a basis for force measuring techniques. Note that the measurement of forces in separate engineering applications is very application specific, and care must be taken in the selection of the measuring techniques outlined below.

Basic Methods of Force Measurement

An unknown force may be measured by the following means:

1. Balancing the unknown force against a standard mass through a system of levers.
2. Measuring the acceleration of a known mass.
3. Equalizing it to a magnetic force generated by the interaction of a current-carrying coil and a magnet.
4. Distributing the force on a specific area to generate pressure, and then measuring the pressure.
5. Converting the applied force into the deformation of an elastic element.

The aforementioned methods used for measuring forces yield a variety of designs of measuring equipment. The challenge involved with the task of measuring force resides primarily in sensor design. The basics of sensor design can be resolved into two problems:

1. Primary geometric, or physical constraints, governed by the application of the force sensor device.
2. The means by which the force can be converted into a workable signal form (such as electronic signals or graduated displacements).

The remaining sections will discuss the types of devices used for force to signal conversion and finally illustrate some examples of applications of these devices for measuring forces.

23.3 Force Sensors

Force sensors are required for a basic understanding of the response of a system. For example, cutting forces generated by a machining process can be monitored to detect a tool failure or to diagnose the causes of this failure in controlling the process parameters, and in evaluating the quality of the surface produced. Force sensors are used to monitor impact forces in the automotive industry. Robotic handling and assembly tasks are controlled by detecting the forces generated at the end effector. Direct measurement of forces is useful in controlling many mechanical systems.

Some types of force sensors are based on measuring a deflection caused by the force. Relatively high deflections (typically, several micrometers) would be necessary for this technique to be feasible. The excellent elastic properties of helical springs make it possible to apply them successfully as force sensors that transform the load to be measured into a deflection. The relation between force and deflection in the elastic region is demonstrated by Hooke's law. Force sensors that employ strain gage elements or piezoelectric (quartz) crystals with built-in microelectronics are common. Both impulsive forces and slowly varying forces can be monitored using these sensors.

Of the available force measuring techniques, a general subgroup can be defined as that of load cells. Load cells are comprised generally of a rigid outer structure, some medium that is used for measuring

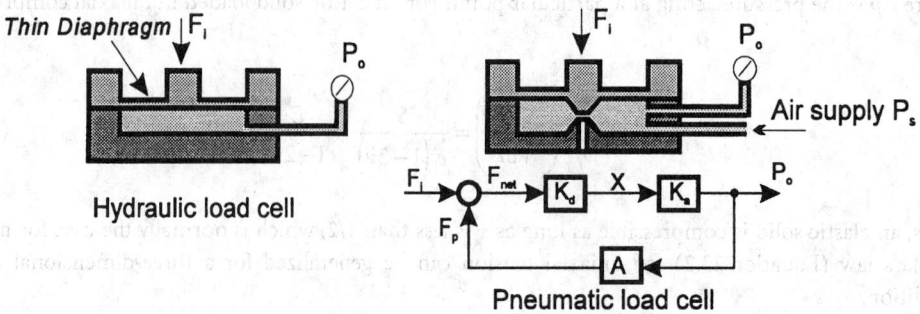

FIGURE 23.3   Different types of load cells [2].

the applied force, and the measuring gage. Load cells are used for sensing large, static or slowly varying forces with little deflection and are a relatively accurate means of sensing forces. Typical accuracies are of the order of 0.1% of the full-scale readings. Various strategies can be employed for measuring forces that are strongly dependent on the design of the load cell. For example, Figure 23.3 illustrates different types of load cells that can be employed in sensing large forces for relatively little cost. The hydraulic load cell employs a very stiff outer structure with an internal cavity filled with a fluid. Application of a load increases the oil pressure, which can be read off an accurate gage.

Other sensing techniques can be utilized to monitor forces, such as piezoelectric transducers for quicker response of varying loads, pneumatic methods, strain gages, etc. The proper sensing technique needs special consideration based on the conditions required for monitoring.

Strain Gage Load Cell

The strain gage load cell consists of a structure that elastically deforms when subjected to a force and a strain gage network that produces an electrical signal proportional to this deformation. Examples of this are beam and ring types of load cells.

Strain Gages

Strain gages use a length of gage wire to produce the desired resistance (which is usually about 120 Ω) in the form of a flat coil. This coil is then cemented (bonded) between two thin insulating sheets of paper or plastic. Such a gage cannot be used directly to measure deflection. It has to be first fixed properly to a member to be strained. After bonding the gage to the member, they are baked at about 195°F (90°C) to remove moisture. Coating the unit with wax or resin will provide some mechanical protection. The resistance between the member under test and the gage itself must be at least 50 MΩ. The total area of all conductors must remain small so that the cement can easily transmit the force necessary to deform the wire. As the member is stressed, the resulting strain deforms the strain gage and the cross-sectional area diminishes. This causes an increase in resistivity of the gage that is easily determined. In order to measure very small strains, it is necessary to measure small changes of the resistance per unit resistance ($\Delta R/R$). The change in the resistance of a bonded strain gage is usually less than 0.5%. A wide variety of gage sizes and grid shapes are available, and typical examples are shown in Figure 23.4.

The use of strain gages to measure force requires careful consideration with respect to rigidity and environment. By virtue of their design, strain gages of shorter length generally possess higher response frequencies (examples: 660 kHz for a gage of 0.2 mm and 20 kHz for a gage of 60 mm in length). The environmental considerations focus mainly on the temperature of the gage. It is well known that resistance is a function of temperature and, thus, strain gages are susceptible to variations in temperature. Thus, if it is known that the temperature of the gage will vary due to any influence, temperature compensation is required in order to ensure that the force measurement is accurate.

A Wheatstone bridge (Figure 23.5) is usually used to measure this small order of magnitude. In Figure 23.5, no current will flow through the galvanometer (G) if the four resistances satisfy a certain

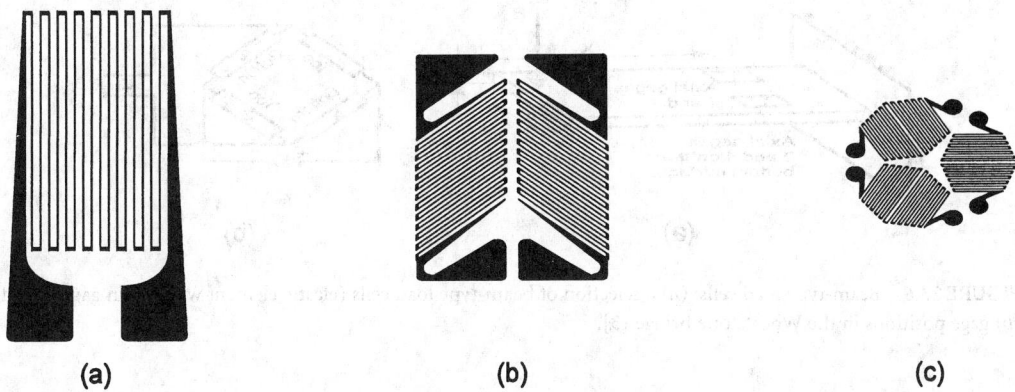

FIGURE 23.4 Configuration of metal-foil resistance strain gages: (*a*) single element; (*b*) two element; and (*c*) three element.

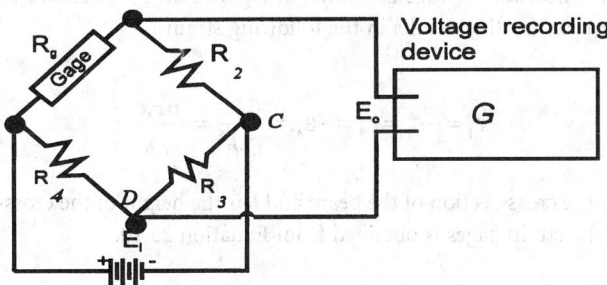

FIGURE 23.5 The Wheatstone bridge.

condition. In order to demonstrate how a Wheatstone bridge operates [3], a voltage scale has been drawn at points C and D of Figure 23.5. Assume that R_1 is a bonded gage and that initially Equation 23.7 is satisfied. If R_1 is now stretched so that its resistance increases by one unit ($+\Delta R$), the voltage at point D will be increased from zero to plus one unit of voltage ($+\Delta V$), and there will be a voltage difference of one unit between C and D that will give rise to a current through C. If R_4 is also a bonded gage, and at the same time that R_1 changes by $+\Delta R$, R_4 changes by $-\Delta R$, the voltage at D will move to $+2\Delta V$. Also, if at the same time, R_2 changes by $-\Delta R$, and R_3 changes by $+\Delta R$, then the voltage of point C will move to $-2\Delta V$, and the voltage difference between C and D will now be $4\Delta V$. It is then apparent that although a single gage can be used, the sensitivity can be increased fourfold if two gages are used in tension while two others are used in compression.

$$\frac{R_1}{R_4} = \frac{R_2}{R_3} \tag{23.7}$$

The grid configuration of the metal-foil resistance strain gages is formed by a photo-etching process. The shortest gage available is 0.20 mm; the longest is 102 mm. Standard gage resistance are 120 Ω and 350 Ω A strain gage exhibits a resistance change $\Delta R/R$ that is related to the strain in the direction of the grid lines by the expression in Equation 23.8 (where S_g is the gage factor or calibration constant for the gage).

$$\frac{\Delta R}{R} = S_g \varepsilon \tag{23.8}$$

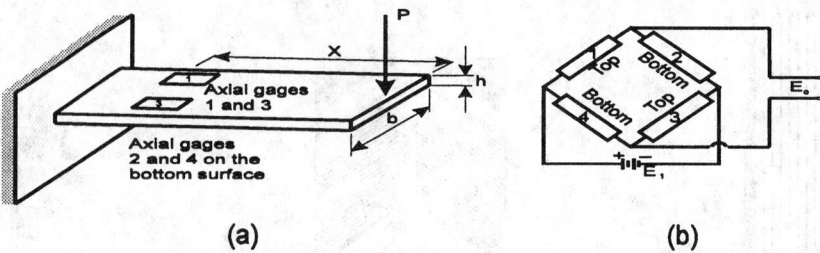

FIGURE 23.6   Beam-type load cells: (*a*) a selection of beam-type load cells (elastic element with strain gages); and (*b*) gage positions in the Wheatstone bridge [3].

Beam-Type Load Cell

Beam-type load cells are commonly employed for measuring low-level loads [3]. A simple cantilever beam (see Figure 23.6(*a*)) with four strain gages, two on the top surface and two on the bottom surface (all oriented along the axis of the beam) is used as the elastic member (sensor) for the load cell. The gages are wired into a Wheatstone bridge as shown in Figure 23.6(*b*). The load *P* produces a moment $M = Px$ at the gage location (*x*) that results in the following strains:

$$\varepsilon_1 = -\varepsilon_2 = \varepsilon_3 = -\varepsilon_4 = \frac{6M}{Ebh^2} = \frac{6Px}{Ebh^2} \tag{23.9}$$

where *b* is the width of the cross-section of the beam and *h* is the height of the cross-section of the beam. Thus, the response of the strain gages is obtained from Equation 23.10.

$$\frac{\Delta R_1}{R_1} = -\frac{\Delta R_2}{R_2} = \frac{\Delta R_3}{R_3} = -\frac{\Delta R_4}{R_4} = \frac{6S_g Px}{Ebh^2} \tag{23.10}$$

The output voltage E_o from the Wheatstone bridge, resulting from application of the load *P*, is obtained from Equation 23.11. If the four strain gages on the beam are assumed to be identical, then Equation 23.11 holds.

$$E_o = \frac{6S_g PxE_1}{Ebh^2} \tag{23.11}$$

The range and sensitivity of a beam-type load cell depends on the shape of the cross-section of the beam, the location of the point of application of the load, and the fatigue strength of the material from which the beam is fabricated.

Ring-Type Load Cell

Ring-type load cells incorporate a proving ring (see Figure 23.7) as the elastic element. The ring element can be designed to cover a very wide range of loads by varying the diameter *D*, the thickness *t*, or the depth *w* of the ring. Either strain gages or a linear variable-differential transformer (LVDT) can be used as the sensor.

The load *P* is linearly proportional to the output voltage E_o. The sensitivity of the ring-type load cell with an LVDT sensor depends on the geometry of the ring (*R*, *t*, and *w*), the material from which the ring is fabricated (*E*), and the characteristics of the LVDT (*S* and E_i). The range of a ring-type load cell is controlled by the strength of the material used in fabricating the ring.

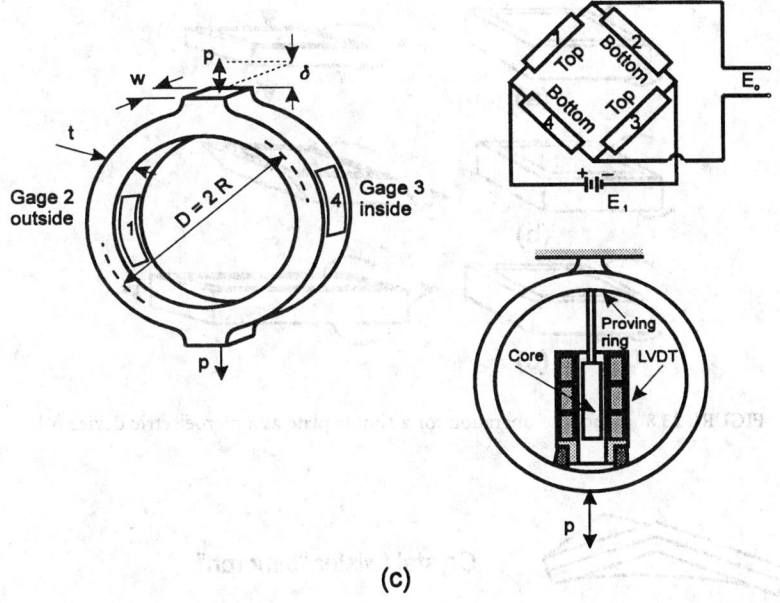

FIGURE 23.7 Ring-type load cells: (*a*) elastic element with strain-gage sensors; (*b*) gage positions in the Wheatstone bridge; and (*c*) elastic element with an LVDT sensor [3].

Piezoelectric Methods

A piezoelectric material exhibits a phenomenon known as the *piezoelectric effect*. This effect states that when asymmetrical, elastic crystals are deformed by a force, an electrical potential will be developed within the distorted crystal lattice. This effect is reversible. That is, if a potential is applied between the surfaces of the crystal, it will change its physical dimensions [4]. Elements exhibiting piezoelectric qualities are sometimes known as electrorestrictive elements.

The magnitude and polarity of the induced surface charges are proportional to the magnitude and direction of the applied force [4]:

$$Q = dF \qquad (23.12)$$

where d is the charge sensitivity (a constant for a given crystal) of the crystal in C/N. The force F causes a thickness variation Δt meters of the crystal:

$$F = \frac{aY}{t} \Delta t \qquad (23.13)$$

where a is area of crystal, t is thickness of crystal, and Y is Young's modulus.

$$Y = \frac{stress}{strain} = \frac{Ft}{a\Delta t} \qquad (23.14)$$

The charge at the electrodes gives rise to a voltage $E_0 = Q/C$, where C is capacitance in farads between the electrodes and $C = \varepsilon a/t$ where ε is the absolute permittivity.

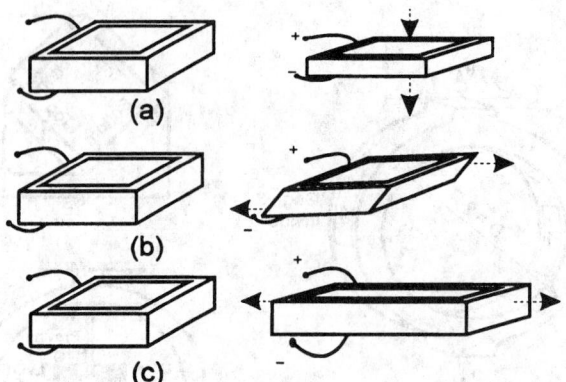

FIGURE 23.8 Modes of operation for a simple plate as a piezoelectric device [4].

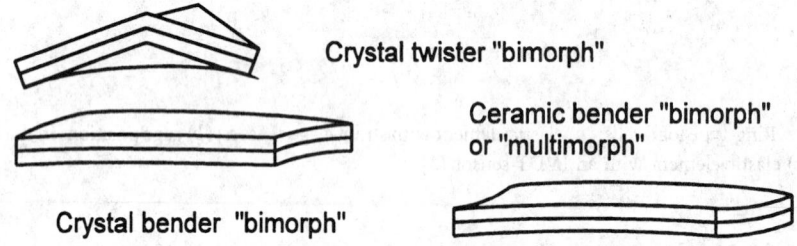

FIGURE 23.9 Curvature of "twister" and "bender" piezoelectric transducers when voltage applied [4].

$$E_o = \frac{dF}{C} = \frac{d}{\varepsilon}\frac{tF}{a} \qquad (23.15)$$

The voltage sensitivity = $g = d/\varepsilon$ in volt m/N can be obtained as:

$$E_o = g\frac{t}{a}F = gtP \qquad (23.16)$$

The piezoelectric materials used are quartz, tourmaline, Rochelle salt, ammonium dihydrogen phosphate (ADP), lithium sulfate, barium titanate, and lead zirconate titanate (PZT) [4]. Quartz and other earthly piezoelectric crystals are naturally polarized. However, synthetic piezoelectric materials, such as barium titanate ceramic, are made by baking small crystallites under pressure and then placing the resultant material in a strong dc electric field [4]. After that, the crystal is polarized, along the axis on which the force will be applied, to exhibit piezoelectric properties. Artificial piezoelectric elements are free from the limitations imposed by the crystal structure and can be molded into any size and shape. The direction of polarization is designated during their production process.

The different modes of operation of a piezoelectric device for a simple plate are shown in Figure 23.8 [4]. By adhering two crystals together so that their electrical axes are perpendicular, bending moments or torque can be applied to the piezoelectric transducer and a voltage output can be produced (Figure 23.9) [4]. The range of forces that can be measured using piezoelectric transducers are from 1 to 200 kN and at a ratio of 2×10^5.

Piezoelectric crystals can also be used in measuring an instantaneous change in the force (dynamic forces). A thin plate of quartz can be used as an electronic oscillator. The frequency of these oscillations will be dominated by the natural frequency of the thin plate. Any distortion in the shape of the plate caused by an external force, alters the oscillation frequency. Hence, a dynamic force can be measured by the change in frequency of the oscillator.

Resistive Method

The resistive method employs the fact that when the multiple contact area between semiconducting particles (usually carbon) and the distance between the particles are changed, the total resistance is altered. The design of such transducers yields a very small displacement when a force is applied. A transducer might consist of 2 to 60 thin carbon disks mounted between a fixed and a movable electrode. When a force is applied to the movable electrode and the carbon disks move together by 5 to 250 μm per interface, the transfer function of their resistance against the applied force is approximately hyperbolic, that is, highly nonlinear. The device is also subject to large hysteresis and drift together with a high transverse sensitivity.

In order to reduce hysteresis and drift, rings are used instead of disks. The rings are mounted on an insulated rigid core and prestressed. This almost completely eliminates any transverse sensitivity error. The core's resonant frequency is high and can occur at a frequency as high as 10 kHz. The possible measuring range of such a transducer is from 0.1 kg to 10 kg. The accuracy and linear sensitivity of this transducer is very poor.

Inductive Method

The inductive method utilizes the fact that a change in mechanical stress of a ferromagnetic material causes its permeability to alter. The changes in magnetic flux are converted into induced voltages in the pickup coils as the movement takes place. This phenomenon is known as the *Villari effect* or *magneto-striction*. It is known to be particularly strong in nickel–iron alloys.

Transducers utilizing the Villari effect consist of a coil wound on a core of magnetostrictive material. The force to be measured is applied on this core, stressing it and causing a change in its permeability and inductance. This change can be monitored and used for determining the force.

The applicable range for this type of transducer is a function of the cross-sectional area of the core. The accuracy of the device is determined by a calibration process. This transducer has poor linearity and is subject to hysteresis. The permeability of a magnetostrictive material increases when it is subjected to pure torsion, regardless of direction. A flat frequency response is obtained over a wide range from 150 Hz to 15,000 Hz.

Piezotransistor Method

Devices that utilize *anisotropic stress effects* are described as piezotransistors. In this effect, if the upper surface of a *p–n* diode is subjected to a localized stress, a significant reversible change occurs in the current across the junction. These transistors are usually silicon nonplanar type, with an emitter base junction. This junction is mechanically connected to a diaphragm positioned on the upper surface of a typical TO-type can [4]. When a pressure or a force is applied to the diaphragm, an electronic charge is produced. It is advisable to use these force-measuring devices at a constant temperature by virtue of the fact that semiconducting materials also change their electric properties with temperature variations. The attractive characteristic of piezotransistors is that they can withstand a 500% overload.

Multicomponent Dynamometers Using Quartz Crystals As Sensing Elements

The Piezoelectric Effects in Quartz.

For force measurements, the *direct piezoelectric effect* is utilized. The direct longitudinal effect measures compressive force; the direct shear effect measures shear force in one direction. For example, if a disk of crystalline quartz (SiO_2) cut normally to the crystallographic x-axis is loaded by a compression force, it will yield an electric charge, nominally 2.26 pC/N. If a disk of crystalline quartz is cut normally to the

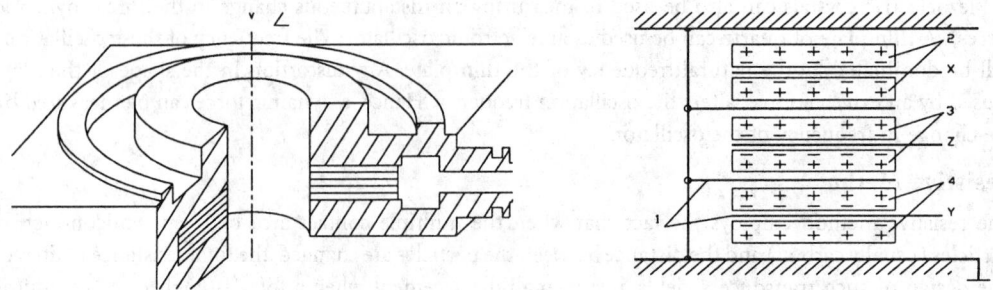

FIGURE 23.10   Three-component force transducer.

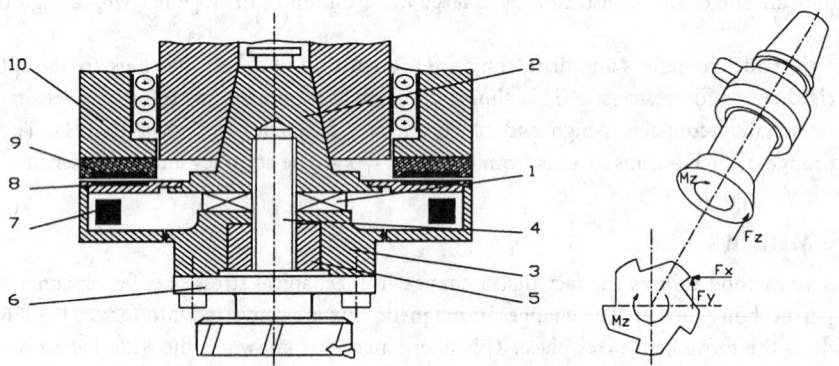

FIGURE 23.11   Force measuring system to determine the tool-related cutting forces in five-axis milling [6].

crystallographic y-axis, it will yield an electric charge (4.52 pC/N) if loaded by a shear force in one specific direction. Forces applied in the other directions will not generate any output [5].

A charge amplifier is used to convert the charge yielded by a quartz crystal element into a proportional voltage. The range of a charge amplifier with respect to its conversion factor is determined by a feedback capacitor. Adjustment to mechanical units is obtained by additional operational amplifiers with variable gain.

The Design of Quartz Multicomponent Dynamometers.
The main element for designing multicomponent dynamometers is the three-component force transducer (Figure 23.10). It contains a pair of X-cut quartz disks for the normal force component and a pair of Y-cut quartz disks (shear-sensitive) for each shear force component.

Three-component dynamometers can be used for measuring cutting forces during machining. Four three-component force transducers sandwiched between a base plate and a top plate are shown in Figure 23.10. The force transducer is subjected to a preload as shear forces are transmitted by friction. The four force transducers experience a drastic change in their load, depending on the type and position of force application. An overhanging introduction of the force develops a tensile force for some transducers, thus reducing the preload. Bending of the dynamometer top plate causes bending and shearing stresses. The measuring ranges of a dynamometer depend not only on the individual forces, but also on the individual bending stresses.

Measuring Signals Transmitted by Telemetry.
Figure 23.11 shows the newly designed force measuring system RCD (rotating cutting force dynamometer). A ring-shaped sensor (1) is fitted in a steep angle taper socket (2) and a base ring (3) allowing sensing of the three force components F_x, F_y and F_z at the cutting edge as well as the moment M_z. The

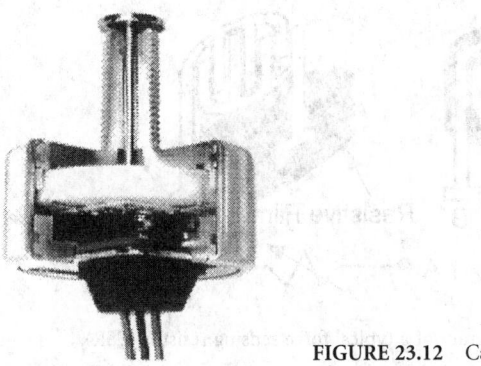

FIGURE 23.12 Capacitive force transducer [7].

physical operating principle of this measuring cell is based on the piezoelectric effect in quartz plates. The quartz plates incorporated in the sensor are aligned so that the maximum cross-sensitivity between the force components is 1%. As a result of the rigid design of the sensor, the resonant frequencies of the force measuring system range from 1200 Hz to 3000 Hz and the measuring ranges cover a maximum of 10 kN [6].

Force-proportional charges produced at the surfaces of the quartz plates are converted into voltages by four miniature charge amplifiers (7) in hybrid construction. These signals are then filtered by specific electrical circuitry to prevent aliasing effects, and digitized with 8 bit resolution using a high sampling rate (pulse-code modulation). The digitized signals are transmitted by a telemetric unit consisting of a receiver and transmitter module, an antenna at the top of the rotating force measuring system (8), as well as a fixed antenna (9) on the splash cover of the two-axis milling head (10). The electrical components, charge amplifier, and transmitter module are mounted on the circumference of the force measuring system [6].

The cutting forces and the moment measured are digitized with the force measuring system described above. They are modulated on an FM carrier and transmitted by the rotating transmitter to the stationary receiver. The signals transmitted are fed to an external measured-variable conditioning unit.

Measuring Dynamic Forces.

Any mechanical system can be considered in the first approximation as a weakly damped oscillator consisting of a spring and a mass. If a mechanical system has more than one resonant frequency, the lowest one must be taken into consideration. As long as the test frequency remains below 10% of the resonant frequency of the reference transducer (used for calibration), the difference between the dynamic sensitivity obtained from static calibration will be less than 1%. The above considerations assume a sinusoidal force signal. The static calibration of a reference transducer is also valid for dynamic calibration purposes if the test frequency is much lower (at least 10 times lower) than the resonant frequency of the system.

Capacitive Force Transducer

A transducer that uses capacitance variation can be used to measure force. The force is directed onto a membrane whose elastic deflection is detected by a capacitance variation. A highly sensitive force transducer can be constructed because the capacitive transducer senses very small deflections accurately. An electronic circuit converts the capacitance variations into dc-voltage variations [7].

The capacitance sensor illustrated in Figure 23.12 consists of two metal plates separated by an air gap. The capacitance C between terminals is given by the expression:

$$C = \varepsilon_o \varepsilon_r \frac{A}{h} \qquad (23.17)$$

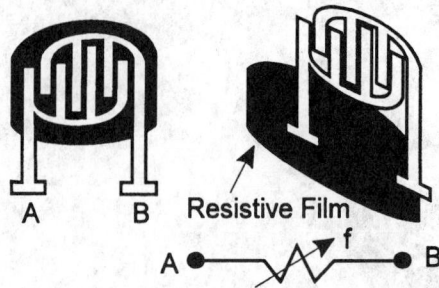

FIGURE 23.13   Diagram of a typical force sensing resistor (FSR).

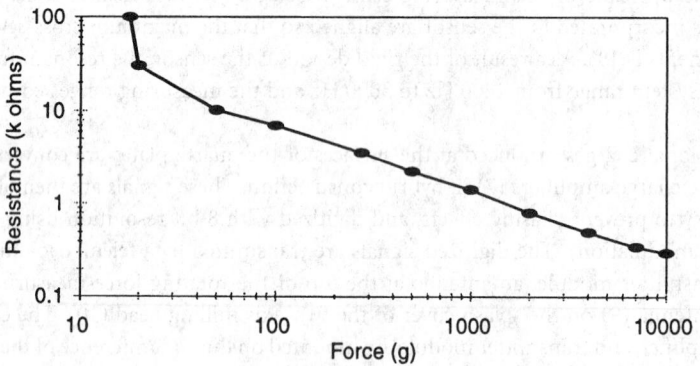

FIGURE 23.14   Resistance as a function of force for a typical force sensing resistor.

where C = Capacitance in farads (F)
   ε_0 = Dielectric constant of free space
   ε_r = Relative dielectric constant of the insulator
   A = Overlapping area for the two plates
   h = Thickness of the gap between the two plates

The sensitivity of capacitance-type sensors is inherently low. Theoretically, decreasing the gap h should increase the sensitivity; however, there are practical electrical and mechanical conditions that preclude high sensitivities. One of the main advantages of the capacitive transducer is that moving of one of its plate relative to the other requires an extremely small force to be applied. A second advantage is stability and the sensitivity of the sensor is not influenced by pressure or temperature of the environment.

Force Sensing Resistors (Conductive Polymers)

Force sensing resistors (FSRs) utilize the fact that certain polymer thick-film devices exhibit decreasing resistance with the increase of an applied force. A force sensing resistor is made up of two parts. The first is a resistive material applied to a film. The second is a set of digitating contacts applied to another film. Figure 23.13 shows this configuration. The resistive material completes the electrical circuit between the two sets of conductors on the other film. When a force is applied to this sensor, a better connection is made between the contacts; hence, the conductivity is increased. Over a wide range of forces, it turns out that the conductivity is approximately a linear function of force. Figure 23.14 shows the resistance of the sensor as a function of force. It is important to note that there are three possible regions for the sensor to operate. The first abrupt transition occurs somewhere in the vicinity of 10 g of force. In this region, the resistance changes very rapidly. This behavior is useful when one is designing switches using force sensing resistors.

FSRs should not be used for accurate measurements of force because sensor parts may exhibit 15% to 25% variation in resistance between each other. However, FSRs exhibit little hysteresis and are considered far less costly than other sensing devices. Compared to piezofilm, the FSR is far less sensitive to vibration and heat.

Magnetoresistive Force Sensors

The principle of *magnetoresistive force sensors* is based on the fact that metals, when cooled to low temperatures, show a change of resistivity when subjected to an applied magnetic field. Bismuth, in particular, is quite sensitive in this respect. In practice, these devices are severely limited because of their high sensitivity to ambient temperature changes.

Magnetoelastic Force Sensors

Magnetoelastic transducer devices operate based on the Joule effect; that is, a ferromagnetic material is dimensionally altered when subjected to a magnetic field. The principle of operation is as follows: Initially, a current pulse is applied to the conductor within the waveguide. This sets up a magnetic field circumference-wise around the waveguide over its entire length. There is another magnetic field generated by the permanent magnet that exists only where the magnet is located. This field has a longitudinal component. These two fields join vectorally to form a helical field near the magnet which, in turn, causes the waveguide to experience a minute torsional strain or twist only at the location of the magnet. This twist effect is known as the *Wiedemann effect* [8].

Magnetoelastic force transducers have a high frequency response (on the order of 20 kHz). Some of the materials that exhibit magnetoelastic include Monel metal, Permalloy, Cekas, Alfer, and a number of nickel–iron alloys. Disadvantages of these transducers include: (1) the fact that excessive stress and aging may cause permanent changes, (2) zero drift and sensitivity changes due to temperature sensitivity, and (3) hysteresis errors.

Torsional Balances

Balancing devices that utilize the deflection of a spring may also be used to determine forces. *Torsional balances* are equal arm scale force measuring devices. They are comprised of horizontal steel bands instead of pivots and bearings. The principle of operation is based on force application on one of the arms that will deflect the torsional spring (within its design limits) in proportion to the applied force. This type of instrument is susceptible to hysteresis and temperature errors and therefore is not used for precise measurements.

Tactile Sensors

Tactile sensors are usually interpreted as a touch sensing technique. Tactile sensors cannot be considered as simple touch sensors, where very few discrete force measurements are made. In tactile sensing, a force "distribution" is measured using a closely spaced array of force sensors.

Tactile sensing is important in both grasping and object identification operations. Grasping an object must be done in a stable manner so that the object is not allowed to slip or damaged. Object identification includes recognizing the shape, location, and orientation of a product, as well as identifying surface properties and defects. Ideally, these tasks would require two types of sensing [9]:

1. Continuous sensing of contact forces
2. Sensing of the surface deformation profile

These two types of data are generally related through stress–strain relations of the tactile sensor. As a result, almost continuous variable sensing of tactile forces (the sensing of the tactile deflection profile) is achieved.

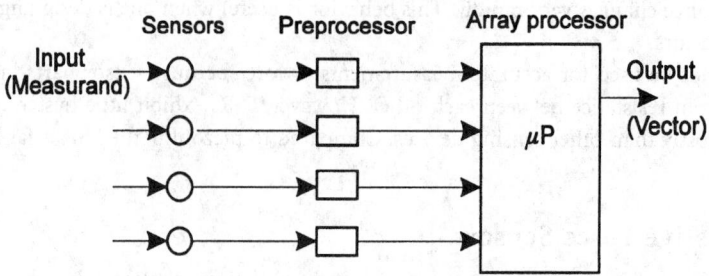

FIGURE 23.15   Tactile array sensor.

Tactile Sensor Requirements.

Significant advances in tactile sensing are taking place in the robotics area. Applications include automated inspection of surface profiles, material handling or parts transfer, parts assembly, and parts identification and gaging in manufacturing applications and fine-manipulation tasks. Some of these applications may need only simple touch (force–torque) sensing if the parts being grasped are properly oriented and if adequate information about the process is already available.

Naturally, the main design objective for tactile sensing devices has been to mimic the capabilities of human fingers [9]. Typical specifications for an industrial tactile sensor include:

1. Spatial resolution of about 2 mm
2. Force resolution (sensitivity) of about 2 g
3. Maximum touch force of about 1 kg
4. Low response time of 5 ms
5. Low hysteresis
6. Durability under extremely difficult working conditions
7. Insensitivity to change in environmental conditions (temperature, dust, humidity, vibration, etc.)
8. Ability to monitor slip

Tactile Array Sensor.

Tactile array sensors (Figure 23.15) consist of a regular pattern of sensing elements to measure the distribution of pressure across the finger tip of a Robot. The 8×8 array of elements at 2 mm spacing in each direction, provides 64 force sensitive elements. Table 23.1 outlines some of the characteristics of early tactile array sensors. The sensor is composed of two crossed layers of copper strips separated by strips of thin silicone rubber. The sensor forms a thin, compliant layer that can be easily attached to a variety of finger-tip shapes and sizes. The entire array is sampled by computer.

A typical tactile sensor array can consist of several sensing elements. Each element or taxel (Figure 23.16) is used to sense the forces present. Since tactile sensors are implemented in applications where sensitivity providing semblance to human touch is desired, an elastomer is utilized to mimic the human skin. The elastomer is generally a conductive material whose electrical conductivity changes locally when pressure is applied. The sensor itself consists of three layers: a protective covering, a sheet of conductive elastomer, and a printed circuit board. The printed circuit board consists of two rows of two "bullseyes," each with conductive inner and outer rings that compromise the taxels of the sensor. The outer rings are connected together and to a column-select transistor. The inner rings are connected to diodes (D) in Figure 23.16. Once the column in the array is selected, the current flows through the diodes, through the elastomer, and thence through a transistor to ground. As such, it is generally not possible to excite just one taxel because the pressure applied causes a local deformation in neighboring taxels. This situation is called *crosstalk* and is eliminated by the diodes [10].

Tactile array sensor signals are used to provide information about the contact kinematics. Several feature parameters, such as contact location, object shape, and the pressure distribution, can be obtained.

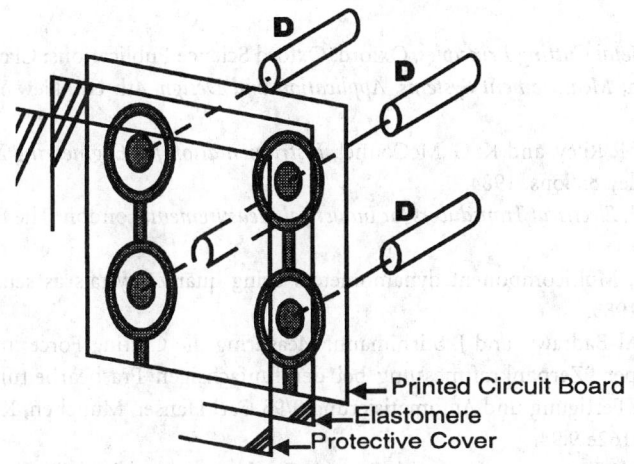

FIGURE 23.16 Typical taxel sensor array.

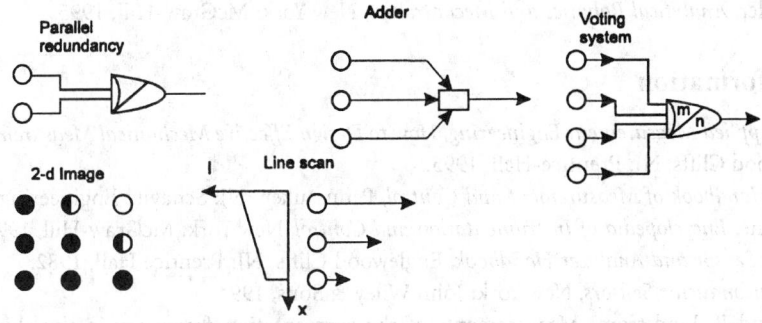

FIGURE 23.17 General arrangement of an intelligent sensor array system [9].

TABLE 23.1 Summary of Some of the Characteristics
of Early Tactile Arrays Sensors

	Size of array		
Device parameter	(4 × 4)	(8 × 8)	(16 × 16)
Cell spacing (mm)	4.00	2.00	1.00
Zero-pressure capacitance (fF)	6.48	1.62	0.40
Rupture force (N)	18.90	1.88	0.19
Max. linear capacitance (fF)	4.80	1.20	0.30
Max. output voltage (V)	1.20	0.60	0.30
Max. resolution (bit)	9.00	8.00	8.00
Readout (access) time (μs)	—	<20	—

©IEEE 1985.

The general layout of a sensor array system can be seen in Figure 23.17. An example of this is a contact and force sensing finger. This tactile finger has four contact sensors made of piezoelectric polymer strips on the surface of the fingertip that provide dynamic contact information. A strain gage force sensor provides static grasp force information.

References

1. M. C. Shaw, *Metal Cutting Principles*, Oxford: Oxford Science Publications: Clarendon Press, 1989.
2. E. O. Doebelin, *Measurement Systems, Application and Design*, 4th ed., New York: McGraw-Hill, 1990.
3. J. W. Dally, W. F. Riley, and K. G. McConnel, *Instrumentation for Engineering Measurements*, New York: John Wiley & Sons, 1984.
4. P. H. Mansfield, *Electrical Transducers for Industrial Measurement*, London: The Butterworth Group, 1973.
5. K. H. Martini, Multicomponent dynamometers using quartz crystals as sensing elements, *ISA Trans.*, 22(1), 1983.
6. G. Spur, S. J. Al-Badrawy, and J. Stirnimann, Measuring the Cutting Force in Five-Axis Milling, Translated paper "Zerpankraftmessung bei der funfachsigen Frasbearbeitung", Zeitschrift fur wirtschaftliche Fertigung und Automatisierung 9/93 Carl Hanser, Munchen, Kistler Piezo-Instrumentation, 20.162e 9.94.
7. C. L. Nachtigal, *Instrumentation and Control, Fundamentals and Applications*, Wiley Series in Mechanical Engineering Practice, New York: Wiley Interscience, John Wiley & Sons, 1990.
8. C. W. DeSilva, *Control Sensors and Actuators*, Englewood Cliffs, NJ: Prentice-Hall, 1989.
9. J. W. Gardner, *Microsensors Principles and Applications*, New York: John Wiley & Sons, 1995.
10. W. Stadler, *Analytical Robotics and Mechatronics*, New York: McGraw-Hill, 1995.

Further Information

C. P. Wright, *Applied Measurement Engineering, How to Design Effective Mechanical Measurement Systems*, Englewood Cliffs, NJ: Prentice-Hall, 1995.

E. E. Herceg, *Handbook of Measurement and Control*, Pennsauken, NJ: Schavitz Engineering, 1972.

D. M. Considine, *Encyclopedia of Instrumentation and Control*, New York: McGraw-Hill, 1971.

H. N. Norton, *Sensor and Analyzer Handbook*, Englewood Cliffs, NJ: Prentice Hall, 1982.

S. M. Sze, *Semiconductor Sensors*, New York: John Wiley & Sons, 1994.

B. Lindberg and B. Lindstrom, Measurements of the segmentation frequency in the chip formation process, *Ann. CIRP*, 32(1), 1983.

J. Tlusty and G. C Andrews, A critical review of sensors for unmanned machining, *Ann. CIRP*, 32(2), 1983.

24

Torque and Power Measurement

Torque, speed, and power are the defining mechanical variables associated with the functional performance of rotating machinery. The ability to accurately measure these quantities is essential for determining a machine's efficiency and for establishing operating regimes that are both safe and conducive to long and reliable services. On-line measurements of these quantities enable real-time control, help to ensure consistency in product quality, and can provide early indications of impending problems. Torque and power measurements are used in testing advanced designs of new machines and in the development of new machine components. Torque measurements also provide a well-established basis for controlling and verifying the tightness of many types of threaded fasteners. This chapter describes the basic concepts as well as the various methods and apparati in current use for the measurement of torque and power; the measurement of speed, or more precisely, angular velocity, is discussed elsewhere in this handbook [1].

24.1 Fundamental Concepts

Angular Displacement, Velocity, and Acceleration

The concept of *rotational* motion is readily formalized: all points within a rotating rigid body move in parallel or coincident planes while remaining at fixed distances from a line called the *axis*. In a perfectly rigid body, all points also remain at fixed distances from each other. Rotation is perceived as a change in the angular position of a reference point on the body, i.e., as its *angular displacement*, $\Delta\theta$, over some time interval, Δt. The motion of that point, and therefore of the whole body, is characterized by its

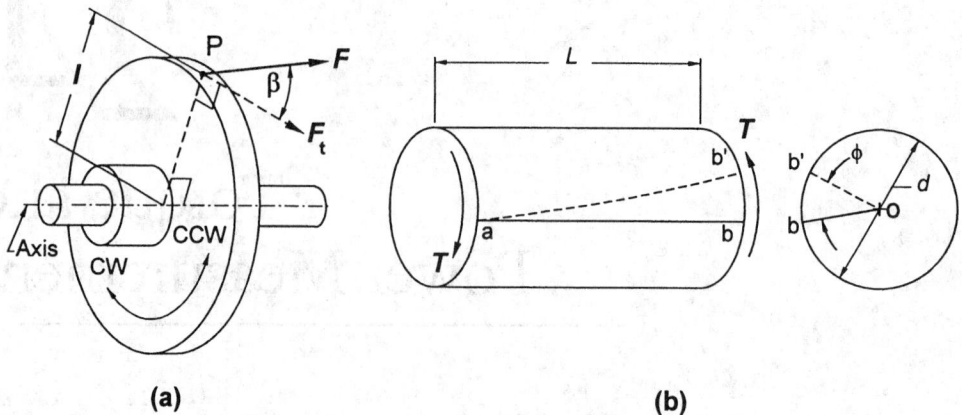

FIGURE 24.1 (*a*) The off-axis force *F* at P produces a torque $T = (F \cos \beta)l$ tending to rotate the body in the CW direction. (*b*) Transmitting torque *T* over length *L* twists the shaft through angle ϕ.

clockwise (CW) or counterclockwise (CCW) *direction* and by its *angular velocity*, $\omega = \Delta\theta/\Delta t$. If during a time interval Δt, the velocity changes by $\Delta\omega$, the body is undergoing an *angular acceleration*, $\alpha = \Delta\omega/\Delta t$. With angles measured in radians, and time in seconds, units of ω become radians per second (rad s^{-1}) and of α, radians per second per second (rad s^{-2}). Angular velocity is often referred to as *rotational speed* and measured in numbers of complete revolutions per minute (rpm) or per second (rps).

Force, Torque, and Equilibrium

Rotational motion, as with motion in general, is controlled by *forces* in accordance with Newton's laws. Because a force directly affects only that component of motion in its line of action, forces or components of forces acting in any plane that includes the axis produce no tendency for rotation about that axis. Rotation can be initiated, altered in velocity, or terminated only by a *tangential force* F_t acting at a finite radial distance *l* from the axis. The effectiveness of such forces increases with both F_t and *l*; hence, their product, called a *moment*, is the activating quantity for rotational motion. A moment about the rotational axis constitutes a *torque*. Figure 24.1(*a*) shows a force *F* acting at an angle β to the tangent at a point P, distant *l* (the moment arm) from the axis. The torque *T* is found from the *tangential component* of *F* as:

$$T = F_t l = \left(F \cos \beta \right) l \qquad (24.1)$$

The combined effect, known as the *resultant*, of any number of torques acting at different locations along a body is found from their *algebraic sum*, wherein torques tending to cause rotation in CW and CCW directions are assigned opposite signs. Forces, hence torques, arise from physical contact with other solid bodies, motional interaction with fluids, or via gravitational (including inertial), electric, or magnetic force fields. The *source* of each such torque is subjected to an equal, but oppositely directed, *reaction* torque. With force measured in newtons and distance in meters, Equation 24.1 shows the unit of torque to be a Newton meter (N·m).

A nonzero resultant torque will cause the body to undergo a proportional angular acceleration, found, by application of Newton's second law, from:

$$T_r = I\alpha \qquad (24.2)$$

where *I*, having units of kilogram meter^2 (kg m^2), is the moment of inertia of the body around the axis (i.e., its *polar* moment of inertia). Equation 24.2 is applicable to any body regardless of its state of motion.

When $\alpha = 0$, Equation 24.2 shows that T_r is also zero; the body is said to be in *equilibrium*. For a body to be in equilibrium, there must be either more than one *applied* torque, or none at all.

Stress, Rigidity, and Strain

Any portion of a rigid body in equilibrium is also in equilibrium; hence, as a condition for equilibrium of the portion, any torques applied thereto from *external* sources must be balanced by equal and directionally opposite *internal* torques from adjoining portions of the body. Internal torques are *transmitted* between adjoining portions by the collective action of *stresses* over their common cross-sections. In a solid body having a round cross-section (e.g., a typical shaft), the *shear stress* τ varies linearly from zero at the axis to a maximum value at the surface. The shear stress, τ_m, at the surface of a shaft of diameter, d, transmitting a torque, T, is found from:

$$\tau_m = \frac{16T}{\pi d^3} \tag{24.3}$$

Real materials are not *perfectly* rigid but have instead a *modulus of rigidity*, G, which expresses the finite ratio between τ and *shear strain*, γ. The maximum strain in a solid round shaft therefore also exists at its surface and can be found from:

$$\gamma_m = \frac{\tau_m}{G} = \frac{16T}{\pi d^3 G} \tag{24.4}$$

Figure 24.1(*b*) shows the manifestation of shear strain as an angular displacement between axially separated cross-sections. Over the length L, the solid round shaft shown will be *twisted* by the torque through an angle ϕ found from:

$$\phi = \frac{32LT}{\pi d^4 G} \tag{24.5}$$

Work, Energy, and Power

If during the time of application of a torque, T, the body rotates through some angle θ, mechanical work:

$$W = T\theta \tag{24.6}$$

is performed. If the torque acts in the same CW or CCW sense as the displacement, the work is said to be done *on* the body, or else it is done *by* the body. Work done *on* the body causes it to accelerate, thereby appearing as an increase in *kinetic energy* (KE = $I\omega^2/2$). Work done *by* the body causes deceleration with a corresponding decrease in kinetic energy. If the body is not accelerating, any work done on it at one location must be done by it at another location. Work and energy are each measured in units called a joule (J). Equation 24.6 shows that 1 J is equivalent to 1 N·m rad, which, since a radian is a dimensionless ratio, $\equiv$ 1 N·m. To avoid confusion with torque, it is preferable to quantify mechanical work in units of m·N, or better yet, in J.

The *rate* at which work is performed is termed *power*, P. If a torque T acts over a small interval of time Δt, during which there is an angular displacement $\Delta\theta$, work equal to $T\Delta\theta$ is performed at the rate $T\Delta\theta/\Delta t$. Replacing $\Delta\theta/\Delta t$ by ω, power is found simply as:

$$P = T\omega \tag{24.7}$$

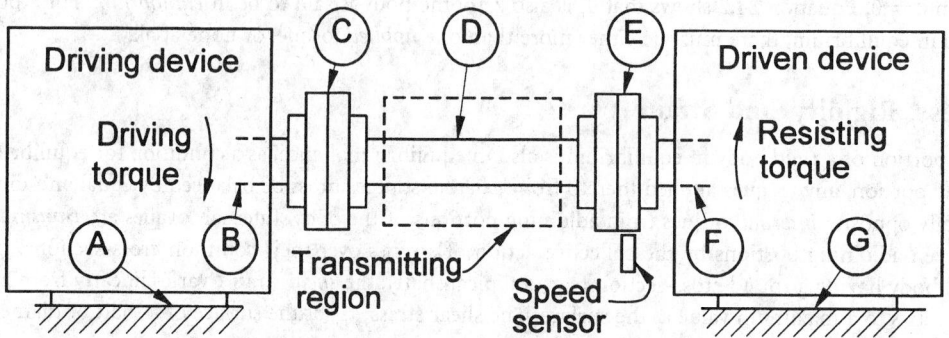

FIGURE 24.2 Schematic arrangement of devices used for the measurement of torque and power.

The unit of power follows from its definition and is given the special name watt (W). 1 W = 1 J s⁻¹ = 1 m·N s⁻¹. Historically, power has also been measured in horsepower (Hp), where 1 Hp = 746 W. Rotating bodies effectively transmit power between locations where torques from external sources are applied.

24.2 Arrangements of Apparatus for Torque and Power Measurement

Equations 24.1 through 24.7 express the physical bases for torque and power measurement. Figure 24.2 illustrates a generalized measurement arrangement. The actual apparatus used is selected to fulfill the specific measurement purposes. In general, a driving torque originating within a device at one location (B in Figure 24.2), is resisted by an opposing torque developed by a different device at another location (F). The driving torque (from, e.g., an electric motor, a gasoline engine, a steam turbine, muscular effort, etc.) is coupled through connecting members C, transmitting region D, and additional couplings E, to the driven device (an electric generator, a pump, a machine tool, mated threaded fasteners, etc.) within which the resisting torque is met at F. The torque at B or F is the quantity to be measured. These torques may be *indirectly* determined from a correlated physical quantity, e.g., an electrical current or fluid pressure associated with the operation of the driving or driven device, or more directly by measuring either the *reaction* torque at A or G, or the *transmitted* torque through D. It follows from the cause-and-effect relationship between torque and rotational motion that most interest in transmitted torque will involve rotating bodies.

To the extent that the frames of the driving and driven devices and their mountings to the "Earth" are *perfectly* rigid, the reaction at A will *at every instant* equal the torque at B, as will the reaction at G equal the torque at F. Under equilibrium conditions, these equalities are independent of the compliance of any member. Also under equilibrium conditions, and except for usually minor *parasitic* torques (due, e.g., to bearing friction and air drag over rapidly moving surfaces), the driving torque at B will equal the resisting torque at F.

Reaction torque at A or G is often determined, using Equation 24.1, from measurements of the forces acting at known distances fixed by the apparatus. Transmitted torque is determined from measurements, on a suitable member within region D, of τ_m, γ_m, or ϕ and applying Equations 24.3, 24.4, or 24.5 (or analogous expressions for members having other than solid round cross-sections [2]). *Calibration*, the measurement of the stress, strain, or twist angle resulting from the application of a *known* torque, makes it unnecessary to know any details about the member within D. When $\alpha \neq 0$, and is measurable, T may also be determined from Equation 24.2. Requiring only noninvasive, observational measurements, this method is especially useful for determining transitory torques; for example those associated with firing events in multicylinder internal combustion engines [3].

Equations 24.6 and 24.7 are applicable *only* during rotation because, in the absence of motion, no work is done and power transfer is zero. Equation 24.6 can be used to determine *average* torque from calorimetric measurements of the heat generated (equal to the mechanical work W) during a totalized number of revolutions ($\equiv \theta/2\pi$). Equation 24.7 is routinely applied in power measurement, wherein T is determined by methods based on Equations 24.1, 24.3, 24.4, or 24.5, and ω is measured by any suitable means [4].

F, T, and ϕ are sometimes measured by simple mechanical methods. For example, a "torque wrench" is often used for the controlled tightening of threaded fasteners. In these devices, torque is indicated by the position of a needle moving over a calibrated scale in response to the elastic deflection of a spring member, in the simplest case, the bending of the wrench handle [5]. More generally, instruments, variously called *sensors* or *transducers*, are used to convert the desired (torque or speed related) quantity into a linearly proportional electrical signal. (Force sensors are also known as *load cells*.) The determination of P most usually requires multiplication of the two signals from separate sensors of T and ω. A transducer, wherein the amplitude of a *single* signal proportional to the power being transmitted along a shaft, has also been described [6].

24.3 Torque Transducer Technologies

Various physical interactions serve to convert F, τ, γ, or ϕ into proportional electrical signals. Each requires that some axial portion of the shaft be dedicated to the torque sensing function. Figure 24.3 shows typical features of sensing regions for four sensing technologies in present use.

Surface Strain

Figure 24.3(*a*) illustrates a sensing region configured to convert surface strain (γ_m) into an electric signal proportional to the transmitted torque. Surface strain became the key basis for measuring both force and torque following the invention of bonded wire strain gages by E. E. Simmons, Jr. and Arthur C. Ruge in 1938 [7]. A modern strain gage consists simply of an elongated electrical conductor, generally formed in a serpentine pattern in a very thin foil or film, bonded to a thin insulating carrier. The carrier is attached, usually with an adhesive, to the surface of the load carrying member. Strain is sensed as a change in gage resistance. These changes are generally too small to be accurately measured directly and so it is common to employ two to four gages arranged in a Wheatstone bridge circuit. Independence from axial and bending loads as well as from temperature variations are obtained by using a four-gage bridge comprised of two diametrically opposite pairs of matched strain gages, each aligned along a *principal strain* direction. In round shafts (and other shapes used to transmit torque), tensile and compressive principal strains occur at 45° angles to the axis. Limiting strains, as determined from Equation 24.4 (with τ_m equal to the shear proportional limit of the shaft material), rarely exceed a few parts in 10^3. Typical practice is to increase the compliance of the sensing region (e.g., by reducing its diameter or with hollow or specially shaped sections) in order to attain the limiting strain at the highest value of the torque to be measured. This maximizes the measurement sensitivity.

Twist Angle

If the shaft is *slender* enough (e.g., $L > 5\ d$) ϕ, at limiting values of τ_m for typical shaft materials, can exceed 1°, enough to be resolved with sufficient accuracy for practical torque measurements (ϕ at τ_m can be found by manipulating Equations 24.3, 24.4, and 24.5). Figure 24.3(*b*) shows a common arrangement wherein torque is determined from the difference in tooth-space phasing between two identical "toothed" wheels attached at opposite ends of a compliant "torsion bar." The phase displacement of the periodic electrical signals from the two "pickups" is proportional to the peripheral displacement of salient features on the two wheels, and hence to the twist angle of the torsion bar and thus to the torque. These features are chosen to be sensible by any of a variety of noncontacting magnetic, optical, or capacitive techniques. With more elaborate pickups, the relative angular position of the two wheels appears as the amplitude of a *single* electrical signal, thus providing for the measurement of torque even on a stationary shaft (e.g., [13–15]). In still other constructions, a shaft-mounted variable displacement transformer or a related type of electric device is used to provide speed independent output signals proportional to ϕ.

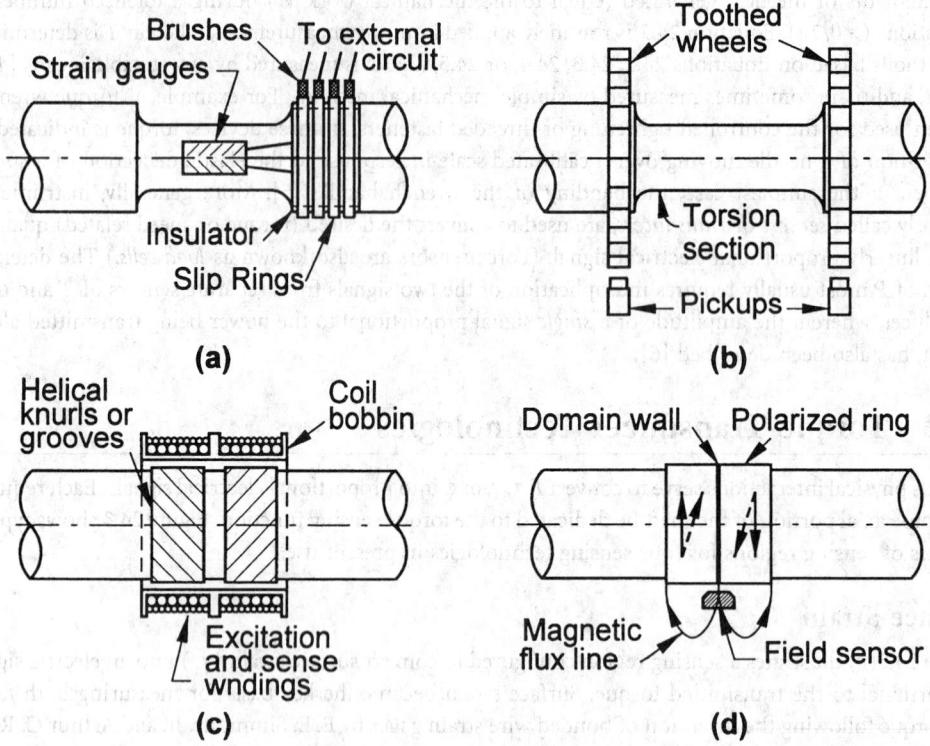

FIGURE 24.3 Four techniques in present use for measuring transmitted torque. (*a*) Torsional strain in the shaft alters the electrical resistance for four strain gages (two not seen) connected in a Wheatstone bridge circuit. In the embodiment shown, electrical connections are made to the bridge through slip rings and brushes. (*b*) Twist of the torsion section causes angular displacement of the surface features on the toothed wheels. This creates a phase difference in the signals from the two pickups. (*c*) The permeabilities of the two grooved regions of the shaft change oppositely with torsional stress. This is sensed as a difference in the output voltages of the two sense windings. (*d*) Torsional stress causes the initially circumferential magnetizations in the ring (solid arrows) to tilt (dashed arrows). These helical magnetizations cause magnetic poles to appear at the domain wall and ring ends. The resulting magnetic field is sensed by the field sensor.

Stress

In addition to elastic strain, the stresses by which torque is transmitted are manifested by changes in the magnetic properties of ferromagnetic shaft materials. This "magnetoelastic interaction" [8] provides an inherently noncontacting basis for measuring torque. Two types of magnetoelastic (sometimes called magnetostrictive) torque transducers are in present use: Type 1 derive output signals from torque-induced variations in magnetic circuit permeances; Type 2 create a magnetic field in response to torque. Type 1 transducers typically employ "branch," "cross," or "solenoidal" constructions [9]. In branch and cross designs, torque is detected as an imbalance in the permeabilities along orthogonal 45° helical paths (the principal stress directions) on the shaft surface or on the surface of an *ad hoc* material attached to the shaft. In solenoidal constructions torque is detected by differences in the *axial* permeabilities of two adjacent surface regions, preendowed with symmetrical magnetic "easy" axes (typically along the 45° principal stress directions). While branch and cross type sensors are readily miniaturized [10], local variations in magnetic properties of typical shaft surfaces limit their accuracy. Solenoidal designs, illustrated in Figure 24.3(*c*), avoid this pitfall by effectively averaging these variations. Type 2 transducers are generally constructed with a ring of magnetoelastically active material rigidly attached to the shaft. The ring is magnetized during manufacture of the transducer, usually with each axial half polarized in an

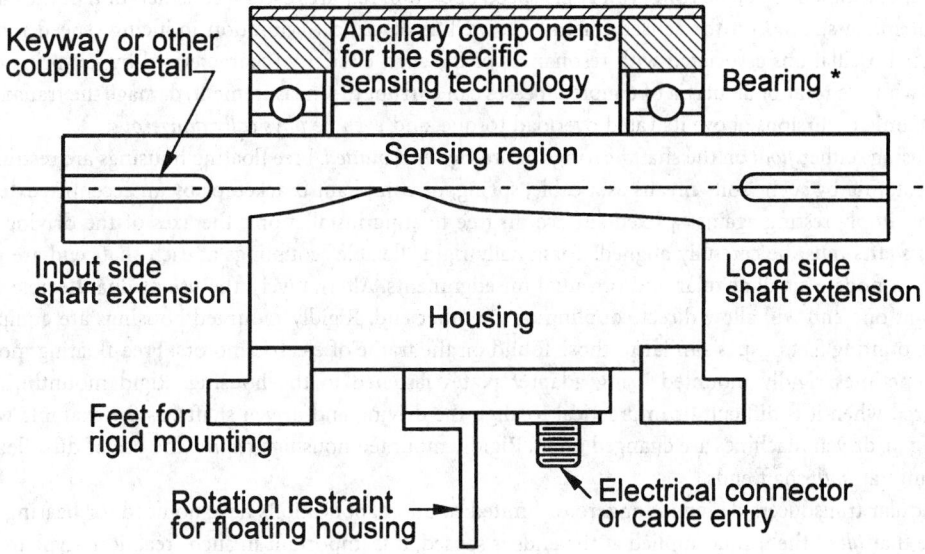

FIGURE 24.4 Modular torque transducer showing generic features and alternative arrangements for free floating or rigid mounting. Bearings* are used only on rotational models. Shaft extensions have keyways or other features to facilitate torque coupling.

opposite circumferential direction as indicated by the solid arrows in Figure 24.3(*d*) [11]. When torque is applied, the magnetizations tilt into helical directions (dashed arrows), causing magnetic poles to develop at the central domain wall and (of opposite polarity) at the ring end faces. Torque is determined from the output signal of one or more magnetic field sensors (e.g., Hall effect, magnetoresistive, or flux gate devices) mounted so as to sense the intensity and polarity of the magnetic field that arises in the space near the ring.

24.4 Torque Transducer Construction, Operation, and Application

Although a torque sensing region can be created directly on a desired shaft, it is more usual to install a preassembled *modular* torque transducer into the driveline. Transducers of this type are available with capacities from 0.001 N·m to 200,000 N·m. Operating principle descriptions and detailed installation and operating instructions can be found in the catalogs and literature of the various manufactures [12–20]. Tradenames often identify specific type of transducers; for example, *Torquemeters* [13] refers to a family of noncontact strain gage models; *Torkducer®* [18] identifies a line of Type 1 magnetoelastic transducers; *Torqstar™* [12] identifies a line of Type 2 magnetoelastic transducers; *Torquetronic* [16] is a class of transducers using wrap-around twist angle sensors; and *TorXimitor™* [20] identifies optoelectronic based, noncontact, strain gage transducers. Many of these devices show generic similarities transcending their specific sensing technology as well as their range. Figure 24.4 illustrates many of these common features.

Mechanical Considerations

Maximum operating speeds vary widely; upper limits depend on the size, operating principle, type of bearings, lubrication, and dynamic balance of the rotating assembly. Ball bearings, lubricated by grease, oil, or oil mist, are typical. Parasitic torques associated with bearing lubricants and seals limit the accuracy of low-end torque measurements. (Minute capacity units have no bearings [15]). Forced lubrication can allow operation up to 80,000 rpm [16]. High-speed operation requires careful consideration of the effects of centrifugal stresses on the sensed quantity as well as of critical (vibration inducing) speed ranges. Torsional oscillations associated with resonances of the shaft elasticity (characterized by its spring constant) with the rotational inertia of coupled masses can corrupt the measurement, damage the transducer by dynamic excursions above its rated overload torque, and *even be physically dangerous.*

Housings either *float* on the shaft bearings or are *rigidly mounted.* Free floating housings are restrained from rotating by such "soft" means as a cable, spring, or compliant bracket, or by an eccentric external feature simply resting against a fixed surface. In free floating installations, the axes of the driving and driven shafts must be carefully aligned. Torsionally rigid "flexible" couplings at each shaft end are used to accommodate small angular and/or radial misalignments. Alternatively, the use of dual flexible couplings at one end will allow direct coupling of the other end. Rigidly mounted housings are equipped with mounting feet or lugs similar to those found on the frame of electric motors. Free-floating models are sometimes rigidly mounted using adapter plates fastened to the housing. Rigid mountings are preferred when it is difficult or impractical to align the driving and driven shafts, as for example when driving or driven machines are changed often. Rigidly mounted housings *require* the use of dual flexible couplings at *both* shaft ends.

Modular transducers designed for zero or limited rotation applications have no need for bearings. To ensure that *all* of the torque applied at the ends is sensed, it is important in such "reaction"-type torque transducers to limit attachment of the housing to the shaft to only one side of the sensing region. Whether rotating or stationary, the external shaft ends generally include such torque coupling details as flats, keyways, splines, tapers, flanges, male/female squares drives, etc.

Electrical Considerations

By their very nature, transducers require some electrical input power or *excitation.* The "raw" output signal of the actual sensing device also generally requires "conditioning" into a level and format appropriate for display on a digital or analog meter or to meet the input requirements of data acquisition equipment. Excitation and signal conditioning are supplied by electronic circuits designed to match the characteristics of the specific sensing technology. For example, strain gage bridges are typically powered with 10 V to 20 V (dc or ac) and have outputs in the range of 1.5 mV to 3.0 mV per volt of excitation at the rated load. Raising these millivolt signals to more usable levels requires amplifiers having gains of 100 or more. With ac excitation, oscillators, demodulators (or rectifiers) are also needed. Circuit elements of these types are normal when inductive elements are used either as a necessary part of the sensor or simply to implement noncontact constructions.

Strain gages, differential transformers, and related sensing technologies require that electrical components be mounted *on* the torqued member. Bringing electrical power to and output signals from these components on rotating shafts require special methods. The most direct and common approach is to use conductive means wherein brushes (typically of silver graphite) bear against (silver) slip rings. Useful life is extended by providing means to lift the brushes off the rotating rings when measurements are not being made. Several "noncontacting" methods are also used. For example, power can be supplied via inductive coupling between stationary and rotating transformer windings [12–15], by the illumination of shaft mounted photovoltaic cells [20], or even by batteries strapped to the shaft [21] (limited by centrifugal force to relatively low speeds). Output signals are coupled off the shaft through rotary transformers, by frequency-modulated (infrared) LEDs [19, 20], or by radio-frequency (FM) telemetry [21]. Where shaft rotation is limited to no more than a few full rotations, as in steering gear, valve actuators or oscillating mechanisms, hard wiring both power and signal circuits is often suitable. Flexible cabling minimizes incidental torques and makes for a long and reliable service life. All such wiring considerations are avoided when noncontact technologies or constructions are used.

Costs and Options

Prices of torque transducers reflect the wide range of available capacities, performance ratings, types, styles, optional features, and accessories. In general, prices of any one type increase with increasing capacity. Reaction types cost about half of similarly rated rotating units. A typical foot-mounted, 565 N·m capacity, strain gage transducer with either slip rings or rotary transformers and integral speed sensor, specified nonlinearity and hysteresis each within ±0.1%, costs about $4000 (1997). Compatible instrumentation providing transducer excitation, conditioning, and analog output with digital display of torque and speed costs about $2000. A comparable magnetoelastic transducer with ±0.5% accuracy costs about $1300. High-capacity transducers for extreme speed service with appropriate lubrication options can cost more than $50,000. Type 2 magnetoelastic transducers, mass produced for automotive power steering applications, cost approximately $10.

24.5 Apparatus for Power Measurement

Rotating machinery exists in specific types without limit and can operate at power levels from fractions of a watt to some tens of megawatts, a range spanning more than 10^8. Apparatus for power measurement exists in a similarly wide range of types and sizes. Mechanical power flows from a *driver* to a *load*. This power can be determined *directly* by application of Equation 24.7, simply by measuring, in addition to ω, the output torque of the driver or the input torque to the load, whichever is the device under test (DUT). When the DUT is a driver, measurements are usually required over its full service range of speed and torque. The test apparatus therefore must act as a controllable load and be able to *absorb* the delivered power. Similarly, when the DUT is a pump or fan or other type of load, or one whose function is simply to alter speed and torque (e.g., a gear box), the test apparatus must include a *driver* capable of supplying power over the DUT's full rated range of torque and speed. Mechanical power can also be determined *indirectly* by conversion into (or from) another form of energy (e.g., heat or electricity) and measuring the relevant calorimetric or electrical quantities. In view of the wide range of readily available methods and apparatus for accurately measuring both torque and speed, indirect methods need only be considered when special circumstances make direct methods difficult.

Dynamometer is the special name given to the power-measuring apparatus that includes absorbing or/and driving means and wherein torque is determined by the reaction forces on a stationary part (the *stator*). An effective dynamometer is conveniently assembled by mounting the DUT in such a manner as to allow measurement of the reaction torque on its frame. Figure 24.5 shows a device designed to facilitate such measurements. Commercial models (Torque Table® [12]) rated to support DUTs weighing 222 N to 4900 N are available with torque capacities from 1.3 N·m 226 to N·m. "Torque tubes" [4] or other DUT mounting arrangements are also used. Other than for possible rotational/elastic resonances, these systems have no speed limitations. More generally, and especially for large machinery, dynamometers include a specialized driving or absorbing machine. Such dynamometers are classified according to their function as *absorbing* or *driving* (sometimes *motoring*). A *universal dynamometer* can function as either a driver or an absorber.

Absorption Dynamometers

Absorption dynamometers, often called *brakes* because their operation depends on the creation of a controllable *drag* torque, convert mechanical work into heat. A drag torque, as distinguished from an active torque, can act only to restrain and not to initiate rotational motion. Temperature rise within a dynamometer is controlled by carrying away the heat energy, usually by transfer to a moving fluid, typically air or water. Drag torque is created by inherently dissipative processes such as: friction between rubbing surfaces, shear or turbulence of viscous liquids, the flow of electric current, or magnetic hysteresis. Gaspard Riche de Prony (1755–1839), in 1821 [22], invented a highly useful form of a friction brake to meet the needs for testing the steam engines that were then becoming prevalent. Brakes of this

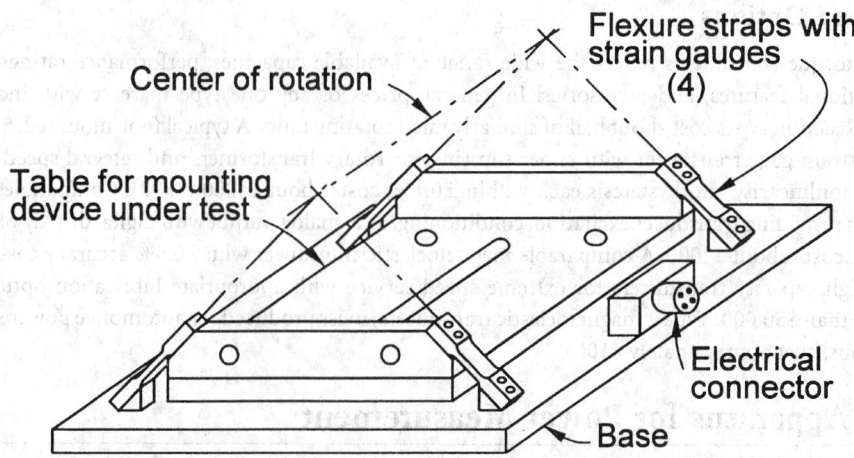

FIGURE 24.5 Support system for measuring the reaction torque of a rotating machine. The axis of the machine must be accurately set on the "center of rotation." The holes and keyway in the table facilitate machine mounting and alignment. Holes in the front upright provide for attaching a lever arm from which calibrating weights may be hung [4, 11].

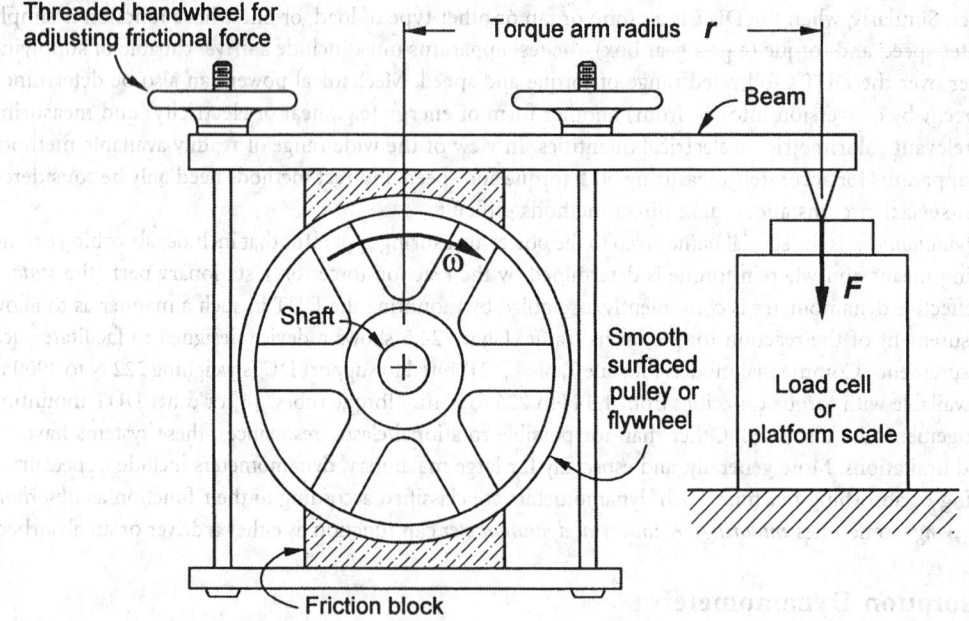

FIGURE 24.6 A classical prony brake. This brake embodies the defining features of all absorbing dynamometers: conversion of mechanical work into heat and determination of power from measured values of reaction torque and rotational velocity.

type are often used for instructional purposes, for they embody the general principles and major operating considerations for all types of absorption dynamometers. Figure 24.6 shows the basic form and constructional features of a *prony brake*. The power that would normally be delivered by the shaft of the driving engine to the driven load is (for measurement purposes) converted instead into heat via the work done by the frictional forces between the friction blocks and the flywheel rim. Adjusting the tightness of the clamping bolts varies the frictional drag torque as required. Heat is removed from the inside surface of the rim by arrangements (not shown) utilizing either a continuous flow or evaporation of water. There is no need to know the magnitude of the frictional forces nor even the radius of the flywheel (facts recognized by Prony), because, while the drag torque tends to rotate the clamped-on apparatus, it is held stationary by the equal but opposite reaction torque Fr. F at the end of the torque arm of radius r (a fixed dimension of the apparatus) is monitored by a scale or load cell. The power is found from Equations 24.1 and 24.7 as $P = Fr\omega = Fr2\pi N/60$ where N is in rpm.

Uneven retarding forces associated with fluctuating coefficients of friction generally make rubbing friction a poor way to generate drag torque. Nevertheless, because they can be easily constructed, *ad hoc* variations of prony brakes, often using only bare ropes or wooden cleats connected by ropes or straps, find use in the laboratory or wherever undemanding or infrequent power measurements are to be made. More sophisticated prony brake constructions are used in standalone dynamometers with self-contained cooling water tanks in sizes up to 746 kW (1000 Hp) for operation up to 3600 rpm with torques to 5400 N·m [23]. Available in stationary and mobile models, they find use in testing large electric motors as well as engines and transmissions on agricultural vehicles. Prony brakes allow full drag torque to be imposed down to zero speed.

William Froude (1810–1879) [24] invented a *water brake* (1877) that does not depend on rubbing friction. Drag torque within a *Froude brake* is developed between the rotor and the stator by the momentum imparted by the rotor to water contained within the brake casing. Rotor rotation forces the water to circulate between cup-like pockets cast into facing surfaces of both rotor and stator. The rotor is supported in the stator by bearings that also fix its axial position. Labyrinth-type seals prevent water leakage while minimizing frictional drag and wear. The stator casing is supported in the dynamometer frame in cradle fashion by *trunnion* bearings. The torque that prevents rotation of the stator is measured by reaction forces in much the same manner as with the prony brake. Drag torque is adjusted by a valve, controlling either the back pressure in the water outlet piping [25] or the inlet flow rate [26] or sometimes (to allow very rapid torque changes) with two valves controlling both [27]. In any case, the absorbed energy is carried away by the continuous water flow. Other types of cradle-mounted water brakes, while externally similar, have substantially different internal constructions and depend on other principles for developing the drag torque (e.g., smooth rotors develop viscous drag by shearing and turbulence). Nevertheless, all *hydraulic dynamometers* purposefully function as *inefficient* centrifugal pumps. Regardless of internal design and valve settings, maximum drag torque is low at low speeds (zero at standstill) but can rise rapidly, typically varying with the square of rotational speed. The irreducible presence of some water, as well as windage, places a speed-dependent lower limit on the *controllable* drag torque. In any one design, wear and vibration caused by cavitation place upper limits on the speed and power level. Hydraulic dynamometers are available in a wide range of capacities between 300 kW and 25,000 kW, with some portable units having capacities as low as 75 kW [26]. The largest ever built [27], absorbing up to about 75,000 kW (100,000 Hp), has been used to test propulsion systems for nuclear submarines. Maximum speeds match the operating speeds of the prime movers that they are built to test and therefore generally decrease with increasing capacity. High-speed gas turbine and aerospace engine test equipment can operate as high as 30,000 rpm [25].

In 1855, Jean B. L. Foucault (1819–1868) [22] demonstrated the conversion of mechanical work into heat by rotating a copper disk between the poles of an electromagnet. This simple means of developing drag torque, based on *eddy currents*, has, since circa 1935, been widely exploited in dynamometers. Figure 24.7 shows the essential features of this type of brake. Rotation of a toothed or spoked steel rotor through a spatially uniform magnetic field, created by direct current through coils in the stator, induces locally circulating (eddy) currents in electrically conductive (copper) portions of the stator. Electromagnetic forces between the rotor, which is magnetized by the uniform field, and the field arising from the eddy currents, create the drag torque. This torque, and hence the mechanical input power, are controlled by adjusting the *excitation* current in the stator coils. Electric input power is less than 1% of the rated capacity. The dynamometer is effectively an internally short-circuited generator because the power associated with the resistive losses from the generated eddy currents is dissipated *within* the machine.

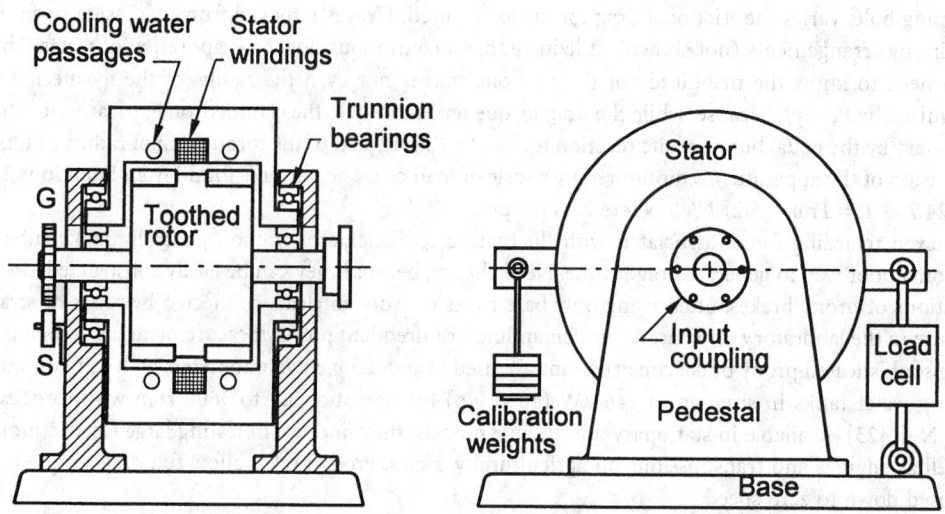

FIGURE 24.7 Cross-section (left) and front view (right) of an eddy current dynamometer. G is a gear wheel and S is a speed sensor. Hoses carrying cooling water and cable carrying electric power to the stator are not shown.

Being heated by the flow of these currents, the stator must be cooled, sometimes (in smaller capacity machines) by air supplied by blowers [23], but more often by the continuous flow of water [25, 27, 28]. In *dry gap* eddy current brakes (the type shown in Figure 24.7), water flow is limited to passages within the stator. Larger machines are often of the *water in gap* type, wherein water also circulates around the rotor [28]. Water in contact with the moving rotor effectively acts as in a water brake, adding a nonelectromagnetic component to the total drag torque, thereby placing a lower limit to the controllable torque. Windage limits the minimum value of controllable torque in dry gap types. Since drag torque is developed by the motion of the rotor, it is zero at standstill for any value of excitation current. Initially rising rapidly, approximately linearly, with speed, torque eventually approaches a current limited saturation value. As in other cradled machines, the torque required to prevent rotation of the stator is measured by the reaction force acting at a fixed known distance from the rotation axis. Standard model eddy current brakes have capacities from less than 1 kW [23, 27] to more than 2000 kW [27, 28], with maximum speeds from 12,000 rpm in the smaller capacity units to 3600 rpm in the largest units. Special units with capacities of 3000 Hp (2238 kW) at speeds to 25,000 rpm have been built [28].

Hysteresis brakes [29] develop drag torque via magnetic attractive/repulsive forces between the magnetic poles established in a reticulated stator structure by a current through the field coil, and those created in a "drag cup" rotor by the stator field gradients. Rotation of the special steel rotor, through the spatial field pattern established by the stator, results in a cyclical reversal of the polarity of its local magnetizations. The energy associated with these reversals (proportional to the area of the hysteresis loop of the rotor material) is converted into heat within the drag cup. Temperature rise is controlled by forced air cooling from a blower or compressed air source. As with eddy current brakes, the drag torque of these devices is controlled by the excitation current. In contrast with eddy current brakes, rated drag torque is available down to zero speed. (Eddy current effects typically add only 1% to the drag torque for each 1000 rpm). As a result of their smooth surfaced rotating parts, hysteresis brakes exhibit low parasitic torques and hence cover a dynamic range as high as 200 to 1. Standard models are available having continuous power capacities up to 6 kW (12 kW with two brakes in tandem cooled by two blowers). Intermittent capacities per unit (for 5 min or less) are 7 kW. Some low-capacity units are convection cooled; the smallest has a continuous rating of just 7 W (35 W for 5 min). Maximum speeds range from 30,000 rpm for the smallest to 10,000 rpm for the largest units. Torque is measured by a strain gage bridge on a moment arm supporting the machine stator.

Driving and Universal Dynamometers

Electric generators, both ac and dc, offer another means for developing a controllable drag torque and they are readily adapted for dynamometer service by cradle mounting their stator structures. Moreover, electric machines of these types can also operate in a motoring mode wherein they can deliver controllable *active* torque. When configured to operate selectively in either driving or absorbing modes, the machine serves as a universal dynamometer. With dc machines in the absorbing mode, the generated power is typically dissipated in a convection-cooled resistor bank. Air cooling the machine with blowers is usually adequate, since *most* of the mechanical power input is dissipated externally. Nevertheless, *all* of the mechanical input power is accounted for by the product of the reaction torque and the rotational speed. In the motoring mode, torque and speed are controlled by adjustment of both field and armature currents. Modern ac machines utilize regenerative input power converters to allow braking power to be returned to the utility power line. In the motoring mode, speed is controlled by high-power, solid-state, adjustable frequency inverters. Internal construction is that of a simple three-phase induction motor, having neither brushes, slip rings, nor commutators. The absence of rotor windings allows for higher speed operation than dc machines. Universal dynamometers are "four-quadrant" machines, a term denoting their ability to produce torque in the same or opposite direction as their rotational velocity. This unique ability allows the effective drag torque to be reduced to zero at any speed. Universal dynamometers [25, 28] are available in a relatively limited range of capacities (56 to 450 kW), with commensurate torque (110 to 1900 N·m) and speed (4500 to 13,500 rpm) ranges, reflecting their principal application in automotive engine development. Special dynamometers for testing transmissions and other vehicular drive train components insert the DUT between a diesel engine or electric motor prime mover and a hydraulic or eddy current brake [30].

Measurement Accuracy

Accuracy of power measurement (see discussion in [4]) is generally limited by the torque measurement ($\pm 0.25\%$ to $\pm 1\%$) since rotational speed can be measured with almost any desired accuracy. Torque errors can arise from the application of extraneous (i.e., not indicated) torques from hose and cable connections, from windage of external parts, and from miscalibration of the load cell. Undetected friction in the trunnion bearings of cradled dynamometers can compromise the torque measurement accuracy. Ideally, well-lubricated antifriction bearings make no significant contribution to the restraining torque. In practice, however, the unchanging contact region of the balls or other rolling elements on the bearing races makes them prone to brinelling (a form of denting) from forces arising from vibration, unsupported weight of attached devices, or even inadvertently during the alignment of connected machinery. The problem can be alleviated by periodic rotation of the (primarily outer) bearing races. In some bearing-in-bearing constructions, the central races are continuously rotated at low speeds by an electric motor while still others avoid the problem by supporting the stator on hydrostatic oil lift bearings [28].

Costs

The wide range of torque, speed, and power levels, together with the variation in sophistication of associated instrumentation, is reflected in the very wide range of dynamometer prices. Suspension systems of the type illustrated in Figure 24.5 (for which the user must supply the rotating machine) cost $4000 to $6000, increasing with capacity [12]. A 100 Hp (74.6 kW) *portable* water brake equipped with a strain gage load cell and a digital readout instrument for torque, speed, and power costs $4500, or $8950 with more sophisticated data acquisition equipment [26]. Stationary (and some *transportable* [23]) hydraulic dynamometers cost from $113/kW in the smaller sizes [25], down to $35/kW for the very largest [27]. Transportation, installation, and instrumentation can add significantly to these costs. Eddy current dynamometers cost from as little as $57/kW to nearly $700/kW, depending on the rated capacity, type of control system, and instrumentation [24, 25, 28]. Hysteresis brakes with integral speed sensors cost from $3300 to $14,000 according to capacity [29]. Compatible controllers, from manual to fully programmable for PC test control and data acquisition via an IEEE-488 interface, vary in price from $500 to $4200. The flexibility and high performance of ac universal dynamometers is reflected in their comparatively high prices of $670 to $2200/kW [25, 28].

References

1. Pinney, C. P. and Baker, W. E., Velocity Measurement, *The Measurement, Instrumentation and Sensors Handbook*, Webster, J. G., ed., Boca Raton, FL: CRC Press, 1999.
2. S. Timoshenko, *Strength of Materials*, 3rd ed., New York: Robert E. Kreiger, Part I, 281–290; Part II, 235–250, 1956.
3. S. J. Citron, *On-line engine torque and torque fluctuation measurement for engine control utilizing crankshaft speed fluctuations*, U. S. Patent No. 4,697,561, 1987.
4. Supplement to ASME Performance Test Codes, Measurement of Shaft Power, ANSI/ASME PTC 19.7-1980 (Reaffirmed 1988).
5. See, for example, the catalog of torque wrench products of Consolidated Devices, Inc., 19220 San Jose Ave., City of Industry, CA 91748.
6. I. J. Garshelis, C. R. Conto, and W. S. Fiegel, A single transducer for non-contact measurement of the power, torque and speed of a rotating shaft, SAE Paper No. 950536, 1995.
7. C. C. Perry and H. R. Lissner, *The Strain Gage Primer*, 2nd ed., New York: McGraw-Hill, 1962, 9. (This book covers all phases of strain gage technology.)
8. B. D. Cullity, *Introduction to Magnetic Materials*, Reading, MA: Addison-Wesley, 1972, Section 8.5, 266–274.
9. W. J. Fleming, Magnetostrictive torque sensors—comparison of branch, cross and solenoidal designs, SAE Paper No. 900264, 1990.
10. Y. Nonomura, J. Sugiyama, K. Tsukada, M. Takeuchi, K. Itoh, and T. Konomi, Measurements of engine torque with the intra-bearing torque sensor, SAE Paper No. 870472, 1987.
11. I. J. Garshelis, *Circularly magnetized non-contact torque sensor and method for measuring torque using same*, U.S. Patent 5,351,555, 1994 and 5,520,059, 1996.
12. Lebow® Products, Siebe, plc., 1728 Maplelawn Road, Troy, MI 48099, Transducer Design Fundamentals/Product Listings, Load Cell and Torque Sensor Handbook No. 710, 1997, also: Torqstar™ and Torque Table®.
13. S. Himmelstein & Co., 2490 Pembroke, Hoffman Estates, IL 60195, MCRT® Non-Contact Strain Gage Torquemeters and Choosing the Right Torque Sensor.
14. Teledyne Brown Engineering, 513 Mill Street, Marion, MA 02738-0288.
15. Staiger, Mohilo & Co. GmbH, Baumwasenstrasse 5, D-7060 Schorndorf, Germany (In the U.S.: Schlenker Enterprises Ltd., 5143 Electric Ave., Hillside, IL 60162), Torque Measurement.
16. Torquemeters Ltd., Ravensthorpe, Northampton, NN6 8EH, England (In the U.S.: Torquetronics Inc., P.O. Box 100, Allegheny, NY 14707), Power Measurement.
17. Vibrac Corporation, 16 Columbia Drive, Amherst, NH 03031, Torque Measuring Transducer.
18. GSE, Inc., 23640 Research Drive, Farmington Hills, MI 48335-2621, Torkducer®.
19. Sensor Developments Inc., P.O. Box 290, Lake Orion, MI 48361-0290, 1996 Catalog.
20. Bently Nevada Corporation, P.O. Box 157, Minden, NV 89423, TorXimitor™.
21. Binsfield Engineering Inc., 4571 W. MacFarlane, Maple City, MI 49664.
22. C. C. Gillispie (ed.), *Dictionary of Scientific Biography*, Vol. XI, New York: Charles Scribner's Sons, 1975.
23. AW Dynamometer, Inc., P.O. Box 428, Colfax, IL 61728, Traction dynamometers: Portable and stationary dynamometers for motors, engines, vehicle power take-offs.
24. Roy Porter (ed.), *The Biographical Dictionary of Scientists*, 2nd ed., New York: Oxford University Press, 1994.

25. Froude-Consine Inc., 39201 Schoolcraft Rd., Livonia, MI 48150, F Range Hydraulic Dynamometers, AG Range Eddy Current Dynamometers, AC Range Dynamometers.

26. Go-Power Systems, 1419 Upfield Drive, Carrollton, TX 75006, Portable Dynamometer System, Go-Power Portable Dynamometers.

27. Zöllner GmbH, Postfach 6540, D-2300 Kiel 14, Germany (In the U.S. and Canada: Roland Marine Inc., 90 Broad St., New York, NY 10004), Hydraulic Dynamometers Type P, High Dynamic Hydraulic Dynamometers.

28. Dynamatic Corporation, 3122 14th Ave., Kenosha, WI 53141-1412, Eddy Current Dynamometer—Torque Measuring Equipment, Adjustable Frequency Dynamometer.

29. Magtrol, Inc., 70 Gardenville Parkway, Buffalo, NY 14224-1322, Hysteresis Absorption Dynamometers.

30. Hicklin Engineering, 3001 NW 104th St., Des Moines, IA 50322, Transdyne™ (transmission test systems, brake and towed chassis dynamometers).

25. Himmelstein & Co., 2490 Pembroke, Hoffman Estates, IL 60195, MCRT Non-Contact Strain Gage Torquemeters, S. Himmelstein and Company.

26. Magtrol, Inc., 70 Gardenville Pkwy., Buffalo, NY 14224-1322, Hysteresis Absorption Dynamometers.

27. High Speed Engineering, 3001 NW 104th St., Des Moines, IA 50322, Transducer and transmission testing systems (brake and load cells as dynamometers).

25
Tactile Sensing

R. E. Saad
University of Toronto

A. Bonen
University of Toronto

K. C. Smith
University of Toronto

B. Benhabib
University of Toronto

Robots in industrial settings perform repetitive tasks, such as machine loading, parts assembly, painting, and welding. Only in rare instances can these autonomous manipulators modify their actions based on sensory information. Although, thus far, a vast majority of research work in the area of robot sensing has concentrated on computer vision, contact sensing is an equally important feature for robots and has received some attention as well. Without tactile-perception capability, a robot cannot be expected to effectively grasp objects. In this context, robotic tactile sensing is the focus of this chapter.

25.1 Sensing Classification

Robotic sensing can be classified as either of the noncontact or contact type [1]. *Noncontact sensing* involves interaction between the robot and its environment by some physical phenomenon, such as acoustic or electromagnetic waves, that interact without contact. The most important types of robotic sensors of the noncontact type are vision and proximity sensors. *Contact sensing*, on the other hand, implies measurement of the general interaction that takes place when the robot's end effector is brought into contact with an object. Contact sensing is further classified into force and tactile sensing.

Force sensing is defined as the measurement of the global mechanical effects of contact, while *tactile sensing* implies the detection of a wide range of local parameters affected by contact. Most significant among those contact-based effects are contact stresses, slippage, heat transfer, and hardness.

The properties of a grasped object that can be derived from tactile sensing can be classified into geometric and dynamometric types [2]. Among the geometric properties are presence, location in relation to the end-effector, shape and dimensions, and surface conditions [3–7]. Among the dynamometric parameters associated with grasping are: force distribution, slippage, elasticity and hardness, and friction [8–12].

Tactile sensing requires sophisticated transducers; yet the availability of these transducers alone is not a sufficient condition for successful tactile sensing. It is also necessary to accurately control the modalities through which the tactile sensor interacts with the explored objects (including contact forces, as well as end-effector position and orientation) [13–15]. This leads to active tactile sensing, which requires a high degree of complexity in the acquisition and processing of the tactile data [16].

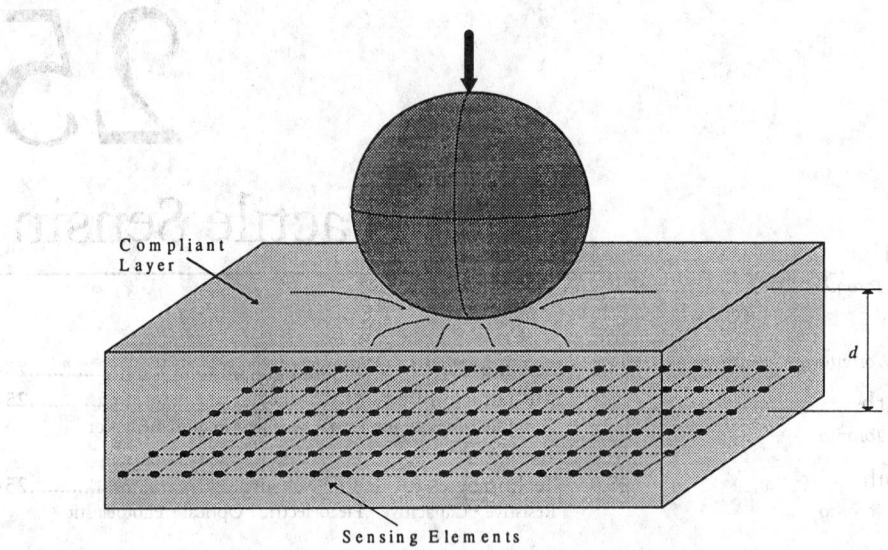

FIGURE 25.1   An object indenting a compliant layer, where an array of force-sensing elements is placed at a distance d from the surface.

25.2   Mechanical Effects of Contact

Tactile sensing normally involves a rigid object indenting the compliant cover layer of a tactile sensor array [17], Figure 25.1. The indentation of a compliant layer due to contact can be analyzed from two conceptually different points of view [1]. The first one is the measurement of the actual contact stresses (force distribution) in the layer, which is usually relevant to controlling manipulation tasks. The second one is the deflection profile of the layer, which is usually important for recognizing geometrical object features. Depending on the approach adopted, different processing and control algorithms must be utilized.

There exists a definite relationship between the local shape of a contacting body and a set of subsurface strains (or displacements); however, this relationship is quite complex. Thus, it requires the use of the Theory of Elasticity and Contact Mechanics to model sensor–object interaction [18], and the use of Finite Element Analysis (FEA) as a practical tool for obtaining a more representative model of the sensor [19].

In general, the study of tactile sensors comprises two steps: (1) the *forward analysis*, related to the acquisition of data from the sensor (changes on the stress or strains, induced by the indentation of an object on the compliant surface of the transducer); and, (2) the *inverse problem*, normally related to the recovery of force distribution or, in some cases, the recovery of the indentor's shape.

Simplified Theory for Tactile Sensing

For simplicity, the general two-dimensional tactile problem is reduced herein to a one-dimensional one. Figure 25.2 shows a one-dimensional transducer that consists of a compliant, homogeneous, isotropic, and linear layer subjected to a normal stress $q_v(x)$ created by the indentation of an object.

For modeling purposes, it is assumed that the compliant layer is an elastic half-space. This simplification yields closed-form equations for the analysis and avoids the formation of a more complex problem, in which the effect of the boundary conditions at x_{min} and x_{max} must be taken into account. It has been proven that the modeling of the sensor by an elastic half-space represents a reasonable approximation to the real case [18]. Under these conditions, it can be shown that the normal strain, at a depth $y = d$, due to the normal stress $q_v(y)$ is given by [20]:

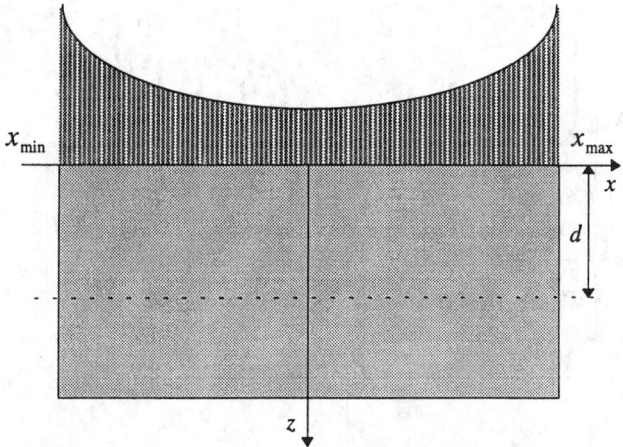

FIGURE 25.2 Ideal one-dimensional transducer subjected to a normal stress.

$$\varepsilon_z(x) = \int_{-\infty}^{\infty} q_v(x - x_0) h_z(x_0, d) \, dx_0 \tag{25.1}$$

where ε_z is the strain at x and $z = d$ due to the normal stress on the surface, and

$$h_z(x) = -\frac{2d(1+v)\left[d^2(1-v) - vx^2\right]}{\pi r E\left(x^2 + d^2\right)^2} \tag{25.2}$$

E and v are, respectively, the modulus of elasticity and the Poisson's coefficient of the compliant layer. In obtaining Equation 25.2, it is assumed that the analysis is performed under *planar strain* conditions. It should be noted that a similar analysis can be performed for tangential contact stresses or strains.

The normal displacement at the surface, w, is given by:

$$w(x) = \int_{-\infty}^{\infty} q_v(x - x_0) k(x_0) \, dx_0 \tag{25.3}$$

where

$$k(x) = \frac{-2(1 - v^2)}{\pi E} \log\left|\frac{x}{x_a}\right| \tag{25.4}$$

The singularity at $x = 0$ is expected due to the singularity of stress at that point. Note that, $k(x)$ is the deformation of the surface when a singular load of 1 N is applied at $x = 0$. The constant x_a should be chosen such that at $x = x_a$, the deformation is zero. In this case, zero deformation should occur at $x \to \infty$ (note that it has been assumed that the sensor is modeled by an elastic half space), namely $x_a \to \infty$. This problem is associated with the two-dimensional deformation of an elastic half-space. To eliminate this difficulty, the boundary conditions of the transducer must be taken into account (i.e., a finite transducer must be analyzed), which requires, in general, the use of FEA.

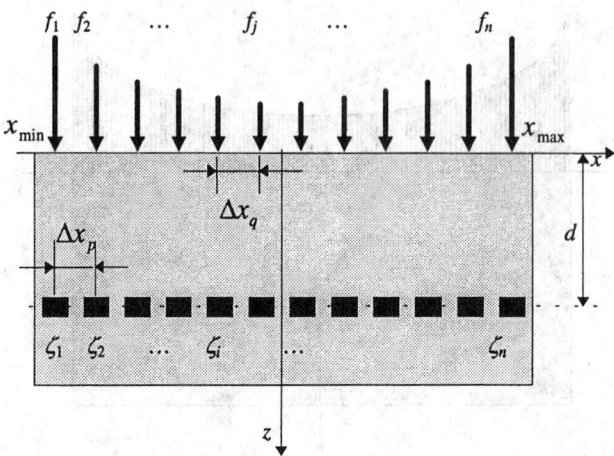

FIGURE 25.3   One-dimensional transducer with discrete sensing elements located at $z = d$.

Since measurements of strain (or stress) are usually done by a discrete number of sensing elements, Equation 25.2 must be discretized (Figure 25.3). Correspondingly, the force distribution must be reconstructed at discrete positions as shown in Figure 25.3. Let Δx_q be the distance between points, where the force distribution must be reconstructed from strain (or stress) measurements carried out by strain (or stress) sensing elements uniformly distributed at intervals Δx_p, at $z = d$. Also assume, even though it is not necessary, that $\Delta x_q = \Delta x_p = \Delta x$ and that the forces are applied at positions immediately above the sensor elements. One can now define the strain (stress)-sample vector, ζ, whose components are given by $\zeta_i = \varepsilon_x(x_i)$, $i = 1, 2, ..., n$, and the force distribution vector, F, whose components are given by $f_i = q_v(x_j)$, $j = 1, 2, ..., $ n. Then, the discrete form of Equation 25.1 is given by:

$$\zeta = \mathbf{T}\mathbf{F} \tag{25.5}$$

where the elements of the matrix $\mathbf{T}$ are given by $T_{ij} = k_v(x_i - x_j)$, $i = 1, 2, ..., n$ and $j = 1, 2, ..., n$ [23]. A similar relation to Equation 25.5 can be obtained discretizing Equation 25.3. In the general case, where $\Delta x_q \neq \Delta x_p$, $\mathbf{T}$ is not square. Furthermore, in the general case, the vector $\mathbf{F}$ comprises both vertical and tangential components.

Equations 25.1 and 25.3 represent the regular *forward problem*, while Equation 25.5 represents the discretized version of the forward problem. The *inverse problem*, in most cases, consists of recovering the applied force profile from the measurements of strain, stress, or deflection. (Note that the surface displacement can also be used to recover the indentor's shape.)

In [20], it was shown that the inverse problem is ill-posed because the operators h and k, of Equations 25.1 and 25.3, respectively, are ill-conditioned. Consequently, the inverse problem is susceptible to noise. To solve this problem, regularization techniques must be utilized [20].

It has been proven that, in order to avoid aliasing in determining the continuous strain (stress) at a depth d using a discretized transducer, the elements have to be separated by one tenth of the compliant layer's thickness. However, good results were obtained, without much aliasing, by separating the sensing elements by a distance equal to the sensor's depth [18].

Requirements for Tactile Sensors

In 1980, Harmon conducted a survey to determine general specifications for tactile sensors [21]. Those specifications have been used subsequently as guidelines by many tactile sensor designers:

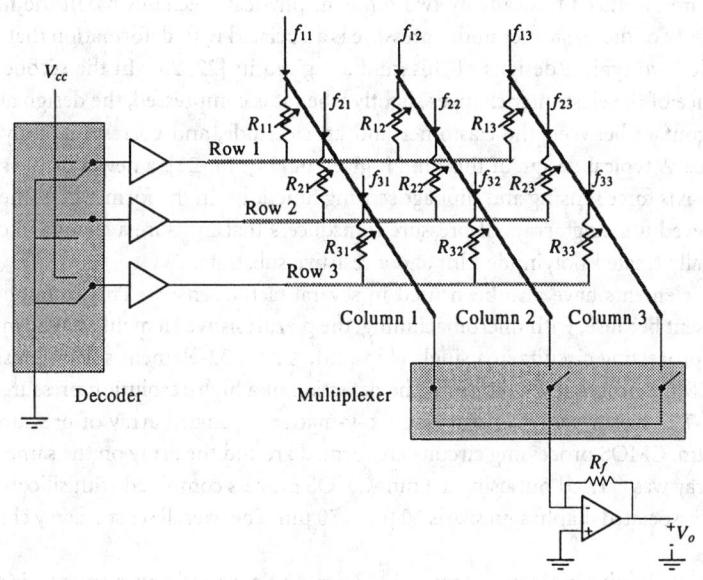

FIGURE 25.4 General configuration of a resistive transducer.

1. Spatial resolution of 1 to 2 mm
2. Array sizes of 5×10 to 10×20 points
3. Sensitivity of 0.5×10^{-2} to 1×10^{-2} N for each force-sensing element (tactel)
4. Dynamic range of 1000:1
5. Stable behavior and with no hysteresis
6. Sampling rate of 100 Hz to 1 kHz
7. Monotonic response, though not necessarily linear
8. Compliant interface, rugged and inexpensive

While properties (5), (7), and (8) above should apply to any practical sensor, the others are merely suggestions, particularly with respect to the number of array elements and spatial resolution.

Developments on tactile sensing following [21] have identified additional desirable qualities; namely, reliability, modularity, speed, and the availability of multisensor support [16].

25.3 Technologies for Tactile Sensing

The technologies associated with tactile sensing are quite diverse: extensive surveys of the state-of-the-art of robotic-tactile-transduction technologies have been presented in [2, 3, 16, 17]. Some of these technologies will be briefly discussed.

Resistive

The transduction method that has received the most attention in tactile sensor design is concerned with the change in resistance of a conductive material under applied pressure. A basic configuration of a resistive transducer is shown in Figure 25.4. Each resistor, whose value changes with the magnitude of the force, represents a resistive cell of the transducer. Different materials have been utilized to manufacture the basic cell.

Conductive elastomers were among the first resistive materials used for the development of tactile sensors. They are insulating, natural or silicone-based rubbers made conductive by adding particles of conductive or semiconductive materials (e.g., silver or carbon). The changes in resistivity of the elastomers under pressure are produced basically by two different physical mechanisms. In the first approach, the change in resistivity of the elastomer under pressure is associated with deformation that alters the particle density within it. Two typical designs of this kind are given in [22, 23]. In the second approach, while the bulk resistance of the elastomer changes slightly when it is compressed, the design allows the increase of the area of contact between the elastomer and an electrode, and correspondingly a change in the contact resistance. A typical design of this kind is given in [24]. In [25], a newer tactile sensor is reported with both three-axis force sensing and slippage sensing functions. In the former case, the pressure sensing function is achieved utilizing arrays of pressure transducers that measure a change in contact resistance between a specially treated polyimide film and a resistive substrate.

Piezoresistive elements have also been used in several tactile sensors. This technology is specifically attractive at present because, with micromachining, the piezoresistive elements can be integrated together with the signal-processing circuits in a single chip [26]. A 32 × 32-element silicon pressure sensor array incorporating CMOS processing circuits for the detection of a high-resolution pressure distribution was reported in [8]. The sensor array consists of an x–y-matrix-organized array of pressure cells with a cell spacing of 250 μm. CMOS processing circuits are formed around the array on the same chip. Fabrication of the sensor array was carried out using a 3 mm CMOS process combined with silicon micromachining techniques. The associated diaphragm size is 50 μm × 50 μm. The overall sensor-array chip size is 10 mm × 10 mm.

In Figure 25.4, a circuit topology, to scan a 3 × 3 array of piezoresistive elements, is shown. The basic idea was originally proposed in [24] and adapted on several occasions by different researchers. Using this method, the changes in resistance are converted into voltages at the output. With the connections as shown in Figure 25.4, the resistance R_{21} can be determined from:

$$V_0 = \frac{R_f}{R_{21}} V_{cc} \tag{25.6}$$

where V_0 is the output voltage, V_{cc} is the bias voltage, and R_f is the feedback resistance of the output amplifier stage.

One problem with the configuration shown in Figure 25.4 is the difficulty in detecting small changes in resistance due to the internal resistance of the multiplexer as well changes in the voltage of power source, which have a great influence at the output. Other methods utilized to scan resistive transducer arrays are summarized in [3].

When piezoresistors and circuits are fabricated on the same silicon substrate, the sensor array can be equipped with a complex switching circuit, next to the sensing elements, that allows a better resolution in the measurements [9].

Capacitive

Tactile sensors within this category are concerned with measuring capacitance, which varies under applied load. The capacitance of a parallel-plate capacitor depends on the separation of the plates and their areas. A sensor using an elastomeric separator between the plates provides compliance such that the capacitance will vary according to the applied normal load, Figure 25.5(*a*).

Figure 25.5(*b*) shows the basic configuration of a capacitive tactile sensor. The intersections of rows and columns of conductor strips form capacitors. Each individual capacitance can be determined by measuring the corresponding output voltage at the selected row and column. To reduce cross-talk and electromagnetic interference, the rows and columns that are not connected are grounded. Figure 25.5(*c*) shows an equivalent circuit when the sensor is configured to measure the capacitance formed at the intersection of row *i* and row *j*, C_{ij}. R_d is the input resistance of the detector and C_d represents the effects of the stray capacitances, including the detector-amplifier input capacitance, the stray capacitance due

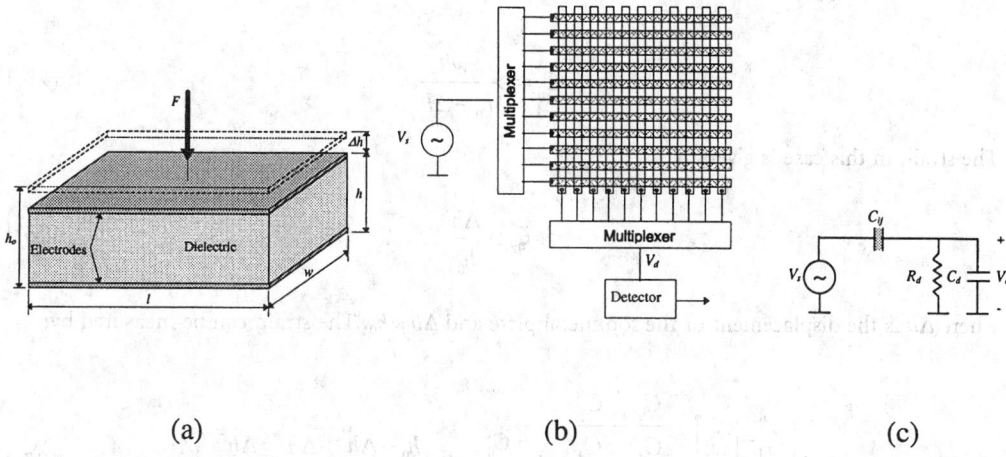

FIGURE 25.5 (*a*) Basic cell of a capacitor tactile sensor. (*b*) Typical configuration of a capacitive tactile sensor. (*c*) Equivalent circuit for the measurement of the capacitance C_{ij}.

to the unselected rows and columns, and the capacitance contributed by the cable that connects the transducer to the detector. Since the stray capacitance due to the unselected rows and columns changes with the applied forces, the stray capacitance due to the cable is designed to be predominant [18].

The magnitude of voltage at the input of the detector, $|V_d|$ is given by:

$$|V_d| = \frac{C_{ij} R_d \omega}{\sqrt{1+\left[\omega R_d \left(C_{ij}+C_d\right)\right]^2}}|V_s| \qquad (25.7)$$

Assuming that $C_d \gg C_{ij}$ and ω is sufficiently large,

$$|V_d| \cong \frac{C_{ij}}{C_d}|V_s| \qquad (25.8)$$

When a load is applied to the transducer, the capacitor is deformed as shown in Figure 25.5(*a*). For modeling purposes, it is assumed that the plate capacitor is only under compression. When no load is applied, the capacitance due to the element in the *i*th row and the *j*th column, C_{ij}^0, is given by:

$$C_{ij}^0 = \varepsilon \frac{wl}{h_0} \qquad (25.9)$$

where ε is the permittivity of the dielectric, w and l are the width and the length of the plate capacitor, respectively, and h_0 is the distance between plates when no load is applied. The voltage at the input of the detector for this particular case is indicated by V_{d0}; then from Equation 25.8, one obtains:

$$|V_{d0}| \cong \frac{C_{ij}^0}{C_d}|V_s| \qquad (25.10)$$

When a load is applied, the capacitor is under compression and the capacitance is given by:

$$C_{ij} = \varepsilon \frac{wl}{h_0 - \Delta h} \tag{25.11}$$

The strain in this case is given by:

$$\zeta_z \cong \frac{\Delta h}{h_0} \tag{25.12}$$

where Δh is the displacement of the top metal plate and $\Delta h \ll h_0$. The strain can be measured by:

$$\frac{|V_d| - |V_{d0}|}{|V_d|} = \frac{\dfrac{C_{ij}}{C_d} - \dfrac{C_{ij}^0}{C_d}}{\dfrac{C_{ij}}{C_d}} = 1 - \frac{C_{ij}^0}{C_{ij}} = 1 - \frac{h_0 - \Delta h}{h_0} = \frac{\Delta h}{h_0} = \frac{\Delta h}{h_0} \cong \zeta_z \tag{25.13}$$

Consequently, the strain at each tactel can be determined by measuring the magnitudes of V_d and V_{d0} for each element.

Note that the presence of a tangential force would offset the plates tangentially and change the effective area of the capacitor plates. An ideal capacitive pressure sensor can quantify basic aspects of touch by sensing normal forces, and can detect slippage by measuring tangential forces. However, distinguishing between the two forces at the output of a single sensing element is a difficult task and requires a more complex transducer than the one presented in Figure 25.5(a) [27].

Micromachined, silicon-based capacitive devices are especially attractive due to their potential for high accuracy and low drift. A sensor with 1024 elements and a spatial resolution of 0.5 mm was reported in [28]. Several possible structures for implementing capacitive high-density tactile transducers in silicon have been reported in [29]. A cylindrical finger-shaped transducer was reported in [18].

The advantages of capacitive transducers include: wide dynamic range, linear response, and robustness. Their major disadvantages are susceptibility to noise, sensitivity to temperature, and the fact that capacitance decreases with physical size, ultimately limiting the spatial resolution. Research is progressing toward the development of electronic processing circuits for the measurement of small capacitances using charge amplifiers [30], and the development of new capacitive structures [29].

Piezoelectric

A material is called piezoelectric, if, when subjected to a stress or deformation, it produces electricity. Longitudinal piezoelectric effect occurs when the electricity is produced in the same direction of the stress, Figure 25.6. In Figure 25.6(a), a normal stress σ (= F/A) is applied along the Direction 3 and the charges are generated on the surfaces perpendicular to Direction 3. A transversal piezoelectric effect occurs when the electricity is produced in the direction perpendicular to the stress.

The voltage V generated across the electrodes by the stress σ is given by:

$$V = d_{33} \frac{h}{\varepsilon} \sigma \tag{25.14}$$

where d_{33} = Piezoelectric constant associated with the longitudinal piezoelectric effect
$\quad\;\; \varepsilon$ = Permittivity
$\quad\;\; h$ = Thickness of the piezoelectric material

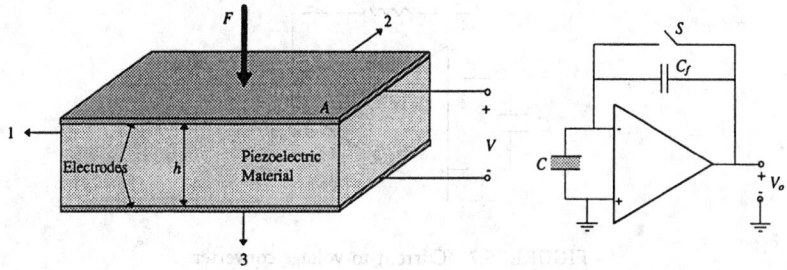

(a)   (b)

FIGURE 25.6 (a) Basic cell of a pizoelectric transducer. (b) Charge amplifier utilized for the measurement of the applied force.

Since piezoelectric materials are insulators, the transducer shown in Figure 25.6(a), can be considered as a capacitor, from an electrical point of view. Consequently,

$$V = \frac{Q}{C} = \frac{Q}{\varepsilon A} h \qquad (25.15)$$

where Q = Charge induced by the stress σ
           C = Capacitance of the parallel capacitor
           A = Area of each electrode

A comparison of Equations 25.14 and 25.15 leads to:

$$Q = d_{33} A \sigma \qquad (25.16)$$

It is concluded that the force applied to the photoelastic material can be determined by finding the charge Q. Charge amplifiers are usually utilized for determining Q. The basic configuration of a charge amplifier is shown in Figure 25.6(b). The charge generated in the transducer is transferred to the capacitor C_f and the output voltage, V_o is given by:

$$V_o = -\frac{Q}{C_f} \qquad (25.17)$$

The circuit must periodically discharge the feedback capacitor C_f to avoid saturation of the amplifier by stray charges generated by the offset voltages and currents of the operational amplifier. This is achieved by a switch as shown in Figure 25.6(b) or by a resistor parallel to C_f.

The piezoelectric material most widely used in the implementation of tactile transducers is PVF2. It shows the largest piezoelectric effect of any known material. Its flexibility, small size, sensitivity, and large electrical output offer many advantages for sensor applications in general, and tactile sensors in particular. Examples of tactile sensors implemented with this technology can be found in [1, 31].

The major advantages of the piezoelectric technology are its wide dynamic range and durability. Unfortunately, the response of available materials does not extend down to dc and therefore steady loads cannot be measured directly. Also, the PVF2 material produces a charge output that is prone to electrical interference and is temperature dependent.

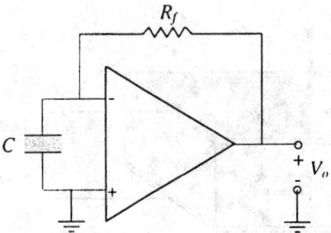

FIGURE 25.7 Current-to-voltage converter.

The possibility of measuring transient phenomenon using piezoelectric material has recently encouraged some researchers to use the piezoelectric effect for detecting vibrations that indicate incipient slip, occurrence of contact, local change in skin curvature, and estimating friction and hardness of the object [7, 10, 11]. If the piezoelectric transducer shown in Figure 25.6(*a*) is connected to an FET-input operational amplifier configured as a current-to-voltage converter as shown in Figure 25.7, the output voltage is given by:

$$V_o = \frac{dQ}{dt} R_f = AR_f d_{33} \frac{d\sigma}{dt} \tag{25.18}$$

where R_f is the feedback resistor. Correspondingly, the circuit configuration provides the mean to measure of changes in the contact stress. A detailed explanation of the behavior of this sensor can be found in [7].

Optical

Recent developments in fiber optic technology and solid-state cameras have led to numerous novel tactile sensor designs [32, 33]. Some of these designs employ flexible membranes incorporating a reflecting surface, Figure 25.8. Light is introduced into the sensor via a fiber optic cable. A wide cone of light propagates out of the fiber, reflects back from the membrane, and is collected by a second fiber. When an external force is applied onto the elastomer, it shortens the distance between the reflective side of the

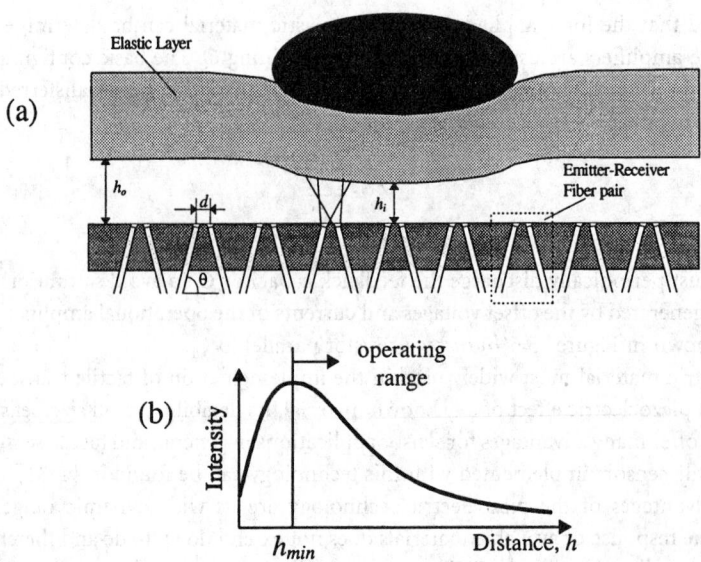

FIGURE 25.8 (*a*) Reflective transducer. (*b*) Light-intensity as a function of the distance *h*.

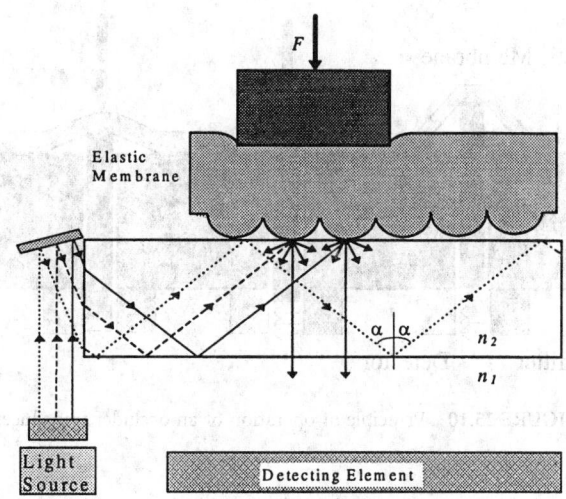

FIGURE 25.9   Tactile transducer based on the principle of internal reflection.

membrane and the fibers, h. Consequently, the light gathered by the receiving fiber changes as a function of h, Figure 25.8(b). To recover univocally the distance from the light intensity, a monotonic function is needed. This can be achieved by designing the transducer such it operates for $h > h_{min}$, where h_{min} is indicated in Figure 25.8(b). (The region $h > h_{min}$ is preferred to the $h < h_{min}$ for dynamic range reasons.)

Another optical effect that can be used is that of frustrated total internal reflection [5, 34]. With this technique, an elastic rubber membrane covers, without touching, a glass plate (waveguide); light entering the side edge of the glass is totally reflected by the top and bottom surfaces and propagates along it, Figure 25.9.

The condition for total internal reflection occurs when:

$$n_2 \sin \alpha \leq n_1 \tag{25.19}$$

where n_1 = Index of refraction of the medium surrounding the waveguide (in this case air, $n_1 \cong 1$)

n_2 = Index of refraction of the waveguide

α = Angle of incidence at the interface glass-air

Objects in contact with the elastic membrane deform it and induce contact between the bottom part of the membrane and the top surface of the waveguide, disrupting the total internal reflection. Consequently, the light in the waveguide is scattered at the contact location. Light that escapes through the bottom surface of the waveguide can be detected by an array of photodiodes, a solid-state sensor, or, alternatively, transported away from the transducer by fibers [3]. The detected imaged is stored in a computer for further analysis. A rubber membrane with a flat surface yields a high-resolution binary (contact or noncontact) image [5]. If the rubber sheet is molded with a textured surface (Figure 25.9), then an output proportional to the area of contact is obtained and, consequently, the applied forces can be detected [3]. Shear forces can also be detected using special designs [35]. Sensors based on frustrated internal reflection can be molded into a finger shape [5] and are capable of forming very high-resolution tactile images. Such sensors are commercially available. An improved miniaturized version of a similar sensor was proposed in [34].

Other types of optical transducers use "occluder" devices. One of the few commercially available tactile sensors uses this kind of transducer [36]. In one of the two available designs, the transducer's surface is made of a compliant material, which has on its underside a grid of elongated pins. When force is applied to the compliant surface, the pins on the underside undergo a mechanical motion normal to the surface,

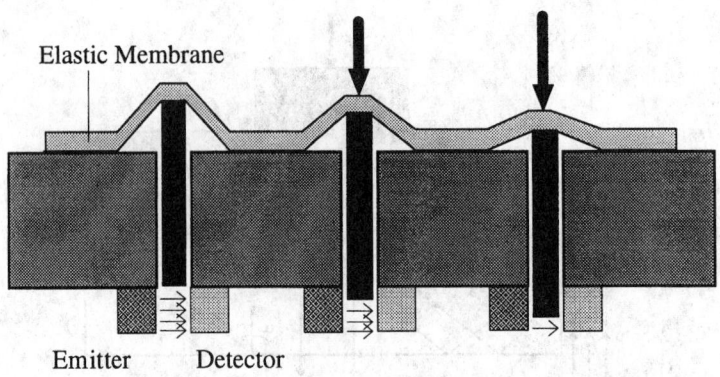

FIGURE 25.10   Principle of operation of an occluder transducer.

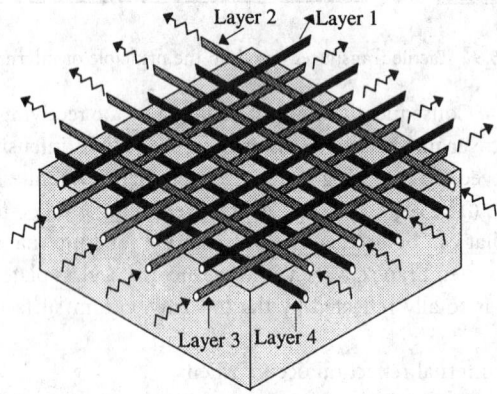

FIGURE 25.11   A four-layer tactile transducer.

blocking the light path of a photoemitter–detector pair. The amount of movement determines the amount of light reaching the photoreceiver. Correspondingly, the more force applied, the less amount of light is collected by the photoreceiver, Figure 25.10. The major problems with this specific device are associated with creep, hysteresis, and temperature variation. This scheme also requires individual calibration of each photoemitter–photodetector pair.

Fibers have also been used directly as transducers in the design of tactile sensors. Their use is based on two properties of fiber optic cables: (1) if a fiber is subjected to a significant amount of bending, then the angle of incidence at the fiber wall can be reduced sufficiently for light to leave the core [37]; and (2) if two fibers pass close to one another and both have roughened surfaces, then light can pass between the fibers. Light coupling between adjacent fibers is a function of their separation [3].

An example of an optical fiber tactile sensor, whose sensing mechanism is based on the increase of light attenuation due to the microbend in the optical fibers, is shown in Figure 25.11 [37]. The transducer consists of a four-layer, two-dimensional fiber optic array constructed by using two layers of optical fibers as a corrugation structure, through which microbends are induced in two orthogonal layers of active fibers. Each active fiber uses an LED as the emitter and a PIN photodiode as a detector. When an object is forced into contact with the transducer, a light distribution is detected at each detector. This light distribution is related to the applied force and the shape of the object. Using complex algorithms and active sensing (moving the object in relation to the transducer), the object position, orientation, size, and contour information can be retrieved [37]. However, the recovery of the applied force profiles was not reported in [37].

Photoelastic

An emerging technology in optical tactile sensing is the development of photoelastic transducers. When a light ray propagates into an optically anisotropic medium, it splits into two rays that are linearly polarized at right angles to each other and propagate at different velocities. This splitting of a ray into two rays that have mutually perpendicular polarizations results from a physical property of crystalline material that is called *optical birefringence* or simply *birefringence*. The direction in which light propagates with the higher velocity is called the *fast axis*; and the one in which it propagates more slowly is called the *slow axis*. Some optically isotropic materials — such as glass, celluloid, bakelite, and transparent plastics in general — become birefringent when they are subjected to a stress field. The birefringent effect lasts only during the application of loads. Thus, this phenomenon is called *temporary* or *artificial birefringence* or, more commonly, the *photoelastic phenomenon*.

Figure 25.12(*a*) shows a photoelastic transducer proposed in [38]. It consists of a fully supported two-layer beam with a mirrored surface sandwiched in between. Normal line forces are applied to the top surface of the beam at discrete tactels, separated by equal distances, *s*, along the beam. The upper compliant layer is for the protection of the mirror, while the lower one is the photoelastic layer.

Circularly polarized monochromatic light, incident along the *z*-axis, illuminates the bottom surface of the transducer. The light propagates parallel to the *z*-axis, passes through the photoelastic layer, and then reflects back from the mirror. If no force is applied to the transducer, the returning light is circularly polarized because unstressed photoelastic material is isotropic. If force is applied, stresses are induced in the photoelastic layer, making the material birefringent. This introduces a certain phase difference between the components of the electric field associated with the light-wave propagation. The two directions of polarization are in the plane perpendicular to the direction of propagation (in this case, the *x–y* plane). As a consequence of this effect, the output light is elliptically polarized, creating a phase difference distribution, *p*, between the input light ant the output light at each point in the *x–y* plane. The phase difference distribution carries the information of the force distribution applied to the transducer.

A *polariscope* is a practical method to observe the spatial variation on light intensity (fringes) due to the effect of induced phase difference distribution. Polariscopes can be either linear or circular, depending on the required polarization of the light. They can also be characterized as a reflective or a transparent type, depending on whether the photoelastic transducer reflect or transmits the light.

A circular, reflective polariscope, shown in Figure 25.12(*b*), is utilized to illuminate the transducer shown in Figure 25.12(*a*). The input light is linearly polarized and is directed toward the photoelastic transducer by a beam splitter. Before reaching the transducer, the light is circularly polarized by a quarter-wave plate. The output light is elliptically polarized when a force is applied. This light is directed toward a detector passing through the quarter-wave plate, the beam splitter, and an analyzer. Finally, it is detected by a camera linked to a frame grabber connected to a PC, for further data processing. The light that illuminates the camera consists of a set of fringes from where the force distribution applied to the transducer must be recovered. A technique for the recovery of the forces from the fringes is described in [38]. A model of the transducer using FEA is reported in [39].

One of the earlier applications of photoelasticity to tactile sensing dates back to the development phase of the Utah/MIT dexterous hand [40]. The researchers proposed the use of the photoelastic phenomenon as a transduction method for the recovery of the force profile applied to the fingers of the hand. They limited their application to the development of a single-touch transducer, although they claimed that an array of such devices could be implemented. However, the construction of a large array of their devices would be difficult. To overcome this difficulty, another research group proposed a different transducer [41]. Although an analytical model was developed for the sensor, a systematic method for recovering the two-dimensional force profile from the light intensity distribution was not reported. Thus, the sensor was used mainly for the study of the forward analysis, namely, observing the light intensity distribution for different touching objects brought into contact with the sensor. This sensor could eventually be used for determining some simple geometric properties of a touching object.

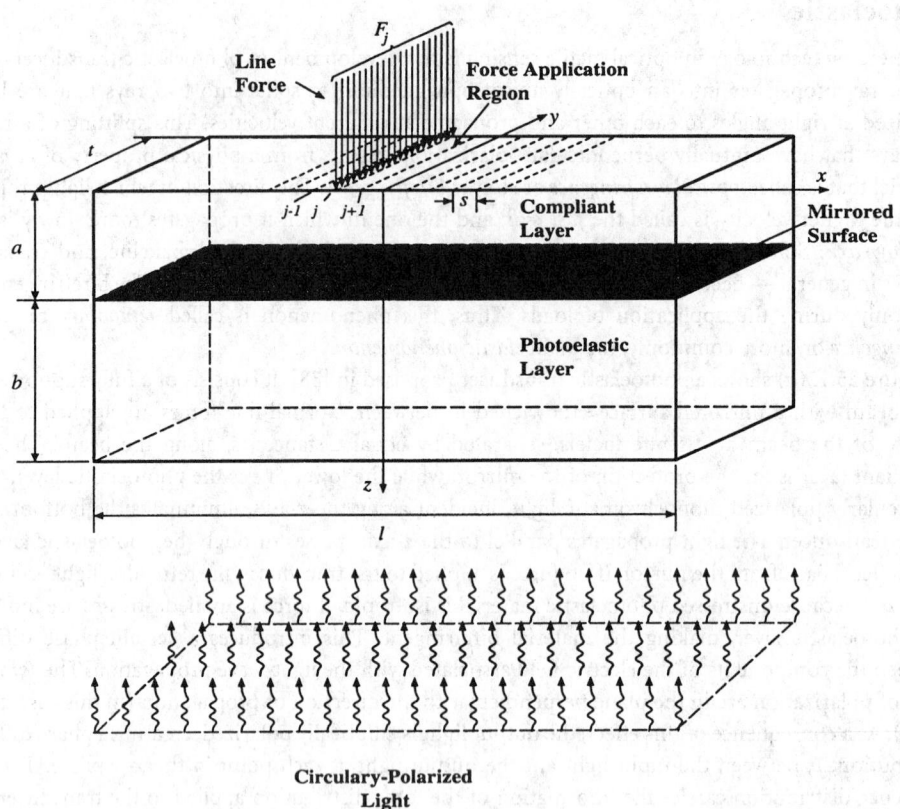

(a)

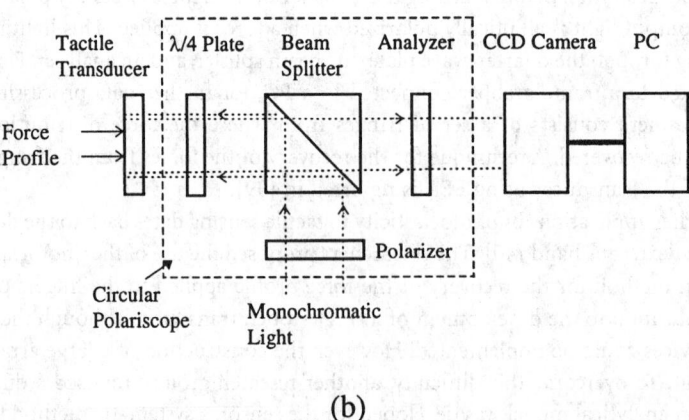

(b)

FIGURE 25.12   (*a*) Photoelastic transducer. (*b*) Circular reflective polariscope.

A tactile sensor reported in [42] is capable of detecting slippage. The output light intensity (the fringe pattern) is captured by a camera interfaced to a PC. When an object moves across the surface of the transducer, the light intensity distribution changes. A direct analysis of the fringes is used to detect movement of the grasped object; a special technique was reported to optimize the comparison process for detecting differences between two fringe patterns occurring due to the slippage of the object in contact with the sensor [42]. It is important to note that such an analysis of the fringes does not require the recovery of the applied force profile.

Photoelasticity offers several attractive properties for the development of tactile sensors: good linearity, compatibility with vision-base sensing technologies, and high spatial resolution associated with the latter, that could lead to the development of high-resolution tactile imagers needed for object recognition and fine manipulation. Also, photoelastic sensors are compatible with fiber optic technology that allows remote location of electronic processing devices and avoidance of interference problems.

Other technologies for tactile sensing include acoustic, magnetic, and microcavity vacuum sensors [43, 44].

References

1. P. Dario, Contact sensing for robot active touch, in *Robotic Science*, M. Brady (ed.), Cambridge, MA: MIT Press, 1989, chap. 3, 138-163.
2. P. P. L. Regtien, Tactile imaging, *Sensors and Actuators, A*, 31, 83-89, 1992.
3. R. A. Russell, *Robot Tactile Sensing*, Brunswick, Australia: Prentice-Hall, 1990.
4. A. D. Berger and P. K. Khosla, Using tactile data for real-time feedback, *Int. J. Robotics Res.*, 10(2), 88-102, 1991.
5. S. Begej, Planar and finger-shaped optical tactile sensors for robotic applications, *IEEE J. Robotics Automation*, 4, 472-484, 1988.
6. R. A. Russell and S. Parkinson, Sensing surface shape by touch, *IEEE Int. Conf. Robotics Automation*, Atlanta, GA, 1993, 423-428.
7. R. D. Howe, A tactile stress rate sensor for perception of fine surface features, *IEEE Int. Conf. Solid-State Sensors Actuators*, San Francisco, CA, 1991, 864-867.
8. S. Sugiyama, K. Kawahata, H. Funabashi, M. Takigawa, and I. Igarashi, A 32 × 32 (1K)-element silicon pressure-sensor array with CMOS processing circuits, *Electron. Commun. Japan*, 75(1), 64-76, 1992.
9. J. S. Son, E. A. Monteverde, and R. D. Howe, A tactile sensor for localizing transient events in manipulation, *IEEE Int. Conf. Robotics Automation*, San Diego, CA, 1994, 471-476.
10. M. R. Tremblay and M. R. Cutkosky, Estimating friction using incipient slip sensing during manipulation task, *IEEE Int. Conf. Robotics Automation*, Atlanta, GA, 1993, 429-434.
11. S. Omata and Y. Terubuna, New tactile sensor like the human hand and its applications, *Sensors and Actuators, A*, 35, 9-15, 1992.
12. R. Bayrleithner and K. Komoriya, Static friction coefficient determination by force sensing and its applications, *IROS'94*, Munich, Germany, 1994, 1639-1646.
13. M. A. Abidi and R. C. Gonzales, The use of multisensor data for robotic applications, *IEEE Trans. Robotics Automation*, 6, 159-177, 1990.
14. A. A. Cole, P. Hsu, and S. S. Sastry, Dynamic control of sliding by robot hands for regrasping, *IEEE Trans. Robotics Automation*, 8, 42-52, 1992.
15. P. K. Allen and P. Michelman, Acquisition and interpretation of 3-D sensor data from touch, *IEEE Trans. Robotics Automation*, 6, 397-404, 1990.
16. H. R. Nicholls (ed.), *Advanced Tactile Sensing for Robotics*, Singapore: World Scientific Publishing, 1992.

17. J. G. Webster (ed.), *Tactile Sensors for Robotics and Medicine*, New York: John Wiley & Sons, 1988.
18. R. S. Fearing, Tactile sensing mechanism, *Int. J. Robotics Res.*, 9(3), 3-23, 1990.
19. T. H. Speeter, Three-dimensional finite element analysis of elastic continua for tactile sensing, *Int. J. Robotics Res.*, 11(1), 1-19, 1992.
20. Y. C. Pati, P. S. Krishnaprasad, and M. C. Peckerar, An analog neural network solution to the inverse problem of early taction, *IEEE Trans. Robotics Automation*, 8(2), 196-212, 1992.
21. L. D. Harmon, Automated tactile sensing, *Int. J. Robotics Res.*, 1(2), 3-32, 1982.
22. W. E. Snyder and J. St. Clair, Conductive elastomers as a sensor for industrial parts handling equipment, *IEEE Trans. Instrum. Meas.*, 27(1), 94-99, 1991.
23. M. Shimojo, M. Ishikawa, and K. Kanaya, A flexible high resolution tactile imager with video signal output, *IEEE Int. Conf. Robotics Automation*, Sacramento, CA, 1991, 384-391
24. W. D. Hillis, A high resolution imaging touch sensor, *Int. J. Robotic Res.*, 1(2), 33-44, 1982.
25. Y. Yamada and M. R. Cutkosky, Tactile sensor with 3-axis force and vibration sensing functions and its applications to detect rotational slip, *IEEE Int. Conf. Robotics Automation*, San Diego, CA, 1994, 3550-3557.
26. K. Njafi and C. H. Mastrangelo, Solid-state microsensors and smart structure, *Ultrasonic Symp.*, Baltimore, MD, 1993, 341-350.
27. F. Zhu and J. W. Spronck, A capacitive tactile sensor for shear and normal force measurements, *Sensors and Actuators, A*, 31, 115-120, 1992.
28. K. Suzuki, K. Najafi, and K. D. Wise, A 1024-element high-performance silicon tactile imager, *IEEE Trans. Electron Devices*, 17(8), 1852-1860, 1990.
29. M. R. Wolffenbuttel and P. L. Regtien, The accurate measurement of a micromechanical force using force-sensitive capacitances, *Conf. Precision Electromagnetic Meas.*, Boulder, CO, 1994, 180-181.
30. M. R. Wolffenbuttel, R. F. Wolffenbuttel, and P. P. L. Regtien, An integrated charge amplifier for a smart tactile sensor, *Sensors and Actuators, A*, 31, 101-109, 1992.
31. E. D. Kolesar, Jr. and C. S. Dyson, Object imaging with piezoelectric robotic tactile sensor, *J. Microelectromechanical Syst.*, 4(2), 87-96, 1995.
32. J. L. Scheiter and T. B. Sheridan, An optical tactile sensor for manipulators, *J. Robot Computer-Integrated Manufacturing*, 1, 65-71, 1989.
33. R. Ristic, B. Benhabib, and A. A. Goldenberg, Analysis and design of a modular electrooptical tactile sensor, *IEEE Trans. Robotics Automation*, 5(3), 362-368, 1989.
34. H. Maekawa, K. K. Tanie, K. Komoriya, M. Kaneko, C. Horiguchi, and T. Sugawara, development of a finger shaped tactile sensor and it evaluation by active touch, *IEEE Int. Conf. Robotics Automation*, Nice, France, 1992, 1327-1334.
35. M. Ohka, Y. Mitsurya, S. Takeuchi, and O. Kamekawa, A three-axis optical tactile sensor (fem contact analyses and sensing experiments using a large-sized tactile sensor), *IEEE Int. Conf. Robotics Automation*, Nagoya, Aichi, Japan, 1995, 817-824.
36. J. Rebman and K. A. Morris, A tactile sensor with electro-optical transduction, in *Robots Sensors, Tactile and Non-Vision*, Vol. 2, A. Pugh (ed.), EFS Publications, 1986, 145-155.
37. S. R. Emge and C. L. Chen, Two dimensional contour imaging with a fiber optic microbend tactile sensor array, *Sensors and Actuators, B*, 3, 31-42, 1991.
38. R. E. Saad, A. Bonen, K. C. Smith, and B. Benhabib, Distributed-force recovery for a planar photoelastic tactile sensor, *IEEE Trans. Instrum. Meas.*, 45, 541-546, 1996.
39. R. E. Saad, A. Bonen, K. C. Smith, and B. Benhabib, Finite-element analysis for photoelastic tactile sensors, *Proc. IEEE Int. Conf. Industrial Electronics, Control, and Instrumentation*, Orlando, FL, 1995, 1202-1207.
40. S. C. Jacobsen, J. E. Wood, D. F. Knutti, and B. Biggers, The Utah/MIT dexterous hand: work in progress, in *Robotics Research: The First International Symposium*, M. Brady and R. Paul (eds.), Cambridge: MIT Press, 1983, 601-653.
41. A. Cameron, R. Daniel, and H. Durrant-Whyte, Touch and motion, *IEEE, Int. Conf. Robotics Automation*, Philadelphia, PA, 1988, 1062-1067.

42. S. H. Hopkins, F. Eghtedari, and D. T. Pham, Algorithms for processing data from a photoelastic slip sensor, *Mechatronics*, 2(1), 15-28, 1992.
43. S. Ando and H. Shinoda, Ultrasonic emission tactile sensing, *IEEE Trans. Control Syst.*, 15(1), 61-69, 1996.
44. J. C. Jiang, V. Faynberg, and R. C. White, Fabrication of micromachined silicon tip transducer for tactile sensing, *J. Vacuum Sci. Technol.*, B, 11, 1962-1967, 1993.

V

Mechanical Variables Measurement — Fluid

26
Pressure and Sound Measurement

Kevin H.-L. Chau
Analog Devices, Inc.

Ron Goehner
The Fredericks Company

Emil Drubetsky
The Fredericks Company

Howard M. Brady
The Fredericks Company

William H. Bayles, Jr.
The Fredericks Company

Peder C. Pedersen
Worcester Polytechnic Institute

26.1 Pressure Measurement

Kevin H.-L. Chau

Basic Definitions

Pressure is defined as the normal force per unit area exerted by a fluid (liquid or gas) on any surface. The surface can be either a solid boundary in contact with the fluid or, for purposes of analysis, an imaginary plane drawn through the fluid. Only the component of the force normal to the surface needs to be considered for the determination of pressure. Tangential forces that give rise to shear and fluid motion will not be a relevant subject of discussion here. In the limit that the surface area approaches zero, the ratio of the differential normal force to the differential area represents the pressure at a point on the surface. Furthermore, if there is no shear in the fluid, the pressure at any point can be shown to be independent of the orientation of the imaginary surface under consideration. Finally, it should be noted that pressure is not defined as a vector quantity and is therefore nondirectional.

Three types of pressure measurements are commonly performed:

Absolute pressure is the same as the pressure defined above. It represents the pressure difference between the point of measurement and a perfect vacuum where pressure is zero.

Gage pressure is the pressure difference between the point of measurement and the ambient. In reality, the ambient (atmospheric) pressure can vary, but only the pressure difference is of interest in gage pressure measurements.

TABLE 26.1 Pressure Unit Conversion Table

Units	kPa	psi	in H$_2$O	cm H$_2$O	in. Hg	mm Hg	mbar
kPa	1.000	0.1450	4.015	10.20	0.2593	7.501	10.00
psi	6.895	1.000	27.68	70.31	2.036	51.72	68.95
in. H$_2$O	0.2491	3.613×10^{-2}	1.000	2.540	7.355×10^{-2}	1.868	2.491
cm H$_2$O	0.09806	1.422×10^{-2}	0.3937	1.000	2.896×10^{-2}	0.7355	0.9806
in. Hg	3.386	0.4912	13.60	34.53	1.000	25.40	33.86
mm Hg	0.1333	1.934×10^{-2}	0.5353	1.360	3.937×10^{-2}	1.000	1.333
mbar	0.1000	0.01450	0.04015	1.020	0.02953	0.7501	1.000

Key:
(1) kPa = kilopascal;
(2) psi = pound force per square inch;
(3) in. H$_2$O = inch of water at 4°C;
(4) cm H$_2$O = centimeter of water at 4°C;
(5) in. Hg = inch of mercury at 0°C;
(6) mm Hg = millimeter of mercury at 0°C;
(7) mbar = millibar.

Differential pressure is the pressure difference between two points, one of which is chosen to be the reference. In reality, both pressures can vary, but only the pressure difference is of interest here.

Units of Pressure and Conversion

The SI unit of pressure is the *pascal* (Pa), which is defined as the newton per square meter (N·m^{-2}); 1 Pa is a very small unit of pressure. Hence, decimal multiples of the pascal (e.g., kilopascals [kPa] and megapascals [MPa]) are often used for expressing higher pressures. In weather reports, the hectopascal (1 hPa = 100 Pa) has been adopted by many countries to replace the millibar (1 bar = 10^5 Pa; hence, 1 millibar = 10^{-3} bar = 1 hPa) as the unit for atmospheric pressure. In the United States, pressure is commonly expressed in pound force per square inch (psi), which is about 6.90 kPa. In addition, the absolute, gage, and differential pressures are further specified as psia, psig, and psid, respectively. However, no such distinction is made in any pressure units other than the psi. There is another class of units e.g., millimeter of mercury at 0°C (mm Hg, also known as the *torr*) or inch of water at 4°C (in H$_2$O), which expresses pressure in terms of the height of a static liquid column. The actual pressure *p* referred to is one that will be developed at the base of the liquid column due to its weight, which is given by Equation 26.1.

$$p = \rho\, g\, h \qquad\qquad (26.1)$$

where ρ is the density of the liquid, g is the acceleration due to gravity, and h is the height of the liquid column. A conversion table for the most popular pressure units is provided in Table 26.1.

Sensing Principles

Sensing Elements

Since pressure is defined as the force per unit area, the most direct way of measuring pressure is to isolate an area on an elastic mechanical element for the force to act on. The deformation of the sensing element produces displacements and strains that can be precisely sensed to give a calibrated measurement of the pressure. This forms the basis for essentially all commercially available pressure sensors today. Specifically, the basic requirements for a pressure-sensing element are a means to isolate two fluidic pressures (one to be measured and the other one as the reference) and an elastic portion to convert the pressure difference into a deformation of the sensing element. Many types of pressure-sensing elements are currently in use. These can be grouped as diaphragms, capsules, bellows, and tubes, as illustrated in Figure 26.1. Diaphragms

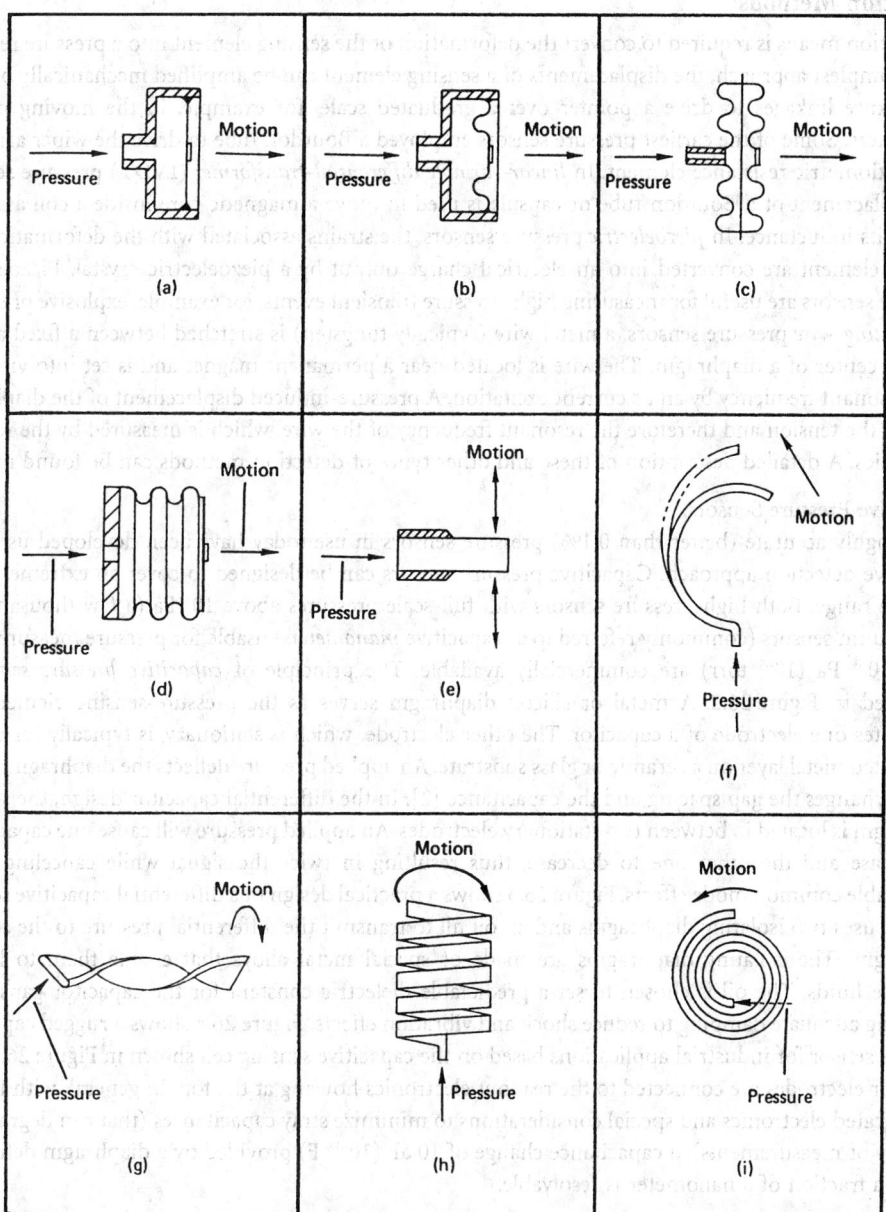

FIGURE 26.1 Pressure-sensing elements: (*a*) flat diaphragm; (*b*) corrugated diaphragm; (*c*) capsule; (*d*) bellows; (*e*) straight tube; (*f*) C-shaped Bourdon tube; (*g*) twisted Bourdon tube; (*h*) helical Bourdon tube; (1) spiral Bourdon tube. (From Norton, H. N., *Handbook of Transducers,* Englewood Cliffs, NJ: Prentice-Hall, 1989, 294–330. Reprinted with permission.)

are by far the most widely used of all sensing elements. A special form of tube, known as the *Bourdon tube,* is curved or twisted along its length and has an oval cross-section. The tube is sealed at one end and tends to unwind or straighten when it is subjected to a pressure applied to the inside. In general, Bourdon tubes are designed for measuring high pressures, while capsules and bellows are usually for measuring low pressures. A detailed description of these sensing elements can be found in [1].

Detection Methods

A detection means is required to convert the deformation of the sensing element into a pressure readout. In the simplest approach, the displacements of a sensing element can be amplified mechanically by lever and flexure linkages to drive a pointer over a graduated scale, for example, in the moving pointer barometers. Some of the earliest pressure sensors employed a Bourdon tube to drive the wiper arm over a potentiometric resistance element. In *linear-variable differential-transformer* (LVDT) pressure sensors, the displacement of a Bourdon tube or capsule is used to move a magnetic core inside a coil assembly to vary its inductance. In *piezoelectric* pressure sensors, the strains associated with the deformation of a sensing element are converted into an electrical charge output by a piezoelectric crystal. Piezoelectric pressure sensors are useful for measuring high-pressure transient events, for example, explosive pressures. In *vibrating-wire* pressure sensors, a metal wire (typically tungsten) is stretched between a fixed anchor and the center of a diaphragm. The wire is located near a permanent magnet and is set into vibration at its resonant frequency by an ac current excitation. A pressure-induced displacement of the diaphragm changes the tension and therefore the resonant frequency of the wire, which is measured by the readout electronics. A detailed description of these and other types of detection methods can be found in [1].

Capacitive Pressure Sensors.

Many highly accurate (better than 0.1%) pressure sensors in use today have been developed using the capacitive detection approach. Capacitive pressure sensors can be designed to cover an extremely wide pressure range. Both high-pressure sensors with full-scale pressures above 10^7 Pa (a few thousand psi) and vacuum sensors (commonly referred to as capacitive *manometers*) usable for pressure measurements below 10^{-3} Pa (10^{-5} torr) are commercially available. The principle of *capacitive pressure sensors* is illustrated in Figure 26.2. A metal or silicon diaphragm serves as the pressure-sensing element and constitutes one electrode of a capacitor. The other electrode, which is stationary, is typically formed by a deposited metal layer on a ceramic or glass substrate. An applied pressure deflects the diaphragm, which in turn changes the gap spacing and the capacitance [2]. In the differential capacitor design, the sensing diaphragm is located in between two stationary electrodes. An applied pressure will cause one capacitance to increase and the other one to decrease, thus resulting in twice the signal while canceling many undesirable common mode effects. Figure 26.3 shows a practical design of a differential capacitive sensing cell that uses two isolating diaphragms and an oil fill to transmit the differential pressure to the sensing diaphragm. The isolating diaphragms are made of special metal alloys that enable them to handle corrosive fluids. The oil is chosen to set a predictable dielectric constant for the capacitor gaps while providing adequate damping to reduce shock and vibration effects. Figure 26.4 shows a rugged capacitive pressure sensor for industrial applications based on the capacitive sensing cell shown in Figure 26.3. The capacitor electrodes are connected to the readout electronics housing at the top. In general, with today's sophisticated electronics and special considerations to minimize stray capacitances (that can degrade the accuracy of measurements), a capacitance change of 10 aF (10^{-18} F) provided by a diaphragm deflection of only a fraction of a nanometer is resolvable.

Piezoresistive Pressure Sensors.

Piezoresistive sensors (also known as *strain-gage* sensors) are the most common type of pressure sensor in use today. *Piezoresistive effect* refers to a change in the electric resistance of a material when stresses or strains are applied. Piezoresistive materials can be used to realize strain gages that, when incorporated into diaphragms, are well suited for sensing the induced strains as the diaphragm is deflected by an applied pressure. The sensitivity of a strain gage is expressed by its *gage factor*, which is defined as the fractional change in resistance, $\Delta R/R$, per unit strain:

$$\text{Gage factor} = \left(\Delta R/R\right)\big/\varepsilon \qquad (26.2)$$

where strain ε is defined as $\Delta L/L$, or the extension per unit length. It is essential to distinguish between two different cases in which: (1) the strain is parallel to the direction of the current flow (along which

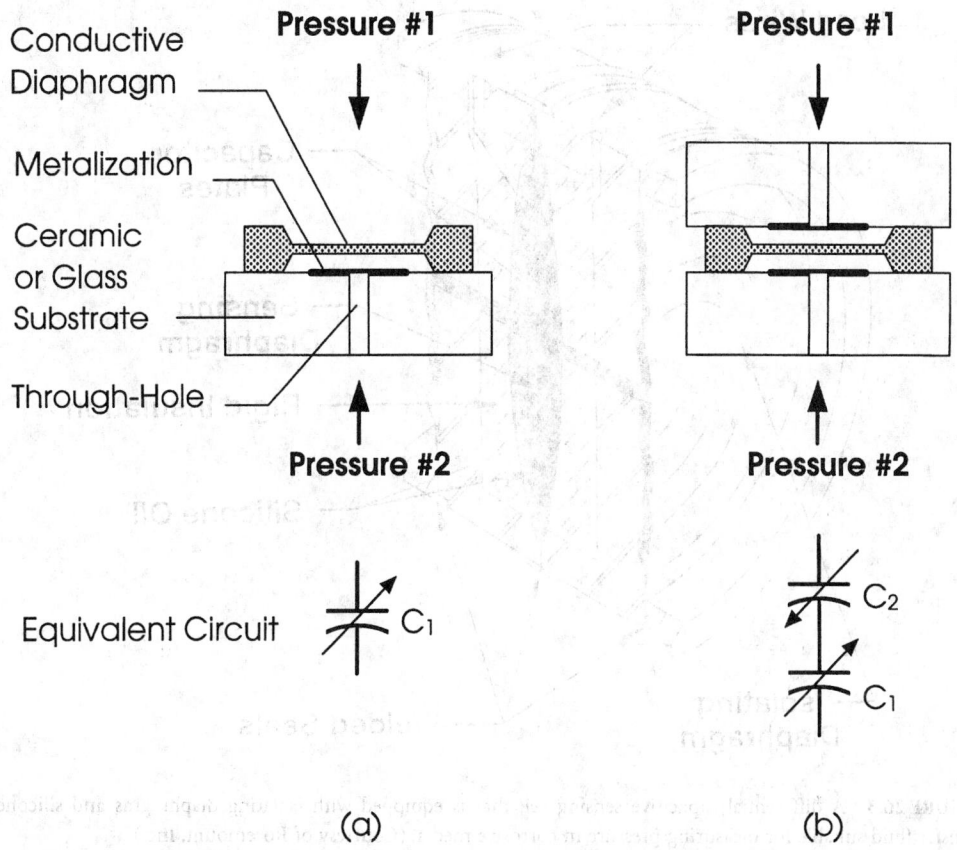

FIGURE 26.2 Operating principle of capacitive pressure sensors. (*a*) Single capacitor design; and (*b*) differential capacitor design.

the resistance change is to be monitored); and (2) the strain is perpendicular to the direction of the current flow. The gage factors associated with these two cases are known as the *longitudinal gage factor* and the *transverse gage factor*, respectively. The two gage factors are generally different in magnitude and often opposite in sign. Typical longitudinal gage factors are ~2 for many useful metals, 10 to 35 for polycrystalline silicon (polysilicon), and 50 to 150 for single-crystalline silicon [3–5]. Because of its large piezoresistive effect, silicon has become the most commonly used material for strain gages. There are several ways to incorporate strain gages into pressure-sensing diaphragms. For example, strain gages can be directly bonded onto a metal diaphragm. However, hysteresis and creep of the bonding agent are potential issues. Alternatively, the strain gage material can be deposited as a thin film on the diaphragm. The adhesion results from strong molecular forces that will not creep, and no additional bonding agent is required. Today, the majority of piezoresistive pressure sensors are realized by integrating the strain gages into the silicon diaphragm using integrated circuit fabrication technology. This important class of silicon pressure sensors will be discussed in detail in the next section.

Silicon Micromachined Pressure Sensors

Silicon micromachined pressure sensors refer to a class of pressure sensors that employ integrated circuit batch processing techniques to realize a thinned-out diaphragm sensing element on a silicon chip. Strain gages made of silicon diffused resistors are typically integrated on the diaphragm to convert the pressure-induced diaphragm deflection into an electric resistance change. Over the past 20 years, silicon micro-machined pressure sensors have gradually replaced their mechanical counterparts and have captured over

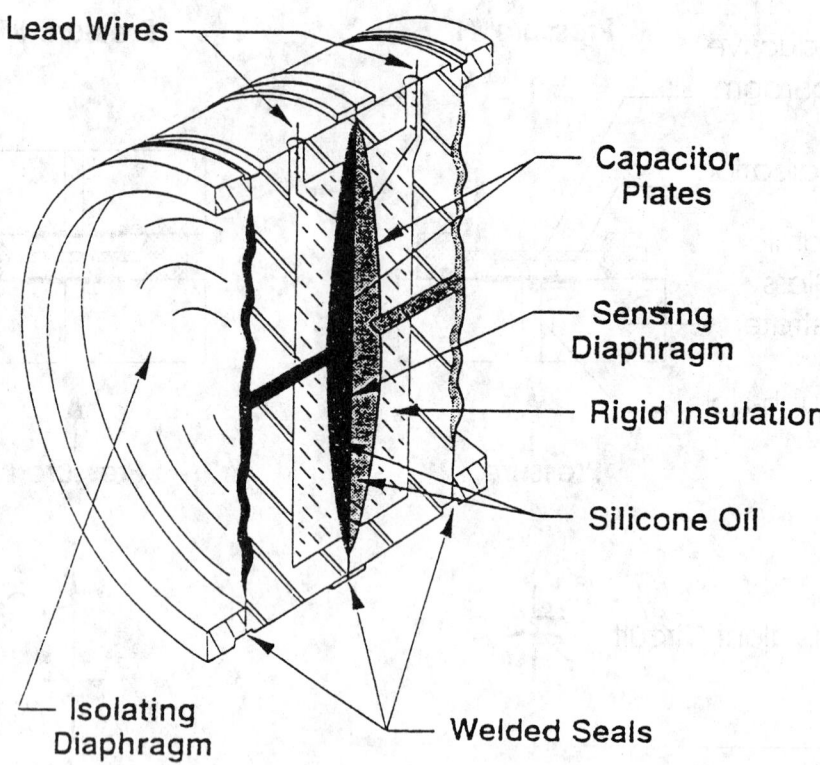

FIGURE 26.3 A differential capacitive sensing cell that is equipped with isolating diaphragms and silicone oil transfer fluid suitable for measuring pressure in corrosive media. (Courtesy of Rosemount, Inc.)

80% of the pressure sensor market. There are several unique advantages that silicon offers. Silicon is an ideal mechanical material that does not display any hysteresis or yield and is elastic up to the fracture limit. It is stronger than steel in yield strength and comparable in Young's modulus [6]. As mentioned in the previous section, the piezoresistive effect in single-crystalline silicon is almost 2 orders of magnitude larger than that of metal strain gages. Silicon has been widely used in integrated circuit manufacturing for which reliable batch fabrication technology and high-precision dimension control techniques have been well developed. A typical silicon wafer yields hundreds of identical pressure sensor chips at very low cost. Further, the necessary signal conditioning circuitry can be integrated on the same sensor chip no more than a few millimeters in size [7]. All these are key factors that contributed to the success of silicon micromachined pressure sensors.

Figure 26.5 shows a typical construction of a silicon piezoresistive pressure sensor. An array of square or rectangular diaphragms is "micromachined" out of a (100) oriented single-crystalline silicon wafer by selectively removing material from the back. An anisotropic silicon etchant (e.g., potassium hydroxide) is typically employed; it etches fastest on (100) surfaces and much slower on (111) surfaces. The result is a pit formed on the backside of the wafer bounded by (111) surfaces and a thinned-out diaphragm section on the front at every sensor site. The diaphragm thickness is controlled by a timed etch or by using suitable etch-stop techniques [6, 8]. To realize strain gages, *p*-type dopant, typically boron, is diffused into the front of the *n*-type silicon diaphragm at stress-sensitive locations to form resistors that are electrically isolated from the diaphragm and from each other by reverse biased *p–n* junctions. The strain gages, the diaphragm, and the rest of the supporting sensor chip all belong to the same single-crystalline silicon. The result is a superb mechanical structure that is free from creep, hysteresis, and thermal expansion coefficient mismatches. However, the sensor die must still be mounted to a sensor housing, which typically has mechanical properties different from that of silicon. It is crucial to ensure

FIGURE 26.4   A rugged capacitive pressure sensor product for industrial applications. It incorporates the sensing cell shown in Figure 26.3. Readout electronics are contained in the housing at the top. (Courtesy of Rosemount, Inc.)

a high degree of stress isolation between the sensor housing and the sensing diaphragm that may otherwise lead to long-term mechanical drifts and undesirable temperature behavior. A common practice is to bond a glass wafer or a second silicon wafer to the back of the sensor wafer to reinforce the overall composite sensor die. This way, the interface stresses generated by the die mount will also be sufficiently remote from the sensing diaphragm and will not seriously affect its stress characteristics. For gage or differential pressure sensing, holes must be provided through the carrier wafer prior to bonding that are aligned to the etch pits of the sensor wafer leading to the back of the sensing diaphragms. No through holes are necessary for absolute pressure sensing. The wafer-to-wafer bonding is performed in a vacuum to achieve a sealed reference vacuum inside the etch pit [6, 9]. Today's silicon pressure sensors are available in a large variety of plastic, ceramic, metal can, and stainless steel packages (some examples are shown in Figure 26.6). Many are suited for printed circuit board mounting. Others have isolating diaphragms and transfer fluids for handling corrosive media. They can be readily designed for a wide range of industrial, medical, automotive, aerospace, and military applications.

Silicon Piezoresistive Pressure Sensor Limitations

Despite the relatively large piezoresistive effects in silicon strain gages, the full-scale resistance change is typically only 1% to 2% of the resistance of the strain gage (which yields an unamplified voltage output of 10 mV/V to 20 mV/V). To achieve an overall accuracy of 0.1% of full scale, for example, the combined effects of mechanical and electrical repeatability, hysteresis, linearity, and stability must be controlled or compensated to within a few parts per million (ppm) of the gage resistance. Furthermore, silicon strain gages are also very temperature sensitive and require careful compensations. There are two primary sources of temperature drifts: (1) the temperature coefficient of resistance of the strain gages (from 0.06%/°C to 0.24%/°C); and (2) the temperature coefficient of the gage factors (from –0.06%/°C to

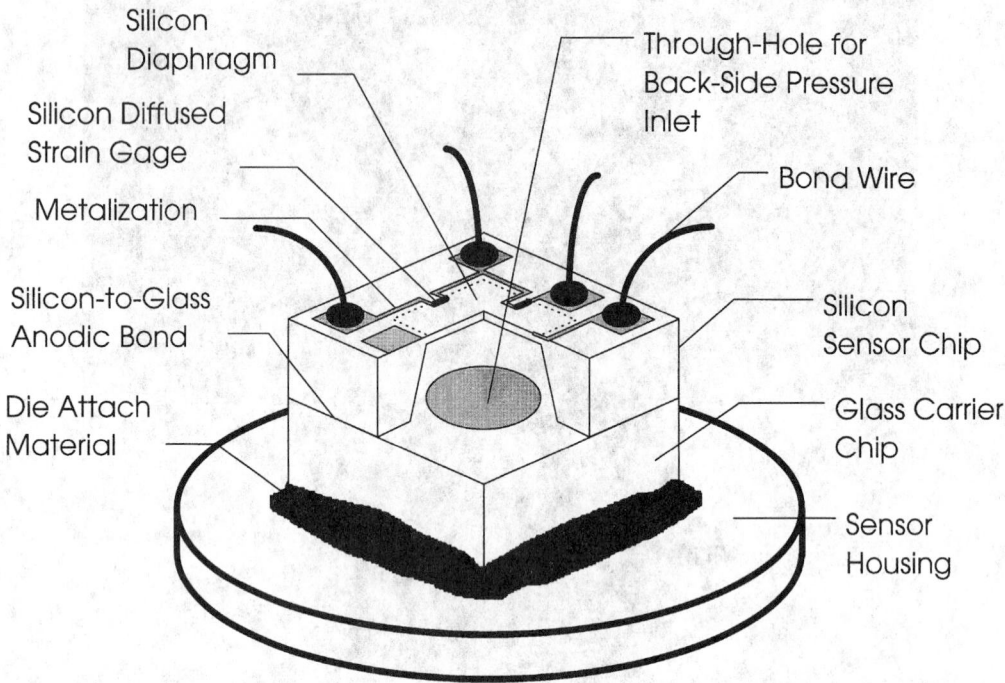

Silicon
Diaphragm

Silicon Diffused
Strain Gage

Metalization

Silicon-to-Glass
Anodic Bond

Die Attach
Material

Through-Hole for
Back-Side Pressure
Inlet

Bond Wire

Silicon
Sensor Chip

Glass Carrier
Chip

Sensor
Housing

FIGURE 26.5   A cut-away view showing the typical construction of a silicon piezoresistive pressure sensor.

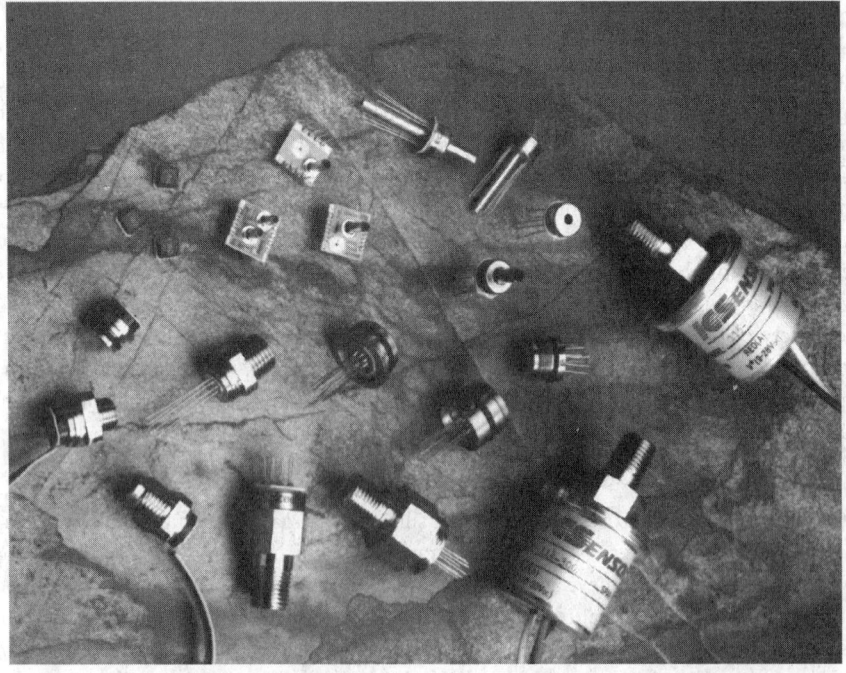

FIGURE 26.6   Examples of commercially available packages for silicon pressure sensors. Shown in the photo are surface-mount units, dual-in-line (DIP) units, TO-8 metal cans, and stainless steel units with isolating diaphragms. (Courtesy of EG&G IC Sensors.)

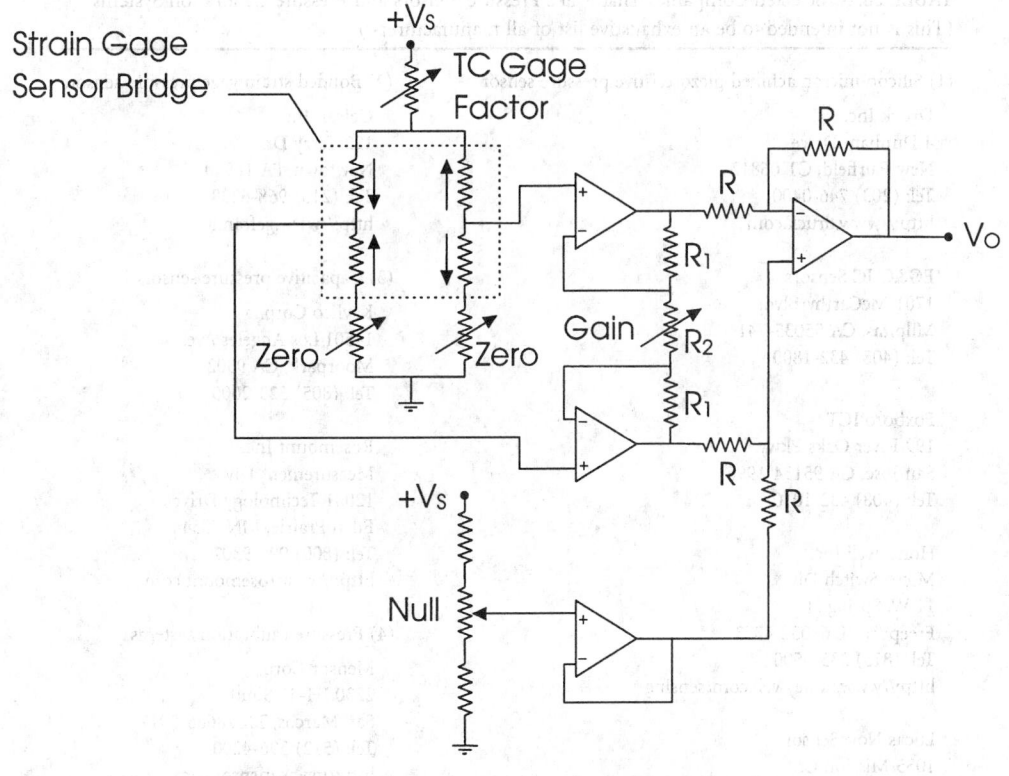

FIGURE 26.7 A signal-conditioning circuit for silicon piezoresistive pressure sensor.

−0.24%/°C), which will cause a decrease in pressure sensitivity as the temperature rises. Figure 26.7 shows a circuit configuration that can be used to achieve offset (resulting from gage resistance mismatch) and temperature compensations as well as providing signal amplification to give a high-level output. Four strain gages that are closely matched in both their resistances and temperature coefficients of resistance are employed to form the four active arms of a Wheatstone bridge. Their resistor geometry on the sensing diaphragm is aligned with the principal strain directions so that two strain gages will produce a resistance increase and the other two a resistance decrease on a given diaphragm deflection. These two pairs of strain gages are configured in the Wheatstone bridge such that an applied pressure will produce a bridge resistance imbalance while the temperature coefficient of resistance will only cause a common mode resistance change in all four gages, keeping the bridge balanced. As for the temperature coefficient of the gage factor, because it is always negative, it is possible (e.g., with the voltage divider circuit in Figure 26.7) to utilize the positive temperature coefficient of the bridge resistance to increase the bridge supply voltage, compensating for the loss in pressure sensitivity as temperature rises. Another major limitation in silicon pressure sensors is the nonlinearity in the pressure response that usually arises from the slight nonlinear behavior in the diaphragm mechanical and the silicon piezoresistive characteristics. The nonlinearity in the pressure response can be compensated by using analog circuit components. However, for the most accurate silicon pressure sensors, digital compensation using a microprocessor with correction coefficients stored in memory is often employed to compensate for all the predictable temperature and nonlinear characteristics. The best silicon pressure sensors today can achieve an accuracy of 0.08% of full scale and a long-term stability of 0.1% of full scale per year. Typical compensated temperature range is from −40°C to 85°C, with the errors of compensation on span and offset both around 1% of full scale. Commercial products are currently available for full-scale pressure ranges from 10 kPa to 70 MPa (1.5 psi to 10,000 psi). The 1998 prices are U.S.$5 to $20 for the most basic uncompensated sensors; $10 to $50 for the compensated (with additional laser trimmed resistors either integrated on-chip or on a ceramic substrate)

TABLE 26.2 Selected Companies That Make Pressure Sensors and Pressure Calibration Systems (This is not intended to be an exhaustive list of all manufacturers.)

(1) Silicon micromachined piezoresistive pressure sensor	(2) Bonded strain gage pressure sensors
Druck Inc. 4 Dunham Drive New Fairfield, CT 06812 Tel: (203) 746-0400 http://www.druck.com	Gefran Inc. 122 Terry Dr. Newtown, PA 18940 Tel: (215) 968-6238 http://www.gefran.it
EG&G IC Sensors 1701 McCarthy Blvd. Milpitas, CA 95035-7416 Tel: (408) 432-1800	(3) Capacitive pressure sensors Kavlico Corp. 14501 Los Angeles Ave. Moorpark, CA 93021 Tel: (805) 523-2000
Foxboro ICT 199 River Oaks Pkwy. San Jose, CA 95134-1996 Tel: (408) 432-1010	Rosemount Inc. Measurement Div. 12001 Technology Drive Eden Prairie, MN 55344 Tel: (800) 999-9307 http://www.rosemount.com
Honeywell Inc. Micro Switch Div. 11 W. Spring St. Freeport, IL 61032-4353 Tel: (815) 235-5500 http://www.honeywell.com/sensing	(4) Pressure calibration systems Mensor Corp. 2230 IH-35 South San Marcos, TX 78666-5917 Tel: (512) 396-4200 http://www.mensor.com
Lucas NovaSensor 1055 Mission Ct. Fremont, CA 94539 Tel: (800) 962-7364 http://www.novasensor.com	Ruska Instrument 10311 Westpark Drive Houston, TX 77042 Tel: (713) 975-0547 http://www.ruska.com
Motorola, Inc. Sensor Products Div. 5005 E. McDowell Rd. Phoenix, AZ 85008 Tel: (602) 244-3381 http://mot-sps.com/senseon	
SenSym, Inc. 1804 McCarthy Blvd. Milpitas, CA 95035 Tel: (408) 954-1100 http://www.sensym.com	

or signal-conditioned (compensated with amplified output) sensors; and $60 to $300 for sensors with isolating diaphragms in stainless steel housings. Table 26.2 provides contact information for selected companies making pressure sensors.

References

1. H. N. Norton, *Handbook of Transducers*, Englewood Cliffs, NJ: Prentice-Hall, 1989, 294-330.
2. W. H. Ko, Solid-state capacitive pressure transducers, *Sensors and Actuators*, 10, 303-320, 1986.
3. C. S. Smith, Piezoresistance effect in germanium and silicon, *Phys. Rev.*, 94, 42-49, 1954.
4. O. N. Tufte and E. L. Stelzer, Piezoresistive properties of silicon diffused layers, *J. Appl. Phys.*, 34, 313-318, 1963.

5. D. Schubert, W. Jenschke, T. Uhlig, and F. M. Schmidt, Piezoresistive properties of polycrystalline and crystalline silicon films, *Sensors and Actuators*, 11, 145-155, 1987.
6. K. E. Petersen, Silicon as a mechanical material, *IEEE Proc.*, 70, 420-457, 1982.
7. R. F. Wolffenbuttel (ed.), *Silicon Sensors and Circuits: On-Chip Compatibility*, London: Chapman & Hall, 1996, 171-210.
8. H. Seidel, The mechanism of anisotropic silicon etching and its relevance for micromachining, Tech. Dig., *Transducers '87*, Tokyo, Japan, June 1987, 120-125.
9. E. P. Shankland, Piezoresistive silicon pressure sensors, *Sensors*, 22-26, Aug. 1991.

Further Information

R. S. Muller, R. T. Howe, S. D. Senturia, R. L. Smith, and R. M. White (eds.), *Microsensors*, New York: IEEE Press, 1991, provides an excellent collection of papers on silicon microsensors and silicon micromachining technologies.

R. F. Wolffenbuttel (ed.), *Silicon Sensors and Circuits: On-Chip Compatibility*, London: Chapman & Hall, 1996, provides a thorough discussion on sensor and circuit integration.

ISA Directory of Instrumentation On-Line (http://www.isa.org) from the Instrument Society of America maintains a list of product categories and active links to many sensor manufacturers.

26.2 Vacuum Measurement

Ron Goehner, Emil Drubetsky, Howard M. Brady, and William H. Bayles, Jr.

Background and History of Vacuum Gages

To make measurements in the vacuum region, one must possess a knowledge of the expected pressure range required by the processes taking place in the vacuum chamber as well as the accuracy and/or repeatability of the measurement required for the process. Typical vacuum systems require that many orders of magnitude of pressures must be measured. In many applications, the pressure range may be 8 orders of magnitude, or from atmospheric (1.01×10^5 Pa, 760 torr) to 1×10^{-3} Pa (7.5×10^{-6} torr).

For semiconductor lithography, high-energy physics experiments and surface chemistry, ultimate vacuum of 7.5×10^{-9} torr and much lower are required (a range of 11 orders of magnitude below atmospheric pressure). One gage will not give reasonable measurements over such large pressure ranges. Over the past 50 years, vacuum measuring instruments (commonly called gages) have been developed that used transducers (or sensors) which can be classified as either direct reading (usually mechanical) or indirect reading [1] (usually electronic). Figure 26.8 shows vacuum gages typically in current use. When a force on a surface is used to measure pressure, the gages are mechanical and are called *direct reading* gages, whereas when any property of the gas that changes with density is measured by electronic means, they are called *indirect reading* gages. Figure 26.9 shows the range of operating pressure for various types of vacuum gages.

Direct Reading Gages

A subdivision of direct reading gages can be made by dividing them into those that utilize a liquid wall and those that utilize a solid wall. The force exerted on a surface from the pressure of thermally agitated molecules and atoms is used to measure the pressure.

Liquid Wall Gages

The two common gages that use a liquid wall are the manometer and the McLeod gage. The liquid column *manometer* is the simplest type of vacuum gage. It consists of a straight or U-shaped glass tube

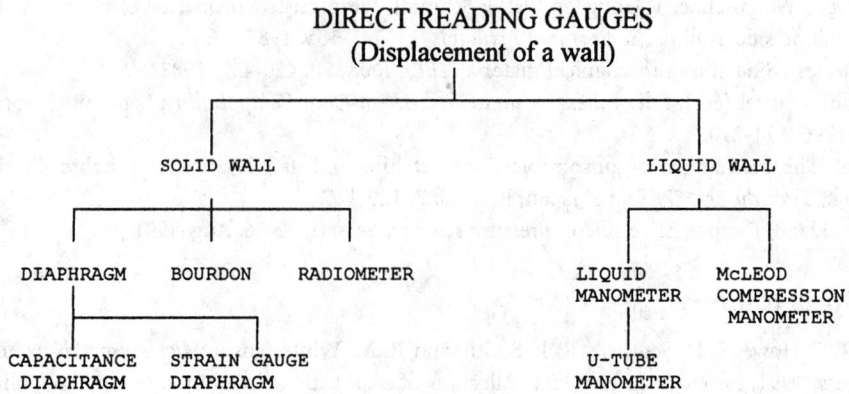

DIRECT READING GAUGES
(Displacement of a wall)

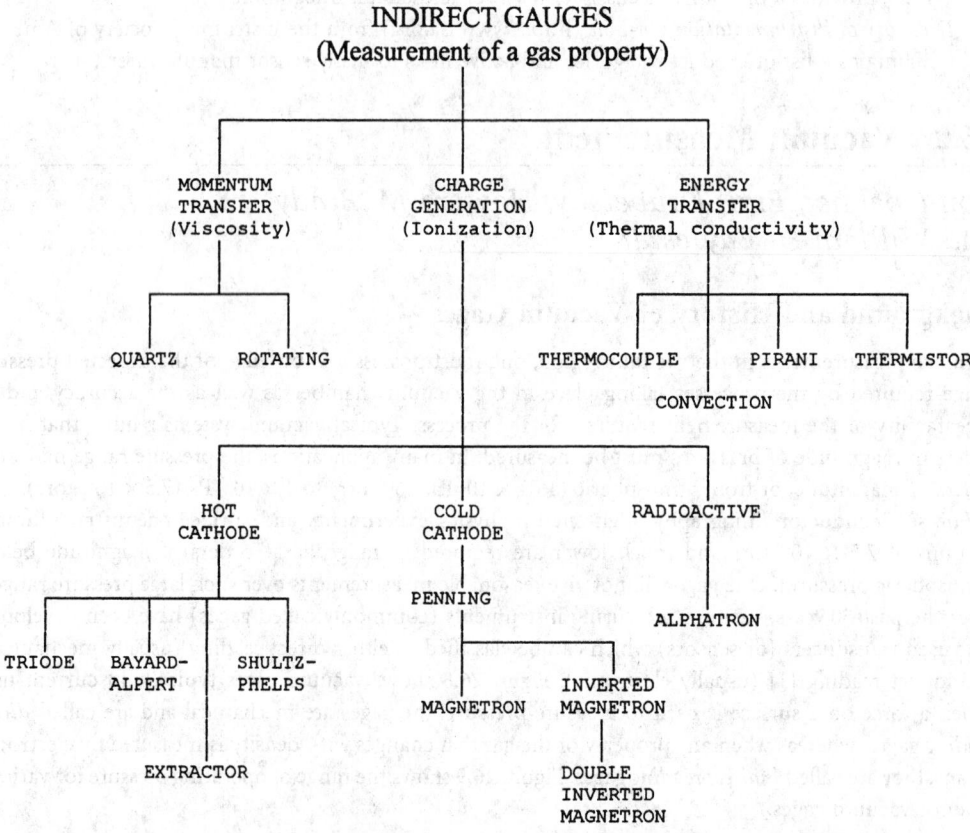

INDIRECT GAUGES
(Measurement of a gas property)

FIGURE 26.8 Classification of pressure gages. (From D.M. Hoffman, B. Singh, and J.H. Thomas, III (eds.), *The Handbook of Vacuum Science and Technology,* Orlando, FL: Academic Press, 1998. With permission.)

evacuated and sealed at one end and filled partly with mercury or a low vapor pressure liquid such as diffusion pump oil (See Figure 26.10). In the straight tube manometer, as the space above the mercury is evacuated, the length of the mercury column decreases. In the case of the U-tube, as the free end is evacuated, the two columns approach equal height. The pressure at the open end is measured by the difference in height of the liquid columns. If the liquid is mercury, the pressure is directly measured in mm of Hg (torr). The manometer is limited to pressures equal to or greater than ~1 torr (133 Pa). If

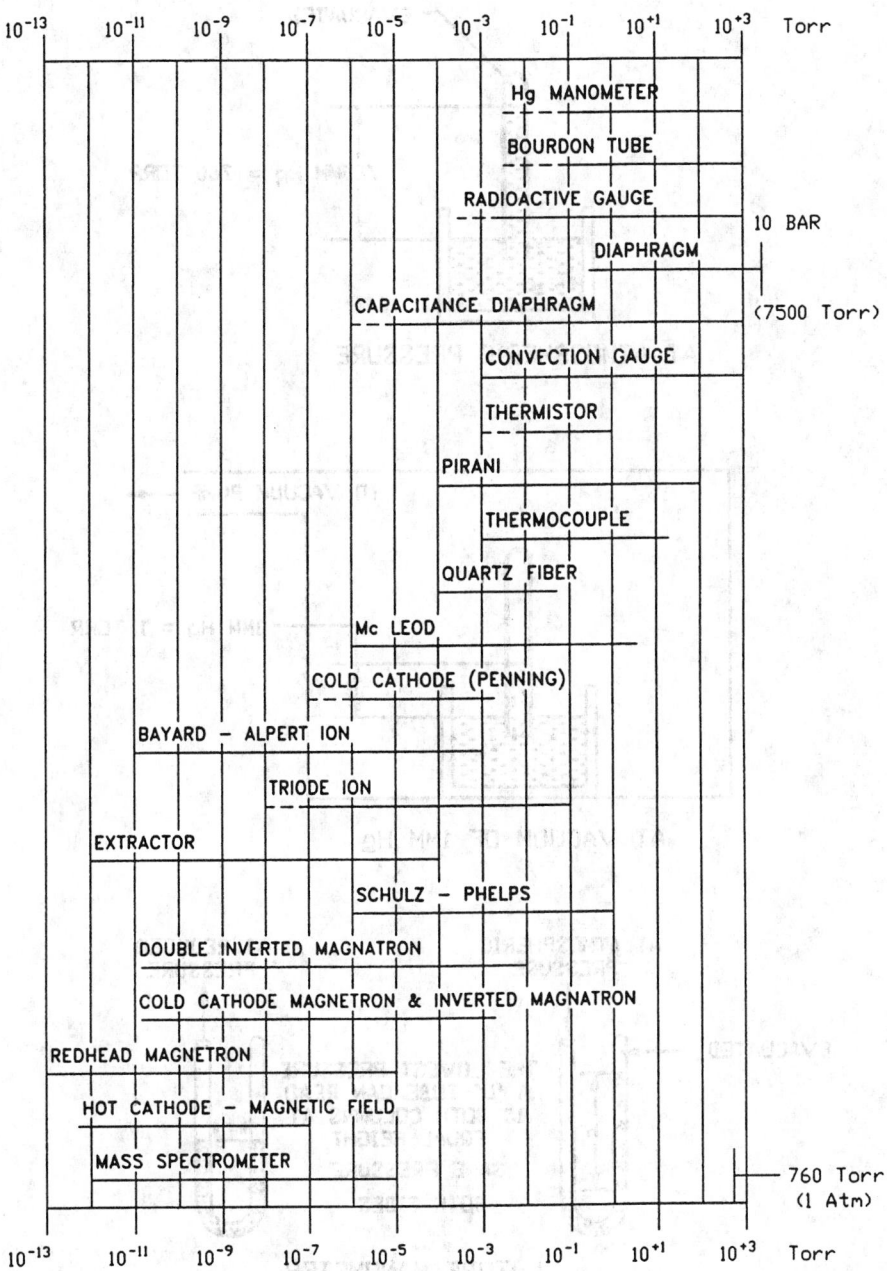

FIGURE 26.9 Pressure ranges for various gages. (From D.M. Hoffman, B. Singh, and J.H. Thomas, III (eds.), *The Handbook of Vacuum Science and Technology,* Orlando, FL: Academic Press, 1998. With permission.)

the liquid is a low density oil, the U-tube is capable of measuring a pressure as low as ~0.1 torr. This is an absolute, direct reading gage but the use of mercury or low density oils that will in time contaminate the vacuum system preclude its use as a permanent vacuum gage.

Due to the pressure limitation of the manometer, the McLeod gage [2, 3] was developed to significantly extend the range of vacuum measurement (see Figure 26.11). This device is essentially a mercury manometer in which a volume of gas is compressed before measurement. This can be used as a primary standard device when a liquid nitrogen trap is used on the vacuum system. Figure 26.11 shows gas at 10^{-6} torr and

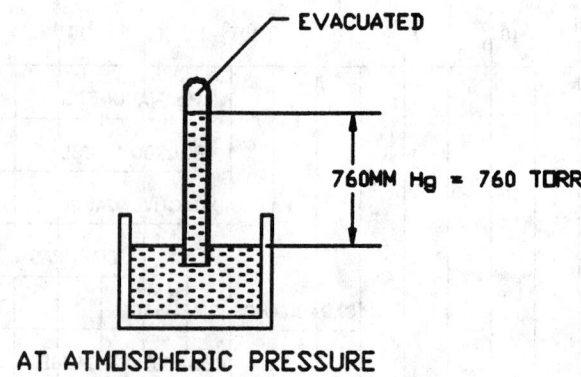

AT ATMOSPHERIC PRESSURE

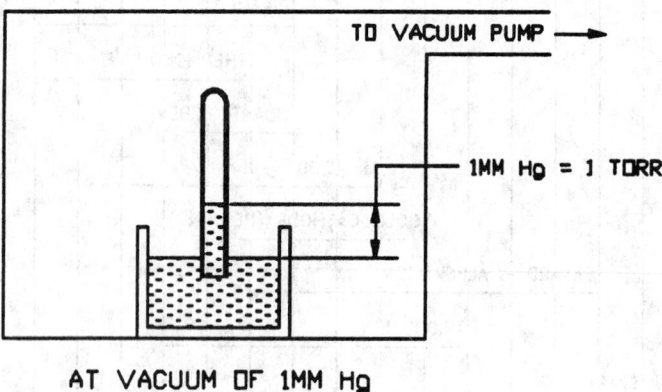

AT VACUUM OF 1MM Hg

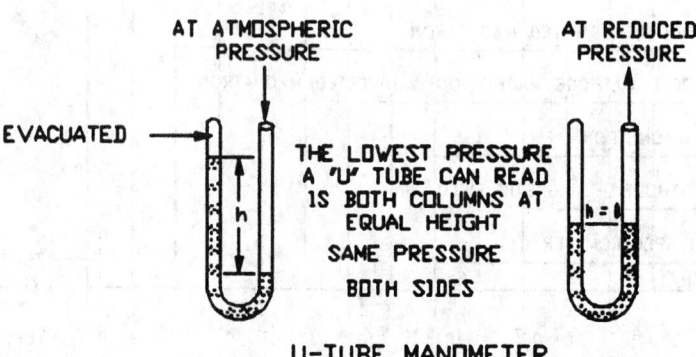

U-TUBE MANOMETER

FIGURE 26.10 Mercury manometers. (From W.H. Bayles, Jr., Fundamentals of Vacuum Measurement, Calibration and Certification, Industrial Heating, October 1992. With permission.)

a compression ratio of 10^{+7}. In this example, the difference of the columns will be 10 mm. Extreme care must be taken not to break the glass and expose the surroundings to the mercury. The McLeod Gage is an inexpensive standard but should only be used by skilled and careful technicians. The gage will give a false low reading unless precautions are taken to ensure that any condensible vapors present are removed by liquid nitrogen trapping.

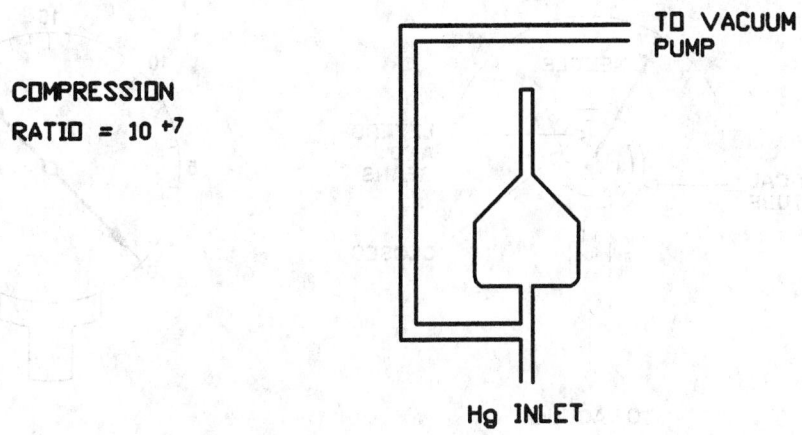

COMPRESSION
RATIO = 10^{+7}

TO VACUUM
PUMP

Hg INLET

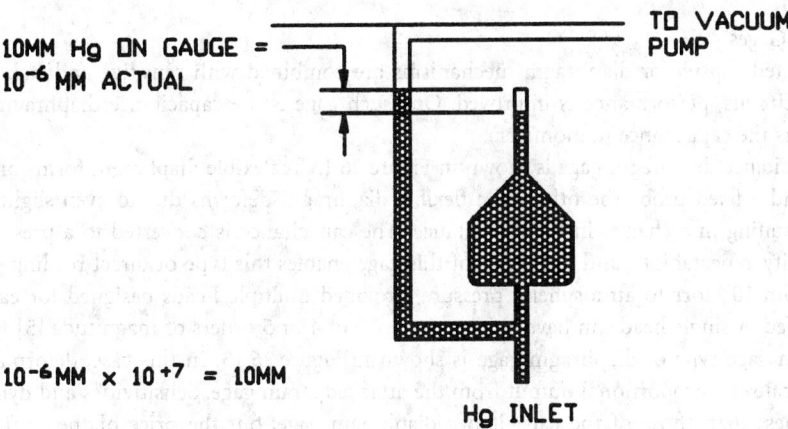

10MM Hg ON GAUGE =
10^{-6} MM ACTUAL

TO VACUUM
PUMP

10^{-6}MM × 10^{+7} = 10MM

Hg INLET

FIGURE 26.11 McLeod gage. (From D.M. Hoffman, B. Singh, and J.H. Thomas, III (eds.), *The Handbook of Vacuum Science and Technology,* Orlando, FL: Academic Press, 1998. With permission.)

Solid Wall Gages

There are two major mechanical solid wall gage types: capsule and diaphragm.

Bourdon Gages.
The capsule-type gages depend on the deformation of the capsule with changing pressure and the resultant deflection of an indicator. Pressure gages using this principle measure pressures above atmospheric to several thousand psi and are commonly used on compressed gas systems. This type of gage is also used at pressures below atmospheric, but the sensitivity is low. The Bourdon gage (Figure 26.12), is used as a moderate vacuum gage. In this case, the capsule is in the form of a thin-walled tube bent in a circle, with the open end attached to the vacuum system with a mechanism and a pointer attached to the other end. The atmospheric pressure deforms the tube; a linear indication of the pressure is given that is independent of the nature of the gas. Certain manufacturers supply capsule gages capable of measuring pressures as low as 1 torr. These gages are rugged, inexpensive, and simple to use and can be made of materials inert to corrosive vapors. Since changing atmospheric pressure causes inaccuracies in the readings, compensated versions of the capsule and Bourdon gage have been developed that improve the accuracy [4].

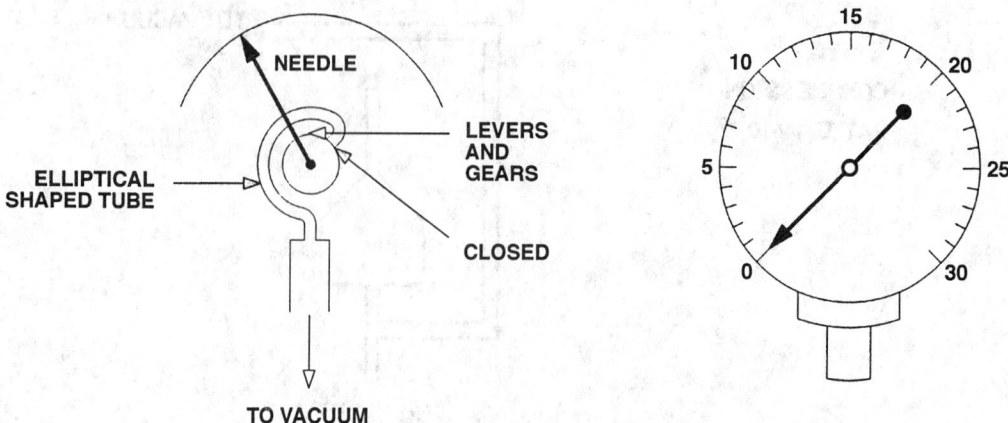

FIGURE 26.12   Bourdon gage. (From Varian Associates, Basic Vacuum Practice, Varian Associates, Inc., Lexington, MA, 1992. With permission.)

Diaphragm Gages.

If compensated capsule or diaphragm mechanisms are combined with sensitive and stable electronic measuring circuits, performance is improved. One such gage is the capacitance diaphragm gage (also referred to as the capacitance manometer).

The capacitance diaphragm gage is shown in Figure 26.13. A flexible diaphragm forms one plate of a capacitor and a fixed probe the other. The flexible diaphragm deforms due to even slight changes in pressure, resulting in a change in the capacitance. The capacitance is converted to a pressure reading. The sensitivity, repeatability, and simplicity of this gage enables this type of direct reading gage to be a standard from 10^{-6} torr to atmospheric pressure, provided multiple heads designed for each pressure range are used. A single head can have a dynamic range of 4 or 5 orders of magnitude [5].

The strain gage type of diaphragm gage is shown in Figure 26.13. In this case, deformation of the diaphragm causes a proportional output from the attached strain gage. Sensitivities and dynamic range tend to be less than those of the capacitance diaphragm gage, but the price of the strain gage type diaphragm gage is usually lower.

Both of these gages are prone to errors caused by small temperature changes due to the inherent high sensitivity of this gage type. Temperature-controlled heads or correction tables built into the electronics have been used to minimize this problem. Other sources of error in all solid wall gages are hysteresis and metal fatigue.

Indirect Reading Gages

Indirect reading gages measure some property of the gas that changes with the density of the gas and usually produces an electric output. Electronic devices amplify and compensate this output to provide a pressure reading.

Thermal Conductivity Gages

Thermal conductivity gages utilize the property of gases in which reduced thermal conductivity corresponds to decreasing density (pressure). The thermal conductivity decreases from a nearly constant value above ~1 torr to essentially 0 at pressures below 10^{-2} torr. The gage controllers are designed to work with a specific sensor tube, and substitutions are limited to those that are truly functionally identical. Heat transfer at various pressures is related to the Knudsen number, as is shown in Figure 26.14 for various heat transfer regimes. The Knudsen number can then be related to pressure through the geometry of the sensor, providing a relationship of heat transfer to pressure for a particular design thermal conductivity gage.

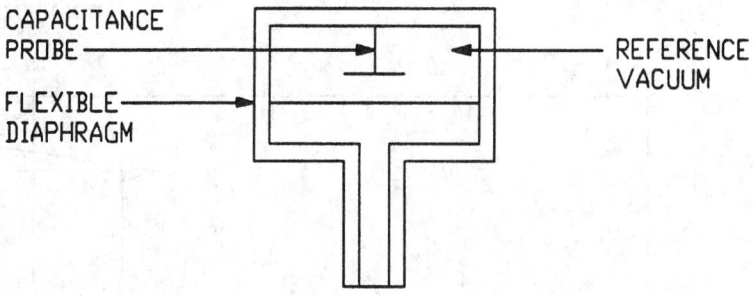

CAPACITANCE DIAPHRAGM GAUGE

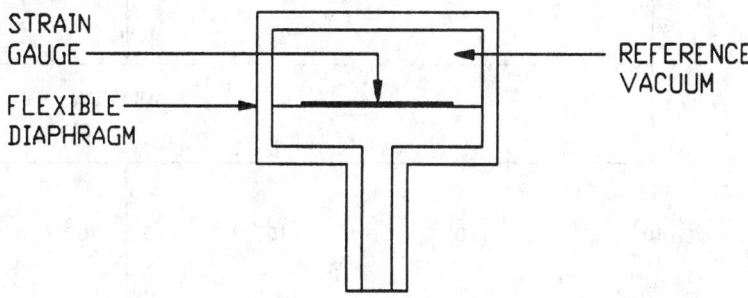

STRAIN GAUGE DIAPHRAGM GAUGE

FIGURE 26.13 Diaphragm gage. (From W.H. Bayles, Jr., Fundamentals of Vacuum Measurement, Calibration and Certification, Industrial Heating, October 1992. With permission.)

Pirani Gages.

The Pirani gage is perhaps the oldest indirect gage that is still used today. In operation, a sensing filament carrying current and producing heat is surrounded by the gas to be measured. As the pressure changes, the thermal conductivity changes, thus varying the temperature of the sensing filament. The temperature change causes a change in the resistance of the sensing filament. The sensing filament is usually one leg of a Wheatstone bridge. The bridge can be operated so that the voltage is varied to keep the bridge balanced; that is, the resistance of the sensing filament is kept constant.

This method is called the constant temperature method and is deemed the fastest, most sensitive, and most accurate. To reduce the effect of changing ambient temperature, an identical filament sealed off at very low pressure is placed in the leg adjacent to the sensing filament as a balancing resistor. Because of its high thermal resistance coefficient, the filament material is usually a thin tungsten wire. It has been demonstrated that a 10 W light bulb works quite well [6]. (see Figure 26.15.)

A properly designed, compensated Pirani gage with sensitive circuitry is capable of measuring to 10^{-4} torr. However, the thermal conductivity of gases varies with the gas being measured, causing a variation in gage response. These variations can be as large as a factor of 5 at low pressures and as high as 10 at high pressures (see Figure 26.16). Correction for these variations can be made on the calibration curves supplied by the manufacturer if the composition of the gas is known. Operation in the presence of high partial pressures of organic molecules such as oils is not recommended.

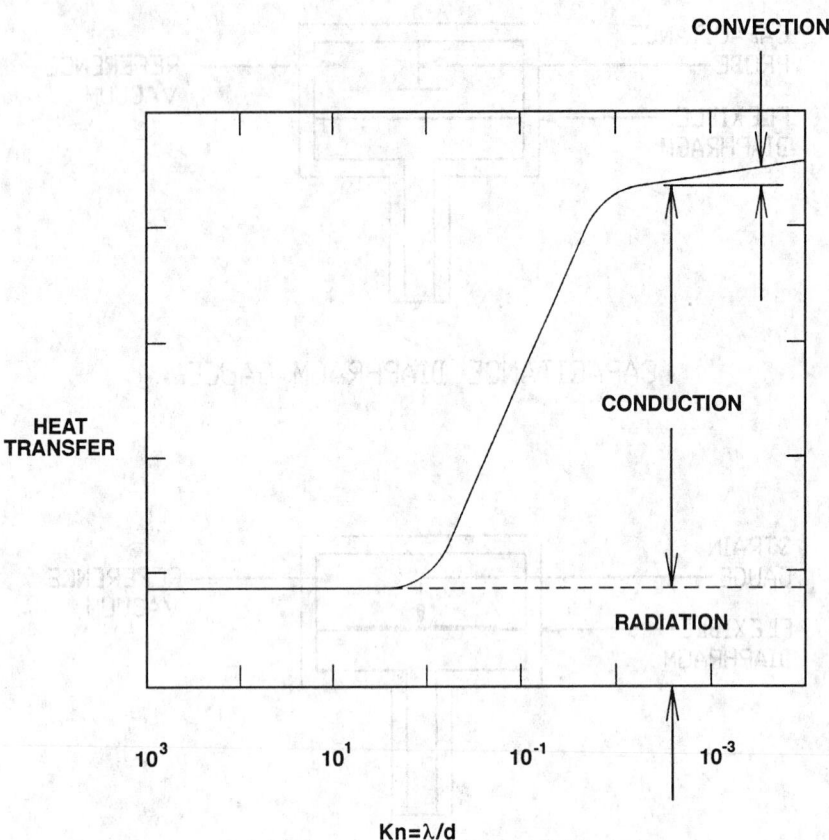

FIGURE 26.14   Heat transfer regimes in a thermal conductivity gage. (From J.F. O'Hanlon, *A User's Guide to Vacuum Technology*, New York: John Wiley & Sons, 1980, 47. With permission.)

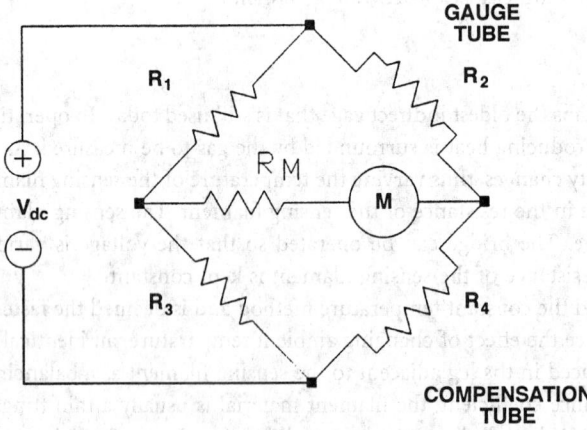

FIGURE 26.15   Pirani gage. (From J.F. O'Hanlon, *A User's Guide to Vacuum Technology*, New York: John Wiley & Sons, 1980, 47. With permission.)

Thermistor Gages.

In the thermistor gage, a thermistor is used as one leg of a bridge circuit. The inverse resistive characteristics of the thermistor element unbalances the bridge as the pressure changes, causing a corresponding change in current. Sensitive electronics measure the current and are calibrated in pressure units. The

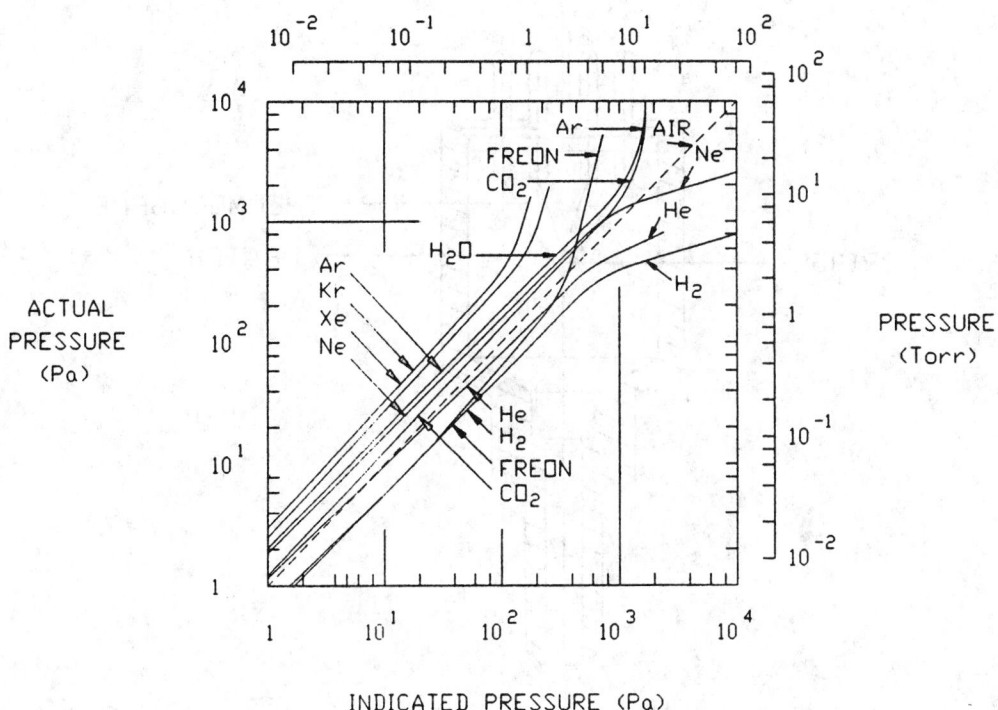

FIGURE 26.16 Calibration curves for the Pirani gage. (Reprinted with permission from Leybold-Herqeus GMblt, Köhn, Germany.)

thermistor gage measures approximately the same pressure range as the thermocouple. The exact calibration depends on the the gas measured. In a well-designed bridge circuit, the plot of current vs. pressure is practically linear in the range 10^{-3} to 1 torr [7]. Modern thermistor gages use constant-temperature techniques.

Thermocouple Gages.
Another example of an indirect reading thermal conductivity gage is the thermocouple gage. This is a relatively inexpensive device with proven reliability and a wide range of applications. In the thermocouple gage, a filament of resistance alloy is heated by the passage of a constant current (see Figure 26.17). A thermocouple is welded to the midpoint of the filament or preferably to a conduction bridge at the center of the heated filament. This provides a means of directly measuring the temperature. With a constant current through the filament, the temperature increases as the pressure decreases as there are fewer molecules surrounding the filament to carry the heat away. The thermocouple output voltage increases as a result of the increased temperature and varies inversely with the pressure. The thermocouple gage can also be operated in the constant-temperature mode.

   Gas composition effects apply to all thermal conductivity gages. The calibration curves for a typical thermocouple gage are shown in Figure 26.18. The thermocouple gage can be optimized for operation in various pressure ranges. Operation of the thermocouple gage in high partial pressures of organic molecules such as oils should be avoided. One manufacturer pre-oxidizes the thermocouple sensor for stability in "dirty" environments and for greater interchangeability in clean environments.

Convection Gages.
Below 1 torr a significant change in thermal conductivity occurs as the pressure changes. Thus, the thermal conductivity gage is normally limited to 1 torr.

   At pressures above 1 torr, there is, in most gages, a small contribution to heat transfer caused by convection. Manufacturers have developed gages that utilize this convection effect to extend the usable

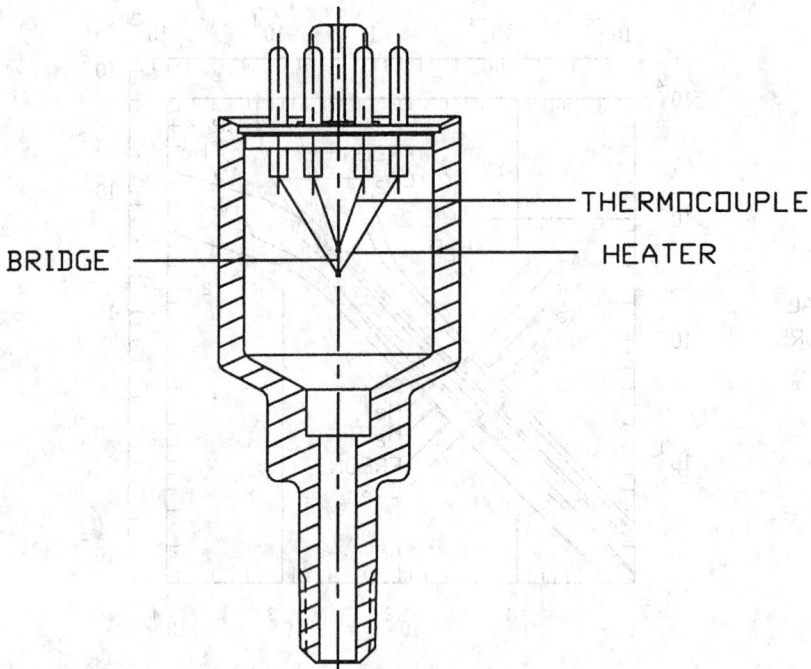

FIGURE 26.17   Thermocouple gage. (Reprinted with permission of Televac Division, The Fredericks Co., Hunting-don Valley, PA.)

range to atmospheric pressure and slightly above [8–12]. Orientation of a convection gage is critical because this convection heat transfer is highly dependent on the orientation of the elements within the gage.

The Convectron™ uses the basic structure of the Pirani with special features to enhance convection cooling in the high-pressure region [13]. To utilize the gage above 1 torr (133 Pa), the sensor tube must be mounted with its major axis in a horizontal position. If the only area of interest is below 1 torr, the tube can be mounted in any position. As mentioned above, the gage controller is designed to be used with a specific model sensor tube; because extensive use is made of calibration curves and look-up tables stored in the controller, no substitution is recommended.

The Televac convection gage uses the basic structure of the thermocouple gage except that two thermocouples are used [14]. As in any thermocouple gage, the convection gage measures the pressure by determining the heat loss from a fine wire maintained at constant temperature. The response of the sensor depends on the gas type. A pair of thermocouples is mounted a fixed distance from each other (see Figure 26.19). The one mounted lower is heated to a constant temperature by a variable current power supply. Power is pulsed to this lower thermocouple and the temperature is measured between heating pulses. The second (upper) thermocouple measures convection effects and also compensates for ambient temperature. At pressures below ~2 torr (270 Pa) the temperature in the upper thermocouple is negligible. The gage tube operates as a typical thermocouple in the constant-temperature mode. Above 2 torr, convective heat transfer causes heating of the upper thermocouple. The voltage output is subtracted from that of the lower thermocouple, thus requiring more current to maintain the wire temperature. Consequently, the range of pressure that can be measured (via current change) is extended to atmospheric pressure (see Figure 26.20). Orientation of the sensor is with the axis vertical.

The use of convection gages with process control electronics allows for automatic pump-down with the assurance that the system will neither open under vacuum nor be subject to over-pressure during backfill to atmospheric pressure. These gages, with their controllers, are relatively inexpensive. In oil-free systems, they afford long life and reproducible results.

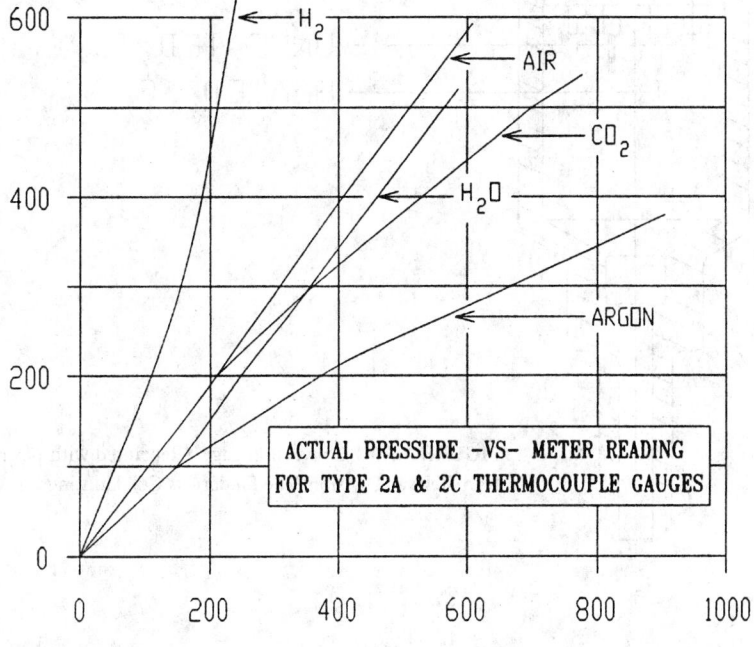

FIGURE 26.18   Calibration curves for the thermocouple gage. (Reprinted with permission from Televac Division, The Fredericks Co., Huntingdon Valley, PA.)

Hot Cathode Ionization Gages

Hot cathode ionization gage designs consist of triode gages, Bayard-Alpert gages, and others.

Triode Hot Cathode Ionization Gages.

For over 80 years, the triode electron tube has been used as an indirect way to measure vacuum [15, 16]. A typical triode connection is as an amplifier, as is shown in Figure 26.21. A brief description of its operation is given here, but more rigorous treatment of triode performance is given in [17–20]. However, if the triode is connected as in Figure 26.22 so that the grid is positive and the plate is negative with respect to the filament, then the ion current collected by the plate for the same electron current to the grid is greatly increased [21].

Today, the triode gage is used in this higher sensitivity mode. Many investigators have shown that a linear change in molecular density (pressure) results in a linear change in ion current [15, 21, 22]. This linearity allows a sensitivity factor S to be defined such that:

$$I_i = S \times I_e \times P \qquad (26.3)$$

where I_i = Ion current (A)

$\quad I_e$ = Electron current (A)

$\quad P$ = Pressure

$\quad S$ = Sensitivity (in units of reciprocal pressure)

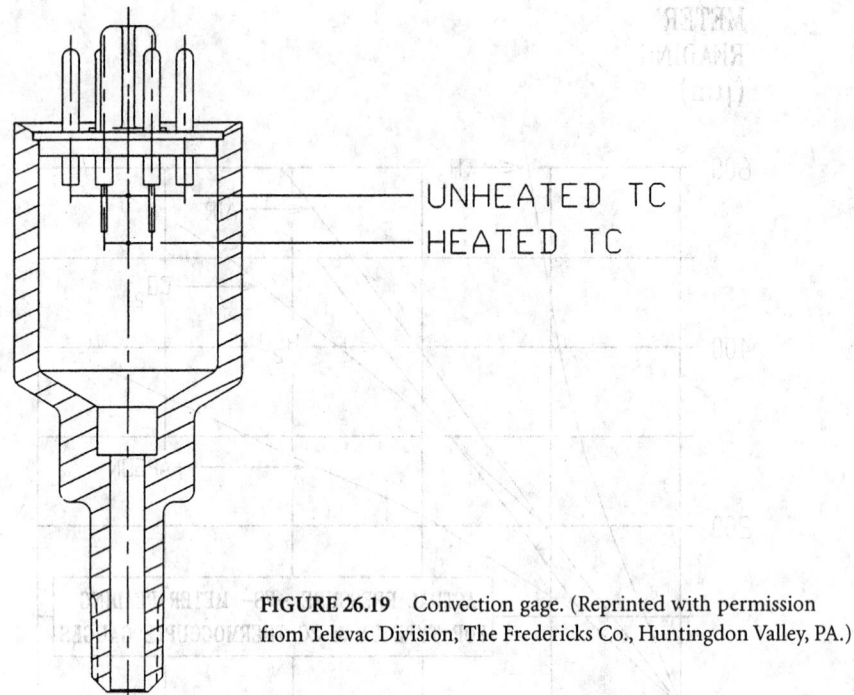

FIGURE 26.19   Convection gage. (Reprinted with permission from Televac Division, The Fredericks Co., Huntingdon Valley, PA.)

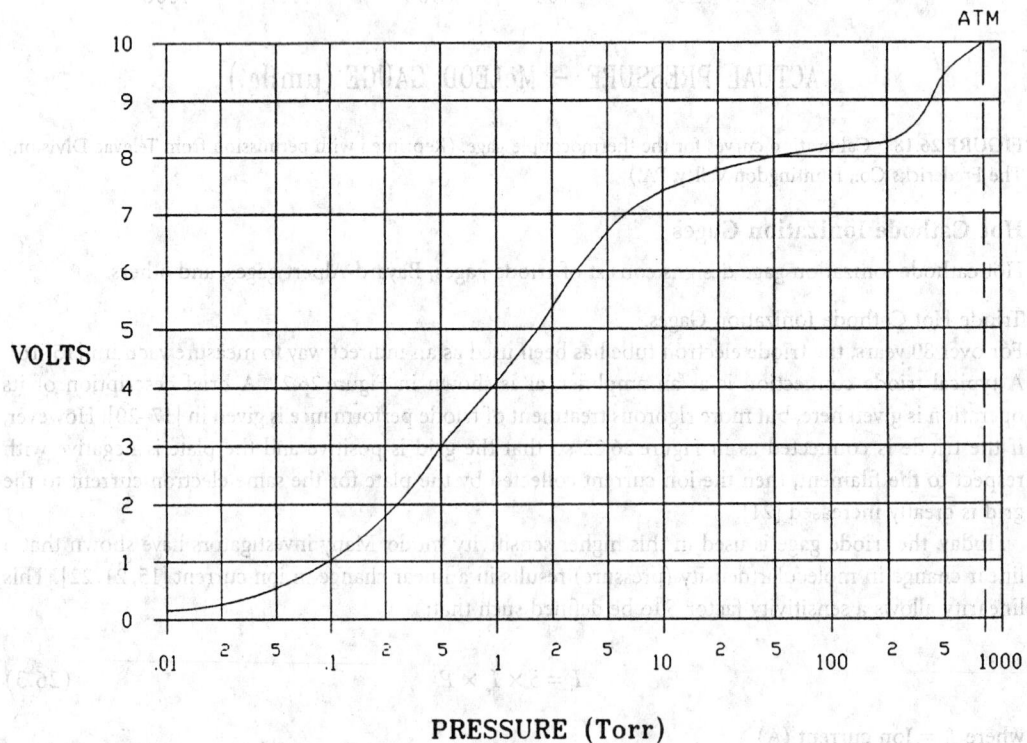

FIGURE 26.20   Output curve for the convection gage. (Reprinted with permission from Televac Division, The Fredericks Co., Huntingdon Valley, PA.)

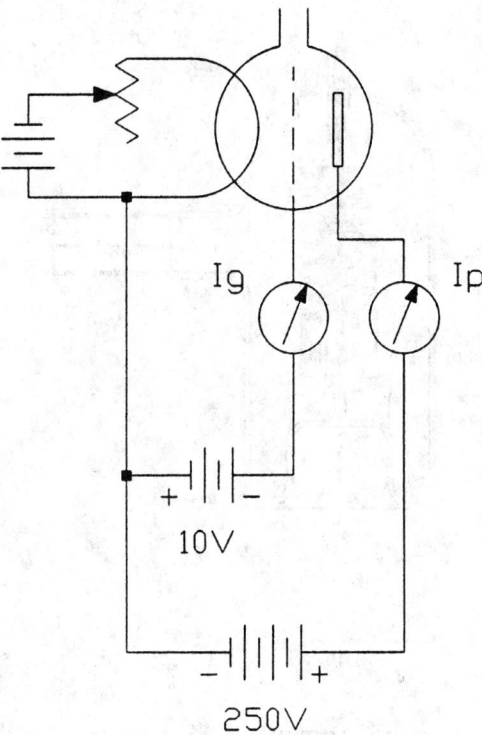

FIGURE 26.21 Typical triode connection.

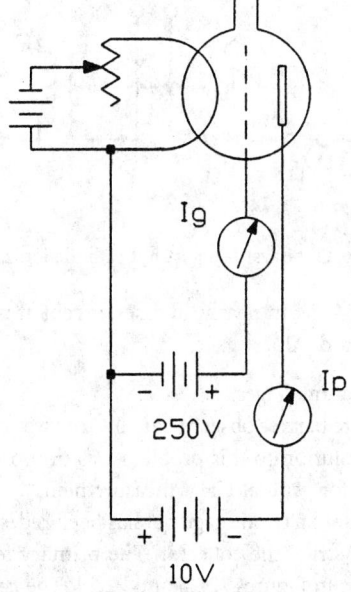

FIGURE 26.22 Alternative triode connection.

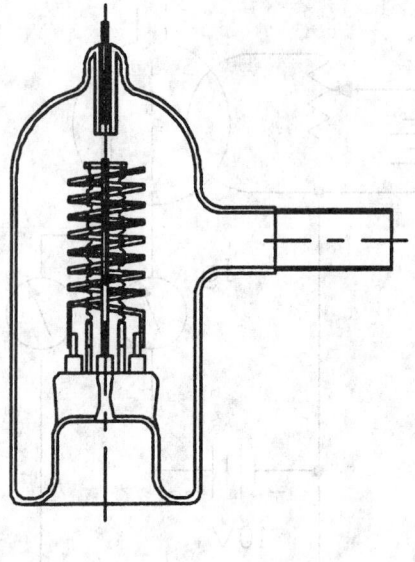

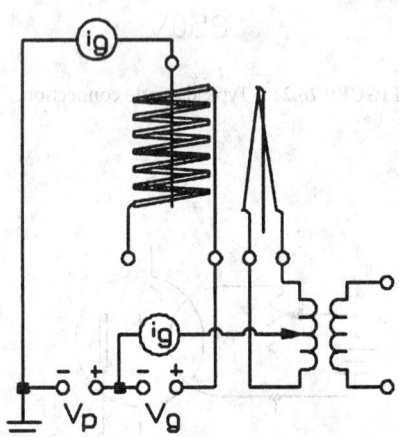

FIGURE 26.23   Bayard–Alpert hot cathode ionization gage.

Additional details are found in [23–25]. In nearly all cases, except at relatively high pressures, the triode gage has been replaced by the Bayard–Alpert gage.

Bayard–Alpert Hot Cathode Ionization Gages.
It became apparent that the pressure barrier observed at 10^{-8} torr was caused by a failure in measurement rather than pumping [26, 27]. A solution to this problem was proposed by Bayard and Alpert [28] that is now the most widely used gage for general UHV measurement.

The Bayard–Alpert gage is similar to a triode gage but has been redesigned so that only a small quantity of the internally generated X-rays strike the collector. The primary features of the Bayard–Alpert gage and its associated circuit are shown in Figures 22.23 and 22.24. The cathode has been replaced by a thin collector located at the center of the grid, and the cathode filament is now outside and several millimeters away from the grid. The Bayard–Alpert design utilizes the same controller as the triode gage, with

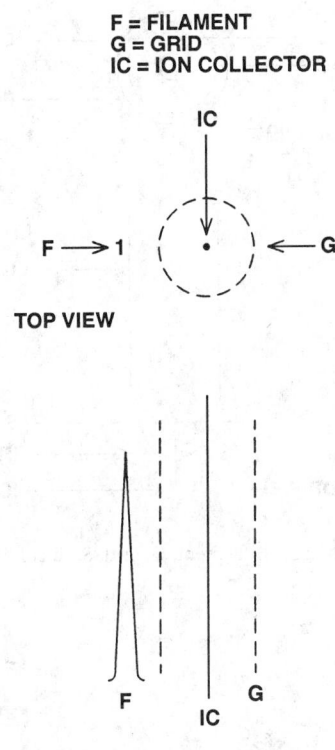

FIGURE 26.24 Bayard–Alpert gage configuration.

corrections for sensitivity differences between the gage designs. When a hot filament gage is exposed to high pressures, burn-out of the tungsten filaments often occurs. To prevent this, platinum metals were coated with refractory oxides to allow the gage to withstand sudden exposure to atmosphere with the filament hot [29, 30]. Typical materials include either thoria or yttria coatings on iridium. Bayard–Alpert and triode gages of identical structure and dimensions but with different filaments (i.e., tungsten vs. thoria iridium)were observed to have different sensitivities, with the tungsten filament versions being 20% to 40% more sensitive than the iridium of the same construction.

The lowest pressure that can be measured is limited by low energy X-rays striking the ion collector and emitting electrons. Several methods to reduce this X-ray limit were developed. Gage designs with very small diameter collectors have been made that extend the high vacuum range down to 10^{-12} torr, but accuracy was lost at the high pressures [31, 32].

The modulated gage was designed by Redhead [33] with an extra electrode near the ion collector. In this configuration, the X-ray current could be subtracted by measuring the ion current at two modulator potentials, thus increasing the range to 5×10^{-12}. Other gages use suppressor electrodes in front of the ion collector [34, 35].

The extractor gage (Figure 26.25) is the most widely used UHV hot cathode gage for those who need to measure 10^{-12} torr [36]. In this gage, the ions are extracted out of the ionizing volume and deflected or focused onto a small collector. More recent designs have been developed [37, 38]. The use of a channel electron multiplier [39] has reduced the low pressure limit to 10^{-15} torr.

The Bayard–Alpert gage suffers from some problems, however. The ion current is geometry dependent. Investigators have reported on the sensitivity variations, inaccuracy, and instability of Bayard-Alpert gages with widely differing results [40–46, 49, 50, 56]. Investigators have developed ways to reduce or eliminate some of these problems [47, 48].

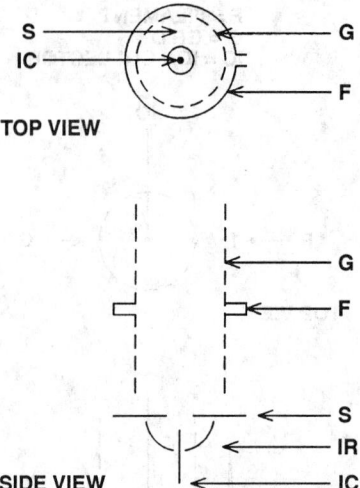

FIGURE 26.25 Extractor ionization gage. F, filament; G, grid; S, shield; IR, iron reflector; IC, ion collector.

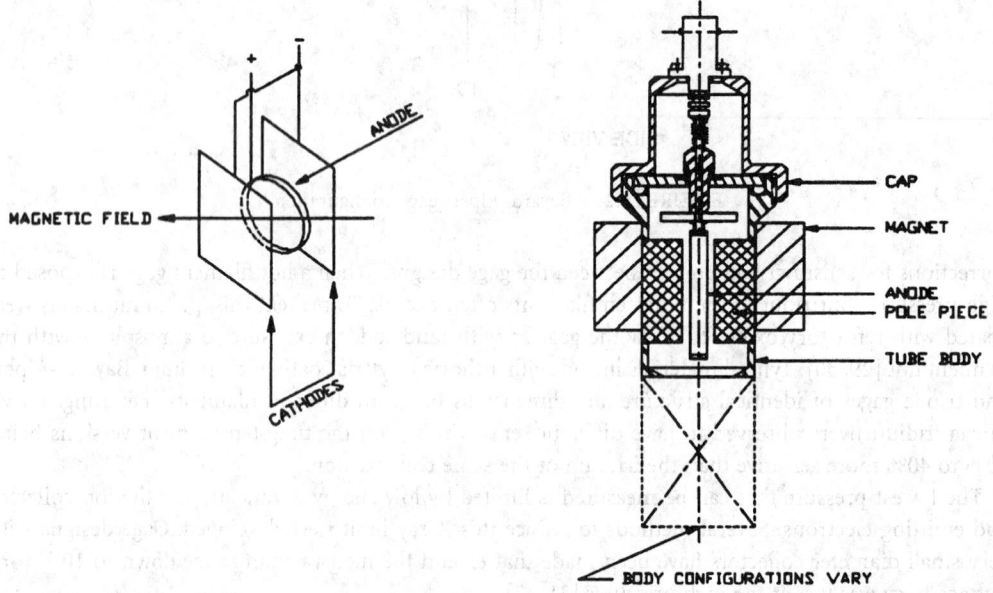

FIGURE 26.26 Penning gages. ([Left] From J.F. O'Hanlon, *A User's Guide to Vacuum Technology*, New York: John Wiley & Sons, 1980, 47. With permission. [Right] Reprinted with permission from Televac Division, The Fredericks Co., Huntingdon Valley, PA.)

Cold Cathode Ionization Gages

To measure pressures below 10^{-3} torr, Penning [51] developed the cold cathode discharge gage. Below 10^{-3} torr, the mean free path is so high that little ionization takes place. The probability of ionization was increased by placing a magnetic field parallel to the paths of ions and electrons to force these particles into helical trajectory.

This gage consists of two parallel cathodes and an anode, which is placed midway between them (see Figure 26.26). The anode is a circular or rectangular loop of metal wire whose plane is parallel to that of the cathodes. A few kilovolts potential difference is maintained between the anode and the cathodes.

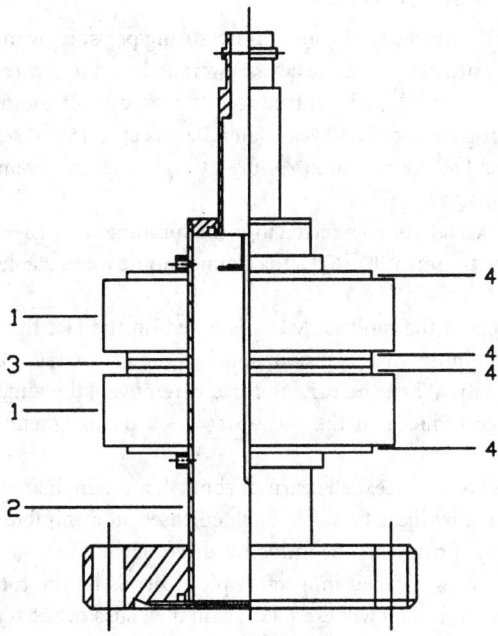

FIGURE 26.27 Double inverted magnetron. (Reprinted with permission from Televac Division, The Fredericks Co., Huntingdon Valley, PA.)

Furthermore, a magnetic field is applied between the cathodes by a permanent magnet usually external to the gage body. Electrons emitted from either of the two cathodes must travel in helical paths due to the magnetic field eventually reaching the anode, which carries a high positive charge. During the travel along this long path, many electrons collide with the molecules of the gas, thus creating positive ions that travel more directly to the cathodes. The ionization current thus produced is read out on a sensitive current meter as pressure.

This is a rugged gage used for industrial applications such as in leak detectors, vacuum furnaces, electron beam welders, and other industrial processes. The Penning gage is rugged, simple, and inexpensive. Its range is typically 10^{-3} torr to 10^{-6} torr, and some instability and lack of accuracy has been observed [52]. The magnetron design [53] and the inverted magnetron design [54] extended the low pressure range to 10^{-12} Torr [55] or better. These improvements produced a better gage but instability, hysterisis, and starting problems remain [57, 58]. Magnetrons currently in use are simpler and do not use an auxiliary cathode.

More recently, a double inverted magnetron was introduced [59]. This gage has greater sensitivity (amp/torr) than the other types (see Figure 26.27). It has been operated successfully at $\sim 1 \times 10^{-11}$ torr. The gage consists of two axially magnetized, annular-shaped magnets (**1**) placed around a cylinder (**2**) so that the north pole of one magnet faces the north pole of the other one. A nonmagnetic spacer (**3**) is placed between the two magnets and thin shims (**4**) are used to focus the magnetic fields. This gage has been operated to date at 10^{-11} torr, and stays ignited and reignites quickly when power is restored at this pressure. Essentially instantaneous reignition has been demonstrated by use of radioactive triggering [60].

Resonance Gages

One example of a resonance-type vacuum gage is the quartz friction vacuum gage [61]. A quartz oscillator can be built to measure pressure by a shift in resonance frequency caused by static pressure of the surrounding gas or by the increased power required to maintain a constant amplitude. Its range is from near atmospheric pressure to about 0.1 torr. A second method is to measure the resonant electrical impedance of a tuning fork oscillator. Test results for this device show an accuracy within ±10% for pressure from 10^{-3} torr to 10^{3} torr. There is little commercial use to date for these devices.

Molecular Drag (Spinning Rotor) Gages

Meyer [62] and Maxwell [63] introduced the idea of measuring pressure by means of the molecular drag of rotating devices in 1875. The rotors of these devices were tethered to a wire or thin filament. The gage was further enhanced by Holmes [64], who introduced the concept of the magnetic rotor suspension, leading to the spinning rotor gage. Nearly 10 years later, Beams et al. [65] disclosed the use of a magnetically levitated, rotating steel ball to measure pressure at high vacuum. Fremerey [66] reported on the historical development of this gage.

The molecular drag gage (MDG), often referred to as the spinning rotor gage, received wider acceptance after its commercial introduction in 1982 [67]. It is claimed to be more stable than other gages at lower pressures [68].

The principle of operation of the modern MDG is based on the fact that the rate of change of the angular velocity of a freely spinning ball is proportional to the gas pressure and inversely proportional to the mean molecular velocity. When the driving force is removed, the angular velocity is determined by measuring the ac voltage induced in the pickup coils by the magnetic moment of the ball (see Figure 26.28).

In current practice, a small rotor (steel ball bearing) about 4.5 mm in diameter is magnetically levitated and spun up to about 400 Hz by induction. The ball, enclosed in a thimble connected to the vacuum system, is allowed to coast by turning off the inductive drive. Then, the time of a revolution of the ball is measured by timing the signal induced in a set of pickup coils by the rotational component of the ball's magnetic moment. Gas molecules will exert a drag on the ball, slowing it at a rate set by the pressure P, its molecular mass m, temperature T, and the coefficient of momentum transfer σ, between the gas and the ball. A perfectly smooth ball would have a value of unity. There is also a pressure-independent residual drag (RD) caused by eddy current losses in the ball and surrounding structure. There will also be temperature effects that will cause the ball diameter and moment of inertia to change.

The pressure in the region of molecular flow is given by:

$$P = \frac{\pi \rho\, a\, \bar{c}}{10\, \sigma_{\text{eff}}} \left(\frac{-\omega' - \text{RD} - 2\,\alpha\, T'}{\omega} \right) \tag{26.4}$$

Note: Some sources include the term $\frac{(8kT')}{(\pi m)}$

where ρ = Density of the rotor
   a = Radius of the rotor
   ω'/ω = Fractional rate of slowing of the rotor
   $\bar{c}$ = Mean gas molecular velocity
   α = Linear coefficient of expansion of the ball
   T' = Rate of change of the ball's temperature

All of the terms in the first part of the equation can be readily determined except for the accommodation coefficient σ, which depends on the surface of the ball and the molecular adhesion between the gas and the surface of the ball. The accommodation coefficient σ must be determined by calibration of the MDG against a known pressure standard or, if repeatability is more important than the highest accuracy, by assuming a value of 1 for σ. Measurements of σ on many balls over several years have been repeatedly performed by Dittman et al. [68]. The values obtained ranged from 0.97 to 1.06 for 68 visually smooth balls, so using a value of 1 for σ would not introduce a large error and would allow the MDG to be considered a primary standard (Fremerey [66]).

The controller [68] contains the electronics to power and regulate the suspension and drive, detect and amplify the signal from the pickup coils, and then time the rotation of the ball. It also contains a data processor that stores the calibration data and computes the pressure.

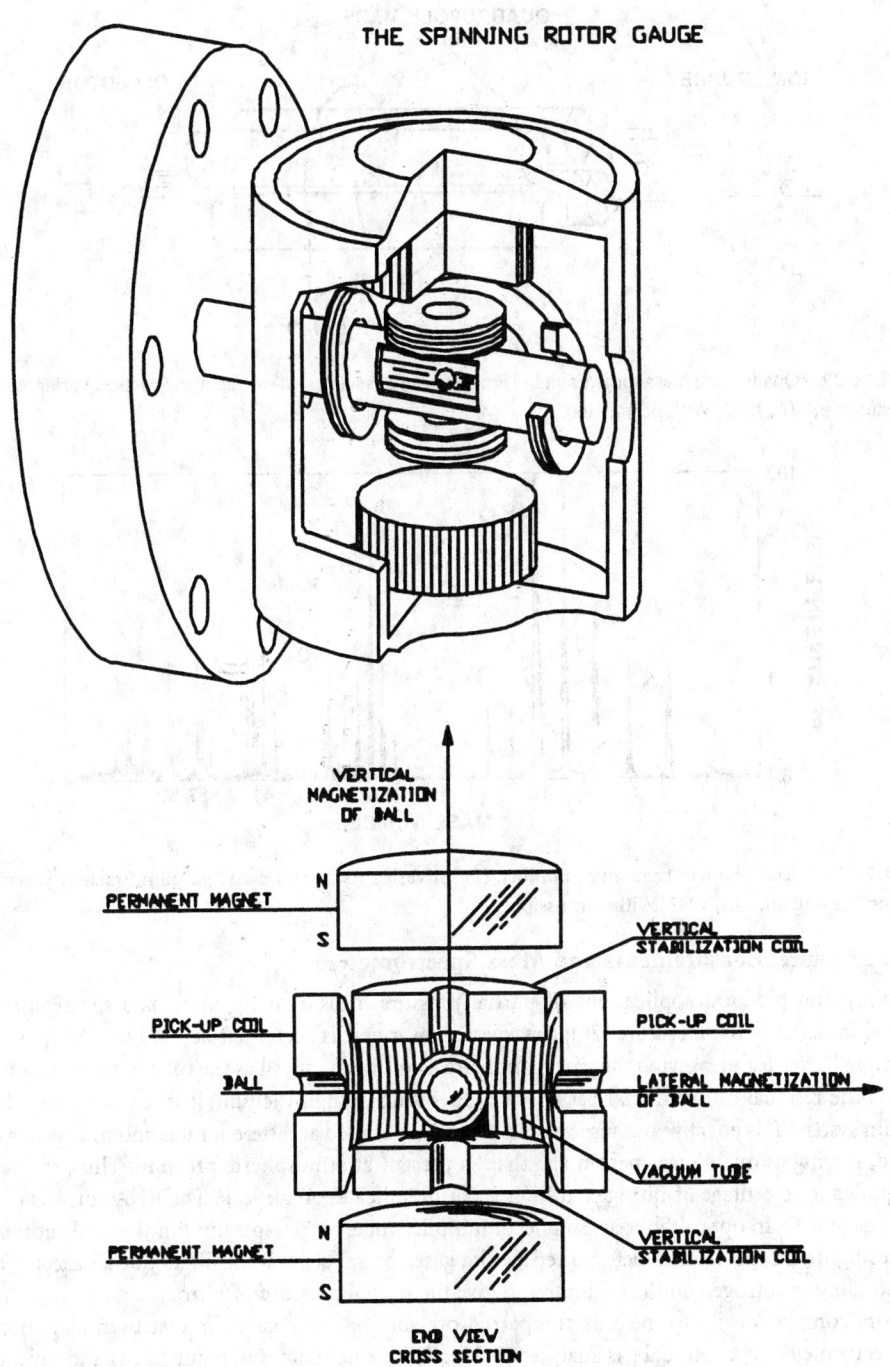

FIGURE 26.28   Molecular drag gage (spinning rotor).

The MDG is perhaps the best available transfer standard for the pressure range of 10^{-2} torr to 10^{-7} torr (1 Pa to 10^{-5} Pa) because it is designed for laboratory use in controlled, relatively vibration-free environments [69, 70].

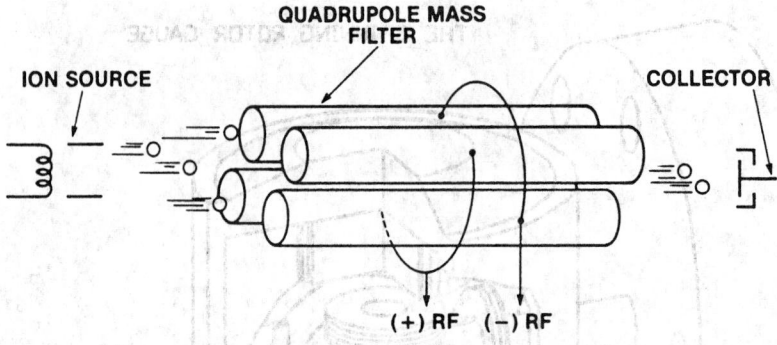

FIGURE 26.29 Quadrupole mass spectrometer. (From Varian Associates, Basic Vacuum Practice, Varian Associates, Inc., Lexington, MA, 1992. With permission.)

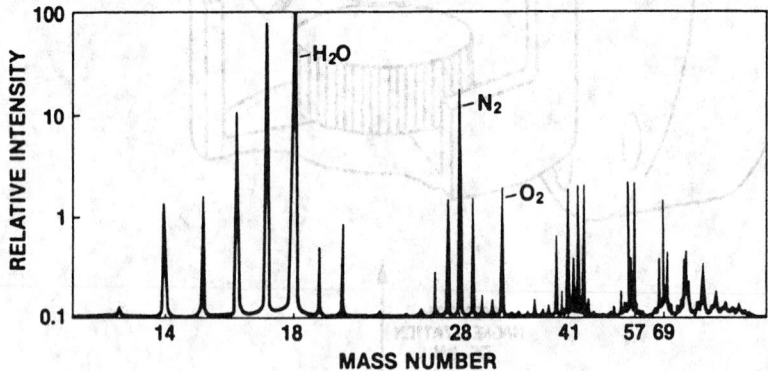

FIGURE 26.30 Relative intensity vs. mass number. (From Varian Associates, Basic Vacuum Practice, Varian Associates, Inc., Lexington, MA, 1992. With permission.)

Partial Pressure Measurements and Mass Spectrometers

The theory and practical applications of partial pressure measurements and mass spectrometers are discussed in detail in the literature [76]; however, an overview is presented herein [78].

A simple device for measuring the partial pressure of nitrogen as well as the total pressure in a vacuum system is the residual nitrogen analyzer (RNA). Operating in high vacuum, it is used to detect leaks in a vacuum system. It is effective because various gases are pumped at different rates and nitrogen is readily pumped, leaving a much lower percentage than is present at atmospheric pressure. Thus, the presense of a significant percentage of nitrogen at high vacuum indicates an air leak. The RNA consists of a cold cathode gage with an optical filter and a photomultiplier tube. Since ionization in the cold cathode tube produces light and the color is determined by the gases present, the RNA filters out all except for that corresponding to nitrogen and is calibrated to give the partial pressure of nitrogen.

A more complex device to measure the partial pressure of many gases in a vacuum chamber is the mass spectrometer-type residual gas analyzer (RGA). This device comes in many forms and several sizes. The quadrupole mass spectrometer is shown in Figure 26.29. The sensing head consists of an ion source, a quadrupole mass filter, and a Faraday cup collector. The quadrupole mass filter consists of two pairs of parallel rods having equal and opposite RF and dc voltages. For each combination of voltages, only ions of a specific mass will pass through the filter. The mass filter is tuned to pass only ions of a specific mass-to-charge ratio at a given time. As the tuning is changed to represent increasing mass numbers, a display such as Figure 26.30 is produced, showing the relative intensity of the signal vs. the mass number. This display can then be compared electronically with similar displays for known gases to determine the composition of the gases in the vacuum chamber. The head can operate only at high vacuum. However, by maintaining the head at high vacuum and using a sampling technique, the partial pressures of gases at higher pressures can be determined.

Calibration of mass spectrometers can be accomplished by equating the integral (the total area under all the peaks, taking into account the scale factors) to the overall pressure as measured by another gage (cold cathode, BA, or MDG). Once calibrated, the mass spectrometer can be used as a sensitive monitor of system pressure. It is also important when monitoring system pressure to know what gases are present. Most gages have vastly different sensitivities to different gas species.

Additional references on the details of the MDG and other types of gages available and on calibration are found in the literature [69–73, 75, 76]. The material in this chapter was summarized from an article on the fundamentals of vacuum measurement, calibration, and certification [77] from the authors' contribution to *The Handbook of Vacuum Technology* [76] and from other referenced sources.

References

1. J.F. O'Hanlon, *A User's Guide to Vacuum Technology,* New York: John Wiley & Sons, 1980, 47.
2. H. McLeod, *Phil. Mag.,* 47, 110, 1874.
3. C. Engleman, Televac Div. The Fredericks Co. 2337 Philmont Ave., Huntingdon Valley, PA, 19006, private communication.
4. Wallace and Tiernan Div., Pennwalt Corp., Bellville, NJ.
5. R.W. Hyland and R.L. Shaffer, Recommended practices of calibration and use of capacitance diaphragm gage for a transfer standard, *J. Vac. Sci. Tech., A,* 9(6), 2843, 1991.
6. K.R. Spangenberg, *Vacuum Tubes,* New York: McGraw-Hill, 1948, 766.
7. S. Dushman, *Scientific Foundations of Vacuum Technique,* 2nd ed., J.M. Lafferty (ed.), New York: John Wiley & Sons, 1962.
8. W. Steckelmacher and B. Fletcher, *J. Physics E.,* 5, 405, 1972.
9. W. Steckelmacher, *Vacuum,* 23, 307, 1973.
10. A. Beiman, *Total Pressure Measurement in Vacuum Technology,* Orlando, FL: Academic Press, 1985.
11. Granville-Phillips, 5675 Arapahoe Ave., Boulder CO, 80303.
12. Televac Div. The Fredericks Co. 2337 Philmont Ave., Huntingdon Valley, PA, 19006.
13. Granville-Phillips Data Sheet 360127, 3/95.
14. Televac U.S. Patent No. 5351551.
15. O.E. Buckley, *Proc. Natl. Acad. Sci.,* 2, 683, 1916.
16. M.D. Sarbey, *Electronics,* 2, 594, 1931.
17. R. Champeix, *Physics and Techniques of Electron Tubes,* Vol. 1, New York: Pergamon Press, 1961, 154-156.
18. N. Morgulis, *Physik Z. Sowjetunion,* 5, 407, 1934.
19. N.B. Reynolds, *Physics,* 1, 182, 1931.
20. J.H. Leck, *Pressure Measurement in Vacuum Systems,* London: Chapman & Hall, 1957, 70-74.
21. S. Dushman and C.G. Found, *Phys. Rev.,* 17, 7, 1921.
22. E.K. Jaycock and H.W. Weinhart, *Rev. Sci. Instr.,* 2, 401, 1931.
23. G.J. Schulz and A.V. Phelps, *Rev. Sci. Instr.,* 28, 1051, 1957.
24. Japanese Industrial Standard (JIS-Z-8570), Method of Calibration for Vacuum Gages,
25. J.W. Leck, op. cit., 69.
26. W.B. Nottingham, *Proc. 7th Annu. Conf. Phys. Electron.,* M.I.T., Cambridge, MA, 1947.
27. H.A. Steinhertz and P.A. Redhead, *Sci. Am.,* March, 2, 1962.
28. R.T. Bayard and D. Alpert, *Rev. Sci. Instr.,* 21, 571, 1950.
29. O.A. Weinreich, *Phys. Rev.,* 82, 573, 1951.
30. O.A. Weinreich and H. Bleecher, *Rev. Sci. Instr.,* 23, 56, 1952.
31. H.C. Hseuh and C. Lanni, *J. Vac. Sci. Technol.,* A 5, 3244, 1987.
32. T.S. Chou and Z.Q. Tang, *J. Vac. Sci. Technol.,* A4, 2280, 1986.
33. P.A. Redhead, *Rev. Sci. Instr.,* 31, 343, 1960.

34. G.H. Metson, *Br. J. Appl. Phys.*, 2, 46, 1951.
35. J.J. Lander, *Rev. Sci. Inst.*, 21, 672, 1950.
36. P.A. Redhead, *J. Vac. Sci. Technol.*, 3, 173, 1966.
37. J. Groszkowski, *Le Vide*, 136, 240, 1968.
38. L.G. Pittaway, *Philips Res. Rept.*, 29, 283, 1974.
39. D. Blechshmidt, *J. Vac Sci. Technol.*, 10, 376, 1973.
40. P.A. Redhead, *J. Vac. Sci. Technol.*, 6, 848, 1969.
41. S.D. Wood and C.R. Tilford, *J. Vac. Sci. Technol.*, A3, 542, 1985.
42. C.R. Tilford, *J. Vac. Sci. Technol.*, A3, 546, 1985.
43. P.C. Arnold and D.G. Bills, *J. Vac. Sci. Technol.*, A2, 159, 1984.
44. P.C. Arnold and J. Borichevsky, *J. Vac. Sci. Technol.*, A12, 568, 1994.
45. D.G. Bills, *J. Vac. Sci. Technol.*, A12, 574, 1994.
46. C.R. Tilford, A.R. Filippelli, et al., *J. Vac. Sci. Technol.*, A13, 485, 1995.
47. P.C. Arnold, D.G. Bills, et al., *J. Vac. Sci.Technol.*, A12, 580, 1994.
48. ETI Division of the Fredericks Co., Gage Type 8184.
49. T.A. Flaim and P.D. Owenby, *J. Vac. Sci. Technol.*, 8, 661, 1971.
50. J.F. O'Hanlon, op. cit., 65.
51. F.M. Pennin, *Physica*, 4, 71, 1937.
52. F.M. Penning and K. Nienhauis, *Philips Tech. Rev.*, 11, 116, 1949.
53. P.A. Redhead, *Can. J. Phys.*, 36, 255, 1958.
54. J.P. Hobson and P.A. Readhead, *Can. J. Phys.*, 33, 271, 1958.
55. NRC type 552 data sheet.
56. N. Ohsako, *J. Vac. Sci.Technol.*, 20, 1153, 1982.
57. D. Pelz and G. Newton, *J. Vac. Sci. Technol.*, 4, 239, 1967.
58. R.N. Peacock, N.T. Peacock, and D.S. Hauschulz, *J. Vac. Sci. Technol.*, A9, 1977 1991.
59. E. Drubetsky, D.R. Taylor, and W.H. Bayles, Jr., *Am. Vac. Soc., New Engl. Chapter, 1993 Symp.*
60. B.R. Kendall and E. Drubetsky, *J. Vac. Sci. Technol.*, A14, 1292, 1996.
61. M. Ono, K. Hirata, et al., Quartz friction vacuum gage for pressure range from 0.001 to 1000 torr, *J. Vac. Sci. Technol.*, A4, 1728, 1986.
62. O.E. Meyer, *Pogg. Ann.*, 125, 177, 1865.
63. J.C. Maxwell, *Phil. Trans. R. Soc.*, 157, 249, 1866.
64. F.T. Holmes, *Rev. Sci. Instrum.*, 8, 444, 1937.
65. J.W. Beams, J.L. Young, and J.W. Moore, *J. Appl. Phys.*, 17, 886, 1946.
66. J.K. Fremery, *Vacuum*, 32, 685, 1946.
67. NIST, Vacuum Calibrations Using the Molecular Drag Gage, Course Notes, April 15-17, 1996.
68. S. Dittman, B.E. Lindenau, and C.R. Tilford, *J. Vac. Sci. Technol.*, A7, 3356, 1989.
69. K.E. McCulloh, S.D. Wood, and C.R. Tilford, *J. Vac. Sci. Technol.*, A3, 1738, 1985.
70. G. Cosma, J.K. Fremerey, B. Lindenau, G. Messer, and P. Rohl, *J. Vac. Sci. Technol.*, 17, 642, 1980.
71. C.R. Tilford, S. Dittman, and K.E. McCulloh, *J. Vac. Sci. Technol.*, A6, 2855, 1988.
72. S. Dittman, NIST Special Publication 250-34, 1989.
73. National Conference of Standards Laboratories, Boulder, CO.
74. M. Hirata, M. Ono, H. Hojo, and K. Nakayama, *J. Vac. Sci. Technol.*, 20(4), 1159, 1982.
75. H. Gantsch, J. Tewes, and G. Messer, *Vacuum*, 35(3), 137, 1985.
76. D.M. Hoffman, B. Singh, and J.H. Thomas, III (eds.), *The Handbook of Vacuum Science and Technology,* Orlando, FL: Academic Press, 1998.
77. W.H. Bayles, Jr., Fundamentals of Vacuum Measurement, Calibration and Certification, Industrial Heating, October 1992.
78. Varian Associates, Basic Vacuum Practice, Varian Associates, Inc., 121 Hartwell Ave., Lexington, MA 02173, 1992.

26.3 Ultrasound Measurement

Peder C. Pedersen

Applications of Ultrasound

Medical

Ultrasound has a broad range of applications in medicine, where it is referred to as *medical ultrasound*. It is widely used in obstetrics to follow the development of the fetus during pregnancy, in cardiology where images can display the dynamics of blood flow and the motion of tissue structures (referred to as real-time imaging), and for locating tumors and cysts. 3-D imaging, surgical applications, imaging from within arteries (intravascular ultrasound), and contrast imaging are among the newer developments.

Industrial

In industry, ultrasound is utilized for examining critical structures, such as pipes and aircraft fuselages, for cracks and fatigue. Manufactured parts can likewise be examined for voids, flaws, and inclusions. Ultrasound has also widespread use in process control. The applications are collectively called *Non-Destructive Testing* (NDT) or *Non-Destructive Evaluation* (NDE). In addition, *acoustic microscopy* refers to microscopic examinations of internal structures that cannot be studied with a light microscope, such as an integrated circuit or biological tissue.

Underwater

Ultrasound is likewise an important tool for locating structures in the ocean, such as wrecks, mines, submarines, or schools of fish; the term SONAR (SOund Navigation And Ranging) is applied to these applications.

There are many other usages of ultrasound that lie outside the scope of this handbook: ultrasound welding, ultrasound cleaning, ultrasound hyperthermia, and ultrasound destruction of kidney stones (lithotripsy).

Definition of Basic Ultrasound Parameters

Ultrasound refers to acoustic waves of frequencies higher than 20,000 cycles per second (20 kHz), equal to the assumed upper limit for sound frequencies detectable by the human ear. As acoustic waves fundamentally are mechanical vibrations, a medium (e.g., water, air, or steel) is required for the waves to travel, or propagate, in. Hence, acoustic waves cannot exist in vacuum, such as outer space. If a single frequency sound wave is produced, also termed a *continuous wave* (CW), the fundamental relationship between frequency, f, in Hz, the sound speed of the medium, c_0, in m s^{-1}, and the wavelength, λ, in meters, is given as:

$$\lambda = \frac{c_0}{f}$$

(26.5)

The wavelength λ describes the length, in the direction of propagation, of one period of the sound wave. The wavelength determines, or influences, the behavior of many acoustic functions: The sound field emitted from an acoustic radiator (e.g., a transducer or loudspeaker) is determined by the radiator's size measured in wavelengths; the ability to differentiate between closely spaced reflectors is a function of the separation measured in wavelengths. Even when a sound pulse, rather than a CW sound, is transmitted, the wavelength concept is still useful, as the pulse typically contains a dominant frequency.

The vibrational activity on the surface of the sound source transfers the acoustic energy into the medium. If one were able to observe a very small volume, referred to as a *particle*, of the medium during transmission of sound energy, one would see the particle moving back and forth around a fixed position. Associated with the particle motion is an acoustic pressure, which refers to the pressure variation around the mean pressure (which is typically the atmospheric pressure). This allows the introduction of two important — and closely related — acoustic quantities: the *particle velocity*, $\vec{\mathbf{u}}(\vec{r},t)$, and the *acoustic pressure*, $p(\vec{r},t)$. In this notation, the arrow above a symbol in bold indicates a vector. The symbol $\vec{r}$ represents the position vector, which simply defines a specific location in space. Thus, both particle velocity and pressure are functions of three spatial variables, x, y, and z, and the time variable, t.

To characterize a medium acoustically, the most important parameter is the *specific acoustic impedance, z*. For a lossless medium, z is given as follows:

$$z = \rho_0 c_0 \tag{26.6}$$

In Equation 26.6, ρ_0 is the density of the medium, measured in kg m^{-3}. When a medium absorbs acoustic energy (which all media do to a greater or smaller extent), the expression for acoustic impedance also contains a small imaginary term; this will be ignored in the discussions presented in this chapter. The acoustic impedance relates the particle velocity to the acoustic pressure:

$$z = \frac{p(\vec{r},t)}{u(\vec{r},t)} \tag{26.7}$$

Note that the relationship in Equation 26.7 uses the scalar value of the particle velocity (a scalar is a quantity, such as temperature, that does not have a direction associated with it). Equation 26.7 is exact for plane wave fields and a very good approximation for arbitrary acoustic fields. In a plane wave, all points in a plane normal to (i.e., which forms a 90° angle with respect to) the direction of propagation have the same pressure and particle velocity.

The acoustic impedances on either side of an interface (boundary between different media) determine the acoustic pressure reflected from the interface. Let a plane wave traveling in a medium with the acoustic impedance z_1 encounter a planar, smooth interface with another medium having the acoustic impedance z_2. Assume that the plane wave propagates directly toward the interface; this is commonly referred to as insonification under normal incidence. In this case, the pressure and the intensity reflection coefficients, R and R_I, respectively, are as follows:

$$R = \frac{p_r}{p_i} = \frac{z_2 - z_1}{z_2 + z_1}$$

$$R = \frac{I_r}{I_i} = \left(\frac{z_2 - z_1}{z_2 + z_1} \right)^2 \tag{26.8}$$

Intensity is a measure of the mean power transmitted through unit area and is measured in watts per square meter. The corresponding pressure and intensity transmission coefficients, T and T_I, respectively, are:

$$T = \frac{p_t}{p_i} = \frac{2 z_2}{z_2 + z_1}$$

$$T_I = \frac{I_t}{I_i} = \frac{4 z_1 z_2}{\left(z_2 + z_1 \right)^2} \tag{26.9}$$

The subscripts "i," "r," and "t" in Equations 26.8 and 26.9 refer to incident, reflected, and transmitted, respectively. The expressions in these two equations can be considered approximately valid for nonplanar waves under near-normal incidence. However, when the incident angle (angle between direction of propagation and the normal to the surface) becomes large, the reflection and transmission coefficients can change dramatically. In addition, the reflected and transmitted signals will also change if the surface is rough to the extent that the rms (root mean square) height exceeds a few percent of the wavelength.

The wave propagation can take several forms. In fluids and gases, only *longitudinal* (or *compressional*) waves exist, meaning that the direction of wave propagation is equal to the direction of the particle velocity vector. In solids, both longitudinal and *shear* waves exist which propagate in different directions and with different sound speeds. Transverse waves can exist on strings where the particle motion is normal to the direction of propagation. (Strictly speaking, shear waves can propagate a short distance in liquids [fluids] if the viscosity is sufficiently high.)

Conceptual Description of Ultrasound Imaging and Measurements

Most ultrasound measurements are based on the generation of a short ultrasound pulse that propagates in a specified direction and is partly reflected wherever there is an abrupt change in the acoustic properties of the medium and detection of the resulting echoes (pulse-echo ultrasound). A change in properties can be due to a cyst in liver tissue, a crack in a high-pressure pipe, or reflection from layers in the sea bottom. The degree to which a pulse is reflected at an interface is determined by the change in acoustic impedance as described in Equation 26.8. An image is formed by mapping echo strength vs. travel time (proportional to distance) and beam direction, as illustrated in Figure 26.31. This is referred to as *B-mode* imaging (Brightness-mode). Further signal processing can be applied to compensate for attenuation (the damping out of the pressure pulse as it propagates) of the medium or to control focusing. Signal processing can also be applied to analyze echoes for information about the structure of materials or about the surface characteristics of rough surfaces.

A block diagram of a simplified pulse-echo ultrasound measurement system is shown in Figure 26.32. The pulser circuit can generate a large voltage spike for exciting the transducer in B-mode applications, or the arbitrary function source can produce a short burst for Doppler measurements, or a coded waveform, such as a linear sweep. The amplifier brings the driving voltage to a level where the transducer can generate an adequate amount of acoustic energy.

The transducer is made from a piezoelectric ceramic that has the property of producing mechanical vibrations in response to an applied voltage pulse and generating a voltage when subjected to mechanical stress. When an image is required, the transducer can be a mechanical sector probe that produces a fan-shaped image by means of a single, mechanically steered, focused transducer element. Alternatively, the transducer can be a linear array transducer (described later) that produces a rectangular image. In the case of an array transducer, the pulser/amplifier must contain a driving circuit for each element in the array, in addition to delay control. To achieve a short pulse and good sensitivity, the transducer is equipped with backing material and matching layers (to be discussed later).

The receiver block contains a low-noise amplifier with time-varying gain to correct for medium attenuation and often a circuit for logarithmic compression. In the case of array transducers, the receiver circuitry is a complex system of amplifiers, time-varying delay elements, and summing circuits. The signal processing block can be part analog and part digital. The echo signals are envelope detected and digitized; the *envelope* of a signal is a curve that follows the amplitude of the received signal. A scan converter changes the signal into a format suitable for display on a gray-scale monitor. Information about ultrasound pulser-receiver instrumentation for NDE measurements can be found in [1] and for medical imaging in [2].

Not all ultrasound measurement systems are based on the pulse-echo concept. For material characterization, *transmission measurements* are frequently used, as illustrated in Figure 26.33. The main difference between the pulse-echo system and the transmission system is that two transducers are used in the transmission system; the description of the individual blocks for the pulse-echo system applies generally here also. Imaging is generally not possible, although tomographic imaging of either attenuation

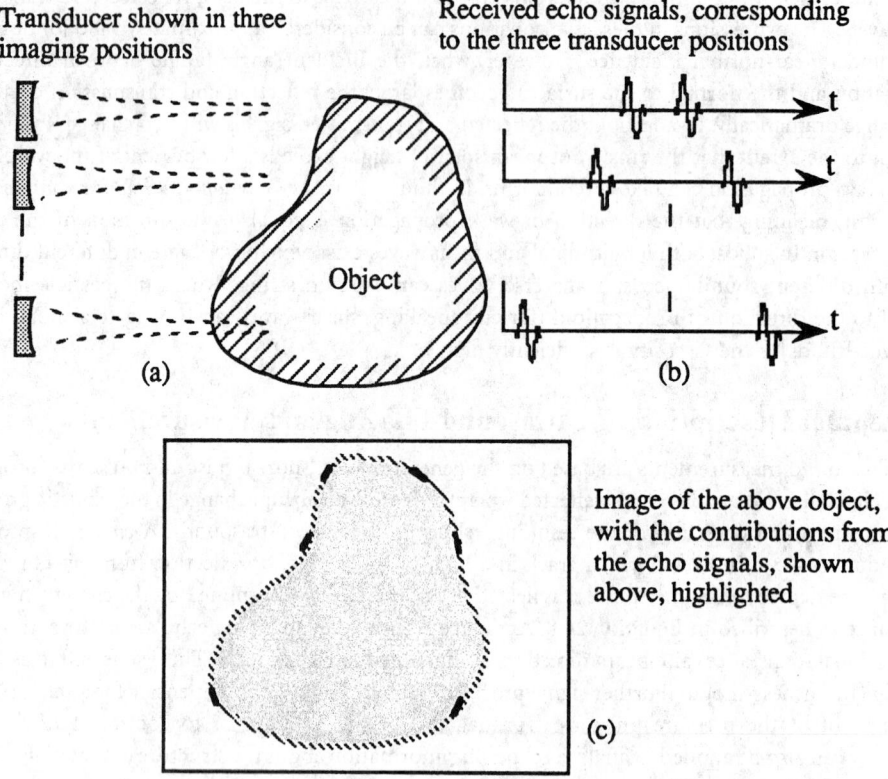

FIGURE 26.31   (*a*) A focused transducer insonifies the irregular object from different positions out of which only three are shown. The different transducer positions can readily be obtained by the use of an array transducer (to be described later). (*b*) Received echoes from the front and back of the object are displayed vs. travel time. It is here assumed that the structure is only weakly reflecting and that the attenuation has only a minimal effect. (*c*) An image is formed, based on the echo strengths and the echo arrival times.

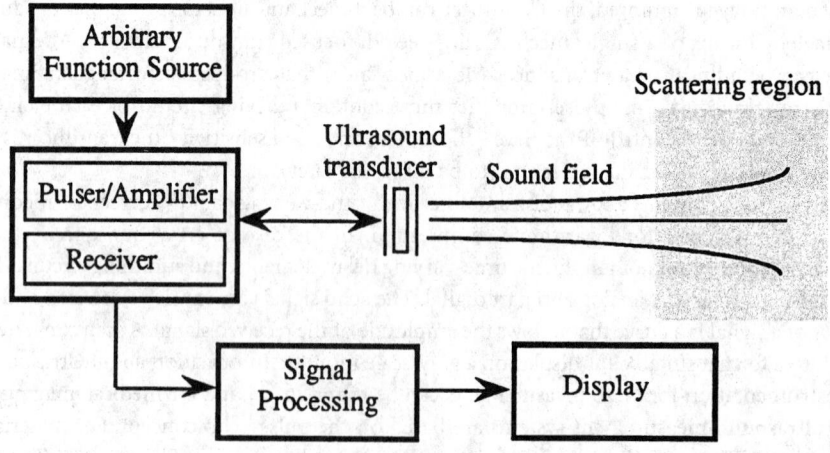

FIGURE 26.32   Simplified block diagram of a pulse-echo ultrasound system.

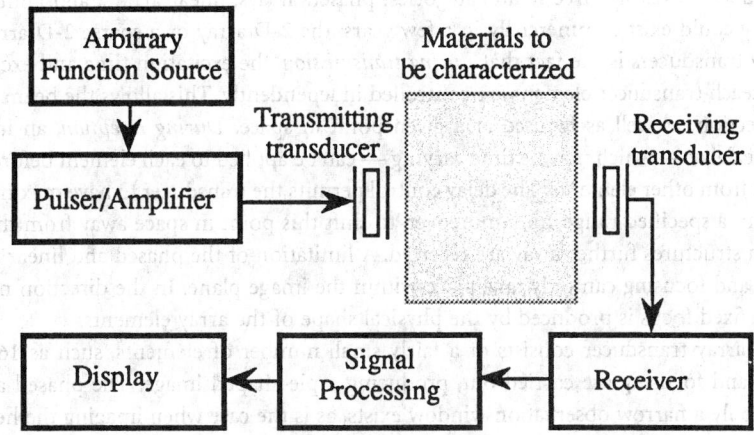

FIGURE 26.33 Simplified block diagram of ultrasound transmission system.

or velocity has been attempted. If the two transducers are moved together, voids and inclusions located in the transmission path can be detected.

Transmission measurements require, of course, that the structure or the medium of interest is accessible from opposite sides. This system allows accurate characterization of the attenuation and the sound speed of the medium from which a variety of material properties can be derived. Flow in a pipe or a channel can also be determined with a transmission system, by determining the difference between upstream and downstream propagation times.

Single-Element and Array Transducers

The device that converts electric energy into acoustic energy and vice versa is termed the *transducer* or, more specifically, the ultrasound transducer. Piezoelectric ceramics can be used for all frequencies in the ultrasound range; however, for ultrasound frequencies in the 20 kHz to 200 kHz range, magnetostrictive devices are occasionally used, which are based on materials that exhibit mechanical deformation in response to an applied magnetic field. Ultrasound transducers are commercially available for many applications over a wide range of frequencies.

Single-element transducers are used for basic measurements and material characterization, while array transducers with many individual transducer elements are used for imaging purposes. The former type costs from a few hundreds dollars and up, whereas the latter type costs in the thousands of dollars and requires extensive electronics for beam control and signal processing. A third alternative is the mechanical sector scanner where a single transducer is mechanically rotated over a specified angle. A single-element transducer consists of a piezoelectric element, a backing material that enables the transducer to respond to a fairly wide range of frequencies (but with reduced sensitivity), and a matching layer and faceplate that provide improved coupling into the medium and protect the transducer.

Although broadband transducers are very desirable, the energy conversion characteristics of the practical transducer correspond to that of a bandpass filter, i.e., the ultrasound pulse has most of its energy distributed around one frequency. This is the frequency referred to when, for example, one orders a 3.5 MHz transducer. The 6 dB bandwidth can be from 50% to 100% of the transducer frequency; the 6 dB bandwidth is the frequency range over which the transducer can produce an acoustic pressure that is at least 50% of the acoustic pressure at the most efficient frequency. The radiation characteristics of a transducer are determined by the geometry of the transducer (square, circular, plane, focused, etc.) and by the dimensions measured in wavelengths. Hence, a 10 MHz transducer tends to be smaller than a 5 MHz transducer. Focusing can also be achieved by means of an acoustic lens.

Array transducers exist in three main categories: phased arrays, linear arrays, and annular arrays. A fourth category could exist commercially in a few years: the 2-D array or a sparse 2-D array. Common for these array transducers is the fact that *during transmission*, the excitation time and excitation signal amplitude for each transducer element are controlled independently. This allows the beam to be steered in a given direction, as well as focused at a given point in space. *During reception*, an independently controlled time delay — which may be time varying — can be applied to each element before summation with the signal from other elements. The delay control permits the transducer to have maximal sensitivity to an echo from a specified range, and, moreover, to shift this point in space away from the transducer as echoes from structures further away are received. A limitation of the phased and linear arrays is that beam steering and focusing can only take place within the image plane. In the direction normal to the image plane, a fixed focus is produced by the physical shape of the array elements.

The phased array transducer consists of a fairly small number of elements, such as 16 to 32. Both beam steering and focusing are carried out, producing a pie-shaped image. The phased array is most suitable when only a narrow observation window exists, as is the case when imaging the heart from the space between the ribs. The linear array transducer has far more elements, typically between 128 and 256. It activates only a subset of all the elements for a given measurement, and produces focusing, but generally not beam steering, so that the beam direction is normal to the array surface. Consider a linear array transducer with 128 elements, labeled 1 to 128. If one assumes that the first pulse-echo measurement is made with elements 1 to 16, the next measurement with elements 2 to 17, etc., the beam will have been *linearly* translated along the long dimension of the linear array (hence the name), producing a rectangular image format. The annular array transducer consists of a series of concentric rings. As such, it cannot steer the beam, but can focus both in transmit and receive mode, just as the phased and linear arrays. Thus, the annular array transducer is not suitable for imaging unless the transducer is moved mechanically. Its specific advantage is uniform focusing.

Selection Criteria for Ultrasound Frequencies

The discussion so far has often mentioned *frequency* as an important variable, without providing any guidance as to what ultrasound frequencies should be used. There is generally no technological limitation with respect to choice of ultrasound frequency, as frequencies even in the gigahertz (GHz, 10^9 Hz) range can be produced. The higher the frequency, the better is the control over the direction and the width of the ultrasound beam, leading to improved axial and lateral resolution. However, the attenuation increases with higher frequency. For many media, the attenuation varies with the frequency squared; while for biological soft tissue, the attenuation varies nearly linearly with frequency.

Generally, the ultrasound frequency is chosen as high as possible while still allowing a satisfactory signal-to-noise ratio (SNR) of the received signal. This rule does not always hold; for example, for grain size estimation in metals and for rough surface characterization, the optimal frequency has a specific relationship to the mean grain dimension or the rms roughness.

Basic Parameters in Ultrasound Measurements

Ultrasound technology today has led to a very wide range of applications, and the following overview mentions only the basic parameters. For an in-depth review, see [3]. A discussion of applications is given at the end of the Ultrasound Theory and Applications section.

Reflection (or Transmission) Detected or Not

The most basic measurement consists of determining whether a reflected or transmitted signal is received or not. This can be used to monitor liquid level where the absence of liquid prevents ultrasound transmission from taking place, or to detect the existence of bubbles in a liquid (e.g., for dialysis purposes, where the bubbles are the reflectors) or for counting objects, such as bottles, on a conveyer belt.

Travel Time

Travel time is the elapsed time from transmission of an ultrasound pulse to the detection of a received pulse in either a pulse-echo or a transmission system. If the sound speed is known, the thickness of an object can be found from the time difference between the front surface echo and the back surface echo. Conversely, sound speed can be determined when the thickness is known. By measuring with a broadband pulse (a short pulse containing a broad range of frequencies), velocity dispersion (frequency dependence of velocity) can be found, which has applications for materials characterization. Elastic parameters can also be found from velocity measurements. Imaging applications are generally made based on the assumption of a known, constant velocity; thus, round-trip travel times from the transducer to the reflecting structures and back in a known direction gives the basis for the image formation.

Attenuation

The attenuation and its frequency dependence are important materials parameters, as they can be used to differentiate between normal and pathological biological tissue, measure porosity of ceramics, the grain size of metals, and the structure of composite materials. Attenuation can be found by means of transmission measurements by determining the signal amplitude with the test object first absent and then present; here, transmission losses must also be considered, as quantified in Equation 26.9. Attenuation can be obtained from pulse-echo measurements, by comparing the strength of echoes from the front and back surfaces of the specimen under test. Diffraction effects (described later in the *Advanced Topics in Ultrasound* section) might need to be considered. Attenuation can be considered as a bulk parameter for the medium as a whole, or it can be determined for small regions of the medium.

Reflection Coefficient

The strength of a reflection at an interface between two media is determined by the change in acoustic impedance across the interface, as shown in Equation 26.8. This allows, at least in principle, the impedance of one medium to be determined from the measured reflection coefficient and the impedance of the other medium. In fact, the term *impediography* refers to the determination of the impedance profile of a layered medium by means of this concept. In practice, the measurement is not easy to carry out, as the transducer must be aligned very accurately at normal incidence to the medium surface. By measuring the reflection coefficient vs. incident angle at a planar, smooth surface, a velocity estimate of the reflecting medium can be made; in an alternative application, reflection measurements vs. frequency permit the evaluation of rough surface parameters.

Parameters Obtained Through Signal Processing

A number of object parameters can be obtained from analysis of the received signals. Material properties can be extracted from speckle analysis (analysis of statistical fluctuations in the image) and other forms of statistical analysis. Recognition of objects with complex shapes can, in some cases, be done by extracting specific features of the received signals. Doppler processing for velocity estimation is another example of information that is only attainable by means of signal processing. By combining several "stacked" images, 3-D reconstruction is possible, from which volume estimation and rate of change of volume are possible, such as for determining the dynamics of the left ventricle of the heart.

Ultrasound Theory and Applications

Sound Velocity

The propagation speed, c_0, is generally determined by the density and the elastic properties of the medium. For an ideal gas, c_0 may be expressed as [4]:

$$c_0 = \sqrt{\gamma P_0/\rho_0} = \sqrt{\gamma r T_K} \qquad (26.10)$$

where $\gamma = C_P/C_V$, the ratio of specific heats; for air, $\gamma = 1.402$. P_0 is the static pressure of the gas, which at 1 atm is 1.013×10^5 N m^{-2}. ρ_0 is the specific density of the gas, equal to 1.293 kg m^{-3} for air at 1 atm and 0°C. The constant r is the ratio of the universal gas constant and the molecular weight of the gas, and T_K is the temperature in kelvin. Substituting the values for air into the first equation in Equation 26.10 gives $c_0 = 331.6$ m s^{-1} at 0°C. As the ratio P_0/ρ_0 is constant for varying pressure, but constant temperature, c_0 is likewise pressure independent. (For a real gas, c_0 exhibits, in fact, a small dependence on pressure.)

The temperature dependence of an ideal gas can be obtained by rewriting the second expression in Equation 26.10 as follows

$$c_0 = c_{\mathrm{ref}} \sqrt{1 + T/273} \qquad (26.11)$$

where c_{ref} is the sound speed at 0°C, and T is the temperature in °C.

In liquids, the expression for c_0 is given as

$$c_0 = \sqrt{K_s/\rho_0} \qquad (26.12)$$

where K_s is the adiabatic bulk modulus. Expressions for actual liquids as a function of temperature and pressure are not easily derived, and are often empirically determined. For pure water, the sound speed as a function of temperature and pressure can be given as:

$$c_{\mathrm{H_2O}} = 1402.4 + 5.01T - 0.055T^2 + 0.00022T^3 + 1.6 \times 10^{-6} P_0 \qquad (26.13)$$

In Equation 26.13, T is temperature in °C, and P_0 is the static pressure in N m^{-2}; the expression is valid for a pressure range from 1 atm to 100 atm.

Finally, for solids, the sound speed for longitudinal waves, c_L, is [5]:

$$c_L = \left(\frac{c_{11}}{\rho_0}\right)^{1/2} = \left(\frac{\lambda + 2\mu}{\rho_0}\right)^{1/2} \qquad (26.14)$$

In Equation 26.14, c_{ij} is the the elastic stiffness constant, such that c_{11} refers to longitudinal stress over longitudinal strain, and $\lambda = c_{12}$ and $\mu = c_{44}$ are the Lamé elastic constants.

Wave Propagation in Homogeneous Media

In order to effectively use ultrasound for measurement purposes, it is essential to be able to describe the behavior of acoustic fields. This includes the general behavior, and maybe even a detailed understanding, of the radiated field from ultrasound transducers; the transmission, reflection, and refraction of an acoustic field at boundaries and layers; and the diffraction of of acoustic field at finite-sized reflectors.

The wave equation is a differential equation that formulates how an acoustic disturbance (acoustic pressure or particle velocity) propagates through a homogeneous medium. For an arbitrary wave field, the wave equation takes the following general form:

$$\nabla^2 p = \frac{1}{c^2} \frac{\partial^2 p}{\partial t^2} \qquad (26.15)$$

where ∇^2 is the Laplacian operator. Equation 26.15 is also valid when particle velocity is substituted for pressure. If the condition of a plane wave field at a single frequency (also called a harmonic plane wave field) is imposed, the solution to Equation 26.15 is:

$$p(\vec{r}, t) = A \exp\left[j\left(\omega t - \vec{k} \times \vec{r}\right)\right] + B \exp\left[j\left(\omega t + \vec{k} \times \vec{r}\right)\right] \qquad (26.16)$$

In Equation 26.16, $p(\vec{r},t)$ consists of a plane wave propagating in the direction defined by the propagation vector $\vec{k}$ and a plane wave in the direction $-\vec{k}$. As before, $\vec{r}$ is a position vector that, in a Cartesian coordinate system, can be expressed as $\vec{r} = x\hat{x} + y\hat{y} + z\hat{z}$ where $\hat{x}$, $\hat{y}$, and $\hat{z}$ are unit vectors. The magnitude of $\vec{k}$ is called the wavenumber:

$$k = \omega/c_0 = 2\pi/\lambda \tag{26.17}$$

Expressing $\vec{k}$ in the Cartesian coordinate system gives:

$$\vec{k} = k_x\hat{x} + k_y\hat{y} + k_z\hat{z} \tag{26.18}$$

When k, k_x, and k_y are specified, k_z is defined as well:

$$k_z = \sqrt{\left(\left(\omega/c_0\right)^2 - k_x^2 - k_y^2\right)} \tag{26.19}$$

Applying the expressions for $\vec{r}$ and $\vec{k}$ to Equation 26.16 and considering only the plane wave in the direction of $+\vec{k}$ gives

$$p(\vec{r},t) = A \exp\left[j\left(\omega t - \left(k_x x + k_y y + k_z z\right)\right)\right] \tag{26.20}$$

From Equation 26.20, one sees that the amplitude of a plane wave is constant and equal to A, but that the phase varies with both time and space. When the direction of $\vec{k}$ is specified, k_x, k_y, and k_z are found from projection onto the coordinate axes.

Solving Equation 26.15 under the assumption of spherical waves at a single frequency, with the source placed at the origin of the coordinate system, gives:

$$p(r,t) = \frac{A}{r}\exp\left[j\left(\omega t - kr\right)\right] + \frac{B}{r}\exp\left[j\left(\omega t + kr\right)\right] \tag{26.21}$$

The first term is a diverging spherical wave, and the second term is a converging spherical wave. The reason that the vector dot product does not appear in Equation 26.21 is that $\vec{k}$ and $\vec{r}$ always point in the same direction for spherical waves. While Equation 26.21 gives the complete solution, in most cases only a diverging spherical wave exists. If spherical waves are produced by a spherical source of radius a, then:

$$p(r,t) = \frac{A}{r}\exp\left[j\left(\omega t - kr\right)\right], \quad r > a \tag{26.22}$$

Acoustic Intensity and Sound Levels

The acoustic intensity, I, describes the mean power transported through a unit area normal to the direction of propagation. A general expression for I for a CW pressure function is:

$$I = \frac{1}{T}\int_0^T p(t)u(t)\,dt \tag{26.23}$$

In Equation 26.23, T represents one full cycle of the pressure function, and $u(t)$ is the particle velocity function. For a plane wave, the intensity is:

$$I = \frac{P^2}{2\rho_0 c_0} = \frac{1}{2} PU \tag{26.24}$$

where P and U are the magnitudes of the pressure and the particle velocity functions, respectively. The expression for I in Equation 26.24 can also be used as an approximation for nonplanar waves, such as spherical waves, as long as (kr) is large.

The intensity levels for pulse-echo measurements are often described by the SPTA (*Spatial Peak, Temporal Average*) value, which therefore refers to the mean intensity at the point in space where the intensity is the highest. Although the guidelines for exposure levels for medical imaging, set by the FDA, vary for different parts of the body, a general upper limit is 100 mW cm^{-2}. Given that the duty cycle (pulse duration/pulse interval) is typically in the order of 0.0001 for ultrasound imaging, the temporal peak intensity is much higher, and is even approaching the level where nonlinear effects can begin to be observed. Intensity levels used for NDE are application dependent, but are generally in the same range as in medical ultrasound.

Sound levels are logarithmic expressions (expressions in dB) of the pressure level or the intensity level. The term SPL stands for sound pressure level and is given as:

$$SPL = 20 \log\left(P_e / P_{ref}\right) \ \left[dB\right] \tag{26.25}$$

In Equation 26.25, P_e is the effective pressure of the acoustic wave and P_{ref} is the reference effective pressure. Correspondingly, IL is the intensity level defined as:

$$IL = 10 \log\left(I / I_{ref}\right) \ \left[dB\right] \tag{26.26}$$

where I_{ref} is the reference acoustic intensity.

Several reference levels are commonly used [6]. In air, $P_{ref} = 20 \ \mu Pa$ and $I_{ref} = 10^{-12}$ W m^{-2} are nearly equivalent reference levels. For water, typical pressure reference levels are $P_{ref} = 1 \ \mu Pa$, $P_{ref} = 20 \ \mu Pa$ or $P_{ref} = 1 \ \mu bar = 0.1$ Pa, with corresponding intensity reference levels of $I_{ref} = 6.76 \times 10^{-19}$ W m^{-2}, $I_{ref} = 2.70 \times 10^{-16}$ W m^{-2}, and $I_{ref} = 6.76 \times 10^{-9}$ W m^{-2}.

Wave Propagation Across Layers and Boundaries

The reflection and transmission coefficients for a plane wave impinging at normal incidence on the interface between two half spaces were given in Equations 26.8 and 26.9. It is also important to consider the transmission and reflection of waves at non-normal incidence and across a layer.

When a longitudinal wave impinges at a liquid–liquid interface, both the transmitted and the reflected waves are longitudinal waves, as illustrated in Figure 26.34(a). However, when a longitudinal wave propagating in a liquid medium encounters a liquid–solid interface under oblique incidence, both a longitudinal and a shear transmitted wave and a longitudinal reflected wave will result, as shown in Figure 26.34(b). The change of direction of the transmitted waves relative to the incident wave is referred to as *refraction*, which is caused by the difference in sound speed between the two media.

The magnitude and direction of the reflected and transmitted waves are determined by two simple boundary conditions: (1) the *pressure* on both sides of the boundary must be equal at all times, and (2) the *normal particle velocity* on both sides of the boundary must be equal at all times; the normal particle velocity refers to the component of the particle velocity that is normal to the boundary.

The boundary conditions lead to the following two relationships:

$$\sin\theta_i = \sin\theta_r ; \quad \frac{\sin\theta_i}{c_1} = \frac{\sin\theta_t}{c_2} \tag{26.27}$$

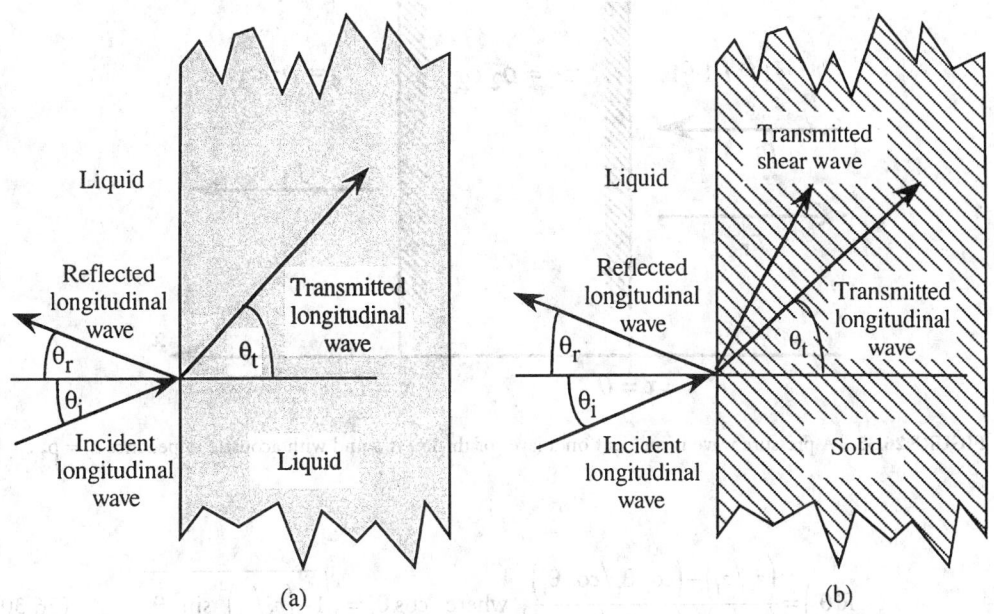

FIGURE 26.34 (*a*). An incident longitudinal wave at a liquid–liquid interface produces a reflected and a transmitted longitudinal wave. (*b*) An incident longitudinal wave at a liquid–solid interface produces a reflected longitudinal wave, transmitted longitudinal, and shear waves. The incident, reflected, and transmitted angles are indicated.

where c_1 and c_2 are the sound speeds of the two media, and θ_i, θ_r, and θ_t are the incident, reflected and transmitted angles, respectively. From Equation 26.27, one sees that

$$\theta_r = \theta_i; \quad \theta_t = \arcsin\left(\frac{c_2}{c_1}\sin\theta_i\right) \tag{26.28}$$

The second expression in Equation 26.28 is referred to as Snell's law and quantifies the degree of refraction. Refraction affects the quality of ultrasound imaging because image formation is based on the assumption that the ultrasound beam travels in a straight path through all layers and inhomogeneities, and that the sound speed is constant throughout the medium. When the actual beam travels along a path that deviates to some extent from a straight path and passes through some parts of the medium faster than it does other parts, the resulting image is a distorted depiction of reality. Correcting this distortion is a very complex problem and, in the near future, one should only expect image improvement in the simplest cases.

It can be seen from Equation 26.28 that angle θ_t is not defined if the argument to the arcsin function exceeds unity. This defines a critical incident angle as follows:

$$\sin\theta_c = \frac{c_1}{c_2} \tag{26.29}$$

at or above which the reflection coefficient is 1 and the transmission coefficient is 0. A critical angle only exists when $c_1 < c_2$. The transmitted shear and longitudinal velocities each correspond to a different critical angle of incidence.

From the boundary conditions, the reflection coefficient as a function of incident angle can be determined [7]:

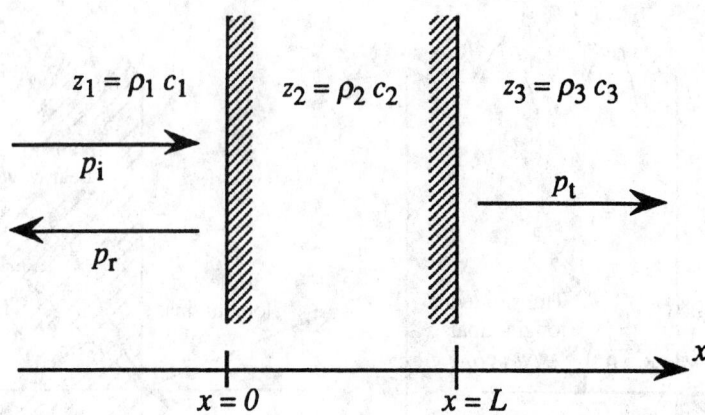

FIGURE 26.35 A pressure wave is incident on a layer of thickness L and with acoustic impedance $z_2 = \rho_2 c_2$.

$$R(\theta_i) = \frac{(z_2/z_1) - (\cos\theta_t/\cos\theta_i)}{(z_2/z_1) + (\cos\theta_t/\cos\theta_i)}, \text{ where } \cos\theta_t = \sqrt{1 - (c_2/c_1)^2 \sin^2\theta_i} \qquad (26.30)$$

The result in Equation 26.30 is referred to as the Rayleigh reflection coefficient.

The transmission of plane waves under normal incidence through a single layer is often of interest. The dimensions and the medium parameters are defined in Figure 26.35.

By applying the boundary conditions to both interfaces of the layer, both reflection and transmission coefficients are obtained [7], as given in Equations 26.31 and 26.32.

$$R = \frac{p_r}{p_i} = \frac{(1 - z_1/z_3)\cos k_2 L + j(z_2/z_3 - z_1/z_2)\sin k_2 L}{(1 + z_1/z_3)\cos k_2 L + j(z_2/z_3 + z_1/z_2)\sin k_2 L} \qquad (26.31)$$

$$T = \frac{p_t}{p_i} = \frac{2}{(1 + z_1/z_3)\cos k_2 L + j(z_2/z_3 + z_1/z_2)\sin k_2 L} \qquad (26.32)$$

A number of special conditions for Equation 26.31 can be considered, such as: (1) $z_1 = z_3$, which simplifies the numerator in Equation 26.31; and (2) the layer thickness is only a small fraction of a wavelength, that is, $k_2 L \ll 1$ which makes $\cos k_2 L \approx 1$ and $\sin k_2 L \approx k_2 L$. One case is of particular interest: the choice of thickness and acoustic impedance for the layer that makes $R = 0$, and therefore gives 100% energy transmission. This is fulfilled for:

$$k_2 L = \pi/2 + n\pi; \quad z_2 = \sqrt{z_1 z_3} \qquad (26.33)$$

The result in Equation 26.33 states that the layer must be a quarter of a wavelength thick (plus an integer number of half wavelengths) and must have an acoustic impedance that is the geometric mean of the impedances of the media on either side. Among several applications of the *quarter wavelength impedance matching* is the impedance matching between a transducer and the medium, such as water. A drawback with a single matching layer is that it only works effectively over a narrow frequency range, while the actual acoustic pulse contains a fairly broad spectrum of frequencies. A better matching is achieved by using more than one matching layer, and it has been shown that a matching layer that has a continuously varying acoustic impedance across the layer provides broadband impedance matching.

Attenuation: Its Origin, Measurement, and Applications

Attenuation refers to the damping of a signal, here specifically an acoustic signal, with travel time or with travel distance. Attenuation is typically expressed in dB, i.e., on a logarithmic scale. Attenuation is an important parameter to measure in many types of materials characterization, but also the parameter that sets an upper limit for the ultrasound frequency that can be used for a given measurement. In NDE, attenuation is used for grain size estimation [8, 9], for characterization of composite materials, and for determination of porosity [10]. In medical ultrasound, attenuation can be used for tissue characterization [11], such as differentiating between normal and cirrhotic liver tissue and for classification of malignancies. In flowmeters, attenuation caused by vortices can be used to measure the frequency at which they are shed; this frequency is proportional to the flow velocity.

Attenuation represents the combined effect of *absorption* and *scattering*, where absorption refers to the conversion of acoustic energy into heat due to the viscosity of the medium, and scattering refers to the spreading of acoustic energy into other directions due to inhomogeneities in the medium. The absorption can, in part, be due to *classical absorption*, which varies with frequency squared, and *relaxation absorption*, which can result in a complicated frequency dependence of absorption. Gases (except for noble gases), liquids, and biological tissue exhibit mainly relaxation absorption, whereas classical absorption is most prominent in solids, which can also have a significant amount of scattering attenuation.

In general, absorption dominates in homogeneous media (e.g., liquids, gases, fine-grained metals, polymers), whereas scattering dominates in heterogeneous media (e.g., composites, porous ceramics, large-grained materials, bone). The actual attenuation and its frequency dependence can be specified fairly unambiguously for gases and liquids, while for solids it is very dependent on the manufacturing process, which determines the microstructure of the material, such as the grain structure.

Measurement of attenuation can be carried out for at least two purposes: (1) to measure the *bulk attenuation* of a given homogeneous medium; and (2) to measure the *spatial distribution of attenuation* over a plane in an inhomogeneous medium. The former approach is most common in materials characterization, whereas the latter approach is found mainly in medical ultrasound. Bulk attenuation can be performed either with transmission measurements or with pulse-echo measurements, as illustrated in Figure 26.36(*a*) and (*b*), respectively.

For the measurement of attenuation, in dB/cm, of a medium of thickness d, by transmission measurements, define $v_1(t)$ as the received signal without medium present and $v_2(t)$ as the received signal with medium present, as shown in Figure 26.36(*a*). The attenuation is then determined from the ratio of the energies of the two signals, corrected for the transmission losses, as:

$$\text{Att}\left[\text{dB}/\text{cm}\right] = \frac{1}{d} \, 10 \log \left\{ \frac{\displaystyle\int_0^\infty \left[v_1(t)\right]^2 dt}{\displaystyle\int_0^\infty \left[v_2(t)\right]^2 dt} \right\} - 20 \log \left\{ \frac{4 z_1 z_2}{\left(z_1 + z_2\right)^2} \right\} \tag{26.34}$$

Correction for transmission losses (2nd term) can be avoided by alternatively measuring the incremental attenuation due to an incremental thickness increase.

For the measurement of attenuation by *pulse-echo measurements*, the front and the back wall echoes are termed $v_f(t)$ and $v_b(t)$, respectively. Based on an *a priori* knowledge of the pulse duration, a time window ΔT is defined. The attenuation is most accurately measured when based on the energies of $v_f(t)$ and $v_b(t)$, but may alternatively be based on the amplitudes of $v_f(t)$ and $v_b(t)$, as stated in Equation 26.35.

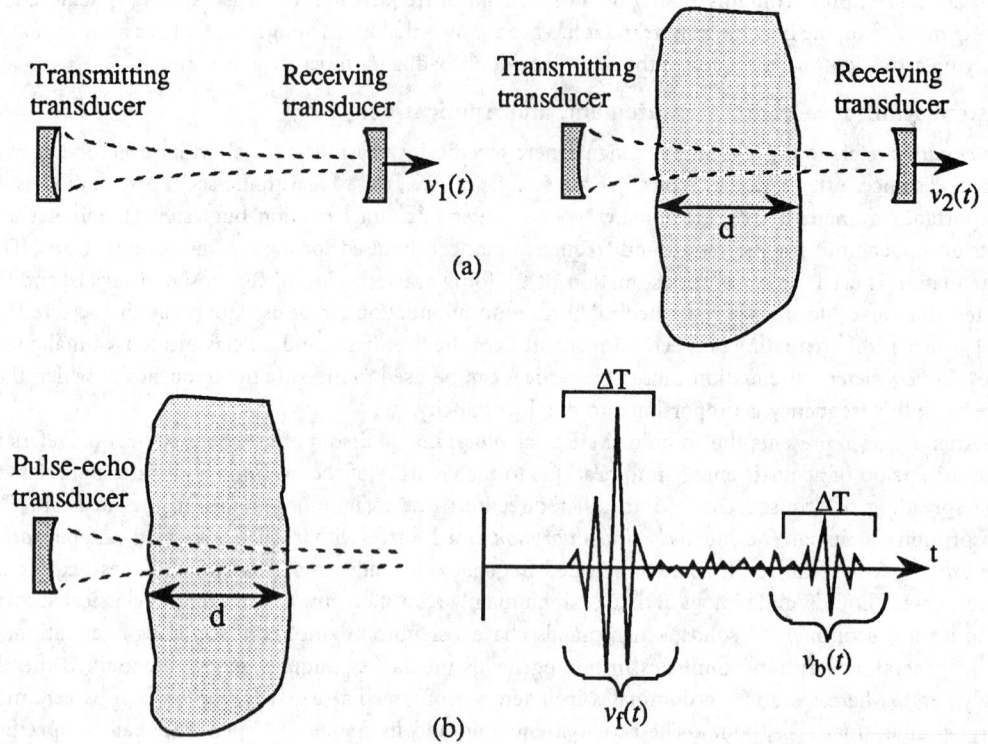

FIGURE 26.36 Measurement of bulk attenuation. (*a*) Measurement of attenuation by transmission measurement. (*b*) Measurement of attenuation by pulse-echo measurements.

$$
\text{Att}\left[\text{dB/cm}\right] = \frac{1}{2d}\, 10\log\left\{ \frac{\displaystyle\int_0^{\Delta t}\left[v_f(t)\right]^2 dt}{\displaystyle\int_0^{\Delta T}\left[v_b(t)\right]^2 dt} \right\} \cong \frac{1}{2d}\, 20\log\left\{ \frac{\text{peak ampl., } v_f(t)}{\text{peak ampl., } v_b(t)} \right\} \tag{26.35}
$$

Accurate attenuation measurements require attention to several potential pitfalls: (1) diffraction effects (even in the absence of attenuation, echoes from different ranges vary in amplitude, due to beam spreading); (2) misalignment effects (if the reflecting surface is not normal to the transducer axis, there is a reduction in detected signal amplitude, due to phase cancellation at the transducer surface); and (3) transmission losses whose magnitude it is not always easy to determine.

The spatial distribution of attenuation must be measured with pulse-echo measurements, where it is assumed that a backscatter signal of sufficient amplitude can be received from all regions of the medium. Use is made of the frequency dependence of attenuation, which has the effect that the shift in mean frequency of a given received echo relative to the mean frequency of the transmitted signal varies proportional to the total absorption. The *rate of shift* in mean frequency is thus proportional to the local attenuation.

CW Fields from Planar Piston Sources

In discussing the acoustic fields generated by acoustic radiators (transducers), a clear distinction must be made between fields due to CW excitation and due to pulse excitation. Although the overall field

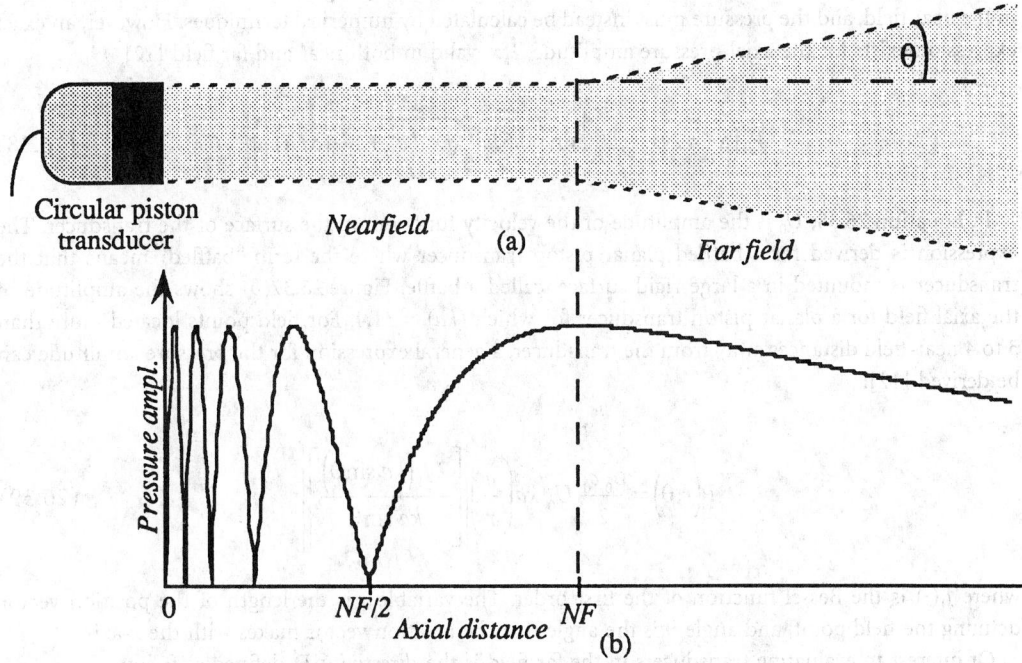

FIGURE 26.37 Pressure fields from a circular planar piston transducer operating with CW excitation. NF = Near Field. (*a*) Approximate field distribution in near field and far field. (*b*) Axial pressure in near field and far field for (*ka*) = 31.4.

patterns for these two cases are quite similar, the detailed field structure is very different. In this section, only CW fields are discussed.

When a CW excitation voltage is applied to a planar piston transducer, the resulting acoustic field can be divided into a *near field* region and a *far field* region. This division is particularly distinct when the transducer has a circular geometry. A piston transducer simply refers to a transducer with the same velocity amplitude at all points on the surface. The length of the near field, NF, is given as:

$$NF = \frac{a^2}{\lambda} \tag{26.36}$$

where *a* is the radius of the transducer and λ is the wavelength. As shown in Figure 26.37(*a*), the near field is approximately confined while the far field is diverging. The angle of divergence, θ, is approximately:

$$\theta = \arcsin\left(0.61\frac{\lambda}{a}\right) \tag{26.37}$$

Additional comments to the simplified representation in Figure 26.37(*a*) are in order:

1. The actual field has no sharp boundaries, in contrast to what is shown in Figure 26.37.
2. While the beam diameter is roughly constant in the near field, it is not as regular as shown; and at the same time, the near field structure is very complex.
3. The angle of divergence, θ, is only clearly established well into the far field.
4. The depiction of the far field shows only the main lobe; in addition, there are side lobes, which are directions of significant pressure amplitude, separated by *nulls* (i.e., directions with zero pressure amplitude).

In general, analytical expressions do not exist for the pressure magnitude at an arbitrary field point in the near field, and the pressure must instead be calculated by numerical techniques. However, an exact expression exists for the axial pressure amplitude, $P_{ax}x$, valid in both near and far field [12].

$$P_{ax}(x) = 2\rho_0 c_0 U_0 \left| \sin\left(0.5ka\left[\sqrt{x^2/a^2 + 1} - (x/a) \right] \right) \right|$$ (26.38)

In Equation 26.38, U_0 is the amplitude of the velocity function on the surface of the transducer. The expression is derived for a baffled planar piston transducer where the term "baffled" means that the transducer is mounted in a large rigid surface, called a baffle. Figure 26.37(b) shows the amplitude of the axial field for a planar piston transducer for which $(ka) = 31.4$. For field points located more than 3 to 4 near-field distances away from the transducer, a general expression for the pressure amplitude can be derived [12]:

$$P(r,\theta) = \frac{\rho_0 c_0}{2} U_0 \, ka\left(\frac{a}{r}\right) \left| \left[\frac{2J_1(ka\sin\theta)}{ka\sin\theta} \right] \right|$$ (26.39)

where $J_1(\cdot)$ is the Bessel function of the first order. The variable r is the length of the position vector defining the field point, and angle θ is the angle that the position vector makes with the x-axis.

Of interest in evaluating transducers in the far field is the *directivity*, D, defined as follows:

$$D(x) = \frac{I_{ax}(x)\left[\text{given source}\right]}{I_{ax}(x)\left[\text{simple source}\right]}$$ (26.40)

Thus, the directivity gives the factor with which the axial intensity of the given source is increased over that of a simple source (omnidirectional radiator), radiating the same total energy. For a baffled, circular planar piston transducer, the directivity in the far field can be calculated to be [12]:

$$D = \frac{(ka)^3}{ka - J_1(2ka)}$$ (26.41)

When $(2ka) \gg 1$, Equation 26.41 can be approximated to $D = (ka)^2$. Directivity values in the 1000s are common. For example, a 3.5 MHz transducer with 1 cm diameter radiating into water has a (ka) value of 73.3 and a directivity of 5373.

Often, focused transducers are used where the focusing is either created by the curvature of the piezoelectric element or by an acoustic lens in front of the transducer. The degree of focusing is determined by the (ka) value of the transducer.

Generation of Ultrasound: Piezoelectric and Magnetostrictive Phenomena

Ultrasound transducers today are available over a wide frequency range, in many sizes and shapes, and for a wide variety of applications. The behavior of the ultrasound transducer is determined by several parameters: the transduction material, the backing material, the matching layer(s), and the geometry and dimension of the transducer. A good overview of ultrasound transducers is available in [13].

The transduction material is most commonly a piezoelectric material, but can for some applications be a magnetostrictive material instead. These materials are inherently narrowband, meaning that they work efficiently only over a narrow frequency range. This is advantageous for CW applications such as

TABLE 26.3 List of Most Significant Piezoelectric Parameters for Common Piezoelectric Materials

Parameter	Barium titanate (BaTiO$_3$)	Lead zirconate titanate, PZT-5	Lead meta-niobate, PbNb$_2$O$_6$	Polyvinylidene fluoride (PVDF)
d_{33}	149 (10^{-12} m/V)	374 (10^{-12} m/V)	75 (10^{-12} m/V)	22 (10^{-12} m/V)
g_{33}	14 (10^{-3} Vm/N)	25 (10^{-3} Vm/N)	35 (10^{-3} Vm/N)	339 (10^{-3} Vm/N)
k_{33}	0.50	0.70	0.38	k_{31} = 0.12
k_T	0.38	0.68	0.40	0.11
Q_m	600	75	5	19

ultrasound welding and ultrasound hyperthermia, but is a problem for imaging applications, as impulse excitation will produce a long pulse with poor resolving abilities. To overcome this deficiency, a *backing material* is tightly coupled to the back side of the transducer for the purpose of damping the transducer and shortening the pulse. However, the backing material also reduces the sensitivity of the transducer. Some of this reduced sensitivity can be regained by the use of a *matching layer*, specifically selected for the medium of interest. As seen in Equation 26.33, a quarter wavelength matching layer can provide 100% efficient coupling to a medium, albeit only at one frequency. A combination of several matching layers can provide a more broadband impedance matching. The field is determined by both the geometry (planar, focused, etc.) of the transducer and by the frequency content of the velocity function of the surface of the transducer. In this section, unique aspects of the transduction material itself are described.

Piezoelectric Materials.

A piezoelectric material exhibits a mechanical strain (relative deformation) due to the presence of an electric field, and generates an electric field when subjected to a mechanical stress. A detailed review of piezoelectricity is given in [14]. Piezoelectric materials can either be: (1) natural material such as quartz; (2) man-made ceramics (e.g., barium titanate (BaTi), lead zirconate titanate (PZT), or lead meta-niobate); (3) man-made polymers (e.g., polyvinylidene fluoride (PVDF)). The piezoelectric ceramics are the most commonly used materials for ultrasound transducers. These ceramics are made piezoelectric by a so-called poling process in which the material is subjected to a strong electric field while at the same heating it to above the material's Curie temperature.

Several material constants determine the behavior of a given piezoelectric material, the most important of which are listed in Table 26.3 and defined below. An extensive list of parameter values for various piezoelectric materials is available in [15].

d_{33} Transmission constant
g_{33} Receiving constant
k_{33} Piezoelectric coupling coefficient
k_T Piezoelectric coupling coefficient for a transverse clamped material
Q_M Mechanical Q

The transmission constant, d_{33}, gives the mechanical deformation of piezoelectric materials for frequencies well below the resonance frequency. A transducer with a large d_{33} value will therefore become an efficient transmitter. If an electric field, E_3, is applied in the polarized direction of a piezoelectric rod or disk, the strain, S_3, in that direction is approximately:

$$S_3 = d_{33} E_3 = d_{33} \frac{V_{appl}}{l} \tag{26.42}$$

where V_{appl} is the applied voltage and l is the thickness. The total deformation, Δl, becomes

$$\Delta l = d_{33} V_{appl} \tag{26.43}$$

The receiving constant, g_{33}, defines the sensitivity of a transducer element as a receiver when the frequency of the applied pressure is well below the resonance frequency of the transducer. If the applied stress (force/area) in the direction of polarization is T_3, the output voltage from a rod or disk is approximately:

$$v_{out} = g_{33} T_3 l \qquad (26.44)$$

The coupling coefficient describes the power conversion efficiency of a piezoelectric transducer, *operating at or near resonance*. Specifically, k_{33} is the coupling coefficient for an unclamped rod; that is, the rod is allowed to deform in the directions orthogonal to the direction of applied force or voltage. In contrast, k_T is the coupling coefficient for a *clamped* disk. Finally, Q_M gives the mechanical Q, which is a measure for how narrowband the transducer material inherently is.

Whereas expressions of the type given in Equations 26.42 to 26.44 are adequate for describing static or low-frequency behavior, the behavior near resonance where most transducers operate requires more complex models, which are beyond the scope of this chapter. The Mason model is adequate for narrowband modeling of transducers, whereas the KLM or Redwood models better describe the transducers for broadband applications [13, 15].

Magnetostrictive Materials.
The magnetostrictive phenomenon refers to a magnetically induced contraction or expansion in ferroelectric media, such as in nickel or alfenol, and was discovered by Joule in 1847. Magnetostrictive materials are generally used for ultrasound frequencies below 100 kHz, and are therefore relevant mainly for underwater applications. Eddy current losses influence the performance of magnetostrictive materials, but the losses can be reduced by constructing the magnetostrictive transducer from thin laminations.

Transducer Specifications.
Ultrasound transducers are generally specified by their diameter, center frequency, focal distance, and type of focusing (if applicable). The transducer can be designed as a contact transducer, a submersible transducer, an air transducer, etc.; in addition, the type of connector or cabling can be specified. In many cases, measurement data for the actual transducer can be supplied by the vendor in the form of a measured pressure pulse and the corresponding frequency spectrum, recorded with a hydrophone at a specific field point. Similarly, the beam profile can be measured in the form of pulse amplitude or pulse energy as a function of lateral position.

Detailed information about the acoustic field from a radiating transducer can be obtained with either the Schlieren technique or the optical interferometric technique. Such instruments are quite expensive, falling in the $50K to $120K range.

Display Formats for Pulse-Echo Measurements

The basic description of ultrasound imaging was presented earlier in this chapter. Based on the information presented so far in this section, more specific aspects of pulse-echo ultrasound imaging will now be described. Different display formats are used; the simplest of these is the *A-mode* display.

A-mode.
When a pulse-echo transducer has emitted a pulse in a given direction and has been switched to receive mode, an output signal from the transducer is produced based on detected echoes from different ranges, as illustrated conceptually in Figure 26.31(*a*). In this signal, distance to the reflecting structure has been converted into time from the start of signal. This signal is often referred to as the RF signal. Demodulating this signal (i.e., generating the envelope of the signal) produces the A-mode (amplitude mode) display, or the A-line signal. This signal can be the basis for range measurements, attenuation measurements, and measurement of reflection coefficient.

M-mode.
If the A-mode signal from a transducer in a fixed position is used to *intensity* modulate a cathode-ray tube (CRT), such as a monitor or oscilloscope, along a straight vertical line, a line of dots with brightness

according to echo strengths would appear on the screen. Moving the display electronically across the screen results in a set of straight horizontal lines. Now consider the case where the reflecting structures are moving in a direction toward or away from the transducer while pulse-echo measurements were being performed. This results in variations in the arrival time of echoes in the A-line signal, and the resulting lines across the screen are no longer straight, but curved, as determined by their velocity. Such a display is called *M-mode*, or motion mode, display. An application for this would be measurement of the diameter variation of a flexible tube, or blood vessel, due to a varying pressure inside the tube.

B-mode.

If pulse-echo measurements are repeatedly being performed, while the transducer scans the object of interest, an image of the object can be generated, as illustrated in Figure 26.31. Specifically, each A-line signal is used to intensity modulate a line on a CRT corresponding to the location of ultrasound beam, which produces an image that maps the reflectivity of the structures in the object. The resulting image is called a *B-mode* (or brightness mode) image. If the transducer is moved linearly, a rectangular image is produced, whereas a rotated transducer generates a pie-shaped, or sector image. This motion is typically done electronically or electromechanically, as described earlier under single-element and array transducers. When the scanning is done rapidly (say, 30 scans/s), the result is a real-time image.

Many forms of signal processing and image enhancement can be applied in the process of generating the B-mode image. Echoes from deeper lying structures will generally be of smaller amplitude, due to the attenuation of the overlying layers of the medium. Attenuation correction is made especially in medical imaging, so that echoes are displayed approximately with their true strength. This correction consists of a time-varying gain, called *time-gain control*, or TGC, such that the early arriving echoes experience only a low gain and later echoes from deeper lying structures experience a much higher gain. Various forms for signal compressions or nonlinear signal transfer function can selectively enhance the weaker or the stronger echoes.

C-mode.

In a C-mode display, only echoes from a specific depth will be imaged. To generate a complete image, the transducer must therefore be moved in a raster scan fashion over a plane. C-scan imaging is slow and cannot be used for real-time imaging, but has several applications in NDE for examining a given layer, or interface, in a composite structure, or the inner surface of a pipe.

Flow Measurements by Doppler Signal Processing

Flow velocity can be obtained by various ultrasonic methods (Chapter 28, Section 7), e.g., by measuring the Doppler frequency or Doppler spectrum. The Doppler frequency is the difference between the frequency (or pitch) of a moving sound source, as measured by a *stationary* observer, and the actual frequency of the sound source. The change in frequency is determined by the speed and direction of the moving source. The classical example is a moving locomotive with its whistle blowing; the pitch is increased when the train moves toward the observer, and vice versa. With respect to ultrasound measurements, only the reflecting (or scattering) gas, fluid, or structure is moving and not the the sound source, yet the Doppler phenomenon is present here as well. In order for ultrasound Doppler to function, the gas or fluid must contain scatterers that can reflect some of the ultrasound energy back to the receiver. The Doppler frequency, f_d, is given as follows:

$$f_d = \frac{2v}{c_0} \cos \theta \qquad (26.45)$$

where v is the velocity of the moving scatterers, and θ is the angle between the velocity vector and the direction of the ultrasound beam. Doppler flowmeters require only access to the moving gas, fluid, or object from one side. For industrial use, when the fluids or gases often do not contain scatterers, *transmission* methods are preferred; these methods, however, are not based on the Doppler principle (Chapter 28, Section 7).

Two main categories of Doppler systems exist: the CW Doppler system and the PW (pulsed wave) Doppler system. The CW Doppler system transmits a continuous signal and does not detect the distance to the moving structure. It is therefore only applicable when there is just one moving structure or cluster of scatterers in the acoustic field. For example, a CW Doppler is appropriate for assessing the pulsatility and nature of blood flow in an arm or leg. CW Doppler systems are small and relatively inexpensive.

The PW Doppler system transmits a short burst at precise time intervals; it is therefore inherently a sampled system, and, as such, is subject to aliasing. It measures changes from one transmitted pulse to the next in the received signal from moving structures at one or several selected ranges, and is able to determine both velocity and range. Assuming that the ultrasound beam is narrow and thus very directional, the PW Doppler can create a *flow image*; that is, it can indicate the magnitude and direction of flow over an image plane. Commonly, color and color saturation are used to indicate direction and speed in a flow image, respectively. In contrast to CW Doppler systems, the PW Doppler systems require much signal processing and are therefore typically expensive and often part of a complete imaging system. When a distribution of velocities, rather than a single velocity, is encountered, as is often the case with fluid flow, a Doppler spectrum rather than a Doppler frequency is determined. An FFT (Fast Fourier Transform) routine is then used to reveal the Doppler frequencies and thus the velocity components present.

Review of Common Applications of Ultrasound and Their Instrumentation

Range Measurements, Air

Ultrasound range measurements are used in cameras, in robotics, for determining dimensions of rooms, etc. Measurement frequencies are typically around 50 kHz to 60 kHz. The measurement concept is pulse-echo, but with burst excitation rather than pulse excitation. Special electronic circuitry and a thin low-acoustic-impedance air transducer is most commonly used. Rugged solid or composite piezoelectric-based transducers, however, can also be used, sometimes up to about 500 kHz.

Thickness Measurement for Testing, Process Control, Etc.

Measurement of thickness is a widely used application of ultrasound. The measurements can be done with direct coupling between the transducer and the object of interest, or — if good surface contact is difficult to establish — with a liquid or another coupling agent between the transducer and the object. Ultrasound measurements of thickness have applications in process control, quality control, measuring build-up of ice on an aircraft wing, detecting wall thickness in pipes, as well as medical applications. The instrumentation involves a broadband transducer, pulser-receiver, and display or, alternatively, echo detecting circuitry and numerical display.

Detection of Defects, such as Flaws, Voids, and Debonds

The main ultrasound application in NDE is inspection for the localization of voids, crack, flaws, debonding, etc. [3]. Such defects can exist immediately after manufacturing, or were formed due to stresses, corrosion, etc. Various types of standard or specialized flaw detection equipment are available from ultrasound NDE vendors.

Doppler Flow Measurements

The flow velocity of a liquid or a moving surface can be determined through Doppler measurements, provided that the liquid or the surface scatters ultrasound back in the direction of the transducer, and that the angle between the flow direction and the ultrasound beam is known. Further details are given in the section about Doppler processing. CW and PW Doppler instruments are commercially available, with CW instrumentation being by far the least expensive.

Upstream/Downstream Volume Flow Measurements

When flow velocity is measured in a pipe with access to one or both sides, an ultrasound transmission technique can be used in which transducers are placed on the same or opposite sides of the pipe, with one transducer placed further upstream than the other transducer. From the measured *difference* in travel time between the upstream direction and the downstream direction, and knowledge about the pipe geometry, the volume flow can be determined. Special clamp-on transducers and instrumentation are available. An overview of flow applications in NDE is given in [16].

Elastic Properties of Solids

Since bulk sound speed varies with the elastic stiffness of the object, as given in Equation 26.15, sound speed measurements can be used to estimate elastic properties of solids under different load conditions and during solidification processes. Such measurements can also be used for measurement of product uniformity and for quality assurance. The measurements can be performed on bulk specimens or on thin rods, using either pulse-echo or transmission instrumentation [17]. Alternatively, measurements of the material's own resonance frequencies can be performed for which commercial instruments, such as the *Grindo-sonics*, are available.

Porosity, Grain Size Estimation

Measurement of ultrasound attenuation can reveal several materials parameters. By observing the attenuation in metals as a function of frequency, the grain size and grain size distribution can be estimated. Attenuation has been used for estimating porosity in composites. In medical ultrasound, attenuation is widely used for tissue characterization, that is, for differentiating between normal and pathological tissues. Pulse-echo instrumentation interfaced with a digitizer and a computer for data analysis is required.

Acoustic Microscopy

The measurement approaches utilized in acoustic microscopy are similar to other ultrasound techniques, in that A-scan, B-scan, and C-scan formats are used. It is in the applications and the frequency ranges where acoustic microscopy differs from conventional pulse-echo techniques. Although acoustic microscopes have been made with transducer frequencies up to 1 GHz, the typical frequency range is 20 MHz to 100 MHz, giving spatial resolutions in the range from 100 μm to 25 μm. Acoustic microscopy is used for component failure analysis, electronic component packaging, and internal delaminations and disbonds in materials, and several types of acoustic microscopes are commercially available.

Medical Ultrasound

Medical imaging is a large and diverse application area of ultrasound, especially in obstetrics, cardiology, vascular studies, and for detecting lesions and abnormalities in organs. The display format is either B-mode, using gray scale image, or a combination of Doppler and B-mode, with flow presented in color and stationary structures in gray scale. A wide variety of instruments and scanners for medical ultrasound are available.

Selected Manufacturers of Ultrasound Products

Table 26.4 contains a representative list of ultrasound equipment and manufacturers.

Advanced Topics in Ultrasound

Overview of Diffraction

The ultrasound theory presented thus far has emphasized basic concepts, and the applications that have been discussed tacitly assume that the field from the transducer is a plane wave field. This simplifying assumption is acceptable for applications such as basic imaging and measurements based on travel time. However, the plane wave assumption introduces errors when materials parameters (e.g., attenuation, surface roughness, and object shape) are sought to be measured with ultrasound. Therefore, to use ultrasound as a quantitative tool, an understanding is needed of the structure of the radiated acoustic field from a given transducer with a given excitation, and — equally important — the ability to calculate the actual radiated field. This leads to the topic of diffraction, which is the effect that accounts for the complex structure of both radiated and scattered fields. Not surprisingly, there are direct parallels between optical diffraction and acoustic diffraction. (As a separate issue, it should be noted that the ultrasound

TABLE 26.4 List of Products for and Manufacturers of Ultrasound Measurements

Product type	Manufacturer
Ultrasound transducers	Panametrics, 221 Crescent St., Waltham, MA 02154. (800) 225-8330
Ultrasound transducers	Krautkramer Branson Inc., 50 Industrial Park Rd., Lewistown, PA 17044. (717) 242-0327
Range measurements, air	Polaroid Corporation, Ultrasonics Components Group, 119 Windsor Street, Cambridge, MA 02139. (800) 225-1618
Pulser-receivers	Panametrics, 221 Crescent St., Waltham, MA 02154. (800) 225-8330
Pulser-receivers	JSR Ultrasonics, 3800 Monroe Ave., Pittsfield, NY 14534. (716) 264-0480
Ultrasound power ampl.	Amplifier Research, 160 School House Rd., Souderton, PA 18964. (800) 254-2677
Ultrasound power ampl.	Ritec, 60 Alhambra Rd., Suite 5, Warwick, RI 02886. (401) 738-3660
NDE instrumentation	Panametrics, 221 Crescent St., Waltham, MA 02154. (800) 225-8330
NDE instrumentation	Krautkramer Branson Inc., 50 Industrial Park Rd., Lewistown, PA 17044. (717) 242-0327
Acoustic microscopy	Sonoscan, 530 E. Green St., Bensenville, IL 60106. (708) 766-4603
Medical Imaging	Hewlett Packard, Andover, MA; ATL, Bothell, WA; Diasonics, Milpitas, CA; Siemens Ultrasound, Issaquah, WA.
Schlieren based imaging of acoustic fields	Optison, 568 Weddell Drive, Suite 6, Sunnyvale, CA 94089. (408) 745-0383
Optical based imaging of acoustic fields	UltraOptec, 27 de Lauzon, Boucherville, Quebec, Canada J4B 1E7. (514) 449-2096

theory presented here assumes that the wave amplitudes [pressure, displacement] are small enough so that nonlinear effects can be disregarded.)

Diffraction is basically an edge effect. Whereas a plane wave incident on a large planar interface is reflected in a specific direction, the plane wave incident on an edge results in waves scattered in many directions. Similar considerations hold for the field produced by a transducer: The surface of the transducer produces a so-called *geometric wave* that has the shape of the transducer itself; the edge of the transducer, however, generates an *edge wave* with the shape of an expanding torus. The actual pressure field is a combination of the two wave fields. Very close to a large transducer, the geometric wave dominates, and diffraction effects might not need to be considered. Over small regions far away from the transducer, the field can be approximated by a plane wave field, and diffraction does not need to be considered. However, in many practical cases, diffraction must be considered if detailed information about the pressure field is desired, and numerical methods must be employed for the calculations.

The structure of the axial field, shown in Figure 26.37(*b*), is a direct result of diffraction. Numerical evaluation of the diffracted field from a transducer can be done in several ways: (1) use of the Fresnel or Fraunhofer integrals (not applicable close to the transducer) to calculate the field at a single frequency at a specified plane normal to the transducer axis; (2) calculation of the pressure function at any point of interest in space, based on a specified velocity function, $u(t)$, on the surface of the transducer, using Rayleigh integral; (3) decomposing the velocity field in the plane of the transducer into its plane wave components, using a 2-D Fourier transform technique, followed by a forward propagation of the plane waves to the plane of interest and an inverse Fourier transform to give the diffracted field; or (4) use of finite element methods to calculate the diffracted field at any point or plane of interest. In the following, methods (1) and (2) will be described.

Fresnel and Fraunhofer Diffraction.
Let the velocity function on the surface of the transducer, $u(x,y)$, be specified for a particular frequency, ω. Assume that the transducer is located in the $(x,y,0)$ plane and that one is interested in the pressure field in the (x_0,y_0,z) plane. The Fresnel diffraction formulation [18] assumes that the paraxial approximation is fulfilled, requiring z to be at least 5 times greater than the transducer radius, in which case the *Fresnel diffraction integral* applies:

$$p\left(x_0, y_0, z, \omega\right) = \frac{A_0}{\lambda_z} \iint_S u\left(x, y, \omega\right) \exp\left[-jk\frac{x^2+y^2}{2z}\right] \exp\left[jk\frac{x_0x+y_0y}{z}\right] dx\, dy \quad (26.46)$$

where S is the surface of the transducer and A_0 is a constant. If one defines the two first terms of the integrand as some complex spatial function, $\Gamma(x,y)$, then Equation 26.46 is a scaled Fourier transform of $\Gamma(x,y)$.

If $k(x^2 + y^2)/2z \ll 1$, or, equivalently, $z > 10\ a^2/\lambda$, the second term in Equation 26.46 can be ignored, and the resulting equation is called the *Fraunhofer diffraction integral*:

$$p\left(x_0,\ y_0,\ z,\ \omega\right) = \frac{A_0}{\lambda z} \iint_S u\left(x,\ y,\ \omega\right) \exp\left[jk\frac{x_0 x + y_0 y}{z} \right] dx\ dy \qquad (26.47)$$

Thus, one can only use the Fraunhofer integral for calculating the far field diffraction. From Equation 26.47, one can make the interesting observation that the far field of a transducer is a scaled version of the Fourier transform of the source.

Pressure Function at a Given Field Point, Based on Rayleigh Integral.
While the Fresnel and Fraunhofer diffraction methods are CW methods, calculation of pressure from the Rayleigh integral is fundamentally an impulse technique, and is as such better suited for analysis of pulse-echo measurements. The basis for the calculation is the *velocity potential impulse response*, $h(\vec{r},t)$, obtained from the Rayleigh integral:

$$h\left(\vec{r},t\right) = \frac{1}{2\pi} \iint_S \frac{\delta\left(t - r'/c\right)}{r'} dS \qquad (26.48)$$

In Equation 26.48, r' is the distance from dS on the surface of the transducer to the field point, defined by the position vector $\vec{r}$, and $\delta(t)$ is the Dirac delta function. As can be seen, $h(\vec{r},t)$ is the result of an impulsive velocity excitation on the surface of transducer and is a function of both time and a spatial location. It is important to note that $h(\vec{r},t)$ exists in analytical form for several transducer geometries, and, by extension, for annular and linear array transducers [19].

For the case of an arbitrary velocity function, $u(t)$, on the transducer surface, the corresponding velocity potential, $\phi(\vec{r},t)$, is obtained as:

$$\phi\left(\vec{r},t\right) = u\left(t\right) \otimes h\left(\vec{r},t\right) \qquad (26.49)$$

where $\otimes$ refers to time domain convolution. Both particle velocity, $u(\vec{r},t)$, and pressure, $p(\vec{r},t)$, can be found from $\phi(\vec{r},t)$, as follows:

$$u\left(\vec{r},t\right) = \nabla \phi\left(\vec{r},t\right) \qquad (26.50)$$

$$p\left(\vec{r},t\right) = -\rho_0 \frac{\partial}{\partial t} \phi\left(\vec{r},t\right) \qquad (26.51)$$

where ∇ is the gradient operator.

Thus, from the expressions above, the pressure can be calculated for any field point, $\vec{r}$, when $u(t)$ and the transducer geometry are defined. In this calculation, all diffraction effects are included. However, given the high frequency content in $h(\vec{r},t)$ and in particular in the time derivative of $h(\vec{r},t)$, care must be taken to avoid aliasing errors, as described in [19].

Received Signal in Pulse-Echo Ultrasound.
The expression in Equation 26.51 allows for quantitative evaluation of the pressure field for an arbitrary point, line, or plane. However, it does not describe the calculation of the received signal in a pulse-echo

system. Consider a small planar reflector, placed in a homogeneous medium, and referred to as dR. The reflector has the area dA. The dimensions of dR must be small with respect to the shortest wavelength in the insonifying pulse. The location and the orientation of the planar reflector is given by $\vec{r}$ and $\hat{n}$, respectively, where $\hat{n}$ is a unit normal vector to the small reflector.

The voltage from the receiving transducer in a pulse-echo system due to dR is termed $d\,v(\vec{r},t)$ and can be determined when $u(t)$ is specified. The electro-acoustic transfer function for both the transmitting and the receiving transducer is assumed to be unity for all frequencies. For the case when the acoustic impedance of dR is much higher than that of the medium, $d\,v(\vec{r},t)$ is given as [20]:

$$d\,v\!\left(\vec{r},t\right)=A_0\,\rho_0\,\frac{\cos\!\left[\psi\!\left(\vec{r}\right)\right]}{c_0}\left[h\!\left(\vec{r},t\right)\otimes h\!\left(\vec{r},t\right)\otimes\frac{\partial^2}{\partial t^2}u\!\left(t\right)\right]dA$$

$$=A_0\cos\!\left[\psi\!\left(\vec{r}\right)\right]u\!\left(t\right)\otimes\frac{\rho_0}{c_0}\left[\frac{\partial^2}{\partial t^2}\left(h\!\left(\vec{r},t\right)\otimes h\!\left(\vec{r},t\right)\right)\right]dA$$

$$(26.52)$$

In Equation 26.52, A_0 is determined by the reflection coefficient of the reflector. The term $\cos[\psi(\vec{r})]$ is a correction term (obliquity factor) where $\psi(\vec{r})$ is the angle between $\hat{n}$ and the propagation direction of the wave field at $\vec{r}$. For an extended surface, the received voltage can be found by decomposing the surface into small reflectors and calculating the total received signal as the sum of the contributions from all the small reflectors. An efficient numerical technique for this type of integration has been developed [21].

References

1. C. M. Fortunko and D. W. Fitting, Appropriate ultrasonic system components for NDE of thick polymer-composites, *Review of Progress in Quantitative Nondestructive Evaluation, Vol. 10B.* New York: Plenum Press, 1991, 2105-2112.
2. C. R. Hill, Medical imaging and pulse-echo imaging and measurement, in C.R. Hill (ed.) *Physical Principles of Medical Ultrasound,* New York: Halsted Press, 1986, chaps. 7 and 8, 262-304.
3. L. C. Lynnworth, *Ultrasonic Measurements for Process Control,* San Diego: Academic Press, 1989, 53-89.
4. L. E. Kinsler, A. R. Frey, A. B. Coppens, and J. V. Sanders, *Fundamentals of Acoustics,* 3rd. ed., New York: John Wiley, 1982, 106.
5. J. D. Achenbach, *Wave Propagation in Elastic Solids,* 1st ed., New York: Elsevier Science, 1975, 123.
6. L. E. Kinsler, A. R. Frey, A. B. Coppens, and J. V. Sanders, *Fundamentals of Acoustics,* 3rd. ed., New York: John Wiley, 1982, 115-117.
7. L. E. Kinsler, A. R. Frey, A. B. Coppens, and J. V. Sanders, *Fundamentals of Acoustics,* 3rd. ed., New York: John Wiley, 1982, 127-133.
8. J. Saniie and N. M. Bilgutay, Quantitative grain size evaluation using ultrasonic backscattered echoes, *J. Acoust. Soc. Am.,* 80, 1816-1824, 1986.
9. E. P. Papadakis, Scattering in polycrystalline media, in P. D. Edmonds (ed.) *Ultrasonics,* New York: Academic Press, 1981, 237-298.
10. S. M. Handley, M. S. Hughes, J. G. Miller, and E. I. Madaras, Characterization of porosity in graphite/epoxy laminates with polar backscatter and frequency dependent attenuation, *1987 Ultrasonics Symp.,* 1987, 827-830.
11. J. C. Bamber, Attenuation and absorption, in C.R. Hill (ed.), *Physical Principles of Medical Ultrasound,* New York: Halsted Press, 1986, 118-199.
12. L. E. Kinsler, A. R. Frey, A. B. Coppens, and J. V. Sanders, *Fundamentals of Acoustics,* 3rd. ed., New York: John Wiley, 1982, 176-185.
13. M. O'Donnell, L. J. Busse, and J. G. Miller, Piezoelectric transducers, in P. D. Edmonds (ed.), *Ultrasonics,* New York: Academic Press, 1981, 29-65.

14. IEEE Standard on Piezoelectricity, *IEEE Trans. Sonics Ultrasonics*, 31, 8–55, 1984.

15. G.S. Kino, *Acoustic Waves*, Englewood Cliffs, NJ: Prentice-Hall, 1987, 17-83 and 554-557.

16. L. C. Lynnworth, *Ultrasonic Measurements for Process Control*, San Diego, CA: Academic Press, 1989, 245-368.

17. L. C. Lynnworth, *Ultrasonic Measurements for Process Control*, San Diego, CA: Academic Press, 1989, 537-557.

18. V. M. Ristic, *Principles of Acoustic Devices*, New York: John Wiley & Sons, 1983, 316-320.

19. D. P. Orofino and P. C. Pedersen, Multirate digital signal processing algorithm to calculate complex acoustic pressure fields, *J. Acoust. Soc. Am.*, 92, 563-582, 1992.

20. A. Lhemery, Impulse-response method to predict echo responses from targets of complex geometry. I. Theory, *J. Acoust. Soc. Am.*, 90, 2799-2807, 1991.

21. S. K. Jespersen, P. C. Pedersen, and J. E. Wilhjelm, The diffraction response interpolation method, *IEEE Trans. Ultrasonics, Ferroelectrics, and Frequency Control*, 45, Nov. 1998.

Further Information

L. C. Lynnworth, *Ultrasonic Measurements for Process Control*, San Diego, CA: Academic Press, 1989, an excellent overview of industrial applications of ultrasound.

L. E. Kinsler, A. R. Frey, A. B. Coppens, and J. V. Sanders, *Fundamentals of Acoustics*, 3rd. ed., New York: John Wiley & Sons, 1982, a very readable introduction to acoustics.

P. D. Edmonds (ed.), *Ultrasonics*, (Vol. 19 in the series: *Methods of Experimental Physics*). New York: Academic Press, 1981, in-depth description of ultrasound interaction with many types of materials, along with discussion of ultrasound measurement approaches.

J. A. Jensen, *Estimation of Blood Velocities Using Ultrasound*, Cambridge, UK: Cambridge University Press, 1996, a very up-to-date book about ultrasound Doppler measurement of flow and the associated signal processing.

E. P. Papadakis (ed.), *Ultrasonic Instruments and Devices: Reference for Modern Instrumentations, Techniques, and Technology*, in the series *Physical Acoustics*, Vol. 40, New York: Academic Press, 1998.

F. W. Kremkau, Diagnostic Ultrasound: Principles and Instruments, 5th ed., Philadelphia, PA: W. B. Saunders Co., 1998, a very readable and up-to-date introduction to medical ultrasound.

27

Acoustic Measurement

Per Rasmussen
G.R.A.S. Sound and Vibration

Sound is normally defined as vibration of a solid, liquid, or gaseous medium in the frequency range of the human ear, i.e., between about 20 Hz and 20 kHz. Here, the definition is further limited and only vibrations in liquids and gaseous media are considered. In contrast to solid media, a liquid or gaseous medium cannot transmit shear forces, so sound waves are always longitudinal waves, in which the particles moves in the direction of propagation of the wave. The wave propagation in gaseous and liquid media can be described by the three variables: the pressure p, the particle velocity u, and the density ρ. The relation between these is described by the wave equation [1], and this can be derived from three basic equations: the Euler equation (this is essentially Newton's second law applied to a fluid), the Continuity equation, and the State equation. Although the wave equation in principle can be used to describe and calculate all sound waves in all situations, it will in practice often be impossible to perform the necessary calculations. In some special cases, it is possible to get analytical results directly from the wave equation and these cases are therefore of special interest. The cases most often encountered in acoustics is the *free field,* the *diffuse* (or reverberant) *field,* and the *closed coupler.* The free field is, in principle, an infinite, empty (except for the medium and the source) space, with no reflections. Here, the waves are allowed to radiate freely in all directions without reflections. In practice, the free field is implemented in an anechoic chambers, where all walls have been made nearly 100% absorptive. The diffuse field is obtained in a reverberation room where all walls have been made, in principle, 100% reflective. At the same time, the walls are made nonparallel and the result is a sound field with sound waves in all directions. The closed coupler is a small chamber, with dimensions small compared to the wavelength of the sound. A special case of this is the standing wave tube. This is a tube with a diameter smaller than the wavelength and with a sound source in one end. With a suitable loudspeaker as a source, the wave propagation in the tube can be assumed to be one-dimensional. This simplifies the mathematical description so that it is possible to calculate the sound field.

In practice, almost the only parameter measured directly in acoustics is the sound pressure, and all other parameters like sound power, particle velocity, reverberation time, directivity, etc. are derived from pressure measurements. These are performed with measurement microphones in gaseous media and hydrophones in liquid media. The measurement microphones are all of the condenser type to ensure precision, long-term stability, and sensitivity. Hydrophones are usually made with a rubber coating over a sensitive element of piezoelectric material.

The traditional frequency range from 20 Hz to 20 kHz for acoustic measurements is selected because this is the range audible to the human ear. Sound waves exist outside this range in the form of infrasound (below 20 Hz) and ultrasound (above 20 kHz). As the basic equations (the wave equation) and measurement principles are the same for both infrasound and ultrasound, many of the principles from the frequency range from 20 Hz to 20 kHz can be extended to these ranges.

27.1 The Wave Equation

Sound wave propagation cannot take place in a vacuum, but is always associated with some kind of medium. For simplicity, assume that this medium is air, although the same equations are valid also for all gaseous and fluid media. In this medium, the concept of an air particle can be introduced. An *air particle* is a small volume of air in which the acoustical parameters like pressure, density, etc. can be considered constant. On the other hand, the air volume must be large enough to include a very large number of air molecules, so that the air volume can be considered to be a continuous medium and not a collection of molecules. The *Euler equation* for such an air particle is given by:

$$-\text{grad } p = \rho_0 \left(\frac{\partial \vec{v}}{\partial t} \right) \tag{27.1}$$

where p is the pressure, ρ_0 is the density, and v is the particle velocity. This equation can be considered as Newton's second law ($F = ma$) applied to a fluid. Here, the gradient of the pressure equals the force F acting on the air particle, the density equals the mass m, and the time differentiated particle velocity $\partial v/\partial t$ equals the acceleration.

The second equation necessary to derive the wave equation is the *continuity equation*. This simply states that if you have a small volume of air and you bring in some extra air, the density (or the mass) will increase. Mathematically, this can be formulated as:

$$\text{div } \vec{v} = -\frac{1}{\rho_0} \cdot \frac{\partial \rho}{\partial t} \tag{27.2}$$

where c is the sound velocity. The sound velocity depends on the composition of the air and the temperature. For normal air at 0°C, the velocity is 314 m s^{-1} while at 20°C, the velocity is 340 m s^{-1}.

The third equation is the *state equation*, which relates pressure changes to changes in the density, that is, if a small volume of air is compressed, the density will increase:

$$\frac{\partial \rho}{\partial t} = \frac{1}{c^2} \times \frac{\partial p}{\partial t} \tag{27.3}$$

Now we have three equations relating the three variables: pressure, particle velocity, and density. By eliminating the particle velocity and the density, we obtain one differential equation for the sound pressure:

$$\nabla^2 p = \frac{1}{c^2} \frac{\partial^2 p}{\partial t^2} \tag{27.4}$$

This is the *wave equation* for acoustic waves in gaseous and fluid media. In principle, this allows one to calculate the sound pressure anywhere in a sound field, if some suitable boundary conditions are given. In practice, however, it is only possible to find solutions in a few simple cases.

27.2 Plane Sound Waves

In a free space, at great distance from the sound source, sound waves are approximately plane waves. This means that the wave equation only depends on one of the coordinates in the wave equation. If the direction of propagation of the wave fronts is in the *x*-direction, the solution to the wave equation reduces to:

$$p = A \cos\left[\left(\omega t - k\right)\left(x + \varphi_a\right)\right] \tag{27.5}$$

where ω is the frequency. Similarly, the particle velocity in the free field is given by:

$$v = \frac{A}{\rho c} \cos\left[\left(\omega t - k\right)\left(x + \varphi_a\right)\right] = \frac{p}{\rho c} \tag{27.6}$$

This means that in a plane wave, the particle velocity is equal to the pressure divided by the constant ρc and the pressure and particle velocity are in phase. The constant ρc is the *acoustic impedance;* and for air at 20°C, the density is 1.29 kg m^{-3} and the sound velocity is 340 m s^{-1}, giving an acoustic impedance of 438.6 kg m^{-2} s^{-1}.

The plane wave transmits energy in the direction of propagation. The power transmitted per unit area is the intensity in the direction of the propagation (in many older textbooks, terms like "the intensity of the sound" were mistakenly used for the magnitude of the sound pressure). In general, the intensity is given by the product of the sound pressure and the sound velocity; thus, in the case of the plane wave, the intensity (*I*) can be calculated from Equations 27.5 and 27.6:

$$I = vp = \frac{p^2}{\rho c} \tag{27.7}$$

Thus, in the plane wave, the intensity transmitted by the wave can be calculated from the sound pressure and, as the sound power is the intensity per unit area, the sound power can be calculated by multiplying the intensity by the area.

27.3 Spherical Waves

Another simple solution to the wave equation can be found for the radiation from a point source into free space. The point source is an infinitely small sphere whose surface is pulsating radially. In practice, for small sound sources (i.e., where the dimensions of the sound source is small compared to the wavelength of the sound), the point source is a good approximation for the real physical source that makes the spherical wave solution of special interest.

The wave equation in Equation 27.5 is transformed into the spherical coordinates r, θ, and ψ. As the point source radiates equally in all directions, the solution depends only on the distances r from the center of the point source:

$$p = \frac{P_0}{r} \cos\left(\omega t - kr\right) \tag{27.8}$$

It can be seen that the sound pressure is inversely proportional to the distance from the sound source. The particle velocity can be divided into a near-field contribution v_n and a far-field contribution v_f:

$$v_f = \frac{P_0}{r\rho c} \cos(\omega t - kr) \qquad (27.9)$$

$$v_n = \frac{P_0}{\omega \rho r^2} \sin(\omega t - kr) \qquad (27.10)$$

The far-field contribution in Equation 27.9 can be seen to be in-phase with the pressure in Equation 27.8 and also the particle velocity is inversely proportional to the distance r from the point source. The near-field contribution is inversely proportional to the square of the distance to the source and therefore dies away rapidly as the distance to the source increases.

As in the case of the plane wave, the intensity in the spherical wave is the product of the pressure and particle velocity. For the near-field contribution, one obtains:

$$I = v_n p = \frac{P_0}{\omega \rho r^2} \sin(\omega t - kr) \frac{P_0}{r} \cos(\omega t - kr) = 0 \qquad (27.11)$$

That is, the near-field part of the particle velocity does not contribute to the radiated power as the particle velocity is 90° out of phase with the pressure.

The far-field contribution is given by:

$$I = v_f p = \frac{P_0}{r\rho c} \cos(\omega t - kr) \frac{P_0}{r} \cos(\omega t - kr) = \frac{P_0^2}{r^2 \rho c} \cos^2(\omega t - kr) \qquad (27.12)$$

It can be seen that the intensity decreases with the square of the distance to the source and by combining Equations 27.8 and 27.12, one obtains:

$$I = \frac{P_0^2}{r^2 \rho c} \cos^2(\omega t - kr) = \frac{p^2}{\rho c} \qquad (27.13)$$

which is identical to Equation 27.7 for the plane wave. Thus, as for the plane progressive wave, the intensity in the spherical wave can be calculated from the pressure.

27.4 Acoustic Measurements

As can be seen from the wave equation, the full acoustic field can in principle be described from only pressure measurements. This means that all other acoustic parameters can be derived from pressure measurements and, in practice, pressure is often the only parameter measured. There have been a few attempts to make transducers for particle velocity measurements based on, for example, transmission of ultrasonic waves; but the absolute dominating transducers for acoustic measurements are the condenser-type microphones, Figure 27.1. These have proven to be superior with respect to temperature stability, long-term stability, and insensitivity to rough handling. While measurement microphones are designed and produced to ensure well-defined and accurate measurements, a wide range of other microphones are available for other purposes. These can, for example, be for incorporation in telephones, where price is a very decisive factor, or for studio recordings, where a subjective evaluation is more important than the objective performance.

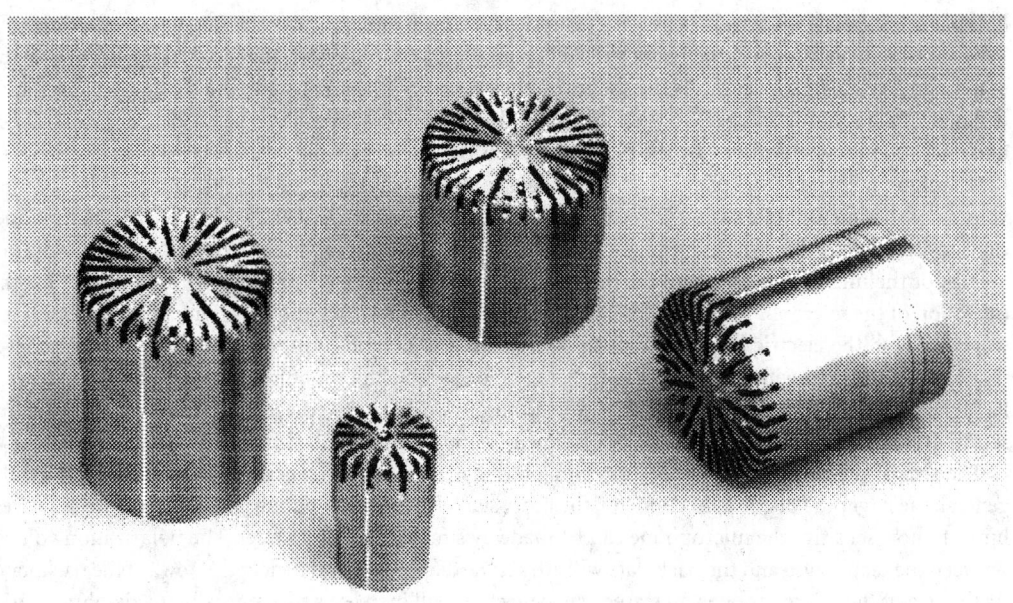

FIGURE 27.1   Measurement microphones: ½″ and ¼″.

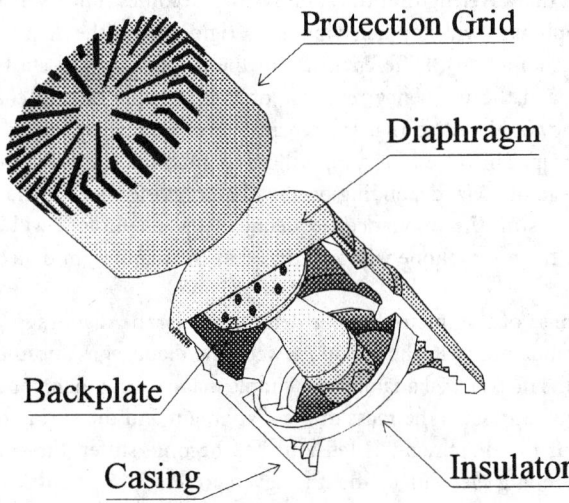

FIGURE 27.2   Basic elements of a measurement microphone.

Condenser Microphones

The condenser microphone consists basically of five elements: protection grid, microphone casing, diaphragm, backplate, and insulator; see Figure 27.2. The diaphragm and the backplate form the parallel plates of an air capacitor. This capacitor is polarized with a charge from an external voltage supply (externally polarized type) or by an electric charge injected directly into an insulating material on the backplate (prepolarized type). When the sound pressure in the sound field fluctuates, the distances between the diaphragm and the backplate will change, and consequently change the capacitance of the diaphragm/backplate capacitor. As the charge on the capacitor is kept constant, the change in capacitance will generate an output voltage on the output terminal of the microphone. The acoustical performance of a microphone is determined by the physical dimensions such as diaphragm area, the distance between the diaphragm and the backplate, the stiffness and mass of the suspended diaphragm, and the internal volume of the microphone casing. These factors will determine the frequency range of the microphone, the sensitivity, and the dynamic range. The sensitivity of the microphone is described as the output voltage of the microphone for a given sound pressure excitation, and is in itself of little interest for the operation of the microphone, except for calibration purposes. However, the sensitivity of the microphone (together with the electric impedance of the cartridge) also determines the lowest sound pressure level that can be measured with the microphone. For example, with a microphone with a sensitivity of 2.5 mV Pa^{-1}, the lowest level that can be measured is around 40 dB (re. 20 µPa), while a microphone with a sensitivity of 50 mV Pa^{-1} can measure levels down to approximately 15 dB (re. 20 µPa).

The size of the microphone is the first parameter determining the sensitivity of the microphone. In general, the larger the diaphragm diameter, the more sensitive the microphone will be. There are, however, limits to how sensitive the microphone can be made by simply making it larger. The polarization voltage between the diaphragm and the backplate will attract the diaphragm and deflect this toward the backplate. As the size of the microphone is increased, the deflection will increase and eventually the diaphragm will be deflected so much that it will touch the backplate. To avoid this, the distance between the diaphragm and the backplate can be increased or the polarization voltage can be decreased. Both of these actions will, however, decrease the sensitivity, so that the optimum size of a practical measurement microphone for use up to 20 kHz is very close to ½″ (12.6 mm).

As the size of the microphone is decreased, the useful frequency range of the microphone is increased. The frequency range, which can be obtained, is determined in part by the size of the microphone. At high frequencies, when the wavelength of the sound waves becomes much smaller than the diameter of the diaphragm, the diaphragm will stop behaving like a rigid piston (the diaphragm "breaks up" — this is not a destructive phenomenon). Different parts of the diaphragm will start to move with different magnitude and phase, and the frequency response of the microphone will change. To avoid this, the upper limiting frequency is placed so that the sensitivity of the microphone drops off before the diaphragm starts to break up. This gives, for a typical 0.5 in. microphone, an upper limiting frequency in the range from 20 kHz to 40 kHz, depending on the diaphragm tension. If the diaphragm is tensioned so that it becomes more stiff, the resonance frequency of the diaphragm will be higher; on the other hand, the sensitivity of the microphone will be reduced as the diaphragm deflection by a certain sound pressure level decreases.

The frequency response of the microphone is determined by the diaphragm tension, the diaphragm mass, and the acoustical damping in the airgap between the diaphragm and the backplate. This system can be represented by the mechanical analogy of a simple mass–spring–damper system as in Figure 27.3. The mass in the analogy represents the mass of the diaphragm and the spring represents the tension in the diaphragm. Thus, if the diaphragm is tensioned to become stiffer, the corresponding spring will become stiffer. The damping element in the analogy represents the acoustical damping between the diaphragm and the backplate. This can be adjusted by, for example, drilling holes in the backplate. This will make it easier for the air to move away from the airgap when the diaphragm is deflected, and therefore decrease damping.

The frequency response of the simple mechanical model of the microphone is given in Figure 27.4, together with the influence of the different parameters. At low frequencies (below the resonant frequency), the response of the microphone is determined by the diaphragm tension, and as described above, the sensitivity will increase if the tension is decreased. The resonant frequency is determined by the diaphragm tension and the diaphragm mass, with an increased tension giving an increased resonant frequency, and an increased mass giving a decreased resonant frequency. The response around the

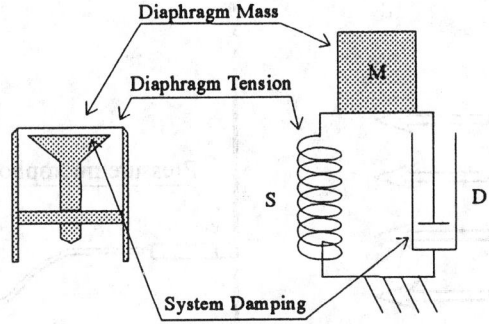

FIGURE 27.3 Mechanical analogy of a microphone.

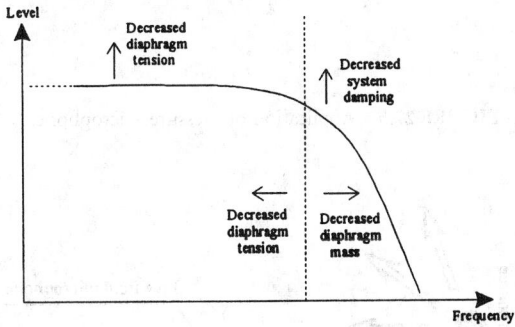

FIGURE 27.4 Influence of microphone parameters on frequency response.

resonant frequency is determined by the acoustical damping, where an increase in the damping will decrease the response.

Although the material selection and assembling techniques have changed during the last few years, the basic types of microphones remain unchanged. The basic types are the free-field microphones, the pressure microphones, and the random incidence microphones. They have been constructed with different frequency characteristics, corresponding to the different requirements.

The *pressure microphone* is meant to measure the actual sound pressure as it exists on the diaphragm. A typical application could be the measurement of the sound pressure in a closed coupler or as in Figure 27.5, the measurement of the sound pressure at a boundary. In this case, the microphone forms part of the wall and measures the sound pressure on the wall itself. The frequency response of this microphone should be flat in a frequency range as wide as possible, taking into account that the sensitivity will decrease as the frequency range is increased. The acoustical damping in the airgap between the diaphragm and the backplate is adjusted so that the frequency response is flat up to and a little beyond the resonant frequency.

The *free-field microphone* is designed to essentially measure the sound pressure as it existed before the microphone was introduced into the sound field. At higher frequencies, the presence of the microphone itself in the sound field will change the sound pressure. In general, the sound pressure around the microphone cartridge will increase due to reflections and diffraction. The free-field microphone is designed so that the frequency characteristics compensate for this pressure increase. The resulting output of the free-field microphone is a signal proportional to the sound pressure as it existed before the microphone was introduced into the sound field. The free-field microphone should always be pointed toward the sound source (0° incidence), as in Figure 27.6. In this situation, the presence of the microphone diaphragm in the sound field will result in a pressure increase in front of the diaphragm, see

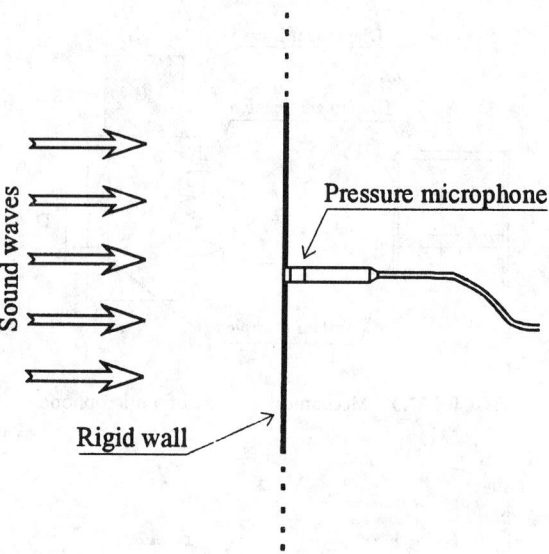

FIGURE 27.5   Application of pressure microphones.

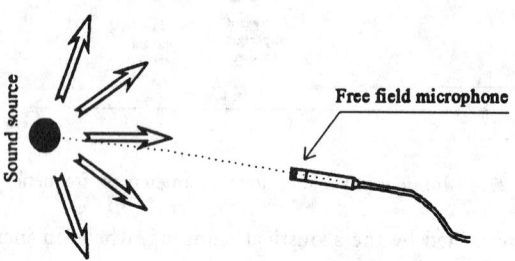

FIGURE 27.6   Application of a free-field microphone.

Figure 27.7(*a*), depending on the wavelength of the sound waves and the microphone diameter. For a typical ½″ microphone, the maximum pressure increase will occur at 26.9 kHz, where the wavelength of the sound (λ = 342 ms^{-1}/26.9 kHz ≈ 12.7 mm ≈ 0.5 in.) coincides with the diameter of the microphone. The microphone is then designed so that the sensitivity of the microphone decreases by the same amount as the acoustical pressure increases in front of the diaphragm. This is obtained by increasing the internal acoustical damping in the microphone cartridge, to obtain a frequency response as in Figure 27.7(*b*). The result is an output from the microphone, Figure 27.7(*c*), which is proportional to the sound pressure as it existed before the microphone was introduced into the sound field. The curve in Figure 27.7(*a*) is also called the "free-field correction curve" for the microphone, as this is the curve that must be added to the frequency response of the microphone cartridge to obtain the acoustical characteristic of the microphone in the free field.

The free-field microphone is required in principle, to be pointed toward the sound source and that the sound waves travel in essentially one direction. In some cases, (e.g., when measuring in a reverberation room or other highly reflecting surroundings), the sound waves will not have a well-defined propagation direction, but will arrive at the microphone from all directions simultaneously. The sound waves arriving at the microphone from the front will cause a pressure increase, as described for the free-field microphone, while the waves arriving from the back of the microphone will be decrease to a certain extent due to the shadowing effects of the microphone cartridge. The combined influence of the waves coming from different directions therefore depends on the distribution of sound waves from different directions. For

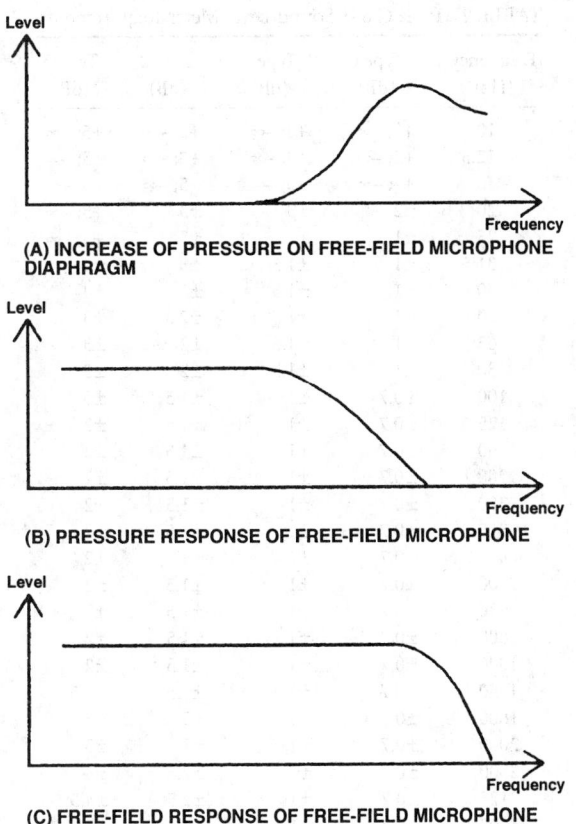

FIGURE 27.7 Frequency response of free-field microphone: (*a*) pressure increase in front of diaphragm; (*b*) microphone pressure response; (*c*) resulting microphone output.

measurement microphones, a standard distribution has been defined, based on statistical considerations, resulting in a standardized random incidence microphone.

As mentioned previously, measurement microphones can be either of the externally polarized type or the prepolarized type. The externally polarized types are by far the most stable and accurate microphones and should be preferred for precision measurements. The prepolarized microphones are, however, preferred in some cases, in that they do not require the external polarization voltage source. This is typically the case when the microphone will be used on small hand-held devices like sound level meters, where a power supply for polarization voltage would add excessively to cost, weight, and battery consumption. Still, it should be realized that prepolarized microphones in general are much less stable to environmental changes than externally polarized microphones.

27.5 Sound Pressure Level Measurements

The human ear basically hears the sound pressure, but the sensitivity varies with the frequency. The human ear is most sensitive to sound in the frequency range from 1 kHz to 5 kHz, while the sensitivity drops at higher and lower frequencies. This has led to the development of several frequency weighting functions, which attempt to replicate the sensitivity of the human ear. Also, the response of the human ear to time-varying signals and impulses has led to the development of instruments with well-defined time weighting functions. The resulting measurement instrumentation is the *sound level meter*, as defined in for example by the IEC International Standard 651, "Sound Level Meters" [2]. The standard defines four classes of sound level meters for different accuracy's (Table 27.1). Type 0 is the most accurate,

TABLE 27.1   IEC 651 Sound Level Meter Requirements

Frequency (Hz)	Type 0 (dB)	Type 1 (dB)	Type 2 (dB)	Type 3 (dB)
10	+2; −∞	+3; −∞	+5; −∞	+5; −∞
12.5	+2; −∞	+3; −∞	+5; −∞	+5; −∞
16	+2; −∞	+3; −∞	+5; −∞	+5; −∞
20	±2	±3	±3	+5; −∞
25	±1.5	±2	±3	+5; −∞
31.5	±1	±1.5	±3	±4
40	±1	±1.5	±2	±4
50	±1	±1.5	±2	±3
63	±1	±1.5	±2	±3
80	±1	±1.5	±2	±3
100	±0.7	±1	±1.5	±3
125	±0.7	±1	±1.5	±2
160	±0.7	±1	±1.5	±2
200	±0.7	±1	±1.5	±2
250	±0.7	±1	±1.5	±2
315	±0.7	±1	±1.5	±2
400	±0.7	±1	±1.5	±2
500	±0.7	±1	±1.5	±2
630	±0.7	±1	±1.5	±2
800	±0.7	±1	±1.5	±2
1000	±0.7	±1	±1.5	±2
1250	±0.7	±1	±1.5	±2.5
1600	±0.7	±1	±2	±3
2000	±0.7	±1	±2	±3
2500	±0.7	±1	±2.5	±4
3125	±0.7	±1	±2.5	±4.5
4000	±0.7	±1	±3	±5
5000	±1	±1.5	±3.5	±6
6300	+1; −1.5	+1.5; −2	±4.5	±6
8000	+1; −2	+1.5; −3	±5	±6
10000	+2; −3	+2; −4	+5; −∞	+6; −∞
12500	+2; −3	+3; −6	+5; −∞	+6; −∞
16000	+2; −3	+3; −∞	+5; −∞	+6; −∞
16000	+2; −3	+3; −∞	+5; −∞	+6; −∞

intended for precision laboratory measurements, while Type 1 is most widely used for general-purpose measurements, see Figure 27.8. Type 2 is used where low price is of importance, while Type 3 is not used in practice because of the wide tolerances, making the results too unreliable. The output of the sound level meter is, in principle, assumed to be an approximate measure of the impression perceived by the human ear.

The sound level meter can be functionally divided into four parts: microphone and preamplifier, A-weighting filter, rms detector and display (Figure 27.9). The microphone should ensure the correct measurement of the sound pressure within the frequency range for the given class. Also, the standard gives requirements for the directionality of the microphone. The frequency response of the instrument, including the weighting filter, is given for sound waves arriving at the microphone along the reference direction. For sound waves arriving from other directions, the standard allows wider tolerances at higher frequencies, taking into account the inevitable reflections and diffraction occurring at higher frequencies.

The preamplifier converts the high-impedance output signal from the microphone to a low-impedance signal, but has in itself no or even negative voltage amplification. The signal from the preamplifier is then passed through an A-weighting filter. This is a standardized filter which, in principle, resembles the sensitivity of the human ear, so that a measurement utilizing this filter will give a result which correlates with the subjective response of an average listener. The filter, with the filter characteristic as in

NA-27

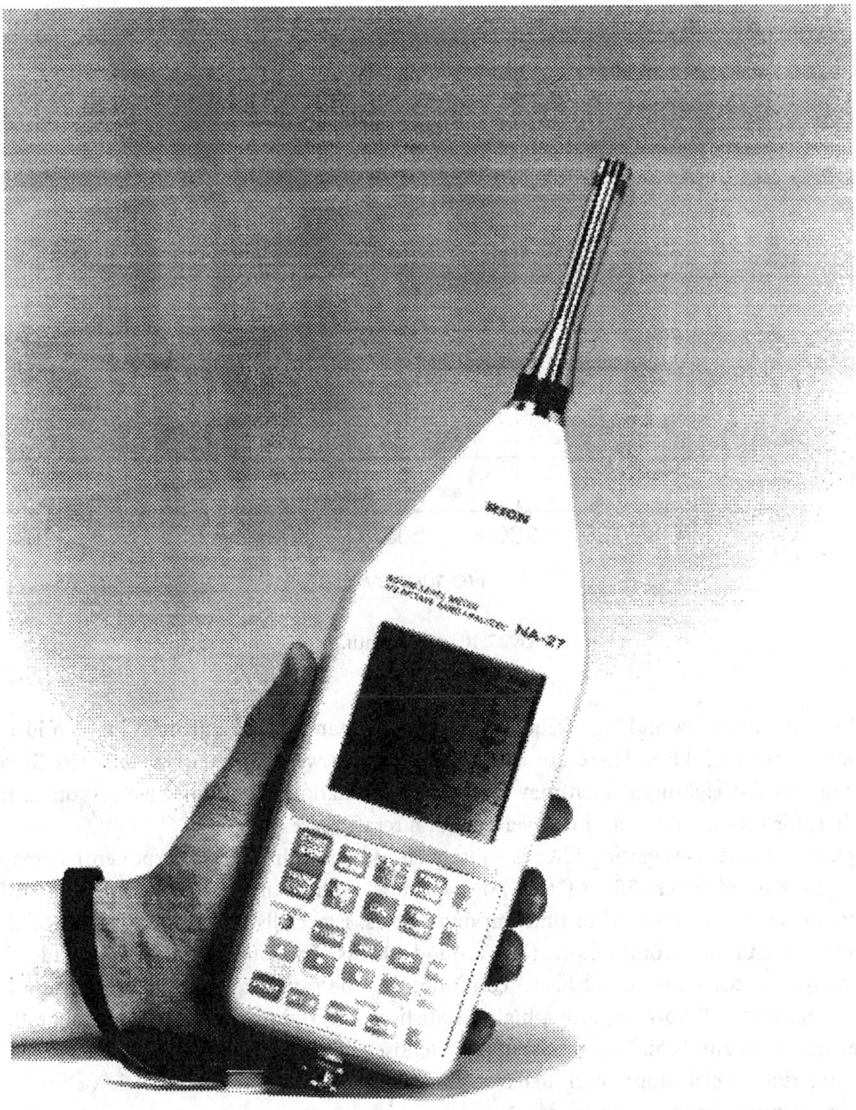

FIGURE 27.8 Modern Type 1 sound level meter with built-in frequency analyzer.

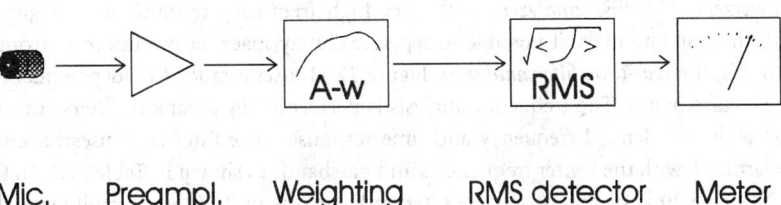

A-w
RMS
Mic.
Preampl.
Weighting
RMS detector
Meter

FIGURE 27.9 Functional parts of a sound level meter.

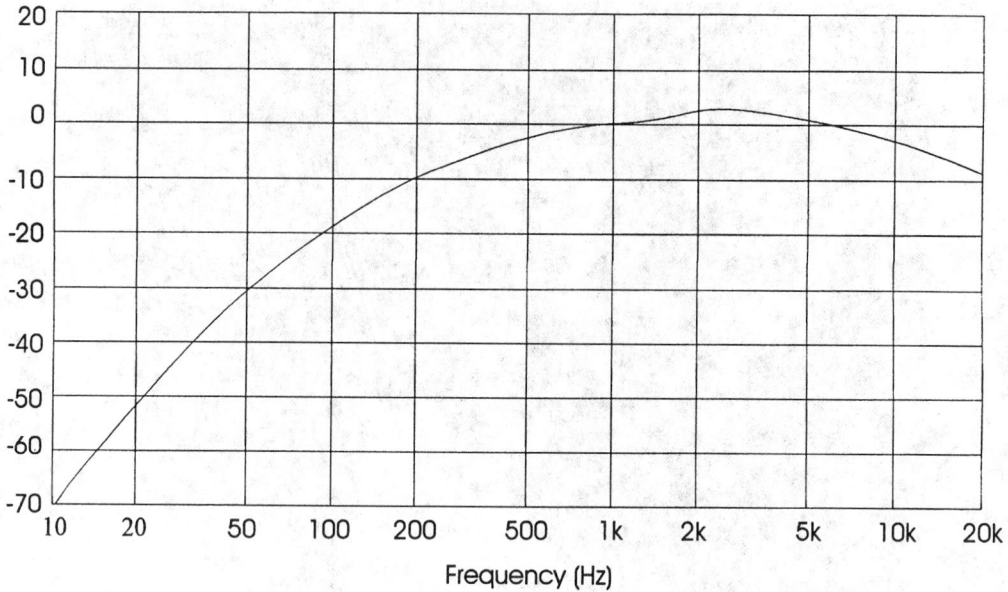

FIGURE 27.10   A-weighting curve.

Figure 27.10, attenuates low and high frequencies and slightly amplifies frequencies in the mid-frequency range from 1 kHz to 5 kHz. There are a number of other weighting curves, denoted B-weighting, C-weighting and D-weighting, which may give better correlation with subjective responses in special cases, such as for very high or very low levels, or for aircraft noise.

The signal from the A-weighting filter is subsequently passed through an exponential rms detector, with a time constant of either 125 ms ("fast") or 1 s ("slow"). These time constants simulate the behavior of the human ear when subjected to time-varying signals. Especially when the duration of the sound stimuli to the human ear becomes shorts (e.g., around 200 ms), the sound is subjectively judged as being lower compared to the same sound heard continuously. The same effect is obtained using the "fast" averaging time. This will however give a higher statistical uncertainty on the level estimate than when using the time constant "slow," so this should be chosen if the sound signal is continuously. Other standards describe special sound level meters such as integrating sound level meters or impulse sound level meters intended for special purposes.

27.6   Frequency Analyzers

While the sound level meter gives a single reading for the sound level in the frequency range from 20 Hz to 20 kHz, it is often desirable to have more detailed information about the frequency content of the signal. Two types of frequency analyzers are commonly used in acoustic measurements: FFT analyzers and real-time filter analyzers. The *FFT analyzers*, with very high frequency resolutions, can give a wealth of frequency information and make it possible to separate closely space harmonics (e.g., from a gear box). In contrast to this, the *real-time filter analyzers*, Figure 27.11, uses a much broader frequency resolution, usually in 1/3-octave bands. The frequency analysis is performed by a bank of filters (nowadays mostly digital filters) with well-defined frequency and time responses. The filter responses have been internationally standardized, with the center frequencies and passbands as shown in Table 27.2. In the frequency range from 20 Hz to 20 kHz, the 1/3-octave filterbank consists of 31 filters, simultaneously measuring the input signal. The resulting 1/3-octave spectrum resembles the subjective response of the human ear.

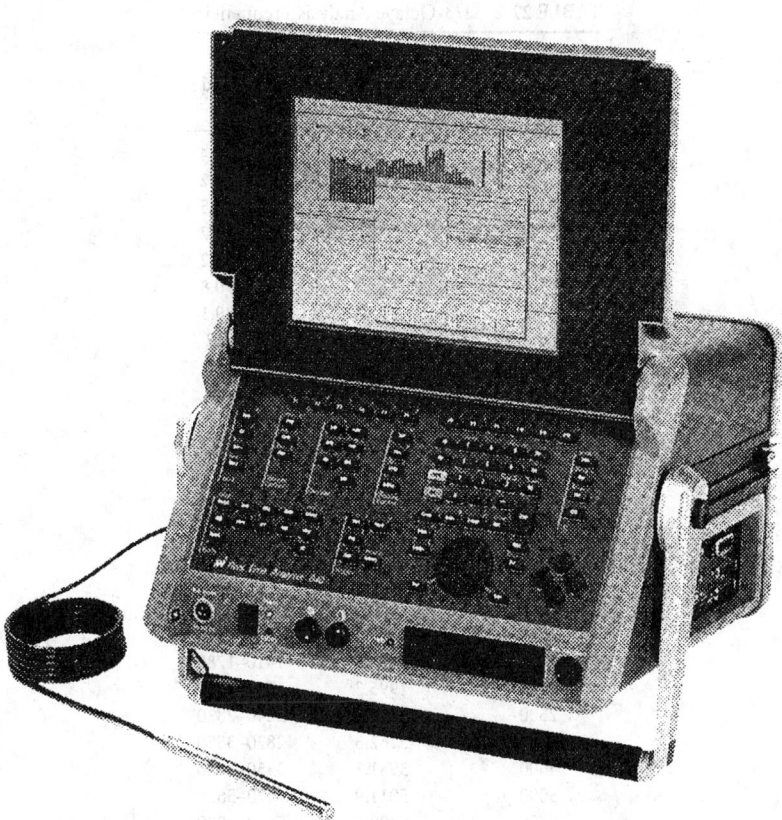

FIGURE 27.11   Real-time frequency analyzer for acoustic measurements.

27.7   Pressure-Based Measurements

The result of sound pressure measurements will be influenced by many factors: source, source operating conditions, surroundings, measurement position, etc. Depending on the goal of the measurement, these parameters can be controlled in different manners. If the goal of the measurement is to quantify the noise exposure to an operator's ear in a noisy environment, it is important that the microphone is in the same position as the operator's ear would normally be in, and that the environment is equal to the normal operating environment. If, on the other hand, the task is to describe the sound source as a noise-emitting machine, it is important to minimize the influence of the environment on the measurement result.

If the aim is to describe the measuring object as a noise source, it is customary to state the radiated sound power for the source. This is a global parameter quantifying the total noise radiation from the source, and to a certain extent independent of the environment. The sound power can be measured in a number of different ways: in a free field, in a reverberation room, using a substitution technique, or using sound intensity technique. A free field is a sound field in which the sound is radiated freely in all directions, with no restricting walls or reflections. This is most often obtained in a semi-anechoic chamber, where all walls and ceiling have been covered by nearly 100% absorptive material, with only the floor made of reflecting material. When the sound source is placed in the semi-anechoic chamber, the emitted sound waves will radiate freely away from the source and, in the far field, the waves can be considered to be plane waves or spherical waves. Therefore, the sound intensity can be calculated from pressure measurements using Equation 27.7.

TABLE 27.2   1/3-Octave Analysis Frequencies

Nominal center frequency (Hz)	Exact center frequency (Hz)	Passband (Hz)
20	19.95	17.8–22.4
25	25.12	22.4–28.2
31.5	31.62	28.2–35.5
40	39.81	35.5–44.7
50	50.12	44.7–56.2
63	63.1	56.2–70.8
80	79.43	70.8–89.1
100	100.0	89.1–112
125	125.89	112–141
160	158.49	141–178
200	199.53	178–224
250	251.19	224–282
315	316.23	282–355
400	398.11	355–447
500	501.19	447–562
630	630.96	562–708
800	794.33	708–891
1000	1000.0	891–1120
1250	1258.9	1120–1410
1600	1584.9	1410–1780
2000	1995.3	1780–2240
2500	2511.9	2240–2820
3150	3162.3	2820–3550
4000	3981.1	3550–4470
5000	5011.9	4470–5620
6300	6309.6	5620–7080
8000	7943.3	7080–8910
10000	10000.0	8910–11200
12500	12589.3	11200–14100
16000	15848.9	14100–17800
20000	19952.6	17800–22400

As real sound sources seldom radiate equally in all directions, a number of measurements around the test object are averaged. ISO Standard 3745 "Acoustics—Determination of sound power levels of noise sources—Precision method for anechoic and semi-anechoic rooms" [3] specifies an array of microphone positions on a hemisphere over the test object, as in Figure 27.12, with the coordinates as in Table 27.3. As all points are associated with the same area, and as the sound power is intensity times the area, the total radiated sound power can be calculated as:

$$P = A \sum I_n = \frac{2\pi r}{\rho c} \sum p_n^2 \tag{27.14}$$

where A is the area of the test hemisphere with radius r, and p_n is the pressure measured in point number n.

27.8 Sound Intensity Measurements

The calculation in Equation 27.14 of the sound power from sound pressure measurements is based on Equation 27.7. This equation, which gives the intensity based on a pressure measurement, is however only valid in a free field, in the direction of propagation. In general, in the presence of background noise or with reflections from walls, etc., it is not possible to calculate the sound intensity from a single pressure

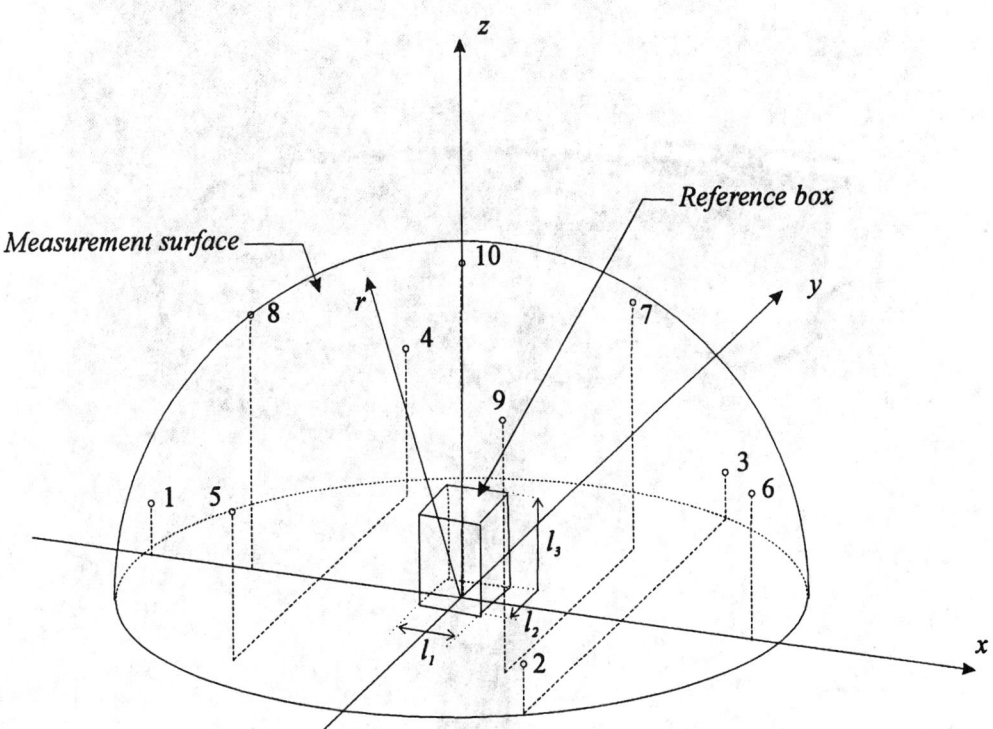

FIGURE 27.12 Measurement points for sound power determination.

TABLE 27.3. Coordinates of Measurement Points for Hemisphere with Radius r

Measurement point no.	x/r	y/r	z/r
1	−0.99	0	0.15
2	0.5	−0.86	0.15
3	0.5	0.86	0.15
4	−0.45	0.77	0.45
5	−0.45	−0.77	0.45
6	0.89	0	0.45
7	0.33	0.57	0.75
8	−0.66	0	0.75
9	0.33	−0.57	0.75
10	0	0	1.0

measurement. In these cases, it is however possible to measure directly the sound intensity with a two-microphone intensity probe, Figure 27.13.

Sound intensity I is the product of the pressure and the particle velocity:

$$I = pv \tag{27.15}$$

While the pressure p is a scalar and independent of the direction, the particle velocity is a vector quantity and directionally dependent. When the particle velocity is stated as in Equation 27.15, it is implicit that the velocity is in a certain direction and that the resulting intensity is calculated in the same direction.

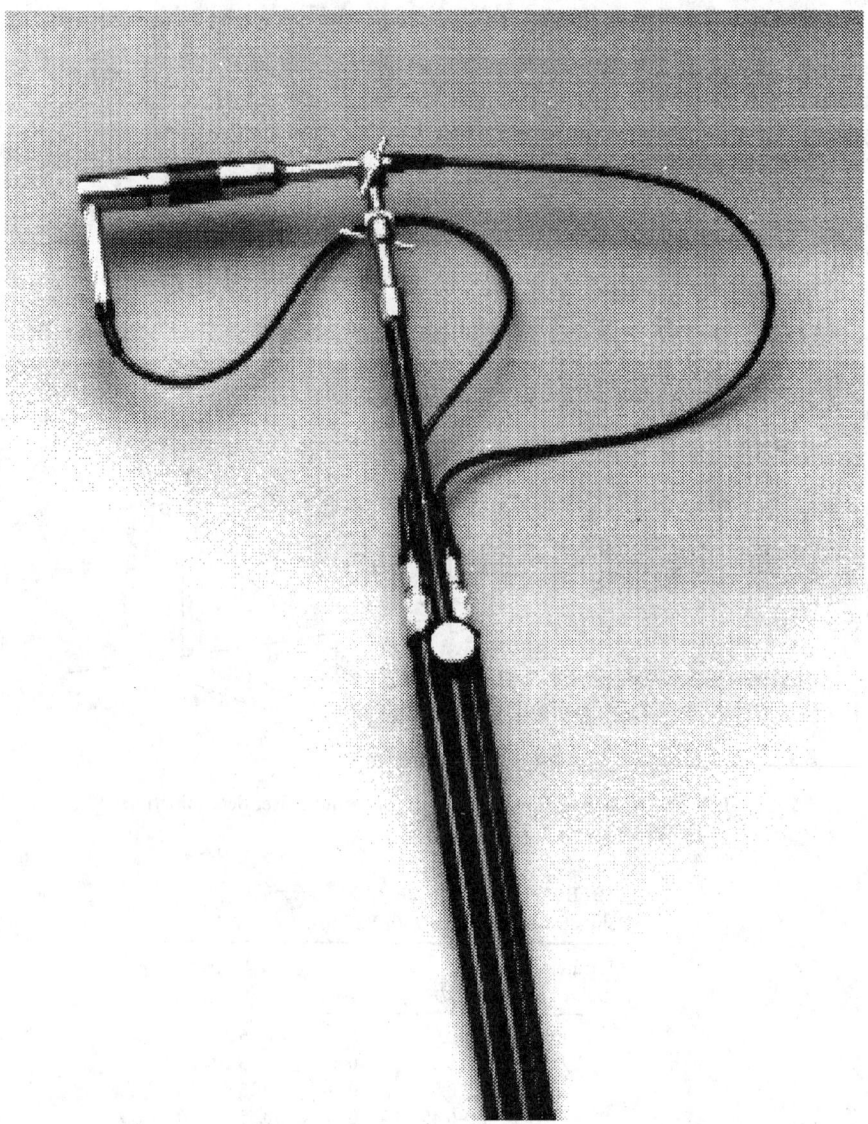

FIGURE 27.13 Two-microphone sound intensity probe.

For example, the particle velocity v in the direction of propagation, Figure 27.14(a), gives the intensity radiation away from the point source, while the particle velocity perpendicular to the propagation direction, Figure 27.14(b), is zero. The intensity calculated from Equation 27.15 will therefore be zero in the direction perpendicular to the propagation direction even though the sound pressure is the same. This means that the sound energy flows away radially from the point source and no energy is flowing tangentially.

The measurement of the sound intensity according to Equation 27.15 requires the measurement of the sound pressure and the particle velocity. With the two-microphone intensity probe, the pressure in a position in between the two microphones is calculated as the mean pressure measured by the two microphones:

$$p = \frac{p_1 + p_2}{2} \qquad (27.16)$$

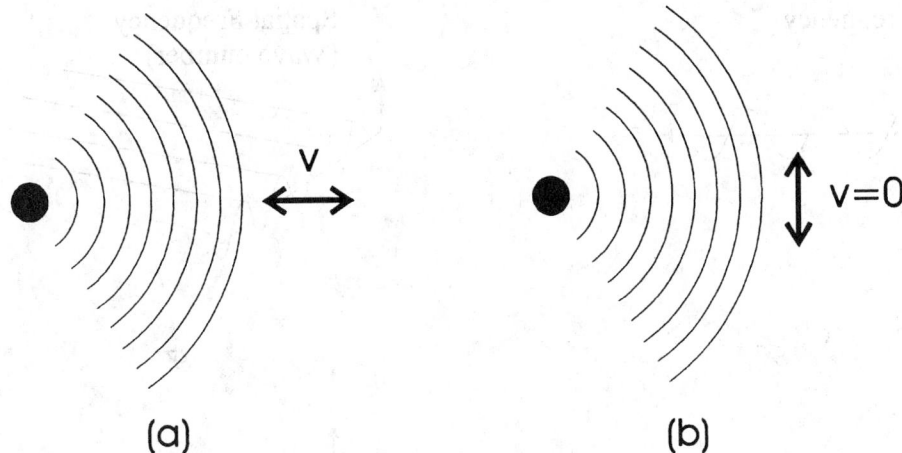

FIGURE 27.14 Particle velocity (*a*) along direction of propagation, and (*b*) perpendicular to direction of propagation.

The air particle velocity *v*, in the direction of the intensity probe, can be calculated from the pressure differences between the two microphone measurements:

$$v = \int \frac{\left(p_2 - p_1\right)}{\rho \Delta r} \, \partial \tau \qquad (27.17)$$

where ρ is the density of the air and Δr is the distance between the microphones. The intensity *I* is then obtained by multiplying the pressure and the velocity:

$$I = pv = \frac{p_1 + p_2}{2} \int \frac{\left(p_2 - p_1\right)}{\rho \Delta r} \, \partial \tau \qquad (27.18)$$

The intensity measurement technique is a powerful tool to localize acoustical noise source and to determine the sound power radiated from a sound source, even in the presence of other strong sound sources.

27.9 Near-Field Acoustic Holography Measurements

The term *acoustic holography* comes from the analogy to optical holographs. It is well known how holography, as opposed to a normal photo, enables one to reconstruct the full image of an object. This is obtained by "recording" information about both the magnitude and the phase of the light, while a normal photo only "records" the magnitude of the light. Similarly, with *acoustic holography*, both the magnitude and the phase of the sound field are measured over a plane surface. These measurements result in a complete description of the sound field where both magnitude and phase are known at all points. It is then possible to calculate acoustic quantities, including sound intensity distribution, particle velocity, sound power, radiation pattern, etc.

The basic assumption behind *near-field acoustic holography* (NAH) is that the sound field can be decomposed into two simple wave types: plane waves and evanescent waves. The *plane waves* describe the part of the sound field that is propagated away from the near field toward the far field, and the *evanescent waves* describe the complicated sound field existing in the near field. Any sound field can be described as a combination of plane waves and evanescent waves with different magnitude and directions.

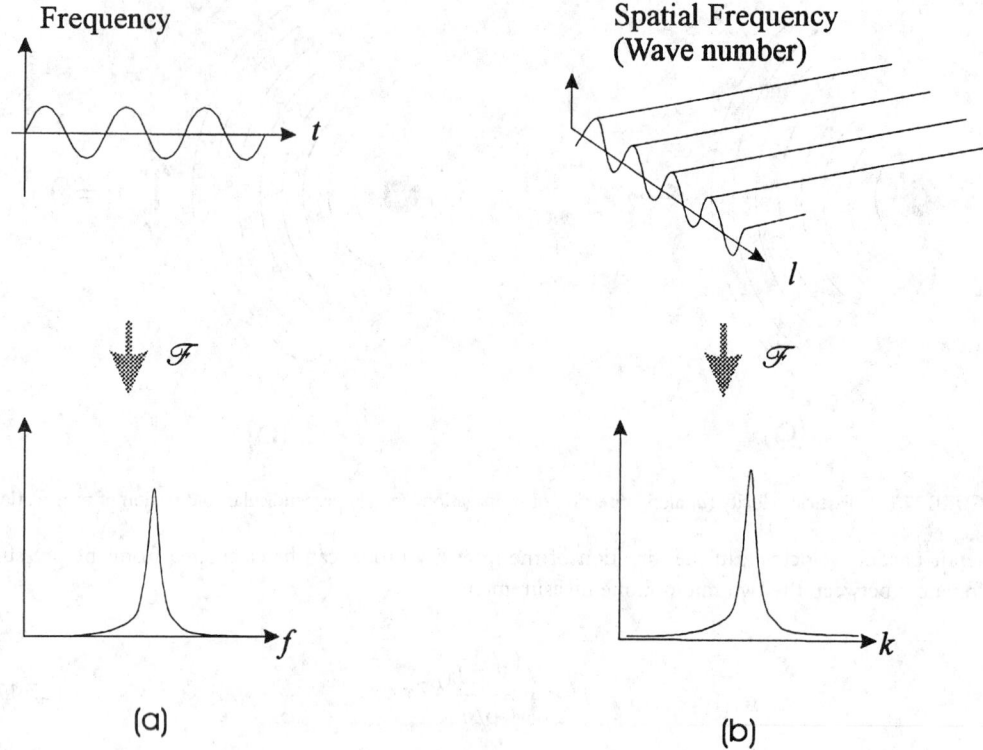

FIGURE 27.15 Temporal and spatial frequency of plane wave.

The magnitude and direction of the individual waves can be described by their spatial frequencies or wave numbers. For a simple plane wave propagating in a certain direction, this can be described in terms of its temporal frequency as well as by its spatial frequency. The temporal frequency, Figure 27.15(a), is obtained by looking at the pressure changes with time at a certain point in the sound field. This gives the temporal frequency in hertz or radians per second. Similarly, the spatial frequency, Figure 27.15(b), is obtained by looking at the pressure changes at a certain time. At that instant in time, the pressure will be different in different positions in space. If one moves in a certain direction in space, one will see a certain change in the pressure, corresponding to a spatial frequency, measured with the unit cycles per meter or radians per meter. As the temporal frequency gives information about how often the pressure changes with time at a certain point, the spatial frequency gives information about how often the pressure changes with position at a certain time. In the example of Figure 27.15(b), the propagation direction of the plane wave was identical to the direction of the axis along which the spatial frequency was measured. In this case, shown again in Figure 27.16(a), the relationship between the spatial frequency k_0 (i.e., the wave number) and the temporal frequency f is given by the speed of sound c:

$$k_0 = \frac{2\pi f}{c} = \frac{\omega}{c} = \frac{2\pi}{\lambda} \tag{27.19}$$

where λ is the wavelength. If, however, the axis along which the spatial frequency is measured is not the same as the propagation direction, see the example in Figure 27.16(b), this simple relationship is not valid. In this case, although the temporal frequency is the same as in Figure 27.16(a), the spatial frequency is lower. For one particular temporal frequency, the spatial frequencies will thus give information about the propagation directions. Therefore, if the sound field is made up of several plane waves with the same temporal frequency, but with different propagation directions, this will be shown in the spatial spectrum

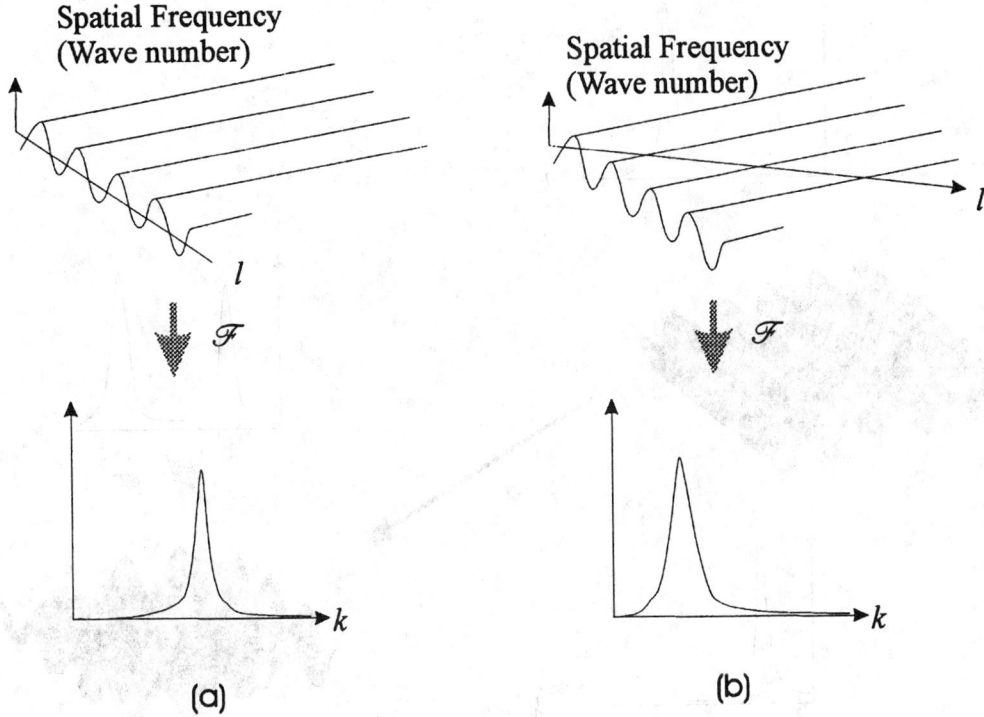

FIGURE 27.16   Spatial frequencies of waves propagating in different directions.

as several spatial frequency components. If, for example, the sound field is a sum of two waves, Figure 27.17, where one wave is traveling along the axis of measurement and the other at an angle of 45° relative to the first wave, the spatial spectrum will contain two spatial frequencies. One spatial frequency will be k, corresponding to a wave in the direction along the axis, and the other frequency will be k^* cos (45°). Thus far, the spatial frequencies have been defined along a single axis corresponding to a one-dimensional Fourier transformation. In the NAH technique, the sound field is sampled not only along a single axis, but over a plane. Therefore, a two-dimensional Fourier transformation is used instead. This gives as a result a two-dimensional spatial frequency spectrum, but otherwise the information is the same as before: namely, information about the direction and magnitude of the simple wave types.

The sound field from a point source cannot be explained by simple plane waves such as those in Figures 27.16 and 27.17, as the amplitude decreases with the distance from the origin. The plane waves retain the same magnitude over the full plane. Thus, to described the near-field phenomenon, one must introduce evanescent waves. In the one-dimensional Fourier spectrum, the evanescent waves can be identified as spatial frequencies higher than $k_0 = 2\pi f/c$. Similarly, in the two-dimensional spatial frequency spectrum, the evanescent waves can be identified as having spatial frequencies or wavenumbers higher than k_0.

The individual spatial frequencies in the two-dimensional spatial frequency spectrum correspond to simple plane waves or evanescent waves in the scan plane (i.e., the measurement plane). For each of these simple wave types, it is easy to calculate the pressure in other planes, see Figure 27.18. For the plane waves, a simple phase shift of the wave is required to calculate the result in a new plane. For the evanescent waves, the changes in amplitude must be taken into account; but in principle, this is also a simple transfer function applied to the two-dimensional spatial frequency spectrum. In this way, the two-dimensional spatial frequency spectrum in a new plane can be calculated from the original data by applying simple transfer function operations. The new two-dimensional spatial frequency spectrum is then an inverse Fourier transform (in two dimensions) to get the sound field in the new plane.

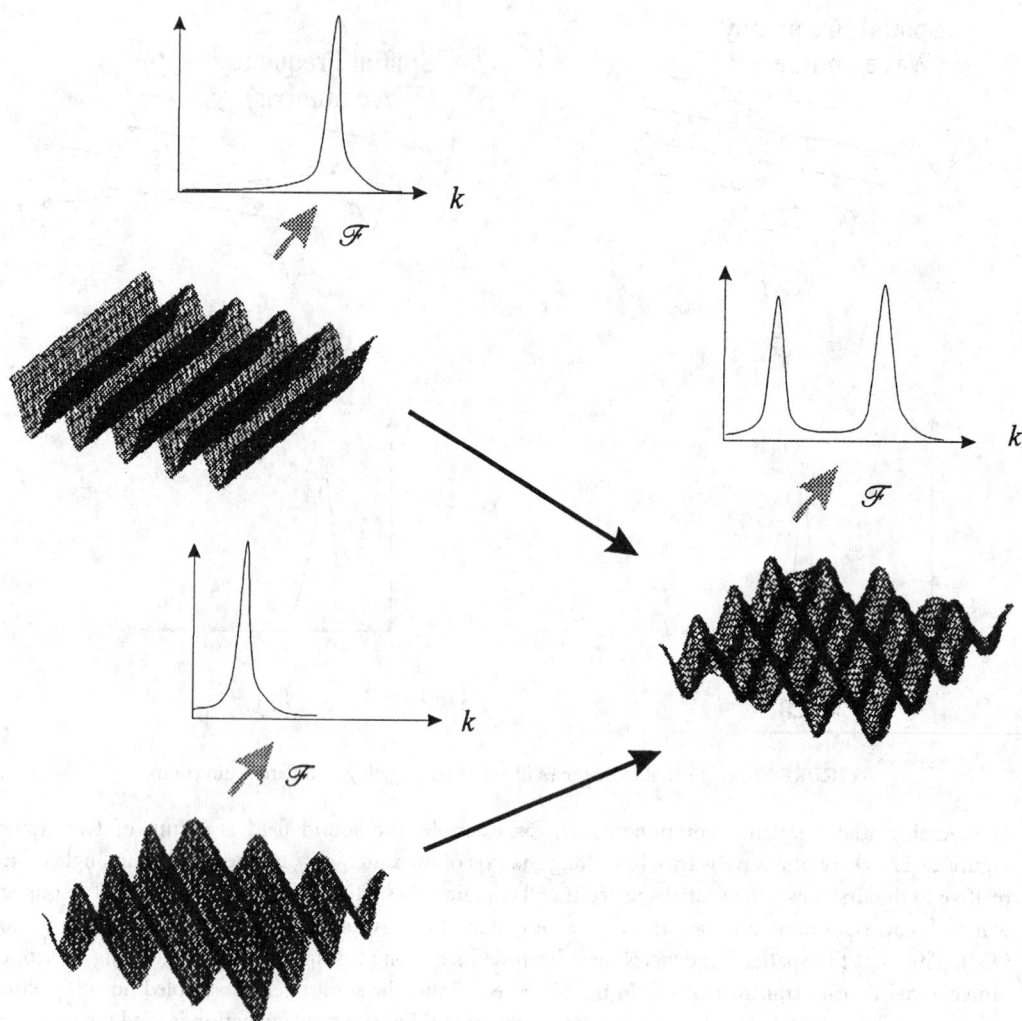

FIGURE 27.17 Spatial spectrum of two waves propagating in different directions.

The overall principle of near-field acoustic holography can be simplified as in Figure 27.19. The sound field is scanned in a plane close to the measuring object. This gives an array of temporal spectra, one for each scan position. Looking at one temporal frequency at a time, one takes out the information from each of the spectra corresponding to the actual frequency of interest. This generates a new array with information about only one temporal frequency. A Fourier transform (in two dimensions) is then applied to the array to generate a two-dimensional spatial frequency spectrum. This can then be transformed to new planes using simple transfer function operations. When the two-dimensional spatial frequency spectrum in the new plane has been calculated, an inverse Fourier transform is used to obtain the new pressure distribution in the new plane. In principle, the NAH technique requires that all cross-spectra between all the scan positions are given; that is, in each of the scan positions, all the cross-spectra to all other scan positions must be determined. A simple scan of a sound field with 2540 scan positions, defining $N = 1000$ scan positions, would result in $\frac{1}{2}N(N + 1) = 500,500$ cross-spectra. Instead of measuring all these cross-spectra, the system uses a set of reference transducers to reduce the amount of cross-spectra. The number of necessary reference transducers to give a complete description of the sound field without measuring the full amount of data is determined by the complexity of the sound field. A measurement

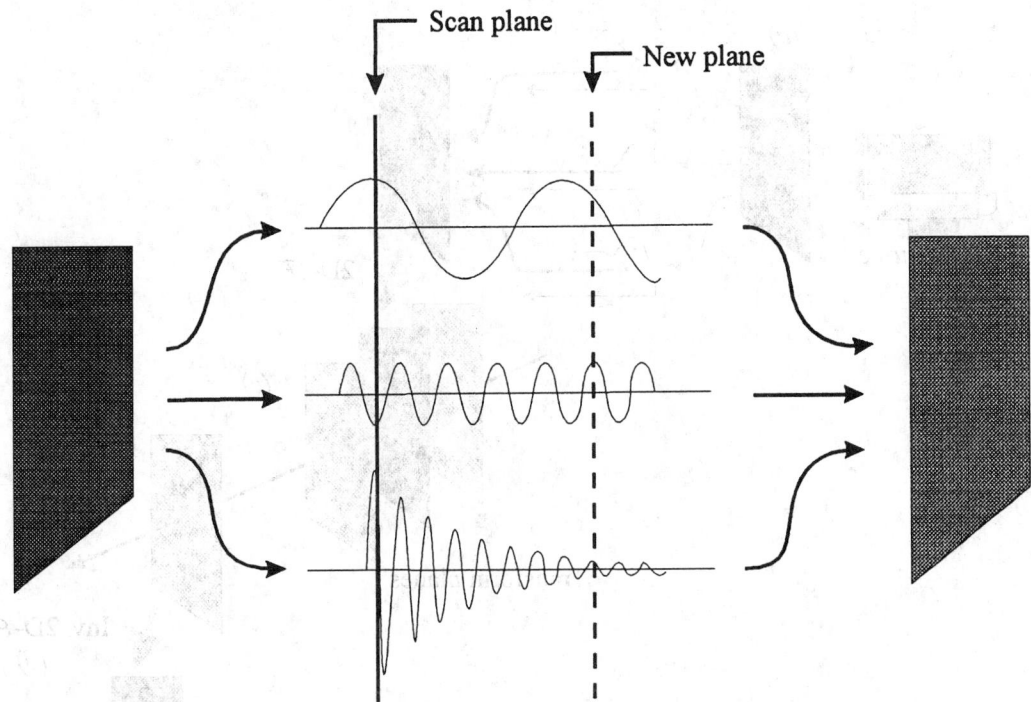

FIGURE 27.18   Transformation of simple wave types in a spatial spectrum, from one measurement plane to another plane.

with, for example, four reference transducers and 2540 scan positions will then be reduced to $4N = 4000$ cross-spectrum measurements.

27.10   Calibration

In order to make accurate and reliable measurements, the microphone and connected instruments must be properly calibrated. The calibration of measurement microphones can be divided into two parts: a level calibration and a frequency response calibration. The *level calibration* establishes the output signal of the microphone for a given acoustic input signal at a given frequency, while the *frequency response* gives the output at other frequencies relative to the level calibration frequency. The level calibration can be performed by a number of different methods with different accuracies.

The most accurate method is the *reciprocity calibration method*. This method utilizes the fact that a condenser microphone is a reciprocal transducer; that is, it can be used as a microphone (to convert an acoustical signal to a voltage signal) and as a loudspeaker (to convert a voltage signal into an acoustical signal). By measuring the relationship between three test microphones driven as both transmitters and receivers, one obtains a set of three equations with the three microphone sensitivities as the unknowns. By solving these three equations, one obtains the sensitivity of the three microphones. The reciprocity calibration method is very accurate but rather tedious and requires well-controlled environmental conditions and is therefore seldom used in practical situations.

The comparison or substitution methods are essentially identical in that they are based on measuring the differences between the test microphone and a reference microphone with known sensitivity. In this case, the reference microphone is often calibrated at an accredited national acoustical laboratory like NIST, NPL, or PTB, whereby the traceability is ensured. In the substitution method, the acoustical output of a sound source is measured with the reference microphone. Afterward, the reference microphone is replaced with a test microphone and the output is measured again. Provided that the sound source has

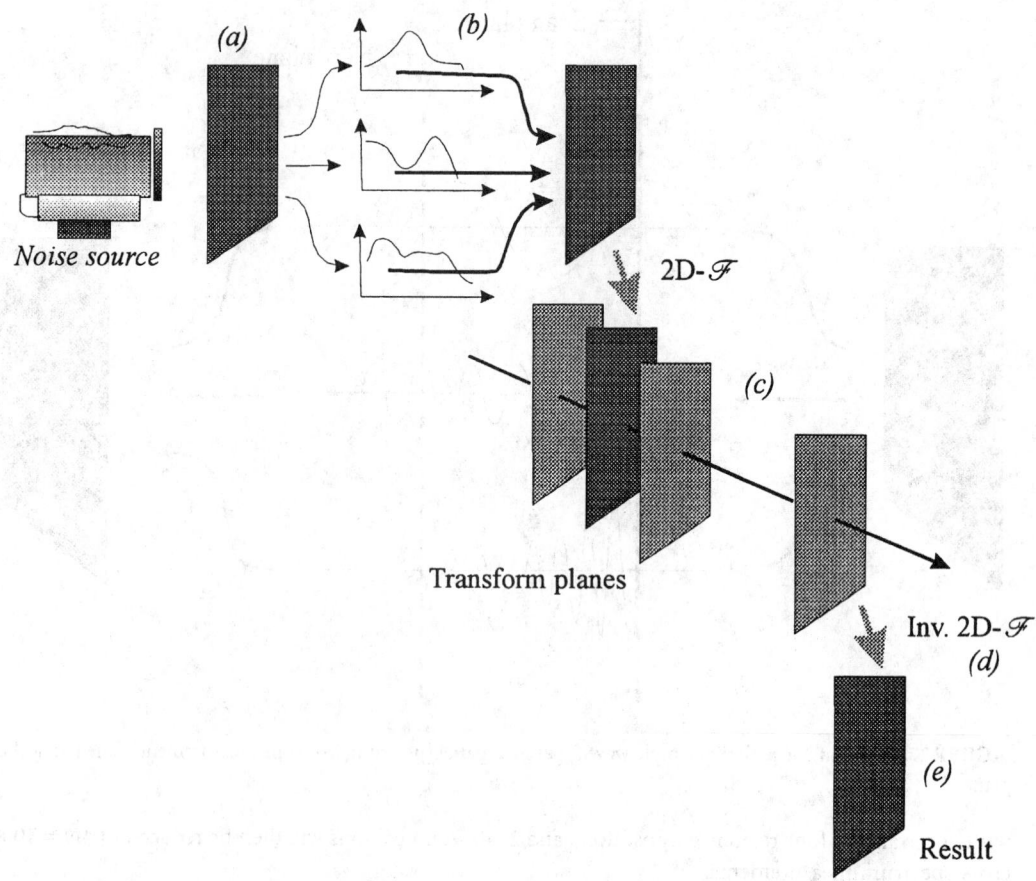

Noise source

2D-$\mathcal{F}$

Transform planes

Inv. 2D-$\mathcal{F}$
(d)

(e)

Result

FIGURE 27.19 Overall principle of NAH: (*a*) measurement of cross-spectra in the scan plane; (*b*) calculation for one temporal frequency at a time; (*c*) 2-D spatial Fourier transformation; (*d*) transformation of simple wave types; (*e*) inverse 2-D transformation; (*f*) to obtain the sound field in the new plane.

been stable, the sensitivity of the test microphone can then be calculated. The comparison method is similar to the substitution method, except that the reference microphone and the test microphone are subjected to the same sound pressure simultaneously and therefore the requirements to the sound source stability are less important.

An often-used method for microphone calibration is the *pistonphone method*. The pistonphone, Figure 27.20, is a very stable sound source, which produces a well-defined sound pressure level inside a closed coupler. It works by volume displacements, Figure 27.21, with a well-defined velocity, usually at 250 Hz. As the piston is moving in and out, the volume of the closed coupler is changed and this will result in pressure variations. The actual pressure level obtained in the pistonphone depends on the volume of the coupler, the volume displacement of the pistons, the barometric pressure, and—to a lesser degree—on other factors such as humidity, heat dissipation, etc. As the pistonphone is based on a relatively simple mechanical system, it is very reliable and easy to use in practice, with an accuracy around 0.1 dB. Also, the pistonphone is often used as the stable sound source for calibrations using comparisons or substitution methods.

A *sound pressure calibrator* is basically a small self-contained comparison calibration device. The test microphone is inserted into a small, closed volume and a small loudspeaker produces a single frequency signal, usually at 1 kHz. The output level of the loudspeaker is controlled in a feedback system with a signal from a reference microphone. Provided that the reference microphone and the feedback gain are

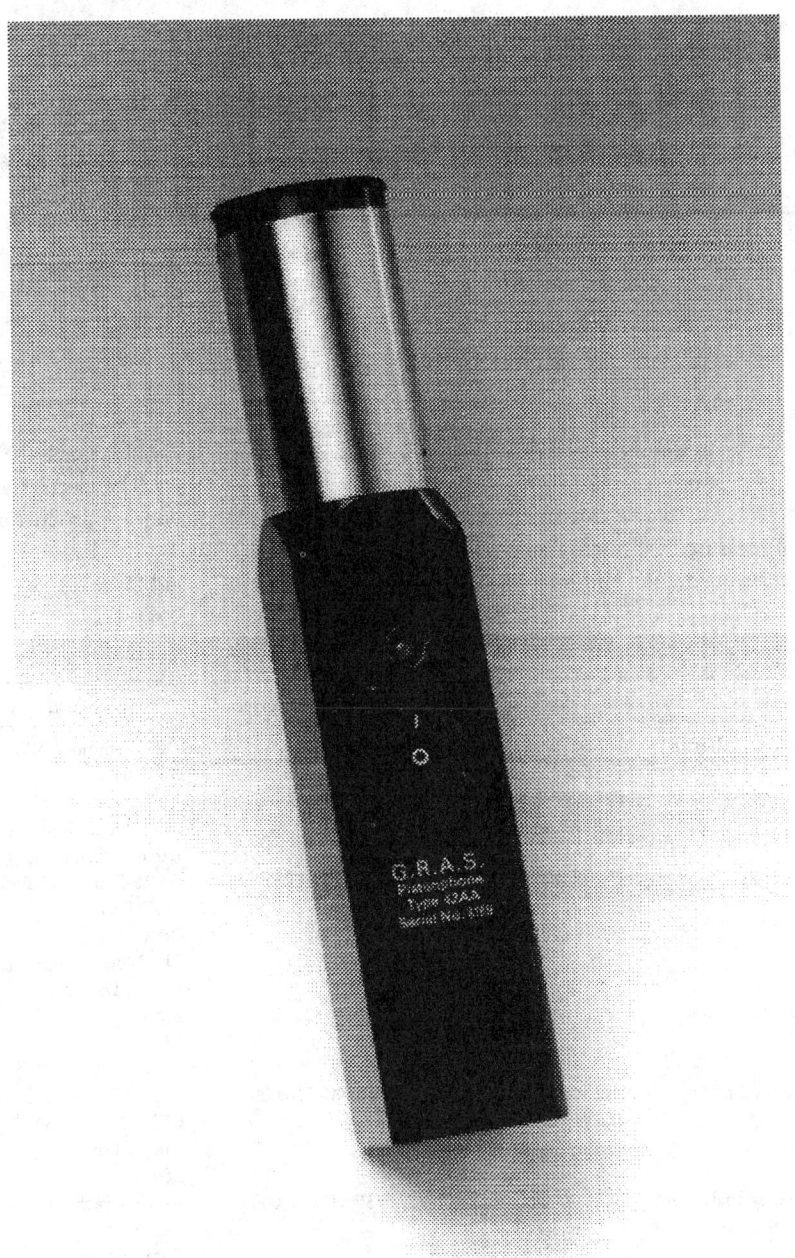

FIGURE 27.20 Pistonphone for microphone calibration.

stable, the sound level at the test microphone will be well-defined and the sensitivity can be determined. The sound level calibrators are normally not used to make accurate microphone calibrations, but rather to make field checks of the integrity of a complete measurement system.

The frequency response of a microphone is most often determined by the *electrostatic actuator method*. A conducting grid is placed close to and parallel to the microphone diaphragm. An electric field is established between the actuator and the diaphragm by applying 800 V dc to the actuator. A test signal of 50 to 150 V ac is superimposed on the dc signal, and the electrostatic forces will push and pull the

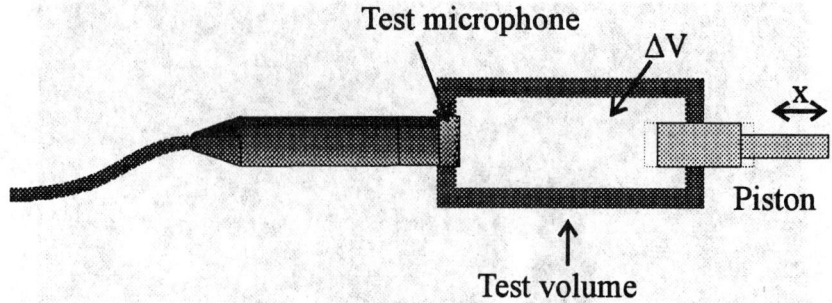

FIGURE 27.21   Principle of a pistonphone.

diaphragm, similar to a sound pressure of 1 to 10 Pa. By sweeping the test signal through the frequencies of interest, the pressure response of the test microphone can be recorded. The electrostatic actuator technique is widely used as a convenient and accurate test method, both during production and final calibration of measurement microphones.

Available instrumentation and manufacturers are given in Tables 27.4 and 27.5.

TABLE 27.4

Instrumentation	Types available	Approx. price	Manufacturers
Meas. microphones	½″ Free field	$750–$825	GRAS Sound & Vibration
	½″ Pressure		ACO Pacific
	¼″ Free field		B&K
	¼″ Pressure		The Modal Shop
			Larson Davies
Preamplifiers	½″ and 0.25 in.	$600–$850	GRAS Sound & Vibration
			ACO Pacific
			B&K
			The Modal Shop
			Larson Davies
Sound level meters	Simple type 1 SLM	$800–$2000	Rion
			CEL
			B&K
Sound level meters	Advanced SLM with freq. analysis and data storage	$2000–$10,000	Rion
			CEL
			Larson Davies
			B&K
Frequency analyzers	Real-time frequency analyzers/FFT	$5000–$50,000	Hewlett Packard
			Norsonic
			Data Physics
			01dB
Near-field acoustical holography	Complete system with 16–64 channel acquisition and postprocessing	$100,000–$200,000	LMS
			B&K

TABLE 27.5   Companies That Makes Acoustical Measurement Instruments

G.R.A.S. Sound & Vibration Skelstedet 10B 2950 Vedbaek Denmark Tel: +45 45 66 40 46	Larson Davies Inc. 1681 West 820 North Provo, UT 84601 Tel: (801) 375 0177
LMS International Interleuvenlaan 68 B-3001 Leuven Belgium Tel: +32 16 384 571	Brüel & Kjær Spectris Technologies Inc. 2364 Park Central Blvd. Decatur, GA 30035-3987 Tel: (800) 332 2040
Hewlett-Packard Co. P.O. Box 95052-8059 Santa Clara, CA 95052 Tel: (206) 335 2000	Rion Scantek, Inc. 916 Gist Avenue Silver Springs, MD 20910 Tel: (301) 495 7738
Norsonic AS P.O. Box 24 N-3420 Lierskogen Norway Tel: +47 32 85 20 80	ACO Pacific, Inc. 2604 Read Avenue Belmont, CA 94002 Tel: (415) 595 8588
The Modal Shop Inc. 1776 Mentor Avenue, Suite 170 Cincinnati, OH 45212-3521 Tel: (513) 351 9919	CEL Instruments 1 Westchester Drive Milford, NH 03055 Tel: (800) 366 2966
01dB 111 rue du 1er Mars F69100 Villeurbanne France Tel: +33 4 78 53 96 96	

References

1. E. Skudrzyk, *The Foundation of Acoustics,* New York: Springer-Verlag, 1971.
2. International Electrotechnical Commission, *Publication 651: Sound Level Meters,* Genève, Switzerland, IEC, 1971.
3. International Organization for Standardization, *Standard 3745 "Acoustics—Determination of sound power levels of noise sources—Precision method for anechoic and semi-anechoic rooms,* Genève, Switzerland, ISO, 1981.

Richard Thorn
University of Derby

Adrian Melling
Universitaet Erlangen-Nuember

Herbert Köchner
Universitaet Erlangen-Nuember

Reinhard Haak
Universitaet Erlangen-Nuember

Zaki D. Husain
Daniel Flow Products, Inc.

Donald J. Wass
Daniel Flow Products, Inc.

David Wadlow
Sensors Research Consulting, Inc.

Harold M. Miller
Data Industrial Corporation

Halit Eren
Curtin University of Technology

Hans-Peter Vaterlaus
Rittmeyer Ltd.

Thomas Hossle
Rittmeyer Ltd.

Paolo Giordano
Rittmeyer Ltd.

Christophe Bruttin
Rittmeyer Ltd.

Wade M. Mattar
The Foxboro Company

James H. Vignos
The Foxboro Company

Nam-Trung Nguyen
University of California at Berkeley

Jesse Yoder
Automation Research Corporation

Rekha Philip-Chandy
Liverpool John Moores University

Roger Morgan
Liverpool John Moores University

Patricia J. Scully
Liverpool John Moores University

28

Flow Measurement

0-8493-8347-1/99/$0.00+$.50
© 1999 by CRC Press LLC

28.1 Differential Pressure Flowmeters

Richard Thorn

Flow measurement is an everyday event. Whether you are filling up a car with petrol (gasoline) or wanting to know how much water the garden sprinkler is consuming, a flowmeter is required. Similarly, it is also difficult to think of a sector of industry in which a flowmeter of one type or another does not play a part. The world market in flowmeters was estimated to be worth $2500 million in 1995, and is expected to grow steadily for the foreseeable future. The value of product being measured by these meters is also very large. For example, in the U.K. alone, it was estimated that in 1994 the value of crude oil produced was worth $15 billion.

Given the size of the flowmeter market, and the value of product being measured, it is somewhat surprising that both the accuracy and capability of many flowmeters are poor in comparison to those instruments used for measurement of other common process variables such as pressure and temperature. For example, the orifice plate flowmeter, which was first used commercially in the early 1900s and has a typical accuracy of ±2% of reading, is still the only flowmeter approved by most countries for the fiscal measurement of natural gas. Although newer techniques such as Coriolis flowmeters have become increasingly popular in recent years, the flow measurement industry is by nature conservative and still dominated by traditional measurement techniques. For a review of recent flowmeter developments, refer to [1].

Over 40% of all liquid, gas, and steam measurements made in industry are still accomplished using common types of *differential pressure flowmeter*; that is, the orifice plate, Venturi tube, and nozzle. The operation of these flowmeters is based on the observation made by Bernoulli that if an annular restriction is placed in a pipeline, then the velocity of the fluid through the restriction is increased. The increase in velocity at the restriction causes the static pressure to decrease at this section, and a pressure difference is created across the element. The difference between the pressure upstream and pressure downstream of this obstruction is related to the rate of fluid flowing through the restriction and therefore through the pipe. A differential pressure flowmeter consists of two basic elements: an obstruction to cause a pressure drop in the flow (a *differential producer*) and a method of measuring the pressure drop across this obstruction (a *differential pressure transducer*).

One of the major advantages of the orifice plate, Venturi tube, or nozzle is that the measurement uncertainty can be predicted without the need for calibration, if it is manufactured and installed in accordance with one of the international standards covering these devices. In addition, this type of differential pressure flowmeter is simple, has no moving parts, and is therefore reliable. The main disadvantages of these devices are their limited range (typically 3:1), the permanent pressure drop they produce in the pipeline (which can result in higher pumping costs), and their sensitivity to installation effects (which can be minimized using straight lengths of pipe before and after the flowmeter). The combined advantages of this type of flowmeter are still quite hard to beat, and although it has limitations, these have been well investigated and can be compensated for in most circumstances. Unless very high accuracy is required, or unless the application makes a nonintrusive device essential, the differential flowmeter should be considered. Despite the predictions of its demise, there is little doubt that the differential pressure flowmeter will remain a common method of flow measurement for many years to come.

Important Principles of Fluid Flow in Pipes

There are a number of important principles relating to the flow of fluid in a pipe that should be understood before a differential pressure flowmeter can be used with confidence. These are the difference between laminar and turbulent flow, the meaning of Reynolds number, and the importance of the flow's velocity profile.

Fluid motion in a pipe can be characterized as one of three types: laminar, transitional, or turbulent. In *laminar flow*, the fluid travels as parallel layers (known as streamlines) that do not mix as they move in the direction of the flow. If the flow is turbulent, the fluid does not travel in parallel layers, but moves in a haphazard manner with only the average motion of the fluid being parallel to the axis of the pipe. If the flow is *transitional*, then both types may be present at different points along the pipeline or the flow may switch between the two.

In 1883, Osborne Reynolds performed a classic set of experiments at the University of Manchester that showed that the flow characteristic can be predicted using a dimensionless number, now known as the Reynolds number. The Reynolds number Re is the ratio of the inertia forces in the flow ($\rho \bar{v} D$) to the viscous forces in the flow (η) and can be calculated using:

$$Re = \frac{\rho \bar{v} D}{\eta} \tag{28.1}$$

where ρ = Density of the fluid
$\bar{v}$ = Mean velocity of the fluid
D = Pipe diameter
η = Dynamic viscosity of the fluid

If Re is less than 2000, viscous forces in the flow dominate and the flow will be laminar. If Re is greater than 4000, inertia forces in the flow dominate and the flow will be turbulent. If Re is between 2000 and 4000, the flow is transitional and either mode can be present. The Reynolds number is calculated using mainly properties of the fluid and does not take into account factors such as pipe roughness, bends, and valves that also affect the flow characteristic. Nevertheless, the Reynolds number is a good guide to the type of flow that can be expected in most situations.

The fluid velocity across a pipe cross-section is not constant, and depends on the type of flow present. In laminar flow, the velocity profile is parabolic since the large viscous forces present cause the fluid to move more slowly near the pipe walls. Under these conditions, the velocity at the center of the pipe is twice the average velocity across the pipe cross-section. The laminar flow profile is unaffected by the roughness of the pipe wall. In turbulent flow, inertia forces dominate, pipe wall effects are less, and the flow's velocity profile is flatter, with the velocity at the center being about 1.2 times the mean velocity. The exact flow profile in a turbulent flow depends on pipe wall roughness and Reynolds number. Figure 28.1 shows the "fully developed" flow profiles for laminar and turbulent flow. These are the flow profiles that would be obtained at the end of a very long pipe, thus ensuring that any changes to the flow profile due to pipe bends and fittings are no longer present. To have confidence in the performance of a differential pressure flowmeter, both the characteristic and velocity profile of the flow passing through the flowmeter should be stable and known.

Bernoulli's Equation

The Bernoulli equation defines the relationship between fluid velocity (*v*), fluid pressure (*p*), and height (*h*) above some fixed point for a fluid flowing through a pipe of varying cross-section, and is the starting point for understanding the principle of the differential pressure flowmeter. For the inclined, tapered pipe shown in Figure 28.2, Bernoulli's equation states that:

$$\frac{p_1}{\rho g} + \frac{v_1^2}{2g} + h_1 = \frac{p_2}{\rho g} + \frac{v_2^2}{2g} + h_2 \tag{28.2}$$

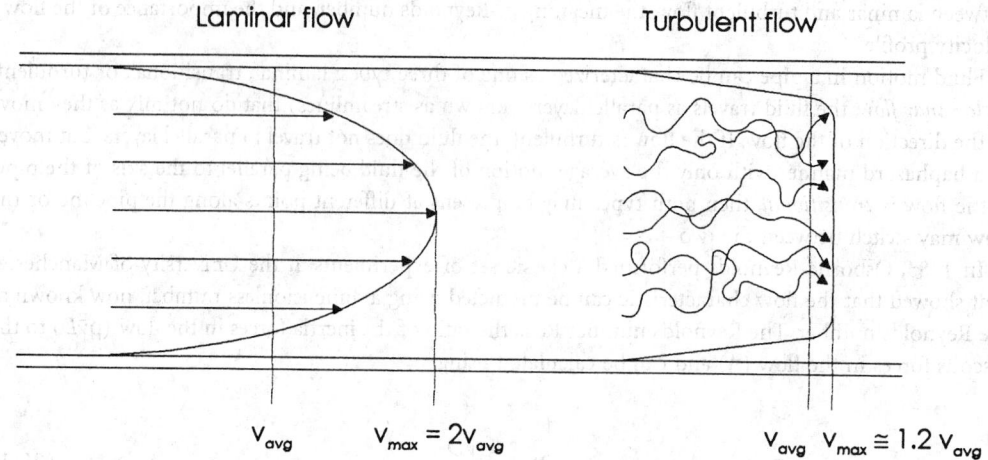

FIGURE 28.1 Velocity profiles in laminar and turbulent flow.

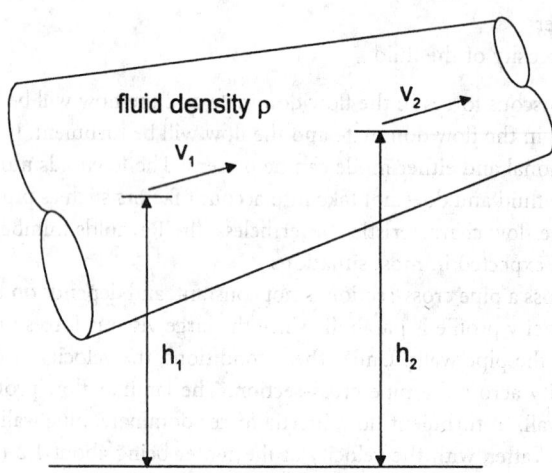

FIGURE 28.2 Flow through an inclined, tapered pipe.

Thus, the sum of the pressure head ($p/\rho g$), the velocity head ($v/2g$), and potential head (h) is constant along a flow streamline. The term "head" is commonly used because each of these terms has the unit of meters. Equation 28.2 assumes that the fluid is frictionless (zero viscosity) and of constant density (incompressible). Further details on the derivation and significance of Bernoulli's equation can be found in most undergraduate fluid dynamics textbooks (e.g., [2]).

Bernoulli's equation can be used to show how a restriction in a pipe can be used to measure flow rate. Consider the pipe section shown in Figure 28.3. Since the pipe is horizontal, $h_1 = h_2$, and Equation 28.2 reduces to:

$$\frac{p_1 - p_2}{\rho} = \frac{v_1^2 - v_2^2}{2}$$ (28.3)

The conservation of mass principle requires that:

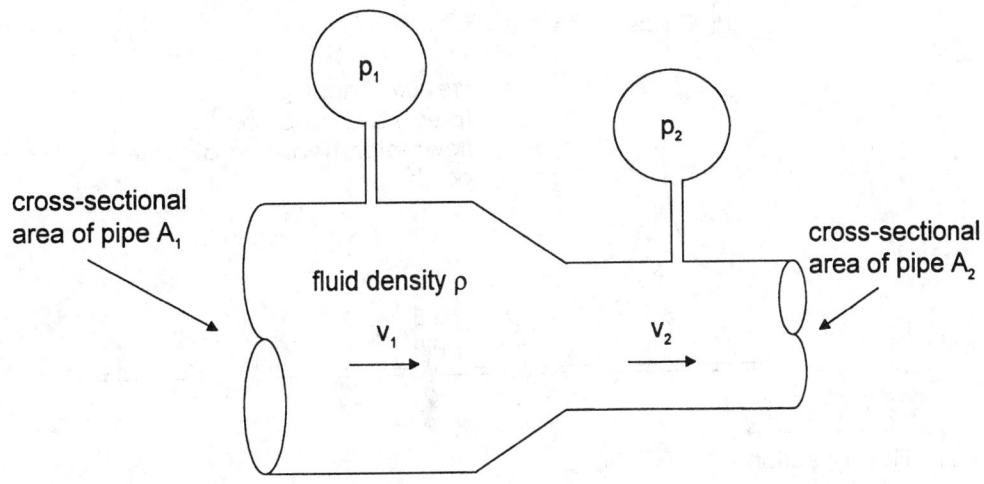

FIGURE 28.3   Using a restriction in a pipe to measure fluid flow rate.

$$v_1 A_1 \rho = v_2 A_2 \rho \tag{28.4}$$

Rearranging Equation 28.4 and substituting for v_2 in Equation 28.3 gives:

$$Q = v_1 A_1 = \frac{A_2}{\sqrt{1 - \left(\dfrac{A_2}{A_1}\right)^2}} \sqrt{\frac{2\left(p_1 - p_2\right)}{\rho}} \tag{28.5}$$

This shows that the volumetric flow rate of fluid Q can be determined by measuring the drop in pressure $(p_1 - p_2)$ across the restriction in the pipeline — the basic principle of all differential pressure flowmeters. Equation 28.5 has limitations, the main ones being that it is assumed that the fluid is incompressible (a reasonable assumption for most liquids), and that the fluid has no viscosity (resulting in a flat velocity profile). These assumptions need to be compensated for when Equation 28.5 is used for practical flow measurement.

Common Differential Pressure Flowmeters

The Orifice Plate

The orifice plate is the simplest and cheapest type of differential pressure flowmeter. It is simply a plate with a hole of specified size and position cut in it, which can then clamped between flanges in a pipeline (Figure 28.4). The increase that occurs in the velocity of a fluid as it passes through the hole in the plate results in a pressure drop being developed across the plate. After passing through this restriction, the fluid flow jet continues to contract until a minimum diameter known as the vena contracta is reached. If Equation 28.5 is used to calculate volumetric flow rate from a measurement of the pressure drop across the orifice plate, then an error would result. This is because A_2 should strictly be the area of the vena contracta, which of course is unknown. In addition, turbulence between the vena contracta and the pipe wall results in an energy loss that is not accounted for in this equation.

To overcome the problems caused by the practical application of Equation 28.5, two empirically determined correction factors are added. After some reorganization Equation 28.5 can be written as:

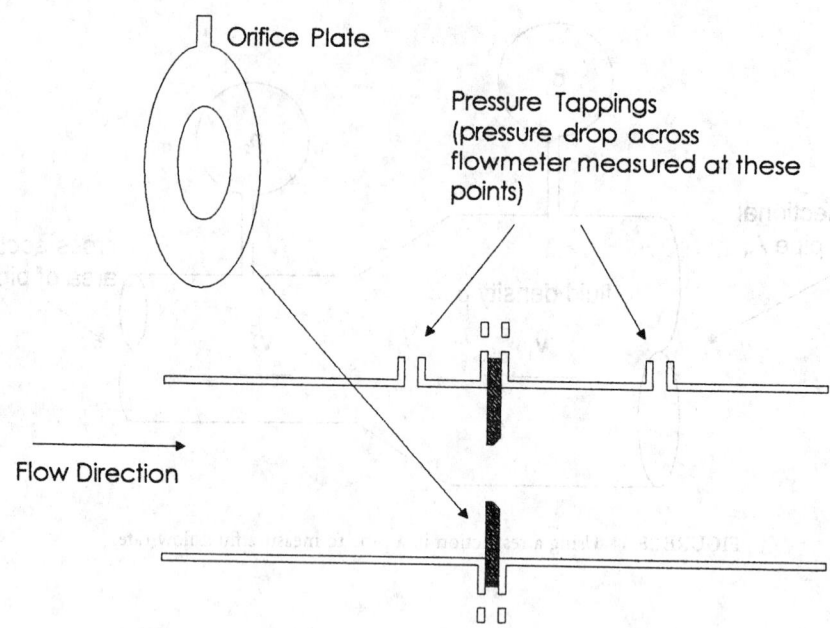

FIGURE 28.4 A square-edged orifice plate flowmeter.

$$Q = \frac{C}{\sqrt{1-\beta^4}} \, \varepsilon \frac{\pi}{4} d^2 \sqrt{\frac{2(p_1 - p_2)}{\rho}} \qquad (28.6)$$

where ρ = Density of the fluid upstream of the orifice plate

d = Diameter of the hole in the orifice plate

β = Diameter ratio d/D, where D is the upstream internal pipe diameter

The two empirically determined correction factors are C the discharge coefficient, and ε the expansibility factor. C is affected by changes in the diameter ratio, Reynolds number, pipe roughness, the sharpness of the leading edge of the orifice, and the points at which the differential pressure across the plate are measured. However, for a fixed geometry, it has been shown that C is only dependent on Reynolds number and so this coefficient can be determined for a particular application. ε is used to account for the compressibility of the fluid being monitored. Both C and ε can be determined from equations and tables in a number of internationally recognized documents known as standards. These standards not only specify C and ε, but also the geometry and installation conditions for the square-edged orifice plate, Venturi tube, and nozzle, and are essentially a design guide for the use of the most commonly used types of differential pressure flowmeter. Installation recommendations are intended to ensure that fully developed turbulent flow conditions exist within the measurement section of the flowmeter. The most commonly used standard in Europe is ISO 5167-1 [3], while in the U.S., API 2530 is the most popular [4]. There are differences between some of the recommendations in these two standards (e.g., the minimum recommended length of straight pipe upstream of the flowmeter), but work is underway to resolve these.

Equation 28.6 illustrates perhaps the greatest strength of the orifice plate, which is that measurement performance can be confidently predicted without the need for calibration if the device is manufactured, installed, and operated in accordance with one of the international standards. In addition, the device is cheap to manufacture, has no moving parts, is reliable, and can be used for metering most clean gases, liquids, and steam.

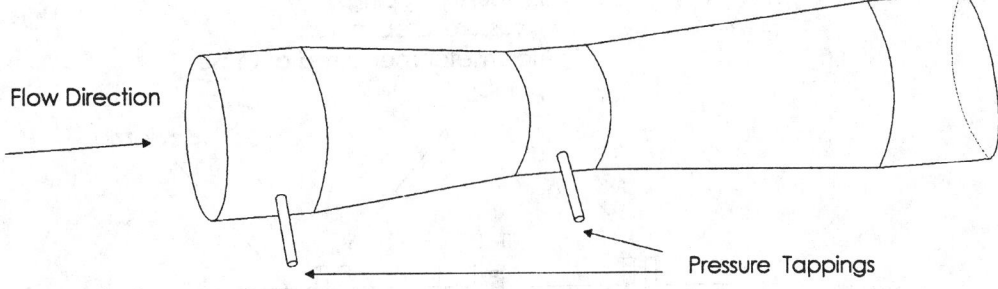

Flow Direction

Pressure Tappings

FIGURE 28.5 A Venturi tube flowmeter.

The major disadvantages of the orifice plate are its limited range and sensitivity to flow disturbances. The fact that fluid flow rate is proportional to the square root of the measured differential pressure limits the range of a one plate/one differential pressure transmitter combination to about 3:1. The required diameter ratio (also known as beta ratio) of the plate depends on the maximum flow rate to be measured and the range of the differential pressure transducer available. Sizing of the orifice plate is covered in most of the books in the further reading list, and nowadays computer programs are also available to help perform this task. The flow measurement range can be increased by switching; ways in which this may be achieved are described in [5]. Equation 28.6 assumes a fully developed and stable flow profile, and so installation of the device is critical, particularly the need for sufficient straight pipework upstream of the meter. Wear of the leading edge of the orifice plate can severely alter measurement accuracy; thus; this device is normally only used with clean fluids.

Only one type of orifice plate, the square-edged concentric, is covered by the standards. However, other types exist, having been designed for specific applications. One example is the eccentric orifice plate, which is suited for use with dirty fluids. Details of these other types of orifice plate can be found in [6].

The Venturi Tube

The classical or Herschel Venturi tube is the oldest type of differential pressure flowmeter, having first been used in 1887. As Figure 28.5 shows, a restriction is introduced into the flow in a more gradual way than for the orifice plate. The resulting flow through a Venturi tube is closer to that predicted in theory by Equation 28.5 and so the discharge coefficient C is much nearer unity, being typically 0.95. In addition, the permanent pressure loss caused by the Venturi tube is lower, but the differential pressure is also lower than for an orifice plate of the same diameter ratio. The smooth design of the Venturi tube means that it is less sensitive to erosion than the orifice plate, and thus more suitable for use with dirty gases or liquids. The Venturi tube is also less sensitive to upstream disturbances, and therefore needs shorter lengths of straight pipework upstream of the meter than the equivalent orifice plate or nozzle. Like the orifice plate and nozzle, the design, installation, and use of the Venturi tube is covered by a number of international standards.

The major disadvantages of the Venturi tube flowmeter are its size and cost. It is more difficult, and therefore more expensive to manufacture than the orifice plate. Since a Venturi tube can be typically 6 diameters long, it can become cumbersome to use with larger pipe sizes, with associated maintenance of upstream and downstream pipe lengths also becoming a problem.

The Nozzle

The nozzle (Figure 28.6) combines some of the best features of the orifice plate and Venturi tube. It is compact and yet, because of its curved inlet, has a discharge coefficient close to unity. There are a number of designs of nozzle, but one of the most commonly used in Europe is the ISA-1932 nozzle, while in the U.S., the ASME long radius nozzle is more popular. Both of these nozzles are covered by international standards.

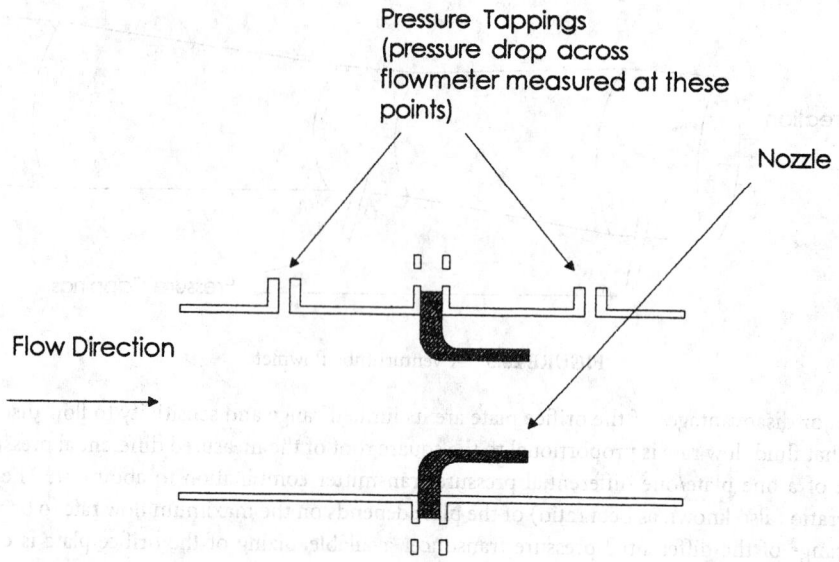

FIGURE 28.6   A nozzle flowmeter.

The smooth inlet of the nozzle means that it is more expensive to manufacture than the orifice plate as the curvature of the inlet changes with diameter ratio, although it is cheaper than the Venturi tube. The device has no sharp edges to erode and cause changes in calibration, and thus is well suited for use with dirty and abrasive fluids. The nozzle is also commonly used for high-velocity, high-temperature applications such as steam metering.

A variation of the nozzle is the sonic (or critical flow Venturi) nozzle, which has been used both as a calibration standard for testing gas meters and a transfer standard in interlaboratory comparisons [7].

Other Differential Pressure Flowmeters

There are many other types of differential pressure flowmeter, including the segmental wedge, V-cone, elbow, and Dall tube. Each of these has advantages over the orifice plate, Venturi tube, and nozzle for specific applications. For example, the segmental wedge can be used with flows having a low Reynolds number, and a Dall tube has a lower permanent pressure loss than a Venturi tube. However, none of these instruments are yet covered by international standards and, thus, calibration is needed to determine their accuracy. Further information on these, and other less-common types of differential pressure flowmeter, can be found in [8].

Performance and Applications

Table 28.1 shows the performance characteristics and main application areas of the square-edged orifice plate, Venturi tube, and nozzle flowmeters. Compared to other types of flowmeters on the market, these differential pressure flowmeters only have moderate accuracy, typically ±2% of reading; but of course, this can be improved if the device is calibrated after installation. Although in some circumstances these flowmeters can be used with dirty gases or liquids, usually only small amounts of a second component can be tolerated before large measurement errors occur. When calculating the cost and performance of a differential flowmeter, both the primary element and the differential pressure transducer should be taken into account. Although the orifice plate is the cheapest of the primary elements, the cost of the fitting needed to mount it in the pipeline, particularly if on-line removal is required, can be significant.

Choosing which flowmeter is best for a particular application can be very difficult. The main factors that influence this choice are the required performance, the properties of the fluid to be metered, the

TABLE 28.1   The Performance and Application Areas of Common Differential Pressure Flowmeters

	Performance					Applications				
	Typical uncalibrated accuracy	Typical range	Typical pipe diameter (mm)	Permanent pressure loss	Comparative cost	Clean gas	Dirty gas	Clean liquid	Slurry	Steam
Orifice plate	±2%	3:1	10–1000	High	Low	Yes	No	Yes	No	Yes
Venturi tube	±2%	3:1	25–500	Low	High	Yes	Maybe	Yes	Maybe	Maybe
Nozzle	±2%	3:1	25–250	High	Medium	Yes	Maybe	Yes	No	Yes

TABLE 28.2   A Selection of Companies That Supply Differential Pressure Flowmeters

ABB Kent-Taylor
Oldens Lane, Stonehouse
Gloucestershire, GL10 3TA
England
Tel: + 44 1453 826661
Fax: + 44 1453 826358

Daniel Industries Inc.
9720 Katy Road
P.O. Box 19097
Houston, TX 77224
Tel: (713) 467-6000
Fax: (713) 827-3880

Hartmann & Braun (UK) Ltd.
Bush Beach Engineering Division
Stanley Green Trading Estate, Cheadle Hulme
Cheshire SK8 6RN
England
Tel: + 44 161 4858151
Fax: + 44 161 4884048

ISA Controls Ltd.
Hackworth Industrial Park
Shildon
County Durham DL4 1LH
England
Tel: + 44 1388 773065
Fax: + 44 1388 774888

Perry Equipment Corporation
Wolters Industrial Park
P.O. Box 640
Mineral Wells, TX 76067
Tel: (817) 325-2575
Fax: (817) 325-4622

installation requirements, the environment in which the instrument is to be used, and, of course, cost. There are two standards that can be used to help select a flowmeter: BS 1042: Section 1.4, which is a guide to the use of the standard differential pressure flowmeters [9]; and BS 7405, which is concerned with the wider principles of flowmeter selection [10].

Because all three flowmeters have similar accuracy, one strategy for selecting the most appropriate instrument is to decide if there are any good reasons for not using the cheapest flowmeter that can be used over the widest range of pipe sizes: the orifice plate. Where permanent pressure loss is important, the Venturi tube should be considered, although the high cost of this meter can only usually be justified where large quantities of fluid are being metered. For high-temperature or high-velocity applications, the nozzle should be considered because under these conditions, it is more predictable than the orifice plate. For metering dirty fluids, either the Venturi tube or the nozzle should be considered in preference to the orifice plate, the choice between the two depending on cost and pressure loss requirements. Table 28.2 lists some suppliers of differential pressure flowmeters.

Installation

Correct installation is essential for successful use of a differential pressure flowmeter because the predicted uncertainty in the flow rate/differential pressure relationship in Equation 28.6 assumes a steady flow,

TABLE 28.3   The Minimum Straight Lengths of Pipe Required between Various Fittings and an Orifice Plate or Venturi Tube (as recommended in ISO 5167-1) to Ensure That a Fully Developed Flow Profile Exists in the Measurement Section. All Lengths Are Multiples of the Pipe Diameter

Diameter Ratio β	Single 90° bend	Two 90° bends in the same plane	Two 90° bends in different planes	Globe valve fully open	For any of the fittings shown to the left
			Upstream of the flowmeter		Downstream of the flowmeter
0.2	10	14	34	18	4
0.4	14	18	36	20	6
0.6	18	26	48	26	7
0.8	46	50	80	44	8

with a fully developed turbulent velocity profile, is passing through the flowmeter. Standards contain detailed recommendations for the minimum straight lengths of pipe required before and after the flowmeter, in order to ensure a fully developed flow profile. Straight lengths of pipe are required after the flowmeter because disturbances caused by a valve or bend can travel upstream and thus also affect the installed flowmeter. Table 28.3 gives examples of installation requirements taken from ISO 5167-1. If it is not possible to fit the recommended lengths of straight pipe before and after the flowmeter, then the flowmeter must be calibrated once it has been installed.

The other major problem one faces during installation is the presence of a rotating flow or swirl. This condition distorts the flow velocity profile in a very unpredictable way, and is obviously not desirable. Situations that create swirl, such as two 90° bends in different planes, should preferably be avoided. However, if this is not possible, then swirl can be removed by placing a flow conditioner (also known as a flow straightener) between the source of the swirl and the flowmeter. There are a wide range of flow conditioner designs, some of which can be used to both remove swirl and correct a distorted velocity profile [11]. Because they obstruct the flow, all flow conditioners produce an unrecoverable pressure loss, which in general increases with their capability (and complexity).

Differential Pressure Measurement

Apart from the differential producer, the other main element of a differential pressure flowmeter is the *transducer* needed to measure the pressure drop across the producer. The correct selection and installation of the differential pressure transducer plays an important part in determining the accuracy of the flow rate measurement.

The main factors that should be considered when choosing a differential pressure transducer for a flow measurement application are the differential pressure range to be covered, the accuracy required, the maximum pipeline pressure, and the type and temperature range of the fluid being metered.

Most modern differential pressure transducers consist of a pressure capsule in which either capacitance, strain gage, or resonant wire techniques are used to detect the movement of a diaphragm. Using these techniques, a typical accuracy of ±0.1% of full scale is possible. See Chapter 26, Section 1 for further details of pressure transducers. The transducer is usually part of a unit known as a transmitter, which converts differential pressure, static pressure, and ambient temperature measurements into a standardized analog or digital output signal. "Smart" transmitters use a local, dedicated microprocessor to condition signals from the individual sensors and compute volumetric or mass flow rate. These devices can be remotely configured, and a wide range of diagnostic and maintenance functions are possible using their built-in "intelligence."

As far as installation is concerned, the transmitter should be located as close to the differential producer as possible. This helps ensure a fast dynamic response and reduces problems caused by vibration of the connecting tubes. The position of the pressure tappings is also important. If liquid flow in a horizontal pipe is being measured, then the pressure tappings should be located at the side of the pipe so that they cannot be blocked with dirt or filled with air bubbles. For horizontal gas flows, if the gas is clean, the pressure tappings should be vertical; if steam or dirty gas is being metered, then the tappings should be

TABLE 28.4   Standards Related to Differential Pressure Flow Measurement

American National Standards Institute, New York	
ANSI/ASHRAE 41.8	Standard methods of measurement of flow of liquids in pipes using orifice flowmeters.
ANSI/ASME MFC-7M	Measurement of gas flow by means of critical flow Venturi nozzles.
ANSI/ASME MFC-14M	Measurement of fluid flow using small bore precision orifice meters.
American Petroleum Institute, Washington, D.C.	
API 2530	Manual of Petroleum Measurement Standards, Chapter 14 — Natural gas fluids measurement, Section 3 — Orifice metering of natural gas and other related hydrocarbon fluids.
American Society of Mechanical Engineers, New York	
ASME MFC-3M	Measurement of fluid flow in pipes using orifice, nozzle and Venturi.
ASME MFC-8M	Fluid flow in closed conduits — Connections for pressure signal transmissions between primary and secondary devices.
British Standards Institution, London	
BS 1042	Measurement of fluid flow in closed conduits.
International Organization for Standardization, Geneva	
ISO 2186	Fluid flow in closed conduits — connections for pressure transmissions between primary and secondary elements.
ISO TR 3313	Measurement of pulsating fluid flow in a pipe by means of orifice plates, nozzles or Venturi tubes.
ISO 5167-1	Measurement of fluid flow by means of pressure differential devices.
ISO 9300	Measurement of gas flow by means of critical flow Venturi nozzles.

located at the side of the pipe. These general guidelines show that considerable care must be taken with the installation of the differential pressure transmitter if large measurement errors are to be avoided. For further details on the installation of differential pressure transmittters, see ISO 2186 [12].

Standards

International standards that specify the design, installation, and use of the orifice plate, Venturi tube, and nozzle, and allow their accuracy to be calculated without the need for calibration, are probably the main reason for the continuing use of this type of flowmeter. Table 28.4 gives details of the most common standards related to differential pressure flowmeters. There are still inconsistencies in the various standards. For example, ISO 5167-1 states that any flow conditioner should be preceded by at least 22 diameters of straight pipe and followed by at least 20 diameters of straight pipe. This would seem to contradict one application of a flow conditioner, which is to reduce the length of straight pipe required upstream of a flowmeter.

Despite the occasional inconsistancy and difference between the standards, they are the internationally accepted rules for the installation and use of the square-edged orifice plate, Venturi tube, and nozzle.

Future Developments

In spite of the vast amount of published data available on differential pressure flowmeters, continued research is needed to improve the understanding of the effect of flow conditions, and flowmeter geometry, on the uncalibrated accuracy of these devices. For example, work has been recently undertaken to derive an improved equation for the discharge coefficient of an orifice plate [13].

The metering of multiphase flow is an area of increasing importance. In addition to developing new measurement techniques, many people are investigating ways in which traditional flowmeters can be used to meter multiphase flows. A good review of the use of differential pressure flowmeters for multiphase flow measurement can be found in [14].

The development of "smart" differential pressure transmitters has overcome the limitations of differential pressure flowmeters in some applications. For example, these devices are being used to linearize and extend the range of differential pressure flowmeters.

The above developments should help to ensure the continued popularity of the differential pressure flowmeter for the forseeable future, despite increasing competition from newer types of instrument.

Defining Terms

Differential pressure flowmeter: A flowmeter in which the pressure drop across an annular restriction placed in the pipeline is used to measure fluid flow rate. The most common types use an orifice plate, Venturi tube, or nozzle as the primary device.

Orifice plate: Primary device consisting of a thin plate in which a circular aperture has been cut.

Venturi tube: Primary device consisting of a converging inlet, cylindrical mid-section, and diverging outlet.

Nozzle: Primary device consisting of a convergent inlet connected to a cylindrical section.

Differential pressure transmitter: Secondary device that measures the differential pressure across the primary device and converts it into an electrical signal.

References

1. R. A. Furness, Flowmetering: evolution or revolution, *Measurement and Control*, 27 (8), 15-18, 1994.
2. B. S. Massey, *Mechanics of Fluids*, 6th ed., London: Chapman and Hall, 1989.
3. International Organization for Standardization, ISO 5167-1, Measurement of Fluid Flow by Means of Pressure Differential Devices — Part 1 Orifice plates, nozzles and Venturi tubes inserted in circular cross-section conduits running full, Geneva, Switzerland, 1991.
4. American Petroleum Institute, API 2530, Manual of Petroleum Measurement Standards Chapter 14 — Natural Gas Fluids Measurement, Section 3 — Orifice Metering of Natural Gas and Other Related Hydrocarbon Fluids, Washington, 1985.
5. E. L. Upp, *Fluid Flow Measurement*, Houston: Gulf Publishing, 1993.
6. H. S. Bean, *Fluid Meters Their Theory and Application*, 6th ed., New York: American Society of Mechanical Engineers, 1983.
7. P. H. Wright, The application of sonic (critical flow) nozzles in the gas industry, *Flow Meas. Instrum.*, 4 (2), 67-71, 1993.
8. D. W. Spitzer, *Industrial Flow Measurement*, 2nd ed., Research Triangle Park, NC: ISA, 1990.
9. British Standards Institution, BS 1042, Measurement of Fluid Flow in Closed Conduits — Part 1 Pressure differential devices — Section 1.4 Guide to the use of devices specified in Sections 1.1 and 1.2, London, 1992.
10. British Standards Institution, BS7405, Guide to the Selection and Application of Flowmeters for Measurement of Fluid Flow in Closed Conduits, London, 1991.
11. E. M. Laws and A. K. Ouazzane, Compact installations for differential pressure flow measurement, *Flow Meas. Instrum.*, 5 (2), 79-85, 1994.
12. International Organization for Standardization, ISO 2186, Fluid Flow in Closed Conduits — Connections for Pressure Signal Transmissions Between Primary and Secondary Elements, Geneva, Switzerland, 1973.
13. M. J. Reader-Harris, J. A. Slattery, and E. P. Spearman, The orifice plate discharge coefficient equation — further work, *Flow. Meas. Instrum.*, 6 (2), 101-114, 1995.
14. F. C. Kinghorn, Two-phase flow measurement using differential pressure meters, *Multi- Phase Flow Measurement Short Course*, London, 17th-18th June 1985.

Further Information

R. C. Baker, *An Introductory Guide to Flow Measurement*, London: Mechanical Engineering Publications, 1989, a good pocket sized guide on the choice and use of flowmeters commonly used in industry.

A. T. J. Hayward, *Flowmeters — A Basic Guide and Source-Book for Users*, London: Macmillan, 1979, an overview of the important areas of flow measurement which is a joy to read.

R. W. Miller, *Flow Measurement Engineering Handbook*, 3rd ed., New York: McGraw-Hill, 1996, a thorough reference book particularly on differential pressure flowmeters, covers European and U.S. standards with calculations using both U.S. and SI units.

D. W. Spitzer, *Flow Measurement: Practical Guides for Measurement and Control*, Research Triangle Park, NC: ISA, 1991, intended for practicing engineers, this book covers most aspects of industrial flow measurement.

E. L. Upp, *Fluid Flow Measurement*, Houston: Gulf Publishing, 1993, contains a lot of practical advice on the use of differential pressure flowmeters.

Flow Measurement and Instrumentation, UK: Butterworth-Heinemann, a quarterly journal covering all aspects of flowmeters and their applications. A good source of information on current research activity.

28.2 Variable Area Flowmeters

Adrian Melling, Herbert Köchner, and Reinhard Haak

The term *variable area flowmeters* refers to those meters in which the minimum cross-sectional area available to the flow through the meter varies with the flow rate. Meters of this type that are discussed in this section include the rotameter and the movable vane meter used in pipe flows, and the weir or flume used in open-channel flows. The measure of the flow rate is a geometrical quantity such as the height of a bob in the rotameter, the angle of the vane, or the change in height of the free surface of the liquid flowing over the weir or through the flume.

Most of the discussion here is devoted to the rotameter, firstly because the number of installed rotameters and movable vane meters is large relative to the number of weirs and flumes, and secondly because the movable vane is often used simply as a flow indicator rather than as a meter.

The following section includes basic information describing the main constructional features and applications of each type of meter. In the third section, the principles of measurement and design of rotameters and open-channel meters are described in some detail; the movable vane meter is not considered further because most design aspects are similar to those of rotameters. Then, the contribution of modern computational and experimental methods of fluid mechanics to flowmeter design is discussed, using the results of a detailed investigation of the internal flow.

Details of manufacturers of rotameters and movable vane meters, together with approximate costs of these meters, are also tabulated in Tables 28.5 through 28.7.

General Description of Variable Area Flowmeters

Rotameter

The *rotameter* is a robust and simple flowmeter for gases and liquids, and holds a large share of the market for pipe diameters smaller than about 100 mm. In its basic form, the rotameter consists of a conical transparent vertical glass tube containing a "bob" (Figure 28.7), which rises in the tube with increasing flow rate until a balance is reached between gravitational, buoyancy, and drag forces on the bob. Within the range of a particular flowmeter (depending on the bob shape and density, the tube shape and the fluid density and viscosity), the flow rate is linearly proportional to the height of the bob in the tube and is determined simply by reading the level of the upper edge of the bob. The rotameter is also

TABLE 28.5   Approximate Rotameter Prices

Type	Size	Price (plastic)	Price (stainless steel, borosilicate glass)
Glass	< 1/2 in. (12.7 mm)	$50	$200
Plastic	1/2 in. (12.7 mm)	$50	$200
	1 in. (25.4 mm)	$70	$150
	2 in. (50.8 mm)	$200	$500
Metal	1/2 in. (12.7 mm)		$500
	1 in. (25.4 mm)		$600
	2 in. (50.8 mm)		$700
	3 in. (76.2 mm)		$1100
	4 in. (101.6 mm)		$1400

Additional costs

Electric limit switch	$100
Current transducer	$500
Pneumatic	$1200

TABLE 28.6   Approximate Movable Vane Meter Prices

Type	Price
Plastic versions (small scale)	From $10
Direct coupled pointer versions incl. electric flow switch: Metal, screw connections 1/2 in. (12.7 mm) to 2 in. (50.8 mm)	$400 (brass) to $600 (stainless steel)
Flanged versions	Add $200 (1/2 in., 12.7 mm) to $400 (2 in., 50.8 mm)
Magnetic coupled indicator, stainless steel	$1200 (1/2 in., 12.7 mm) to $2100 (8 in., 203 mm)

known as "floating element flowmeter," although the buoyancy force on the "floating element" is not sufficient to make it float.

The bob is commonly formed of a combination of cylindrical and conical sections, but a spherical bob is often used in small diameter tubes (see Figures 28.8). More complicated bob geometries can reduce the sensitivity to viscosity of the fluid. Frequently, the bob has several shallow inclined grooves around the upper rim that induce a slow rotation (frequency about 1 Hz) of the bob, which helps to maintain a stable position of the bob. In larger rotameters where the additional friction is acceptable, the bob can be allowed to slide up and down a rod on the tube axis to prevent any sideways motion.

Various tube geometries are in use. The basic requirement for an increase in the cross-sectional area up the height of the tube leads to the conical tube. An alternative form uses three ribs arranged circumferentially around the tube to guide the bob; the flow area between the ribs increases along the tube height. For spherical bobs, a triangular tube in which the bob remains in contact with three surfaces over the whole height of the tube prevents lateral movement of the bob with minimum friction. Commonly, the tube is made of glass to facilitate the reading of the flow rate from a scale engraved on the tube. For general-purpose applications, the scale can be marked in millimeters; the flow rate is then determined by a conversion factor depending on the tube dimensions, the mass of the bob, the pressure and temperature, and the properties of the fluid. For specific application to a single fluid under controlled conditions, it is more convenient to have the flow rate directly marked on the tube.

For laboratory use, medical equipment, and other applications with small flow rates, the glass tube is almost universal. Laboratory meters are often equipped with a flow valve that allows direct flow setting and reading. For many small rotameters, integrated flow controllers are offered; these hold the pressure drop and the flow setting constant, even with changing upstream or downstream pressure.

For most rotameters, electric limit switches or analog electric signal transducers are available. For glass and plastic meters, inductive or optical switches are used that can be positioned to the desired level of

TABLE 28.7 Manufacturers of Rotameters and Movable Vane Meters

Tokyo Keiso Co. Ltd. Shiba Toho Building 1-7-24, Shibakoen Minato-ku Tokyo 105, Japan Tel: +81-3-3431-1625 Fax: +81-3-3433-4922	Porter Instruments Co. 245 Township Line Road Hatfield, PA 19440 Tel: (215) 723-4000 Fax: (215) 723-2199
Brooks Instrument Division Emerson Electric 407 W. Vine St. Hatfield, PA 19440 Tel: (215) 362-3500 Fax: (215) 362-3745	Wallace & Tiernan, Inc. 25 Main St. St. Belleville, NJ 07109 Tel: (201) 759-8000 Fax: (201) 759-9333
Krohne America Inc. 7 Dearborn Rd. Peabody, MA 01960 Tel: (508) 535-6060, (800) 356-9464 Fax: (508) 535-1720	KDG-Mobrey Ltd. Crompton Way Crawley West Sussex RH10 2YZ, Great Britain Tel: +44-1293-518632 Fax: +44-1293-533095
Bailey Fischer & Porter Co. County Line Rd. Warminster, PA 18974 Tel: (215) 674-6000 Fax: (215) 674-7183	CT Platon Jays Close Viables Basingstoke, Hampshire RG22 4BS, Great Britain Tel: +44-1256-470456 Fax: +44-1256-363345
Bopp & Reuther Heinrichs Messtechnik GmbH Stolberger Str. 393 D-50933 Köln, Germany Tel: +49-221-49708-0 Fax: +49-221-497088	Kobold Messring GmbH Nordring 22-24 D-65719 Hofheim/Taunus, Germany Tel: +49 6192 2990 Fax: +49 6192 23398
Rota Yokogawa GmbH & Co. KG Rheinstrasse 8 D-79664 Wehr, Germany Tel: +49-7761-567-0 Fax: +49-7761-567-126	

the bob. While some units just detect the bob within their active area, other switches have bistable operation, wherein the switch is toggled by the float passing the switch.

For industrial applications, the metering of corrosive fluids or fluids at high temperature or pressure leads to the use of stainless steel tubes or a wide range of other materials chosen to suit the application requirements (e.g., temperature, pressure, corrosion). Most metal rotameters use a permanent magnetic coupling between the float and the pointer of the indicator, enabling a direct analog flow reading without electric supply. Electric switches can, however, be added to the indicator, and electric and pneumatic transmitters for the flow reading are offered by most producers. Some units are equipped with options such as flow totalizers, mass flow calculation units for gas applications in conjunction with temperature and pressure sensors, or energy calculation units. Due to the frequent application of the rotameter in the chemical and petrochemical industry, electric options are often designed for use in hazardous areas.

The rotameter is characterized by:

- Simple and robust construction
- High reliability
- Low pressure drop

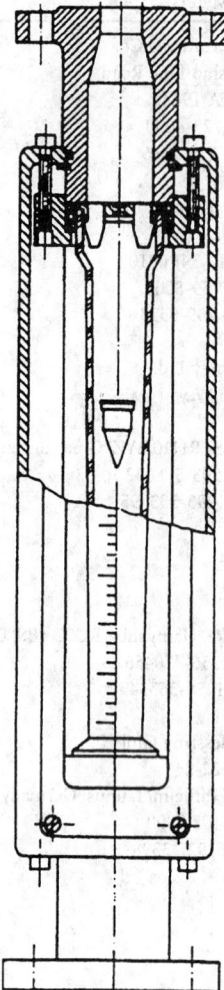

FIGURE 28.7 Cross-section of a rotameter.
The level of the bob rises linearly with increasing
flow rate.

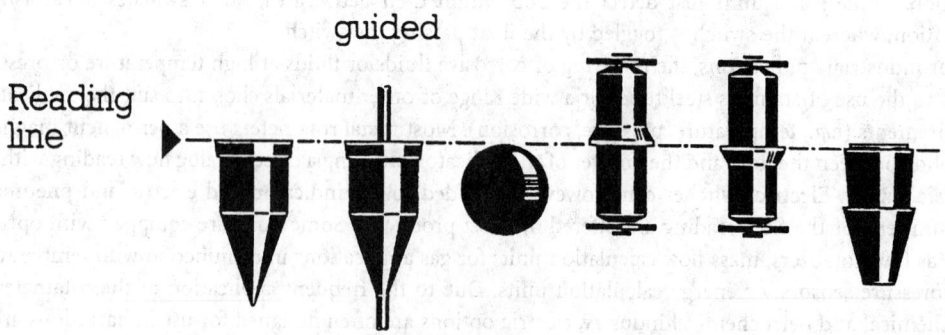

FIGURE 28.8 Typical rotameter bob geometries.

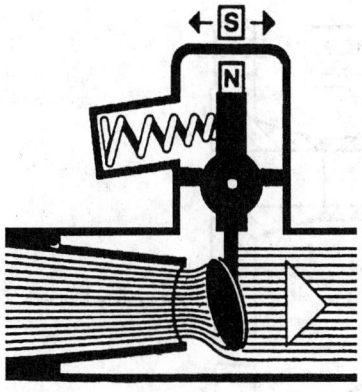

FIGURE 28.9   Movable vane meter. The magnet (N,S) transmits the vane position to an indicator.

- Applicable to a wide variety of gases and liquids
- Flow range typically 0.04 L h^{-1} to 150 m^3 h^{-1} for water
- Flow range typically 0.5 L h^{-1} to 3000 m^3 h^{-1} for air
- 10:1 flow range for given bob-tube combination
- Uncertainty 0.4% to 4% of maximum flow
- Insensitivity to nonuniformity in the inflow (no upstream straight piping needed)
- Typical maximum temperature 400°C
- Typical maximum pressure 4 MPa (40 bar)
- Low investment cost
- Low installation cost

Movable Vane Meter

The movable vane meter is a robust device suitable for the measurement of high flow rates where only moderate requirements on the measurement accuracy are made. Dirty fluids can also be metered. It contains a flap that at zero flow is held closed by a weight or a spring (Figure 28.9). A flow forces the vane open until the dynamic force of the flow is in balance with the restoring force of the weight or the spring. The angle of the vane is thus a measure of the flow rate, which can be directly indicated by a pointer attached to the shaft of the vane on a calibrated scale. The resistance provided by the vane depends on the vane position and hence on the flow rate or Reynolds number; a recalibration is therefore necessary when the fluid is changed. An important application is the metering of the air flow in automotive engines with fuel injection.

In low-cost flow indicators, a glass or plastic window allows a direct view of the flap (Figure 28.10). Sometimes, optical or reed switches in combination with a permanent magnet attached to the flap are used as electric flow switches. All-metal units use a magnetic coupling between the vane and the pointer of the indicator, thus avoiding most of the friction and material selection problems. Many applications use the movable vane meter as a flow switch with appropriate mechanical, magnetic, inductive, or optical switches. The setting of the switch is done once for all during the installation. Since a continuous flow indication is not needed, flow reading uncertainties are not of concern.

Most aspects of design and application of rotameters can be applied to movable vane meters, although the latter are characterized by a higher uncertainty of the flow reading. General features of both types of variable area flowmeter are summarized, for example, in reference [1]. Since the basic construction of the two meter types is very similar, the unit costs also lie in the same range.

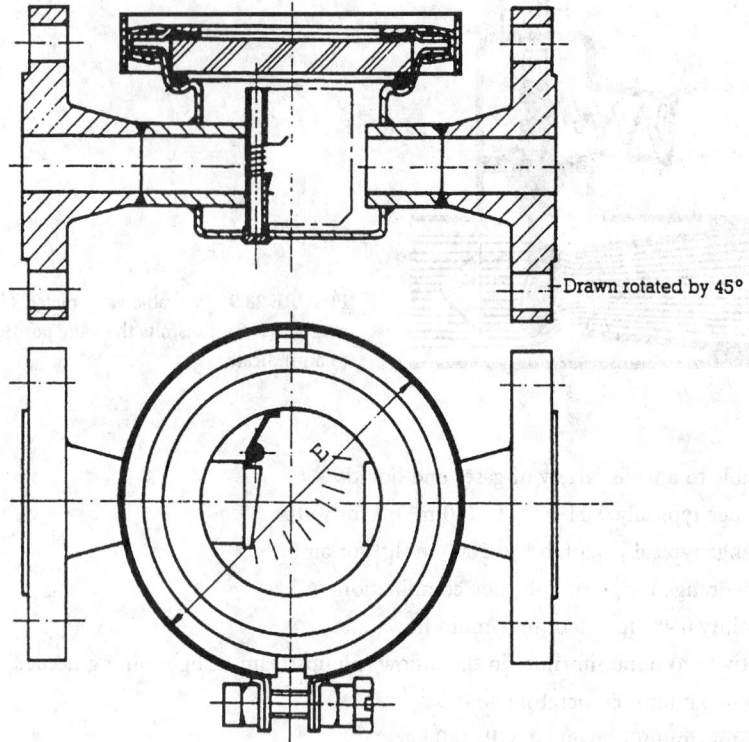

Drawn rotated by 45°

FIGURE 28.10   Flow indicator. The angle of the vane provides a measure of the flow rate.

Weir, Flume

Of the methods available for the metering of a liquid (generally water) in open-channel flow, the weir and the flume fall within the scope of this discussion on variable area flowmeters. In each case, the flow metering depends on measurement of the difference in height h of the water surface over an obstruction across the channel and the surface sufficiently far upstream. There is a wide variety of geometries in use (see reference [2]), but most of these can be described as variants of three basic types. In the sharp crested weir, also known as the thin plate weir (Figure 28.11), the sill or crest is only about 1 mm to 2 mm thick. The sheet of liquid flowing over the weir, called the nappe or vein, separates from the weir body after passing over the crest. An air-filled zone at atmospheric pressure is formed underneath the outflowing jet, and the streamlines above the weir are strongly curved. If the width of the weir is less than that of the upstream channel (Figures 28.12 and 28.13), the term "notch" is frequently used. A broad crested weir (Figure 28.14) has a sill which is long enough for straight, parallel streamlines to form above the crest. In this review, the term "weir" is applied generally to both the weir and the notch.

The resistance to flow introduced by the weir causes the water level upstream to rise. If it is assumed that the flow velocity some distance upstream of the weir is zero, then a simple measure of the upstream water level suffices to determine the discharge over a weir of known geometry. As in the case of an orifice in pipe flow, there will be a contraction of the nappe and a frictional resistance at the sides as water flows over a weir. The actual discharge is less than the theoretical discharge according to an empirically determined coefficient of discharge C_d. In the analysis of rectangular notches and weirs, it is frequently assumed that the channel width remains constant so that a contraction of the nappe occurs only in the vertical direction as the water accelerates over the obstacle. When the weir is narrower than the upstream channel, there is an additional horizontal contraction. The effect of a significant upstream velocity is accounted for empirically.

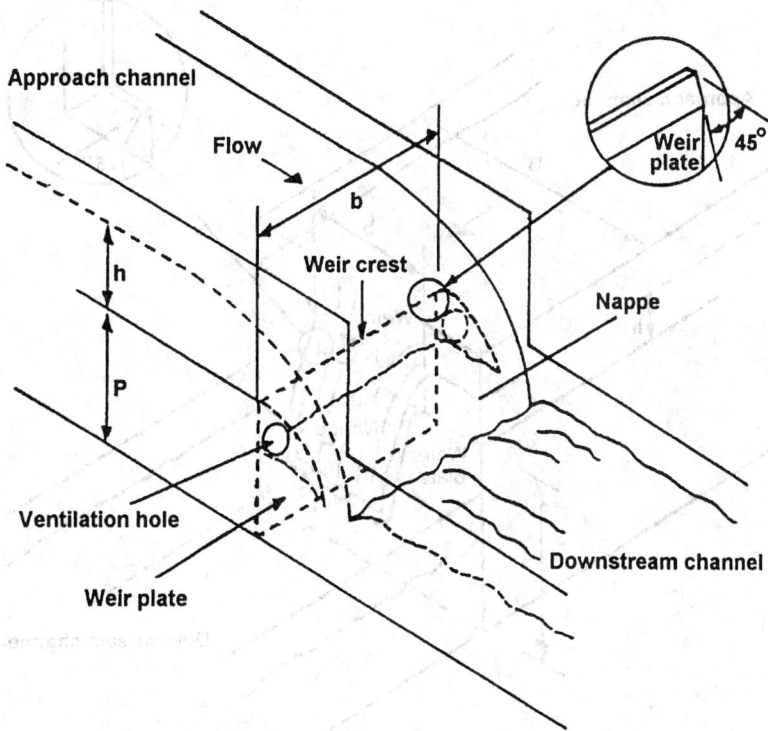

FIGURE 28.11 Full-width thin plate weir. Flow rate $\dot{Q}$ varies with water level h above the weir crest ($\dot{Q} \sim h^{3/2}$) and with a flow-dependent discharge coefficient C_d.

As an alternative to the weir, the flume (Figure 28.15) provides a method for flow metering with relatively low pressure loss. By restricting the channel width, analogous to the Venturi tube used in pipe flow, the flow velocity in the narrow portion of the channel is increased and the water level sinks accordingly. Most of the head of water is recovered in the diffusing section of the weir. The water levels upstream and in the throat of the weir can be determined by simple floats and recorded on a chart by pens driven mechanically from the floats. For remote monitoring, the use of echo sounders is advantageous.

The weir is preferentially used in natural river beds and the flume in canalized water courses. The flume must be used in streams with sediment transport to avoid the accumulation of deposits that would occur at the approach to a weir. Weirs and flumes are characterized by

- Simple measurement of the water level
- Simple maintenance
- Reliable measurement of large flow rates at low stream velocity
- Limited measurement accuracy (at best about 2%)
- High installation costs, particularly for flumes

Recommendations for installation of weirs and flumes, for the location of the head measuring station upstream, and for determining the discharge from the measured head are given, for example, in [2] and [3] as well as in appropriate standards (e.g., [4–6]).

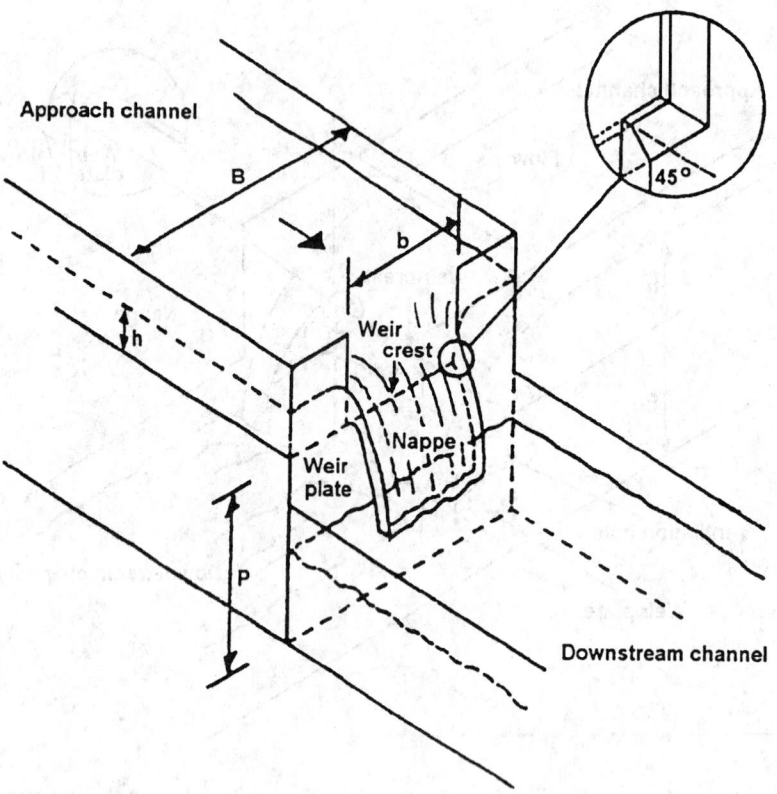

FIGURE 28.12   Thin plate rectangular notch weir: discharge coefficient C_d is flow dependent.

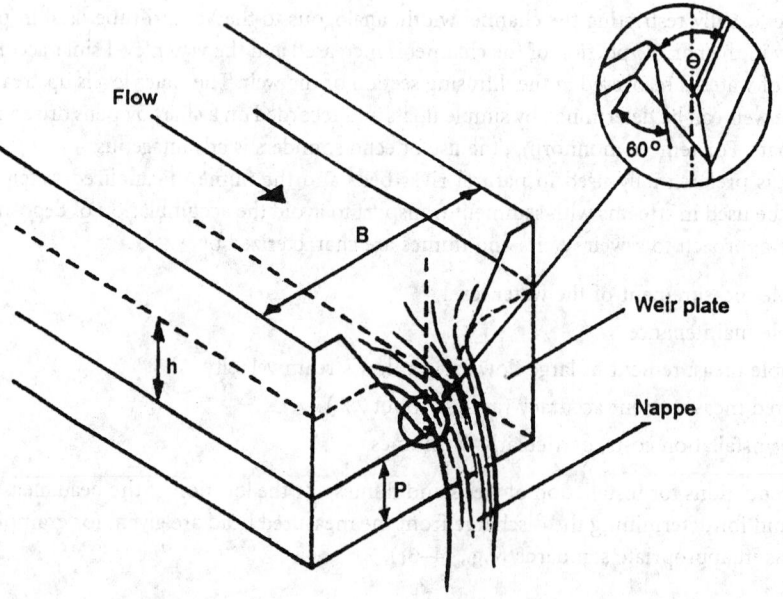

FIGURE 28.13   Thin plate V-notch weir: discharge coefficient C_d is almost independent of the flow.

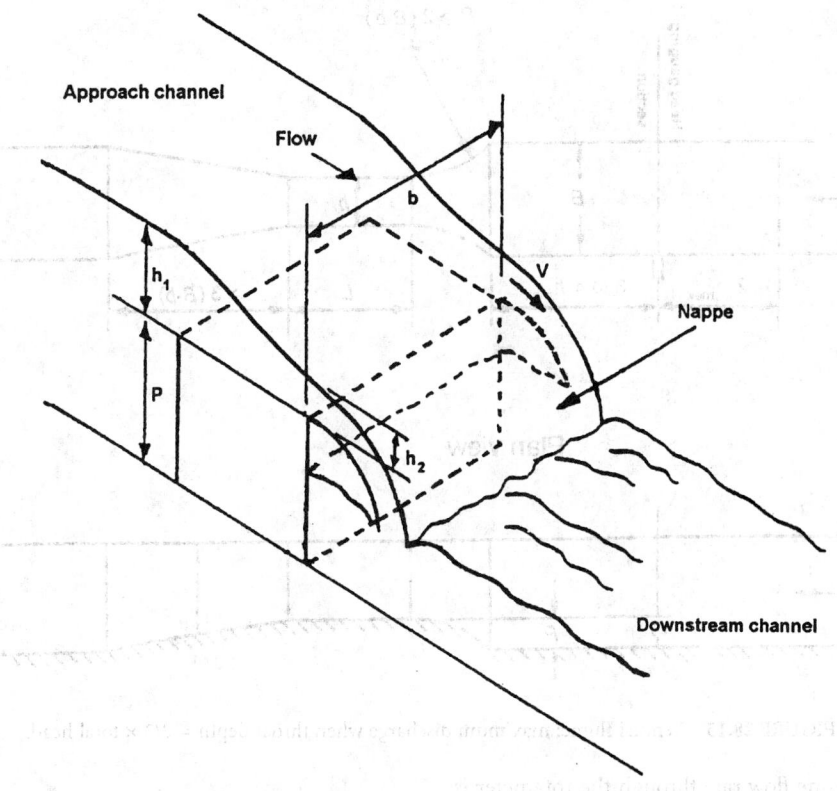

FIGURE 28.14 Broad-crested weir: maximum discharge when $h_2 = 2h_1/3$.

Measuring Principles of Variable Area Flowmeters

Rotameter

Flow Rate Analysis.

The forces acting on the bob lead to equilibrium between the weight of the bob $\rho_b g V_b$ acting downwards and the buoyancy force $\rho g V_b$ and the drag force F_d acting upwards, where V_b is the volume and ρ_b is the density of the bob, ρ is the density of the fluid, and g is the gravitational acceleration:

$$\rho_b g V_b = \rho g V_b + F_d \tag{28.7}$$

The drag force results from the flow field surrounding the bob and particularly from the wake of the bob. In flow analyses based on similarity principles, these influences are accounted for by empirical coefficient C_L or C_T in the drag law for:

$$\text{Laminar flow} \quad F_d = C_L \mu D_b U \tag{28.8}$$

$$\text{Turbulent flow} \quad F_d = C_T \rho D_b^2 U^2 \tag{28.9}$$

where μ = Fluid viscosity
D_b = Maximum bob diameter
U = Velocity in the annular gap around the bob at the minimum cross-section

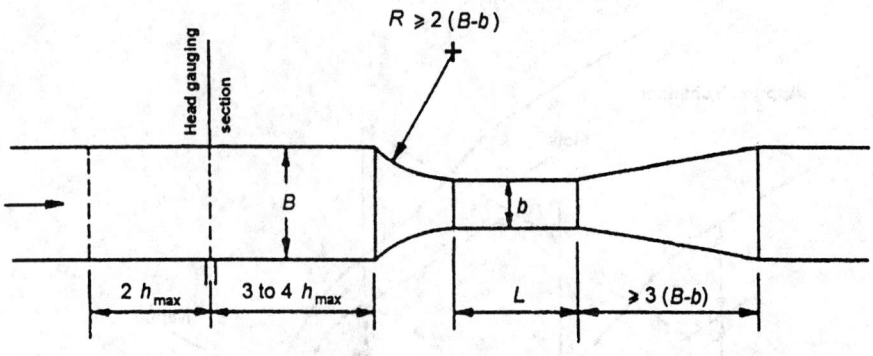

Plan view

FIGURE 28.15   Venturi flume: maximum discharge when throat depth = 2/3 × total head.

The volume flow rate through the rotameter is:

$$\dot{Q} = \frac{\pi}{4}\left(D^2 - D_b^2\right)U \tag{28.10}$$

or

$$\dot{Q} = m\frac{\pi}{4}D_b^2 U \tag{28.11}$$

where m is the open area ratio, defined as:

$$m = \frac{D^2 - D_b^2}{D_b^2} \tag{28.12}$$

and D is the tube diameter at the height of the bob.

Combining Equations 28.7, 28.8, and 28.10 gives for laminar flow:

$$\dot{Q}_L = \alpha D_b^4 \frac{\left(\rho_b - \rho\right)g}{\mu} \tag{28.13}$$

where the parameter α is defined in terms of a constant $K = V_b/D_b^3$ characteristic of the shape of the bob:

$$\alpha = \frac{\pi m K}{4C_L} \tag{28.14}$$

Using Equation 28.9 instead of Equation 28.8 yields for turbulent flow:

$$\dot{Q}_T = \beta D_b^{5/2} \sqrt{\frac{\left(\rho_b - \rho\right)g}{\rho}} \tag{28.15}$$

where

$$\beta = \frac{\pi m}{4} \sqrt{\frac{K}{C_T}} \tag{28.16}$$

With either laminar or turbulent flow through the rotameter, it is clear from Equations 28.13 and 28.15 that the flow rate is proportional to m. If the cross-sectional area of the tube is made to increase linearly with length, i.e.,

$$D = D_b \left(1 + h \tan \phi\right) \tag{28.17}$$

then since the cone angle ϕ of the tube is small, Equation 28.12 can be written as:

$$m = 2h \tan \phi \tag{28.18}$$

and the flow rate is directly proportional to the height h of the bob.

Similarity Analysis.
In early studies of floating element flowmeters, Ruppel and Umpfenbach [7] proposed the introduction of characteristic dimensionless quantities, to permit the use of experimentally determined flow coefficients in flowmeter analysis. Lutz [8] extended these ideas by showing that the transfer of flow coefficients from one flowmeter to another is possible if geometrical similarity exists. More recent works [9–12] have used these principles to produce graphical or computer-based design schemes and have proposed general guidelines for laying out practical flow metering systems.

The basic scaling parameter for flow is the Reynolds number, defined as:

$$Re = \frac{\rho U_{IN} D_b}{\mu} \tag{28.19}$$

where U_{IN} is the velocity at the rotameter inlet, and the tube diameter D is represented by its value at the inlet, equal to the bob diameter D_b. Through the Reynolds number regimes of laminar or turbulent flow, and particularly important for the rotameter flow regimes with strong or weak viscosity dependence, can be distinguished. Originating in the work [7], it has been found to be practical for rotameters to use an alternative characteristic number, the Ruppel number, defined as:

$$Ru = \frac{\mu}{\sqrt{m_b g \rho \left(1 - \rho/\rho_b\right)}} \tag{28.20}$$

where $m_b = \rho_b D_b^3$ is the mass of the bob. By combining Equations 28.15, 28.16, and 28.20, the mass flow $\dot{m}$ through the rotameter can be written as:

$$\dot{m} = \frac{\pi}{4} \frac{mD_b\mu}{\sqrt{C_T Ru}} \qquad (28.21)$$

Alternatively, from the definition:

$$\dot{m} = \rho\dot{Q} = \rho U_{IN} \frac{\pi}{4} D_b^2 \qquad (28.22)$$

and Equation 28.19, the flow rate is:

$$\dot{m} = \frac{\pi}{4} D_b\mu Re \qquad (28.23)$$

Equations 28.21 and 28.23 give the following relationship between the Ruppel number and the Reynolds number:

$$Ru = \frac{m}{\sqrt{C_T Re}} \qquad (28.24)$$

An analysis for laminar flow leads to the relationship:

$$Ru = \sqrt{\frac{m}{C_L Re}} \qquad (28.25)$$

The advantage of the Ruppel number is its independence of the flow rate. Since the Ruppel number contains only fluid properties and the mass and the density of the bob, it is a constant for a particular instrument.

At low Ruppel numbers the linear resistance law assumed in Equation 28.8 applies and α is a constant for given m, as shown in Figure 28.16. At higher Ruppel numbers, the flow is transitional or turbulent, and log α decreases linearly with log Ru. In Figure 28.17, curves for β against Ruppel number show a linear increase of log β with log Ru in the laminar region, followed by a gradual transition to horizontal curves in fully turbulent flow.

Similarity analysis of rotameters allows the easy calculation of the flow reading with changing density and viscosity of the fluid. For lower viscosities, only the density effect has to be taken into account; hence, for most gas measurements, simple conversion factors can be used. With higher viscosity, manufacturers offer either Ruppel number-related conversion factors or two-dimensional tables of conversion factors to be applied to different heights of the bob. Some manufacturers offer recalibration services, allowing the user to order new scales for instruments to be used in changed applications and environments. For large-order users, simple computer programs are available from some manufacturers for scale calculations.

Theories based on similarity considerations can predict the variation of the flow coefficient with the Ruppel number in the laminar and turbulent flow regimes but not in laminar-turbulent transitional flow. Detailed experimental and computational studies can assist the flow analysis of floating element flowmeters. An example is given in the section "Rotameter Internal Flow Analysis."

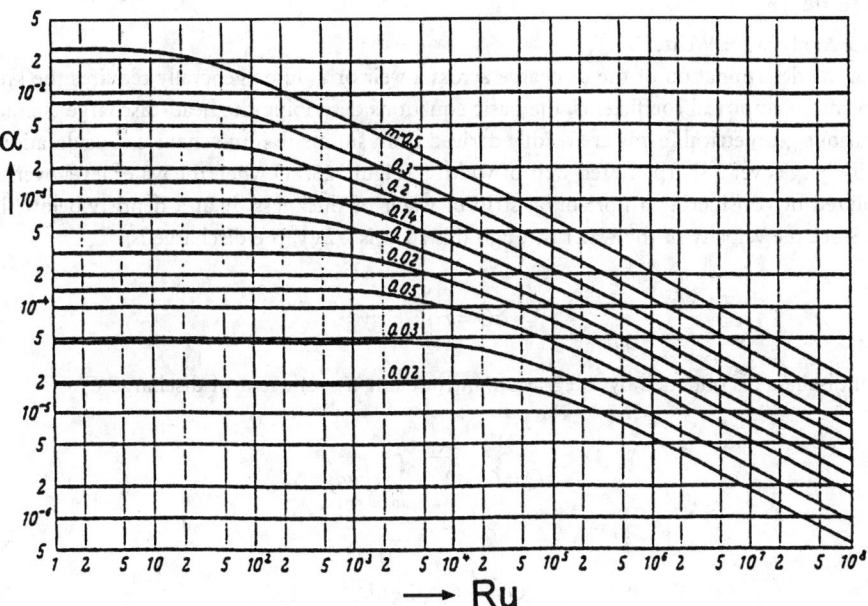

FIGURE 28.16 Rotameter flow coefficient α as a function of Ruppel number *Ru* and open area ratio *m*. For a given geometry, α is constant for low *Ru* (laminar flow).

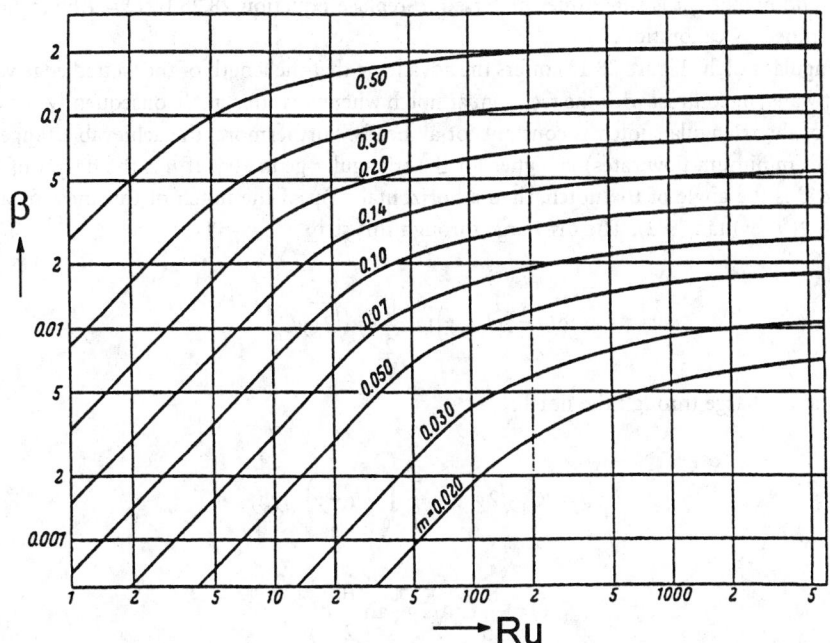

FIGURE 28.17 Rotameter flow coefficient β as a function of Ruppel number *Ru* and open area ratio *m*. For a given geometry, β is constant for high *Ru* (turbulent flow).

Weir, Flume

Flow Rate Analysis for Weirs.

Although the determination of the discharge across a weir or a flume generally requires the knowledge of one or more empirical coefficients, the basic equations describing the head-discharge characteristics of the various geometrical forms are readily derived from simple fluid mechanics considerations.

The discharge over a sharp crested weir of width b (Figure 28.11) when the water level over the sill is h is analyzed by considering a horizontal strip of water of thickness δy at a depth y below the water surface. Since the velocity of the water through this strip is $\sqrt{2gy}$, the discharge is:

$$\delta \dot{Q} = b \delta y \sqrt{2gy} \tag{28.26}$$

The total discharge is obtained by integration, introducing a coefficient of discharge C_d:

$$\dot{Q} = C_d \sqrt{2g} b \int_0^h \sqrt{y}\, dy \tag{28.27}$$

$$\dot{Q} = \frac{2}{3} C_d \sqrt{2g} b h^{3/2} \tag{28.28}$$

Empirical corrections to Equation 28.28 to account for the weir height p are given, for example, in [3], but provided h/p does not exceed 0.5, it is adequate to use Equation 28.28 with a discharge coefficient of 0.63 to achieve a flow rate tolerance of 3%.

For the rectangular notch, Equation 28.28 also applies with a discharge coefficient of 0.59 and a flow rate tolerance of 3%. Since in practice the discharge coefficient varies slightly with the head of water for both weirs and notches, it is sometimes preferred to replace Equation 28.28 by $\dot{Q} = K b h^n$, where K and n are determined by calibration.

The triangular notch (Figure 28.13) offers the advantage that the length of the wetted edge varies with the head of water, in contrast with the rectangular notch where it is constant. Consequently, the discharge coefficient of the triangular notch is constant for all heads. Furthermore, the achievable range (ratio of maximum to minimum flow rates) is higher for the triangular geometry. If h is the height of the water surface and θ is the angle of the notch, then a horizontal strip of the notch of thickness δy at depth y has a width $2(h - y) \tan (\theta/2)$. The discharge through this strip is

$$\delta \dot{Q} = 2\left(h - y\right) \tan \frac{\theta}{2} \delta y \sqrt{2gy}\, C_d \tag{28.29}$$

giving a total discharge through the notch:

$$\dot{Q} = 2 C_d \sqrt{2g}\, \tan \frac{\theta}{2} \int_0^h \left(h - y\right) \sqrt{y}\, dy \tag{28.30}$$

$$\dot{Q} = \frac{8}{15} C_d \sqrt{2g}\, \tan \frac{\theta}{2}\, h^{5/2} \tag{28.31}$$

The discharge coefficient is about 0.58.

Commonly used notch angles are 90° (tan $\theta/2$ = 1, C_d = 0.578), 53.13° (tan $\theta/2$ = 0.5, C_d = 0.577), and 28.07° (tan $\theta/2$ = 0.25, C_d = 0.587). Notches with θ = 53.13° and θ = 28.07° deliver, respectively, one half and one quarter of the discharge of the 90° notch at the same head.

The above derivations assume that the water is discharging from a reservoir with cross-sectional area far exceeding the flow area at the weir, so that the velocity upstream is negligible. When the weir is built into a channel of cross-sectional area A, the water will have a finite velocity of approach $v_1 = \dot{Q}/A$. As a first approximation, $\dot{Q}$ is obtained from Equation 28.28 assuming zero velocity of approach. Assuming further that v_1 is uniform over the weir, there will be an additional head $v_1^2/2g$ acting over the entire weir. Referring to Figure 28.11, the total discharge is then:

$$\dot{Q}=C_d \sqrt{2g}\, b \int_{v_1^2/2g}^{h+v_1^2/2g} \sqrt{y}\, dy \qquad (28.32)$$

$$\dot{Q}=\frac{2}{3} C_d \sqrt{2g}\, b\left[\left(h+v_1^2/2g\right)^{3/2}-\left(v_1^2/2g\right)^{3/2}\right] \qquad (28.33)$$

From Equation 28.33, a corrected value of v_1 can be determined. Further iterations converge rapidly to the final value of the discharge $\dot{Q}$. Equation 28.33 can be written as:

$$\dot{Q}=\frac{2}{3} C_v C_d \sqrt{2g}\, b h^{3/2} \qquad (28.34)$$

where

$$C_v =\left(1+\frac{v_1^2}{2gh}\right)^{3/2}-\left(\frac{v_1^2}{2gh}\right)^{3/2} \qquad (28.35)$$

is the coefficient of velocity, which is frequently determined empirically.

Over a broad-crested weir (Figure 28.14), the discharge depends on the head h_1, the width b, and the length l of the sill. There is also a dependence on the roughness of the sill surface and the viscosity. Consequently, there is a loss of head as the water flows over the sill. It is assumed that the sill is of sufficient length to allow the velocity to be uniform at a value v throughout the depth h_2 of the water at the downstream edge of the sill. Neglecting losses

$$v=\sqrt{2g\left(h_1 - h_2\right)}$$

and the discharge is:

$$\dot{Q}=C_d b h_2 v = C_d b \sqrt{2g\left(h_1 h_2^2 - h_2^3\right)} \qquad (28.36)$$

From Equation 28.36, the discharge is a maximum when $(h_1 h_2^2 - h_2^3)$ reaches a maximum, i.e., when $h_2 = 2h_1/3$. This discharge is then:

$$\dot{Q}=\left(\frac{2}{3}\right)^{3/2} C_d b \sqrt{g}\, h_1^{3/2} \qquad (28.37)$$

The stable condition of the weir lies at the maximum discharge.

Flow Rate Analysis for Flumes.
The discharge through a flume depends on the water level h, the channel width b, and velocity v at each of the stations 1 (upstream) and 2 (at the minimum cross-section) in Figure 28.15. Applying Bernoulli's equation to the inlet and the throat, neglecting all losses, one obtains:

$$\rho g H = \rho g h_1 + \frac{1}{2}\rho v_1^2 = \rho g h_2 + \frac{1}{2}\rho v_2^2 \tag{28.38}$$

where H is the total head. The analysis is then formally the same as that for the broad-crested weir, giving the same expression for the discharge as in Equation 28.37. The flow through the flume is a maximum when the depth at the throat is two thirds of the total head. Normally, a coefficient of velocity is introduced so that the discharge equation is:

$$\dot{Q} = \left(\frac{2}{3}\right)^{3/2} C_v C_d b \sqrt{g} h_1^{3/2} \tag{28.39}$$

The combined coefficient is determined empirically. In [3], a value $C_v C_d = 1.061 \pm 0.085$ is quoted.

The validity of the empirical coefficients quoted above, or of other coefficients given in the various standards for metering of open-channel flows, can only be guaranteed with a given confidence level for meter installations satisfying certain geometrical constraints. Limits on quantities such as b, h, h/b, etc. are given in [3] and the standards.

Rotameter: Internal Flow Analysis

Computation of Internal Flow.
Improvement of rotameter design could be assisted by detailed knowledge of the internal flow field, which is characterized by steep velocity gradients and regions of separated flow. Measurements of the internal flow field are complicated by the small dimensions of the gap around the bob and the strongly curved glass tube. Bückle et al. [13] successfully used laser Doppler anemometry [14] for velocity measurements in a rotameter. The working fluid was a glycerine solution with an index of refraction (1.455) close to that of glass (1.476). Problems with refraction of the laser beams at the curved tube wall were thus avoided, but the high viscosity of glycerine restricted the experiments to laminar flow at Reynolds numbers $Re = \rho U_{IN} D_b / \mu < 400$.

The application of computational fluid dynamics to the flow in a rotameter [13] involves the finite volume solution of the conservation equations for mass and momentum. For a two-dimensional laminar flow, these equations can be written in a cylindrical polar coordinate system as:

$$\frac{\partial \rho}{\partial t} + \frac{\partial(\rho u)}{\partial z} + \frac{1}{r}\frac{\partial(\rho r v)}{\partial r} = 0 \tag{28.40}$$

$$\frac{\partial(\rho u)}{\partial t} + \frac{\partial}{\partial z}\left(\rho u u - 2\mu\frac{\partial u}{\partial z}\right) + \frac{1}{r}\frac{\partial}{\partial r}\left(\rho r u v - \mu r\left(\frac{\partial u}{\partial r} + \frac{\partial v}{\partial z}\right)\right) = -\frac{\partial p}{\partial z} \tag{28.41}$$

$$\frac{\partial(\rho v)}{\partial t} + \frac{\partial}{\partial z}\left(\rho u v - \mu r\left(\frac{\partial u}{\partial r} + \frac{\partial v}{\partial z}\right)\right) + \frac{1}{r}\frac{\partial}{\partial r}\left(\rho r v v - 2\mu r\frac{\partial v}{\partial r}\right) = -\frac{\partial p}{\partial r} \tag{28.42}$$

where ρ is the density, u, v and r, z are the velocity components and coordinate directions in the axial and radial directions, respectively, μ is the dynamic viscosity, and p is the pressure.

For the numerical solution, the governing equations for a generalized transport variable ϕ (i.e., u or v) are formally integrated over each control volume (CV) of the computational grid. The resulting flux balance equation can in turn be discretized as an algebraic equation for ϕ at the center P of each CV in terms of the values ϕ_{nb} at the four nearest neighbors of point P and known functions A:

$$A_{p\phi p} + \sum_{nb} A_{nb}\,\phi_{nb} = S_\phi \qquad (28.43)$$

For the whole solution domain, a system of equations results that can be solved by a suitable algorithm (e.g. [15]).

For a rotameter, the symmetry of the problem allows the computational solution domain to be chosen as one half diametral plane of the rotameter. The boundary conditions are:

$$u = v = 0 \qquad (28.44)$$

along the walls, and

$$\frac{\partial u}{\partial r} = v = 0 \qquad (28.45)$$

along the axis of symmetry. At the input boundary, the initial profile for the u-velocity is taken from the experiment. At the outlet boundary, zero gradient is assumed for all dependent variables.

Computed Flow Field.
Representative calculations for Reynolds number 220 are shown as velocity vectors (Figure 28.18) and streamlines (Figure 28.19). Experimental and computed velocity profiles over a radius of the flowmeter tube are shown in Figure 28.20.

At $z = 45$ mm, the computations indicate a stagnation point at the bob tip, but the measurements show a low but finite axial velocity. The deviation from zero is attributable to slight unsteadiness in the position of the bob and to smearing of the steep radial velocity gradient along the measuring volume. This effect is also apparent at radii up to about 5 mm, where the computations show a notably steeper velocity gradient than the measurements. The measured and computed peak velocities agree well, although the measured peak is located at a larger radius.

In the strongly converging annulus between the bob and the tube wall ($z = 65$ mm) and in the plane $z = 88$ mm, the computations reproduce well the trend of the measured results, particularly near the flowmeter wall. Discrepancies in the region adjacent to the bob are attributable to asymmetry in measured profiles arising from the piping upstream of the rotameter tube.

Above the bob ($z = 110$ mm), there is a strong upward flow in an annular region near the tube wall and a recirculation region around the axis occupying almost half the tube cross-section. The forward velocities show a good match between computations and measurements, but there is a very marked discrepancy in the recirculation zone. At $z = 135$ mm, the recirculation zone is finished, but the wake of the bob is still very evident. Except for a discrepancy of about 10% on the axis, the results from the two methods agree remarkably well.

The effect of rotation of the bob on the flow field was considered by computations, including an equation for the azimuthal velocity component w.

$$\frac{\partial(\rho w)}{\partial t} + \frac{\partial}{\partial z}\left(\rho u w - \mu \frac{\partial w}{\partial z}\right) + \frac{1}{r}\frac{\partial}{\partial r}\left(\rho r v w - \mu r \frac{\partial w}{\partial r}\right) = -\rho\frac{vw}{r} - \mu\frac{w}{r^2} \qquad (28.46)$$

Calculations for 1 Hz rotation frequency showed the highest swirl velocities close to the axis above the bob. There was no observable influence of the rotation on the u- and v-velocity components [16].

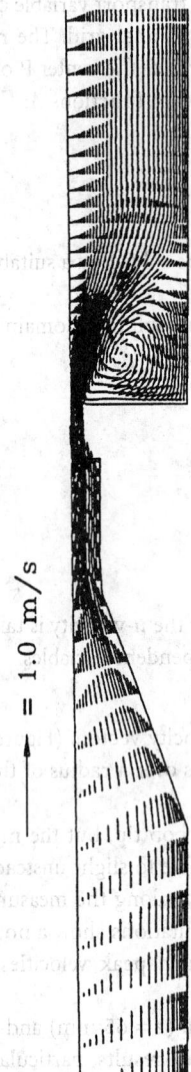

FIGURE 28.18 Computed velocity vectors for laminar flow through a rotameter. Computations in a half diametral plane for axisymmetric flow at Reynolds number 220.

Summary

For pipe flows, variable area flowmeters are most suitable for low flow rates of gases or liquids at moderate temperatures and pressures. Favorable features include rugged construction, high reliability, low pressure drop, easy installation, and low cost. Disadvantages include measurement uncertainty of 1% or more, limited range (10:1), slow response, and restrictions on the meter orientation. A generally good price/performance ratio has led to widespread use of these meters in numerous scientific and medical instruments and in many industrial applications for flow monitoring.

Variable area flowmeters in open-channel flows have applications for flow measurements in waste water plants, waterworks, rivers and streams, irrigation, and drainage canals. Hydrological applications of weirs with adjustable sill or crest height include flow regulation, flow measurement, upstream water level control, and discharge of excess flow in streams, rivers, and canals. Ruggedness, simplicity, and low maintenance costs are favorable characteristics for field applications of these meters.

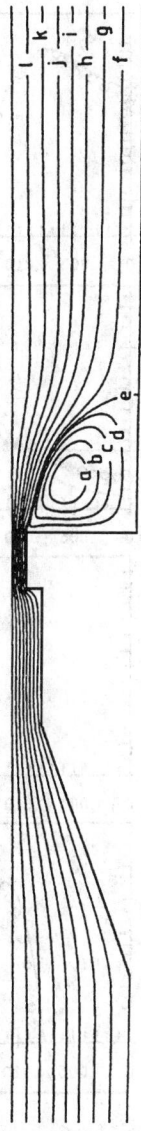

FIGURE 28.19 Computed streamlines for laminar flow through a rotameter (Reynolds number 220).

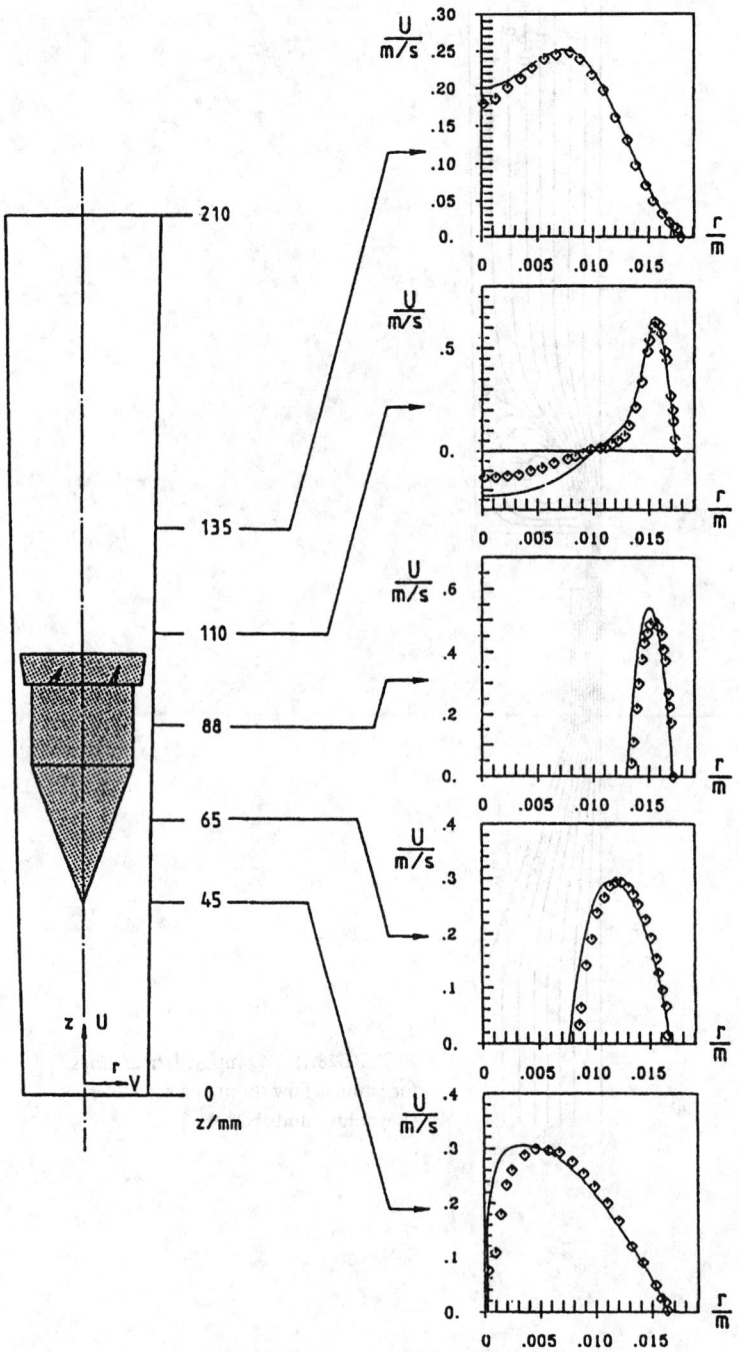

FIGURE 28.20 Comparison of measured and computed velocity profiles for laminar flow through a rotameter (Reynolds number 220). U = axial component (m/s), z = axial position (mm), r = radial position (m).

References

1. R. A. Furness, BS7045: the principles of flowmeter selection, *Flow Meas. Instrum.*, 2, 233–242, 1991.
2. W. Boiten, Flow-measuring structures, *Flow Meas. Instrum.*, 4, 17–24, 1993.
3. R. Hershey, General purpose flow measurement equations for flumes and thin plate weirs, *Flow Meas. Instrum.*, 6, 283–293, 1995.
4. ISO 1438, Thin Plate Weirs, International Standards Organisation, Geneva, 1980.
5. ISO 4359, Rectangular, Trapezoidal and U-shaped Flumes, International Standards Organisation, Geneva, 1983.
6. ISO 8368, Guidelines for the Selection of Flow Gauging Structures, International Standards Organisation, Geneva, 1985.
7. G. Ruppel and K. J. Umpfenbach, Strömungstechnische Untersuchungen an Schwimmermessern, *Technische Mechanik und Thermodynamik*, 1, 225–233, 257–267, 290–296, 1930.
8. K. Lutz, Die Berechnung des Schwebekörper-Durchflußmessers, *Regelungstechnik*, 10, 355–360, 1959.
9. D. Bender, Ähnlichkeitsparameter und Durchflußgleichungen für Schwebekörperdurchflußmesser, *ATM-Archiv für Technisches Messen*, 391, 97–102, 1968.
10. VDE/VDI-Fachgruppe Meßtechnik 3513, *Schwebekörperdurchflußmesser, Berechnungsverfahren*, VDE/VDI 3513, 1971.
11. H. Nikolaus, Berechnungsverfahren für Schwebekörperdurchflußmesser, theoretische Grundlagen und Aufbereitung für die Datenverarbeitungsanlage, *ATM-Archiv für Technisches Messen*, 435, 49–55, 1972.
12. H. Nikolaus and M. Feck, Graphische Verfahren zur Bestimmung von Durchflußkennlinien für Schwebekörperdurchflußmeßgeräte, *ATM-Archiv für Technisches Messen*, 1247-5, 171–176, 1974.
13. U. Bückle, F. Durst, B. Howe, and A. Melling, Investigation of a floating element flow meter, *Flow Meas. Instrum.*, 3, 215–225, 1992.
14. F. Durst, A. Melling, and J. H. Whitelaw, *Principles and Practice of Laser-Doppler Anemometry*, 2nd ed., London: Academic Press, 1981.
15. M. Perić, M. Schäfer, and E. Schreck, Computation of fluid flow with a parallel multigrid solver, *Proc. Conf. Parallel Computational Fluid Dynamics*, Stuttgart, 1991.
16. U. Bückle, F. Durst, H. Köchner, and A. Melling, Further investigation of a floating element flowmeter, *Flow Meas. Instrum.*, 6, 75–78, 1995.

28.3 Positive Displacement Flowmeters

Zaki D. Husain and Donald J. Wass

A *positive displacement flowmeter*, commonly called a PD meter, measures the volume flow rate of a continuous flow stream by momentarily entrapping a segment of the fluid into a chamber of known volume and releasing that fluid back into the flow stream on the discharge side of the meter. By monitoring the number of entrapments for a known period of time or number of entrapments per unit time, the total volume of flow or the flow rate of the stream can be ascertained. The total volume and the flow rate can then be displayed locally or transmitted to a remote monitoring station.

The positive displacement flowmeter has been in use for many decades to measure both liquid and gas flows. A PD meter can be viewed as a hydraulic motor with high volumetric efficiency that generally absorbs a small amount of energy from the flowing fluid. The energy absorption is primarily to overcome the internal resistance of the moving parts of the meter and its accessories. This loss of energy is observed as the pressure drop across the meter. The differential pressure across the meter is the driving force for the internals of the PD meter.

Design and Construction

A positive displacement meter has three basic components: an outer housing, the internal mechanism that creates the dividing chamber through its repetitive motion, and the display or counter accessories that determines the number of entrapments of fluid in the dividing chamber and infers the flow rate and the total volume of flow through the meter.

The external housing acts as the holding chamber of the flowing fluid before and after the entrapments by the internals. For low line pressures, the housing is usually single walled while, for higher operating pressures, the housing is double walled where the inner wall is the containment wall for the entrapment chamber and the outer wall is the pressure vessel. For the double-walled housing, the entrapment chamber walls are subjected to the differential pressure across the meter while the external housing is subjected to the total line pressure. This allows fabrication of a thin-walled inner chamber that can retain its precise shape and dimensions independent of line pressure.

The measuring mechanism consists of precise metering elements that require tight tolerances on the mating parts. The metering elements consist of the containment wall of the metering chamber and the moving components of the meter that forms the entrapment volume of the flowing fluid by cyclic or repetitive motion of those elements. The most common types of positive displacement meters are the oscillating piston, nutating disk, oval gear, sliding vane, birotor, trirotor, and diaphragm designs.

The counter or output mechanism converts the motion of the internal measuring chamber of a PD meter and displays the flow rate or total flow by correlating the number of entrapments and each entrapped volume. Many positive displacement meters have a mechanical gear train that requires seals and packing glands to transmit the motion of the inner mechanism to the outside counters. This type of display requires more driving power to overcome resistance of the moving parts and the seals, which results in an additional pressure drop for the meter. Many PD meters transmit the motion of the inner mechanism to the counters through switch output utilizing electromechanical, magnetic, optical, or purely electronic techniques in counting the entrapments and displaying the flow rate and total flow volume. The latter processing techniques normally have less pressure drop than the all-mechanical or part-mechanical transmission methods. All-mechanical drive counters do not require external power and, through proper selection of gear train, can display the actual flow volume or the flow rate. Thus, meters can be installed at remote locations devoid of any external power source. Meters with mechanical display cannot easily correct for the changes in volume due to thermal expansion or contraction of the measuring chamber due to flow temperature variation and, in the case of single-walled housing, the changes in the entrapment volume due to variations of the line pressure. An electronically processed output device could monitor both the pressure and temperature of the meter and provide necessary corrections to the meter output. With constantly changing and improving electronics technology, many PD meters are now installed with solar or battery-powered electronic output for installations at remote locations with no external power source. Many electronic displays allow access to meter data from a central monitoring station via radio or satellite communication.

Some Commercially Available PD Meter Designs

Commercially available PD meters have many noticeably different working mechanisms and a few designs are unique and proprietary. Although each design and working mechanism can be noticeably different from another, all positive displacement meters have a stationary fluid retaining wall and a mechanism that momentarily entraps inlet fluid into a partitioned chamber before releasing it to the downstream side of the meter. This entrapment and release of the flowing fluid occur with such repetitive and sweeping motion that, for most practical purposes, the flow rate appears to be uniform and steady, even though, in reality, the exit flow does have some pulsation. The flow pulsation out of the meter may be more pronounced for some designs of positive displacement meters than others. These flow pulsations are more pronounced at the lower flow rates for all designs. Some designs are more suitable for either liquid or gas flows, while some designs can measure both gas and liquid. For liquid applications, PD meters

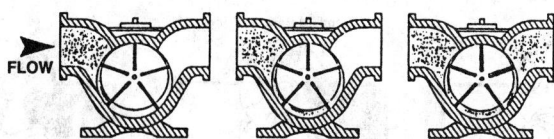

FIGURE 28.21   Sliding-vane type PD meter. (Courtesy of Daniel Industries, Inc.)

FIGURE 28.22   Trirotor type PD meter. (Courtesy of Liquid Controls LLC.)

work best for liquids with heavy viscosities. Almost all PD meters require precisely machined, high-tolerance mating parts; thus, measured fluid must be clean for longevity of the meter and to maintain the measurement precision.

Sliding-Vane Type Positive Displacement Meter

Figure 28.21 shows a working cycle of a sliding-vane type PD meter where vanes are designed to move in and out of the rotating inner mechanism. The position of each sliding vane relative to specific angular rotation of the rotor is usually maintained by a mechanical cam. In the design shown in Figure 28.21, each blade is independently retained in the slots. There are designs with even numbers of blades where diametrically opposite blades are one integral unit. High-pressure application would utilize dual wall construction.

Tri-Rotor Type PD Meter

This design has three rotating parts that entrap fluid between the rotors and an outer wall. The working cycle of this type of meter is shown Figure 28.22. In this design, for one rotation of the top mechanism, two blades rotate twice. The driving mechanism and rotation of each blade with respect to the others is maintained by a three-gear assembly where each rotating blade shaft is connected to one gear of the three-gear assembly to maintain relative rotational speed. This design is used to measure liquid flows.

Birotor PD Meter

The measuring unit of the birotor PD meter has two spiral type rotors kept in perfect timing by a set of precision gears and is used to measure liquid flows. Flow can enter the meter either perpendicular or parallel to the axis of rotation of the mating rotors, as shown in Figure 28.23. The axial design birotor is identical to the standard model in component parts and in principle; however, it utilizes a measuring unit mounted parallel, rather than perpendicular, to flow. Meter output is registered mechanically through a gear train located outside the measuring chamber or, electronically, using a pickup assembly mounted with a timing gear. Axial orientation results in compact installation, improved accuracy, and low pressure loss. The axial design is ideal for high flow rate applications.

Piston Type PD Meter

A typical design of a piston type PD meter is shown in Figure 28.24. A centrally located rotating part generates the reciprocating motion for each of the four pistons of the meter. The reciprocating motion of each of the piston is timed such that the discharge from an individual piston cylinder occurs in a cycle to generate a semicontinuous discharge from the meter. These PD meters are used for very low flow rates of liquid flows. A piston-cylinder design can withstand large differential pressures across the meter, so high viscous liquids can be measured by this type of meters and measurements are very precise. Mechanical

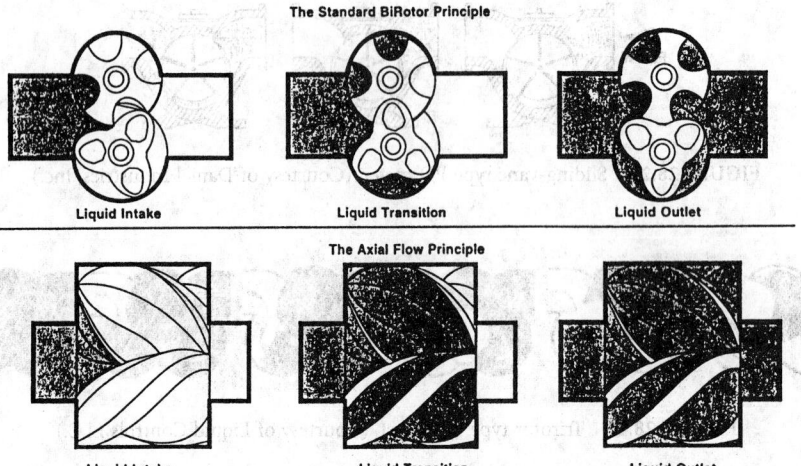

FIGURE 28.23 Birotor type PD meter. (Courtesy of Brooks Instruments Division, Emerson Electric Company.)

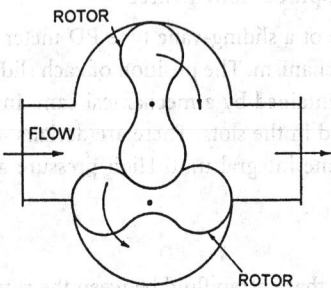

FIGURE 28.24 Piston-type PD meter. (Courtesy of Pierburg Instruments, Inc.)

FIGURE 28.25 Oval Gear Meter. (Courtesy of Daniel Industries, Inc.)

components of this type of meter require very precise mechanical tolerances, which increase the product cost.

Oval Gear PD Meter

The measurement of volumetric flow of an oval gear meter is obtained by separating partial volumes formed between oval gears and the measuring chamber wall, as shown in Figure 28.25. The rotation of the oval gears results from the differential pressure across the flowmeter. During one revolution of the oval gears, four partial volumes are transferred. The rotation of the gears is transmitted from the measuring chamber to the output shaft directly with mechanical seals or via a magnetic coupling. This type of meter is used to measure liquids having a wide range of viscosity.

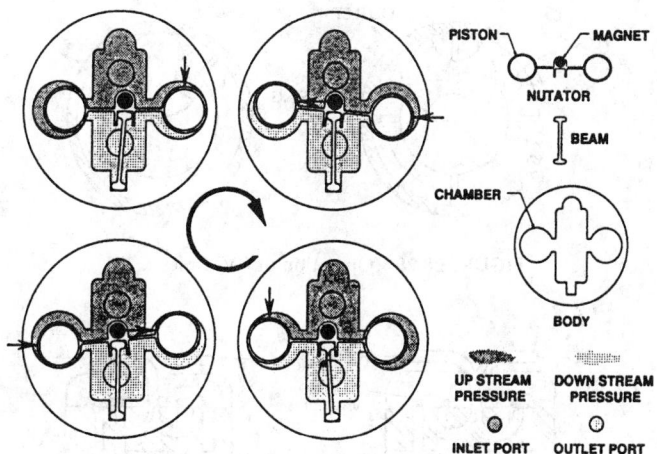

FIGURE 28.26 Nutating-disk type PD meter. (Courtesy of DEA Engineering Company.)

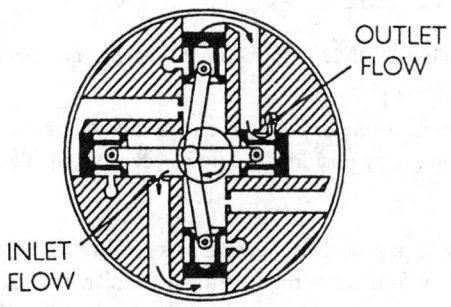

FIGURE 28.27 Schematic diagram of roots meter.

Nutating-Disk Type PD Meters

A disk placed within the confines of a boundary wall at a specific orientation can induce a flow instability that can generate a wobbling or nutating motion to the disk. The operating cycle of one such design is shown in Figure 28.26. The entrapment and discharge of the fluid to and from the two chambers occur during different phases of the repetitive cycle of the nutating disk. This type of meter is used to measure liquids. The design provides economic flow measurement where accuracy is not of great importance and is often used in water meters.

Meter designs described hereafter are for gas flows. PD meters for gas flows require very tight tolerances of the moving parts to reduce leak paths and, therefore, the gas must be clean because the meters are less tolerant to solid particles.

Roots PD Meter

The roots meter design is shown in Figure 28.27. This is the most commonly used PD meter to measure gas flows. This is one of the oldest designs of a positive displacement meter. The mechanical clearance between the rotors and the housing requires precisely machined parts. Roots meters are adversely affected if inlet flow to the meter has a relatively high level of pulsation. The main disadvantage of this meter is that it introduces cyclic pulsations into the flow. This meter cannot tolerate dirt and requires filtering upstream of the meter. A roots meter in single-case design is used at near-ambient line pressures. This

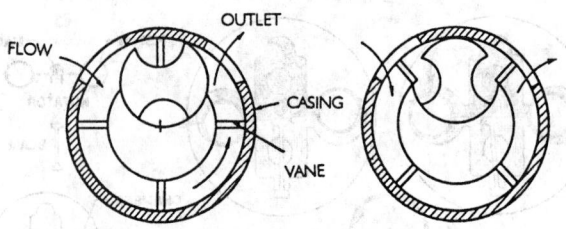

FIGURE 28.28 Operation of CVM meter.

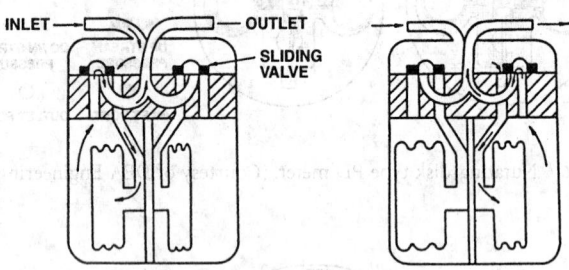

FIGURE 28.29 Operation of a diaphragm meter.

meter design is very widely used in the low-pressure transmission and distribution market of natural gas. If this meter ever seizes up at any position, the flow is completely blocked.

The CVM Meter

The CVM meter (Figure 28.28) is a proprietary design of a positive displacement meter used to measure gas flows. This meter has a set of four vanes rotating in an annular space about the center of the circular housing. A gate similar in shape to the center piece of the trirotor design (Figure 28.22) allows the vanes to pass — but not the gas. The measurement accuracy of the CVM meter is similar to the roots meter, but the amplitude of exit pulsation is lower than the roots meter for similar meter size and flow rate. This design is also not as sensitive to inlet flow fluctuations, and is widely used in the low-pressure, natural gas distribution market in Europe.

Diaphragm Meter

The diaphragm meter, also known as a bellows-type meter, is simple, relatively inexpensive, has reliable metering accuracy, and is widely used as a domestic gas meter. This design, shown in Figure 28.29, acts like a reciprocating piston meter, where the bellows act as the piston. The operation is controlled by a double slide valve. In one position, the slide valve connects the inlet to the inside of the bellow, while the outside of the bellow is open to the outlet. In the other position, the outside is connected to the inlet while the inside of the bellow is open to the outlet. This design can measure extremely low flow rates (e.g., pilot light of a gas burner).

Advantages and Disadvantages of PD Meters

High-quality, positive displacement meters will measure with high accuracy over a wide range of flow rates, and are very reliable over long periods. Unlike most flowmeters, a positive displacement meter is insensitive to inlet flow profile distortions. Thus, PD meters can be installed in close proximity to any upstream or downstream piping installations without any loss of accuracy. In general, PD meters have minimal pressure drop across the meter; hence, they can be installed in a pipeline with very low line pressures. Until the introduction of electronic correctors and flow controls on other types of meters, PD meters were most widely used in batch loading and dispensing applications. All mechanical units can be installed in remote locations.

Positive displacement meters are generally bulky, especially in the larger sizes. Due to the tight clearance necessary between mating parts, the fluid must be clean for measurement accuracy and longevity of the meter. More accurate PD meters are quite expensive. If a PD meter ever seizes up, it would completely block the flow. Many PD meters have high inertia of the moving parts; therefore, a sudden change in the flow rate can damage the meter. PD meters are normally suitable over only limited ranges of pressure and temperature. Some designs will introduce noticeably high pulsations into the flow. Most PD meters require a good maintenance schedule and are high repair and maintenance meters. Recurring costs in maintaining a positive displacement flowmeter can be a significant factor in overall flowmeter cost.

Applications

Liquid PD meters are capable of measuring fluids with a wide range of viscosity and density. Minor changes in viscosity of the fluid have minimal influence on the accuracy of a PD meter. However, no one specific design of PD meter is capable of such a broad application range. The physical properties of the fluid, especially the fluid viscosity, must be reviewed for proper selection of a meter for the application. Many PD meter designs accurately measure flow velocities over a wide range.

Careful dimensional control is required to produce well-regulated clearances around the moving parts of the PD meter. The clearances provide a leakage path around the flow measurement mechanism, and the amount of leakage depends on the relative velocity of the moving parts, the viscosity of the fluid, and the clearances between parts. In general, the leakage flow rate follows the Poiseuille equation; that is, leakage is directly proportional to the differential pressure and inversely proportional to the absolute viscosity and the length of the leakage path.

The leakage flow, in percentage of the total flow rate, is useful information for error measurement. At low flow rates, percent leakage error can be significant; while with modest flow rates, leakage can be insignificant. However, the differential pressure across the PD meter increases exponentially with increasing flow rates. Therefore, at very high flow rates, leakage flow can again be a significant portion of the total flow. As a result, PD meters tend to dispense more fluid than is indicated by the register at the very low and very high flow rates. The amount of error due to leakage depends on the meter design and on the viscosity of the fluid.

PD meters are driven by the differential pressure across the meter. The primary losses within the meter can be attributed to the friction losses and the viscous drag of the fluid on the moving parts. At very low flow rates, the bearing losses are predominant, while at high flow rates, the viscous drags predominate. The viscous drag is directly proportional to the fluid viscosity and the relative velocity of the moving parts and inversely proportional to the clearance. Within each design, the differential pressure is limited to a predetermined value. Excessive differential pressures can damage the meter. High differential pressure can be avoided by limiting the maximum flow rate, lowering viscosity by heating the fluid, increasing clearances, or by various combinations of these techniques. In general, if the differential pressure limit can be addressed, PD meters measuring high-viscosity fluids have less measurement error over a wide flow rate range.

The viscosity of most gases is too low to cause application problems relating to viscous drag of the form discussed for liquid PD meters. Gas PD meters utilize smaller clearances than liquid PD meters; therefore, gas must be very clean. Even small particles of foreign material can damage gas PD meters. PD meters used in gas flow measurement are primarily in the low line pressure application.

Accessories

Many accessories are available for PD meters. Among the most common and useful accessories for liquid PD meters are the automatic air eliminator and the direct-coupled shut-off valve for batch loading operations. The automatic air eliminator consists of a pressure-containing case surrounding a float valve.

When gas in the liquid flow enters the chamber, the float descends by gravity and allows the float valve to open and purge the gas. An air eliminator improves the precision of flow measurement in many applications and is an absolute necessity in applications where air or gas is trapped in the flow stream. The direct-coupled shut-off valve is directly linked to the mechanical register. A predetermined fluid quantity can be entered into the register. When the dispensed amount equals the amount entered in the register, the mechanical linkage closes the valve. Two-stage shut-off valves are often used to avoid line shock in high flow rate dispensing applications. Similarly, all-electronic units can control batch operations. The all-electronic units provide great convenience in the handling of data. However, the register-driven mechanical valves have advantages in simplicity of design, field maintenance or repair, and initial cost.

Gas PD meters can be equipped with mechanical registers that correct the totals for pressure and temperature. The pressure and temperature compensation range is limited in the mechanical units but with appropriate application; the range limitation is not a problem. Electronic totalizers for gas PD meters can correct for broad ranges of temperature and pressure. Electronic accessories can provide various data storage, logging, and remote data access capabilities.

Price and Performance

In general, the accuracy of the PD meter is reflected in the sales price. However, several additional factors influence the cost. The intended application, the materials of construction, and the pressure rating have a strong influence on the cost. Although they provide excellent accuracy, small PD meters for residential water or residential natural gas service are very inexpensive. Meters produced for the general industrial market provide good accuracy at a reasonable cost, and meters manufactured for the pipeline or petro-chemical markets provide excellent accuracy and are expensive.

Very inexpensive PD meters are available with plastic cases or plastic internal mechanisms. Aluminum, brass, ductile iron, or cast iron PD meters are moderately priced. Steel PD meters are relatively expensive, especially for high-pressure applications. If stainless steel or unusual materials are necessary for corrosive fluids or special service requirements, the costs are very high. PD meters are manufactured in a wide range of sizes, pressure ratings, and materials, with a multitude of flow rate ranges, and with an accuracy to match most requirements. The cost is directly related to the performance requirements.

Further Information

D. W. Spitzer, Ed., *Flow Measurement: Practical Guides for Measurement and Control*, Research Triangle Park, NC: ISA, 1991.

R. W. Miller, *Flow Measurement Engineering Handbook*, New York: McGraw-Hill, 1989.

Fluid Meters: Their Theory and Application, 6th ed., Report of ASME Research Committee on Fluid Meters, American Society of Mechanical Engineers, 1971.

V. L. Streeter, *Fluid Mechanics, 4th ed.*, New York: McGraw-Hill, 1966.

V. L. Streeter, *Handbook of Fluid Dynamics*, New York: McGraw-Hill, 1961.

Measurement of Liquid Hydrocarbons by Displacement Meters, Manual of Petroleum Measurement Standard, Chapter 5.2, American Petroleum Institute, 1992.

28.4　Turbine and Vane Flowmeters

David Wadlow

This section describes a range of closed-conduit flowmeters that utilize rotating vaned transduction elements, with particular emphasis on axial turbine flowmeters. Single jet and insertion tangential turbines, also known as paddlewheel flowmeters, are described in another section. The various vaned flowmeters used for open-channel and free-field flow measurement are not included in this section.

Axial Turbine Flowmeters

The modern axial turbine flowmeter, when properly installed and calibrated, is a reliable device capable of providing the highest accuracies attainable by any currently available flow sensor for both liquid and gas volumetric flow measurement. It is the product of decades of intensive innovation and refinements to the original axial vaned flowmeter principle first credited to Woltman in 1790, and at that time applied to measuring water flow. The initial impetus for the modern development activity was largely the increasing needs of the U.S. natural gas industry in the late 1940s and 1950s for a means to accurately measure the flow in large-diameter, high-pressure, interstate natural gas lines. Today, due to the tremendous success of this principle, axial turbine flowmeters of different and often proprietary designs are used for a variety of applications where accuracy, reliability, and rangeability are required in numerous major industries besides water and natural gas, including oil, petrochemical, chemical process, cryogenics, milk and beverage, aerospace, biomedical, and others.

Figure 28.30 is a schematic longitudinal section through the axis of symmetry depicting the key components of a typical meter. As one can see, the meter is an in-line sensor comprising a single turbine rotor, concentrically mounted on a shaft within a cylindrical housing through which the flow passes. The shaft or shaft bearings are located by end supports inside suspended upstream and downstream aerodynamic structures called diffusers, stators, or simply cones. The flow thus passes through an annular region occupied by the rotor blades. The blades, which are usually flat but can be slightly twisted, are inclined at an angle to the incident flow velocity and hence experience a torque that drives the rotor. The rate of rotation, which can be up to several $\times\ 10^4$ rpm for smaller meters, is detected by a pickup, which is usually a magnetic type, and registration of each rotor blade passing infers the passage of a fixed volume of fluid.

General Performance Characteristics

Axial turbines perform best when measuring clean, conditioned, steady flows of gases and liquids with low kinematic viscosities (below about 10^{-5} m^{2}s^{-1}, 10 cSt, although they are used up to 10^{-4} m^{2}s^{-1}, 100 cSt), and are linear for subsonic, turbulent flows. Under these conditions, the inherent mechanical stability of the meter design gives rise to excellent repeatability performance. Not including the special case of water meters, which are described later, the main performance characteristics are:

- Sizes (internal diameter) range from 6 mm to 760 mm, (1/4 in. to 30 in.).
- Maximum measurement capacities range from 0.025 Am^3 h^{-1} to 25,500 Am^3 h^{-1}, (0.015 ACFM to 15,000 ACFM), for gases and 0.036 m^3 h^{-1} to 13,000 m^3 h^{-1}, (0.16 gpm to 57,000 gpm or 82,000 barrels per hour), for liquids, where A denotes actual.
- Typical measurement repeatability is ±0.1% of reading for liquids and ±0.25% for gases with up to ±0.02% for high-accuracy meters. Typical linearities (before electronic linearization) are between ±0.25% and ±0.5% of reading for liquids, and ±0.5% and ±1.0% for gases. High-accuracy meters have linearities of ±0.15% for liquids and ±0.25% for gases, usually specified over a 10:1 dynamic range below maximum rated flow. Traceability to NIST (National Institute of Standards and Technology) is frequently available, allowing one to estimate the overall absolute accuracy performance of a flowmeter under specified conditions. Under ideal conditions, absolute accuracies for optimum designs and installations can come close to the accuracy capabilities at the NIST, which are stated as ±0.13% for liquid flows and ±0.25% for air.
- Rangeability, when defined as the ratio of flow rates over which the linearity specification applies, is typically between 10:1 and 100:1.
- Operating temperature ranges span –270°C to 650°C, (–450°F to 1200°F).
- Operating pressure ranges span coarse vacuum to 414 MPa (60,000 psi).
- Pressure drop at the maximum rated flow rate ranges from around 0.3 kPa (0.05 psi) for gases to in the region of 70 kPa (10 psi) for liquids.

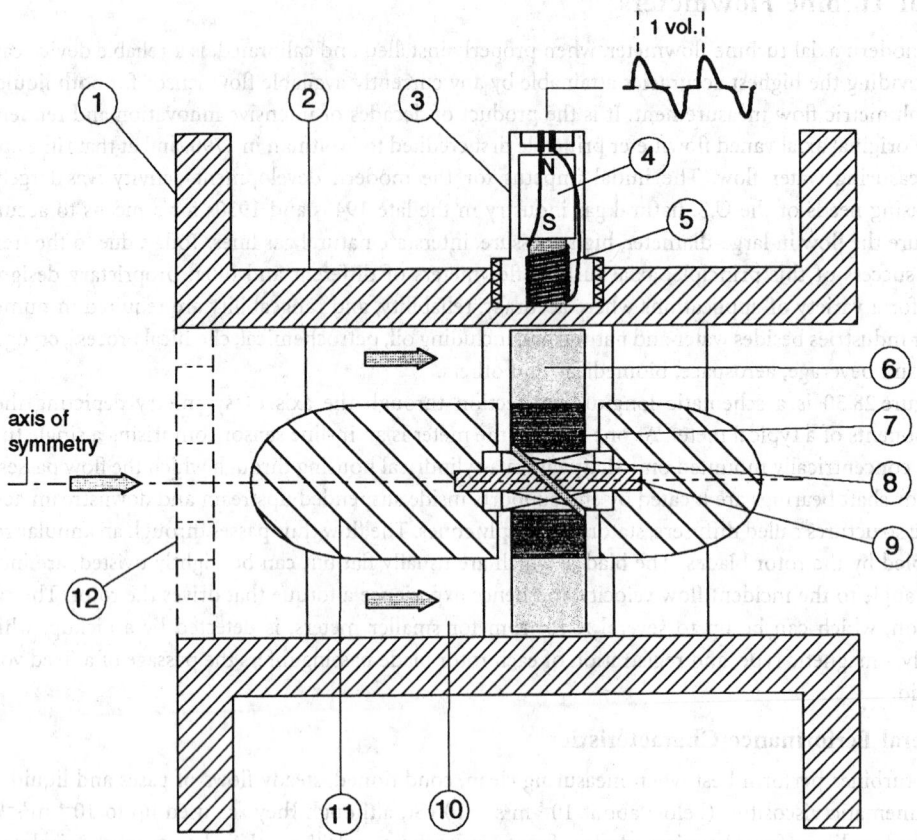

FIGURE 28.30 Longitudinal section of an axial turbine flowmeter depicting the key components. The flowmeter body is usually a magnetically transparent stainless steel such as 304. Common end-fittings include face flanges (depicted), various threaded fittings and tri-clover fittings. The upstream and downstream diffusers are the same in bidirectional meters, and generally supported by three or more flat plates, or sometimes tubular structures, aligned with the body, which also act as flow straighteners. The relative size of the annular flow passage at the rotor varies among different designs. Journal rotor bearings are frequently used for liquids, while ball bearings are often used for gases. Magnetic reluctance pickups (depicted) are frequently used. Others types include mechanical and modulated carrier pickups. (1) End fitting — flange shown; (2) flowmeter body; (3) rotation pickup — magnetic, reluctance type shown; (4) permanent magnet; (5) pickup cold wound on pole piece; (6) rotor blade; (7) rotor hub; (8) rotor shaft bearing — journal type shown; (9) rotor shaft; (10) diffuser support and flow straightener; (11) diffuser; (12) flow conditioning plate (dotted) — optional with some meters.

Theory

There are two approaches described in the current literature for analyzing axial turbine performance. The first approach describes the fluid driving torque in terms of momentum exchange, while the second describes it in terms of aerodynamic lift via airfoil theory. The former approach has the advantage that it readily produces analytical results describing basic operation, some of which have not appeared via airfoil analysis. The latter approach has the advantage that it allows more complete descriptions using fewer approximations. However, it is mathematically intensive and leads rapidly into computer-generated solutions. One prominent pioneer of the momentum approach is Lee [1] who, using this approach, later went on to invent one of the few, currently successful, dual rotor turbine flowmeters, while Thompson and Grey [2] provided one of the most comprehensive models currently available using the airfoil

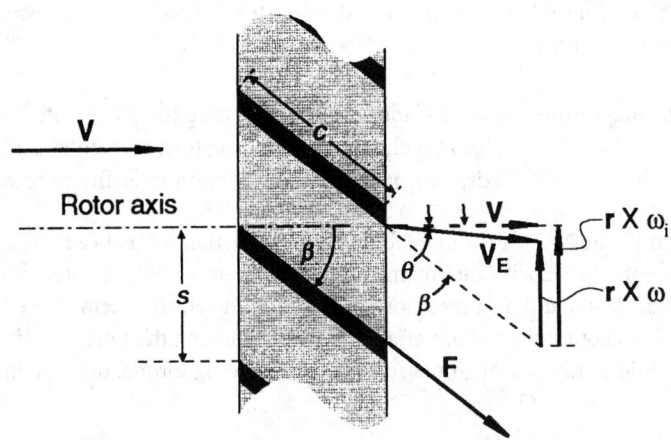

FIGURE 28.31 Vector diagram for a flat-bladed axial turbine rotor. The difference between the ideal (subscript i) and actual tangential velocity vectors is the rotor slip velocity and is caused by the net effect of the rotor retarding torques. This gives rise to linearity errors and creates swirl in the exit flow. V, incident fluid velocity vector; V_E, exit fluid velocity vector; θ, exit flow swirl angle due to rotor retarding torques; β, blade pitch angle, same as angle of attack for parallel flow; ω, rotor angular velocity vector; r, rotor radius vector; F, flow-induced drag force acting on each blade surface; c, blade chord; s, blade spacing along the hub; c/s, rotor solidity factor.

approach, which for example, took into account blade interference effects. In the following, the momentum exchange approach is used to highlight the basic concepts of the axial turbine flowmeter.

In a hypothetical situation, where there are no forces acting to slow down the rotor, it will rotate at a speed that exactly maintains the fluid flow velocity vector at the blade surfaces. Figure 28.31 is a vector diagram for a flat-bladed rotor with a blade pitch angle equal to β. Assuming that the rotor blades are flat and that the velocity is everywhere uniform and parallel to the rotor axis, then referring to Figure 28.31, one obtains:

$$r\omega_i = \frac{\tan\beta}{V} \qquad (28.47)$$

When one introduces the total flow rate, this becomes:

$$\frac{\omega_i}{Q} = \frac{\tan\beta}{\bar{r}\,A} \qquad (28.48)$$

where ω_i = "Ideal" rotational speed
 Q = Volumetric flow rate
 A = Area of the annular flow cross-section
 $\bar{r}$ = Root-mean-square of the inner and outer blade radii, (R, a)

Eliminating the time dimension from the left-hand-side quantity reduces it to the number of rotor rotations per unit fluid volume, which is essentially the flowmeter K factor specified by most manufacturers. Hence, according to Equation 28.48, in the ideal situation, the meter response is perfectly linear and determined only by geometry. (In some flowmeter designs, the rotor blades are helically twisted to improve efficiency. This is especially true of blades with large radius ratios, (R/a). If the flow velocity profile is assumed to be flat, then the blade angle in this case can be described by $\tan\beta$ = constant $\times$ r. This is sometimes called the "ideal" helical blade.) In practice, there are instead a number of rotor retarding torques of varying relative magnitudes. Under steady flow, the rotor assumes a speed that satisfies the following equilibrium:

Fluid driving torque = rotor blade surfaces fluid drag torque + rotor hub
and tip clearance fluid drag torque + rotation sensor   (28.49)
drag torque + bearing friction retarding torque

Referring again to Figure 28.31, the difference between the actual rotor speed, $r\omega$, and the ideal rotor speed, $r\omega_i$, is the rotor slip velocity due to the combined effect of all the rotor retarding torques as described in Equation 28.49, and as a result of which the fluid velocity vector is deflected through an exit or swirl angle, θ. Denoting the radius variable by r, and equating the total rate of change of angular momentum of the fluid passing through the rotor to the retarding torque, one obtains:

$$\int_a^R \frac{\rho Q 2\pi r^2 \left(r\omega_i - r\omega\right)}{\pi\left(R^2 - a^2\right)} \, dr = N_T \qquad (28.50)$$

which yields:

$$\bar{r}^2 \rho Q\left(\omega_i - \omega\right) = N_T \qquad (28.51)$$

where ρ is the fluid density and N_T is the total retarding torque. Combining Equations 28.47 and 28.51 and rearranging, yields:

$$\frac{\omega}{Q} = \frac{\tan\beta}{\bar{r} A} - \frac{N_T}{\bar{r}^2 \rho Q^2} \qquad (28.52)$$

The trends evident in Equation 28.52 reflect the characteristic decline in meter response at very low flows and why lower friction bearings and lower drag pickups tend to be used in gas vs. liquid applications and small diameter meters. In most flowmeter designs, especially for liquids, the latter three of the four retarding torques described in Equation 28.49 are small under normal operating conditions compared with the torque due to induced drag across the blade surfaces. As shown in Figure 28.31, the force, F, due to this effect acts in a direction along the blade surface and has a magnitude given by:

$$F = \frac{\rho V^2}{2} C_D S \qquad (28.53)$$

where C_D is the drag coefficient and S is the blade surface area per side. Using the expression for drag coefficient corresponding to turbulent flow, selected by Pate et al. [3] and others, this force can be estimated by:

$$F = \rho V^2 0.074 \, \mathrm{Re}^{-0.2} S \qquad (28.54)$$

where Re is the flow Reynolds number based on the blade chord shown as dimension c in Figure 28.31. Assuming θ is small compared with β, then after integration, the magnitude of the retarding torque due to the induced drag along the blade surfaces of a rotor with n blades is found to be:

$$N_D = n\left(R + a\right)\rho V^2 \, 0.037 \, \mathrm{Re}^{-0.2} S \sin\beta \qquad (28.55)$$

Combining Equations 28.55 and 28.52, and rearranging yields:

$$\frac{\omega}{Q} = \frac{\tan\beta}{\bar{r}A} - \frac{0.036n\left(R+a\right)SA^2\,\text{Re}^{-0.2}\sin\beta}{\bar{r}^2} \qquad (28.56)$$

Equation 28.56 is an approximate expression for the K factor because it neglects the effects of several of the rotor retarding torques, and a number of important detailed meter design and aerodynamic factors, such as rotor solidity and flow velocity profile. Nevertheless, it reveals that linearity variations under normal, specified operating conditions are a function of certain basic geometric factors and Reynolds number. These results reflect general trends that influence design and calibration. Additionally, the marked departure from an approximate ρV^2 (actually $\rho^{0.8}V^{1.8}\mu^{-0.2}$ via Re in Equation 28.54) dependence of the fluid drag retarding torque on flow properties under turbulent flow, to other relationships under transitional and laminar flow, gives rise to major variations in the K factor vs. flow rate and media properties for low-flow Reynolds numbers. This is the key reason why axial turbine flowmeters are generally recommended for turbulent flow measurement.

Calibration, Installation, and Maintenance

Axial turbine flowmeters have a working dynamic range of at least 10:1 over which the linearity is specified. The maximum flow rate is determined by design factors related to size vs. maximum pressure drop and maximum rotor speed. The minimum of the range is determined by the linearity specification itself. Due to small, unavoidable, manufacturing variances, linearity error curves are unique to individual meters and are normally provided by the manufacturer. However, although recommended where possible, the conditions of the application cannot usually and need not necessarily duplicate those of the initial or even subsequent calibrations. This has pivotal importance in applications where actual operating conditions are extreme or the medium is expensive or difficult to handle. Figure 28.32 depicts a typically shaped calibration curve of linearity vs. flow rate expressed in terms of multiple alternative measures, various combinations of which can be found in current use. The vertical axis thus represents either the linearity error as a percentage of flow rate, a K factor expressed in terms of the number of pulses from the rotation sensor output per volume of fluid, or the deviation from 100% registration; the latter only applies to flowmeters with mechanical pickups. The horizontal axis can be expressed in terms of flow rate in volume units/time, Reynolds number (Re), or pulse frequency (from the rotation sensor for nonmechanical) divided by kinematic viscosity, (f/v), in units of Hz per $m^2 s^{-1}$, (Hz/cSt or Hz/SSU; $10^{-6}\ m^2 s^{-1} = 1$ centistoke $= 31.0$ seconds, Saybolt Universal), and where kinematic viscosity is the ratio of absolute viscosity (μ) to density. Calibrations are preferably expressed vs. Re or f/v, which is proportional to Re. The hump shown in the curve is a characteristic frequently observed at lower Re and is due to velocity profile effects. K factor vs. f/v calibration curves are specifically called universal viscosity curves (UVC) and, for most meters, are available from the manufacturer for an extra charge. A key utility of UVC is that where media type and properties differ significantly from those of the original calibration, accuracies much greater than the overall linearity error can still readily be obtained via the flowmeter UVC if the kinematic viscosity of the application is known. An alternative, advanced calibration technique [4] is to provide response in terms of Strouhal number vs. Re or Roshko number. This approach is not widely adopted, but it is particularly relevant to high-accuracy and extreme temperature applications because it further allows correct compensation for flowmeter thermal expansion errors.

The accuracy of axial turbine flowmeters is reduced by unconditioned flow, especially swirl. An installation incorporating flow conditioners along with specific upstream and downstream straight pipe lengths is generally recommended [5]. Some axial turbine flowmeters can be purchased with additional large flow straighteners that mount directly ahead of the flowmeter body or conditioning plates that are integral to the body. The manufacturer is the first source of information regarding installation. Errors due to flow velocity pulsations are another concern, particularly in certain gas installations. However, no standard technique for effectively counteracting this source of error has yet been adopted. Periodic

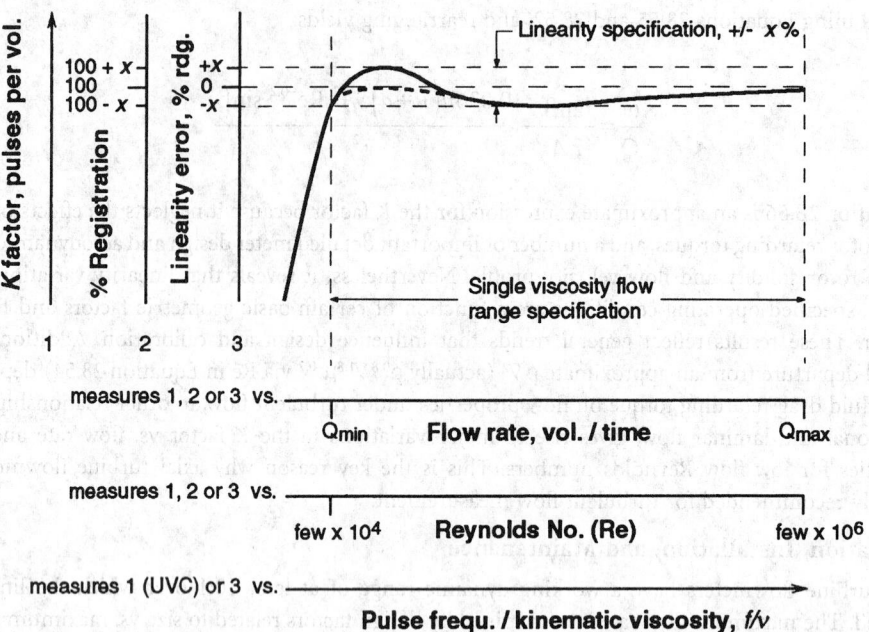

FIGURE 28.32 A typical single rotor axial turbine linearity error, or calibration, curve for a low-viscosity fluid showing the main alternative presentations in current use. Higher accuracy specifications usually correspond to a 10:1 flow range down from Q_{max}, while extended operating ranges usually correspond to reduced accuracies. The hump in the depicted curve is a characteristic feature caused by flow velocity profile changes as Re approaches the laminar region. This feature varies in magnitude between meters. Sensitivity and repeatability performance degrade at low Re. Percent registration is only used with meters that have mechanical pickups. All other meters have a K factor. Universal viscosity curve (UVC) and Re calibrations remain in effect at different known media viscosities provided Re or f/v stays within the specified range. Re is referenced to the connecting conduit diameter and is less within the flowmeter. The Re range shown is therefore approximate and can vary by an order of magnitude, depending on the meter. Linearity error can also be expressed in terms of Strouhal number (fD/V) vs. Re (VD/υ) or Roshko number (fD^2/υ), when instead D is a flowmeter reference diameter [4]. UVC, Universal Viscosity Curve; – – –, the effect of a rotor shroud in a viscosity compensated flowmeter.

maintenance, testing, and recalibration is required because the calibration will shift over time due to wear, damage, or contamination. For certain applications, especially those involving custody transfer of oil and natural gas, national standards, international standards, and other recommendations exist that specify the minimum requirements for turbine meters with respect to these aspects [6–10].

Design and Construction

There are numerous, often proprietary, designs incorporating variations in rotors, bearings, pickups, and other components in format and materials that are tailored to different applications. Meter bodies are available with a wide range of standard end-fittings. Within application constraints, the primary objective is usually to optimize the overall mechanical stability and fit in order to achieve good repeatability performance. Design for performance, application, and manufacture considerations impacts every internal component, but most of all the rotor with respect to blade shape and pitch, blade count, balance and rigidity vs. drag, stress, and inertia, bearings with respect to precision vs. friction, speed rating and durability, and rotation pickup vs. performance and drag.

Most low-radius ratio blades are machined flat, while high-ratio blades tend to be twisted. The blade count varies from about 6 to 20 or more, depending on the pitch angle and blade radius ratio so that the required rotor solidity is achieved. Rotor solidity is a measure of the "openness" to the flow such that higher solidity rotors are more highly coupled to the flow and achieve a better dynamic range. The pitch angle, which primarily determines the rotor speed, is typically 30° to 45° but can be lower in flowmeters designed for low-density, gas applications. Rotor assemblies are usually a close fit to the inside of the housing. In large-diameter meters, the rotor often incorporates a shroud around the outer perimeter for enhanced stability. Also, since large meters are often used for heavy petroleum products, via selection of a suitable wall clearance, the fluid drag resulting from this clearance gap is often designed to offset the tendency at high media viscosities for the meter to speed up at lower Reynolds numbers. The materials of construction range from nonmagnetic to magnetic steels to plastics.

Stainless steel ball bearings tend to be used for gas meters and low lubricity liquids such as cryogenic liquids and freon, while combination tungsten carbide or ceramic journal and thrust bearings are often considered best for many other liquid meters, depending on the medium lubricity. Fluid bearings (sometimes called "bearingless" designs) are often used in conjunction with the latter, but also sometimes with gases, for reducing the drag. They operate by various designs that use flow-induced forces to balance the rotor away from the shaft ends. Bearing lubrication is either derived from the metered medium or an internal or external system is provided. The more fragile, jeweled pivot bearings are also used in certain gas applications and small meters. Sanitary meters can incorporate flush holes in the bearing assembly to meet 3A crack and crevice standards.

The most common types of rotation sensor are magnetic, modulated carrier, and mechanical, while optical, capacitive, and electric resistance are also used. In research, a modulated nuclear radiation flux rotation sensor for use in certain nuclear reactors has also been reported [11, 12]. Mechanical pickups, which sometimes incorporate a magnetic coupling, are traditional in some applications and can have high resolution; one advantage is that they require no electric power. However, the pickup drag tends to be high. The magnetic and modulated carrier types utilize at least a coil in a pickup assembly that screws into the meter housing near the rotor. In magnetic inductance types, which are now less common, the blades or shroud carry magnetized inserts, and signals are induced in the coil by the traversing magnetic fields. In the more prevalent magnetic reluctance type, an example of which is schematically depicted in Figure 28.30, the coil is wrapped around a permanent magnet or magnet pole piece in the pickup assembly which is mounted next to a high magnetic permeability bladed rotor (or machined shroud). The latter is then typically made of a magnetic grade of stainless steel such as 416, 430, or 17-4Ph. As the rotor turns, the reluctance of the magnetic circuit varies, producing signals at the coil. In the more expensive modulated carrier types, the rotor need only be electrically conductive. The coil is part of a radio frequency (RF) oscillator circuit and the proximity of the rotor blades changes the circuit impedance, giving rise to modulation at a lower frequency that is recovered. The RF types have much lower drag, higher signal levels at low flow, and can operate at temperatures above the Curie point of typical ferromagnetic materials. They are preferred for wide dynamic range and high-temperature applications. Bidirectional flowmeters usually have two magnetic pickups to determine flow direction. This is useful, for example, in the monitoring of container filling and emptying operations often encountered in sanitary applications. Multiple magnetic pickups are also used in some designs to provide increased measurement resolution. Regarding output, various pulse amplifiers, totalizers, flow computers for gas pressure and temperature correction, along with 4–20 mA and other standard interface protocols, are available to suit particular applications. As an example of advanced transmitters, at least one manufacturer (EG&G Flow Technology, Inc.) provides a real-time, miniature, reprogrammable, "smart" transmitter that is integrated into the pickup housing along with a meter body temperature sensor, for full viscosity compensation and UVC linearization. These are for use in dedicated applications, such as airborne fuel management, where the medium viscosity–temperature relationship is known.

Certain applications have uniquely different design requirements and solutions, and two are discussed separately in the following.

Propeller Meters.
Propeller meters are used in either municipal, irrigation, or wastewater measurement. Although in some designs, propeller and turbine meters look almost identical and operate on the same axial rotor principle, this type of flowmeter is currently commercially and officially [13, 14] distinguished as a separate category

distinct from the axial turbine. Diameters up to 2440 mm (96 in.) are available. The flow rate capacity of a 1800 mm (72 in.) diameter propeller meter is up to about 25,000 $m^3\,h^{-1}$, (110,000 gpm). Typical accuracies are ±2% of reading. A primary requirement is ruggedness, and it is in the designs most suited to harsh environments that the formats are most distinctive. Rotor and pickup assemblies are generally flanged to the housing and removable. The rotors have large clearances, are often cantilevered into the flow, and supported via a sealed bearing without stators. The rotors are typically made of plastic or rubber and carry as few as three highly twisted, high radius ratio blades. Pickups are always mechanical and frequently have magnetic couplings.

Spirometers.

Monitoring spirometers measure the volumes of gas flows entering and leaving the lungs and can also be incorporated in ventilator circuits. Diagnostic spirometers are used to monitor the degree and nature of respiration. With these, a clinician can determine patient respiratory condition by various measures and clinical maneuvers. Low cost, light weight, speed of response, and patient safety are major considerations. Measurement capabilities include the gas volume of a single exhalation and also the peak expiratory flow for diagnostic types, measured in liters and liters per second, respectively. Various technologies are used. However, the Wright respirometer, named after the original inventor [15], today refers to a type of hand-held monitoring spirometer that utilizes a special type of tangential turbine transducer with a two-bladed rotor connected to a mechanical pickup and a dial readout for the volume. These particular spirometers are routinely used by respiratory therapists for patient weaning and ventilator checking. Other axial turbine-based flowmeters are available for ventilation measurements involving, for example, patient metabolics measurements. One axial turbine-based diagnostic spirometer made by Micro Medical, Ltd., currently claims most of the European market. This device utilizes an infrared, optical pickup and has a battery-powered microprocessor-controlled display. In these medical devices, rotors tend to be plastic with a large blade radius ratio. Flow conditioning is minimal or absent. The meters are typically accurate to ±a few percent of reading. In the U.S., spirometers are designated as class 2 medical devices and as such certain FDA approvals are required concerning manufacture and marketing. In the EU they are class IIb medical devices under a different system, and other approvals are required.

Dual-Rotor Axial Turbines

Dual-rotor axial turbines have performance features not found in single rotor designs. In 1981, Lee et al. [16] were issued a U.S. patent for a self-correcting, self-checking dual-rotor turbine flowmeter that is currently manufactured exclusively by Equimeter, Inc. and sold as the Auto-Adjust. This is a high-accuracy flowmeter primarily intended for use on large natural gas lines where even small undetected flow measurement errors can be costly. It incorporates two closely coupled turbine rotors that rotate in the same direction. The upstream rotor is the main rotor and the second rotor, which has a much shallower blade angle, is the sensor rotor. Continuous and automatic correction of measurement errors due to varying bearing friction is achieved by calculating the flow rate based on the difference between the rotor speeds. As shown in Figure 28.31 and discussed in the theory section, the flow exit angle is due to the net rotor retarding torque. If this torque increases in the main rotor, thereby reducing its speed, the exit angle increases and the speed of the sensor rotor is then also reduced. The meter is also insensitive to inlet swirl angle because the swirl affects both rotor speeds in the same sense and the effect is then subtracted in the flow calculation. The meter also checks itself for wear and faults by monitoring the ratio of the two rotor speeds and comparing this number with the installation value [17].

A dual-rotor liquid flowmeter, invented by Ruffner et al. [18], was recently introduced by Exact Flow, LLC. It is being offered as a high-accuracy flowmeter (up to ±0.1% accuracy and ±0.02% repeatability), which has an extraordinarily wide dynamic range of 500:1 with a single-viscosity liquid. This flowmeter has had early commercial success in fuel flow measurement in large jet engine test stands where the wide dynamic range is particularly useful [19]. The meter comprises two, closely and hydraulically coupled rotors that rotate in opposite directions. Due to the exit angle generated by the first rotor, the second rotor continues to rotate to much lower flow rates compared with the first.

Two-Phase Flow Measurement Using Axial Turbines

A differential pressure producing flowmeter such as a venturimeter in series with a turbine is known to be a technically appropriate and straightforward method for measuring the volumetric and mass flow rates of some fine, solid aerosols. However, this section highlights a current research area in the application of axial rotor turbine meters to a range of industrial flow measurement problems where gas/liquid, two-phase flows are encountered. Customarily, turbine meters are not designed for and cannot measure such flows accurately. Errors of the order 10% arise in metering liquids with void fractions of around 20%. Such flows are normally measured after gas separators. Although this problem is not restricted to these industries, the current main impetuses for research are the direct measurement of crude oil in offshore multiphase production systems, the measurement of water/steam mixtures in the cooling loops of nuclear reactors, and the measurement of freon liquid–vapor flows in refrigeration and air conditioning equipment. Several techniques investigated thus far use an auxiliary sensor. This can either be a void fraction sensor or a pressure drop device such as a venturimeter or drag disk, of which the pressure drop approach appears to be technically more promising [20, 21]. Also, from a practical standpoint, gamma densitometers for measuring void fraction are additional and expensive equipment and not, for example, well adapted for use in undersea oil fields. Two techniques currently being studied do not require an auxiliary in-line sensor. The first uses the turbine meter itself as the drag body and combines the output of the turbine with that of a small differential pressure sensor connected across the inlet and outlet regions. This technique requires a homogenizer ahead of the turbine, and measurement accuracies of ±3% for the volumetric flow rates of both phases have recently been reported for air/water mixtures up to a void fraction of 80% [22]. The second technique is based entirely on analysis of the turbine output signal and has provided significant correlations of the signal fluctuations with void fraction. Accuracies of water volumetric flow rate measurement of ±2% have been reported when using air/water mixtures with void fractions of up to 25% [23].

Insertion Axial Turbine Flowmeters

These flowmeters comprise a small axial rotor mounted on a stem that is inserted radially through the conduit wall, often through a shut-off valve. They measure the flow velocity at the rotor position from which the volumetric flow rate is inferred. They are an economical solution to flow measurement problems where pipe diameters are high and accuracy requirements are moderate, and also may be technically preferred where negligible pressure drop is an advantage, as in high-speed flows. They are typically more linear than insertion tangential turbine flowmeters and compete also with magnetic and vortex shedding insertion flowmeters. They are available for the measurement of a range of liquids and gases, including steam, similar to the media range of full bore axial turbines, and have a similarly linear response. Flow Automation, Inc., is currently the leading manufacturer. The rotors, which are usually metal but can be plastic, typically have diameters of 25 mm to 51 mm (1 in. to 2 in.). They can be inserted into pipes with diameters ranging from 51 mm to 2032 mm (2 in. to 80 in.). Velocity measurement ranges cover 0.046 ms^{-1} to 91 ms^{-1}, (9 fpm to 18,000 fpm) for gases and 0.03 ms^{-1} to 30 ms^{-1} (6 fpm to 6000 fpm) for liquids. Dynamic ranges vary between 10:1 and 100:1. The maximum flow rate measurement capacity in a 1836 mm (72 in.) diameter pipe can be as high as nearly 56,500 m^3 h^{-1}, (about 250,000 gpm). Since these devices are local velocity sensors, calculating the volumetric flow rate requires a knowledge of the area velocity profile and the actual flow area. Flow conditioning is therefore particularly important for accurate volumetric measurements, while radial positioning, which is a further responsibility of the user, must be according to the manufacturer's recommendation, which can either be centerline, one third of the diameter, 12% of the diameter, or determined by "profiling." Quick [24] discusses operation and installation for natural gas measurement. Although linearities or "accuracies" can be quoted up to ±1% of velocity, achieving the same accuracy for the volumetric flow rate, although possible, can be difficult or impractical. In this respect, a unique dual rotor design, exclusive to Onicon, Inc., and primarily used for chilled-water flow measurement in HVAC systems, requires less flow conditioning than single rotor designs. An Onicon dual rotor turbine assembly is depicted in Figure 28.33. It comprises two rotors that

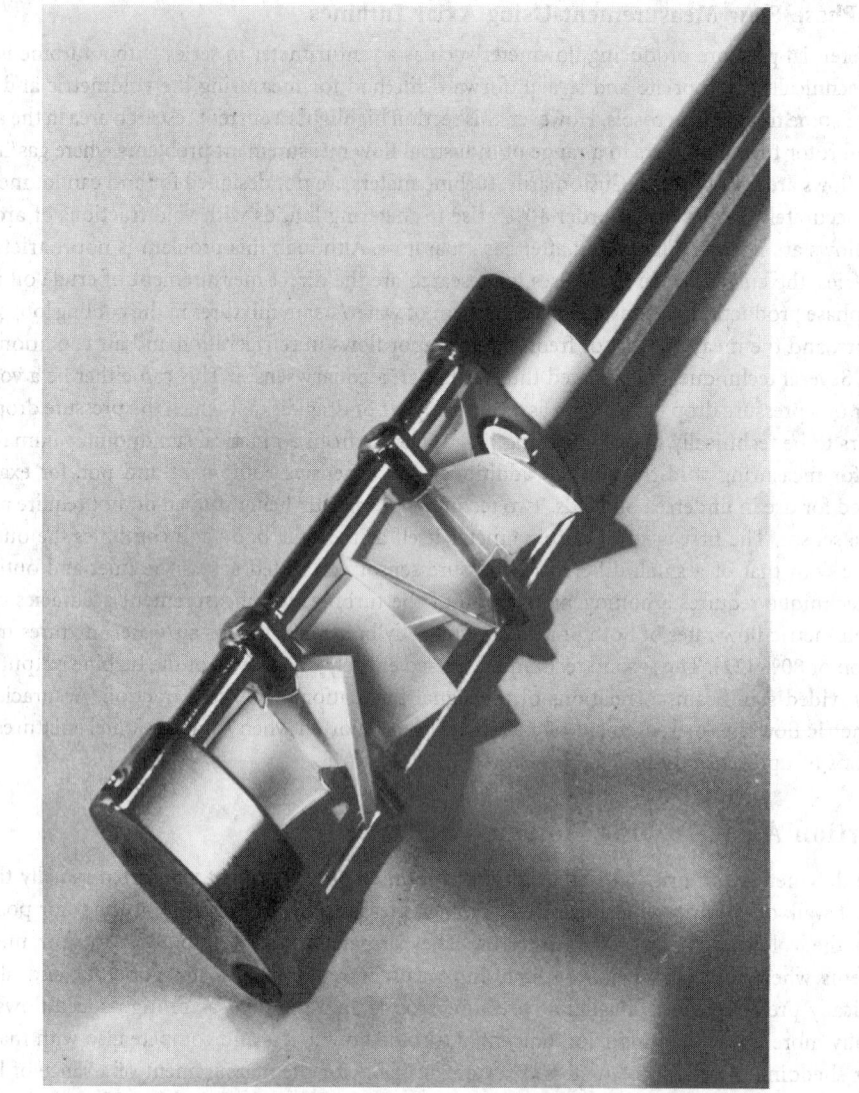

FIGURE 28.33   The rotor assembly of a dual rotor, insertion axial turbine flowmeter for water flow measurement. This patented design renders the flowmeter insensitive to errors due to flow swirl; an important source of potential error in single rotor axial turbine flowmeters. The rotations are sensed by two separate, electric impedance sensors. (Courtesy of Onicon Incorporated.)

rotate in opposite directions. The output is based on the average rotor speed. Any flow swirl present due to poor flow conditioning changes the speed of rotation of each rotor by the same but opposite amounts. Swirl-induced error is thus virtually absent in the averaged output. Also, flow profile sampling is improved over that of a single rotor. The devices are calibrated using a volumetric prover and the specified accuracy of ±2% of reading is for volumetric flow rate rather than velocity. This is the total error and includes an allowance for dimensional variations in industry standard pipes.

Angular Momentum Flowmeters

These are accurate, linear, liquid mass flowmeters that utilize vaned components and are used in aerospace applications. They are currently the instrument of choice for airborne, jet engine fuel flow measurement for large commercial aircraft and some military aircraft for afterburner applications. They are more expensive than the equivalent turbine flowmeters for this application, but they provide mass flow rate measurements directly and are unaffected by fuel density variations. Typical accuracies lie between ±0.5% and ±1.0% over a 40:1 or greater dynamic range. Measurement ranges are available for fuel flows from about 0.01 to 6 kg s^{-1} (70 to 46,000 PPH).

The principle of operation is long established and based on imparting angular momentum to the fluid flow using a driven, flat-bladed impeller. The force required to drive the impeller at constant speed is monitored as a proportional indication of mass flow rate as this quantity varies. Some designs use an electric motor to drive the impeller. However, the current trend in design is motorless. In such a device, a constant driving speed in the region of 100 rpm to 200 rpm is provided by one or more turbine rotors driven by the flow. A variable shunt metering valve assembly adjacent to the turbine rotor mechanically opens and closes in response to flow dynamic pressure and thereby automatically maintains a constant speed of rotation provided that the flow rate is above the minimum of the range. The driven impeller carries vanes that are parallel to the flow, and resides on a common axis with the turbine rotor. A flow straightener ensures parallel flow past the impeller. There is a carefully engineered constant rate spring connection between the turbine shaft and the impeller so that the angular deflection between the two is proportional to the applied torque, and this quantity is directly proportional to the mass flow rate. Two pickup coils sense the rotations of the turbine and the vaned impeller, and only the time difference between the pulses in these two signals is measured and used to calculate the flow rate. Chiles et al. [25] provide a detailed illustration and explanation of the intricate mechanism.

Multijet Turbine Flowmeters

These are linear, volumetric flowmeters designed for liquids measurements and comprise a single, radial-vaned impeller, vertically mounted on a shaft bearing within a vertically divided flow chamber, sometimes called a distributor. The impeller is often plastic and can even be neutrally buoyant in water. There are various designs, but typically, both chambers access a series of radially distributed and angled jets. The lower chamber jets connect to the flowmeter input port and distribute the flow tangentially onto the lower region of the impeller blades, while the upper series, which is angled oppositely, allow the flow to exit. The flow pattern within the flow chamber is thus a vertical spiral and the dynamic pressure drives the impeller to track the flow. This design gives the meters good sensitivity at low flow rates. Due to the distribution principle, the meters are also insensitive to upstream flow condition. Impeller rotation pickups are always mechanical, often magnetically coupled, and frequently also connect with electric contact transmitters. They are primarily used in water measurement, including potable water measurement for domestic and business billing purposes and in conjunction with energy management systems such as hot water building or district heating, and to a much lesser extent in some chemical and pharmaceutical industries for dosing and filling systems involving solvents, refrigerants, acids and alkalis with absolute viscosities less than 4.5 mPa s, (0.045 Poise). Available sizes range from 15 mm to 50 mm. Dynamic ranges lie between 25:1 and 130:1 and flow measurement ranges cover 0.03 to 30 m^3 h^{-1}, (0.13 to 130 gpm). Measurement linearities range between ±1% to ±2%, with typical repeatabilities of ±0.3%. Operating temperatures range from normal to 90°C (200°F) and maximum operating pressures are available up to 6.9 MPa (1000 psi). A number of potable water measurement systems come with sophisticated telemetry options that allow remote interrogation by radio or telephone. For potable water applications in the U.S., these meters normally comply with the applicable AWWA standard [26], while in Europe EEC, DIN, and other national standards apply.

In the author's opinion, there is also another type of vaned flowmeter that could be classified as a type of multijet turbine. This type comprises an axially mounted, vaned impeller with an upstream element that imparts a helical swirl to the flow. The transducer is typically a small, low-cost, sometimes disposable, plastic component, and is usually designed for liquids (but also to lower accuracies, gases), low-flow rate measurements (down to 50 mL min^{-1}). The dynamic range is high and accuracies range up to ±0.5%.

Goss [27] describes one particular design in current use. Specialized applications cover the pharmaceutical, medical, and beverage industries.

Cylindrical Rotor Flowmeter

Instead of coupling to a turbulent flow using the fluid dynamic pressure or momentum flux, as in most axial and tangential turbines, a demonstrated research device due to Wadlow et al. [28] provides a linear, volumetric, low gas flow rate measurement using a single, low inertia, smooth cylindrical tangential rotor that couples to a laminar flow via surface friction. The geometry is conceptually that of a single-surfaced rotating vane. A plane Poiseuille flow is created that passes azimuthally over most of the curved rotor surface in a narrow annular passage between the rotor and a concentric housing. The rotor is motor driven via a feedback control loop connected to an error signal producing, differential pressure sensor connected across the gas ports so as to maintain the meter pressure drop equal to zero. Under this condition, the response, as indicated by motor shaft speed, is exactly linear, determined only by geometry and is independent of gas density and viscosity. The demonstration device reported has a 40 mm diameter rotor and 15 mm diameter gas ports. It measures up to a maximum of 25 Lpm bidirectionally, has a linear dynamic range greater than 100:1, is insensitive to upstream flow condition, and has a $1/e$ step response time of 42 ms, limited only by the external motor torque and combined motor and rotor inertia.

Manufacturers

There is significant and dynamic competition among the numerous manufacturers of the different types of flowmeters described in this section. In all flowmeter types, most manufacturers have exclusive patent rights concerning one or more detailed design aspects that make their products perform differently from those of the competition. Every few years, one or more major turbine flowmeter company can be identified that has changed ownership and name or formed a new partnership. Identifying the competition and selecting the manufacturer are important and sometimes time-consuming parts of the flowmeter selection and specification process. To assist with this, Table 28.8 gives a few selected examples from different manufacturers, of all of the different flowmeter types described in this section. Table 28.9 gives the corresponding contact information for those selected manufacturers, along with the general types offered by each.

Conclusion

However anachronistic intricate mechanical sensors might appear amid current everyday high technology, there are fundamental reasons why axial turbines are likely to experience continued support and development rather than obsolescence, especially for in-line applications requiring in the region of tenth percent volumetric accuracy. Mechanical coupling is the most direct *volume* interaction for a flowing fluid, which is why mechanical meters historically developed first and continue to be the most accurate and reliable types of flowmeter for so many different fluids. Other nonmechanical, perhaps higher technology, or newer approaches thus face high demands for accurate compensation to render such less directly volume-coupled techniques as generally accurate, or more accurate. This is because the error in each corrected factor or assumption in an indirect technique contributes to the overall error. The technology of high-accuracy flowmeters continues to be driven by applications, such as the custody transfer of valuable oil and natural gas, which demand high accuracy and reliability. There is a continuing demand for accurate and reliable water flowmeters. By reason of long proven field experience, turbine and other vane type devices have become one of a few broadly accepted techniques in many major applications such as these where the demands for flow sensors is significant or growing.

TABLE 28.8   Examples of Turbine and Vaned Flowmeters

Type	Size(s), in.	Description	Example application(s)	Manufacturer	Approx. price range
Axial	0.5	FT 4-8, with SIL smart transmitter	Helicopters — fuel	EG&G Flow Technology	$2,500
Axial	1	6700 series, type 60	Raw milk, de-ionized water	Flow Automation	$1,500
Axial	16	Parity series	Custody transfer oil	Fisher-Rosemount Petroleum	$30,000–$38,000
Axial dual rotor	12	Auto-Adjust Turbo-Meter	Custody transfer gas	Equimeter	$42,000
Axial dual rotor	0.25–2	DR Series	Fuel — large jet engine test stands	Exact Flow	$1,300–$1,800
Propeller	48	FM182 (150 PSI)	Municipal water	Sparling	$6,500
Propeller	6	FM102 (150 PSI)	Irrigation	Sparling	$850
Special tangential	—	Wright Mark 8	Monitoring spirometer	Ferraris Medical	$800
Axial	—	MS03/MS04 MicroPlus	Diagnostic spirometer	Micro Medical	$600
Axial	4	WTX802	Chemical dosing for water treatment	SeaMetrics	$450–$550
Axial insertion	1.5–10	TX101	Municipal water, water — HVAC	SeaMetrics	$450
Axial insertion	2+	VTS-300	High pressure steam	Flow Automation	$2,900
Axial insertion	3+	VL-150-LP	Flare stack control	Flow Automation	$2,500
Axial insertion dual rotor	2.5–72	F-1200	Water — HVAC	Onicon	$900
Angular momentum	—	Model 9-217 True Mass Fuel Flowmeter	Large jet aircraft, e.g., Airbus A320, A321	Eldec	Not available
Multijet	1	1720	Domestic water	ABB Kent Messtechnik	$170
Multijet	1.5	AMD3000	Chemical liquid filling and dosing	Aquametro AG	$950
Multijet axial	9/32	DFS-2W	Pharmaceutical filling lines, kidney dialysis dialyte, beverage dispensers, OEM	Digiflow Systems	$42 or less, transducer; $235 electronics

References

1. W. F. Z. Lee and H. J. Evans, Density effect and Reynolds number effect on gas turbine flowmeters, *Trans. ASME, J. Basic Eng.,* 87 (4): 1043-1057, 1965.
2. R. E. Thompson and J. Grey, Turbine flowmeter performance model, *Trans. ASME, J. Basic Eng.,* 92(4), 712-723, 1970.
3. M. B. Pate, A. Myklebust, and J. H. Cole, A computer simulation of the turbine flow meter rotor as a drag body, *Proc. Int. Comput. in Eng. Conf. and Exhibit 1984,* Las Vegas: 184-191, New York: ASME, 1984.
4. P. D. Olivier and D. Ruffner, Improved turbine meter accuracy by utilization of dimensionless data, *Proc. 1992 National Conf. Standards Labs.* (NCSL) Workshop and Symp., Boulder, CO: NCSL, 1992, 595-607.
5. ISA-RP 31.1, *Specification, Installation and Calibration of Turbine Flowmeters,* Research Triangle Park, NC: ISA, 1977.

TABLE 28.9 Manufacturer Contact Information

Manufacturer	U.S. distributor	Relevant types
EG&G Flow Technology, Inc. 4250E Broadway Road Phoenix, AZ 85040 Tel: (602) 437-1315	Not applicable	Axial turbines — liquid and gas, including 3A sanitary Insertion axial turbines — liquid and gas
Flow Automation, Inc. 9303 W. Sam Houston Pkwy S. Houston, TX 77099 Tel: (713) 272-0404	Not applicable	Insertion axial turbines — liquid and gas Axial turbines — liquid and gas, including 3A sanitary
Fisher-Rosemount Petroleum Highway 301 North P.O. Box 450 Statesboro, GA 30459 Tel: (912) 489-0200	Not applicable	Axial turbines — liquid
Equimeter, Inc. 805 Liberty Blvd. DuBois, PA 15801 Tel: (814) 371-8000	Not applicable	Axial turbines — gas Dual rotor axial turbines — gas
Exact Flow, LLC P.O. Box 14515 Scottsdale, AZ 85267-4545 Tel: (602) 922-7446	Not applicable	Dual rotor axial turbines — liquid
Sparling Instruments Co., Inc. 4097 North Temple City Blvd. P.O. Box 5988 El Monte, CA 91734 Tel: (818) 444-0571	Not applicable	Propeller meters
Ferraris Medical, Ltd Ferraris House Aden Road Enfield Middlesex EN3 7SE U.K. Tel: 44 (0)1818059055	Ferraris Medical, Inc. P.O. Box 344 9681 Wagner Road Holland, NY 14080 (716) 537-2391	Tangential turbine monitoring spirometers
Micro Medical, Ltd. The Admiral's Offices The Chatham Historic Dockyard Chatham, Kent ME4 4TQ U.K. 44 (0)163 843383	Micro Direct, Inc. P.O. Box 239 Auburn, ME 04212 (800) 588-3381	Axial turbine diagnostic spirometers
SeaMetrics Inc. P.O. Box 1589 Kent, WA 98035 Tel: (206) 872-0284	Not applicable	Axial turbines — liquid Insertion axial turbines — liquid Multijet turbines
Onicon, Inc. 2161 Logan St. Clearwater, FL 34625 Tel: (813) 447-6140	Not applicable	Axial turbines — liquid Insertion axial turbines, single and dual rotor — liquid

TABLE 28.9 (continued)   Manufacturer Contact Information

Manufacturer	U.S. distributor	Relevant types
Eldec Corporation 16700 13th Avenue W. P.O. Box 97027 Lynnwood, WA 98046 Tel: (206) 743-8499	Not applicable	Angular momentum flowmeters
ABB Kent Messtechnik GmbH Otto-Hahn-Strasse 25 D-68623 Lampertheim Germany 49 62069330	ISTEC Corporation 415 Hope Avenue Roselle, NJ 07203 (908) 241-8880	Multijet turbines
Aquametro AG Ringstrasse 75 CH-4106 Therwil 061 725 11 22	ISTEC Corporation 415 Hope Avenue Roselle, NJ 07203 (908) 241-8880	Multijet turbines
Digiflow Systems B.V. Postbus 46 6580 AA Malden The Netherlands 31 243 582929	Digiflow Systems 781 Clifton Blvd. Mansfield, OH 44907 (419) 756-1746	'Multijet' axial turbines — liquid

6. ANSI/ASME MFC-4M-1986 (R1990), *Measurement of Gas Flow by Turbine Meters*, New York: ASME.

7. API MPM, *Measurement of Liquid Hydrocarbons by Turbine Meters*, 3rd ed., Washington, D.C.: API (Amer. Petroleum Inst.), 1995, chap. 5.3.

8. AGA Transmission Meas. Committee Rep. No. 7, Measurement of fuel gas by turbine meters, Arlington, VA: AGA (Amer. Gas Assoc.), 1981.

9. Int. Recommendation R32, Rotary piston gas meters and turbine gas meters, Paris: OIML (Int. Organization of Legal Metrology), 1989.

10. ISO 9951:1993, Measurement of gas flow in closed conduits — turbine meters, Geneva, Switzerland: Int. Organization for Standardization, (also available ANSI), 1993.

11. T. H. J. J. Van Der Hagen, Proof of principle of a nuclear turbine flowmeter, *Nucl. Technol.*, 102(2), 167-176, 1993.

12. K. Termaat, W. J. Oosterkamp, and W. Nissen, *Nuclear turbine coolant flow meter*, U.S. Patent No. 5,425,064, 1995.

13. AWWA C704-92, Propeller-type meters for waterworks applications, Denver, CO: Amer. Water Works Assoc., 1992.

14. ANSI/AWWA C701-88, Cold water meters — turbine type, for customer service, Denver, CO: Amer. Water Works Assoc., 1988.

15. B. M. Wright and C. B. McKerrow, Maximum forced expiratory flow rate as a measure of ventilatory capacity, *Br. Med. J.*, 1041-1047, 1959.

16. W. F. Z. Lee, R. V. White, F. M. Sciulli, and A. Charwat, *Self-correcting self-checking turbine meter*, U.S. Patent No. 4,305,281, 1981.

17. W. F. Z. Lee, D. C. Blakeslee, and R. V. White, A self-correcting and self-checking gas turbine meter, Trans. ASME, *J. Fluids Eng.*, 104, 143-149, 1982.

18. D. F. Ruffner, and P. D. Olivier, *Wide range, high accuracy flow meter*, U.S. Patent No. 5,689,071, 1997.

19. D. F. Ruffner, Private communication, 1996.

20. A. Abdul-Razzak, M. Shoukri, and J. S. Chang, Measurement of two-phase refrigerant liquid-vapor mass flow rate. III. Combined turbine and venturi meters and comparison with other methods, *ASHRAE Trans.: Research,* 101(2), 532-538, 1995.

21. W. J. Shim, T. J. Dougherty, and H. Y. Cheh, Turbine meter response in two-phase flows, *Proc. Int. Conf. Nucl. Eng. — 4,* 1 part B: 943-953, New York: ASME, 1996.

22. K. Minemura, K. Egashira, K. Ihara, H. Furuta, and K. Yamamoto, Simultaneous measuring method for both volumetric flow rates of air-water mixture using a turbine flowmeter, *Trans. ASME, J. Energy Resources Technol.,* 118, 29-35, 1996.

23. M. W. Johnson and S. Farroll, Development of a turbine meter for two-phase flow measurement in vertical pipes, *Flow Meas. Instrum.,* 6(4), 279-282, 1995.

24. L. A. Quick, Gas measurement by insertion turbine meter, *Proc. 70th Int. School Hydrocarbon Meas.,* OK, 1995. (Available E. Blanchard, Arrangements Chair, Shreveport, LA, (318) 868–0603.)

25. W. E. Chiles, L. E. Vetsch, and J. V. Peterson, *Shrouded flowmeter turbine and improved fluid flowmeter using the same,* U.S. Patent No. 4,012,957, 1977.

26. ANSI/AWWA C708-91, *Cold-water meters, multi-jet-type,* Denver, CO: Amer. Water Works Assoc., 1991.

27. J. Goss, *Flow meter,* U.S. Patent No. 5,337,615, 1994.

28. D. Wadlow, and L. M. Layden, *Controlled flow volumetric flowmeter,* U.S. Patent No. 5,284,053, 1994.

29. C. R. Sparks, private communication, 1996.

Further Information

D. W. Spitzer (ed.), *Flow measurement,* Research Triangle Park, NC: ISA, 1991, is a popular 646 page practical engineering guide of which four chapters concern turbine flowmeters, sanitary flowmeters, insertion flowmeters and custody transfer issues.

A. J. Nicholl, Factors affecting the performance of turbine meters, *Brown Bov. Rev.,* 64(11), 684-692, 1977, describes some basic design factors not commonly discussed elsewhere.

J. W. DeFeo, Turbine flowmeters for measuring cryogenic liquids, *Adv. Instrum. Proc.,* 47 pt.1, 465-472, ISA, 1992, provides guidance for a less frequently discussed application in which axial turbines perform well.

ATS Standardization of spirometry, 1994 Update, *Am. J. Respir. Care Med.,* 152, 1107-1136, 1995, is the latest version of the official U.S. guideline for spirometry generated by the American Thoracic Society.

M. D. Lebowitz, The use of expiratory flow rate measurements in respiratory disease, *Ped. Pulmonol.,* 11, 166-174, 1991, provides a review of the diagnostic usefulness of PEFR using portable spirometers.

W. M. Jungowski and M. H. Weiss, Effects of flow pulsation on a single-rotor turbine meter, *Trans. ASME, J. Fluids Eng.,* 118(1), 198-201, 1996.

C. R. Sparks and R. J. McKee, *Method and apparatus for assessing and quantifying pulsation induced error in gas turbine flow meters,* U.S. Patent No. 5,481,924, 1996, assigned to the Gas Research Institute, Chicago, is a potential solution to the accurate measurement of pulsating gas flows which will require engineering development and a high performance rotation sensor [29].

K. Ogawa, S. Ito, and C. Kuroda, Laminar-turbulent velocity profile transition for flows in concentric annuli, parallel plates and pipes, *J. Chem. Eng. Japan,* 13(3), 183-188, 1990, provides mathematical descriptions of velocity profiles.

J. Lui and B. Huan, Turbine meter for the measurement of bulk solids flow rate, *Powder Technol.,* 82, 145-151, 1995, describes theory and experiments relating to a very simple design for a unique application, namely the volumetric measurement of plug flows of sands in a pipe.

28.5 Impeller Flowmeters

Harold M. Miller

Impeller flowmeters, sometimes called *paddlewheel* meters, are one of the most commonly used flowmeter variety. The impeller flow sensor is a direct offshoot of the old-fashioned undershot waterwheel. They are a cost-effective alternative to turbine meters, and can be used in applications that are difficult to handle with other types of flow metering instruments. In their mechanical construction, they are usually very simple, with any sophistication residing in the electronics used to detect the rotation rate of the impeller, and in the choice of materials of construction for chemical corrosion attributes of the metered fluids. They are related to turbine meters in that both types use a rotating mechanical element to produce the output signal. They differ in the fact that the impeller meter has its rotary axis transverse to the flow stream, as opposed to the turbine meter axis, which is parallel thereto.

These devices are available in two basic types. The insertion style is the most common. (See Figures 28.34(*b*) and 28.35.) The sensor is directly installed into a hole in a pipe, with saddle or welded fitting installed at the entry to seal the sensor to the pipe. The sensor can also be preinstalled in appropriate Tees in the pipeline. Most suppliers have designs that can be installed in operating pipe systems, with little, if any, loss of the fluid during the installation process. The impeller design is also supplied in in-line (through-flow) sensors for those applications where such use is desirable (Figure 28.34(*a*)). The in-line meters are commonly of somewhat higher accuracy than the insertion style.

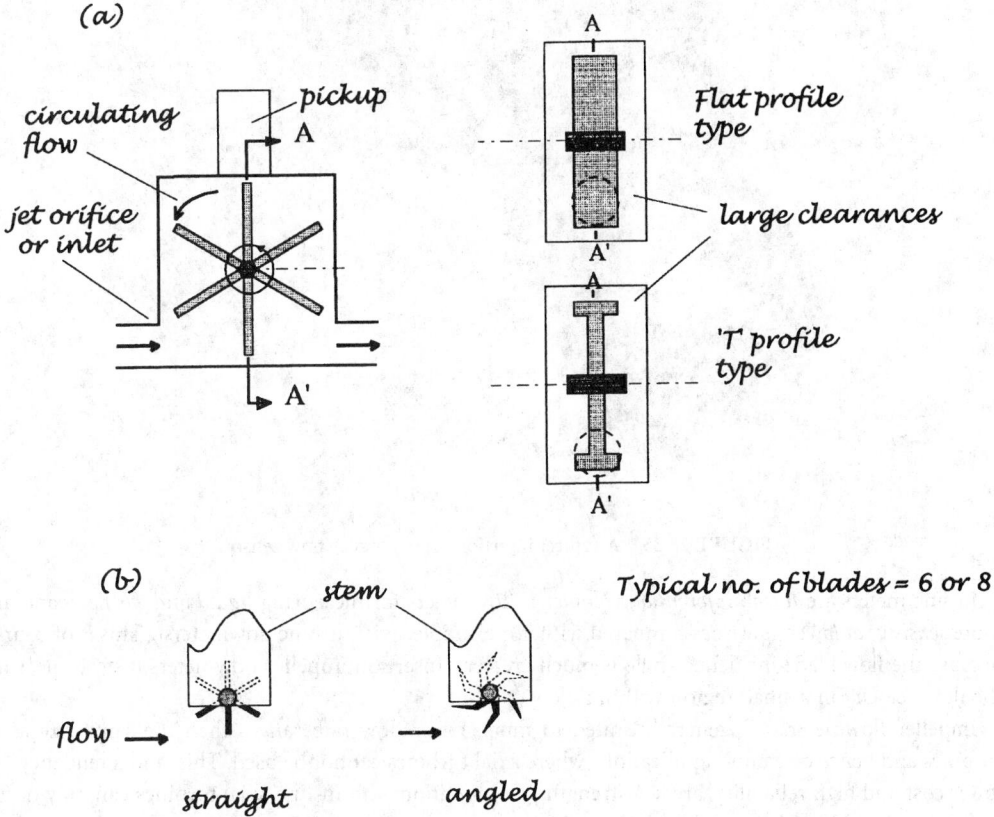

FIGURE 28.34 Impeller flowmeter rotor design variations (*a*) in-line meters; (*b*) insertion meters. (Figure courtesy of David Wadlow.)

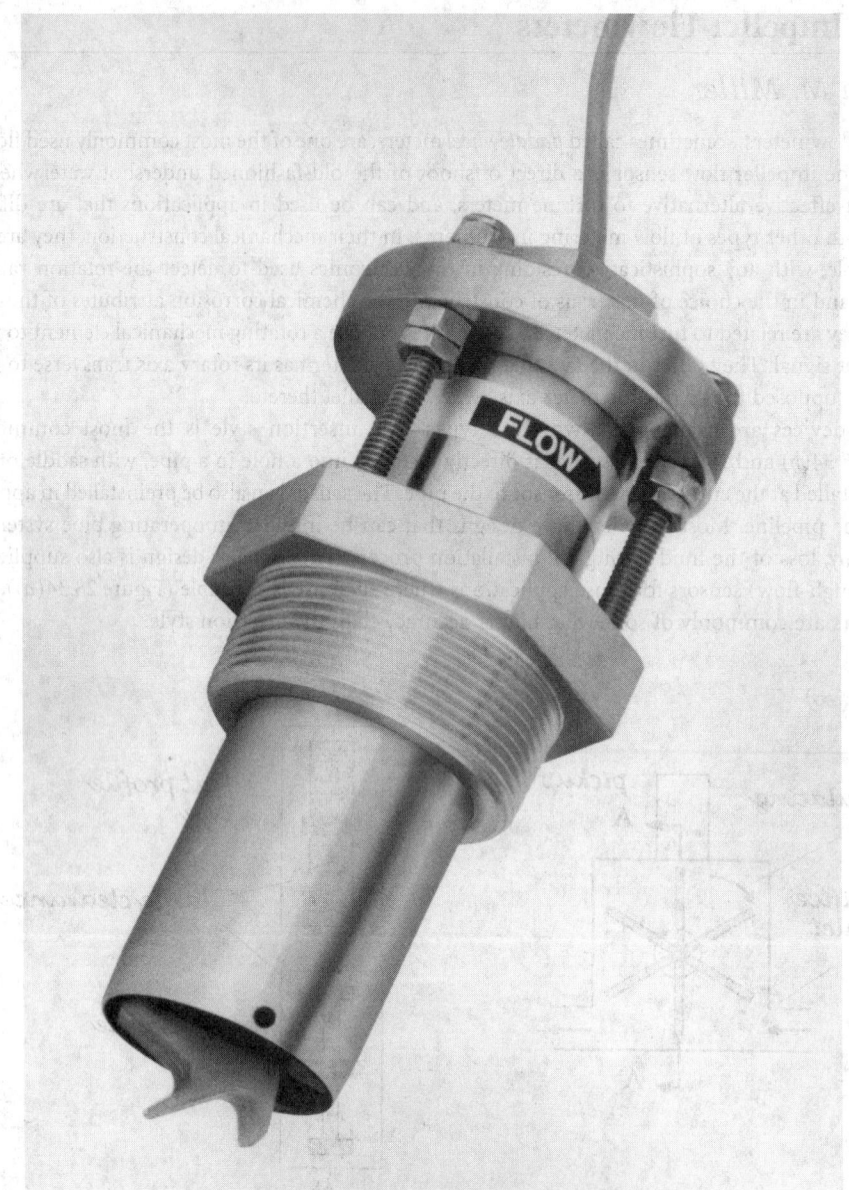

FIGURE 28.35   A typical insertion type impeller flow sensor.

In-line meters are *linear, inferential, volumetric* flowmeters for measuring *liquid and gas flows* and are more *sensitive at lower flow rates* compared with, for example, axial turbine flowmeters. (This is of course because the flow blade incidence angle is much greater.) Insertion impeller flowmeters instead measure the flow velocity in a small region within a flow conduit.

Impeller flowmeters are generally suited to much lower flow rate ranges than the same size axial turbines and hence often find applications where axial turbines cannot be used. This, and a tendency for lower cost and high reliability, are key strengths. Competition with in-line axial turbines can only occur in overlapping design flow ranges.

Some in-line meters have interchangeable orifice sizes, allowing the same body to be used over different flow ranges. The orificed in-line meters are typically insensitive to flow condition. Insertion meters are, of course, very sensitive to flow condition if volumetric measurements are inferred. This is an important distinction and can extend to differences between requirements for specific in-line versions. Installation is important. It affects the user when he or she is designing an installation, selecting a flowmeter, and deciding on how much confidence to place in the measurement reading.

Paddlewheel flowmeters are *never* used in liquid hydrocarbon custody transfer applications. The flow measurement capacity is insufficient to compete in monitoring large volume transfers, and accuracy is not sufficient for valuable fluids.

These meters go by a variety of names, depending on which name the manufacturer selects and also the application. One often-quoted, historical root is also the impulse turbine invented by Pelton in the 19th century and has given rise to the "Pelton wheel turbine" description currently used by some suppliers. A device like this was originally used to drive milling wheels directly about the vertical axis, rather than through a right angle gear as in the "undershot" wheel described. Another common name, besides paddlewheel and impeller-type, based at least on suppliers' descriptions for the in-line variety, is single-jet tangential, or simply tangential, turbine. The former description distinguishes this design from that of multijet tangential turbine flowmeters and therefore deserves mention. There are thus a confusing variety of names in current use, all relating to essentially the same impeller-like, vaned flowmetering principle:

1. Impeller type: insertion and in-line
2. Paddlewheel type: insertion and in-line
3. Pelton turbine wheel type: in-line
4. Tangential turbine flowmeter: insertion and in-line
5. Single-jet tangential turbine flowmeter: in-line
6. Impulse turbine flowmeter: in-line

Historically, the impeller sensor is based on early electronic speedometers used in pleasure boating. Signet, followed quickly by Data Industrial, moved to modify the basic design of such marine instruments to meet the significantly more demanding service life requirements of industrial flow measurement. The resulting flowmeters, and their descendants, have been widely used since 1975. Significant engineering efforts have resulted in highly reliable instruments used in many extremely demanding applications.

Sensing Principles

All impeller flow sensors must detect the rotation of the impeller, and in their usual form transmit a pulse train, at a frequency related to the rotational velocity of the impeller. Being essentially a digital output, impeller sensors can typically transmit signals over quite long distances, up to 1 km when so required. Detection principles used include:

1. One or more magnets retained in the impeller or mechanically connected thereto, using the zero crossing of an induced ac field to generate the pulse train.
2. One or more coupling devices contained within the impeller, modulating a transmitted frequency that is processed to produce the pulse train.
3. One or more metallic targets installed within the sensor, sensed by any of the proximity pickup techniques commercially available, to produce output pulses.
4. One or more magnets retained in the impeller, used to switch a Hall effect device, producing the output pulse train.
5. Optical devices, both transmissive and reflective, have been used to sense the passage of the impeller blades to produce the output pulse train.
6. Measurement of the change in electric reactance due to the passage of impeller vanes through the measurement field area, conditioned to produce the output pulse train.

Any given supplier can produce several types of impeller flowmeter, each type using a different detection method depending on market requirements. Since the impeller can operate to rotational velocities of 4000 rpm or higher, output frequencies can be as high as 500 Hz. At low flow rates, the frequencies can be as low as 0.2 Hz. This factor should be addressed early in the selection procedure because the chosen output device must be capable of the frequency output range of the sensor.

Most sensor manufacturers can either supply or specify an appropriate meter to relate sensor output to flow rate in Engineering units, either U.S. Customary or SI. These meters are usually capable of displaying both flow rate and accumulated flow for the sensor to which they are connected. Additional outputs are available, either stand-alone or in combination with the meter, providing periodic pulse outputs at definable flow increments or analog outputs scaled to flow rate. In addition, certain control functions, alarms, and other special features required by the various markets served are often incorporated in these meters.

Flow Ranges

The insertion style of impeller meter, even when Tee-installed, is a local sensing device, measuring the flow velocity in only a part of the flow stream. The manufacturer calibrates the meter for average flow across the entire cross-section of the pipe. The paddlewheel location is usually close to the inner diameter of the pipe in a region of flow with a velocity significantly below the average flow velocity in the pipe. Proper calibration practices by the manufacturer make the meters effective at flow rates as low as 10 cm s^{-1} average velocity in spite of the low local velocities at the impeller. Generally speaking, flows at Reynolds numbers as low as 5000 can be run with no requirement for special calibration, and frequently lower numbers can be handled with no difficulty. The usual range specification is a flow rate equivalent to 0.3 to 10 m s^{-1} average flow velocity. The diversity of products available allows operation to velocities considerably lower than the 0.3 m s^{-1} range, indeed to as low as 0.07 m s^{-1} in certain specialty impeller flowmeters.

Installation

Pipe sizes: Pipe sizes in which these sensors have been installed run the gamut from small bore tubing to 2.3 m outside diameter. The larger pipe sizes are those that show the greatest installed cost savings over alternative metering systems. The impeller meter, in fact, can be cost-effective in any flowmeter application that is consistent with the accuracy of the instrument, particularly if the application involves pump or valve control.

Piping system restrictions: Most suppliers require at least 10 pipe diameter lengths of straight pipe upstream of the installed meter, and 5 downstream. These conditions are required to minimize the asymmetry of the flow stream in the neighborhood of the impeller, in the installed piping, which can be caused by elbows, tees, and valves. More is better; no supplier is likely to complain that there is too great a straight pipe length upstream. Less can adversely affect the calibration of the sensor due to the local variations of velocity resulting from flow disturbances.

Operating pressure: Manufacturers' standard offerings are usually consistent with the pressure limitations of the materials of construction of the piping system with plastic piping systems, and are commonly as high as 2.7 MPa with steel or brass piping. Higher pressures are available, but are usually somewhat more expensive. Pressure drop generated by the installed flowmeter is usually low. Manufacturers can usually supply information on the anticipated pressure drop.

Calibration: Calibration of the sensors is usually specified by the manufacturer. Some manufacturers provide a calibration factor in terms of gallons per pulse or in pulses per gallon. Others, such as Data Industrial, provide data relating frequency to flow rate in GPM or other volumetric rate units. For insertion-style sensors, any such instrument must be field calibratable to accommodate the variation in the relation of impeller rotational velocity with average flow rate with pipe size variation.

Accuracy: Accuracies for in-line impeller flowmeters vary considerably but can be high, ranging from ±0.2% reading for liquids for one manufacturer to several percent of full scale for several others. The difference between accuracies specified as % full scale and % reading should be particularly noted.

Manufacturers use both, particularly with this class of meter. A ±1% full-scale device is often far less accurate than, for example, a ±2% reading device. For example, it is in error by 10% at the minimum of its flow range if the turndown is 10:1, whereas the ±2% reading device is still only in error by ±2%.

Output accuracy is usually specified as ±1% of full rated flow. This specification is broad enough to handle the dimensional and frictional variations in the sensors as manufactured. Some producers will custom-calibrate the sensors to meet special needs, but this task should be performed only when the mating pipe entry and exit can be shipped to, and accommodated by, the manufacturer. Alternatively, the manufacturer can usually provide new calibration values if the meter used is identified, the meter reading is known, and the actual flow is known for at least two points on the flow curve. When calibration in place is required, the anticipated accuracy is in the range of ±0.5% of full scale or 1% of indicated flow, whichever is greater.

Repeatability of readings is usually on the order of 0.5%. Linearity, except at the extremes of the flow range, is also expected to be no worse than 0.5%, and over the full design range of the meter is normally accurate within the ±1% of full rated flow.

To achieve the accuracies noted above, careful attention must be paid to proper installation, particularly with regard to insertion depth.

Design features: These flow sensors are offered in materials of construction compatible with a broad range of aqueous solutions, of both high and low pH and with deionized water. For this latter service, the history of the sensor, particularly for particle generation, should be reviewed. Flow streams with high concentrations of solids, and particularly of fibrous solids, should be carefully reviewed when the impeller sensor is under consideration. Construction is reasonably forgiving with particulates, but caution in such applications is warranted. Heat transfer fluids are also compatible with sensors from at least two of the suppliers. Basic application limits should be discussed with the supplier, who should have a broad history of successful applications to back up recommendations, as well as sensor materials with test results specific to the measured fluid. So-called "hot tap" or "wet tap" versions are available for installation in operating pipelines and in-line or on-line service. A submersible version capable of metering in flooded well pits for extended periods of time is available from some suppliers. At least two manufacturers provide "intrinsically safe" sensors for application in hazardous environments.

Areas of application: Impeller flow sensors are widely used in the following fields with considerable success:

- Agricultural and horticultural irrigation
- Deionized water systems, including silicon wafer fabrication
- Heating, ventilating, and air conditioning, energy management
- Industrial waste treatment
- Industrial filtration systems
- Chemical reagent metering and batching
- Municipal water systems (potable water)

The sensors can be supplied with analog output for control applications, if desired.

Some suppliers have developed specialty impeller meters adapted to provide long life, ruggedness, and maintainability required in more demanding applications. The history of acceptable operation in applications similar to that proposed for the metering system should be reviewed with the supplier. Custodial transfer applications are normally unsuitable for the impeller sensor.

Manufacturers

The following is a selection of U.S. manufacturers. Their locations and phone numbers are listed in Table 28.10.

TABLE 28.10   U.S. Manufacturers of Impeller Flowmeters

Manufacturer	City, State	Phone number
Data Industrial Corp.	Mattapoisett, MA	(508) 758-6390
G. Fischer (Signet)	Tustin, CA	(714) 731-8800
SeaMetrics	Kent, WA	(206) 872-0284
Roper Flow Technology, Inc.	Phoenix, AZ	(602) 437-1315
Blancett	Altus, OK	(405) 482-0036
Hoffer Flow Controls, Inc.	Elizabeth City, NC	(800) 628-4584; (919) 331
McMillan Co.	Georgetown, TX	(512) 863-0231
Flowmetrics, Inc.	Canoga Park, CA	(818) 346-4492
Proteus Industries, Inc.	Mountain View, CA	(415) 964-4163

Data Industrial

In-line and Insertion Meters: 0.07 to 18 m/s (0.25 to 60 fps) fluid velocity. Available materials to accommodate most industrial fluids. Temperatures from −29°C to 152°C (−20°F to 305°F). Energy monitoring. Digital and/or analog meter outputs. NEMA 4, 4X, and 6P constructions. FM, CSA Approvals.

G. Fischer (Signet)

In-line and Insertion Meters: 0.1 to 7 m/s (0.3 to 20 fps) fluid velocity. Available materials to accommodate most industrial fluids. Temperatures to 149°C (300°F). Energy monitoring. Digital and/or analog meter ouputs. NEMA 4, 4X constructions. FM, CE Approvals.

SeaMetrics

In-line meter. Sizes 3/8 in. to 2 in. Plastic, brass. Clean water applications. Flows from 0.05 to 40.0 gpm. Accuracy ±1% of full scale, (FS). Insertion. Range of IP probes. Plastic, brass, stainless. Sapphire bearings and carbide shafts. Up to 250°F. 0.3 to 30 fps. Accuracy ±1% to ±2% FS.

Roper Flow Technology

In-line meters: Omniflow®, liquids and gases, highly precise, range 7.6 to 5677 mL min^{-1} (liquid), 0.0025 to 0.68 Am^{3}h^{-1} (gases). Cryogenics to 593°C, pressure to 60,000 psi, viscosity compensated.

Optiflo®, plastic construction for liquids only with some reduction in performance capabilities.

Hoffer

MF Series (Miniflow). High-performance, in-line low flowmeters using a "Pelton rotor" for liquids and gases. Liquid flow ranges from 57 mL min^{-1} to 11.5 rpm. Linearity ±1% reading over 10:1 range. Repeatability ±0.1% typical. Gas flow ranges depending on density from 0.02 ACFM to 1.0 ACFM. Linearity ±1.5% reading. Repeatability ±0.2%. Various bearings available. Ball bearings used for high accuracy applications. UVC curves available. "Smart" transmitter available for temperature–viscosity correction + linearization.

McMillan

Low viscosity liquids: In-line meters. Ranges from 13 mL min^{-1} to 10 L min^{-1}. Accuracies ±1% to ±3% full scale. IR rotation sensors. Have a Teflon version for corrosives.

Gases: Teflon version for chlorine, fluorine, etc. Accuracy ±3% full scale. Ranges: 25 AmL min^{-1} to 5 AmL min^{-1}.

Flowmetrics

Series 600: Stainless steel, in-line tangential turbine for gases and liquids. Ranges: 0.001 to 2.0 GPM liquids; 0.001 to 2.0 ACFM gases. Linearities vary according to range from ±1% FS to ±5% FS. Repeatabilities ± 0.05%, traceable NIST.

Proteus

In-line paddlewheel flowmeters and flow switches — liquids only. Typically used for water cooling lines on electrical or vacuum equipment. Quoted for use with liquids up to 120 cSt. Media include: water, treated water, deionized water, ethylene glycol, light oils, etc. Range 0.08 to 60 gpm. Accuracies vary between models in the ±3% to ±4% FS range. Meters bodies can incorporate optional integral temperature and pressure transducers. These meters are also available with a FluidTalk™ computer interface protocol and PC software for constructing embedded control systems for flow, temperature, and pressure.

References

1. N. P. Cheremisinoff, *Applied Fluid Flow Measurement*, New York: Marcel Dekker, 1979, 106.
2. H. S. Bean (ed.), *Fluid Meters, 6th ed.*, New York: American Society of Mechanical Engineers, 1971, 99.
3. Flow Meter Finder™ Impeller-Type Flow Meters, http://www.seametrics.com/flowmeter-finder/impeller.htm.
4. R. Koch and D. Palmer, Revisiting Measurement Technologies, Multijet vs. Positive Displacement Meters in 1996, http://www.wateronline.com/companies/mastermeterinc/tech.html.
5. L. C. Kjelstrom, Methods of measuring pumpage through closed conduit irrigation systems, *J. Irrigation Drainage Eng.-ASCE*, 117, 748-757. 1991.

28.6 Electromagnetic Flowmeters

Halit Eren

Magnetic flowmeters have been widely used in industry for many years. Unlike many other types of flowmeters, they offer true noninvasive measurements. They are easy to install and use to the extent that existing pipes in a process can be turned into meters simply by adding external electrodes and suitable magnets. They can measure reverse flows and are insensitive to viscosity, density, and flow disturbances. *Electromagnetic flowmeters* can rapidly respond to flow changes and they are linear devices for a wide range of measurements. In recent years, technological refinements have resulted in much more economical, accurate, and smaller instruments than the previous versions.

As in the case of many electric devices, the underlying principle of the electromagnetic flowmeter is Faraday's law of electromagnetic induction. The induced voltages in an electromagnetic flowmeter are linearly proportional to the mean velocity of liquids or to the volumetric flow rates. As is the case in many applications, if the pipe walls are made from nonconducting elements, then the induced voltage is independent of the properties of the fluid.

The accuracy of these meters can be as low as 0.25% and, in most applications, an accuracy of 1% is used. At worst, 5% accuracy is obtained in some difficult applications where impurities of liquids and the contact resistances of the electrodes are inferior as in the case of low-purity sodium liquid solutions.

Faraday's Law of Induction

This law states that if a conductor of length l (m) is moving with a velocity v (m s^{-1}), perpendicular to a magnetic field of flux density B (Tesla), then the induced voltage e across the ends of conductor can be expressed by:

$$e = B\,l\,v \tag{28.57}$$

The principle of application of Faraday's law to an electromagnetic flowmeter is given in Figure 28.36. The magnetic field, the direction of the movement of the conductor, and the induced emf are all perpendicular to each other.

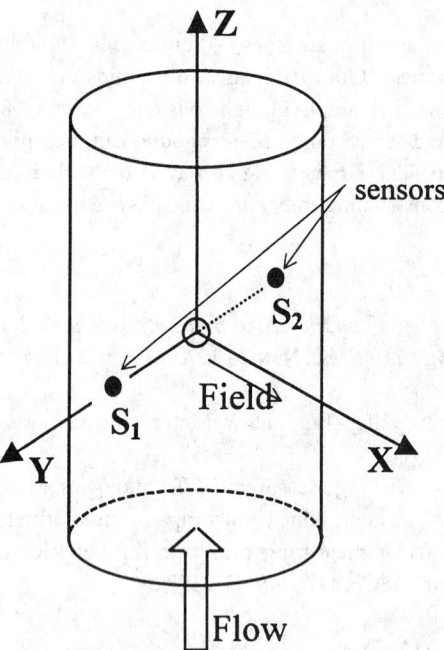

FIGURE 28.36 Operational principle of electromagnetic flowmeters. Faraday's law states that a voltage is induced in a conductor moving in a magnetic field. In electromagnetic flowmeters, the direction of movement of the conductor, the magnetic field, and the induced emf are perpendicular to each other on x, y, and z axes. Sensors S1 and S2 experience a virtual conductor due to liquid in the pipe.

Figure 28.37 illustrates a simplified electromagnetic flowmeter in greater detail. Externally located electromagnets create a homogeneous magnetic field passing through the pipe and the liquid inside it. When a conducting flowing liquid cuts through the magnetic field, a voltage is generated along the liquid path between two electrodes positioned on the opposite sides of the pipe.

In the case of electromagnetic flowmeters, the conductor is the liquid flowing through the pipe, and the length of the conductor is the distance between the two electrodes, which is equal to the tube diameter. The velocity of the conductor is proportional to the mean flow velocity of the liquid. Hence, the induced voltage becomes:

$$e = B D v \tag{28.58}$$

where D (m) is the diameter of pipe. If the magnetic field is constant and the diameter of the pipe is fixed, the magnitude of the induced voltage will only be proportional to the velocity of the liquid. If the ends of the conductor, in this case the sensors, are connected to an external circuit, the induced voltage causes a current, i, to flow, which can be processed suitably as a measure of the flow rate. The resistance of the moving conductor can be represented by R to give the terminal voltage v_T of the moving conductor as $v_T = e - iR$.

Electromagnetic flowmeters are often calibrated to determine the volumetric flow of the liquid. The volume of liquid flow, Q (L s^{-1}), can be related to the average fluid velocity as:

$$Q = A v \tag{28.59}$$

Writing the area, A (m^2), of the pipe as:

$$A = \pi D^2 / 4 \tag{28.60}$$

gives the induced voltage as a function of the flow rate.

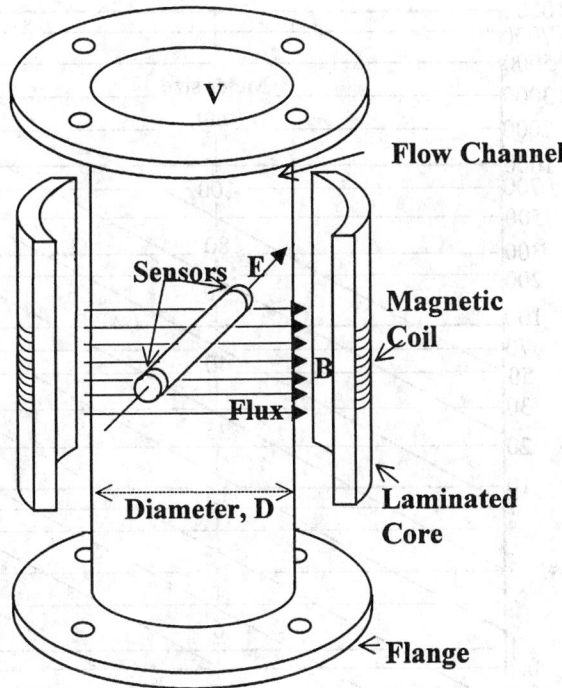

FIGURE 28.37 Construction of practical flowmeters. External electromagnets create a homogeneous magnetic field that passes through the pipe and the liquid inside. Sensors are located 90° to the magnetic field and the direction of the flow. Sensors are insulated from the pipe walls. Flanges are provided for fixing the flowmeter to external pipes. Usually, manufacturers supply information about the minimum lengths of the straight portions of external pipes.

$$e = \frac{4BQ}{\pi D} \tag{28.61}$$

Equation 28.61 indicates that in a carefully designed flowmeter, if all other parameters are kept constant, then the induced voltage is linearly proportional to the liquid flow only.

Based on Faraday's law of induction, there are many different types of electromagnetic flowmeters available, such as ac, dc, and permanent magnets. The two most commonly used ones are the ac and dc types. This section concentrates mainly on ac and dc type flowmeters.

Although the induced voltage is directly proportional to the mean value of the liquid flow, the main difficulty in the use of electromagnetic flowmeters is that the amplitude of the induced voltage is small relative to extraneous voltages and noise. Noise sources include:

- Stray voltage in the process liquid
- Capacitive coupling between signal and power circuits
- Capacitive coupling in connection leads
- Electromechanical emf induced in the electrodes and the process fluid
- Inductive coupling of the magnets within the flowmeter

Construction and Operation of Electromagnetic Flowmeters

Common to both ac and dc electromagnetic flowmeters, the magnetic coils create a magnetic field that passes through the flow tube and process fluid. As the conductive fluid flows through the flowmeter, a voltage is induced between the electrodes in contact with the process liquid. The electrodes are placed at positions where maximum potential differences occur. The electrodes are electrically isolated from the

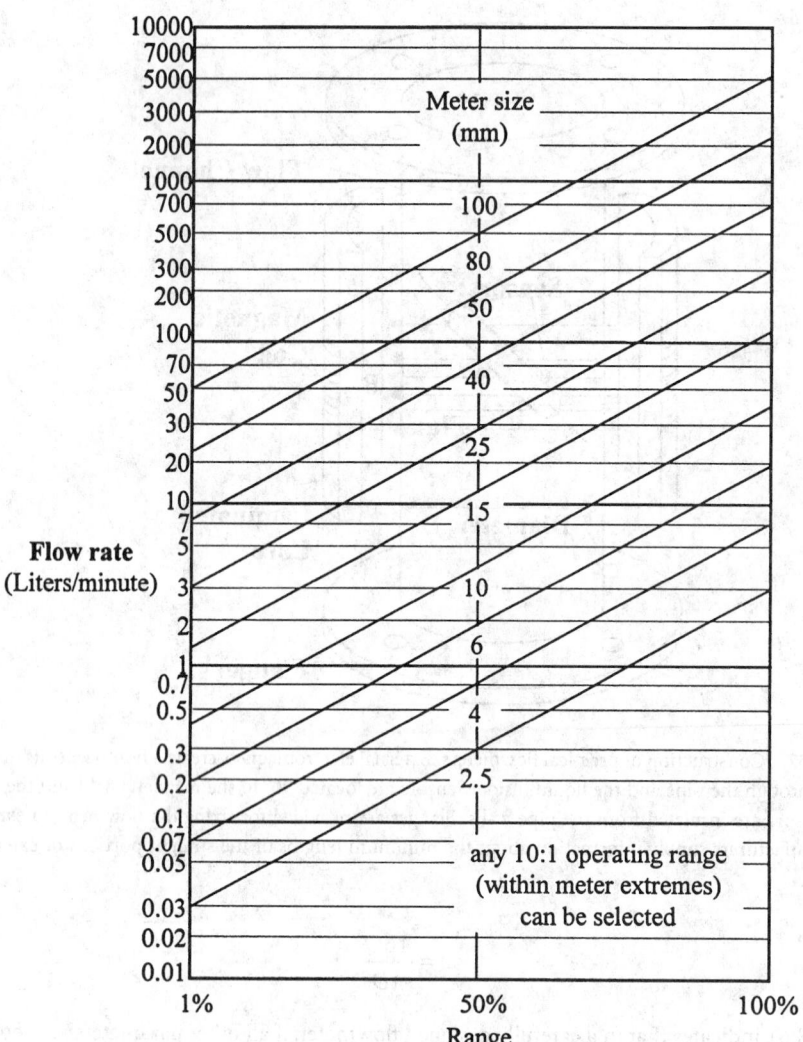

FIGURE 28.38 Selection of flowmeters. In the selection of a suitable flowmeter for a particular application, care must be exercised in handling the anticipated liquid velocities. The velocity of liquid must be within the linear range of the device. For example, a flowmeter with 100 mm internal diameter can handle flows between 50 L min⁻¹ to 4000 L min⁻¹. An optimum operation will be achieved at a flow rate of 500 L min⁻¹.

pipe walls by nonconductive liners to prevent short-circuiting of electrode signals. The liner also serves as protection to the flow tube to eliminate galvanic action and possible corrosion due to metal contacts. Electrodes are held in place by holders that also act as sealing.

Dimensionally, magnetic flowmeters are manufactured from 2 mm to 1.2 m in diameter. In a particular application, the determination of the size of the flowmeter is a matter of selecting the one that can handle the anticipated liquid velocities. The anticipated velocity of the liquid must be within the linear range of the device. As an example of a typical guide for selection, the capacities of various size flowmeters are given in Figure 28.38.

Some electromagnetic flowmeters are made from replaceable flow tubes whereby the field coils are located external to the tubes. In these flowmeters, the flanges are located far apart in order to reduce their adverse effects on the accuracy of measurements; hence, they are relatively larger in dimensions. Whereas in others, the field coils are located closer to the flow tube or even totally integrated together.

In this case, the flanges could be located closer to the magnets and the electrodes, thus giving relatively smaller dimensions. On the other hand, the miniature and electrodeless magnetic flowmeters are so compact in size that face-to-face dimensions are short enough to allow them to be installed between two flanges.

The wetted parts of a magnetic flowmeter include the liners, electrodes, and electrode holders. Many different materials such as rubber, teflon, polyurethane, polyethylene, etc. are used in the construction to suit process corrosivity, abrasiveness, and temperature constraints. The main body of a flowmeter and electrodes can be manufactured from stainless steel, tantalum, titanium, and various other alloys. Liners are selected mainly to withstand the abrasive and corrosive properties of the liquid. The electrodes must be selected such that they cannot be coated with insulating deposits of the process liquid during long periods of operation.

The pipe between the electromagnets of a flowmeter must be made from nonmagnetic materials to allow the field to penetrate the fluid without any distortion. Therefore, the flow tubes are usually constructed of stainless steel or plastics. The use of steel is a better option because it adds strength to the construction. Flanges are protected with appropriate liners and do not make contact with the process fluid.

In some electromagnetic flowmeters, electrodes are cleaned continuously or periodically by ultrasonic or electric means. Ultrasonics are specified for ac and dc type magnetic flowmeters when frequent severe insulating coating is expected on the electrodes that might cause the flowmeter to cease to operate in an anticipated manner.

The operation of a magnetic flowmeter is generally limited by factors such as linear characteristics, pressure ratings of flanges, and temperatures of the process fluids. The maximum temperature limit is largely dependant on the liner material selection and usually is set to around 200°C. For example, ceramic liners can withstand high temperatures, but are subject to cracking in case of sudden changes in temperatures of the process fluid.

During the selection of electromagnetic flowmeters, the velocity constraints should be evaluated carefully to secure accurate performance over the expected range. The full-scale velocity of the flowmeter is typically 0.3 m s^{-1} to 10 m s^{-1}. Some flowmeters can measure lower velocities with somewhat poorer accuracy. Generally, employment of electromagnetic flowmeters over a velocity of 5 m s^{-1} should be considered carefully because erosion of the pipe and damages to liners can be significant.

The accuracy of a conventional magnetic flowmeter is usually expressed as a function of full scale (FS), typically 0.5% to 1% FS. However, dc flowmeters have a well-defined zero due to an automatic zeroing capabilities; therefore, they have a percentage rate of accuracy better than ac types, typically 0.5% to 2%.

Types of Electromagnetic Flowmeters

AC Magnetic Flowmeters

In many commercial electromagnetic flowmeters, an alternating current of 50 Hz to 60 Hz in coils creates the magnetic field to excite the liquid flowing within the pipe. A voltage is induced in the liquid due to Faraday's law of induction, as explained above. A typical value of the induced emf in an ac flowmeter fixed on a 50 mm internal diameter pipe carrying 500 L min^{-1} is about 2.5 mV.

Historically, ac magnetic flowmeters were the most commonly used types because they reduced polarization effects at the electrodes. In general, they are less affected by the flow profiles of the liquid inside the pipes. They allow the use of high Z_{in} amplifiers with low drift and high pass filters to eliminate slow and spurious voltage drifts emanating mainly from thermocouple and galvanic actions. These flowmeters find many applications as diverse as the measurement of blood flow in living specimens. Miniaturized sensors allow measurements on pipes and vessels as small as 2 mm in diameter. In these applications, the excitation frequencies are higher than industrial types, 200 Hz to 1000 Hz.

A major disadvantage of the ac flowmeter is that the powerful ac field induces spurious ac signals in the measurement circuits. This necessitates periodic adjustment of zero output at zero velocity conditions — more frequently than for dc counterparts. Also, in some harsh industrial applications, currents in the magnetic field can vary, due to voltage fluctuations and frequency variations in the power

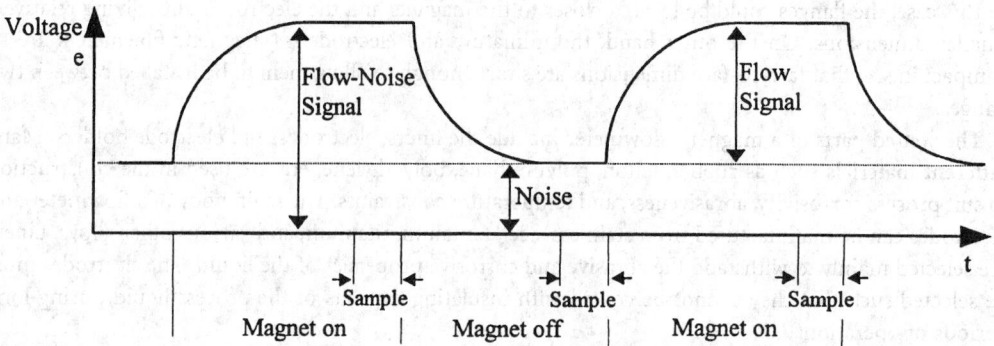

FIGURE 28.39   The signals observed at the electrodes represent the sum of the induced voltage and the noise. When the current in the magnetic coils is turned off, the signal across the electrodes represents only the noise. Subtracting the measurement of the flowmeter when no current flows through the magnet from the measurement when current flows through the magnet effectively cancels out the effect of noise.

lines. The effect of fluctuations in the magnetic field can be minimized by the use of a reference voltage proportional to the strength of the magnetic field to compensate for these variations. To avoid the effects of noise and fluctuations, special cabling and calibration practices recommended by the manufacturers must be used to ensure accurate operation. Usually, the use of two conduits is required — one for signals and one for power. The cable lengths should also be set to certain levels to minimize noise and sensitivity problems.

Ac flowmeters operating at 50, 60, or 400 Hz are readily available. In general, ac flowmeters can operate from 10 Hz to 5000 Hz. High frequencies are preferred in determining the instantaneous behavior of transients and pulsating flows. Nevertheless, in applications where extremely good conducting fluids and liquid metals are used, the frequency must be kept low to avoid skin effects. On the other hand, if the fluid is a poor conductor, the frequency must not be so high such that dielectric relaxation is not instantaneous.

Dc Magnetic Flowmeters

Unlike ac magnetic flowmeters, direct current or pulsed magnetic flowmeters excite the flowing liquid with a field operating at 3 Hz to 8 Hz. As the current to the magnet is turned on, a dc voltage is induced at the electrodes. The signals observed at the electrodes represent the sum of the induced voltage and the noise, as illustrated in Figure 28.39. When the current in the magnetic coils is turned off, the signal represents only the noise. Subtracting the measurement of the flowmeter when no current flows through the magnet from the measurement when current flows through the magnet effectively cancels out the effect of noise.

If the magnetic field coils are energized by normal direct current, then several problems can occur: polarization, which is the formation of a layer of gas around the measured electrodes, as well as electro-chemical and electromechanical effects. Some of these problems can be overcome by energizing the field coils at higher frequencies or ac. However, higher frequencies and ac generate transformer action in the signal leads and fluid path. Therefore, the coils are excited by dc pulses at low repetition rates to eliminate the transformer action. In some flowmeters, by appropriate sampling and digital signal processing techniques, the zero errors and the noise can be rejected substantially.

The zero compensation inherent in dc magnetic flowmeters eliminates the necessity of zero adjustment. This allows the extraction of flow signals regardless of zero shifts due to spurious noise or electrode coating. Unlike ac flowmeters, a larger insulating electrode coating can be tolerated that could shift the effective conductivity significantly without affecting performance. If the effective conductivity remains high enough, a dc flowmeter will operate satisfactorily. Therefore, dc flowmeters are less susceptible to drifts, electrode coatings, and changes in process conditions in comparison to conventional ac flowmeters.

TABLE 28.11   List of Manufacturers of Electromagnetic Flowmeters

ABB K-Flow Inc. P.O. Box 849 45 Reese Rd. Millville, NJ 08332 Tel: (800) 294-8116 Fax: (609) 825-1678	Marsh-McBirney, Inc. 4539 Metropolitan Court Frederick, MD 21704 Tel: (301) 879-5599 Fax: (301) 874-2172
Control Warehouse Shores Industrial park Ocala, FL 34472 Tel: (800) 633-0319 Fax: (352) 687-8925	Nusonics Inc. 11391 E. Tecumseh St. Tulsa, OK 74116-1606 Tel: (918) 438-1010 Fax: (918) 438-6420
Davis Instruments 4701 Mount Hope Dr. Baltimore, MD 21215 Tel: (410) 358-3900 Fax: (410) 358-0252	Rosemount Inc. Dept. MCA 15 12001 Technology Dr. Eden Prairie, MN 55344 Tel: (612) 828-3006 Fax: (612) 828-3088
Fischer Porter 50 Northwestern Dr. P.O. Box 1167T Salem, NH 03079-1137 Tel: (603) 893-9181 Fax: (603) 893-7690	Sparling Instruments Co., Inc. 4097 Temple City Blvd. P.O. Box 5988 El Monte, CA 91734-1988 Tel: (800) 423-4539
Johnson Yokogawa Dept. P, Dart Rd. Newman, GA 30265 Tel: (800) 394-9134 Fax: (770) 251-6427	Universal Flow Monitors, Inc. 1751 E. Nine Mile Rd. Hazel Park, MI 48030 Tel: (313) 542-9635 Fax: (313) 398-4274

Dc magnetic flowmeters do not have good response times due to the slow pulsed nature of operations. However, as long as there are not rapid variations in the flow patterns, zero to full-scale response times of a few seconds do not create problems in the majority of applications. Power requirements are also much less as the magnet is energized part of the time. This gives an advantage in power saving of up to 75%.

If the dc current to the magnet is constant, the proportional magnetic field can be kept steady. Therefore, the amplitudes of the dc voltages generated at the electrodes will be linearly proportional to the flow. However, in practice, the current to the magnet varies slightly due to line voltage and frequency variations. As in the case of ac flowmeters, voltage and frequency variations could necessitate the use of a reference voltage. Because the effect of noise can be eliminated more easily, the cabling requirements are not as stringent.

To avoid electrolytic polarization of the electrodes, bipolar pulsed dc flowmeters have been designed. Also, modification of dc flowmeters led to the development of miniature dc magnetic flowmeters with wafer design for a limited range of applications. The wafer design reduces the weights as well as the power requirements.

Table 28.11 provides a listing of several manufacturers of electromagnetic flowmeters.

Installation and Practical Applications of Electromagnetic Flowmeters

Conventional ac and dc magnetic flowmeters have flanges at the inlet and the outlet that need to be bolted to the flanges of the pipe. Face-to-face dimensions of magnetic flowmeters differ between manufacturers; therefore, once the flowmeter is installed, new piping arrangements could be necessary if a flowmeter is replaced by one from a different manufacturer.

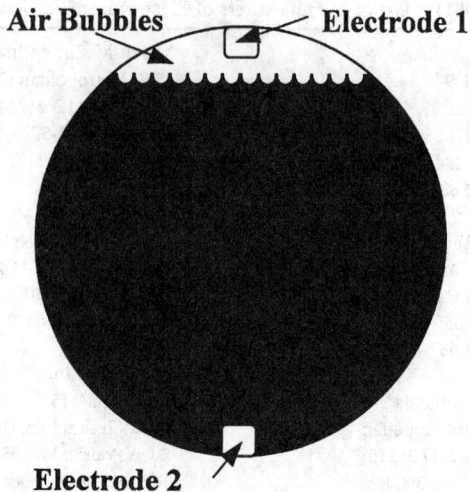

FIGURE 28.40   The pipes of electromagnetic flowmeters must be full of liquid at all times for accurate measurement. If the liquid does not make full contact with electrodes, the high impedance prevents the current flow; hence, measurements cannot be taken. Also, if the pipe is not full, even if contact is maintained between the liquid and electrodes, the empty portions of the pipe will lead to miscalculated flow rates.

Upstream and downstream straight piping requirements can vary from one flowmeter to another, depending on the manufacturer's specifications. As a rule of thumb, the straight portion of the pipe should be at least 5D/2D from the electrodes and 5D/5D from the face of the flowmeter in upstream and downstream directions, respectively. For good accuracy, one should adhere carefully to the recommendations of manufacturers for piping requirements. In some magnetic flowmeters, coils are used in such a way that the magnetic field is distributed in the coil to minimize the piping effect.

For accurate measurements, magnetic flowmeters must be kept full of liquid at all times. If the liquid does not contact the electrodes, measurements cannot be taken. Figure 28.40 illustrates this point. If the measurements are made in other than vertical flows, the electrodes should be located in horizontal directions to eliminate the possible adverse effect of the air bubbles, because the air bubbles tend to concentrate on the top vertical part of the liquid.

In the selection of magnetic flowmeters, a number of considerations must be taken into account, such as:

- Cost, simplicity, precision, and reproducibility
- Metallurgical aspects
- Velocity profiles and upstream disturbances

Most processes employ circular piping that adds simplicity to the construction of the system. The flowmeters connected to circular pipes give relatively better results compared to rectangular or square-shaped pipes, and velocity profiles of the liquid are not affected by the asymmetry. However, in circular pipes, the fringing of the magnetic field can be significant, making it necessary to employ empirical calibrations.

Selection of materials for constructing the channel of the magnetic flowmeter demands care. If the fluid is nonmetallic, a nonconducting or an insulated channel should be sufficient. In this case, wetted electrodes must be used. Electrodes must be designed to have sufficiently large dimensions to keep the output impedance at acceptable levels. Also, careful handling of electrode signals must be observed because, in many cases, malfunctioning of reference signal electronics is the main cause of flowmeter failure.

Magnetic flowmeters do not require continuous maintenance, except for periodic calibrations. Nevertheless, electrode coating, damage to the liners, and electronic failures can occur. Any modification or repair must be treated carefully because, when installed again, some accuracy can be lost. After each modification or repair, recalibration is usually necessary.

Often, magnetic flowmeter liners are damaged by the presence of debris and solids in the process liquid. Also, the use of incompatible liquid with the liners, wear due to abrasion, excess temperature, and installation and removals can contribute to the damage of liners. The corrosion in the electrodes can also be a contributing factor for the damage. In some cases, magnetic flowmeters can be repaired on-site even if severe damage occurs; in other cases, they must be shipped to the manufacturer for repairs. Usually, manufacturers supply spare parts for electrodes, liners, flow tubes and electronic components.

Calibration of electromagnetic flowmeters is achieved with a magnetic flowmeter calibrator or by electronic means. The magnetic flowmeter calibrators are precision instruments that inject simulated output signals of the primary flowmeter into the transmitter. Effectively, this signal is used to check correct operation of electronic components and make adjustments to the electronic circuits. Alternatively, calibrations can also be made by injecting suitable test signals to discrete electronic components. In some cases, empirical calibrations must be performed at zero flow while the flowmeter is filled with the stationary process liquid.

Application of magnetic flowmeters can only be realized with conductive liquids such as acids, bases, slurries, foods, dyes, polymers, emulsions, and suitable mixtures that have conductivities greater than the minimum conductivity requirements. Generally, magnetic flowmeters are not suitable for liquids containing organic materials and hydrocarbons. As a rule of thumb, magnetic flowmeters can be applied if the process liquids constitute a minimum of about 10% conductive liquid in the mixture.

The lack of any direct Reynolds number constraints and the obstructionless design of magnetic flowmeters make it practical for applications that involve conductive liquids that have high viscosity which could plug other flowmeters. They also measure bidirectional flows.

Despite the contrary belief, magnetic flowmeters demonstrate a certain degree of sensitivity to flow profiles. Another important aspect is the effect of turbulence. Unfortunately, there is very little information available on the behavior of turbulent flows when they are in transverse magnetic fields. Figure 28.41 shows an example of a flow profile in which the velocity profile is perturbed. The fluid is being retarded near the center of the channel, and accelerated at the top and bottom near the electrodes.

An important point in electromagnetic flowmeters is the effect of magneto-hydrodynamics, especially prominent in fluids with magnetic properties. Hydrodynamics is the ability of a magnetic field to modify the flow pattern. In some applications, the velocity perturbation due to the magneto-hydrodynamic effect can be serious enough to influence the accuracy of operations, as in the case of liquid sodium and its solutions.

Effects of Electric Conductivity of Fluid

For electromagnetic flowmeters to operate accurately, the process liquid must have minimum conductivity of $1\ \mu S\ cm^{-1}$ to $5\ \mu S\ cm^{-1}$. Most common applications involve liquids with conductivities greater than $5\ \mu S\ cm^{-1}$. Nevertheless, for accurate operation, the requirement for the minimum conductivity of liquid can be affected by length of leads from sensors to transmitter electronics.

Ac flowmeters are sensitive to the nonconductive coating of the electrodes that may result in calibration shift or complete loss of signals. The dc flowmeters, on the other hand, should not be affected by nonconductive coating to such a great extent, unless the conductivity between electrodes is less than the minimum required value. In some flowmeters, electrodes can be replaced easily; while in others, electrodes can be cleaned by suitable methods. If coating is a continual problem, ultrasonic and other cleaning methods should be considered.

Zero adjustment of ac magnetic flowmeters requires compensation for noise. If the zero adjustment is performed with any fluid other than the process fluid, serious errors can result because of possible

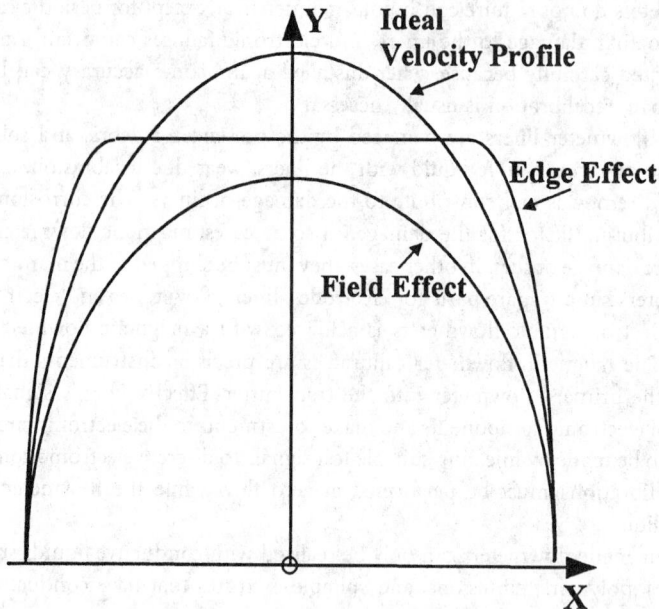

FIGURE 28.41 Flow profiles in the pipes. Magnetic flowmeters demonstrate a certain degree of sensitivity to flow profiles. The ideal velocity profile can be distorted due to edge effects and also field effects known as magneto-hydrodynamics. In some applications, the velocity perturbation due to the magnetohydrodynamic effect can be serious enough to severely influence the accuracy of operations.

differences in conductivities. Similarly, if the electrodes are coated with an insulating substance, the effective conductivity of the electrodes can be altered, thereby causing a calibration shift. If the coating changes in time, the flowmeter can continually require calibration for repeatable readings.

The resistance between electrodes can be approximated by $R = 1/\delta d$, where δ is the fluid conductivity and d is the electrode diameter. For tap water, $\delta = 200\ \mu S\ cm^{-1}$; for gasoline, $\delta = 0.01\ \mu S\ cm^{-1}$; and for alcohol, $\delta = 0.2\ \mu S\ cm^{-1}$. A typical electrode with a 0.74 cm diameter in contact with tap water results in a resistance of 6756 Ω.

Signal Pickup and Demodulation Techniques

Magnetic flowmeters are four-wire devices that require an external power source for operations. Particularly in ac magnetic flowmeters, the high-voltage power cables and low-voltage signal cables must run separately, preferably in different conduits; whereas, in dc magnetic flowmeters, the power and signal cables can be run in one conduit. This is because in dc-type magnetic flowmeters, the voltage and the frequency of excitation of the electromagnets are relatively much lower. Some manufacturers supply special cables along with their flowmeters.

In ac flowmeters, the electrode signals can be amplified much more readily compared to their dc counterparts. That is the reason why ac flowmeters have been used successfully to measure very low flow rates, as well as the flow of very weakly conducting fluids. Nevertheless, ac flowmeters tend to be more complicated, bulky, expensive, and they require electromagnets with laminated yokes together with stabilized power supplies. In some magnetic flowmeters, it is feasible to obtain sufficiently large flow signal outputs without the use of a yoke by means of producing a magnetic field by naked coils. In this case, the transformer action to the connecting leads can be reduced considerably.

One of the main drawbacks of ac-type flowmeters is that it is difficult to eliminate the signals due to transformer action from the useful signals. The separation of the two signals is achieved by exploiting the fact that the flow-dependent signal and the transformer signal are in quadrature. That is, the useful signal is proportional to the field strength, and the transformer action is proportional to the time derivative of the field strength. The total voltage v_T can be expressed as:

$$v_T = v_F + v_t = V_F \sin(\omega t) + V_t \cos(\omega t) \tag{28.62}$$

where v_F = Induced voltage due to liquid flow
$\quad\quad v_t$ = Voltage due to transformer action on wires, etc.

Phase-sensitive demodulation techniques can be employed to eliminate the transformer action voltage. The coil magnetizing current, $i_m = I_m \sin(\omega t)$, is sensed and multiplied by the total voltage v_T, giving:

$$v_T i_m = \left[V_F \sin(\omega t) + V_t \cos(\omega t) \right] I_m \sin(\omega t) \tag{28.63}$$

Integration of Equation 28.63 over one period between 0 and 2π eliminates the transformer voltage, yielding only the voltage that is proportional to the flow.

$$V_f = V_F I_m \pi \tag{28.64}$$

Where V_f is the voltage after integration. This voltage is proportional to the induced voltage modified by constants I_m and π.

In reality, this situation can be much more complicated because of phase shift due to eddy currents in nearby solids and conductors. Other reasons for complexity include: the harmonics because of non-linearity such as hysteresis, and capacitive pickup.

A good electrical grounding of magnetic flowmeters, as illustrated in Figure 28.42, is required to isolate relatively high common mode potential. The sources of ground potential can be in the liquid or in the pipes. In practice, if the pipe is conductive and makes contact with the liquid, the flowmeter should be grounded to the pipe. If the pipe is made from nonconductive materials, the ground rings should be installed to maintain contact with the process liquid.

If the flowmeter is not grounded carefully relative to the potential of the fluid in the pipe, then the flowmeter electrodes could be exposed to excessive common mode voltages that can severely limit the accuracy. In some cases, excessive ground potential can damage the electronics because the least-resistance path to the ground for any stray voltage in the liquid would be via the electrodes.

Some commercial magnetic flowmeters have been developed that can operate on saw-tooth or square waveforms. Universally standardized magnetic flowmeters and generalized calibration procedures still do not exist, and manufacturers use their own particular design of flow channels, electromagnets, coils, and signal processors. Most manufacturers provide their own calibration data.

Further Information

J. P. Bentley, *Principles of Measurement Systems,* 2nd ed., New York: Longman Scientific and Technical, 1988.

E. O. Doebelin, *Measurement Systems: Application and Design,* 4th ed., New York: McGraw-Hill, 1990.

J. P. Holman, *Experimental Methods for Engineers,* 5th ed., New York: McGraw-Hill, 1989.

J. A. Shercliff, *Electromagnetic Flow-Measurements,* New York: Cambridge University Press, 1987.

D. W. Spitzer, *Industrial Flow Measurement,* Research Triangle Park, NC: Instrument Society of America, 1990.

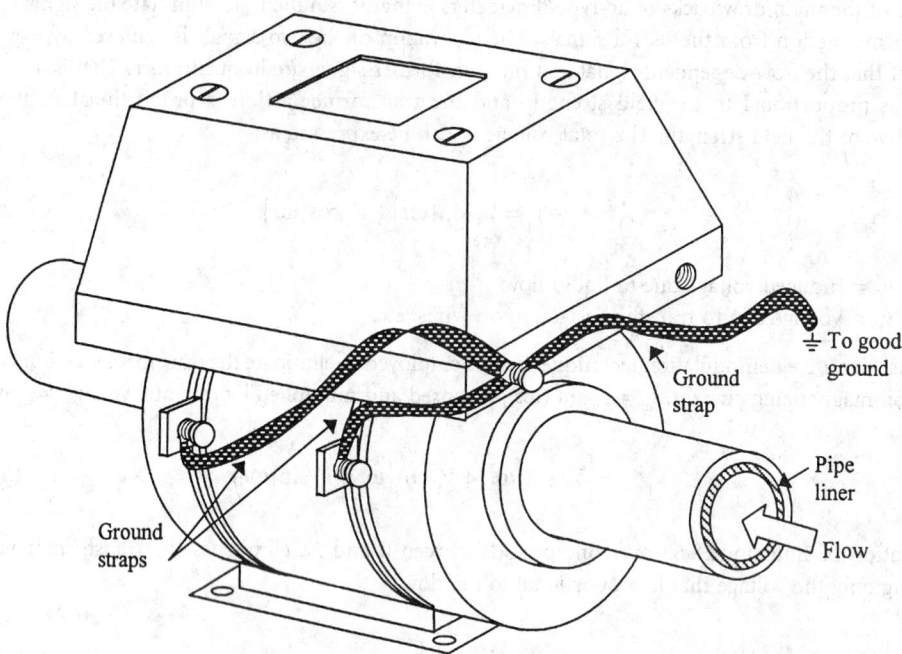

FIGURE 28.42 Grounding of electromagnetic flowmeters. A good grounding is absolutely essential to isolate noise and high common mode potential. If the pipe is conductive and makes contact with the liquid, the flowmeter should be grounded to the pipe. If the pipe is made from nonconductive materials, ground rings should be installed to maintain contact with the process liquid. Improper grounding results in excessive common mode voltages that can severely limit the accuracy and also damage the processing electronics.

28.7 Ultrasonic Flowmeters

Hans-Peter Vaterlaus, Thomas Hossle, Paolo Giordano,
and Christophe Bruttin

Flow is one of the most important physical parameters measured in industry and water management. There are various kinds of flowmeters available, depending on the requirements defined by the different market segments. For many years, differential pressure types of flowmeters have been the most widely applied flow measuring device for fluid flows in pipes and open channels that require accurate measurement at reasonable cost. In markets like waterpower, water supply, irrigation, etc., however, flow must be measured without any head losses or any pressure drop. This means no moving parts, no secondary devices, nor are any restrictions allowed. Two types of flowmeters presently fulfill this requirement: Electromagnetic and ultrasonic flowmeters. Whereas *ultrasonic flowmeters* can be applied in nearly any kind of flowing liquid, *electromagnetic* flowmeters require a minimum electric conductivity of the liquid for operation. In addition, the cost of ultrasonic flowmeters is nearly independent of pipe diameter, whereas the price of electromagnetic flowmeters increases drastically with pipe diameter.

There are various types of ultrasonic flowmeters in use for discharge measurement: (1) *Transit time:* This is today's state-of-the-art technology and most widely used type, and will be discussed in this chapter section. This type of ultrasonic flowmeter makes use of the difference in the time for a sonic pulse to travel a fixed distance, first against the flow and then in the direction of flow. Transmit time flowmeters are sensitive to suspended solids or air bubbles in the fluid. (2) *Doppler:* This type is more popular and less expensive, but is not considered as accurate as the transit time flowmeter. It makes use of the Doppler

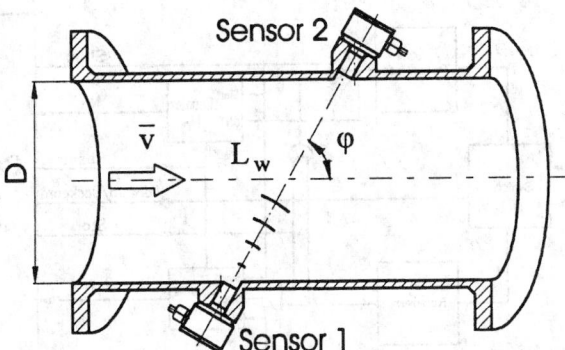

FIGURE 28.43 Principle of transit time flowmeters. Transmitting an ultrasonic pulse upstream and downstream across the flow: the liquid is moving with velocity $\bar{v}_a$ and with angle φ to the ultrasonic pulse.

frequency shift caused by sound reflected or scattered from suspensions in the flow path and is therefore more complementary than competitive to transit time flowmeters. (3) *Cross-correlation:* Two measuring sections are installed with a certain distance to each other. Both measure the energy absorption of the ultrasonic signal. A cross-correlation calculates the flow velocity. (4) *Phase shift:* The phase position of the transmitting and receiving signal is measured in the direction of the flow and against it. The resulting phase shift angle is directly proportional to the flow velocity. (5) *Drift:* The drift of an ultrasonic signal crossing the flow is measured by signal attenuation.

Transit Time Flowmeter

Principle of Operation

The acoustic method of discharge measurement is based on the fact that the propagation velocity of an acoustic wave and the flow velocity are summed vectorially. This type of flowmeter measures the difference in transit times between two ultrasonic pulses transmitted upstream t_{21} and downstream t_{12} across the flow, as shown in Figure 28.43. If there are no transverse flow components in the conduit, these two transmit times of acoustic pulses are given by:

$$t_{12} = \frac{L_w}{c + v_a \cos \varphi} \quad \text{and} \quad t_{21} = \frac{L_w}{c - v_a \cos \varphi} \tag{28.65}$$

where L_w = Distance in the fluid between the two transducers
$\quad c$ = Speed of sound at the operating conditions
$\quad \phi$ = Angle between the axis of the conduit and the acoustic path
$\quad \bar{v}_a$ = Axial low velocity averaged along the distance L_w

Since the transducers are generally used both as transmitters and receivers, the difference in travel time can be determined with the same pair of transducers. Thus, the mean axial velocity $\bar{v}_a$ along the path is given by:

$$\bar{v}_a = \frac{L_w}{2 \cos \varphi} \left(\frac{1}{t_{21}} - \frac{1}{t_{12}} \right) = \frac{D}{2 \cos \varphi \sin \varphi} \left(\frac{1}{t_{21}} - \frac{1}{t_{12}} \right) \tag{28.66}$$

The following example shows the demands on the time measurement technique: assuming a closed conduit with diameter $D = 150$ mm, angle $\phi = 60°$, flow velocity $\bar{v}_a = 1$ m·s^{-1}, and water temperature = 20°C. This results in transmit times of about 116 s and a time difference Δt ($\Delta t = t_{12} - t_{21}$) on the order

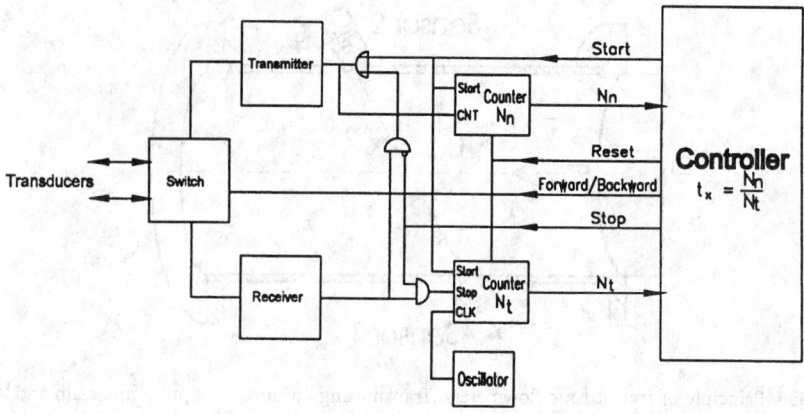

FIGURE 28.44 Block diagram of a transit time ultrasonic flowmeter using oversampling for higher resolution.

of 78 ns. To achieve an accuracy of 1% of the corresponding full-scale range, Δt has to be measured with a resolution of at least 100 ps (1×10^{-10} s).

Standard time measurement techniques are not able to meet such requirements so that special techniques must be applied. The advantage of the approach of state-of-the-art real digital measurement is to process the measured value directly by a microcomputer. The most difficult problem is to reach the required resolution and to cope with the jitter of digital logic gates. It is well known in the measurement technique that, if signals are sampled multiple times (N_n), the resolution increases with the number of samples ("oversampling") [1]. This knowledge is not only applied for analog signals but also for transit time signals and is used in today's technology of flowmeters. The propagation time t (t_{12} t_{21}), depending on the distance the sound pulse has to travel through the fluid, is measured several times. Due to this fact, one obtains the following relation [2]:

$$t = \frac{1}{N_n} \int_0^\tau \frac{1}{f} dt \qquad (28.67)$$

where τ is integration time, and f is the frequency.

According to the block diagram in Figure 28.44, two counters are used. One counter N_t is clocked during the measuring period by a stable quartz oscillator; the other one counts the number of samples N_n. The measurement stops after a certain period of some milliseconds and after having reached an integer value for N_n. The two counts of N_t and N_n are used to calculate propagation time t_{12} or t_{21} by dividing N_t by N_n. The resolution r is calculated by:

$$r = \frac{1}{N_n \cdot f} \qquad (28.68)$$

Advantages of this measurement method for transit time flowmeters are:

1. The resolution of the velocity measurement is constant (typical value 0.8 mm s^{-1}).
2. The accuracy depends almost only on the stability and the temperature coefficient of the quartz oscillator.
3. Due to the multiple sampling, the jitter of the digital logic is averaged.

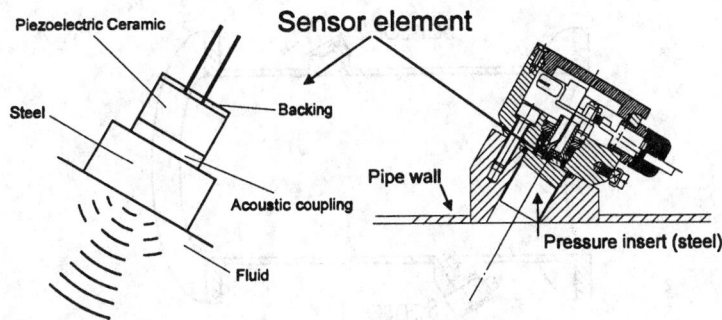

FIGURE 28.45 Ultrasonic transducer in principle and as an example of an existing version. The sensor element can be changed even under pressure.

FIGURE 28.46 Axial sensor type. The ultrasonic pulse passes directly down the axis of the pipe.

Sensors.

The transducer comprises the piezoelectric element that converts electric to acoustic energy and the basic structure for supporting the piezoceramic and providing electric connections. Transducer design entails choice of the piezoelectric element, determination of suitable dimensions and resonant frequency, and construction to withstand thermal and mechanical stress, see Figure 28.45. Considering the transmission line model [3, 4], the optimal electroacoustic response and the best matching of acoustic impedances $Z = \rho c$ of the different transducer elements is important. Not only must suitable waveforms to be detected by the electronics be obtained, but also the energy loss must be minimized when crossing several interface boundaries. Because of the wide versatility of pipe sizes and flow conditions, there are a number of different sensor configurations for transit time flowmeters.

Axial.

Due to the small differences in small diameter pipes, it is necessary to pass the sonic pulse directly down the axis of the pipe to ensure that there is sufficient path length. Figure 28.46 shows an example of an axial sensor pipe section. The upper pipe size limit for this kind of sensor is about 0.075 m [5].

Radial.

Many manufacturers supply complete metering sections with built-in sonic transducers on either side of a spool section. Such sensors, shown in Figure 28.47, are generally called "radial" because of the transducer placement. A lower pipe size limit of radial type sensors is about 100 mm.

Field Mounting.

Radial-type sensors are often used to instrument an existing line, where it is desirable to make the installation without cutting out a section of the pipe. To meet this requirement, field-mountable transducers, cemented, drilled, or welded into an existing pipe section, can be used. Metering sections with diameters up to 13 m or more can be achieved in this way.

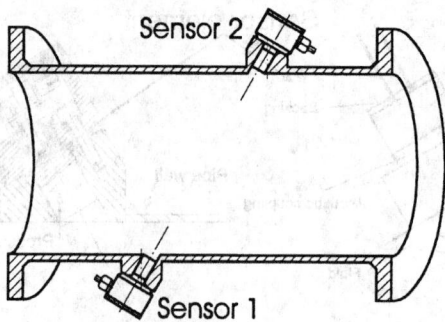

FIGURE 28.47   Radial sensor type. Manufacturers provide them as complete metering sections or as field-mounting sensors for existing conduits.

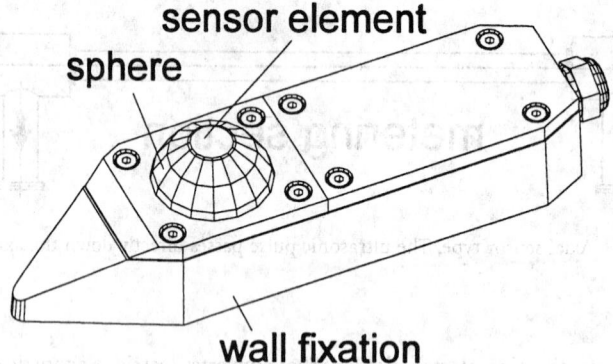

FIGURE 28.48   Open-channel sensor with sensor element placed on a movable sphere. Alignment of two sensors can be executed by laser equipment.

Clamp-on.

If there is a need for an installation where the pipe wall it not penetrated by the transducers, clamp-on systems are the right choice. Achieving a lower accuracy and being somewhat more complex to calibrate, clamp-on systems have their entitlement in applications where an easy movement of the metering section is an important requirement or where an existing process cannot be interrupted. The transducers are mounted on a calibration device and acoustically coupled to the pipe wall with grease and/or epoxy.

Open Channels and Special Applications.

In open channels, the transducers are normally mounted on or dug into the channel walls. Figure 28.48 shows an example of an open-channel sensor. The piezoceramic element is placed on a sphere to achieve a wide range of mounting possibilities. Sometimes, existing pipe sections are completely enclosed with rock or concrete. In these cases, the transducers can be fixed in the wall of the conduit. For small diameters, the resulting protrusion of the transducers into the metering section must be taken into account.

Measurement of Flow in Closed Conduits.

The most important issue in applying ultrasonic flowmeters is an understanding of the effects of the velocity profile of the flowing fluid within the conduit. The flow profile depends on the fluid, the Reynolds number Re, the relative roughness and shape of the conduit, upstream and downstream disturbances, and other factors. Transit time flowmeters give an average flow velocity v along the sonic path. The acoustic flow rate Q_{ADM} is therefore calculated through $Q = \bar{v}A$, with $\bar{v}$ the area-averaged flow velocity and A the cross-section of the conduit. In order to obtain the area-averaged flow velocity $\bar{v}$, the measured velocity $\bar{v}_a$ must be corrected by a hydraulic coefficient k_h that depends on the type of the conduit and

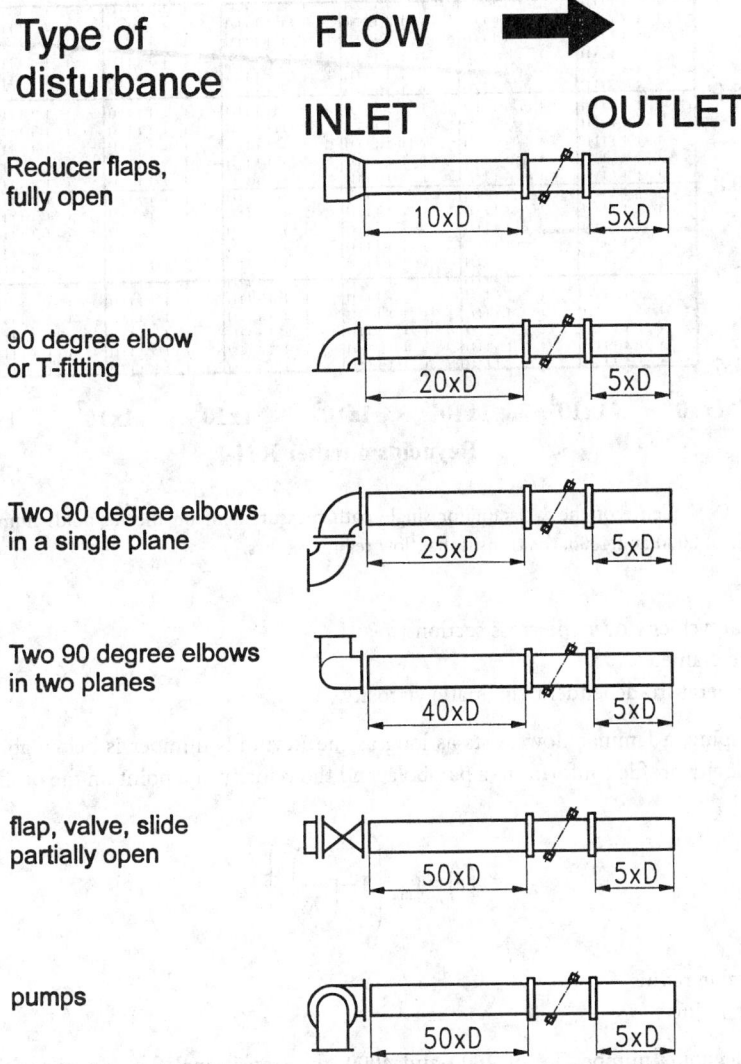

FIGURE 28.49 Minimum straight run requirements for a 1% accuracy of a single-path transit time flowmeter.

the Reynolds number. In order to achieve maximum performance and accuracy of ultrasonic flowmeters, one has to keep to sufficient straight run requirements as shown for some examples in Figure 28.49. By doing so, a typical accuracy of 1% of reading or better can be achieved, even when applying a single-path measurement system. Reduced straight runs lead to reduced accuracy. In some applications, this reduced accuracy is acceptable; if not, a multipath ultrasonic flowmeter must be installed. These flowmeters provide averaging of the various error-producing flow components. Accuracy of 0.5% of reading can be achieved even under nonideal conditions or insufficient straight runs. Figure 28.50 shows four examples of possible sonic path arrangements in a closed conduit.

Single Path with Circular and Rectangular Cross-sections.
The Reynolds number Re is given by:

$$Re = \frac{\bar{v} \cdot D}{\nu} \qquad (28.69)$$

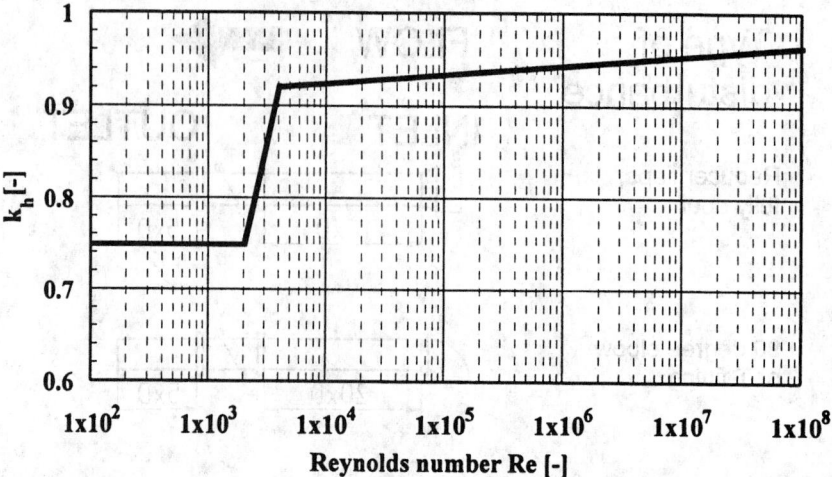

FIGURE 28.50 Dependence of the k_h factor for single-path measurement on the Reynolds number. Between a Reynolds number of 2000 and 4000, the transitional flow regime occurs.

where $\bar{v}$ = Mean velocity over the cross section [m·s^{-1}]
  D = Pipe diameter
  v = Temperature-dependent cinematic viscosity

In normal piping, a laminar flow exists as long as the Reynolds number is below about 2600. The shape of the velocity profile conforms to a parabola, and the velocity of a point on the profile is given by:

$$v(r) = v_{max}\left(1 - \left(\frac{r}{R}\right)^2\right) \tag{28.70}$$

where r = Variable radius
  R = Pipe radius

Between a Reynolds number Re of 2600 and 4000, the transitional flow regime with continuous switching between laminar and turbulent velocity profile comes into existence. When the Reynolds number exceeds 4000, the velocity profile enters the turbulent flow regime. Nikuradse [6] showed that the turbulent velocity profile of an axis-symmetrical flow in a closed conduit without swirl and sufficient inlet and outlet sections and smooth walls can be expressed by:

$$v(r) = v_{max}\left(\frac{R-r}{R}\right)^{\frac{1}{n}} \tag{28.71}$$

where, according to Nikuradse, n is a Reynolds number-dependent exponent given by:

$$n = \frac{1}{\left(0.2525 - 0.0229 \times \log(\text{Re})\right)} \tag{28.72}$$

With the definition of a hydraulic corrective coefficient k_h given by:

$$k_h = \frac{\bar{v}}{\bar{v}_a}$$ (28.73)

where $\bar{v}$ = Mean velocity over the cross-section
$\bar{v}_a$ = Mean velocity along the sonic path

and integrating over the cross-section according to:

$$k_h = \frac{\dfrac{1}{A}\displaystyle\int_0^R\int_0^{2\pi} v(r)\, r\, dr\, d\theta}{\dfrac{1}{2R}\displaystyle\int_{-R}^{R} v(r)\, dr}$$ (28.74)

one obtains for laminar flow a hydraulic corrective coefficient of $k_h = 0.75$.

In the turbulent flow regime, according to Nikuradse [6], the hydraulic corrective coefficient k_h, dependent on the Reynolds number Re, can be expressed by:

$$k_h = \frac{1}{1.125 - 0.011 \times \log(\text{Re})}$$ (28.75)

for a circular cross-section, and

$$k_h = 0.79 + 0.02 \times \log(\text{Re})$$ (28.76)

for a rectangular cross-section.

In hydropower applications, the Reynolds number Re generally exceeds the value of 4000. Figure 28.50 shows the dependence of the k_h factor in closed conduits with circular cross-section on the Reynolds number. The Reynolds number not only changes its value as a function of the flow velocity v for a given diameter D, but is also strongly dependent on the temperature-dependent cinematic viscosity $v(T)$. Not taking into account for correct value of Re can easily lead to errors of the k_h factor and thus to the flow-rate Q on the order of 2% to 3%. Modern transit time meters with microprocessors update the k_h factor at a rate of 4 times a second and by measuring the temperature T of the fluid at the same rate. A correct k_h factor is obtained and hence a temperature and Reynolds number-compensated flow rate Q.

Multipath Integration in Circular Cross-sections.
In reality, however, the straight run requirements as defined in Figure 28.49 cannot always be kept. In addition, cross flow errors occur when nonaxial components of velocity in a pipe alter the transit times of a pulse between the sensors. Nonaxial velocities are caused by such disturbances in closed conduits as bends, asymmetric intake flows, and discontinuities in the pipe wall or pumps. It has become accepted practice to eliminate the sensitivity of an acoustic flowmeter to velocity distributions by increasing the number of acoustic paths n. Additional paths not only decrease substantially the velocity distribution error, but also reduce the numerical integration error as well as errors due to path misalignment. This suggests that for accurate measurements for which a few tenths of a percent error are significant, the cost of installing more sensors for a four-path arrangement according to Figure 28.51 can be justified. In a

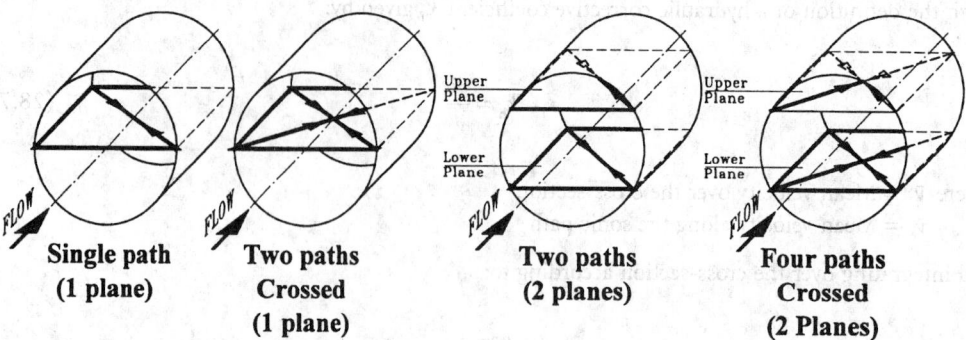

Single path	**Two paths**	**Two paths**	**Four paths**
(1 plane)	Crossed	(2 planes)	Crossed
	(1 plane)		(2 Planes)

FIGURE 28.51 Possible sonic path arrangements for transit time flowmeters in closed conduits. Using multiple paths leads to reduced straight run requirements.

multipath measurement system, according to the integration method described by IEC 41, Appendix J1 [7], the flowrate Q_{ADM} can be expressed by:

$$Q_{ADM} = k\frac{D}{2}\sum_{i=1}^{n}\bar{v}_{ai}\,W_i\,L_{wi}\,\sin\varphi_i \qquad (28.77)$$

where $\bar{v}_{ai}$ = Velocities along the acoustic path i
W_i = Corresponding weighting factor
L_{wi} = Corresponding path length
φ_i = Corresponding path angle

The method described in IEC 41 needs a very accurate transducer positioning due to the fact that the weighting factors W_i obtained by mathematical analysis are only valid when the sensors are positioned at the correct locations. Misalignment of the acoustic paths in conjunction with fixed weights can lead to considerable errors.

Measurement of Flow in Open Channels

Open-channel flow measurement is used in many applications, including water supply networks, hydrography, allocation of water for irrigation and agricultural purposes, sewage treatment plants, etc. Discharge measurements in rivers and open channels are often computed by means of a rating curve, used to convert records of water level readings into flow rates. The rating curve is developed using a set of discharge measurements and water level in the stream, and must be checked periodically to ensure that the level-discharge relationship has remained constant; many phenomena can cause the rating curve to change so that the same recorded water level produces a different discharge. This is the case of open channels under changeable hydraulic conditions due, for example, to backwater effects, gates, and where an univocal stage-discharge relationship does not exist. In this context, acoustic flowmeter application is extremely interesting and is currently experiencing wide success in water management. In fact, while different flow rate values can correspond to a given water level in relation to the hydraulic characteristics, there is always an univocal relationship between the acoustic wave propagation velocity in a flowing fluid and the flowing fluid itself.

By means of the ultrasonic technique, discharge through open channels can be determined using single- and multipath technology. In a single-path configuration, particular attention must be paid to define the vertical velocity distribution (Figure 28.52) in order to achieve a good level of accuracy by a single "line" velocity reading $\bar{v}_{az}$ at the distance z above the bottom. On the other hand, in a multipath configuration, the mean line velocity profile is well described. In this case, special attention must be paid to the integration method used to determine the flow rate from the acoustic path readings.

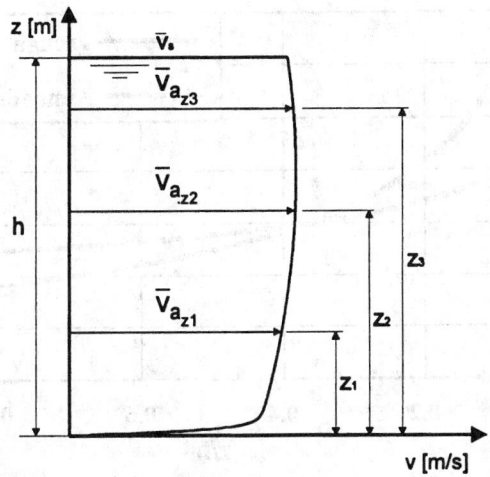

FIGURE 28.52 Open channel: possible vertical velocity distribution.

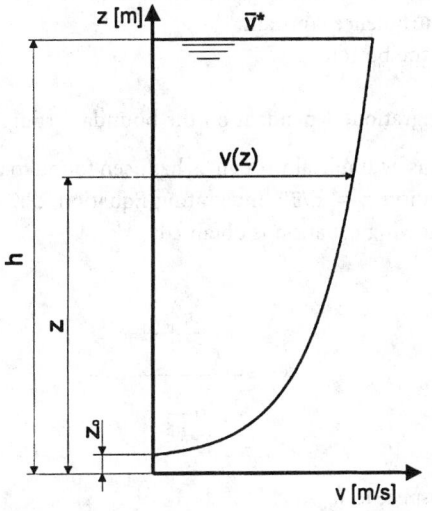

FIGURE 28.53 Logarithmic law describing open-channel velocity profile.

Single Path.
In single-path measurements, using the area-velocity method, the flow rate Q is calculated through $Q = \bar{v}A$ with $\bar{v}$ (m s^{-1}) the area-averaged flow velocity and A the cross-section. To obtain the velocity average $\bar{v}$ over the entire cross-section, the mean path velocity $\bar{v}_{az}$, measured by the acoustic flowmeter at a given depth z, must be corrected by a dimensionless hydraulic corrective coefficient k_h according to the relation $\bar{v} = k_h \bar{v}_{az}$. In general, the coefficient k_h reflects the influence of the horizontal and vertical velocity profile. It mainly depends on the water level, on the cross-section shape, and on the boundary roughness. The mean vertical velocity profile can be described by the well-known logarithmic law (Figure 28.53) given by [8].

$$\bar{v}(z) = \left(\frac{\bar{v}^*}{k} \right) \ln \frac{z}{z_0} \tag{28.78}$$

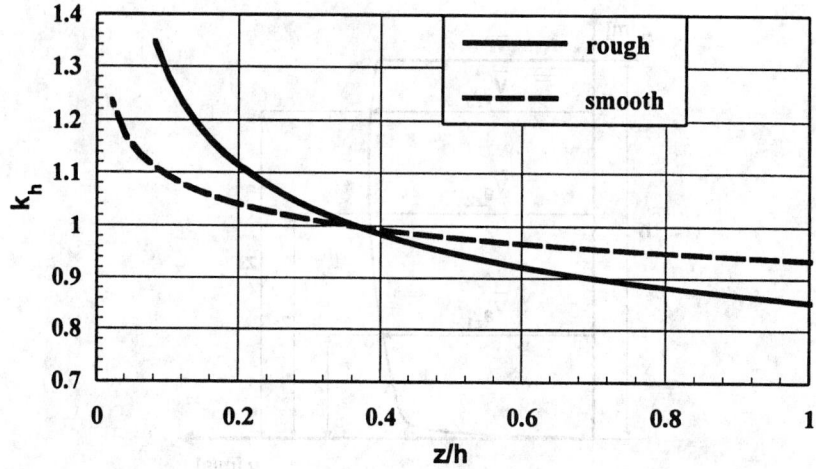

FIGURE 28.54   Logarithmic k_h factor in open channels for two different roughnesses k_s.

where $\bar{v}(z)$ = Mean flow velocity at a distance z above the bottom
  k = Von Kàrmàn's turbulence constant
  z = Distance above the bottom
  $\bar{v}^*$ = Shear velocity
  z_0 = Constant of integration, dependent on the boundary roughness

When the boundary surface is hydraulically rough, z_0 has been found to depend solely on the roughness height k_s according to the relation $z_0 = k_s/30$. Integrating Equation 28.78 over the total water height h and substituting for $\bar{v}$, the following equation is obtained:

$$k_h = \frac{\ln\left(\dfrac{h}{k_s}\right) - 1}{\ln\dfrac{z}{k_s}} \tag{28.79}$$

assuming z_0 negligible with respect to h.

In Figure 28.54, the logarithmic k_h factor is represented for two different bottom roughness. However, in many practical applications, it is often difficult to define correct values for k_s. For this reason, another k_h model has been developed on the basis of the power law [9, 10]. In this formula, the exponent $1/m$ is not constant anymore, but depends on the roughness and the hydraulic radius to take into account the influence of the channel shape. The mean vertical velocity profile $\bar{v}(z)$, according to the power law, can be expressed by the following formula (Figure 28.55):

$$\bar{v}(z) = \bar{v}_s \left(\frac{z}{h}\right)^{\frac{1}{m}} \tag{28.80}$$

where h = Current water level
  $\bar{v}$ = Mean line velocity at the free surface (maximum value)
  $1/m$ = Exponent

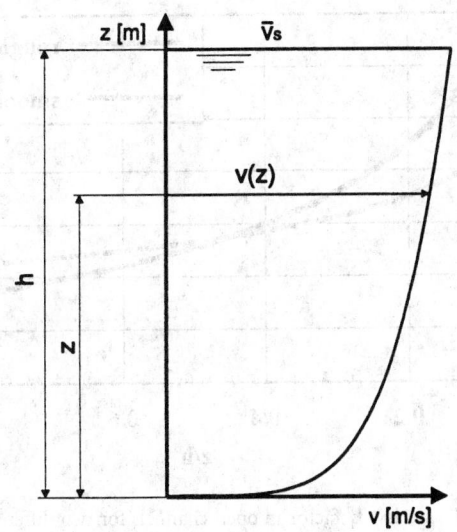

FIGURE 28.55 Power law describing open-channel velocity profile.

From the integration of Equation 28.80 over the total water depth, one obtains the following expression for the k_h factor:

$$k_h = \frac{m}{m+1}\left(\frac{h}{z}\right)^{\frac{1}{m}} \tag{28.81}$$

The value of m, depending on the roughness, can be expressed using the a dimensional friction factor f of the Darcy–Weisbach formula [8], according to:

$$m = k\sqrt{\frac{8}{f}} \tag{28.82}$$

where k = Von Kàrmàn constant, varying from 0.2 to 0.4

k depends on the suspended load (low values for high turbidity), or can be expressed using the Manning formula by the relation [11]:

$$m = \left(\frac{k}{\sqrt{g}}\right)\frac{R_h^{\frac{1}{6}}}{n} \tag{28.83}$$

where g = Gravity acceleration
    n = Manning's roughness coefficient
    R_h = Hydraulic radius

In Figure 28.56, the corrective factor is represented for smooth and rough surfaces. By means of this relation, an easier field application has been obtained due to wide familiarity with the Manning formula in open-channel flow computation because of its simplicity of form, high versatility, and satisfactory results.

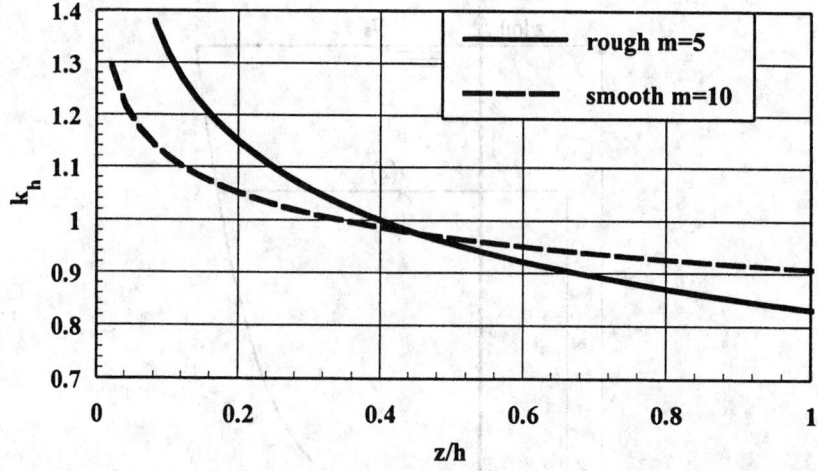

FIGURE 28.56   Power law k_h factor in open channels for two different roughnesses.

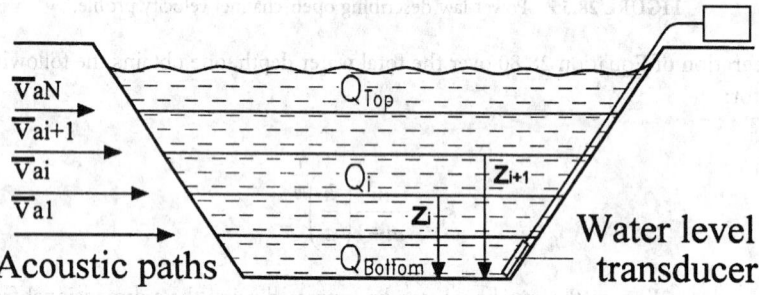

FIGURE 28.57   Multipath measurement in open channels by the mean section method. The flow velocity is measured by several levels.

Multipath.

The foregoing equations for the mean vertical velocity profile in open channels predict that the maximum mean velocity occurs at the free surface. Field and laboratory measurements, however, demonstrate that the maximum mean velocity occurs below the free surface, strongly depending on the ratio B/h, with B being the channel width. These observations show that a one-dimensional velocity distribution law cannot always completely describe flow profiles in open channels. Therefore, to reduce uncertainties in velocity profile description and to achieve high accuracies even under unfavorable hydraulic conditions, a multipath configuration must be used.

In multipath measurement, using the "mean section method," the flow velocity is measured at several levels between the free surface and the channel bottom (Figure 28.57). The total discharge Q_{ADM} is performed by the relation [12]:

$$Q_{ADM} = Q_b + Q_t + \sum_{i=1}^{n} \left(\frac{\bar{v}_{ai} + \bar{v}_{ai+1}}{2} \left[A(z_{i+1}) - A(z_i) \right] \right) \qquad (28.84)$$

where Q_b = Flow rate in the bottom section with the bottom velocity obtained from the lowest path velocity by correction for bottom friction

Q_t = Flow rate at the highest active section with velocity v_{top}, interpolated from the velocity profile

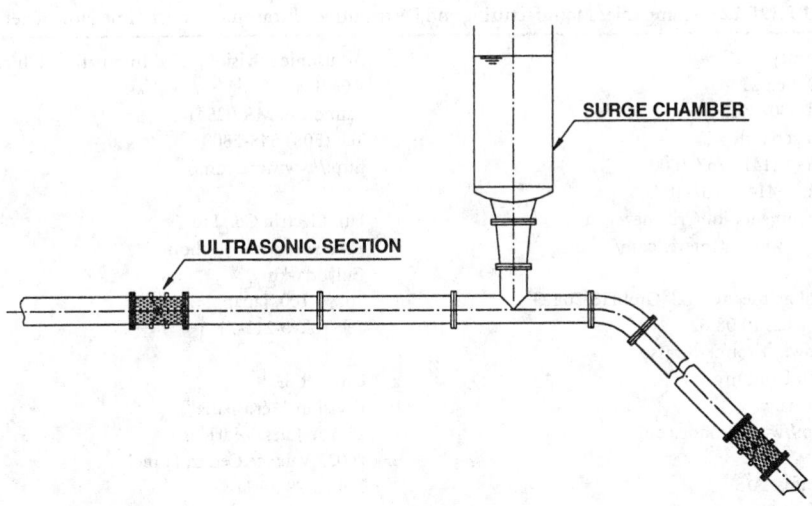

FIGURE 28.58 Penstock leak detection by transit time flowmeter. The penstock contains a surge chamber that causes mass oscillation in case of load changes of the turbine.

$\bar{v}_{ai}$ = Mean velocity along the ith acoustic path
$A(z_i)$ = Cross-section below the ith path

Modern microprocessor-controlled ultrasonic flowmeters can cope with either single-path or multi-path measurement in open channels, using the logarithmic or the power law for single-path measurement. In addition, up-to-date completely modular systems are able to use the major part of the equipment for measuring both in open channels and closed conduits.

Application: Penstock Leak Detection with Surge Chambers

Penstock leak detection is a typical acoustic flowmeter application [13]. It is used for immediate recognition of pipe rupture and leak losses in penstocks. Two flowmeters are installed at opposite ends of the penstock: one for measuring Q_{up} as near as possible to the intake, the other for measuring Q_{down} at the powerhouse entrance. In this description, the upstream flow is compared to the downstream flow and the flow difference. $\Delta Q = Q_{up} - Q_{down}$ is calculated and supervised. If the difference ΔQ exceeds a given threshold, the system enunciates alarms or valve closure contacts.

In hydropower applications another problem arises due to the presence of surge chambers (Figure 28.58). This hydraulic structure causes mass oscillations due to load changes of the turbine. Up to now, if there was a surge tank in a penstock, one had two possibilities. On the one hand, one could divide the penstock into two parts: one before the surge chamber, one after it, and protect them separately. This solution is expensive and very often impractical because the surge chamber is built into the rock and inaccessible. On the other hand, one could develop something like a "leak detection algorithm," which includes the oscillatory behavior of a penstock. Such a system, however, needs extensive field tests and the knowledge for exact settings of resonant frequency, damping factor, and thresholds. The latest technology in ultrasonic flowmeters, combined with accurate water level sensors, can offer a solution to this problem. By applying the formula:

$$\Delta Q = Q_{up} - Q_{down} - \frac{\Delta V_{chamber}}{\Delta t} \qquad (28.85)$$

changes of the volume of the surge chamber due to mass oscillations are taken into account, leading to a better and more realistic behavior of penstock leak detection in the presence of surge tanks.

TABLE 28.12   Companies Manufacturing and Distributing Ultrasonic Transit Time Flowmeters

Rittmeyer AG	Accusonic Division, ORE International, Inc.
P.O. Box 2143	P.O. Box 709
CH-6302 Zug	Falmouth, MA 02541
Switzerland	Tel: (508) 548-5800
Tel: (+4141)-767-1000	http://www.ore.com/
Fax: (+4141)-767-1075	
instrumentation@rittmeyer.ch	Fuji Electric Co., Ltd.
http://www.rittmeyer.com/	12-1 Yurakucho 1-chome
	Chiyoda-ku
Krohne Messtechnik GmbH&Co.KG	Tokyo 100, Japan
Postfach 10 08 62	Tel: Tokyo 211-7111
Ludwig-Krohne-Strasse 5	
4100 Duisburg 1	Crouzet SA
Germany	Division "Aérospatial"
http://www.krohne.com/	25, rue Jules-Védrines
	26027 Valence Cedex, France
Danfoss A/S	Tel: 75 79 85 11
DK-6430	
Nordborg	Nusonics Inc.
Denmark	11391 E. Tecumseh St.
Tel: (+45) 74 88 22 22	Tulsa, OK 74116-1602
	Tel: (918) 438-1010
Panametrics, Inc.	
221 Crescent Street	Ultraflux
Waltham, MA 02254	le technoparc
Tel: (617) 899-2719	17, rue Charles Edouard Jeanneret
http://www.panametrics.com/	78306 Poissy Cedex, France
	Tel: 33(1)39 79 26 40

Instrumentation and Manufacturers/Distributors

Table 28.12 gives an overview of some companies manufacturing and distributing transit time flowmeters. Prices of transit time flowmeters have a very wide range due to the wide versatility of different applications and are therefore difficult to list accurately. Generally, ultrasonic flowmeters are seldom sold as a "prepacked" instrument. For this reason, the price for a metering section, including installation, varies from a few $1000 to nearly $100,000.

References

1. M. Barmettler and P. Gruber, Anwendung von Oversampling-Verfahren zur Erhöhung der Auflösung digital erfasster Signale, *Technisches Messen*, Oldenbourg Verlag, 1992.
2. D. Hoppe, Kombinierte Zählung und Abstandsbestimmung von Impulssignalen, *Technisches Messen*, Oldenbourg Verlag, 10, 1991.
3. R. Krimholtz, D. Leedom, and G. Matthaei, New equivalent circuits for elementary piezoelectric transducers, *Electronics Lett.*, 6, 398, 1970.
4. P. D. Edmonds, *Methods of Experimental Physics, Ultrasonics*, New York: Academic Press, 1981.
5. D. W. Spitzer, *Flow Measurement, Practical Guides for Measurement and Control*, Research Triangle Park, NC: Instrument Society of America, 1991.
6. J. Nikuradse, Gesetzmässigkeiten der turbulenten Strömung in glatten Rohren, *VDI Verlag GmbH*, 1932.
7. F. L. Brand, Akustische Verfahren zur Durchflussmessung, *Messen, Prüfen Automatisieren*, April 1987.
8. *International Standard IEC 41*, 3rd ed., 1991.
9. R. H. French, *Open Channel Hydraulics*, New York: McGraw-Hill, 1985, 30.

10. Chen-Iung Chen, *J. Hydraulic Eng.*, 379, 117, 1990.
11. M. F. Karim and J. F. Kennedy, *J. Hydraulic Eng.*, 162, 113, 1987.
12. G. Grego, M. Baldin, et al., *Application of an Acoustic Flowmeter for Discharge Measurement in the Po.*
13. G. Grego and M. Baldin, *Energia Elettrica*, 1, 52, 72, 1995.
14. H. P. Vaterlaus and H. Gabler, A new intelligent ultrasonic flowmeter for hydropower applications, *Int. Water Power & Dam Construction*, 1994.

28.8 Vortex Shedding Flowmeters

Wade M. Mattar and James H. Vignos

The *vortex shedding flowmeter* first emerged 25 to 30 years ago and has steadily grown in acceptance since then to be a major flow measurement technique. Its appeal is due, in part, to the fact that it has no moving parts yet produces a frequency output that varies linearly with flow rate over a wide range of Reynolds numbers. The vortex meter has a very simple construction, provides accuracy (1% or better) comparable to higher priced and/or more maintenance-intensive techniques, and works equally well on liquids and gases. In addition, it is powered primarily by the fluid and lends itself more readily than other linear flow devices to two-wire operation. Comparing the vortex shedding flowmeter to an orifice plate, the former has higher accuracy and rangeability, does not require complex pressure impulse lines, is less sensitive to wear and, for volumetric flow measurement, does not require the need to compensate for fluid density.

Industrial vortex shedding flowmeters are normally available in pipe sizes ranging from 15 mm to 300 mm (1/2 in. to 12 in.), with some manufacturers offering sizes up to 400 mm (16 in.). Flow ranges covered depend on fluid properties and meter design. Typical ranges for a 15 mm meter are:

- Water at 21°C (70°F); 0.06 to 2.2 L s⁻¹ (1 to 35 gallons per minute)
- Air at 16°C (60°F) and 101 kPa (14.7 psia); 1.1 to 15.7 L s⁻¹ (140 to 2000 cubic feet per hour)
- Dry saturated steam at 689 kPa (100 psig); 4.5 to 225 kg h⁻¹ (10 to 500 pounds per hour)

Typical ranges for a 300 mm (12 in.) meter are:

- Water at 21°C (70°F); 5.4 to 5400 L s⁻¹ (85 to 8500 gallons per minute)
- Air at 16°C (60°F) and 101 kPa (14.7 psia); 157 to 12500 L s⁻¹ (20,000 to 1,600,000 cubic feet per hour)
- Dry saturated steam at 689 kPa (100 psig); 1240 to 124000 kg h⁻¹ (2750 to 275,000 pounds per hour)

Temperature capability ranges from cryogenic temperatures up to 427°C (800°F). Pressure capability as high as 20.7 MPa (3000 psig) is available.

Principle of Operation

Probably the first time, ages ago, that anyone placed a blunt obstacle in a flowing fluid, he or she observed the whirlpools or vortices that naturally form and shed downstream. In everyday life, examples of vortex shedding are numerous. The undulation of a flag is due to vortex shedding from the pole, and the singing of telephone wires in a strong wind is due to shedding from the wires. Analysis by Theodore von Karman in 1911 described the stability criterion for the array of shed vortices. Consequently, when a stable array of vortices form downstream from an obstacle, it is often referred to as the von Karman vortex street (Figure 28.59).

Very early on, it was noted that, for a large class of obstacles, as the velocity increased, the number of vortices shed in a given time (or frequency of vortex shedding) increased in direct proportion to the velocity. The dimensionless Strouhal number, St, is used to describe the relationship between vortex shedding frequency and fluid velocity and is given by:

FIGURE 28.59   Von Karman vortex street.

FIGURE 28.60   Vortex shedding in a pipe.

$$St = \frac{f \times d}{U} \tag{28.86}$$

where f = Vortex shedding frequency
d = Width of shedding body
U = Fluid velocity

Alternatively,

$$U = \frac{f \times d}{St} \tag{28.87}$$

Although early studies were conducted in unconfined flow, it was later observed that vortex shedding also occurred in confined flow, such as exists in a pipe (see Figure 28.60). For this case, the average fluid velocity, $\overline{U}$, and the meter Strouhal number, St', replace the fluid velocity and Strouhal number, respectively, in Equation 28.87 to give:

$$\overline{U} = \frac{f \times d}{St'} \tag{28.88}$$

Since the cross-sectional area, A, of the pipe is fixed, it is possible to define a flowmeter K factor, K, that relates the volumetric flow rate (Q) to the vortex shedding frequency. Given that:

$$Q = A \times \overline{U} \tag{28.89}$$

From Equation 28.88, one obtains:

$$Q = \left(\frac{(A \times d)}{St'}\right) \times f \tag{28.90}$$

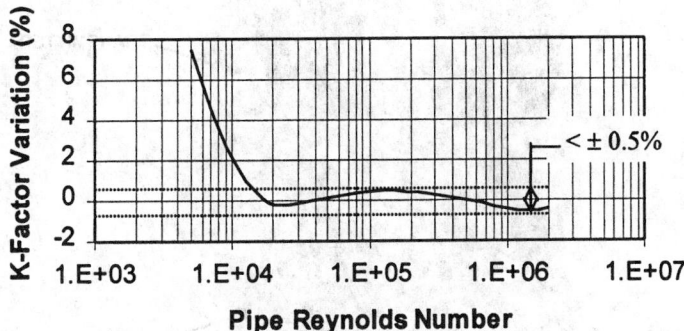

FIGURE 28.61   Typical *K* factor curve.

Defining:

$$K = \frac{\mathrm{St}'}{\left(\mathrm{A} \times d\right)} \tag{28.91}$$

results in:

$$Q = \frac{f}{K} \tag{28.92}$$

Vortex shedding frequencies range from less than 1 Hz to greater than 3000 Hz, the former being for large meters at low velocities and the latter for small meters at high velocities.

For a vortex shedding flowmeter, an obstacle is chosen that will produce a constant *K* factor over a wide range of pipe Reynolds numbers. Thus, simply counting the vortices that are shed in a given amount of time and dividing by the *K* factor will give a measurement of the total volume of fluid that has passed through the meter. A typical *K* factor vs. Reynolds number curve is shown in Figure 28.61.

The variation in *K* factor over a specified Reynolds number range is sometimes referred to as *linearity*. For the example in Figure 28.61, it can be seen that between Reynolds numbers from 15,000 to 2,000,000, the *K* factor is the most linear. This is referred to as the *linear range* of the shedder. The wider the linear range a shedder exhibits, the more suitable the device is as a flowmeter.

At Reynolds numbers below the linear range, linearization is possible but flowmeter uncertainty can increase.

Calculation of Mass Flow and Standard Volume

Although the vortex flowmeter is a volumetric flowmeter, it is often combined with additional measurements to calculate or infer mass flow or standard volume.

To determine mass flow, $\dot{M}$:

$$\dot{M} = \rho_f \times Q = \rho_f \times \frac{f}{K} \tag{28.93}$$

where ρ_f = Fluid density at flowing conditions.

It is often desirable to know what the volumetric flow rate would be at standard process conditions with respect to pressure and temperature. This is referred to as *standard volume*. In different parts of the world and for different industries, the standard temperature and pressure can be different. The fluid density at standard conditions is referred to as the *base density*, ρ_b.

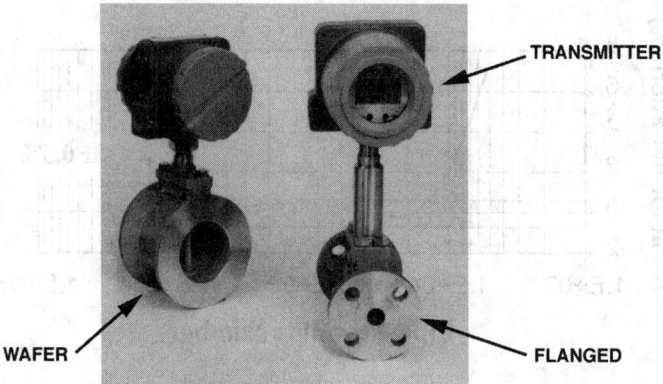

FIGURE 28.62   Vortex flowmeter construction.

FIGURE 28.63   Shedder cross-sections.

To calculate *standard volume,* Q_V:

$$Q_V = \left(\frac{\rho_f}{\rho_b}\right) \times \frac{f}{K} \qquad (28.94)$$

Flowmeter Construction

The vortex shedding flowmeter can be described as having two major components: the *flow tube* and the *transmitter*. Both are described below.

Flow Tube

The flow tube is composed of three functional parts: the flowmeter *body*, which contains the fluid and acts as a housing for the hydraulic components; the *shedder*, which generates the vortices when the fluid passes by; and the *sensor(s)*, which by some transducing means detects the vortices and produces a usable electric signal.

Flowmeter Body.
The pressure-containing portion of the vortex flowmeter, sometimes referred to as the *flowmeter body*, is available in two forms: wafer or flanged (see Figure 28.62). The wafer design, which is sandwiched between flanges of adjacent pipe, is generally lower in cost than the flanged design, but often its use is limited by piping codes.

Shedder.
The *shedder* spans the flowmeter body along the diameter and has a constant cross-section along its length. Typical cross-sections are shown in Figure 28.63.

Sensors.
When shedding is present, both the pressure and velocity fields in the vicinity of the shedder will oscillate at the vortex shedding frequency. *Pressure or velocity sensors* are used to transform the oscillating fields to an electric signal, current or voltage, from which the vortex shedding frequency can be extracted.

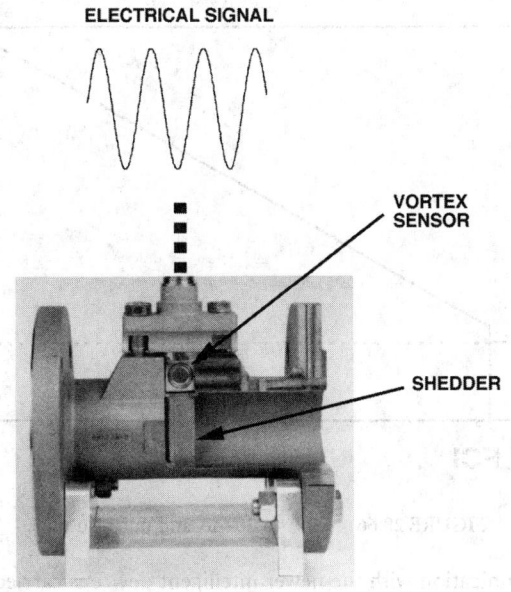

FIGURE 28.64 Example of vortex sensor.

FIGURE 28.65 Isolation manifold for sensor replacement.

Figure 28.64 shows an example of a piezoelectric differential pressure sensor located in the upper portion of the shedder, which converts the oscillating differential pressure that exists between the two sides of the shedder into an electric signal. Some other sensing means utilized to detect vortex shedding are capacitive, thermal, and ultrasonic. Often times the flowmeter electronics are mounted remotely from the flow tube. This might require a local preamplifier to power the sensor or boost its signal strength.

Since the sensor is the most likely mechanical component in a vortex flowmeter to fail, most designs have a provision for replacement of the sensors. In some cases, they can be replaced without removing the flowmeter from the pipeline. An example of an isolation manifold for sensor replacement under process conditions is shown in Figure 28.65.

Transmitter

Vortex flowmeter *transmitters* can be classified into two broad groups: analog and intelligent (or smart). Communication with the analog transmitter is carried out via local manual means, including switches

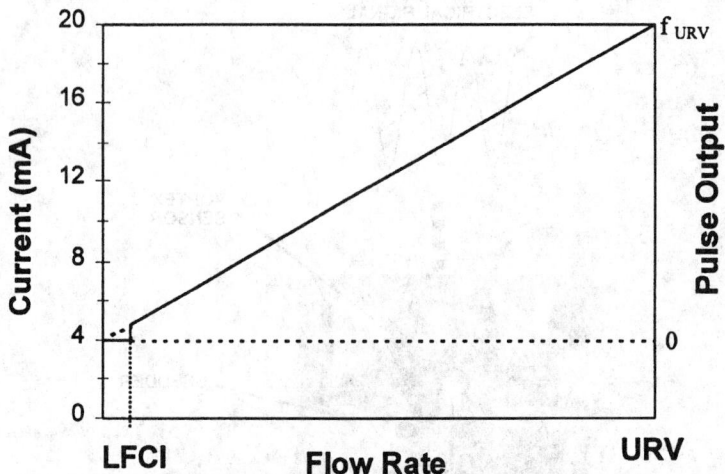

FIGURE 28.66   4 mA to 20 mA and pulse outputs.

and jumper wires. Communication with the newer intelligent device is carried out via digital electronic techniques. Both types of transmitters are two-wire devices. One of the more important features of the intelligent transmitter is that it allows application-specific information to be loaded into the transmitter. This information is then used internally to automatically tailor the transmitter to the application, including calibration of the 4 mA to 20 mA output. Before describing these devices, it is useful to consider the three most common forms of flow measurement signals provided by vortex transmitters.

Measurement Output Signals.
The three most common ways for a transmitter to communicate measurement information to the outside world are 4 mA to 20 mA, digital, and pulse signals.

The *4 mA to 20 mA signal* is the dc current flowing in the power leads to the transmitter. This current is directly proportional to the vortex shedding frequency and, hence, is also linear with flow (see Figure 28.66). 4 mA corresponds to zero flow and 20 mA to the maximum flow rate, i.e., the upper range value (URV) of the meter. A frequency-to-analog converter in the analog meter and a digital-to-analog converter in the intelligent meter produce this output.

The *digital signal* is a digitized numeric value of the measured flow rate in engineering units transmitted over the two wires powering the meter.

The *pulse signal* is a squared-up version of the raw vortex signal coming from the sensor (see Figure 28.67), and is accessible via a pair of electric terminals inside the transmitter housing. The frequency of the pulse signal is either identical to the vortex shedding frequency (raw pulse) or some multiple thereof (scaled pulse). As discussed in the Principle of Operation section, in either case, the frequency of the pulse signal is linearly proportional to flow rate, going from zero to the frequency at the URV, f_{URV} (see Figure 28.66).

As shown in Figure 28.66, at a low but nonzero flow rate, the frequency and mA signals drop to 0 Hz and 4 mA, respectively. The flow rate at which this abrupt change takes place is normally referred to as the low flow cut-in, LFCI, or cut-off. The reason for this forced zero is to avoid erroneous flow measurements at low flow, which result from process noise, including hydrodynamic fluctuations, mechanical vibration, and electrical interference. The digital signal also drops to zero below the LFCI flow rate.

Analog Transmitter.
Originally, *analog transmitters* were constructed entirely of analog electronic components. Today, they are built around a combination of analog and digital electronic components. In either case, the measurement output is in the form of a raw pulse and/or a 4 mA to 20 mA signal. Depending on the particular

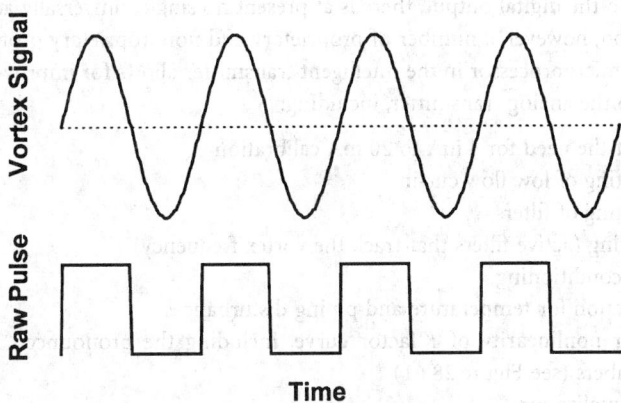

FIGURE 28.67 Raw pulse output.

transmitter, one or more of the following functions are available for tailoring the device via mechanical means to the specific application.

1. Signal output selection: If the transmitter provides both raw pulse and 4 mA to 20 mA signals, but not simultaneously, a means is available for selecting the one desired.

2. 4 mA to 20 mA calibration: Use of the this signal requires that 20 mA correspond to the desired URV. This is accomplished by inputting, via a signal or pulse generator, a periodic signal whose frequency corresponds to the upper range frequency (URF), and adjusting the output current until it reads 20 mA. The URF, which is the frequency corresponding to the vortex shedding frequency at the desired URV, is calculated using the equation URF = $K \times$ URV.

   In order to achieve the accuracy specified by the manufacturer, the K used in the above calculation must be corrected for process temperature and piping effects according to the manufacturer's instructions. The temperature effect is a result of thermal expansion of the flow tube, and is described by:

$$\Delta K(\%) = -300 \times \alpha \times (T - T_0) \qquad (28.95)$$

   where α is the thermal expansion coefficient of the flow tube material, and T_0 is the fluid temperature at which the meter was calibrated. If the shedder and meter body materials are different, α must be replaced by $(2\alpha_1 + \alpha_2)/3$, where α_1 is the thermal expansion coefficient of the meter body material and α_2 that of the shedder.

   Piping disturbances also affect the K factor because they alter the flow profile within the flow tube. This will be discussed in more detail in the section entitled "Adjacent Piping."

3. LFCI: For optimum measurement performance, the low flow cut-in should be set to fit the specific application. The goal is to set it as low as possible, while at the same time avoiding an erroneous flow measurement output.

4. Filter settings: To reduce noise present on the signal from the sensor, electronic filters are built into the transmitter. Normally, means are provided for adjusting these filters, that is, setting the frequencies at which they become active. By attenuating frequencies outside the range of the vortex shedding frequency, which varies from one application to another, better measurement performance is achieved.

Intelligent Transmitters.
Intelligent transmitters, which are microprocessor-based digital electronic devices, have measurement outputs that usually include two or more of the following: raw pulse, scaled pulse, 4 mA to 20 mA and

digital. With regard to the digital output, there is at present no single, universally accepted protocol for digital communication; however, a number of proprietary and nonproprietary protocols exist.

The presence of a microprocessor in the intelligent transmitter allows for improved functionality and features compared to the analog transmitter, including:

- Elimination of the need for 4 mA to 20 mA calibration
- Automatic setting of low flow cut-in
- Automatic setting of filters
- Adaptive filtering (active filters that track the vortex frequency)
- Digital signal conditioning
- K factor correction for temperature and piping disturbances
- Correction for nonlinearity of K factor curve, including the pronounced nonlinearity at low Reynolds numbers (see Figure 28.61)
- Integral flow totalization
- Digital measurement output in desired engineering units

Configuring — that is, tailoring the transmitter to a specific application — is carried out by one or more of the following digital communicators:

- Local configurator: a configurator, built into a transmitter, that has a display and keypad
- Hand-held terminal: a palm-size digital device programmed for configuration purposes
- PC configurator: a personal computer containing configuration software
- System configurator: a digital measurement and control system with imbedded configuration software

Using one of these configurators, the dataset of parameters that defines the configuration can be modified to fit the application in question. The details of this dataset vary, depending on the specific transmitter; however, the general categories of information listed below apply.

- Flowtube parameters (e.g., tube bore, K factor, serial no.)
- User identification parameters (e.g., tag no., location)
- Transmitter options (e.g., measurement units, function selections)
- Process fluid parameters (e.g., fluid density and viscosity, process temperature)
- Application parameters (e.g., K factor corrections, URV, LFCI level)
- Output options (e.g., measurement output modes, damping, fail-safe state)

Application Considerations

Meter Selection

From a safety viewpoint, it is essential that the vortex flowmeter satisfy the appropriate electrical safety requirements and be compatible with the process, that is, be able to withstand the temperature, pressure, and chemical nature of the process fluid. From a mechanical viewpoint, it must have the proper end connections and, if required for critical applications, have a sensor that can be replaced without shutting down the process. Meter size and measurement output signal type are also very important selection factors.

Size.
Contrary to what one might expect, the required *meter size* is not always the same as the nominal size of the piping in which it is to be installed. In some applications, selecting the size based on adjacent piping will not allow the low end of the required flow range to be measured. The appropriate criteria for selecting meter size is that the meter provides a reliable and accurate measurement over the entire required flow range. This could dictate a meter size that is less than the adjacent piping.

Pressure drop is a competing sizing criteria to that described above. This drop is given by:

$$\Delta P = C \times \rho_f \times Q^2 / D^4 \tag{28.96}$$

where C is a constant dependent on meter design, and D is the bore diameter of the flow tube. The tendency is to pick a flow tube with the same nominal diameter as the adjacent piping to eliminate the extra pressure drop introduced by a smaller-sized meter. However, in the majority of cases, this added drop is of little consequence.

The meter manufacturer can provide the needed information for making the proper selection. In some cases, sizing programs from manufacturers are available on the Internet in either an interactive or downloadable form.

Measurement Output Options.

As mentioned previously, three types of *measurement outputs* are in current use: a 4 mA to 20 mA analog signal, a pulse train, and a digital signal. Some vortex meters will provide all three of these outputs, but not always simultaneously. It is essential that the meter has the output(s) required by the application.

Meter Installation

The performance specifications of a vortex flowmeter are normally established under the following conditions: (1) the flow tube is installed in a pipeline running full with a single-phase fluid; (2) the piping adjacent to the flow tube consists of straight sections of specified schedule pipe (normally Schedule 40), typically a minimum of 30 PD (pipe diameters) in length upstream and 5 PD downstream of the flow tube with no flow-disturbing elements located within these sections; and (3) the meter is located in a vibration free and electrical interference free environment. As a consequence, certain constraints are placed on where and how the meter is installed in process piping if published performance specifications are to be achieved. These constraints are discussed below. Because meters from different suppliers differ in their sensitivity to the above influences, the statements made are of a qualitative nature. The specific manufacturer should be consulted for more quantitative information.

Location.

The flowmeter should be located in a place where vibration and electrical interference levels are low. Both of these influences can decrease the signal-to-noise ratio at the input to the transmitter. This reduction can degrade the ability of the meter to measure low flows.

The meter should not be installed in a vertical line in which the fluid is flowing down because there is a good possibility that the pipe will not be full.

Adjacent Piping.

Recommended practice is to mount the flowmeter in the process piping according to the manufacturer's stated upstream and downstream minimum straight-length piping requirements. These are typically 15 to 30 PD and 5 PD, respectively. Piping elements such as elbows or reducers upstream of the meter normally affect its K factor, but not its linearity. This allows a bias correction to be applied to the K factor. Many manufacturers provide bias factors for common upstream piping arrangements. Some who offer intelligent flowmeters make the corrections internally once the user has selected the appropriate configuration from a picklist. For piping elements and arrangements where the bias correction is not available, an *in situ* calibration should be run if the manufacturer's specified uncertainty is to be achieved. If this is not possible, calibration in a test facility with an identical configuration should be run.

The same situation as above applies if the pipe schedule adjacent to the meter differs from that under which the meter was calibrated.

To avoid disturbance to the flow, flange gaskets should never protrude into the process fluid.

The following recommendations apply if a control valve is to be situated near a vortex flowmeter. In liquid applications, the control valve should be located a minimum of 5 PD downstream of the flowmeter. This not only prevents disturbance to the flow profile in the flow tube, but also aids in preventing flashing and cavitation (see below). In gas applications, the control valve should be installed upstream of the meter, typically a minimum of 30 PD upstream of the meter to ensure an undisturbed flow profile. Having the pressure drop across the valve upstream of the meter results in a decreased density and subsequent increased velocity at the flowmeter. This helps in achieving good measurements at low flows.

For condensable gases, such as steam, it also helps to reduce the amount of condensate that might otherwise be present at the flowmeter.

Orientation.

In general, meter orientation is not an issue for vortex flowmeters, particularly for vertical pipe installations. However, for meters having electronics at the flow tube, it is recommended in high-temperature horizontal pipe applications that the flow tube be oriented with the electronics beneath the meter. Although vortex flowmeters are not recommended for multiphase applications, they do operate with somewhat degraded performance with dirty fluids (i.e., small amounts of gas bubbles in liquid, solid particles in liquid, or liquid droplets in gas). The degree of degradation in horizontal pipe applications depends to some extent on the specific meter design. Orienting the flow tube according to manufacturer's recommendations for the dirty fluid in question can help to alleviate this problem.

Pressure and Temperature Taps.

The placement of pressure and temperature taps for determining gas densities, if required, is also an important consideration. Recommendations for location of these taps vary, depending on the manufacturer. The temperature probe is inserted typically 6 PD downstream of the flow tube. This prevents any flow disturbance in the meter, and at the same time gets the probe as close to the meter as possible. The pressure tap is made typically 4 PD downstream of the meter. Although a pressure tap does not significantly affect the flow, its placement is critical for achieving an accurate density measurement.

Process Conditions.

Flashing and cavitation can occur in a liquid application just downstream of the shedder if the pressure drop across the meter results in the downstream pressure being below the vapor pressure of the liquid. These phenomena lead to undefined measurement errors and possibly to structural damage, and hence should be avoided. This is usually accomplished by increasing the inlet pressure or inserting a backpressure valve downstream of the meter. To avoid flashing and cavitation, the downstream pressure after recovery (approximately 5 PD downstream) must be equal to or greater than P_{dmin}, where:

$$P_{dmin} = c_1 \times \Delta P + c_2 \times P_{vap} \tag{28.97}$$

where P_{dmin} = Minimum absolute downstream pressure after recovery
　　　　P_{vap}　= Vapor pressure of the liquid at the flowing temperature
　　　　ΔP　 = Overall pressure drop
　　　　c_1, c_2 = Empirical constants for a specific meter (normally available from the meter manufacturer)

Pulsating flow can also, in some circumstances, lead to measurement errors. It is best to avoid placing the meter in process lines where noticeable pulsation exists.

Meter Configuration

It is important when installing an analog or intelligent vortex flowmeter that it be configured for the specific application (see section above on Flowmeter construction). This is often done by the supplier prior to shipping if the user supplies the relevant information at the time the order is placed. If this is not the case, the user must carry out the configuration procedures provided by the manufacturer.

Recent Developments

Recent efforts have been made to make the vortex flowmeter into a real-time mass flow measurement device. As was demonstrated in the "Principle of Operation" section, the output of the meter, based on the frequency of vortex shedding, is related to actual volumetric flow (see Equation 28.92). In intelligent transmitters, the flowing density (the density at flowing conditions) and the base density can be entered into the transmitter's database. Based on these values, mass flow or standard volumetric flow can be computed (see Equations 28.93 and 28.94). This procedure is valid if the flowing density does not vary in time. If this is not the case, an on-line, real-time measure of the density must be provided. Two different approaches have been used. One (multisensor) employs sensors in addition to the vortex sensor; the other (single sensor) relies on additional information being extracted from the vortex shedding signal.

Multisensor

In this method, temperature and pressure measurements are made in addition to the vortex frequency. This approach is similar to that used in orifice-d/p mass flowmetering, in which case temperature and pressure ports are located in the pipe normally downstream of the orifice plate. However, for the multisensor vortex, the temperature and pressure sensors are incorporated into the flowmeter rather than located in the adjacent piping. Using these two additional measurements, the flowing density is calculated from the equation of state for the process fluid.

Single Sensor

This approach takes advantage of the fact that, in principle, for a force- or pressure-based vortex shedding sensor, the amplitude of the vortex shedding signal is directly proportional to the density times the square of the fluid velocity; that is:

$$\text{Signal amplitude} \propto \rho_f \times U^2 \tag{28.98}$$

The fluid velocity can be determined from the vortex frequency; that is:

$$\text{Frequency} \propto U \tag{28.99}$$

Hence,

$$\frac{\text{Signal amplitude}}{\text{Frequency}} \propto \rho_f \times U \propto \text{Mass flow} \tag{28.100}$$

This approach, in principle, is independent of the process fluid, and requires no additional sensors.

Further Information

R. W. Miller, *Flow Measurement Engineering Handbook,* 3rd ed., New York: McGraw-Hill, 1996, chap. 14.

W. C. Gotthardt, Oscillatory flowmeters, in *Practical Guides for Measurement and Control: Flow Measurement,* D. W. Spitzer (ed.), Research Triangle Park, NC: Instrument Society of America, 1991, chap. 12.

J. P. DeCarlo, *Fundamentals of Flow Measurement,* Research Triangle Park, NC: Instrument Society of America, 1984, chap. 8.

ASME MFC-6M, Measurement of Fluid Flow in Closed Conduits Using Vortex Flowmeters, American Society of Mechanical Engineeers, 1998.

28.9 Thermal Mass Flow Sensors

Nam-Trung Nguyen

This chapter section deals with thermal mass flowmeters. The obvious question that arises is, what is actually meant by mass flow and thermal mass flowmeter? To answer this question, one should first understand what is the flow and what is the physical quantity measured by the meter. The flow can be understood here by the motion of a continuum (fluid) in a closed structure (channel, orifice), and is the

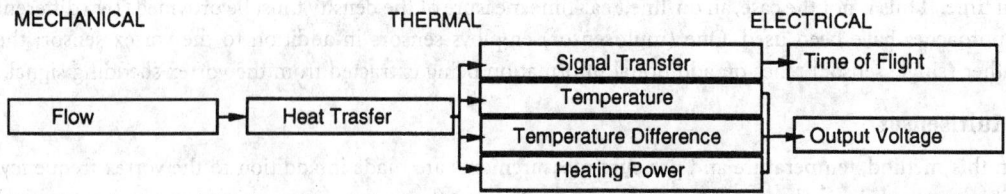

FIGURE 28.68   The three signal domains and the signal transfer process of a thermal flow sensor.

measured object. The associated physical quantity measured by the meter is the mass flux that flows through a unit cross-section. The equation for the volume flow rate Φ_v is given by:

$$\Phi_v = dV/dt = vA \qquad (28.101)$$

where V = Volume through in the time t
   v = Average velocity over the cross-section area A of the channel

With the relation between volume V, mass M, and the density of fluid ρ:

$$V = M/\rho \qquad (28.102)$$

the mass flow rate Φ_m can be derived by using Equations 28.101 and 28.102 to obtain:

$$\Phi_m = \left(\Phi_v \rho\right) + \left(V\frac{d\rho}{dt}\right) \qquad (28.103)$$

For time invariable fluid density, one obtains:

$$\Phi_m = \Phi_v \rho = Av\rho \qquad (28.104)$$

A thermal mass flow sensor will generally output a signal related to the mass flux:

$$\phi_m = \Phi_m / A = v\rho \qquad (28.105)$$

and convert the mechanical variable (mass flow) via a thermal variable (heat transfer) into an electrical signal (current or voltage) that can be processed by, for example, a microcontroller. Figure 28.68 illustrates this working principle. The working range for any mass flux sensor is somewhat dependent on the fluid properties, such as thermal conductivity, specific heat, and density, but not on the physical state (gas or liquid) of the fluid.

Principles of Conventional Thermal Mass Flowmeters

With two heater control modes and two evaluation modes, there are six operational modes shown in Table 28.13 and three types of thermal mass flowmeters:

- Thermal mass flowmeters that measure the effect of the flowing fluid on a hot body (increase of heating power with constant heater temperature, decrease of heater temperature with constant heating power). They are usually called hot-wire, hot-film sensors, or hot-element sensors.
- Thermal mass flowmeters that measure the displacement of temperature profile around the heater, which is modulated by the fluid flow. These are called calorimetric sensors.
- Thermal mass flowmeters that measure the passage time of a heat pulse over a known distance. They are usually called time-of-flight sensors.

TABLE 28.13 Operational Modes of Thermal Mass Flow Sensors

Heater controls	Constant heating power		Constant heater temperature	
Evaluation	Heater temperature	Temperature difference	Heating power	Temperature difference
Operational	Hot-wire and	Calorimetric type	Hot-wire and	Calorimetric type
Modes	hot-film type		hot-film type	
	Time-of-flight type		Time-of-flight type	

TABLE 28.14 Typical Arrangements of Flow Channel

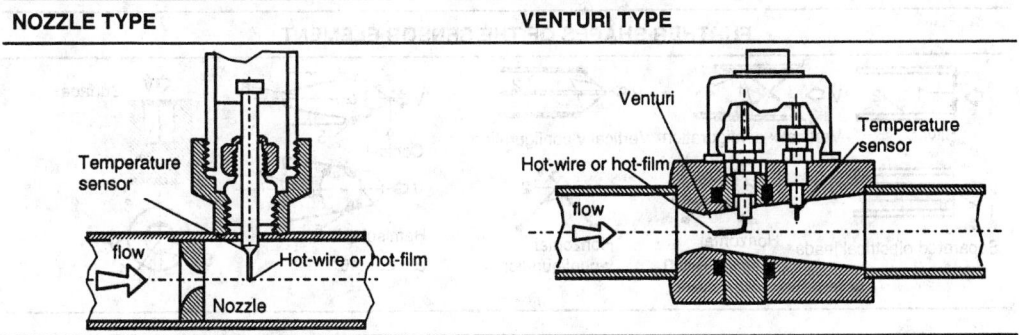

Hot-Wire and Hot-Film Sensors

The dependence of the heat loss between a fine wire as well as a thin film and the surrounding fluid has traditionally been the most accepted method for measuring a fluid flow: the hot-wire method. The hot-film method uses film sensors for detecting the flow. The basic elements of this sensor type are discussed.

Flow channel: In contrast to the thermal anemometer described in Chapter 29 for point measurement, thermal mass flow sensors of the hot-wire and hot-film type have a flow channel defining the mass flow. Table 28.14 shows two typical arrangements.

Sensor element: Hot-wire sensors are fabricated from platinum, platinum-coated tungsten, or a platinum–iridium alloy. Since the wire sensor is extremely fragile, hot-wire sensors are usually used only for clean air or gas applications. On the other hand, hot-film sensors are extremely rugged; therefore, they can be used in both liquid and contaminated-gas environments. In the hot-film sensor, the high-purity platinum film is bonded to the rod. The thin film is protected by a thin coating of alumina if the sensor will be used in a gas, or of quartz if the sensor will be used in a liquid. The alumina coatings have a high abrasion resistance and high thermal conductivity. Quartz coatings are less porous and can be used in heavier layers for electrical insulation. Typical hot-wire and hot-film sensors are shown in Table 28.15.

The sensor element, whether it is a wire or a film, should be a resistor that has a resistance with a high temperature coefficient α. For most sensor materials, the temperature dependence can simply be expressed by a first-order function:

$$R = R_r \left[1 + \alpha \left(T - T_r \right) \right] \qquad (28.106)$$

where R = Resistance at operating temperature T
R_r = Resistance at reference temperature T_r
α = Temperature coefficient

For research applications, cylindrical sensors are most common, either a fine wire (typical diameters from 1 to 15 μm) or a cylindrical film (typical diameters from 25 to 150 μm). Industrial sensors are often a resistance wire wrapped around a ceramic substrate that has typical diameters from 0.02 mm to 2 mm. Table 28.16 shows some typical parameters of industrial hot-wire and hot-film sensors.

TABLE 28.15 Typical Hot-Wire and Hot-Film Sensors (**1**, hot-wire; **2**, sensor supports; **3**, electric leads; **4**, hot-film; **5**, contact caps; **6**, quartz rod)

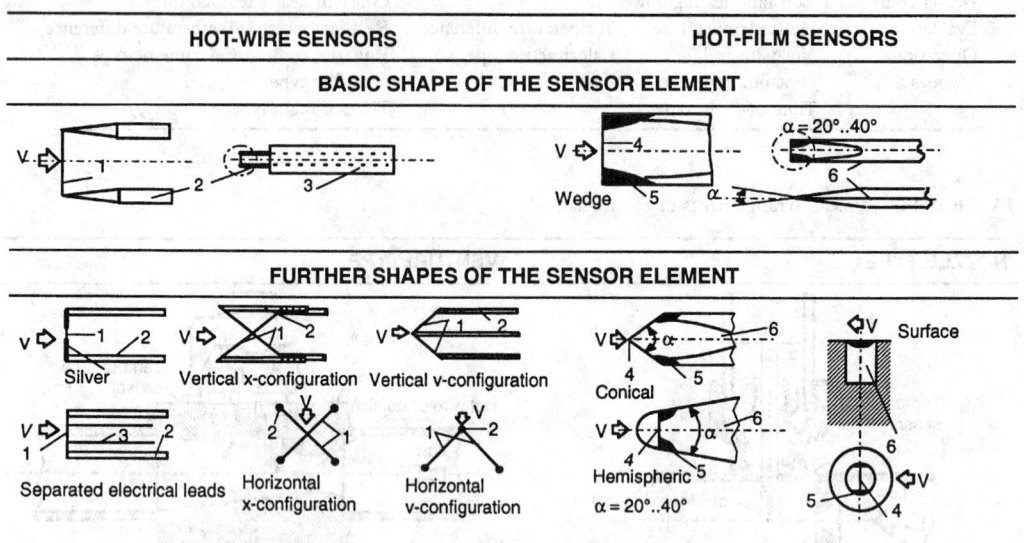

TABLE 28.16 Typical Parameters of Hot-Wire and Hot-Film Sensors [3]

Parameter	Hot-wire	Hot-film
Sensor element	Platinum hot-wire (diameter 70 μm)	Platinum hot-film (alumina coated)
Operational mode	Constant heater temperature in air	
Working temperature range	−30 to 200°C	
Characteristics	Nonlinear	
Accuracy in %	±4	±2
Time response in ms	<5	12
Sensitivity in mV kg^{-1} h^{-1}	1	5

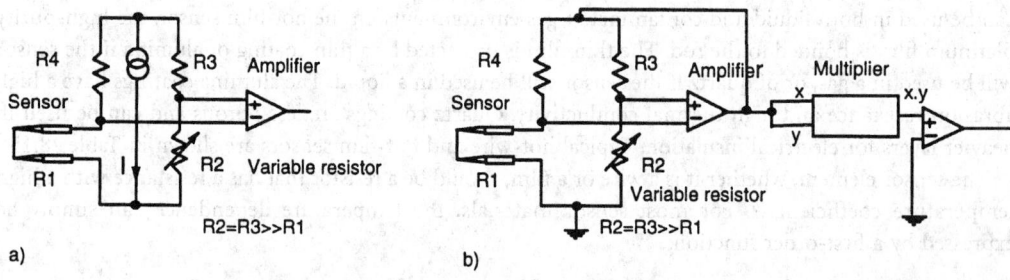

FIGURE 28.69 Control and evaluation circuit of heat-wire and heat-film sensors: (*a*) constant-current bridge; (*b*) constant-temperature bridge.

Control and evaluation circuit: The constant-current and constant-temperature bridge are conventional circuits for control and evaluation of heat-wire or heat-film sensors.

A constant-current Wheatstone bridge with a hot-wire sensor is shown schematically in Figure 28.69(*a*). In this circuit, resistors $R3$ and $R4$ are much larger than sensor resistor $R1$. Therefore, current through $R1$ is essentially independent of changes in the sensor resistor $R1$. Any flow in the channel cools the hot wire, decreases its resistance as given by Equation 28.106, and unbalances the bridge. The

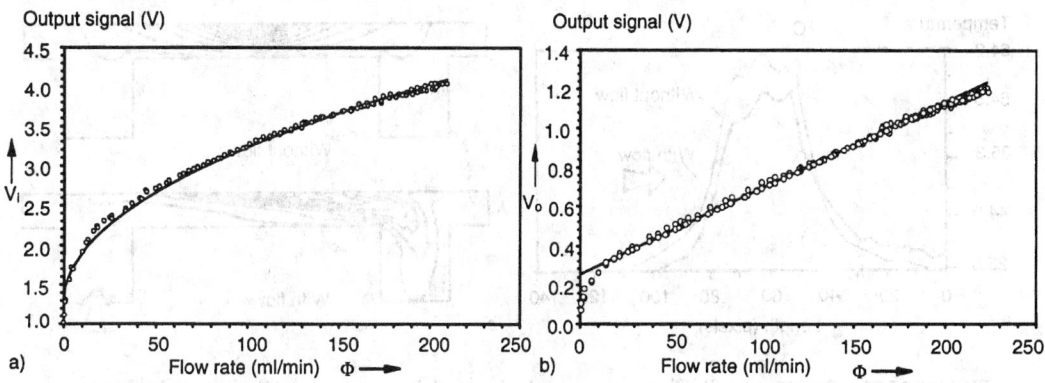

FIGURE 28.70   Sensor characteristics of the constant-temperature mode before (*a*) and after (*b*) linearization.

TABLE 28.17   Typical Calorimetric Sensors

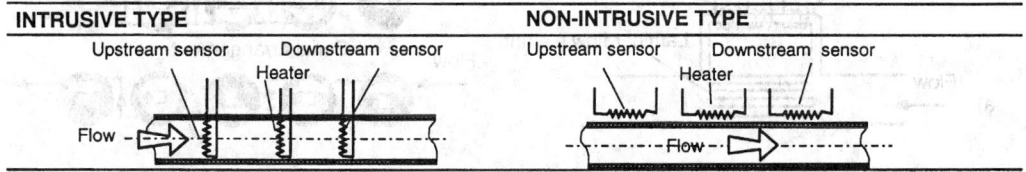

unbalanced bridge produces an output voltage V_o, which is related to the mass flow. Because the output voltage V_o from the bridge is small, it must be amplified before it is recorded. The value of the thermal coefficient α for $R1$ and $R2$ should be equal in order to eliminate signal errors due to changes in ambient temperature. Similarly, thermal coefficients for $R3$ and $R4$ should also be equal.

The constant-temperature Wheatstone bridge is shown in Figure 28.69(*b*). The bridge is balanced under no-flow conditions with the variable resistor $R2$. The flow cools the hot wire, and its resistance decreases and unbalances the bridge. A differential amplifier balances the bridge with the feedback voltage. The output signal can be linearized before recording.

Because the output signal has a square-root-like characteristic, the linearizer can be realized easily using a multiplier (i.e., AD534 of Analog Devices) with two equal input signals (a squarer). Figure 28.70 illustrates the results. With this method, there is linearization error in the low flow range.

Calorimetric Sensors

The displacement of the temperature profile caused by the fluid flow around a heating element can be used for measuring very small mass flow. Depending on the location of the heating and sensing elements, there are two types of calorimetric sensors: the intrusive sensors that lie in the fluid, and the nonintrusive sensors that are located outside the flow. Table 28.17 illustrates the two typical calorimetric sensors.

The intrusive type has many limitations. The heater and the temperature sensors must protrude into the fluid. Therefore, corrosion and erosion damage these elements easily. Furthermore, the integrity of the piping is sacrificed by the protrusions into the flow, thus increasing the danger of leakage.

In the nonintrusive sensor type, the heater and the temperature sensors essentially surround the flow by being located on the outside of the tube that contains the flow. The major advantage of this sensor type is the fact that no sensor is exposed to the flowing fluid, which can be very corrosive. This technique is generally applied to flows in the range of 1 mL min⁻¹ to 500 L min⁻¹. The larger flows are measured using the bypass arrangement. Figure 28.71(*a*) shows the measured shift of the temperature distribution around the heater. The asymmetricity of the temperature profile increases with flow. The measurement

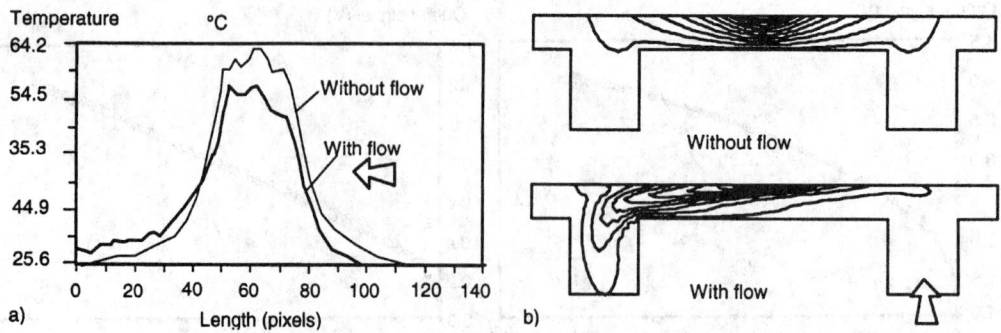

FIGURE 28.71 Temperature distribution around the heater: (*a*) measurement (*b*) numerical simulation.

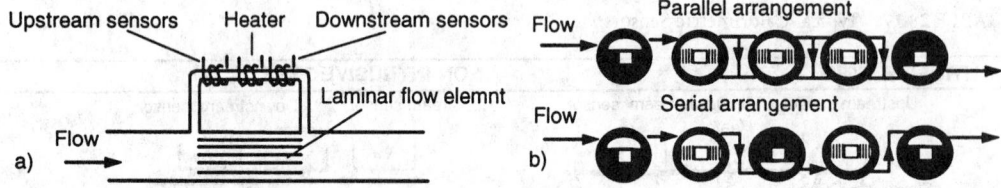

FIGURE 28.72 Bypass arrangement for large flow range: (*a*) principle (*b*) a solution for the laminar flow element [4].

TABLE 28.18 Typical Parameters of Calorimetric Sensors [4]

Parameter	Gases	Liquids
Working temperature	0°C to 70°C	0°C to 70°C
Accuracy in %	±1%	±1%
Linearity	±0.2%	±0.2%
Flow range 1:50	min. 5 mL min⁻¹	min. 5 g h⁻¹
	max. 100 L min⁻¹	max. 1000 g h⁻¹

was carried out using a thermography system [10]. To understand the working principle, the effect of fluid mechanics and heat transfer should be reviewed. The mathematical theory for this problem is discussed later in this chapter. Figure 28.71(*b*) illustrates the influence of the flow over the temperature distribution.

Because the calorimetric mass flow sensors are sensitive in low-flow ranges, bypass designs have been introduced in order to make the sensors suitable for the measurement of larger flow ranges. The sensor element is a small capillary tube (usually less than 3 mm in diameter). They ensure laminar flow over the full measurement range. The laminar flow elements are located parallel to the sensor element as a bypass (Figure 28.72(*a*)). They are usually a small tube bundle, a stack of disks with etched capillary channels [4] (Figure 28.72(*b*)), or a machined annular channel.

Table 28.18 shows typical parameters of calorimetric flow sensors. Compared to the hot-wire or hot-film sensors, this sensor type has good linearity and is only limited by signal noise at low flows and saturation at high flows. While the linear range may exceed a 100:1 ratio, the measurable range may be as large as 10,000:1 [21]. The small size of the capillary sensor tube is advantageous in minimizing the electric power requirement and also in increasing the time of response. Because of the small size of the tube, it necessitates the use of upstream filters to protect against plugging of dust particles. With the bypass arrangement, a relatively wide flow range is possible.

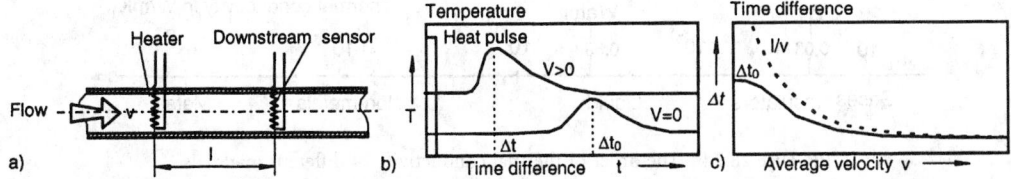

FIGURE 28.73 Time-of-flight sensors: (*a*) principle; (*b*) temperature at downstream sensor; (*c*) sensor characteristic.

Time-of-Flight Sensors

The time-of-flight sensor consists of a heater and one or more temperature sensors downstream, Figure 28.73(*a*). The heater is activated by a current pulse. The transport of the generated heat is a combination of diffusion and forced convection. The resulting temperature field can be detected by temperature sensors located downstream. The detected temperature output signal of the temperature sensor is a function of time and flow velocity. The sensor output is the time difference between the starting point of the generated heat pulse and the point in time at which a maximum temperature at the downstream sensor is reached, Figure 28.73(*b*). At the relatively low flow rates, the time difference depends mainly on the diffusivity of the fluid medium. At relatively high flow rates, the time difference tends to relate to the ratio of the heater–sensor distance and the average flow velocity [5].

Because of the arrangement shown in Figure 28.73(*a*), the time-of-flight sensors have the same limitations as the intrusive type of calorimetric sensors: corrosion, erosion, and leakage. Since the signal processing needs a while to measure the time difference, this sensor type is not suitable for dynamic measurement. The advantage of this type of volumetric flow sensor is the independence of fluid properties as well as fluid temperature in the higher flow range. The influence of fluid properties on the mass flow sensor output is described in [21], as well as an approach to compensate for changes in these properties, which is valid for both hot-element and calorimetric sensors.

Mass and Heat Transfer

The most important signal of the transfer process shown in Figure 28.68 is the thermal signal. There are different kinds of thermal signals: temperature, heat, heat capacity, and thermal resistance. In the following, the transfer of heat and the interaction between heat and temperature will be explained by three mean heat transfer processes: conduction, convection, and radiation. The first two processes can be described by the general equation of a transfer process. The transfer variables in the equation can be the momentum (momentum equation), the temperature (energy equation), or the mass (mass equation).

A transfer process consists of four elements: accumulation, conduction, induction, and convection. The *accumulation process* describes the time dependence of the transfer variable. The *conduction* presents the molecular transfer. The *convection* is the result of the interaction between the flow field and the field of the transfer variable. The *induction* describes the influence of external fields and sources.

Conduction

When there is a temperature gradient in a substance, the heat will flow from the hotter to the colder region, and this heat flow q (in W m^{-2})will be directly proportional to the value of the temperature gradient:

$$q = -\lambda \frac{dT}{dx} \qquad (28.107)$$

where T is the temperature. The above expression is called the Fourier's law of heat conduction and defines the material constant λ (in W K^{-1} m^{-1}), the thermal conductivity. Figure 28.74 shows the order of the thermal conductivity λ of different materials.

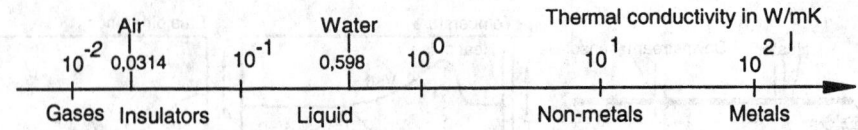

FIGURE 28.74 The order of thermal conductivity of different materials.

The differential form of the heat-conduction equation is a special case of the energy equation (see next subsection on convection). The transfer equation only consists of the accumulative, conductive, and inductive terms:

$$\frac{\partial T}{\partial t} = \frac{\lambda}{\rho c}\left(\frac{\partial^2 T}{\partial x^2} + \frac{\partial^2 T}{\partial y^2} + \frac{\partial^2 T}{\partial z^2}\right) + \frac{q'}{\rho c} \tag{28.108}$$

where ρ is the density, c is the specific heat at constant pressure and q' (in W m^{-3}) is the amount of heat (in joules) per unit of volume and time that can be generated inside the material itself, either through the action of a separate heat source, or through a change in phase of matter.

In the steady state without internal heat sources, the equation of conduction reduces to:

$$\frac{\partial^2 T}{\partial x^2} + \frac{\partial^2 T}{\partial y^2} + \frac{\partial^2 T}{\partial z^2} = 0 \tag{28.109}$$

Convection

In general, there are two kinds of convection: forced convection and natural convection. The first one is caused by a fluid flow, the other one by itself because of the temperature dependency of fluid density and the buoyancy forces. To describe convection, three conservation equations are required:

- Conservation of mass: continuity equation

$$\frac{\partial \rho}{\partial t} + \nabla(\rho v) = 0 \tag{28.110}$$

- Conservation of momentum: Navier–Stokes equation

$$\frac{\rho \partial v}{\partial t} + v\nabla v = -\nabla p + \eta\nabla^2 v + \rho g \tag{28.111}$$

- Conservation of energy: energy equation

$$\frac{\partial T}{\partial t} + v\nabla T = \left(\frac{\lambda}{\rho c}\right)\nabla^2 T + \frac{q'}{\rho c} \tag{28.112}$$

where η is the dynamic viscosity of the fluid. The temperature field and the heat power can be found by solving these three equations. For designing and understanding the thermal flow sensor, the convective heat transfer can be expressed in the simplest form:

$$Q = \varepsilon A \Delta T \tag{28.113}$$

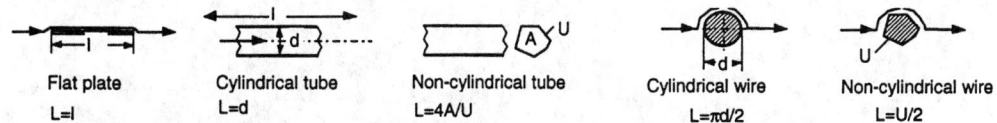

FIGURE 28.75 Definition of the characteristic length of different objects [6].

where Q (in W) = heat transfer rate (or the heat power)

ε = heat transfer coefficient between the heated surface A and the fluid

ΔT = temperature difference between the heated body and ambient

The dimensionless Nusselt number describes the heat transfer. The relationship between the heat transfer coefficient ε and the Nusselt number Nu can be expressed as follows:

$$\varepsilon = \mathrm{Nu}\frac{\lambda}{L} \tag{28.114}$$

where L is the characteristic length (the length L of a flat plate, the hydraulic diameter D_h of a tube, and the half of the perimeter of a wire, Figure 28.75). The hydraulic diameter D_h can be calculated using the wetted perimeter U and the cross-sectional area A of the tube:

$$D_h = \frac{4A}{U} \tag{28.115}$$

The relevant dimensionless number which describes the flow is the Reynolds number Re:

$$Re = \frac{vL}{\nu} \tag{28.116}$$

where v is the average flow velocity and ν the kinematic viscosity of the fluid, which is defined by the density ρ and the dynamic viscosity η:

$$\nu = \frac{\eta}{\rho} \tag{28.117}$$

Table 28.19 shows a collection of formulae for calculating the Nusselt number. The fluid properties (kinematic viscosity ν and Prandtl number Pr) should be chosen at the average temperature T_{av} between the heater temperature $T + \Delta T$ and the fluid temperature T:

$$T_{av} = T + \frac{\Delta T}{2} \tag{2.118}$$

In the case of natural convection, the Nusselt number depends on the Grashof number Gr, which describes the influence of buoyancy forces:

$$Gr = \frac{g\beta L^3 \Delta T}{\nu^2} \tag{28.119}$$

TABLE 28.19 Nusselt Number (Nu) of Forced Convection [6]

Object	Nu_{lam} for laminar regime	Nu_{turb} for turbulent regime	Average Nusselt number Nu
Flat plate	$Nu_{lam} = 0.664\sqrt{Re^3}\sqrt{Pr}$ $Re<10^5$; $0.6<Pr<2000$	$Nu_{turb} = \dfrac{0.037\ Re^{0.8}\ Pr}{1+2.443\ Re^{-0.1}\left(Pr^{2/3}-1\right)}$ $5\cdot10^5<Re<10^7$; $0.6<Pr<2000$	$Nu = \sqrt{Nu_{lam}^2 + Nu_{turb}^2}$ $10<Re<10^7$; $0.6<Pr<2000$
Cylindrical tube	$Nu_{lam} = 3.65 + \dfrac{0.19\left(Re\ Pr\ d/l\right)^{0.8}}{1+0.117\left(Re\ Pr\ d/l\right)^{0.467}}$ $Re<2300$; $0.1<(Re\ Pr\ d/l)<10^4$	$Nu_{turb} = \dfrac{\xi/8\,(Re-1000)\,Pr}{1+12.7\sqrt{\xi/8}\left(Pr^{2/3}-1\right)}\left[1+\left(\dfrac{d}{l}\right)^{2/3}\right]$ where $\xi = \left(1.28\ \log_{10}Re - 1.64\right)^{-2}$	
Short cylindrical tube $0.1<d/l<1$	$Nu_{lam} = 0.664\sqrt[3]{Pr}\sqrt{Re\,d/l}$ $Re<2300$; $0.1<(Re\ Pr\ d/l)<10^4$	Thermal entrance, fully developed flow, $2300<Re<10^4$ $Nu_{turb} = \sqrt[3]{3.66^3 + 1.61^3\ Re\ Pr\ d/l}$ Thermal entrance, developing flow, $2300<Re<10^6$	
Wire	$Nu_{lam} = 0.664\sqrt{Re^3}\sqrt{Pr}$ $10<Re<10^7$; $0.6<Pr<1000$	$Nu_{turb} = \dfrac{0.037\ Re^{0.8}\ Pr}{1+2.443\ Re^{-0.1}\left(Pr^{2/3}-1\right)}$	$Nu = 0.3 + \sqrt{Nu_{lam}^2 + Nu_{turb}^2}$

TABLE 28.20 The Average Nusselt Number of Free Convection in Some Special Cases

Cases	Equation	Ref.
Vertical flat plate or wire	$Nu = 0.55(Gr\ Pr)^{1/4}$	[7]
Horizontal flat plate	For the upper surface: $Nu = 0.76(Gr\ Pr)^{1/4}$	[8]
	For the lower surface: $Nu = 0.38(Gr\ Pr)^{1/4}$	

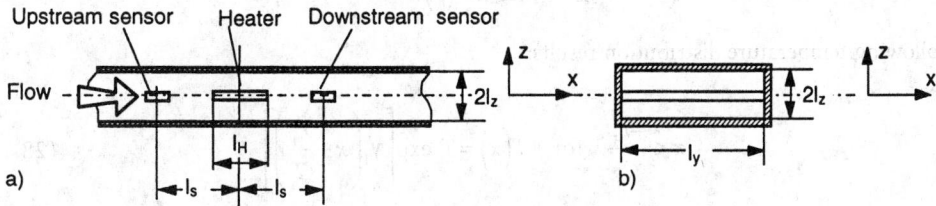

FIGURE 28.76 Analytical model for the intrusive type of calorimetric sensors: (*a*) length cut, (*b*) cross-section.

where g is the acceleration due to the gravity (9.81 m s^{-2}), β the thermal expansion coefficient of the fluid, and ΔT is the temperature difference between the hot fluid and the ambient. The average Nusselt number can be calculated for a laminar flow ($10^4 < GrPr < 10^8$) in Table 28.20.

Radiation

A body can either emit or absorb thermal radiation. Radiation is not important for the operational principle of thermal mass flow because of its relatively low magnitude.

Following, the physical and mathematical backgrounds of these three sensor types are discussed in detail. The working principle and the influence of fluid properties on the sensor signal can be determined using these mathematical models. However, the mathematical models only describe the relationship between thermal variables (heat power, temperature) and the average velocity. Further relationships between mass flow, mass flux, thermal, and electrical variables can be derived using Equations 28.101 to 28.106.

Analytical Models for Calorimetric Flowmeters

Model for the Intrusive Type of Calorimetric Sensors

In a quasi-static situation, the incoming heat at a certain point in the fluid must be equal to the outgoing heat. The heat is transported either by conduction in the fluid and/or supporting beams, or by convection through the thermal mass of the fluid. Ultimately, the heat is transported to the walls of the tube; see Figure 28.76 and Table 28.18. A heat balance equation results in a differential equation for T in x. The temperature profile in the y and z directions is assumed to be constant and linear, respectively [9].

Referring to Figure 28.76 and using A as the cross-section area of the flow channel ($A = l_y 2l_z$), ρ as the fluid density, c as the fluid heat capacity (at constant pressure), v as the average fluid velocity, and λ as the fluid thermal conductivity, one finds that:

$$\frac{\partial^2 T}{\partial x^2} - v\left(\frac{\rho c}{\lambda}\right)\frac{\partial T}{\partial x} - \left(\frac{T}{l_z^2}\right) = 0 \qquad (28.120)$$

or

$$\frac{\partial^2 T}{\partial x^2} - \left(\frac{v}{a}\right)\frac{\partial T}{\partial x} - \frac{T}{l_z^2} = 0 \tag{28.121}$$

where $a = \lambda/\rho c$ is the thermal diffusivity of the fluid. Equation 28.121 is linear in T. Solving the differential equation using a heater length l_H, a heater power Q, and the boundary condition:

$$\lim_{x \to \pm\infty} T(x) = 0 \tag{28.122}$$

the following temperature distribution results:

$$x < \frac{-l_H}{2} \quad \text{for} \quad T(x) = T_0 \exp\left[\gamma_1\left(x + \frac{l_H}{2}\right)\right] \tag{28.123}$$

$$\frac{l_H}{2} < x < \frac{l_H}{2} \quad \text{for} \quad T(x) = T_0 \tag{28.124}$$

$$x > \frac{l_H}{2} \quad \text{for} \quad T(x) = T_0 \exp\left[\gamma_2\left(x - \frac{l_H}{2}\right)\right] \tag{28.125}$$

where

$$\gamma_{1,2} = \frac{\left(v \pm \sqrt{v^2 + 2a^2/l_z^2}\right)}{(2a)} \tag{28.126}$$

$$T_0 = \frac{P}{\left[\left(\frac{2\lambda l_y l_H}{l_z}\right) + A\lambda(\gamma_1 - \gamma_2)\right]} \tag{28.127}$$

The temperature difference between the two sides, upstream (at $x = l_s$) and downstream (at $x = -l_s$) can be then calculated as:

$$\Delta T(v) = T_0\left\{\exp\left[\gamma_2\left(\frac{l_s - l_H}{2}\right)\right] - \exp\left[\gamma_1\left(\frac{-l_s + l_H}{2}\right)\right]\right\} \tag{28.128}$$

Model for the Nonintrusive Type of Calorimetric Sensors

A simple, one-dimensional model is used to show the working principle of the nonintrusive type with capillary-tube and heater wire winding around it. Geometric parameters and assumptions are given in Figure 28.77. Because of the symmetry, only half of the capillary-tube will be considered for the calculation model. The conservation of thermal energy in a lumped element (Figure 28.77(c)) can be given in the following equation:

$$Q_{\text{cond.,x,fluid}} + Q_{\text{conv.,x,fluid}} + Q_{\text{cond.,x,wall}} = Q_{\text{cond.,y,fluid}} \tag{28.129}$$

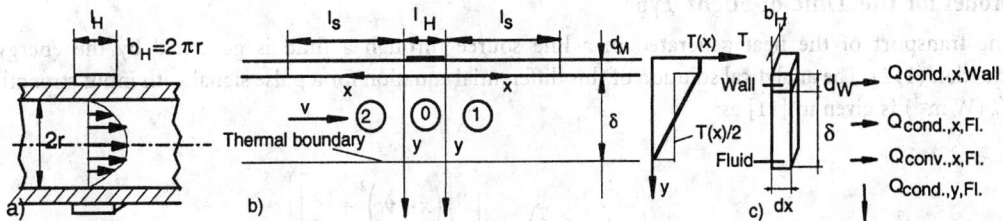

FIGURE 28.77 Analytical model for the nonintrusive type of calorimetric sensors: (*a*) heater and channel geometry, (*b*) model geometry, and (*c*) model of a lumped element.

The indices define the conduction or convection in the *x*- or *y*-axis in the fluid as well as in the heated wall. Defining the temperature along the *x*-axis as $T(x)$, the average flow velocity as v, the thermal conductivities of wall material as λ_w and of fluid as λ, the thermal diffusivity of fluid $a = \lambda/(\rho c)$, and the thickness of the average thermal boundary layer as δ one finds the heat balance equation:

$$\left[\frac{1}{2}+\left(\frac{\lambda_w d_w}{\lambda\delta}\right)\right]\frac{\partial^2 T}{\partial x^2}-\frac{v}{(2a)}\frac{\partial T}{\partial x}-\frac{T(x)}{\partial^2}=0 \tag{28.130}$$

The thickness of the average thermal boundary layer δ depends on the flow velocity [15]. For gases with a small Prandtl number (Pr < 1) or liquids with a low Reynolds number, one can assume that:

$$\delta = r \tag{28.131}$$

After solving Equation 28.130 in the local coordinate systems 1 and 2 (Figure 28.77(*b*)), one obtains the temperature difference $\Delta T(v)$ between the temperature sensors:

$$\Delta T(v)=\vartheta_0\left[\exp(\gamma_2 l_s)-\exp(\gamma_1 l_s)\right] \tag{28.132}$$

with:

$$\gamma_{1,2}=\frac{\left(v\pm\sqrt{v^2+16a^2\kappa/\delta^2}\right)}{(4a\kappa)} \tag{28.133}$$

The dimensionless factor:

$$\kappa=\frac{1}{2}+\frac{(\lambda_w d_w)}{(\lambda\delta)} \tag{28.134}$$

describes the influence of the wall on the heat balance. If the wall is neglected, we get $\kappa = 1/2$ as in the similar case of Equation 28.121. The heater temperature T_0 can be calculated for the constant heat power Q:

$$T_0=\frac{Q}{\left\{2\pi r\lambda\left[\dfrac{l_H}{\delta}+\sqrt{\dfrac{(v^2\delta^2)}{(4a^2)}+4\kappa}\right]\right\}} \tag{28.135}$$

Model for the Time-of-Flight Type

The transport of the heat generated in a line source through a fluid is governed by the energy equation (112). The analytical solution of this differential equation for a pulse signal with input strength q'_0 (W m^{-1}) is given in [11] as:

$$T(x, y, t) = \left(\frac{q'_0}{4\pi\lambda t}\right)\exp\left\{-\frac{\left[(x-vt)^2 + y^2\right]}{4at}\right\}\tag{28.136}$$

where a denotes the thermal diffusivity. By measuring the top time τ at which the signal passes the detection element ($y = 0$), in other words differential Equation 28.136 with respect to time, one can obtain the basic equation for the so-called "time-of-flight" of the heat pulse:

$$v = \frac{x}{t}\tag{28.137}$$

For Equation 28.136 to be valid, the term $4at$ must be much smaller than the heater-sensor distance x. This assumes that forced convection by the flow is dominating over the diffusive component. In other words, Equation 28.136 is true at high flow velocities. When the diffusive effect is taken into account, the time-of-flight is given by:

$$\Delta t = \tau = \frac{\left[-2a + \left(4a^2 + v^2 x^2\right)^{1/2}\right]}{v^2} \qquad v \neq 0\tag{28.138}$$

$$\tau = \frac{x^2}{4a} \qquad v = 0\tag{28.139}$$

Principles of Microflow Sensors

In research papers, the first reference to thermal mass flow sensor normally cited is that of King in 1914. Since then, microsystems technology has been developed. The development of thermal flow sensor can be realized in micron-size using the three current technologies: bulk micromachining, surface micromachining, epimicromachining and LIGA-techniques (LIGA: German description of "Lithographie, Galvanoformung, Abformung"). These fabrication techniques (except LIGA) are compatible with conventional microelectronic processing technology. Thermal flow sensors developed using these technologies will be called "microflow sensors" in this section. The operational modes are similar to the conventional thermal flow sensor. In Table 28.21, the microflow sensors are classified after their transducing principle. With these new sensors, very small flows in the nanoliter and microliter range can be measured. Table 28.22 shows some realized examples of microflow sensors.

Smart Thermal Flow Sensors and Advanced Evaluation Circuits

Conventional sensors usually have separate electronics, which causes high cost and prevents large serial production. An integrated smart thermal flow sensor is defined as a chip that contains one or more sensors, signal conditioning, A/D conversion, and a bus output [15]. Therefore, there is a need for advanced evaluation methods that convert the thermal signal directly into a frequency and duty-cycle output.

TABLE 28.21 The Transducing Principle of Microflow Sensors

Transducing principle	Realization	Application
Thermoresistive	Metal film (platinum), polycrystalline silicon, single crystalline silicon or metal alloys.	Measurement of temperature, temperature difference, and heat power
Thermoelectric	pSi-Al (bipolar-technology), polySi-Al (CMOS-technology) or pPolySi-nPolySi thermopiles	Measurement of temperature and temperature difference
Thermoelectronic	Transistors, diodes	Measurement of temperature and temperature difference
Pyroelectric	Pyroelectric materials (LiTaO$_3$) with metal or silicon resistors as heater and electrodes	Measurement of heat power
Frequency analog	SAW oscillators	Measurement of temperature

The Duty-Cycle Modulation for the Hot-Wire Sensors

The constant heater temperature can be controlled by modulation of the amplitude of the heat voltage (conventional principle) or by modulation of the duty-cycle. The heater is activated when the output signal is high. This heater controlling output goes low when the temperature level $T_0 + \Delta T$ is reached. During low output, the heating temperature decreases to the temperature level $T_0 + \Delta T$, where the output goes high again and restarts the heating cycle. The temperature level is determined by a reference resistor and a variable resistor. An increase in the flow rate increases the convective cooling of the heater, and it needs a longer time to reach the temperature level. That results in the higher output time t_H. Figure 28.78 shows the working principle of the duty-cycle modulation, where t_d is the time delay in the switching action. Defining the maximum heating power as Q_{max} (the output signal is always high), one obtains the relation:

$$Q = Q_{max}\left(\frac{t_H}{t_{total}}\right) \tag{28.140}$$

The Electrical Sigma–Delta Modulation for Calorimetric Sensors

Conventional signal conditioning circuits use an analog-to-digital converter (ADC) to get digital sensor readout, which can be regarded as an amplifier or a voltmeter. They read the information signal from the sensor, but do not interact with the sensor. In contrast, sigma–delta converters are a part of the sensor function since they act as a feedback amplifier for the sensor output. Hence, sigma–delta conversion normally results in a much more robust sensor signal than is provided by conventional ADCs. Following, the principle of sigma–delta conversion applied to calorimetric flow sensors is explained.

The transistors T_1 and T_2 represent two switchers that feed the constant current I_0 into the RC network. It is assumed in this example that the temperature on the resistor R_{s2} is higher than the temperature on R_{s1}. Therefore, the resistance of R_{s2} is larger than R_{s1}. It results in a larger time constant for the charging of C_2. With the help of the comparators and the D-flip-flop, the constant current I_0 can be switched on and off. The switching signals $f1$ and $f2$ have, in the same time period, different pulse numbers $N + S$ and N:

$$\frac{R_{s2}}{R_{s1}} = \frac{(N+S)}{N} = 1 + \frac{\Delta R}{R_{s1}} \tag{28.141}$$

Thus:

$$\frac{\Delta R}{R_{s1}} = \frac{S}{N} \tag{28.142}$$

TABLE 28.22 Examples of Microthermal Mass Flow Sensors

TRANSDUCING PRINCIPLE	HOT-WIRE OR HOT-FILM	CALORIMETRIC
Thermoresistive [13, 14]		
Thermoelectric [15,16]		
Thermoelectronic [17, 18]		
Pyroelectric [19]		
Frequency analog [20]		

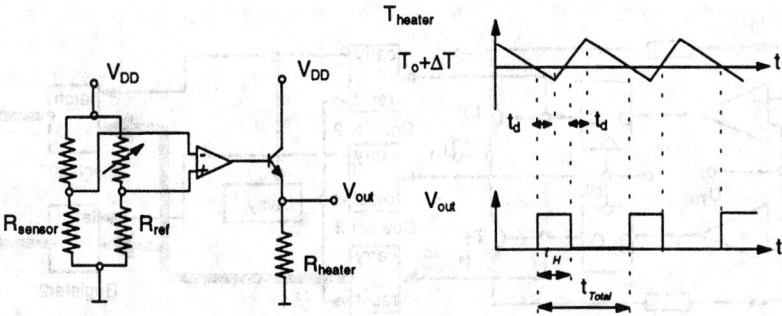

FIGURE 28.78 The duty-cycle modulation for hot wire sensors: (*a*) basic circuit; (*b*) the heater temperature detected by R_{sensor} and the output voltage V_{out} vs. the time.

with:

$$R_{s1} = \frac{\left(V_{\text{ref}} N\right)}{\left(I_0 G\right)} \tag{28.143}$$

The resistance difference as well as the temperature difference can be calculated:

$$\Delta R = \frac{\left(S V_{\text{ref}}\right)}{G I_0} \tag{28.144}$$

The counting and recording of S and G are shown in Figure 28.79.

The Thermal Sigma–Delta Modulation

The principle of thermal sigma–delta modulation is based on the conventional electrical sigma–delta: the thermal sigma-delta converter uses a thermal integrator (thermal R/C-network) instead of an electric integrator (electric R/C-network). Figure 28.80 shows the principle of thermal sigma–delta modulation. The comparator output modulates the flip-flop. The flip-flop synchronizes the heating signal by its clock. Therefore, the heating periods are chopped into further small pulses that depend on the clock frequency. The heater is actuated step-by-step until the comparator switches again, and one obtains a frequency analog output at the flip-flop [22].

Calibration Conditions

Calibration of Hot-Wire and Hot-Film Sensors

The hot-wire and hot-film sensors are based on the point velocity measurement (see Table 28.14). Therefore, the measurement results depend on the velocity distribution inside the flow channel. Achieving a high signal-to-noise ratio can thus require spatial arrays of hot-wire and hot-film sensors that give more information about the velocity field and thus more accurate results of the mass flow in channel. With the use of a nozzle, a Venturi, or a flow conditioner, the flow profile is preconditioned, which leads to an acceptable accuracy.

Temperature Dependence of Fluid Properties

Most fluid properties depend on the working temperature. The heat transfer process depends on the fluid properties. The measurement of fluid temperature (see Table 28.14) keeps the heater on a constant

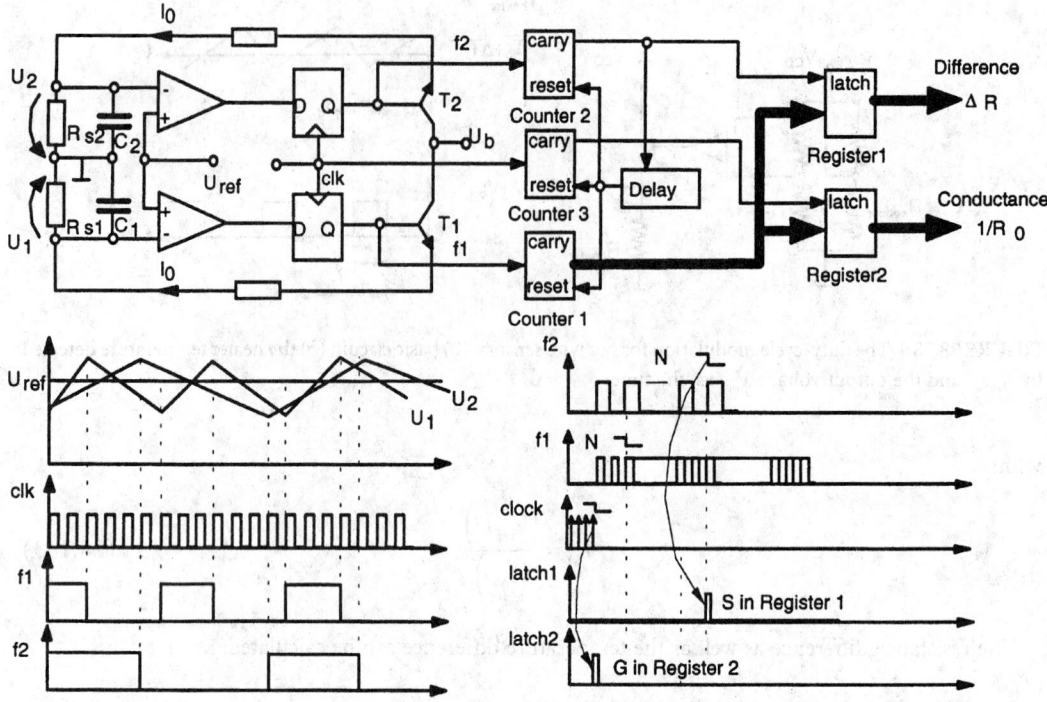

FIGURE 28.79 Sigma–delta conversion for calorimetric sensors.

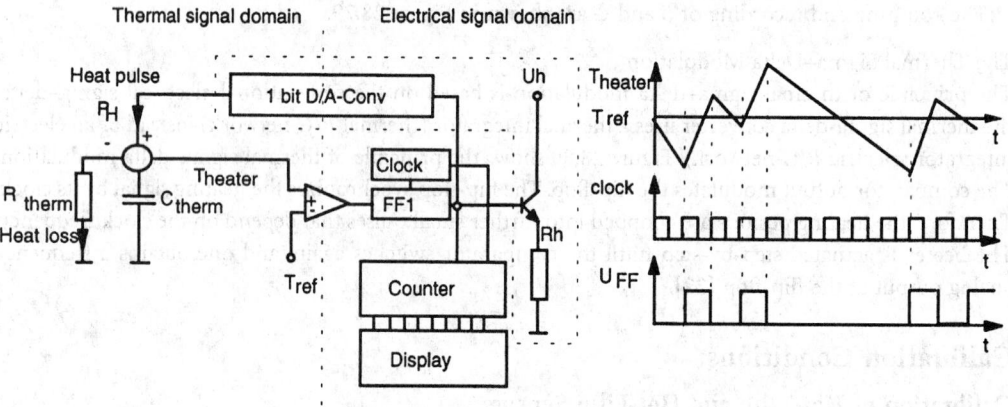

FIGURE 28.80 Principle of thermal sigma–delta modulation.

temperature difference to the fluid and can also be used for compensation of variations in temperature. For these reasons, accurate thermal mass flow sensors require both flow calibration and calibration for temperature compensation. Furthermore, the influence of temperature and/or fluid composition can be derived utilizing the relationships developed in the sections on modeling (Equations 28.120–28.135).

Instrumentation and Components

Table 28.23 lists some companies that manufacture and market thermal mass flowmeters.

TABLE 28.23 Manufactures of Thermal Mass Flow Sensors

Manufacturers	Data	Approximate price
KOBOLD Instruments Inc. 1801 Parkway View Drive Pittsburg, PA 15205	Calorimetric type: MAS-Series; air; min. range 0–10 mL min; max. range 0–40 L min. T max. 50°C; max. pressure 10 bar; accuracy ±2%	~ US$850
	Hot wire type: ANE-Series; air; range 0–20 m/s; working temperature 20–70°C	~ US$1200
HÖNTZSCH GmbH Box 1324, Robert-Bosch Str. 8 D-7050 Waiblingen, Germany	Hot wire type: range 0.05–20 m/s	
BRONKHORST HI-TEC Nijverheidstraat 1A 7261 AK Ruurlo, Netherlands	Calorimetric type: gases and liquids; min. range (gas) 0–5 mL min; max. range (liquid) 0–1000 ml/min; max. pressure 400 bar	~ US$1800
HONEYWELL, MicroSwitch, Freeport, IL	Calorimetric type: Gases; range 0–1000 ml/min; max. pressure 1.75 bar	
SIERRA INSTRUMENTS INC. 5 Harris Ct., Bldg. L Monterey, CA 93924	Calorimetric type: all gases from 1 ml/min–10,000 l/min; –40–100°C; 30 bar max.; 1% accuracy	~ US$400–1200
	Hot-wire type in stainless-steel sheath: gases from 0–100 m/s; –40–400°C; 100 bar max.; 2% accuracy	~ US$60–2000
BROOKS INSTRUMENT DIVISION Emerson Electric Co. 407 W. Vine Street Hatfield, PA 19440	Calorimetric type: all gases from 1 ml/min to 10,000 l/min; –40–100°C; 100 bar max.; 1% accuracy	~ US$400–1200

References

1. Béla G. Lipták, *Flow Measurement*, Radnor, PA: Chilton Book, 1993.
2. H. Strickert, *Hitzdraht und Hitzfilmanemometrie* (Hot-wire and hot-film anemometry), Berlin: Verlag Technik, 1974.
3. G. Schnell, Sensoren in der Automatisierungstechnik (Sensors in the automation techniques), Brauschweing: Verlag Viehweg, 1993.
4. *Mass Flow and Pressure Meters/Controllers*, The Netherlands: Bronkhorst Hi-Tec B. V., 1994.
5. T. S. J. Lammerink, F. Dijkstra, Z. Houkes, and J. van Kuijk, Intelligent gas-mixture flow sensor, *Sensors and Actuators A*, 46/47, 380-384, 1995.
6. VDI-Wärmeatlas, VDI-Verlag, 1994.
7. H. Schlichting, *Boundary Layer Theory*, 7th ed., New York: McGraw-Hill, 1979.
8. A. J. Chapman, *Heat Transfer*, 4th ed., New York: Macmillan, 1984.
9. T. S. J. Lammerink, Niels R. Tas, Miko Elwenspoek, and J. H. J. Fluitman, Micro-liquid flow sensor, *Sensors and Actuators A*, 37/38, 45-50, 1993.
10. N.T. Nguyen and W. Dötzel, A novel method for designing multi-range electrocaloric mass flow sensors: asymmetrical locating with heater- and sensor arrays, *Sensors and Actuators A*, 62, 506-512, 1997.
11. J. van Kuijk, T. S. T. Lammerink, H.-E. de Bree, M. Elwenspoek, and J. H. J. Fluitman, Multi-parameter detection in fluid flows, *Sensors and Actuators A*, 46/47, 380- 384, 1995.
12. S. M. Sze, *Semiconductor Sensors*, New York: John Wiley & Sons, 1994.
13. You-Chong Tai and Richard S. Muller, Lightly-doped polysilicon bridge as a flow meter, *Sensors and Actuators A*, 15, 63-75, 1988.
14. R. G. Johnson and R. E. Higashi, A highly sensitive silicon chip microtransducer for air flow and differential pressure sensing applications, *Sensors and Actuators A*, 11, 63-67, 1987.
15. H. J. Verhoeven and J. H. Huijsing, An integrated gas flow sensor with high sensitivity, low response time and pulse-rate output, *Sensors and Actuators A*, 41/42, 217-220, 1994.

16. F. Mayer, G. Salis, J. Funk, O. Paul, and H. Baltes, Scaling of thermal CMOS gas flow microsensors experiment and simulation, *MEMS '96*, San Diego, 1996, 116-121.

17. Göran Stemme, A CMOS integrated silicon gas-flow sensor with pulse-modulated output, *Sensors and Actuators A*, 14, 293-303, 1988.

18. Canqian Yang and Heinrik Soeberg, Monolithic flow sensor for measuring millilitre per minute liquid flow, *Sensors and Actuators A*, 33, 143-153, 1992.

19. Dun Yu, H. Y. Hsieh, and J. N. Zemel, Microchannel pyroelectric anemometer, *Sensors and Actuators A*, 39, 29-35, 1993.

20. S. G. Joshi, Flow sensors based on surface acoustic waves, *Sensors and Actuators A*, 44, 63-72, 1994.

21. U. Bonne, Fully compensated flow microsensor for electronic gas metering, *Proc. Int. Gas Research Conf.*, 16-19 Nov. '92, Orlando, FL, 3, 859, 1992.

22. J. H. Huijsing, F. R. Riedijk, and G. van der Horn, Developments in integrated smart sensors, *Proc. of Transducer 93*, Yokohama, Japan, 1993, 320-326.

28.10   Coriolis Effect Mass Flowmeters

Jesse Yoder

Coriolis flowmeters were developed in the 1980s to fill the need for a flowmeter that measures mass directly, as opposed to those that measure velocity or volume. Because they are independent of changing fluid parameters, Coriolis meters have found wide application. Many velocity and volumetric meters are affected by changes in fluid pressure, temperature, viscosity, and density. Coriolis meters, on the other hand, are virtually unaffected by these types of changes. By measuring mass directly as it passes through the meter, Coriolis meters make a highly accurate measurement that is virtually independent of changing process conditions. As a result, Coriolis meters can be used on a variety of process fluids without recalibration and without compensating for parameters specific to a particular type of fluid. Coriolis flowmeters are named after Gaspard G. Coriolis (1792–1843), a French civil engineer and physicist for whom the Coriolis force is named.

Coriolis meters have become widely used in industrial environments because they have the highest accuracy of all types of flowmeters. They measure mass directly, rather than inferentially. Coriolis meters do not have moving parts like turbine and positive displacement meters, which have parts that are subject to wear. Maintenance requirements for Coriolis meters are low, and they do not require frequent calibration. Wetted parts can be made from a variety of materials to make these meters adaptable to many types of fluids. Coriolis meters can handle corrosive fluids and fluids that contain solids or particulate matter. While these meters were used mainly for liquids when they were first introduced, they have recently become adaptable for gas applications.

Theory of Operation

Coriolis meters typically consist of one or two vibrating tubes with an inlet and an outlet. While some are U-shaped, most Coriolis meters have some type of complex geometric shape that is proprietary to the manufacturer. Fluid enters the meter in the inlet, and mass flow is determined based on the action of the fluid on the vibrating tubes. Figure 28.81 shows flow tube response to Coriolis acceleration.

Common to Coriolis meters is a central point that serves as the axis of rotation. This point is also the peak amplitude of vibration. What is distinctive about this point is that fluid behaves differently, depending on which side of the axis of rotation, or point of peak amplitude, it is on. As fluid flows toward this central point, the fluid takes on acceleration due to the vibration of the tube. As the fluid flows away from the amplitude of peak vibration, it decelerates as it moves toward the tube outlet. On the inlet side of the tube, the accelerating force of the flowing fluid causes the tube to lag behind its no-flow position. On the outlet side of the tube, the decelerating force of the flowing fluid causes the tube to lead ahead of its no-flow position. As a result of these forces, the tube takes on a twisting motion as it passes through

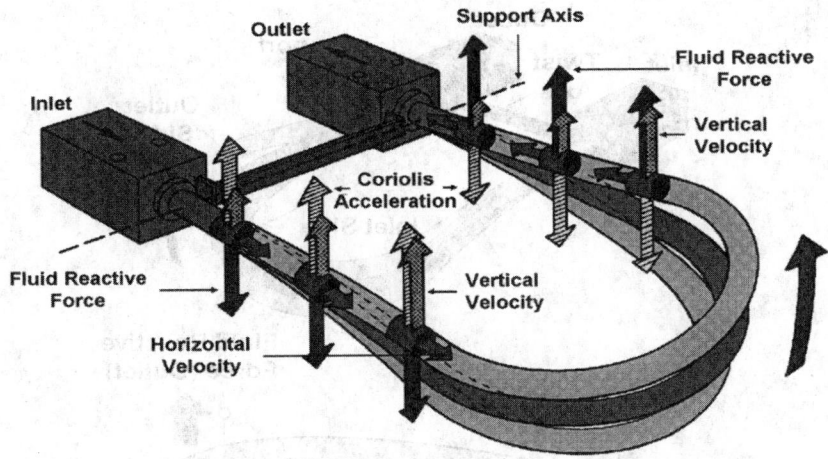

FIGURE 28.81 Flow tube response to Coriolis acceleration.

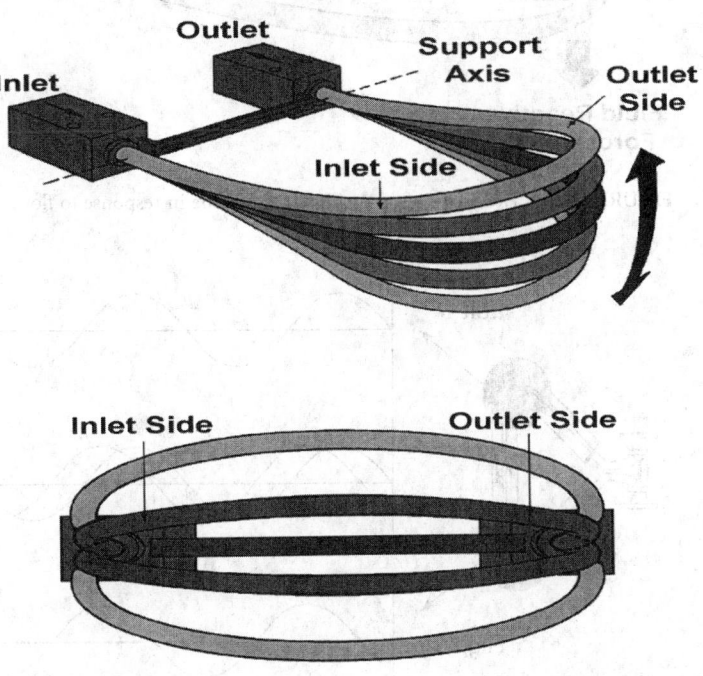

FIGURE 28.82 Two views of an oscillating flow tube with no flow.

each vibrational cycle; the amount of twist is directly proportional to the mass flow through the tube. Figure 28.82 shows the Coriolis flow tube in a no flow situation, and Figure 28.83 shows Coriolis tube response to flow.

The Coriolis tube (or tubes, for multitube devices) is vibrated through the use of electromagnetic devices. The tube has a drive assembly, and has a predictable vibratory profile in the no-flow position. As flow occurs and the tube twists in response to the flow, it departs from this predictable profile. The degree of tube twisting is sensed by the Coriolis meter's detector system. At any point on the tube, tube motion represents a sine wave. As mass flow occurs, there is a phase shift between the inlet side and the outlet side. This is shown in Figure 28.84.

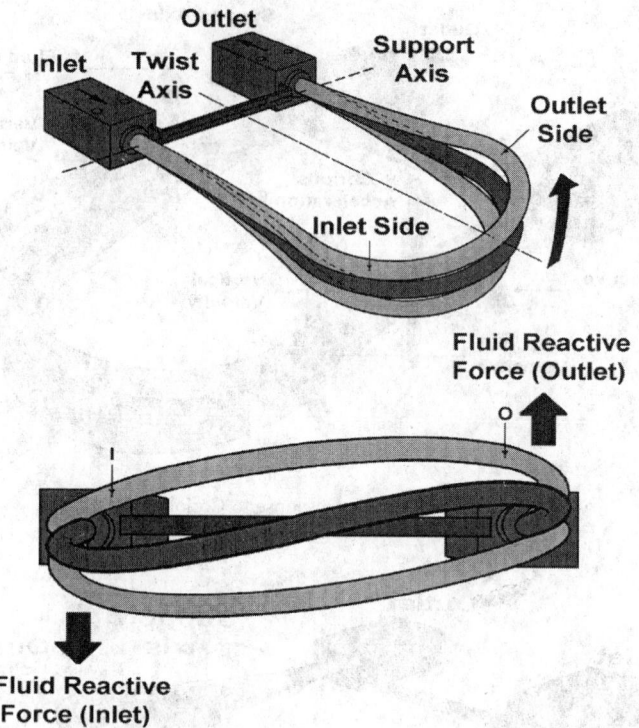

FIGURE 28.83   Two views of an oscillating flow tube in response to flow.

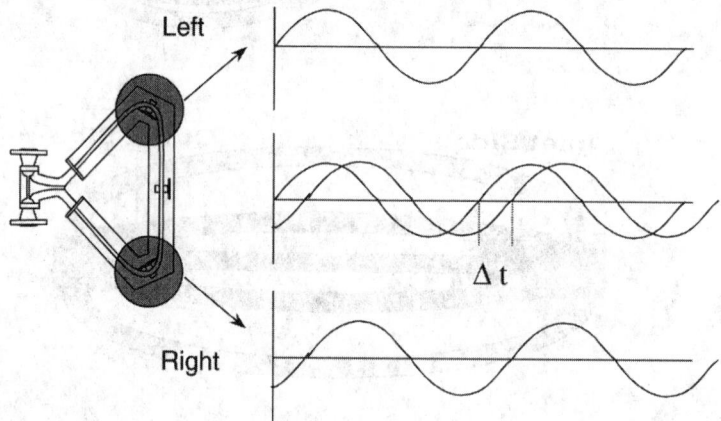

FIGURE 28.84   Phase shift between inlet side and outlet side.

The Coriolis force induced by flow is described by an equation that is equivalent to Newton's second law for rotational motion. This equation is as follows:

$$F = 2m\omega\bar{v} \tag{28.145}$$

In Equation 28.145, F is force, m is the mass to be applied to a known point at a distance L from the axis 0-0, ω is a vector representing angular motion, and $\bar{v}$ is a vector that represents average velocity.

Construction

The internal part of the Coriolis tube is the only part of the meter that is wetted. A typical material of construction is stainless steel. Other corrosion-resistant metals such as Hastelloy are used for tube construction. Some meters are lined with Teflon.

Some designs have thin-wall as well as standard tubes. A thin-wall design makes the meter more useful for gas and low-velocity liquid applications, where the amount of twist by the tube is reduced. It is important to be aware of the extent to which the fluid degrades or attacks the tube wall or lining. If the fluid eats away at the wall, this can reduce the accuracy of the meter.

Advantages

The most significant advantage of Coriolis meters is high accuracy under wide flow ranges and conditions. Because Coriolis meters measure mass flow directly, they have fewer sources of errors. Coriolis meters have a high turndown, which makes them applicable over a wide flow range. This gives them a strong advantage over orifice plate meters, which typically have low turndown. Coriolis meters are also insensitive to swirl effects, making flow conditioning unnecessary. Flow conditioners are placed upstream from some flowmeters to reduce swirl and turbulence for flowmeters whose accuracy or reliability is affected by these factors.

Coriolis meters have a low cost of ownership. Unlike turbine and positive displacement meters, Coriolis meters have no moving parts to wear down over time. The only motion is due to the vibration of the tube, and the motion of the fluid flowing inside the tube. Because Coriolis flowmeters are designed not to be affected by fluid parameters such as viscosity, pressure, temperature, and density, they do not have to be recalibrated for different fluids. Installation is simpler than installation for many other flowmeters, especially orifice plate meters, because Coriolis meters have fewer components.

Coriolis meters can measure more than one process variable. Besides mass flow, they can also measure density, temperature, and viscosity. This makes them especially valuable in process applications where information about these variables reduces costs. It also makes it unnecessary to have a separate instrument to measure these additional variables.

Disadvantages

The chief disadvantage of Coriolis meters is their initial cost. While some small meters have prices as low as $4000, the base price for most Coriolis meters is $6000 and up. The cost of Coriolis meters rises significantly as line sizes increase. The physical size of Coriolis meters increases substantially with the increase in line size, making 150 mm (6 in.) the upper line size limit on Coriolis meters today. The large size of some Coriolis meters makes them difficult to handle, and can also make installation difficult in some cases.

The lack of an established body of knowledge about Coriolis meters is a substantial disadvantage. Because Coriolis meters were recently invented, not nearly as much data are available about them as are for differential pressure-based flowmeters. This has made it difficult for Coriolis meters to gain approvals from industry associations such as the American Petroleum Institute. This will change with time, as more manufacturers enter the market and users build up a larger base of experience.

Applications

Coriolis meters have no Reynolds number constraints, and can be applied to almost any liquid or gas flowing at a sufficient mass flow to affect vibration of the flowmeter. Typical liquid applications include foods, slurries, harsh chemicals, and blending systems. The versatility of Coriolis meters in handling multiple fluids makes them very useful for plants where the flow of multiple fluid types must be measured.

There are an increasing number of gas applications for Coriolis meters. While gas applications are still very much in the minority, the use of this meter to measure gas is likely to increase as more is learned about its use for this purpose.

A Look Ahead

There are some important areas of research for Coriolis meters. While most Coriolis meters have been bent, several manufacturers have recently introduced straight-tube designs. Manufacturers will continue to fine-tune the single-tube vs. double-tube design, and to work on tube geometry. As noted above, the use of Coriolis meters for gas and also for steam applications is another area for future development.

References

E. O. Doebelin, *Measurement Systems: Application and Design*, 4th ed., New York: McGraw Hill, 1990, 603–605.

R. S. Figliola and D. E. Beasley, *Theory and Design for Mechanical Measurements*, 2nd ed., New York: John Wiley & Sons, 1995, 475–478.

K. O. Plache, Coriolis/Gyroscopic Flowmeter, *Mechanical Engineering*, MicroMotion Inc., Boulder, CO, March 1979, 36–41.

L. Smith and J. R. Ruesch, *Flow Measurement*, D. W. Spitzer (ed.), Research Triangle Park, NC: Instrument Society of America, 1996, 221-247.

28.11   Drag Force Flowmeters

Rekha Philip-Chandy, Roger Morgan, and Patricia J. Scully

In a *target flowmeter* a solid object known as a *drag element* is exposed to the flow of fluid that is to be measured. The force exerted by the fluid on the drag element is measured and converted to a value for speed of flow.

The flow-sensing element has no rotating parts, and this makes the instrument suitable for conditions where abrasion, contamination, or corrosion make more conventional instruments unsuitable. An important application of such flowmeters involves environmental monitoring in areas such as meteorology, hydrology, and maritime studies to measure speeds of air or water flow and turbulence close to the surface. In these applications, the fluid flows are sporadic and multidirectional.

A further advantage of the instrument is that it can be made to generate a measurement of flow *direction* in two dimensions, or even in three dimensions, as well as of flow speed. To implement this feature, the drag element must be symmetrical in the appropriate number of dimensions, and it is necessary to measure the force on it vectorially, again in the appropriate number of dimensions. Provided that the deflecting forces are independent in the sensing directions, the resulting outputs can be added vectorially to generate independent values for flow speed and direction. Target flowmeters using strain gage technology have been used by industry, utilities, aerospace, and research laboratories. They have been used to successfully measure the flow of uni- and bidirectional liquids (including cryogenic), gases, and steam (both saturated and superheated) for almost half a century.

Despite these advantages, the target flowmeter appears to have been neglected in favor of more complex and sophisticated devices. The authors have sought to remedy this neglect by developing a sensor suitable for measuring multidirectional flows in two dimensions, instead of measuring only bidirectional flows in a single dimension.

Design of the Flow Sensor

The sensor described in this chapter section is ideally suited for environmental flow measurement. The operation is based on strain measurement of deformation of an elastic rubber cantilever, to which a force is applied by a spherically symmetrical drag element (Figure 28.85). This sensor has many advantages, including compactness and a simple construction requiring no infrastructure other than a rigid support, and it can cope with fluids containing solid matter such as sludge and slurries provided that they do not tangle with the drag element or the rubber beam.

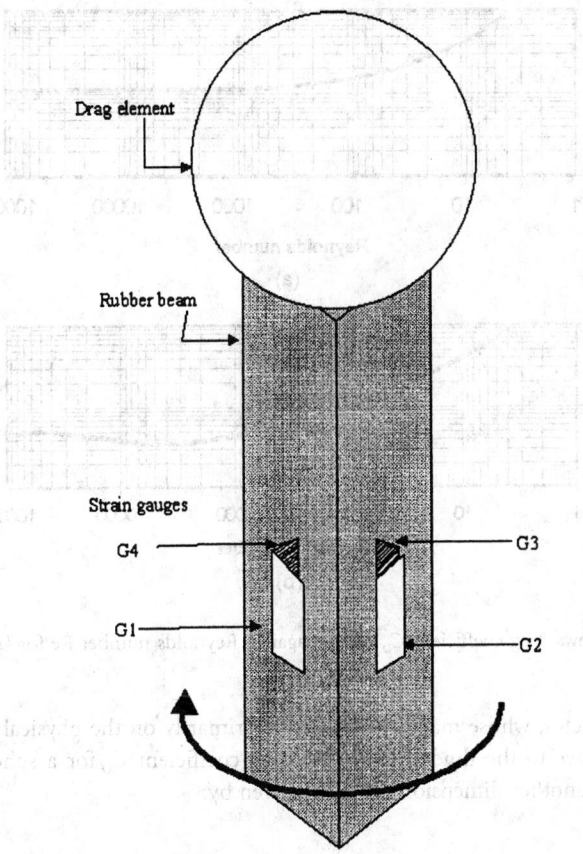

FIGURE 28.85 Schematic of the electric resistance strain gage drag force flow sensor.

According to Clarke [1], the ideal drag element is a flat disk, because this configuration gives a drag coefficient independent of flow rate. Using a spherical drag element, which departs from the ideal of a flat disk [1], the drag coefficient is likely to vary with flow speed, and therefore the gage must be calibrated and optimized for the conditions of intended use. In this discussion, a gage is developed for air flows in the range normally encountered in the natural environment.

The strain measurement can be performed with conventional strain gages, but this limits the applications of the device to conditions where corrosion of the metal-resistive track of the strain gage can be avoided. Therefore, an optical fiber strain gage has been developed as an alternative.

Principle of Fluid Flow Measurement

The drag force F_D exerted by a fluid on a solid object exposed to it is given by the *drag equation*, which from incompressible fluid dynamics is:

$$F_D = \frac{C_D \, \rho \, A V^2}{2} \tag{28.146}$$

where ρ = Fluid density
  V = Fluid's velocity at the point of measurement
  A = Projected area of the body normal to the flow
  C_D = Overall drag coefficient

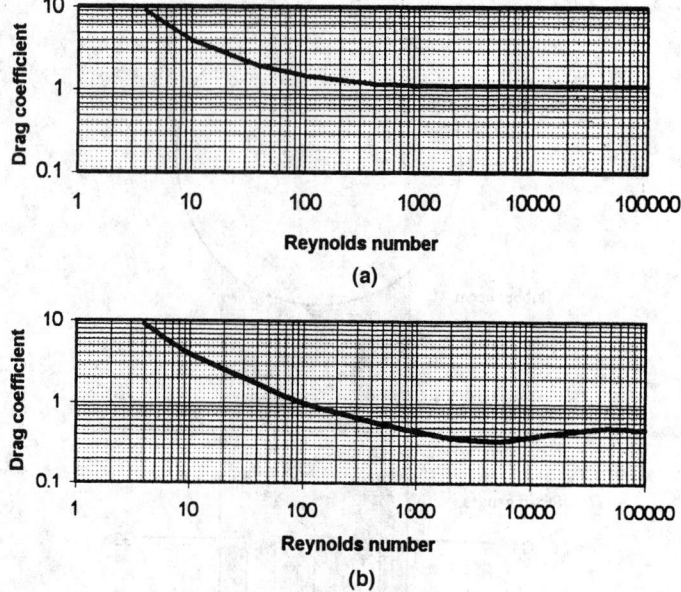

FIGURE 28.86 Graph shows drag coefficient C_D plotted against Reynolds number Re for (*a*) a flat disk and (*b*) a sphere.

C_D is a dimensionless factor, whose magnitude depends primarily on the physical shape of the object and its orientation relative to the fluid stream. The drag coefficient C_D for a sphere is related to the Reynolds number (Re), another dimensionless factor, given by:

$$Re = \rho V A / \eta \qquad (28.147)$$

where η = viscosity of the fluid.

A graph of C_D against Reynolds number (derived from Clarke [1]) is shown in Figure 28.86(*a*) for a flat disk and Figure 28.86(*b*) for a sphere, from which it is evident that the value of C_D, although not constant, is not subject to wide variation over the range of the graph.

Now consider the effect of the force F_D on an elastic beam to which the drag element is attached (as in Figure 28.85). If the mass of the beam can be ignored, the deflection of the beam will be due only to force exerted on the drag element by the fluid. From the theory of elasticity for a cantilever beam of length L with point load at its end [2], the shear force P will be constant along the beam. The bending moment, M at a point x along the beam will be $P(L-x)$ so that it varies linearly from PL at $x = 0$ to 0 at $x = L$. The distance y is measured from the neutral plane, which for a rectangular section is the mid-plane.

The inverse of the radius of curvature is given by:

$$\frac{1}{\rho} = \frac{M}{EI} \qquad (28.148)$$

where E is the modulus of elasticity or the Young's modulus and I is the moment of inertia of the cross-section where it is assumed that the force is applied perpendicular to the broad face of width b.

$$I = \frac{b\,a^3}{12} \qquad (28.149)$$

Through the thickness a of the beam, the strain is ε_x.

$$\varepsilon_x = \frac{\sigma_x}{E} = \frac{Y}{\rho} \tag{28.150}$$

Substituting Equation 28.148 in Equation 28.150 gives:

$$\varepsilon_x = \frac{YM}{EI} \tag{28.151}$$

The strain at the surfaces is:

$$\varepsilon_x = \frac{a}{2\rho} = \frac{YM}{2EI} \tag{28.152}$$

Substituting Equation 28.149 and the bending moment into Equation 28.151 gives:

$$\varepsilon_x = \frac{6\,P\left(L-x\right)}{E\,a^2\,b} \tag{28.153}$$

or, the shear force P is:

$$P = \frac{\varepsilon\,E\,a^2\,b}{6\left(L-x\right)} \tag{28.154}$$

Consequently,

$$\frac{C_D\,\rho\,AV^2}{2} = \frac{\varepsilon\,E\,a^2\,b}{6\left(L-x\right)} \tag{28.155}$$

Therefore, the strain is:

$$\varepsilon = \frac{3C_D\,\rho\,AV^2\left(L-x\right)}{E\,a^2\,b} \tag{28.156}$$

From Equation 28.156, the strain is a square law function of fluid speed.

For most fluid flows in the natural environment, a two-dimensional measurement is necessary as the flow in the natural environment is almost always two-dimensional. For a measurement of flow speed and direction, it is necessary to relate wind speed to strain measured in two orthogonal directions, on the assumption (which has been justified by experiment) that the velocity vector has a zero component along (parallel to) the rubber beam support (i.e., $U_z = 0$). The wind speed U has orthogonal components U_x and U_y that are proportional to the strain measured in the x (strain_x) and y directions (strain_y), respectively. Since U_x is proportional to $\sqrt{\text{strain}_x}$ and U_y is proportional to $\sqrt{\text{strain}_y}$, then the velocity magnitude $|U|$ can be written as

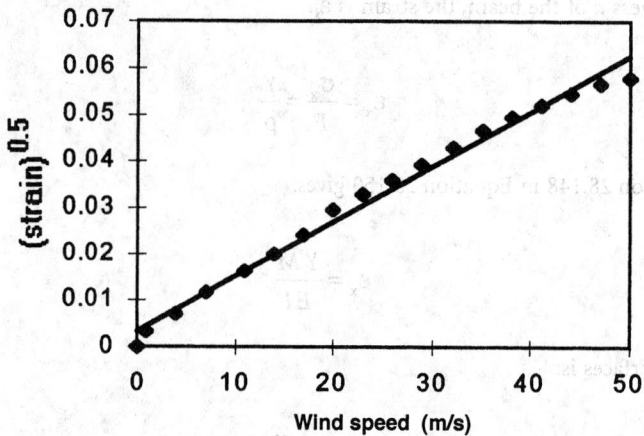

FIGURE 28.87   Root strain vs. wind speed for one dimensional air flow measurement.

$$|U| = \sqrt{U_x^2 + U_y^2} \qquad (28.157)$$

thus,

$$|U| \, \alpha \, \sqrt{strain_x + strain_y} \qquad (28.158)$$

Therefore, the magnitude of the velocity, $|U|$, is proportional to the square root of the sum of the orthogonal strain components, $strain_x$ and $strain_y$. The velocity direction, θ, is calculated using the relation:

$$\theta = \tan^{-1}\left[\frac{\sqrt{strain_y}}{\sqrt{strain_x}}\right] \qquad (28.159)$$

Implementation Using Resistive Strain Gages

A sensor was constructed for measuring one-dimensional flows by bonding two strain gages onto the opposite sides of a square-sectioned elastic beam, 165 mm long and 13.5 mm square, made from poly-butadiene polymer (unfilled vulcanized rubber) supplied by the Malaysian Rubber Products Research Association. The modulus of the beam, at approximately 1.2×10^6 N m^{-2}, was relatively low in order to achieve good sensitivity. The drag element was a table tennis ball of diameter 20 mm, glued to the end of the rubber beam. The strain gages were cupro-nickel alloy on a polyimide film, and were bonded to the rubber beam with epoxy resin. The two gage elements were connected in a half-bridge configuration, and the output was taken to a digital strain-gage indicator reading directly in microstrain units.

The sensor was mounted in a calibrated wind tunnel, in such a way that one gage underwent compression and the other gage underwent extension. The wind speed was set at different values and the corresponding strain readings were noted. Results were plotted as wind speed against square root of strain (Figure 28.87). The graph obtained was linear with a correlation co-efficient of 0.98. The linearity shows that the strain is a square-law function of speed, which confirms the theory above.

In this first prototype of the sensor, the range was limited by beam oscillations at speeds greater than 23 m s^{-1}, causing spurious signals. In principle, however, much greater speeds up to 50 m s^{-1} can be monitored by this sensor, if the construction is suitable.

A second sensor was constructed for measuring two-dimensional flows, in which a square-sectioned rubber beam similar to the previous one was set up with four strain gages attached to the four longitudinal

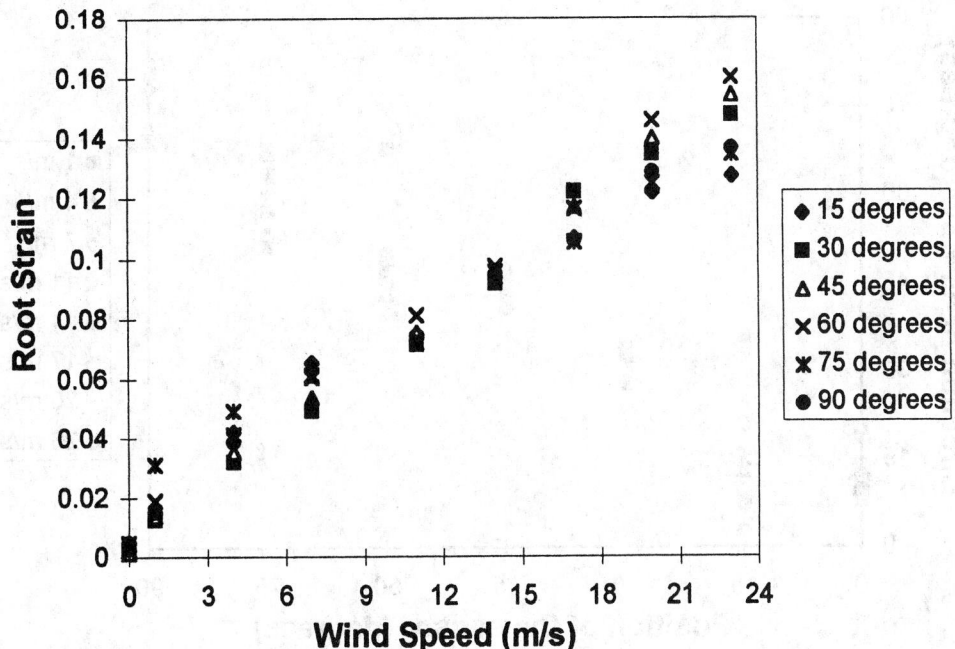

FIGURE 28.88 Root strain plotted against wind speed with the flowmeter oriented at various angles to wind flow.

surfaces of the beam. Each opposite pair of strain gage elements was connected in a half-bridge configuration as before. The two outputs were taken to instrumentation amplifiers and then to a data acquisition
board interfaced to the LabView instrumentation software package (National Instruments Corporation).
After appropriate signal processing by the software, the values of $strain_x$ and $strain_y$ from both the channels
are added and then the square root of the absolute value of this sum is found.

The device was clamped on a turntable and this was rotated about its longitudinal axis from 0° to 90°
at 15° intervals. At each angle, the x and y strain gage outputs were recorded as a function of velocity,
over a range from 0 to 23 m s^{-1}. The graph of speed vs. $(strain_x + strain_y)^{1/2}$ for different angles is shown
in Figure 28.88. The data presented in Figure 28.88 can be used to obtain a speed calibration curve for
the sensor, which will be valid for multidirectional air flow.

The direction of wind flow was calculated according to Equation 28.159 for the different wind speeds.
Figure 28.89 shows the plots of the wind flow direction at different wind speeds for various sensor
positions. Several authors have published the inability of calculating the wind direction accurately at low
wind speeds. The situation is similar in this case and the readings get more accurate as the wind speed
increases, especially beyond 11 m s^{-1}.

For the measurement of gusts, the time response of the sensor becomes a critical parameter. Experiments were performed to measure the response time by dropping known weights on the free end of the
sensor and using a specially written program in Labview to acquire and record the response. The sensor
indicated a 95% response time of 50 ms and a time constant of 30 ms.

Optical Fiber Strain Gage Drag Force Flowmeter

Optical fibers have been applied to measurement of fluid flow, and strain measurement, using interferometric techniques and bending losses [3–5]. The instrument described in the previous sections can be
adapted to use optical strain measurement instead of resistive strain gages. In this way, the advantages
of the instrument can be preserved; in addition, the flowmeter has the usual benefits of optical fiber
sensors such as immunity to electromagnetic interference and intrinsic safety in hazardous environments.

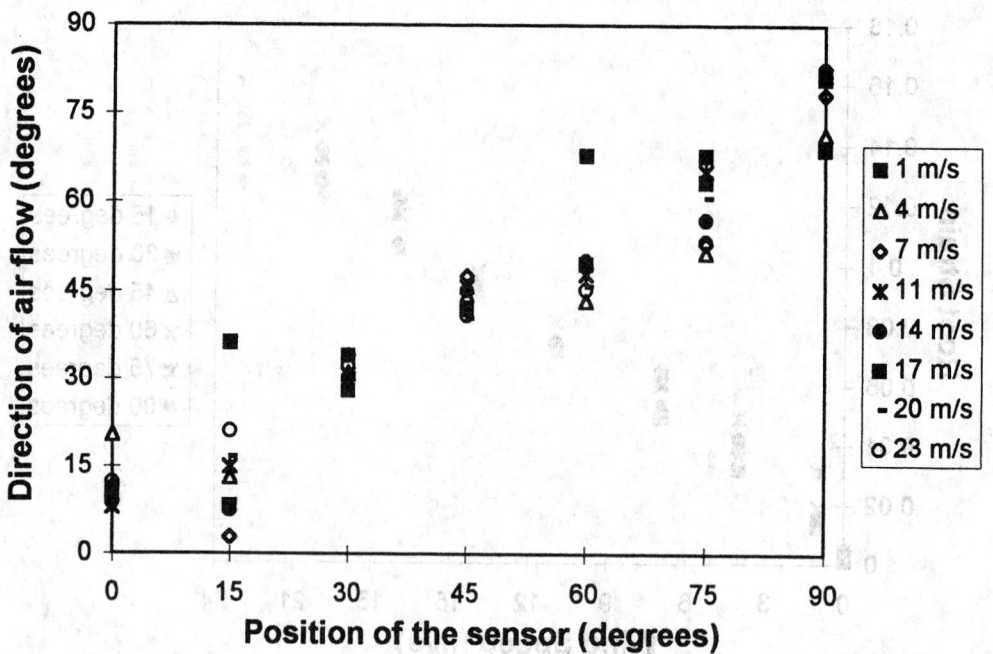

FIGURE 28.89 Calculated direction of wind velocity at different wind speeds.

The principle of an optical fiber strain gage is based on the effect of one or more grooves inserted radially into a 1 mm diameter PMMA fiber, which cause a loss in light transmission. As the optical fiber is bent, as in a cantilever, the angle of the grooves varies. These changes of angle cause light to be attenuated at each groove. The intensity variation can be related to the change of the angle of the groove caused by the bending of the cantilever.

To develop an analogy between the optical strain measurement and the resistive strain measurement, the optical strain was calculated using the formula:

$$Strain_{opt} = \frac{\text{Change in power output}}{\text{Original power output}} = \frac{\Delta P}{P} \qquad (28.160)$$

Hence,

$$Strain_x = \frac{\Delta P_x}{P_x} \quad \text{and} \quad strain_y = \frac{\Delta P_y}{P_y} \qquad (28.161)$$

Therefore, the magnitude of the velocity, $|U|$ is given by:

$$|U| = \sqrt{U_x^2 + U_y^2} = \sqrt{strain_x + strain_y}$$

Substituting Equation 28.161 in $|U|$, the optical root strain magnitude is:

$$|U| = \sqrt{\left(\frac{\Delta P_x}{P_x} + \frac{\Delta P_y}{P_y} \right)} \qquad (28.162)$$

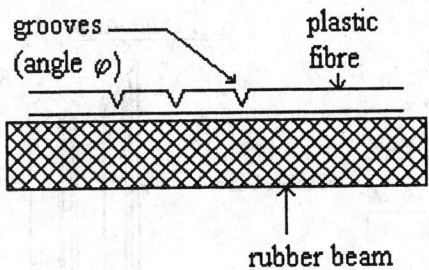

FIGURE 28.90 Principle of operation of the fiber optic drag force flow sensor.

and the direction of wind velocity, θ:

$$\theta = \tan^{-1}\left[U_y/U_x\right] = \tan^{-1}\sqrt{\frac{\text{Strain}_y}{\text{Strain}_x}} \qquad (28.163)$$

Substituting Equation 28.161 in θ, the optical strain direction is:

$$\theta = \tan^{-1}\frac{\sqrt{\left(\dfrac{\Delta P_y}{P_y}\right)}}{\sqrt{\left(\dfrac{\Delta P_x}{P_x}\right)}} \qquad (28.164)$$

Two types of the optical fiber flow sensor are described here. In the first type, the rubber beam is used as the deflected device with optical fiber strain gages attached to the deflected beam. The second version of this flow sensor uses an unsupported sensitised 1 mm diameter plastic optical fiber that undergoes deflection in the airflow.

Fiber Optic Flow Sensor WITH Rubber Beam

The sensor has been built using a rectangular cross-sectioned rubber beam and a spherical drag element with a 1 mm plastic optical fiber glued to the beam with epoxy. The fiber is looped inside the drag element to enhance the sensitivity of the device. Grooves have been made in the fiber surface that extends into the core of the fiber to increase the losses as a function of the bending of the rubber beam (Figure 28.90).

The grooves are normal to the rubber beam on its outside face, and their depths affect only the cladding of the fiber. As the cantilever bends due to the force exerted by the air flow, the angle of the grooves vary. The groove angle increases when the air flow is facing the sensor, and vice versa. These changes of groove angle cause an intensity modulation of the light transmitted through the fiber because light is lost at each groove. Changes in intensity can be related to changes in the angle caused by the force inducing the bending of the cantilever, and therefore to the velocity of the fluid.

The orthogonal components of strain were measured by attaching two grooved optical fibers on adjacent sides of the rubber beam, orthogonal to each other, as illustrated in Figure 28.91. These two fibers measured the x and y components of optical strain.

Light from a 1 mW helium-neon laser of peak wavelength 633 nm was split into equal components using a cubic beam splitter, and coupled into each fiber. The transmitted intensity through each of the two fibers was monitored using a power meter. The signals from the power meter were sent to a 486 DX2 laptop computer via a data acquisition card to be acquired and processed by Labview. Experiments indicated that this version of the flow sensor could measure wind speeds up to 30 m s^{-1} with a resolution of 1.3 m s^{-1}.

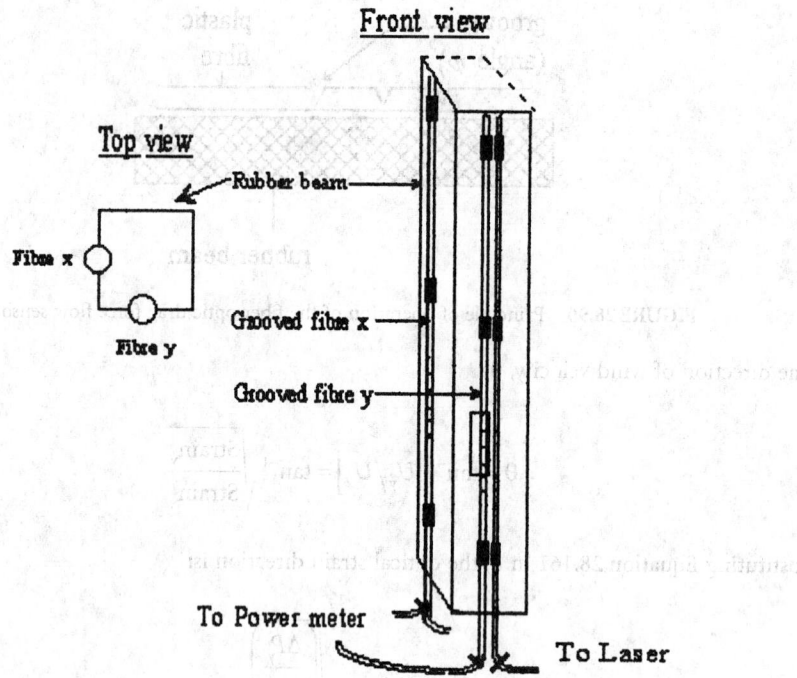

FIGURE 28.91   Front and top views of a section of the fiber optic drag force flow sensor used to measure the two dimensional fluid flow.

Fiber Optic Flow Sensor WITHOUT Rubber Beam

This version of the sensor uses 1 mm diameter polymer fiber as the deflection element, with a core diameter of 0.980 mm, and a thin cladding layer of approximately 20 μm [6]. Multiple grooves were etched radially into the fiber surface to a depth of 0.5 mm, extending into the core of the fiber, using a hot scalpel, and a manufactured V-groove template to ensure uniformity of grooves. Six grooves were determined as the optimum number to achieve a compromise between insertion losses and strain sensitivity, spaced 0.4 cm apart, over a length of 2.5 cm.

In order to measure strain in two orthogonal directions, perpendicular to the longitudinal axis of the fiber, two fibers were used, as shown in Figure 28.92, with the grooves oriented at 90° to each other. This was achieved by positioning the fibers on two adjacent faces of a beam of square cross-section. This beam was short enough to prevent any restriction to the deflection and yet long enough to hold and support the optical fibers. The fibers were looped around so that the looped ends acted as drag elements. The grooved portion of the fiber was unsupported and free to deflect in the air stream. Wind tunnel calibration indicated that the sensor could measure two-dimensional flow up to 35 m s^{-1} with a resolution of 0.96 m s^{-1}.

Conclusion

The measurement of one- and two-dimensional fluid velocity, using both optical fiber and conventional strain gages on a deflected beam, for a range of 0 to 30 m s^{-1}, has been demonstrated experimentally. Although flow visualization and modeling techniques are well advanced in engineering, there is still a need for real measurements, especially in the natural environment. The sensors described in this study are particularly suited for such measurements, which are almost always two-dimensional. The outputs

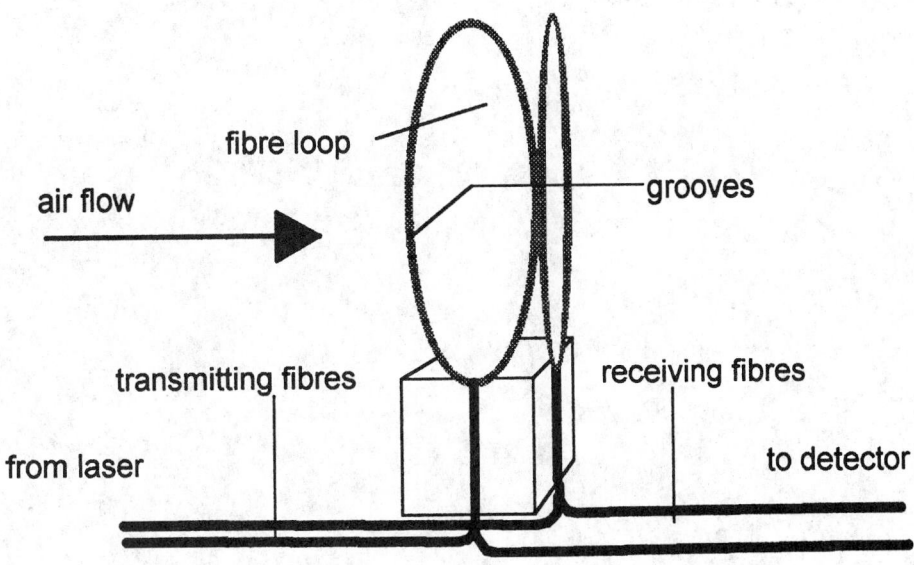

FIGURE 28.92 Optical fiber drag force flow sensor without rubber beam.

of the sensor representing speed and direction of fluid flow are independent of each other, so the sensor is suitable for environmental applications such as wind measurement or river flow, where the fluid forms gusts and can change direction as well as speed. One noteworthy feature of this sensor is its quick response time of 50 ms, which easily enables the measurement of gusts. The dimensions and materials of the sensor must be chosen to suit the fluid. In terms of sensitivity and resolution, the resistive strain gage sensor is a better option, but replacing the conventional strain gages with optical fiber strain gages ensures electrical noise immunity and intrinsic safety for use in hazardous environments.

References

1. T. Clarke, Design and operation of target flowmeters, *Encyclopedia of Fluid Mechanics*, Vol 1, Houston, TX: Gulf Publishing Company, 1986.
2. F. H. Newman and V. H. L. Searle, The general properties of matter. Edward Arnold, 1948.
3. S. Webster, R. McBride, J. S. Barton, and J. D. C. Jones, Air flow measurement by vortex shedding from multimode and monomode fibres. *Measurement, Science and Technology*, 3, 210-216, 1992.
4. J. S. Barton and M. Saudi, A fibre optic vortex flowmeter. *J. Phys. E: Sci. Instrum.* 19, 64-66, 1986.
5. N. Narendran, A. Shukla, and S. Letcher, Optical fibre interferometric strain sensor using a single fibre, *Experimental Techniques*, 16(2), 33-36, 1992.
6. R. Philip-Chandy, Ph.D. thesis, *Fluid flow measurement using electrical and optical fibre strain gauges*, Liverpool John Moores University, UK, 1997.

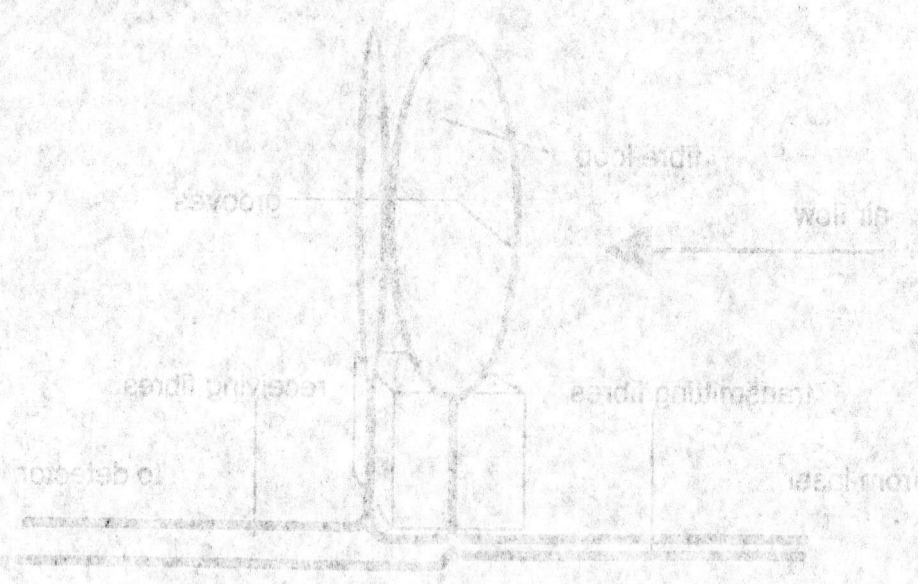

29

Point Velocity Measurement

John A. Kleppe
University of Nevada

John G. Olin
Sierra Instruments, Inc.

Rajan K. Menon
TSI Inc.

29.1   Pitot Probe Anemometry

John A. Kleppe

Theory

It is instructive to review briefly the principles of fluid dynamics in order to understand *Pitot tube theory* and applications. Consider, for example, a constant-density fluid flowing steadily without friction through the simple device shown in Figure 29.1. If it is assumed that there is no heat being added and no shaft work being produced by the fluid, a simple expression can be developed to describe this flow:

$$\frac{p_1}{w} + \frac{v_1^2}{2g} + z_1 = \frac{p_2}{w} + \frac{v_2^2}{2g} + z_2 \qquad (29.1)$$

where p_1, v_1, z_1 = Pressure, velocity, and elevation at the inlet
$\qquad p_2, v_2, z_2$ = Pressure, velocity, and elevation at the outlet
$\qquad w \qquad$ = ρg, the specific weight of the fluid
$\qquad \rho \qquad$ = Density
$\qquad g \qquad$ = 9.80665 m s^{-2}

Equation 29.1 is the well known Bernoulli equation. The following example will demonstrate the use of Equation 29.1 and lead to a discussion of the theory of Pitot tubes.

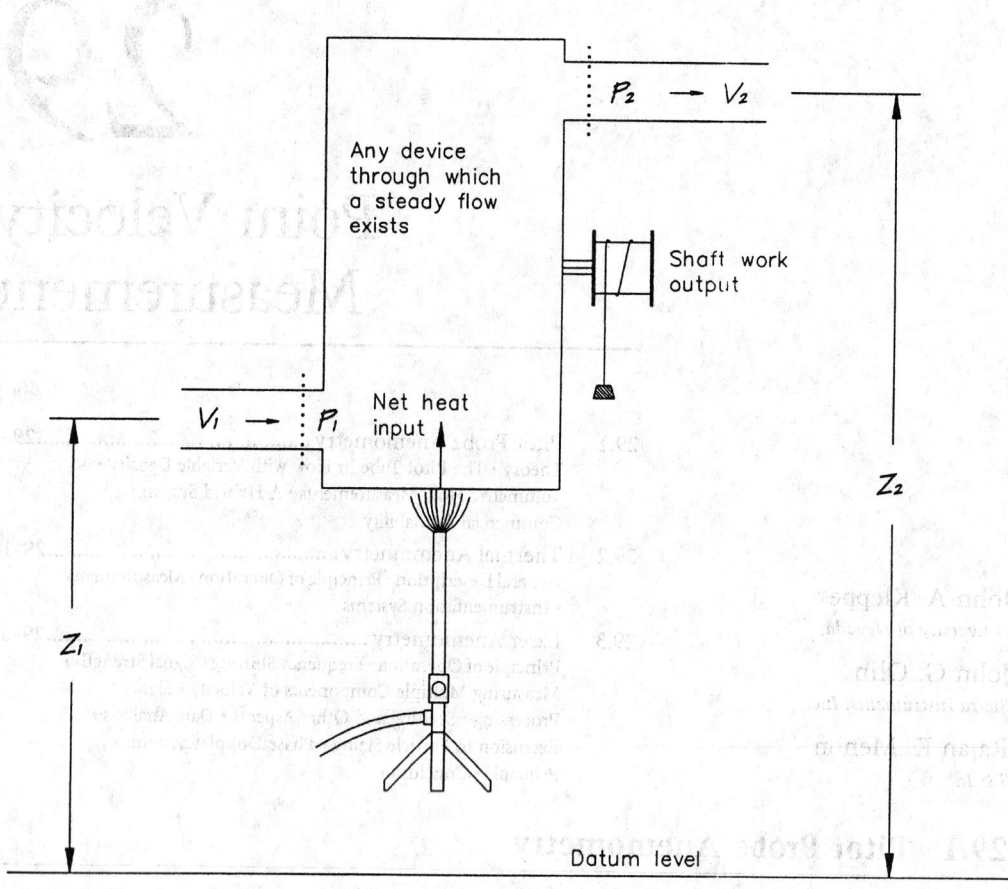

FIGURE 29.1 A device demonstrating Bernoulli's equation for steady flow, neglecting losses. (From [1].)

Example

A manometer [2] is used to measure the dynamic pressure of the tube assembly shown in Figure 29.2 [3]. The manometer fluid is mercury with a density of 13,600 kg m^{-3}. For a measured elevation change, Δh, of 2.5 cm, calculate the flow rate in the tube if the flowing fluids is (a) water, (b) air. Neglect all losses and assume STP conditions for the air flowing in the tube and $g = 9.81$ m s^{-2}.

Solution

Begin by writing expressions for the pressure at point 3.

$$p_3 = h_1\, w_{Hg} + \left(h_3 - h_1\right) w + p_1 \qquad (29.2)$$

and

$$p_3 = h_2\, w_{Hg} + \left(h_3 - h_2\right) w + p_2 \qquad (29.3)$$

Subtracting these equations and rearrangement yields an expression for the pressure difference.

$$p_2 - p_1 = \Delta h \left(w_{Hg} - w\right) \qquad (29.4)$$

where w is the specific weight for water or air, etc.

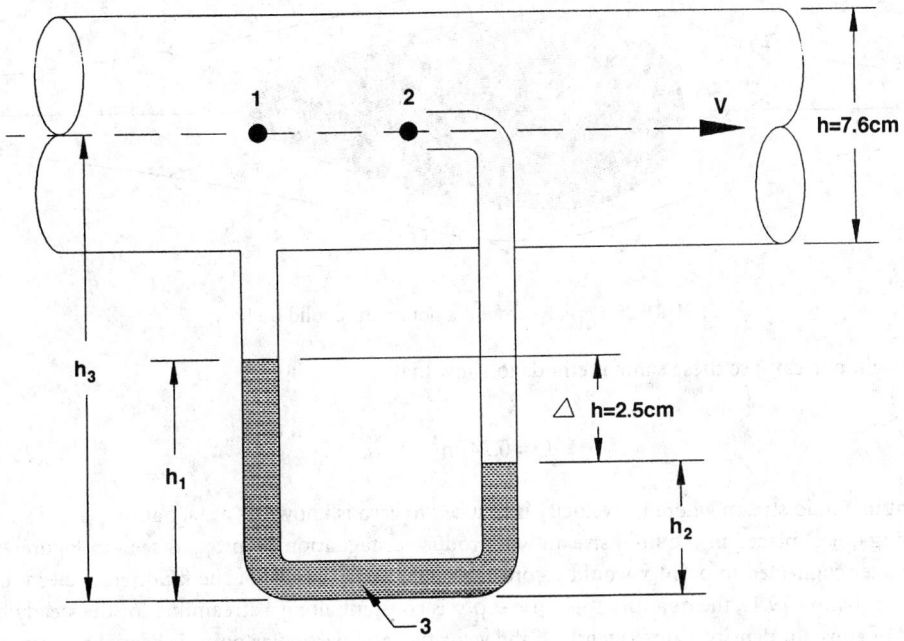

FIGURE 29.2 Using a manometer to measure a Pitot-static tube type assembly - Example (1).

Also using Equation 29.1, one can show that for $z_1 = z_2$ and $v_2 = 0$:

$$p_2 - p_1 = \frac{w \, v_1^2}{2g} \tag{29.5}$$

(a) For water,

$$p_2 - p_1 = \Delta h \left(w_{\text{Hg}} - w_{\text{H}_2\text{O}} \right)$$

$$= 0.025 \left[13{,}600 \, (9.81) - 998 \, (9.81) \right] \tag{29.6}$$

$$= 3090.6 \text{ Pa}$$

Then,

$$3090.6 = \frac{(998)(9.81) v_1^2}{2(9.81)} \tag{29.7}$$

or

$$v_1 = 2.5 \text{ m s}^{-1} \tag{29.8}$$

The flow Q is then calculated to be:

$$Q = A_1 v_1 = \frac{\pi d^2 v_1}{4} = \frac{\pi (.076)^2 (2.5)}{4} = 0.011 \text{ m}^3 \text{ s}^{-1} \tag{29.9}$$

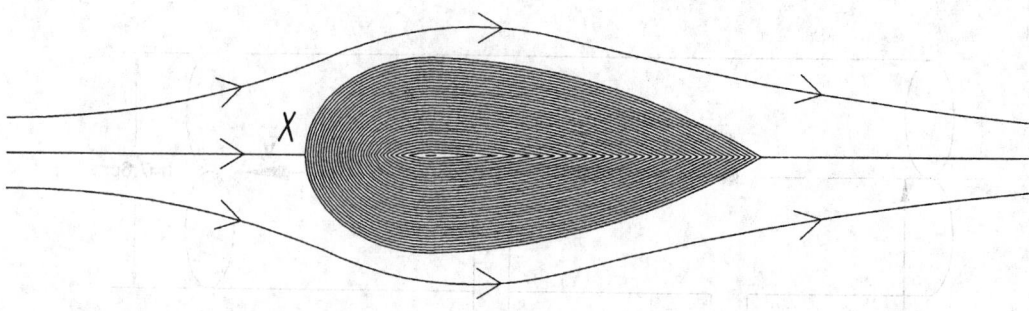

FIGURE 29.3 Flow around a nonrotating solid body.

(b) For air, one can use these same methods to show that:

$$Q = 0.34 \text{ m}^3 \text{ s}^{-1} \tag{29.10}$$

A point in a fluid stream where the velocity is reduced to zero is known as a stagnation point [1]. Any nonrotating object placed in the fluid stream will produce a stagnation point, x, as seen in Figure 29.3. A manometer connected to point x would record the stagnation pressure of the fluid. From Bernoulli's equation (Equation 29.1), the quantity $p + \frac{1}{2}\rho v^2 + \rho gz$ is constant along a streamline for the steady flow of a fluid of constant density. Consequently, if the velocity v at a particular point is brought to zero, the pressure there is increased from p to $p + \frac{1}{2}\rho v^2$. For a constant-density fluid, the quantity $p + \frac{1}{2}\rho v^2$ is known as the stagnation pressure p_0 of that streamline, while the term $\frac{1}{2}\rho v^2$ — that part of the stagnation pressure due to the motion — is termed the dynamic pressure. A manometer connected to point x would measure the stagnation pressure and, if the static pressure p were also known, then $\frac{1}{2}\rho v^2$ could be obtained. One can show that:

$$p_t = p + p_v \tag{29.11}$$

where p_t = Total pressure, which is the sum of the static and dynamic pressures which can be sensed by
 a probe that is at rest with respect to the system boundaries when it locally stagnates the fluid
 isentropically

p = The actual pressure of the fluid whether in motion or at rest and can be sensed by a probe
 that is at rest with respect to the fluid and does not disturb the fluid in any way

p_v = The dynamic or velocity pressure equivalent of the directed kinetic energy of the fluid

Using Equation 29.11, one can develop an expression that relates to the velocity of the fluid:

$$p_t = p + 1/2\rho v^2 \tag{29.12}$$

or, solving for v:

$$v = \sqrt{\frac{2\left(p_t - p\right)}{\rho}} \tag{29.13}$$

Consider as an example the tube arrangement shown in Figure 29.4. A right-angled tube, large enough to neglect capillary effects, has one end A facing the flow. When equilibrium is attained, the fluid at A is stationary and the pressure in the tube exceeds that of the surrounding stream by $\frac{1}{2}\rho v^2$. The liquid is forced up the vertical part of the tube to a height:

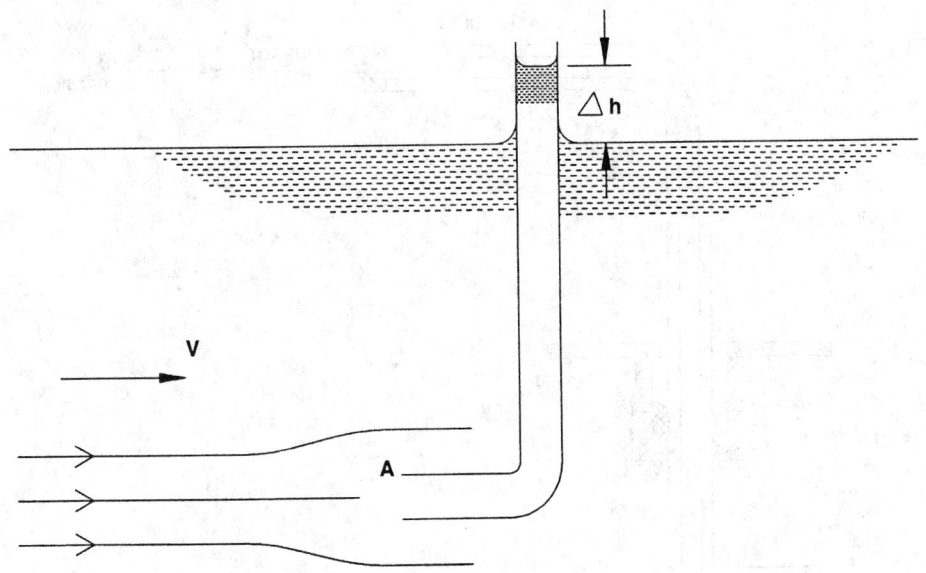

FIGURE 29.4   Right-angle tube in a flow system.

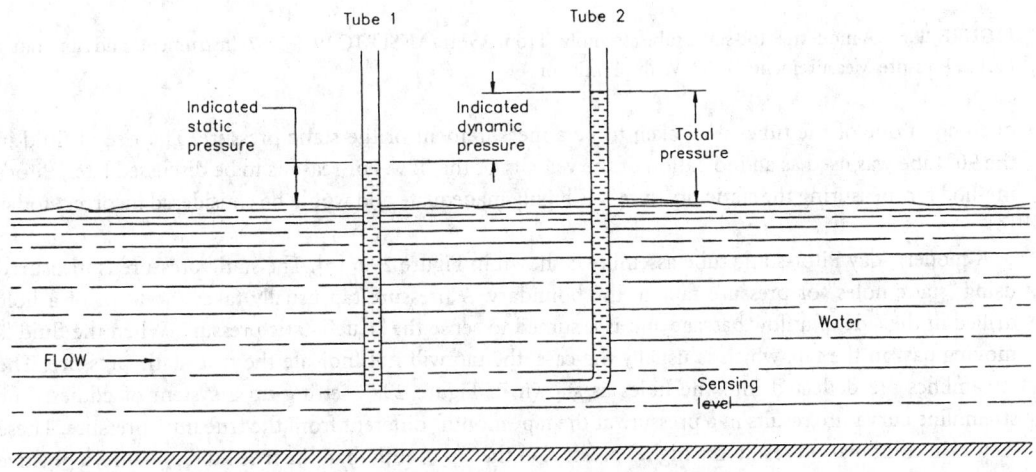

FIGURE 29.5   Basic Pitot tube method of sensing static, dynamic, and total pressure. (From R. P. Benedict, *Fundamentals of Temperature, Pressure and Flow Measurements*, 3rd ed., New York: John Wiley & Sons, 1984. With permission.)

$$\Delta h = \frac{\Delta p}{w} = \frac{v^2}{2g} \tag{29.14}$$

This relationship was used in the example given earlier to solve for *v*. It must be remembered that the total pressure in a fluid can be sensed only by stagnating the flow isentropically; that is, when its entropy is identical at all points in the flow. Such stagnation can be accomplished by a Pitot tube, as first developed by Henri de Pitot in 1732 [4]. In order to obtain a velocity measurement in the River Seine (in France), Pitot made use of two tubes immersed in water. Figure 29.5 shows his basic Pitot tube method. The lower

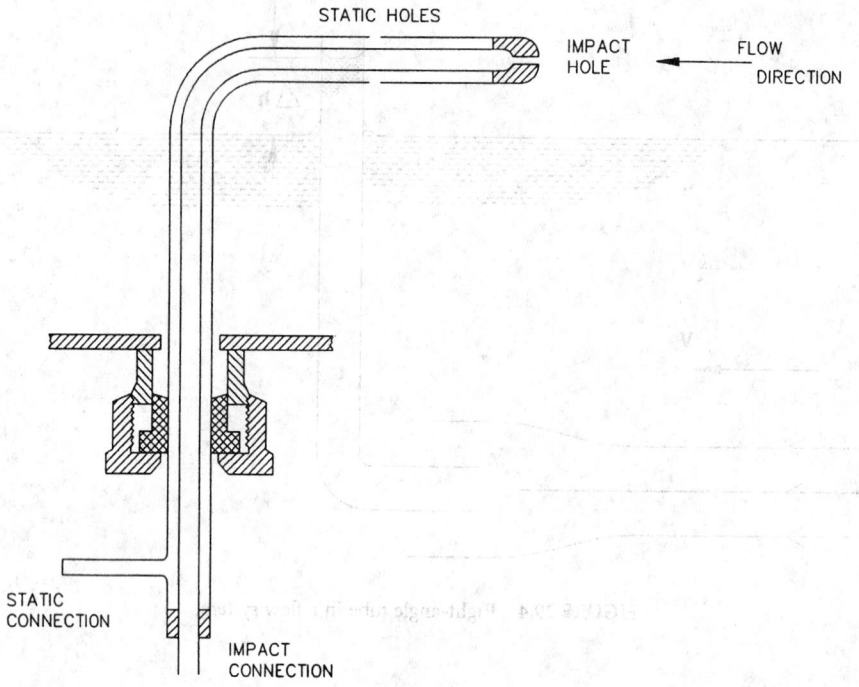

FIGURE 29.6 A modern Pitot-static tube assembly. (From ASME/ANSI PTC 19.2-1987, Instruments and Apparatus, Part 2, Pressure Measurements, 1987. With permission.)

opening in one of the tubes was taken to be a measurement of the static pressure. The rise of fluid in the 90° tube was used as an indication of the velocity of the flow. For reasons to be discussed later, Pitot's method for measuring the static pressure was highly inadequate and would be considered incorrect today [4].

A modern-day Pitot-static tube assembly is shown in Figure 29.6 [5]. The static pressure is measured using "static holes" or pressure taps in the boundary. A pressure tap usually takes the form of a hole drilled in the side of a flow passage and is assumed to sense the "true" static pressure. When the fluid is moving past in the tap, which is usually the case, the tap will not indicate the true static pressure. The streamlines are deflected into the holes as shown in Figure 29.7, setting up a system of eddies. The streamline curvature results in a pressure at the tap "mouth" different from the true fluid pressure. These factors in combination result in a higher pressure at the tap mouth than the true fluid pressure, a positive pressure error. The magnitude of this pressure error is a function of the Reynolds number based on the shear velocity and the tap diameter [5]. Larger tap diameters and high velocities give larger errors [5]. The effect of compressibility on tap errors is not well understood or demonstrated, although correlations for this effect have been suggested [5]. It is possible to reduce tap errors by moving the location of the tap to a nonaccelerating flow location, or use pressure taps of smaller diameter. The effect of edge burrs is also noteworthy. All burrs must be removed. There is also an error that results with the angle of attack of the Pitot tube with the flow direction. Figure 29.8 shows the variation of total pressure indications as a function of the angle of attack. It can be seen that little error results if the angle of attack is less than ±10°.

A widely used variation of the Pitot-static tube is the type S Pitot tube assembly shown in Figure 29.9. It must be carefully designed and fabricated to ensure it will properly measure the static pressure. The "static" tube faces backwards into the wake behind the probe where the pressure is usually somewhat lower than the undisturbed static pressure. The type S Pitot tube therefore requires the application of a correction factor (usually in the range of 0.84). This correction factor will be valid only over a limited

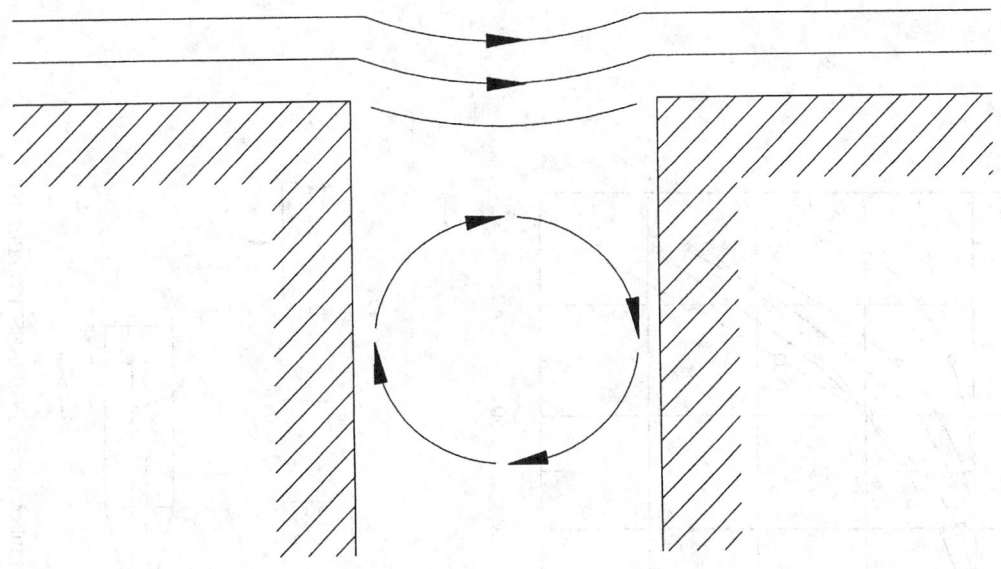

FIGURE 29.7 Pressure tap flow field.

range of velocity measurement. The type S Pitot tube does, however, have the advantage of being compact and relatively inexpensive. A type S Pitot tube can be traversed across a duct or stack to determine the velocity profile and hence total volumetric flow. This is discussed later.

The Pitot Tube in Flow with Variable Density

When a Pitot-static tube is used to determined the velocity of a constant-density fluid, the stagnation pressure and static pressure need not be separately measured: It is sufficient to measure their difference. A high-velocity gas stream, however, can undergo an appreciable change of density in being brought to rest at the front of the Pitot-static tube; under these circumstances, stagnation and static pressures must be separately measured. Moreover, if the flow is initially supersonic, a shock wave is formed ahead of the tube, and, thus, results for supersonic flow differ essentially from those for subsonic flow. Consider first the Pitot-static tube in uniform subsonic flow, as in Figure 29.10.

The process by which the fluid is brought to rest at the nose of the tube is assumed to be frictionless and adiabatic. From the energy equation for a perfect gas, it can be shown that [1]:

$$\frac{v^2}{2} = C_p \left(T_0 - T \right) = C_p T_0 \left\{ 1 - \left(\frac{p}{p_0} \right)^{(\gamma-1)/\gamma} \right\} \tag{29.15}$$

where v = Velocity
C_p = Specific heat at constant pressure
T = Absolute temperature of the gas
T_0 = Absolute temperature at stagnation conditions
p = Total pressure
γ = Ratio of specific heats

For measuring T_0, it is usual to incorporate in the instrument a small thermocouple surrounded by an open-ended jacket. If T_0 and the ratio of static to stagnation pressure are known, the velocity of the stream can then be determined from Equation 29.15.

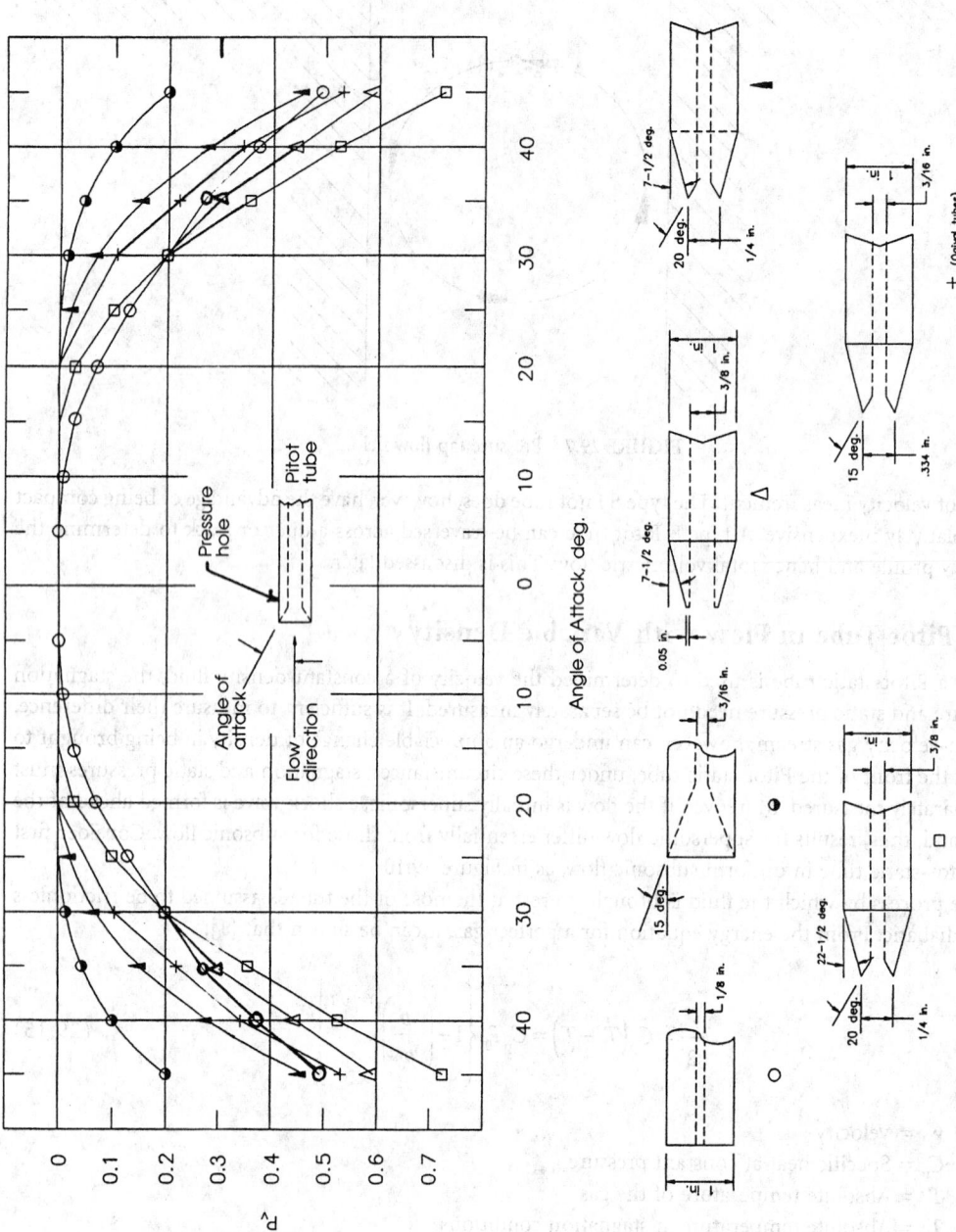

FIGURE 29.8 Variation of total pressure indication with angle of attach and geometry for Pitot tubes. (From ASME/ANSI PTC 19.2-1987, Instruments and Apparatus, Part 2, Pressure Measurements, 1987. With permission.)

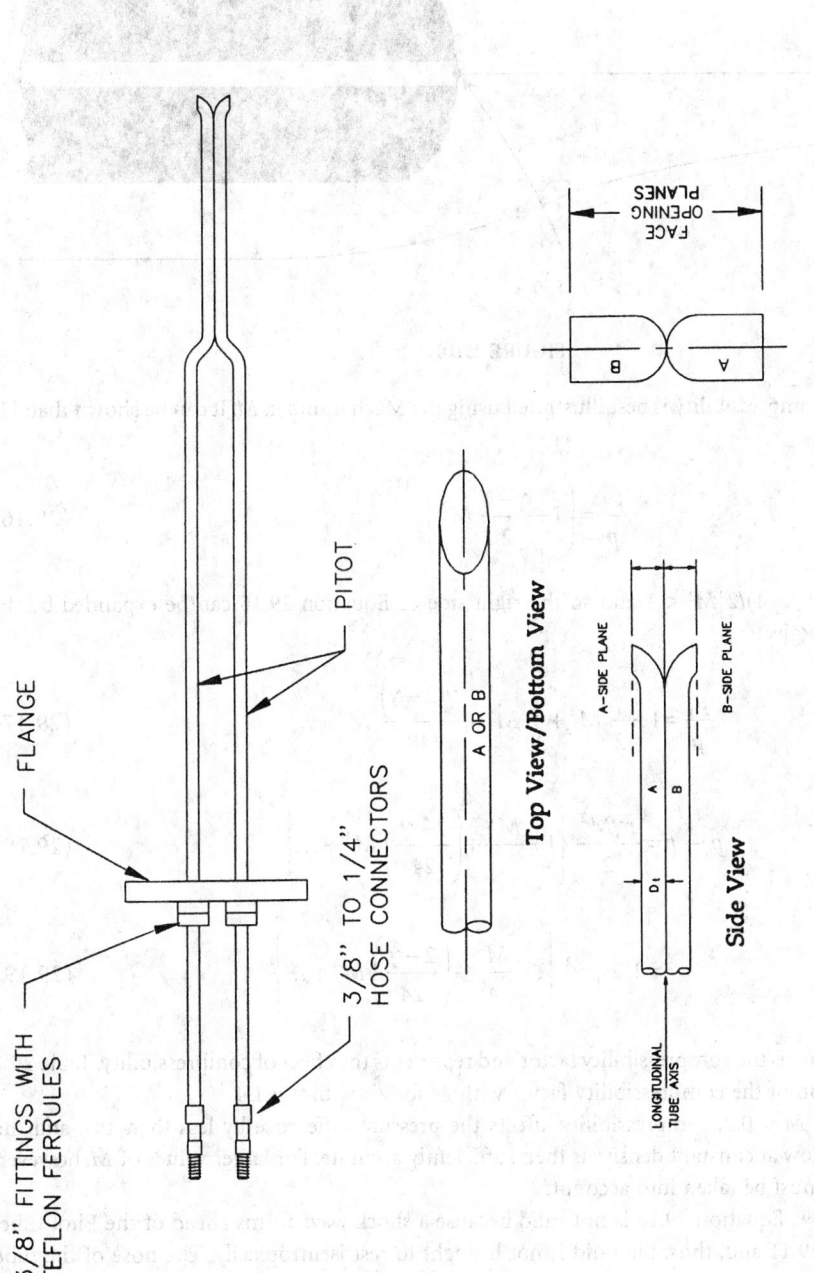

FIGURE 29.9 An S type Pitot tube for use in gas flow measurement will have specific design parameters. For example, the diameter of the tubing D_t, a gas probe will be between 0.48 and 0.95 cm. There should be equal distances from the base of each leg of the Pitot tube to its face opening plane, dimensions d_1, d_2. This distance should be between 1.05 and 1.50 times the external tubing diameter, D_t. The face openings of the Pitot tube should be aligned as shown. This configuration of the type S Pitot tube results in a correction coefficient of approximately 0.84. (From EPA, CFR 40 Part 60, Appendix A—Test Methods, 1 July 1995.)

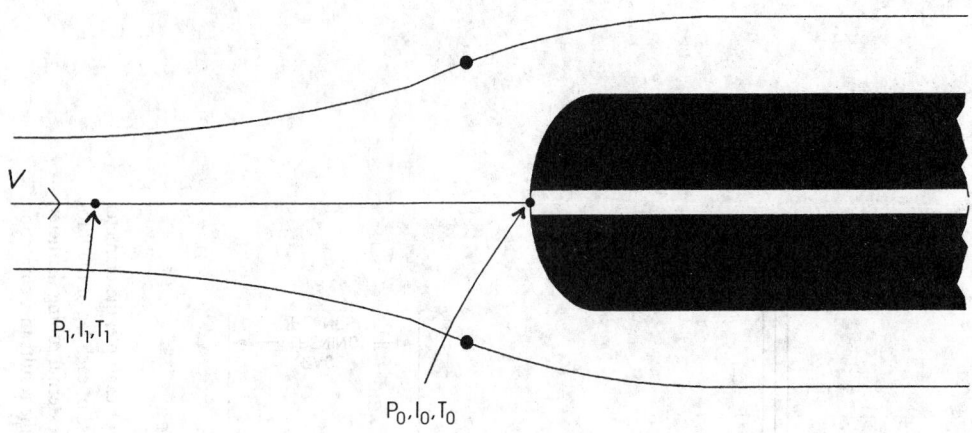

FIGURE 29.10

The influence of compressibility is best illustrated using the Mach number, M. It can be shown that: [1]

$$\frac{p_0}{p} = \left(1 + \frac{\gamma-1}{2} M^2\right)^{\gamma/(\gamma-1)} \tag{29.16}$$

For subsonic flow, $[(\gamma - 1)/2]M^2 < 1$ and so the right side of Equation 29.16 can be expanded by the binomial theorem to give:

$$\frac{p_0}{p} = 1 + \frac{\gamma}{2} M^2 + \frac{\gamma}{8} M^4 + \frac{\gamma(2-\gamma)}{48} M^6 + \ldots \tag{29.17}$$

$$p_0 - p = \frac{p\gamma M^2}{2}\left\{1 + \frac{M^2}{4} + \left(\frac{2-\gamma}{24}\right) M^4 + \ldots\right\} \tag{29.18}$$

$$= 1/2\ \rho v^2\left\{1 + \frac{M^2}{4} + \left(\frac{2-\gamma}{24}\right) M^4 + \ldots\right\} \tag{29.19}$$

The bracketed quantity is the compressibility factor and represents the effect of compressibility. Table 29.1 indicates the variation of the compressibility factor with M for air with $\gamma = 1.4$

It is seen that for $M < 0.2$, compressibility affects the pressure difference by less than 1%, and the simple formula for flow at constant density is then sufficiently accurate. For larger values of M, however, the compressibility must be taken into account.

For supersonic flow, Equation 29.16 is not valid because a shock wave forms ahead of the Pitot tube, as shown in Figure 29.11 and, thus, the fluid is not brought to rest isentropically. The nose of the tube must be shaped so that the shock wave is detached, i.e., the semiangle must be greater than 45.6° [1].

If the axis of the tube is parallel to the oncoming flow, the wave can be assumed normal to the streamline leading to the stagnation point. The pressure rise across the shock can therefore be given by:

$$\frac{p_2}{p_1} = \frac{1 + \gamma M_1^2}{1 + \gamma M_2^2} \tag{29.20}$$

TABLE 29.1 Variation of
"Compressibility Factor" for Air

M	$\dfrac{P_0 - P}{\frac{1}{2}\rho v^2}$
0.1	1.003
0.2	1.010
0.3	1.023
0.4	1.041
0.5	1.064
0.6	1.093
0.7	1.129
0.8	1.170
0.9	1.219
1.0	1.276

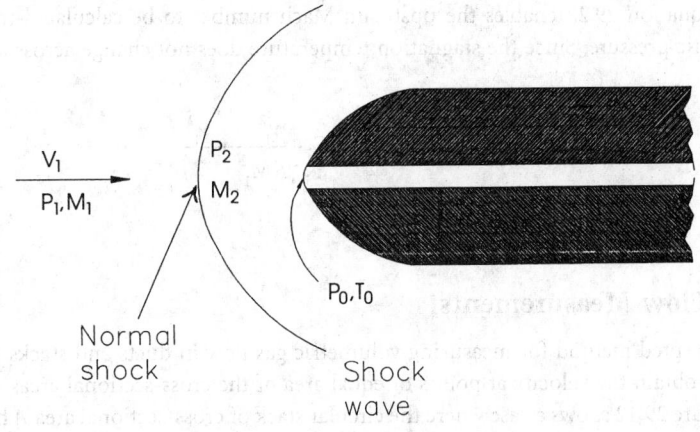

FIGURE 29.11

In the subsonic region downstream of the shock, there is a gradual isentropic pressure rise that can be represented as:

$$\frac{p_0}{p_1} = \frac{p_0}{p_2}\frac{p_2}{p_1} = \left(1 + \frac{\gamma-1}{2}M_2^2\right)^{\gamma/(\gamma-1)}\frac{1+\gamma M_1^2}{1+\gamma M_2^2} \qquad (29.21)$$

Finally, one obtains Rayleigh's formula:

$$\frac{p_0}{p_1} = \left\{\frac{(\gamma+1)^{\gamma+1}}{2\gamma M_1^2 - \gamma + 1}\left(\frac{M_1^2}{2}\right)^{\gamma}\right\}^{\gamma/(\gamma-1)} \qquad (29.22)$$

This expression for air reduces to:

$$\frac{p_0}{p_1} = \frac{166.9 M_1^7}{\left(7M_1^2 - 1\right)^{2.5}} \quad \text{when } \gamma = 1.4 \qquad (29.23)$$

Although a conventional Pitot-static tube gives satisfactory results at Mach numbers low enough for no shock waves to form, it is unsuitable in supersonic flow because its "static holes" or "pressure taps", being in the region downstream of the shock, do not then register p_1; nor do they register p_2 since this is found only on the central streamline, immediately behind the normal part of the shock wave. Consequently, p_1 is best determined independently — for example, through an orifice in a boundary wall well upstream of the shock. Where independent measurement of p_1 is not possible, a special Pitot-static tube can be used, in which the static holes are much further back (about 10 times the outside diameter of the tube) from the nose. The oblique shock wave on each side of the tube has by then degenerated into a Mach wave across which the pressure rise is very small.

When $M_1 = 1$, the pressure rise across the shock is infinitesimal and, thus, Equations 29.16 and 29.22 both give:

$$\frac{p_0}{p_1} = \left\{ (\gamma+1)/2 \right\}^{\gamma/(\gamma-1)} = 1.893 \left(\text{for air} \right)$$
(29.24)

A small value of p_0/p therefore indicates subsonic flow, a larger value supersonic flow.

Notice that Equation 29.22 enables the upstream Mach number to be calculated from the ratio of stagnation to static pressure. Since the stagnation temperature does not change across a shock wave:

$$C_p T_0 = C_p T_1 + \frac{v_1^2}{2} = C_p \frac{v_1^2}{\gamma R M_1^2} + \frac{v_1^2}{2}$$
(29.25)

Thus, v_1 can also be calculated if T_0 is determined.

Volumetric Flow Measurements

The currently accepted method for measuring volumetric gas flow in ducts and stacks involves the use of Pitot tubes to obtain the velocity at points of equal area of the cross-sectional areas of the stack [7]. For example, Figure 29.12 shows a case where the circular stack of cross-sectional area A has been divided into twelve (12) equal areas. An estimate of the average volumetric flow velocity is determined using the following relationship:

$$\bar{v}_n \approx \frac{\Sigma\, v_n A_n}{A} = \frac{A_i\, \Sigma\, v_n}{N A_i} = \frac{1}{N} \Sigma\, v_n$$
(29.26)

where A_i = One segment of the equal area segments
N = Number of equal area segments
v_n = Velocity measured at each point of equal area segment

This relationship shows that one can estimate the average volumetric flow velocity by taking velocity measurements at each point of equal area and then calculate the arithmetic mean of these measurements. It is clearly seen that a different result would be obtained if one were to simply take velocity measurements at equidistant points across the measurement plane and then take the arithmetic mean of these measurements. What would result in this case would be the path-averaged velocity, $\bar{v}_p$, which would be in error.

The sampling site and the number of traverse points designated will affect the quality of the volumetric flow measurement. The acceptability of the sampling procedure is generally determined by the distances from the nearest upstream and downstream disturbances (obstruction or change in direction) to gas flow. The minimum requirements for an acceptable sampling procedure can be found in the literature [7].

An automated system for accomplishing this measurement is shown in Figure 29.13 [8].

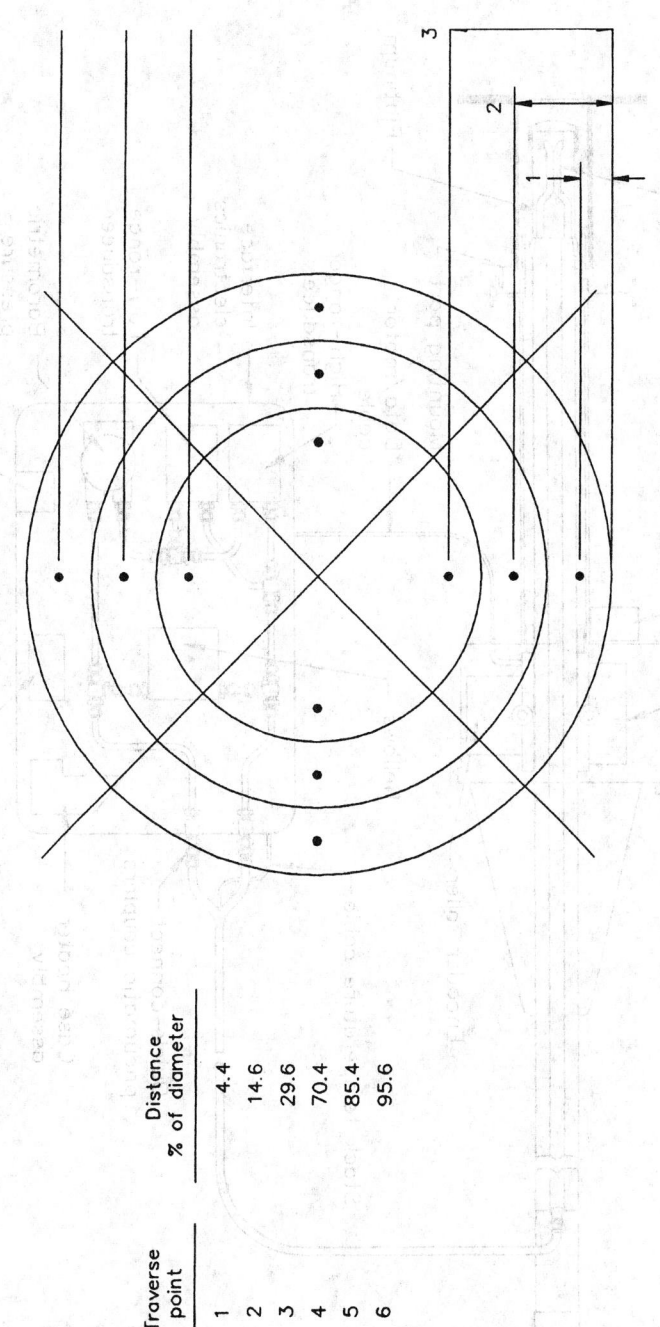

Traverse point	Distance % of diameter
1	4.4
2	14.6
3	29.6
4	70.4
5	85.4
6	95.6

FIGURE 29.12 Example showing circular stack cross-section divided into 12 equal areas, with location of traverse points indicated.

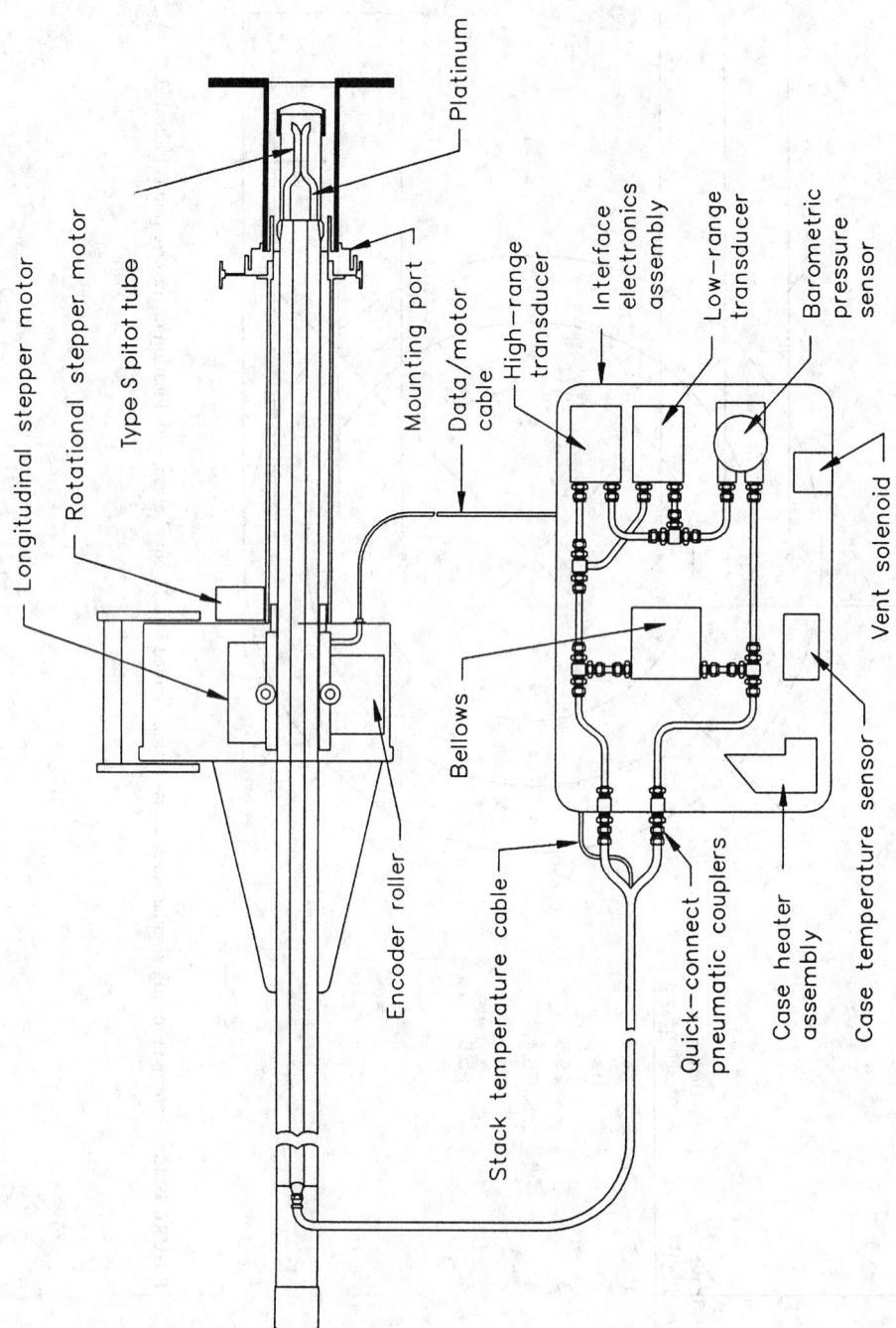

FIGURE 29.13   Automated probe consists of a type S Pitot tube and a platinum RTD mounted onto a type 319 stainless steel probe. (From T. C. Elliott, CEM System: Lynchpin Holding CAA Compliance Together, *Power*, May 1995, 31–40. With permission.)

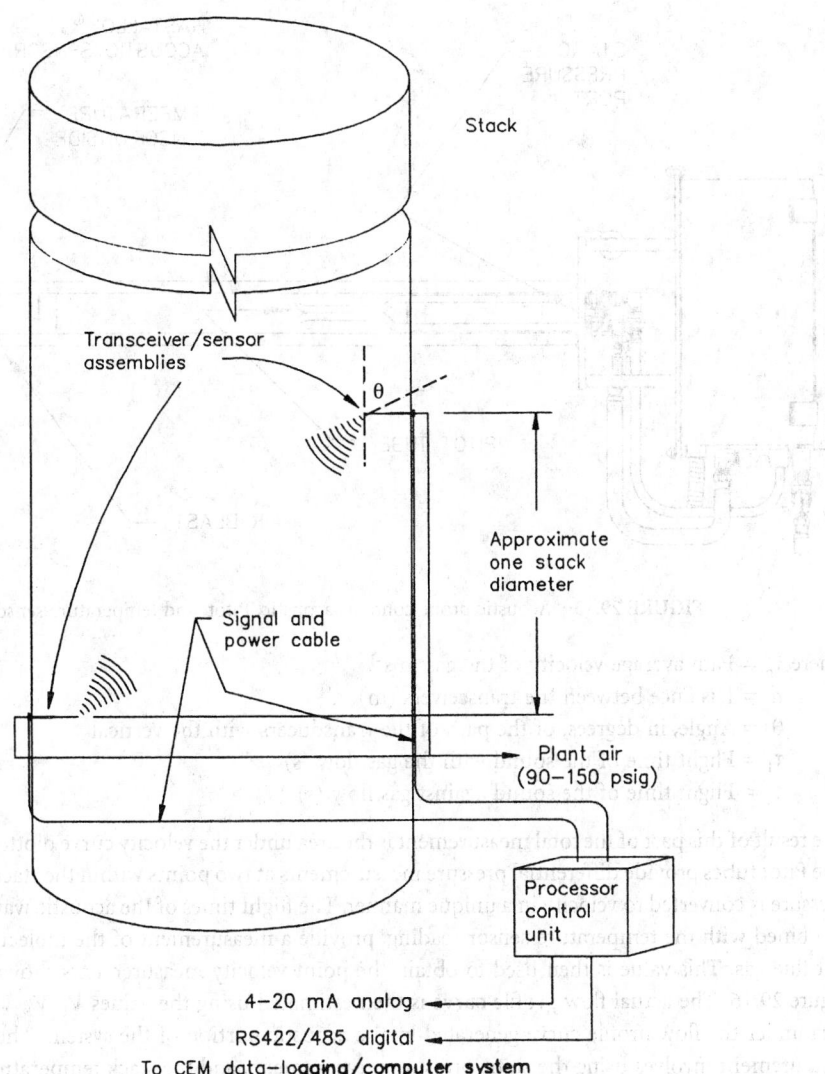

FIGURE 29.14 Block diagram of the hybrid system.

A Hybrid System

A hybrid system that combines sonic (acoustic) and Pitot tube technology has been developed to measure volumetric flow in large ducts and stacks [9–12]. A block diagram of this system is shown in Figure 29.14. The sensors (Figure 29.15) are mounted on opposite sides of the stack or duct at an angle θ to the flow direction. The acoustic portion of the sensor measures the flight time of the sound waves with and against the gas flow. It can easily be shown [9] that by transmitting and receiving the sound waves in opposite directions, the path average velocity of the gaseous medium can be determined from:

$$\bar{v}_\mathrm{p} = \frac{d}{2\cos\theta}\left(\frac{\tau_2 - \tau_1}{\tau_1\,\tau_2}\right)\mathrm{m\ s}^{-1} \tag{29.27}$$

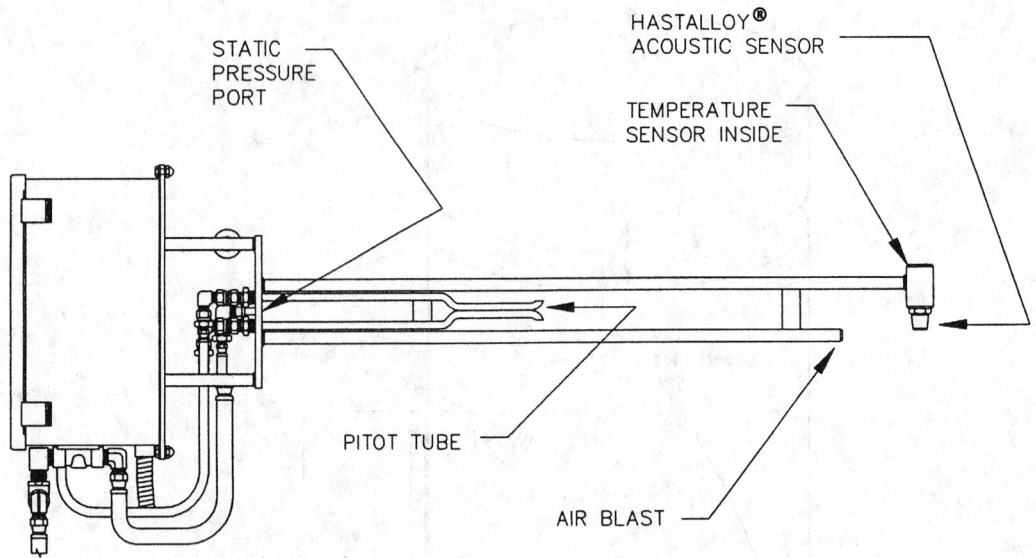

FIGURE 29.15   Acoustic probe contains acoustic, Pitot, and temperature sensors.

where $\bar{v}_p$ = Path average velocity of the gas m s^{-1}
    d = Distance between the transceivers (m)
    θ = Angle, in degrees, of the path of the transducers with the vertical
    τ_1 = Flight time of the sound with the gas flow (s)
    τ_2 = Flight time of the sound against gas flow (s)

The result of this part of the total measurement is the area under the velocity curve plotted in Figure 29.16. The Pitot tubes provide differential pressure measurements at two points within the stack. The differential pressure is converted to velocity in a unique manner. The flight times of the acoustic wave, when properly combined with the temperature sensor reading, provide a measurement of the molecular weight of the wet flue gas. This value is then used to obtain the point velocity measurements shown as V_2 and V_3 in Figure 29.16. The actual flow profile curve is then estimated using the values V_1, V_2, V_3, and V_4 and the area under the flow profile curve generated by the acoustic portion of the system. The final part of the measurement involves using the static pressure measurements and the stack temperature measurements to calculate the total standard volumetric flow in scfh (wet).

Commercial Availability

There are a variety of material used to construct Pitot tubes. The reasons for this are that Pitot tubes are used to measure a wide range of fluids. For example, to use a type S Pitot tube in a large power plant stack with a wet scrubber where the environment is extremely hostile and corrosive, stainless steel 316 or C276 (Hastaloy®) must be used. This, of course, makes the price of the Pitot tube as varied as its application. Many of the basic type S Pitot tube probes themselves are manufactured by a few small companies who, in turn, supply them on an OEM basis to others.

A typical type S Pitot tube assembly, such as that shown in Figure 29.9, constructed using stainless steel 316 can be purchased (in small quantities) for $310 each. They are available from:

EEMC/EMRC
3730 North Pellegrino Drive
Tucson, AZ 85749
Tel: (520) 749-2167
Fax: (520) 749-3582

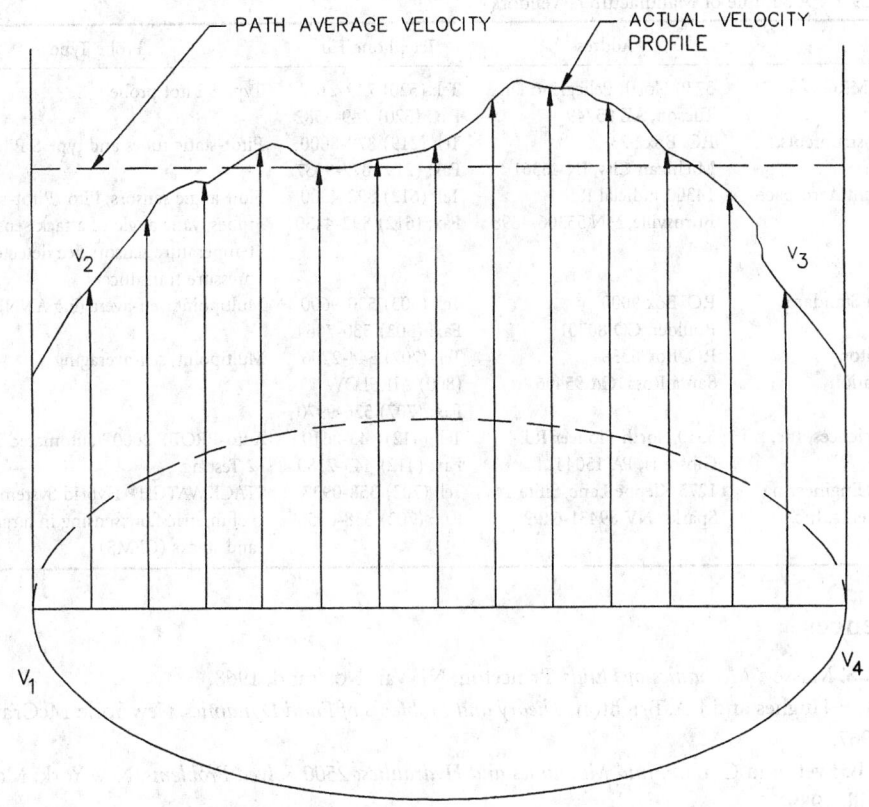

PATH AVERAGE VELOCITY

ACTUAL VELOCITY PROFILE

V_2

V_3

V_1

V_4

$V_1 = V_4 = 0$ AT WALLS
V_2 = VELOCITY AT POINT, MEASURED BY PITOT TUBE
V_3 = VELOCITY AT DIAMETRIC POINT, MEASURED BY PITOT TUBE

FIGURE 29.16 The velocity profile in a typical large duct or stack can vary greatly, thus changing the total volumetric flow. The hybrid system assumes V_1 and V_4 to be zero; measures V_2 and V_3 using the Pitot tubes; and provides the path average (area under the curve) using the acoustic portions of this sensor.

A typical modern Pitot-static assembly, such as that shown in Figure 29.6, can be purchased (in small quantities) for $34 and are available from:

Dwyer Instruments, Inc.
P.O. Box 373
Michigan City, IN 46361
Tel: (219) 879-8000
Fax: (219) 872-9057

More complex, custom-designed and fabricated Pitot-static probes for use on aircraft are available from:

Rosemount Aerospace Inc.
14300 Judicial Road
Burnsville, MN 55306-4898
Tel: (612) 892-4300
Fax: (612) 892-4430

Table 29.2 lists a number of manufactures/vendors that sell Pitot tube and general differential pressure measurement instrumentation.

TABLE 29.2   A Sample of Manufacturers/Vendors

Name	Address	Telephone/Fax	Probe Type
EEMC/EMRC	3730 North Pellegrino Dr. Tucson, AZ 85749	Tel: (520) 749-2167 Fax: (520) 749-3582	Type S Pitot probe
Dwyer Instruments, Inc.	P.O. Box 373 Michigan City, IN 46361	Tel: (219) 879-8000 Fax: (219) 872-9057	Pitot-static tubes and type S Pitot probe
Rosemount Aerospace, Inc.	14300 Judicial Rd. Burnsville, MN 55306-4898	Tel: (612) 892-4300 Fax: (612) 892-4430	Flow angle sensors, Pitot/Pitot-static tubes, vane angle of attack sensors, temperature sensors, ice detectors, and pressure transducers
Dieterich Standard	P.O. Box 9000 Boulder, CO 80301	Tel: (303) 530-9600 Fax: (303) 530-7064	Multipoint, self-averaging ANNUBAR®
Air Monitor Corporation	P.O. Box 6358 Santa Rosa, CA 95406	Tel: (707) 544-2706 (800) AIRFLOW Fax: (707) 526-9970	Multipoint, self-averaging
United Sciences, Inc.	5310 North Pioneer Rd. Gibsonia, PA 15044	Tel: (412) 443-8610 Fax: (412) 443-7180	Auto-PROBE 2000® automated Method 2 Testing
Scientific Engineering Instruments, Inc.	1275 Kleppe Lane, Suite 14 Sparks, NV 89431-6499	Tel: (702) 358-0937 Fax: (702) 358-0956	STACKWATCH® Hybrid System for volumetric flow sensing in large ducts and stacks (CEMS)

References

1. B. S. Massey, *Mechanics of Fluids,* Princeton, NJ: Van Nostrand, 1968.
2. W. F. Hughes and J. A. Brighton, *Theory and Problems of Fluid Dynamics,* New York: McGraw-Hill, 1967.
3. J. B. Evett and C. Liu, *Fluid Mechanics and Hydraulics: 2500 Solved Problems,* New York: McGraw-Hill, 1989.
4. R. P. Benedict, *Fundamentals of Temperature, Pressure and Flow Measurements,* 3rd ed., New York: John Wiley & Sons, 1984.
5. ASME/ANSI PTC 19.2 - 1987, *Instruments and Apparatus, Part 2, Pressure Measurement,* 1987.
6. S. P. Parker, *Fluid Mechanics Source Book,* New York: McGraw-Hill, 1988.
7. EPA, *CFR 40 Part 60, Appendix A—Test Methods,* 1 July 1995.
8. T. C. Elliott, CEM System: Lynchpin Holding CAA Compliance Together, *Power,* May 1995, 31–40.
9. J. A. Kleppe, Principles and Applications of Acoustic Sensors Used for Gas Temperature and Flow Measurement, *Proc. SENSOR EXPO,* Boston, May 1995, 337–374.
10. J. A. Kleppe, Acoustic Gas Flow Measurement in Large Ducts and Stacks, *Sensors J.,* 12(5), 18–24 and 85–87, 1995.
11. *Guidelines for Flue Gas Flow Rate Monitoring,* EPRI TR-104527, Project 1961-13 Final Report, June 1995.
12. A. Mann and J. A. Kleppe, A Report on the Performance of a Hybrid Flow Monitor Used for CEMS and Heat Rate Applications, *Proc. EPRI 1996 Heat Rate Improvement Conf.,* Dallas, TX, May 1996, Part 33, 1–13.

29.2   Thermal Anemometry

John G. Olin

General Description

A thermal anemometer measures the velocity at a point in a flowing fluid — a liquid or a gas. Figure 29.17 shows a typical industrial thermal anemometer used to monitor velocity in gas flows. It has two sensors — a velocity sensor and a temperature sensor — that automatically correct for changes in gas temperature.

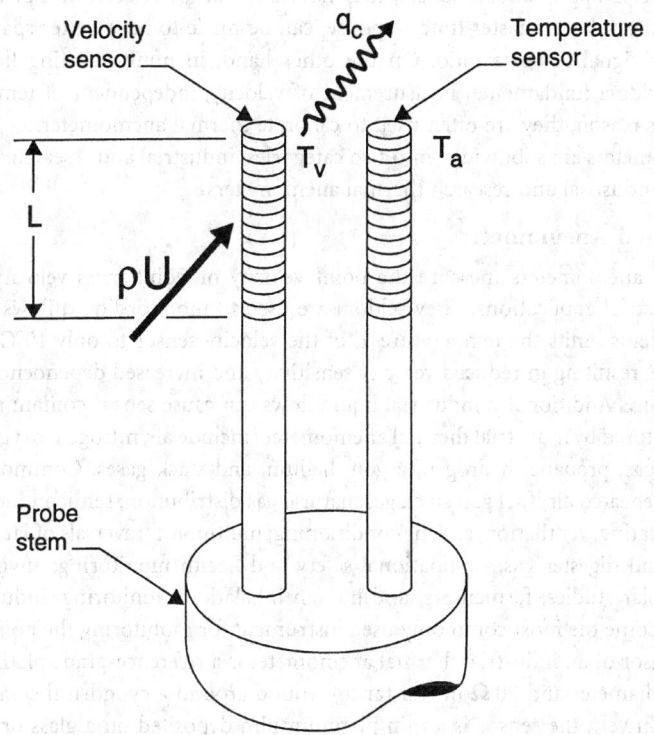

FIGURE 29.17 The principle of operation of a typical industrial thermal anemometer. T_v is the temperature of the heated velocity sensor; T_a is the gas temperature measured by the temperature sensor; ρ is the gas mass density; U is the gas velocity; q_c is the heat carried away by the flowing gas stream; and L is the length of the heated tip of the sensor. (Reprinted with the permission of Sierra Instruments, Inc.)

Both sensors are reference-grade platinum resistance temperature detectors (RTDs). The electric resistance of RTDs increases as temperature increases. For this reason, they are one of the most commonly used sensors for accurate temperature measurements. The electronics circuit passes current through the velocity sensor, thereby heating it to a constant temperature differential $(T_v - T_a)$ above the gas temperature T_a and measures the heat q_c carried away by the cooler gas as it flows past the sensor. Hence, it is called a "constant-temperature thermal anemometer."

Because the heat is carried away by the gas molecules, the heated sensor directly measures gas mass velocity (mass flow rate per unit area) ρU. The mass velocity is typically expressed as U_s in engineering units of normal meters per second, or *normal* m s^{-1}, referenced to normal conditions of 0°C or 20°C temperature and 1 atm pressure. If the fluid's temperature and pressure are constant, then the anemometer's measurement can be expressed as *actual* meters per second, or m s^{-1}. When the mass velocity is multiplied by the cross-sectional area of a flow channel, the mass flow rate through the channel is obtained. Mass flow rate, rather than volumetric flow rate, is the direct quantity of interest in most practical and industrial applications, such as any chemical reaction, combustion, heating, cooling, drying, mixing, fluid power, human respiration, meteorology, and natural convection.

The thermal anemometer is often called an *immersible* thermal mass flowmeter because it is immersed in the flow stream, in contrast to the *capillary-tube* thermal mass flowmeter, another thermal methodology commonly configured as an in-line mass flowmeter for low gas flows. The thermal anemometer has some advantages and disadvantages when compared with the two other common point-velocity instruments — Pitot tubes and laser Doppler anemometers. Compared with Pitot tubes, the thermal anemometer measures lower velocities, has much wider rangeability, and can be made smaller, but it generally has a higher cost and is not recommended for nonresearch liquid flows. When thermal anemometers are compared with laser Doppler anemometers, they have a much lower cost, do not require seeding the flow with particles, can have a faster time response, can be made to have better spatial resolution, and can have a higher signal-to-noise ratio. On the other hand, in nonfluctuating flows, laser Doppler anemometers provide a fundamental measurement of velocity, independent of temperature and fluid properties. For this reason, they are often used to calibrate thermal anemometers.

Thermal anemometers are subdivided into two categories: industrial and research. Figure 29.18 shows typical sensors of industrial and research thermal anemometers.

Industrial Thermal Anemometers

Industrial thermal anemometers measure the point velocity or point mass velocity of gases in most practical and industrial applications. They seldom are used to monitor liquid flows because avoidance of cavitation problems limits the temperature T_v of the velocity sensor to only 10°C to 20°C above the liquid temperature, resulting in reduced velocity sensitivity and increased dependence on small changes in liquid temperature. Additionally, industrial liquid flows can cause sensor contamination and fouling. Typical gases monitored by industrial thermal anemometers include air, nitrogen, oxygen, carbon dioxide, methane, natural gas, propane, hydrogen, argon, helium, and stack gases. Common applications are: combustion air; preheated air; fuel gas; stack gas; natural gas distribution; semiconductor manufacturing gas distribution; heating, ventilation, and air conditioning; multipoint traversals of large ducts and stacks; drying; aeration and digester gas; occupational safety and health monitoring; environmental, natural convection, and solar studies; fermentors; and human inhalation monitoring. Industrial thermal anemometers have become the most commonly used instrument for monitoring the point velocity of gases.

The velocity sensor of an industrial thermal anemometer is a reference-grade platinum wire (approximately 25 μm in diameter and 20 Ω in resistance) wound around a cylindrical ceramic mandrel, such as alumina. Alternatively, the sensor is a thin platinum film deposited on a glass or ceramic substrate. To withstand the harsh environment encountered in many industrial applications, the cylindrical platinum RTD is tightly cemented into the tip of a thin-walled, stainless-steel, Hastelloy, or Inconel tube (typically 3 mm outside diameter and 2 cm to 6 cm long). Because the gas temperature usually varies in industrial applications, industrial thermal anemometer probes almost always have a separate, but integrally mounted, unheated platinum RTD sensor for measuring the local gas temperature T_a. When operated in the constant-temperature anemometer mode, the temperature difference $(T_v - T_a)$ is usually in the 30°C to 100°C range. The temperature sensor is constructed just like the velocity sensor, but has a resistance in the 300 Ω to 1000 Ω range. As shown in Figure 29.17, the dual-sensor probe has the velocity and temperature sensor mounted side-by-side on a cylindrical probe stem (usually 6 mm to 25 mm in diameter and 0.1 m to 3 m long). A shield usually is provided to prevent breakage of the sensing head. The spatial resolution of this industrial thermal anemometer is 1 cm to 2 cm. The electronics for the industrial thermal anemometer is usually mounted directly on the probe stem in an explosion-proof housing. Industrial thermal anemometer systems like this measure gas velocity over the range of 0.5 normal m s^{-1} to 150 normal m s^{-1}.

In use, the industrial thermal anemometer probe is inserted through a sealed compression fitting or flanged stub in the wall of a duct, pipe, stack, or other flow passage. In this case, it is usually called an *insertion* thermal mass flowmeter. In another common configuration, the dual-sensor probe is permanently fitted into a pipe or tube (typically 8 mm to 300 mm in diameter) with either threaded or flanged gas connections. This configuration is called an *in-line* thermal mass flowmeter. In-line meters are directly calibrated for the total gas mass flow rate flowing through the pipe. The several flow body sizes facilitate mass flow monitoring over the range of 10 mg s^{-1} to 10 kg s^{-1}.

Research Thermal Anemometers

Research thermal anemometers measure the point velocity and/or turbulence of clean gases and liquids in research, product development, and laboratory applications. Because of their more fragile nature, they are not used for industrial applications. Typically, the gas is ambient air. Constant-temperature, filtered, degasified water is the primary liquid application, but the technique has also been applied to clean

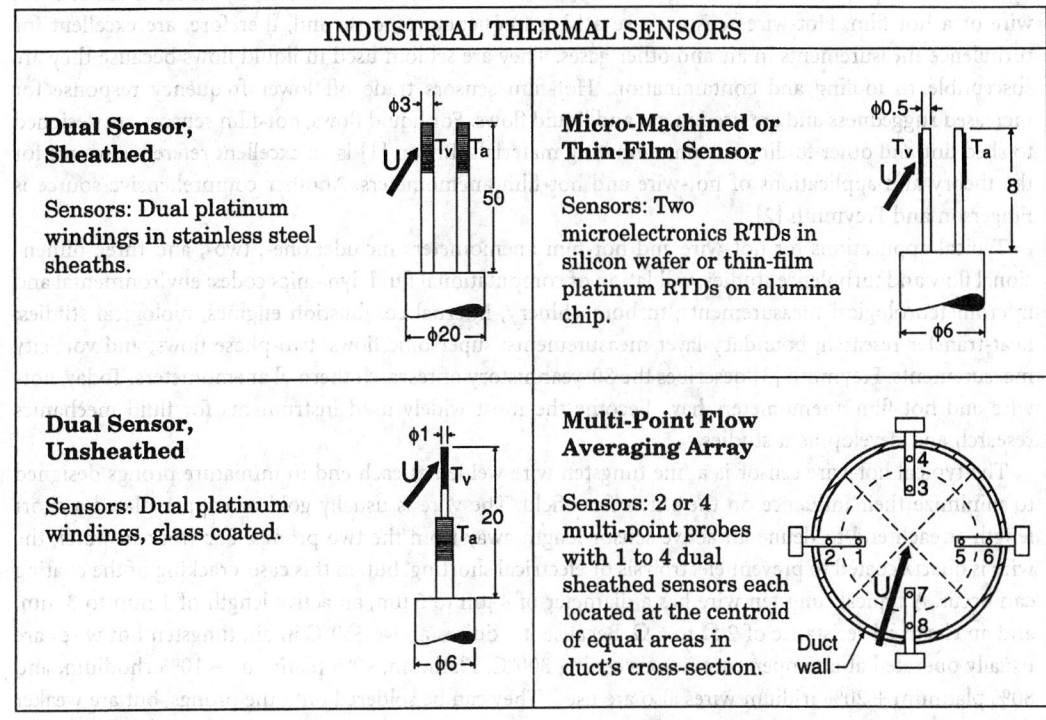

INDUSTRIAL THERMAL SENSORS

Dual Sensor, Sheathed

Sensors: Dual platinum windings in stainless steel sheaths.

Micro-Machined or Thin-Film Sensor

Sensors: Two microelectronics RTDs in silicon wafer or thin-film platinum RTDs on alumina chip.

Dual Sensor, Unsheathed

Sensors: Dual platinum windings, glass coated.

Multi-Point Flow Averaging Array

Sensors: 2 or 4 multi-point probes with 1 to 4 dual sheathed sensors each located at the centroid of equal areas in duct's cross-section.

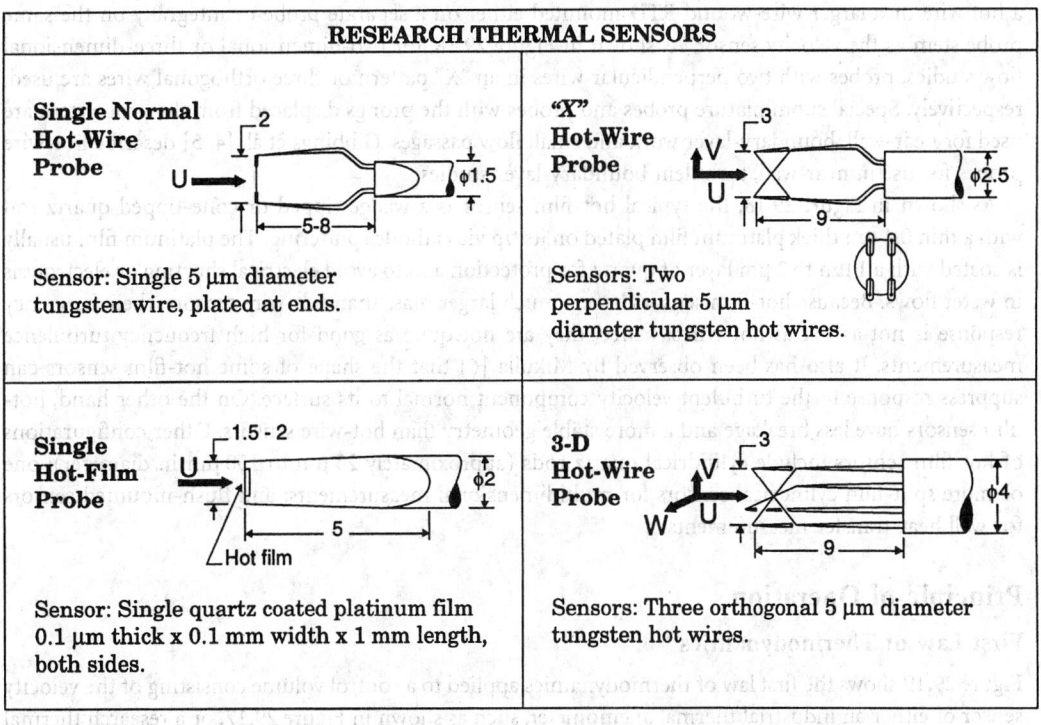

RESEARCH THERMAL SENSORS

Single Normal Hot-Wire Probe

Sensor: Single 5 µm diameter tungsten wire, plated at ends.

"X" Hot-Wire Probe

Sensors: Two perpendicular 5 µm diameter tungsten hot wires.

Single Hot-Film Probe

Sensor: Single quartz coated platinum film 0.1 µm thick x 0.1 mm width x 1 mm length, both sides.

3-D Hot-Wire Probe

Sensors: Three orthogonal 5 µm diameter tungsten hot wires.

FIGURE 29.18 Typical industrial and research thermal anemometer sensors. All dimensions are in millimeters. T_v indicates the heated velocity sensor; T_a indicates the temperature sensor; U is the major velocity component in the x-direction; V is the transverse velocity component in the y-direction; and W is the transverse velocity component in the z-direction.

hydrocarbon liquids. As shown in Figure 29.18, the research anemometer's velocity sensor is either a hot wire or a hot film. Hot-wire sensors have a high frequency response and, therefore, are excellent for turbulence measurements in air and other gases. They are seldom used in liquid flows because they are susceptible to fouling and contamination. Hot-film sensors trade off lower frequency response for increased ruggedness and are used in gas and liquid flows. For liquid flows, hot-film sensors are designed to shed lint and other fouling or contaminating materials. Bruun [1] is an excellent reference source for the theory and applications of hot-wire and hot-film anemometers. Another comprehensive source is Fingerson and Freymuth [2].

Typical applications for hot-wire and hot-film anemometers include: one-, two-, and three-dimensional flow and turbulence studies; validation of computational fluid dynamics codes; environmental and micrometeorological measurements; turbomachinery; internal combustion engines; biological studies; heat-transfer research; boundary-layer measurements; supersonic flows; two-phase flows; and vorticity measurements. Freymuth [3] describes the 80-year history of research thermal anemometers. Today, hot-wire and hot-film anemometers have become the most widely used instruments for fluid mechanics research and development studies.

The typical hot-wire sensor is a fine tungsten wire welded at each end to miniature prongs designed to minimize their influence on the wire's flow field. The wire is usually gold or copper plated a short length at each end to define an active sensor length away from the two prongs. For work in water, the wire is quartz coated to prevent electrolysis or electrical shorting, but, in this case, cracking of the coating can occur. A typical tungsten wire has a diameter of 4 μm to 5 μm, an active length of 1 mm to 3 mm, and an electrical resistance of 2 Ω to 6 Ω. Because it oxidizes above 350°C in air, tungsten hot wires are usually operated at a temperature not exceeding 300°C. Platinum, 90% platinum + 10% rhodium, and 80% platinum + 20% iridium wires also are used. They can be soldered onto the prongs, but are weaker than tungsten. In cases where the fluid temperature T_a changes enough to cause measurement errors, a separate sensor is used to measure T_a and make temperature corrections. The temperature sensor is either a hot wire or a larger wire-wound RTD mounted either on a separate probe or integrally on the same probe stem as the velocity sensor. As shown in Figure 29.18, for two-dimensional or three-dimensional flow studies, probes with two perpendicular wires in an "X" pattern or three orthogonal wires are used, respectively. Special subminiature probes and probes with the prongs displaced from the probe stem are used for near-wall, boundary-layer work and small flow passages. Gibbings et al. [4, 5] describe hot-wire probes for use in near-wall, turbulent boundary-layer studies.

As shown in Figure 29.18, the typical hot-film sensor is a wedge-tipped or cone-tipped quartz rod with a thin 0.1 μm thick platinum film plated on its tip via cathode sputtering. The platinum film usually is coated with a 1 μm to 2 μm layer of quartz for protection and to avoid electrical shorting or electrolysis in water flows. Because hot-film sensors have a much larger mass than hot-wire sensors, their frequency response is not as flat as hot wires; hence, they are not quite as good for high frequency turbulence measurements. It also has been observed by Mikulla [6] that the shape of some hot-film sensors can suppress response to the turbulent velocity component normal to its surface. On the other hand, hot-film sensors have less breakage and a more stable geometry than hot-wire sensors. Other configurations of hot-film sensors include cylindrical quartz rods (approximately 25 μm to 150 μm in diameter); one or more split-film cylindrical sensors for multidimensional measurements; and flush-mounted sensors for wall heat-transfer measurements.

Principle of Operation

First Law of Thermodynamics

Figure 29.19 shows the first law of thermodynamics applied to a control volume consisting of the velocity sensor of either an industrial thermal anemometer, such as shown in Figure 29.17, or a research thermal anemometer. Application of the first law to thermal anemometer sensors provides the basis for determining point velocity. Applied to Figure 29.19, the first law states that the energy into the control volume

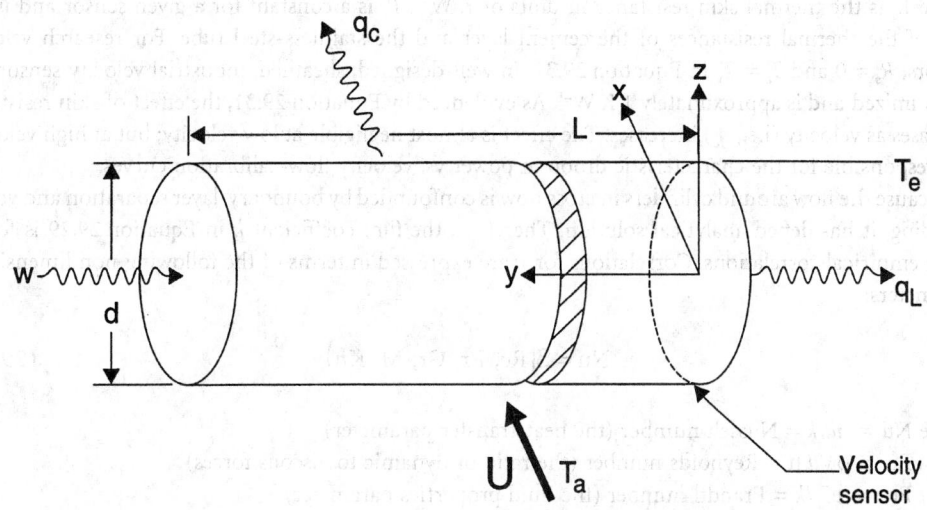

FIGURE 29.19 First law of thermodynamics applied to a thermal anemometer velocity sensor. The term w is the electric power (Watts) supplied to the sensor; q_c is the heat convected away from the sensor by the flowing fluid having a velocity U and temperature T_a; q_L is the conductive heat lost; T_e is the average surface temperature of the sensor over its length L; and d is the sensor's outside diameter.

equals the energy out plus the energy stored. Making the practical simplifying assumptions of steady-state operation (i.e., no energy stored) and no heat transfer via radiation, one obtains:

$$w = q_c + q_L \qquad (29.28)$$

The heat transfer q_c due to natural and forced convection normally is expressed in terms of the heat transfer coefficient h as:

$$q_c = hA_v \left(T_e - T_a \right) \qquad (29.29)$$

where $A_v = \pi dL$ is the external surface area of the velocity sensor. The electric power w usually is expressed as:

$$w = E_v^2 / R_v \qquad (29.30)$$

where E_v is the voltage across the sensor, and R_v is its electric resistance.

For the industrial velocity sensor shown in Figure 29.17, q_L is the heat conducted from the end of the heated velocity sensor of length L to the remainder of the sensor's length. Most of this heat is convected away by the flowing fluid, and a small fraction is conducted to the probe stem. In the case of research hot-wire or cylindrical hot-film sensors, q_L is conducted to the two prongs, of which a major fraction is convected away and a minor fraction enters the probe stem. In well-designed velocity sensors, q_L is at most 10% to 15% of w, a fraction that decreases as velocity increases.

For research velocity sensors, the surface temperature T_e is identical to the wire or film temperature T_v. However, the surface temperature T_e of industrial velocity sensors with stainless-steel sheaths is slightly less than the temperature T_v of the platinum winding because a temperature drop is required to pass the heat q_c through the intervening "skin" — the cement layer and the stainless-steel tube. This is expressed as:

$$T_e = T_v - q_c R_s \qquad (29.31)$$

where R_s is the thermal skin resistance in units of K W^{-1}. R_s is a constant for a given sensor and is the sum of the thermal resistances of the cement layer and the stainless-steel tube. For research velocity sensors, $R_s = 0$ and $T_e = T_v$ in Equation 29.31. In well-designed, sheathed, industrial velocity sensors, R_s is minimized and is approximately 1 K W^{-1}. As evidenced by Equation 29.31, the effect of skin resistance increases as velocity (i.e., q_c) increases. The effect is almost negligible at low velocity; but at high velocity, it is responsible for the characteristic droop in power vs. velocity flow-calibration curves.

Because the flow around cylinders in cross flow is confounded by boundary-layer separation and vortex shedding, it has defied analytical solution. Therefore, the film coefficient h in Equation 29.29 is found using empirical correlations. Correlations for h are expressed in terms of the following nondimensional parameters:

$$\text{Nu} = \Im\left(\text{Re, Pr, Gr, M, Kn}\right) \tag{29.32}$$

where Nu = hd/k = Nusselt number (the heat-transfer parameter)

Re = $\rho V d/\mu$ = Reynolds number (the ratio of dynamic to viscous forces)

Pr = $\mu C_p/k$ = Prandtl number (the fluid properties parameter)

M = Mach number (the gas compressibility parameter)

Kn = Knudsen number (the ratio of the gas mean free path to d)

In the above, k is the fluid's thermal conductivity; μ is its viscosity; and C_p is its coefficient of specific heat at constant pressure. If one takes the practical case where: (1) natural convection is embodied in Re and Pr, (2) the velocity is less than one third the fluid's speed of sound (i.e., <100 m s^{-1} in ambient air), and (3) the flow is not in a high vacuum, then one can ignore the effects of Gr, M, and Kn, respectively. Thus,

$$\text{Nu} = \Im\left(\text{Re, Pr}\right) \tag{29.33}$$

Over the years, many attempts have been made to find universal correlations for the heat transfer from cylinders in cross flow. For an isothermal fluid at constant pressure, King [7] expresses Equation 29.33 as:

$$\text{Nu} = A + B\text{Re}^{0.5} \tag{29.34}$$

where A and B are empirical calibration constants that are different for each fluid and each temperature and pressure. Kramers [8] suggests the following correlation:

$$\text{Nu} = 0.42\text{Pr}^{0.2} + 0.57\text{Pr}^{0.33}\text{Re}^{0.50} \tag{29.35}$$

This correlation accounts for the variation in fluid properties (k, μ, and Pr) with temperature. Kramers [8] evaluates these properties at the so-called "film" temperature $(T_v + T_a)/2$, rather than at T_a itself. Another comprehensive correlation is given by Churchill and Bernstein [9]. Several other correlations are similar to Equation 29.35, but have exponents for the Reynolds number ranging from 0.4 to 0.6. Others have 0.36 and 0.38 for the exponent of the Prandtl number. Equations 29.34 and 29.35 are strictly valid only for hot-wire sensors with very high L/d ratios, in which case q_L and R_s are zero. The following universal correlation is suggested for real-world velocity sensors with variable fluid temperature and nonzero q_L and R_s:

$$\text{Nu} = A + B\text{Pr}^{0.33}\text{Re}^n \tag{29.36}$$

where constants A, B, and n are determined via flow calibration. Equation 29.36 is applicable to most commercial industrial and research velocity sensors.

Combining Equations 29.28, 29.29, 29.30, and 29.36, and recognizing that $h = kNu/d$, one obtains:

$$E_v^2/R_v = \left(Ak + Bk\mathrm{Pr}^{0.33}\mathrm{Re}^n\right)\left(T_v - T_a\right) \tag{29.37}$$

where A and B are new constants. A, B, and n are determined via flow calibration and account for all nonidealities, including end conduction and skin resistance. Equation 29.37 is applicable to most commercial industrial and research velocity sensors. Manufacturers of industrial thermal anemometers can add other calibration constants to Equation 29.37 to enhance its correlation with flow-calibration data. The presence of end conduction means that the temperature of the velocity sensor varies with the axial coordinate y in Figure 29.19. The temperature T actually sensed by the velocity sensor is the *average* temperature over length L, or:

$$T_v = \left(1/L\right)\int_o^L T_v(y)\,dy \tag{29.38}$$

Bruun [1] presents an analytical solution for $T_v(y)$ for hot-wire sensors. Equation 29.38 is the correct expression for T_v in Equation 29.37 and is so defined hereafter.

For fluid temperatures less than 200°C, the electric resistance of the RTD velocity and temperature sensors is usually expressed as:

$$R_v = R_{v0}\left[1 + \alpha_v\left(T_v - T_0\right)\right] \tag{29.39}$$

$$R_T = R_{T0}\left[1 + \alpha_T\left(T_T - T_0\right)\right] \tag{29.40}$$

where R_{v0} and R_{T0} are, respectively, the electric resistances of the velocity sensor and the temperature sensor at temperature T_0 (usually 0°C or 20°C), and α_v and α_T are the temperature coefficients of resistivity at temperature T_0. Additional terms are added to Equations 29.39 and 29.40 when fluid temperatures exceed 200°C. When evaluated at the fluid temperature T_a, the resistance R_a of the velocity sensor is:

$$R_a = R_{v0}\left[1 + \alpha_v\left(T_a - T_0\right)\right] \tag{29.41}$$

For applications with wide excursions in fluid temperature, additional terms are added to Equations 29.39–29.41. At 20°C, α_v and α_T are approximately 0.0036°C^{-1} for tungsten wire; 0.0038°C^{-1} for pure platinum wire; 0.0016°C^{-1} for 90% platinum + 10% rhodium wire; 0.0024°C^{-1} for platinum film; and 0.0040°C^{-1} for tungsten film. R_v and R_a are called the "hot" and "cold" resistances of the velocity sensor, respectively. The ratio R_v/R_a is called the "overheat ratio." For gas flows, sheathed industrial velocity sensors are operated at overheat ratios from 1.1 to 1.4 ($T_v - T_a = 30$°C to 100°C). For gas flows, the overheat ratio of tungsten hot-wire and hot-film sensors are usually set to approximately 1.8 ($T_v - T_a = 200$°C to 300°C) and 1.4 ($T_v - T_a = 150$°C to 200°C), respectively. For water flows, the overheat ratio of hot-film sensors is approximately 1.05 to 1.10 ($T_v - T_a = 10$°C to 20°C). Mikulla [6] shows the importance of the effect of overheat ratio on frequency response.

Combining Equations 29.39 and 29.41, one obtains:

$$T_v - T_a = \frac{R_v - R_a}{\alpha_v R_{v0}} \tag{29.42}$$

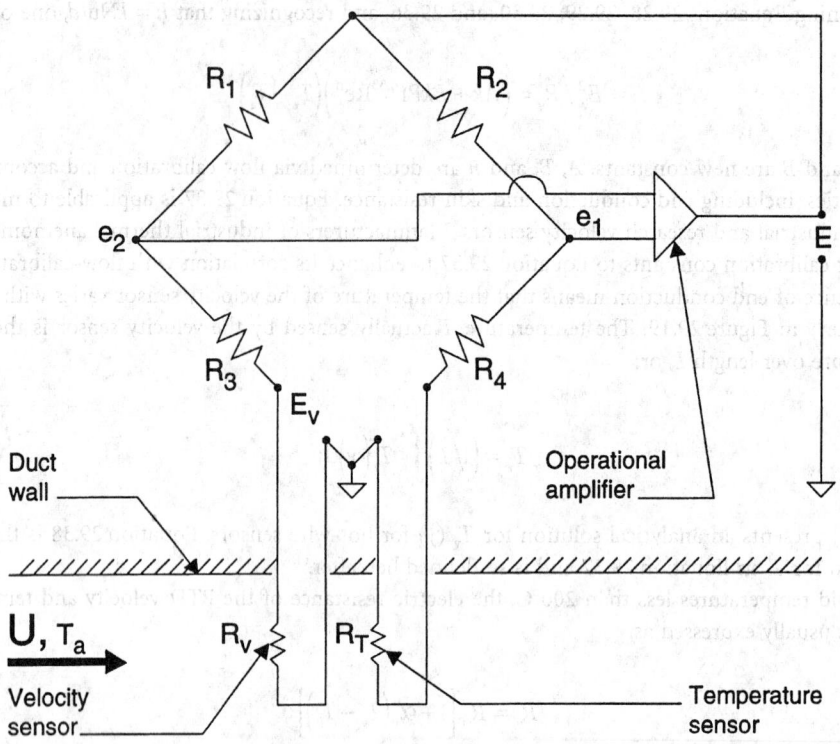

FIGURE 29.20   Constant-temperature thermal anemometer bridge circuit with automatic temperature compensation. R_1, R_2, and R_4 are fixed resistors selected to achieve temperature compensation; R_3 is the probe and cable resistance; R_v is the velocity sensor's resistance; R_T is the temperature sensor's resistance; and E is the bridge voltage output signal. For research anemometers operating in isothermal flows, the temperature sensor is eliminated and replaced with a variable bridge resistor. Some temperature compensation circuits have an additional resistor in parallel with R_T.

Inserting this into Equation 29.37 obtains:

$$\frac{E_v^2}{R_v\left(R_v - R_a\right)} = Ak + Bk\text{Pr}^{0.33}\text{Re}^n \qquad (29.43)$$

where new constants A and B have absorbed the constants α_v and R_{v0}.

Figures 29.20 to 29.22 show three typical electronic drives for thermal anemometer sensors. Figure 29.20 shows the commonly used constant-temperature anemometer Wheatstone bridge circuit described by Takagi [10]. Figure 29.21 is similar, but is controlled and operated via a personal computer. In the constant-temperature mode, the hot resistance R_v, and hence the velocity sensor's temperature, remains virtually constant, independent of changes in velocity. With the addition of the temperature sensor shown in Figure 29.20, the bridge circuit also compensates for variations in fluid temperature T_a, as described later. Another common analog sensor drive is the constant-current anemometer. In this mode, a constant current is passed through the velocity sensor, and the sensor's temperature decreases as the velocity increases. Because the entire mass of the sensor must participate in this temperature change, the sensor is slower in responding to changes in velocity. Because the constant-temperature anemometer has a flatter frequency response, excellent signal-to-noise ratio [2], and is easier to use, it is favored over constant-current anemometers by most researchers and manufacturers for velocity and turbulence measurements. The constant-current anemometer with a very low overheat ratio is often used

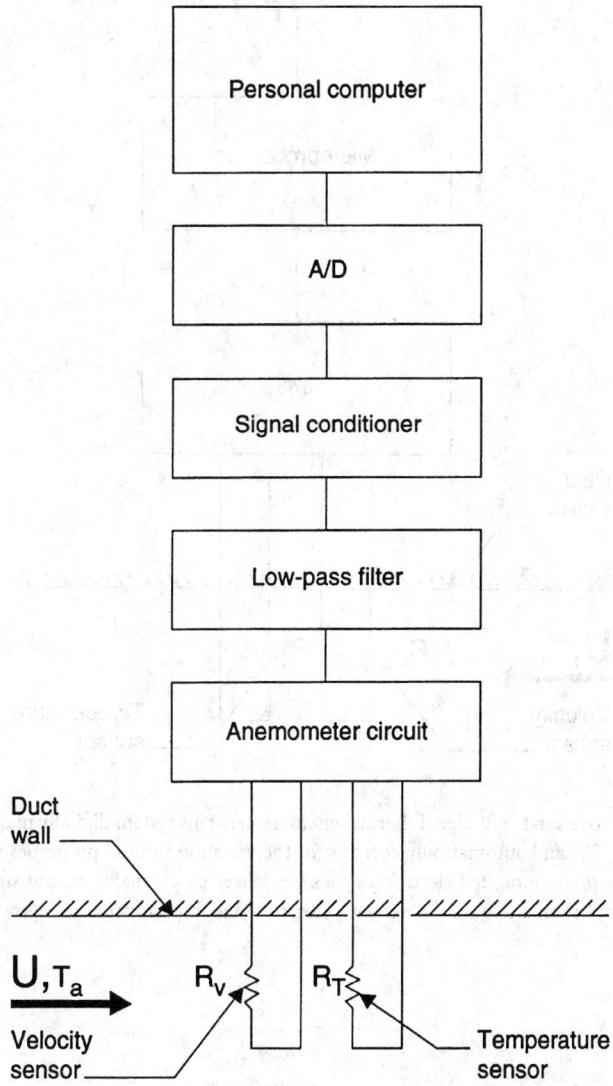

FIGURE 29.21 Personal computer-based digital thermal anemometer system. The signal conditioner matches the anemometer circuit's output to the ADC. For isothermal flows, the temperature sensor is eliminated.

as the temperature sensor. Subsequently, references made herein to sensor electronics will be based on the constant-temperature anemometer.

In the constant-temperature anemometer drive shown in Figure 29.20, the resistances R_1 and R_2 are chosen to: (1) maximize the current on the velocity-sensor side of the bridge so it becomes self-heated and (2) minimize the current on the temperature-sensor side of the bridge so it is not self-heated and is independent of velocity. Additionally, the temperature sensor must be sufficiently large in size to avoid self-heating. The ratio R_2/R_1 is called the "bridge ratio." A bridge ratio of 5:1 to 20:1 is normally used; but for optimum frequency response and compensation for long cable length, a bridge ratio of 1:1 can be used. In Figure 29.20, the operational amplifier, in a feedback control loop, senses the error voltage $(e_2 - e_1)$ and feeds the exact amount of current to the top of the bridge necessary to make $(e_2 - e_1)$ approach zero. In this condition, the bridge is balanced; that is,

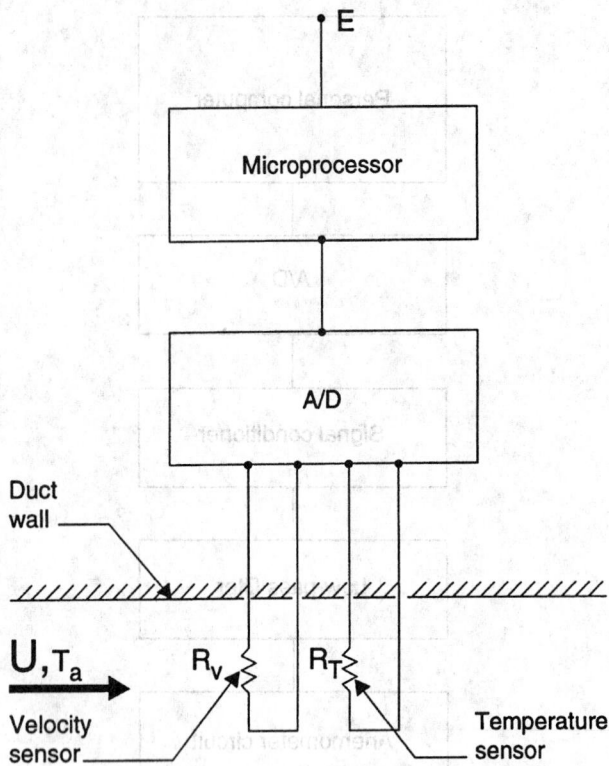

FIGURE 29.22 Microprocessor-based digital thermal anemometer. This system digitally maintains a constant temperature difference $(T_v - T_T)$ and automatically corrects for the variation in fluid properties with temperature. The manufacturer provides a probe-mounted electronics package delivering an analog output signal E and/or a digital RS485 signal linearly proportional to gas mass velocity. (Reprinted with permission of Sierra Instruments, Inc.)

$$\frac{R_1}{R_v + R_3} = \frac{R_2}{R_T + R_4} \qquad (29.44)$$

or

$$R_v = \frac{R_1}{R_2}\left(R_T + R_4\right) - R_3 \qquad (29.45)$$

From Equation 29.45, one sees that R_v is a linear function of R_T. This relationship forms the basis for analog temperature compensation.

Expressing the voltage E_v across the velocity sensor in terms of the bridge voltage E, one obtains:

$$E_v = \frac{ER_v}{R_1 + R_3 + R_v} \qquad (29.46)$$

Inserting this into Equation 29.43, one arrives at the generalized expression for the first law of thermodynamics for the thermal anemometer velocity sensor:

$$E^2 = G\left[Ak + Bk\left(\frac{\rho_s}{\mu}\right)^n \mathrm{Pr}^{0.33} U_s^n\right] \tag{29.47}$$

where $G = (R_1 + R_3 + R_v)^2 (R_v - R_a)/R_v$, and where A and B again are new constants. In Equation 29.47, one recognizes that conservation-of-mass considerations require that $\rho U = \rho_s U_s$, where ρ and U are referenced to the actual fluid temperature and pressure, and ρ_s and U_s are referenced to normal conditions of 0°C or 20°C temperature and 1 atm pressure. To write Equation 29.47 in terms of U, one simply replaces ρ_s by ρ and U_s by U.

Temperature Compensation

The objective of temperature compensation is to make the bridge voltage E in Equation 29.47 independent of changes in the fluid temperature T_a. This is accomplished if: (1) the term G in Equation 29.47 is independent of T_a and (2) compensation is made for the change in fluid properties (k, μ, and Pr) with T_a. Since these fluid properties have a weaker temperature dependence than G in Equation 29.47, for small temperature changes (less than ±10°C) in gas flows, only G requires compensation.

The two-temperature method is a typical procedure for compensating for both G and fluid properties. In this method, fixed-bridge resistors R_1, R_2, and R_4 in Figure 29.20 are selected so that E is identical at two different temperatures, but at the *same* mass flow rate. This procedure is accomplished during flow calibration and has variations among manufacturers.

The two-temperature method adequately compensates for temperature variations less than approximately ±50°C. In higher temperature gas flow applications, such as the flow of preheated combustion air and stack gas, temperature variations typically are higher. The microprocessor-based digital sensor drive in Figure 29.22 provides temperature compensation for temperature variations ranging from ±50°C to ±150°C. This sensor drive has no analog bridge. Instead, it has a virtual digital bridge that maintains $(T_v - T_a)$ constant within 0.1°C and has algorithms that automatically compensate for temperature variations in k, μ, and Pr. For this digital sensor drive, the first law of thermodynamics is found from Equation 29.37 as:

$$w = \left[Ak + Bk\left(\frac{\rho_s}{\mu}\right)^n \mathrm{Pr}^{0.33} U_s^n\right]\Delta T \tag{29.48}$$

where $\Delta T = (T_v - T_a)$ is now a known constant.

Flow Calibration

Figure 29.23 shows a typical flow calibration curve for the digital electronics drive shown in Figure 29.22. The curve is nonlinear of a logarithmic nature. The nonlinearity is disadvantageous because it requires linearization circuitry, but is advantageous because it provides rangeabilities up to 1000:1 for a single sensor. Additionally, the high-level output of several volts provides excellent repeatability and requires no amplification other than that for spanning. Since the critical dimensions of thermal anemometer sensors are so small, current manufacturing technology is incapable of maintaining sufficiently small tolerences to ensure sensor reproducibility. Therefore, each thermal anemometer must be flow calibrated, for example as in Figure 29.23, over its entire velocity range, either at the exact fluid temperature of its usage or over the range of temperatures it will encounter if it is to be temperature compensated. A 10 to 20 point velocity calibration is required to accurately determine the calibration constants A, B, and n in Equation 29.47. A least-squares curve-fitting procedure usually is applied. Proper flow calibration requires two critical elements: (1) a stable, reproducible, flow-generating facility and (2) an accurate velocity transfer standard. Bruun [1] and Gibbings et al. [4] provide more insight into curve fitting.

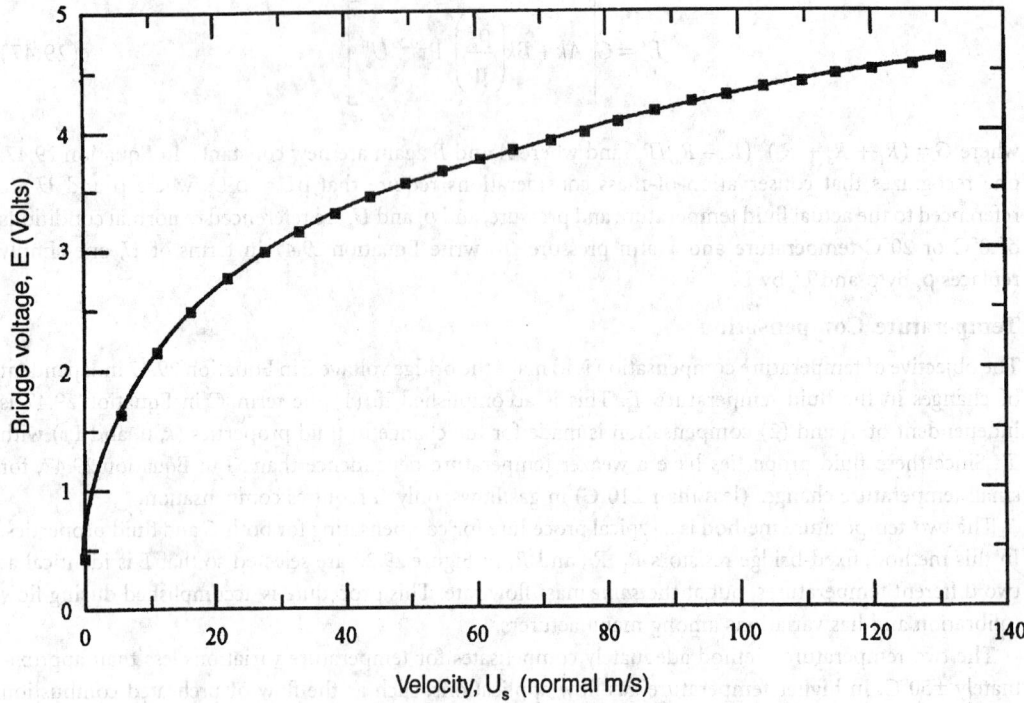

FIGURE 29.23 Typical flow calibration curve for an industrial thermal anemometer. The electronics drive is that shown in Figure 29.22. The constant temperature differential $(T_v - T_T)$ is 50.0°C. The cold resistances R_{v0} and R_{T0} of the velocity and temperature sensors at 20°C are approximately 20 Ω and 200 Ω, respectively. (Reprinted with permission of Sierra Instruments, Inc.)

Flow-generating facilities are of two types — open loop and closed loop. An open-loop facility consists of: (1) a flow source such as a fan, pump, elevated tank, or compressed gas supply; (2) a flow-quieting section, such as a plenum with flow straighteners, screens, or other means to reduce swirling, turbulence, or other flow nonuniformities; (3) a nozzle to accelerate the flow and further flatten, or uniformize, the velocity profile; (4) a test section or free jet into which the thermal anemometer probe is inserted; and (5) a means for holding and sealing the thermal anemometer probe and velocity transfer standard. The test section or free jet must have: a velocity profile which is uniform within approximately 0.5% to 1.0% in its central portion; a turbulence intensity less than about 0.5%; and an area large enough so that the projected area of the velocity probe is less than 5% to 10% of the cross-sectional area. Manufacturers of small open-loop flow calibrators often determine the calibration flow velocity by measuring the pressure drop across the nozzle.

The closed-loop flow-generating facility, or wind tunnel, has the same components, but the exit of the test section is connected via ductwork to the inlet of the fan or pump so that the air mass inside the facility is conserved. Open-loop facilities are less expensive than closed-loop tunnels and are far more compact, making them suitable for flow calibrations in the field. But, a laboratory open-loop air-flow calibrator with a fan as the flow generator actually is *closed loop*, with the loop closing within the laboratory. For air velocities less than about 5 m s⁻¹, open-loop calibrators can experience shifts due to changing pressure, temperature, or other conditions in the laboratory. Properly designed closed-loop wind tunnels generate precise, reproducible air velocities from about 0.5 m s⁻¹ to 150 m s⁻¹. When fitted with water chillers, they remove compression heating and provide a constant-temperature air flow within ±2°C. When fitted with an electric heater and proper thermal insulation, they provide air temperatures up to 300°C. Gibbings [4] describes a water box displacement rig for flow calibration at very low velocities in the range of 0.1 m s⁻¹ to 4 m s⁻¹.

Pitot tubes and laser Doppler anemometers are the two most common velocity transfer standards used to calibrate thermal anemometers. Both have detailed descriptions earlier in this chapter. The Pitot tube usually has the classical "L" shape and an outside diameter of about 3 mm. Its tip is located in the same plane in the test section as the thermal anemometer probe but is no closer than approximately 3 cm. The focal volume of the laser Doppler anemometer is similarly located. The Pitot tube is far less expensive and easier to operate, but is difficult to use if air velocities are less than about 3 m s⁻¹. A proper Pitot-tube flow transfer standard should have its calibration recertified every 6 months by an accredited standards laboratory. On the other hand, the laser Doppler anemometer is a fundamental standard that accurately measures air velocity from approximately 0.5 m s⁻¹ to 100 m s⁻¹. Since it provides noncontact anemometry, it is usable at high temperatures. Its primary disadvantages are high expense and complications associated with properly seeding the flow with particles.

Measurements

Point Velocity

Based on the first law of thermodynamics expressed by Equation 29.47, one now can solve for the desired quantity — either the actual point velocity U (m s⁻¹) or the point mass velocity U_s (normal m s⁻¹). Here, one assumes that the velocity vector is normal to the flow sensor. Two- and three-dimensional velocity measurements are discussed later. In the following, A, B, and n are constants, but are different for each case.

The simplest case is isothermal flow with a hot-wire sensor having a very high length-to-diameter ratio (L/d). In this case, the exponent n in Equation 29.47 is 0.5, as shown by Equation 29.34. The applicable first law and velocity expressions are:

$$E^2 = A + BU^{0.5} \tag{29.49}$$

and

$$U = \left[\frac{E^2 - A}{B}\right]^2 \tag{29.50}$$

In the case of a real-world sensor in an isothermal flow having either end loss only or both end loss and skin resistance, one obtains:

$$E^2 = A + BU^n \tag{29.51}$$

and

$$U = \left[\frac{E^2 - A}{B}\right]^{1/n} \tag{29.52}$$

Often, Equation 29.52 is replaced with a polynomial of the form $U = F(E^2)$, where the function $F(\ )$ is a fourth-order polynomial whose coefficients are determined from flow calibration data using least-squares curve-fitting software. For the same case as above, but with nonisothermal flow, the first law is expressed by Equation 29.47, and the velocity is expressed as:

$$U_s = \frac{\mu}{\rho_s}\left[\frac{E^2/G - Ak}{Bk\mathrm{Pr}^{0.33}}\right]^{1/n} \tag{29.53}$$

For the digital sensor drive of Figure 29.22, the first law is given by Equation 29.48, and the velocity by:

$$U_s = \frac{\mu}{\rho_s}\left[\frac{w/\Delta T - Ak}{Bk Pr^{0.33}}\right]^{1/n}$$
(29.54)

Current commercial industrial thermal anemoneter systems have temperature-compensation and "linearization" electronics that automatically calculate U_s as a linear function of E or w, based on the foregoing relationships.

Turbulence

Turbulence measurements are the second most common application of research thermal anemometers. This measurement requires the high-freqency response of hot-wire and hot-film research anemometers operated in the constant-temperature mode. The vast majority of fluid flows are turbulent. Only flows with very low Reynolds numbers are nonturbulent, or laminar. Turbulent flows are time variant and usually are separated as follows into time-mean and fluctuating parts:

$$U(t) = U + u$$

$$V(t) = \overline{V} + v$$

$$W(t) = \overline{W} + w$$
(29.55)

$$T_a(t) = \overline{T}_a + \theta$$

$$E = E + e$$

where $U(t)$, $V(t)$, $W(t)$ are the orthogomal components in the x, y, and z directions, respectively, such as shown in Figure 29.18 for the 3-D hot-wire probe. $T_a(t)$ is the fluid temperature, and $E(t)$ is the bridge voltage. $\overline{U}$, $\overline{W}$, $\overline{V}$, $\overline{T}_a$, and $\overline{E}$ are the time-mean parts, and $u(t)$, $v(t)$, $w(t)$, $\theta(t)$, and $e(t)$ are the time-dependent fluctuating parts. The time-mean parts are averaged sufficiently long to become independent of turbulent fluctuations, yet respond to changes with time in the main flow. In the previous subsection, the expressions given were for the time-mean velocity. In the study of turbulence, one is primarily interested in the time average of the product of two fluctuating velocity components (turbulence correlations) because these terms appear in the time-averaged Navier–Stokes equation. Two important turbulence correlations are $\overline{u^2}$ and $\overline{uv}$. The correlation $(\sqrt{\overline{u^2}})/\overline{U}$ is called the *turbulence intensity*. Manufacturers of research anemometer systems provide electronics for automatically computing turbulence correlations.

For a fluid with changes in temperature sufficiently small that fluid properties are essentially constant, one can write Equation 29.47 in the following form:

$$E^2 = \left(A + BU^n\right)\left(T_v - T_a\right)$$
(29.56)

where A, B, and n are constant and where R_v is virtually constant because the anemometer is in the constant-temperature mode. Elsner [11] shows that the fluctuating voltage e is found by taking the total derivative of Equation 29.56, as follows:

$$e = S_u u + S_\theta \theta$$
(29.57)

where

$$S_u = \frac{\delta E}{\delta U} = \frac{nBU^{n-1}}{2} \left[\frac{(T_v - T_a)}{A + BU^n} \right]^{1/2} = \text{Velocity sensitivity} \tag{29.58}$$

$$S_\theta = \frac{\delta E}{\delta T_a} = -\frac{1}{2} \left[\frac{(A + BU^n)}{(T_v - T_a)} \right]^{1/2} = \text{Temperature sensitivity} \tag{29.59}$$

It is seen from Equations 29.58 and 29.59 that increasing $(T_v - T_a)$, i.e., operating the sensor as hot as possible, maximizes the velocity sensitivity and minimizes the sensitivity to temperature fluctuations. This is why tungsten hot wires are operated at high temperatures (typically 200°C to 300°C).

The fluctuating components of velocity have a broad frequency spectrum, ranging from 10^{-2} Hz to 10^5 Hz, and sometimes even higher. Therefore, it is imperative that the frequency response of constant-temperature research anemometers have a flat frequency response, i.e., minimized attenuation and phase shift at higher frequencies. Blackwelder [12] and several other investigators have studied the frequency response of hot-wire anemometers. For turbulence measurements, Borgos [13] describes commercial research anemometer systems with features such as: low-pass filters to decrease electronics noise; a subcircuit for determining and setting overheat ratio; a square-wave generator for frequency response testing; and two or more controls to optimize the frequency response to fast fluctuations. Recent systems have electronics that compensate for frequency attenuation. When used with 5 μm diameter hot-wire sensors in air, commercial systems are capable of nearly flat frequency response and very small phase lag from 0 Hz to approximately 10^4 Hz. As reported by Nelson and Borgos [14], wedge and conical hot-film sensors in water have a relatively flat response from 0 Hz to 10 Hz for velocities above 0.3 m s^{-1}.

Two- and three-component velocity and turbulence measurements are made using hot-wire or hot-film research anemometers, such as shown in Figure 29.18. As described by Müller [15], hot-wire or cylindrical hot-film probes in the "X"-configuration are used to measure the U and V velocity components. In a three-sensor orthogonal array, they measure U, V, and W. Döbbeling, Lenze, and Leuckel [16] and other investigators have developed four-wire arrays for measurement of U, V, and W. Olin and Kiland [17] describe an orthogonal array of three cylindrical split hot-film sensors. Each of the three sensors in this array has two individually operated hot-films separated by two axial splits 180° apart along its entire length. The two split films take advantage of the nonuniform heat-transfer distribution around a cylinder in cross flow.

In multisensor arrays, the velocity vector is not necessarily normal to a cylindrical sensor. If the discussion is limited to isothermal flows, the first law expressed by Equation 29.47 becomes:

$$E^2 = A + BV_e^n \tag{29.60}$$

where V_e is the effective velocity sensed by a single cylindrical sensor in the array, and A, B, and n are constants. Jörgenson [18] describes V_e as follows:

$$V_e^2 = U_N^2 + a^2 U_T^2 + b^2 U_B^2 \tag{29.61}$$

where U_N = velocity component normal to the sensor
U_T = tangential component
U_B = component perpendicular to both U_N and U_T (i.e., binormal)

The constants a and b in Equation 29.61 are referred to as the sensor's yaw and pitch coefficients, respectively, and are determined via flow calibration. Typical values for a and b for a plated hot-wire sensor are 0.2 and 1.05, respectively. Inserting Equation 29.61 into Equation 29.60, we get the following expression for the output signal of a single sensor in the array:

$$E^2 = A + B\left(U_N^2 + a^2 U_T + b^2 U_B^2\right)^{n/2} \tag{29.62}$$

Expressions like this, or similar ones such as given by Lekakis, Adrian, and Jones [19], are written for all sensors in the array. These expressions and trigonometry are then used to solve for the components of velocity U, V, and W in the x, y, z spatially fixed reference frame.

Channel Flows

Based on the following relationship, a single-point industrial *insertion* thermal anemometer monitors the mass flow rate $\dot{m}$ (kg s^{-1}) in ducts, pipes, stacks, or other flow channels by measuring the velocity $U_{s,c}$ at the channel's centerline:

$$\dot{m} = \rho_s \gamma \, U_{s,c} A_c \tag{29.63}$$

where $U_{s,c}$ is the velocity component parallel to the channel's axis measured at the channel's centerline and referenced to *normal* conditions of 0°C or 20°C temperature and 1 atmosphere pressure; ρ_s, a constant, is the fluid's mass density at the same normal conditions; A_c, another constant, is the cross-sectional area of the channel; and γ is a constant defined as $\gamma = U_{s,ave}/U_{s,c}$, where $U_{s,ave}$ is the average velocity over area A_c. The velocity in channel flows is seldom uniform and therefore γ is not unity. If the flow channel has a length-to-diameter ratio of 40 to 60, then its flow profile becomes unchanging and is called "fully developed." In fully developed flows, the fluid's viscosity has retarded the velocity near the walls, and hence γ is always less than unity. If the channel's Reynolds number is less than 2000, the flow is laminar; the fully developed profile is a perfect parabola; and γ is 0.5. If the Reynolds number is larger than 4000, the flow is turbulent; the fully developed profile has a flattened parabolic shape; and for pipes with typical rough walls, γ is 0.79, 0.83, and 0.83 for Reynolds numbers of 10^4, 10^5, and 10^6, respectively. If the Reynolds number is between 2000 and 4000, the flow is transitioning between laminar and turbulent flows, and γ ranges between 0.5 and 0.8.

Unfortunately, in most large ducts and stacks, 40 to 60 diameters of straight run preceding the flow monitoring location does not exist. Instead, the flow profile usually is highly nonuniform, swirling, and, in air-preheater ducts and in stacks, is further confounded by temperature nonuniformities. In these cases, single-point monitoring is ill-advised. Fortunately, multipoint monitoring with industrial thermal anemometer flow-averaging arrays, such as shown in Figure 29.18, have proven successful in these applications. As described by Olin [20], this method consists of a total of N (usually, $N = 4$, 8, or 12) industrial thermal anemometer sensors, each similar to that shown in Figure 29.17, located at the centroid of an equal area A_c/N in the channel's cross-sectional area A_c. The individual mass flow rate $\dot{m}_i$ monitored by each sensor is $\rho_s\, U_{s,i}\,(A_c/N)$, where $U_{s,i}$ is the individual velocity monitored by the sensor at point i. The desired quantity, the total mass flow rate $\dot{m}$ through the channel, is the sum of the individual mass flow rates, or:

$$\dot{m} = \sum_{i=1}^{N} \dot{m}_i = \rho_s A_c U_{s,ave} \tag{29.64}$$

where $U_{s,ave}$ is the arithmetic average of the N individual velocities $U_{s,i}$. As described by Olin [21], industrial multipoint thermal anemometers are used as the flow monitor in stack continuous emissions monitoring systems required by governmental air-pollution regulatory agencies.

Table 29.3 Typical Commercial Thermal Anemometer Systems

Product Description	Average 1997 U.S. List Price
Industrial systems	
Insertion mass flow transducer	$1,900
50 mm (2 in.) NPT in-line mass flowmeter	$2,500
8-point smart industrial flow averaging array	$15,000
Research systems	
Single-channel hot-wire or hot-film anemometer system	$10,000
Three-component hot-wire anemometer system	$21,000
Portable air velocity meter	$1,000

Note: Prices listed are the average of the manufacturers listed in Table 29.4. Insertion probe is 25 cm in length. Insertion and in-line mass flowmeters have: probe-mounted FM/CENELEC approved, explosion-proof housing; ac line voltage input power; 5-0 V dc output signal; 316 SS construction; and ambient air calibration. In-line industrial mass flowmeter has built-in flow conditioning. Industrial flow averaging array has four 1 m long probes, 2 points per probe, 316 SS construction, line voltage input power, 0 to 5 V dc output signal, and smart electronics mounted on probe. Research anemometer systems have standard hot-wire probes, most versatile electronics, and include ambient air calibrations.

TABLE 29.4 Manufacturers of Thermal Anemometer Systems

Industrial Systems and Portable Air Velocity Meters	Research Systems and Portables
Sierra Instruments, Inc.	TSI Inc.
5 Harris Court	500 Cardigan Road
Building L	St. Paul, MN 55164
Monterey, CA 93940	Tel: (612) 490-2811
Tel: (831) 373-0200	Fax: (612) 490-3824
Fax: (831) 373-4402	
	Dantec Measurement Technology, Inc.
Fluid Components, Inc.	Denmark
1755 La Costa Meadows Drive	Tel: (45) 4492 3610
San Marcos, CA 92069	Fax: (45) 4284 6136
Tel: (619) 744-6950	
Fax: (619) 736-6250	
Kurz Instruments, Inc.	
2411 Garden Road	
Monterey, CA 93940	
Tel: (831) 646-5911	
Fax: (831) 646-8901	

Instrumentation Systems

Table 29.3 lists examples of typical commercial thermal anemometer systems. Table 29.4 lists their major manufacturers. Thermal anemometer systems include three elements: sensors, probe, and electronics. Sensors and probes have been described in previous sections. The electronics of industrial systems are enclosed in an explosion-proof or other industrial-grade housing mounted either directly on the probe or remotely (usually within 30 m). The electronics is powered with a 24 V dc source or with 100, 115, or 230 V ac line voltage. The output signal typically is 0 to 5 V dc, 4 to 20 mA, RS232, or RS485 linearly proportional to gas mass velocity U_s over the range of 0.5 normal m s^{-1} to 150 normal m s^{-1}. In-line mass flowmeters have the same output-signal options and are calibrated directly in mass flow rate $\dot{m}$ (kg s⁻¹). In-line meters are now available with built-in flow conditioners that eliminate errors associated with upstream disturbances, such as elbows, valves, and pipe expansions. Systems are available either with lower cost analog electronics or with smart microprocessor-based electronics. The repeatability of these systems is ±0.2% of full scale. The typical accuracy of a smart industrial system is ±2% of reading over 10 to 100% of full scale and ±0.5% of full scale below 10% of full scale. Automatic temperature compensation facilitates temperature coefficients of ±0.04% of reading per °C within ±20°C of calibration temperature and ±0.08% of reading per °C within ±40°C. High-temperature applications have temperature compensation over a range of ±150°C. Pressure effects are negligible within ±300 kPa of calibration pressure.

Research thermal anemometer systems usually are coupled with a personal computer, as shown in Figure 29.21. The PC provides system set-up and control, as well as data display and analysis. Modern systems feature low-noise circuits, together with smart bridge optimization technology that eliminates tuning and automatically provides flat frequency response up to 300,000 Hz. Lower cost units provide flat response up to 10,000 Hz. A built-in thermocouple circuit simplifies temperature measurement. The PC's windows-based software provides near real-time displays of velocity, probability distribution, and turbulence intensity. Post-processing gives additional statistics, including: mean velocity; turbulence intensity; standard deviation; skewness; flatness; normal stress for one-, two-, and three-component probes; as well as shear stress, correlation coefficients and flow-direction angle for two- and three-dimensional probes. In addition, power spectrum, auto correlations, and cross correlations can be displayed. The software automatically handles flow calibration set-up and calculates calibration velocity. Systems are available in 1-, 2-, 8-, and 16-channel versions.

Commercial industrial and research thermal anemometer systems were first introduced in the early 1960s. At first, industrial thermal anemometers were not considered sufficiently durable for the rigors of industrial use. With the advent of stainless-steel sheathed sensors and microprocessor-based electronics, industrial thermal anemometers now enjoy the credibility formerly attributed to only traditional flow-meter approaches. Initial research systems required a high level of user knowledge and considerable involvement in operation. In contrast, current research systems have nearly flat frequency response, high accuracy, and easy-to-use controls providing the flexibility researchers require. Research systems based on personal computers have graphical user interfaces that enhance both performance and simplicity of operation.

References

1. H. H. Bruun, *Hot-Wire Anemometry: Principles and Signal Analysis,* Oxford: Oxford University Press, 1995.
2. L. M. Fingerson and P. Freymuth, Thermal anemometers, in R. J. Goldstein (ed.), *Fluid Mechanics Measurements,* Washington, D.C.: Hemisphere, 1983.
3. P. Freymuth, History of thermal anemometry, in N.P. Cheremisinoff and R. Gupta (ed.), *Handbook of Fluids in Motion,* Ann Arbor, MI: Ann Arbor Science Publishers, 1983.
4. J. C. Gibbings, J. Madadnia, and A.H. Yousif, The wall correction of the hot-wire anemometer, *Flow Meas. Instrum.,* 6(2), 127-136, 1995.
5. J. C. Gibbings, J. Madadnia, S. Riley, and A.H. Yousif, The proximity hot-wire probe for measuring surface shear in air flows, *Flow Meas. Instrum.,* 6(3), 201-206, 1995.
6. V. Mikulla, *The Measurement of Intensities and Stresses of Turbulence in Incompressible and Compressible Air Flow,* Ph.D. Thesis, University of Liverpool, 1972.
7. L. V. King, On the convection of heat from small cylinders in a stream of fluid: determination of the convection constants of small platinum wires with application to hot-wire anemometry, *Phil. Trans. Roy. Soc.,* A214, 373-432, 1914.
8. H. Kramers, Heat transfer from spheres to flowing media. *Physica,* 12, 61-80, 1946.

9. S. W. Churchill and M. Bernstein, A correlating equation for forced convection from gases and liquids to a circular cylinder in crossflow, *J. Heat Transfer*, 99, 300-306, 1997.
10. S. Takagi, A hot-wire anemometer compensated for ambient temperature variations, *J. Phys. E.: Sci. Instrum.*, 19, 739-743, 1986.
11. J. W. Elsner, An analysis of hot-wire sensitivity in non-isothermal flow, *Proc. Dynamics Flow Conf., Marseille*, 1972.
12. R. F. Blackwelder, Hot-wire and hot-film anemometers, in R.J. Emrich (ed.), *Methods of Experimental Physics: Fluid Dynamics*, New York: Academic Press, 18A, 259-314, 1981.
13. J. A. Borgos, A review of electrical testing of hot-wire and hot-film anemometers, *TSI Quart.*, VI(3), 3-9, 1980.
14. E.W. Nelson and J. A. Borgos, Dynamic response of conical and wedge type hot films: comparison of experimental and theoretical results, *TSI Quart.*, IX(1), 3-10, 1983.
15. U. R. Müller, Comparison of turbulence measurements with single, X and triple hot-wire probes, *Exp. in Fluids*, 13, 208-216, 1992.
16. K. Döbbeling, B. Lenze, and W. Leuckel, Four-sensor hot-wire probe measurements of the isothermal flow in a model combustion chamber with different levels of swirl, *Exp. Thermal and Fluid Sci.*, 5, 381-389, 1992.
17. J.G. Olin and R. B. Kiland, Split-film anemometer sensors for three-dimensional velocity-vector measurement, *Proc. Symp. on Aircraft Wake Turbulence*, Seattle, Washington, 1970, 57-79.
18. F. E. Jörgenson, Directional sensitivity of wire and fibre-film probes, *DISA Info.*, (11), 31-37, 1971.
19. I. C. Lekakis, R. J. Adrian, and B. G. Jones, Measurement of velocity vectors with orthogonal and non-orthogonal triple-sensor probes, *Experiments in Fluids*, 7, 228-240, 1989.
20. J. G. Olin, A thermal mass flow monitor for continuous emissions monitoring systems (CEMS), *Proc. ISA/93 Int. Conf. Exhibition & Training Program*, (93-404), 1993, 1637-1653.
21. J. G. Olin, Thermal flow monitors take on utility stack emissons, *Instrumentation and Control Systems*, 67(2), 71-73, 1994.

Further Information

J. A. Fay, *Introduction to Fluid Mechanics*, Cambridge, MA: MIT Press, 1994.
A. Bejan, *Convection Heat Transfer*, New York: John Wiley & Sons, 1995.

29.3   Laser Anemometry

Rajan. K. Menon

Laser anemometry, or *laser velocimetry*, refers to any technique that uses lasers to measure velocity. The most common approach uses the Doppler shift principle to measure the velocity of a flowing fluid at a point and is referred to as Laser Doppler Velocimetry (LDV) or Laser Doppler Anemometry (LDA). This technique (also known as dual beam, differential Doppler or fringe mode technique), incorporating intersecting (focused) laser beams, is also used to measure the motion of surfaces [1]. In some special flow situations, another approach using two *nonintersecting*, focused laser beams known as *dual focus* (also known as L2F) technique is used to measure flow velocity at a point [2]. More recently, laser illumination by light sheets is used to make global flow measurements and is referred to as *particle image velocimetry* (PIV) [3]. The strength of PIV (including particle tracking velocimetry) lies in its ability to capture turbulence structures within the flow and transient phenomena, and examine unsteady flows [4]. The development of this technique to obtain both spatial and temporal information about flow fields is making this a powerful diagnostic tool in fluid mechanics research [5, 6]. Other approaches to measure global flow velocities come under the category of molecular tagging velocimetry [7] or Doppler global velocimetry [8, 9].

The noninvasive nature of the LDV technique and its ability to make accurate velocity measurements with high spatial and temporal resolution, even in highly turbulent flows, have led to the widespread use of LDV for flow measurement. Flow velocities ranging from micrometers per second to hypersonic speeds have been measured using LDV systems. Measurements of highly turbulent flows [10], flows in rotating machinery [11], especially in the interblade region of rotors [12, 13], very high [14] or very low [15] velocity flows, flows at high temperatures [16] and in other hostile environments [17, 18], and flows in small spaces [19] have been performed using the LDV technique. The versatility and the widespread use of the LDV approach to measure flows accurately has resulted in referring to this technique as *laser velocimetry* or *laser anemometry*. Many details of the technique, including some of the early developments of the hardware, are provided in the book by Durst [20]. A bibliography of the landmark papers in LDV has been compiled by Adrian [21].

For the case of spherical scatterers, the technique has also been extended to measure size of these particles. In this case, the scattered light signal from a suitably placed receiver system is processed to obtain the diameter of the particle, using the phase Doppler technique [22].

The first reported fluid flow measurements using LDV principles was by Yeh and Cummins [23]. Although in this case an optical arrangement referred to as the reference beam system was used to measure the Doppler shift, in almost all measurement applications, what is referred to as the dual beam or differential Doppler arrangement [24] is used now. This arrangement, also referred to as the "fringe" mode of operation, uses two intersecting laser beams (Figure 29.24) to measure one velocity component.

The advantages of the LDV technique in measuring flows include (1) a small measuring region (i.e., point measurement), (2) high measurement accuracy, (3) the ability to measure any desired velocity component, (4) accurate measurement of high turbulence intensities, including flow reversals, (5) a large dynamic range, (6) no required velocity calibration, (7) no probe in the flow (does not disturb the flow; measures in hostile environments), and (8) good frequency response.

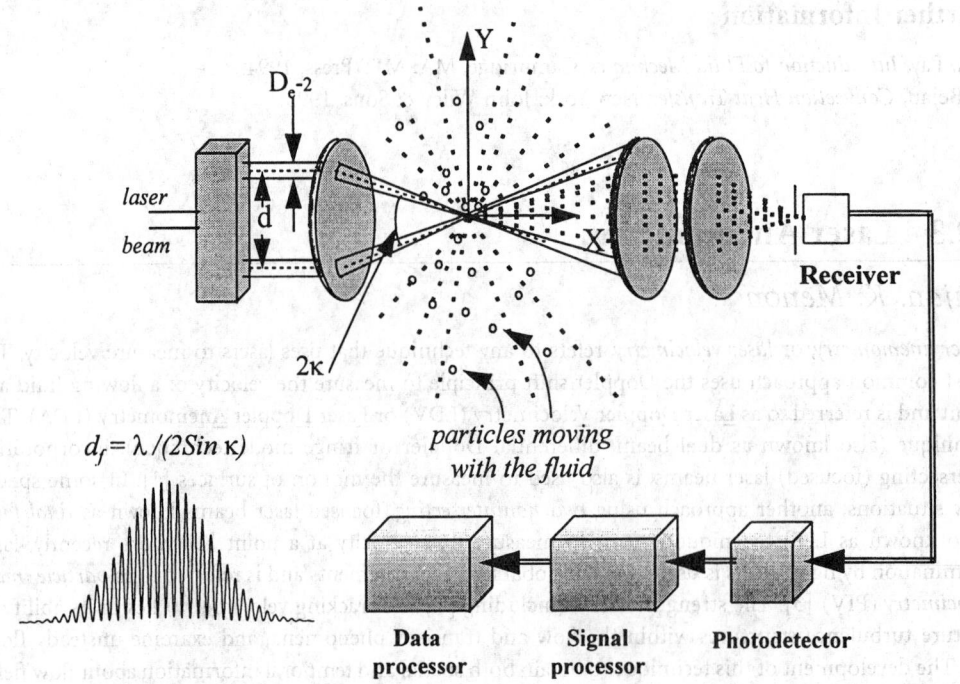

FIGURE 29.24 Schematic of a dual-beam system.

The LDV technique relies on the light scattered by scattering centers in the fluid to measure flow velocity. These scattering centers will also be referred to as *particles*, with the understanding that bubbles or anything else that has a refractive index different from that of the fluid could be the source of scattered light. The particles, whose velocities are measured, must be small enough (generally in the micron range) to follow the flow variations and large enough to provide signal strength adequate for the signal processor to give velocity measurements. It should be noted that the signal exists only when a "detectable" particle is in the measuring volume and, hence, is discontinuous. This, along with other properties of the signal, adds special requirements on the signal processing and the subsequent data analysis systems. The scattered light signal is processed to obtain the Doppler shift frequency and from that the velocity of the particle. Hence, the rate at which the velocity measurements are made depends on the rate of particle arrival. It is desirable to have a high particle concentration to obtain a nearly continuous update of velocity. In carefully controlled experiments, the LDV system can provide very high accuracy (0.1% or better) measurements in mean velocity. Thermal anemometer systems are generally able to measure lower turbulence levels compared to that by an LDV system [25]. While the direct measurement of the Doppler-shifted frequency from a single laser beam caused by a moving particle is possible [26], most LDV systems employ the heterodyne principle to obtain and process only the Doppler shift (difference) frequency.

Principle of Operation

Dual-Beam Approach

The dual-beam approach is the most common optical arrangement used for LDV systems for flow measurement applications. The schematic (Figure 29.24) shows the basic components of a complete LDV system to measure one component of velocity. The transmitting optics include an optical element to split the original laser beam into two parallel beams and a lens system to focus and cross the two beams. The intersection region of the two beams becomes the measuring region. The receiving optics (shown to be set up in the forward direction) collect a portion of the light scattered by the particles, in the fluid stream, passing through the beam-crossing region (measuring volume) and direct this light to a photodetector, which converts the scattered light intensity to an analog electrical signal. The frequency of this signal is proportional to the velocity of the particle. A signal processor extracts the frequency information from the photodetector output and provides this as a digital number corresponding to the instantaneous velocity of the particle. The data processing system obtains the detailed flow properties from these instantaneous velocity measurements. The idealized photodetector signal, for a particle passing through the center of the measuring volume, is shown in the lower left side of Figure 29.24. Actual signals will have noise superimposed on them; and the signal shape will vary, depending on the particle trajectory through the measuring volume [27].

Fringe Model Description.
While there are several ways to describe the features of a dual-beam system, the description based on a fringe model is, perhaps, the simplest. For simplicity, the diameter and the intensity of both the beams are assumed to be the same. After the beams pass through the transmitting lens, the diameter of each beam continuously decreases to a minimum value (beam waist) at the focal point of the lens, and then increases again. Thus, the beam waists cross where the two laser beams intersect (at the focal point of the lens), and the wavefronts in the beams interfere with each other, creating a fringe pattern [28]. In this pattern, assuming equal intensity beams and other needed qualities of the beams, the light intensity varies from zero (dark fringe) to a maximum (bright fringe), and the fringes are equally spaced. The particles in the flow passing through the intersection region (measuring region) scatter light in all directions. An optical system, including a receiving lens (to collimate the scattered light collected) and a focusing lens, is used to collect the scattered light and focus it onto the receiver. The aperture in front of the receiver is used to block out stray light and reflections and collect only the light scattered from the measuring region.

As a particle in the flow, with velocity *u*, moves across the fringes, the intensity pattern of the light scattered by the particle resembles that shown in the lower left of Figure 29.24. The velocity component,

u_y (perpendicular to the optical axis and in the plane of the incident beams) can be obtained from the ratio of the distance between fringes (or fringe spacing, d_f), and the time t (= $1/f_D$) for the particle to cross one pair of fringes, where f_D is the frequency of the signal. The amplitude variation of the signal reflects the Gaussian intensity distribution across the laser beam. Collection (receiving) optics for the dual-beam system can be placed at any angle, and the resulting signal from the receiving system will still give the same frequency. However, signal quality and intensity will vary greatly with the collection optics angle.

Doppler Shift Explanation.
The description of the dual-beam system using the Doppler shift principle is as follows. At the receiver, the frequencies of the Doppler-shifted light scattered by a particle from beam one and beam two are given by:

$$v_{D1} = v_{01} + \frac{\vec{u}}{\lambda}\left(\hat{r} - \hat{S}_1\right); \quad v_{D2} = v_{02} + \frac{\vec{u}}{\lambda}\left(\hat{r} - \hat{S}_2\right) \tag{29.65}$$

where v_{01} and v_{02} are the frequencies of laser beam 1 and laser beam 2; $\hat{r}$ is the unit vector directed from the measuring volume to the receiving optics; $\hat{S}_1$ and $\hat{S}_2$ are the unit vectors in the direction of incident beam 1 and incident beam 2; $\vec{u}$ is the velocity vector of the particle (scattering center); and λ is the wavelength of light. The frequency of the net (heterodyne) signal output from the photodetector system is given by the difference between v_{D1} and v_{D2}.

$$f_D = f_s + \frac{\vec{u}}{\lambda}\left(\hat{S}_2 - \hat{S}_1\right) \tag{29.66}$$

where $f_s = v_{01} - v_{02}$ is the difference in frequency between the two incident beams. This difference frequency is often intentionally imposed (see section on frequency shifting) to permit unambiguous measurement of flow direction and high-turbulence intensities. Assuming $f_s = 0$, the frequency detected by the photodetector is:

$$f_D = \frac{\vec{u}}{\lambda}\left(\hat{S}_2 - \hat{S}_1\right) = 2u_y \sin\kappa \tag{29.67}$$

Hence,

$$u_y = \frac{f_D \lambda}{2\sin\kappa} = f_D \, d_f \tag{29.68}$$

This is the equation for u_y and shows that the signal frequency f_D is directly proportional to the velocity u_y. The heterodyning of the scattered light from the two laser beams at the photodetector actually gives both the sum and difference frequency. However, the sum frequency is too high to be detected and so only the difference frequency ($v_{D1} - v_{D2}$) is output from the photodetector as an electrical signal. The frequency f_D is often referred to as the Doppler frequency of the output signal, and the output signal is referred to as the Doppler signal.

It can be seen from Equation 29.68 that the Doppler frequency is independent of the receiver location ($\hat{r}$). Hence, the receiver system location can be chosen based on considerations such as signal strength, ease of alignment, and clear access to the measuring region. The expressions for the other optical configurations can be reduced similarly [29], giving the identical equation for the Doppler shift frequency f_D. It should be noted that the fringe description does not involve a "Doppler shift" and is, in fact, not always appropriate. The fringe model is convenient and gives the correct expression for the frequency. However, it can be misleading when studying the details of the Doppler signal (e.g., signal-to-noise ratio) and other important parameters e.g., modulation depth or visibility ($\overline{V}$) of the signal [30].

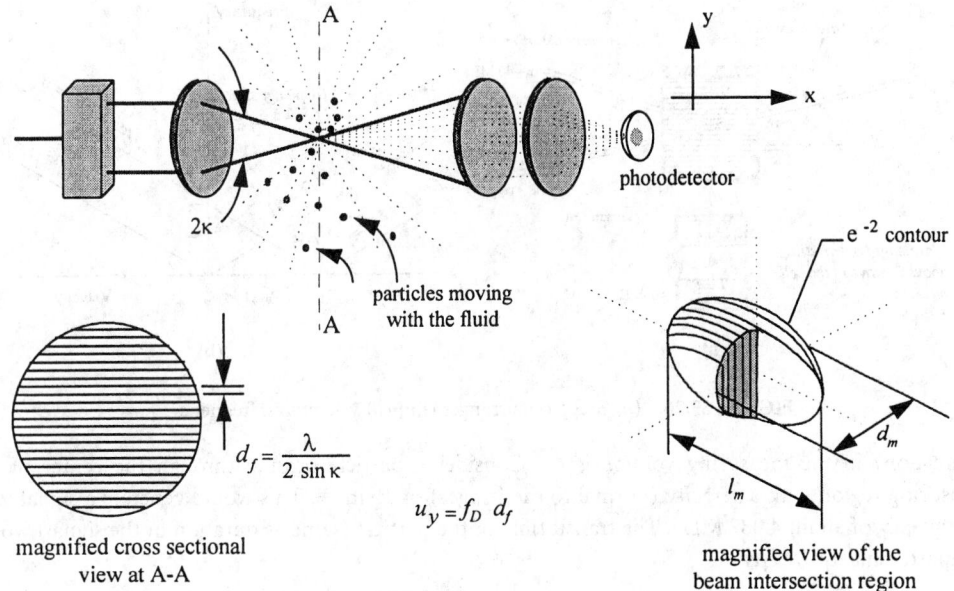

FIGURE 29.25 Details of the beam crossing.

The time taken by the particle to cross the measuring volume is referred to as *transit time, residence time,* or *total burst time,* τ_B, and corresponds to the duration of the scattered light signal. The number of cycles (N) in the signal (same as the number of fringes the particle crosses) is given by the product of the transit time (τ_B) and the frequency, f_D, of the signal.

It should also be noted that the fringe spacing (d_f), depends only on the wavelength of the laser light (λ) and the angle (2κ) between the two beams. It can be shown that the effect of the fluid refractive index on these two terms tends to cancel out and, hence, the value of fringe spacing is independent of the fluid medium [31]. The values of λ and κ are known for any dual-beam system and, hence, an actual velocity calibration is not needed. In some cases, an actual velocity calibration using the rim of a precisely controlled rotating wheel has been performed to overcome the errors in measuring accurately the angle between the beams.

The intensity distribution in a laser beam operating in the TEM_{00} mode is Gaussian [32]. Using wave theory and assuming diffraction-limited optics, the effective diameter of the laser beam and the size of the measurement region can be defined. The conventional approach to the definition of laser beam diameter and measuring volume dimensions is based on the locations where the light intensity is $1/e^2$ of the maximum intensity (at the center of the beam). This definition of the dimensions is analogous to that of the boundary layer thickness. The dimensions d_m and l_m of the ellipsoidal measuring volume (Figure 29.25) are based on the $1/e^2$ criterion and are given by:

$$d_m = 4f\lambda/\pi D_{e^{-2}}; \quad l_m = d_m/\tan\kappa; \quad N_{FR} = d_m/d_f \qquad (29.69)$$

N_{FR} is the maximum number of fringes in the ellipsoidal measuring region. Note that as the value of $D_{e^{-2}}$ increases, the measuring volume becomes smaller. In flow measurement applications, this relationship is exploited to arrive at the desired size of the measuring volume.

The measuring volume parameters for the following sample situation are wavelength, $\lambda = 514.5$ nm (green line of argon-ion laser), $D_{e^{-2}} = 1.1$ mm, $d = 35$ mm, and $f = 250$ mm. Then, $\kappa = 4°$, $d_m = 149$ μm, and $l_m = 2.13$ mm. The fringe spacing, d_f, is 3.67 μm and the maximum number (N_{FR}) of fringes (number of cycles in a signal burst for a particle going through the center of the measuring volume in the

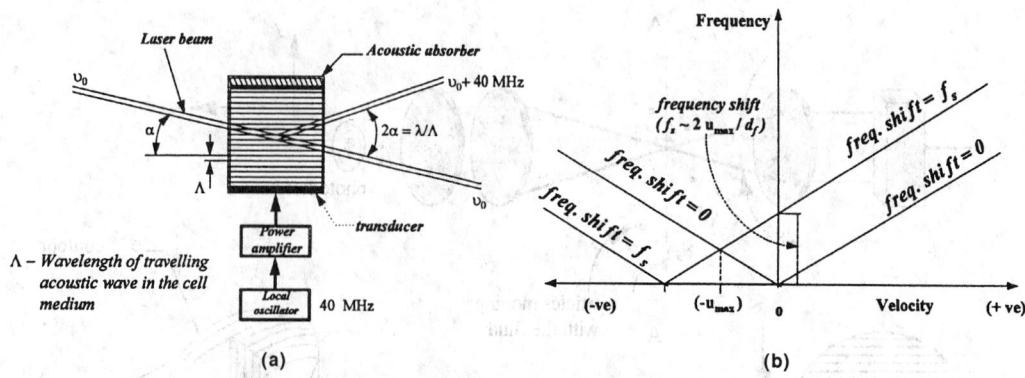

FIGURE 29.26   (*a*) Bragg cell arrangement; (*b*) velocity vs. frequency.

y-direction) in the measuring volume is 40. Consider a particle passing through the center of the measuring region with a velocity (normal to the fringes) of 15 m s⁻¹. This would generate a signal with a frequency of about 4.087 MHz. The transit time of the particle (same as duration of the signal) would be approximately 9.93 μs!

Frequency Shifting

The presence of high turbulence intensity and recirculating or oscillatory flow regions is common in most flow measuring situations. In the fringe model and the Doppler shift (with $f_3 = 0$) descriptions of the dual-beam system, the Doppler signal does not indicate the influence of the sign (positive or negative) of the velocity. Further, a particle passing through the measuring volume parallel to the fringes would not cross any fringes and, hence, not generate a signal having the cyclic pattern resulting in the inability to measure the zero normal (to the fringes) component of velocity. In addition, signal processing hardware used to extract the frequency information often requires the signals to have a minimum number of cycles. This, as well as the ability to measure flow reversals, is achieved by a method of frequency offsetting referred to as *frequency shifting*. Frequency shifting is also used to measure small velocity components perpendicular to the dominant flow direction and to increase the effective velocity measuring range of the signal processors [31].

By introducing a phase or frequency offset (f_s) to one of the two beams in a dual-beam system, the directional ambiguity can be resolved. From the fringe model standpoint, this situation corresponds to a moving (instead of a stationary) fringe system. A stationary particle in the measuring volume will provide a continuous signal at the photodetector output whose frequency is equal to the difference in frequency, f_s, between the two incident beams. In other words, as shown in Figure 29.26(*b*), the linear curve between velocity and frequency is offset along the positive frequency direction by an amount equal to the frequency shift, f_s. Motion of a particle in a direction opposite to fringe movement would provide an increase in signal frequency, while particle motion in the direction of fringe motion would provide a decrease in frequency. To create a signal with an adequate number of cycles even while measuring negative velocities (e.g., flow reversals, recalculating flows), a convenient "rule-of-thumb" approach for frequency shifting is often used. The approach is to select the frequency shift ($f_s \sim 2 u_{max}/d_f$) to be approximately twice the frequency corresponding to the magnitude of the maximum negative velocity (u_{max}) expected in the flow. This provides approximately equal probability of measurement for all particle trajectories through the measuring volume [33, 34].

Frequency shifting is most commonly achieved by sending the laser beam through a Bragg cell (Figure 29.26(*a*)), driven by an external oscillator [35]. Typically, the propagation of the 40 MHz acoustic wave (created by a 40 MHz drive frequency) inside the cell affects the beam passing through the cell to yield a frequency shift of 40 MHz for that beam. By properly adjusting the angle the cell makes with the incoming beam and blocking off the unwanted beams, up to about 80% of input light intensity is recovered in the shifted beam. The Bragg cell approach will provide a 40 MHz frequency shift in the photodetector output signal. To improve the measurement resolution of the signal processor, the resulting photodetector signal is often "downmixed" to have a more appropriate frequency shift (based on the rule-of-thumb shift value) for the flow velocities being measured. Frequency shifting using two Bragg cells (one for each beam of a dual-beam system) operating at different frequencies is attractive to systems where the bandwidth of the photodetector is limited. However, the need to readjust the beam crossing with a change in frequency shift has not made this approach (double Bragg cell technique) attractive for applications where frequency shift needs to be varied [31].

More recently, Bragg cells have been used in a multifunctional mode to split the incoming laser beam into two equal intensity beams, with one of them having the 40 MHz frequency shift. This is accomplished by adjusting the Bragg cell angle differently. In addition to Bragg cells, rotating diffraction gratings and other mechanical approaches have been used for frequency shifting. However, limits on rotational speed and other mechanical aspects of these systems make them limited in frequency range [20]. Other frequency shifting techniques have been suggested for use with laser diodes [36, 37]. Because so many flow measurement applications involve recirculating regions and high turbulence intensities, frequency shifting is almost always a part of an LDV system used for flow measurement.

Signal Strength

Understanding the influence of various parameters of an LDV system on the signal-to-noise ratio (SNR) of the photodetector signal provides methods or approaches to enhance signal quality and hence improve the performance of the measuring system. The basic equation for the ratio of signal power to noise power (SNR) of the photodetector signal can be written as [38]:

$$\text{SNR} = A_1 \frac{\eta_q P_0}{\Delta f} \left[\frac{D_a}{r_a} \frac{D_{e^{-2}}}{f} \right]^2 d_p^2 \overline{G} \overline{V}^2 \tag{29.70}$$

Equation 29.70 shows that higher laser power (P_0) provides better signal quality. The quantum efficiency of the photodetector, η_q depends on the type of photodetector used and is generally fixed. The SNR is inversely proportional to the bandwidth, Δf, of the Doppler signal. The term in brackets relates to the optical parameters of the system; the "f-number" of the receiving optics, D_a/r_a, and the transmitting optics, $D_{e^{-2}}/f$. The square dependence of SNR on these parameters makes them the prime choice for improving signal quality and, hence, measurement accuracy. The focal length of the transmitting (f) and receiving (r_a) lenses are generally decided by the size of the flow facility. Using the smallest possible values for these would increase the signal quality. The first ratio (D_a is the diameter of the receiving lens) determines the amount of the scattered light that is collected, and the second ratio determines the diameter of (and hence the light intensity in) the measuring volume. The last three terms are the diameter, d_p, of the scattering center and the two terms (scattering gain $\overline{G}$, visibility $\overline{V}$) relating to properties of the scattered light. These need to be evaluated using the Mie scattering equations [38] or the generalized Lorentz–Mie theory [39].

Measuring Multiple Components of Velocity

A pair of intersecting laser beams is needed to measure (Figure 29.24) one component of velocity. This concept is extended to measure two components of velocity (perpendicular to the optical axis) by having two pairs of beams that have an overlapping intersection region. In this case, the plane of each pair of beams is set to be orthogonal to that of the other. The most common approach to measure two components of velocity is to use a laser source that can generate multiwavelength beams so that the wavelength of one pair of beams is different from the other pair. The Doppler signals corresponding to the two components of velocity are separated by wavelength [31].

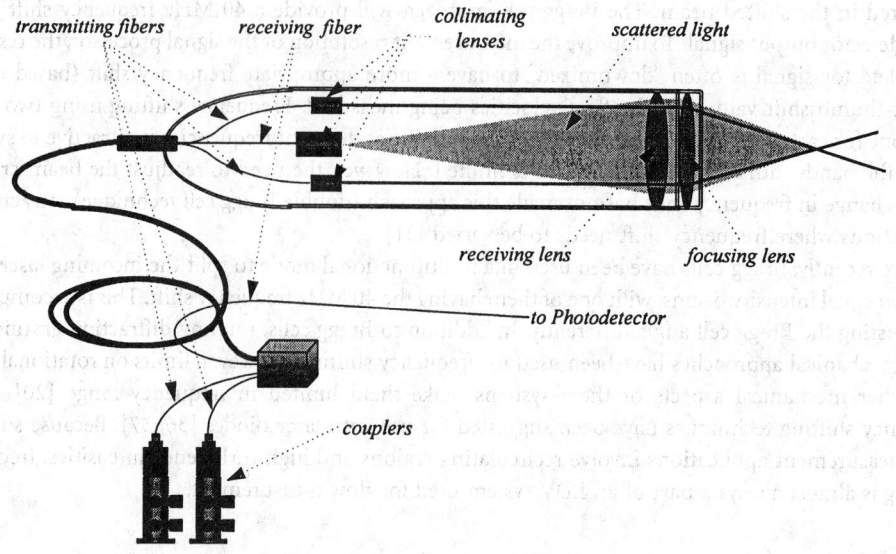

FIGURE 29.27   Schematic arrangement of a fiberoptic system.

Historically, LDV systems were assembled by putting together a variety of optical modules. These modules included beam splitters, color separators, polarization rotators, and scattered light collection systems. The size of such a modular system depended on the number of velocity components to be measured.

The use of optical fibers along with multifunctional optical elements has made the systems more compact, flexible, and easier to make measurements. The laser, optics to generate the necessary number of beams (typically, one pair per component of velocity to be measured), photodetectors, and electronics can be isolated from the measurement location [40]. The fibers carrying the laser beams thus generated are arranged in the probe to achieve the desired beam geometry for measuring the velocity components. Hence, flow field mapping is achieved by moving only the fiber-optic probes, while keeping the rest of the system stationary. To achieve maximum power transmission efficiency and beam quality, special single-mode, polarization-preserving optical fibers along with precision couplers are used. In most cases, these fiber probes also have a receiving system and a separate fiber (multimode) to collect (in back scatter) the scattered light and carry that back to the photodetector system.

A schematic arrangement of a fiber probe system to measure one component of velocity is shown in Figure 29.27. In flow measurement applications, LDV systems using these types of fiber-optic probes have largely replaced the earlier modular systems.

The best way to make three-component of velocity measurements is to use an arrangement using two probes [13]. In this case, the optical axis of the system to measure the third component of velocity (u_x) is perpendicular to that of the two-component system. Unfortunately, access and/or traversing difficulties often make this arrangement impractical or less attractive. In most practical situations, the angle between the two probes is selected to be less than 90°. Such an arrangement using two fiber-optic probes to measure three components of velocity simultaneously is shown in Figure 29.28.

Signal Processing

Nature of the Signal

Every time a particle passes through the measuring region, the scattered light signal level (Figure 29.29) suddenly increases ("burst"). The characteristics of the burst signal are (1) amplitude in the burst not constant, (2) lasts for only a short duration, (3) amplitude varies from burst to burst, (4) presence of noise, (5) high frequency, and (6) random arrival.

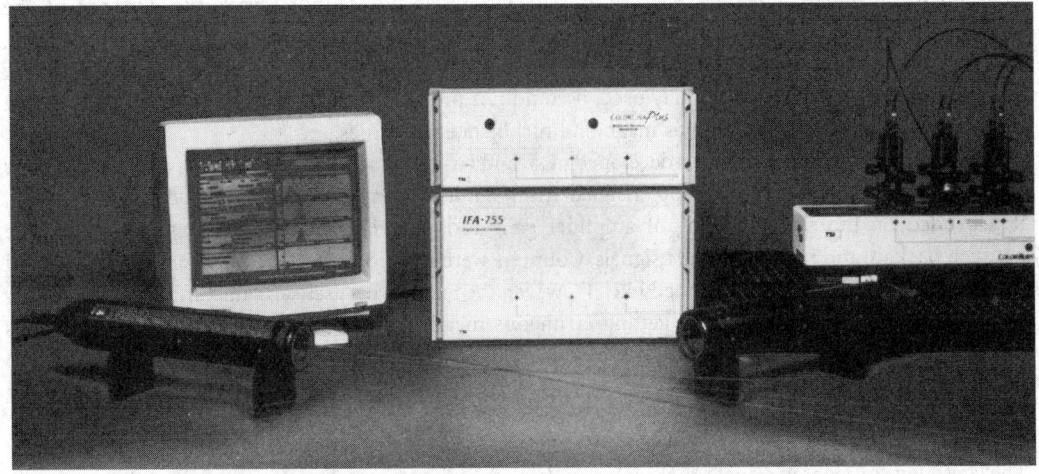

FIGURE 29.28 Three-component LDV system with fiber-optic probes.

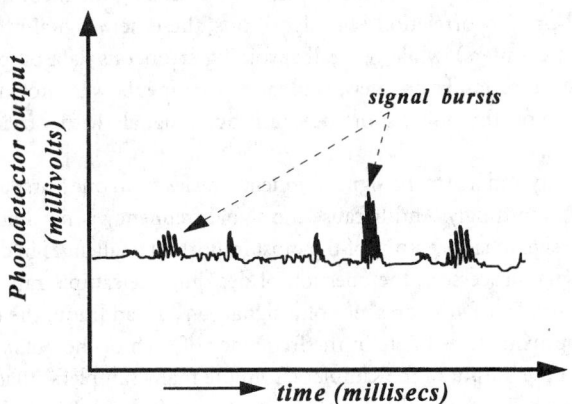

FIGURE 29.29 Time history of the photodetector signal.

The primary task of the signal processor is to extract the frequency information from the burst signal generated by a particle passing through the measuring volume, and provide an analog or digital output proportional to the frequency of the signal. The unique nature of the signals demands the use of a special signal processing system to extract the velocity information.

A variety of techniques has been used for processing Doppler signals. Signal processors have been based on spectrum analysis, frequency tracking, photon correlation, frequency counting, Fourier transform, and autocorrelation principles. The evolution of the signal processing techniques shows the improvement in their ability to handle more difficult measuring situations (generally implies noisier signals), give more accurate measurements, and have higher processing speed.

The traditional instrument to measure signal frequency is a spectrum analyzer. The need to measure individual particle velocities, and to obtain the time history and other properties of the flow, has eliminated the use of "standard" spectrum analyzers [29]. The "tracker" can be thought of as a fixed bandwidth filter that "tracked" the Doppler frequency as the fluid velocity changed. This technique of "tracking flow" worked quite well at modest velocities and where the concentration of scattering centers was high enough to provide an essentially continuous signal. However, too frequently these conditions could not be met in the flows of most interest [29].

When the scattered light level is very low, the photodetector output reveals the presence of the individual photon pulses. By correlating the actual photon pulses from a wide bandwidth photodetector, the photon correlator was designed to work in situations where the attainable signal intensity was very low (low SNR). However, as normally used, it could not provide the velocity of individual particles but only the averaged quantities, such as mean and turbulence intensities.

The "counter" type processor was developed next, and basically measured the time for a certain number (typically, eight) of cycles of the Doppler signal. Although it measured the velocity of individual particles, it depended on the careful setting of amplifier gain and, especially, threshold levels to discriminate between background noise and burst signals. Counters were the processors of choice for many years, and excellent measurements were obtained [42]. However, the reliance on user skill, the difficulty in handling low SNR signals, the possibility of getting erroneous measurements, the inclination to ignore signals from small particles, and the desire to make measurements close to surfaces and in complex flows led to the need for a better signal processor.

Digital Signal Processing

The latest development in signal processing is in the area of digital signal processors. Recent developments in high-speed digital signal processing now permit the use of these techniques to extract the frequency from individual Doppler bursts fast enough to actually follow the flow when the seeding concentration is adequate in a wide range of measurement situations. By digitizing the incoming signal and using the Fourier transform [43] or autocorrelation [44] algorithms, these new digital processors can work with lower SNR signals (than counters), while generally avoiding erroneous data outputs. While instruments using these techniques are certainly not new, standard instruments were not designed to make rapid individual measurements on the noisy, short-duration burst signals with varying amplitudes that are typical of Doppler bursts.

Because the flow velocity and hence the signal frequency varies from one burst to the next, the sampling rate needs to be varied accordingly. And because the signal frequency is not known *a priori*, the ability to optimally sample the signal has been one of the most important challenges in digital signal processing. In one of the digital signal processors, the question of deciding the sample rate is addressed by a burst detector that uses SNR to identify the presence of a signal [44]. In addition, the burst detector provides the duration and an approximate estimate of the frequency of each of the burst signals. This frequency estimate is used to select the output of the sampler (from the many samplers) that had sampled the burst signal at the optimum rate. Besides optimizing the sample rate for each burst, the burst detector information is also used to focus on and process the middle portion of the burst where the SNR is maximum. These optimization schemes, followed by digital signal processing, provide an accurate digital output that is proportional to the signal frequency, and hence the fluid velocity.

Seeding and Other Aspects

The performance of an LDV system can be significantly improved by optimizing the source of the signal, the scattering particle. The first reaction of many experimentalists is to rely on the particles naturally present in the flow. There are a few situations (e.g., LDV systems operating in forward scatter to measure water or liquid flows) where the particles naturally present in the flow are sufficient in number and size to provide good signal quality and hence good measurements. In most flow measurement situations, particles are added to the flow (generally referred to as seeding the flow) to obtain an adequate number of suitable scatterers. Use of a proper particle can result in orders of magnitude increase in signal quality (SNR), and hence can have greater impact on signal quality than the modification of any other component in the LDV system. Ideally, the seed particles should be naturally buoyant in the fluid, provide adequate scattered light intensity, have large enough number concentration, and have uniform properties from particle to particle. While this ideal is difficult to achieve, adequate particle sources and distribution systems have been developed [29, 45–47].

LDV measurements of internal flows such as in channels, pipes and combustion chambers result in the laser beams (as well as the scattered light) going through transparent walls or "windows." In many cases, the window is flat and, hence, the effect of light refraction can be a simple displacement of the measuring region. In the case of internal flows with curved walls, each beam can refract by different amounts and the location of the measuring region needs to be carefully estimated [48]. For internal flows in models with complex geometries, the beam access needs to be carefully selected so that the beams do cross inside. Further, to make measurements close to the wall in an internal flow, the refraction effect of the wall material on the beam path needs to be minimized. One of the approaches is to use a liquid [49] that has the same refractive index as that of the wall material.

Data Analysis

The flow velocity is "sampled" by the particle passing through the measuring volume, and the velocity measurement is obtained only when the Doppler signal, created by the particle, is processed and output as a data point by the signal processor. While averaging the measurements to get, for example, mean velocity would seem reasonable, this method gives the wrong answer. This arises from the fact that the number of particles going through the measuring region per unit time is higher at high velocities than at low velocities. In effect, there is a correlation between the measured quantity (velocity) and the sampling process (particle arrival). Hence, a simple average of the data points will bias the mean value (and other statistical parameters) toward the high-velocity end and is referred to as *velocity bias* [50]. The magnitude of the bias error depends on the magnitude of the velocity variations about the mean. If the variations in velocity are sufficiently small, the error might not be significant.

If the actual data rate is so high that the output data is essentially able to characterize the flow (time history), then the output can be sampled at uniform time increments. This is similar to the procedure normally used for sampling a continuous analog signal using an ADC. This will give the proper value for both the mean and the variance when the data rate is sufficiently high compared to the rates based on the Taylor microscale for the temporal variation of velocity. This is referred to as a high data density situation [29].

In many actual measurement situations, the data rate is not high enough (low data density) to actually characterize the flow. Here, sampling the output of the signal processor at uniform time increments will not work because the probability of getting an updated velocity (new data point) is higher at high velocity than at low velocity (velocity bias). The solution to the velocity bias problem is to weight the individual measurements with a factor inversely proportional to the probability of making the measurement.

$$\overline{U} = \frac{\Sigma u_j \tau_{Bj}}{\Sigma \tau_{Bj}} \tag{29.71}$$

where u_j = Velocity of particle j

$\quad\quad \tau_{Bj}$ = Transit time for particle j

Similar procedures can be used to obtain unbiased estimators for variance and other statistical properties of the flow [29]. Modern signal processors provide the residence time and the time between data points along with the velocity data. A comparison of some of the different approaches to do bias correction has been presented by Gould and Loseke [51]. Some of the other types of biases associated with LDV have been summarized by Edwards [52].

A variety of techniques to obtain spectral information of the flow velocity from the random data output of the signal processors have been tried. The goal of all these techniques has been to get accurate and unbiased spectral information to as high a frequency as possible. Direct spectral estimation of the digital output of the processors [53] exhibit the spectrum estimates at high frequency to be less reliable. The "slotting" technique [54, 55] of estimating the autocorrelation of the (random) velocity data followed by Fourier transform continues to be attractive from a computational standpoint. To obtain reliable spectrum estimates at high frequencies, a variety of methods aimed at interpolation of measured velocity values have been attempted. These are generally referred to as *data* or *signal reconstruction* techniques. A review article [37] emphasizes the need to correct for velocity bias in the spectrum estimates. It also covers some of the recent reconstruction algorithms and points out the difficulties in coming up with a general-purpose approach.

Extension to Particle Sizing

In LDV, the frequency of the scattered light signal provides the velocity of the scatterer. Processing the scattered light to get information about the scatterer other than velocity has always been a topic of great interest in flow and particle diagnostics. One of the most promising developments is the extension of the LDV technique to measure the surface curvature and, hence, the diameter of a spherical scatterer [22]. This approach (limited to spherical particles) uses the phase information of the scattered light signal to extract the size information. To obtain a unique and, preferably, monotonic relation between phase of the signal and the size of the particle, the orientation and the geometry (aperture) of the scattered light collection system needs to be carefully chosen. In the following, unless otherwise mentioned, the particles are assumed to be spherical.

The light scattered by a particle, generally, contains contributions from the different scattering mechanisms — reflection, refraction, diffraction, and internal reflection(s). It can be shown that, by selecting the position of the scattered light collection set-up, contributions from one scattering mechanism can be made dominant over the others. The aim in phase Doppler measurements is to have the orientation of the receiver system such that the scattered light collected is from one dominant scattering mechanism.

The popularity of the technique is evidenced by its widespread use for measuring particle diameter and velocity in a large number of applications, especially in the field of liquid sprays [56]. The technique has also been used in diagnosing flow fields associated with combustion, cavitation, manufacturing processes, and other two-phase flows.

Phase Doppler System: Principle

The phase Doppler approach, outlined as an extension to an LDV system, was first proposed by Durst and Zare [57] to measure velocity and size of spherical particles. The first practical phase Doppler systems using a single receiver were proposed by Bachalo and Houser [22].

A schematic arrangement of a phase Doppler system is shown in Figure 29.30(*a*). This shows a receiver system arrangement that collects, separates, and focuses the scattered light onto multiple photodetectors. In general, the receiving system aperture is divided into three parts and the scattered light collected through these are focused into three separate photodetectors. For simplicity, in Figure 29.30(*a*), the output of two detectors are shown. The different spatial locations of the detectors (receiving apertures) results in the signals received by each detector having a slightly different phase. In general, the difference in phase between the signals from the detectors is used to obtain the particle diameter whereas the signal frequency provides the velocity of the particle.

Fringe Model Explanation

The fringe model provides an easy and straightforward approach to arrive at the expressions for Doppler frequency and phase shift created by a particle going through the measuring volume. As the particle moves through the fringes in the measuring volume, it scatters the fringe pattern (Figure 29.30(*b*)). The phase shift in the signals can be examined by looking at the scattered fringe pattern. If the particle acts like a spherical mirror (dominant reflection) or a spherical lens (dominant refraction), it projects fringes from the measuring volume into space all around as diverging bands of bright and dark light, known as *scattered fringes*. Scattered fringes as seen on a screen placed in front of the receivers are shown in Figure 29.30(*b*). The spacing between the scattered fringes at the plane of the receiver is s_f. The receiver system shown in Figure 29.30(*b*) shows two apertures. The distance between (separation) the centroids

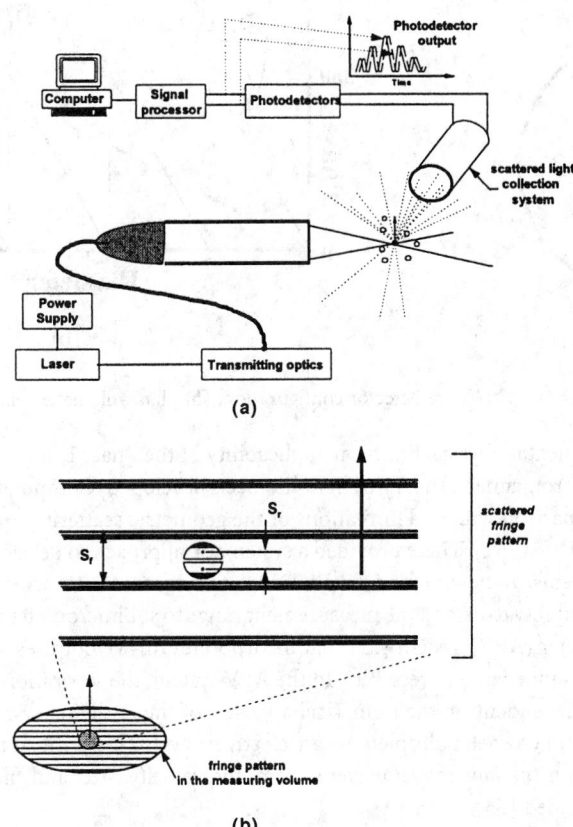

FIGURE 29.30 (*a*) Phase/Doppler system: schematic. (*b*) phase/Doppler System: fringe model.

of the two receiving apertures is s_r. Scattered fringes move across the receivers as the particle moves in the measuring volume, generating temporally fluctuating signals. The two photodetector output signals are shifted in phase by s_r/s_f times 360° [31]. Large particles create a scattered fringe pattern with a smaller fringe spacing (compared to that for small particles), i.e., particle diameter is inversely proportional to s_f, while s_f is inversely proportional to phase difference. Thus, the fringe model shows the particle diameter to be directly proportional to the phase difference. It can also be seen that the sensitivity (degrees of phase difference per micrometer) of the phase Doppler system can be increased by increasing the separation (s_r) between the detectors.

The phase Doppler system shown above measures the phase difference between two detectors in the receiver system to obtain particle diameter. This brings in the limitation that the maximum value of phase that could be measured is 2π. A three-detector arrangement in the receiver system is used to overcome this 2π ambiguity. Figure 29.31 shows the three-detector (aperture) arrangement. Scattered light collected through apertures 1, 2, and 3 are focused into detectors 1, 2, and 3. Φ_{13} is the phase difference between the detectors 1 and 3 and provides the higher phase sensitivity because of their greater separation compared to detectors 1 and 2. As Φ_{13} exceeds 2π, the value of Φ_{12} is below 360° and is used to keep track of Φ_{13}. It should be noted that the simplified approach in terms of geometrical scattering provides a linear relationship between the phase difference and diameter of the particle.

It has been pointed out that significant errors in measured size can occur due to trajectory-dependent scattering [58]. These errors could be minimized by choosing the appropriate optical configuration of the phase Doppler system [59]. An intensity-based validation technique has also been proposed to reduce the errors [60].

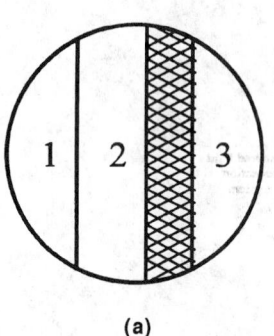
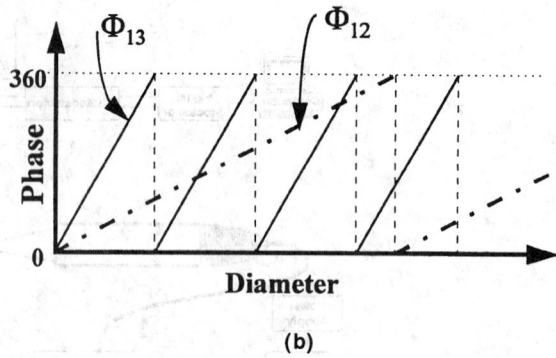

(a)   (b)

FIGURE 29.31   (*a*) Three-detector configuration; (*b*) phase–diameter relationship.

To explore the fundamental physical limits on applicability of the Phase Doppler technique, a rigorous model based on the electromagnetic theory of light has been developed. Computational results based on Mie scattering and comparison with and limitations of the geometric scattering approach have also been outlined by Naqwi and Durst [61]. These provided a systematic approach to develop innovative solutions to particle sizing problems. A new approach (PLANAR arrangement) to achieve high measurement resolution provided the ability to extend the measurement range to submicrometer particles. The Adaptive Phase Doppler Velocimeter (APV) system [59] that incorporates this layout uses a scattered light collection system that employs independent receivers. In the APV system, the separation between the detectors is selectable and is not dependent on the numerical aperture of the receiving system. Such a system was used for measuring submicrometer droplets in an electrospray [62]. By integrating a phase Doppler velocimeter system with a *rainbow refractometer* system, the velocity, size, and the refractive index of a droplet could be determined [63].

The velocity and diameter information is obtained by processing the photodetector output signals. The frequency of the photodetector output signal provides the velocity information. In general, the signal processing system for velocity measurements is expanded to measure the phase difference between two photodetector signals. The digital signal processing approaches described earlier have been complimented by the addition of accurate phase measurement techniques [64, 65].

Although the phase Doppler technique is limited to spherical particles, there has always been an interest in extending the technique to nonspherical particles. In the past, symmetry checks [66] and other similar techniques have been used to check on the sphericity of particles. An equivalent sphere approach has been used to describe these nonspherical particles. Sizing irregular particles is a more complex problem because the local radius of curvature concept is not meaningful in these cases. An innovative stochastic modeling approach has been used to study irregular particles using a phase Doppler system [67].

Conclusion

LDV has become the preferred technique for measuring flow velocity in a wide range of applications. The ability to measure noninvasively the velocity, without calibration, of any transparent flowing fluid has made it attractive for measuring almost any type of flow. Velocity measurement of moving surfaces by LDV is used to monitor and control industrial processes. Use of laser diodes, fiber optics, and advances in signal processing and data analysis are reducing both the cost and complexity of measuring systems. The extension of LDV to the phase Doppler technique provides an attractive, noncontact method for measuring size and velocity of spherical particles. Recent developments in the phase Doppler technique have generated a method to size submicrometer particles as well. These ideas have been extended to examine irregular particles also.

Acknowledgments

The input and comments from Dr. L. M. Fingerson and Dr. A. Naqwi of TSI Inc. have been extremely valuable in the preparation of this chapter section. The author is sincerely grateful to them for the help.

References

1. D. Niccum, A new tool for fiber spinning process control and diagnostics, *Int. Fiber J.*, 10(1), 48-57, 1995.
2. R. Schodl, On the extension of the range of applicability of LDA by means of a the laser-dual-focus (L-2-F) technique, *The Accuracy of Flow Measurements by Laser Doppler Methods*, Skovulunde, Denmark: Dantec Measurement Technology, 1976, 480-489.
3. R. J. Adrian, Particle imaging techniques for experimental fluid mechanics, *Annu. Rev. Fluid Mech.*, 23, 261-304, 1991.
4. I. Grant, *Selected Papers in Particle Image Velocimetry*, SPIE Milestone Series, MS 99, Bellingham, WA: SPIE Optical Engineering Press, 1994.
5. R. J. Adrian, *Bibliography of Particle Velocimetry Using Imaging Methods: 1917–1995*, TAM Report, University of Illinois Urbana-Champaign, Produced and distributed in cooperation with TSI Inc., March 1996. (Also available in electronic format.)
6. W. T. Lai, Particle image velocimetry: a new approach to experimental fluid research, in *Three Dimensional Velocity and Vorticity Measuring and Image Analysis Techniques*, Th. Dracos (ed.), Boston: Kluwer Academic, 1996, 61-92.
7. M. M. Koochesfahani, R. K. Cohn, C. P. Gendrich, and D. G. Nocera, Molecular tagging diagnostics for the study of kinematics and mixing in liquid phase flows, *8th Int. Symp. Appl. Laser Techniques Fluid Mechanics*, Lisbon, 1996.
8. H. Komine, S. J. Brosnan, A. B. Litton, and E. A. Stappaerts, Real time Doppler global velocimetry, *AIAA 29th Aerospace Sciences Meeting*, Paper No. AIAA-91-0337, January 1991.
9. R. L. McKenzie, Measurement capabilities of planar Doppler velocimetry using pulsed lasers, *Appl. Opt.*, 35, 948-964, 1996.
10. C. Berner, Supersonic base flow investigation over axisymmetric bodies, *Proc. 5th Inc. Conf. Laser Anemometry and Applications*, Netherlands, SPIE, 2052, 1993.
11. K. Jaffri, H. G. Hascher, M. Novak, K. Lee, H. Schock, M. Bonne, and P. Keller, Tumble and Swirl Quantification within a Four-valve SI Engine Cylinder Based on 3D LDV Measurements, SAE Paper No. 970792, Feb. 1997.
12. G. G. Podboy and M. J. Krupar, Laser Velocimeter Measurements of the Flow Field Generated by a Forward-Swept Propfan During Flutter, NASA Technical Memorandum 106195, 1993.
13. Y. O. Han, J. G. Leishman, and A. J. Coyne, Measurements of the velocity and turbulence structure of a rotor tip vortex, *AIAA J.*, 35, 477-485, 1997.
14. T. Mathur and J. C. Dutton, Velocity and turbulence measurements in a supersonic base flow with mass bleed, *AIAA J.*, 34, 1153-1159, 1996.
15. E. J. Johnson, P. V. Hyer, P. W. Culotta, and I. O. Clark, Laser velocimetry in nonisothermal CVD systems, *Proc. 4th Int. Conf. Laser Anemometry*, Cleveland, OH, August 1991.
16. R. W. Dibble, V. Hartmann, R. W. Schefer, and W. Kollmann, Conditional sampling of velocity and scalars in turbulent flames using simultaneous LDV-Raman scattering, *Exp. Fluids*, 5, 103-113, 1987.
17. D. V. Srikantiah and W. W. Wilson, Detection of a pulsed flow in an MHD environment by laser velocimetry, *Exp. Fluids*, 6, 500-503, 1988.
18. P. O. Witze, Velocity measurements in end-gas region during homogeneous-charge combustion in a spark ignition engine, *Laser Techniques and Applications in Fluid Mechanics*, Adrian, et al. (eds.), Lisbon: Ladoan, 1992, 518-534.

19. G. L. Morrison, M. C. Johnson, R. E. DeOtte, H. D. Thames, and B. J. Wiedner, An experimental technique for performing 3D LDA measurements inside whirling annular seals, *Flow Meas. Instrum.*, 5, 43-49, 1994.

20. F. Durst, A. Melling, and J. H. Whitelaw, *Principles and Practice of Laser Doppler Anemometry*, 2nd ed., New York: Academic Press, 1981.

21. R. J. Adrian (ed.), *Selected Papers on Laser Doppler Velocimetry*, SPIE Milestone Series, MS 78, Bellingham, WA: SPIE Optical Engineering Press, 1993.

22. W. D. Bachalo and M. J. Houser, Phase Doppler spray analyzer for simultaneous measurements of drop size and velocity distributions, *Opt. Eng.*, 23, 583-590, 1984.

23. Y. Yeh and H. Z. Cummins, Localized fluid flow measurements with an He-Ne laser spectrometer, *Appl. Phys. Lett.*, 4, 176-178, 1964.

24. C. M. Penney, Differential Doppler velocity measurements, *IEEE J. Quantum Electron.*, QE-5, 318, 1969.

25. L. M. Fingerson and P. Freymuth, Thermal anemometers, in *Fluid Mechanics Measurements*, R. J. Goldstein (ed.), New York: Hemisphere, 1983, 99-154.

26. G. Smeets and A. George, Michelson spectrometer for instantaneous Doppler velocity measurements, *J. Phys. E: Sci. Instrum.*, 14, 838-845, 1981.

27. D. Brayton, Small particle signal characteristics of a dual scatter laser velocimeter, *J. Appl. Opt.*, 13, 2346-2351, 1974.

28. F. Durst and W. H. Stevenson, Moiré patterns to visually model laser Doppler signals, *The Accuracy of Flow Measurements by Laser Doppler Methods*, Skovulunde, Denmark: Dantec Measurement Technology, 1976, 183-205.

29. R. J. Adrian, Laser Velocimetry, in *Fluid Mechanics Measurements*, R. J. Goldstein (ed.), New York: Hemisphere, 1983, 155-240.

30. R. J. Adrian and K. L. Orloff, Laser anemometer signal: visibility characteristics and application to particle sizing, *Appl. Opt.*, 16, 677-684, 1977.

31. L. M. Fingerson, R. J. Adrian, R. K. Menon, S. L. Kaufman, and A. Naqwi, Data Analysis, Laser Doppler Velocimetry and Particle Image Velocimetry, TSI Short Course Text, TSI Inc., St. Paul, MN, 1993.

32. H. Kogelnik and T. Li, Laser beams and resonators, *Appl. Opt.*, 5, 1550-1567, 1966.

33. M. C. Whiffen, Polar response of an LV measurement volume, *Minnesota Symp. Laser Anemometry*, University of Minnesota, 1975.

34. C. Tropea, A practical aid for choosing the shift frequency in LDA, *Exp. Fluids*, 4, 79-80, 1986.

35. M. K. Mazumder, Laser Doppler velocity measurement without directional ambiguity by using frequency shifted incident beams, *Appl. Phys. Lett.*, 16, 462-464,1970.

36. H. Muller, V. Tobben, V. Arndt, V. Strunck, H. Wang, R. Kramer, and D. Dopheide, New frequency shift techniques in laser anemometry using tunable semiconductor lasers and solid state lasers, *Proc. 2nd Int. Conf. Fluid Dynamic Measurement Applications*, Beijing, Oct. 1994, 3-19.

37. E. Muller, H. Nobach, and C. Tropea, LDA signal reconstruction: application to moment and spectral estimation, *Proc. 7th Int. Symp. Applications Laser Techniques Fluid Mechanics*, Lisbon, 1994b.

38. R. J. Adrian and W. L. Early, Evaluation of laser Doppler velocimeter performance using Mie scattering theory, *Proc. Minnesota Symp. Laser Anemometry*, University of Minnesota, 1975, 429-454.

39. G. Grehan, G. Gouesbet, A. Naqwi, and F. Durst, Trajectory ambiguities in phase Doppler systems: study of a new forward and a near-backward geometry, *Part. Part. Syst. Charact.*, 11, 133-144, 1994.

40. D. J. Fry, Model submarine wake survey using internal LDV probes, *Proc. ASME Fluids Engineering Meeting*, T. T. Huang, J. Turner, M. Kawahashi, and M. V. Otugen (eds.), FED- Vol. 229, August 1995, 159-170.

41. P. A. Chevrin, H. L. Petrie, and S. Deutsch, Accuracy of a three-component laser Doppler velocimeter system using a single lens approach, *J. Fluids Eng.*, 115, 142-147, 1993.

42. R. I. Karlsson and T. G. Johansson, LDV measurements of higher order moments of velocity fluctuations in a turbulent boundary layer, in *Laser Anemometry in Fluid Mechanics III*, Ladoan-Instituto Superior Technico, 1096 Lisbon Codex, Portugal, 1988, 273-289.

43. K. M. Ibrahim, G. D. Werthimer, and W. D. Bachalo, Signal processing considerations for laser Doppler and phase Doppler applications, *Proc. 5th Int. Symp. Applications Laser Techniques Fluid Mechanics*, Lisbon, 1990.

44. L. Jenson, LDV digital signal processor based on Autocorrelation, *Proc. 6th Int. Symp. Applications Laser Techniques Fluid Mechanics*, Lisbon, 1992.

45. W. W. Hunter and C. E. Nichols (compilers), Wind Tunnel Seeding Systems for Laser Velocimeters, NASA Conference Publication 2393, 1985.

46. A. Melling, Seeding gas flows for laser anemometry, AGARD CP-339, 1986, 8-1–8-11.

47. R. K. Menon and W. T. Lai, Key considerations in the selection of seed particles for LDV measurements, *Proc. 4th Int. Conf. Laser Anemometry*, Cleveland, OH, August 1991.

48. M. L. Lowe and P. H. Kutt, Refraction through cylindrical tubes, *Exp. Fluids*, 13, 315-320, 1992.

49. R. Budwig, Refractive index matching methods for liquid flow investigations, *Exp. Fluids*, 17, 350-355, 1994.

50. D. K. McLaughlin and W. G. Tiederman, Biasing correction for individual realization of laser anemometer measurements in turbulent flows, *Phys. Fluids*, 16, 2082-2088, 1973.

51. R. D. Gould and K. W. Loseke, A comparison of four velocity bias correction techniques in laser Doppler velocimetry, *J. Fluids Eng.*, 115, 508–514, 1993.

52. R. V. Edwards (ed.), Report on the special panel on statistical particle bias problems in laser anemometry, *J. Fluids Eng.*, 109, 89-93, 1987.

53. J. B. Roberts, J. Downie, and M. Gaster, Spectral analysis of signals from a laser Doppler anemometer operating in the burst mode, *J. Physics, E: Sci. Instrum.*, 13, 977-981, 1980.

54. W. T. Mayo, Spectrum measurements with laser velocimeters, *Proc. Dynamic Flow Conf. Dynamic Measurements in Unsteady Flows*, DISA Electronik A/S, Denmark, 1978, 851-868.

55. H. L. Petrie, Reduction of noise effects on power spectrum estimates using the discretized lag product method, *ASME Fluids Engineering Meeting*, FED-229, 139-144, 1995.

56. W. D. Bachalo, A. Brena de la Rosa, and S. V. Sankar, Diagnostics for fuel spray characterization, *Combustion Measurements*, N. Chigier (ed.), New York: Hemisphere, 1991, chap. 7.

57. F. Durst and M. Zare, Laser Doppler measurements in two-phase flows, *The Accuracy of Flow Measurements by Laser Doppler Methods*, Skovulunde, Denmark: Dantec Measurement Technology, 1976, 480-489.

58. M. Saffman, The use of polarized light for optical particle sizing, *Laser Anemometry in Fluid Mechanics III*, Adrian, et al. (eds.), Lisbon: Ladoan, 1988, 387-398.

59. A. Naqwi, Innovative phase Doppler systems and their applications, *Part. Part. Syst. Charact.*, 11, 7-21, 1994.

60. S. V. Sankar, D. A. Robart, and W. D. Bachalo, An adaptive intensity validation technique for minimizing trajectory dependent scattering errors in phase Doppler interferometry, *Proc. 4th Int. Congr. Optical Particle Sizing*, Nuremberg, Germany, March 1995.

61. A. Naqwi and F. Durst, Light scattering applied to LDA and PDA measurements. 2. Computational results and their discussion, *Part. Part. Syst. Charact.*, 9, 66-80, 1992.

62. A. Naqwi, *In-situ* measurement of submicron droplets in electrosprays using a planar phase Doppler system, *J. Aerosol Sci.*, 25, 1201-1211, 1994.

63. S. V. Sankar, D. H. Buermann, D. A. Robart, and W. D. Bachalo, An advanced rainbow signal processor for improved accuracy of droplet temperature measurements in spray flames, *Proc. 8th Int. Symp. Applications Laser Techniques Fluid Mechanics*, Lisbon, 1996.

64. J. Evenstad, A. Naqwi, and R. Menon, A device for phase shift measurement in an advanced phase Doppler velocimeter, *Proc. 8th Int. Symp. Applications Laser Techniques Fluid Mechanics*, Lisbon, 1996.

65. K. M. Ibrahim and W. D. Bachalo, A novel architecture for real-time phase measurement, *Proc. 8th Int. Symp. Applications of Laser Techniques to Fluid Mechanics*, Lisbon, 1996.
66. M. Saffman, P. Buchave, and H. Tanger, Simultaneous measurement of size, concentration and velocity of spherical particles by a laser Doppler method, in *Laser Anemometry in Fluid Mechanics II*, R. J. Adrian, et al. (eds.), Lisbon: Ladoan, 1986, 85-104.
67. A. Naqwi, Sizing of irregular particles using a phase Doppler system, *Proc. ASME Heat Transfer and Fluid Engineering Divisions*, FED-Vol. 233, 1995.

Further Information

C. A. Greated and T. S. Durrani, *Laser Systems and Flow Measurement*, New York: Plenum, 1977.

L. E. Drain, *The Laser Doppler Technique*, New York: John Wiley & Sons, 1980.

Proc. Int. Symp. (1 to 8) *on Applications of Laser Techniques to Fluid Mechanics*, Lisbon, Portugal, 1982, 1984, 1986, 1988, 1990, 1992, 1994, 1996.

P. Buchave, W. K. George, and J. L. Lumley, The measurement of turbulence with the laser Doppler anemometer, *Annu. Rev. Fluid Mech.*, 11, 443-504, 1979.

Proc. 5th Int. Conf. Laser Anemometry and Applications, Netherlands, SPIE, Vol. 2052, 1993.

Proc. ASME Fluids Engineering Meeting, T. T. Huang, J. Turner, M. Kawahashi, and M. V. Otugen, eds., FED- Vol. 229, August 1995.

L. H. Benedict and R. D. Gould, Experiences using the Kalman reconstruction for enhanced power spectrum estimates, *Proc. ASME Fluids Engineering Meeting*, T. T. Huang, J. Turner, M. Kawahashi, and M. V. Otugen (eds.), FED 229, 1-8, 1995.

D. Dopheide, M. Faber, G. Reim, and G. Taux, Laser and avalanche diodes for velocity measurement by laser Doppler anemometry, *Exp. Fluids*, 6, 289-297, 1988.

F. Durst, R. Muller, and A. Naqwi, Measurement accuracy of semiconductor LDA systems, *Exp. Fluids*, 10, 125-137, 1990.

A. Naqwi and F. Durst, Light scattering applied to LDA and PDA measurements. 1. Theory and numerical treatments, *Particle and Particle System Characterization*, 8, 245-258, 1991.

30

Viscosity Measurement

G.E. Leblanc
McMaster University

R.A. Secco
The University of Western Ontario

M. Kostic
Northern Illinois University

30.1 Shear Viscosity

An important mechanical property of fluids is *viscosity*. Physical systems and applications as diverse as fluid flow in pipes, the flow of blood, lubrication of engine parts, the dynamics of raindrops, volcanic eruptions, planetary and stellar magnetic field generation, to name just a few, all involve fluid flow and are controlled to some degree by fluid viscosity. *Viscosity* is defined as the internal friction of a fluid. The microscopic nature of internal friction in a fluid is analogous to the macroscopic concept of mechanical friction in the system of an object moving on a stationary planar surface. Energy must be supplied (1) to overcome the inertial state of the interlocked object and plane caused by surface roughness, and (2) to initiate and sustain motion of the object over the plane. In a fluid, energy must be supplied (1) to create viscous flow units by breaking bonds between atoms and molecules, and (2) to cause the flow units to move relative to one another. The resistance of a fluid to the creation and motion of flow units is due to the viscosity of the fluid, which only manifests itself when motion in the fluid is set up. Since viscosity involves the transport of mass with a certain velocity, the viscous response is called a *momentum transport process*. The velocity of flow units within the fluid will vary, depending on location. Consider a liquid between two closely spaced parallel plates as shown in Figure 30.1. A force, *F*, applied to the top plate causes the fluid adjacent to the upper plate to be dragged in the direction of *F*. The applied force is communicated to neighboring layers of fluid below, each coupled to the driving layer above, but with diminishing magnitude. This results in the progressive decrease in velocity of each fluid layer, as shown by the decreasing velocity vector in Figure 30.1, away from the upper plate. In this system, the applied force is called a *shear* (when applied over an area it is called a *shear stress*), and the resulting deformation rate of the fluid, as illustrated by the *velocity gradient dU_x/dz*, is called the *shear strain rate, $\dot{\gamma}_{zx}$*. The mathematical expression describing the viscous response of the system to the shear stress is simply:

$$\tau_{zx} = \frac{\eta dU_x}{dz} = \eta \dot{\gamma}_{zx} \tag{30.1}$$

where τ_{zx}, the shear stress, is the force per unit area exerted on the upper plate in the x-direction (and hence is equal to the force per unit area exerted by the fluid on the upper plate in the x-direction under the assumption of a no-slip boundary layer at the fluid–upper plate interface); dU_x/dz is the gradient of the x-velocity in the z-direction in the fluid; and η is the *coefficient of viscosity*. In this case, because one is concerned with a shear force that produces the fluid motion, η is more specifically called the *shear*

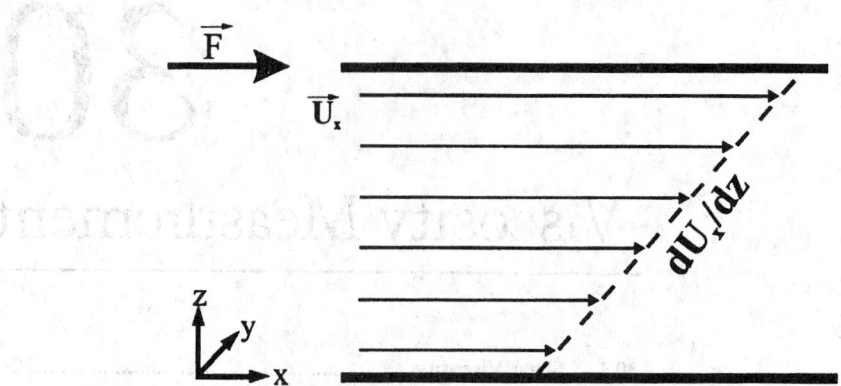

FIGURE 30.1 System for defining Newtonian viscosity. When the upper plate is subjected to a force, the fluid between the plates is dragged in the direction of the force with a velocity of each layer that diminishes away from the upper plate. The reducing velocity eventually reaches zero at the lower plate boundary.

dynamic viscosity. In fluid mechanics, diffusion of momentum is a more useful description of viscosity where the motion of a fluid is considered without reference to force. This type of viscosity is called the *kinematic viscosity*, ν, and is derived by dividing dynamic viscosity by ρ, the mass density:

$$\nu = \frac{\eta}{\rho} \tag{30.2}$$

The definition of viscosity by Equation 30.1 is valid only for *laminar* (i.e., layered or sheet-like) or streamline flow as depicted in Figure 30.1, and it refers to the molecular viscosity or *intrinsic viscosity*. The molecular viscosity is a property of the material that depends microscopically on bond strengths, and is characterized macroscopically as the fluid's resistance to flow. When the flow is turbulent, the diffusion of momentum is comprised of viscous contributions from the motion, sometimes called the *eddy viscosity*, in addition to the intrinsic viscosity. Viscosities of turbulent systems can be as high as 10^6 times greater than viscosities of laminar systems, depending on the Reynolds number.

Molecular viscosity is separated into *shear viscosity* and bulk or *volume viscosity*, η_v, depending on the type of strain involved. Shear viscosity is a measure of resistance to isochoric flow in a shear field, whereas volume viscosity is a measure of resistance to volumetric flow in a three-dimensional stress field. For most liquids, including hydrogen bonded, weakly associated or unassociated, and polymeric liquids as well as liquid metals, $\eta/\eta_v \approx 1$, suggesting that shear and structural viscous mechanisms are closely related [1].

The shear viscosity of most liquids decreases with temperature and increases with pressure, which is opposite to the corresponding responses for gases. An increase in temperature usually causes expansion and a corresponding reduction in liquid bond strength, which in turn reduces the internal friction. Pressure causes a decrease in volume and a corresponding increase in bond strength, which in turn enhances the internal friction. For most situations, including engineering applications, temperature effects dominate the antagonistic effects of pressure. However, in the context of planetary interiors where the effects of pressure cannot be ignored, pressure controls the viscosity to the extent that, depending on composition, it can cause fundamental changes in the molecular structure of the fluid that can result in an anomalous viscosity decrease with increasing pressure [2].

Newtonian and Non-Newtonian Fluids

Equation 30.1 is known as Newton's law of viscosity and it formulates Sir Isaac Newton's definition of the viscous behavior of a class of fluids now called Newtonian fluids.

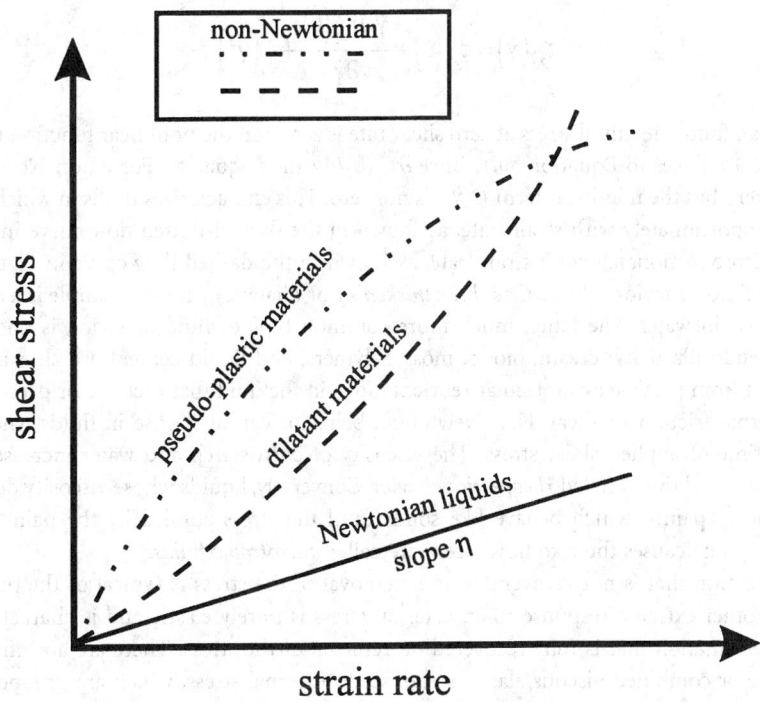

FIGURE 30.2 Flow curves illustrating Newtonian and non-Newtonian fluid behavior.

If the viscosity throughout the fluid is independent of strain rate, then the fluid is said to be a *Newtonian fluid*. The constant of proportionality is called the coefficient of viscosity, and a plot of stress vs. strain rate for Newtonian fluids yields a straight line with a slope of η, as shown by the solid line flow curve in Figure 30.2. Examples of Newtonian fluids are pure, single-phase, unassociated gases, liquids, and solutions of low molecular weight such as water. There is, however, a large group of fluids for which the viscosity is dependent on the strain rate. Such fluids are said to be non-Newtonian fluids and their study is called *rheology*. In differentiating between Newtonian and non-Newtonian behavior, it is helpful to consider the time scale (as well as the normal stress differences and phase shift in dynamic testing) involved in the process of a liquid responding to a shear perturbation. The velocity gradient, dU_x/dz, in the fluid is equal to the shear strain rate, $\dot{\gamma}$, and therefore the time scale related to the applied shear perturbation about the equilibrium state is t_s, where $t_s = \dot{\gamma}^{-1}$. A second time scale, t_r, called the *relaxation time*, characterizes the rate at which the relaxation of the strain in the fluid can be accomplished and is related to the time it takes for a typical flow unit to move a distance equivalent to its mean diameter. For Newtonian water, $t_r \sim 10^{-12}$ s and, because shear rates greater than 10^6 s^{-1} are rare in practice, the time required for adjustment of the shear perturbation in water is much less than the shear perturbation period (i.e., $t_r \ll t_s$). However, for non-Newtonian macromolecular liquids like polymeric liquids, for colloidal and fiber suspensions, and for pastes and emulsions, the long response times of large viscous flow units can easily make $t_r > t_s$. An example of a non-Newtonian fluid is liquid elemental sulfur, in which long chains (polymers) of up to 100,000 sulfur atoms form flow units that are easily entangled, which bind the liquid in a "rigid-like" network. Another example of a well-known non-Newtonian fluid is tomato ketchup.

With reference to Figure 30.2, the more general form of Equation 30.1 also accounts for the nonlinear response. In terms of an initial shear stress required for flow to start, $\tau_{xy}(0)$, an initial linear term in the Newtonian limit of a small range of strain rate, $\dot{\gamma}\partial\tau_{xy}(0)/\partial\gamma$, and a nonlinear term $O(\dot{\gamma}^2)$, the shear stress dependence on strain rate, $\tau_{xy}(\dot{\gamma})$ can be described as:

$$\tau_{xy}\left(\dot{\gamma}\right)=\tau_{xy}\left(0\right)+\frac{\dot{\gamma}\partial\tau_{xy}\left(0\right)}{\partial\dot{\gamma}}+O\left(\dot{\gamma}^2\right) \tag{30.3}$$

For a Newtonian fluid, the initial stress at zero shear rate is zero and the nonlinear function $O(\dot{\gamma}^2)$ is zero, so Equation 30.3 reduces to Equation 30.1, since $\partial\tau_{xy}(0)/\partial\dot{\gamma}$ then equals η. For a non-Newtonian fluid, $\tau_{xy}(0)$ may be zero but the nonlinear term $O(\dot{\gamma}^2)$ is nonzero. This characterizes fluids in which shear stress increases disproportionately with strain rate, as shown in the dashed-dotted flow curve in Figure 30.2, or decreases disproportionately with strain rate, as shown in the dashed flow curve in Figure 30.2. The former type of fluid behavior is known as *shear thickening* or dilatancy, and an example is a concentrated solution of sugar in water. The latter, much more common type of fluid behavior, is known as *shear thinning* or pseudo-plasticity; cream, blood, most polymers, and liquid cement are all examples. Both behaviors result from particle or molecular reorientations in the fluid that increase or decreases, respectively, the internal friction to shear. Non-Newtonian behavior can also arise in fluids whose viscosity changes with time of applied shear stress. The viscosity of corn starch and water increases with time duration of stress, and this is called *rheopectic behavior*. Conversely, liquids whose viscosity decreases with time, like nondrip paints, which behave like solids until the stress applied by the paint brush for a sufficiently long time causes them to flow freely, are called *thixotropic fluids*.

Fluid deformation that is not recoverable after removal of the stress is typical of the purely viscous response. The other extreme response to an external stress is purely elastic and is characterized by an equilibrium deformation that is fully recovered on removal of the stress. There are an infinite number of intermediate or combined viscous/elastic responses to external stress, which are grouped under the behavior known as *viscoelasticity*. Fluids that behave elastically in some stress range require a limiting or yield stress before they will flow as a viscous fluid. A simple, empirical, constitutive equation often used for this type of rheological behavior is of the form:

$$\tau_{yx}=\tau_y+\dot{\gamma}^n\eta_p \tag{30.4}$$

where τ_y is the *yield stress*, η_p is an *apparent viscosity* called the plastic viscosity, and the exponent n allows for a range of non-Newtonian responses: $n=1$ is pseudo-Newtonian behavior and is called a *Bingham fluid*; $n<1$ is shear thinning behavior; and $n>1$ is shear thickening behavior. Interested readers should consult [3–9] for further information on applied rheology.

Dimensions and Units of Viscosity

From Equation 30.1, the dimensions of dynamic viscosity are $M\,L^{-1}\,T^{-1}$ and the basic SI unit is the Pascal second (Pa·s), where 1 Pa·s $=1$ N s m^{-2}. The c.g.s. unit of dyn s cm^{-2} is the poise (P). The dimensions of kinematic viscosity, from Equation 30.2, are $L^2\,T^{-1}$ and the SI unit is m^2 s^{-1}. For most practical situations, this is usually too large and so the c.g.s. unit of cm^2 s^{-1}, or the stoke (St), is preferred. Table 30.1 lists some common fluids and their shear dynamic viscosities at atmospheric pressure and 20°C.

TABLE 30.1 Shear Dynamic Viscosity of Some Common Fluids at 20°C and 1 atm

Fluid	Shear dynamic viscosity (Pa·s)
Air	1.8×10^{-4}
Water	1.0×10^{-3}
Mercury	1.6×10^{-3}
Automotive engine oil (SAE 10W30)	1.3×10^{-1}
Dish soap	4.0×10^{-1}
Corn syrup	6.0

TABLE 30.2 Viscometer Classification and Basic Characteristics

Drag Flow Types:
Flow set by motion of instrument boundary/surface using external or gravity force.

Type/Geometry	Basic characteristics/Comments
Rotating concentric cylinders (Couette)	Good for low viscosity, high shear rates; for $R_2/R_1 \cong 1$, see Figure 30.3; hard to clean thick fluids
Rotating cone and plate	Homogeneous shear, best for non-Newtonian fluids and normal stresses; need good alignment, problems with loading and evaporation
Rotating parallel disks	Similar to cone-and-plate, but inhomogeneous shear; shear varies with gap height, easy sample loading
Sliding parallel plates	Homogeneous shear, simple design, good for high viscosity; difficult loading and gap control
Falling body (ball, cylinder)	Very simple, good for high temperature and pressure; need density and special sensors for opaque fluids, not good for viscoelastic fluids
Rising bubble	Similar to falling body viscometer; for transparent fluids
Oscillating body	Needs instrument constant, good for low viscous liquid metals

Pressure Flow Types:
Fluid set in motion in fixed instrument geometry by external or gravity pressure

Type/Geometry	Basic characteristics/Comments
Long capillary (Poiseuille flow)	Simple, very high shears and range, but very inhomogeneous shear, bad for time dependency, and is time consuming
Orifice/Cup (short capillary)	Very simple, reliable, but not for absolute viscosity and non-Newtonian fluids
Slit (parallel plates) pressure flow	Similar to capillary, but difficult to clean
Axial annulus pressure flow	Similar to capillary, better shear uniformity, but more complex, eccentricity problem and difficult to clean

Others/Miscellaneous:

Type/geometry	Basic characteristics/Comments
Ultrasonic	Good for high viscosity fluids, small sample volume, gives shear and volume viscosity, and elastic property data; problems with surface finish and alignment, complicated data reduction

Adapted from C. W. Macosko, *Rheology: Principles, Measurements, and Applications,* New York: VCH, 1994.

Viscometer Types

The instruments for viscosity measurements are designed to determine "a fluid's resistance to flow," a fluid property defined above as viscosity. The fluid flow in a given instrument geometry defines the strain rates, and the corresponding stresses are the measure of resistance to flow. If strain rate or stress is set and controlled, then the other one will, everything else being the same, depend on the fluid viscosity. If the flow is simple (one dimensional, if possible) such that the strain rate and stress can be determined accurately from the measured quantities, the absolute dynamic viscosity can be determined; otherwise, the relative viscosity will be established. For example, the fluid flow can be set by dragging fluid with a sliding or rotating surface, falling body through the fluid, or by forcing the fluid (by external pressure or gravity) to flow through a fixed geometry, such as a capillary tube, annulus, a slit (between two parallel plates), or orifice. The corresponding resistance to flow is measured as the boundary force or torque, or pressure drop. The flow rate or efflux time represents the fluid flow for a set flow resistance, like pressure drop or gravity force. The viscometers are classified, depending on how the flow is initiated or maintained, as in Table 30.2.

The basic principle of all viscometers is to provide as simple flow kinematics as possible, preferably one-dimensional *(isometric) flow,* in order to determine the shear strain rate accurately, easily, and independent of fluid type. The resistance to such flow is measured, and thereby the shearing stress is

TABLE 30.3 Different Causes of Viscometers Errors

Error/Effect	Cause/Comment
End/edge effect	Energy losses at the fluid entrance and exit of main test geometry
Kinetic energy losses	Loss of pressure to kinetic energy
Secondary flow	Energy loss due to unwanted secondary flow, vortices, etc.; increases with Reynolds number
Nonideal geometry	Deviations from ideal shape, alignment, and finish
Shear rate non-uniformity	Important for non-Newtonian fluids
Temperature variation and viscous heating	Variation in temperature, in time and space, influences the measured viscosity
Turbulence	Partial and/or local turbulence often develops even at low Reynolds numbers
Surface tension	Difference in interfacial tensions
Elastic effects	Structural and fluid elastic effects
Miscellaneous effects	Depends on test specimen, melt fracture, thixotropy, rheopexy

TABLE 30.4 Viscometer Manufacturers

Manufacturers	Model	Description
Brookfield Eng. Labs Inc.	DV-I+	Concentric cylinder
Custom Scientific Inst. Inc.	CS245	Concentric cylinder
Reologica Inst.	Various	Falling sphere, capillary, rotational
Haake GmbH	Various	Falling sphere, rotational
Cannon Inst. Co.	Various	Extensive variety of capillary viscometers
Toyo Seikl Seisaku-Sho Ltd.	Capirograph	Capillary
Gottfert Werkstoff-Prufmaschinen GmbH	Various	Extensive variety of capillary viscometers
Cole-Palmer Inst. Co.	GV2100	Falling sphere
Paar Physica U.S.A. Inc.	Various	Concentric cylinder, falling sphere, capillary
Monsanto Inst. & Equipment	ODR 2000	Oscillating viscometer
Nametre Co.	Vibrational viscometer	Oscilating viscometer
Rheometric Scientific Inc.	RM180	Cone-and-plate, parallel plate
	RM265	Concentric cylinders
T.A. Instruments Inc.	Various	Concentric cylinder

Note: All the above manufacturers can be found via the Internet (World Wide Web), along with the most recent contact information, product description and in some cases, pricing.

determined. The shear viscosity is then easily found as the ratio between the shearing stress and the corresponding shear strain rate. Practically, it is never possible to achieve desired one-dimensional flow nor ideal geometry, and a number of errors, listed in Table 30.3, can occur and need to be accounted for [4–8]. A list of manufacturers/distributors of commercial viscometers/rheometers is given in Table 30.4.

Concentric Cylinders

The main advantage of the rotational as compared to many other viscometers is its ability to operate continuously at a given shear rate, so that other steady-state measurements can be conveniently performed. That way, time dependency, if any, can be detected and determined. Also, subsequent measurements can be made with the same instrument and sampled at different shear rates, temperature, etc. For these and other reasons, *rotational viscometers* are among the most widely used class of instruments for rheological measurements.

Concentric cylinder-type viscometers/rheometers are usually employed when absolute viscosity needs to be determined, which in turn, requires a knowledge of well-defined shear rate and shear stress data. Such instruments are available in different configurations and can be used for almost any fluid. There are models for low and high shear rates. More complete discussion on concentric cylinder viscometers/rheometers is given elsewhere [4–9]. In the *Couette-type viscometer*, the rotation of the outer cylinder, or cup, minimizes centrifugal forces, which cause Taylor vortices. The latter can be present in the *Searle-type viscometer* when the inner cylinder, or bob, rotates.

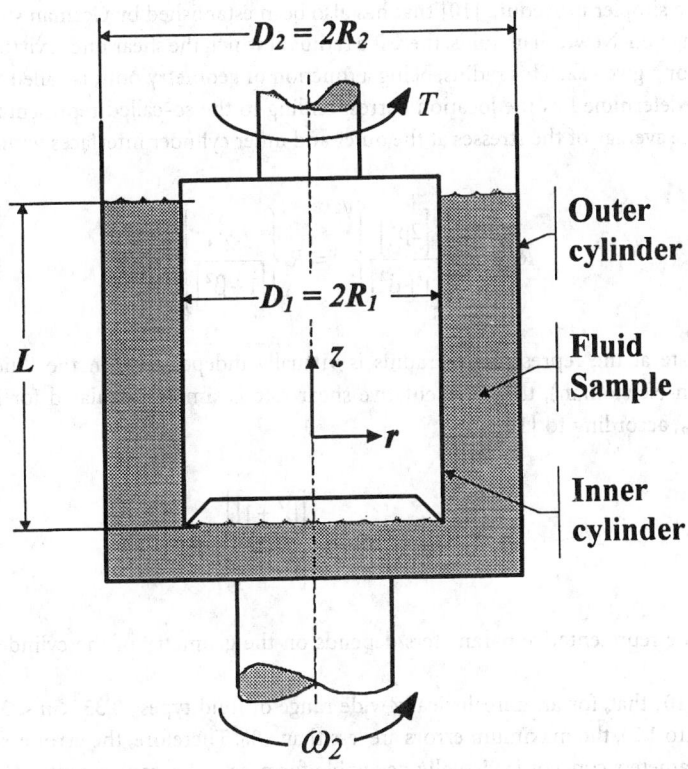

FIGURE 30.3   Concentric cylinders viscometer geometry.

Usually, the torque on the stationary cylinder and rotational velocity of the other cylinder are measured for determination of the shear stress and shear rate, which is needed for viscosity calculation. Once the torque, T, is measured, it is simple to describe the fluid shear stress at any point with radius r between the two cylinders, as shown in Figure 30.3.

$$\tau_{r\theta}(r) = \frac{T}{2\pi r^2 L_e} \qquad (30.5)$$

In Equation 30.5, $L_e = (L + L_c)$, is the effective length of the cylinder at which the torque is measured. In addition to the cylinder's length L, it takes into account the end-effect correction L_c [4–8].

For a narrow gap between the cylinders ($\beta = R_2/R_1 \cong 1$), regardless of the fluid type, the velocity profile can be approximated as linear, and the shear rate within the gap will be uniform:

$$\gamma(r) \cong \frac{\Omega \bar{R}}{(R_2 - R_1)} \qquad (30.6)$$

where $\Omega = (\omega_2 - \omega_1)$ is the relative rotational speed and $\bar{R} = (R_1 + R_2)/2$, is the mean radius of the inner (1) and outer (2) cylinders. Actually, the shear rate profile across the gap between the cylinders depends on the relative rotational speed, radii, and the unknown fluid properties, which seems an "open-ended" enigma. The solution of this complex problem is given elsewhere [4–8] in the form of an infinite series, and requires the slope of a logarithmic plot of T as a function of Ω in the range of interest. Note that for a stationary inner cylinder ($\omega_1 = 0$), which is the usual case in practice, Ω becomes equal to ω_2.

However, there is a simpler procedure [10] that has also been established by German standards [11]. For any fluid, including non-Newtonian fluids, there is a radius at which the shear rate is virtually independent of the fluid type for a given Ω. This radius, being a function of geometry only, is called the *representative radius*, R_R, and is determined as the location corresponding to the so-called representative shear stress, $\tau_R = (\tau_1 + \tau_2)/2$, the average of the stresses at the outer and inner cylinder interfaces with the fluid, that is:

$$R_R = R_1 \left\{ \frac{\left[2\beta^2 \right]}{\left[1 + \beta^2 \right]} \right\}^{1/2} = R_2 \left\{ \frac{2}{\left[1 + \beta^2 \right]} \right\}^{1/2} \tag{30.7}$$

Since the shear rate at the representative radius is virtually independent on the fluid type (whether Newtonian or non-Newtonian), the representative shear rate is simply calculated for Newtonian fluid ($n = 1$) and $r = R_R$, according to [10]:

$$\dot{\gamma}_R = \dot{\gamma}_{r=R_R} = \omega_2 \left\{ \frac{\left[\beta^2 + 1 \right]}{\left[\beta^2 - 1 \right]} \right\} \tag{30.8}$$

The accuracy of the representative parameters depends on the geometry of the cylinders (β) and fluid type (n).

It is shown in [10] that, for an unrealistically wide range of fluid types ($0.35 < n < 3.5$) and cylinder geometries ($\beta = 1$ to 1.2), the maximum errors are less than 1%. Therefore, the error associated with the representative parameters concept is virtually negligible for practical measurements.

Finally, the (apparent) fluid viscosity is determined as the ratio between the shear stress and corresponding shear rate using Equations 30.5 to 30.8, as:

$$\eta = \eta_R = \frac{\tau_R}{\dot{\gamma}_R} = \left\{ \frac{\left[\beta^2 - 1 \right]}{\left[4\pi\beta^2 R_1^2 L_e \right]} \right\} \frac{T}{\omega_2} = \left\{ \frac{\left[\beta^2 - 1 \right]}{\left[4\pi R_2^2 L_e \right]} \right\} \frac{T}{\omega_2} \tag{30.9}$$

For a given cylinder geometry (β, R_2, and L_e), the viscosity can be determined from Equation 30.8 by measuring T and ω_2.

As already mentioned, in Couette-type viscometers, the Taylor vortices within the gap are virtually eliminated. However, vortices at the bottom can be present, and their influence becomes important when the Reynolds number reaches the value of unity [10, 11]. Furthermore, flow instability and turbulence will develop when the Reynolds number reaches values of 10^3 to 10^4. The Reynolds number, Re, for the flow between concentric cylinders is defined [11] as:

$$\text{Re} = \left\{ \frac{\left[\rho \omega_2 R_1^2 \right]}{2\eta} \right\} \left[\beta^2 - 1 \right] \tag{30.10}$$

Cone-and-Plate Viscometers

The simple cone-and-plate viscometer geometry provides a uniform rate of shear and direct measurements of the first normal stress difference. It is the most popular instrument for measurement of non-Newtonian fluid properties. The working shear stress and shear strain rate equations can be easily derived in spherical coordinates, as indicated by the geometry in Figure 30.4, and are, respectively:

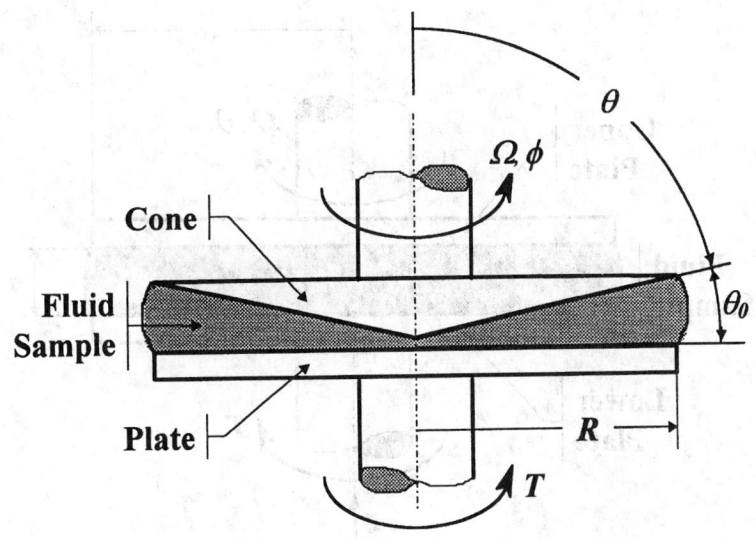

FIGURE 30.4 Cone-and-plate viscometer geometry.

$$\tau_{\theta\phi} = \frac{3T}{\left[2\pi R^3\right]} \tag{30.11}$$

and

$$\dot{\gamma} = \frac{\Omega}{\theta_0} \tag{30.12}$$

where R and $\theta_0 < 0.1$ rad ($\approx 6°$) are the cone radius and angle, respectively. The viscosity is then easily calculated as:

$$\eta = \frac{\tau_{\theta\phi}}{\dot{\gamma}} = \frac{\left[3T\theta_0\right]}{2\pi\Omega R^3} \tag{30.13}$$

Inertia and secondary flow increase while shear heating decreases the measured torque (T_m). For more details, see [4, 5]. The torque correction is given as:

$$\frac{T_m}{T} = 1 + 6 \cdot 10^{-4} Re^2 \tag{30.14}$$

where

$$Re = \frac{\left\{\rho\left[\Omega\theta_0 R\right]^2\right\}}{\eta} \tag{30.15}$$

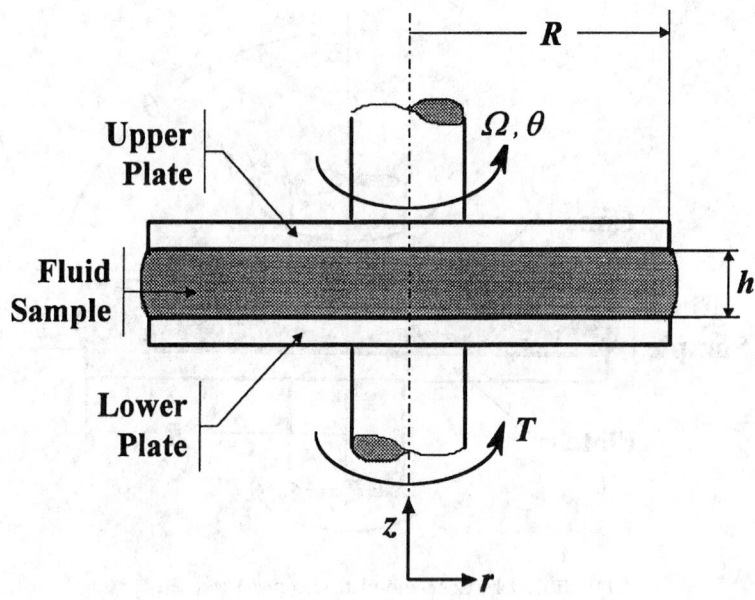

FIGURE 30.5   Parallel disks viscometer geometry.

Parallel Disks

This geometry (Figure 30.5), which consists of a disk rotating in a cylindrical cavity, is similar to the cone-and-plate geometry, and many instruments permit the use of either one. However, the shear rate is no longer uniform, but depends on radial distance from the axis of rotation and on the gap h, that is:

$$\dot{\gamma}(r) = \frac{r\Omega}{h} \tag{30.16}$$

For Newtonian fluids, after integration over the disk area, the torque can be expressed as a function of viscosity, so that the latter can be determined as:

$$\eta = \frac{2Th}{\left[\pi\Omega R^4\right]} \tag{30.17}$$

Capillary Viscometers

The *capillary viscometer* is based on the fully developed laminar tube flow theory (*Hagen–Poiseuille flow*) and is shown in Figure 30.6. The capillary tube length is many times larger than its small diameter, so that entrance flow is neglected or accounted for in more accurate measurement or for shorter tubes. The expression for the shear stress at the wall is:

$$\tau_w = \left[\frac{\Delta P}{L}\right] \cdot \left[\frac{D}{4}\right] \tag{30.18}$$

and

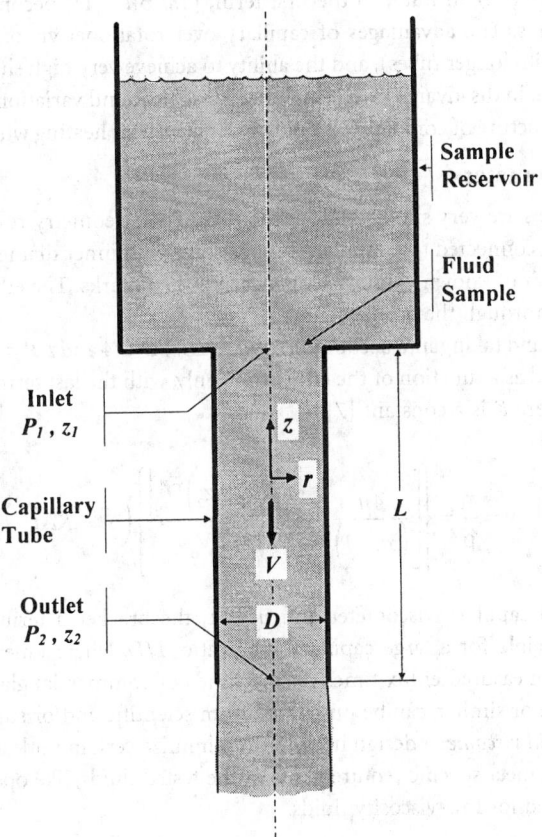

FIGURE 30.6 Capillary viscometer geometry.

$$\Delta P = \left(P_1 - P_2\right) + \left(z_1 - z_2\right) - \frac{\left[CpV^2\right]}{2} \tag{30.19}$$

where, $C \cong 1.1$, P, z, $V = 4Q/[\pi D^2]$, and Q are correction factor, pressure, elevation, the mean flow velocity, and the fluid volume-flow rate, respectively. The subscripts 1 and 2 refer to the inlet and outlet, respectively.

The expression for the shear rate at the wall is:

$$\dot{\gamma} = \left\{ \left[\frac{3n+1}{4n}\right] \right\} \circ \left\{ \frac{8V}{D} \right\} \tag{30.20}$$

where $n = d[\log \tau_w]/d[\log (8V/D)]$ is the slope of the measured $\log(\tau_w) - \log (8V/D)$ curve. Then, the viscosity is simply calculated as:

$$\eta = \frac{\tau_w}{\dot{\gamma}} = \left\{ \frac{4n}{[3n+1]} \right\} \circ \left\{ \left[\frac{\Delta P D^2}{32LV}\right] \right\} = \left\{ \frac{4n}{[3n+1]} \right\} \circ \left\{ \left[\frac{\Delta P D^4 \pi}{128QL}\right] \right\} \tag{30.21}$$

Note that $n = 1$ for a Newtonian fluid, so the first term, $[4n/(3n + 1)]$ becomes unity and disappears from the above equations. The advantages of capillary over rotational viscometers are low cost, high accuracy (particularly with longer tubes), and the ability to achieve very high shear rates, even with high-viscosity samples. The main disadvantages are high residence time and variation of shear across the flow, which can change the structure of complex test fluids, as well as shear heating with high-viscosity samples.

Glass Capillary Viscometers

Glass capillary viscometers are very simple and inexpensive. Their geometry resembles a U-tube with at least two reservoir bulbs connected to a capillary tube passage with inner diameter D. The fluid is drawn up into one bulb reservoir of known volume, V_0, between etched marks. The efflux time, Δt, is measured for that volume to flow through the capillary under gravity.

From Equation 30.21 and taking into account that $V_0 = (\Delta t)VD^2\pi/4$ and $\Delta P = \rho g(z_1 - z_2)$, the kinematic viscosity can be expressed as a function of the efflux time only, with the last term, $K/\Delta t$, added to account for error correction, where K is a constant [7]:

$$\nu = \frac{\eta}{\rho} = \left\{ \left[\frac{4n}{(3n+1)} \right] \cdot \frac{\left[\pi g\left(z_1 - z_2\right)D^4 \right]}{128LV_0} \right\} \left(\Delta t - K\Delta t \right) \tag{30.22}$$

Note that for a given capillary viscometer and $n \cong 1$, the bracketed term is a constant. The last correction term is negligible for a large capillary tube ratio, L/D, where kinematic viscosity becomes linearly proportional to measured efflux time. Various kinds of commercial glass capillary viscometers, like Cannon-Fenske type or similar, can be purchased from scientific and/or supply stores. They are the modified original *Ostwald viscometer* design in order to minimize certain undesirable effects, to increase the viscosity range, or to meet specific requirements of the tested fluids, like opacity, etc. Glass capillary viscometers are often used for low-viscosity fluids.

Orifice/Cup, Short Capillary: Saybolt Viscometer

The principle of these viscometers is similar to glass capillary viscometers, except that the flow through a short capillary ($L/D \ll 10$) does not satisfy or even approximate the Hagen–Poiseuille, fully developed, pipe flow. The influences of entrance end-effect and changing hydrostatic heads are considerable. The efflux time reading, Δt, represents relative viscosity for comparison purposes and is expressed as "viscometer seconds," like the Saybolt seconds, or Engler seconds or degrees. Although the conversion formula, similar to glass capillary viscometers, is used, the constants k and K in Equation 30.23 are purely empirical and dependent on fluid types.

$$\nu = \frac{\eta}{\rho} = k\Delta t - \frac{K}{\Delta t} \tag{30.23}$$

where $k = 0.00226, 0.0216, 0.073$; and $K = 1.95, 0.60, 0.0631$; for Saybolt Universal ($\Delta t < 100$ s), Saybolt Furol ($\Delta t > 40$ s), and Engler viscometers, respectively [12, 13]. Due to their simplicity, reliability, and low cost, these viscometers are widely used for Newtonian fluids, like in oil and other industries, where the simple correlations between the relative properties and desired results are needed. However, these viscometers are not suitable for absolute viscosity measurement, nor for non-Newtonian fluids.

Falling Body Methods

Falling Sphere

The falling sphere viscometer is one of the earliest and least involved methods to determine the absolute shear viscosity of a Newtonian fluid. In this method, a sphere is allowed to fall freely a measured distance through a viscous liquid medium and its velocity is determined. The viscous drag of the falling sphere results in the creation of a restraining force, F, described by Stokes' law:

$$F = 6\pi\eta r_s U_t \qquad (30.24)$$

where r_s is the radius of the sphere and U_t is the *terminal velocity* of the falling body. If a sphere of density ρ_2 is falling through a fluid of density ρ_1 in a container of infinite extent, then by balancing Equation 30.24 with the net force of gravity and buoyancy exerted on a solid sphere, the resulting equation of absolute viscosity is:

$$\eta = 2gr_s^2 \frac{\left(\rho_2 - \rho_1\right)}{9U_t} \qquad (30.25)$$

Equation 30.25 shows the relation between the viscosity of a fluid and the terminal velocity of a sphere falling within it. Having a finite container volume necessitates the modification of Equation 30.25 to correct for effects on the velocity of the sphere due to its interaction with container walls (W) and ends (E). Considering a cylindrical container of radius r and height H, the corrected form of Equation 30.25 can be written as:

$$\eta = 2gr_s^2 \frac{\left(\rho_2 - \rho_1\right)W}{\left(9U_t E\right)} \qquad (30.26)$$

where

$$W = 1 - 2.104\left(\frac{r_s}{r}\right) + 2.09\left(\frac{r_s}{r}\right)^3 - 0.95\left(\frac{r_s}{r}\right)^5 \qquad (30.27)$$

$$E = 1 + 3.3\left(\frac{r_s}{H}\right) \qquad (30.28)$$

The wall correction was empirically derived [15] and is valid for $0.16 \leq r_s/r \leq 0.32$. Beyond this range, the effects of container walls significantly impair the terminal velocity of the sphere, thus giving rise to a false high viscosity value.

Figure 30.7 is a schematic diagram of the falling sphere method and demonstrates the attraction of this method — its simplicity of design. The simplest and most cost-effective approach in applying this method to transparent liquids would be to use a sufficiently large graduated cylinder filled with the liquid. With a distance marked on the cylinder near the axial and radial center (the region least influenced by the container walls and ends), a sphere (such as a ball bearing or a material that is nonreactive with the liquid) with a known density and sized to within the bounds of the container correction, free falls the length of the cylinder. As the sphere passes through the marked region of length d at its terminal velocity, a measure of the time taken to traverse this distance allows the velocity of the sphere to be calculated. Having measured all the parameters of Equation 30.26, the shear viscosity of the liquid can be determined.

This method is useful for liquids with viscosities between 10^{-3} Pa·s and 10^5 Pa·s. Due to the simplicity of design, the falling sphere method is particularly well suited to high pressure–high temperature viscosity studies.

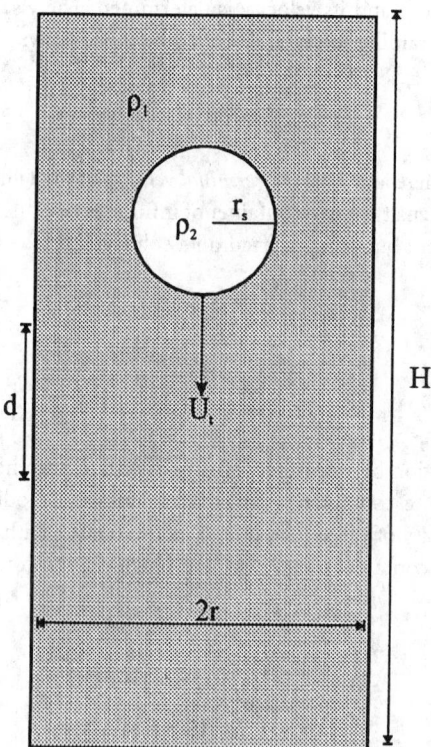

FIGURE 30.7 Schematic diagram of the falling sphere viscometer. Visual observations of the time taken for the sphere to traverse the distance d, is used to determine a velocity of the sphere. The calculated velocity is then used in Equation 30.24 to determine a shear viscosity.

Falling Cylinder

The *falling cylinder method* is similar in concept to the falling sphere method except that a flat-ended, solid circular cylinder freely falls vertically in the direction of its longitudinal axis through a liquid sample within a cylindrical container. A schematic diagram of the configuration is shown in Figure 30.8. Taking an infinitely long cylinder of density ρ_2 and radius r_c falling through a Newtonian fluid of density ρ_1 with infinite extent, the resulting shear viscosity of the fluid is given as:

$$\eta = g r_c^2 \frac{\left(\rho_2 - \rho_1\right)}{2U_t} \tag{30.29}$$

Just as with the falling sphere, a finite container volume necessitates modifying Equation 30.29 to account for the effects of container walls and ends. A correction for container wall effects can be analytically deduced by balancing the buoyancy and gravitational forces on the cylinder, of length L, with the shear force on the sides and the compressional force on the cylinder's leading end and the tensile force on the cylinder's trailing end. The resulting correction term, or geometrical factor, $G(k)$ (where $k = r_c/r$), depends on the cylinder radius and the container radius, r, and is given by:

$$G(k) = \frac{\left[k^2\left(1 - \ln k\right) - \left(1 + \ln k\right)\right]}{\left(1 + k^2\right)} \tag{30.30}$$

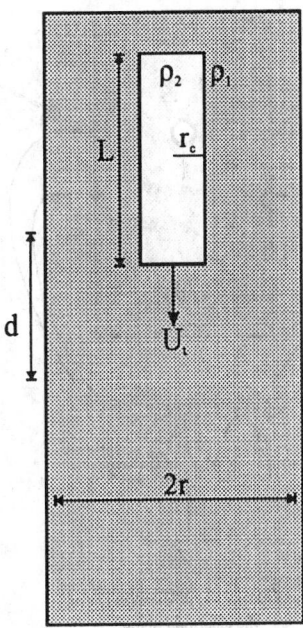

FIGURE 30.8 Schematic diagram of the falling cylinder viscometer. Using the same principle as the falling sphere, the velocity of the cylinder is obtained, which is needed to determine the shear viscosity of the fluid.

Unlike the fluid flow around a falling sphere, the fluid motion around a falling flat-ended cylinder is very complex. The effects of container ends are minimized by creating a small gap between the cylinder and the container wall. If a long cylinder (here, a cylinder is considered long if $\psi \geq 10$, where $\psi = L/r$) with a radius nearly as large as the radius of the container is used, then the effects of the walls would dominate, thereby reducing the end effects to a second-order effect. A major drawback with this approach is, however, if the cylinder and container are not concentric, the resulting inhomogeneous wall shear force would cause the downward motion of the cylinder to become eccentric. The potential for misalignment motivated the recently obtained analytical solution to the fluid flow about the cylinder ends [16]. An analytical expression for the end correction factor (ECF) was then deduced [17] and is given as:

$$\frac{1}{\text{ECF}} = 1 + \left(\frac{8k}{\pi C_w}\right)\left(\frac{G(k)}{\psi}\right) \tag{30.31}$$

where $C_w = 1.003852 - 1.961019k + 0.9570952k^2$. C_w was derived semi-empirically [17] as a disk wall correction factor. This is based on the idea that the drag force on the ends of the cylinder can be described by the drag force on a disk. Equation 30.31 is valid for $\psi \leq 30$ and agrees with the empirically derived correction [16] to within 0.6%.

With wall and end effects taken into consideration, the working formula to determine the shear viscosity of a Newtonian fluid from a falling cylinder viscometer is:

$$\eta = \frac{\left[gr_c^2\left(\rho_2 - \rho_1\right)G(k)\right]}{\left(\dfrac{2U_t}{\text{ECF}}\right)} \tag{30.32}$$

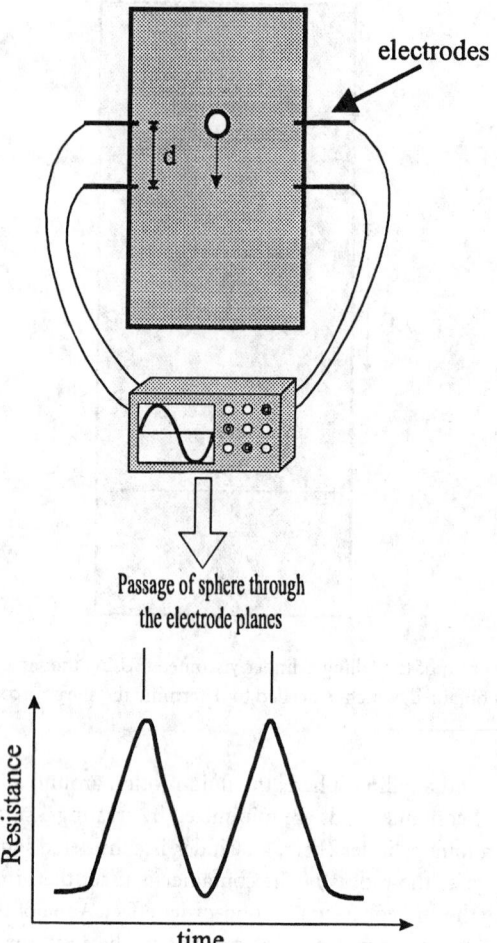

FIGURE 30.9 Diagram of one type of apparatus used to determine the viscosity of opaque liquids *in situ*. The electrical signal from the passage of the falling sphere indicates the time to traverse a known distance (d) between the two sensors.

In the past, this method was primarily used as a method to determine relative viscosities between transparent fluids. It has only been since the introduction of the ECF [16, 17] that this method could be rigorously used as an absolute viscosity method. With a properly designed container and cylinder, this method is now able to provide accurate absolute viscosities from 10^{-3} Pa·s to 10^7 Pa·s.

Falling Methods in Opaque Liquids

The falling body methods described above have been extensively applied to transparent liquids where optical (often visual) observation of the falling body is possible. For *opaque liquids*, however, falling body methods require the use of some sensing technique to determine, often *in situ*, the position of the falling body with respect to time. Techniques have varied but they all have in common the ability to detect the body as it moves past the sensor. A recent study at high pressure [18] demonstrated that the contrast in electric conductivity between a sphere and opaque liquid could be exploited to dynamically sense the moving sphere if suitably placed electrodes penetrated the container walls as shown in the schematic diagram in Figure 30.9. References to other similar *in situ* techniques are given in [18].

Rising Bubble/Droplet

For many industrial processes, the rising bubble viscometer has been used as a method of comparing the relative viscosities of transparent liquids (such as varnish, lacquer, and beer) for decades. Although its use was widespread, the actual behavior of the bubble in a viscous liquid was not well understood until long after the method was introduced [19]. The rising bubble method has been thought of as a derivative of the falling sphere method; however, there are primary differences between the two. The major physical differences are (1) the density of the bubble is less that of the surrounding liquid, and (2) the bubble itself has some unique viscosity. Each of these differences can, and do, lead to significant and extremely complex rheological problems that have yet to be fully explained. If a bubble of gas or droplet of liquid with a radius, r_b, and density, ρ', is freely rising in some enclosing viscous liquid of density ρ, then the shear viscosity of the enclosing liquid is determined by:

$$\eta = \left(\frac{1}{\varepsilon}\right)\frac{\left[2gr_b^2\left(\rho-\rho'\right)\right]}{9U_t} \tag{30.33}$$

where

$$\varepsilon = \frac{\left(2\eta+3\eta'\right)}{3\left(\eta+\eta'\right)} \tag{30.34}$$

where η' is the viscosity of the bubble. It must be noted that when the value of η' is large (solid spheres), $\varepsilon = 1$, which reduces Equation 30.33 to Equation 30.25. For small values of η' (gas bubbles), ε becomes 2/3, and the viscosity calculated by Equation 30.33 is 1.5 times greater than the viscosity calculated by Equation 30.25. It is apparent from Equation 30.33 and 30.25 that if the density of the bubble is less than the density of the enclosing liquid, and the terminal velocity of the sphere is negative, which indicates upward motion since the downward direction is positive.

During the rise, great care must be taken to avoid contamination of the bubble and its surface with impurities in the surrounding liquid. Impurities can diffuse through the surface of the bubble and combine with the fluid inside. Because the bubble has a low viscosity, the upward motion in a viscous medium induces a drag on the bubble that is responsible for generating a circulatory motion within it. This motion can efficiently distribute impurities throughout the whole of the bubble, thereby changing its viscosity and density. Impurities left on the surface of the bubble can form a "skin" that can significantly affect the rise of the bubble, as the skin layer has its own density and viscosity that are not included in Equation 30.33. These surface impurities also make a significant contribution to the inhomogeneous distribution of interfacial tension forces. A balance of these forces is crucial for the formation of a spherical bubble. The simplest method to minimize the above effects is to employ minute bubbles by introducing a specific volume of fluid (gas or liquid), with a syringe or other similar device, at the lower end of the cylindrical container. Very small bubbles behave like solid spheres, which makes interfacial tension forces and internal fluid motion negligible.

In all rising bubble viscometers, the bubble is assumed to be spherical. Experimental studies of the shapes of freely rising gas bubbles in a container of finite extent [20] have shown that (to 1% accuracy) a bubble will form and retain a spherical shape if the ratio of the radius of the bubble to the radius of the confining cylindrical container is less than 0.2. These studies have also demonstrated that the effect of the wall on the terminal velocity of a rising spherical bubble is to cause a large decrease (up to 39%) in the observed velocity compared to the velocity measured within an unbounded medium. This implies that the walls of the container influence the velocity of the rising bubble sooner than its geometry. In this method, end effects are known to be large. However, a rigorous, analytically or empirically derived

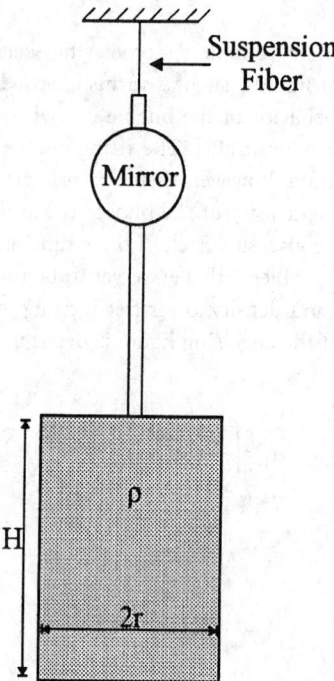

FIGURE 30.10 Schematic diagram of the oscillating cup viscometer. Measurement of the logarithmic damping of the amplitude and period of vessel oscillation are used to determine the absolute shear viscosity of the liquid.

correction factor has not yet appeared. To circumvent this, the ratio of container length to sphere diameter must be in the range of 10 to 100. As in other Stokian methods, this allows the bubble's velocity to be measured at locations that experience negligible end effects.

Considering all of the above complications, the use of minute bubbles is the best approach to ensure a viscosity measurement that is least affected by the liquid to be investigated and the container geometry.

Oscillating Method

If a liquid is contained within a vessel suspended by some torsional system that is set in oscillation about its vertical axis, then the motion of the vessel will experience a gradual damping. In an ideal situation, the damping of the motion of the vessel arises purely as a result of the *viscous coupling* of the liquid to the vessel and the viscous coupling between layers in the liquid. In any practical situation, there are also frictional losses within the system that aid in the damping effect and must be accounted for in the final analysis. From observations of the amplitudes and time periods of the resulting oscillations, a viscosity of the liquid can be calculated. A schematic diagram of the basic set-up of the method is shown in Figure 30.10. Following initial oscillatory excitation, a light source (such as a low-intensity laser) can be used to measure the amplitudes and periods of the resulting oscillations by reflection off the mirror attached to the suspension rod to give an accurate measure of the logarithmic decrement of the oscillations (δ) and the periods (T).

Various working formulae have been derived that associate the oscillatory motion of a vessel of radius r to the absolute viscosity of the liquid. The most reliable formula is the following equation for a cylindrical vessel [21]:

$$\eta = \left[\frac{I\delta}{(\pi r^3 H Z)}\right]^2 \left[\frac{1}{\pi \rho T}\right] \qquad (30.35)$$

where

$$Z = \left(\frac{1+r}{4H}\right)a_0 - \frac{\left(\frac{3}{2}+\frac{4r}{\pi H}\right)1}{p} + \frac{\left(\frac{3}{8}+\frac{9r}{4H}\right)a_2}{2p^2} \qquad (30.36)$$

$$p = \left(\frac{\pi\rho}{\eta T}\right)^{1/2} r \qquad (30.37)$$

$$a_0 = 1 - \left(\frac{\delta}{4\pi}\right) - \left(\frac{3\delta^2}{32\pi^2}\right) \qquad (30.38)$$

$$a_2 = 1 + \left(\frac{\delta}{4\pi}\right) + \left(\frac{\delta^2}{32\pi^2}\right) \qquad (30.39)$$

I is the mass moment of inertia of the suspended system and ρ is the density of the liquid.

A more practical expression of Equation 30.35 is obtained by introducing a number of simplifications. First, it is a reasonable assumption to consider δ to be small (on the order of 10^{-2} to 10^{-3}). This reduces a_0 and a_2 to values of 1 and -1, respectively. Second, the effects of friction from the suspension system and the surrounding atmosphere can be experimentally determined and contained within a single variable, δ_0. This must then be subtracted from the measured δ. A common method of obtaining δ_0 is to observe the logarithmic decrement of the system with an empty sample vessel and subtract that value from the measured value of δ. With these modifications, Equation 30.35 becomes:

$$\frac{(\delta - \delta_0)}{\rho} = \left[A\left(\frac{\eta}{\rho}\right)^{1/2} - B\left(\frac{\eta}{\rho}\right) + C\left(\frac{\eta}{\rho}\right)^{3/2} \right] \qquad (30.40)$$

where

$$A = \left(\frac{\pi^{3/2}}{I}\right)\left[1 + \left(\frac{r}{4H}\right)Hr^3T^{1/2}\right] \qquad (30.41)$$

$$B = \left(\frac{\pi}{I}\right)\left[\left(\frac{3}{2}\right) + \frac{4r}{\pi H}\right]Hr^2T \qquad (30.42)$$

$$C = \left(\frac{\pi^{1/2}}{2I}\right)\left[\left(\frac{3}{8}\right) + \frac{9r}{4H}\right]HrT^{3/2} \qquad (30.43)$$

It has been noted [22] that the analytical form of Equation 30.40 needs an empirically derived, instrument-constant correction factor (ζ) in order to agree with experimentally measured values of η. The discrepancy between the analytical form and the measured value arises as a result of the above assumptions. However, these assumptions are required as there are great difficulties involved in solving the differential equations of motion of this system. The correction factor is dependent on the materials, dimensions, and densities of each individual system, but generally lies between the values of 1.0 and 1.08. The correction factor is obtained by comparing viscosity values of calibration materials determined by an individual system (with Equation 30.35) and viscosity values obtained by another reliable method such as the capillary method.

With the above considerations taken into account, the final working Roscoe's formula for the absolute shear viscosity is:

$$\frac{(\delta - \delta_0)}{\rho} = \zeta \left[A \left(\frac{\eta}{\rho} \right)^{1/2} - B \left(\frac{\eta}{\rho} \right) + C \left(\frac{\eta}{\rho} \right)^{3/2} \right] \qquad (30.44)$$

The *oscillating cup method* has been used, and is best suited for use with low values of viscosity within the range of 10^{-5} Pa s to 10^{-2} Pa·s. Its simple closed design and use at high temperatures has made this method very popular when dealing with liquid metals.

Ultrasonic Methods

Viscosity plays an important role in the absorption of energy of an *acoustic wave* traveling through a liquid. By using ultrasonic waves (10^4 Hz $< f < 10^8$ Hz), the elastic, viscoelastic, and viscous response of a liquid can be measured down to times as short as 10 ns. When the viscosity of the fluid is low, the resulting time scale for structural relaxation is shorter than the ultrasonic wave period and the fluid is probed in the relaxed state. High-viscosity fluids subjected to ultrasonic wave trains respond as a stiff fluid because structural equilibration due to the acoustic perturbation does not go to completion before the next wave cycle. Consequently, the fluid is said to be in an unrelaxed state that is characterized by dispersion (frequency-dependent wave velocity) and elastic moduli that reflect a much stiffer liquid. The frequency dependence of the viscosity relative to some reference viscosity (η_0) at low frequency, η/η_0, and of the absorption per wavelength, $\alpha\lambda$, where α is the absorption coefficient of the liquid and λ is the wavelength of the compressional wave, for a liquid with a single relaxation time, t, is shown in Figure 30.11. The maximum absorption per wavelength occurs at the *relaxation frequency* when $\omega\tau = 1$

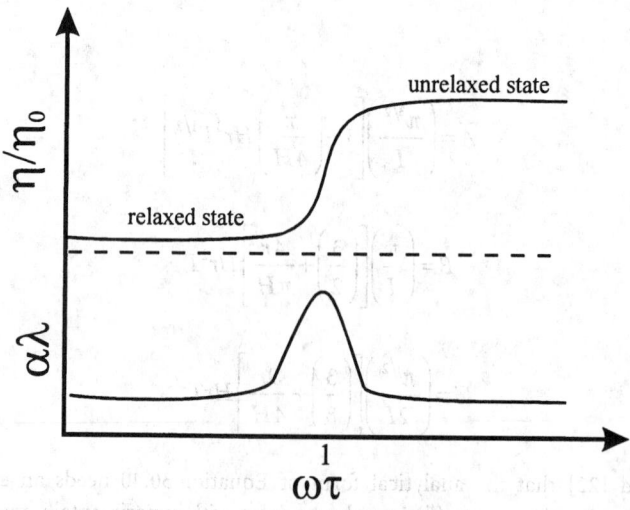

FIGURE 30.11 Effects of liquid relaxation (relaxation frequency corresponds to $\omega\tau = 1$ where $\omega = 2\pi f$) on relative viscosity (upper) and absorption per wavelength (lower) in the relaxed elastic ($\omega\tau < 1$) and unrelaxed viscoelastic ($\omega\tau > 1$) regimes.

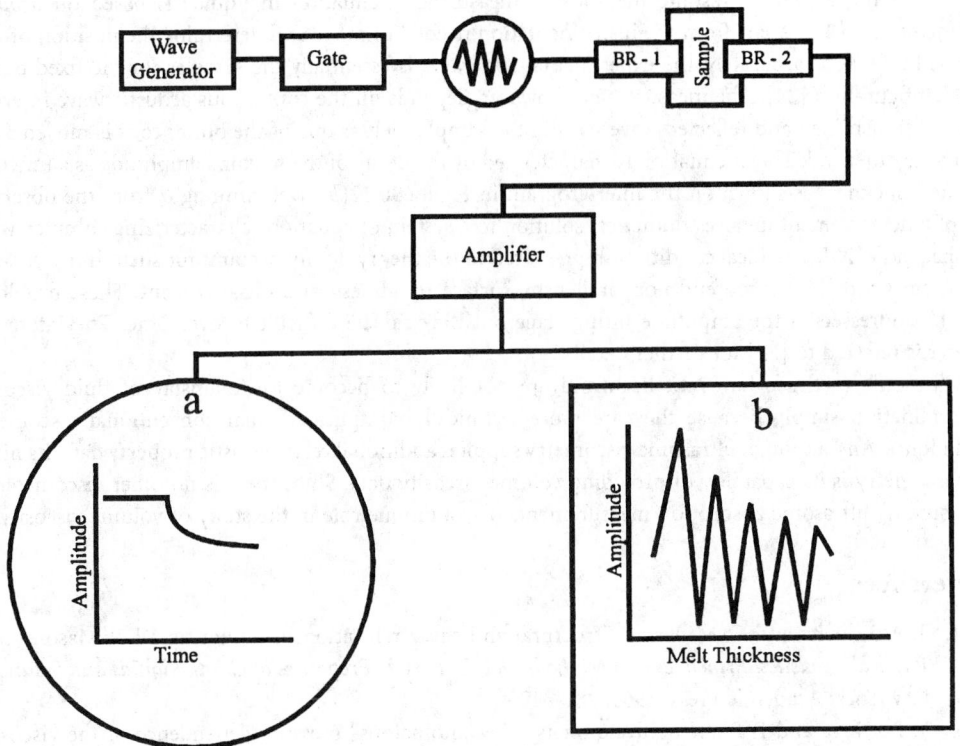

FIGURE 30.12   Schematic diagram of apparatus for liquid shear and volume viscosity determination by ultrasonic wave attenuation measurement showing the recieved signal amplitude through the exit buffer rod (BR-2) using (*a*) a fixed buffer rod (BR-2) using (*a*) a fixed buffer rod configuration, and (*b*) an interferometric technique with moveable buffer rod.

and is accompanied by a step in η/η_0, as well as in other properties such as velocity and compressibility. Depending on the application of the measured properties, it is important to determine if the liquid is in a relaxed or unrelaxed state.

A schematic diagram of a typical apparatus for measuring viscosity by the ultrasonic method is shown in Figure 30.12. Mechanical vibrations in a *piezoelectric* transducer travel down one of the buffer rods (BR-1 in Figure 30.12) and into the liquid sample and are received by a similar transducer mounted on the other buffer rod, BR-2. In the fixed buffer rod configuration, once steady-state conditions have been reached, the applied signal is turned off quickly. The decay rate of the received and amplified signal, displayed on an oscilloscope on an amplitude vs. time plot as shown in Figure 30.12(*a*), gives a measure of α. The received amplitude decays as:

$$A = A_0\, e^{-\left(b + \alpha c\right) t'} \tag{30.45}$$

where A is the received decaying amplitude, A_0 is the input amplitude, b is an apparatus constant that depends on other losses in the system such as due to the transducer, container, etc. that can be evaluated by measuring the attenuation in a standard liquid, c is the compressional wave velocity of the liquid, and t' is time. At low frequencies, the absorption coefficient is expressed in terms of volume and shear viscosity as:

$$\left(\eta_v + \frac{4\eta}{3}\right) = \frac{\alpha\rho c^3}{2\pi^2 f^2} \tag{30.46}$$

One of the earliest ultrasonic methods of measuring attenuation in liquids is based on *acoustic interferometry* [23]. Apart from the instrumentation needed to move and determine the position of one of the buffer rods accurately, the experimental apparatus is essentially the same as for the fixed buffer rod configuration [24]. The measurement, however, depends on the continuous acoustic wave interference of transmitted and reflected waves within the sample melt as one of the buffer rods is moved away from the other rod. The attenuation is characterized by the decay of the maxima amplitude as a function of melt thickness as shown on the interferogram in Figure 30.12(*b*). Determining α from the observed amplitude decrement involves numerical solution to a system of equations characterizing complex wave propagation [25]. The ideal conditions represented in the theory do not account for such things as wave front curvature, buffer rod end nonparallelism, surface roughness, and misalignment. These problems can be addressed in the amplitude fitting stage but they can be difficult to overcome. The interested reader is referred to [25] for further details.

Ultrasonic methods have not been and are not likely to become the mainstay of fluid viscosity determination simply because they are more technically complicated than conventional viscometry techniques. And although ultrasonic viscometry supplies additional related elastic property data, its niche in viscometry is its capability of providing volume viscosity data. Since there is no other viscometer to measure η_v, ultrasonic absorption measurements play a unique role in the study of volume viscosity.

References

1. T. A. Litovitz and C. M. Davis, Structural and shear relaxation in liquids, in W. P. Mason (ed.), *Physical Acoustics: Principles and Methods, Vol. II. Part A, Properties of Gases, Liquids and Solutions*, New York: Academic Press, 1965, 281-349.
2. Y. Bottinga and P. Richet, Silicate melts: The "anomalous" pressure dependence of the viscosity, *Geochim. Cosmochim. Acta*, 59, 2725-2731, 1995.
3. J. Ferguson and Z. Kemblowski, *Applied Fluid Rheology*, New York: Elsevier, 1991.
4. R. W. Whorlow, *Rheological Techniques*, 2nd ed., New York: Ellis Horwood, 1992.
5. K. Walters, *Rheometry*, London: Chapman and Hall, 1975.
6. J. M. Dealy, *Rheometers for Molten Plastics*, New York: Van Nostrand Reinhold, 1982.
7. J. R. Van Wazer, J. W. Lyons, K. Y. Kim, and R. E. Colwell, *Viscosity and Flow Measurement*, New York: Interscience, 1963.
8. C. W. Macosko, *Rheology: Principles, Measurements, and Applications*, New York: VCH, 1994.
9. W. A. Wakeham, A. Nagashima, and J. V. Sengers (eds.), *Measurement of the Transport Properties of Fluids*, Oxford, UK: Blackwell Scientific, 1991.
10. J. A. Himenez and M. Kostic, A novel computerized viscometer/rheometer, *Rev. Sci. Instrum.*, 65(1), 229-241, 1994.
11. DIN 53018 (Part 1 and 2), 53019, German National Standards.
12. Marks' *Standard Handbooks for Mechanical Engineers*, New York: McGraw-Hill, 1978.
13. ASTM D445-71 standard.
14. W. D. Kingery, *Viscosity in Property Measurements at High Temperatures*, New York: John Wiley & Sons, 1959.
15. H. Faxen, Die Bewegung einer Starren Kugel Langs der Achsee eines mit Zaher Flussigkeit Gefullten Rohres: Arkiv for Matematik, *Astronomi och Fysik*, 27(17), 1-28, 1923.
16. F. Gui and T. F. Irvine Jr., Theoretical and experimental study of the falling cylinder viscometer, *Int. J. Heat and Mass Transfer*, 37(1), 41-50, 1994.
17. N. A. Park and T. F. Irvine Jr., Falling cylinder viscometer end correction factor, *Rev. Sci. Instrum.*, 66(7), 3982-3984, 1995.
18. G. E. LeBlanc and R. A. Secco, High pressure stokes' viscometry: a new *in-situ* technique for sphere velocity determination, *Rev. Sci. Instrum.*, 66(10), 5015-5018, 1995.
19. R. Clift, J. R. Grace, and M. E. Weber, *Bubbles, Drops, and Particles*, San Diego: Academic Press, 1978.

20. M. Coutanceau and P. Thizon, Wall effect on the bubble behavior in highly viscous liquids, *J. Fluid Mech.*, 107, 339-373, 1981.
21. R. Roscoe, Viscosity determination by the oscillating vessel method I: theoretical considerations, *Proc. Phys. Soc.*, 72, 576-584, 1958.
22. T. Iida and R. I. L. Guthrie, *The Physical Properties of Liquid Metals*, Oxford, UK: Clarendon Press, 1988.
23. H. J. McSkimin, Ultrasonic methods for measuring the mechanical properties of liquids and solids, in W.P. Mason (ed.), *Physical Acoustics: Principles and Methods, Vol. I Part A, Properties of Gases, Liquids and Solutions*, New York: Academic Press, 1964, 271-334.
24. P. Nasch, M. H. Manghnani, and R. A. Secco, A modified ultrasonic interferometer for sound velocity measurements in molten metals and alloys, *Rev. Sci. Instrum.*, 65, 682-688, 1994.
25. K. W. Katahara, C. S. Rai, M. H. Manghnani, and J. Balogh, An interferometric technique for measuring velocity and attenuation in molten rocks, *J. Geophys. Res.*, 86, 11779-11786, 1981.

Further Information

M. P. Ryan and J. Y. K. Blevins, *The Viscosity of Synthetic and Natural Silicate Melts and Glasses at High Temperatures and 1 Bar (10^5 Pascals) Pressure and at Higher Pressures*, U.S. Geological Survey Bulletin 1764, Denver, CO, 1987, 563, an extensive compilation of viscosity data in tabular and graphic format and the main techniques used to measure shear viscosity.

M. E. O'Neill and F. Chorlton, *Viscous and Compressible Fluid Dynamics*, Chichester: Ellis Horwood, 1989, mathematical methods and techniques and theoretical description of flows of Newtonian incompressible and ideal compressible fluids.

J. R. Van Wazer, J. W. Lyons, K. Y. Kim, and R. E. Colwell, *Viscosity and Flow Measurement: A Laboratory Handbook of Rheology*, New York: Interscience Publishers Div. of John Wiley & Sons, 1963, A comprehensive overview of viscometer types and simple laboratory measurements of viscosity for liquids.

31

Surface Tension Measurement

David B. Thiessen
California Institute of Technology

Kin F. Man
California Institute of Technology

The effect of surface tension is observed in many everyday situations. For example, a slowly leaking faucet drips because the force of surface tension allows the water to cling to it until a sufficient mass of water is accumulated to break free. Surface tension can cause a steel needle to "float" on the surface of water although its density is much higher than that of water. The surface of a liquid can be thought of as having a skin that is under tension. A liquid droplet is somewhat analogous to a balloon filled with air. The elastic skin of the balloon contains the air inside at a slightly higher pressure than the surrounding air. The surface of a liquid droplet likewise contains the liquid in the droplet at a pressure that is slightly higher than ambient. A clean liquid surface, however, is not elastic like a rubber skin. The tension in a piece of rubber increases as it is stretched and will eventually rupture. A clean liquid surface can be expanded indefinitely without changing the surface tension.

The mechanical model of the liquid surface is that of a skin under tension. Any given patch of the surface thus experiences an outward force tangential to the surface everywhere on the perimeter. The force per unit length of the perimeter acting perpendicular to the perimeter is defined as the *surface tension*, γ. Molecules in the interfacial region have a higher potential energy than molecules in the bulk phases because of an imbalance of intermolecular attractive forces. This results in an excess free energy per unit area associated with the surface that is numerically equivalent to the surface tension, as shown below. Consider a flat rectangular patch of fluid interface of width W and length L. In order to expand the length to $L + \Delta L$, an amount of work $\gamma W \Delta L$ must be done at the boundary. The product $W \Delta L$ is just the change in area ΔA of the surface. The work done to increase the area is thus $\Delta A \gamma$, which corresponds to the increase in surface free energy. Thus, the surface tension γ is seen to be equivalent to the surface free energy per unit area. Room-temperature organic liquids typically have surface tensions in the range of 20 mN m^{-1} to 40 mN m^{-1}, while pure water has a value of 72 mN m^{-1} at 25°C. The interface between two immiscible liquids, such as oil and water, also has a tension associated with it, which is generally referred to as the *interfacial tension*.

The surface energy concept is useful for understanding the shapes adopted by liquid surfaces. An isolated system in equilibrium is in a state of minimum free energy. Because the surface free energy contributes to the total free energy of a multiphase system, the surface free energy is minimized at

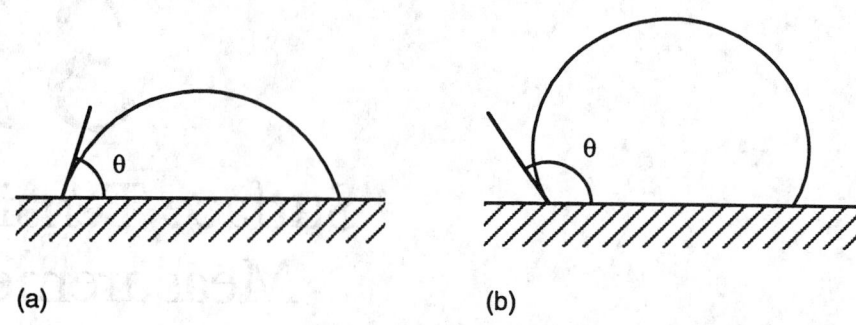

FIGURE 31.1   Illustration of contact angles and wetting. The liquid in (*a*) wets the solid better than that in (*b*).

equilibrium subject to certain constraints. Also, because the surface free energy is directly proportional to the surface area, surface area is also minimized. In the absence of gravity, a free-floating liquid droplet assumes a spherical shape because, for a given volume of liquid, the sphere has the least surface area. However, a droplet suspended from a needle tip on Earth does not form a perfect sphere because the minimum free energy configuration involves a trade-off between a reduction of the surface energy and a reduction of the gravitational potential. The droplet elongates to reduce its gravitational potential energy.

The surface energy concept is also useful for understanding the behavior of so-called surface active agents or *surfactants*. A two-component liquid mixture in thermodynamic equilibrium exhibits preferential adsorption of one component at the surface if the adsorption causes a decrease in the surface energy. The term surfactant is reserved for molecular species that strongly adsorb at the surface even when their concentration in the bulk liquid is very low. Surfactants are common in natural waters and are very important in many biological and industrial processes.

The interface between a solid and a fluid also has a surface free energy associated with it. Figure 31.1(*a*) shows a liquid droplet at rest on a solid surface surrounded by air. This system contains three different types of interfaces: solid–gas, solid–liquid, and liquid–gas, each with a characteristic surface free energy per unit area. The state of minimum free energy for the system then involves trade-offs in the surface area for the various interfaces. The region of contact between the gas, liquid, and solid is termed the *contact line*. The liquid–gas surface meets the solid surface with an angle θ measured through the liquid, which is known as the *contact angle*. The contact angle attains a value that minimizes the free energy of the system and is thus a characteristic of a particular solid–liquid–gas system. The system shown in Figure 31.1(*a*) has a smaller contact angle than that shown in Figure 31.1(*b*). The smaller the contact angle, the better the liquid is said to wet the solid surface. For θ = 0, the liquid is said to be perfectly wetting.

Measurement of surface tension is important in many fields of science and engineering, as well as in medicine. A number of standard methods exist for its measurement. In many systems of interest, the surface tension changes with time, perhaps, for example, because of adsorption of surfactants. Several standard methods can be used to measure dynamic surface tension if it changes slowly with time. Special techniques have been developed to measure dynamic surface tensions for systems that evolve very rapidly.

31.1   Mechanics of Fluid Surfaces

Some methods of measuring surface tension depend on the mechanics at the line of contact between a solid, liquid, and gas. When the system is in static mechanical equilibrium, the contact line is motionless, meaning that the net force on the line is zero. Forces acting on the contact line arise from the surface tensions of the converging solid–gas, solid–liquid, and liquid–gas interfaces, denoted by γ_{SG}, γ_{SL}, and γ_{LG}, respectively (Figure 31.2). The condition of zero net force along the direction tangent to the solid surface gives the following relationship between the surface tensions and contact angle θ:

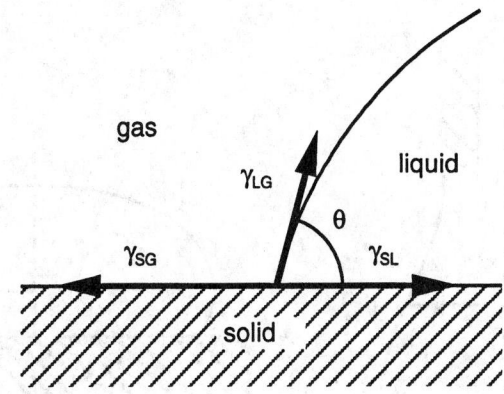

FIGURE 31.2 Surface tension forces acting on the contact line.

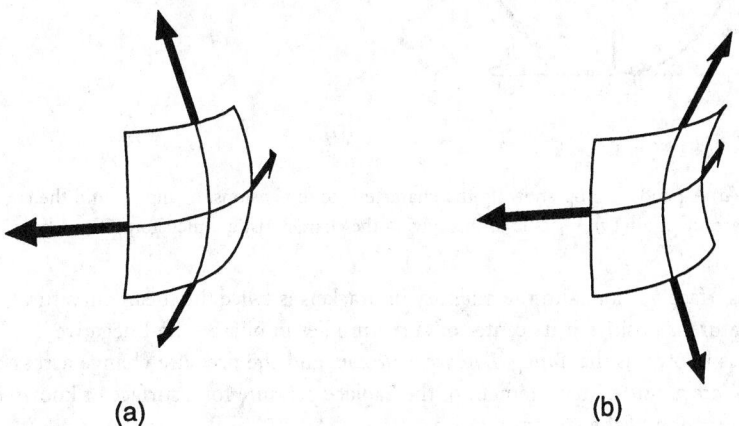

FIGURE 31.3 Mechanics of curved surfaces that have principal radii of curvature of: (*a*) the same sign, and (*b*) the opposite sign.

$$\gamma_{SG} = \gamma_{SL} + \gamma_{LG} \cos\theta \tag{31.1}$$

This is known as *Young's equation.* The contact angle is thus seen to be dependent on the surface tensions between the various phases present in the system, and is therefore an intrinsic property of the system.

As discussed in the introduction, the surface tension of a droplet causes an increase in pressure in the droplet. This can be understood by considering the forces acting on a curved section of surface as illustrated in Figure 31.3(*a*). Because of the curvature, the surface tension forces pull the surface toward the concave side of the surface. For mechanical equilibrium, the pressure must then be greater on the concave side of the surface. Figure 31.3(*b*) shows a saddle-shaped section of surface in which surface tension forces oppose each other, thus reducing or eliminating the required pressure difference across the surface. The mean curvature of a two-dimensional surface is specified in terms of the two principal radii of curvature, R_1 and R_2, which are measured in perpendicular directions. A detailed mechanical analysis of curved tensile surfaces shows that the pressure change across the surface is directly proportional to the surface tension and to the mean curvature of the surface:

$$P_A - P_B = \gamma\left(\frac{1}{R_1} + \frac{1}{R_2}\right) \tag{31.2}$$

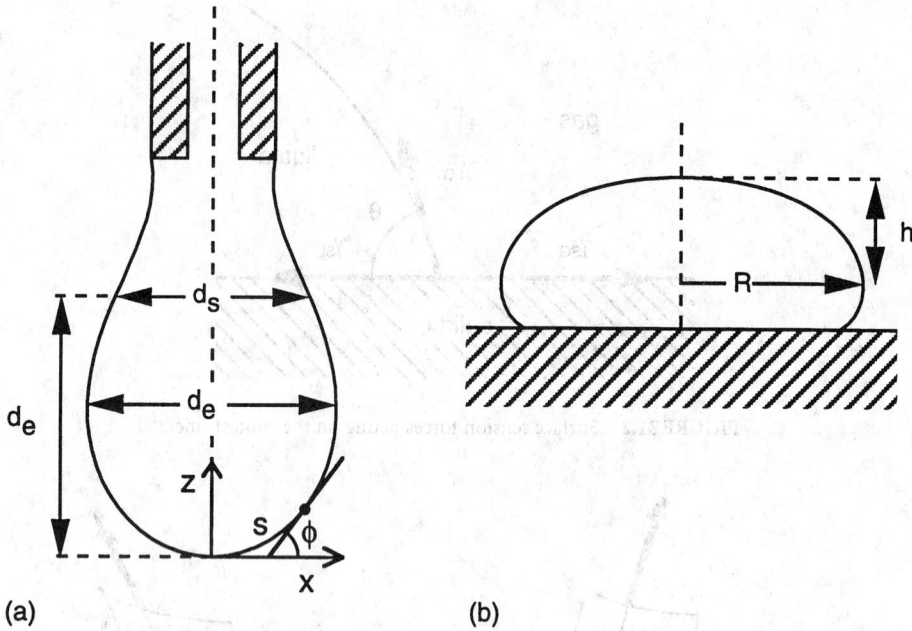

FIGURE 31.4 (*a*) A pendant drop showing the characteristic dimensions, d_e and d_s, and the coordinates used in the Young–Laplace equation. (*b*) A sessile drop showing the characteristic dimensions R and h.

where γ is the surface tension, and the quantity in brackets is twice the mean curvature. The sign of the radius of curvature is positive if its center of curvature lies in phase A and negative if it lies in phase B. Equation 31.2 is known as the *Young–Laplace equation*, and the pressure change across the interface is termed the *Laplace pressure*. Measurement of the Laplace pressure for a surface of known curvature then allows a determination of the surface tension.

Several methods of surface tension measurement are based on the measurement of the static shape of an axisymmetric drop or bubble or on the point of mechanical instability of such drops or bubbles. In a gravitational field a drop or bubble that is attached to a solid support assumes a nonspherical shape. Figure 31.4(*a*) shows the shape of a hanging droplet, also known as a pendant drop, and Figure 31.4(*b*) shows a so-called sessile drop. Axisymmetric air bubbles in water attain the same shapes as water drops in air, except that they are inverted. A bubble supported from below is thus called a hanging or pendant bubble, and a bubble supported from above is called a captive or sessile bubble. The reason for the deviation of the shape from that of a sphere can be understood from Equation 31.2. The hydrostatic pressure changes with depth more rapidly in a liquid than in a gas. The pressure difference across the surface of a pendant drop in air therefore increases from top to bottom, requiring an increase in the mean curvature of the surface according to Equation 31.2. The drop in Figure 31.4(*a*) has a neck at the top, which means that the two principal radii of curvature have opposite signs and cancel to some extent. At the bottom of the drop, the radii of curvature have the same sign, thus making the mean curvature larger. The Young–Laplace equation can be written as coupled first-order differential equations in terms of the coordinates of the interface for an axisymmetric surface in a gravitational field as:

$$\frac{dx}{ds} = \cos\phi$$

$$\frac{dz}{ds} = \sin\phi$$

(31.3)

$$\frac{d\phi}{ds} = \frac{2}{b} + \left(\frac{\Delta\rho g}{\gamma}\right)z - \frac{\sin\phi}{x}$$

$$x(0) = z(0) = \phi(0) = 0$$

where x and z are the horizontal and vertical coordinates, respectively, with the origin at the drop apex; s is the arc-length along the drop surface measured from the drop apex; and ϕ is the angle between the surface tangent and the horizontal (Figure 31.4(a)). The parameter b is the radius of curvature at the apex of the drop or bubble, $\Delta\rho$ is the density difference between the two fluid phases, and g is the acceleration of gravity. Numerical integration of Equation 31.3 allows one to compute the shape of an axisymmetric fluid interface. Comparison of computed shapes with experimentally measured shapes of drops or bubbles is a useful method of measuring surface tension. If all lengths in Equation 31.3 are made dimensionless by dividing them by b, the resulting equation contains only one parameter, $\beta = \Delta\rho g b^2/\gamma$, which is called the Bond number (or shape factor). The shape of an axisymmetric drop, bubble, or meniscus depends only on this one dimensionless parameter. The Bond number can also be written as $\beta = 2b^2/a^2$ where $a = \sqrt{2\gamma/\Delta\rho g}$ is known as the capillary constant and has units of length.

Several dynamic methods of measuring surface tension are based on capillary waves. Capillary waves result from oscillations of the liquid surface for which surface tension is the restoring force. The frequency of the surface oscillation is thus dependent on the surface tension and wavelength. Very low amplitude capillary waves with a broad range of frequencies are always present on liquid surfaces owing to thermal fluctuations. Larger amplitude capillary waves can be excited by purposely perturbing the surface.

31.2 Standard Methods and Instrumentation

A number of commonly used methods of measuring surface tension exist. The choice of a method depends on the system to be studied, the degree of accuracy required, and possibly on the ability to automate the measurements. In the discussion that follows, these methods are grouped according to the kind of instruments used in the measurements. Because the information presented for each method is necessarily brief, readers who are interested in constructing their own apparatus should consult the more detailed treatises in [1–4]. A list of commercially available instruments is given in Table 31.1, together with manufacturer names and approximate prices. Vendors can be contacted at the addresses given in Table 31.2.

TABLE 31.1 Commercially Available Instruments

Method	Instrument type	Manufacturer/Product name	Approximate price (range)
Capillary rise	Manual	Fisher	$79
Wilhelmy plate/du Noüy ring	Manual, mechanical balance	CSC, Fisher, Kahl	$2000–$4000
Wilhelmy plate/du Noüy ring	Manual, electrobalance	KSV, Lauda, NIMA	$4000–$11,000
Wilhelmy plate/du Noüy ring	Automatic, electrobalance	Cahn, Krüss, KSV, NIMA	$9000–$24,000
Maximum bubble pressure	Automatic	Krüss, Lauda, Sensa Dyne	$5000–$23,000
Pendant/sessile drop	Manual	Krüss, Rame-Hart	$7000–$10,000
Pendant/sessile drop	Automatic	ADSA, AST, FTA, Krüss, Rame-Hart, Temco	$10,000–$100,000
Drop weight/volume	Automatic	Krüss, Lauda	$16,000–$21,000
Spinning drop	Manual	Krüss	$20,100

Note: Price ranges reflect differences in degree of automation, the number of accessories included, or variation in price between manufacturers.

TABLE 31.2   Manufacturers and Suppliers of Instruments for Surface Tension Measurement

AST Products 9 Linnell Circle Billerica, MA 01821-3902 Tel: (508) 663-7652	Fisher Scientific 711 Forbes Ave. Pittsburgh, PA 15219-4785 Tel: (800) 766-7000
Applied Surface Thermodynamics Research Associates (ASTRA) (distributor of ADSA instrumentation) 15 Brendan Rd. Toronto, Ontario Canada, M4G 2W9 Tel: (416) 978-3601	Kahl Scientific Instrument Corp. P.O. Box 1166 El Cajon, CA 92022-1166 Tel: (619) 444-2158 Krüss U.S.A. 9305-B Monroe Road Charlotte, NC 28270-1488
Brinkmann Instruments, Inc. (distributor for Lauda tensiometers) One Catiaque Road P.O. Box 1019 Westbury, NY 11590-0207 Tel: (800) 645-3050	Tel: (704) 847-8933 KSV Instruments U.S.A. P.O. Box 192 Monroe, CT 06468 Tel: (800) 280-6216
Cahn Instruments 5225 Verona Rd., Bldg. 1 Madison, WI 53711 Tel: (800) 244-6305	Rame-Hart, Inc. 8 Morris Ave. Mountain Lakes, NJ 07046 Tel: (201) 335-0560
CSC Scientific Company, Inc. 8315 Lee Highway Fairfax, VA 22031 Tel: (800) 458-2558	Sensa Dyne Instrument Div. Chem-Dyne Research Corp. P.O. Box 30430 Mesa, AZ 85275-0430 Tel: (602) 924-1744
CTC Technologies, Inc. (distributor for NIMA tensiometers) 7925-A North Oracle Road, Suite 364 Tucson, AZ 85704-6356 Tel: (800) 282-8325	Temco, Inc. 4616 North Mingo Tulsa, OK 74117-5901 Tel: (918) 834-2337
First Ten Angstroms (FTA) 465 Dinwiddie Street Portsmouth, VA 23704 Tel: (800) 949-4110	

Capillary Rise Method

If a glass capillary tube is brought into contact with a liquid surface, and if the liquid wets the glass with a contact angle of less than 90°, then the liquid is drawn up into the tube as shown in Figure 31.5(*a*). The surface tension is directly proportional to the height of rise, *h*, of the liquid in the tube relative to the flat liquid surface in the larger container. By applying Equation 31.2 to the meniscus in the capillary tube, the following relationship is obtained:

$$\Delta \rho g h = \frac{2\gamma}{b} \tag{31.4}$$

where *b* is the radius of curvature at the center of the meniscus and $\Delta \rho$ is the density difference between liquid and gas. For small capillary tubes, *b* is well approximated by the radius of the tube itself, assuming that the contact angle of the liquid on the tube is zero. For larger tubes or for increased accuracy, the

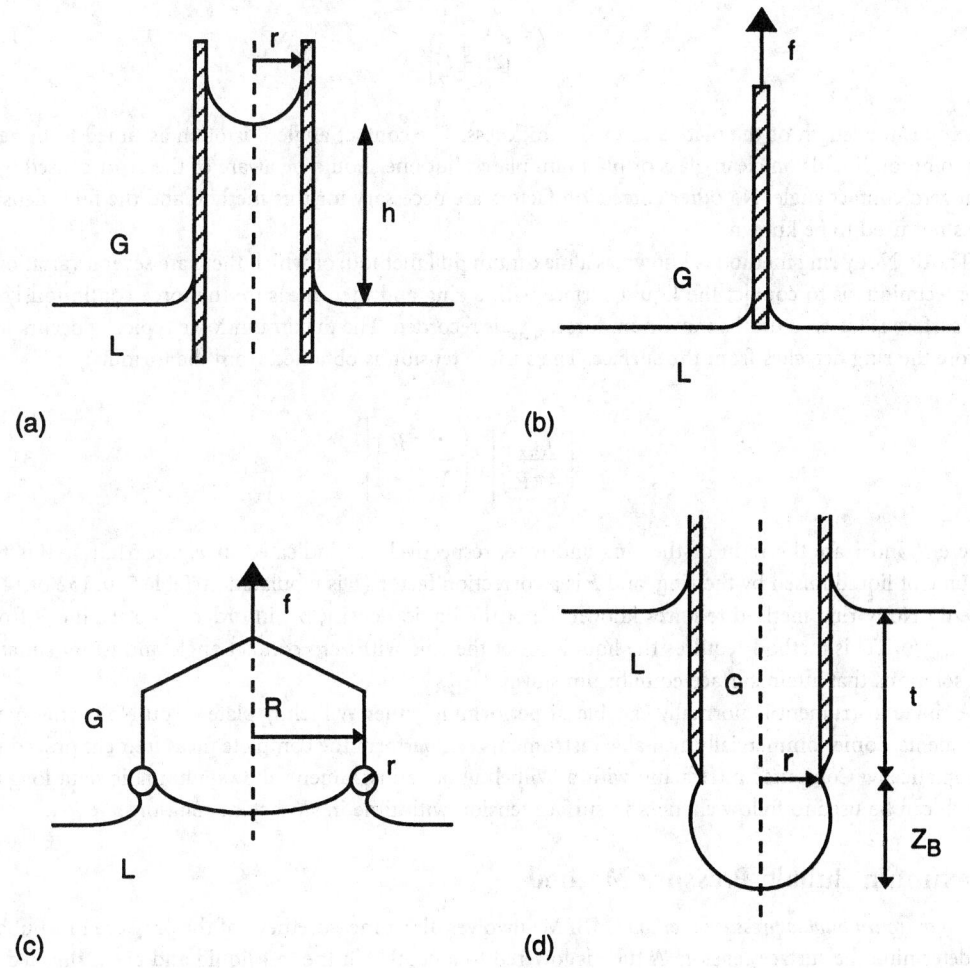

FIGURE 31.5 Geometries for: (*a*) capillary rise method, (*b*) Wilhelmy plate method, (*c*) du Noüy ring method, and (*d*) maximum bubble pressure method.

value of *b* must be corrected for gravitational deformation of the meniscus (p. 12 of [1]). Obtaining accurate results with the capillary rise method requires using a thoroughly clean glass capillary tube with a very uniform diameter of less than 1 mm. The container for the liquid should be at least 8 cm in diameter and the liquid must wet the capillary tube with a contact angle of zero. This method is primarily useful for pure liquids and is capable of high accuracy at relatively low cost.

Wilhelmy Plate and du Noüy Ring Methods

Measurement of the pull of a liquid surface directly on a solid object is the basis for two of the standard methods discussed here. In the Wilhelmy plate method, the solid object is a flat, thin plate that the test liquid should wet with a zero contact angle. The plate is suspended vertically from a delicate balance that is zeroed with the plate suspended in air. The test liquid is brought into contact with the bottom of the plate, causing the plate to be pulled down into the liquid by the surface tension force. The force applied to the plate from above is then increased to bring the bottom edge of the plate level with the flat surface of the liquid (Figure 31.5(*b*)). This avoids the necessity to make buoyancy corrections to the measurement. The surface tension is computed from the force measurement, *f*, using:

$$\gamma = \frac{f \cos\theta}{\left[2(l+t)\right]}$$ (31.5)

where l is the length of the plate and t is its thickness. The contact angle θ is often assumed to be zero for common liquids on clean glass or platinum plates, but one should be aware of the error caused by a non-zero contact angle. No other correction factors are necessary for this method and the fluid density does not need to be known.

The du Noüy ring method is known as a maximum pull method, of which there are several variations. The technique is to contact the liquid surface with a ring and then measure the force continuously as the surface is lowered until a maximum force, f_{max}, is recorded. The maximum force typically occurs just before the ring detaches from the surface. The surface tension is obtained from the formula:

$$\gamma = \left(\frac{f_{max}}{4\pi R}\right)\left[F\left(\frac{R^3}{V}, \frac{R}{r}\right)\right]$$ (31.6)

where R and r are the radii of the ring and wire, respectively, as indicated in Figure 31.5(c), V is the volume of liquid raised by the ring, and F is a correction factor (F is tabulated in Table 5, p. 132 of [4]). The du Noüy ring method requires knowledge of the liquid density, ρ_L, in order to determine V from $V = f_{max}/\rho_L$. This method requires the liquid to wet the ring with zero contact angle and is not suitable for solutions that attain surface equilibrium slowly.

A single instrument is normally capable of performing either Wilhelmy plate or du Noüy ring measurements. Some commercially available instruments can perform the complete measurement procedure automatically. Computer interfacing with a Wilhelmy plate instrument allows automatic data logging which can be used to follow changes in surface tension with time in surfactant solutions.

Maximum Bubble Pressure Method

The *maximum bubble pressure method* (MBPM) involves direct measurement of the pressure in a bubble to determine the surface tension. A tube is lowered to a depth t in the test liquid and gas is injected to form a bubble of height Z_B at the tip of the tube as shown in Figure 31.5(d). The increase in bubble pressure, P_b, over ambient pressure, P_a, arising from the interface is given by the sum of a hydrostatic pressure and Laplace pressure:

$$\delta p = P_b - P_a - \Delta\rho g t = \Delta\rho g Z_B + \frac{2\gamma}{b}$$ (31.7)

As a new bubble begins to form, Z_B increases while b, the radius of curvature at the bubble apex, decreases, resulting in an increase in pressure in the bubble. Ultimately, b increases as the bubble grows larger, thus reducing the pressure. The pressure in the bubble thus reaches a maximum when δp reaches a maximum, which in turn can be theoretically related to the surface tension. For $\delta p = \delta p_{max}$, Equation 31.7 can be rewritten in dimensionless form as follows:

$$\frac{r}{X} = \frac{r}{b} + \frac{r}{a}\frac{Z_B}{b}\left(\frac{\beta}{2}\right)^{1/2}$$ (31.8)

where r is the tube radius, X is a length defined as $X = 2\gamma/\delta p_{max}$, a is the capillary constant, and β is the Bond number. The dimensionless quantity r/X depends only on r/a, the relationship being determined by Equation 31.8 combined with numerical solutions of Equation 31.3. Tabulations of this relationship are used to calculate the surface tension by an iterative procedure (p. 18 of [1]). The standard MBPM requires a knowledge of the fluid densities, tube radius, and depth of immersion of the tube.

A differential MBPM uses two tubes of different diameters immersed to the same depth. The difference in the maximum bubble pressure for the two tubes, ΔP, is measured, eliminating the need to know the immersion depth and making the method less sensitive to errors in the knowledge of the liquid density. For the differential MBPM, surface tension is computed from (see [5]):

$$\gamma = A\Delta P \left[1 + \left(\frac{0.69 r_2 \rho_L}{\Delta P} \right) \right] \tag{31.9}$$

where r_2 is the radius of the larger tube, ρ_L is the liquid density, and A is an apparatus-dependent constant that is determined by calibration with several standard liquids [6]. Automated MBPM units are commercially available (see Table 31.1). Sensa Dyne manufactures differential MBPM units that allow for on-line process measurements under conditions of varying temperature and pressure.

Pendant Drop and Sessile Drop Methods

The shape of an axisymmetric pendant or sessile drop (Figure 31.4) depends on only a single parameter, the Bond number, as discussed above. The Bond number is a measure of the relative importance of gravity to surface tension in determining the shape of the drop. For Bond numbers near zero, surface tension dominates and the drop is nearly spherical. For larger Bond numbers, the drop becomes significantly deformed by gravity. In principle, the method involves obtaining an image of the drop and comparing its shape and size to theoretical profiles obtained by integrating Equation 31.3 for various values of β and b. Once β and b have been determined from shape and size comparison, the surface tension is calculated from:

$$\gamma = \frac{\Delta \rho g b^2}{\beta} \tag{31.10}$$

In practice, the drop shape and size has traditionally been determined by the manual measurement of several characteristic dimensions (see Figure 31.4) of the drop from a photographic print. For pendant drops, the ratio d_s/d_e is correlated to a shape factor H from which surface tension is calculated (p. 27 of [3], [7]) according to:

$$\gamma = \frac{\Delta \rho g d_e^2}{H} \tag{31.11}$$

For sessile drops, various analytical formulae are available for computation of surface tension directly from the characteristic dimensions (p. 36 of [3]). Drop shape methods based on characteristic dimensions require very accurate measurement of the dimensions for good results. For more accurate results, methods that fit the entire shape of the edge of the drop to the Laplace equation are recommended.

In recent years, the entire procedure has been automated using digital imaging and computer image analysis [8, 9]. Typically, several hundred coordinates on the edge of the drop are located with subpixel resolution by computer analysis of the digital image. The size, shape, and horizontal and vertical offsets of the theoretical profile given by Equation 31.3 are varied by varying four parameters: b, β, and the pixel coordinates of the drop apex, x_0 and z_0. A best fit of the theoretical profile to the measured edge coordinates is obtained by minimizing an objective function. A digital image of a pendant drop can be analyzed for surface tension on a desktop computer in 1 or 2 s [10]. The speed of algorithms for pendant drop analysis on modern desktop computers has allowed this method to be used to track changes in surface tension for surfactant-covered surfaces by analyzing a sequence of images. The algorithms can simultaneously track the surface area and volume of the drop or bubble. Both soluble and insoluble surfactants have been studied using the pendant drop, sessile drop, pendant bubble, and captive bubble configurations [11–13]. Table 31.1 lists several manufacturers that can provide software for automated analysis of surface tension from drops or bubbles in pendant or sessile configurations. The increased accuracy and simplicity of the automated pendant drop procedure makes it a very flexible method that has been applied to measure ultralow interfacial tensions, pressure, temperature and time dependence of interfacial tension, relaxation of adsorption layers, measurement of line tensions, and film-balance measurements [14].

Drop Weight or Volume Method

A pendant drop will become unstable and detach from its support if it grows too large. The weight of the detached portion is related to the surface tension of the fluid by:

$$\gamma = \left(\frac{mg}{r}\right)\left[F\left(\frac{r}{V^{1/3}}\right)\right] \tag{31.12}$$

where mg is the weight of the detached drop, r is the radius of the tip from which the drop hangs, and V is the volume of the detached drop. An empirical correction factor, F, is tabulated as a function of $r/V^{1/3}$ (p. 50 of [3]). For Equation 31.12 to apply, drops must be formed slowly. Measurements typically involve weighing the accumulated liquid from a large number of drops to determine the average weight per drop. The density of the fluid must be known in order to determine the drop volume and then obtain the factor F. Another method involves measuring the volumetric flow rate of liquid to the tip while counting the drops. The density of the fluid must be known in order to determine the drop weight. The latter method allows for automation of measurements [15].

Spinning Drop Method

The spinning drop method is a shape-measurement method similar to the pendant and sessile drop methods. However, the deformation of the drop in this case is caused by radial pressure gradients in a rapidly spinning tube. This method is normally used for measuring interfacial tensions between immiscible liquids. A horizontal glass tube with sealed ends is filled with the more dense liquid through a filling port. The tube is then spun about its axis while a drop of the lower density liquid is injected into the tube. The pressure in the outer liquid increases from the center of the tube toward the walls as a result of the spinning motion. The pressure gradient forces the drop to move to the center of the tube and causes it to elongate, while surface tension opposes elongation. Measurement of the maximum drop diameter, $2r_{max}$, and length, $2h_{max}$, together with the angular velocity of rotation, Ω, allows for calculation of the surface tension according to:

$$\gamma = \frac{1}{2}\left(\frac{r_{max}}{r_{max}^*}\right)^3 \Delta\rho\Omega^2 \tag{31.13}$$

where r_{max}^* is correlated to the aspect ratio r_{max}/h_{max} [16]. The spinning drop method is particularly suited for measuring ultralow interfacial tensions (10^{-2} mN m^{-1} to 10^{-4} mN m^{-1}).

31.3 Specialized Methods

Dynamic Surface Tension

In an aqueous solution of soluble surfactant, the surface tension decreases following creation of new surface area because of adsorption of surfactant molecules. Surfactant adsorption kinetics can be studied by measuring the change in surface tension with time. For a dilute solution, the rate of change of surface tension is often slow enough that automated versions of static methods such as the Wilhelmy plate or pendant drop [17] methods can be used to follow the changes in surface tension. In concentrated solutions in which large changes in surface tension can occur within a fraction of a second following surface creation, a dynamic method must be used. A liquid jet emerging from an elliptical orifice has stationary waves on its surface, the wavelengths of which are related to the surface tension. The oscillating jet method has been used to measure surface tension for surface ages as low as 0.6 ms [18]. A dynamic version of the maximum bubble pressure method has been used to measure dynamic surface tension at surface ages down to 0.1 ms [19].

Surface Viscoelasticity

A liquid surface covered by a monolayer of surfactant exhibits viscoelastic behavior. In addition to surface tension, the surface rheology is characterized in terms of dilatational and shear elasticities as well as dilatational and shear viscosities. The dilatational properties in particular are important in a variety of situations from foam stability to the functioning of the human lung. The surface dilatational modulus is proportional to the change in surface tension for a given change in surface area. This modulus depends on the rate of change of surface area for both soluble and insoluble surfactant monolayers, which indicates that relaxation processes are active. These relaxation processes give rise to the surface dilatational viscosity. For the case of soluble surfactants, one of the relaxation processes is the adsorption or desorption of molecules at the surface. The equilibrium dilatational elasticity of an insoluble monolayer can be measured by slowly expanding or compressing the monolayer in a Langmuir trough while monitoring the surface tension with a Wilhelmy plate apparatus [20]. Studies of surface rheology at high rates of surface expansion or compression are of interest for both soluble and insoluble surfactants.

Surface tension relaxation following sudden expansion or compression of the surface for a soluble surfactant has been studied by the automated pendant drop method [21]. A method known as oscillating bubble tensiometry has been applied to measure the kinetics of adsorption and desorption for soluble surfactants [22, 23]. Other methods for studying dynamic dilatational viscoelastic properties are reviewed in [24], including transverse and longitudinal capillary wave methods, a modified maximum bubble pressure method, and an oscillating bubble method.

Measurements at Extremes of Temperature and Pressure

Several of the standard methods described in this chapter can be adapted to make surface or interfacial tension measurements at extreme temperatures and/or pressures. The most common methods used to measure the surface tension of high-temperature molten metals, alloys, and semiconductors are the maximum bubble pressure method [25] and the pendant or sessile drop method [26–28]. Measurement of the interfacial tension between oil and a second immiscible phase at high pressure and elevated temperature is of interest for understanding aspects of enhanced oil recovery. The pendant drop method has been applied under pressures of 82 MPa at 449 K [29], while a capillary wave method has been applied at 136 MPa and 473 K [30]. A pendant drop apparatus capable of measurements to 10,000 psi (69 MPa) and 350°F (450 K) is commercially available from Temco Inc. (Table 31.2).

Interfacial Tension

Measurement of the interfacial tension between two immiscible liquids can present special difficulties. Measurement by the capillary rise, du Noüy ring, or Wilhelmy plate method is problematic in that the contact angle is often nonzero. The pendant drop [7] and drop weight [31] methods can both be applied, provided the densities of the two liquids are sufficiently different. The pendant drop method, in particular, is widely used for interfacial tension measurement. Interfacial tension can be measured by a modified maximum bubble pressure method in which one measures the maximum pressure in a liquid drop injected into a second immiscible liquid [32]. The modified maximum bubble pressure method [32] and a liquid bridge method [33] have been used to measure interfacial tension between two liquids of equal density. Ultralow values of interfacial tension can be measured by the spinning drop [34], pendant drop [35], and capillary wave methods [34].

Defining Terms

Surface tension: A force per unit length that acts tangential to a liquid surface and perpendicular to any line that lies within the surface.

Surface energy: The excess free energy per unit area associated with a surface between two phases. For a liquid–fluid surface, the surface energy is numerically equivalent to the surface tension.

Acknowledgments

During the preparation of this chapter, one of us (DBT) was supported in part by the National Aeronautics and Space Administration (NASA) and by the Office of Naval Research. The work by one of us (KFM) was performed at the Jet Propulsion Laboratory, California Institute of Technology, under contract with NASA.

References

1. A. W. Adamson, *Physical Chemistry of Surfaces*, 5th ed., New York: John Wiley & Sons, 1990.
2. A. E. Alexander and J. B. Hayter, Determination of surface and interfacial tension, in A. Weissberger and B. W. Rossiter (eds.), *Physical Methods of Chemistry, Part V*, 4th ed., New York: John Wiley & Sons, 1971.
3. A. Couper, Surface tension and its measurement, in B. W. Rossiter and R. C. Baetzold (eds.), *Physical Methods of Chemistry, Vol. 9A*, 2nd ed., New York: John Wiley & Sons, 1993.
4. J. F. Padday, Surface tension. II. The measurement of surface tension, in E. Matijevic (ed.), *Surface and Colloid Science, Vol. 1*, New York: John Wiley & Sons, 1969.
5. S. Sugden, The determination of surface tension from the maximum pressure in bubbles. Part II, *J. Chem. Soc.*, 125, 27-31, 1924.
6. ASTM Standard D3825-90, Standard test method for dynamic surface tension by the fast-bubble technique, *1996 Annual Book of ASTM Standards, Vol. 05.02*, West Conshohocken, PA: ASTM, 1996, 575-579.
7. D. S. Ambwani and T. Fort, Jr., Pendant drop technique for measuring liquid boundary tensions, in R. J. Good and R. R. Stromberg (eds.), *Surface and Colloid Science, Vol. 11*, New York: Plenum Press, 1979.
8. Y. Rotenberg, L. Boruvka, and A. W. Neumann, Determination of surface tension and contact angle from the shapes of axisymmetric fluid interfaces, *J. Colloid Interface Sci.*, 93, 169-183, 1983.
9. P. Cheng, D. Li, L. Boruvka, Y. Rotenberg, and A. W. Neumann, Automation of axisymmetric drop shape analysis for measurements of interfacial tensions and contact angles, *Colloids Surf.*, 43, 151-167, 1990.
10. D. B. Thiessen, D. J. Chione, C. B. McCreary, and W. B. Krantz, Robust digital image analysis of pendant drop shapes, *J. Colloid Interface Sci.*, 177, 658-665, 1996.
11. S. Lin, K. McKeigue, and C. Maldarelli, Diffusion-controlled surfactant adsorption studied by pendant drop digitization, *AIChE J.*, 36, 1785-1795, 1990.
12. D. Y. Kwok, D. Vollhardt, R. Miller, D. Li, and A. W. Neumann, Axisymmetric drop shape analysis as a film balance, *Colloids Surf., A*, 88, 51-58, 1994.
13. W. M. Schoel, S. Schurch, and J. Goerke, The captive bubble method for the evaluation of pulmonary surfactant: surface tension, area, and volume calculations, *Biochim. Biophys. Acta*, 1200, 281-290, 1994.

14. S. Lahooti, O. I. Del Rio, P. Cheng, and A. W. Neumann, Axisymmetric drop shape analysis (ADSA), in A. W. Neumann and J. K. Spelt (eds.), *Applied Surface Thermodynamics*, New York: Marcel Dekker, 1996.

15. M. L. Alexander and M. J. Matteson, The automation of an interfacial tensiometer, *Colloids Surf.*, 27, 201-217, 1987.

16. J. C. Slattery and J. Chen, Alternative solution for spinning drop interfacial tensiometer, *J. Colloid Interface Sci.*, 64, 371-373, 1978.

17. D. Y. Kwok, M. A. Cabrerizo-Vilchez, Y. Gomez, S. S. Susnar, O. Del Rio, D. Vollhardt, R. Miller, and A. W. Neumann, Axisymmetric drop shape analysis as a method to study dynamic interfacial tensions, in V. Pillai and D. O. Shah (eds.), *Dynamic Properties of Interfaces and Association Structures*, Champaign, IL: AOCS Press, 1996.

18. W. D. E. Thomas and L. Potter, Solution/air interfaces. I. An oscillating jet relative method for determining dynamic surface tensions, *J. Colloid Interface Sci.*, 50, 397-412, 1975.

19. V. B. Fainerman and R. Miller, Dynamic surface tension measurements in the sub-millisecond range, *J. Colloid Interface Sci.*, 175, 118-121, 1995.

20. G. L. Gaines Jr., *Insoluble Monolayers at Liquid-Gas Interfaces*, New York: John Wiley & Sons, 1966, 44.

21. R. Miller, R. Sedev, K.-H. Schano, C. Ng, and A. W. Neumann, Relaxation of adsorption layers at solution/air interfaces using axisymmetric drop-shape analysis, *Colloids Surf.*, 69, 209-216, 1993.

22. D. O. Johnson and K. J. Stebe, Oscillating bubble tensiometry: a method for measuring the surfactant adsorptive-desorptive kinetics and the surface dilatational viscosity, *J. Colloid Interface Sci.*, 168, 21-31, 1994.

23. D. O. Johnson and K. J. Stebe, Experimental confirmation of the oscillating bubble technique with comparison to the pendant bubble method: the adsorption dynamics of 1-decanol, *J. Colloid Interface Sci.*, 182, 526-538, 1996.

24. D. A. Edwards, H. Brenner, and D. T. Wasan, *Interfacial Transport Processes and Rheology*, Boston: Butterworth-Heinemann, 1991.

25. C. Garcia-Cordovilla, E. Louis, and A. Pamies, The surface tension of liquid pure aluminium and aluminium-magnesium alloy, *J. Mater. Sci.*, 21, 2787-2792, 1986.

26. B. C. Allen, The surface tension of liquid transition metals at their melting points, *Trans. Metall. Soc. AIME*, 227, 1175-1183, 1963.

27. D. B. Thiessen and K. F. Man, A quasi-containerless pendant drop method for surface tension measurements on molten metals and alloys, *Int. J. Thermophys.*, 16, 245-255, 1995.

28. S. C. Hardy, The surface tension of liquid silicon, *J. Cryst. Growth*, 69, 456-460, 1984.

29. V. Schoettle and H. Y. Jennings, Jr., High-pressure high-temperature visual cell for interfacial tension measurement, *Rev. Sci. Instrum.*, 39, 386-388, 1968.

30. R. Simon and R. L. Schmidt, A system for determining fluid properties up to 136 MPa and 473K, *Fluid Phase Equilib.*, 10, 233-248, 1983.

31. K. Hool and B. Schuchardt, A new instrument for the measurement of liquid-liquid interfacial tension and the dynamics of interfacial tension reduction, *Meas. Sci. Technol.*, 3, 451-457, 1992.

32. A. Passerone, L. Liggieri, N. Rando, F. Ravera, and E. Ricci, A new experimental method for the measurement of the interfacial tension between immiscible fluids at zero Bond number, *J. Colloid Interface Sci.*, 146, 152-162, 1991.

33. G. Pétré and G. Wozniak, Measurement of the variation of interfacial tension with temperature between immiscible liquids of equal density, *Acta Astronaut.*, 13, 669-672, 1986.

34. D. Chatenay, D. Langevin, J. Meunier, D. Bourbon, P. Lalanne, and A. M. Bellocq, Measurement of low interfacial tension, comparison between a light scattering technique and the spinning drop technique, *J. Dispersion Sci. Technol.*, 3, 245-260, 1982.

35. D. Y. Kwok, P. Chiefalo, B. Khorshiddoust, S. Lahooti, M. A. Cabrerizo-Vilchez, O. Del Rio, and A. W. Neumann, Determination of ultralow interfacial tension by axisymmetric drop shape analysis, in R. Sharma (ed.), *Surfactant Adsorption and Surface Solubilization (ACS Symp. Ser. 615)*, Washington, D.C.: ACS, 1995, 374-386.

VI

Mechanical Variables Measurement — Thermal

32

Temperature
Measurement

Robert J. Stephenson
University of Cambridge

Armelle M. Moulin
University of Cambridge

Mark E. Welland
University of Cambridge

Jim Burns
Burns Engineering Inc.

Meyer Sapoff
MS Consultants

R. P. Reed
Proteun Services

Randy Frank
Motorola, Inc.

Jacob Fraden
Advanced Monitors Corporation

J.V. Nicholas
Industrial Research Limited

Franco Pavese
*CNR Instituto di Metrologia
"G. Colonnetti"*

Jan Stasiek
Technical University of Golansk

Tolestyn Madaj
Technical University of Golansk

Jaroslaw Mikielewicz
Institute of Fluid Flow Machinery

Brian Culshaw
University of Strathclyde

32-2   *The Measurement, Instrumentation, and Sensors Handbook*

32.8   Liquid-in-Glass Thermometers.................................32-117
        General Description • Liquid Expansion • Time-Constant
        Effects • Thermal Capacity Effects • Separated Columns •
        Immersion Errors • Organic Liquids • Storage • High
        Accuracy • Defining Terms

32.9   Manometric Thermometers.....................................32-129
        Vapor Pressure • Gas Thermometry

32.10   Temperature Indicators 32-136
        Melting and Shape/Size Changing Temperature Indicators •
        Color-Change Temperature Indicators

32.11   Fiber-Optic Thermometers 32-152
        Fiber Optic Temperature Sensors • Fiber Optic Point
        Temperature Measurement Systems • Distributed and Quasi-
        distributed Optical Fiber Temperature Measurement
        Systems • Applications for Optical Fiber Temperature Probes

32.1   Bimaterials Thermometers

Robert J. Stephenson, Armelle M. Moulin, and Mark E. Welland

The first known use of differential thermal expansion of metals in a mechanical device was that of the English clockmaker John Harrison in 1735. Harrison used two dissimilar metals in a clock escapement to account for the changes in temperature on board a ship. This first marine chronometer used a gridiron of two metals that altered the flywheel period of the clock through a simple displacement. This mechanical actuation, resulting from the different thermal expansivities of two metals in contact, is the basis for all bimetallic actuators used today.

The bimetallic effect is now used in numerous applications ranging from domestic appliances to compensation in satellites. The effects can be used in two ways: either as an actuator or as a temperature measuring system. A bimetallic actuator essentially consists of two metal strips fixed together. If the two metals have different expansivities, then as the temperature of the actuator changes, one element will expand more than the other, causing the device to bend out of the plane. This mechanical bending can then be used to actuate an electromechanical switch or be part of an electrical circuit itself, so that contact of the bimetallic device to an electrode causes a circuit to be made. Although in its simplest form a bimetallic actuator can be constructed from two flat pieces in metal, in practical terms a whole range of shapes are used to provide maximum actuation or maximum force during thermal cycling.

As a temperature measuring device, the bimetallic element, similar in design to that of the actuator above, can be used to determine the ambient temperature if the degree of bending can be measured. The advantage of such a system is that the amount of bending can be mechanically amplified to produce a large and hence easily measurable displacement.

The basic principle of a bimetallic actuator is shown in Figure 32.1. Here, two metal strips of differing thermal expansion are bonded together. When the temperature of the assembly is changed, in the absence

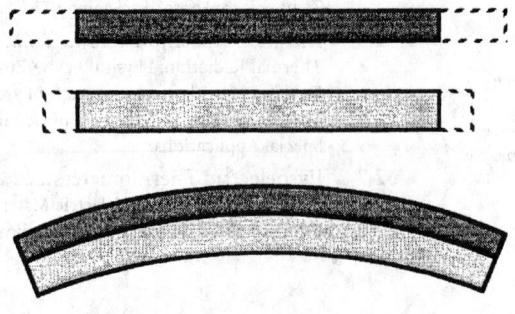

FIGURE 32.1   Linear bimetallic strip.

of external forces, the bimetallic actuator will take the shape of an arc. The total displacement of the actuator out of the plane of the metal strips is much greater than the individual expansions of the metallic elements. To maximize the bending of the actuator, metals or alloys with greatly differing coefficients of thermal expansion are normally selected. The metal having the largest thermal expansitivity is known as the *active element*, while the metal having the smaller coefficient of expansion is known as the *passive element*. For maximum actuation, the passive element is often an iron–nickel alloy, Invar, having an almost zero thermal expansivity (actually between 0.1 and 1×10^{-6} K^{-1}, depending upon the composition). The active element is then chosen to have maximum thermal expansivity given the constraints of operating environment and costs.

In addition to maximizing the actuation of the bimetallic element, other constraints such as electrical and thermal conductivity can be made. In such cases, a third metallic layer is introduced, consisting of either copper or nickel sandwiched between the active and passive elements so as to increase both the electrical and thermal conductivity of the actuator. This is especially important where the actuator is part of an electrical circuit and needs to pass current in addition to being a temperature sensor.

Linear Bimaterial Strip

Basic Equations

The analysis of the stress distribution and the deflection of an ideal bimetallic strip was first deduced by Timoshenko [1], who produced a simple derivation from the theory of elasticity. Figure 32.2 shows the internal forces and moments that induce bending in a bimetallic strip followed by the ideal stress distribution in the beam. This theory is derived for bimetallic strips, but is equally applicable to bimaterial strips.

The general equation for the curvature radius of a bimetallic strip uniformly heated from T_0 to T in the absence of external forces is given by [1]:

$$\frac{1}{R} - \frac{1}{R_0} = \frac{6(1+m)^2 (\alpha_2 - \alpha_1)(T - T_0)}{t\left[3(1+m)^2 + (1+mn)\left(m^2 + 1/mn\right)\right]} \tag{32.1}$$

where $1/R_0$ = Initial curvature of the strip at temperature T_0
    α_1 and α_2 = Coefficients of expansion of the two elements: (1) low expansive material and (2) high expansive material
    n = E_1/E_2, with E_1 and E_2 their respective Young's moduli
    m = t_1/t_2, with t_1 and t_2 their respective thicknesses
    t = $t_1 + t_2$ thickness of the strip

The width of the strip is taken as equal to unity.

Equation 32.1 applies for several strip configurations, including the simply supported strip and a strip clamped at one end (i.e., a cantilever). For a given configuration, the deflection of a strip can be determined by its relationship with curvature, $1/R$.

An example of a calculation of deflection is a bimetallic strip simply supported at its two ends. The initial curvature $1/R_0$ is assumed to be zero. Figure 32.3 shows the geometrical relationship between the radius R of the strip and the deflection d at its mid-point and is given by:

$$\left(R - t_2\right)^2 = \left(R - d - t_2\right)^2 + \left(\frac{L}{2}\right)^2 \tag{32.2}$$

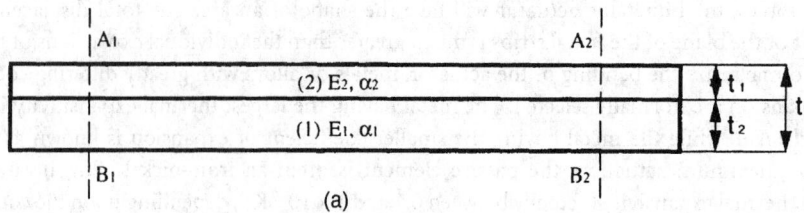

(a)

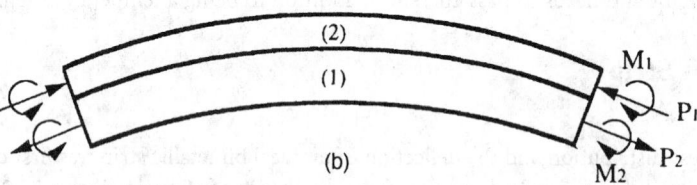

(b)

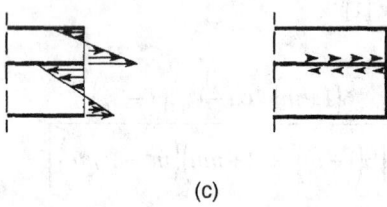

(c)

FIGURE 32.2 Bending of bimetallic strip uniformly heated with $\alpha_2 \geq \alpha_1$. (a) Bimetallic strip. A_1B_1–A_2B_2 is an element cut out from the strip. (b) Bending of the element A_1B_1–A_2B_2 when uniformly heated. Assuming $\alpha_2 > \alpha_1$, the deflection is convex up. The total force acting over the section of (1) is an axial tensile force P_1 and bending moment M_1, whereas over the section of (2) it is an axial compressive force P_2 and bending moment M_2. (c) Sketch of the internal resulting stress distribution. (Left): normal stresses over the cross section of the strip. The maximum stress during heating is produced at the interface between the two components of the strip. This stress is due to both axial force and bending. (Right): shearing stresses at the ends of the strip.

Hence,

$$\frac{1}{R} = \frac{8d}{L^2 + 4d^2 + 8dt_2} \tag{32.3}$$

Making the assumption that the deflection and the thickness are less than 10% of the length of the strip (which is true in most practical cases) means the terms $8dt_2$ and $4d^2$ are therefore negligible and the expression reduces to:

$$d = \frac{L^2}{8R} \tag{32.4}$$

or

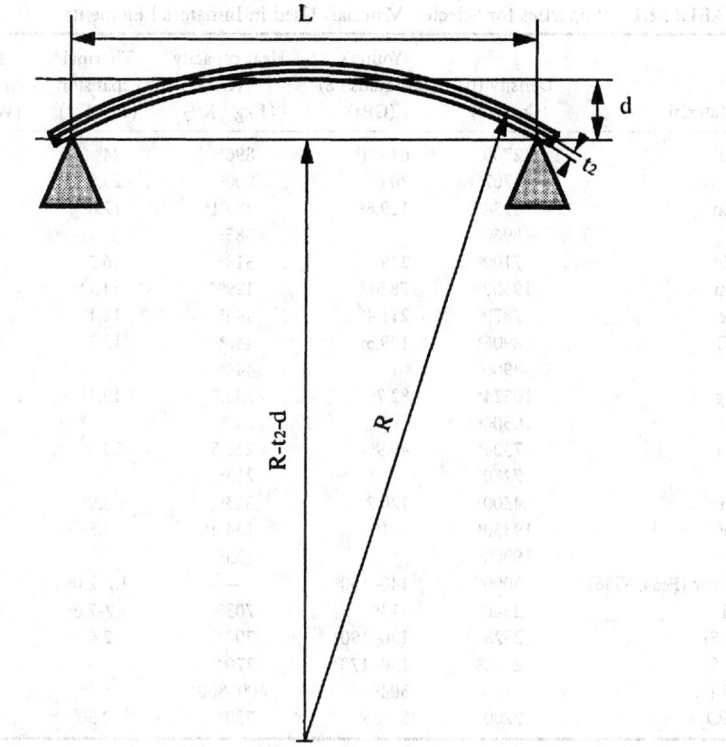

FIGURE 32.3 Bending of a strip simply supported at its ends.

$$d = L^2 \frac{3(1+m)^2}{4t\left[3(1+m)^2 + (1+mn)(m^2 + 1/mn)\right]}(\alpha_2 - \alpha_1)(T - T_0) \qquad (32.5)$$

If a 100-mm strip is composed of two layers of the same thickness (0.5 mm) with the high-expansive layer being made of iron (from Table 32.1, E_2 = 211 GPa and α_2 = 12.1 × 10^{-6} K^{-1}), the low-expansive layer made of Invar (from Table 32.1, E_1 = 140 GPa and α_1 = 1.7 × 10^{-6} K^{-1}), and the temperature increases from 20°C to 120°C, then the theoretical bending at the middle of the strip will be 1.92 mm.

As a second example, consider the calculation of the deflection of the free end of a bimetallic cantilever strip as illustrated in Figure 32.4. In this case, the geometrical relation is:

$$(R + t_1)^2 = (R + t_1 - d)^2 + L^2 \qquad (32.6a)$$

or

$$\frac{1}{R} = \frac{2d}{L^2 + d^2 - 2dt_1} \qquad (32.6b)$$

Making the same assumptions as before, that is, $d^2 \ll L^2$ and $dt_1 \ll L^2$, then the deflection of the free end is given by:

$$d = \frac{L^2}{8R} \qquad (32.7)$$

TABLE 32.1 Properties for Selected Materials Used in Bimaterial Elements

Material	Density (ρ) (kg m^{-3})	Young's Modulus (E) (GPa)	Heat capacity (C) (J kg^{-1} K^{-1})	Thermal expansion (10^{-6} K^{-1})	Thermal conductivity (W m^{-1} K^{-1})
Al	2700[c]	61–71[b]	896[a]	24[b]	237[c]
	2707[a]	70.6[c]	900[c]	23.5[c]	204[a]
Cu	8954[a]	129.8[c]	383.1[a]	17.0[c]	386[a]
	8960[c]		385[c]		401[c]
Cr	7100[c]	279[c]	518[c]	6.5[c]	94[c]
Au	19300[b,c]	78.5[b,c]	129[b,c]	14.1[b,c]	318[b,c]
Fe	7870[c]	211.4[c]	444[c]	12.1[c]	80.4[c]
Ni	8906[a]	199.5[c]	446[a]	13.3[c]	90[a]
	8900[c]		444[c]		90.9[c]
Ag	10524[a]	82.7[c]	234.0[a]	19.1[c]	419[a]
	10500[c]		237[c]		429[c]
Sn	7304[a]	49.9[c]	226.5[a]	23.5[c]	64[a]
	7280[c]		213[c]		66.8[c]
Ti	4500[c]	120.2[c]	523[c]	8.9[c]	21.9[c]
W	19350[a]	411[c]	134.4[a]	4.5[c]	163[a]
	19300[c]		133[c]		173[c]
Invar (Fe64/Ni36)	8000[c]	140–150[c]	—	1.7-2.0[c]	13[c]
Si	2340[c]	113[c]	703[c]	4.7-7.6[c]	80–150[c]
n-Si	2328[b]	130–190[b]	700[b]	2.6[b]	150[b]
p-Si	2300[b]	150–170[b]	770[b]	—	30[b]
Si$_3$N$_4$	3100[a]	304[b]	600–800[b]	3.0[b]	9–30[b]
SiO$_2$	2200[b]	57-85[b]	730[b]	0.50[b]	1.4[b]

[a] From Reference [13], Table A1 at 20°C.
[b] From Reference [13], Table A2 at 300K.
[c] From Goodfellow catalog 1995/1996 [14].

and combining this with Equation 32.1 yields:

$$d = L^2 \frac{3(1+m)^2}{t\left[3(1+m)^2 + (1+mn)(m^2 + 1/mn)\right]}(\alpha_2 - \alpha_1)(T - T_0) \qquad (32.8)$$

If an aluminum and silicon nitride bimaterial microcantilever as used for sensor research [2] is considered, then $L = 200$ μm, $t_1 = 0.6$ μm, $t_2 = 0.05$ μm, $E_1 = 300$ GPa, $E_2 = 70$ GPa, $\alpha_1 = 3 \times 10^{-6}$ K^{-1}, $\alpha_2 = 24 \times 10^{-6}$ K^{-1} (see Table 32.1). In this situation, a temperature difference of 1°C gives a theoretical deflection of the cantilever of 0.103 μm.

Terminology and Simplifications

For industrial purposes, bimetallic thermostatic strips and sheets follow a standard specification — ASTM [3] in the U.S. and DIN [4] in Europe. Important parameters involved in this specification are derived directly from the previous equations, in which simplifications are made based on common applications.

It can be seen that the magnitude of the ratio $E_1/E_2 = n$ has no substantial effect on the curvature of the strip, and taking $n = 1$ implies an error less than 3%. Assuming again that the initial curvature is zero, Equation 32.1 can be simplified to:

$$\frac{1}{R} = \frac{6m}{t(m+1)^2}(\alpha_2 - \alpha_1)(T - T_0) \qquad (32.9)$$

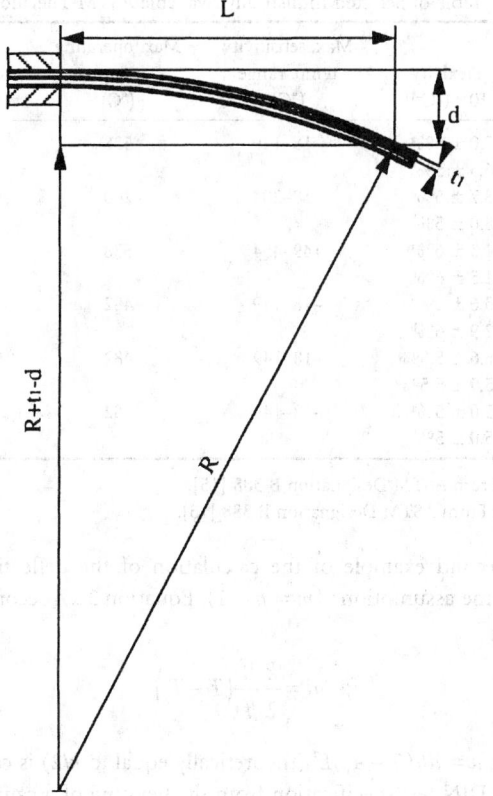

FIGURE 32.4 Bending of a strip fixed at one end.

In most industrial applications involving bimetallic elements, the thicknesses of the two component layers are taken to be equal ($m = 1$), thus Equation 32.6 becomes:

$$\frac{1}{R} = \frac{3}{2}\frac{(\alpha_2 - \alpha_1)(T - T_0)}{t} \tag{32.10}$$

The constant $\frac{3}{2}(\alpha_2 - \alpha_1)$ is known as flexivity in the U.S. and as specific curvature in Europe. Introducing the flexivity, k, and rearranging Equation 32.10 gives:

$$k = \frac{\frac{t}{R}}{T - T_0} \tag{32.11}$$

Flexivity can be defined as "the change of curvature of a bimetal strip per unit of temperature change times thickness" [5]. The experimental determination of the flexivity for each bimetallic strip has to follow the test specifications ASTM B388 [3] and DIN 1715 [4]. The method consists of measuring the deflection of the midpoint of the strip when it is simply supported at its ends. Using Equation 32.4 derived above and combining with Equation 32.11 gives:

$$k = \frac{8dt}{(T - T_0)L^2} \tag{32.12}$$

TABLE 32.2 Table of Selected Industrially Available ASTM Thermostatic Elements

Type (ASTM)	Flexivity 10^{-6} (°C^{-1})	Max. sensitivity temp. range (°C)	Max. operating temp. (°C)	Young's Modulus (GPa)
TM1	$27.0 \pm 5\%^a$ $26.3 \pm 5\%^b$	−18–149	538	17.2
TM2	$38.7 \pm 5\%^a$ $38.0 \pm 5\%^b$	−18–204	260	13.8
TM5	$11.3 \pm 6\%^a$ $11.5 \pm 6\%^b$	149–454	538	17.6
TM10	$23.6 \pm 6\%^a$ $22.9 \pm 6\%^b$	−18–149	482	17.9
TM15	$26.6 \pm 5.5\%^a$ $25.9 \pm 5.5\%^b$	−18–149	482	17.2
TM20	$25.0 \pm 5\%^a$ $25.0 \pm 5\%^b$	−18–149	482	17.2

[a] 10–93°C. From ASTM Designation B 388 [15].
[b] 38–149°C. From ASTM Designation B 388 [15].

Coming back to the second example of the calculation of the deflection (cantilever case), using Equation 32.10 and the same assumptions ($m = n = 1$), Equation 32.7 becomes:

$$d = \frac{k}{2} \frac{L^2}{t}\left(T - T_0\right) \tag{32.13}$$

In Europe, the constant $a = dt/(T - T_0)L^2$ (theoretically equal to $k/2$) is called specific deflection and is measured following the DIN test specification from the bending of a cantilever strip. It can be noted that the experimental value differs from the theoretical value as it takes into account the effect of the external forces suppressing the cross-curvature where the strip is fastened (i.e., the theory assumes that the curvature is equal along the strip; whereas in reality, the fact that the strip is fastened implies that the radius is infinite at its fixed end).

Tables 32.2 and Table 32.4 present a selection of bimetallic elements following ASTM and DIN standards, respectively. Flexivity (or specific curvature), linear temperature range, maximum operating temperature, and specific deflection (DIN only) are given. The details of the chemical composition of these elements are specified in Tables 32.3 and Table 32.5.

Industrial Applications

The mechanical thermostat finds a wide range of applications in temperature control in industrial processes and everyday life. This widespread use of thermostats is due to the discovery of Invar, a 36% nickel alloy that has a very low thermal expansion coefficient, and was so named because of its property of invariability [6].

There are two general classes of bimetallic elements based on their movement in response to temperature changes. Snap-action devices jump from one position to another at a specific temperature depending on their design and construction. There are numerous different shapes and sizes of snap-action elements and they are typically ON/OFF actuators. The other class of elements, creep elements, exhibit a gradual change in shape in response to a change in temperature and are employed in temperature gauges and other smooth movement applications. Continuous movement bimetals will be considered first. A linear configuration was covered previously, so the discussion will focus on coiled bimetallic elements.

Spiral and Helical Coil Configurations

For industrial or commercial measurements, a spiral or helical coil configuration is useful for actuating a pointer on a dial as the thermal deflection is linear within a given operating range. Linearity in this

TABLE 32.3 Composition of Selected Industrially Available ASTM Thermostatic Elements Given in Table 32.2

	Element	TM1	TM2	TM5	TM10	TM15	TM20
High-expansive material chemical composition (% weight)	Nickel	22	10	25	22	22	18
	Chromium	3	72	8.5	3	3	11.5
	Manganese	—	18	—	—	—	—
	Copper	—	—	—	—	—	—
	Iron	75	—	66.5	75	75	70.5
	Aluminum	—	—	—	—	—	—
	Carbon	—	—	—	—	—	—
Intermediate nickel layer		No	No	No	Yes	Yes	No
Low-expansive material chemical composition (% weight)	Nickel	36	36	50	36	36	36
	Iron	64	64	50	64	64	64
	Cobalt	—	—	—	—	—	—
Component ratio (% of thickness)	High	50	53	50	34	47	50
	Intermediate	—	—	—	32	6	—
	Low	50	47	50	34	47	50

From ASTM Designation B 388 [15].

TABLE 32.4 Table of Selected Industrially Available DIN Thermostatic Elements

Type (DIN)	Specific deflection (10^{-6} K^{-1})	Specific curvature (10^{-6} K^{-1}) ± 5%	Linear range (°C)	Max. operating temperature (°C)
TB0965	9.8	18.6	−20–425	450
TB1075	10.8	20.0	−20–200	550
TB1170A	11.7	22.0	−20–380	450
TB1577A	15.5	28.5	−20–200	450
TB20110	20.8	39.0	−20–200	350

Note: From DIN 1715 standard [4]. Specific deflection and curvature are for the range 20°C to 130°C.

TABLE 32.5 Composition of Selected Industrially Available DIN Thermostatic Elements Given in Table 32.4

	Element	TB0965	TB1075	TB1170A	TB1577A	TB20110
High-expansive chemical composition (% mass)	Nickel	20	16	20	20	10-16
	Chromium	—	11	—	—	—
	Manganese	6	—	6	6	Remainder
	Copper	—	—	—	—	18-10
	Iron	Remainder	Remainder	Remainder	Remainder	0.5
	Carbon	—	—	—	—	—
Low-expansive chemical composition (% mass)	Nickel	46	20	42	36	36
	Iron	Remainder	Remainder	Remainder	Remainder	Remainder
	Cobalt	—	26	—	—	—
	Chromium	—	8	—	—	—

From DIN 1715 Standard [4].

case means that the deflection does not vary by more than 5% of the deflection, as calculated from the flexivity [4]. The basic bimaterial ideas from the previous section still apply, with some additional equations relating the movement of a bimetal coil to a change in temperature. As in the previous section, standard methods for testing the deflection rate of spiral and helical coils exist and can be found in [7]. The following equations have been taken from the Kanthal Thermostatic Bimetal Handbook [8], with some change in notation. The angular rotation of a bimetal coil is given by (see Figure 32.5):

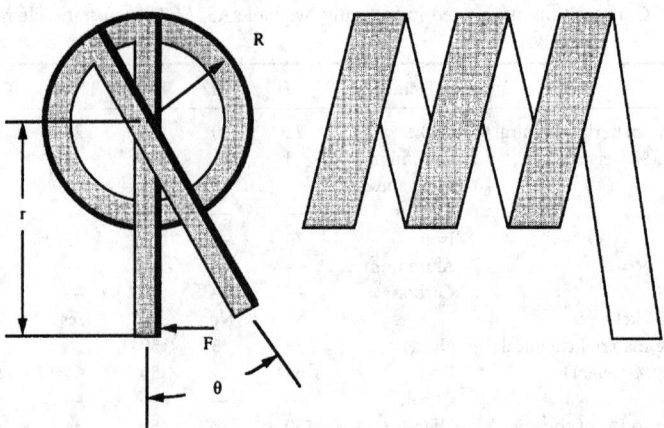

FIGURE 32.5 Helical coiled bimetal element.

$$\theta = \left(\frac{1}{R} - \frac{1}{R_0}\right)L \tag{32.14}$$

where L = length of the strip
R_0 and R = initial and final bending radii (assumed to be constant along the strip), respectively.

In terms of the specific deflection a, this can be written as:

$$\theta = \frac{2aL}{t}\left(T - T_0\right)\frac{360}{2\pi} \tag{32.15}$$

where t = thickness of the device
T_0 and T = initial and final temperatures, respectively.

An example would be a helical bimetal coil inside a steel tube with one end of the coil fixed to the end of the tube and the other connected to a pointer. The accuracy of a typical commercial product is 1% to 2% of full-scale deflection with an operating range of 0°C to 600°C [9].

If a change in temperature is required to both move a pointer and produce a driving force, then the angular rotation is reduced and is given by:

$$\theta = \left(\frac{2a\left(T - T_0\right)L}{t} - \frac{12\left(F - F_0\right)Lr}{wt^3 E}\right)\frac{360}{2\pi} \tag{32.16}$$

where w = width of the element
r = distance from the center of the coil to the point of applied force, F.

This can be rewritten as:

$$T - T_0 = \frac{\theta t}{2aL}\frac{2\pi}{360} + \frac{6\left(F - F_0\right)r}{wt^2 Ea} \tag{32.17}$$

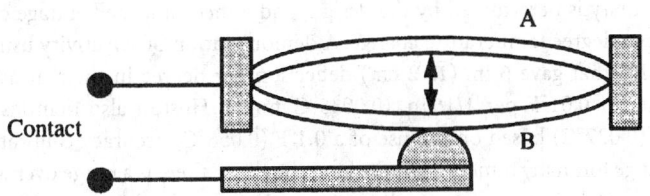

FIGURE 32.6 Snap-action bimetallic element.

where the first term represents the temperature associated with the angular rotation of the strip and the second term represents the temperature associated with the force generated by the strip. In general, the strip is designed so that the two are equal, as this leads to the minimum volume for the strip and consequently less weight and cost for the device.

Finally, if the coil is prevented from moving then the change in torque is given by

$$\left(F - F_0\right)r = \frac{1}{6}wt^2 Ea\left(T - T_0\right) \tag{32.18}$$

Example: Consider a bimetal element that measures a temperature change from 20°C to 100°C and moves a lever 50 mm away with a force of 1 N. A dial reading range of 180° is required.

Choosing thermostatic bimetal TM2 gives the largest deflection per degree temperature change and TM2 meets the operating temperature requirements. Both a force and a movement are involved, so use Equation 32.7. Choosing each term equal to half the temperature change gives the minimum volume for the strip as:

$$\frac{1}{2}\left(T - T_0\right) = \frac{\theta t}{2aL}\frac{2\pi}{360} = \frac{6\left(F - F_0\right)r}{wt^2 Ea} \tag{32.19}$$

Selecting a thickness of 1.0 mm and using a specific deflection of 19×10^{-6} °C^{-1} ($a = k/2$ from Table 32.2) gives a width of 29 mm. Similarly, the length of the bimetal strip is obtained from the second term in the equality, giving $L = 2.1$ m.

Snap-Action Configurations

Snap-action bimetal elements are used in applications where an action is required at a threshold temperature. As such, they are not temperature measuring devices, but rather temperature-activated devices. The typical temperature change to activate a snap-action device is several degrees and is determined by the geometry of the device (Figure 32.6). When the element activates, a connection is generally made or broken and in doing so, a gap between the two contacts exists for a period of time. For a mechanical system, there is no problem; however, for an electrical system, the gap can result in a spark that can lead to premature aging and corrosion of the device. The amount and duration of spark is reduced by having the switch activate quickly, hence the use of snap-action devices.

Snap-action elements also incorporate a certain amount of hysteresis into the system, which is useful in applications that would otherwise result in an oscillation about the set-point. It should be noted, however, that special design of creep action bimetals can also lead to different ON/OFF points, such as in the reverse lap-welded bimetal [8].

Sensitivity and Accuracy

Modern techniques are more useful where sensitivity and accuracy are concerned for making a temperature measurement; however, bimetals find application in commercial and industrial temperature control where an action is required without external connections. Evidently, geometry is important for bimetal systems as the sensitivity is determined by the design, and a mechanical advantage can be used to yield a large movement per degree temperature change. A demonstration of sensitivity using a helical coil was made by Huston [10] that gave 6 in. (15.2 cm) deflection per degree in their measurement system — yielding a sensitivity of 0.01°F per 1/16 in. (0.0035°C mm^{-1}). Huston also demonstrated a repeatable accuracy of 0.05°F (0.027°C) based on the use of a 0.1°F (0.056°C) accuracy calibration instrument.

The operating range for many bimetals is quite large; however, there is a range over which the sensitivity is a maximum. A bimetal element is generally chosen to operate in this range and specific details are provided by manufacturers in their product catalogs. Of particular note is that, despite extended thermal cycling, bimetal strips reliably return to the same position (i.e., show no hysteresis) at a given temperature and are very robust as long as they are not subjected to temperatures outside their specified operating range.

Advanced Applications

Thermostatic valves are a ready and robust means of measuring temperature and controlling heating and cooling in industrial settings. The basic designs have been around for many years and are the mainstay of many commercial temperature-control systems. New applications of bimaterials are being found in microactuators and microsensors.

Besides operating as temperature-measuring instruments, bimaterial devices can be used for a variety of applications where temperature is the controlling or triggering phenomenon, or indeed, other material properties are inferred from the temperature response. One such example is a nickel–silicon actuator developed by engineers at HP labs in Palo Alto, CA; the actuator operates by heating a thin nickel resistor on the silicon side of a bilayer device[11]. Both the silicon and nickel layers expand due to heating; however, the nickel layer expands more, thereby curling the device, which leads to the actuation of a tiny valve. The device can control gas flow rates from 0.1 to 600 standard cm^3 per minute with pressures ranging from 5 psi to 200 psi (34.5 kPa to 1379 kPa).

The use of micromachined thermal sensors compatible with modern silicon integrated circuit fabrication methods has recently received significant attention. One way of achieving highly sensitive thermal measurements is by using micromechanical bimetallic cantilevers and measuring the deflection as a result of thermal fluctuations. Rectangular or triangular cantilevers, typically 100 μm long made of silicon or silicon nitride, are coated with a thin high-expansive metallic layer (e.g., aluminum or gold). Precise measurement of the deflection of the end of the cantilever is achieved using an optical sensing arrangement commonly used in atomic force microscopes. The micromechanical nature of the cantilever-based sensor leads to significant advantages in the absolute sensitivity achievable. In this way, the device is capable of measuring temperature, heat, and power variations of 10^{-5} K, 150 fJ, or 100 pW, respectively [2]. In addition to their use as thermal sensors, the bimetallic cantilever systems have been used to investigate physical phenomena where heat is produced by the sample. Examples include photothermal spectroscopy studies of amorphous silicon [2] and the observation of oscillations in the catalyzed reaction of hydrogen and oxygen on platinum [12]. Thus, bimaterial sensors are becoming an increasingly important area of development.

Defining Terms

Linear coefficient of thermal expansion: The change in length of a material per degree change in temperature expressed as a fraction of the total length ($\Delta L/L$).

Flexivity: The change in radius of curvature of a bimaterial per degree change in temperature times the width of the element (See Equation 32.11).

Specific curvature: The European term for flexivity.

Specific deflection: Theoretically equal to half the flexivity. Specific deflection is measured by mounting the test element as a cantilever — supported at one end and free to move at the other.

References

1. S.P. Timoshenko, *The Collected Papers*, New York: McGraw-Hill, 1953.
2. J.R. Barnes, R.J. Stephenson, M.E. Welland, C. Gerber, and J.K. Gimzewski, Photothermal spectroscopy with femtojoule sensitivity using a micromechanical device. *Nature*, 372, 79-81, 1994.
3. ASTM Designation B 388.
4. DIN 1715. Part 1. Thermostat Metals. 1983.
5. ASTM Designation B 106.
6. M. Kutz, *Temperature Control*, New York: John Wiley & Sons, 1968.
7. ASTM Designation B 389.
8. *The Kanthal Thermostatic Bimetal Handbook*. Box 502. S-73401 Hallstammar, Sweden, 1987.
9. Bourdon Sedeme, F-41103 Vendome Cedex, France.
10. W.D. Huston, The accuracy and reliability of bimetallic temperature measuring elements, in C.M. Herzfeld and A.I. Dahl (eds.), *Temperature — Its Measurement and Control in Science and Industry*, New York: Reinhold, 1962.
11. L. O'Connor, A bimetallic silicon microvalve. *Mechanical Engineering*, 117(1), 1, 1995.
12. J.K. Gimzewski, C. Gerber, E. Meyer, and R. Schlittler, Observation of a chemical reaction using a micromechanical sensor. *Chem. Phys. Lett.* 217, 589-594, 1994.
13. G.C.M. Meijer and A.W. van Herwaarden, *Thermal Sensors*, Bristol, U.K.: Institute of Physics Publishing, 1994.
14. Goodfellow Cambridge Limited. Cambridge Science Park. U.K. CB4 4DJ.
15. American Society for Testing and Materials, *Annual Book of ASTM Standards*, Philadelphia, 1991.

Further Information

V.C. Miles, *Thermostatic Control — Principles and Practice*, Liverpool: C. Tinling and Co., 1965.

32.2 Resistive Thermometers

Jim Burns

Introduction to Resistance Temperature Detectors

One common way to measure temperature is by using Resistive Temperature Detectors (RTDs). These electrical temperature instruments provide highly accurate temperature readings: simple industrial RTDs used within a manufacturing process are accurate to ±0.1°C, while Standard Platinum Resistance Thermometers (SPRTs) are accurate to ±0.0001°C.

The electric resistance of certain metals changes in a known and predictable manner, depending on the rise or fall in temperature. As temperatures rise, the electric resistance of the metal increases. As temperatures drop, electric resistance decreases. RTDs use this characteristic as a basis for measuring temperature.

The sensitive portion of an RTD, called an element, is a coil of small-diameter, high-purity wire, usually constructed of platinum, copper, or nickel. This type of configuration is called a wire-wound element. With thin-film elements, a thin film of platinum is deposited onto a ceramic substrate.

Platinum is a common choice for RTD sensors because it is known for its long-term stability over time at high temperatures. Platinum is a better choice than copper or nickel because it is chemically inert, it withstands oxidation well, and works in a higher temperature range as well.

In operation, the measuring instrument applies a constant current through the RTD. As the temperature changes, the resistance changes and the corresponding change in voltage is measured. This measurement is then converted to thermal values by a computer. Curve-fitting equations are used to define this resistance vs. temperature relationship. The RTD can then be used to determine any temperature from its measured resistance.

A typical measurement technique for industrial thermometers involves sending a constant current through the sensor (0.8 mA to 1.0 mA), and then measuring the voltage generated across the sensor using digital voltmeter techniques. The technique is simple and few error-correcting techniques are applied.

In a laboratory where measurement accuracies of 10 ppm or better are required, specialized measurement equipment is used. High-accuracy bridges and digital voltmeters with special error-correcting functions are used. Accuracies of high-end measurement equipment can reach 0.1 ppm (parts per million). These instruments have functions to compensate for errors such as thermoelectric voltages and element self-heating.

In addition to temperature, strain on and impurities in the wire also affect the sensor's resistance vs. temperature characteristics. The Matthiessen rule states that the resistivity (ρ) of a metal conductor depends on temperature, impurities, and deformation. ρ is measured in (Ω cm):

$$\rho\left(\text{total}\right) = \rho\left(\text{temperature}\right) + \rho\left(\text{impurities}\right) + \rho\left(\text{deformation}\right) \tag{32.20}$$

Proper design and careful material selection will minimize these effects so that resistivity will only vary with a change in temperature.

Resistance of Metals

Whether an RTD's element is constructed of platinum, copper, or nickel, each type of metal has a different sensitivity, accuracy, and temperature range. Sensitivity is defined as the amount of resistance change of the sensor per degree of temperature change. Figure 32.7 shows the sensitivity for the most common metals used to build RTDs.

Platinum, a noble metal, has the most stable resistance-to-temperature relationship over the largest temperature range −184.44°C (−300°F) to 648.88°C (1200°F). Nickel elements have a limited temperature

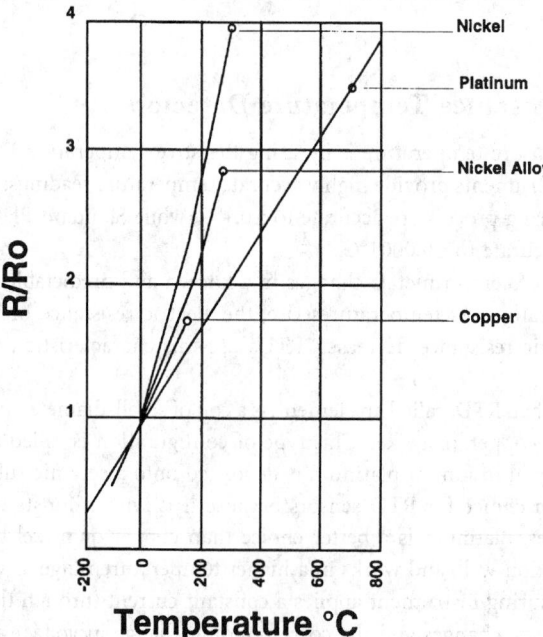

FIGURE 32.7   Of the common metals, nickel has the highest sensitivity.

TABLE 32.6

Probe	Basic application	Temperature	Cost	Probe style[a]	Handling
SPRT	Calibration of Secondary SPRT	−200 to 1000°C (−328 to 1832°F)	$5000	I	Very fragile
Secondary SPRT	Lab use	−200 to 500°C (−328 to 932°F)	$700	I, A	Fragile
Wirewound IPRT	Industrial field use	−200 to 648°C (−328 to 1200°F)	$60–$180	I, S, A	Rugged
Thin-film IPRT	Industrial field use	−50 to 260°C (−200 to 500°F)	$40–$140	I, S, A	Rugged

[a] I = immersion; A = air; S = surface.

range because the amount of change in resistance per degree of change in temperature becomes very nonlinear at temperatures above 300°C (572°F). Copper has a very linear resistance-to-temperature relationship. However, copper oxidizes at moderate temperatures and cannot be used above 150°C (302°F).

Platinum is the best metal for RTD elements for three reasons. It follows a very linear resistance-to-temperature relationship; it follows its resistance-to-temperature relationship in a highly repeatable manner over its temperature range; and it has the widest temperature range among the metals used to make RTDs. Platinum is not the most sensitive metal; however, it is the metal that offers the best long-term stability.

The accuracy of an RTD is significantly better than that of a thermocouple within an RTD's normal temperature range of −184.44°C (−300°F) to 648.88°C (1200°F). RTDs are also known for high stability and repeatability. They can be removed from service and recalibrated for verifiable accuracy and checked for any possible drift.

Who Uses RTDs? Common Assemblies and Applications

Different applications require different types of RTDs. A direct-immersion Platinum Resistance Thermometer (PRT) and a connection head can be used for low-velocity pipelines, tanks, or air temperature measurement. A spring-loaded PRT, thermowell, and connection head are often used in pipelines or storage tanks. An averaging temperature element senses and measures temperatures along its entire sheath, which can range from 1 to 20 m in length. A heavy-duty underwater temperature sensor is designed for complete submersion under rivers, cooling ponds, or sewers. These are just a few examples of RTD configurations and applications.

Overview of Platinum RTDs

There are three main classes of Platinum Resistance Thermometers (PRTs): Standard Platinum Resistance Thermometers (SPRTs), Secondary Standard Platinum Resistance Thermometers (Secondary SPRTs), and Industrial Platinum Resistance Thermometers (IPRTs). Table 32.6 presents information about each.

Temperature Coefficient of Resistance

Each of the different metals used for sensing elements (platinum, nickel, copper) has a different amount of relative change in resistance per unit change in temperature. A measure of a resistance thermometer's sensitivity is its temperature coefficient of resistance. It is defined as the element's change in resistance per degree C change in temperature per ohm of sensor resistance over the range of 0°C to 100°C.

The alpha value is the average change in resistance per degree C per ohm resistance. The actual change in resistance per degree C per ohm is largest at −200°C and decreases steadily as the use temperatures increase.

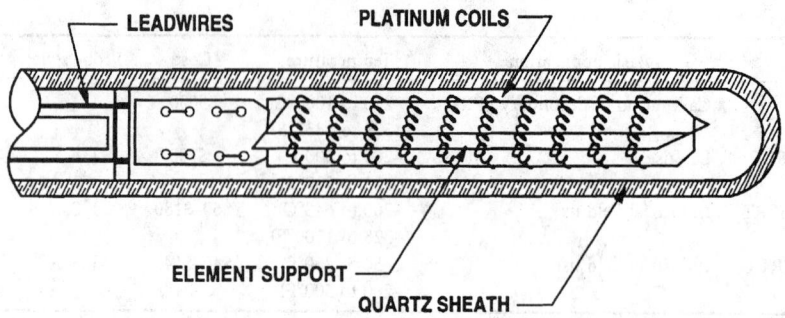

FIGURE 32.8   The Standard Platinum Resistance Thermometer is fragile and used only in laboratory environments.

The units for the coefficient are $\Omega/\Omega^{-1}/°C^{-1}$. This is called the alpha value and is commonly denoted by the Greek letter α. The larger the temperature coefficient, the greater the change in resistance for a given change in temperature. Of the commonly used RTD metals, nickel has the highest temperature coefficient, 0.00672, while that of copper is 0.00427. The α value of the sensor is calculated using the equation:

$$\alpha = \frac{R_{100} - R_0}{100°C \times R_0} \qquad (32.21)$$

where R_0 = the resistance of the sensor at 0°C
      R_{100} = the resistance of the sensor at 100°C

Three primary temperature coefficients are specified for platinum:

1. ITS-90, the internationally accepted temperature scale, requires a minimum temperature coefficient of 0.003925 for SPRTs. This is achieved using high-purity wire (99.999% or better) wound in a strain-free configuration.
2. With reference-grade platinum wire used in industrial elements, the temperature coefficient is 0.003902.
3. IEC 751 [1] and ASTM 1137 [2] have standardized the temperature coefficient of 0.0038500 for platinum.

RTD Construction

Standard Platinum Resistance Thermometers (SPRTs), the highest-accuracy platinum thermometers, are fragile and used in laboratory environments only (Figure 32.8). Fragile materials do not provide enough strength and vibration resistance for industrial environments. SPRTs feature high repeatability and low drift, but they cost more because of their materials and expensive production techniques.

SPRT elements are wound from large-diameter, high-purity platinum wire. Internal leadwires are usually made from platinum and internal supports from quartz or fused silica. SPRTs are used over a very wide range, from −200°C (−328°F) to above 1000°C (1832°F). For SPRTs used to measure temperatures up to 660°C (1220°F), the ice point resistance is typically 25.5 Ω. For high-temperature thermometers, the ice point resistance is 2.5 Ω or 0.25 Ω. SPRT probes can be accurate to ±0.001°C (0.0018°F) if properly used.

Secondary Standard Platinum Resistance Thermometers (Secondary SPRTs) are also intended for laboratory environments (Figure 32.9). They are constructed like the SPRT, but the materials are less expensive, typically reference-grade, high-purity platinum wire, metal sheaths, and ceramic insulators. Internal leadwires are usually a nickel-based alloy. The secondary grade sensors are limited in temperature

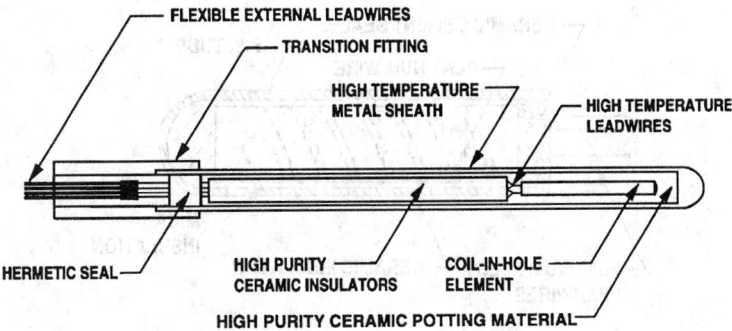

FIGURE 32.9   The Secondary Standard Platinum Resistance Thermometer is intended for laboratory environments.

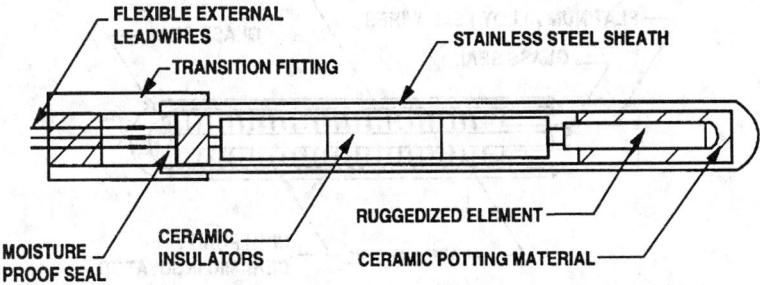

FIGURE 32.10   Industrial Platinum Resistance Thermometers are almost as durable as thermocouples.

range — –200°C (–328°F) to 500°C (932°F) — and are accurate to ±0.03°C (±0.054°F) over their temperature range.

Secondary standard thermometers can withstand some handling, although they are still quite strain-free. Rough handling, vibration, and shock will cause a shift in calibration. The nominal resistance of the ice point is most often 100 Ω. This simplifies calibration procedures when calibrating other 100-Ω RTDs. The temperature coefficient for secondary standards using reference-grade platinum wire is usually 0.00392 Ω Ω⁻¹ °C⁻¹ or higher.

Industrial Platinum Resistance Thermometers (IPRTs) are designed to withstand industrial environments and are almost as durable as thermocouples (Figure 32.10). IEC 751 [1] and ASTM 1137 [2] standards cover the requirements for industrial platinum resistance thermometers. The most common temperature range is –200°C (–328°F) to 500°C (932°F). Standard models are interchangeable to an accuracy of ±0.25°C (±0.45°F) to ±2.5°C (±4.5°F) over their temperature range.

Several element designs are available for different applications. One common configuration is the wirewound element (Figure 32.11). This durable design was developed as a substitute for the fragile SPRT. The small platinum sensing wire (usually within 7 to 50 µm (0.0003 in. to 0.002 in. diameter) is noninductively wound around a cylindrical ceramic mandrel, and usually covered with a thin layer of material that provides electrical insulation and mechanical protection. Because the sensing element wire is firmly supported, it cannot expand and contract as freely as the SPRT's relatively unsupported platinum wire. This type of element offers higher durability than SPRTs and secondary standards, and very good accuracy for most industrial applications.

In another wirewound design, the coil suspension, a coil of fine platinum wire is assembled into small holes in a cylindrical ceramic mandrel (Figure 32.12). The coils are supported by ceramic powder or cement, and sealed at both ends. When ceramic powder is loosely packed in the bores of the mandrel,

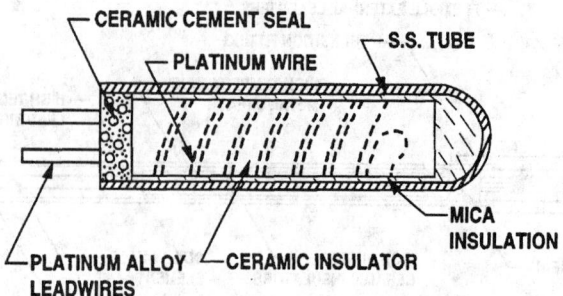

FIGURE 32.11 The wirewound RTD is noninductively wound around a cylindrical ceramic mandrel.

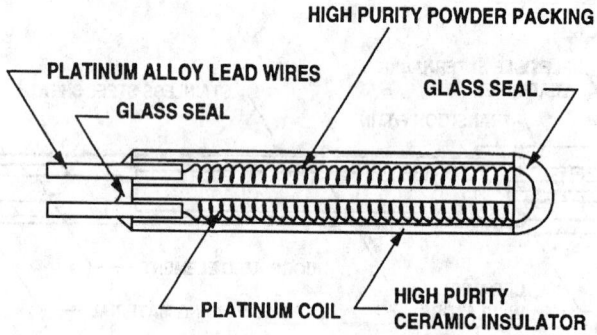

FIGURE 32.12 The coil suspension RTD has a coil of wire assembled into small holes.

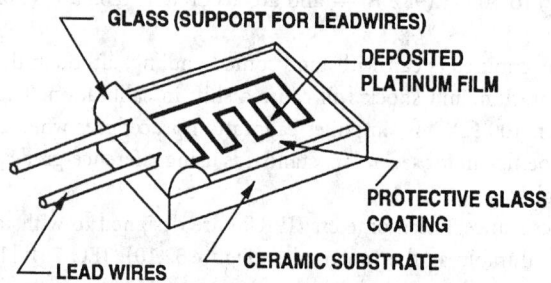

FIGURE 32.13 Thin-film elements have a thin film of platinum deposited onto a ceramic substrate.

the element can expand and contract freely. This reduces the effects of strain on the resistance characteristics, resulting in very high accuracy and stability for use in secondary temperature standards and
docile industrial applications (with little or no vibration or shock). Recent improvements in ceramic
materials give the sensing coil more stability — it will be capable of maintaining accuracies of 0.03°C
after thousands of hours at temperatures of 500°C. These ceramic powders support the coils in the
mandrel bores and hold them firmly in place with minimum strain.

Thin-film elements are extremely small, often less than 1.6 mm (1/16 in.) square (Figure 32.13). They
are manufactured by similar techniques employed to make integrated circuits. First, a thin film of
platinum is deposited onto a ceramic substrate. Some manufacturers use photolithography to etch the
deposited platinum, leaving the element pattern on the ceramic substrate. Then, the element's surfaces
are covered with glass material to protect the elements from humidity and contaminants.

The temperature range of thin film platinum elements is –50°C (–58°F) to 400°C (752°F); accuracy is from 0.5°C (0.9°F) to 2.0°C (3.6°F). The most common thin-film element has a 100-Ω ice point resistance and a temperature coefficient of 0.00385°C.

Thin-film RTDs can be extremely durable if the small-diameter leadwires and the thin element are properly protected. The accuracy and stability might not be as good as some wirewound elements due to hysteresis, long-term stability errors, and self-heating errors.

Self-heating Errors

The current that measures sensor resistance also heats the sensor. This is known as I^2R* heating or Joule heating. Because of this effect, the sensor's indicated temperature is somewhat higher than the actual temperature. This inconsistency is commonly called self-heating error. Self-heating errors, which are dependent on the application, can range from negligible values to 1°C. The greatest heating errors occur because of poor heat transfer between the sensing element and application, or excessive current used in measuring resistance.

The following are methods for reducing the self-heating error.

1. Minimize the power dissipation in the sensor. There is a tradeoff between the signal level and the self-heating of the sensor. Typically, 1 mA of current is used as the sensing current.
2. Use a sensor with a low thermal resistance. The lower the thermal resistance of the sensor, the better the sensor can dissipate the I^2R power and the lower the temperature rise in the sensor. Small time constants indicate a sensor with a low thermal resistance.
3. Maximize thermal contact between the sensor and the application.

Calibration

Testing programs are essential to verify the accuracy of PRTs. Some IPRTs are factory-calibrated to a temperature such as at the ice point, but PRT users might want to calibrate them at other temperatures, depending on their application.

The calibration results can be compared to prior calibrations from the same instrument. This will determine if it is necessary to repair or replace the instrument, or if calibration is required more often.

Since frequent repairs and recalibration are usually costly, purchasers and specifiers of PRTs might want to investigate various RTDs before installation by referring to ASTM Standard #E1137-95 [2], published by The American Society for Testing and Materials (ASTM).

Frequency. An RTD's stability depends on its working environment. High temperatures can cause drift and contamination of the platinum wire. The higher the temperature, the faster the drift occurs. Below 400°C, the high-temperature drift is not a significant problem, but temperatures reaching 500°C to 600°C are the most significant causes of drift — up to several degrees per year. Severe shock can damage a sensor instantly and cause failure. Shock, vibration, and rough handling will put strains in the platinum wire and change its characteristics, and ultimately damage the entire unit. If a sensor is not properly sealed, humidity can get into the sensor and cause some problems with the insulation resistance. Since the water in humidity is conductive, it will get between the lead wires and the sensing element, and basically shunt off the resistance of the element's wires. Under extreme operating conditions, a sensor should be calibrated on a monthly or bimonthly basis. If five or more calibrations are completed without a significant change, then the time between calibrations can be doubled; at least once a year is recommended, however.

Techniques. Two common calibration methods are the fixed point method and the comparison method.

Fixed point calibration, used for the highest accuracy calibrations, uses the triple point, freezing point or melting point of pure substances such as water, zinc, tin, and argon to generate a known and repeatable temperature (Figure 32.14). The fixed point cells are sealed to prevent contamination and to prevent atmospheric conditions from affecting the cell's temperature. These cells allow the user to reproduce actual conditions of the ITS-90 temperature scale.

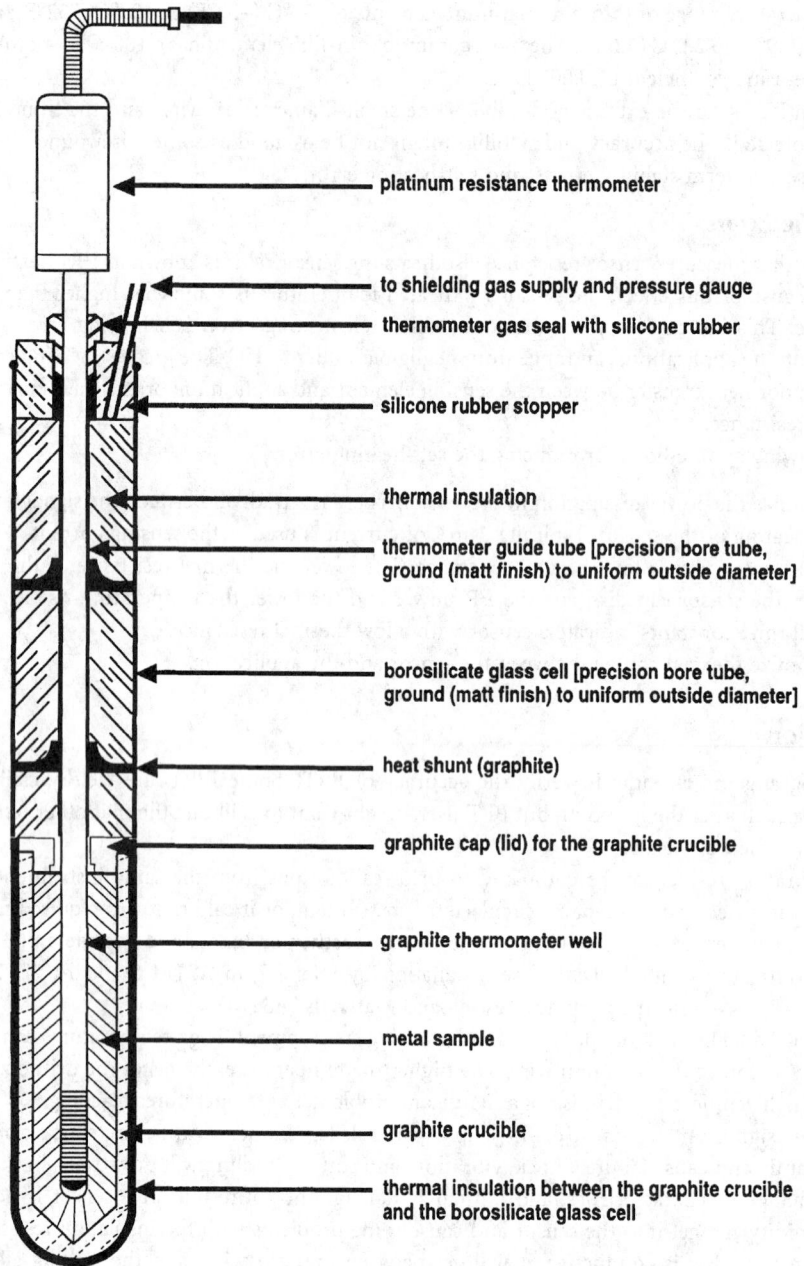

platinum resistance thermometer

to shielding gas supply and pressure gauge

thermometer gas seal with silicone rubber

silicone rubber stopper

thermal insulation

thermometer guide tube [precision bore tube,
ground (matt finish) to uniform outside diameter]

borosilicate glass cell [precision bore tube,
ground (matt finish) to uniform outside diameter]

heat shunt (graphite)

graphite cap (lid) for the graphite crucible

graphite thermometer well

metal sample

graphite crucible

thermal insulation between the graphite crucible
and the borosilicate glass cell

FIGURE 32.14   Fixed point calibration uses the triple point, freezing point, or melting point of water, zinc, tin, and argon.

Fixed point calibrations provide extremely accurate calibrations (within ±0.001°C), but the cells are time-consuming to use and can only accommodate one sensor at a time. For this reason, they are not widely used in calibrating industrial sensors. Each fixed point cell has a unique procedure for achieving the fixed point.

A generalized procedure for fixed point calibration is as follows:

1. Prepare the cell. Different procedures exists for each fixed point.
2. Insert the thermometer to be calibrated.

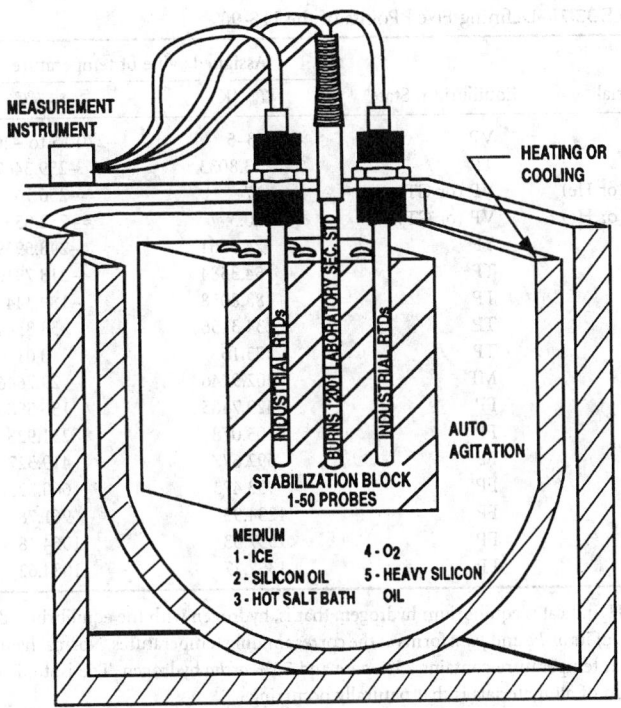

FIGURE 32.15   An isothermal bath permits calibration of industrial RTDs compared with a secondary standard.

3. Allow the system to stabilize. Stabilization times depend on thermometer/fixed point cells. Usually, 15 to 30 min is sufficient.
4. Measure the resistance of the thermometer. For the highest accuracy measurements, special resistance bridges are used. They have accuracies in the range of 10 ppm to 0.1 ppm.

A common fixed point calibration method for industrial-grade probes is the ice bath. The equipment is inexpensive, easy to use, and can accommodate several sensors at once. The ice point is designated as a secondary standard because its accuracy is ±0.005°C (±0.009°F), compared to ±0.001°C (±0.0018°F) for primary fixed points.

In *comparison calibrations*, commonly used with secondary SPRTs and industrial RTDs, the thermometers being calibrated are compared to calibrated thermometers by means of an isothermal bath whose temperature is uniformly stable (Figure 32.15). Unlike fixed point calibrations, comparisons can be made at any temperature between −100°C (−148°F) and 500°C (932°F). This method might be more cost-effective since several sensors can be calibrated simultaneously with automated equipment.

These isothermal baths, electrically heated and well-stirred, use silicone oils as the medium for temperatures ranging from −100°C (−148°F) to 200°C (392°F), and molten salts for temperatures above 200°C (392°F). At temperatures above 500°C (932°F), air furnaces or fluidized beds are used, but are significantly less uniformly stable.

The procedure for comparison calibration is as follows:

1. Insert the standard thermometer and thermometers being calibrated into the bath.
2. Allow the bath to stabilize.
3. Measure the resistance of the standard to determine the temperature of the bath.
4. Measure the resistance of each thermometer under calibration.

Deriving the resistance vs. temperature relationship of a PRT. After determining the PRT's resistance, the calibration coefficients can be determined. By plugging these values into an equation, temperature

TABLE 32.7 Defining Fixed Points of the ITS-90

Material[a]	Equilibrium State[b]	Assigned value of temperature	
		T_{90} (K)	t_{90} (°C)
He	VP	3–5	−270.15 to −268.15
e-H$_2$	TP	13.8033	−259.3467
e-H$_2$ (or He)	VP (or GT)	≈17	≈−256.16
e-H$_2$ (or He)	VP (or GT)	≈20.3	≈−252.85
Ne	TP	24.5561	−248.5939
O$_2$	TP	54.3584	−218.7916
Ar	TP	83.8058	−189.3442
Hg	TP	234.3156	−38.8344
H$_2$O	TP	273.16	0.01
Ga	MP	302.9146	29.7646
In	FP	429.7485	156.5985
Sn	FP	505.078	231.928
Zn	FP	692.677	419.527
Al	FP	933.473	660.323
Ag	FP	1234.93	961.78
Au	FP	1337.33	1064.18
Cu	FP	1357.77	1084.62

[a] e-H$_2$ indicates equilibrium hydrogen; that is, hydrogen with the equilibrium distribution of its *ortho* and *para* forms at the corresponding temperatures. Normal hydrogen at room temperature contains 25% *para* and 75% *ortho* hydrogen. The isotopic composition of all materials is that naturally occurring.

[b] VP indicates vapor pressure point or equation; GT indicates gas thermometer point; TP indicates triple point; FP indicates freezing point; MP indicates melting point.

from any measured resistance can be derived. The two most common curve-fitting techniques are the ITS-90 and Callendar–Van Dusen equations.

On January 1, 1990, the International Temperature Scale of 1990 (ITS-90) became the official international temperature scale [3]. ITS-90 extends upward from 0.65 K (−272.5°C or −458.5°F) and defines temperatures of 0.65 K (0.65°C above absolute zero) and up by fixed points (see Table 32.7).

Two reference functions are used to define the temperature coefficient for an ideal SPRT: one for temperatures below 0°C and the other for temperatures above 0°C. When a PRT is calibrated on the ITS-90, the coefficients determined in the calibration are used to describe a deviation function that represents the difference between the resistance of the standard PRT and the reference function at all temperatures within the range. Using the calibration coefficients and the deviation functions, the SPRT can be used to determine any temperature from its measured resistance. Because ITS-90 equations are complex, computer software is necessary for accurate calculations.

ITS-90 affects:

- Standards and temperature calibration laboratories
- Users of standard and secondary SPRTs with traceability to standards laboratories
- Users of temperature measurement and control systems within companies concerned with verifiable total quality management

The National Institute for Standards and Technology (NIST) has published Technical Note 1265, *Guidelines for Realizing the International Temperature Scale of 1990* [4]. Not all PRT users need to follow the complex equations and computer programs associated with ITS-90. As a rule of thumb: if the minimum required uncertainty of measurement is less than 0.1°C, one probably will want to use ITS-90. For uncertainty of measurements greater than 0.1°C, the effect of the change in scales is relatively small and one will not be affected.

Callendar–Van Dusen equations are interpolation equations that describe the temperature vs. resistance relationship of industrial PRTs. These simple-to-use second- and fourth-order equations can be programmed easily into many electronic controllers. The equation for the temperature range of 0°C (32°F) to 850°C (1562°F) is:

$$R(t) = R_0 \left(1 + At + Bt^2\right) \tag{32.22}$$

For the temperature range −200°C (−392°F) to 0°C (32°F):

$$R(t) = R_0 \left[1 + At + Bt^2 + C(t - 100)t^3\right] \tag{32.23}$$

where $R(t)$ = Resistance of the PRT at temperature t
  t = Temperature in °C
  R_0 = Nominal resistance of the PRT at 0°C
  α, δ, β = Calibration coefficients

To determine the temperature from a measured resistance, a different set of equations and calibration coefficients is required. For temperatures greater than 0°C (measured resistances greater than the known ice point resistance of the PRT):

$$t(°C) = \left[(R_t - R_0)/(\alpha R_0)\right] + \delta\left[(t/100) - 1\right](t/100) \tag{32.24}$$

For temperatures less than 0°C (measured resistances less than the known ice point resistance of the PRT):

$$t(°C) = \left[(R_t - R_0)/(\alpha R_0)\right] + \delta\left[(t/100) - 1\right](t/100) + \beta\left[(t/100) - 1\right](t/100)^3 \tag{32.25}$$

where t = Temperature to be calculated
  $R(t)$ = Measured resistance at unknown temperature
  R_0 = Resistance of the sensor at 0°C
  α, δ, β = Coefficients

To correctly determine the temperature from a given resistance with these equations, one must iterate the equations a minimum of five times. After each calculation, the new value of temperature (t) is plugged back into the equations. The calculated temperature value will converge on its true value. After five iterations, the calculated temperature should be within ±0.001°C of the true value.

For industrial sensors, an alternative method would be to use nonlinear least squares curve fits to produce the temperature/resistance relationship. However, these methods should not be used for secondary and primary level thermometers as they cannot sufficiently match the defined ITS-90 scale. Curve fitting errors of up to 0.05°C are possible.

Usage of RTDs Today

Throughout the industry, usage of RTDs is increasing for many reasons. With the advent of the computer age, industries recognize the need for better temperature measurement, and an electrical signal to accompany advances in computerized process instrumentation. RTDs produce an electrical signal; and because of automatic control in industrial plants, it makes it simpler and easier to interface with process controllers. With the focus on reengineering, companies are searching for ways to improve processes. Improved temperature measurement and control is one good way to save energy, reduce material waste, reduce expenses and improve overall operating efficiencies.

Governmental regulations are another reason why RTDs are gaining popularity. In the pharmaceutical industry, the FDA [5] requires validation; among them the verification that temperature measurement is accurate. New regulations are currently being written for the food industry as well [6].

The growing worldwide acceptance of ISO 9000 standards has forced companies to calibrate their temperature measurement systems and instrumentation. In addition, the calibration must be documented, and must be traceable to a recognized national, legal standard.

RTDs are safer for the environment. With mercury thermometers, disposal of mercury is a problem. In many industries, the mere presence of mercury thermometers presents a risk.

Examples of Advanced Applications for Critical Temperature Measurement

RTD technology allows for custom design in a wide range of applications and industries. In many of these cases, the RTD becomes an integral part of an advanced application when temperature is critical.

Power plants use RTD sensors to monitor fuel and coolant temperatures entering and exiting heat exchangers. Accurate temperature measurement is also critical for nuclear power plants to perform pressure leak tests on the containment vessel surrounding the reactor core.

Microprocessor manufacturers require precise temperature control throughout their clean room areas. Air temperature is critical to production; many temperature measurement points need to be accurate ±0.028°C within (±0.05°F). To achieve this, an RTD is calibrated with a temperature transmitter. This matched pair ensures the highest level of system accuracy and eliminates the interchangeability error of the RTD.

The Future of RTD Technology

The future of RTD technology is driven by end-user needs and unsolved problems. For example, the need for high-temperature industrial RTDs exists for applications above 600°C (1200°F). In order to function at high temperatures, the RTD's platinum element must be protected from contamination. However, the sheath material can be a problem because, at high temperatures, it will react with the oxygen in the air and give off metal particles that can attach themselves to the platinum.

RTDs must be mechanically strong enough to survive higher temperatures as well. A high-temperature industrial RTD would require a thermally resistant sheath, perhaps Inconel 600. In addition, sensor components must be designed and manufactured to resist ultra-high temperatures. Some RTDs are specified to operate above 600°C (1200°F), but when tested, are not always reliable. Drift and nonrepeatability are problem areas that are in need of further attention.

Advances in RTD calibration provide significant improvements for temperature measurement and control. New measurement technologies combined with powerful computational techniques, have simplified the calibration process and made it more reliable.

Another current area of growth is in-house calibration and calibration baths. Governmental validation requirements and ISO-9000 standards are the driving forces in this area. Because third-party calibration services are expensive, companies want to be more productive and more cost-efficient. Wong [7] discusses the benefits of setting up an in-house calibration lab in addition to traceability concepts. Traceability refers to an unbroken chain of comparisons, linking the temperature measurement to a recognized national, legal standard. In the U.S., this national standard is maintained by the National Institute of Standards and Technology (NIST). All RTD manufacturers, laboratories, and calibration labs must adjust their standards to meet NIST standards.

Defining Terms

Accuracy: The degree of agreement between an actual measurement and its reference standard.

Alpha (α):   The temperature coefficient of resistance of a PRT over the range 0°C to 100°C. For example, α for a standard platinum resistance thermometer (SPRT) is 0.003925.

Calibrate: To check, adjust or determine an RTD's accuracy by comparing it to a standard.

DIN (Deutsche Industrial Norm):   A German organization that develops technical, scientific, and dimensional standards that are recognized worldwide.

Error: The difference between a correct value and the actual reading taken.

Primary Standard (or Standard PRT): A platinum resistance thermometer that meets the requirements for establishing calibrations according to the ITS-90. This highly accurate instrument is intended for laboratory environments and is accurate to 0.001°C.

Reliability: Used to designate precision for measurements made within a very restricted set of conditions.

Repeatability: The ability to give the same measurement under repeated, matching conditions.

Stability: The state of being resistant to change or deterioration.

Sensitivity: The amount of resistance change of the sensor per degree temperature change.

References

1. IEC, Industrial platinum resistance thermometer sensors, IEC International Standard 751.1995-07, Genéve, Suisse: Bureau Central de la Commission Electrotechnique Internationale, 1995.
2. ASTM Standards, Standard Specification for Industrial Platinum Resistance Thermometers, Standard E 1137-95, 1995.
3. H. Preston-Thomas, The International Temperature Scale of 1990 (ITS-90), *Metrologia*, 27(1), 3-10, 1990. For errata, see *ibid*, 27(2), 107, 1990.
4. NIST Technical Note 1265, Guidelines for Realizing the International Temperature Scale of 1990 (ITS-90), National Institute of Standards and Technology, 1990.
5. FDA validation for the pharmaceutical industry.
6. FDA validation for the food industry.
7. W. Wong, Traceability tops in-house calibration, *InTech*, 41(8), 27-29, 1994.

32.3 Thermistor Thermometers

Meyer Sapoff

A thermistor is a thermally sensitive resistor whose primary function is to exhibit a change in electric resistance with a change in body temperature. Unlike a wirewound or metal film resistance temperature detector (RTD), a thermistor is a ceramic semiconductor. An RTD exhibits a comparatively low temperature coefficient of resistance on the order of 0.4 to 0.5% °C^{-1}. Depending on the type of material system used, a thermistor can have either a large positive temperature coefficient of resistance (PTC device) or a large negative temperature coefficient of resistance (NTC device).

Two types of PTC thermistors are available. Silicon PTC thermistors rely on the bulk properties of doped silicon and exhibit resistance–temperature characteristics that are approximately linear. They have temperature coefficients of resistance of about 0.7 to 0.8% °C^{-1}. The most common application of silicon PTC thermistors is compensation of silicon semiconductor devices and circuits [1]. The materials used for switching-type PTC thermistors are compounds of barium, lead, and strontium titanates. Figure 32.16 shows the resistance–temperature characteristic of a typical switching-type PTC thermistor [2]. At low temperatures, from below 0°C to R_{min}, the resistance value is low, and R_T vs. T exhibits a small negative temperature coefficient of resistance on the order of −1% °C^{-1}. As the temperature increases, the temperature coefficient of resistance becomes positive and the resistance begins to rise. At a threshold or switching temperature, the rate of rise becomes very rapid and the PTC characteristic becomes very steep. Within its switching range, the temperature coefficient of resistance can be as high as 100% °C^{-1} and the device exhibits a high resistance value. At temperatures above the switching range, the resistance reaches a maximum value beyond which the temperature coefficient becomes negative again. The switching temperature can be varied between 80°C and 240°C by altering the chemical composition of the ceramic. Typical applications for switching-type PTC thermistors are over-temperature protection, current limiting, and self-regulated heating. The temperature coefficient of resistance of a unit used as a heating element typically is about 25% °C^{-1} at a switching temperature of 240°C [1, 2].

NTC thermistors consist of metal oxides such as the oxides of chromium, cobalt, copper, iron, manganese, nickel, and titanium. Such units exhibit a monotonic decrease in electric resistance with an

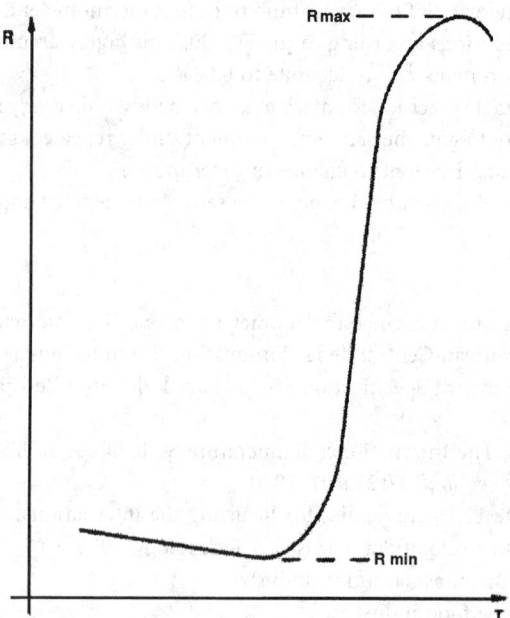

FIGURE 32.16 Resistance–temperature characteristic of a typical swtiching-type PTC thermistor. At low temperatures, from below 0°C to R_{min}, the resistance value is low and R_T vs. T exhibits a small negative temperature coefficient of resistance on the order of −1% °C⁻¹. As the temperature increases, the temperature coefficient of resistance becomes positive and the resistance begins to rise. At a threshold or switching temperature, the rate of rise becomes very rapid and the PTC characteristic becomes very steep. Within its switching range, the device exhibits a high resistance value. At temperatures above the switching range, the resistance reaches a maximum value beyond which the temperature coefficient becomes negative again.

increase in temperature. The resistance–temperature characteristics of NTC thermistors are nonlinear and approximate the characteristics exhibited by intrinsic semiconductors for which the temperature dependence of resistance is due to the excitation of carriers across a single energy gap. As such, the logarithm of resistance of an NTC thermistor is approximately a linear function of its inverse absolute temperature. Below room temperature, the slope of the function decreases and the thermistor behaves more like an extrinsic semiconductor. The actual conduction mechanism is comparable to the "hopping" mechanism observed in ferrites and manganites that have a spinel crystal structure. Conduction occurs when charge carriers hop from one ionic site in the spinel lattice to an adjacent site. Such hopping can occur when ions of the same element, with valences differing by 1, are present on equivalent lattice sites.

Because of its nonlinear resistance–temperature characteristic, the temperature coefficient of resistance of an NTC thermistor changes with temperature. Depending on the material system used, the temperature coefficient at 25°C typically is in the range of −3 to −5% °C⁻¹. At −60°C, it is in the range of −6.4 to −11.3% °C⁻¹; and at 100°C, it varies between −2.1 and −3.7% °C⁻¹. The corresponding resistance ratios with respect to 25°C, R_T/R_{25}, are 41 to 228 at −60°C and 0.13 to 0.03 at 100°C. The slope of the log R vs. $1/T$ characteristic β is relatively constant. The material systems for which the above data are presented exhibit β values of 2930 to 5135 K over the range of 25 to 125°C [1].

Although the resistance–temperature characteristic of an NTC thermistor is nonlinear, it is possible to achieve good linearity of the conductance–temperature and resistance–temperature characteristics using thermistor–resistor networks. The use of a resistor in series with a thermistor results in a linear conductance network, such as a voltage divider. The current obtained in response to a voltage applied to the network is linear with temperature. Consequently, the output of the voltage divider (voltage across the resistor) is linear with temperature. The parallel combination of a resistor and thermistor is a linear resistance network. Techniques exist for optimizing the linearity of single thermistor networks [3, 4].

Such networks exhibit S-shaped curves for their voltage–temperature or resistance–temperature characteristics. For temperature spans of 20°C to 30°C, the maximum deviation is small between any point on the curve and the best straight-line approximation of the curve. As the span increases beyond 30°C, the maximum deviation from linearity increases rapidly from about 0.15°C for a 30°C span to 0.7°C for a 50°C span. Using a two-thermistor network, a maximum linearity deviation of 0.22°C can be obtained over a range of 0°C to 100°C, as compared with a 2°C deviation obtained with a single-thermistor network [5]. Applications for NTC thermistors include temperature measurement, control, and compensation.

When current flows through a thermistor, there is a self-heating effect caused by the power dissipated in it. The thermistor temperature rises until the rate of heat loss from the thermistor to its surrounding environment equals the power dissipated. For this condition of thermal equilibrium, the temperature rise is directly proportional to the power dissipated in the thermistor. The constant of proportionality δ is the dissipation constant of the thermistor. By definition, δ is the ratio — at a specified ambient temperature, mounting condition, and environment — of the change in power dissipation in a thermistor to the resultant body temperature change. Factors that affect the dissipation constant are the thermistor surface area, the thermal conductivity and relative rate of motion of the medium surrounding the thermistor, the heat loss due to free convection in a still medium, the heat transfer between the thermistor and its mount through the thermistor lead wires, and heat loss due to radiation (significant for gases at low pressure). Because the thermal conductivities of fluids vary with temperature and free convection depends on the temperature difference between the thermistor and its ambient, δ is not a true constant. For gases in particular, the dissipation constant varies both with the thermistor body temperature and the amount of self-heating.

The thermal time constant τ is the 63.2% response to a step-function change in the thermistor temperature when the power dissipated in the thermistor is negligible. The thermal time constant only has meaning when there is a single exponential response. In such a case, an elapsed time of τ results in a 63.2% change between the initial and final steady-state temperatures, while an elapsed time of 5τ results in a 99.3% change. For a more complex structure, such as a thermistor encapsulated in a sensor housing, multiple exponentials can exist and one cannot predict that the 99% response will occur after an interval of five time constants. As with the dissipation constant, the thermal time constant is dependent on the rate of heat transfer between the thermistor and its environment. Consequently, all of the factors that increase the dissipation constant decrease the time constant.

The most common types of silicon PTC devices are the glass diode package having diameters of 1.8 mm to 2.5 mm and lengths of 3.8 mm to 7.5 mm, and the molded epoxy package with diameters of 3.6 mm to 6.0 mm and lengths of 10.4 mm to 15.0 mm. Such units have axial leads. Epoxy-coated chips and disks with radial leads also are available, with diameters of approximately 3 mm. Silicon PTC thermistors that comply with MIL-T-23648 have an operating temperature range of –55°C to 125°C. Commercial versions can be obtained with a range of –65°C to 150°C [6].

The most common configuration for switching-type PTC thermistors is the radial lead disk, with and without an insulating coating. Such units are available in diameters of 4 mm to 26 mm. The thickness dimension is in the range between 0.5 mm to 6.5 mm. They also are available as surface-mount devices, disks without leads, and in glass diode packages. Switching-type PTC disks have a storage temperature range of –25°C to 155°C and an operating range of 0°C to 55°C [1].

The emphasis of this text will be on NTC thermistors. Most thermistor applications use such devices. Two major categories of NTC thermistors exist. Bead-type thermistors have platinum alloy lead wires sintered into the body of the ceramic. Chip, disk, surface-mount, flake, and rod-type thermistors have metallized surface electrodes. The latter category includes glass diode packages in which dumet leads are compression-bonded to disks or chips.

Glass-coated beads include both adjacent and opposite lead configurations with diameters of 0.125 mm to 1.5 mm. Glass probes with diameters of 0.4 mm to 2.5 mm have lengths ranging from 1.5 mm to 6.35 mm. Probes with diameters of 1.5 mm to 2.5 mm have lengths of 3 mm to 12.7 mm, while larger probes with diameters of 2 mm to 2.5 mm have lengths of up to 50 mm. Glass rods typically are 6.3 mm long with diameters of 1.5 mm to 2.5 mm. A glass probe consists of a bare-bead-type thermistor sealed at the tip of a solid glass rod and radial dumet leads. A glass rod, by comparison, has its bead sealed in the center of a solid glass rod and has axial dumet leads. The temperature range specified in MIL-T-23648 is −55°C to 200°C for glass-coated beads and −55°C to 275°C for glass probes and glass rods. Commercial versions of such glass-enclosed devices typically have a range of −80°C to 300°C. Some units are rated for intermittent operation at 600°C, while special cryogenic devices are rated for operation in the range of −196°C to 25°C.

Chip thermistors with radial leads are available with cross-sections of 0.25 mm × 0.25 mm to 13 mm × 13 mm and a thickness range of 0.175 mm to 1.5 mm. Disks having diameters of 1 mm to 25 mm are available in both radial and axial lead configurations, with a thickness range of 0.25 mm to 6.35 mm. Chips and disks also are available in glass diode packages with axial dumet leads. Surface-mount chips are available in sizes ranging from 2 mm to 3.2 mm long × 1.1 mm to 1.6 mm wide × 0.36 mm to 1.3 mm thick. Flake thermistors with both adjacent and opposite leads are available in a thickness range of 0.025 mm to 0.125 mm, with cross-sections of 0.5 mm × 0.5 mm to 3 mm × 3 mm. The temperature range specified for chips, disks, rods, and glass diode packages in MIL-T-23648 is −55°C to 125°C. Commercial chips and disks are available with an operating temperature range of −80°C to 155°C [1]. Rods are available for use over the range of −60°C to 150°C [7], and glass diode packages are available with an operating range of −60°C to 300°C [8].

Thermal Properties of NTC Thermistors

The energy dissipated as heat in a thermistor connected to an electric circuit causes the thermistor body temperature to rise above the ambient temperature of its environment. At any instant, the applicable heat transfer equation is:

$$\frac{dH}{dt} = P = E_T I_T = \delta\left(T - T_a\right) + cm\frac{dT}{dt} \tag{32.26}$$

where $dH/dt = P = E_T I_T$ = Rate of thermal energy or heat supplied to the thermistor
$\quad\quad \delta$ = Dissipation constant
$\quad\quad T_a$ = Ambient temperature
$\quad\quad \delta\,(T - T_a)$ = Rate of heat loss to the surrounding environment
$\quad\quad c$ and m = Respectively, specific heat and mass of the thermistor
$\quad\quad cm(dT/dt)$ = Rate of heat absorbed by the thermistor

The solution of Equation 32.26 when P is constant is:

$$T = T_a + \frac{P}{\delta}\left[1 - \exp\left(-\frac{\delta}{cm}t\right)\right] \tag{32.27}$$

The transient solution of Equation 32.27 is the basis for the current–time characteristic of a thermistor. When, $t \gg cm/\delta$, then $dT/dt \to 0$, and the steady state solution of Equation 32.27 becomes:

$$P = E_T I_T = \delta\left(T - T_a\right) \tag{32.28}$$

where E_T and I_T are the steady-state thermistor voltage and current, respectively. Equation 32.28 is the basis for the voltage–current characteristic of a thermistor.

By reducing the thermistor power to a value that results in negligible self-heating, $P \to 0$, Equation 32.26 becomes:

$$\frac{dT}{dt} = -\frac{\delta}{cm}\left(T - T_a\right) \tag{32.29}$$

TABLE 32.8 Thermal Properties of Hermetically Sealed Beads and Probes

Style	Diameter (mm) Still air	Dissipation constant (mW °C⁻¹) Still water	Time constant (s) Still air	Water plunge	
Glass-coated bead	0.13	0.045	0.45	0.12	0.005
Glass-coated bead	0.25	0.09	0.9	0.5	0.010
Glass-coated bead	0.36	0.10	0.98	1.0	0.015
Ruggedized bead	0.41	0.12	1.1	1.2	0.016
Glass probe	0.63	0.19	1.75	1.9	0.020
Glass-coated bead	0.89	0.35	3.8	4.5	0.10
Glass-coated bead	1.1	0.40	4.0	5.5	0.14
Ruggedized bead	1.4	0.51	4.3	7.0	0.20
Glass probe	1.5	0.72	4.4	12.0	0.30
Glass probe	2.16	0.90	4.5	16.0	0.40
Glass probe	2.5	1.0	4.5	22.0	0.65

From Reference [1].

TABLE 32.9 Thermal Properties of Thermistors with Metallized Surface Electrodes

Style	Diameter (mm)	Dissipation constant (mW °C⁻¹)	Time constant (s)
Chip or disk in glass diode package	2	2–3	7–8
Interchangeable epoxy coated chp or disk	2.4	1	10
Disk with radial or axial leads	2.5	3–4	8–15
Disk with radial or axial leads	5.1	4–8	13–50
Disk with radial or axial leads	7.6	7–9	35–85
Disk with radial or axial leads		10.28–11	28–150
Disk with radial or axial leads		12.75–16	50–175
Disk with radial or axial leads		19.615–20	90–300
Disk with radial or axial leads		25.424–40	110–230
Disk with radial or axial leads	1.3	25–3	16–20
Rod with radial or axial leads	1.8	4–10	35–90
Rod with radial or axial leads	4.4	8–24	80–125

From References [1, 6, 9, 10].

Equation 32.29 is a mathematical statement of Newton's law of cooling and has the solution:

$$T = T_a + \left(T_i - T_a\right)\exp\left(-\frac{t}{\tau}\right)$$

where T_i = Initial thermistor body temperature
T_a = Ambient temperature
$\tau = cm/\delta$ = Thermal time constant

Table 32.8 gives dissipation and time constant data for glass-coated beads, glass probes, and glass rods. Table 32.9 lists similar data for chip, disk, and rod-type thermistors with metallized surface contacts. Data for the dissipation and time constants are for the thermistor suspended by its leads at 25°C using the procedure specified in MIL-T-23648A. The temperature increment used for computing the dissipation constant results from self-heating the thermistors to 75°C from an ambient of 25°C. The time constant data result from allowing the thermistors to cool from 75°C to an ambient of 25°C. The water-plunge time constant data result from rapidly immersing the thermistors from room temperature air into still water. The transit time was negligible. The air temperature was approximately 25°C and the water temperature was about 5°C. The diameters specified for epoxy-coated interchangeable units are maximum diameters. To illustrate the effect of the environment on the thermal properties of these units, the

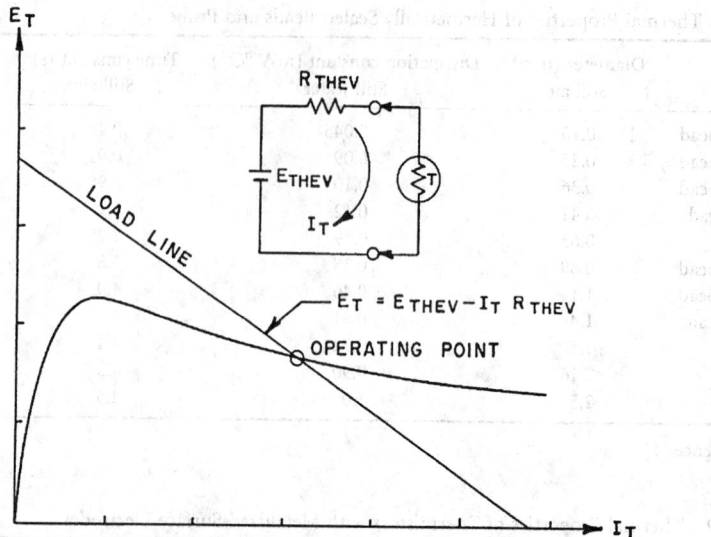

FIGURE 32.17 Typical voltage–current characteristic of an NTC thermistor. Due to self-heating of the thermistor, the slope of the E_T vs. I_T curve decreases with increasing current. This continues until a maximum value of E_T is reached for which the slope is equal to zero. Beyond this value, the slope continues to decrease and the thermistor exhibits a negative resistance characteristic. The Thevenin equivalent circuit with respect to the thermistor terminals provides a straight-line relationship between E_T and I_T that can be plotted as a load line. The intersection of the load line with the E_T vs. I_T curve is the operating point.

dissipation and time constants are 8 mW °C⁻¹ and 1 s, respectively, when tested in stirred oil, as compared with 1 mW °C⁻¹ and 10 s, respectively, when tested in still air. The dissipation and time constants for disks depend on the disk thickness, lead wire diameter, quantity of solder used for lead attachment, distance between the thermistor body and its mount, and the thermistor material. The dissipation and time constants for rods depend on the rod length as well as the variables specified above for disks.

Electrical Properties of NTC Thermistors

Thermistor applications depend on three fundamental electrical characteristics. These are the voltage–current characteristic, the current–time characteristic, and the resistance–temperature characteristic.

Voltage–Current Characteristics

Figure 32.17 shows a typical voltage–current curve. At low currents, the power dissipated is small compared with the dissipation constant and the self-heating effect is negligible. Under these conditions, the resistance is constant, independent of the current, and the voltage is proportional to the current. Consequently, at low currents, the characteristic curve is tangent to a constant resistance line equal to the zero-power resistance of the thermistor. As the current increases, the effects of self-heating become more evident; the thermistor temperature rises, and its resistance begins to decrease. For each subsequent incremental increase in current, there is a corresponding decrease in resistance. Hence, the slope of the voltage–current characteristic decreases with increasing current. This continues until the current reaches a value at which the slope becomes zero. Beyond this point, at which the voltage exhibits its maximum value, the slope continues to decrease and the thermistor exhibits a negative resistance characteristic. The temperature, voltage, and current corresponding to the peak of the curve are:

$$T_P = \frac{\beta - \sqrt{\beta^2 - 4\beta T_a}}{2}$$

(32.30)

TABLE 32.10 Applications Based on the Voltage–Current Characteristic of Thermistors

1. Variation in dissipation constant (fixed load line on a family of curves)
    - Vacuum manometers
    - Anemometers, flowmeters, fluid velocity
    - Thermal conductivity analysis, gas detectors, gas chromotography
    - Liquid level measurement, control, and alarm

2. Variation in circuit parameters (rotation and/or translation of load line on a fixed *E–I* curve)
    - Oscillator amplitude and/or frequency regulation
    - Gain or level stabilization and equalization
    - Volume limiters
    - Voltage compression and expansion
    - Switching devices

3. Variation in ambient temperature (fixed or variable load line on a family of curves)
    - Temperature control or alarm

4. Microwave power measurement
    - Bolometers

From References [1, 14].

$$E_P = \sqrt{R_P \delta \left(T_P - T_a \right)} \tag{32.31}$$

$$I_P = \sqrt{\delta \left(T_P - T_a \right) / R_P} \tag{32.32}$$

where T_P = Absolute temperature (in K) at which the peak occurs
$\quad\ T_a$ = Absolute ambient temperature (in K)
$\quad\ R_P$ = Thermistor resistance at T_P

Applications based on the voltage–current characteristics involve changes in the operating point on a single curve or family of curves. Such changes result from changes in the environmental conditions or variations in circuit parameters. Figure 32.17 shows a graphical solution for obtaining the operating point. The curve represents a nonlinear relationship between the thermistor voltage and current. The Thevenin equivalent circuit with respect to the thermistor terminals provides the relationship given by Equation 32.33.

$$E_T = E_{THEV} - I_T R_{THEV} \tag{32.33}$$

where E_T and I_T are the thermistor voltage and current, respectively. The straight-line relationship of Equation 32.33 is the load line of the self-heated thermistor. Its intersection with the voltage–current characteristic is the operating point. Table 32.10 categorizes the more familiar applications into four groups distinguished by the form of thermistor excitation [1, 14].

Current–Time Characteristics

The voltage–current characteristic discussed above deals with a self-heated thermistor operated under steady-state conditions. This condition, for which a decrease in thermistor resistance results from a current sufficiently high to cause self-heating, does not occur instantaneously. A transient condition exists in a thermistor circuit from the time at which power is first applied ($t = 0$) until the time equilibrium occurs ($t \gg \tau$). The relationship between the thermistor current and the time required to reach thermal equilibrium is the current–time characteristic. Generally, the excitation is a step function in voltage through a Thevenin equivalent source. Figure 32.18 shows the current–time characteristics for several

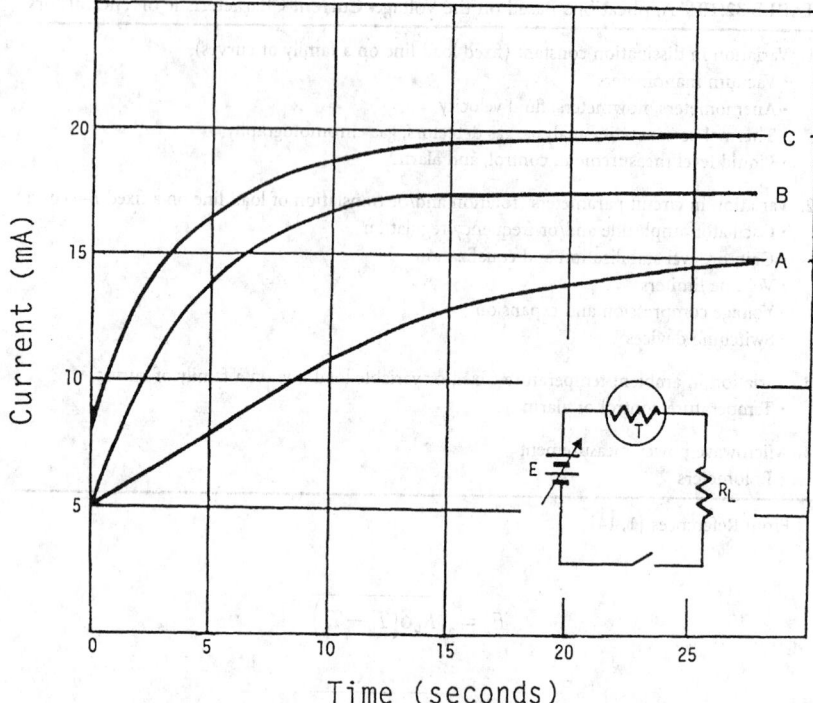

FIGURE 32.18 Current–time characteristics of Thermometrics, Inc. epoxy-coated chip thermistors. Curves A and C are for a DC95F402 having a zero-power resistance at 25°C of 4 kΩ. Curve B is for a DC9F802 having a zero-power resistance at 25°C of 8 kΩ. The excitation is a step function in voltage through a Thevenin equivalent source. The source voltage and resistance values for curve A are 24 V and 1 kΩ, respectively. For curves B and C, the source voltage is 48 V and source resistance is 2 kΩ.

thermistor disks and voltage sources. For any given characteristic, the source voltage, source resistance, and thermistor zero-power resistance determine the initial current. The source voltage, source resistance, and voltage–current characteristic determine the final equilibrium value. The curve between the initial and final values depends on the circuit design parameters, as well as the dissipation constant and heat capacity of the thermistor. The proper choice of thermistor and circuit design results in a transient time range of a few milliseconds to several minutes.

Applications based on the current–time characteristic are time delay, surge suppression, filament protection, overload protection, and sequential switching.

Resistance–Temperature Characteristics

The term zero-power resistance applies to thermistors operated with negligible self-heating. This characteristic describes the relationship between the zero-power resistance of a thermistor and its ambient temperature. Over small temperature spans, the approximately linear relationship between the logarithm of resistance and inverse absolute temperature is given by:

$$\ln R_T \cong A + \frac{\beta}{T} \tag{32.34}$$

where T = Absolute temperature (K)
β = Material constant of the thermistor

If one lets $R_T = R_{T_0}$ at a reference temperature $T = T_0$ and solves for R_T, one obtains:

$$R_T \cong R_{T_0} \exp\left[\frac{\beta(T_0 - T)}{TT_0}\right] \tag{32.35}$$

The temperature coefficient of resistance α is defined as:

$$\alpha \equiv \frac{1}{R_T}\frac{dR_T}{dT} \tag{32.36}$$

Solving Equation 32.34 for α yields:

$$\alpha = -\frac{\beta}{T^2} \tag{32.37}$$

In Equation 32.34, the deviation from linearity results in temperature errors of 0.01°C, 0.1°C, and 0.3°C for temperature spans of 10°C, 30°C, and 50°C, respectively, within the range of 0°C to 50°C. Using a polynomial for $\ln R_T$ vs. $1/T$ reduces the error considerably. The degree of the polynomial required depends on the temperature range and material system used. The use of a third-degree polynomial is adequate for most applications. Hence, more accurate expressions for the resistance–temperature characteristic are:

$$\ln R_T = A_0 + \frac{A_1}{T} + \frac{A_2}{T^2} + \frac{A_3}{T^3} \tag{32.38a}$$

$$\ln R_T = A_0 + \frac{A_1}{T} + \frac{A_2}{T^2} + \frac{A_3}{T^3} \tag{32.38b}$$

$$\frac{1}{T} = \alpha_0 + \alpha_1 \ln R_T + \alpha_2 \left(\ln R_T\right)^2 + \alpha_3 \left(\ln R_T\right)^3 \tag{32.39}$$

Steinhart and Hart proposed the use of Equation 32.39 for the oceanographic range of −2°C to 30°C [11]. Their analysis showed that no significant loss in accuracy occurred by eliminating the square term $a_2(\ln R_T)^2$. Consequently, they proposed the use of Equation 32.40.

$$\frac{1}{T} = b_0 + b_1 \ln R_T + b_3 \left(\ln R_T\right)^3 \tag{32.40}$$

Mangum reported that the use of Equation 32.40 resulted in interpolation errors of approximately 0.001°C over the range of 0°C to 70°C [12].

The technical staff at Thermometrics, Inc. evaluated Equation 32.38 over the range of −80°C to 260°C using 17 different thermistor materials. The glass probes investigated encompassed a span of specific resistance values of 2 Ω cm to 300 kΩ cm and a resistance range at 25°C of 10 Ω to 2 MΩ. The results were presented at the *Sixth International Symposium on Temperature* and show that the interpolation errors do not exceed the total measurement uncertainties. For temperature spans of 100°C within the range of −80°C to 260°C, 150°C within the range of −60°C to 260°C, and 150°C to 200°C within the

TABLE 32.11 Interpolation Errors for $\ln R_T = C_0 + C_1/T + C_3/T^3$

Temperature range (°C)	Temperature span (°C)	Interpolation error (°C)
−80 to 0	50	0.002–0.01
0 to 200	50	0.001–0.003
−80 to 0	100	0.02–0.03
0 to 200	100	0.01
−60 to 90	150	0.1
0 to 150	150	0.045
50 to 200	150	0.015
0 to 200	200	0.08

range of 0°C to 260°C, the interpolation errors are 0.005°C to 0.01°C [13]. Lowering the temperature span to 50°C within the range of 0°C to 260°C reduces the interpolation error to 0.001°C to 0.003°C.

Sapoff and Siwek [14] evaluated the loss in accuracy introduced by eliminating the quadratic term in Equation 32.38. For this condition, Equation 32.38 reduces to Equation 32.41.

$$\ln R_T = C_0 + \frac{C_1}{T} + \frac{C_3}{T^3} \tag{32.41}$$

The interpolation errors introduced by Equations 32.40 and 32.41 depend on the material system, temperature range (nonlinearity increases at low temperatures), and the temperature span considered. Table 32.11 summarizes the errors associated with the use of Equation 32.41 for various temperature spans and ranges.

Equations 32.38 through 32.41 can be rewritten as:

$$R_T = \exp\left(A_0 + \frac{A_1}{T} + \frac{A_2}{T^2} + \frac{A_3}{T^3} \right) \tag{32.42}$$

$$T = \left[a_0 + a_1 \ln R_T + a_2 \left(\ln R_T \right)^2 + a_3 \left(\ln R_T \right)^3 \right]^{-1} \tag{32.43}$$

$$T = \left[b_0 + b_1 \ln R_T + b_3 \left(\ln R_T \right)^3 \right]^{-1} \tag{32.44}$$

$$R_T = \exp\left(C_0 + \frac{C_1}{T} + \frac{C_3}{T^3} \right) \tag{32.45}$$

The solutions for Equations 32.38, 32.39, 32.42, and 32.43 require four calibration points and the use of simultaneous equations. Similarly, the solutions for Equations 32.40, 32.41, 32.44, and 32.45 require three calibration points. The use of a polynomial regression analysis involving additional calibration points can minimize the effects of calibration uncertainties.

Applications that depend on the resistance–temperature characteristic are temperature measurement, control, and compensation. There also are applications for which the thermistor temperature depends on some other physical phenomenon. For example, an hypsometer is an instrument in which the temperature of a boiling liquid provides an indication of the liquid vapor pressure. Another example involves the use of thermistor-type cardiac catheters for thermodilution analysis. A saline or dextrose

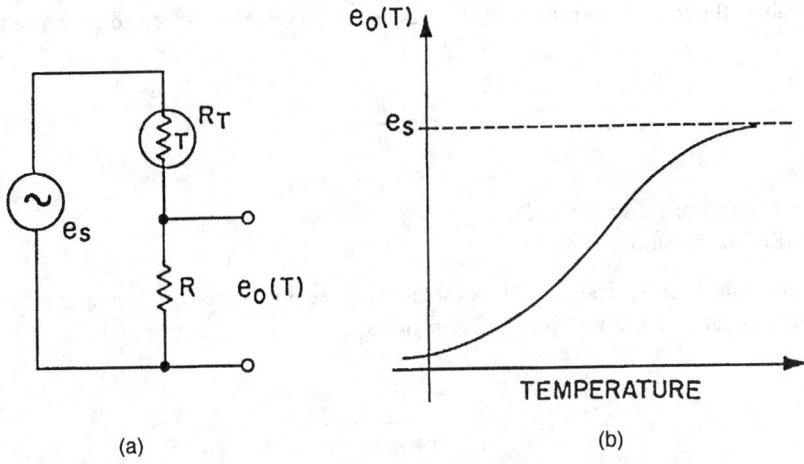

FIGURE 32.19 (a) Thermistor voltage divider. The resistor R represents the parallel combination of the load resistance and the fixed divider resistor. (b) The output, e_0 vs. T, is an S-shaped curve that is linear over a portion of the temperature range.

solution of known volume and temperature is injected into the bloodstream through one of the catheter lumens. The mixing of the solution with the blood dilutes the solution and decreases the temperature of the blood as it flows downstream past a thermistor located at the surface of another lumen in the catheter. The cardiac output computed from the temperature–time response data measured by the thermistor provides an indication of the heart pump efficiency.

Linearization and Signal Conditioning Techniques

The measured resistance and computed constants for Equations 32.43 and 32.44 can be used with a computer to determine temperature without the need for a linear network. This is useful for obtaining high accuracy over relatively wide temperature ranges. Most applications based on the resistance–temperature characteristics of thermistors, however, use some form of linearization or signal conditioning. The use of a constant current source and a linear resistance network provides a voltage output that is linear with temperature. By using the proper combination of current and resistance level, a digital voltmeter connected across the network provides a direct display of temperature. The use of a constant voltage source results in a linear conductance network with a current that is linear with temperature.

Linear Conductance Networks

Voltage dividers, ohmmeter circuits and Wheatstone bridge circuits are linear conductance networks. Consider the voltage divider circuit shown in Figure 32.19. Using the fixed resistor for the output eliminates the effect of the load resistance. For the purpose of analysis, the resistor denoted by R is the parallel combination of the load and fixed divider resistors. Another advantage is that the output voltage increases with temperature with this arrangement. The ratio of output voltage to input voltage is given by:

$$\frac{e_0}{e_s} = \frac{R}{R+R_T} = \frac{1}{1+R_T/R} \qquad (32.46)$$

where e_0 = Output voltage
 e_s = Source voltage
 R = Parallel combination of the load and fixed resistors
 R_T = Thermistor resistance at a specified temperature T

If one normalizes the thermistor resistance with respect to its value at a specified reference temperature T_0, then:

$$r_T = \frac{R_T}{R_{T_0}}$$

(32.47)

where R_{T_0} = Thermistor resistance at T_0
r_T = Resistance ratio

Thermistor manufacturers typically supply resistance ratio–temperature characteristics in their catalogs. Substituting Equation 32.47 in Equation 32.46 yields:

$$\frac{e_0}{e_s} = \frac{1}{1 + r_T R_{T_0}/R}$$

(32.48)

If the circuit constant $\sigma = R_{T_0}/R$, then Equation 32.48 becomes:

$$F(T) = \frac{e_0}{e_s} = \frac{1}{1 + r_T \sigma}$$

(32.49)

The output for the voltage divider of Figure 32.19a is the S-shaped curve shown in Figure 32.19b. The value of the circuit constant σ determines the temperature range for which good linearity exists between e_0 and T. References [3, 4] include families of S-shaped curves over the range $0.01 \leq \sigma \leq 20.0$ for the three basic material systems of MIL-T-23648. A good criterion for achieving optimum linearity is to equate the slopes of the function $F(T)$ at the end-points of the specified temperature range $T_L \leq \sigma \leq T_H$. Specifying that $dF(T_L)/dT = dF(T_H)/dT$ results in the following:

$$\sigma = \frac{X - Y}{Yr_{T_L} - Xr_{T_H}}$$

(32.50)

where X and Y are determined by the end-point conditions and the equation constants for $\ln R_T$ vs. T. When Equation 32.41 is used, X and Y are given by:

$$X = T_H \sqrt{r_{T_L}\left(C_1 + 3\frac{C_3}{T_H^2}\right)}$$

(32.51)

$$Y = T_L \sqrt{r_{T_H}\left(C_1 + 3\frac{C_3}{T_L^2}\right)}$$

(32.52)

When Equation 32.38 is used, X and Y are given by:

$$X = T_H \sqrt{r_{T_L}\left(A_1 + 2\frac{A_2}{T_L} + 3\frac{A_3}{T_L^2}\right)}$$

(32.53)

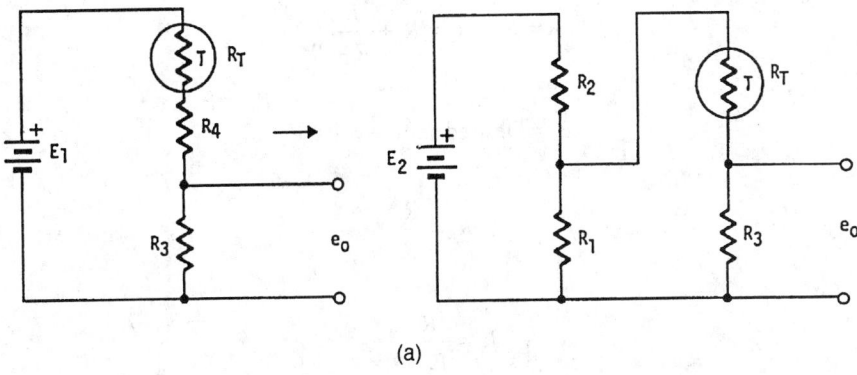

(a)

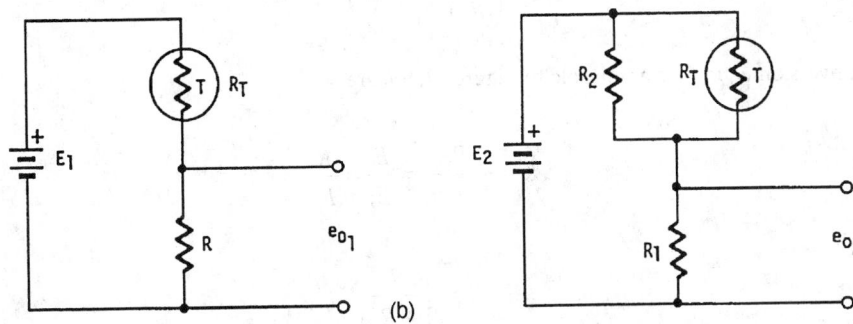

(b)

FIGURE 32.20 Equivalent thermistor voltage divider circuits for converting a network with a nonstandard voltage source to a network that uses a standard voltage source. Only a portion of the output is available in (*a*). Ohmmeter-type thermometers use such circuits. The conversion circuit of (*b*) provides the full available output voltage in a series with a bias voltage. (*b*) is used with bridge circuit-type thermometers.

$$Y = T_\mathrm{L} \sqrt{r_{T_\mathrm{H}}\left(A_1 + 2\frac{A_2}{T_\mathrm{H}} + 3\frac{A_3}{T_\mathrm{H}^2}\right)} \tag{32.54}$$

By allowing e_s and R to be the Thevenin equivalent voltage and resistance with respect to the thermistor terminals, respectively, the voltage divider analysis applies to any more complex circuit such as a Wheatstone bridge [3, 4].

Thermistor self-heating constraints typically result in nonstandard voltage sources for thermometer circuit designs. The modified divider circuits of Figure 32.20 provide equivalent circuits with standard voltage sources. Only a portion of the available output voltage appears across the output detector in Figure 32.20*a*. Ohmmeter-type thermometers use such circuits. The circuit of Figure 32.20*a* provides the full available output voltage. However, the conversion results in a bias voltage in series with the output. Bridge circuit-type thermometers use the circuit of Figure 32.20*b*. The conversion equations applicable to Figure 32.20*a* are:

$$K = \frac{R_1}{R_1 + R_2} = \frac{E_1}{E_2} = \frac{R_4}{R_2} \tag{32.55}$$

$$R = R_3 + R_4 = R_3 + \frac{R_1 R_2}{R_1 + R_2} \tag{32.56}$$

$$E_2 = \text{Desired source voltage} \tag{32.57}$$

$$R_2 = \frac{R_4}{K} \tag{32.58}$$

$$R_1 = \frac{R_2 R_4}{R_2 - R_4} \tag{32.59}$$

$$\left(\frac{R_3}{R_3 + R_4}\right) F(T) \tag{32.60}$$

The conversion equations applicable to Figure 32.20*b* are:

$$K = \frac{R_2}{R_1 + R_2} = \frac{E_1}{E_2} = \frac{R}{R_1} \tag{32.61}$$

$$R = \frac{R_1 R_2}{R_1 + R_2} \tag{32.62}$$

$$E_2 = \text{Desired source voltage} \tag{32.63}$$

$$R_1 = \frac{R}{K} \tag{32.64}$$

$$R_2 = \frac{R R_1}{R_1 - R} \tag{32.65}$$

$$e_{0_2} = \left(E_2 - E_1\right) + e_{0_1} \tag{32.66}$$

Reference [15] includes design examples that use these conversion equations.

Temperature Controllers.

Thermistor temperature controllers frequently use voltage divider and bridge circuits. The use of a thermistor in such a circuit results in much higher sensitivity than that obtainable with a thermocouple or RTD. The most sensitive standard thermocouples exhibit output voltage slopes of 50 to 55 μV $^\circ$C^{-1} in the range of 0°C to 300°C. Thermistor voltage dividers typically exhibit output slopes of about 8 to 10 mV $^\circ$C^{-1} per volt applied to the divider. Since the input to a thermistor voltage divider typically is in the range of 1 to 5 V, a thermistor provides about 200 to 1000 times the sensitivity of a thermocouple. The temperature coefficient of resistance of a thermistor is about 10 times that of an RTD. However, the temperature span about the control point of a thermistor controller is small compared with that available with a thermocouple or RTD.

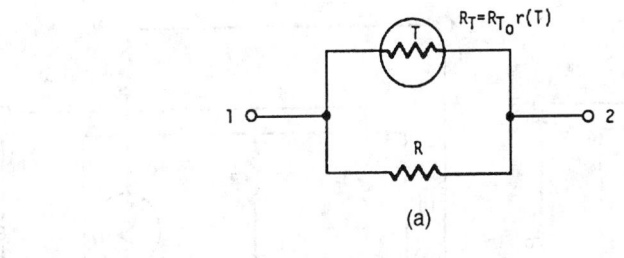

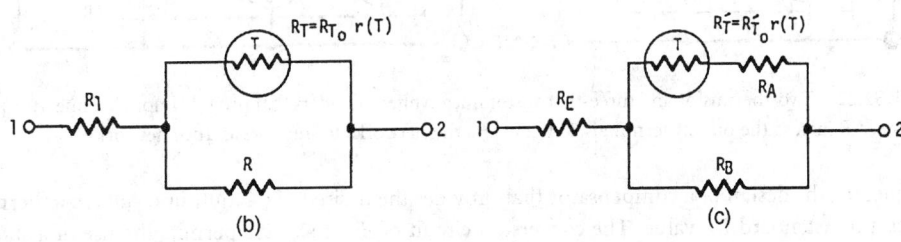

FIGURE 32.21 Equivalent linear thermistor networks for converting a network with a nonstandard R_{T_0} to a network with a standard catalog value for R'_{T_0}. A requirement is that both thermistors have the same resistance ratio–temperature characteristic.

Reference [1] includes some typical low-cost thermistor temperature controllers. Thermistor temperature controllers are available from Hart Scientific, Inc. that provide control stabilities of better than 0.001°C [16].

Linear Resistance Networks

A common technique for designing a thermometer circuit is to apply a constant-current source to a linear resistance network. This results in a voltage across the network that is linear with temperature. Temperature compensation for resistance changes that occur in coil windings, instruments, relays, motors, and generators also use such networks. For example, a copper coil has a positive temperature coefficient of resistance of approximately 0.39% °C^{-1}. The thermistor network has a negative temperature coefficient. The compensator is designed to provide a slope that is equal in magnitude, but opposite in sign to that of the copper coil. Hence, the Ω vs. °C slope of the compensator is equal to the Ω/°C change of the coil. This results in a current through the coil that is independent of temperature. Additional applications include compensation of drift in silicon strain gages, infrared detectors, and circuits that contain both passive and active components.

For the basic linear resistance network of Figure 32.21a:

$$R_{12} = R_C = \frac{RR_T}{R + R_T} = R\left(1 - \frac{1}{1 + R_T/R}\right) \tag{32.67}$$

Normalizing with respect to R and substituting $r_T = R_T/R_{T_0}$ and $R_{T_0}/R = \sigma$ in Equation 32.67 yields:

$$R_n = \frac{R_C}{R} = 1 - \frac{1}{1 + r_T\sigma} = 1 - F(T) \tag{32.68}$$

Consequently, the techniques used for optimizing the linearity of voltage dividers also apply to linear resistance networks. The use of the series resistor R_1 in Figure 32.21b, translates the curve to a higher resistance level and does not affect the Ω/°C output.

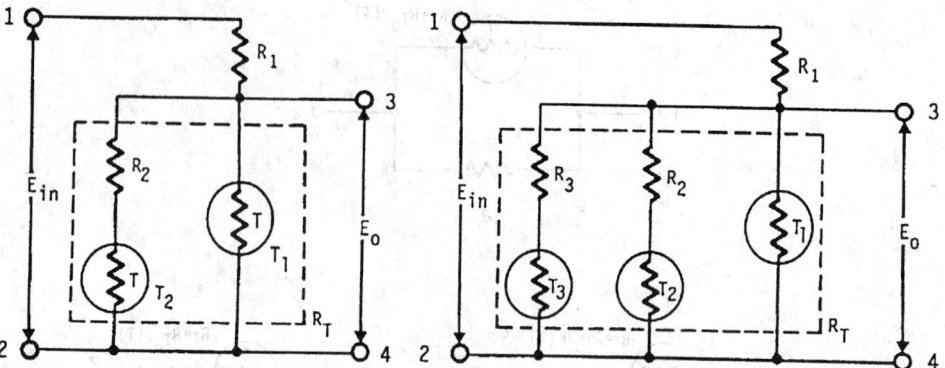

FIGURE 32.22 Two-thermistor and three-thermistor linear voltage dividers that provide improved linearity. Placing the resistors R_1 across the output terminals 3–4 converts the networks to linear resistance networks.

Frequently, the design of a compensator that provides the desired $\Omega/°C$ output results in a thermistor that has a nonstandard R_{T_0} value. The conversion circuit of Figure 32.21c permits the use of a standard R_{T_0} value. The conversion equations are as follows:

$$K_1 = \sqrt{\frac{R'_{T_0}}{R_{T_0}}} \tag{32.69}$$

$$R_E = R_1 - R(K_1 - 1) \tag{32.70}$$

$$R_A = RK_1(K_1 - 1) \tag{32.71}$$

$$R_B = RK_1 \tag{32.72}$$

where R_{T_0} is the nonstandard value and R'_{T_0} is the standard value. A requirement is that both thermistors have the same resistance ratio–temperature characteristic. Setting $R_1 = R(K_1 - 1)$ in Figure 32.21b results in $R_E = 0$ for Figure 32.21c. Consequently, the conversion of Figure 32.21a to that of Figure 32.21c requires the use of the minimum insertion resistance $R_1 = R(K_1 - 1)$.

The network provides good linearity for small temperature spans of 10°C to 30°C. However, the error increases rapidly from about 0.15°C for a 30°C span to 0.7°C for a 50°C span (0°C to 50°C). The use of the two-thermistor network shown in Figure 32.22 results in a maximum linearity error of 0.22°C over the range of 0°C to 100°C, while the three-thermistor network shown reduces the error to 0.04°C for the same 0°C to 100°C range [5]. The networks shown in Figure 32.22 are linear voltage dividers. Placing the resistor R_1 across the output terminals 3–4 converts these networks to linear resistance networks. Many manufacturers sell an interchangeable, three-wire, dual thermistor that is suitable for use in the two-thermistor circuit of Figure 32.2.

Interchangeable thermistors and the linearization techniques described above have been available for many years. In addition, the stability of NTC thermistors is better than that of thermocouples and frequently better than or equal to that of commercial RTDs [17]. However, the nonlinear resistance–temperature characteristics of thermistors continue to limit their use in industrial temperature measurement and control applications. The availability of low-cost microprocessors has eliminated this limitation. Such devices can use the equation constants of Equations 32.38 or 32.39 to compute and display temperature directly. They also can be used to compute lookup tables to provide interpolation uncertainties in the range of 0.001°C to 0.01°C. There are instruments described in the literature that utilize such microprocessor chips [18, 19]. Thermometrics, Inc. sells a commercial instrument that reads the equation constants from a chip in the connector of each probe supplied for use with the instrument [1]. The use of thermistors for industrial applications will continue to increase as the cost of microprocessor chips continues to fall.

References

1. Thermometrics, Inc., *Thermometrics NTC & PTC Thermistors*, Edison, NJ, 1993.
2. J. Fabien, Heating with PTC thermistors, *EDN Products Edition*, 41(12A), 10, 1996.
3. M. Sapoff and R. M. Oppenheim, The design of linear thermistor networks, *IEEE International Convention Record*, Part 8, 12, 1964.
4. M. Sapoff, Thermistors: Part 4, Optimum linearity techniques, *Measurements & Control*, 14(10), 1980.
5. C. D. Kimball and R. W. Harruff, Thermistor thermometry design for biological systems, *Temperature, Its Measurement and Control in Science and Industry*, Vol. 4, Pittsburgh, PA: Instrument Society of America, 1972, Part 2.
6. Ketema, Rodan Division, *Thermistor Product Guide*, Anaheim, CA, 1995.
7. Cesiwid Inc., *Alphalite® Bulk Ceramic NTC Thermistors*, Niagara Falls, NY, 1996.
8. Fenwal Electronics, Inc., *Standard Products Catalog*, Milford, MA, 1994.
9. Fenwal Electronics, Inc., *Thermistor Manual*, Milford, MA, 1974.
10. Victory Engineering Corporation, *Technical Corporation of Thermistors & Varistors*, Springfield, NJ, 1962.
11. J. S. Seinhart and S. R. Hart, Calibration curves for thermistors, *Deep Sea Research*, 15, 497, 1968.
12. B. W. Mangum, The triple point of succinonitrile and its use in the calibration of thermistor thermometers, *Rev. Sci. Instrum.*, 54(12), 1687, 1983.
13. M. Sapoff, W. R. Siwek, H. C. Johnson, J. Slepian, and S. Weber, The exactness of fit of resistance–temperature data of thermistors with third-degree polynomials, in J. F. Schooley (ed.), *Temperature, Its Measurement and Control in Science and Industry*, Vol. 5, New York, NY: American Institute of Physics, 1982, 875.
14. M. Sapoff and R. M. Oppenheim, Theory and application of self-heated thermistors, *Proc. IEEE*, 51, 1292, 1963.
15. M. Sapoff, Thermistors: Part 5, Applications, *Measurements & Control*, 14(12), 1980.
16. Hart Scientific, Inc., *Hart Scientific Calibration Equipment*, Pleasant Grove, UT, 1995.
17. W. R. Siwek, M. Sapoff, A. Goldberg, H. C. Johnson, M. Botting, R. Lonsdorf, and S. Weber, Stability of NTC thermistors, in J. F. Schooley (ed.), *Temperature, Its Measurement and Control in Science and Industry*, Vol. 6, New York, NY: American Institute of Physics, 1992, 497.
18. R. L. Berger, T. Clem, C. Gibson, W. Siwek, and M. Sapoff, A digitally linearized thermistor thermometer referenced to IPTS—26(13), 68, 1980.
19. W. R. Siwek, M. Sapoff, A. Goldberg, H. C. Johnson, M. Botting, R. Lonsdorf, and S. Weber, A precision temperature standard based on the exactness of fit of thermistor resistance–temperature data using third degree polynomials, in J. F. Schooley (ed.), *Temperature, Its Measurement and Control in Science and Industry*, Vol. 6, New York, NY: American Institute of Physics, 1992, 491.

32.4   Thermocouple Thermometers

R. P. Reed

The Simplest Thermocouple

Despite an increasing variety of temperature sensors, the self-generating thermocouple remains the most generally used sensor for thermometry because of its versatility, simplicity, and ease of use. Any pair of

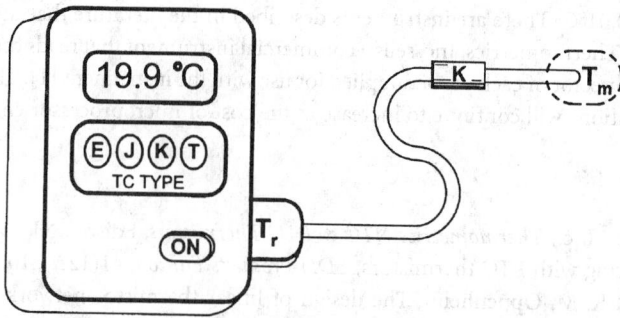

FIGURE 32.24 The simple modern digital thermocouple thermometer. Modern digital electronics has made casual thermometry very easy, but has obscured the continuing need to have an authentic understanding of thermoelectric principles for accurate thermometry with more complicated circuits and more important measurements.

electrically conducting and thermoelectrically dissimilar materials coupled at an interface is a *thermocouple* [1]. The legs are *thermoelements*. The *Seebeck effect* produces a voltage in all such thermoelements where they are not at a uniform temperature. Any electric interface between dissimilar electric conductors is a *real thermoelectric junction*. A free end of a thermoelement is a *terminus*, not a junction. Couplings between *identical* thermoelements are *splices* or *joins*, not junctions.

It is the thermoelements that determine thermocouple *sensitivity* and calibration; but, it is the temperatures of the end-points of thermoelements (i.e., junction temperatures) that determine the *net* emf observed in thermometry. The Seebeck effect, which converts temperature to voltage, is used for thermoelectric thermometry but is also a primary low-frequency noise source in all low-level electronic circuits [2].

Simple Thermocouple Thermometry

In the simplest applications, thermocouple thermometry now is as easy to use as is a multimeter to measure resistance. In fact, many present-day digital multimeters (in addition to voltage, current, and resistance) do provide a thermocouple temperature probe, and temperature measurement is just another button-selectable function. These thermocouple thermometers consist of an indicator that digitally displays the temperature of the tip of a plug-in thermocouple probe (Figure 32.24). The simpler of such versatile multimeters can be purchased for less than U.S.$60 [3]. Interchangeable thermocouple probes of different standard thermocouple material types and specialized sensing tips of widely varying designs are adapted to measurement from surface temperatures to internal temperatures in foodstuffs [4].

With some probes, temperatures up to 1370°C can be indicated merely by (1) ensuring that the selected thermocouple types of the indicator and probe correspond, (2) pressing a power-on button, and (3) applying the probe tip at the point where temperature is to be measured. Promptly, and without calibration, the present temperature of the probe tip is digitally displayed selectably in °C or °F, usually to a resolution of 0.1°C or 1°C. Some specifications claim "accuracy" of 0.3% (4°C at 1370°C) of reading. Some even offer certified *calibration traceability* to NIST or other national standards laboratory.

Modern digital electronics has created the illusion that very accurate thermocouple thermometry is no different than other routine electrical measurements. It has encouraged the perception that arbitrarily fine accuracy can be always be accomplished by calibration and guaranteed by certification. The remarkable simplification of instruments that is now commonplace and an abundance of misleading tutorials obscure the real need for a sound understanding of principles of thermocouple circuits for anything other than inconsequential thermometry.

Manufacturers can shelter users from many problems inherent in thermocouple application. Unfortunately, there are many pitfalls from which the manufacturer cannot isolate the user by design or construction. In less simple applications, it is necessary for the user to avoid problems by carefully learning and applying the true principles of thermoelectric circuits and thermocouple thermometry. Even the simplest indicators and probes can easily be misused and produce unrecognized substantial error.

Thermometry errors of only a few °C or even much less in energy, process, manufacturing, and research fields annually cost many millions of dollars in lost yield, fuel cost, performance bonuses, etc. Consequence of error can also be incurred as incorrect interpretation of data, failure of objective, equipment damage, personal injury, or even loss of life.

Unusual thermocouple circuits, installations, and special applications often produce *inconspicuous* error or else *peculiar* results that are very puzzling if an authentic model of thermoelectric circuits is not understood. This chapter section presents the factual principles of thermoelectric circuits that equip the user to easily apply thermocouples for reliable and critical thermometry, even in unusual circumstances with justified confidence. These principles are very simple, yet they justify study even by experienced thermocouple users as thermoelectric circuits are very often misrepresented or misunderstood in subtle ways that, although unrecognized, degrade measurement. The aim of the chapter section is to allow thermometry with all *practical* simplicity while avoiding possibly costly measurement errors.

Thermoelectric Effects

The three thermoelectric phenomena are the Seebeck, Peltier, and Thomson effects [1, 2, 5-7]. Of these, *only* the Seebeck effect converts thermal energy to electric energy and results in the thermocouple voltage used in thermometry. The current-dependent Peltier and Thomson effects are insignificant in practical thermometry. Neither produces a voltage, contrary to common misconceptions. The Peltier and Thomson effects only transport heat by electric current and redistribute it around a circuit. Thermocouple thermometry is properly conducted by *open-circuit* measurement. The Seebeck emf occurs even without current where Peltier and Thomson effects *necessarily* vanish. Related *thermo-magneto-electric* effects are significant only in the presence of large magnetic fields and infrequently degrade applied thermoelectric thermometry [2, 5, 6].

The Seebeck Effect

The Seebeck effect is the occurrence of a net source emf, the *absolute Seebeck emf*, between pairs of points in any individual electrically conducting material due to a difference of temperature between them [1, 2, 5, 7, 8]. The Seebeck emf occurs *without* dissimilar materials. It is *not* a junction phenomenon, nor is it related to Volta's contact potential.

Absolute Seebeck Properties

The *absolute Seebeck coefficient* expresses the measurement sensitivity (volts per unit of temperature) of the Seebeck effect. It is defined over any *thermoelectrically homogeneous* region of a slender individual conducting material by:

$$\sigma(T) = \lim_{\Delta T \to 0} \Delta E / \Delta T = dE/dT, \text{ or} \qquad (32.73)$$

$$dE = \sigma(T)dT \qquad (32.74)$$

The Seebeck coefficient is a transport property of *all* electrically conducting materials. Equation 32.74 will be acknowledged later as the functional law that governs thermoelectric emf. From Equation 32.74,

$$\Delta E = \int_{T_1}^{T_2} \sigma(T)dT = E(T_2) - E(T_1) \qquad (32.75)$$

where ΔE is the increment of emf between a pair of points, separated by *any* distance, between which the temperature difference is $\Delta T = (T_2 - T_1)$.

From Equation 32.75, the net Seebeck effect for a particular material depends only on the temperatures at the two points and *not* on the values of temperature gradients between the two points. The Seebeck coefficient is a nonlinear function of temperature. It is not a constant. For accurate thermometry, the Seebeck coefficient must remain dependent on temperature alone. $\sigma(T)$ cannot vary along a thermoelement, nor can it vary significantly during the time interval of use. Although vulnerable to such environmental effects, for accurate thermometry it must not depend during measurement on such environmental variables as strain, pressure, or magnetic field.

The coefficient, $\sigma_M(T)$, is an *absolute Seebeck coefficient* for an individual material M [1, 2, 7, 8]. A corresponding *source* voltage within a single material is an *absolute Seebeck emf*, $E_M(T)$. The absolute Seebeck emf does physically exist but it is not simply observable. The absolute Seebeck coefficient can be determined indirectly by measuring the *Thomson coefficient*, τ, of the individual material and applying a Kelvin relationship,

$$\sigma = \int_0^{T_{abs}} \left(\tau/T_{abs}\right) dT \tag{32.76}$$

to deduce the thermodynamically related Seebeck coefficient [1, 2, 5, 7, 8].

Relative Seebeck Properties

The difference between the Seebeck emfs of two thermoelements, of materials A and B, of a thermocouple with their shared junction at temperature, T_m, and both their termini at a *physical reference temperature*, T_r, is their *relative Seebeck emf*, $E_{AB}(T)$, expressed as:

$$E_{AB}\left(T_m, T_r\right) = E_A\left(T_m, T_r\right) - E_B\left(T_m, T_r\right) \tag{32.77}$$

The corresponding relative Seebeck coefficient for the pair is:

$$\sigma_{AB}\left(T_m, T_r\right) = \sigma_A\left(T_m, T_r\right) - \sigma_B\left(T_m, T_r\right) \tag{32.78}$$

It is this relative Seebeck coefficient that has been called by the anachronistic and inept term "thermopower" [1]. It is these *relative* voltage or coefficient values that are directly observable and usually used in thermometry. These relative properties, defined for convenience in the *series* circuits of thermometry, have no general meaning for electrically paralleled thermoelements [2]. These are the relative values that are presented in the familiar tables of thermocouple emf vs. measuring junction temperature, T_m, referred to as a *designated reference temperature* $T_0 = T_r$ [1, 4, 9-12]. For convenience, T_0 is usually taken as 0°C, but the value is arbitrary.

Realistic Thermocouple Circuits

The thermocouple is often represented as only a pair of dissimilar thermoelements joined by two junctions in a closed circuit. One junction, at temperature T_m, is the *measuring junction*; the other, at temperature T_r, is the *reference junction*. The net Seebeck emf is proportional to the temperature difference between the two junctions and to a relative coefficient for the paired materials.

The Seebeck phenomenon has wrongly been characterized as the occurrence of current in the closed loop. The true nature of the Seebeck phenomenon is the occurrence of a *source emf* that, for accurate thermometry, must be measured in open-circuit mode that suppresses current. In practical thermometry, no realistic thermocouple circuit has only two dissimilar materials. Some have many and several of these can be expected to contribute some Seebeck emf. The most common thermometry circuits have two separate reference junctions, not one. Valid, but uncommon, circuits can simultaneously have more than one reference temperature [1].

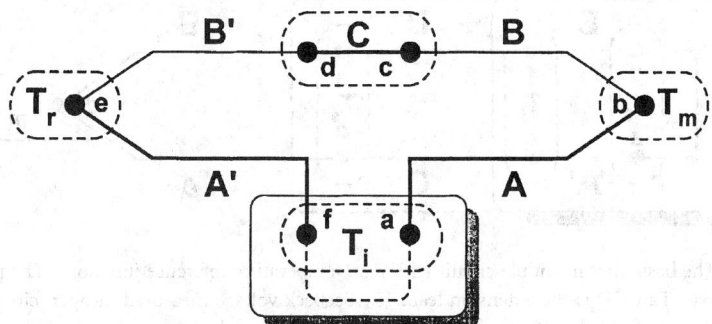

FIGURE 32.25 The basic thermocouple circuit with a *single* temperature reference junction, **e**. The Seebeck voltage measured in open-circuit mode at terminals **a** and **f** is proportional to the temperature difference between thermocouple *measuring junction* **b** and the necessary temperature *reference junction* **e**. For convenience, T_r is usually made to be 0°C. For thermometry, the zones at temperatures T_r and T_i must be isothermal.

Reference Temperature

The *physical* reference temperature, T_r, can be different from the *designated* reference temperature, T_0, of the characterizing relation. If T_r is not identical to T_0, then, to use standard scaling functions, the observed thermocouple emf must be corrected for the temperature difference by adding an emf equivalent to $E_{AB}(T_r) - E_{AB}(T_0)$ to the observed thermocouple emf [1, 7, 9-12]. This is often accomplished by separately monitoring T_r and applying a correction, either numerically or electrically, using a fixed $E_{AB}(T)$ relation that only *approximates* that of the actual thermocouple and the standardized characteristic over a limited range in the vicinity of T_0 and T_r. The error due to the slight discrepancy between the approximation and the actual $E_{AB}(T)$ is small if the two temperatures are similar.

Special thermocouple extension leads are used in most applied thermometry. Many industrial principal thermocouples are inflexibly metal sheathed [1, 4, 10-12]. Others have bare thermoelements separated by bead insulators. Often, these kinds of assemblies are housed in protective wells, have measuring junctions of complex construction, or are distant from the monitoring instrument. Short "pigtails" and extension leads can be of larger wire size, lower resistance, greater flexibility, and very different rugged cable construction than the principal thermocouple (Figure 32.24). All thermoelements must be very well electrically isolated except at measuring and reference junctions. Most modern thermometry is conducted with variants of two basic circuits.

Circuit with Single Reference Junction.

Figure 32.25 shows a thermometry circuit now used mostly in calibration laboratories. This form is convenient when a *fixed point temperature reference* such as an ice point bath or water triple point cell is used to impose the known *physical* reference temperature, T_r [1, 7, 9-12]. The circuit of the thermocouple indicator between **a** and **f** inconspicuously includes many incidental materials and complex circuitry within the instrument. It is necessary that $T_i = T_a = T_f$. Unless instrument temperatures are constant throughout measurement, any nonisothermal portion of the circuit within the instrument can contribute Seebeck emf as noise. This emf can be correctly offset *only* if it is constant.

Often, in circuits like Figure 32.25, the measuring junction, **b** of the principal thermocouple, and reference junction, **e**, are provided by *separately* acquired thermocouples **a-b-c** and **d-e-f**. The two might have significantly different calibrations although they are nominally of the same thermocouple material type. With T_m at 40°C and T_0 at 0°C the reference, thermocouple **d-e-f** contributes about half the Seebeck emf.

This commonplace circuit has at least four distinct thermoelement materials — A, A′, B, and B′ — each of which must be homogeneous. Also, they can be joined, as between **c** and **d**, by an intermediate uncalibrated linking material, C (unless $T_c = T_d$, material C contributes unwanted Seebeck emf). If B

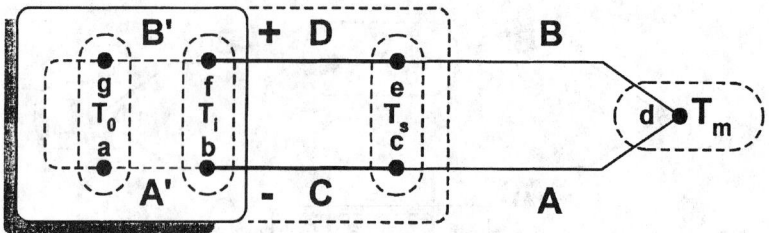

FIGURE 32.26 The basic thermocouple circuit with *dual* temperature reference junctions. The principal thermo-couple is the AB pair. Pair CD is the extension lead. The Seebeck voltage measured in open-circuit mode between terminals **b** and **f** is proportional to the temperature difference between thermocouple *measuring junction* **d** and the necessary temperature *reference junctions*. Depending on the type of extension leads, the reference junctions might be either **c** and **e**, or else **b** and **f**. For thermometry, the zones at temperatures T_s and T_i must be isothermal.

and B′ are not *identical* in Seebeck characteristics, then **c** (and/or **d**) are *real* junctions. If so, it is also necessary that $T_i = T_c = T_d$. The termini **a** and **f**, when connected to a monitor, also become real junctions. They must be controlled so that both stay at the same temperature.

In Figure 32.25, for thermometry the physical reference temperature is intended to be $T_r = T_e$. However, monitoring instruments that internally compensate for reference junction temperature presume (*incorrectly for this circuit*) that $T_r = T_a = T_f$. Therefore, the *single-reference circuit of Figure 32.25 cannot be used directly with thermocouple instruments that automatically apply reference junction temperature compensation.*

Circuits with Dual Reference Junctions.
In the simple "black box" thermocouple thermometer (Figure 32.24), as well as in most applied thermo-electric thermometry, the most common circuit is that of Figure 32.26. This circuit is now the most commonly used in modern thermocouple thermometry. The thermocouple **c-d-e** with thermoelements joined at the measuring junction, **d**, is the *principal thermocouple*. This circuit might have only the principal thermocouple with a plug and jack at the indicator input. More often, as in Figure 32.24, the relatively inflexible principal thermocouple probe also has flexible *extension leads, pair C-D*, which can reside unseen within the indicator. Dashed pair A′-B′ schematically represents internal reference junction temperature compensation.

Circuits like Figure 32.26 have two separate *reference junctions* that must be held at the same *known* temperature. Net Seebeck terminal voltage is measured between **b** and **f**. When thermocouple leads *C* and *D* are connected to the monitor, the input terminals, **b-f**, might be intended to be reference junctions. If so, their reference temperature is accurately monitored and emf reference correction corresponding to the difference between T_r and T_0 is applied. Accurate thermocouple measurements cannot be made with ordinary voltmeters in which the temperatures of **b** and **f** are not deliberately controlled to be equal and known.

Extension Leads

Paired thermoelements *C* and *D* are thermoelectric *extension leads*. Extension leads are of three kinds: (1) *neutral*, (2) *matching*, and (3) *compensating*. All types are readily available; some are proprietary. Which of the junctions in Figure 32.26 must be reference junctions varies with the type of extension. Depending on the extension type, materials *A*, *B*, *C*, and *D* might, intentionally, all be of very different materials.

Neutral Extension Leads.
In the simplest application of Figure 32.26, legs *C* and *D* are thermoelectrically homogeneous and are carefully matched to have the same Seebeck coefficient ($\sigma_C = \sigma_D$). Such pairs are *neutral extension leads*. Provided that $T_c = T_e$ and $T_b = T_f$, they contribute no *net* Seebeck emf, so the like pair function effectively serve only as "passive" leads. Therefore, the leads could be made of any electrically conducting material. Such extension leads should be made of solid unplated copper, for which the Seebeck coefficient is small and uniform.

Neutral extension leads require that junctions **c** and **e** be the reference junctions at temperature $T_r = T_s$. *Thermocouple monitors that provide reference junction temperature compensation at input terminals* **b** *and* **f** *cannot be used with neutral extension leads.*

The reference temperature can be controlled on both junctions by a fixed point physical temperature reference such as an ice bath external to the voltage monitor [1, 9-12] Ice baths, very carefully made and maintained, can routinely approximate 0°C to within 0.1°C to 0.2°C [9, 11]. Carelessly applied and maintained, they can deviate from the "icepoint temperature" by up to 4°C [10]. Peltier thermoelectric refrigeration is also used. More often, in thermometry, that reference temperature is not physically imposed. Instead, the actual temperature of the reference junctions is accurately measured and electrical compensation is applied (schematically by A′-B′) for the difference between T_r and T_0. The particular value of the temperature at **b** and **f** does not matter if $T_b = T_f$ is compensated.

Complementary Extension Leads.
Two kinds of complementary extension leads are allowed to contribute to the circuit Seebeck emf: *matched* and *compensating* extensions. The legs C and D are thermoelectric extension leads that are usually uncalibrated, and possibly are of broader or even unknown tolerance. It is expected that T_s and T_i are similar so that the uncalibrated contribution is small and the error is negligible. For very accurate thermometry, this assumption might not be justified.

Complementary extension leads can be used directly with thermocouple indicators that internally compensate the reference junction temperature for the difference between the physical reference junction temperature, T_r, and the designated reference temperature, T_0. For both matched and compensating complementary extensions, the reference temperature is $T_r = T_i$ over the isothermal reference zone around the terminals **b** and **f**.

Matching Extension Leads.
Matching leads have $\sigma_C = \sigma_A$ and also $\sigma_D = \sigma_B$. It is essential that $T_b = T_e$, but it is not essential (although it is desirable) that $T_c = T_e$. Near room temperature, thermoelements are intended to have σ_{CD}, as a pair, be nearly the same as σ_{AB} for the temperature span near T_r. Error due to slightly mismatched and unproven calibration of the extension is minimized if $T_s \cong T_i$. Matched extensions are used with base metal thermocouples, not with refractory or precious metal thermocouples.

Compensating Extension Leads.
The third variation of the circuit in Figure 32.26 is often used for economy and convenience with expensive precious and refractory metal principal thermocouples that are intended for use at very high temperatures and in special environments [1, 7, 9-12]. Usually, only a portion of the principal thermo-couple need be exposed to the adverse environment. Extension leads C and D need only survive a more benign environment. Therefore, they can be made of a less expensive or more conveniently handled material, use lower temperature insulation, be more flexible and of lower resistance, add bulky shielding and mechanical protection, and extend a great distance to a recording facility.

Compensating leads have σ_{CD} of the extension C–D, only as a pair, match σ_{AB} of the A–B pair. It is not necessary that material A be like C, nor that C and D be alike. A practical circuit can have four very dissimilar materials. However, to allow this, it is essential that $T_c = T_e$. Therefore, this circuit with compensating leads is more vulnerable to error from improperly matched temperatures of incidental junctions than with matching extension leads. The reference junction temperature is $T_r = T_b = T_f$, and only this temperature must be independently known. Reference junction compensation is the same as for the matching leads.

Modular Signal Conditioning Components

The myriad of diverse and capable thermocouple indicators and recorders now commercially available, ranging from simple and very inexpensive to sophisticated and versatile yet reasonably priced, makes it unnecessary for most applied thermometry to assemble special signal conditioning circuits. Some, seeking economy of hardware, have built custom systems. Too often, these have not achieved accurate measurement and have proved to be very costly because the critical distinctions between thermoelectric circuits and ordinary electric circuits were not appreciated. For special situations where thermocouples are incorporated into special measurement or operational components or systems, several manufacturers now offer modular and integrated circuit components that simplify special application and do protect the unaware from some pitfalls [13].

Thermocouple Signal Conditioning on a Chip.
As thermocouple signals are low level, it is sometimes desired to amplify them for improved resolution, recording, or control. It might be desirable to incorporate thermocouple sensing as a functional component in other instrumentation packages. There are now several miniature and inexpensive integrated circuit modules for thermocouples that provide reference junction compensation, linearization, isolation, open input indication, amplification, and set point control. Properly applied, these make the integration of thermocouple sensing into other devices, such as computer data acquisition boards, very simple [13].

When a user applies modular components such as these, it is particularly important to understand and to apply the authentic thermoelectric circuit model and principles introduced in the section "The Functional Model of Thermoelectric Circuits." Some precautions normally provided in commercial thermocouple monitors must be provided by the user.

Reference Temperature Sensors.
Thermocouples are self-generating. For casual approximate temperature and differential temperature measurement, they require neither excitation nor reference temperature. For accurate thermometry, unless a known reference temperature is physically imposed, an accurate measurement of the reference junction temperature by independent means is necessary. This sensing is usually by powered resistance temperature detectors, thermistors, transistors, or integrated circuit sensors [1, 4].

Reference Temperature Compensators.
Proper reference compensation requires (1) the establishment of an isothermal temperature zone that includes the terminals of the thermocouples, (2) sensing of the temperature of this zone, and (3) application of a complementary physical or numerical voltage to the thermocouple terminal voltage before scaling the total voltage to temperature.

An analog method includes the resistive zone temperature sensor in a bridge that nonlinearly modulates a supplied voltage according to an approximated nonlinear curve of the $E(T)$ characteristic of the thermocouple. This method is adapted to compensation of only a single thermocouple type. An alternate method numerically converts the sensed reference temperature to the appropriate compensating voltage value. The numerical approach allows applying a separate compensating voltage to each individual thermocouple and for different standard types. In principle, for accuracy, a numerical compensating voltage could be programmed by the user to conform to the specific calibration of the individual thermocouple, whether or not of standardized type.

Reference temperature compensators are internal to the more advanced thermocouple monitors. In advanced units, eight to ten thermocouple inputs are grouped on a separately removable isothermal terminal assembly so they can be removed for reference sensor calibration or for replacement. Like thermocouples, reference sensors occasionally drift, causing significant temperature error. For most thermocouple types, a 1° error in temperature reference produces a similar error in the measured temperature. A variety of external battery- or line-powered, single- or multichannel reference junction compensating units are also commercially available [4, 14].

Grounding, Shielding, and Noise

The technically well-founded principles of noise control for electric circuits apply also to thermoelectric circuits [15, 16]. There are, however, some additional considerations that are necessary for thermocouple circuits. Some noise control problems stem from the nature of thermoelectricity; others from commercial thermocouple design practices.

Noise Problems Peculiar to Thermocouples

Temperature control of all deliberate and incidental junctions and components, and the distributed nature of Seebeck source emf, cannot be ignored with impunity in thermocouple application. The unavoidable requirements of such control are best visualized using methods such as given later in "The Functional Model of Thermoelectric Circuits." Incidental circuit components such as balancing or swamping resistors, feedthrus, splices, and terminal strips, not intended to contribute Seebeck emf, must deliberately be maintained *isothermal*. Shields, unavoidably nonisothermal, made electrically common with the thermocouple must be explicitly recognized as latent sources of random dc noise which contributes Seebeck emf if not properly connected.

Electromagnetic (EM) Noise

In thermocouple probes of mineral-insulated metal-sheathed (MIMS) construction, ceramic-bead-insulated thermocouples, and in some paired insulated thermocouple wires, the paired thermoelements are not twisted and are well separated. Lengthy and larger diameter thermocouple probes present a significant circuit loop area to couple magnetic noise fields. Even if probes are entirely metal-sheathed, the standardized sheaths are very thin and of low magnetic permeability. They are scarcely effective for electromagnetic (EM) shielding. These features prevent the use of some classic techniques for the rejection of EM noise [15, 16]. To reduce troublesome EM noise, the principal thermocouple probe should be of the smallest practical diameter, and of minimum length. In instances where the EM source is localized and identified, the orientation of the probe relative to the source can be arranged to minimize EM coupling. For rejection of EM noise, extension leads should always be of twisted-pair construction, with a pitch small enough to reject high-frequency noise [15].

Electrostatic (ES) Noise

Electrostatic noise is more easily reduced. Shields of low resistance, though thin, can be effective if properly connected [15, 16]. Plated copper braid is commonly used and is effective for noise of moderate frequency. Continuous shields, such as the MIMS sheaths and aluminized mylar film, are effective for low-frequency ES noise and are more effective than braid shields for very-high-frequency noise. Electrostatic shields must be continuous (without gaps and holes as in braid) for maximum effectiveness. Optimum benefits from shielding require use with thermocouple monitors that provide three-wire input and multiplexers that, individually for each thermocouple, switch both the signal pair and their separate shield lead.

Grounding

Shielding for ES noise and pair twisting for EM noise are important for reduction of high-frequency noise. A secondary overall shield, isolated from inner shields and separately grounded, can improve the rejection of EM noise. Appropriate circuit grounding is also particularly important for both high-frequency EM and ES control, but it is even more significant for low-frequency and dc noise, which are often more consequential in thermocouple measurements. Appropriate grounding is complicated where intimate thermal contact of the measuring junction with an earth-grounded conducting test subject is needed for rapid transient thermometry or for accuracy in the presence of static temperature gradients.

Figure 32.27 illustrates appropriate grounding for several of many possible situations. Low-level thermocouple circuits should be grounded only at a single electric reference point. Because the measuring junction is not a localized site of source emf, the point of grounding must be carefully considered. Grounding should never be at any point of the thermocouple circuit other than measuring or reference junctions. Electric contact, and particularly shunting or shorting, on a thermoelement at any point between measuring and reference junctions usually introduces spurious Seebeck emf from incidental unrecognized thermoelements.

Allowable grounding might be dictated by the internal design of monitors and data loggers. Each instrument design addresses noise control in a distinctive way. Simple and inexpensive line-powered thermocouple thermometers might have only single-ended inputs with the negative lead internally

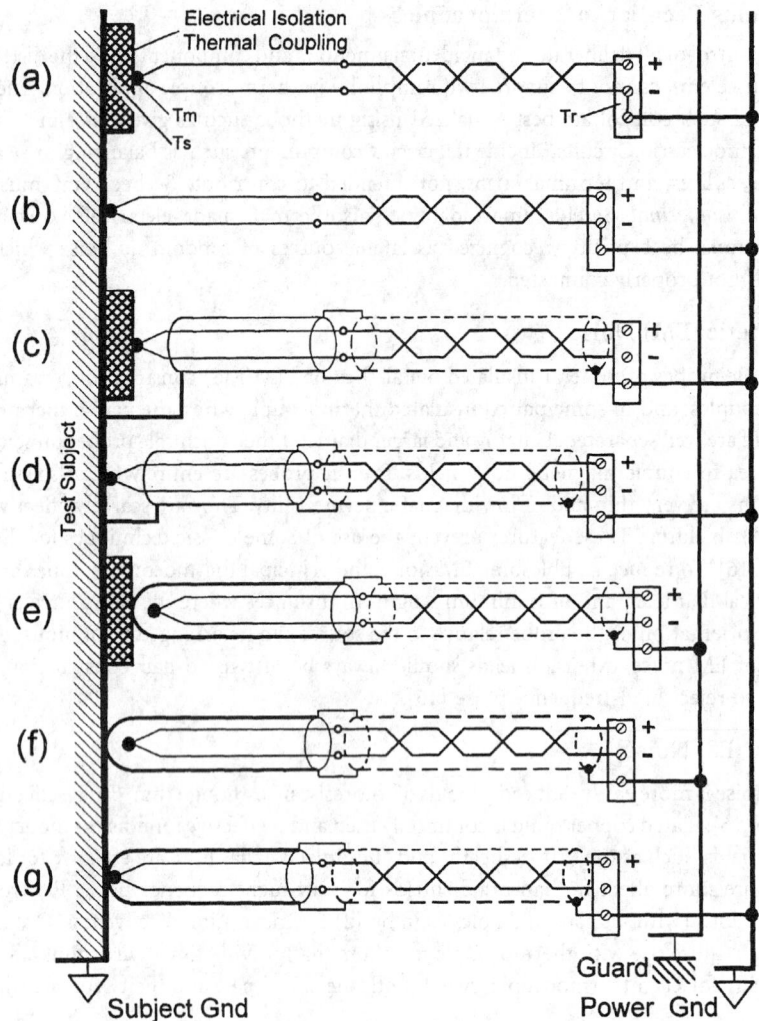

FIGURE 32.27 Preferred grounding and shielding for several thermometry situations. Electromagnetic (EM) and electrostatic (ES) noise must be controlled by different means. The design of the thermocouple monitoring instrument may dictate the grounding and shielding scheme that can be used.

connected to the power ground. This forces the sensor grounding to also refer to power ground. A grounded-junction arrangement might also be required for fast transient thermometry. In use, initially isolated thermocouples can short to the protective sheath or test subject. Although this does change the grounding configuration, resulting errors due to other effects usually are then predominant.

Balanced Thermocouple Circuits

The more sophisticated multichannel data acquisition systems specialized to thermocouple thermometry usually have high-resistance balanced inputs for high common-mode signal rejection. Some differential input amplifiers require significant input resistance for stability. Most have an internal chassis that provides a common signal reference guard surface that is well-isolated from power ground. In a few designs, the guard is electrically driven to a particular reference point. Because many variations are used, it is prudent to study and follow the specific grounding instructions recommended in the user's manual for the instrument. Most accommodate alternative grounding arrangements.

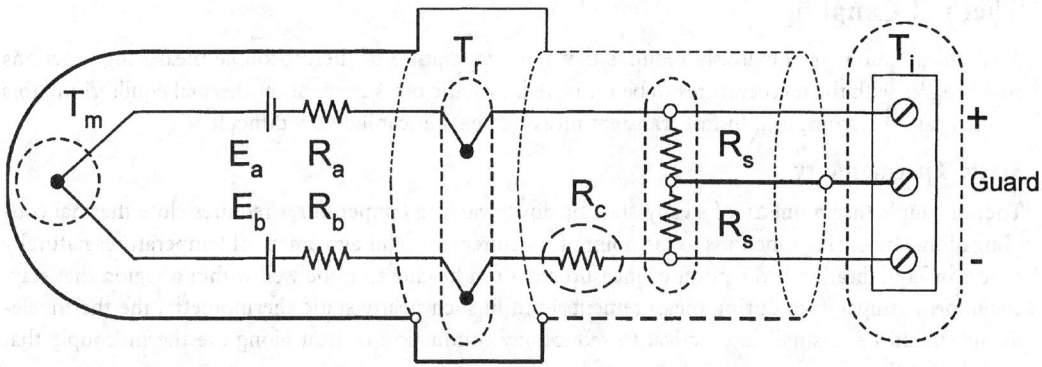

FIGURE 32.28 A circuit arrangement for balanced input recorders to improve common-mode noise rejection. The thermoelement balancing resistor, $R_c = |R_a - R_b|$, and the matched pair of high-resistance input resistors R_s must be held isothermal.

In sheathed thermocouple probes of standard MIMS dimensions, the thermoelements are of the same cross-section, parallel and well spaced, and located within the cylindrical sheath symmetrically [10]. Noise tends to be coupled equally to the paired thermoelements in common mode. However, lengthy or fine gage thermocouples have legs of high resistance compared to copper circuits. The ratio between the resistances of the dissimilar legs can be up to 25:1. Noise introduced in common mode adversely tends to convert to normal mode.

In addition, unlike most voltage sources, the Seebeck emf is not localized to the measuring junction (a material interface). The emf can arise anywhere along the thermoelements. In transient thermometry, it first occurs adjacent to, and then spreads away from the measuring junction. In the more commonplace *static* thermometry, the site of source emf is usually distant from the measuring junction.

Most specialized thermocouple recorders provide reference temperature compensation. They do not allow convenient user access to the circuit beyond the reference junctions. This restricts the noise control methods available to the user. For externally referenced thermocouples, some effective simple circuit modifications are possible. Figure 32.28 shows the addition of three resistors that, used with monitors of balanced input, improve noise rejection. Isothermal balancing resistor, $R_c = |R_a - R_b|$ is added in series with the thermoelement of lower resistance. Making the resistance of the two circuit branches equal reduces the undesired conversion of noise from common mode to normal mode. The matched pair of shunting resistors, R_s, presents a significant definite minimum source resistance to the amplifier input for stability and symmetrically couples the signals and sheath to the common reference point. The resistors must be large enough, $R_s \geq 1000\ (R_a + R_b)$, to avoid significant shunting resistive attenuation of the thermocouple source Seebeck emf. With great care to maintain these resistors strictly isothermal, the balancing and shunt resistors could be placed just at the input terminals *external* to the instrument-provided reference junction temperature compensation.

Filtering

For low-frequency thermometry, to complement the user-provided passive rejection of EM and ES noise by proper grounding and shielding, internal filtering provided by data acquisition monitor manufacturers can effectively reject high-frequency noise. Filtering before recording is preferred. This control can be performed by conventional filters; but for quasi-static thermocouple measurements, other very effective techniques such as "double-slope integration" are provided in some designs. These techniques effectively average the signal over several measured samples. The noise reduction and stability under real-life conditions is remarkable. Routine resolution and stability to 0.1 µV variation over a period of days is achieved by many modern thermocouple data loggers. However, this noise rejection method restricts the sampling speed so that samples can be limited to intervals no shorter than 10 ms to 20 ms that are unsuited to very fast transient thermometry.

Thermal Coupling

Accurate thermometry obviously requires that the temperature of thermocouple measuring junctions closely agree with the temperatures to be measured. In static measurements at thermal equilibrium, this is often easy to accomplish. In fast transient measurements, it can be very difficult.

Static Thermometry

Thermocouple thermometry of steady-state or slowly varying temperatures requires close thermal coupling of the measuring junctions to the point of measurement. This agreement of temperatures naturally occurs by equilibration if the point of measurement can be chosen to be well within a region that stays at uniform temperature during measurement. Even in such nearly static thermometry, the thermoelements should be as small as practical to reduce any conduction of heat along the thermocouple that could affect the temperature being measured.

In typical situations, however, the vicinity of the measuring point is not isothermal and, even at steady state, static temperature gradients can cause the measuring junction to be at a significantly different temperature from the subject [18]. This source of error is more severe for surface, liquid, and gas measurements than for internal solid measurements, and is aggravated for the measurement of subjects of low thermal mass. Efficient thermal coupling of the thermocouple to the subject, as in Figure 32.27, is essential.

Noise or safety might require electrical isolation of the thermocouple. Commonplace electrical insulating materials are poor thermal conductors. There are now special greases, coatings, and thin sheet materials available designed to provide high electrical isolation, yet good thermal contact [17]. Where needed, they can improve accuracy by efficient thermal coupling.

Transient Thermometry

Monitoring very rapid variations of temperature is most effectively accomplished by thermocouples because of the low sensing mass and small size of measuring junction that they allow. Fast thermal response requires the use of special thermoelement configurations in the vicinity of the measuring junction, special circuitry, special thermocouple monitoring instruments, and characterizing methods [16, 18]. Some suppliers offer very thin (10 μm thick) foil and fine wire thermocouples that are especially well adapted to such measurement [19, 20].

A thermocouple alone has no characteristic response time. The Seebeck effect occurs at electronic speed. The thermocouple source emf always promptly corresponds to the temperatures of all junctions in the circuit. However, the transient temperature response of a thermocouple installation is very much slower, and is primarily governed by the thermal interaction of the thermocouple components adjacent to the junctions and the adjacent mass of the test subject [18, 22, 23]. That slow conductive heat transfer is retarded many-fold by even a thin film of air, vacuum, or an insulating solid layer. In fast transient thermometry, the electrical parameters of the thermocouple and associated circuitry and dynamic response of the recording system can further degrade response time.

Intimate thermal coupling by special materials required for static measurement in a thermal gradient is even more critical to transient than to static thermometry [17]. Even the special coupling materials slow response. They should be as thin as feasible. For very fast thermometry, such coupling might not be tolerable. It may be necessary to place the bare measuring junction in direct electric contact with the subject (Figure 32.27).

Intrinsic Thermocouples.

Where the subject of thermometry is an electric conductor, it is possible to electrically fuse the two thermoelements separately to the test subject. This is an "intrinsic" thermocouple junction arrangement. Fine wires or thin foils spot-welded to the subject allow the fastest possible electrothermal response and reduce the influence of the thermocouple on the temperature being measured.

Observe two special precautions in the use of intrinsic junctions. First, the two thermoelements should be fused to the subject side-by-side, close together (not one over the other.) With intrinsic junctions, the bridging segment of the subject becomes a part of the series thermoelectric circuit. (There are two intrinsic junctions coupled by the intermediate subject material.) This produces the fastest response; but in very fast transients, the temperatures of the two junctions can very briefly be slightly different and thus introduce a momentary error, as the unknown Seebeck coefficient of the bridging subject material is different from that of one or both of the principal thermoelements [21].

Second, for transient thermometry, the fine thermoelements leading to the intrinsic junctions should be as short and fine as feasible to reduce their thermal loading; however, this results in a relatively high electric lead resistance. A strain-relieved transition from the fine filaments to more substantial low-resistance compensating lead wires should be made near the measuring junction to enhance electric response. The electrical effect of capacitive coupling of the circuit on electric transient response must also be minimized.

Numerical Correction.

All feasible physical techniques to achieve fast transient temperature response when applied might not be sufficient for a very fast measurement. An inadequate transient response can be further enhanced by numerical analysis [21]. Two additional steps must be taken.

First, an *authentic* experimental or *correctly* modeled transient response characterization of the overall thermal-electrical thermometry system must be made. The usual simplistic first-order representations of *in situ* thermocouple response might be inadequate [21–23].

Second, this authentic characterization must be applied by mathematical *deconvolution* to better estimate the true temperature history of data indicated by a system that had inadequate physical transient response [16, 19]. Critical and reliable improvements of effective response time by factors of four or more are often possible. If uncorrected, indicated transient peak temperatures can be in error by a much greater percentage than the error that results from static thermocouple calibration error. *No general uncertainty can be assigned to a transient measurement without such response characterization.*

Thermocouple Materials

Thermocouple Standardization

Many materials are in regular use as thermocouples for thermometry. Some pairs are *standardized;* some are not. The distinction is a matter of formal consensus group approval by balanced standards committees of experienced users, producers, and standards laboratory staff [10, 24]. The most extensive application data is available for standardized thermocouples. In the U.S., eight materials presently are letter-designated (Types *B, E, J, K, N, R, S,* and *T*) [9–12, 24]. Some properties of these are summarized in Table 32.12. The standardized *E(T)* characteristics of these are now *defined exactly* by polynomial functions of high degree rather than by tabulated values [1, 9, 10]. Thermoelements should be used as selectively paired for thermocouples by the producer, as randomly paired thermoelements of the same type from the same or different manufacturers are not assured to conform to the standard pair values or tolerances [1, 9–12].

The properties of a few other popular pairs, such as precious metal and tungsten refractory alloys, have also been committee-proposed, but letter designations and color codes have not been formally assigned [10]. Infrequently, other popular materials are considered for standardization. Limited Seebeck and application information is available for a multitude of nonstandard materials [25, 26].

Low Temperature Thermometry

Most thermometry is at elevated temperatures. Thermocouple measurement of temperatures below the ice point requires special consideration. Cryogenic thermometry has been very loosely defined as measurement below 280 K (7°C) [1]. More restrictively, the defined cryogenic range has been limited to below 90 K (–183°C, the boiling point temperature of liquid oxygen at 1 atm) [1]. The former range broadly overlaps the measure of atmospheric temperatures (down to –50°C) below the ice point that, along with higher ambient temperatures, often are measured by a single thermocouple system. The latter definition favors characterizing as cryogenic only the extremely low-temperature regime over which thermometry involves distinctive severe problems and different techniques.

TABLE 32.12 Characteristics of U.S. Letter-Designated Thermocouples

Type	Common name	Color code	M.P. (°C)	Recommended range, (°C)[d]	emf at 400°C, (mV)	Uncertainty, +/− Special tolerance Normal tolerance	ρ (μΩ-cm)
B	—	Brown[a]	1810	870 to 1700	0.787	0.25%	34.4
BX	—	Gray[a]	—	—	—	0.50%	—
BP	Pt30Rh	Gray	1910	—	—	—	18.6
BN	Pt6Rh	Red	1810	—	—	—	15.8
E	—	Brown[a]	1270	−200 to 870	28.946	1.0°C or 0.40%	127
EX	—	Purple[a]	—	—	—	1.7°C or 0.50%	—
EP	Chromel[b]	Purple	1430	—	—	—	80
EN	Constantan	Red	1270	—	—	—	46
J	—	Brown[a]	1270	0 to 760	21.848	1.1°C or 0.40%	56
JX	—	White[a]	—	—	—	2.2°C or 0.75%	—
JP	Iron	White	1536	—	—	—	10
JN	Constantan	Red	1270	—	—	—	46
K	—	Brown[a]	1400	−200 to 1260	16.397	1.1°C or 0.40%	112
KX	—	Yellow[a]	—	—	—	2.2°C or 0.75%	—
KP	Chromel	Yellow	1430	—	—	—	80
KN	Alumel[b]	Red	1400	—	—	—	31
N	—	Brown[a]	—	0 to 1260	12.974	1.1°C or 0.40%	
NX	—	Orange[a]	—	—	—	2.2°C or 0.75%	—
NP	Nisil	Orange		—	—	—	
NN	Nicrosil	Red					
R	—	Brown[a]	1769	0 to 1480	3.408	0.6°C or 0.10%	29
RX	—	Green[a]	—	—	—	1.5°C or 0.25%	—
RP	Pt13Rh	Green	1840	—	—	—	19
RN	Pt	Red	1769	—	—	—	10
S	—	Brown[a]	1769	0 to 1480	3.259	0.6°C or 0.10%	30
SX	—	Green[a]	—	—	—	1.5°C or 0.25%	—
SP	Pt10Rh	Green	1830	—	—	—	20
SN	Pt	Red	1769	—	—	—	10
T	—	Brown[a]	1083	−200 to 370	20.810	0.5°C or 0.40%	48
TX	—	Blue[a]	—	—	—	1.0°C or 0.75%	—
TP	Copper	Blue	1083	—	—	—	2
TN	Constantan	Red	1270	—	—	—	46

From References [1, 4, 9, 10]

[a] Overall jacket color.

[b] Chromel and Alumel are trademarks of Hoskins Mfg. Co.

[c] Initial tolerances are for material as manufactured and used within recommended temperature range, Table 32.15, protected in a benign environment.

[d] Recommended temperature range is a guideline for service in compatible environments and for short durations.

The Seebeck coefficient of all conductors is insignificant at 0 K and common materials progressively decrease in thermoelectric sensitivity below the ice point. A few natural superconductor metal elements experience an abrupt drop in Seebeck coefficient to zero at a characteristic superconducting threshold that is below 10 K for most unalloyed metal superconductors [27]. Special alloys have recently been developed to raise the superconducting threshold to about 120 K — well above the formal cryogenic range. Superconducting transitions complicate thermoelectric thermometry at the lowest cryogenic temperatures [7, 8].

Standard Seebeck characteristics are defined for Types *E, J, K, N,* and *T* down to –270°C. The characteristics for Types *R* and *S* extend only down to –50°C, and Type *B* is not characterized below 0°C. The standard polynomials that define the Seebeck properties of letter-designated thermocouples and the production tolerances for commercial materials are different for temperatures below and above the ice point. Materials manufactured for thermometry at elevated temperatures might conform less well to the standard cryogenic characteristics than alloys of the same type especially furnished for such use [1]. This quality issue should be discussed with the thermocouple supplier. Special alloys are available for cryogenic thermometry.

For the lower cryogenic range, of the letter-designated thermocouple materials, Type *E* is preferred for use down to –233°C (40 K) because of its higher relative Seebeck coefficient [1]. The less-sensitive Type *K* and Type *T* are also used in this range. Below 40 K, special alloy combinations, such as Type *KP* vs. Au/0.07 Fe, are recommended. Special thermoelectric relations apply to these materials in the cryogenic range [1].

As Peltier heating at junctions and Thomson heating along thermoelements are current-dependent thermoelectric effects, neither is a significant problem if thermometry is properly conducted by open-circuit measurement. There is no significant thermocouple self-heating as with resistance thermometers. Some thermoelement alloys experience grain growth and incur serious inhomogeneity under prolonged exposure to deep cryogenic temperatures. More sensitively at cryogenic temperatures than at elevated temperatures, the Seebeck coefficient of most thermocouple alloys is very strongly dependent on magnetic field. Strong magnetic fields are often involved in cryogenic experiments, so thermo-magneto-electric effects become significant in studies of superconductivity [2, 27].

Sensitivity

The need for large thermocouple output has been drastically reduced by the enhanced sensitivity, stability, and noise suppression of modern solid-state digital thermocouple indicators. These instruments can routinely indicate temperature to 0.1°C resolution and stability for all letter-designated types. The standard pairs differ significantly in their sensitivity (Table 32.12). The sensitivity of the Type *E* thermocouple is 10 times that of the Type *B* thermocouple at 1000°C. Because the Type *B* thermocouple has extremely small sensitivity around room temperature, it is intended for use only with the measuring junction above 870°C. The most sensitive standardized thermocouple, Type *E*, has a maximum Seebeck coefficient of 81 µV °C^{-1} and, referenced to 0°C, has a maximum Seebeck emf of 76 mV. While initial tolerances for both normal- and special-grade material have been standardized, the difference between *commercial* and *premium* grades is small and, in use, special-grade materials can degrade to exceed the initial tolerances of the normal-grade material.

Letter Designations

The U.S. ANSI standard letter designations and color codes for eight particular thermocouple types (*B, E, J, K, N, R, S,* and *T*) were first established by the ISA in Standard MC-96.1 [28]. The same conventions are followed by Standards of the ASTM and ANSI [10, 24]. A first suffix to the type letter designator, *P* or *N*, as in types *KP* or *KN*, designates the positive or negative thermoelement of a thermocouple pair. A final "*X*" suffix designates an extension wire material, as in *KPX* for a positive type *K* extension thermoelement. For non-standardized material pairs, producers and vendors often apply their own unofficial letter designations, color codes, and trade names. These *commercial* identifiers of individual manufacturers have no universal meaning.

Color Codes

Intended to ease identification, the standards of many nations have assigned color codes to the different letter-designated thermocouples and to thermoelements and extensions. As the individual thermoelements determine both polarity and sensitivity, it is very important to properly identify each leg. Color codes now used in the U.S. are shown in Table 32.12.

TABLE 32.13 International Thermocouple Color Codes

Type	U.S.	IEC	England	China	France	Germany	Japan	Russia
B	Brown	—	—	—	—	—	—	— B
BX	Gray	—	—	—	—	Gray	Gray	— BX
BP	Gray	—	—	—	—	Red	Red	— BP
BN	Red	—	—	—	—	Gray	Gray	— BN
E	Brown	—	—	—	—	—	—	— E
EX	Purple	Purple	Brown	—	Purple	Black	Purple	— EX
EP	Purple	Purple	Brown	Red	Yellow	Red	Red	Purple or black EP
EN	Red	White	Blue	Brown	Purple	Black	White	Yellow or orange EN
J	Brown	—	Red	—	—	—	—	— J
JX	Black	Black	Black	—	Black	Blue	Yellow	— JX
JP	White	Black	Yellow	Red	Yellow	Red	Red	White JP
JN	Red	White	Blue	Purple	Black	Blue	White	Yellow or orange JN
K	Brown	—	—	—	—	—	—	— K
KX	Yellow	Green	Red	—	Yellow	Green	Blue	— KX
KP	Yellow	Green	Brown	Red	Yellow	Red	Red	Red KP
KN	Red	White	Blue	Blue	Purple	Green	White	Brown KN
N	Brown	—	—	—	—	—	—	— N
NX	Orange	—	—	—	—	—	—	— NX
NP	Orange	—	—	—	—	—	—	— NP
NN	Red	—	—	—	—	—	—	— NN
R	Brown	—	—	—	—	—	—	— R
RX	Black	Orange	Green	—	Green	White	Black	— RX
RP	Black	Orange	White	Red	Yellow	Red	Red	— RP
RN	Red	White	Blue	Green	Green	White	White	— RN
S	Brown	—	—	—	—	—	—	— S
SX	Black	Orange	Green	—	Green	White	Black	— SX
SP	Black	Orange	White	Red	Yellow	Red	Red	Red or pink SP
SN	Red	White	Blue	Green	Green	White	White	Green SN
T	Brown	—	—	—	—	—	—	— T
TX	Blue	Brown	Blue	—	Blue	Brown	Brown	— TX
TP	Blue	Brown	White	Red	Yellow	Red	Red	Red or pink TP
TN	Red	White	Blue	White	Blue	Brown	White	Brown TN
Std.	ANSI	IEC	BS	NMI	NFC42	DIN	JIS	
No.	MC96.1	584-3	1843		42-323	43714	1620	

From References [1, 4, 9, 10].

U.S.-standardized color codes have remained uniform for several decades; thus color code confusion of material types in the U.S. is mostly due to user carelessness. The present globally discordant color codes can cause costly misinterpretation in multinational use, particularly outside the U.S. where neighboring countries and trading partners have unlike or multiple color codes. Clearly, a single universal international color code accepted by all nations would be beneficial. Such an international color code is embodied in standard *IEC 584* that is being considered by several nations [28].

Unlike the standardized Seebeck characteristics that are fairly uniform worldwide, the uncoordinated color codes of different national standards are very inconsistent. The unfortunate Babel of national color codes that existed in 1998 is displayed in Table 32.13 [1, 4].

Unfortunately, and uniquely, in present U.S. thermocouple standards, the *negative* thermoelement is always *red*, contrary to customary U.S. electrical and instrument practices. This is also contrary to the historic national thermocouple color codes of China, Germany, and Japan, in which *red* designates the *positive* thermoelement. In English standards, the *negative leg* of all types is *blue*.

In France, the *positive* thermoelement is always coded *yellow*. However, a *yellow positive* leg in the U.S. standard designates Type *KP* material. The wire leads of some U.S. electric blasting caps use yellow insulated wire with a parallel red tracer that has been confused with Type *KX* thermocouple extension wire. In England, yellow denotes Type *JP* material. The *white positive* and *black negative* of the present U.S. ANSI color code for Type *J* are *transposed* in international standard *IEC* 584.

Despite the clear desirability for a uniform color code, there remains a huge quantity of material of different color codes in stock and in use worldwide. For any nation that switches to any new color standard, there would be, immediately and over an unavoidably lengthy transition period following acceptance, the new color-coded thermocouple material intermixed with the multitude of inconsistent legacy color codes. The immediate possibility for confusion of material type would greatly increase rather than decrease. Also, even the present color codes can become indistinct on long-installed material, colors can fade, and/or colors may have been incorrectly applied in manufacture.

Identifying Thermoelements

The color codes apply directly to extension lead wires and effectively to the principal thermocouple wire. Many principal thermoelements and thermocouples are not color coded. A user might not correctly recall the color code. The prudent user will, before use, *confirm* material type identification independent of the color code. *Definite* type identification must be by a combination of methods. Any single identification method can be indefinite, and no method is adaptable to all materials or circumstances.

Visual Identification.
TP thermoelements are of copper and thus are distinguished by their distinctive reddish color. *JP* thermoelements are iron and have a distinctive matte gray cast. Other base metal alloys and platinum and its alloys, if bare, all have a very similar bright silvery appearance unless, if bared from compressed mineral insulation, fabrication has effected a roughened gray matte surface appearance.

Magnetic Identification.
Type *JP* (iron) is strongly magnetic. Type *KN* (Alumel) is slightly magnetic. All other standard thermoelement materials are nonmagnetic. *JP* and *KN* thermoelements can be distinguished from the others by testing the attraction to them of a small magnet.

Resistive Identification.
Resistivities of thermoelements are given in Table 32.12. Although assembled thermocouples have a measuring junction that cannot generally be removed for testing, a bare junction can be directly accessible or it can be electrically accessible at a probe tip for resistance measurement of each leg if the junction is of the type made common at the measuring junction to a conductive sheath. Thermocouple assemblies or cable usually have paired thermoelements of the same length and cross-section. The resistance of each leg distinguishes the material if the wire size and length are known. In these instances, *with both ends of the cable at the same temperature*, the resistance of each thermoelement can be directly measured. The ratio between positive and negative leg resistances can aid type confirmation. For assembled thermocouples with inaccessible measuring junctions, only the loop resistance can be measured and compared with calculated loop resistances.

Thermoelectric Identification.
Less conveniently, a pair can be identified by the output for a known temperature of measuring junction and reference junction. *Complementary extension cables* can be thermally identified by temporarily forming a junction between a pair at one end. Identification can be definite using a less precise procedure than necessary for formal calibration. For identification, both reference and measuring junction temperatures must be independently known or measured. The approximate temperature of the reference junction can be determined by momentarily shorting the input directly at the indicator input terminals of an instrument that provides reference compensation. Because the thermocouple material might not correspond to the compensation applied, the temperature must be determined separately.

The temperature applied to the measuring junction for identification must be at least 200°C because uncertainty of the thermocouple calibration and of the imposed temperature makes unreliable the emf distinction of thermocouple pairs, such as *E* from *J*, and *K* from *N* or *T* using either ice or boiling water baths. Very similar Types *R* and *S* can be reliably distinguished only at much higher temperatures or by formal calibration.

The Functional Model of Thermoelectric Circuits

For simplicity, the relation between junction temperatures required for accurate measurement was merely asserted in the section on "Practical Thermocouple Circuits," without any physical explanation. Some subtle problems of realistic thermoelectric circuits are difficult to visualize without an explicit circuit model. A simple, practical, and general-purpose model of thermoelectric circuits now explains why those temperature structures are appropriate. More significantly, it makes clear the consequences of deviation from these temperatures. It illuminates the common problems of calibration and inhomogeneity. It explains why the commonplace misconception that the Seebeck emf is localized to junctions can lead to serious error in general thermocouple circuits.

Real thermocouple circuits involve several materials and incidental real junctions, often many more than in Figures 32.24 or 32.25. The incidental uncontrolled materials of feedthrus, terminals, splices, etc. might not be recognized as source elements, yet they can, unnoticed, contribute significant unwanted Seebeck emf to the measurement. Therefore, for practical thermocouple thermometry, it is essential to understand and use a descriptive circuit model that is simple to apply and that forces the attention to locations of potential error so that problems can be avoided.

One such authentic model, the *Functional Model of Thermoelectric Circuits* [1, 2, 5, 7, 29–31], combines (1) a basic thermoelectric circuit element, (2) a single fundamental law that describes the sensitivity of that element, (3) a set of practical corollaries from that law that illuminate its practical implications, and (4) a graphic tool for circuit visualization to simplify analysis. This very simple but nontraditional model is crafted for practical thermometry and is worth studying.

The Basic Thermoelectric Circuit Element

Any thermoelectrically homogeneous nonisothermal segment of arbitrary length of material M within any thermoelement is a *Seebeck cell*. Each such segment across which a net temperature difference exists (Figure 32.29) is a *non-ideal voltage source* with internal resistance $R(T)$. The Seebeck *source emf* must be observed in an "open-circuit" (null current) mode for the most accurate thermometry. Any iR voltage

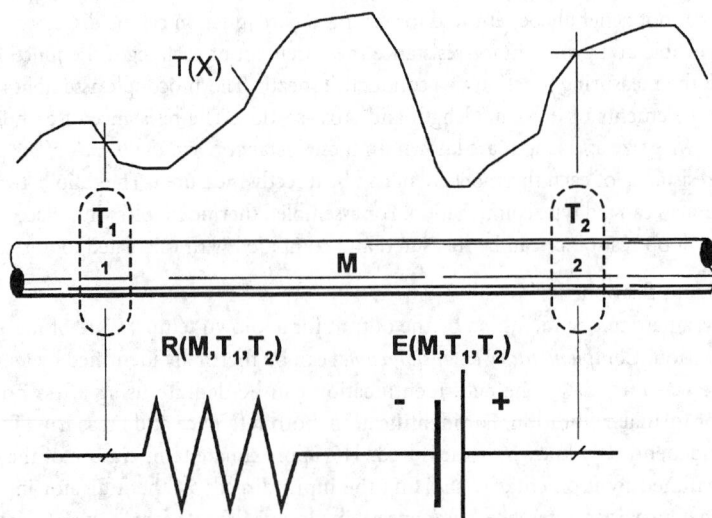

FIGURE 32.29 The *Seebeck emf cell. Every* homogeneous, nonisothermal, electrically conducting material is a source of Seebeck emf. The basic cell, a nonideal voltage source with internal resistance, contributes a Seebeck emf that depends only on the material M and the temperatures at the segment end-points 1 and 2.

drop due to current allowed by a low input resistance of the voltage monitor reduces the Seebeck source emf to a lower terminal Seebeck voltage. For thermometry, that voltage difference causes a temperature error unless corrected.

In *static* thermometry, as in calibration and in typical process measurement, the measuring junction and a substantial length of the adjacent thermoelements are immersed in a stationary and somewhat isothermal zone so that most of the Seebeck emf occurs well apart from the measuring junction in thermoelements where they cross remote temperature transition regions. In *transient* thermometry, the zone of principal temperature difference initially is adjacent to the measuring junction so that the emf arises across a region of spreading extent adjacent to, but not in, the measuring junction (a material interface).

The Law of Seebeck emf

Equation 32.74, $dE = \sigma(T)dT$, is the functional law that governs the emf of the Seebeck cell and, thus, the net voltage of the most complex thermoelectric circuits. It is the *Law of Seebeck emf*. Every thermo-electric aspect of circuit behavior follows from only this simple relation. Physical details of the process that leads to the Seebeck effect are very complex [5, 6]. Nevertheless, this one simple law is entirely consistent with all physical theory and is experimentally confirmed. If this simple relationship *does* apply to the basic Seebeck cell then accurate thermometry is possible. If it does not, then accurate and reliable thermoelectric thermometry is *not* possible.

The source emf of an individual cell of material M, from Equation 32.75, is:

$$\Delta E_M\left(T\right) = E_M\left(T_2\right) - E_M\left(T_1\right) \qquad (32.79)$$

and, for *thermally paired* segments of materials A and B that happen *at any instant* to share a pair of end-point temperatures, *regardless of their residence or proximity in a circuit,*

$$\Delta E_{AB}\left(T\right) = E_{AB}\left(T_2\right) - E_{AB}\left(T_1\right) \qquad (32.80)$$

They need not be directly joined at a junction (nor, indeed, be electrically joined). The values of $E_{AB}(T)$ are obtained directly from the standard thermocouple polynomial defining equations, simpler representations of those equations, tables, or graphs [1, 9–12]. Absolute Seebeck properties for many materials are also available, but are less commonly reported [1, 5, 7, 8]. As is evident from this model, absolute properties could always be used in thermoelectric analysis. They *must* be used in some realistic circumstances where the conventional relative properties are meaningless or where the relative properties are not known.

Either the absolute Seebeck coefficient, the temperature increment across the segment, or both of them, can be either positive or negative. Therefore, the polarity of a cell within a circuit depends both on the material and, unlike the electrochemical emf cell, on the *momentary* sense of temperature difference across the segment. *Polarity, and even function, changes with temperature distribution.*

Corollaries from the Seebeck Law

From the single Law of Seebeck emf (Equation 32.74), five particularly instructive practical corollaries that aid thermoelectric circuit analysis have been recognized. These are the corollaries of: (1) functional roles, (2) functional determinacy, (3) temperature determinacy, (4) emf determinacy, and (5) Seebeck emf [1, 2, 7, 29–31]. These are stated in Table 32.14, abbreviated from [29]. These revealing corollaries relate more directly to practical thermometry than do the three familiar thermocouple "laws" [32] that actually are only oblique alternative corollaries to the sole physical law (Equation 32.74).

The T/X Visualization Sketch

The practical significance of the fundamental law and its corollaries for realistic thermoelectric circuits is revealed by a simple graphic sketch (Figure 32.30). The *T/X sketch* is used for visualization and numerical analysis only. It is *not* drawn to scale. It is *not* used for graphic solution. It illuminates essential

TABLE 32.14   Corollaries from the Law of Seebeck emf

In any circuit of electrically conducting materials that have an absolute Seebeck coefficient $\sigma(T)$, that are each *thermoelectrically homogeneous*, and which follow the Seebeck Law, $dE(T) = \sigma(T)dT$:

1. **The *Corollary of Functional Roles***

   There are three thermoelectric functional roles: *junctions*, "*conductors*," and *Seebeck emf sources*:
   • *Real thermoelectric junctions* are interfaces that ohmically couple dissimilar materials,
   • "*Conductors*" are segments that, *in effect*, individually or in combination, contribute no *net* Seebeck emf, and
   • *All* other segments are sources of Seebeck emf.

2. **The *Corollary of Functional Determinacy***

   Instantaneous thermoelectric roles around a circuit *cannot* be predetermined by physical construction, material choice, or circuit arrangement alone; they are governed by temperature distribution.

3. **The *Corollary of Temperature Determinacy***

   In a circuit with multiple junctions, the temperature of a single junction can be determined from the net Seebeck emf only if the temperatures of all other real junctions are defined.

4. **The *Corollary of emf Determinacy***

   Seebeck emf is produced *only* by thermoelements, but the net Seebeck emf is governed by the temperatures only of *real junctions*.

5. **The *Corollary of Seebeck emf***

   The Seebeck emf of any segment of material M with end-point temperatures T_1 at segment endpoint X_1 and T_2 at segment endpoint X_2 is independent of *temperature distribution, temperature gradient,* or *cross-section* as it is determined by:

   $$E_M(T) = \int_{T_1(x_1)}^{T_2(x_2)} \sigma_M(T)dT = E_M\left(T_2\left(X_2\right)\right) - E_M\left(T_1\left(X_1\right)\right)$$

From References [1, 7, 29, 30].

facts that are not obvious from a conventional electrical schematic or $E(T)$ plot. The T/X sketch shows the temperatures of *all* real junctions in the sequence in which they occur around the circuit. Real junctions are indicated by closed circles. Junctions are joined by thermoelements. The sketch is not drawn to scale, so slopes do not represent temperature gradients.

The T/X sketch reveals that segments that span a temperature interval always occur in pairs — but significantly only in *series* circuits. It is this fact that allows the convenient use of relative Seebeck properties. This conventional simplification does *not* apply to circuits with paralleled branches [2]. The T/X sketch, applied to the thermocouple circuit of Figure 32.26, depicts the significant circuit elements (the junctions and thermoelements) in a way that focuses on their unavoidable thermoelectric *functions* (*Corollary 1*). Figure 32.30 shows a circuit temperature distribution with junction temperatures $T_b \neq T_e$, that were shown (intentionally improperly different) to illustrate a principle and the benefit of the sketch.

Temperatures of Incidental Junctions.

The *reference junction* temperatures must independently be accurately known for measurement. The unknown temperature of the *measuring junction* is to be deduced from the Seebeck voltage. The specific temperatures of all incidental real junctions of a circuit are rarely known accurately in practice. For use in the T/X sketch, *the actual values need not be accurately known*. Nevertheless, if the *relative* values of all incidental junctions are not properly controlled, as described in the section "Practical Thermocouple Circuits," and if some essential junction temperatures are not known well enough to draw the sketch, then *accurate* thermometry cannot be ensured! *The revealing T/X sketch requires no more information than is essential for the physical measurement.*

In the estimate of measurement consequences and to visualize how junction temperatures must be controlled in circuits of many materials, it is sufficient to assume, for qualitative analysis, plausible *relative*

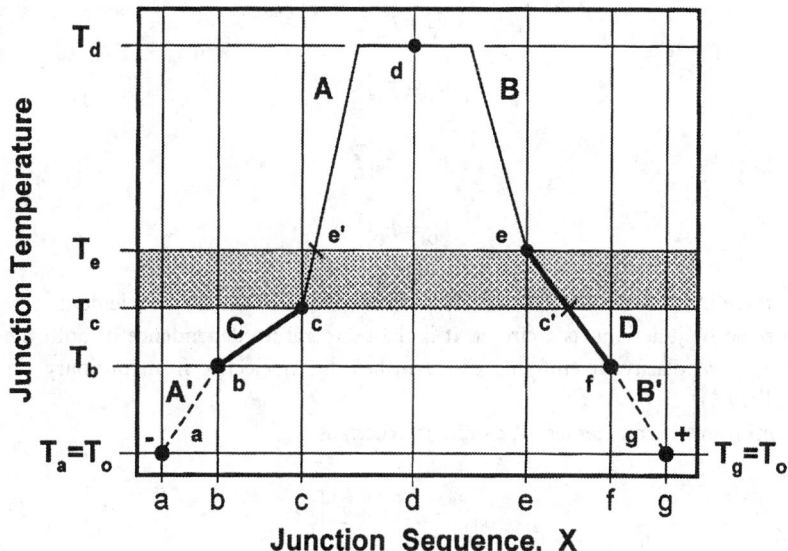

FIGURE 32.30 The T/X sketch for thermoelectric circuit visualization. The temperature of the reference junction(s) and the relation between (not necessarily the specific values of) the temperatures of *all* incidental real junctions in a circuit must be known for accurate thermometry, just as for the use of the T/X sketch. The simple sketch is an aid in recognizing temperatures of incidental junctions that must be controlled to the same temperature. It also makes clear which thermoelements are *thermally paired*, even if not directly joined in the series circuit.

temperature levels and their consequences. The sketch is most often used for visualizing consequences by inspection [1, 2, 29–31]. Nevertheless, it also can aid in quantitative analysis of error for *plausible* temperature distributions.

Virtual Junctions and Thermoelements.
On the T/X sketch, isotherms through *real* junctions intersect some thermoelements. It is useful for analysis to view these intercepts (e.g., c′ and e′), each marked by a tic, as *virtual junctions*. For inspection and analysis, they conveniently delineate the arbitrary temperature end-points of segments. Also, on the diagram, virtual thermoelements, **b-c** and **f-g** are indicated by dashed lines. These segments represent the imaginary thermoelectric source of complementary emf that must be supplied to extend the physical reference temperature, T_r, to the designated reference temperature value, T_0. The virtual thermoelements represent the reference junction compensation.

The T/X sketch aids in assigning segment bounding temperatures and thus a polarity and an emf to each thermoelement in the circuit. The circuit is traversed on the sketch from one instrument terminal, conveniently the negative, to the other terminal. Then, if the absolute Seebeck coefficient of a thermoelement is positive, the emf contribution of the segment adds emf if the temperature increases in proceeding across the segment from the negative toward the positive instrument terminal, etc. For inspection and analysis, it is efficient to consider segments as thermally paired over a temperature zone bounded by isotherms.

Examples.
On the T/X sketch, consider the absolute Seebeck emfs supplied by the four thermoelements of Figure 32.26. As connected in this series circuit, proceeding from negative to positive terminals, the net Seebeck emf of the physical circuit between terminals b and f is:

$$E_{net} = E_C + E_A + E_B + E_D \qquad (32.81)$$

where

$$E_{\mathrm{C}} = E_{\mathrm{C}}\Big|_{T_{\mathrm{b}}}^{T_{\mathrm{c}}}$$

$$E_{\mathrm{A}} = E_{\mathrm{A}}\Big|_{T_{\mathrm{c}}}^{T_{\mathrm{d}}}$$

$$E_{\mathrm{B}} = E_{\mathrm{B}}\Big|_{T_{\mathrm{d}}}^{T_{\mathrm{e}}} \qquad (32.82)$$

$$E_{\mathrm{D}} = E_{\mathrm{D}}\Big|_{T_{\mathrm{e}}}^{T_{\mathrm{f}}}$$

Recall that these individual Seebeck emfs *do* physically exist in the thermoelements whether or not they are connected by junctions as a circuit. It is the temperature dependence of both the magnitude and the momentary polarity of emf that distinguishes thermoelectric from ordinary electric circuit analysis (Corollary 5).

Note that emf from thermoelement *A* can be rewritten as:

$$E_{\mathrm{A}}\Big|_{T_{\mathrm{c}}}^{T_{\mathrm{d}}} = E_{\mathrm{A}}\Big|_{T_{\mathrm{c}}}^{T_{\mathrm{e'}}} + E_{\mathrm{A}}\Big|_{T_{\mathrm{e'}}}^{T_{\mathrm{d}}} \qquad (32.83)$$

This arbitrarily breaks thermoelement *A* into two segments joined at *virtual* junction e'. Thermoelement *D* can be segmented as well. Now, across the shaded temperature zone between T_{c} and T_{e}, there are improperly thermally paired segments c-e' and c'-e. The net emf from these thermally paired segments is:

$$E_{\mathrm{net}}\Big|_{T_{\mathrm{c}}}^{T_{\mathrm{e}}} = E_{\mathrm{A}}\Big|_{T_{\mathrm{c}}}^{T_{\mathrm{e'}}} - E_{\mathrm{D}}\Big|_{T_{\mathrm{c}}}^{T_{\mathrm{e}}} \qquad (32.84)$$

From Equation 32.75, recognize that this is the *relative* Seebeck emf for the *unintended A-D* pair over that arbitrary temperature zone even though they are not directly joined in the circuit. That improper pairing of segments clearly is avoided only if $T_{\mathrm{c}} = T_{\mathrm{e}}$, whatever the temperature. In the instance that Figure 32.26 represents a principal thermocouple *A-B* with matching extension leads *C-D*, materials *A* and *C* are alike and *B* and *D* are alike. If the legs of extension leads *C-D* each exactly match the corresponding legs of thermocouple *AB*, then

$$E_{\mathrm{CD}}\Big|_{T_{\mathrm{c}}}^{T_{\mathrm{e}}} = E_{\mathrm{AB}}\Big|_{T_{\mathrm{c}}}^{T_{\mathrm{e}}} \qquad (32.85)$$

and (*for matching extension leads only*) the accidental pairing is benign.

Otherwise, with *compensating extension leads* ($\sigma_{\mathrm{A}} = \sigma_{\mathrm{C}}$ and $\sigma_{\mathrm{B}} = \sigma_{\mathrm{D}}$), error occurs even if σ_{CD} closely matches σ_{AB} as a pair but not individually. In the instance of *neutral extension leads* where *C* and *D* are alike, the net emf from the pair is *null* over the zone from T_{b} to T_{c}. In this instance, the physical reference junction is recognized as necessarily $T_{\mathrm{r}} = T_{\mathrm{c}} = T_{\mathrm{e}}$, rather than $T_{\mathrm{r}} = T_{\mathrm{b}} = T_{\mathrm{f}}$.

If T_{c} and T_{e} were interchanged, the error would be of different magnitude, not merely of opposite sign, simply because of the temperature distribution (Corollary 2). The unknown temperature of only one junction can be determined; the others, including incidental junctions, must be defined by value or, indirectly, as being isothermal (Corollary 4).

Note that such relative contributions (the null net contribution from two opposed like segments or the inappropriate thermal mispairing between *B* and *D*) are immediately evident simply *by inspection* of the informal *T/X* sketch without tedious algebra. These critical facts of thermoelectric circuits are not evident from the usual electrical schematic or from $E(T)$ plots.

Most real circuits include several incidental uncalibrated materials such as connectors, terminals, splices, feedthrus, short pigtails, or extension leads that each have their own (usually indefinite) Seebeck properties. In some circuits, some thermoelements might accidentally be paralleled. It is important to include them in the sketch to recognize the unnoticed potential error that could be contributed by such circuit elements if they are not held isothermal. Also, multiple extension circuit elements might be improperly connected with crossed polarity. The specific voltage or temperature consequence of these situations is easily calculated for any plausible temperature distribution. The T/X plot was designed specifically to aid in circuit visualization to avoid these very common practical problems and to easily assess their possible impact.

Inhomogeneity

The Nature of Inhomogeneity

A slender thermoelement is inhomogeneous if σ_M (T) varies along its length. The environment during application can introduce irregular inhomogeneity in one or both thermoelements of a pair. The effect is as if one or more additional dissimilar materials had been added to the circuit.

The Significance of Inhomogeneity

Adequate thermoelectric homogeneity is the most critical assumption of thermocouple application. It usually is presumed; rarely is it confirmed. Inhomogeneity is a real but phantom problem. Inhomogeneity often remains undetected even while producing substantial error. It rarely is discovered by even the most careful conventional calibration [1, 7, 11, 29–31, 33, 34]. In physical thermometry, thermocouple *drift* is *invariably* a symptom of progressing *inhomogeneity*. Such change is progressive, often insidious, and usually is misinterpreted [34].

Rather than envisioned correctly as localized degradation of Seebeck coefficient, *drift* is often viewed improperly as a uniform "black-box" decalibration of the thermocouple rather than progressing inhomogeneity. For this reason, it is a far more commonplace problem than recognized by most experienced users. It can occur in use, in fabrication, or in calibration because of mechanical strain, thermal phase change, surface contamination, chemical interaction between materials, evaporation or migration of alloy constituents, transmutation under radiation, and a variety of other realistic causes.

Inhomogeneity error in thermocouples, as manufactured, is intended to be covered by standard tolerances [1, 9, 10, 28]. Troublesome inhomogeneity is most common in used and abused thermocouples, but can sometimes occur in new thermocouples that have been individually calibrated to high temperature. It can occur within and through apparently impervious protective metal sheaths or thermal wells and between bead insulators.

The example, Figure 32.30, illustrated the effect of improperly controlled junction temperatures that resulted in the subtle introduction of relative Seebeck emf from an unintended thermal pairing of segments of homogeneous thermoelements. The analogous *inhomogeneity* problem, best visualized with the T/X sketch, arises when a portion of one or both thermoelements locally changes in Seebeck coefficient over some nonisothermal span of the thermocouple. In effect, this introduces *phantom* dissimilar segments of indefinite graduated property and unrecognized location [35]. This most often occurs over a lengthy region near the measuring junction where the thermoelements are exposed to damaging environments in an oven or process.

The *maximum possible* error of inhomogeneity is determined by the location and magnitude of greatest deviation from normal of the Seebeck coefficient. The *actual* error depends on the momentary distribution of temperature during use. In thermometry and in calibration, the *likely* error introduced by inhomogeneity is moderated. While present only within an isothermal region, inhomogeneities introduce *no* error. Under unfavorable temperature distributions, inhomogeneity error can be extreme and results in peculiar puzzling responses [34]. Changes of relative Seebeck coefficient by more than 60% over 25-cm spans have been observed in individually calibrated, certified, premium-grade fine wire Type R MIMS thermocouples entirely enclosed within intact platinum sheaths and exposed to temperature within the tabulated temperature range [18].

Testing for Inhomogeneity

Many authentic sensitive and accurate tests for thermoelectric inhomogeneity have been developed, and their practical need has been demonstrated; however, they are rarely used [33]. All true inhomogeneity tests require moving an abrupt *step* of temperature progressively along thermoelements.

The commonplace application of a *very narrow symmetric* temperature pattern is the antithesis of an inhomogeneity test. It is *not* a test for inhomogeneity. Its popular use has misled many to discount inhomogeneity as a real and significant problem in accurate thermoelectric thermometry.

Valid tests range from simple methods of low resolution to advanced methods that can accurately resolve inhomogeneity with spatial resolution of a few millimeters [21, 33]. Regrettably, as inhomogeneity errors usually are not recognized, inhomogeneity is not authentically tested in commercial practice nor by calibration laboratories. Nevertheless, invalid though certified NIST-traceable certification to great *precision* is possible on a thermocouple that can be accurately measured to be *severely* inhomogeneous and of indefinite uncertainty [29–31, 33, 34].

Calibration

The measurement uncertainty of most kinds of sensors can be reduced by *individual* calibration. Initial calibration, periodic recalibration of unused thermocouples, and even of degraded *individual* thermocouples, although commonplace, is often less beneficial (or even harmful) and more costly. Such ill-advised thermocouple calibration has been mandated by some "quality assurance" programs.

Surveys reveal a trend for customers to demand progressively higher accuracy of thermometry [36]. In some applications, a 1°F (0.56°C) error now is deemed too large, and 0.1°F tolerance may be specified independent of temperature level. (Compare with tolerances in Table 32.12.) There are a few industrial applications that truly require such accuracy. More often, the specification merely presumes that such accuracy in thermocouple thermometry is routinely attainable merely by calibration. Calibration is an opiate of quality assurance. The illusion of accuracy provided by NIST-traceable certified calibration and purported *in situ* calibration is counterproductive if it is not authentic.

Consideration of the details of thermocouple calibration suggests why authentic accuracy at the 0.1°C level is unlikely in industrial thermoelectric thermometry. The approach to achieving authentic calibration of thermocouple system accuracy (*ignoring* thermal coupling and transient response errors) is illustrated in Figure 32.31. Figure 32.31(a) represents the thermocouple system to be calibrated. Calibration for thermoelectric thermometry must distinguish three system components: (1) the thermocouple circuit, (2) the reference junction temperature compensator, and (3) the monitoring instrument. Each separately affects temperature uncertainty.

The Principal Thermocouple

Calibration of principal thermocouples is performed by immersing the vicinity of the measuring junction in the isothermal region of a bath or oven [37]. Several compact dry-well calibrators available are convenient for calibration at the job site [38]. Fixed-point cells, liquid baths, and fluidized solid beds can be more accurate and are widely used in the calibration laboratory [39]. Sufficient depth of immersion into an isothermal calibration zone, usually at least 10 to 20 times the probe diameter, ensures that conduction of heat along the thermocouple from the environment does not affect the junction temperature.

The appropriate concern for the effect of longitudinal heat conduction on measuring junction temperature has led to the misperception that it is the *junction* that is being calibrated [30]. Clearly, thermocouple calibration is *not* of the measuring junction; it *is* of unidentified segments of thermoelements, remote from the measuring junction, only where they enter the isothermal calibration zone through a temperature transition.

Segments of service-induced inhomogeneity that seriously degrade measurement accuracy are often placed, during unsuccessful attempted recalibration, within the isothermal region where they contribute no Seebeck emf, so inhomogeneity is not discovered. The act of calibration at temperatures above 400°C can actually induce inhomogeneity and degrade accuracy [9, 10]. Not even costly "NIST-traceable"

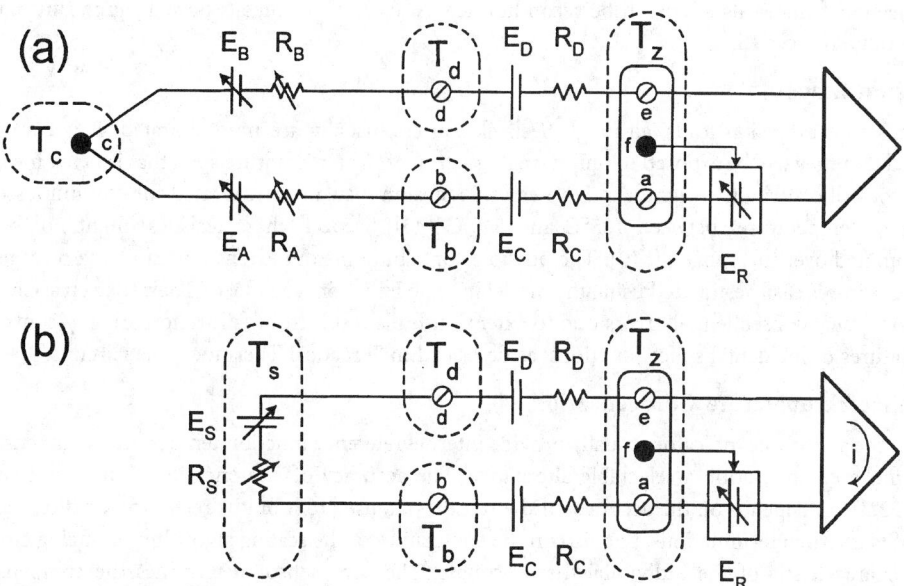

FIGURE 32.31 Contributions of Seebeck and compensating emf in system calibration: (a) the circuit to be calibrated or simulated, (b) a general calibration simulation of a system that includes internal reference junction temperature compensation.

calibration and certification by competent calibration laboratories, without specific assurance of homogeneity, are a guarantee of accuracy at *any* level.

Commercial Tolerances. Standardized tolerances for new thermocouple materials are established by the formal consensus judgment of many experienced producers, users, and calibration standards laboratories staff [10, 24, 28]. The tolerances include not only deviation from the overall Seebeck properties, but also cover inevitable uncertainty from inhomogeneity and irregular deviations from the smoothed standardized characteristic over small spans of temperature. They apply to material as delivered and not exposed to excessive temperature. The conservative tolerances for standard letter-designated thermocouple types as delivered are usually reliable until the thermocouples are exposed too long to excessive temperatures or adverse environments. Recalibration of used thermocouples without separate assurance of homogeneity can be misleading [31].

To address calibration problems presented by undiscovered inhomogeneity, two practices are common. For inexpensive base metal thermocouples, it is presumed that unused thermocouples are homogeneous and uniform within standardized tolerances. The typical Seebeck property of a production batch is characterized by the manufacturer or user by calibrating one or more expended surrogate samples. These are discarded after calibration. Their first-cycle calibration is taken as representative of other thermocouples of the same batch. Used principal thermocouples that have experienced drift are discarded.

A commonplace second approach is recalibration. This is usually in response to drift observed in service. Drift signals progressing inhomogeneity. Apart from an authentic inhomogeneity prescreening, recalibration is not recommended. Even widely promoted *in situ* recalibration of degraded thermocouples is ineffective if temperature distribution will vary in use [31].

Platinum-based thermocouple wire can be annealed full length by electrically heating in air [9–12]. Such annealed thermoelements can be presumed to be free of reversible strain-induced inhomogeneity and, thus, accurate recalibration might be justified. However, annealing cannot reverse decalibration from the migration, absorption, or evaporation of alloy constituents or other chemical contamination; therefore, recalibration, even of used precious metal thermocouples, should be performed only where the likelihood of homogeneity is factually based. Unlike base metal thermoelements, precious metal thermoelement materials often can be reconditioned. Also, precious metals have a significant material salvage value if recycled.

Extension Leads

Compensating extension leads, although Seebeck sources, usually are not calibrated. It is commonly assumed that they will be exposed to only a small fraction of the temperature span that is being measured and, if so, will contribute insignificantly to error. Extension insulations are rated for continuous use to maximum temperatures between 105°C and 540°C [43]. Seebeck characteristics might not be well approximated over this range [1, 10]. The possible contribution from extensions can be very large. The plausible error is easily estimated using the model in "The Functional Model of Thermoelectric Circuits." The more usual consequential errors due to extension leads result from failure to correctly control the temperatures of incidental splice junctions, as described in "Practical Thermocouple Circuits."

Reference Temperature Compensation

Modern thermocouple indicators usually provide internal reference junction temperature compensation. A few make compensation a selectable alternative. The accuracy of reference junction compensation, Figure 32.31(a), depends on the accuracy of the sensor(s), **f**, used to monitor reference junction temperature of isothermal terminations, but also on the conformity of the scaling algorithm or analog circuitry to the characteristic of the individual thermocouple. The zone sensor determines the compensating voltage, E_R. Some monitors have replaceable isothermal terminal blocks that include the zone temperature sensor. Most monitor designs unfortunately do not allow separate thermal calibration of the reference temperature sensor. Simply shorting the indicator input terminals, **b-d**, should produce a temperature indication near, and usually slightly above, ambient temperature. Specifications claim reference uncertainty on the order of 0.1°C to 0.5°C and error contributions usually *are* small but *infrequently* they have been *very* large and insidiously progressive up to *many* times the standard thermocouple tolerances [35].

Monitoring Instruments

The accuracy of conversion of the Seebeck voltage at terminals **b-d** to deduce the temperature T_c requires calibration of the monitoring instrument. Both the accuracy of voltage measurement and of linearization are sources of uncertainty. The scaling accuracy is based on an approximation of the defining $E(T)$, and *not* on the individual thermocouple characteristic.

Instrument designs are varied. The general principles of *authentic* calibration required to approach 0.1°C uncertainty for a thermocouple indicator that provides internal reference temperature compensation are illustrated with the circuit of Figure 32.31(b). A thermocouple simulator/calibrator supplies a well-known voltage E_S that corresponds, for the standard thermocouple type, to a desired temperature calibration point. The simulator must remain in thermal equilibrium during calibration, with irrelevant Seebeck voltage properly nulled .

If the input resistance of the indicator is very *large*, resistance matching of the simulator to the thermocouple is not required. If the indicator has a *low* input resistance or it internally produces current **i** when its terminals b–d are shorted, then a temperature error proportional to the indicator current and the loop resistance will be experienced. For accuracy with such indicators, the simulator R_S must also mimic the source resistance of the individual thermocouple ($R_A + R_B$). However, that thermocouple loop resistance may vary considerably in application as it depends on the temperature distribution around the thermocouple over a broad range of temperatures.

The input terminals of the indicator are **b** and **d**. For accurate calibration, it is essential that $T_b = T_d = T_z$. Some indicators and reference temperature compensators connect the isothermal zone block to the terminals with compensating thermocouple leads. For these, if $T_b \neq T_d \neq T_z$, an indefinite error results that cannot be overcome by calibration.

The simulator voltage must imitate the behavior of only the thermocouple, **b-c-d**. The voltage it must provide depends on whether or not the indicator supplies internal reference temperature compensation. If $T_b = T_d = T_z$, the proper total Seebeck voltage is $E_{AB}\big|_{T_0=0°C}^{T_c} = E_{AB}\big|_{T_z}^{T_c} + E_{AB}\big|_{0°C}^{T_z}$. If the indicator *does not* apply reference temperature compensation, E_R, then the simulator must provide $E_s = E_{AB}\big|_{0°C}^{T_c}$. If the indicator *does* apply E_R, then the simulator must supply *only* $E_s = E_{AB}\big|_{T_r}^{T_c}$.

This requires that the internal T_z of the indicator be accurately known by the simulator. Some simulators allow setting a *presumed* T_z in calibration. Those that do not *cannot* be used directly without error for calibration of the commonplace reference temperature compensating indicators. Many convenient thermocouple instrument calibrators allow setting the desired calibration by temperature rather than by voltage for standardized thermocouple types. This convenience introduces an additional nonlinear scaling and only approximate conformity to the standard characteristic.

Thermocouple Failure and Validation

A thermocouple measurement has *failed* when its indications are beyond uncertainty limits required for a measurement. Thermocouples sometimes fail "open" as junctions separate or thermoelements corrode, yield, or melt. Explicit "open circuit" indication is a promoted feature of many modern thermocouple indicators. This popular indication is useful, but is not an adequate indicator of thermocouple integrity.

The more common but less apparent circuit failures are by inobvious shunting, shorting, or progressing inhomogeneity of thermoelements. These are not detected by the "open circuit" indication. The continual indication of a *plausible* temperature is not proof of *authentic* temperature measurement.

A deliberate or accidental electrical shunting or direct short between thermoelements at ambient temperature local to an electronic compensating reference junction should result in an indication near *ambient* temperature, not 0°C. A thermocouple failure resulting from direct shorting between thermoelements at some distance from the intended measuring or reference junctions can go undetected because it occurs in a temperature region that is very different from the reference temperature and where a valid or plausible temperature measurement can continue to be made *but at an irrelevant and unexpected and unrecognized location* [35]. In some situations, as in predictably hostile environments, such failure of some thermocouples in service is anticipated. Where the failure is because of thermocouple circuit damage, it may not be evident immediately (or ever) without special circuit diagnostics. For such critical situations, special thermoelectric circuits and monitoring methods have been developed to assess the continued circuit integrity, although not the accuracy, of thermometry [35].

Environmental Compatibility

A primary consideration in thermocouple selection is compatibility of the thermoelements with their protective enclosures and of both thermoelements with the environment of measurement. Thermoelements must be protected from corrosive environments and electrically conductive fluid or solid shunts. Plastic, ceramic, or fiber insulators, metallic sheaths, and thermowells are intended to serve this function but may fail in service [1, 9–12, 25]. The compressed granular insulation of mineral-insulated, metal-sheathed thermocouples, if exposed, rapidly absorbs moisture that can seriously degrade resistive isolation [1, 10–12]. When cut, ends of MIMS thermocouple sheaths must be quickly resealed to avoid moisture absorption. Fiber and bead insulators are easily contaminated and can degrade accuracy [1, 10].

Temperature Exposure

The *unavoidable* environmental variables in thermometry are *temperature* and *duration*. Although thermoelements may not be visibly affected by an environment, the Seebeck coefficient might be substantially degraded. Even if only a very thin surface layer of a thermoelement is modified, the Seebeck coefficient could be changed. The properties can be degraded by use for a long period of time, at extreme temperatures, and in adverse environments [1, 9–12]. For these reasons, application should be limited to within the suggested temperature limits (Table 32.15). Calibration should be performed quickly, allowing time only for equilibration, and only to the maximum temperature of intended use. Degradation, observed in use as *instability* or *drift*, is necessarily a symptom of progressing *inhomogeneity* [30, 31].

Sustained excessive temperature alone can quickly degrade a thermoelement. Melting points define the absolute upper temperature range of thermocouple use. However, below this definite value, there are

TABLE 32.15 Temperature Upper Limits For Different Wire Diameters

Dia., mm	0.025	0.127	0.254	0.406	0.813	1.600	3.175	
Dia., in.	0.001	0.005	0.010	0.016	0.032	0.063	0.125	
Dia., AWG	50	35	30	26	20	14	8	
Type	Temperature limit,°C							Type
E	290	325	370	400	510	775	855	E
J	230	275	305	350	460	600	750	J
K, N	690	730	790	840	950	1095	1250	K, N
T	90	110	150	185	270	370	375	T

Note: Recommended limits are guidelines for continual use of bead-insulated thermocouples in closed-end protection tubes in compatible environments. Mineral insulated metal-sheathed thermocouples can have slightly higher limits and tolerate longer exposure.
From References [1, 9, 13].

TABLE 32.16 Environmental Tolerance of Letter Designated Thermocouples

Type	Oxygen rich	Oxygen poor	Reducing	Vacuum	Humid	Below 0°C	Sulfur traces	Neutron radiation
B	Good	Good	Poor	Fair	Good	Poor	Poor	Fair
E	Good	Poor	Poor	Poor	Good	Good	Poor	Poor
J	Fair	Good	Good	Good	Poor	Poor	Fair	Poor
K	Good	Poor	Poor	Poor	Good	Fair	Poor	Good
N	Good	Fair	Poor	Poor	Good	Good	Fair	Good
R,S	Good	Good	Poor	Poor	Good	Fair	Poor	Poor
T	Fair	Fair	Good	Good	Good	Good	Fair	Poor

(Environment spans Oxygen rich, Oxygen poor, Reducing, Vacuum, Humid, Below 0°C, Sulfur traces, Neutron radiation)

From References [1, 4, 9, 10, 43].

indefinite application limits at which stability might be substantially reduced [1, 9–12]. Standard thermocouple tables extend only to the greatest temperature of *recommended* use for benign protected environments, short durations, and for wire materials of 3 mm or greater diameter [1, 9, 10]. Significantly reduced temperature limits apply for materials of smaller cross-section, Table 32.14 [1]. The thermocouples of smaller cross-section degrade more quickly. The standard tolerances apply only to material as manufactured.

Chemical Environment

There are broad guidelines for environmental compatibility. These recommendations are conditional and critically depend on many specifics of exposure. A concise summary of environmental compatibility characteristics is given in Table 32.16. The references should be consulted for detailed compatibility information [1, 4, 9, 22, 23]. Many thermocouple catalogs include tables of *usually* tolerable chemical environments for thermoelements or thermowell materials [19, 20]. Service experience in working with customers in a wide variety of process environments sometimes enables manufacturers to advise users on special problems of compatibility. Often, suitability in a particular service can be assured only by trial.

Metallurgical Change

Alloyed thermoelements can locally change composition by evaporation of constituents when exposed to vacuum. Alloy constituents evaporated from a sheath or from one thermoelement can deposit and coat adjacent thermoelements even through insulators [1, 9–12, 24, 30, 31]. Alloy components can migrate from one thermoelement to the other through the measuring junction. Traces of contaminants can penetrate through pinholes and hairline cracks and can actually diffuse through intact protective sheaths and affect apparently isolated thermoelements. Minute trace impurities in sheaths or insulating materials can interact with thermoelements. Appropriate MIMS construction and material selection usually extends life and increases temperature limits, but damage can occur even within apparently fully sheathed assemblies [34]. Excessive strain can locally substantially modify the Seebeck coefficient of one or both thermoelements resulting in inhomogeneity. Very localized strain, as introduced by sharp bends, usually has little effect on practical thermometry.

Data Acquisition

Thermocouple Indicators

The instrument designer can shelter the user from many troublesome details that complicated measurement for thermocouple pioneers. Microvolt-level signal resolution, high input resistance, stability, noise reduction, nonlinear scaling, and reference temperature compensation are now routinely provided to allow the user to focus on the measurement rather than on details of its indication or recording.

Some thermocouple indicators are hand-held and accommodate only a single thermocouple and only of a single type. A few provide dual thermocouple inputs and allow direct differential temperature measurement for which reference junction compensation *must not* be directly applied. Other bench-top units accommodate several thermocouple inputs of the same or of intermixed thermocouple and other sensor types.

Claimed instrument accuracy could mislead the casual reader. Although many instruments are supplied complete with a thermocouple probe, some accuracy specifications describe only the accuracy of input terminal voltage measurement. The actual instrument accuracy should be determined by electrical calibration (see "Calibration" section). Occasionally, the error of the indicating instrument becomes substantially greater than specified due to drift of reference compensation, offset, or gain. The additional uncertainty of thermocouple calibration, Table 32.13, and the often larger discrepancy between measuring junction temperature and the actual temperature of the object being measured, cannot be included because the latter are governed by details of heat transfer and transient thermal response.

Thermocouple Transmitters

A less familiar form of thermocouple signal conditioning, the *thermocouple transmitter*, is commonplace in process industries [4, 39]. Most two-wire thermocouple transmitters (not an RF wireless transmitter) are single-channel devices that perform reference junction compensation, linearizing, and isolation, and transmit the conditioned information to a remote monitoring site [1]. The transmitter converts the temperature-scaled Seebeck emf to an analogous signal for wire transmission over long lines to a remote recording or monitoring location. Such inexpensive and compact single-sensor signal conditioners are available specialized to most process variables. Thermocouple transmitters are well suited for monitoring slowly changing temperatures at monitoring sites distant from the measuring junction. Many are electrically isolated for safety. Some are adapted for use with the DIN rail mounting system. A transmitter can be installed near each thermocouple. Some are small enough to install in the connection head of a thermowell.

The transmitted signal can take either current-modulating or voltage-modulating form following ISA standards. The current-style transmitter temperature modulates, in proportion to temperature, a dc current supplied from the monitoring site. The signal ranges between 4 mA and 20 mA. The current-style transmitter conveys information over great distances without the voltage loss attendant to long-line voltage transmission. Some of the voltage-style transmitters modulate a supplied voltage over the span from 1 Vdc to 5 Vdc. Offset and scaling for some are adjusted at the transmitter. Some transmitters now have a local digital readout, and the scaling and offset can be remotely adjusted [4, 39].

Recording

Some multichannel thermocouple data recorders are now battery powered for stand-alone use and small enough to be hand held [41]. Some simpler recorders now cost less than U.S.$200. More sophisticated units accommodate multiple thermocouples of the same or intermixed type. Larger digital data loggers with multiplexers that sequentially sample the output from individual thermocouples can accommodate up to several hundred inputs and can record a mix of both thermocouple and other signals at sample intervals programmable from a few milliseconds to hours or days. Desirable three-wire switching between thermocouples is commonplace.

Personal computers now accommodate inexpensive internal digital data acquisition boards that can convert the computer to a high-resolution temperature recording system. Laptop computers can accept special palm-sized PCMCIA plug-in cards for multiple thermocouple input. Digital computer based data acquisition systems with software specialized for data acquisition, analysis, and presentation, allow the user to apply special reference temperature compensation, customized linearization of thermocouple scaling functions for either standard or individual calibrations and of special thermocouple types. Computer-based thermocouple systems with suitable commercial data acquisition and other software, provide for experiment design, prediction, data acquisition, information management, analysis, reporting, archiving, and communication within a single compact fieldable laptop computer [42].

Signal Transmission

The length of the principal thermocouple should be as short as feasible. The thermoelements should be continuous between measuring junction and reference junctions for best accuracy, but lengthy extension cables can be used if necessary (see "Practical Thermocouple Circuits" section). In field experiments and in process monitoring, recording might be separated from the measuring junction by hundreds of meters. There are several satisfactory possibilities for signal transmission over great distances.

Extension Cable

Matching and compensating thermoelectric extension cables (see "Practical Thermocouple Circuits" section) are available in a wide variety of constructions and insulations [4, 43]. An extension cable, as a transmission line, can be a single twisted thermoelectric pair or a multipair cable designed for suppression of both electrostatic and electromagnetic noise and for environmental protection. While inexpensive unshielded insulated extension wire is available, in the common situation where electric noise is likely, each extension should be an insulated, individually shielded, and twisted pair. Where several extension pairs from a cluster of nearby thermocouples extend to a remote recorder, multipair thermoelectric cable is available that adds an overall electrostatic shield, mechanical strengthening reinforcement, and a robust environmental overjacket to protect from mechanical damage and chemical intrusion in harsh process environments [4, 42].

Such elaborate thermoelectric extension is more expensive than copper instrumentation cable; thus, alternative transmission systems should be considered. The proper use of twisted pair copper instrumentation cable as neutral extension leads, extending from external reference junction temperature compensation, can be sufficient and less costly than thermoelectric extensions.

Higher quality twisted-shielded-pair instrumentation cable can also improve signal fidelity in transient thermometry. For such neutral extension leads, insulated solid *unplated* copper instrumentation pair only should be used because nonuniform plating thickness can result in irregularly distributed inhomogeneity and cause Seebeck emf noise. Conventional *coaxial* instrumentation cable should never be used as lengthy neutral thermoelectric extension leads because the outer braid and center conductors have very different Seebeck coefficients. If not at uniform temperature, the cable can be a significant source of Seebeck noise emf.

Thermoelectric extensions and "pigtailed" assemblies should be checked full-length for hidden junctions between slightly mismatched spliced thermoelements and for deliberate, but hidden, pair splices incorrectly made in crossed polarity [10]. Pass a narrow heat source along the full length of the cable, while monitoring for local jumps of output local to such unintended junctions, using a thermocouple indicator. This test is very effective to pinpoint hidden junctions. It does not test for inhomogeneity [10, 30].

Optic fiber cable with electro-optic transmitters can be used for electric noise-free cable transmission over a long distance between signal conditioning that is local to the thermocouple and to a remote recorder or monitor.

Wireless Transmission

In several situations, copper wire transmission is less effective, more expensive, or less convenient than wireless methods. Several low-cost wireless (not thermocouple transmitter) data acquisition systems are now adapted to combine thermocouple data acquisition and remote signal transmission [44]. *Radio modems* communicate data from a measuring location to recording sites as much as 5 km away. Most provide spread spectrum transmission that does not require a communications license. Wireless data transmission can be particularly economical where the distance between the measuring junction and recorder is great, when many thermocouples must be recorded, when little setup time is available for field cabling, and where the measuring setup must be moved occasionally. Some systems combine data logging and radio communication in a single unit. Others require the use of a thermocouple transmitter, a data logger, and the radio modem to form a system.

Sources of Thermocouple Application Information

Technical Papers

The most concentrated and broadest sources of refereed technical papers addressing thermocouple thermometry are the serial proceedings of a decennial international symposium on temperature [25]. The six volumes to date are a rich and reliable source of application data. Proceedings of annual symposia of the ISA and NCSL also have a few papers describing current developments in thermometry. Reference [26] has thermoelectric data on the widest variety of common and nonstandard materials. Current papers devoted specifically, not incidentally, to thermocouples are infrequent and are distributed broadly across trade magazines and journals of professional societies. Essential details of thermocouple thermometry are more often (poorly) described only incidentally in reports of experimental studies. Some manufacturers publish subscription technical journals specializing in thermometry [45].

One measurement journal frequently publishes articles on applied thermometry and, annually in the June issue, publishes an extensive directory of manufacturers of thermocouple materials and related instrumentation [46]. The directory includes, for several dozen listed manufacturers, concise descriptions of selected products and current prices.

Books, Reports, and Standards

A very detailed and authoritative NIST monograph is the primary source for Seebeck properties and physical characteristics of letter-designated thermocouples [9]. Standards of the ASTM and ISA adopt the values and complement the NIST report [10, 24]. U.S. standards are reviewed and updated at intervals of 5 years or less. The ASTM also publishes interim tables of Seebeck emf for a few popular materials that have not been standardized [10]. Many standards are collected in a single, annually revised volume specializing in thermometry [10]. The hardcopy thermocouple tables from these volumes are now available as functional computer programs that calculate $E(T)$ and $T(E)$ of the NIST 175 document over ranges, at intervals, and in units selected by the user [46].

A comprehensive thermocouple application manual is published by the ASTM [1]. Many commercial publishers issue new reference and textbooks addressing measurement. Most of these include a brief obligatory section or chapter describing the bare elements of thermocouple thermometry. A few devote an entire volume to thermometry. There are many deeper scientific treatments of thermoelectricity, mostly in the historic literature [1, 11, 12].

Many national and international consensus standards organizations publish thermoelectric test methods and material characterizations [10, 21, 25]. Standards are instructive as well as being formalizations of procedures and materials. ASTM Standards and indexes to them are now available in computer CD-ROM file format. Some standards now are available on the World Wide Web or by FAX. Societies with committees that specialize in thermoelectric standards now have Internet Web sites (such as *http://www.isa.org* and *astm.org*) that provide information about available standards and solicit on-line technical questions from users.

Workshops

Annually, there are several excellent workshops of a few hours or days duration devoted to measurement, thermometry, and even thermocouples. Professional societies such as the ISA, in association with special symposia or annual meetings, present special tutorial workshops. Some are offered without fee by manufacturers. Other measurement courses are offered by independent measurement specialists. Announcements of these workshops are published in the technical and trade journals.

Trade Literature

Abundant, free, commercial catalog literature, handbooks, and Web site advertising describe thermocouple hardware and data acquisition products. Some manufacturers publish elaborate catalogs of thermocouple-related hardware, software, and books [4]. Most include tutorial material. A few include technical reprints, discussion of principles and practices, and present extensive tables of thermoelectric characteristics and physical properties of thermocouple materials. Demonstration programs and product information are available from manufacturers on CD-ROM and on the Internet. There are several trade journals that relate to measurement, explore current issues and developments, and are distributed free to qualified recipients.

Caveat

The reader is cautioned that the extensive current and historic literature of thermoelectric thermometry, tutorial articles (*even this one*), standards, specifications, and advice from "experts," all must be studied very critically with an authentic thermoelectric model in mind. At every level of sophistication, from the promotional to the most esteemed esoteric and sophisticated mathematical and physical thermoelectric theory, innocently propagated misconceptions concerning the thermoelectric effects are very commonplace.

Summary

The Seebeck effect can be used to measure temperature with finer spatial and time resolution, over a broader temperature range, in more diverse geometries, and at lower cost than most other electric temperature sensors. The unique versatility of the thermocouple ensures that it will continue as the thermometry means of choice for many applications despite competition from an increasing variety and abundance of alternative special-purpose thermometers. However, the misleading seeming simplicity of the idealized thermocouple and the convenience and apparent accuracy afforded by modern signal conditioning instruments can be misleading. Correct application is simple.

Properly used within its limitations, the thermocouple is capable of accurate reliable thermometry. Improperly understood, the thermocouple is subject to misapplication and inconspicuous, but very significant, error. This precautionary overview described pitfalls and their avoidance in thermocouple practice. This can provide the receptive reader with an authentic basis for the knowledgeable use of even the most complex series and generalized thermoelectric circuits in circuit design, diverse applications, particularly in thermometry. The simple Functional Model presented here (and perhaps requiring some study) was deliberately crafted to aid the user in clearly distinguishing the authentic information from the misleading, and to aid in evaluating the designs and specifications of manufacturers of thermoelectric products.

References

1. R. M. Park (ed.), *Manual on the Use of Thermocouples in Temperature Measurement*, MNL 12, 4th ed., Philadelphia, PA, American Society for Testing and Materials, 1993.
2. R. P. Reed, Thermal effects in industrial electronics circuits, in J. D. Irwin (ed.), *CRC Industrial Electronics Handbook*, Boca Raton, FL, CRC Press, 1996, 57–70.

3. *Catalog*, TRANSCAT/EIL, Rochester, NY, 1998.

4. *Catalog, The Temperature Handbook*, Issue 29, Omega Engineering, Inc., Stamford, CT, 1995.

5. D. D. Pollock, *Physics of Engineering Materials*, Englewood Cliffs, NJ, Prentice-Hall, 1990.

6. D. M. Rowe (ed.), *CRC Handbook of Thermoelectrics*, Boca Raton, FL, CRC Press, 1995.

7. R. P. Reed, Principles of thermoelectric thermometry, in R. M. Park (ed.), *Manual on the Use of Thermocouples in Temperature Measurement*, MNL 12, 4th ed., Philadelphia, PA, American Society for Testing and Materials, 1993, chap. 2, 4–42.

8. R. P. Reed, Absolute Seebeck thermoelectric characteristics — principles, significance, and applications, in J. F. Schooley (ed.), *Temperature, Its Measurement and Control in Science and Industry*, Vol. 6, Part 1, New York, American Institute of Physics, 1992, 503-508.

9. G. W. Burns, M. G. Scroger, G. F. Strouse, M. C. Croarkin, and W. F. Guthrie, *Temperature-Electromotive Force Reference Functions and Tables for the Letter-Designated Thermocouple Types Based on the ITS-90*, NIST, Monograph 175, Washington, D.C., Department of Commerce, 1993.

10. *Annual Book of ASTM Standards, Temperature Measurement*, 14.03, Philadelphia, PA, American Society for Testing and Materials, 1998.

11. J. V. Nicholas and D. R. White, *Traceable Temperatures — An Introduction to Temperature Measurement and Calibration*, New York, John Wiley & Sons, 1994.

12. T. W. Kerlin and R. L. Shepard, *Industrial Temperature Measurement*, Philadelphia, PA, Instrument Society of America, 1982.

13. *Catalog*, Analog Devices, Inc., Norwood, MA.

14. *Catalog, Equipment for Temperature Measurement and Control Systems.*, Hades Manufacturing Corp., Farmingdale, NY.

15. R. Morrison, *Noise and Other Interfering Signals*, New York, John Wiley & Sons, 1992.

16. H. W. Markenstein, Proper shielding reduces EMI, *Electron. Packaging & Production*, 37, 72-78, 1997.

17. *Catalog*, Thermagon, Inc., Cleveland, OH.

18. N. R. Keltner and J. V. Beck, Surface temperature measurement errors, *J. Heat Transfer*, 105, 312-318, 1983.

19. *Catalog*, RdF Corporation, Hudson, NH.

20. *Catalog, Temperature Measurement Handbook*, Vol. IX, NANMAC Corporation, Framingham, MA.

21. R. P. Reed, Convolution & deconvolution in measurement and control, professional course, *Measurements & Control*, (178-188), 1997, 1998.

22. H. M. Hashemian, K. M. Peterson, D. W. Mitchell, M. Hashemian, and D. D. Beverly, In situ response time testing of thermocouples, *ISA Trans.*, 29, 1986.

23. R. P. Reed, The transient response of embedded thin film temperature sensors, *Temperature, Its Measurement and Control in Science and Industry*, Vol. 4, Part 3, New York, Instrument Society of America, 1972.

24. *Standard, Temperature Measurement Thermocouples*, ISA/ANSI Standard MC96.1-1982, Research Triangle Park, NC, ISA International Society for Measurement and Control, 1982. (Standard withdrawn 1993.

25. *Temperature, Its Measurement and Control in Science and Industry*, Vols. 2–6 (Serial), New York, American Institute of Physics, 1942–1992.

26. P. A. Kinzie, *Thermocouple Temperature Measurement*, New York, Wiley-Interscience, 1973.

27. H. L. Anderson (ed.), *Physics Vade Mecum*, New York, American Institute of Physics, 1981.

28. *Standard, Norme Internationale/International Standard*, IEC 584, Parts 1–3, Thermocouples, International Electrotechnical Commission, Geneva, Switzerland, 1995.

29. R. P. Reed, Thermoelectric thermometry: A functional model, in J. F. Schooley (ed.), *Temperature, Its Measurement and Control in Science and Industry*, vol. 5, Part 2, New York, American Institute of Physics, 1982, 915-922.

30. R. P. Reed, Ya can't calibrate a thermocouple junction! *Measurements & Control*, Part 1. *Why not?*, 178, 137–145; Part 2. *So What?*, 179, 93–100, 1996.

31. R. P. Reed, Thermocouples: calibration, traceability, instability, and inhomogeneity, *Isotech J. Thermometry*, 7(2), 91-114, 1996.

32. W. F. Roeser, Thermoelectric thermometry, *J. Appl. Phys.*, 11, 213-232, 1940.

33. R. P. Reed, Thermoelectric inhomogeneity testing. Part I: Principles; Part II: Advanced methods, in J. F. Schooley (ed.), *Temperature, Its Measurement and Control in Science and Industry*, Vol. 6, Part 1, New York, American Institute of Physics, 1992, 519-530.

34. W. Rosch, A. Fripp, W. Debnum, S. Sorokach, and R. Simchick, Damage of fine diameter platinum sheathed Type R thermocouples at temperatures between 950 and 1100°C, in J. F. Schooley (ed.), *Temperature, Its Measurement and Control in Science and Industry*, Vol. 6, Part 1, New York, American Institute of Physics, 1992, 569-574.

35. R. P. Reed, Validation diagnostics for defective thermocouple circuits, in J. F. Schooley (ed.), *Temperature, Its Measurement and Control in Science and Industry*, 5, Part 2, New York, American Institute of Physics, 1982, 915-922.

36. *Survey, Measurement Needs Tracking Study — 1997*, Keithley Instruments, Cleveland, OH, 1997.

37. J. P. Tavener, Temperature calibration, *Measurements and Control*, Sept., 160-164, 1986.

38. T. B. Fisher, Selecting a dry well calibrator, *Measurements and Control*, (185), 105, 1997.

39. *Catalog, Reference Manual for Temperature Products and Services*, 1st ed., Isothermal Technology Ltd., Merseyside, England, 1997.

40. *Catalog*, Moore Industries, Sepulveda, CA, 1997.

41. *Catalog*, DCC Corporation, Pennsauken, NJ, 1997.

42. *Catalog, Instrumentation Reference and Catalogue*, National Instruments, Austin, TX, 1997.

43. *Catalog, Temperature Sensors, Wire and Cable*, Watlow Gordon, Richmond, IL, 1997.

44. *Catalog*, ENCOM Radio Services, Calgary, Alberta, Canada.

45. *Isotech Journal of Thermometry*, ISSN 0968-347X, Isothermal Technology, Ltd., Merseyside, England.

46. *Measurements and Control* magazine, ISSN 0148-0057, Measurements & Data Corporation, Pittsburgh, PA.

47. R. P. Reed, A comparison of programs that convert thermocouple properties to the 1990 international temperature and voltage scales, *Measurements and Control*, 30(177), 105-109, 1996.

* *Note:* Mention of *representative* products is to introduce the reader to the variety of available thermocouple-related hardware and is not an endorsement.

32.5 Semiconductor Junction Thermometers

Randy Frank

Temperature sensors can be easily produced with semiconductor processing technology by using the temperature characteristics of the *pn junction*. The batch processing and well-defined manufacturing processes associated with semiconductor technology can provide low cost and consistent quality temperature sensors. The temperature sensitivity of the *pn junction* is part of the transistor's defining equations and is quite predictable over the typical semiconductor operating range of −55°C to +150°C.

Most semiconductor junction temperature sensors use a diode-connected bipolar transistor (short-circuited collector-base junction) [1]. A constant current passed through the base-emitter junction produces a junction voltage between the base and emitter (V_{be}) that is a linear function of the absolute temperature (Figure 32.32). The overall forward voltage drop has a temperature coefficient of approximately 2 mV °C⁻¹.

When compared to a thermocouple or a resistive temperature device (RTD), the temperature coefficient of a semiconductor sensor is larger but still quite small. Also, the semiconductor sensor's forward voltage has an offset that varies significantly from unit to unit. However, the semiconductor junction voltage vs. temperature is much more linear than that of a thermocouple or RTD. In addition to the

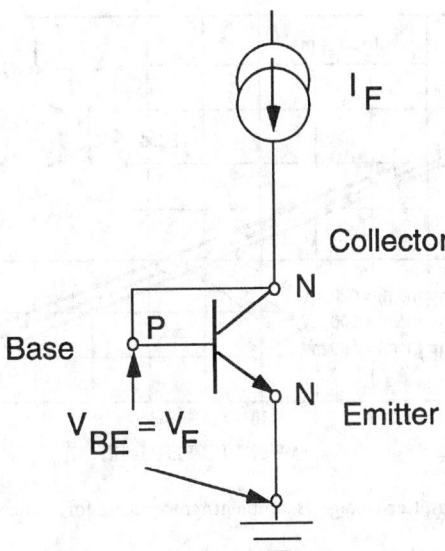

FIGURE 32.32 Bipolar transistor configured as a temperature sensor. The base of the transistor is shorted to the collector. A constant current flowing in the remaining *pn* (base to emitter) junction produces a forward voltage drop V_F proportional to temperature.

temperature-sensing element, circuitry is easily integrated to produce a monolithic temperature sensor with an output that can be easily interfaced to a microcontroller and to provide features that are useful in specific applications. For example, by using an *embedded temperature sensor* with additional circuitry, protection features can be added to integrated circuits (ICs). A temperature sensor becomes an embedded item in a semiconductor product when it has a secondary or supplemental purpose instead of the primary function.

The Transistor as a Temperature Sensor

A common semiconductor product for temperature sensing is a small-signal transistor such as a 2N2222 or a 2N3904. By selecting a narrow portion of the overall distribution of the V_{be} for these devices, a temperature sensor with a lower variation in characteristics can be obtained. The lower variation can provide a part-for-part replacement when a tolerance of only a few percent is acceptable. This device (formerly offered as an MTS102 but no longer in production) demonstrates the performance characteristics of the transistor used as a temperature sensor [2].

As shown in Figure 32.33 [2], a silicon temperature sensor has a nominal output of 730 mV at –40°C and an output of 300 mV at 150°C. The narrowly specified V_{be} ranges between 580 mV and 620 mV at 25°C. The linearity error, or variation from a straight line, of this device is shown in Figure 32.34 [2]. The total accuracy is within ±3.0 mV including nonlinearity which is typically within ±1°C in the range of –40°C to 150°C. These readings are made with a constant (collector) current of 0.1 mA, passing through the device to minimize the effect of self-heating of the junction. When the constant current applied is larger than 0.1 mA, the effect of self-heating in the device must be taken into account. The variation of the V_{be} with current is shown in Figure 32.35 [2].

Thermal Properties of Semiconductors: Defining Equations

A constant forward current supplied through an ideal silicon *pn* junction produces a forward voltage drop, V_F [3, 4]:

$$V_F = V_{be} = (kT/q)\ln(I_F/I_S)$$

(32.86)

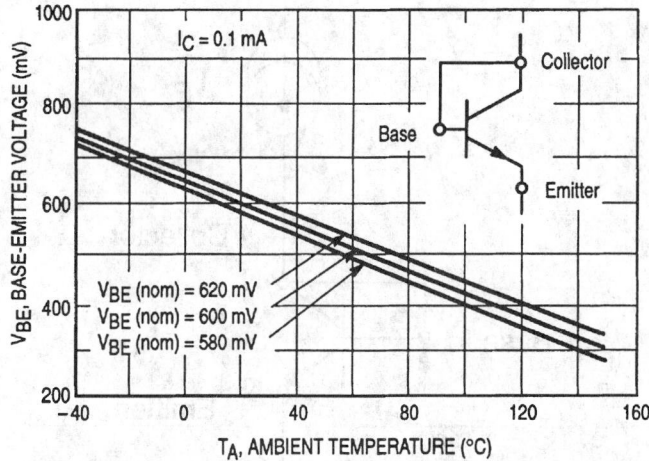

FIGURE 32.33   Base–emitter voltage vs. ambient temperature for a silicon temperature sensor.

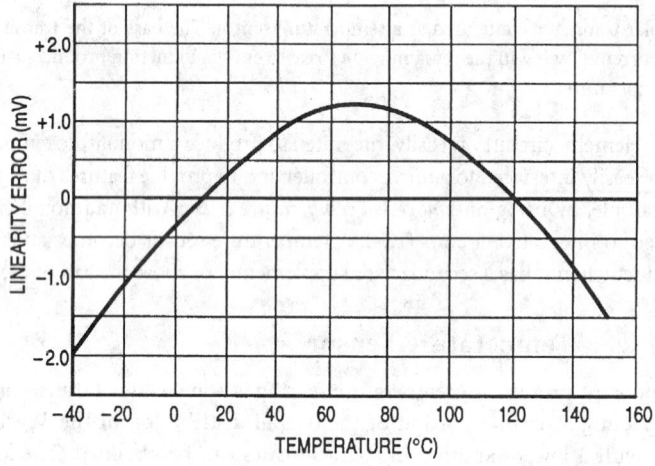

FIGURE 32.34   Linearity error (in mV) vs. temperature for a silicon temperature sensor.

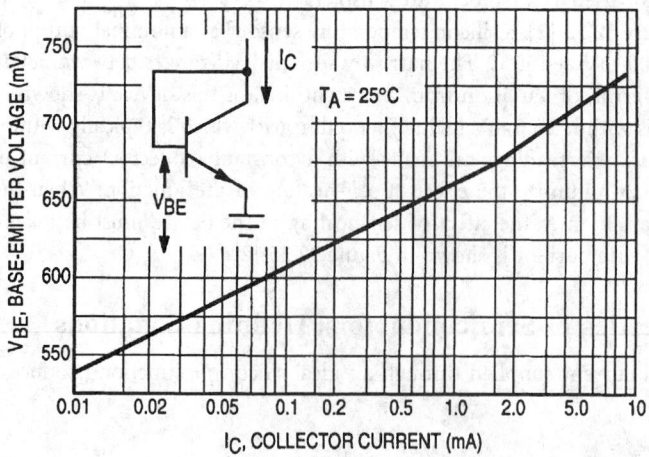

FIGURE 32.35   Base–emitter voltage vs. collector–emitter current.

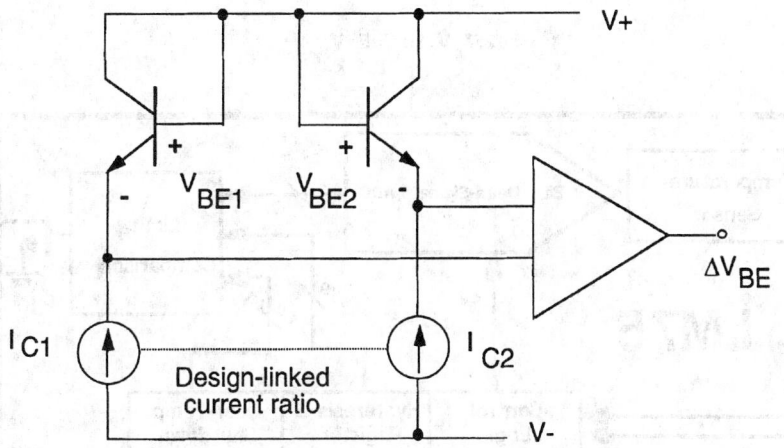

FIGURE 32.36 Differential pair formed by two *pn* junctions. The transistors are diode connected to form a temperature sensor independent of variations in source current.

where k = Boltzmann's constant (1.38×10^{-23} J K^{-1})

T = Temperature (K)

q = Charge of electron (1.6×10^{-19} C)

I_F = Forward current (A)

I_S = Junction's reverse saturation current (A)

For constant I_S, the junction voltage (V_{be}) would be directly proportional to absolute temperature. Unfortunately, I_S is temperature dependent and varies with the cube of absolute temperature. As a result, V_F has an overall temperature coefficient of approximately -2mV °C^{-1}.

To reduce the temperature variation, a *band gap reference* is formed based on two adjacent and essentially identical-behavior transistors with proportional emitter area designed in an integrated circuit process. The two base–emitter junctions are biased with different current densities (I/A), but the ratio of current densities is essentially constant over the operating temperature range (-55°C to $+150$°C). The following equation shows how the differential voltage (ΔV_{be}) is related to the current (I) and emitter area (A) of the respective transistors:

$$V_{be1} - V_{be2} = \Delta V_{be} = \left(kT/q\right)\ln\left(\left(I_1/A_1\right)/\left(I_2/A_2\right)\right) \qquad (32.87)$$

The differential voltage appearing at the output can be amplified as shown in Figure 32.36 and used as a direct indication of absolute temperature. Additional circuitry can eliminate the offset voltage at 0°C and provide an output in degrees Celsius or degrees Fahrenheit.

The ability to obtain temperature sensing using semiconductor processing techniques has two significant consequences: (1) semiconductor processing and integrated circuit design can be used to improve the temperature sensor's performance for specific applications, and (2) temperature sensors can be integrated within other integrated circuits to obtain additional features. The next two sections explain these approaches.

Integrated Temperature Sensors

Once a temperature sensor can be manufactured using semiconductor processing techniques, a number of shortcomings of the sensor can be corrected by additional circuitry integrated into the sensor or by using circuit techniques external to the sensor. The linearity improvement, addition of precision voltage references, precision voltage amplifiers, and digital output for direct interface to a microcontroller (MCU)

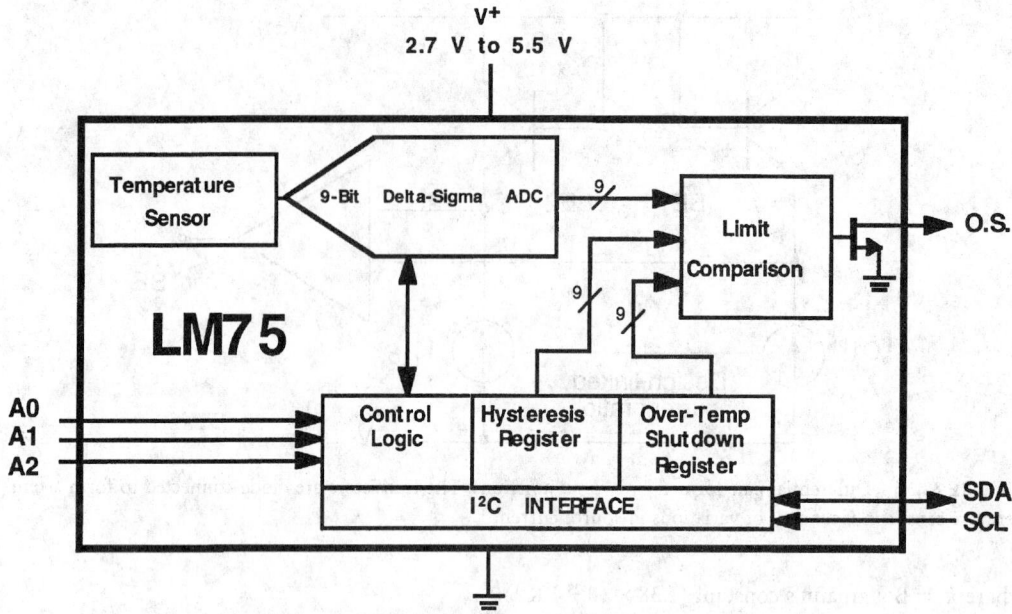

FIGURE 32.37   Block diagram of a monolithic digital-output temperature sensor. (Courtesy National Semiconductor Corp.)

are among the enhancements possible. Furthermore, resistance-measuring circuitry (i.e., RTD sensors) or cold junction compensation (i.e., thermocouple sensors) are not required. Three integrated circuits and one external circuit example are discussed.

Integrated Digital Temperature Sensor

A *monolithic* (one piece of silicon) semiconductor junction temperature sensor (LM75 from National Semiconductor) that incorporates several features, including a digital output, is shown in Figure 32.37 [3]. The analog signal of the temperature sensor is converted to digital format by an on-board sigma–delta converter [3]. Digital communication is provided directly to a host microcontroller through a serial two-wire interface. The sensor has a software-programmable setpoint that can be used to terminate the operation of the controller or implementing protection [5]. To avoid false triggering, a user-programmable number of comparisons (up to six successive over-temperature occurrences) can be implemented.

Eight different sensors can be operated on the bus. Resolution is ±1/2°C and the accuracy is ±2% from −25°C to +100°C. The sensor consumes 250 µA during operation and only 10 µA in sleep mode.

Analog Output Integrated Temperature Sensor

Another approach to integrated temperature sensing is shown in Figure 32.38 [4].The TMP-1 resistor-programmable temperature controller features a 5-mV °C⁻¹ output high and low set points and over- and under-temperature output. A low-drift voltage reference is also included in the 70 mil × 78 mil (2.76 mm × 3.07 mm) design. Figure 32.39 shows a photomicrograph of the silicon die. The TMP-1 is specified for operation between −55°C and +125°C, with ±1°C accuracy over the entire range.

Digital Output Temperature Sensor

A direct-to-digital temperature sensor has been designed for multi-drop temperature sensing applications [6]. A unique serial number is etched onto each device. The 64-bit read-only memory (ROM) identifies the temperature of a particular sensor in a measurement system, with several sensors providing readings from different locations. The signal can be transmitted for distances up to 300 m.

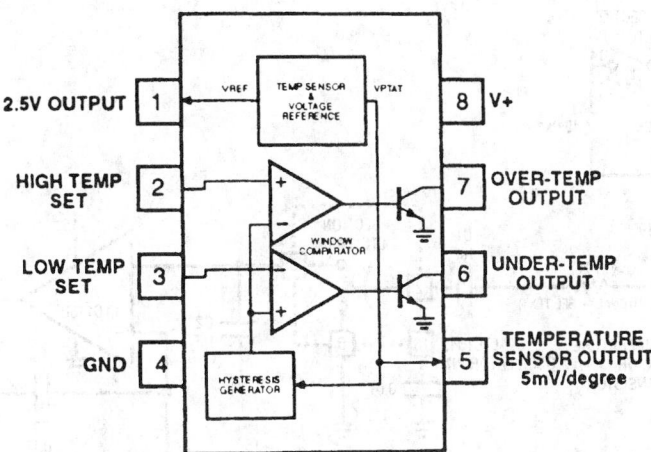

FIGURE 32.38 Block diagram and pinout of TMP-1 monolithic, programmable temperature controller. (Courtesy of Analog Devices, Inc.)

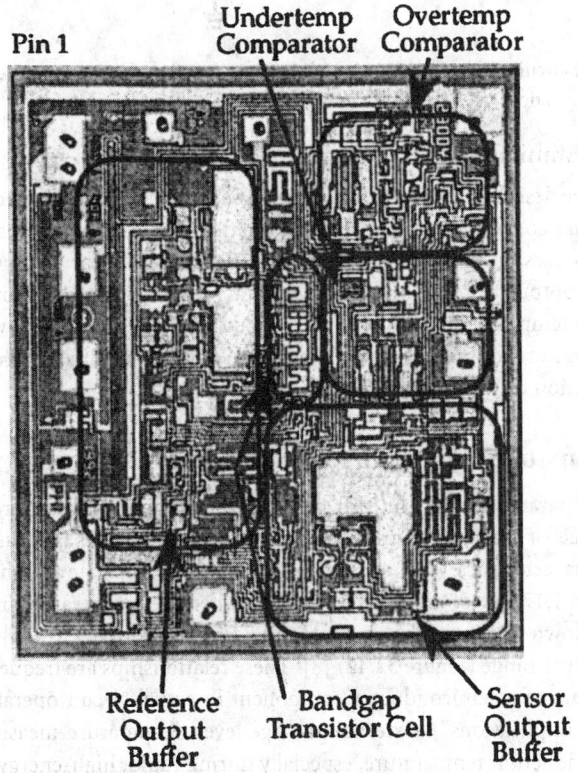

FIGURE 32.39 Detail of TMP-1 die. Note the relative size of the bandgap transistor cell to the other circuitry included in the monolithic device. (Courtesy of Analog Devices, Inc.)

The temperature sensor operates from −55°C to 125°C with the power supplied from the data line. The measurement is resolved in 0.5°C increments as a 9-bit digital value, with the conversion occurring within 200 ms. User-definable alarm settings are included in the device.

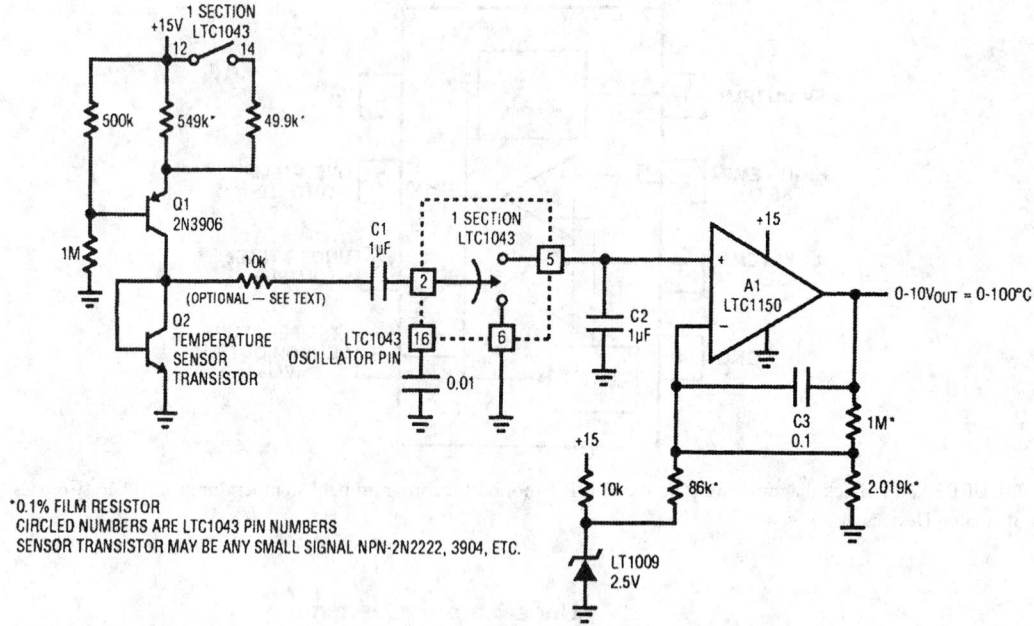

FIGURE 32.40   External circuitry provides ΔV_{be}-based thermometer that does not require calibration. (Courtesy of Linear Technologies Corporation.)

External Circuitry Eliminates Calibration

External circuitry can be designed to utilize the inherent sensing capability of low-cost transistors without incurring the additional cost of integrated circuitry or the need to calibrate each device. The circuit shown in Figure 32.40 uses an integrated circuit (LTC1043 from Linear Technology Corporation) to provide a 0 V to 10 V output from 0°C to 100°C, with an accuracy of ±1°C using any common small signal transistor as the temperature sensor [7]. The circuitry establishes a ΔV_{be} vs. current relationship that is constant regardless of the V_{be} diode's absolute value. Substituting different transistors from multiple sources showed a variation of less than 0.4°C.

Other Applications of Semiconductor Sensing Techniques

Several semiconductor parameters vary linearly over the operating temperature range. Power MOSFETs used to switch high levels of current (typically several amperes) at voltages that can exceed 500 V provide an example of these characteristics. As shown in Figure 32.41 [8], the gate threshold voltage of a power MOSFET changes from 1.17 to 0.65 times its 25°C value when the temperature increases from −40°C to 150°C. Also, the breakdown voltage of the power MOSFET varies from 0.9 to 1.18 times its value at 25°C over the same temperature range (Figure 32.42) [8]. These relationships are frequently used to determine the junction temperature of a semiconductor component in actual circuit operation during the design phase (see "Reliability Implications"). External package level temperature measurements can be many degrees lower than the junction temperature, especially during rapid, high-energy switching events. The actual junction temperature and the resulting effect on semiconductor parameters must be taken into account for the proper application of semiconductor devices.

Polysilicon diodes (and resistors) that are isolated from the power MOSFET can be produced as part of the semiconductor manufacturing process with minor process modifications. The diodes can be used as temperature sensing elements in an actual application [9]. The thermal sensing that is performed by the polysilicon elements is a significant improvement over power device temperature sensing that is performed by an external temperature sensing element. By sensing with polysilicon diodes, the sensor can be located close to the center of the power device near the source bond pads where the current

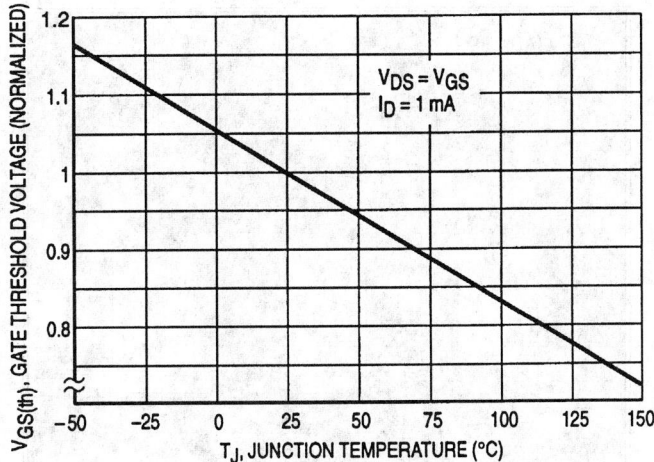

FIGURE 32.41 Power MOSFET's gate threshold variation vs. temperature.

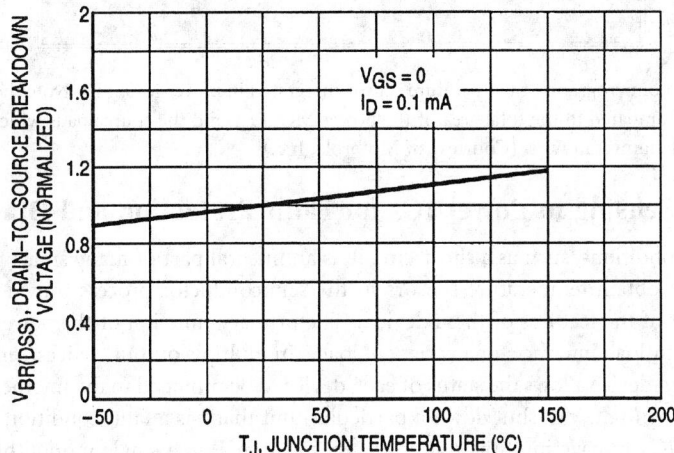

FIGURE 32.42 Power MOSFET's breakdown voltage variation vs. temperature.

density is the highest and, consequently, the highest die temperature occurs. The thermal conductivity of the oxide that separates the polysilicon diodes from the power device is 2 orders of magnitude less than that of silicon. However, because the layer is thin, the polysilicon element offers an accurate indication of the actual peak junction temperature.

A power FET that incorporates temperature sensing diodes is shown in Figure 32.43 [9]. By monitoring the output voltage when a constant current is passed through the integrated polysilicon diode(s), an accurate indication of the maximum die temperature is obtained. A number of diodes are actually provided in the design. A single diode in this design has a temperature coefficient of 1.90 mV °C^{-1}. Two or more can be placed in series if a larger output is desired. For greater accuracy, the diodes can be trimmed during wafer-level testing by blowing fusible links made from polysilicon. The response time of the diodes is less than 100 μs, which has allowed the device to withstand a direct connection across an automobile battery with external circuitry providing shutdown prior to device failure. The sensing capability also allows the output device to provide an indication (with additional external circuitry) if the heatsinking is not proper when the unit is installed in a module or if a change occurs in the application that would ultimately cause a failure.

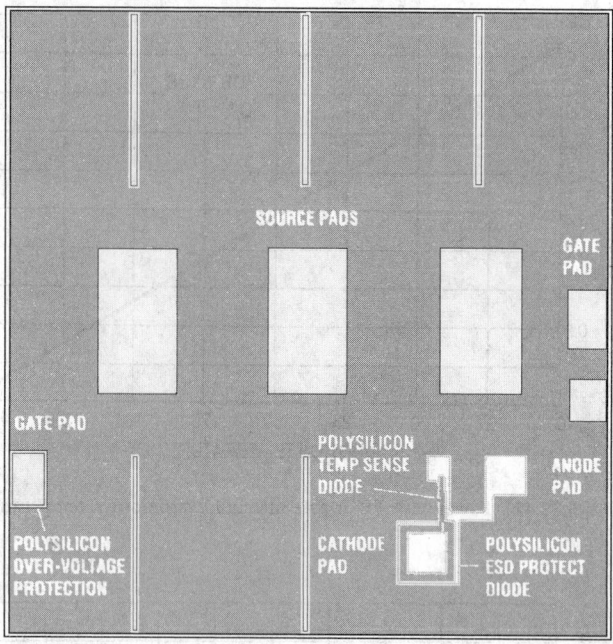

FIGURE 32.43   Photomicrograph of temperature sensor integrated in power MOSFET. Note the relative size of the temperature sensor compared to the total area of the power MOSFET and the source pads which allow attachment of 15 mil (0.60 mm) aluminum wire. (Courtesy of Motorola, Inc.)

Temperature Sensing in Power ICs for Fault Protection and Diagnostics

Sensing for fault conditions, such as a short-circuit, is an integral part of many smart power (or power) ICs. The ability to obtain temperature sensors in the semiconductor process provides protection and diagnostics as part of the features of these devices. The primary function of the power IC is to provide a microcontroller-to-load interface for a variety of loads. In multiple output devices, sensing the junction temperature of each device allows the status of each device to be provided to the microcontroller (MCU), and, if necessary, the MCU can shut down a particular unit that has a fault condition.

A *smart power IC* can have multiple power drivers integrated on a single monolithic piece of silicon [10]. Each of these drivers can have a temperature sensor integrated to determine the proper operating status and shut off only a specific driver if a fault occurs. Figure 32.44 shows an eight-output driver that independently shuts down the output of a particular driver if its temperature is excessive (i.e., between 155°C and 185°C) [10].

The octal serial switch (OSS) adds independent thermal sensing through over-temperature detection circuitry to the protection features. Faults can be detected for each output device, and individual shut-down can be implemented. In a multiple output power IC, it is highly desirable to shut down only the device that is experiencing a fault condition and not all of the devices that are integrated on the power IC. With outputs in various physical locations on the chip, it is difficult to predict the thermal gradients that could occur in a fault situation. Local temperature sensing at each output, instead of a single global temperature sensor, is required.

As shown in Figure 32.45, the eight outputs of the device with individual temperature sensors can be independently shut down when the thermal limit of 170°C is exceeded [10]. All of the outputs were connected to a 16-V supply at a room temperature ambient. A total current of almost 30 A initially flowed through the device. Note that each device turns off independently. The hottest device turns off first. Variations can result from differences in current level and thermal resistance. As each device turns off, the total power dissipation in the chip decreases and the devices that are still on can dissipate heat more effectively.

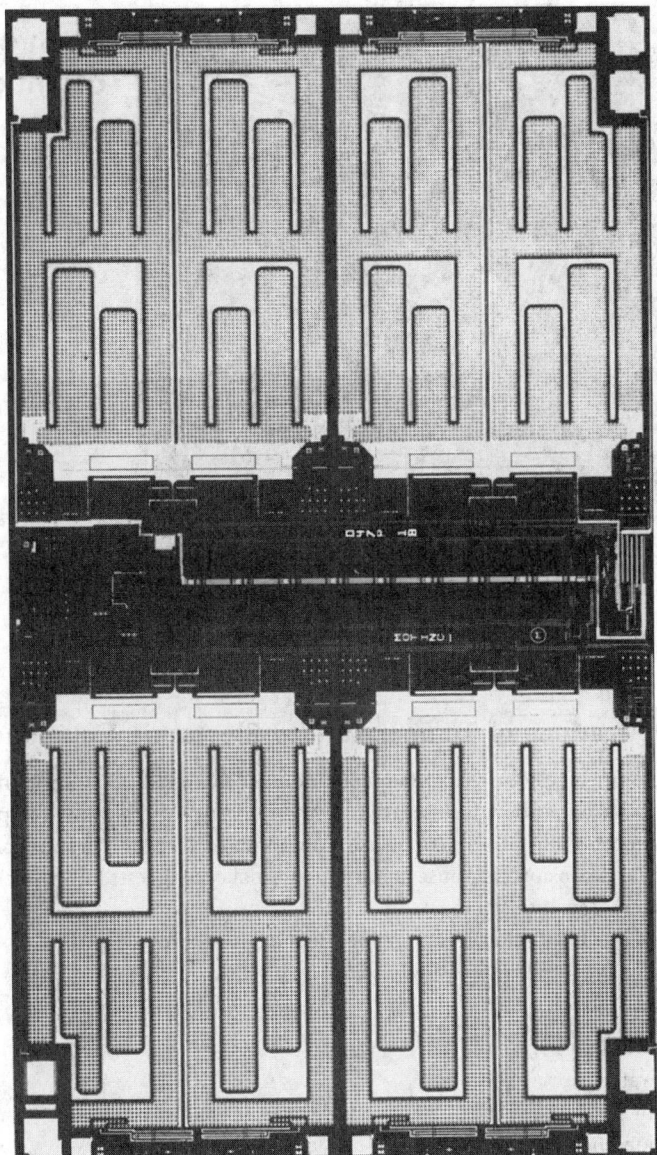

FIGURE 32.44 Photomicrograph of eight-output power IC. Note that the area of the eight output devices (two located at each corner of the die) are considerably larger than the circuitry in the center, and top and bottom that provides the temperature sensing, signal conditioning, and other control features. (Courtesy of Motorola, Inc.)

Connecting directly to the battery is a hard short that could have been detected by current limit circuitry. However, a soft short is below the current limit, but exceeds the power-dissipating capability of the chip, and can be an extremely difficult condition to detect. Soft shorts require over-temperature sensing to protect the IC from destructive temperature levels.

The over-temperature condition sensed by the power IC could mean that the device turns itself off to prevent failure in one case; and in another situation, a fault signal provides a warning to the MCU but no action is taken, depending on the fault circuit design. The remaining portion of the system is allowed to function normally. With the fault conditions supplied to the MCU, an orderly system shutdown can be implemented. Integrated temperature sensing is essential to provide this type of protection in a multiple-output power IC.

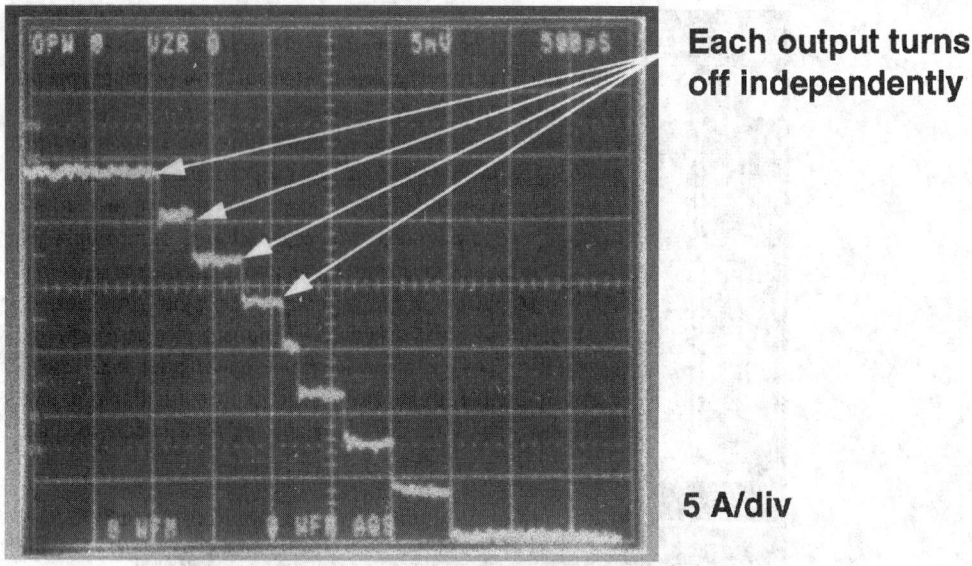

Each output turns off independently

5 A/div

10 ms

FIGURE 32.45   Independent thermal shutdown of an 8-output power IC. (Courtesy of Motorola, Inc.)

Reliability Implications of Temperature to Electronic Components

The effect of temperature on electronic components and their successful application in electronic systems is one of the issues that must be addressed during the design of the system. Temperature affects the performance and expected life of semiconductor components. Mechanical stress created by different coefficients of thermal expansion can cause failures in thermal cycling tests (air-to-air) or during thermal shock (water-to-water) transitions.

The typical failure rate for semiconductor component can be expressed by the Arrhenius equation [9]:

$$\lambda = A e^{-\varnothing/KT} \tag{32.88}$$

where λ = Failure rate
   A = Constant
   $\varnothing$ = Activation energy (eV)
   K = Boltzmann's constant (8.62×10^{-5} eV K^{-1})
   T = Junction temperature (K)

The failure rate of semiconductor components is typically stated to double for every 10 to 15°C increase in operating (i.e., junction) temperature. However, increased testing and design improvements have minimized the failures due to specific failure mechanisms.

One of the temperature-related parameters that must be taken into account during the design phase of a power switch is the transient thermal response, which is designated as $r(t)\ R\varnothing_{JC}$, where $r(t)$ represents the normalized transient thermal resistance. The value of $r(t)$ is determined from the semiconductor manufacturer's data sheet using duty cycle and pulse duration used in the application. This reduced level of the thermal resistance (see "Junction Temperature Measurement"), based on the transistor operating in a switching mode and being off for a period of time, approaches the dc level within a second. Excessive temperatures can be generated quickly and must be detected within milliseconds to prevent failure.

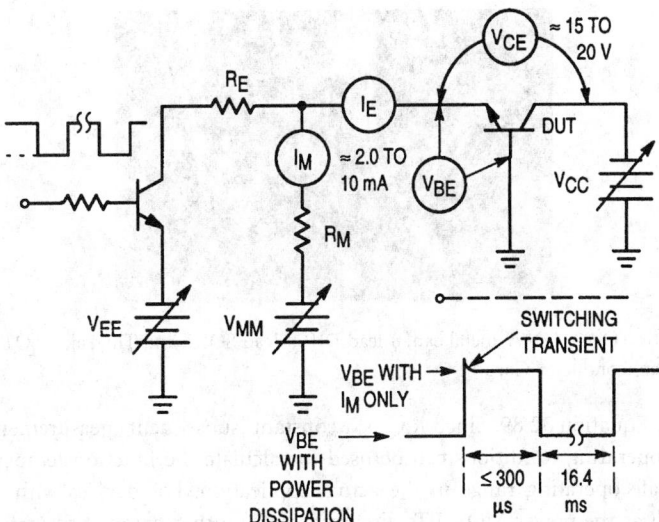

FIGURE 32.46 Example of steady-state thermal resistance test circuit for a bipolar power transistor. Power is applied for 16.4 ms and interrupted for ≤300 μs to measure the V_{BE}.

Junction Temperature Measurement

In a semiconductor, the change in temperature is directly related to the power dissipated through the thermal resistance. The steady-state dc thermal resistance junction-to-case, $R\varnothing_{JC}$, is defined as the temperature rise per unit power above an external reference point (typically the case). The relationship is shown in Equation 32.89 [8].

$$R\varnothing_{JC} = \Delta T / P_D \qquad (32.89)$$

where ΔT = Junction temperature minus the case temperature (°C)
   P_D = Power dissipated in the junction (W)

The semiconductor device or silicon die is typically enclosed in a package that prevents a direct measurement of the junction temperature. The junction temperature is measured indirectly by measuring the case temperature, T_C; the heatsink temperature, T_S; for those higher-power applications that require a heatsink; the ambient temperature, T_A; and a temperature-sensitive electrical parameter of the device.

The first step of the process requires calibrating the temperature-sensitive parameter. Using a bipolar power transistor as an example, the base-emitter forward voltage is measured and recorded with a low calibration current (I_M) flowing through the device that is low enough to avoid self-heating (typically between 2 mA and 10 mA) and yet sufficiently high to be in the linear range of the forward voltage curve. The procedure is performed at room and elevated temperatures, typically 100°C.

After calibration, a power switching fixture (such as Figure 32.46) is used to alternately apply and interrupt the power to the device [8]. The on portion is long (typically several milliseconds) and the off portion is short (only a few 100 μs), so the temperature of the case is stabilized and junction cooling is minimal. The transistor is operated in its active region and the power dissipation is varied by adjusting the I_E and/or V_{CE} until the junction is at the calibration temperature. This point is known by measuring V_{BE} during the time that I_M is the only current flowing.

When the V_{BE} value equals the value on the calibration curve, the junction temperature is at the calibration temperature. Measurements of V_{BE}, T_C, and I_E allow the thermal resistance for the device to

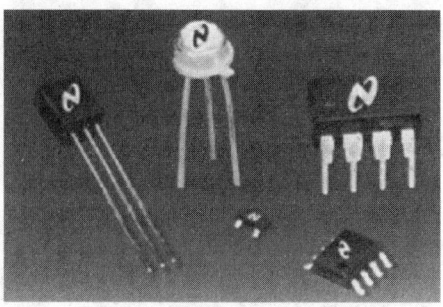

FIGURE 32.47   Plastic TO-92, TO-99 metal can, 8-lead DIP, 8-lead SOIC, and TinyPak™ SOT-23 plastic packages. (Courtesy National Semiconductor Corp.)

be calculated using Equation 32.89. Since $R\phi_{JC}$ is a constant, subsequent measurements of T_C, V_{CE}, and I_E under different operating conditions can be used to calculate the junction temperature to keep the device within its safe operating range in the actual application. For devices with different electrical characteristics, such as the power MOSFETs discussed earlier, other parameters that have a linear relationship to temperature are used for calibration and measurement.

Semiconductor Temperature Sensor Packaging

Temperature sensors that are manufactured using semiconductor technology are typically packaged in packages common to the semiconductor industry. These include metal can (TO-99), ceramic, and more commonly available plastic (SOT-23, 8-lead DIP, TO- 92, 8-lead SOIC, etc.) packages. These packages are designed for circuit board solder attachment that can be either through-hole or surface-mount technology. As a result, package form factors can be considerably different from packages for temperature sensors manufactured using other technologies. Figure 32.47 shows examples of five available silicon temperature sensor packages.

Defining Terms

Bandgap reference: Forward-biased emitter junction characteristics of adjacent transistors used to provide an output voltage with zero temperature coefficient.

Die: An unpackaged semiconductor chip separated from the wafer.

Embedded sensor: A sensor included within an integrated circuit.

Integrated circuit: A multiplicity of transistors, as well as diodes, resistors, capacitors, etc., on the same silicon die.

Junction: The interface at which the conductivity type of a material changes from p type to n type.

Junction voltage: The voltage drop across a forward-biased pn interface in a transistor (V_{be}) or diode.

Junction temperature: The temperature of the pn interface in a transistor or diode.

Monolithic (integrated circuit):   Constructed from a single piece of silicon.

Power IC or smart power IC: Hybrid or monolithic (semiconductor) device that is capable of being conduction-cooled, performs signal conditioning, and includes a power control function such as fault management and/or diagnostics.

Self-heating: Temperature rise within a (semiconductor) device caused by current flowing in the device.

Soft short: An excessive load condition that causes excessive temperature but is below the current limit of a device.

Thermal resistance: The steady-state dc thermal resistance junction-to-case, $R\phi_{JC}$, is the temperature rise per unit power above an external reference point (typically the case).

TinyPak: A trademark of National Semiconductor Corp.

References

1. J. Carr, *Sensors and Circuits,* Englewood Cliffs, NJ, PTR Prentice-Hall, 1993.
2. *Pressure Sensor Device Data* DL200/D Rev. 1, Phoenix, AZ, Motorola, 1994.
3. K. Lacanette, "Silicon Temperature Sensors: Theory and Applications," *Measurements and Control,* pp. 120-126, April 1996.
4. R. Wegner and H. Hulsemann, "New Family of Monolithic Temperature Sensor and Controller Circuits Present Challenges In Maintaining Temperature Measurement Accuracy," *Proceedings of Sensors Expo West,* Anaheim, CA, Feb. 8-10, 1994.
5. W. Schweber, "Temperature sensors fill different needs," *EDN,* p. 20, 3/14/96.
6. "Digital Thermometer IC simplifies distributed sensing," *Electronic Products,* p. 56, 12/95.
7. J. Williams, "High Performance Signal Conditioning for Transducers," *Proceedings of Sensors Expo West,* San Jose, CA, March 2-4, 1993.
8. *TMOS Power MOSFET Transistor Data* DL135/D Rev. 4, Phoenix, AZ, Motorola Semiconductor.
9. R. K. Jurgen (ed.), *Automotive Electronics Handbook,* New York, McGraw-Hill, 1994.
10. R. Frank, *Understanding Smart Sensors,* Boston, MA, Artech House, 1995.

32.6   Infrared Thermometers

Jacob Fraden

Thermal Radiation: Physical Laws

In any material object, every atom and every molecule exist in perpetual motion. When an atom moves, it collides with other atoms and transfers to them part of its kinetic energy, thus losing some of its own energy in this perpetual bouncing. On the other hand, an atom having a smaller kinetic energy, after a collision gains some energy. Afterward, the material body consisting of such agitated atoms reaches the energetic equilibrium where all atoms, while not vibrating with exactly the same intensity, still can be described by an average kinetic energy. Such an average kinetic energy of agitated particles is represented by the *absolute temperature,* which is measured in degrees kelvin. In other words, what is commonly called temperature, is a measure of the atomic motion.

According to the laws of electrodynamics, a moving electric charge (all atoms are made of electric charges) is associated with a variable electric field. The field, in turn, produces an alternating magnetic field. And again, when the magnetic field changes, it results in a coupled with it variable electric field, and so on. Thus, a moving particle becomes a source of electromagnetic field that propagates outwardly with the speed of light and is called *thermal radiation.* This radiation is governed by the laws of optics — it can be reflected, filtered, focused, etc. Also, it can be used to measure the object's temperature.

Electromagnetic waves originating from mechanical movement of particles can be characterized by their intensities and wavelengths. Both of these characteristics relate to temperature; that is, the hotter the object, the shorter the wavelength. Very hot objects radiate electromagnetic energy in the visible portion of the spectrum — wavelengths between 0.4 μm (blue) and 0.7 μm (red). For example, a filament in an incandescent lamp is so hot that it radiates bright visible light. If such a lamp is controlled by a dimmer, the light intensity can be reduced by turning the knob and observing that the dimmed light becomes more yellowish, reddish, and finally disappears. Near the end of the dimmer control, the filament is still quite hot, yet one cannot see it because it emanates light in the invisible infrared spectral range — wavelengths greater than 0.8 μm. Cooler objects radiate light in the near-, mid-, and far-infrared spectral ranges, which one cannot see. For example, electromagnetic radiation emanating from human skin primarily is situated at wavelengths between 5 μm and 15 μm — in the mid- and far-infrared ranges and is not visible to human eyes; otherwise, we all would glow in the dark (sick people with fever would look even brighter). If one imagines that all atomic vibration stopped for some mysterious reason, no electromagnetic radiation would be emanated. Such an imaginable but impossible event is characterized by infinitely cold temperature, which is called *absolute zero*.

Because temperature is a measure of the average atomic kinetic energy, it is logical to assume that one can determine the object's temperature by measuring the intensity of the emanated electromagnetic radiation or its spectral characteristics. This presumption is the basis for noncontact temperature measurements that are known by various names, depending on the application: infrared thermometry, optical pyrometry, radiation thermometry, etc. *Pyrometry* is derived from the Greek word *pyr*, which means fire, and thus is more appropriate for measuring hot temperatures. For lower temperatures, *infrared thermometry* is used interchangeably with term *radiation thermometry*.

Planck's Law

A relationship between the magnitude of radiation at a particular wavelength λ and absolute temperature T is rather complex and is governed by Planck's law, which was discovered in 1901. It establishes radiant flux density W_λ as power of electromagnetic radiation per unit of wavelength:

$$W_\lambda = \frac{\varepsilon(\lambda)C_1}{\pi\lambda^5\left(e^{C_2/\lambda T}-1\right)} \tag{32.90}$$

where $\varepsilon(\lambda)$ = Emissivity of an object
$\quad C_1$ = 3.74×10^{-12} Wcm² and $C_2 = 1.44$ cmK are Constants
$\quad e$ = Base of natural logarithms

Spectral densities for different temperatures are shown in Figure 32.48.

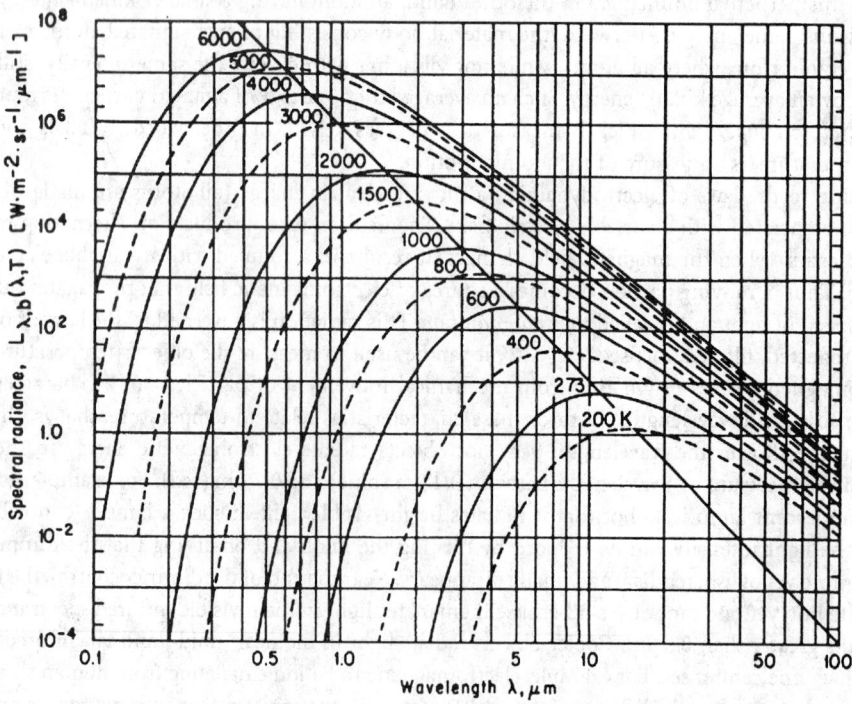

FIGURE 32.48 Spectral densities calculated within a solid angle of 1 steradian for blackbody source ($\varepsilon = 1$) radiation toward infinitely cold space (at absolute zero). (From J. C. Richmond and D.P. DeWitt (Eds). Application of radiation thermometry. ASTM PCN 04-895000-40. Used with permission.)

Wien's Law

Equation 32.90 does not lend itself to a simple mathematical analysis and thus is approximated by a simplified version, which is known as Wien's law:

$$W_\lambda = \frac{C_1}{\pi}\,\varepsilon(\lambda)\lambda^{-5}\,e^{\frac{C_2}{\lambda T}} \tag{32.91}$$

Because temperature is a statistical representation of an average kinetic energy, it determines the highest probability for the particles to vibrate with a specific frequency and to have a specific wavelength. This most probable wavelength follows from Wien's law by equating to zero a first derivative of Equation 32.91. The result of the calculation is a wavelength near which most of the radiant power is concentrated when light is emanated toward infinitely cold space at absolute zero:

$$\lambda_m = \frac{2898}{T} \tag{32.92}$$

where λ_m is in μm and T in kelvin. Wien's law states that the higher the temperature, the shorter the wavelength. This formula also defines the midpoint of spectral response of a pyrometer or infrared thermometer.

Stefan–Boltzmann Law

Theoretically, a thermal radiation bandwidth is infinitely wide. Yet, most of the emanated power is situated within quite a limited bandwidth. Also, one must account for the filtering properties of the real world windows used in instruments. In order to determine the total radiated power limited within a particular bandwidth, Equation 32.90 or 32.91 must be integrated within the limits from λ_1 to λ_2:

$$\Phi_{bo} = \frac{1}{\pi}\int_{\lambda_1}^{\lambda_2} \frac{\varepsilon(\lambda)C_1\lambda^{-5}}{e^{C_2/\lambda T}-1} \tag{32.93}$$

This integral can be resolved only numerically or by approximation. For a narrow bandwidth (λ_1 and λ_2 are close to one another), the solution can be approximated by:

$$\Phi_{bo} = kT^x \tag{32.94}$$

where k is constant and $x \approx (12/\lambda_2)(1200/T)$. For example, in the visible portion of spectrum at $\lambda_2 \approx$ 0.7 μm and for temperatures near 2000 K, the approximation is a 10th-order parabola. An approximation for a very broad bandwidth ($\lambda_2 \to \infty$ or practically, when the range between λ_1 and λ_2 embrace well over 50% of the total radiated power) is a 4th-order parabola, which is known as the *Stefan–Boltzmann law*:

$$\Phi_{bo} = A\varepsilon\sigma T^4 \tag{32.95}$$

where $\sigma = 5.67 \times 10^{-8}$ W/m^{2}K^4 (Stefan–Boltzmann constant), and ε is assumed to be wavelength independent. It is seen that with an increase in temperature, the intensity of electromagnetic radiation Φ_{bo} grows very fast due to the 4th power of T.

Kirchhoff's Law

While wavelengths of the radiated light are temperature dependent, the magnitude of radiation also is a function of the surface property. That property is called *emissivity*, ε. Emissivity is measured on a scale

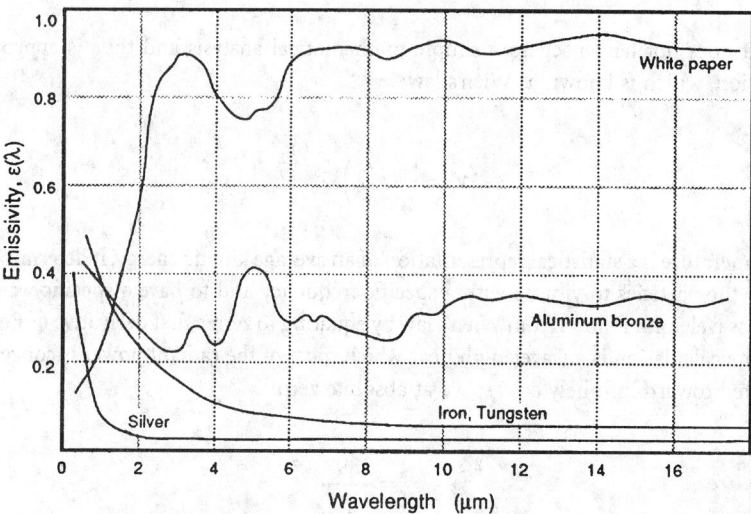

FIGURE 32.49 Wavelength dependence of emissivities. (From J. Fraden, *Handbook of Modern Sensors*, AIP Press, 1997. Used with permission.)

from 0 to 1. It is a ratio of electromagnetic flux that is emanated from a surface to the flux that would be emanated from the ideal emitter having the same temperature. *Reflectivity*, ρ, and *transparency*, γ, also on a scale from 0 to 1, show what portion of incident light is reflected and passed through, respectively. There is a fundamental equation that connects these three characteristics:

$$\varepsilon + \gamma + \rho = 1 \tag{32.96}$$

Equation (32.96) indicates that any one of the three properties of the material can be changed only at the expense of the others. As a result, for an opaque object ($\gamma = 0$), reflectivity ρ and emissivity ε are connected by a simple relationship: $\rho = 1 - \varepsilon$, which, for example, makes a mirror a good reflector but a poor emitter.

Emissivity

The emissivity of a material is a function of its dielectric constant and, subsequently, refractive index n. It should be noted, however, that emissivity is generally wavelength dependent (Figure 32.49). For example, a white sheet of paper is very much reflective in the visible spectral range and emits no visible light. In the far-infrared spectral range, its reflectivity is low and emissivity is high (about 0.92), thus making paper a good emitter of thermal radiation. However, for many practical purposes in infrared thermometry, emissivity can be considered constant.

For nonpolarized far-infrared light in normal direction, emissivity can be expressed by the equation:

$$\varepsilon = \frac{4n}{\left(n+1\right)^2} \tag{32.97}$$

As a rule, emissivities of dielectrics are high and of metals are low. Due to the high emissivity of dielectrics, they lend themselves to easy and accurate noncontact temperature measurement. On the other hand, such measurements from nonoxidized metals are difficult, due to small amounts of emanated infrared flux. Table 32.17 gives typical emissivities of some opaque materials in a temperature range between 0°C and 100°C.

TABLE 32.17 Typical Emissivities of Different Materials (from 0 to 100°C)

Material	Emissivity	Material	Emissivity
Blackbody (ideal)	1.00	Green leaves	0.88
Cavity radiator	0.99–1.00	Ice	0.96
Aluminum (anodized)	0.70	Iron or steel (rusted)	0.70
Aluminum (oxidized)	0.11	Nickel (oxidized)	0.40
Aluminum (polished)	0.05	Nickel (unoxidized)	0.04
Aluminum (rough surface)	0.06–0.07	Nichrome (80Ni-20Cr) (oxidized)	0.97
Asbestos	0.96	Nichrome (80Ni-20Cr) (polished)	0.87
Brass (dull tarnished)	0.61	Oil	0.80
Brass (polished)	0.05	Silicon	0.64
Brick	0.90	Silicone rubber	0.94
Bronze (polished)	0.10	Silver (polished)	0.02
Carbon-filled latex paint	0.96	Skin (human)	0.93–0.96
Carbon lamp black	0.96	Snow	0.85
Chromium (polished)	0.10	Soil	0.90
Copper (oxidized)	0.6–0.7	Stainless steel (buffed)	0.20
Copper (polished)	0.02	Steel (flat rough surface)	0.95–0.98
Cotton cloth	0.80	Steel (ground)	0.56
Epoxy resin	0.95	Tin plate	0.10
Glass	0.95	Water	0.96
Gold	0.02	White paper	0.92
Gold-black	0.98–0.99	Wood	0.93
Graphite	0.7–0.8	Zinc (polished)	0.04

Source: J. Fraden, *Handbook of Modern Sensors,* AIP Press, 1997. Used with permission.

Unlike most solid bodies, gases in many cases are transparent to thermal radiation. When they absorb and emit radiation, they usually do so only in certain narrow spectral bands. Some gases, such as N_2, O_2, and others of nonpolar symmetrical molecular structure, are essentially transparent at low temperatures, while CO_2, H_2O, and various hydrocarbon gases radiate and absorb to an appreciable extent. When infrared light enters a layer of gas, its absorption has an exponential decay profile, governed by *Beer's law*:

$$\frac{\Phi_x}{\Phi_0} = e^{-\alpha_\lambda x} \tag{32.98}$$

where Φ_0 = Incident thermal flux
Φ_x = Flux at thickness x
α_λ = Spectral coefficient of absorption

The above ratio is called a monochromatic transmissivity γ_λ at a specific wavelength λ. If gas is nonreflecting, then its emissivity is defined as:

$$\varepsilon_\lambda = 1 - \gamma_\lambda = 1 - e^{-\alpha_\lambda x} \tag{32.99}$$

It should be emphasized that since gases absorb only in narrow bands, emissivity and transmissivity (transparency) must be specified separately for any particular wavelength. For example, water vapor is highly absorptive at wavelengths of 1.4, 1.8, and 2.7 µm, and is very transparent at 1.6, 2.2, and 4 µm.

All non-metals are very good diffusive emitters of thermal radiation with a remarkably constant emissivity defined by Equation 32.97 within a solid angle of about ±70°. Beyond that angle, emissivity begins to decrease rapidly to zero with the angle approaching 90°. Near 90°, emissivity is very low. A typical calculated graph of the directional emissivity of non-metals into air is shown in Figure 32.50A. It should be emphasized that the above considerations are applicable only to wavelengths in the far

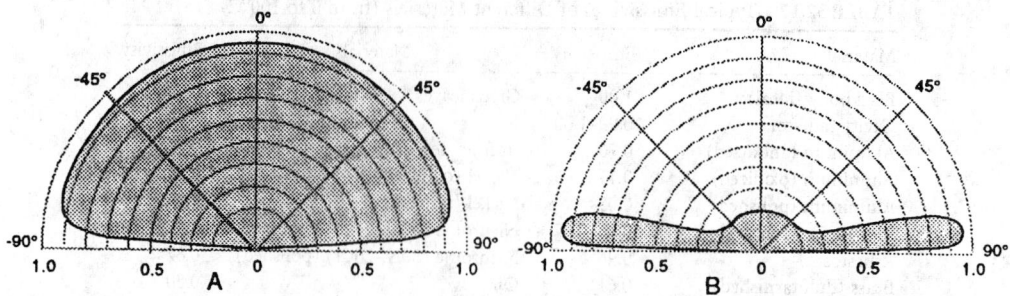

FIGURE 32.50 Spatial emissivities for non-metal (A) and a polished metal (B). (From J. Fraden, *Handbook of Modern Sensors*, AIP Press, 1997. Used with permission.)

infrared spectral range and are not true for visible light, since emissivity of thermal radiation is a result of electromagnetic effects that occur at an appreciable depth.

Metals behave quite differently. Their emissivities greatly depend on surface finish. Generally, polished metals are poor emitters within the solid angle of $\pm70°$, while their emissivity increases at larger angles (Figure 32.50B). Oxidized metals start behaving more and more like dielectrics with increasing thickness of oxides.

Blackbody

By definition, the highest possible emissivity is unity. It is attributed to the so-called blackbody — an ideal emitter of electromagnetic radiation. If the object is opaque ($\gamma = 0$) and nonreflective ($\rho = 0$) according to Equation 32.96, it becomes an ideal emitter and absorber of electromagnetic radiation. The name blackbody implies its appearance at normal room temperatures — indeed, it does look black because it is not transparent and not reflective at any wavelength. In reality, a blackbody does not exist, and any object with a nonunity emissivity often is called a *graybody*. A practical blackbody (ε is about 0.99 or higher) is an essential tool for calibrating and verifying the accuracy of infrared thermometers.

Cavity Effect

To make a practical blackbody, a cavity effect is put to work. The effect appears when electromagnetic radiation is measured from a cavity of an object. For this purpose, a cavity means an opening in a concave void of a generally irregular shape whose inner wall temperature is uniform over an entire surface. The emissivity of a cavity opening dramatically increases, approaching unity at any wavelength, as compared with a flat surface. The cavity effect is especially pronounced when its inner walls have relatively high emissivity. Consider a non-metal cavity. All non-metals are diffuse emitters. Also, they are diffuse reflectors. It is assumed that the temperature and surface emissivity of the cavity are homogeneous over an entire area. The ideal emitter (blackbody) would emanate from area a, the infrared photon flux $\Phi_0 = a\sigma T_b^4$. However, the object has the actual emissivity ε_b and, as a result, the flux radiated from that area is smaller: $\Phi_r = \varepsilon_b\Phi_0$ (Figure 32.51). Flux emitted by other parts of the object toward area a is also equal to Φ_r (because the object is thermally homogeneous, one can disregard the spatial distribution of flux). A substantial portion of that incident flux Φ_r is absorbed by the surface of area a, while a smaller part is diffusely reflected:

$$\Phi_\rho = \rho\Phi_r = \left(1-\varepsilon_b\right)\varepsilon_b\Phi_0 \qquad (32.100)$$

and the combined radiated and reflected flux from area a is:

$$\Phi = \Phi_r + \Phi_\rho = \varepsilon_b\Phi_0 + \left(1-\varepsilon_b\right)\varepsilon_b\,\Phi_0 = \left(2-\varepsilon_b\right)\varepsilon_b\Phi_0. \qquad (32.101)$$

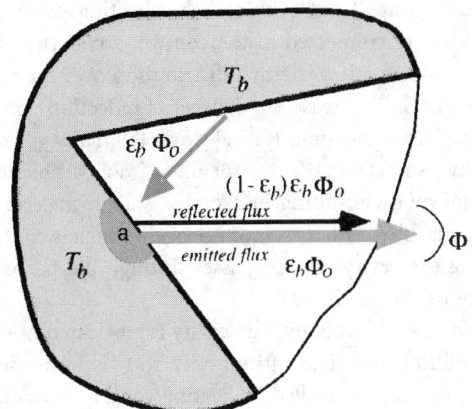

FIGURE 32.51   Cavity effect enhances emissivity. Note that $\varepsilon_e > \varepsilon_b$. (From J. Fraden, *Handbook of Modern Sensors*, AIP Press, 1997. Used with permission.)

As a result, for a single reflection, the *effective emissivity* can be expressed as:

$$\varepsilon_e = \frac{\Phi}{\Phi_0} = \left(2 - \varepsilon_b\right)\varepsilon_b \qquad (32.102)$$

It follows from the above that due to a single reflection, a perceived (effective) emissivity of a cavity at its opening (aperture) is equal to the surface emissivity magnified by a factor of $(2 - \varepsilon_b)$. Of course, there may be more than one reflection of radiation before it exits the cavity. In other words, the incident on area *a* flux could already be a result of a combined effect from the reflectance and emittance at other parts of the cavity's surface. For a cavity effect to work, the effective emissivity must be attributed only to the cavity opening (aperture) from which radiation escapes. If a sensor is inserted into the cavity facing its wall directly, the cavity effect could disappear and the emissivity would be close to that of a wall surface.

Practical Blackbodies

A practical blackbody can be fabricated in several ways. Copper is the best choice for the cavity body material, thanks to its high thermal conductivity, which helps to equalize temperatures of the cavity walls. As an example, Figure 32.52 shows two practical blackbodies. One is a solid-state blackbody having

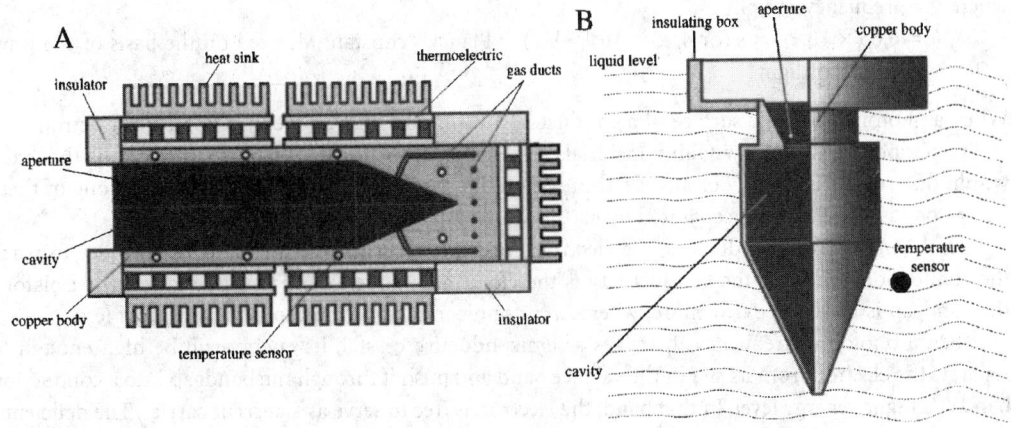

FIGURE 32.52   (A): Solid-state blackbody with thermoelectric elements. (B): Immersed blackbody.

thermoelectric elements (heat pumps) that provide either heating or cooling to the cavity body. The embedded temperature sensors are connected to the control circuit (not shown). The function of the multiple temperature sensors is to monitor thermal distribution over the length of the cavity. The inner shape of the cavity is partly conical to increase the number of reflections, and the entire surface is treated to provide it with as high emissivity as possible, typically over 0.9. This type of a blackbody has a relatively wide aperture that potentially can act as an entry for undesirable ambient air. The air can disturb the thermal uniformity inside the cavity, resulting in excessively high uncertainty of the radiated flux. To reduce this problem, the cavity is filled with dry air or nitrogen, which is continuously pumped in through the gas ducts. Before entering the cavity, the gas passes through the narrow channels inside the cavity and acquires the temperature of the blackbody.

Another example is an immersed blackbody. The cavity is fabricated of copper or aluminum and has relatively thin walls (a few millimeters). The entire cavity body is immersed into a stirred liquid bath, the temperature of which is precisely controlled by heating/cooling devices. The liquid assures uniform temperature distribution around the cavity with a typical thermal instability on the order of ±0.02°C. The inner surface of the cavity is coated with high-emissivity paint. The aperture of the cavity is relatively small. The ratio of the inner surface of the cavity to the aperture area should be at least 100, and preferably close to 1000.

Detectors for Thermal Radiation

Classification

Generally speaking, there are two types of sensors (detectors) known for their capabilities to respond to thermal radiation within the spectral range from the near-infrared to far-infrared; that is, from approximately 0.8 μm to 40 μm. The first type is quantum detectors; and the second type is thermal detectors. The latter, in turn, can be subdivided into passive (PIR) and active (AFIR) detectors.

Quantum Detectors

Quantum detectors (photovoltaic and photoconductive devices) relay on the interaction of individual photons with a crystalline lattice of semiconductor materials. Their operations are based on the photoeffect that was discovered by Einstein, and brought him the Nobel Prize. In 1905, he made a remarkable assumption about the nature of light: that at least under certain circumstances, its energy was concentrated into localized bundles, later named photons. The energy of a single photon is given by:

$$E = h\nu, \tag{32.103}$$

where ν = frequency of light

$h = 6.63 \times 10^{-34}$ J × s (or 4.13×10^{-15} eV s) = Planck's constant, derived on the basis of the wave theory of light

When a photon strikes the surface of a conductor, it can result in the generation of a free electron.

The periodic lattice of crystalline materials establishes allowed energy bands for electrons that exist within that solid. The energy of any electron within the pure material must be confined to one of these energy bands, which can be separated by gaps or ranges of forbidden energies.

In isolators and semiconductors, the electron must first cross the energy bandgap in order to reach the conduction band and the conductivity is therefore many orders of magnitude lower. For isolators, the bandgap is usually 5 eV or more; whereas, for semiconductors, the gap is considerably less.

When a photon of frequency ν_1 strikes a semiconductive crystal, its energy will be high enough to separate the electron from its site in the valence band and push it through the bandgap into a conduction band at a higher energy level. In that band, the electron is free to serve as a current carrier. The deficiency of an electron in the valence band creates a hole that also serves as a current carrier. This is manifested

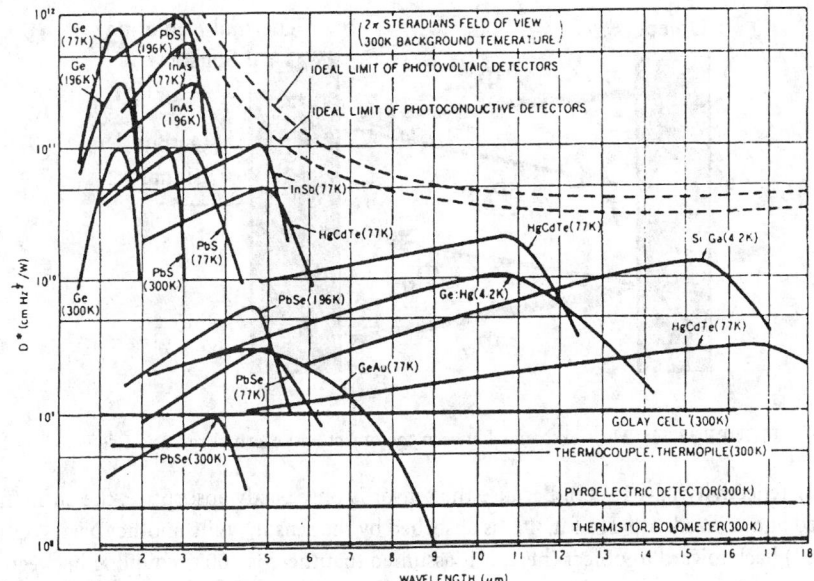

FIGURE 32.53   Operating ranges for some infrared detectors.

in the reduction of specific resistivity of the material. The energy gap serves as a photon energy threshold, below which the material is not light sensitive.

For measurements of objects emanating photons in the range of 2 eV or greater, quantum detectors having room temperature are generally used. For the smaller energies (longer wavelengths), narrower bandgap semiconductors are required. However, even if a quantum detector has a sufficiently small energy bandgap, at room temperature, its own intrinsic noise is much greater than a photoconductive signal. Noise level is temperature dependent; therefore, when detecting long-wavelength photons, a signal-to-noise ratio can become so small that accurate measurement becomes impossible. This is the reason why, for the operation in the near- and far-infrared spectral ranges, a detector not only should have a sufficiently narrow energy gap, but its temperature must be lowered to the level where intrinsic noise is reduced to an acceptable level. Depending on the required sensitivity and operating wavelength, the following crystals are typically used for the cryogenically cooled sensors (Figure 32.53): lead sulfide (PbS), indium arsenide (InAs), germanium (Ge), lead selenide (PbSe), and mercury-cadmium-telluride (HgCdTe).

The sensor cooling allows responses to longer wavelengths and increases sensitivity. However, response speeds of PbS and PbSe become slower with cooling. Methods of cooling include dewar cooling using dry ice, liquid nitrogen, liquid helium, or thermoelectric coolers operating on the Peltier effect.

Thermal Detectors

Another class of infrared radiation detectors is called *thermal detectors*. Contrary to quantum detectors that respond to individual photons, thermal detectors respond to heat resulting from absorption of thermal radiation by the surface of a sensing element. The heat raises the temperature of the surface, and this temperature increase becomes a measure of the net thermal radiation.

The Stefan–Boltzmann law (Equation 32.95) specifies radiant power (flux) which would emanate from a surface of temperature, T, toward an infinitely cold space (at absolute zero). When thermal radiation is detected by a thermal sensor, the opposite radiation from the sensor toward the object must also be taken in account. A thermal sensor is capable of responding only to a net thermal flux, i.e., flux from the object minus flux from itself. The surface of the sensor that faces the object has emissivity ε_s (and,

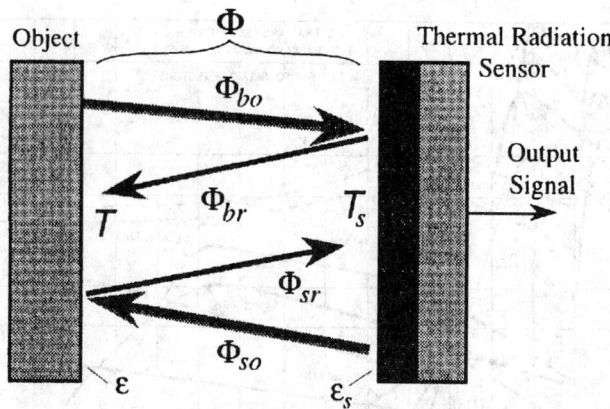

FIGURE 32.54 Heat exchange between the object and thermal radiation detector.

subsequently reflectivity $\rho_s = 1 - \varepsilon_s$). Because the sensor is only partly absorptive, the entire flux, Φ_{bo}, is not absorbed and utilized. A part of it, Φ_{ba}, is absorbed by the sensor, while another part, Φ_{br}, is reflected (Figure 32.54) back toward to object (here, it is assumed that there is 100% coupling between the object and the sensor and there are no other objects in the sensor's field of view). The reflected flux is proportional to the sensor's coefficient of reflectivity:

$$\Phi_{br} = -\rho_S \Phi_{bo} = -A\varepsilon \left(1 - \varepsilon_S\right)\sigma T^4 \qquad (32.104)$$

A negative sign indicates an opposite direction with respect to flux Φ_{bo}. As a result, the net flux originated from the object is:

$$\Phi_b = \Phi_{bo} + \Phi_{br} = A\varepsilon\varepsilon_s\sigma T^4 \qquad (32.105)$$

Depending on its temperature T_s, the sensor's surface radiates its own net thermal flux toward the object in a similar way:

$$\Phi_s = -A\varepsilon\varepsilon_s\sigma T_s^4 \qquad (32.106)$$

Two fluxes propagate in the opposite directions and are combined into a final net flux existing between two surfaces:

$$\Phi = \Phi_b + \Phi_s = A\varepsilon\varepsilon_s\sigma\left(T^4 - T_s^4\right) \qquad (32.107)$$

This is a mathematical model of a net thermal flux that is converted by a thermal sensor into the output signal. It establishes a connection between thermal power, Φ, absorbed by the sensor, and the absolute temperatures of the object and the sensor. It should be noted that since the net radiation exists between the two bodies, the spectral density will have the maximum not described by Equation 32.92; but depending on the temperature gradient, it will be somewhat shifted toward the shorter wavelengths.

Dynamically, the temperature T_s of a thermal element in a sensor, in general terms, can be described by the first-order differential equation:

$$cm\frac{dT_s}{dt} = P - P_L - \Phi \qquad (32.108)$$

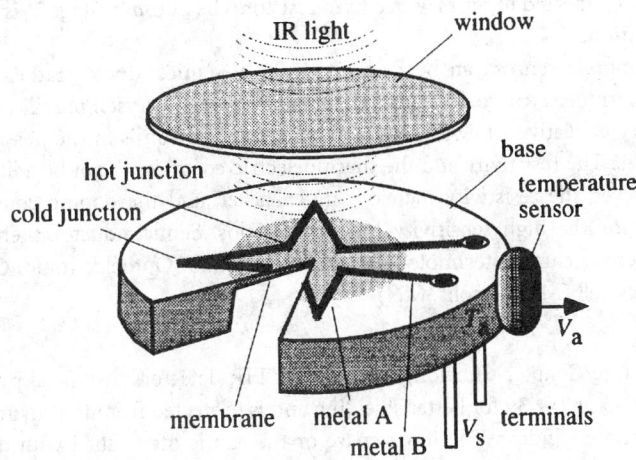

FIGURE 32.55 Thermopile sensor. "Hot" junctions are deposited on a membrane and "cold" junctions on the supporting ring.

where P is the power supplied to the element from a power supply or an excitation circuit (only in AFIR sensors; see below); P_L is a nonradiative thermal loss attributed to thermal conduction and convection; m and c are the sensor's mass and specific heat, respectively; and $\Phi = \Phi_\eta + \Phi_b$ is the net radiative thermal flux. We select a positive sign for power P when it is directed toward the element.

In the PIR detector (thermopiles, pyroelectric, and bolometers), no external power is supplied ($P = 0$); hence, the speed response depends only on the sensor's thermal capacity and heat loss, and is a first-order function that is characterized by a thermal time constant τ_T.

Thermopile Sensors.
Thermopiles belong to a class of PIR detectors. Their operating principle is the same as that of thermocouples. In effect, a thermopile can be defined as serially connected thermocouples. Originally, it was invented by Joule to increase the output signal of a thermoelectric sensor; he connected several thermocouples in series and thermally joined together their hot junctions. Presently, thermopiles have a different configuration. Their prime application is detection of thermal radiation.

A cut-out view of a thermopile sensor is shown in Figure 32.55. The sensor consists of a base having a relatively large thermal mass, which is the place where the "cold" junctions are positioned. The base can be thermally coupled with a reference temperature sensor or attached to a thermostat having a known temperature. The base supports a thin membrane whose thermal capacity and thermal conductivity are small. The membrane is the surface where the "hot" junctions are positioned.

The best performance of a thermopile is characterized by high sensitivity and low noise, which can be achieved by the junction materials having high thermoelectric coefficient, low thermal conductivity, and low volume resistivity. Besides, the junction pairs should have thermoelectric coefficients of opposite signs. This dictates the selection of materials. Unfortunately, most of metals having low resistivity (i.e., gold, copper, silver) have only very poor thermoelectric coefficients. The higher resistivity metals (especially bismuth and antimony) possess high thermoelectric coefficients and they are the prime selection for designing thermopiles. By doping these materials with Se and Te, the thermoelectric coefficient has been improved up to 230 μV K^{-1} [1].

Methods of construction of metal junction thermopiles can differ to some extent, but all incorporate vacuum deposition techniques and evaporation masks to apply the thermoelectric materials, such as bismuth and antimony on thin substrates (membranes). The number of junctions varies from 20 to several hundreds. The "hot" junctions are often blackened (e.g., with goldblack or organic paint) to improve their absorptivity of the infrared radiation. A thermopile is a dc device with an output voltage that follows its "hot" junction temperature quite well. It can be modeled as a thermal flux-controlled voltage source that is connected in series with a fixed resistor. The output voltage V_s is nearly proportional to the incident radiation.

An efficient thermopile sensor can be designed using a semiconductor rather than double-metal junctions [2]. The thermoelectric coefficients for crystalline and polycrystalline silicon are very large and the volume resistivity is relatively low. The advantage of using silicon is in the possibility of employing standard IC processes. The resistivity and the thermoelectric coefficients can be adjusted by the doping concentration. However, the resistivity increases much faster, and the doping concentration must be carefully optimized for the high sensitivity–low noise ratios. Semiconductor thermopile sensors are produced with a micromachining technology by EG&G Heimann Optoelectronics GmbH (Wiesbaden, Germany) and Honeywell (Minneapolis, MN).

Pyroelectrics.
Pyroelectric sensors (see Chapter 6) belong to a class of PIR detectors. A typical pyroelectric sensor is housed in a metal TO-5 or TO-39 for better shielding and is protected from the environment by a silicon or any other appropriate window. The inner space of the can is often filled with dry air or nitrogen. Usually, two sensing elements are oppositely, serially or in parallel, connected for better compensation of rapid thermal changes and mechanical stresses resulting from acoustic noise and vibrations [3].

Bolometers.
Bolometers are miniature RTDs or thermistors, which are mainly used for measuring rms values of electromagnetic signals over a very broad spectral range from microwaves to near-infrared. An external bias power is applied to convert resistance changes to voltage changes. For the infrared thermometers, the bolometers are often fabricated in the form of thin films having relatively large area. The operating principle of a bolometer is based on a fundamental relationship between the absorbed electromagnetic signal and dissipated power [3].

The sensitivity of the bolometer to the incoming electromagnetic radiation can be defined as [4]:

$$\beta = \frac{1}{2}\varepsilon\alpha_0 \sqrt{\frac{R_0 Z_T \Delta T}{\left(1+\alpha_0 \Delta T\right)\left[1+\left(\omega\tau\right)^2\right]}} \tag{32.109}$$

where $\alpha = (dR/dT)/R$ = TCR (temperature coefficient of resistance) of the bolometer
ε = Surface emissivity
Z_T = Bolometer thermal resistance, which depends on its design and the supporting structure
τ = Thermal time constant, which depends on Z_T and the bolometer's thermal capacity
ω = Angular frequency
ΔT = Bolometer's temperature increase

Bolometers are relatively slow sensors and are used primarily when no fast response is required. For thermal imaging, bolometers are available as two-dimensional arrays of about 80,000 sensors [5].

Active Far-Infrared Sensors.
In the active far-infrared (AFIR) sensor, a process of measuring thermal radiation flux is different from previously described passive (PIR) detectors. Contrary to a PIR sensing element — the temperature of which depends on both the ambient and the object's temperatures — the AFIR sensor's surface is actively controlled by a special circuit to have a defined temperature T_s that, in most applications, is maintained constant during an entire measurement process [6]. To control the sensor's surface temperature, electric power P is provided by a control (or excitation) circuit (Figure 32.56). To regulate T_s, the circuit measures the element's surface temperature and compares it with an internal reference.

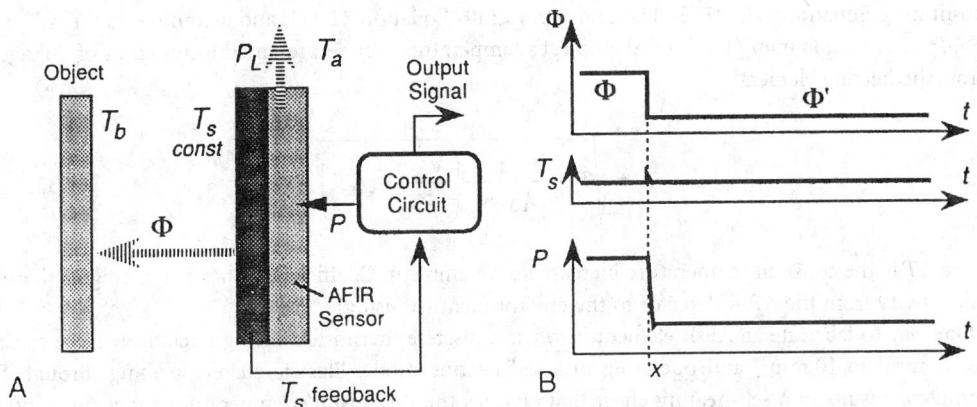

FIGURE 32.56 (A) AFIR element radiates thermal flux Φ_n toward its housing and absorbs flux Φ_b from the object. (B) Timing diagrams for radiative flux, surface temperature, and supplied power. (From J. Fraden, *Handbook of Modern Sensors*, AIP Press, 1997. Used with permission.)

Obviously, the incoming power maintains T_s higher than ambient, practically by just several tenths of a degree Celsius. Since the element's temperature is above ambient, the sensing element loses thermal energy toward its surroundings, rather than passively absorbing it, as in a PIR detector. Part of the heat loss is in the form of a thermal conduction; part is a thermal convection; and the other part is thermal radiation. The third part is the one that must be measured. Of course, the radiative flux is governed by the fundamental Stefan–Boltzmann law for two surfaces (Equation 32.99).

Some of the radiation power goes out of the element to the sensor's housing, while the other is coming from the object (or goes to the object). What is essential is that the net thermal flow (coductive + convective + radiative) must always come out of the sensor; that is, it must have a negative sign.

In the AFIR element, after a warm-up period, the control circuit forces the element's surface temperature T_s to stay constant; thus,

$$\frac{dT_s}{dt} = 0 \tag{32.110}$$

and Equation 32.108 becomes algebraic:

$$P = P_L + \Phi \tag{32.111}$$

It follows from the above that, under idealized conditions, its response does not depend on thermal mass and is not a function of time, meaning that practical AFIR sensors are quite fast. If the control circuit is highly efficient, since P_L is constant at given ambient conditions, the electronically supplied power P should track changes in the radiated flux Φ with high fidelity. Nonradiative loss P_L is a function of ambient temperature T_a and a loss factor α_s:

$$P_L = \alpha_s \left(T_s - T_a \right) \tag{32.112}$$

To generate heat in the AFIR sensor, it may be provided with a heating element having electrical resistance R. During the operation, electric power dissipated by the heating element is a function of voltage V across that resistance

$$P = V^2 / R \tag{32.113}$$

Substituting Equations 32.107, 32.112, and 32.113 into Equation 32.111, and assuming that $T = T_b$ and $T_s > T_a$, after simple manipulations, the object's temperature can be presented as function of voltage V across the heating element:

$$T_b = \sqrt{T_s^4 - \frac{1}{A\sigma\varepsilon_s\varepsilon_b}\left(\frac{V^2}{R} - \alpha_s\Delta T\right)} \qquad (32.114)$$

where ΔT is the constant temperature increase above ambient. Coefficient α_s has a meaning of thermal conductivity from the AFIR detector to the environment (housing).

One way to fabricate an AFIR element is to use a discrete thermistor having a relatively large surface area (3 mm² to 10 mm²) and operating in a self-heating mode. Electric current passing through the thermistor results in a self-heating effect that elevates the thermistor's temperature above ambient. In effect, the thermistor operates as both the heater and a temperature sensor.

Contrary to a PIR detector, an AFIR sensor is active and can generate a signal only when it works in orchestra with a control circuit. A control circuit must include the following essential components: a reference to preset a controlled temperature, an error amplifier, and a driver stage for the heater. In addition, it may include an RC network for correcting a loop response function and for stabilizing its operation; otherwise, an entire system could be prone to oscillations [7].

It can be noted that an AFIR sensor, along with its control circuit, is a direct converter of thermal radiative power into electric voltage and a quite efficient one. Its typical responsivity is in the range of 3000 V W⁻¹, which is much higher as compared with a thermopile, whose typical responsivity is in the range of 100 V W⁻¹. More detailed description of an AFIR sensor can be found in [3, 6].

Pyrometers

Disappearing Filament Pyrometer

Additional names for the disappearing filament pyrometer include: *optical pyrometer* and *monochromatic-brightness radiation thermometer*. This type of pyrometer is considered the most accurate radiation thermometer for temperatures over 700°C. This limitation is a result of human-eye sensitivity within a specific wavelength. The operating principle of this thermometer is based on Planck's law (Equation 32.90 and Figure 32.48) which states that intensity and color of the surface changes with temperature. The idea behind the design is to balance a radiation from an object having a known temperature against unknown temperature from a target. The pyrometer has a lens through which the operator views the target (Figure 32.57A). An image of a tungsten filament is superimposed on the image of the target. The filament

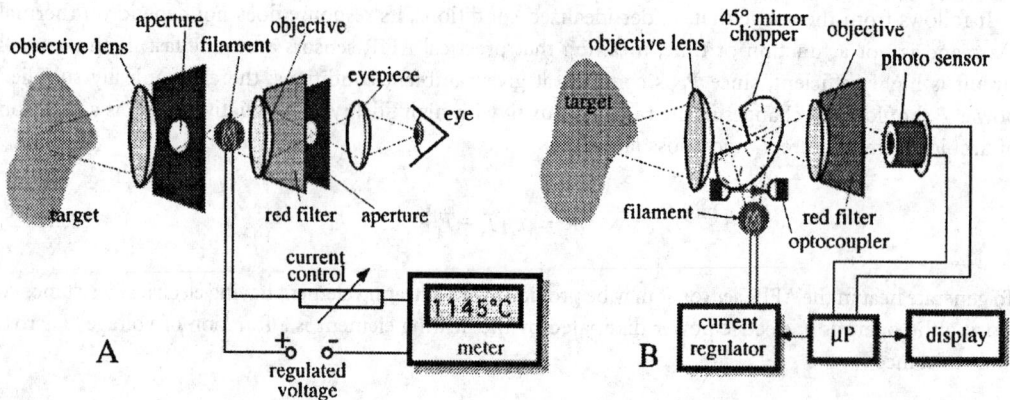

FIGURE 32.57 Disappearing filament optical pyrometer: (A): Manual, and (B): automatic versions.

is warmed up by electric current to glow. During the calibration, the relationship between the current and the filament temperature was established by measuring brightness of a blackbody of known temperature. The operator views the target through the eyepiece and manually adjusts the heating current to the level when an image of the glowing filament visible in the foreground disappears — that is, when both the target and the filament have the same brightness and color. A color component complicates the measurement somewhat; so to remove this difficulty, a narrow-band red filter ($\lambda_f = 0.65\ \mu m$) is inserted in front of an eyepiece. Therefore, the operator has to balance only the brightness of an image visible in red color. Another advantage of the filter is that the emissivity ε_{bf} of the target needs to be known only at λ_f. The error in temperature measurement is given by [8]:

$$\frac{dT_b}{T_b} = -\frac{\lambda_f T_b}{C_2}\frac{d\varepsilon_{bf}}{\varepsilon_{bf}} \tag{32.115}$$

Thus, for a target at $T_b = 1000$ K, a 10% uncertainty in knowing emissivity of the target at the filter's wavelength results in only $\pm 0.45\%$ (± 4.5 K) uncertainty in measured temperature.

The instrument can be further improved by removing an operator from the measurement loop. Figure 32.57B shows an automatic version of the pyrometer where a rotating mirror tilted by 45° has a removed sector that allows the light from a target to pass through to the photosensor. Such a mirror serves as a chopper, which alternately sends light to a photosensor, either from a target or from the filament. The microprocessor (μP) adjusts current through the filament to bring the optical contrast to zero. The optocoupler provides a synchronization between the chopper and the microprocessor.

Two-color Pyrometer

Since emissivities of many materials are not known, measurement of a surface temperature can become impractical, unless the emissivity is excluded from the calculation. This can be accomplished by use of a ratiometric technique: the so called "two-color radiation thermometer" or *ratio thermometer*. In such a thermometer, the radiation is detected at two separate wavelengths λ_x and λ_y for which emissivities of the surface can be considered nearly the same. The coefficients of transmission of the optical system at each wavelength respectively are γ_x and γ_y, then the ratio of two equations (2) calculated for two wavelengths is:

$$\phi = \frac{W_x}{W_y} = \frac{\gamma_x \varepsilon(\lambda_x)\dfrac{C_1}{\pi}\varepsilon(\lambda_x)\lambda_x^{-5} e^{-\frac{C_2}{\lambda_x T}}}{\gamma_y \varepsilon(\lambda_y)\dfrac{C_1}{\pi}\varepsilon(\lambda_y)\lambda_y^{-5} e^{-\frac{C_2}{\lambda_y T}}} \tag{32.116}$$

Because the emissivities $\varepsilon(\lambda_x) \approx \varepsilon(\lambda_y)$, after manipulations, Equation 32.116 can be rewritten for the displayed temperature T_c, where ϕ represents the ratio of the thermal radiation sensor outputs at two wavelengths:

$$T_c \approx C_2\left(\frac{1}{\lambda_y} - \frac{1}{\lambda_x}\right)\left(\ln\phi\frac{\gamma_y}{\gamma_x}\frac{\lambda_x^5}{\lambda_y^5}\right)^{-1} \tag{32.117}$$

It is seen that the equation for calculating temperature does not depend on the emissivity of the surface. Equation 32.117 is the basis for calculating temperature by taking the ratio of the sensor outputs at two different wavelengths. Figure 32.58 shows a block diagram of an IR thermometer where an optical modulator is designed in the form of a disk with two filters.

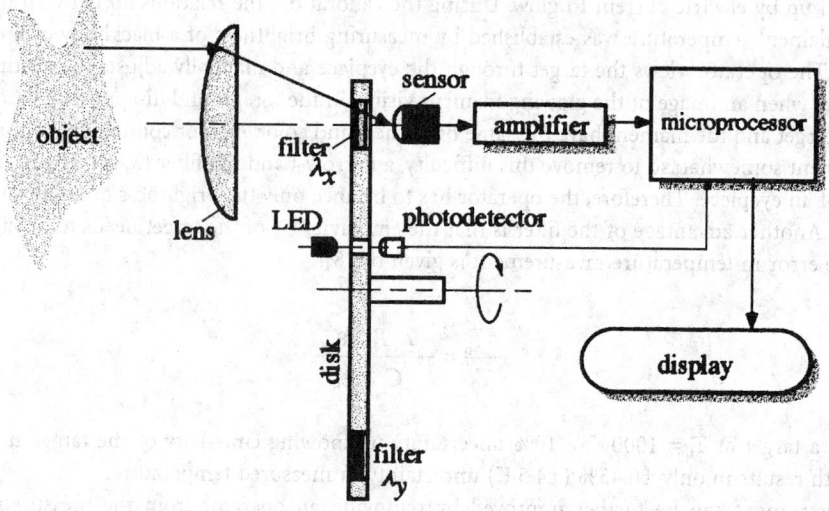

FIGURE 32.58 Two-color, non-contact thermometer. LED and photodetector are for synchronizing the disk rotation with the sensor response.

IR Thermometers

Operating Principle

Infrared (IR) Thermometer with PIR Sensors.

The operating principle of a noncontact infrared (IR) thermometer is based on the fundamental Stefan-Boltzmann law (Equation 32.118). For the purposes of calculating the object's temperature, when using the PIR sensor (thermopile, bolometer, or pyroelectric), the equation can be manipulated as:

$$T_c = \sqrt[4]{T_s^4 + \frac{\Phi}{A\sigma\varepsilon\varepsilon_s}} \tag{32.118}$$

where T_c is the calculated object's temperature in kelvin. Hence, to calculate the temperature of an object, one should first determine the magnitude of net thermal radiation flux Φ and the sensor's surface temperature T_s. The other parts of Equation 32.118 are considered as constants and must be determined during the calibration of the instrument from a blackbody. Emissivity of the object ε also must be known before the calculation. In practice, it is sometimes difficult to determine the exact temperature T_s of the sensor's surface, due to changing ambient conditions, drifts, handling of the instrument, etc. In such cases, the IR thermometer can be supplied with a reference target. Then, the calculation can still be done with the use of Equation 32.118; however, the value of Φ will have a meaning of a flux differential between the object and the reference target, and the value of T_s will represent the reference target temperature.

IR Thermometer with AFIR Sensor.

The object's temperature can be calculated from Equation 32.118; but first the sensing element surface temperature should be determined by measuring the temperature of the sensor's housing: $T_s = T_a + \Delta T$, where ΔT is the constant. In addition, the value of V must be measured. The AFIR sensor allows for continuous monitoring of temperature in a manner similar to thermopile sensors. Its prime advantage is simplicity and low cost.

Continuous Monitoring of Temperature

Depending on the type of thermal radiation sensor employed, a non-contact infrared thermometer for continuous monitoring can incorporate different components. Thus, if a sensor with a dc response is

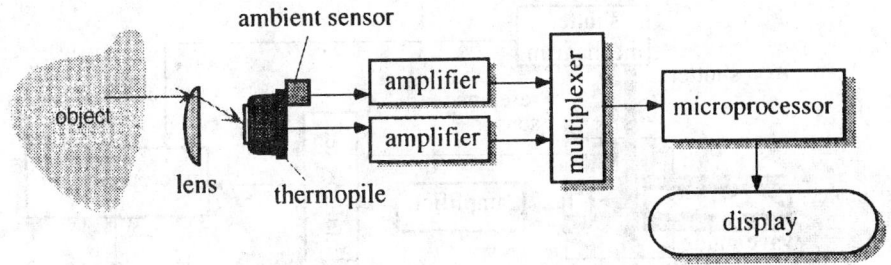

FIGURE 32.59   Infrared thermometer with a dc type of thermal radiation sensor.

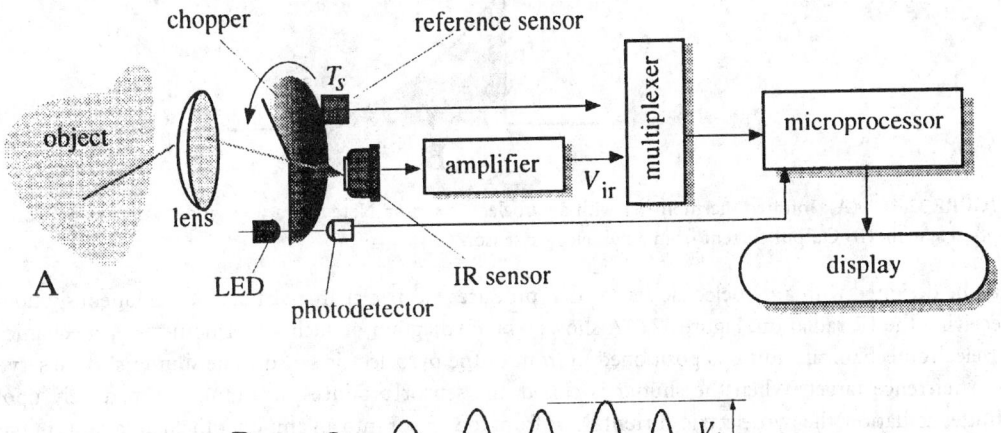

FIGURE 32.60   Infrared thermometer with a chopper.

used, the circuit might look like the one shown in Figure 32.59. It has the following essential components: the optical system (the lens), the IR sensor (a thermopile in this example), the reference (ambient) sensor, and a conventional data acquisition circuit. The dc-type sensors are the thermopiles, bolometers, and AFIR. When the AFIR sensor is used, it should be supplied with an additional control circuit, as described above.

In the case when an ac-type of IR sensor is used (pyroelectric), the thermometer needs a chopper that breaks the IR radiation into series pulses, usually with a 50% duty cycle (Figure 32.60). The output of an amplifier is an alternate signal that has a magnitude dependent on both the chopping rate and the net IR radiation. The rotation of the chopper is synchronized with the data processing by the microprocessor by means of an optocoupler (LED and photodetector). It should be noted that the chopper not only converts the IR flux into an ac signal, but it also serves as a reference target, as described above. Use of a reference target can significantly improve IR thermometer accuracy; thus, the chopper is often employed, even with the dc-type sensors, although it is not essential for operation. Naturally, the chopper's surface emissivity will be high and known.

Intermittent Monitoring of Temperature

Obviously, any infrared thermometer capable of continuous monitoring of temperature can be made operational in the intermittent mode. In other words, the continuous temperature can be processed to extract a single value of temperature that might be of interest to the user. For example, it may be a maximum, a minimum, or an averaged over time temperature; yet it is possible to design a low-cost IR thermometer that takes only a "snapshot" of the temperature at any specific moment. Such a thermometer

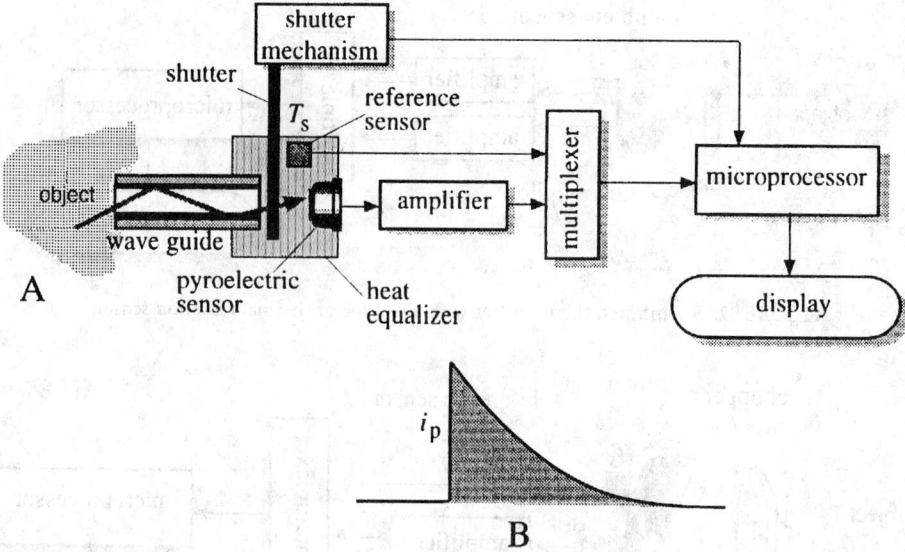

FIGURE 32.61   (A): Infrared thermometer with a pyroelectric sensor. Note that a waveguide is used as part of an optical system. (B) Output current from a pyroelectric sensor.

can be designed with a pyroelectric sensor that produces electric charge virtually instantaneously upon receiving the IR radiation. Figure 32.61A shows a block diagram of such a thermometer. A mechanical or electromechanical shutter is positioned in front of the pyroelectric sensor. The shutter surface serves as a reference target. When the shutter is closed, the sensor produces no output. Immediately upon shutter actuation, the pyroelectric current flows from the sensor into an amplifier that contains a charge-to-voltage converter. The sensor's current response has a shape close to the exponential function (Figure 32.61B). The magnitude of the spike is nearly proportional to the IR flux at the moment of the shutter opening. The pyroelectric thermometer is widely used for medical purposes and is dubbed the "instant thermometer."

Response Speed

All infrared thermometers are relatively fast; their typical response time is on the order of a second. Along with the non-contact way of taking temperature, this makes these devices very convenient whenever fast tracking of temperature is essential. However, an IR thermometer, while being very fast, still can require a relatively long warm-up time, and may not be accurate whenever it is moved from one environment to another without having enough time to adapt itself to new ambient temperature. The reason for this is that the reference and IR sensors (or the IR sensor and the shutter or chopper) must be in thermal equilibrium with one another; otherwise, the calculated temperature is erroneous.

Components of IR Thermometers

Optical Choppers and Shutters

The shutters or choppers must possess the following properties: (1) they should have high surface emissivity that does not change with time; (2) the opening and closing speed must be high; (3) the thermal conductivity between the front and back sides of the blade should be as small as possible; (4) the blade should be in good thermal contact with the reference sensor; and (5) while operating, the blade should not wear off or produce microscopic particles of dust that could contaminate the optical components of the IR thermometer.

For operation in the visible and near-infrared portions of the spectrum, when measured temperatures are over 800 K, the shutter can be fabricated as a solid-state device without the use of moving components

TABLE 32.18 Materials Useful for Infrared Windows and Lenses

Material	n	ρ	Wavelength (μm)	Note
AMTIR-1 ($Ge_{33}As_{12}Se_{55}$)	2.6	0.330	1	Amorphous glass
	2.5	0.310	10	
AMTIR-3 ($Ge_{28}Sb_{12}Se_{60}$)	2.6	0.330	10	Amorphous glass
As_2S_3	2.4	0.3290	8.0	Amorphous glass
CdTe	2.67	0.342	10.6	
Diamond	2.42	0.292	0.54	Best IR material
Fused silica (SiO_2)	1.46	0.067	3.5	Near-IR range
GaAs	3.13	0.420	10.6	
Germanium	4.00	0.529	12.0	Windows and lenses
Irtran 2 (ZnS)	2.25	0.258	4.3	
KRS-6	2.1	0.224	24	Toxic
Polyethylene	1.54	0.087	8.0	Low-cost IR windows; lenses
Quartz	1.54	0.088		Near-IR range
Sapphire (Al_2O_3)	1.59	0.100	5.58	Chemically resistant. Near- and mid-IR ranges
Silicon	3.42	0.462	5.0	Windows in IR sensors
ZnSe	2.4	0.290	10.6	IR windows; brittle

Note: n is the refractive index and ρ is the coefficient of reflection from two surfaces in air.

(e.g., employing liquid crystals). However, for longer wavelengths, only the mechanical blades are useful. Special attention should be paid with regard to the prevention of reflection by the shutter or chopper of any spurious thermal radiation that originates from within the IR thermometer housing.

Filters and Lenses

The IR filters and lenses serve two purposes: they selectively pass specific wavelengths to the sensing components, and they protect the interior of the instrument from undesirable contamination by outside pollutants. In addition, lenses — due to their refractive properties — divert light rays into specific direction [3]. In IR thermometry, the selection of filters and lenses is limited to a relatively short list. Table 32.18 lists some materials that are transparent in the infrared range. Note that most of these materials have a relatively high refractive index, which means that they have high surface reflectivity loss. For example, silicon (the cheapest material for the IR windows) reflects over 46% of incoming radiation which along with the absorptive loss, amounts to an even higher value of combined loss. To a certain degree, this problem can be solved by applying special antireflective (AR) coatings on both surfaces of the window or lens. These coatings are geared to specific wavelengths; thus, for a broad bandwidth, a multilayer coating may need to be deposited by a sputtering process.

Another problem with IR refractive materials is the relatively high absorption of light. This becomes a serious limitation for the lenses, as they need to be produced with appreciable thickness. The solution is to select materials with low absorption in the spectral range of interest. Examples are zinc selenide (ZnSe) and AMTIR. Another solution is the use of polyethylene Fresnel lenses, which are much thinner and less expensive [3]. Any window or lens that is absorptive will also emanate thermal radiation according to its own temperature. Hence, to minimize this effect on the overall accuracy of an IR thermometer, the refractive devices should be kept at the same temperature as the IR sensor, the shutter (chopper), and the reference sensor.

Waveguides

Waveguides are optical components intended for channeling IR radiation from one point to another. Usually, they are used when an IR sensor cannot be positioned close to the source of thermal radiation, yet must have a wide field of view. These components employ light reflection and are not focusing, even if they are designed with refractive materials. If focusing is required, the waveguides can be combined with lenses and curved or tilted mirrors. A waveguide has relatively wide entry and exit angles. A typical application is in a medical IR thermometer, where the waveguide is inserted into an ear canal (Figure 32.61).

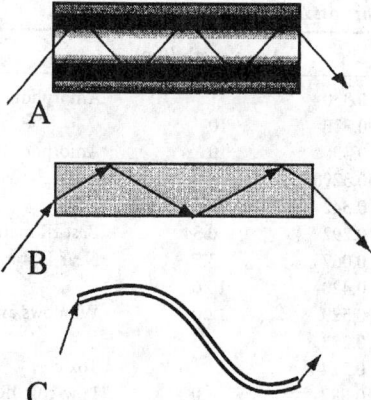

FIGURE 32.62   Light inside the barrel (A), rod (B) and fiber-optic (C) propagates on a zigzag path.

There are three types of waveguides: hollow tubes (barrels), optical fibers, and rods [3, 9, 10]. The latter two are made of IR-transparent materials, such as ZnSe or AMTIR, and use the angle of total internal reflection to propagate light inside in a zigzag pattern (Figure 32.62). The barrels are hollow tubes, polished and gold-plated on the inner surface.

Error Sources in IR Thermometry

Any error in detection of either radiated flux (Φ) or reference temperature (T_a) will result in inaccuracy in the displayed temperature. According to Equation 32.118, the emissivities of both the sensor (ε_s) and the object (ε) must be known for the accurate detection of thermal flux. Emissivity of the sensor usually remains constant and is taken care of during the calibration. However, uncertainty in the value of emissivity of the object can result in significant uncertainty in temperatures measured by non-contact infrared thermometers (Figure 32.63).

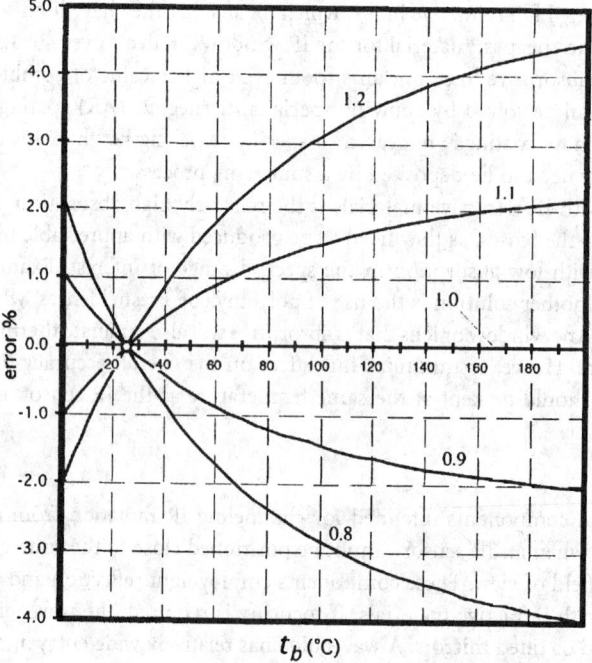

FIGURE 32.63   Error in temperature measurement resulted from uncertainty in the value of object's emissivity.

Another source of an error is spurious heat sources, which can emanate their thermal radiation either directly into the optical system of an IR thermometer, or by means of reflection from the measured surface [11]. Since no surface is an ideal emitter (absorber), its reflectivity may have significant value. For example, an opaque object with emissivity 0.9 has reflectivity of 0.1 (Equation 32.54); thus, about 10% of the radiation from a hot object in the vicinity of a measured surface is reflected and can greatly disturb the measurement.

Since emissivities of metals are quite small and one never can be sure about their actual values, it is advisable, whenever practical, to coat a portion of a metal surface with a dielectric material, such as metal oxide or organic paint having known emissivity. Alternatively, temperatures should not be taken from a flat surface, but rather from a hole or cavity that inside has a nonmetallic surface.

Another source of error is the thermal instability of the thermal sensing element. For example, in a thermopile or pyroelectric sensor, upon exposure to thermal radiation, the element's temperature might increase above ambient by just a few millidegrees. Hence, if for some reason, the element's temperature varies by the same amount, the output signal becomes indistinguishable from that produced by the thermal flux. To minimize these errors, the infrared sensors are encapsulated into metal bodies having high thermal capacity and poor coupling to the environment (high thermal resistance), which helps to stabilize the temperature of the sensor. Another powerful technique is to fabricate the IR sensor in a dual form; that is, with two identical sensing elements. One element is subjected to an IR signal, while the other is shielded from it. The signals from the elements are subtracted, thus canceling the drifts that affect both elements [3].

Some Special Applications

Semiconductor Materials

Temperature measurement of semiconductor material during processing, such as growth of epitaxial films in vacuum chambers, always has been a difficult problem. Various process controls require accurate temperature monitoring, often without physical contact with the substrates. As a rule, heating of a substrate is provided by resistive heaters. The substrates are often loaded into the chambers through various locks and rotated during processing. Therefore, it is very difficult to attach a contact sensor, such as a thermocouple, to the wafer. And, even if one does so, the thermal gradient between the sensor and the substrate may be so large that one should never trust the result. The attractive alternative is optical pyrometry; however, this presents another set of difficulties. The major problem is that semiconductors are substantially transparent in the spectral region where thermal radiation would be emanated. In other words, the emissivity of a semiconductor is negligibly small and the amount of thermal radiation from a semiconductor is also not only small, but due to wafer transparency, the radiation from the underlying devices (e.g., heater) will go through to the IR thermometer.

One relatively simple method is to coat a small portion of the semiconductor with a material having high emissivity in the IR spectral range. Then, the thermal radiation can be measured from that patch. An example of such a material is nichrome (see Table 32.17).

An attractive method of temperature monitoring, when no emissive patch can be deposited, is the use of temperature dependence of bandgaps of common semiconductors. The bandgap is determined from the threshold wavelength at which the radiation from the heaters behind the substrate is transmitted through the substrate [12]. Another method is based on the temperature dependence of diffuse reflection of a semiconductor. In effect, this method is similar to the former; however, it relies on reflection, rather than on transmission of the semiconductor. An external lamp is used for the measurement of a threshold wavelength from a front polished and backside textured substrate [13]. The temperature measurement arrangement is shown in Figure 32.64 where the diffused light is detected by cryogenically cooled quantum detector. The monochromator has resolution of 3 nm and scans through the threshold area at a rate of 100 nm/s.

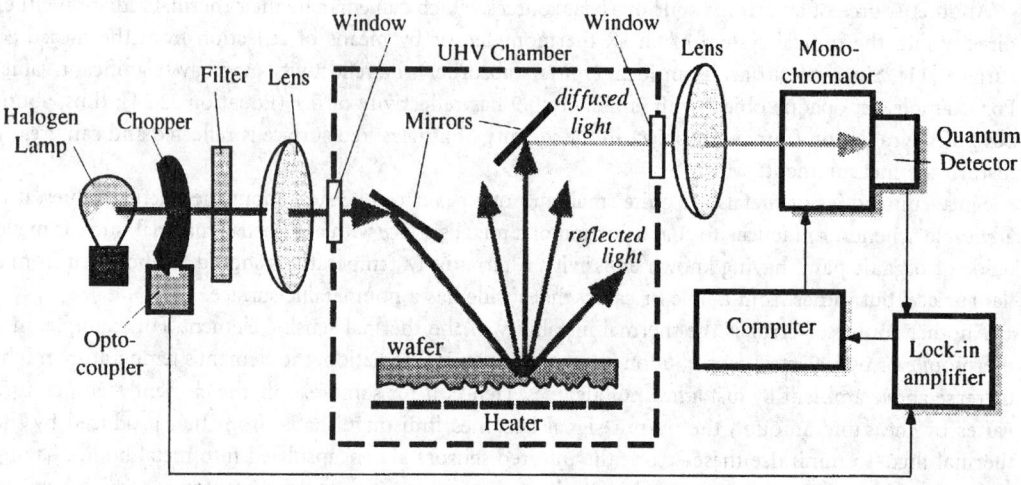

FIGURE 32.64   Diffused-light thermometer for measuring temperature of GaAs wafers. (After [12].)

Medical Applications

Medical infrared thermometry has two distinct types of measurements: *body* temperature measurement and *skin* surface temperature measurement. Skin temperature measurements have specific applications in determining surface temperature of a human body. That temperature greatly depends on both the skin blood perfusion and environmental conditions. Therefore, skin temperature cannot be independently correlated with the internal body temperature.

Now, it is customary to measure the internal body temperature by placing the probe of an IR thermometer into the ear canal aiming it at a tympanic membrane [14]. The tympanic membrane has temperature close to that of blood. As a rule, the probe is designed with a gold-plated waveguide (Figure 32.61) which is protected either by a semiconductor or polymer window. Because a medical IR thermometer is used in direct contact with patient tissues, it is imperative to protect the probe from becoming a carrier of soiling compounds and a transmitter of infection from one patient to another (cross-contamination) or even from re-infecting the same patient (recontamination). Thus, a probe of a medical IR thermometer is protected by a special probe cover made of a polymer film (polyethylene, polypropylene, etc.) which is transparent in the spectrum range of thermal radiation. In effect, the probe cover becomes a part of the optical system. This demands that covers be produced with very tight tolerances so that they will not significantly affect transmission of IR signal.

References

1. A. Völklein, A. Wiegand, and V. Baier, *Sensors and Actuators A*, 29, pp: 87-91, 1991.
2. J. Schieferdecker, R. Quad, E. Holzenkämpfer, and M. Schulze, Infrared thermopile sensors with high sensitivity and very low temperature coefficient. *Sensors and Actuators* A, 46-47, 422-427,1995.
3. J. Fraden, *Handbook of Modern Sensors*. 2nd ed., Woodbury, NY: AIP Press, 1996.
4. J-S. Shie and P.K. Weng, Fabrication of micro-bolometer on silicon substrate by anizotropic etching technique. *Transducers'91, Int. Conf. Solid-state Sensors and Actuators.* 627-630, 1991.
5. R.A. Wood, Uncooled thermal imaging with silicon focal plane arrays. *Proc. SPIE* 2020, *Infrared Technology XIX*, pp: 329-36, 1993.
6. J. Fraden, Active far infrared detectors. In *Temperature, Its Measurement and Control in Science and Industry*. Vol. 6, Part 2, New York: American Institute of Physics, 1992, 831-836.

7. C.J. Mastrangelo and R.S. Muller, Design and performance of constant-temperature circuits for microbridge-sensor applications. *Transducers'91. Int. Conf. Solid-state Sensors and Actuators.* 471-474, 1991.

8. E. O. Doebelin, *Measurement Systems. Application and Design,* 4th ed., New York: McGraw-Hill Co., 1990.

9. A. R. Seacord and G. E. Plambeck. *Fiber optic ear thermometer.* U.S. Patent No. 5,167,235.

10. J. Fraden, *Optical system for an infrared thermometer.* U.S. Patent No. 5,368,038.

11. D.R. White and J.V. Nicholas, Emissivity and reflection error sources in radiation thermometry, in *Temperature, Its Measurement and Control in Science and Industry.* Vol. 6, Part 2, New York: American Institute of Physics, 1992, 917-922.

12. E.S. Hellman et al., *J. Crystal Growth*, 81, 38, 1986.

13. M.K. Weilmeier et al., A new optical temperature measurement technique for semiconductor substrates in molecular beam epitaxy. *Can. J. Phys.*, 69, 422-426, 1991.

14. J. Fraden, Medical infrared thermometry (review of modern techniques), in *Temperature, Its Measurement and Control in Science and Industry.* Vol. 6, Part 2, New York: American Institute of Physics, 1992, 825-830.

32.7 Pyroelectric Thermometers

Jacob Fraden

Pyroelectric Effect

The pyroelectric materials are crystalline substances capable of generating an electric charge in response to heat flow [1]. The pyroelectric effect is very closely related to the piezoelectric effect. The materials belong to a class of ferroelectrics. The name was given in association with ferromagnetics and is rather misleading because most such materials have nothing to do with iron.

A crystal is considered to be pyroelectric if it exhibits a spontaneous temperature-dependent polarization. Of the 32 crystal classes, 21 are noncentrosymmetric and 10 of these exhibit pyroelectric properties. Besides pyroelectric properties, all these materials exhibit some degree of piezoelectric properties as well — they generate an electric charge in response to mechanical stress.

Pyroelectricity was observed for the first time in tourmaline crystals in the 18th century (some claim that the Greeks noticed it 23 centuries ago). Later, in the 19th century, Rochelle salt was used to make pyroelectric sensors. A large variety of materials became available after 1915: KDP (KH_2PO_4), ADP ($NH_4H_2PO_4$), $BaTiO_3$, and a composite of $PbTiO_3$ and $PbZrO_3$ known as PZT. Presently, more than 1000 materials with reversible polarization are known. They are called ferroelectric crystals. Most important among them are triglycine sulfate (TGS) and lithium tantalate ($LiTaO_3$).

A pyroelectric material can be considered a composition of a large number of minute crystallites, where each behaves as a small electric dipole. All these dipoles are randomly oriented, however, along a preferred direction. Above a certain temperature, known as the Curie point, the crystallites have no dipole moment.

When temperature of a pyroelectric material changes, the material becomes polarized. In other words, an electric charge appears on its surface. It should be clearly understood that the polarization occurs not as a function of temperature, but only as function of a *change in temperature* of the material. There are several mechanisms by which changes in temperature will result in pyroelectricity. Temperature changes can cause shortening or elongation of individual dipoles. It can also affect the randomness of the dipole orientations due to thermal agitation. These phenomena are called *primary pyroelectricity*. There is also *secondary pyroelectricity*, which, in a simplified way, can be described as a result of the piezoelectric effect; that is, a development of strain in the material due to thermal expansion.

The dipole moment, M, of the bulk pyroelectric sensor is:

$$M = \mu A h \qquad (32.119)$$

TABLE 32.19　Physical Properties of Pyroelectric Materials

Material	Curie temperature °C	Thermal conductivity W mK^{-1}	Relative permittivity ε_r	Pyroelectric charge coeff. C (m^{2}K)$^{-1}$	Pyroelectric voltage coeff. V (mK)$^{-1}$	Coupling k_p^2 (%)
Single Crystals						
TGS	49	0.4	30	3.5×10^{-4}	1.3×10^6	7.5
LiTaO$_3$	618	4.2	45	2.0×10^{-4}	0.5×10^6	1.0
Ceramics						
BaTiO$_3$	120	3.0	1000	4.0×10^{-4}	0.05×10^6	0.2
PZT	340	1.2	1600	4.2×10^{-4}	0.03×10^6	0.14
Polymers						
PVDF polycrystalline layers	205	0.13	12	0.4×10^{-4}	0.40×10^6	0.2
PbTiO$_3$	470	2 (monocrystal)	200	2.3×10^{-4}	0.13×10^6	0.39

Note: The above figures may vary depending on manufacturing technologies.
From Reference [2].

where μ = Dipole moment per unit volume
　　　A = Sensor's area
　　　h = Thickness

The charge, Q_a, which can be picked up by the electrodes, develops the dipole moment across the material:

$$M_0 = Q_a h \tag{32.120}$$

M must be equal to M_0, so that:

$$Q_a = \mu A \tag{32.121}$$

As the temperature varies, the dipole moment also changes, resulting in an induced charge.

Thermal absorption can be related to a dipole change, so that μ must be considered a function of both temperature, T_a, and an incremental thermal energy, ΔW, absorbed by the material:

$$\Delta Q_a = A\mu\left(T_a, \Delta W\right) \tag{32.122}$$

The above equation shows the magnitude of electric charge resulting from absorption of thermal energy. To pick up the charge, the pyroelectric materials are fabricated in the shapes of a flat capacitor with two electrodes on opposite sides and the pyroelectric material serving as a dielectric.

Pyroelectric Materials

To select the most appropriate pyroelectric material, energy conversion efficiency should be considered. It is, indeed, the function of the pyroelectric sensor to convert thermal energy into electrical. "How effective is the sensor?" — is a key question in the design practice. A measure of efficiency is: k_p^2 which is called the pyroelectric coupling coefficient [2, 3]. It shows the factor by which the pyroelectric efficiency is lower than the Carnot limiting value $\Delta T/T_a$. Numerical values for k_p^2 are shown in Table 32.19.

Table 32.19 shows that triglycine sulfate (TGS) crystals are the most efficient pyroelectric converters. However, for a long time they were quite impractical for use in sensors because of a low Curie temperature. If the sensor's temperature is elevated above that level, it permanently loses its polarization. In fact, TGS sensors proved to be unstable even below the Curie temperature, with the signal being lost quite spontaneously [4]. It was discovered that doping of TGS crystals with L-alanine (LATGS process patented by

Philips) during its growth stabilizes the material below the Curie temperature. The Curie temperature was raised to 60°C, which allows its use at the upper operating temperature of 55°C, which is sufficient for many applications.

Other materials, such as lithium tantalate and pyroelectric ceramics, are also used to produce pyroelectric sensors. Polymer films (KYNAR from AMP Inc.) have become increasingly popular for a variety of applications. During recent years, deposition of pyroelectric thin films has been intensively researched. Especially promising is the use of lead titanate ($PbTiO_3$), which is a ferroelectric ceramic having both a high pyroelectric coefficient and a high Curie temperature of about 490°C. This material can be easily deposited on silicon substrates by the so called sol-gel spin casting deposition method [5].

In 1969, Kawai discovered strong piezoelectricity in the plastic materials, polyvinyl fluoride (PVF) and polyvinylidene fluoride (PVDF) [6]. These materials also possess substantial pyroelectric properties. PVDF is a semicrystalline polymer with an approximate degree of crystallinity of 50% [7]. Like other semicrystalline polymers, PVDF consists of a lamellar structure mixed with amorphous regions. The chemical structure contains the repeat unit of doubly fluorinated ethene CF_2-CH_2:

$$\left[\begin{array}{cc} H & F \\ | & | \\ -C & -C- \\ | & | \\ H & F \end{array} \right]_n$$

The molecular weight of PVDF is about 10^5, which corresponds to about 2000 repeat units. The film is quite transparent in the visible and near-IR regions, and is absorptive in the far-infrared portion of the electromagnetic spectrum. The polymer melts at about 170°C. Its density is about 1780 kg m^{-3}. PVDF is mechanically durable and flexible. In piezoelectric applications, it is usually drawn, uniaxially or biaxially, to several times its length. Elastic constants, for example, Young modulus, depend on this draw ratio. Thus, if the PVDF film was drawn at 140°C to the ratio of 4:1, the modulus value is 2.1 GPa; while for the draw ratio of 6.8:1, it was 4.1 GPa. Resistivity of the film also depends on the stretch ratio. For example, at low stretch, it is about $6.3 \times 10^{15} \Omega$ cm, while for the stretch ratio 7:1 it is $2 \times 10^{16} \Omega$ cm.

Since silicon does not possess pyroelectric properties, such properties can be added on by depositing crystalline layers of pyroelectric materials. The three most popular materials are zinc oxide (ZnO), aluminum nitride (AlN), and the so-called solid solution system of lead-zirconite-titanium oxides $Pb(Zr,Ti)O_3$ known as PZT ceramic, which is basically the same material used for fabrication of discrete piezoelectric and pyroelectric sensors. One of the advantages of using zinc oxide is the ease of chemical etching. The zinc oxide thin films are usually deposited on silicon by employing sputtering technology. Note, however, that silicon has a large coefficient of thermal conductivity. That is, its thermal time constant is very small (see below), so the pyroelectric sensors made with silicon substrates possess low sensitivity yet are capable of fast response.

Manufacturing Process

Manufacturing of ceramic PZT elements begins with high-purity metal oxides (lead oxide, zirconium oxide, titanium oxide, etc.) in the form of fine powders having various colors. The powders are milled to a specific fineness, and mixed thoroughly in chemically correct proportions. In a process called "calcining," the mixtures are then exposed to an elevated temperature, allowing the ingredients to react to form a powder, each grain of which has a chemical composition close to the desired final composition. At this stage, however, the grain does not yet have the desired crystalline structure.

The next step is to mix the calcined powder with solid and/or liquid organic binders (intended to burn out during firing) and mechanically form the mixture into a "cake" that closely approximates a shape of the final sensing element. To form the "cakes" of desired shapes, several methods can be used. Among them are pressing (under force of a hydraulic-powered piston), casting (pouring viscous liquid

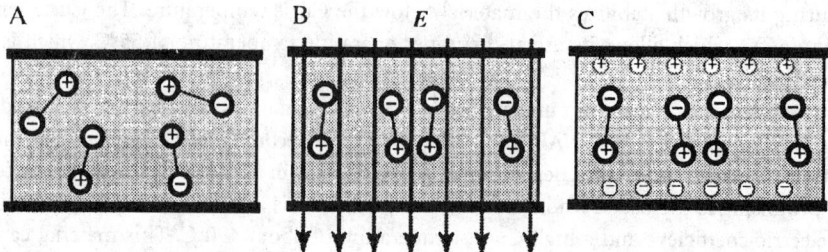

FIGURE 32.65 Poling of a pyroelectric crystal in a strong electric field. The sensor must be stored and operated below the Curie temperature.

into molds and allowing to dry), extrusion (pressing the mixture through a die, or a pair of rolls to form thin sheets), and tape casting (pulling viscous liquid onto a smooth moving belt).

After the "cakes" have been formed, they are placed into a kiln and exposed to a very carefully controlled temperature profile. After burning out of organic binders, the material shrinks by about 15%. The "cakes" are heated to a red glow and maintained at that state for some time, which is called the "soak time," during which the final chemical reaction occurs. The crystalline structure is formed when the material is cooled down. Depending on the material, the entire firing may take 24 h.

When the material is cold, contact electrodes are applied to its surface. This can be done by several methods. The most common are: fired-on silver (a silk-screening of silver-glass mixture and refiring), electroless plating (a chemical deposition in a special bath), and sputtering (an exposure to metal vapor in partial vacuum).

Crystallities (crystal cells) in the material can be considered electric dipoles. In some materials, like quartz, these cells are naturally oriented along the crystal axes, thus giving the material sensitivity to stress. In other materials, the dipoles are randomly oriented and the materials need to be "poled" to possess piezoelectric properties. To give a crystalline material pyroelectric properties, several poling techniques can be used. The most popular poling process is thermal poling, which includes the following steps:

1. A crystalline material (ceramic or polymer film) that has randomly oriented dipoles (Figure 32.65A) is warmed to slightly below its Curie temperature. In some cases (for a PVDF film), the material is stressed. High temperature results in stronger agitation of dipoles and permits one to more easily orient them in a desirable direction.
2. The material is placed in strong electric field, E, (Figure 32.65B) where dipoles align along the field lines. The alignment is not total. Many dipoles deviate from the filed direction quite strongly; however, statistically predominant orientation of the dipoles is maintained.
3. The material is cooled down while the electric field across its thickness is maintained.
4. The electric field is removed and the poling process is complete. As long as the poled material is maintained below the Curie temperature, its polarization remains permanent. The dipoles stay "frozen" in the direction that was given to them by the electric field at high temperature (Figure 32.65C). The above method is used to manufacture ceramic and plastic pyroelectric materials.

Another method, called a corona discharge poling, can be used to produce polymer piezo/pyroelectric films. The film is subjected to a corona discharge from an electrode at several million volts per centimeter of film thickness for 40 s to 50 s [8, 9]. Corona polarization is uncomplicated to perform and can be easily applied before electric breakdown occurs, making this process useful at room temperature.

The final operation in preparing the sensing element is shaping and finishing. This includes cutting, machining, and grinding. After the piezo (pyro) element is prepared, it is installed into a sensor's housing, where its electrodes are bonded to electrical terminals and other electronic components.

After poling, the crystal remains permanently polarized; however, it remains electrically charged for a relatively short time. There is a sufficient amount of free carriers that move in the electric field setup

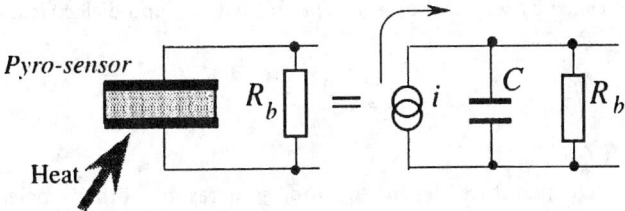

FIGURE 32.66 Pyroelectric sensor and its equivalent circuit.

inside the bulk material and there are plenty of charged ions in the surrounding air. The charge carriers move toward the poled dipoles and neutralize their charges (Figure 32.65C). Hence, after a while, the poled piezoelectric material becomes electrically discharged as long as it remains under steady-state conditions. When temperature changes and thermally induced stress develops, the balanced state is degraded and the pyroelectric material develops an electric charge. If the stress is maintained for a while, the charges again will be neutralized by the internal leakage. Thus, a pyroelectric material is responsive only to a changing temperature rather than to a steady level of it. In other words, a pyroelectric sensor is an ac device, rather than a dc device. Sometimes, it is called a *dynamic* sensor, which reflects the nature of its response.

Pyroelectric Sensors

To make sensors, the pyroelectric materials are used in the form of thin slices or films with electrodes deposited on the opposite sides to collect the thermally induced charges (Figure 32.66). The pyroelectric detector is essentially a capacitor that can be charged by an influx of heat. The detector does not require any external electrical bias (excitation signal). It needs only an appropriate electronic interface circuit to measure the charge. Contrary to thermoelectrics (thermocouples), which produce a steady voltage when two dissimilar metal junctions are held at steady but different temperatures, pyroelectrics generate charge in response to a change in temperature. Since a change in temperature essentially requires propagation of heat, a pyroelectric device is a heat flow detector rather than a heat detector. Figure 32.66 shows a pyroelectric detector (pyro-sensor) connected to a resistor R_b that represents either the internal leakage resistance or a combined input resistance of the interface circuit connected to the sensor. The equivalent electrical circuit of the sensor is shown on the right. It consists of three components: (1) the current source generating a heat induced current, i, (remember that a current is a movement of electric charges), (2) the sensor's capacitance, C, and (3) the leakage resistance, R_b. Since the leakage resistance is very high and often unpredictable, an additional bias resistor is often connected in parallel with the pyroelectric material. The value of that resistor is much smaller than the leakage resistance, yet its typical value is still on the order of 10^{10} Ω (10 GΩ).

The output signal from the pyroelectric sensor can be taken in the form of either charge (current) or voltage, depending on the application. Being a capacitor, the pyroelectric device is discharged when connected to a resistor, R_b (Figure 32.66). Electric current through the resistor and voltage across the resistor represent the heat flow-induced charge. It can be characterized by two pyroelectric coefficients [2]:

$$P_Q = \frac{dP_s}{dT} \qquad \text{Pyroelectric charge coefficient}$$

(32.123)

$$P_V = \frac{dE}{dT} \qquad \text{Pyroelectric voltage coefficient}$$

where P_s = Spontaneous polarization (which is the other way to say "*electric charge*")
  E = Electric field strength
  T = Temperature in K

Both coefficients are related by way of the electric permittivity, ε_r, and dielectric constant, ε_0:

$$\frac{P_Q}{P_V} = \frac{dP_s}{dE} = \varepsilon_r \varepsilon_0 \qquad (32.124)$$

The polarization is temperature dependent and, as a result, both pyroelectric coefficients in Equation 32.123 are also functions of temperature.

If a pyroelectric material is exposed to a heat source, its temperature rises by ΔT, and the corresponding charge and voltage changes can be described by the following equations.

$$\Delta Q = P_Q A \Delta T \qquad (32.125)$$

$$\Delta V = P_V h \Delta T \qquad (32.126)$$

Remembering that the sensor's capacitance can be defined as:

$$C_e = \frac{\Delta Q}{\Delta V} = \varepsilon_r \varepsilon_0 \frac{A}{h} \qquad (32.127)$$

then, from Equations 32.124, 32.126, and 32.127, it follows that:

$$\Delta V = P_Q \frac{A}{C_e} \Delta T = P_Q \frac{\varepsilon_r \varepsilon_0}{h} \Delta T \qquad (32.128)$$

It is seen that the peak output voltage is proportional to the sensor's temperature rise and pyroelectric charge coefficient and inversely proportional to its thickness.

Figure 32.67 shows a pyroelectric sensor whose temperature, T_0, is homogeneous over its volume. Being electrically polarized, the dipoles are oriented (poled) in such a manner as to make one side of the material positive and the opposite side negative. However, under steady-state conditions, free charge carriers (electrons and holes) neutralize the polarized charge and the capacitance between the electrodes does not appear to be charged. That is, the sensor generates zero charge. Now, assume that heat is applied to the bottom side of the sensor. Heat can enter the sensor in a form of thermal radiation that is absorbed by the bottom electrode and propagates toward the pyroelectric material via the mechanism of thermal conduction. The bottom electrode can be given a heat-absorbing coating, such as goldblack or organic paint. As a result of heat absorption, the bottom side becomes warmer (the new temperature is T_1), which causes the bottom side of the material to expand. The expansion leads to flexing of the sensor, which, in turn, produces stress and a change in dipole orientation. Being piezoelectric, stressed material generates electric charges of opposite polarities across the electrodes. Hence, one can regard secondary pyroelectricity as a sequence of events: a thermal radiation — a heat absorption — a thermally induced stress — an electric charge.

The temperature of the sensor T_s is a function of time. That function is dependent on the sensing element: its density, specific heat, and thickness. If the input thermal flux has the shape of a step function of time, for the sensor freely mounted in air, the output current can be approximated by an exponential function, so that:

$$i = i_0 e^{-t/\tau_T} \qquad (32.129)$$

where i_0 = Peak current

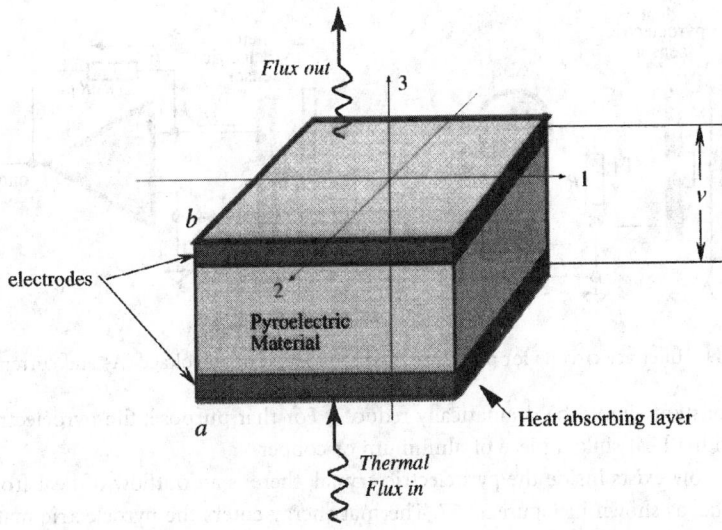

FIGURE 32.67 Pyroelectric sensor has two electrodes at the opposite sides of the crystalline material. Thermal radiation is applied along axis 3.

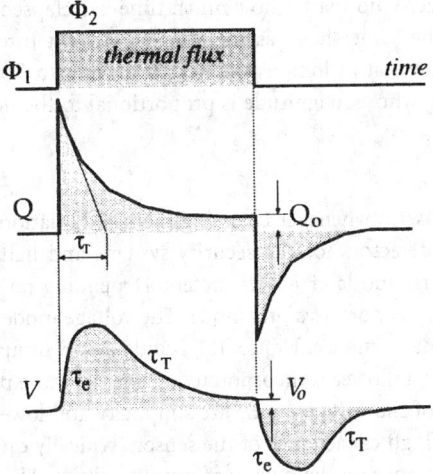

FIGURE 32.68 Response of a pyroelectric sensor to a thermal step function. The magnitudes of charge Q_0 and voltage v_0 are exaggerated for clarity.

Figure 32.68 shows timing diagrams for a pyroelectric sensor when it is exposed to a step function of heat. It is seen that the electric charge reaches its peak value almost instantaneously, and then decays with a *thermal time constant*, τ_T. This time constant is a product of the sensor's thermal capacitance, C, and thermal resistance, r, which defines a thermal loss from the sensing element to its surroundings:

$$\tau_T = Cr = cAhr \qquad (32.130)$$

where c = Specific heat of the sensing element.

The thermal resistance r is a function of all thermal losses to the surroundings through convection, conduction, and thermal radiation. For low-frequency applications, it is desirable to use sensors with τ_T as large as practical; while for the high-speed applications (e.g., to measure the power of laser pulses), a

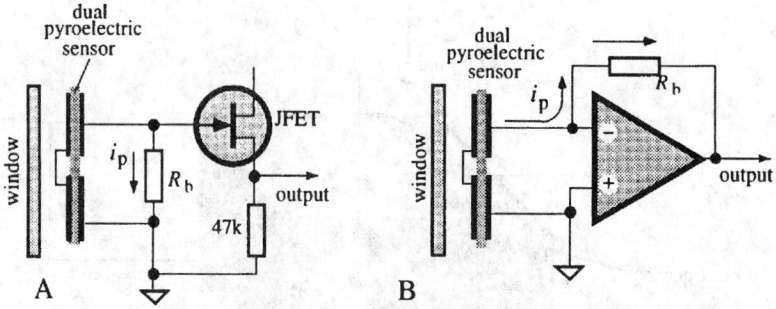

FIGURE 32.69   Interface circuits for pyroelectric sensors operating in voltage (A) and current (B) modes.

thermal time constant should be dramatically reduced. For that purpose, the pyroelectric material can be laminated with a heat sink: a piece of aluminum or copper.

When a heat flow exists inside the pyroelectric crystal, there is an outflow of heat from the opposite side of the crystal, as shown in Figure 32.67. Thermal energy enters the pyroelectric material from side *a*. Since the other side *b* of the sensor faces a cooler environment, part of the thermal energy leaves the sensor and is lost to its surroundings. Because the sides *a* and *b* face objects of different temperatures (one is the temperature of a target and the other is the temperature of the environment), a continuous heat flow exists through the pyroelectric material. As a result, in Figure 32.68, charge Q and voltage V do not completely return to zero, no matter how much time has elapsed. Electric current generated by the pyroelectric sensor has the same shape as the thermal current through its material. An accurate measurement can demonstrate that as long as the heat continues to flow, the pyroelectric sensor will generate a constant voltage V_0 whose magnitude is proportional to the heat flow.

Applications

The pyroelectric sensors are useful whenever changing thermal radiation or heat flow need to be measured. Examples are motion detectors for the security systems and light control switches [1], instant medical infrared thermometers, and laser power meters. Depending on the application, a pyroelectric sensor can be used either in *current* or in *voltage* mode. The voltage mode (Figure 32.69A) uses a voltage follower with a very high input resistance. Hence, JFET and CMOS input stages are essential. As a rule, in the voltage mode sensor, the follower is incorporated inside the same package along with the element and bias resistor. Advantages of the voltage mode are simplicity and lower noise. The disadvantages are slower speed response due to high capacitance of the sensor (typically on the order of 30 pF) and other influences of the sensor capacitance on the quality of output voltage. The output voltage of the follower is shown in Figure 32.68 (V). It is seen that it rises slowly with electric time constant τ_e and decays with thermal time constant τ_T.

The current mode uses an electronic circuit having a "virtual ground" as its input (Figure 32.69B). An advantage of this circuit is that the output signal is independent of the sensor's capacitance and, as a result, is much faster. The signal reaches its peak value relatively fast and decays with thermal time constant τ_T. The output voltage V_0 has the same shape of charge Q in Figure 32.68. The disadvantages of the circuit are higher noise (due to wider bandwidth) and higher cost.

Note that Figure 32.69 shows dual pyroelectric sensors, where two sensing elements are formed on the same crystalline plate by depositing two pairs of electrodes. The electrodes are connected in a serial-opposite manner. If both sensors are exposed to the same magnitude of far-infrared radiation, they will produce nearly identical polarizations and, due to the opposite connection, the voltage applied to the input of the transistor or the current passing through resistor R_b will be nullified. This feature allows for cancellation of undesirable common-mode input signals in order to improve stability and reduce noise. Signals that arrive only to one of the elements will not be canceled.

References

1. J. Fraden, *Handbook of Modern Sensors,* 2nd ed., Woodbury, NY: AIP Press, 1997.
2. H. Meixner, G. Mader, and P. Kleinschmidt, Infrared sensors based on the pyroelectric polymer polyvinylidene fluoride (PVDF), *Siemens Forsch. Entwickl. Ber.,* 15(3), 105-114, 1986.
3. P. Kleinschmidt, Piezo- und pyroelektrische Effekte. Heywang, W., ed., in *Sensorik,* Kap. 6: New York: Springer, 1984.
4. Semiconductor Sensors, *Data Handbook,* Philips Export B.V, 1988.
5. C. Ye, T. Tamagawa, and D.L. Polla, Pyroelectric PbTiO$_3$ thin films for microsensor applications, *Transducers'91. Int. Conf. Solid-State Sensors and Actuators.* 904-907, 1991.
6. H. Kawai, The piezoelectricity of poly(vinylidene fluoride), *Jap. J. Appl. Phys.,* 8, 975-976, 1969.
7. A. Okada, Some electrical and optical properties of ferroelectric lead-zirconite-lead-titanate thin films, *J. Appl. Phys.,* N. 48, 2905, 1977.
8. P.F. Radice, *Corona Discharge poling process,* U.S. Patent No. 4,365,283; 1982.
9. P.D. Southgate, *Appl. Phys. Lett.,* 28, 250, 1976.

32.8   Liquid-In-Glass Thermometers

J.V. Nicholas

The earliest form of thermometer, known as a thermoscope, was a liquid-in-glass thermometer that was open to the atmosphere and thus responded to pressure. By sealing a thermoscope so that it responded only to temperature, the modern form of a liquid-in-glass thermometer resulted. As a temperature sensor, its use dominated temperature measurement for at least 200 years. Liquid-in-glass thermometers had a profound effect on the development of thermometry and, in popular opinion, they are the only "real" thermometers! Liquid-in-glass thermometer sensors were developed in variety to fill nearly every niche in temperature measurement from −190°C to +600°C, including the measurement of temperature differences to a millikelvin. In spite of the fragile nature of glass, the popularity of these thermometers continues because of the chemical inertness of the glass, as well as the self-contained nature of the thermometer.

Measurement sensor designers are unlikely to develop their own liquid-in-glass thermometers, but many will use them to check the performance of a new temperature sensor. The emphasis in this chapter section will therefore be on the use of mercury-in-glass thermometers — the most successful liquid-in-glass thermometer — as calibration references. Mercury-in-glass thermometers provide a stable temperature reference to an accuracy of 0.1°C, provided they are chosen and used with care. The extra requirements to achieve higher accuracy are indicated, but are beyond the scope of this section.

The trend is, however, to move away from mercury-in-glass thermometers for specialized uses (in particular, where the risk from glass or mercury contamination is not acceptable; for example, in the food or aviation industries). Other forms of temperature sensors described in this handbook are be more suitable.

General Description

A common form of mercury-in-glass is a solid-stem glass thermometer illustrated in Figure 32.70. The other major form of liquid-in-glass thermometer is the enclosed scale thermometer, for which the general principles discussed will also apply.

There are four main parts to the liquid-in-glass thermometer:

- The bulb is a thin glass container holding the bulk of the thermometric liquid. The glass must be of suitable type and properly annealed. The thinness is essential for good thermal contact with the medium being measured, but it can result in instabilities due to applied stress or sudden shock.

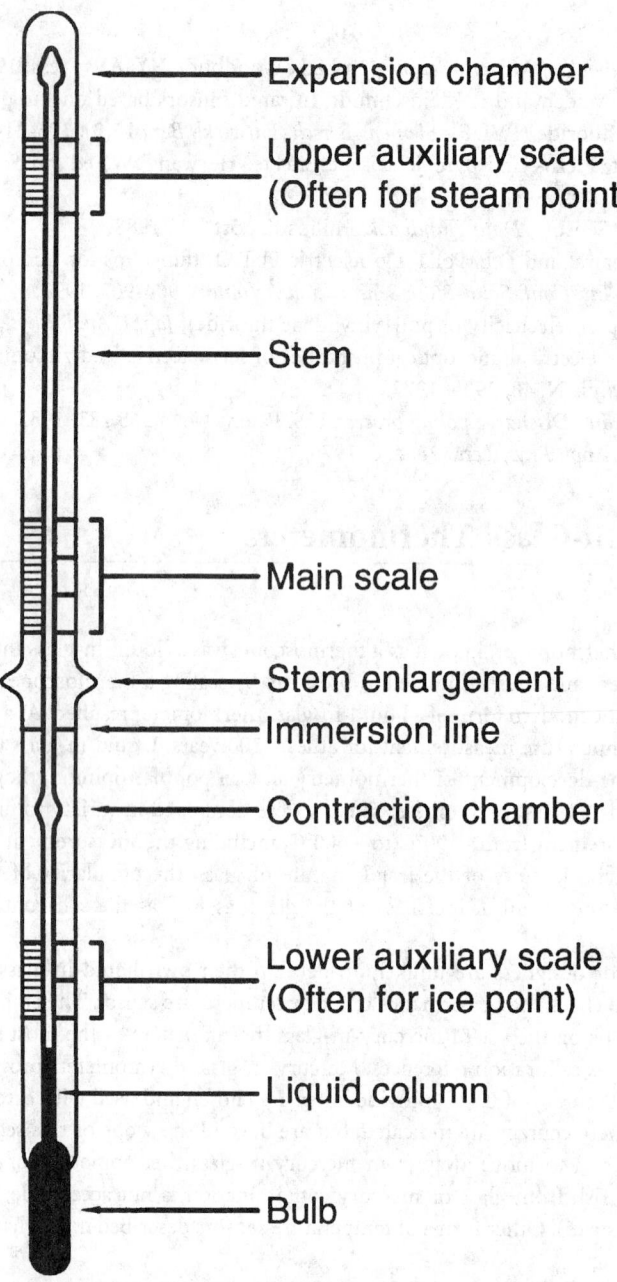

FIGURE 32.70   The main features of a solid-stem glass thermometer. The thermometer can have an enlargement in the stem or an attachment at the end of the stem to assist in the positioning of the thermometer. Thermometers will display several of these features, but seldom all of them.

Some lower-accuracy, general-purpose thermometers are made with a thicker glass bulb to lower the risk of breakage.
- The stem is a glass capillary. Again, a suitable glass is necessary and may differ from that of the bulb. The bore can be gas-filled or vacuous. The volume of the bore must be somewhat smaller than the volume of the bulb for good sensitivity. In addition, the bore must be smooth, with a uniform cross section.

- The liquid is usually mercury for best precision, or an organic liquid for lower temperature ranges.
- The markings are usually etched or printed onto the stem. The markings include the scale, to allow direct reading of the temperature, as well as other important information.

Figure 32.70 illustrates the main parts of a mercury-in-glass thermometer, along with a nomenclature for other features commonly found. Not all of the features will be found on all thermometers.

The operation of liquid-in-glass thermometers is based on the expansion of the liquid with temperature; that is, the liquid acts as a transducer to convert thermal energy into a mechanical form. As the liquid in the bulb becomes hotter, it expands and the liquid is forced up the capillary stem. The temperature of the bulb is indicated by the position of the top of the mercury column with respect to the marked scale. The flattest part of the liquid meniscus is used as the indicator: for mercury, this is the top of the curve; for organic liquids, the bottom.

The thermometers appear to have a simplicity about them, but this is lost when accurate measurements are required. By accuracy, we mean any reading where the temperature needs to be known to within 1°C or better. The chief cause of inaccuracy is that not all of the liquid is at the required temperature due to its necessary presence in the stem. Thus, the thermometer is also sensitive to the stem temperature. The main advantage of a liquid-in-glass thermometer is that it is self-contained; but this means that the thermometer stem has to be seen to read the scale. Even in a well-designed apparatus, a good part of the stem will not be at the temperature of the bulb. For example, with a bulb immersed in boiling water and the entire stem outside, an error of 1°C results from the cooler stem. Correction for this error can be incorporated in the scale of a partial immersion thermometer, or the error can be corrected with a chart of stem corrections.

The next most significant cause of error comes from the glass, a substance with complex mechanical properties. Like mercury, it expands rapidly on heating but does not contract immediately on cooling. This produces a hysteresis which, for a good glass, is about 0.1% of the temperature change. A good, well-annealed thermometer glass will relax back over a period of days. An ordinary glass might never recover its shape. Besides this hysteresis, the glass bulb undergoes a secular contraction over time; that is, the thermometer reading increases with age, but fortunately the effect is slow and calibration checks at the ice point, 0°C, can correct for it.

A bewildering number of types of liquid-in-glass thermometers are available, with many of the variations being designed with different dimensions and temperature ranges to suit specific applications.

For best performance, solid-stem mercury-in-glass thermometers should be restricted to operation over the maximum range −38 °C to 250°C. The purchase should be guided by a specification as published by a recognized standards body. Such bodies include the International Standards Organisation (ISO) [1]; the American Society for Testing and Materials (ASTM) [2]; the British Standards Institute (BS) [3]; or the Institute for Petroleum (IP). Be aware that some type numbers are the same, yet may refer to different thermometers, such as in the IP and ASTM ranges. Make sure the specification body is referred to; for example, an order for a 16C thermometer could result in either an ASTM 10C, the equivalent of an IP 16C, or an IP 61C, the equivalent of ASTM 16C.

One's choice of thermometer will most probably be a compromise between the best range, scale division, and length for the purpose. If good precision is required, then the thermometer range will be constrained to avoid extremely long and unwieldy thermometers. Table 32.20 gives the specification for precision thermometers based on the compromise as seen by the ASTM [2]. The cost of these thermometers depends on the range and varies from $50 to $180 at 1996 prices. The best precision for the ASTM liquid-in-glass thermometers is around 0.1°C, with the thermometers supplied being accurate to one scale division. Consult the references at the end of this chapter section if higher-resolution thermometers are required. As a rule, choose thermometers subdivided to the accuracy desired, and do not rely on visual interpolation to increase the accuracy. If relying heavily on interpolation, then a better choice of thermometer should be made.

Table 32.20 has thermometers with an ice point either in the main scale or as an auxiliary scale. The ice point is a very convenient way to check on the on-going performance of a thermometer, and without it, more expensive time-consuming procedures may be needed.

TABLE 32.20 Summary of Requirements for ASTM Precision Thermometers

ASTM Thermometer Number	Range (°C)	Maximum length (mm)	Graduation at each (°C)	Maximum error (°C)
62C	−38 to +2	384	0.1	0.1
63C	−8 to +32	384	0.1	0.1
64C	−0.5 to +0.5 and 25 to 55	384	0.1	0.1
65C	−0.5 to +0.5 and 50 to 80	384	0.1	0.1
66C	−0.5 to +0.5 and 75 to 105	384	0.1	0.1
67C	−0.5 to +0.5 and 95 to 155	384	0.2	0.2
68C	−0.5 to +0.5 and 145 to 205	384	0.2	0.2
69C	−0.5 to +0.5 and 195 to 305	384	0.5	0.5
70C	−0.5 to +0.5 and 295 to 405	384	0.5	0.5

TABLE 32.21 Working Range of Some Thermometric Liquids and Their Apparent Thermal Expansion Coefficient in Thermometer Glasses Around Room Temperature

Liquid	Typical apparent expansion coefficient (°C⁻¹)	Possible temperature range (°C)
Mercury	0.00016	−35 to 510
Ethanol	0.00104	−80 to 60
Pentane	0.00145	−200 to 30
Toluene	0.00103	−80 to 100

Liquid Expansion

The equation that best describes the expansion of the mercury volume is:

$$V = V_0\left(1 + \alpha t + \beta t^2\right) \tag{32.131}$$

where V_0 = Volume of mercury at 0°C
  α and β = Coefficients of thermal expansion of mercury, with

$$\alpha = 1.8 \times 10^{-4}\,°C^{-1}$$

$$\beta = 5 \times 10^{-8}\,°C^{-2}$$

See Table 32.21 for the expansion coefficients of other liquids and their range of use.

Equation 32.131 is the ideal equation for a mercury-in-glass thermometer. In practice, several factors modify the ideal behavior because of the way in which the thermometers are constructed.

Because the glass of a mercury-in-glass thermometer also expands, it is the apparent expansion coefficient due to the differential expansion of the mercury with respect to the glass that is of interest. Glass used in a typical thermometer has a value of $\alpha = 2 \times 10^{-5}\,°C^{-1}$, about 10% that of mercury. Hence, both the glass and the mercury act as temperature transducers and thus justify the description "mercury-in-glass."

The mercury also serves as the temperature indicator in the stem and consequently might not be at the same temperature as the mercury in the bulb. While this effect is small for mercury, where the bulb volume is 6250 times the volume of the mercury in a 1°C length of the capillary stem, thermometers used in partial immersion often need correcting.

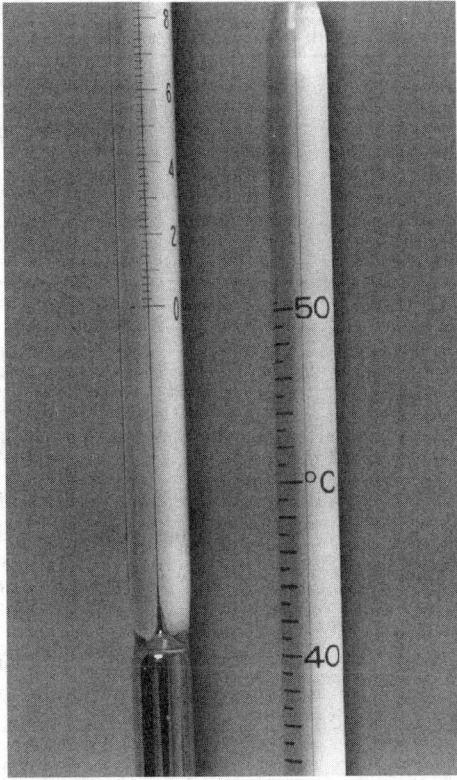

FIGURE 32.71 Calibration marks are usually scratched on at both ends of a thermometer's scale to locate the ruling of the scale. Left: A good quality thermometer. The calibration mark is immediately alongside the 0°C mark. Right: A general-purpose thermometer. Here, the calibration mark is about ¼ scale division above the 50°C mark. Since it is a cheaper thermometer, the manufacturer is content to locate the scale within the ¼ scale division, and this would vary from thermometer to thermometer in the same batch. Readings could be expected to be accurate to about one scale division, 0.5°C in this instance.

The bore in the stem needs to be smooth and uniform. An allowed departure from uniformity is a contraction chamber that, by taking up a volume of the expanding mercury, allows the overall length of the thermometer to be kept a reasonable size. The chamber shape must be very smooth to prevent bubbles of gas being trapped. An auxiliary scale is usually added for the ice point if a contraction chamber is used.

The marked scale allows the user to read the column length as a temperature. For a well-made thermometer, the change in length is proportional to the change in volume and hence to the temperature, as per Equation 32.131. In order to make the scale, the manufacturer first places "calibration" marks on the thermometer stem, as shown in Figure 32.71. Depending on the range and accuracy, more than two calibration marks can be used and, thus, the thermometer stem is divided into segments. A ruling engine is then used to rule a scale between each pair of marks, with careful alignments between the adjacent segments if they occur. The scale rulings will be spaced to approximate Equation 32.131 to the accuracy expected for the thermometer type. A good indicator of the quality of a thermometer is how close these calibration scratches are to the scale markings.

Because of the segmented ruling, it pays to check that the scale markings are uniform in appearance with no obvious glitches. Quite marked discontinuities in the scale are sometimes found. The markings are individual to each thermometer and the total scale length can vary from thermometer to thermometer. This can be an inconvenience if a thermometer has to be replaced; but fortunately, most quality thermometer specifications restrict the amount of variation permissible.

TABLE 32.22   Time Constants for a Mercury-In-Glass
Thermometer with a 5-mm Diameter Bulb

Medium	Still (s)	0.5 m s⁻¹ flow (s)	Infinite flow velocity (s)
Water	10	2.4	2.2
Oil	40	4.8	2.2
Air	190	71	2.2

Time-Constant Effects

The time constant is determined almost entirely by the diameter of the bulb because heat must be conducted from the outside to the center of the bulb. A typical bulb of diameter 5 mm has a relatively short time constant. The length of the bulb is then determined by the sensitivity required of the thermometer, given that there is a minimum useful diameter for the capillary bore.

The choice of bore diameter is a compromise involving several error effects. A large-diameter bore requires a larger volume bulb to achieve a given resolution, thus increasing the thermal capacity. A small-diameter bore not only becomes difficult to read but also suffers from stiction — the mercury moving in fits and starts due to the surface tension between the mercury and the bore wall. Stiction should be kept less than 1/5 of a scale division.

Table 32.22 gives the $1/e$ time constants in various media for a 5-mm diameter bulb. Time constants for other diameters can be estimated by scaling the time in proportion to the diameter. The table clearly indicates that the thermometer is best used with flowing (or stirred) fluids.

Thermal Capacity Effects

Glass thermometers are bulky and have considerable mass, especially high-precision thermometers that have a long bulb. The high thermal mass or heat capacity can upset temperature measurements, making high precision difficult. Inappropriate use of liquid-in-glass thermometers occurs when the thermometer is too massive to achieve the precision required. Preheating the thermometer can alleviate the worst of the problem. For higher precision and low mass, choose a platinum resistance thermometer or thermistor instead.

Simple estimates of the heat requirements are made by measuring the volume of thermometer immersed, and assuming that 2 J are required to raise 1 cm³ of the thermometer volume (glass or mercury) by 1°C.

Separated Columns

A common problem is for a part of the thermometric liquid in the stem to become separated from the main volume. While this will show as an ice point shift, it is still important to make a simple visual check when using the thermometer.

With organic liquids, the problem might be more difficult to identify because liquid adheres to the surface of the capillary and may not be visible. Spirit thermometers need to be held vertically to allow the thermometric liquid to drain down. Warm the top of the thermometer to prevent condensation of any vapor. Allow time for drainage of the liquid in the thermometer if the temperature is lowered quickly (approximately 3 min per centimeter). Cool the bulb first in order to keep the viscosity of the liquid in the stem low for better drainage.

For mercury, the separation is usually visible. Two causes can be identified: boil-off and mechanical separation (Figure 32.72).

To help retard the boil-off of mercury vapor at high temperatures (e.g., above 150°C), the capillary tube is filled with an inert gas when manufactured. Usually, dry nitrogen is used under pressure to prevent

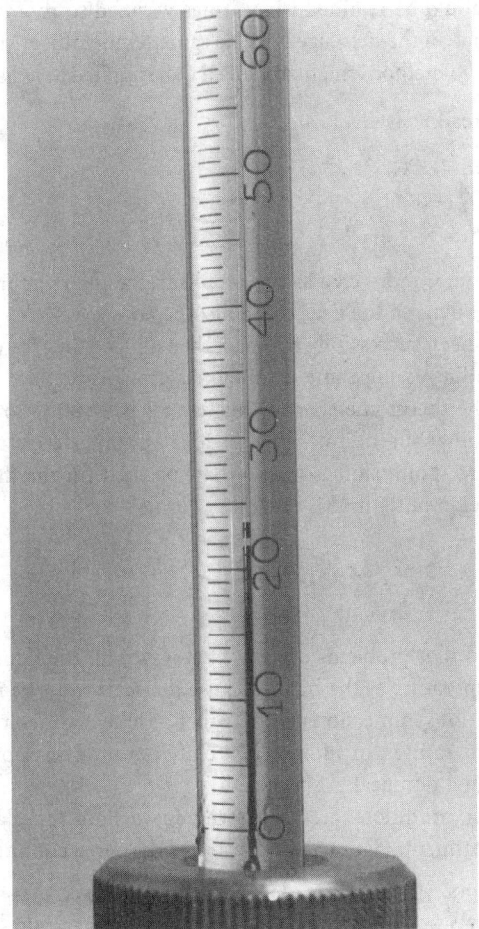

FIGURE 32.72 A typical break in the mercury column of a thermometer.

oxidation of the mercury. The expansion chamber must be kept cooler than the bulb to prevent a high pressure build-up. The high pressure can permanently distort the bulb even if rupture does not occur.

Mechanical separation of the liquid column is, unfortunately, a common occurrence, particularly after shipment. A gas fill will help prevent this separation but, conversely, the gas makes it more difficult to rejoin. There is also a risk of trapped gas bubbles in the bulb or expansion chambers and careful inspection is needed to locate them. A vacuum in the capillary tube will give rise to more breaks, but they are easily rejoined.

With care, it is often possible to rejoin the column and still have a viable thermometer. However, it must be realized that attempts to join a broken column could also result in the thermometer being discarded if the procedure is not successful. Column breaks that occur only when the thermometer is heated often require that the thermometer be discarded.

Various procedures for joining a broken mercury column can be tried. The procedures below are given in order of preference.

- Lightly tap the thermometer while holding it vertically. This may work for a vacuous thermometer.
- Apply centrifugal force, but avoid a flicking action, and be careful to avoid striking anything. This can be best done by holding the bulb alongside the thumb, protecting it with the fingers, and with the stem along the arm. Raise the arm above the head and bring it down quickly to alongside the leg.
- If both the above are unsuccessful, a cooling method can be attempted. This method relies on sufficient cooling of the bulb for all the mercury to contract into the bulb, leaving none in the

stem. The column should be rejoined when it has warmed to room temperature. Carry out the warming slowly so that all the mercury is at the same temperature. More than one attempt might be needed. The first two methods might also need to be applied to assist movement of the mercury.

Three cooling mediums readily available are:

1. Salt, ice, and water (to −18°C)
2. Dry ice, i.e., solid CO_2 (−78°C)
3. Liquid nitrogen (−196°C)

The last two refrigerants require more care as they could freeze the mercury. An excessive cooling rate could stress the glass. **Cold burns to the user could also occur.**

If the broken column has been successfully rejoined, then an ice point (or other reference point) check should be made. If the reading is the same as obtained previously, then the join can be considered completely successful and the thermometer ready for use. (It is essential to keep written records here.) However, a significant ice-point shift indicates that the join was not successful and that the thermometer should be discarded. If the ice-point shift is within that specified for the thermometer type, then treat the thermometer with suspicion until there is evidence of long-term stability, i.e., no significant ice-point changes after use.

Immersion Errors

It was previously mentioned that problems are expected if not all the liquid in a liquid-in-glass thermometer is at the same temperature as the bulb. Because the scale must be read visually, liquid-in-glass thermometers are used at various immersion depths, which results in different parts of the thermometric liquid being at different temperatures. In addition, the clutter around an apparatus often necessitates the thermometer being placed in a nonideal position.

Three distinct immersion conditions are recognized for a liquid-in-glass thermometer, and each requires a different error treatment. Figure 32.73 illustrates the three conditions.

- *Complete immersion*: By definition, if the complete bulb and stem are immersed at the same temperature, the thermometer is completely immersed. This condition is not common, except at room temperature, and should be avoided at higher temperatures. High pressure build-up in the thermometer can cause it to rupture, spreading deadly mercury vapor throughout the laboratory. Laboratories where there is a danger of mercury exposure to high temperatures should be kept well ventilated. In other words, DO NOT put a mercury thermometer completely inside an oven to measure the temperature. Specialized applications that use complete immersion, take into account pressure effects on the thermometers.
- *Total immersion*: Total immersion applies to the situation where all the thermometric liquid, i.e., all the mercury in the bulb, the contraction chamber, and the stem, is at the temperature of interest. The remaining portion of the stem will have a temperature gradient to room temperature (approximately). Especially at high temperatures, the expansion chamber should be maintained close to room temperature to avoid pressure build-up. A very small part of the mercury column can be outside the region of interest, to allow visual readings to be made. The error introduced by this can be estimated by the procedures given below for partial-immersion thermometers. Obviously, the thermometer will have to be moved to allow a range of temperatures to be measured. Total-immersion thermometers are generally calibrated at total immersion and therefore do not need additional corrections.
- *Partial immersion*: One way around the problem of scale visibility and the need to move the thermometer is to immerse the thermometer to some fixed depth so that most, but not all, of the mercury is at the temperature of interest. The part of the mercury column not immersed is referred to as the emergent column. Corrections will be needed to compensate for the error arising from the emergent column not being at the same temperature as the bulb. Many thermometers are

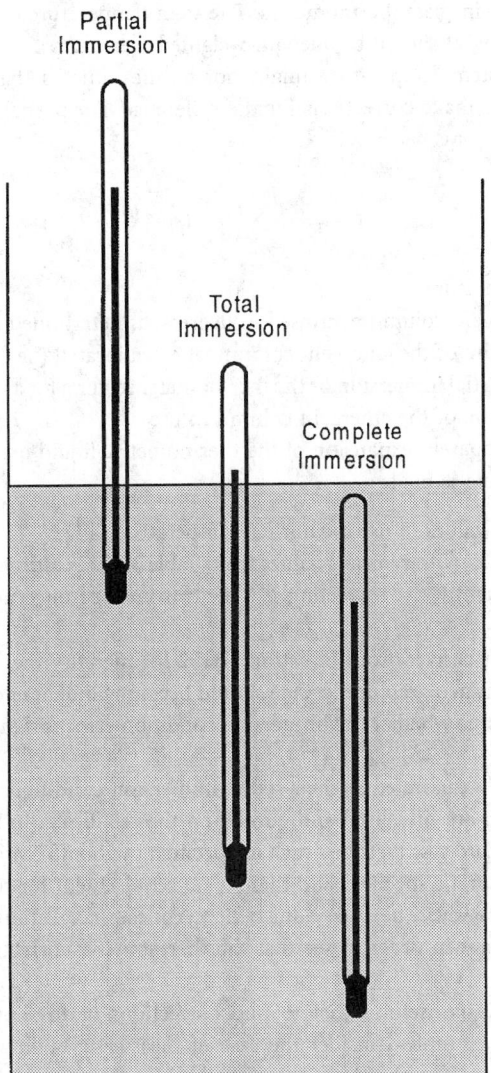

FIGURE 32.73 Three types of immersion conditions used for liquid-in-glass thermometers. The preferred immersion condition is usually marked as a line or distance on the stem for partial-immersion thermometers or is given by the thermometer specification.

designed and calibrated for partial immersion and are marked accordingly on the stem with an immersion depth or an immersion line (see Figure 32.70).

A partial-immersion thermometer is properly defined when the temperature profile of the emergent column is also specified. Usually, an average stem temperature is quoted to represent the temperature profile of the emergent column. Thermometer specifications can define the expected stem temperature for a set of test temperatures, but they do not usually define stem temperatures for all possible readings.

A measure of the stem temperature is required if the accuracy of the thermometer reading is to be assessed. The traditional way to measure the stem temperature is with a Faden thermometer. These are mercury-in-glass thermometers with a very long bulb, and various bulb lengths available. The bulb is mounted alongside the part of the stem containing the emergent column with the bottom of the bulb in the fluid. An average stem temperature is obtained as indicated by the Faden thermometer. Other ways of measuring the temperature profile are to use thermocouples along the length of the thermometer, or even several small mercury-in-glass thermometers. The stem temperature can be calculated as a simple average; but strictly speaking, it should be a length-weighted average.

Because the measured stem temperature might not be the same as that given on the calibration certificate, it is necessary to make corrections for the difference. For partial immersion thermometers, the true temperature reading t is given by:

$$t = t_i + N \times \left(t_2 - t_1\right) \times k \qquad (32.132)$$

where t_i = Indicated temperature

N = Length of emergent column expressed in degrees, as determined by the thermometer scale

t_2 = Mean temperature of the emergent column when calibrated (i.e., the stem temperature on a certificate for partial immersion or the thermometer reading for a total-immersion certificate)

t_1 = Mean temperature of the emergent column in use

k = Coefficient of apparent expansion of the thermometric liquid used in the glass of which the thermometer stem is made

See Table 32.21 for suitable values to use for normal temperature ranges.

The use of Equation 32.132 with typical k values from Table 32.21 is estimated to give a 10% accuracy for the correction. Consequently, the correction is a major source of uncertainty for large temperature differences.

Figure 32.74 gives a chart derived from Equation 32.132 for mercury thermometers that enables the stem correction to be determined graphically. One should become familiar enough with it to make quick estimates in order to determine whether the immersion condition error is significant and therefore needs correction.

Thermometers are usually calibrated at their stated immersion conditions and the actual stem temperatures during calibration are measured and quoted on the certificate. In many applications, a thermometer is used for a standard test method (such as specified by the ASTM or IP). For these instances, the expected stem temperature is specified and there is no requirement for the stem temperature to be measured. The user will, however, need to adjust the certificate correction terms to the immersion conditions of the specification in order to see that the thermometer corrections meet the appropriate quality criteria.

The chart of Figure 32.74 is useful, either to find corrections or to show faults with a particular measurement method. For example, consider the case of measuring boiling water in a beaker with a total-immersion thermometer. The thermometer is too long to immerse and the water level is around the 20°C mark. The emergent column is therefore 80°C long and one assumes that the stem is close to room temperature of 20°C, resulting in an 80°C temperature difference from calibration conditions. On the chart, one finds that the intersection of the 80°C emergent line and the 80°C difference line gives a correction value of just over 1°C. If this value is unacceptable, then clearly a redesign of the measurement method is warranted. More detailed examples can be found in the text of Nicholas and White [5].

Organic Liquids

Thermometers with organic liquids have three possible uses:

- To measure temperatures below −38°C
- In situations where mercury is to be avoided
- For inexpensive thermometers

The utility of spirit thermometers is limited because of the lower achievable accuracy, the high nonlinearities, and the volatile nature of the liquids. Organic-liquid thermometers are also difficult to read because of the very clear liquid and concave meniscus. However, the use of a suitable dye and wide bore

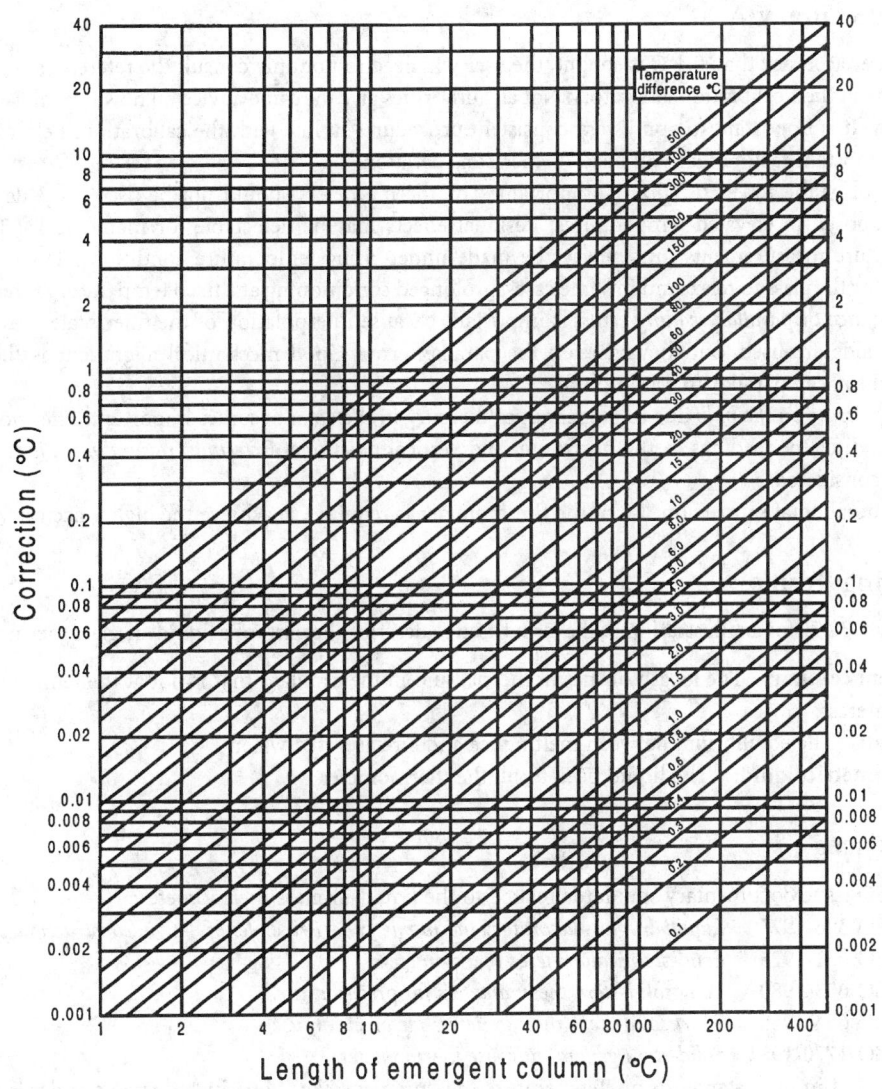

Length of emergent column (°C)

FIGURE 32.74 Chart of stem exposure corrections for mercury-in-glass thermometers with $k = 0.00016°C^{-1}$.

can give them as good a readability as mercury. Follow the recommendations of the section on Separated Columns and the section on Storage to get the best result from organic-liquid thermometers.

Storage

Most mercury-in-glass thermometers can be stored horizontally on trays in cabinets, care being taken to avoid any weight or pressure on the bulbs (one reason for the horizontal position). Avoid vibration. Corrugated cardboard, or similar material, can be used as a liner for a tray to prevent the thermometers from rolling.

Thermometers whose main range is below 0°C are better stored vertically, bulb down, in a cool place, but do not rest the thermometer on its bulb. This particularly applies to organic-liquid thermometers, which also should be shielded from light sources, as ultraviolet radiation can often degrade the liquid. If the top of the bore of a spirit thermometer is kept at a slightly higher temperature than the rest of the thermometer, then the volatile liquid will not condense in the expansion chamber.

High Accuracy

If a higher accuracy than 0.1°C is sought, the user will need to not only consult the references [4, 5], but also consult their calibration laboratory. Not all authorities give the same advice on how to achieve higher accuracy. It is important to apply very consistent procedures in line with the calibration. Below are the more important factors that will need further consideration.

Control of the *hysteresis effect* is important. The thermometers should not be used for 3 days after being exposed to elevated temperatures. Residual effects may be detectable for many weeks. That is, temperature measurements must always be made under rising temperature conditions. Three days is needed for the glass to relax, and in some cases, prolonged conditioning at a fixed temperature is required.

Avoidance of *parallax reading errors* is important because interpolation of the finer scale is essential. Optical aids are used, but they increase the parallax error. Good mechanical alignment is therefore required to keep parallax to a minimum.

The *pressure on the bulb* due to the length of mercury in the stem becomes important. Thermometers will give different readings in the horizontal and vertical positions. *External pressure variations* should also be considered.

In general, total immersion use of the thermometer is required to achieve the higher accuracy.

Defining Terms

See Figure 32.70 for an illustration of the terms used to describe the parts of a liquid-in-glass thermometer.

Emergent column: The length of thermometric fluid in the capillary that is not at the temperature of interest.

Ice point: The equilibrium between melting ice and air-saturated water.

Thermometric liquid: The liquid used to fill the thermometer.

References

1. ISO issue documentary standards related to the liquid-in-glass thermometer:
   ISO 386-1977   *Liquid-in-glass laboratory thermometer — Principles of design, construction and use.*
   ISO 651-1975   *Solid-stem calorimeter thermometers.*
   ISO 653-1980   *Long solid-stem thermometers for precision use.*
   ISO 654-1980   *Short solid-stem thermometers for precision use.*
   ISO 1770-1981 *Solid-stem general purpose thermometer.*
2. ASTM in their standards on Temperature Measurement, Vol. 14.03 include two standards related to liquid-in-glass thermometers:
   E1-95   *Specification for ASTM Thermometers.*
   E77-92 *Test Method for Inspection and Verification of Liquid-in-glass Thermometers.*
3. BSI publish a series of documentary specifications for thermometers, including:
   BS 593:1989   *Laboratory Thermometers.*
   BS 791:1990   *Thermometers for Bomb Calorimeters.*
   BS 1704:1985   *General Thermometers.*
   BS 1900:1976   *Secondary Reference Thermometers.*
4. J. A. Wise, *Liquid-in-glass Thermometer Calibration Service*, Natl. Inst. Stand. Technol. Spec. Publ., 250-23, 1988. A good treatment of calibration practice for liquid-in-glass thermometers, with a wider coverage than given here.
5. J. V. Nicholas and D. R. White, *Traceable Temperatures*, Chichester: John Wiley & Sons, 1990. The present chapter section was extracted and adapted from this text. The text explains how to make traceable calibrations of various temperature sensors to meet international requirements.

32.9 Manometric Thermometers

Franco Pavese

Manometric thermometers are defined in this Handbook as those thermometers that make use of the pressure of a *gaseous* medium as the physical quantity to obtain temperature. Very seldom are they available from commercial sources; for example, the temperature control of a home freezer is often of this kind. Consequently, instead of simply buying one, every user must build his own if this kind of thermometer is needed. They can be a quite useful choice since, in the era of electronic devices and sensors, it is still possible to make a totally nonelectronic thermometer, which in addition keeps its calibration indefinitely, as long as the quantity of substance sealed in it remains unchanged. The range of temperatures that can be covered by this kind of thermometer depends on the principle and on the substance used. When the thermodynamic equilibrium between the *condensed* phase of a substance (either liquid or solid) and its vapor is used, one has a "vapor-pressure thermometer" and the temperature range spanned by each substance is generally narrow. In addition, only substances that are gaseous at room temperature (i.e., condensed at temperatures lower than 0°C) are normally used, confining the working range to below room temperature; however, some substances that are liquid at room temperature and have a high vapor pressure (i.e., which easily evaporate) have bee used, but do not result in a sizable extension of the working range much above room temperature. A special case of vapor pressure being used at high temperature is the device called a "heat pipe," which is not used as a thermometer, but instead as an accurate thermostat [1]; using sodium, the working range is pushed up to ~1100°C. When a pure substance is used only in its gaseous state, one has a "gas thermometer," whose temperature range can be very wide, especially for moderate accuracy, depending mainly on the manometer; on the other hand, its fabrication is somewhat more complex and its use less straightforward.

Both thermometers can be built to satisfy the state-of-the-art accuracy of national standards (uncertainty better than ±0.001 K) or for lower accuracies, down to an uncertainty of ±1% or higher. Both thermometers require the measurement of pressure, in the range from less than 1 Pa up to 100 bar. Directions about the choice of the manometer can be found in Chapter 5.1 of this Handbook. A complete and specialized treatment on both vapor-pressure and gas thermometers up to room temperature and on pressure measurement instruments and techniques for gaseous media can be found in [2]. Gas thermometry above room temperature is treated in [3, 4].

In consideration of the fact that these kinds of thermometers typically must be built by the users, the following will concentrate on the basic guidelines for their design and fabrication.

Vapor Pressure

Figures 32.75 and 32.76 show the pressure values and the sensitivities in the temperature range allowed for each of the most common gaseous substances, considering also, in addition to the liquid phase, the use of the solid phase (where vapor pressure is lower) below the triple point. The lower end of the range is determined by the manometer uncertainty (for a given accuracy), the upper end by the full-scale pressure of the manometer (or by the full evaporation of the condensed phase).

Table 32.23 reports "certified" vapor pressure equations, linking the measured pressure p to the unknown temperature T. The reader might prefer them to the plethora of equations found in the literature, since they have been checked by official bodies and T is expressed in the ITS-90, the International Temperature Scale [5, text in 2]. More checked equations can be found in [6].

Figure 32.77 shows the general layout of a vapor-pressure thermometer. The fabrication of a vapor-pressure thermometer is not exceedingly difficult if a few guidelines are followed. Table 32.24 summarizes the most critical ones — design criteria and filling information — in a compact form [2]. Much more constructional details can be found in [1, 3]. In most cases, the manometer is located at room temperature, and the bulb is connected to it via a small-bore tube (the "capillary tube") without critical drawbacks. The accuracy of these thermometers ranges from ± 0.0001 K using very pure substances in calorimeters and precision mercury manometers, to ≈±1% using dial manometers.

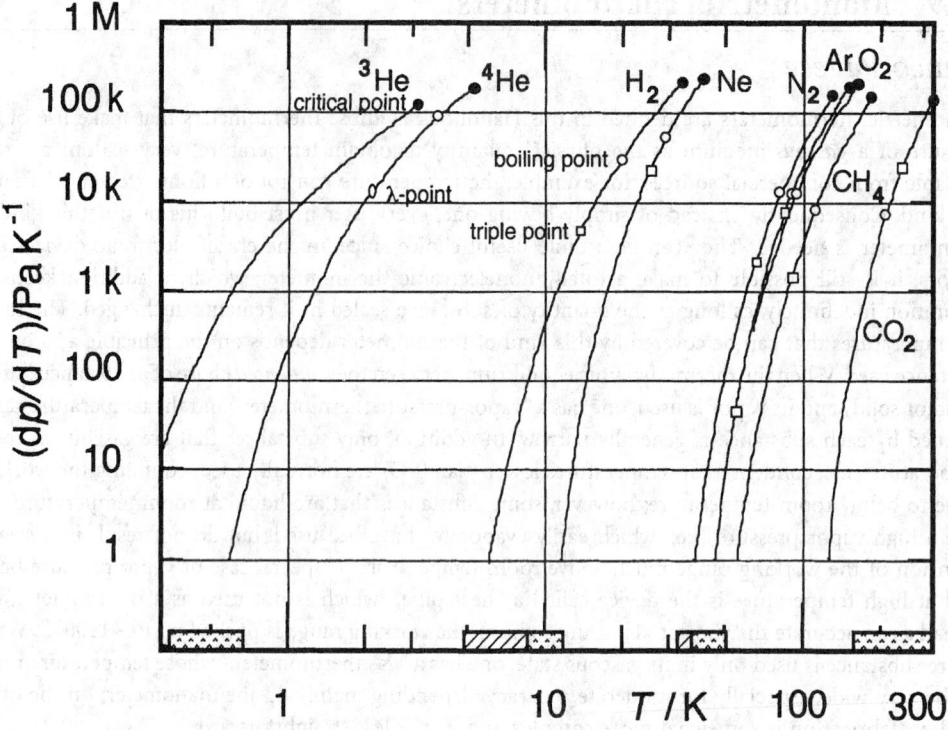

FIGURE 32.75 Range for vapor-pressure thermometry of various gases. The shaded parts indicate regions where it is less common or less accurate. ▨▨▨, not available; ▨▨▨, lower accuracy; ●, critical point; ○, triple point; 0, lambda-point.

Gas Thermometry

The layout of Figure 32.77 also applies to the design of a *constant-volume* gas thermometer (more common than the constant-pressure type), with the differences indicated in the relevant caption. The lower temperature end of the working range of a gas thermometer is stated, well before condensation of the substance takes place, by the increase of the uncertainty due to the increase in the nonideality of the gas — i.e., deviation from linearity of the relationship $p(T)$ — which takes place when approaching the condensed state or for increasing gas densities, or due to excessive absorption of gas on the bulb surface, thereby changing the quantity of the thermometric substance. All these conditions act at the lower end of the range. The upper end is stated by technological reasons or by the manometer full-scale capability. The best substances are, as listed, helium (either isotopes), hydrogen, and nitrogen.

From a design point of view, Table 32.25 summarizes the most critical issues. The major problem, apart from gas purity and ideality, is meeting the constant-volume requirement. Being that the manometer is generally at room temperature, the fraction of gas filling the connecting "capillary" tube is subtracted from the total amount of thermometric substance amount filling the system, and since this fraction is not constant, but depends on temperature and on technical conditions, it tends to increase the measurement uncertainty, which is contrary to the case of the vapor-pressure thermometer; this error is called the *dead-volume error*. Also, the bulb volume itself changes with temperature, due to thermal expansion and, to a much smaller extent, to the change in internal pressure. Design and fabrication criteria and measuring procedures are given in great detail in [2]. The case where the gas thermometer is *calibrated* at a number of fixed points is also described, with a discussion of the simplification in the use of the gas thermometer introduced with this instrument (called the *interpolating gas thermometer*, defined in the ITS-90 for use between 3 K and 26 K).

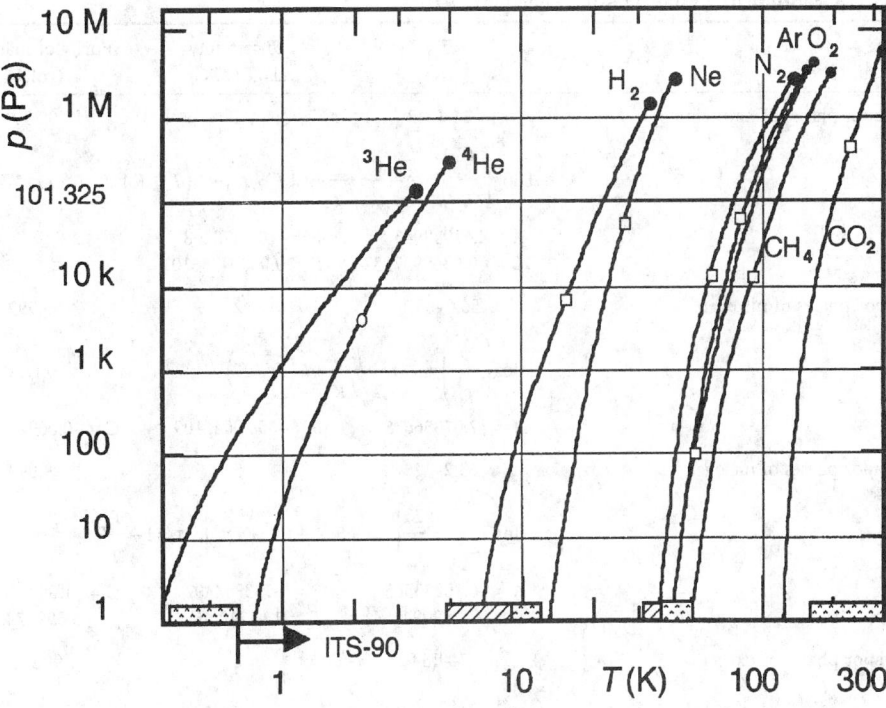

FIGURE 32.76 Sensitivity dp/dT of vapor-pressure thermometry for selected gases. The shaded parts indicate regions where it is less common or less accurate. ▨, not available; ▨, lower accuracy; ●, critical point; ○, triple point; □, lambda-point.

TABLE 32.23 Vapor Pressure Equations

Equilibrium state	T_{90} (K)	Uncertainty $\pm\delta T$ (mK)	Purity of material[1] (vol%)
Liquid-vapor phases of helium-4	1.25–2.1768	0.1	99.9999

$$T_{90}/K = A_0 + \sum_{i=1}^{9} A_i\left[\left(\ln\left(p/\text{Pa}\right) - B\right)/C\right]^i$$

$A_0 = 1.392408$	$A_1 = 0.527153$	$A_2 = 0.166756$
$A_3 = 0.050988$	$A_4 = 0.026514$	$A_5 = 0.001975$
$A_6 = -0.017976$	$A_7 = 0.005409$	$A_8 = 0.013259$
$B = 5.6$	$C = 2.9$	

2.1768–5.0	0.1	99.9999
$A_0 = 3.146631$	$A_1 = 1.357655$	$A_2 = 0.413923$
$A_3 = 0.091159$	$A_4 = 0.016349$	$A_5 = 0.001826$
$A_6 = -0.004325$	$A_7 = -0.004973$	$B = 10.3$
$C = 1.9$		

Equilibrium state	T_{90} (K)	Uncertainty	Purity
Liquid-vapor phases of equilibrium hydrogen	13.8–20.3	1[b]	99.99

$$p/\text{Pa} = \left(p_0/\text{Pa}\right)\exp\left[A + \frac{B}{T_{90}/K} + C\, T_{90}/K\right] + \sum_{i=0}^{5} b_i\left(T_{90}/K\right)^i$$

$A = 4.037592968$	$B = -101.2775246$
$C = 0.0478333313$	

$b_0 = 1902.885683$	$b_1 = -331.2282212$	$b_2 = 32.25341774$
$b_3 = -2.106674684$	$b_4 = 0.060293573$	$b_5 = -0.000645154$

TABLE 32.23 (continued) Vapor Pressure Equations

Equilibrium state	T_{90} (K)	Uncertainty $\pm\delta T$ (mK)	Purity of material[1] (vol%)
Liquid-vapor phases of natural neon[c]	24.6–40	2	99.99

$$\log\left(\frac{p}{p_0}\right) = A + \frac{B}{T_{90}/K} + C\left(T_{90}/K\right) + D\left(T_{90}/K\right)^2$$

$A = 4.61948943$ $\qquad$ $B = -106.478268$
$C = -0.0369937132$ $\qquad$ $D = 0.00004256101$

Solid-vapor phases of nitrogen	56.0–63.1	2	99.999

$$\log\left(\frac{p}{p_0}\right) = A + \frac{B}{T_{90}/K} + C\left(T_{90}/K\right)$$

$A = 12.07856655$ $\qquad$ $B = -858.0046109$ $\qquad$ $C = -0.009224098$

Liquid-vapor phases of nitrogen	63.2–125	5	99.999

$$\ln\left(\frac{p}{p_c}\right) = \frac{T_c}{T_{90}}\left[A\tau + B\tau^{0.5} + C\tau^3 + D\tau^6\right]; \quad \tau = 1 - \frac{T_{90}}{T_c}$$

$A = -6.10273365$ $\qquad$ $B = 1.153844492$ $\qquad$ $C = -1.087106903$
$D = -1.759094154$ $\qquad$ $T_c = 126.2124$ K $\qquad$ $p_c = 3.39997$ MPa

Liquid-vapor phases of oxygen	54–154	2	99.999

$$\ln\left(\frac{p}{p_c}\right) = \frac{T_c}{T_{90}}\left[A\tau + B\tau^{1.5} + C\tau^3 + D\tau^7 + D\tau^9\right]$$

$\tau = 1 - T_{90}/T_c$ $\qquad$ $A = -6.044437278$
$B = 1.176127337$ $\qquad$ $C = -0.994073392$ $\qquad$ $D = -3.449554987$
$E = 3.343141113$ $\qquad$ $T_c = 154.5947$ K $\qquad$ $p_c = 5.0430$ MPa

Liquid-vapor phases of argon	83.8–150	5	99.999

$$\ln\left(\frac{p}{p_c}\right) = \frac{T_c}{T_{90}}\left[A\tau + B\tau^{1.5} + C\tau^3 + D\tau^6\right]; \quad \tau = 1 - \frac{T_{90}}{T_c}$$

$A = -5.906852299$ $\qquad$ $B = 1.132416723$ $\qquad$ $C = -0.7720072001$
$D = -1.671235815$ $\qquad$ $T_c = 150.7037$ K $\qquad$ $p_c = 4.8653$ MPa

Liquid-vapor phases of methane	90.7–190	5[d]	99.99

$$\ln\left(\frac{p}{p_c}\right) = \frac{T_c}{T_{90}}\left[A t + B\tau^{1.5} + C\tau^{2.5} + D\tau^5\right]; \quad \tau = 1 - \frac{T_{90}}{T_c}$$

$A = -6.047641425$ $\qquad$ $B = 1.346053934$ $\qquad$ $C = -0.660194779$
$D = -1.304583684$ $\qquad$ $T_c = 190.568$ K $\qquad$ $p_c = 4.595$ MPa

Liquid-vapor phases of carbon dioxide	216.6–304	15	99.99

$$\ln\left(\frac{p}{p_c}\right) = A_0\left(1 - \frac{T_{90}}{T_c}\right)^{1.935} + \sum_{i=1}^{4} A_i\left(\frac{T_c}{T_{90}} - 1\right)^i$$

$p_c = 7.3825$ MPa $\qquad$ $T_c = 304.2022$ K $\qquad$ $A_0 = 11.37453929$
$A_1 = -6.886475614$ $\qquad$ $A_2 = -9.589976746$ $\qquad$ $A_3 = 13.6748941$
$A_4 = -8.601763027$

Note: For the relevant references and more gases, see [6]. $p_0 = 101325$ Pa, except when otherwise indicated.
[a] Minimum purity of the material to which the listed values of temperature and uncertainty apply.
[b] The summation term in the equation adds to the value of p a pressure amounting to the equivalent of 1 mK maximum.
[c] These values are for neon with an isotopic composition close to that specified in the ITS-90 definition.
[d] Above 100 K. It increases to 15 mK at 91 K, and to 10 mK near the critical point.

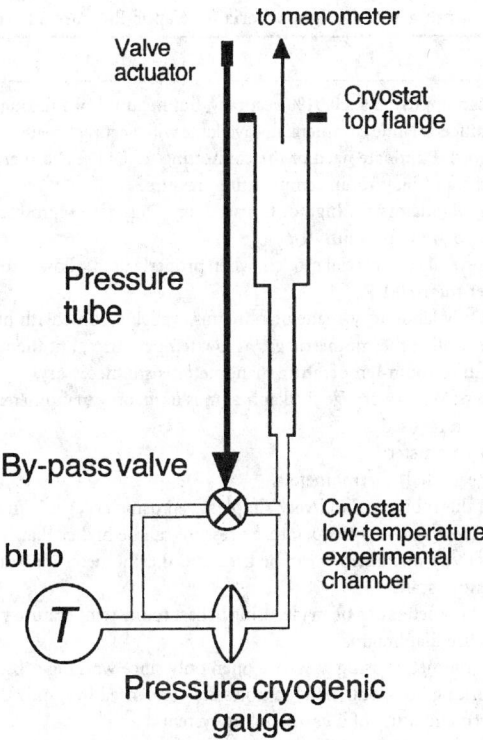

to manometer

Valve
actuator

Cryostat
top flange

Pressure
tube

By-pass valve

bulb

Cryostat
low-temperature
experimental
chamber

T

Pressure cryogenic
gauge

FIGURE 32.77 The general layout of a manometric thermometer. It is shown with a cryogenic diaphragm pressure transducer; when the transducer is placed instead at room temperature, the bypass valve is also placed at room temperature. *Vapor-pressure thermometer:* the diameter of the pressure tube increases in steps when pressures lower than ≈10 Pa must be measured, in order to decrease the thermomolecular pressure drop. *Gas thermometer* (constant-volume): the diameter of the pipes connecting the bulb to the cryogenic pressure transducer must be small in order to reduce the so-called "dead-volume." This requirement is much more stringent when the pressure transducer is moved up to room temperature. In this case, in order to reduce the error due to the "dead-volume," the bulb volume must be increased significantly.

TABLE 32.24 Summary of Design Criteria for Vapor-Pressure Thermometers

	Example
1. Choice of working substance:	$T_{max} = p_c$
• Temperature range: Each substance spans only a narrow temperature interval. $T_{max}/T_{min} <2-3$	$T_{min} = 100$ Pa K^{-1}
(including solid-vapor range), except helium. The limit:	$T_{max}/T_{min} =$
– T_{max} set by maximum manometer pressure.	^{3}He ≈ 10
– T_{min} set by manometer sensitivity.	^{4}He ≈ 9
• Accuracy:	H$_2$ ≈ 3
– Manometer: No single manometer spans whole range from ≈ 1 Pa (dp/dT ≈ 100 Pa K^{-1}) and	Ne ≈ 3
critical point ($p_c > 10^6$ Pa, except helium) with high or constant accuracy, or with sufficient	N$_2$ ≈ 2.5
sensitivity.	O$_2$ ≈ 2.5
– Substance: Not all substances allow for maximum accuracy, due to purity or to thermal	Ar ≈ 2.5
problems related to a low thermal diffusivity value.	CO$_2$ ≈ 2
2. Choice of pressure measuring system:	(solid ≈ 1.5)
Sensitivity and accuracy must be matched to the range of dp/dT and of p, i.e., T, to be measured.	
• *Without* separating diaphragm: Can be used only for low to medium accuracy, as thermometric gas also fills the entire manometric apparatus, with problems of contamination and increases in vapor volume.	
– Dial manometers: Used only for accuracy > ±1%.	
– Metal diaphragm or bellows (electronic) manometers: Can achieve a ±0.1–0.03% accuracy.	

TABLE 32.24 (continued)   Summary of Design Criteria for Vapor-Pressure Thermometers

	Example

- – Quartz bourdon gages: can approach a ±0.01% accuracy, but helium leaks through quartz.
- – Cryogenic pressure transducers: None commercially available with accuracy better than ±0.1% (after cryogenic calibration). Eliminate need of the connecting tube in sealed thermometers, but transducer must withstand high room-temperature pressure.
- • *With* separation diaphragm: Mandatory for high or top accuracy. Only zero reproducibility and a moderate linearity near zero are important.
  - – Capacitive diaphragms: Several commercial models, when properly used, allow zero sensitivity and reproducibility better than ±0.1 Pa.
  - – Cryogenic diaphragms: Only laboratory-made diaphragms available, some with high zero reproducibility. Allow to confine thermometric gas at low temperatures, but the tube connecting the diaphragm to room-temperature manometer is still necessary.
  Room-temperature manometers: When a cryogenic diaphragm is used, only manometers allowing helium as manometric gas can be used.
3. Choice of sealed vs. "open" thermometer:
  - • Sealed: Low-accuracy only (e.g., dial) thermometers.
    - – Medium-accuracy sealed thermometers still very simple when using cryogenic manometer and reducing vapor volume, but room-temperature pressure can be higher than 10 MPa. Therefore, only low-sensitivity manometers can be used and thermometer measures only upper part of vapor-pressure scale.
    - – High-accuracy sealed thermometers can be made, using ballast room-temperature volume and precision room-temperature diaphragm.
  - • "Open": Vapor-pressure thermometers using gases are open only since working substance does not stay permanently in working bulb, but (new) samples are condensed in it only during measurements. Requires permanent use of a gas-handling system.
4. Gas purity, isotopic composition and spin equilibrium:
  - • Purity: Must be known, and possibly checked, e.g., by performing a triple-point temperature measurement. Dew-boiling point difference measurement must also routinely be performed, before sealing in the case of sealed devices.
  - • Isotopic composition: Some gases show irreproducibility in results due to sample-to-sample changes in isotopic composition. It is impossible to obtain top accuracy with these substances, unless pure isotopes are used.
  - • Spin equilibrium: With some gases, showing different spin species, equilibrium must be ensured with use of a suitable catalyst.

Problems only for high accuracy Kr, Xe H_2, D_2

5. Thermometer filling:
  - • Amount of substance n_{max} at $T_{min} \rightarrow V^L \approx V_b$:

$$n_{max} \leq \frac{p_{min}}{R\,T_r}\left[\frac{2V_c T_r}{T_r + T_{min}} + V_r\right] + \frac{V_b}{M}\,\rho_{min}$$

  - • Amount of substance n_{min} at $T_{max} \rightarrow V^L = V_e^{;L} \approx 0$:

$$n_{min} \geq \frac{V_e^L \rho_{max}}{M} + \frac{p_{max}}{R\,T_r}\left[\frac{2V_c T_r}{T_r + T_{min}} + V_r + \frac{T_r}{T_{max}}\left(V_{max} - V_e^L\right)\right]$$

  - • Bulb volume V_b:

$$V_b\left[\frac{\rho_{min}}{M} - \frac{p_{max}}{R\,T_{max}}\right] \leq \frac{V_r}{R\,T_r}\left[p_{max} - p_{min}\right] + V^L\left[\frac{\rho_{max}}{M} - \frac{p_{max}}{R\,T_{max}}\right]$$

$$+ \frac{2V_c}{R}\left[\frac{p_{max}}{T_r + T_{max}} - \frac{p_{min}}{T_r + V_{min}}\right]$$

to a first approximation the terms in **bold** can be omitted.

^{4}He thermometer
T_{min} = 2.2 K
p_{min} = 5.263 kPa
ρ_{min} = 146 kg m^{-3}
T_{max} = 5.2 K
p_{max} = 227.5 kPa
ρ_{max} = 67.5 kg m^{-3}
T_f = 4.2 K
p_f = 100 kPa
T_r = 300 K
p_r = 200 kPa
V_r = 220 cm^3
V_c = 16 cm^3
V_T = 500 cm^3
M = 4 g mol^{-1}

It follows:
$V_b \geq \approx 2$ cm^3
Taking the minimum volume
0.074 $\geq n$
$n \geq 0.034$

TABLE 32.24 (continued)   Summary of Design Criteria for Vapor-Pressure Thermometers

	Example
• Calculation of the amount of substance n to condense in the thermometer: the gas is stored at p_r in the room-temperature reservoir of volume V_T. When the substance is condensed in the thermometer bulb at a temperature T_f, a residual $n_o = p_f V_T / R\,T_r$ remains in V_T. Therefore, in order to condense a quantity N, one must have in the system:	In order to seal-in 0.05 mol, the filling system must contain 0.069 mol

$$n' = \frac{p_r V_T}{R\,T_r}\left[1 - \frac{p_f}{p_r}\right] + \left(V_b + V_c + V_r\right)$$

Symbol caption: V^L = volume of the liquid phase; ρ = density; p = pressure; V = volume; subscript r = room temperature, c = capillary, b = bulb.

TABLE 32.25   Summary of Design Criteria for an Absolute Constant-Volume Gas Thermometer (CVGT) in the Low-Temperature Range ($T < 273.16$ K)

1. Choice of temperature range and of span $T_{min} \leftrightarrow T_{max}$:
   This choice is preliminary to the choice of most of the design parameters.
   - Below 273.16 K, ^{4}He gas thermometry is limited down to 2.5 K. With ^{3}He, accurate virial corrections available down to 1.5 K.
   - Only CVGTs of special design can be used in full span. Being that $p \propto T$, the 2.5 K to 273.16 K range corresponds to $p_{max}/p_{min} > 100$. For top accuracy, $\delta p/p < 0.01\%$, corresponding at p_{min} to $\delta p < 10^{-6}$, p_{max}, generally not achievable.
   - Being that $p \propto n/V$, molar density must generally be changed over the range to optimize accuracy, but n/V must be limited to avoid third virial correction, especially below ≈ 2 K.
   - In general, a CVGT is designed for work only below or only above a temperature between 25 K and 100 K.
2. Choice of reference temperature T_0:
   - Truly absolute thermometer: only one choice possible — 273.16 K.
     - Two-bulb CVGT: Avoids necessity to bring up to T_0 the bulb measuring $T_{min} > T < T_{max}$. Useful with thermometers designed for use at $T \ll T_0$.
     - Single-bulb CVGT: Same bulb spans the entire range up to T_0.
   - Low-temperature reference temperature T_0^* ($\approx$ from 25 K to 90 K):
     - Single-bulb CVGT commonly used. T_0^* value assigned by an independent experiment, and, therefore, not exact by definition. However, the additional uncertainty is a minor inconvenience with respect to the advantage of limiting bulb temperature within the span $T_{min} \leftrightarrow T_{max}$.
3. Choice of thermometric gas and filling density:
   - Thermometric gas:
     - Nitrogen: Low-medium accuracy.
     - e-Hydrogen: Not used for over 50 years, but still suitable for low-medium accuracy and temperature range above ≈ 20 K.
     - Helium-4: Commonly employed in recent gas thermometers. Use limited to above 2.5 K.
     - Helium-3: Considered more in modern gas thermometry. Use presently limited to above 1.5 K; potential for use down to <1 K.
   - Filling density: $p \propto n/V$ and $dp/dT \propto n/V$ (1 kPa K^{-1} $\triangleq$ 121 mol m^{-3}). Always advantageous increasing n/V, up to an upper boundary set by need of third virial correction. As a rule, $n/V < 250$ mol m^{-3} above ≈ 2.5 K, $n/V < 160$ mol m^{-3}, down to 1.2 K and $n/V < 30$ mol m^{-3} at 0.8 K.
4. Choice of the pressure measuring system:
   - See Table 32.24.
5. CVGT parameter design:
   - A. Room-temperature pressure transducer
     - Bulb: Top accuracy, 1 L volume typical; low accuracy, as low as 50 cm^3.
     - Dead-volume: Top accuracy, <10 cm^3; low accuracy: up to 10% of bulb volume.
   - B. Cryogenic pressure transducer
     No difference with respect to a vapor-pressure thermometer.
6. Bulb design:
   - Volume may not be constant, because of:
     - Compression modulus: Walls must be thick to limit deflection due to pressure, or bulb must be enclosed in a guard chamber kept at bulb pressure. Stress in bulb material must be relieved by annealing after machining.

TABLE 32.25 (continued)   Summary of Design Criteria for an Absolute Constant-Volume Gas Thermometer (CVGT) in the Low-Temperature Range ($T < 273.16$ K)

- Thermal expansion: Nothing can be done to suppress this effect, except using glass; must be corrected for. Small effect below ≈30 K.
  - Amount of "active" gas might not be constant, because of:
    - Gas adsorption: Physicochemical interaction of bulb walls with the gas determines the amount adsorbed. Copper often gold-plated to limit adsorption: this prevents heating the bulb above 50–70°C.
    - Impurity molecules on walls and leaks: Clean machining used for metal bulbs, followed by physicochemical cleaning. The bulb sealing gaskets must be stable in shape and leak-proof at working temperatures.
7. Dead-volume design:
   Dead-volume effect comes from combination of geometrical volume, working pressure, and gas density distribution, i.e., from the amount of substance contained in it.
   - Room-temperature dead-volume: Consists of all volumes of the gas measuring system at room temperature. Must be kept at uniform temperature (except diaphragm, often thermostated at ≈40°C), to be measured within 0.1–1°C.
   - Low-temperature dead-volume: (Part of) capillary tube between room and bulb temperature. Temperature and density change from one end to the other. Tube diameter is a tradeoff between geometrical volume and thermomolecular pressure effect: typical values between 0.5 mm and 3 mm. Advantageous keeping the parts of tube where temperature variations occur as short as possible. For medium-high accuracy, temperature distribution must be known accurately.
8. Gas handling and measuring system (for non-sealed CVGTs):
   - Handling system: Must ensure purity, checked on-line with a mass spectrometer for the highest accuracy, and include gas recovery with cryogenic pumps and clean storage (or purification).
   - Measuring system (case A): Separating diaphragm requires valve system for zero check, including constant-value valves and provisions to avoid (or to restore) thermometric gas losses and contamination from the manometric gas. For this purpose, a second diaphragm separator can be used.

References

1. R. E. Bedford, G. Bonnier, H. Maas, and F. Pavese, *Techniques for approximating the ITS-90,* Monograph 90/1 of the Bureau International des Poids et Mesures, Sèvres: BIPM, 1990.
2. F. Pavese and G. F. Molinar, *Modern gas-based temperature and pressure measurements,* International Monograph Series on Cryogenic Engineering, New York: Plenum Publishing, 1992, and references therein.
3. J. F. Schooley, *Thermometry,* Boca Raton, FL: CRC Press, 1986.
4. T. J. Quinn, *Temperature,* London: Academic Press, 1983.
5. R. E. Bedford, G. Bonnier, H. Maas, and F. Pavese, Recommended values of temperature on the ITS-90 for a selected set of secondary reference points, *Metrologia,* 33, 133-154, 1996.
6. F. Pavese, Recalculation on ITS-90 of accurate vapour-pressure equations for e-H_2, Ne, N_2 O_2, Ar, CH_4 and CO_2, *J. Chem. Thermodynam.,* 25, 1351-1361, 1993.

32.10 Temperature Indicators

Jan Stasiek, Tolestyn Madaj, and Jaroslaw Mikielewicz

Temperature indicators serve for approximate determination of bodies' temperatures and are used to control a variety of temperature treatment processes. The temperatures are determined based on knowledge of characteristic rated temperatures, which are mean critical temperatures of the indicator. However, it should be stressed that the accuracy of these measurements is satisfactory only if the measurement conditions are similar to the standard conditions for which the temperature indicators were calibrated. Otherwise, the critical temperatures of the indicators can be different from their rated temperatures listed in the standards and the measurements can have considerable errors.

The temperature indicators can be classified into two groups, each group using different physical properties for the determination of the temperature. The indicators belonging to the first group melt at certain temperatures. For some of these indicators, such as pyrometric cones, thermoscope bars and rings, the process of melting manifests itself as a shape/size deformation for which the temperature is determined by measuring the degree/rate of deformation of the indicator. For others, such as melting pellets, liquids, crayons, and monitors, the melting means turning entirely into a liquid smear. This can also be accompanied by color changing. The second group consists of color-change indicators containing pigments that at different temperatures, show different colors by selectively reflecting incident white light. Among this group are reversible and irreversible paints, color-change crayons, and liquid-crystal indicators.

Melting and Shape/Size Changing Temperature Indicators

The latest British Standard BS 1041, Part 7, 1988 [1] lists the following temperature indicators: Seger cones, thermoscope bars, and Bullers rings. Some temperature indicators used in the past — such as Watkin cylinders and Holcdorft bars — are no longer used and are of historical value only. In the U.S., melting pellets, crayons, liquids and monitors are available on the market and widely used.

The rated temperature for Seger cones (pyrometric cones) and thermoscope bars is defined by appropriate shape deformation resulting from the transformation of a certain amount of the indicator substance from the solid to liquid state. For chemically pure elements and compounds at a constant pressure, the temperature during the entire process of phase change remains constant. If the pressure changes within the range of changes for atmospheric pressure, then the temperature changes are insignificant and can be neglected even during precise measurements. The pyrometric cones and thermoscope bars are prepared from complex mixes of frits, fluxes, clays, calcium and magnesium compounds, silica, etc. The melting temperatures of the indicators, also referred to as the critical or rated temperatures, vary with the proportions of the above compounds. Therefore, a set of indicators differing in the proportions of the compounds is capable of covering a required range of rated temperatures. The melting temperatures can also change, to a degree, with the proportion of phases. For mixes that constitute pyrometric cones and thermoscope bars, the temperature difference between the beginning and the end of the melting process can be as large as 25 to 40°C. At the rated temperature, one can assume that the temperature is either at the beginning or at the end of the melting process. In practice, an intermediate value is assumed, referring to an expected shape deformation of the temperature indicator.

For Bullers rings, the rated temperature is determined by a shape deformation that can be described as a temperature shrinkage. An indication of the required temperature is a proper contraction of the outer diameter of the ring made of special clay (a mix of appropriate materials) that contracts uniformly with the increase in temperature throughout the operating range.

For melting pellets, crayons, liquids, and monitors, the rated temperature is that of the beginning of the melting process when the indicator turns entirely into a liquid smear. Usually, on cooling, the liquid mark solidifies and becomes glossy-transparent or translucent in appearance. The entire process can be accompanied by a change in color — mostly because the color of the workpiece surface or the back of an adhesive label, which enables the contact of the indicator with the surface, will show up from under the transparent mark. However, the moment of melting — not a color change — is the temperature signal.

Seger Cones

The pyrometric cones are typically slender, truncated, trihedral pyramids, about 25 mm to 60 mm in height. The base of the pyramid is a regular triangle of side 7 mm to 16 mm. One edge of the pyramid is vertical or slightly leaned outward (see Figure 32.77a). The recommended height of the standard cones is 60 mm; the laboratory cones are 30 mm high.

The pyramids are manufactured by pressing a powder mixture of a number of minerals mixed in different proportions throughout the required range of rated temperatures. The main components are silicon oxide (SiO_2), aluminium oxide (Al_2O_3) with additives in the form of oxides (MgO, K_2O, Na_2O, CaO, B_2O_3, PbO), and an organic binder. The following equation is an example of chemical constitution of the pyrometric cone for the temperature range of 600°C to 900°C:

$$X\left(2SiO_2 + Al_2O_3\right) + \left(1-X\right)\left(0.5Na_2O + 0.5PbO + B_2O_3 + 2SiO_2\right)$$

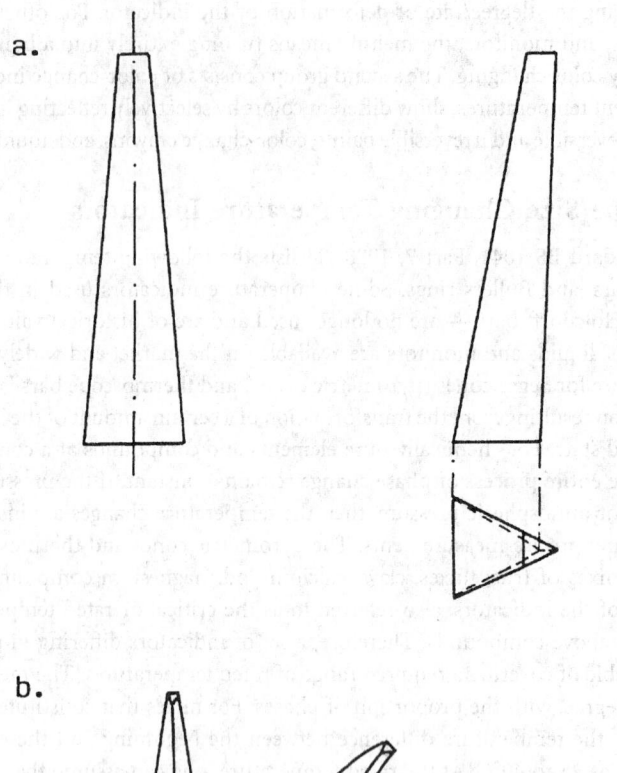

FIGURE 32.77   Pyrometric cones (a) in cross-sections; (b) on a plaque during firing.

where X is a mass unit. The pyramids designed for higher rated temperatures are prepared based on similar equations.

Touch-down Temperatures for Seger Cones.

As the heating progresses, the cone used for the measurement begins to soften and bends until its tip touches down on the surface on which it was placed. The rated temperature referring to this deformation is called the *touch-down temperature*. The touch-down temperatures for the Seger cones are determined in an electric kiln with clean atmospheric air at a heating rate of 60°C h^{-1}.

According to the earlier German standard DIN 51063 [2], the range of touch-down temperatures from 600°C to 2000°C at 10 to 50°C steps is covered by a series of Seger cones denoted traditionally by numbers from 022 to 42 (see Table 32.26a).

According to the latest British Standard BS 1041, Part 7, 1988, the touch-down temperatures within the range of 600°C to 1535°C at temperature intervals of 15 to 35°C are realized by a series of Seger cones numbered from 022 to 20 (see Table 32.26b).

The precision of determination of the touch-down temperatures for the industrial cones should be better than ±15°C; for the laboratory cones, better than ±10°C. If the heating rate undergoes change

TABLE 32.26a Approximate Touch-Down Temperatures of Pyrometric Cones (DIN 51063)

Cone no.	Temperature (°C)	Cone no.	Temperature (°C)	Cone no.	Temperature (°C)	Cone no.	Temperature (°C)
022	600	07a	960	9	1280	29	1650
021	650	06a	980	10	1300	30	1670
020	670	05a	1000	11	1320	31	1690
019	690	04a	1020	12	1350	32	1710
018	710	03a	1040	13	1380	33	1730
017	730	02a	1060	14	1410	34	1750
016	750	01a	1080	15	1435	35	1770
015a	790	1a	1100	16	1460	36	1790
014a	815	2a	1120	17	1480	37	1825
013a	835	3a	1140	18	1500	38	1850
012a	855	4a	1160	19	1520	39	1880
011a	880	5a	1180	20	1530	40	1920
010a	900	6a	1200	26	1580	41	1960
09a	920	7	1230	27	1610	42	2000
08a	940	8	1250	28	1630		

TABLE 32.26b Approximate Touch-Down Temperatures of Pyrometric Cones (BS 1041)

Cone no.	Temperature (°C)	Cone no.	Temperature (°C)	Cone no.	Temperature (°C)	Cone no.	Temperature (°C)
022	600	011	880	1	1135	11	1310
021	615	010	900	2	1150	12	1330
020	630	09	925	3	1165	13	1350
019	665	08	950	4	1180	14	1380
018	700	07	975	5	1195	15	1410
017	730	06	1000	6	1210	16	1435
016	760	05	1030	7	1230	17	1460
015	790	04	1060	8	1250	18	1485
014	810	03	1085	9	1270	19	1510
013	830	02	1105	10	1290	20	1535
012	860	01	1120				

Note: 1. Each temperature given in the table is that at which the tip of a cone will bend sufficiently to touch the base in an electric kiln with a heating rate of 60°C h⁻¹. 2. The touch-down temperature depends on the rate of heating: reports on firing behavior should quote the cone number, not the temperature taken from the above table. 3. Intermediate degrees of bending can be referred to the hands of a clock, e.g., 3 o'clock would represent a cone bent halfway to the stand.

within the range of 20 to 150°C h⁻¹, then the rated temperatures can change by −40°C for the above lower limiting value of the heating rate up to 60°C for the upper value. A 0.35% content of SO_2 in the atmosphere increases the rated temperatures by about 35°C. Also, the presence of soot in the atmosphere slightly raises the rated temperatures.

How to Use the Materials.

While single cones are sometimes used, usually three or four consecutively numbered cones, including a cone whose rated temperature is equal to the required temperature of the heat treatment and two cones of neighboring numbers (one less and one more) are employed for the temperature determination (see Figure 32.77b). They are installed into specially unfired plaques with tapered holes and protrusions that hold the cones firmly. The plaques are mounted to a workpiece surface to allow observations. A cone can be set up in other ways, such as inserting its base into refractory clay. However, it is necessary to assure a correct angle and firm hold of the cones during the firing cycle. Failure in these respects will

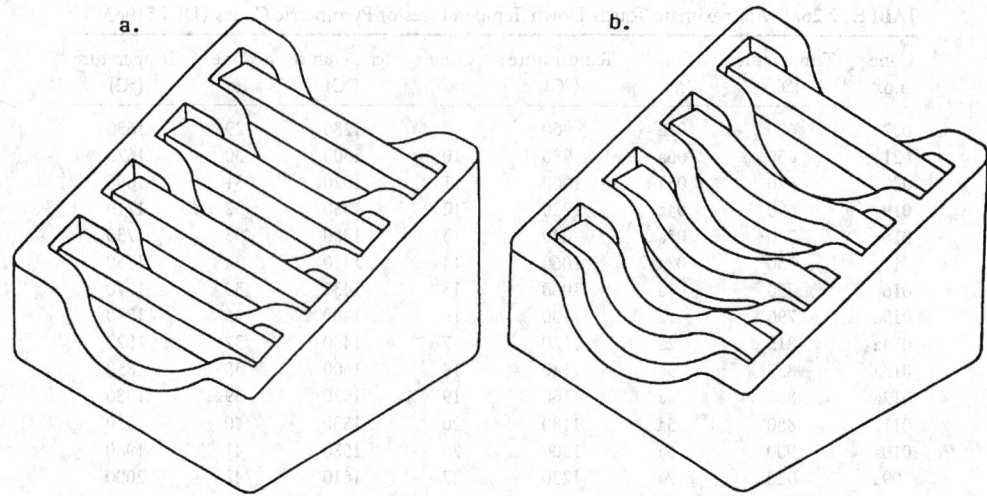

FIGURE 32.78   Thermoscope bars on a stand before and after firing.

cause the cone to bend unpredictably and give incorrect assessment of the heat treatment. If the process of heating takes place with a standard heating rate, then the rated temperature is reached when the tip of the central cone touches the base of the plaque. With further prolongation of the firing cycle, the cone will melt completely to form a blob on the plaque. The process of reaching the rated temperature is signaled in advance by the cone of one-less number. The cone of one-more number is there to prove that the required temperature value is not exceeded. Placing a series of cones with lower numbers (lower rated temperatures) provides the opportunity to carry out the process of heating at a required rate.

Typical Application.
Seger cones are used for the control of firing processes in the ceramics industry and artistry.

Thermoscope Bars

These indicators have the shape of bars of rectangular cross-sections. The typical dimensions of the bars are: length, 57 mm; width, 8 mm; and height, 6 mm. Bars of consecutive numbers (rated temperatures) are placed horizontally on a refractory stand as in Figure 32.78. The set of thermoscope bars is a more convenient and slightly modified form of Holcdorft bars. The bars are made of the same composites (mineral mixes and organic binder) as the pyrometric cones. The mixed powders are pressed and can be hardened by prefiring at relatively low temperatures, below those at which bending should occur.

Bending Temperatures of Thermoscope Bars.
The rated temperatures of thermoscope bars, referred to as the bending temperatures, are found during the calibration in an electric kiln with a heating rate of 60°C h^{-1} when the bars start to exhibit deformation (i.e., begin to bend).

According to the British Standard BS 1041, the range of rated temperatures from 590°C to 1525°C at temperature intervals of 15 to 35°C is covered by 42 thermoscope bars (see Table 32.27).

The precision of determination of the bending is about ±15°C; the other properties of the thermoscope bars referring to changes of the standard conditions are the same as for the pyrometric cones.

How to Use the Materials.
Four thermoscope bars of consecutive numbers — the first two having lower bending temperatures, the third one having the bending temperature equal or close to the required temperature, the fourth one having a higher bending temperature — are placed in sequence on a special refractory stand as in Figure 32.78a. The set is mounted to the workpiece surface where observations take place. If the process of heating takes place with a standard heating rate, then the beginning of deformation (bending) of the

TABLE 32.27 Approximate Bending Temperatures of Thermoscope Bars (BS 1041)

Bar no.	Temperature (°C)	Cone no.	Temperature (°C)	Cone no.	Temperature (°C)	Cone no.	Temperature (°C)
1	590	12	870	23	1130	33	1300
2	610	13	890	24	1145	34	1320
3	625	14	915	25	1160	35	1340
4	650	15	940	26	1175	36	1365
5	685	16	965	27	1190	37	1395
6	715	17	990	28	1205	38	1425
7	745	18	1015	29	1220	39	1450
8	775	19	1045	30	1240	40	1475
9	800	20	1075	31	1260	41	1500
10	820	21	1095	32	1280	42	1525
11	845	22	1115				

Note: 1. Each temperature given in the table is that at which the bar starts to bend in an electric kiln with a heating rate of 60°C h^{-1}. 2. The bending temperature depends on the rate of heating: reports on firing behavior should quote the bar number, not the temperature taken from the above table. 3. The bar can be expected to bend sufficiently to touch the stand at a temperature of 10°C to 30°C higher than the values given in the table, depending on the composition of the bar.

third bar indicates that the required temperature is reached. The process of reaching the rated temperature for the third bar is signaled in advance by the deformation of the proceeding bars whose behavior allows the evaluation of the heating rate. The unbent fourth bar testifies that the required temperature is not exceeded (see Figure 32.78b).

Typical Application.
The application of the thermoscope bars is identical to that of the Seger cones.

Bullers Rings

These temperature indicators in the form of rings have the following dimensions: outer diameter, 63 mm; inner diameter, 22 mm; and width, 8 mm. The appropriate measuring unit consists of a Bullers ring and a specially prescaled device for measurement of the temperature shrinkage of the ring. This contraction gage measures the outer diameter of the heated ring, based on which the heating work is assessed. The full measuring range of rated temperatures from 960°C to 1440°C is covered by four types of rings manufactured by pressing powders of ceramics mixes, with a binder, and without prefiring.

1. Rings denoted as 55 of brown color, suitable for temperatures from 960°C to 1100°C are used in the firing of glost ware and common building bricks where the finishing temperatures are relatively low.
2. Rings numbered 27/84, colored green, suitable for temperatures from 960°C to 1250°C are used for firing earthenware at the medium finishing temperatures, as well as tiles and bricks refractory with respect to low temperatures.
3. Rings numbered 75/84, colored natural, recommended for firing temperatures from 960°C to 1320°C, allow for higher finishing temperatures and are used for firing electrical porcelain, china, grinding wheels, and bricks refractory with respect to higher temperatures.
4. Rings numbered 73, colored yellow, recommended for temperatures from 1280°C to 1440°C for slow firing conditions as used in the manufacture of high-temperature ceramics and heavy refractories.

Approximate rated temperatures for the Bullers rings and corresponding readings of the contraction gage according to the British Standard BS 1041 are presented in Table 32.28.

How to Use the Materials.
One or more rings of the same type are placed vertically in a prefired stand and mounted to a workpiece surface. To determine the temperature as the firing progresses, the heated rings are withdrawn from their

TABLE 32.28   Approximate Rated Temperatures for Bullers Rings

Temperature (°C)	Gage readings			
	Ring no. 55	Ring no. 27/84	Ring no. 75/84	Ring no. 73
960	3	0	0	
970	7	1	1	
980	11	2.5	2	
990	15	4	3	
1000	18	5.5	4	
1010	21	7	5	
1020	24	8.5	6	
1030	27	10	7	
1040	30	11.5	8.5	
1050	32	13	10	
1060	34	14	11	
1070	36	15.5	12.5	
1080	37	17	14	
1090	38	18.5	15.5	
1100	39	20	17	
1110		21.5	18	
1120		23	20	
1130		24.5	21	
1140		26	22	
1150		27	23	
1160		28.5	24.5	
1170		30	26	
1180		31.5	27	
1190		33	28	
1200		34.5	29	
1210		36	30	
1220		37.5	31	
1230		38.5	32	
1240		40	33	
1250		41.5	34.5	
1260			36.5	
1270			38.5	
1280			40	29.5
1290			42	30
1300			44	31
1320			46	34
1340				37
1360				40.5
1380				44
1400				48
1420				51
1440				54

Note: These values should be used with caution because they are dependent on the firing cycle to which the rings are subjected.

stands, cooled to the ambient temperature, and then measured for contraction. This measurement is carried out on a gage consisting of a brass base plate on which a radial arm with a pointer moving over a scale and two steel dowel pins, against which the ring is pressed by the movable arm, are mounted (see Figure 32.79). A contraction of the ring gives rise to an amplified movement of the pointer over the scale. The divisions on the scale are numbered from −5 to 60. More heavily fired rings contract more and give higher readings on the gage. The divisions below 0 indicate expansion of the ring; above 0 indicates contraction. Rings should be measured across several diameters by turning them around in the gage so as to find the mean value to which the temperature can be assigned with the help of Table 32.28. Placing several rings in the stand in a manner that allows their easy withdrawal gives the possibility of measuring

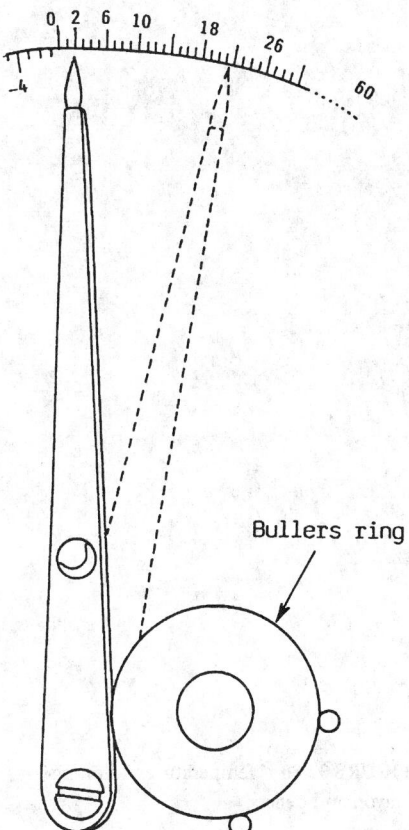

FIGURE 32.79   Contraction gage.

the heating rate. In a similar way, distribution of a number of rings throughout the furnace enables the determination of the temperature field in the furnace.

Usually, one or more test pieces from a series of rings are picked out for the sake of calibration so as to compare the obtained readings with the standard values enclosed in Table 32.28. Intermediate measurements are also carried out to evaluate the effect of the heating rate on the temperature shrinkage of the rings. The accuracy of the temperature determination for the standard heating conditions is ±0.5 of a single division of the scale.

Typical Application.
The application of the Bullers rings is similar to the Seger cones and thermoscope bars. An inconvenience is the fact that the Bullers rings require gage measurements and the temperature cannot be solely determined based on naked-eye observations.

Temperature-Indicating Pellets, Liquids, Crayons, and Monitors

Temperature-Indicating Pellets.
Temperature-indicating pellets are manufactured by pressing powders of mineral mixes of certain melting temperatures and an indifferent binder. Melting pellets are recommended as standard tablets $\phi 7/16 \times 1/8$ and miniature tablets $\phi 1/8 \times 1/8$ (see Figure 32.80). There are 112 different pellets which cover the temperature range from 40°C (100°F) to 1650°C (3000°F); see Table 32.29. The accuracy of the temperature determination is ±1% of the rated temperature. There are also available pellets for temperature control in strongly reducing atmospheres.

How to Use the Materials.
A pellet of the rated temperature equal to the required temperature of heat treatment is placed on the investigated surface before the heating starts. When the heating progresses, the beginning of melting

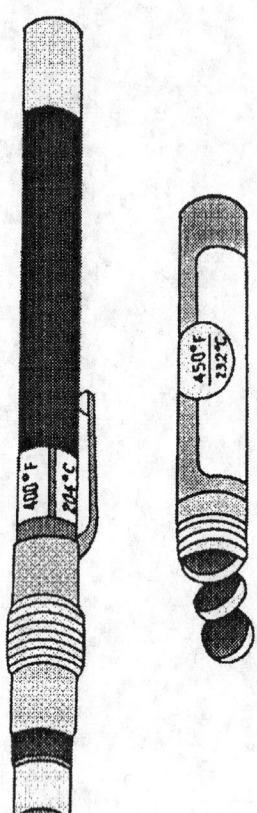

FIGURE 32.80   Temperature indicating crayon and pellets.

signals that the rated temperature is reached. Placing more pellets with the rated temperatures lower and higher than the required temperature enables more precise control of the heating process.

Typical Application.
Typical applications are checking furnace temperatures, control of heat treating of large units, as well as other applications involving long-duration heating.

Temperature-Indicating Liquids.
Temperature-indicating liquids are solutions of powdered mineral mixes in indifferent highly volatile solvents. They are available for use by brushing or spraying. There are over 100 different liquids with the rated temperatures from 40°C (100°F) to 1371°C (2500°F); see Table 32.29. The accuracy of the temperature determination is ±1% of the rated temperature.

How to Use the Materials.
A thin coating of the liquid is put on the clean and dry surface by brushing or spraying before the heating starts. It dries almost instantly to a dull opaque mark. When the required temperature is reached, this mark liquefies. The melted coating does not revert to its original opaque appearance but remains glossy-transparent on cooling. It should be noted that color changes do not signal the required temperature. The melting, not the color change, is the temperature signal.

Typical Application.
They are recommended for temperature control on fabrics, rubber, plastics, on smooth surfaces such as glass or polished metals, as well as for monitoring critical temperatures in electronic fields.

TABLE 32.29 Rated Temperatures for Temperature-Indicating Pellets, Liquids, and Crayons

°F	°C	°F	°C	°F	°C
100	38	325	163	1200[a]	649
103	39	331	166	1250[a]	677
106	41	338	170	1300[a]	704
109	43	344	173	1350[a]	732
113	45	350	177	1400[a]	760
119	48	363	184	1425	774
125	52	375	191	1450[a]	788
131	55	388	198	1480	804
138	59	400	204	1500[a]	816
144	62	413	212	1550	843
150	66	425	218	1600	871
156	69	438	226	1650	899
163	73	450	232	1700	927
169	76	463	239	1750[a]	954
175	79	475	246	1800	982
182	83	488	253	1850	1010
188	87	500	260	1900[a]	1032
194	90	525	274	1950	1066
200	93	550	288	2000	1093
206	97	575	302	2050	1121
213	101	600	316	2100	1149
219	104	625	329	2150[a]	1177
225	107	650[a]	343	2200[a]	1204
231	111	675	357	2250[a]	1232
238	114	700	371	2300[a]	1260
244	118	725	385	2350[a]	1288
250	121	750[a]	399	2400	1316
256	124	800[a]	427	2450	1343
263	128	850[a]	454	2500[a]	1371
269	132	900	482	2550[b]	1390
275	135	932	500	2600[b]	1427
282	139	950	510	2650[b]	1454
288	142	977	525	2700[b]	1482
294	146	1000	538	2800[b]	1538
300	149	1022	550	2900[b]	1593
306	152	1050[a]	566	3000[b]	1649
313	156	1100	593		
319	159	1150	621		

[a] Series "R" pellets for use in strongly reducing atmospheres.
[b] Available in pellets only.

Temperature-Indicating Crayons.

Temperature-indicating crayons are sticks manufactured from powders of mineral mixes of certain melting temperatures and an indifferent binder. The crayons are put in specially adjustable metal holders with labels saying their rated temperatures; see Figure 32.80. Similar to the temperature-indicating liquids, there are over 100 different crayons that cover the temperature range from 40°C (100°F) to 1371°C (2500°F); see Table 32.29. The accuracy of the temperature determination is also ±1% of the rated temperature.

How to Use the Materials.

During heating, the workpiece should be struck repeatedly by the crayon. Below its rated temperature, the crayon leaves a dry opaque mark. When the rated temperature is reached, the crayon leaves a liquid smear. On cooling, the liquid mark will solidify with a transparent or translucent appearance. For temperatures below 700°F, the mark can be put on the workpiece surface before the heating process. The mark will liquefy when the rated temperature is reached. It should be remembered that the moment of melting, not any change in color, is the temperature signal.

Typical Application.

The crayons can be applied in welding, forging, heat treating and fabrication of metals, molding of rubber and plastics, wherever the workpiece is accessible during the heating process. Very smooth surfaces are excluded.

Temperature Monitors (Labels).

These temperature indicators are adhesive-backed labels with one or more heat-sensitive indicators under transparent circular windows. The indicators turn black (show black paper backing) when the rated temperature is reached. The rated temperatures are from 40°C (100°F) to 320°C (600°F). They are available as single temperature or multi-temperature indicators with 10°, 25°, or 50° steps. The tolerance of the temperature determination is ±1°C (±1.8°F) below 100°C, and ±1% of the rated temperature above 100°C. Exemplary rated temperatures for a series of 4-temperature (4-dot) indicators are presented in Table 32.30. A 4-dot temperature monitor is displayed in Figure 32.81.

TABLE 32.30   Rated Temperatures for 4-Temperature (4-dot) Labels

Model no.	°F	°C	°F	°C	°F	°C	°F	°C
4A-100	100	38	110	43	120	49	130	54
4A-110	110	43	120	49	130	54	140	60
4A-120	120	49	130	54	140	60	150	66
4A-130	130	54	140	60	150	66	160	71
4A-140	140	60	150	66	160	71	170	77
4A-150	150	66	160	71	170	77	180	82
4A-160	160	71	170	77	180	82	190	88
4A-170	170	77	180	82	190	88	200	93
4A-180	180	82	190	88	200	93	210	99
4A-190	190	88	200	93	210	99	220	104
4A-200	200	93	210	99	220	104	230	110
4A-210	210	99	220	104	230	110	240	116
4A-220	220	104	230	110	240	116	250	121
4A-230	230	110	240	116	250	121	260	127
4A-240	240	116	250	121	260	127	270	132
4A-250	250	121	260	127	270	132	280	138
4A-260	260	127	270	132	280	138	290	143
4A-270	270	132	280	138	290	143	300	149

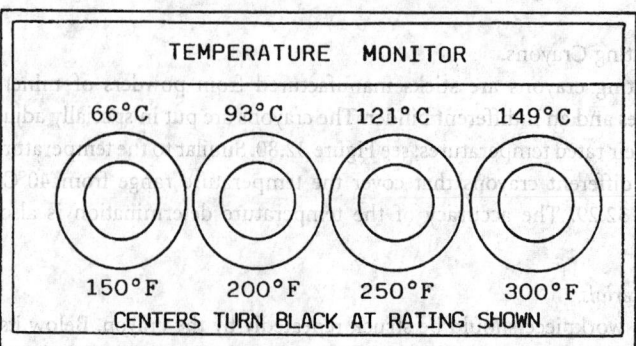

FIGURE 32.81   Four-dot label.

How to Use the Materials.

After removing the backing, the label is pressed firmly to the dry and clean workpiece surface. A change in color to black is the temperature signal. Application of multitemperature-indicating labels allows more precise temperature determination.

Typical Application.

They are especially applied for monitoring the safe operating temperature of equipment and processes, safeguarding temperature-sensitive materials during storage and transport.

The melting temperature indicators are described in the catalogs of their manufacturers [3, 4]. As the melting indicators are widely used in the U.S., temperatures in Fahrenheit are also given.

Color-Change Temperature Indicators

Color-change indicators comprise temperature-indicating paints, crayons, as well as liquid crystal indicators.

Temperature-Indicating Paints and Crayons

Temperature-indicating paints are basically acrylic lacquers containing finely dispersed temperature-sensitive inorganic pigments. The principle of operation of these indicators draws on the change in color of incident light reflected from the surface of the paint due to chemical reactions which the dispersed pigments undergo and creation of new compounds at specific transition temperatures. The color-change temperatures are also determined by the heating time. According to the British Standard, it is assumed that the rated temperatures of the paints correspond to the change in color at a heating interval of 10 min. To make the characteristics of temperature-indicating paints complete, the manufacturers also provide, together with the paints, graphs of trigger temperature vs. heating time relationships.

Temperature-indicating paints and crayons can be divided into two groups:

- Irreversible indicators: where the change of color becomes permanent
- Reversible indicators: after cooling and some time, they revert to previous colors.

Irreversible color-change indicators are complex compounds containing various metals, including cobalt, chromium, molybdenum, nickel, copper, vanadium, or uranium. However, they are lead- and sulphur-free. They are available on the market in the form of paints and crayons.

Irreversible Color-Change Paints.

The irreversible paints can change color once or several times during the heating process. Thus, another division can be made on single-change and multichange paints. The range of rated temperatures is from 40°C to 1350°C at 10 to 200°C steps. At the standard conditions, the tolerance of measurements is ±5°C for lower temperature values and ±1% for higher temperatures. Exemplary single-change paints with two critical temperatures — the initial trigger temperature for which the paint changes color after 10 min heating, and the cut-off temperature being the lowest temperature for which the color change is achieved for long-duration heating — are collected in Table 32.31. Color changes and critical temperatures for some multichange paints (changing color 2, 3, 3 or 6 times throughout the heating cycle) are presented in Table 32.32.

How to Use the Materials.

A thin layer of the paint is applied to a workpiece surface by brushing or spraying like an ordinary paint and allowed to dry before the heating starts. During the heating, when a point of the surface reaches or exceeds the critical temperature, a color change will take place. To determine the distribution of temperature, a multichange paint can be applied. With a nonuniform temperature rise, a number of colored bands separated by isothermal lines will appear on the workpiece surface, allowing the thermal record to be made of the temperature gradient across the surface.

Typical Application.

The temperature-indicating paints are widely used in industrial applications for observing heat patterns, detecting high and low temperature points on surfaces of heat engines, pipelines, and refrigeration fins.

TABLE 32.31   Single-Change Paints

Original color	Signal color	Initial trigger temperature[a] (°C)	Cut-off temperature (°C)
Pink	Blue	48	30
Pink	Blue	135	110
Mauve pink	Blue	148	120
Blue	Dark green	155	46
Yellow	Red	235	180
Blue	Fawn	275	150
Mauve red	Grey	350	220
Mauve	White	386	290
Green	Salmon pink	447	312
Green	White	458	312
Orange	Yellow	555	482
Red	White	630	450

[a] Color-change temperature for 10-min heating.

TABLE 32.32   Multichange Paints

Original color	Signal color	Initial trigger temperature[a] °C	Cut-off temperature °C
Light tan	Bronze green	160	150
Bronze green	Pale indian red	230	210
Reddish orange	Dark gray	242	193
Dark gray	Medium gray	255	211
Medium gray	Dirty white	338	228
Purple	Pink	395	355
Pink	Fawn	500	386
Fawn	Blue	580	408
Red	Dusty gray	420	310
Dusty gray	Yellow	555	328
Yellow	Orange	610	450
Orange	Green	690	535
Green	Brown	820	621
Brown	Green/gray	1050	945

[a] Color-change temperature for 10-min heating.

They can be also used for controlling temperatures of powered elements and surfaces that are inaccessible or revolve at high speeds.

Color-Change Crayons.

Color-change crayons, available in more than 10 distinct colors, similar in shape to regular crayons for drawing, cover the temperature range from 65°C to 670°C at 10 to 100°C temperature intervals. Exemplary single-change crayons are presented in Table 32.33. The accuracy of the temperature determination is the same as for the temperature-sensitive paints. They can be used for evaluating the temperature on already heated surfaces. They change color 2 min after reaching the rated temperature. Easy to use and inexpensive, they are invaluable for occasional temperature control in auto repairs, soldering, welding, electrical wiring, enameling, and for any operation involving boiling, baking, and other forms of heating.

TABLE 32.33 Single-Change Crayons

Original color	Signal color	Initial trigger temperature (°C)
Ivory	Light green	65
Yellow & green	Light green	75
Light pink	Blue	100
Gray & white	Light blue	120
Light ivory	Pink	150
Light blue	Black	200
Green	Black	280
Light green	Gray & brown	300
Blue	White	320
Brown	Red orange	350
White	Yellow	410
Light pink	Black	450
Ochre	Black	500
Blue	White	600
Green	White	670

Reversible Color-Change Indicators.
Reversible color-change indicators are available on the market as paints and in label form. The thermal pigments of these temperature indicators are mercury-based complexes. Therefore, they cannot be applied directly to metal surfaces as this causes decomposition. They also tend to decompose after long exposure to heat, but the decomposition can be retarded by using a clear over-lacquer. The pigments find their most successful application when encapsulated into labels.

The rated temperatures for the reversible color-change paints do not exceed 170°C. For temperatures up to 70°C under standard conditions, the tolerance of measurements is ±1°C; for 70 to 150°C, ±2°C; and for 150 to 170°C, ±3°C.

How to Use the Materials.
A thin layer of a reversible paint is applied to a workpiece surface by brushing or spraying, or a label is pressed to the surface. During the heating, a color change will take place when the temperature of the surface reaches or exceeds the critical temperature.

Typical Application.
The reversible color-change paints are widely used in the electrical industry, especially on busbars, live conductors, and connectors in high-current switches and in electronic fault-finding. They also find application as warning and indicating devices of domestic appliances. They are invaluable for controling lower temperatures when it is necessary to detect undesirable temperature excursions, correct faults, and revert to normal conditions.

Thermochromic Liquid Crystals

Liquid crystals constitute a class of matter unique in exhibiting mechanical properties of liquids (fluidity and surface tension) and optical properties of solids (anisotropy to light, birefringence). Certain liquid crystals are thermochromic and react to changes in temperature by changing color. They can be painted on a surface or suspended in a fluid and used to make the distribution of temperature visible. Normally clear, or slightly milky in appearance, liquid crystals change in appearance over a narrow range of temperatures called the color-play bandwidth (the temperature interval between first red and last blue), centered around the nominal event temperature (midgreen temperature). The displayed color is red at the low temperature margin of the color-play interval and blue at the high end. Within the color-play interval, the colors range smoothly from red to blue as a function of temperature; see Figure 32.82. Liquid

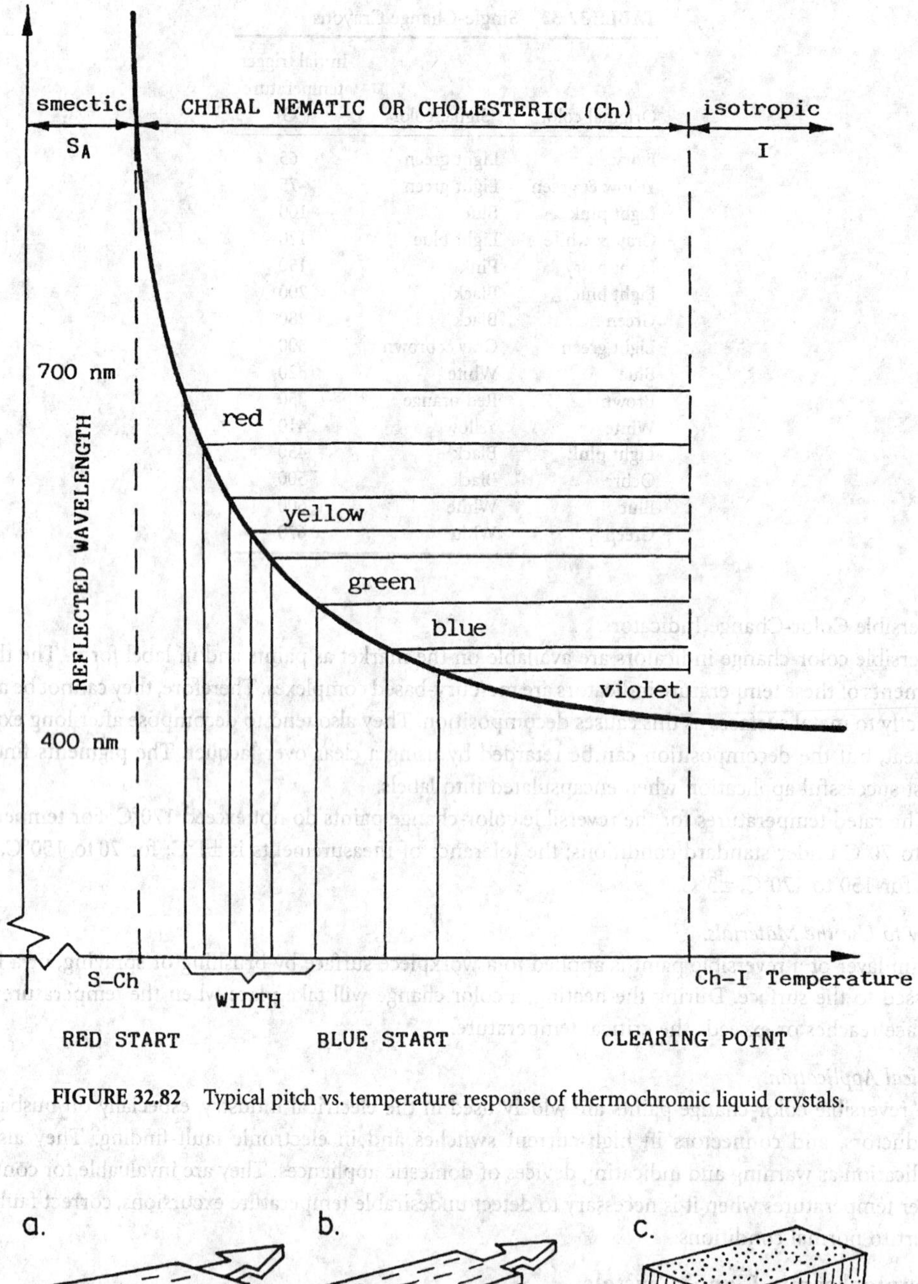

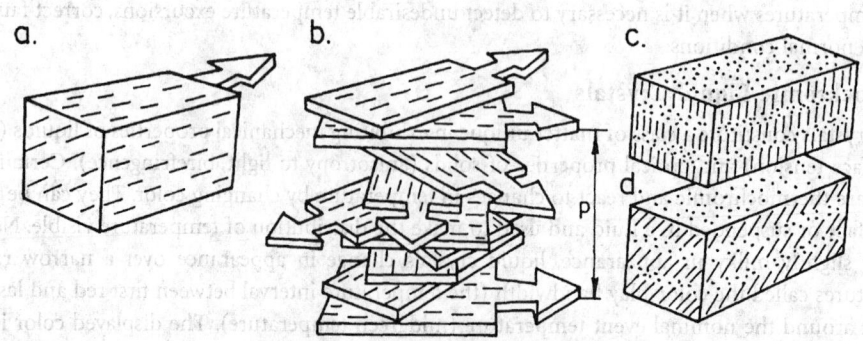

FIGURE 32.82 Typical pitch vs. temperature response of thermochromic liquid crystals.

FIGURE 32.83 Structures of liquid crystals (a) nematic; (b) choleteric; (c) smectic A; (d) smectic B.

crystals or mesophases have been classified as smectic, chiral nematic, cholesteric, and blue. The structure of liquid crystals is shown schematically in Figure 32.83.

Temperature-Sensitive and Shear-Sensitive Formulations.
Temperature-sensitive liquid crystals show colors by selectively reflecting incident white light. Conventional temperature-sensitive mixtures turn from colorless (or black against a black background) to red at a given temperature and, as the temperature is increased, pass smoothly through the other colors of the visible spectrum in sequence (orange, yellow, green, blue, violet) before turning colorless (or black) again in the ultraviolet at a higher temperature. The color changes are reversible and on cooling the color change sequence is reversed.

Temperature-insensitive (sometimes called shear-sensitive) formulations can also be made. These mixtures show just a single color below a given transition temperature (called the clearing point) and change to colorless (black) above it. The working temperature range is thus below the clearing point. Both reversible and hysteretic (memory) formulations can be made. All liquid crystal mixtures should be viewed against nonreflecting backgrounds (ideally black, totally absorbing) for best visualization of the colors.

Color-Play Properties and Resolution.
Temperature-sensitive thermochromic mixtures have a characteristic red start or midgreen temperature and color-play bandwidth. The bandwidth is defined as the blue start temperature minus the red start temperature. The color play is defined by specifying either the red start or midgreen temperature and the bandwidth. For example, R35C1W describes a liquid crystal with a red start at 35°C and a bandwidth of 1°C, i.e., a blue start 1°C higher, at 36°C; G100F2W describes a liquid crystal with a midgreen temperature at 100°F and a bandwidth of 2°F.

Both the color-play bandwidth and the event temperature of a liquid crystal can be selected by its proper chemical composition. The event temperatures of liquid crystals range from −30°C to 115°C with color-play bands from 0.5°C to 20°C, although not all combinations of event temperature and color-play bandwidth are available. Liquid crystals with color-play bandwidths of 1°C or less are called narrow-band materials, while those whose bandwidth exceeds 5°C are referred to as wide-band. The type of material to be specified for temperature indicating should depend very much on the type of available image interpretation technique — human observers, intensity-based image processing, or true-color image processing systems (see [7]). The uncertainty associated with direct visual inspection is about 1/3 the color-play bandwidth, given an observer with normal color vision — about ±0.2°C to 0.5°C. The uncertainty of true-color image processing interpreters using wide-band liquid crystals is of the same order as the uncertainty assigned to human observers using narrow-band materials, and depends on the pixel-to-pixel uniformity of the applied paint and the size of the area averaged by the interpreter (about ±0.05°C can be achieved). Using a multiply filtered, intensity-based system, the resolution is better than ±0.1°C.

How to Use the Materials.
Liquid-crystal indicators can be used in a number of different forms: as unsealed liquids (also in solutions), in the microencapsulated form (as aqueous slurries or coating formulations), and as coated (printed) sheets. The different forms of the materials have selective advantages and suit different temperatures and flow visualization applications. Individual products are described in more detail in relevant booklets issued by the manufacturers of liquid crystals [5, 6].

Typical Application.
Liquid-crystal indicators are ideal for monitoring temperatures of electronic parts, transformers, relays, and motors. They are invaluable for a fast visual indication of temperatures.

References

1. BS 1041: Part 7. Temperature Measurement.
2. DIN 51063: Part 1. Testing of Ceramic Raw and Finished Materials, Pyrometric cone of Seger. Part 2. Testing of Ceramic Materials.
3. OMEGA International Corp. P.O. Box 2721, Stanford, CT 06906 (The Temperature Handbook).
4. TEMPIL Division, Big Three Industries, Inc. South Plainfield, NJ 07080 (Catalog GC-75).
5. HALLCREST Products Inc. 1820 Pickwick Lane, Glenview, IL 60025.
6. MERC Industrial Chemicals, Merc House, Poole Dorset, BH15 1TD, U.K.
7. Moffat, R.J., Experimental heat transfer, *Proc. 9th Int. Heat Transfer Conf.*, Jerusalem, Israel, 1990.

32.11   Fiber-Optic Thermometers

Brian Culshaw

Optical fiber sensing is a remarkably versatile approach to measurement. A fiber sensor guides light to and from a measurement zone where the light is modulated by the measurand of interest and returned along the same or a different optical fiber to a detector at which the optical signal is interpreted. The measurement zone can be intrinsic within the fiber that transports the optical signal, or can be extrinsic to the optical waveguide. The versatility of the fiber sensing medium arises in part because of the range of optical parameters that can be modulated and in part because of the diversity of physical phenomena that involve environmentally sensitive interactions with light.

For example, highly coherent light from a laser source can be introduced into a fiber and its phase modulated by a parameter of interest. The resulting phase changes can then be detected interferometrically. The phase change is simply a modification to the optical path length within the fiber, and can be modulated by shifts in temperature, strain, external pressure field or inertial rotation. A well-designed interferometer can detect 10^{-7} radians — equivalent to 10^{-14} m!

The laser light could also be Doppler shifted through reflection from a moving object. Its state of polarization can be changed. Its throughput intensity can be modified or the light can be used to stimulate some secondary emissions, which in turn can be monitored to produce the relevant optical signal. If the light is incoherent, then its wavelength distribution (color) can be modified, in addition, of course, to the possibilities for polarization changes and intensity changes.

The physical phenomena capable of imposing this modulation are again many and varied. They include, for example, periodically bending an optical fiber to introduce a localized loss that depends on the sharpness of the bend (usually referred to as microbend loss); changing the relative refractive indices of the core and the cladding of the optical fiber and thereby changing the guiding properties and again introducing a loss; modifying an optical phase delay by introducing a change in refractive index or a change in physical length; examining changes in birefringence introduced through modifications to physical stress and/or temperature; using external indicators to color modulate a broadband source and relate the color distribution to temperature, chemical activity, etc. These are all linear effects where the input optical frequency is the same as the output optical frequency (regarding Doppler shift as a rate of change of phase of an optical carrier) and where, for a given system setting, the output at all frequencies is directly proportional to the input.

Nonlinear effects are also widely exploited. Of these, the most important are fluorescence, observed usually in fluorophores external to the optical fiber, and Raman and Brillouin scattering, usually observed within the fiber itself. In all these phenomena, the light is absorbed within a material and re-emitted as a different optical wavelength from the one that was observed. The difference in optical wavelengths depends on the material and usually on strain and temperature fields to which the material is subjected. These major features of optical fiber sensors are encapsulated in Figure 32.84.

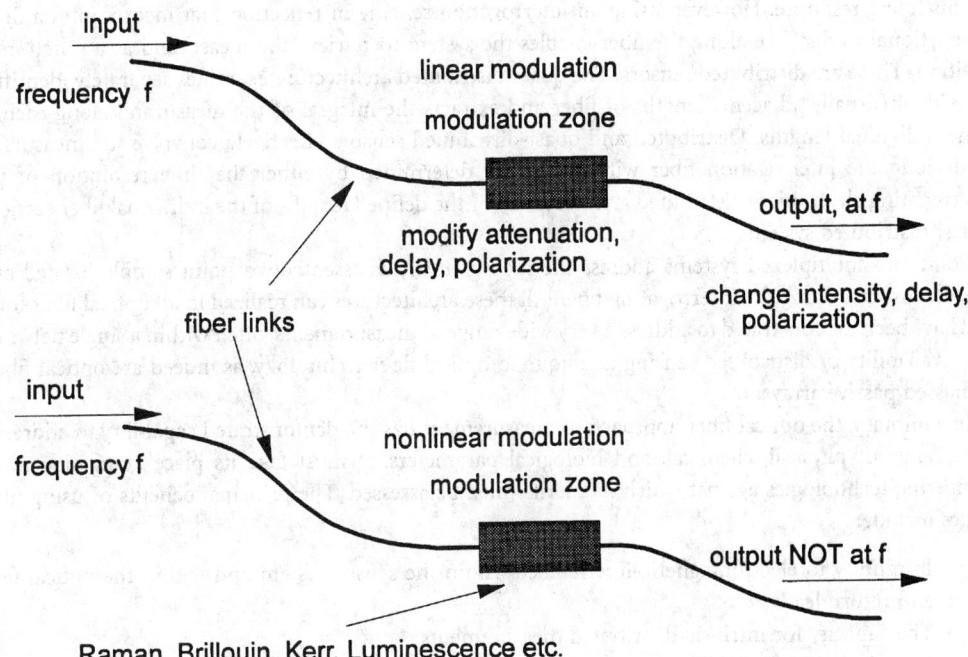

FIGURE 32.84 Linear and nonlinear optical processes for measurement using optical fiber sensors.

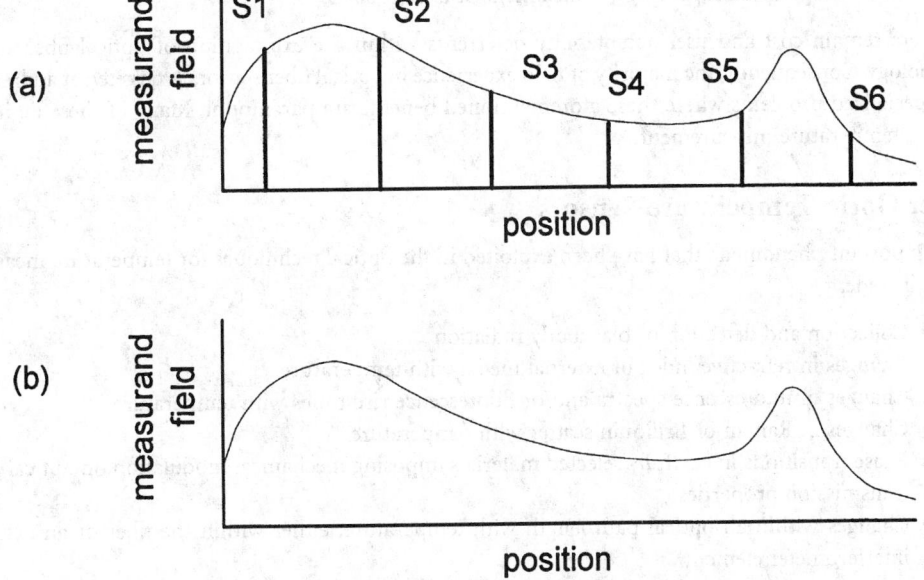

FIGURE 32.85 Sensor system outputs for (a) point array and (b) distributed sensor systems.

Optical fiber sensors have an additional feature that is unique to the medium — namely, the abilities for intrinsic networking in either distributed, quasi-distributed/multiplexed, or discrete (point) configurations. The essential features of these achitectures are sketched in Figure 32.85. For intrinsic sensors, the fiber responds to the measurand throughout its length, and the output in transmission is an integral of this linear response. However, using an interrogation scheme in reflection that incorporates a delay proportional to distance along the fiber enables the system to retrieve the measurand as a function of position. These are distributed sensors. The quasi-distributed architecture examines separately identified individual (usually adjacent) lengths of fiber and extracts the integral of the measurand along each of these individual lengths. Distributed and quasi-distributed sensors effectively convolve the measurand field along the interrogation fiber with a window determined by either the time resolution of the interrogating electronics (distributed architectures) or the defined lengths of the individual fiber sections (quasi-distributed systems).

Point and multiplexed systems address the measurement as essentially a point sample located at a specific distance along the interrogating fiber. All these architectures can realized in all optical fiber form and have been demonstrated to address a very wide range of measurements, often within a single network. The availability of distributed sensing is unique to optical fiber technology, as indeed are optical fiber-addressed passive arrays.

In summary, the optical fiber approach to measurement has the demonstrated capability to address a wide range of physical, chemical, and biological parameters. It must take its place along side other competing technologies against which its merits must be assessed. The principal benefits of using fiber optics include:

- Immunity to electromagnetic interference within the sensor system and within the optical feed and return leads
- The capacity for intrinsic distributed measurements
- Chemical passivity within the sensor system itself and inherent immunity to corrosion
- Small size, providing a physically, chemically, and electrically noninvasive measurement system
- Mechanical ruggedness and flexibility: optical fibers are exceptionally strong and elastic — they can withstand strains of several percent
- High temperature capability — silica melts at over 1500°C

There remain cost and user acceptability deterrents within the exploitation of optical fiber sensor technology. Consequently, the majority of field experience in optical fiber sensors is targeted at addressing the specialized problems where these aforementioned benefits are paramount. Many of these lie in the area of temperature measurement.

Fiber Optic Temperature Sensors

The important phenomena that have been exploited in the optical techniques for temperature measurement include:

- Collection and detection of blackbody radiation
- Changes in refractive index of external media with temperature
- Changes in fluorescence spectra and/or fluorescence rise times with temperature
- Changes in Raman or Brillouin scatter with temperature
- Phase transitions in carefully selected materials imposing mechanical modulation on optical fiber transmission properties
- Changes within an optical path length with temperature, either within the fiber or an external interferometer element

Within these phenomena, Brillouin and Raman scatter and mechanical phase transitions have been primarily used in distributed measurement systems. Some distributed measurement/quasi distributed measurement systems based on modulated to phase delay have also been evaluated, although they have yet to reach commercial reality. The remaining phenomena are almost exclusively used in point sensor systems.

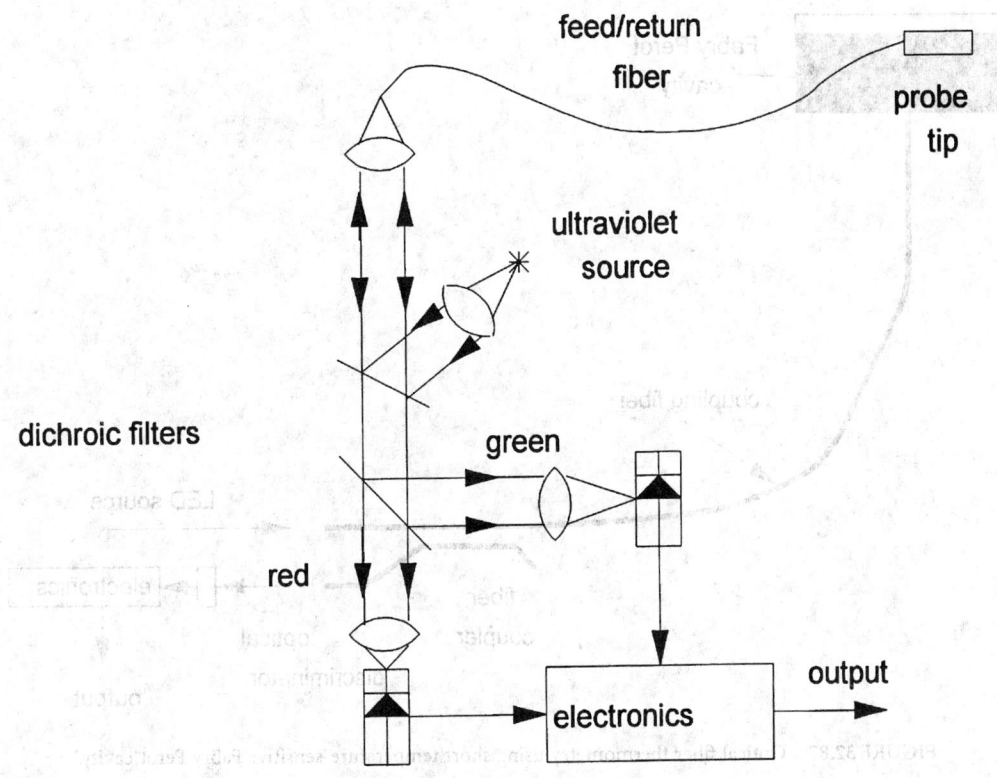

FIGURE 32.86   Optical fiber fluorescent thermometer.

Fiber Optic Point Temperature Measurement Systems

One of the first commercial optical fiber sensors was a fluorescence-based temperature probe introduced in the early 1980s by the Luxtron Corporation of Mountain View, CA. Successors to these early sensors are still commercially available and are a very effective, but expensive, approach to solving specific measurement problems. These include monitoring temperature profiles in both domestic and industrial microwave ovens, examining temperatures in power transformer oils, motor/generator windings, and similar areas where (primarily) the issue is the operation of a reasonably precise temperature probe within very high electromagnetic fields. In such circumstances, a metallic probe either distorts the electromagnetic field significantly (e.g., in microwave ovens) or is subjected to very high levels of interference, producing spurious readings. Other applications sectors exploit the small size or chemical passivity of the device, including operation within corrosive solvents or examination of extremely localized phenomena such as laser heating or in determining the selectivity of radiation and diathermy treatments.

The principles of the probe are quite simple and are shown in Figure 32.86. The rare earth phosphor is excited by an ultraviolet light source (which limits the length of the silica-based feed fiber to a few tens of meters) and the return spectrum is divided into "red" and "green" components, the intensity ratios of which are a simple single-valued function of phosphor temperature. For precision measurement, the detectors and feed fiber require calibration and, especially for the detectors, the calibration is a function of ambient temperature. However, this can be resolved through curve fitting and interrogation of a thermal reference. The instrument, which has now gone through several generations to improve upon the basic concept, is capable of accuracies of about ±0.1°C within subsecond integration times over a temperature range extending from approximately –50°C to ±200°C. Since its introduction, this particular

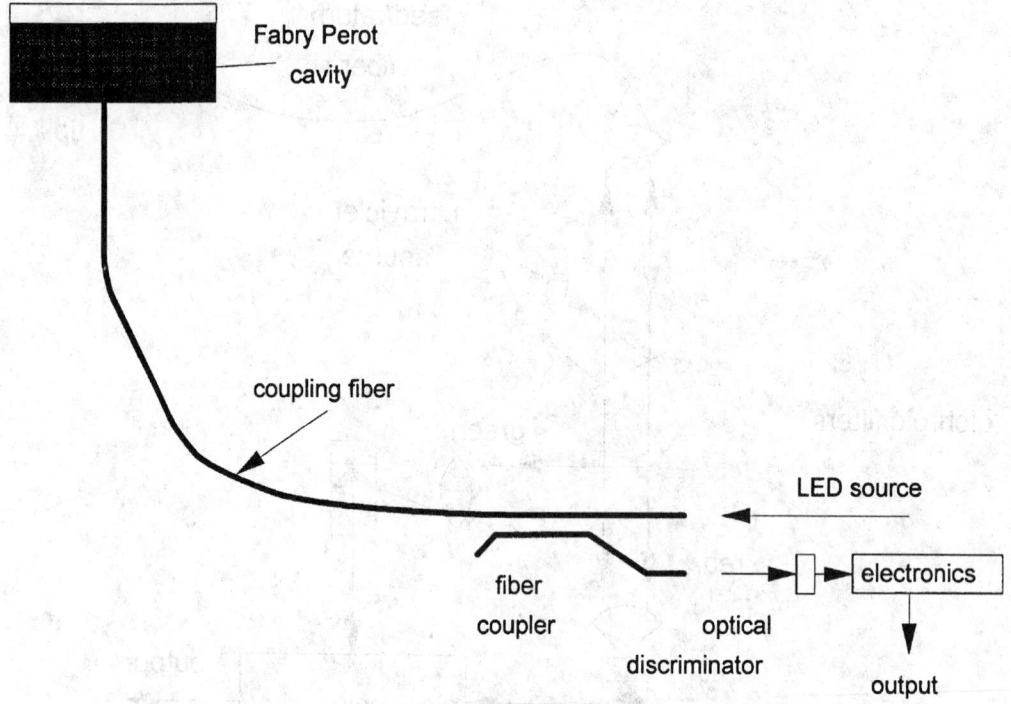

FIGURE 32.87   Optical fiber thermometry using short temperature-sensitive Fabry Perot cavity.

probe has accumulated extensive field experience in a wide variety of applications and remains among the most widely exploited fiber optic sensor concepts.

A number of temperature probes based on fluorescence decay time measurements have also been demonstrated. The level of commercial activity exploiting these concepts has, to date, been very modest, partly because the accurate measurement of decay times can be problematic.

Measuring the temperature response of dyes and other thermally sensitive color-selective materials can afford a very simple approach to temperature measurements. Among the most successful of these has been the temperature probe examining the bandedge of gallium arsenide introduced by ASEA (now ABB), again in the early 1980s and now transferred to Takaoka. The bandedge can either be monitored through examining the spectra of induced fluorescence or through interrogating the absorption characteristics of the material when subjected to a constant spectrum excitation. The accuracy and temperature range of this probe are comparable with those of the Luxtron system, and this particular version of the bandedge probe has the additional benefit of operating primarily in the near-infrared range of the spectrum, thereby accessing the best transmission characteristics of the optical fiber medium. The probe was originally conceived to address ASEA's internal needs in monitoring electrical power system components. Similar bandedge probes have also been demonstrated based on absorption edge detection in materials such as ruby.

Refractometry and interferometry are potential extremely sensitive thermal probes. Several have been demonstrated, some of which achieve microkelvin resolution. Interferometric detection or exploitation of sensitive mode coupling phenomena is the source of this very high sensitivity, although rarely is such high sensitivity required in practice. The relatively simple Fabry Perot probe shown in Figure 32.87 has been introduced commercially with simplified spectral analysis and a semiconductor source, although as yet its market penetration has been relatively modest.

Optical pyrometry is a well-established approach to measuring temperatures in the hundreds to thousands of degrees Centigrade. The disappearing filament pyrometer has been used in this fashion for over half a century. The optical fiber equivalent has also found a few niche applications. The general

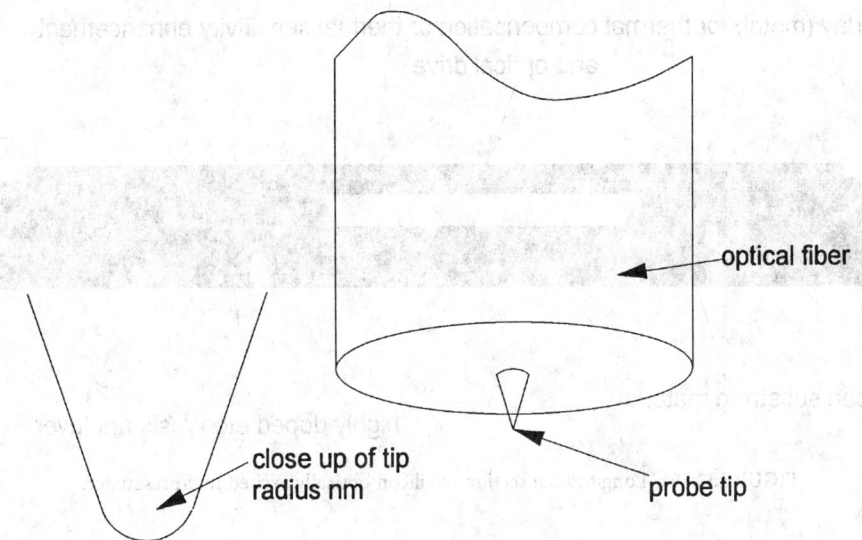

FIGURE 32.88 Probe for photon tunneling microscope and nano optrode.

form of such a sensor is to place the black body at the end of the fiber and place it with the fiber into the hot zone. The consequent radiation within the transmission spectrum of the fiber is then monitored using a semiconductor photodetector that can be based on III-V materials or silicon. The received radiation is then primarily within the red and near-infrared from about 600 nm to, depending on the detector, 1.8 μm. Blackbodies radiate significantly in this range at temperatures in the hundreds of degrees Centigrade and above. The most significant success of this approach has been in the fabrication of the reference standard temperature probe at NIST for temperatures above 1200°C. This uses a sapphire rather than silica collection fiber because of its superior optical and thermal properties within the temperature and the wavelength ranges of interest. It defines these high temperatures with subdegree precision.

Optical fibers also have the capacity to make unique nanoprobes — the opposite end of the scale by orders of ten from the distributed sensors discussed below. These (Figure 32.88) are tapered optical fibers with the end reduced in diameter to tens of nanometers. The tapered region is coated with a metal, often aluminium or silver, to confine the optical field. This produces an intense spot of light at the fiber tip which irradiates an area nanometers in dimensions. The tip can be coated with the dye or the fluorescent thermally sensitive material and used to monitor temperature over extremely small areas. This enables thermal profiles within cellular dimensions to be assessed. In other formats, the same probe can also be used to address chemical activity and chemical composition.

Fiber optic temperature sensing can be realized using a wide variety of techniques primarily, but not exclusively, based on the variation of optical properties of materials with temperature. An example of the exceptions is the optical excited vibrating element probe shown in Figure 32.89. This probe has been primarily used for pressure measurement and is now available commercially for pressure assessment down-hole in oil wells. It can also be configured to exhibit extremely high temperature sensitivity with accuracies and resolutions in the millikelvin region. It uses the beneficial features of mechanical resonators and the consequential frequency read-out in parallel with optomechanical excitation and direct optical interrogation to produce a probe that can be reliably exploited over interrogation lengths of tens of kilometers.

Fiber optic point sensors for temperature measurement are now a relatively mature technology. Most of the devices mentioned above were first introduced and characterized 10 or more years ago and have since been refined to address specific applications sectors. They remain expensive, especially when compared to the ubiquitous thermocouple, but their unique capability for noninvasive electrically passive interference immune measurement give them a very specific market address that cannot be accessed using alternative technologies. Within these market areas, the probes have been extremely successful.

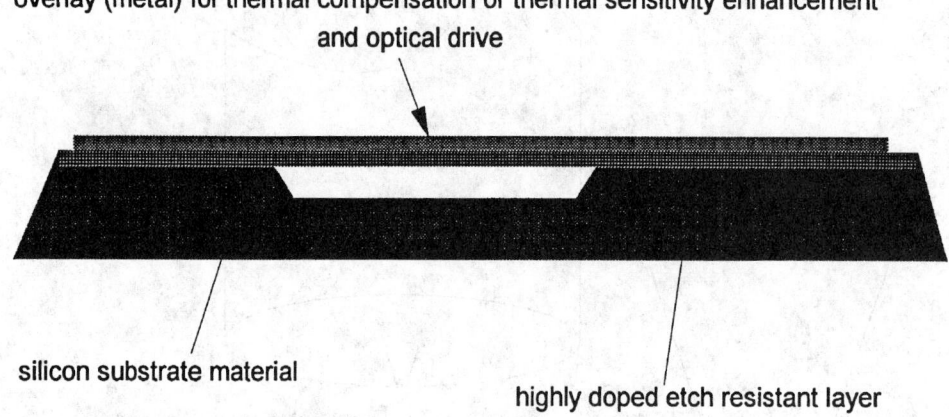

overlay (metal) for thermal compensation or thermal sensitivity enhancement
and optical drive

silicon substrate material

highly doped etch resistant layer

FIGURE 32.89   Longitudinal section of silicon optically excited microresonator.

Distributed and Quasi-distributed Optical Fiber Temperature Measurement Systems

These systems all exploit the unique capability for optical fibers to measure and resolve environmental parameters as a function of position along the fiber length. This generic technology is unique to optical fiber systems and, while there are a few commercial distributed temperature sensor systems available, the research in this sector continues.

The stimulated Raman scatter (SRS) distributed temperature probe is the most well established of these and, in common with many of the point sensors, was originally introduced commercially in the late 1980s. The principle (Figure 32.90) is quite simple. Within the Raman backscatter process (and also within the spontaneous Brillouin backscatter process), the amplitudes of the Stokes and anti-Stokes lines are related to the energy gap between these lines by a simple $\exp(-\Delta E/kT)$ relationship. Therefore, measuring this ratio immediately produces the temperature. Furthermore, this ratio is uniquely related to temperature and cannot be interfered with by the influence of other external parameters. The system block diagram is shown in Figure 32.91. The currently available performance from such systems enables resolutions of around 1 K in integration times of the order of 1 min, with resolution lengths of one to a few meters over total interrogation lengths of kilometers. The interrogation can extend to tens of kilometers if either the interrogation times are increased or the temperature and/or spatial resolutions are relaxed. The system is available from both European and Japanese manufacturers. The applications are very specific, as indeed they must be to accommodate an instrument price that is typically in tens of thousands of dollars. The instruments have been used in a variety of highly specific areas, ranging from monitoring temperature profiles in long process ovens to observing the thermal characteristics within large volumes of concrete during the curing process.

Distributed temperature alarms triggering on and locating the presence of either hot or cold spots along the fiber can be realized at significantly lower costs and have been modestly successful as commercial systems. The first of these — and probably the simplest — was originally conceived in the 1970s. This uses a simple step index fiber in which the refractive index of the core material has a different temperature coefficient than that of the cladding material. The temperature coefficient is designed such that at a particular threshold temperature, the two indices become equal and thereafter that of the cladding exceeds that of the core. Within this section, light is no longer guided. Simple intensity transmittance measurement is then very sensitive to the occurrence of this threshold temperature at a particular point along the fiber. If used with an optical time domain reflectometer, the position at which this first event occurs can be located. This system is now in use as a temperature alarm on liquefied natural gas storage tanks. Here, the core and cladding indices for a plastic-clad silica fiber cross at a temperature in the region of

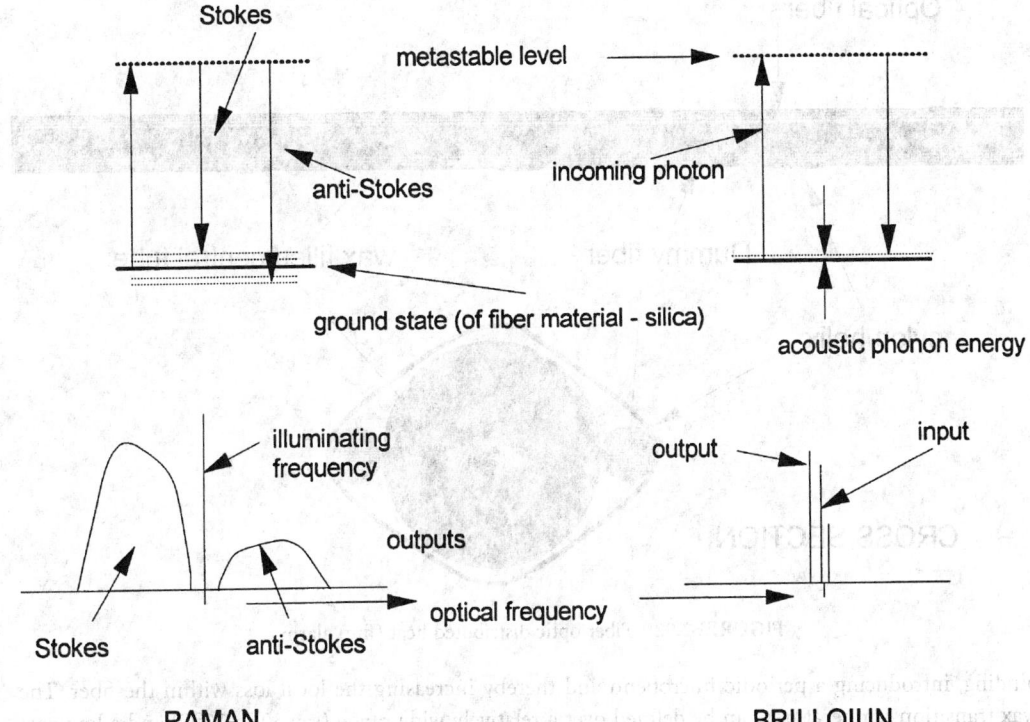

FIGURE 32.90 Thermally sensitive nonlinear scattering processes in optical fibers.

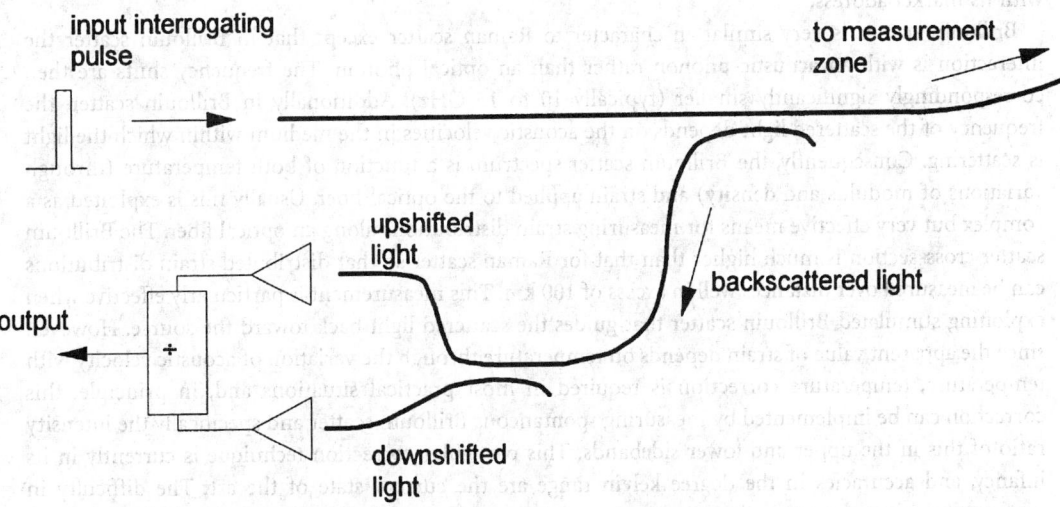

FIGURE 32.91 Raman distributed temperature probe: basic schematic.

50°C. Such temperatures can only be achieved when a leak occurs. Further, the system has the obvious benefit of intrinsic safety and total compatibility with use within potentially explosive atmospheres.

A heat — as opposed to cold — alarm system that has also been introduced commercially is shown in Figure 32.92. In this system, the central tube is filled with a wax that expands by typically 20% when passing through its melting point. This expansion in turn forces the optical fiber against the helical

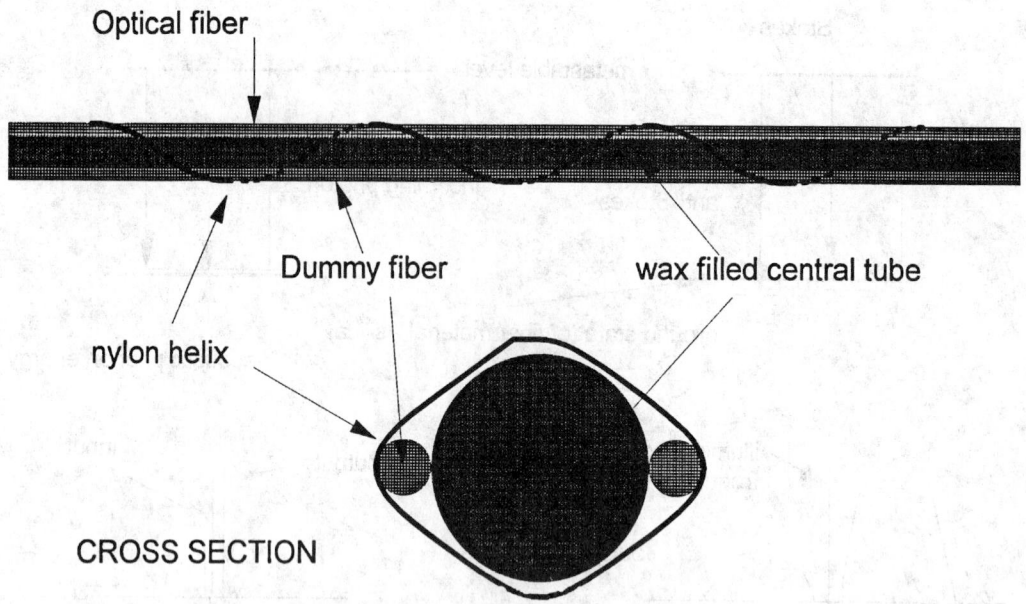

FIGURE 32.92   Fiber optic distributed heat (fire) alarm.

binding, introducing a periodic microbend and thereby increasing the local loss within the fiber. The wax transition temperatures can be defined over a relatively wide range (say 30 to 70°C) and a low-cost OTDR system enables location of the hot spot to within a few meters. This system presents a very cost-effective over-heat or fire alarm when such systems are required and are in intrinsically safe areas or in regions of very high electromagnetic interference. Again, it is the unique properties of the optical fiber sensing medium — especially intrinsic safety and electromagnetic immunity — which provide this system with its market address.

Brillouin scatter is very similar in character to Raman scatter except that in Brillouin scatter the interaction is with an acoustic phonon rather than an optical phonon. The frequency shifts are then correspondingly significantly smaller (typically 10 to 15 GHz). Additionally in Brillouin scatter, the frequency of the scattered light depends on the acoustic velocities in the medium within which the light is scattering. Consequently, the Brillouin scatter spectrum is a function of both temperature (through variations of modulus and density) and strain applied to the optical fiber. Usually this is exploited as a complex but very effective means for measuring strain distributions along an optical fiber. The Brillouin scatter cross-section is much higher than that for Raman scatter so that distributed strain distributions can be measured over distances well in excess of 100 km. This measurement is particularly effective when exploiting stimulated Brillouin scatter that guides the scattered light back toward the source. However, since the apparent value of strain depends on temperature through the variation of acoustic velocity with temperature, temperature correction is required in most practical situations and, in principle, this correction can be implemented by measuring spontaneous Brillouin scatter and specifically the intensity ratio of this in the upper and lower sidebands. This particular correction technique is currently in its infancy, and accuracies in the degree kelvin range are the current state of the art. The difficulty in temperature measurement is that the energy gap between the two sidebands is very small so that the ratio of the amplitudes is close to unity but must be measured very accurately in order to invert the exponential.

The optical Kerr effect manifests itself as an intensity-dependent refractive index. Consequently, this nonlinearity gives rise to either second harmonic generation or sum and difference frequencies. It has been investigated for distributed temperature sensing using pump:probe configurations and birefringent fiber from which the beat length is a function of temperature and strain. This beat length in turn

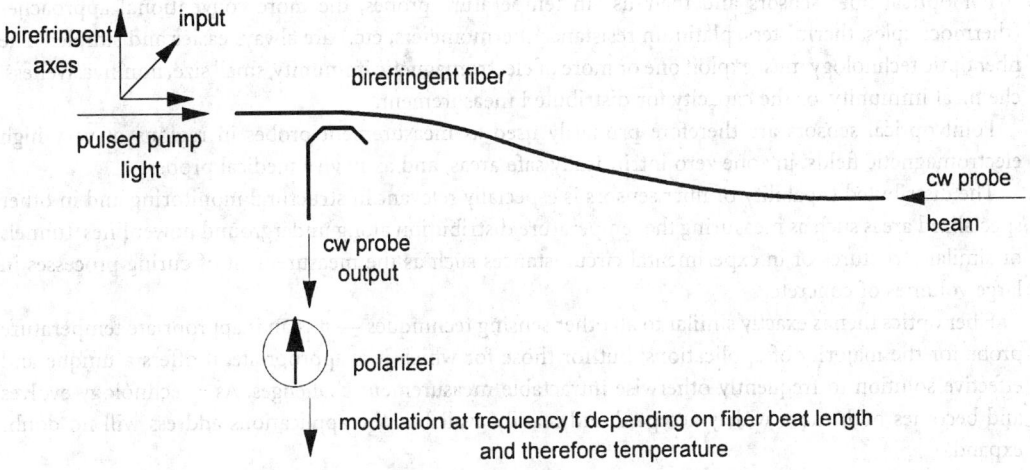

FIGURE 32.93 Distributed Kerr effect probe for temperature or strain field measurements.

determines the frequency offset through phase matching conditions of the mixed pump and probe signal (Figure 32.93). The overall situation is conceptually similar — this offset frequency depends on temperature and strain although in principle, dual measurements and adequate calibration can retrieve both, or alternatively the probe can measure a temperature field in the absence of strain. Again, the actual experimental results that have been achieved remain in the laboratory and the accuracies and resolutions are modest.

In quasi-distributed sensing and point multiplexed systems, temperature probes have, as yet, been but little exploited. There are many variations on the basic theme of a marked optical fiber within which the optical interrogation system measures the optical path length between the marks. These marks can be introduced using partially reflective gratings, mode coupling Bragg gratings, partially reflective splices or connectors, low reflectivity directional couplers, or a multitude of other arrangements. Similarly, the optical delay between the markings can be measured directly as an optical or subcarrier phase or indirectly through monitoring dispersion between adjacent modes typically in low moded or birefringent fibers. Yet again, the different delays depend on both temperature and strain so that for temperature measurement, a strain-free mounting is ideal. The context within which most, if not all, of this class of system has been evaluated is that of smart structures and here the technique does offer some promise as a means for deconvolving strain, mechanical, and thermal effects, and assessing structural integrity. It could also function as a temperature measurement probe, but to date has been minimally addressed in this application.

In multiplexed systems, the current fashion, again primarily for combined strain and temperature measurement, is to incorporate arrays of Bragg gratings used here as wavelength filters rather than as broadband reflectors. In this configuration, the Bragg grating presents a combined temperature/strain field at its location encoded within the reflection wavelength. Multiple addressing schemes can deconvolve temperature and strain sensitivities. There have also been a few demonstrations of discrete temperature-sensitive elements inserted at strategic points along an optical fiber. Of these the use of bandedge shifting in ruby crystals interrogated using a pulsed source observed in reflection has probably been the most successful. In this arrangement, the reflectors are replaced by the crystals and sample the temperature field at these points. The receiver then determines the return to spectrum as a function of time.

Applications for Optical Fiber Temperature Probes

Instrumentation is a very applications-specific discipline and, in particular for sensors, a particular technology is usually only relevant in a limited number of application sectors. As the technology becomes more and more specialized and expensive, these applications niches become much more tightly defined.

For optical fiber sensors and their use in temperature probes, the more conventional approaches (thermocouples, thermisters, platinum resistance thermometers, etc.) are always easier and simpler. The fiber optic technology must exploit one or more of electromagnetic immunity, small size, noninvasiveness, chemical immunity, or the capacity for distributed measurement.

Point optical sensors are therefore primarily used as measurement probes in regions of very high electromagnetic fields, in zone zero intrinsically safe areas, and as *in vivo* medical probes.

The distributed capability of fiber sensors is especially relevant in structural monitoring and in other specialized areas such as measuring the temperature distribution along underground power lines, tunnels or similar structures or in experimental circumstances such as the measurement of curing processes in large volumes of concrete.

Fiber optics then is exactly similar to all other sensing techniques — it is an inappropriate temperature probe for the majority of applications; but for those for which it is appropriate, it offers a unique and effective solution to frequently otherwise intractable measurement challenges. As a technology evolves and becomes both more widely accepted and readily available, the applications address will no doubt expand.

Further Information

Additional information on optical fiber temperature measurements can be obtained from the following.

B. Culshaw and J. P. Dakin, *Optical Fiber Sensors, Vol. I–IV,* Boston: Artech House, Vol. 1, 1988; Vol. II, 1989; Vols. III and IV, 1997.

B. Culshaw, *Optical Fiber Sensing and Signal Processing,* Stevenage, UK: IEE/Peter Perigrinus, 1983.

International Conferences on Optical Fiber Sensors (OFS) are regarded as the principal forum for the dissemination of research results OFS(1), London 1983 to OFS(12) Williamsburg, 1997, various publishers.

Proceedings of series of *Distributed and Multiplexed Optical Fiber Sensors and of Laser and Fiber Sensors Conf.* available from SPIE, Bellingham, Washington.

E. Udd (ed.), *Optical Fiber Sensors,* New York: John Wiley & Sons, 1991.

33

Thermal Conductivity Measurement

William A. Wakeham
Imperial College

Marc J. Assael
Aristotle University of Thessaloniki

There are three mechanisms whereby energy can be transported from one region of space to another under the influence of a temperature difference. One is by transmission in the form of electromagnetic waves (radiation); the second is the process of convection, in which a bulk or local motion of the material effects the transport; and the final process is that of thermal conduction, when energy is transported through a medium. In most practical situations, energy transport is accomplished by all three processes to some extent, but the relative importance of each contribution varies markedly. For example, within an evacuated region, radiation is the sole mechanism of transport; whereas, in an opaque solid, conduction is the only mechanism possible.

These processes of heat transfer are often very important in a wide variety of scientific and industrial applications. In the cooling of cast or crystalline materials (e.g., metals, semiconductors, or polymers) from a molten state to a solid state, the heat transfer within the material can have a profound effect on the final properties of the solid. Equally, the heat transfer in a foodstuff is a determinant of its cooking, freezing, or processing time, while the size of equipment needed to heat or cool the liquid or gas stream in a chemical plant depends sensitively on the heat transfer within and between one stream and another.

For these reasons, there has been great interest in the understanding and description of all three heat transfer processes. Among the three, that of thermal conduction is the simplest to describe in principle, since the empirical law of Fourier simply states that the heat transported by conduction per unit area in a particular direction is proportional to the gradient of the temperature in that direction. The coefficient of proportionality in this law is known as the thermal conductivity and denoted here by the symbol λ. Many important materials, whether made of pure chemical components or mixtures, are of uniform composition throughout and for them the thermal conductivity is a true physical property of the material, depending often only on the temperature, pressure, and composition of the sample. However, particularly in the solid state, the thermal conductivity can depend on the direction of the heat flow, for example, in the case of a molecular crystal.

It is also conventional to speak of the thermal conductivity of various types of composite materials such as bricks, glass-fiber insulation, carbon-fiber composites, or polymer blends. In this case, the thermal conductivity is taken to be the empirical constant of proportionality in the linear relationship between a measured heat transport per unit area and the temperature difference over a prescribed distance in the material. The thermal conductivity is not then, strictly, a property of the material, since it can often depend on a large number of parameters, including the history of the material, its method of manufacture, and even the character of its surface. However, this distinction between homogeneous and inhomogeneous materials is often ignored and leads to more than a little confusion, especially where intercomparisons among measurements are concerned.

The fact that in most practical situations all three heat transfer mechanisms are present greatly complicates the process of measurement of the thermal conductivity. Thus, much early work in the field is substantially in error, and it has been really quite difficult to devise methods of measurement that unequivocally determine the thermal conductivity. For that reason, the instruments to be described in the following sections often seem to be rather far removed from the apparent simplicity implied by Fourier's Law.

33.1 Fundamental Equations

The essential constitutive equation for thermal conduction relates the heat flux in a material to the temperature gradient by the equation:

$$Q = -\lambda \nabla T \tag{33.1}$$

It is not possible to measure local heat fluxes and gradients; thus, all experimental techniques must make use of an integrated form of the equation, subject to certain conditions at the boundaries of the sample. All experiments are designed so that the mathematical problem of the ideal model is reduced to an integral of the one-dimensional version of Equation 33.1, which yields, in general:

$$Q_a = G \lambda \Delta T \tag{33.2}$$

in which G is constant for a given apparatus and depends on the geometric arrangement of the boundaries of the test sample. Typical arrangements of the apparatus, which have been employed in conjunction with Equation 33.2, are two flat, parallel plates on either side of a sample, concentric cylinders with the sample in the annulus and concentric spheres.

Techniques that make use of Equation 33.2 are known as steady-state techniques and they have found wide application, some of which are discussed below. They are operated usually by measuring the temperature difference ΔT that is generated by the application of a measured heat input Q_a at one of the boundaries. The absolute determination of the thermal conductivity, λ, of the sample contained between the boundaries then requires a knowledge of the geometry of the cell contained in the quantity G. In practice, it is impossible to arrange an exactly one-dimensional heat flow in any finite sample so that great efforts have to be devoted to approaching these circumstances and then there must always be corrections to Equation 33.2 to account for departures from the ideal situation.

If the application of heat to one region of the test sample is made in some kind of time-dependent fashion, then the temporal response of the temperature in any region of the sample can be used to determine the thermal conductivity of the fluid. In these transient techniques, the fundamental differential equation that is important for the conduction process is:

$$\rho C_p \frac{\partial T}{\partial t} = \nabla \cdot \left(\lambda \nabla T \right) \tag{33.3}$$

which arises from an elementary energy balance in the absence of any other processes and in which ρ is the density of the material and C_p its isobaric heat capacity. In most, but not all, circumstances, it is acceptable to ignore the temperature dependence of the thermal conductivity in this equation and to write:

$$\frac{\partial T}{\partial t} = \frac{\lambda}{\rho C_p} \nabla^2 T = a \, \nabla^2 T \tag{33.4}$$

in which a is known as the thermal diffusivity.

Experimental techniques for the measurement of the thermal conductivity based on Equation 33.4 generally take the form of the application of heat at one surface of the sample in a known time-dependent manner, followed by detection of the temperature change in the material at the same or a different location. In most applications, every effort is made to ensure that the heat conduction is unidirectional so that the integration of Equation 33.4 is straightforward. This is never accomplished in practice, so some corrections to the integrated form of Equation 33.4 are necessary. The techniques differ among each other by virtue of the method of generating the heating, of measuring the transient temperature rise, and of the geometric configuration. Interestingly, in one geometric configuration only, is it possible to determine the thermal conductivity essentially independently of a knowledge of ρ and C_p, which has evident advantages. More usually, it is the thermal diffusivity, a, that is the quantity measured directly, so that the evaluation of the thermal conductivity requires further, independent measurements.

In the following sections, brief descriptions of the specific applications of these general principles are given. The examples chosen for study are intended to cover the full spectrum of materials and thermodynamic states and, in each case, attention is concentrated on a method that has proved most accurate and is widely used. The steady-state and transient techniques are considered separately.

33.2 Measurement Techniques

Steady-State Methods

The steady-state methods employed for the measurement of the thermal conductivity of fluids and solids have most often employed the geometry of parallel plates so that it is that configuration described here in two variants. Coaxial cylinder equipment has largely been used within the preserve of the research laboratory, with the apparatus of Tufeu and Le Neindre [1] an excellent example of the genre.

Parallel-Plate Instrument

A schematic diagram of a guarded parallel-plate instrument is shown in Figure 33.1. The sample is contained in the gap between two plates (upper and lower) maintained a distance d apart by spacers. A small amount of heat, Q_a, is generated electrically in the upper plate and is transported through the sample to the lower plate. Around the upper plate, and very close to it, is placed a guard plate. This plate is, in many instruments, maintained automatically at the same temperature as the upper plate so as to reduce heat losses from the upper surface of the upper plate and to most nearly secure a one-dimensional heat flow at the edges of the sample.

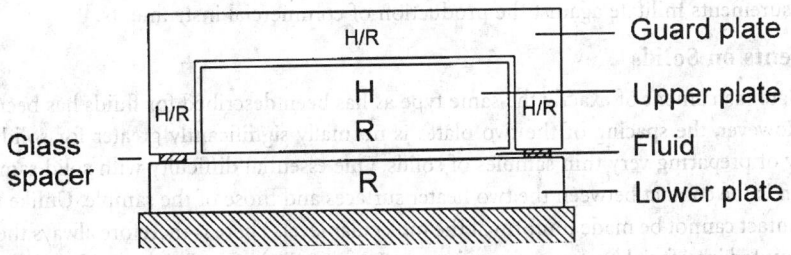

FIGURE 33.1 Schematic diagram of a guarded parallel-plate instrument. (H = heater, R = resistance thermometer).

The temperatures at the surfaces of the upper and lower plates are measured very precisely, as is the electric input of energy, so that the thermal conductivity can, in principle, be evaluated from the equation:

$$Q_a = \frac{A\lambda\Delta T}{d} \tag{33.5}$$

where ΔT is the measured temperature difference, and A is the area of the upper plate.

Whether the sample is a fluid or a solid, the electric energy generated in the upper plate is not all conducted to the lower plate. Thus, it is necessary in all cases to account for spurious heat losses and for all except opaque materials for the radiative transfer between the two surfaces. When the sample is transparent to radiation, this correction is straightforward and can be reduced by means of surface coating the plates to reduce their emissivity; but when the material adsorbs radiation, the problem is much more complicated and has been the subject of some controversy in the past, which has since been resolved (p.147 of [2]). In many cases, the effort of performing absolute measurements cannot be justified so that the ratio A/d is determined by calibration with a material of known thermal conductivity.

Measurement on Fluids

Measurements with parallel-plate instruments on fluids have been performed for a considerable period of time. The technique has particular advantages in some special regions of thermodynamic space but requires great attention to detail if accurate results are to be obtained. In the most accurate instruments for fluids, the gap between upper and lower plates is kept as small as possible (perhaps as small as 0.2 mm). This has the benefit of reducing the effect of heat flows that are not normal to the heat surfaces, but more importantly the small gap contributes to the reduction of the heat transferred by bulk convective motion of the fluid. Indeed, if very small temperature differences are employed by heating the top plate (to have a stable density gradient) *and* considerable care is taken to align the parallel plates normal to the earth's gravitational acceleration, then the effects of convective heat transfer can be rendered negligible (p.154 of [2]). It is a fact of history that the necessary care with this instrument has been taken by only a few workers so that despite the fact that the instrument has been used in the temperature range from 4 K to 800 K, and for pressures up to 250 MPa, only some of the measurements are reliable.

An example of what can be achieved with a parallel-plate instrument is provided by the work of Mostert [3] and Sakonidou [4] at the van der Waals laboratory in Amsterdam. They have used the technique near the critical point of a pure fluid or mixture where the extreme values of the compressibility make the fluid exceedingly prone to convection. In such circumstances, the small vertical extent of the fluid layer required for this technique avoids large density variations in the test layer and, combined with the fact that very small temperature differences (0.3 mK) can be employed, has enabled measurements of the thermal conductivity to be conducted to within 100 mK of the critical temperature at the critical density. Under these circumstances, the thermal conductivity of a pure fluid reveals an enhancement that is, in principle, infinite at the critical point itself.

The unique characteristics of the parallel-plate instrument mean that it is the method of choice for work near the critical region of a material. The arguments above pertaining to the care required for reliable measurements militate against the production of commercial instruments.

Measurements on Solids

A parallel-plate instrument of exactly the same type as has been described for fluids has been employed for solids. However, the spacing of the two plates is normally significantly greater for solids, owing to the difficulty of preparing very thin samples of solids. One essential difficulty with solid samples in this configuration is the contact between the two heater surfaces and those of the sample. Unlike the case for fluids, the contact cannot be made uniform at the molecular level. There is therefore always the possibility of an unaccounted interfacial heat transfer resistance. It seems likely that these considerations contribute to the wide differences between values reported for the same sample by different authors.

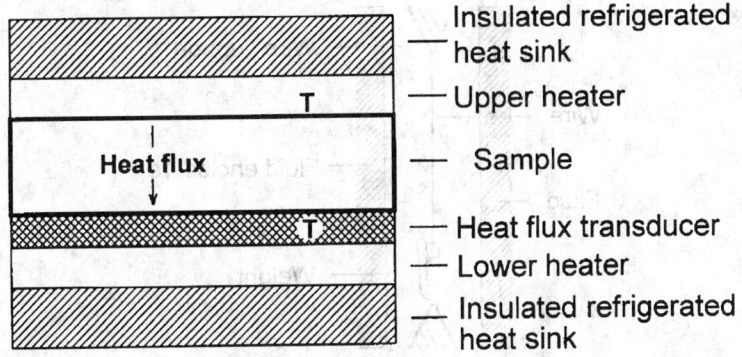

FIGURE 33.2   Schematic diagram of a heat flow-meter instrument. (T = thermocouple).

A more popular implementation of the parallel-plate configuration for solids is the so-called heat flow-meter instrument. When applied to materials such as building insulation, the dimensions of this type of instrument can be very large.

In such instruments (see Figure 33.2), the upper heater plate is set at a higher temperature than the lower one. The hot and cold surface temperatures of the sample are measured with the two thermocouples permanently installed on the adjacent surface plates, while a precalibrated heat flow transducer on the lower plate measures the magnitude of the heat flux through the sample. The thermal conductivity is calculated directly from Equation 33.5. In some cases, contact resistances (i.e., insulation) can be characterized and controlled by employing a pressurized gas in the sample chamber. Commercial instruments of this type are included in the listing of Table 33.1.

Transient Methods

There are rather more transient techniques that have achieved popularity than steady-state instruments. This is because transient techniques generally require much less precise alignment and dimensional knowledge and stability. Furthermore, some of the techniques have distinct advantages that arise from the speed of the measurement. Here we have space to describe only one technique in detail which has, in a variety of ways, far greater applicability.

TABLE 33.1   Companies That Make Thermal Conductivity Instruments

Type of instrument	Temperature range	Supplier	Approximate price (U.S.$)
Transient hot disk			
Thermal Analyser TPS	290–1,000 K	K-analys AB	$20,000
Guarded parallel plate			
Thermatest GHP-300	290–1,000 K	Holometrix Inc.	Variable
TCT 416	290–340 K	NETZSCH	$30,000
Heat flow meter			
Rapid-k RK-70	290–500 K	Holometrix Inc.	Variable
Unitherm 2021	290–500 K	ANTER	$18,000
Radial heat-flow			
Orton D.C.A.	290–1400 K	Orton	$50,000
Laser flash			
Thermaflash 1100	290–1300 K	Holometrix Inc.	Variable

Note: Prices are only indicative.

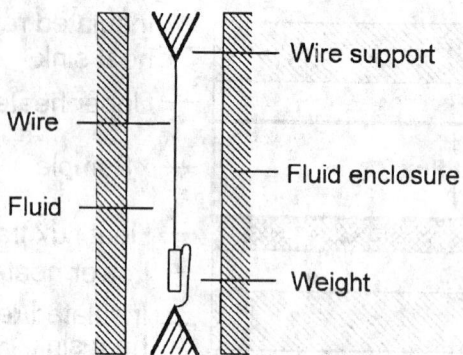

FIGURE 33.3 Schematic diagram of a transient hot-wire instrument for fluids. (Note that the hot wire is kept under constant tension by a weight.)

Transient Hot-Wire Technique

In this technique, the thermal conductivity of a material is determined by observing the temporal evolution of the temperature of a very thin metallic wire (see Figure 33.3) after a step change in voltage has been applied to it. The voltage applied results in the creation of a line source of nearly constant heat flux in the fluid. As the wire is surrounded by the sample material, this produces a temperature field in the material that increases with time. The wire itself acts as the temperature sensor and, from its resistance change, its temperature change is evaluated and this is related to the thermal conductivity of the surrounding material.

According to the ideal model of this instrument, an infinitely long, line source of heat possessing zero heat capacity and infinite thermal conductivity is immersed in an infinite isotropic material, with physical properties independent of temperature and in thermodynamic equilibrium with the line source at time $t = 0$ at a temperature T_0. The heat transferred from the line source to the sample when a stepwise heat flux, q, per unit length is applied, is assumed to be entirely conductive. Then the temperature rise of the material at a radial distance, r_0, which it transpires, is the same as the temperature rise at the surface of a wire of radius r_0, is $\Delta T_i(r_0, t)$ is given by:

$$\Delta T_i\left(r_0,t\right)=T\left(r_0,t\right)-T_0=\frac{q}{4\pi\lambda}\left[\ln\left(\frac{4at}{r_0^2C}\right)+\frac{r_0^2}{4at}+\cdots\right] \tag{33.6}$$

In the above equation, C is a known constant [5]. The equation suggests that, provided the radius of the wire is chosen small enough so that the second term on the right-hand side of Equation 33.6 is negligible, the thermal conductivity of the fluid can be obtained from the slope of the line ΔT_i vs. ln t. Any practical implementation of this method of measurement inevitably deviates from this ideal model. The success of the technique, however, rests on the fact that by proper design, it is possible to construct an instrument that can operate very closely to the ideal model, while at the same time small departures can be treated by a first-order analysis [5].

The transient hot-wire technique was first developed in the 1930s to measure the effective thermal conductivity of powders. However, its application to other materials was somewhat slower until in the late 1960s the new technology associated with electronics made it possible to measure small, transient resistance changes with high accuracy in a period of less than 1 s. This development, pioneered by Haarman [6], made it possible to complete the transient heating process so quickly that, despite the inevitability of convective motion in the fluid from time zero, the inertia of the fluid ensures that the fluid velocity is sufficiently small that there is no significant contribution to heat transfer. This fact prompted a rapid development of the measurement technique for fluids — first in gases and then in liquids — that was then followed by further developments in solids. The differences in the technique between solids and fluids are rather small; thus, some aspects of the instrumentation for liquids are briefly discussed and the differences for solids are just outlined.

In the case of fluids, the instrumentation generally involves a wire some 7 μm to 25 μm in diameter (in order to reduce the correction owing to its heat capacity) and some 150 mm long. The wire is mounted vertically in a cylindrical cell containing the test sample. Often, a second wire differing only in length is employed to compensate automatically for effects at the ends of the wires via the electrical measurement system, but this can also be accomplished with potential taps [5]. Whenever possible, platinum is used for the wire material because its resistance/temperature characteristics are well known and it can be readily obtained in the form of wires with a diameter as small as 5 μm. When the material under test is electrically conducting, it is necessary to insulate electrically the wire from the fluid. A variety of techniques have been employed for this purpose that enjoy different degrees of success depending on the range of conditions to be studied. Near ambient temperature over a range of pressures, it has been found adequate to use a tantalum wire as the sensor that is electrolytically anodized to cover the wire with an insulating layer of tantalum pentoxide 100 nm thick [7]. Under more aggressive conditions, it has been necessary to employ ion-plating of the wire with a ceramic to secure the isolation [8]. In either case, the theory has been modified by the addition of a small correction.

In the case of solids, the need for the wire to be straight and vertical is removed by virtue of the rigidity of the material. Thus, Bäckström and colleagues [9] were able to employ a wire embedded as an arc within the compressed solid matrix of the material under study, particularly at very high pressures (up to 4 GPa).

The transient hot-wire technique has a unique advantage among transient methods that the thermal conductivity of the test material can be evaluated directly from the slope of the line relating the temperature rise of the wire to the logarithm of time. The heat capacity and density of the test material are required only to evaluate small corrections. Furthermore, the exact dimensions of the heating element and the cell are also unimportant so that the method avoids the intricate alignment problems of the parallel-plate technique while securing absolute measurements of the property. Despite these advantages and its wide application to measurements in gases, solids, and liquids, there has been no commercial development of an instrument of this kind, presumably because of the delicacy of the long, thin wire in the case of devices for fluids, and the difficulty of sample preparation for solids.

Hot-Disk Instrument

A transient technique for which there is a commercial version suitable for solid materials is the transient hot-disk instrument shown in Figure 33.4.

The sensor in this case comprises a thin metal strip, often of nickel, wound in the form of a double-spiral in a plane. It is printed on, and embedded within, a thin sandwich formed by two layers of a material that is a poor electric conductor but a good thermal conductor. This disk heater is then, in turn, placed either between two halves of a disk-shaped sample of solid material or affixed to the outside of the sample.

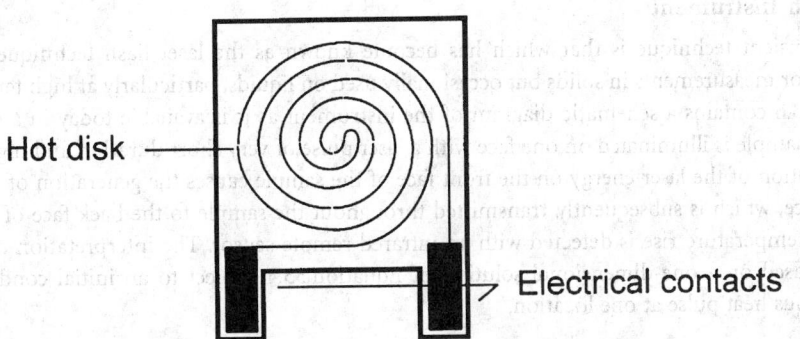

FIGURE 33.4 Schematic diagram of a transient hot-disk instrument.

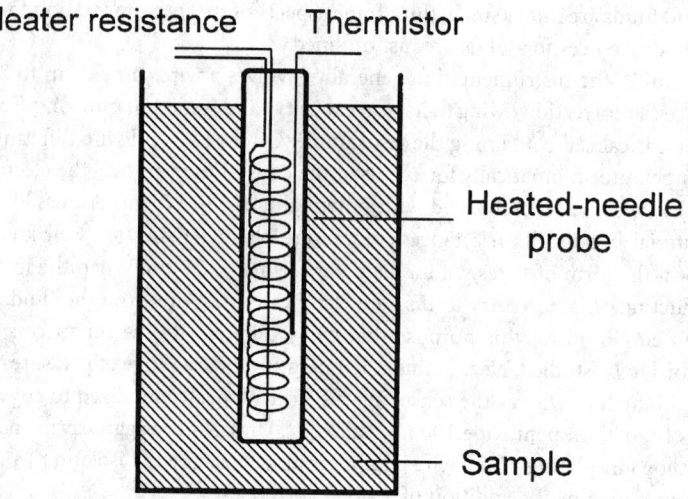

FIGURE 33.5   Schematic diagram of a transient heated-needle probe.

However it is configured, the essential measurement performed is the same as that for the hot-wire technique, and the temperature history of the sensor when subject to known electrical dissipation is inferred from its resistance charge. In the most recent version of the instrument, developed by Gustafsson [10], and also available commercially, the interpretation of the data is accomplished via a numerical solution of the differential equation rather than by some analytical approximation to it. The technique is used frequently for studies of polymer composites, glasses, superconductors, and insulating materials.

Heated-Needle Probe

A further commercial device exists for the measurement of the thermal conductivity of granular materials such as powders and soils, natural materials such as rock and concrete and, indeed, of food. The probe is shown schematically in Figure 33.5 where it is seen that it consists of a thin, hollow, metallic needle (diameter 3 mm) containing an electric heater and a separate thermistor as a probe of the temperature history of the needle following initiation of a heat pulse [11]. The temperature history of the probe is generally interpreted with the aid of the equation appropriate to a transient hot-wire instrument but in a relative manner whereby its response is calibrated against known standards. This rather simplistic approach to the analysis of a somewhat complex cell inevitably restricts the accuracy that can be achieved, but does provide a measurement capability where no other technique is viable. It is often employed for measurements in inhomogeneous samples such as rocks or soils where it is simply the effective thermal conductivity that is required.

Laser-flash Instrument

A final transient technique is that which has become known as the laser-flash technique developed originally for measurements in solids but occasionally used on liquids, particularly at high temperatures.

Figure 33.6 contains a schematic diagram of the instrument as it is available today in a commercial form. The sample is illuminated on one face with a laser pulse of very short duration and high intensity. The absorption of the laser energy on the front face of the sample causes the generation of heat at that front surface, which is subsequently transmitted throughout the sample to the back face of the sample where the temperature rise is detected with an infrared remote sensor. The interpretation of measurements is based on a one-dimensional solution of Equation 33.4 subject to an initial condition of an instantaneous heat pulse at one location.

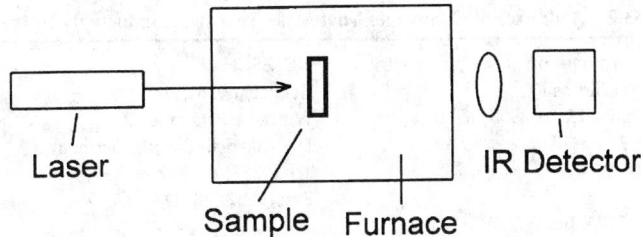

FIGURE 33.6 Schematic diagram of a laser-flash instrument.

The temperature rise at the back face of a sample of thickness l and radius r, is therefore given by [12]:

$$\Delta T(l,t) = \frac{Q}{\rho C_p l \pi r^2}\left[1 + 2\sum_{n=1}^{\infty}(-1)^n \exp\left(-\frac{n^2 \pi^2 a t}{l^2}\right)\right] \tag{33.7}$$

where Q is the energy absorbed at the front surface at time zero. The thermal diffusivity of the sample, a, is then often deduced from the measurement of the time taken for the back face of the sample to reach one half of its maximum value. The technique has the very distinct advantage that it does not require physical contact between the test sample and the heater or detector. For this reason, it is a particularly appropriate technique for use at high temperatures or in aggressive environments.

However, there are a number of precautions that must be taken to ensure accurate results. First, the theory should be modified to account for non-unidirectional heat flow. Secondly, care must be taken to ensure that no radiation incident on the front face penetrates to the back face for transparent samples. Due care must also be taken to match the laser power to the system being studied so that there is neither fusion nor ablation at the front face that can distort the results. Finally, when the fluid state is studied, due care should be taken to eliminate convective heat transport. Seldom are all of these precautions adopted in routine work, so that some results obtained with the technique are of dubious validity. Nevertheless, the method has seen widespread application to a wide range of materials, including composites, polymers, glasses, metals, refracting materials, insulating solids, and coatings.

Finally, the radial heat-flow method should also be mentioned. This is a transient technique in which the sample is heated and cooled continuously [13]. From the recording of the temperature gradient via thermocouples, the thermal diffusivity is obtained and thus the thermal conductivity is calculated. The advantage of the radial heat-flow method is that the measurements are fast and only small temperature gradients are necessary in the sample. A commercially available instrument operating according to this technique is listed in Table 33.1.

33.3 Instrumentation

Table 33.1 lists various instruments for the measurement of the thermal conductivity of solids, while the addresses of the suppliers are shown in Table 33.2. As already mentioned, to our knowledge, there is no company that produces instruments specifically for the measurement of the thermal conductivity of fluids.

33.4 Appraisal

Naturally, the technique to be employed for the measurement of thermal conductivity depends on the type of sample to be studied and the range of conditions to be employed. For fluid samples under most conditions, a variant of the transient hot-wire method must be the preferred technique. Under favorable

TABLE 33.2   Addresses of Companies That Make Thermal Conductivity Instruments

ANTER Corporation 1700 Universal Road Pittsburgh, PA 15235-3998 Tel: (412) 795-6410 Fax: (412) 795-8225	NETZSCH Gerätebau GmbH Wittelsbacherstraße 42 D-95100 Selb/Bayern, Germany Tel: (9287) 88136 Fax: (9287) 88144
HOLOMETRIX Inc. 25 Wiggins Avenue Bedford, MA 0170-2323 Tel: (617) 275-3300 Fax: (617) 275-3705	ORTON The Edward Orton JR. Ceramic Foundation P.O. Box 460 Westerville, OH 43081 Tel: (614) 895-2663 Fax: (614) 895-5610
K-ANALYS AB Seminariegatan 33 H S-752 28 Uppsala, Sweden Tel: (46) 18 50 01 66 Fax: (46) 18 54 36 38	

conditions, an accuracy of ±0.3% can be achieved and a level of ±1% is possible under all but the most aggressive circumstances. Near the critical region of fluids, a parallel-plate instrument is essentially the only viable method. For molten materials at high temperature, while a variant of the hot-wire system has advantages, the laser-flash technique is very attractive but an accuracy of no better than 10% is then to be expected.

For solids, the hot-disk or laser-flash techniques have many features that recommend them when the sample is amenable to appropriate preparation. In those cases, an accuracy of a few percent should be possible but is rarely attained. For samples such as rocks, the heated needle-probe is undoubtedly the only viable technique.

It should be emphasized again here that when a sample is inhomogeneous, the quantity determined is not the thermal conductivity of any element of it, but rather an effective value suitable for engineering purposes. It is not an intrinsic thermophysical property of the material.

References

1. B. Le Neindre, *Contribution à l' étude expérimentale de la conductivité thermique de quelques fluides a haute température et à haute pression*, Ph.D. thesis, Université de Paris, 1969.
2. M. Sirota, Steady-State Measurements for Thermal Conductivity, in A. Nagashima, G. V. Sengers, and W. A. Wakeham (eds.), *Experimental Thermodynamics. Vol. III. Measurement of the Transport Properties of Fluids*, London: Blackwell Scientific Publications, 1991, chap. 6.
3. R. Mostert, H. R. van der Berg, and P. S. van der Gulik, The thermal conductivity of ethane in the critical region, *J. Chem. Phys.*, 92, 5454-5462, 1990.
4. E. Sakonidou, *The thermal conductivity of methane and ethane mixtures around the critical point*, Ph.D. thesis, University of Amsterdam, 1996.
5. M. J. Assael, C. A. Nieto de Castro, H. M. Roder, and W. A. Wakeham, Transient Methods for Thermal Conductivity, in A. Nagashima, J. V. Sengers, and W. A. Wakeham (eds.), *Experimental Thermodynamics. Vol. III. Measurement of the Transport Properties of Fluids*, London: Blackwell Scientific Publications, 1991, chap. 7.
6. J. W. Haarman, *Thermal conductivity measurements with a transient hot-wire method*, Ph.D. thesis, Technische Hogeschool Delft, 1969.
7. A. Alloush, W. B. Gosney, and W. A. Wakeham, A transient hot-wire instrument for thermal conductivity measurements in electrically conducting liquids at elevated temperatures, *Int. J. Thermophys.*, 3, 225-234, 1982.

8. Y. Nagasaka and A. Nagashima, Absolute measurements of the thermal conductivity of electrically conducting liquids by the transient-hot wire method, *J. Phys., E,* 14, 1435, 1981.

9. P. Andersson and G. Bäckström, Measurement of the thermal conductivity under high pressures, *Rev. Sci. Instrum.,* 47, 205, 1976.

10. S. E. Gustafsson, E. Karawacki, and M.N. Khan, Transient hot-strip method for simultaneously measuring thermal conductivity and thermal diffusivity of solids and fluids, *J. Phys. D: Appl. Phys.,* 12, 1411, 1979.

11. J. Nicolas, Ph. André, J. F. Rivez, and V. Debaut, Thermal conductivity measurements in soil using an instrument based on the cylindrical probe method, *Rev. Sci. Instrum.,* 64, 774-780, 1993.

12. W. J. Parker, R. J. Jenkins, C. P. Butler, and G. L. Abbott, Flash method for determining thermal diffusivity, heat capacity and thermal conductivity, *J. Appl. Phys.,* 32, 1679-1684, 1960.

13. G. S. Sheffield and J. R. Schorr, Comparison of thermal diffusivity and thermal conductivity methods, *Ceram. Bull.,* 70, 102, 1991.

34

Heat Flux

Thomas E. Diller
Virginia Polytechnic Institute

Thermal management of materials and processes is becoming a sophisticated science in modern society. It has become accepted that living spaces should be heated and cooled for maximum comfort of the occupants. Many industrial manufacturing processes require tight temperature control of material throughout processing to establish the desired properties and quality control. Examples include control of thermal stresses in ceramics and thin films, plasma deposition, annealing of glass and metals, heat treatment of many materials, fiber spinning of plastics, film drying, growth of electronic films and crystals, and laser surface processing.

Temperature control of materials requires that energy be transferred to or from solids and fluids in a known and controlled manner. Consequently, the proper design of equipment such as dryers, heat exchangers, boilers, condensers, and heat pipes becomes crucial. The constant drive toward higher power densities in electronic, propulsion, and electric generation equipment continually pushes the limits of the associated cooling systems.

Although the measurement of temperature is common and well accepted, the measurement of heat flux is often given little consideration. Temperature is one of the fundamental properties of a substance. Moreover, because temperature can be determined by human senses, most people are familiar with its meaning. Conversely, heat flux is a derived quantity that is not easily sensed. It is not enough, however, to only measure the temperature in most thermal systems. How and where the thermal energy goes is often equally or more important than the temperature. For example, the temperature of human skin can indicate the comfort level of the person, but has little connection with the energy being dissipated to the surroundings, particularly if evaporation is occurring simultaneously. Wind chill factor is another common example of the importance of convection heat transfer in addition to air temperature.

Maximizing or minimizing the thermal energy transfer in many systems is crucial to their optimum performance. Consequently, sensors that can be used to directly sense heat flux can be extremely important. The subsequent material in this chapter is intended to help individuals understand and implement heat flux measurements that are best suited for the required applications.

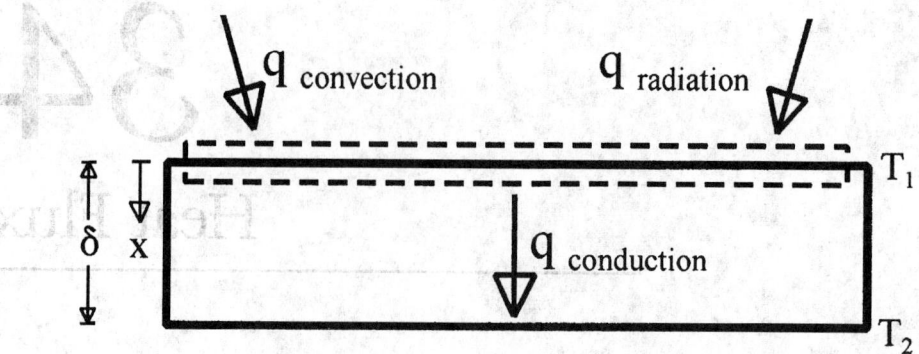

FIGURE 34.1   Illustration of energy balance.

34.1 Heat Transfer Fundamentals

The movement of thermal energy is known as "heat" and the rate of this transfer is commonly called "heat transfer." It is given the symbol q and has the units of watts. The heat transfer per unit area is termed the "heat flux" and is given the symbol q'' with the units of W m^{-2}. Although in some cases only the overall heat transfer from a system is required, often the spatial and temporal variation of the heat flux is important to performance enhancement. Methods for measuring the spatial *or* temporal distribution of heat flux are identified and discussed in this chapter. Detailed simultaneous measurements of *both* spatial and temporal distributions of heat flux, however, are generally not feasible at this time.

One of the most important principles concerning heat transfer is the first law of thermodynamics, which states that the overall energy transfer to and from a system is conserved. It includes all types of energy transfer across the system boundary, including the three modes of heat transfer — conduction, convection, and radiation. For the simple example shown in Figure 34.1, the transient energy balance on the control volume marked can be expressed as:

$$mC\frac{\partial T}{\partial t} = q_{\text{convection}} + q_{\text{radiation}} - q_{\text{conduction}} \qquad (34.1)$$

where m is the mass of the system and C is the corresponding specific heat. The effect of the thermal capacitance (mC) of the material causes a time lag in the temperature response of the material to a change in heat transfer. A short summary of important heat transfer principles follows, with many engineering textbooks available in the field that give additional details.

Conduction

Conduction encompasses heat transfer through stationary materials by electrons and phonons. It is related to the temperature distribution by Fourier's law, which states that the heat flux vector is proportional to and in the opposite direction of the temperature gradient:

$$\vec{q}'' = -k\vec{\nabla}T \qquad (34.2)$$

The constant k is the thermal conductivity of the material. Measuring this temperature gradient is one of the basic methods for determining heat flux.

For a homogeneous material in Cartesian coordinates Equation 34.1 becomes:

$$\frac{\partial T}{\partial t} = a\left(\frac{\partial^2 T}{\partial x^2} + \frac{\partial^2 T}{\partial y^2} + \frac{\partial^2 T}{\partial z^2}\right) \qquad (34.3)$$

where a is the thermal diffusivity of the material with a density of ρ, $a = k/(\rho C)$. Measuring the temperature response of the system according to this equation is the second major method for determining the heat transfer. Because of the complexity of solutions to Equation 34.3, this method can be complicated if multidimensional effects are present.

If steady-state one-dimensional heat transfer can be assumed throughout the planar solid in Figure 34.1, the temperature distribution in the direction of heat flux q'' is linear. Equation 34.2 becomes simply:

$$q'' = k\frac{T_1 - T_2}{\delta} \tag{34.4}$$

where the temperatures are specified on either side of the material of thickness δ.

As illustrated in Figure 34.1, convection and radiation are the other modes of heat transfer typically present at the surface of a solid. These are usually the quantities of interest to measure with a heat flux sensor. Both are present at least to some extent in virtually all cases, although the effects of one or the other are often purposely minimized to isolate the effects of the other.

Convection

Although heat transfer by convection occurs by the same physical mechanisms as conduction, the fluid is free or forced to move relative to the surface. The fluid motion greatly complicates the analysis by coupling the heat transfer problem with the fluid mechanics. Particularly when the flow is turbulent, the fluid equations are generally impossible to solve exactly. Consequently, the heat transfer and fluid mechanics are commonly isolated by introduction of a heat transfer coefficient, which encompasses all of the fluid flow effects.

$$q'' = h_T\left(T_r - T_s\right) \tag{34.5}$$

The temperature of the fluid is represented by T_r, which for low-speed flows is simply the fluid temperature away from the surface. The recovery temperature is used for high-speed flows because it includes the effect of frictional heating in the fluid. The subscript T on the heat transfer coefficient, h_T, implies that the boundary condition on the surface is a constant temperature, T_s. Although other surface temperature conditions can be encountered, it is then important to carefully document the surface temperature distribution because it can have a profound effect on the values of h and q'' [1, 2].

Radiation

Heat transfer by radiation occurs by the electromagnetic emission and adsorption of photons. Because this does not rely on a medium for transmission of the energy, radiation is very different from conduction and convection. Radiation has a spectrum of wavelengths dependent on the temperature and characteristics of the emitting surface material. Moreover, the surface properties are often dependent on the wavelength and angular direction of the radiation. One classic example is material that looks black to the visible spectrum, but is transparent to the longer wavelengths of the infrared spectrum. Consequently, special coatings are sometimes applied to surfaces to control the absorption characteristics. For example, the surface of a radiation sensor is often coated with a high absorptivity paint or graphite.

Because the power emitted from a surface is proportional to the fourth power of the absolute temperature, radiation detectors are usually cooled sufficiently for the power emitted from the detector itself to be negligible. In this case, the temperature distribution of the sensor is not important. This is a big advantage over convection measurements where the temperature distribution on the surface has a big influence on the measurement.

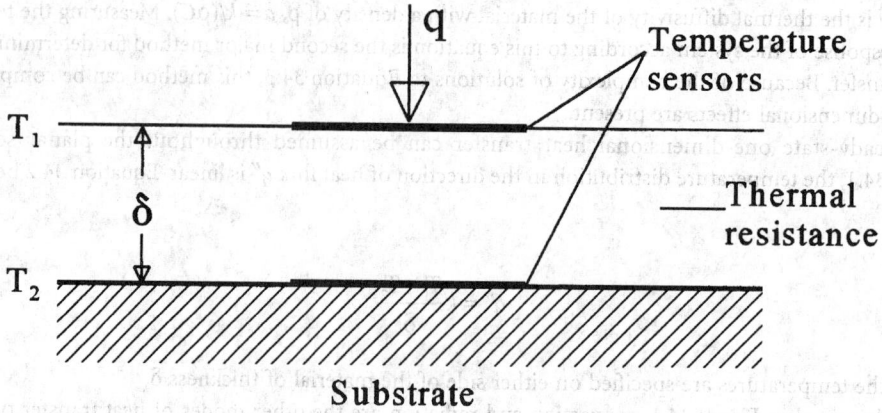

FIGURE 34.2   One-dimensional planar sensor concept.

34.2 Heat Flux Measurement

Most of the many methods for measuring heat flux are based on temperature measurements on the
surface or close to the surface of a solid material. Usually this involves insertion of a device either onto
or into the surface, which has the potential to cause both a physical disruption and a thermal disruption
of the surface. As with any good sensor design, the goal for good measurements must be to minimize
the disruption caused by the presence of the sensor. It is particularly important to understand the thermal
disruption caused by the sensor because it cannot be readily visualized and because all heat flux sensors
have a temperature change associated with the measurement. Consequently, wise selection of the sensor
type and operating range is important for good heat flux measurements. The following sections emphasize
important factors in using the currently available heat flux sensors, followed by short summaries of
sensors used in research and possible future developments. They are grouped by the general type of
sensor action.

34.3 Sensors Based on Spatial Temperature Gradient

The heat flux at the material surface can be found at a location if the temperature gradient can be
determined at that position, as indicated in Equation 34.2. Because it is difficult to position temperature
sensors with the requisite accuracy inside the material, sensors to measure heat flux are either applied
on the surface or mounted in a hole in the material. In the following sections, the different types of
commercially available sensors are discussed first and listed in a table by manufacturer. Shorter sections
briefly describing sensors used in research labs or currently being developed follow.

One-Dimensional Planar Sensors

The simplest heat flux sensor in concept is illustrated in Figure 34.2. The one-dimensional heat flux
perpendicular to the surface is found directly from Equation 34.4 for steady-state conditions:

$$q'' = \frac{k}{\delta}\left(T_1 - T_2\right) \tag{34.6}$$

The thickness of the sensor δ and thermal conductivity k are not known with sufficient accuracy for any
particular sensor to preclude direct calibrations of each sensor. An adhesive layer may also be required
between the sensor and surface to securely attach the sensor, which adds an additional thermal resistance

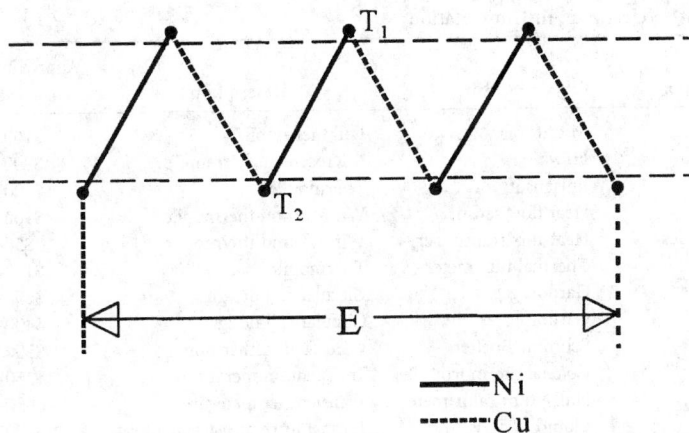

FIGURE 34.3 Thermopile for differential temperature measurement [2].

and increases the thermal disruption. Temperature measurements on the sensor and on the surrounding undisturbed material are recommended to quantify this disruption.

Although the temperature difference can be measured by any number of methods, the most commonly used are thermocouples. Thermocouples have the advantage that they generate their own voltage output corresponding to the temperature difference between two junctions. Consequently, they can be connected in series to form a thermopile that amplifies the output from a given temperature difference. An illustration of a thermopile for measuring a temperature difference is given in Figure 34.3. Most any pair of conductors (e.g., copper–constantan) can be used for the legs of the thermopile, but the output leads should be of the same material so that additional thermocouple junctions are not created. The voltage output, E, is simply:

$$E = N S_T \left(T_1 - T_2 \right) \tag{34.7}$$

where N represents the number of thermocouple junction pairs, and S_T is the Seebeck coefficient or thermoelectric sensitivity of the materials, expressed in volts per degree Centigrade. The corresponding sensitivity of the heat flux sensor is:

$$S = \frac{E}{q''} = \frac{N S_T \delta}{k} \tag{34.8}$$

Although the sensitivity is determined in practice from a direct calibration, the last part of the equation can be used to determine the effects of different parameters for design purposes.

One successful design using a thermopile was described by Ortolano and Hines [3] and is currently manufactured by RdF Corp., as listed in Table 34.1. Thin pieces of two types of metal foil are alternately wrapped around a thin plastic (Kapton) sheet and butt-welded on either side to form thermocouple junctions, as illustrated in Figure 34.4. A separate thermocouple is included to provide a measure of the sensor temperature. The flexible, micro-foil sensors are 75 μm to 400 μm thick and can be glued to a variety of surface shapes, but are limited to temperatures below (250°C) and heat fluxes less than 100 kW m⁻². This covers many general-purpose industrial and research applications. The time response can be as fast as 20 ms, but transient signals can be attenuated unless the frequency of the disturbance is less than a few hertz. First-order systems, such as these sensors, give 70% response to a sinusoidal input with a period six times the exponential time constant.

TABLE 34.1 Heat Flux Instrumentation

Manufacturer	Sensor	Description	Approximate price (U.S.$)
RdF	Micro-foil	Foil thermopile	$100
Vatell	HFM	Microsensor thermopile	$900
Vatell	Episensor	Thermopile	$100–250
Concept	Heat flow sensor	Wire-wound thermopile	$100–300
Thermonetics	Heat flux transducer	Wire-wound thermopile	$50–900
ITI	Thermal flux meter	Thermopile	$150–350
Vatell	Gardon gage	Circular foil design	$250–500
Medtherm	Gardon gage	Circular foil design	$400–800
Medtherm	Schmidt-Boelter	Wire-wound thermopile	$500–800
Medtherm	Coaxial thermocouple	Transient temperature	$250–450
Medtherm	Null-point calorimeter	Transient temperature	$650–800
Hallcrest	Liquid crystals	Temperature measurement kit	$200
Image Therm Eng.	TempVIEW	Liquid crystal thermal system	$30k–50k

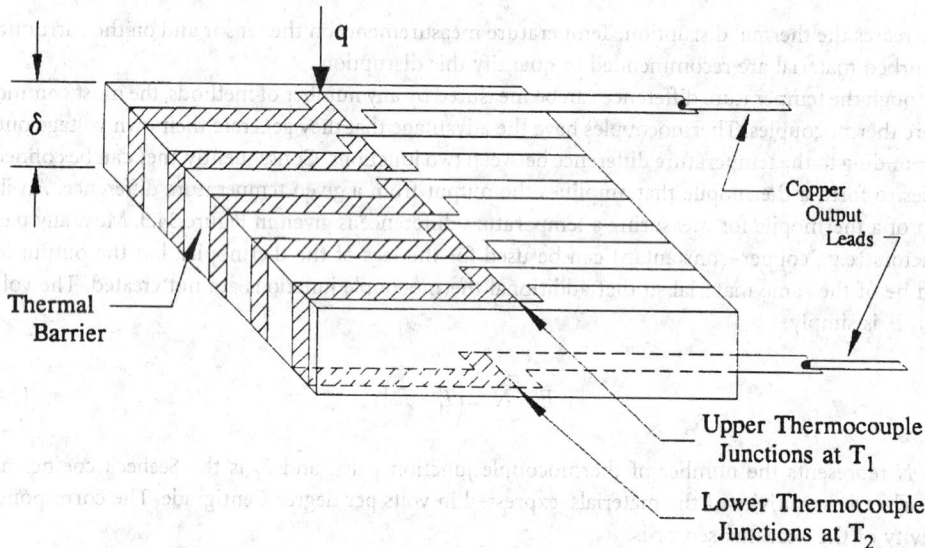

FIGURE 34.4 Thermopile heat flux sensor [3].

A similar design uses welded wire to form the thermopile across a sensor about 1 mm thick. This gives a higher sensitivity to heat flux, but also a larger thermal resistance. Time constants are on the order of 1 s and the upper temperature limit is 300°C. These are manufactured in a range of sizes by International Thermal Instrument Co., as listed in Table 34.1. Applications include heat transfer in buildings and physiology. Sensors with higher sensitivity are made with semiconductor thermocouple materials for geothermal applications. Lower sensitivity sensors are made for operating temperatures up to 1250°C.

A much thinner thermopile sensor called the Heat Flux Microsensor (HFM) was described by Hager et al. [4] and is manufactured by Vatell Corp., as listed in Table 34.1. Because it is made with thin-film sputtering techniques, the entire sensor is less than 2-μm thick. The thermal resistance layer of silicon monoxide is also sputtered directly onto the surface. The resulting physical and thermal disruption of the surface due to the presence of the sensor is extremely small. Use of high-temperature thermocouple materials allows sensor operating temperatures to exceed 800°C for the high-temperature models. They are best suited for heat flux values above 1 kW m^{-2}, with no practical upper limit. Because the sensor is so thin, the thermal response time is less than 10 μs [5], giving a good frequency response well above 1 kHz. A temperature measurement that is integrated into the sensor is very useful for checking the heat flux calibration [6] and determining the heat transfer coefficient. The high temperature and fast time response capabilities are useful for aerodynamic applications, combusting flows in engines and propulsion systems, and capturing high-speed events such as shock passage.

Terrell [7] describes a similar sensor design made with screen printing techniques of conductive inks. A copper/nickel thermocouple pair was used with a dielectric ink for the thermal resistance layer. The inks were printed onto anodized aluminum shim stock for the substrate. Although the entire package is 350-μm thick, the thermal resistance is low because of the high thermal conductivity of all of the materials. These are currently offered commercially by Vatell Corp., as listed in Table 34.1. Because of the large number of thermocouple pairs (up to 10,000), sensitivities are sufficient to measure heat fluxes as low as 0.1 W m^{-2}. The thermal time constant is about 1 s, and the upper temperature limit is approximately 150°C. The aluminum base allows some limited conformance to a surface. Applications include the low heat fluxes typical of building structures, biomedicine, and fire detection.

Another technique for measuring the temperature difference across the thermal resistance layer is to wrap wire and then plate one side of it with a different metal. A common combination is constantan wire with copper plating. The resulting wire-wound sensor looks similar to the sensor shown in Figure 34.4. The difference is that the constantan wire is continuous all around the sensor, so it does not form discrete thermocouple junctions. A summary of the theory is given by Hauser [8] and a general review is given by van der Graaf [9]. Concept Engineering offers a range of these type of sensors at moderate cost. Because of the hundreds of windings on these sensors around 2-mm thick plastic strips, the sensitivity to heat flux is high. The corresponding thermal resistance is also large and time constants are around 1 s. Temperatures are limited to about 150°C. Thermonetics also makes a plated wire-wound heat flux sensor. Thicknesses range from 0.5 mm to 3 mm, with time constants greater than 20 s. Some of the units are flexible and can be wrapped around objects. The normal temperature limit is 200°C, but ceramic units are available for operation above 1000°C. The main use for these sensors is to measure heat flux levels less than 1 kW m^{-2}, with applications including building structures, insulation, geothermal, and medicine.

One popular version of plated wire sensors uses a small anodized piece of aluminum potted into a circular housing, commonly known as a Schmidt-Boelter gage. Kidd [10] has performed extensive analyses on these gages to determine the effect of the potting on the measured heat flux. Neumann [11] discusses applications in aerodynamic testing. The sensors are commercially available from Medtherm Corp. in sizes as small as 1.5-mm diameter. There is also some ability to contour the surface of the sensor to match a curved model surface for complex test article shapes.

Circular Foil Gages

The circular foil or Gardon gage consists of a hollow cylinder of one thermocouple material with a thin foil of a second thermocouple material attached to one end. A wire of the first material is attached to the center of the foil to complete a differential thermocouple pair between the center and edge of the foil as illustrated in Figure 34.5. The body and wire are commonly made of copper with the foil made of constantan. Heat flux to the gage causes a radial temperature distribution along the foil as illustrated in Figure 34.5. The circular foil gage was originated by Robert Gardon [12] to measure radiation heat transfer. For a uniform heat flux typical of incoming radiation the center to edge temperature difference is proportional to the heat flux (neglecting heat losses down the center wire):

$$T_o - T_s = \frac{q'' R^2}{4k\delta} \tag{34.9}$$

The thickness of the foil is δ and the active radius of the foil is R. The temperature difference produced from center to edge of the foil is measured by a single thermocouple pair, typically copper-constantan. The output voltage is proportional to the product of the temperature difference in Equation 34.9 and the thermoelectric sensitivity of the differential thermocouple.

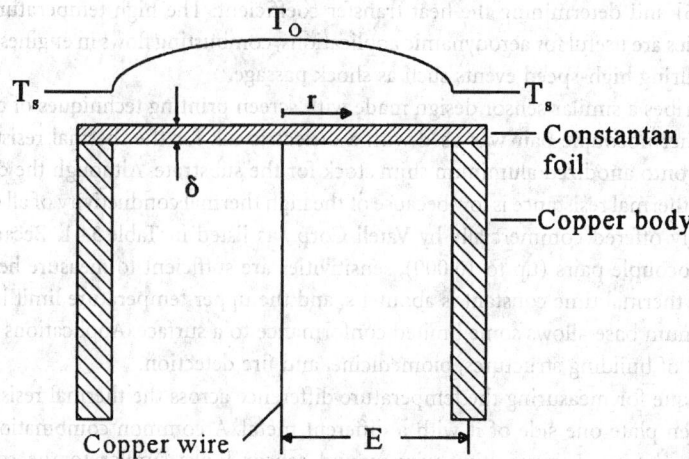

FIGURE 34.5   Circular foil heat flux gage.

These sensors are reasonably rugged and simple. They are manufactured by two companies at moderate cost (Medtherm and Vatell), and are often used as secondary standards for measurement of radiation. One important application is the measurement of the heat flux occurring during fire tests to check flammability of materials. The biggest problems with the circular foil gages arise when they are used with any type of convection heat transfer. It has been shown analytically and experimentally that the output is incorrect for convective heat transfer because of the distortion of the temperature profile in the foil from the assumed radially symmetric, parabolic profile of radiation [13]. Because the amount of error is a function of the gage geometry, the fluid flow, and the heat transfer coefficient, it is difficult to reliably correct. The errors become particularly large when the sensor is used in a flow that has a shear flow component [14], which encompasses most convection situations. Consequently, great care must be used to keep the temperature difference across the gage $T_o - T_s$ small if Gardon gages are used to measure convective heat transfer.

When the gages are used in high heat flux situations, such as combustors, water cooling is usually supplied through the body of the sensor to keep the temperature from exceeding material limits. Because of the resulting temperature mismatch of the gage and surrounding material in which it is mounted, a water-cooled gage is not recommended for convection heat transfer measurements. It is also important for a water-cooled gage to ensure that condensation does not occur on the sensor face.

Although most heat flux sensors are designed to measure the total heat flux, sensors have been developed to separate convection from radiation. The most common method is to put a transparent window over the sensor to eliminate convection to the sensor face. Because the resulting sensor only measures radiation, it is termed a radiometer. The field of view is limited, however, in these radiometers and must be included in the interpretation of results. In a dirty environment where the transmission of the window could be degraded, air is blown across the face of the window to keep the particles away from the sensor. Both manufacturers of the circular foil gages (Medtherm and Vatell) make these radiometer versions. Applications include use in high-ash boilers and gas turbine combustors.

Research Sensors

Although not commercially available, heat flux sensors using RTDs (resistance temperature devices) have been developed to measure the required temperature difference across a heat flux sensor. They are not as convenient for measuring small temperature differences as thermocouples, however, because RTDs require two individual temperature readings to be subtracted. Conversely, a thermocouple pair reads the temperature difference between the two junctions directly and allows the formation of thermopiles with many pairs of junctions to amplify the signal.

Researchers at MIT have developed a sensor like the Micro-Foil thermopile sensor using a nickel resistance element on each side of the plastic sheet [15]. One advantage of knowing the individual temperatures rather than the temperature difference is that the time response of the sensor can be analytically enhanced up to 100 kHz with appropriate modeling. Hayashi et al. [16] used vacuum deposition to create thin-film heat flux sensors like the HFM sputtered thermopile sensors except using a nickel RTD on either side of the silicon monoxide thermal resistance layer. The frequency response was estimated from shock-tunnel experiments to be 600 Hz.

Future Developments

The most exciting recent advances in the field of heat flux measurement have been provided by thin-film technology. Continued improvements in size, sensitivity, price, and time response are anticipated. As the size of sensors continues to decrease, the deposition of thin-film heat flux sensors directly on parts as they are manufactured should become a reality.

34.4 Sensors Based on Temperature Change with Time

Equations 34.1 to 34.3 give the form of the relationship between the unsteady response of temperature and surface heat transfer. If the thermal properties of the wall material are known along with sufficient detail about the temperature history and distribution, the heat transfer as a function of time can, in principle, be determined. Although temperature sensors are available from manufacturers, the necessary data manipulation must be done by the user to obtain heat flux. There are two types of solutions used to reduce the temperature history to heat flux. These are discussed separately as the semi-infinite solution and calorimeter methods. In addition, a variety of methods for measuring the required temperature history are discussed.

Semi-Infinite Surface Temperature Methods

An important technique for short-duration heat flux tests is to measure the surface temperature history on a test object with a fast-response temperature sensor. For short enough times and sufficiently thick material, it can be assumed that the transfer is one-dimensional and that the thermal effects do not reach the back surface of the material. Equation 34.3 reduces to a one-dimensional, semi-infinite solution, which is simple to implement for this case. For example, the surface temperature for a step change of heat flux at time zero is

$$T_s - T_i = \frac{2q_o'' \sqrt{t}}{\sqrt{\pi k \rho C}} \tag{34.10}$$

The substrate properties are the thermal conductivity, density, and specific heat, represented by the product $k\rho C$. T_i is the uniform initial temperature of the substrate and T_s is the surface temperature as a function of time.

A good criteria for the time limit on the test is the time before 1% error occurs in the heat flux [17]:

$$t = 0.3 \frac{L^2}{a} \tag{34.11}$$

Here, a represents the thermal diffusivity of the substrate material and L is the substrate thickness. For a typical ceramic substrate (MACOR), the corresponding minimum thickness for 1 s of test time is 1.6 mm. High conductivity materials, such as metals, have much larger required thicknesses.

Data analysis of the measured temperature record can be performed by several methods. The simplest is to use the analytical solution with each sampled data point to recreate the heat flux signal. The most popular equation for this conversion is that attributed to Cook and Felderman [18] for uniformly sampled data:

$$q''(t_n) = \frac{2\sqrt{k\rho C}}{\sqrt{\pi \Delta t}} \sum_{j=1}^{n} \frac{T_j - T_{j-1}}{\sqrt{n-j} + \sqrt{n+1-j}} \tag{34.12}$$

This can easily be implemented for digital data with a short computer program to perform the summation for the measured points. Modifications are also available to provide more solution stability [17]. More complex techniques include the use of parameter estimation techniques [19] and numerical solutions to account for changes in property values with the changing temperature [20]. Because of the noise amplification inherent in the conversion from temperature to heat flux, analog methods have been developed to convert the temperature signal electronically before digitizing the signal [21].

There are many methods for measuring surface temperature that can be used to determine heat flux. Two broad categories are point measurements using thermocouples or RTDs and optical methods that allow for simultaneous measurement of temperatures over the entire surface. They all require substantial effort from the user to initiate the test procedure and reduce the data to find heat flux. Places for additional information and a few sources for temperature sensors are given.

Point temperature measurements for determining convective heat flux are often made with thin-film RTDs. A metallic resistance layer is sputtered, painted and baked, or plated onto the surface. Because the resulting thickness of the sensor is less than 0.1 μm, the response time is a fraction of 1 μs and there should be no physical or thermal disruption of the measured temperature due to the sensor. Most researchers develop techniques for instrumenting models themselves. Transient flow facilities provide an easy method for quickly initiating the flow and heat transfer, as required by this transient method. However, the model can also be injected into the flow or the flow can be quickly diverted to provide the fast initiation of heat flux. The method is used for basic measurements applicable to gas turbine engines, rockets, internal combustion engines, and high-speed aerodynamics [2].

A special type of thermocouple is made for surface temperature measurement, called coaxial thermocouples [22]. It has one thermocouple wire inside the second thermocouple material with an insulating layer in between. One end is mounted into a metal sheath for press fitting into the surface material for testing. A thin thermocouple is formed by combining the two metals right at the end of the assembly. Response times are typically 1 ms or less, which although slower than the thin-film sensors is sufficient for most applications [11]. The cost per sensor is moderate and they are available from Medtherm Corp., as listed in Table 34.1.

A new approach to measure the transient surface temperature at a point is being developed using a fiber optic probe embedded in the surface [23]. A Fabry-Perot interferometer is the basis for the technique, which has the advantages of high spatial resolution and no electrical connections.

Null-point calorimeters [24] are a further extension of the semi-infinite surface temperature method. They are designed for measurement of extremely high levels of heat flux (over 1000 kW m^{-2}). To protect the thermocouple and wires, they are mounted in a cavity behind the surface. The geometry of the null-point calorimeter is designed, however, for the thermocouple to measure a temperature that would match the surface temperature of a semi-infinite material so that Equation 34.14 can be used for data reduction [24]. Medtherm is the current supplier of null-point calorimeters, as indicated in Table 34.1.

Optical methods give the opportunity to measure the entire temperature field over a section of the surface. Consequently, much data can be collected over each test, but interpretation to obtain quantitative heat flux values is more challenging than measurements with point sensors. The visual display of the temperature over the surface can be very qualitatively informative, however.

The most popular optical method for measuring temperature is to record the color change using liquid crystals. These are specially prepared molecules that reversibly change their color reflection through several distinct colors as a function of temperature, typically in the range of 25°C to 40°C. The best for transient measurements are the chiral-nematic form that have been microencapsulated to stabilize their properties. A variety of types can be obtained from Hallcrest (in Table 34.1), which can be used over the temperature range from 5°C to 150°C. They can easily be spray-painted onto a blackened surface for testing. Setting the lighting for reproducible color, temperature calibration, image acquisition, and accurately establishing the starting temperature are crucial steps. Detailed procedures for accurate measurements have been established by several groups [25–28]. The basic materials are cheap, but the associated equipment is expensive. As listed in Table 34.1, Image Therm Engineering offers a complete system for temperature measurement, including a high-quality video camera, lighting system, calibration system, computer hardware and software for image processing, and a liquid crystal kit.

As previously discussed for radiation heat transfer, all surfaces emit radiation with an intensity and wavelength distribution that can be related to the surface temperature. The advent of high-speed infrared scanning radiometers has made it feasible to record the transient temperature field for determination of the heat flux distribution [29]. A coating is usually applied to establish a known, high absorptivity surface. To convert the measured radiation emission to surface temperature, the radiation field of the surroundings is also required. The camera and associated equipment are quite expensive.

Thermographic phosphors emit radiation in the visible spectrum when illuminated with ultraviolet light. The intensity of emission at specific wavelengths can be related to the temperature over a wide range of surface temperatures. The potential high-temperature applications are particularly appealing [30]. A CCD camera is required to record the transient optical images and calibration is challenging.

Calorimeter Methods

A *calorimeter* is a device for measuring the amount of absorbed thermal energy. The slug calorimeter [31] assumes that the temperature throughout the sensor is uniform while it changes with time. When exposed to a fluid at a temperature T_r and heat transfer coefficient h over an active area of A, the solution for the temperature change is simply an exponential:

$$\frac{T - T_r}{T_i - T_r} = e^{-t/\tau} \tag{34.13}$$

where T_i is the initial temperature and the time constant is:

$$\tau = \frac{mC}{hA} \tag{34.14}$$

with the active surface area represented by A. The thermal capacitance is the product of the mass of the sensor and the specific heat. The time constant can be found from the temperature response of the system, which can then be used in Equation 34.14 to quantify the heat transfer coefficient, h. Although these calorimeters are simple in principle, it is often difficult to obtain reliable results because of heat losses and nonuniform temperatures.

A more useful device, called the plug-type heat flux gage, was developed by Liebert [32] at NASA Lewis. An annulus is created on the backside of the surface by electrical discharge machining. Four thermocouples are attached along the remaining plug to estimate both the temperature gradient and the change in thermal energy content in the plug. This gives a better estimate of the heat flux than the slug calorimeter. An additional advantage is that the measurement surface is physically undisturbed.

Another calorimeter technique, called the *thin skin method*, uses the entire test article as the sensor. Models are constructed of thin metals and instrumented with thermocouples on the back surface. The temperature is assumed constant throughout the material at any location, but varies with time and position around the model. The main errors to be avoided are: (1) lateral conduction along the surface material, (2) heat loss by conduction down the thermocouple wires, and (3) heat loss from the back surface, which is usually considered adiabatic. Because of the recent advances in thin-film and optical surface temperature measurement, the thin-skin method is considered outdated for most modern aerodynamic testing [11].

34.5   Measurements Based on Surface Heating

For research on convective heat transfer, electric heating provides an easy method of controlling and measuring the heat flux to the surface. A combination of guard heaters and proper insulation allows control of the heat losses to give an accurate knowledge of the heat flux leaving the surface based on the total electrical power supplied once steady-state conditions have been established. The temperature of the surface and fluid are used in Equation 34.5 to give the heat transfer coefficient for the surface:

$$h = \frac{q''}{T_r - T_s} \tag{34.15}$$

As with the transient temperature methods, it requires considerable experimental design and expertise by the user to begin making measurements.

34.6   Calibration and Problems to Avoid

Calibration of heat flux is a complicated issue because some heat flux sensors respond to different modes and conditions of heat flux differently. For example, a sensor calibrated by radiation can have a substantially different response to the same amount of heat flux in convection.

For the low heat fluxes seen in building applications, the guarded hot-plate method has been well established [33]. The National Institute of Standards and Technology (NIST) maintains calibration devices for this range of conduction heat transfer. Calibrated insulation samples are readily available to check other guarded hot-plate calibrators. Calibration of sensors at elevated temperatures has demonstrated that there is a dependence of the heat flux sensitivity on sensor temperature [34].

There have been several important advances in heat flux calibration for more general industrial applications within the past few years. NIST is completing three heat flux calibration facilities. A blackbody radiation facility operating to 100 kW m^{-2}, a laminar flow convection facility, and a helium conduction facility are currently being completed and tested [35–37]. This combination of facilities will allow comparison of sensor response under the different modes of heat transfer. In addition, Arnold Engineering Development Center (AEDC) has recently acquired a radiation calibration facility for elevated sensor temperatures (up to 800°C). The temperature dependence of the heat flux sensitivity is thought to be substantial for many sensors. Because in the past most all calibrations have only been performed with the sensors at room temperature, this is an important new facility.

As with many other measurements, the major problem with heat flux measurement is the error caused in the heat flux by the disruption of the sensor itself. For sensors based on the spatial temperature gradient methods, a larger signal implies a larger temperature difference and a larger temperature disruption. For the second type of sensors based on the transient temperature change, the surface temperature is changing while the measurement occurs. The larger the temperature change, the easier the determination of the sensor heat flux, but the larger the error from the sensor temperature disruption.

The error caused by the thermal disruption of the sensor can be estimated for conduction [38]. In convection, however, the effect of the surface temperature disruption on the developing thermal boundary layer is much more difficult to estimate and the effect on the heat flux can be much larger than the percentage change of the temperature [2]. Therefore, it is imperative in convection measurements to keep the thermal disruption of the sensor to a minimum.

34.7   Summary

There are a large number of off-the-shelf heat flux sensors available. Those commercially available have been listed in Table 34.1, and the information for contacting the manufacturers is given in Table 34.2.

TABLE 34.2 Companies That Make Sensors for Heat Flux Measurement

Concept Engineering	RdF Corporation
43 Ragged Rock Road	P.O. Box 490
Old Saybrook, CT 06475	Hudson, NH 03051-9981
(860) 388-5566	(603) 882-5195
Hallcrest Liquid Crystal Division	Thermonetics Corp.
1820 Pickwick Lane	Box 9112
Glenview, IL 60025	San Diego, CA 92109
(312) 998-8580	(619) 488-2242
International Thermal Instrument Co.	Vatell Corporation
P.O. Box 309	P.O. Box 66
Del Mar, CA 92014	Christiansburg, VA 24073
(619) 755-4436	(540) 961-3576
Medtherm Corporation	Image Therm Eng., Inc.
P.O. Box 412	159 Summer St.
Huntsville, AL 35804	Waltham, MA 02154
(205) 837-2000	(781) 893-7793

The differential temperature devices provide a direct readout of the heat flux over the surface of the sensor. With the proper choice of sensor for the application and care in measurement method, the results are easily interpreted and used. Alternatively, the transient temperature methods can provide more surface details, but the output is a surface temperature history that must be analyzed to obtain the corresponding heat flux. Although certain components of these systems are off-the-shelf, more work is required of the user to interpret the results. Issues of calibration and errors have been addressed briefly here. More detail on all aspects of heat flux measurement can be obtained from the manufacturers and the references listed.

References

1. R. J. Moffat, Experimental heat transfer, in G. Hetsroni (ed.), *Heat Transfer 1990,* Vol. 1, New York: Hemisphere, 1990, 187-205.
2. T. E. Diller, Advances in heat flux measurement, in J. P. Hartnett et al. (eds.), *Advances in Heat Transfer,* Vol. 23, Boston: Academic Press, 1993, 279-368.
3. D. J. Ortolano and F. F. Hines, A simplified approach to heat flow measurement. *Advances in Instrumentation,* Vol. 38, Part II, Research Triangle Park: ISA, 1983, 1449-1456.
4. J. M. Hager, S. Onishi, L. W. Langley, and T. E. Diller, High temperature heat flux measurements, *AIAA J. Thermophysics Heat Transfer,* 7, 531-534, 1993.
5. D. G. Holmberg and T. E. Diller, High-frequency heat flux sensor calibration and modeling, *ASME J. Fluids Eng.,* 117, 659-664, 1995.
6. J. M. Hager, J. P. Terrell, E. Sivertson, and T. E. Diller, *In-situ* calibration of a heat flux microsensor using surface temperature measurements, *Proc. 40th Int. Instrum. Symp.,* Research Triangle Park, NC: ISA, 1994, 261-270.
7. J. P. Terrell, New high sensitivity, low thermal resistance surface mounted heat flux transducer, *Proc. 42nd Int. Instrum. Symp.,* Research Triangle Park, NC: ISA, 1996, 235-249.
8. R. L. Hauser, Construction and performance of *in situ* heat flux transducers, in E. Bales et al. (eds.), *Building Applications of Heat Flux Transducers,* ASTM STP 885, 1985, 172-183.
9. F. Van der Graaf, Heat flux sensors, in W. Gopel et al. (eds.), *Sensors,* Vol. 4, New York: VCH, 1989, 295-322.
10. C. T. Kidd and C. G. Nelson, How the Schmidt-Boelter gage really works, *Proc. 41st Int. Instrum. Symp.,* Research Triangle Park, NC: ISA, 1995, 347-368.

11. D. Neumann, Aerothermodynamic instrumentation, AGARD Report No. 761, 1989.
12. R. Gardon, An instrument for the direct measurement of intense thermal radiation, *Rev. Sci. Instrum.*, 24, 366-370, 1953.
13. C. H. Kuo and A. K. Kulkarni, Analysis of heat flux measurement by circular foil gages in a mixed convection/radiation environment, *ASME J. Heat Transfer*, 113, 1037-1040, 1991.
14. N. R. Keltner, Heat flux measurements: theory and applications, Ch. 8, in K. Azar (ed.) *Thermal Measurements in Electronics Cooling*, Boca Raton, FL: CRC Press, 1997, 273-320.
15. A. H. Epstein, G. R. Guenette, R. J. G. Norton, and Y. Cao, High-frequency response heat-flux gauge, *Rev. Sci. Instrum.*, 57, 639-649, 1986.
16. M. Hayashi, S. Aso, and A. Tan, Fluctuation of heat transfer in shock wave/turbulent boundary-layer interaction, *AIAA J.*, 27, 399-404, 1989.
17. T. E. Diller and C. T. Kidd, Evaluation of Numerical Methods for Determining Heat Flux With a Null Point Calorimeter, in *Proc. 43rd Int. Instrum. Symp.*, Research Triangle Park, NC: ISA, 357-369, 1997.
18. W. J. Cook and E. M. Felderman, Reduction of data from thin film heat-transfer gages: a concise numerical technique, *AIAA J.*, 4, 561-562, 1966.
19. D. G. Walker and E. P. Scott, One-dimensional heat flux history estimation from discrete temperature measurements, in R. J. Cochran et al. (eds.), *Proc. ASME Heat Transfer Division*, Vol. 317-1, New York: ASME, 1995, 175-181.
20. W. K. George, W. J. Rae, P. J. Seymour, and J. R. Sonnenmeier, An evaluation of analog and numerical techniques for unsteady heat transfer measurement with thin film gages in transient facilities, *Exp. Thermal. Fluid Sci.*, 4, 333-342, 1991.
21. D. L. Schultz and T. V. Jones, Heat transfer measurements in short duration hypersonic facilities, AGARDograph 165, 1973.
22. C. T. Kidd, C. G. Nelson, and W. T. Scott, Extraneous thermoelectric EMF effects resulting from the press-fit installation of coaxial thermocouples in metal models, *Proc. 40th Int. Instrum. Symp.*, Research Triangle Park, NC: ISA, 1994, 317-335.
23. S. R. Kidd, P. G. Sinha, J. S. Barton, and J. D. C. Jones, Wind tunnel evaluation of novel interferometric optical fiber heat transfer gages, *Meas. Sci. Technol.*, 4, 362-368, 1993.
24. ASTM Standard E598-77, Standard method for measuring extreme heat-transfer rates from high-energy environments using a transient null-point calorimeter, *Annu. Book of ASTM Standards*, 15.03, 381-387, 1988.
25. D. J. Farina, J. M. Hacker, R. J. Moffat, and J. K. Eaton, Illuminant invariant calibration of thermochromic liquid crystals, *Exp. Thermal Fluid Sci.*, 9, 1-12, 1994.
26. J. W. Baughn, Liquid crystal methods for studying turbulent heat transfer, *Int. J. Heat Fluid Flow*, 16, 365-375, 1995.
27. Z. Wang, P. T. Ireland, and T. V. Jones, An advanced method of processing liquid crystal video signals from transient heat transfer experiments, *ASME J. Turbomachinery*, 117, 184-189, 1995.
28. C. Camci, K. Kim, S. A. Hippensteele, and P. E. Poinsatte, Evaluation of a hue capturing based transient liquid crystal method for high-resolution mapping of convective heat transfer on curved surfaces, *ASME J. Heat Transfer*, 115, 311-318, 1993.
29. G. Simeonides, J. P. Vermeulen, and H. L. Boerrigter, Quantitative heat transfer measurements in hypersonic wind tunnels by means of infrared thermography, *IEEE Trans. Aerosp. Electron. Syst.*, 29, 878-893, 1993.
30. D. J. Bizzak and M. K. Chyu, Use of laser-induced fluorescence thermal imaging system for local jet impingement heat transfer measurement, *Int. J. Heat Mass Transfer*, 38, 267-274, 1995.
31. ASTM Standard E457-72, Standard method for measuring heat-transfer rate using a thermal capacitance (slug) calorimeter, *Annual Book of ASTM Standards*, 15.03, 299-303, 1988.
32. C. H. Liebert, Miniature convection cooled plug-type heat flux gages, *Proc. 40th Int. Instrum. Symp.*, Research Triangle Park, NC: ISA, 1994, 289-302.

33. M. Bomberg, A workshop on measurement errors and methods of calibration of a heat flow meter apparatus, *J. Thermal Insulation and Building Environments*, 18, 100-114, 1994.

34. M. A. Albers, Calibration of heat flow meters in vacuum, cryogenic, and high temperature conditions, *J. Thermal Insulation and Building Environments*, 18, 399-410, 1995.

35. W. Grosshandler and D. Blackburn, Development of a high flux conduction calibration apparatus, in M. E. Ulucakli et al. (eds.), *Proc. ASME Heat Transfer Division*, Vol. 3, New York: ASME, 1997, 153-158.

36. A. V. Murthy, B. Tsai, and R. Saunders, Facility for calibrating heat flux sensors at NIST: an overview, in M. E. Ulucakli et al. (eds.), *Proc. ASME Heat Transfer Division*, Vol. 3, New York: ASME, 1997, 159-164.

37. D. Holmberg, K. Steckler, C. Womeldorf, and W. Grosshandler, Facility for calibrating heat flux sensors in a convective environment, in M. E. Ulucakli et al. (eds.), *Proc. ASME Heat Transfer Division*, Vol. 3, New York: ASME, 1997, 165-171.

38. S. N. Flanders, Heat flux transducers measure *in-situ* building thermal performance, *J. Thermal Insulation and Building Environments*, 18, 28-52, 1994.

35

Thermal Imaging

Herbert M. Runciman
Pilkington Optronics

All objects at temperatures above absolute zero emit electromagnetic radiation. Radiation thermometry makes use of this fact to estimate the temperatures of objects by measuring the radiated energy from selected regions. Thermal imaging takes the process one stage further and uses the emitted radiation to generate a picture of the object and its surroundings, usually on a TV display or computer monitor, in such a way that the desired temperature information is easily interpreted by the user.

Thermal imagers require no form of illumination to operate, and the military significance of this, together with their ability to penetrate most forms of smoke, has been largely responsible for driving thermal imager development. Although thermal imagers intended for military or security applications can be used for temperature measurement, they are not optimized for this purpose since the aim is to detect, recognize, or identify targets at long ranges by their shape; thus, resolution and sensitivity are favored over radiometric accuracy.

35.1 Essential Components

All thermal imagers must have a detector or array of detectors sensitive to radiation in the required waveband, and optics to form an image of the object on the detector. In modern thermal imagers, the detector array might have a sufficient number of sensitive elements to cover the focal plane completely (a "staring array"), in the same way as a CCD television camera. Some of the most recent staring arrays can give good performance without cooling. In other imagers, the detector might take the form of a single row or column of elements, in which case a scanning mechanism is required to sweep the image across the detector array. If a single-element detector, or a very small detector array, is used, a means of

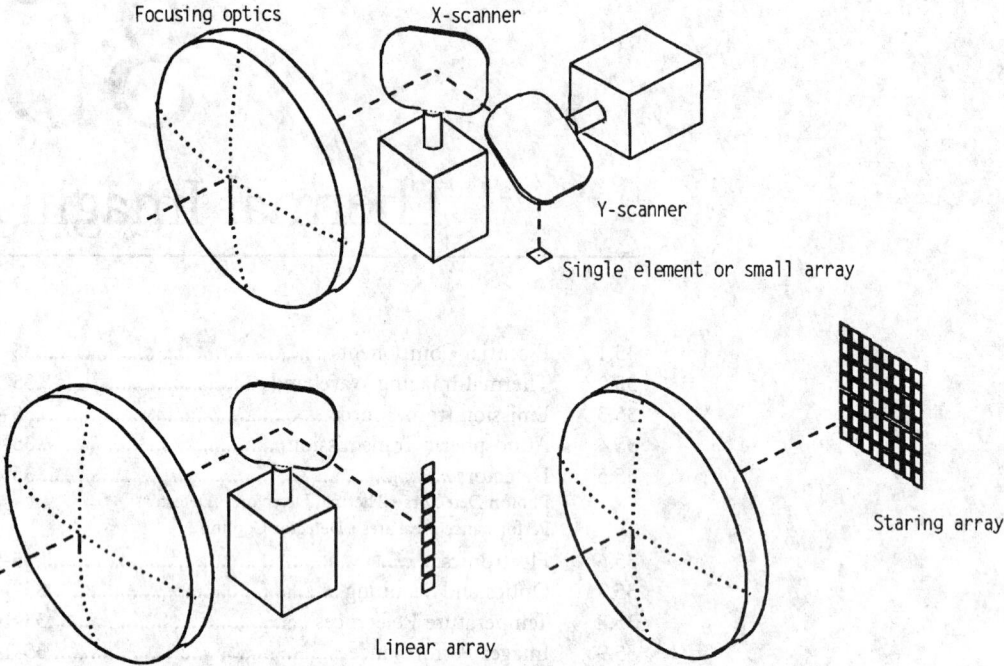

FIGURE 35.1 Thermal imaging options: (a) 2-D scanning for small detector array or single element; (b) 1-D scanning with linear detector array; (c) staring array without scanning.

providing a two-dimensional scan is required. (Figure 35.1 shows these options schematically.) For scanning imagers, it is necessary to cool the detectors (usually to about 80 K to 120 K) to achieve adequate performance.

Although in principle it would be possible to deduce target temperature from the absolute value of the detector signal, it is necessary in practice to estimate temperature by comparison with one or more reference bodies of known temperature. The temperature references are usually internal to the equipment, and accessed by mechanical movement of the reference (which may take the form of a rotating chopper) or by deflecting the optical path using a mirror.

35.2   Thermal Imaging Wavebands

The optimum waveband for thermal imaging is determined partly by the wavelength distribution of the emitted radiation, partly by the transmission of the atmosphere, and partly by the chosen detector technology.

The power radiated from a given area of an object depends only on its temperature and the nature of its surface. If the surface absorbs radiation of all wavelengths completely, it is referred to as a "blackbody." It then also emits the maximum amount possible, which can be calculated using the Planck equation (given later). Figure 35.2(a) shows the way in which blackbody emission varies with wavelength for several temperatures. It will be seen that for objects near normal ambient temperature, maximum output occurs at a wavelength of about 10 μm, or about 20 times the wavelength of visible light. At wavelengths below about 3 μm, there is generally insufficient energy emitted to allow thermal imaging of room-temperature objects. The emissivity at any wavelength is defined as the ratio of the energy emitted at that wavelength to the energy that would be emitted by a blackbody at the same wavelength.

It is important that the atmosphere should have sufficient transparency to permit the target to be observed. There are two important "atmospheric windows" — one between 3 μm and 5 μm (with a notch at 4.2 μm due to carbon dioxide absorption) and one between 7.5 μm and 14 μm. These are commonly

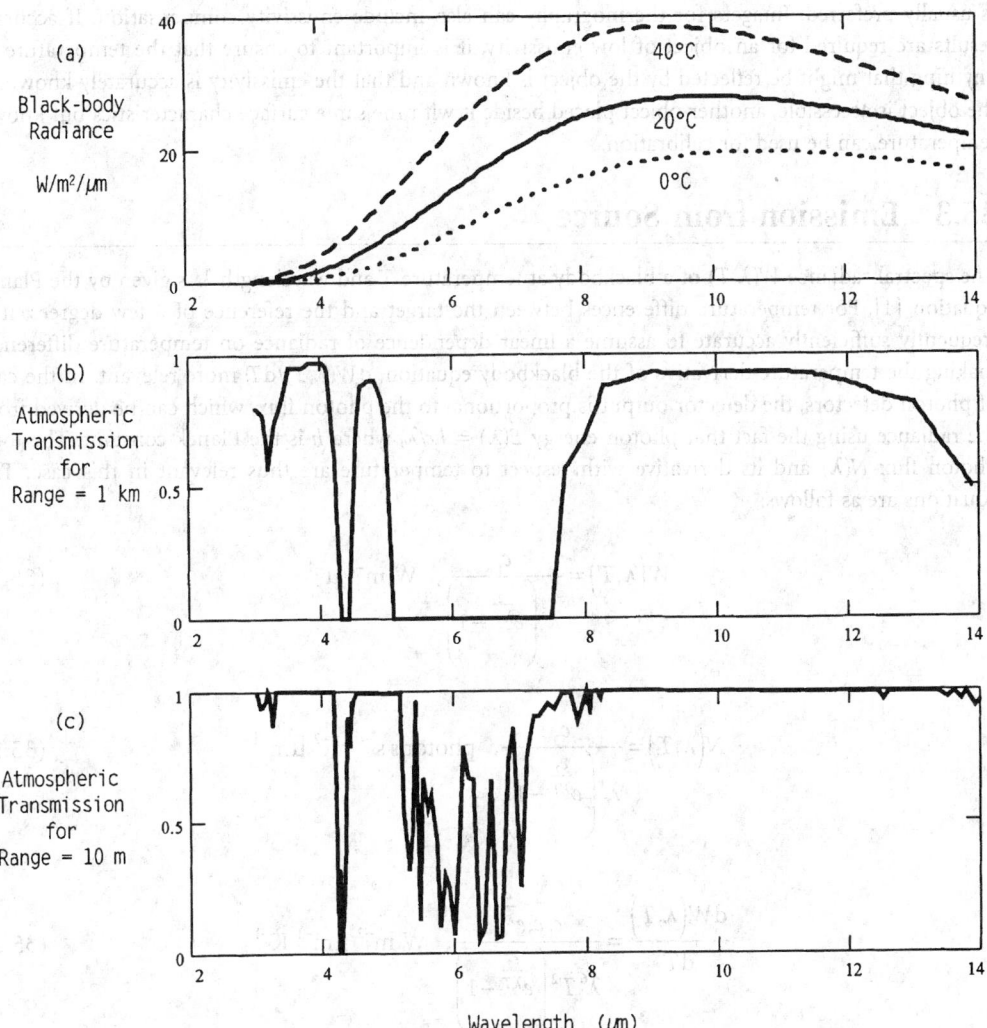

FIGURE 35.2 Factors determining thermal imaging wavebands. Imager must operate in regions where radiance is sufficiently high (a) and atmospheric transmission is good (b) and (c).

referred to as the medium-wave infrared (MWIR) and long-wave infrared (LWIR) windows, respectively. For thermal measurement over short ranges in the laboratory, it is possible to work outside these bands, but most instruments are optimized for either the MWIR or LWIR. Typical transmissions through 1 km and 10 m of a clear U.S. Standard Atmosphere are shown in Figures 35.2(b) and 35.2(c).

Emissivity for most naturally occurring objects and organic paints is high (>0.8) in the LWIR, but is lower and more variable in the MWIR. Metallic surfaces have low emissivity in both bands. Solar radiation in the MWIR is significant, and can cause errors in measurements made outdoors. These considerations favor the LWIR for quantitative imaging, but the band chosen can also be influenced by the chosen detector technology, the latter frequently being determined by cost. Scanning imagers can be used in either band, but are more sensitive for a given detector architecture in the LWIR. Cooled staring arrays give similar sensitivity in either band, but are currently more readily available in the MWIR. Uncooled staring arrays work well only in the LWIR band.

For temperature measurement, the electronics can be used to encode signal level as false color, a color scale derived from the thermal references being injected into the display to allow the user to identify the temperature of the object under examination. For general surveillance, a conventional gray-scale image is usually preferred. Imagers for thermography can also include emissivity compensation. If accurate results are required for an object of low emissivity, it is important to ensure that the temperature of anything that might be reflected by the object is known and that the emissivity is accurately known. If the object is accessible, another object placed beside it with the same surface characteristics but known temperature can be used for calibration.

35.3 Emission from Source

The spectral radiance $W(\lambda, T)$ of a blackbody at temperature T and wavelength λ is given by the Planck equation [1]. For temperature differences between the target and the reference of a few degrees, it is frequently sufficiently accurate to assume a linear dependence of radiance on temperature difference, making the temperature derivative of the blackbody equation, $dW(\lambda, T)/dT$, more relevant. In the case of photon detectors, the detector output is proportional to the photon flux, which can be derived from the radiance using the fact that photon energy $E(\lambda) = hc/\lambda$, where h is the Planck constant. The total photon flux $N(\lambda)$ and its derivative with respect to temperature are thus relevant in this case. The equations are as follows:

$$W(\lambda, T) = \frac{c_1}{\lambda^5 \left(e^{\frac{c_2}{\lambda T}} - 1 \right)}, \quad \mathrm{W\ m^{-2}\ \mu^{-1}} \tag{35.1}$$

$$N(\lambda, T) = \frac{c_3}{\lambda^4 \left(e^{\frac{c_2}{\lambda T}} - 1 \right)}, \quad \mathrm{photons\ s^{-1}\ m^{-2}\ \mu m^{-1}} \tag{35.2}$$

$$\frac{dW(\lambda, T)}{dT} = \frac{c_1 c_2 e^{\frac{c_2}{\lambda T}}}{\lambda^6 T^2 \left(e^{\frac{c_2}{\lambda T}} - 1 \right)}, \quad \mathrm{W\ m^{-2}\ \mu m^{-1}\ K^{-1}} \tag{35.3}$$

$$\frac{dN(\lambda, T)}{dT} = \frac{c_3 c_2 e^{\frac{c_2}{\lambda T}}}{\lambda^5 T^2 \left(e^{\frac{c_2}{\lambda T}} - 1 \right)}, \quad \mathrm{photons\ s^{-1}\ m^{-2}\ \mu m^{-1}\ K^{-1}} \tag{35.4}$$

where: Numerical values of the constants are:
$c_1 = 3.742 \times 10^8$
$c_2 = 1.439 \times 10^4$
$c_3 = 1.884 \times 10^{27}$

The unit of wavelength is chosen for convenience to be the micrometer (μm).

   The above values are for radiation into a hemisphere. The intensities (watts per steradian, photons per steradian, etc.) are obtained by dividing the above values by π. The actual radiances for real targets are obtained by multiplying by the spectral emissivity $\varepsilon(\lambda)$; but since target reflectivity $\rho(\lambda) = 1 - \varepsilon(\lambda)$, some caution is required. For example, a target at temperature T surrounded by a background of temperature T_b will appear to emit $W(\lambda, T)\varepsilon(\lambda) + W(\lambda, T_b)\rho(\lambda) = [W(\lambda, T) - W(\lambda, T_b)]\varepsilon(\lambda) + W(\lambda, T_b)$.

Provided that the background surrounds the target and that the target is reasonably small, the background acts as an isothermal enclosure, which can be shown [2] to behave as an ideal blackbody (i.e., $\varepsilon(\lambda) = 1$). The differential spectral radiance against background $\Delta W(\lambda)$ is thus $[W(\lambda, T) - W(\lambda, T_b)]\varepsilon(\lambda)$, which for a small temperature difference ΔT is simply:

$$\Delta W(\lambda) = \varepsilon(\lambda) \frac{dW(\lambda, T)}{dT} \Delta T \qquad (35.5)$$

The spectral emissivity of a wide variety of natural and man-made objects is also given in [2].

A major difference between thermal imaging and visual imaging is the very low contrast. In the MWIR, the contrast calculated from Equation 35.1 due to 1 K at the target is about 4%, falling to about 2% in the LWIR.

35.4 Atmospheric Transmission

Provided that the absorption bands shown in Figures 35.2(b) and 35.2(c) are avoided, atmospheric transmission can frequently be ignored in the laboratory or industrial context. For longer ranges, an atmospheric transmission model must be used or calibrating sources must be placed at the target range. The standard atmospheric transmission model is LOWTRAN [3], currently at version 7. The atmospheric transmission $T_a(\lambda)$ reduces the differential signal from the target proportionally, but has no effect on the background flux if the atmosphere is at background temperature. Atmospheric transmission in the LWIR is severely affected by high humidity, making the MWIR the band of choice for long-range operation in the Tropics. (Many gases and vapors such as methane or ammonia have very strong absorptions in the infrared, making thermal imaging a possible means of leak detection and location.)

35.5 Detectors

There are two main types of detector — photon (or quantum) and thermal. A more detailed discussion of photon detectors is given in this handbook in Chapter 8.1.1 and 8.1.2, and of thermal detectors in Chapter 6.1.8, so only aspects unique to thermal imaging are discussed here.

Photon Detectors

In photon detectors, the response is caused by photons of radiation that generate free carriers in a semiconductor, which in turn increase the conductivity (for photoconductive detectors) or generate a potential difference across a junction (for photovoltaic detectors). Photovoltaic devices have the advantage of not requiring a bias current (important to reduce the heat load on the cooling system), and they have 40% lower noise because the electric field at the junction separates the carriers, thereby eliminating recombination noise. Whether or not the lower noise is achieved in practice depends on the read-out electronics. The photon energy in the LWIR is only about 1/20th of that of a photon in the visible region of the spectrum, so a semiconductor with a much smaller bandgap than silicon must be used. The most widely used material is a compound of mercury, cadmium, and tellurium (MCT or CMT) since not only is the quantum efficiency excellent (70% or more), but the bandgap can be tuned to the desired wavelength (in either waveband) by altering the composition. Cooling of the detector to about 80 K is desirable for the LWIR, but about 120 K is acceptable for the MWIR. For the MWIR, indium antimonide (InSb) is also an excellent material; and since it is a true stoichiometric compound, it is easier to achieve good uniformity of response, but cooling to 80 K is required.

Detectors for use in scanning systems are frequently arranged so that several elements are scanned over the same part of the scene in rapid succession, the output of each element being delayed and added

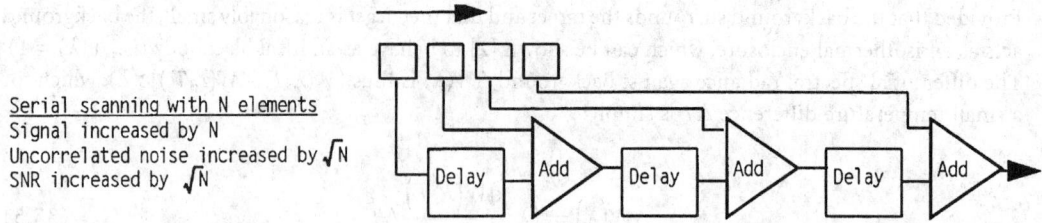

Serial scanning with N elements
Signal increased by N
Uncorrelated noise increased by $\sqrt{N}$
SNR increased by $\sqrt{N}$

FIGURE 35.3 Use of serial scanning to enhance signal-to-noise ratio. The delay times are chosen to match the speed at which the image is swept along the detector array. Serial scanning is usually combined with parallel scanning using a detector matrix.

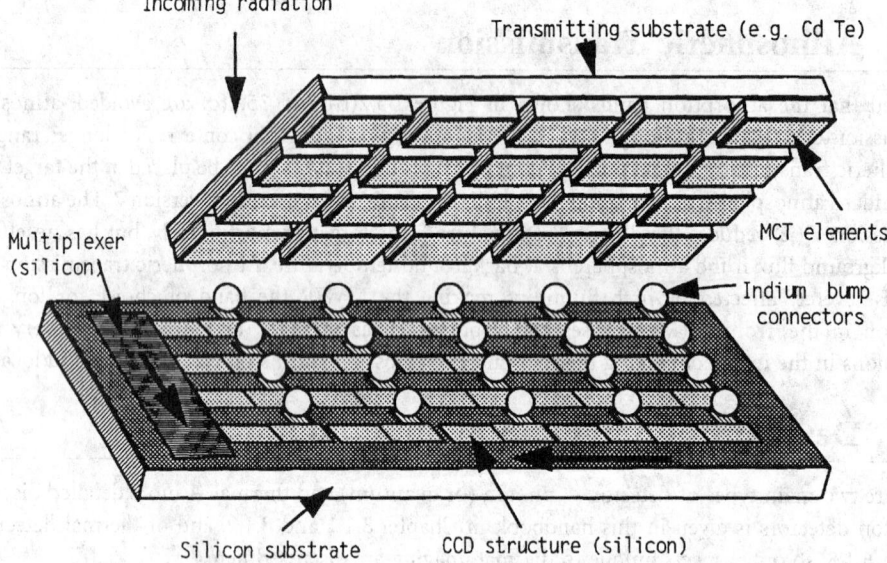

FIGURE 35.4 Typical hybrid detector construction. A typical element size is 30 μm. A large array of this type to match U.S. TV standard would have 640 × 480 elements.

to the previous one to enhance the signal-to-noise ratio (SNR). This approach (Figure 35.3) is termed serial scanning or "time-delay and integrate" (TDI) mode, and gives a theoretical gain in the SNR equal to the square root of the number of elements in TDI. It is also possible to perform TDI in the detector material itself. In the SPRITE detector (Signal Processing In The Element), the sensitive element is an elongated strip of CMT. Photons incident on the device generate carriers that drift toward a read-out electrode near one end. If the image is scanned along the detector at the same speed as the carrier drift, the signal builds up along the length of the device. The useful length is limited by carrier recombination, while diffusion of the carriers limits spatial resolution.

Large arrays of photon detectors are generally of hybrid construction, the sensitive elements being bonded to a silicon CCD or CMOS addressing circuit using indium "bumps" (Figure 35.4). An exception is the Schottky barrier detector (e.g., platinum silicide), which can be manufactured by a monolithic process, and thus tends to be lower in cost, but quantum efficiency is much lower and operation is usually limited to the MWIR band. Detector arrays and read-out architectures are discussed in depth in [2] Vol. 3, p. 246–341 and [4].

The detector assembly is encapsulated in a Dewar as shown in Figure 35.5. In front of the detector is a "cold shield" that limits the acceptance angle of the radiation to match that of the optics.

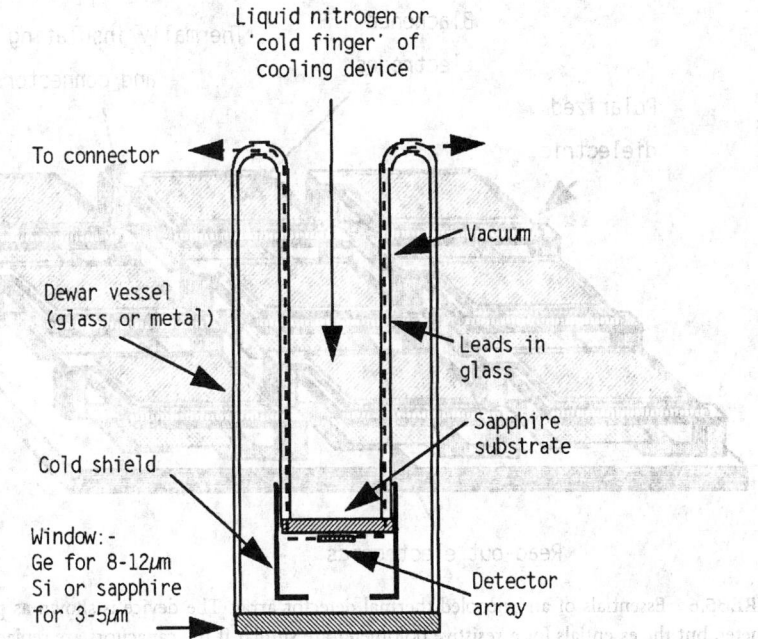

FIGURE 35.5 Construction of typical cooled detector. Cooling can be by liquid nitrogen, Joule-Thomson expansion of compressed gas, or a cooling engine.

Thermal Detectors

Thermal detectors rely on the heating effect of the incoming radiation, the change in temperature causing a change in resistance, capacitance, or electrical polarization that might be detected electrically. They are generally fairly slow in response (several milliseconds) but have the advantage that cooling is not essential (although it can be of considerable benefit with some types). The detectivity of uncooled thermal detectors is typically 1/100 that of cooled photon detectors, so real-time imaging requires the use of staring arrays. The essentials of a pyroelectric array are shown in Figure 35.6. Incoming radiation is absorbed by the blackened electrode, and the heat generated is transferred to the pyroelectric layer, which comprises a dielectric material that has been polarized by means of a high electric field during manufacture. The change in electrical polarization with temperature gives the electric signal. One of the most important parameters is the thermal isolation of the sensitive elements, so some kind of insulating support structure is necessary; and for good performance, the device must be evacuated to prevent convection. In a variant of this approach, the dielectric bolometer, the rapid variation of the dielectric constant at temperatures near the Curie point, causes the capacitance of the sensitive element to change, and hence the voltage for a constant charge. A detailed description of this approach is given in [5]. In both techniques, the detector responds only to change in temperature, so it is necessary to modulate the incoming radiation using a chopper. In the technique used initially by Wood [6] (now licensed to several manufacturers), the sensitive elements are vanadium dioxide coatings that undergo a large change in resistivity for a small temperature change. The elements are supported by silicon strips that are micromachined from the substrate and give excellent thermal insulation of the element. Changes in resistivity are read out by circuitry on the substrate, and no chopping is required, but the array must be maintained at a precise and uniform temperature.

Detector Performance Measures

The wavelength-dependent power responsivity of a detector $R(\lambda)$ is defined as the output potential or current that would result from 1 W of radiation at wavelength λ, assuming that linearity was maintained at such a high flux level. The units are $V\,W^{-1}$ or $A\,W^{-1}$. Photon responsivities in $V\,\text{photon}^{-1}\,s^{-1}$ and $A\,\text{photon}^{-1}\,s^{-1}$ are similarly defined.

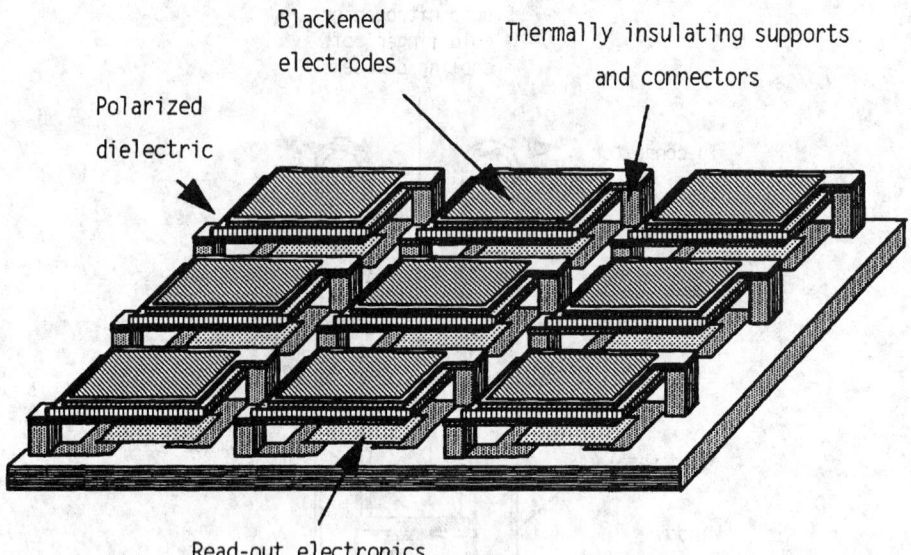

FIGURE 35.6 Essentials of an uncooled thermal detector array. The device is shown as pyroelectric or dielectric bolometer, but the essentials for a resistive bolometer are similar if the capacitors are replaced by resistors.

A thermal detector has a power responsivity that is essentially independent of wavelength, the limits of response being determined by the transparency of the window and the absorption spectrum of the element or the material used to blacken it.

In an ideal photon detector, the quantum efficiency η (defined as the number of carriers generated per photon) would be constant at all wavelengths for which the photon energy is greater than the bandgap, i.e., the photon responsivity is independent of wavelength up to the chosen cut-off wavelength. Since a given number of Watts corresponds to a number of photons proportional to the wavelength, the power responsivity (V W^{-1} or A W^{-1}) would increase linearly with wavelength until the cut-off. In practice, the cut-off is spread over about 0.5 μm and shortwave performance is modified by window transmission and antireflection coatings.

The sensitivity of a detector is limited by noise that may be due to the detector itself or due to the background radiation (as is discussed later). Noise-equivalent power $\text{NEP}(\lambda)$ is defined as the power incident on the detector at wavelength λ, which gives a signal equal to the rms noise when the measurement is made with a 1-Hz bandwidth. The NEP depends also on the modulation frequency, the latter effect being large for thermal detectors, but frequently negligible for quantum detectors. For many types of detector, the noise is proportional to the square root of the sensitive area A_d and the electrical bandwidth B, so that $\sqrt{(A_d B)}/\text{NEP}$ is constant. A performance figure that is proportional to sensitivity can then be defined as specific detectivity $D^*(\lambda) = \sqrt{(A_d B)}/\text{NEP}(\lambda)$. For historical reasons, specific detectivity is usually given in units of cm$\sqrt{\text{Hz}}$ W^{-1}, so it is important to remember to convert this to SI units or to measure detector area in square centimeters. Since noise is an electrical quantity particular to the detector under the conditions for which D^* is defined and is independent of wavelength, $\text{NEP}(\lambda)$ is proportional to $1/R(\lambda)$; so if the value of detectivity D_p^* at the wavelength of peak responsivity R_p is known, for other wavelengths $D^*(\lambda) = D_p^*(R(\lambda)/R_p)$. Sometimes, "blackbody $D^*(T)$" figures are quoted rather than D_p^*. If the blackbody temperature is T, the value of D_p^* is given by:

$$D_p^* = \frac{D^*(T)R_p \int_0^\infty W(\lambda,T)\,d\lambda}{\int_0^\infty W(\lambda,T)R(\lambda)\,d\lambda} \tag{35.6}$$

The rms noise voltage V_n is simply the NEP multiplied by the responsivity. Since the ratio $D^*(\lambda)/R(\lambda)$ is independent of wavelength, one obtains:

$$V_n = \sqrt{A_d B}\,\frac{R_p}{D_p^*} \tag{35.7}$$

The detector also affects the modulation transfer function (MTF) of the imager, defined as the ratio of the modulation depth of the signal due to a target with sinusoidally varying brightness of spatial frequency f cycles per milliradian to the modulation depth from a similar target at very low frequency. If there are no limitations due to time constant, the MTF of a detector is due to its instantaneous field of view (IFOV), which is given by IFOV = x/F where F is the focal length of the optics. For example, a lens of 500-mm focal length used with a 50-μm square detector would give an IFOV of 0.1 mrad. Then,

$$MTF_d = \frac{\sin(\pi \times f \times IFOV)}{\pi \times f \times IFOV} \tag{35.8}$$

For staring arrays, frequencies above 1/2 cycle per element pitch (the Nyquist frequency) give aliasing, and it is undesirable to rely on performance above this frequency.

Detector Cooling

The method with the lowest cost is to use liquid nitrogen poured directly into the detector Dewar, and many detector manufacturers supply detectors in Dewars with sufficient capacity for many hours of use per filling. Outside the laboratory, this technique is seldom practical, and a cooling engine (usually based on the Stirling thermodynamic cycle) is commonly used. The cost of such an engine has now fallen to a level where it no longer dominates the cost of the instrument, and power consumption is only a few watts to give cooling to 80 K. Dimensions vary widely, but a typical low-power cooling engine is about 40 mm × 40 mm × 60 mm excluding the length of the cold finger that lies inside the detector Dewar. A disadvantage is the relatively slow cool-down time (several minutes typically) and, where this is critical (as in military or security applications), Joule-Thomson cooling can be used. This method operates by expansion of air or nitrogen compressed to about 15 MPa through a nozzle, and can give cool-down times of a few seconds. Thermoelectric cooling (using the Peltier effect) can be used for temperatures down to about 200 K. Some detectors (e.g., lead selenide) have been designed to operate at this temperature in the 3 μm to 5 μm band, but thermoelectric cooling is inadequate for most types of photon detectors. A survey of cooling methods and devices is given in [2], Vol. 3, p. 345–431.

35.6 Electronics

The electronics architecture depends on the type of detector and the application, but typically electronics are required to provide bias and clocking signals to the detector, to amplify the low-level signals from the detector, to equalize the responses of the outputs from different detector elements, to provide scan conversion to a form suitable for display, and to provide image processing suitable to the application.

In scanning imagers with a small number of detector elements, the output of each detector element can be amplified continuously, the amplifier outputs then being multiplexed to give the required display. The bandwidth (important for estimating sensitivity) must be at least sufficient to accommodate the data rate, but too wide a bandwidth gives excess high-frequency noise. An approach frequently used for the measurement of sensitivity (e.g., [1], p. 167) is to make the electronics response equivalent to that of a single-pole filter with the 3-dB point placed at a frequency corresponding to $1/(2 \times$ dwell time), in which case the bandwidth is:

$$B = \frac{\pi \times \text{FOV}_h \times \text{FOV}_v \times f_f}{4E \times N \times \text{IFOV}^2} \qquad (35.9)$$

where FOV_h and FOV_v = Horizontal and vertical fields of view, respectively
$\quad\quad f_f$ = Field rate
$\quad\quad E$ = Ratio of the active scan period to the total period
$\quad\quad N$ = Number of parallel detector channels

Often high-frequency boost filters are used to compensate for optics and detector MTFs, in which case the noise bandwidth can be much increased. A commonly used criterion for the electronic filtering is to make the perceived noise independent of frequency up to the cut-off frequency of the detector.

To maintain good spatial resolution, it is desirable that the detector output be sampled at least twice during the dwell time so that the Nyquist frequency is not a major limitation [7]. Frequencies above the Nyquist frequency are changed by the sampling process into lower frequencies, a process known as aliasing. Thus, the fidelity of the image is affected so that noise that might be expected to be of sufficiently high frequency to be filtered out can appear within the passband. To avoid aliasing, a steep-cut filter is used to eliminate frequencies above the Nyquist frequency before sampling. In staring systems, the Nyquist sampling frequency is 1/2 cycle per element pitch, so a 256 × 256 element array would be limited to a resolution of 128 cycles per line. This can be overcome by using "microscan" (or μscan), in which the image is collected over a number of fields with an image shift performed optically between each field, the commonest patterns being diagonal (low implementation cost) and 2 × 2 with four fields per frame.

For large detector arrays, charge is usually accumulated on a capacitor associated with each pixel for an integration time τ (which can often be controlled independently of the frame rate or dwell time to prevent saturation), the capacitor being discharged when the pixel is read out. The effective bandwidth of such an integrator is:

$$B = \frac{1}{2\tau} \qquad (35.10)$$

With state-of-the-art amplifiers, it is generally possible to make amplifier and read-out noise less than the detector noise, the exceptions being systems of small aperture that are "photon starved" and uncooled systems.

In multichannel systems, a crucial role of the electronics is to provide channel equalization. The importance of this is due to the very low contrast of the target against the background. If the full sensitivity of an imager with a typical NETD of 0.1 K is to be realized, the difference in response between adjacent detector channels must be less than 0.4% in the MWIR or 0.2% in the LWIR. In a real-time imager where the eye integrates over several frames, the requirement is 2 to 3 times more stringent, since nonuniformity, unlike signal-to-noise ratio, is not improved by eye integration.

35.7 Optics and Scanning

The materials commonly used for visual optics are opaque in the thermal wavebands. In the LWIR, germanium is by far the most widely used material. It has a refractive index of 4, and chromatic dispersion is sufficiently low that it is frequently unnecessary to use a second material for color correction. These properties allow high performance to be achieved with few optical components, largely offsetting the relatively high cost of the material. In the MWIR, germanium has fairly high dispersion, but silicon/germanium doublets give highly achromatic performance. Zinc selenide and zinc sulfide are commonly used for color correction, some forms of the latter being transparent also in the visible band. The high refractive indices make anti-reflection coating essential to reduce surface losses — the transmission of a thin piece

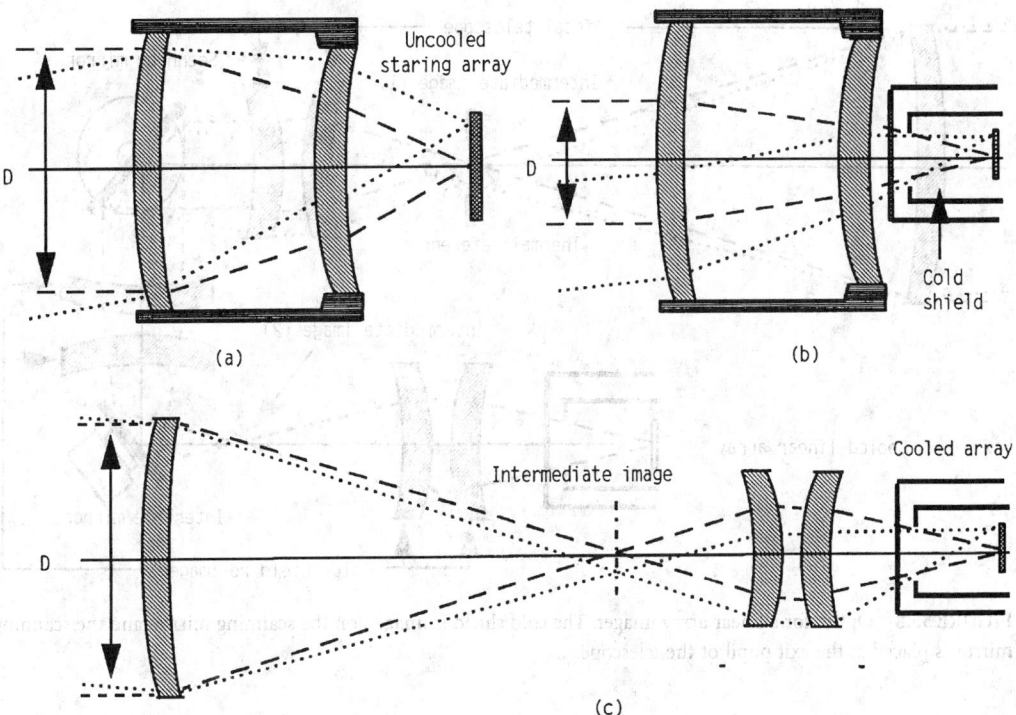

FIGURE 35.7 Optics for staring arrays. The same lens as is used for uncooled operation (a) can be used as in (b) for a cooled device, but re-imaging optics (c) are needed for a cooled imager to use the full aperture. In practice, the optics in (c) would usually have more components than shown.

of uncoated germanium is only 40%, rising to better than 99% when coated. Front-surface mirrors coated with aluminum or gold perform well in the thermal bands, and optics based on parabolic mirrors can be used if the detector array is small. For most applications, it is necessary to seal the imager to prevent ingress of dust or moisture, in which case a window is required, making reflecting optics less attractive than it first appears, since the cost of a mirror and window can be a little less than that of a lens. Plastic materials in general have poor transmission, although a thin polythene or "cling-wrap" window might be acceptable for laboratory use.

For uncooled staring arrays, the sole function of the optics is to focus an image of the scene on the detector. Good performance in the 8 μm to 12 μm band can be achieved with a two-element Petzval lens [8] with aspheric surfaces (Figure 35.7(a)). With a cooled array, it is desirable that the cold shield inside the detector should form the aperture stop of the system, since any radiation from the interior of the instrument will add to the system noise and might give shading effects. If the field of view is reasonably small, it is sometimes possible to use the same type of lens in the manner shown in Figure 35.7(b), but it can be seen that the beam diameter that can be accepted is now much smaller than the lens diameter; thus, for long-range applications requiring large beam diameters, the lens becomes very costly. The solution is to use re-imaging optics as shown in simplified form in Figure 35.7(c). The relay stage not only re-images the scene on the detector, but images the cold shield on the objective lens so that the latter need be no larger than the input beam. The intermediate image can be useful to allow insertion of temperature references or microscanning.

Scanning, when required, is most commonly performed by moving reflective surfaces since electro-optic and acousto-optic techniques either provide insufficient deflection angle or are highly wavelength dependent, causing smearing over the thermal wavebands. Rotating refractive polygons are now less used than previously. For fast line scanners, rotating reflective polygons (sometimes with curved facets) are

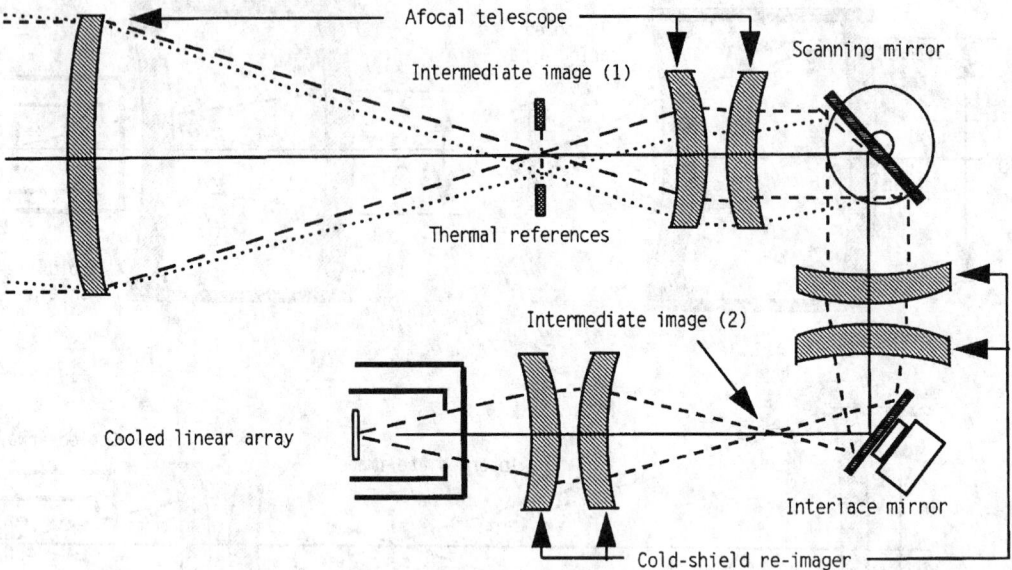

FIGURE 35.8 Optics for a linear array imager. The cold shield is imaged on the scanning mirror, and the scanning mirror is placed at the exit pupil of the telescope.

used, in some cases using gas-bearing motors operating in helium to reduce windage. For frame scanning, a plane mirror driven by a powerful galvanometer is customary. This can give very linear sweeps at 60 Hz with scan efficiencies of about 80% when operated in a closed loop. For microscan or interlace, a small image movement can be achieved by tilting a mirror using piezoelectric actuators or by tilting a refractive plate using a galvanometer. For scanning systems, it is generally necessary to re-image the detector cold shield at each scan mechanism, since otherwise the optics must be enlarged to accommodate pupil movements. Figure 35.8 shows how this is done in a typical imager using a linear array. The movement of the interlace mirror is sufficiently small that pupil re-imaging at this mirror is not required.

Figure 35.9 shows a typical arrangement for a 2-D scanner, a high-speed polygon rotor being used to generate the line scan.

The main optical parameters to be specified are transmission T_o, focal length F, and f/number $F_\#$ or numerical aperture (NA). Since IFOV = x/F, the focal length determines the spatial resolution for a given detector. The NA is defined as the sine of the semi-cone angle of the output beam from the optics. It can be shown that if a diffuse (Lambertian) source emits W W m^{-2} into a hemisphere, the radiance due to the source in the focal plane is $WT_o(\text{NA})^2$. $F_\#$ is defined as the ratio of focal length to diameter, and for a distant object NA = $1/(2\,F_\#)$, so the ratio of the irradiance in the focal plane to the source radiance is:

$$\frac{\text{Irradiance in focal plane}}{\text{Radiance from extended object}} = \frac{T_o\!\left(\lambda\right)T_a\!\left(\lambda\right)}{4F_\#^2} \tag{35.11}$$

Transmission of thermal imaging optics is typically 60% to 90%, depending on complexity. The transmission of any optics between the temperature reference and the target must be known for quantitative measurement.

The imaging performance of a lens can be indicated by the size of the spot generated in the focal plane by a distant point object, and accurate results can be obtained if the intensity as a function of the angle α from the center of the image (point spread function, PSF) is known. A more usual approach is to use

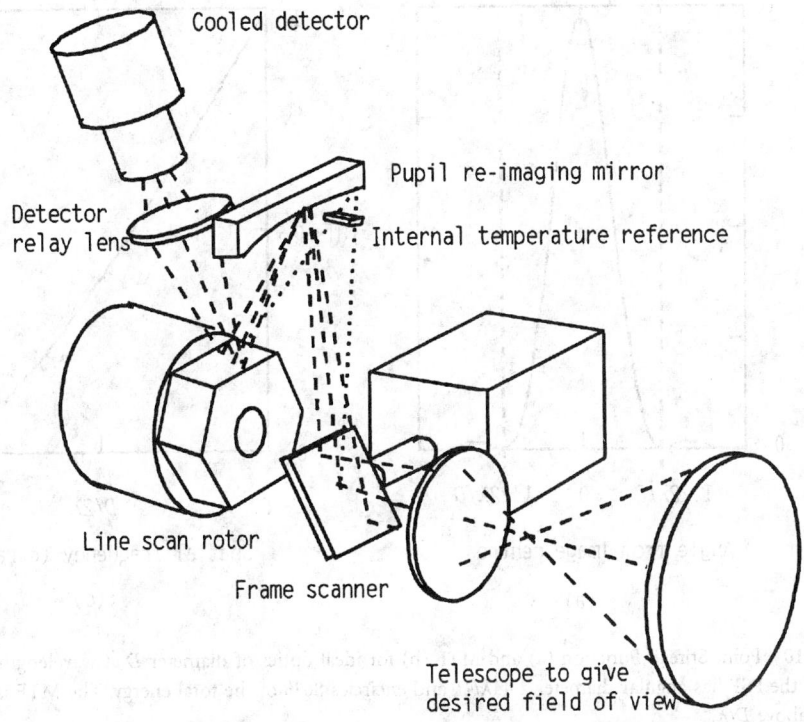

FIGURE 35.9 Simplified layout of imager using 2-D scanning (based on Pilkington Optronics HDTI). The line scan uses a high-speed polygon, and the concave strip mirror images the line scan pupil on the frame scanner.

the modulation transfer function (MTF), defined as the ratio of the contrast of the image of a target with a sinusoidal spatial variation in intensity to the contrast of the target itself. The MTF depends on the spatial frequency f of the target (expressed in cycles per radian) and on the wavelength of the radiation. For perfect optics of diameter D, diffraction gives the following values for PSF and MTF, which are also plotted in Figure 35.10:

$$\mathrm{PSF}(a) = 4\left[\left(\frac{\lambda}{\pi Da}\right)J_1\left(\frac{\pi Da}{\lambda}\right)\right]^2 \tag{35.12}$$

where J_1 is the Bessel function.

$$\mathrm{MTF_o}(f) = \frac{2}{\pi}\left\{a\cos\left(\frac{\lambda f}{D}\right) - \frac{\lambda f}{D}\sqrt{1 - \left(\frac{\lambda f}{D}\right)^2}\right\} \tag{35.13}$$

In practice, optics are frequently far from the diffraction limit, so manufacturers' figures must be used. If a thermal imager is to be used for measurement, correction for MTF will be required unless the IFOV and the PSF are both smaller than the region over which the temperature is to be measured. The signal level for a subresolution point source can be obtained by integrating the PSF over the detector area. For diffraction-limited optics, a rule of thumb for the LWIR band is that the diffraction spot diameter in mrad is the reciprocal of the lens diameter in inches (or $25/D$ when D is in millimeters).

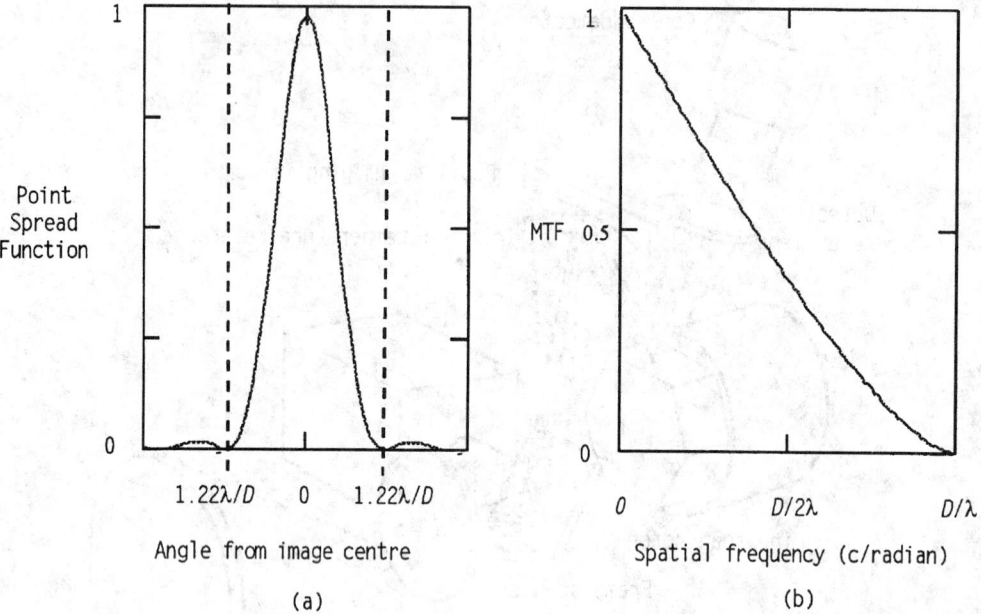

FIGURE 35.10   Point Spread Function (a) and MTF (b) for ideal optics of diameter D at wavelength λ. The first dark ring of the PSF has angular diameter $2.44\lambda/D$, and encircles 86% of the total energy. The MTF is zero for all frequencies above D/λ.

35.8   Temperature References

If a thermal imager is to be used for measurement, it is generally necessary to compare the signal from the target with that from one or more bodies at known temperature. The most precise results are obtained if two bodies at different known temperatures and having the same emissivity as the target (e.g., painted with the same paint type) are placed adjacent to the target and the target temperature is deduced by interpolation. There is then no dependence on emissivity, atmospheric transmission, and optical transmission. Outside the laboratory this is seldom practical, however, and it is necessary to use thermal references within the imager itself. The fewer optical components there are between the reference and the scene, the better will be the accuracy. If the interior of a cavity of uniform temperature is viewed through a small hole, the emission from the hole follows the Planck equation accurately, irrespective of the emissivity of the interior surface of the cavity. Although very accurate, such a blackbody cavity is usually inconveniently large, and it is more usual to use a blackened surface with deep grooves or pits to give high emissivity.

Unless the detector has only one element, thermal references are also desirable if not essential to allow the outputs of the different detector elements to be equalized.

35.9   Imager Performance

SNR and NETD

If the target is larger than the IFOV and the PSF, the signal level due to a small temperature difference is found by multiplying the source differential output, the atmospheric and optical transmissions, the geometric attenuation due to the *f*/number, the detector area, and the responsivity as derived or defined in the previous sections, and integrating over the imager passband to give:

$$V = \frac{A_d \Delta T}{4F_\#^2} \int R(\lambda) T_t(\lambda) T_a(\lambda) \varepsilon(\lambda) \frac{dW(\lambda,T)}{dT} d\lambda \tag{35.14}$$

The noise voltage V_n is given by Equation 35.7, so the signal-to-noise ratio is obtained. If the number of serial detector elements is N_s, the signal level is increased by this factor; but since the noise is uncorrelated between the elements, it increases only as $\sqrt{N_s}$, so SNR improves as $\sqrt{N_s}$. The SNR also improves as the square root of the number of parallel channels, since the dwell time is increased, giving a reduction in bandwidth B.

The sensitivity of a thermal imager for large targets is normally defined in terms of its noise equivalent temperature difference (NETD) — the temperature difference between a large blackbody at zero range and its background — which gives a signal equal to the rms noise. This is given by V_n/V when $\Delta T = 1$, $T_a = 1$ and $\varepsilon(\lambda) = 1$, and is found to be:

$$\text{NETD} = \frac{4F_\#^2}{\int_0^\infty D^*(\lambda) T_t(\lambda) \frac{dW(\lambda,T)}{dT} d\lambda} \sqrt{\frac{B}{A_d N_s}} \tag{35.15}$$

A good indication of signal-to-noise ratio is given if it is assumed that T_a and ε are constant within the imager passband, in which case SNR $= \varepsilon \, T_a \, \Delta T/\text{NETD}$.

Minimum Resolvable Temperature Difference

The performance of a thermal imager is frequently defined by its minimum resolvable temperature difference (MRTD). MRTD(f) is the temperature difference between a four-bar square test pattern of frequency f c mrad^{-1} and its background required for an observer to count the imaged bars. The test is subjective, but has the advantage of characterizing the complete system and display, including any effects of nonuniformity. MRTD is proportional to NETD, and inversely proportional to the MTF and the square root of the number of frames presented within the integration time of the eye τ_e. The constant of proportionality depends on the degree of overlap between the scan lines. Discussion of the full MRTD model is beyond the scope of this chapter, but a simple expression (based on [1] p.167) — which gives an indication of performance for an imager with square detector elements without overlap and at least two samples per IFOV, and in which the electronics bandwidth is the same as that used for NETD calculation — is:

$$\text{MRTD}(f) = \frac{3 \cdot \text{NETD} \cdot f \cdot \text{IFOV}}{\text{MTF}_o \text{MTF}_d \sqrt{\tau_e f_f}} \tag{35.16}$$

Calculation of MRTD is best performed using standard models, the most widely adopted being FLIR 92 [9] .

35.10 Available Imagers

Recent advances in detector technology are only now beginning to be incorporated in commercially available systems. The result is that product ranges are currently changing rapidly, and prices are very unstable. The prices for uncooled systems in particular are likely to drop significantly in the near future as the market size increases. An indication of cost at present is:

Military high-performance imagers $100,000–300,000
Medium-performance imagers for measurement $30,000–100,000
Uncooled imagers $10,000–30,000 (but falling rapidly)

TABLE 35.1 Typical Commercially Available Thermal Imagers

Manufacturer	Model	Data	Description
AGEMA	880 LWB	LWIR, CMT NETD = 0.07	Thermal measurement system 175 pixels (50% MTF)
Amber	Radiance1	MWIR, InSb NETD = 0.025	Compact imager for measurement 256 × 256 pixels, InSb
Amber	Sentinel	LWIR NETD = 0.07K	Uncooled compact imager 320 × 240 pixels
Cincinnati	IRRIS-160ST	MWIR, InSb NETD = 0.025	Compact imager 160 pixels/line
FLIR Systems	2000F	LWIR, CMT NETD = 0.1K	Surveillance imager >350 pixels/line
GEC Sensors	Sentry	LWIR	Uncooled low-cost imager 100 × 100 pixels
Hamamatsu	THEMOS 50	MWIR NETD = 0.2 K	Microscope, 4-μm resolution 256 × 256 pixels
Inframetrics	ThermaCAM	MWIR, PtSi NETD<0.1K	Thermal measurement system 256 × 256 pixels
Mitsubishi	IR-M600	MWIR, PtSi NETD = 0.08K	High-definition imager 512 × 512 pixels
Nikon	LAIRD-3	MWIR, PtSi NETD = 0.1K	High-definition imager 768 × 576 pixels
Pilkington Optronics	LITE	LWIR NETD = 0.2K	Hand held surveillance imager 350 × 175 pixels
Quest	TAM200	MWIR NETD = 0.05K	Microscope with probe facility Bench system with 12.5-μm resolution

Prices reflect not only performance, but ruggedness, environmental survivability, and image processing software. Military imagers in many countries are based on "common modules" that are sometimes multisourced, and that must be configured for specific applications. Table 35.1 lists some typical commercial imagers, which were selected to emphasize the wide variety of imager types, and the list must not be taken as a comprehensive survey. Some compact imagers weigh less than 2 kg, while some of the bench systems weigh over 100 kg. Accuracy of temperature measurement is not generally specified, but ±2 K or ±2% is a good figure for a calibrated imager. The information is based on published brochures, and there is every likelihood that improved models will be available by the time this book is published. It is strongly recommended that prospective purchasers should contact as many manufacturers as possible to obtain specifications and prices.

35.11 Performance Trade-offs

The expression for NETD presented earlier is appropriate for evaluation of existing systems where D^* is known. To predict performance of future systems based on photon detectors, and to carry out design trade-offs, it is important to appreciate that D^* is very dependent on conditions of use and on waveband. One reason is that the photons from an object of given temperature, although having a well-defined average flux, are emitted at random times. Statistical theory shows that if on average N photons are collected within a given time interval, the standard deviation is $\sqrt{N}$ if N is large enough to make the distribution Gaussian (as is almost always the case in the infrared due to the high background flux). This "photon noise" frequently predominates, in which case the detector is referred to as BLIP (background-limited photodetector). The D^* is then determined more by the conditions of use than by the detector itself. If the efficiency of the cold shield η_c is defined as the ratio of the effective f/number for the background flux to that of the signal, the number of background electrons generated within an integration time τ is given by:

$$N_e = \frac{\tau A_d}{4 F_\#^2 \eta_c^2} \int_0^\infty N(\lambda, T) \eta(\lambda) d\lambda \qquad (35.17)$$

For the reasons discussed above, the rms electron noise for a BLIP device within time τ is simply the square root of the above figure. If the detector also has read-out noise of N_n electrons rms, this could be regarded as the noise that would be caused by a background flux that caused the generation of N_n^2 photoelectrons, so the noise can be written as:

$$\text{Noise} = \sqrt{N_e + N_n^2} \text{ electrons rms} \qquad (35.18)$$

The signal due to a temperature difference of 1 K between target and background expressed in electrons within the integration time is:

$$\text{Signal} = \frac{\tau A_d}{4 F_\#^2} \int_0^\infty T_t(\lambda) T_a(\lambda) \eta(\lambda) \frac{dN(\lambda, T)}{dT} d\lambda \qquad (35.19)$$

Thus, the signal-to-noise ratio for $\Delta T = 1$ K is found, NETD being simply 1/SNR when $T_a = 1$.

For staring systems, and some scanning systems, the integration time is limited by saturation. If the maximum number of electrons that can be stored is N_m, we find that the maximum integration time is:

$$\tau_m = \frac{4 F_\#^2 \eta_c^2 N_m}{A_d \int_0^\infty N(\lambda) \eta(\lambda) d\lambda} \qquad (35.20)$$

When the integration time is controlled to restrict the number of electrons generated within the sampling interval to τ_m, the noise is given by letting $N_e = N_m$ in Equation 35.18, so is independent of *f*/number. If one also substitutes τ_m for τ in Equation 35.19, and divides Equation 35.19 by Equation 35.18, to obtain SNR, one obtains:

$$\text{SNR} = \frac{\Delta T \, \eta_c^2 \, N_m \int T_o(\lambda) T_a(\lambda) \eta(\lambda) \frac{dN(\lambda, T)}{dT} d\lambda}{\sqrt{N_m + N_n^2} \int N(\lambda) \eta(\lambda) d\lambda} \qquad (35.21)$$

The ratio of integrals is effectively the contrast of the scene behind the optics, which is further reduced by the inefficiency of the cold shield; thus, performance depends mainly on image contrast at the detector and the number of electrons stored. (Quantum efficiency essentially cancels out; and for a saturated device, $\sqrt{N_m}$ is usually greater than N_n.)

The above formulae are sufficiently general to allow trade-offs for photon detectors to be carried out against aperture, waveband, range, and frame rate, the main uncertainty being the read-out noise. For staring array detectors with processing on the focal plane, a typical figure is $N_n = 1000$, but much better (and worse) values are possible. The SNR and NETD values obtained in this way are for a single pixel. If the samples overlap spatially, the SNR is improved by the square root of the overlap factor. In practice, nonuniformity of the detector array will introduce spatial noise that will affect MRTD; and even after

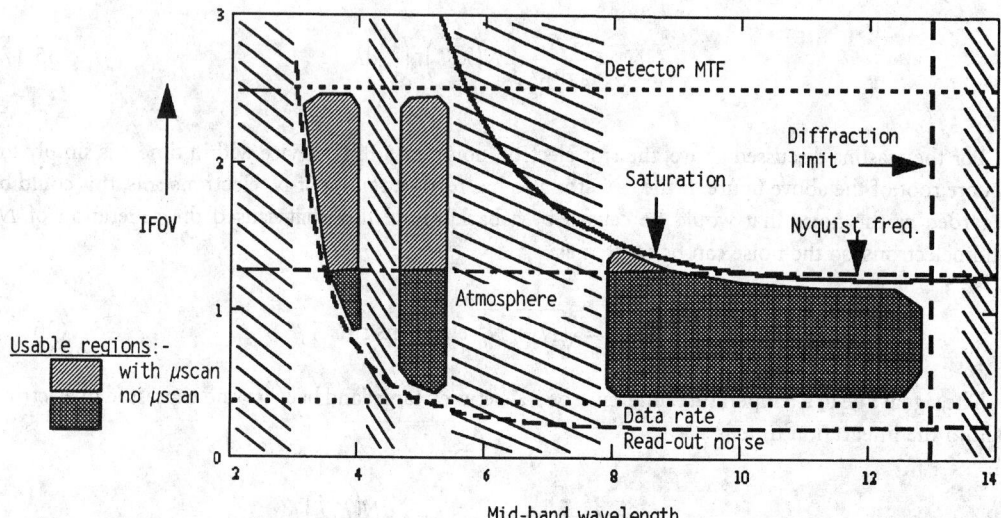

FIGURE 35.11 Physical limitations to thermal imaging when a specified spatial frequency must be resolved. The shaded regions indicate the combinations of wavelength and IFOV for which the task can be achieved. All the lines on the graph depend on system parameters such as aperture, quantum efficiency, read-out noise, and electron storage, but can be calculated from the equations given in the text.

electronic correction, this is frequently the main limitation on performance. Nonuniformity effects are discussed in [10] and [11].

The resulting trade-offs are discussed at some length in [12]. It is observed that if the integration time is controlled to prevent saturation, NETD is independent of *f*/number and quantum efficiency, but depends strongly on image contrast. This tends to favor the MWIR band, and makes high cold-shield efficiency crucial. The problem can be overcome in the LWIR by operating at a high frame rate, which has no effect on NETD but improves MRTD due to eye integration. For long-range applications, it is frequently necessary to use long focal lengths, giving large *f*/numbers if the optics diameter is to be kept within reasonable bounds. Under these conditions, saturation is less likely, and the high photon flux in the LWIR is necessary to overwhelm the read-out noise. If high spatial resolution rather than sensitivity is required, the MWIR band has a much better diffraction limit. Figure 35.11 indicates the ways in which the various limitations combine to define the combination of IFOV and waveband that gives optimum performance when it is necessary to resolve a given spatial frequency; the positions of the lines on this chart are of course peculiar to the system under investigation.

35.12 Future Trends in Thermal Imaging

At the high-performance end of the market, the developments are most likely to be driven initially by military requirements. Many countries already have thermal imaging common module programs, and many developments will be aimed at productionizing these modules to reduce cost and enhance performance. Multispectral instruments to aid in camouflage penetration and to broaden the conditions of operability through the atmosphere can be taken out of the laboratory into service. Currently, these are mainly scanning devices using adjacent long linear arrays of LWIR and MWIR detectors. For spectral agility, detector arrays based on multiquantum-well (MQW) technology can be used, since they can be tuned to some extent by varying the electrical bias. Multispectral refracting optics are complex and have poor transmission, so reflecting optics are generally used. Multispectral systems will allow more accurate compensation for emissivity, but their high cost might limit their use except for high-value applications such as earth resources surveys in aircraft or satellites. Techniques employed include frame-sequential filtering and imaging Fourier transform spectroscopy.

TABLE 35.2 Companies That Make Commercial Thermal Imagers

AGEMA Infrared Systems Inc.
550 County Avenue
Secaucus, NJ 07094
Tel: (201) 867-5390

Amber Engineering Inc.
57566 Thornwood Drive
Goleta, CA 93117-3802
Tel: (800) 232-6237, Fax: (805) 964-2185

Cincinnati Electronics Corporation
Detector and Microcircuit Devices Laboratories
7500 Innovation Way
Mason, OH 45040-9699
Tel: (513) 573-6275, Fax: (513) 573-6290

Hamamatsu Photonic Systems
360 Foothill Rd.
Bridgewater, NJ 08807-0910
Tel: (908) 231-1116, Fax: (908) 231-0852

Inframetrics Imaging Radiometer Group
12 Oak Park Drive
Bedford, MA 01730
Tel: (617) 275-8990, TWX: (710) 326-0659

Mitsubishi Electronics America Inc.
5665 Plaza Drive, P.O. Box 6007
Cypress, CA 90630-0007
Tel: (800) 843-2515

Nikos Corporation (Nikon)
1502 West Campo Bello Drive
Phoenix, AZ 85023
Tel: (602) 863-6182

Pilkington Optronics Inc
7550 Chapman Ave.
Garden Grove, CA 92841
Tel: (714) 373-6061, Fax: (714) 373-6074

Quest Integrated Inc.
21414-68th Avenue South
Kent, WA 98032
Tel: (206) 872-9500, Fax: (206) 872-8967

For commercial thermal imaging, the main thrust is likely to be cost reduction via the use of uncooled imagers, and in packaging to provide increased ease of use and functionality using in-built software. Optics cost can be reduced or performance improved by the use of hybrid aspheric components, which are now practical due to improvements in diamond turning. A diffractive surface is generated on one surface of a refracting lens. Because the power of a diffractive component is proportional to wavelength, a very low diffractive power can compensate for the chromatic aberration of the lens. This allows the use of materials such as zinc sulfide, which would otherwise be unacceptable due to chromatic aberration, but which have other desirable properties such as low cost or a low thermal coefficient of refractive index.

References

1. J.M. Lloyd, *Thermal Imaging Systems*, New York, Plenum Press, 1975, 20-21.
2. J.S. Acetta and D.L. Shumaker (eds.), *The Infrared & Electro-Optical Systems Handbook*, Bellingham, WA: SPIE Optical Engineering Press, 1993, Vol. 1, 52, 251-254.
3. F.X. Kneizys, E.P. Shettle, L.W. Abreu, G.P. Anderson, J.H. Chetwynd, W.O. Gallery, J.E.A. Selby, and S.A. Clough, *Users' Guide to LOWTRAN 7*, Report no. AFGL-TR-88-0177 Hanscom, Air Force Geophysics Laboratory, 1988.
4. M. Bass (ed.), *Handbook of Optics*, New York, McGraw-Hill, 1995, Vol. 1, chap. 23.
5. C.M. Hanson, Uncooled ferroelectric thermal imaging, in *Proc. SPIE 2020, Infrared Technology XIX*, 1993.
6. R.A. Wood, Uncooled thermal imaging with monolithic silicon focal plane arrays, *Proc. SPIE 2020, Infrared Technology XIX*, 1993, 329.
7. G.C. Holst, *Testing and Evaluation of Infrared Imaging Systems*, Winter Park, FL, JCD Publishing, 1993, 36-42.
8. M.J. Riedl, *Optical Design Fundamentals for Infrared Systems*, Bellingham, WA, SPIE Optical Engineering Press, 1995, 55.

9. L. Scott and J. D'Agostino, NVEOD FLIR92 thermal imaging systems performance model, *Proc. SPIE, Vol. 1689, Infrared Imaging*, 1992, 194-203.

10. A.F. Milton, F.R. Barone, and M.R. Kruer, Influence of nonuniformity on infrared focal plane performance, *Opt. Eng.*, 24, 855-862, 1985.

11. H.M. Runciman, Impact of FLIR parameters on waveband selection, *Infrared Phys. Technol.*, 37, 581-593, 1996.

Further Information

G. Gaussorgues, *Infrared Thermography*, London, Chapman and Hall, 1994.

F. Grum and R.J. Becherer (eds.), *Optical Radiation Measurements*, New York, Academic Press, 1979, Vol. 1.

G.C. Holst, *Electro-optical Imaging System Performance*, Winter Park, FL, JCD Publishing, 1993, 36-42.

C.L. Wyatt, *Radiometric Calibration: Theory and Methods*, New York, Academic Press, 1978.

36

Calorimetry Measurement

Sander van Herwaarden

Xensor Integration

36.1 Heat and Other Thermal Signals

Calorimetry is the science of measuring heat or thermal signals. This chapter begins by explaining thermal and other signals. For a correct understanding of calorimetry, knowledge of thermodynamics is necessary; however, space permits an introduction only. An overview of the most important types of calorimeters, followed by some of the many applications, will then be presented. For those who want to do measurements without having to make their own devices, an overview of the most important calorimeters, and their vendors and approximate prices will be given. The chapter concludes with some hints on present and future developments, and on further reading, since there is a large amount of literature on calorimeters.

Signal Domains

In measurement science, physical quantities can be distinguished by six different so-called signal domains [1]. All signals in calorimeters are either thermal signals or other signals that are transduced into thermal signals. The thermal signal (heat, for example) is then transduced into an electrical signal. In addition to the domains of thermal and electric signals, there is also the domains of the chemical, the mechanical, the magnetic, and the radiant signals. Calorimetric measurement of signals occurs from all domains [2]. Many of them are described in different sections of this Handbook. Calorimeters usually measure the thermal effects of (bio)chemical or mechanical processes, or they measure the thermal effect of temperature changes on matter.

Heat and Temperature

One of the forms in which energy can be present in a system is the random, internal kinetic energy of the particles (molecules or atoms) of a system, which can intuitively be called "thermal energy." This is to be distinguished from the average, external movement of a system of particles as a whole, which can be called the "mechanical energy" of the system. For gases, thermal energy is closely related to the random velocity of the molecules; and in the case of multi-atom molecules, the rotations and vibrations of the atoms within the molecules. The zeroeth law of thermodynamics states that, if two systems are each in equilibrium with a third system, they will be in equilibrium with each other. They are said to have the same temperature. For gases, statistical mechanics shows the direct relation between the thermal energy (or heat) stored in the system and temperature [3]. This law, however, also applies to liquids and solids, although the quantitative relation between thermal energy (agitation of the particles) and temperature is less straightforward and not so easy to calculate as for gases. Temperature and thermal energy distribution can be viewed as the result of statistical processes, such as diffusion. Diffusion ensures that if ever there is a surplus of thermal energy (i.e., of fast molecules or electrons, or a higher density of phonons) in some area, some of it will flow toward areas with a lower thermal energy density until thermal equilibrium has been established. This flow of energy P (in $W = J\,s^{-1}$) is called heat flow. Heat flow is therefore the transfer of (thermal) energy from one (part of a) system to another. Note that heat is not conserved, because a change in thermal energy of a system may be achieved by heat exchange with the environment, but also by mechanical interaction, and heat can be used to do mechanical work (motor!) just as well as heating up a system. Heat can be viewed as "thermal energy on the move," but whether it ends up as thermal, mechanical or another form of energy depends on the circumstances.

Work and Enthalpy

The first law of thermodynamics is also of importance for calorimetry. In a simple form, in the absence of other forms of energy exchange, it states that the change in internal energy ΔU of a system is equal to the heat Q supplied by the ambient to the system, minus the work (mechanical energy) $p\Delta V$ done by the system on the ambient:

$$\Delta U = Q - p\Delta V \tag{36.1}$$

This is of importance when measuring the specific heat capacity c (in $J\,kg^{-1}\,K^{-1}$) of matter. This measurement can be performed under two conditions: at constant volume V and at constant pressure p. At constant volume, no work will be done by the system, since $p\Delta V = 0$. So, the specific heat capacity at constant volume c_v is simply the change in internal energy. The specific heat capacity of a system at constant pressure c_p is (usually) higher, since additional energy (heat) is needed to perform work on the ambient: $p\Delta V$. The quantity combining these contributions to the energy at constant pressure is called

TABLE 36.1 Overview of the Classification Criteria for Calorimeters

Relation to surroundings	Heat measurement
Isothermal	Phase-transition compensation of heat
	Thermoelectric compensation of heat
Adiabatic	Measurement of temporal temperature difference
Isoperibol	Measurement of spatial temperature difference

enthalpy H, and the change in enthalpy vs. temperature (in J K^{-1}) at constant pressure is the specific heat capacity C_p of a sample at constant pressure: $C_p = (\mathrm{d}H/\mathrm{d}T)_p$.

36.2 Calorimeters Differ in How They Relate to Their Surroundings

In essence, a calorimeter performs three functions: it encloses a chamber in which a thermal experiment is carried out; it measures the heat exchange between the sample under test and the calorimeter (and often other quantities are being measured as well, such as temperature and amount of substance); and it thermally separates the experimental chamber from its surroundings. There exist many types of calorimeters. They all have an experimental chamber. Apart from this, four essential criteria can be used to classify calorimeters [4, 5]. The first criterion is, what does the calorimeter do with the heat that is generated (or absorbed) by the experiment? The second criterion is, how does the calorimeter relate to its surroundings? The third criterion is, does it measure a single experimental sample, or is it a twin design with a reference compensating for common mode errors? The fourth criterion is, does the calorimeter function at a fixed temperature, or can it scan a temperature range? Table 36.1 lists the various possibilities for the relation to the surroundings, and for the heat measurement. In principle, almost every combination is possible. In practice, some combinations naturally go together because they compensate for their strengths and weaknesses. For example, calorimeters in which the experimental temperature is scanned often use a twin configuration to eliminate the common mode errors arising from the continually changing temperature.

Isothermal Calorimeters

In the isothermal calorimeter, the experiment is always kept at a fixed temperature. This is attained by instantly removing (or supplying) any heat that the experiment releases (or absorbs). The isothermal calorimeter was the first to be developed. In 1780, Lavoisier and Laplace made the "ice calorimeter" in which the heat generated by the experiment is used to melt ice into water. If enough ice is available, the calorimeter will remain at 0°C, regardless of the progress of the experiment. The experimentally generated heat Q (J) is calculated by weighing the mass m (kg) of the melt water and multiplying by the heat of the ice-water transition q_{fus} (J kg^{-1}):

$$Q = m\, q_{\text{fus}} \tag{36.2}$$

The experiment with the melting ice is enveloped within a thermostat, which consists of a double-walled vessel with melting ice between the walls, which is always at 0°C as well. Thus, the isothermal calorimeter has a perfect thermal isolation between experiment and surroundings in the form of a second guard-vessel that buffers all the heat coming from the surroundings.

Phase-Transition or Thermoelectric Compensation of the Heat

In the isothermal calorimeter, the heat generated by the experiment is immediately absorbed by the calorimeter. This can be accomplished by phase-transition compensation of the heat, e.g., by melting of solids or evaporation of liquids. But nowadays, compensation by electrical means is preferred because it can be measured so easily. Thus, heat Q (J) absorbed by the experiment is replaced using Joule heating (dissipation of heat by a current I (A) through an electric heater with resistance R_h (Ω)), while heat generated by the experiment is absorbed by Peltier coolers:

$$Q = \int I^2 R_h dt \tag{36.3}$$

36.3 Adiabatic Calorimeters Often Measure Time-Dependent Temperature Differences

In the adiabatic calorimeter, no heat exchange with the surroundings is allowed, and all the heat generated by the experiment is used to increase the temperature of the calorimeter. The amount of heat Q (J) generated follows from the temperature increase ΔT (K), divided by the heat capacity of the calorimeter C_c (K J^{-1}):

$$Q = \Delta T / C_c \tag{36.4}$$

The absence of heat exchange with the surroundings of the calorimeter is obtained by immersing the experimental chamber of the adiabatic calorimeter in an outer vessel; see Figure 36.1. The temperature of the outer vessel is kept at the same (increasing) temperature as the experimental chamber by means of electronic feedback, heating the outer vessel to maintain a practically zero temperature difference. In the adiabatic calorimeter, one must wait a few minutes after the experiment has finished to allow the heat to spread itself uniformly over the chamber and to obtain the final temperature. In Figure 36.1, the experimental chamber is a so-called "calorimetric bomb" in which fuel is fully burned to measure its heat of reaction. The bomb is immersed in a vessel filled with water — the inner vessel. A stirrer assures fast heat exchange between the bomb and the inner vessel, thermometers measure the temperature of the inner vessel and the outer guard vessel. The outer vessel is regulated to the inner-vessel temperature by means of electrical heaters and refrigerator coolers.

Isoperibol and Heat Flux Calorimeters

The term *isoperibol* was devised to indicate a calorimeter having "uniform surroundings." In this calorimeter, the outer shell provides a reference temperature, and customarily, the experiment starts at the same temperature. The experimental chamber is connected to the outer shell by a well-defined thermal conductance. Any heat generated by the experiment will cause a well-defined temperature difference ΔT (K) across the thermal conductance G_{th} (W K^{-1}), and this temperature difference is subsequently measured as a "local temperature difference." Calorimeters utilizing this way of measuring the heat are referred to as "heat flux calorimeters," and often measure the power P (J s^{-1} or W) generated by the experiment, rather than the energy:

$$P = \Delta T \, G_{th} \tag{36.5}$$

Scanning Calorimetry Sweeps the Experiment Temperature

While calorimeters are over 200 years old, a recent innovation is that of the *scanning calorimeter*. In the scanning calorimeter, the experimental chamber is not kept at (approximately) one temperature, but it is swept over a temperature range. The temperature is increased at given rate (e.g., 10 K min^{-1}) by electric heating, or decreased by, for example, cooling with liquid nitrogen. To achieve the temperature sweep, the experiment is placed inside a computer-controlled oven. To compensate for common-mode errors

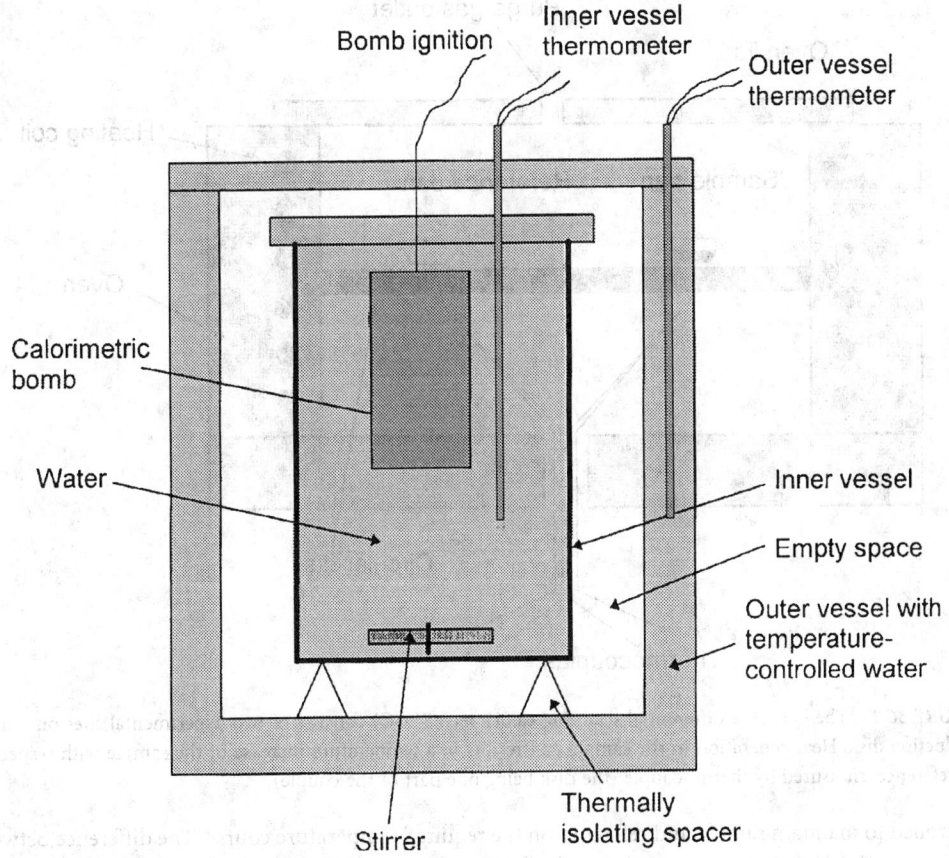

FIGURE 36.1 Adiabatic calorimeter, with inner experimental vessel being heated by a chemical reaction in the calorimetric bomb, and outer guard vessel electronically adjusted to the temperature of the inner vessel to prevent heat loss of the inner vessel. The temperature increase of the inner vessel is a measure of the heat of burning of fuel in the bomb.

(i.e., back-logging or deviation of the sample temperature from the oven temperature, and imperfections in the oven temperature profile with respect to time and location), scanning calorimeters are often built with twin experimental sites for which the difference is measured. One site is for the sample under test; the other site serves as reference, which is either left empty or contains material resembling the sample under test as much as possible, except for the phenomena to be measured. Such calorimeters are called *differential scanning calorimeters*, DSCs (see "Further Information" for a general book on DSC). Three different types of DSCs are common. The first type is the DSC based on heat flux measurement; see Figure 36.2. In this, the reference and the sample are both heated by the oven through a thermal resistance (gaseous or a solid circular disk). In Figure 36.2, the heating block is the oven, while the (constantan) disk is the path from oven to sample and reference, assuring both heat conduction from oven to sample/reference. The disk also forms a well-defined heat resistance from sample to oven, to measure the heat generation and absorption in the sample using thermocouples that measure the resulting temperature differences. Because of the small size of the experimental site and sample (typically less than 1 cm diameter and 0.1 mL volume), the DSC is usually very fast, and the temperature curve is an accurate representation of the momentaneous heat production in the sample. In the second type of DSC — the power compensation DSC — the temperature of the sample and the reference are both measured with platinum resistors. Sample and reference each have an individual heating source, which is electronically

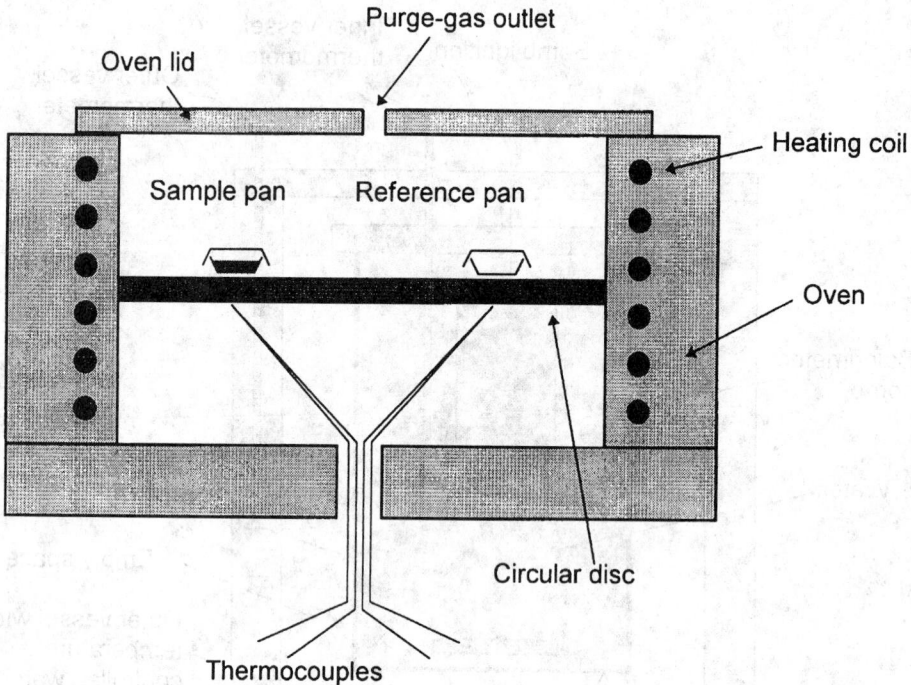

FIGURE 36.2 The heat flux differential scanning calorimeter (DSC) consists of two experimental sites on a heat-conducting disk. Heat generation in the sample pan results in a temperature increase of the sample with respect to the reference, measured by thermocouple (the disk being one part of the couple).

controlled to maintain sample and reference on the required temperature course. The difference between reference and sample heating power immediately gives the experimental heat. In the third (less accurate) type, only the temperature at which phenomena occur is registered. This type can be used for bigger samples (2 to 3 mL against 0.1 mL for the heat-flux DSC), or at very high temperatures up to 2000°C.

Converting the Measured Curve to the Actual Progress of the Experiment

Due to the thermal resistance and the heat capacitance inside the experimental chamber of a calorimeter, there is a time constant associated with distributing the experimental heat over the chamber, and reaching the final temperature. A heat pulse does not create a temperature pulse, but rather a smeared-out curve. In a DSC, this smearing-out is usually not significant because of the small time constant (1 to 3 s). For some other instruments, so-called *curve desmearing* has to be carried out using convolution integrals to extract all the information from the measurement curve. With commercially available instruments, software is often supplied that does the job for you, but the accuracy of this software is not satisfactory in all cases, so one might need to do some further study on desmearing [4].

Calibration of Calorimeters

The inaccuracy of the calorimeter can be reduced by *calibration*. Heat production can easily be simulated electrically. Unfortunately, the thermal leakage of the electric leads reduces the accuracy of this method. More common, therefore, is the use of reference materials with a known heat of transition or reaction for calibration. For bomb calorimeters, the response can be calibrated using the heat of reaction of benzoic acid, which has been carefully determined to be 26,457 J g^{-1} [6]. For a DSC, a range of materials can be used for calibration. This is due to the wide range of applications, and also to the wide range of temperatures used. Generally, one or more metals are used for calibration of both the power and the temperature scale; for example, indium, which has a melting point of 156.6°C, and an enthalpy of melting of 28.6 J g^{-1}. The uncertainty of the value for the heat of fusion (in this case, of melting) is about 0.5%. This is the limit when calibrating in this manner. Basically, DSC inaccuracy is around a few percent. For up-to-date details on calibration and the calibration materials, please consult [7, 8], the references given there, and the most recent data published, since calibration procedures for the DSC are still improving. Specific heat capacity measurements are often performed in three steps. First, the baseline-offset of the DSC is measured with empty reference and empty sample sites (one measures the difference between sample and reference baseline, i.e., the systematic asymmetry of the instrument). Then, a measurement of a reference material with accurately determined specific heat capacity is made; for this, sapphire (crystalline Al_2O_3) is often used. Finally, a measurement of the unknown sample is carried out. By correcting for the baseline-offset, and division of the measurements of the known and the unknown samples, the specific heat capacity of the unknown sample can be determined. Currently, these corrections are all made by the computer controlling the DSC [9].

36.4 Typical Applications of Calorimeters

Specific Heat, Transition Heat and Temperatures

Specific heat capacity can be measured by accurate point-wise measurements, but measurement with a DSC is much more efficient. With the DSC, one also obtains the temperature and heat of phase transitions by enthalpy measurement at the transition temperature. Many materials also exhibit changes in crystallization at some given temperature (e.g., glass transition points). Also, these are transition points with their specific heat and temperature, and these can be measured with DSC as well.

Analysis of Chemical and Physical Reactions

Calorimetry is very well suited for analysis of chemical reactions. In particular, the heat of exothermic and endothermic reactions can be determined. An important example is the calorific value of fuels. Households and factories buying fuels are primarily interested in the calorimetric value of their purchases. The so-called oxygen bomb calorimeter is indispensable for primary calibration in this application. The oxygen bomb is usually an adiabatic calorimeter, in which a sample of the fuel (such as solid coal, liquid oil, or gaseous methane) is brought together with an excessive amount of oxygen (e.g., at a pressure of 30 bar). Then, the mixture is ignited, and the heat of burning will spread over the bomb, until equilibrium is reached. The temperature increase of the bomb, divided by the heat capacity of the device, gives the heat of reaction. In general, corrections have to be made for volume and pressure changes (work!) as liquids or solids are burned and converted to gases and liquids, and also for the additional heat capacity and for the transition heat of the reaction products. Of course, chemical reactions other than burning of fuel can be analyzed in this way as well. In a slightly modified version, the adiabatic calorimeter can also serve to analyze, for example, weak bases and weak acids that do not easily respond to other analysis methods. Similarly, the heat of mixing two solutions, of dissolving a solid in a solvent, of diluting a mixture, and even the heat of wetting can be determined in this so-called *solution calorimeter*.

36.5 Thermal Analysis of Materials and Their Behavior with Temperature

Apart from directly measuring chemical reactions, one can also analyze materials by exposing them to a temperature sweep, either in an inert atmosphere (nitrogen or helium) or in an oxidizing atmosphere.

Schematic Diagram STA 409 EP

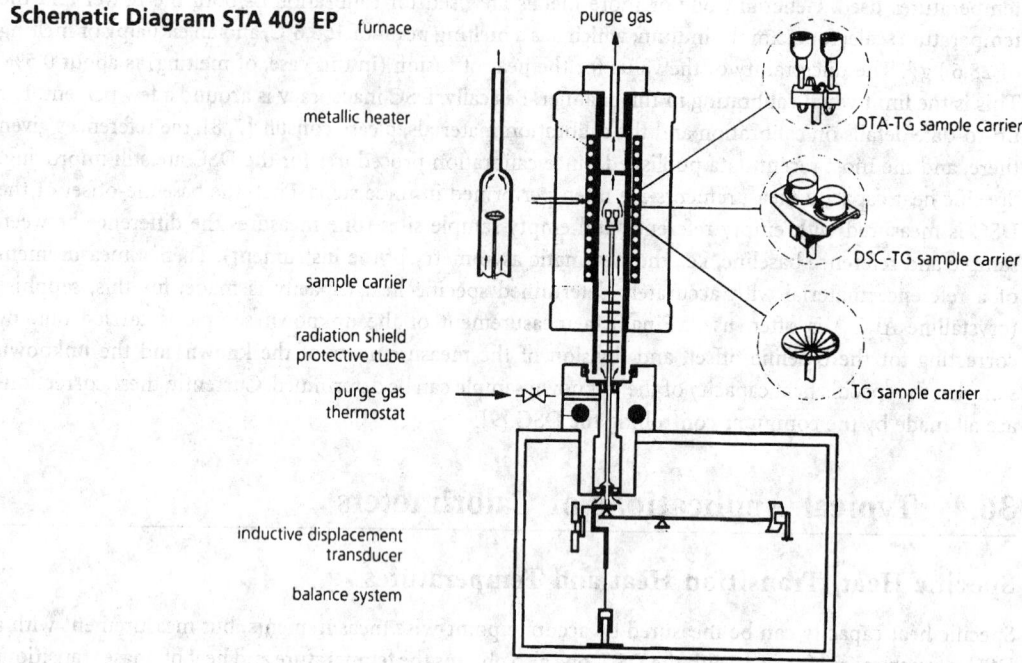

FIGURE 36.3   In a DSC-TG combined thermal analysis instrument, heat is transferred to the sample from the oven by the enveloping purge gas, with radiation shields to diminish heat loss by infrared radiation. The entire sample carrier is placed on a balance to enable thermogravimetric analysis. (Drawing courtesy of Netzsch GmbH.)

Then, all effects — such as glass transition, crystallization, melting, evaporation, decomposition, and even oxidation — can be detected. For this, a DSC is again utilized. In the more expensive analysis systems, DSC instruments often offer the possibility of incorporating other techniques. Mass changes due to oxidation and evaporation are detected by thermogravimetry (TG). This is accomplished by placing the entire experimental chamber on a balance, that continuously measures the sample mass. These data are available as a function of temperature as well, parallel to the heat data. The DSC inner gas atmosphere is usually refreshed continuously using purge gas. This makes it possible to collect the gases coming from the calorimeter furnace and subject them to further analysis, such as mass spectrometry (MS), gas chromatography (GC), Fourier transform infrared analysis (FTIR), and other analysis methods. With sensitive DSC, it is thus possible to detect almost any structural change in matter as a function of temperature. Figure 36.3 shows a drawing of a combined DSC and TG instrument, capable of analyses up to 2000°C. The DSC-TG sample carrier is accurate up to 1500°C, using heat transfer from oven to samples carrier by the surrounding purge gas. Radiation shields are required to reduce heat losses by infrared radiation, which is significant at high temperatures.

Biological Analysis from Cells to Entire Human Beings

Calorimetry is not just the measurement of thermal effects in 1-mL samples. Calorimeters receiving entire human beings are available as well. In practice, there is a wide range of application of calorimeters to biology. This is not surprising, since all forms of live produce heat in activity and often also in rest. The first calorimeters were already used to measure the heat produced by animals. But also, the study of cells (with their heat production of about 1 pW [10]) and the efficiency of enzymatic conversions can be studied using calorimetry. Microcalorimetry is used for very small effects, sometimes using micro-technology to fabricate very small and very sensitive calorimeters [11, 12]; see [10] for an overview and

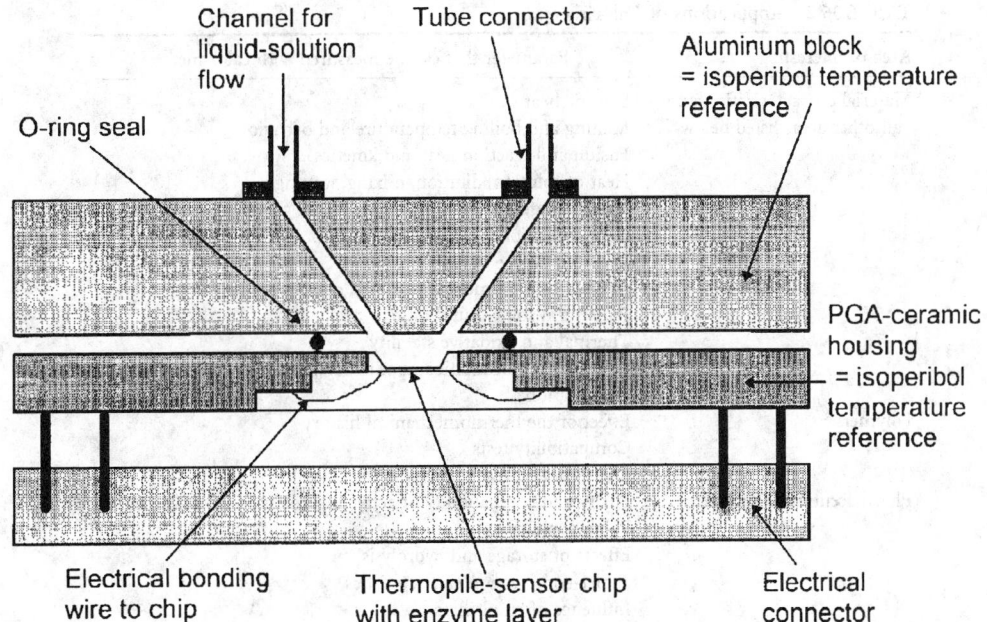

FIGURE 36.4 In microcalorimetry for (bio)chemical analyses, flow-through systems allow continuous, on-line measurement of (bio)chemical quantities in microliter reaction chambers, using 5 × 5-mm silicon microchips coated with enzymes or living cells.

many references. Some microcalorimeters are already being commercialized, using integrated-circuits technology to make very sensitive sensors. Figure 36.4 shows such microcalorimeters encapsulated in a ceramic housing. So far, the use of these sensors is restricted to isoperibol operation around room temperature [11], using aluminum heat sinks to provide the reference temperature. Applications of calorimetry are also found in the food industry, for routine analyses, ecology, plants, etc. See [13, 14] for collections of papers on these subjects.

Summary

The applications of thermal analysis are overwhelming in number. In many cases, when wanting to learn more about materials, thermal analysis can add insight. Some applications were mentioned above. In Table 36.2 some of the often-encountered applications for various disciplines are listed.

36.6   Choosing the Proper Calorimeter for an Application

When faced with a practical analysis problem, one can choose between many instruments. The three main categories are the DSC, the calorimetric bomb or solution calorimeter, and the large reaction or fermentation calorimeter.

What Do You Want To Measure?

If one wants to measure specific heats of reaction, of oxidation, solution, etc., at room temperature, a calorimetric bomb or a solution calorimeter might be the first choice. They are accurate and economical. If one wants to optimize a chemical or biological process, reaction calorimeters or fermentation calorimeters might be the choice. In almost all other cases, a DSC will be the most effective instrument. A

TABLE 36.2   Applications of Calorimeters

Area of interest	Parameter that can be measured with calorimeters
Material characterization and all other areas listed below	Specific heat
	Melting and boiling temperature and behavior
	Fusion and reaction heat and kinetics
	Heat of solution, dilution, mixing, wetting
	Thermal safety
	Glass transition
	Rate and degree of cure
	Crystallization time, temperature, and percentage
	Purity and solid–liquid ratios
	Thermal and oxidative stability
	Identification of multicomponent systems
	Dehydration
Polymers	Effect of the thermomechanical history
	Compatibility tests
	Effects of additives
Pharmaceuticals/Cosmetics	Purity and compatibility of active ingredients
	Polymorphism
	Effects of storage and hydrolysis
	Tablet-compression characteristics
	Influence of emulsifiers
	Concentration of medicines in polymers
	Melting and crystallization behavior of waxes
Foods	Melting, solidification behavior, and specific heat of fats and oils
	Polymophism
	Denaturation of proteins
	Gelatinization of starch
	Freezing-thawing behavior
Biology	Metabolism of cells, organs, animals, and human beings
	Influence of nutrition, toxins, and others on organisms
	Enzymatic efficiency and selectivity
	Concentration of solutions using enzymes or organisms
	Environmental monitoring

DSC is especially useful when one wants to obtain the thermal behavior of materials as a function of temperature. In turn, the temperature-dependent behavior of materials can tell a lot about their structure, their properties, and even their thermomechanical history. Important is the scanning range and rate of the DSC. In some cases, the measurement problem cannot be solved by a standard available calorimeter, and one is forced to either use an experimental calorimeter or develop a special-purpose calorimeter.

Budget, and the Need of Pre- and After-Sales Service

A cost-effective DSC will cost about $25,000, the most expensive models can cost up to $150,000 and include thermogravimetric measurement. Calorimetric bombs are somewhat cheaper, starting at $18,000. The large reaction calorimeters are more costly, at about $120,000 and above. Finally, one can buy experimental instruments or make one's own system. An adiabatic system consisting of a Dewar in a polystyrene housing, using a commercially available stirrer and platinum resistance (measured with any 5½ digit DMM) will put you in business, although at reduced accuracy. However, the budget for this will not need to exceed much more than about $1000 (excluding the DMM). It is less advisable to make one's own DSC (for cost reasons) or calorimetric bomb (for cost and safety reasons).

36.7 Can the Instrument of Choice Measure the Signals Desired?

Here, various parameters must be considered. Accuracy, the degree to which an instrument reading approaches the true value, lies around a few percent for DSCs and parts-of-percent for adiabatic calorimeters. Repeatability of the major calorimeter measurement results — such as heat peak area and peak starting temperature — depends, among other things, on such matters as baseline noise/drift, influence of sample preparation, and positioning in the DSC sample site. Resolution should be considered for two parameters: heat and temperature. First, there is the sensitivity of the instrument for generated heat or power, related to its noise. For DSCs, resolution for power is usually around 0.1 μW to 10 μW; for bomb calorimeters, resolution for heat is around 0.1 J to 10 J. But there is also the resolution of a DSC for separating two heat pulses at two nearby temperatures. This can be designated as the temperature resolution, and depends on time constant and heating rate. Finally, there is the point of linearity and time constant. In case the time constant of the instrument is much larger than that of the process, good linearity is required if one wants to "desmear" the measured curve and obtain the actually produced heat as a function of time. But, a current DSC is a very fast instrument, and desmearing is not really necessary anymore. So, linearity is less important for fast DSCs.

Instrument Control and Data Management

Software is becoming more and more important. Since it will control the measurement, its user-friendliness determines how easy and how error-free one can operate the instrument. It also can take care of analysis of the measurement results, and graphical and numerical presentation of the results. The software can also take care of quality-control aspects of analysis, such as writing to file all the measurement details (not only what was measured, but also how it was measured). Presently, top models of all three types described above will perform these functions. With DSC instruments, software can often be used to control all kinds of thermal analysis instruments apart from the DSC, such as TG, DMA, etc., and merge results obtained with DSCs, TG, DMA, and other analysis techniques. This facilitates interpretation of measurements.

36.8 Commercially Available Calorimeters

Tables 36.3 and 36.4 describe the different instruments and vendors. Table 36.3 gives an overview of instruments and some characteristics, while Table 36.4 gives vendor information. Tables 36.3 and 36.4 are (necessarily) incomplete since only the major vendors have been listed; thus, if looking for a calorimeter, please complete the list with local (and up-to-date) information.

36.9 Advanced Topic: Modulated or Dynamic DSC Operation

Modulating the Temperature Scan Improves Insight into the Measurement

One of the recent developments in DSC is the use of a nonuniform temperature scan. On the standard temperature increase (for example, 1 K min^{-1}), an alternating fast temperature change is superimposed, which can be a sinusoidal signal. Alternatively, a stepped temperature profile can be used. Here, the temperature is increased during, for example, 0.5 min with 2°C, and then stabilized for 0.5 min, resulting in an overall scan rate of 2°C min^{-1}. In fact, the scan is now modulated with a block wave of amplitude 2°C min^{-1}, see Figure 36.5(a). For an instrument with small time constant, such as the DSC-7 of Perkin

TABLE 36.3   Some Commercially Available Calorimeters and Their Specifications

Type	Instrument	Vendor	Cost[a] ($1000)	Range[b] (°C)	Scan rate[c] (K min^{-1})	Resolution[d]
PC DSC[e]	Pyris	Perkin Elmer	55	−170 to +725	500	0.2 µW
CHF DSC[f]	DSC 12E	Mettler	20	−40 to 400	20	10 µW
	DSC 821	Mettler	30	−150 to 700	100	0.7 µW
	DSC 200	Netzsch	45	−170 to 530	40	4 µW
	DSC 6	Perkin Elmer	25	−120 to 450	50	
	Exstar 6000	Seiko	40	−150 to 1500	100	0.2 µW
	DSC 141	Setaram	40	−150 to 600	100	10 µW
	DSC 50	Shimadzu	30	20 to 725	100	10 µW
	DSC 2920	TA Instr	45	−180 to 725	200	0.2 µW
	DSC 2010	TA Instr	30	−180 to 725	200	1 µW
FFHF DSC[g]	DSC 404	Netzsch	60	−120 to 1500	50	8 µW
FFHC DSC + TG	STA 409	Netzsch	70	−160 to 2000	100	8 µW
	Labsys	Setaram	45	20 to 1600	100	10 µW
C DSC[h]	DSC VII	Setaram	55	−45 to 120	1.2	1 µW
	DSC 111	Setaram	80	−120 to 830	30	5 µW
Bomb	1425	Parr	18	Ambient	—	4 J
	1271	Parr	45	Ambient	—	2 J
	C 5000	IKA	33	Ambient	—	6 J
	C 7000	IKA	35	Ambient	—	0.5 J
Solution	1455	Parr	18	0 to 70		0.4 J
Calvet	MS 80D	Setaram	95	20 to 200	—	0.1 µW
	HT 1000	Setaram	220	20 to 1000	1	10 µW
Process	RC1	Mettler	120	−50 to 300	30	
	BFK	Berghof	150	20 to 60		40 mW
Micro	LCM-2526	Xensor	—	Ambient	—	0.1 µW

a Cost: simplest complete system.
b Range: addition of the widest ranges available.
c Scan Rate: the highest controlled scan rate (usually for heating).
d Resolution: vendor specification or 2x rms noise, for the most accurate system.
e PC DSC: Power-Compensated DSC.
f CHF DSC: Circular-Disk Heat-Flux DSC.
g FFHF DSC: Floating-Foil Heat-Flux DSC.
h C DSC: Calvet DSC.

Elmer at about 1 to 2 s, this means that the heat flow for increasing the experiment temperature in the stabilization time will be completely stopped. Any heat flows remaining result from processes in the sample itself; see Figure 36.5(a). Proper interpretation of the measurement results obtained by this method is still under discussion by Reading [15], Schawe [16], and many others.

PET Is Often Used as Example

Figure 36.5(b) gives the analysis of PET (polyethylene terephthalate), a polymer often used as a reference material because of its convenient and exemplary behavior in polymer analysis [17]. The curve is that of shock-cooled PET, i.e., the PET (from an ordinary beverage bottle) is heated to 300°C in a nitrogen atmosphere, and then quench-cooled in 1 min to maintain an amorphous structure. Six regions can be seen. From 50°C to about 75°C, the heat capacity of the PET sample requires power to achieve the temperature increase. At about 75°C, the so-called "glass transition" takes place, where reordering of the PET molecules takes place, increasing the specific heat of PET (an endothermic process), as can be seen from the lifting of the baseline (even at zero temperature increase, heat has to be supplied to the PET). Then follows a region with increased specific heat, caused by the higher freedom of movement of the

TABLE 36.4   Companies That Sell Calorimeters

Berghof GmbH Harretstrasse 1 D-72800 Eningen/Reutlingen, Germany Tel: + 49-7121-8940, Fax: + 49-7121-894100	Perkin-Elmer Corp. 761 Main Ave. Norwalk, Connecticut 06859-0012 Tel: + 1-203-762-1000, Fax: + 1-203-762-6000
IKA Analysentechnik GmbH P.O. Box 1240 D-79420 Heitersheim, Germany Tel: + 49-7633-8310, Fax: + 49-7633-83198	Seiko Instruments 1-8, Nakase, Mihami-Ku, Chiba-shi Chiba 261, Japan Tel: + 81-43-211-1340, Fax: + 81-43-211-8067
Linseis GmbH P.O. Box 1404 D-95088 Selb, Germany Tel: + 49-9287-8800, Fax: + 49-9287-70488	Setaram 7, rue de l'Oratoire, BP 34 F-69641 Caluire Cedez, France Tel: + 33-72 10 2525, Fax: + 33-78 28 6355
Mettler Toledo AG Analytical Sonnenbergstrasse 74 CH-8603 Schwerzenbach, Switzerland Tel: + 41-1-806-7711, Fax: + 41-1-806-7350	Shimadzu Corp. 3. Kanda-Nishikicho 1-chome, Chiyoda-ku Tokyo 101, Japan Tel: + 81-3-3219-5641, Fax: + 81-3-3219-5710
Netzsch Gerätebau GmbH P.O. Box 1460 D-95088 Selb/Bavaria, Germany Tel: + 49-9287-8810, Fax: + 49-9287-88144	TA Instruments Inc New Castle, Delaware 19720 Tel: + 1-302-427-4000, Fax: + 1-302-427-4001
Parr Instrument Company 211 Fifty-Third Street Moline, Illinois 61265 Tel: + 1-309-762-7716, Fax: + 1-309-762-9453	Xensor Integration P.O. Box 3233 2601 DE Delft, the Netherlands Tel. + 31-15-2578040, Fax: + 31-15-2578050

PET molecules compared to the structure below glass transition. Around 140°C, cold crystallization of the material occurs: again a reordering of the material. This is a strongly exothermic process, which again displaces the baseline. The power to be supplied to the sample then steadily increases, and also, with rising temperature, the baseline starts to fall again until a maximum just below melting at 250°C. This is the result of further crystallization, which is facilitated by the higher energy in the sample, and the much better mobility of the molecules just before melting. In the modulated DSC, recrystallization and melting are two concurring phenomena that can be distinguished. Similarly, the hope is that glass transition and cold crystallization can be better distinguished in materials where they overlap (in PET, they are clearly distinct).

Acknowledgments

The author wishes to thank Dr. P. J. van Ekeren of Utrecht University and Dr. G. W. H. Höhne of Universität Ulm for their suggestions in improving this chapter.

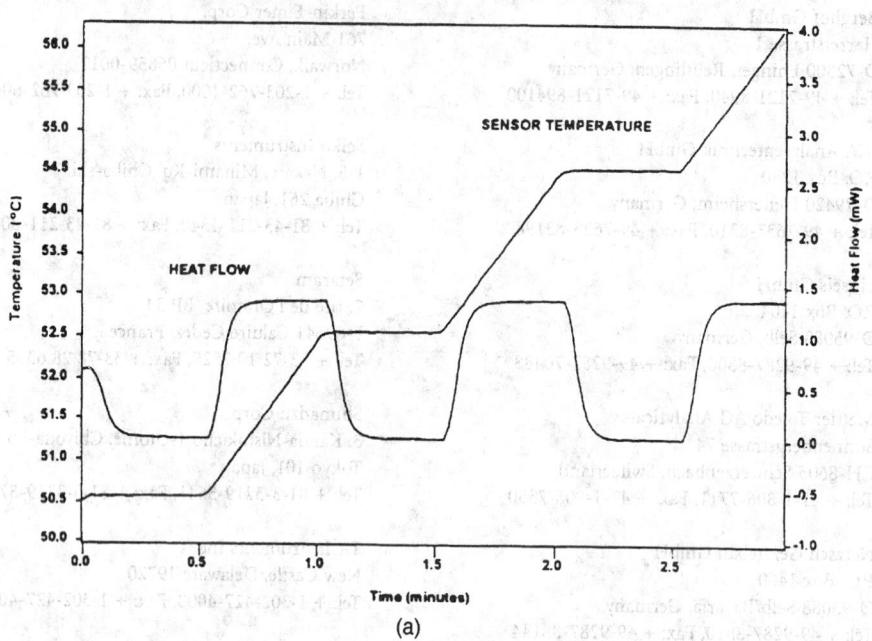

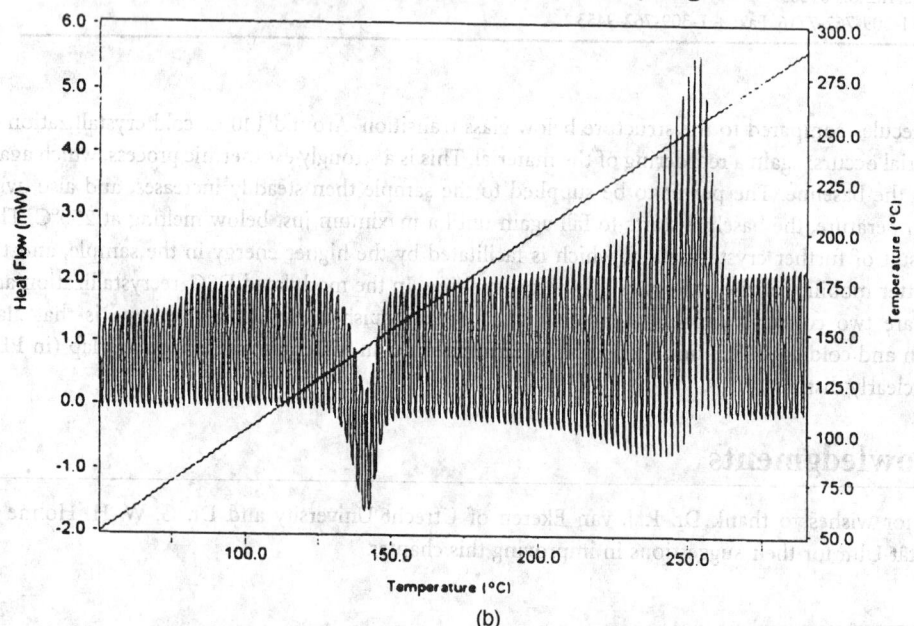

FIGURE 36.5 In modulated or dynamic DSCs, the temperature is not increased at a given rate, but modulated with a sinusoidal or stepwise deviation (a) to separate heat transfer due to temperature increase (specific heat), and heat transfer due to phase-changes in the material (glass transition, crystallization, melting). The heat flow in dynamic a DSC for shock-cooled PET (b) clearly shows the absence of heat flow in regions without material change (only temperature increase), and also the residual heat flow when not increasing temperature due to glass transition ($\approx 75°C$), cold crystallization ($\approx 140°C$), and melting ($\approx 250°C$). (Drawings courtesy of Perkin Elmer.)

References

1. S. Middelhoek and S.A. Audet, *Silicon Sensors*, London, Academic Press, 1989.
2. A.W. van Herwaarden, Physical principles of thermal sensors, *Sensors and Materials*, 8, 373-387, 1996.
3. H.B. Callen, *Thermodynamics and an Introduction to Thermostatistics*, 2nd ed., New York, John Wiley & Sons, 1985.
4. W. Hemminger and G.W.H. Höhne, *Calorimetry — Fundamentals and Practice*, Weinheim, Verlag Chemie, 1984.
5. W. Hemminger, Calorimetric methods, in V.B.F. Mathot (ed.), *Calorimetry and Thermal Analysis of Polymers*, Munich, Hanser Publishers, 1994.
6. K.N. Marsh (ed.), *Recommended Reference Materials for Realization of Physicochemical Properties*, Oxford, Blackwell Scientific, 1987.
7. G.W.H.Höhne, Fundamentals of differential scanning calorimetry and differential thermal analysis, in V.B.F. Mathot (ed.), *Calorimetry and Thermal Analysis of Polymers*, Munich, Hanser Publishers, 1994.
8. E. Gmelin and St.M. Sarge, Calibration of differential scanning calorimeters, *Pure Appl. Chem.*, 67, 1789-1800, 1995.
9. T.M.V.R. de Barros, R.C. Santos, A.C. Fernandes, and M.E. Minas da Piedade, Accuracy and precision of heat capacity measurements using a heat flux differential scanning calorimeter, *Thermochim. Acta*, 269/272, 51-60, 1995.
10. P. Bataillard, Calorimetric sensing in bioanalytical chemistry: principles, applications and trends, *Trends in Anal. Chem.*, 12, 387-394, 1993.
11. A.W. van Herwaarden, P.M. Sarro, J.W. Gardner, and P. Bataillard, Liquid and gas micro-calorimeters for (bio)chemical measurements, *Sensors and Actuators*, A43, 24-30, 1994.
12. G.W.H. Höhne, A.E. Bader, and St. Höhnle, Physical properties of a vacuum-deposited thermopile for heat measurements, *Thermochim. Acta*, 251, 307-317, 1995.
13. J. Lamprecht, W. Hemminger, and G.W.H. Höhne (eds.), Calorimetry in the biological sciences, *Thermochim. Acta*, 193, 1991.
14. R.B. Kemp and B. Schaarschmidt (eds.), Calorimetric and thermodynamic studies in biology, *Thermochim. Acta*, 251, 1995.
15. M. Reading, A. Luget, and R. Wilson, Modulated differential scanning calorimetry, *Thermochim. Acta*, 238/239, 295-307, 1994.
16. J.E.K. Schawe, Principles for the interpretation of modulated temperature DSC measurement. Part 1. Glass transition, *Thermochim. Acta*, 261, 183-194, 1995.
17. Characterization of Amorphous Polyethylene Terephtalate by Dynamic Differential Scanning Calorimetry, *Perkin Elmer Thermal Analysis Newsletter* 69, Perkin-Elmer Corp., Norwalk, CT.

Further Information

Much is being published on Calorimetry each year. Some very good books to obtain a basic understanding of calorimetry are:

W. Hemminger and G.W.H. Höhne, *Calorimetry — Fundamentals and Practice*, Weinheim, Verlag Chemie, 1984.

V.B.F. Mathot (ed.), *Calorimetry and Thermal Analysis of Polymers*, Munich, Hanser Publishers, 1994.

B. Wunderlich, *Thermal Analysis*, San Diego, CA, Academic Press, 1990.

M. Brown, *Introduction to Thermal Analysis*, London, Chapman and Hall, 1988.

On DSC, an up-to-date book is:

G.W.H. Höhne, W. Hemminger, and H.-J. Flammersheim, *Differential Scanning Calorimetry: An Introduction for Practitioners*, Berlin, Springer-Verlag, 1996.

Apart from these, many relevant papers appear in:

Thermochimica Acta, Amsterdam, Elsevier Science Publishers.
Journal of Thermal Analysis, Chichester, U.K., John Wiley & Sons.

    Leafing through the latest volumes will bring one up to date on calorimetry research. Regular conferences on Calorimetry and Thermal Analysis are being held, organized by local and global institutes, such as the ICTAC (International Confederation on Thermal Analysis and Calorimetry). Proceedings of these conferences are often published in *Thermochimica Acta* or the *Journal of Thermal Analysis.*

    Very interesting is the publication of ICTA from 1991, which contains significant information on nomenclature, literature, suppliers of instrumentation, etc.:

J.O. Hill (ed.), *For Better Thermal Analysis and Calorimetry, Edition III*, ICTA, 1991.

VII

Electromagnetic Variables Measurement

37
Voltage Measurement

Alessandro Ferrero
Politecnico di Milano

Jerry Murphy
Hewlett Packard Company

Cipriano Bartoletti
University of Rome "La Sapienza"

Luca Podestà
University of Rome "La Sapienza"

Giancarlo Sacerdoti
University of Rome "La Sapienza"

37.1 Meter Voltage Measurement

Alessandro Ferrero

Instruments for the measurement of electric voltage are called *voltmeters*. Correct insertion of a voltmeter requires the connection of its terminals to the points of an electric circuit across which the voltage has to be measured, as shown in Figure 37.1. To a first approximation, the electric equivalent circuit of a voltmeter can be represented by a resistive impedance Z_v (or a pure resistance R_v for dc voltmeters). This means that any voltmeter, once connected to an electric circuit, draws a current I_v given by:

$$I_v = \frac{U}{Z_v}$$

(37.1)

where U is the measured voltage. The higher the value of the internal impedance, the higher the quality of the voltmeter, since it does not significantly modify the status of the electric circuit under test.

Different operating principles are used to measure an electric voltage. The mechanical interaction between currents, between a current and a magnetic field, or between electrified conductors was widely adopted in the past to generate a mechanical torque proportional to the voltage or the squared voltage to be measured. This torque, balanced by a restraining torque, usually generated by a spring, causes the instrument pointer, which can be a mechanical or a virtual optical pointer, to be displaced by an angle proportional to the driving torque, and hence to the voltage or the squared voltage to be measured. The value of the input voltage is therefore given by the reading of the pointer displacement on a graduated scale. The thermal effects of a current flowing in a conductor are also used for measuring electric voltages, although they have not been adopted as widely as the previous ones. More recently, the widespread diffusion of semiconductor devices led to the development of a completely different class of voltmeters: *electronic* voltmeters. They basically attain the required measurement by processing the input signal by means of electronic semiconductor devices. According to the method, analog or digital, the input signal is processed, the electronic voltmeters can be divided into *analog* electronic voltmeters and *digital*

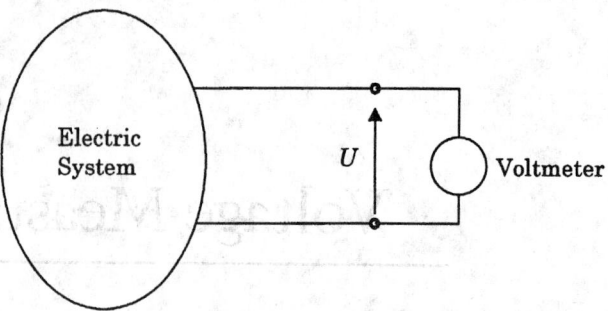

FIGURE 37.1   Voltmeter insertion.

TABLE 37.1   Classification of Voltage Meters

Class	Operating principle	Subclass	Application field
Electromagnetic	Interaction between currents and magnetic fields	Moving magnet	Dc voltage
		Moving coil	Dc voltage
		Moving iron	Dc and ac voltage
Electrodynamic	Interactions between currents	—	Dc and ac voltage
Electrostatic	Electrostatic interactions	—	Dc and ac voltage
Thermal	Current's thermal effects	Direct action	Dc and ac voltage
		Indirect action	Dc and ac voltage
Induction	Magnetic induction	—	Ac voltage
Electronic	Signal processing	Analog	Dc and ac voltage
		Digital	Dc and ac voltage

electronic voltmeters. Table 37.1 shows a rough classification of the most commonly employed voltmeters, according to their operating principle and their typical application field.

This chapter section briefly describes the most commonly employed voltmeters, both electromechanical and electronic.

Electromechanical Voltmeters

Electromechanical voltmeters measure the applied voltage by transducing it into a mechanical torque. This can be accomplished in different ways, basically because of the interactions between currents (*electrodynamic voltmeters*), between a current and a magnetic field (*electromagnetic voltmeters*), between electrified conductors (*electrostatic voltmeters,* or *electrometers*), and between currents induced in a conducting vane (*induction voltmeters*). According to the different kinds of interactions, different families of instruments can be described, with different application fields. Moving-coil electromagnetic voltmeters are restricted to the measurement of dc voltages; moving-iron electromagnetic, electrodynamic, and electrostatic voltmeters can be used to measure both dc and ac voltages; while induction voltmeters are restricted to ac voltages.

The most commonly employed electromechanical voltmeters are the electromagnetic and electrodynamic ones. Electrostatic voltmeters have been widely employed in the past (and are still employed) for the measurement of high voltages, both dc and ac, up to a frequency on the order of several megahertz. Induction voltmeters have never been widely employed, and their present use is restricted to ac voltages.

Therefore, only the electromagnetic, electrodynamic, and electrostatic voltmeters will be described in the following.

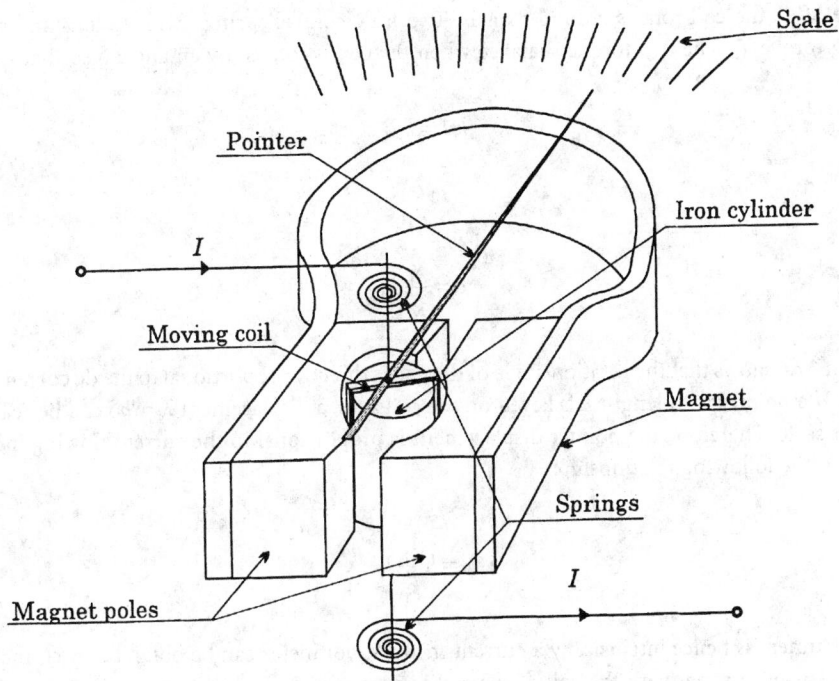

FIGURE 37.2 Dc moving-coil meter.

Electromagnetic Voltmeters

Dc Moving-Coil Voltmeters.
The structure of a dc moving-coil meter is shown in Figure 37.2. A small rectangular pivoted coil is wrapped around an iron cylinder and placed between the poles of a permanent magnet. Because of the shape of the poles of the permanent magnet, the induction magnetic field B in the air gap is radial and constant.

Suppose that a dc current I is flowing in the coil, the coil has N turns, and that the length of the sides that cut the magnetic flux (active sides) is l; the current interacts with the magnetic field B and a force F is exerted on the conductors of the active sides. The value of this force is given by:

$$F = NBlI \tag{37.2}$$

Its direction is given by the right-hand rule. Since the two forces applied to the two active sides of the coil are directed in opposite directions, a torque arises in the coil, given by:

$$T_i = Fd = NBldI \tag{37.3}$$

where d is the coil width. Since N, B, l, d are constant, Equation 37.3 leads to:

$$T_i = k_i I \tag{37.4}$$

showing that the mechanical torque exerted on the coil is directly proportional to the current flowing in the coil itself.

Because of T_i, the coil rotates around its axis. Two little control springs, with k_r constant, provide a restraining torque T_r. The two torques balance when the coil is rotated by an angle δ so that:

$$k_i I = k_r \delta \tag{37.5}$$

which leads to:

$$\delta = \frac{k_i}{k_r} I \tag{37.6}$$

Equation 37.6 shows that the rotation angle of the coil is directly proportional to the dc current flowing in the coil. If a pointer with length h is keyed on the coil axes, a displacement $\lambda = h\delta$ can be read on the instrument scale. Therefore, the pointer displacement is proportional to the current flowing in the coil, according to the following relationship:

$$\lambda = h \frac{k_i}{k_r} I \tag{37.7}$$

This instrument is hence intrinsically a current meter. A voltmeter can be obtained by connecting an additional resistor in series with the coil. If the coil resistance is R_c, and the resistance of the additional resistor is R_a, the current flowing in the coil when the voltage U is applied is given by:

$$I = \frac{U}{R_a + R_c} \tag{37.8}$$

and therefore the pointer displacement is given by:

$$\lambda = h\delta = h \frac{k_i}{k_r} I = h \frac{k_i}{k_r (R_a + R_c)} U \tag{37.9}$$

and is proportional to the applied voltage. Because of this proportionality, moving-coil dc meters show a proportional-law scale, where the applied voltage causes a proportional angular deflection of the pointer.

Because of the operating principle expressed by Equation 37.3, these voltmeters can measure only dc voltages. Due to the inertia of the mechanical part, ac components typically do not cause any coil rotation, and hence these meters can be also employed to measure the dc component of a variable voltage. They have been widely employed in the past for the measurement of dc voltages up to some thousands volts with a relative measurement uncertainty as low as 0.1% of the full-scale value. At present, they are being replaced by electronic voltmeters that feature the same or better accuracy at a lower cost.

Dc Galvanometer.

General characteristics. A galvanometer is used to measure low currents and low voltages. Because of the high sensitivity that this kind of measurement requires, galvanometers are widely employed as null indicators in all dc balance measurement methods (like the bridge and potentiometer methods) [1, 2].

A dc galvanometer is, basically, a dc moving coil meter, and the relationship between the index displacement and the current flowing in the moving coil is given by Equation 37.7. The instrument constant:

$$k_a = h \frac{k_i}{k_r} \qquad (37.10)$$

is usually called the galvanometer *current constant* and is expressed in mm μA^{-1}. The galvanometer *current sensitivity* is defined as $1/k_a$ and is expressed in μA mm^{-1}.

According to their particular application field, galvanometers must be chosen with particular care. If k_a is taken into account, note that once the full-scale current and the corresponding maximum pointer displacement are given, the value of the ratio hk_i/k_r is also known. However, the single values of h, k_i, and k_r can assume any value and are usually set in order to reduce the friction effects. In fact, if the restraining friction torque T_f is taken into account in the balance equation, Equation 37.5 becomes:

$$k_i I = k_r \frac{\lambda}{h} \pm T_f \qquad (37.11)$$

where the ± sign shows that the friction torque does not have its own sign, but always opposes the rotation.

The effects of T_f can be neglected if the driving torque $hk_i I$ and the restraining torque $k_r \lambda$ are sufficiently greater than T_f. Moreover, since the galvanometer is employed as a null indicator, a high sensitivity is needed; hence, k_a must be as high as possible. According to Equations 37.10 and 37.11, this requires high values of hk_i and low values of k_r. A high value of h means a long pointer; a high value of k_i means a high driving torque, while a low value of k_r means that the inertia of the whole moving system must be low.

The pointer length can be increased without increasing the moving system inertia by employing virtual optical pointers: a little, light concave mirror is fixed on the moving coil axis and is lit by an external lamp. The reflected light hits a translucid, graduated ruler, so that the mirror rotation can be observed (Figure 37.3). In this way, a virtual pointer is obtained, whose length equals the distance between the mirror and the graduated ruler.

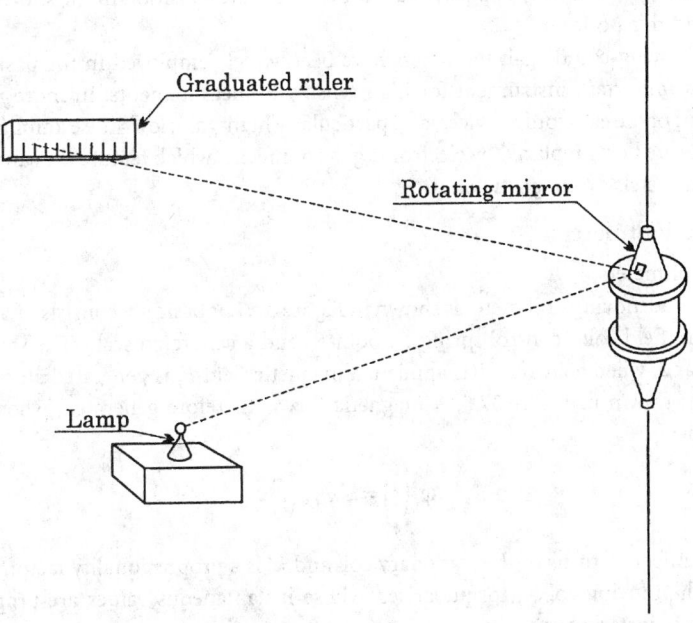

FIGURE 37.3 Virtual optical pointer structure in a dc galvanometer.

The reduction of the moving system inertia is obtained by reducing the weight and dimension of the moving coil, and reducing the spring constant. This is usually done by suspending the moving coil with a thin fiber of conducting material (usually bronze). Thus, the friction torque is practically removed, and the restraining spring action is given by the fiber torsion.

According to Equations 37.3 and 37.4, the driving torque can be increased by increasing the coil flux linkage. Three parameters can be modified to attain this increase: the induction field B, the coil section ld, and the number of turns N of the coil winding.

The induction field B can be increased by employing high-quality permanent magnets, with high coercive force, and minimizing the air gap between the magnet's poles. This minimization prevents the use of moving coils with a large section. Moreover, large coil sections lead to heavier coils with greater inertia, which opposes the previous requirement of reduced inertia. For this reason, the coil section is usually rectangular (although a square section maximizes the flux linkage) and with $l > d$.

If the galvanometer is used to measure a low voltage U, the *voltage sensitivity*, expressed in $\mu V\ mm^{-1}$ is the inverse of:

$$k_v = \frac{\lambda}{U} \tag{37.12}$$

where k_v is called the galvanometer's *voltage constant* and is expressed in $mm\ \mu V^{-1}$.

Mechanical characteristics. Due to the low inertia and low friction, the galvanometer moving system behaves as an oscillating mechanical system. The oscillations around the balance position are damped by the electromagnetic forces that the oscillations of the coil in the magnetic field exert on the coil active sides. It can be proved [1] that the oscillation damping is a function of the coil circuit resistance: that is, the coil resistance r plus the equivalent resistance of the external circuit connected to the galvanometer. In particular, the damping effect is nil if the coil circuit is open, and maximum if the coil is short-circuited.

In practical situations, a resistor is connected in series with the moving coil, whose resistance is selected in such a way to realize a critical damping of the coil movement. When this situation is obtained, the galvanometer is said to be *critically damped* and reaches its balance position in the shortest time, without oscillations around this position.

Actual trends. Moving-coil dc galvanometers have been widely employed in the past when they represented the most important instrument for high-sensitivity measurements. In more recent years, due to the development of the electronic devices, and particularly high-gain, low-noise amplifiers, the moving-coil galvanometers are being replaced by electronic galvanometers, which feature the same, or even better, performance than the electromagnetic ones.

Electrodynamic Voltmeters

Ac Moving-coil Voltmeters.

The structure of an ac moving-coil meter is shown in Figure 37.4. It basically consists of a pivoted moving coil, two stationary field coils, control springs, a pointer, and a calibrated scale. The stationary coils are series connected and, when a current i_f is applied, a magnetic field B_f is generated along the axis of the stationary coils, as shown in Figure 37.5. A magnetic flux is therefore generated, whose instantaneous values are given by:

$$\varphi_f(t) = k' m_f i_f(t) \tag{37.13}$$

where m_f is the number of turns of the stationary coil and k' is a proportionality factor. When a current i_m is applied to the moving coil, a torque arises, whose instantaneous values are proportional to the product of φ_f and i_m instantaneous values:

$$T_i(t) = k'' \varphi_f(t) i_m(t) = k i_f(t) i_m(t) \tag{37.14}$$

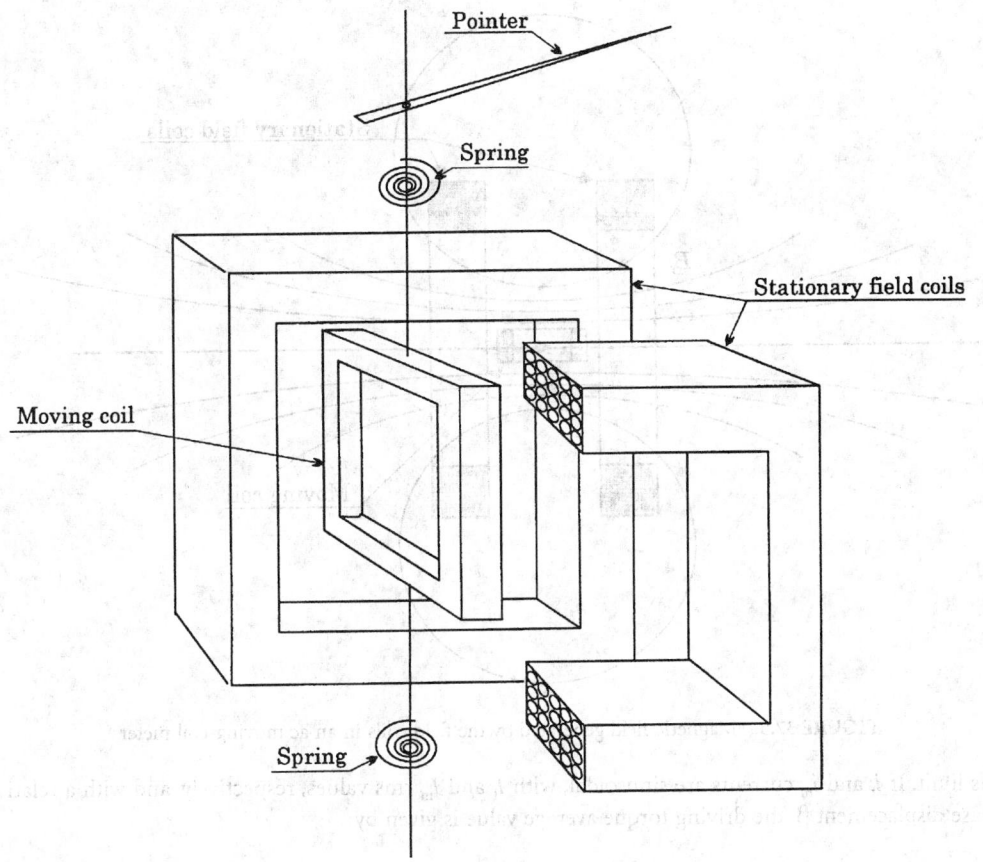

FIGURE 37.4 Ac moving-coil meter.

The driving torque is therefore proportional to the instantaneous product of the currents flowing in the two coils. Due to this driving torque, the moving element is displaced by an angle (δt), until the spring restraining torque $T_s(t) = k_s\delta(t)$ balances the driving torque. The moving element rotation is thus given by:

$$\delta(t) = \frac{k}{k_s}i_f(t)i_m(t) \tag{37.15}$$

and, if the pointer length is h, the following pointer displacement can be read on the scale:

$$\lambda(t) = h\frac{k}{k_s}i_f(t)i_m(t) \tag{37.16}$$

The proportionality factor k is generally not constant, since it depends on the mutual inductance between the two coils, and thus on their number of turns, shape and relative position. However, if the two coils are carefully designed and placed, the magnetic field can be assumed to be constant and radial in the rotation area of the moving coil. Under this condition, k is virtually constant.

Because the bandwidth of the moving element is limited to a few hertz, due to its inertia, the balance position is proportional to the average value of the driving torque when the signal bandwidth exceeds

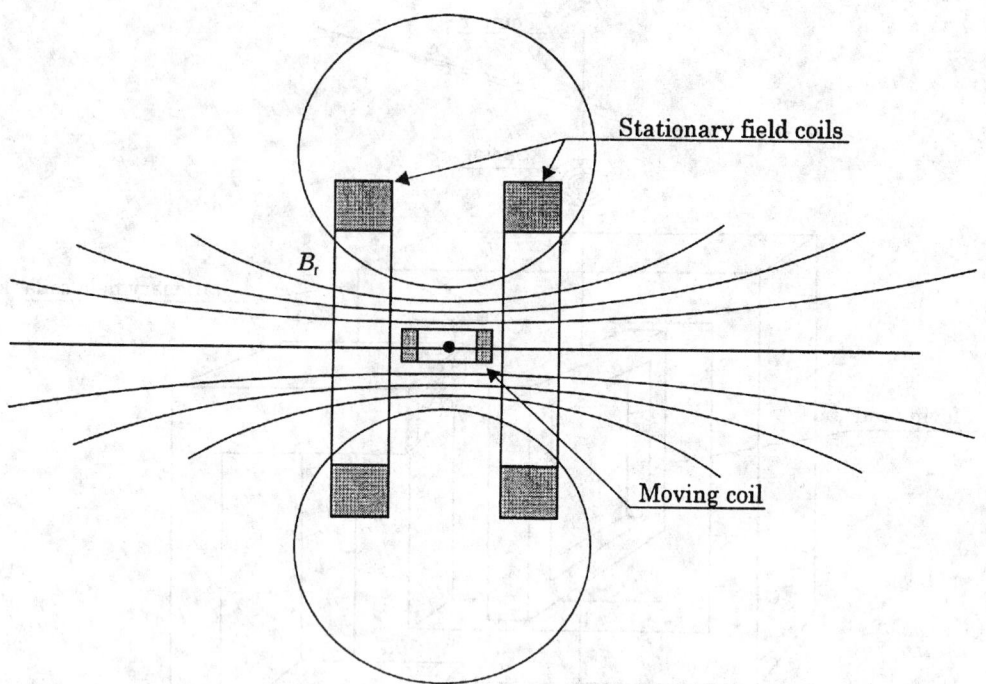

FIGURE 37.5 Magnetic field generated by the field coils in an ac moving-coil meter.

this limit. If i_f and i_m currents are sinusoidal, with I_f and I_m rms values, respectively, and with a relative phase displacement β, the driving torque average value is given by:

$$\overline{T_i} = k I_f I_m \cos\beta \tag{37.17}$$

and thus, the pointer displacement in Equation 37.16 becomes:

$$\lambda = h \frac{k}{k_s} I_f I_m \cos\beta \tag{37.18}$$

In order to realize a voltmeter, the stationary and moving coils are series connected, and a resistor, with resistance R, is also connected in series to the coils. If R is far greater than the resistance of the two coils, and if it is also far greater than the coil inductance, in the frequency operating range of the voltmeter, the rms value of the coils' currents is given by:

$$I_f = I_m = \frac{U}{R} \tag{37.19}$$

U being the applied voltage rms value. From Equation 37.18, the pointer displacement is therefore given by:

$$\lambda = h \frac{k}{k_s} \frac{U^2}{R^2} = k_v U^2 \tag{37.20}$$

Because of Equation 37.20, the voltmeter features a square-law scale, with k_v constant, provided that the coils are carefully designed, and that the coils' inductance can be neglected with respect to the resistance

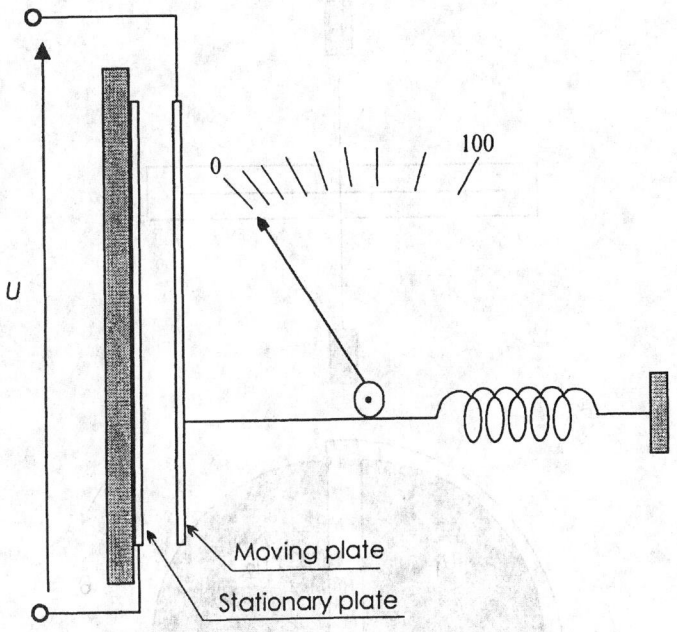

FIGURE 37.6   Basic structure of an electrostatic voltmeter.

of the coils themselves and the series resistor. This last condition determines the upper limit of the input voltage frequency.

These voltmeters feature good accuracy (their uncertainty can be as low as 0.2% of the full-scale value), with full-scale values up to a few hundred volts, in a frequency range up to 2 kHz.

Electrostatic Voltmeters

The action of electrostatic instruments is based on the force exerted between two charged conductors. The conductors behave as a variable plate air capacitor, as shown in Figure 37.6. The moving plate, when charged, tends to move so as to increase the capacitance between the plates. The energy stored in the capacitor, when the applied voltage is U and the capacitance is C, is given by:

$$W = \frac{1}{2}CU^2 \qquad (37.21)$$

This relationship is valid both under dc and ac conditions, provided that the voltage rms value U is considered for ac voltage.

When the moving plate is displaced horizontally by ds, while the voltage is held constant, the capacitor energy changes in order to equal the work done in moving the plate. The resulting force is:

$$F = \frac{dW}{ds} = \frac{U^2}{2}\frac{dC}{ds} \qquad (37.22)$$

For a rotable system, Equation 37.21 leads similarly to a resulting torque:

$$T = \frac{dW}{d\vartheta} = \frac{U^2}{2}\frac{dC}{d\vartheta} \qquad (37.23)$$

If the action of a control spring is also considered, both Equations 37.22 and 37.23 show that the balance position of the moving plate is proportional to the square of the applied voltage, and hence electrostatic

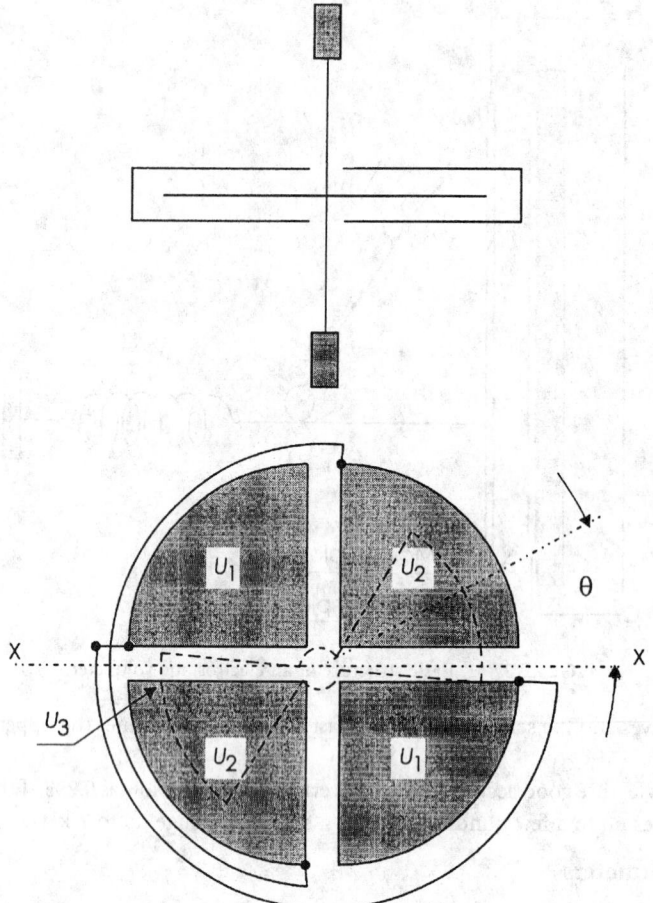

FIGURE 37.7   Quadrant electrometer structure.

voltmeters have a square-law scale. These equations, along with Equation 37.21, show that these instruments can be used for the measurement of both dc and ac rms voltages. However, the force (or torque) supplied by the instrument schematically represented in Figure 37.6 is generally very weak [2], so that its use is very impractical.

The Electrometer.

A more useful configuration is the quadrant electrometer, shown in Figure 37.7. Four fixed plates realize four quadrants and surround a movable vane suspended by a torsion fiber at the center of the system. The opposite quadrants are electrically connected together, and the potential difference $(U_1 - U_2)$ is applied. The moving vane can be either connected to potential U_1 or U_2, or energized by an independent potential U_3.

Let the zero torque position of the suspension coincide with the symmetrical X-X position of the vane. If $U_1 = U_2$, the vane does not leave this position; otherwise, the vane will rotate.

Let C_1 and C_2 be the capacitances of quadrants 1 and 2, respectively, relative to the vane. They both are functions of ϑ and, according to Equation 37.23, the torque applied to the vane is given by:

$$T = \frac{\left(U_3 - U_1\right)^2}{2}\frac{dC_1}{d\vartheta} + \frac{\left(U_3 - U_2\right)^2}{2}\frac{dC_2}{d\vartheta} \qquad (37.24)$$

Since the vane turns out of one pair of quadrants as much as it turns into the other, the variations of C_1 and C_2 can be related by:

$$-\frac{dC_1}{d\vartheta} = \frac{dC_2}{d\vartheta} = k_1 \qquad (37.25)$$

Taking into account the suspension restraining torque $T_r = k_2\vartheta$, the balance position can be obtained by Equations 37.24 and 37.25 as:

$$\vartheta = \frac{k_1}{2k_2}\left[\left(U_3 - U_2\right)^2 - \left(U_3 - U_1\right)^2\right] \qquad (37.26)$$

If the vane potential U_3 is held constant, and is large compared to the quadrant potentials U_1 and U_2, Equation 37.26 can be simplified as follows:

$$\vartheta = \frac{k_1}{k_2}U_3\left(U_1 - U_2\right) \qquad (37.27)$$

Equation 37.27 shows that the deflection of the vane is directly proportional to the voltage difference applied to the quadrants. This method of use is called the *heterostatic* method.

If the vane is connected to quadrant 1, $U_3 = U_1$ follows, and Equation 37.26 becomes

$$\vartheta = \frac{k_1}{2k_2}\left(U_1 - U_2\right)^2 \qquad (37.28)$$

Equation 37.28 shows that the deflection of the vane is proportional to the square of the voltage difference applied to the quadrants, and hence this voltmeter has a square-law scale. This method of use is called the *idiostatic* method, and is suitable for the direct measurement of dc and ac voltages without an auxiliary power source.

The driving torque of the electrometer is extremely weak, as in all electrostatic instruments. The major advantage of using this kind of meter is that it allows for the measurement of dc voltages without drawing current by the voltage source under test. Now, due to the availability of operational amplifiers with extremely high input impedance, they have been almost completely replaced by electronic meters with high input impedance.

Electronic Voltmeters

Electronic meters process the input signal by means of semiconductor devices in order to extract the information related to the required measurement [3, 4]. An electronic meter can be basically represented as a three-port element, as shown in Figure 37.8.

The input signal port is an input port characterized by high impedance, so that the signal source has very little load. The measurement result port is an output port that provides the measurement result (in either an analog or digital form, depending on the way the input signal is processed) along with the power needed to energize the device used to display the measurement result. The power supply port is an input port which the electric power required to energize the meter internal devices and the display device flows through.

One of the main characteristics of an electronic meter is that it requires an external power supply. Although this may appear as a drawback of electronic meters, especially where portable meters are

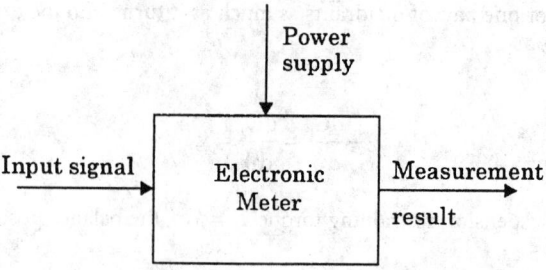

FIGURE 37.8   Electronic meter.

concerned, note that, this way, the energy required for the measurement is no longer drawn from the signal source.

The high-level performance of modern electronic devices yields meters that are as accurate (and sometime even more accurate) as the most accurate electromechanical meters. Because they do not require the extensive use of precision mechanics, they are presently less expensive than electromechanical meters, and are slowly, but constantly, replacing them in almost all applications.

Depending on the way the input signal is processed, electronic meters are divided into *analog* and *digital* meters. Analog meters attain the required measurement by analog, continuous-time processing of the input signal. The measurement result can be displayed both in analog form using, for example, an electromechanical meter; or in digital form by converting the analog output signal into digital form. Digital meters attain the required measurement by digital processing of the input signal. The measurement result is usually displayed in digital form. Note that the distinction between analog and digital meters is not due to the way the measurement result is displayed, but to the way the input signal is processed.

Analog Voltmeters

An electronic analog voltmeter is based on an electronic amplifier and an electromechanical meter to measure the amplifier output signal. The amplifier operates to make a dc current, proportional to the input quantity to be measured, flow into the meter. This meter is hence a dc moving-coil milliammeter.

Different full-scale values can be obtained using a selectable-ratio voltage divider if the input voltage is higher than the amplifier dynamic range, or by selecting the proper amplifier gain if the input voltage stays within the amplifier dynamic range.

The main features of analog voltmeters are high input impedance, high possible gain, and wide possible bandwidth for ac measurements. The relative measurement uncertainty can be lower than 1% of full-scale value. Because of these features, electronic analog voltmeters can have better performance than the electromechanical ones.

Dc Analog Voltmeters.

Figure 37.9 shows the circuit for an electronic dc analog voltmeter. Assuming that the operational amplifier exhibits ideal behavior, current I_m flowing in the milliammeter A is given by:

$$I_m = I_o + I_2 = \frac{U_o}{R_o} + \frac{U_o}{R_2} = -U_i \frac{R_2}{R_1} \frac{R_2 + R_o}{R_2 R_o} = -\frac{U_i}{R_1}\left(1 + \frac{R_2}{R_o}\right) \qquad (37.29)$$

If $R_1 = R_2$, and the same resistances are far greater than R_o, Equation 37.29 can be simplified to:

$$I_m = -\frac{U_i}{R_o} \qquad (37.30)$$

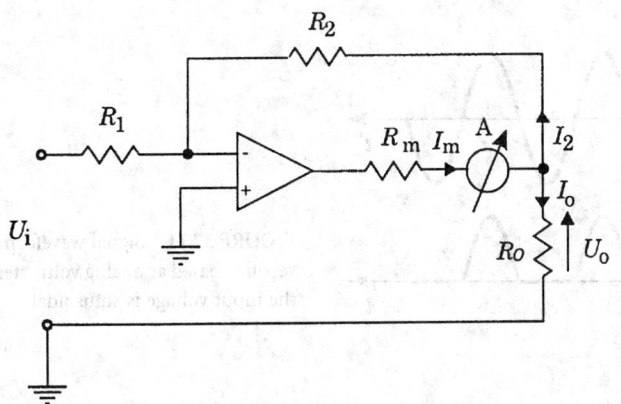

FIGURE 37.9 Electronic dc analog voltmeter schematics.

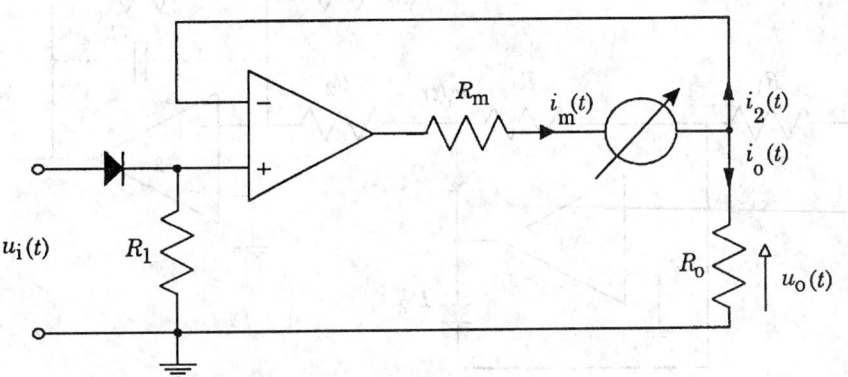

FIGURE 37.10 Electronic, rectifier-based ac analog voltmeter schematics.

Equation 37.30 shows that the milliammeter reading is directly proportional to the input voltage through resistance R_o only. This means that, once the milliammeter full-scale value is set, the voltmeter full-scale value can be changed, within the dynamic range of the amplifier, by changing the R_o value. This way, the meter full-scale value can be changed without changing its input impedance.

Rectifier-Based ac Analog Voltmeters.
Analog meters for ac voltages can be obtained starting from the dc analog voltmeters, with a rectifying input stage. Figure 37.10 shows how the structure in Figure 37.9 can be modified in order to realize an ac voltmeter.

Because of the high input impedance of the electronic amplifier, $i_2(t) = 0$, and the current $i_m(t)$ flowing in the milliammeter A is the same as current $i_o(t)$ flowing in the load resistance. Since the amplifier is connected in a voltage-follower configuration, the output voltage is given by:

$$u_o\left(t\right) = u_i\left(t\right) \qquad (37.31)$$

Due to the presence of the input diode, current $i_m(t)$ is given by:

$$i_m\left(t\right) = \frac{u_i\left(t\right)}{R_o} \qquad (37.32)$$

when $u_i(t) > 0$, and

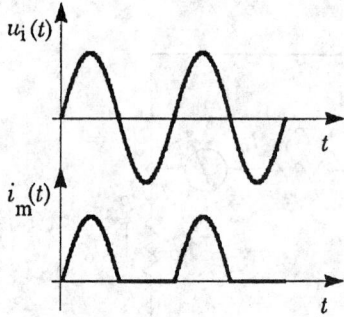

FIGURE 37.11 Signal waveforms in a rectifier-based ac analog voltmeter when the input voltage is sinusoidal.

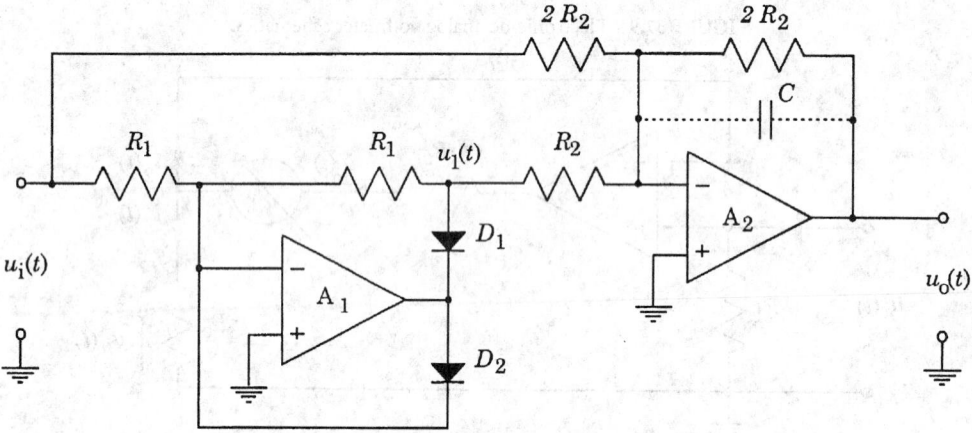

FIGURE 37.12 Electronic, full-wave rectifier-based ac analog voltmeter schematics.

$$i_m\left(t\right)=0 \tag{37.33}$$

when $u_i(t) \leq 0$. If $u_i(t)$ is supposed to be a sine wave, the waveform of $i_m(t)$ is shown in Figure 37.11.

The dc moving-coil milliammeter measures the average value $\bar{I}_m$ of $i_m(t)$, which, under the assumption of sinusoidal signals, is related to the rms value U_i of $u_i(t)$ by:

$$\bar{I}_m = \frac{2\sqrt{2}}{\pi R_o} U_i \tag{37.34}$$

The performance of the structure in Figure 37.10 can be substantially improved by considering the structure in Figure 37.12 which realizes a full-wave rectifier. Because of the presence of diodes D_1 and D_2, the output of amplifier A1 is given by:

$$u_1\left(t\right)=\begin{cases}-u_i\left(t\right) & \text{for } u_i\left(t\right)\geq 0\\ 0 & \text{for } u_i\left(t\right)<0\end{cases} \tag{37.35}$$

where $u_i(t)$ is the circuit input voltage.

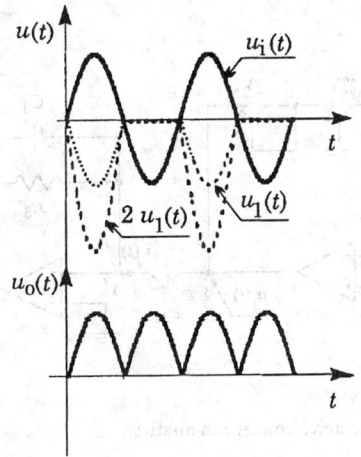

FIGURE 37.13 Signal waveforms in a full-wave rectifier-based ac analog voltmeter when the input voltage is sinusoidal.

If capacitor C is supposed to be not connected, amplifier A_2 output voltage is:

$$u_o(t) = -\left[u_i(t) + 2u_1(t)\right] \qquad (37.36)$$

which gives:

$$u_o(t) = \begin{cases} u_i(t) & \text{for } u_i(t) \geq 0 \\ -u_i(t) & \text{for } u_i(t) < 0 \end{cases} \qquad (37.37)$$

thus proving that the circuit in Figure 37.12 realizes a full-wave rectifier.

If $u_i(t)$ is a sine wave, the waveforms of $u_i(t)$, $u_1(t)$ and $u_o(t)$ are shown in Figure 37.13.

Connecting capacitor C in the feedback loop of amplifier A2 turns it into a first-order low-pass filter, so that the circuit output voltage equals the average value of $u_o(t)$:

$$\overline{U}_o = \overline{\left|u_i(t)\right|} \qquad (37.38)$$

In the case of sinusoidal input voltage with rms value U_i, the output voltage is related to this rms value by:

$$\overline{U}_o = \frac{2\sqrt{2}}{\pi}U_i \qquad (37.39)$$

$\overline{U}_o$ can be measured by a dc voltmeter.

Both meters in Figures 37.10 and 37.12 are actually average detectors. However, due to Equations 37.34 and 37.39, their scale can be labeled in such a way that the instrument reading gives the rms value of the input voltage, provided it is sinusoidal. When the input voltage is no longer sinusoidal, an error arises that depends on the signal form factor.

True rms Analog Voltmeters.

The rms value U_i of a periodic input voltage signal $u_i(t)$, with period T, is given by:

$$U_i = \sqrt{\frac{1}{T}\int_0^T u_i^2(t)\,dt} \qquad (37.40)$$

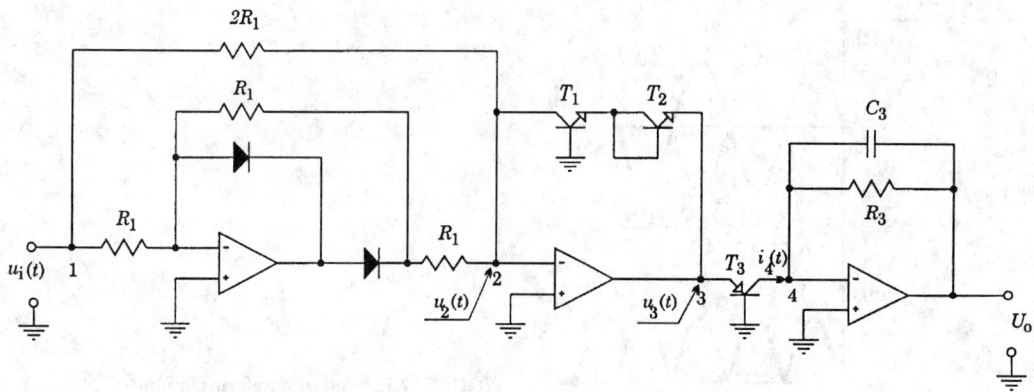

FIGURE 37.14   True rms electronic ac voltmeter schematics.

The electronic circuit shown in Figure 37.14 provides an output signal U_o proportional to the squared rms value of the input signal $u_i(t)$. The circuit section between nodes 1 and 2 is a full-wave rectifier. Hence, node 2 potential is given by:

$$u_2(t) = |u_i(t)| \tag{37.41}$$

The circuit section between nodes 2 and 4 is a log multiplier. Because of the logarithmic characteristic of the feedback path due to the presence of T1 and T2, node 3 potential is given by:

$$u_3(t) = 2k_1 \log\left[u_2(t)\right] = k_1 \log\left[u_2^2(t)\right] = k_1 \log\left[\left|u_i(t)\right|^2\right] = k_1 \log\left[u_i^2(t)\right] \tag{37.42}$$

and, due to the presence of T_3, the current flowing in node 4 is given by:

$$i_4(t) = k_2 \exp\left[u_3(t)\right] = k_3 u_i^2(t) \tag{37.43}$$

The circuit section after node 4 is a low-pass filter that extracts the dc component of the input signal. Therefore, the circuit output voltage is given by:

$$U_o = \frac{k}{T} \int_0^T u_i^2(t) dt = k U_i^2 \tag{37.44}$$

thus providing an output signal proportional to the squared rms value of the input signal $u_i(t)$ in accordance with Equation 37.40. Quantities k_1, k_2, and k depend on the values of the elements in the circuit in Figure 37.14. Under circuit operating conditions, their values can be considered constant, so that k_1, k_2, and k can be considered constant also.

If carefully designed, this circuit can feature an uncertainty in the range of ±1% of full scale, for signal frequencies up to 100 kHz.

Digital Voltmeters

A digital voltmeter (DVM) attains the required measurement by converting the analog input signal into digital, and, when necessary, by discrete-time processing of the converted values. The measurement result is presented in a digital form that can take the form of a digital front-panel display, or a digital output signal. The digital output signal can be coded as a decimal BCD code, or a binary code.

The main factors that characterize DVMs are speed, automatic operation, and programmability. In particular, they presently offer the best combination of speed and accuracy if compared with other available voltage-measuring instruments. Moreover, the capability of automatic operations and programmability make DVMs very useful in applications where flexibility, high speed and computer controllability are required. A typical application field is therefore that of automatically operated systems.

When a DVM is directly interfaced to a digital signal processing (DSP) system and used to convert the analog input voltage into a sequence of sampled values, it is usually called an analog-to-digital converter (ADC).

DVMs basically differ in the following ways: (1) number of measurement ranges, (2) number of digits, (3) accuracy, (4) speed of reading, and (5) operating principle.

The basic measurement ranges of most DVMs are either 1 V or 10 V. It is however possible, with an appropriate preamplifier stage, to obtain full-scale values as low as 0.1 V. If an appropriate voltage divider is used, it is also possible to obtain full-scale values as high as 1000 V.

If the digital presentation takes the form of a digital front-panel display, the measurement result is presented as a decimal number, with a number of digits that typically ranges from 3 to 6. If the digital representation takes the form of a binary-coded output signal, the number of bits of this representation typically ranges from 8 to 16, though 18-bit ADCs are available.

The accuracy of a DVM is usually correlated to its resolution. Indeed, assigning an uncertainty lower than the 0.1% of the range to a three-digit DVM makes no sense, since this is the displayed resolution of the instrument. Similarly, a poorer accuracy makes the three-digit resolution quite useless. Presently, a six-digit DVM can feature an uncertainty range, for short periods of time in controlled environments, as low as the 0.0015% of reading or 0.0002% of full range.

The speed of a DVM can be as high as 1000 readings per second. When the ADC is considered, the conversion rate is taken into account instead of the speed of reading. Presently, the conversion rate for 12-bit, successive approximation ADCs can be on the order of 10 MHz. It can be in the order of 100 MHz for lower resolution, flash ADCs [5].

DVMs can be divided into two main operating principle classes: the *integrating* types and the *nonintegrating* types [3]. The following sections give an example for both types.

Dual Slope DVM.
Dual slope DVMs use a counter and an integrator to convert an unknown analog input voltage into a ratio of time periods multiplied by a reference voltage. The block diagram in Figure 37.15 shows this operating principle. The switch S1 connects the input signal to the integrator for a fixed period of time t_f. If the input voltage is positive and constant, $u_i(t) = U_i > 0$, the integrator output represents a negative-slope ramp signal (Figure 37.16). At the end of t_f, S1 switches and connects the output of the voltage reference U_R to the integrator input. The voltage reference output is negative for a positive input voltage. The integrator output starts to increase, following a positive-slope ramp (Figure 37.16). The process stops when the ramp attains the 0 V level, and the comparator allows the control logic to switch S1 again. The period of time t_v the ramp takes to increase to 0 V is variable and depends on the ramp peak value attained during period t_f.

The relationship between the input voltage U_i and the time periods t_v and t_f is given by:

$$\frac{1}{RC}\int_0^{t_f} U_i \, dt = \frac{t_v}{RC} U_R \tag{37.45}$$

that, for a constant input voltage U_i, leads to:

$$U_i = U_R \frac{t_v}{t_f} \tag{37.46}$$

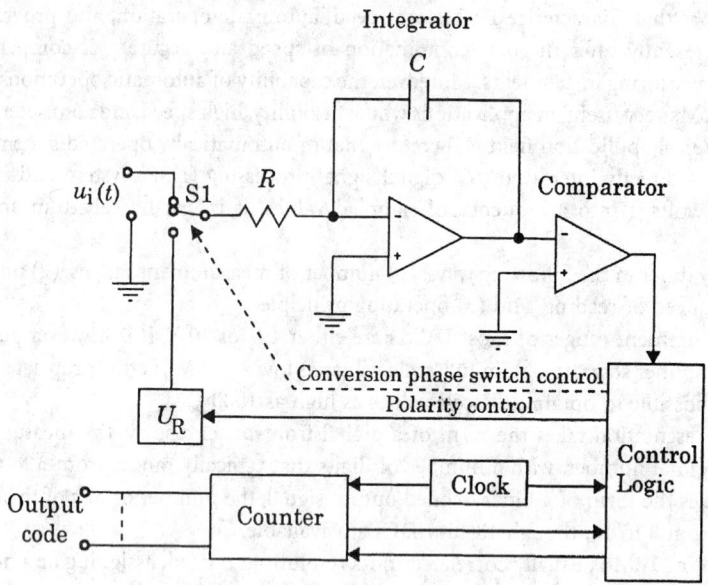

FIGURE 37.15   Dual slope DVM schematics.

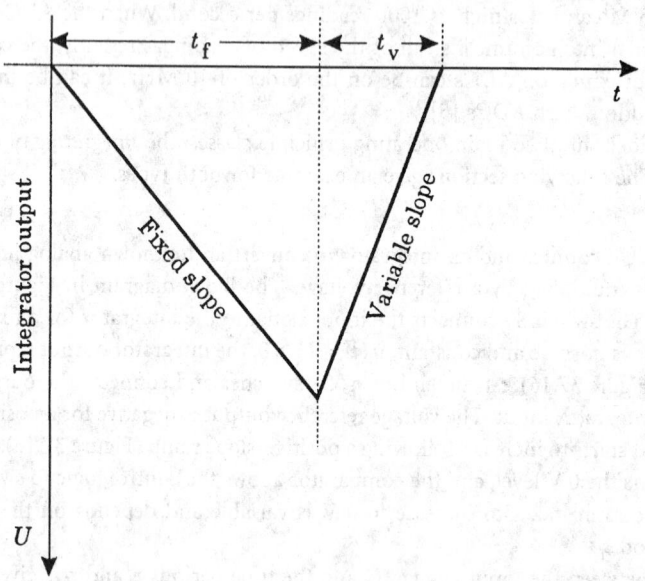

FIGURE 37.16   Integrator output signal in a dual slope DVM.

Since the same integrating circuit is used, errors due to comparator offset, capacitor tolerances, long-term counter clock drifts, and integrator nonlinearities are eliminated. High resolutions are therefore possible, although the speed of reading is low (in the order of milliseconds).

Slowly varying voltages can be also measured by dual slope DVMs. However, this requires that the input signal does not vary for a quantity greater than the DVM resolution during the reading time. For high-resolution DVMs, this limits the DVM bandwidth to a few hertz.

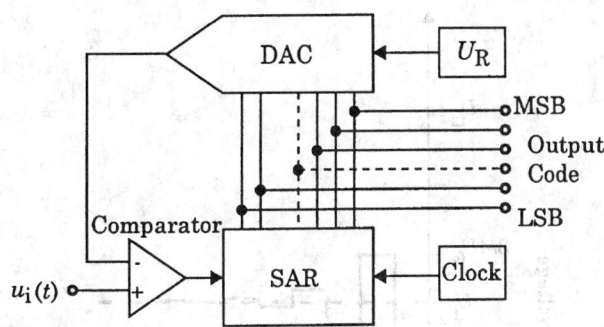

FIGURE 37.17 Successive approximation ADC schematics.

Successive Approximation ADC.

The successive approximation technique represents the most popular technique for the realization of ADCs. Figure 37.17 shows the block diagram of this type of converter. The input voltage is assumed to have a constant value U_i and drives one input of the comparator. The other comparator's input is driven by the output of the digital-to-analog converter (DAC), which converts the binary code provided by the successive approximation register (SAR) into an analog voltage. Let n be the number of bits of the converter, U_R the voltage reference output, and C the code provided by the SAR. The DAC output voltage is then given by:

$$U_c = \frac{C}{2^n} U_R \tag{37.47}$$

When the conversion process starts, the SAR most significant bit (MSB) is set to logic 1. The DAC output, according to Equation 37.47, is set to half the reference value, and hence half the analog input full-scale range. The comparator determines whether the DAC output is above or below the input signal. The comparator output controls the SAR in such a way that, if the input signal is above the DAC output, as shown in Figure 37.18, the SAR MSB is retained and the next bit is set to logic 1.

If now the input signal is below the DAC output (Figure 37.18), the last SAR bit set to logic 1 is reset to logic 0, and the next one is set to logic 1. The process goes on until the SAR least significant bit (LSB) has been set. The entire conversion process takes time $t_c = nT_c$, where T_c is the clock period. At the end of conversion, the SAR output code represents the digitally converted value of the input analog voltage U_i.

According to Equation 37.47, the ADC resolution is $U_R/2^n$, which corresponds to 1 LSB. The conversion error can be kept in the range $\pm\frac{1}{2}$ LSB. Presently, a wide range of devices is available, with resolution from 8 to 16 bits, and conversion rates from 100 μs to below 1 μs.

Varying voltages can be sampled and converted into digital by the ADC, provided the input signal does not vary by a quantity greater than $U_R/2^n$ during the conversion period t_c. The maximum frequency of an input sine wave that satisfies this condition can be readily determined starting from given values of n and t_c.

Let the input voltage of the ADC be an input sine wave with peak-to-peak voltage $U_{pp} = U_R$ and frequency f. Its maximum variation occurs at the zero-crossing time and, due to the short conversion period t_c, is given by $2\pi f t_c U_{pp}$. To avoid conversion errors, it must be:

$$2\pi f t_c U_{pp} \leq \frac{U_R}{2^n} \tag{37.48}$$

Since $U_{pp} = U_R$ is assumed, this leads to:

$$f \leq \frac{1}{2^n 2\pi t_c} \tag{37.49}$$

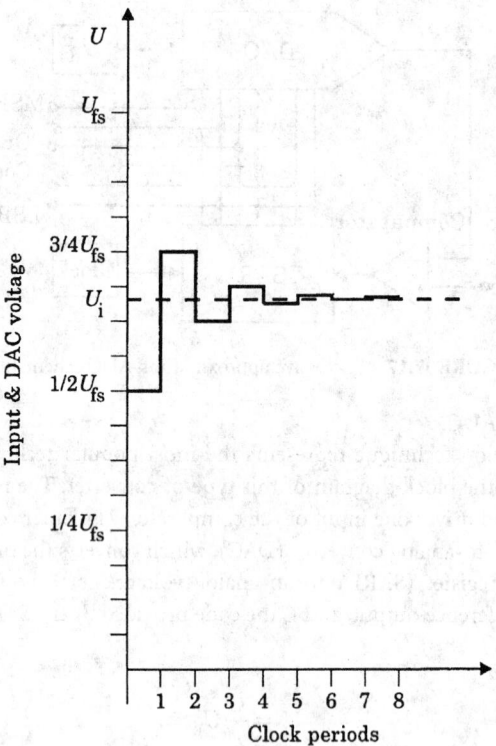

FIGURE 37.18   DAC output signal in a successive approximation ADC.

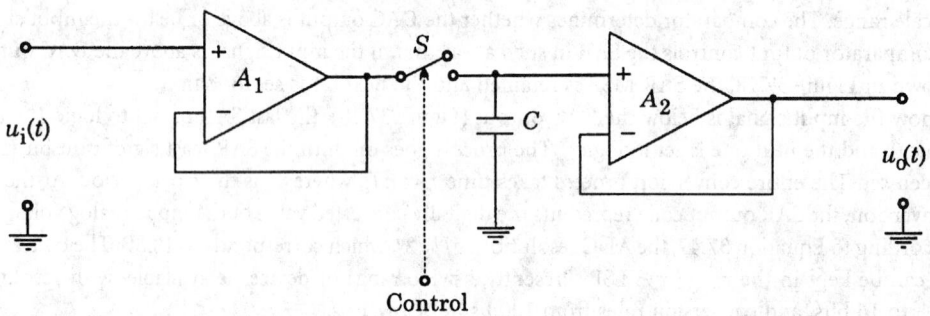

FIGURE 37.19   Sample and Hold schematics

If $t_c = 1$ μs and $n = 12$, Equation 37.49 leads to $f \leq 38.86$ Hz. However, ADCs can still be employed with input signals whose frequency exceeds the value given by Equation 37.49, provided that a *Sample and Hold* circuit is used to keep the input voltage constant during the conversion period.

The *Sample and Hold* circuit is shown in Figure 37.19. When the electronic switch S is closed, the output voltage $u_o(t)$ follows the input voltage $u_i(t)$. When switch S is open, the output voltage is the same as the voltage across capacitor C, which is charged at the value assumed by the input voltage at the time the switch was opened. Due to the high input impedance of the operational amplifier A_2, if a suitable value is chosen for capacitor C, its discharge transient is slow enough to keep the variation of the output voltage below the ADC resolution.

Ac Digital Voltmeters.

True rms ac voltmeters with digital reading can be obtained using an electronic circuit like the one in Figure 37.14 to convert the rms value into a dc voltage signal, and measuring it by means of a DVM.

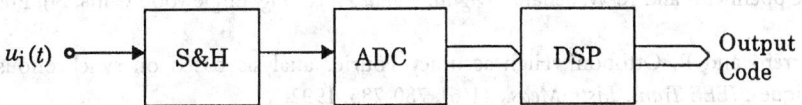

FIGURE 37.20 Block diagram of a modern digital meter.

However, this structure cannot actually be called a digital structure, because the measurement is attained by means of analog processing of the input signal.

A more modern approach, totally digital, is shown in Figure 37.20. The input signal $u_i(t)$ is sampled at constant sampling rate f_s, and converted into digital by the ADC. The digital samples are stored in the memory of the digital signal processor (DSP) and then processed in order to evaluate Equation 37.40 in a numerical way. Assuming that the input signal is periodic, with period T, and its frequency spectrum is upper limited by harmonic component of order N, the sampling theorem is satisfied if at least $(2N + 1)$ samples are taken over period T in such a way that $(2N + 1)T_s = T$, $T_s = 1/f_s$ being the sampling period [6, 7]. If $u_i(kT_s)$ is the kth sample, the rms value of the input signal is given by, according to Equation 37.40:

$$U^2 = \sqrt{\frac{1}{2N+1}\sum_{k=0}^{2N} u_i^2(kT_s)} \qquad (37.50)$$

This approach can feature a relative uncertainty as low as ±0.1% of full scale, with an ADC resolution of 12 bits. The instrument bandwidth is limited to half the sampling frequency, according to the sampling theorem. When modern ADCs and DSPs are employed, a 500-kHz bandwidth can be obtained. Wider bandwidths can be obtained, but with a lower ADC resolution, and hence with a lower accuracy.

Frequency Response of ac Voltmeters.
When the frequency response of ac voltmeters is taken into account, a distinction must be made between the analog voltmeters (both electromechanical and electronic) and digital voltmeters, based on DSP techniques.

The frequency response of the analog meters is basically a low-pass response, well below 1 kHz for most electromechanical instruments, and up to hundreds of kilohertz for electronic instruments.

When digital, DSP-based meters are concerned, the sampling theorem and aliasing effects must be considered. To a first approximation, the frequency response of a digital meter can be considered flat as long as the frequency-domain components of the input signal are limited to a frequency band narrower than half the sampling rate. If the signal components exceed this limit (the so-called Nyquist frequency), the aliasing phenomenon occurs [6]. Because of this phenomenon, the signal components at frequencies higher than half the sampling rate are folded over the lower frequency components, changing them. Large measurement errors occur under this situation.

To prevent the aliasing, a low-pass filter must be placed at the input stage of any digital meter. The filter cut-off frequency must ensure that all frequency components above half the sampling rate are negligible. If the low-pass, anti-aliasing filter is used, the digital DSP-based meters feature a low-pass frequency response also.

References

1. M. B. Stout, *Basic Electrical Measurements*, Englewood Cliffs, NJ, Prentice-Hall, 1960.
2. I. F. Kinnard, *Applied Electrical Measurements*, New York, John Wiley & Sons, Chapman & Hall, Ltd. London, 1956.
3. B. M. Oliver and J. M. Cage, *Electronic Measurements and Instrumentation*, London, McGraw-Hill, Inc. 1975.
4. T. T. Lang, *Electronics of Measuring Systems*, New York, John Wiley & Sons, 1987.
5. Analog Devices, *Analog-Digital Conversion Handbook*, Englewood Cliffs, NJ, Prentice-Hall, 1986.

6. A. V. Oppenheim and R. W. Schafer, *Digital Signal Processing*, Englewood Cliffs, NJ, Prentice-Hall, 1975.
7. A. Ferrero and R. Ottoboni, High-accuracy Fourier analysis based on synchronous sampling techniques. *IEEE Trans. Instr. Meas.*, 41(6), 780-785, 1992.

37.2 Oscilloscope Voltage Measurement

Jerry Murphy

Engineers, scientists, and other technical professionals around the world depend on oscilloscopes as one of the primary voltage measuring instruments. This is an unusual situation because the oscilloscope is not the most accurate voltage measuring instrument usually available in the lab. It is the graphical nature of the oscilloscope that makes it so valued as a measurement instrument — not its measurement accuracy.

The oscilloscope is an instrument that presents a graphical display of its input voltage as a function of time. It displays voltage waveforms that cannot easily be described by numerical methods. For example, the output of a battery can be completely described by its output voltage and current. However, the output of a more complex signal source needs additional information such as frequency, duty cycle, peak-to-peak amplitude, overshoot, preshoot, rise time, fall time, and more to be completely described. The oscilloscope, with its graphical presentation of complex waveforms, is ideally suited to this task. It is often described as the "screwdriver of the electronic engineer" because the oscilloscope is the most fundamental tool that technical professionals apply to the problem of trying to understand the details of the operation of their electronic circuit or device. So, what is an oscilloscope?

> The oscilloscope is an electronic instrument that presents a high-fidelity graphical display of the rapidly changing voltage at its input terminals.

The most frequently used display mode is voltage vs. time. This is not the only display that could be used, nor is it the display that is best suited for all situations. For example, the oscilloscope could be called on to produce a display of two changing voltages plotted one against the other, such as a Lissajous display. To accurately display rapidly changing signals, the oscilloscope is a high bandwidth device. This means that it must be capable of displaying the high-order harmonics of the signal being applied to its input terminals in order to correctly display that signal.

The Oscilloscope Block Diagram

The oscilloscope contains four basic circuit blocks: the vertical amplifier, the time base, the trigger, and the display. This section treats each of these in a high-level overview. Many textbooks exist that cover the details of the design and construction of each of these blocks in detail [1]. This discussion will cover these blocks in enough detail so that readers can construct their own mental model of how their operation affects the application of the oscilloscope for their voltage measurement application. Most readers of this book have a mental model of the operation of the automatic transmission of an automobile that is sufficient for its successful operation but not sufficient for the overhaul or redesign of that component. It is the goal of this section to instill that level of understanding in the operation of the oscilloscope. Those readers who desire a deeper understanding will get their needs met in later sections.

Of the four basic blocks of the oscilloscope, the most visible of these blocks is the display with its *cathode-ray tube* (CRT). This is the component in the oscilloscope that produces the graphical display of the input voltage and it is the component with which the user has the most contact. Figure 37.21 shows the input signal is applied to the vertical axis of a cathode ray tube. This is the correct model for an analog oscilloscope but it is overly simplified in the case of the digital oscilloscope. The important thing to learn from this diagram is that the input signal will be operated on by the oscilloscope's vertical axis circuits so that it can be displayed by the CRT. The differences between the analog and digital oscilloscope are covered in later sections.

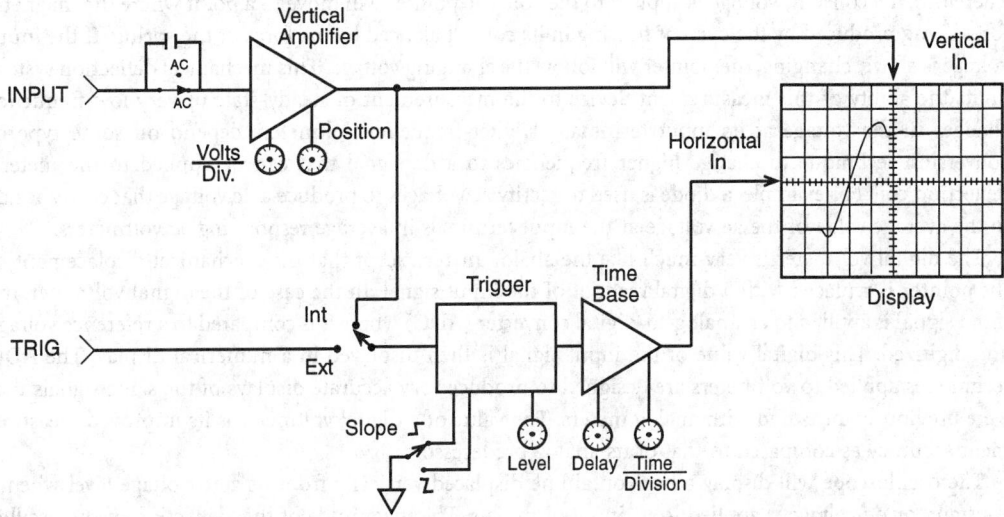

FIGURE 37.21 Simplified oscilloscope block diagram that applies to either analog or digital oscilloscopes. In the case of the digital oscilloscope, the vertical amplifier block will include the ADC and high-speed waveform memory. For the analog scope the vertical block will include delay lines with their associated drivers and a power amplifier to drive the CRT plates.

The *vertical amplifier* conditions the input signal so that it can be displayed on the CRT. The vertical amplifier provides controls of volts per division, position, and coupling, allowing the user to obtain the desired display. This amplifier must have a high enough bandwidth to ensure that all of the significant frequency components of the input signal reach the CRT.

The *trigger* is responsible for starting the display at the same point on the input signal every time the display is refreshed. It is the stable display of a complex waveform that allows the user of an oscilloscope to make judgments about that waveform and its implications as to the operation of the device under test.

The final piece of the simplified block diagram is the *time base*. This circuit block is also known as the horizontal system in some literature. The time base is the part of the oscilloscope that causes the input signal to be displayed as a function of time. The circuitry in this block causes the CRT beam to be deflected from left to right as the input signal is being applied to the vertical deflection section of the CRT. Controls for time-per-division and position (or delay) allow the user of the oscilloscope to adjust the display for the most useful display of the input signal. The time-per-division controls of most oscilloscopes provide a wide range of values, ranging from a few nanoseconds (10^{-9} s) to seconds per division. To get a feeling for the magnitude of the dynamic range of the oscilloscope's time base settings, keep in mind that light travels about 1 m in 3 ns.

The Oscilloscope As a Voltage Measurement Instrument

That the oscilloscope's vertical axis requires a wide bandwidth amplifier and its time base is capable of displaying events that are as short as a few nanoseconds apart, indicates that the oscilloscope can display rapidly changing voltages. Voltmeters, on the other hand, are designed to give their operator a numeric readout of steady-state or slowly changing voltages. Voltmeters are not well suited for displaying voltages that are changing levels very quickly. This can be better understood by examination of the operation of a voltmeter as compared to that of an oscilloscope. The analog voltmeter uses the magnetic field produced by current flowing through a coil to move the pointer against the force of a spring. This nearly linear deflection of the voltmeter pointer is calibrated by applying known standard voltages to its input.

Therefore, if a constant voltage is applied to the coil, the pointer will move to a point where the magnetic force being produced by the current flowing in its coil is balanced by the force of the spring. If the input voltage is slowly changing, the pointer will follow the changing voltage. This mechanical deflection system limits the ability of this measurement device to the measurement of steady-state or very low-frequency changes in the voltage at its input terminals. Higher-frequency voltmeters depend on some type of conversion technique to change higher frequencies to a dc signal that can be applied to the meter's deflection coil. For example, a diode is used to rectify ac voltages to produce a dc voltage that corresponds to the average value of the ac voltage at the input terminals in average responding ac voltmeters.

The digital voltmeter is very much like the analog meter except that the mechanical displacement of the pointer is replaced with a digital readout of the input signal. In the case of the digital voltmeter, the input signal is applied to an analog-to-digital converter (ADC) where it is compared to a reference voltage and digitized. This digital value of the input signal is then displayed in a numerical display. The ADC techniques applied to voltmeters are designed to produce very accurate displays of the same signals that were previously measured with analog meters. The value of a digital voltmeter is its improved measurement accuracy as compared to that of its analog predecessors.

The oscilloscope will display a horizontal line displaced vertically from its zero-voltage level when a constant, or dc voltage is applied to its input terminals. The magnitude of this deflection of the oscilloscope's beam vertically from the point where it was operating with no input being applied is how the oscilloscope indicates the magnitude of the dc level at its input terminals. Most oscilloscopes have a graticule as a part of their display and the scope's vertical axis is calibrated in volts per division of the graticule. As one can imagine, this is not a very informative display of a dc level and perhaps a voltmeter with its numeric readout is better suited for such applications.

There is more to the scope–voltmeter comparison than is obvious from the previous discussion. That the oscilloscope is based on a wide-bandwidth data-acquisition system is the major difference between these two measurement instruments. The oscilloscope is designed to produce a high fidelity display of rapidly changing signals. This puts additional constraints on the design of the oscilloscope's vertical system that are not required in the voltmeter. The most significant of these constraints is that of a constant group delay. This is a rather complex topic that is usually covered in network analysis texts. It can be easily understood if one realizes the effect of group delay on a complex input signal.

Figure 37.22 shows such a signal. The amplitude of this signal is a dc level and the rising edge is made up of a series of high-frequency components. Each of these high-frequency components is a sine wave of specific amplitude and frequency. Another example of a complex signal is a square wave with a frequency of 10 MHz. This signal is made up of a series of odd harmonics of that fundamental frequency. These harmonics are sine waves of frequencies of 10 MHz, 30 MHz, 50 MHz, 70 MHz, etc. So, the oscilloscope must pass all of these high-frequency components to the display with little or no distortion. Group delay is the measure of the propagation time of each component through the vertical system. A constant group delay means that each of these components will take the same amount of time to propagate through the vertical system to the CRT, independent of their frequencies. If the higher-order harmonics take more or less time to reach the scope's deflection system than the lower harmonics, the resulting display will be a distorted representation of the input signal. Group delay (in seconds) is calculated by taking the first derivative of an amplifier's phase-vs.-frequency response (in radians/(l/s)). If the amplifier has a linearly increasing phase shift with frequency, the first derivative of its phase response will be a horizontal line corresponding to the slope of the phase plot (in seconds). Amplifier systems that have a constant group delay are known as Gaussian amplifiers. They have this name because their pass band shape resembles that of the bell curve of a Gaussian distribution function (Figure 37.23). One would think that the oscilloscope's vertical amplifier should have a flat frequency response, but this is not the case because such amplifiers have nonconstant group delay [1].

The oscilloscope's bandwidth specification is based on the frequency where the vertical deflection will be −3 dB (0.707) of the input signal. This means that if a constant 1-V sine wave is applied to the oscilloscope's input, and the signal's frequency is adjusted to higher and higher frequencies, the oscilloscope's bandwidth will be that frequency where its display of the input signal has been reduced to be

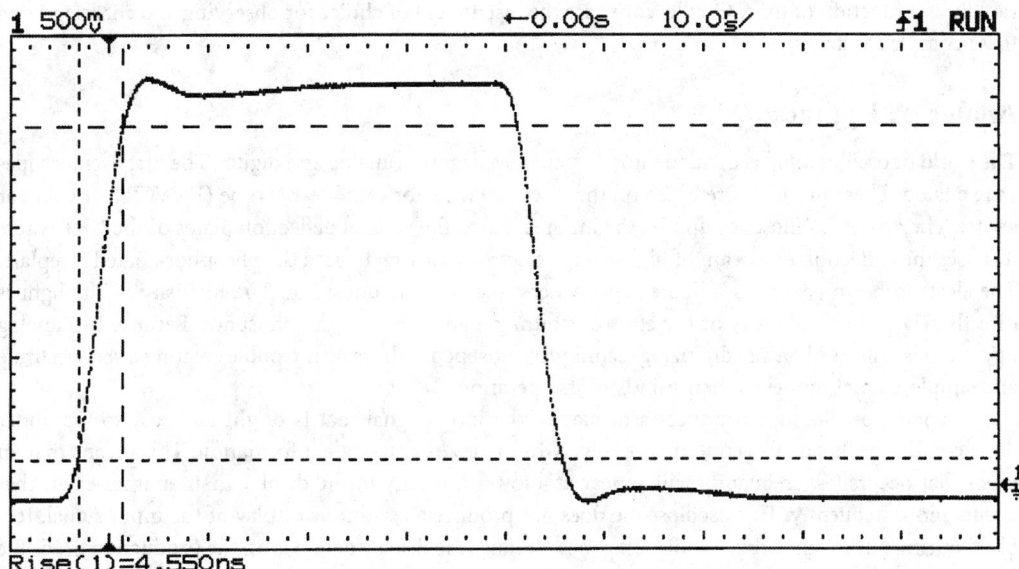

FIGURE 37.22   A typical complex waveform. This waveform is described by measurements of its amplitude, offset, risetime, falltime, overshoot, preshoot, and droop.

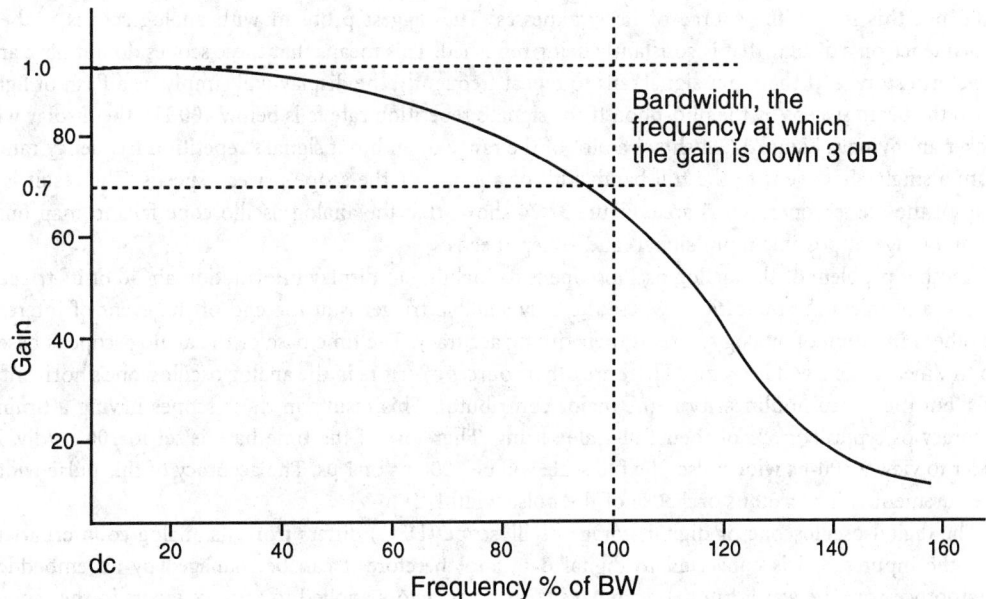

FIGURE 37.23   The Gaussian frequency response of the oscilloscope's vertical system which is not flat in its pass band. Amplitude measurements made at frequencies greater than 20% of the scope's bandwidth will be in error.

0.707 V. Noticable errors in amputude measurements will start at 20% of the scope's bandwidth. The oscilloscope's error-free display of complex waveforms gives it poor voltage accuracy. For the measurement of dc and single frequency signals such as sine waves, other instruments can produce more accurate measurements.

Conclusion: The voltmeter makes the most accurate measurements of voltages that are dc, slowly changing, or can be converted to a dc analog of their ac content. The oscilloscope is not the most accurate voltage measurement instrument, but it is well suited to measurements of voltages that are changing very rapidly as a function of time. Oscilloscopes are the instrument of choice for observing and characterizing these complex voltages.

Analog or Digital

The world of oscilloscopes is divided into two general categories: analog and digital. The first oscilloscopes were analog. These products are based on the direct-view vector cathode-ray tube (DVVCRT or CRT for short). The analog oscilloscope applies the input signal to the vertical deflection plates of the CRT where it causes the deflection of a beam of high-energy electrons moving toward the phosphor-coated faceplate. The electron beam generates a lighted spot where it strikes the phosphor. The intensity of the light is directly related to the density of the electrons hitting a given area of the phosphor. Because this analog operation is not based on any digitizing techniques, most people have little trouble creating a very accurate and simple mental model in their minds of its operation.

The analog oscilloscope produces a display of the input signal that is bright and easy to see under most conditions. It can also contain as many as 16 shades of gray-scale information. This means that an event that occurs less frequently will appear at a lower intensity in the display than another event that occurs more frequently. This oscilloscope does not produce a continous display of the input signal. It is blind during retrace and trugger hold-off times. Because the display depends on the production of visible light from the phosphor being excited by an electron beam, the display must be refreshed frequently. This makes the analog oscilloscope a low-dead-time display system that can follow rapidly changing signals. Also, there is little lag time in front panel control settings.

The analog oscilloscope is not without its shortcomings. The strength of the analog oscilloscope is its CRT, but this is also the source of its weaknesses. The biggest problem with analog scopes is their dependence on a display that is constantly being refreshed. This means that these scopes do not have any waveform storage. If the input signal fails to repeat frequently, the display will simply be a flash of light when the beam sweeps by the phosphor. If the signal's repetition rate falls below 100 Hz, the display will flicker annoyingly. Figure 37.24 shows a plot of the range of an input signal's repetition frequency range from a single-shot event to the full bandwidth of a scope vs. the scope's sweep speeds. The result is a map of the scope's operational area. Figure 37.24 shows that the analog oscilloscope fails to map onto the full range of possible input signals and sweep speeds.

Another problem of the analog oscilloscope is its inability to display information ahead of its trigger. This is a problem in applications where the only suitable trigger is at the end of the event of interest. Another limitation of analog scopes is their timing accuracy. The time base of the analog scope is based on the linearity of a voltage ramp. There are other sources of errors in the analog oscilloscope's horizontal axis, but the sweep nonlinearity is the major contributor. This results in these scopes having a timing accuracy of typically ±3% of their full-scale setting. Therefore, if the time base is set to 100 ns/div, in order to view a 100-ns wide pulse, the full scale will be 1000 ns or 1 μs. The accuracy of this pulse width measurement will be ±30 ns or ±30% of the pulse width!

The digital oscilloscope or digital storage oscilloscope (DSO) differs from its analog counterpart in that the input signal is converted to digital data and therefore it can be managed by an embedded microprocessor. The waveform data can have correction factors applied to remove errors in the scope's acquisition system and can then be stored, measured, and/or displayed. That the input signal is converted from analog to digital and manipulations are performed on it by a microprocessor results in people not having a good mental model of the digital oscilloscope's operation. This would not be a problem except for the fact that the waveform digitizing process is not totally free from errors, and a lack of a correct mental model of the scope's operation on the part of its user can increase the odds of a measurement error. To make matters worse, various manufacturers of these products make conflicting claims, making it easy to propagate incorrect mental models of the digital scope's operation. It is the intention of this presentation to give the information needed to create a mental model of the operation of these devices that will enable the user to perform error-free measurements with ease.

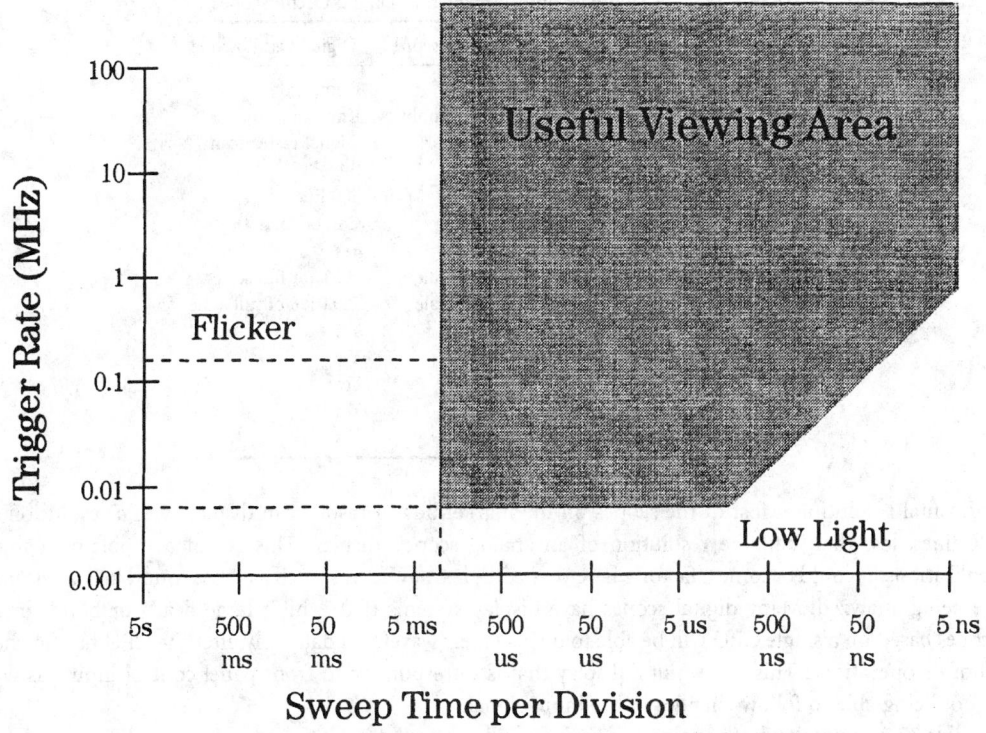

FIGURE 37.24 The operating range of the analog oscilloscope. This is a plot of input signal repetition rate from the lower limit of single shot to the full bandwidth of the scope plotted against sweep speed. The shaded area is the area where the analog oscilloscope will produce a usable display.

The digital storage oscilloscope offers many advantages over its analog counterpart. The first is accuracy. The voltage measurement accuracy of the digital oscilloscope is better than that of an analog scope because the microprocessor can apply correction factors to the data to correct for errors in the calibration of the scope's vertical system. The timing accuracy of a digital oscilloscope is an order of magnitude better than that of an analog scope. The digital scope can store the waveform data for comparison to other test results or uploading to a computer for analysis or project documentation. The digital oscilloscope does not depend on the input signal being continuously updated to produce an easily viewable display. A single-shot event is displayed at the same brightness level as a signal that repeats in time periods corresponding to the full bandwidth of the scope.

The disadvantages of the digital oscilloscope are its more complex operation, aliasing, and display performance. The analog-to-digital conversion process [1] is used to convert the input signal into a series of discrete values, or samples, uniformly spaced in time, which can be stored in memory. Voltage resolution is determined by the total number of codes that can be produced. A larger number permits a smoother and more accurate reproduction of the input waveform but increases both the cost and difficulty in achieving a high sample frequency. Most digital oscilloscopes provide 8-bit resolution in their ADC. As the ADC's sampling speed is increased, the samples will be closer together, resulting smaller gaps in the waveform record.

All digital scopes are capable of producing an aliased display. Some models are more prone to this problem than others, but even the best will alias under the right conditions. An alias is a lower frequency false reproduction of the input signal resulting from under-sampling, i.e., sampling less than the Nyquist frequency. The display of the digital scope is based on computer display technology. This results in a display that is very bright and easy to see, even under conditions where an analog scope would have difficulty in producing a viewable display. The disadvantage of the digital scope's display is its lower

TABLE 37.2 A Comparison of Analog and Digital Oscilloscopes

	Analog Oscilloscope	Digital Oscilloscope
Operation	Simple	Complex
Front panel controls	Direct access knobs	Knobs and menus
Display	Real-time vector	Digital raster scan
Gray scales	>16	>4
Horizontal resolution	>1000 lines	500 lines
Dead-time	Short	Can be long
Aliasing	No	Yes
Voltage accuracy	±3% of full scale	±3% of full scale
Timing accuracy	±3% of full scale	±0.01% of full scale
Single shot capture	None	Yes
Glitch capture	Limited	Yes
Waveform storage	None	Yes
Pretrigger viewing	None	Yes
Data out to a computer	No	Yes

horizontal resolution. Most of the scopes on the market have a raster scan display with a resolution of 500 lines, less than half the resolution of an analog scope's display. This is not a problem in most applications. It could become a factor where very complex waveforms, such as those found in TV systems, are being analyzed. Many digital scopes have display systems that exhibit large dead- or blind-times. Scopes based on a single CPU will be able to display their waveform data only after the CPU has finished all of its operations. This can result a display that is unresponsive to front panel control inputs as well as not being able to follow changes in the input signal.

Table 37.2 shows that both analog and digital oscilloscopes have relative advantages and disadvantages. All the major producers of oscilloscopes are pushing the development of digital scopes in an attempt to overcome their disadvantages. All the major producers of these products believe that the future is digital. However, a few manufacturers produce scopes that are both analog and digital. These products appear to have the best of both worlds; however, they have penalties with respect to both cost and complexity of operation.

One of the driving forces making scope manufacturers believe that the future of the digital oscilloscope is bright is that modern electronic systems are becoming ever more digital in nature. Digital systems place additional demands on the oscilloscope that exceed the capabilities of the analog scope. For example, often in digital electronic systems, there is a need to view fast events that occur at very slow or infrequent rates. Figure 37.24 shows that these events fail to be viewable on analog scopes. Another common problem with digital systems is the location of trigger events. Often the only usable trigger is available at the end of the event being viewed. Analog scopes can only display events that occur after a trigger event. The rapid growth of digital electronics that in the late 1990s is being attributed to the lowering of the cost of single-chip microcontrollers. These devices, which contain a complete microprocessor on one integrated circuit, are responsible for the "electronics everywhere" phenomenon, where mechanical devices are becoming electronic as well as those devices that were previously electrical in nature. In 1996, Hewlett Packard introduced a new class of oscilloscope designed to meet the unique needs of the microcontroller-based applications. This new class of oscilloscope is known as the mixed signal oscilloscope or MSO [2].

Voltage Measurements

Voltage measurements are usually based on comparisons of the waveform display to the oscilloscope's graticule. Measurements are made by counting the number of graticule lines between the end-points of the desired measurement and then multiplying that number by the sensitivity setting. This was the only measurement available to most analog scope users, and it is still used by those performing troubleshooting with their digital scope as a time-saving step. (Some late-model analog oscilloscopes incorporate cursors

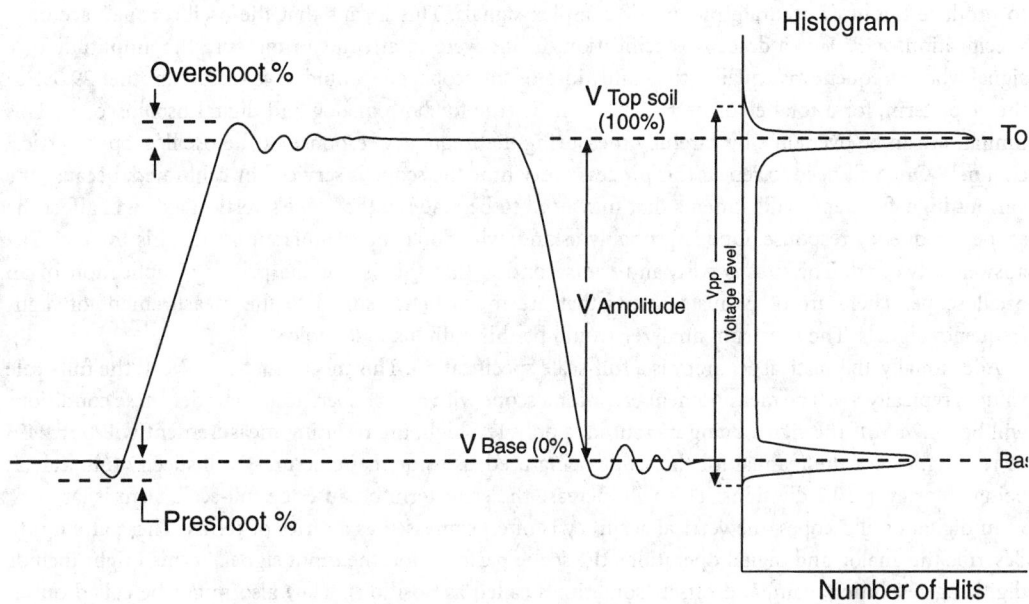

FIGURE 37.25 Voltage histograms as applied by a digital oscilloscope. The complex waveform is measured by use of the voltage histogram. This histogram is a plot of each voltage level in the display and the number of data points at that level.

to enhance their measurement ability.) For example, a waveform that is 4.5 divisions high at a vertical sensitivity of 100 mV/div would be 450 mV high.

Switching the scope's coupling between ac and dc modes will produce a vertical shift in the waveform's position that is a measurement of its dc component. This technique can be applied to either analog or digital scopes. Simply note the magnitude of the change in waveform position and multiply by the channel's sensitivity.

Additional measurements can be performed with an analog oscilloscope but they usually require more skill on the part of the operator. For example, if the operator can determine the location of the top and base of a complex waveform, its amplitude can be measured. Measurements based on percentages can be made using the scope's vernier to scale the waveform so that its top and bottom are 5 divisions apart. Then, each division represents 20% of the amplitude of the waveform being studied. The use of the vernier, which results in the channel being uncalibrated, prevents performance of voltage measurements. Many analog scopes have a red light to warn the operator that the scope is uncalibrated when in vernier mode.

The digital oscilloscope contains an embedded microprocessor that automates the measurement. This measurement automation is based on a histogramming technique, where a histogram of all the voltages levels in the waveform are taken from the oscilloscope's waveform data. The histogram is a plot of the voltage levels in the waveform plotted against the number of samples found at each voltage level. Figure 37.25 shows the histogramming technique being applied to the voltage measurements of complex waveforms.

Understanding the Specifications

The oscilloscope's vertical accuracy is one place that a person's mental model of the scope's operation can lead to measurement trouble. For example, the oscilloscope's vertical axis has a frequency response that is not flat across its pass band. However, as noted above, the scope has a Gaussian frequency response to produce the most accurate picture of complex signals. This means that the oscilloscope's accuracy specification of ±3% is a dc-only specification. If one were to attempt to measure the amplitude of a signal whose frequency is equal to the bandwidth of the scope, one would have to add another 29.3% to the error term, for a total error of ±32.3%. This is true for both analog and digital oscilloscopes. This limitation can be overcome by carefully measuring the frequency response of the oscilloscope's vertical channels. One will need to repeat this process every time the scope is serviced or calibrated, because the various high-frequency adjustments that may need to be made in the scope's vertical axis will affect the scope's frequency response. One is probably asking, why don't the manufacturers do this for me? The answer is twofold. The first is cost, and the second is that this is not the primary application of an oscilloscope. There are other instruments that are much better suited to the measurement of high-frequency signals. The spectrum analyzer would be this author's first choice.

Additionally, the vertical accuracy is a full-scale specification. This means that at 1 V/div, the full-scale value is typically 8 V. The measurement error for a scope with a ±3% specification under these conditions will be ±0.24 V. If the signal being measured is only 1 V high, the resulting measurement will be ±24% of reading. Check the manual for the scope being used, as some manufacturers will specify full-scale as being 10 or even 10.2 divisions. This will increase the error term because the full-scale term is larger.

In digital oscilloscopes, the vertical accuracy is often expressed as a series of terms. These attempt to describe the analog and digital operations the scope performs on the input signal. Terms might include digitizing resolution, gain, and offset (sometimes called as position). They also might be called out as single and dual cursor accuracies. The single cursor accuracy is a sum of all three terms. In the dual cursor case, where the voltage measurement is made between two voltage cursors, the offset term will cancel out, leaving only the digitizing resolution and gain errors. For example, the Hewlett Packard model 54603B has a single cursor accuracy specification of ±1.2% of full scale, ±0.5% of position value, and a dual cursor specification of ±0.4% of full scale.

HINT: Always try to make the voltage measurements on the largest possible vertical and widest possible display of the signal.

The horizontal accuracy specifications of analog and digital scopes are very different; however, both are based on a full-scale value. In the analog scope, many manufacturers limit accuracy specifications to only the center eight divisions of their display. This means that a measurement of a signal that starts or ends in either the first or ninth graticule, will be even more error prone than stated in the scope's specifications. To the best of this author's knowledge, this limitation does not apply to digital scopes. The horizontal specifications of digital scopes are expressed as a series of terms. These might include the crystal accuracy, horizontal display resolution, and trigger placement resolution. These can be listed as cursor accuracy. For example, the Hewlett Packard model 54603B has a horizontal cursor accuracy specification of ±0.01% ±0.2% full-scale ±200 ps. In this example, the first term is the crystal accuracy, the second is the display resolution (500 lines), and the final term is twice the trigger placement error. By comparing the analog and digital scopes' horizontal specifications, it can be seen that in either case, the measurement is more accurate if it can be made at full screen. The digital scope is more accurate than its analog counterpart.

Digital scopes also have acquisition system specifications. Here is another place where the operator's mental model of the operation of a digital scope can produce measurement errors. All manufacturers of digital scopes specify the maximum sampling speed of their scope's acquisition system as well as its memory depth and number of bits. The scope's maximum sampling speed does not apply to all sweep speeds, only memory depth and number of bits applies to all sweep speeds. The scope's maximum sampling speed applies only to its fastest sweep speeds.

The complexity of the digital scope results from the problem of having to sample the input. There is more to be considered than Nyquist's Sampling Theorem in the operation of a digital scope. For example, how does the scope's maximum sampling rate relate to the smallest time interval that the scope can capture and display? A scope that samples at 100 MSa s^{-1} takes a sample every 10 ns; therefore, in principle, it cannot display any event that is less than 10 ns wide because that event will fall between the samples. In practice, however, this limit can — under certain circumstances — be extended. If the scope is operating in an "equivalent time" or "random repetitive" mode and if the signal is repetitive, even if very

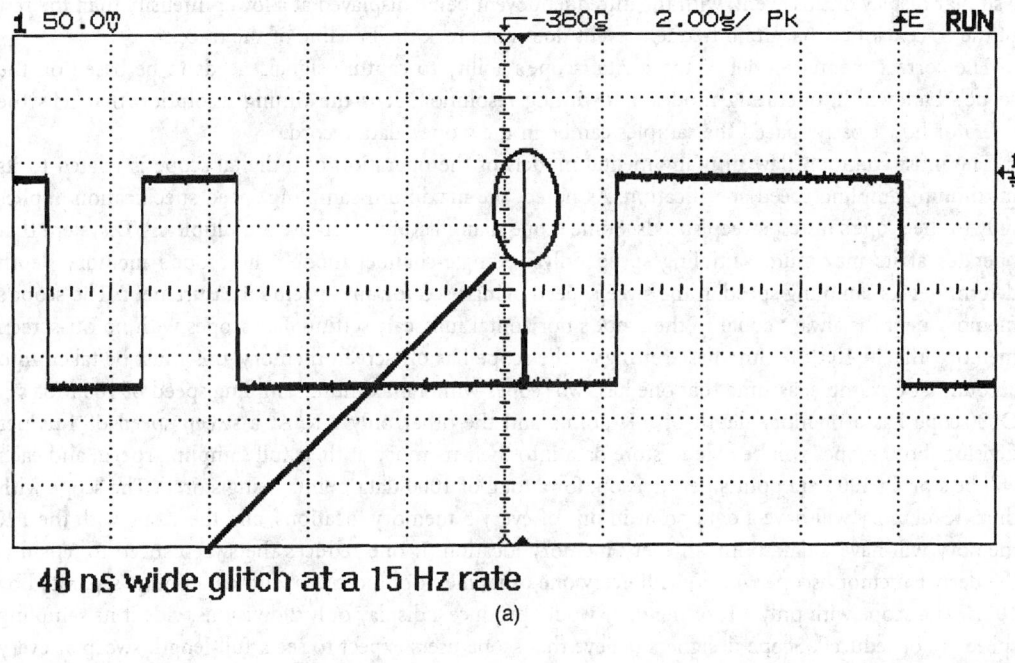

48 ns wide glitch at a 15 Hz rate

(a)

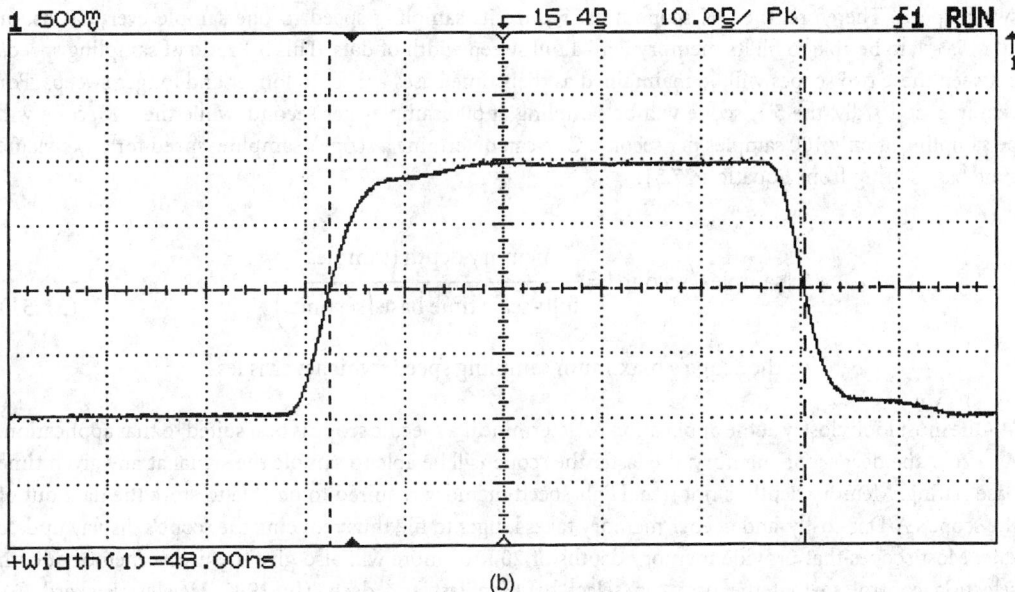

+Width(1)=48.00ns

(b)

FIGURE 37.26 An infrequently occurring event as displayed on a digital oscilloscope with random repetitive sampling. (a): the event embedded in a pulse train. (b): shows the same event at a faster sweep speed. The fact that the waveform baseline is unbroken under the narrow pulse indicates that it does not occur in every sweep. The width of this pulse is less than half the scope's sampling period in (b). Both traces are from a Hewlett Packard model 54603B dual channel 60-MHz scope.

infrequently, the scope will be able to capture any event that is within its vertical system bandwidth. Figure 37.26 shows an infrequently occurring pulse that is 25 ns wide embedded into a data stream being captured and displayed on an oscilloscope with a maximum sampling speed of 20 MSa s^{-1} (sampling interval of 50 ns). Figure 37.26(b) shows this pulse at a faster sweep speed. An analog scope would produce a similar display of this event, with the infrequent event being displayed at a lower intensity than the rest of the trace. Notice that the infrequent event does not break the baseline of the trace.

The correct mental model of the digital scope's ability to capture signals needs to be based on the scope's bandwidth, operating modes, and timing resolution. It is the timing resolution that tells the operator how closely spaced the samples can be in the scope's data record.

The most common flaw in many mental models of the operation of a digital scope is related to its maximum sampling speed specification. As noted, the maximum sampling speed specification applies only to the scope's fastest sweep speeds. Some scope manufacturers will use a multiplex A/D system that operates at its maximum sampling speed only in single-channel mode. The scope's memory depth determines its sampling speed at the sweep speed being used for any specific measurement. The scope's memory depth is always equal to the scope's horizontal full-scale setting. For scopes with no off-screen memory, this in 10× the time base setting. If the scope has off-screen memory, this must be taken into account. For example, assume that one has two scopes with a maximum sampling speed of 100 MSa s⁻¹. One scope has a memory depth of 5 K points and the other only 1 K. At a sweep speed of 1 μs per division, both scopes will be able to store data into their memory at their full sampling speed, and each will be storing 100 data points per division, for a total of 1000 data points being stored. The scope with the 5 K memory will have a data point in one of every 5 memory locations, and the scope with the 1 K memory will have a data point in every memory location. If one reduces the sweep speed to 5 μs/div, the deeper memory scope will now fill every one of its memory locations with data points separated by 10 ns. The scope with only 1 K of memory would produce a display only 2 divisions wide if its sampling speed is not reduced. Scope designers believe that scope users expect to see a full-length sweep at every sweep speed. Therefore, the 1 K scope must reduce its sampling speed to one sample every 50 ns, or 20 MSa s⁻¹, to be able to fill its memory with a full sweep width of data. This 5:1 ratio of sampling speeds between these two scopes will be maintained as their time bases are set to longer and longer sweeps. For example, at 1 s/div, the 5 K scope will be sampling at 500 samples per second, while the 1 K scope will be sampling at only 100 samples per second. One can determine a scope's sampling speed for any specific time base setting from Equation 37.51.

$$S\,(\text{samples}/\text{second}) = \frac{\text{memory depth}\,(\text{samples})}{\text{full-scale time base}\,(\text{seconds})},\tag{37.51}$$

or the scope's maximum sampling speed, whichever is less

One must look closely at the application to determine if a specific scope is best suited to that application. As a rule, the deeper the memory, the faster the scope will be able to sample the signal at any given time base setting. Memory depth is not free. High-speed memory required to be able to store the data out of the scope's A/D is costly, and deeper memory takes longer to fill, thus reducing the scope's display update rate. Most scopes that provide memory depths of 20 K or more will also give the user a memory depth selection control so that the user can select between fast and deep. (In 1996, Hewlett Packard Co. introduced two scopes based on an acquisition technology known as MegaZoom (TM) [10] that removes the need for a memory depth control.) A correct mental model for the sampling speed of a digital scope is based on Equation 37.51 and not just on the scope's maximum performance specifications.

Some digital oscilloscopes offer a special sampling mode known as *peak detection*. Peak detection is a special mode that has the effect of extending the scope's sampling speed to longer time records. This special mode can reduce the possibility of an aliased display. The performance of this special mode is specified as the minimum pulse width that the peak detection system can capture. There are several peak detection systems being used by the various manufacturers. Tektronix has an analog-based peak detection system in some of its models, while Hewlett Packard has a digital system in all of its models. Both systems perform as advertised, and they should be evaluated in the lab to see which system best meets one's needs. There is a down side to peak detection systems and that is that they display high-frequency noise

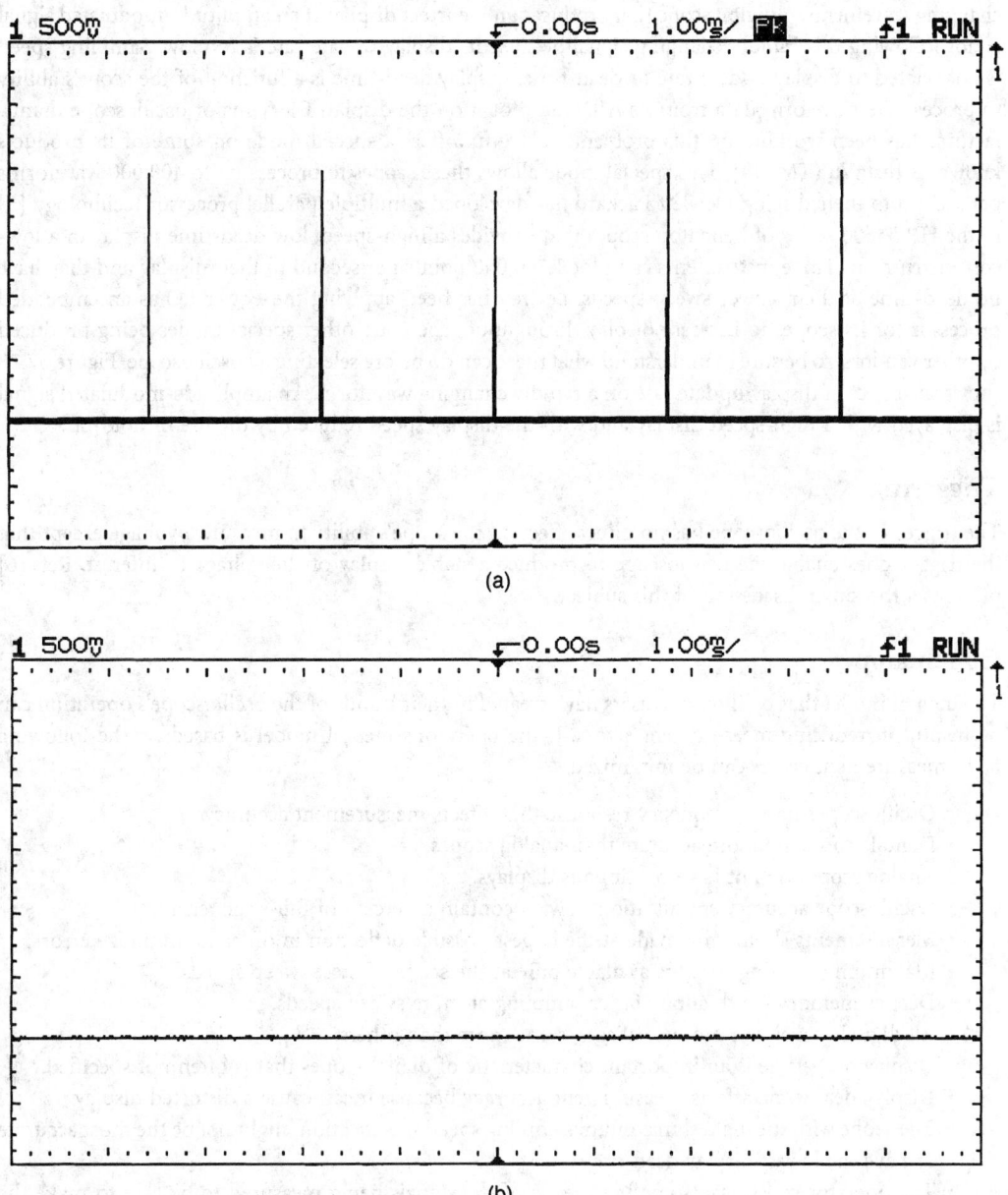

FIGURE 37.27 Peak detection. This special mode has the effect of increasing the scope's sampling speed at time base settings where it would be decimated. In operation, each memory location contains either the maximum or minimum value of the waveform at that location in time. (a): a series of 300-ns wide pulses being captured at a slow sweep speed; (b): the same setup with peak detection disabled. These narrow pulses would appear as intermittent pulses if the scope could be seen in operation with peak detection disabled.

that might not be within the bandwidth of the system under test. Figure 37.27 shows a narrow pulse being captured by peak detection and being missed when the peak detection is off.

What effect does display dead-time have on the oscilloscope's voltage measurement capabilities? Display dead-time applies to both analog and digital oscilloscopes, and it is that time when the oscilloscope is not capturing the input signal. This is also a very important consideration in the operation of a digital scope because it determines the scope's ability to respond to front-panel control commands and to follow changing waveforms. A digital scope that produces an incorrect display of an amplitude-modulated signal is not following this rapidly changing signal because its display update rate is too low. Sampling speed is not related to display update rate or dead-time. Display dead-time is a function of the scope's ability to process the waveform data from its A/D and plot it on the display. Every major oscilloscope manufacturer has been working on this problem. Tektronix offers a special mode on some of its products known as InstaVu (TM) [4]. This special mode allows these scopes to process up to 400,000 waveforms per second to their display. Hewlett Packard has developed a multiple parallel processor technology [5] in the HP 54600 series of benchtop scopes that provides a high-speed, low dead-time display in a low-cost instrument. These instruments can plot 1,500,000 points per second to their display and they have no dead-time at their slower sweep speeds. LeCroy has been applying the Power PC as an embedded processor for its scopes to increase display throughput. There are other special modes being produced by other vendors, so be sure to understand what these can do before selecting an oscilloscope. Figure 37.28 shows the effect of display update rate on a rapidly changing waveform. An amplitude-modulated signal is displayed with a high-speed display and with the display speed reduced by the use of hold-off.

Triggering

The trigger of the oscilloscope has no direct effect on the scope's ability to measure a voltage except that the trigger does enable the oscilloscope to produce a stable display of the voltage of interest. Ref. [6] presents a thorough discussion of this subject.

Conclusion

The mental model that oscilloscope users have created in their minds of the oscilloscope's operation can be helpful in reducing measurement errors. If the operator's mental model is based on the following facts, measurement errors can be minimized:

- Oscilloscopes have a frequency response that affects measurement accuracy.
- Digital scopes are more accurate than analog scopes.
- Analog scopes do not have continuous displays.
- Oscilloscope accuracy specifications always contain a percent of full-scale term.
- Measurements should me made at the largest possible deflection in order to minimize errors.
- Maximum sampling speed is available only at the scope's fastest sweep speeds.
- Deeper memory depth allows faster sampling at more sweep speeds.
- All digital scopes can produce aliases, some more than others.
- Display dead-time is an important characteristic of digital scopes that is often not specified.
- Display dead-time affects measurement accuracy because it can cause a distorted display.
- The scope with the highest maximum sampling speed specification might not be the most accurate or have the lowest display dead-time.
- The operator must have some knowledge of the signals being measured to be able to make the best possible measurements.

The person who has the mental model of the oscilloscope that takes these factors into account will be able to purchase the scope that is best suited to his/her application and not spend too much money on unnecessary performance. In addition, that person will be able to make measurements that are up to the full accuracy capabilities of the scope.

Selecting the Oscilloscope

There are ten points to consider when selecting an oscilloscope. This author has published a thorough discussion of these points [7] and they are summarized as follows:

1. **Analog or Digital?** There are a few places where the analog scope might be the best choice, and the reader can make an informed selection based on the information presented here.

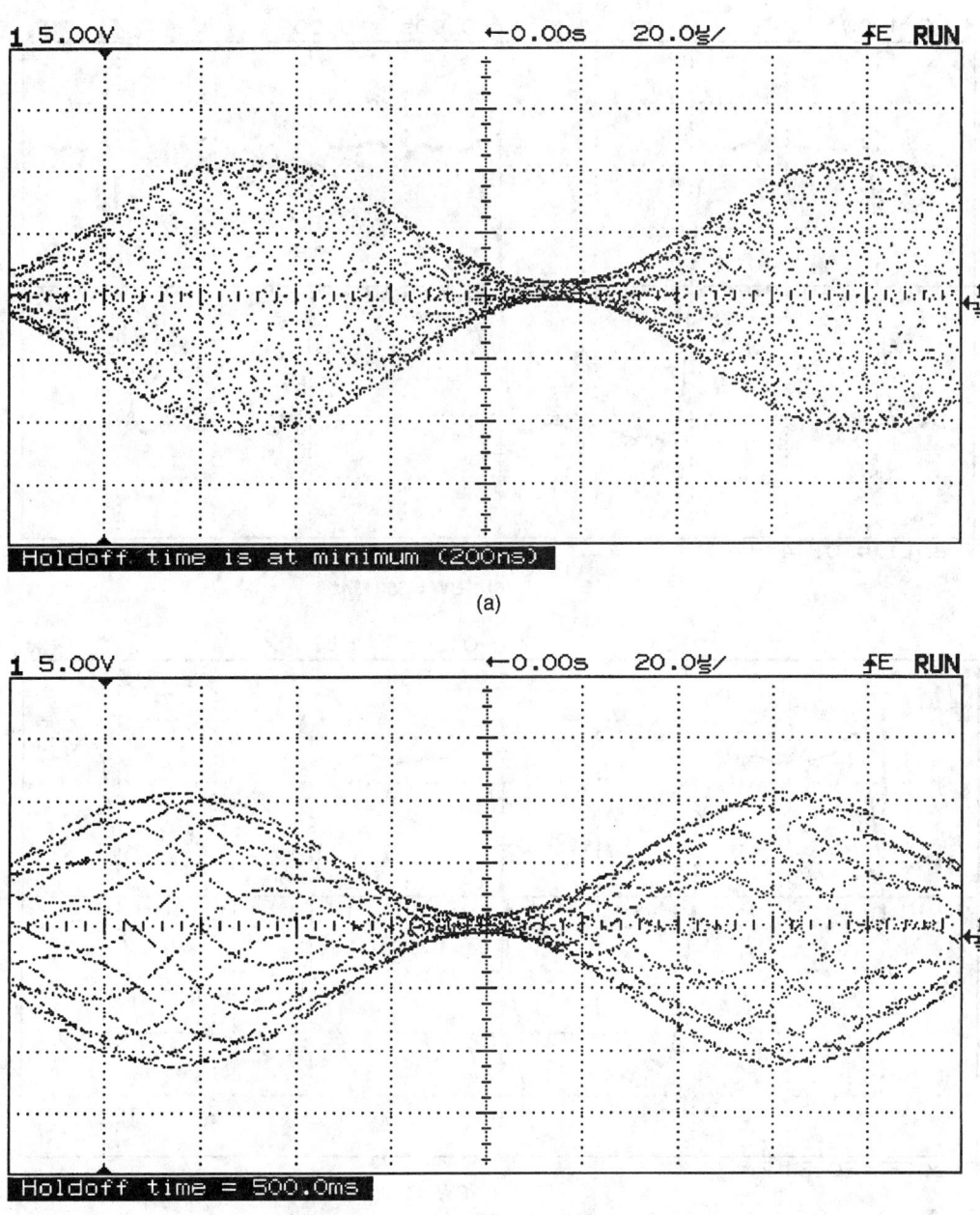

FIGURE 37.28 Display dead-time. The time that an oscilloscope is blind to the input signal has an effect on the scope's ability to correctly display rapidly changing signals. (a): an amplitude-modulated signal with a high-speed display; (b): the same signal with the dead-time increased by use of hold-off.

2. **How much bandwidth?** This is a place where the person selecting an oscilloscope can save money by not purchasing more bandwidth than is needed. When analog oscilloscopes were the only choice, many people were forced to purchase more bandwidth than they needed because they needed to view infrequent or low repetition signals. High bandwidth analog scopes had brighter CRTs so that they were able to display high-frequency signals at very fast time base settings. At a sweep speed of 5 ns/div, the phosphor is being energized by the electron beam for 50 ns, so the electron beam had

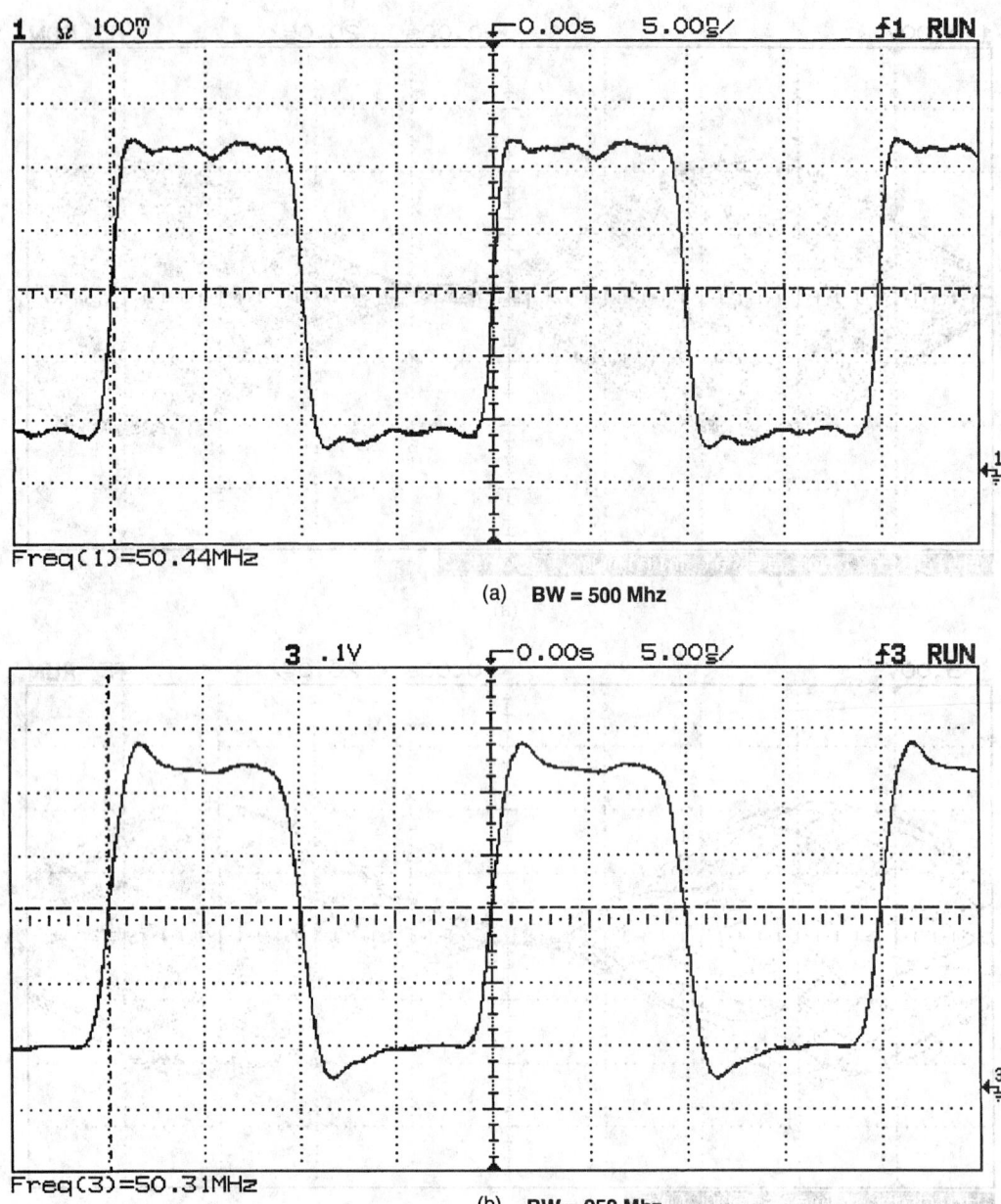

(a)   **BW = 500 Mhz**

(b)   **BW = 250 Mhz**

FIGURE 37.29   The effect of the scope's bandwidth is shown in this set of waveforms. The same 50-MHz square wave is shown as it was displayed on scopes of 500 MHz in Figure 37.28(a) all the way down to 20 MHz in Figure 37.29(e). Notice that the 100-MHz scope produced a usable display although it was missing the high-frequency details of the 500-MHz display. The reason that the 100-MHz scope looks so good is the fact that its bandwidth is slightly greater than 100 MHz. This performance, which is not specified on any data sheet, is something to look for in any evaluation.

to be very high energy to produce a visible trace. This situation does not apply to digital scopes. Now, one needs to be concerned only with the bandwidth required to make the measurement. Figure 37.29 shows the effect of oscilloscope bandwidth on the display of a 50-MHz square wave.

The oscilloscope's bandwidth should be >2× the fundamental highest frequency signal to be measured.

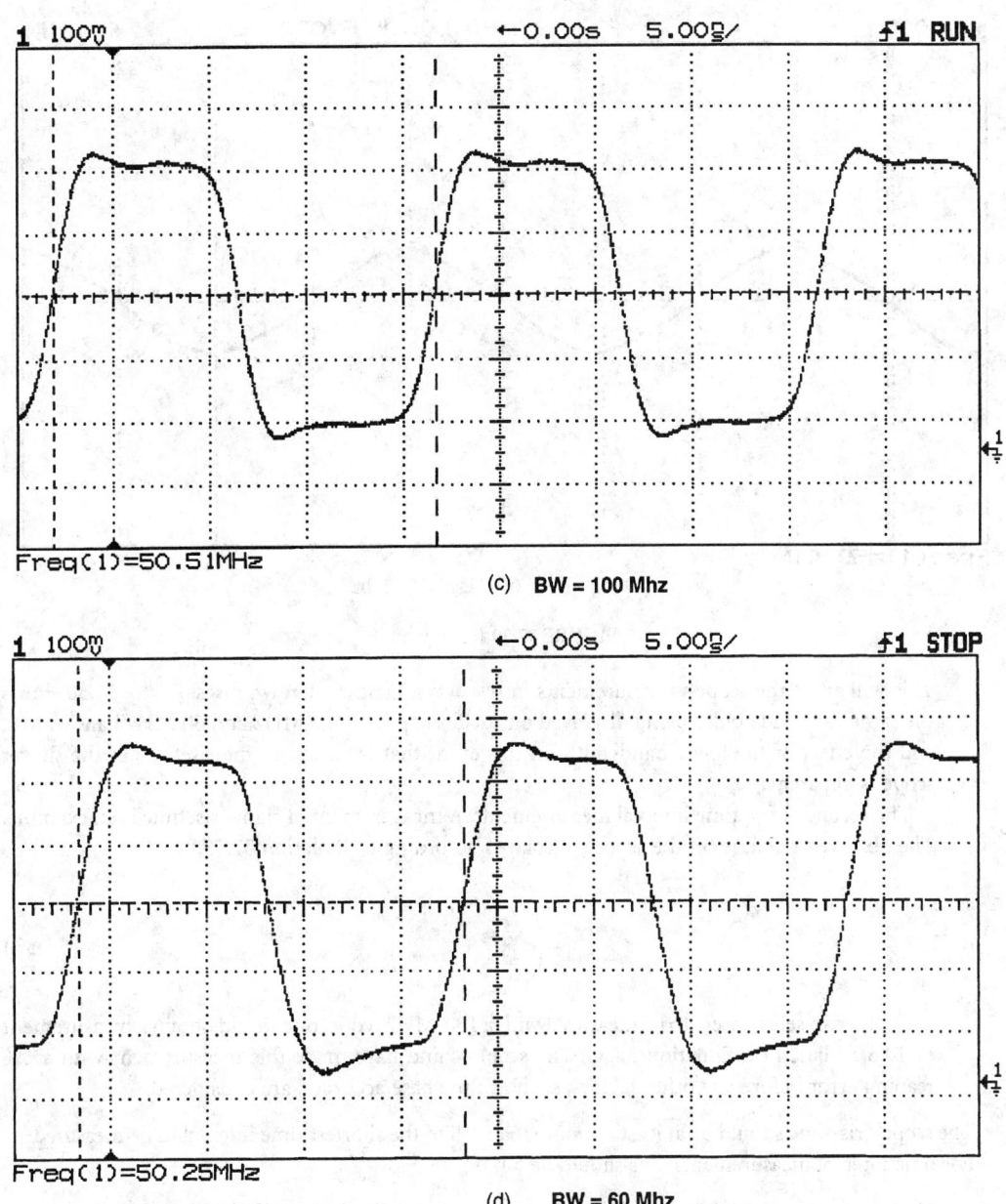

FIGURE 37.29 (continued)

The bandwidth of the scope's vertical system can affect the scope's ability to correctly display narrow pulses and to make time interval measurements. Because of the scope's Gaussian frequency response, one can determine its ability to correctly display a transient event in terms of risetime with Equation 37.52.

$$t_r = 0.35/\text{BW} \qquad (37.52)$$

Therefore, a 100-MHz scope will have a rise time of 3.5 ns. This means that if the scope were to have a signal at its input with zero rise time edges, it would be displayed with 3.5-ns edges.

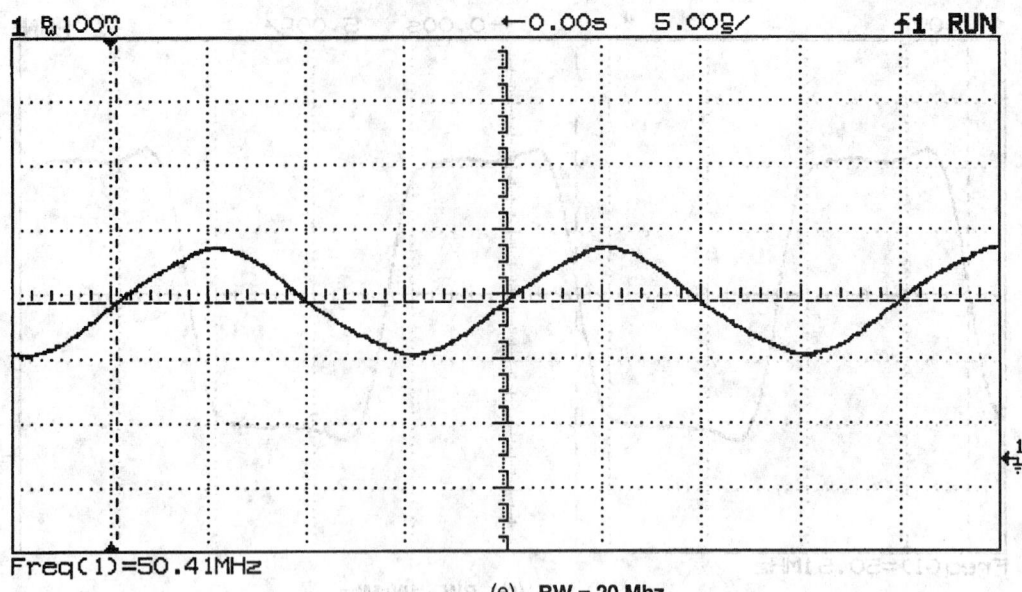

(e) **BW = 20 Mhz**

FIGURE 37.29 (continued)

This will affect the scope's measurements in two ways. First is narrow pulses. Figure 37.30 shows the same 5-ns wide pulse being displayed on oscilloscopes of 500 MHz and 60 MHz bandwidths, and the effect of the lower bandwidth on this event that is closest to the risetime of the slower scope is apparent.

The second is fast time interval measurements. A measurement of signal risetime is an example. The observed risetime on the scope's display is according to Equation 37.53.

$$t_{\text{observed}} = \left(t_{\text{signal}}{}^2 + t_{\text{scope}}{}^2 \right)^{1/2} \tag{37.53}$$

If a 10-ns risetime were to be measured with a 100-MHz scope, one would obtain a measurement of 10.6 ns based on Equation 37.53. The scope would have made this measurement with a 6% reading error before any other factors, such as time base accuracy, are considered.

The scope's risetime should be at least no more than 1/5 of the shortest time interval to be measured. For time interval measurements, this should be >1/10.

3. **How many channels?** Most oscilloscopes in use today are dual-channel models. In addition, there are models described as being 2+2 and four channels. This is one time where 2+2 is not equal to 4. The 2+2 models have limited features on two of their channels and cost less than 4-channel models. Most oscilloscope suppliers will hold the 4-channel description only for models with four full-featured channels, but user should be sure to check that the model under consideration so as to be sure if it is a 4- or 2+2 model. Either of the four channel classes is useful for applications involving the testing and development of digital-based systems where the relationship of several signals must be observed.

Hewlett Packard introduced a new class of oscilloscopes that is tailored for the applications involving both analog and digital technologies, or mixed-signal systems. The mixed signal oscilloscope (MSO) [4] provides two scope channels and 16 logic channels so that it can display both the analog and digital operation of a mixed-signal system on its display.

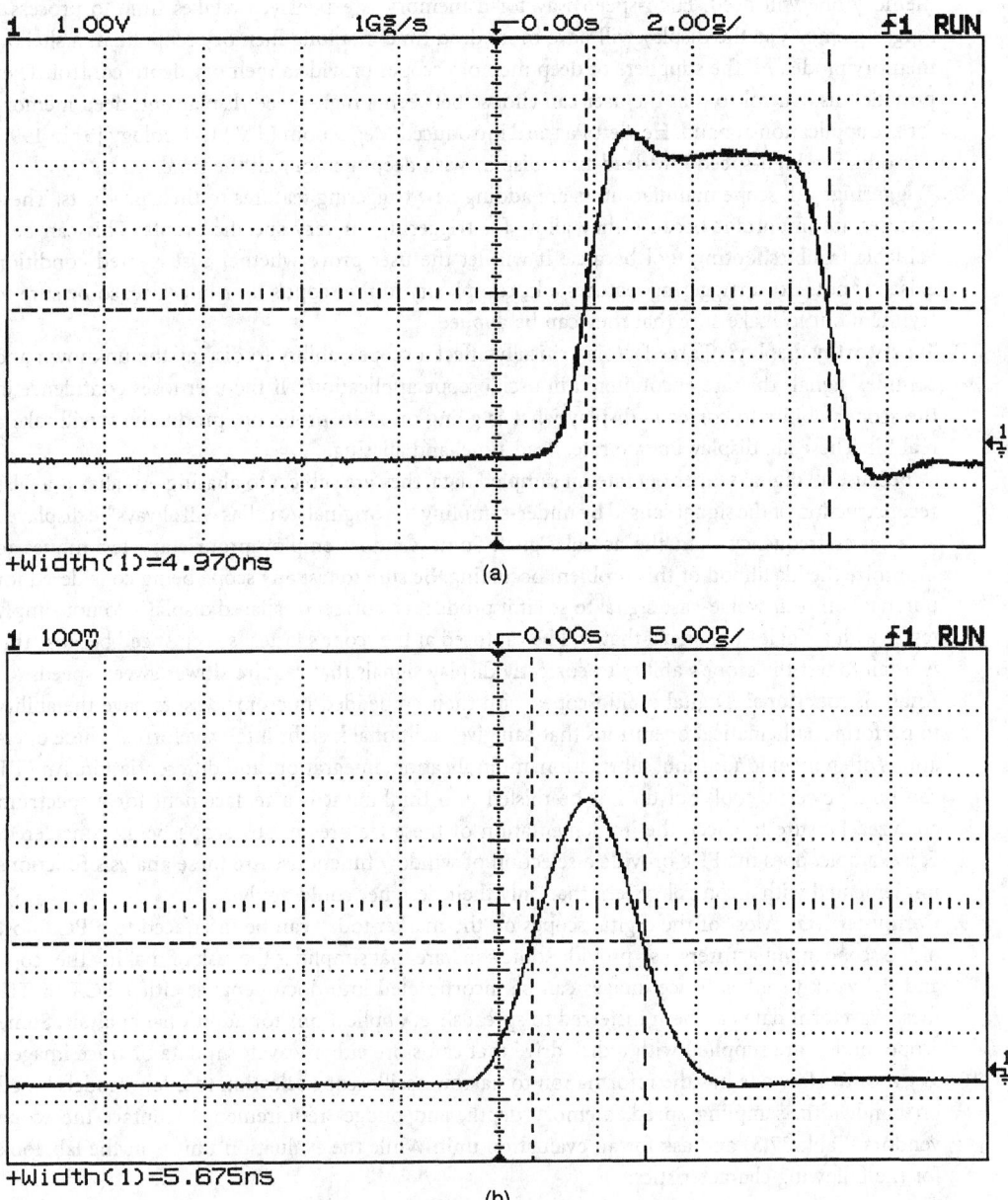

FIGURE 37.30 Bandwidth and narrow events. (a): a 5-ns wide pulse as displayed on a 500-MHz scope; (b): the same pulse displayed on a 60-MHz scope. The 60-MHz scope has a risetime of 5.8 ns, which is longer than the pulse width. This results in the pulse shape being incorrectly displayed and its amplitude being in error.

4. **What sampling speed?** Do not simply pick the scope with the highest banner specification. One needs to ask, what is the sampling speed at the sweep speeds that my application is most likely to require? As observed in Equation 37.51 the scope's sampling speed is a function of memory depth and full-scale time base setting. If waveforms are mostly repetitive, one can save a lot of money by selecting an oscilloscope that provides equivalent time or random repetitive sampling.

5. **How much memory?** As previously discussed, memory depth and sampling speed are related. The memory depth required depends on the time span needed to measure and the time resolution required. The longer the time span to be captured and the finer the resolution required, the more

memory one will need. High-speed waveform memory is expensive. It takes time to process a longer memory, so the display will have more dead time in a long memory scope than a shallow memory model. All the suppliers of deep memory scopes provide a memory depth control. They provide this control so that the user can choose between a high-speed display and deep memory for the application at hand. Hewlett Packard introduced MegaZoom (TM) technology [3] in 1996; it produces a high-speed low dead-time display with deep memory all the time.

6. **Triggering?** All scope manufacturers are adding new triggering features to their products. These features are important because they allow for triggering on very specific events. This can be a valuable troubleshooting tool because it will let the user prove whether a suspected condition exists or not. Extra triggering features add complexity to the scope's user interface; so be sure to try them out to make sure that they can be applied.

7. **Trustworthy display?** Three factors critically affect a scope's ability to display the unknown and complex signals that are encountered in oscilloscope applications. If the user loses confidence in the scope's ability to correctly display what is going on at its probe tip, productivity will take a real hit. These are display update rate, dead-time, and aliasing.

   Because all digital scopes operate on sampled data, they are subject to aliasing. An alias is a false reconstruction of the signal caused by under-sampling the original. An alias will always be displayed as a lower frequency than the actual signal. Some vendors employ proprietary techniques to minimize the likelihood of this problem occurring. Be sure to test any scope being considered for purchase on your worse-case signal to see if it produces a correct or aliased display. Do not simply test it with a single-shot signal that will be captured at the scope's fastest sweep speed because this will fail to test the scope's ability to correctly display signals that require slower sweep speeds.

8. **Analysis functions?** Digital oscilloscopes with their embedded microprocessors have the ability to perform mathematical operations that can give additional insight into waveforms. These operations often include addition, subtraction, multiplication, integration, and differentiation. An FFT can be a powerful tool, but do not be misled into thinking it is a replacement for a spectrum analyzer. Be sure to check the implementation of these features in any scope being considered. For example, does the FFT provide a selection of window functions? Are these analysis functions implemented with a control system that only their designer could apply?

9. **Computer I/O?** Most of the digital scopes on the market today can be interfaced to a PC. Most of the scope manufacturers also provide some software that simplifies the task of making the scope and PC work together. Trace images can be incorporated into documents as either PCX or TIF files. Waveform data can be transferred to spreadsheet applications for additional analysis. Some scope models are supplied with a disk drive that can store either waveform data or trace images.

10. **Try it out?** Now one has the information to narrow oscilloscope selection to a few models based on bandwidth, sampling speed, memory depth, and budget requirements. Contact the scope vendors (Table 37.3) and ask for an evaluation unit. While the evaluation unit is in the lab, look for the following characteristics:

    • Control panel responsiveness: Does the scope respond quickly to inputs or does it have to think about it for a while?
    • Control panel layout: Are the various functions clearly labeled? Does the user have to refer to the manual even for simple things?
    • Display speed: Turn on a couple of automatic measurements and check that the display speed remains fast enough to follow changing signals.
    • Aliasing: Does the scope produce an alias when the time base is reduced from fast to slow sweep speeds? How does the display look for the toughest signal?

The oscilloscope is undergoing a period of rapid change. The major manufacturers of oscilloscopes are no longer producing analog models and the digital models are evolving rapidly. There is confusion in the oscilloscope marketplace because of the rapid pace of this change. Hopefully, this discussion will prove valuable to the user in selecting and applying oscilloscopes in the lab in the years to come.

TABLE 37.3   Major Suppliers of Oscilloscopes and their Web Addresses

Vendor	Description	Web address
B&K Precision 6460 W. Cortland St. Chicago, IL 60635	Analog and digital scopes and Metrix scopes in France	http://bkprecision.com
Boonton Electronics Corp. 25 Estmans Road P.O. Box 465 Parsippany, NJ 07054-0465	U.S. importer for Metrix analog, mixed analog, and digital scopes from France	http://www.boonton.com
Fluke P.O. Box 9090 Everett, WA 98206-9090	Hand-held, battery-powered scopes (ScopeMeter), analog scopes, and CombiScopes(R)	http://www.fluke.com
Gould Roebuck Road, Hainault, Ilford, Essex IG6 3UE, England	200-MHz DSO products	http://www.gould.co.uk
Hewlett Packard Co. Test & Measurement Mail Stop 51LSJ P.O. Box 58199 Santa Clara, CA 95052-9952	A broad line of oscilloscopes and the Mixed Signal oscilloscope for technical professionals	http://www.tmo.hp.com/tmo/pia search on "oscilloscopes"
LeCroy Corp. 700 Chestnut Ridge Road Chestnut Ridge, NY 10977	Deep memory oscilloscopes for the lab	http://www.lecroy.com
Tektronix Inc. Corporate Offices 26600 SW Parkway P.O. Box 1000 Watsonville, OR 97070-1000	The broad line oscilloscope supplier with products ranging from hand-held to high-performance lab scopes	http://www.tek.com/measurement search on "oscilloscopes"
Yokogawa Corp. of America Corporate offices Newnan, GA 1-800-258-2552	Digital oscilloscopes for the lab	http://www.yca.com

References

1. A. DeVibiss, Oscilloscopes, in C. F. Coombs, Jr. (ed.), *Electronic Instrument Handbook,* 2nd ed., New York, McGraw-Hill, 1995.
2. R. A. Witte, A family of instruments for testing mixed-signal circuits and systems, *Hewlett Packard J.,* April 1996, Hewlett Packard Co., Palo Alto, CA.
3. M.S. Holcomb, S.O. Hall, W.S. Tustin, P.J. Burkart, and S.D. Roach, Design of a mixed signal oscilloscope, *Hewlett Packard J.,* April 1996, Hewlett Packard Co., Palo Alto, CA.
4. InstaVu acquisition mode, *Tektronix Measurement Products Catalog,* Tektronix Inc., Beaverton, OR, 1996, 69.
5. M.S. Holcomb and D.P. Timm, A high-throughput acquisition architecture for a 100 MHz digitizing oscilloscope, *Hewlett Packard J.,* February 1992, Hewlett Packard Co., Palo Alto,CA.
6. R.A. Witte, *Elecronic Test Instruments, Theory and Applications,* Englewood Cliffs, NJ, Prentice-Hall, 1993.
7. J. Murphy, Ten points to ponder in picking an oscilloscope, *IEEE Spectrum,* 33(7), 69-77, 1996.

37.3 Inductive and Capacitive Voltage Measurement

Cipriano Bartoletti, Luca Podestà, and Giancarlo Sacerdoti

This chapter section addresses electrical measurements where the voltage range to be measured is very large — from 10^{-10} V to 10^7 V. The waveform can be continuous, periodic, or impulsive. If it is periodic, the spectrum components can vary for different situations, and within the same electric power network there may be subharmonic components. In impulsive voltage measurement, it is often important to get maximum value, pulse length, etc. Capacitive and inductive voltage sensors are mainly utilized in low-frequency electric measurements.

Capacitive Sensors

The voltage to be measured can be reduced by means of capacitive dividers (Figure 37.31). Capacitive dividers are affected by temperature and frequency and therefore are not important, at least in Europe. Capacitive sensors detect voltage by different methods:

1. Electrostatic force (or torque)
2. Kerr or Pockels effect
3. Josephson effect
4. Transparency through a liquid crystal device
5. Change in refractive index of the optic fiber or in light pipe

1. The relations that rule the listed capacitive voltage sensors are reported below. The force between two electrodes is (Figure 37.32):

$$F = \varepsilon_0 \frac{S}{d} \left(V_1 - V_2 \right)^2 \qquad (37.54)$$

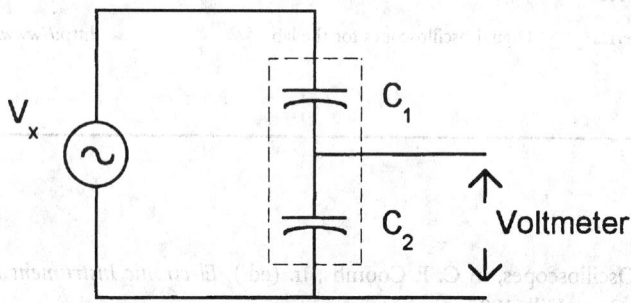

FIGURE 37.31 Schematic arrangement of a capacitive divider.

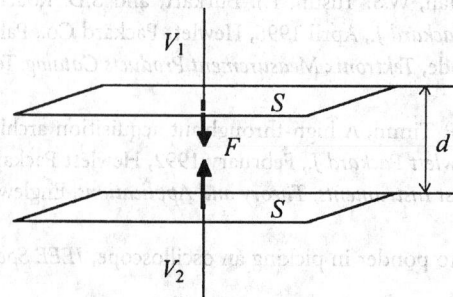

FIGURE 37.32 Force between two electrodes with an applied voltage.

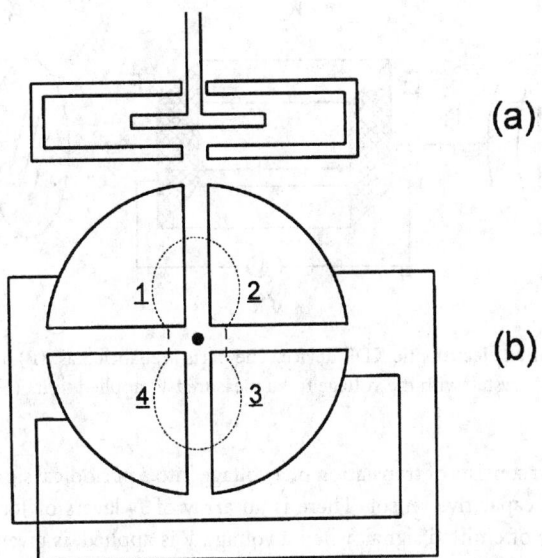

FIGURE 37.33 Scheme of an electrostatic voltmeter. (a) lateral view, (b) top view: (1), (2), (3), (4) are the static electrodes; the moving vane is shown in transparency.

where ε_0 = Dielectric constant
S = Area of the electrode
d = Distance
V_1, V_2 = Potentials of the electrodes

The *torque* between electrostatic voltmeter quadrants (Figure 37.33) is given by:

$$T = \frac{1}{2}\frac{\partial C}{\partial \theta}\left(V_1 - V_2\right)^2 \tag{37.55}$$

where C = Capacitance
θ = Angle between electrodes

To get the torque from the rate of change (derivative) of electrostatic energy vs. the angle is easy. Obtaining the torque by mapping the electric field is difficult and requires long and complex field computing.

2. The rotation of the polarization plane of a light beam passing through a KDP crystal under the influence of an electric field (*Pockels effect*) is expressed by (Figure 37.34):

$$\theta = k_p l\left(V_1 - V_2\right) \tag{37.56}$$

where k_p = Electro-optic constant
l = Length of crystal

One obtains a rotation of $\pi/2$ by applying a voltage of the order of 1 kV to a KDP crystal of a few centimeters in length.

If a light beam passes through a light pipe that performs the *Kerr effect*, one observes a quadratic dependence of the rotation vs. *V.*

$$\theta \equiv kE^2 \equiv k'V^2 \tag{37.57}$$

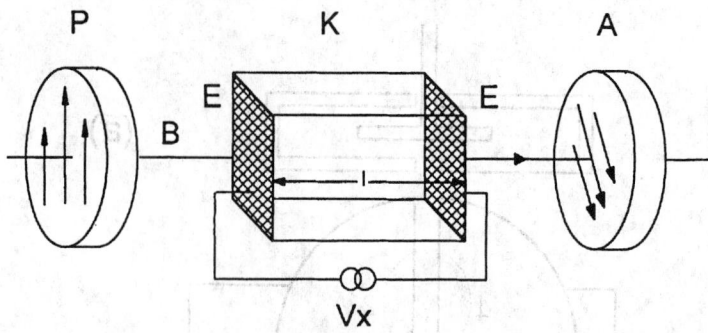

FIGURE 37.34 Scheme of an electrooptic KDP device. The parts are labeled as: (B) a light beam, (P) a polarizer, (A) an analyzer, (K) a KDP crystal, with the voltage to be measured V_x applied to its (E) transparent electrodes.

3. The *Josephson effect* consists of translation of a voltage into a periodical signal of a certain frequency, carried out by a special capacitive sensor. There is an array of N layers of Josephson superconducting junctions; the frequency of emitted signal, when a voltage V is applied, is given by:

$$v = \frac{2eV}{Nh} \tag{37.58}$$

4. The *transparency* of a liquid crystal device depends on the difference of potential applied. There are liquid crystal devices working in transmission or in reflection. A change in transparency is obtained when a difference of potential of a few volts is applied.

5. The *change in refractive index* due to the presence of an electric field can be detected by:

   - Interferometric methods (where the velocity of light is equal to c/n)
   - Change in light intensity in a beam passing through an optical wave guide device like Li-Nb (Figure 37.35).

By means of method 1, many kinds of instruments (voltmeters) can be realized. Methods 2 through 5 are used in research laboratories but are not yet used in industrial measurements.

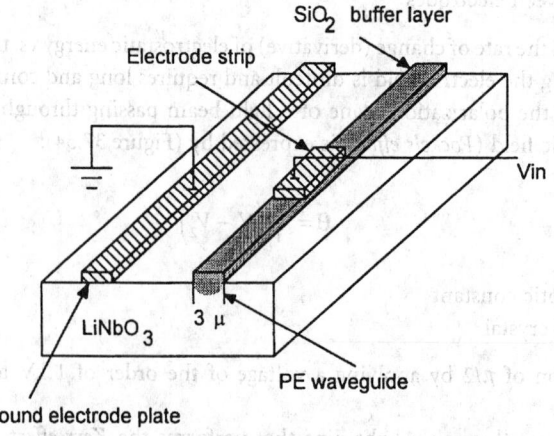

FIGURE 37.35 Li-Nb optical wave guide device.

Inductive Sensors

Voltage Transformers (VTs)

Voltage transformers have two different tasks:

- Reduction in voltage values for meeting the range of normal measuring instruments or protection relays
- Insulation of the measuring circuit from power circuits (necessary when voltage values are over 600 V)

Voltage transformers are composed of two windings — one primary and one secondary winding. The primary winding must be connected to power circuits; the secondary to measuring or protection circuits. Electrically, these two windings are insulated but are connected magnetically by the core.

One can define:

$$Nominal\ ratio = K_n = \frac{V_{1n}}{V_{2n}} \qquad (37.59)$$

as the ratio between the magnitude of primary and secondary rated voltages.

$$Actual\ ratio = K = \frac{V_1}{V_2} \qquad (37.60)$$

as the ratio between the magnitudes of primary and secondary actual voltages.

Burden is the value of the apparent power (normally at $\cos\varphi = 0.8$) that can be provided on the secondary circuit (instruments plus connecting cables).

Burden limits the maximum value of secondary current and then the minimum value of impedance of the secondary circuit is:

$$Z_{min} = \frac{V_{2n}^2}{A_n} \qquad (37.61)$$

where A_n = VT burden

For example, if $A_n = 25$ VA and $V_{2n} = 100$ V, one obtains:

$$Z_{min} = \frac{100}{0.25} = 400\ W \qquad (37.62)$$

There are two kinds of errors:

1. *Ratio error* = $h_\% = \dfrac{K_n - K}{K}$ $\qquad (37.63)$

2. *Angle error* = the phase displacement between the primary voltage and the secondary voltage (positive if the primary voltage lags the secondary one).

Voltage transformers are subdivided into accuracy classes related to the limits in ratio and angle error (according to CEI and IEC normative classes 0.1, 0.2, 0.5, 1, 3; see Table 37.4). To choose the voltage transformer needed, the following technical data must be followed:

TABLE 37.4 Angle and Ratio Error Limit Table Accepted by CEI-IEC Standards

Class	Percentage voltage (ratio) error (±)	Phase displacement Minutes (±)	Centiradians (±)
0.1	0.1	5	0.15
0.2	0.2	10	0.3
0.5	0.5	20	0.6
1	1	40	1.2
3	3	—	—
3P	3	120	3,5
6P	6	240	7

- Primary and secondary voltage (rated transformation ratio). Normally, the secondary value is 100 V.
- Accuracy class and rated burden in VA: e.g., cl. 0.5 and A_n = 10 VA.
- Rated working voltage and frequency
- Insulation voltage
- Voltage factor: the ratio between maximum operating voltage permitted and the rated voltage. The standard voltage factor is 1.2 V_n (i.e., the actual primary voltage) for an unlimited period of time (with VT connected with phases), and is 1.9 V_n for a period of 8 h for VT connected between phase and neutral.
- Thermal power is the maximum burden withstood by VT (errors excluded).

For extremely high voltage values, both capacitive dividers and voltage transformers are normally used, as shown in Figure 37.36. The capacitive impedance must compensate for the effect of the transformer's internal inductive impedance at the working frequency.

Other Methods

The ac voltage inductive sensors act by interaction between a magnetic field (by an electromagnet excited by voltage to be measured) and the eddy current induced in an electroconductive disk, producing a force or a torque. This can be achieved by the scheme shown in Figure 37.37. The weight of many parts of the indicator can be some tens of grams. The power absorbed is on the order of a few watts. The precision is not high, but it is possible to get these sensors or instruments as they are similar to the widely produced induction energy meters. They are quite robust and are priced between $50 and $100, but they are not widely used. The relation between torque and voltage is quadratic:

$$T = k_i V^2 \tag{37.64}$$

The proportionality factor k_i depends on magnet characteristics and disk geometry.

G.E.C., Landys & Gyr, A.B.B., Schlumberger, etc. are the major companies that furnish components and instruments measuring voltage by inductive and capacitive sensors.

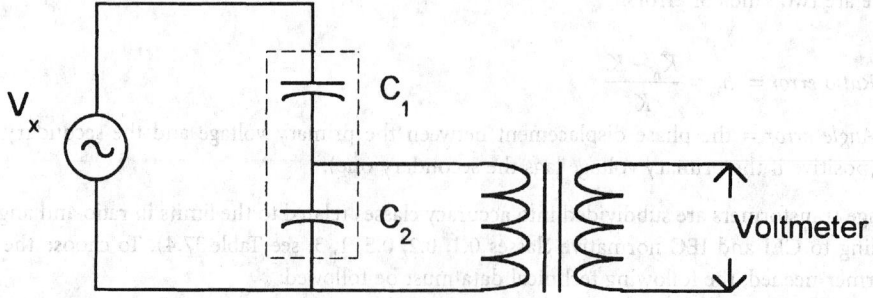

FIGURE 37.36 Capacitive divider and voltage transformer device for extremely high voltage.

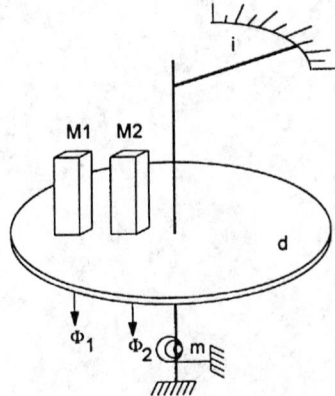

FIGURE 37.37 Schematic inductive voltmeter. The parts are labeled as: (i) index, (d) metallic disk, (M1) and (M2) electromagnets, (m) spring, (Φ1) and (Φ2) generated fluxes.

Defining Terms

CEI: Comitato Elettrotecnico Italiano.
IEC: International Electric Committee.
KDP: Potassium dihydrogen phosphate.
Li-Nb: ($LiNbO_3$) lithium niobate

Further Information

J. Moeller and G. Rosenberger, Instrument transformers for HV Systems, Siemens Power Engineering III
      (1981) Special Issue, *High-Voltage Technology.*
G. Sacerdoti, O. Jappolo, and R. Paggi, *Misure Elettriche, Vol. I Strumenti*, Bologna, Zanichelli, 1994.
G. Zingales, *Metodi e Strumenti per le Misure Elettriche*, Bologna, UTET, 1976.

38

Current Measurement

Douglas P. McNutt
The MacNauchtan Laboratory

Current measuring devices are selected for:

- Precision
- Stability
- Frequency response, including dc
- Galvanic isolation
- Presentation of the data
- Effect on measured circuit

A few of the most common sensors will be introduced in this chapter. Details and some less common sensors will follow after definitions and a bit of magnetic theory, which can be skipped. Any magnetic field sensor can be used as a current sensor and there are some exotic examples, such as quantum effects in low-temperature superconductors used to measure currents in neurons within the brain. This discussion is limited to measurement of currents in wires with commercially practical devices.

An isolated current sensor is free of any metallic connection to the circuit being measured. It is also essentially free of capacitive coupling so that it is safe to use with grounded amplifiers and other equipment. The quality of the isolation is measured in volts and is usually the breakdown potential of an insulator. 5 kV is typical for personal safety around light industrial power.

By far, the simplest current-to-voltage converter is the resistor. In current measuring service, it is called a shunt although it is typically placed in series with the load. That is because shunts are sometimes used to increase the range of another current-measuring device using a connection that bypasses part of the current around a meter. Frequency response of a shunt is good and includes dc. Shunts produce a voltage output that can be presented by a variety of secondary meters, including analog meters, digital meters, oscilloscopes, and 4- to 20-mA converters. Shunts provide no isolation and have a potentially unacceptable effect on the circuit being measured. Shunts used for dc are as accurate as the resistance and the associated voltmeter.

The common moving-coil meter, the D'Arsonval movement [1, 2], probably with an internal shunt and/or rectifier, is an easily used device and still available. Its isolation is by means of the human eye since that is the only way to read out the result. It is useful for power panels where an operator needs quick data. Accuracy is no better than 2%.

For power frequency, 25 Hz to 400 Hz service, the current transformer, called a "donut" transformer, or CT in the trade, is commonly employed. The current-carrying conductor is passed through the hole in a toroid of magnetic material. A shorted secondary winding of n turns carries current, which is $1/n$ times the measured current, and is typically passed to another ammeter or used as the current input to a power measuring device. Isolation is as good as the insulation on the primary conductor; frequency response is fair but does not include dc; there is minimal effect on the measured circuit, and the cost is low. Operational safety is an issue; see the note below.

A variety of noncontact sensors is available for dc sensing. Most depend on the Hall effect and all require a source of operating power. Frequency response from dc to 200 kHz is advertised. Because operating power is available, output to later processing can be voltage, current, or digital. Accuracy is whatever one wants to pay for. Long-term stability depends on dc-coupled operational amplifiers and can exhibit zero drift. Externally, these look much like CTs.

CTs, Hall devices, and other similar noncontact schemes are available in wrap-around form so they can be installed without disconnecting power. The wrapping process always involves breaking a magnetic path, and consistency of reassembly becomes a limit on precision. Everything from current-sensing probes for oscilloscopes to CTs for 10000-A circuits can be found as wrap-arounds.

38.1 Definition of the Ampere

There is perpetual argument about the number of "fundamental" quantities required to describe our environment. Is there a fundamental unit of electricity or are the electrical units derived from more basic things such as mass, length, and time? Standards laboratories such as the U.S. National Institute of Standards and Technology, NIST, measure the force between current carrying wires to compare the ampere to the meter, the kilogram, and the second, and provide a standard stated as "that constant current which, if maintained in two straight parallel conductors of infinite length, of negligible circular cross-section, and placed 1 m apart in vacuum, would produce between these conductors a force equal to 2×10^{-7} newton per meter of length" [3]. For the rest of this discussion, it is assumed that instrumentation is "traceable to the NIST," meaning that one way or another, an electrical unit as measured is compared to a standard ampere maintained by NIST. Absolute calibrations using the quantum properties of the Hall effect and the Josephson junction are not practical in the field.

For practical reasons, it is often easier to distribute a voltage standard than a current standard. Chemical cells and, more recently, semiconductor voltage references are quite stable and do not depend on the length of the wires used to connect them together during calibration and use. As a result, current measuring is usually a matter of conversion of current to an equivalent voltage.

An exception is current measurement by comparison to magnetic forces provided by a permanent magnet and that might be why the older unrationalized centimeter-gram-second, the cgs units, with no separate electrical unit is still found in specifications of magnetic quantities. The gauss and the oersted are particular examples of these now deprecated units. It is this confusion of units that frightens many who would measure current away from even attempting calculations involving magnetic devices.

38.2 Magnetics

Magnetic current sensors have advantages over shunts. To understand them, one should delve into the interaction between currents and magnetic fields. Following Maxwell by way of Sommerfeld [4], it is convenient to describe magnetic effects in terms of two vector fields, **B** and **H**. **H** is the field created by an electric current, and **B** is the field that acts on a moving charge or a current carrying wire. **B** and **H** are related by characteristics of the material in which they coexist. Strictly speaking, **B** is the flux density and **H** is the field, but they are both called the field in less than precise usage.

The SI units of **B** and **H** are the tesla (T) and the ampere per meter (A m^{-1}). They are called rationalized because a ubiquitous 4π has been suppressed in the underlying equations.

For practical engineering, it is still necessary to understand the unrationalized equivalents, the gauss and the oersted, because they are universally used in specifications for magnetic materials. To convert from gauss to tesla, divide by 10^4. To convert from oersted to amperes per meter, multiply by $1000/(4\pi)$, a number that is commonly approximated as simply 80; but strictly speaking, the units of **H** in the two systems are dimensionally different and cannot be converted.

The relationship between **B** and **H** is most generally a tensor that reflects spatial anisotropy in the material, but for common magnetic materials used in current sensing, a scalar value μ applies. In SI units, **B** and **H** have different physical dimensions and are not equal in vacuum. The "permeability of free space" μ_0 is defined so that the μ for a material is a dimensionless constant. Thus,

$$\mathbf{B} = \mu\mu_0\mathbf{H} \tag{38.1}$$

where μ_0 is exactly $4\pi \times 10^{-7}$ H m^{-1}. (H here is the henry, the SI unit of inductance, base units m^2 kg s^{-2} A^{-2}). For many problems, μ will be a constant of the magnetic material, but when magnetic saturation needs to be considered it will be a variable. For engineering calculations, μ is often treated as though it depends on frequency. Values of μ range from less than 100 for high-frequency ferrite to 5000 for transformer iron, to 10^5 for carefully annealed magnetic alloys.

The field due to a long straight wire is shown in Figure 38.1 and Equation 38.2. The **H** vector obeys the right-hand rule and is everywhere perpendicular to the wire. The amplitude of **H** is proportional to the current and falls off linearly with the radial distance from the wire. This is the field that makes noncontact sensing of current possible.

$$\mathbf{H} = \frac{1}{2\pi r} \tag{38.2}$$

The field at the center of a circular loop of wire is shown in Figure 38.2 and Equation 38.3. It obeys the right-hand rule, is proportional to the current, and inversely proportional to the radius of the loop.

$$\mathbf{H} = \frac{1}{r} \tag{38.3}$$

A magnetic field can often be calculated by direct application of Maxwell's relations in integral form [4]. The line integral over any closed path of **H·dl** is equal to the current passing through a surface delimited by the closed path.

Magnetic flux passing through a surface, usually denoted by ϕ, is the integral of **B·n** dA over the surface with normal vector **n**. The SI unit of ϕ is the weber. The unrationalized unit, the line of force, is best relegated to history except that the lines, which form continuous loops, dramatize the fact that it does not matter how a surface is drawn to close a bounding line. The flux is the same. It is convenient to think of magnetic lines, forming closed loops, that can be bent but not interrupted or removed.

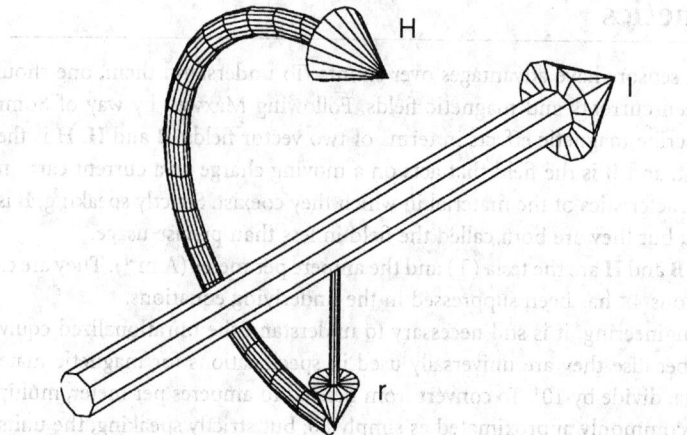

FIGURE 38.1   The magnetic field associated with a current-carrying long wire. It is this field that makes contactless sensing of current possible.

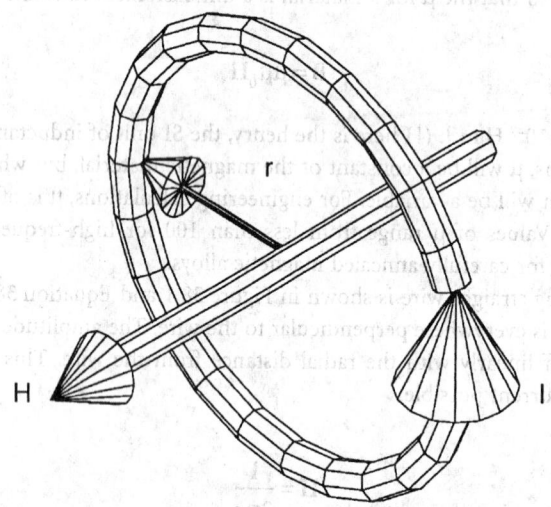

FIGURE 38.2   The magnetic field at the center of a loop of wire. This is the starting point for many calculations involving inductors and transformers.

A carrier of charge, an electron in a wire, a particle in vacuum, or a hole in a semiconductor, moving in a magnetic field is acted on by a force that is perpendicular to the field and the velocity. For positive charge, the force obeys the right-hand rule. The magnitude of the force is proportional to the magnitude of **B**, the velocity **V**, and the sine of the angle between them.

$$\mathbf{F} = \mathbf{V} \times \mathbf{B} \tag{38.4}$$

A carrier, moving or not, is affected in a similar way by a changing magnetic field. The result is an electromotive force, *EMF,* in a loop of wire through which a changing magnetic flux passes. The *EMF* is equal to the rate of change of the flux enclosed by the loop with a change of sign. That is Faraday's law of induction.

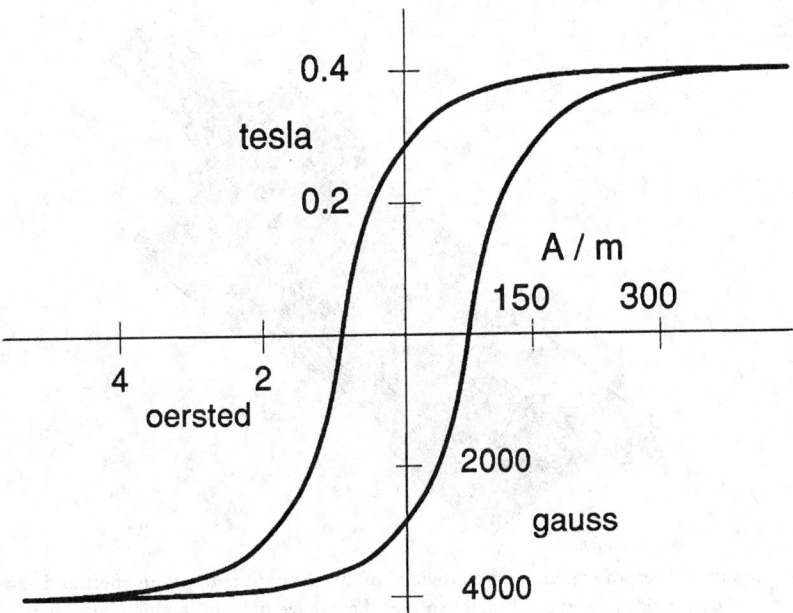

FIGURE 38.3 A hysteresis curve for some rather poor transformer iron. Unrationalized units are shown below and left of the origin because that remains standard practice in the industry.

$$\text{EMF} = -\frac{d\phi}{dt} \qquad (38.5)$$

Most magnetic materials exhibit hysteresis. That is, the relationship between **B** and **H** depends on the history of the applied **H**. A plot of **B** vs. **H** is called a hysteresis loop and a sample is shown as Figure 38.3. The area of the loop represents an energy. If a magnetic core is repeatedly driven around its hysteresis loop, there is a power loss due to hysteresis that is proportional to the area of the loop and the frequency. A material with very large hysteresis is a permanent magnet.

Magnetic materials which are also electric conductors have free carriers which are affected by alternating magnetic fields. As they move back and forth they encounter electric resistance and dissipate energy in the form of heat. These eddy currents can be minimized by use of high resistance ferrites, powdered iron, by laminating cores, or by keeping the flux small.

38.3 Shunts

Shunts were introduced above. They dissipate power as heat and the resistance changes in response to the rising temperature. The dissipation is proportional to the voltage across the shunt and a design compromise must be made because low voltage implies less accuracy in the voltmeter. A standard of 50 mV has evolved. Shunts do not provide galvanic isolation between the measured circuit and the measurement device.

Measurement of alternating current using shunts is also affected by skin effect and the inductance of the shunt. Skin effect can be minimized in the design of the shunt by the use of several parallel sheets of thin metal, Figure 38.4, a feature that also improves heat dissipation. There is not much to be done about inductance except to minimize size.

Safety is enhanced if shunts are placed in the ground leg of a circuit; that way, the output leads, usually only at 50 mV, are near ground. However, that introduces a resistance in the ground path and can interfere

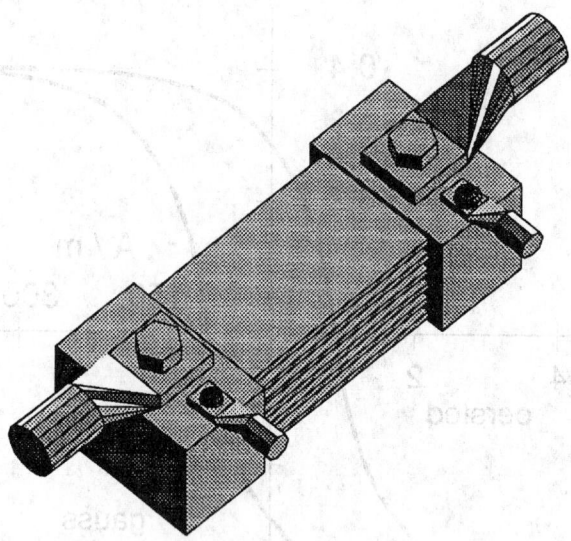

FIGURE 38.4 Multiple sheets of conductor are provided in this shunt to reduce skin effect and allow air cooling. Kelvin connections are provided for the voltmeter so that the voltage drop in the high current connectors is not inadvertently included in the measurement.

with common mode requirements of interdevice signal connections. If the shunt is placed in the high side, care must be taken to protect the wiring and the meter to which it is attached from accidental grounds.

Sometimes, sufficient accuracy can be obtained by measuring the voltage drop along a length of conductor that is otherwise required in the installation. In vehicular service, it is common to sense the voltage drop in a battery cable using a millivoltmeter. Such installations are always high side and should be protected with fuses installed near the points of measurement. It is also wise to provide a connection means that is independent of contact resistance where the current-carrying conductor is installed — a Kelvin connection. Including one lead of an in-line fuse holder in the terminals before they are crimped is one such technique.

38.4 The Moving Magnet Meter

The simplest current indicator balances the force on a permanent magnet created by current in a wire against a spring. A magnetic material is usually placed around the conductor to concentrate the field and reduce the effect of the magnetic field of the Earth. Use is limited to low-precision indicators such as a battery charging meter for a vehicle. It is a dc device.

38.5 The D'Arsonval Meter

This indicator balances the force on a current-carrying wire due to a permanent magnet against a spring. The measured current flows in a coil of wire supported in bearings. It is described in elementary texts [1, 2]. It is generally a dc instrument, but chart recorders have been built with frequency response in the kilohertz range using mirrors on the moving coil. They are then called galvanometers. For ac service, these meters are often equipped with internal copper oxide rectifiers and a nonlinear scale to correct for the diode drop. For current ranges above a few milliamperes, they will have an internal shunt.

The moving magnet meter and the D'Arsonval movement are the only current sensors that do not convert current to voltage and then depend on other devices to read out the voltage.

38.6   The Electrodynamometer

A variation of the D'Arsonval meter for ac service can be built by replacing the permanent magnet with an electromagnet. The force on the moving coil becomes proportional to both the current being measured and the voltage applied to the coil of the electromagnet. It is sensitive to the relative phase of the voltage and current in just the right way to be useful for measurement of power in a circuit with correction for power factor. An electrodynamometer in a power panel is often the load for a current transformer.

38.7   The RF Ammeter and True rms

Current to a radio transmitting antenna is commonly passed through a small resistor, a shunt, that is thermally connected to a thermocouple or other thermometer and mounted in a thermally insulating blanket. The rise in temperature is a measure of the current and is often sensed with a thermocouple.

This is an example of true rms indication. Root mean square current is the square root of the integral of the square of the instantaneous current over an unspecified time divided by that time. It is intended to represent a stationary ac waveform by a single value that is equal to the direct current which would dissipate the same power in a resistive load. The RF ammeter does that precisely. The indication is not particularly linear, but it can easily be calibrated by applying dc to the input.

Other schemes for measuring rms current depend on analog multipliers and subsequent integration. They are limited by crest factor, the ratio of highest instantaneous peak to the rms over a measuring period. Inexpensive meters simply measure the peak, assume a sinusoidal waveform, and scale to rms.

38.8   The Current Transformer

Consider the magnetics of a toroidal core of high-μ material through which a current-carrying conductor passes. Include a secondary winding of n turns as shown in Figure 38.5. The secondary winding is connected to a low-resistance load.

In this current transformer, universally referred to as a CT, alternating current in the single-turn primary attempts to magnetize the core but in so doing, creates an emf and current in the secondary that tend to cancel the field. If the secondary truly has zero resistance, the current in it exactly cancels the field due to the primary. The result is a secondary current equal to the primary current divided by the number of secondary turns. The secondary current is in phase with the primary current. Because of the tightly closed magnetic loop, there is little effect from nearby conductors or position of the primary wire in the hole.

The secondary circuit can now be connected to a low-resistance current- or power-sensing device with assurance of calibration. But the secondary resistance is never really zero and the magnetic coupling is never perfect, so there are other considerations.

First, the concept of *secondary burden* is introduced. It is called that to avoid calling it a "load" because it behaves differently; the best burden is a short-circuit. Burden is sometimes expressed as a secondary resistance in ohms, but more often as an equivalent power in kVA for a defined current without consideration of phase. When the burden is not a perfect short-circuit, energy is dissipated and the magnetic field present in the core is no longer zero. The secondary current leads the primary current with a phase that depends on frequency.

Manufacturers of CTs have techniques to optimize accuracy of the CT when specified for a particular burden. The finished units might not have the number of secondary turns one would expect, but will nonetheless provide results accurate to a percent or so. They have laminations selected to minimize heating of the core. One will see ratings like 100:5, meaning 100 A in the primary will produce 5 A in the secondary rather than "20 turns." They should be installed in a circuit that provides the burden for which they were calibrated. The voltage across a burden resistor is commonly amplified and passed to a data collection device. CTs for large currents need to be large to avoid magnetic saturation when burdened.

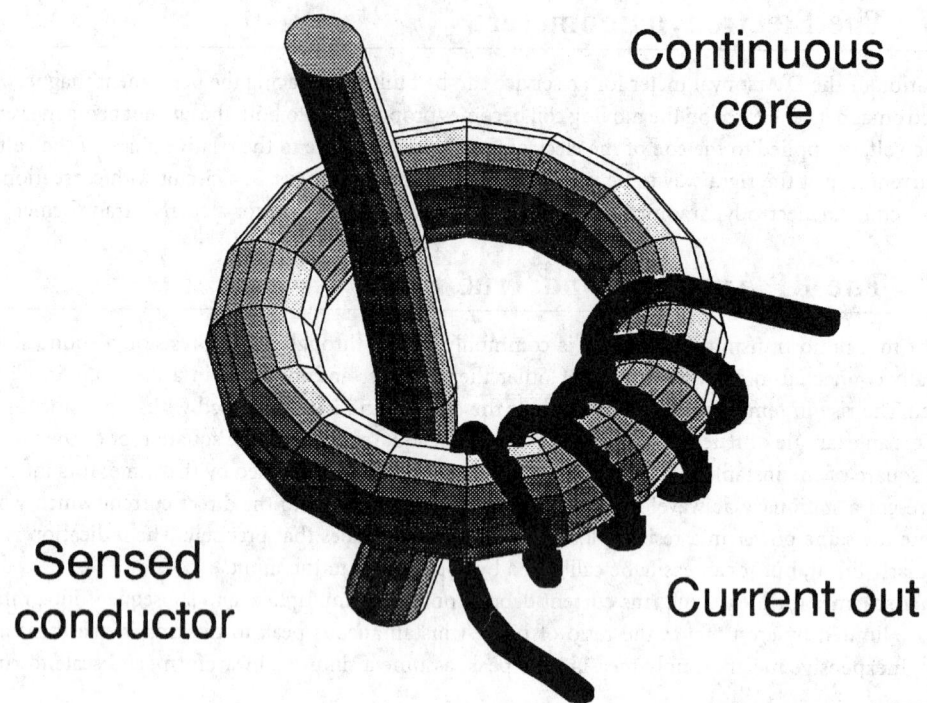

Continuous core

Sensed conductor

Current out

FIGURE 38.5   The ideal current transformer or CT is tightly coupled with no magnetic gap. Current in the secondary exactly balances current in the single-turn primary so that the magnetic flux in the core is zero.

Cores are prepared with laminates of silicon iron in the form of disks, concentric rings, or tape that is wound on a bobbin. Even with the best of materials, eddy current and hysteresis loss are present. When power dissipation is unacceptable, another choice of sensor might be preferable.

Most CTs are used for measurement of low-frequency power and energy. They are found at the front end of kilowatt-hour meters used by power providers. Radio frequency current in transmitting antennas can be measured with suitable core material. Ferrite cores with appropriate compensation are used for sensing pulse width modulated current in switching power supplies. Very large cores are used to sense pulsing beams of high-energy particles. Some oscilloscope probes are highly compensated CTs with a core that can be opened to allow a current-carrying wire to be introduced. With modern winding equipment for toroids, it is possible to put 2000 or more turns on a secondary [Coilcraft]. The CT then begins to look more like a current to voltage converter in its own right without need for very small values of the burden resistor and subsequent voltage amplification.

Safety Note

The secondary of a CT should always remain shorted while there is any possibility of current flow in the primary. With an open secondary, the core can be driven back and forth between saturation in opposite directions, resulting in high-speed changes in the internal B field. The result is very high, dangerous to life, voltages on the open secondary. Insulation in the secondary circuit can be damaged by arcing. Many CTs are made with a provision for shorting the secondary if the circuit must be opened. Use it.

38.9   Gapped Inductive Sensors

It is common practice in the design of transformers to introduce a small gap in the magnetic path. For even very small gaps, the magnetic properties of the magnetic loop become almost completely determined

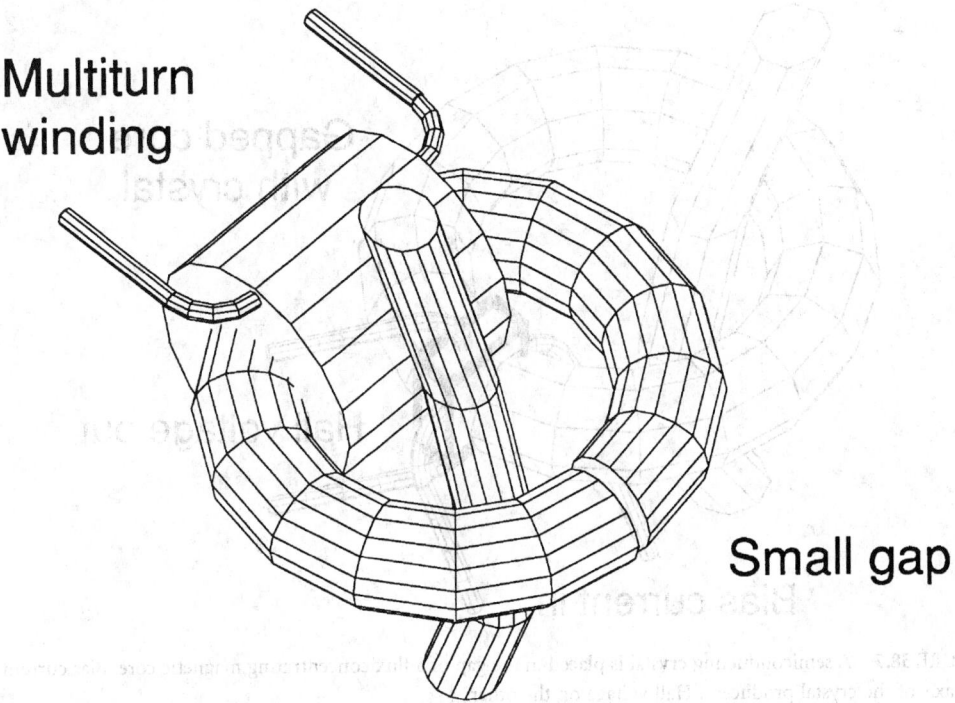

Multiturn
winding

Small gap

FIGURE 38.6 Placing a gap in the core and dramatically increasing the secondary turns count results in a current-to-voltage converter with high output voltage.

by the length of the gap, the rest of the material serving only to contain the lines of flux. Analysis of such a device begins with understanding that the B field is continuous in the core and through the gap. The H field is not, but it still satisfies the relation that the line integral of **H·dl** around the core is equal to the linked current. For a magnetic path of length s in material of permeability μ with a gap g, the ratio B/H, the effective permeability is given by:

$$\mu_{eff} = \frac{s}{g + \dfrac{s}{\mu}}$$

(38.6)

which applies for g much smaller than s. Note that when μ is sufficiently large, the effective permeability becomes independent of μ.

Introducing a gap into what would otherwise be a CT and drastically increasing the secondary turns count to 10000 or more results in a current sensor that is intrinsically safe because the core cannot saturate: Figure 38.6 [SRT]. Because the B field is always small, the heating effect of eddy currents is less important than in the CT. When loaded with an appropriate resistor, the high inductance of the secondary causes the sensor to act like a current source that generates a voltage across the load proportional to the primary current with better than 1% linearity. Useful bandwidths of a sensor can be over 3 decades. The output impedance is high and requires the use of electronic voltmeters. Power dissipation is low, even for very high current models. Output voltage is high enough that simple diode rectifiers can be used to provide for dc output to further processing. In many cases, such a sensor can be used without any special electronics other than a voltmeter.

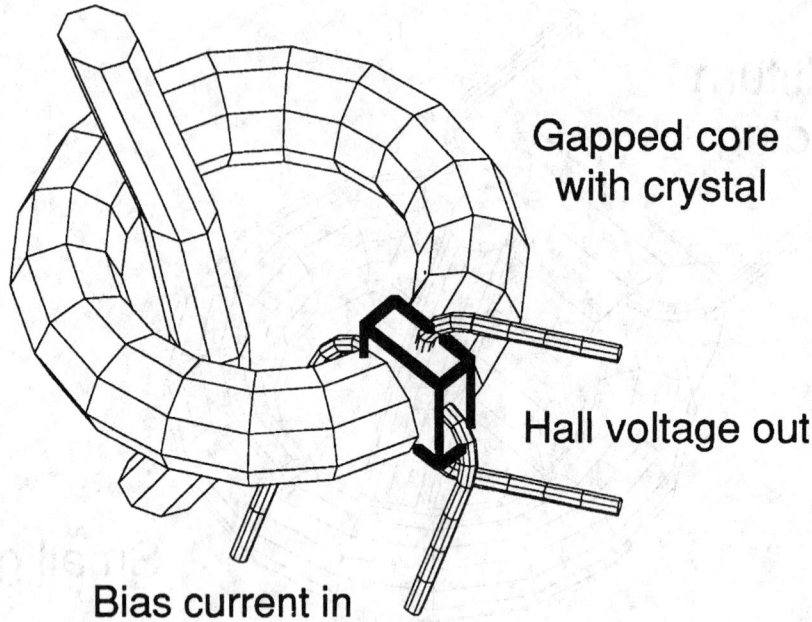

Gapped core with crystal

Hall voltage out

Bias current in

FIGURE 38.7   A semiconducting crystal is placed in the gap of a flux concentrating magnetic core. Bias current on one axis of the crystal produces a Hall voltage on the other.

38.10 Hall Effect Sensor

The *Hall effect* as a sensor for magnetic fields is described in Chapter 48 of this volume and in [5]. It depends on a semiconductor crystal selected for its high carrier mobility and is placed in a magnetic field. A current is passed through the crystal along an axis perpendicular to the field. The carriers assume a mean velocity that causes them to be acted upon by the field and they move toward the other axis perpendicular to the field. The result is an emf at the faces of the crystal that can be measured. The emf is proportional to the field, the bias current, and the mobility.

In principle, such a field sensor could be placed near a current-carrying wire and oriented to sense the field created by the current, Figure 38.1, but the sensitivity is insufficient and there would always be interfering fields from currents in other nearby wires. A flux concentrator that looks like a CT with a large gap is always used. See Figure 38.7.

The device is sensitive to direct current and the polarity is preserved. The Hall voltage is a few millivolts and amplification is always required. Zero drift in the amplifiers must be properly compensated although this is not so important for ac service. The bias current must be carefully controlled and it can be used to provide intrinsic analog multiplication and metering of power if it is made proportional to the circuit voltage.

Sensitivity is best with the smallest gap but there must be room for the crystal so gaps are larger than in the gapped inductive sensors. Fringing of the field in the larger gap reduces the natural shielding of the toroid from unwanted magnetic fields.

The accuracy and linearity of the Hall effect sensor can be improved in closed-loop mode. A feedback winding is added to the core and driven by a servo amplifier. The emf from the Hall device is used to drive the servo amplifier until the field is zero. The output is then the feedback current which is less than the sensed current by the number of turns in the feedback winding. The frequency response of the closed-loop system is surprisingly good, hundreds of kilohertz [F. W. Bell].

38.11 Clamp-on Sensors

It is often desirable to measure current in an existing system without removing power in order to install a device; thus, most of the magnetic devices are available in a clamp-on configuration. The variety ranges from tiny oscilloscope probes to clamps for 3000-A power buses.

Accuracy is always reduced in clamp-on mode because the clamp itself constitutes a gap that is uncontrollable and subject to wear. Some manufacturers provide highly polished surfaces that slide together. Others have iron fingers that interlace as the clamp is closed. Still others do not worry about it because the instrument is intended for field use where accuracy is not so critical.

Some units have handles for one-hand operation; others require a wrench or other tool. An interesting variation is the flexible core by [Flexcorp]. Installation is by bending the core and allowing it to spring back to shape.

38.12 Magnetoresistive Sensors

Most of the features of a Hall effect sensor are available if the Hall crystal is replaced by a device whose resistance changes with magnetic field. The discovery of giant magnetoresistive devices has recently made this idea attractive. [NVE]

Such devices still exhibit rather small resistance change and are sensitive to other effects such as temperature, so it is imperative that they be used in self-compensating bridge circuits in the manner of a strain gage. They are also insensitive to the polarity of the field.

Zetex has delivered magnetoresistive current sensors using thin-film permalloy in a variety of printed circuit packages. They are constructed in the form of a bridge and require bias and a differential amplifier. Measured current up to 20 A passes through the chip via its solder pins.

38.13 The Magnetic Amplifier

The efficiency of a transformer can be adjusted by a dc or low-frequency current that moves the operating point on a hysteresis curve. Excitation, a pump, is required at a higher frequency; it is passed through the transformer and then synchronously rectified and filtered into a higher power representation of the low-frequency signal. The magnetic configuration must be designed so that the pump does not put power into the signal circuit. Figure 38.8 shows one such arrangement. The pump coils are phased in series so that the associated flux cancels in the center leg of the E-E transformer core. The output coils are also in series and phased to add. When dc is applied to the center winding, it saturates both sides and reduces the signal on the output coils. More complicated arrangements can preserve polarity in the phase of the output.

As a current measuring device the magnetic amplifier leaves much to be desired in linearity and frequency response, but it does provide isolation and is limited in sensitivity only by the number of turns placed on the center winding. The 20 mA dc off-hook current in a telephone is one such application.

38.14 Fluxgates

In its simplest form, a *fluxgate magnetometer* uses a driving coil to drive a high-permeability rod into saturation, first in one direction and then in the other. A second winding observes the rate of change of the B field inductively. In the absence of an external magnetic field, the observed signal is symmetric; but when an external field shifts the hysteresis loop to the right or left, the symmetry is lost. An amplifier tuned and phase-locked to the second harmonic of the driving frequency can be used to determine the amplitude of the external field.

In principle, such a fluxgate could be used to sense the field in the gap of a flux concentrator as with the Hall sensor, but it is too big and the linearity would be unacceptable. It is better to think of a way

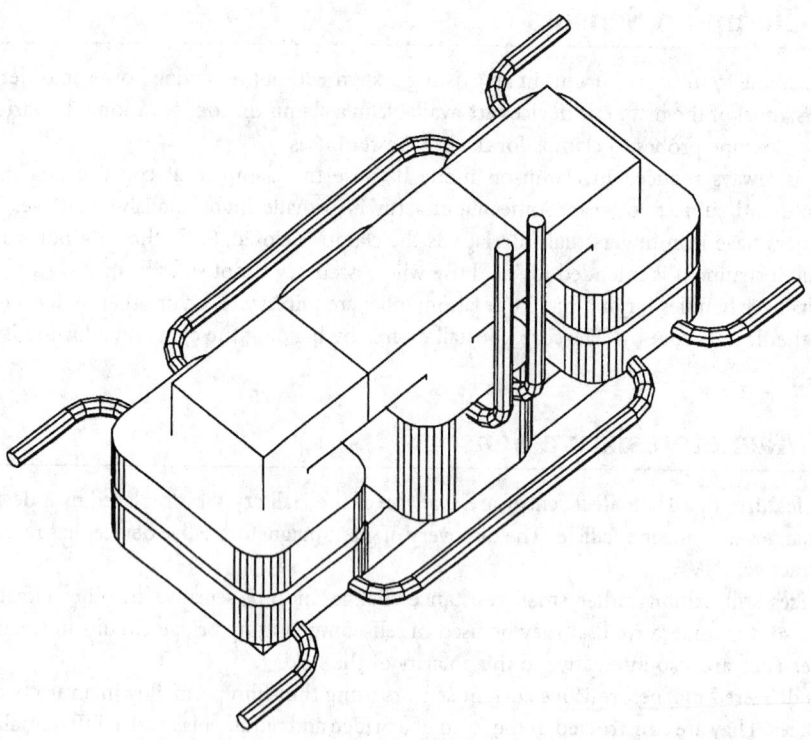

FIGURE 38.8 The windings of a magnetic amplifier. For simple current detection, the coils can be as little as two turns on ferrite toroids using an RF pump.

to drive the whole flux concentrator with a pumping frequency similar to that used in a magnetic amplifier.

Driving a toroid this way has the undesirable effect of coupling the pump energy into the circuit being measured, but one can pump two toroids in opposite directions. Now a current to be sensed passes through both cores and biases one in one direction and one in the other relative to the pump. A pickup winding senses second harmonic, which is demodulated with a phase-locked sensor to preserve the direction of the sensed current.

Linearity is improved by adding one more winding through both toroids; it is driven by a servo amplifier that is satisfied only when the amplitude of the second harmonic is zero; the flux at the measured frequency is zero. This is now the ideal current transformer. At the frequency of the sensed current, which might be dc, there is zero average flux in the toroids and the output of the servo amplifier is a true representation of the current being measured, reduced by the turns count of the servo winding.

Unfortunately, complicated electronics and considerable power are required to accomplish all of this, but the resulting accuracy is significantly better than anything other than precision shunts. Two suppliers — GMW and Holec — provide equipment that uses similar principles. Many of the details are either patented or proprietary.

38.15 Optical Sensors

The Faraday effect is a rotation of the plane of polarization of light as it passes in a transparent medium parallel to a magnetic field. Optical fibers that do not randomize polarization are available with a small Faraday coefficient. Winding such a fiber on a form to be installed around a wire so that the light propagates parallel to the field produced by current in the wire, Figure 38.1, results in a sensor. A measurable rotation proportional to the current comes about because of the long optical path. When the myriad problems of preparation are solved, the result is a sensor that looks externally like a current transformer but has no wires. Using a reflecting ring made of yttrium-iron-garnet with a large Faraday coefficient, NIST reports sensitivity of 220 nA [6]. Matsushita Electric Industrial Company makes sensors using thin garnet films.

Analysis of the polarization requires a polarized light source and a polarization analyzer at the other. An advantage of such a sensor is that the fiber leading to and from can be long enough to allow isolation for very high voltages.

Winding and annealing of fiber sensors without destroying the polarization-preserving properties of the fiber and temperature sensitivity of the Faraday coefficient must be addressed. Further information on these sensors can be found in [7–9]. ABB, Sweden, reports a maximum detectable current of >23 kA, a sensitivity of about 2 A, and a relative error of ±0.15% [7].

38.16 Fault Indicators

Latching indicators which save an indication of high pulse current which was present sometime in the past are needed for power distribution systems subject to lightning strikes. Such indicators are often a bar magnet that moves into a latched position when a current pulse occurs. Readout is visual and they can be reset and installed on live wires using high-voltage tools.

38.17 Other Schemes

Cantor of Ford Aerospace has described a flux reset transformer scheme for measurement of direct current [10]. An oscillator and semiconductor switch periodically drive a core into reverse saturation. When the drive current is switched off, the flux rises due to dc in the sense winding and produces a voltage on the drive winding that is sensed synchronously with the oscillator. No commercial device based on this principle is available.

The double-balanced mixer, familiar to RF engineers, is also a sensor for low-frequency current. The IF port is usually dc-coupled through diodes in a ring configuration. dc applied there will modulate an RF pump applied to one of the RF ports and recovered on the other. There are no known products that use this principle.

38.18 Some Generalities and Warnings

Except for very simple loads, the waveform of the current drawn by a load does not resemble the voltage waveform. Besides the well-known phase shift and power factor, current flows in harmonics and sub-harmonics of the power frequency. Bridge rectifiers with capacitor input filters draw pulses of current near the peaks of the voltage waveform. Triacs cause a phase shift of the fundamental and introduce odd harmonics. Triacs in full-cycle mode with zero current switching can draw full current for a few cycles, followed by zero current for a few more introducing frequency components below the power frequency. Frequency changers are likely to draw current in pulse width modulated bursts.

A classic error is to measure true rms voltage and true rms current and multiply the two to get power. Even after correcting for phase, this is usually wrong. In short, accurate measurement demands some knowledge of the characteristics of the load. Beware of sensors labeled "true rms" for they can be anything but.

38.19 Current Actuated Switches and Indicators

Another form taken by current sensors is the current actuated switch. There was a time when contacts were placed on the pointer of a D'Arsonval movement to establish upper and lower limits for a process variable. The modern way is to configure a sensor so that it operates a solid-state switch.

When operating power is available, any of the current sensors can be configured as a switch; but when the switch must be controlled solely by the power that can be extracted from the current being measured, the gapped toroid is superior. The high voltage available can be rectified to control MOSFETs or the base of an open collector Darlington transistor [SRT].

One company — CR Magnetics — markets a light-emitting diode, LED, indicator that shows the presence of alternating current without any connection or external source of power.

Circuit breakers can be implemented with a magnetic coil or with a bimetallic thermal element. For completeness, there is also the fuse.

Ground fault breakers use a toroidal CT through which the line current is passed twice: once on the way to the load and again on the way back to the neutral wire in such a way as to cancel. Any fault current to ground associated with the load causes the return current to be less than the source and induces voltage on the secondary of the toroid. This is amplified and used to trip a switch. This application demands absolute insensitivity to the position of a current-carrying wire in the hole of the toroid. Gapped sensors typically are not suitable because of magnetic leakage.

38.20 Where to Get Current Sensors

CTs, shunts, and some other sensors are commodity items available from a variety of manufacturers and distributors. Well-known manufacturers of switchgear — General Electric and Westinghouse — have their own CTs for use with their power meters. Test equipment manufacturers — Simpson, Tektronix, Fluke, Extech, and the like — offer hand-held current meters. Tables 38.1 and 38.2 show distributors and manufacturers of current sensing equipment. There is no way that a list such as this can be complete. It has selected itself from those companies willing to contribute to the author's library.

TABLE 38.1 Purchasing Current Sensors, Sample Prices in $U.S.

Newark Electronics catalog 114 (1996)	
1365 Wiley Road	
Schaumburg, IL, 60173-4325	
Tel: (708) 310-8980, Fax: (708) 310-0275, (800) 298-3133	
Crompton Instruments CTs, 2 VA burden 50 A to 200 A, 60 Hz	$23
Crompton shunts, 5 A to 200 A, 50 mV, 0.25%	$25
Carlo Gavazzi CTs with 4 mA to 20 mA output circuits	$154
F. W. Bell Hall effect sensors, 100 A to 3000 A AC or DC, ±15 V power required	$285
Honeywell Hall effect sensor using the null balance technique, 1 µS response time, ±15 Vdc required	$40
Amprobe digital clamp-on AC/DC to 600 A	$141
Amprobe harmonic analyzer	$1058
Simpson 0-1 mA D'Arsonval panel meter	$64
The Grainger catalog 387 (1996)	
Tel: (708) 498-1920, (800) 323-0620	
http://www.grainger.com	
Simpson CTs ("donut" transformers), 50 A to 1000 A	$30
Simpson 50-mV shunts	$40
Jensen Tools catalog for 1996	
7815 S. 46th St.	
Phoenix, AZ 85044-5399	
Tel: (602) 968-6231, Fax: (602) 438-1690	
Tel: (800) 426-1194, Fax: (800) 366-9662	
http://www.jensentools.com	
Various Fluke and Extech products	
Yokogawa digital clamp-on AC/DC to 200 A	$279
Tektronix "Pen meter" with clamp-on to 300 A, 1.7%	$104

TABLE 38.1 (continued)   Purchasing Current Sensors, Sample Prices in $U.S.

Techni-Tool, catalog 59 (1997)
5 Apollo Road
Plymouth Meeting, PA 19462
Tel: (610) 941-2400
http://www.techni-tool.com, techtool@interserv.com

Fluke clamp-on AC/DC oscilloscope probes, 100 A, dc to 100 kHz	$395
Fluke clamp-on Hall effect current probe, 700 kA, 2%	$297
Fluke clamp-on inductive current probe, 150 A	$79
Extech clamp-on true rms AC/DC complete digital meter	$249

Omega, Data Acquisition Guide (1996)
One Omega Drive
P.O. Box 4047
Stamford, CT 06907-0047
Tel: (800) 848-4297
http://www.industry.net/omega

ac current transducer, 0–5 A for use with CT, 4–20 mA output	$240
CT, 1000:5, 52-mm diam. window, 10 VA burden, 0.6%	$32

TABLE 38.2   Selected Manufacturers of Current Sensors

AEMC Instruments
99 Chauncy St.
Boston, MA 02111-9990
Tel: (617) 451-0227, (800) 343-1391, Fax: (617) 423-2952
Manufactures a selection of clamp-on current transformers suitable for extending the range of other alternating current meters.

ABB Inc., with numerous outlets, offers an optical sensor using the Faraday effect.

Amecon
1900 Chris Lane
Anaheim, CA 92805
Tel: (714) 634-2220, Fax: (714) 634-0905
Offers a line of smaller CTs

American Aerospace Controls
570 Smith St.
Farmingdale, NY 11735
Tel: (516) 694-5100, Fax: (516) 694-6739
Has a line of ac and dc current sensors for mounting in remote locations with wiring to collect data.

Coilcraft
1102 Silver Lake Rd.
Cary, IL
Tel: (708) 639-6400
Manufactures smaller current transformers usable up to 100 A for mounting on printed circuit boards.

CR Magnetics Inc.
304 Axminister Dr.
Fenton, MO 63026
Tel: (314) 343-8518, Fax: (314) 343-5119
http://www.crmagnetics.com
Provides a full line of current transformers and current actuated switches. Their catalog contains a useful discussion of accuracy classes and selection criteria.

Dranetz Technologies, Inc.
1000 New Durham Rd., P.O. Box 4019
Edison, NJ 08818-4019
Tel: (908) 287-3680, (800) DRANTEC, Fax: (908) 248-9240
Offers a line of equipment for analysis of power and is included here although it does not make instruments dedicated to current sensing.

TABLE 38.2 (continued)   Selected Manufacturers of Current Sensors

EIL Instruments
10946 Golden West Drive
Hunt Valley, MD 21031-1320
Tel: (410) 584-7400, Fax: (410) 584-7561
Offers a complete line of CTs and other current transducers in a variety of accuracy classes.

Extech Instruments Corporation
335 Bear Hill Road
Waltham, MA 02154-1020
Tel: (617) 890-7440, Fax: (617) 890-7864
Builds a line of hand-held meters suitable for current measurement to 2000 A.

F. W. Bell
6120 Hanging Moss Rd.
Orlando, FL 32807
Tel: (407) 677-6900, (800) 775-2550, Fax: (407) 677-5765
mbuechin@belltechinc.com
Offers noncontact sensors mostly using the Hall effect with a variety of output connections. In closed-loop mode, a 100 Adc
  to 100 kHz unit can be bought for $52 plus your cost of the ±15 V to operate it.

Fluke, numerous sales offices and distributors, has a line of sensors for their hand-held instruments.

Flex-Core division of Morlan and Associates
6625 McVey Blvd.
Worthington, OH 43235
Tel: (614) 889-6152, (614) 876-8308, (614) 876-8538
Makes a line of sensors that cover the range of CTs. Of particular interest is their flexible iron core, which can be installed
  over an existing conductor by bending it.

GMW Danfysik
P.O. Box 2378
Redwood City, CA 94064
Tel: (415) 802-8292, Fax: (415) 802-8298
Offers a line of laboratory grade equipment for precision noncontact measurement. Precision in the ppm range and prices
  to match. Up to 10000 Adc to more than 200 kHz. Large-area CTs suitable for beam current in accelerators under the
  Bergoz tradename.

Holec Power Protection B.V.
P.O. Box 4
7550 GA Hengelo
The Netherlands
Tel: +31.74.246.28.50, Fax: +31.74.246.28.00
hppbv@pi.net
Offers "zero-Flux" current transformers using servo amplifiers to drive the secondary actively. Their document 5.20.2 helps
  to understand the principles involved that operate from dc to 500 kHz.

LEM U.S.
6643 West Mill Rd.
Milwaukee, WI
Tel: (414) 353-0711
Offers Hall effect sensors in both open- and closed-loop form for ac and dc currents up to 10000 A and up to 200 kHz.

Microswitch Division of Honeywell Inc.
11 W. Spring St.
Freeport, IL 61032-9945
Tel: (815) 235-6600
Offers current sensors mostly using the Hall effect, along with their line of magnetic sensors and switches.

Neilson-Kuljian
P.O. box 328
Palo Alto, CA 94302
Tel: (415) 856-3555

TABLE 38.2 (continued) Selected Manufacturers of Current Sensors

Nonvolatile Electronics (NVE)
11409 Valley View Road
Eden Prairie, MN 55344-3617
Tel: (612) 829-9217, (800) 467-7141, Fax: (612) 996-1600
info@nve.com, http://www.nve.com
Has a giant magnetoresistive sensor in an SO8 package that could be used to make a sensor. Less than $3. Other companies
are working with NVE on current sensors, but they prefer not to be known at this time.

Ohio Semitronics
4242 Reynolds Drive
Hilliard, OH
Tel: (614) 777-4611, Fax: (614) 777-4511
http://www.ohiosemi.com
Offers a wide range of current, voltage, and power transducers for ac and dc. They will soon be introducing network-capable
digital output devices, including such things as harmonic analysis.

Pearson Electronics Inc.
1860 Embarcadero Road
Palo Alto, CA 94303
Tel: (415) 494-6444
Provides high-speed compensated current transformers for pulse work with oscilloscopes.

Smith Research & Technology, Inc. (SRT)
3109 S. Cascade Ave., #201
Colorado Springs, CO 80907-5190
Tel: (719) 634-2259, Fax: (719) 634-2601
http://www.srt_inc.com
Offers gapped inductive sensors with ac, dc, and 4–20-mA output up to 1000A. ac current actuated switches which control
isolated 200-W loads using no external power for the switch are offered. A combined voltage and current sensor with
integrated programmable microprocessor is available for harmonic analysis of power systems with network connectivity.

Sprague Electric
San Diego, CA
Tel: (619) 575-9353
Offers their model 97Z current sense transformer, which is for feedback in switching power supplies above 50 kHz.

SSAC
8220 Loop Road
Baldwinsville, NY 13027
Tel: (315) 638-1300, Fax: (305) 638-0333
Has a line of less-expensive switching products that are often found in catalogs issued by distributors.

Tektronix, numerous sales offices and distributors, has announced current sensors and hand-held instruments.

Zetex division of Telemetrix PLC
Fields New Road
Chadderton, Oldham, OL9 8NP, U.K.
Tel: 0161-627-5105, Fax: 0161-627-5467
Tel: (516) 543-7100, Fax: (516) 864-7630 in the U.S.
This supplier of integrated circuits and transistors offers magnetoresistive current and magnetic field sensors for use on
printed circuit boards.

References

1. I. Genzer and P. Youngner, *Physics*, Chapter 10-6, Morristown, NJ: General Learning Corporation, 1973. A high school physics text.
2. H. E. White, *Modern College Physics*, Chapter 51, New York: D. Van Nostrand, 1952.
3. B. N. Taylor, *Guide for the Use of the International System of Units* (SI), NIST Special Publication 811, Washington, D.C.: U.S. Government Printing Office, 1995, Appendix A.5.

4. A. Sommerfeld, *Electrodynamics, Lectures on Theoretical Physics, Vol. III*, New York: Academic Press, 1952.

5. R. C. Dorf (ed.), *The Electrical Engineering Handbook*, Chapter 49, The Hall Effect, Boca Raton, FL, CRC Press, 1993.

6. K. B. Rochford, A. H. Rose, M. N. Deeter, and G. W. Day, Faraday effect current sensor with improved sensitivity-bandwidth product, *Optics Lett.*, 19, 1903-1905, 1994.

7. S. R. Forrest, (ed.), *JTEC Panel Report on Optoelectronics in Japan and the United States*, Baltimore, MD: International Technology Research Institute, Loyola College, February 1996. NTIS PB96-152202. Available *http://itri.loyola.edu/opto/c6_s3.htm.*

8. M. N. Deeter, Domain effects in Faraday effect sensors based on iron garnets, *Appl. Opt.*, 34, 655, 1995.

9. M. N. Deeter, A. H. Rose, and G. W. Day, Faraday effect magnetic field sensors based on substituted iron garnets, in *Fiber Optic and Laser Sensors VIII*, R. P. DePaula and E. Udd, eds., Proc. Soc. Photo-Opt. Instrumentation. Eng. 1367, 243-248, 1990.

10. S. Cantor, NASA Tech Briefs, November 1993, p. 54.

39

Pasquale Arpaia
Università di Napoli Federico II

Francesco Avallone
Università di Napoli Federico II

Aldo Baccigalupi
Università di Napoli Federico II

Claudio De Capua
Università di Napoli Federico II

Carmine Landi
Università de L'Aquila

Power Measurement

In this chapter, the concept of electric power is first introduced, and then the most popular power measurement methods and instruments in *dc, ac* and *pulse* waveform circuits are illustrated.

Power is defined as the *work performed per unit time*. So, dimensionally, it is expressed as joules per second, $J s^{-1}$. According to this general definition, electric power is the electric work or energy dissipated per unit time and, dimensionally, it yields:

$$J s^{-1} = J C^{-1} \times C s^{-1} = V \times A \tag{39.1}$$

where J = Joules
     s = Seconds
     C = Coulombs
     V = Volts
     A = Amperes

The product *voltage times current* gives an electrical quantity equivalent to *power*.

39.1 Power Measurements in dc Circuits

Electric power (P) dissipated by a load (L) fed by a dc power supply (E) is the product of the voltage across the load (V_L) and the current flowing in it (I_L):

$$P = V_L \times I_L \tag{39.2}$$

Therefore, a power measurement in a dc circuit can be generally carried out using a voltmeter (V) and an ammeter (A) according to one of the arrangements shown in Figure 39.1. In the arrangement of Figure 39.1(a), the ammeter measures the current flowing into the voltmeter, as well as that into the load; whereas in the arrangement of Figure 39.1(b), this error is avoided, but the voltmeter measures the voltage drop across the ammeter in addition to that dropping across the load. Thus, both arrangements give a surplus of power measurement absorbed by the instruments. The corresponding measurement errors are generally referred to as insertion errors.

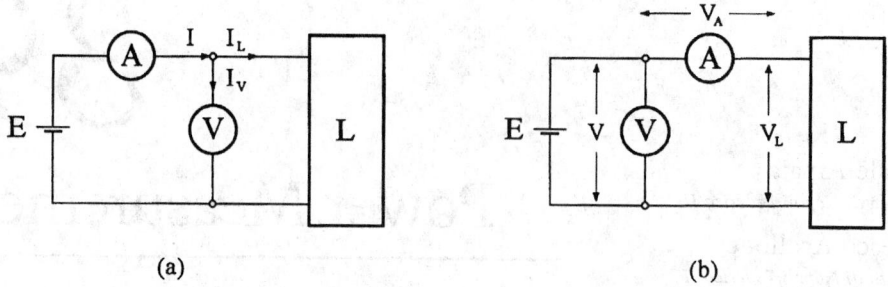

FIGURE 39.1   Two arrangements for dc power measurement circuits.

According to the notation:

- I,   Current measured by the ammeter (Figure 39.1(a))
- V,   Voltage measured by the voltmeter (Figure 39.1(b))
- R_V, R_A, Internal resistance of the voltmeter and the ammeter, respectively
- R_L,   Load resistance
- I_V Current flowing into the voltmeter (Figure 39.1(a))
- V_A,   Voltage drop across the ammeter (Figure 39.1(b))

the following expressions between the measurand electric power P and the measured power $V \times I$ are derived by analyzing the circuits of Figures 39.1(a) and 39.1(b), respectively:

$$P = V_L \times I_L = V \times I \times \left(\frac{R_V - R_L}{R_V} \right) \tag{39.3}$$

$$P = V_L \times I_L = V \times I \times \left(\frac{R_L - R_A}{R_L} \right) \tag{39.4}$$

If:

- I_V, compared with I
- V_A, compared with V

are neglected for the arrangements of Figure 39.1(a) and 39.1(b), respectively, it approximately yields:

$$\frac{I_V}{I} = \frac{R_L}{R_V + R_L} \cong \frac{R_L}{R_V} \cong 0; \quad \frac{V_A}{V} \cong \frac{R_A}{R_A + R_L} \cong \frac{R_A}{R_L} \cong 0; \tag{39.5}$$

consequently, measured and measurand power will be coincident.

On this basis, from Equations 39.3, 39.4, and 39.5, analytical corrections of the insertion errors can be easily derived for the arrangement of Figures 39.1(a) and 39.1(b), respectively.

The instrument most commonly used for power measurement is the *dynamometer*. It is built by (1) two fixed coils, connected in series and positioned coaxially with space between them, and (2) a moving coil, placed between the fixed coils and equipped with a pointer (Figure 39.2(a)).

The torque produced in the dynamometer is proportional to the product of the current flowing into the fixed coils times that in the moving coil. The fixed coils, generally referred to as *current coils*, carry the load current while the moving coil, generally referred to as *voltage coil*, carries a current that is proportional, via the multiplier resistor R_V, to the voltage across the load resistor R_L. As a consequence, the deflection of the moving coil is proportional to the power dissipated into the load.

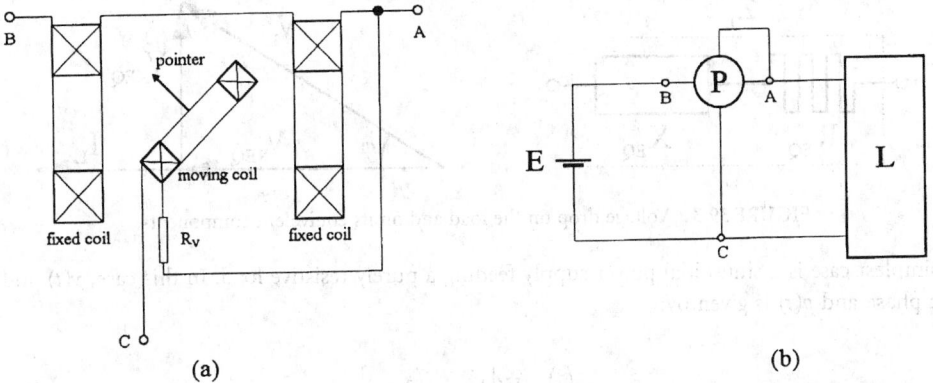

FIGURE 39.2   Power measurement with a dynamometer. (a) Working principle; (b) measurement circuit.

As for the case of Figure 39.1, insertion errors are also present in the dynamometer power measurement. In particular, by connecting the voltage coil between A and C (Figure 39.2(b)), the current coils carry the surplus current flowing into the voltage coil. Consequently, the power P_L dissipated in the load can be obtained by the dynamometer reading P as:

$$P_L = P - \frac{V^2}{R'_v} \tag{39.6}$$

where R'_v is the resistance of the voltage circuit ($R'_v = R_v + R_{vc}$, where R_{vc} is the resistance of the voltage coil). By connecting the moving coil between B and C, this current error can be avoided, but now the voltage coil measures the surplus voltage drop across the current coils. In this case, the corrected value is:

$$P_L = P - I^2 R_C \tag{39.7}$$

where R_C is the resistance of the current coil.

39.2   Power Measurements in ac Circuits

Definitions

All the above considerations relate to *dc* power supplies. Now look at power dissipation in *ac* fed circuits. In this case, electric power, defined as voltage drop across the load times the current flowing through it, is the function:

$$p(t) = v(t) \times i(t) \tag{39.8}$$

referred to as the *instantaneous power*. In ac circuits, one is mainly interested in the mean value of instantaneous power for a defined time interval. In circuits fed by periodic ac voltages, it is relevant to define the mean power dissipated in one period T (*active power P*):

$$P = \frac{1}{T} \int_0^T p(t) dt \tag{39.9}$$

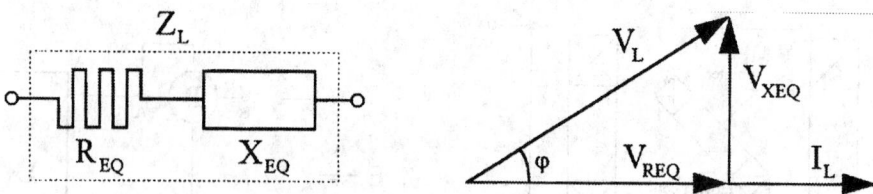

FIGURE 39.3   Voltage drop on the load and on its equivalent components.

The simplest case is a sinusoidal power supply feeding a purely resistive load. In this case, $v(t)$ and $i(t)$ are in phase and $p(t)$ is given by:

$$p(t) = VI\left[1 - \cos(2\omega t)\right] \tag{39.10}$$

where V and I = rms values of $v(t)$ and $i(t)$, respectively
    ω = power supply angular frequency

Therefore, the instantaneous power is given by a constant value VI plus the *ac* quantity oscillating with twice the angular frequency of the power supply; thus, the active power is simply the product VI. In this case, all the above considerations referring to active power for *dc* circuits are still correct, but voltages and currents must be replaced by the corresponding rms values.

The case of purely reactive loads is the opposite; the voltage drop across the load and current flowing through it are out of phase by 90°. Instantaneous power $p(t)$ is given by:

$$p(t) = VI\cos(2\omega t) \tag{39.11}$$

Thus, the active power dissipated by a reactive load is zero, owing to the phase introduced by the load itself between voltage and current.

The simplest cases of sinusoidal power sources supplying purely resistive and purely reactive loads have been discussed. In these cases, the load is expressed by a real or a pure imaginary number. In general, the load is represented by a complex quantity (the impedance value). In this case, load impedance can be represented by its equivalent circuit (e.g., a pure resistance and a pure reactance in series). With this representation in mind, the electric power dissipated in the load Z_L (Figure 39.3) can be expressed by the sum of power components separately dissipated by resistance R_{EQ} and reactance X_{EQ} of the equivalent circuit Z_L.

Considering that no active power is dissipated in the reactance X_{EQ}, it yields:

$$P = V_{REQ}I_L = V_L I_L \cos\varphi \tag{39.12}$$

The term $\cos\varphi$ appearing in Equation 39.12 is referred to as the *power factor*. It considers that only a fraction of voltage V_L contributes to the power; in fact, its component V_{XEQ} (the drop across the reactance) does not produce any active power, as it is orthogonal to the current I_L flowing into the load.

Figure 39.4 plots the waveforms of instantaneous power $p(t)$, voltage $v(t)$, and current $i(t)$. The effect of the power factor is demonstrated by a dc component of $p(t)$ that varies from a null value (i.e., $v(t)$ and $i(t)$ displaced by 90°) toward the value VI (i.e., $v(t)$ and $i(t)$ in phase).

The term:

$$P_A = V_L I_L \tag{39.13}$$

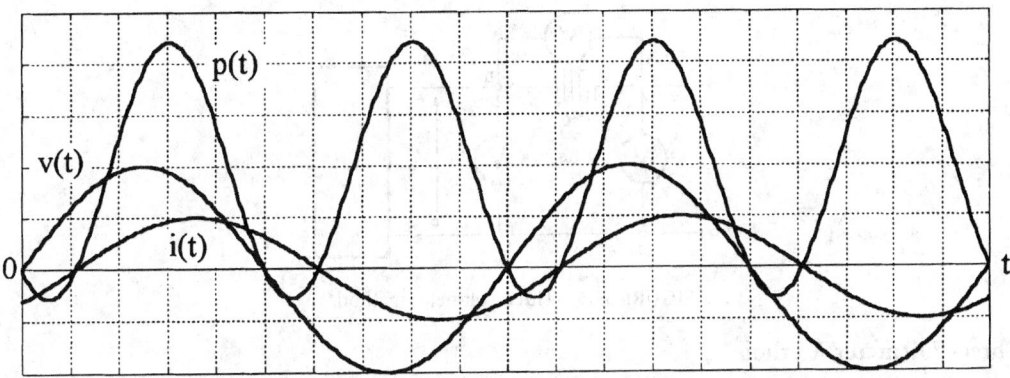

FIGURE 39.4 Waveforms of instantaneous power (*p*), voltage (*v*), and current (*i*).

is called the *apparent power*, while the term:

$$Q = V_{XEQ} I_L = V_L I_L \sin \varphi \tag{39.14}$$

is called the *reactive power* because it represents a quantity that is dimensionally equivalent to power. This is introduced as a consequence of the voltage drop across a pure reactance and, therefore, does not give any contribution to the active power. From Figure 39.3, the relationship existing between *apparent power*, *active power*, and *reactive power* is given by:

$$P_A = \sqrt{P^2 + Q^2} \tag{39.15}$$

Dynamometers working in ac circuits are designed to integrate instantaneous power according to Equation 39.9. Insertion errors can be derived by simple considerations analogous to the dc case. However, in ac, a phase uncertainty due to the not purely resistive characteristic of voltage circuit arises. In sinusoidal conditions, if ε_w (in radians) is the phase of the real coil impedance, and $\cos\varphi$ is the load power factor, the relative uncertainty in active power measurements can be shown to be equal to $-\varepsilon_w \mathrm{tg}\varphi$. The phase uncertainty depends on the frequency. By using more complex circuits, the frequency range of the dynamometer can be extended to a few tens of kilohertz.

The above has presented the power definitions applied to *ac* circuits with the restrictions of sinusoidal quantities. In the most general case of distorted quantities, obviously symbolic representation can no longer be applied. In any case, active power is always defined as the mean power dissipated in one period.

As far as methods and instruments for *ac* power measurements are concerned, some circuit classification is required. In fact, the problems are different, arising in circuits as the frequency of power supply increases. Therefore, in the following, *ac* circuits will be classified into (1) line-frequency circuits, (2) low- and medium-frequency circuits (up to a few megahertz) and (3) high-frequency circuits (up to a few gigahertz). Line-frequency circuits will be discussed separately from low-frequency circuits, principally because of the existence of problems related specifically to the three-phase power supply of the main.

Low- and Medium-Frequency Power Measurements

In the following, the main methods and instruments for power measurements at low and medium frequencies are considered.

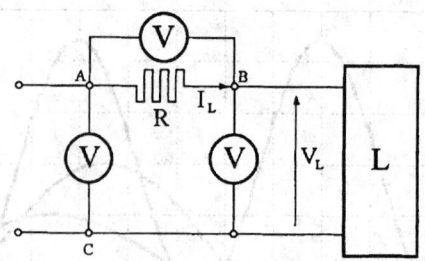

FIGURE 39.5   Three-voltmeter method.

Three-Voltmeter Method

The power dissipation in the load L can be measured using a noninductive resistor R and measuring the three voltages shown in Figure 39.5 [1]. Although one of the voltages might appear redundant on a first analysis of the circuit, in actual fact, three independent data are needed in order to derive power from Equation 39.12. In particular, from voltage drops v_{AB} and v_{BC}, the load current and load voltage can be directly derived; instead, v_{AC} is used to retrieve information about their relative phase.

If currents derived by voltmeters are neglected and the current i_L flowing into the load L is assumed to be equal to that flowing into the resistor R, the statement can be demonstrated as follows:

$$v_{AC} = v_L + Ri_L$$
$$v_{AC}^2 = R^2 i_L^2 + v_L^2 + 2Rv_L i_L \tag{39.16}$$

where the small characters indicate instantaneous values. By computing rms values (indicated as capital characters), one obtains the power P_L:

$$\frac{1}{T}\int_0^T v_{AC}^2 dt = \frac{1}{T}\int_0^T R^2 i_L^2 dt = \frac{1}{T}\int_0^T v_L^2 dt + \frac{1}{T}\int_0^T 2Rv_L i_L dt$$

$$V_{AC}^2 = RI_L^2 + V_L^2 + 2RP_L \tag{39.17}$$

$$P_L = \frac{V_{AC}^2 - R^2 I_L^2 - V_L^2}{2R} = \frac{V_{AC}^2 - V_{AB}^2 - V_{BC}^2}{2R}$$

Equation 39.17 is also the same in dc by replacing rms values with dc values. Since the result is obtained as a difference, problems arise from relative uncertainty when the three terms have about sum equal to zero.

Such a method is still used for high-accuracy applications.

Thermal Wattmeters

The working principle of thermal wattmeters is based on a couple of twin thermocouples whose output voltage is proportional to the square of the rms value of the currents flowing into the thermocouple heaters [2].

The principle circuit of a thermal wattmeter is shown in Figure 39.6(a). Without the load, with the hypothesis $S \ll r_1$ and $S \ll r_2$, the two heaters are connected in parallel and, if they have equal resistance r $(r_1 = r_2 = r)$, they are polarized by the same current i_p

$$i_1 = i_2 = \frac{i_p}{2} = \frac{v}{2R+r} \tag{39.18}$$

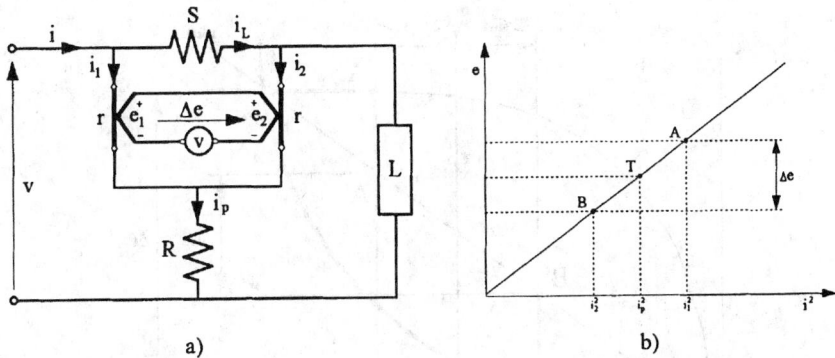

FIGURE 39.6 Thermal wattmeter based on twin thermocouples (*a*); working characteristic in ideal conditions (*b*).

In this case, the output voltages of the two thermocouples turn out to be equal ($e_1 = e_2$); thus, the voltage Δe measured by the voltmeter is null. In Figure 39.6(b), this situation is highlighted by the working point T equal for both thermocouples. By applying a load L with a corresponding current i_L, a voltage drop across S arises, causing an imbalance between currents i_1 and i_2. With the hypothesis that r << R, the two heaters are in series; thus, the current imbalance through them is:

$$i_1 - i_2 = \frac{S i_L}{2R} \tag{39.19}$$

This imbalance increases the current i_1 and decreases i_2. Therefore, the working points of the two thermocouples change: the thermocouple polarized by the current i_1 operates at A, and the other thermocouple operates at B (Figure 39.6(b)). In this situation, with the above hypotheses, the voltmeter measures the voltage imbalance Δe proportional to the active power absorbed by the load (except for the surplus given by the powers dissipated in R, S, r_1, and r_2):

$$\Delta e = k\left(\left\langle i_1^2 \right\rangle - \left\langle i_2^2 \right\rangle\right) = k\left(\left\langle \left(i_p + i_L\right)^2 \right\rangle - \left\langle \left(i_p - i_L\right)^2 \right\rangle\right)$$

$$= k\left\langle 4 i_p i_L \right\rangle = k_1 \left\langle v(t) i(t) \right\rangle = k_1 P \tag{39.20}$$

where the notation $\langle i \rangle$ indicates the time average of the quantity i.

If the two thermocouples cannot be considered as twins and linear, the power measurement accuracy will be obviously compromised. This situation is shown in Figure 39.7 where the two thermocouples are supposed to have two quite different nonlinear characteristics. In this case, the voltage measured by voltmeter will be Δe_n instead of Δe.

Wattmeters based on thermal principle allow high accuracy to be achieved in critical cases of highly distorted wide-band spectrum signals.

Wattmeters Based on Multipliers

The multiplication and averaging processes (Figure 39.8) involved in power measurements can be undertaken by electronic means.

Electronic wattmeters fall into two categories, depending on whether multiplication and averaging operations are performed in a continuous or discrete way. In continuous methods, multiplications are mainly carried out by means of analog electronic multipliers. In discrete methods, sampling wattmeters take simultaneous samples of voltage and current waveforms, digitize these values, and provide multiplication and averaging using digital techniques.

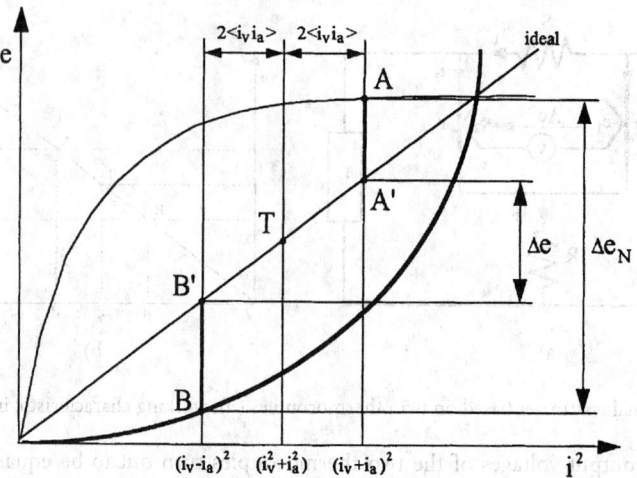

FIGURE 39.7 Ideal and actual characteristics of thermal wattmeter thermocouples.

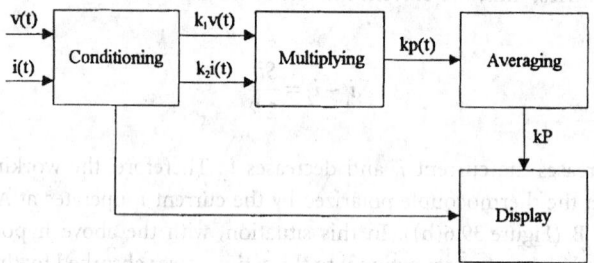

FIGURE 39.8 Block diagram of a multiplier-based wattmeter.

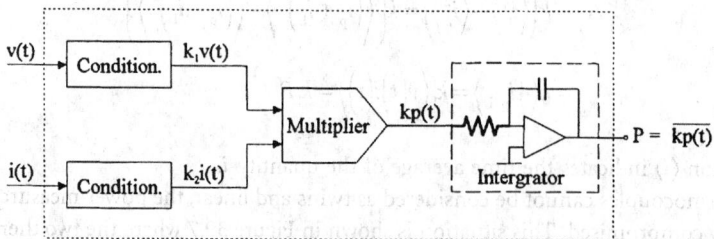

FIGURE 39.9 Block diagram of a four-quadrant, multiplier-based wattmeter.

Analogous to the case of dynamometers, the resistances of the voltage and current circuits have to be taken into account (see Equations 39.6 and 39.7). Also, phase errors of both current ε_{wc} and voltage ε_{wv} circuits increase the relative uncertainty of power measurement, e.g., in case of sinusoidal conditions increased at $(\varepsilon_{wc}-\varepsilon_{wv})T_g\varphi$.

Wattmeters Based on Analog Multipliers

The main analog multipliers are based on a transistor-based popular circuit such as a four-quadrant multiplier [3], which processes voltage and current to give the instantaneous power, and an integrator to provide the mean power (Figure 39.9). More effective solutions are based on (1) Time Division Multipliers (TDMs), and (2) Hall effect-based multipliers.

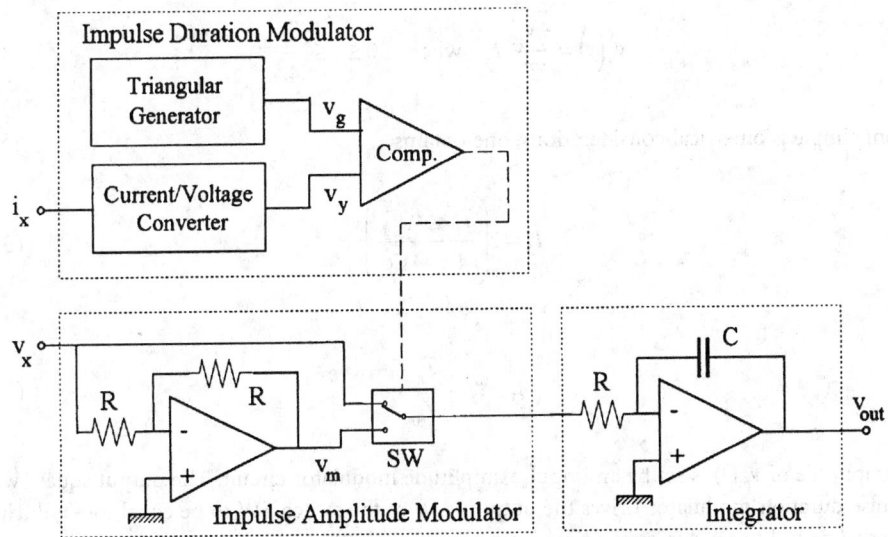

FIGURE 39.10 Block diagram of a TDM-based wattmeter.

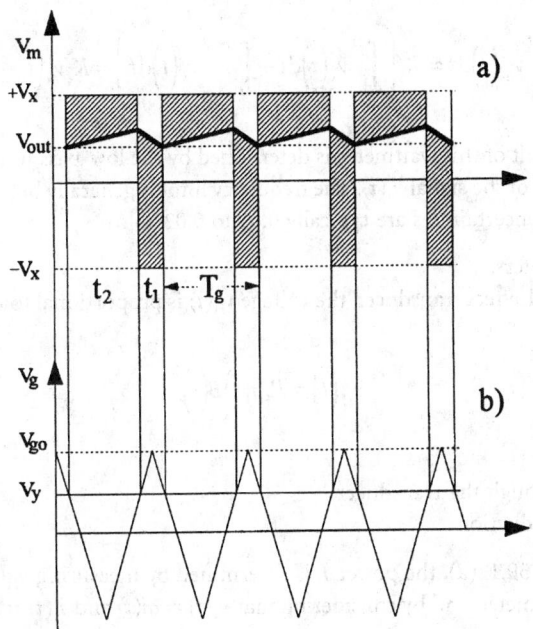

FIGURE 39.11 Waveform of the TDM-based power measurement: (*a*) impulse amplitude modulator output, (*b*) ramp generator output.

TDM-Based Wattmeters.

The block diagram of a wattmeter based on a TDM is shown in Figure 39.10 [4]. A square wave v_m (Figure 39.11(a)) with constant period T_g, and duty cycle and amplitude determined by $i(t)$ and $v(t)$, respectively, is generated. If T_g is much smaller than the period of measurands $v_x(t)$ and $v_y(t)$, these voltages can be considered as constant during this time interval.

The duty cycle of v_m is set by an impulse duration modulator circuit (Figure 39.10). The ramp voltage $v_g(t)$ (Figure 39.11(b)) is compared to the voltage $v_y(t)$ proportional to $i(t)$, and a time interval t_2, whose duration is proportional to $v_y(t)$, is determined. If

$$v_g(t) = \frac{4V_{g0}}{T_g} t \quad \text{when} \quad 0 \le t \le \frac{T_g}{4} \tag{39.21}$$

then from simple geometrical considerations, one obtains:

$$t_2 = 2\left(\frac{T_g}{4} - \frac{v_y T_g}{4V_{g0}}\right) \tag{39.22}$$

and

$$t_1 - t_2 = \frac{T_g}{V_{g0}} v_y \tag{39.23}$$

The amplitude of $v_m(t)$ is set by an impulse amplitude modulator circuit. The output square wave of the impulse duration modulator drives the output $v_m(t)$ of the switch SW to be equal to $+v_x$ during the time interval t_1, and to $-v_x$ during the time interval t_2 (Figure 39.11(a)).

Then, after an initial transient, the output voltage $v_{out}(t)$ of the low-pass filter (integrator) is the mean value of $v_m(t)$:

$$V_{out} = \frac{1}{RC}\int_0^t v_m(t)\,dt = K'\left(\int_0^{t1} v_x(t)\,dt - \int_{t1}^{t1+t2} v_x(t)\,dt\right) = K'v_x(t_1 - t_2) = Kv_x v_y \tag{39.24}$$

The high-frequency limit of this wattmeter is determined by the low-pass filter and it must be smaller than half of the frequency of the signal $v_g(t)$. The frequency limit is generally between 200 Hz and 20 kHz, and can reach 100 kHz. Uncertainties are typically 0.01 to 0.02% [5].

Hall Effect-Based Wattmeters.

As is well known, in a Hall-effect transducer, the voltage $v_H(t)$ is proportional to the product of two time-dependent quantities [6]:

$$v_H(t) = R_H i(t) B(t) \tag{39.25}$$

where R_H = Hall constant
$i(t)$ = Current through the transducer
$B(t)$ = Magnetic induction

In the circuit of Figure 39.12(a), the power P is determined by measuring $v_H(t)$ through a high-input impedance averaging voltmeter, and by considering that $v_x(t) = ai(t)$ and $i_x(t) = bB(t)$, where a and b are proportionality factors:

$$P = \frac{1}{T}\int_0^T v_x(t) \cdot i_x(t)\,dt = ab\frac{1}{T}\int_0^T i(t) \cdot B(t)\,dt = abR_H V_H \tag{39.26}$$

where T is the measurand period, and V_H the mean value of $v_H(t)$.

In the usual realization of the Hall multiplier (0.1% up to a few megahertz), shown in Figure 39.12(a), the magnetic induction is proportional to the load current and the optimal polarizing current i_v is set by the resistor R_v.

For the frequency range up to megahertz, an alternative arrangement is shown in Figure 39.12(b), in which the load current I_L flows directly into the Hall device, acting as a polarizing current, and the

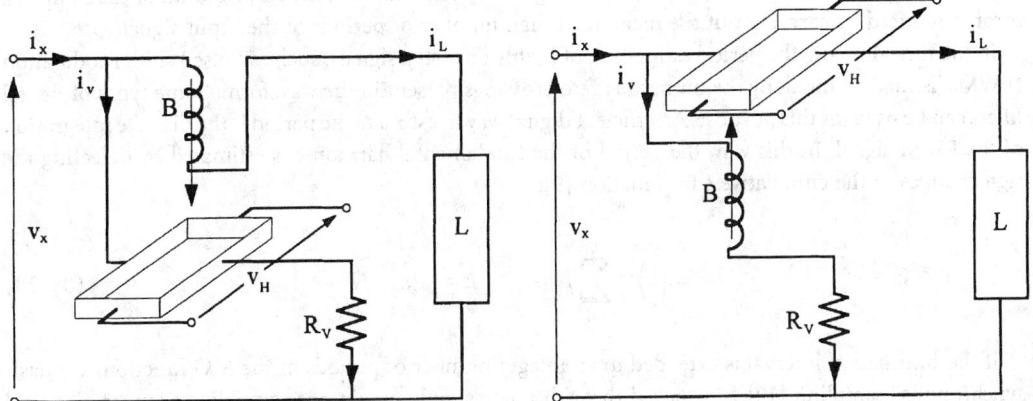

FIGURE 39.12   Configurations of the Hall effect-based wattmeter.

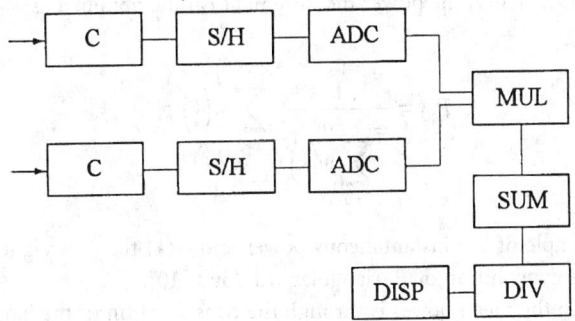

FIGURE 39.13   Block diagram of the sampling wattmeter.

magnetic field is generated by the voltage *v*. In this same way, the temperature influence is reduced for line-frequency applications with constant-amplitude voltages and variable load currents.

In the megahertz to gigahertz range, standard wattmeters use probes in waveguides with rectifiers.

Wattmeters Based on Digital Multipliers

Sampling Wattmeters.

The most important wattmeter operating on discrete samples is the sampling wattmeter (Figure 39.13). It is essentially composed of two analog-to-digital input channels, each constituted by (1) a conditioner (C), (2) a sample/hold (S/H), (3) an analog-to-digital converter (ADC), (4) a digital multiplier (MUL), and (5) summing (SUM), dividing (DIV), and displaying units (DISP). The architecture is handled by a processing unit not shown in Figure 39.13.

If samples are equally spaced, the active power is evaluated as the mean of the sequence of instantaneous power samples $p(k)$:

$$\bar{p} = \frac{1}{N}\sum_{k=0}^{N-1} p(k) = \frac{1}{N}\sum_{k=0}^{N-1} v(k)\,i(k) \tag{39.27}$$

where N^* represents the number of samples in one period of the input signal, and $v(k)$ and $i(k)$ are the kth samples of voltage and current, respectively. A previous estimation of the measurand fundamental period is made to adjust the summation interval of Equation 39.27 and/or the sampling period in order to carry out a synchronous sampling [7]. The sampling period can be adjusted by using a frequency multiplier with PLL circuit driven by the input signal [8]. Alternatively, the contribution of the sampling error is reduced by carrying out the mean on a high number of periods of the input signal.

In the time domain, the period estimation of highly distorted signals, such as Pulse Width Modulation (PWM), is made difficult by the numerous zero crossings present in the waveform. Some types of digital filters can be used for this purpose. An efficient digital way to estimate the period is the discrete integration of the PWM signal. In this way, the period of the fundamental harmonic is estimated by detecting the sign changes of the cumulative sum function [9]:

$$S(k) = \sum_{i=1}^{k} p_i \qquad k = 1, 2, \ldots, N \qquad (39.28)$$

If the summation interval is extended to an integer number of periods of the $S(k)$ function, a "quasi-synchronous" sampling [10] is achieved through a few simple operations (cumulative summation and sign detection) and the maximum synchronization error is limited to a sampling period. Through relatively small increases in computational complexity and memory size, the residual error can be further reduced through a suitable data processing algorithm; that is, the multiple convolution in the time domain of triangular windows [9]. Hence, the power measurement can be obtained as:

$$P_{(B)} = \frac{1}{\displaystyle\sum_{k=0}^{2B(N^*-1)} w(k)} \sum_{k=0}^{2B(N^*-1)} w(k)\, p(k) \qquad (39.29)$$

where $p(k)$ is the kth sample of the instantaneous power and $w(k)$ the kth weight corresponding to the window obtained as the convolution of B triangular windows [10].

Another way to obtain the mean power is through the consideration of the harmonic components of voltages and currents in the frequency domain using the Discrete Fourier Transform [11]. In particular, a Fast Fourier Transform algorithm is used in order to improve efficiency. Successively, a two-step research of the harmonic peaks is carried out: (1) the indexes of the frequency samples corresponding to the greatest spectral peaks provide a rough estimate of the unknown frequencies when the wide-band noise superimposed onto the signal is below threshold; (2) a more accurate estimate of harmonic frequencies is carried out to determine the fractional bin frequency (i.e., the harmonic determination under the frequency resolution); to this aim, several approaches such as zero padding, interpolation techniques, and flat-top window-based technique can be applied [12].

Line-Frequency Power Measurements

For line applications where the power is directly derived by the source network, the assumption of infinite power source can be reliably made, and at least one of the two quantities voltage or current can be considered as sinusoidal. In this case, the definition of the power as the product of voltage and current means that only the power at the fundamental frequency can be examined [13].

Single-Phase Measurements

Single-phase power measurements at line frequency are carried out by following the criteria previously mentioned. In practical applications, the case of a voltage greater than 1000 V is relevant; measurements must be carried out using voltage and current transformers inserted as in the example of Figure 39.14. The relative uncertainty is equal to:

$$\frac{\Delta P}{P} = \left(\eta_w + \eta_a + \eta_v \right) + \left(\varepsilon_w + \varepsilon_a + \varepsilon_v \right) tg\varphi_c \qquad (39.30)$$

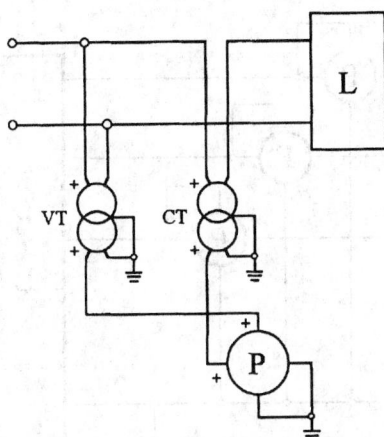

FIGURE 39.14 Single-phase power measurement with voltage (VT) and current (CT) transformers.

where η_w and ε_w are the instrumental and phase uncertainty of the wattmeter, η_a and η_v are the ratio uncertainties of current (CT) and voltage (VT) transformers, and ε_a and ε_v their phase uncertainties, respectively.

If the load current exceeds the current range of the wattmeter, a current transformer must be used, even in the case of low voltages.

Polyphase Power Measurements

Three-phase systems are the polyphase systems most commonly used in practical industrial applications. In the following, power measurements on three-phase systems will be derived as a particular case of polyphase systems (systems with several wires) and analyzed for increasing costs: (1) balanced and symmetrical systems, (2) three-wire systems, (3) two wattmeter-based measurements, (4) unbalanced systems, (5) three wattmeter-based measurements, and (6) medium-voltage systems.

Measurements on Systems with Several Wires

Consider a network with sinusoidal voltages and currents composed by n wires. For the currents flowing in such wires, the following relation is established:

$$\sum_{1}^{n} \dot{I}_i = 0 \tag{39.31}$$

The network can be thought as composed of $n-1$ single-phase independent systems, with the common return on any one of the wires (e.g., the sth wire). Then, the absorbed power can be measured as the sum of the readings of $n-1$ wattmeters, each one inserted with the current circuit on a different wire and the voltmeter circuit between such a wire and the sth one (Figure 39.15):

$$P = \sum_{1}^{n-1} \left(\dot{V}_{is} \times \dot{I}_i \right) \tag{39.32}$$

The absorbed power can be also measured by referring to a generic point O external to the network. In this case, the absorbed power will be the sum of the readings of n wattmeters, each inserted with the ammeter circuit on a different wire and the voltmeter circuit connected between such a wire and the point O:

$$P = \sum_{1}^{n} \left(\dot{V}_{io} \times \dot{I}_i \right) \tag{39.33}$$

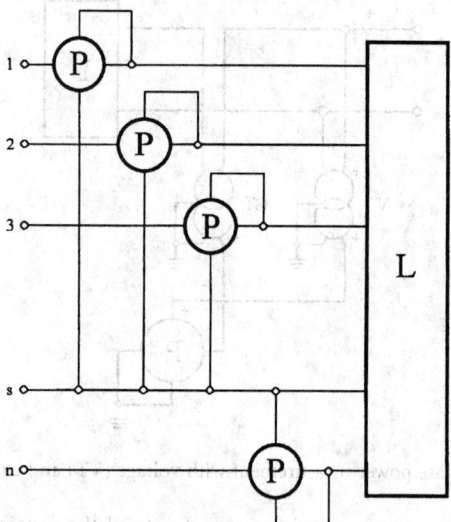

FIGURE 39.15 Power measurement on systems with several wires.

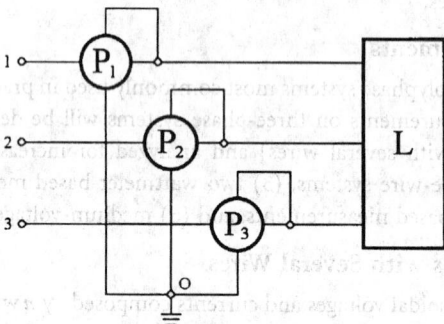

FIGURE 39.16 Power measurement on three-wire systems.

Power Measurements on Three-Wire Systems

Active power in a three-phase power system can generally be evaluated by three wattmeters connected as shown in Figure 39.16.

For each power meter, the current lead is connected on a phase wire and the voltmeter lead is connected between the same wire and an artificial neutral point O, whose position is fixed by the voltmeter impedance of power meters or by suitable external impedances.

Under these conditions, absorbed power will be the sum of the three wattmeter indications:

$$P = \sum_{1}^{3} \left(\dot{V}_{io} \times \dot{I}_i \right) \tag{39.34}$$

If the three-phase system is provided by four wires (three phases with a neutral wire), the neutral wire is utilized as a common wire.

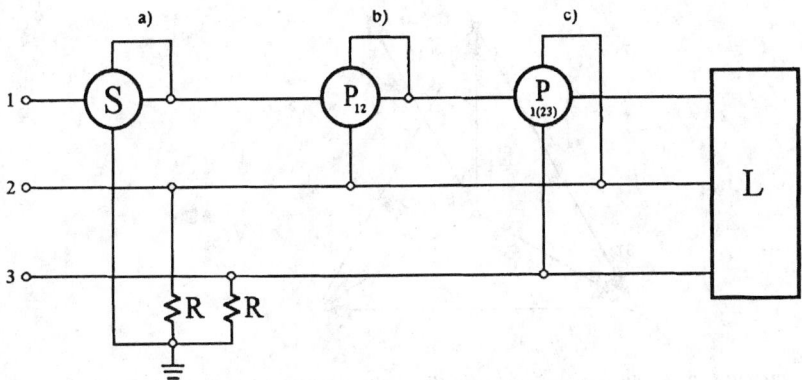

FIGURE 39.17 The three kinds of insertion of a power meter.

Symmetrical and Balanced Systems

The supply system is symmetrical and the three-phase load is balanced; that is:

$$\begin{cases} V_1 = V_2 = V_3 \\ I_1 = I_2 = I_3 \end{cases}$$ (39.35)

In Figure 39.17, the three possible kinds of insertion of an instrument S (an active power or a reactive power meter) are illustrated. The first (*a* in Figure 39.17) was described in the last subsection; if S is a wattmeter; the overall active power is given by three times its indication, and similarly for the reactive power if S is a reactive power meter. Notice that a couple of twin resistors with the same resistance R of the voltage circuit of S are placed on the other phases to balance the load.

The other two insertions are indicated by the following convention: S_{ijk} indicates a reading performed with the current leads connected to the line "*i*" and the voltmeter leads connected between the phases "*j*" and "*k*." If "*i*" is equal to "*j*", one is omitted (e.g., the notation P_{12} (*b* in Figure 39.17)). The active power absorbed by a single phase is usually referred to as P_1.

The wattmeter reading corresponding to the (c) case in Figure 39.17 is equal to the reactive power Q_1 involved in phase 1, save for the factor $\sqrt{3}$. Hence, in the case of symmetrical and balanced systems, the overall reactive power is given by:

$$Q = 3Q_1 = 3 P_{1(23)} \big/ \sqrt{3} = \sqrt{3}\, P_{1(23)}$$ (39.36)

In fact, one has:

$$P_{1(23)} = \dot{I}_1 \times \dot{V}_{23}$$ (39.37)

but:

$$\dot{V}_{12} + \dot{V}_{23} + \dot{V}_{31} = 0 \quad \Rightarrow \quad P_{1(23)} = \dot{I}_1 \times \left(-\dot{V}_{12} - \dot{V}_{31} \right)$$

$$\dot{V}_{13} = -\dot{V}_{31} \quad\quad\quad \Rightarrow \quad P_{1(23)} = -\dot{I}_1 \times \dot{V}_{12} + \dot{I}_1 \times \dot{V}_{13}$$

$$\begin{cases} \dot{I}_1 \times \dot{V}_{12} = P_{12} \\ \dot{I}_1 \times \dot{V}_{13} = P_{13} \end{cases} \quad \Rightarrow \quad P_{1(23)} = P_{13} - P_{12}$$

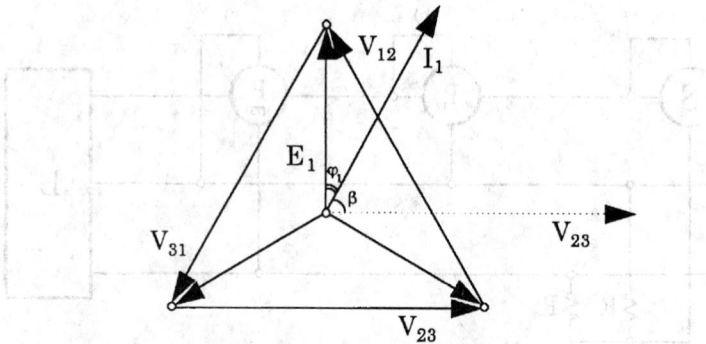

FIGURE 39.18   Phasor diagram for a three-phase symmetrical and balanced system.

In the same manner, the following relationships, which are valid for any kind of supply and load, can be all proved:

$$P_{1(23)} = P_{13} - P_{12}$$

$$P_{2(31)} = P_{21} - P_{23} \tag{39.38}$$

$$P_{3(12)} = P_{32} - P_{31}$$

If the supply system is symmetrical, $P_{1(23)} = \sqrt{3}Q_1$.

In fact, moving from the relationship (Figure 39.18):

$$P_{1(23)} = \vec{I}_1 \times \vec{V}_{23} = I_1 V_{23} \cos\beta \tag{39.39}$$

where $\beta = 90° - \varphi_1$, one obtains $P_{1(23)} = \sqrt{3}E_1 I_1 \sin\varphi_1 = \sqrt{3}Q_1$.

In the same manner, the other two corresponding relationships for $P_{2(31)}$ and $P_{3(12)}$ are derived. Hence:

$$P_{1(23)} = \sqrt{3}Q_1 = P_{13} - P_{12}$$

$$P_{2(31)} = \sqrt{3}Q_2 = P_{21} - P_{23} \tag{39.40}$$

$$P_{3(12)} = \sqrt{3}Q_3 = P_{32} - P_{31}$$

Power Measurements Using Two Wattmeters

The overall active power absorbed by a three-wire system can be measured using only two wattmeters. In fact, Aron's theorem states the following relationships:

$$P = P_{12} + P_{32}$$

$$P = P_{23} + P_{13} \tag{39.41}$$

$$P = P_{31} + P_{21}$$

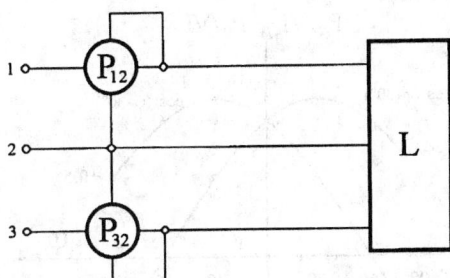

FIGURE 39.19 Power measurements using two wattmeters.

Analogously, the overall reactive power can be measured by using only two reactive power meters:

$$Q = Q_{12} + Q_{32}$$
$$Q = Q_{23} + Q_{13} \tag{39.42}$$
$$Q = Q_{31} + Q_{21}$$

Here one of the previous statements, that is:

$$P = P_{12} + P_{32}$$

is proved. The two wattmeters connected as shown in Figure 39.19 furnish P_{12}, P_{32}:
Hence, the sum of the two readings gives:

$$P_{12} + P_{32} = \dot{I}_1 \times \dot{V}_{12} + \dot{I}_3 \times \dot{V}_{32} = \dot{I}_1 \times \left(\dot{E}_1 - \dot{E}_2\right) + \dot{I}_3 \times \left(\dot{E}_3 - \dot{E}_2\right)$$

$$= \dot{I}_1 \times \dot{E}_1 - \dot{I}_1 \times \dot{E}_2 + \dot{I}_3 \times \dot{E}_3 - \dot{I}_3 \times \dot{E}_2 = \dot{I}_1 \times \dot{E}_1 + \dot{I}_3 \times \dot{E}_3 - \left(\dot{I}_1 + \dot{I}_3\right) \times \dot{E}_2 \tag{39.43}$$

$$= \dot{I}_1 \times \dot{E}_1 + \dot{I}_3 \times \dot{E}_3 + \dot{I}_2 \times \dot{E}_2 = P_1 + P_2 + P_3 = P$$

Provided that the system has only three wires, Aron's theorem applies to any kind of supply and load. In the case of symmetrical and balanced systems, it also allows the reactive power to be evaluated:

$$Q = \sqrt{3} \cdot \left(P_{32} - P_{12}\right) \tag{39.44}$$

Using Equations 39.41 and 39.44, the power factor is:

$$\cos\varphi = \frac{P_{12} + P_{32}}{\sqrt{\left(P_{12} + P_{32}\right)^2 + 3\left(P_{32} - P_{12}\right)^2}} = \frac{P_{12} + P_{32}}{\sqrt{4P_{12}^2 + 4P_{32}^2 - 4P_{12}P_{32}}} = \frac{1 + \dfrac{P_{12}}{P_{32}}}{2\sqrt{\left(\dfrac{P_{12}}{P_{32}}\right)^2 - \left(\dfrac{P_{12}}{P_{32}}\right) + 1}} \tag{39.45}$$

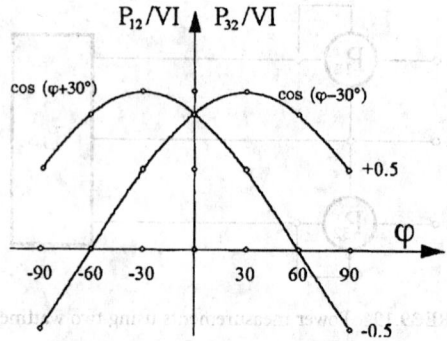

FIGURE 39.20   Sign of powers in Aron insertion.

Aron's insertion cannot be utilized when the power factor is low. In fact, if the functions:

$$\cos\!\left(\varphi+30\right)=\frac{P_{12}}{VI}$$

$$\cos\!\left(\varphi-30\right)=\frac{P_{32}}{VI}$$

(39.46)

are considered (Figure 39.20), it can be argued that: (1) for $\varphi \leq 60°$, P_{12} and P_{32} are both greater than zero; (2) for $\varphi > 60°$, $\cos(\varphi - 30)$ is still greater than zero, and $\cos(\varphi + 30)$ is lower than zero.

The absolute error in the active power is:

$$\Delta P=\frac{\partial\!\left(P_{12}+P_{32}\right)}{\partial P_{12}}\Delta P_{12}+\frac{\partial\!\left(P_{12}+P_{32}\right)}{\partial P_{32}}\Delta P_{32}=\Delta P_{12}+\Delta P_{32}$$

(39.47)

This corresponding relative error is greater as P_{12} and P_{32} have values closer to each other and are opposite in polarity; in particular, for $\cos\varphi = 0$ ($\varphi = 90°$), the error is infinite.

If η_w and ε_w are the wattmeter amplitude and phase errors, respectively, then the error in the active power is:

$$\frac{\Delta P}{P}=\frac{\left(\eta_w+\varepsilon_w tg\varphi_{12}\right)P_{12}+\left(\eta_w+\varepsilon_w tg\varphi_{32}\right)P_{32}}{P_{12}+P_{32}}=\eta_w+\varepsilon_w\,\frac{Q}{P}$$

(39.48)

Let two wattmeters with nominal values V_0, I_0, $\cos\varphi_0$, and class c be considered; the maximum absolute error in each reading is:

$$\Delta P=\frac{cV_0I_0\cos\varphi_0}{100}$$

(39.49)

Therefore, the percentage error related to the sum of the two indications is:

$$\frac{\Delta P}{P}=\frac{cV_0I_0\cos\varphi_0}{\sqrt{3}VI\cos\varphi}=1.11\frac{cV_0I_0\cos\varphi_0}{VI\cos\varphi}$$

(39.50)

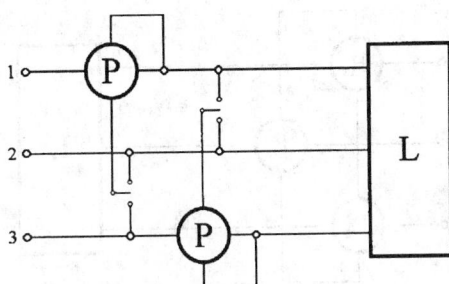

FIGURE 39.21 Barbagelata insertion for symmetrical and unbalanced systems.

equal to approximately the error of only one wattmeter inserted in a single-phase circuit with the same values of *I, V,* and cosφ. Consequently, under the same conditions, the use of two wattmeters involves a measurement uncertainty much lower than that using three wattmeters.

If the Aron insertion is performed via current and voltage transformers, characterized by ratio errors η_a and η_v, and phase errors ε_a and ε_v, respectively, the active power error is:

$$\frac{\Delta P}{P} = \frac{\left(\eta_{TOT} + \varepsilon_{TOT} tg\varphi_{12}\right)P_{12} + \left(\eta_{TOT} + \varepsilon_{TOT} tg\varphi_{32}\right)P_{32}}{P_{12} + P_{32}} = \eta_{TOT} + \varepsilon_{TOT}\frac{Q}{P} = \eta_{TOT} + \varepsilon_{TOT} tg\Phi_c \quad (39.51)$$

where $\cos\Phi_c$ = Conventional power factor

$$\left.\begin{array}{l}\eta_{TOT} = \eta_w + \eta_a + \eta_v \\ \varepsilon_{TOT} = \varepsilon_w + \varepsilon_a + \varepsilon_v\end{array}\right\}$$ the error sums with η_w and ε_w being the wattmeter errors.

Symmetrical Power Systems Supplying Unbalanced Loads

If the load is unbalanced, the current amplitudes are different from each other and their relative phase is not equal to 120°. Two wattmeters and one voltmeter have to be connected as proposed by Barbagelata [13] (Figure 39.21). The first wattmeter can provide P_{12} and P_{13}, and the second one gives P_{31} and P_{32}.

From the Aron theorem, the active power is:

$$P = P_{12} + P_{32} \quad (39.52)$$

and then the reactive power *Q* is:

$$Q = Q_1 + Q_2 + Q_3 = \frac{1}{\sqrt{3}}\left[P_{13} - P_{12} + \underline{P_{21}} - \underline{P_{23}} + P_{32} - P_{31}\right] \quad (39.53)$$

For the underlined terms, from Aron's theorem it follows that:

$$P = P_{13} + P_{23} = P_{12} + P_{32} = P_{21} + P_{31} \quad (39.54)$$

then:

$$P_{13} + P_{23} = P_{21} + P_{31} \quad \Rightarrow \quad P_{21} - P_{23} = P_{13} - P_{31}$$

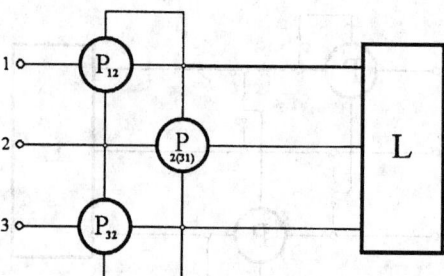

FIGURE 39.22   Righi insertion for symmetrical and unbalanced systems.

Thus, one obtains:

$$Q = \frac{1}{\sqrt{3}}\left[2\left(P_{13} - P_{31}\right) + P_{32} - P_{12}\right]$$ (39.55)

Therefore, using only four power measurements, the overall active and reactive powers can be obtained.

The main disadvantage of this method is that the four measurements are not simultaneous; therefore, any load variations during the measurement would cause a loss in accuracy. In this case, a variation proposed by Righi [13] can be used. In this variation, three wattmeters are connected as shown in Figure 39.22 and give simultaneously P_{12}, P_{32}, and $P_{2(31)}$. Reactive power is:

$$Q = \frac{1}{\sqrt{3}}\left[P_{13} - P_{12} + P_{21} - P_{23} + P_{32} - P_{31}\right].$$ (39.56)

Analogously as above, from the Aron theorem it follows that:

$$P_{21} - P_{23} = P_{13} - P_{31} \quad \Rightarrow \quad P_{2(31)} = P_{21} - P_{23} = P_{13} - P_{31}$$ (39.57)

then:

$$Q = \frac{1}{\sqrt{3}}\left[P_{32} - P_{12} + 2P_{2(31)}\right]$$ (39.58)

For symmetrical and unbalanced systems, another two-wattmeter insertion can be carried out (Figure 39.23). The wattmeters give:

$$P_{1(30)} = \dot{E}_3 \times \dot{I}_1 = j\frac{\dot{V}_{12}}{\sqrt{3}} \times \dot{I}_1 = -\frac{Q_{12}}{\sqrt{3}}$$

$$P_{3(10)} = \dot{E}_1 \times \dot{I}_3 = j\frac{\dot{V}_{23}}{\sqrt{3}} \times \dot{I}_3 = -\frac{Q_{32}}{\sqrt{3}}$$ (39.59)

Hence, the overall reactive power is:

$$Q = Q_{12} + Q_{32} = \sqrt{3}\left[-P_{1(30)} + P_{3(10)}\right]$$ (39.60)

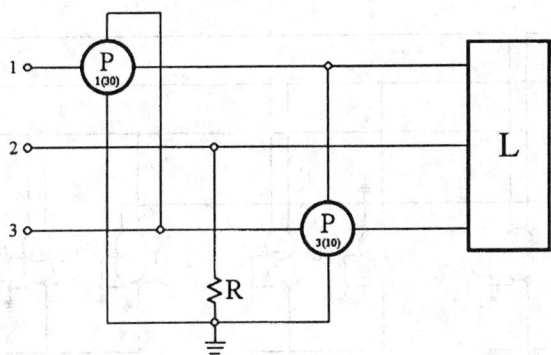

FIGURE 39.23 Two wattmeters-based insertion for symmetrical and unbalanced systems.

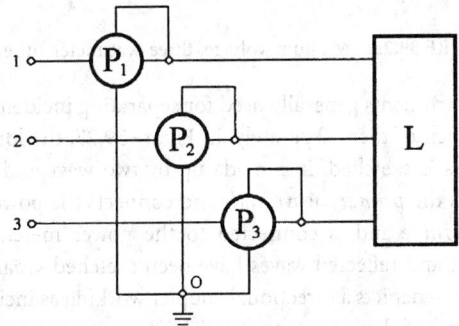

FIGURE 39.24 Three wattmeters-based insertion for three-wire, three-phase systems.

Three-Wattmeter Insertion

A three-wire, three-phase system can be measured by three wattmeters connected as in Figure 39.24. The artificial neutral point position does not affect the measurement; it is usually imposed by the impedance of the voltmeter leads of the wattmeters.

Medium-Voltage, Three-Wattmeter Insertion

Analogously to the single-phase case, for medium-voltage circuits, the three-wattmeter insertion is modified as in Figure 39.25.

Method Selection Guide

For three-wire systems, the flow chart of Figure 39.26 leads to selecting the most suitable method according to system characteristics.

High-Frequency Power Measurements

Meters used for power measurements at radio or microwave frequencies are generally classified as *absorption type* (containing inside their own load, generally 50 Ω for RF work) and *transmitted* or *through-line type* (where the load is remote from the meter). Apart from the type, power meters are mainly based on thermistors, thermocouples, diodes, or radiation sensors. Therefore, to work properly, the sensor should sense all the RF power (P_{LOAD}) incoming into the sensor itself. Nevertheless, line-to-sensor impedance mismatches cause partial reflections of the incoming power (P_{INCIDENT}) so that a meter connected to a sensor does not account for the total amount of reflected power ($P_{\text{REFLECTED}}$). The relationship existing among power dissipated on the load, power incident, and power reflected is obviously:

$$P_{\text{LOAD}} = P_{\text{INCIDENT}} - P_{\text{REFLECTED}} \qquad (39.61)$$

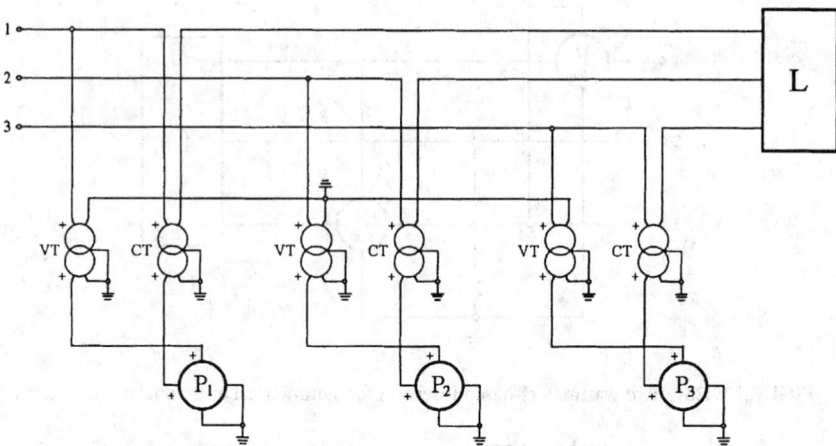

FIGURE 39.25   Medium-voltage, three-wattmeters insertion.

Directional couplers are instruments generally used for separating incident and reflected signals so that power meters can measure each of them separately. In Figure 39.27, the longitudinal section of a directional coupler for waveguides is sketched. It is made up by two waveguides properly coupled through two holes. The upper guide is the *primary waveguide* and connects the power source and load; the lower guide is the *secondary waveguide* and is connected to the power meter. To explain the working of directional couplers, incident and reflected waves have been sketched separately in Figure 39.27(*a*) and 39.27(*b*). In particular, section *a* depicts a directional coupler working as incident wave separator, whereas section *b* shows the separation of the reflected wave. The correct working is based on the assumption that the distance between the holes matches exactly *one quarter of the wave length* (λ). In fact, in the secondary waveguide, each hole will give rise to two waves going in opposite directions (one outside and the other inside the waveguide); consequently, in front of each hole, two waves are summed with their own phases. The assumption made on the distance between the holes guarantees that, in front of one hole, (1) the two waves propagating outside the waveguide will be in phase, causing an enforcing effect in that direction; (2) while, in front of the other hole, the two waves (always propagating outside) will be in opposition, causing a canceling effect in that direction. The enforcing and canceling effects for incident and reflected waves are opposite. In particular, according to the directions chosen in Figure 39.27, incident power propagates on the right side and is canceled on the left side (Figure 39.27(*a*)), while reflected power propagates on the left side and is canceled on the right side (Figure 39.27(*b*)). Therefore, directional couplers allow separate measurement of incident and reflected power by means of power meters applied, respectively, on the right and on the left side of the secondary waveguide.

In any case, the secondary waveguide must be correctly matched from the impedance point of view at both sides (by adaptive loads and/or a proper choice of the power meter internal resistance) in order to avoid unwanted reflections inside the secondary waveguide.

Directional couplers are also used to determine the *reflection coefficient* ρ of the sensor, which takes into account mismatch losses and is defined as:

$$P_{\text{REFLECTED}} = \rho^2 \times P_{\text{INCIDENT}} \tag{39.62}$$

In order to take into account also the absorptive losses due to dissipation in the conducting walls of the sensor, leakage into instrumentation, power radiated into space, etc., besides the reflection coefficient, the *effective efficiency* η_C of the sensor should also be considered. Generally, the reflection coefficient and effective efficiency are included into the *calibration factor K*, defined as:

$$K = \eta_C \left(1 - \rho^2\right) \times 100 \tag{39.63}$$

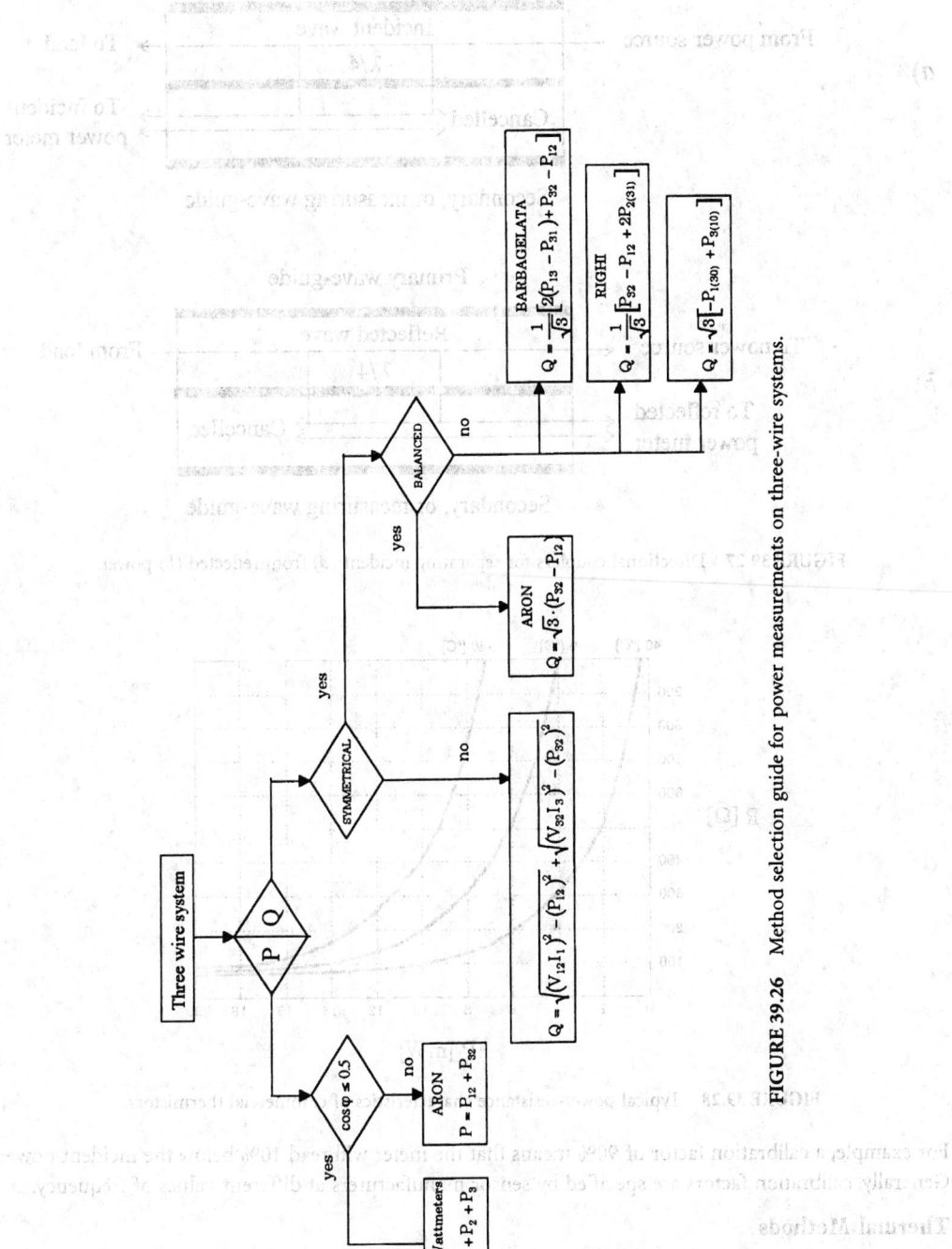

FIGURE 39.26 Method selection guide for power measurements on three-wire systems.

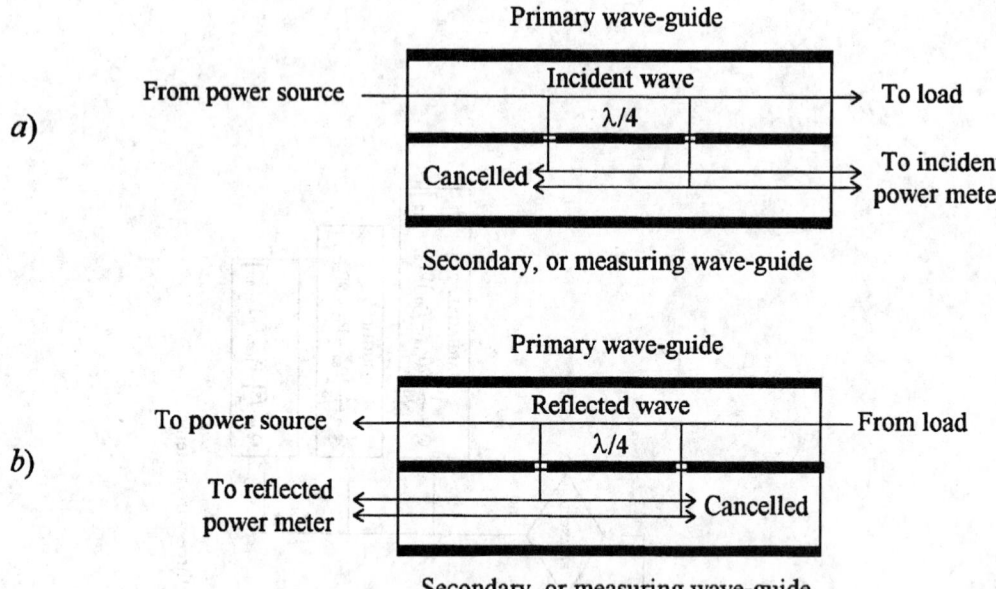

FIGURE 39.27 Directional couplers for separating incident (*a*) from reflected (*b*) power.

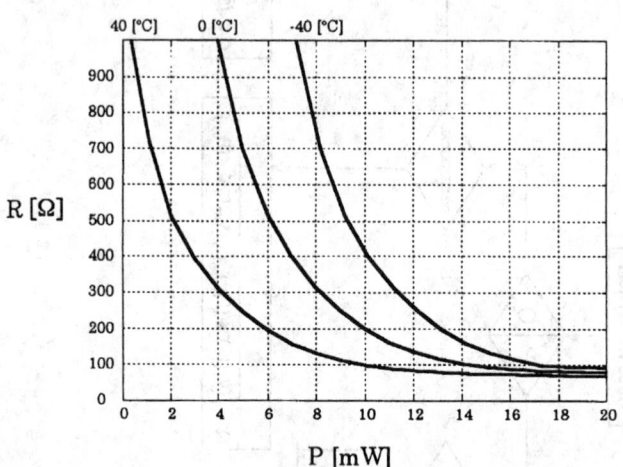

FIGURE 39.28 Typical power-resistance characteristics of commercial thermistors.

For example, a calibration factor of 90% means that the meter will read 10% below the incident power. Generally, calibration factors are specified by sensor manufacturers at different values of frequency.

Thermal Methods

In this section, the main methods based on power dissipation will be examined, namely: (1) thermistor-based, (2) thermocouple-based, and (3) calorimetric.

Thermistor-Based Power Meters

A thermistor is a resistor made up of a compound of highly temperature-sensitive metallic oxides [14]. If it is used as a sensor in a power meter, its resistance becomes a function of the temperature rise produced by the applied power. In Figure 39.28, typical power-resistance characteristics are reported for several values of the operating temperature.

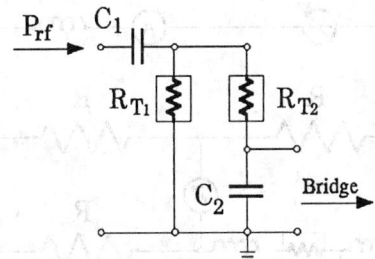

FIGURE 39.29 Working principle of the thermistor-based power meter.

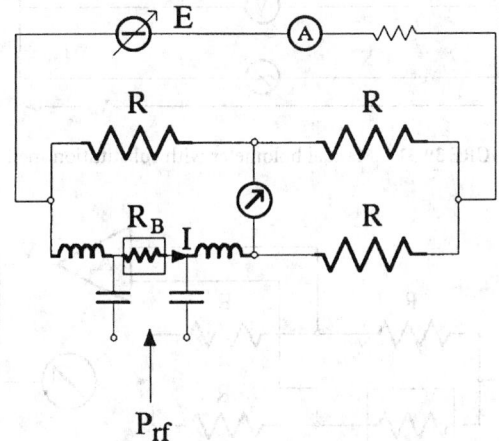

FIGURE 39.30 The manual bolometer.

The working principle of the thermistor power meter is illustrated in Figure 39.29 [15]: two thermistors (R_{T1} and R_{T2}) are connected (1) in parallel, for measurand signals appearing at the RF input (P_{RF}); and (2) in series, for the following measuring circuit (e.g., a bridge). The capacitance C_1 prevents the power dc component from flowing to the thermistors; the C_2 stops the RF power toward the bridge.

A bridge with a thermistor or a barretter in one arm is called a *bolometer*. Bolometer-based measurements can be performed with (1) a manual bolometer with variation of the bias current, (2) a manual bolometer with substitution method, or (3) a self-balancing bolometer.

The *manual bolometer with a variation of the bias current* is illustrated in Figure 39.30. Its working principle consists of two steps. In the first, no RF power is applied to the sensor; the equilibrium is obtained by varying the dc power supply E until the sensor resistance R_B, related to the dc power flowing in it, is equal to R. In this condition, let the current I flowing into the sensor be equal to I_1. In the second step, an RF power P_{RF} is fed to the sensor; the power increase must be compensated by a dc power decrease, which is performed by lowering the bridge dc supply voltage E; in this case, let I be equal to I_2.

Since the power dissipated in the sensor has been maintained constant in both steps, the power P_{RF} can be evaluated as:

$$P_{RF} = \frac{R}{4}\left(I_1^2 - I_2^2\right) \tag{39.64}$$

The *manual bolometer with substitution method* (Figure 39.31) consists of two sequential steps; in the first, both RF power (P_{RF}) and dc power (P_{dc}) are present, and the power (P_d) necessary to lead the bridge to the equilibrium is:

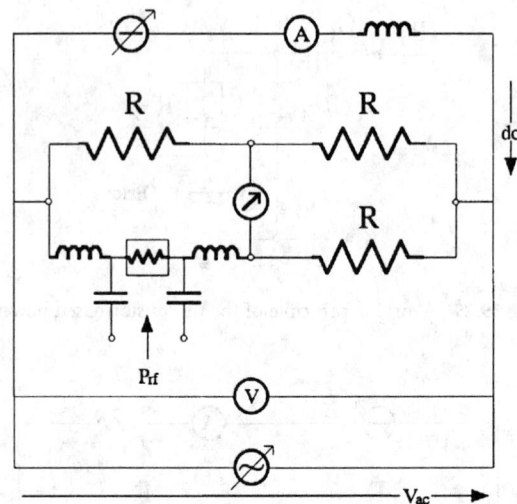

FIGURE 39.31   Manual bolometer with substitution method.

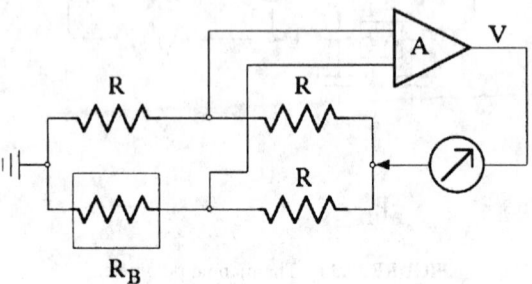

FIGURE 39.32   Self-balancing bolometer.

$$P_d = P_{dc} + P_{RF} \tag{39.65}$$

During the second step, P_{RF} is set to zero and an alternative voltage V_{ac} is introduced in parallel to the dc power supply. In this case, the power P_d necessary to balance the bridge:

$$P_d = P_{dc} + P_{ac} \tag{39.66}$$

is obtained by varying v_{ac}.

Since P_d is the same in both cases, the power supplied by the alternative generator is equal to P_{RF}:

$$P_{RF} = P_{ac} = \frac{V_{ac}^2}{4R} \tag{39.67}$$

Equation 39.66 implies that the RF power can be obtained by a voltage measurement. The *self-balancing bolometer* (Figure 39.32) automatically supplies a dc voltage V to balance the voltage variations due to sensor resistance R_B changes for an incident power P_{RF}. At equilibrium, R_B is equal to R and the RF power will then be:

$$P_{RF} = \frac{V^2}{4R} \tag{39.68}$$

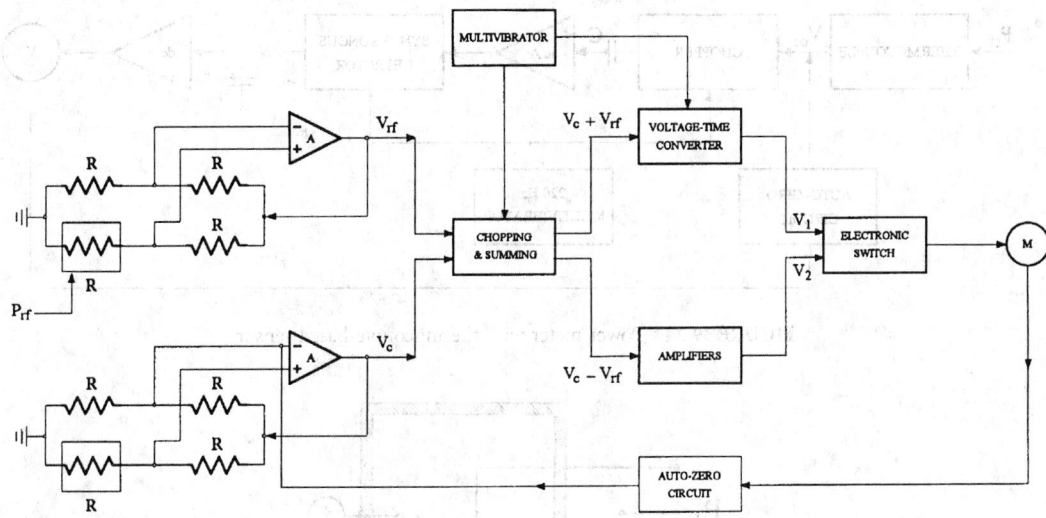

FIGURE 39.33 Power meter based on two self-balancing bridges.

As mentioned above, the thermistor resistance depends on the surrounding temperature. This effect is compensated in an instrument based on two self-balancing bridges [15]. The RF power is input only to one of these, as shown in Figure 39.33.

The equilibrium voltages V_c and V_{RF} feed a chopping and summing circuit, whose output $V_c + V_{RF}$ goes to a voltage-to-time converter. This produces a pulse train V_1, whose width is proportional to $V_c + V_{RF}$. The chopping section also generates a signal with an amplitude proportional to $V_c - V_{RF}$, and a frequency of a few kilohertz, which is further amplified. The signals V_1 and V_2 enter an electronic switch whose output is measured by a medium value meter M. This measure is proportional to the RF power because:

$$P_{RF} = \frac{\left(V_c + V_{RF}\right)\left(V_c - V_{RF}\right)}{4R} \Rightarrow P_{RF} = \frac{V_c^2 - V_{RF}^2}{4R} \tag{39.69}$$

Owing to the differential structure of the two bolometers, this device is capable of performing RF power measurements independent of the surrounding temperature. In addition, an offset calibration can be carried out when P_{RF} is null and V_c is equal to V_{RF}.

These instruments can range from 10 mW to 1 μW and utilize sensors with frequency bandwidths ranging from 10 kHz to 100 GHz.

Thermocouple-Based Power Meters

Thermocouples [14] can be also used as RF power meters up to frequencies greater than 40 GHz. In this case, the resistor is generally a thin-film type. The sensitivity of a thermocouple can be expressed as the ratio between the dc output amplitude and the input RF power. Typical values are 160 μV mW⁻¹ for minimum power of about 1 μW.

The measure of voltages of about some tens of millivolts requires strong amplification, in that the amplifier does not have to introduce any offset. With this aim, a chopper microvoltmeter is utilized [16], as shown in the Figure 39.34.

The thermocouple output voltage V_{dc} is chopped at a frequency of about 100 Hz; the resulting square wave is filtered of its mean value by the capacitor C and then input to an ac amplifier to further reduce offset problems. A detector, synchronized to the chopper, and a low-pass filter transform the amplified square voltage in a dc voltage finally measured by a voltmeter.

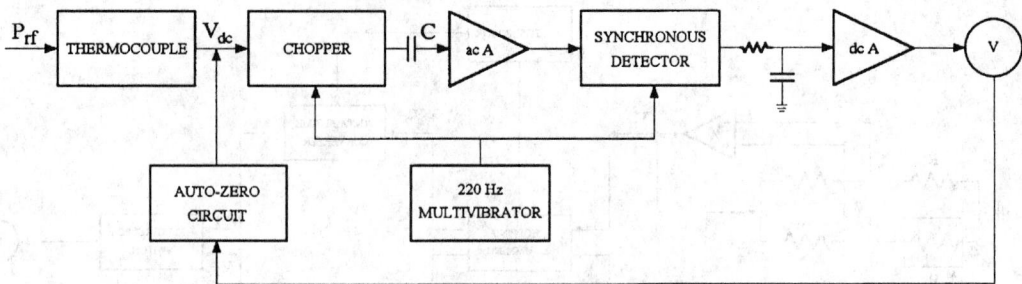

FIGURE 39.34   Power meter with thermocouple-based sensor.

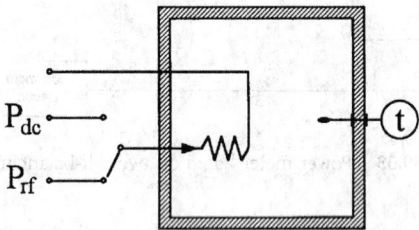

FIGURE 39.35   Calorimetric method based on a substitution technique.

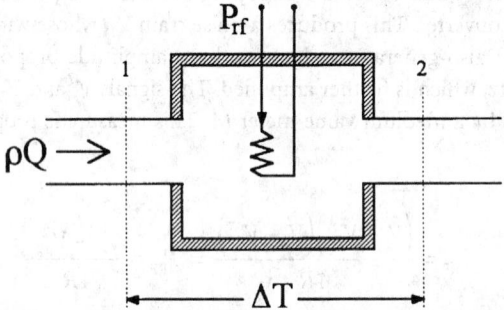

FIGURE 39.36   Calorimetric method based on a comparison technique.

Calorimetric Method

For high frequencies, a substitution technique based on a calorimetric method is utilized (Figure 39.35) [17]. First, the unknown radio frequency power P_{RF} is sent to the measuring device t, which measures the equilibrium temperature. Then, once the calorimetric fluid has been cooled to its initial temperature, a dc power P_{dc} is applied to the device and regulated until the same temperature increase occurs in the same time interval. In this way, a thermal energy equivalence is established between the known P_{dc} and the measurand P_{RF}.

A comparison version of the calorimetric method is also used for lower frequency power measurements (Figure 39.36). The temperature difference ΔT of a cooling fluid between the input (1) and the output (2) sections of a cooling element where the dissipated power to be measured P is determined. In this case, the power loss will correspond to P:

$$P = C_p \rho Q \Delta T \tag{39.70}$$

where C_p is the specific heat, ρ the density, and Q the volume flow, respectively, of the refreshing fluid.

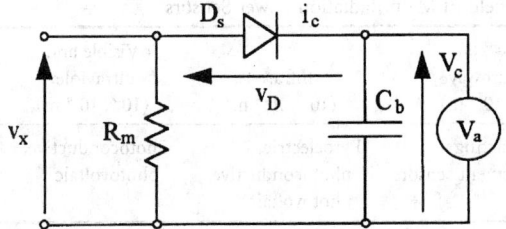

FIGURE 39.37   Circuit of the diode sensor-based power measurement.

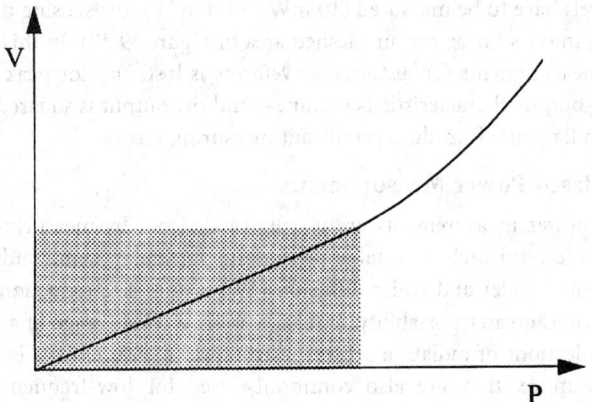

FIGURE 39.38   Characteristic of a low-barrier Schottky diode.

Diode Sensor-Based Power Measurements

Very sensitive (up to 0.10 nW, −70 dBm), high-frequency (10 MHz to 20 GHz) power measurements are carried out through a diode sensor by means of the circuit in Figure 39.37 [18]. In particular, according to a suitable selection of the components in this circuit, (1) true-average power measurements, or (2) peak power measurements can be performed.

The basic concept underlying *true-average power measurements* exploits the nonlinear squared region of the characteristic of a low-barrier Schottky diode (non-dashed area in Figure 39.38). In this region, the current flowing through the diode is proportional to the square of the applied voltage; thus, the diode acts as a squared-characteristic sensor.

In the circuit of diode sensor-based wattmeters shown in Figure 39.37, the measurand v_x, terminated on the matching resistor R_m, is applied to the diode sensor D_s working in its squared region in order to produce a corresponding output current i_c in the bypass capacitor C_b. If C_b has been suitably selected, the voltage V_c between its terminals, measured by the voltmeter amplifier V_a, is proportional to the average of i_c, i.e., to the average of the squares of instantaneous values of the input signal v_x, and, hence, to the true average power.

In the true-average power measurement of nonsinusoidal waveforms having the biggest components at low frequency, such as radio-frequency AM (Amplitude Modulated), the value of C_b must also satisfy another condition. The voltage v_d on the diode must be capable of holding the diode switched-on into conduction even for the smallest values of the signal. Otherwise, in the valleys of the modulation cycle, the high-frequency modulating source is disconnected by the back-biased diode and the measurement is therefore misleading.

On the other hand, for the same signal but for a different selection of the C_b value, the circuit can act as a peak detector for *peak power measurements*. As a matter of fact, the voltage v_c on the bypass capacitor C_b during the peak of the modulation cycle is so large that in the valleys, the high-frequency peaks are

TABLE 39.1   Operating Field of Main Radiation Power Sensors

Operating field (Wavelength)	Microwave $(1, 10^{-3}$ m)	Infrared $(10^{-3}, 10^{-6}$ m)	Visible and ultraviolet $(10^{-6}, 10^{-9}$ m)	Nuclear rays $(10^{-8}, 10^{-15}$ m)
Sensors	Noncontacting displacement sensors	Pyroelectric, photoconductive, photovoltaic	Photoconductive, photovoltaic	Scintillation counters, plastic films, solid-state, thermolum

not capable of switching on the diode into conduction; thus, these peaks do not contribute to the measured power level.

If higher power levels have to be measured (10 mW to 100 mW), the sensing diode is led to work out of the squared region into its linear region (dashed area in Figure 39.38). In this case, the advantage of true-average power measurements for distorted waveforms is lost; and for peak power measurements, since the diode input-output characteristic is nonlinear and the output is squared, spectral components different from the fundamental introduce significant measuring errors.

Radiation Sensor-Based Power Measurements

Very high-frequency power measurements are usually carried out by measuring a radiant flux of an electromagnetic radiation through a suitable sensor. In particular, semiconductor-based radiation microsensors have gained wider and wider diffusion [19], in that size reduction involves several well-known advantages such as greater portability, fewer materials, a wider range of applications, etc. One of the most familiar applications of radiation sensor-based power measurements is the detection of object displacement. Furthermore, they are also commonly used for low-frequency power noninvasive measurements.

Radiation sensors can be classified according to the measurand class to which they are sensitive: nuclear particles or electromagnetic radiations. In any case, particular sensors capable of detecting both nuclear particles and electromagnetic radiations, such as gamma and X-rays, exist and are referred to as nucleonic detectors.

In Table 39.1, the different types of radiation sensors utilized according to the decrease of the measurand wavelength from microwaves up to nuclear (X, gamma, and cosmic) rays are indicated.

In particular, *microwave* power radiation sensors are mainly used as noncontacting detectors relying on ranging techniques using microwaves [20]. Shorter and longer (radar) wavelength microwave devices are employed to detect metric and over-kilometer displacements, respectively.

Beyond applications analogous to microwave, power radiation *infrared* sensors also find use as contact detectors. In particular, there are two types of infrared detectors: thermal and quantum. The thermal type includes contacting temperature sensors such as thermocouples and thermopiles, as well as noncontacting pyroelectric detectors. On the other hand, the quantum type, although characterized by a strong wavelength dependence, has a faster response and includes photoconductive (spectral range: 1 μm to 12 μm) and photovoltaic (0.5 μm to 5.5 μm) devices.

The main power radiation *visible and ultraviolet* sensors are photoconductive cells, photodiodes, and phototransistors. Photodiodes are widely used to detect the presence, the intensity, and the wavelength of light or ultraviolet radiations. Compared to photoconductive cells, they are more sensitive, smaller, more stable and linear, and have lower response times. On the other hand, phototransistors are more sensitive to light.

At very low light levels, rather than silicon-based microsensors, *nuclear radiation* power microsensors are needed. In this case, the most widespread devices are scintillation counters, solid-state detectors, plastic films, and thermoluminescent devices. The scintillation counter consists of an active material that converts the incident radiation to pulses of light, and a light-electric pulse converter. The active material can be a crystal, a plastic fluorine, or a liquid. The scintillator size varies greatly according to the radiation energy, from thin solid films to large tanks of liquid to detect cosmic rays. A thin (5 μm) plastic

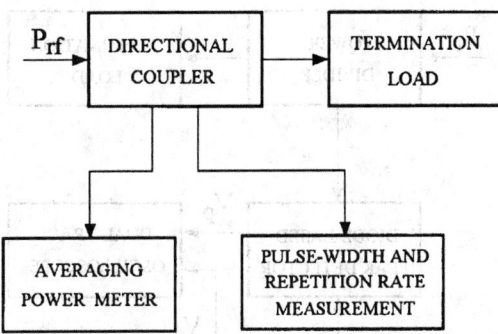

FIGURE 39.39 Block diagram of an instrument measuring average power per duty cycle.

polycarbonate film or a thermoluminescent material (e.g., LiF) can measure the radiation power falling on a surface. The film is mechanically damaged by the propagation of highly α–ionizing particles. Consequent etching of the film reveals tracks that can be observed and counted.

39.3 Pulse Power Measurements

Pulse waveforms are becoming more and more diffused in several fields such as telecommunications, power source applications, etc. The *pulse power* P_p is defined as the average power P_m in the pulse width:

$$P_p = \frac{P_m}{\tau_d} \qquad (39.71)$$

where τ_d is the duty cycle of the pulse waveform (i.e., the pulse width divided by the waveform period). If the pulse width cannot be accurately defined (e.g., nonrectangular pulses in the presence of noise), the pulse power P_p becomes unmeaningful. In this case, the *peak envelope power* is introduced as the maximum of the instantaneous power detected on a time interval, including several periods of the pulse waveform (but negligible with respect to the modulation period, in the case of PWM waveforms).

Several techniques are used to measure pulse power [21]. In particular, they can be classified according to the *pulse frequency* and the necessity for *constraining real-time applications* (i.e., measuring times, including a few of the modulation periods). For real-time, low-frequency applications (up to 100 kHz), the algorithms mentioned in the above sections on wattmeters based on digital multipliers can be applied [9].

If constraining limits of real-time do not have to be satisfied, either digital or analog techniques can be utilized. As far as the digital techniques are concerned, for high-frequency applications, if the measurand pulse waveform is stationary over several modulation periods, digital wattmeters based on equivalent sampling can be applied, with accuracies increasing according to measuring times. As far as the analog techniques are concerned, three traditional methods are still valid: (1) average power per duty cycle, (2) integration-differentiation, and (3) dc/pulse power comparison.

A block diagram of an instrument measuring *average power per duty cycle* is illustrated in Figure 39.39. At first, the mean power of the measurand pulse signal, terminated on a suitable load, is measured by means of an average power meter; then, the pulse width and the pulse waveform period are measured by a digital counter. Finally, the power is obtained by means of Equation 39.71.

The *integration-differentiation* technique is based on a barretter sensor capable of integrating the measurand, and on a conditioning and differentiating circuit to obtain a voltage signal proportional to the measurand power. The signal is input to the barretter sensor, having a thermal constant such that the barretter resistance will correspond to the integral of the input. The barretter is mounted as an arm of a conditioning Wheatstone bridge; in this way, the barretter resistance variations are transformed into voltage variations, and an integrated voltage signal is obtained as an output of the bridge detecting arm.

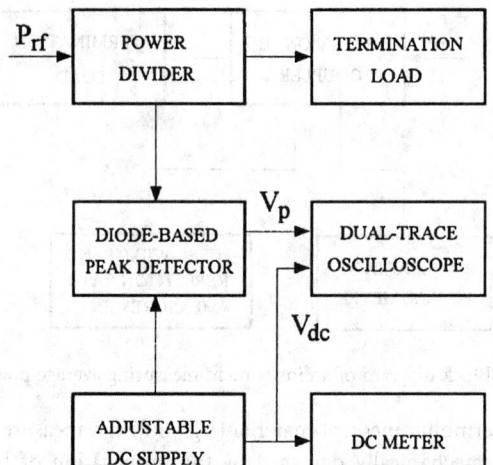

FIGURE 39.40 Block diagram of an instrument based on dc/pulse power comparison technique.

This signal, suitably integrated to reach a voltage signal proportional to the output, is detected by a peak voltmeter calibrated in power. Analogously to the selection of the time constant of an *RC* integrating circuit, attention must be paid to the thermal constant selection of the barretter in order to attain the desired accuracy in the integration. With respect to the measurand pulse period, a very long thermal constant will give rise to insufficient sensitivity. On the other hand, a very short constant approaching the pulse duration will give rise to insufficient accuracy.

The *dc/pulse power comparison* technique is based on the concept of first revealing the peak envelope power through a diode sensor, and then comparing the peak to a known dc source with a dual trace scope. A block diagram of an instrument based on this concept is illustrated in Figure 39.40. The peak of the measurand pulse signal terminated on a suitable load is sensed by a peak detector by obtaining a proportional signal V_p. This signal is input to a channel of a dual trace oscilloscope and the envelope peak is displayed. An adjustable dc source is input to the other channel of the scope to obtain a signal V_{dc} to be compared to the envelope peak signal. When the two signals are made equal, a dc voltage meter directly calibrated in peak power measures the output power.

References

1. G. Zingales, Power measurements on single-phase ac circuits (Chap. VI, 6.2), in *Methods and Instruments for Electrical Measurements*, (in Italian), Torino: UTET, 1980.
2. G. Korànyi, Measurement of power and energy, in L. Schnell (ed.), *Technology of Electrical Measurements*, Chichester: John Wiley & Sons, 1993.
3. J. Milmann and C.C. Halkias, *Integrated Electronics: Analog and Digital Circuits and Systems*, New York: McGraw-Hill, 1972.
4. F.F. Mazda, AC analogue instruments (Chap. VI, 6.3), in *Electronic Instruments and Measurement Techniques*, Cambridge, U.K.: Cambridge University Press, 1987.
5. P.S. Filipski, A TDM wattmeter with 0.5 MHz carrier frequency, *IEEE Trans. Instrum. Meas.*, IM39, 15-18, 1990.
6. J.R. Carstens, *Electrical Sensors and Transducers*, Englewood Cliffs, NJ: Prentice-Hall, 1992.
7. Lu Zu-Liang, An error estimate for quasi-integer-period sampling and an approach for improving its accuracy, *IEEE Trans. Instrum. Meas.*, IM-23, 337-341, 1984.
8. V. Haasz, The error analysis of digital measurements of electrical power, *Measurement*, 6, 1986.
9. P. Arpaia, F. Avallone, A. Baccigalupi, and C. De Capua, Real-time algorithms for active power measurements on PWM-based electric drives, *IEEE Trans. Instrum. Meas.*, IM-45, 462-466, 1996.

10. X. Dai and R. Gretsch, Quasi-synchronous sampling algorithm and its applications, *IEEE Trans. Instrum. Meas.*, IM-43, 204-209, 1994.

11. M. Bellanger, *Digital Processing of Signals: Theory and Practice*, Chichester: John Wiley & Sons, 1984.

12. M. Bertocco, C. Offelli, and D. Petri, Numerical algorithms for power measurements, *Europ. Trans. Electr. Power*, ETEP 3, 91-101, 1993.

13. G. Zingales, Measurements on steady-state circuits (Chap. VI), in *Methods and Instruments for Electrical Measurements*, (In Italian), Torino: UTET, 1980.

14. H.N. Norton, Thermometers (Chap. 19), in *Handbook of Transducers*, Englewood Cliffs, NJ: Prentice-Hall, 1989.

15. Anonymous, Thermistor mounts and instrumentation (Chap. II), in *Fundamental of RF and Microwave Power Measurements*, Application Note 64-1, Hewlett Packard, 1978.

16. R.E. Pratt, Power measurements (15.1–15.16), in C.F. Coombs, (ed.), in *Handbook of Electronic Instruments*, New York: McGraw-Hill, 1995.

17. F. F. Mazda, High-frequency power measurements (Chap. VIII, 8.4), in *Electronic Instruments and Measurement Techniques*, Cambridge: Cambridge University Press, 1987.

18. Anonymous, Diode detector power sensors and instrumentation (Chap. IV), in *Fundamental of RF and Microwave Power Measurements*, Application Note 64-1, Hewlett Packard, 1978.

19. J.W. Gardner, *Microsensors: Principles and Applications*, Chichester: John Wiley & Sons, 1994.

20. H.N. Norton, Radiation pyrometers (Chap. 20), in *Handbook of Transducers*, Englewood Cliffs, NJ: Prentice-Hall, 1989.

21. F.F. Mazda, Pulse power measurement (Chap. VIII, 8.5), in *Electronic Instruments and Measurement Techniques*, Cambridge, U.K.: Cambridge University Press, 1987.

Further Information

F.K. Harris, The Measurement of Power (Chap. XI), in *Electrical Measurements*, Huntington, NY: R.E. Krieger Publishing, 1975; a clear reference for line-frequency power measurements.

Anonymous, *Fundamental of RF and Microwave Power Measurements*, Application Note 64-1, Hewlett Packard, 1978; though not very recent, is a valid and comprehensive reference for main principles of high-frequency power measurements.

J.J. Clarke and J.R. Stockton, Principles and theory of wattmeters operating on the basis of regularly spaced sample pairs, *J. Phys. E. Sci. Instrum.*, 15, 645-652, 1982; gives basics of synchronous sampling for digital wattmeters.

T.S. Rathore, Theorems on power, mean and RMS values of uniformly sampled periodic signals, *IEE Proc. Pt. A*, 131, 598-600, 1984; provides fundamental theorems for effective synchronous sampling wattmeters.

J.K. Kolanko, Accurate measurement of power, energy, and true RMS voltage using synchronous counting, *IEEE Trans. Instrum. Meas.*, IM-42, 752–754, 1993; provides information on the implementation of a synchronous dual-slope wattmeter.

G.N. Stenbakken, A wideband sampling wattmeter, *IEEE Trans. Power App. Syst.*, PAS-103, 2919-2925, 1984; gives basics of asynchronous sampling-based wattmeters and criteria for computing uncertainty in time domain.

F. Avallone, C. De Capua, and C. Landi, Measurement station performance optimization for testing on high efficiency variable speed drives, *Proc. IEEE IMTC/96 (Brussels, Belgium)*, 1098–1103, 1996; proposes an analytical model of uncertainty arising from power measurement systems working under highly distorted conditions.

F. Avallone, C. De Capua, and C. Landi, Metrological performance improvement for power measurements on variable speed drives, *Measurement*, 21, 1997, 17-24; shows how compute uncertainty of measuring chain components for power measurements under highly distorted conditions.

F. Avallone, C. De Capua, and C. Landi, Measurand reconstruction techniques applied to power measurements on high efficiency variable speed drives, *Proc. of XIV IMEKO World Congress (Tampere,
Fi)*, 1997; proposes a technique to improve accuracy of power measurements under highly distorted
conditions.

F. Avallone, C. De Capua, and C. Landi, A digital technique based on real-time error compensation for
high accuracy power measurement on variable speed drives, *Proc. of IEEE IMTC/97 (Ottawa,
Canada)*, 1997; reports about a real-time technique for error compensation of transducers working
under highly distorted conditions.

J.C. Montano, A. Lopez, M. Castilla, and J. Gutierrrez, DSP-based algorithm for electric power measurement, *IEE Proc. Pt.A*, 140, 485–490, 1993; describes a Goertzel FFT-based algorithm to compute
power under nonsinusoidal conditions.

S.L. Garverick, K. Fujino, D.T. McGrath, and R.D. Baertsch, A programmable mixed-signal ASIC for
power metering, *IEEE J. Solid State Circuits*, 26, 2008–2015, 1991; reports about a programmable
mixed analog-digital integrated circuit based on six first-order sigma-delta ADCs, a bit serial DSP,
and a byte-wide static RAM for power metering.

G. Bucci, P. D'Innocenzo, and C. Landi, A modular high-speed dsp-based data acquisition apparatus for
on-line quality analysis on power systems under non-sinusoidal conditions, *Proc. of 9th IMEKO
TC4 Int. Sym.* (Budapest, Hungary), 286–289, 1996; shows the strategy of measurement system
design for power measurements under non-sinusoidal conditions.

K.K. Clarke and D.T. Hess, A 1000 A/20 kV/25 kHz-500 kHz volt-ampere-wattmeter for loads with power
factors from 0.001 to 1.00, *IEEE Trans. Instrum. Meas.*, IM-45, 142-145, 1996; provides information
on the implementation of a instrument to perform an accurate measurement of currents (1 A to
1000 A), voltages (100 V to 20 kV), and powers (100 W to 20 MW) over the frequency range from
25 kHz to 500 kHz.

P. Arpaia, G. Betta, A. Langella, and M. Vanacore, An Expert System for the Optimum Design of
Measurement Systems, *IEE Proc. (Pt. A)*, 142, 330–336, 1995; reports about an Artificial Intelligence
tool for the automatic design of power measuring systems.

In any case, *IEEE Transactions on Instrumentation and Measurement* and *Measurement* journals provide
current research on power measurements.

40
Power Factor Measurement

Michael Z. Lowenstein
Harmonics Limited

Electricity can be thought of as a means of delivering power from one place to another to do work. The laws and relationships for delivering power were originally developed for direct current. Power delivered, expressed in *watts*, was calculated by multiplying the voltage and current as shown in Equation 40.1

$$P_{dc} = EI \tag{40.1}$$

The situation becomes more complex when alternating current is used to deliver power. Figure 40.1 shows a sine wave representing either ac current or voltage. Since the instantaneous value of the wave is continually changing, a numerical quantity is defined to represent an average property of this wave. This quantity, the root-mean-square, or **rms** value, calculated by squaring the instantaneous value, integrating it during one cycle, dividing by the period, and taking the square root of the result, is equal to the peak value of the ac wave divided by the square root of 2, or, for ac current, $I_{rms} = i_{peak}/\sqrt{2}$. In the physical world, a sine wave ac current having an rms value of 1 A (A = ampere), passed through a resistive load, produces the same heating effect as 1 A of dc current. Thus, one might expect delivered ac power to be easily calculated in watts using Equation 40.1 and inserting rms values for current and voltage. While this simple relationship holds true for the instantaneous voltage and current values as shown in Equation 40.1a in general, it is not true for the rms quantities except for the special case when the ac current and voltage are restricted to perfect sine waves *and* the load is a pure resistance.

$$p_{inst.} = ei \tag{40.1a}$$

In real-world situations where current and/or voltage waveforms are not perfectly sinusoidal and/or the loads are other than resistive, the relationships are no longer simple and the power delivered, or *active power*, is usually less than the product of rms voltage and current, as shown in Equation 40.2.

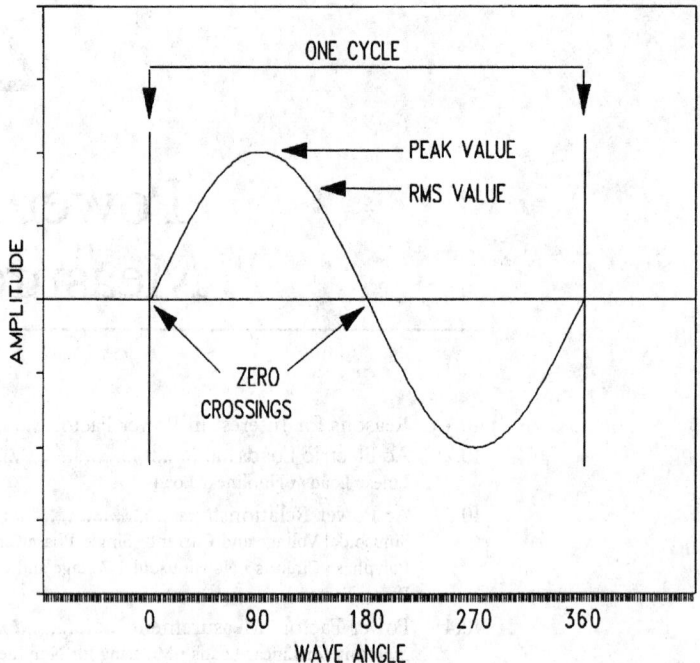

FIGURE 40.1   Sine wave characteristics. One cycle of a continuous wave is shown. The wave begins at a zero-crossing, reaches a positive peak, continues through zero to a negative peak, and back to zero. The wave repeats every 360°. The wave angle can also be expressed as a function of frequency *f* and time *t*. For a given frequency, the wave angle is related to the expression, $2\pi f t$.

$$P_{ac} \leq E_{rms} I_{rms} \tag{40.2}$$

The product of rms voltage and rms current does, however, define a quantity termed *apparent power*, *U*, as shown in Equation 40.3.

$$U = E_{rms} I_{rms} \tag{40.3}$$

A derived term, the *power factor*, F_p, used to express the relationship between delivered or active power, *P*, and apparent power, *U*, is defined by Equation 40.4.

$$F_p = \frac{P}{U} \tag{40.4}$$

From Equations 40.2, 40.3, and 40.4, it is clear that the value of F_p must lie in a range between zero and one.

This chapter focuses on: (1) ac power relationships and the calculation of power factor; (2) the physical meaning of these relationships; and (3) measurement techniques and instrumentation for determining these relationships and calculating power factor.

40.1   Reasons for Interest in Power Factor

Power factor is a single number that relates the active power, *P*, to the apparent power, *U*. Electric components of a utility distribution system are designed on a kVA basis; i.e., they are designed to operate at a given voltage and carry a rated current without undue temperature rise. Transformer and conductor sizes are chosen on this basis. While active power does useful work, reactive and harmonic powers do no useful work, absorb system capacity, and increase system losses; but reactive and harmonic powers are needed to provide magnetic fields or nonlinear currents. *The capacity of electric systems is limited by apparent power, not active power. Power factor expresses, with a single value, the extent to which an electrical distribution system is efficiently and effectively utilized.* A low value for the power factor means that much of the system capacity is not available for useful work. From a utility viewpoint, this means reduced ability to deliver revenue-producing active power; from a user viewpoint, a low power factor reduces the available active power or requires increased system size to deliver needed power.

40.2   ac Electric Loads

Linear Loads

Electric loads in ac power systems with sinusoidal voltages are categorized by the way they draw current from the system. Loads that draw sinusoidal currents, i.e., the current waveshape is the same as the voltage waveshape, are defined as *linear loads*. Historically, a high percentage of electric loads have been linear. Linear loads include: (1) induction motors; (2) incandescent lighting; and (3) heaters and heating elements. Linear loads use ac electric power directly to accomplish their functions.

Nonlinear Loads

Electric loads that draw nonsinusoidal currents, i.e., the current waveshape differs from the voltage waveshape, are defined as *nonlinear loads*. As energy savings and efficient use of electricity are emphasized, an increased percentage of nonlinear electric devices, both new and replacement, are being installed. Nonlinear loads include: (1) adjustable-speed motor drives; (2) fluorescent and arc-discharge lighting; (3) computers and computerized controls; and (4) temperature-controlled furnaces and heating elements. Nonlinear loads, rather than using ac electric power directly, often convert ac power into direct current before it is used to accomplish their functions. A common element in nonlinear loads is some kind of rectifier to accomplish this ac to dc conversion. Rectifiers do not draw sinusoidal currents.

40.3   ac Power Relationships

Sinusoidal Voltage and Current

Power calculations for sinusoidal ac electric systems require knowledge of the rms voltage, the rms current, and the phase relationships between the two. Figure 40.2 illustrates possible phase relationships between voltage and current. If the positive-going zero-crossing of the voltage is considered the reference point, then the nearest positive-going zero-crossing of the current can occur at a wave angle either less than or greater than this reference. If the current zero-crossing occurs before the reference, the current is said to *lead* the voltage. If the current zero-crossing occurs after the reference, the current is *lagging*. If the zero-crossing for the current coincides with the reference, the two waves are said to be *in phase*. The wave angle, θ, by which the current leads or lags the voltage is called the *phase angle*, in this case, 30°.

Single-Phase Circuits

Power Calculations

The power delivered to do work is easily calculated [1]. Given a sinusoidal voltage of rms magnitude E and sinusoidal current of rms magnitude I, displaced by angle θ, at time t,

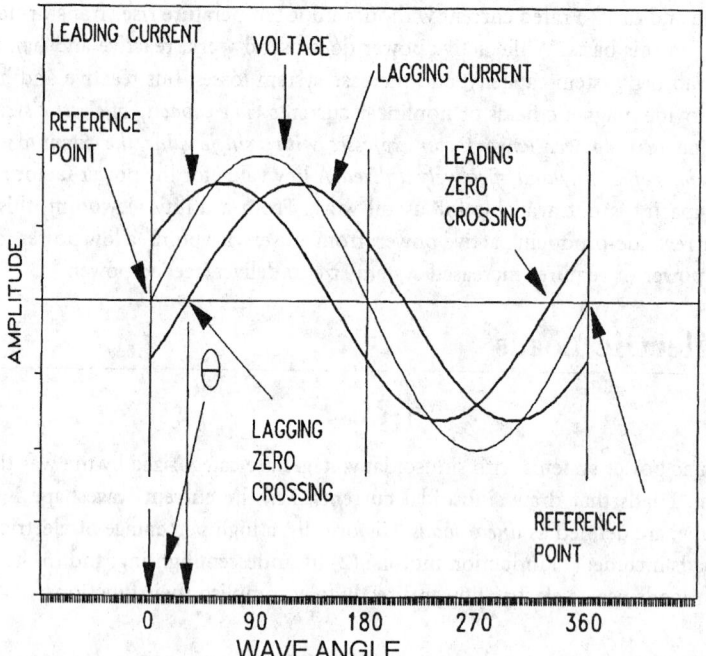

FIGURE 40.2   Sine wave phase angle. Two waves with the same zero-crossing are *in phase*. A sine wave that crosses zero before the reference wave is *leading*, and one that crosses zero after the reference wave is *lagging*. The *phase angle* θ illustrated is 30° lagging.

$$\text{Instantaneous voltage} = e\sqrt{2}E\sin(2\pi ft)$$

$$\text{Instantaneous current} = i = \sqrt{2}I\sin(2\pi ft - \theta)\ \big(\text{note that } \theta \text{ can have a positive or negative}$$

$$\text{value}\big)$$

(40.5)

$$\text{Instantaneous power} = p = ei$$

$$p = 2EI\sin(2\pi ft)\sin(2\pi ft - \theta)$$

$$p = EI\cos(\theta) - EI\cos(4\pi ft - \theta)$$

$$\text{Average power over an integral number of cycles}\quad P = EI\cos(\theta)$$

(40.5)

$$\text{Power factor } F_p = \frac{P}{U} = \frac{EI\cos(\theta)}{EI} = \cos(\theta)$$

(40.5)

Equation 40.5 is the fundamental equation that defines power for systems in which the current and voltage are sinusoidal. The application of this equation is illustrated for three cases: (1) the current and voltage are in phase; (2) the current and voltage are out of phase by an angle less than 90°; and (3) the current and voltage are out of phase by exactly 90°.

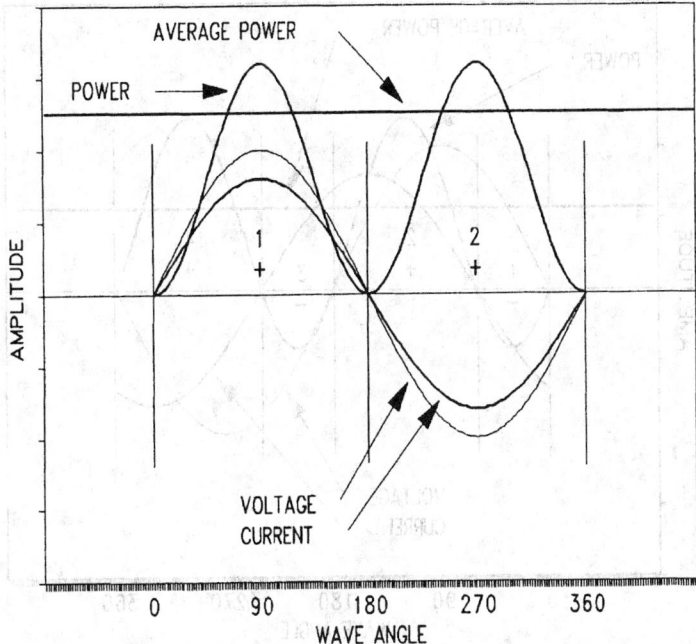

FIGURE 40.3 Voltage, current, and power for sine waves in phase. The vertical scales for voltage and current are equal. The scale for power is relative and is selected to permit the display of all three waves on a single graph. The voltage and current are in phase and both are sinusoidal. The power is everywhere positive, and average power delivered to do work is the maximum power.

Ac Power Examples

Figure 40.3 shows voltage current and power when the voltage and current are in phase and the current displacement angle is zero (0). (An example would be a resistance space heater.) The power curve is obtained by multiplying together the instantaneous values of voltage and current as the wave angle is varied from 0° to 360°. Instantaneous power, the product of two sine waves, is also a sine wave. There are two zero-crossings per cycle, dividing the cycle into two regions. In region (1) both the voltage and current are positive and the resultant product, the power, is positive. In region (2) both the voltage and current are negative and the power is again positive. The average power in watts, given by Equation 40.5, $EI\cos(0°) = EI$, is the maximum power that can be delivered to do work. *When sinusoidal voltage and current are in phase, the delivered power in watts is the same as for dc and is the maximum that can be delivered.* The power factor, $F_p = \cos(0°) = 1$, or *unity*.

Figure 40.4 shows voltage, current, and power when the current lags the voltage by 60°. (An example might be a lightly loaded induction motor.) The power sine wave again is obtained by multiplying together the instantaneous values of voltage and current. There are now four zero-crossings per cycle, dividing the cycle into four regions. In regions (2) and (4), voltage and current have the same sign and the power is positive. In regions (1) and (3), voltage and current have opposite signs, resulting in a negative value for the power. The average power in watts, given by Equation 40.5, $EI\cos(60°) = EI(0.5)$, is less than the maximum that could be delivered for the particular values of voltage and current. *When voltage and current are out of phase, the delivered power in watts is always less than the maximum.* In this example, $F_p = \cos(60°) = 0.5$.

Figure 40.5 shows voltage, current, and power when the current lags the voltage by 90°. (This situation is not attainable in the real world.) The power sine wave again is obtained by multiplying together the instantaneous values of voltage and current. Again, four zero-crossings divide the cycle into four regions.

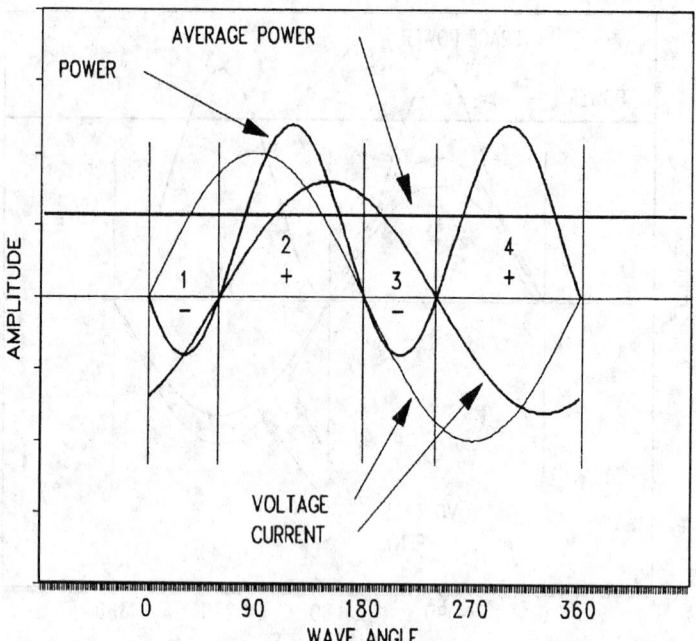

FIGURE 40.4 Voltage, current, and power for sine waves 60° out of phase. The vertical scales for voltage and current amplitudes are the same as those for Figure 40.3. Current lags voltage by 60° and both are sinusoidal. The power is positive in regions (2) and (4), and negative in regions (1) and (3). Average power delivered to do work is less than the maximum power.

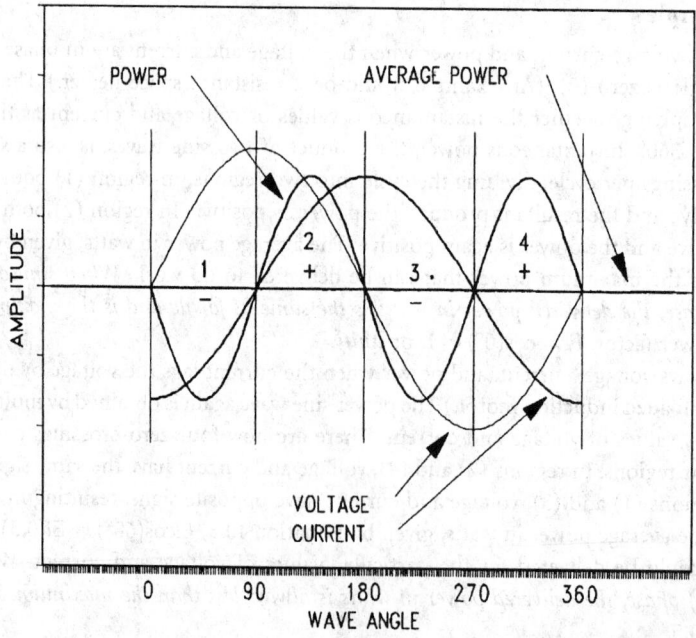

FIGURE 40.5 Voltage, current, and power for sine waves 90° out of phase. The vertical scales for voltage and current amplitudes are the same as those for figure 40.3. Current lags voltage by 90° and both are sinusoidal. The power is positive in regions (2) and (4), negative in regions (1) and (3), and is of equal absolute magnitude in all four regions. Average power delivered to do work is zero.

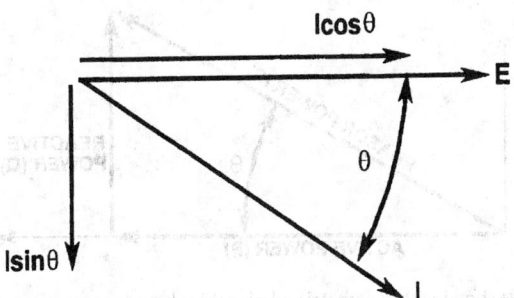

FIGURE 40.6 Phasor diagram for current and voltage. Voltage is the reference phasor. The current *I* has been resolved into orthogonal components *I*cosθ and *I*sinθ. (From Reference [1].)

In regions (2) and (4), the power is positive, while in regions (1) and (3), the power is negative. The average power in watts is given by Equation 40.5, $EI\cos(90°) = 0$. No matter what the values of voltage and current, *when voltage and current are exactly 90° out of phase, the delivered power in watts is always zero.* The power factor, $F_p = \cos(90°) =$ zero.

Power Factor

Resolving the current into orthogonal components on a phasor diagram illustrates how delivered power can vary from a maximum to zero, depending on the phase angle between the voltage and the current sine waves. Figure 40.6 shows the voltage vector along with the current resolved into orthogonal components [1]. The current *I* at an angle θ relative to the voltage can be resolved into two vectors: $I\cos(\theta)$ and $I\sin(\theta)$. The in-phase component $I\cos(\theta)$ multiplied by the voltage gives average power in watts. The current component that is 90° out of phase with the voltage, $I\sin(\theta)$, is not associated with delivered power and does not contribute to work. For want of a better name, this was often termed the *wattless component* of the current. Since this wattless current could be associated with magnetic fields, it was sometimes termed *magnetizing current* because, while doing no work, this current interacts through the inductive reactance of an ac motor winding to provide the magnetic field required for such a motor to operate.

Three types of power have been defined for systems in which both the voltage and current are sinusoidal. Throughout the years, a number of different names have been given to the three power types. The names in present usage will be emphasized.

Active power is given the symbol *P* and is defined by Equation 40.5:

$$P = EI\cos\left(\theta\right) \tag{40.5}$$

Other names for active power include: (1) real power and (2) delivered power. Active power is the power that does work. Note that while all power quantities are volt-ampere products, only *active power is expressed in watts.*

Reactive power is given the symbol *Q* and is defined by the equation:

$$Q = EI\sin\left(\theta\right) \tag{40.6}$$

Other names for reactive power include: (1) imaginary power; (2) wattless power; (3) and magnetizing power. Reactive power is expressed in *voltamperes*$_{reactive}$ or *vars*. If the load is predominantly inductive, current lags the voltage and the reactive power is given a positive sign. If the load is predominantly capacitive, current leads the voltage and the reactive power is given a negative sign.

Phasor power is given the symbol *S* and is defined by the equation:

$$S = \sqrt{P^2 + Q^2} \tag{40.7}$$

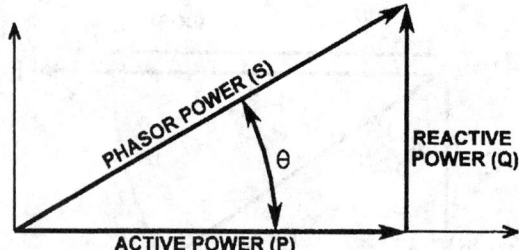

FIGURE 40.7   Power triangle showing the geometric relationships between active, reactive, and phasor power. Power factor is defined geometrically as cosθ.

Phasor power was called apparent power for many years, and it will be seen in a later section that phasor power S, *for sinusoidal voltages and currents*, is identical to what is now called apparent power U. Phasor power is expressed in *voltamperes* or *VA*.

Figure 40.7 is a phasor diagram, often called a *power triangle*, which illustrates the relationships among the three types of power defined above. Reactive power is orthogonal to active power, and is shown as positive for lagging current. It is clear that the definition of phasor power, Equation 40.7, is geometrically derived from active and reactive power.

Power factor is given the symbol F_p and *for sinusoidal quantities* is defined by the equation:

$$F_p = \frac{\text{ACTIVE POWER}}{\text{PHASOR POWER}} = \frac{P}{S} = \frac{\text{WATTS}}{\text{VOLTAMPS}} = \cos\theta \qquad (40.8)$$

Since the power factor can be expressed in reference to the displacement angle between voltage and current, power factor so defined should be termed *displacement power factor*, and the symbol is often written $F_{p\ displacement}$. Values for displacement power factor range from one (unity) to zero as the current displacement angle varies from 0° (current and voltage in phase) to 90°. Since the cosine function is positive in both the first and fourth quadrants, the power factor is positive for both leading and lagging currents. To completely specify the voltage–current phase relationship, the words *leading* or *lagging* must be used in conjunction with power factor. Power factor can be expressed as a decimal fraction or as percent. For example, the power factor of the case shown in Figure 40.4 is expressed either as 0.5 lagging or 50% lagging.

Polyphase Circuits

Power Calculations

The power concepts developed for single-phase circuits with sinusoidal voltages and currents can be extended to polyphase circuits. Such circuits can be considered to be divided into a group of two-wire sets, with the neutral conductor (or a resistively derived neutral for the case of a delta-connected, three-wire circuit) paired with each other conductor. Equations 40.3–40.5 can be rewritten to define power terms equivalent to the single-phase terms. In these equations, k represents a phase number, m is the total number of phases, and α and β are, respectively, the voltage and current phase angles with respect to a common reference frame.

$$P = \sum_{k=1}^{m} E_k I_k \cos(\alpha - \beta) \qquad (40.9)$$

$$Q = \sum_{k=1}^{m} E_k I_k \sin(\alpha - \beta) \qquad (40.10)$$

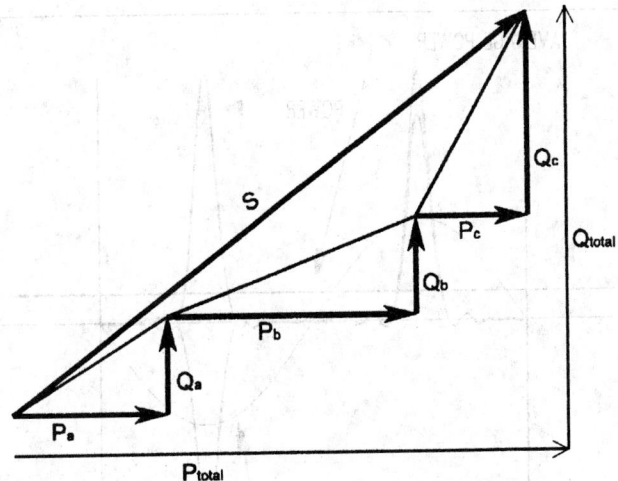

FIGURE 40.8 Phasor diagram for a sample polyphase sinusoidal service in which each phase has a different phase angle. Power factor cannot be defined as the cosine of the phase angle in this case. (From Reference [3].)

and, restating Equation 40.7:

$$S = \sqrt{P^2 + Q^2}$$

For example, a three-phase sinusoidal power distribution service, with phases a, b, and c:

$$P = E_a I_a \cos(\alpha_a - \beta_a) + E_b I_b \cos(\alpha_b - \beta_b) + E_c I_c \cos(\alpha_c - \beta_c)$$

$$Q = E_a I_a \sin(\alpha_a - \beta_a) + E_b I_b \sin(\alpha_b - \beta_b) + E_c I_c \sin(\alpha_c - \beta_c)$$

$$S = \sqrt{P^2 + Q^2}$$

Power Factor

Power factor is defined by Equation 40.11. Note that it is no longer always true to say that power factor is equal to the cosine of the phase angle. In many three-phase balanced systems, the phase angles of all three phases are equal and the cosine relationship holds. In unbalanced systems, such as that represented by the phasor diagram Figure 40.8, each phase has a different phase angle, the phase voltages and currents are not equal, and the cosine relationship fails [3].

$$F_p = \frac{\text{TOTAL ACTIVE POWER}}{\text{PHASOR POWER}} = \frac{P_{Eq.9}}{S} = \frac{\text{WATTS}}{\text{VOLTAMPS}} \; often \neq \cos\theta \qquad (40.11)$$

Nonsinusoidal Voltage and Current

Fourier Analysis

Figure 40.9 shows voltage, current, and power for a typical single-phase nonlinear load, a computer switch-mode power supply. Due to the nature of the bridge rectifier circuit in this power supply, current is drawn from the line in sharp spikes. The current peak is only slightly displaced from the voltage peak and the power is everywhere positive. However, power is delivered to the load during only part of the cycle and the average power is much lower than if the current had been sinusoidal. The current waveshape

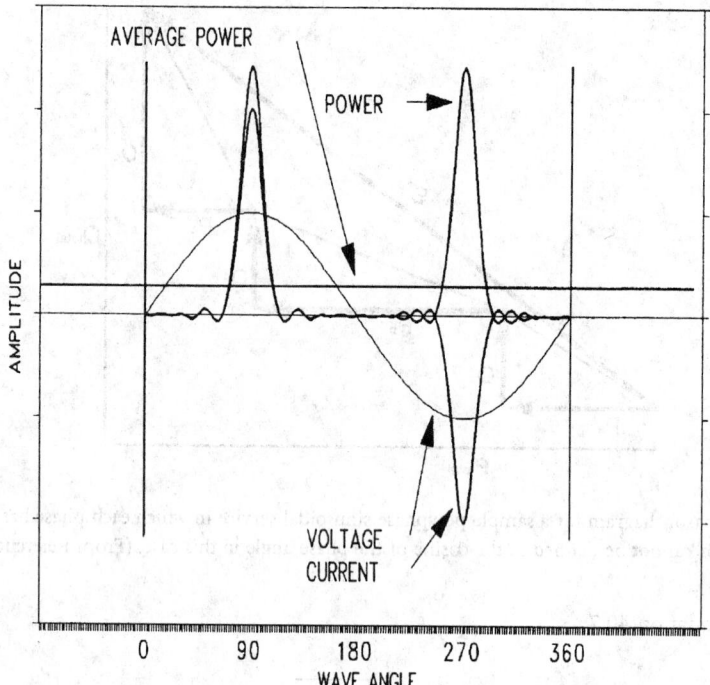

FIGURE 40.9 Voltage, current, and power for a single-phase, switch-mode computer power supply, a typical nonlinear load. The current is no longer sinusoidal.

required by the load presents a problem to the ac power system, which is designed to deliver only sine wave current. The solution to this problem is based on mathematical concepts developed in 1807 for describing heat flow by Jean Baptiste Joseph Fourier, a French mathematician [4]. Fourier's theorem states that any periodic function, however complex, can be broken up into a series of simple sinusoids, the sum of which will be the original complex periodic variation. Applied to the present electrical problem, Fourier's theorem can be stated: *any periodic nonsinusoidal electrical waveform can be broken up into a series of sinusoidal waveforms, each a harmonic of the fundamental, the sum of which will be the original nonsinusoidal waveform.*

Harmonics

Harmonics are defined as continuous integral multiples of the fundamental waveform. Figure 40.10 shows a fundamental sine wave and two harmonic waves — the 3rd and 5th harmonics. The harmonic numbers 3 and 5 express the number of complete cycles for each harmonic wave per cycle of the fundamental (or 1st harmonic). Each harmonic wave is defined by its harmonic number, its amplitude, and its phase relationship to the fundamental. Note that the fundamental frequency can have any value without changing the harmonic relationships, as shown in Table 40.1.

Power Calculations

Calculating power delivered to do work for a nonlinear load is somewhat more complicated than if the current were sinusoidal. If the fundamental component of the voltage at frequency f is taken as a reference (the a-phase fundamental for a polyphase system), the subscript "1" means the fundamental, and E denotes the peak value of the voltage; then the voltage can be expressed as:

$$\varepsilon_{a1(t)} = E_{a1} \sin(2\pi f t + 0°)$$

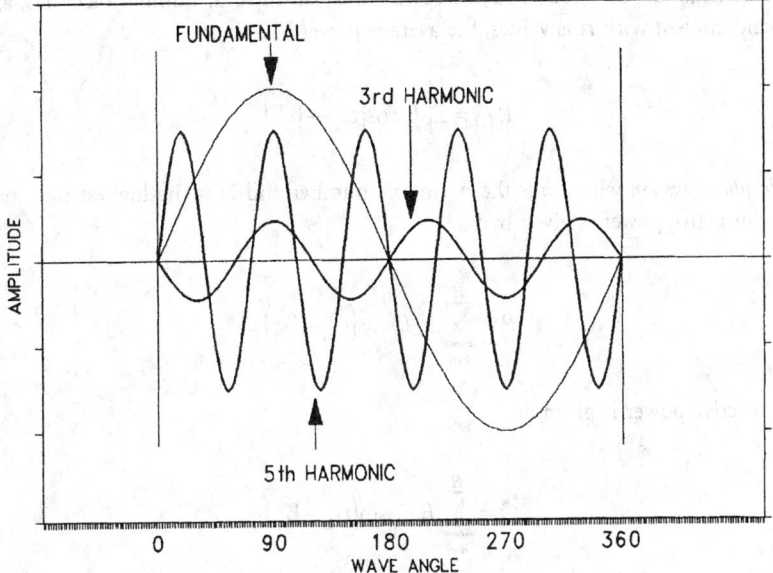

FIGURE 40.10 Harmonics are continuous integral multiples of the fundamental frequency. The 5th harmonic is shown, in phase with the fundamental, while the 3rd harmonic is 180° out of phase.

TABLE 40.1 Harmonics and Their Relationship to the Fundamental Frequency

Harmonic number	Frequency (Hz)	Frequency (Hz)	Frequency (Hz)
1	60	50	400
2	120	100	800
3	180	150	1200
4	240	200	1600
5	300	250	2000

The voltage fundamental will then have an amplitude E_{a1} and pass through zero in the positive going direction at time $t = 0$. If h = harmonic number, and E_h and I_h are peak amplitudes of the harmonic voltage and current, respectively, then general expressions for any harmonic will be:

$$e_{h(t)} = E_h \sin\left(2\pi f h t + \alpha_h^\circ\right)$$

$$i_{h(t)} = I_h \sin\left(2\pi f h t + \beta_h^\circ\right)$$

To compute the power associated with a voltage and current waveform, take advantage of the fact that products of harmonic voltages and currents of different frequency have a time average of zero. Only products of voltages and currents of the same frequency are of interest, giving a general expression for harmonic power as:

$$P_{h(t)} = E_h I_h \sin\left(2\pi f h t + \alpha_h^\circ\right)\sin\left(2\pi f h t + \beta_h^\circ\right)$$

Simplifying with trigonometric identities, evaluating over an integral number of cycles, and replacing peak voltage and current with rms values, the average power becomes:

$$P_{h(t)} = E_h I_h \cos\left(\alpha_h^\circ - \beta_h^\circ\right)$$

For a *single-phase system* where h is the harmonic number and H is the highest harmonic, the total average power or active power is given by:

$$P = \sum_{h=1}^{H} E_h I_h \cos\left(\alpha_h - \beta_h\right) \qquad (40.12)$$

Total average reactive power is given by:

$$Q = \sum_{h=1}^{H} E_h I_h \sin\left(\alpha_h - \beta_h\right) \qquad (40.13)$$

It should be noted that in the real world, the actual contribution of harmonic frequencies to active and reactive power is small (usually less than 3% of the total active or reactive power). The major contribution of harmonic frequencies to the power mix comes as distortion power, which will be defined later.

For a *polyphase system* wherein r is the phase identification and N is the number of conductors in the system, including the neutral conductor, total average power for a polyphase system is given by:

$$P = \sum_{r=1}^{N-1} \sum_{h=1}^{H} E_{rh} I_{rh} \cos\left(\alpha_{rh} - \beta_{rh}\right) \qquad (40.14)$$

Total average reactive power is given by:

$$Q = \sum_{r=1}^{N-1} \sum_{h=1}^{H} E_{rh} I_{rh} \sin\left(\alpha_{rh} - \beta_{rh}\right) \qquad (40.15)$$

Power Factor

Single-Phase Systems

For a single-phase system, phasor power is again given by Equation 40.7 and illustrated by Figure 40.7, where P is the algebraic sum of the active powers for the fundamental and all the harmonics (Equation 40.12), and Q is the algebraic sum of the reactive powers for the fundamental and all the harmonics (Equation 40.13). Therefore, phasor power is based on the fundamental and harmonic active and reactive powers. It is found, however, that phasor power S is no longer equal to apparent power U and a new power phasor must be defined to recognize the effects of waveform distortion. A phasor representing the distortion, termed *distortion power* and given the symbol D, is defined by:

$$D = \pm\sqrt{U^2 - S^2} \qquad (40.16)$$

Without further definite information as to the sign of distortion power, its sign is selected the same as the sign of the total active power. The relationships among the various power terms are displayed in

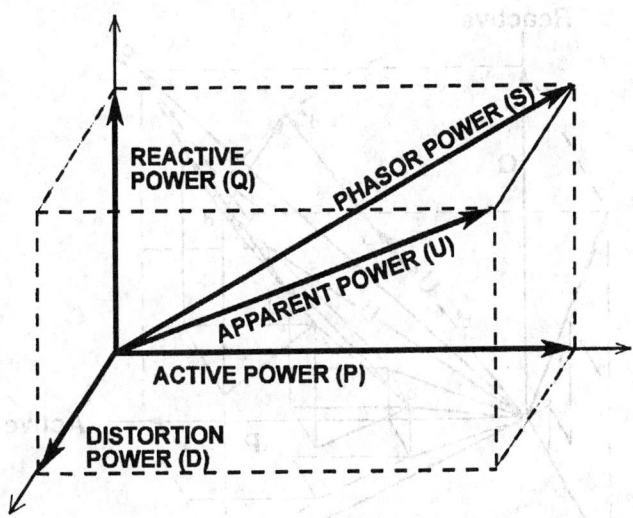

FIGURE 40.11 Phasor diagram for a single-phase, nonsinusoidal service in which the voltage and current contain harmonics. Geometric relationships are shown between active, reactive, phasor, distortion, and apparent powers.

Figure 40.11, a 3-dimensional phasor diagram. Power factor, in direct parallel with sinusoidal waveforms, is defined by the equation:

$$F_p = \frac{\text{TOTAL ACTIVE POWER}}{\text{APPARENT POWR}} = \frac{P}{U} = \frac{\text{WATTS}}{\text{VOLTAMPS}} \tag{40.17}$$

From Equations 40.7 and 40.16 we obtain:

$$S = \sqrt{P^2 + Q^2} \tag{40.7}$$

$$U = \sqrt{S^2 + D^2} = \sqrt{P^2 + Q^2 + D^2} \tag{40.18}$$

It is clear that when waveforms are sinusoidal, i.e., linear loads are drawing current, that there is no distortion power and Equation 40.18 reduces to Equation 40.7. Likewise as shown in Figure 40.13, as the distortion power vector goes to zero, the figure becomes two-dimensional and reduces to Figure 40.7, and U becomes equal to S. When, however, nonlinear loads are drawing harmonic currents from the system, U will be greater than S. As already noted, the contribution of the harmonics to the total power quantities is small and one is frequently interested mainly in the fundamental quantities.

The power factor associated only with the fundamental voltage and current components was termed the displacement power factor $F_{p\,\text{displacement}}$ where Equations 40.7 and 40.8 are written [5]:

$$S_{60} = \sqrt{P_{60}^2 + Q_{60}^2}$$

and

$$F_{p\,\text{displacement}} = \frac{P_{60}}{S_{60}}$$

When harmonics are present, F_p is always smaller than $F_{p\,\text{displacement}}$.

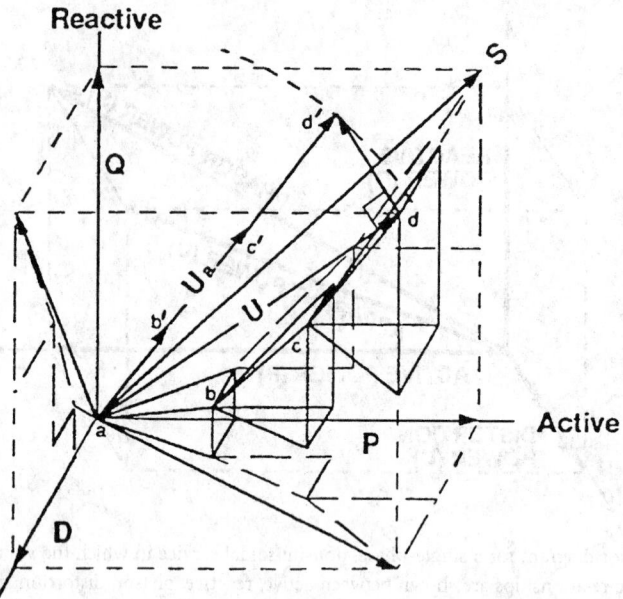

FIGURE 40.12 Phasor diagram for a three-phase, nonsinusoidal service in which the voltage and current contain harmonics. Arithmetic apparent power U_a is the length of the segmented line *abcd* and is a scaler quantity U_a can be represented by the line *ab'c'd'*. The diagonal *ad*, a vector quantity, is the apparent power U [6].

Polyphase Systems

For a polyphase system, phasor power, S, is again given by Equation 40.7, but one must now use the total values for P and Q calculated using Equations 40.14 and 40.15. One can then define the apparent power U in one of two ways.

- *Arithmetic apparent power.* The *arithmetic apparent power* is given the symbol U_a, and is defined by Equation 40.19, where E_r and I_r are the rms values for the respective phases and M equals the number of phases. U_a is a scalar quantity.

$$U_A = \sum_{r=1}^{M-1} E_r I_r \tag{40.19}$$

- *Apparent power.* Apparent power is given the symbol U and is defined by Equation 40.18 using total values for P and Q as defined by Equations 40.14 and 40.15, and a total value for D determined using Equation 40.16 for each phase. Figure 40.12 illustrates the two alternative concepts for polyphase apparent power [6]. Note that U_a uses arithmetic addition of vector magnitudes and is equal to apparent power U only if the polyphase voltages and currents have equal magnitudes and equal angular spacings, a situation that often exists in balanced three-phase systems. The two alternative definitions of apparent power, U and U_a, give rise to two possible values for power factor: (1) $F_p = P/U$; and (2) $F_{pa} = P/U_a$. Apparent power U and power factor F_p are the preferred definitions since using U_a can give unexpected results with some nonsymmetric service arrangements such as four-wire delta, and, with extremely unbalanced resistive loads, F_{pa} can exceed 1.0. Despite these shortcomings, arithmetic apparent power has become quite widely used due to the comparative simplicity of its measurement and calculation. With the advent of sophisticated digital meters, there is no longer any advantage to using arithmetic apparent power and its use will surely decrease.

40.4 Power Factor "Measurement"

There are no instruments that measure power factor directly. (Power stations and large substations often use phase angle meters with a power factor scale representing cos (θ) to display power factor. Such meters are accurate only for sinusoidal balanced polyphase systems.) One must remember that, of all the ac power quantities discussed, the only ones that can be directly measured are voltages, currents, and their time relationships (phase angles). All other ac power quantities are derived mathematically from these measured quantities. The only one of these derived values that has physical reality is the active power P (the quantity that does work); the others are mathematical constructs. Therefore, correct determination of power factor requires accurate measurement of voltage and current, and proficient mathematics.

Metering for Linear Loads

By the early 1920s, the concepts of active, reactive, and apparent (since renamed phasor) power, and power factor were known, and metering capabilities had been developed to enable their determination. Energy meters utilizing voltage and current coils driving an induction disk inherently measured active power P ($EI \cos\theta$), which was displayed on a mechanical register. Using the trigonometric identity $EI \sin\theta = EI \cos(90° + \theta)$, with voltage delayed 90°, a similar energy meter displayed reactive power Q and, with these two values, displacement power factor was calculated. Voltage delay (phase shifting) was accomplished using specially wound transformers.

  Through the years, the method of obtaining the 90° phase shift has been updated. Analog electronic meters are available that provide the 90° phase shift electronically within the meter. More recently, digital meters have been developed that sample voltages and currents at regular intervals and digitize the results. Voltages and currents are multiplied as they are captured to compute active power. Past voltage samples delayed by a time equal to a quarter cycle (90°) are multiplied by present current values to obtain reactive power. Active–reactive metering of this type is a widely utilized method for determining displacement power factor for utility billing. These meters do not accurately measure the effect of harmonic currents because the delay of the voltage samples is based on the fundamental frequency and is incorrect for the harmonics. (The important 5th harmonic, which is frequently the predominant harmonic component, is accurately measured because it is delayed 450° ($5 \times 90°$), which is equivalent to the correct 90° delay).

Metering for Nonlinear Loads

With the application of high-speed digital computing techniques to measurement of ac currents and voltages, together with digital filtering, the quantities necessary for accurate and correct calculation of power factor are susceptible to direct computation. In practice, the ac waveforms are sampled at a frequency greater than twice the highest frequency to be measured, in compliance with well-known sampling theories. Data obtained can be treated using Fourier's equations to calculate rms values for voltage, current, and phase angle for the fundamental and each harmonic frequency. Power quantities can be obtained with digital filtering in strict accordance with their *ANSI/IEEE STD* 100 definitions. Power quantities can be displayed for the fundamental only (displacement power factor), or for fundamental plus all harmonics (power factor for nonsinusoidal waveforms).

Metering Applications

Instruments with the capabilities described above are often called *harmonic analyzers*, and are available in both single-phase and polyphase versions. They can be portable, in which case they are often used for power surveys, or panel mounted for utility and industrial revenue metering. Polyphase analyzers can be connected as shown in Figures 40.13 and 40.14. Care must be taken to connect the instrument properly. Both voltage leads and current transformers must be connected to the proper phases, and the current transformers must also be placed with the correct polarity. Most instruments use color-coded voltage connectors. Correct polarity is indicated on the current transformers by arrows or symbols, and complete

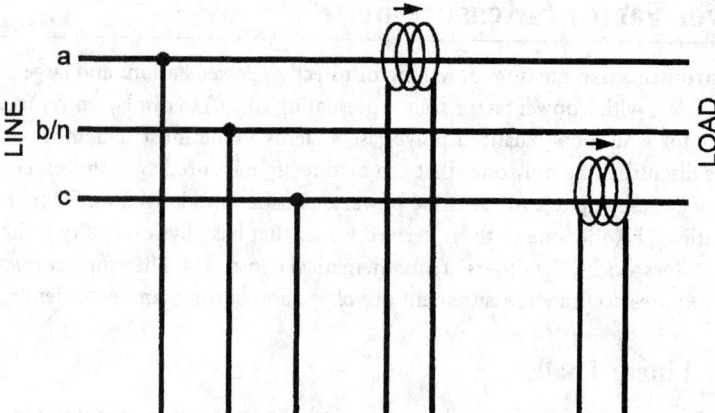

FIGURE 40.13 Meter connections for a one-, two-, or three-phase, three-wire service. Voltage leads are connected to phases a, b, and c, *or* to a, c, and neutral as the system dictates. Care must be taken to connect the voltage lead for each phase to the input corresponding to the input for the current transformer reading that phase, and directional characteristics of the transformers must be observed. Only two current connections are required.

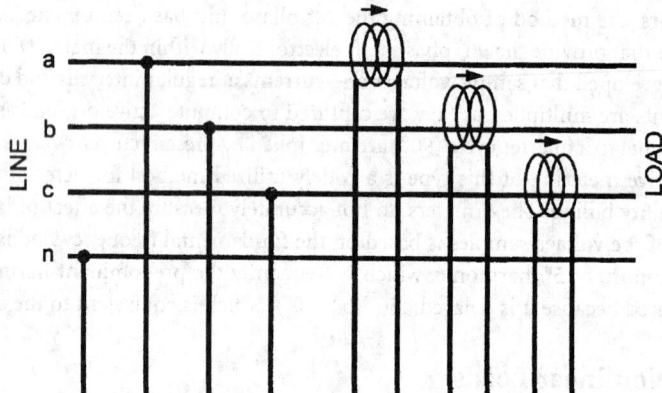

FIGURE 40.14 Meter connections for a three-phase, four-wire wye or delta connected service. Voltage leads are connected to phases a, b, and c, and to the neutral. Care must be taken to connect the voltage lead for each phase to the input corresponding to the input for the current transformer reading that phase, and directional characteristics of the transformers must be observed. In situations where measurement of neutral currents is desired, a fourth current transformer can be used to read the neutral. A fourth voltage connection might be used to read neutral-ground voltages.

hook-up and operating instructions are included. When single-phase instruments are used, the same precautions must be followed for making connections.

40.5 Instrumentation

Table 40.2 lists a sampling of harmonic and power factor measuring instruments available from major manufacturers. All instruments listed use some type of Fourier calculations and/or digital filtering to determine power values in accordance with accepted definitions. Many can be configured to measure non-harmonic-related power quality concerns. Unless otherwise noted, all instruments require the purchase of one current probe per input. Probes are available for measuring currents from 5 A to several thousand amperes. For comparison purposes, priced probes will be those with a 600-A range. Voltage leads are usually supplied as standard equipment. Table 40.3 contains addresses of these manufacturers.

TABLE 40.2 Selected Instrumentation for Harmonic and Power Factor Measurement

Manufacturer	Model	V/I inputs	Display type	Communication	Special features	List price (US$)
Hand-held units						
Amprobe	HA2000	1/1	Visual	RS232	Hand-held, probe included, 21 nonvolatile memories	990
BMI	155	1/1	Visual	Optional printer, RS232	Hand-held	1145 + 550 (printer) + 380 (probe)
BMI	355	4/3	Visual	Optional printer, RS232	Hand-held	1995 + 550 (printer) + 380/probe
Dranetz	4300	4/4	Panel for data and graphs	RS232	Hand-held, battery or ac power, optional PCMCIA memory card	5000 + 450/probe
Fluke	39	1/1	Visual	None	Probe included	895
Fluke	41b	1/1	Visual	RS232	Probe included, 8 memories, logging with computer and supplied software	1795
Portable units						
BMI	3030A	4/4	Built-in printer for data and graphs optional internal modem		Portable, optional built-in disk drive for storage, long-term monitoring, optional PQ configurations	6800 + 600 (modem) + 1895 (disk storage) + 590/probe
Dranetz	PP1-R	4/4	Panel for data and graphs	PCMCIA card slot	Long-term monitoring, optional PQ configurations, remote control by software	12,000 + 450/probe
PC-based units						
BMI	7100	4/4	PC-based (not included)	PC connection cord	Portable, uses computer for storage, optional software, long-term monitoring, optional PQ configurations	5294 + PC + 395 (software) + 415/probe
Cooper	V-Monitor II	4/4	PC-based (not included)	Serial port	Portable, software and signal interface and data acquisition board, long-term monitoring	12,000 + 500/probe

TABLE 40.2 (continued)   Selected Instrumentation for Harmonic and Power Factor Measurement

Manufacturer	Model	V/I inputs	Display type	Communication	Special features	List price (US$)
RPM	1650	4/5	PC-based, not included	Ethernet long-term monitoring, optional PQ configurations, remote control by software	6250 + 750 software + 490/probe	
			Panel-mounted			
Cutler-Hammer/ Westinghouse	4/4	Panel	Optional IMPACC	Panel-mount to replace meters, monitoring	3690 + input devices	
General Electric	kV	Vector Electricity Meter	Socket	Programmable multifunction LCD display pulse output for measured power quantities replaces industrial revenue meters, calculates and accumulates all power and revenue data	595 + 495 (software)	
Square D	Powerlogic PM620	3/3	LCD panel	RS485	Panel meter replacement, calculates and accumulates power data	1583 + probes
Square D	Powerlogic CM2350	3/3	Panel	RS485	Connect to network, remote controller, monitoring	4290 + probes

TABLE 40.3   Manufacturers of Power Factor Measuring Harmonic Analyzers

Amprobe Instruments
630 Merrick Road
Lynbrook, NY 11563
Tel: (516) 593-5600

BMI
3250 Jay Street
Santa Clara, CA 95054
Tel: (408) 970-3700

Cooper Power Systems Division
11131 Adams Road
P.O. Box 100
Franksville, WI 53126-0100
Tel: (414) 835-2921

Cutler Hammer Inc.
Westinghouse & Cutler-Hammer Products
Five Parkway Center
Pittsburgh, PA 15220
Tel: (412) 937-6100

Dranetz Technologies, Inc.
1000 New Durham Road
Edison, NJ 08818-4019
Tel: (800) DRANTEC

Fluke Corporation
P.O. Box 9090
Everett, WA 98206
Tel: (800) 44FLUKE

GE Meter
130 Main Street
Somersworth, NH 03878
Tel: (603) 749-8477

Reliable Power Meters, Inc.
400 Blossom Hill Road
Los Gatos, CA 95032
Tel: (408) 358-5100

Square D Power Logic
330 Weakley Road
Smyrna, TN 37167-9969
Tel: (615) 459-8552

Defining Terms

Active power: A term used to express the real power delivered to do work in an ac distribution system.
Reactive power: A term used to express the imaginary power that does no work but provides magnetization to enable work in an ac distribution system.
Phasor power: A term used to express the product of volts and amperes in an ac distribution system in which voltage and current are sinusoidal.
Harmonic power: A term used to express the power due to harmonic frequencies in an ac distribution system in which voltage and/or current are nonsinusoidal.
Apparent power: A term used to express the product of volts and amperes in an ac distribution system in which voltage and/or current are nonsinusoidal.
Power factor: A single number, calculated by dividing active power by either the phasor power or the apparent power, which describes the effective utilization of ac distribution system capacity.

References

1. D. F. Bullock, Methods of measuring apparent power, GE Meter, Missouri Valley Electrical Assoc., *Annu. Eng. Conf.*, Kansas City, MO, 1996.
2. *ANSI/IEEE Std 100, Dictionary of Electrical and Electronic Terms, 1992*, New York: IEEE, 1993.
3. D. F. Bullock, Phase Angle Relationships, Part 1: Theory, GE Meter Arkansas Electric Meter School, 1995.
4. I. Assimov, *Assimov's Biographical Encyclopedia of Science and Technology, New Rev. Ed.*, Garden City, NY: Doubleday & Co., 1972.
5. D. F. Bullock, private communication.
6. D. F. Bullock and D. D. Elmore, *MIND UR PS & QS, GE Meter EEI-AEIC Meter and Service Committee Meeting*, Dallas, TX, 1994.

Further Information

H. L. Curtis and F. B. Silsbee, Definitions of power and related quantities, *AIEE Summer Conf.*, 1935.

41

Phase Measurement

Peter O'Shea
Royal Melbourne Institute of Technology

The notion of "phase" is usually associated with *periodic* or repeating signals. With these signals, the waveshape perfectly repeats itself every time the *period* of repetition elapses. For periodic signals one can think of the phase at a given time as the fractional portion of the period that has been completed. This is commonly expressed in degrees or radians, with full cycle completion corresponding to 360° or 2π radians. Thus, when the cycle is just beginning, the phase is zero. When the cycle is half completed, the phase is half of 360°, or 180° (See Figure 41.1). It is important to note that if phase is defined as the portion of a cycle that is completed, the phase depends on where the beginning of the cycle is taken to be. There is no universal agreement on how to specify this beginning. For a sinusoidal signal, probably the two most common assumptions are that the start of the cycle is (1) the point at which the maximum value is achieved, and (2) the point at which the negative to positive zero-crossing occurs. Assumption (1) is common in many theoretical treatments of phase, and for that reason is adopted in this chapter. It should be noted, however, that assumption (2) has some benefits from a measurement perspective, because the zero-crossing position is easier to measure than the maximum.

The measurement of phase is important in almost all applications where sinusoids proliferate. Many means have therefore been devised for this measurement. One of the most obvious measurement techniques is to directly measure the fractional part of the period that has been completed on a cathode-ray oscilloscope (CRO). Another approach, which is particularly useful when a significant amount of noise is present, is to take the Fourier transform of the signal. According to Fourier theory, for a sinusoidal signal, the energy in the Fourier transform is concentrated at the frequency of the signal; the initial phase of the signal (i.e., the phase at time, $t = 0$) is the phase of the Fourier transform at the point of this energy concentration. The measurements of initial phase and frequency obtained from the Fourier transform can then be used to deduce the phase of the signal for any value of time.

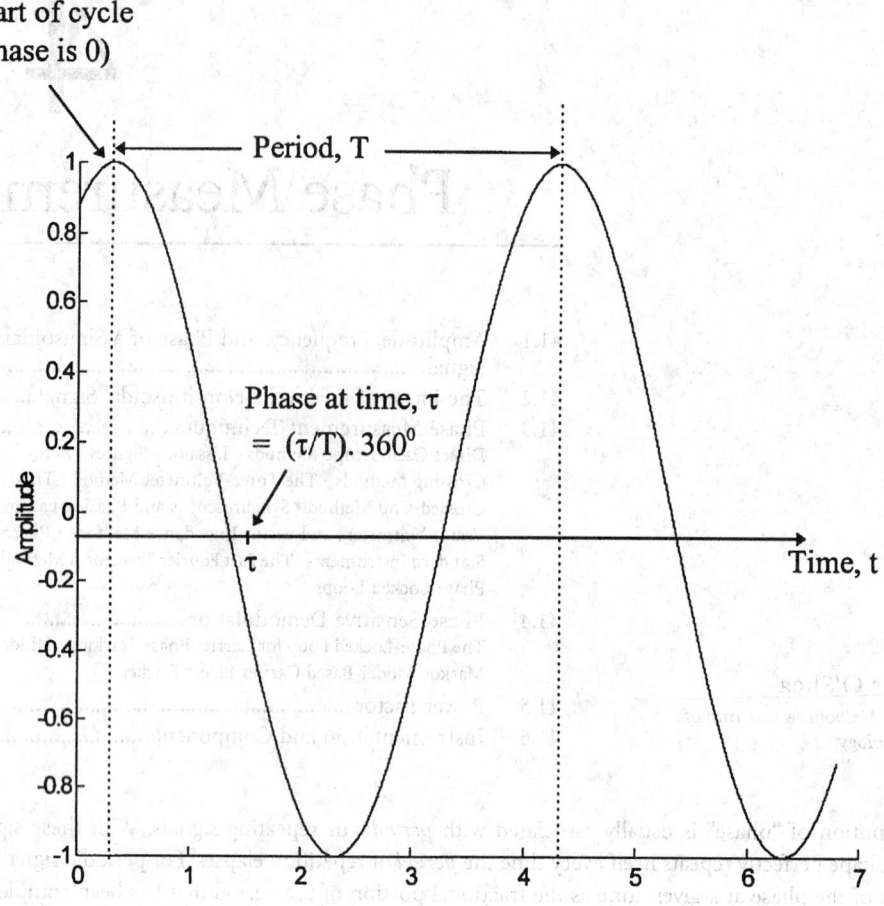

Start of cycle (phase is 0)

Period, T

Phase at time, τ
= (τ/T). 360⁰

Amplitude

Time, t

FIGURE 41.1   The phase of a periodic sinusoidal signal. The time scale is arbitrary.

Frequently what is needed in practice is a measurement of the *phase difference* between two signals of the same frequency; that is, it is necessary to measure the *relative phase* between two signals rather than the *absolute phase* of either one (see Figure 41.2). Often, in the measurement of the relative phase between two signals, both signals are derived from the same source. These signals might, for example, be the current and voltage of a power system; the relative phase, φ, between the current and voltage would then be useful for monitoring power usage, since the latter is proportional to the cosine of φ.

Several techniques are available for the measurement of "relative phase." One crude method involves forming "Lissajous figures" on an oscilloscope. In this method, the first of the two signals of interest is fed into the vertical input of a CRO and the other is fed into the horizontal input. The result on the oscilloscope screen is an ellipse, the intercept and maximum height of which can be used to determine the relative phase. Other methods for determining relative phase include the crossed-coil meter (based on electromagnetic induction principles), the zero-crossing phase meter (based on switching circuitry for determining the fractional portion of the period completed), the three-voltmeter method (based on the use of three signals and trigonometric relationships), and digital methods (based on analog-to-digital conversion and digital processing).

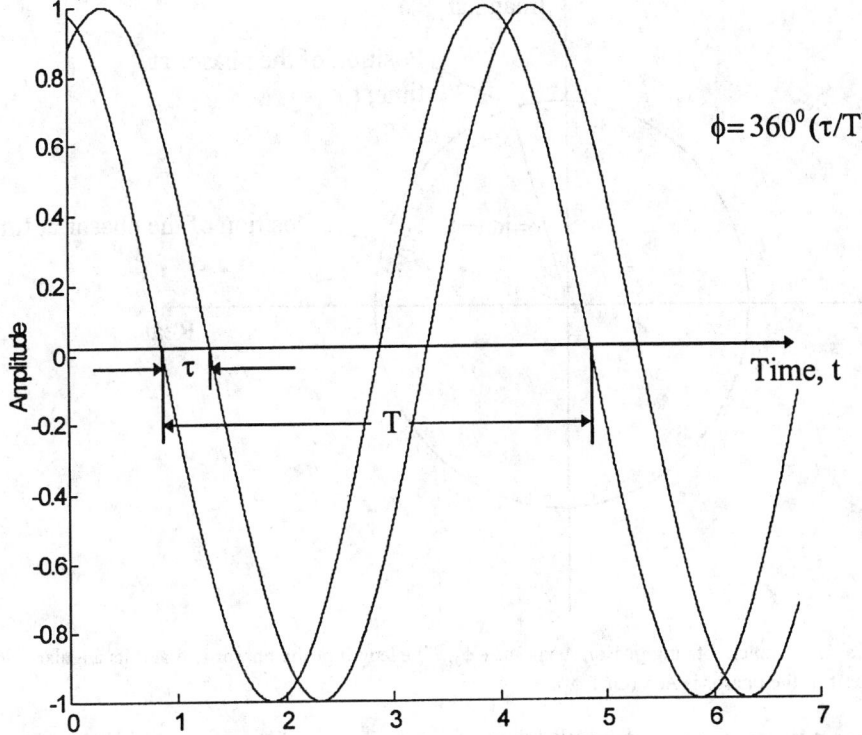

$$\phi = 360^0 \left(\tau/T\right)$$

FIGURE 41.2 Two signals with a relative phase difference of ϕ between them. The time scale is arbitrary.

41.1 Amplitude, Frequency, and Phase of a Sinusoidal Signal

An arbitrary sinusoidal signal can be written in the form:

$$s(t) = A\cos\left(2\pi ft + \phi_0\right) = A\cos\left(\omega t + \phi_0\right) \tag{41.1}$$

where A = Peak amplitude
  f = Frequency
  ω = Angular frequency
  ϕ_0 = Phase at time $t = 0$

This signal can be thought of as being the real part of a complex phasor that has amplitude, A, and which rotates at a constant angular velocity $\omega = 2\pi f$ in the complex plane (see Figure 41.3).

Mathematically, then, $s(t)$ can be written as:

$$s(t) = \Re\left\{Ae^{j2\pi ft + \phi_0}\right\} = \Re\left\{z(t)\right\} \tag{41.2}$$

where $z(t)$ is the complex phasor associated with $s(t)$, and $\Re\{.\}$ denotes the real part. The "phase" of a signal at any point in time corresponds to the angle that the rotating phasor makes with the real axis. The initial phase (i.e., the phase at time $t = 0$) is ϕ_0. The "frequency" f of the signal is $1/2\pi$ times the phasor's angular velocity.

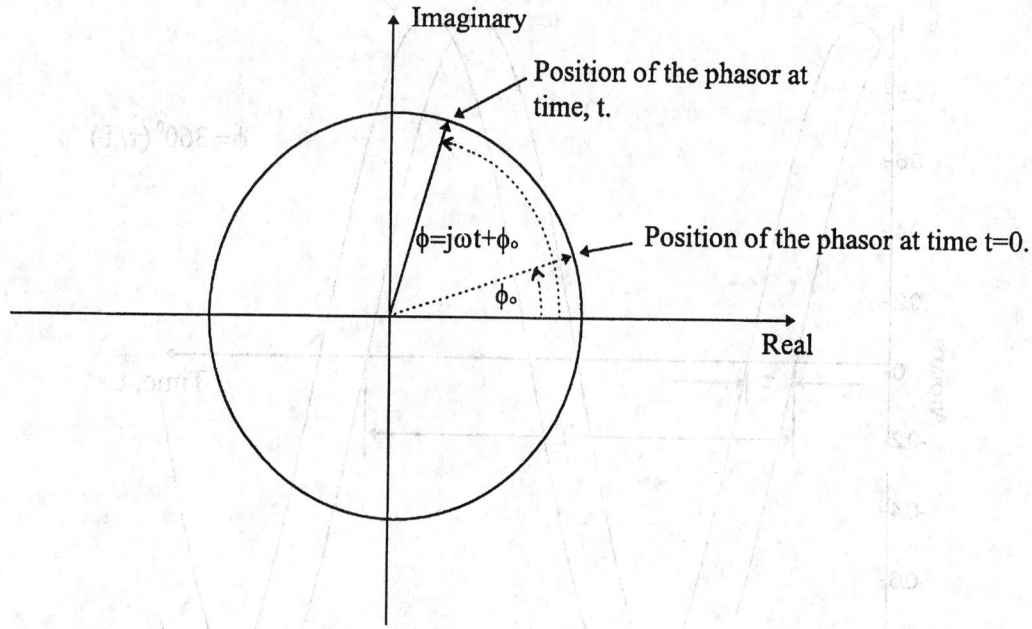

FIGURE 41.3   A complex rotating phasor, $A\exp(j\omega t + \phi_0)$. The length of the phasor is A and its angular velocity is ω. The real part of the phasor is $A\cos(\omega t + \phi_0)$.

There are a number of ways to define the phase of a real sinusoid with unknown amplitude, frequency, and initial phase. One way, as already discussed, is to define it as the fractional part of the period that has been completed. This is a valid and intuitively pleasing definition, and one that can be readily generalized to periodic signals that contain not only a sinusoid, but also a number of harmonics. It cannot, however, be elegantly generalized to allow for slow variations in the frequency of the signal, a scenario that occurs in communicatons with phase and frequency modulation. Gabor put forward a definition in 1946 that can be used for signals with slowly varying frequency. He proposed a mathematical definition for generating the complex phasor, $z(t)$, associated with the real signal, $s(t)$. The so-called *analytic signal* $z(t)$ is defined according to the following definition [1]:

$$z\!\left(t\right) = s\!\left(t\right) + j\mathcal{H}\!\left\{s\!\left(t\right)\right\}$$

(41.3)

where $\mathcal{H}\{\ \}$ denotes the *Hilbert transform* and is given by:

$$\mathcal{H}\!\left\{s\!\left(t\right)\right\} = p.v.\left[\int_{-\infty}^{+\infty} \frac{s\!\left(t-\tau\right)}{\pi\tau}\,d\tau\right]$$

(41.4)

with *p.v.* signifying the Cauchy principal value of the integral [2].

The imaginary part of the analytic signal can be generated practically by passing the original signal through a "Hilbert transform" filter. From Equations 41.3 and 41.4, it follows that this filter has impulse response given by $1/\pi t$. The filter can be implemented, for example, with one of the HSP43xxx series of ICs from Harris Semiconductors. Details of how to determine the filter coefficients can be found in [2].

Having formally defined the analytic signal, it is possible to provide definitions for phase, frequency, and amplitude as functions of time. They are given below.

$$\text{Phase: } \phi(t) = \arg\{z(t)\} \tag{41.5}$$

$$\text{Frequency: } f(t) = \frac{1}{2\pi} \frac{d\left[\arg\{z(t)\}\right]}{dt} \tag{41.6}$$

$$\text{Amplitude: } A(t) = \text{abs}\left[z(t)\right] \tag{41.7}$$

The definitions for phase, frequency, and amplitude can be used for signals whose frequency and/or amplitude vary slowly with time. If the frequency and amplitude do vary with time, it is common to talk about the "instantaneous frequency" or "instantaneous amplitude" rather than simply the frequency or amplitude.

Note that in the analytic signal, the imaginary part lags the real part by 90°. This property actually holds not only for sinusoids, but for the real and imaginary parts of all frequency components in "multicomponent" analytic signals as well. The real and imaginary parts of the analytic signal then correspond to the "in-phase (I)" and "quadrature (Q)" components used in communications systems.

In a balanced three-phase electrical power distribution system, the analytic signal can be generated by appropriately combining the different outputs of the electrical power signal; that is, it can be formed according to:

$$z(t) = v_a(t) + \frac{j}{\sqrt{3}}\left[v_c(t) - v_b(t)\right] \tag{41.8}$$

where $v_a(t)$ = Reference phase
$\quad\quad v_b(t)$ = Phase that leads the reference by 120°
$\quad\quad v_c(t)$ = Phase that lags the reference by 120°.

41.2 The Phase of a Periodic Nonsinusoidal Signal

It is possible to define "phase" for signals other than sinusoidal signals. If the signal has harmonic distortion components present in addition to the fundamental, the signal will still be periodic, but it will no longer be sinusoidal. The phase can still be considered to be the fraction of the period completed. The "start" of the period is commonly taken to be the point at which the initial phase of the fundamental component is 0, or at a zero-crossing. This approach is equivalent to just considering the phase of the fundamental, and ignoring the other components. The Fourier method provides a very convenient method for determining this phase — the energy of the harmonics in the Fourier transform can be ignored.

41.3 Phase Measurement Techniques

Direct Oscilloscope Methods

Cathode-ray oscilloscopes (CROs) provide a simple means for measuring the phase difference between two sinusoidal signals. The most straightforward approach to use is direct measurement; that is, the signal of interest is applied to the vertical input of the CRO and an automatic time sweep is applied to the horizontal trace. The phase difference is the time delay between the two waveforms measured as a fraction of the period. The result is expressed as a fraction of 360° or of 2π radians; that is, if the time

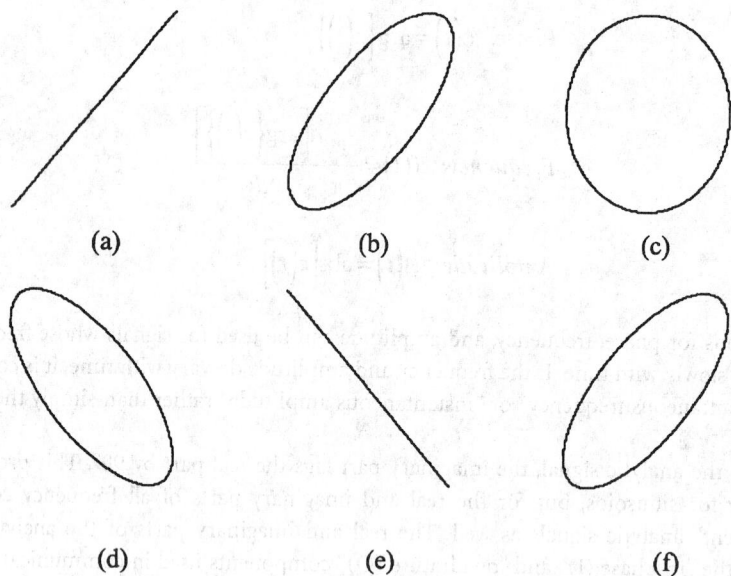

FIGURE 41.4 Lissajous figures for two equal-amplitude, frequency-synchronized signals with a relative phase difference of (a) O, (b) π/4, (c) π/2, (d) 3π/4, (e) π, (f) –π/4.

delay is 1/4 of the period, then the phase difference is 1/4 of 360° = 90° (see Figure 41.2). If the waveforms are not sinusoidal but are periodic, the same procedure can still be applied. The phase difference is just expressed as a fraction of the period or as a fractional part of 360°.

Care must be taken with direct oscilloscope methods if noise is present. In particular, the noise can cause triggering difficulties that would make it difficult to accurately determine the period and/or the time delay between two different waveforms. The "HF reject" option, if available, will alleviate the triggering problems.

Lissajous Figures

Lissajous figures are sometimes used for the measurement of phase. They are produced in an oscilloscope by connecting one signal to the vertical trace and the other to the horizontal trace. If the ratio of the first frequency to the second is a rational number (i.e., it is equal to one small integer divided by another), then a closed curve will be observed on the CRO (see Figures 41.4 and 41.5). If the two frequencies are unrelated, then there will be only a patch of light observed because of the persistance of the oscilloscope screen.

If the two signals have the same frequency, then the Lissajous figure will assume the shape of an ellipse. The ellipse's shape will vary according to the the phase difference between the two signals, and according to the ratio of the amplitudes of the two signals. Figure 41.6 shows some figures for two signals with synchronized frequency and equal amplitudes, but different phase relationships. The formula used for determining the phase is:

$$\sin(\phi) = \pm \frac{Y}{H} \tag{41.9}$$

where H is half the maximum vertical height of the ellipse and Y is the intercept on the *y-axis*. Figure 41.7 shows some figures for two signals that are identical in frequency and have a phase difference of 45°, but with different amplitude ratios. Note that it is necessary to know the direction that the Lissajous trace is moving in order to determine the sign of the phase difference. In practice, if this is not known *a priori*,

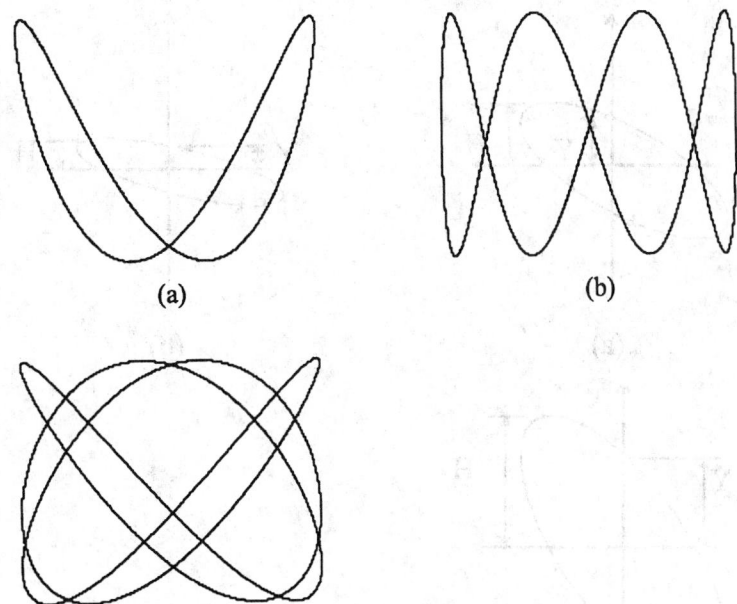

FIGURE 41.5 Lissajous figures for two signals with vertical frequency: horizontal frequency ratios of (a) 2:1, (b) 4:1, (c) 4:3.

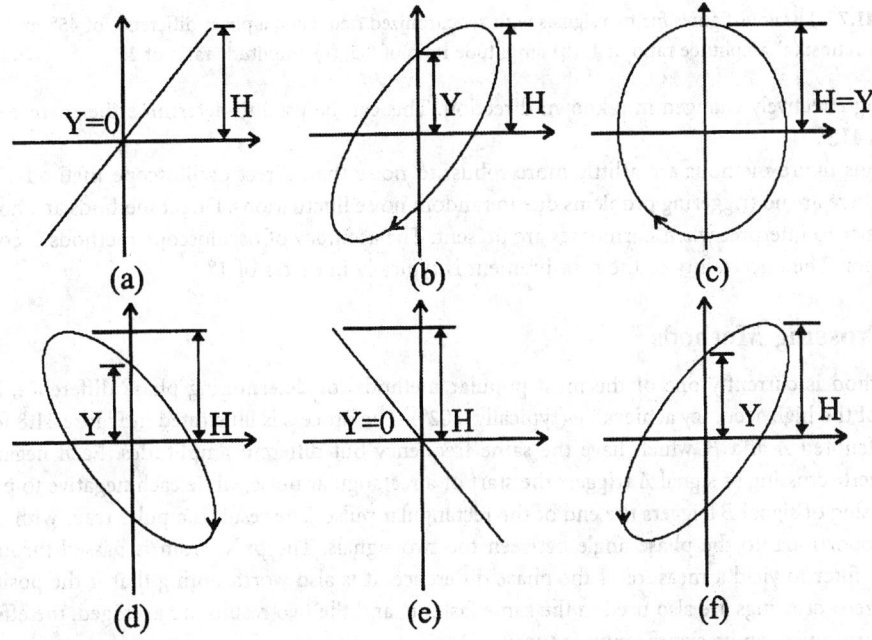

FIGURE 41.6 Lissajous figures for two signals with synchronized frequency and various phase differences: (a) phase difference = 0°, (b) phase difference = 45°, (c) phase difference = 90°, (d) phase difference = 135°, (e) phase difference = 180°, (f) phase difference = 315°.

then it can be determined by testing with a variable frequency signal generator. In this case, one of the signals under consideration is replaced with the variable frequency signal. The signal generator is adjusted until its frequency and phase equal that of the other signal input to the CRO. When this happens, a straight line will exist. The signal generator frequency is then increased a little, with the relative phase

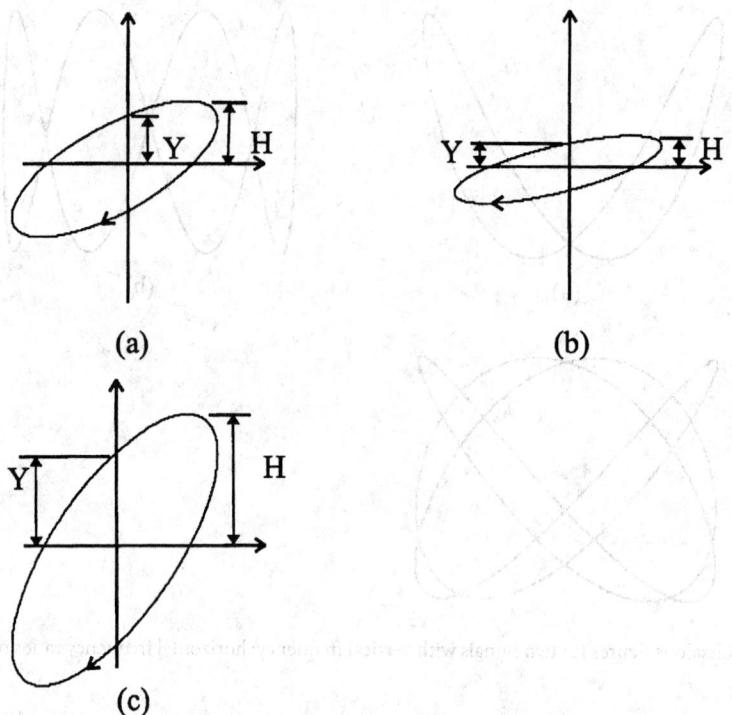

FIGURE 41.7 Lissajous figures for two signals with synchronized frequency, a phase difference of 45°, and various amplitude ratios: (a) amplitude ratio of 1, (b) amplitude ratio of 0.5, (c) amplitude ratio of 2.

thus being effectively changed in a known direction. This can be used to determine the correct sign in Equation 41.9.

Lissajous figure methods are a little more robust to noise than direct oscilloscope methods. This is because there are no triggering problems due to random noise fluctuations. Direct methods are, however, much easier to interpret when harmonics are present. The accuracy of oscilloscope methods is comparatively poor. The uncertainty of the measurement is typically in excess of 1°.

Zero-Crossing Methods

This method is currently one of the most popular methods for determining phase difference, largely because of the high accuracy achievable (typically 0.02°). The process is illustrated in Figure 41.8 for two signals, denoted *A* and *B*, which have the same frequency but different amplitudes. Each negative to positive zero-crossing of signal *A* triggers the start of a rectangular pulse, while each negative to positive zero-crossing of signal *B* triggers the end of the rectangular pulse. The result is a pulse train with a pulse width proportional to the phase angle between the two signals. The pulse train is passed through an averaging filter to yield a measure of the phase difference. It is also worth noting that if the positive to negative zero-crossings are also used in the same fashion, and the two results are averaged, the effects of dc and harmonics can be significantly reduced.

To implement the method practically, the analog input signals must first be converted to digital signals that are "high" if the analog signal is positive, and "low" if the analog signal is negative. This can be done, for example, with a Schmitt trigger, along with an *RC* stabilizing network at the output. Chapter 81 provides a circuit to do the conversion. In practice, high-accuracy phase estimates necessitate that the switching of the output between high and low be very sharp. One way to obtain these sharp transitions is to have several stages of "amplify and clip" preceding the Schmitt trigger.

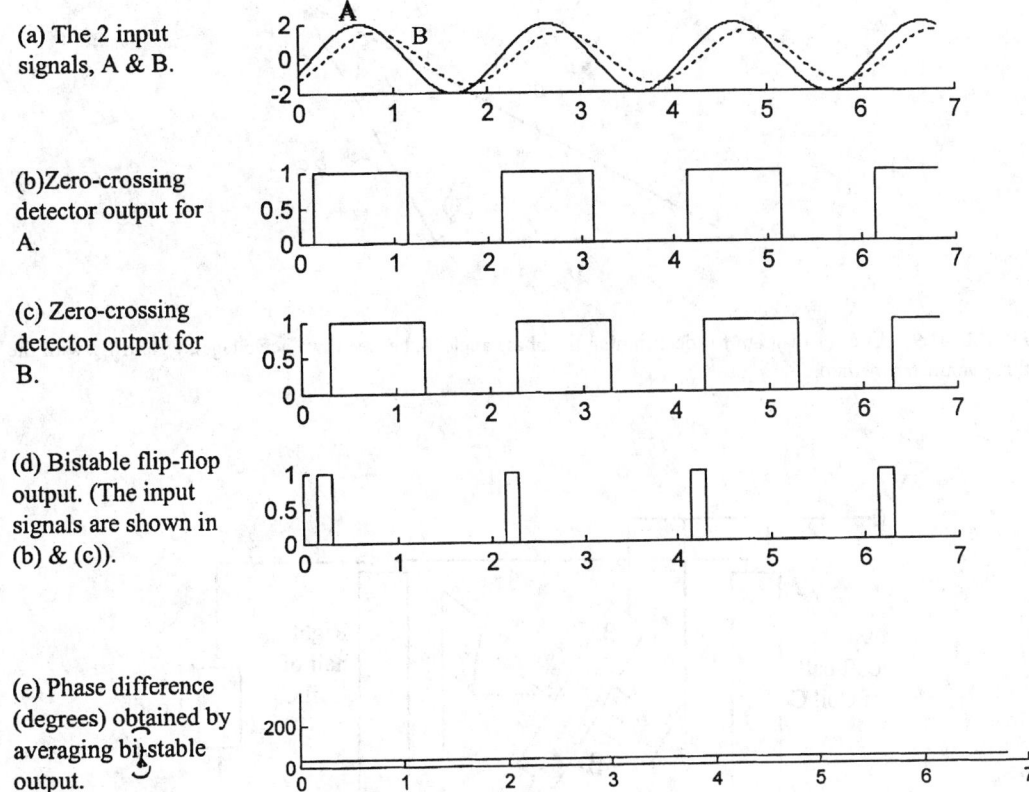

(a) The 2 input signals, A & B.

(b) Zero-crossing detector output for A.

(c) Zero-crossing detector output for B.

(d) Bistable flip-flop output. (The input signals are shown in (b) & (c)).

(e) Phase difference (degrees) obtained by averaging bistable output.

FIGURE 41.8 Input, output and intermediate signals obtained with the zero-crossing method for phase measurement. Note that the technique is not sensitive to signal amplitude.

The digital portion of the zero-crossing device can be implemented with an edge-triggered RS flip-flop and some ancillary circuitry, while the low-pass filter on the output stage can be implemented with an RC network. A simple circuit to implement the digital part of the circuit is shown in Chapter 81.

A simpler method for measuring phase based on zero-crossings involves the use of an exclusive or (XOR) gate. Again, the analog input signals must first be converted to digital pulse trains. The two inputs are then fed into an XOR gate and finally into a low-pass averaging filter. The circuit is illustrated in Chapter 81. A disadvantage with this method is that it is only effective if the duty cycle is 50% and if the phase shift between the two signals is betwen 0 and π radians. It is therefore not widely used.

The Three-Voltmeter Method

The measurement of a phase difference between two voltage signals, v_{ac}, and v_{bc}, can be expedited if there is a common voltage point, c. The voltage between points b and a (v_{ba}), the voltage between points b and c (v_{bc}), and the voltage between points c and a (v_{ca}) are measured with three different voltmeters. A vector diagram is constructed with the three measured voltages as shown in Figure 41.9. The phase difference between the two vectors, v_{ac} and v_{bc}, is determined using a vector diagram (Figure 41.9) and the cos rule. The formula for the phase difference, ϕ, in radians is given by:

$$\pi - \phi = \cos^{-1}\left(\frac{v_{ca}^2 + v_{bc}^2 - v_{ba}^2}{2v_{ca}v_{bc}}\right) \tag{41.10}$$

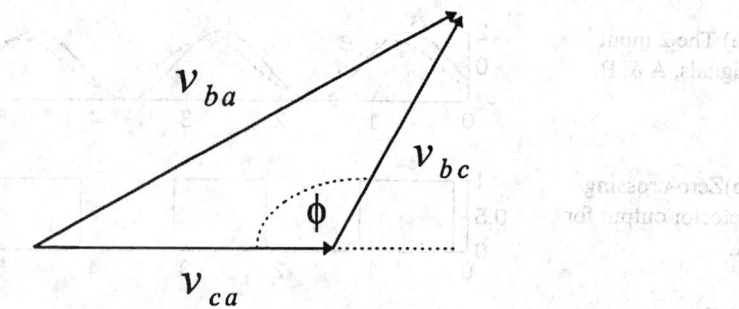

FIGURE 41.9 A vector diagram for determining the phase angle, ϕ, between two ac voltages, v_{ac} and v_{bc}, with the three-voltmeter method.

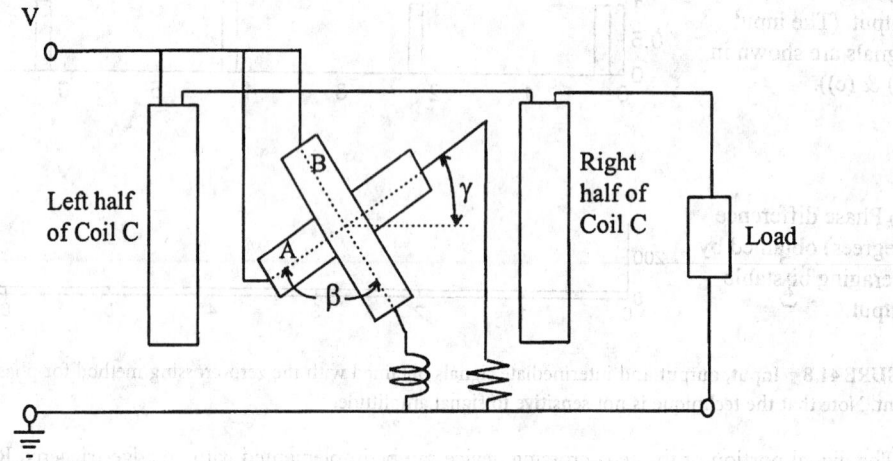

FIGURE 41.10 Diagram of a crossed-coil device for measuring phase. Coils A and B are the rotating coils. Coil C (left and right parts) is the stationary coil.

The Crossed-Coil Method

The crossed-coil phase meter is at the heart of many analog power factor meters. It has two crossed coils, denoted A and B, positioned on a common shaft but aligned at different angles (see Figure 41.10). The two coils move as one, and the angle between them, β, never changes. There is another independent nonrotating coil, C, consisting of two separate parts, "enclosing" the rotating part (see Figure 41.10). The separation of coil C into two separate parts (forming a Helmholtz pair) allows the magnetic field of coil C to be almost constant in the region where the rotating A and B coils are positioned.

Typically the system current, I, is fed into coil C, while the system voltage, V, is applied to coil A via a resistive circuit. The current in coil A is therefore in phase with the system voltage, while the current in coil C is in phase with the system current. Coil B is driven by V via an inductive circuit, giving rise to a current that lags V (and therefore the current in coil A) by 90°. In practice, the angle between the currents in coils A and B is not quite 90° because of the problems associated with achieving purely resistive and purely inductive circuits. Assume, then, that this angle is β. If the angle between the currents in coil B and in coil C is ϕ, then the angle between the currents in coils A and C is $\beta + \phi$. The average torque induced in coil A is proportional to the product of the average currents in coils A and C, and to the cosine of the angle between coil A and the perpendicular to coil C. The average torque induced in coil A is therefore governed by the equation:

$$\overline{T_A} \propto I_A I_C \cos(\phi+\beta)\cos(\gamma) = k_A \cos(\phi+\beta)\cos(\gamma) \qquad (41.11)$$

where I_A and I_C = Constants
$\quad \omega \qquad$ = Angular frequency
$\quad \phi + \beta \qquad$ = Relative phase between the currents in coils A and C
$\quad \gamma \qquad$ = Angle between coil A and the perpendicular to coil C
$\quad \propto \qquad$ = Signifies "is proportional to"

Assuming that the current in coil B lags the current in coil A by β, then the average torque in coil B will be described by:

$$\overline{T_B} \propto I_B I_C \cos(\phi)\cos(\gamma+\beta) = k_B \cos(\phi)\cos(\gamma+\beta) \qquad (41.12)$$

where I_B is a constant, ϕ is the relative phase between the currents in coils B and C, and the other qunatities are as in Equation 41.11.

Now, the torques due to the currents in coils A and B are designed to be in opposite directions. The shaft will therefore rotate until the two torques are equal; that is, until:

$$k_A \cos(\phi+\beta)\cos(\gamma) = k_B \cos(\phi)\cos(\gamma+\beta) \qquad (41.13)$$

If $k_A = k_B$, then Equation 41.13 will be satisfied when $\phi = \gamma$. Thus, the A coil will be aligned in the direction of the phase shift between the load current and load voltage (apart from errors due to the circuits of the crossed coils not being perfectly resistive/inductive). Thus, a meter pointer attached to the A plane will indicate the angle between load current and voltage. In practice, the meter is usually calibrated to read the cosine of the phase angle rather than the phase angle, and also to allow for the errors that arise from circuit component imperfections.

The accuracy of this method is limited, due to the heavy use of moving parts and analog circuits. Typically, the measurement can only be made accurate to within about 1°.

Synchroscopes and Phasing Lamps

The crossed-coil meter described above is used as the basis for *synchroscopes*. These devices are often used in power generation systems to detemine whether two generators are phase and frequency synchronized before connecting them together. In synchroscopes, the current from one generator is fed into the fixed coil and the current from the other generator is fed into the movable crossed coils. If the two generators are synchronized in frequency, then the meter needle will move to the position corresponding to the phase angle between the two generators. If the generators are not frequency synchronized, the meter needle will rotate at a rate equal to the difference between the two generator frequencies. The direction of rotation will indicate which generator is rotating faster.

When frequency synchronization occurs (i.e., the meter needle rotation ceases) and the phase difference is zero, the generators can be connected together. Often in practice, the generators are connected before synchronization occurs; the generator coming on-line is deliberately made a little higher in frequency so that it can provide extra power rather than be an extra drain on the system. The connection is still made, however, when the instantaneous phase difference is zero.

Phasing lamps are sometimes used as a simpler alternative to synchroscopes. A lamp is connected between the two generators, and any lack of frequency synchronization manifests as a flickering of the lamp. A zero phase difference between the two generators corresponds to maximum brightness in the lamp.

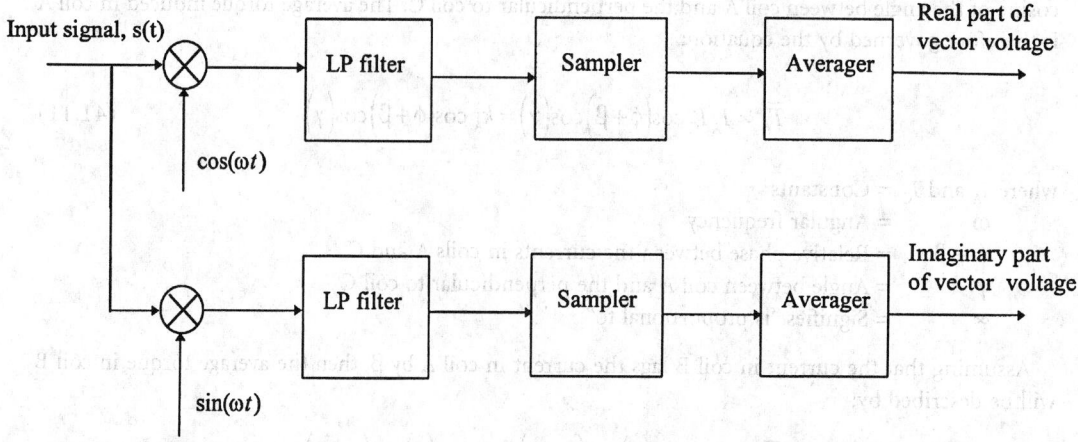

FIGURE 41.11 Vector voltmeter block diagram. The vector voltmeter determines the voltage (amplitude and phase or real and imaginary parts) of the component of the input signal at frequency f.

Vector Voltmeters and Vector Impedance Methods

Alternating voltages (and currents) are often characterized as vectors consisting of a magnitude and a phase, with the phase being measured relative to some desired reference. Many instruments exist that can display the voltage amplitude and phase of a signal across a wide range of frequencies. These instruments are known as *vector voltmeters* or *network analyzers*. The phase and amplitude as a function of frequency can be obtained very simply in principle by taking the Fourier transform of the signal and simply reading the amplitude and phase across the continuum of frequencies. To achieve good accuracy, this is typically done with down-conversion and digital processing in the baseband region. The down-conversion can be analog, or it can be digital. The procedure is described more fully in the succeeding paragraphs.

To determine the real part of the voltage vector at a given frequency f, the signal is first down-converted by mixing with a local oscillator signal, $\cos(2\pi ft)$. This mixing of the signal recenters the frequency component of interest at 0 Hz. The resultant signal is low-pass filtered, digitally sampled (if not in the digital domain already), and averaged. The digital sampling and averaging enables the amplitude of the newly formed 0-Hz component to be evaluated. The imaginary part is obtained in similar fashion by mixing the signal with $\sin(2\pi ft)$, low-pass filtering, digitally sampling, and again averaging the samples. The amplitude and phase of the voltage vector, V, are obtained from the real and imaginary parts using the standard trigonometric relationships:

$$Magnitude = Abs\big(V\big) = \sqrt{\left[\Re e\{V\}\right]^2 + \left[\Im m\{V\}\right]^2} \qquad (41.14)$$

$$Phase = \arg\big(V\big) = \arctan\big(\Im m\{V\}\big/\Re e\{V\}\big) \qquad (41.15)$$

where $\Re\{.\}$ and $\Im m\{.\}$ denote the real and imaginary parts, respectively.

The procedure for forming the vector voltage is summarized in the block diagram in Figure 41.11. In practice, the down-conversion can be carried out in more than one step. For high-frequency signals, for example, the first stage might shift a large band of frequencies to the audio region, where further down-conversion is carried out. Alternatively, the first stage might shift a band of frequencies to the intermediate frequency (IF) band, and the second stage to the audio band. More details on the physics of the down-conversion process are available in the article on "Modulation" in Chapter 81.

Just as it is possible to analyze a voltage signal and produce a magnitude and phase across any given frequency band, so it is possible to obtain a frequency profile of the magnitude and phase of a *current* signal. If current vectors and voltage vectors can be obtained for a given impedance, then it is possible to obtain a "vector impedance." This impedance is defined simply as the result of the complex division of voltage by current:

$$Z = \frac{V}{I} \tag{41.16}$$

The calculation of vector impedances are useful for such applications as impedance matching, power factor correction, and equalization.

Typically, much of the current processing for vector voltmeters and vector impedance meters is done digitally. One of the great advantages of this type of processing is the high accuracy achievable. Accuracies of 0.02° are common, but this figure is improving with developing technology. The high sampling rates that can be employed (typically beyond 1 GHz) cause the errors in the A/D conversion to be spread out over very large bandwidths. Since the ultimate measurement of a vector voltage or impedance is usually made over a very narrow bandwidth, the errors are substantially eliminated. The developments in technology that enable greater accuracy are (1) sampling rate increases, (2) word-length increases, and (3) increased A/D converter fidelity.

Phase Standard Instruments

For high-precision phase measurements and calibration, "phase standard" instruments can be used. These instruments provide two sinusoidal signal outputs, whose phase difference can be controlled with great accuracy. They typically use crystal-controlled timing to digitally synthesize two independent sinusoids with a variable user-defined phase difference. The Clarke-Hess 5500 Digital Phase Standard is one such instrument. For this standard, the sinusoids can have frequencies ranging from 1 Hz to 100 kHz, while amplitudes can vary between 50 mV and 120 V rms. The phase can be set with a resolution of 0.001°, with a typical accuracy of about 0.003°.

The Fast Fourier Transform Method

This method is one in which virtually all the processing is done in the digital domain. It operates on the pulse code modulated (PCM) digital samples of a signal. This and other similar methods are very promising means for measuring phase. This is because of the digital revolution that has resulted in cheap, fast, accurate, and highly versatile digital signal processors (DSPs). The latter are small computer chips capable of performing fast additions and multiplications, and which can be programmed to emulate conventional electronic functions such as filtering, coding, modulation, etc. They can also be programmed to perform a wide variety of functions not possible with analog circuitry. Up until the end of the 1980s, digital measurement was limited by the relatively inaccurate analog-to-digital (A/D) conversion process required before digital processing could be performed. Developments in the early 1990s, however, saw the introduction of oversampling analog-to-digital converters (ADCs), which can achieve accuracies of about one part in 100,000 [3]. ADC speeds as well as DSP chips are now running reliably at very high speeds.

In the fast Fourier transform (FFT) method, the digital signal samples are Fourier transformed with an FFT [2]. If the signal is sinusoidal, the initial phase is estimated as that value of the phase where the Fourier transform is maximized [4]. The frequency of the signal is estimated as that value of frequency where the Fourier transform is maximized. Once measurements of the frequency f and initial phase ϕ_0 have been obtained, the phase at any point in time can be calculated according to:

$$\phi = 2\pi ft + \phi_0 \tag{41.17}$$

One important practical issue in the measurement of the frequency and initial phase with an FFT arises because the FFT yields only *samples* of the Fourier transform; that is, it does not yield a continuous Fourier transform. It is quite possible that the true maximum of the Fourier transform will fall *between* samples of the FFT. For accurate measurement of frequency and initial phase, then, it is necessary to *interpolate* between the FFT samples. An efficient algorithm to do this is described in [5].

The FFT method is particularly appealing where there is significant noise present, as it is effective down to quite low signal-to-noise ratios (SNRs). Furthermore, it provides an optimal estimate of the frequency and initial phase, providing the background noise is white and Gaussian, and that no harmonic components are present [4]. If harmonics are present, the estimate of the phase is commonly taken as the phase at the FFT peak corresponding to the fundamental; this is not an optimal estimate, but serves well in many appplications. An optimal estimate in the case when harmonics are present can be obtained, if necessary, with the algorithms in [6], [7], and [8]. DSP chips such as the Texas Instruments TMS320C3x family, the Analog Devices ADSP21020 chip, or the Motorola MC5630x series can be used to implement the real-time FFTs.

If long word-lengths are used (say 32 bits) to perform the arithmetic for the FFTs, then determination of the phase from the samples of the FFT is virtually error-free. The only significant inaccuracy incurred in determining the phase is due to the ADC errors. Moreover, the error due to the digitization will typically be spread out over a large bandwidth, only a small amount of which will be "seen" in the phase measurement. With a high-quality ADC, accuracies of less than 0.001° are possible.

Phase-Locked Loops

If the frequency of a signal changes significantly over the period of the measurement, the FFT method described above will provide inaccurate results. If the signal's frequency does change substantially during measurement, one means to estimate the phase of the signal is to use a phase-locked loop (PLL). In this case, the signal, $s(t) = A \sin(\omega t + \phi(t))$, can be thought of as a constant frequency component, $A \sin(\omega t)$, which is phase modulated by a time-varying phase component, $\phi(t)$. The problem then reduces largely to one of demodulating a phase-modulated (PM) signal. A PLL can be used to form an estimate of the "phase-modulating" component, $\hat{\phi}(t)$, and the overall phase of the signal, ϕ_{oa}, can be estimated according to:

$$\phi_{oa}(t) = \omega t + \hat{\phi}(t) \tag{41.18}$$

Either analog or digital PLLs can be used, although higher accuracy is attainable with digital PLLs. Analog PLLs for demodulating a frequency-modulated (FM) signal are discussed in Chapter 81 and in [12]. The digital PLL (DPLL) was developed as an extension of the conventional analog PLL and is therefore similar in structure to its analog counterpart. The DPLL is discussed in [9] and [10]. The equation to demodulate the digital modulated signal with a first order DPLL is a simple recursive equation [9].

A block diagram of the DPLL for demodulating a PM signal is shown in Figure 41.12. In this diagram, n represents the discrete-time equivalent of continuous time t. It can be seen that there are strong similarities between Figure 41.12 and the analog PLL-based FM demodulator in Chapter 81. Both have phase comparators (implemented by the multiplier in Figure 41.12), both have loop filters, and both have modulators (either PM or FM) in the feedback loop.

The DPLL can easily be extended to measure the phase of a signal consisting not just of a fundamental component, but also of harmonically related components. Details are provided in [11]. The DPLL is near optimal for phase esitmation in white Gaussian background noise down to a signal power-to-noise power ratio of about 8 dB [10].

The DPLL will function effectively whether the phase is constant or time-varying. Unlike the FFT, the DPLL is a recursive algorithm, with the feedback involved in the recursion creating a vulnerability to quantization errors. However, with proper precautions and long word-lengths, the DPLL will introduce

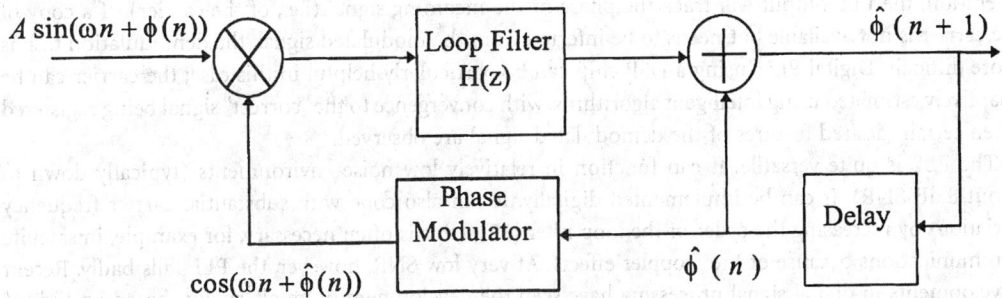

FIGURE 41.12   Block diagram of a digital phase-locked loop to implement phase demodulation.

minimal processing error. The main error would then arise from the inaccuracy of the ADC. With appropriate conditioning, one could expect the DPLL to provide accuracies approaching 0.001°.

41.4   Phase-Sensitive Demodulation

It is frequently necessary to track the phase of a carrier that "jitters" in some uncertain manner. This tracking of the carrier phase is necessary for synchronous demodulation schemes, where the phase of the demodulating signal must be made equal to the phase of the carrier. This need is explained in Chapter 81, and is briefly re-explained here. Consider, for example, double sideband (DSB) amplitude modulation. In DSB, the modulated signal is given by $f_s(t) = A[k + \mu m(t)]\cos(\omega_c t)$, where $m(t)$ is the message signal, A is the amplitude of the unmodulated carrier, μ is the modulation index, k is the proportion of modulating signal present in the modulated signal, and $\cos(\omega_c t)$ is the carrier. Demodulation is typically carried out by multiplying $f_s(t)$ by the carrier, and then low-pass filtering so that the demodulated signal is given by:

$$f_d(t) = \frac{A_c\left[k + \mu m(t)\right]}{2}$$ (41.19)

However, if because of carrier uncertainty, one multiplies the modulated signal by $\cos(\omega_c t + \phi)$, then the demodulated signal is given by:

$$f_d(t) = \frac{A_c\left[k + \mu m(t)\right]\cos(\phi)}{2}$$ (41.20)

It can be seen from Equation 41.20 that the error in the carrier phase can affect both the amplitude and the sign of the demodulated signal. The phase errors can thus yield substantial errors in system output. The following sections outline important techniques used for tracking the phase of carriers, and thus reducing phase errors.

The Phase-Locked Loop for Carrier Phase Tracking

The phase-locked loop (PLL) is well known as a means for demodulating frequency-modulated signals. It is also frequenctly used for tracking the phase of a carrier in noise, so that a copy of the carrier with correct phase is available for demodulation. This tracking is simple enough if a (noisy) copy of the carrier is directly available; either a digital or analog PLL can be used. In either case, the input can be assumed to have the form, $A\sin(\omega t + \phi(t))$, where $\phi(t)$ is the carrier phase. The PLL consists of a multiplier (effectively a phase comparator), a phase modulator, and a loop filter, arranged as shown in Chapter 81. The design of the loop filter is critical if noise is to be optimally removed. In proper operation, the PLL output will track the phase of the incoming signal (i.e., of the carrier). If a copy of the carrier is not available but needs to be inferred from the modulated signal, the demodulation task is more difficult. Digital PLLs using a DSP chip can be particularly helpful in this case; the carrier can be adaptively estimated using intelligent algorithms, with convergence to the "correct" signal being registered when certain desired features of the demodulated signal are observed.

The PLL is quite versatile. It can function in relatively low noise environments (typically down to about 8 dB SNR). It can be implemented digitally. It can also cope with substantial carrier frequency variations by increasing the order of the loop filter [12]; (this is often necessary, for example, in satellite communications because of the Doppler effect). At very low SNR, however, the PLL fails badly. Recent developments in digital signal processing have seen the development of an alternative based on hidden Markov models (HMMs), which will function down to about –5 dB SNR [13]. The HMM method is discussed in the next section.

Hidden Markov Model-Based Carrier Phase Tracker

In the HMM method, the problem of estimating the phase and frequency of a noisy waveform is couched as a "state estimation" problem. The phase of the signal at any point in time can go from 0 to 360°. The 0 to 360° value range is divided into a finite number of intervals or "states," so that the phase at any time occupies a particular (though unknown) state. Similarly, the angular frequency normalized by the sampling frequency at any time in a digital system must be between $-\pi$ to $+\pi$. This value range is also divided into a number of states, so that the frequency at any time has a (hidden or unknown) state associated with it. The frequency is assumed to be a first-order Markov process and probabilities are assigned to the possibility of the frequency changing from one state to another for successive values of time, i.e., frequency transition probabilities are assigned. Large frequency changes are assigned low probabilities, while small changes are assigned high probabilities. The problem of estimating the true phase and frequency states underlying the noisy signal then reduces to one of estimating which states the phase and frequency occupy as time evolves, given the observed noisy signal and the transition probabilities. Computationally efficient optimal algorithms have been developed to estimate these "optimal state sequences" for both the phase and frequency. Details are provided in [13].

41.5 Power Factor

Of particular interest in many applications is the phase angle between the current and voltage of a system. This angle is important because it is a measure of the power which is dissipated in the system. The following paragraphs discuss this angle, its cosine (the system power factor), and its measurement.

In a linear electric circuit that is fed by a current of peak amplitude, I_M, with an angular frequency of ω, the current will have the form, $I_M \cos(\omega t)$. The system voltage will be given by $V_M \cos(\omega t + \phi)$, where V_M is the peak voltage and ϕ is the phase difference between the current and voltage. Then the average power dissipated in the circuit will be given by:

$$P_{av} = \frac{1}{2} V_M I_M \cos\left(\phi\right) = V_{rms} I_{rms} \cos\left(\phi\right) \qquad (41.21)$$

where V_{rms} and I_{rms} are the root mean square (rms) values of the voltage and current respectively. The term $\cos(\phi)$ is known as the *power factor*. It may alternatively be expressed as the ratio of real average power to the product of the rms values of voltage and current, respectively:

$$PF = \frac{P_{av}}{V_{rms} I_{rms}} \qquad (41.22)$$

TABLE 41.1 Integrated Circuits Used in Phase Measurement

Function	Designation	Manufacturer	Approximate price
Phase-locked loop	LM566	National, Motorola, Phillips	$2.70
Phase-locked loop	74HC4046	Harris, Motorola	$2
Phase/frequency detector	MC4044P	Motorola	$18.29
Pair of retriggerable monostables (one-shot)	74HC4538	Motorola, Harris	$2
DSP Chip	TMS320C32	Texas Instruments	$50
DSP Chip	TMS320C31	Texas Instruments	$80
DSP Chip	MC56303	Motorola	$60
DSP Chip	ADSP21020	Analog Devices	$110

The above expression is, in fact, a general definition of power factor for any current and voltage waveforms. For the special case of sinusoidal voltage and currents, *PF* reduces to $\cos(\phi)$.

There are a number of ways to measure the power factor. One way is to use a wattmeter to measure the real average power and a voltmeter and an ammeter to measure the rms voltage and current, respectively. The power factor is then determined according to Equation 41.22. This is probably the most effective way when the currents and/or voltages are nonsinusoidal. This procedure can easily be implemented with "digital power meters." The power is measured by time-averaging the product of the instantaneous voltage and current, while the rms values are calculated by taking the square root of the time averaged value of the square of the parameter of interest — current or voltage. Some digital power meters also provide an analysis of the individual harmonics via FFT processing. These meters are accurate and versatile, and consequently very popular.

A more direct method is based on the crossed-coil meter, the operation of which was described earlier in this chapter. Note that this meter is a "single-phase meter," which is accurately designed for one frequency only. Errors will occur at other frequencies because of the dependance of the crossed-coil meter method on a constant and known phase angle between the currents in the crossed coils.

With balanced polyphase circuits, it is possible to use a single-phase meter applied to one of the phases. Alternatively, one can use specially designed polyphase meters. In a three-phase meter, for example, one phase is connected to the fixed coil, while the other two phases are connected to the crossed coils on the rotating shaft. The crossed coils are constructed with a 60° angle between them. With four-phase systems, consecutive lines are 90° out of phase. Two of these consecutive lines are connected to the two crossed-coils and the angle between the coils is made equal to 90°.

With unbalanced polyphase circuits amid the presence of harmonics, each of the harmonic components has its own power factor, and so it is likely to be misleading to use a meter that measures a single angle. These methods based on the crossed-coil meter are thus much more limited than their digital counterparts.

41.6 Instrumentation and Components

Table 41.1 lists some integrated circuits and DSP chips that can be used in the various techniques for measuring phase. The list is really only illustrative of what is available and prices are approximate costs in U.S. dollars for small quantities at the end of 1996. Table 41.2 lists some companies that manufacture these products. An extensive (and indeed rapidly increasing) product range exists for DSP chip-based products, with details being available from the companies listed in Table 41.2. Table 41.3 lists instruments used for phase measurement. These instruments include CROs, vector voltage meters, vector impedance meters, crossed-coil meters and digital power meters, zero-crossing meters, and phase standards. Again, the table is only representative, as the full range of available instruments is enormous. Addresses of some of the relevant companies are provided in Table 41.4.

TABLE 41.2   Companies Making Integrated Circuits and DSP Chips Which Can Be Used for Phase Measurement

Analog Devices, Inc. One Technology Way Box 9106, Norwood, MA 02062 Tel: (617) 329-4700.	National Semiconductor Corp. 2900 Semiconductor Dr. P.O. Box 58090 Santa Clara, CA 95052-8090
Harris Semiconductor Products Division P.O. Box 883 Melbourne, FL 37902 Tel: (407) 724-3730	Texas Instruments Incorporated P.O. Box 1443 Houston, Texas 77251-1443
Motorola, Semiconductor Products Sector 3102 N. 56th St. Phoenix, AZ 85018 Tel: (602) 952-3248	

TABLE 41.3   Instruments for Measuring Phase

Description	Model number	Manufacturer	Approximate price
CRO	HP54600B	Hewlett Packard	$2,495
CRO	HP54602B	Hewlett Packard	$2,995
CRO	HP54616	Hewlett Packard	$5,595
CRO	TDS220	Tektronix	$1,695
CRO	TDS510A	Tektronix	$9,956
Vector signal analyzer	HP89410A	Hewlett Packard	$29,050
Vector signal analyzer	HP89440A	Hewlett Packard	$52,500
Vector signal analyzer	HP89441A	Hewlett Packard	$58,150
Gain/phase Impedance meter	HP4193A	Hewlett Packard	$13,700
Zero-crossing phase meter	KH6500	Krohn-Hite	$1,300
Digital power analyzer (with power factor & phase)	NANOVIP	Elcontrol	$660
Digital analyzing vector voltmeter	NA2250	North Atlantic Instruments	
Digital power analyzer (with power factor & phase, FFT analysis)	3195	Hioki	$25,000
Crossed-coil meter	246-425G	Crompton Industries	$290
Digital phase standard	5500	Clarke-Hess	$11,000

TABLE 41.4   Companies Making Instruments for Measuring Phase

Hewlett-Packard Co. Test and Measurement Sector P.O. Box 58199 Santa Clara, CA 95052-9943 Tel: (800) 452-4844	Krohn-Hite Corporation Bodwell St., Avon Industrial Park Avon, MA	Hioki 81 Koizumi Veda, Nagano 386-11 Japan
	Crompton Instruments Freebournes Road, Witham Essex, CM83AH England	Clarke-Hess Comm. Research Corporation 220 W. 19 Street
Tektronix Inc. Corporate Offices 26600 SW Parkway P.O. Box 1000 Wilsonville, OR 97070-1000 Tel: (503) 682-3411, (800) 426-2200	Elcontrol Via San Lorenzo 1/4 - 40037 Sasso Marconi Bologna, Italy	New York, NY North Atlantic Instruments htttp://www.naii.com

References

1. D. Gabor, The theory of communication, *J. Inst. Elec. Eng.*, 93(III), 429-457, 1946.
2. A. V. Oppenheim and R. W. Schafer, *Discrete-Time Signal Processing*, Engelwood-Cliffs, NJ: Prentice-Hall, 1989.
3. K. C. Pohlmann (ed.), *Advanced Digital Audio*, Carmel, IN: Howard Sams and Co., 1991.
4. D. Rife and R. Boorstyn, Single tone parameter estimation from discrete-time observations, *IEEE Trans. Inf. Theory*, 20, 591-598, 1974.
5. T. Abotzoglou, A fast maximum likelihood algorithm for estimating the frequency of a sinusoid based on Newton's algorithm, *IEEE Trans. Acoust., Speech Signal Process.*, 33, 77-89, 1985.
6. D. McMahon and R. Barrett, ML estimation of the fundamental frequency of a harmonic series, *Proc. of ISSPA 87*, Brisbane, Australia, 1987, 333-336.
7. A. Nehorai and B. Porat, Adaptive comb filtering for harmonic signal enhancement, *IEEE Trans. Acoust., Speech Signal Process.*, 34, 1124-1138, 1986.
8. L. White, An iterative method for exact maximum likelihood estimation of the parameters of a harmonic series, *IEEE Trans. Automat. Control*, 38, 367-370 1993.
9. C. Kelly and S. Gupta, Discrete-time demodulation of continuous time signals, *IEEE Trans. Inf. Theory*, 18, 488-493, 1972.
10. B. D. O. Anderson and J. B. Moore, *Optimal Filtering*, Englewood Cliffs, NJ: Prentice-Hall, 1979.
11. P. Parker and B. Anderson, Frequency tracking of periodic noisy signals, *Signal Processing*, 20(2), 127-152, 1990.
12. R. E. Best, *Phase-Locked Loops; Theory, Design and Applications*, 2nd ed., New York: McGraw-Hill, 1993.
13. L. White, Estimation of the instantaneous frequency of a noisy signal, in B. Boashash (ed.), *Time-Frequency Signal Analysis, Methods and Applications*, Melbourne, Australia: Longman-Cheshire; New York: Halsted Press, 1992.

Further Information

A. D. Helfrick and W. D. Cooper, *Modern Electronic Instrumentation and Measurement Techniques*, Englewood Cliffs, NJ: Prentice-Hall, 1990.

McGraw-Hill Encyclopedia of Science and Technology, 8th ed., New York: McGraw-Hill, 1997.

H. Taub and D. L. Schilling, *Principles of Communication Systems*, 2nd ed., New York: McGraw-Hill, 1986.

J. D. Lenk, *Handbook of Oscilloscopes: Theory and Application*, Englewood Cliffs, NJ: Prentice-Hall, 1982.

42

Energy Measurement

Arnaldo Brandolini
Politecnico di Milano

Alessandro Gandelli
Politecnico di Milano

Energy is one of the most important physical quantities in any branch of science and engineering and especially in electrical engineering. Energy exchange processes lead to the study of electric networks from the physical point of view and allow an in-depth knowledge of power transfer within the electrical world and between electric and other forms of energy.

The definitions of energy and power represent the starting point for any successive study.

1. Energy is the amount of work that the system is capable of doing.
2. Power is the time rate of doing work.

Energy can be mathematically defined as the definite integral of the power over a given time interval Δt.

The power available in a two-terminal section of an electric circuit is given by the product of the voltage across the terminals and the current flowing through the section itself ($p = vi$). The electric energy (E) flowing through the same section is defined by the integral of the power over the observation interval:

$$E(\Delta t) = \int_{t_0}^{\Delta t + t_0} p \, dt \qquad (42.1)$$

For this reason, energy measurement is a dynamic measurement, which means it varies with time. The energy measurement unit is the Joule (J); but for the electric energy, the Watthour (Wh) is most common.

The electrostatic energy is defined as the product of the electric charge and the difference of electric potential.

Electricity is generated from different forms of energy (thermal, hydraulic, nuclear, chemical, etc.); after electric transfer and distribution processes, it is converted to other forms of energy. The main feature of electric energy is the simplicity by which one can transfer it over long distances, control the distribution, and measure energy consumption.

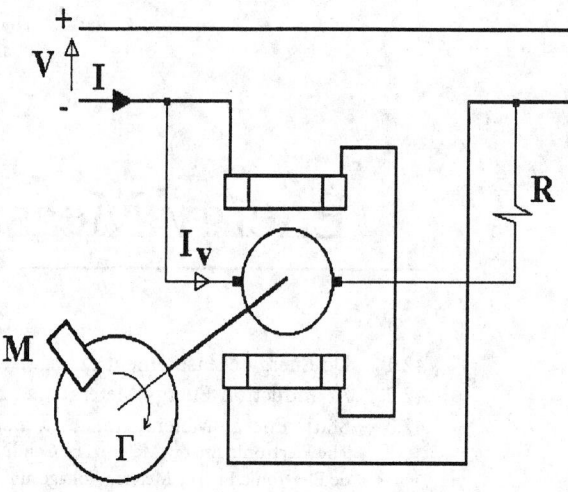

FIGURE 42.1   The electrodynamic dc energy meter; M = permanent magnet.

42.1 Dc Energy Measurement

The simplest way to perform this measurement is to measure voltage and current and then compute the product:

$$E = VI\Delta t \tag{42.2}$$

where Δt is the observation interval measured by means of a chronometer or a time counter.

Note that dc systems are limited to a restricted number of applications in power systems, as, for example: electric traction, electric drives, electrochemical power plants, and for HVDC transmission system in limited operating conditions. All these cases, nevertheless, allow energy measurement either on the dc or ac side of the network.

The dc energy measurement has been performed in the past by means of different methodologies and instruments such as electrodynamic measurement devices (Electrodynamics dc Energy Meter) operating as an integrating wattmeter (Figure 42.1). This measuring instrument is built using a small dc motor without iron, whose magnetic field is generated by the line current flowing through a coil arranged as the fixed part of the system. The rotor is connected in series with an additional resistor and is powered by the line voltage (V). Because of the lack of the iron in the magnetic circuit, the rotor magnetic flux ϕ is strictly proportional to the current I.

The rotor current (derived from the line voltage) is:

$$I_V = \left(V - E\right)\big/ R \tag{42.3}$$

where $E = k_1\Gamma\phi$ is the emf induced by the angular speed Γ, and R is the total resistance of the voltage circuit. It is possible to make the emf E negligible because of low angular speed Γ, limited amplitude of the flux ϕ, and a significant resistance R. In this way, Equation 42.3 becomes:

$$I_V \approx V\big/ R \tag{42.4}$$

The torque C_m provided by the motor can be written:

$$C_m = k_2\phi I_V \approx k_3 IV\big/ R = k_4 P \tag{42.5}$$

C_m is therefore approximately proportional to the power P flowing through the line. It is necessary, however, to remember that this torque could create a high angular speed to the rotor, because of constantly incrementing speed. In order to maintain dynamic equilibrium, a simple aluminum disk mounted on the rotor axis and placed in a constant magnetic field provided by a permanent magnet M, is added to the dc motor system. In this way, the induced currents in the disk introduce a damped torque proportional to the angular speed Γ, so, at equilibrium, there is a linear dependence of Γ on the power P. Thus,

$$E = \int_{\Delta t} P \, dt = k_5 \int_{\Delta t} \Gamma \, dt \qquad (42.6)$$

A mechanical counter transfers the rotating motion into a digital representation of the total energy consumed during a specific time interval Δt in the power system.

42.2 ac Induction Energy Meters

The most traditional and widely used ac energy meter is the *induction meter.* This device is built by means of three electric circuits, magnetically coupled, two of them fixed and one rotating around the mechanical axis of the system. Figure 42.2 shows the two fixed circuits, (1) and (2), which are the voltage and the current coils. The third circuit is the rotating disk (3), generally made of aluminum, mounted on a rigid axis (4) transmitting the disk rotation to a mechanical counter (6), which provides the energy display.

The fixed circuits (1) and (2) provide magnetic fluxes interacting with the rotating disk. Fixed circuits (1) and (2) form a C shape and the disk is placed in their iron gaps. Another similar structure, arranged using a permanent magnet (5), is placed over the disk as well. The magnetic fluxes generated by the voltage and current circuits are at the same frequency and are sinusoidal. They induce currents in the rotating disk that, by means of a cross-interaction with the two generating fluxes, provide a mechanical torque acting on the disk. The torque is given by:

$$C_m = KVI \sin(\alpha) \qquad (42.7)$$

where C_m = Mechanical torque
  K = System constant
  V = rms of the value of the applied voltage
  I = rms of the value of the applied current
  α = Phase angle between the fluxes generated by V and I

The acting torque causes the disk to rotate around its axis. This rotation reaches a dynamic equilibrium by balancing the torque C_m of the voltage and current coils and the reacting torque generated by permanent magnet. The resulting angular speed, Γ, is therefore proportional to the flowing power if:

- The angular speed Γ of the disk is much smaller than the voltage and current frequency ω
- The phase difference between the voltage and current fluxes is equal to $\alpha = \pi - \varphi$, where φ is the phase difference between the voltage and current signals

The angular speed of the rotating disk can be written as:

$$\Gamma = (1/k)\omega \left(R_3/Z_3^2\right)\left(M_1 I\right)\left(M_2 V/Z_2\right)\cos(\phi) = K P \qquad (42.8)$$

where Γ = Angular speed of the rotating circuit (conductor disk), in rad s^{-1}
  K = Instrument constant, in rad s^{-1} W^{-1}
  P = Mean power in the circuit, in W
  $1/k$ = Constant, in Ω V^{-2} s^{-2}

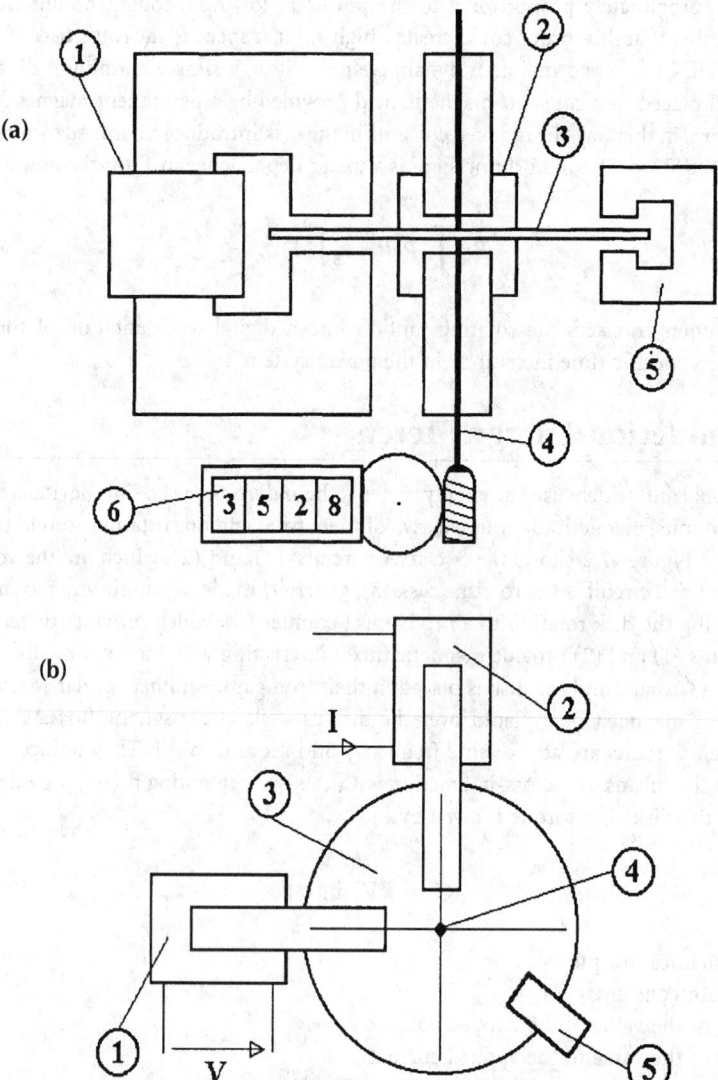

FIGURE 42.2   (a) Side View of an ac induction energy meter: (1) voltage coil and magnetic circuit; (2) current coil and magnetic circuit; (3) aluminum rotating disk; (4) disk axis; (5) permanent magnet; (6) mechanical display. (b) Top view of an ac induction energy meter: (1) voltage coil and magnetic circuit; (2) current coil and magnetic circuit; (3) aluminum rotating disk; (4) disk axis; (5) permanent magnet.

ω	= Voltage and current frequency, in rad s^{-1}
R_3	= Equivalent resistance of the rotating disk, relative to the induced current fields, in Ω
Z_3	= Equivalent impedance of the rotating disk, relative to the induced current fields, in Ω
$(M_2 \, V/Z_2)$	= rms value of the common flux related to the circuits n. 1 and 3, in Wb
$(M_1 I)$	= rms value of the common flux related to the circuits n. 2 and 3, in Wb
Z_2	= Impedance of the voltage circuit (n. 1), in Ω
V	= rms value of the applied voltage, in V
I	= rms value of the applied current, in A
ϕ	= Phase difference between current and voltage signals

The integral of Γ over a defined period Δt is proportional (with enough accuracy) to the energy flowing in the power circuit. Thus, it is true that the instrument constant K is strictly related (but not proportional) to the signal frequency ω .

42.3 Static Energy Meters

The development of electronic multipliers led to their use in energy meters that directly multiply voltage by current. In their first version, electronic multipliers used analog components (operational amplifiers, resistors, capacitors, etc.), while recent devices use digital components and programmable logic systems. Voltage and current signals are processed to obtain a signal proportional to the real power flowing into the line. The result is integrated over the observation time in order to calculate the *measured* energy. The devices based on these components are completely static (i.e., they do not have any moving parts). Moreover, because these electronic components have a frequency range from dc to high frequencies, instruments based on them can be applied to dc, ac, or distorted power systems (some care must be taken in order to provide a correct sampling of signals in all-digital systems).

There are many different prototypes in this class of energy meters. The first realizations were based on analog multipliers and, even if they were not able to replace the traditional induction energy meters, they represented a good solution for all those applications where an increased accuracy was required (up to 0.2%). Now, more sophisticated digital instruments are under design and development, based on dedicated structures mainly implementing DSPs (digital signal processors) as powerful tools for numerical computation and sigma-delta analog-to-digital converters (ADCs) in order to optimize the conversion process.

Many of these instruments can be analyzed by means of the following functional descriptions.

The Electronic Energy Meter

Figure 42.3 shows the block diagram of an electronic energy meter. The main feature of this type of instrument is the presence of voltage inputs on both voltage and current channels, because the electronic circuitry accepts only voltage signals. It has negligible current consumption from the system under measurement, due to high input impedance. Moreover, the maximum amplitude level of the input signal must be limited to around 5 V to 15 V. For this reason, the conditioning apparatus must guarantee the correct current-to-voltage transformation and the proper voltage reduction. This type of instrument can work at dc (which omits voltage and current transformers) or ac power systems and can also measure energy from distorted signals.

The Conditioning System for dc Electronic Energy Meters

The basic blocks of the conditioning system for a dc energy meter are formed from a voltage divider for the voltage input, and a shunt for the current input. After these passive components, two preamplifiers are usually introduced before the processing system. The current preamplifier is very important because:

1. The voltage output level of the current shunt is very low, even at full scale (≤ 1 V).
2. Many times, the current input has to support overloaded signals; the presence of a variable gain amplifier allows acceptable working conditions for the system.
3. It can be used to implement an active filter before signal processing.

Voltage and Current Adapters for ac Electronic Energy Meters

The most common devices to process ac signals for static energy meters are the traditional voltage and current transformers. They must be made with proper components to achieve the right amplitude of the voltage inputs (by nonreactive shunts for the current transformers, and nonreactive voltage dividers for

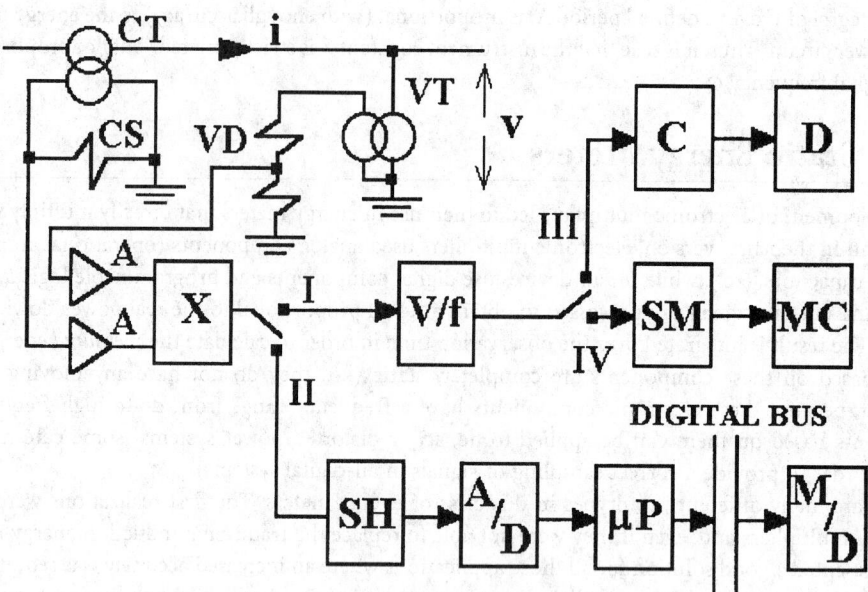

FIGURE 42.3 Electronic energy meter. Mechanical display option (I to IV). Electronic display option (I to III). Electronic display option and digital processing of the power signal (II). CT, current transformer; VT, voltage transformer; CS, current shunt; VD, voltage divider; A, analog signal processing block; X, multiplier; V/f, voltage-to-frequency converter; SM, step motor; MC, mechanical counter; C, electronic counter; D, display; SH, sample and hold; A/D, analog-to-digital converter; μP, microprocessor (CPU); M/D, memory and display.

the voltage transformers). After the transformers, and related devices, a second block, based on electronic amplifiers, provides the final analog processing of the input signals, as for the dc conditioning systems. It is useful to introduce this second processing element because analog filters are generally required when the input signals need to be digitally processed.

Electronic-Analog Energy Meters with Digital Output

These instruments provide the product of the two input signals (both voltages) through an analog multiplier that evaluates a voltage output proportional to the power of the input signals. This output can be followed by a filtering block.

The output signal is proportional to the instantaneous electric power flowing through the line. To calculate the energy, it is now necessary to complete the process by integrating over the observation time. This last procedure can be performed in two different ways.

1st procedure: The power signal at the output of the analog multiplier is applied to the input of a voltage frequency converter. Thus, the power information is converted from a voltage level to the frequency pulse sequence, for which the counting process performs the integration of the power in the observation interval, i.e., the measurement of energy.

The final measurement can be performed by means of an electronic counter with digital display or using a dc step motor incrementing the rotor angular position every pulse by a fixed angular increment. The rotor position is shown by a mechanical counter (similar to the system mounted on the induction energy meters) indicating the total number of complete rotations performed by the system, proportional to the energy of the system under measurement. This second arrangement is normally adopted because it allows a permanent record of the energy information, which is not subject to possible lack of electric energy as in the first case.

2nd procedure: This arrangement is based on an analog-to-digital converter (ADC) connected to the output of the analog multiplier. The sampling process is driven by an internal clock. Thus, the ADC

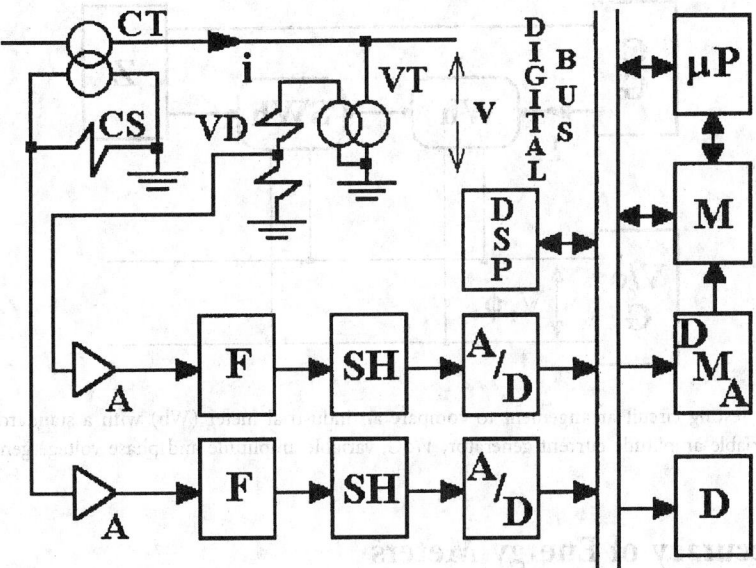

FIGURE 42.4 All-digital energy meter. CT, current transformer; VT, voltage transformer; CS, current shunt; VD, voltage divider; A, analog signal processing block; F, analog electronic filter; SH, sample and hold; A/D, analog-to-digital converter; μP, microprocessor (CPU); M, memory; DSP, digital signal processor; DMA, direct memory access circuit; D, display.

provides uniform sampling over the signal period and, under the condition imposed by the sampling theorem, the sum of the samples is proportional to the integral of the power signal, i.e., to the energy during the observation interval.

The calculation is performed by means of a dedicated CPU and then the results are sent to the digital memory to be stored and displayed. They can also be used to manage any other automatic processes based on the energy measurement. For this purpose, data are available on a data bus (serial or parallel) connecting the measuring system with other devices.

The sampling process is performed by a Sample & Hold circuit.

All-Digital Energy Meters

The most advanced solution for energy measurement can be found in all-digital meters (Figure 42.4), where both the voltage and current signals are sampled before any other processing. Thus, the data bus presents the sampled waveforms in digital form, giving the opportunity to perform a wide choice of digital signal processing on the power and energy information. Both sampling devices are driven by a CPU, providing synchronized sampling signals.

Sometimes, the system is equipped with a DSP capable of providing hardware resources to implement real-time evaluation of complex parameters (i.e., signal transforms) of the signal and energy measurement. Dedicated hardware and software performing instrument testing are also integrated into the meter to complete the device with the most advanced features.

Filters able to meet the sampling theorem requirements, programmable gain amplifiers, and Sample & Hold circuits generally precede the ADCs.

Data management is arranged in two possible ways: sending the sampled data directly to the processing system for calculations, or accessing the memory using DMA procedures, so the data string for a specific time period is first stored and then used for computation of energy and related parameter values. Final results of this computation are then available on the system bus to be sent to the other system resources or to be displayed.

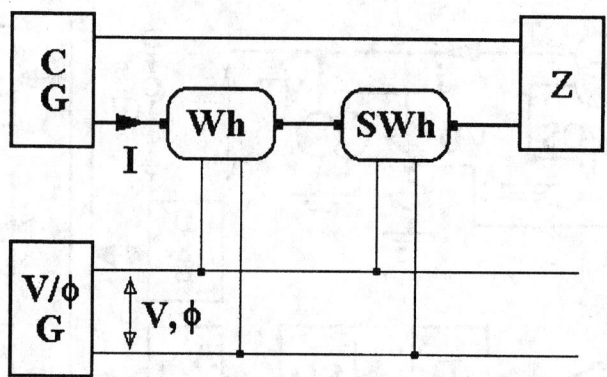

FIGURE 42.5 Testing circuit arrangement to compare an industrial meter (Wh) with a standard energy meter (SWh). CG, variable-amplitude current generator; V/φG, variable-amplitude and phase voltage generator; Z, load impedance.

42.4 Accuracy of Energy Meters

Accuracy of energy meters is defined by means of relative parameters (in percent) obtained from a testing process by powering the instrument with a constant (nominal) voltage signal and a variable current signal (5, 10, 20, 50, 100, 120% of the nominal value). The testing procedures are performed by comparing the meter under test with a standard meter (Figure 42.5), or using equivalent methods.

The accuracy of commercial electromechanical (induction) energy meters is generally around 2%. Energy meters with accuracies of 1% have also been built. Electronic energy meters have a better accuracy, generally between 0.5% and 0.2%.

Further Information

S. Kusui and T. Nagai, A single-phase 3-wire watt-to-pulse frequency-converter using simple PWM and its accuracy analysis, *IEEE Trans. Instrum. Meas.*, 43, 770–774, 1994.

P. Bosnjakovic and B. Djokic, Reactive energy measurement using frequency controlled analog-to-pulse-rate converter, *Archiv fur Elektrotechnik*, 75, 131–135, 1992.

B. Djokic, P. Bosnjakovic, and M. Begovic, New method for reactive power and energy measurement, *IEEE Trans. Instrum. Meas.*, 41, 280–285, 1992.

J. C. Montano, A. Lopez, M. Castilla, and J. Gutierrez, DSP-based algorithm for electric-power measurement, *IEE Proc. A, Sci. Meas. Technol.*, 140, 485–490, 1993.

C. V. Drisdale and A. C. Jolley, *Electrical Measuring Instruments*, 2nd ed., New York: John Wiley & Sons, 1952.

F. K. Harris, *Electrical Measurements*, New York: John Wiley & Sons, 1952.

L. Schnell (ed.), *Technology of Electrical Measurements*, New York: John Wiley & Sons, 1993.

L. Finkelstein and K. T. V. Grattan (eds.), *Concise Encyclopedia of Measurement & Instrumentation*, New York: Pergamon Press, 1994.

43

Electrical Conductivity and Resistivity

Michael B. Heaney
Huladyne Research

Electrical resistivity is a key physical property of all materials. It is often necessary to accurately measure the resistivity of a given material. The electrical resistivity of different materials at room temperature can vary by over 20 orders of magnitude. No single technique or instrument can measure resistivities over this wide range. This chapter describes a number of different experimental techniques and instruments for measuring resistivities. The emphasis is on explaining how to make practical measurements and avoid common experimental errors. More theoretical and detailed discussions can be found in the sources listed at the end of this chapter.

43.1 Basic Concepts

The *electrical resistivity* of a material is a number describing how much that material resists the flow of electricity. Resistivity is measured in units of ohm·meters (Ω m). If electricity can flow easily through a material, that material has low resistivity. If electricity has great difficulty flowing through a material, that material has high resistivity. The electrical wires in overhead power lines and buildings are made of copper or aluminum. This is because copper and aluminum are materials with very low resistivities (about 20 nΩ m), allowing electrical power to flow very easily. If these wires were made of high resistivity material like some types of plastic (which can have resistivities about 1 EΩ m (1×10^{18} Ω m)), very little electric power would flow.

Electrical resistivity is represented by the Greek letter ρ. Electrical conductivity is represented by the Greek letter σ, and is defined as the inverse of the resistivity. This means a high resistivity is the same as a low conductivity, and a low resistivity is the same as a high conductivity:

$$\sigma \equiv \frac{1}{\rho}$$

(43.1)

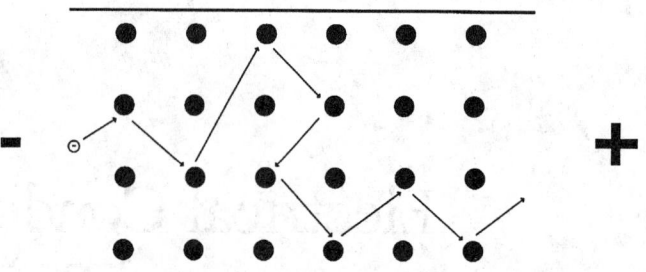

FIGURE 43.1   Simple model of electricity flowing through a material under an applied voltage. The white circle is an electron moving from left to right through the material. The black circles represent the stationary atoms of the material. Collisions between the electron and the atoms slow down the electron, causing electrical resistivity.

This chapter will discuss everything in terms of resistivity, with the understanding that conductivity can be obtained by taking the inverse of resistivity. The electrical resistivity of a material is an intrinsic physical property, independent of the particular size or shape of the sample. This means a thin copper wire in a computer has the same resistivity as the Statue of Liberty, which is also made of copper.

43.2   Simple Model and Theory

Figure 43.1 shows a simple microscopic model of electricity flowing through a material [1]. While this model is an oversimplification and incorrect in several ways, it is still a very useful conceptual model for understanding resistivity and making rough estimates of some physical properties. A more correct understanding of the electrical resistivity of materials requires a thorough understanding of quantum mechanics [2].

On a microscopic level, electricity is simply the movement of electrons through a material. The smaller white circle in Figure 43.1 represents one electron flowing through the material. For ease of explanation, only one electron is shown. There are usually many electrons flowing through the material simultaneously. The electron tends to move from the left side of the material to the right side because an external force (represented by the large minus and plus signs) acts on it. This external force could be due to the voltage produced by an electrical power plant, or a battery connected to the material. As the electron moves through the material, it collides with the "stationary" atoms of the material, represented by the larger black circles. These collisions tend to slow down the electron. This is analogous to a pinball machine. The electron is like the metal ball rolling from the top to the bottom of a pinball machine, pulled by the force of gravity. The metal ball occasionally hits the pins and slows down. Just like in different pinball machines, the number of collisions the electron has can be very different in different materials. A material that produces lots of collisions is a high-resistivity material. A material that produces few collisions is a low-resistivity material.

The resistivity of a material can vary greatly at different temperatures. The resistivity of metals usually increases as temperature increases, while the resistivity of semiconductors usually decreases as temperature increases. The resistivity of a material can also depend on the applied magnetic field.

The discussion thus far has assumed that the material being measured is homogeneous and isotropic. Homogeneous means the material properties are the same everywhere in the sample. Isotropic means the material properties are the same in all directions. This is not always a valid assumption. A more exact definition of resistivity is the proportionality coefficient ρ relating a local applied electric field to the resultant current density:

$$E \equiv \rho J \qquad (43.2)$$

where E is the electric field (V/m), J is the current density (A m^{-2}), and ρ is a proportionality coefficient (Ω m). Equation 43.2 is one form of Ohm's law. Note that E and J are vectors, and ρ is, in general, a

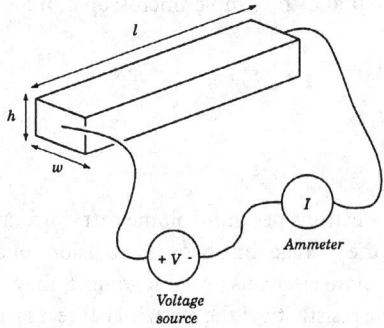

FIGURE 43.2 A two-point technique for measuring the resistivity of a bar of material. The voltage source applies a voltage across the bar, and the ammeter measures the current flowing through the bar.

tensor. This implies that the current does not necessarily flow in the direction of the applied electric field. In this chapter, isotropic and homogeneous materials are assumed, so ρ is a scalar (a single number).

Now consider the bar-shaped sample shown in Figure 43.2. The electric field E is given by the voltage V divided by the distance l over which the voltage is applied:

$$E \equiv \frac{V}{l} \tag{43.3}$$

The current density J is given by the current I, divided by the cross-sectional area A through which the current flows:

$$J \equiv \frac{I}{A} \tag{43.4}$$

where the area A in Figure 43.2 is equal to the width w times the height h. Combining Equations 43.2, 43.3, and 43.4 and rearranging gives:

$$V = \frac{I\rho l}{A} \tag{43.5}$$

Now define a new quantity called "resistance" R with the definition:

$$R \equiv \frac{\rho l}{A} \tag{43.6}$$

Combining Equations 43.5 and 43.6 then gives:

$$I = \frac{V}{R} \tag{43.7}$$

where I is the current in amps (A) flowing through the sample, V is the voltage in volts (V) applied across the sample, and R is the resistance in ohms (Ω) of the sample. Equation 43.7 is another form of Ohm's law.

Note that the resistance R can depend on the size and shape of the sample, while ρ is independent of the size or shape of the sample. For example, if the length l of the sample bar is doubled, the resistance will double but the resistivity will remain constant.

The quantitative relationship between the resistivity ρ and the simple microscopic model shown in Figure 43.1 is given by:

$$\rho = \frac{m}{ne^2\tau} \tag{43.8}$$

where m is the mass of an electron, n is the number of electrons per unit volume carrying current in the material, e is the electric charge on an electron, and τ is the average time between collisions of an electron with the stationary atoms of the material. If there were more electrons per unit volume, they could carry more current through the material. This would lower the resistivity. If the electric charge on the electrons were greater, then the applied voltage would pull harder on the electrons, speeding them up. This would lower the resistivity. If the average time between collisions with the stationary atoms were longer, then the electrons could get through the material quicker. This would lower the resistivity. If electrons could be made more massive, they would move slower and take longer to get through the material. This would increase the resistivity.

43.3 Experimental Techniques for Measuring Resistivity

Two-Point Technique

The resistivity of a material can be obtained by measuring the resistance and physical dimensions of a bar of material, as shown in Figure 43.2. In this case, the material is cut into the shape of a rectangular bar of length l, height h, and width w. Copper wires are attached to both ends of the bar. This is called the two-point technique, since wires are attached to the material at two points. A voltage source applies a voltage V across the bar, causing a current I to flow through the bar. (Alternatively, a current source could force current through the sample bar, while a voltmeter in parallel with the current source measures the voltage induced across the sample bar.) The amount of current I that flows through the bar is measured by the ammeter, which is connected in series with the bar and voltage source. The voltage drop across the ammeter should be negligible. The resistance R of the bar is given by Equation 43.8a:

$$R = \frac{V}{I} \tag{43.8a}$$

where R = Resistance in Ω
       V = Voltage in volts
       I = Current in amps

The physical dimensions can be measured with a ruler, a micrometer, or other appropriate instrument. Consult earlier sections of this Handbook for guidance on measuring spatial dimensions. The two-point resistivity of the material is then:

$$\rho \equiv \frac{Rwh}{l} \tag{43.9}$$

where ρ is the resistivity in Ω m, R is the measured resistance in Ω, and w, h, and l are the measured physical dimensions of the sample bar in meters.

In practice, measuring resistivity with a two-point technique is often not reliable. There is usually some resistance between the contact wires and the material, or in the measuring equipment itself. These additional resistances make the resistivity of the material measure higher than it really is. A second potential problem is modulation of the sample resistivity due to the applied current. This is often a

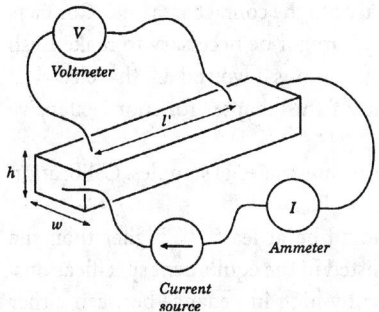

FIGURE 43.3 A four-point technique for measuring the resistivity of a bar of material. The current source forces a current through the bar, which is measured by a separate ammeter. The voltmeter measures the voltage across the middle of the bar.

possibility for semiconducting materials. A third problem is that contacts between metal electrodes and a semiconducting sample tend to have other electrical properties that give wrong estimates for the actual sample resistivity. The four-point measurement technique overcomes many of these problems.

Four-Point Technique

Figure 43.3 shows the four-point measurement technique on a bar of material. Four wires are attached to the sample bar as shown. A current source forces a constant current through the ends of the sample bar. A separate ammeter measures the amount of current I passing through the bar. A voltmeter simultaneously measures the voltage V produced across the inner part of the bar. (Alternatively, a voltage source could apply a voltage across the outer contacts, while an ammeter in series with this voltmeter measures the current flowing through the sample bar.)

The four-point resistivity of the material is then:

$$\rho = \frac{V\,w\,h}{I\,l^{l}} \tag{43.10}$$

where ρ = Resistivity in Ω m
V = Voltage measured by the voltmeter in volts
w = Width of the sample bar measured in meters
h = Height of the sample bar measured in meters
I = Current the ammeter measures flowing through the sample in amperes
l^{l} = Distance between the two points where the voltmeter wires make contact to the bar, measured in meters

Note that the total length l of the bar is not used to calculate the four-point resistivity: the length l^{l} between the two inner contacts is used.

Common Experimental Errors

There are many experimental pitfalls to avoid when making resistivity measurements. The most common sources of error arise from doing a two-point measurement on a material that has any of the contact problems discussed earlier. For this reason, it is advisable to do four-point measurements whenever possible. This section describes experimental techniques to avoid errors in measuring resistivity:

1. The most difficult part of making resistivity measurements is often making good electric contacts to the sample. The general technique for making good electric contacts is to clean the areas of the sample where contacts are to be made with alcohol or an appropriate solvent, and then apply the contacts. If this does not work, try scraping the surface with a razor blade where contact is to be made, or cutting the sample to expose a fresh surface. Contacts can be made in many ways, such as using alligator clips, silver-paint, squeezing a wire against the material, soldering wires to the

material, pressing small pieces of indium or a similar soft metal onto the contact areas, etc. Contacts can age: a good contact can become a bad contact over time. It might be necessary to make fresh contacts to a sample that has aged. There are many complications involved in the electrical properties of contacts. Refer to the sources listed at the end of this chapter for more extensive discussions.

2. The measurement system should be calibrated before measuring any material samples. Calibration procedures are usually described in the equipment manuals.

3. The input resistance (or "impedance") of the voltmeter should be at least 10^5 higher than the resistance of the sample bar. The input impedance is usually listed in the equipment specifications. Note that some voltmeters and electrometers have a sufficiently high impedance between either of the inputs and ground, but not between the two inputs. In this case, it is necessary to use two voltmeters/electrometers (each with one input connected to ground and the other input connected to the sample bar). Measure the difference between them to obtain the voltage across the sample.

4. The measurement system should be tested before measuring any material samples. First test "short" with a thick copper wire or sheet in place of the sample. Then test "open" with nothing in place of the sample. Finally, test with a known, calibrated resistor whose resistance is within an order of magnitude of the sample resistance.

5. The geometry of the sample and electric contacts can be important. Contacts are often made by painting silver-paint or applying metal electrodes to the sample. If these contact areas are large or close to each other, this could reduce the accuracy of the resistivity measurement. It is best to make the two voltage contacts in a four-point measurement as small or thin as possible, and make the distance between inner electrodes much larger than the sample thickness. This also allows a more accurate estimate of the effective volume of the sample being probed.

6. It is critical that the four contacts to the sample bar in a four-point measurement are completely independent; there should be nothing other than the material of the bar connecting each of the four wires at the bar. For example, when pieces of indium are used to attach wires to a small sample, it is easy to smudge two adjacent indium pieces into one another. Those two contacts are no longer independent, and could easily cause errors. Visually inspect the contacts for this condition. If visual inspection is impractical, measure the resistance between the wires going to adjacent contacts. An unusually low resistance might indicate that two contacts are touching each other.

7. The applied voltage or current can cause heating of the material, which can change its resistivity. To avoid this problem, start with very small voltages or currents, and increase until the measured voltages and currents are at least 10 times larger than the random fluctuations of the meters. Then make sure the measured resistance is constant with time: the average resistance should not drift more than 10% in a few minutes.

8. Even if heating of the sample is not a problem, Ohm's law is not always obeyed. Many materials have a resistance that varies as the applied voltage varies, especially at higher voltages. Test for a linear relationship between current and voltage by measuring the resistance at several voltages on both sides of the measurement voltage. Whenever possible, make measurements in the linear (ohmic) region, where resistance is constant as voltage changes.

9. If one or both of the contacts to the voltmeter are bad, the voltmeter may effectively be disconnected from the material. In this situation, the voltmeter might display some random voltage unrelated to the voltage in the material. It might not be obvious that something is wrong, since this random voltage could accidentally appear to be a reasonable value. Check for this by setting the current source to zero amps and seeing if the voltmeter reading drops to zero volts. If it does not, try remaking the two inner contacts.

10. A critical check of a four-point measurement is to reverse the leads and remeasure the resistance. First, turn the current source to zero amps. Without disturbing any of the four contacts at the sample, swap the two sets of wires going to the voltmeter and the current source/ammeter. The two wires that originally plugged into the voltmeter should now plug into one terminal of the current source and one terminal of the ammeter. The two wires that originally plugged into the

current source and ammeter should now plug into the voltmeter. Turn the current source on and remeasure the resistance. Note that current is now being forced to flow between the two inner contact points on the sample, while the voltage is being measured between the two outer contacts on the sample. The two measured resistances should be within 10% of each other.

11. The resistivity of some materials can depend on how much light is hitting the material. This is especially a problem with semiconductors. If this is a possibility, try blocking all light from the sample during measurement.

Sheet Resistance Measurements

It is often necessary to measure the resistivities of thin films or sheets of various materials. If the material can be made into the form of a rectangle, then the resistivity can be measured just like the bar samples in Figure 43.2:

$$\rho \equiv \frac{Vwh}{Il} \tag{43.11}$$

where ρ = sample resistivity in Ω m
V = Voltage measured by the voltmeter in volts
w = Width of the sample measured in meters
h = Thickness of the sample measured in meters
I = Current the ammeter measures flowing through the sample in amperes
l = Length of the film measured in meters

For the special case of a square film, the width w is equal to the length l, and Equation 43.11 becomes:

$$\rho\left(of\ square\ film\right) \equiv \frac{Vh}{I} \tag{43.12}$$

The resistivity of a square film of material is called the "sheet resistivity" of the material, and is usually represented by the symbol ρ_s. The "sheet resistance" R_s is defined by:

$$R_s \equiv R\left(of\ square\ film\right) = \frac{V}{I} \tag{43.13}$$

where V = Voltage measured by the voltmeter in volts
I = Current the ammeter measures flowing through the sample in amps

The units for sheet resistance are Ω, but people commonly use the units "Ω per square" or $\Omega/\square$. The sheet resistance is numerically equal to the measured resistance of a square piece of the material. Note that sheet resistance is independent of the size of the square measured, and it is not necessary to know the film thickness to measure sheet resistance. This makes sheet resistance a useful quantity for comparing different thin films of materials.

It is usually more convenient to measure thin-film samples of arbitrary shape and size. This is usually done by pressing four collinear, equally spaced contacts into a film whose length and width are both much greater than the spacing between contacts. In this situation, the sheet resistance is [3]:

$$R_s = 4.532\frac{V}{I} \tag{43.14}$$

where V = Voltage measured across the two inner contacts
I = Current applied through the two outer contacts

In many practical cases, the size of the thin-film sample will not be much greater than the spacing between the four-point contacts. In other cases, it might be necessary to measure a thin film near a corner or edge. In this situation, use geometric correction factors to accurately estimate the sheet resistance. These correction factors are available for the most commonly encountered sample geometries [3].

Instrumentation for Four-Point Resistivity Measurements

The resistivities of thin films of materials are often measured using commercial four-point probes. These probes generally have four equally spaced, collinear metal points that are pressed against the surface of the film. A current is applied between the outer two points, while the voltage is measured across the inner two points. These probes can also be used to measure the resistivity of bulk samples. Some companies that make probes and systems specifically for four-point resistivity measurements are listed in Table 43.1.

Instrumentation for High-Resistivity Measurements

Many materials such as rocks, plastics, and paper have very high resistivities, up to 1 EΩ m. The techniques described earlier for measuring resistivity are usually not reliable for these materials. In particular, it is often not possible to make a four-point measurement. One problem is that high voltages are needed to get any measurable current flowing through these materials. A second problem is that very long time constants prevent making steady-state measurements. A third problem is that the surfaces of these materials can often have significantly lower resistivity than the bulk, due to defects or other contamination. Measurements using the techniques described above then give falsely low values for the bulk resistivity. The best way to measure the resistivity of these materials is to use a specialized commercial instrument. These are designed to separate out the bulk resistivity from the surface resistivity, and to minimize the many other problems encountered when measuring very high resistivities. Table 43.2 lists some companies that make high-resistivity measurement systems.

van der Pauw Technique

The four-point measurement technique described earlier has assumed the material sample has the shape of a rectangular thin film or a bar. There is a more general four-point resistivity measurement technique that allows measurements on samples of arbitrary shape, with no need to measure all the physical dimensions of the sample. This is the van der Pauw technique [4]. There are four conditions that must be satisfied to use this technique:

1. The sample must have a flat shape of uniform thickness.
2. The sample must not have any isolated holes.
3. The sample must be homogeneous and isotropic.
4. All four contacts must be located at the edges of the sample.

In addition to these four conditions, the area of contact of any individual contact should be at least an order of magnitude smaller than the area of the entire sample. For small samples, this might not be possible or practical. If sufficiently small contacts are not achievable, it is still possible to do accurate van der Pauw resistivity measurements, using geometric correction factors to account for the finite size of the contacts. See Ref. [5] for further details.

The inset illustration of Figure 43.4 illustrates one possible sample measurement geometry. A more common geometry is to attach four contacts to the four corners of a square-shaped sheet of the material.

The procedure for doing a van der Pauw measurement is as follows:

1. Define a resistance $R_{ij,kl} \equiv V_{kl}/I_{ij}$, where $V_{kl} \equiv V_k - V_l$ is the voltage between points k and l, and I_{ij} is the current flowing from contact i to contact j.
2. Measure the resistances $R_{21,34}$ and $R_{32,41}$. Define $R_>$ as the greater of these two resistances and $R_<$ as the lesser of these two resistances.

TABLE 43.1   Companies That Make Four-Point Resistivity Measurement Probes and Systems

Company and comments

Creative Design Engineering, Inc.
20565 Elves Drive
Cupertino, CA 95014
Tel: (408) 736-7273
Fax: (408) 738-3912

Creative Design Engineering makes manual and automatic four-point resistivity systems specially designed for both small
and large semiconductor wafers.

Four Dimensions, Inc.
3138 Diablo Ave.
Hayward, CA 94545
Tel: (510) 782-1843
Fax: (510-786-9321
http://www.4dimensions.com

Four Dimensions makes a variety of manual and automatic four-point probe systems for measurement of resistivity and
resistivity mapping of flat samples such as semiconductor wafers.

Hewlett-Packard Company
Test and Measurement Organization
5301 Stevens Creek Blvd.
Santa Clara, CA 95052-8059
Tel: (800) 452-4844
Fax: (303) 754-4801
http://www.hp.com

Hewlett-Packard makes a variety of high-quality instruments useful for four-point measurements.

Jandel Engineering, Ltd.
Grand Union House
Leighton Road
Linslade, Leighton Buzzard
LU7 7LA
U.K.
Tel: (01525)-378554
Fax: (01525)-381945
http://www.getnet.com/~bridge/jandel.html

Jandel makes four-point probes useful for flat samples such as semiconductor wafers. They will build custom four-point
probes for your particular needs. They also make a combined constant current source and digital voltmeter for resistivity
measurements.

Keithley Instruments, Inc.
28775 Aurora Road
Cleveland, OH 44139-1891
Tel: (440) 248-0400
Fax: (440) 248-6168
http://www.keithley.com

Keithley makes a wide variety of four-point measurement systems. They also have useful, free literature detailing techniques
for making accurate resistivity measurements.

KLA-Tencor Corp.
1 Technology Drive
Milpitas, CA 95035
Tel: (408) 875-3000
Fax: (408) 875-3030
http://www.kla-tencor.com

KLA-Tencor makes automated sheet resistance mapping systems designed for semiconductor wafers.

TABLE 43.1 (continued)   Companies That Make Four-Point Resistivity Measurement Probes and Systems

Company and comments

Lucas-Signatone Corp.
393-J Tomkins Ct.
Gilroy, CA 95020
Tel: (408) 848-2851
Fax: (408) 848-5763
http://www.signatone.com

Signatone makes four-point resistivity measurement systems and a variety of four-point probe heads. They make a high-temperature, four-point probe head for temperatures up to 670 K.

Miller Design and Equipment, Inc.
2231-C Fortune Drive
San Jose, CA 95131-1806
Tel: (408) 434-9544
Fax: (408) 943-1491

Miller Design makes semi-automatic resistivity probe systems, designed for semiconductor wafers.

Mitsubishi Chemicals Corp./Yuka Denshi Co., Ltd.
Kyodo Bldg., 1-5 Nihonbashi Muromachi 4-chome
Chuo-kyu, Tokyo 103
Japan
Tel: 03-3270-5033
Fax: 03-3270-5036

Yuka Denshi makes a low-resistivity meter and a variety of four-point probe heads.

MMR Technologies, Inc.
1400 North Shoreline Blvd., # A5
Mountain View, CA 94043
Tel: (650) 962-9620
Fax: (650) 962-9647
http://www.mmr.com

MMR makes systems for four-point resistivity, Hall mobility, and Seebeck potential measurements over the temperature range 80 K to 400 K.

Napson Corporation
Momose Bldg. 7F
2-3-6 Kameido
Koto-kyu
Tokyo 136
Japan

QuadTech, Inc.
100 Nickerson Rd., Suite 3
Marlborough, MA 01752-9605
Tel: (800) 253-1230
Fax: (508) 485-0295
http://www.quadtechinc.com

QuadTech makes a four-point ohmmeter capable of measuring resistances from 1 μΩ to 2 MΩ.

Quantum Design
11578 Sorrento Valley Rd.
San Diego, CA 92121-1311
Tel: (800) 289-6996
Fax: (619) 481-7410
http://www.quandsn.com

Quantum Design makes an automated system for measuring four-point resistivity, Hall mobility, and other properties over the temperature range 2 K to 400 K in magnetic fields up to 14 T.

TABLE 43.2 Companies That Make High-Resistivity Measurement Probes and Systems

---

Company and comments

Hewlett-Packard Company
Test and Measurement Organization
5301 Stevens Creek Blvd.
Santa Clara, CA 95052-8059
Tel: (800) 452-4844
Fax: (303) 754-4801
http://www.hp.com

Hewlett-Packard makes high resistance meters and specially designed resistivity test chambers.

Keithley Instruments, Inc.
28775 Aurora Road
Cleveland, OH 44139-1891
Tel: (440) 248-0400
Fax: (440) 248-6168
http://www.keithley.com

Keithley makes special meters and resistivity test chambers for measuring high resistivities. They also have
useful, free literature detailing techniques for making accurate resistivity measurements.

Mitsubishi Chemicals Corp./Yuka Denshi Co., Ltd.
Kyodo Bldg., 1-5 Nihonbashi Muromachi 4-chome
Chuo-kyu, Tokyo 103
Japan
Tel: 03-3270-5033
Fax: 03-3270-5036

Yuka Denshi makes high-resistance meters and a variety of probes and resistivity test chambers.

Monroe Electronics, Inc.
100 Housel Avenue
Lyndonville, NY 14098
Tel: (800) 821-6001
Fax: (716) 765-9330
http://www.monroe-electronics.com

Monroe Electronics makes portable and hand-held instruments for measuring surface resistivity, designed
for testing antistatic materials.

QuadTech, Inc.
100 Nickerson Rd. Suite 3
Marlborough, MA 01752-9605
Tel: (800) 253-1230
Fax: (508) 485-0295
http://www.quadtechinc.com

QuadTech makes a high-resistance ohmmeter.

---

3. Calculate the ratio $R_>/R_<$ and find the corresponding value of the function $f(R_>/R_<)$ from Figure 43.4. Be careful to use the appropriate horizontal scale!
4. Calculate the resistivity ρ_a using:

$$\rho_a = \frac{\pi d \left(R_> + R_< \right) f\left(R_> / R_< \right)}{\ln 4} \qquad (43.15)$$

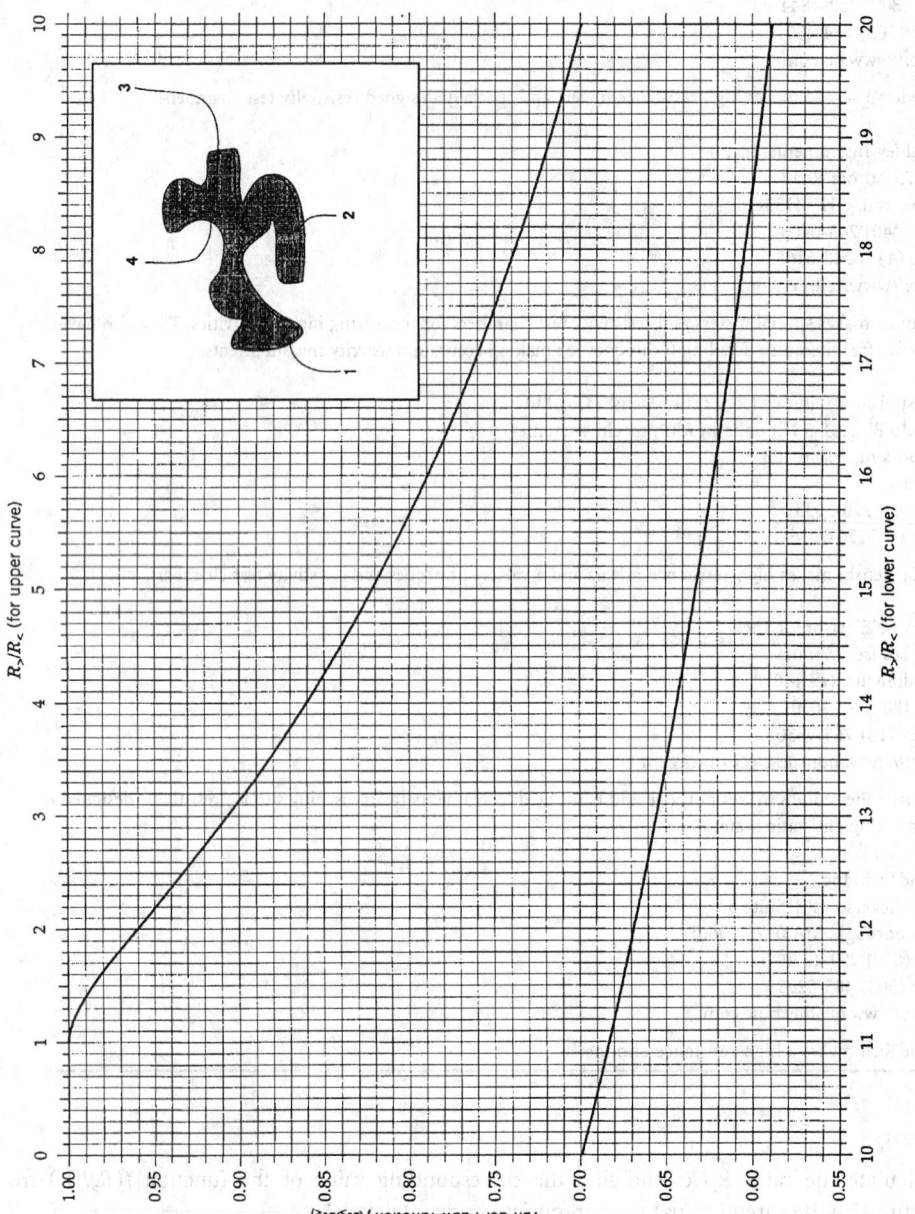

FIGURE 43.4 The van der Pauw technique. The inset shows one possible measurement geometry. The graph shows the function $f(R_>/R_<)$ needed to find the resistivity in Equation 43.15.

where ρ_a = Resistivity in Ω m

$\quad\quad d$ = Thickness of the sample in m

$\quad\quad$ Resistances $R_>$ and $R_<$ are measured in Ω

$\quad\quad$ ln4 = Approximately 1.3863

It is not necessary to measure the width or length of the sample.

5. Switch the leads to measure $R_{43,12}$ and $R_{14,23}$. Repeat steps 3 and 4 to calculate ρ_b using these new values for $R_>$ and $R_<$. If the two resistivities ρ_a and ρ_b are not within 10% of each other, either the contacts are bad, or the sample is too nonuniform to measure reliably. Try making new contacts. If the two resistivities ρ_a and ρ_b are within 10% of each other, the best estimate of the material resistivity ρ is the average:

$$\rho = \frac{\left(\rho_a + \rho_b\right)}{2} \tag{43.16}$$

Note: The function $f(R_>/R_<)$ plotted in Figure 43.4 is defined by the transcendental equation:

$$f\left(R_>/R_<\right) \equiv \frac{-\ln 4\left(R_>/R_<\right)}{\left[1+\left(R_>/R_<\right)\ln\left\{1-4^{-\left[\left(1+R_>/R_<\right)f\right]^{-1}}\right\}\right]} \tag{43.17}$$

Defining Terms

Conductance: The inverse of resistance.

Conductivity: The inverse of resistivity.

Contact resistance: The resistance between the surface of a material and the electric contact made to the surface.

Four-point technique: A method for measuring the resistivity of a material, using four electric contacts to the material, which avoids many contact resistance problems.

Resistance: The physical property of a particular piece of a material, quantifying the ease with which electricity can flow through it. Resistance will depend on the size and shape of the piece of material.

Resistivity: The intrinsic physical property of a material quantifying the ease with which electricity can flow through it. Resistivity will not depend on the size and shape of the piece of material. Higher resistivity means the flow of electricity is more difficult.

Sheet resistance: The resistance of a square thin film or sheet of material.

Two-point technique: A method for measuring the resistivity of a material, using two electric contacts to the material.

Van der Pauw technique: A method of measuring the four-point resistivity of an arbitrarily shaped material sample.

Acknowledgments

I thank Alison Breeze, John Clark, Kirsten R. Daehler, James M. E. Harper, Linda D. B. Kiss, Heidi Pan, and Shukri Souri for many useful suggestions.

References

1. P. Drude, Zur elektronentheorie der metalle, *Annalen der Physik,* 1, 566–613, 1900; 3, 369–402, 1900. See Ref. [2] for modern discussions of the Drude model and electrical conductivity.

2. N. W. Ashcroft and N. D. Mermin, *Solid State Physics,* Philadelphia, PA: Saunders College, 1976; C. Kittel, *Introduction to Solid State Physics,* 7th ed., New York: John Wiley & Sons, 1996.

3. L. B. Valdes, Resistivity measurements on germanium for transistors, *Proc. I.R.E.*, 42, 420–427, 1954.
4. L. J. van der Pauw, A method of measuring specific resistivity and Hall effect of discs of arbitrary shape, *Philips Res. Rep.*, 13, 1–9, 1958.
5. R. Chwang, B. J. Smith, and C. R. Crowell, Contact size effects on the van der Pauw method for resistivity and Hall coefficient measurement, *Solid-State Electron.*, 17, 1217–1227, 1974.

Further Information

H. H. Wieder, *Laboratory Notes on Electrical and Galvanomagnetic Measurements*, New York: Elsevier, 1979.

L. I. Maissel, Electrical properties of metallic thin films, 13-1 to 13-33, in L. I. Maissel and R. Glang (eds.), *Handbook of Thin Film Technology*, San Francisco: McGraw-Hill, 1970.

D. C. Look, *Electrical Characterization of GaAs Materials and Devices*, New York: John Wiley & Sons, 1989.

D. C. Look, Bulk and contact electrical properties by the magneto-transmission-line method: application to GaAs, *Solid-State Electron.*, 30, 615–618, 1987.

44

Charge Measurement

Saps Buchman
Stanford University

John Mester
Stanford University

T. J. Sumner
Imperial College

Electric charge, a basic property of elementary particles, is defined by convention as negative for the electron and positive for the proton. In 1910, Robert Andrews Millikan (1868–1953) demonstrated the quantization and determined the value of the elementary charge by measuring the motion of small charged droplets in an adjustable electric field. The SI unit of charge, the *coulomb* (*C*), is defined in terms of base SI units as:

$$1 \text{ coulomb} = 1 \text{ ampere} \times 1 \text{ second} \qquad (44.1a)$$

In terms of fundamental physical constants, the coulomb is measured in units of the elementary charge *e*:

$$1\,C = 1.60217733 \times 10^{19}\,e \qquad (44.1b)$$

where the relative uncertainty in the value of the elementary charge is 0.30 ppm [1].

   Charge measurement is widely used in electronics, physics, radiology, and light and particle detection, as well as in technologies involving charged particles or droplets (as for example, toners used in copiers). Measuring charge is also the method of choice for determining the average value for small and/or noisy electric currents by utilizing time integration. The two standard classes of charge-measuring devices are the electrostatic voltmeters and the charge amplifiers.

   Electrostatic instruments function by measuring the mechanical displacement caused by the deflecting torques produced by electric fields on charged conductors [2,3]. Electrostatic voltmeters also serve as charge-measurement devices, using the fact that charge is a function of voltage and instrument capacitance. This class of instruments can be optimized for a very wide range of measurements, from about 100 V to 100 kV full-scale, with custom devices capable of measuring voltages in excess of 200 kV. The accuracy of electrostatic voltmeters is about 1% of full scale, with typical time constants of about 3 s. Their insulation resistance is between $10^{10}\,\Omega$ and $10^{15}\,\Omega$, with instrument capacitances in the range of 1 pF to 500 pF. Figure 44.1 gives a schematic representation of several types of electrostatic voltmeters.

   Modern electronic instruments have replaced in great measure the *electrostatic voltmeters* as devices of choice for the measurement of charge. The charge amplifier is used for the measurement of charge or charge variation [4]. Figure 44.2 shows the basic configuration of the charge amplifier. The equality of charges on C_1 and C_f results in:

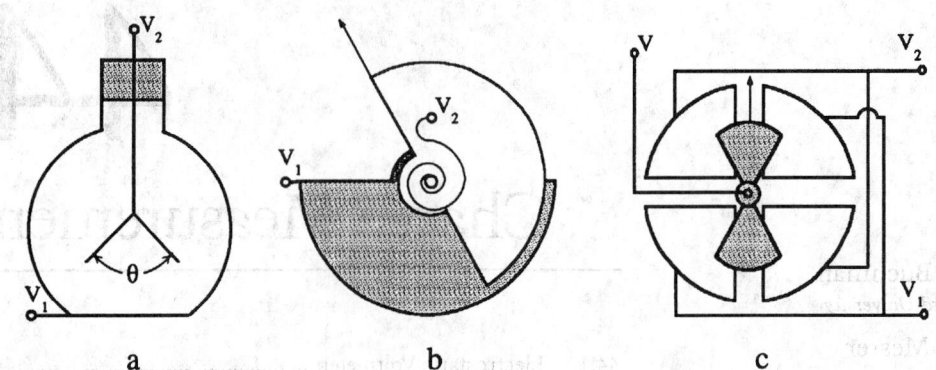

FIGURE 44.1 Examples of the repulsion, attraction, and symmetrical mechanical configurations of electrostatic voltmeters: (a) gold-leaf electroscope, (b) schematic representation of an attraction electrostatic voltmeter, (c) a symmetric quadrant electrostatic voltmeter [2].

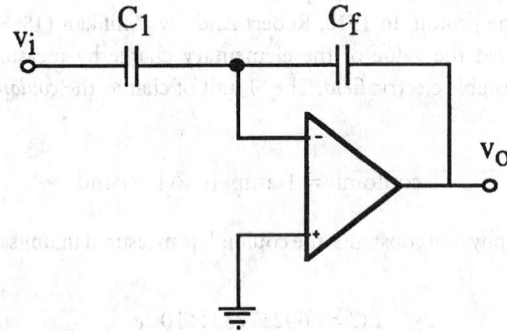

FIGURE 44.2 Basic concept of the charge amplifier. The output voltage is $v_0 = C_1/C_f \times v_i$.

$$v_0 = \frac{C_1}{C_f} v_i \quad \text{or} \quad \Delta v_0 = \frac{C_1}{C_f} \Delta v_i \qquad (44.2)$$

This same measurement principle is realized in the *electrometer*. The charge, Q, to be measured is transferred to the capacitor, C, and the value, V, of the voltage across the capacitor is measured: $Q = CV$. Figure 44.3 shows the block diagram for the typical digital electrometer [5]. Charge is measured in the coulomb mode, in which a capacitor C_f is connected across the operational amplifier, resulting in the input capacitance AC_f. Typical gain A for these amplifiers is in the range 10^4 to 10^6, making AC_f very large, and thus eliminating the problem of complete charge transfer to the input capacitor of the coulombmeter. Electrometers have input resistances in the range 10^{14} Ω to 10^{16} Ω, resulting in very long time constants, and thus minimizing the discharging of the capacitor. Typical leakage currents are 5×10^{-14} A to 5×10^{-16} A, again minimizing the variation in the charge. In the coulombmeter mode, electrometers can measure charges as low as 10^{-15} C and currents as low as 10^{-17} A.

Errors in charge-measuring instruments are caused by extraneous currents [5]. These currents are generated as thermal noise in the shunt resistance, by resistive leakage, and by triboelectric, piezoelectric, pyroelectric, electrochemical, and dielectric absorption effects. The coulombmeter function of the electrometers does not use internal resistors, thus eliminating this thermal noise source. Triboelectric charges due to friction between conductors and insulators can be minimized by using low-noise triaxial cables, and by reducing mechanical vibrations in the instrument. Under mechanical stress, certain insulators

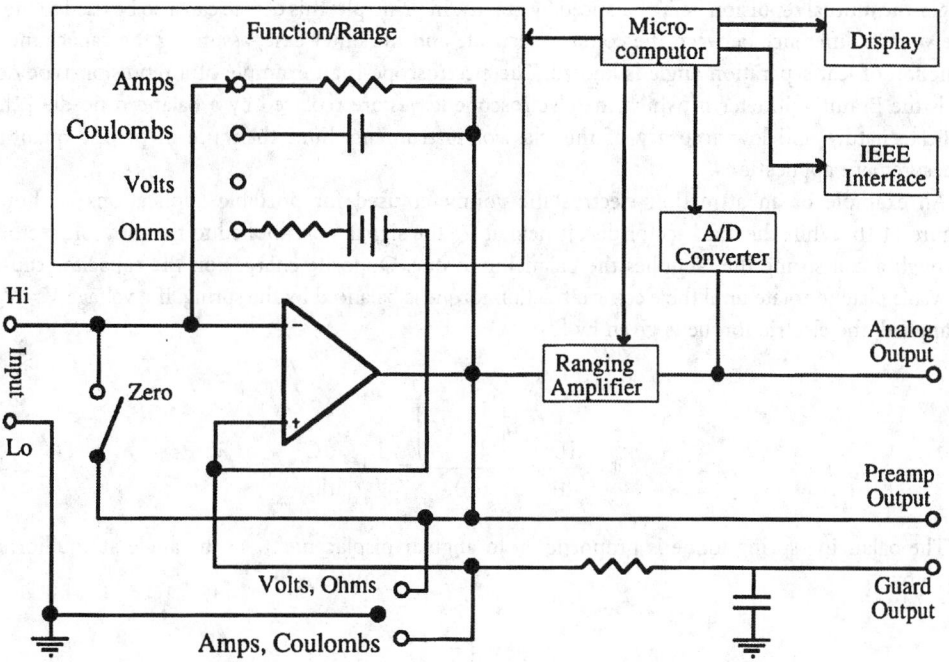

FIGURE 44.3 Conceptual block diagram of the digital electrometer. In the coulombs function, the charge to be determined is transferred to the corresponding capacitor, and the voltage across this capacitor is measured.

will generate electric charge due to piezoelectric effects. Judicious choices of materials and reduction of stress and mechanical motion can significantly reduce this effect.

Trace chemicals in the circuitry can give rise to electrochemical currents. It is therefore important to thoroughly clean and dry chemicals of all sensitive circuitry. Variations in voltages applied across insulators cause the separation and recombination of charges, and thus give rise to dielectric absorption parasitic currents. The solution is to limit the voltages applied to insulators used for high-sensitivity charge measurements to less than about 5 V.

Dielectric materials used in sensitive charge-measurement experiments should be selected for their high resistivity (low resistive leakage), low water absorptivity, and minimal piezoelectric, pyroelectric, triboelectric, and dielectric absorption effects. Sapphire and polyethylene are two examples of suitable materials. *Guarding* is used to minimize both shunt currents and errors associated with the capacitance of cables and connectors. The block diagram in Figure 44.3 shows typical guarding arrangements for modern electrometers.

44.1 Electrostatic Voltmeters

Electrostatic voltmeters and the more sensitive mechanical electrometers use an indicator to readout the position of a variable capacitor. Depending on their mechanical configuration, the electrostatic voltmeters can be categorized into three types: repulsion, attraction, and symmetrical [2, 3]. The moving system in the high-sensitivity instruments is suspended from a torsion filament, or pivoted in precision bearings to increase ruggedness. A wide variety of arrangements is used for the capacitive elements, including parallel plates, concentric cylinders, hinged plates, and others. Motion damping of the moving parts is provided by air or liquid damping vanes or by eddy current damping.

One of the oldest devices used to measure charge is the *gold leaf electroscope*, shown in Figure 44.1a. Thin leaves of gold are suspended from a conductive contact that leads out of a protective case through an insulator. As charge applied to the contact is transferred to the leaves, the leaves separate by a certain angle, the mutual repulsion being balanced by gravity. In principle, this device can also be used to measure the voltage difference between the contact electrode and the outer case, assuming the capacitance as a function of leaf separation angle is known. The electroscope is an example of a repulsion-type device, as is the Braun voltmeter in which the electroscope leaves are replaced by a balanced needle [2]. The delicate nature and low accuracy of this class of instruments limit their use in precise quantitative measurement applications.

An example of an attraction electrostatic voltmeter used for portable applications is shown in Figure 44.1b. While the fixed sector disk is held at V_1, the signal V_2 is applied to the movable sector disk through a coil spring that supplies the balancing torque. Opposite charges on the capacitor cause the movable plate to rotate until the electric attraction torque is balanced by the spring. If a voltage $V = V_1 - V_2$ is applied, the electric torque is given by [3]:

$$\tau = \frac{dU}{d\theta} = \frac{d\left(\frac{1}{2}CV^2\right)}{d\theta} = \frac{1}{2}V^2\frac{dC}{d\theta} \tag{44.3}$$

The balancing spring toque is proportional to angular displacement, so the angle at equilibrium is given by:

$$\frac{1}{2}V^2\frac{dC}{d\theta} = K\theta \tag{44.4}$$

Since the rotation is proportional to V^2, such an instrument can be used to measure ac voltages as well.

Symmetrical instruments are used for high-sensitivity, low-voltage measurements. The voltage is applied to a mobile element positioned between a symmetrical arrangement of positive and negative electrodes. Common mode displacement errors are thus reduced, and the measurement accuracy increased. One of the first devices sensitive enough to be called an "electrometer," was the quadrant electrometer shown schematically in Figure 44.1c. As a voltage difference, $V_1 - V_2$, is applied across the quadrant pairs, the indicator is attracted toward one pair and repelled by the other. The indicator is suspended by a wire allowing the stiffness of the suspension to be controlled by the potential V, so that the displacement is given by [2]:

$$\theta = K\left[\left(V_1 - V_2\right)\left(V - \frac{1}{2}\left(V_1 - V_2\right)\right)\right] \tag{44.5}$$

where K is the unloaded spring constant of the suspension.

The advantage of electrostatic instruments is that the only currents they draw at dc are the leakage current and the current needed to charge up the capacitive elements. High-performance symmetrical electrostatic instruments have leakage resistances in excess of 10^{16} Ω, sensitivities of better than 10 μV, and capacitances of 10 pF to 100 pF. They are capable of measuring charges as small as 10^{-16} C, and are sensitive to charge variations of 10^{-19} C.

Historically, as stated above, the symmetrical electrostatic voltmeters have been called "electrometers." Note that this can give rise to some confusion, as the term *electrometer* is presently also used for the electronic electrometer. This is a high-performance dc multimeter with special input characteristics and high sensitivity, capable of measuring voltage, current, resistance, and charge.

Modern noncontacting electrostatic voltmeters have been designed for voltage measurements up to the 100-kV range. An advantage of these instruments is that no physical or electric contact is required between the instrument and test surface, ensuring that no charge transfer takes place. Kawamura, Sakamoto, and

TABLE 44.1 Instruments Used in Charge Measurement Applications

Instrument manufacturer	Model #	Description	Approximate price
Advantest	TR8652	Electrometer	$2500.00
	R8340/8340A	Electrometer	$5400.00
	R8240	Digital electrometer	$2300.00
	TR8601	Micro current meter	$3500.00
	TR8641	Pico ammeter	$2500.00
Amptek	A101	Charge preamplifier	$300.00
	A111	Charge preamplifier	$375.00
	A203	Charge preamplifier/shaper	$300.00
	A225	Charge preamplifier/shaper	$395.00
	A250	Charge preamplifier	$420.00
EIS	ESH1-33	Electrostatic voltmeter	$1650.00–$5890.00[a]
	ESD1-11	Electrostatic voltmeter	$1465.00–$1740.00[a]
	CRV	Electrostatic peak voltmeter	$2100.00
Jennings	J-1005	RF kilovoltmeter	$5266.00
Keithley	610C	Electrometer	$4990.00
	614	Digital electrometer	$2490.00
	617	Programmable electrometer	$4690.00
	642	Digital electrometer	$9990.00
	6512	Electrometer	$2995.00
	6517	Electrometer	$4690.00
Kistler	5011B	Charge amplifier	$2700.00
	5995	Charge amplifier	$1095.00
	5395A	Charge calibrator	$11655.00
Monroe	168-3	Electrostatic voltmeter	$4975.00
	174-1	Electrostatic voltmeter	$5395.00
	244AL	Electrostatic millivoltmeter	$3695.00
	253-1	Nanocoulomb meter/Faraday cup	$1765.00
Nuclear Associates	37-720FW	Digital electrometer for dosimetry	$1234.00
Trek	320B	Electrostatic voltmeter	$1930.00
	341	Electrostatic voltmeter	$6900.00
	344	Electrostatic voltmeter	$2070.00
	362A	Electrostatic voltmeter	$2615.00
	368	Electrostatic voltmeter	$2440.00–$9160.00[a]

[a] Available in a range of specifications.

Noto [6] report the design of an attraction-type device that uses a strain gage to determine the displacement of a movable plate electrode. Hsu and Muller [7] have constructed a micromechanical shutter to modulate the capacitance between the detector electrode and the potential surface to be measured. Trek Inc. [8] electrostatic voltmeters achieve a modulated capacitance to the test surface by electromechanically vibrating the detector electrode. Horenstein [9], Gunter [10], and MacDonald and Fallone [11] have employed noncontacting electrostatic voltmeters to determine the charge distributions on semiconductor and insulator surfaces. Tables 44.1 and 44.2 contain a selection of available commercial devices and manufacturers.

44.2 Charge Amplifiers

The conversion of a charge, Q, into a measurement voltage involves at some stage the transfer of that charge onto a reference capacitor, C_r. The voltage, V_r, developed across the capacitor gives a measure of the charge as $Q = V_r / C_r$. There are two basic amplifier configurations for carrying out such measurements using the reference capacitor in either a *shunt* or *feedback* arrangement.

TABLE 44.2   Instrument Manufacturers

Advantest Corporation	Kistler Intrumente AG
Shinjuku-NS Building 4-1	CH-8408 Winterthur, Switzerland
Nishi-Shinjuku 2-Chome, Shinjuku-ku	
Tokyo 163-08 Japan	Monroe Electronics
	100 Housel Ave.
Amptek Inc.	Lyndonville, NY 14098
6 De Angelo Drive	
Bedford, MA 01730	Nuclear Associates
	Div. of Victoreen, Inc.
Electrical Instrument Service Inc.	100 Voice Rd.
Sensitive Research Instruments	P.O. Box 349
25 Dock St.	Carle Place, NY 11514-0349
Mount Vernon, NY 10550	
	Trek Incorporated
Jennings Technology Corporation	3922 Salt Works Rd.
970 McLaughlin Ave.	P.O. Box 728
San Jose, CA 95122	Medina, NY 14103
Keithley Instruments Inc.	
28775 Aurora Road	
Cleveland, OH 44139	

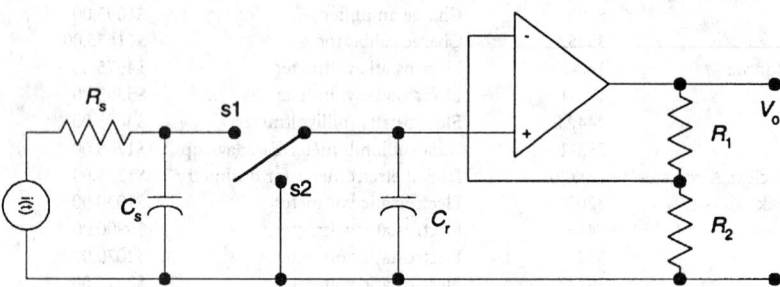

FIGURE 44.4   Schematic representation of a charge amplifier using a shunt reference capacitor. With the switch in position s1, the measurement circuit is connected and the charge is proportional to the output voltage and to the sum $C_s + C_r$. Note the significant sensitivity to C_s.

Shunt Amplifiers

Figure 44.4 shows a typical circuit in which the reference capacitor is used in a shunt mode. In this example, it is assumed that the charge that is to be measured is the result of the integrated current delivered by a time-dependent current source, $i(t)$. With the measurement circuit disconnected (switch in position s2), the charge on the source capacitor, C_s, at time τ will be $Q = \int_0^c i(t)dt$ (assuming Q starts from zero at $t = 0$) and the output voltage, V_o will be zero, as the input voltage to the (ideal) operational amplifier is zero. On closing the switch in position s1, the charge on C_s will then be shared between it and C_r and:

$$V_0 = \left(\frac{R_1 + R_2}{R_2} \right) \frac{Q}{C_s + C_r} \qquad (44.6)$$

In order to accurately relate the output voltage to the charge Q not only does the gain of the noninverting amplifier and the reference capacitance need to be known, which is relatively straightforward,

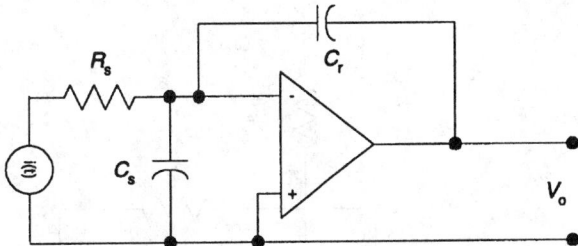

FIGURE 44.5 Schematic representation of a charge amplifier with reference feedback capacitor. The charge is proportional to the output voltage and to the sum $C_s + AC_r$, where A is the amplifier gain. Note the reduced sensitivity to C_s.

but it is also necessary to know the source capacitance. This is not always easy to determine. The effect of any uncertainty in the value of C_s can be reduced by increasing the value of the reference capacitor to the point where it dominates the total capacitance. However, in so doing, the output voltage is also reduced and the measurement becomes more difficult. The dependence of the measurement on C_s is one of the main limitations to this simple method of charge measurement. In addition, any leakage currents into the input of the operational amplifier, through the capacitors, or back into the source circuitry during the measurement period will affect the result. For the most accurate measurements of low charge levels, *feedback amplifiers* are more commonly used.

Feedback Amplifiers

Figure 44.5 shows a circuit where the reference capacitor now provides the feedback path around the operational amplifier. The output voltage from this configuration for a given charge Q transfer from the source is then:

$$V_0 = \frac{AQ}{C_S + AC_r} \qquad (44.7)$$

where A is the open-loop gain of the operational amplifier. For most situations, $AC_r > C_s$ and the charge measurement becomes independent of the source capacitance. In addition, the inverting input to the operational amplifier is kept close to ground potential, reducing the magnitude of leakage currents in that part of the circuit. However, in contrast to these two benefits is the new problem that the input bias current for the operational amplifier is integrated by the feedback capacitor, producing a continual drift in the output voltage. Several solutions have been used to overcome this problem, including the use of a parallel feedback resistor, R_f, which suppresses the integrating behavior at low frequencies (periods longer than $R_f C_r$), balancing the bias current with another externally provided current, and incorporating a reset switch that discharges the circuit each time the output voltage ramps beyond a set trigger level.

The sensitivity of feedback amplifiers depends on the noise sources operating within any specific application. The most impressive performance is obtained by amplifiers integrated into CCD chips (charge coupled devices) that can, under the right operational conditions, provide sensitivities measured in terms of a few electron charges. To illustrate the important parameters involved in the design of ultra-low noise charge preamplifiers for CCD-type applications, consider the circuit shown in Figure 44.6. The source (detector) is now shown as a biased photodiode, which is assumed to be producing individual bursts of charge each time a photon (an X-ray, for example) interacts in it. In this example, the photodiode is coupled to the amplifier using a large value capacitor, C_c. This blocks the direct current path from the diode bias supply, V_b, but provides a low impedance path for the short-duration charge deposits. The preamplifier is a variant on that shown in Figure 44.5, in which there is now a parallel feedback resistor to provide baseline restoration on long time scales and an FET transistor to reduce the effect of the operational amplifier input bias current by virtue of its high current gain factor, β.

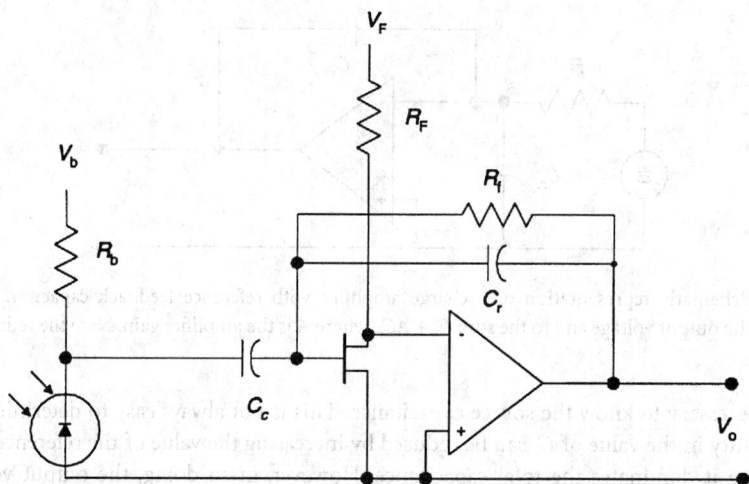

FIGURE 44.6 Typical ultralow noise charge preamplifier configuration for charge pulse readout from ionization type radiation detectors (e.g., X-ray detection using photodiodes or CCDs). The large capacitor C_c provides a low impedance path for the short charge deposit pulses, while the parallel feedback resistor provides baseline restoration on long time scales. An FET transistor reduces the effect of the operational amplifier input bias current.

In practice, the dominant noise contributions in most applications of this type come from Johnson (current) noise in the bias and feedback resistors, shot noise on the photodiode bias current, voltage noise across the FET, and finally the inevitable $1/f$ component. The two resistors effectively feed thermal current noise into the input of the integrator. Similarly, the shot noise associated with the photodiode bias current feeds into the input. Together, these components are known as *parallel* noise and the total parallel noise charge is given by:

$$q_p = \sqrt{\left(\frac{4kT}{R_f + R_b} + 2eI_b(T)\right)\frac{1}{\Delta B}} \tag{44.8}$$

where k is Boltzmann's constant, e is the charge on the electron, T is absolute temperature, and ΔB is the bandwidth associated with the measurement that will depend on the details of subsequent shaping amplifier stages [12]. Voltage noise across the FET (and hence operational amplifier inputs) will arise from junction noise in the FET itself and from Johnson noise in its bias resistor, R_F. In practice, the FET junction noise usually dominates, in which case this *series* noise contribution is given by:

$$q_s = \sqrt{\varepsilon_n^2 C_{in}^2 \Delta B} \tag{44.9}$$

where ε_n is the junction voltage noise for the FET in $V\sqrt{Hz^{-1}}$ and C_{in} is the total capacitance seen at the gate of the FET. This will include both the source capacitance, the gate capacitance of the FET, and any stray capacitance. The total noise is then the quadrature sum of Equations 44.8 and 44.9. The different dependencies on the bandwidth for Equations 44.8 and 44.9 imply there will be some optimum bandwidth for the measurement and this will depend on the relative contributions from each. $1/f$ noise manifests itself as a bandwidth-independent term that again, must be added in quadrature. The Johnson noise associated with the resistors and FET junction will show a temperature dependence decreasing with $\sqrt{T}$. For the FET, this reduction does not continue indefinitely and there is usually an optimum temperature for the FET around 100 K. Photodiode bias currents also fall with decreasing temperature and, for silicon devices, this is about a factor of 2 for every 10-K drop in temperature. Most ultralow noise applications thus operate at reduced temperature, at least for the sensitive components. Bias resistors and feedback resistors are kept as high as possible (typically > 100 MΩ) and FETs are specially selected for low junction voltage noise (typically 1 nV$\sqrt{\text{Hz}^{-1}}$). Ideally, photodiode capacitances should be kept as low as possible and there is also an interplay between the FET junction noise, ε_n, and the FET gate capacitance that is affected by altering the FET bias current, which can be used to fine-tune the series noise component.

Finally, there is another noise component that can often be critical and difficult to deal with. This is from microphonics. There are two effects. First, the feedback reference capacitor is typically made as small as possible (<1 pF) to reduce the effect of noise in the following shaping amplifier stages. This makes it sensitive to any stray capacitances and, if there are vibrations in the system that alter the local geometry, then this can change the feedback capacitance that changes the "gain" of the preamplifier. Second, the photodiode will be operating with some applied bias voltage (often several tens of volts) and any change in its apparent capacitance through mechanical movement of components will result in charge being moved around. These charge movements will be sensed by the charge amplifier. Ultralow noise applications that ignore mechanical stability in their design phase are in peril. Tables 44.1 and 44.2 contain a selection of available commercial devices and manufacturers.

44.3 Applications

The Millikan technique of measuring charges on particles suspended in electric fields continues to be developed for various applications. Kutsuwada et al. [13] use a modified Millikan experiment to measure the charge distribution on electrophotographic toner particles of various sizes. They show a comparison of the results obtained using an ordinary Millikan apparatus, and a modified system in which an additional ac electrode has been inserted in the hyperbolic quadrupole electrode assembly. The two methods agree to within a small multiplicative calibration factor. A different method of measuring the charge of toner particles uses the q/d meter [14] in which q and d refer, respectively, to the charge and the diameters of the particles. In the q/d meter, the charged particles are transported horizontally in a steady laminar air flow, and move vertically in an electric field until deposited on a registration electrode. The position at which the particle is deposited on the registration electrode defines the charge-to-diameter ratio. The size of the deposited particle is then measured, thus completing the determination of the charge distribution for various particle sizes.

The experiments searching for fractional charges [15] make use of superconducting niobium spheres 0.25 mm in diameter, suspended in vacuum at 4.2 K in a magnetic field. The vertical position of the spheres is modulated by an alternating electric field and measured with an ultra-sensitive magnetometer. Positrons and electrons generated by radioactive sources are used to cancel all integer charges on the spheres. Fractional charges are detected and measured as that residual charging of the niobium spheres that cannot be neutralized by the integral charges from the radioactive sources. Although this experiment is sensitive to about 0.01 electron charges (10^{-21} C), it has produced no conclusive evidence of fractional charges.

A similar approach, using force modulation, is used for the noncontact measurement of charge on gyroscopes [16]. Out-of-phase equal forces are applied to an electrostatically suspended gyroscope at a frequency well within the suspension control bandwidth. The charge of the gyroscope is then proportional to modulation frequency component of the suspension control effort. The sensitivity of this method is about 10^{-12} C gyroscope charge, limited by the allowable modulation force and position sensor noise.

Noncontact measurement of charge on liquid drops in a microgravity environment can be also performed using field mill instruments [17]. A mechanical chopper is used to modulate the electric field induced by the spherical charge on a grounded sensing plate. The resulting alternating current from sensor to ground is a measure of the charge on the drop. In a refinement of this method, the modulation is achieved by varying the distance between charge and sensor. The authors [18] claim that this system provides an increase in sensitivity of 2 to 3 orders of magnitude over the original technique.

Optical sensors based on the Pockels effect [19] are used to measure the space-charge field in gaseous dielectrics. The Pockels effect involves the change in the birefringence of certain crystalline materials on the application of an electric field. In the measuring system, a circularly polarized beam is detected after passing through the Pockels sensor and a polarizing plate. The detected intensity varies linearly with the intensity of the electric field applied to the sensor. This system is capable of performing a vector measurement of the electric field produced by the space charge, determining both intensity and direction.

Optical methods of charge measurement are also used for particles whose physical structure depends on their charge. An example is the degree of dissociation of the end groups of polystyrene particles in colloidal solutions [20]. In this application, the intensities of the Raman scattering spectrum lines depend on the degree of dissociation of the polystyrene end groups, and thus determine the charge of these particles.

A widely used application of charge measurement is as an integral element of radiation dosimetry. Radiation is passed through ionization chambers, where the generated ions are collected and the charge measured with electrometers. Ionization chambers used in dosimetry for radiation therapy have typical sensitivities in the range 0.02 nC R^{-1} to 0.2 nC R^{-1}. Coulombmeter electrometers with sensitivities of 1 pC to 10 pC are therefore required for this application.

Defining Terms

Charge, also **Electric charge:** A basic property of elementary particles, defined by convention as negative for the electron and positive for the proton. The SI unit of charge is the coulomb (**C**), defined as 1 ampere × 1 second.

Electrostatic instrument: An instrument that functions by measuring the mechanical displacement or strain caused by electric fields.

Electrostatic voltmeter: Electrostatic instrument used to measure charge. The charge is determined as a function of voltage and instrument capacitance.

Charge amplifier: Charge-measuring instrument. The charge is transferred to a reference capacitor and the resulting voltage across the capacitor is measured. Shunt and feedback versions of the charge amplifier have the reference capacitor used in shunt and feedback mode, respectively.

Electrometer: *Historic usage*, type of electrostatic voltmeter. *Modern usage*, electronic electrometer, a high-performance dc multimeter with special input characteristics and high sensitivity, capable of measuring voltage, current, resistance, and charge.

References

1. E. Cohen and B. Taylor, The 1986 Adjustment of the Fundamental Physical Constants, *Rev. Modern Phys.*, 59(4), 1121-1148, 1987.
2. F. Harris, *Electrical Measurements*, New York: John Wiley & Sons, 1952.
3. W. Michels, *Electrical Measurements and Their Applications*, Princeton, NJ: D. Van Nostrand, 1969.
4. A. J. Diefenderfer, *Principles of Electronic Instrumentation*, 3rd ed., Philadelphia: Saunders College Publishing, 1994.
5. Keithley Instruments, Inc., *Low Level Measurements*, 4th ed., Cleveland: Keithley Instruments, Inc., 1993.
6. K. Kawamura, S. Sakamoto, and F. Noto, Design and Development of New Electrostatic Voltmeter Using Strain Gauge, *IEEE Trans. Ind. Appl.*, 25, 563-568, 1989.
7. C. Hsu and R. Muller, Micromechanical Electrostatic Voltmeter, *Transducers '91. Int. Conf. Solid-State Sensors Actuators*, 1991, 659-662.
8. Trek Incorporated, 3922 Salt Works Rd., P.O. Box 728, Medina, NY 14103.
9. M. M. Horenstein, Measuring surface charge with a noncontacting voltmeter, *Proc. IEEE Industry Appl. Soc. Annu. Meeting*, Toronto, Ontario, Canada IAS'93, 3, 1811-1816, 1993.

10. P. Gunther, Determination of charge density and charge centroid location in electrets with semi-conducting substrates, *IEEE Trans. Electrical Insul.*, 27, 698-701, 1992.

11. B. MacDonald and B. Fallone, Surface-charge distributions on radiation-charged electrets. *7th Int. Symp. Electrets* (ISE 7). Berlin, Germany, 1991, 798-803.

12. V. Radeka, Signal, noise and resolution in position-sensitive detectors, *20th Nuclear Science Symp. 5th Nuclear Power Syst. Symp.*, San Francisco, CA, *IEEE Trans. Nucl. Sci.*, 21, 51-64, 1974.

13. N. Kutsuwada, T. Shohdohji, N. Okada, H. Izawa, T. Sugai, Y. Nakamura, and T. Murata, Measurement of electric charge of electrophotographic toner, *J. Imaging Sci. Technol.*, 37(5), 480-484, 1993.

14. M. Mehlin and R. M. Hess, Electrical charge measurement of toner particles using the q/d meter, *J. Imaging Sci. Technol.*, 36, 142-150, 1992.

15. A. F. Hebard, G. S. LaRue, J. D. Phillips, and C. R. Fisel, *Search for Fractional Charge in Near Zero: New Frontiers of Physics*, New York: W. H. Freeman and Company, 1988, 511.

16. Saps Buchman, T. Quinn, G. M. Keiser, Dale Gill, and T. J. Sumner, Charge measurement and control for the Gravity Probe B gyroscopes, *Rev. Sci. Instrum.*, 66, 120-129, 1995.

17. M. N. Horenstein, Peak sampled vibrating-reed for the measurement of electric fields in the presence of space charge, *Rev. Sci. Instrum.*, 54, 591-593, 1983.

18. Kuan-Chan Lin and Taylor G. Wang, Noncontact charge measurement, *Rev. Sci. Instrum.*, 63, 2040-2043, 1992.

19. Kunihido Hidaka, Progress in Japan of space charge field measurement in gaseous dielectrics using a Pockels sensor, *IEEE Electrical Insul. Mag.*, 12(1), 17-28, 1996.

20. R. Kesavamoorthy, T Sakuntala, and Akhilesh K. Arora, *In situ* measurement of charge on polystyrene particles in colloidal suspension, *Meas. Sci. Technol.*, 1(5), 440-445, 1990.

10. R. Jonckheer, Determination of charge density and charge centroid location in electrets with semi-conducting substrates, *IEEE Trans. Electr. Insul.*, 27, 695–701, 1992.

11. J. MacDonald and B. Fallone, Surface-charge distributions on radiation-charged electrets, *Am. Inst. Phys. Conf. Proc.* (353), Berlin, Germany, 1991, 79–302.

12. V. Radeka, Sigma, noise and resolution in position-sensitive detectors, 20th Nuclear Science Symp. on Nuclear Detectors, San Francisco, CA, *IEEE Trans. Nucl. Sci.*, 21, 51–64, 1974.

13. S. Kusuwada, J. Johnohi, N. Okada, H. Ikawa, T. Sugai, Y. Ihjimma, and T. Murata, Measurement of electric charge of electrophotographic toner, *Denppy. Soc. Zudai Zhaol.*, 37(6), 450–457, 1991.

14. M. Machlin and T. M. Hess, Electrical charge measurement of toner particles using the grid method, *J. Imaging Sci. Technol.*, 36(2), 1490–1500, 1992.

15. A. P. Hebard, Q. S. Santen, J. U. Phillips, and C. R. Fisch, *Search for Fractional Charge in Near-Zero: New Frontiers of Physics*, New York, W. H. Freeman and Company, 1988, 311.

16. Igor Blichman, T. Ouana, O. M. Kelson, Dale Gill, and T. J. Sumner, Charge measurement apparatus, control for the Gravity Probe I gyroscopes, *Rev. Sci. Instrum.*, 68, 190–176, 1995.

17. M. N. Horenstein, Peak sampled vibrating-reed for the measurement of electric fields in the presence of space charge, *Rev. Sci. Instrum.*, 54, 591–593, 1983.

18. Kuan-chun Lin and Taylor C. Wang, Noncontact charge measurement, *Rev. Sci. Instrum.*, 65, 2640–2643, 1992.

19. Kunihido Hidaka, Progress in Japan of space-charge field measurement in gaseous dielectrics using a Pockels sensor, *IEEE Electrical Insul. Mag.*, 12(1), 19–38, 1996.

20. R. Kesavamoorthy, T. Sivkumar, and A. V. Vrsh, R. A. ota, In situ measurement of charge or polystyrene particles in colloidal suspension, *Meas. Sci. Technol.*, 3(5), 490–495, 1930.

45

Capacitance and Capacitance Measurements

Halit Eren
Curtin University of Technology

James Goh
Curtin University of Technology

A capacitor is a system of two conducting electrodes, having equal and opposite charges separated by a dielectric. The capacitance C of this system is equal to the ratio of the absolute value of the charge Q to the absolute value of the voltage between bodies as:

$$C = Q/V \qquad (45.1)$$

where C = Capacitance in farad (F)
Q = Charge in coulomb (C)
V = Voltage (V)

The unit of capacitance, the farad, is a large unit; practical capacitors have capacitances in mirofarads (μF or 10^{-6} F), nanofarads (nF or 10^{-9} F), and picofarads (pF or 10^{-12} F). The unit conversions are shown in Table 45.1.

TABLE 45.1 Capacitance Unit Conversions

microfarads (μF)	nanofarads (nF)	picofarads (pF)
10^{-6} F	10^{-9} F	10^{-12} F
0.000001 μF	0.001 nF	1.0 pF
0.001 μF	1.0 nF	1000 pF
1.0 μF	1000 nF	1,000,000 pF

The capacitance C depends on the size and shape of charged bodies and their relative positions; examples are shown in Table 45.2. Generally, capacitance is inherent wherever an electrostatic field appears. In many electronic systems, it is necessary to deal with capacitances that are designed within the circuits and also with unwanted interference and stray capacitances that are introduced externally or internally at various stages of the circuits. For example, some sensors operate on capacitive principles, giving useful signals; in others, capacitance is inherent but undesirable. In many cases, cables and external

TABLE 45.2

Capacitances of various electrode systems

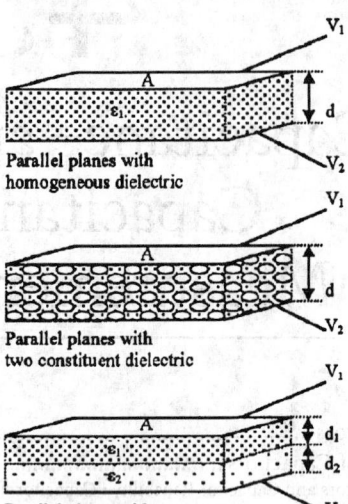

Parallel planes with
homogeneous dielectric

$$C = \frac{\varepsilon A}{d}$$

Parallel planes with
two constituent dielectric

$$C = \frac{(\varepsilon_1 r + \varepsilon_2)A}{(1+r)d} \quad \text{where } r \text{ value is the volumetric ratio}$$

Parallel planes with
two different dielectrics

$$C = \frac{\varepsilon A}{\dfrac{d_1}{\varepsilon_1} + \dfrac{d_2}{\varepsilon_2}}$$

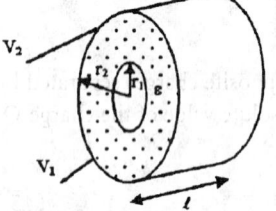

Coaxial cylinders with
single dielectric

$$C = \frac{4\pi\varepsilon\ell}{\ln(r_2 / r_1)}$$

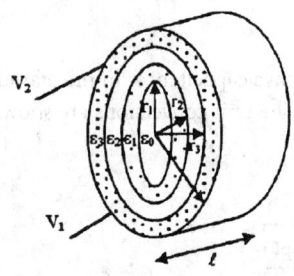

Coaxial cylinders with
three dielectrics

$$C = \frac{2\pi\varepsilon_0\ell}{\ln\left[\left(\dfrac{r_2}{r_1}\right)^{\frac{\varepsilon_0}{\varepsilon_1}} \times \left(\dfrac{r_3}{r_2}\right)^{\frac{\varepsilon_0}{\varepsilon_2}} \times \left(\dfrac{r_4}{r_3}\right)^{\frac{\varepsilon_0}{\varepsilon_3}}\right]}$$

$$= \left(\frac{1}{C_1} + \frac{1}{C_2} + \frac{1}{C_3}\right)^{-1}$$

Cylinder parallel with plane

$$C = \frac{\pi\varepsilon\ell}{\ln\left[\dfrac{d}{r} + \sqrt{\left(\dfrac{d}{r}\right)^2 - 1}\right]} = \frac{2\pi\varepsilon\ell}{\text{Cosh}^{-1}\left(\dfrac{d}{r}\right)}$$

TABLE 45.2 (continued)

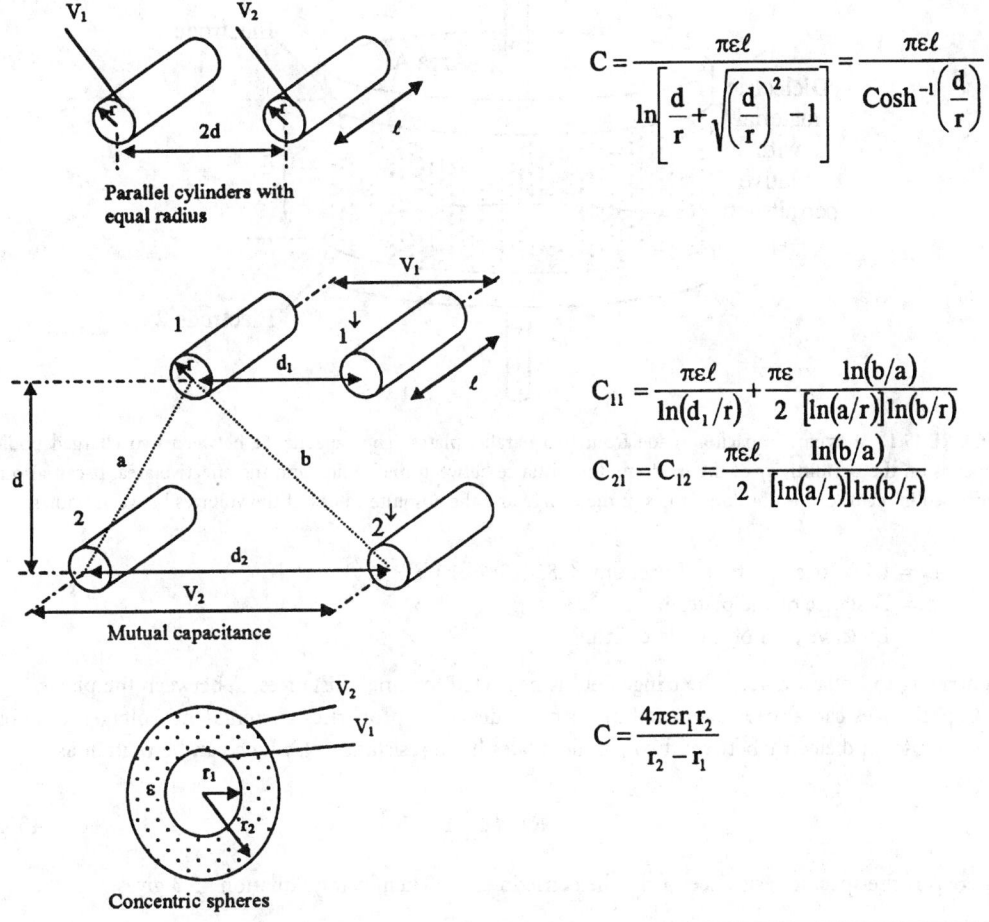

$$C = \frac{\pi \varepsilon \ell}{\ln\left[\dfrac{d}{r} + \sqrt{\left(\dfrac{d}{r}\right)^2 - 1}\right]} = \frac{\pi \varepsilon \ell}{\text{Cosh}^{-1}\left(\dfrac{d}{r}\right)}$$

Parallel cylinders with equal radius

$$C_{11} = \frac{\pi \varepsilon \ell}{\ln(d_1/r)} + \frac{\pi \varepsilon}{2} \frac{\ln(b/a)}{[\ln(a/r)]\ln(b/r)}$$

$$C_{21} = C_{12} = \frac{\pi \varepsilon \ell}{2} \frac{\ln(b/a)}{[\ln(a/r)]\ln(b/r)}$$

Mutual capacitance

$$C = \frac{4\pi \varepsilon r_1 r_2}{r_2 - r_1}$$

Concentric spheres

circuits introduce additional capacitances that need to be accounted for the desirable operation of the system. In these cases, Table 45.2 is useful to identify and analyze possible sources of capacitances where charged bodies are involved.

In general, the capacitance can be determined by solving Laplace's equations $\nabla^2 V(x,y,z) = 0$ with appropriate boundary conditions. One type of boundary condition specifies the electrode voltages V_1 and V_2 of the plates. Laplace's equation yields to V and the electric field $E(x,y,z) = -\nabla V(x,y,z)$ between the electrodes. The charge of each electrode can also be obtained by integration of the flux density over each electrode surface as:

$$Q = \int \varepsilon(x, y, z) \, E(x, y, z) \, dA \tag{45.2}$$

If the capacitor is made from two parallel plates, as shown in Figure 45.1, the capacitance value in terms of dimensions can be expressed by:

$$C = \varepsilon \, A/d = \varepsilon_r \varepsilon_0 \, A/d \tag{45.3}$$

where ε = Dielectric constant or permittivity
ε_r = Relative dielectric constant (in air, $\varepsilon_r = 1$)

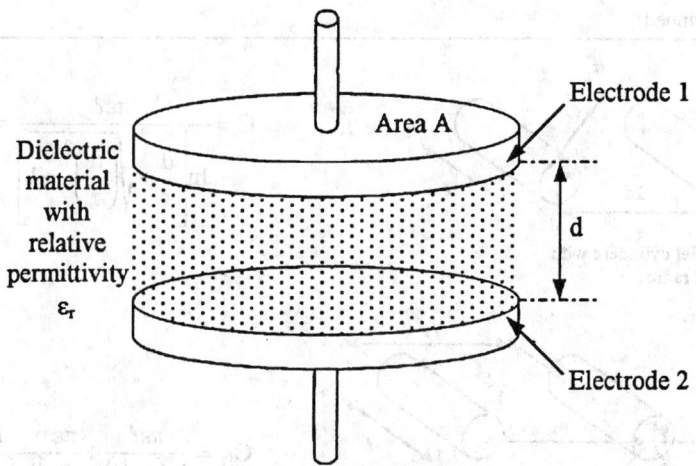

FIGURE 45.1   A typical capacitor made from two parallel plates. The capacitance between two charged bodies depends on the permittivity of the medium, the distance between the bodies, and the effective area. It can also be expressed in terms of the absolute values of the charge and the absolute values of the voltages between bodies.

ε_0 = Dielectric constant of vacuum (8.854188×10^{-12} F m^{-1})
d = Distance of the plates in m
A = Effective area of the plates in m^2

In arriving to Equation 45.3, the fringe field is neglected for small distances, d, between the plates.

Capacitances can also be expressed in terms of dielectric properties currents and voltages. Suppose that a uniform dielectric between two parallel plates has a resistance, R, which can be written as:

$$R = d\rho/A \qquad (45.4)$$

where ρ is the specific resistance of the dielectric in Ω m. Then, using Equation 45.3 gives:

$$C = \varepsilon\rho/R \qquad (45.5)$$

A voltage V across the capacitor causes a leakage current $I_l = V/R$ such that:

$$C = \varepsilon\rho\, I_l/V \qquad (45.6)$$

This indicates that the leakage current of a capacitor is proportional to its capacitance value.

As seen in Equations 45.3 and 45.4, the value of the capacitance is proportional to the permittivity of the dielectric material used. In the construction of capacitors, the permittivity of commonly used materials is given in Table 45.3.

45.1   Types of Capacitors

Commonly used fixed capacitors are constructed with air, paper, mica, polymers, and ceramic dielectric materials. A comprehensive list of common capacitors and their characteristics are given in Table 45.4(a) and Table 45.4(b) and a list of manufacturers is given in Table 45.5. Variable capacitors are generally made with air or ceramic dielectric materials. The capacitors used in electronic circuits can be classified as: low-loss, medium-loss, and high-tolerance capacitors.

TABLE 45.3 Permittivity (Dielectric Constants of Materials Used in Capacitors)

Material	Permittivity
Vacuum	1.0
Air	1.0006
Teflon	2.1
Polyethylene, etc.	2.0–3.0
Impregnated paper	4.0–6.0
Glass and mica	4.0–7.0
Ceramic (low K)	≤20.0
Ceramic (medium K)	80.0–100.0
Ceramic (high K)	≥1000.0

- *Low-loss capacitors* such as: mica, glass, low loss ceramic and low-loss plastic film capacitors generally have a good capacitance stability. These capacitors are expensive and often selected in precision applications, e.g., telecommunication filters.
- *Medium-loss capacitors* have medium stability in a wide range of ac and dc applications. These are paper, plastic film, and medium-loss ceramic capacitors. Their applications include, coupling, decoupling, bypass, energy storage, and some power electronic applications (e.g., motor starter, lighting, power line applications, and interference suppressions).
- *High-tolerance capacitors* such as aluminum and tantalum electrolytic capacitors deliver high capacitances. Although these capacitors are relatively larger in dimension, they are reliable and have longer service life. They are used in polarized voltage applications, radios, televisions and, other consumer goods, as well as military equipment and harsh industrial environments.

There are also specially designed capacitors (e.g., mica, glass, oil, gas and vacuum). These capacitors are used particularly in high-voltage (35 kV) and high-current (200 A) applications.

In the majority of cases, the manufacturing process of the capacitors begins by forming one plate using metallization of one side of a flexible dielectric film. A foil such as aluminum is used as the other plate. The film/foil combination is rolled on a suitable core with alternate layers slightly extended and then heat-treated. In some cases, two-foil layers are divided by a dielectric film or paper impregnated with oil.

Generally, capacitors are two-terminal devices with one electrode as the ground terminal. However, if both terminals are separated from the common terminal, the additional capacitances between ground and electrodes might have to be taken into account. Usually, capacitance between electrodes and ground are small compared to the dominant capacitance between plates. Three-terminal capacitances exist and are manufactured in many different ways, as illustrated in Table 45.2.

As far as construction and materials and construction techniques are concerned, the capacitors can broadly be classified as: electrolytic, ceramic, paper, polymer, mica, variable capacitors, or integrated circuit capacitors.

Paper capacitors: Usually, paper capacitors are made with thin (5- to 50-μm in thickness) wood pulp. A number of sheets are used together to eliminate possible chemical and fibrous defects that may exist in each sheet. The paper sheets are placed between thin aluminum foils and convolutely wound, as shown in Figure 45.2. The moisture of the paper is removed at high-temperature vacuum drying before the capacitor is vacuum impregnated with oil, paraffin, or wax. The losses and self-inductance are sizeable and frequency dependent. The applications are usually restricted to low frequency and high voltages. When impregnated with silicone-oil, they can withstand voltages up to 300 kV.

Electrolytic capacitors: This describes any capacitor in which the dielectric layer is formed by an electrolytic method. The electrolytic capacitors in dry foil form may be similar to construction to paper film capacitors; that is, two foil layers separated by an impregnated electrolyte paper spacer are rolled together. In this case, one of the plates is formed using metallization of one side of a flexible dielectric

TABLE 45.4a Characteristics of Common Capacitors

Capacitor types	Range (F)	Tolerance (%)	Voltage range (V)	Temperature range (°C)	Temperature Coefficient (ppm/°C)	Frequency range (Hz)	Permittivity ($\varepsilon/\varepsilon_0$ C²/Nm²)	Dielectric strength (C/C_0)	Dissipation factor (%)	Insulation resistance (MΩ/μF)	Typical average failure rates (fail per 10^6 h)
Mica, glass, porcelain and Teflon											
Mica	5p–0.01μ	5	100–600	–55/125	–50	100Hz–10GHz	7.0	1000	0.001	2.5×10^4	0.0133
Glass	5p–1000p	5	100–600	–55/125	40	100Hz–10GHz	6.6	2500	0.001	10^6	
Porcelain	100p–0.1μ	5	50–400	–55/125	120				0.1	5×10^5	
Teflon	1000p–2μ	10	50–200	–70/250	–200				0.04	5×10^6	
Ceramic											
Low-Loss Disk Multilayer Plate	100p–1μ	10	50–400	–55/125	±30	100Hz–10GHz	5.7	200–300	0.02	5×10^3	0.11–0.008
High Permittivity Disk Multilayer Plate	10p–1μ		50–30000	–55/125		1kHz–1GHz	1000–7000	100			
Paper											
Paper Metalized paper	0.1μ–10μ	10	200–1600	–55/125	±800	D.C.–1MHz	4.5	500–1000	1.0	5×10^3	0.002
Plastic											
Kapton	1000p–1μ	10		–55/220	100				0.3	10^5	
Polyester/Mylar	1000p–50μ	10	50–600	–55/125	400	D.C.–10GHz	2.3	1000	0.75	10^5	0.05

Parylene	5000p–1μ	10		–55/125	±100	D.C.–10GHz		1000	0.1	10^5	
Polysulfone	1000p–1μ	5		–55/150	80				0.3	10^5	
Polycarbonate											
Axial	100p–30μ	10	50–800	–55/125	±100	D.C.–10GHz	2.8		0.2	5×10^5	
Can											
Radial											
Polypropylene											
Axial	100p–50μ	10	100–800	–55/105	–200	D.C.–10GHz			0.2	10^5	
Can											
Radial											
Polystyrene											
Axial	10p–2.7μ	10	100–600	–55/85	–100	D.C.–10GHz	10		0.05	10^6	
Can											
Radial											0.04
Electrolytic and solid											
Aluminum	0.1μ–1.6	–10/100	3–600	–40/85	2500		8–10		10	100	
Tantalum											
Axial	0.1μ–1000μ	–10/100	6–100	–55/85	800	D.C.–1kHz	25–27		4.0	20	
Can											
Radial											
TiTiN Film	10p–200p	10	6–30	–55/125	100	D.C.–1MHz	2.4	1000	0.01	10^6	
Oil	0.1μ–20μ		200–10000				1.0		0.5		
Air/Vacuum	1p–100p		2000–3600								

TABLE 45.4b (continued)   Capacitor Specifications and Applications

Capacitor types	Typical commercial specifications			Applications	Samples/codes
	Voltage A.C. V	Capacitance F	Tolerance %		
Mica, glass, porcelain and Teflon					
Mica	350	2.2 p–1000 p	1	High temperature, low absorption, good in RF applications, and circuit requiring long-term stability	350 V DC
Glass					
Porcelain					
Teflon					
Ceramic					
Low-loss					
Disk	100	1.8 p–470 p	±2	Active filters, and high-density PCB applications, power-tuned circuits, coupling and decoupling of high-frequency circuits (PCB versions available)	Colour code of temperature coefficient; Capacitance value (27 pF); 27 pF
Multilayer	63/50	10 p–1 µ	10		
Plate	100	390 p–4700 p	10		
High permittivity					
Disk					
Multilayer					
Plate					
Paper					
Paper				General-purpose motor applications	ABCDE
Metalized paper					

Colour	α ppm/°C
Red Purple	100
Black	0
Brown	-33
Red	-75
Orange	-150
Yellow	-220
Green	-330
Blue	-470
Purple	-750

Colour	Significant figures		Multiplier	Tolerance %	Voltage V
	A	B	C	D	E
Black	-	0	1	±20	
Brown	1	1	10		125
Red	2	2	10^2		160
Orange	3	3	10^3		250
Yellow	4	4	10		
Green	5	5	10^5		
Blue	6	6	10^6		
Purple	7	7	10^7		
Gray	8	8	-		
White	9	9	-	±10	

Colour	Significant figures		Multiplier pF	Tolerance %	Voltage V (D.C.)
	A	B	C	D	E
Black	0	0	1	±20	-
Brown	1	1	10		100
Red	2	2	10^2		250
Orange	3	3	10^3		-
Yellow	4	4	10^4		400
Green	5	5	10^5		-
Blue	6	6			630
Purple	7	7	-		-
Gray	8	8	10^{-2}	±10	-
White	9	9	10^{-1}		-

Colour	Significant figures		Multiplier µF	Voltage V
	A	B	C	D
Black	-	0	1	10
Brown	1	1	-	1.6
Red	2	2	-	4
Orange	3	3	-	40
Yellow	4	4	-	6.3
Green	5	5	-	16
Blue	6	6	-	-
Purple	7	7	10^{-3}	-
Gray	8	8	10^{-2}	25
White	9	9	10^{-1}	2.5

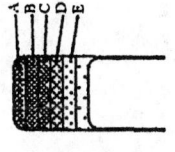

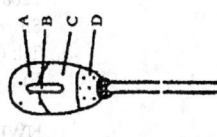

Plastic

Type	Range			Applications
Kapton polyester/mylar	0.1 µ–68 µ	63	10	Filters, timing and other high-stability applications, high quality, small low TC, tuned circuits, timing networks, stable oscillator circuits, resonance circuits and other high-performance pulse handling applications, phase shifting, pulse applications
Parylene				
Polysulfone				
Polycarbonate				
Axial	1 p–10 p	280	10	
Can	0.1 µ–68 µ	100	10	
Radial				
Polypropylene				
Axial	100 p–2200 p	63	5	
Can	150 p–1000 p	400	1	
Radial	1 n–470 n	1000	20	
Polystyrene				
Axial	47 p–680 p	450	1	
Can	1 n–470 n	1000	20	
Radial	10 p–10000 p	160	2.5	

Electrolytic and solid

Type	Range			Applications
Aluminum	680 p–6800 p	25	20	General-purpose to high-performance applications, power supply filters, motor capacitors, switching circuits, high-voltage filter transmitters and long life applications
Tantalum				
Axial	6.8 µ–150 µ	6.3	20	
Can	6.8 µ–150 µ	35	20	
Radial	2.2 µ–68 µ	16	20	

TABLE 45.5   List of Manufacturers

Bycap Inc.	Magnetek
5115 N. Ravenswood, Dept. T	902 Crescent Avenue
Chicago, IL 60640	Bridgeport, CT 06607
Tel: (312) 561-4976	Tel: (800) 541-9997
Fax: (312) 561-5095	Fax: (203) 335-2820
Chenelex	Maxwell Laboratories Inc.
Barr Road, P.O. Box 82	8888 Balboa Avenue
Norwich, NY 13815	San Diego, CA, 92123
Tel: (607) 344-3777	Tel: (619) 576-7545
Fax: (607) 334-9076	Fax: (619) 576-7545
Chicago Condenser	Metuchen Capacitors Inc.
2900-T W. Chicago Ave.	139 White Oak Lane
Chicago, IL 60622	Old Bridge, NJ 08857
Tel: (312) 227-7070	Tel: (800) 679-0514
Fax: (312) 227-6646	Fax: (800) 679-9959
Comet North America Inc.	Murata Electronics
11 Belden Avenue	Marketing Communications
Norwalk, CT 06850	2200 Lake Park Drive
Tel: (203) 852-1231	Smyrna, GA 30080
Fax: (203) 838-3827	Tel: (800) 394-5592
	Fax: (800) 4FAXCAT
Condenser Products	
2131 Broad Street	NWL Capacitors
Brooksville, FL 34609	204 Caroline Drive, P.O. Box 97
Tel: (800) 382-6874	Snow Hill, NC 28580
Fax: (904) 799-0221	Tel: (919) 747-5943
	Fax: (919) 747-8979
CSI Capacitors	
810 Rancheros Drive	Okaya Electric America Inc.
San Marcos, CA 92069-3009	503 Wall Street
Tel: (619) 747-4000	Valparaiso, IN 46383
Fax: (619) 743-5094	Tel: (219) 477-4488
	Fax: (219) 477-4856
HVCO Inc.	
P.O. Drawer 223, 7137 Sycamore Ln.	
Cedarburg, WI 53012	
Tel: (414) 375-0172	
Fax: (414) 275-0173	

film. A foil (e.g., aluminum) is used as the plate. The capacitor is then hermetically sealed in an aluminum or plastic can, as shown in Figure 45.3. These capacitors can be divided into two main subgroups.

Tantalum electrolytic: The anode consists of sintered tantalum powder and the dielectric is Ta_2O_5, which has a high value of ε_r. A semiconductor layer MnO_2 surrounds the dielectric. The cathode made from graphite is deposited around MnO_2 before the capacitor is sealed. The form of a tantalum electrolytic capacitor includes a porous anode slug to obtain a large active surface. These capacitors are highly stable and reliable, with good temperature ranges, and are suitable for high-frequency applications.

Aluminum electrolytic capacitors: Aluminum foil is oxidized on one side as Al_2O_3. The oxide layer is the dielectric having a thickness of about 0.1 μm and a high electric field strength (7×10^5 Vmm^{-1}). A second layer acting as the cathode, made from etched Al-foil, is inserted. The two layers are separated by a spacer when the layers are rolled and mounted.

Electrolytic capacitors must be handled with caution, since in these capacitors the electrolytic is polarized. That is, the anode should always be positive with respect to the cathode. If not connected correctly, hydrogen gas will form; this damages the dielectric layer, causing a high leakage current or

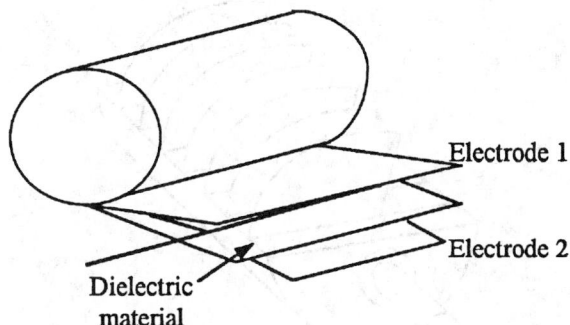

FIGURE 45.2   Construction of a typical capacitor. Dielectric material sheets are placed between electrode foils and convolutely wound. The moisture of the dielectric material is removed at high temperatures by vacuum drying before the capacitor is impregnated with oil, paraffin, or wax.

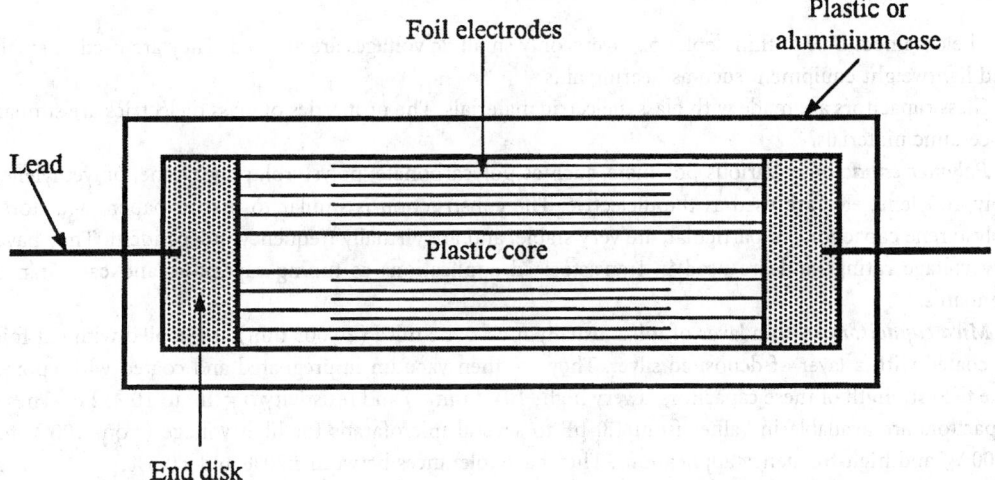

FIGURE 45.3   Construction of an electrolytic capacitor. The two foil layer electrodes are separated by an impregnated electrolyte paper spacer and rolled together on a plastic core. Usually, a flexible metallized dielectric film is used as one of the plates and an ordinary foil is used for the other. The capacitor is then hermetically sealed in an aluminum or plastic can.

blow-up. These capacitors can be manufactured in values up to 1 F. They are used in not so critical applications such as coupling, bypass, filtering, etc. However, they are not useful at frequencies above 1 kHz or so.

Ceramic and glass capacitors: The dielectric is a ceramic material with deposited metals. They are usually rod or disk shaped. They have good temperature characteristics and are suitable in many high-frequency applications. There are many different types, such as (1): *Low K ceramic:* These capacitors are made with materials that contains a large fraction of titanium dioxide (TiO_2). The relative permittivity of these materials varies from 10 to 500, with negative temperature coefficient. The dielectric is TiO_2 + MgO + SiO_2, suitable in high-frequency applications in filters, tuned circuits, coupling and bypass circuits, etc. (2): *High K ceramic:* The dielectric contains a large fraction of barium titanate, $BaTiO_3$, mixed with $PbTiO_3$ or $PbZrO_3$ giving relative permittivity of 250 to 10,000. They have high losses, and also have high-voltage time dependence with poor stability. (3): *Miniature ceramic capacitors:* These are used in critical high-frequency applications. They are made in the ranges of 0.25 pF to 1 nF. (4): *Dielectric ceramic capacitors:* The material is a semiconducting ceramic with deposited metals on both sides. This arrangement results in two depletion layers that make up the very thin dielectric. In this way, high capacitances

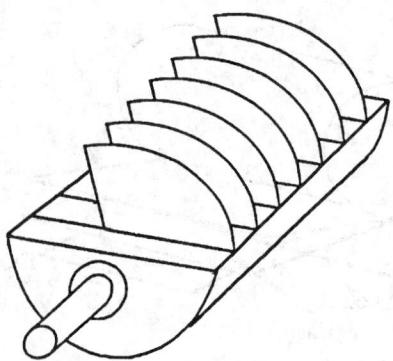

FIGURE 45.4 A variable capacitor. It consists of two assemblies of spaced plates positioned together by insulation members such that one set of plates can be rotated. The majority of variable capacitors have air as the dielectric. They are used mainly in adjustment of resonant frequency of tuned circuits in receivers and transmitters. By shaping the plates suitably, they can be made to be linear or logarithmic.

can be obtained. Due to thin depletion layers, only small dc voltages are allowed. They are used in small and lightweight equipment such as hearing aids.

Glass capacitors are made with glass dielectric materials. The properties of glass dielectrics are similar to ceramic materials.

Polymer capacitors: Various polymers, such as polycarbonate, polystyrol, polystyrene, polyethylene, polypropylene, etc., are used as the dielectric. The construction is similar to that of paper capacitors. Polystyrene capacitors, in particular, are very stable, and are virtually frequency independent. They have low voltage ratings and are used in transistorized applications as tuning capacitors and capacitance standards.

Mica capacitors: A thin layer of mica, usually muscovite mica (≥ 0.003 mm) are stapled with Cu-foil or coated with a layer of deposited silver. They are then vacuum impregnated and coated with epoxy. The field strength of these capacitors is very high (10^5 V mm^{-1}) and resistivity $\rho = 10^6$ to 10^{15} Ω m. These capacitors are available in values from 1.0 pF to several microfarads for high voltage (from 100 V to 2000 V) and high-frequency applications. They have tolerances between $\pm 20\%$ and $\pm 0.5\%$.

Variable capacitors: These capacitors usually have air as the dielectric and consist of two assemblies of spaced plates positioned together by insulation members such that one set of plates can be rotated. A typical example of variable capacitors is given in Figure 45.4. Their main use is the adjustment of resonant frequency of tuned circuits in receivers and transmitters, filters, etc. By shaping the plates, various types of capacitances can be obtained, such as: *linear capacitance*, in which capacitance changes as a linear function of rotation, and *logarithmic capacitance*.

Variable capacitors can be grouped as: precision types, general-purpose types, transmitter types, trimmer types, and special types such as phase shifters.

Precision-type variable capacitors are used in bridges, resonant circuits, and many other instrumentation systems. The capacitance swing can be from 100 pF to 5000 pF. They have excellent long-term stability with very tight tolerances.

General-purpose type variable capacitors are used as tuning capacitors in radio and other broadcasting devices. They are available in many laws such as, straight line frequency, straight line wavelength, etc. The normal capacitance swing is from 400 pF to 500 pF. In some cases, a swing of 10 pF to 600 pF are available.

Transmitter-type variable capacitors are similar to general-purpose variable capacitors, but they are specially designed for high-voltage operations. The vanes are rounded and spaced wider to avoid flashover and excessive current leakages. The swing of these capacitors can go from few picofarads up to 1000 pF. In some cases, oil filling or compressed gases are used to increase operating voltages and capacitances.

Trimmer capacitors are used for coil trimming at intermediate radio frequencies. They can be air-spaced rotary types (2 pF to 100 pF), compression types (1.5 pF to 2000 pF), ceramic-dielectric rotary types (5 pF to 100 pF), and tubular types (up to 3 pF).

Sometimes, special type variable capacitors are produced for particular applications, such as differential and phase shift capacitors in radar systems. They are used for accurate measurement of time intervals, high-speed scanning circuits, transmitters and receivers, etc.

Integrated circuit capacitors: These are capacitors adapted for use in microelectronic circuits. They include some miniature ceramic capacitors, tantalum oxide solid capacitors, and tantalum electrolyte solid capacitors. The ceramic and tantalum oxide chips are unencapsulated and are fitted with end caps for direct surface mounting onto the circuit board. The beam-leaded tantalum electrolytic chips are usually attached by pressure bonding. Typical values of these capacitors are: 1 pF to 27 nF for temperature compensating ceramic, (100 to 3000 pF) for tantalum oxide, 390 pF to 0.47 µF for general-purpose ceramic, and (0.1 to 10 µF) for tantalum electrolyte. Operating voltages ranges from 25 to 200 V for ceramic, 12 to 35 V for tantalum electrolyte and 12 to 25 V for tantalum oxide.

Integrated circuit capacitors are made mostly within MOS integrated circuits as monolayer capacitors containing tantalum or other suitable deposits. The plates of the capacitors of the integrated circuits are generally formed by two heavily doped polysilicon layers formed on a thick layer of oxide. The dielectric is usually made from a thin layer of silicon oxide. These capacitors are temperature stable, with a temperature coefficient of about 20 ppm/°C. Integrated circuit capacitive sensors are achieved by incorporating a dielectric sensitive to physical variables. Usually, the metallization layer formed on top of the dielectric forms a shape to provide access to measured physical variable to the dielectric.

Voltage variable capacitors: These capacitors make use of the capacitive effect of the reversed-biased p-n junction diode. By applying different reverse bias voltages to the diode, the capacitance can be changed. Hence, the name, varicap or varactor diodes is given to theses devices. Varactors are designed to provide various capacitance ranges from a few picofarads to more than 100 pF. It is also possible to make use of high-speed switching silicon diodes as voltage variable capacitors. However, they are limited by the very low maximum capacitance available. Typical applications of these varactor diodes are in the tuning circuits in radio frequency receivers. Present-day varactor diodes operate into the microwave part of the spectrum. These devices are quite efficient as frequency multipliers at power levels as great as 25 W. The efficiency of a correctly designed varactor multiplier can exceed 50% in most instances. It is also worth noting that some Zener diodes and selected silicon power-supply rectifier diodes can work effectively as varactors at frequencies as high as 144 MHz. In the case of the Zener diode, it should be operated below its reverse breakdown voltage.

45.2 Characteristics of Capacitors

Capacitors are characterized by dielectric properties, break-down voltages, temperature coefficients, insulation resistances, frequency and impedances, power dissipation and quality factors, reliability and aging, etc. Typical characteristics of common capacitors are given in Table 45.4.

Dielectric properties: Dielectrics of capacitors can be made from polar or nonpolar materials. Polar materials have dipolar characteristics; that is, they consist of molecules whose ends are oppositely charged. This polarization causes oscillations of the dipoles at certain frequencies, resulting in high losses.

In general, capacitor properties are largely determined by the dielectric properties. For example, the losses in the capacitors occur due to the current leakage and the dielectric absorption. These losses are frequency dependent, as typified by Figure 45.5.

The dielectric absorption introduces a time lag during the charging and discharging of the capacitor, thus reducing the capacitance values at high frequencies and causing unwanted time delays in pulse circuits. The leakage current, on the other hand, prevents indefinite storage of energy in the capacitor. An associated parameter to leakage currents is the leakage resistance, which is measured in megohms, but usually expressed in megohm-microfarads or ohms-farads. The leakage resistance and capacitance

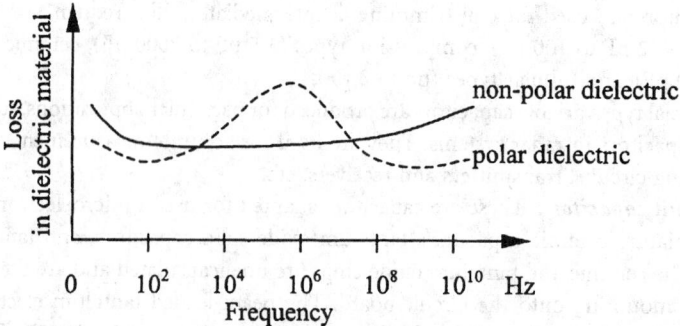

FIGURE 45.5   Frequency dependence of dielectric loss. The dielectric material of capacitors can be polar or nonpolar. In polar materials, polarization causes oscillations at certain frequencies, resulting in high losses. The dielectric losses introduce a time lag during the charging and discharging of the capacitor, thus reducing the capacitance values at high frequencies.

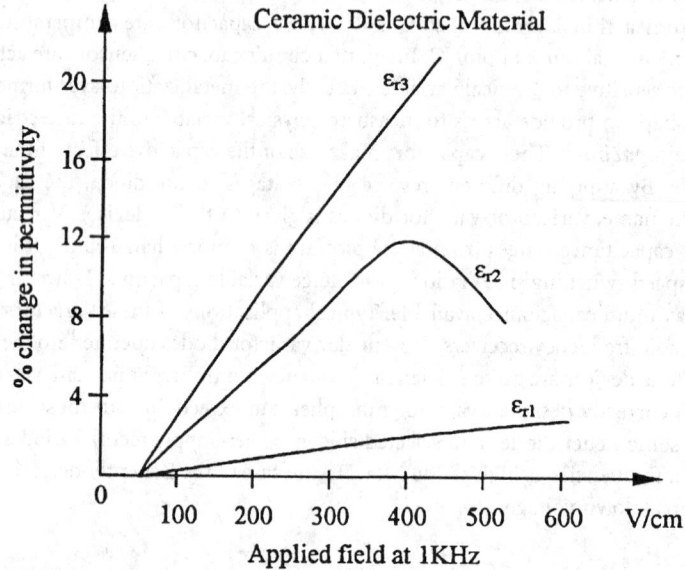

FIGURE 45.6   Changes in relative permittivity vs. field strength. The dielectric strength depends on the temperature, frequency, and applied voltage. Increases in the applied voltage cause higher changes in the dielectric strength. If the capacitor is subjected to high operating voltages, the electric field in the dielectric exceeds the breakdown value which can damage the dielectric permanently.

introduces time constants that can vary from a few days for polystyrene to several seconds in some electrolytic capacitors. It is important to mention that the leakage current does not only depend on the properties of the dielectric materials, but also depends on the construction and structure of capacitors. This is particularly true for capacitors having values less than 0.1 µF, having very thin dielectric materials between the electrodes.

     Breakdown voltage: If the capacitor is subjected to high operating voltages, the electric field in the dielectric exceeds the breakdown value, which damages the dielectric permanently. The dielectric strength, which is the ability to withstand high voltages without changing properties, depends on the temperature, frequency, and applied voltage. An example of this dependence on the applied voltage is given in Figure 45.6. It is commonly known that the use of capacitors below their rated values increases the reliability and the expected lifetime. The standard voltage ratings of most capacitors are quoted by the

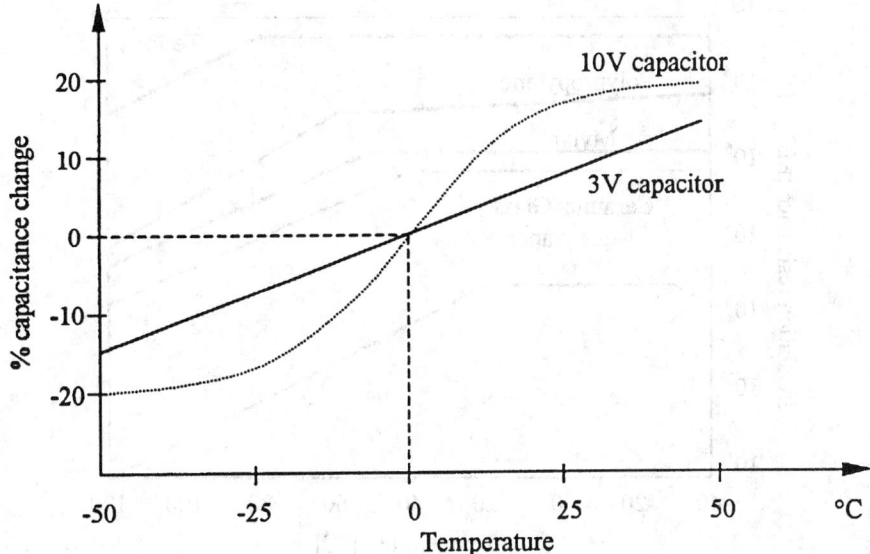

FIGURE 45.7 Temperature dependence of capacitors. The temperature characteristics of capacitors are largely dependent on the temperature properties of the dielectric materials used. The variations in capacitance due to temperature also depends on the type of capacitor and the operational voltage. The temperature coefficient of glass, teflon, mica, and polycarbonate are very small, and relatively high in ceramic capacitors.

manufacturers as 50, 100, 200, 400, and 600 V. Tantalum and electrolytic capacitors have ratings of 6, 10, 12, 15, 20, 25, 35, 50, 75, 100 V and higher.

Usually, values for surge voltages are given to indicate the ability of capacitors to withstand high transients. Typically, the surge voltages for electrolytic capacitors is 10% above the rated voltage, 50% for aluminum capacitors and about 250% for ceramic and mica capacitors.

The rated reverse voltages of electrolytic capacitors are limited to 1.5 V and, in some cases, to 15% of the rated forward voltages.

Temperature coefficient: The temperature characteristics of capacitors largely dependent on the temperature properties of the dielectric materials used, as given in Figure 45.7. The temperature coefficients of glass, teflon, mica, polycarbonate, etc. are very small, whereas in ceramic capacitors, they can be very high.

Insulation resistance: The insulation resistance of capacitors is important in many circuits. The insulation resistance is susceptible to temperature and humidity. For example, unsealed capacitors show large and rapid changes against temperature and humidity. For most capacitors, under high temperature conditions, the change in insulation resistance is an exponential function of temperature ($R_{T1} = R_{T2}e^{K(T1-T2)}$). The temperature dependence of insulation resistance of common capacitors is shown in Figure 45.8.

Frequency and impedance: Practical capacitors have increases in losses at very low and very high frequencies. At low frequencies, the circuit becomes entirely resistive and the dc leakage current becomes effective. At very high frequencies, the current passes through the capacitance and the dielectric losses become important. Approximate useable frequency ranges of capacitors are provided in Table 45.4.

An ideal capacitor should have an entirely negative reactance, but losses and inherent inductance prevents ideal operation. Depending on the construction, capacitors will resonate at certain frequencies due to unavoidable construction-based inductances. A typical impedance characteristic of a capacitor is depicted in Figure 45.9.

Power dissipation and quality factors: Ideally, a capacitor should store energy without dissipating any power. However, due to equivalent resistances, R_{eq}, some power will be dissipated. The power factor of a capacitor can be expressed as:

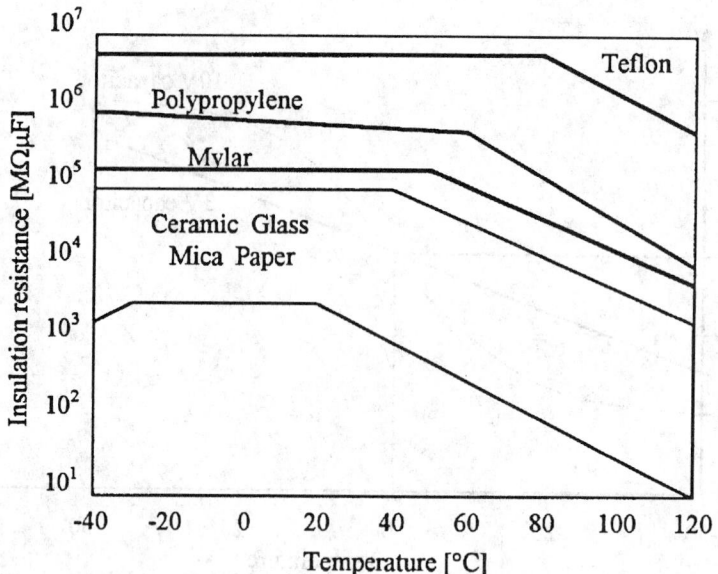

FIGURE 45.8 Temperature dependence of insulation resistance. The insulation resistance of many capacitors is not affected at low temperatures. However, under high temperature conditions, the change in insulation resistance can be approximated by an exponential relation. The insulation resistance is also susceptible to variations in humidity.

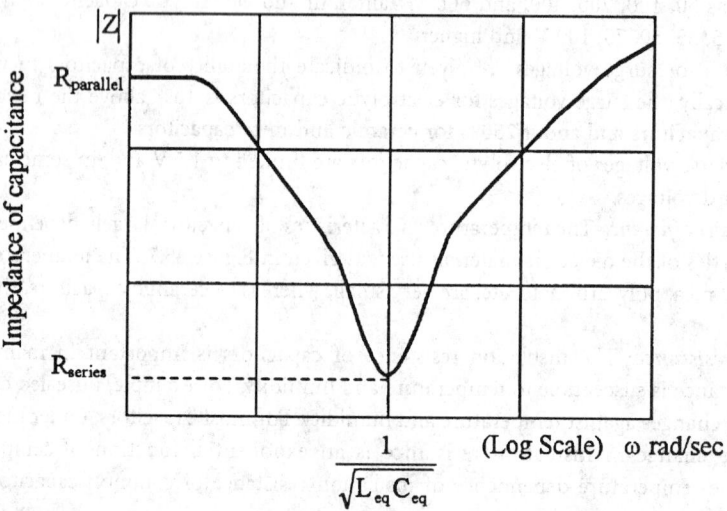

FIGURE 45.9 Frequency and impedance relation of capacitors. The losses and inherent inductance affects the ideal operation of capacitors and the capacitance impedance becomes a function of frequency. Depending on the construction, all capacitors will resonate at a certain frequency.

$$PF = \cos\theta = R_{eq} / |Z_{eq}| \qquad (45.7)$$

where θ = Phase angle

Z_{eq} = Equivalent total impedance

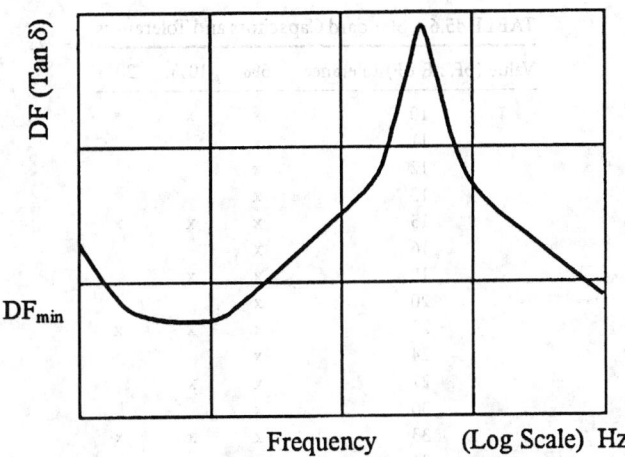

FIGURE 45.10 Power dissipation factors. In ideal operations, capacitors should store energy without dissipating power. Nevertheless, due to resistances, some power will be dissipated. The dissipation depends on frequency. The standard measurement of dissipation factor δ is determined by applying 1.0 Vrms at 1 kHz.

An important characteristic, the *dissipation factor* of capacitors, is expressed as:

$$DF = \tan\delta = R_{eq}/X_{eq} \tag{45.8}$$

where δ = Angle of loss

X_{eq} = Equivalent reactance

The dissipation factor depends on the frequency. Capacitors are designed such that this dependence is minimal. The measurement of dissipation factor δ is made at 1 kHz and 1.0 Vrms applied to the capacitor. A typical dissipation factor curve is depicted in Figure 45.10.

Selection of Capacitors and Capacitor Reliability

Capacitor Selection

Experience shows that a substantial part of component failures in electronic equipment is due to capacitors. The major cause of this can be attributed to the improper selection and inappropriate applications of capacitors. The following factors are therefore important criteria in the selection of capacitors in circuit applications: (1) the capacitance *values and tolerances* are determined by operating frequencies or by the value required for timing, energy storage, phase shifting, coupling, or decoupling; (2) the *voltages* are determined by the type and nature of the source, ac, dc, transient, surges, and ripples; (3) the *stability* is determined by operating conditions like temperature, humidity, shock, vibration, and life expectancy; (4) the *electrical properties* are determined by life expectancy, leakage current, dissipation factor, impedance, and self-resonant frequency; (5) the mechanical properties are determined by the types and construction, e.g., size, configuration, and packaging; and (6) the cost is determined by the types and physical dimensions of capacitors and the required tolerance.

Capacitor Reliability

Some of the common causes of capacitor failure are due to voltage and current overloads, high temperature and humidity, shock, vibration pressure, frequency effects, and aging. The voltage overload produces an excessive electric field in the dielectric that results in the breakdown and destruction of the dielectric. The current overload caused by rapid voltage variations results in current transients. If these currents are of sufficient amplitude and duration, the dielectric can be deformed or damaged, resulting in drastic

TABLE 45.6   Standard Capacitors and Tolerances

Value (pF, nF, uF)\tolerance	5%	10%	20%
10	x	x	x
11	x		
12	x		
13	x		
15	x	x	x
16	x		
18	x	x	
20	x		
22	x	x	x
24	x		
27	x	x	
30	x		
33	x	x	x
36	x		
39	x		
43	x		
47	x	x	x
51	x		
56	x	x	x
62	x		
68	x	x	x
75	x		
82	x		
91	x		

changes in capacitance values, and thus leading to equipment malfunction. The high temperatures are mainly due to voltage and current overloads. The overheating and high temperatures accelerate the dielectric aging. This causes the plastic film to be brittle and also introduces cracks in the hermetic seals. The moisture and humidity due to severe operating environments cause corrosion, reduce the dielectric strength, and lower insulation resistances. The mechanical effects are mainly the pressure, variation, shock, and stress, which can cause mechanical damages of seals that result in electrical failures. Aging deteriorates the insulation resistance and affects the dielectric strength. The aging is usually determined by shelf-life; information about aging is supplied by the manufacturers.

Capacitor Standard Values and Tolerances

General-purpose capacitors values tend to be grouped close to each other in a bimodal distribution manner within their tolerance values. Usually, the tolerances of standard capacitors are 5%, 10%, and 20% of their values, as shown in Table 45.6. Nevertheless, tolerances of precision capacitors are much tighter — in the range of 0.25%, 0.5%, 0.625%, 1%, 2%, and 3% of the values. These capacitors are much more expensive than the standard range.

For capacitors in the small pF range, the tolerances can be given as ±1.5, ±1, ±0.5, ±0.25, and ±0.1 pF. Usually, low tolerance ranges are achieved by selecting manufactured items.

Standard capacitors are constructed from interleaved metal plates using air as the dielectric material. The area of the plates and distance between them are determined and constructed with precision. National Bureau of Standards maintains a bank of primary standard air capacitors that can be used to calibrate the secondary and working standards for laboratories and industry. Generally, smaller capacitance working standards are obtained from air capacitors, whereas larger working standards are made from solid dielectric materials. Usually, silver-mica capacitors are selected as working standards. These capacitors are very stable, have very low dissipation factors and small temperature coefficients, and have very little aging effect. They are available in decade mounting forms.

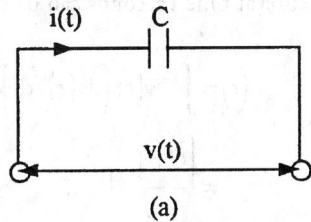

(a)

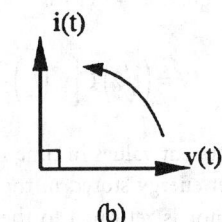

(b)

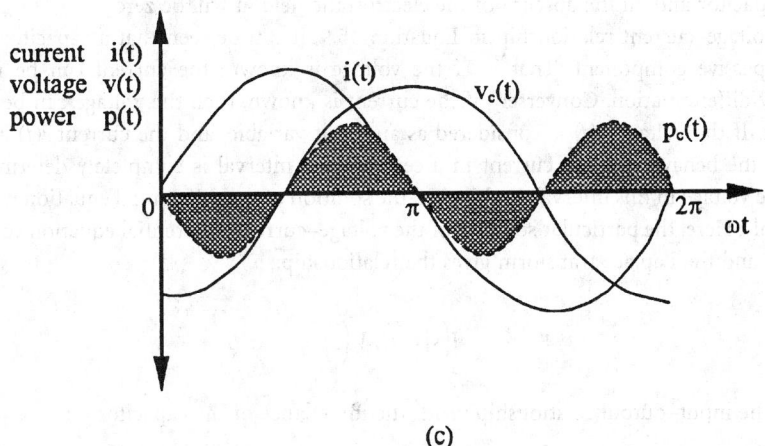

(c)

FIGURE 45.11 A capacitor as a two-terminal circuit element: (a) connection of a capacitor in electric circuits; (b) current–voltage relationship under sinusoidal operations; and (c) the power, voltage, and current relationships. The power has positive and negative values with twice the frequency of the applied voltage.

Capacitors as Circuit Components

The capacitor is used as a two-terminal element in electric circuits with the current–voltage relationship given by:

$$i(t) = C\, dv(t)/dt \tag{45.9}$$

where C is the capacitance.

This element is represented by the circuit shape as shown in Figure 45.11(a).

From the $v(t)$, $i(t)$ relationship, the instantaneous power of this element can then be given by:

$$p(t) = v(t)\, i(t) \tag{45.10}$$

The stored energy in the capacitor at time t seconds can be calculated as:

$$w(t) = \int C\,v(t)\left\{dv(t)/dt\right\}dt$$
$$= \frac{\left[C\,v^2(t)\right]}{2} \tag{45.11}$$

If the voltage across the capacitor is changing in time, the energy stored in the capacitor in the time interval t_1 to t_2 can be found using Equation 45.11 as:

$$W = \left(1/2\right)C\left[v^2(t_2) - v^2(t_1)\right] \tag{45.12}$$

Although the voltage assumed different values in time interval t_1 to t_2, if the initial and final voltages are equal (e.g., $v(t_1) = v(t_2)$), the net energy stored in the capacitor will be equal to zero. This implies that the energy stored in the capacitor is returned to the circuit during the time interval; that is, the capacitor transforms the energy without dissipation. The stored energy is associated with the electrostatic field of the capacitor and, in the absence of the electrostatic field, it will be zero.

From the voltage–current relationship in Equation 45.9, it can be seen that a capacitor as a circuit element is a passive component. That is, if the voltage is known, the current can be immediately determined by differentiation. Conversely, if the current is known, then the voltage can be determined by integration. If the voltage $v(t)$ is considered as an input variable, and the current $i(t)$ as an output variable, then the behavior of the current in a certain time interval is completely determined by the behavior of the voltage in this interval. In this case, the solution of the differential equation has the forced component only. Here, the particular solution of the voltage–current differential equation coincides with a full solution and the Laplace transform gives the relationship:

$$I(s) = sC\,V(s) \tag{45.13}$$

From this, the input–output relationship yields the impedance of the capacitor as:

$$Z(s) = \frac{1}{sC}$$

or

$$Y(s) = sC \tag{45.14}$$

In the stationary condition, $s \to 0$, and $Z \to \infty$, and $Y \to 0$.

In the sinusoidal condition, $s = j\omega = 2\pi f$, where f = frequency and, hence:

$$Z(j\omega) = \frac{1}{\left\{j\omega C\right\}}$$
$$= \frac{-j}{\left\{\omega C\right\}} \tag{45.15}$$

and

$$Y(j\omega) = j\omega C \tag{45.16}$$

The capacitor can then be characterized under sinusoidal conditions, by a reactance of $X_C = 1/\omega C$ measured in ohms with the current leading the voltage by 90°, as shown in the phasor diagram in Figure 45.11(b).

In sinusoidal operations, the instantaneous power $p(t) = v(t)\, i(t)$ can be calculated as:

$$v(t) = V_{max} \cos\omega t = \sqrt{2}\, V \cos\omega t \tag{45.17}$$

Using the relationship given by Equation 45.9, the current can be written as:

$$i(t) = C\, dv/dt = -\omega C \sqrt{2}\, V \sin\omega t \tag{45.18}$$

giving:

$$p(t) = v(t)\, i(t) = -2\,\omega C\, V^2 \sin\omega t \cos\omega t = \frac{V^2 \sin 2\omega t}{X_C} \tag{45.19}$$

This indicates that the average power is zero because of the $\sin 2\omega t$ term, but there is a periodic storage and return of energy and the amplitude of that power is V^2/X_c. The power, voltage, and current relationship in a capacitor is given in Figure 45.11(c).

Series and Parallel Connection of Capacitors

The formulae for series and parallel connection of capacitors can be obtained from the general consideration of series and parallel connection of impedances as shown in Figures 45.12(a) and (b), respectively. For the series connection, the impedances are added such that:

$$\frac{1}{sC} = \frac{1}{sC_1} + \frac{1}{sC_2} + \cdots + \frac{1}{sC_n} \tag{45.20}$$

where $C_1, C_2, \ldots C_n$ are the capacitance of the capacitors connected in series as in Figure 45.12(a). The equivalent capacitance is then given by:

$$C = \left\{ \frac{1}{C_1} + \frac{1}{C_2} + \cdots + \frac{1}{C_n} \right\}^{-1} \tag{45.21}$$

The final capacitance value will always be smaller than the smallest value.

In a similar way, the equivalent capacitance of parallel connected capacitors is

$$C = C_1 + C_2 + \ldots + C_n \tag{45.22}$$

and the final value of C is always larger than the largest capacitance in the circuit.

Distributed Capacitances in Circuits

Since capacitance is inherent whenever an electric potential exists between two conducting surfaces, its effect will be most noticeable in coils and in transmission lines at high frequencies. In the case of coils,

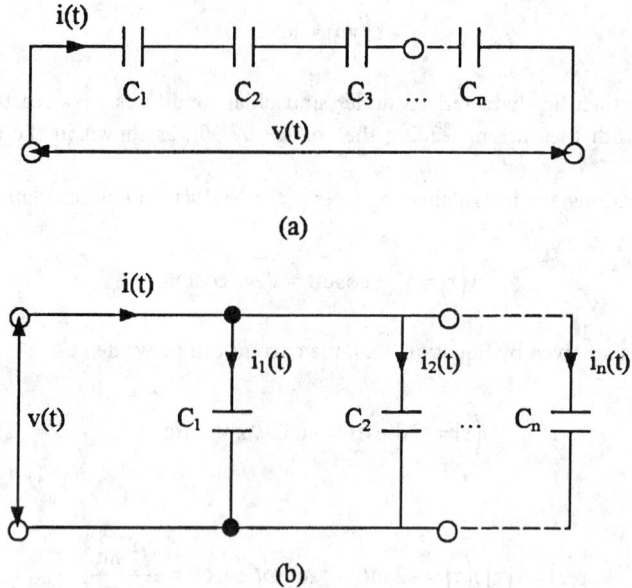

FIGURE 45.12 Series and parallel connection of capacitors: (a) series connection, and (b) parallel connection. In a series connection, the final capacitance value will always be smaller than the smallest value of the capacitor at the circuit element, whereas in parallel connection the final value is greater than the largest capacitance.

there are small capacitances between adjacent turns, between turns that are not adjacent, between terminal leads, between turns and ground, etc. Each of the various capacitance associated with the coil stores a quantity of electrostatic energy that is determined by the capacitance involved and the fraction of the total coil voltage that appears across it. The total effect is that the numerous small coil capacitances can be replaced by a single capacitor of appropriate size shunted across the coil terminals. This equivalent capacitance is called either the *distributed capacitance* or the *self-capacitance* of the coil, and it causes the coil to show parallel resonance effects under some conditions. In the case of a mismatched or unterminated transmission line, the distributed capacitance, together with the inductive effect, will create a phase difference between the voltage and current in the line. This phase difference depends on the type of termination and the electrical length of the line and, as a result, the input impedance of the line can effectively be an equivalent capacitor when its electrical length is less than a quarter wavelength for an open-circuit termination, or between a quarter wavelength and half a wavelength for a short-circuit termination.

Capacitor Equivalent Circuits

The electric equivalent circuit of a capacitor consists of a pure capacitance (C_p), plate inductances (L_1, L_2), plate resistances (R_1, R_2), and a parallel resistance R_p that represents the resistance of the dielectric or leakage resistance, as shown in Figure 45.13. The capacitors that have high leakage currents flowing through the dielectric have relatively low R_p values. Very low leakage currents are represented by extremely large R_p values. Examples of these two extremes are electrolytic capacitors that have high leakage current (low R_p), and plastic film capacitors, which have very low leakage current (high R_p). Typically, an electrolytic capacitor might easily have several microamperes of leakage current ($R_p < 1$ MΩ), while a plastic film capacitor could have a resistance greater than 100,000 MΩ.

It is usual to represent a low leakage capacitor (high R_p) by a series RC circuit, while those with high leakage (low R_p) are represented by parallel RC circuits. However, when a capacitor is measured in terms of the series C and R quantities, it is desirable to resolve them into parallel equivalent circuit quantities. This is because the (parallel) leakage resistance best represents the quality of the capacitor dielectric.

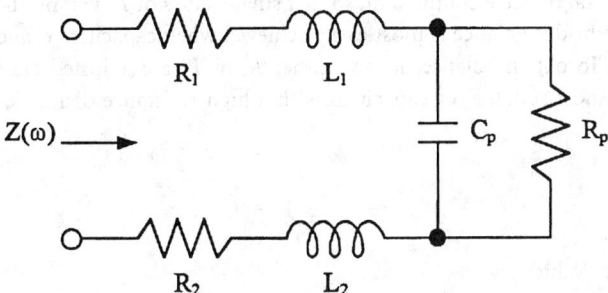

FIGURE 45.13 Capacitor equivalent circuit. A practical capacitor has resistance and inductances. Often, the electrical equivalent circuit of a capacitor can be simplified by a pure capacitance C_p and a parallel resistance R_p by neglecting resistances R_1, R_2 and inductances L_1, L_2. In low-leakage capacitors where R_p is high, the equivalent circuit can be represented by a series RC circuit.

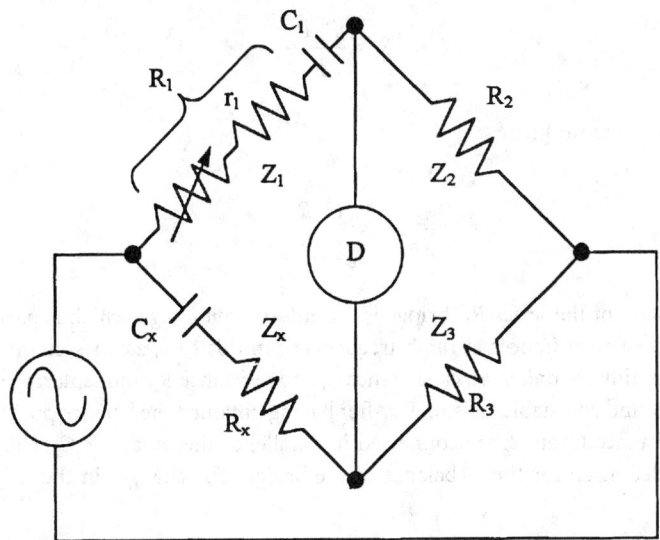

FIGURE 45.14 A series RC bridge. In these bridges, the unknown capacitance is compared with a known capacitance. The voltage drop across R_1 balances the resistive voltage drop in branch Z_2 when the bridge is balanced. The bridge balance is most easily achieved when capacitive branches have substantial resistive components. The resistors R_1 and either R_3 or R_4 are adjusted alternately to obtain the balance. This type of bridge is found to be most suitable for capacitors with a high-resistance dielectric, hence very low leakage currents.

Capacitive Bridges and Measurement of Capacitance

Bridges are used to make precise measurements of unknown capacitances and associated losses in terms of some known external capacitances and resistances. Most commonly used bridges are: series-resistance-capacitance bridge, parallel-resistance-capacitance bridge, Wien bridge, and Schering bridge.

Series-Resistance-Capacitance Bridge

Figure 45.14 is a series-resistance-capacitance (RC) bridge, which is used for the comparison of a known capacitance with an unknown capacitance. The unknown capacitance is represented by C_x and R_x. A standard adjustable resistance R_1 is connected in series with a standard capacitor C_1. The voltage drop across R_1 balances the resistive voltage drop when the bridge is balanced. The additional resistor in series with C_x increases the total resistive component, so that small values of R_1 will not be required to achieve balance. Generally, the bridge balance is most easily achieved when capacitive branches have substantial resistive components. To obtain balance, R_1 and either R_3 or R_4 are adjusted alternately. This type of bridge is found to be most suitable for capacitors with a high-resistance dielectric and hence very low leakage currents.

At balance:

$$Z_1 Z_3 = Z_2 Z_x \qquad (45.23)$$

Substituting impedance values gives:

$$\left(R_1 - \frac{j}{\omega C_1}\right) R_3 = \left(R_x - \frac{j}{\omega C_2}\right) R_2 \qquad (45.24)$$

Equating the real terms gives:

$$R_x = \frac{R_1 R_3}{R_2} \qquad (45.25)$$

and equating *imaginary* terms gives:

$$C_x = \frac{C_1 R_2}{R_3} \qquad (45.26)$$

An improved version of the series *RC* bridge is the *substitution bridge*, which is particularly useful to determine the values of capacitances at radio frequencies. In this case, a series-connected *RC* bridge is balanced by disconnecting the unknown capacitance C_x and resistance R_x, and replacing it by an adjustable standard capacitor C_s and adjustable resistor R_s. After having obtained the balance position, the unknown capacitance and resistance C_x an R_x are connected in parallel to the capacitor C_s. The capacitor C_s and resistor R_s are adjusted again for the re-balance of the bridge. The changes in the ΔC_s and ΔR_s lead to unknown values as:

$$C_x = \Delta C_s \quad \text{and} \quad R_x = \Delta R_s \left(C_{s1}/C_x\right)^2 \qquad (45.27)$$

where C_{s1} is the value of C_s in the initial balance condition.

The Parallel-Resistance-Capacitance Bridge

Figure 45.15 illustrates a parallel-resistance-capacitance bridge. In this case, the unknown capacitance is represented by its parallel equivalent circuit C_x in parallel with R_x. The Z_2 and Z_3 impedances are pure resistors with either or both being adjustable. The Z_1 is balanced by a standard capacitor C_1 in parallel with an adjustable resistor R_1. The bridge balance is achieved by adjustment of R_1 and either R_2 or R_3. The parallel-resistance-capacitance bridge is found to be most suitable for capacitors with a low-resistance dielectric, hence relatively high leakage currents. At balance:

$$\frac{1}{\left(\dfrac{1}{R_1} + j\omega C_1\right)} R_3 = \frac{1}{\left(\dfrac{1}{R_x} + j\omega C_x\right)} R_2 \qquad (45.28)$$

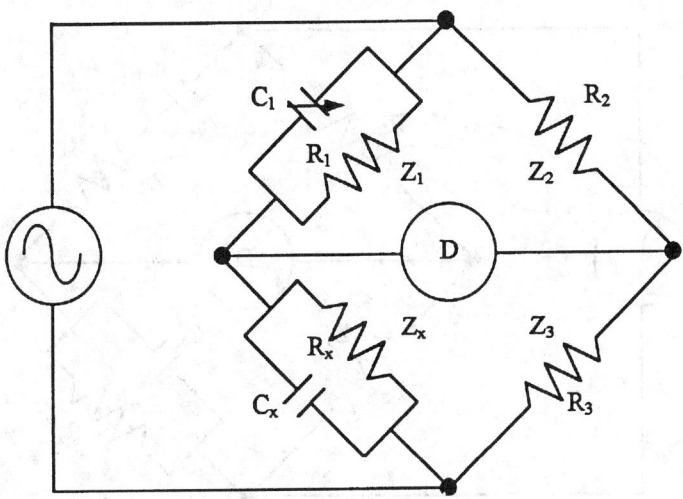

FIGURE 45.15 A parallel-resistance-capacitance bridge. The unknown capacitance is represented by its parallel equivalent circuit; C_x in parallel with R_x. The bridge balance is achieved by adjustment of R_1 and either R_3 or R_4. The parallel-resistance-capacitance bridge is found to be most suitable for capacitors with a low-resistance dielectric, hence relatively high leakage currents.

Equating *real* terms gives:

$$R_x = \frac{R_3\,R_1}{R_2} \tag{45.29}$$

and equating *imaginary* terms gives:

$$C_x = \frac{C_1\,R_2}{R_3} \tag{45.30}$$

The Wien Bridge

Figure 45.16 shows a Wien bridge. This is a special resistance-ratio bridge that permits two capacitances to be compared once all the resistances of the bridge are known. At balance, it can be proven that the unknown resistance and the capacitance are:

$$R_x = \frac{R_3\left(1+\omega^2 R_1^2 C_1^2\right)}{\omega^2 R_1 R_2 C_1^2} \tag{45.31}$$

and

$$C_x = \frac{C_1 R_2}{\left[R_3\left(1+\omega^2 R_1^2 C_1^2\right)\right]} \tag{45.32}$$

It can also be shown that:

$$\omega^2 = \frac{1}{R_1 C_1 R_x C_x} \tag{45.33}$$

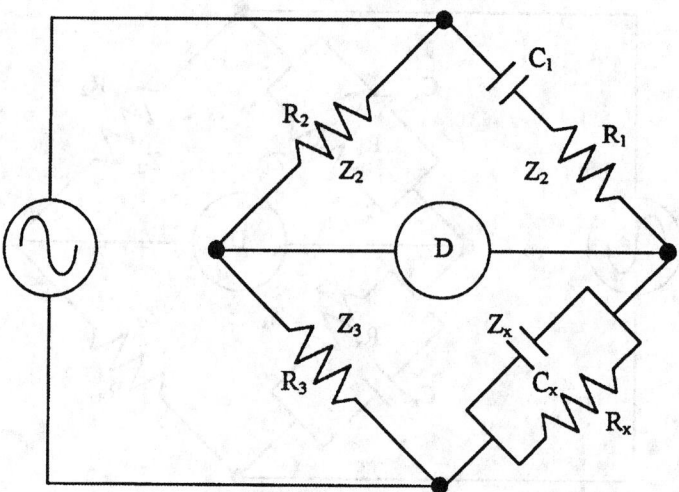

FIGURE 45.16 The Wien bridge. This bridge is used to compare two capacitors directly. It finds applications particularly in determining the frequency in *RC* oscillators. In some cases, capacitors C_1 and C_x are made equal and ganged together so that the frequency at which the null occurs varies linearly with capacitance.

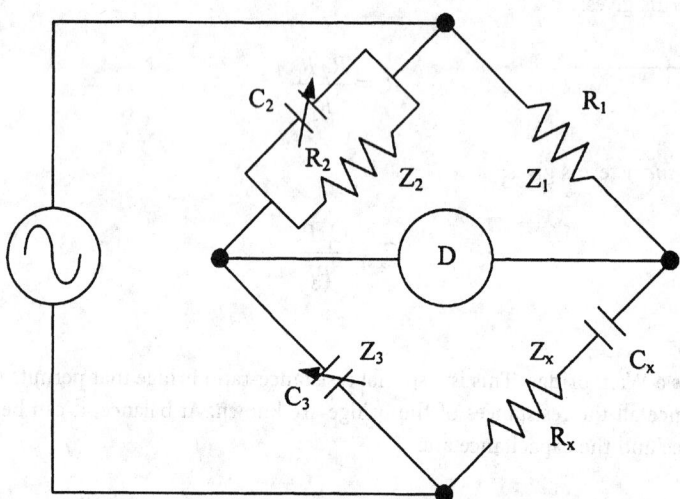

FIGURE 45.17 The Schering bridge. This bridge is particularly useful for measuring the capacitance, associated dissipation factors, and the loss angles. The unknown capacitance is directly proportional to the known capacitance C_1. The Schering bridge is frequently used as a high-voltage bridge with a high-voltage capacitor as C_1.

As indicated in Equation 45.33, the Wien bridge has an important application in determining the frequency in *RC* oscillators. In frequency meters, C_1 and C_x are made equal and the two capacitors are ganged together so that the frequency at which the null occurs varies linearly with capacitances.

The Schering Bridge

Figure 45.17 illustrates the configuration of the Schering bridge. This bridge is used for measuring the capacitance, the dissipation factors, and the loss angles. The unknown capacitance is directly proportional to the known capacitance C_3. That is, the bridge equations are:

$$C_x = \frac{C_3 R_2}{R_1} \quad \text{and} \quad R_x = \frac{C_2 R_1}{C_3} \tag{45.34}$$

Usually, R_2 and R_3 are fixed C_2 and C_3 are made variable. Schering bridges are frequently used in high-voltage applications with high-voltage capacitor C_3. They are also used as high-frequency bridges since the use of two variable adjustment capacitors are convenient for precise balancing.

Further Information

J. C. Whitaker, *The Electronics Handbook,* Boca Raton, FL: CRC Press & IEEE Press, 1996.

D. G. Fink and D. Christiansen, *Electronics Engineers' Handbook,* New York: McGraw-Hill, 1982.

L. J. Giacoletto, *Electronics Designers' Handbook,* 2nd ed., New York: McGraw-Hill, 1977.

F. F. Mazda, *Electronics Engineers' Reference Book,* 5th ed., London: Butterworth, 1983.

46

Permittivity Measurement

Devendra K. Misra
University of Wisconsin

Dielectric materials possess relatively few free charge carriers. Most of the charge carriers are bound and cannot participate in conduction. However, these bound charges can be displaced by applying an external electric field. In such cases, the atom or molecule forms an electric dipole that maintains an electric field. Consequently, each volume element of the material behaves as an electric dipole. The dipole field tends to oppose the applied field. Dielectric materials that exhibit nonzero distribution of such bound charge separations are said to be *polarized*. The volume density of polarization $\vec{P}$ describes the volume density of those electric dipoles. When a material is linear and isotropic in nature, the polarization density is related to applied electric field intensity, $\vec{E}$, as follows:

$$\vec{P} = \varepsilon_0 \chi_e \vec{E} \qquad (46.1)$$

where ε_0 (= 8.854×10^{-12} F m^{-1}) is the permittivity of free-space and χ_e is called the electric susceptibility of the material.

The electric flux density, or displacement, $\vec{D}$ is defined as follows:

$$\vec{D} = \varepsilon_0 \vec{E} + \vec{P} = \varepsilon_0 \left(1 + \chi_o\right)\vec{E} = \varepsilon_0 \varepsilon_r \vec{E} = \varepsilon \vec{E} \qquad (46.2)$$

where ε is called the permittivity of the material and ε_r is its relative permittivity or dielectric constant. Electric flux density is expressed in coulombs per meter (C m^{-1}).

Equation 46.2 represents a relation between the electric flux density and electric field intensity in frequency domain. It will hold well in time-domain only if the permittivity is independent of frequency. A material is called *dispersive* if its characteristics are frequency dependent. The product of Equation 46.2 in frequency domain will be replaced by a convolution integral for the time-domain fields.

Assuming that the fields are time-harmonic as $e^{j\omega t}$, the generalized Ampere's law can be expressed in phasor form as follows:

$$\nabla \times \vec{H} = \vec{J}^e + \vec{J} + j\omega \vec{D} \qquad (46.3)$$

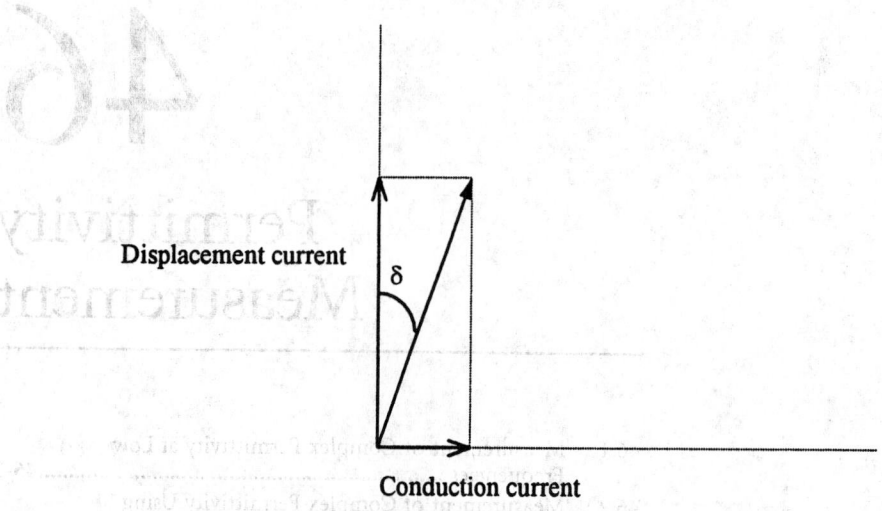

FIGURE 46.1 A phasor diagram representing displacement and loss currents.

where H is the magnetic field intensity in A m^{-1} and J^e is current-source density in A m^{-2}. J is the conduction current density in A m^{-2} and the last term represents the displacement current density. J^e will be zero for a source-free region.

The conduction current density is related to the electric field intensity through Ohm's law as follows:

$$\vec{J} = \sigma \vec{E} \qquad (46.4)$$

where σ is the conductivity of material in S m^{-1}.

From Equations 46.2 through 46.4, one obtains:

$$\nabla \times \vec{H} = \vec{J}^e + \sigma \vec{E} + j\omega\varepsilon\vec{E} \qquad (46.5)$$

Conduction current represents the loss of power. There is another source of loss in dielectric materials. When a time-harmonic electric field is applied, the dipoles flip back and forth constantly. Because the charge carriers have finite mass, the field must do work to move them and they might not respond instantaneously. This means that the polarization vector will lag behind the applied electric field. This factor shows up at high frequencies. Therefore, Equation 46.5 is modified as follows:

$$\nabla \times \vec{H} = \vec{J}^e + \sigma \vec{E} + \omega\kappa''\vec{E} + j\omega\varepsilon\vec{E} = \vec{J}^e + j\omega\left(\varepsilon - j\frac{\sigma + \omega\kappa''}{\omega}\right)\vec{E} = \vec{J}^e + j\omega\varepsilon^*\vec{E} \qquad (46.6)$$

The complex relative permittivity of a material is defined as follows:

$$\varepsilon_r^* = \frac{\varepsilon^*}{\varepsilon_0} = \frac{1}{\varepsilon_0}\left(\varepsilon - j\frac{\sigma + \omega\kappa''}{\omega}\right) = \varepsilon_r' - j\varepsilon_r'' = \varepsilon_r\left(1 - j\tan\delta\right) \qquad (46.7)$$

where ε_r' and ε_r'' represent real and imaginary parts of the complex relative permittivity. The imaginary part is zero for a lossless material. The term $\tan\delta$ is called the *loss tangent*. It represents the tangent of angle between the displacement phasor and total current, as shown in Figure 46.1. Thus, it will be close to zero for a low-loss material.

TABLE 46.1 Dielectric Dispersion Parameters for Some Liquids at Room Temperature

Substance	ε_∞	ε_s	α	τ (picoseconds)
Water	5	78	0	8.0789
Methanol	5.7	33.1	0	53.0516
Ethanol	4.2	24	0	127.8545
Acetone	1.9	21.2	0	3.3423
Ethylene glycol	3	37	0.23	79.5775
Propanol	3.2	19	0	291.7841
Butanol	2.95	17.1	0.08	477.4648
Chlorobenzene	2.35	5.63	0.04	10.2920

TABLE 46.2 Complex Permittivity of Some Substances at Room Temperature

Substance	60 Hz	1 MHz	10 GHz
Nylon	3.60–j 0.06	3.14–j 0.07	2.80–j 0.03
Plexiglas	3.45–j 0.22	2.76–j 0.04	2.5–j 0.02
Polyethylene	2.26–j 0.0005	2.26–j 0.0005	2.26–j 0.0011
Polystyrene	2.55–j 0.0077	2.55–j 0.0077	2.54–j 0.0008
Styrofoam	1.03–j 0.0002	1.03–j 0.0002	1.03–j 0.0001
Teflon	2.1–j 0.01	2.1–j 0.01	2.1–j 0.0008
Glass (lead barium)	6.78–j 0.11	6.73–j 0.06	6.64–j 0.31

Dispersion characteristics of a large class of materials can be represented by the following empirical equation of Cole-Cole.

$$\varepsilon_r^* = \varepsilon_\infty + \frac{\varepsilon_s - \varepsilon_\infty}{1 + \left(j\omega\tau \right)^{1-\alpha}} \tag{46.8}$$

where ε_∞ and ε_s are the relative permittivities of material at infinite and zero frequencies, respectively. ω is the signal frequency in radians per second, and τ is the characteristic relaxation time in seconds. For α equal to zero, Equation 46.8 reduces to the Debye equation. Dispersion parameters for a few liquids are given in Table 46.1.

Complex permittivity of a material is determined using lumped circuits at low frequencies, and distributed circuits or free-space reflection and transmission of waves at high frequencies. Capacitance and dissipation factor of a lumped capacitor are measured using a bridge or a resonant circuit. The complex permittivity is calculated from this data. Complex permittivities for some substances are presented in Table 46.2.

At high frequencies, the sample is placed inside a transmission line or a resonant cavity. Propagation constants of the transmission line or resonant frequency and quality factor of the cavity resonator are used to calculate the complex permittivity. Propagation characteristics of electromagnetic waves are influenced by the complex permittivity of that medium. Therefore, a material can be characterized by monitoring the reflected and transmitted wave characteristics as well.

46.1 Measurement of Complex Permittivity at Low Frequencies [1, 2]

A parallel-plate capacitor is used to determine the complex permittivity of dielectric sheets. For a separation d between the plates of area A in vacuum, the capacitance is given by:

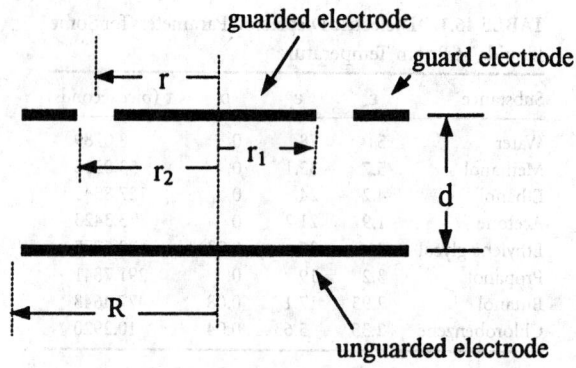

FIGURE 46.2   Geometry of a guarded capacitor.

$$C_0 = 8.854 \frac{A}{d} \ \text{pF} \tag{46.9}$$

where all dimensions are measured in meters. If the two plates have different areas, then the smaller one is used to determine C_0. Further, it is assumed that the field distribution is uniform and perpendicular to the plates. Obviously, the fringing fields along the edges do not satisfy this condition. The guard electrodes, as shown in Figure 46.2, are used to ensure that the field distribution is close to the assumed condition. For best results, the width of the guard electrode must be at least $2d$, and the unguarded plate must extend to outer edge of the guard electrode. Further, the gap between the guarded and guard electrodes must be as small as possible.

The radius of guarded electrode is r_1, and the inner radius of guard electrode is r_2. It is assumed that $R - r_2 \geq 2d$. The area A for this parallel plate capacitor is πr^2, where r is defined as follows:

$$r = r_1 + \Delta \tag{46.10}$$

$$\Delta = \frac{1}{2}(r_2 - r_1) - \frac{2d}{\pi}\ln\left(\cosh\frac{\pi(r_2 - r_1)}{4d}\right) = \frac{1}{2}(r_2 - r_1) - 1.4659d\ln\left(\cosh 0.7854\frac{r_2 - r_1}{d}\right) \tag{46.11}$$

Using the Debye model (i.e., $\alpha = 0$ in Equation 46.8), an equivalent circuit for a dielectric-filled parallel plate capacitor can be drawn as shown in Figure 46.3. If a step voltage V is applied to it, then the current I can be found as follows [2].

$$I = \varepsilon_\infty C_0 V \delta(t) + \frac{V C_0(\varepsilon_0 - \varepsilon_\infty)}{\tau}\exp\left(-\frac{t}{\tau}\right) \tag{46.12}$$

where $\tau = RC_0(\varepsilon_0 - \varepsilon_\infty)$

The first term in Equation 46.12 represents the charging current of capacitor $\varepsilon_\infty C_0$ in the upper branch. This current is not measured because it disappears instantaneously. In practice, it needs to be bypassed at short times to protect the detector from overloading or burning. The second term of Equation 46.12 represents charging current of the lower branch of an equivalent circuit. The time constant, τ, is determined following the decay characteristics of this current. Further, the resistance R can be found after

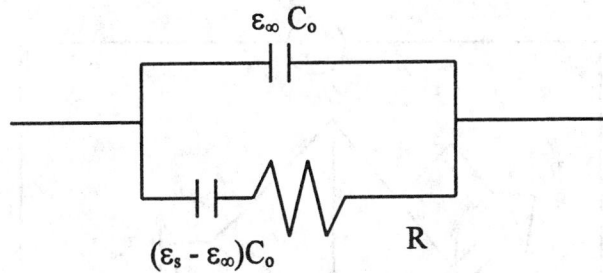

FIGURE 46.3 Equivalent circuit of a parallel-plate capacitor based on the Debye model.

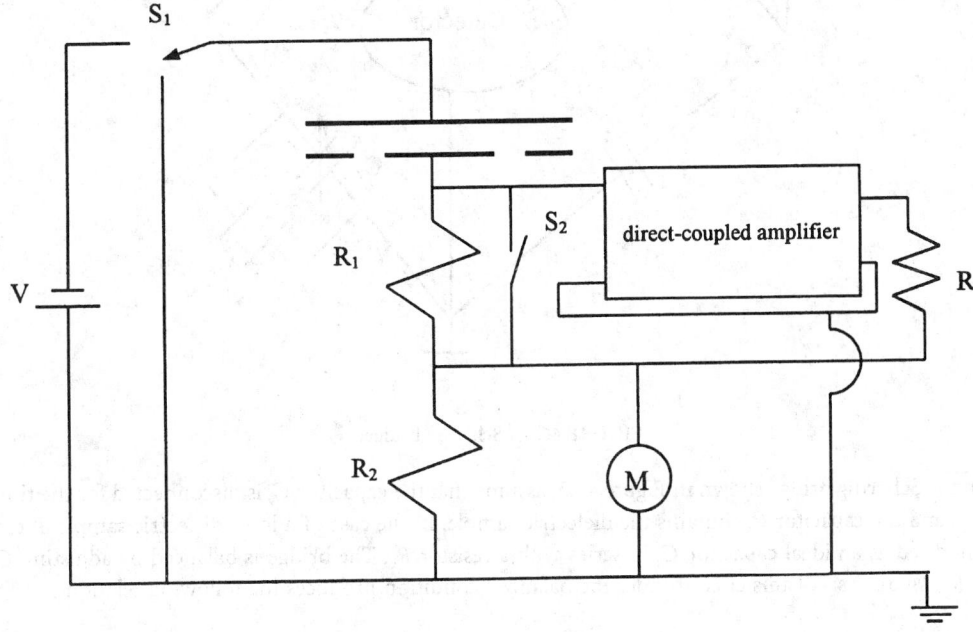

FIGURE 46.4 Circuit arrangement for the characterization of dielectric materials using a step voltage.

extrapolating this current-time curve to $t = 0$. The discharging current characteristics are used to remove V at $t = 0$.

A typical circuit arrangement for the characterization of dielectric materials using a step voltage is shown in Figure 46.4. A standard resistor R_1 of either 10 GΩ or 1 TΩ is connected between the guarded electrode and the load resistor R_2. A feedback circuit is used that forces the voltage drop across R_1 equal in magnitude but opposite in polarity to that of across R_2. It works as follows. Suppose that the node between capacitor and R_1 has a voltage V_1 with respect to ground. It is amplified but reversed in polarity by the amplifier. Therefore, the current through R_1 will change. This process continues until the input to the amplifier is zero. The junction between R_1 and the capacitor will then be at the ground potential. Thus, the meter M measures voltage across R_2 that is negative of the voltage across R_1. Since R_1 is known, the current through it can be calculated. This current also flows through the sample. S_1 is used to switch from charging to discharging mode while S_2 is used to provide a path for surge currents.

Capacitance and dissipation factor of the dielectric-loaded parallel-plate capacitor are used in the medium frequency range to determine the complex permittivity of materials. A substitution method is generally employed in a Schering bridge circuit for this measurement.

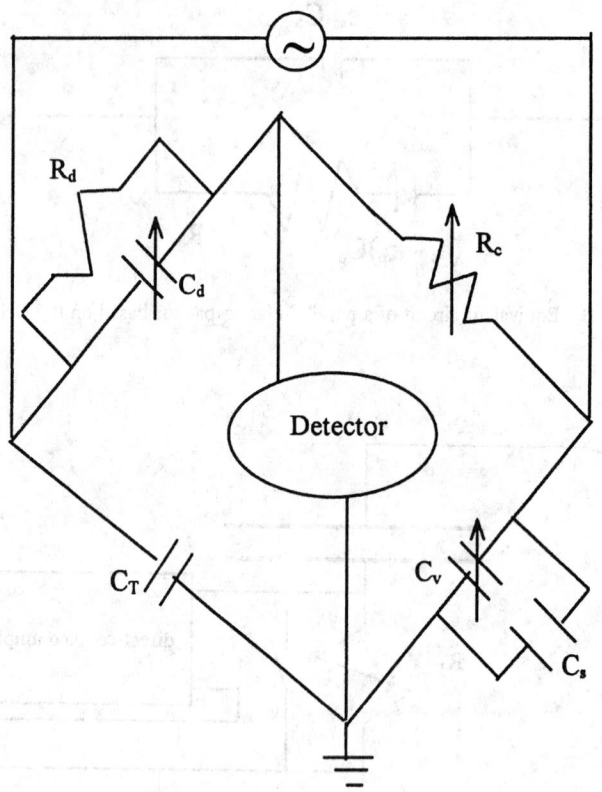

FIGURE 46.5   Schering bridge.

In the Schering bridge shown in Figure 46.5, assume that the capacitor C_v is disconnected for the time being, and the capacitor C_s contains the dielectric sample. In the case of a lossy dielectric sample, it can be modeled as an ideal capacitor C_x in series with a resistor R_x. The bridge is balanced by adjusting C_d and R_c. An analysis of this circuit under the balanced condition produces the following relations.

$$R_x = \frac{C_d R_c}{C_T} \tag{46.13}$$

and

$$C_x = \frac{C_T R_d}{R_c} \tag{46.14}$$

Quality factor Q of a series RC circuit is defined as the tangent of its phase angle, while the inverse of Q is known as the dissipation factor D. Hence,

$$Q = \frac{X_x}{R_x} = \frac{1}{\omega C_x R_x} = \frac{1}{D} \tag{46.15}$$

For a fixed R_d, the capacitor C_d can be calibrated directly in terms of dissipation factor. Similarly, the resistor R_c can be used to determine C_x. However, an adjustable resistor limits the frequency range. A substitution method is preferred for precision measurement of C_x at higher frequencies. In this technique,

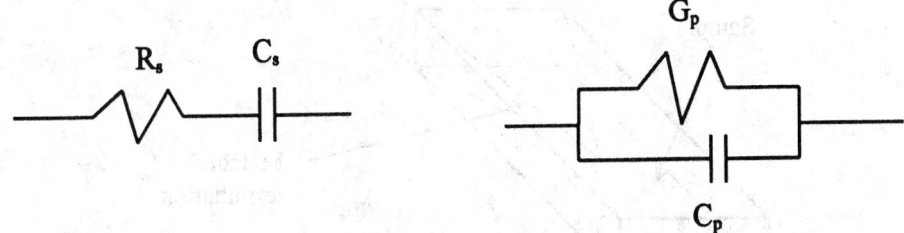

FIGURE 46.6 Series and parallel equivalent circuits of a dielectric loaded capacitor.

a calibrated precision capacitor C_v is connected in parallel with C_s as shown in Figure 46.5 and the bridge is balanced. Assume that the settings of two capacitors at this condition are C_{d1} and C_{v1}. The capacitor C_s is then removed and the bridge is balanced again. Let the new settings of these capacitors be C_{d2} and C_{v2}, respectively. Equivalent circuit parameters of the dielectric loaded capacitor C_s are then found as follows.

$$C_x = C_{v2} - C_{v1} \tag{46.16}$$

$$D_x = \frac{C_{v2}}{C_x}\delta D \tag{46.17}$$

where $\delta D = \omega R_d (C_{d1} - C_{d2})$

Complex permittivity of the specimen is calculated from these data as follows:

$$\varepsilon'_r = \frac{C_x}{C_0} \tag{46.18}$$

and

$$\varepsilon''_r = \frac{C_x D_x}{C_0} \tag{46.19}$$

Thus far, a series *RC* circuit equivalent model is used for the dielectric-loaded capacitor. As illustrated in Figure 46.6, an equivalent parallel *RC* model can also be obtained for it. The following equations can be used to switch back and forth between these two equivalent models.

$$G_p = \frac{R_s}{R_s^2 + \dfrac{1}{\omega^2 C_s^2}} = \frac{1}{R_s}\left(\frac{1}{1+Q^2}\right) \tag{46.20}$$

$$C_p = \frac{C_s}{1+\left(\omega R_s C_s\right)^2} = \frac{C_s}{1+D^2} \tag{46.21}$$

$$R_s = \frac{G_p}{G_p^2 + \omega^2 C_p^2} = \frac{1}{G_p}\left(\frac{1}{1+Q^2}\right) \tag{46.22}$$

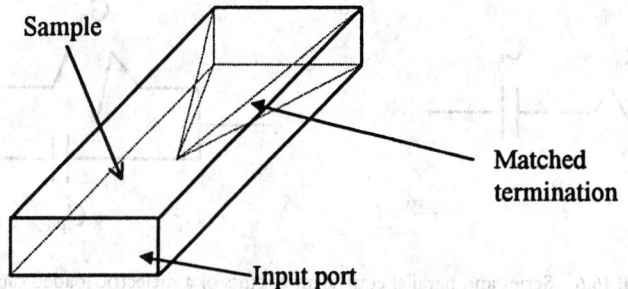

FIGURE 46.7   A waveguide termination filled with liquid or powder sample.

$$C_s = \frac{G_p^2 + \omega^2 C_p^2}{\omega^2 C_p} = C_p\left(1 + D^2\right) \tag{46.23}$$

and

$$Q = \frac{1}{D} = \frac{\omega C_p}{G_p} = \frac{1}{\omega R_s C_s} \tag{46.24}$$

Proper shielding and grounding arrangements are needed for a reliable measurement, especially at higher frequencies. Grounding and edge capacitances of the sample holder need to be taken into account for improved accuracy. Further, a guard point needs to be obtained that may require balancing in some cases. An inductive-ratio-arm capacitance bridge can be another alternative to consider for such application [1].

46.2   Measurement of Complex Permittivity Using Distributed Circuits

Measurement techniques based on the lumped circuits are limited up to the lower end of the VHF band. Characterization of materials at microwave frequencies requires the distributed circuits. A number of techniques have been developed on the basis of wave reflection and transmission characteristics inside a transmission line or in free space. Some other methods employ a resonant cavity that is loaded with the sample. Cavity parameters are measured and the material characteristics are deduced from that. A number of these techniques, described in [3, 4], can be used for a sheet material. These techniques require cutting a piece of sample to be placed inside a transmission line or a cavity. In case of liquid or powder samples, a so-called modified infinite sample method can be used. In this technique, a waveguide termination is filled completely with the sample, as shown in Figure 46.7. Since a tapered termination is embedded in the sample, the wave incident on it will be dissipated with negligible reflection and it will look like the sample is extending to infinity. The impedance at its input port will depend on the electrical properties of filling sample. Its VSWR S and location of first minimum d from the load plane are measured using a slotted line. The complex permittivity of sample is then calculated as follows [5].

$$\varepsilon_r' = \left(\frac{\lambda}{\lambda_c}\right)^2 + \frac{\left[1 - \left(\frac{\lambda}{\lambda_c}\right)^2\right] \times \left[S^2 \sec^4\left(\beta d\right) - \left(1 - S^2\right)^2 \tan^2\left(\beta d\right)\right]}{\left[1 + S^2 \tan^2\left(\beta d\right)\right]^2} \tag{46.25}$$

and

$$\varepsilon''_r = \frac{\left[1-\left(\lambda \big/ \lambda_c\right)^2\right] \times \left[2S\left(1-S^2\right)^2 \sec^4\left(\beta d\right)\tan\left(\beta d\right)\right]}{\left[1+S^2\tan^2\left(\beta d\right)\right]^2} \tag{46.26}$$

where λ = Free-space wavelength
λ_c = Cut-off wavelength for the mode of propagation in empty guide
β = Propagation constant in the feeding guide

It is assumed that the waveguide supports TE_{10} mode only.

Resonant Cavity Method

A cavity resonator can be used to determine the complex permittivity of materials at microwave frequencies. If a cavity can be filled completely with the sample, then the following procedure can be used.

Measure the resonant frequency f_1 and the quality factor Q_1 of an empty cavity. Next, fill that cavity with the sample material and measure its new resonant frequency f_2 and quality factor Q_2. The dielectric parameters of the sample are then calculated from the following formulae [3].

$$\varepsilon_r = \left(1+\frac{f_1-f_2}{f_2}\right)^2 \tag{46.27}$$

and

$$\tan\delta = \frac{1}{Q_2}-\frac{1}{Q_1}\sqrt{\frac{f_1}{f_2}} \tag{46.28}$$

On the other hand, a cavity perturbation technique will be useful for smaller samples [4]. If the sample is available in a circular cylindrical form, then it may be placed inside a TE_{101} rectangular cavity through the center of its broad face where the electric field is maximum. Its resonant frequency and quality factor with and without sample are then measured. Complex permittivity of sample is calculated as follows.

$$\varepsilon'_r = 1+\frac{1}{2}\frac{f_1-f_2}{f_2}\frac{V}{v} \tag{46.29}$$

and

$$\varepsilon''_r = \frac{V}{4v}\frac{Q_2-Q_1}{Q_1Q_2} \tag{46.30}$$

where V and v are cavity and sample volumes, respectively.

Similarly, for a small spherical sample of radius r that is placed in a uniform field at the center of the rectangular cavity, the dielectric parameters are as follows.

$$\varepsilon'_r = \frac{abd}{8\pi r^3}\frac{f_1-f_2}{f_2} \tag{46.31}$$

and

$$\varepsilon_r'' = \frac{abd}{16\pi r^3}\left(\frac{Q_2 - Q_1}{Q_1 Q_2}\right) \tag{46.32}$$

Where a, b, and d are the width, height, and length of the rectangular cavity, respectively. For best accuracy in cavity perturbation method, the shift in frequency $(f_1 - f_2)$ must be very small.

Free-Space Method for Measurement of Complex Permittivity

When a plane electromagnetic wave is incident on a dielectric interface, its reflection and transmission depend on the contrast in the dielectric parameters. Many researchers have used it for determining the complex permittivity of dielectric materials placed in free space. An automatic network analyzer and phase-corrected horn antennas can be used for such measurements [6]. The system is calibrated using the TRL (through, reflect, and line) technique. A time-domain gating is used to minimize the error due to multiple reflections. The sample of thickness d is placed in front of a conducting plane and its reflection coefficient S_{11} is measured. A theoretical expression for this reflection coefficient is found as follows.

$$S_{11} = \frac{jZ_d\,\tan(\beta_d d) - 1}{jZ_d\,\tan(\beta_d d) + 1} \tag{46.33}$$

Where:

$$Z_d = \frac{1}{\sqrt{\varepsilon_r^*}} \tag{46.34}$$

$$\beta_d = \frac{2\pi}{\lambda}\sqrt{\varepsilon_r^*} \tag{46.35}$$

λ = Free-space wavelength of electromagnetic signal

Equation 46.33 is solved for ε_r^* after substituting the measured S_{11}. Since it represents a nonlinear relation, an iterative numerical procedure can be used.

A Nondestructive Method for Measuring the Complex Permittivity of Materials

Most of the techniques described thus far require cutting and placing a part of sample in the test fixture. Sometimes, it may not be permissible to do so. Further, the dielectric parameters can change in that process. It is especially important in the case of a biological specimen to perform *in vivo* measurements. In one such technique, an open-ended coaxial line is placed in close contact with the sample and its input reflection coefficient is measured using an automatic network analyzer [7, 8]. As recommended by the manufacturers, the network analyzer is calibrated initially using an open-circuit, a short-circuit, and a matched load. The reference plane is then moved to the measuring end of the coaxial line using a short-circuit.

Assume that a and b are inner and outer radii of the coaxial line, respectively. ω is the angular frequency; μ_0 and ε_0 are the permeability and permittivity of the free space, respectively, $k = \omega\sqrt{\mu_0\varepsilon_0\varepsilon_r^*}$ is the wavenumber in material medium. Admittance of the coaxial aperture in contact with material medium is as follows.

$$Y_L = \frac{2}{\int_a^b E_\rho(\rho',0)d\rho'} - \frac{2\pi}{\left[\sqrt{\frac{\mu_0}{\varepsilon_0\varepsilon_1}}\ln(b/a)\right]} \tag{46.36}$$

where $E_\rho(\rho',0)$ is radial electric field intensity over the aperture. It is evaluated from the following integral equation.

$$\frac{1}{\pi\rho} + j\omega\varepsilon_1\varepsilon_0 \int_a^b E_\rho(\rho',0)K_c(\rho,\rho')\rho'\,d\rho' = \frac{j\omega\varepsilon_r^*\varepsilon_0}{\pi}\int_a^b E_\rho(\rho',0)\rho'\,d\rho' \int_0^\pi \cos(\phi')\frac{\exp(-jkr)}{r}d\phi' \tag{46.37}$$

Where:

$$r = \sqrt{\rho^2 + \rho'^2 - 2\rho\rho'\cos(\phi')} \tag{46.38}$$

$$K_c(\rho,\rho') = j\sum_{n=0}^{\infty}\frac{\phi_n(\rho)\phi_n(\rho')}{A_n^2\beta_n} \tag{46.39}$$

$$\phi_n = Y_0(\gamma_n a)J_1(\gamma_n\rho) - J_0(\gamma_n a)Y_1(\gamma_n\rho) \tag{46.40}$$

$$\beta_n = \begin{cases} \sqrt{k_1^2 - \gamma_n^2} & k_1 > \gamma_n \\ -j\sqrt{\gamma_n^2 - k_1^2} & k_1 < \gamma_n \end{cases} \tag{46.41}$$

$$A_n^2 = \frac{2}{\pi^2\gamma_n^2}\left[\frac{J_0^2(\gamma_n a)}{J_0^2(\gamma_n b)} - 1\right] \quad n > 0; \quad A_0^2 = \ln\left(\frac{b}{a}\right) \tag{46.42}$$

The eigenvalues γ_n are solutions to the following characteristic equation:

$$J_0(\gamma_n b)Y_0(\gamma_n a) = J_0(\gamma_n a)Y_0(\gamma_n b) \tag{46.43}$$

J_n and Y_n are Bessel functions of the first and second kind of order n, respectively. ε_1 is the dielectric constant of the insulator and k_1 is wavenumber inside the coaxial line.

Equation 46.37 is solved numerically using the method of moments. A numerical root-finding procedure, such as the Muller's method, is used to solve Equation 46.36 for the complex wavenumber k. Complex permittivity, in turn, is determined from the following relation.

$$\varepsilon_r^* = \frac{k^2}{\omega^2\mu_0\varepsilon_0} \tag{46.44}$$

Defining Terms

Electric dipole: A pair of equal and opposite electric charges separated by a small distance.

Isotropic material: A material in which the electrical polarization has the same direction as the applied electric field.

Electric polarization density: The average electric dipole moment per unit volume.

Electric susceptibility: A dimensionless parameter that relates the polarization density in a material with electric field intensity.

Electric flux density: A fundamental electric field quantity that is related to volume density of free charges. It is also known as the electric displacement.

Time domain field: A field expressed as a function of time. It is a real function that is dependent on time and space coordinates.

Frequency domain field: A phasor quantity (a complex function in general) that depends on space coordinates. The time dependency is assumed to be sinusoidal.

Displacement current density: It represents the time rate of change of electric flux density.

Conduction current density: Current per unit area caused by conduction of charge carriers.

Dielectric constant: A dimensionless constant that represents the permittivity of a material relative to the permittivity of free space.

Loss tangent: A ratio of the imaginary part to the real part of the complex permittivity of a material.

Relaxation time: It represents the time taken by a charge placed inside a material volume to decay to about 37% of its initial value.

Quality factor: A dimensionless quantity that represents the time average energy stored in an electrical circuit relative to energy dissipated in one period.

Dissipation factor: It is the inverse of the quality factor.

Voltage standing wave ratio (VSWR): Defined as a ratio of maximum voltage to the minimum voltage on a transmission line.

Reflection coefficient: Defined as a ratio of reflected phasor voltage to that of incident phasor voltage at a point in the circuit.

References

1. A. R. Von Hippel, *Dielectric Materials and Applications*, Cambridge, MA: MIT Press, 1961.
2. N. E. Hill, W. E. Vaughan, A. H. Price, and M. Davies, *Dielectric Properties and Molecular Behaviour*, London: Van Nostrand Reinhold, 1969.
3. M. Sucher and J. Fox (eds.), *Handbook of Microwave Measurements*, Vol. II, Brooklyn, NY: Polytechnic Press, 1963.
4. R. Chatterjee, *Advanced Microwave Engineering*, Chichester, U.K.: Ellis Horwood Limited, 1988.
5. D. K. Misra, Permittivity measurement of modified infinite samples by a directional coupler and a sliding load, *IEEE Trans. Microwave Theory Technol.*, 29, 65-67, 1981.
6. D. K. Ghodgaonkar, V. V. Varadan, and V. K. Varadan, A free-space method for measurement of dielectric constants and loss tangents at microwave frequencies, *IEEE Trans. Instrum. Meas.*, 38, 789-793, 1989.
7. A. P. Gregory, R. N. Clarke, T. E. Hodgetts, and G. T. Symm, RF and microwave dielectric measurements upon layered materials using a reflectometric coaxial sensor, *NPL Rep.*, DES 125, UK, March 1993.
8. D. Misra, On the measurement of the complex permittivity of materials by an open-ended coaxial probe, *IEEE Microwave Guided Wave Lett.*, 5, 161-163, 1995.

47
Electric Field Strength[1]

David A. Hill
National Institute of Standards and Technology

Motohisa Kanda
National Institute of Standards and Technology

Electric field strength is defined as the ratio of the force on a positive test charge at rest to the magnitude of the test charge in the limit as the magnitude of the test charge approaches zero. The units of electric field strength are volts per meter (V m^{-1}). Electric charges and currents are sources of electric and magnetic fields, and Maxwell's equations [1] provide the mathematical relationships between electromagnetic (EM) fields and sources.

The electric field at a point in space is a vector defined by components along three orthogonal axes. For example, in a rectangular coordinate system, the electric field $\vec{E}$ can be written as:

$$\vec{E} = \hat{x}E_x + \hat{y}E_y + \hat{z}E_z \qquad (47.1)$$

where $\hat{x}$, $\hat{y}$, and $\hat{z}$ are unit vectors and E_x, E_y, and E_z are scalar components. For electrostatic fields, the components are real scalars that are independent of time. For steady-state, time-harmonic fields, the components are complex phasors that represent magnitude and phase. The time dependence, $e^{j\omega t}$, is suppressed.

47.1 Electrostatic Fields

Electrostatic fields are present throughout the atmosphere, and there are strong electrostatic fields near high-voltage dc power lines. The commonly used electrostatic field meters generate an ac signal by periodic conductor motion (either rotation or vibration). This ac signal is proportional to the electric field strength, and field meter calibration is performed in a known electrostatic field.

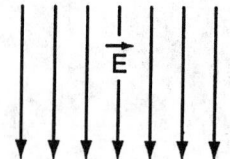

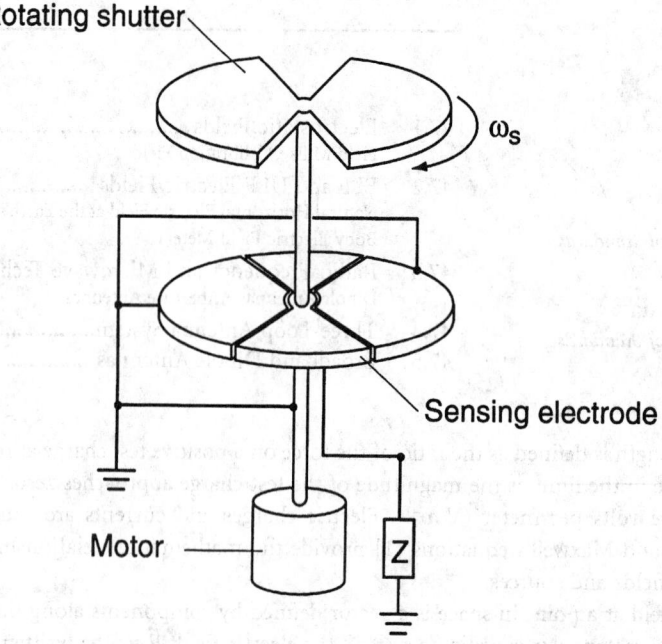

FIGURE 47.1　Shutter-type electric field mill for measurement of the polarity and magnitude of an electrostatic field.

Field Mills

Field mills (also called generating voltmeters) determine electric field strength by measuring modulated, capacitively induced charges or currents on metal electrodes. Two types of field mills — the shutter type and the cylindrical type — are described in the technical literature [2]. The shutter type is more common; a simplified version is shown in Figure 47.1. The sensing electrode is periodically exposed to and shielded from the electric field by a grounded, rotating shutter. The charge q_s induced on the sensing electrode and the current i_s between the sensing electrode and ground are both proportional to the electric field strength E normal to the electrode:

$$q_s\left(t\right)=\epsilon_0 E a_s\left(t\right) \quad \text{and} \quad i_s\left(t\right)=\epsilon_0 E \frac{d a_s\left(t\right)}{dt} \tag{47.2}$$

where ϵ_0 is the permittivity of free space [1] and $a_s(t)$ is the effective exposed area of the sensing electrode at time t.

Thus, the field strength can be determined by measuring the induced charge or current (or voltage across the impedance Z). If the induced signal is rectified by a phase-sensitive detector (relative to the shutter motion), the dc output signal will indicate both the polarity and magnitude of the electric field [3].

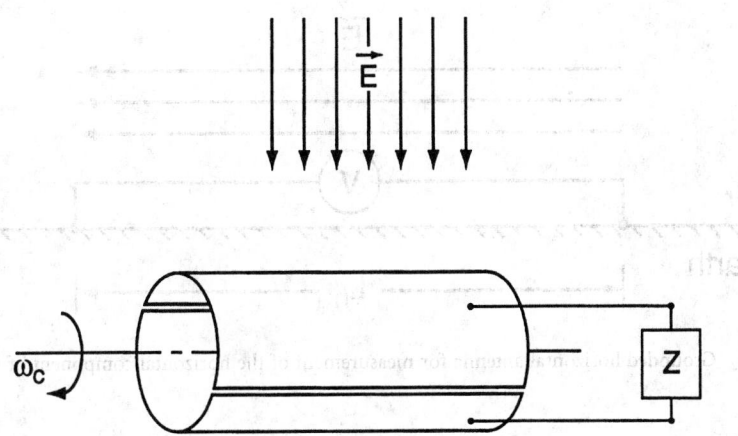

FIGURE 47.2 Cylindrical field mill for measurement of electrostatic field strength.

Shutter-type field mills are typically operated at the ground or at a ground plane, but a cylindrical field mill can be used to measure the electric field at points removed from a ground plane. A cylindrical field mill consists of two half-cylinder sensing electrodes as shown in Figure 47.2. Charges induced on the two sensing electrodes are varied periodically by rotating the sensing electrodes about the cylinder axis at a constant angular frequency ω_c. The charge q_c induced on a half- cylinder of length L and the current i_c between the half-cylinders are given by:

$$q_c = 4\epsilon_0 r_c LE \sin \omega_c t \quad \text{and} \quad i_c = 4\epsilon_0 r_c LE \omega_c \cos \omega_c t \tag{47.3}$$

where r_c is the cylinder radius. Equation 47.3 is based on the two-dimensional solution for a conducting cylinder in an electric field and neglects end effects for finite L. Equation 47.3 shows that the electric field strength E can be determined from a measurement of the induced charge or current.

A third type of electric field meter uses a vibrating plate [4] to generate an ac signal that is proportional to the electric field strength. With any type of electric field strength meter, the observer should be at a sufficient distance from the measurement location to avoid perturbing the electric field.

Calibration Field

A known uniform field for meter calibration can be produced between a pair of parallel plates [2]. If a potential difference V is applied between a pair of plates with a separation d_p, the field strength away from the plate edges is V/d_p. The plate dimensions should be much larger than d_p to provide an adequate region of uniform field. Also, d_p should be much larger than the field meter dimensions so that the charge distribution on the plates is not disturbed. The parallel plates can be metal sheets or tightly stretched metal screens.

The field meter should be located in the type of environment in which it will be used. Shutter-type field mills that are intended to be located at a ground plane should be located at one of the plates. Cylindrical field mills that are not intended to be used at a ground plane should be located in the center of the region between the plates.

For simplicity, only field mills and calibration in the absence of space charge were mentioned. However, near power lines or in the upper atmosphere [5], the effects of space charge can be significant and require modifications in field mill design. A field mill for use in a space charge region and a calibration system with space charge are described in [6].

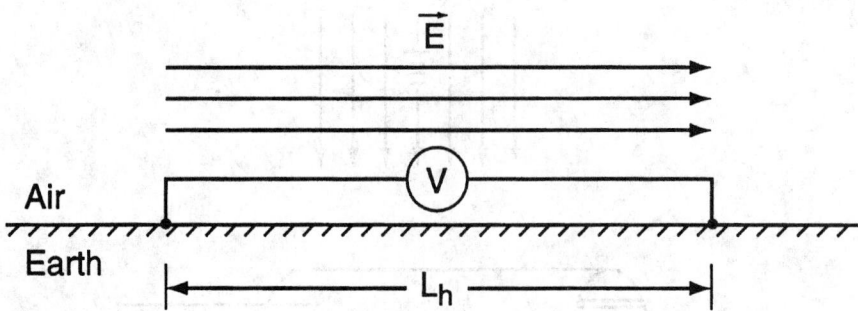

FIGURE 47.3   Grounded horizontal antenna for measurement of the horizontal component of the electric field.

47.2   ELF and ULF Electric Fields

In this section, measurement techniques for extremely low frequency (ELF, 3 Hz to 3 kHz) and ultralow frequency (ULF, below 3 Hz) electric fields are considered. Natural ELF fields are produced by thunderstorms, and natural ULF fields are produced by micropulsations in the earth's magnetic field [7]. Geophysicists make use of these natural fields in the magnetotelluric method for remote sensing of the Earth's crust [8]. ac power lines are dominant sources of fields at 50 Hz or 60 Hz and their harmonics.

An ac electric field strength meter [9] includes two essential parts: (1) an antenna and (2) a detector (receiver). Other possible features are a transmission line or optical link, frequency-selective circuits, amplifying and attenuating circuits, an indicating device, and a nonconducting handle. The antenna characteristics can be calculated for simple geometries or determined by calibration. For example, linear antennas are often characterized by their effective length L_{eff} [10], which determines the open-circuit voltage V_{oc} induced at the antenna terminals:

$$V_{oc} = L_{eff}\, E_{inc} \tag{47.4}$$

where E_{inc} is the component of the incident electric field parallel to the axis of the linear antenna. The detector could respond to the terminal voltage or current or to the power delivered to the load.

Natural Horizontal Electric Field at the Earth's Surface

Magnetotelluric sounding of the Earth's crust requires measurement of the horizontal electric and magnetic fields at the Earth's surface [8]. The magnetic field is measured with a horizontal axis loop, and the electric field is measured with a horizontal wire antenna as shown in Figure 47.3. The antenna wire is insulated since it lies on the ground, but it is grounded at its end points. Nonpolarizing grounding electrodes should be used to avoid polarization potentials between the electrodes and the ground.

Since the natural electric field strength to be measured is on the order of 1 μV m^{-1}, the antenna length L_h needs to be on the order of 1 km to produce a measurable voltage. Since the effective length of a grounded antenna is equal to the physical length ($L_{eff} = L_h$), the horizontal component E_h of the electric field parallel to the antenna is equal to the open-circuit voltage divided by the antenna length:

$$E_h = V_{oc}/L_h \tag{47.5}$$

The frequencies used in typical magnetotelluric sounding range from approximately 0.1 mHz to 10 Hz. If both horizontal components of the electric field are needed, a second orthogonal antenna is required.

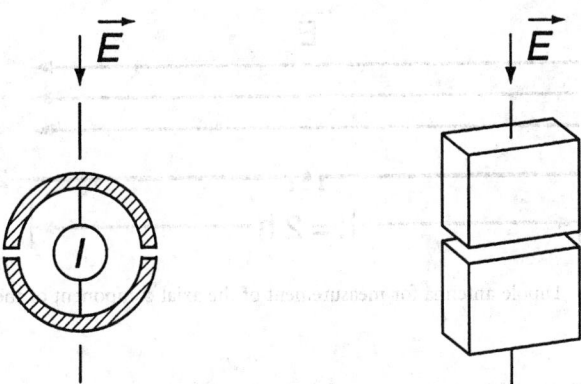

FIGURE 47.4 Electric field meters for measurement of the axial component of the electric field: (a) spherical geometry and (b) rectangular geometry.

Free-Body Electric Field Meters

ELF electric fields in residential and industrial settings are most conveniently measured with free-body field meters [11, 12], which measure the steady-state current or charge oscillating between two halves of a conducting body in free space. (Ground reference meters [13] are also available for measuring the electric field normal to the ground or some other conducting surface.) Geometries for free-body electric field meters are shown in Figure 47.4. Commercial field meters are usually rectangular in shape, and typical dimensions are on the order of 10 cm. A large dynamic range (1 V m^{-1} to 30 kV m^{-1}) is required to cover the various field sources (ac power lines, video display terminals, mass transportation systems, etc.) of interest. (Electro-optic field meters with less sensitivity are described in [12].) A long, nonconducting handle is normally attached perpendicular to the field-meter axis for use in measurement surveys.

The charge Q on half of the field meter is proportional to the incident electric field E along the meter axis:

$$Q = A \epsilon_0 E \tag{47.6}$$

where ϵ_0 is the permittivity of free space [1] and A is a constant proportional to the surface area. For the spherical geometry in Figure 47.4(a), $A = 3\pi a^2$, where a is the sphere's radius. Since the current I between the two halves is equal to the time derivative of the charge, for time-harmonic fields it can be written:

$$I = j\omega A \epsilon_0 E \tag{47.7}$$

This allows E to be determined from the measured current. For commercial field meters that are not spherical, the constant A needs to be determined by calibration. A known calibration field can be generated between a pair of parallel plates where the plates are sufficiently large compared to the separation to produce a uniform field with small edge effects. This technique produces a well-characterized field with an uncertainty less than 0.5% [11]. However, the presence of harmonic frequencies can cause less accurate meter readings in field surveys.

47.3 Radio-Frequency and Microwave Techniques

Dipole Antennas

A thin, linear dipole antenna of length L is shown in Figure 47.5. Its effective length is approximately [10]:

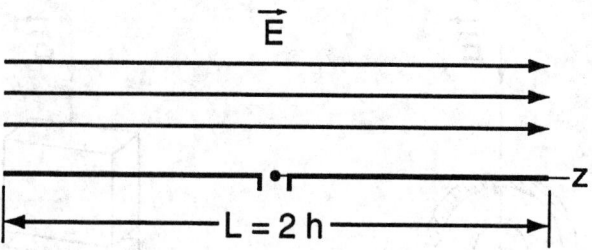

FIGURE 47.5 Dipole antenna for measurement of the axial component of the electric field.

$$L_{eff} = \frac{\lambda}{\pi} \tan\left(\frac{\pi L}{2\lambda}\right) \qquad (47.8)$$

where λ is the free-space wavelength. Resonant half-wave dipoles ($L = \lambda/2$) have an effective length of $L_{eff} = 2L/\pi$ and are of convenient length for frequencies from 30 to 1000 MHz. The physical length of a dipole at resonance is actually slightly shorter than $\lambda/2$ to account for the effect of a finite length-to-diameter ratio. Resonant dipoles are used as standard receiving antennas to establish a known standard field in the *standard antenna method* [9]. Commercial antennas and field meters are calibrated in such standard fields.

For $L < \lambda/2$, Equation 47.8 must be used to determine L_{eff}. For very short dipoles ($L/\lambda \ll 1$), the current distribution is approximately linear, and the effective length is approximately one half the physical length ($L_{eff} \approx L/2$). Short dipoles are frequently used as electric field probes.

Aperture Antennas

Aperture antennas are commonly used for receiving and transmitting at microwave frequencies (above 1 GHz). As receiving antennas, they are conveniently characterized by their on-axis gain g or effective area A_{eff}. Effective area is defined as the ratio of the received power P_r to the incident power density S_{inc} and can also be written in terms of the gain [9]:

$$A_{eff} = \frac{P_r}{S_{inc}} = \frac{g\lambda^2}{4\pi} \qquad (47.9)$$

Equation 47.9 applies to the case where the incident field is polarization-matched to the receiving antenna. The incident power density in a plane wave is $S_{inc} = E^2/\eta_0$, where E is the rms electric field strength and η_0 is the impedance of free space ($\approx 377 \ \Omega$). Thus, the electric field strength can be determined from the received power:

$$E = \sqrt{P_r\eta_0/A_{eff}} = \lambda^{-1}\sqrt{4\pi\eta_0 P_r/g} \qquad (47.10)$$

In general, the gain can be measured using the two-antenna method [9]. For a pyramidal horn antenna as shown in Figure 47.6, the gain can be calculated accurately from [14]:

$$g = \frac{32 \, ab}{\pi\lambda^2} R_E R_H \qquad (47.11)$$

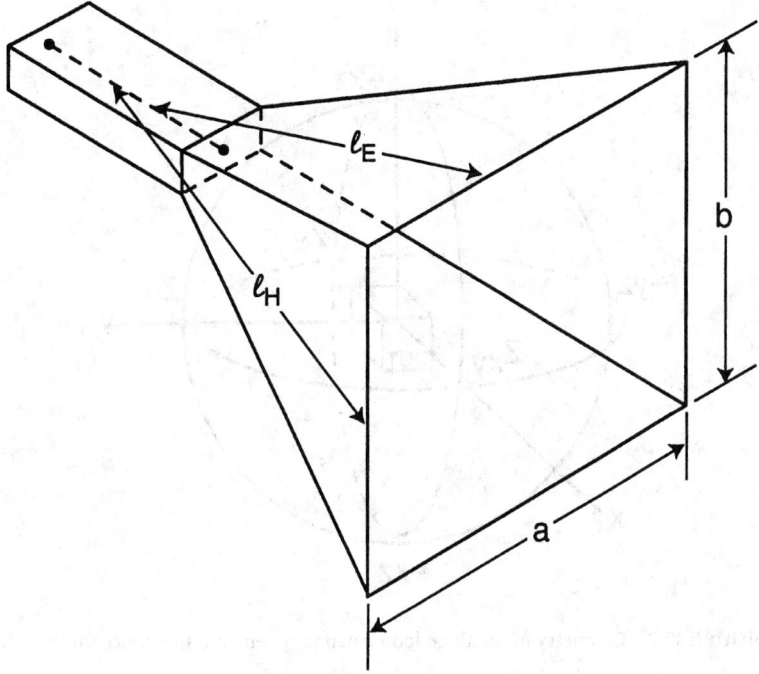

FIGURE 47.6 Pyramidal horn for measuring power density or electric field strength.

where R_E and R_H are gain reduction factors due to the E and H plane flare of the horn. The gain reduction factors are

$$R_E = \frac{C^2(w)+S^2(w)}{w^2} \quad \text{and} \quad R_H = \frac{\pi^2\left\{\left[C(u)-C(v)\right]^2+\left[S(u)-S(v)\right]^2\right\}}{4(u-v)^2}, \quad (47.12)$$

Where:

$$w = \frac{b}{\sqrt{2\lambda l_E}} \quad \text{and} \quad \left.\begin{matrix} u \\ v \end{matrix}\right\} = \frac{\sqrt{\lambda l_H/2}}{a} \pm \frac{a}{\sqrt{2\lambda l_H}}$$

The Fresnel integrals C and S are defined as [15]:

$$C(w) = \int_0^w \cos\left(\frac{\pi}{2}t^2\right)dt \quad \text{and} \quad S(w) = \int_0^w \sin\left(\frac{\pi}{2}t^2\right)dt \quad (47.13)$$

Well-characterized aperture antennas are also used to generate standard fields [16] for calibrating commercial antennas and field strength meters. This method of calibration is called the *standard field method* [9]. The electric field strength E at a distance d from the transmitting antenna is:

$$E = \sqrt{\eta_0 P_{del}g/(4\pi)}/d, \quad (47.14)$$

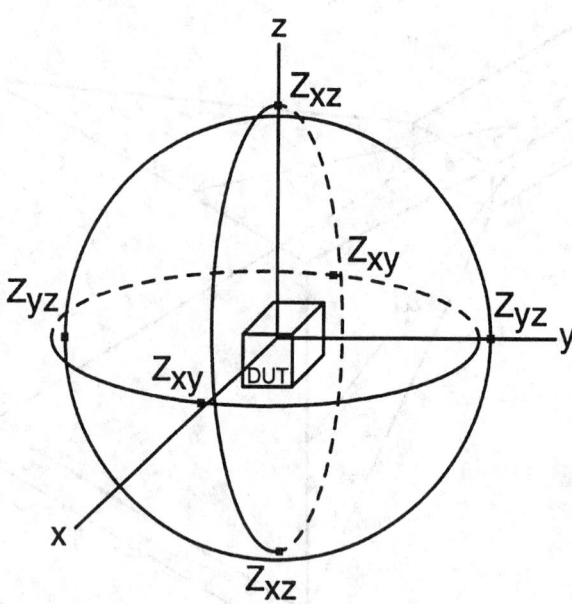

FIGURE 47.7   Geometry of the three-loop antenna system and the device under test.

where P_{del} is the net power delivered to the transmitting antenna and is typically measured with a directional coupler [16]. The gain g for a pyramidal horn can be calculated from Equation 47.11. The National Institute of Standards and Technology (NIST) uses rectangular open-ended waveguides from 200 MHz to 500 MHz and a series of pyramidal horns from 500 MHz to 40 GHz to generate standard fields in an anechoic chamber [16]. The uncertainty of the field strength is less than 1 dB over the entire frequency range of 200 MHz to 40 GHz.

47.4   Three-Loop Antenna System

Electronic equipment can emit unintentional electromagnetic radiation that can interfere with other electronic equipment. If the radiating source is electrically small (as is a video display terminal), then it can be characterized by equivalent electric and magnetic dipole moments. The three-loop antenna system (TLAS), shown in Figure 47.7, consists of three orthogonal loop antennas that are terminated at diametrically opposite points. The unique feature of loop antennas with double terminations [17] is that they can measure both electric and magnetic fields. For electromagnetic interference (EMI) applications, a device under test (DUT) is placed at the center of the TLAS. On the basis of six terminal measurements, the TLAS determines three equivalent electric dipole components and three equivalent magnetic dipole moments of the DUT and hence its radiation characteristics.

Here, the theory [18] is summarized for one of the three loops. The DUT in Figure 47.7 is replaced by an electric dipole moment $\vec{m}_e$ and a magnetic dipole moment $\vec{m}_m$, both located at the origin of the coordinate system. The dipole moments can be written in terms of their rectangular components:

$$\vec{m}_e = \hat{x}m_{ex} + \hat{y}m_{ey} + \hat{z}m_{ez} \text{ and } \vec{m}_m = \hat{x}m_{mx} + \hat{y}m_{my} + \hat{z}m_{mz} \tag{47.15}$$

The loop in the xy plane has radius r_0 and has impedance loads Z_{xy} located at the intersections with the x axis ($\phi = 0, \pi$).

The solution for the current induced in the loop is based on Fourier series analysis [19]. The incident azimuthal electric field $E_\phi^i(\phi)$ tangent to the loop is:

$$E_\phi^i(\phi) = A_0 + A_1 \cos\phi + B_1 \sin\phi \qquad (47.16)$$

Where:

$$A_0 = m_{mx} G_m$$

$$A_1 = m_{ey} G_e$$

$$B_1 = -m_{ex} G_e$$

$$G_m = \frac{\eta_0}{4\pi}\left(\frac{k^2}{r_0} - \frac{jk}{r_0^2}\right)e^{-jkr_0}$$

$$G_e = \frac{-\eta_0}{4\pi}\left(\frac{jk}{r_0} + \frac{1}{r_0^2} + \frac{1}{jkr_0^2}\right)e^{-jkr_0}$$

and $k = 2\pi/\lambda$ is the free-space wavenumber. An approximate solution [17] for the loop current $I(\phi)$ yields the following results for the load currents $I(0)$ and $I(\pi)$:

$$I(0) = 2\pi r_0 \left(\frac{m_{mz} G_m Y_0}{1 + 2Y_0 Z_{xy}} + \frac{m_{ey} G_e Y_1}{1 + 2Y_1 Z_{xy}}\right)$$

$$\qquad (47.17)$$

$$I(\pi) = 2\pi r_0 \left(\frac{m_{mz} G_m Y_0}{1 + 2Y_0 Z_{xy}} - \frac{m_{ey} G_e Y_1}{1 + Y_1 Z_{xy}}\right)$$

where Y_0 and Y_1 are the admittances for the constant and $\cos\phi$ currents [17].

One can solve Equation 47.17 for the magnetic and electric dipole components:

$$m_{mz} = \frac{I_\Sigma(1 + 2Y_0 Z_{xy})}{2\pi r_0 G_m Y_0} \quad \text{and} \quad m_{ey} = \frac{I_\Delta(1 + 2Y_1 Z_{xy})}{2\pi r_0 G_e Y_1} \qquad (47.18)$$

Where:

$$I_\Sigma = \left[I(0) + I(\pi)\right]/2$$

$$I_\Delta = \left[I(0) - I(\pi)\right]/2$$

Thus, the sum current I_Σ can be used to measure the magnetic dipole moment, and the difference current I_Δ can be used to measure the electric dipole moment. The four remaining dipole components can be obtained in an analogous manner. The loop in the xz plane can be used to measure m_{my} and m_{ex}, and the loop in the yz plane can be used to measure m_{mx} and m_{ez}.

The total power P_T radiated by the source can be written in terms of the magnitudes of the six dipole components:

$$P_T = \frac{2\pi\eta_0}{3\lambda^2}\left[|m_{ex}|^2 + |m_{ey}|^2 + |m_{ez}|^2 + k^2\left(|m_{mx}|^2 + |m_{my}|^2 + |m_{mz}|^2\right)\right] \qquad (47.19)$$

The expression for the power pattern is more complicated and involves the amplitudes and phases of the dipole moments [20]. The TLAS has been constructed with 1-m diameter loops and successfully tested from 3 kHz to 100 MHz [21]. It is currently being used to measure radiation from video display terminals and other inadvertent radiators.

47.5 Broadband Dipole Antennas

The EM environment continues to grow more severe and more complex as the number of radiating sources increases. Broadband antennas are used to characterize the EM environment over a wide frequency range. For electric-field measurements, electrically short dipole antennas with a high capacitive input impedance are used with a capacitive load, such as a field-effect transistor (FET). The transfer function S of frequency f is defined as the ratio of the output voltage V_L of the antenna to the incident electric field E_i [22]:

$$S(f) = \frac{V_L(f)}{E_i(f)} = \frac{h\alpha/2}{1 + C/C_a} \tag{47.20}$$

Where:

$$C_a = \frac{4\pi h}{c\eta_0\left(\Omega_a - 2 - \ln 4\right)}$$

$$\alpha = \frac{\Omega_a - 1}{\Omega_a - 2 + \ln 4}$$

$$\Omega_a = 2\ln\left(2h/r_a\right)$$

r_a = Antenna radius
C = Load capacitance
C_a = Antenna capacitance
h = Half the physical length of the dipole antenna, as shown in Figure 47.5
c = Free-space speed of light
Ω_a = Antenna thickness factor

Since the input impedance of an electrically short dipole is predominantly a capacitive reactance, a very broadband frequency response can be achieved with a high-impedance capacitive load. However, given the present state of the art, it is not possible to build a balanced, high-input impedance FET with high common-mode rejection above 400 MHz. For this reason, it is more common practice to use a high-frequency, beam-lead Schottky-barrier diode with a very small junction capacitance (less than 0.1 pF) and very high junction resistance (greater than several M3) for frequencies above 400 MHz.

The relationship between the time-dependent diode current $i_d(t)$ and voltage $v_d(t)$ is

$$i_d(t) = I_s\left[e^{\alpha_d v_d(t)} - 1\right] \tag{47.21}$$

where I_s and α_d are constants of the diode. For very small incident fields $E_i(t)$, the output detected dc voltage v_0 is:

$$v_0 = \frac{-b_d^2}{2\,\alpha_d}\left\langle \bar{v}_i^2 \right\rangle \tag{47.22}$$

$$v_i = E_i L_e$$

Where:

$$\tilde{v}_i(t) = v_i(t) - \langle v_i \rangle$$

$$b_d = \frac{C_a \alpha_d}{C_a + C_d}$$

C_a = Dipole capacitance
C_d = Diode capacitance
L_e = Effective length of the dipole antenna
$< >$ Indicates time average

Thus, the dc detected voltage is frequency independent and is directly proportional to the average of $(E_i - \langle E_i \rangle)^2$.

For large incident fields $E_i(t)$, the output detected dc voltage is:

$$v_0 = -\frac{b_d \tilde{V}_i}{\alpha_d} \tag{47.23}$$

where $\tilde{V}_i$ is the peak value of $v_i(t)$. Consequently, for a large incident field, v_0 is also frequency independent and is directly proportional to the peak field.

Conventional dipole antennas support a standing-wave current distribution; thus, the useful frequency range of this kind of dipole is usually limited by its natural resonant frequency. In order to suppress this resonance, a resistively loaded dipole (traveling-wave dipole) has been developed. If the internal impedance per unit length $Z_i(z)$ as function of the axial coordinate z (measured from the center of the dipole) has the form:

$$Z_i(z) = \frac{60 \, \psi}{h - |z|} \tag{47.24}$$

then the current distribution $I_z(z)$ along the dipole is that of a traveling wave. Its form is

$$I_z(z) = \frac{V_0}{60 \, \psi (1 - j/kh)} \left[1 - \frac{|z|}{h} \right] e^{-jk|z|} \tag{47.25}$$

where $2h$ = Total physical length of the dipole
V_0 = Driving voltage

$$\psi = 2 \left[\sinh^{-1} \frac{h}{a_d} - C(2ka_d, 2kh) - jS(2ka_d, 2kh) \right] + \frac{j}{kh} \left(1 - e^{-j2kh} \right) \tag{47.26}$$

$C(x,y)$ and $S(x,y)$ = Generalized cosine and sine integrals
a_d = Dipole radius

This type of resistively tapered dipole has a fairly flat frequency response from 100 kHz to 18 GHz [23].

Defining Terms

Electric field strength: The ratio of the force on a positive test charge to the magnitude of the test charge in the limit as the magnitude of the test charge approaches zero.

Electrostatic field: An electric field that does not vary with time.

Field mill: A device used to measure an electrostatic field.

Antenna: A device designed to radiate or to receive time-varying electromagnetic waves.

Microwaves: Electromagnetic waves at frequencies above 1 GHz.

Power density: The time average of the Poynting vector.

Aperture antenna: An antenna that radiates or receives electromagnetic waves through an open area.

Dipole antenna: A straight wire antenna with a center feed used for reception or radiation of electromagnetic waves.

References

1. J.A. Stratton, *Electromagnetic Theory*, New York: McGraw-Hill, 1941.
2. ANSI/IEEE Std. 1227-1990, *IEEE Guide for the Measurement of DC Electric-Field Strength and Ion Related Quantities.*
3. P.E. Secker and J.N. Chubb, Instrumentation for electrostatic measurements, *J. Electrostatics*, 16, 1–19, 1984.
4. R.E. Vosteen, DC electrostatic voltmeters and fieldmeters, *Conference Record of Ninth Annual Meeting of the IEEE Industrial Applications Society*, October 1974.
5. P.J.L. Wildman, A device for measuring electric field in the presence of ionisation, *J. Atmos. Terr. Phys.*, 27, 416–423, 1965.
6. M. Misakian, Generation and measurement of dc electric fields with space charge, *J. Appl. Phys.*, 52, 3135–3144, 1981.
7. G.V. Keller and F.C. Frischknecht, *Electrical Methods in Geophysical Prospecting*, Oxford, U.K.: Pergamon Press, 1966.
8. A.A. Kaufman and G.V. Keller, *The Magnetotelluric Sounding Method*, Amsterdam: Elsevier, 1981.
9. IEEE Std. 291-1991, *IEEE Standard Methods for Measuring Electromagnetic Field Strength of Sinusoidal Continuous Waves, 30 Hz to 30 GHz.*
10. E.C. Jordan and K.G. Balmain, *Electromagnetic Waves and Radiating Systems*, 2nd ed., Englewood Cliffs, NJ: Prentice-Hall, 1968.
11. ANSI/IEEE Std. 644-1987, *IEEE Standard Procedures for Measurement of Power Frequency Electric and Magnetic Fields from AC Power Lines.*
12. IEEE Std. 1308-1994, *IEEE Recommended Practice for Instrumentation: Specifications for Magnetic Flux Density and Electric Field Strength Meters — 10 Hz to 3 kHz.*
13. C.J. Miller, The measurements of electric fields in live line working, *IEEE Trans. Power Apparatus Sys.*, PAS-16, 493–498, 1967.
14. E.V. Jull, *Aperture Antennas and Diffraction Theory*, Stevenage, U.K.: Peter Peregrinus, 1981.
15. M. Abramowitz and I.A. Stegun, *Handbook of Mathematical Functions*, Nat. Bur. Stand. (U.S.), Spec. Pub. AMS 55, 1968.
16. D.A. Hill, M. Kanda, E.B. Larsen, G.H. Koepke, and R.D. Orr, Generating standard reference electromagnetic fields in the NIST anechoic chamber, 0.2 to 40 GHz, Natl. Inst. Stand. Technol. Tech. Note 1335, 1990.
17. M. Kanda, An electromagnetic near-field sensor for simultaneous electric and magnetic-field measurements, *IEEE Trans. Electromag. Compat.*, EMC-26, 102–110, 1984.
18. M. Kanda and D.A. Hill, A three-loop method for determining the radiation characteristics of an electrically small source, *IEEE Trans. Electromag. Compat.*, 34, 1–3, 1992.
19. T.T. Wu, Theory of the thin circular antenna, *J. Math. Phys.*, 3, 1301–1304, 1962.

20. I. Sreenivasiah, D.C. Chang, and M.T. Ma, Emission characteristics of electrically small radiating sources from tests inside a TEM cell, *IEEE Trans. Electromag. Compat.*, EMC-23, 113–121, 1981.

21. D.R. Novotny, K.D. Masterson, and M. Kanda, An optically linked three-loop antenna system for determining the radiation characteristics of an electrically small source, *IEEE Int. EMC Symp.*, 1993, 300–305.

22. M. Kanda, Standard probes for electromagnetic field measurements, *IEEE Trans. Antennas Propagat.*, 41, 1349–1364, 1993.

23. M. Kanda and L.D. Driver, An isotropic electric-field probe with tapered resistive dipoles for broadband use, 100 kHz to 18 GHz, *IEEE Trans. Microwave Theory Techniques*, MTT-35, 124–130, 1987.

48

Magnetic Field Measurement

Steven A. Macintyre
Macintyre Electronic Design

Magnetic field strength is measured using a variety of different technologies. Each technique has unique properties that make it more suitable for particular applications. These applications can range from simply sensing the presence or change in the field to the precise measurements of a magnetic field's scalar and vector properties. A very good and exhaustive fundamental description of both mechanical and electrical means for sensing magnetic fields can be found in Lion [1]. Less detailed but more up-to-date surveys of magnetic sensor technologies can be found in [2, 3]. It is not possible to adequately describe all of these technologies in the space available in a Handbook. This chapter concentrates on sensors that are commonly used in magnetic field measuring instruments.

As shown in Figure 48.1, magnetic field sensors can be divided into vector component and scalar magnitude types. The vector types can be further divided into sensors that are used to measure low fields (<1 mT) and high fields (>1 mT). Instruments that measure low fields are commonly called *magnetometers*. High-field instruments are usually called *gaussmeters*.

The induction coil and fluxgate magnetometers are the most widely used vector measuring instruments. They are rugged, reliable, and relatively less expensive than the other low-field vector measuring instruments. The fiber optic magnetometer is the most recently developed low-field instrument. Although it currently has about the same sensitivity as a fluxgate magnetometer, its potential for better performance is large. The optical fiber magnetometer has not yet left the laboratory, but work on making it more rugged and field worthy is under way. The superconducting quantum interference device (SQUID) magnetometers are the most sensitive of all magnetic field measuring instruments. These sensors operate at temperatures near absolute zero and require special thermal control systems. This makes the SQUID-based magnetometer more expensive, less rugged, and less reliable.

The Hall effect device is the oldest and most common high-field vector sensor used in gaussmeters. It is especially useful for measuring extremely high fields (>1 T). The magnetoresistive sensors cover the middle ground between the low- and high-field sensors. Anisotropic magnetoresistors (AMR) are currently being used in many applications, including magnetometers. The recent discovery of the giant

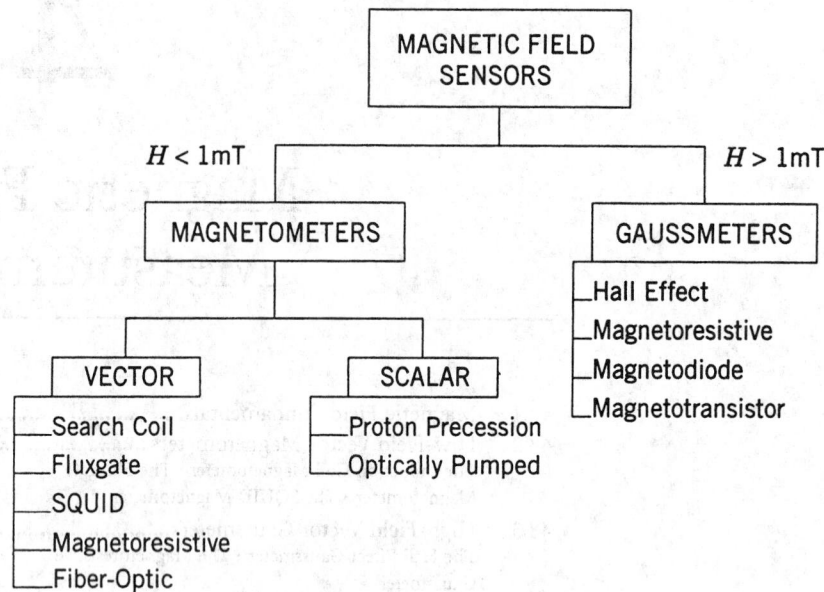

FIGURE 48.1 Magnetic field sensors are divided into two categories based on their field strengths and measurement range: magnetometers measure low fields and gaussmeters measure high fields.

TABLE 48.1   Field Strength Instrument Characteristics

Instrument	Range (mT)	Resolution (nT)	Bandwidth (Hz)	Comment
Induction coil	10^{-10} to 10^6	Variable	10^{-1} to 10^6	Cannot measure static fields
Fluxgate	10^{-4} to 0.5	0.1	dc to 2×10^3	General-purpose vector magnetometer
SQUID	10^{-9} to 0.1	10^{-4}	dc to 5	Highest sensitivity magnetometer
Hall effect	0.1 to 3×10^4	100	dc to 10^8	Best for fields above 1T
Magnetoresistance	10^{-3} to 5	10	dc to 10^7	Good for mid-range applications
Proton precession	0.02 to 0.1	0.05	dc to 2	General-purpose scalar magnetometer
Optically pumped	0.01 to 0.1	0.005	dc to 5	Highest resolution scalar magnetometer

magnetoresistive (GMR) effect, with its tenfold improvement in sensitivity, promises to be a good competitor for the traditional fluxgate magnetometer in medium-sensitivity applications.

The proton (nuclear) precession magnetometer is the most popular instrument for measuring the scalar magnetic field strength. Its major applications are in geological exploration and aerial mapping of the geomagnetic field. Since its operating principle is based on fundamental atomic constants, it is also used as the primary standard for calibrating magnetometers. The proton precession magnetometer has a very low sampling rate, on the order of 1 to 3 samples per second, so it cannot measure fast changes in the magnetic field. The optically pumped magnetometer operates at a higher sampling rate and is capable of higher sensitivities than the proton precession magnetometer, but it is more expensive and not as rugged and reliable.

Table 48.1 lists various magnetic field strength instruments and their characteristics.

48.1   Magnetic Field Fundamentals

An understanding of the nature of magnetic fields is necessary in order to understand the techniques used for measuring magnetic field strength. The most familiar source of a magnetic field is the bar

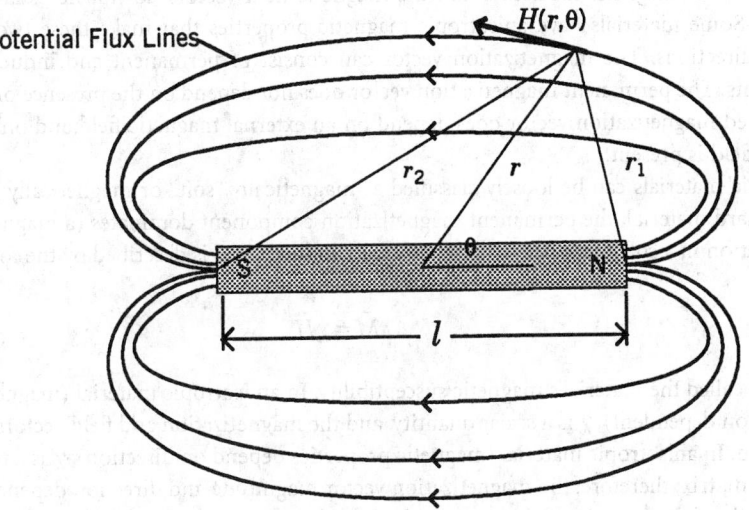

FIGURE 48.2 Magnets produce magnetic fields. A magnetic field is a vector quantity with both magnitude and direction properties.

magnet. The field it produces is shown in Figure 48.2. Magnetic field is a vector quantity; that is, it has both a magnitude and a direction. The field of a bar magnet or any other magnetized object, when measured at a distance much greater than its longest dimension, is described by Equation 48.1:

$$\vec{H} = \frac{3(\vec{m} \times \hat{a}_r)\hat{a}_r - \vec{m}}{r^3} \tag{48.1}$$

where $\hat{a}_r$ is a unit vector along r, r is the distance between the magnetic field source and the measurement point, and $\vec{m}$ is called the magnetic dipole moment. The derivation of this equation can be found in many textbooks on electromagnetics. This is a very convenient equation for estimating the field produced by many magnetized objects.

The strength or intensity of a magnetized object depends on the density of its volume-distributed moments. This intensity is called its magnetization M, which is defined as the moments per unit volume:

$$\vec{M} = \frac{\vec{m}}{volume} \tag{48.2}$$

Like magnetic field, magnetization is a vector quantity. Magnetization is a material property that can arise from internal magnetic sources as well as be induced by an external magnetic field.

There is a third magnetic vector $\vec{B}$ called magnetic induction or flux density. In free space, magnetic field and magnetic induction are proportional to one another by a constant factor μ_0.

$$\vec{B} = \mu_0 \vec{H} \tag{48.3}$$

Things are different in matter. Equation 48.4 describes the relationship among the magnetic field, magnetic induction, and magnetization vectors in matter:

$$\vec{B} = \mu_0 \left(\vec{H} + \vec{M} \right) \tag{48.4}$$

In this case, the magnetic induction and the magnetic field vectors do not necessarily have the same direction. Some materials have anisotropic magnetic properties that make these two vectors point in different directions. The magnetization vector can consist of permanent and induced magnetization components. The permanent magnetization vector does not depend on the presence of an external field. The induced magnetization vector does depend on an external magnetic field and only exists while the inducing field is present.

Magnetic materials can be loosely classified as magnetically "soft" or magnetically "hard." In a magnetically hard material, the permanent magnetization component dominates (a magnet is an example). Magnetization in a soft magnetic material is largely induced and is described by the following equation:

$$\vec{M} = \chi \vec{H} \tag{48.5}$$

where χ is called the material's magnetic susceptibility. In an isotropic material (magnetic properties are not direction dependent), χ is a scalar quantity, and the magnetization and field vectors are proportional and aligned. In anisotropic material (magnetic properties depend on direction), χ is a tensor represented by a 3×3 matrix; therefore, the magnetization vector magnitude and direction depend on the direction and strength of the inducing field. As a result, the magnetization vector will not always align with the magnetization inducing field vectors. Equation 48.5 can be modified for magnetically "soft" material to the following:

$$\vec{B} = \mu_0 \left(1 + \chi\right) \vec{H} = \mu_0 \mu \vec{H} \tag{48.6}$$

where μ is called the relative permeability of the material.

A magnetized object with a magnetic moment $\vec{m}$ will experience torque $\vec{T}$ in the presence of a uniform magnetic field $\vec{H}$. Equation 48.7 expresses this relationship.

$$\vec{T} = \vec{m} \times \vec{H} \tag{48.7}$$

Torque is the cross-product of the magnetic moment and field vectors. The magnitude equation is:

$$T = mH \sin\theta \tag{48.8}$$

where θ is the angle between the direction of $\vec{m}$ and $\vec{H}$.

There is an intimate relationship between electric and magnetic fields. Oersted discovered that passing a current through a wire near a compass causes the compass needle to rotate. The compass was the first magnetic field strength sensor. Faraday found that he could produce an electric voltage at the terminals of a loop of wire if he moved a magnet near it. This led to the induction or search coil sensor.

Magnetic fields are produced by the flow of electric charge (i.e., electric currents). In effect, a magnetic field is a velocity-transformed electric field (through a Lorentz transformation). Current flowing through a straight wire, a loop of wire, or a solenoid will also produce a magnetic field as illustrated in Figure 48.3.

Units are always a problem when dealing with magnetic fields. The Gaussian cgs (centimeter, gram, and second) system of units was favored for many years. Since $\mu_0 = 1$ in the cgs system, magnetic field and flux density have the same numeric value in air, and their units (oerstedt for field and gauss for flux density) are often indiscriminately interchanged. This has led to great confusion. The cgs system has now been replaced by the International System of Units (SI). The SI system uses, among others, the meter (m), kilogram (kg), second (s) and ampere (A) as the fundamental units. Payne [4] gives a very good explanation of the differences between these systems of units as they relate to magnetic fields. Table 48.2 summarizes the relationships between the two systems of units.

The SI system of units is used throughout this chapter.

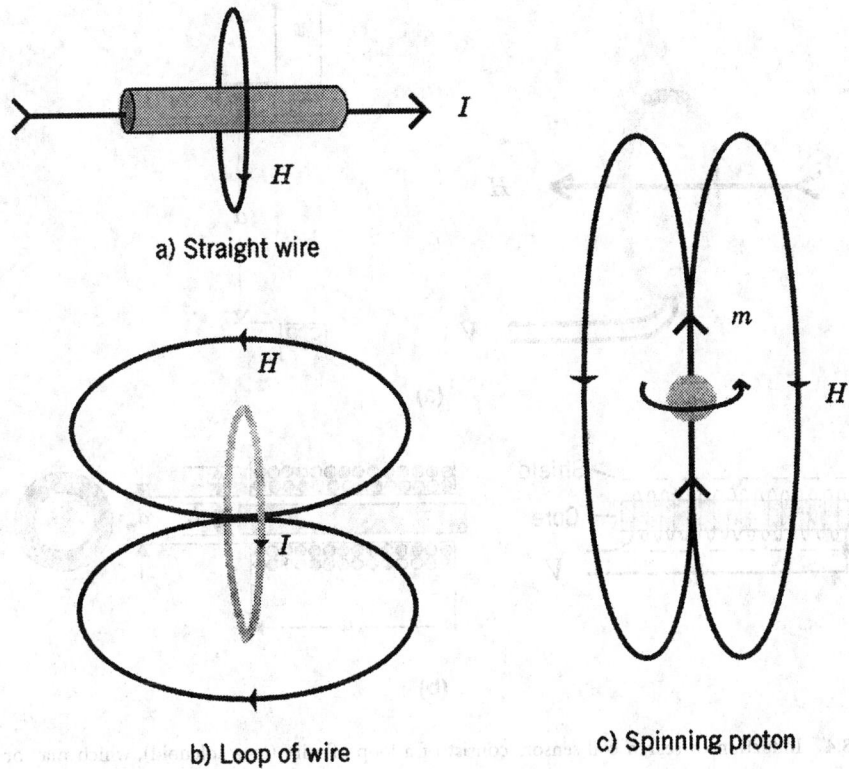

a) Straight wire

b) Loop of wire

c) Spinning proton

FIGURE 48.3 Magnetic fields are also produced by electric currents.

TABLE 48.2 Factors for Converting from cgs to SI Magnetic Field Units

Description	Symbol	SI unit	Gaussian cgs unit	Multiply by
Magnetic induction	B	Tesla	gauss	10^4
Magnetic field strength	H	A m^{-1}	oerstedt (oe)	$4\pi \times 10^{-3}$
Magnetization	M	A m^{-1}	emu m^3	10^{-3}
Magnetic dipole moment	m	A m^2	emu	10^3
Magnetic flux	ϕ	Weber (Wb)	maxwell	10^8
Magnetic pole strength	p	A m	emu	
Permeability of free space	μ_0	H m^{-1}	$4\pi \times 10^{-7}$	1

48.2 Low-Field Vector Magnetometers

The Induction Coil Magnetometer

The induction or search coil, which is one of the simplest magnetic field sensing devices, is based on Faraday's law. This law states that if a loop of wire is subjected to a changing magnetic flux, ϕ, through the area enclosed by the loop, then a voltage will be induced in the loop that is proportional to the rate of change of the flux:

$$e(t) = -\frac{d\phi}{dt} \tag{48.9}$$

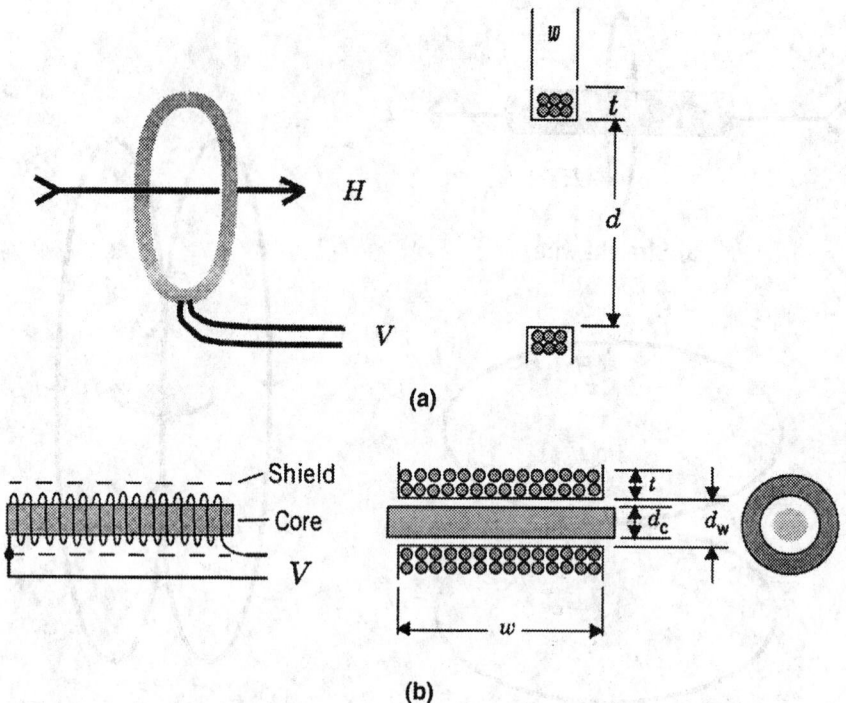

FIGURE 48.4 Induction or search coil sensors consist of a loop of wire (or a solenoid), which may or may not surround a ferromagnetic core. (a) Air core loop antenna; (b) solenoid induction coil antenna with ferromagnetic core.

Since magnetic induction $\vec{B}$ is flux density, then a loop with cross-sectional area $\vec{A}$ will have a terminal voltage:

$$e(t) = -\frac{d(\vec{B} \bullet \vec{A})}{dt} \tag{48.10}$$

for spatially uniform magnetic induction fields.

Equation 48.10 states that a temporal change in $\vec{B}$ or the mechanical orientation of $\vec{A}$ relative to $\vec{B}$ will produce a terminal voltage. If the coil remains fixed with respect to $\vec{B}$, then static fields cannot be detected; but if the loop is rotated or the magnitude of $\vec{A}$ is changed, then it is possible to measure a static field. The relationship described by Equation 48.10 is exploited in many magnetic field measuring instruments (see [1]).

Figure 48.4 shows the two most common induction coil configurations for measuring field strength: the air core loop antenna and the rod antenna. The operating principle is the same for both configurations. Substituting $\mu_0\mu_e H(t)$ for B in Equation 48.10 and, assuming the loop is stationary with respect to the field vector, the terminal voltage becomes:

$$e(t) = -\mu_0\mu_e nA \frac{dH(t)}{dt} \tag{48.11}$$

where n is the number of turns in the coil, and μ_e is the effective relative permeability of the core. The core of a rod antenna is normally constructed of magnetically "soft" material so one can assume the flux density in the core is induced by an external magnetic field and, therefore, the substitution above is valid. With an air (no) core, the effective relative permeability is one. The effective permeability of an induction

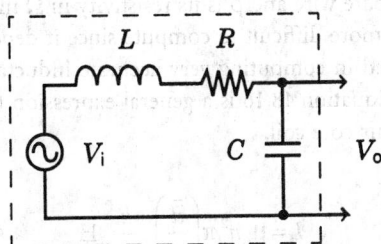

FIGURE 48.5   The induction coil equivalent circuit is a frequency-dependent voltage source in series with an inductor, resistor, and lumped capacitor.

coil that contains a core is usually much greater than one and is strongly dependent on the shape of the core and, to some extent, on the configuration of the winding.

Taking the Laplace transform of Equation 48.11 and dividing both sides by H, one obtains the following transfer function $T(s)$ for an induction coil antenna:

$$T(s) = -\mu_0 \mu_e nAs = -Ks \quad \left(\text{VmA}^{-1} \right) \tag{48.12}$$

where $E(s) = T(s) H(s)$, $E(s)$ and $H(s)$ are the Laplace transforms of $e(t)$ and $H(t)$, and s is the Laplace transform operator. Inspection of Equation 48.12 reveals that the magnitude of the coil voltage is proportional to both the magnitude and frequency of the magnetic field being measured. The coil constant or sensitivity of the loop antenna is:

$$K = \mu_0 \mu_e nA \quad \left(\text{VsmA}^{-1} \right) \tag{48.13}$$

Figure 48.5 is the equivalent circuit for an induction coil antenna. The actual voltage measured at the terminals of the loop is modified by the inductance L, resistance R, and the distributed stray and shield capacitances represented by the lumped capacitor C. These circuit parameters depend on the geometry of the core, coil, and winding.

The electrostatic shield made of nonmagnetic material shown in Figure 48.4 is an important element in the design of an induction coil. It prevents coupling of electric fields into the coil, thereby assuring that the signal seen at the coil terminals is only that due to a magnetic field. The shield should not be placed too close to the winding since it contributes to coil capacitance and noise.

The Air Core Loop Antenna

The air core loop antenna consists of a circular or rectangular loop containing one or more turns of wire and no magnetic core. The diameter of the loop is usually much greater than the dimensions of the winding cross-section. The sensitivity of a circular loop antenna with a winding inside diameter d and rectangular cross-section is approximately:

$$K = \mu_0 n\pi \frac{d^2}{4} \left[1 + 2\left(\frac{t}{d}\right) + \frac{3}{4}\left(\frac{t}{d}\right)^2 \right] \tag{48.14}$$

where t is the thickness of the winding and n is the number of turns.

The resistance of the coil is:

$$R = 4n \frac{d}{d_w^2} \left(1 + \frac{t}{d}\right) \rho \ \Omega \tag{48.15}$$

where d_w is the diameter of the bare wire and ρ is its resistivity in Ω m (1.7×10^{-8} Ω m for copper).

The inductance of the coil is more difficult to compute since it depends heavily on the geometry of the coil. Those who are interested in computing very accurate inductance values for a wide variety of coil shapes should consult [5]. Equation 48.16 is a general expression that gives a good approximation for the inductance of a circular air core coil.

$$L = \mu_0 n^2 \pi \left(\frac{\bar{d}}{2}\right)^2 \frac{k}{w} \text{ H} \tag{48.16}$$

where w is the width of the winding, $\bar{d}$ is the average diameter, and k is Nagaoka's constant:

$$k = \frac{1}{1 + 0.45 \dfrac{\bar{d}}{w} + 0.64 \dfrac{t}{d} + 0.84 \dfrac{t}{w}} \tag{48.17}$$

The distributed capacitance of the coil contributes the most to the overall antenna capacitance. The parasitic capacitances can usually be ignored. Equation 48.18 can be used to estimate the distributed capacitance of a coil.

$$C_d = \left[\frac{\varepsilon_w \varepsilon_1}{\varepsilon_w t_1 + \varepsilon_1 t_w}\right] \frac{0.018544 \bar{d} w (n_1 - 1)}{n_1^2} \tag{48.18}$$

where ε_w is the dielectric constant of the wire insulation, ε_1 is the dielectric constant of the interlayer insulation if any, t_w is the thickness of the wire insulation, t_1 is the thickness of the interlayer insulation, and n_1 is the number of layers. Adding a second layer to a single-layer coil significantly increases the capacitance but, as the number of layers increases, the capacitance decreases.

The air core loop antenna is particularly useful for measuring magnetic fields with frequencies from 100 Hz to several megahertz. Because it has a linear response to magnetic field strength, it has virtually no intermodulation distortion. On the negative side, the size of the sensor can get quite large for applications that require high sensitivities at low frequencies.

The Rod Antenna

The rod antenna is a good alternative to an air core loop antenna. It is smaller in size than a loop antenna with the same sensitivity, and it can be designed to operate at lower frequencies. Unfortunately, its response to magnetic field strength can be nonlinear and the core adds noise.

Figure 48.4(b) is a typical configuration for a rod antenna. It is basically a solenoid with a magnetic core. The core can have a circular or rectangular cross-section and can be made from a ferrite, a nickel-iron alloy, an amorphous metal glass alloy, or some other material with high relative permeability. The winding can be wound directly on the core or on a form through which the core is inserted. Insulation is sometimes placed between layers of the winding to reduce distributed capacitance. An electrostatic shield is placed around the winding to attenuate any electric field coupling into the signal. The shield has a gap that runs the length of the winding. This prevents circulating currents in the shield from attenuating the magnetic field within the coil.

The most common rod antenna configuration is a core with a circular cross-section and a tightly coupled winding that runs most of the length of the core. The sensitivity of the rod antenna is computed by substituting μ_e in Equation 48.13 with the following:

$$\mu_e = 1 + \left(\frac{d_c}{d + t}\right)^2 (\bar{\mu} - 1) \tag{48.19}$$

TABLE 48.3 Demagnetizing Factors, N for Rods and Ellipsoids Magnetized Parallel to Long Axis

Dimensional ratio (length/diameter)	Rod	Prolate ellipsoid	Oblate ellipsoid
0	1.0	1.0	1.0
1	0.27	0.3333	0.3333
2	0.14	0.1735	0.2364
5	0.040	0.0558	0.1248
10	0.0172	0.0203	0.0696
20	0.00617	0.00675	0.0369
50	0.00129	0.00144	0.01472
100	0.00036	0.000430	0.00776
200	0.000090	0.000125	0.00390
500	0.000014	0.0000236	0.001576
1000	0.0000036	0.0000066	0.000784
2000	0.0000009	0.0000019	0.000392

where d_c is the core diameter and $\bar{\mu}$ is the core average effective permeability. The core effective or apparent permeability depends on its geometry and initial permeability, as well as the winding length relative to the core length. A rod becomes magnetized when a magnetic field is applied to it. In response, a magnetic field is created within the rod that opposes the externally applied field and reduces the flux density. The demagnetizing field is proportional to the magnetization and the net field H in the core is:

$$H = H' - NM \tag{48.20}$$

where H' is the applied external field, N is the demagnetizing factor, and M is the magnetization. The apparent relative permeability of a core is the ratio of the flux density B in the middle of the core to the flux density in air:

$$\frac{B}{\mu_0 H'} = \mu_a = \frac{\mu_i}{1 + N(\mu_i - 1)} \tag{48.21}$$

where μ_i is the initial relative permeability of the core material. Initial relative permeability is the slope of the B–H magnetization curve near zero applied field for a closed magnetic path.

The value of N is shape dependent. As the length-to-diameter ratio m of a rod increases, N decreases and the apparent relative permeability approaches the initial permeability. Table 48.3, which is reproduced from [6], lists demagnetizing factors for a rod, prolate ellipsoid (cigar shape), and oblate ellipsoid (disk shape).

Equation 48.22 can be used to approximate the value of N for cylindrical rods with $m > 10$ and $\mu_i > 1000$:

$$N = \frac{2.01 \times \log_{10} m - 0.46}{m^2} \tag{48.22}$$

The apparent permeability of a rod with a small m and large μ_i is almost exclusively determined by m alone. Table 48.4 lists the magnetic properties of several ferromagnetic materials that can be used to construct a core.

Bozorth [7] found that the apparent permeability of a rod is not constant throughout the length of the rod. It reaches a maximum at the center of the rod and continuously drops in value until the ends of the rod are reached. The variation in permeability can be approximated by:

TABLE 48.4   Magnetic Properties of Typical Core Material

Name	Composition	Manufacturer	μ_i	μ_{max}
Mild steel	0.2 C, 99 Fe		120	2000
Silicon iron	4.0 Si, 96 Fe		500	7000
CN20	Ni-Zn Ferrite	Ceramic Magnetics	800	4500
MN60	Mn-Zn Ferrite	Ceramic Magnetics	5000	10,500
"49" Alloy	48 Ni, 52 Fe	Carpenter	6500	75,000
2605S-2	Fe-based amorphous alloy	Allied-Signal	10,000	600,000
4-79 Permalloy	4 Mn, 79 Ni, 17 Fe	Magnetics	20,000	100,000
Mumetal	5 Cu, 2 Cr, 77 Ni, 16 Fe	Magnetics	20,000	100,000
HyMu "80"	4.2 Mo, 80 Ni, 15 Fe	Carpenter	50,000	200,000
2826MB	NiFe-based amorphous alloy	Allied-Signal	100,000	800,000

Note: μ_i is the slope of the magnetization curve at the origin. μ_{max} is the maximum incremental slope of the magnetization curve.

$$\mu(l) = \mu_a \left[1 - F \left(\frac{l}{l_0} \right)^2 \right] \tag{48.23}$$

where l is the distance from the center of the rod to the measurement point, l_0 is the half length of the rod, and F is a constant that varies from 0.72 to 0.96. The average permeability seen by the coil is the integral of Equation 48.23 over the length of the coil:

$$\overline{\mu} = \mu_a \left[1 - F \left(\frac{l_w}{l_c} \right)^2 \right] \tag{48.24}$$

where l_w is the length of the winding and l_c is the length of the core. Equation 48.24 is substituted into Equation 48.19 to compute the rod's effective relative permeability which is used in Equation 48.13 to compute sensitivity.

The inductance of the rod antenna can be computed using the following equations:

$$L = \frac{\mu_0 \mu_e n^2 \pi (d+t)^2 l_w}{4 l_c} \tag{48.25}$$

$$\mu_e = 1 + \left(\frac{d_c}{d+t} \right)^2 \left[\mu_a f(l_w/l_c) - 1 \right] \tag{48.26}$$

$$f(l_w/l_c) = 1.9088 - 0.8672(l_w/l_c) - 1.1217(l_w/l_c)^2 + 0.8263(l_w/l_c)^3 \tag{48.27}$$

The function $f(l_w/l_c)$ accounts for the variation in flux density from the middle of the winding to its ends and assumes the winding is centered about the middle of the core.

Equations 48.15 and 48.16 can be used to compute the resistance and capacitance of a rod antenna.

Signal Conditioning

To be useful, the induction coil signal must be conditioned using either a voltage or a current amplifier. Figure 48.6 illustrates the circuit configurations for both of these signal conditioning methods. The voltage

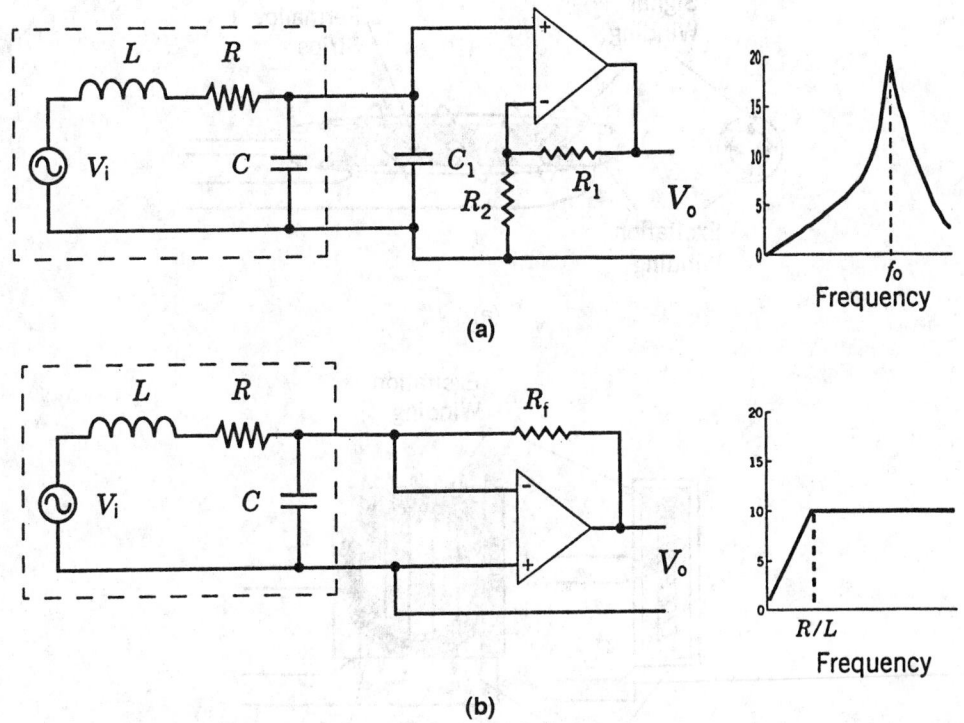

(a)

(b)

FIGURE 48.6 (a) The amplitude of a voltage-amplified induction coil signal is proportional to the frequency and strength of the field. (b) The amplitude of a current-amplified induction coil signal is only proportional to field strength beyond its L/R corner frequency.

amplifier can have either a single-ended or differential input and it can be tuned or untuned. The signal output of the voltage amplifier is proportional to the magnitude and frequency of the field for frequencies well below resonance. Its output will peak at the resonant frequency of the coil or at the tuning frequency. Because its output signal depends on both the frequency and strength of the magnetic field, the voltage amplifier is more suited to narrow band or tuned frequency applications.

In the current amplifier configuration, the induction coil terminals are connected to a virtual ground. As long as the product of the amplifier forward gain and the coil ohmic resistance is much greater than the feedback resistor, the output signal magnitude is independent of the frequency of the magnetic field beyond the R/L (rad s^{-1}) corner of the coil. This remains true up to the coil's resonant frequency. For this reason, the current amplifier configuration is particularly suited to broadband magnetic field strength measurements. The current amplifier configuration minimizes intermodulation distortion in induction coils with magnetic cores. The current flowing through the coil produces a magnetic field that opposes the ambient field. This keeps the net field in the core near zero and in a linear region of the *B–H* curve.

Current-amplifier-based induction coil magnetometers have been built that have a flat frequency response from 10 Hz to over 200 kHz. Some magnetometers designed for geophysical exploration applications have low frequency corners that extend down to 0.1 Hz. For further information on this subject, see [8, 9].

The Fluxgate Magnetometer

The fluxgate magnetometer has been and is the workhorse of magnetic field strength instruments both on Earth and in space. It is rugged, reliable, physically small, and requires very little power to operate. These characteristics, along with its ability to measure the vector components of magnetic fields over a

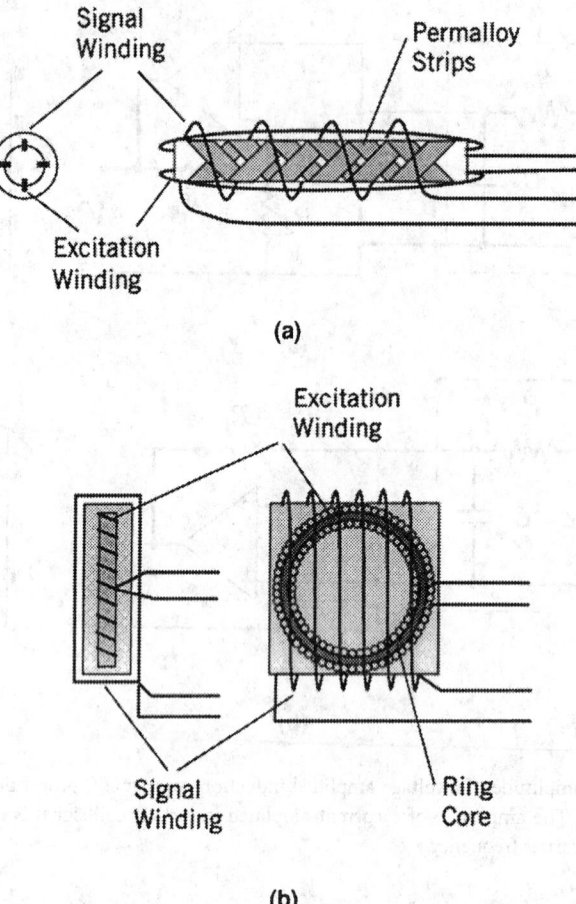

(a)

(b)

FIGURE 48.7 In Schonstedt (a) and ring core (b) fluxgate sensors, the excitation field is at right angles to the signal winding axis. This configuration minimizes coupling between the excitation field and the signal winding.

0.1 nT to 1 mT range from dc to several kHz, make it a very versatile instrument. Geologists use them for exploration and geophysicists use them to study the geomagnetic field (about 20 μT to 75 μT on the Earth's surface). Satellite engineers use them to determine and control the attitude of spacecraft, scientists use them in their research, and the military uses them in many applications, including mine detection, vehicle detection, and target recognition. Some airport security systems use them to detect weapons.

The Fluxgate

The heart of the magnetometer is the *fluxgate*. It is the transducer that converts a magnetic field into an electric voltage. There are many different fluxgate configurations. Two of the more popular ones are shown in Figure 48.7. A very comprehensive explanation of the fluxgate principle and the different fluxgate configurations is given in [10].

The ring core fluxgate is constructed from a thin ribbon of easily saturable ferromagnetic material, such as 4-79 Permalloy wrapped around a bobbin to form a ring or toroid. As shown in Figure 48.8, an alternating current is applied through a coil that is wound about the toroid. This creates a magnetic field that circulates around the magnetic core. This magnetic field causes the flux in the ferrous material to periodically saturate first clockwise and then counterclockwise. A pick-up (signal) winding is wrapped around the outside of the toroid. While the ferrous material is between saturation extremes, it maintains an average permeability much greater than that of air. When the core is in saturation, the core permeability

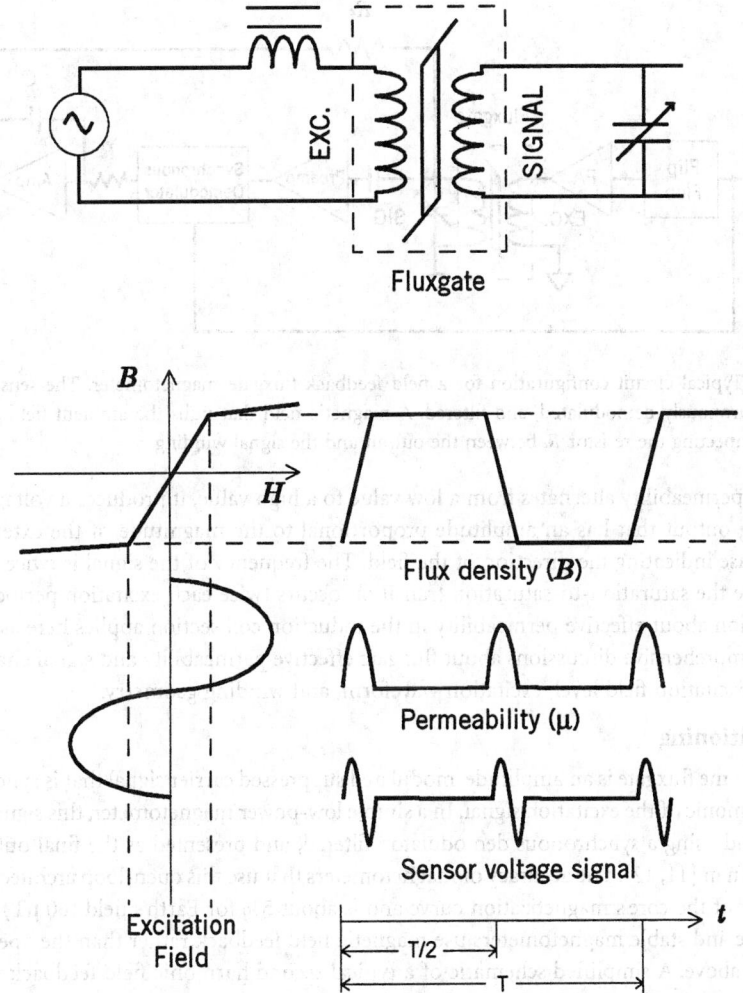

FIGURE 48.8 The excitation field of a fluxgate magnetometer alternately drives the core into positive or negative saturation, causing the core's effective permeability to switch between 1 and a large value twice each cycle.

becomes equal to that of air. If there is no component of magnetic field along the axis of the signal winding, the flux change seen by the winding is zero. If, on the other hand, a field component is present along the signal winding axis, then each time the ferrous material goes from one saturation extreme to the other, the flux within the core will change from a low level to a high level. According to Faraday's law, a changing flux will produce a voltage at the terminals of the signal winding that is proportional to the rate of change of flux. For dc and low-frequency magnetic fields, the signal winding voltage is:

$$e(t) = nA \frac{d(\mu_0 \mu_e H)}{dt} = nA\mu_0 H \frac{d\mu_e(t)}{dt} \tag{48.28}$$

where H = Component of the magnetic field being measured
n = Number of turns on the signal winding
A = Cross-sectional area of the signal winding
$\mu_e(t)$ = Effective relative permeability of the core

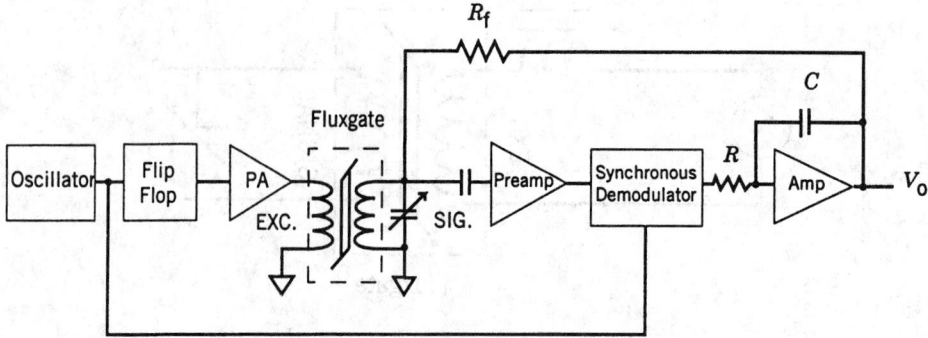

FIGURE 48.9 Typical circuit configuration for a field feedback fluxgate magnetometer. The sensor output is ac amplified, synchronously demodulated, and filtered. A magnetic field that nulls the ambient field at the sensor is produced by connecting the resistor R_f between the output and the signal winding.

As the core permeability alternates from a low value to a high value, it produces a voltage pulse at the signal winding output that has an amplitude proportional to the magnitude of the external magnetic field and a phase indicating the direction of the field. The frequency of the signal is twice the excitation frequency since the saturation-to-saturation transition occurs twice each excitation period.

The discussion about effective permeability in the induction coil section applies here as well. Consult [10, 13] for comprehensive discussions about fluxgate effective permeability and signal characteristics as they relate to excitation field level, excitation waveform, and winding geometry.

Signal Conditioning

The signal from the fluxgate is an amplitude-modulated suppressed carrier signal that is synchronous with the second harmonic of the excitation signal. In a simple low-power magnetometer, this signal is converted to the base band using a synchronous demodulator, filtered, and presented as the final output. Example circuits are given in [11, 12]. The accuracy of magnetometers that use this open-loop architecture is limited by the linearity of the core's magnetization curve and is about 5% for Earth's field (60 μT) applications.

More precise and stable magnetometers use magnetic field feedback rather than the open-loop structure described above. A simplified schematic of a typical second harmonic field feedback fluxgate magnetometer is shown in Figure 48.9. The circuitry to the left of the fluxgate is called the excitation circuit. It consists of an oscillator tuned to twice the excitation frequency, a flip-flop that divides the oscillator frequency by two, and a power amplifier driven by the flip-flop and, in turn, provides the excitation current to the excitation winding.

The circuitry to the right of the fluxgate is called the signal channel circuit. It amplifies the output from the fluxgate signal winding, synchronously demodulates the ac signal using the oscillator signal as a reference, integrates and amplifies the base band output, and then feeds back the output through a resistor to the signal winding. The fed-back signal produces a magnetic field inside the sensor that opposes the external field. This keeps the field inside the sensor near zero and in a linear portion of the magnetization curve of the ferromagnetic core.

The flow diagram for the magnetometer is given in Figure 48.10. The external field H_a is opposed by the feedback field H_f, and the difference is converted into a voltage signal (K_s represents the transfer function from field to voltage). This signal is amplified (A), and the amplified signal is converted into a current I_f and then into the feedback field (K_c represents the transfer function from current to field). The overall transfer function for the magnetometer is:

$$\frac{V_0}{H_a} = \frac{AK_s}{1 + \frac{K_c AK_s}{R_f}} \tag{48.29}$$

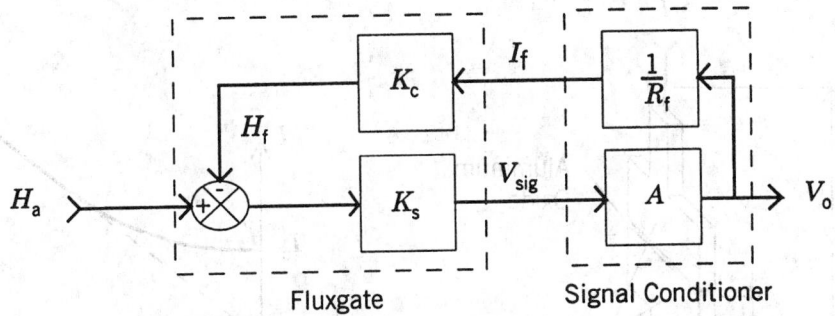

Fluxgate   Signal Conditioner

FIGURE 48.10 Block diagram of a field feedback fluxgate magnetometer. K_c is the current-to-field constant for the coil. K_s is the field-to-voltage transduction constant for the sensor. The feedback field H_f opposes the ambient field H_a, thus keeping the net sensor field very small.

The amplifier gain is normally very high such that the second term in the denominator is much larger than one, and Equation 48.29 reduces to

$$\frac{V_0}{H_a} = \frac{R_f}{K_c} \tag{48.30}$$

Under these circumstances, the transfer function becomes almost completely determined by the ratio of R_f (the feedback resistor) to K_c (the current-to-field coil constant of the sensor winding). Both of these constants can be very well controlled. The consequence of this circuit topology is a highly stable and accurate magnetometer that is insensitive to circuit component variations with temperature or time. An accuracy of 1% over a temperature range of −80°C to 80°C is easily achievable. Accuracy and stability can be improved using a current feedback circuit, like the one described in [13], that compensates for the resistance of the signal winding or by using a separate feedback winding and a high-quality voltage-to-current converter instead of a simple feedback resistor.

The SQUID Magnetometer

Brian D. Josephson in 1962, while a graduate student at Cambridge University, predicted that superconducting current could flow between two superconductors that are separated by a thin insulation layer. The magnitude of the superconductor (critical) current through this "Josephson junction" is affected by the presence of a magnetic field and forms the basis for the SQUID magnetometer.

Figure 48.11 illustrates the general structure of a Josephson junction and the voltage–current (V–I) relationship. Two superconductors (e.g., niobium) are separated by a very thin insulating layer (e.g., aluminum oxide). The thickness of this layer is typically 1 nm. When the temperature of the junction is reduced to below 4.2 K (−269°C), a superconductor current will flow in the junction with 0 V across the junction. The magnitude of this current, called the critical current I_c, is a periodic function of the magnetic flux present in the junction. Its maximum magnitude occurs for flux values equal to $n\phi_0$, where ϕ_0 is one flux quantum (2 fW), and its minimum magnitude occurs for flux values equal to $(n + \frac{1}{2})\phi_0$. The period is one flux quantum. This phenomenon is called the "dc Josephson effect" and is only one of the "Josephson effects."

Magnetometers based on the Superconducting Quantum Interference Device (SQUID) are currently the most sensitive instruments available for measuring magnetic field strength. SQUID magnetometers measure the change in the magnetic field from some arbitrary field level; they do not intrinsically measure the absolute value of the field. Biomedical research is one of the more important applications of SQUID magnetometers. SQUID magnetometers and gradiometers (measure spatial variation in the magnetic field) have the high sensitivities needed to measure the weak magnetic fields generated by the body [15].

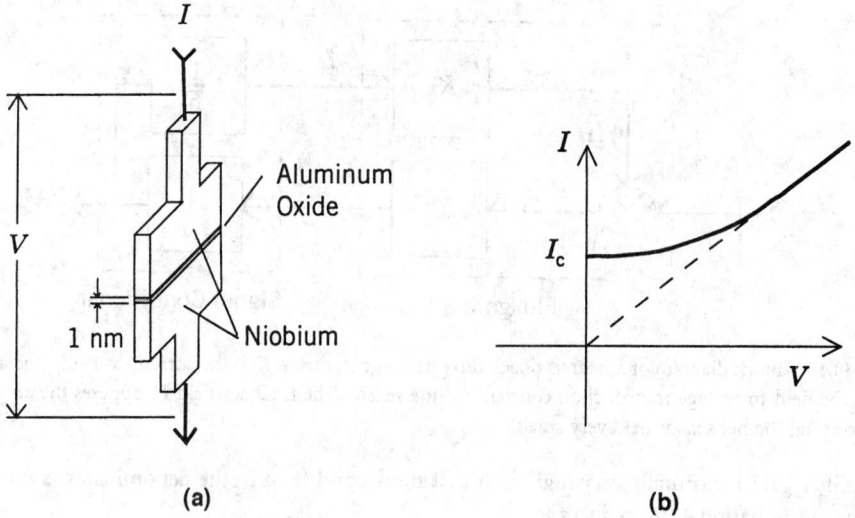

FIGURE 48.11   The Josephson junction in (a) consists of a superconductor such as niobium separated by a thin insulation layer. The voltage (V) vs. current (I) curve in (b) shows that a superconducting current flows through the junction with zero volts across the junction.

Other application areas include paleomagnetics (measuring remnant magnetism in rocks) and magnetotellurics (Earth resistivity measurements). Descriptions of these applications as well as the general theory of SQUIDs can be found in [16]. Clark [17], one of the pioneers in SQUID magnetometers, provides a good contemporary overview of SQUID technology and applications.

A dc SQUID magnetometer uses two Josephson junctions in the two legs of a toroid as shown in Figure 48.12(a). The toroid is biased with a constant current that exceeds the maximum critical current of the junctions. When the flux through the toroid is an integral multiple of ϕ_0, the voltage across the junctions is determined by the intersection of I_b and the $n\phi_0$ V–I curve (point A). As the flux increases, the critical current decreases. The V–I curve and thus the intersection point move to the right (the junction voltage increases). The critical current reaches a minimum when the flux has increased by $\frac{1}{2}\phi_0$ and the junction voltage is at its maximum (point B). As the flux continues to increase, the critical current increases back toward its maximum value and the junction voltage decreases. Thus, the period of the flux cycle is ϕ_0.

Signal Conditioning

Figure 48.13 is a block diagram of one implementation of a dc SQUID magnetometer that can be used for wide dynamic range field measurements. A large superconducting loop, which is exposed to the magnetic field being measured, is connected to a multiturn signal winding that is magnetically coupled directly to the SQUID. At cryogenic temperatures, the loop and signal winding effectively form a dc induction coil. External flux applied to the coil will generate a current in the loop that keeps the net flux within the loop constant, even for dc magnetic fields. The signal winding magnifies the flux that is applied to the SQUID.

The SQUID is magnetically biased at an optimal sensitivity point. A small ac magnetic field at 100 kHz to 500 kHz is superimposed on the bias field. The output of the SQUID is a suppressed carrier amplitude modulated signal where the amplitude indicates the change in magnetic field from the bias point, and the phase indicates the polarity of the change. The output signal is amplified and then synchronously demodulated down to the base band. The resulting dc signal is amplified and fed back through a resistor to a coil coupled to the SQUID. The current through the coil generates a magnetic field at the SQUID that opposes the applied field. This keeps the SQUID operating point very near the bias point. The scale

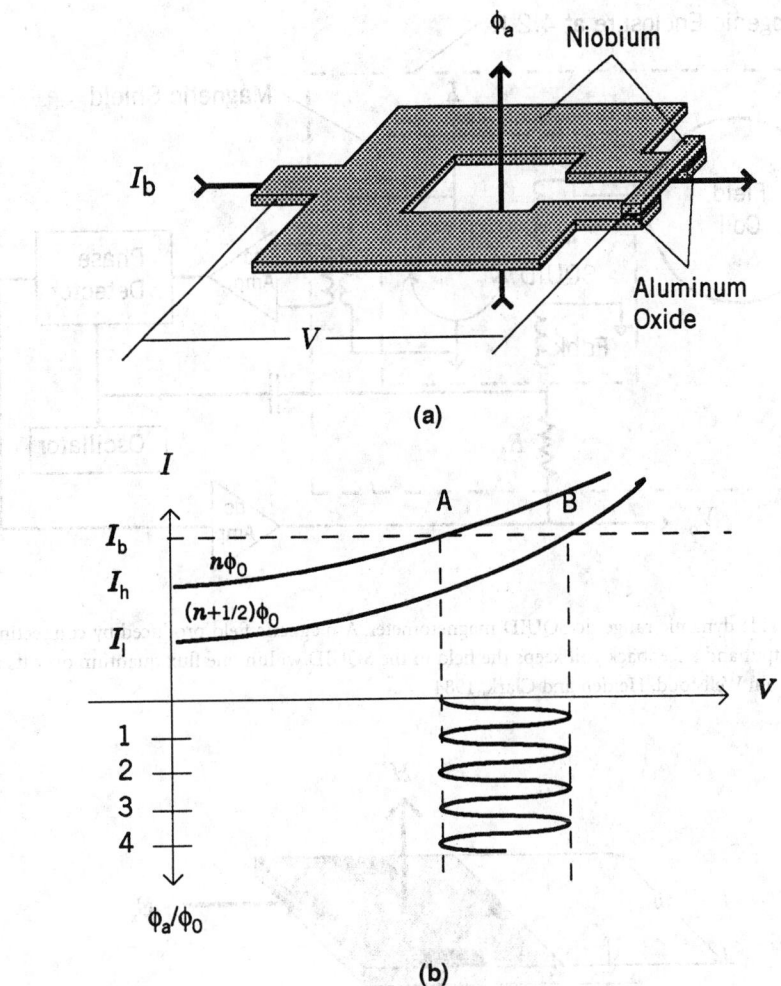

FIGURE 48.12 Use of a dc SQUID to measure magnetic flux. The dc SQUID in (a) consists of a superconductor loop and two Josephson junctions with a bias current that is greater than the maximum critical current I_h. The V–I curve in (b) illustrates how the voltage across the SQUID oscillates with a period equal to one flux quantum ϕ_0.

factor of the magnetometer depends on the feedback resistor and the coil constant of the feedback winding in the same manner that it does for a field feedback fluxgate magnetometer.

The pick-up loop, signal coil, SQUID, feedback coil and feedback resistor are kept in a cryogenic temperature chamber and, except for the pick-up coil, are magnetically shielded. The rest of the circuit is operated at room temperature.

48.3 High-Field Vector Gaussmeters

The Hall Effect Gaussmeter

The Hall effect device, which is probably the most familiar and widely used sensor for measuring strong magnetic fields, is based on the discovery of the Hall effect by Edwin H. Hall in 1897. The Hall effect is a consequence of the Lorentz force law, which states that a moving charge q, when acted upon by a magnetic induction field $\vec{B}$, will experience a force $\vec{F}$ that is at right angles to the field vector and the velocity vector v of the charge as expressed by the following equation:

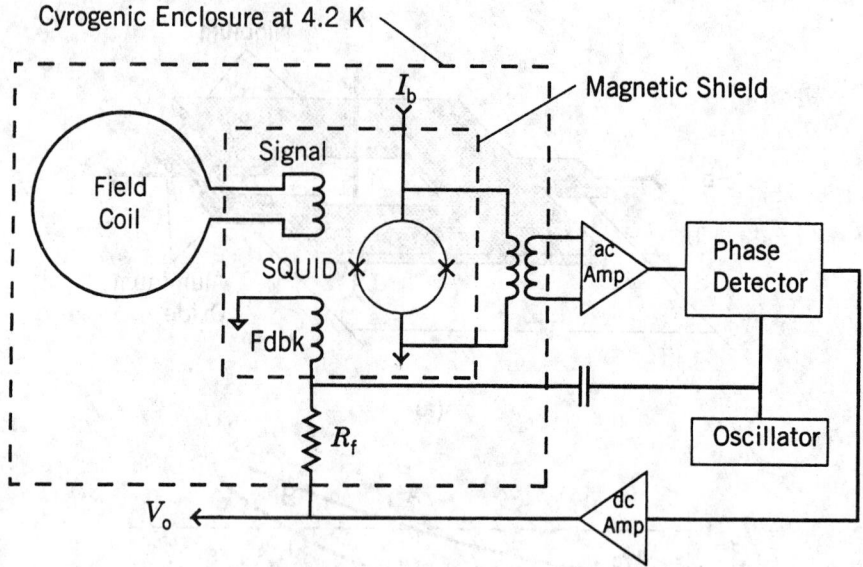

FIGURE 48.13 Wide dynamic range dc SQUID magnetometer. A magnetic field produced by connecting resistor R_f between the output and a feedback coil keeps the field in the SQUID within one flux quantum over its operating range. (Adapted from Wellstood, Heiden and Clark, 1984.)

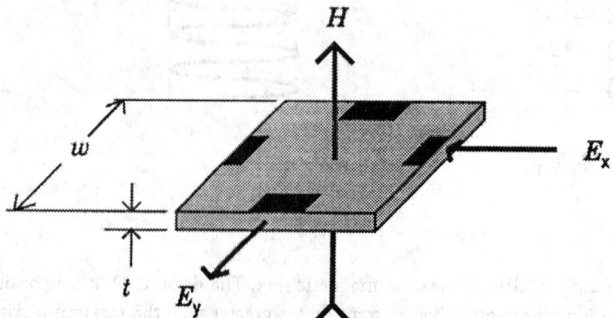

FIGURE 48.14 Hall effect sensor. A magnetic field H applied normal to the surface of the sensor, which is conducting current along the x-direction, will generate a voltage along the y-direction. E_x is the applied electric field along the x-direction, and E_y is the Hall effect electric field along the y-direction.

$$\vec{F} = -q\left(\vec{E} + \vec{v} \times \vec{B}\right) \tag{48.31}$$

The Hall effect device consists of a flat, thin rectangular conductor or semiconductor with two pairs of electrodes at right angles to one another as illustrated in Figure 48.14. An electric field E_x is applied along the x or control axis. When a magnetic field B_z is applied perpendicular to the surface of the device, the free charge, which is flowing along the x-axis as a result of E_x, will be deflected toward the y or Hall voltage axis. Since current cannot flow in the y-axis under open-loop conditions, this will cause a buildup of charge along the y-axis that will create an electric field which produces a force opposing the motion of the charge:

$$E_y = v_x B_z \tag{48.32}$$

where v_x is the average drift velocity of the electrons (or majority carriers). In a conductor that contains n free charges per unit volume having an average drift velocity of v_x, the current density is:

$$J_x = qnv_x \tag{48.33}$$

and

$$E_y = \frac{J_x B_z}{qn} = R_H J_x B_z \tag{48.34}$$

where R_H is called the Hall coefficient.

A semiconductor is treated in terms of the mobility μ (drift velocity/field) of the majority carrier (electron or hole) and conductivity σ. In this case,

$$E_y = \mu E_x B_z \text{ and } E_x = \frac{J_x}{\sigma} \tag{48.35}$$

Therefore,

$$E_y = \frac{\mu}{\sigma} J_x B_z \text{ and } R_H = \frac{\mu}{\sigma} \tag{48.36}$$

The value of R_H varies substantially from one material to another and is both temperature and field magnitude dependent. Its characteristics can be controlled to a certain extent by doping the base material with some impurities. For example, doping germanium with arsenic can reduce the temperature dependence at the expense of magnitude.

The voltage measured across the y-axis terminals is the integral of the electric field along the y-axis. If a constant control current I is flowing along the x axis, then:

$$J_x = \frac{I}{wt} \tag{48.37}$$

and the measured output voltage is:

$$e_y = \frac{R_H I B_z}{t} \tag{48.38}$$

where t is thickness (m) and w is the distance between the y-axis terminals.

Another characteristic specified by manufacturers of Hall effect devices is the magnetic sensitivity γ_b at the rated control current I_c:

$$\gamma_b = \frac{e_y}{B_z} = \frac{R_H I_c}{t} \tag{48.39}$$

Although conductors such as copper (Cu) can be used to make a Hall effect device, semiconductor materials, such as gallium arsenide (GaAs), indium antimonide (InSb), and indium arsenide (InAs), produce the highest and most stable Hall coefficients. InAs, because of its combined low temperature coefficient of sensitivity (<0.1%/°C), low resistance, and relatively good sensitivity, is the material favored by commercial manufacturers of Hall effect devices.

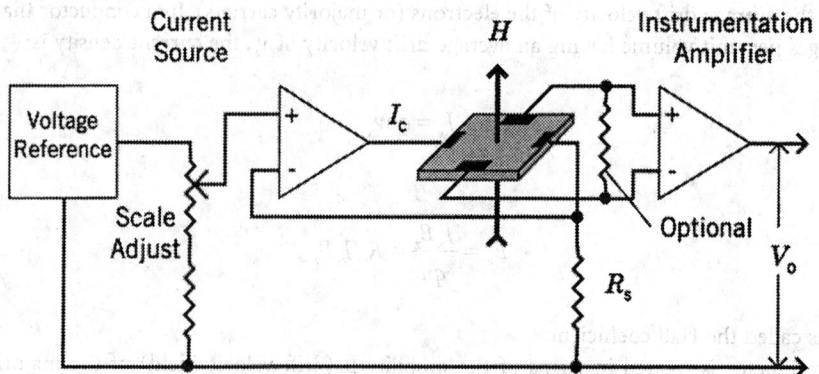

FIGURE 48.15 Example of how to construct a Hall effect gaussmeter. The operational amplifier and resistor R_s form a stable constant-current source for the Hall effect sensor. An instrumentation or differential amplifier amplifies and scales the Hall voltage. A load resistor is sometimes required across the Hall voltage output terminals.

The typical control current for Hall effect devices is 100 mA, but some do operate at currents as low as 1 mA. Sensitivities range from 10 mV/T to 1.4 V/T. Linearity ranges from ¼% to 2% over their rated operating field range. The control input and the voltage output resistance are typically in the range of 1 Ω to 3 Ω The sensor element is usually tiny (on the order of 10 mm square by 0.5 mm thick), and a three-axis version can be housed in a very small package. These devices are most effective for measuring flux densities ranging from 50 µT to 30 T.

Signal Conditioning

A simple Hall effect gaussmeter can be constructed using the signal conditioning circuit shown in Figure 48.15. The voltage reference, operational amplifier, and sense resistor R_s form a precision constant-current source for the Hall effect device control current I_c. For best performance, the voltage reference and R_s should be very stable with temperature and time. A general-purpose operational amplifier can be used for low control currents. A power amplifier is required for control currents above 20 mA.

The Hall voltage can be conditioned and amplified by any high input impedance (>1 kΩ) differential amplifier. A precision instrumentation amplifier is a good choice because it has adequate input impedance, its gain can be determined by a stable resistor, and the amplifier zero offset trim resistor can be used to cancel the zero offset of the Hall effect device. Some devices require a load resistor across the Hall voltage terminal to achieve optimum linearity.

The zero offset and $1/f$ noise of the Hall voltage amplifier limit the performance of a Hall effect gaussmeter for low field strength measurements. Sometimes, these effects can be reduced by using an ac precision current source. The ac amplitude modulated Hall voltage can then be amplified in a more favorable frequency band and synchronously detected to extract the Hall voltage signal. If the field to be measured requires this amount of signal conditioning, it probably is better to use a fluxgate magnetometer for the application.

The Magnetoresistive Gaussmeter

The magnetoresistance effect was first reported by William Thomson (Lord Kelvin) in the middle of the 19th century. He found that a magnetic field applied to a ferromagnetic material caused its resistivity to change. The amount of change depends on the magnetization magnitude and the direction in which the current used to measure resistivity is flowing. Nickel–iron alloys show the greatest change in resistivity (about 5% maximum). Figure 48.16 illustrates how the resistivity changes in Permalloy (a nickel–iron alloy) for a field applied parallel to the current flow. As magnetic field is increased, the change in resistivity increases and asymptotically approaches its maximum value when the material approaches saturation. Bozorth [6] points out that the shape of the curve and the magnitude of the change depend on the

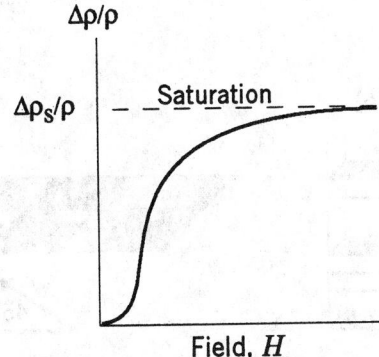

FIGURE 48.16 Change in resistivity in a ferromagnetic material. As field is applied, the resistivity changes rapidly at first. As the material approaches magnetic flux saturation, the resistivity change approaches its maximum value.

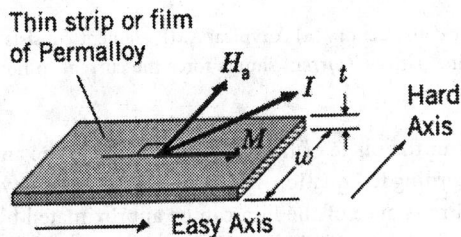

FIGURE 48.17 An AMR resistor element. During fabrication, a magnetic field is applied along the strip's length to magnetize it and establish its easy axis. Current I is passed through the film at 45° to the easy or anisotropic axis. A magnetic field H_a applied at right angles to the magnetization vector M causes the magnetization vector to rotate and the magnetoresistance to change.

composition of the alloy. Permalloy with 80% Ni and 20% Fe provides a high magnetoresistance effect with near-zero magnetostriction and is a favorite material for magnetoresistors.

The change in resistivity in permalloy film [18] is also a function of the angle θ between the magnetization direction and the current direction:

$$\rho(\theta) = \rho_0 + \Delta\rho_m \cos^2\theta \qquad (48.40)$$

where $\Delta\rho_m$ is the magnetoresistivity anisotropy change and ρ_0 is the resistivity for $\theta = \pi/2$.

It was mentioned earlier that magnetic materials have anisotropic magnetic properties (their magnetic properties are direction dependent). The physical shape of an object (see the discussion on demagnetizing factor above) and the conditions that exist during fabrication strongly determine its anisotropic characteristics. A thin long film of permalloy can be made to have highly uniaxial anisotropic properties if it is exposed to a magnetizing field during deposition. This characteristic is exploited in the anisotropic magnetoresistance (AMR) sensor.

The basic resistor element in an AMR is a thin rectangular shaped film as shown in Figure 48.17. One axis, called the anisotropy or easy axis, has a much higher susceptibility to magnetization than the other two. The easy axis is normally along the length of the film. Because of its thinness, the axis normal to the film has virtually no magnetic susceptibility. The axis transverse to the easy axis (across the width) has very little susceptibility as well.

A bias field H_b is used to saturate the magnetization along the easy axis and establish the magnetization direction for zero external field. For a simplified analysis, the film can be modeled as a single domain.

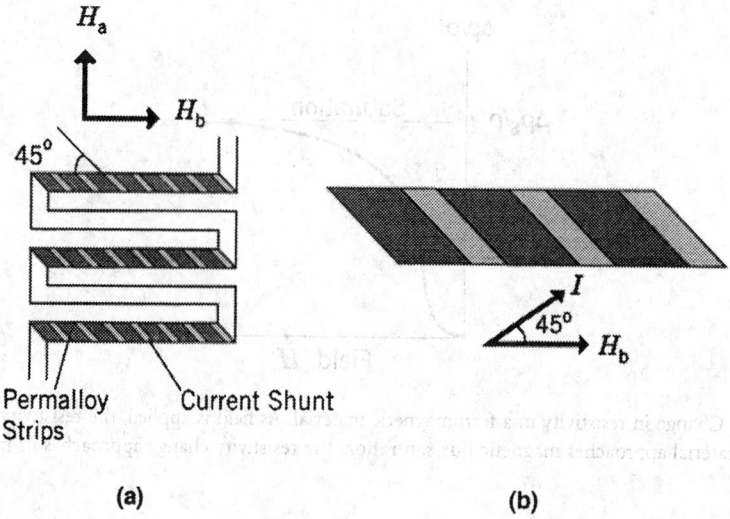

FIGURE 48.18 Magnetoresistor construction. (a) A typical AMR element consists of multiple strips of permalloy connected together in a serpentine pattern. Current shunts force the current to flow through the permalloy at 45° to the easy axis. (b) A close-up view.

The effect of an external field in the plane of the film and normal to the anisotropy axis is to rotate the magnetization vector and, according to Equation 48.40, change the resistivity. Kwiatkowski and Tumanski [19] stated that the change in resistance of the film can be approximated by Equation 48.41:

$$\Delta R \approx R_s \frac{\Delta \rho_m}{\rho} \left(h_a^2 \cos 2\theta + h_a \sqrt{1 - h_a^2} \sin 2\theta - \frac{1}{2} \cos 2\theta \right) \qquad (48.41)$$

where h_a is the normalized externally applied field (i.e., $h_a = H_a/H_k$), R_s is the nominal resistance, and $\Delta \rho_m/\rho$ is the maximum resistivity change. H_k is the anisotropy field. Optimum linear performance is obtained when $\theta = \pi/4$ and Equation 48.41 reduces to:

$$\Delta R \approx R_s \frac{\Delta \rho_m}{\rho} \frac{1}{H_k + H_b} H_a \qquad (48.42)$$

The anisotropy field is given by:

$$H_k = \sqrt{H_{k0}^2 + \left(N M_s \right)^2} \qquad (48.43)$$

where H_{k0} is the film anisotropy field, N is the demagnetizing factor ($\approx$ thickness(t)/width(w)) and M_s is the saturation magnetization.

An AMR is constructed using long thin film segments of deposited permalloy. During deposition, a magnetic field is applied along the length of the film to establish its easy axis of magnetization. The shape of the film also favors the length as an easy axis. As shown in Figure 48.18, a series of these permalloy films is connected together to form the magnetoresistor. The current is forced to flow at a 45° angle to the easy axis by depositing thin strips of highly conductive material (e.g., gold) across the permalloy film. The level of magnetization of the film is controlled by a bias field that is created through the deposition of a thin layer of cobalt over the resistors, which is then magnetized parallel to the easy axis of the permalloy.

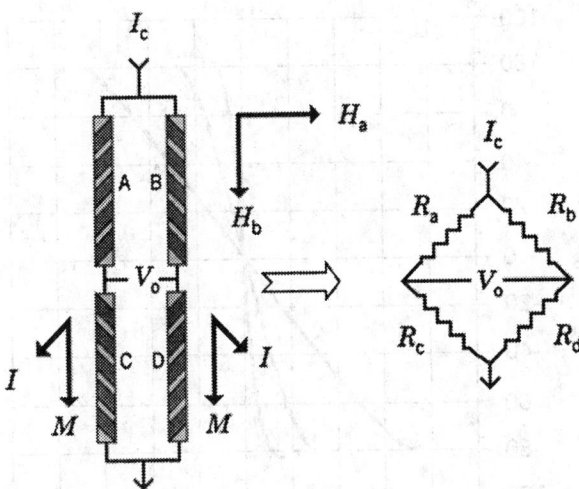

FIGURE 48.19 AMR bridge sensor. In an AMR bridge, the current shunts of resistors A and D are the same and reversed from B and C. Thus, the resistors on diagonal legs of the bridge have the same response to an applied field and opposite that of the other diagonal pair. Bridge leg resistance varies from 1 kΩ to 100 kΩ.

A typical AMR sensor suitable for a gaussmeter or magnetometer consists of four AMRs connected in a Wheatstone bridge as shown in Figure 48.19. The transfer function polarity of the A and D resistors is made to be opposite that of the B and C resistors by rotating the current shunt 90°. This complimentary arrangement enhances the output voltage signal for a given field by a factor of four over a single resistor. Kwiatkowski and Tumanski [19] showed that the transfer function for the bridge configuration is described by:

$$v = IR_s \frac{\Delta \rho_m}{\rho} \cos 2\Delta \varepsilon h_a \sqrt{1 - h_a^2} \qquad (48.44)$$

Where:

$$\cos 2\Delta \varepsilon = \frac{H_{k0}^2 + H_k^2 - \left(NM_s\right)^2}{2H_{k0}H_k} \qquad (48.45)$$

$$h_a = \frac{H_a}{H_k + H_b} \qquad (48.46)$$

For best linearity, $H_a < 0.1\, H_k$. The linearity of the bridge can be controlled during fabrication by adjusting the l/w ratio and H_{k0}. The bias field can also be used to optimize linearity and establish the measurement field range. Some transfer functions for a typical AMR bridge [20] are shown in Figure 48.20. A more comprehensive discussion of AMR theory can be found in [21–23].

Signal Conditioning

Conventional Wheatstone bridge signal conditioning circuits can be used to process the AMR bridge. The bridge sensitivity and zero offset are proportional to the bridge voltage, so it is important to use a well-regulated supply with low noise and good temperature stability.

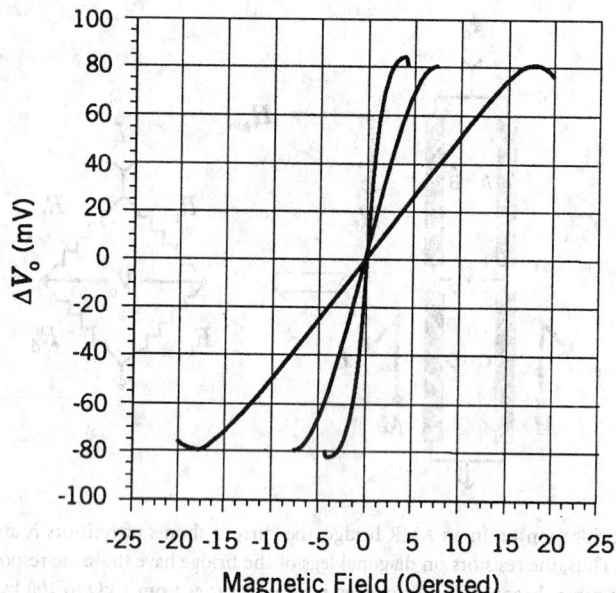

FIGURE 48.20 Typical AMR bridge sensor transfer functions. The sensitivity of an AMR bridge can be adjusted by changing its bias field. Increases in sensitivity are accompanied by corresponding decreases in range.

As Equation 48.44 shows, the polarity of the transfer function is determined by the polarity of (H_k + H_b). If the sensor is exposed to an external field that is strong enough to reverse this field, then the transfer function polarity will reverse. To overcome this ambiguity, the polarity should be established prior to making a measurement. This can be accomplished by momentarily applying a strong magnetic field along the easy axis of the AMR bridge. Some commercial AMR bridges come with a built-in method for performing this action.

Figure 48.21 is a block diagram for a signal conditioner that takes advantage of the bias field polarity flipping property to eliminate zero offset errors and low frequency $1/f$ noise. A square wave oscillator is used to alternately change the direction of the bias field and thus the polarity of the transfer function. The duration of the current used to set the bias field direction should be short in order to minimize power consumption. The amplitude of the ac signal from the bridge is proportional to the field magnitude, and its phase relative to the oscillator gives the field direction. This signal can be amplified and then phase-detected to extract the field-related voltage. Optionally, the output signal can be fed back through a coil that produces a magnetic field opposing the field being measured. This feedback arrangement makes the AMR bridge a null detector and minimizes the influence of changes in its transfer function on overall performance. Of course, the added circuitry increases the size, cost, and complexity of the instrument.

48.4 Scalar Magnetometers

Scalar magnetometers measure the magnitude of the magnetic field vector by exploiting the atomic and nuclear properties of matter. The two most widely used scalar magnetometers are the proton precession and the optically pumped magnetometer. When operated under the right conditions, these instruments have extremely high resolution and accuracy and are relatively insensitive to orientation. They both have several common operating limitations. The instruments require the magnetic field to be uniform throughout the sensing element volume. They have a limited magnetic field magnitude measurement range: typically 20 μT to 100 μT. And they have limitations with respect to the orientation of the magnetic field vector relative to the sensor element.

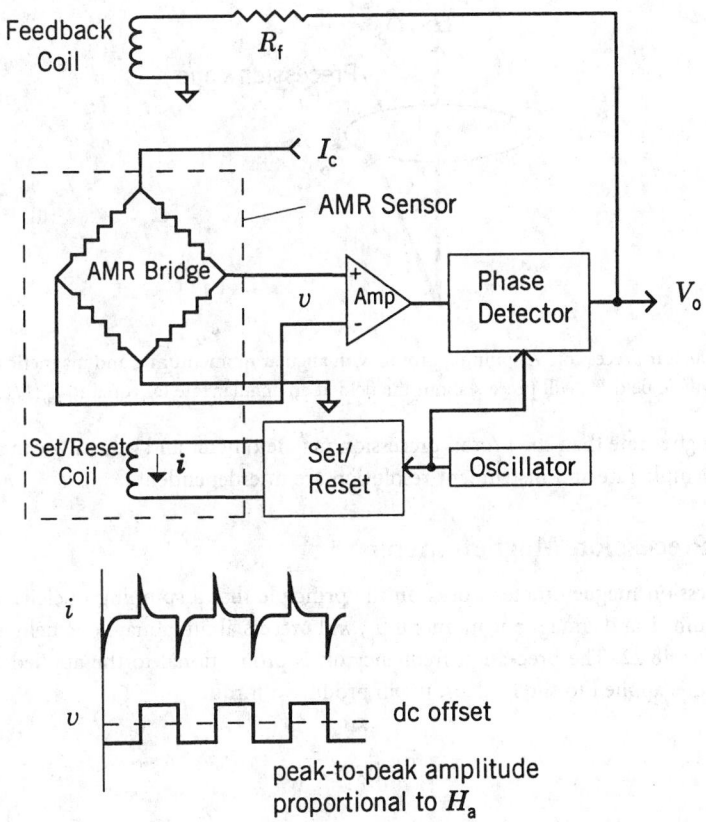

FIGURE 48.21 Example AMR gaussmeter. The magnetization direction can be alternately flipped to eliminate zero offset. The resulting ac signal can then be amplified and synchronously phase-detected to recover the field-related signal. Optionally, the range and stability of the AMR gaussmeter can be increased by connecting the output voltage through a resistor to a feedback coil that produces a field that nulls the applied field.

The proton precession magnetometer uses a strong magnetic field to polarize the protons in a hydrocarbon and then detects the precession frequency of the protons while they decay to the nonpolarized state after the polarizing field is turned off. The precession frequency is proportional to the magnitude of any ambient magnetic field that is present after the polarizing field is removed. This sampling of the magnetic field strength through the polarize-listen sequence makes the proton precession magnetometer response very slow. Maximum rates of only a few samples per second are typical. Because of its dependence on atomic constants, the proton precession magnetometer is the primary standard for calibrating systems used to generate magnetic fields and calibrate magnetometers.

The optically pumped magnetometer is based on the Zeeman effect. Zeeman discovered that applying a field to atoms, which are emitting or absorbing light, will cause the spectral lines of the atoms to split into a set of new spectral lines that are much closer together than the normal lines. The energy-related frequency interval between these hyperfine lines is proportional to the magnitude of the applied field. These energy levels represent the only possible energy states that an atom can possess. The optically pumped magnetometer exploits this characteristic by optically stimulating atoms to produce an overpopulated energy state in one of the hyperfine spectral lines and then causing the energy state to depopulate using an RF magnetic field. The RF frequency required to depopulate the energy state is equal to the spectral difference of the hyperfine lines produced by a magnetic field and, therefore, is proportional to the magnetic field strength. The optically pumped magnetometer can be used to sample the magnetic

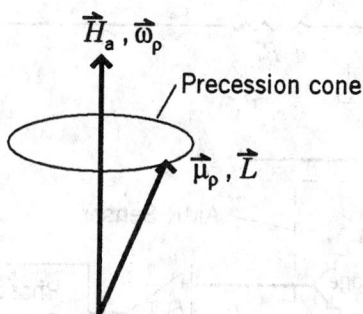

FIGURE 48.22 Nuclear precession. A spinning proton with angular momentum L and magnetic moment μ_p, when subjected to a magnetic field H_a, will precess about the field at an angular rate ω_p equal to $\mu_p H_a/L$.

field at a much higher rate than the proton precession magnetometer and generally can achieve a higher resolution. The sample rate and instrument resolution are interdependent.

The Proton Precession Magnetometer

The proton precession magnetometer works on the principle that a spinning nucleus, which has both angular momentum $\vec{L}$ and a magnetic moment $\vec{\mu}_p$, will precess about a magnetic field like a gyroscope, as shown in Figure 48.22. The precession frequency ω_p is proportional to the applied field. When the magnetic field $\vec{H}_a$ is applied to the nucleus, it will produce a torque:

$$\vec{T} = \vec{\mu}_p \times \vec{H}_a \qquad (48.47)$$

on the nucleus. Because the nucleus has angular momentum, this torque will cause the nucleus to precess about the direction of the field. At equilibrium, the relationship between the torque, precession rate, and angular momentum is:

$$\mu_p \times \vec{H}_a = \vec{\omega}_p \times \vec{L} \qquad (48.48)$$

Solving for the magnitude of the (Larmor) precession frequency, one finds that:

$$\omega_p = \left(\frac{\mu_p}{L}\right) H_a = \gamma H_a \qquad (48.49)$$

where γ is called the gyromagnetic ratio and equals $(2.6751526 \pm 0.0000008) \times 10^{-8}$ T^{-1} s^{-1}.

Figure 48.23 is a block diagram of a proton precession magnetometer. The sensor is a container of hydrocarbon rich in free hydrogen nuclei. A solenoid wrapped around the container is used to both polarize the nuclei and detect the precession caused by the ambient field. Before the polarizing field is applied, the magnetic moments of the nuclei are randomly oriented, and the net magnetization is zero. Application of the polarizing field (typically 3 mT to 10 mT) causes the nuclei to precess about the field. The precession axis can be parallel or antiparallel (nuclear magnetic moment pointing in the direction of the field) to the applied field. From a quantum mechanical standpoint, the antiparallel state is a lower energy level than the parallel state. In the absence of thermal agitation, which causes collisions between atoms, the fluid would remain unmagnetized. When a collision occurs, the parallel precession-axis nuclei lose energy and switch to the antiparallel state. After a short time, there are more nuclei with magnetic

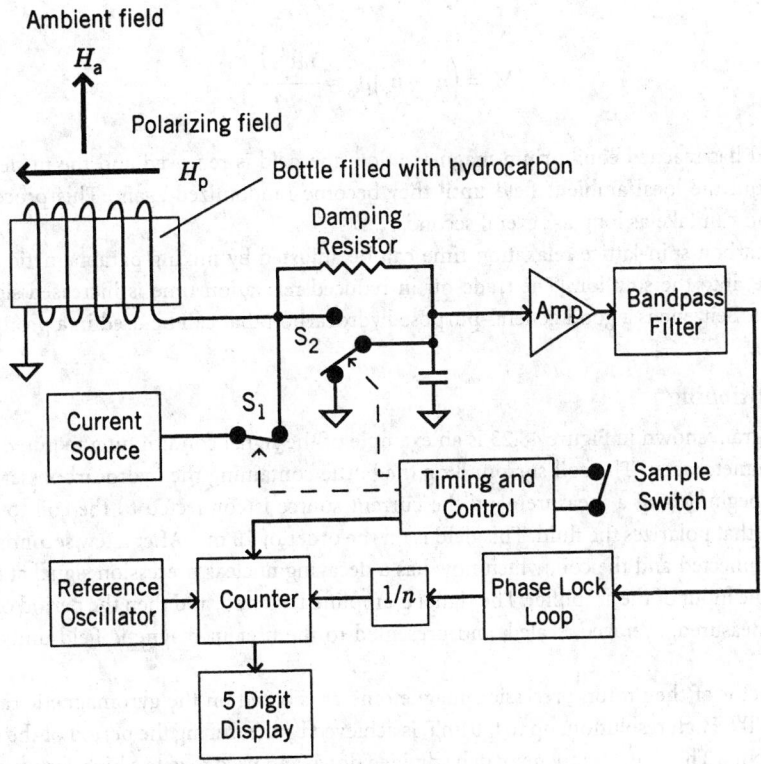

FIGURE 48.23 Typical proton precession magnetometer. A polarizing field is applied to the hydrocarbon when S1 is closed. The amplifier input is shorted to prevent switching transients from overdriving it. After a few seconds, S1 is opened and the coil is connected to the signal processor to measure the Larmor frequency.

moments pointing in the direction of the field than away from it, and the fluid reaches an equilibrium magnetization M_0. The equation that relates magnetization buildup to time is:

$$M(t) = M_0 \left(1 - e^{-t/\tau_e}\right) \tag{48.50}$$

where τ_e is the spin-lattice relaxation time.

The equilibrium magnetization is based on thermodynamic considerations. From Boltzmann statistics for a system with spins of ½:

$$\frac{n_p}{n_a} = e^{2\mu_p H_a / kT} \tag{48.51}$$

where n_p is the number of precession spin axes parallel to H_a, n_a is the number of precession spin axes antiparallel to H_a, k is Boltzmann's constant, and T is temperature (kelvin). If n is the number of magnetic moments per unit volume, then:

$$n = n_p + n_a = n_a \left(1 + e^{2\mu_p H_a / kT}\right) \tag{48.52}$$

and

$$M_0 = \left(n_p - n_a\right)\mu_p \approx \frac{n\mu_p^2 H_a}{kT} \tag{48.53}$$

Once the fluid has reached equilibrium magnetization, the field is removed and the nuclei are allowed to precess about the local ambient field until they become randomized again. This process of excitation–relaxation can take as long as several seconds.

The hydrocarbon spin-lattice relaxation time can be adjusted by mixing paramagnetic salts, such as ferrous nitrate, into the solution. The trade-off in reduced relaxation time is increased signal-to-noise and resolution. Benzene is a good general-purpose hydrocarbon that can be used in a proton precession magnetometer.

Signal Conditioning

The block diagram shown in Figure 48.23 is an example of the signal conditioning required for a proton precession magnetometer. The coil surrounding the bottle containing the hydrocarbon serves two purposes. At the beginning of a measurement, the current source is connected to the coil to generate the magnetic field that polarizes the fluid. This field is on the order of 10 mT. After a few seconds, the current source is disconnected and the coil, which now has a decaying nuclear precession signal at its output, is connected to the input of the amplifier. The signal is amplified, filtered, and then the period of the Larmor frequency is measured, averaged, scaled, and presented to the user in magnetic field units on a digital display.

The scale factor of the proton precession magnetometer is based on the gyromagnetic ratio, which is 0.042579 Hz nT^{-1}. High resolution, up to 0.01 nT, is achieved by measuring the period of the signal rather than the frequency. The signal frequency can be divided down and used to gate a high-frequency oscillator that is driving a counter.

The sampling of the field is controlled manually in many commercially available proton precession magnetometers. Some magnetometers have an internally controlled sample rate. The sample rate and resolution are inversely related to one another. A higher sample rate produces a poorer resolution.

The Optically Pumped Magnetometer

As explained earlier, the optically pumped magnetometer is based on the Zeeman effect. This effect is most pronounced in alkaline vapors (rubidium, lithium, cesium, sodium, and potassium). Figure 48.24 is the hyperfine spectral structure for the valence electrons of rubidium (Rb) 85, which is commonly used in these types of magnetometers. The energy-related frequency interval between these hyperfine lines is proportional to the applied field. The magnetic quantum number m is related to the angular momentum number and specifies the possible component magnitude of the magnetic moment along the applied field. The optically pumped magnetometer takes advantage of this characteristic.

Transitions occur between levels of different m values and obey the rule that the change in m can only have the values 0, 1, and −1. Table 48.5 lists the relationship between the polarization of the light stimulating the transition and the allowable change in m.

When not optically excited, the energy states of the valence electrons will be distributed according to Boltzmann statistics and will be in a state of equilibrium. If the electrons are excited with circularly polarized light at the D1 frequency (794.8 nm wavelength), they will absorb photons and transition from the $^2S_{1/2}$ state to the $^2P_{1/2}$ state according to the transition rules. The excited electrons will then fall back in a random fashion to the lower states, being distributed with an equal probability among all the m states.

But the rules state that the change in m can only be 1 or −1 for polarized light. If one uses right circularly polarized light, then the change in m can only be 1, and the electrons in the $m = 3$ level of the $^2S_{1/2}$ state cannot transition since there is no $m = 4$ level at the $^2P_{1/2}$ state. Therefore, these electrons remain in the $m = 3$ state. All other electrons transition to the higher state and then fall back to the lower state

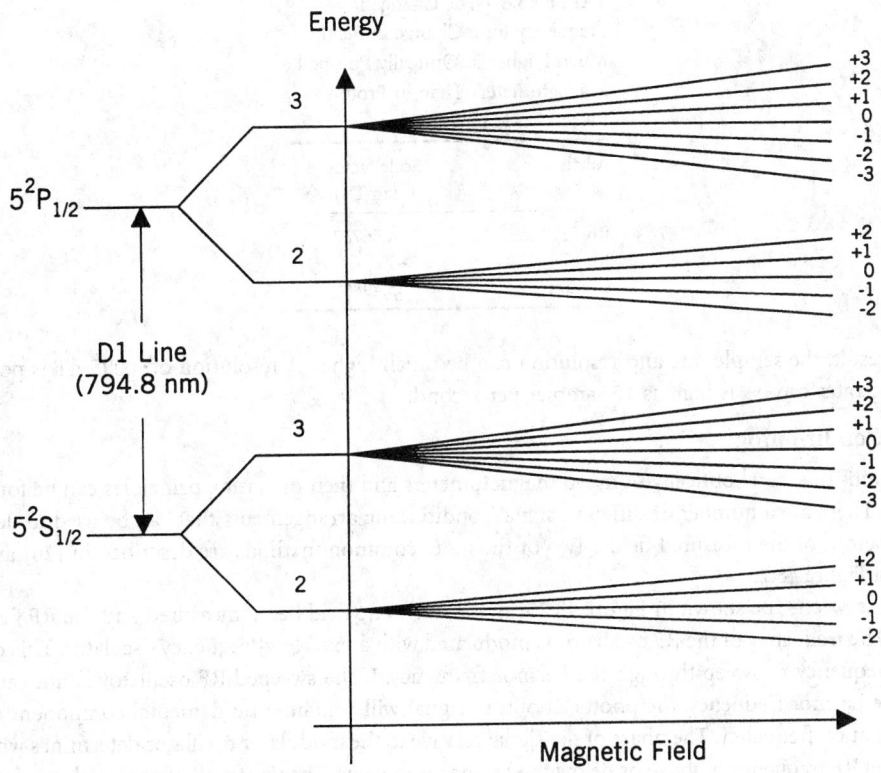

FIGURE 48.24 Rb-85 energy diagram. When a magnetic field is applied, the energy levels split into Zeeman sublevels that diverge as the field increases. Quantum mechanical factors determine the number of sublevels at each primary energy level.

TABLE 48.5 The Allowable Change in m When Jumping from One Energy Level to Another Depends on the Polarization of the Light Causing the Transition

Polarization	m
Left circular	−1
Parallel	0
Right circular	1

with equal probability of arriving at any of the m levels, including $m = 3$. Thus, the $m = 3$ level fills up, and the other levels empty until all the electrons are in the $m = 3$ level, and no more transitions to the higher state can take place. Pumping stops.

When pumping begins, the vapor is opaque. As time goes on, less electrons are available for absorbing photons, and the vapor becomes more transparent until, finally, pumping action stops and the vapor is completely transparent.

If a small RF magnetic field at the Larmor frequency is applied at right angles to the magnetic field being measured, the electrons in the $m = 3$ state will be depumped to the other m levels, making them available for further pumping. The optically pumped magnetometer exploits this situation in a positive feedback arrangement to produce an oscillator at the Larmor frequency.

The scale factors for optically pumped magnetometers are significantly higher than for the proton precession magnetometer. Table 48.6 lists these scale factors for a number of alkali vapors.

TABLE 48.6 The Change in
Frequency for a Change in Field Is
Much Higher in Optically Pumped
Magnetometers Than in Proton
Precession Magnetometers

Alkali	Scale factor (Hz nT^{-1})
Rb-85	4.66737
Rb-87	~7
Cesium	3.4986

As a result, the sample rate and resolution can be much higher. A resolution of 0.005 nT is possible. Sampling rates can be as high as 15 samples per second.

Signal Conditioning

Descriptions of several optically pumped magnetometers and their operating principles can be found in [24–26]. There are a number of different signal conditioning arrangements that can be used to derive a useful readout of the measured fields. Two of the more common methods are described in [26] and are shown in Figure 48.25.

In the servoed type shown in Figure 48.25(a), the magnetic field being measured and the RF field are coaxial. The frequency of the RF oscillator is modulated with a fixed low-frequency oscillator. This causes the RF frequency to sweep through the Larmor frequency. If the sweeped RF oscillator is not centered about the Larmor frequency, the photocell output signal will contain a fundamental component of the RF modulation frequency. The phase of the signal relative to the modulator oscillator determines whether the central RF frequency is above or below the Larmor frequency. The photocell output is phase-detected to produce an error voltage that is used to drive the RF frequency toward the Larmor frequency. The RF frequency can be measured to determine the magnetic field. If a linear voltage controlled oscillator is used as the RF oscillator, its control voltage can also be used as an output since it is a measure of the Larmor frequency.

The auto-oscillating type shown in Figure 48.25(b) is based on the transmission of a polarized beam that is at right angles to the field being measured. The intensity of this cross-beam will be modulated at the Larmor frequency. The photocell signal will be shifted by 90° relative to the RF field. By amplifying the photocell signal, shifting it 90° and feeding it back to drive the RF field coil, an oscillator is created at the Larmor frequency. In practice, only one light source is used, and the field being measured is set at an angle of 45°.

Defining Terms

Anisotropic: The material property depends on direction.

Gaussmeter: An instrument used to measure magnetic fields greater than 1 mT.

Induced magnetization: The object's magnetization is induced by an external magnetic field and disappears when the inducing field is removed.

Initial permeability: The slope at the origin of the magnetization curve.

Isotropic: The material property is the same in all directions.

Magnetic dipole moment: A vector quantity that describes the strength and direction of a magnetic field source, such as a small current loop or spinning atomic nucleus.

Magnetically "hard" material: The material has a significant residual (permanent) magnetization after an external magnetic field is removed.

Magnetically "soft" material: The material's magnetization is induced by an external magnetic field and the material has no significant residual (permanent) magnetization after the field is removed.

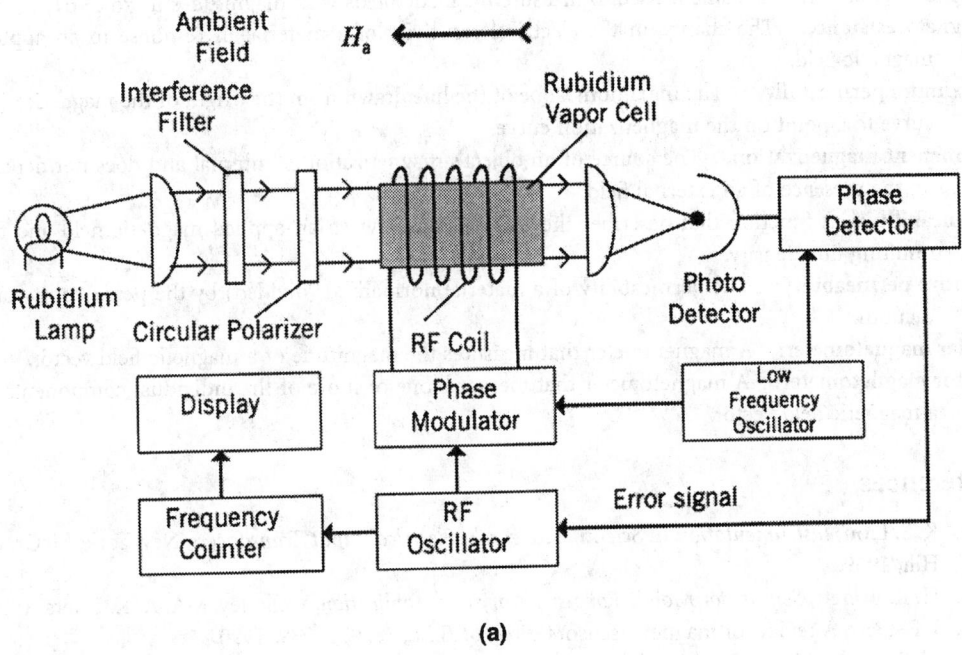

(a)

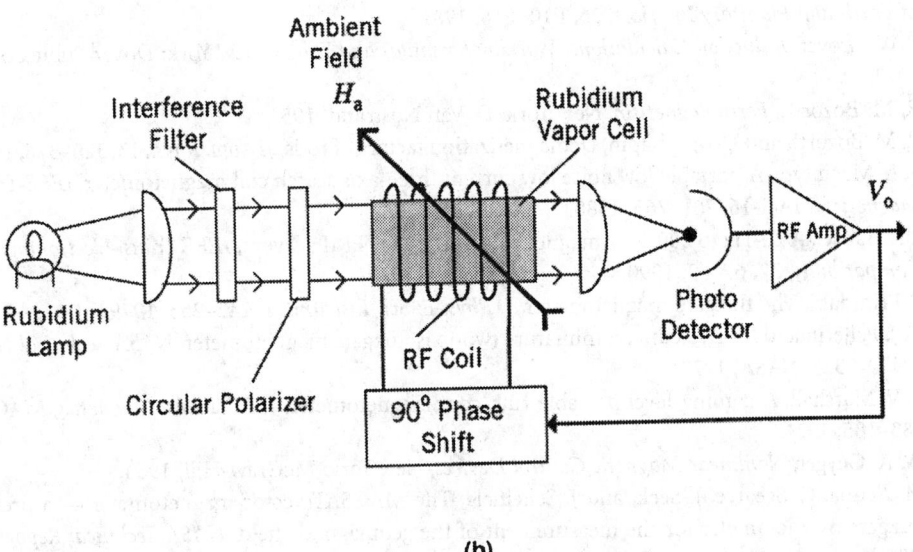

(b)

FIGURE 48.25 Two examples of optically pumped scalar magnetometers. The servoed magnetometer: (a) slightly modulates the RF field at a low frequency, causing the vapor transmissivity to modulate. A phase detector provides an error signal that is used to lock the RF oscillator to the Larmor frequency. (b) A self-oscillating magnetometer: the transmissivity of the vapor, at right angles to the applied field, is made to oscillate at the Larmor frequency by phase-shifting the detected light modulation and feeding it back to the RF field generator. (Adapted from Hartmann, 1972.)

Magnetization curve: A plot of flux density B vs. magnetic field H for an initially unmagnetized ferromagnetic material.

Magnetization: A vector quantity describing the average density and direction of magnetic dipole moments.

Magnetometer: An instrument used to measure magnetic fields with magnitudes up to 1 mT.

Magnetoresistance: The change in the electrical resistivity of a material in response to an applied magnetic field.

Maximum permeability: The maximum slope of the line drawn from the origin of the magnetization curve to a point on the magnetization curve.

Permanent magnetization: The source of an object's magnetization is internal and does not depend on the presence of an external field.

Permeability: A function that describes the relationship between an applied magnetic field and the resulting flux density.

Relative permeability: The permeability of a material normalized (divided) by the permeability of a vacuum.

Scalar magnetometer: A magnetometer that measures the magnitude of a magnetic field vector.

Vector magnetometer: A magnetometer that measures one or more of the individual components of a magnetic field vector.

References

1. K. S. Lion, *Instrumentation in Scientific Research. Electrical Input Transducers*, New York: McGraw-Hill, 1959.
2. H. R. Everett, *Sensors for Mobile Robots: Theory and Application*, Wellesley, MA: A. K. Peters, 1995.
3. J. E. Lenz, A review of magnetic sensors, *Proc. of IEEE*, 78, 973–989, 1990.
4. M. A. Payne, SI and Gaussian cgs units, conversions and equations for use in geomagnetism, *Phys. of Earth and Planitary Interiors*, 26, P10–P16, 1981.
5. F. W. Grover, *Induction Calculations. Working Formulas and Tables*, New York: Dover Publications, 1973.
6. R. M. Bozorth, *Ferromagnetism*, New York: D. Van Nostrand, 1951.
7. R. M. Bozorth and D. M. Chapin, Demagnetization factors of rods, *J. Appl. Phys.*, 13, 320–326, 1942.
8. S. A. Macintyre, A portable low noise low current three-axis search coil magnetometer, *IEEE Trans. Magnetics*, MAG-16, 761–763, 1980.
9. J. P. Hauser, A 20-Hz to 200-kHz magnetic flux probe for EMI surveys, *IEEE Trans. Electromagnetic Compatibility*, 32, 67–69, 1990.
10. F. Primdahl, The fluxgate magnetometer, *J. Phys. E: Sci. Instrum.*, 1, 242–253, 1979.
11. C. J. Pellerin and M. H. Acuna, A miniature two-axis fluxgate magnetometer, *NASA Technical Note*, TN D-5325, NASA, 1970.
12. S. V. Marshall, A gamma-level portable ring-core magnetometer, *IEEE Trans. Magnetics*, MAG-7, 183–185, 1971.
13. W. A. Geyger, *Nonlinear-Magnetic Control Devices*, New York: McGraw-Hill, 1964.
14. M. Acuna, C. Scearce, J. Seek, and J. Schelfiele, The MAGSAT vector magnetometer — a precise fluxgate magnetometer for the measurement of the geomagnetic field, *NASA Technical Report*.
15. D. Cohen, Measurements of the magnetic field produced by the human heart, brain and lung, *IEEE Trans. Magnetics*, MAG-11, 694–700, 1975.
16. C. M. Falco and I. K. Schuller, SQUIDs and their sensitivity for geophysical applications, *SQUID Applications to Geophysics*, The Society of Exploration Geophysicists, 13–18, 1981.
17. J. Clark, SQUIDs, *Sci. Am.*, 46–53, August 1994.
18. T. H. Casselman and S. A. Hanka, Calculation of the performance of a magnetoresistive permalloy magnetic field sensor, *IEEE Trans. Magnetics*, MAG-16, 461–464, 1980.
19. W. Kwaitkawski and S. Tumanski, The permalloy magnetoresistive sensors-properties and applications, *J. Phys. E: Sci. Instrum.*, 19, 502–515, 1986.
20. Permalloy Magnetic Sensors, Honeywell Technical Note.
21. U. Dibbern and A. Petersen, The magnetoresistor sensor-a sensitive device for detecting magnetic field variations, *Electronic Components and Applications*, 5(3), 148–153, 1983.

22. S. Tumanski and M. M. Stabrowski, Optimization of the performance of a thin film magnetore-sistive sensor, *IEEE Trans. Magnetics*, MAG-20, 963–965, 1984.

23. L. W. Parson and Z. M. Wiatr, Rubidium vapor magnetometer, *J. Sci. Instrum.*, 39, 292–299, 1962.

24. W. H. Farthing and W. C. Folz, Rubidium vapor magnetometer for near Earth orbiting spacecraft, *Rev. Sci. Instrum.*, 38, 1023–1030, 1967.

25. F. Hartmann, Resonance magnetometers, *IEEE Trans. Magnetics*, MAG-8, 66–75, 1972.

26. F. Wellstood, C. Heiden, and J. Clark, Integrated dc SQUID magnetometer with high slew rate, *Rev. Sci. Instrum.*, 66, 952–957, 1984.

49

Permeability and Hysteresis Measurement

Jeff P. Anderson
LTV Steel Corporation

Richard J. Blotzer
LTV Steel Corporation

Magnetic fields are typically conceptualized with so-called "flux lines" or "lines of force." When such flux lines encounter any sort of matter, an interaction takes place in which the number of flux lines is either increased or decreased. The original magnetic field therefore becomes amplified or diminished in the body of matter as a result of the interaction. This is true whether the matter is a typical "magnetic" material, such as iron or nickel, or a so-called "nonmagnetic" material, such as copper or air.

The *magnetic permeability* of a substance is a numerical description of the extent to which that substance interacts with an applied magnetic field. In other words, permeability refers to the degree to which a substance can be magnetized.

Different substances possess varying degrees of magnetization. The aforementioned examples of strongly magnetic materials have the ability to strengthen an applied magnetic field by a factor of several thousand. Such highly magnetizable materials are called *ferromagnetic*. Certain other substances, such as Al, only marginally increase an applied magnetic field. Such weakly magnetizable materials are called *paramagnetic*. Still other substances, such as Cu and the rare gases, slightly weaken an applied magnetic field. Such "negatively magnetizable" substances are called *diamagnetic*.

In common parlance, diamagnetic and paramagnetic substances are often called *nonmagnetic*. However, as detailed below, all substances are magnetic to some extent. Only empty space is truly nonmagnetic.

The term *hysteresis* has been used to describe many instances where an effect lags behind the cause. However, Ewing was apparently the first to use the term in science [1] when he applied it to the particular magnetic phenomenon displayed by ferromagnetic materials. Magnetic hysteresis occurs during the cyclical magnetization of a ferromagnet. The magnetization path created while increasing an externally applied field is not retraced on subsequent decrease (and even reversal) of the field. Some magnetization, known as *remanence*, remains in the ferromagnet after the external field has been removed. This remanence, if appreciable, allows for the permanent magnetization observed in common bar magnets.

TABLE 49.1   Conversion Factors Between the mks and cgs Systems for Important Quantities in Magnetism

Quantity	mks	cgs
H, applied field	A/m	$= 4\pi \times 10^{-3}$ Oe
B, flux density	Wb/m²	$= 10^4$ G
M, magnetization	Wb/m²	$= 10^4/4\pi$ emu/cm³
κ, susceptibility	Wb/(A·m)	$= 16\pi^2 \times 10^{-7}$ emu/Oe·cm³

49.1   Definition of Permeability

Let an externally applied field be described by the vector quantity **H**. This field may be produced by a solenoid or an electromagnet. Regardless of its source, **H** has units of ampere turns per meter (A m⁻¹). On passing through a body of interest, **H** magnetizes the body to a degree, **M**, formally defined as the magnetic moment per unit volume. The units of **M** are usually webers per square meter. A secondary coil (and associated electronics) is typically used to measure the combined effects of the applied field and the body's magnetization. This sum total flux-per-unit-area (flux density) is known as the induction, **B**, which typically has units of Wb/m², commonly refered to as a Tesla (T). Because **H**, **M**, and **B** are usually parallel to one another, the vector notation can be dropped, so that:

$$B = \mu_0 H + M \qquad (49.1)$$

where μ_0 is the permeability of free space ($4\pi \times 10^{-7}$ Wb/A⁻¹ m⁻¹).

The absolute permeability, μ, of a magnetized body is defined as the induction achieved for a given strength of applied field, or:

$$\mu = \frac{B}{H} \qquad (49.2)$$

Often, the absolute permeability is normalized by μ_0 to result in the relative permeability, μ_r ($=\mu/\mu_0$). This relative permeability is numerically equal and physically equivalent to the cgs version of permeability. This, unfortunately, is still in common usage, and often expressed in units of gauss per oersted (G Oe⁻¹), although the cgs permeability is actually dimensionless. Conversion factors between the mks and cgs systems are listed in Table 49.1 for the important quantities encountered.

49.2   Types of Material Magnetization

All substances fall into one of three magnetic groups: diamagnetic, paramagnetic, or ferromagnetic. Two important subclasses, antiferromagnetic and ferrimagnetic, will not be included here. The interested reader can find numerous discussions of these subclasses; for example, see [1].

Diamagnetism

Diamagnetic and paramagnetic (see next section) substances are usually characterized by their magnetic susceptibility rather than permeability. Susceptibility is derived by combining Equations 49.1 and 49.2, viz.

$$\mu_r = 1 + \frac{M}{\mu_0 H} = 1 + \frac{\kappa}{\mu_0} \qquad (49.3)$$

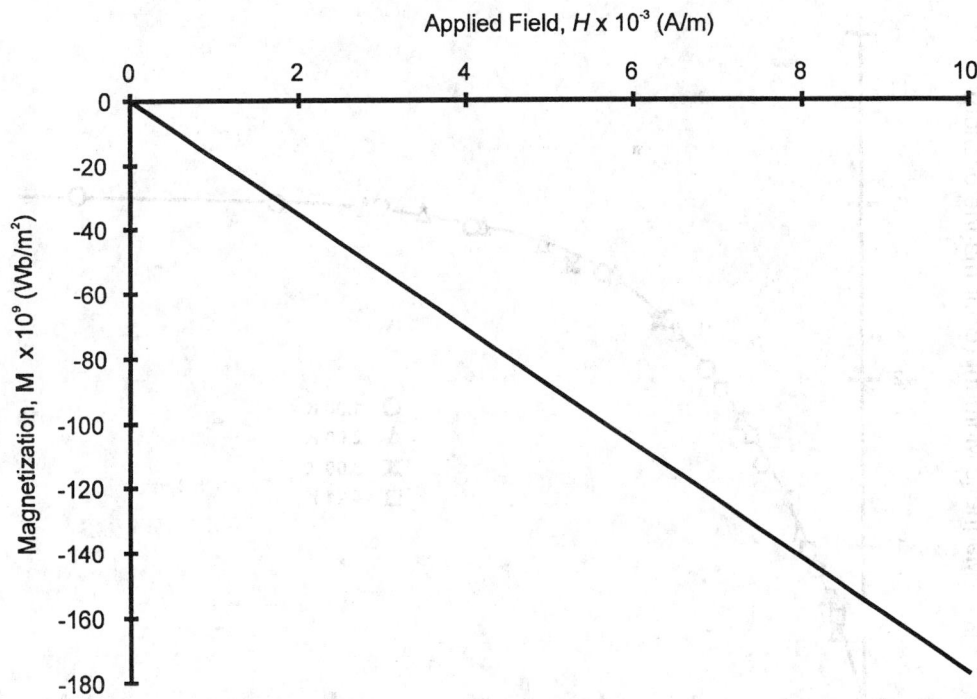

FIGURE 49.1 For diamagnetic substances, magnetization M is small and opposite the applied field H as in this schematic example for graphite ($\kappa = -1.78 \times 10^{-11}$ Wb A^{-1} m^{-1}).

where κ is the susceptibility with units of Wb A^{-1} m^{-1}. This so-called *volume susceptibility* is often converted to a mass susceptibility (χ) or a molar susceptibility (χ_M). Values for the latter are readily available for many pure substances and compounds [2]. Susceptibility is also often called "intrinsic permeability" [3].

In any atom, the orbiting and spinning electrons behave like tiny current loops. As with any charge in motion, a magnetic moment is associated with each electron. The strength of the moment is typically expressed in units of Bohr magnetons.

Diamagnetism represents the special case in which the moments contributed by all electrons cancel. The atom as a whole possesses a net zero magnetic moment. An applied field, however, can induce a moment in the diamagnetic material, and the induced moment opposes the applied field. The magnetization, $\mathbf{M}$, in Equation 49.3 is therefore antiparallel to the applied field, $\mathbf{H}$, and the susceptibility, κ, is negative. For diamagnetic materials, $\mu < 1$. Figure 49.1 shows a schematic M vs. H curve for graphite with $\kappa = -1.78 \times 10^{-11}$ Wb A^{-1} m^{-1}. Note that κ is a constant up to very high applied field values.

Paramagnetism

In a paramagnetic substance, the individual electronic moments do not cancel and the atom possesses a net nonzero moment. In an applied field, the weak diamagnetic response is dominated by the atom's tendency to align its moment parallel with the applied field's direction. Paramagnetic materials have relatively small positive values for κ, and $\mu > 1$.

Thermal energy retards a paramagnet's ability to align with an applied field. Over a considerable range of applied field and temperature, the paramagnetic susceptibility is constant. However, with very high applied fields and low temperatures, a paramagnetic material can be made to approach saturation — the condition of complete alignment with the field. This is illustrated in Figure 49.2 for potassium chromium alum, a paramagnetic salt. Even at a temperature as low as 1.30 K, an applied field in excess of about 3.8×10^6 A m^{-1} is necessary to approach saturation. [Note in Figure 49.2, that 1 Bohr magneton $= 9.27 \times 10^{-24}$ J T^{-1}.]

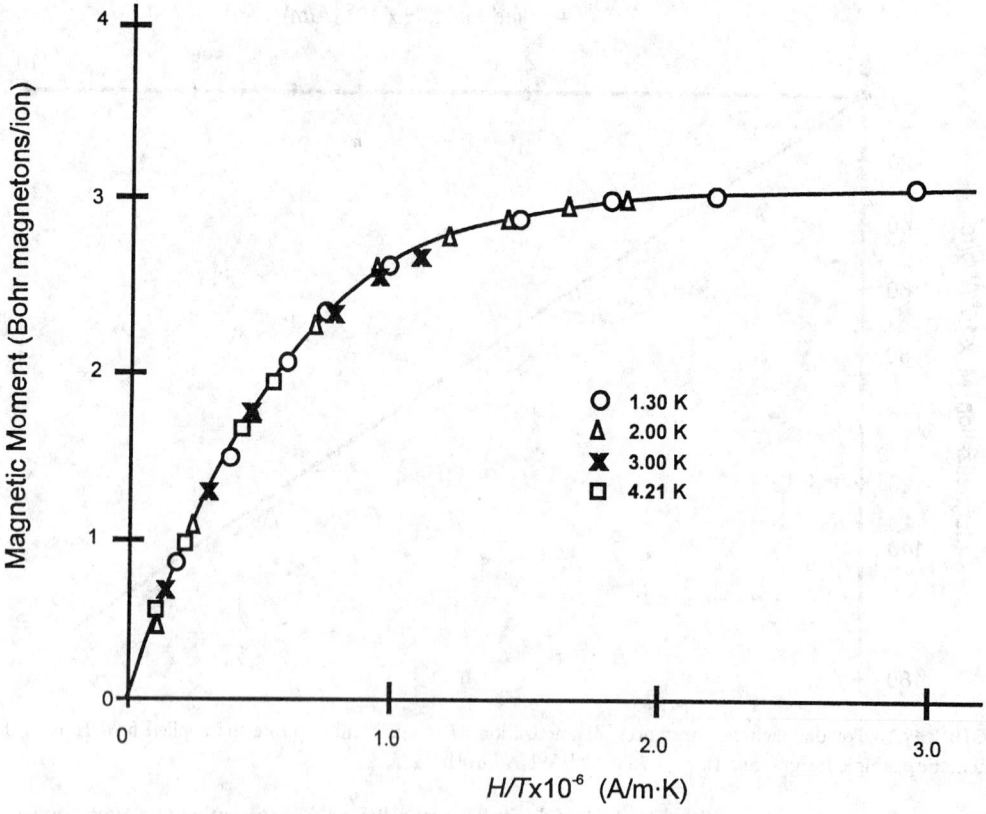

FIGURE 49.2   For paramagnetic substances, the susceptibility is constant over a wide range of applied field and temperature. However, at very high H and low T, saturation can be approached, as in this example for potassium chromium alum. (After W. E. Henry, *Phys. Rev.*, 88, 559-562, 1952.)

Ferromagnetism

Ferromagnetic substances are actually a subclass of paramagnetic substances. In both cases, the individual electronic moments do not cancel, and the atom has a net nonzero magnetic moment that tends to align itself parallel to an applied field. However, a ferromagnet is much less affected by the randomizing action of thermal energy compared to a paramagnet. This is because the individual atomic moments of a ferromagnet are coupled in rigid parallelism, even in the absence of an applied field.

With no applied field, a demagnetized ferromagnet is comprised of several magnetic domains. Within each domain, the individual atomic moments are parallel to one another, or coupled, and the domain has a net nonzero magnetization. However, the direction of this magnetization is generally opposed by a neighboring domain. The vector sum of all magnetizations among the domains is zero. This condition is called the *state of spontaneous magnetization.*

With an increasing applied field, domains with favorable magnetization directions, relative to the applied field direction, grow at the expense of the less favorably oriented domains. This process is schematically illustrated in Figure 49.3. The exchange forces responsible for the ferromagnetic coupling are explained by Heisenberg's quantum mechanical model [4]. Above a critical temperature known as the Curie point, the exchange forces disappear and the formerly ferromagnetic material behaves exactly like a paramagnet.

During magnetization, ferromagnets show very different characteristics from diamagnets and para-magnets. Figure 49.4 is a so-called B–H curve for a typical soft ferromagnet. Note that B is no longer

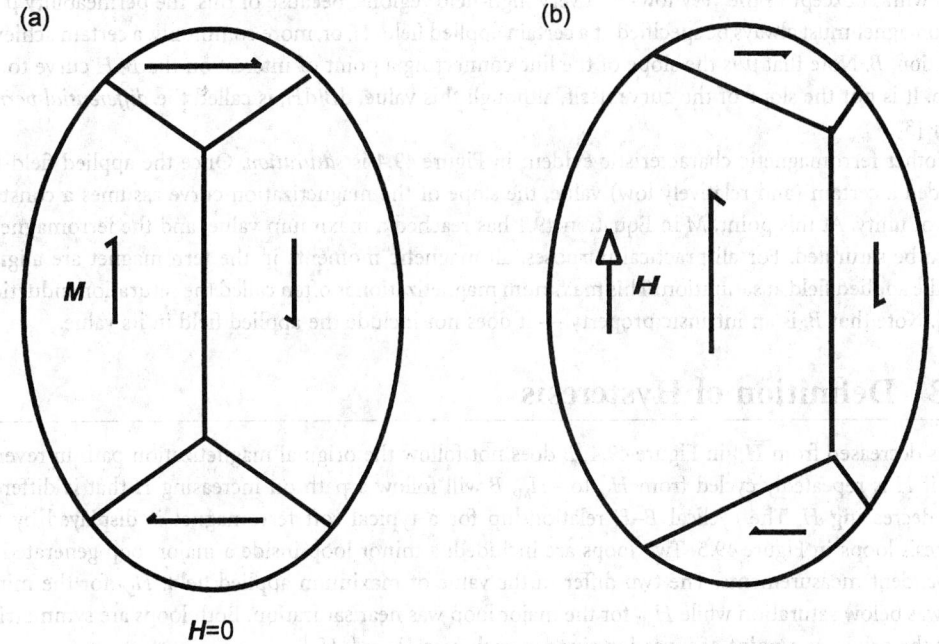

FIGURE 49.3 With no applied field (a) a ferromagnet assumes spontaneous magnetization. With an applied field (b) domains favorably oriented with *H* grow at the expense of other domains.

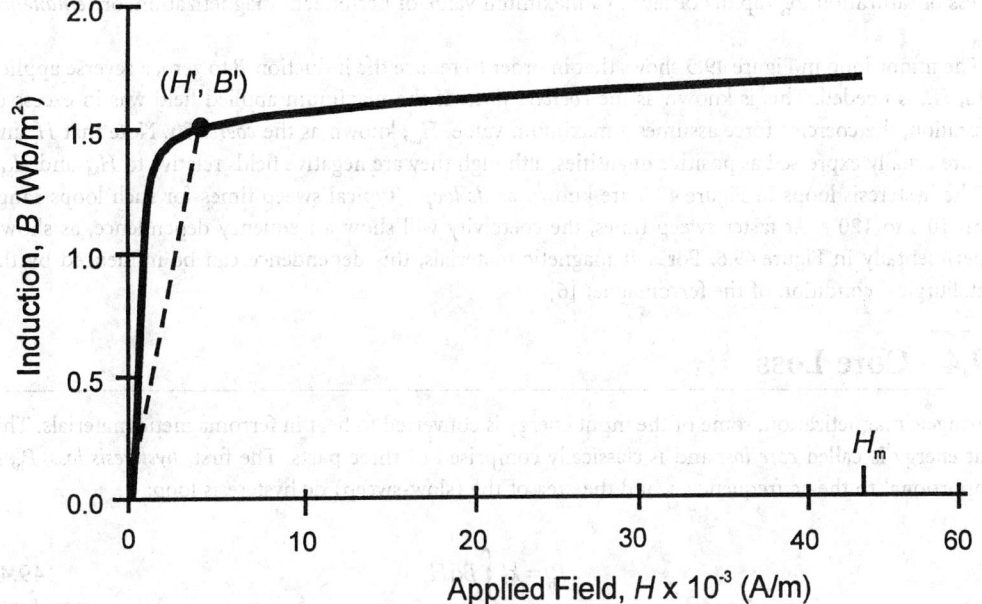

FIGURE 49.4 Magnetization (*B–H*) curve for a typical soft ferromagnet. Permeability at point (H', B') is the slope of the dashed line.

linear with H except in the very low- and very high-field regions. Because of this, the permeability μ for a ferromagnet must always be specified at a certain applied field, H, or, more commonly, a certain achieved induction, B. Note that μ is the slope of the line connecting a point of interest on the B–H curve to the origin. It is not the slope of the curve itself, although this value, dB/dH, is called the *differential permeability* [3].

Another ferromagnetic characteristic evident in Figure 49.4 is *saturation*. Once the applied field has exceeded a certain (and relatively low) value, the slope of the magnetization curve assumes a constant value of unity. At this point, M in Equation 49.1 has reached a maximum value, and the ferromagnet is said to be saturated. For all practical purposes, all magnetic moments in the ferromagnet are aligned with the applied field at saturation. This maximum magnetization is often called the saturation induction, B_s [5]. Note that B_s is an intrinsic property — it does not include the applied field in its value.

49.3   Definition of Hysteresis

If H is decreased from H_M in Figure 49.4, B does not follow the original magnetization path in reverse. Even if H is repeatedly cycled from H_M to $-H_M$, B will follow a path on increasing H that is different from decreasing H. The cyclical B–H relationship for a typical soft ferromagnet is displayed by the hysteresis loops in Figure 49.5. Two loops are included: a minor loop inside a major loop generated by independent measurements. The two differ in the value of maximum applied field: $H_{M'}$ for the minor loop was below saturation while $H_{M''}$ for the major loop was near saturation. Both loops are symmetrical about the origin as a point of inversion since in each case $H_M = \left| -H_M \right|$.

Notice for the minor loop that when the applied field is reduced from $H_{M'}$ to 0, the induction does not also decrease to zero. Instead, the induction assumes the value B_r, known as the *residual induction*. If the peak applied field exceeds the point of saturation, as for the major loop in Figure 49.5, B_r assumes a maximum value known as the *retentivity*, B_{rs}.

Note that B_r and B_{rs} are short-lived quantities observable only during cyclical magnetization conditions. When the applied field is removed, B_r rapidly decays to a value B_d, known as the *remanent induction*. B_d is a measure of the permanent magnetization of the ferromagnet. If the maximum applied field was in excess of saturation, B_{rs} rapidly decays to a maximum value of permanent magnetization, or *remanence*, B_{dm}.

The minor loop in Figure 49.5 shows that in order to reduce the induction B to zero, a reverse applied field, H_c, is needed. This is known as the *coercive force*. If the maximum applied field was in excess of saturation, the coercive force assumes a maximum value, H_{cs}, known as the *coercivity*. Note that H_c and H_{cs} are usually expressed as positive quantities, although they are negative fields relative to $H_{M'}$ and $H_{M''}$.

The hysteresis loops in Figure 49.5 are known as *dc loops*. Typical sweep times for such loops range from 10 s to 120 s. At faster sweep times, the coercivity will show a frequency dependence, as shown experimentally in Figure 49.6. For soft magnetic materials, this dependence can be influenced by the metallurgical condition of the ferromagnet [6].

49.4   Core Loss

During ac magnetization, some of the input energy is converted to heat in ferromagnetic materials. This heat energy is called *core loss* and is classically comprised of three parts. The first, *hysteresis loss*, P_h, is proportional to the ac frequency, f, and the area of the (slow-sweep) dc hysteresis loop:

$$P_h = kf \int B\, dH \qquad (49.4)$$

The second part is the *loss due to eddy current formation*, P_e. In magnetic testing of flat-rolled strips (e.g., the Epstein test; see next section), this core loss component is classically expressed as

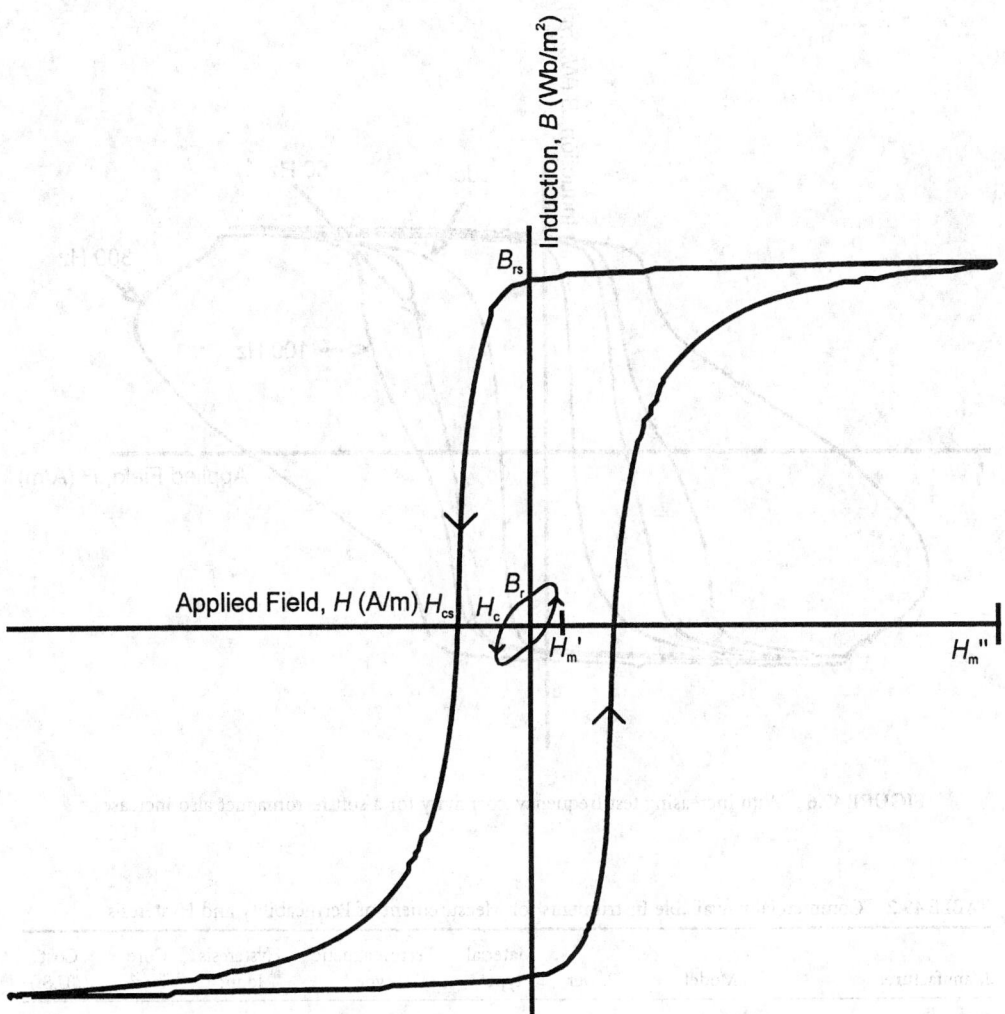

FIGURE 49.5 Major and minor dc hysteresis loops for a typical soft ferromagnet. Labeled points of interest are described in the text.

$$P_e = \frac{\left(\pi B f t\right)^2}{6 d \rho}$$ (49.5)

where B = Peak induction
  t = Strip thickness
  d = Material density
  ρ = Material resistivity

The sum total $P_h + P_e$ almost never equals the observed total core loss, P_t. The discrepancy chiefly originates from the assumptions made in the derivation of Equation 49.5. To account for the additional observed loss, an anomalous loss term, P_a, has often been included, so that

$$P_t = P_h + P_e + P_a$$ (49.6)

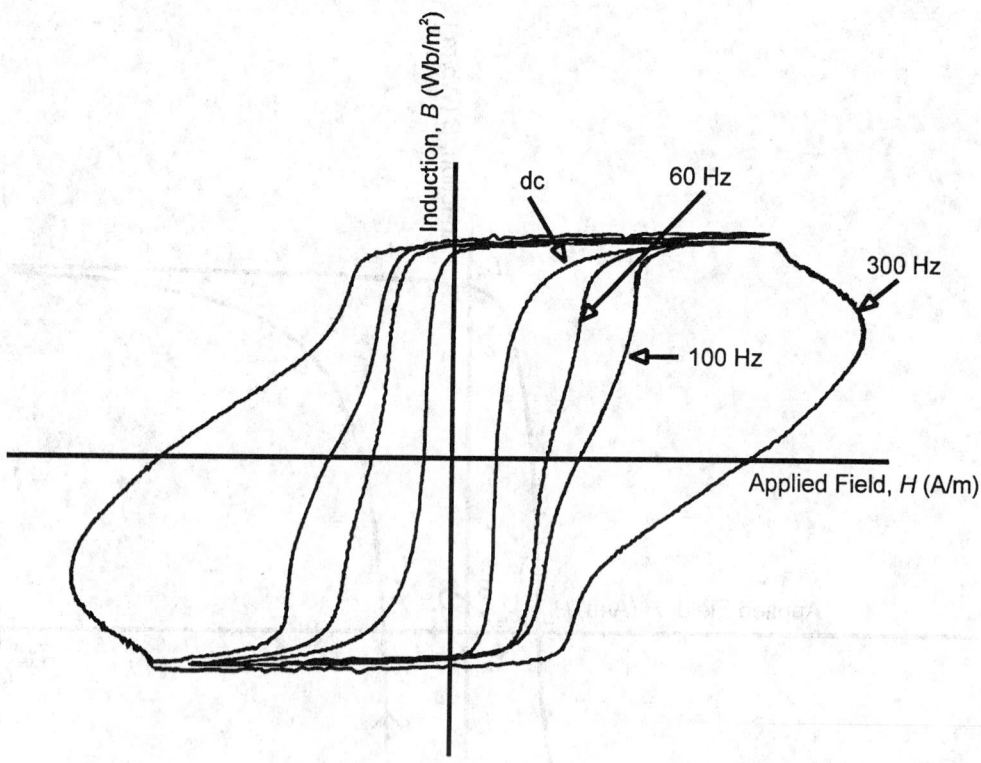

FIGURE 49.6 With increasing test frequency, coercivity for a soft ferromagnet also increases.

TABLE 49.2 Commercially Available Instruments for Measurement of Permeability and Hysteresis

Manufacturer	Model	Power	Material type[a]	Ferromagnetic type	Hysteresis loop?	Core loss?	Cost ($U.S.)
LDJ	3500/5600	ac/dc	F	Soft & hard	Y	Y	30–90k
Troy, MI	5500	dc	F	Soft & hard	Y	N	30–90k
(810) 528-2202	VSM	dc	D, P, F	Soft & hard	Y	N	50–110k
Lakeshore Cryotronics	VSM	dc	D, P, F	Soft & hard	Y	N	45–120k
Westerville, OH	Susceptometer	ac	F	Soft & hard	N	N	50–110k
(614) 891-2243	Magnetometer	dc	F	Soft & hard	N	N	50–110k
Donart Electronics	3401	dc	F	Soft	Y	N	20k+
Pittsburgh, PA	MS-2	ac	F	Soft	N	Y	20k+
(412) 796-5941							
Soken/Magnetech	DAC-BHW-2	ac	F	Soft	Y	Y	38k+
Racine, WI							
(501) 922-6899							

[a] D — diamagnetic, P — paramagnetic, F — ferromagnetic.

49.5 Measurement Methods

Reference [3] is a good source for the various accepted test methods for permeability and hysteresis in diamagnetic, paramagnetic, and ferromagnetic materials. Unfortunately, only a few of the instruments described there are available commercially. Examples of these are listed in Table 49.2.

The instruments in Table 49.2 include hysteresigraphs (LDJ models 3500, 5600, and 5500) and vibrating sample magnetometers (LDJ and Lakeshore Cryotronics VSM models). Also included are two Donart models of Epstein testers. The Epstein test is commonly used to characterize flat-rolled soft ferromagnets such as silicon electrical steels in sheet form. A recent alternative to the Epstein test is the single-sheet test method. The Soken instrument in Table 49.2 is an example of such a tester. This method requires much less sample volume than the Epstein test. It can also accommodate irregular sample geometries. However, the Soken instrument is not yet accepted by the American Society for Testing and Materials (ASTM) for reasons explained in the next section.

Note that all instruments in Table 49.2 can measure permeability (or susceptibility), but not all can provide hysteresis loop measurements. Diamagnetic and paramagnetic materials generally require VSM instruments unless one is prepared to construct their own specialty apparatus [3]. All instruments in Table 49.2 can measure ferromagnetic materials, although only a few can accommodate hard (i.e., permanently magnetizable) ferromagnets.

The price ranges in Table 49.2 account for such options as temperature controls, specialized test software, high-frequency capabilities, etc.

49.6 Validity of Measurements

For a ferromagnet under sinusoidal ac magnetization, the induction will show a waveform distortion (i.e., B is nonsinusoidal) once H_m exceeds the knee of the B–H curve in Figure 49.4. Brailsford [7] has discussed such waveform distortion in detail. With one exception, all ac instruments in Table 49.2 determine H from its sinusoidal waveform and B from its distorted waveform.

The single exception is the Soken instrument. Here, feedback amplification is employed to deliberately distort the H waveform in a way necessary to force a sinusoidal B waveform. In general, this will result in a smaller measured value for permeability compared to the case where feedback amplification is not used. Some suggest this to be the more precise method for permeability measurement, but the use of feedback amplification has prevented instruments such as the Soken from achieving ASTM acceptance to date.

Defining Terms

Permeability: The extent to which a material can be magnetized

Hysteresis: A ferromagnetic phenomenon in which the magnetic induction B is out of phase with the magnetic driving force H.

References

1. B. D. Cullity, *Introduction to Magnetic Materials*, Reading, MA: Addison-Wesley, 1972.
2. D. R. Lide (ed.), *CRC Handbook of Chemistry and Physics*, Boca Raton, FL: CRC Press, 1992-3.
3. Anonymous, 1995 *Annual Book of ASTM Standards*, Philadelphia, PA: ASTM, 3.04, 1995.
4. C. Kittel, *Introduction to Solid State Physics*, 5th ed., New York: John Wiley & Sons, 1976.
5. Page 11 of Ref. [3].
6. R. Thomas and G. Morgan, *Proc. Eleventh Ann. Conf. on Properties and Applications of Magnetic Materials*, Chicago, 1992.
7. F. Brailsford, *Magnetic Materials*, 3rd ed., London: Metuen and New York: Wiley, 1960.

50
Inductance
Measurement

Michał Szyper
University of Mining and Metallurgy

Inductance is an electrical parameter that characterizes electric circuit elements (two- or four-terminal networks) that become magnetic field sources when current flows through them. They are called inductors, although inductance is not a unique property of them. Electric current i (A) and magnetic flux Φ (Wb) are interdependent; that is, they are coupled. Inductance is a measurable parameter; therefore, it has a physical dimension, a measurement unit (the henry), as well as reference standards. Inductance is a property of all electrical conductors. It is found as *self-inductance L* of a single conductor and as *mutual inductance M* in the case of two or more conductors. Inductors can vary in construction, but windings in the form of coils are the most frequent. In this case, they have characteristic geometrical dimensions: surface A, length l, and N number of turns of windings. The path of magnetic flux can be specially shaped by magnetic cores. Figure 50.1 shows examples of different inductive elements of electric circuits.

Figures 50.1*a* and *b* present windings with self-inductance made as a coreless coil (*a*), and wound on a ferromagnetic core that is the concentrator of the magnetic field (*b*). A transformer loaded by impedance Z_L and an electromagnet loaded by impedance of eddy currents in the metal board, shown in Figures 50.1*c* and *d*, will have not only self-inductance of windings, but also mutual inductance between windings (*c*) or between winding and eddy currents (*d*). Self-inductances and mutual inductances can be detected in busbars of electric power stations as shown in Figure 50.1*e*, and also on tracks on printed circuit boards as in Figure 50.1*f*. Couplings between currents and electromagnetic fields can be made deliberately but can also be harmful, e.g., due to energy losses caused by eddy currents, or due to electromagnetic disturbances induced in the tracks of printed circuit boards. Inductors made as windings on ferromagnetic cores have numerous other applications.

The presence of a ferromagnetic core changes the shape of a magnetic field and increases inductance; but in the case of closed cores, it also causes nonlinear relations between inductance and current as well as current frequency. Closed magnetic cores then have a nonlinear and ambiguous magnetization characteristic Φ (i) because of magnetic saturation and hysteresis. Inductances of open magnetic cores with an air gap are mainly dependent on the length of the magnetic path in the air. Real coils made of metal (e.g., copper) wire also show resistance R. However, if resistance is measured at the coil terminals with ac current, it depends not only on the cross-section, length, and resistivity of the wire, but also on losses

0-8493-8347-1/99/$0.00+$.50

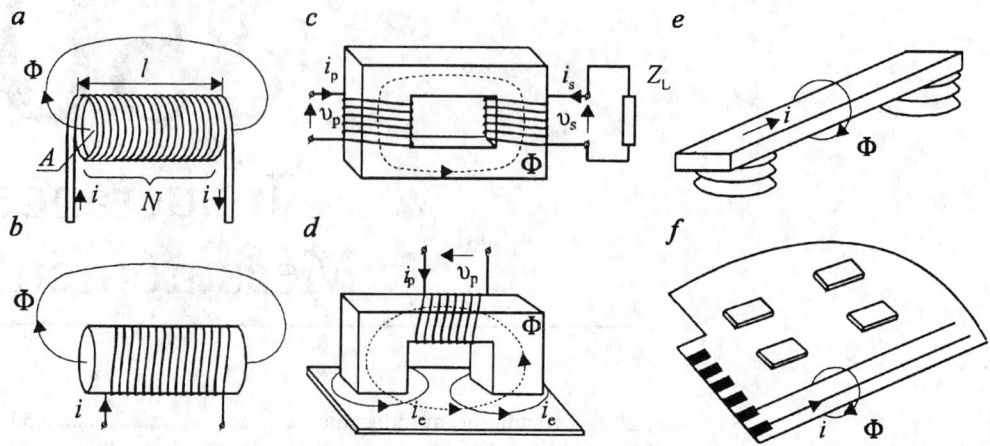

FIGURE 50.1   Examples of different inductive elements: coreless coil (*a*), coil with ferromagnetic concentrating core (*b*), transformer (*c*), electromagnet (*d*), element of electrical power station bus-bars (*e*), and printed circuit board with conductive tracks (*f*).

of active power in the magnetic core. These losses depend on both the current value and frequency. Resistances of inductive elements can also depend on the skin effect. This phenomenon consists of the flow of electric current through a layer of the conductor, near its outer surface, as the result of the effects of the conductor's own magnetic field generated by the current flowing inside the conductor. Notice that the coils have also interturns and stray (to Earth) capacitances C. From its terminals, the inductive elements can then be described by impedances (two-terminal networks) or transmittances (four-terminal networks), values that are experimentally evaluated by current and voltage measurements, or by comparison them with reference impedances. The measured values and the equivalent circuit models of inductive elements are used for evaluation of model parameters: self-inductances, mutual inductances, resistances, and capacitances.

50.1 Definitions of Inductance

Self-inductance is defined as the relation between current i flowing through the coil and voltage v measured at its terminals [1].

$$v = L \frac{di}{dt} \tag{50.1}$$

Using Equation 50.1, the unit of inductance, i.e., the henry, can be defined as follows:

One henry (1 H) is the inductance of a circuit in which an electromotive force of one volt (1 V) is induced, when the current in the circuit changes uniformly by one ampere (1 A) per second (1 s).

The definition of the unit implies the use of uniform-ramp current excitation (the derivative is constant); in practice, however, mainly sinusoidal current excitation is used in inductance measurement.

Mutual inductance M of windings coupled by magnetic flux Φ is a parameter that depends on the coupling coefficient. The coupling coefficient is defined as *perfect coupling* in the case in which the total flux of one winding links the second one; *partial coupling* is the case in which only a fraction of flux links the second one; and *zero coupling* is the case in which no part of the flux of one winding links the second one. Assuming that, as in Figure 50.1c, the second N_s winding is not loaded (i.e., $i_s = 0$) and the flux Φ is a part of the total flux produced by current i_p, then voltage v_s is described by:

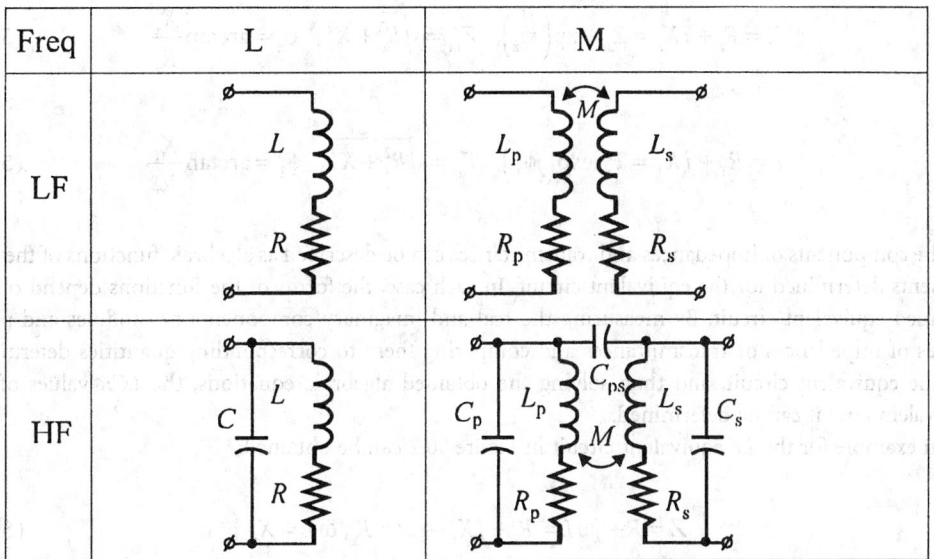

Freq	L	M
LF	L R	L_p M L_s R_p R_s
HF	C L R	C_p L_p C_{ps} L_s C_s M R_p R_s

FIGURE 50.2 Basic equivalent circuits of inductive elements with self-inductance L, mutual inductance M for low (LF) and high (HF) frequencies.

$$v_s = N_s \frac{d\Phi}{dt} = \pm k \sqrt{L_p L_s} \frac{di_p}{dt} = \pm M \frac{di_p}{dt} \quad (50.2)$$

where $\pm k$ is the coupling coefficient of the primary (p) and secondary (s) windings. Its sign depends on the direction of the windings. Because of the similarity between Equations 50.1 and 50.2, mutual inductance is defined similarly to self-inductance. The definitions of self-inductance and mutual inductance described above are correct when L and M can be assumed constant, i.e., not depending on current, frequency, or time.

50.2 Equivalent Circuits and Inductive Element Models

Equivalent circuits of inductive elements and their mathematical models are built on the basis of analysis of energy processes in the elements. They are as follows: energy storage in the parts of the circuits represented by lumped inductances (L) and capacitances (C), and dissipation of energy in the parts of the circuits represented by lumped resistances (R). Essentially, the above-mentioned energy processes are never lumped, so the equivalent circuits and mathematical models of inductive elements only approximate reality. Choosing an equivalent circuit (model), one can influence the quality of the approximation. In Figure 50.2, the basic models of typical coreless inductive elements of electric circuits are presented, using the type of inductance (self-inductance L, mutual inductance M) and frequency band (low LF, high HF) as criteria. Models of inductive elements with ferromagnetic cores and problems concerning model parameter evaluation are beyond the scope of this chapter. The equivalent circuits of coreless inductive elements at low frequencies contain only inductances and series resistances. At high frequencies, parallel and coupling equivalent capacitances are included. Calculations of LCR values of complicated equivalent circuits (with many LCR elements) are very tedious, so often for that purpose a special dedicated processor with appropriate software is provided in the measuring device.

In metrology, complex notation [1] is frequently used to describe linear models of inductive elements. In complex notation, the impedances and transmittances are as follows:

$$Z = R_z + jX_z = Z_m \exp\left(j\phi_z\right), \quad Z_m = \sqrt{R_z^2 + X_z^2}, \quad \phi_z = \arctan\frac{X_z}{R_z} \tag{50.3}$$

$$T = R_T + jX_T = T_m \exp\left(j\phi_T\right), \quad T_m = \sqrt{R_T^2 + X_T^2}, \quad \phi_T = \arctan\frac{X_T}{R_T} \tag{50.4}$$

The components of impedances and transmittances can be described as algebraic functions of the *LCR* elements determined for the equivalent circuit. In each case, the forms of the functions depend on the assumed equivalent circuit. By measuring the real and imaginary components or modules and phase angles of impedances or transmittances and comparing them to corresponding quantities determined for the equivalent circuit, and then solving the obtained algebraic equations, the *LCR* values of the equivalent circuit can be determined.

An example for the *LF* equivalent circuit in Figure 50.2 can be obtained:

$$Z = R + j\omega L = R_z + jX_z \rightarrow R = R_z, \ \omega L = X_z \tag{50.5}$$

$$T = j\omega M = R_T + jX_T \rightarrow R_T = 0, \ \omega M = X_T \tag{50.6}$$

For the HF equivalent circuits shown in Figure 50.2, the models are more complex. More particularly, models of circuits with mutual inductance can represent:

- Ideal transformers, with only their self-inductances and mutual inductances of the coils
- Perfect transformers, without losses in the core
- Real transformers, having a specific inductance, resistance, and capacity of the coils, and also a certain level of losses in the core

The quality factor Q and dissipation factor D are defined for the equivalent circuits of inductive elements. In the case of a circuit with inductance and resistance only, they can be defined as follows:

$$Q = \frac{1}{D} = \frac{\omega L}{R} = \tau\omega \tag{50.7}$$

where parameter τ is a time constant.

The presented models with lumped inductance are not always a sufficiently good approximation of the real properties of electric circuits. This particularly applies to circuits made with geometrically large wires (i.e., of a significant length, surface area, or volume). In such cases, the models applied use an adequately distributed inductance (linearly, over the surface, or throughout the volume). Inductances determined to be the coefficients of such models depend on the geometrical dimensions and, in the case of surface conductivity or conduction by the surface or volume, by the frequencies of the currents flowing in the conductor lines.

A complex formulation is used to represent and analyze circuits with lumped and linear inductances. Sometimes the same thing can be done using simple linear differential equations or the corresponding integral operators.

The analytical methodology for such circuits is described in [1]. In the case of nonlinear inductances, the most frequently used method is the linearized equivalent circuit, also represented in a simplified form using a complex formulation. Circuits with distributed inductance are represented by partial differential equations.

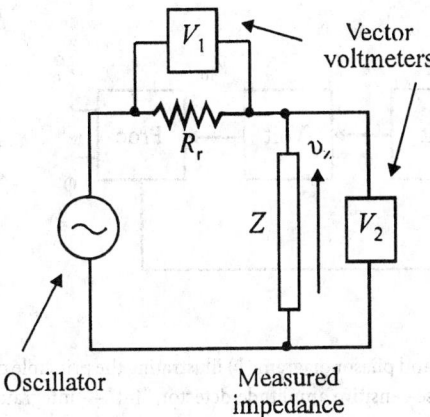

FIGURE 50.3 Circuit diagram for impedance measurement by current and voltage method.

50.3 Measurement Methods

Impedance (or transmittance) measurement methods for inductors are divided into three basic groups:

1. Current and voltage methods based on impedance/transmittance determination.
2. Bridge and differential methods based on comparison of the voltages and currents of the measured and reference impedances until a state of balance is reached.
3. Resonance methods based on physical connection of the measured inductor and a capacitor to create a resonant system.

Current–Voltage Methods

Current–voltage measurement methods are used for all types of inductors. A current–voltage method using vector voltmeters is shown in Figure 50.3. It is based on evaluation of the modules and phase angles of impedances (in the case of self-inductance) or transmittances (in the case of mutual inductance) using Equations 50.8 and 50.9

$$Z = \frac{v_2}{i} = R_r \frac{v_2}{v_1} = R_r \frac{V_{m2} \exp(j\phi_2)}{V_{m1} \exp(j\phi_1)} = R_r \frac{V_{m2}}{V_{m1}} \exp j(\phi_2 - \phi_1) = Z_m \exp(j\phi_z) \tag{50.8}$$

$$T = \frac{v_s}{i_p} = R_r \frac{v_s}{v_{pr}} = R_r \frac{V_{ms} \exp(j\phi_s)}{V_{mpr} \exp(j\phi_{pr})} = R_r \frac{V_{ms}}{V_{mpr}} \exp j(\phi_s - \phi_{pr}) = T_m \exp(j\phi_T) \tag{50.9}$$

where R_r = Sampling resistor used for measurement of current
v_1 = Voltage proportional to the current
v_2 = Voltage across the measured impedance
v_{pr} = Voltage proportional to primary current
v_s = Voltage of the secondary winding of the circuit with mutual inductance

A block diagram illustrating the principle of the vector voltmeter is shown in Figure 50.4a. The system consists of the multiplier or gated synchronous phase-sensitive detector (PSD) [3, 4] of the measured

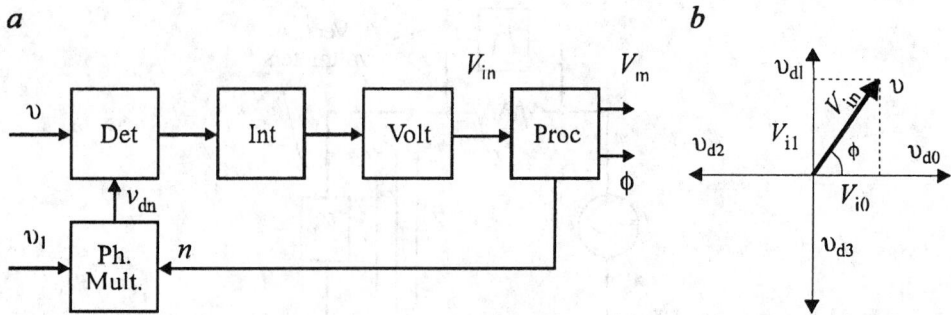

FIGURE 50.4 Block diagram (*a*) and phasor diagram (*b*) illustrating the principle of operation of a vector voltmeter. Block abbreviations: "Det" — phase-sensitive amplitude detector, "Int" — integrator, "Volt" — voltmeter, "Proc" — processor, "Ph. mult" — controlled phase multiplexer.

voltage v with the switching system of the phase of the reference voltage v_{dn}, integrator, digital voltmeter, and processor. The principle of vector voltmeter operation is based on determination of the magnitude V_m and phase angle ϕ of the measured voltage v in reference to voltage v_1, which is proportional to the current i. Assume that voltages v and v_{dn} are in the following forms:

$$v = V_m \sin\left(\omega t + \phi\right) = V_m \left(\sin\omega t \, \cos\phi + \cos\omega t \sin\phi\right) \tag{50.10}$$

$$v_{dn} = V_{md} \sin\left(\omega t + n\frac{\pi}{2}\right), \quad n = 0,1,2,3 \tag{50.11}$$

Phase angle $n\pi/2$ of voltage v_{dn} can take values from the set $\{0, \pi/2, \pi, 3/2\,\pi\}$ by choosing the number n that gives the possibility of detecting the phase angle ϕ in all four quadrants of the Cartesian coordinate system, as is shown in Figure 50.4*b*. A multiplying synchronous phase detector multiplies voltages v and v_{dn} and bilinearly, and the integrator averages the multiplication result during time T_i.

$$V_{in} = \frac{1}{T_i} \int_0^{T_i} v v_{dn} \, dt \tag{50.12}$$

Averaging time $T_i = k\,T$, $k = 1,2,3\ldots$ is a multiple of the period T of the measured voltage. From Equations 50.10 through 50.12, an example for $0 \le \phi \le \pi/2$ (e.g., $n = 0$ and $n = 1$), a pair of numbers is obtained:

$$V_{i0} = 0.5\,V_m\,V_{md}\,\cos\phi, \quad V_{i1} = 0.5\,V_m\,V_{md}\,\sin\phi \tag{50.13}$$

which are the values of the Cartesian coordinates of the measured voltage v. The module and phase angle of voltage are calculated from:

$$V_m = \frac{2}{V_{md}}\sqrt{V_{i0}^2 + V_{i1}^2}, \quad \phi = \arctan\frac{V_{i1}}{V_{i0}} \tag{50.14}$$

Both coordinates of the measured voltage v can be calculated in a similar way in the remaining quadrants of the Cartesian coordinate system. A vector voltmeter determines the measured voltage as vector (phasor) by measurement of its magnitude and angle as shown in Figure 50.4.

a

b

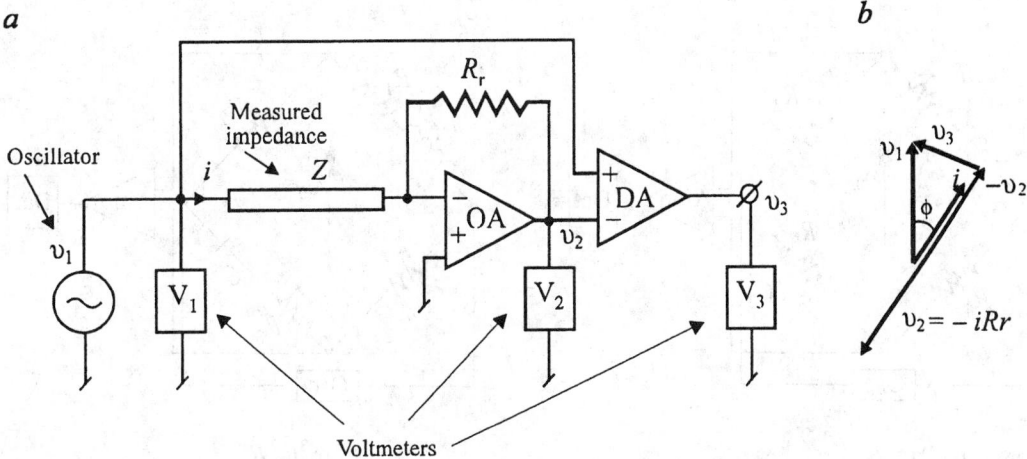

FIGURE 50.5   Block diagram (*a*) and phasor diagram (*b*) of the "three-voltmeter" method. Operational and differential amplifiers are represented by blocks "OA" and "DA."

The current and voltage method of impedance or transmittance measurement is based on measurement of voltages $v_1 = i\,R_r$ and v_2 or $v_{pr} = i\,R_r$ and v_s, and the use of Equations 50.8 or 50.9. Calculation of the voltage measurement results and control of number n is performed by a processor. References [11–13] contain examples of PSD and vector voltmeter applications. Errors of module and phase angle measurement of impedance when using vector voltmeters are described in [11] as being within the range of 1% to 10% and between 10^{-5} rad and 10^{-3} rad, respectively, for a frequency equal to 1.8 GHz. Publications [9] and [10] contain descriptions of applications of comparative methods which have had an important influence on the development of the methods.

Another method of comparative current/voltage type is a modernized version of the "three-voltmeter" method [7]. A diagram of a measurement system illustrating the principle of the method is shown in Figure 50.5*a* and *b*.

The method is based on the properties of an operational amplifier (OA), in which output voltage v_2 is proportional to input voltage v_1 and to the ratio of the reference resistance R_r to measured impedance Z. The phasor difference v_3 of voltages v_1 and v_2 can be obtained using the differential amplifier (DA). The three voltages (as in Figure 50.5*b*) can be used for the module Z_m and phase φ calculation using relations:

$$v_2 = -i\,R_r = -\frac{R_r}{Z}v_1 \rightarrow Z_m = \frac{V_1}{V_2}R_r \tag{50.15}$$

$$v_3 = v_1 - v_2 \rightarrow \phi = \arccos\frac{V_1^2 + V_2^2 - V_3^2}{2V_1 V_2} \tag{50.16}$$

where V_1, V_2, V_3 are the results of rms voltage measurements in the circuit. The advantage of the method lies in limiting the influence of stray capacitances as a result of attaching one of the terminals of the measured impedance to a point of "virtual ground." However, to obtain small measurement errors, especially at high frequencies, amplifiers with very good dynamic properties must be used.

Joint errors of inductance measurement, obtained by current and voltage methods, depend on the following factors: voltmeter errors, errors in calculating resistance R_r, system factors (residual and leakage inductances, resistances and capacitances), and the quality of approximation of the measured impedances by the equivalent circuit.

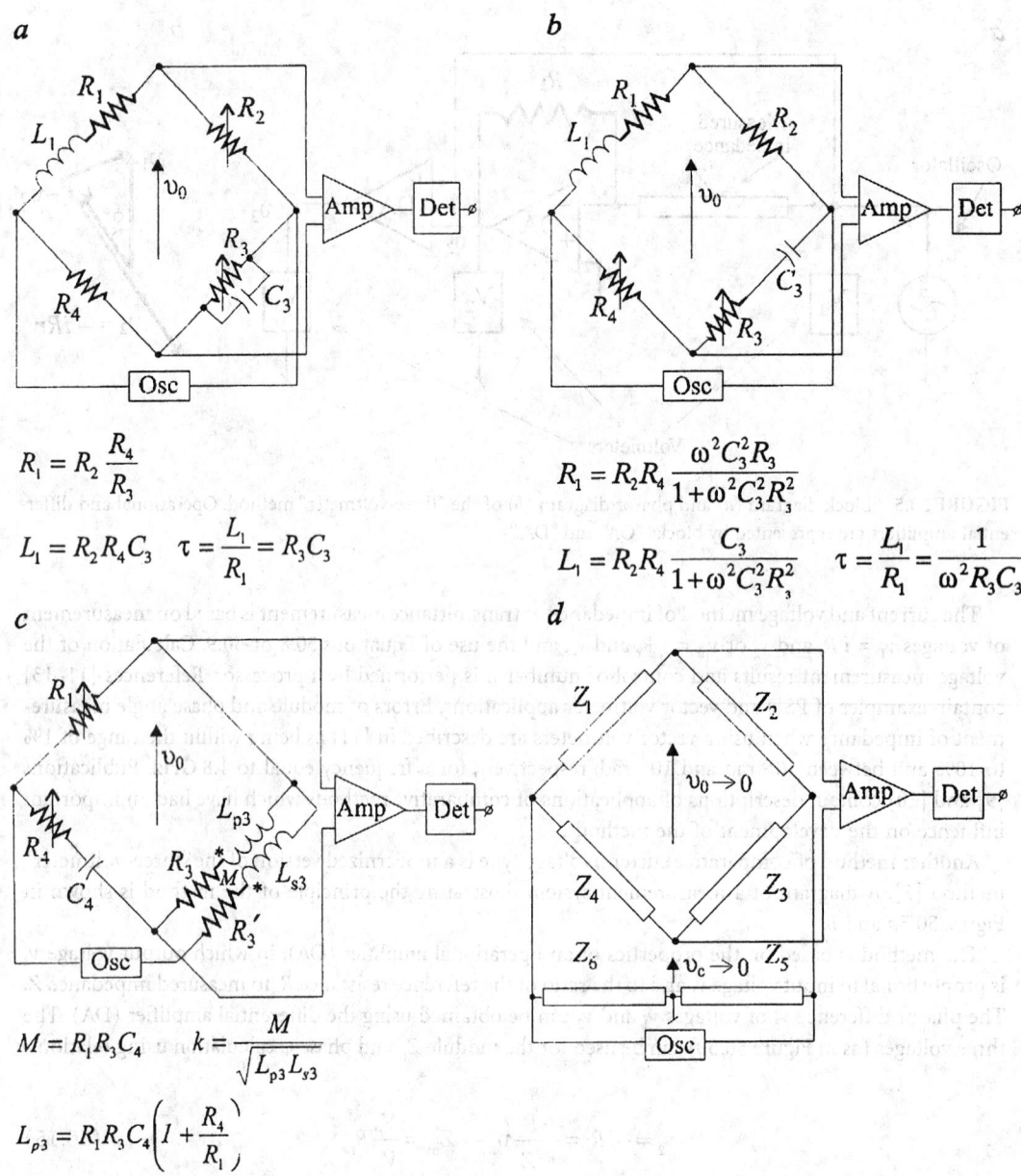

$$R_1 = R_2 \frac{R_4}{R_3}$$

$$L_1 = R_2 R_4 C_3 \quad \tau = \frac{L_1}{R_1} = R_3 C_3$$

$$R_1 = R_2 R_4 \frac{\omega^2 C_3^2 R_3}{1 + \omega^2 C_3^2 R_3^2}$$

$$L_1 = R_2 R_4 \frac{C_3}{1 + \omega^2 C_3^2 R_3^2} \quad \tau = \frac{L_1}{R_1} = \frac{1}{\omega^2 R_3 C_3}$$

$$M = R_1 R_3 C_4 \qquad k = \frac{M}{\sqrt{L_{p3} L_{s3}}}$$

$$L_{p3} = R_1 R_3 C_4 \left(1 + \frac{R_4}{R_1}\right)$$

FIGURE 50.6 Bridge circuits used for inductance measurements: Maxwell-Wien bridge (*a*), Hay bridge (*b*), Carey-Foster bridge (*c*), and ac bridge with Wagner branch (*d*). Block abbreviations: "Osc" — oscillator, "Amp" — amplifier, "Det" — amplitude detector.

Bridge Methods

There are a variety of bridge methods for measuring inductances. Bridge principles of operation and their circuit diagrams are described in [2–5] and [7]. The most common ac bridges for inductance measurements and the formulae for calculating measurement results are shown in Figure 50.6.

The procedure referred to as *bridge balancing* is based on a proper selection of the reference values of the bridge so as to reduce the differential voltage to zero (as referred to the output signal of the balance indicator). It can be done manually or automatically.

The condition of the balanced bridge $v_0 = 0$ leads to the following relation between the impedances of the bridge branches; one of them (e.g., Z_1) is the measured impedance:

$$Z_1 Z_3 = Z_2 Z_4 \rightarrow \left(R_1 + jX_1\right)\left(R_3 + jX_3\right) = \left(R_2 + jX_2\right)\left(R_4 + jX_4\right) \tag{50.17}$$

Putting Equation 50.17 into complex form and using expressions for the impedances of each branch, two algebraic equations are obtained by comparing the real and imaginary components. They are used to determine the values of the equivalent circuit elements of the measured impedance. In the most simple case, they are L and R elements connected in series. More complicated equivalent circuits need more equations to determine the equivalent circuit parameters. Additional equations can be obtained from measurements made at different frequencies.

In self-balancing bridges, vector voltmeters preceded by an amplifier of the out-of-balance voltage v_0 are used as "zero" detectors. The detector output is coupled with variable standard bridge components.

The Maxwell-Wien bridge shown in Figure 50.6a is one of the most popular ac bridges. Its range of measurement values is large and the relative error of measurement is about 0.1% of the measured value. It is used in the wide-frequency band 20 Hz to 1 MHz. The bridge is balanced by varying the R_2 and R_3 resistors or by varying R_3 and capacitor C_3. Some difficulties can be expected when balancing a bridge with inductors with high time constants.

The Hay bridge presented in Figure 50.6b is also used for measurement of inductors, particularly those with high time constants. The balance conditions of the bridge depend on the frequency value, so the frequency should be kept constant during the measurements, and the bridge supply voltage should be free from higher harmonic distortions. The dependence of bridge balance conditions on frequency also limits the measurement ranges. The bridge is balanced by varying R_3 and R_4 resistors and by switching capacitor C_3.

Mutual inductance M of two windings with self-inductances L_p and L_s can be determined by Maxwell-Wien or Hay bridges. For this, two inductance measurements have to be made for two possible combinations of the series connection of both coupled windings: one of them for the corresponding directions of the windings, and one for the opposite directions. Two values of inductances L_1 and L_2 are obtained as the result of the measurements:

$$L_1 = L_p + L_s + 2\,M, \quad L_2 = L_p + L_s - 2\,M \tag{50.18}$$

Mutual inductance is calculated from:

$$M = 0.25\left(L_1 - L_2\right) \tag{50.19}$$

The Carey-Foster bridge described in Figure 50.6c is used for mutual inductance measurement. The self-inductances of the primary and secondary windings can be determined by two consecutive measurements. The expressions presented in Figure 50.6c yield the magnetic coupling coefficient k. The bridge can be used in a wide frequency range. The bridge can be balanced by varying R_1 and R_4 resistances and switching the remaining elements.

For correct measurements when using ac bridges, it is essential to minimize the influence of harmful couplings among the bridge elements and connection wires, between each other and the environment. Elimination of parallel (common) and series (normal) voltage distortions is necessary for high measurement resolution. These couplings are produced by the capacitances, inductances, and parasitic resistances of bridge elements to the environment and among themselves. Because of their appearance, they are called stray couplings. Series voltage distortions are induced in the bridge circuit by varying common electromagnetic fields. Parallel voltage distortions are caused by potential differences between the reference point of the supply voltage and the points of the out-of-balance voltage detector system.

Magnetic shields applied to connection wires and bridge-balancing elements are the basic means of minimizing the influence of parasitic couplings and voltage distortions [8]. All the shields should be connected as a "star" connection; that is, at one point, and connected to one reference ("ground") point of the system. For these reasons, amplifiers of out-of-balance voltage with symmetric inputs are frequently used in ac bridges, as they reject parallel voltage distortions well.

When each of the four nodes of the bridge has different stray impedances to the reference ground, an additional circuit called a Wagner branch is used (see Figure 50.6d). By varying impedances Z_5 and Z_6 in the Wagner branch, voltage v_c can be reduced to zero; by varying the other impedances, voltage v_0 can also be reduced to zero. In this way, the bridge becomes symmetrical in relation to the reference ground point and the influence of the stray impedances is minimized.

The joint error of the inductance measurement results (when using bridge methods) depends on the following factors: the accuracy of the standards used as the bridge elements, mainly standard resistors and capacitors; errors of determining the frequency of the bridge supplying voltage (if it appears in the expressions for the measured values); errors of the resolution of the zero detection systems (errors of state of balance); errors caused by the influence of residual and stray inductances; resistances and capacitances of the bridge elements and wiring; and the quality of approximation of the measured impedances in the equivalent circuit.

The errors of equivalent resistance measurements of inductive elements using bridge methods are higher than the errors of inductance measurements. The number of various existing ac bridge systems is very high. Often the bridge system is built as a universal system that can be configured for different applications by switching elements. One of the designs of such a system is described in [3]. Reference [13] describes the design of an automatic bridge that contains, as a balancing element, a multiplying digital-to-analog converter (DAC), controlled by a microcontroller.

Differential Methods

Differential methods [7] can be used to build fully automatic digital impedance meters (of real and imaginary components or module and phase components) that can also measure inductive impedances. Differential methods of measurement are characterized by small errors, high resolution, and a wide frequency band, and often utilize precise digital control and digital processing of the measurement results. The principle of differential methods is presented through the example of a measuring system with an inductive voltage divider and a magnetic current comparator (Figure 50.7). An inductive voltage divider (IVD) is a precise voltage transformer with several secondary winding taps that can be used to vary the secondary voltage in precisely known steps [7]. By combining several IVDs in parallel, it is possible to obtain a precise voltage division, usually in the decade system. The primary winding is supplied from a sinusoidal voltage source. A magnetic current comparator (MCC) is a precise differential transformer with two primary windings and a single secondary winding. The primary windings are connected in a differential way; that is, the magnetic fluxes produced by the currents in these windings subtract. The output voltage of the secondary winding depends on the current difference in the primary windings. MCCs are characterized by very high resolution and low error but are expensive. In systems in common use, precise control of voltages (IVD) is provided by digitally controlled (sign and values) digital voltage dividers (DVD), and the magnetic comparator is replaced by a differential amplifier.

The algorithms that enable calculation of L and R element values of the series equivalent circuit of an inductive impedance Z in a digital processor result from the mathematical model described in Figure 50.7. When the system is in a state of equilibrium, the following relations occur:

$$v_0 = 0 \rightarrow i_z - i_r = 0 \qquad (50.20)$$

$$i_z = v_1 \frac{1}{R + j\omega L}, \quad i_r = D_r v_1 \left(\frac{b}{R_r} - a j\omega C_r \right) \qquad (50.21)$$

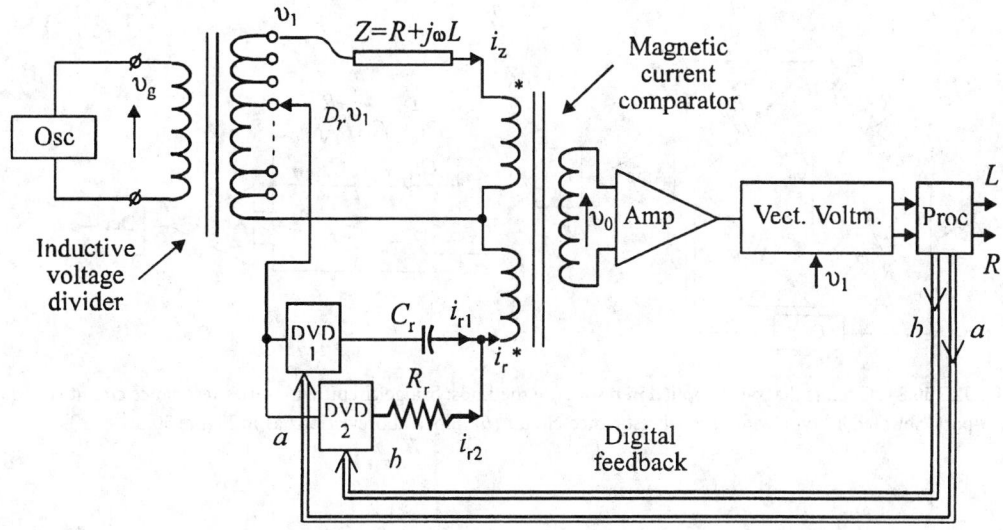

FIGURE 50.7 Scheme of the differential method. Block abbreviations: "Osc", "Amp" — as in Figure 50.6, "Vect. voltm" — vector voltmeter, "Proc" — processor, "DVD" — digital voltage divider.

where $0 < D_r \leq 1$ is the coefficient of v_1 voltage division and the values a and b are equivalent to the binary signals used by the processor to control DVD. Multiple digital-to-analog converters (DACs) are used as digitally controlled voltage dividers. They multiply voltage $D_r v_1$ by negative numbers $(-a)$, which is needed in the case of using a standard capacitor C_r for measurements of inductance. After substituting Equation 50.21 into Equation 50.20 and equating the real and imaginary parts, the following formulae are obtained:

$$L = \frac{a\, R_r^2\, C_r}{D_r \left(b^2 + a^2\, \omega^2\, R_r^2\, C_r^2 \right)} \tag{50.22}$$

$$R = \frac{b\, R_r}{D_r \left(b^2 + a^2\, \omega^2\, R_r^2\, C_r^2 \right)} \tag{50.23}$$

The lengths N of the code words $\{a_n\} \leftrightarrow a$ and $\{b_n\} \leftrightarrow b$, $n = 1, 2, \ldots, N$ determine the range and resolution of the measurement system; that is, the highest measurable inductance and resistance values and the lowest detectable values. The range can be chosen automatically by changing the division coefficients D_r. Achieved accuracy is better than 0.1% in a very large range of impedances and in a sufficiently large frequency range. Measurements can be periodically repeated and their results can be stored and processed.

Resonance Methods

Resonance methods are a third type of inductance measurement method. They are based on application of a series or parallel resonance LC circuits as elements of either a bridge circuit or a two-port (four-terminal) "T"-type network. Examples of both circuit applications are presented in Figure 50.8.

In the bridge circuit shown in Figure 50.8a, which contains a series resonance circuit, the resonance state is obtained by varying the capacitor C_r, and then the bridge is balanced ($v_0 = 0$) using the bridge resistors. From the resonance and balance conditions, the following expressions are obtained:

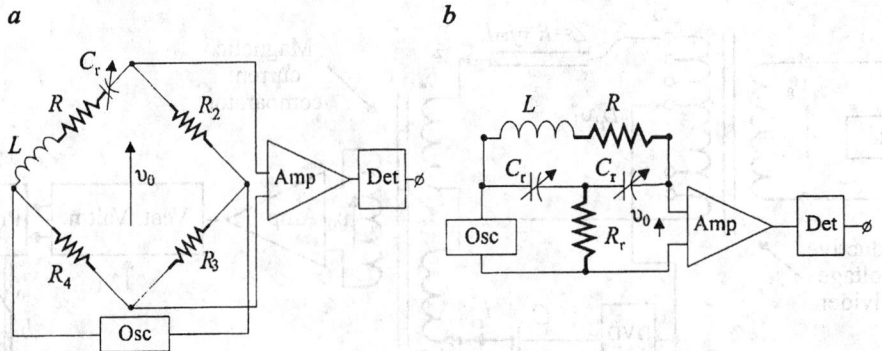

FIGURE 50.8 Circuits diagrams applied in resonance methods: bridge circuit with series resonance circuit (*a*), and two-port "shunted T" type with parallel resonance circuit (*b*). Block abbreviations as in Figure 50.6.

$$L = \frac{1}{\omega^2 C_r}, \qquad R = R_2 \frac{R_4}{R_3} \tag{50.24}$$

To calculate the values of the *LR* elements of the series equivalent circuit of the measured impedance, it is necessary to measure (or know) the angular frequency ω of the supply voltage. The frequency band is limited by the influence of unknown interturn capacitance value. In the "shunted T" network presented in Figure 50.8*b*, the state of balance (i.e., the minimal voltage v_0 value) is achieved by tuning the multiloop *LCR* circuit to parallel resonance. The circuit analysis [7] is based on the "star-delta" transformation of the $C_r R_r C_r$ element loop and leads to the relations for *L* and *R* values:

$$L = \frac{2}{\omega^2 C_r}, \qquad R = \frac{1}{\omega^2 C_r^2 R_r} \tag{50.25}$$

According to reference [7], a "double T" network can be used for inductive impedance measurements at high frequencies (up to 100 MHz).

50.4 Instrumentation

Instruments commonly used for inductance measurements are built as universal and multifunctional devices. They enable automatic (triggered or cyclic) measurements of other parameters of the inductive elements: capacitance, resistance, quality, and dissipation factors. Equipped with interfaces, they can also work in digital measuring systems. Table 50.1 contains a review of the basic features of the instruments called *LCR* meters, specified on the basis of data accessible from the Internet [15]. Proper design of *LCR* meters limits the influence of the factors causing errors of measurement (i.e., stray couplings and distortions). The results of inductance measurements also depend on how the measurements are performed, including how the measured inductor is connected to the meter. Care should be taken to limit such influences as inductive and capacitive couplings of the inductive element to the environment, sources of distortion, and the choice of operating frequency.

TABLE 50.1 Basic Features of Selected Types of LCR Meters

Manufacturer, model (designation)	Measurement range of inductance	Frequency	Basic accuracy	Price
Leader LCR 740 (LCR bridge)	0.1 μH–1100 H	int. 1 kHz ext. 50 Hz–40 kHz	0.5%	$545
Electro Scientific Industries 253 (Digital impedance meter)	200 μH–200 H	1 kHz	(3.5 digit)	$995
Stanford RS SR 715 (LCR meter)	0.1 nH–100 kH	100 Hz–10 kHz	0.2%	$1295 ($1425)
Wayne Kerr 4250	0.01 nH–10 kH	120 Hz–100 kHz	0.1%	$3500
General Radio 1689 (Precision LCR meter)	0.00001 mH–99.999 H	12 Hz–100 kHz	0.02%	$4120
Hewlett-Packard 4284A (Precision LCR meter)	0.01 nH–99.9999 kH	20 Hz–1 MHz	0.05%	$9500

References

1. R. C. Dorf, *Introduction to Electric Circuits*, New York: John Wiley & Sons, 1989.
2. J. P. Bentley, *Principles of Measurement Systems, 2nd ed.*, Harlow: Longman Group U.K., New York: John Wiley & Sons, 1988.
3. A. D. Helfrick and W. D. Cooper, *Modern Electronic Instrumentation and Measurement Techniques*, Englewood Cliffs, NJ: Prentice-Hall, 1990.
4. L. D. Jones and A. F. Chin, *Electronic Instruments and Measurements, 2nd ed.*, Englewood Cliffs, NJ: Prentice-Hall, 1991.
5. M. U. Reissland, *Electrical Measurement*, New York: John Wiley & Sons, 1989.
6. P. H. Sydenham (ed.), *Handbook of Measurement Science, Vol. 1, Theoretical Fundamentals*, New York: John Wiley & Sons, 1982.
7. P. H. Sydenham (ed.), *Handbook of Measurement Science, Vol. 2, Practical Fundamentals*, New York: John Wiley & Sons, 1983.
8. R. L. Bonebreak, *Practical Techniques of Electronic Circuit Design, 2nd ed.*, New York: John Wiley & Sons, 1987.
9. S. Hashimoto and T. Tamamura, An automatic wide-range digital LCR meter, *Hewlett-Packard J.*, 8, 9–15, 1976.
10. T. Wakasugi, T. Kyo, and T. Tamamura, New multi-frequency LCZ meters offer higher-speed impedance measurements, *Hewlett-Packard J.*, 7, 32–38, 1983.
11. T. Yonekura, High-frequency impedance analyzer, *Hewlett-Packard J.*, 5, 67–74, 1994.
12. M. A. Atmanand, V. J. Kumar, and V. G. K. Murti, A novel method of measurement of L and C, *IEEE Trans. Instrum. Meas.*, 4, 898–903, 1995.
13. M. A. Atmanand, V. J. Kumar, and V. G. K. Murti, A microcontroller- based quasi-balanced bridge for the measurement of L, C and R, *IEEE Trans. Instrum. Meas.*, 3, 757–761, 1996.
14. J. Gajda and M. Szyper, Electromagnetic transient states in the micro resistance measurements, *Systems Analysis Modeling Simulation*, Amsterdam: Overseas Publishers Association, 22, 47–52, 1996.
15. Information from *Internet* on *LCR* meters.

51
Immittance Measurement

Achim Dreher
German Aerospace Center

Electronic circuits consist of numerous elements that can be lumped, distributed, or a combination of both. The components are regarded as *lumped* if their size is much smaller than the signal wavelength. This condition holds for resistors, inductors, capacitors, transformers, diodes, transistors, or similar devices operating in printed circuits at frequencies up to a few hundred megahertz or even higher in small integrated circuits. In the microwave or millimeter-wave region, the elements and their connecting transmission lines must be considered as *distributed*. While in lumped circuits a change of voltage or current at one single point immediately affects these quantities at all other points, in distributed circuits the propagation properties now have to be taken into account. The same holds for long connecting cables even at lower frequencies.

To describe the effect of any element within an electronic circuit or of the connection of different circuits, the *immittance* is used as a characteristic quantity. It simply provides a relation of sinusoidal voltage and current at the terminals of the element as a function of frequency. The immittance therefore also characterizes arbitrarily complicated networks considered as one port. This is useful, since in practice the single elements are interconnected to networks. On the other hand, the elements themselves are not ideal. A resistor, for example, made of wound resistive wire, has parasitic components such as capacitance and inductance of winding and terminals. It must be represented by an equivalent circuit forming a complex network [1].

The word "immittance" was proposed by Bode [2] and is a combination of the words "impedance" and the reverse quantity called "admittance." These terms do not only occur in electrodynamics but wherever wave propagation takes place — in acoustics as well as in elasticity. The emphasis of this chapter is on lumped networks and guided electromagnetic waves. Readers interested in more general propagation and scattering phenomena are referred to [3].

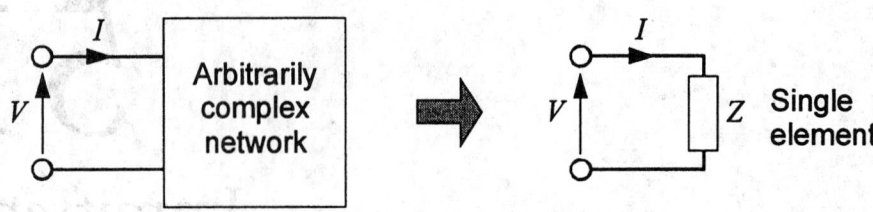

FIGURE 51.1 An arbitrarily complex network can be replaced by its impedance for a given frequency without changing the electrical properties at the terminal.

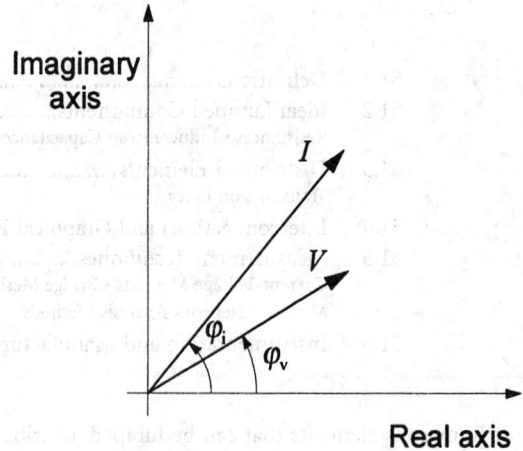

FIGURE 51.2 Voltage and current phasors in the complex plane.

51.1 Definitions

Assume a stable linear and time-invariant (LTI) network with only one port. Linearity and time independence are generally met for combinations of passive elements but also for active devices with small-signal driving under constant physical conditions (temperature, humidity, dimensions, etc.). In the steady state, a voltage $v(t) = V_m \cos(\omega t + \varphi_v)$ with amplitude V_m varying harmonically with the angular frequency $\omega = 2\pi f$ which is applied to the terminal then only produces voltages and currents of the same frequency within the network (Figure 51.1). Using complex notation:

$$v(t) = \mathrm{Re}\{V e^{j\omega t}\} \quad \text{with} \quad V = V_m e^{j\varphi_v} \tag{51.1}$$

the current flowing into the network is given by:

$$i(t) = I_m \cos(\omega t + \varphi_i) = \mathrm{Re}\{I e^{j\omega t}\} \quad \text{with} \quad I = I_m e^{j\varphi_i} \tag{51.2}$$

The phasors V and I are time independent and can be represented in the complex plane (Figure 51.2). Relating voltage and current at the terminal, the network is uniquely described by means of a complex frequency-dependent quantity, the impedance Z:

$$Z = \frac{V}{I} = \frac{V_m}{I_m} e^{j(\varphi_v - \varphi_i)} = |Z| e^{j\varphi_z} \tag{51.3}$$

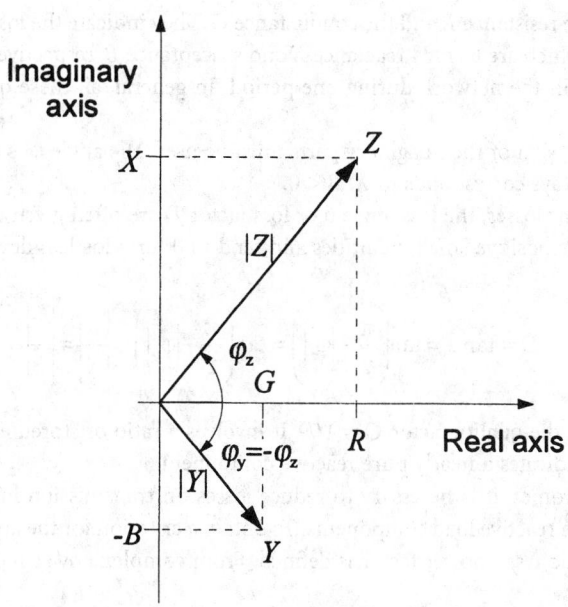

FIGURE 51.3 Representation of impedance and admittance in the complex plane showing the relations between rectangular and polar coordinates. Note that the units are different for each vector.

For a given frequency, an arbitrarily complex network within a circuit thus can be replaced by a single element without changing the electrical properties at the terminals. Sometimes it is more convenient to use the inverse of Z, the admittance Y:

$$Y = \frac{1}{Z} = \frac{I}{V} = |Y| e^{j\varphi_y} \quad \text{with} \quad \varphi_y = \varphi_i - \varphi_v = -\varphi_z \qquad (51.4)$$

Both quantities are combined to form the word "immittance." Figure 51.3 shows their representation in the complex plane. Equations 51.3 and 51.4 give the definition in polar coordinates. In data sheets, they are often written as:

$$|Z| \angle \varphi_z, \quad |Y| \angle \varphi_y \qquad (51.5)$$

Using Euler's identity $e^{j\varphi} = \cos\varphi + j\sin\varphi$, one obtains in rectangular coordinates:

$$Z = |Z|\cos\varphi_z + j|Z|\sin\varphi_z = R + jX$$
$$Y = |Y|\cos\varphi_y + j|Y|\sin\varphi_y = G + jB \qquad (51.6)$$

From Figure 51.3, the following relations between rectangular and polar coordinate representation can be deduced immediately:

$$R = |Z|\cos\varphi_z \quad |Z| = \sqrt{R^2 + X^2} \qquad G = |Y|\cos\varphi_y \quad |Y| = \sqrt{G^2 + B^2}$$
$$X = |Z|\sin\varphi_z \quad \varphi_z = \tan^{-1}\left(\frac{X}{R}\right) \qquad B = |Y|\sin\varphi_y \quad \varphi_y = \tan^{-1}\left(\frac{B}{G}\right) \qquad (51.7)$$

The real parts are the resistance R and the conductance G. They indicate the losses within the network. The imaginary parts, which are termed reactance X and susceptance B, respectively, are a measure of the reactive energy stored in the network during one period. In general, all these quantities are frequency dependent.

Note that the correct sign of the imaginary parts must be used: the angle φ is in the range of $-180° < \varphi \leq 180°$ and $\varphi < 0$ always corresponds to $X, B < 0$.

For elements with low losses, the loss angle δ or loss factor D are often given instead of the phases φ_z and φ_y. They are always positive small quantities and tend to 0 for a lossless device

$$D = \tan\delta = \tan\left(\frac{\pi}{2} - |\varphi_z|\right) = \tan\left(\frac{\pi}{2} - |\varphi_y|\right) = \left|\frac{R}{X}\right| = \left|\frac{G}{B}\right| \tag{51.8}$$

The inverse quantity is the quality factor $Q = 1/D$. It involves a ratio of stored electric energy to power dissipated. A high Q indicates a nearly pure reactive component.

In high-power electronics, it is necessary to reduce losses on transmission lines and therefore avoid currents associated with reactive load components. To obtain a criterion for the application and efficiency of compensation techniques, a power factor is defined. From complex power representation:

$$P = VI^* = |P|\left(\cos\varphi + j\sin\varphi\right) \tag{51.9}$$

(the asterisk indicates the conjugate complex number) follows from Equations 51.3 and 51.4.

$$P = |I|^2 Z = |I|^2 |Z|\left(\cos\varphi_z + j\sin\varphi_z\right) = |V|^2 Y^* = |V|^2 |Y|\left(\cos\varphi_y - j\sin\varphi_y\right) \tag{51.10}$$

and since the effective power is given by the real part of P:

$$P_{eff} = \mathrm{Re}\{P\} = |P|\cos\varphi \tag{51.11}$$

the power factor is:

$$\cos\varphi = \cos\varphi_z = \cos\varphi_y \tag{51.12}$$

In general, rms values are used for the phasors. Otherwise, a factor 1/2 has to be taken into account in Equations 51.9 and 51.10, since $|P| = \frac{1}{2}V_m I_m$ for sinusoidal quantities.

It can also be seen from Equations 51.9 and 51.10 that the immittances are directly related to the apparent power:

$$|P| = |V||I| = |I|^2 |Z| = |V|^2 |Y| \tag{51.13}$$

51.2 Ideal Lumped Components

The immittances of the fundamental passive circuit elements are derived from their instantaneous voltage current relations using Equations 51.1 through 51.4 and the differentiation rules.

Resistances

From Equation 51.14:

$$v(t) = Ri(t) \tag{51.14}$$

it follows $V = RI$ and thus $Z = R$ or $Y = G$. The immittance of a resistance is real and identical to its dc resistance or conductance.

Inductances

Voltage and current are related via the differential equation:

$$v(t) = L\frac{di(t)}{dt} \tag{51.15}$$

with inductance L, from which follows that $V = j\omega LI$ and

$$Z = j\omega L = jX_L, \quad Y = \frac{1}{j\omega L} = -j\frac{1}{X_L} = -jB_L \tag{51.16}$$

Capacitances

From Equation 51.17:

$$i(t) = C\frac{dv(t)}{dt} \tag{51.17}$$

with capacitance C, it follows that $I = j\omega CV$ and

$$Y = j\omega C = jB_C, \quad Z = \frac{1}{j\omega C} = -j\frac{1}{B_C} = -jX_C \tag{51.18}$$

The immittance of ideal inductors and capacitors is purely imaginary with different signs according to the phase shift of ±90° between voltage and current. A general element or network is therefore called inductive or capacitive at a given frequency corresponding to the sign of the imaginary part of its impedance. Note, however, that the frequency dependence can be much more complicated than for these ideal elements and the impedance can even change several times between capacitive and inductive characteristic.

51.3 Distributed Elements

At high frequencies, the size of the elements may no longer be small compared to the signal wavelength. Propagation effects must then be taken into account and the components can no longer be described by means of simple lumped equivalent circuits. If at all possible, they are replaced by transmission line circuits, which are easier to characterize; they realize the required electrical properties more exactly within a defined frequency range.

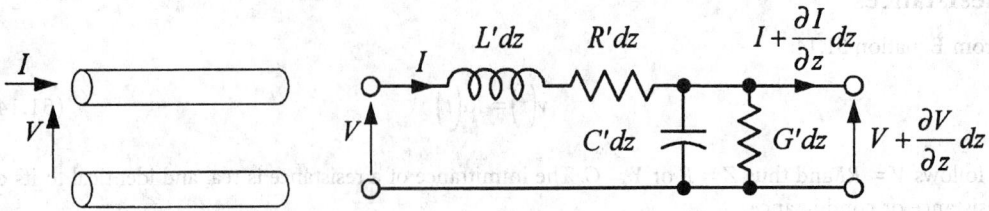

FIGURE 51.4 Equivalent circuit of a differential length of transmission line. The wave equations can be obtained by simply applying Kirchhoff's laws to voltages and currents.

Transmission Lines

Assuming a simplifying transverse electromagnetic wave (TEM mode) with no field components in the propagation direction, voltages and currents can be uniquely defined and are given as solutions of the corresponding wave equations [4]:

$$\frac{d^2V}{dz^2} - \gamma^2 V = 0, \quad \frac{d^2I}{dz^2} - \gamma^2 I = 0 \tag{51.19}$$

They vary along the line in the z-direction according to:

$$V(z) = V_0^+ e^{-\gamma z} + V_0^- e^{\gamma z}, \quad I(z) = I_0^+ e^{-\gamma z} + I_0^- e^{\gamma z} \tag{51.20}$$

These solutions are sums of forward ($e^{-\gamma z}$) and backward ($e^{\gamma z}$) traveling waves with amplitudes V_0^+, I_0^+ and V_0^-, I_0^- and a propagation constant:

$$\gamma = \sqrt{(R' + j\omega L')(G' + j\omega C')} \tag{51.21}$$

The equivalent circuit of the transmission line is shown in Figure 51.4. The energy storage in the electric field is accounted for by the distributed shunt capacitance C' per unit length, while the effect of the magnetic field is represented by the series inductance L' per unit length. The series resistance R' per unit length and the shunt conductance G' per unit length represent the power losses in the conductors and in the dielectric, respectively. The amplitudes of voltage and current are related by means of the characteristic impedance Z_0:

$$Z_0 = \frac{V^+}{I^+} = -\frac{V^-}{I^-} = \sqrt{\frac{R' + j\omega L'}{G' + j\omega C'}} \tag{51.22}$$

Of special interest for the use within a network is the input impedance Z_{in} of the transmission line. It depends also on the termination Z_L at the other end of the line. For a transmission line of length l, it is given by:

$$Z_{in} = Z_0 \frac{Z_L + Z_0 \tanh \gamma l}{Z_0 + Z_L \tanh \gamma l} \tag{51.23}$$

that is, a transmission line transforms the impedance Z_L into Z_{in} at the input.

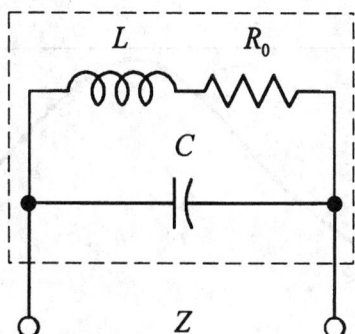

FIGURE 51.5 The simple equivalent circuit of a wire-wound resistor with nominal value R_0, inductance of the winding L, and capacitance of winding and terminal C. It is valid for a wide frequency range.

A quantity more suitable to wave propagation and measurement at high frequencies is the reflection coefficient Γ. It is defined by the relation of the voltages associated with forward and backward traveling waves. At the end of the line, using $V(l) = Z_L I(l)$, one finds:

$$\Gamma = \frac{V_0^- e^{\gamma l}}{V_0^+ e^{-\gamma l}} = \frac{Z_L - Z_0}{Z_L + Z_0} \tag{51.24}$$

For devices that support quasi or strong non-TEM waves like microstrip lines, hollow waveguides, dielectric and optical waveguides, a voltage cannot be uniquely defined. That is why several definitions of the characteristic impedance Z_0 exist [5].

51.4 Interconnections and Graphical Representations

Since Kirchhoff's laws for voltages and currents also holds for complex quantities, the rules for series and parallel connections of resistances and susceptances in the dc case apply as well for immittances.

Series connection:
$$Z = \sum_i Z_i \quad \frac{1}{Y} = \sum_i \frac{1}{Y_i} \tag{51.25}$$

Parallel connection:
$$Y = \sum_i Y_i \quad \frac{1}{Z} = \sum_i \frac{1}{Z_i} \tag{51.26}$$

As an example, consider a simplified equivalent circuit of a resistor with the nominal value R_0 (Figure 51.5). Gradually using the rules for series and parallel connection and the impedances for inductances (Equation 51.16) and capacitances (Equation 51.18), the impedance of the real resistor with parasitic elements as given leads to:

$$Z = \frac{R_0 + j\omega L}{1 - \omega^2 LC + j\omega R_0} \tag{51.27}$$

The magnitude and phase of Z/R_0 as a function of ω/ω_0 are shown in Figure 51.6 with $\omega_0 = 1/\sqrt{LC}$ as the resonant frequency defined by the parasitic elements, which might be caused by the windings of a wire-wound resistor. The network is inductive for low ($\varphi_z > 0$) and capacitve for high frequencies. An

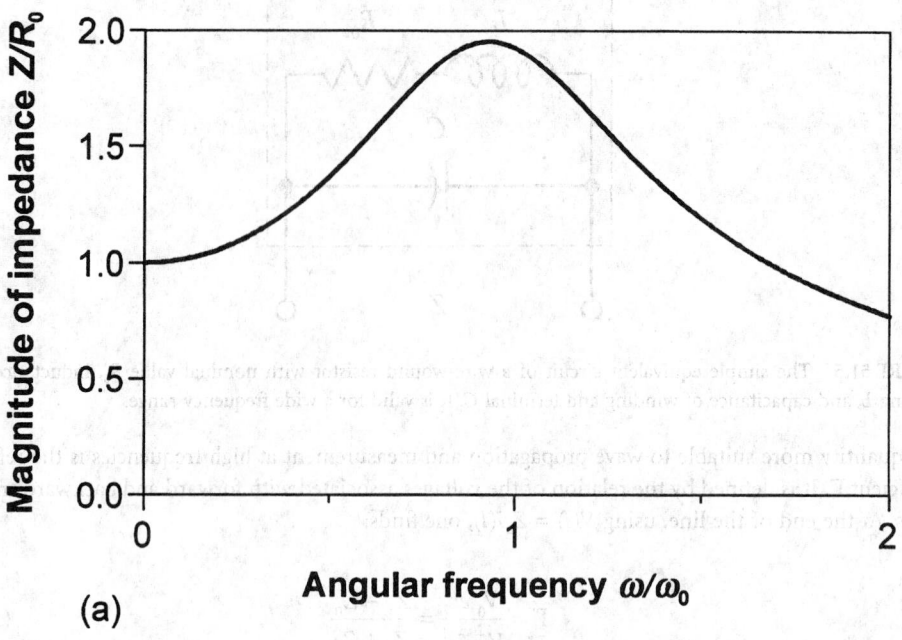

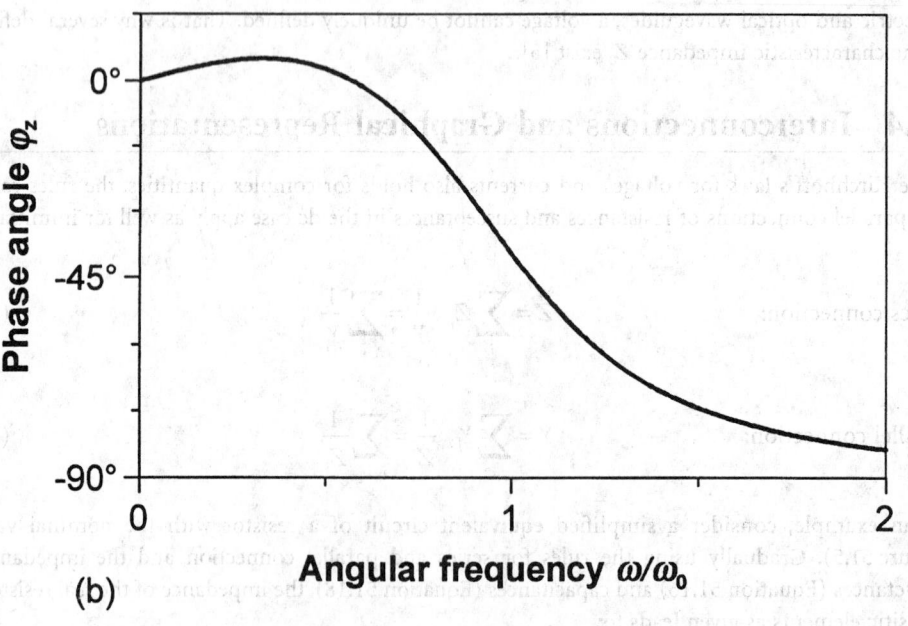

FIGURE 51.6 Normalized magnitude (a) and phase (b) of the impedance of a wire-wound resistor varying with frequency. ω_0 is the resonant frequency defined by the parasitic elements.

alternative representation is to plot real and imaginary parts in the impedance plane with the frequency as parameter as indicated by the labels (Figure 51.7). This version, called the *locus*, is very suitable to see immittance changes caused by parameters like frequency or adjustable elements within the network. Note that both real and imaginary parts are parameter dependent and vary with frequency.

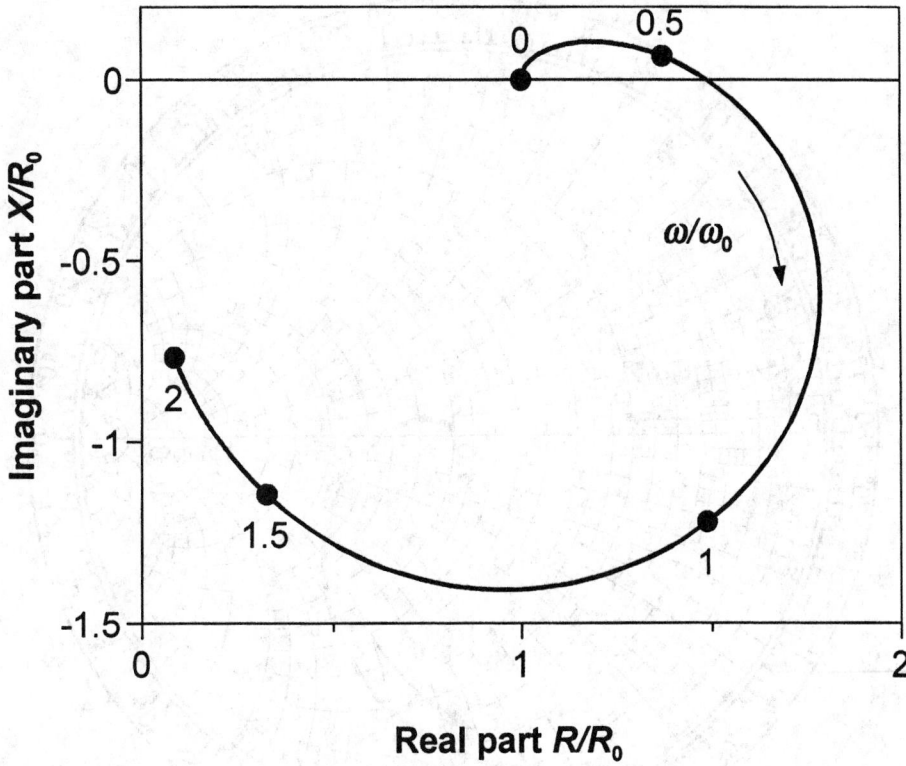

FIGURE 51.7 Normalized impedance of a wire-wound resistor in the complex plane. The arrow indicates the direction of increasing frequency.

In high-frequency applications, one obtains the impedance more easily from the reflection coefficient. Rewriting Equation 51.24 in the form:

$$\Gamma = \frac{\overline{Z}_L - 1}{\overline{Z}_L + 1} \quad \text{with} \quad \overline{Z}_L = \frac{Z_L}{Z_0} \tag{51.28}$$

defines a transformation of which the graphical representation has been called the Smith chart (Figure 51.8). It can be regarded as two coordinate systems lying one on top of the other. The reflection coefficient is given in polar coordinates around the center, the circles give the real and imaginary part of the associated impedance. The Smith chart is very useful for solving transmission line and waveguide impedance matching problems [6].

51.5 Measurement Techniques

Since immittances are complex quantities, one must determine two parameters: magnitude and phase or real and imaginary part, described as vector measurements. There exist several techniques depending on frequency range and required accuracy [7].

Current–Voltage Methods

A simple way to measure immittances follows directly from the defining Equation 51.3. Applying a well-known sinusoidal voltage to the terminal and measuring magnitude and phase of the current gives the

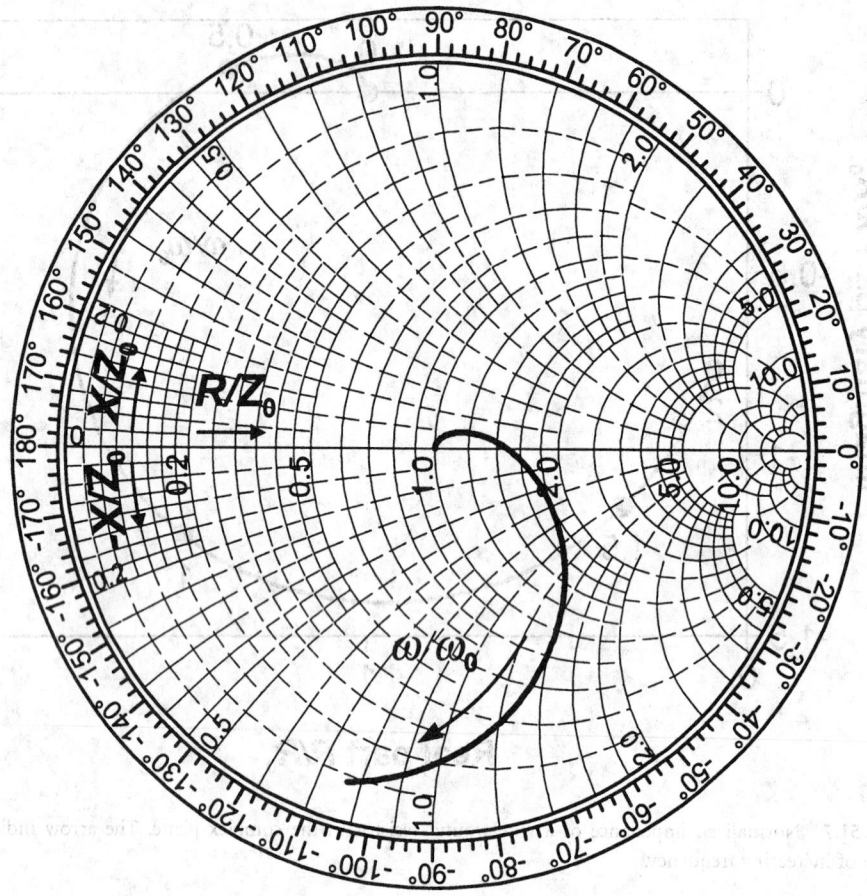

FIGURE 51.8   Smith chart representation of the impedance of a wire-wound resistor.

desired quantity (Figure 51.1). However, the internal impedance Z_A of the ammeter should be known exactly and the unknown impedance is then given by:

$$Z = \frac{V}{I} - Z_A \qquad (51.29)$$

In practical applications, impedances below 1000 Ω are measured by passing a predetermined current through the unknown device and measuring the voltage across it. Phase angle information is obtained by comparing the relative phase between voltage and current by means of a phase detector [8].

A variant on this method using only the better practicable voltage measurements is shown in Figure 51.9. The accurately known resistor R must be small compared to Z_x and to the internal resistance of V_2. One finds that:

$$Z_x = \left(\frac{V_1}{V_2} - 1\right) R, \quad \text{or} \quad Z_x \approx \frac{V_1}{V_2} R \quad \text{if} \quad R \ll |Z_x| \qquad (51.30)$$

The measurement can be enhanced using an operational amplifier with high input and low output resistance in an inverting circuit (Figure 51.10). The unknown is then given by

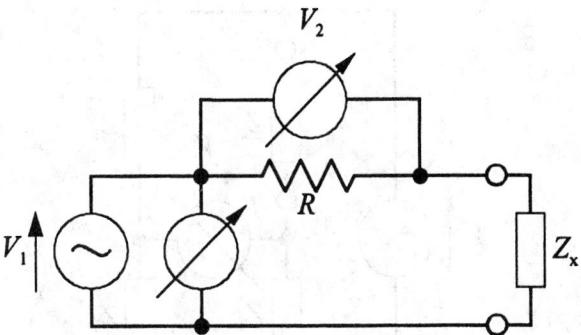

FIGURE 51.9 Determination of an impedance Z_x by phase-sensitive voltage measurements, only using a well-known resistor R.

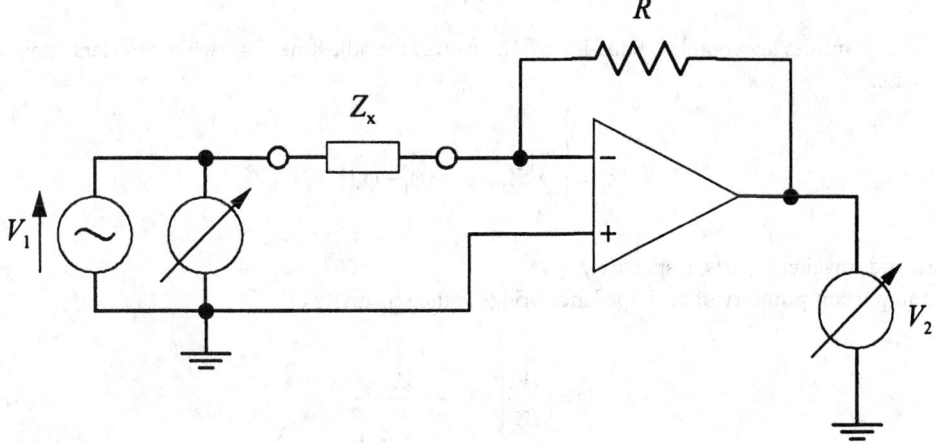

FIGURE 51.10 Impedance measurement with an inverting operational amplifier circuit. Its advantages are high input and low output resistance.

$$Z_x = -\frac{V_1}{V_2} R \qquad (51.31)$$

Practical implementations use operational amplifiers as part of an autobalancing bridge; see [7, 8].

Bridge Methods

Alternating current bridges are low-cost standard laboratory devices to measure impedances over a wide frequency range from dc up to 300 MHz with very high precision (Figure 51.11). A comprehensive survey is given in [1]. Their main advantage is that only a zero indicator in the diagonal branch is necessary. For this reason, the internal impedance does not influence the accuracy and the null point can be detected with a high sensitivity ac galvanometer as well as with headphones in the audio frequency range.

If the bridge is balanced, the unknown immittance is given by:

$$Z_x = \frac{Z_1}{Z_3} Z_2 \quad \text{or} \quad Y_x = \frac{Z_3}{Z_1} Y_2 \qquad (51.32)$$

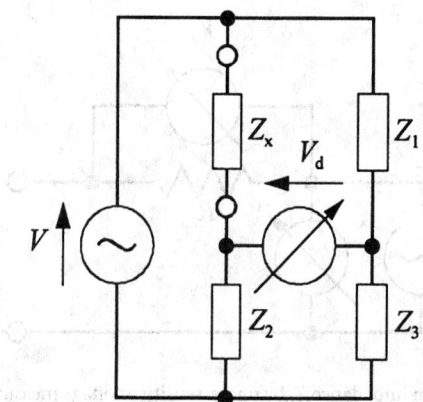

FIGURE 51.11 Impedance measurement by bridge methods. The bridge is balanced when the voltage V_d across the diagonal branch is adjusted to zero by tuning Z_1, Z_2, or Z_3.

Since the quantities are complex, Equation 51.32 involves the adjustment of two parameters: magnitude and phase:

$$|Z_x| = \left|\frac{Z_1}{Z_3}\right| |Z_2|, \quad \varphi_x = \varphi_1 - \varphi_3 + \varphi_2 \tag{51.33}$$

or real and imaginary parts, respectively.

An important property of an impedance bridge is the sensitivity ε:

$$\varepsilon = \left|\frac{\partial V_d}{\partial Z_x}\right| = V \frac{Z_2}{\left(Z_2 + Z_x\right)^2} \tag{51.34}$$

or (independent of Z_x)

$$\varepsilon = V \frac{Z_3^2}{Z_2 \left(Z_1 + Z_3\right)^2} \tag{51.35}$$

in the vicinity of zero crossing when the bridge is balanced.

The precision of the measurement not only depends on the exact zero adjustment, which can be enhanced by choosing the elements and the voltage according to Equation 51.35 to obtain a high sensitivity, but also on the realization of $Z_1 \ldots Z_3$. Mostly, these are connections of resistors and capacitors. Inductors are avoided because they always have a resistive component and it is difficult and expensive to manufacture inductors with exactly defined and reproducible electrical properties. There exist various types of bridges depending on how the elements are designed and interconnected. To choose the correct configuration, it must be known whether the unknown impedance is capacitive or inductive; otherwise, a zero adjustment is not always possible since the balancing condition cannot be fulfilled. Bridges are therefore principally used to measure capacitances and inductances as well as loss and quality factors of capacitors and coils. Since magnitude and phase conditions must be matched simultaneously, two elements must be tuned. To obtain a wide measurement range, the variable elements are designed as combinations of switchable and tunable capacitors and resistors. The sensitivity of the zero indicator can

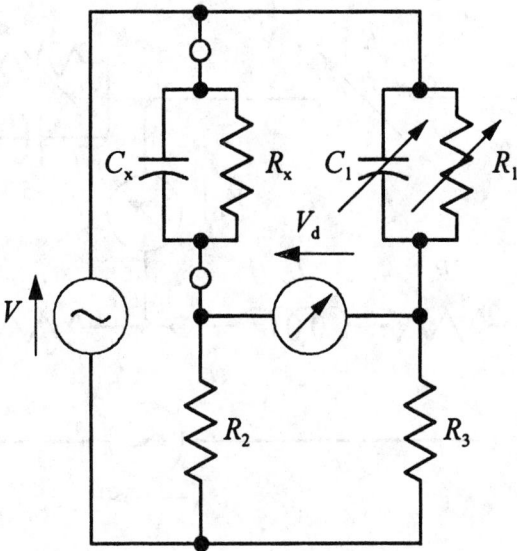

FIGURE 51.12 Wheatstone bridge for the capacitance and dissipation factor measurement of capacitors. The balancing condition is frequency independent. The resistor R_1 and the capacitor C_1 must be tuned successively until the bridge is balanced.

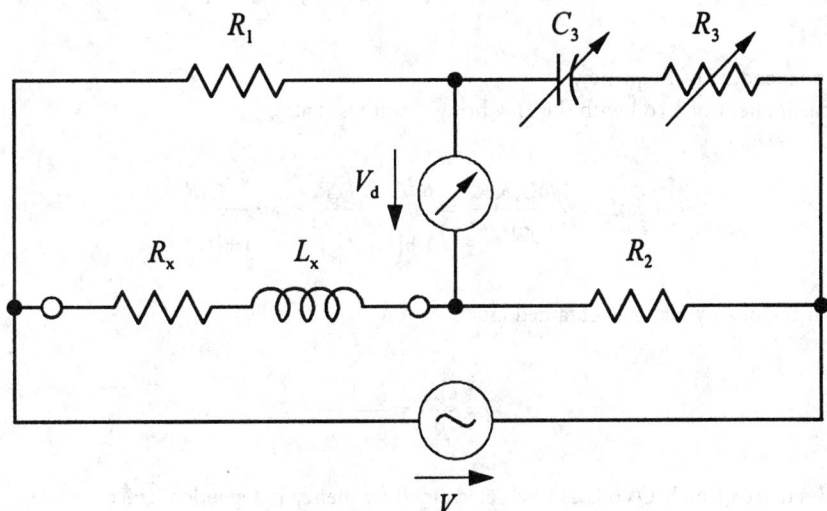

FIGURE 51.13 Hay bridge for the measurement of the inductance and the quality factor of coils. If Q is sufficiently high, the inductance can be determined nearly frequency independent.

be changed for global search and final adjustment. Unfortunately, magnitude and phase cannot be adjusted independently of each other. If the balancing is performed by hand, a suitable strategy is to search the minimum voltage by tuning each element successively.

Frequently used bridges are the Wheatstone bridge (Figure 51.12) for the measurement of lossy capacitances, and the Hay bridge (Figure 51.13) to determine inductivity and quality factor of coils. Because of its symmetrical structure, the balancing condition for the Wheatstone bridge is simply:

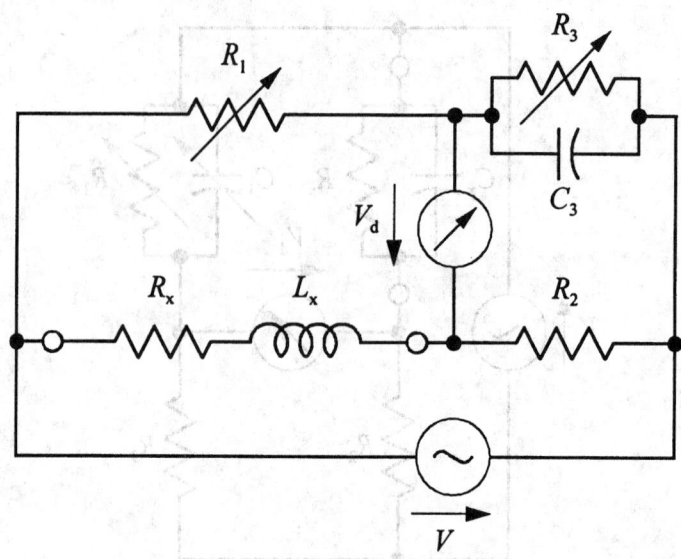

FIGURE 51.14 Maxwell bridge with simple and frequency-independent balancing conditions. Despite these advantages, it is not recommended for high-Q coils because of a very large R_1 value.

$$R_x = \alpha R_1, \quad C_x = \alpha C_1, \quad \alpha = \frac{R_3}{R_2} \tag{51.36}$$

which is independent of frequency.

The measurement of a coil with the Hay bridge requires that:

$$R_x + j\omega L_x = \frac{j\omega C_3 R_1 R_2}{1 + j\omega R_3 C_3} = \frac{\omega^2 C_3^2 R_1 R_2 R_3}{1 + (\omega R_3 C_3)^2} + j\omega \frac{C_3 R_1 R_2}{1 + (\omega R_3 C_3)^2} \tag{51.37}$$

from which the quality factor is obtained as:

$$Q = \frac{\omega L_x}{R_x} = \frac{1}{\omega R_3 C_3} \tag{51.38}$$

The inductance of high-Q coils can be determined frequency independent since

$$L_x \approx R_1 R_2 C_3 \quad \text{if} \quad (\omega R_3 C_3)^2 \ll 1 \tag{51.39}$$

A very interesting alternative is the Maxwell bridge (Figure 51.14), since it requires only resistors as variable elements, which can be manufactured with high precision. The balancing is frequency independent and leads to:

$$R_x = \frac{R_1 R_2}{R_3}, \quad L_x = R_1 R_2 C_3, \quad Q = \omega C R_1 \tag{51.40}$$

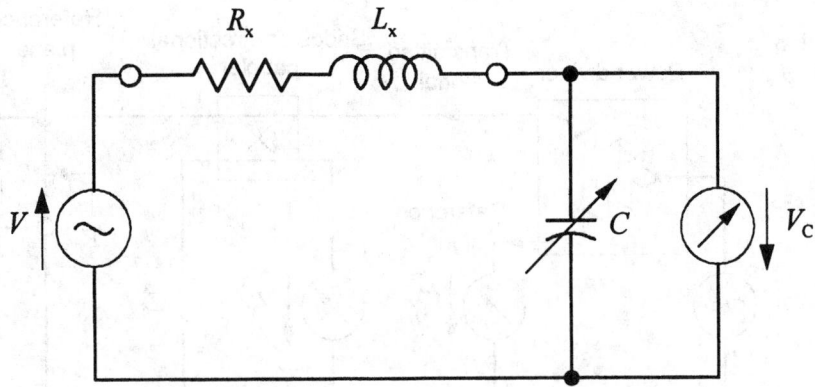

FIGURE 51.15 Coil as part of a resonance circuit to determine inductance and quality factor. The capacitor C is tuned to maximum voltage V_C.

Nevertheless, the Hay bridge is preferred for high-Q coils, because a very large value of R_1 is required for the Maxwell bridge leading to a disadvantageous balancing [9].

Resonant Method

Using the coil as part of a resonance circuit as in Figure 51.15 and tuning C to maximum voltage, the quality factor can be measured directly as:

$$Q = \left| \frac{V_{c,\,max}}{V} \right| = \frac{1}{\omega R_x C} \qquad (51.41)$$

The unknown inductance is then obtained from the test frequency by means of the resonance condition:

$$L_x = \frac{1}{\omega^2 C} \qquad (51.42)$$

If a capacitor with sufficiently low losses is used, Q values as high as 1000 can be measured.

Network Analysis Methods

Frequency Domain

In the case of distributed elements, measurements of currents and voltages depend on the position and are often not directly applicable to high-frequency devices like waveguides or microstrip lines. For that reason the determination of immittances is derived from measuring the reflection coefficient. Equation 51.23 shows the importance of defining a proper measurement plane. This is the cross-section of the line or waveguide perpendicular to the direction of propagation at a definite length l_0, where the reflection coefficient has to be measured. It can then be transformed along the line toward load or source using this relation or the Smith chart. Exact microwave measurements are very sophisticated and need a lot of practical experience. Further details can be found in the literature [5, 10–12].

Automated and precise immittance measurements over a wide frequency range are best carried out with a vector network analyzer [11]. Unfortunately, this is also the most expensive method. The principle of measurement is shown in Figure 51.16. A power divider splits the incident signal into a transmitted and a reference part. The directional bridge or coupler separates forward and backward traveling waves, and the reflected signal now appears in the branch with the phase-sensitive voltmeter V_2. Using a bridge

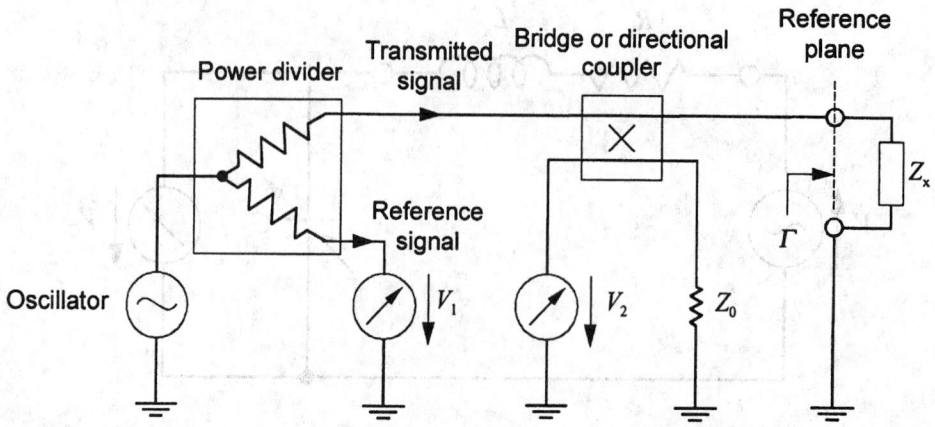

FIGURE 51.16 Schematic of network analyzer measurements. The voltage ratio V_2/V_1 of reflected wave and reference signal is proportional to the reflection coefficient Γ. The impedance Z_x can then be computed.

with impedances matched to the line ($Z_1 = Z_2 = Z_3 = Z_0$), the voltage in the diagonal branch is given by (Figure 51.11):

$$V_d = \frac{Z_x - Z_0}{2(Z_x + Z_0)} V = V_2 \qquad (51.43)$$

and thus the reflection coefficient:

$$\Gamma = \frac{Z_x - Z_0}{Z_x + Z_0} = \alpha \frac{V_2}{V_1} \qquad (51.44)$$

is directly proportional to the voltage ratio.

Network analyzers use an automatic error correction to eliminate the effect of internal and external couplers and junctions. Because of that, a calibration procedure with standard terminations is necessary. These terminations must be manufactured very precisely, since they define the measurement plane and determine the overall measurement error.

Time Domain

It is often necessary to locate an impedance step, whether to find out the distance of a cable defect or to track down the origin of reflections within a connection. To this end, high-performance vector network analyzers have a Fourier transform procedure. But there also exist cheaper time domain reflectometers (TDR) [13, 14]. They use an incident step or impulse signal (Figure 51.17) and the reflected signal is separated by means of a directional coupler and displayed on a CRT in the time domain. From the shape of the signal, the impedance step can be localized by means of the time delay:

$$l = \frac{1}{2} v_g t \qquad (51.45)$$

with v_g as signal or group velocity on the line varying from step to step. Characteristic and magnitude of the impedance can only be estimated, since phase information is usually not available. TDR measurements are restricted to the localization of impedance steps and not to be recommended for exact measurements. Moreover, additional pulse deformations occur in dispersive waveguides.

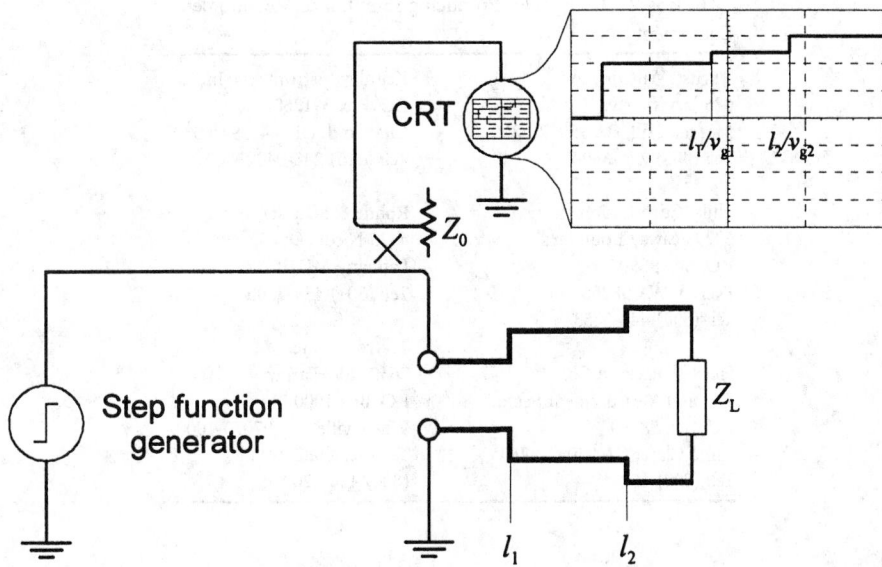

FIGURE 51.17 Detection and measurement of impedance steps on a line or waveguide with a time domain reflectometer (TDR). Since phase information is usually not available, the characteristics and magnitudes of the impedances can only be estimated. Notice that the group or signal velocity v_g varies from step to step.

51.6 Instrumentation and Manufacturers

A broad range of instrumentation for measuring immittance is available. Some of these instruments are included in Table 51.1. Table 51.2 provides the names and addresses of some companies that produce immittance-measuring instrumentation.

TABLE 51.1 Instruments for Immittance Measurements

Manufacturer	Model number	Description
Anritsu Wiltron	360 B	Vector network analyzer 10 MHz–65 GHz
Fluke	PM 6303A	Automatic RCL meter
Fluke	PM 6304	Automatic RCL meter
Hewlett-Packard	HP 4195A	Vector network analyzer 10 Hz–500 MHz
Hewlett-Packard	HP 4396A	Vector network analyzer 100 kHz–1.8 GHz
Hewlett-Packard	HP 8719C	Vector network analyzer 50 MHz–13.5 GHz
Hewlett-Packard	HP 8720C	Vector network analyzer 50 MHz–20 GHz
Hewlett-Packard	HP 8722C	Vector network analyzer 50 MHz–40 GHz
Hewlett-Packard	HP 8510C	Vector network analyzer 45 MHz–110 GHz
Hewlett-Packard	HP 8508A	Vector voltmeter 100 kHz–1 GHz
Hewlett-Packard	HP 4194A	Impedance analyzer 100 Hz–40 MHz
Hewlett-Packard	HP 4191A	HF-impedance analyzer 1 MHz–1 GHz
Hewlett-Packard	HP 4192A	Impedance analyzer 5 Hz–13 MHz
Hewlett-Packard	HP 4193A	Impedance analyzer 400 kHz–110 MHz
Keithley	3321	LCZ meter, 4 test frequencies
Keithley	3322	LCZ meter, 11 test frequencies
Keithley	3330	LCZ meter 40 Hz–100 kHz
Rohde & Schwarz	ZVRL	Vector network analyzer
Rohde & Schwarz	SR 720	LCR meter
Tektronix	CSA 803A	Communications signal analyzer

TABLE 51.2 Companies Producing Immittance Measurement Equipment

Anritsu Wiltron Co.	Keithley Instruments, Inc.
685 Jarvis Drive	P.O. Box 391260
Morgan Hill, CA 95037-2809	Cleveland, OH 44139-9653
Tel: (408) 776-8300	Tel: (216) 248-0400
Fluke Corporation	Rohde & Schwarz, Inc.
6929 Seaway Boulevard	4425 Nicole Dr.
P.O. Box 9090	Lanham, MD 20706
Everett, WA 98206	Tel: (301) 459-8800
Tel: (800) 443-5853	
	Tektronix, Inc.
Hewlett-Packard Co.	26600 SW Parkway
Test and Measurement Sector	P.O. Box 1000
P.O. Box 58199	Wilsonville, OR 97070-1000
Santa Clara, CA 95052-9943	Tel: (503) 682-3411
Tel: (800) 452-4844	(800) 426-2200

Defining Terms

Admittance (Y): The reciprocal of impedance.

Immittance: A response function for which one variable is a voltage and the other a current. Immittance is a general term for both impedance and admittance, used where the distinction is irrelevant.

Impedance (Z): The ratio of the phasor equivalent of a steady-state sine-wave voltage to the phasor equivalent of a steady-state sine-wave current. The real part is the *resistance*, the imaginary part is the *reactance*.

Phasor: A complex number, associated with sinusoidally varying electrical quantities, such that the absolute value (modulus) of the complex number corresponds to either the peak amplitude or root-mean-square (rms) value of the quantity, and the phase (argument) to the phase angle at zero time. The term "phasor" can also be applied to impedance and related complex quantities that are not time dependent.

Reflection coefficient: At a given frequency, at a given point, and for a given mode of propagation, the ratio of voltage, current, or power of the reflected wave to the corresponding quantity of the incident wave.

References

1. B. M. Oliver and J. M. Cage, *Electronic Measurements and Instrumentation*, New York: McGraw-Hill, 1971.
2. H. W. Bode, *Network Analysis and Feedback Amplifier Design*, Princeton: Van Nostrand, 1959.
3. A. T. de Hoop, *Handbook of Radiation and Scattering of Waves*, London: Academic Press, 1995.
4. R. E. Collin, *Foundations for Microwave Engineering*, New York: McGraw-Hill, 1992.
5. P. I. Somlo and J. D. Hunter, *Microwave Impedance Measurement*, London: Peter Peregrinus, 1985.
6. R. L. Thomas, *A Practical Introduction to Impedance Matching*, Dedham, MA: Artech House, 1976.
7. M. Honda, *The Impedance Measurement Handbook*, Yokogawa: Hewlett-Packard, 1989.
8. Anonymous, *4800A Vector Impedance Meter Operating and Service Manual*, Rockaway: Hewlett-Packard, 1967.
9. C. R. Paul, S. A. Nasar, and L. E. Unnewehr, *Introduction to Electrical Engineering*, New York: McGraw-Hill, 1992.
10. T. S. Laverghetta, *Modern Microwave Measurements and Techniques*, Norwood, MA: Artech House, 1988.

11. A. E. Bailey (ed.), *Microwave Measurement,* London: Peter Peregrinus, 1985.

12. T. S. Laverghetta, *Handbook of Microwave Testing,* Dedham, MA: Artech House, 1981.

13. E. K. Miller (ed.), *Time-Domain Measurements in Electromagnetics,* New York: Van Nostrand Reinhold, 1986.

14. Anonymous, *TDR Fundamentals for Use with HP 54120T Digitizing Oscilloscope and TDR,* Application notes AN 62, Palo Alto, CA: Hewlett-Packard, 1988.

Further Information

L. S. Bobrow, *Fundamentals of Electrical Engineering,* New York: Oxford University Press, 1996.

W. H. Roadstrum and D. H. Wolaver, *Electrical Engineering for all Engineers,* New York: John Wiley & Sons, 1994.

A. S. Morris, *Principles of Measurement and Instrumentation,* London: Prentice Hall International (U.K.), 1993.

G. H. Bryant, *Principles of Microwave Measurement,* London: Peter Peregrinus, 1988.

Anonymous, *Low Level Measurements,* Cleveland, OH: Keithley Instruments, 1984.

52

Q Factor Measurement

Q factor is a method of characterizing the rate of dissipation of energy from an oscillating system. Q is defined as 2π times the energy stored in a resonant system divided by the energy dissipated per cycle. The term system used in this context refers to any type of resonance: mechanical, electrical, nuclear, etc. For the purposes of this *Handbook*, Q will be that of an electric circuit. Also, for the discussion of Q, very low values of Q, typically less than 10, will not be considered as these low values of Q produce highly damped oscillations that are more exponential than oscillatory and the concept of Q does not fit.

A common interpretation of Q is quality factor that explains the use of the letter, Q, but this is misleading. A component that has a high Q is not always beneficial and may not be of high quality. In many circuits, a component requires a specific value of Q rather than "higher is better." In other cases, a high Q is an advantage.

The Q factors encountered in common circuit components range from a low of about 50 for many inductors to nearly 1 million found in high-quality quartz crystals. Q can be applied to a resonant circuit or to capacitors and inductors. When Q is applied to a component, such as an inductor, the Q would be that obtained if the inductor were used in a resonant circuit with a capacitor that dissipates no energy. In this case, the value of Q depends on the frequency.

For most LC resonant circuits, the losses in the inductor dominate and the Q of the inductor is, essentially, the Q of the circuit. It is easy to make very low-loss capacitors even in the UHF region. On the other hand, varactor diodes have considerably more loss than fixed capacitors, and a varactor can play a more significant role in setting the circuit Q.

52.1 Basic Calculation of Q

Figure 52.1 shows a simple resonant circuit. The capacitor can store energy in the electric field and the inductor in the magnetic field. The circuit oscillates with the energy transferring between the two elements. For the ideal elements shown, this continues forever. Since the definition of Q has the energy lost per cycle in the denominator — which is zero — the result is an infinite Q.

In practice, circuit elements are not perfect and the energy initially contained within this circuit would be lost by the circuit and the oscillations would decrease in amplitude as the energy diminished. Energy loss in a circuit is represented by that in a resistor which can be included in one of two ways. The first

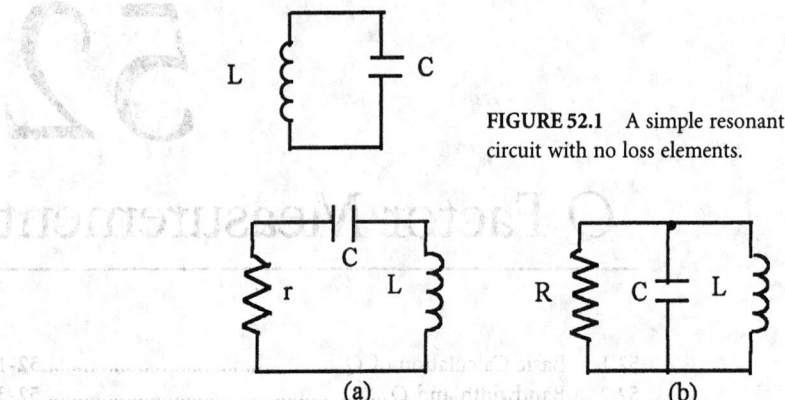

FIGURE 52.1 A simple resonant circuit with no loss elements.

(a)   (b)

FIGURE 52.2   (a) A simple series resonant circuit with the equivalent resistance. (b) A parallel resonant circuit with a parallel equivalent resistance. For the same Q circuit, the values of the resistors are not the same.

way is shown in Figure 52.2(a), where the resistor is in series with the capacitor and inductor. A second representation is a parallel resistor as shown in Figure 52.2(b).

To derive the relationship between the circuit element values and the Q of the circuit, either the current or voltage of the circuit can be used in the equations. Current is the obvious choice for a series circuit, while voltage is the common thread for the parallel circuit. Assume that the amplitude of the current through the series circuit of Figure 52.2(a) is given by:

$$i(t) = I(t)\cos\left(2\pi f_0 t\right) \tag{52.1}$$

where f_0 = Resonant frequency = $f_0 = \dfrac{1}{2\pi\sqrt{LC}}$ $\tag{52.2}$

and $I(t)$ is the amplitude, which is decreasing in some unspecified fashion. The circuit's peak current occurs when the cosine function is equal to 1 and all of the energy is contained in the inductor and is equal to $(1/2)LI^2(t)$.

Assume that a relatively high Q is present in this circuit and that $I(t)$ changes by only a slight amount during the time of one cycle. During this cycle, the peak current is $I(t)$, and the rms value of the current is $(0.707)I(t)$. Therefore, the energy dissipated in one cycle is $(0.5)I^2(t)r/f_0$. Substituting these values in the definition of Q yields:

$$Q = 2\pi\frac{\dfrac{1}{2}LI^2(t)}{\dfrac{1}{2}\dfrac{rI^2(t)}{f_0}} = \frac{2\pi f_0 L}{r} = \frac{X_L}{r} \tag{52.3}$$

where X_L is the reactance of the inductor. The same procedure can be used with the parallel resonant circuit of Figure 52.2(b) using voltage equations to obtain the relationship between a parallel resistance and Q, which is:

$$Q = \frac{R}{X_L} \tag{52.4}$$

It is very important to understand the nature of the circuit resistance in Figure 52.2(a) and (b). This resistance represents all of the losses in the resonant circuit. These losses are from a variety of sources,

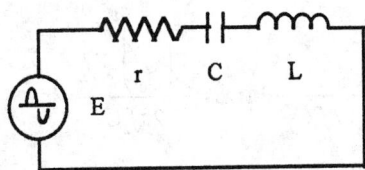

FIGURE 52.3   A series resonant circuit showing a driving source.

such as the resistance of the wire to make an inductor or the leakage current of a capacitor. It can also represent the deliberate addition of resistors to set the Q of the circuit to a specific value. The resistance shown in the circuits of Figure 52.2 represents the equivalent resistance of all of the energy losses. This resistance cannot be measured with an ohmmeter as the value of the equivalent resistor is a function of frequency and other variables such as signal level. Some of the loss in a resonant circuit is due to radiation, which is a function of frequency. The resistance of conductors is mostly due to skin effect, which increases with increasing frequency. The losses in the ferromagnetic materials used for making some inductors are nonlinear; thus, the equivalent resistance is not only a function of frequency but of signal level.

Most resonant circuits are not stand-alone circuits as shown in Figure 52.2, but are a part of other circuits where there are sources and loads. These additional resistances further remove energy from the resonant circuit. The Q of a resonant circuit when there are sources and loads is called the *loaded Q*. In most applications of resonant circuits, the Q of the resonance is set by the external loads rather than the capacitor and inductor that form the resonance.

52.2 Bandwidth and Q

The *bandwidth* of a resonant circuit is a measure of how well a resonant circuit responds to driving signals of a frequency near the resonant frequency and is a function of Q. The relationship between the 3 dB bandwidth and Q will be derived.

Applying a driving signal to a resonant circuit can overcome the losses of the circuit and cause the resonant circuit to oscillate indefinitely. As an example of this, consider the voltage generator in the series resonant circuit shown in Figure 52.3.

When the frequency of the voltage source is equal to the resonant frequency of the circuit, the equivalent impedance of the series resonant circuit is the resistance of the circuit and the current in the circuit, simply E/r.

At frequencies higher or lower than the resonant frequency, the impedance of the circuit is greater because the net reactance is not zero and the circuit current will be less than at resonance.

At what frequency will the circuit current be 3 dB less than at resonance? This frequency is where the circuit impedance is 1.414 that of the impedance at resonance. This is the frequency where the reactive part of the impedance is equal to the real part. This situation occurs at two frequencies. Below the resonant frequency, the net reactance is capacitive and is equal to r, while at a second frequency above the resonant frequency, the reactance is inductive and equal to r. This can be represented by two equations for the two frequencies:

$$\text{For } f_1 > f_0, \qquad \left| X_L - X_C \right| = 2\pi f_1 L - \frac{1}{2\pi f_1 C} = \frac{\left(\dfrac{f_1}{f_0} \right)^2 - 1}{2\pi f_1 C} = r \qquad (52.5)$$

$$\left(\frac{f_1}{f_0} \right)^2 - \frac{1}{Q} \left(\frac{f_1}{f_0} \right) - 1 = 0 \qquad \left(\frac{f_1}{f_0} \right) = \frac{1}{2Q} \pm \sqrt{\frac{1}{4Q^2} + 1}$$

For $f_2 < f_0$,

$$\left| X_L - X_C \right| = \frac{1}{2\pi f_2 C} - 2\pi f_2 L = \frac{-\left(\frac{f_1}{f_0}\right)^2 - 1}{2\pi f_2 C} = r \qquad (52.6)$$

$$\left(\frac{f_2}{f_0}\right)^2 + \frac{1}{Q}\left(\frac{f_2}{f_0}\right) - 1 = 0 \qquad\qquad \left(\frac{f_2}{f_0}\right) = \frac{1}{2Q} \pm \sqrt{\frac{1}{4Q^2} + 1}$$

$$f_1 - f_2 = \frac{f_0}{Q} = \text{bandwidth}$$

Measuring Q

There are a number of methods of measuring Q using a variety of bridges, several of which are described in [1]. One method of measuring a capacitive or inductive Q is to place the component in a resonant circuit. When the Q to be measured of a device that is, in itself, a resonant circuit such as quartz crystal, similar techniques are used except the device's own resonance is used for the measurement. Circuit voltages or currents are measured at the resonance frequency and the Q is determined.

52.3   The Q-Meter

One simple and very popular method of measuring Q is with a device called, appropriately, the Q-meter. Consider the resonant circuit in Figure 52.4 for measuring the Q of inductors. This circuit has a very low-loss capacitor of known value and a constant voltage source.

The usual components measured by the Q-meter are inductors. It was previously mentioned that inductors are the weak link in resonant circuits, and the Q of a circuit is usually set by the inductor. The Q-meter can measure capacitance and capacitor Q. In this theoretical circuit, the circuit resistance is the equivalent series resistance of the inductor under test. This is due to the fact the variable capacitor is assumed to be lossless, the generator has zero resistance, and the voltmeter does not appreciably load the circuit. In a real circuit, it is not possible to achieve this situation, but the instrument is designed to approach this goal.

To measure the Q of an inductor using the Q-meter, the generator is set to the desired frequency while the variable capacitor tunes the circuit to resonance as indicated by the peak reading of the voltmeter.

At resonance, the impedance of the circuit is simply the equivalent series resistance of the inductor. This sets the current of the circuit as:

$$I = E/R_X \qquad (52.7)$$

where E is the generator voltage and R_X is the equivalent resistance of the inductor.

Because the circuit is at resonance, the voltages of the two reactances are equal and of opposite phase. Those voltages are:

$$V = IX_C \quad \text{or} \quad V = IX_L \qquad (52.8)$$

where X_C is the capacitive reactance and X_L is the inductive reactance, which are numerically equal at resonance.

Substituting the relationship of the circuit current, the result is:

$$V = EX_L/R_X = EX_C/R_X = EQ \qquad (52.9)$$

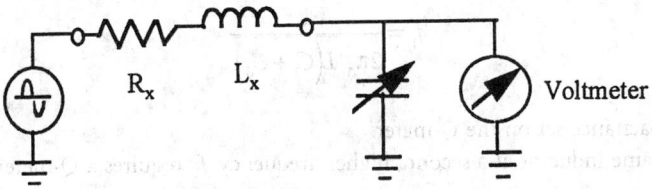

FIGURE 52.4   The basic circuit of a *Q*-meter showing the signal source, the inductor under test, and the voltmeter.

Therefore, the voltage across the reactances is equal to *Q* times the applied voltage. If, as an example, the voltage source were 1 V, the voltmeter would read *Q* directly. *Q* values of several hundred are common and, therefore, voltages of several hundred volts could be measured. Modern circuits do not typically encounter voltages of this magnitude, and many components cannot withstand this potential. Therefore, most *Q*-meters use a much smaller source voltage, typically 20 mV.

If the frequency of the source and the circuit capacitance are known, it is possible to calculate the inductance of the unknown.

52.4   Other Q Measuring Techniques

There are very few *Q*-meters being manufactured today, although there are thousands of old *Q*-meters still in use. Because the *Q*-meter was the only accepted method of measuring *Q* for so many years, it will take decades before alternative methodologies overtake the *Q*-meter.

Measuring *Q* without a *Q*-meter involves a variety of *RLC* measuring instruments that measure the vector impedance and calculate *Q*. The calculated *Q* value is not as valid as that determined with a *Q*-meter unless the *RLC* meter is capable of measuring *Q* at the desired frequency. Only the more sophisticated, and expensive, *RLC* meters allow the use of any test frequency. Despite the new sophisticated *RLC* measuring instruments, there is an adapter for one model *RLC* instrument that allows the classic *Q*-meter-style measurement to be made.

Because vector impedance measurement is covered elsewhere in this Handbook, the remainder of this section will be devoted to measurements using the *Q*-meter.

52.5   Measuring Parameters Other than Q

In addition to *Q*, the *Q*-meter can be used to measure inductance, the *Q* or dissipation factor of a capacitor, and the distributed capacitance, C_d, of an inductor.

If an inductor with capacitance C_d is placed in the *Q*-meter circuit, the total circuit capacitance includes both the *Q*-meter's capacitance plus the additional C_d. Therefore, when resonance is achieved, the actual resonating capacitance is more than what is indicated on the *Q*-meter capacitor's dial. If C_d is not included in the calculation of inductance, the resulting value would be too large.

In many applications, the actual inductance is not the important parameter to be measured by the *Q*-meter. The actual parameter being determined is "what capacitance is required to resonate the particular inductor at a specific frequency," regardless of C_d.

In other applications, such as inductors that are to be used in wide-range oscillators, where C_d will limit the tuning range, C_d is an important parameter.

The *Q*-meter can be used to determine C_d, which will also allow for an accurate calculation of inductance. Determining C_d is a matter of resonating the inductor under test at more than one frequency.

To understand how the two-frequency measurement will allow the determination of C_d, assume an inductor is resonated at a frequency f_1. The relationship between the applied frequency, f_1, and the capacitor of the *Q*-meter to obtain resonance is:

$$f_1 = \frac{1}{2\pi\sqrt{L\left(C_1 + C_d\right)}} \tag{52.10}$$

where C_1 is the capacitance set on the Q-meter.

Resonating the same inductor at a second, higher, frequency, f_2, requires a Q-meter capacitance of C_2 such that:

$$f_2 = \frac{1}{2\pi\sqrt{L\left(C_2 + C_d\right)}} \tag{52.11}$$

This implies that C_2 is a smaller capacitance than C_1. Using these two equations and solving for C_d, the following result is obtained.

$$C_d = \frac{C_2 f_2^2 - C_1 f_1^2}{f_1^2 - f_2^2} \tag{52.12}$$

A convenient relationship between f_1 and f_2 is to set $f_2 = 1.414\, f_1$. With frequencies thus related, the distributed capacitance is:

$$C_d = C_1 - 2C_2 \tag{52.13}$$

C_d causes errors in the measurement of Q because of current through the C_d. The Q measured by the Q-meter is called the "effective Q." Since large inductors with significant C_d are no longer in common use because of the use of active filters, the distinction between effective Q and real Q is seldom considered. For additional information about effective Q and distributed capacitance, see [2].

To measure capacitors on the Q-meter, a relatively high Q inductor is connected to the inductance terminals on the Q-meter and resonated. The capacitor to be measured is connected to the capacitor terminals, which increases the circuit capacitance. The Q-meter variable capacitor is adjusted to regain resonance, which requires that the capacitance be reduced by an amount equal to the unknown capacitor.

The Q of the capacitor can be measured. The addition of the capacitor reduces the circuit Q because of the additional loss introduced by the capacitor. In the description of the Q-meter, the Q-meter's variable capacitor is assumed to have no loss.

Measuring the Q of a capacitor using the Q-meter is seldom done. This is because most capacitors have very high Q values. There are special cases, such as measuring the Q of a transducer or a varactor diode, where low Q values are encountered.

The unknown capacitor is connected to the CAP terminals of the Q-meter, which are simply in parallel with the Q-meter's variable capacitor. The variable capacitor in the Q-meter is set to the minimum capacitance and a suitable inductor is placed across the IND (inductor) terminals. The Q-meter is resonated using the frequency control rather than the variable capacitance.

For best results, the Q of the inductor must be considerably greater than that of the unknown capacitor, and the unknown capacitance must be considerably greater than the internal capacitance. If these criteria are met, the Q of the unknown capacitor can be read from the Q-meter. If these criteria are compromised, corrections can be made but the equations become complex and the accuracy degrades.

Most Q-meters provide measurement ranges to about 500 or to 1000. This is sufficient for measuring inductors, which was the main purpose of the Q-meter. For measuring high-Q devices such as ceramic resonators with Q values greater than 1000 or for quartz resonators with Q values extending into the hundreds of thousands, the Q-meter technique is insufficient. A high-Q circuit implies very little energy loss, and the energy that must be removed to make a measurement must be very small if Q is to be measured accurately.

Defining Terms

Bandwidth: A measurement of the amount of frequency spectrum occupied by a signal or the equivalent spectrum covered by a circuit that passes a finite band of frequencies. There are a number of methods of defining bandwidth, depending on the nature of the spectrum. Relative to resonant circuits, the bandwidth is measured between the –3 dB points of the passband.

Distributed capacitance: The amount of capacitance added to an inductor typically from the capacitance due to adjacent wires in a solenoid-type inductor. The distributed capacitance is given as a single capacitance figure for a specific inductor and can be defined as the equivalent capacitance across the entire coil. This would also allow the inductor to have a self-resonant frequency where the inductor resonates with the distributed capacitance with no external capacitance.

Effective inductance: Due to distributed capacitance, less capacitance than that calculated from an inductance value is required to resonate a circuit. If the actual capacitance required to resonate a circuit is used to calculate an inductance value, the resulting inductance value will be higher than the theoretical inductance value. This higher value is called the "effective inductance." The actual inductor cannot be considered as a pure inductance of a value equal to the effective inductance because the real inductor has a resonant frequency that a pure inductance does not.

Q: A measurement of the rate of energy loss in a resonant circuit.

Q-Meter: An instrument for measuring Q factor by resonating the circuit and measuring the voltage across the reactances.

References

1. *Reference Data for Engineers: Radio Electronics, Computer and Communications*, 8th ed., Sams, Englewood Cliffs, NJ: Prentice-Hall Computer Publishing, 1993.
2. A. Helfrick and W. Cooper, *Modern Instrumentation and Measurement Techniques*, Englewood Cliffs, NJ: Prentice-Hall, 1990.

53

Distortion Measurement

Michael F. Toner
Nortel Networks

Gordon W. Roberts
McGill University

A sine-wave signal will have only a single-frequency component in its spectrum; that is, the frequency of the tone. However, if the sine wave is transmitted through a system (such as an amplifier) having some nonlinearity, then the signal emerging from the output of the system will no longer be a pure sine wave. That is, the output signal will be a distorted representation of the input signal. Since only a pure sine wave can have a single component in its frequency spectrum, this situation implies that the output must have other frequencies in its spectral composition. In the case of *harmonic distortion*, the frequency spectrum of the distorted signal will consist of the fundamental (which is the same frequency as the input sine wave) plus harmonic frequency components that are at integer multiples of the fundamental frequency. Taken together, these will form a Fourier representation of the distorted output signal. This phenomenon can be described mathematically. Refer to Figure 53.1, which depicts a sine-wave input signal $x(t)$ at frequency f_1 applied to the input of a system $A(x)$, which has an output $y(t)$. Assume that system $A(x)$ has some nonlinearity. If the nonlinearity is severe enough, then the output $y(t)$ might have excessive harmonic distortion such that its shape no longer resembles the input sine wave. Consider the example where the system $A(x)$ is an audio amplifier and $x(t)$ is a voice signal. Severe distortion can result in a situation where the output signal $y(t)$ does not represent intelligible speech. The *total harmonic distortion* (THD) is a figure of merit that is indicative of the quality with which the system $A(x)$ can reproduce an input signal $x(t)$. The output signal $y(t)$ can be expressed as:

$$y(t) = a_0 + \sum_{k=1}^{N} a_k \cos\left(2\pi k f_1 t + \theta_k\right) \tag{53.1}$$

where the a_k, $k = 0, 1, …, N$ are the magnitudes of the Fourier coefficients, and θ_k, $k = 0, 1, …, N$ are the corresponding phases. The THD is defined as the percentage ratio of the rms voltage of all harmonics components above the fundamental frequency to the rms voltage of the fundamental. Mathematically, the definition is written:

$$\text{THD} = \frac{\sqrt{\sum_{k=2}^{N} a_k^2}}{a_1} \times 100\% \tag{53.2}$$

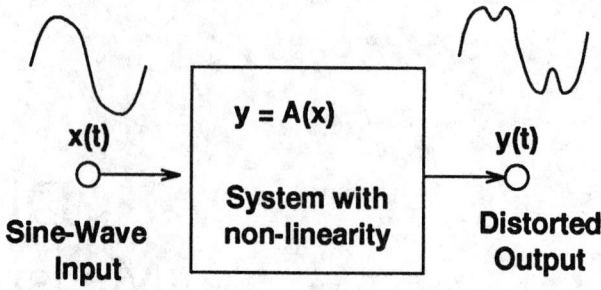

FIGURE 53.1   Any system with a nonlinearity gives rise to distortion.

If the system has good linearity (which implies low distortion), then the THD will be a smaller number than that for a system having poorer linearity (higher distortion). To provide the reader with some feeling for the order of magnitude of a realistic THD, a reasonable audio amplifier for an intercom system might have a THD of about 2% or less, while a high-quality sound system might have a THD of 0.01% or less. For the THD to be meaningful, the bandwidth of the system must be such that the fundamental and the harmonics will lie within the passband. Therefore, the THD is usually used in relation to low-pass systems, or bandpass systems with a wide bandwidth. For example, an audio amplifier might have a 20 Hz to 20 kHz bandwidth, which means that a 1-kHz input sine wave could give rise to distortion up to the 20th harmonic (i.e., 20 kHz), which can lie within the passband of the amplifier. On the other hand, a sine wave applied to the input of a narrow-band system such as a radio frequency amplifier will give rise to harmonic frequencies that are outside the bandwidth of the amplifier. These kinds of narrow-band systems are best measured using *intermodulation distortion*, which is treated elsewhere in this Handbook. For the rest of the discussion at hand, consider the example system illustrated in Figure 53.1 which shows an amplifier system $A(x)$ that is intended to be linear but has some undesired nonlinearities. Obviously, if a linear amplifier is the design objective, then the THD should be minimized.

53.1   Mathematical Background

Let $y = A(x)$ represent the input-output transfer characteristic of the system $A(x)$ in Figure 53.1 containing the nonlinearity. Expanding into a power series yields

$$A(x) = \sum_{k=0}^{\infty} c_k x^k = c_0 + c_1 x + c_2 x^2 + c_3 x^3 + \dots \tag{53.3}$$

Let the input to the system be $x = \cos(2\pi f_0 t)$. Then the output will be

$$y = A(x) = c_0 + c_1 A_0 \cos(2\pi f_0 t) + c_2 A_0^2 \cos^2(2\pi f_0 t) + c_3 A_0^3 \cos^3(2\pi f_0 t) + \dots \tag{53.4}$$

This can be simplified using the trigonometric relationships:

$$\cos^2(\theta) = \frac{1}{2} - \frac{1}{2}\cos(2\theta) \tag{53.5}$$

$$\cos^3(\theta) = \frac{3}{4}\cos(\theta) + \frac{1}{4}\cos(3\theta) \tag{53.6}$$

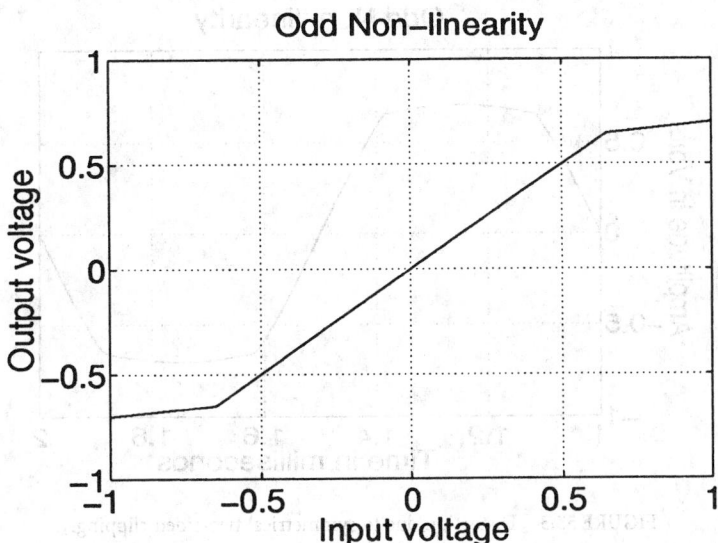

FIGURE 53.2 An off nonlinearity with $f(-x) = -f(x)$.

$$\cos^4(\theta) = \frac{1}{8} - \frac{1}{2}\cos(2\theta) + \frac{1}{8}\cos(4\theta) \tag{53.7}$$

$$\cos^5(\theta) = \frac{5}{8}\cos(\theta) + \frac{5}{16}\cos(3\theta) + \frac{1}{16}\cos(5\theta) \tag{53.8}$$

and so on. Performing the appropriate substitutions and collecting terms results in an expression for the distorted signal $y(t)$ that is of the form shown in Equation 53.1. The THD can then be computed from Equation 53.2.

Closer inspection of Equations 53.6 and 53.8 reveal that a cosine wave raised to an odd power gives rise to only the fundamental and odd harmonics, with the highest harmonic corresponding to the highest power. A similar phenomenon is observed for a cosine raised to even powers; however, the result is only a dc component and even harmonics without any fundamental component. In fact, any nonlinear system that possesses an odd input-output transfer characteristic $A(x)$ (i.e., the function $A(x)$ is such that $-A(x) = A(-x)$) will give rise to odd harmonics only. Consider Figure 53.2, which illustrates an example of two-sided symmetrical clipping. It is an odd function. The application of a sine wave to its input will result in a waveform similar to that shown in Figure 53.3, which has only odd harmonics as shown in Figure 53.4. The majority of physical systems are neither odd nor even. (An even function is one that has the property $A(x) = A(-x)$; for example, a full-wave rectifier.) Consider the enhancement NMOS transistor illustrated in Figure 53.5, which has the square-law characteristic shown. Assume that the voltage V_{GS} consists of a dc bias plus a sine wave such that V_{GS} is always more positive than V_T (the threshold voltage). Then the current flowing in the drain of this NMOS transistor could have the appearance shown in Figure 53.6. It is observed that the drain current is distorted, since the positive-going side has a greater swing than the negative-going side. The equation for the drain current can be derived mathematically as follows. A MOS transistor operating in its saturation region can be approximated as a square-law device:

$$I_{DS} = \frac{\mu C_{ox}}{2} \frac{W}{L} \left(V_{GS} - V_T \right)^2 \tag{53.9}$$

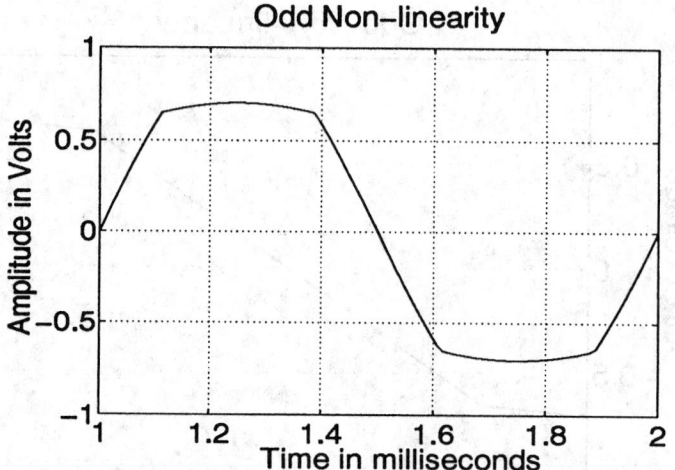

FIGURE 53.3   Distortion due to symmetrical two-sided clipping.

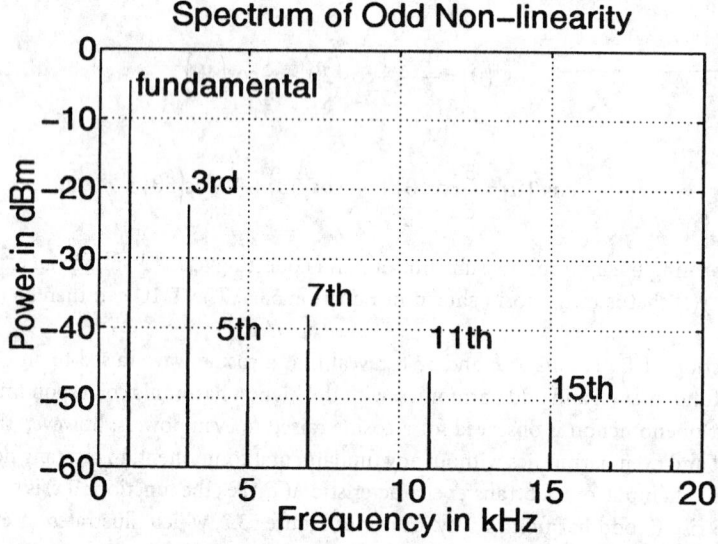

FIGURE 53.4   Frequency spectrum of signal distorted by symmetrical two-sided clipping.

If the gate of the *n*-channel enhancement MOSFET is driven by a voltage source consisting of a sine-wave generator in series with a dc bias, i.e.:

$$V_{GS} = V_B + A_0 \sin(2\pi f_0 t) \qquad (53.10)$$

then the current in the drain can be written as:

$$I_{DS} = \frac{\mu C_{ox}}{2} \frac{W}{L} \left\{ \left[V_B + A_0 \sin(2\pi f_0 t) \right] - V_T \right\}^2 \qquad (53.11)$$

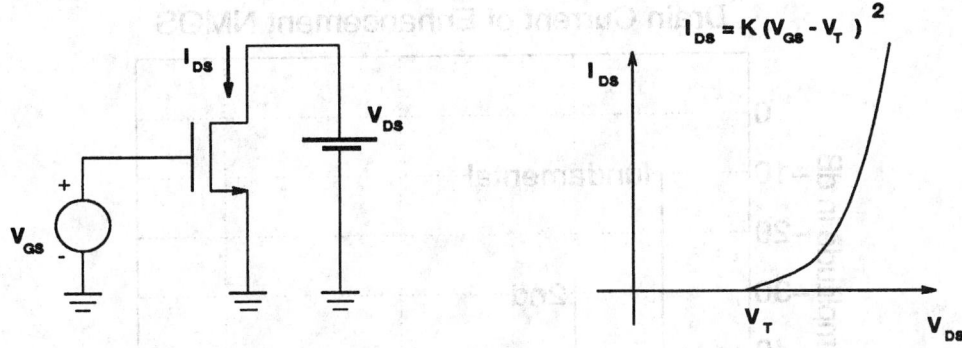

FIGURE 53.5 NMOS enhancement transistor is actually a nonlinear device. It is neither odd nor even in the strict sense.

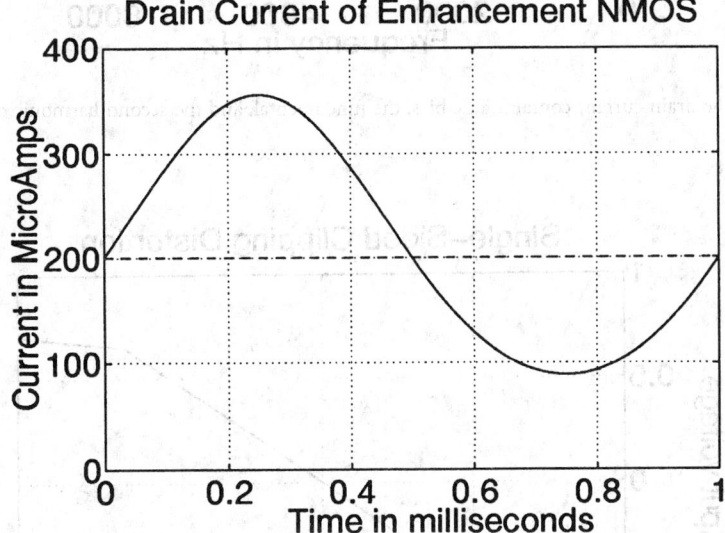

FIGURE 53.6 Showing how the drain current of the enhancement NMOS device is distorted.

Expanding and using the trigonometric relationship:

$$\sin^2(\theta) = \frac{1}{2} - \frac{1}{2}\sin\left(2\theta + \frac{\pi}{2}\right)$$

(53.12)

Equation 53.11 can be rewritten as:

$$I_{DS} = \frac{\mu C_{ox}}{2}\frac{W}{L}\left[\left(V_B - V_T\right)^2 + \frac{A_0^2}{2} + 2\left(V_B - V_T\right)A_0\sin\left(2\pi f_0 t\right) - \frac{A_0^2}{2}\sin\left(4\pi f_0 t\right) + \frac{\pi}{2}\right]$$

(53.13)

which clearly shows the dc bias, the fundamental, and the second harmonic that are visible in the spectrum of the drain current I_{DS} in Figure 53.7. There is one odd harmonic (i.e., the fundamental) and two even harmonics (strictly counting the dc component and the second harmonic). This particular transfer characteristic is neither odd nor even. Finally, for an ideal square-law characteristic, the second harmonic

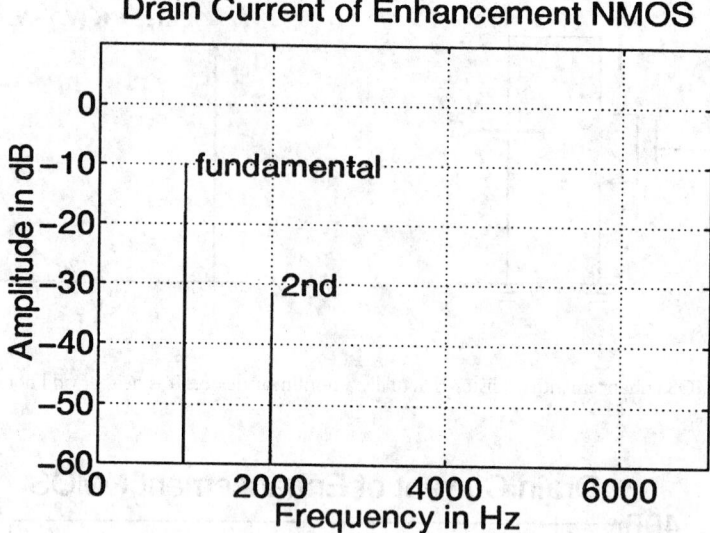

FIGURE 53.7 The drain current contains a dc bias, the fundamental, and the second harmonic only, for an ideal device.

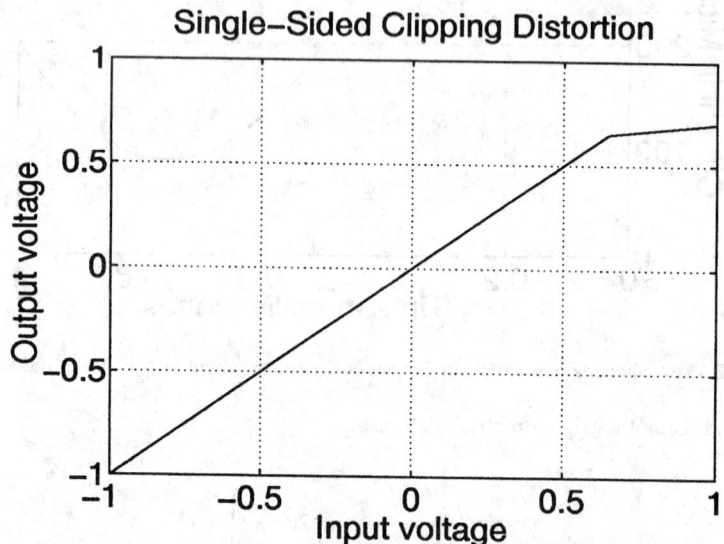

FIGURE 53.8 Single-sided clipping is neither even nor odd.

is the highest frequency component generated in response to a sine-wave input. Another example of a transfer characteristic that is neither even nor odd is single-sided clipping as shown in Figure 53.8, which gives rise to the distortion of Figure 53.9. One last example of an odd input-output transfer characteristic is symmetrical cross-over distortion as depicted in Figure 53.10. The distorted output in response to a 1-kHz sine-wave input is shown in Figure 53.11. The spectrum of the output signal is shown in Figure 53.12. Note that only odd harmonics have been generated.

To round out the discussion, consider a mathematical example wherein the harmonics are derived algebraically. Consider an input-output transfer function $f(x) = c_1 x + c_3 x^3 + c_5 x^5$ that has only odd powers of x. If the input is a cosine $x = A_0 \cos(2\pi f_0 t)$, then the output will be of the form:

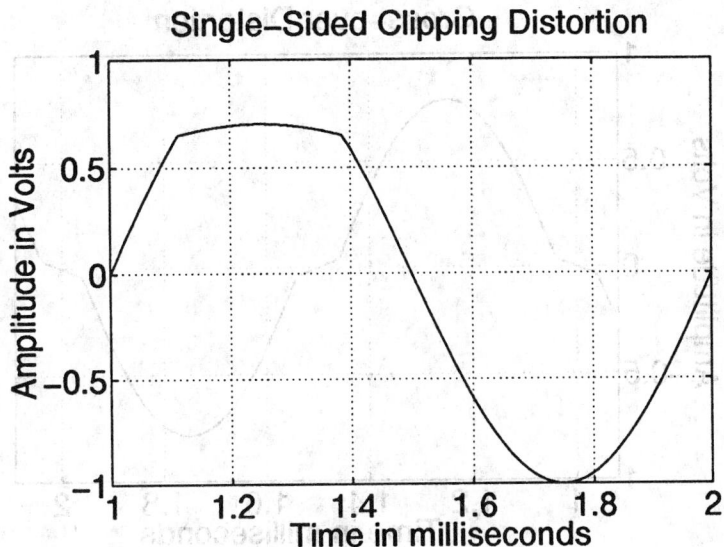

FIGURE 53.9 Distortion due to single-sided clipping.

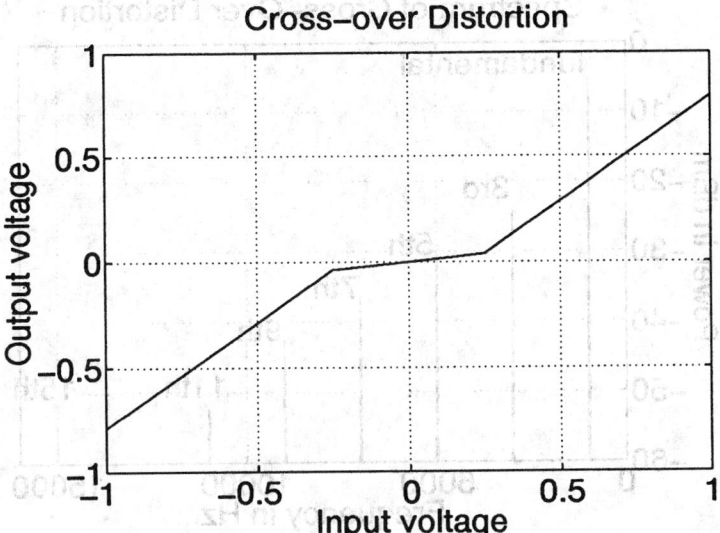

FIGURE 53.10 Symmetrical cross-over distortion is odd.

Clearly only the fundamental plus odd third and fifth harmonics are ... should the exercise be repeated for an input-output transfer characteristic expansion or ... even power of ... then only a dc value ...

$$y(t) = f(x) = c_1 A_0 \cos(2\pi f_0 t) + c_3 A_0^3 \cos^3(2\pi f_0 t) + c_5 A_0^5 \cos^5(2\pi f_0 t) \qquad (53.14)$$

This can be simplified using the trigonometric relationships given in Equations 53.5 through 53.8 with the following result:

$$y(t) = \left(c_1 A_0 + \frac{3c_3 A_0^3}{4} + \frac{5c_5 A_0^5}{8} \right) \cos(2\pi f_0 t) + \left(\frac{c_3 A_0^3}{4} + \frac{5c_5 A_0^5}{16} \right) \cos(2\pi 3 f_0 t) + c_5 A_0^5 \cos^5(2\pi 5 f_0 t) \qquad (53.15)$$

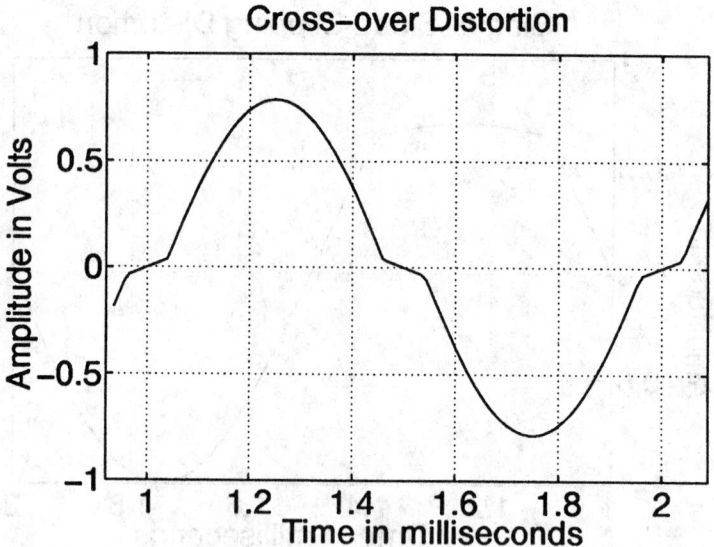

FIGURE 53.11 An example of cross-over distortion.

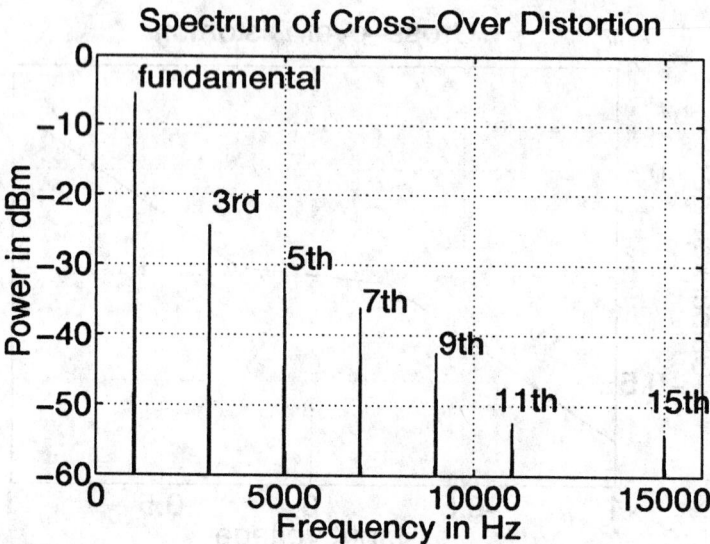

FIGURE 53.12 Symmetrical cross-over distortion gives rise to odd harmonics.

Clearly, only the fundamental plus the third and fifth harmonics are present. Should the exercise be repeated for an input-output transfer function consisting of only even powers of x, then only a dc offset plus even harmonics (not the fundamental) would be present in the output.

53.2 Intercept Points (IP)

It is often desirable to visualize how the various harmonics increase or decrease as the amplitude of the input sine wave $x(t)$ is changed. Consider the example of a signal applied to a nonlinear system $A(x)$ having single-sided clipping distortion as shown in Figure 53.8. The clipping becomes more severe as the amplitude of the input signal $x(t)$ increases in amplitude, so the distortion of the output signal $y(t)$

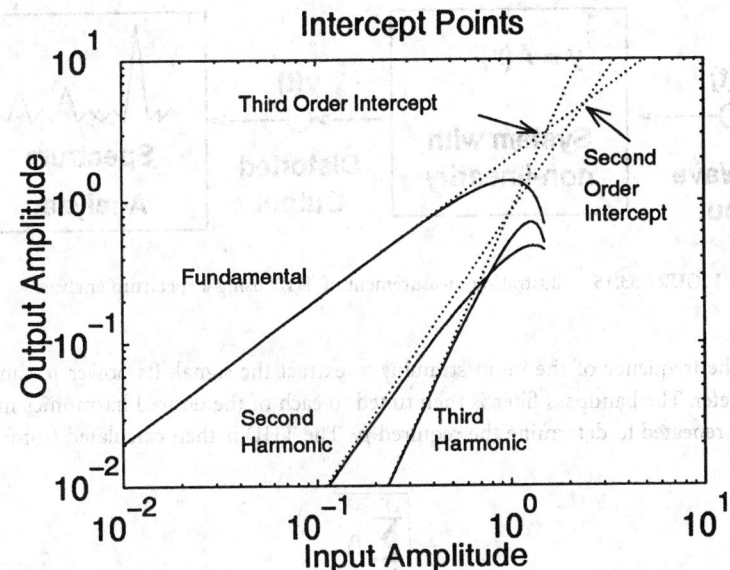

FIGURE 53.13 An example showing the second- and third-order intercept points for a hypothetical system. Both axes are plotted on a logarithmic scale.

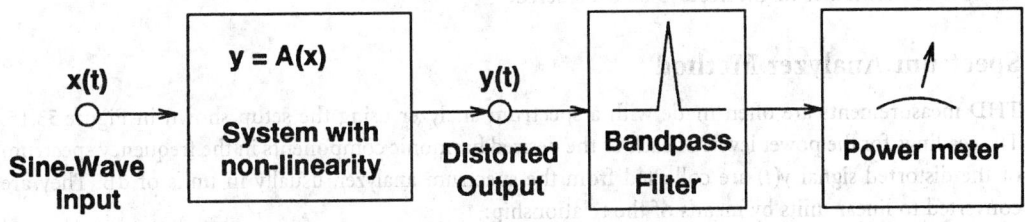

FIGURE 53.14 Illustrating the classical method of measuring THD.

becomes worse. The *intercept point* (IP) is used to provide a figure of merit to quantify this phenomenon. Consider Figure 53.13, which shows an example of the power levels of the first three harmonics of the distorted output $y(t)$ of a hypothetical system $A(x)$ in response to a sine-wave input $x(t)$. It is convenient to plot both axes on a log scale. It can be seen that the power in the harmonics increases more quickly than the power in the fundamental. This is consistent with the observation of how clipping becomes worse as the amplitude increases. It is also consistent with the observation that, in the equations above, the higher harmonics will rapidly become more prominent because they are proportional to higher exponential powers of the input signal amplitude. The intercept point for a particular harmonic is the power level where the extrapolated line for that harmonic intersects with the extrapolated line for the fundamental. The second-order intercept is often abbreviated IP2, the third-order intercept abbreviated IP3, etc.

53.3 Measurement of the THD

Classical Method

The traditional method of measuring THD is shown in Figure 53.14. A sine-wave test stimulus $x(t)$ is applied to the input of the system $A(x)$ under test. The system output $y(t)$ is fed through a bandpass

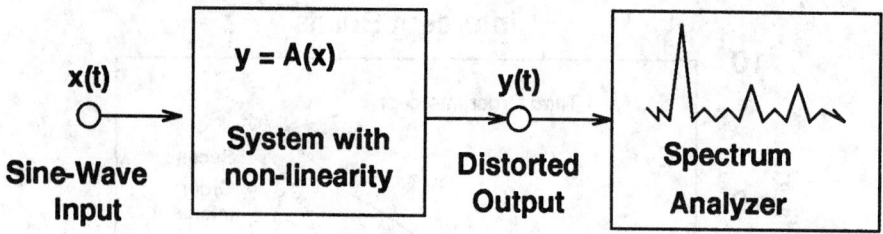

FIGURE 53.15   Illustrating measurement of THD using a spectrum analyzer.

filter tuned to the frequency of the input stimulus to extract the signal. Its power p_1 can be measured with a power meter. The bandpass filter is then tuned to each of the desired harmonics in turn and the measurement is repeated to determine the required p_i. The THD is then calculated from:

$$\text{THD} = \sqrt{\frac{\sum_{k=2}^{N} p_k}{p_1}} \times 100\% \qquad (53.16)$$

In the case of an ordinary audio amplifier, nearly all of the power in the distorted output signal is contained in the first 10 or 11 harmonics. However, in more specialized applications, a much larger number of harmonics might need to be considered.

Spectrum Analyzer Method

THD measurements are often made with a spectrum analyzer using the setup shown in Figure 53.15. The readings for the power levels of each of the desired harmonic components in the frequency spectrum of the distorted signal $y(t)$ are collected from the spectrum analyzer, usually in units of dB. They are converted to linear units by means of the relationship:

$$a_i = 10^{r_i/20} \qquad (53.17)$$

where r_i is the reading for the ith component in dB. The THD is then computed from Equation 53.2. The spectrum analyzer method can be considered as an extension of the classical method described above, except that the spectrum analyzer itself is replacing both the bandpass filter and the power meter.

DSP Method

Digital signal processing (DSP) techniques have recently become popular for use in THD measurement. In this method, the distorted output $y(t)$ is digitized by a precision A/D converter and the samples are stored in the computer's memory as shown in Figure 53.16. One assumes that the samples have been collected with a uniform sample period T_s and that appropriate precautions have been taken with regard to the Nyquist criterion and aliasing. Let $y(n)$ refer to the nth stored sample. A fast Fourier transform (FFT) is executed on the stored data using the relationship:

$$Y(k) = \sum_{n=0}^{N-1} y(n) e^{-j(2\pi/N)kn} \qquad (53.18)$$

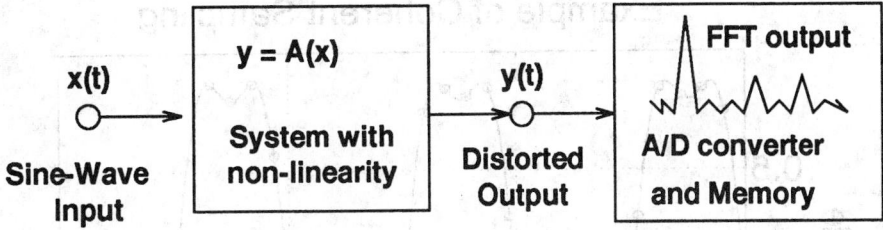

FIGURE 53.16   Illustrating the measurement of THD using FFT.

where N is the number of samples that have been captured. The frequency of the input test stimulus is chosen such that the sampling is coherent. Coherency in this context means that if N samples have been captured, then the input test stimulus is made to be a harmonic of the primitive frequency f_p, which is defined as:

$$f_p = \frac{f_s}{N} = \frac{1}{T_s N}$$ (53.19)

One can view the primitive frequency f_p as the frequency of a sinusoidal signal whose period is exactly equal to the time interval formed by the N-points. Thus, the frequency of the test stimulus can be written as:

$$f_0 = M \times f_p = M \times \frac{f_s}{N} = \frac{M}{N} \times f_s$$ (53.20)

where M and N are integers. To maximize the information content collected by a set of N-points, M and N are selected so that they have no common factors, i.e., relatively prime. This ensures that every sample is taken at a different point on the periodic waveform. An example is provided in Figure 53.17, where $M = 3$ and $N = 32$. The FFT is executed on the distorted signal as per Equation 53.18, and then the THD is computed from:

$$THD = \frac{\sqrt{\sum_{k=2}^{N} |Y(k \times M)|^2}}{|Y(M)|} \times 100\%$$ (53.21)

53.4   Conclusions

The *total harmonic distortion* (THD) is a figure of merit for the quality of the transmission of a signal through a system having some nonlinearity. Its causes and some methods of measuring it have been discussed. Some simple mathematical examples have been presented. However, in real-world systems, it is generally quite difficult to extract all of the parameters c_k in the transfer characteristic. The examples were intended merely to assist the reader's understanding of the relationship between even and odd functions and the harmonics that arise in response to them.

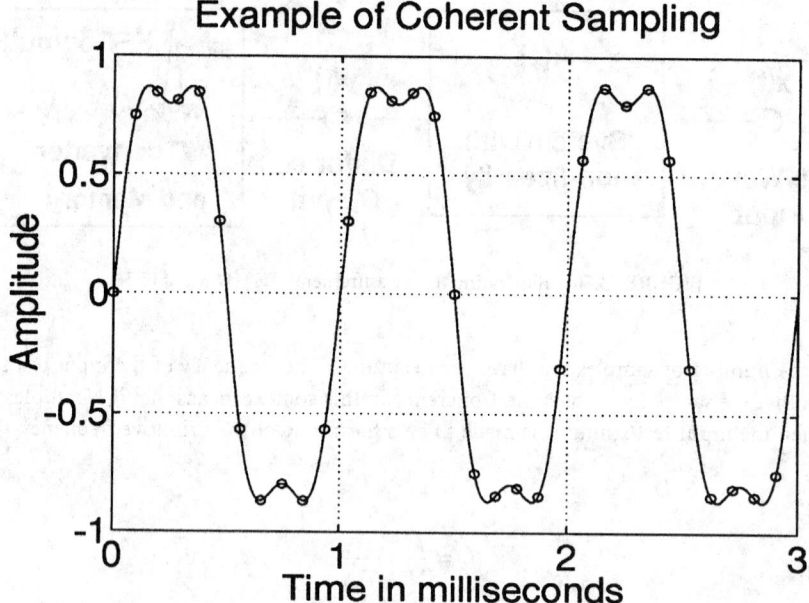

FIGURE 53.17 With coherent sampling, each sample occurs on a unique point of the signal.

Defining Terms

Total harmonic distortion (THD): A numerical figure of merit of the quality of transmission of a signal, defined as the ratio of the power in all the harmonics to the power in the fundamental.

Fundamental: The lowest frequency component of a signal other than zero frequency.

Harmonic: Any frequency component of a signal that is an integer multiple of the fundamental frequency.

Distortion: The effect of corrupting a signal with undesired frequency components.

Nonlinearity: The deviation from the ideal of the transfer function, resulting in such effects as clipping or saturation of the signal.

Further Information

D. O. Pederson and K. Mayaram, *Analog Integrated Circuits for Communications,* New York: Kluwer Academic Press, 1991.

M. Mahoney, *DSP-Based Testing of Analog and Mixed-Signal Circuits,* Los Almos, CA: IEEE Computer Society Press, 1987.

54

Noise Measurement

W. Marshall Leach, Jr.
Georgia Institute of Technology

This chapter describes the principal sources of electric noise and discusses methods for the measurement of noise. The notations for voltages and currents correspond to the following conventions: dc quantities are indicated by an upper-case letter with upper-case subscripts, e.g., V_{BE}. Instantaneous small-signal ac quantities are indicated by a lower-case letter with lower-case subscripts, e.g., v_n. The mean-square value of a variable is denoted by a bar over the square of the variable, e.g., $\overline{v_n^2}$, where the bar indicates an arithmetic average of an ensemble of functions. The root-mean-square or rms value is the square root of the mean-square value. Phasors are indicated by an upper-case letter and lower case subscripts, e.g.,

V_n. Circuit symbols for independent sources are circular, symbols for controlled sources are diamond shaped, and symbols for noise sources are square. In the evaluation of noise equations, Boltzmann's constant is $k = 1.38 \times 10^{-23}$ J K^{-1} and the electronic charge is $q = 1.60 \times 10^{-19}$ C. The standard temperature is denoted by T_0 and is taken to be $T_0 = 290$ K. For this value, $4kT_0 = 1.60 \times 10^{-20}$ J and the thermal voltage is $V_T = kT_0/q = 0.025$ V.

54.1 Thermal Noise

Thermal noise or *Johnson noise* is generated by the random collision of charge carriers with a lattice under conditions of thermal equilibrium [1–7]. Thermal noise in a resistor can be modeled by a series voltage source or a parallel current source having the mean-square values

$$\overline{v_t^2} = 4kTR\Delta f \tag{54.1}$$

$$\overline{i_t^2} = \frac{4kT\Delta f}{R} \tag{54.2}$$

where R is the resistance and Δf is the bandwidth in hertz (Hz) over which the noise is measured. The equation for $\overline{v_t^2}$ is commonly referred to as the *Nyquist formula*. Thermal noise in resistors is independent of the resistor composition.

The *crest factor* for thermal noise is the ratio of the peak value to the rms value. A common definition for the peak value is the level that is exceeded 0.01% of the time. The amplitude distribution of thermal noise is modeled by a gaussian or normal probability density function. The probability that the instantaneous value exceeds 4 times the rms value is approximately 0.01%. Thus, the crest factor is approximately 4.

54.2 Spectral Density

The *spectral density* of a noise signal is defined as the mean-square value per unit bandwidth. For the thermal noise generated by a resistor, the voltage and current spectral densities, respectively, are given by:

$$S_v(f) = 4kTR \tag{54.3}$$

$$S_i(f) = \frac{4kT}{R} \tag{54.4}$$

Because these are independent of frequency, thermal noise is said to have a uniform or flat distribution. Such noise is also called *white noise*. It is called this by analogy to white light, which also has a flat spectral density in the optical band.

54.3 Fluctuation Dissipation Theorem

Consider any system in thermal equilibrium with its surroundings. If there is a mechanism for energy in a particular mode to leak out of that mode to the surroundings in the form of heat, then energy can leak back into that mode from the surrounding heat by the same mechanism. The fluctuation dissipation theorem of quantum mechanics states that the average energy flow in each direction is the same. Otherwise, the system would not be in equilibrium.

Mathematically, the fluctuation dissipation theorem states, in general, that the generalized mean-square force $\overline{\mathfrak{F}^2}$ acting on a system in the frequency band from f_1 to f_2 is given by:

$$\overline{\mathfrak{F}^2} = 4kT \int_{f_1}^{f_2} \text{Re}\left[Z(f)\right] df \qquad (54.5)$$

where Re $[Z(f)]$ is the real part of the system impedance $Z(f)$ and f is the frequency in hertz (Hz). For a mechanical system, the generalized force is the mechanical force on the system and the impedance is force divided by velocity. For an electric system, the generalized force is the voltage and the impedance is the ratio of voltage to current.

Equation 54.1 is a statement of the fluctuation dissipation theorem for a resistor. The theorem can be used to calculate the mean-square thermal noise voltage generated by any two-terminal network containing resistors, capacitors, and inductors. Let $Z(f)$ be the complex impedance of the network. The mean-square open-circuit thermal noise voltage is given by:

$$\overline{v_t^2} = 4kT \int_{f_1}^{f_2} \text{Re}\left[Z(f)\right] df \simeq 4kT \, \text{Re}\left[Z(f)\right] \Delta f \qquad (54.6)$$

where $\Delta f = f_2 - f_1$ and the approximation holds if Re $[Z(f)]$ is approximately constant over the band.

54.4 Equivalent Noise Resistance and Conductance

A mean-square noise voltage can be represented in terms of an *equivalent noise resistance* [8]. Let $\overline{v_n^2}$ be the mean-square noise voltage in the band Δf. The noise resistance R_n is defined as the value of a resistor at the standard temperature $T_0 = 290$ K that generates the same noise. It is given by:

$$R_n = \frac{\overline{v_n^2}}{4kT_0 \Delta f} \qquad (54.7)$$

A mean-square noise current can be represented in terms of an *equivalent noise conductance*. Let $\overline{i_n^2}$ be the mean-square noise current in the band Δf. The noise conductance G_n is defined as the value of a conductance at the standard temperature that generates the same noise. It is given by:

$$G_n = \frac{\overline{i_n^2}}{4kT_0 \Delta f} \qquad (54.8)$$

54.5 Shot Noise

Shot noise is caused by the random emission of electrons and by the random passage of charge carriers across potential barriers [1–7]. The shot noise generated in a device is modeled by a parallel noise current source. The mean-square shot noise current in the frequency band Δf is given by:

$$\overline{i_{sh}^2} = 2qI\Delta f \qquad (54.9)$$

where I is the dc current through the device. This equation is commonly referred to as the *Schottky formula*. Like thermal noise, shot noise is white noise and has a crest formula of approximately 4.

54.6 Flicker Noise

The imperfect contact between two conducting materials causes the conductivity to fluctuate in the presence of a dc current [1–7]. This phenomenon generates what is called *flicker noise* or *contact noise*. It is modeled by a noise current source in parallel with the device. The mean-square flicker noise current in the frequency band Δf is given by:

$$\overline{i_f^2} = \frac{K_f I^m \Delta f}{f^n} \tag{54.10}$$

where K_f is the flicker noise coefficient, I is the dc current, m is the flicker noise exponent, and $n \simeq 1$. Other names for flicker noise are *1/f noise* (read "one-over-f-noise"), *low-frequency noise*, and *pink noise*. The latter comes from the optical analog of pink light, which has a spectral density that increases at lower frequencies.

54.7 Excess Noise

In resistors, flicker noise is caused by the variable contact between particles of the resistive material and is called *excess noise*. Metal film resistors generate the least excess noise, carbon composition resistors generate the most, with carbon film resistors lying between the two. In modeling excess noise, the flicker noise exponent has the value $m = 2$. The mean-square excess noise current is often written as a function of the *noise index NI* as follows:

$$\overline{i_{ex}^2} = \frac{10^{NI/10}}{10^{12} \ln 10} \times \frac{I^2 \Delta f}{f} \tag{54.11}$$

where I is the dc current through the resistor. The noise index is defined as the number of μA of excess noise current in each decade of frequency per A of dc current through the resistor. An equivalent definition is the number of μV of excess noise voltage in each decade of frequency per volt of dc drop across the resistor. In this case, the mean-square excess noise voltage generated by the resistor is given by:

$$\overline{v_{ex}^2} = \frac{10^{NI/10}}{10^{12} \ln 10} \times \frac{V^2 \Delta f}{f} \tag{54.12}$$

where $V = IR$ is the dc voltage across the resistor.

54.8 Burst Noise

Burst noise or *popcorn noise* is caused by a metallic impurity in a *pn* junction [4]. When amplified and reproduced by a loudspeaker, it sounds like corn popping. When viewed on an oscilloscope, it appears as fixed amplitude pulses of randomly varying width and repetition rate. The rate can vary from less than one pulse per minute to several hundred pulses per second. Typically, the amplitude of burst noise is 2 to 100 times that of the background thermal noise.

54.9 Partition Noise

Partition noise occurs when the charge carriers in a current have the possibility of dividing between two or more paths. The noise is generated in the resulting components of the current by the statistical process of partition [9]. Partition noise occurs in BJTs where the current flowing from the emitter into the base can take one of two paths. The recombination of injected carriers in the base region corresponds to the current flow in one path. This current flows in the external base lead. The current carried to the collector corresponds to the current flow in the second path. Because the emitter current exhibits full shot noise, the base and collector currents also exhibit full shot noise. However, the base and collector noise currents are correlated because they have equal and opposite partition components. Partition noise in the BJT can be accounted for if all shot noise is referred to two uncorrelated shot noise current sources, one from base to emitter and the other form collector to emitter [10]. This noise model of the BJT is described here.

54.10 Generation–Recombination Noise

Generation–recombination noise is a semiconductor is generated by the random fluctuation of free carrier densities caused by spontaneous fluctuations in the generation, recombination, and trapping rates [7]. In BJTs, it occurs in the base region at low temperatures. The generation–recombination gives rise to fluctuations in the base spreading resistance which are converted into a noise voltage due to the flow of a base current. In junction FETs, it occurs in the channel at low temperatures. Generation–recombination causes fluctuations of the carrier density in the channel, which gives rise to a noise voltage when a drain current flows. In silicon junction FETs, the effect occurs below 100 K. In germanium junction FETs, it occurs at lower temperatures. The effect does not occur in MOS FETs.

54.11 Noise Bandwidth

The *noise bandwidth* of a filter is the bandwidth of an ideal filter having a constant passband gain which passes the same rms noise voltage, where the input signal is white noise [1–7]. The filter and the ideal filter are assumed to have the same gains. Let $A_v(f)$ be the complex voltage gain transfer function of a filter, where f is the frequency in Hz. Its noise bandwidth B_n is given by:

$$B_n = \frac{1}{A_{vo}^2} \int_0^\infty \left| A_v(f) \right|^2 df \qquad (54.13)$$

where A_{vo} is the maximum value of $\left| A_v(f) \right|$. For a white noise input voltage with the spectral density $S_v(f)$, the mean-square noise voltage at the filter output is $\overline{v_{no}^2} = A_{vo}^2 S_v(f) B_n$.

Two classes of low-pass filters are commonly used in making noise measurements. The first has n real poles, all with the same frequency, having the magnitude-squared transfer function given by:

$$\left| A_v(f) \right|^2 = \frac{A_{vo}^2}{\left[1 + \left(f/f_0 \right)^2 \right]^n} \qquad (54.14)$$

where f_0 is the pole frequency. The second is an n-pole Butterworth filter having the magnitude-squared transfer function given by:

$$\left| A_v(f) \right|^2 = \frac{A_{vo}^2}{1 + \left(f/f_3 \right)^{2n}} \qquad (54.15)$$

TABLE 54.1   Noise Bandwidth B_n of Low-Pass Filters

Number of poles	Slope dB/dec	Real pole B_n		Butterworth B_n
1	20	$1.571\,f_0$	$1.571\,f_3$	$1.571\,f_3$
2	40	$0.785\,f_0$	$1.220\,f_3$	$1.111\,f_3$
3	60	$0.589\,f_0$	$1.155\,f_3$	$1.042\,f_3$
4	80	$0.491\,f_0$	$1.129\,f_3$	$1.026\,f_3$
5	100	$0.420\,f_0$	$1.114\,f_3$	$1.017\,f_3$

where f_3 is the −3 dB frequency. Table 54.1 gives the noise bandwidths as a function of the number of poles n for $1 \le n \le 5$. For the real-pole filter, B_n is given as a function of both f_0 and f_3. For the Butterworth filter, B_n is given as a function of f_3.

54.12   Noise Bandwidth Measurement

The noise bandwidth of a filter can be measured with a white noise source with a known voltage spectral density $S_v(f)$. Let $\overline{v_n^2}$ be the mean-square noise output voltage from the filter when it is driven by the noise source. The noise bandwidth is given by:

$$B_n = \frac{\overline{v_o^2}}{A_{vo}^2 S_v(f)} \tag{54.16}$$

If the spectral density of the source is not known, the noise bandwidth can be determined if another filter with a known noise bandwidth is available. With both filters driven simultaneously, let $\overline{v_{o1}^2}$ and $\overline{v_{o2}^2}$ be the two mean-square noise output voltages, B_{n1} and B_{n2} the two noise bandwidths, and A_{vo1} and A_{vo2} the two maximum gain magnitudes. The noise bandwidth B_{n2} is given by:

$$B_{n2} = B_{n1} \frac{\overline{v_{o2}^2}}{\overline{v_{o1}^2}} \left(\frac{A_{vo1}}{A_{vo2}} \right)^2 \tag{54.17}$$

The white noise source should have an output impedance that is low enough so that the loading effect of the filters does not change the spectral density of the source.

54.13   Spot Noise

Spot noise is the rms noise in a band divided by the square root of the noise bandwidth. For a noise voltage, it has the units $V/\sqrt{Hz}$, which is read "volts per root Hz." For a noise current, the units are $A/\sqrt{Hz}$. For white noise, the spot noise in any band is equal to the square root of the spectral density. Spot noise measurements are usually made with a bandpass filter having a bandwidth that is small enough so that the input spectral density is approximately constant over the filter bandwidth. The spot noise voltage at a filter output is given by $\sqrt{(\overline{v_{no}^2}/B_n)}$, where $\overline{v_{no}^2}$ is the mean-square noise output voltage and B_n is the filter noise bandwidth. The spot noise voltage at the filter input is obtained by dividing the output voltage by A_{vo}, where A_{vo} is the maximum value of $|A_v(f)|$.

A filter that is often used for spot noise measurements is a second-order bandpass filter. The noise bandwidth is given by $B_n = \pi B_3/2$, where B_3 is the −3 dB bandwidth. A single-pole high-pass filter having a pole frequency f_1 cascaded with a single-pole low-pass filter having a pole frequency f_2 is a special case of bandpass filter having two real poles. Its noise bandwidth is given by $B_n = \pi(f_1 + f_2)/2$. The −3 dB bandwidth in this case is $f_1 + f_2$, not $f_2 - f_1$.

54.14   Addition of Noise Voltages

Consider the instantaneous voltage $v = v_n + i_n R$, where v_n is a noise voltage and i_n is a noise current. The mean-square voltage is calculated as follows:

$$\overline{v^2} = \overline{\left(v_n + i_n R\right)^2} = \overline{v_n^2} + 2\rho\sqrt{\overline{v_n^2}}\sqrt{\overline{i_n^2}}R + \overline{i_n^2}R^2 \qquad (54.18)$$

where ρ is the *correlation coefficient* defined by:

$$\rho = \frac{\overline{v_n i_n}}{\sqrt{\overline{v_n^2}}\sqrt{\overline{v_n^2}}} \qquad (54.19)$$

For the case $\rho = 0$, the sources are said to be uncorrelated or independent. It can be shown that $-1 \le \rho \le 1$.

In ac circuit analysis, noise signals are often represented by phasors. The square magnitude of the phasor represents the mean-square value at the frequency of analysis. Consider the phasor voltage $V = V_n + I_n Z$, where V_n is a noise phasor voltage, I_n is a noise phasor current, and $Z = R + jX$ is a complex impedance. The mean-square voltage is given by:

$$\overline{v^2} = \overline{\left(V_n + I_n R\right)\left(V_n^* + I_n^* Z^*\right)}$$

$$= \overline{v_n^2} + 2\sqrt{\overline{v_n^2}}\sqrt{\overline{i_n^2}}\,\mathrm{Re}\left(\gamma Z^*\right) + \overline{i_n^2}\left|Z\right|^2 \qquad (54.20)$$

where the * denotes the complex conjugate and γ is the *complex correlation coefficient* defined by:

$$\gamma = \gamma_r + j\gamma_i = \frac{\overline{V_n I_n^*}}{\sqrt{\overline{v_n^2}}\sqrt{\overline{i_n^2}}} \qquad (54.21)$$

It can be shown that $\left|\gamma\right| \le 1$.

Noise equations derived by phasor analysis can be converted easily into equations for real signals. However, the procedure generally cannot be done in reverse. For this reason, noise formulas derived by phasor analysis are the more general form.

54.15   Correlation Impedance and Admittance

The *correlation impedance* Z_γ and *correlation admittance* Y_γ between a noise phasor voltage V_n and a noise phasor current I_n are defined by [8]:

$$Z_\gamma = R_\gamma + jX_\gamma = \gamma\sqrt{\frac{\overline{v_n^2}}{\overline{i_n^2}}} \qquad (54.22)$$

$$Y_\gamma = G_\gamma + jB_\gamma = \gamma^*\sqrt{\frac{\overline{i_n^2}}{\overline{v_n^2}}} \qquad (54.23)$$

where $\overline{v_n^2}$ is the mean-square value of V_n, $\overline{i_n^2}$ is the mean-square value of I_n, and γ is the complex correlation coefficient between V_n and I_n. With these definitions, it follows that $\overline{V_n I_n^*} = \overline{i_n^2}Z_\gamma = \overline{v_n^2}\,Y_\gamma^*$.

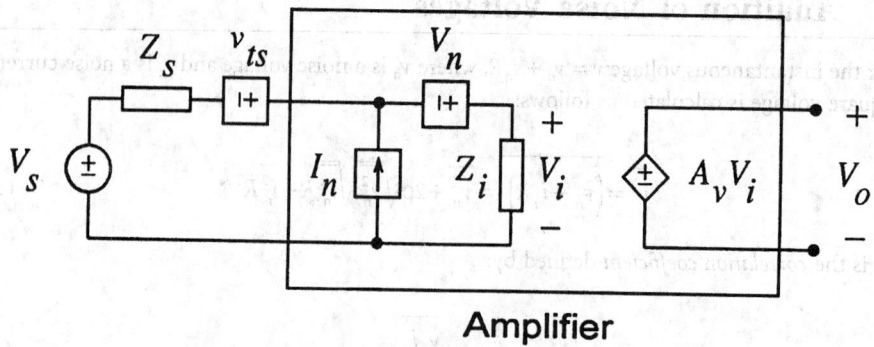

FIGURE 54.1 Amplifier $v_n - i_n$ noise model.

54.16 The $v_n - i_n$ Amplifier Noise Model

The noise generated by an amplifier can be modeled by referring all internal noise sources to the input [1–7], [11]. In order for the noise sources to be independent of the source impedance, two sources are required — a series voltage source and a shunt current source. In general, the sources are correlated.

Figure 54.1 shows the amplifier noise model, where V_s is the source voltage, $Z_s = R_s + jX_s$ is the source impedance, V_{ts} is the thermal noise voltage generated by R_s, $A_v = V_o/V_i$ is the complex voltage gain, and Z_i is the input impedance. The output voltage is given by:

$$V_o = \frac{A_v Z_i}{Z_s + Z_i}\left(V_s + V_{ts} + V_n + I_n Z_s\right) \tag{54.24}$$

The *equivalent noise input voltage* is the voltage in series with V_s that generates the same noise voltage at the output as all noise sources in the circuit. It is given by $V_{ni} = V_{ts} + V_n + I_n Z_s$. The mean-square value is:

$$\overline{v_{ni}^2} = 4kTR_s B_n + \overline{v_n^2} + 2\sqrt{\overline{v_n^2}}\sqrt{\overline{i_n^2}}\,\mathrm{Re}\!\left(\gamma Z_s^*\right) + \overline{i_n^2}\left|Z_s\right|^2 \tag{54.25}$$

where B_n is the amplifier noise bandwidth and γ is the complex correlation between V_n and I_n. For $|Z_s|$ very small, $\overline{v_{ni}^2} \simeq \overline{v_n^2}$ and γ is not important. Similarly, for $|Z_s|$ very large, $\overline{v_{ni}^2} \simeq \overline{i_n^2}\,|Z_s|^2$ and γ is again not important.

When the source is represented by a Norton equivalent consisting of a source current i_s in parallel with a source admittance $Y_s = G_s + jB_s$, the noise referred to the input must be represented by an *equivalent noise input current*. The mean-square value is given by:

$$\overline{i_{ni}^2} = 4kTG_s B_n + \overline{v_n^2}\left|Y_s\right|^2 + 2\sqrt{\overline{v_n^2}}\sqrt{\overline{i_n^2}}\,\mathrm{Re}\!\left(\gamma Y_s\right) + \overline{i_n^2} \tag{54.26}$$

54.17 Measuring $\overline{v_{ni}^2}$, $\overline{v_n^2}$, and $\overline{i_n^2}$

For a given Z_s, the mean-square equivalent noise input voltage can be measured by setting $V_s = 0$ and measuring the mean-square noise output voltage $\overline{v_{no}^2}$. It follows that $\overline{v_{ni}^2}$ is given by:

$$\overline{v_{ni}^2} = \frac{\overline{v_{no}^2}}{|A_v|^2} \times \left|1 + \frac{Z_s}{Z_i}\right|^2 \tag{54.27}$$

To measure $\overline{v_n^2}$ and $\overline{i_n^2}$, $\overline{v_{no}^2}$ is measured with $Z_s = 0$ and with Z_s replaced by a large-value resistor. It follows that $\overline{v_n^2}$ and $\overline{i_n^2}$ are then given by:

$$\overline{v_n^2} = \frac{\overline{v_{no}^2}}{|A_v|^2} \quad \text{for} \quad Z_s = 0 \tag{54.28}$$

$$\overline{i_n^2} = \left|\frac{1}{R_s} + \frac{1}{Z_i}\right|^2 \frac{\overline{v_{no}^2}}{|A_v|^2} \quad \text{for} \quad Z_s = R_s \text{ and } R_s \text{ large} \tag{54.29}$$

54.18 Noise Temperature

The internal noise generated by an amplifier can be expressed as an equivalent *input-termination noise temperature* [12]. This is the temperature of the source resistance that generates a thermal noise voltage equal to the internal noise generated in the amplifier when referred to the input. The noise temperature T_n is given by:

$$T_n = \frac{\overline{v_{ni}^2}}{4kR_s B_n} - T \tag{54.30}$$

where $\overline{v_{ni}^2}$ is the mean-square equivalent input noise voltage in the band B_n, R_s is the real part of the source output impedance, and T is the temperature of R_s.

54.19 Noise Reduction with a Transformer

Let a transformer be connected between the source and the amplifier in Figure 54.1. Let n be the transformer turns ratio, R_1 the primary resistance, and R_2 the secondary resistance. The equivalent noise input voltage in series with the source voltage V_s has the mean-square value:

$$\overline{v_{ni}^2} = 4kT\left(R_s + R_1 + \frac{R_2}{n^2}\right)\Delta f + \frac{\overline{v_n^2}}{n^2}$$

$$\tag{54.31}$$

$$+ 2\sqrt{\overline{v_n^2}}\sqrt{\overline{i_n^2}}\,\text{Re}\left[\gamma\left(Z_s^* + R_1 + \frac{R_2}{n^2}\right)\right] + n^2\overline{i_n^2}\left|Z_s + R_1 + \frac{R_2}{n^2}\right|^2$$

In general, $R_2/R_1 \propto n$, which makes it difficult to specify the value of n that minimizes $\overline{v_{ni}^2}$. For $R_1 + R_2/n^2 \ll |Z_s|$, it is minimized when:

$$n^2 = \frac{1}{|Z_s|}\sqrt{\frac{\overline{v_n^2}}{\overline{i_n^2}}} \tag{54.32}$$

54.20   The Signal-to-Noise Ratio

The *signal-to-noise ratio* of an amplifier is defined by:

$$\text{SNR} = \frac{\overline{v_{so}^2}}{\overline{v_{no}^2}} \tag{54.33}$$

where $\overline{v_{so}^2}$ is the mean-square signal output voltage and $\overline{v_{no}^2}$ is the mean-square noise output voltage. The SNR is often expressed in dB with the equation $\text{SNR} = 10\log\left(\overline{v_{so}^2}/\overline{v_{no}^2}\right)$. In measuring the SNR, a filter should be used to limit the bandwidth of the output noise to the signal bandwidth of interest. An alternative definition of the SNR that is useful in making calculations is:

$$\text{SNR} = \frac{\overline{v_s^2}}{\overline{v_{ni}^2}} \tag{54.34}$$

where $\overline{v_s^2}$ is the mean-square signal input voltage and $\overline{v_{ni}^2}$ is the mean-square equivalent noise input voltage.

54.21   Noise Factor and Noise Figure

The *noise factor F* of an amplifier is defined by [1–8]:

$$F = \frac{\overline{v_{no}^2}}{\overline{v_{nos}^2}} \tag{54.35}$$

where $\overline{v_{no}^2}$ is the mean-square noise output voltage with the source voltage zeroed and $\overline{v_{nos}^2}$ is the mean-square noise output voltage considering the only source of noise to be the thermal noise generated by the source resistance R_s. The *noise figure* is the noise factor expressed in dB and is given by:

$$\text{NF} = 10\log F \tag{54.36}$$

For the $v_n - i_n$ amplifier noise model, the noise factor is given by:

$$F = \frac{\overline{v_{ni}^2}}{4kT\,R_s B_n} = 1 + \frac{\overline{v_n^2} + 2\sqrt{\overline{v_n^2}}\,\sqrt{\overline{i_n^2}}\,\text{Re}\left(\gamma Z_s^*\right) + \overline{i_n^2}\left|Z_s\right|^2}{4kT\,R_s B_n} \tag{54.37}$$

where B_n is the amplifier noise bandwidth. The value of Z_s that minimizes the noise figure is called the *optimum source impedance* and is given by:

$$Z_{so} = R_{so} + jX_{so} = \left[\sqrt{1 - \gamma_i^2} - j\gamma_i\right]\sqrt{\frac{\overline{v_n^2}}{\overline{i_n^2}}} \tag{54.38}$$

where $\gamma_i = \text{Im}(\gamma)$. The corresponding value of F is denoted by F_0 and is given by:

$$F_0 = 1 + \frac{\sqrt{\overline{v_n^2}}\sqrt{\overline{i_n^2}}}{2kTB_n}\left(\gamma_r + j\sqrt{1-\gamma_i^2}\right) \tag{54.39}$$

It follows that F can be expressed in terms of F_0 as follows:

$$F = F_0 + \frac{G_n}{R_{ns}}\left[\left(R_s - R_{so}\right)^2 + \left(X_s - X_{so}\right)^2\right] \tag{54.40}$$

where G_n is the noise conductance of I_n and R_{ns} is the noise resistance of the source. These are given by:

$$G_n = \frac{\overline{i_n^2}}{4kT_0B_n} \tag{54.41}$$

$$R_{ns} = \frac{\overline{v_{ts}^2}}{4kT_0B_n} = \frac{T\,R_s}{T_0} \tag{54.42}$$

When the source is represented by a Norton equivalent consisting of a source current i_s in parallel with a source admittance $Y_s = G_s + jB_s$, the *optimum source admittance* is given by:

$$Y_{so} = G_{so} + jB_{so} = \left[\sqrt{1-\gamma_i^2} + j\gamma_i\right]\sqrt{\frac{\overline{i_n^2}}{\overline{v_n^2}}} \tag{54.43}$$

which is the reciprocal of Z_{so}. The noise factor can be written as:

$$F = F_0 + \frac{R_n}{G_{ns}}\left[\left(G_s - G_{so}\right)^2 + \left(B_s - B_{so}\right)^2\right] \tag{54.44}$$

where R_n is the noise resistance of V_n and G_{ns} is the noise conductance of the source. These are given by:

$$R_n = \frac{\overline{v_n^2}}{4kT_0B_n} \tag{54.45}$$

$$G_{ns} = \frac{\overline{i_{ts}^2}}{4kT_0B_n} = \frac{TG_s}{T_0} \tag{54.46}$$

54.22 Noise Factor Measurement

The noise factor can be measured with a calibrated white noise source driving the amplifier. The source output impedance must equal the value of Z_s for which F is to be measured. The source temperature must be the standard temperature T_0. First, measure the amplifier noise output voltage over the band of interest with the source voltage set to zero. For the amplifier model of Figure 54.1, the mean-square value of this voltage is given by:

$$\overline{v_{nol}^2} = \left|\frac{A_{vo}Z_i}{Z_s + Z_i}\right|^2 \left[4kT_0R_sB_n + \overline{v_n^2} + 2\sqrt{\overline{v_n^2}}\sqrt{\overline{i_n^2}}\,\text{Re}\left(\gamma Z_s^*\right) + \overline{i_n^2}\left|Z_s\right|^2\right] \tag{54.47}$$

The source noise voltage is then increased until the output voltage increases by a factor r. The new mean-square output voltage can be written as:

$$r^2\overline{v_{nol}^2} = \left|\frac{A_{vo}Z_i}{Z_s + Z_i}\right|^2 \left[\left(S_v(f) + 4kT_0R_s\right)B_n + \overline{v_n^2} + 2\sqrt{\overline{v_n^2}}\sqrt{\overline{i_n^2}}\,\text{Re}\left(\gamma Z_s^*\right) + \overline{i_n^2}\left|Z_s\right|^2\right] \tag{54.48}$$

where $S_v(f)$ is the open-circuit voltage spectral density of the white noise source.

The above two equations can be solved for F to obtain:

$$F = \frac{S_s(f)}{\left(r^2 - 1\right)4kT_0R_s} \tag{54.49}$$

A common value for r is $\sqrt{2}$. The gain and noise bandwidth of the amplifier are not needed for the calculation. If a resistive voltage divider is used between the noise source and the amplifier to attenuate the input signal, the source spectral density $S_s(f)$ is calculated or measured at the attenuator output with it disconnected from the amplifier input.

If the noise bandwidth of the amplifier is known, its noise factor can be determined by measuring the mean-square noise output voltage $\overline{v_{no}^2}$ with the source voltage set to zero. The noise factor is given by:

$$F = \left|1 + \frac{Z_s}{Z_i}\right|^2 \frac{\overline{v_{no}^2}}{4kT_0R_sB_nA_{vo}^2} \tag{54.50}$$

This expression is often used with $B_n = \pi B_3/2$, where B_3 is the –3 dB bandwidth. Unless the amplifier has a first-order low-pass or a second-order bandpass frequency response characteristic, this is only an approximation.

54.23 The Junction Diode Noise Model

When forward biased, a diode generates both shot noise and flicker noise [1–7]. The noise is modeled by a parallel current source having the mean-square value:

$$\overline{i_n^2} = 2qI\Delta f + \frac{K_f I^m \Delta f}{f} \tag{54.51}$$

where I is the dc diode current. A plot of $\overline{i_n^2}$ vs. f for a constant Δf exhibits a slope of –10 dB/decade for low frequencies and a slope of zero for higher frequencies.

Diodes are often used as noise sources in circuits. Specially processed zener diodes are marketed as solid-state noise diodes. The noise mechanism in these is called *avalanche noise*, which is associated with

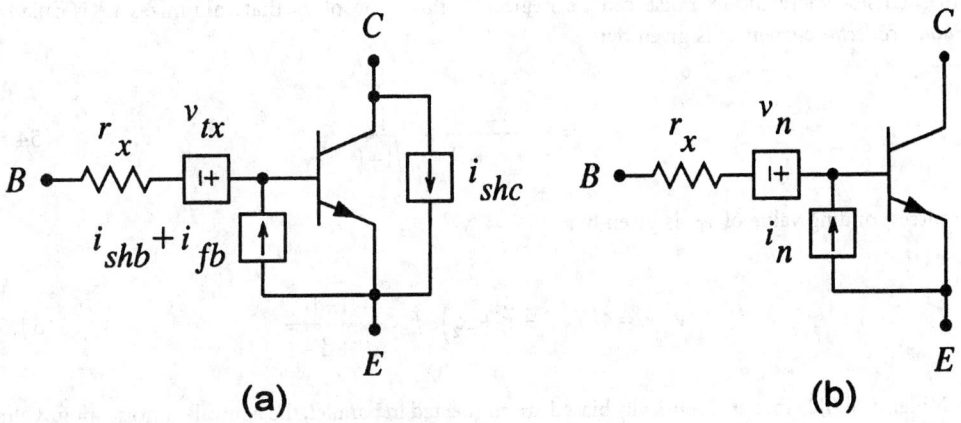

FIGURE 54.2 (a) BJT noise model. (b) BJT $v_n - i_n$ noise model.

the diode reverse breakdown current. For a given breakdown current, avalanche noise is much greater than shot noise in the same current.

54.24 The BJT Noise Model

Figure 54.2(a) shows the BJT noise model [1–7]. The base spreading resistance r_x is modeled as an external resistor; v_{tx} is the thermal noise generated by r_x; i_{shb} and i_{fb}, respectively, are the shot noise and flicker noise in the base bias current I_B; and i_{shc} is the shot noise in the collector bias current I_C. The sources have the mean-square values of:

$$\overline{v_{tx}^2} = 4kTr_x\Delta f \tag{54.52}$$

$$\overline{i_{shb}^2} = 2qI_B\Delta f \tag{54.53}$$

$$\overline{i_{fb}^2} = \frac{K_f I_B^m \Delta f}{f} \tag{54.54}$$

$$\overline{i_{shc}^2} = 2qI_C\Delta f \tag{54.55}$$

Let the resistances to signal ground seen looking out of the base and the emitter, respectively, be denoted by R_1 and R_2. The mean-square equivalent noise input voltages in series with the base or the emitter that generates the same collector noise current is given by:

$$\overline{v_{ni}^2} = 4kT\left(R_1 + r_x + R_2\right)\Delta f + \left(2qI_B\Delta f + \frac{K_f I_B\Delta f}{f}\right)\left(R_1 + r_x + R_2\right)^2 \tag{54.56}$$

$$+ 2qI_C\Delta f\left(\frac{R_1 + r_x + R_2}{\beta} + \frac{V_T}{I_C}\right)^2$$

At frequencies where flicker noise can be neglected, the value of I_C that minimizes $\overline{v_{ni}^2}$ is called the *optimum collector current*. It is given by:

$$I_{C_{opt}} = \frac{V_T}{R_1 + r_x + R_2} \times \frac{\beta}{\sqrt{1+\beta}} \tag{54.57}$$

The corresponding value of $\overline{v_{ni}^2}$ is given by:

$$\overline{v_{ni_{min}}^2} = 4kT\left(R_1 + r_x + R_2\right)\Delta f \times \frac{\sqrt{1+\beta}}{\sqrt{1+\beta}-1} \tag{54.58}$$

If N identical BJTs that are identically biased are connected in parallel, the equivalent noise input voltage is given by Equation 54.56 with r_x replaced with r_x/N, I_B replaced with NI_B, and I_C replaced with NI_C. In this case, R_1 and R_2, respectively, are the resistances to signal ground seen looking out of the parallel connected bases and the parallel connected emitters. For N fixed, the value of I_C that minimizes $\overline{v_{no}^2}$ is given by Equation 54.57 with R_1 replaced with NR_1 and R_2 replaced with NR_2. The corresponding value of $\overline{v_{ni_{min}}^2}$ is given by Equation 54.58 with r_x replaced with r_x/N. It follows that parallel connection of BJTs can be used to reduce the thermal noise of r_x, provided the devices are optimally biased.

The BJT $v_n - i_n$ noise model is given in Figure 54.2(b), where r_x is a noiseless resistor, for its thermal noise is included in v_n. The mean-square values of v_n and i_n and the correlation coefficient are given by:

$$\overline{v_n^2} = 4kTr_x\Delta f + 2kT\frac{V_T}{I_C}\Delta f \tag{54.59}$$

$$\overline{i_n^2} = 2qI_B\Delta f + \frac{K_f I_B^m \Delta f}{f} + \frac{2qI_C\Delta f}{\beta^2} \tag{54.60}$$

$$\rho = \frac{2kT\Delta f}{\beta\sqrt{\overline{v_n^2}}\sqrt{\overline{i_n^2}}} \tag{54.61}$$

where $\beta = I_C/I_B$ is the current gain. An alternative model puts r_x inside the BJT. For this model, the expressions for i_n and ρ are more complicated than the ones given here.

The $v_n - i_n$ noise model of Figure 54.1 does not have a noiseless resistor in series with its input. Before formulae that are derived for this model are applied to the BJT model of Figure 54.2(b), the formulae must be modified to account for the noiseless r_x. For the common-emitter amplifier, for example, the source resistance R_s would be replaced in the expression for $\overline{v_{ni}^2}$ by $R_s + r_x$ in all occurrences except in terms that represent the thermal noise of R_s.

The value of the base spreading resistance r_x depends on the method used to measure it. For noise calculations, r_x should be measured with a noise technique. A test circuit for measuring r_x is shown in Figure 54.3. The emitter bias current I_E and the collector bias voltage V_C are given by:

$$I_E = \frac{-V_{BE} - V_{EE}}{R_E} \tag{54.62}$$

$$V_C = V_{CC} - \alpha I_E R_C \tag{54.63}$$

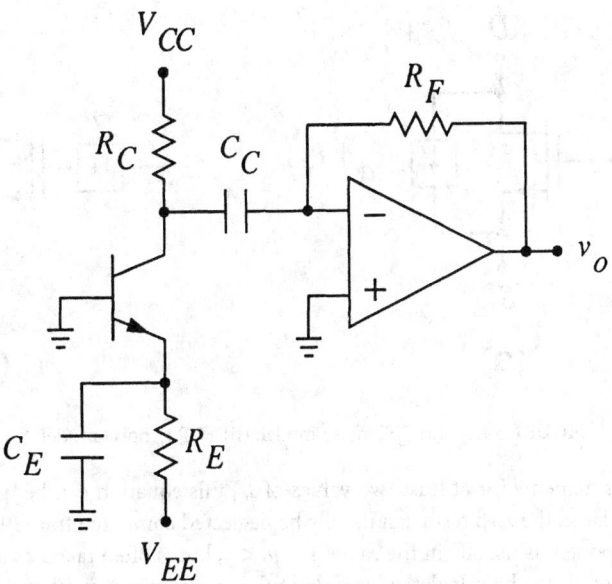

FIGURE 54.3 Test circuit for measuring r_x of a BJT.

where $\alpha = \beta/(1 + \beta)$. Capacitors C_1 and C_2 should satisfy $C_1 \gg I_E/(2\pi f V_T)$ and $C_2 \gg 1/(2\pi f R_C)$, where f is the lowest frequency of interest. To minimize the noise contributed by R_C, R_F, and the op-amp, a low-noise op-amp should be used and R_F should be much larger than R_C. The power supply rails must be properly decoupled to minimize power supply noise.

To prevent flicker noise from affecting the data, the op-amp output voltage must be measured over a noise bandwidth where the spectral density is white. Denote the mean-square op-amp output voltage over the band B_n with the BJT in the circuit by $\overline{v_{no1}^2}$. Denote the mean-square voltage over the band B_n with the BJT removed by $\overline{v_{no2}^2}$. The base spreading resistance r_x can be obtained by solving:

$$r_x^2\left[\frac{A}{\beta^2} - 2qI_B B_n\right] + r_x\left[\frac{2AV_T}{\beta I_C} - 4kT\,B_n\right] + \frac{AV_T^2}{I_C^2} = 0 \tag{54.64}$$

Where:

$$A = \frac{\overline{v_{no1}^2} - \overline{v_{no2}^2}}{R_F^2} - 2qI_C B_n \tag{54.65}$$

The test circuit of Figure 54.3 can be used to measure the flicker noise coefficient K_f and the flicker noise exponent m. The plot of $(\overline{v_{no1}^2} - \overline{v_{no2}^2})$ vs. frequency for a constant noise bandwidth Δf must be obtained, e.g., with a signal analyzer. In the white noise range, the slope of the plot is zero. In the flicker noise range, the slope is -10 dB per decade. The lower frequency at which $(\overline{v_{no1}^2} - \overline{v_{no2}^2})$ is 3 dB greater than its value in the white noise range is the flicker noise corner frequency f_f. It can be shown that:

$$K_f I_B^m = \frac{\left(\overline{v_{no1}^2} - \overline{v_{no2}^2}\right) f_f}{2R_F^2 \Delta f} \times \left(\frac{r_x}{\beta} + \frac{V_T}{I_C}\right)^2 \tag{54.66}$$

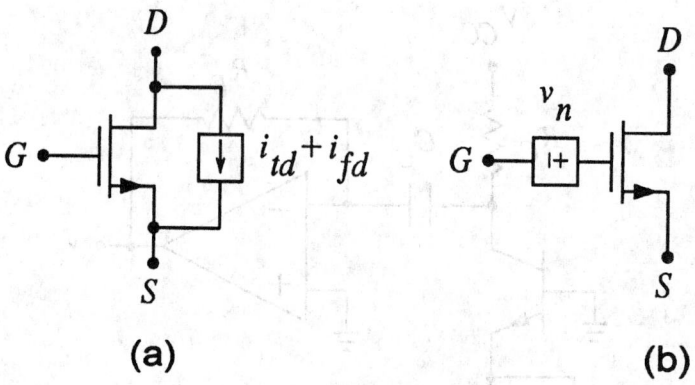

FIGURE 54.4 (a) FET noise model. (b) FET v_n noise model.

By repeating the measurements for at least two values of I_C, this equation can be used to solve for both K_f and m. Unless I_C is large, the r_x/β term can usually be neglected compared to the V_T/I_C term. The value of the flicker noise exponent is usually in the range $1 < m < 3$, but is often taken as unity. If it is assumed that $m = 1$, the value of K_f can be calculated by making the measurements with only one value of I_C.

54.25 The FET Noise Model

Figure 54.4(a) shows the MOSFET noise equivalent circuit, where i_{td} is the channel thermal noise current and i_{fd} is the channel flicker noise current [1–7]. The mean-square values of these currents are given by:

$$\overline{i_{td}^2} = \frac{8kT\Delta f}{3g_m} \tag{54.67}$$

$$\overline{i_{fd}^2} = \frac{K_f\Delta f}{4KfL^2C_{ox}} \tag{54.68}$$

where K is the transconductance parameter, $g_m = 2\sqrt{KI_D}$ is the transconductance, L is the effective length of the channel, and C_{ox} is the gate oxide capacitance per unit area.

Let the resistances to signal ground seen looking out of the gate and the source, respectively, be denoted by R_1 and R_2. The mean-square equivalent noise input voltage in series with either the gate or the source that generates the same drain noise current is given by:

$$\overline{v_{ni}^2} = 4kT\left(R_1 + R_2\right)\Delta f + \frac{4kT\Delta f}{3\sqrt{KI_D}} + \frac{K_f\Delta f}{4KL^2C_{ox}f} \tag{54.69}$$

where it is assumed that the MOSFET bulk is connected to its source in the ac circuit. If N identical MOSFETs that are identically biased are connected in parallel, the equivalent noise input voltage is given by Equation 54.69 with the exception that the second and third terms are divided by N.

The noise sources in Figure 54.4(a) can be reflected into a single source in series with the gate. The circuit is shown in Figure 54.4(b). The mean-square value of v_n is given by:

$$\overline{v_n^2} = \frac{8kT\Delta f}{3g_m} + \frac{K_f\Delta f}{4KfL^2C_{ox}} \tag{54.70}$$

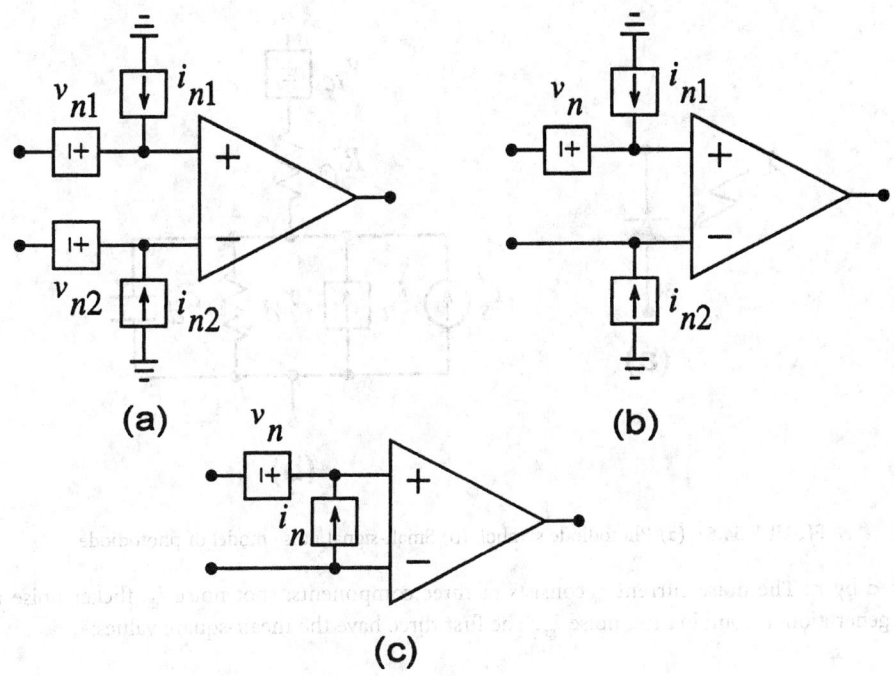

FIGURE 54.5 Op-amp noise models.

The FET flicker noise coefficient K_f can be measured by replacing the BJT in Figure 54.3 with the FET. On a plot of $(\overline{v_{no1}^2} - \overline{v_{no2}^2})$ as a function of frequency for a constant noise bandwidth, the flicker noise corner frequency f_f is the lower frequency at which $(\overline{v_{no1}^2} - \overline{v_{no2}^2})$ is up 3 dB above the white noise level. A signal analyzer can be used to display this plot. The flicker noise coefficient is given by:

$$K_f = \frac{32kT\,K\,f_f L^2 C_{ox}}{3g_m} \tag{54.71}$$

The MOSFET circuits and equations also apply to the junction FET with the exception that the L^2 and C_{ox} terms are omitted from the formulae. This assumes that the junction FET gate-to-channel junction is reverse biased, which is the usual case. Otherwise, shot noise in the gate current must be modeled.

54.26 Operational Amplifier Noise Models

Variations of the $v_n - i_n$ amplifier noise model are used in specifying op-amp noise performance. The three most common models are given in Figure 54.5. In Figures 54.5(b) and (c), v_n can be placed in series with either input [4, 6]. In general, the sources in each model are correlated. In making calculations that use specified op-amp noise data, it is important to use the noise model for which the data apply.

54.27 Photodiode Detector Noise Model

Figure 54.6(a) shows the circuit symbol of a photodiode detector [4]. When reverse biased by a dc source, an incident light signal causes a signal current to flow in the diode. The diode small-signal noise model is shown in Figure 54.6(b), where i_s is the signal current (proportional to the incident light intensity), i_n is the diode noise current, r_d is the small-signal resistance of the reverse-biased junction, c_d is the small-signal junction capacitance, r_c is the cell resistance (typically <50 Ω), and v_{tc} is the thermal noise voltage

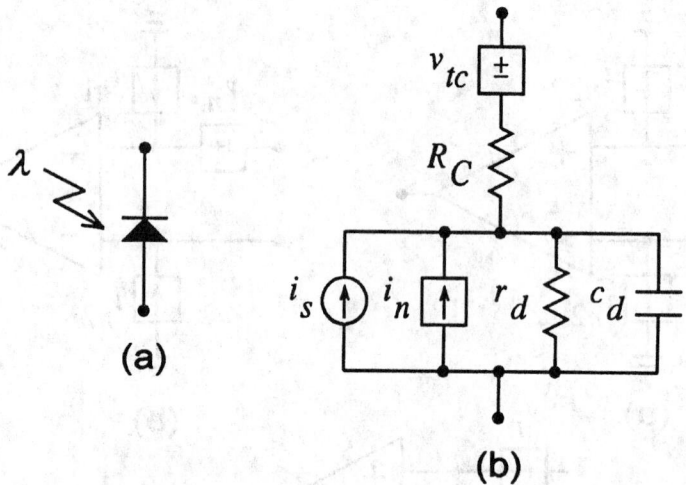

FIGURE 54.6   (a) Photodiode symbol. (b) Small-signal noise model of photodiode.

generated by r_c. The noise current i_n consists of three components: shot noise i_{sh}, flicker noise i_f, and carrier generation–recombination noise i_{gr}. The first three have the mean-square values:

$$\overline{v_{tc}^2} = 4kTr_c\Delta f \tag{54.72}$$

$$\overline{i_{sh}^2} = 2qI_D\Delta f \tag{54.73}$$

$$\overline{i_f^2} = \frac{K_f I_D^m \Delta f}{f} \tag{54.74}$$

where I_D is the reverse-biased diode current. The carrier generation–recombination noise has a white spectral density up to a frequency determined by the carrier lifetime. Because the detector has a large output resistance, it should be used with amplifiers that exhibit a low input current noise.

54.28   Piezoelectric Transducer Noise Model

Figure 54.7(a) shows the circuit symbol of a piezoelectric transducer [4]. This transducer generates an electric voltage when a mechanical force is applied between two of its surfaces. An approximate equivalent circuit that is valid for frequencies near the transducer mechanical resonance is shown in Figure 54.7(b). In this circuit, C_e represents the transducer electric capacitance, while C_s, L_s, and R_s are chosen to have a resonant frequency and quality factor numerically equal to those of the transducer mechanical resonance. The source v_s represents the signal voltage, which is proportional to the applied force. The source v_{ts} represents the thermal noise generated by R_s. It has a mean-square value of:

$$\overline{v_{ts}^2} = 4kTR_s\Delta f \tag{54.75}$$

This noise component is negligible in most applications.

The piezoelectric transducer has two resonant frequencies: a short-circuit resonant frequency f_{sc} and an open-circuit resonant frequency f_{oc} given by:

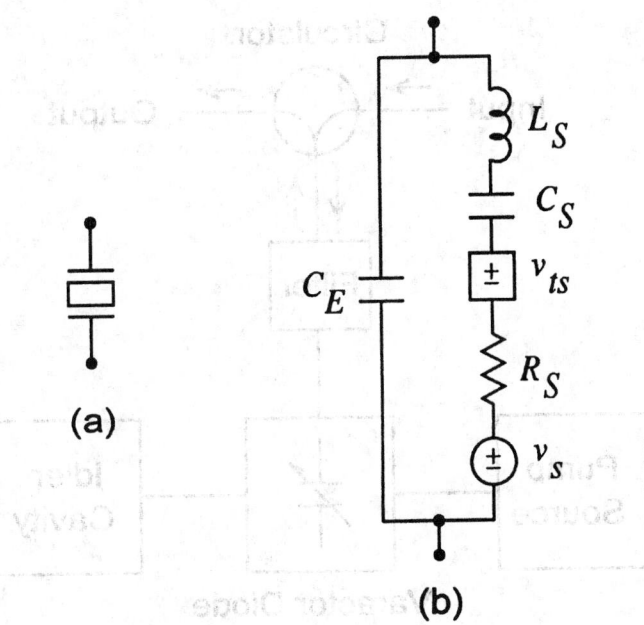

FIGURE 54.7 (a) Piezoelectric transducer symbol. (b) Noise model of piezoelectric transducer.

$$f_{sc} = \frac{1}{2\pi\sqrt{L_s C_s}} \tag{54.76}$$

$$f_{oc} = \frac{1}{2\pi\sqrt{L_s C_1}} \tag{54.77}$$

where $C_1 = C_s C_e/(C_s + C_e)$. It is normally operated at the open-circuit resonant frequency where the transducer output impedance is very high. For this reason, it is should be used with amplifiers that exhibit a low input current noise.

54.29 Parametric Amplifiers

A *parametric amplifier* is an amplifier that uses a time varying reactance to produce amplification [13]. In low-noise microwave parametric amplifiers, a reverse biased *pn* junction diode is used to realize a variable capacitance. Such diodes are called *varactors*, for variable reactance. The depletion capacitance of the reverse-biased junction is varied by simultaneously applying a signal current and a pump current at different frequencies. The nonlinear capacitance causes frequency mixing to occur between the signal frequency and the pump frequency. When the power generated by the frequency mixing exceeds the signal input power, the diode appears to have a negative resistance and signal amplification occurs. The only noise that is generated is the thermal noise of the effective series resistance of the diode, which is very small.

Figure 54.8 shows block diagram of a typical parametric amplifier. The varactor diode is placed in a resonant cavity. A circulator is used to isolate the diode from the input and output circuits. A pump signal is applied to the diode to cause its capacitance to vary at the pump frequency. The filter isolates the pump signal from the output circuit. The idler circuit is a resonant cavity that is coupled to the diode

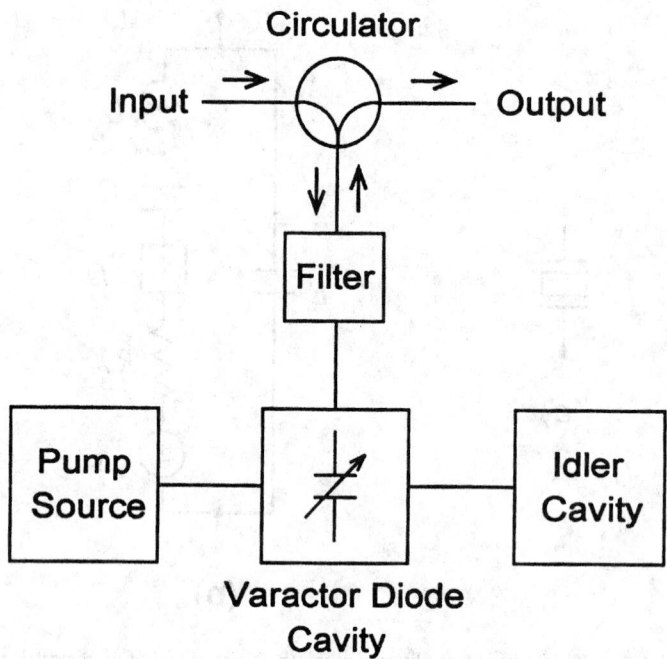

FIGURE 54.8 Block diagram of a typical parametric amplifier.

cavity to reduce the phase sensitivity. Let the signal frequency be f_s, the pump frequency be f_p, and the resonant frequency of the idler cavity be f_i. In cases where $f_p = f_s + f_i$, the varying capacitance of the diode looks like a negative resistance and the signal is amplified. If $f_i = f_s$, the amplifier is called a *degenerate* amplifier. This is the simplest form of the parametric amplifier and it requires the lowest pump frequency and power to operate. For the *nondegenerate* amplifier, $f_p > 2f_s$. In both cases, the input and the output are at the same frequency. In the *up-converter* amplifier, $f_p = f_i - f_s$ and $f_p > 2f_s$. In this case, the varying capacitance of the diode looks like a positive resistance and the signal frequency output is not amplified. However, there is an output at the idler frequency that is amplified. Thus, the output frequency is higher than the input frequency. The conversion gain can be as high as the ratio of the output frequency to the input frequency.

54.30 Measuring Noise

A typical setup for measuring noise is shown in Figure 54.9. To prevent measurement errors caused by signals coupling in through the ground and power supply leads, the circuit under test and the test set must be properly grounded and good power supply decoupling must be used [6]. For measurement schemes requiring a sine-wave source, an internally shielded oscillator is preferred over a function generator. This is because function generators can introduce errors caused by radiated signals and signals coupled through the ground system.

When making measurements on a high-gain circuit, the input signal must often be attenuated. Attenuators that are built into sources might not be adequately shielded, so that errors can be introduced by radiated signals. These problems can be minimized if a shielded external attenuator is used between the source and the circuit under test. Such an attenuator is illustrated in Figure 54.9. When a high attenuation is required, a multi-stage attenuator is preferred. For proper frequency response, the attenuator might require frequency compensation [4]. Unless the load impedance on the attenuator is large compared to its output impedance, both the attenuation and the frequency compensation can be a function of the load impedance.

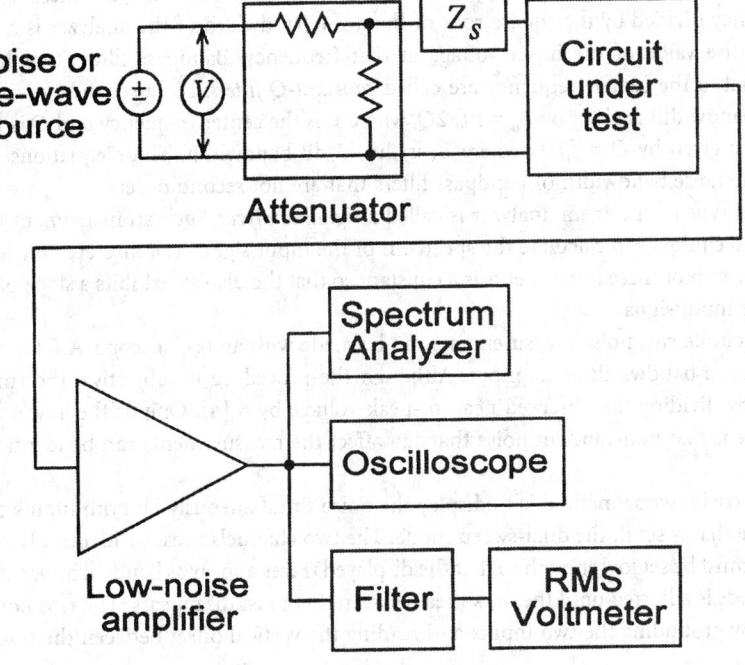

FIGURE 54.9 Noise measuring setup.

Figure 54.9 shows a source impedance Z_s in series with the input to the circuit under test. This impedance is in series with the output impedance of the attenuator. It must be chosen so that the circuit under test has the desired source impedance termination for the noise measurements.

Because noise signals are small, a low-noise amplifier is often required to boost the noise level sufficiently so that it can be measured. Such an amplifier is shown in Figure 54.9. The noise generated by the amplifier will add to the measured noise. To correct for this, first measure the mean-square noise voltage with the amplifier input terminated in the output impedance of the circuit under test. Then subtract this from the measured mean-square noise voltage with the circuit under test driving the amplifier. The difference is the mean-square noise due to the circuit. Ideally, the amplifier should have no effect on the measured noise.

The noise voltage over a band can be measured with either a spectrum analyzer or with a filter having a known noise bandwidth and a voltmeter. The noise can be referred to the input of the circuit under test by dividing by the total gain between its input and the measuring device. The measuring voltmeter should have a bandwidth that is at least 10 times the noise bandwidth of the filter. The *voltmeter crest factor* is the ratio of the peak input voltage to the full-scale rms meter reading at which the internal meter circuits overload. For a sine-wave signal, the minimum voltmeter crest factor is $\sqrt{2}$. For noise measurements, a higher crest factor is required. For gaussian noise, a crest factor of 3 gives an error less than 1.5%. A crest factor of 4 gives an error less than 0.5%. To avoid overload on noise peaks caused by an inadequate crest factor, measurements should be made on the lower one-third to one-half of the voltmeter scale.

A true rms voltmeter is preferred over one that responds to the average rectified value of the input voltage but is calibrated to read rms. When the latter type of voltmeter is used to measure noise, the reading will be low. For gaussian noise, the reading can be corrected by multiplying the measured voltage by 1.13. Noise voltages measured with a spectrum analyzer must also be corrected by the same factor if the spectrum analyzer responds to the average rectified value of the input voltage but is calibrated to read rms.

Noise measurements with a spectrum analyzer require a knowledge of the noise bandwidth of the instrument. For a conventional analyzer, the bandwidth is proportional to frequency. When white noise is analyzed, the display exhibits a slope of +10 dB per decade. However, the measured voltage level at any frequency divided by the square root of the noise bandwidth of the analyzer is a constant equal to the spot-noise value of the input voltage at that frequency. Bandpass filters that have a bandwidth proportional to the center frequency are called *constant-Q filters*. For a second-order constant-Q filter, the noise bandwidth is given by $B_n = \pi f_0/2Q$, where f_0 is the center frequency and Q is the quality factor. The latter is given by $Q = f_0/B_3$, where B_3 is the −3 dB bandwidth. These equations are often used to estimate the noise bandwidth of bandpass filters that are not second order.

A second type of spectrum analyzer is called a *signal analyzer*. Such an instrument uses digital signal processing techniques to calculate the spectrum of the input signal as a discrete Fourier transform. The noise bandwidth of these instruments is a constant so that the display exhibits a slope of zero when white noise is the input signal.

Fairly accurate rms noise measurements can be made with an oscilloscope. A filter should be used to limit the noise bandwidth at its input. Although the procedure is subjective, the rms voltage can be estimated by dividing the observed peak-to-peak voltage by 6 [4]. One of the advantages of using the oscilloscope is that non-random noise that can affect the measurements can be identified, e.g., a 60-Hz hum signal.

Another oscilloscope method is to display the noise simultaneously on both inputs of a dual-channel oscilloscope that is set in the dual-sweep mode. The two channels must be identically calibrated and the sweep rate must be set low enough so that the displayed traces appear as bands. The vertical offset between the two bands is adjusted until the dark area between them just disappears. The rms noise voltage is then measured by grounding the two inputs and reading the vertical offset between the traces.

Defining Terms

Burst noise: Noise caused by a metallic impurity in a *pn* junction that sounds like corn popping when amplified and reproduced by a loudspeaker. Also called *popcorn noise*.

Crest factor: The ratio of the peak value to the rms value.

Equivalent noise input current: The noise current in parallel with an amplifier input that generates the same noise voltage at its output as all noise sources in the amplifier.

Equivalent noise input voltage: The noise voltage in series with an amplifier input that generates the same noise voltage at its output as all noise sources in the amplifier.

Equivalent noise resistance (conductance):   The value of a resistor (conductance) at the standard temperature $T_0 = 290$ K that generates the same mean-square noise voltage (current) as a source.

Excess noise: Flicker noise in resistors.

Flicker noise: Noise generated by the imperfect contact between two conducting materials causing the conductivity to fluctuate in the presence of a dc current. Also called *contact noise, 1/f noise*, and *pink noise*.

Generation–recombination noise: Noise generated in a semiconductor by the random fluctuation of free carrier densities caused by spontaneous fluctuations in the generation, recombination, and trapping rates.

Noise bandwidth: The bandwidth of an ideal filter having a constant passband gain that passes the same rms noise voltage as a filter, where the input signal is white noise.

Noise factor: The ratio of the mean-square noise voltage at an amplifier output to the mean-square noise voltage at the amplifier output considering the thermal noise of the input termination to be the only source of noise.

Noise figure: The noise factor expressed in dB.

Noise index: The number of μA of excess noise current in each decade of frequency per A of dc current through a resistor. Also, the number of μV of excess noise voltage in each decade of frequency per V of dc voltage across a resistor.

Noise temperature: The internal noise generated by an amplifier expressed as an equivalent input-termination noise temperature.

Nyquist formula: Expression for the mean-square thermal noise voltage generated by a resistor.

Optimum source impedance (admittance): The complex source impedance (admittance) that minimizes the noise factor.

Parametric amplifier: An amplifier that uses a time-varying reactance to produce amplification.

Partition noise: Noise generated by the statistical process of partition when the charge carriers in a current have the possibility of dividing between two or more paths.

Shot noise: Noise caused by the random emission of electrons and by the random passage of charge carriers across potential barriers.

Schottky formula: Expression for the mean-square shot noise current.

Signal-to-noise ratio: The ratio of the mean-square signal voltage to the mean-square noise voltage at an amplifier output.

Spectral density: The mean-square value per unit bandwidth of a noise signal.

Spot noise: The rms noise in a band, divided by the square root of the noise bandwidth.

Thermal noise: Noise generated by the random collision of charge carriers with a lattice under conditions of thermal equilibrium. Also called *Johnson noise*.

Varactor diode: A diode used as a variable capacitance.

Voltmeter crest factor: The ratio of the peak input voltage to the full-scale rms meter reading at which the internal meter circuits overload.

White noise: Noise that has a spectral density that is flat, i.e., not a function of frequency.

References

1. P. R. Gray and R. G. Meyer, *Analysis and Design of Analog Integrated Circuits*, New York: John Wiley & Sons, 1993.
2. P. Horowitz and W. Hill, *The Art of Electronics*, 2nd ed., New York: Cambridge Press, 1983.
3. W. M. Leach, Jr., Fundamentals of low-noise analog circuit design, *Proc. IEEE*, 82(10), 1515–1538, 1994.
4. C. D. Motchenbacher and J. A. Connelly, *Low Noise Electronic System Design*, New York: Wiley, 1993.
5. Y. Netzer, The design of low-noise amplifiers, *Proc. IEEE*, 69, 728-741, 1981.
6. H. W. Ott, *Noise Reduction Techniques in Electronic Systems*, 2nd ed., New York: Wiley, 1988.
7. A. Van der Ziel, Noise in solid-state devices and lasers, *Proc. IEEE*, 58, 1178-1206, 1970.
8. H. A. Haus et al., Representation of noise in linear twoports, *Proc. IRE*, 48, 69–74, 1960.
9. R. E. Burgess, Electronic fluctuations in semiconductors, *Br. J. Appl. Phys.*, 6, 185–190, 1955.
10. H. Fukui, The noise performance of microwave transistors, *IEEE Trans. Electron Devices*, ED-13, 329–341, 1966.
11. H. Rothe and W. Dahlke, Theory of noisy fourpoles, *Proc. IRE*, 44, 811–818, 1956.
12. H. A. Haus, et al., IRE standards on methods of measuring noise in linear twoports, 1959, *Proc. IRE*, 48, 60–68, 1960.
13. A. L. Lance, *Introduction to Microwave Theory and Measurements*, New York: McGraw-Hill, 1964.

Further Information

H. Fukui, *Low-Noise Microwave Transistors & Amplifiers*, New York: IEEE Press, 1981.

M. S. Gupta, ed., *Electrical Noise: Fundamentals & Sources*, New York: IEEE Press, 1977.

A. Van der Ziel, *Noise: Sources, Characterization, Measurements*, Englewood Cliffs, NJ: Prentice-Hall, 1970.

J. C. Bertails, Low frequency noise considerations for MOS amplifier design, *IEEE J. Solid-State Circuits*, SC-14, 773–776, 1979.

C. A. Liechti, Microwave field-effect transistors — 1976, *IEEE Trans. Microwave Theory and Tech.*, MTT-24, 279–300, 1976.

M. Steyaert, Z. Y. Chang, and W. Sansen, Low-noise monolithic amplifier design: bipolar vs. CMOS, *Analog Integrated Circuits and Signal Processing*, 1(1), 9–19, 1991.

55

Microwave Measurement

A. Dehé
Institut für Hochfrequenztechnik,
Technische Universität Darmstadt

K. Beilenhoff
Institut für Hochfrequenztechnik,
Technische Universität Darmstadt

K. Fricke
Institut für Hochfrequenztechnik,
Technische Universität Darmstadt

H. Klingbeil
Institut für Hochfrequenztechnik,
Technische Universität Darmstadt

V. Krozer
Institut für Hochfrequenztechnik,
Technische Universität Darmstadt

H. L. Hartnagel
Institut für Hochfrequenztechnik,
Technische Universität Darmstadt

Microwave measurements cover the frequency range from 0.5 GHz to about 20 GHz. Frequencies from 30 GHz to 300 GHz are often referred to as mm-waves. In the following, the most important measurements are described.

55.1 Power Measurement

Exact microwave power measurement is in demand for development, fabrication, and installation of modern telecommunication networks. It is essential for attenuation, insertion, and return loss measurements, as well as for noise measurement and six-port network analysis. This chapter gives a brief overview about the basics of microwave power measurement. Detailed information about this subject can be found in [1–4]. Power detectors usually consist of a sensor that transfers the microwave signal into a dc or low-frequency signal and a power meter to read out the measured power levels. The sensor includes a load impedance for the microwave source (Figure 55.1).

First, several power definitions need to be clarified:

- *Conjugate available power* (P_{CA}): The maximum available power a signal generator can transmit. This power is delivered by the generator if the load impedance is equal to the complex conjugate of the generator's source impedance. Since measurement techniques often require different lengths of waveguides or coaxial lines, the conjugate available power can be achieved only by tuning.

- Z_0 *Available power* (P_{Z0}): The power that is transferred via a coaxial line of characteristic impedance Z_0 into a load impedance Z_L equal to Z_0, while the generator has the impedance Z_G. Consequently, the available power level is reduced due to generator mismatch: $P_{Z0} = P_{CA}(1 - |\Gamma_G|^2)$, where Γ_G is

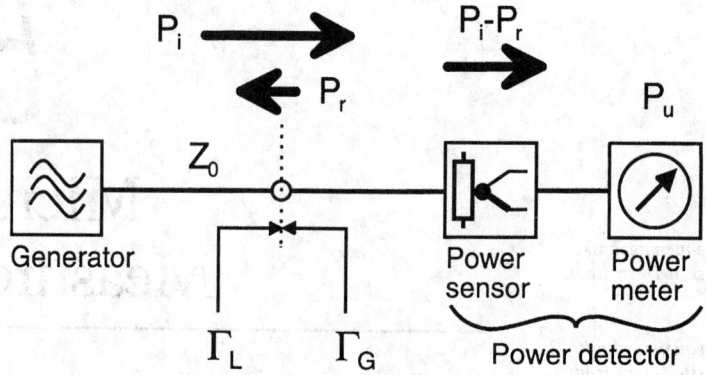

FIGURE 55.1   Setup of a power measurement in load configuration.

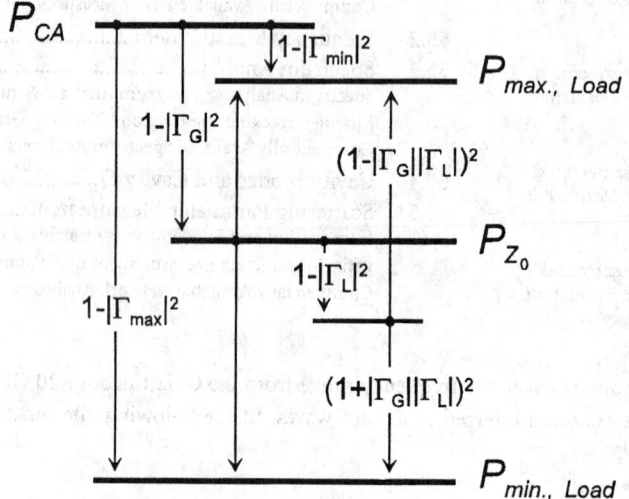

FIGURE 55.2   Relationships between conjugate power P_{CA}, Z_0 available power P_{Z0}, and the power levels available in the load of the power sensor.

the reflection coefficient of the generator. Figure 55.2 shows the relationships between P_{CA}, P_{Z0} and the maximum and minimum power levels that can be measured in the load of a power sensor.

- *Average power* (P_{av}): The result of an averaging over many periods of the lowest modulation frequency of a modulated signal $P(t)$ (Figure 55.3).
- *Envelope power* ($P_e(t)$): The power averaged over the period of the carrier frequency.
- *Peak envelope power* (PEP): The maximum of $P_e(t)$.
- *Pulse power* (P_P): Defined for pulsed signals. If the pulse width τ and the repetition frequency $1/T$ is known, P_P is given by the measured average power:

$$P_P = P_{av} \frac{T}{\tau}$$

Measurement Errors and Uncertainties

Measurement errors occur due to mismatch as well as inside the power sensor and in the power meter. After correcting for these errors, the measurement uncertainties remain. Typically 75% of the total

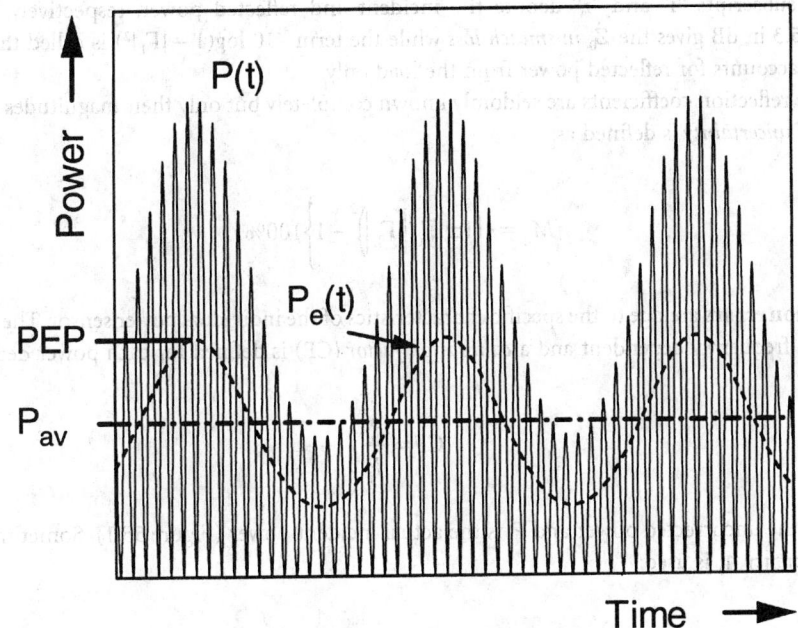

FIGURE 55.3 Power definitions for modulated signals.

uncertainty belongs to mismatch and the smallest part is due to the power meter. The uncertainties and errors can be power dependent, frequency dependent, or independent of both. Of course, the *total uncertainty* must be calculated from the different uncertainties u_i as the root-sum-of-the-squares: rss = $\sqrt{\sum u_i^2}$, provided that the errors are all independent. A pessimistic error definition is the *worst-case uncertainty*, which simply sums up all extreme values of the independent errors.

Mismatch errors occur due to the fact that neither the generator (G), the line, nor the load (L) exhibit exactly the characteristic impedance Z_0. Using the modulus of the reflection coefficient $|\Gamma|$, the available power of the generator P_G can be expressed as [3]:

$$P_G = P_L \frac{\left|1 - \Gamma_G \Gamma_L\right|^2}{\left(1 - \left|\Gamma_G\right|^2\right)\left(1 - \left|\Gamma_L\right|^2\right)} \tag{55.1}$$

The reflection coefficients can also be expressed in terms of the voltage standing wave ratio (VSWR):

$$|\Gamma| = \frac{VSWR - 1}{VSWR + 1} = \frac{Z_L - Z_0}{Z_L + Z_0} \tag{55.2}$$

As mentioned, the knowledge of Z_0 available power P_{Z0} is sufficient in most applications. Then the ratio between the Z_0 available power and the absorbed power is given by:

$$\frac{P_{Z_0}}{P_i - P_r} = \frac{\left|1 - \Gamma_G \Gamma_L\right|^2}{1 - \left|\Gamma_L\right|^2} \tag{55.3}$$

where the subscripts "i" and "r" denote the incident and reflected power, respectively. Expressing Equation 55.3 in dB gives the Z_0 *mismatch loss* while the term $-10 \log(1 - |\Gamma_L|^2)$ is called the *mismatch loss*, which accounts for reflected power from the load only.

Since the reflection coefficients are seldomly known completely but only their magnitudes are known, a *mismatch uncertainty* is defined as:

$$M_u = \left\{\left(1 \pm |\Gamma_G||\Gamma_L|\right)^2 - 1\right\}100\%$$

(55.4)

Conversion errors are due to the specific characteristics of the individual power sensor. The conversion efficiency is frequently dependent and a *calibration factor* (CF) is defined for each power detector:

$$CF = \frac{P_u}{P_i}$$

(55.5)

where P_u is the uncorrected power and P_i is the actual incident power (Figure 55.1). Sometimes also the *effective efficiency* η_e is used:

$$\eta_e = \frac{P_u}{P_i - P_r}$$

(55.6)

Both quantities are related via the reflection coefficient of the load of the power sensor:

$$CF = \eta_e \left(1 - |\Gamma_L|^2\right)$$

(55.7)

The calibration factor is used to correct for efficiency loss and it also accounts for the mismatch loss. Still a remaining *calibration factor uncertainty* has to be taken into account. It is specified for each sensor. The calibration data are usually traceable to a national bureau of standards. Power sensors under test can be compared to the standards using high directivity couplers or power splitters [2].

The next described errors are due to the electronics inside the power meter.

Some power meters exhibit an internal reference oscillator to verify and adjust for the sensitivity of the diode or thermocouple sensor. The *reference power uncertainty* is specified by the manufacturer. Since this reference has its own reflection coefficient, it is related to a *reference oscillator mismatch uncertainty*.

Instrumentation uncertainty depends on the circuit limitations of the power meter and is specified by the manufacturer.

The *±1 Count Error* is for digital output and can be expressed by the relative power reading of the last significant digit.

Autozero can be used on all measurement ranges. Zero-setting immediately prior to the measurement can reduce *drift errors* when measuring in the lower ranges. Still *zero set errors* remain due to noise during the zero-setting operation. This error can be very serious for measurement of low power levels [5]. *Zero carryover* is caused by ADC quantization errors in the zero readings for all measurement ranges except the most sensitive one. ADC quantization also causes a *power quantization error*. If very low power levels have to be measured, averaging can reduce random noise at the expense of measurement speed.

Power Sensors

A large variety of power sensors is available for the frequency range from dc up to 110 GHz and for minimum detectable power levels as low as 100 pW. The sensors differ in measurement principle and hence the correct choice depends on the application. A detailed description of power sensors is given in [2, 3]. Thermal sensors transform the absorbed microwave power into heat that is measured with temperature-sensitive elements:

- *Calorimeters* measure the heat absorbed in a fluid (e.g., water) of well-known specific heat. Applying the substitution principle, their precision can be enhanced. Because of their high stability, they are used in the National Institute of Standards. The manufacturer usually references the sensors to these standards.

- *Bolometers* and *thermistors* [3] make use of the temperature-dependent resistivity change of a resistive load, which is highly nonlinear. Hence, dc power substitution is used to keep the load at constant temperature. The substituted dc power is the measurement. Self-balancing bridges are used for this purpose but need careful ambient temperature compensation. The effective efficiency of bolometers is known very exactly. However, they have only relatively small dynamic range (typical 10 μW to 10 mW). Currently, liquid nitrogen-cooled, high-temperature superconducting bolometers with extremely high sensitivity of several thousands volts per watt are used [6]. A comprehensive overview of bolometers if given in [7].

- *Thermocouple sensors* are based on microwave power conversion into heat via a matched load impedance. Its temperature is controlled by thermocouples utilizing the thermoelectric effect in thin films or semiconductors [8]. One has to distinguish between realizations where the thermocouple itself represents the load impedance [3] and galvanically decoupled thermocouples [9]. The main advantages of the latter sensors are the better flatness of the response from dc to microwave and a lower VSWR. The design of such sensors is simple and a silicon micromachining technique can be applied to enhance the sensitivity [10]. The lowest measurable power levels are 1 μW.

Thermal sensors are well-suited to absolute power measurement, especially with respect to their high linearity. Their drawback is the relatively long response time (>10 ms), which limits these sensors to the measurement of average power. The high speed of Schottky diodes predestines them for the measurement of peak power, envelope power, and peak envelope power.

- *Diode sensors* consist of Schottky barrier diodes that are zero-biased and work in the square-law part of their *I–V* characteristics. For low power levels (<–20 dBm), these devices are very linear in response and measure the average power correctly. Still, diode sensors exhibit nonquadratic contributions [11], which can be important to account for when accurate measurement is required. Minimum detectable power is –70 dBm where the signal level is of the order of 50 nV and requires sophisticated amplification. The diode sensor is part of the sensors described below.

- *Peak power sensors* are specially designed for peak power measurements and account for measurement errors due to waveform and power level, although any diode detector can be used for this purpose if the peak voltages are ≤1 V.

- *Peak envelope analyzers* are designed to detect the envelope power. This is not possible with a simple diode sensor because the electronic setup of the diode must be different.

- *Feedthrough power sensors* are used to measure microwave power in transmission configuration. They have minor losses of approximately 0.5 dB and a limited bandwidth of typically 0.1 GHz to 1 GHz. The limiting device in these systems is a directional coupler with power splitters. Such a measurement can also be implemented with discrete elements. The characteristic figure of merit of these devices is the *directivity* (a_D), relating the read incident power P_i to the read reflected power P_r in case of reflection free load ($\Gamma_L = 0$):

$$a_D = 10 \log \frac{P_i}{P_r} \, [\text{dB}]$$
(55.8)

The directivity should be as high as possible.

The above power sensors are discrete devices. The maturity in microwave monolithic integrated circuit (MMIC) design and fabrication allows the integrated realization of diode power sensors; for example, in an integrated six-port reflectometer [12]. Activities to fabricate thermal power sensors integrable to MMIC-typical processes [13] can presently be implemented on commercial processes, such as the Philips Lemeill HEMT processes with additional postmicromachining [14].

Commercially Available Power Measurement Systems

A collection of different power sensors and corresponding measurement units is shown in Table 55.1. All power meters have GPIB interfaces for easy use in automated measurement.

55.2 Frequency Measurement

Frequency measurement in the microwave regime is usually part of a more complex measurement procedure, for example, determining the scattering parameters and filter characteristics of a DUT. If one is only interested in frequency or a higher accuracy is required, one must use direct frequency measurement systems.

Two different techniques can be distinguished. The rather old-fashioned way is to use mechanically tunable resonators, the so-called *wave meters*. These are not explained in detail here. *Digital frequency counters* are an alternative and are now the state of the art (Table 55.2).

The digital frequency counter measurement system is based on the principle of counting the zero crossovers of a continuous sinusoidal signal. At low frequencies, this method can be used directly; whereas in the microwave region, direct digital counters are not available because of their limited bandwidth. Thus, a modified measurement system must be used.

The standard digital frequency counter usually consists of a mixer, a local oscillator (frequency f_0) in the lower frequency regime, several multipliers, and the digital counter. The principal measurement technique is shown in Figure 55.4.

An extremely stable local oscillator (quartz oscillator) is used to provide the reference signal used in the measurement system. This signal is multiplied by a factor of N and mixed with the RF signal of the DUT. The IF in the low-frequency range can be easily counted and, thus, the frequency of the signal f_s can be calculated according to the following equation.

$$f_s = f_{IF} + Nf_0 \qquad (55.9)$$

For this method, an extremely stable low-frequency oscillator (often temperature controlled) must be provided and, in order to allow a sufficient bandwidth, a high number of multipliers must be implemented in this system.

A pulsed oscillator with extremely short risetime can circumvent this problem. In the frequency domain, this signal is given by spectral lines at $f = if_0$, where f_0 denotes the fundamental frequency of the pulses. Using a bandpass filter, a single frequency can be separated and transferred to the mixer.

55.3 Spectrum Analysis

The expression spectrum analysis subsumes the measurements that are performed to obtain the Fourier transformation $S(f)$ of a given signal $s(t)$. The Fourier transformation of $s(t)$ in the frequency domain is defined by the equation:

$$S(f) = \int_{-\infty}^{+\infty} s(t)e^{-j2\pi ft}dt \qquad (55.10)$$

In practice, the lower and upper bounds of the integral are limited by a finite measurement time that must be fixed by the user.

TABLE 55.1 Available Commercial Power Sensors and Power Meters

Supplier	Power sensor / *Power meter	Frequency range	Dynamic range	VSWR	Remark	~Price $U.S.
			Power Sensors in Matched Load Configuration and Related Power Meter			
Hewlett-Packard	HP8478B	10 MHz–18 GHz	1 μW–100 mW	1.1–1.75	Thermistor	1790
Rohde & Schwarz	NRV-Z52	dc–26.5 GHz	1 μW–100 mW	1.11–1.22	Thermocouple, up to 30 W available	3000/3900
Marconi	6913/6914S	10 MHz–26.5/46 GHz	10 μW–100 μW	1.1–1.4/3.6	Thermocouple, up to 3 W available	
Boonton	51100(9E)	10 MHz–18 GHz	1 μW–100 mW	1.18–1.28	Thermocouple	
Hewlett-Packard	HP8485A	50 MHz–26.5 GHz	1 μW–100 mW	1.10–1.25	Thermocouple	1980
	HP8487A	50 MHz–50 GHz	1 μW–100 mW	1.10–1.50	Thermocouple	
	HPR/Q/W8486A	26.5–40/33–50/75–110 GHz	1 μW–100 mW	1.4/1.5/1.08	Rectangular waveguide, thermocouple	
Rohde & Schwarz	NRV-Z6	50 MHz–26.5 GHz	1 nW–20 mW	1.2–1.4	Diode	4300/5000
Marconi	6923/6924S	10 MHz–26.5/46 GHz	0.1 nW/0.1 μW–10 μW	1.12–1.5/3.6	Diode	
Hewlett-Packard	HP8487D	50 MHz–50 GHz	0.1 nW–10 μW	1.15–1.89	Diode	2710/5290
Rohde & Schwarz	*NRVS/D	dc–26.5 GHz	0.4 nW–30 W		One/two channel	
Boonton	*4230A	10 kHz–100 GHz	0.1 nW–25 W			6900
Marconi	*6960B	30 kHz–46 GHz	0.1 nW–30 W			3800
	*6970	30 kHz–46 GHz	0.1 nW–30 W		Hand portable	
Hewlett-Packard	*HP437B	100 kHz–110 GHz	0.07 nW–25 W			
			Power Analyzer			
Hewlett-Packard	HP84812/13/14A	500 MHz–18/26.5/40 GHz	0.6 μW–100 mW	1.25–1.60	Resolution 100 ps	
	*HP8991A	500 MHz–40 GHz	0.5 μW–100 mW		Rise/fall time 5 ns	
			Peak Power Sensors and Related Power Meter			
Rohde & Schwarz	NRV-Z31/33	0.03–6 GHz	1 μW–20 mW/1 mW–20 W	1.05–1.33	With NRVS	2020/2500
Hewlett-Packard	HP84812/3/4A	500 MHz–18/26.5/40 GHz	1 μW–100 mW	1.25/1.35/1.50		
Boonton	56340	500 MHz–40 GHz		1.25–2.00	Dual diode risetime <15 ns	
	*HP8990A	500 MHz–40 GHz				
Boonton	4500A	1 MHz–40 GHz	0.1 μW–100 mW			
			Feed Through Power Sensor and Related Power Meter			
Rohde & Schwarz	NAS-Z7	1.71–1.99 GHz	0.01–30 W	<1.15	a_D > 26 dB, GSM, DCS1800/1900	1250
	*NAS	0.001–1.99 GHz	10 mW–1200 W			1120

TABLE 55.2   Digital Microwave Frequency Counter

Supplier	Counter	Frequency range	Resolution	Sensitivity	Remark	~Price $US
Hewlett-Packard	HP5351B	26.5 GHz	1 Hz	−40 dBm		7,500
Hewlett-Packard	HP5352B	40(46) GHz	1 Hz	−30 dBm		11,800

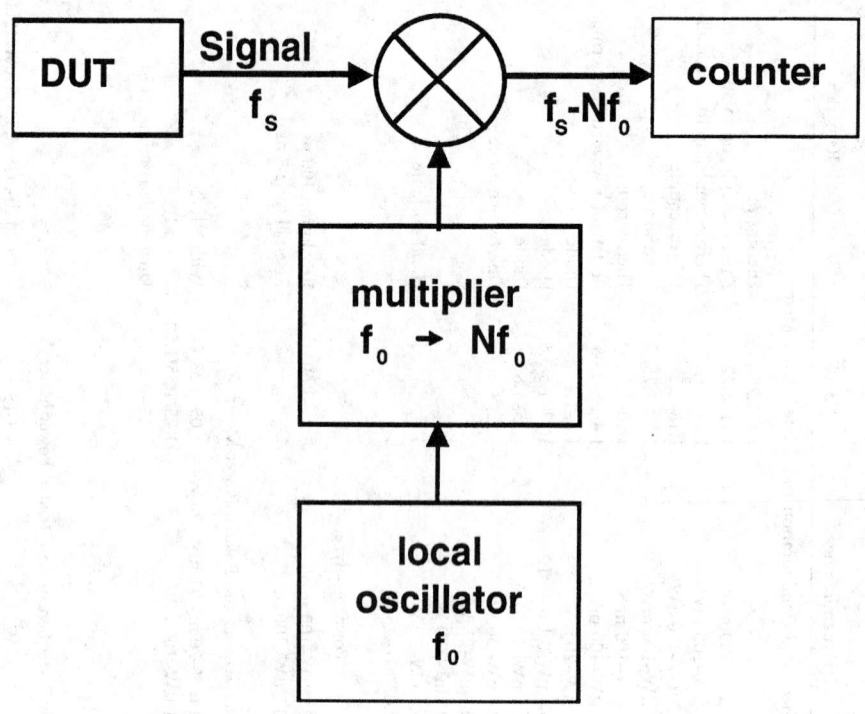

FIGURE 55.4   Block diagram of a digital frequency counter.

For the analysis of an unknown spectrum, different methods can be distinguished:

- *Wave analyzers and selective voltmeters:* These devices utilize a tunable filter for frequency-selective measurements.
- *Spectrum analyzers* rely on the principle of heterodyne mixing with subsequent bandpass filtering.
- Calculation of the spectrum using a *fast Fourier transformation* (FET). This method can be employed only for lower frequencies, since a digital-to-analog converter is needed. For microwave frequencies, the calculation of the spectrum using the FFT is, therefore, difficult to realize.

Since the spectrum analyzer is most suitable for microwave frequencies, it will be described in detail in the following. Brief introductions into the spectrum analyzer measurement techniques are given in [15, 16].

Spectrum Analyzer

The spectrum analyzer is most suitable for the analysis of microwave signals. It is a general-purpose instrument for measurements in the frequency domain and provides the user with the amplitude, power, or noise density of a signal depicted vs. the frequency. The frequency scale is in most cases linear; the vertical axes can be either linear or logarithmic. Spectrum analyzers are available from a few hertz up to more than 100 GHz. They give a quick overview of the spectral power distribution of a signal. Spectrum analyzers have a large dynamic range, a resolution bandwidth of a few hertz, and a reasonable frequency resolution.

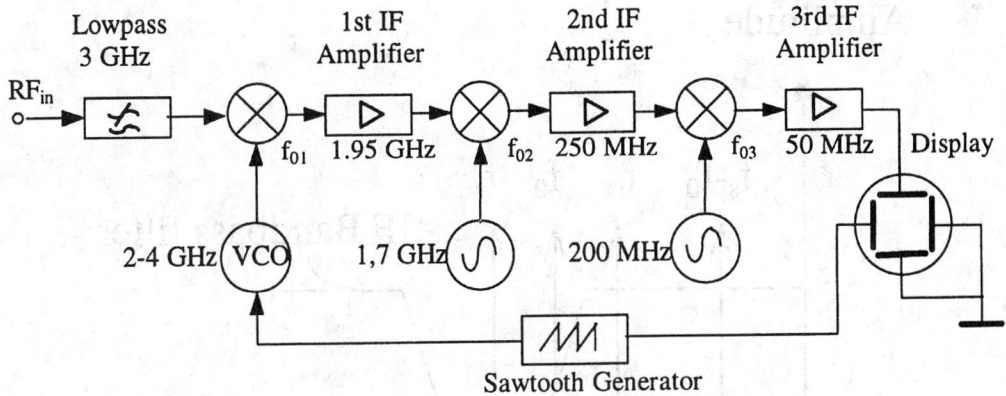

FIGURE 55.5 Simplified setup of a spectrum analyzer.

The spectrum analyzer is suitable for the following measurements:

- *Measurement of absolute and relative frequency:* Frequency drift (unstabilized oscillators), spectral purity, and frequency of harmonics.
- *Absolute and relative amplitude:* Gain of frequency multipliers, harmonics of periodic signals, intermodulation (IM) distortion, and harmonic distortion.
- *Scalar network analysis (if equipped with a tracking generator):* Frequency response of amplifiers and filters.
- *Electromagnetic interference (EMI) measurements:* Broadband spectra.
- *Measurements of modulated signals:* AM, FM, or PM.
- *Noise:* Many spectrum analyzers can be used for noise measurements of active devices.
- *Phase noise:* Phase noise of oscillators can be analyzed with spectrum analyzers [17].

Spectrum Analyzer Setup

The spectrum analyzer is basically an electronically tunable filter that allows the measurement of the amplitude, power or noise at a certain frequency. Using the example shown in Figure 55.5, the principle of operation can be explained as follows.

The tunable filter used to separate the frequencies to be measured is realized using a chain of mixers and IF amplifiers. In this case, three mixers convert a given input signal frequency f_s to the IF passband of the last IF amplifier. At least one of the oscillators is tunable (VCO) in order to scan the input frequency f_s. Sometimes, more than one tunable oscillator is used.

The first mixer and the following IF amplifier with a bandpass center frequency of 1.95 GHz in the given example selects the input frequency to be analyzed according to:

$$f_s = f_{o1} - 1.95\,\text{GHz} \tag{55.11}$$

The input frequency for the spectrum analyzer shown in Figure 55.4 can be due to the scan of the frequency of the tunable oscillator f_{o1}, between 50 MHz and 2.05 GHz. However, the image frequency f_{is} will also be mixed to the IF:

$$f_{is} = f_{o1} + 1.95\,\text{GHz} \tag{55.12}$$

Since f_{is} is in the range of 3.95 GHz to 5.95 GHz, a bandpass filter with a cut-off frequency of 3 GHz is used at the input to reject the image frequency.

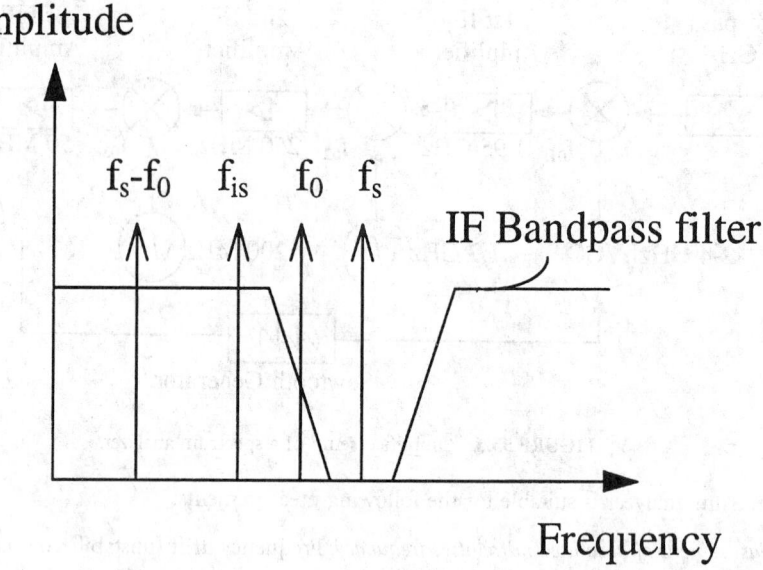

FIGURE 55.6 Example for the blocking of the image frequency by the IF filter in an arrangement according to Figure 55.5.

Because it is difficult to realize a narrow bandpass at 1.95 GHz, the signal is converted to a lower frequency in the megahertz range. At these frequencies, stable quartz filters with high quality factors can be employed. In principle, the RF frequency could be mixed down to the last IF section in one step. However, in such an arrangement, it would be difficult to suppress the image frequencies. In any case, the image frequency of each stage should be rejected by the preceding IF amplifier as shown in Figure 55.6 for the second stage of the given example.

The last IF amplifier is very important for the performance of the complete system. In nearly all spectrum analyzers, its bandwidth — the so-called resolution bandwidth — can be adjusted in steps. For separation of closely spaced spectral components, the bandwidth should be very small. Most spectrum analyzers offer a minimum bandwidth of a few hertz. On the other hand, larger bandwidths are needed, since for a narrow IF filter only a slow scan speed can be allowed (see below) and, therefore, the measurement over several frequency decades would result in a large sweep time. The shape of the IF filter is important for the capability of the spectrum analyzer to separate close spectral components. The performance of the IF filter is described by the shape factor. It is defined by the ratio of the 60 dB to the 3 dB bandwidth of the IF filter. The IF filter can be switched between linear and logarithmic amplification. This is performed numerically in most cases.

A sawtooth generator produces the control voltage for the voltage-controlled oscillator (VCO) and the x-deflection voltage of the screen. A significant error is introduced in the frequency scale by a not-ideal voltage-to-frequency characteristic of the VCO. Therefore, in many cases, synthesizers with a quartz-stabilized phase-locked loop [18] are used.

The detector has to be sensitive either to the amplitude, the power, or the noise (mV $\sqrt{Hz^{-1}}$). In modern spectrum analyzer, digital signal processing is used for this purpose.

It is important to note that there are restrictions on the minimum sweep time T_s. In order to avoid settling errors of the narrow band pass IF amplifier with bandwidth B, the scan time should be for a frequency span S larger than [19, 20]:

$$T_s > 20 S / B^2 \tag{55.13}$$

Most of the spectrum analyzers control the scan time automatically according to this equation.

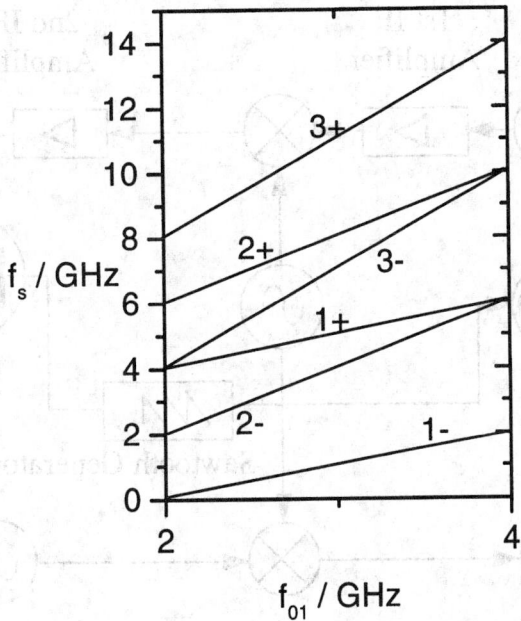

FIGURE 55.7 Measured RF frequency f_s vs. the frequency of the VCO (voltage controlled oscillator) f_{01}. Parameters are n and the plus or the minus sign in Equation 55.4.

Harmonic Mixing

For higher frequencies, the harmonic mixing technique is widely used. If the first 3-GHz low-pass filter in Figure 55.5 is omitted and the VCO produces harmonics, a larger number of input frequencies are converted to the passband of the spectrum analyzer. For the example shown, the possible input signals are depicted in Figure 55.7 vs. the VCO frequency according to:

$$f_s = nf_{01} \pm 1.95 \text{ GHz} \tag{55.14}$$

The notation of the numbers in the figure are the harmonic number n and the plus or the minus sign in the above equation. For a frequency $f_{01} = 3$ GHz of the VCO, the following frequencies will appear at the same frequency location on the screen: 1.05 GHz (1–), 4.95 GHz (1+), 4.05 GHz (2–), 7.95 GHz (2+), 7.05 GHz (3–), 10.95 GHz (3+). A tracking preselection filter (see below), which is scanned with the VCO, can select one of these harmonics.

For further extension of the frequency in the upper mm-wave range, external mixers are used. With such an arrangement, frequencies higher than 500 GHz can be measured. Additionally, equipment for mixing of signals in the optical range is offered by some companies.

Tracking Preselection

For small input signals, the spectrum analyzer can be considered a linear device. However, if the input level increases, harmonics and intermodulation products are generated due to the nonlinearities of a mixer. These products result in spurious signals on the screen of the spectrum analyzer. In addition, image frequencies will appear on the screen, as demonstrated above. To avoid these spurious responses, a tracking preselection filter is employed at the input of the spectrum analyzer. A tracking preselection filter is an electronically tuned bandpass filter usually realized using a YIG filter.

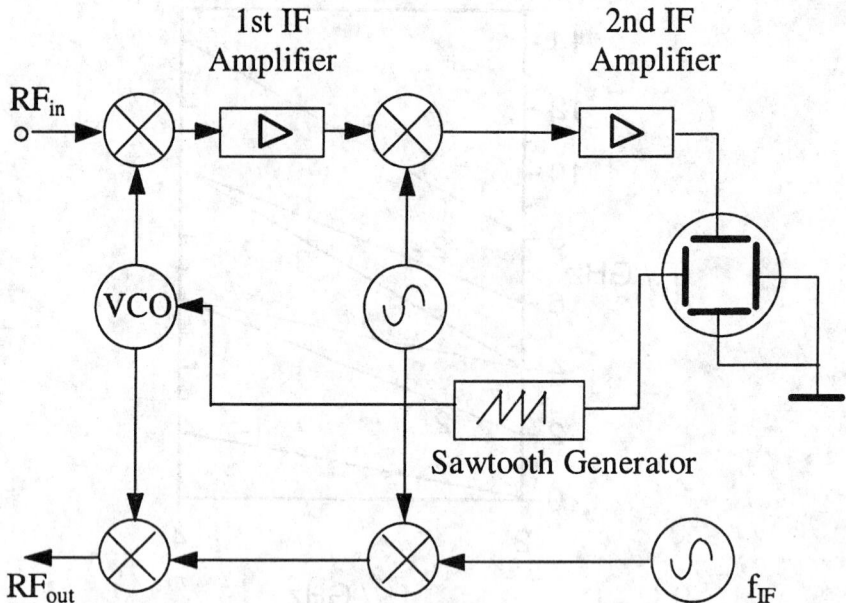

FIGURE 55.8   Principle of a tracking generator, which delivers at the port RF$_{out}$ a signal that is precisely in the passband of the spectrum analyzer.

Tracking Generator

Spectrum analyzers are often equipped with a tracking generator. The principle is shown in Figure 55.8. A local oscillator, with a frequency exactly on the center frequency of the IF amplifier is mixed by an identical setup as in the analyzer path. Using the same local oscillator as in the analyzer path ensures that the frequency of the tracking generator follows precisely the center frequency of the swept window of the analyzed frequency band.

The tracking generator can be used for network analysis. If the tracking frequency is used as an input signal of a two-port, the amplitude of the output can be measured with the spectrum analyzer. Such a network analyzer has the advantage of being sensitive only in a very narrow band. Thus, third-order intermodulation products and noise are suppressed. However, only scalar measurements can be performed.

Commercially Available Spectrum Analyzers

A number of commercially available, general-purpose spectrum analyzers for the microwave and mm-wave range are listed in Table 55.3. Only a small number of spectrum analyzers available on the market are presented. Most of the companies offer special equipment for production quality control. These spectrum analyzers can be computer controlled for fixed measurement routines. On request, spectrum analyzers with special options like fixed frequency operation, multichannel operation, support of external mixers, integrated frequency counters, and digital storage are available.

Typical specifications of microwave spectrum analyzers include:

- The frequency span is several gigahertz.
- With external mixers, the upper frequency limit can be extended to more than 100 GHz.
- The frequency accuracy is between 10^{-5} and 10^{-7}.

TABLE 55.3 Commercially Available Spectrum Analyzers

Company/ model	Frequency range	Min. res. bandw.	Amplitude accuracy	Remarks	Price $U.S.
Anritsu					
MS2602A	100 Hz–8.5 GHz		1.1 dB		$30,600
Avantek					
3365	100 Hz–8 GHz	10 Hz		Portable, tracking	$58,000
3371	100 Hz–26.5 GHz	10 Hz		Portable, tracking	$66,000
R3272	9 kHz–26.5 GHz	300 Hz	1 dB	External mixer 325 GHz	$38,000
Hewlett-Packard					
HP4196A	2 Hz–1.8 GHz	1 Hz	1 dB		
HP8590L	9 kHz–1.8 GHz	1 kHz	1.7 dB		$9080
HP8560E	30 Hz–2.9 GHz	1 Hz	1.85 dB	Portable	$27,530
HP8596E	9 kHz–12.8 GHz	30 Hz	2.7 dB		$25,090
HP8593E	9 kHz–22 GHz	30 Hz	2.7 dB		$27,435
HP8564E	9 kHz–40 GHz	1 Hz	3 dB		$50,890
HP8565E	9 kHz–50 GHz	1 Hz	3 dB		$67,245
Marconi					
2370	30 Hz–1.25 GHz	5 Hz	5 Hz	With frequency extender	
2383	30 Hz–4.2 GHz	3 Hz	1 dB	Tracking	
Rohde & Schwarz					
FSEA30	20 Hz–3.5 GHz	1 Hz	1 dB		$43,000
FSEB30	20 Hz–7 GHz	1 Hz	1 dB		$52,000
FSEM30	20 Hz–26.5 GHz	1 Hz	1 dB	External mixer 110 GHz	$64,000
Tektronix					
2714	9 kHz–1.8 GHz	300 Hz	2 dB	AM/FM demodulation 50 Ω/75 Ω	
2784	100 Hz–40 GHz	3 Hz	1.5 dB	Counter 1.2 THz, external mixer 325 GHz	

- The resolution bandwidth (i.e., the effective bandwidth of the narrow IF filter) can be adjusted in steps between 1 Hz and a few megahertz.
- The resolution bandwidth shape factor is typically 10:1.
- The amplitude accuracy is about 1 dB.
- The noise floor is at about −140 dBm.

55.4 Cavity Modes and Cavity *Q*

Cavities are used in a variety of applications. For example, they can be used to construct filters and they serve as those elements in microwave generators (e.g., klystrons) that determine the operating frequency. Cavities can also be applied in order to measure the frequency or the wavelength of microwaves (wavemeter). The most important parameters of a cavity are its resonant frequency and its q-factor. The latter determines the sharpness of the resonance, or, in case of filters, the bandwidth of the passband.

A cavity can be defined as a volume that is almost completely surrounded by a metallic surface. At one or two positions, coupling elements are applied to the metal surface in order to connect the cavity to other circuit elements. In the case of one coupling element, one speaks of a single-ended cavity; whereas a transmission cavity has two coupling elements. The coupling elements are usually small compared to the dimensions of the cavity. The inside of the cavity is filled with dielectric material (e.g., air: $\varepsilon_r = 1$).

Cavities are used as resonators for microwave applications. Therefore, they are comparable to low-frequency resonant circuits consisting of an inductance L and a capacitance C. In the low-frequency range, lumped elements (inductors and capacitors) are used, which are small in comparison with the wavelength. In contrast, cavities are distributed elements with dimensions comparable to the wavelength. This results in comparatively small losses. Since cavities are distributed elements, it is in general no longer possible to determine L and C directly. Instead of L and C, the resonant frequency $f_0 = \omega_0/2\pi$ is the most important property of a cavity.

Neither lumped elements nor cavities are completely lossless. One must take into account the finite conductivities of the materials, resulting in a resistance R when analyzing resonant circuits. For cavities, however, R cannot be determined directly due to the same reasons that hold for L and C. Therefore, a different parameter, the q-factor Q plays a similar role for cavities. Q is proportional to the stored electric and magnetic energy W, divided by the power loss P:

$$Q = \frac{\omega_0 W}{P} \tag{55.15}$$

Although cavities are distributed elements, one is able to show that their behavior near resonance can be described by a simple parallel resonant circuit consisting of R, L, and C if the reference plane is chosen appropriately. Therefore, the basic properties of such a parallel resonant circuit are analyzed in the following.

Assume that the losses are small ($Q \gg 1$), which is desirable in practice. Therefore, the resistance R of the parallel resonant circuit is comparatively large. In this case, the resonant frequency does not depend on R in a first-order approximation:

$$\omega_0 \approx 1/\sqrt{LC} \tag{55.16}$$

In Equation 55.15 is applied to the analyzed parallel resonant circuit, one obtains:

$$Q \approx \frac{R}{\omega_0 L} \approx R\sqrt{\frac{C}{L}} \tag{55.17}$$

One can characterize the width of a resonance curve by those angular frequencies $\omega_1 = 2\pi f_1$ and $\omega_2 = 2\pi f_2$ where the power has decreased to one-half of its maximum value (–3 dB). The impedance has then decreased to $1/\sqrt{2} \approx 70.7\%$ of its maximum value. This enables calculation of the angular frequencies ω_1 and w_2:

$$\omega_1 \approx \omega_0 - \frac{1}{2RC}, \quad \omega_2 \approx \omega_0 + \frac{1}{2RC} \tag{55.18}$$

Using these relations, one can easily derive the following equation, which is equivalent to Equation 55.17.

$$Q \approx \frac{\omega_0}{\omega_2 - \omega_1} \approx \frac{f_0}{f_2 - f_1} \tag{55.19}$$

This expression shows that Q is a symbol for the sharpness of resonance. Furthermore, it leads to the first principle as to how the q-factor can be measured. This principle, which is often referred to as the bandpass method, is based on the measurement of a resonance curve. For example, one can measure the reflection coefficient S_{11} with a network analyzer. From this curve, the 3 dB-bandwidth $\Delta f = f_2 - f_1$ and the resonant frequency $f_0 = (f_1 + f_2)/2$ can be easily determined. The application of Equation 55.19 yields the desired Q.

A second principle used to measure the q-factor is based on the transient behavior of the cavity after excitation with a pulse. In this case, the energy decays exponentially, and the time constant of the decay is proportional to Q^{-1}. This can again be shown by analyzing the equivalent parallel resonant circuit. Up until now, all signals were time-harmonic, which enabled a solution by complex quantities. This time, the corresponding differential equation must be solved. One can easily show that the amplitude of the voltage is proportional to $e^{-\omega_0 t/2Q}$, which means a $e^{-\omega_0 t/2Q}$ dependency of the energy exists. Measuring the decay time constant, therefore, enables one to determine the q-factor.

The Q, defined here is the unloaded Q of the cavity; that is, no further resistance is connected to the cavity. Sometimes it is desirable to determine the loaded Q (Q_L), which takes into account such resistances. Detailed information about the loaded Q can be found in [21–23]. Although the q-factor is not influenced by the coupling structure if it is lossless, the coupling structure is of great importance. Further information about coupling parameters is presented in [21–23]. Detailed descriptions of measurement methods based on the above mentioned principles can also be found in [21, 22].

Up to now, all analyses were based on the equivalent resonant circuit. There are many other effects that can be explained by this analogy. For example, the energy in both the cavity and in its equivalent circuit oscillates between the magnetic and electric fields. Some properties of cavities, however, can only be seen by examining the electric and magnetic fields themselves. These are governed by Maxwell's equations.

A solution of Maxwell's equations (which can be very complicated for a given cavity) shows that all cavities have an infinite number of resonant frequencies. This could not be seen by analyzing the equivalent parallel resonant circuit. A description of the cavity by a discrete parallel resonant circuit is only valid in the vicinity of *one* of these resonant frequencies.

Furthermore, different modes can exist that have the same resonant frequency. This phenomenon is called *degeneration*. If the resonator will operate at such a frequency, where degenerate modes exist, one has to take care that the companion mode is suppressed. This can be accomplished by an appropriate choice of the positions of the coupling elements.

Even if the desired mode does not have any companion mode, one should take care that no other mode exists in the operating range of the cavity (this can be accomplished with a mode chart [22, 23]); otherwise, these modes must be damped sufficiently in order to be sure that an observed resonance corresponds to the desired mode.

55.5 Scattering Parameter Measurements

Scattering parameters describe multiple port structures in terms of wave variables. The introduction of scattering parameters (S-parameters) arises naturally in microwave circuits and systems, due to the lack of a unique definition for currents and voltages at these frequencies. Most circuits and systems at high frequencies are efficiently described in terms of S-parameters.

This section describes the fundamentals and properties of S-parameters, together with network analysis based on S-parameter calculations and measurements. Measurement procedures are outlined and the most frequently used systems for S-parameter measurement are described. Finally, information on hardware required for the experimental determination of S-parameters is provided, together with the corresponding suppliers.

Introduction and Fundamentals

At high frequencies, the wave variables are a natural extension of the voltages and currents at port terminals. In electric systems where the voltages and currents cannot be uniquely defined, the power flow in a waveguide can be described via wave variables. Whenever a TEM mode of wave propagation cannot be assumed, the currents and voltages are dependent on the integration path. This situation is encountered in all rectangular, circular, and passive waveguide structures, even in the case of lossless wave propagation. It is also true for all guiding structures if losses are to be considered along the path of wave propagation [24–27]. For the case of wave propagation along a transmission line, the wave variables a and b are defined as follows:

$$a(z) = \frac{1}{2}\left(\frac{U(z)}{\sqrt{Z_0}} + I(z)\sqrt{Z_0}\right)$$

$$= \frac{U_+}{\sqrt{Z_0}} = I_+ \sqrt{Z_0}$$

(55.20)

$$b(z) = \frac{1}{2}\left(\frac{U(z)}{\sqrt{Z_0}} - I(z)\sqrt{Z_0}\right)$$

$$= \frac{U_-}{\sqrt{Z_0}} = I_- \sqrt{Z_0}$$

The propagation is along the z-direction. The characteristic impedance of the transmission line is Z_0, and $U(z)$ and $I(z)$ are the voltage and current, respectively at location z along the line. The variables $a(z)$ and $b(z)$ are the complex amplitudes of the modes on the line. The voltage U_+ and U_- and the currents I_+ and I_- denote the voltage and current amplitudes, respectively, in forward and reverse direction. The wave variables are related to the power in the following form:

$$P_+ = \frac{1}{2}\frac{|U_+|^2}{Z_0} = \frac{1}{2}|I_+|^2 Z_0 = \frac{1}{2}|a(z)|^2$$

(55.21)

$$P_- = \frac{1}{2}\frac{|U_-|^2}{Z_0} = \frac{1}{2}|I_-|^2 Z_0 = \frac{1}{2}|b(z)|^2$$

In Equation 55.21, it is assumed that the system is excited by a pure sinusoid and that the characteristic impedance is purely real. The wave variables have the dimensions of $\sqrt{W}$.

Strictly speaking, wave variables and S-parameters can only be applied to linear networks. This is important because many publications are devoted to so-called large-signal S-parameter measurements. The interpretation of such results is not simple and great care must be employed in the correct determination of the characteristic impedance of the system [26]. In the case of analysis of large-signal or nonlinear circuits, two methods exist to introduce the wave variables:

- Volterra series representation [28]
- Harmonic-balance method [26]

Both methods transform the nonlinear circuit into a number of linear circuits at different frequencies, and then change the terminal voltages and currents into wave variables. This situation is sketched in Figure 55.9. Particular attention must be paid to the definition of the characteristic impedance, which can vary between different frequencies. An in-depth treatment of wave variables can be found in [26].

A further utilization of wave variables can be found in the noise analysis of microwave circuits. According to Figure 55.10, a noisy multiport can be analyzed by an associated noiseless two-port with the according noise sources c_i at the corresponding ports of the circuit.

Calculations and Analysis with S-Parameters

The characterization of multiports with S-parameters requires embedding of the multiport into a system consisting of a signal source with a characteristic impedance and appropriate terminations of all ports.

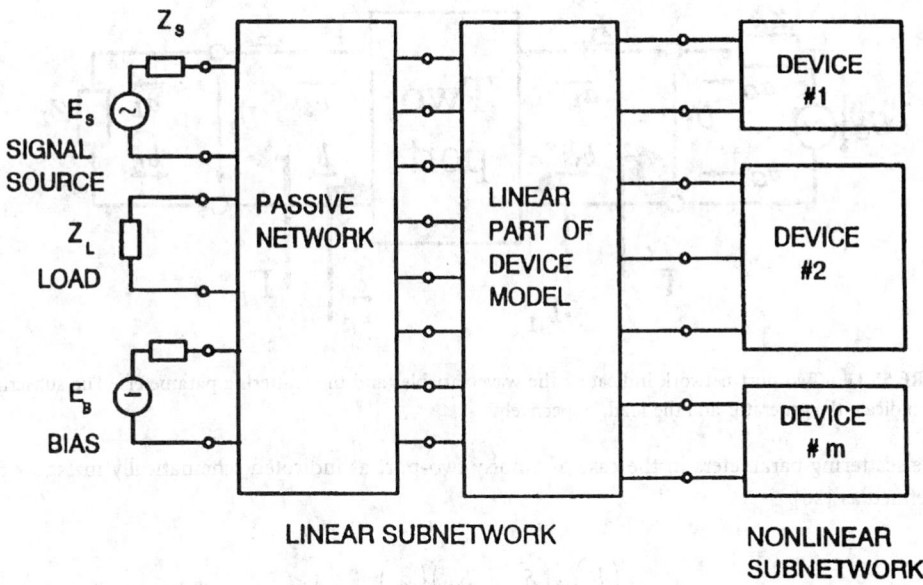

FIGURE 55.9 Schematic illustration of application of wave variables to nonlinear circuits.

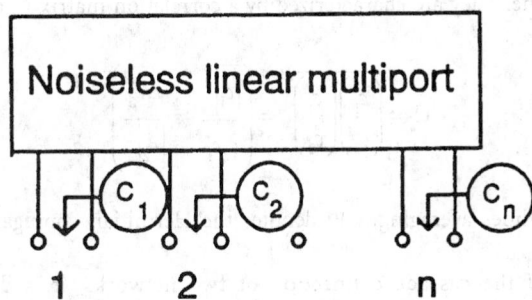

FIGURE 55.10 Schematic illustration of application of wave variables to noisy circuits.

This situation is shown in Figure 55.11. The outgoing wave parameters b are reflections at the corresponding ports. The wave variables are related to the scattering parameters of a two-port in the following manner:

$$\begin{pmatrix} b_1 \\ b_2 \end{pmatrix} = \left(S \right) \begin{pmatrix} a_1 \\ a_2 \end{pmatrix} = \begin{pmatrix} S_{11} & S_{12} \\ S_{21} & S_{22} \end{pmatrix} \begin{pmatrix} a_1 \\ a_2 \end{pmatrix} \tag{55.22}$$

For the determination of the individual scattering matrix elements, all ports of the network must be terminated in their characteristic impedance. The impedances at the corresponding ports need not be equal for all ports. The S-parameters are, in general, complex and are defined with respect to reference planes. These reference planes can be the network terminals, but could also been shifted to other locations in the circuit if this is desirable. The scattering matrix can be transformed into all circuit representations. Table 55.4 indicates the conversion formulae between the S-parameters and $Z\,Y\,h$ parameters for arbitrary characteristic impedances. Additional conversions to $ABCD$ and T parameters can be found in [29].

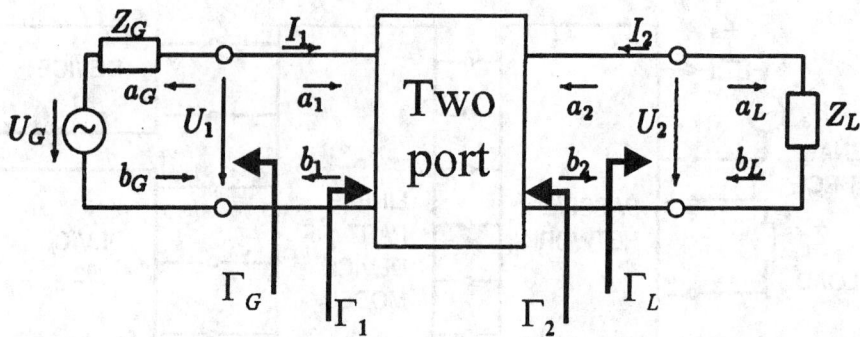

FIGURE 55.11 Two-port network indicating the wave variables and the scattering parameters. The subscripts G and L indicate the generator and the load, respectively.

The scattering parameters in the case of a noisy two-port as indicated schematically in Figure 55.10 are defined as [26]:

$$\begin{pmatrix} b_1 \\ b_2 \end{pmatrix} = \begin{pmatrix} S_{11} & S_{12} \\ S_{21} & S_{22} \end{pmatrix} \begin{pmatrix} a_1 \\ a_2 \end{pmatrix} + \begin{pmatrix} c_1 \\ c_2 \end{pmatrix} \tag{55.23}$$

The noise wave sources c_1 and c_2 represent the noise generated in the circuit and are therefore complex variables varying with time. They are characterized by a correlation matrix C_s as follows:

$$C_s = \overline{\begin{pmatrix} c_1 \\ c_2 \end{pmatrix} \begin{pmatrix} c_1 \\ c_2 \end{pmatrix}^H} = \begin{pmatrix} \overline{|c_1|^2} & \overline{c_1 c_2^*} \\ \overline{c_2 c_1^*} & \overline{|c_2|^2} \end{pmatrix} \tag{55.24}$$

where the bar indicates times averaging, $(\cdot)^H$ denotes the Hermitian conjugate, and * stands for the complex conjugate.

For the calculation of the cascade connection of two networks, it is desirable to convert the S-parameters to T parameters defined in the following way:

$$\begin{pmatrix} b_1 \\ a_1 \end{pmatrix} = \begin{pmatrix} T_{11} & T_{12} \\ T_{21} & T_{22} \end{pmatrix} \cdot \begin{pmatrix} a_2 \\ b_2 \end{pmatrix} \tag{55.25}$$

The conversion between S-parameters and T parameters is given below.

$$T_{11} = S_{12} - \frac{S_{11} S_{22}}{S_{21}} \tag{55.26}$$

$$T_{12} = \frac{S_{11}}{S_{21}} \tag{55.27}$$

$$T_{21} = -\frac{S_{22}}{S_{21}} \tag{55.28}$$

TABLE 55.4 Equations for the Conversion Between Different Two-Port Parameters

$$S_{11} = \frac{\left(X_{11} - Z_{01}^*\right)\left(X_{22} + Z_{02}\right) - X_{12}X_{21}}{\left(X_{11} + Z_{01}\right)\left(X_{22} + Z_{02}\right) - X_{12}X_{21}}$$

$$X_{11} = \frac{\left(Z_{01}^* + S_{11}Z_{01}\right)\left(1 - S_{22}\right) + S_{12}S_{21}Z_{01}}{\left(1 - S_{11}\right)\left(1 - S_{22}\right) - S_{12}S_{21}}$$

$$S_{12} = \frac{2X_{12}\sqrt{R_{01}R_{02}}}{\left(X_{11} + Z_{01}\right)\left(X_{22} + Z_{02}\right) - X_{12}X_{21}}$$

$$X_{12} = \frac{2S_{12}\sqrt{R_{01}R_{02}}}{\left(1 - S_{11}\right)\left(1 - S_{22}\right) - S_{12}S_{21}}$$

$$S_{21} = \frac{2X_{21}\sqrt{R_{01}R_{02}}}{\left(X_{11} + Z_{01}\right)\left(X_{22} + Z_{02}\right) - X_{12}X_{21}}$$

$$X_{21} = \frac{2S_{21}\sqrt{R_{01}R_{02}}}{\left(1 - S_{11}\right)\left(1 - S_{22}\right) - S_{12}S_{21}}$$

$$S_{22} = \frac{\left(X_{11} + Z_{01}\right)\left(X_{22} - Z_{02}^*\right) - X_{12}X_{21}}{\left(X_{11} + Z_{01}\right)\left(X_{22} + Z_{02}\right) - X_{12}X_{21}}$$

$$X_{22} = \frac{\left(1 - S_{11}\right)\left(Z_{02}^* + S_{22}Z_{02}\right) - S_{12}S_{21}Z_{02}}{\left(1 - S_{11}\right)\left(1 - S_{22}\right) - S_{12}S_{21}}$$

$$S_{11} = \frac{\left(1 - Y_{11}Z_{01}^*\right)\left(1 + Y_{22}Z_{02}\right) + Y_{12}Y_{21}Z_{01}^*Z_{02}}{\left(1 + Y_{11}Z_{01}\right)\left(1 + Y_{22}Z_{02}\right) - Y_{12}Y_{21}Z_{01}Z_{02}}$$

$$Y_{11} = \frac{\left(1 - S_{11}\right)\left(Z_{02}^* + S_{22}Z_{02}\right) + S_{12}S_{21}Z_{02}}{\left(Z_{01}^* + S_{11}Z_{01}\right)\left(Z_{02}^* + S_{22}Z_{02}\right) - S_{12}S_{21}Z_{01}Z_{02}}$$

$$S_{12} = \frac{-2Y_{12}\sqrt{R_{01}R_{02}}}{\left(1 + Y_{11}Z_{01}\right)\left(1 + Y_{22}Z_{02}\right) - Y_{12}Y_{21}Z_{01}Z_{02}}$$

$$Y_{12} = \frac{-2S_{12}\sqrt{R_{01}R_{02}}}{\left(Z_{01}^* + S_{11}Z_{01}\right)\left(Z_{02}^* + S_{22}Z_{02}\right) - S_{12}S_{21}Z_{01}Z_{02}}$$

$$S_{21} = \frac{-2Y_{21}\sqrt{R_{01}R_{02}}}{\left(1 + Y_{11}Z_{01}\right)\left(1 + Y_{22}Z_{02}\right) - Y_{12}Y_{21}Z_{01}Z_{02}}$$

$$Y_{21} = \frac{-2S_{21}\sqrt{R_{01}R_{02}}}{\left(Z_{01}^* + S_{11}Z_{01}\right)\left(Z_{02}^* + S_{22}Z_{02}\right) - S_{12}S_{21}Z_{01}Z_{02}}$$

$$S_{22} = \frac{\left(1 + Y_{11}Y_{11}\right)\left(1 - Y_{22}Z_{02}^*\right) + Y_{12}Y_{21}Z_{01}Z_{02}^*}{\left(1 + Y_{11}Z_{01}\right)\left(1 + Y_{22}Z_{02}\right) - Y_{12}Y_{21}Z_{01}Z_{02}}$$

$$Y_{22} = \frac{\left(Z_{01}^* + S_{11}Z_{01}\right)\left(1 - S_{22}\right) - S_{12}S_{21}Z_{01}}{\left(Z_{01}^* + S_{11}Z_{01}\right)\left(Z_{02}^* + S_{22}Z_{02}\right) - S_{12}S_{21}Z_{01}Z_{02}}$$

$$S_{11} = \frac{\left(h_{11} - Z_{01}^*\right)\left(1 + h_{22}Z_{02}\right) - h_{12}h_{21}Z_{02}}{\left(h_{11} + Z_{01}\right)\left(1 + h_{22}Z_{02}\right) - h_{12}h_{21}Z_{02}}$$

$$h_{11} = \frac{\left(Z_{01}^* + S_{11}Z_{01}\right)\left(Z_{02}^* + S_{22}Z_{02}\right) + S_{12}S_{21}Z_{01}Z_{02}}{\left(1 - S_{11}\right)\left(Z_{02}^* + S_{22}Z_{02}\right) - S_{12}S_{21}Z_{02}}$$

$$S_{12} = \frac{2h_{12}\sqrt{R_{01}R_{02}}}{\left(h_{11} + Z_{01}\right)\left(1 + h_{22}Z_{02}\right) - h_{12}h_{21}Z_{02}}$$

$$h_{12} = \frac{2S_{12}\sqrt{R_{01}R_{02}}}{\left(1 - S_{11}\right)\left(Z_{02}^* + S_{22}Z_{02}\right) - S_{12}S_{21}Z_{02}}$$

$$S_{21} = \frac{-2h_{21}\sqrt{R_{01}R_{02}}}{\left(h_{11} + Z_{01}\right)\left(1 + h_{22}Z_{02}\right) - h_{12}h_{21}Z_{02}}$$

$$h_{21} = \frac{-2S_{21}\sqrt{R_{01}R_{02}}}{\left(1 - S_{11}\right)\left(Z_{02}^* + S_{22}Z_{02}\right) - S_{12}S_{21}Z_{02}}$$

$$S_{22} = \frac{\left(h_{11} + Z_{01}\right)\left(h_{22} - Z_{02}^*\right) - h_{12}h_{21}Z_{02}^*}{\left(h_{11} + Z_{01}\right)\left(h_{22}Z_{02}\right) - h_{12}h_{21}Z_{02}}$$

$$h_{22} = \frac{\left(1 - S_{11}\right)\left(1 - S_{22}\right) - S_{12}S_{21}}{\left(1 - S_{11}\right)\left(Z_{02}^* - S_{12}S_{21}Z_{02}\right)}$$

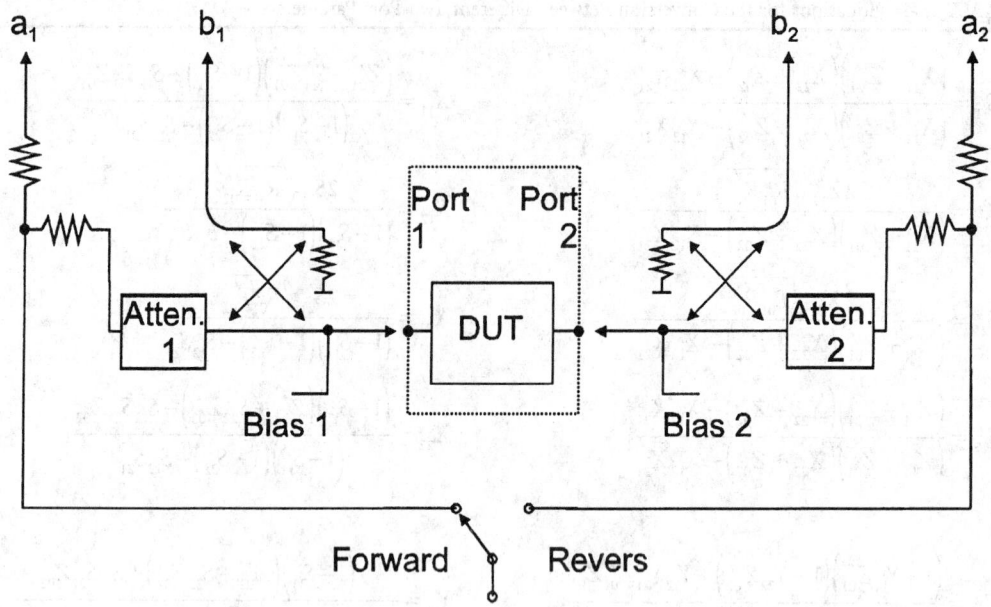

FIGURE 55.12   Schematic illustration of a network analyzer configuration.

$$T_{22} = \frac{1}{S_{21}} \tag{55.29}$$

It should be emphasized that different definitions of the T parameters exist in the literature [2, 24–27, 30]. Power gain, mismatch, insertion loss, etc. can be efficiently described with the help of scattering parameters [2, 30].

Measurement of S-Parameters

Network analyzers are generally used for the measurement of S-parameters. A schematic configuration of a network analyzer is indicated in Figure 55.12. The network analyzer consists of two structures to separate the signals and a heterodyne receiver. According to the definitions, the measurement is performed in two steps as indicated in Figure 55.13. Different error models are employed for the calibration of the network analyzer. The most simple is the one-port error model, which consists of contributions due to directivity, source mismatch, and frequency response. A flow diagram is illustrated in Figure 55.14.

For the characterization of active and passive two-ports, a more sophisticated error model is required. The signal flow graphs of the full two-port model is drawn in Figure 55.15. To determine the S-parameters of the device under test, a de-embedding procedure is required [31–43]. The two-port error model is then divided into two error adapters and the actual device under test (DUT), as depicted in Figure 55.16. The example shown makes reference to the so called "TRL" calibration procedure. This name abbreviates the three calibration standards utilized in this method: a *through* standard with zero length, a *reflecting* standard, and a *line* standard. This method cannot be applied to on-wafer measurements at low frequencies, due to the excessive line length required for a broadband measurement. Other error correction methods are summarized in Table 55.5 [44]. In addition to the measurements given in the table, a known reference impedance and port 1 to port 2 connection are required. Furthermore, at higher frequencies (above ≈15 GHz), a calibration measurement of the isolation should be performed. For this purpose, both ports are terminated by their characteristic impedances and a transmission measurement is performed. This measurement determines the values of C_F and C_R in the calibration flow diagram (see Figure 55.14).

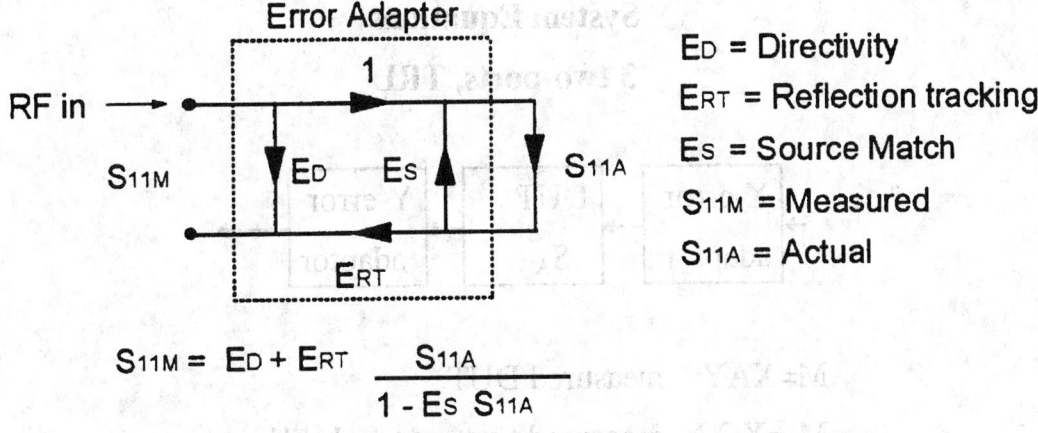

$$S_{11M} = E_D + E_{RT} \frac{S_{11A}}{1 - E_S S_{11A}}$$

FIGURE 55.13 Flow diagram of the measurement procedure.

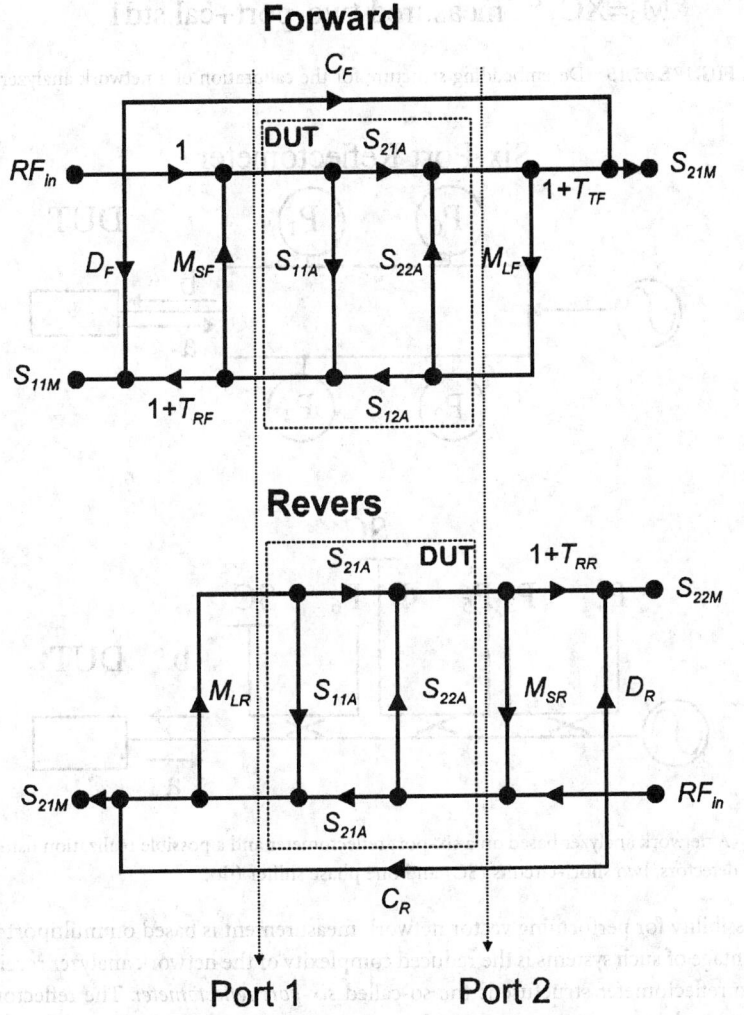

FIGURE 55.14 Flow diagram of the two-port error model.

System Equations

3 two-ports, TRL

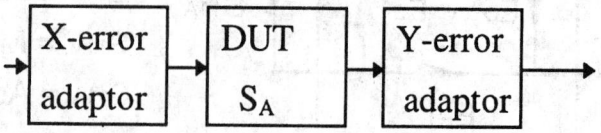

M=XAY   measured DUT

$M_1 = XC_1Y$ measured two-port+cal.std1

$M_2 = XC_2Y$ measured two-port+cal.std1

$M_3 = XC_3Y$ measured two-port+cal.std1

FIGURE 55.15   De-embedding structure for the calibration of a network analyzer.

Six-Port Reflectometer

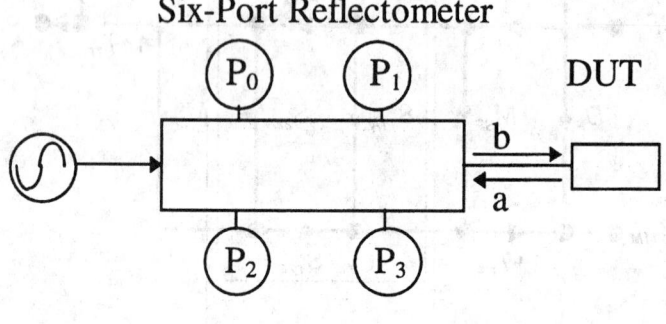

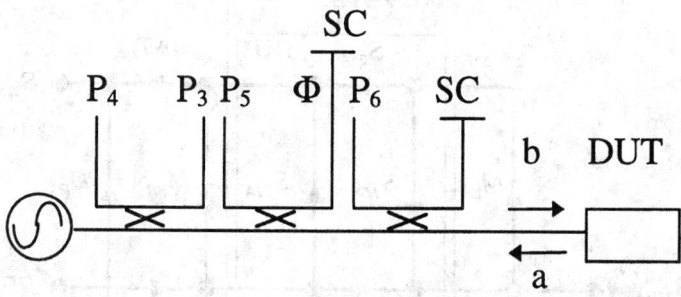

FIGURE 55.16   A network analyzer based on a six-port reflectometer and a possible realization using three couplers and four power detectors, two short-circuits (SC) and one phase shifter (Φ).

Another possibility for performing vector network measurement is based on multiport reflectometers [45]. The advantage of such systems is the reduced complexity of the network analyzer receiver. A possible realization of a reflectometer structure is the so-called *six-port reflectometer*. The reflectometer consists of three couplers and four power sensors. No frequency conversion is required, which simplifies the test equipment.

TABLE 55.5   Summary of Different Calibration Methods

TOSL	Through standard (T) with known length; Fullfills 4 conditions	3 known reflections (OSL) on port 1; Fullfills 3 conditions	3 known reflections (OSL) on port 2; Fullfills 3 conditions
TRL & LRL	Through or line standard (T) or (L) with known length; Fullfills 4 conditions	Unknown equal reflection standard (R) on port 1 and port 2; Fulfills 1 condition	Line (L) with known S_{11} and S_{22}; Fullfills 2 conditions
TRM & LRM	Through or Line standard (T) or (L) with known length; Fullfills 4 conditions	Unknown equal reflection standard (R) on port 1 and port 2; Fullfills 1 condition	Known match (M) on port 1 and port 2; Fullfills 2 conditions
TXYZ & LXYZ	Through or line standard (T) or (L) with known length; Fullfills 4 conditions	3 known reflection standards (XYZ) on port 1 or port 2; Fullfills 3 conditions	
UXYZ	Unknown line standard (U) with S_{11} = S_{21}; Fullfills 1 condition	3 known reflection standards (XYZ) on port 1; Fullfills 3 conditions	3 known reflection standards (XYZ) on port 2; Fullfills 3 conditions

TABLE 55.6   Companies Supplying Network Analyzers for *S*-Parameter Measurements

	Frequency			
Company	min	max	Method	Cal. Methods
Hewlett-Packard HP 8510	45 MHz	110 GHz	Heterodyne	SOLT, TLR, LRL, LRM, TRM
Wiltron	45 MHz	110 GHz	Heterodyne	SOLT, TLR, LRL, LRM, TRM
Rhode & Schwarz	10 Hz	4 GHz	Heterodyne	SOLT, TLR, LTL, LRM, TRM
AB Millimeterique	2 GHz	800 GHz	Heterodyne	TLR, LRL, proprietary

Commercially Available Network Analyzers

Table 55.6 shows some of the current suppliers for network analysis. The frequency range is 10 Hz up to 800 GHz.

References

1. A. Fantom, *Radio Frequency and Microwave Power Measurements,* London: Peter Peregrinus, 1990.
2. A. L. Lance, *Microwave Measurements* in *Handbook of Microwave Optical Components,* Vol. 1, K. Chang (ed.), New York: John Wiley & Sons, 1989.
3. J. Minck, Fundamentals of RF and microwave power measurements, Hewlett-Packard Application Note 64-1A, 1997.
4. G. H. Bryant, *Principles of Microwave Measurements,* London: Peter Peregrinus, 1988.
5. *Microwave Powermate,* Marconi Instruments Ltd., 1989.
6. D. Janik, H. Wolf, and R. Schneider, High-Tc edge bolometer for detecting guided millimeter waves, *IEEE Trans. Appl. Superconduct.,* 3, 2148–2151, 1993.
7. P. L. Richards, Bolometers for infrared and millimeter waves, *J. Appl. Phys.,* 76(1), 1–24, 1994.
8. D. M. Rowe, *Handbook of Thermoelectrics,* Boca Raton, FL: CRC Press, 1995.
9. P. Kopystynski, E. Obermayer, H. Delfs, W. Hohenester, and A. Löser, Silicon power sensor from dc to microwave, in *Micro Systems Technologies 90,* Berlin: Springer, 1990, 605–610.
10. G. C. M. Meijer and A. W. Herwaarden, *Thermal Sources,* Bristol: IOP Ltd., 1994.
11. T. Närhi, Nonlinearity characterisation of microwave detectors for radiometer applications, *Electron. Lett.,* 32, 224–225, 1996.
12. F. Wiedmann, B. Huyart, E. Bergeault, and L. Jallet, New structure for a six-port reflectometer in monolithic microwave integrated-circuit technology, *IEEE Trans. Instrument. Measure.,* 46(2), 527–530, 1997.
13. A. Dehé, V. Krozer, B. Chen, and H. L. Hartnagel, High-sensitivity microwave power sensor for GaAs-MMIC implementation, *Electron. Lett.,* 32(23), 2149–2150, 1996.

14. Circuits Multi Projects, *Information CMP — 42.* Grenoble, France, December 1996.

15. Hints for making better spectrum analyzer measurements, Hewlett-Packard Application Note, No. 1286-1, 1997.

16. C. Brown, Spectrum Analysis Basics, *Hewlett-Packard 1997 Back to Basics Seminar,* 1997.

17. A. Kiiss, Microwave instrumentation and measurements, in *Handbook of Microwave Technology,* T. K. Ishii (ed.), San Diego: Academic Press, 1995, Vol. 2, 562.

18. H. Brand, Spectrum analyzers: precision test instruments, *Microwave J.,* 37(3), 98, 1994.

19. Spectrum analysis: Spectrum analysis basics, Hewlett-Packard Application Note 150, November 1989.

20. T. K. Ishii, Spectrum Analysis: Amplitude and Frequency Modulation, Application Note 150-1, January 1989.

21. M. Sucher and J. Fox, *Handbook of Microwave Measurements,* Vol. II, Polytechnic Press of the Polytechnic Institute of Brooklyn, 1963, chap. VIII.

22. E. L. Ginzton, *Microwave Measurements,* New York: McGraw-Hill, 1957, chap. 7 and 8.

23. F. E. Terman and J. M. Pettit, *Electronic Measurements,* New York: McGraw-Hill, 1952, chapter 4–15.

24. C. A. Lee and G. C. Dalman, *Microwave Devices, Circuits and Their Interaction,* New York: John Wiley & Sons, 1994.

25. G. D. Vendelin, A. M. Pavio, and U. L. Rohde, *Microwave Circuit Design Using Linear and Nonlinear Techniques,* New York: John Wiley & Sons, 1990.

26. J. Dobrowolski, *Microwave Circuit Analysis with Wave Variables,* Norwood, MA: Artech House, 1991.

27. G. Gonzalez, *Microwave Transistor Amplifiers,* Englewood Cliffs, NJ: Prentice-Hall, 1984.

28. D. Weiner and G. Naditch, A scattering variable approach to the Volterro analysis of nonlinear systems, *IEEE Trans. Microwave Theory & Technique,* MTT-24(7), 422–433, 1976.

29. D. A. Frickey, Conversion between S, Z, Y, H, ABSD, and T parameters which are valid for complex source and load impedances, *IEEE Trans. Microwave Theory & Technique,* MTT-42, 205-211, 1994.

30. T. K. Ishii, *Handbook of Microwave Technology,* San Diego, CA: Academic Press, 1995.

31. R. A. Hackborn, An automatic network analyzer systems, *Microwave J.,* 45–52, 1968.

32. S. Rehnmark, On the calibration process of automatic network analyzer systems, *IEEE Trans. Microwave Theory & Technique,* MTT-22(4), 457–458, 1974.

33. J. Fitzpatrick, Error models of vector systems measurements, *Microwave J.,* 21(5), 63–66, 1978.

34. D. Rytting, An analysis of vector measurement accuracy enhancement techniques, *RF & Microwave Measurement Symp. and Exhibition,* Hewlett Packard, 1982.

35. N. R. Franzen and R. A. Speciale, A new procedure for system calibration and error removal in automated S-parameter measurements, *5th European Microwave Conf.,* 1975, 69–73.

36. R. A. Soares and C. A. Hoer, A unified mathematical approach to two-port calibration techniques and some applications, *IEEE Trans. Microwave Theory & Techniques,* MTT-37(11), 1669–1674, 1989.

37. R. A. Soares, *GaAs MESFET Circuit Design,* Norwood, MA: Artech House, 1988.

38. Cascade Microtech, *Microwave Wafer Probe Calibration Standards: HP8510 Network Analyzer Input,* Cascade Microtech Instruction Manual, 1990.

39. Understanding the fundamental principles of vector network analysis, Hewlett-Packard Application Note 1287-1, 1997.

40. B. Donecker, Determining the measurement accuracy of the HP8510 microwave network analyzer, *RF & Microwave Measurement Symp. and Exhibition,* Hewlett-Packard, March 1985.

41. Exploring the architectures of network analyzers, Hewlett-Packard Application Note 1287-2, 1997.

42. Applying error correction to network analyzer measurements, Hewlett-Packard Application Note 1287-3, 1997.

43. D. Ballo, Network analyzer basics, *Hewlett-Packard 1997 Back to Basics Seminar,* 1997.

44. H. J. Eul and B. Schiek, Thru-Match-Reflect: one result of a rigorous theory for deembedding and network analyzer calibration, *18th European Microwave Conf.,* 1988, 909–914.

45. G. F. Engen and C. A. Hoer, Thru-Reflect-Line: an improved technique for calibrating the dual 6-port automatic network analyzer, *IEEE Trans. Microwave Theory & Technique,* MTT-27(12), 993–998, 1979.

VIII

Optical Variables Measurement

Fritz Schuermeyer
Wright Patterson Air Force Base

Thad Pickenpaugh
Wright Patterson Air Force Base

Michael R. Squillante
Radiation Monitoring Devices, Inc.

Kanai S. Shah
Radiation Monitoring Devices, Inc.

J.A. Nousek
Pennsylvania State University

M.W. Bautz
Pennsylvania State University

B.E. Burke
Pennsylvania State University

J.A. Gregory
Pennsylvania State University

R.E. Griffiths
Pennsylvania State University

R.L. Kraft
Pennsylvania State University

H.L. Kwok
Pennsylvania State University

D.H. Lumb
Pennsylvania State University

56
Photometry and Radiometry

56.1 Photoconductive Sensors

Fritz Schuermeyer and Thad Pickenpaugh

Introduction

Photoconduction has been observed, studied, and applied for more than 100 years. In the year 1873, W. Smith [1] noticed that the resistance of a selenium resistor depended on illumination by light. Since that time, photoconduction has been an important tool used to evaluate materials properties, to study semiconductor device characteristics, and to convert optical into electric signals. The Radio Corporation of America (RCA) was a leader in the study and development of photoconductivity and of photoconductive devices. Richard H. Bube of RCA Laboratories wrote the classic book *Photoconductivity in Solids* [2] in 1960. Today, photoconducting devices are used to generate very fast electric pulses using laser pulses with subpicosecond rise and fall times [3]. For optoelectronic communications, photoconducting devices, allow operation in the gigabit per second range.

Photoconductive devices normally have two terminals. Illumination of a photoconductive device changes its resistance. Conventional techniques are used to measure the resistance of the photoconductor. Frequently, small changes in conductivity need to be observed in the study of material or device characteristics. Also, in the measurement of light intensities of faint objects, one encounters small photoconductive signals.

Only solid photoconductors, such as Si, PbS, PbSe, and HgCdTe, will be treated here. Photoconduction has been observed in amorphous, polycrystalline, and single-crystalline materials. During the last decade, major improvements in materials growth have occurred which directly translate in better device performance such as sensitivity and stability. Growth techniques such as molecular beam epitaxy (MBE) and metal organic chemical vapor deposition (MOCVD) allow the growth of single-crystal layers with an accuracy of the lattice constant. Artificially structured materials can be fabricated with these growth techniques for use in novel photoconducting devices.

Absorption of light in semiconductors can free charge carriers that contribute to the conduction process. Figure 56.1 presents the band diagram for a direct bandgap semiconductor where the excitation processes are indicated. Excitation process (a) is a band-to-band transition. The photon energy for this excitation has to exceed the bandgap of the semiconductor. The absorption constant is larger for this process than for any of the other processes shown in this figure. Typical semiconductors used for electronic applications have bandgaps in excess of 1 eV, corresponding to light in the near-infrared region. Special semiconductors have been developed with narrower bandgaps to provide absorption in the mid- and long-wavelength infrared regions. Indium antimonide (InSb) and mercury-cadmium-telluride (HgCdTe) semiconductors provide photosensitivity in the 4- and 10-μm wavelength range, respectively. The photogenerated carriers increase the electron and hole densities in the conduction and valence bands, respectively, which leads to an increase in conductivity [4]. For the simplified case with one type of carrier dominating, the conductivity σ is given by:

$$\sigma = ne\mu \tag{56.1}$$

where n is the density of free carriers, e their charge and μ their mobility. Absorption of light results in a change in free carrier density and a corresponding change in conductivity $\Delta\sigma$:

$$\Delta\sigma = \Delta ne\mu + \Delta\mu en \tag{56.2}$$

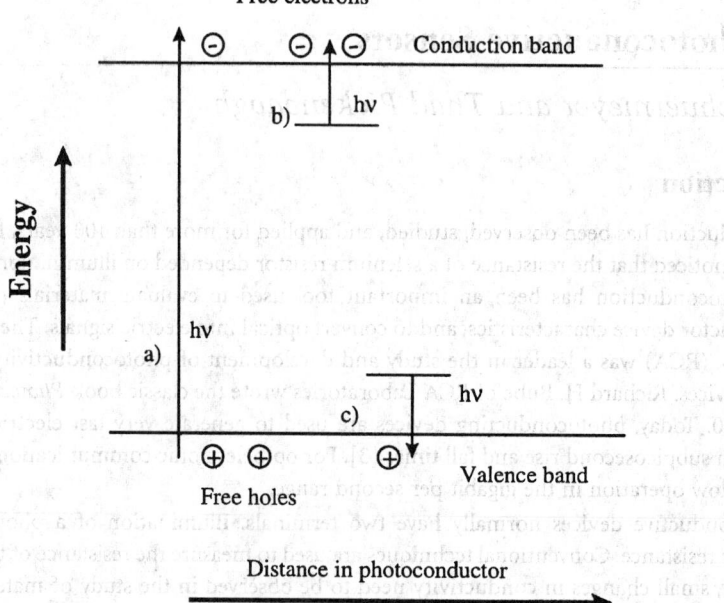

FIGURE 56.1 Example of electronic transitions in a photoconductor: (a) band-to-band excitation, (b) excitation from a trap or a donor, and (c) transition from a trap or an acceptor to the valance band; hν is the energy of the absorbed photon.

$\Delta\sigma$ is the definition for photoconductivity. In Eq. 56.2, one assumes that due to the photon absorption the density of carriers changes. Also, the mobility of the carriers changes due to the modified free carrier density. The latter effect is very small except for special band transitions, as with InSb at very low temperatures.

Figure 56.1 indicates that other excitation processes exist. For example, bound electrons can be excited into the conduction band. This process can lead to persistent photoconductivity. In this example, the trapped holes have a long lifetime while the electrons move freely due to the applied electric field. Charge neutrality requires that the electrons collected at the anode be replenished by electrons supplied by the cathode. This effect leads to an amplification of the photogenerated charge (i.e., more than one electron is collected at the anode of the photoconducting device per absorbed photon). Often, the storage times are long, in the millisecond range. Hence, photoconductive devices with large amplification have a slow signal response.

Small bandgap semiconductors, such as HgCdTe and InSb, are difficult to manufacture. Thus, artificially structured layers of commonly used materials are being developed to replace these. Spatial modulation of doping has been proposed by Esaki and Tsu [5] to achieve a lattice containing a superlattice of *n*-doped, undoped, *p*-doped, and undoped layers *(n-I-p-I)*. Due to the energy configuration of this structure, the effective bandgap is less than that of the undoped material. The effective bandgap depends on the thickness of the layers and their doping concentrations. The quantum-well infrared photodetector (QWIP) [6] is another approach to obtain photoconduction in the far-infrared wavelength range. In this structure, energy wells exist in the conduction band of the material heterostructure due to the energy band discontinuities. Subbands form in the superlattice and electrons in these wells are confined due to

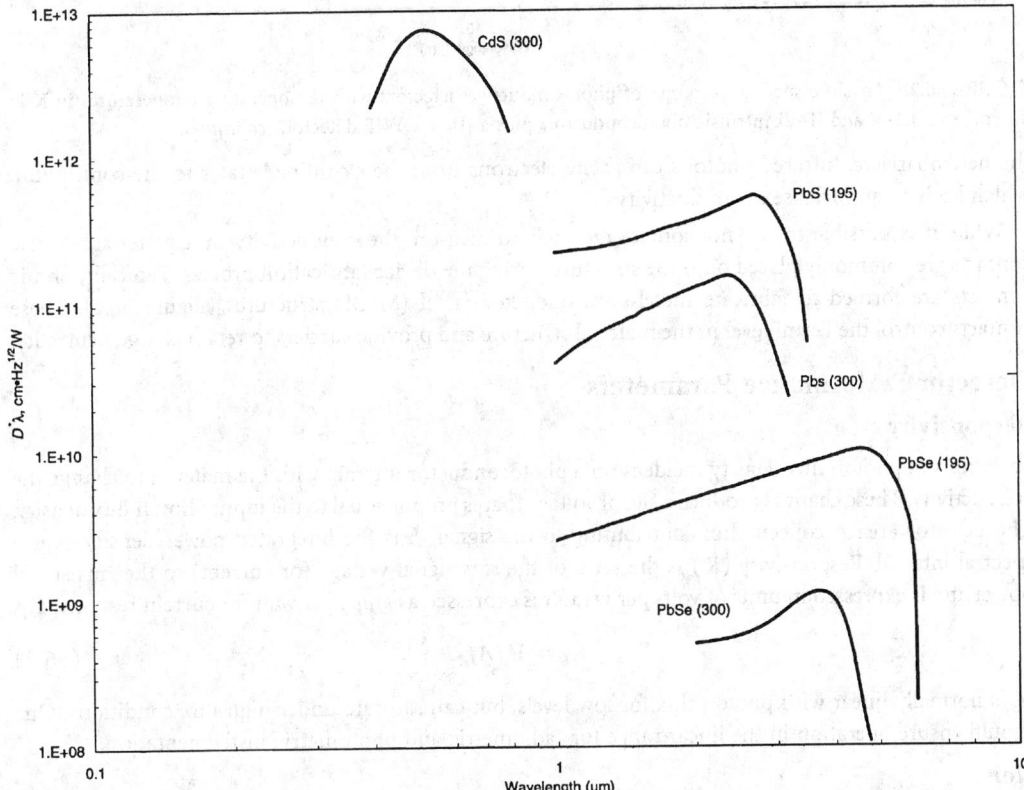

FIGURE 56.2a Absolute spectral response of photoconductive detectors with the operating temperatures in K in parentheses: CdS visible and Pb salt IR detectors. *(continues)*

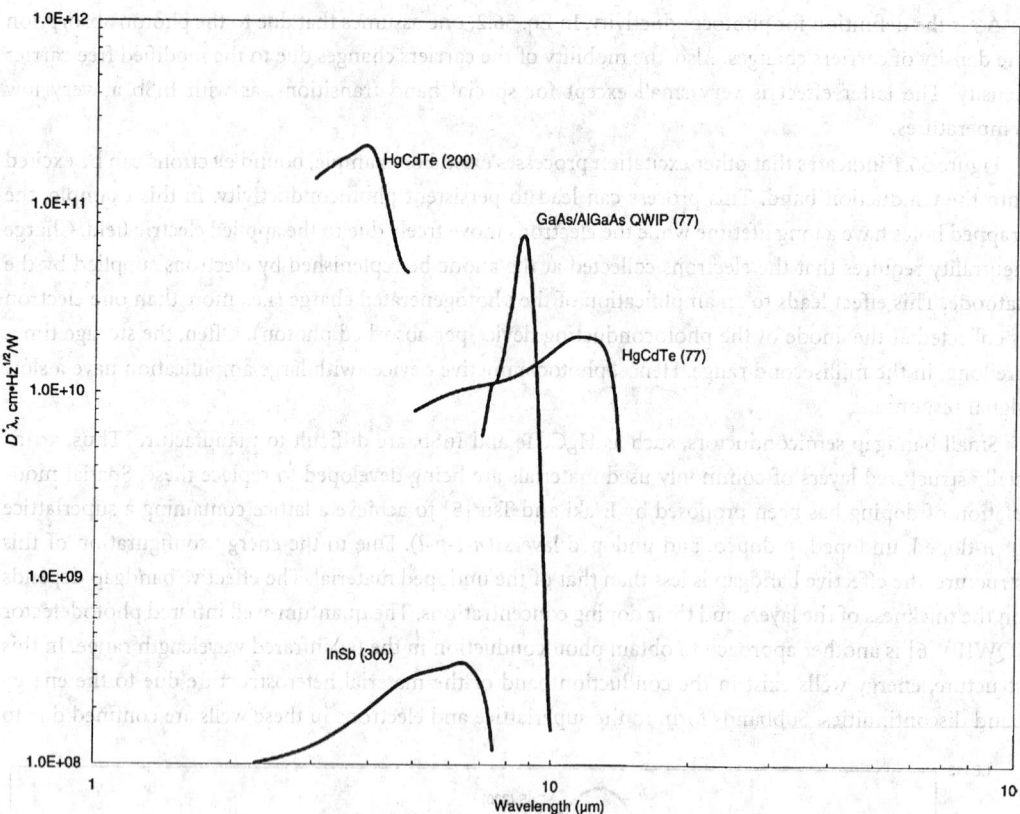

FIGURE 56.2b Absolute spectral response of photoconductive detectors with the operating temperatures in K in parentheses: III–V and II–VI intrinsic photoconductors plus a III–V QWIP detector. *(continues)*

the heterobarriers. Infrared photons can excite electrons from their confined states to the continuum, which leads to an increase in conductivity.

While it is possible to use noncontact methods to measure the conductivity in a material, electric contacts are commonly placed onto the structure during the device fabrication process. Typically, ohmic contacts are formed to fabricate metal-semiconductor-metal (MSM) structures (Figure 56.2). These contacts control the Fermi level in the material structure and provide carriers to retain charge neutrality.

Detector Performance Parameters

Responsivity

Variations in photon flux density incident on a photoconductor interact with the material to change the conductivity. These changes produce a signal voltage that is proportional to the input photon flux density. The detector area A collects flux contributing to the signal. J_s is the integrated power density over a spectral interval. Responsivity (R_v) is the ratio of the rms signal voltage (or current) to the rms signal power and is expressed in units of volts per watt. It is expressed as amps per watt for current responsivity.

$$Rv = V_s/AJ_s \qquad (56.3)$$

V_s is normally linear with photon flux for low levels, but can saturate under high flux conditions. One should ensure operation in the linear range for radiometric and photometric instrumentation.

Noise

The performance of a visible or IR instrument is ultimately limited when the signal-to-noise ratio equals one (SNR = 1). The noise from the instrument's signal processing should be less than the noise from the

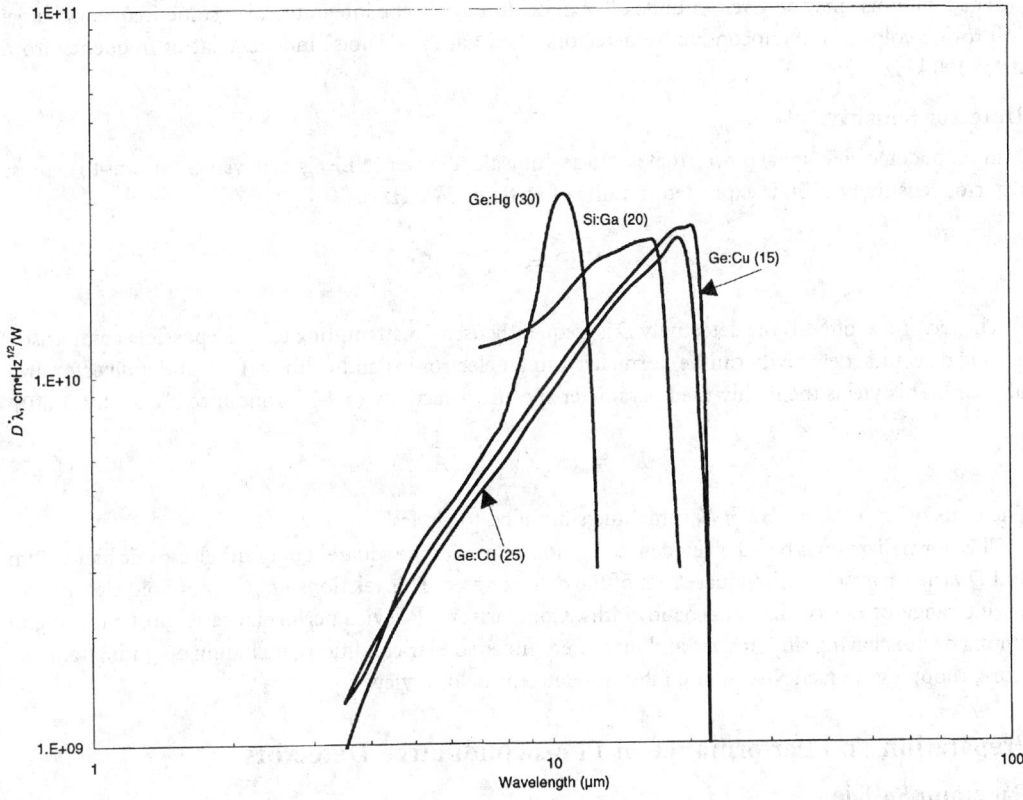

FIGURE 56.2c Absolute spectral response of photoconductive detectors with the operating temperatures in K in parentheses: long-wavelength extrinsic Ge and Si photoconductors.

detector in the ideal case. This means reducing this noise within the restrictions of signal processing design limitations. These may include cost, size, and input power. The detector noise should be minimized.

Johnson noise is the limiting noise in all conductors [7]. It is frequency independent, and independent of the current going through the device. Johnson noise is defined in Equation 56.4, where k is the Boltzmann constant (1.38×10^{-23} J/K), T is the detector temperature (K), R is the resistance (Ω), and Δf is the amplifier bandwidth (Hz).

$$V_{\mathrm{J}} = \sqrt{(4kTR\Delta f)} \tag{56.4}$$

Another type of noise known as $1/f$ noise (V_f) is present in all semiconductor detectors that carry current. The spectrum of this noise varies as $1/f^n$, with n approximately 0.5 [8].

Noise due to fluctuation in generation and recombination of charge carriers [9] varies linearly with current. This noise may be caused by the random arrival of photons from the background (photon noise), fluctuation in the density of charge carriers caused by lattice vibration (g-r noise), by interaction with traps, or between bands.

Excess noise from the amplifier or signal processing (V_{amp}) can also limit photoconductive detector performance.

These uncorrelated noises add in quadrature, giving the total noise (V_{N}):

$$V_{\mathrm{N}}^{2} = V_{\mathrm{J}}^{2} + V_{\mathrm{g-r}}^{2} + V_{\mathrm{amp}}^{2} \tag{56.5}$$

The total noise may be given in units of $V\sqrt{Hz}$. It may also be integrated over some frequency range to provide volts rms. Photoconductive detectors often have a *g-r* noise independent of frequency from dc to 100 kHz.

Detector Sensitivity

Minimum detectable signal power, that is, Noise Equivalent Power (NEP) is a convenient means to express detector sensitivity. NEP is expressed in units of watts or $W\sqrt{Hz}$.

$$NEP = V_N/R_V \qquad (56.6)$$

The reciprocal of NEP, the detectivity *D* is frequently used. In attempting to make possible comparison among detectors, detectivity can be normalized to an electronic bandwidth of 1 Hz and a detector area of 1 cm². This yields the highly used parameter specific detectivity or D^* (pronounced "dee-star") [10]:

$$D^* = (R_V/V_N)\sqrt{(A\Delta f)} \qquad (56.7)$$

The units of D* are cm · Hz$^{1/2}$/W, sometime simplified to "Jones".

This normalization is based on evidence that noise varies as the square root of the electronic bandwidth and *D* varies inversely as the square root of the detector area. This relationship may not hold closely over a wide range of device sizes and bandwidths. Comparison of device performance is most meaningful among devices having similar sizes and measured under similar conditions, including operating temperature, chopping frequency/scanning rate, and detector field of view.

Preparation and Performance of Photoconductive Detectors

Cadmium Sulfide

CdS is normally prepared by vapor deposition or sintering a layer of high-purity CdS powder on a ceramic substrate [11]. It has the largest change in resistance with illumination of any photoconductor. The peak response of this intrinsic detector is at 0.5 µm. Its spectral response is similar to that of the human eye and operates without cooling.

Lead Sulfide

PbS was among the earliest IR detector material investigated. Cashman was one of the earliest researchers in the U.S. [12]. This intrinsic detector material is prepared by deposition of polycrystalline thin films by vacuum sublimation or chemical deposition from a solution. The spectral response extends to approximately 3 µm. PbS operates over the temperature range from 77 K to room temperature. The frequency response slows considerably at the lowest temperatures. The spectral response extends to somewhat longer wavelengths with cooling.

Lead Selenide

PbSe is an intrinsic detector that operates over the temperature range from 77 K to room temperature. Its spectral response extends to longer wavelengths with cooling. Preparation of PbSe is by sublimation or chemical deposition. Noise in PbSe detectors follows a $1/f$ spectrum.

Indium Antimonide

InSb is prepared by melting together stoichiometric quantities of indium and antimony. It operates over the range from 77 K to room temperature. The higher performance and ease of operation with signal processing electronics lead photovoltaic InSb detectors to be much more widely used than photoconductive.

Mercury Cadmium Telluride

HgCdTe is a versatile intrinsic material for IR detectors. CdTe and HgTe are combined to form the alloy semiconductor $Hg_{1-x}Cd_xTe$. For the alloy with $x \approx 0.2$, the bandgap is approximately 0.1 eV, providing a long wavelength cutoff of 12.4 μm. HgCdTe was initially grown into bulk crystals by solid-state crystallization (also called quench and anneal). Currently, thin film growth techniques of liquid phase epitaxy (LPE), MOCVD, and MBE are preferred to obtain larger, more uniform wafers. By appropriately choosing the alloy composition, photoconductive HgCdTe detectors are possible over the 2- to 20-μm range. CdZnTe wafers permit lattice-matched surfaces for HgCdTe thin film growth. Operating temperatures can range from 77K to room temperature, with the lower temperatures necessary for the longer wavelength devices.

Extrinsic Germanium and Silicon

The photoresponse of an extrinsic detector occurs when a photon interacts with an impurity added to a host semiconductor material. With an intrinsic material, the photoresponse is from the interaction with the basic material.

For the extrinsic detector, incident photons may produce free electron-bound hole pairs, or bound electron-free hole pairs. The extrinsic detector's spectral response is achieved using an impurity (or doping element). Intrinsic detection occurs with a detector having the necessary bandgap width for the desired spectral response.

Extrinsic detectors require lower temperatures than do intrinsic and QWIPs, but have the advantage of longer wavelength response.

Ge and Si are zone refined to achieve high purity by making multiple passes of a narrow molten zone from one end to the other of an ingot of the material. Unwanted impurities can be reduced to levels of 10^{12} to 10^{13}/m^3 [13]. Growth of single crystals is by the Czochralski approach of bringing an oriented seed crystal in contact with the melt and withdrawing it slowly while it is rotated, or by applying the horizontal zone refining approach, whereby an oriented seed crystal is melted onto the end of a polycrystalline ingot. A molten zone is started at the meeting of the ingot and seed and moved slowly down the ingot, growing it into a single crystal. An inert atmosphere is required to prevent oxidation.

Hg, Cd, Cu, and Zn are impurities for doping Ge detectors; Ga and As are dopants for Si detectors. See Table 56.1 and Figure 56.3.

TABLE 56.1 Photoconductive Detectors

Material	Cutoff Wavelength (μm)	Temp (K)	Responsivity (V/W)	D* (cm Hz$^{1/2}$/W)
CdS	0.7	300	1×10^6	1×10^{13}
PbS	3	300	$5 \times 10^4 - 1 \times 10^3$	$5 \times 10^{11} - 1 \times 10^{11}$
PbSe	5.8	77–300	$1 \times 10^6 - 1 \times 10^3$	$2 \times 10^{10} - 7 \times 10^8$
InSb	7	300	5	4×10^8
HgCdTe	5	150–220	$1 \times 10^5 - 2 \times 10^4$	
HgCdTe	12	65–100	1×10^5	3×10^{10}
Ge:Hg	13	4–25	8×10^5	2×10^{10}
Ge:Cd	24	20–30	5×10^5	2×10^{10}
Ge:Cu	33	5	5×10^5	3×10^{10}
GaAs/AlGaAs (QWIP)	9	77	780 mA/W	7×10^{10}

From References 14 and 15.

Gallium Arsenide/Aluminum Gallium Arsenide QWIP

QWIP technology uses a quantum-well structure to provide intraband (intersubband) transitions to achieve an effective long-wavelength response in a wide bandgap material. Quantum wells are used to provide states within the conduction or valence bands. Since hυ of the desired spectral region is less than the bandgap of the host material, the quantum wells must be doped. Quantum-well structures are

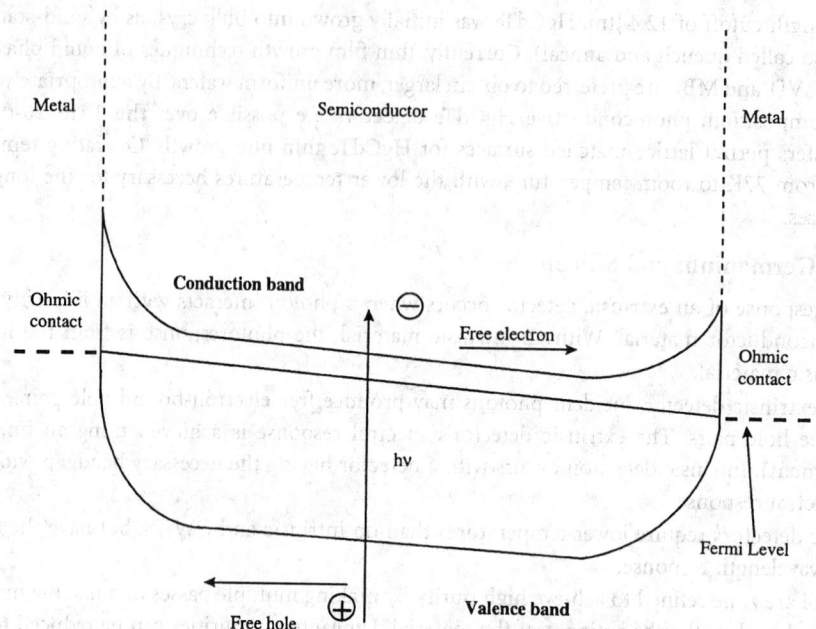

FIGURE 56.3 Energy diagram for a metal-semiconductor-metal (MSM) detector.

designed to permit photoexcited carriers to depart the structure, and be accumulated as signal (photo-current). The QWIP detector is generally comparable to extrinsic photoconductive detectors [16], in that both have lower than desirable quantum efficiency. GaAs/AlGaAs QWIPs have the advantage of higher operating temperatures than extrinsic detectors.

Instrumentation

The Stanford Research Systems SR570 low-noise current preamplifier can be used to amplify the current flowing through a photoconductive device. This preamplifier can be programmed to apply a voltage to the terminals of the photoconducting device. Its output voltage is proportional to the device current. Frequently, the IR radiation or visible light is chopped and the ac component of the device current is detected using lock-in-amplifier techniques. This approach allows the study of very small changes in device conduction. The Stanford Research Systems SR570 and the EG&G Instruments Model 651 are examples of a lock-in amplifier and a mechanical radiation/light chopper, respectively.

References

1. W. Smith, *Nature*, 303 (1873).
2. R.H. Bube, *Photoconductivity of Solids*, New York: John Wiley & Sons, 1960.
3. J.A. Valdmanis, G.A. Mourou, and C.W. Gabel, Pico-second electro-optic sampling system, *Appl. Phys. Lett.*, **41**, 211–212, 1982.
4. R.H. Bube, Photoconductors, in *Photoelectronic Materials and Devices*, S. Larach, Editor, Princeton, NJ, D. Van Nostrand Company, 100–139, 1965.
5. L. Esaki and R. Tsu, Superlattice and negative differential conductivity in semiconductors, *IBM J. Res. Dev.* **14**, 61, 1971.
6. B.F. Levine, Quantum-well Infrared Photodetectors, *J. Appl. Phys.* **74**, R1-R81, 1993.
7. P.W. Kruse, L.D. McGlauchlin, and R.B. McQuistan, *Elements of Infrared Technology*, New York: John Wiley & Sons, 1962.

8. H. Levinstein, Characterization of infrared detectors, in *Semiconductors and Semimetals*, R.K. Willardson and A.C. Beer (Eds.), New York: Academic Press, **5**, 5,1970.

9. K.M. Van Vliet, Noise in semiconductors and photoconductors, *Proc. I.R.E.*, **46**, 1004, 1958.

10. R.C. Jones, Phenomenological description of the response and detecting ability of radiation detectors, *Proc. I.R.E.*, **47**, 1495, 1959.

11. p. 417–418 of Reference 7.

12. R.J. Cashman, Film-type infrared photoconductors, *Proc. I.R.E.*, **47**, 1471, 1959.

13. S.R. Borrello and M.V. Wadsworth, Photodetectors in *Encyclopedia of Chemical Technology*, R.E. Kirk and D.E. Othmer (Eds.), New York: John Wiley & Sons, **18**, 897–898, 1996.

14. p. 862-863 of Reference 13.

15. W.L. Wolfe and G.J. Zissis (Eds.), *The Infrared Handbook, revised ed.*, Ann Arbor, MI: Environmental Research Institute of Michigan, 1985.

16. p. R3 of Reference 6.

17. T.R. Schimert, D.L. Barnes, A.J. Brouns, F.C. Case, P. Mitra, and L.T. Clairborne, Enhanced quantum well infrared photodetector with novel multiple quantum well grating structure, *Appl. Phys. Letts.*, **68** (20), 2846-2848, 1996.

56.2 Photojunction Sensors

Michael R. Squillante and Kanai S. Shah

Introduction

Photojunction sensors (photodiodes and phototransistors) are semiconductor devices that convert the electrons generated by the photoelectric effect into a detectable electronic signal. The *photoelectric effect* is a phenomenon in which photons lose energy to electrons in a material. In the case of a semiconductor, when the energy of an interacting photon ($h\nu$) exceeds the energy of the semiconductor bandgap (E_g), the energy absorbed can promote an electron from the valence band to the conduction band of the material. This causes the formation of an electron-hole pair. In the presence of an electric field, these charges drift toward electrodes on the surface and produce the signal.

The junction in the photojunction device creates a diode that provides a small built-in electric field to propel the charges to the electrodes (photovoltaic mode of operation). In the photovoltaic mode, either the photocurrent or the photovoltage can be measured. This mode of operation provides very high sensitivity because there is no net reverse leakage current, but relatively poor frequency response occurs because of high capacitance and low electric field.

Photodiode devices are most often operated with a bias voltage applied opposing the junction (*reversed bias*) to provide the electric field. The presence of the junction in a diode allows for the application of a relatively large bias to be applied while maintaining a relatively low reverse leakage current and thus relatively low noise. The result of an applied bias on a junction is the increase of the "depletion region," which is the sensitive volume of the detector. Any charges that are generated within this volume are swept toward the electrodes by the field, adding to the reverse leakage current. The *total reverse* current is the sum of the *dark current*, which occurs due to thermal generation of charges in the depletion region, and the *photocurrent*, which is produced due to optical illumination. Thus, the lower the dark current, the higher the sensitivity of the detector to optical illumination.

In an ideal diode, all of the light incident on the photodiode surface is converted to electron-hole pairs and all of the charges drift to the electrodes and are collected. In a real device, there are reflection losses at the surface, additional light is lost in the electrode and/or front layers of the device, and not all of the charges are collected at the electrodes.

There are several fundamental types of junction photodiodes [1], as shown in Figure 56.4. A *Schottky barrier* diode is a device in which the junction is formed at the surface of the semiconductor by the

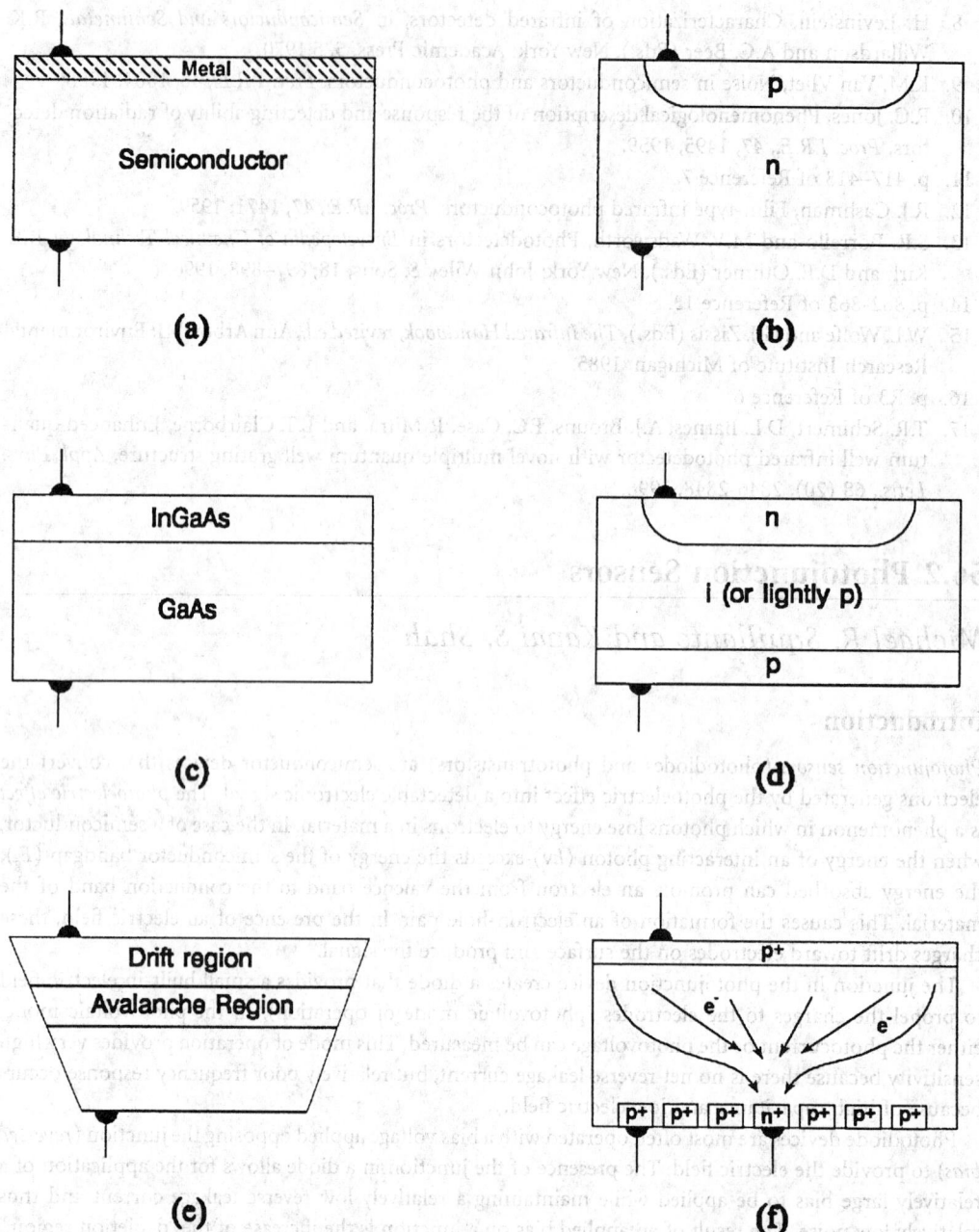

FIGURE 56.4 Schematic of photodiode device structures: (a) Schottky junction, (b) homojunction, (c) heterojunction, (d) *p-i-n*, (e) APD, (f) drift diode.

application of a metal electrode that has a work function that is different from the work function of the semiconductor; a *heterojunction diode* is a device in which two different semiconductor materials with differing work functions are joined; a *homojunction diode* is a device in which the junction is created at an interface in a single material and the difference in work function is created by doping the material *n*-type and *p*-type. Most photodiodes are homojunction devices made using silicon. Other, more complex types of photojunction devices, which are discussed below, include *p-i-n photodiodes*, *avalanche photodiodes* (APD), *drift photodiodes*, and *phototransistors*.

Photodiodes are typically characterized by several properties, including bandwidth, spectral response, operating bias, operating temperature, dark current, junction capacitance, noise equivalent power, and peak wavelength. Other specifications usually provided by manufacturers include size, packaging details, operating temperature range, capacitance, and price.

Photodiodes are used in numerous applications, including CD-ROM systems, television remote control systems, fax machines, copiers, optical scanners, fiber optic telecommunication repeaters, surveillance systems such as motion detectors, certain smoke detectors, light meters, and a wide variety of scientific instrumentation including spectrophotometers, scintillation detectors, optical trackers, laser range finders, LIDAR, LADAR, analytical instrumentation, optical thermometers, nephelometers, densitometers, radiometers laser detectors, shaft encoders, and proximity sensors. Photodiode arrays are available for use as position-sensitive detectors that can either be used for imaging (such as in laser scanners, night vision equipment, spectrophotometers, and edge detection) or alignment systems. Medical imaging applications such as x-ray CT scanners also use large arrays of photodiodes.

A variety of materials are used in the fabrication of photodiodes, but most are fabricated using silicon. Other materials used include CdS, Se, GaAs, InGaAs, HgCdTe, and PbS. In addition, materials with unique properties can be used to solve very specific and unusual problems, including Ge, GaP, HgMnTe, InP, HgI2, and InI.

Photodiodes are an alternative to photomultiplier tubes in many applications. There are a variety of advantages to be gained—including higher quantum efficiency, tailored spectral response, increased ruggedness, reduced power requirements, reduced weight, compact size, elimination of warm-up period, reduced sensitivity to temperature and voltage fluctuations, and insensitivity to magnetic fields. In general, photodiodes are noisier and require more sophisticated readout electronics than photomultiplier tubes, especially at room temperature. Upon cooling, the noise in photodiodes can be reduced significantly due to reduction in dark current.

Figure 56.5(a) shows a simple circuit for operating a photodiode in the photovoltaic mode. In this mode, photocurrent is usually measured because the photocurrent is nearly proportional to the input signal. The output of the photodiodes is typically connected to the input of an op-amp current-to-voltage converter. Figure 56.5(b) shows a simple circuit for operating a photodiode under reverse bias.

Table 56.2 shows a few examples of available photodiodes. These are only a small fraction of the commercially available photodiodes and photodiode manufacturers. Lists of photodiode manufacturers are available [2,3].

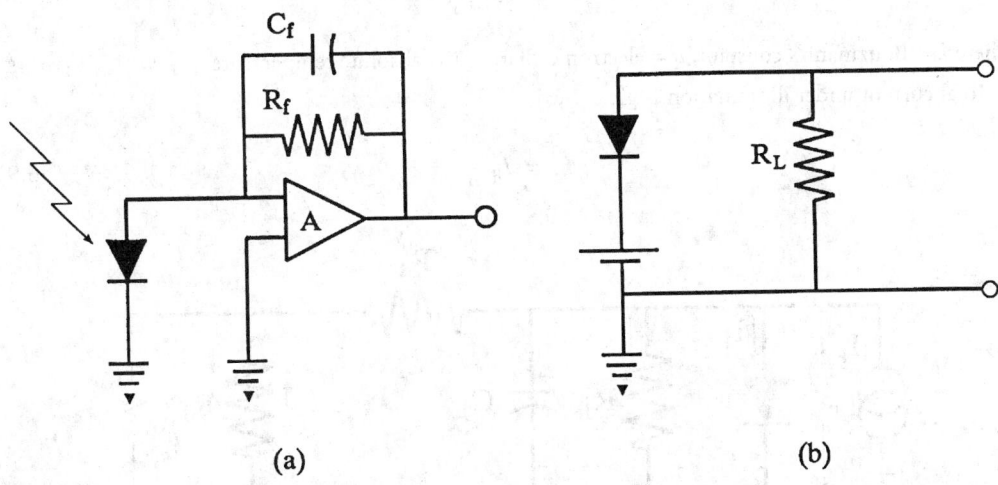

FIGURE 56.5 Typical circuits for operation of a photodiode: (a) circuit for photodiode operation in photovoltaic mode, (b) circuit for photodiode operating under reverse bias.

TABLE 56.2 Examples of the Variety of Commercially Available Photodiodes

Commercial source	Type	Region	Example device	Comments
Hamamatsu	Si	Visible	S2386-44K	13 mm², $13
UDT	Si-*p-i-n*, UV enhanced	To 200 nm	UV50	50 m², $44
UDT	Si-*p-i-n*	Visible	PIH-HS040	0.8 mm², fast, $18
Hamamatsu	Si APD	Visible	S2385	20 mm², $560
RMD	Si APD	Visible	SH8S	169 mm², $2850
UDT	InGaAs	Near IR	InGaAs-300	0.1 mm², $69
Lasertron	InGaAs *p-i-n*	1.3–1.55 μm	QDEP	$200–250
Hamamatsu	GaP	Near UV	G1961	1 mm², $42
Hamamatsu	GaAsP	Near UV and visible	G1125-02	1 mm²
Brimrose	HgMnTe	2 to 12 μm	MMT-212-3-1	0.8 mm², $2240
Komar	HgCdTe	2 to 12 μm	KV104-1-a	1 mm², $3550
Komar	HgCdTe	to 18 μm	KMPC18-1-b1	1 mm², $2000
Komar	InSb	5.1 μm	KISD-1-a	1 mm², $2100

Note: Brimrose Corp. of America, Baltimore, MD; Hamamatsu, Corp. Bridgewater, NJ; Kolmar Technologies, Conyers, GA; Lasertron Corp, Burlington, MA; Loral Lexington, MA; RMD = Radiation Monitoring Devices, Inc., Watertown, MA; UDT = UDT Sensors, Inc., Hawthorne, CA.

Theory

Equivalent Circuit

A simplified version of the equivalent circuit for a photodiode is shown in Figure 56.6, where C_j = junction capacitance, I_d = dark current (current present with no incident photons), I_j = reverse saturation current, I_o = output current, I_p = photocurrent current, R_j = junction shunt resistance (or parallel resistance), R_s = series resistance, V_j = junction voltage, V_o = output voltage.

The dark current from this structure is ideally given by:

$$I_d = I_j\{\exp((qV_j/kT) - 1)\} \qquad (56.8)$$

where k = Boltzmann's constant, q = electronic charge, T = absolute temperature.

Total current under illumination is given by:

$$I_o = I_d + I_p \qquad (56.9)$$

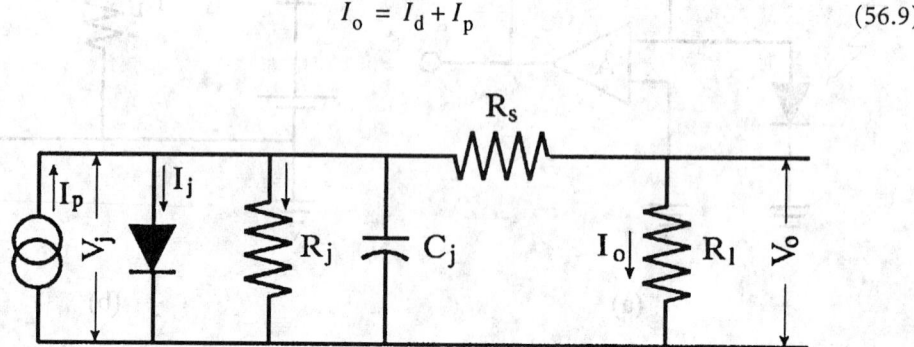

FIGURE 56.6 Equivalent circuit for a photodiode.

For a more rigorous treatment refer to Reference 1, pp. 752–754.

Quantum Efficiency

The *quantum efficiency* of a photodiode is the ratio of the charge pairs generated to the incident photons:

$$\eta = (I_p/q)/(P_i/h\upsilon) \tag{56.10}$$

Where P_i = optical power incident on the photodiode, $h\upsilon$ = energy of the photons.

The responsivity, $\mathcal{R}$ is the ratio of the photocurrent to the incident optical power in amps/watt:

$$\mathcal{R} = I_p/P_i = \eta q/h\upsilon \tag{56.11}$$

The photocurrent is given by rearranging Equation 56.9:

$$I_p = q\eta P_i/h\upsilon \tag{56.12}$$

Noise

There are two main sources of noise when using a photodiode: shot noise in the diode and thermal noise. The *shot noise* is related to the dark current by the formula:

$$I_s = (2qI_dB)^{1/2} \tag{56.13}$$

where B = bandwidth.

Assuming the diode shunt resistance and the input resistance of the measuring circuit to be used to measure the output of the photodiode are high relative to the load resistance, the thermal noise is given by:

$$I_t = (4kTB/R_L)^{1/2} \tag{56.14}$$

where R_L = load resistance. The total noise current, I_n, is the sum of these currents in quadrature.

I–V Characteristics of Photodiodes

With no illumination, photodiodes have I–V curves equivalent standard diodes given by Equation 56.8. Illumination by light causes the current to increase. Figure 56.7 shows a family of I–V curves for a photodiode under illuminations with equally increasing increments of incident light intensity. As the illumination on the device increases, the curve shifts downward by the amount of current generated by the incident light. The lower right-hand quadrant represents the photovoltaic mode of operation. When a photovoltaic device is operated in "current mode" with low or no load resistance (as with an operational amplifier, as in Figure 56.5(a) above), the output is linear with incident light intensity. When operated in "voltage mode" with a high load resistance, there is an exponential relationship between the output and the incident illumination. The lower left-hand quadrant shows the reversed bias mode of operation. Again, in this mode the output is nearly linear with the incident intensity.

Output Current Under Reverse Bias

In a reverse bias p-n junction under bias, the depletion width (W) increases as a function of applied bias (V_b) until the device is fully depleted. The dark leakage current (I_d) under reverse bias conditions can arise from the generation-recombination effects (I_G) and from diffusion (I_D) as well as surface effects. In most cases, the diffusion current is significantly smaller than the generation-recombination component. Thus, it is possible to assume that $I_d \approx I_G$.

The analytical expression for I_G is as follows:

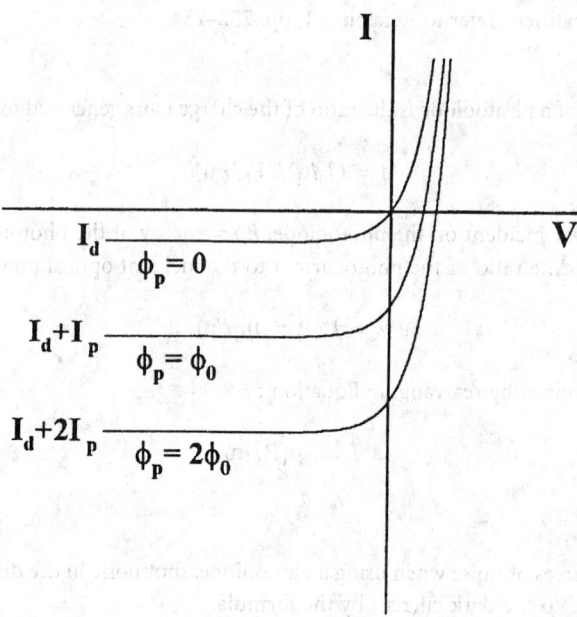

FIGURE 56.7 I–V characteristics of photodiode under illumination.

$$I_G = AWqn_i/2\tau \tag{56.15}$$

where: A = device area, n_i = intrinsic carrier concentration, τ = minority carrier lifetime.

The total diode current under illumination (I_o) is the sum of the dark leakage current (I_d) and the photocurrent (I_p):

$$I_o = I_d + I_p \tag{56.16}$$

Position Sensitive Photodiode Arrays

Many manufacturers offer photodiodes fabricated in a *quadrant* geometry. Four photodiodes are fabricated in a square, 2×2 geometry. When coupled to a lens or a pinhole, they can be operated as position sensitive detectors. In operation, the outputs of the four photodiodes are monitored and the position of the light source can be determined by the projection of the light spot on the detector surface. More recently, manufacturers are offering linear and area arrays of photodiodes that can be used as imaging devices.

Phototransistors

Phototransistors are photojunction devices similar to transistors except that the signal amplified is the charge pairs generated by the optical input. Like transistors, phototransistors can have high gain. Phototransistors can be made on silicon using *p*- and *n*-type junctions or can be heterostructures. Figure 56.8 shows a sketch of the structure of a simple bipolar phototransistor, which is essentially the same as that of a simple bipolar transistor. The main difference is the larger base-collector junction, which is the light-sensitive region. This results in a larger junction capacitance and, although the devices have gain, the capacitance gives phototransistors lower frequency response than photodiodes.

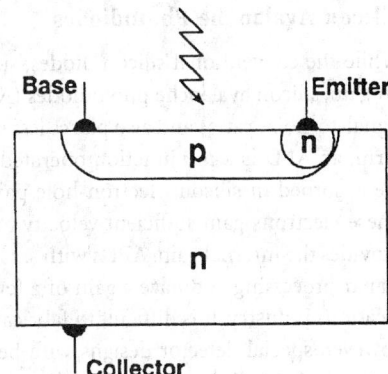

FIGURE 56.8 Schematic representations of a simple bipolar phototransistor. Note that the phototransistor has a large *p-n* junction region that is the photosensitive portion of the device.

Using thin film transistor (TFT) technology developed for flat panel displays, large arrays of phototransistors can be fabricated on amorphous silicon to form imaging devices that can be used in place of other imaging technologies such as vidicon tubes or even film. Examples of this are the very large area detectors (hundreds of square centimeters) being investigated for use in medical radiography by combining the TFT arrays with radiographic phosphor screens [4] or coupled to semiconductor films [5].

Novel Silicon Photojunction Detector Structures

Silicon *p-i-n* Detectors

Silicon *p-i-n* diodes are an extension of the standard *p-n* junction diodes, but are more attractive for low-noise applications due to reduced capacitance in these devices [1]. The reduction in capacitance is achieved by incorporating an intrinsic region between the *p* and *n* regions. This increases the depletion width of the detector and thereby lowers its capacitance. Silicon *p-i-n* detectors can be designed to have higher frequency response than *p-n* junctions and therefore are more popular.

In operation, *p-i-n* detectors are similar to *p-n* junction detectors, but the surface region (either *p* or *n*) is made thin so that the optical photons penetrate this entrance layer and are stopped in the intrinsic (*i*-region) where electron-hole pairs are produced, as shown in Figure 56.4. These electron-hole pairs are swept toward the appropriate electrodes due to applied electric field. For fabrication, *p-i-n* detectors require high resistivity material and typical photodiodes have thickness ranging from 100 to 500 μm. Important applications of *p-i-n* detectors include optical sensing of scintillated light in CT scanners, general scintillation spectroscopy, charged particle spectroscopy, and high-speed sensing applications.

Silicon Drift Detectors

Silicon drift photodiodes are an extension of the *p-i-n* geometry and been extensively studied in recent years to provide very low capacitance (<1 pF for 1 cm² detector with 300 μm thickness) [7]. This is achieved by reducing the area of ohmic electrode (anode in most cases) significantly as compared to the entrance electrode as shown in Fig. 56.4(f). Since the charge sensing electronics is connected to the smaller electrode, the device capacitance is proportional to its size and not to the actual detector area. Thus by exploiting this concept, significantly lower capacitance has been achieved than in comparable *p-i-n* detector.

In drift detectors, it is important to ensure that charges created over the entire active volume will be collected at the anode. In order to achieve this, drift rings are provided around the anode. The outermost ring is biased at highest potential, with the inner rings biased to lower potentials in a successive manner. This arrangement creates a potential minimum at the anode and thereby enables efficient charge collection over the entire detector volume. A variety of device geometries are being explored based on this concept and in some instances these detectors are capable of providing position sensitive detection as well. While these detectors are in the research stage, excellent performance has been demonstrated by prototype detectors. These detectors, when they are commercially available, will have the potential of replacing *p-i-n* diodes in many applications.

Silicon Avalanche Photodiodes

While the conventional silicon diodes, such as *p-n* junction diodes, *p-i-n* diodes and drift diodes, have no gain, silicon avalanche photodiodes (APDs) have internal gain that enables them to operate with high signal-to-noise ratios and also places less stringent requirements on supporting electronics. In its simplest form, an APD is a *p-n* junction operated close to its breakdown voltage in reverse bias. When photons are absorbed in silicon, electron-hole pairs are produced and are accelerated by the high electric field. These electrons gain sufficient velocity to generate additional free carriers by impact ionization, which provides the internal gain. APDs with small areas (few mm diameter) can be manufactured with standard planar processing and have a gain of a few hundred. These detectors are widely used in the telecommunications industry. It is difficult to fabricate high gain detectors with large areas using the planar process; however, special detector designs with beveled edges (see Figure 56.4) have been fabricated to provide high gain (>10,000) in large areas (>1 cm²) [8]. The APD gain versus bias behavior for such a device is shown in Figure 56.9. Recent advances in surface preparation and dead layer reduction have extended the application of these detectors to the UV region. While they are relatively expensive, these detectors are well suited to a number of commercial applications such as medical imaging, astronomy, charged particle and x-ray detection, scintillation spectroscopy, and optical communications.

Amorphous Silicon Detectors

While impressive results have been obtained with various device structures on crystalline silicon such as drift detectors, APDs, and CCDs, they are limited to active areas of only a few square centimeters. As a result, considerable attention has been devoted to development of hydrogenated amorphous silicon (a-Si:H) [4]. This material is produced by an RF plasma technique in large areas (30 cm × 30 cm) on glass substrates. The films are typically a few micrometers in thickness, although films as thick as 200 μm has been reported. Device structures such as *p-n* junctions were developed initially for use in solar cells with lower cost than crystalline silicon devices.

Recently, more complicated devices such as *p-i-n* sensors and thin film transistors (TFTs) have been fabricated from a-Si:H and have been configured in an array format as shown in Fig. 56.10. In these arrays (as large as 20 cm × 20 cm), each pixel consists (200 to 500 μm) of a *p-i-n* sensor connected to a TFT, and the entire array is read out in matrix fashion. These arrays are well suited for high-resolution document imaging and also for medical x-ray imaging applications with phosphors.

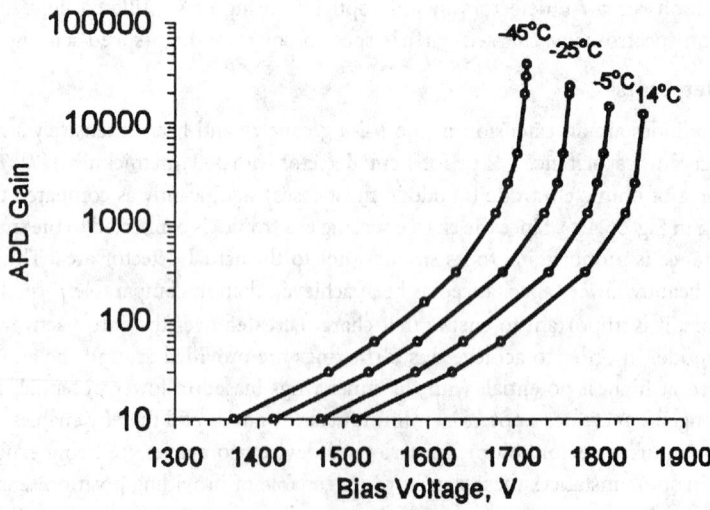

FIGURE 56.9 Gain versus bias relationship for a high-gain APD. Higher gains are achievable at lower voltages as the temperature is decreased.

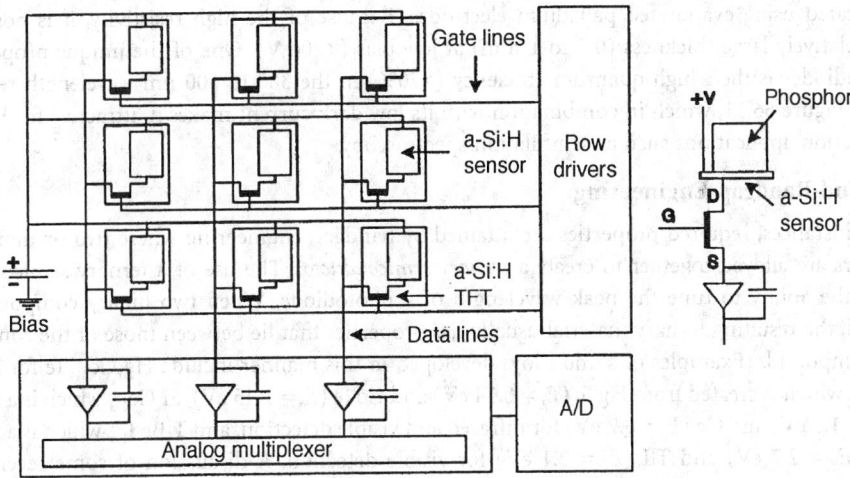

FIGURE 56.10 Schematic representation of a 2-D x-ray imager consisting of a-Si:H *p-i-n* diodes connected to a-Si:H TFTs for readout. The *p-i-n* diodes are coupled to a phosphor layer to increase sensitivity.

Novel Materials for Photodiodes and Bandgap Engineering

Important requirements for photodiodes include high quantum efficiency (QE), good charge collection efficiency, and low noise. The low noise requirement is satisfied by reducing detector capacitance and its dark current. In order to satisfy the QE requirements over a wide range of wavelengths, new semiconductor materials are being extensively investigated [9]. Since most semiconductors show high optical response near their bandgap, special materials are developed for various applications. Furthermore, since the bandgap represents a cut-off point in the optical response of the material, by selecting an appropriate material, it is possible to obtain response in a desired band [10]. For example, materials such as GaN and SiC are being explored to obtain UV detection with no sensitivity in the visible region. Other materials are being studied to exploit their unique properties such as high quantum efficiency, high-temperature operation, and high-speed response.

GaN

GaN is an attractive material for *UV photodiode* fabrication due to its wide bandgap (E_g = 3.4 eV). Due to the difficulty in growing bulk crystals of GaN, much of the work is done with films of GaN prepared by chemical vapor deposition or molecular beam epitaxy [11]. A variety of optical devices (e.g., blue LEDs and lasers, field effect transistors, photoconductive detectors, and photodiodes) have been fabricated using GaN films. GaN photodiodes have the capability of solar blind UV detection and are capable of fast response time due to high electron mobility (as high as 1000 cm^{2}V^{-1}s^{-1}), which is comparable to silicon.

SiC

SiC is another material that has shown promise for UV detection due to its wide bandgap (E_g = 3.0 eV and 3.2 eV for 6H and 4H phases, respectively) [12]. Various optical devices such as blue LEDs, lasers, and UV photodiodes have been developed from 6H-SiC due to the relative ease of doping this material to form *p* and *n* layers by ion implantation or epitaxial methods. SiC devices are also capable of high-temperature operation, and the photodiodes have shown high quantum efficiency (>80%) in the 250 to 280 nm region. Low dark current is another attractive feature of these devices.

InI

Indium iodide is a wide bandgap semiconductor (E_g = 2.0 eV) being developed for detection in visible and near-UV region [13]. The resistivity of the material is quite high (>10^{10} Ω-cm) and Schottky diodes are fabricated using evaporated palladium electrodes. Because of the high resistivity, it is possible to deplete relatively large thickness (0.5 to 1 mm) at low bias (<200 V). One of the unique properties of InI photodiodes is their high quantum efficiency (>70%) in the 300 to 600 nm wavelength region, as shown in Figure 56.11, which in combination with its low dark current makes it attractive for low light level detection applications such as scintillation spectroscopy.

Alloys and Bandgap Engineering

In many instances, required properties are attained by bandgap engineering where two or more semi-conductors are alloyed together to create a *ternary semiconductor*. The use of a ternary semiconductor provides the ability to tune the peak wavelength of a photodiode. When two binary compounds are combined, the resulting ternary material usually has properties that lie between those of the constituent binary compounds. Examples of some alloys developed in this manner include $Hg_xCd_{1-x}Te$ for infrared detection, which is created from HgTe ($E_g = 0.14$ eV) and CdTe ($E_g = 1.45$ eV), Si_xGe_{1-x} which is a mixture of Si ($E_g = 1.1$ eV) and Ge ($E_g = 0.7$ eV) for infrared and visible detection, and $TlBr_xI_{1-x}$ which is a mixture of TlBr ($E_g = 2.7$ eV) and TlI ($E_g = 2.1$ eV) for visible detection. A discussion of some recent novel materials that are being developed is presented in the following section and a compilation of relevant properties of various semiconductor materials is presented in Table 56.3.

III–V Ternary Materials

Ternary alloys of GaN and AlN ($E_g = 6.2$ eV), $Ga_xAl_{1-x}N$, are also being investigated to create optimized UV detectors for desired wavelengths. In the ternary compound, the bandgap depends on the material composition or x and varies almost linearly from 3.4 to 6.2 eV. Such bandgap engineering is desirable to create material with required photoresponse. These devices are also expected to be capable of high

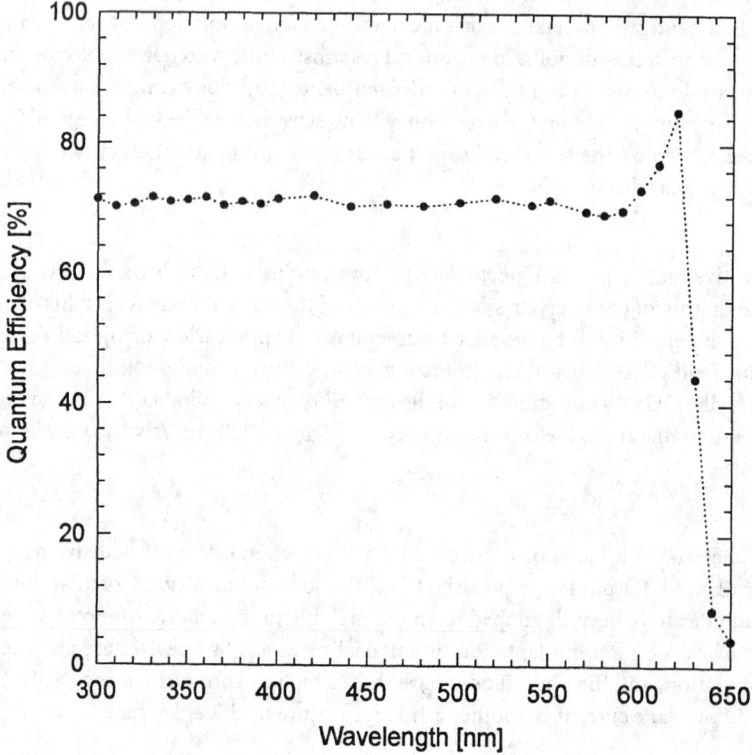

FIGURE 56.11 Quantum efficiency of an InI photodiode. The QE peaks at over 80% near the band edge and has a spectral sensitivity of about 70% into the near-UV.

TABLE 56.3 Properties of Semiconductor Materials Used for Construction of Photodiodes at 25°C

Material	Bandgap	Dielectric constant	Resistivity (Ω-cm)	Electron mobility (cm²/Vs)	Electron lifetime (s)	Hole mobility (cm²/Vs)	Hole lifetime (s)	$\mu\tau(e)$ (cm²/V)	$\mu\tau(h)$ (cm²/V)
HgTe	0.14	6.4		22000		100			
InAs	0.36	12,5		30000		240			
Ge	0.67	16	50	3900	$>10^{-3}$	1900	1×10^{-3}	>1	>1
Si	1.12	11.7	$\leq10^4$	1400	$>10^{-3}$	480	2×10^{-3}	>1	≈1
InP	1.35	12.5	10^7	4600	1.5×10^{-9}	150	$<10^{-7}$	4.8×10^{-6}	$<1.5\times10^{-5}$
GaAs	1.43	12.8	10^7	8000	10^{-8}	400	10^{-7}	8×10^{-5}	4×10^{-6}
CdSe	1.73	10.6	10^8	720	10^{-6}	75	10^{-6}	7.2×10^{-4}	7.5×10^{-5}
a-Si	1.8	11.7	10^{12}	1	6.8×10^{-9}	.005	4×10^{-6}	6.8×10^{-8}	2×10^{-8}
InI	2.01	26	10^{11}					7×10^{-5}	
HgI₂	2.13	8.8	10^{13}	100	10^{-6}	4	10^{-5}	10^{-4}	4×10^{-5}
SiC	2.2								
TlBrI	2.2–2.8		10^{10}					9×10^{-5}	
GaP	2.24								
a-Se	2.3	6.6	10^{12}	.005	10^{-6}	0.14	10^{-6}	5×10^{-9}	1.4×10^{-7}
PbI₂	2.32		10^{12}	8	10^{-6}	2		8×10^{-6}	
CdS	2.5	11.6		300		50			
TlBr	2.68	29.8	10^{12}	6	2.5×10^{-6}			1.6×10^{-5}	1.5×10^{-6}
GaN	3.4	12	$>10^{10}$	300–1000					
Diamond	5.4	5.5		2000	10^{-8}	1600	$<10^{-8}$	2×10^{-5}	$<1.6\times10^{-5}$

temperature operation due to the wide semiconducting bandgap of the material. Similar devices are also been studied from GaP (E_g = 2.1 eV) and AlP (E_g = 2.9 eV) for visible and near UV detection.

Another example is indium gallium arsenide, which is a mixture of InAs (E_g = 0.36 eV) and GaAs (E_g = 1.43 eV) and has been recently commercialized as an infrared detector material in the 1000 to 1700 nm region. InGaAs photodiodes in *p-n* diode, *p-i-n* diode, and avalanche photodiode configurations are available. InGaAs photodiode arrays coupled to amorphous silicon TFTs are being developed for large area infrared imaging. Other ternary III–V materials that have been investigated for similar reasons include GaAsP, GaNP, and BNP.

Heterojunction Photojunction Detectors

A heterojunction is a junction that exists at the interface of two different semiconductors. This concept can be exploited to produce photodiodes with unique properties such as tuned optical response in the region of interest (by adjusting the composition), and reduced optical absorption at the entrance (by irradiating the wider bandgap semiconductor that is transparent to the optical signal). A number of optical sensors have been fabricated using the heterojunction concept using mostly III–V compounds that can be tuned in composition to create heterojunctions with similar lattice constants in both the semiconductors. The research in the heterojunction devices has been aided considerably by the progress in molecular beam epitaxy. One unique application of the heterojunction concept is to fabricate detectors that have capability of distinguishing wavelengths above or below a certain level. This has been accomplished using a multilayer device (see Figure 56.12) that consists of two layers of $Ga_xIn_{1-x}As_yP_{1-y}$ which have different composition and, therefore, different bandgaps. The layer Q1 has a larger bandgap than Q2 and both are grown on InP. The optical response of this device when irradiated through the InP substrate is shown in Figure 56.12 and shows minimal overlap in the desired bands indicating successful wavelength discrimination.

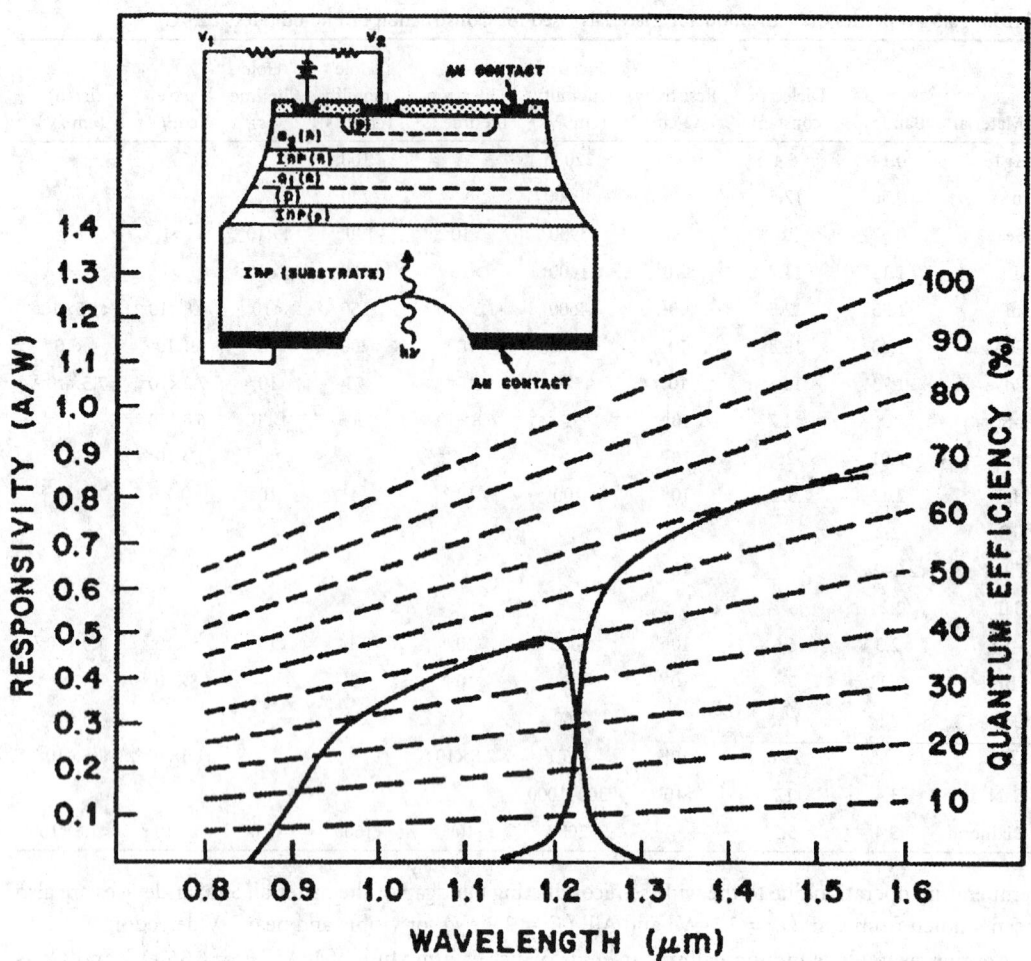

FIGURE 56.12 Responsivity and quantum efficiency of a heterojunction photodiode versus wavelength. The insert shows the cross section of the photodiode (From Reference 1, p. 765.)

Defining Terms

Bandwidth, B: The range of frequencies over which the photodiode operates.

Breakdown voltage, V_b: The reverse bias voltage at which the applied field overcomes ability of the junction to block current and the device acts like a resistor. The reverse leakage current increases abruptly near this voltage.

Dark current, or reverse leakage current, I_d: The leakage current through the device when at the operating voltage with no incident signal.

Depletion region thickness: The depth of the depleted portion of the diode when at the operating voltage. Photodiodes are frequently operated fully depleted.

Junction capacitance, C_j: Capacitance of the photodiode which decreases as the depletion width increases.

Noise equivalent power, NEP: The incident power that generates a signal equal to the noise, i.e., signal-to-noise ratio (S/N or SNR) equals 1.

Operating bias: The applied voltage at which the device operates.

Peak wavelength: The wavelength with the highest quantum efficiency.

Quantum efficiency, η or QE: The efficiency of converting photons incident on the photodiode into electrons that are detected. Reflection of light from the surface and loss of electrons in the semiconductor reduce the efficiency. Reflection losses can be minimized using an antireflection coating on the surface of the device.

Responsivity, (amps/watt): A measure of the signal current produced as a function of the optical power incident on the photodiode.

Spectral response: The quantum efficiency as a function of wavelength.

References

1. S.M. Sze, *Semiconductor Devices: Physics and Technology*, New York: John Wiley & Sons, 1985.

2. *Laser Focus World Buyers Guide*, Pennwell Publishing Co., Nashua, NH, 1997.

3. *Photonics Buyers Guide*, Laurin Publishing Co., Pittsfield, MA, 1997.

4. R.A. Street, Amorphous Silicon Sensor Arrays for Radiation Imaging, *Mat. Res. Soc. Symp. Proc.*, 192, p. 441, 1990.

5. R.A. Street, R.B. Apte, D. Jarad, P. Mei, S. Ready, T. Granberg, T. Rodericks, and R.L. Weisfield, *Amorphous Silicon Sensor Array for X-Ray and Document Imaging*, presented at the Fall Meeting of Materials Research Society Boston, December 1997 and submitted for publication in *Materials Res. Soc.* 478 (1998).

6. K. Shah, L. Cirignano, M. Klugerman, K. Mandal, and L.P. Moy, *Characterization of X-Ray Imaging Properties of PbI_2 Films*, presented at the Fall Meeting of Materials Research Society Boston, December 1997 and submitted for publication in *Materials Res. Soc.* 478 (1998).

7. E. Gatti and P. Rehak, *Semiconductor Drift Chamber on Application of Novel Charge Transport Scheme*, Nucl. Inst. and Meth., A225, p. 608, 1984.

8. R. Farrell, K. Vanderpuye, L. Cirignano, M.R. Squillante, and G. Entine, *Radiation detection performance of very high gain avalanche photodiodes*, Nucl. Inst. and Meth., A353, p. 176, 1994.

9. R.H. Bube, *Photoelectronic Properties of Semiconductors*, Cambridge, UK: Cambridge University Press, 1992.

10. J.I. Pankove, *Optical Processes in Semiconductors*, New York: Dover Publications, Inc., 1971.

11. M.A. Khan, J.N. Kuznia, D.T. Olson, M. Blasingame, and A.R. Bhattarai, *Schottky barrier photodetector based on Mg-doped p-type GaN films*, Appl. Phys. Lett. 63(3), p. 2455, 1993.

12. J.A. Edmond, H.S. Kong, and C.H. Carter, Blue LEDs, UV photodiodes and high temperature rectifiers in 6H-SiC, *Physica B* 185, p. 453, 1993.

13. K.S. Shah, P. Bennett, L.P. Moy, M.M. Misra, and W.W. Moses, Characterization of indium iodide Detectors for Scintillation Studies, *Nucl. Inst. Meth.*, A, 380(1–2), 215–219, 1996.

56.3 Charge-Coupled Devices

J.A. Nousek, M.W. Bautz, B.E. Burke, J.A. Gregory, R.E. Griffiths, R.L. Kraft, H.L. Kwok, and D.H. Lumb

Introduction

Use of CCDs for Precision Light Measurement

Charge-Coupled Devices (CCDs) have become the detector of choice for sensitive, highly precise measurement of light over the electromagnetic spectrum from the near-IR (<1.1 µm) to the x-ray band (up to 10 keV). Key advantages of CCDs over their predecessors (photographic emulsions and vacuum tube, electron beam readout devices such as Vidicons and SIT tubes) are high quantum efficiency, high linearity, large dynamic range, relatively uniform cosmetic response, low noise, and intrinsically digital image capture.

CCDs were initially designed as serial data storage media (an electronic analogy to the magnetic Bubble Memory units) in which charge packets were injected into linked capacitors to store data, and read back by moving the packets back out of the device. When it was found that charge packets could be directly induced in the capacitors by exposing them to light, the CCD as light sensor was born.

Physically, *CCD operation* consists of four critical stages. First, an incident light photon must be photoabsorbed in the sensitive portion of the CCD chip (called the *depletion region*). At optical and infrared wavelengths, the absorption results in a single electron being promoted into the conduction band (leaving a hole in the valence band); at shorter wavelengths, the photon has enough energy to make additional electrons via secondary ionizations by the photoelectron.

Second, the photon-induced electrons must be collected, via an electric field within the silicon, into localized regions near the front surface of the chip. The electric field is shaped by implanted dopants and by electric potentials applied to thin conducting strips (*gates*) that prevent the electrons from diffusing away. The resulting charge distribution corresponds to an electronic analog of the light intensity pattern shone on the CCD. The resolution of this pattern is governed by the size of the potential wells, which are designed to be periodic. Each well is called a *pixel* and corresponds to the minimum picture element detected by the CCD.

Third, after exposure is completed, the CCD charge pattern must be transferred out of the CCD. This is accomplished by modulating the potential applied to the CCD gates in such a way that no charge packets are mixed, but that each packet moves into the next pixel. The end pixel is transferred into a special pixel array called the *serial register.* Each movement of charge resulting from gate potential changes is called a *clock cycle*, and the serial register receives many clock cycles for each cycle of the full pixel array. The net result is a sequence of charge packets emerging from the serial register, each of which is directly proportional to the amount of light striking a particular location on the CCD.

Fourth, the emerging charges are converted into electric signals by a charge-sensitive preamplifier on the CCD chip. These signals are often digitized by electronics in the camera immediately outside the chip, but analog readouts that produce signals compatible with video standards are also used (the popular hand-held video cameras are examples of this). Research-grade camera readouts are able to measure the charge pulses with accuracies as good as one or two electrons rms, if the CCD and electronics are cooled.

Currently available CCDs carry out these steps so well that they are nearly the ideal detector for precision low light level applications, especially in astronomy. Such an ideal detector would have perfect quantum efficiency (i.e., convert every incident photon into detectable signal), no noise, unlimited dynamic range, linearity in response to incident intensity and position, and completely understandable characteristics.

CCDs have high quantum efficiency because photons interact via photoabsorption in the depletion layer, which directly results in one or more electrons promoted into the conduction band of the silicon lattice, and are very efficiently collected by the CCD. The main obstacles to perfect quantum efficiency are absorption of photons by the gate and insulator materials before they ever reach the depletion regions (or optically, by reflection off the front surfaces) or if the photon passes entirely through the depletion region without interacting.

There are many approaches to enhancing *CCD quantum efficiency* for various applications. In soft x-ray and ultraviolet wavelengths, the gate and insulator layers on the front of the CCD absorb too much light. To solve this, CCDs are built with thin gates or thinned substrates and back-side illumination. CCDs are also coated with phosphor coatings that down-convert ultraviolet light to longer wavelengths where the gate transmission is higher.

At hard x-ray and infrared wavelengths, too much light can pass through the depletion region without interacting at all. The depletion region is the part of the CCD pixel that is swept clean of free charges during the readout process. The depletion region gets deeper if higher purity silicon is used, and if higher voltage biases are applied during the readout.

Above 1.1 μm, photons do not have sufficient energy to promote electrons into the silicon conduction band, so other materials, such as germanium or a compound semiconductor such as InAs, InSb, or HgCdTe must be used.

CCD noise results from four major factors: (1) thermal background noise, (2) charge transfer imperfections, (3) charge-to-voltage amplification noise, and (4) cosmetic imperfections in the CCDs due to, for example, microscopic shorts in the insulating layers of the CCD. Factor (1) results from a "dark current" of thermally excited electrons that accumulate in the pixels and can be eliminated by cooling the CCD (typically to −60 to −120°C). Factor (2) results from traps that hold electrons long enough to shift them into following charge packets identified with other pixels. Factor (3) is a fundamental limit related to the temperature and capacitance of the output mode (kT/C), but it can be suppressed by signal processing techniques such as "correlated double sampling" to the equivalent of a few electrons (1–5 in state-of-the-art cameras). Factors (2) and (4) can be greatly reduced by improved manufacturing technique, especially scrupulous contamination control during the process.

CCD dynamic range is set by the maximum charge packet that can be stored in a pixel. Termed "full-well capacity," this is set by the depth of the potential well. When the full-well capacity is exceeded, the image of a point source "blooms" as a result of charge leaking into surrounding pixels, and a trail of brighter pixels forms in the readout direction of the CCD due to charge incompletely transferring from pixel-to-pixel during a clock cycle. Modern CCDs have full-well capacities in excess of 10^5 electrons and can be designed even larger. (Note that larger full-well also requires larger output capacitance, so a trade-off is generally required between blooming and low noise.)

CCD linearity in intensity response and position response is very good because the conduction band in the CCD has so many states that the very small injected photocharge does not affect subsequent photon interactions. The position linearity results from the photolithography of the manufacturing process, which must be accurate to less than 1 μm. The primary limitation on linearity results from imperfect *charge transfer efficiency* (CTE) in the process of clocking charge packets from pixel to pixel. At readout rates below 100,000 pixels per second, CTE imperfections have four causes.

1. Design imperfections: errors in CCD design can leave potential minima that are incompletely drained during clocking.
2. Process-induced traps: random cosmetic defects, presumably due to imperfections in manufacturing.
3. Bulk traps: lattice defects or impurities that introduce local potential minima, which temporarily capture electrons long enough to remove them from the original charge packet, but re-emit them later.
4. Radiation-induced traps: similar to (3) but resulting from lattice defects caused by low-energy protons. This damage is most commonly seen by spacecraft CCD cameras.

CCD Operation and Data Reduction

In order to achieve ultimate *CCD performance* for a given goal, the CCD camera can be operated in special ways, and the postcamera data reduction can be optimized. Typical optical use of CCDs involves timed exposures, where the CCD pixels are exposed to light and the total charge integrated in pixels for a preselected time. At the conclusion of the integration, a shutter closes and the CCD is read out. As noise reduction limits the readout to roughly 100 kpixels/s, a large CCD (2048 × 2048 pixels) readout can take many seconds to complete.

To avoid the deadtime associated with the closed shutter, some CCDs are made with framestore regions. The framestore is a pixel array equal to the integration region which is permanently blocked by a cover from any additional light. The pixels containing the charge pattern resulting from the integration are very rapidly clocked into the framestore region (typically requiring much less than 1 s) and then slowly clocked out into the readout region without moving the integration pixels.

Operationally, CCD reduction requires calibration exposures. These include bias frames, which are readouts with the same integration time but no light striking the CCD, and flat field frames, which have a uniform illumination over the CCD. The bias frames are subtracted from the data frames to set the zero-point corresponding to zero incident radiation. (Note that CCDs will accumulate charge due to thermal electrons and low-level shorts in the gates, even if no light hits the CCD.) The flat field allows correction for pixel-to-pixel sensitivity variations. Proper flat fielding can remove variations of arbitrary amplitude and spatial scale.

The CCD dark current bias can be reduced by cooling the CCD, or by operating it in an inverted phase mode. In inverted phase operation, the gate electrode is given a suitable negative bias that attracts hole carriers to the front surface of the CCD. These holes fill interface states at the Si–SiO_2 boundary between the conducting depletion region and the insulating layer under the gates. Suppression of these interface states dramatically lowers the dark current because the interface states are much more efficient at thermal electron promotion to the conduction band than the bulk material. Not all phases can be operated in inverted mode in a normal CCD because, without the restraining potentials provided by gates held at positive voltage, the pixel charge packets can intermingle. A special CCD called an MPP (multiphase pinned) device has extra implant doping that isolates the pixels even with all three phases inverted, yielding dark current so low that integration times up to minutes become possible in room-temperature MPP CCDs.

Other important uses of CCDs include cases where the CCD is continuously clocked, without any shutter. Suitable for high light level conditions, the effective integration time becomes the time to transfer a pixel charge across the source point spread function on the CCD. This allows sensitive timing of source intensity changes.

A similar technique is called *drift scanning*, where the rate of clocking of pixels equals the rate of motion of the target across the CCD. Such a condition is common in astronomy, where a fixed detector on the Earth sees slow motion of stars in the field of view due to the Earth's rotation.

Drifts and instabilities in the camera electronics can be corrected using a technique called "overclocking." If the serial registers are clocked more times than there are physical pixels in the CCD, then the excess clocks will produce charge pulses corresponding to zero input light and zero dark current. The distribution of the overclock pulse is then a measure of the readnoise of the CCD chip-camera system, and the mean of the distribution sets the zero point of the energy to output voltage curve. Frequently, CCD cameras subtract the mean of the overclock pixels from all output values in a row (called "baseline restoration").

CCD Signal-to-Noise Ratios (SNR)

To see how these characteristics of the CCD relate to measurement, it is instructive to study the SNR predicted for a given exposure time. In a single pixel illuminated by a source that contributes S_o counts (electrons) to the pixel, one also sees contributions from dark current (S_d) and background illumination (S_s, usually called the "sky" in astronomical usages), all in units of counts per pixel per second. The source contribution (S_o) can be expanded into the intensity of light from the source, I; the quantum efficiency of the CCD, Q; and the integration time, t; to provide

$$S_o = I \times Q \times t \qquad (56.17)$$

The camera readout contributes a randomly distributed but fixed Gaussian noise with variance N_r. The SNR of a particular pixel is then:

$$SNR = I \times Q \times t = (I \times Q \times t + N_r^2 + S_d + S_s)^{1/2} \qquad (56.18)$$

If the light from a source is distributed over a number of pixels, n [as might arise from a star viewed through a telescope with a point spread function (PSF) covering n pixels], then if the integral of S_o over the PSF is C_o, and the integral of S_s over the pixels is C_s, then

$$SNR = C_o^{1/2}/(1 + C_sC_o + n + r^2C_o)^{1/2} \qquad (56.19)$$

Clearly, high Q and low r are desirable, and t should be chosen to make C_o greater than both C_s and r. It is worth noting that, in most optical applications (except for extremely faint sources), it is the uncertainty in the flat fielding (i.e., the corrections made for pixel-to-pixel sensitivity variations and background) that ultimately limits the achievable SNR.

CCD Structure and Charge Transport

CCD structure and its potential profile under bias

A CCD is a semiconductor device operating under the principle that charges can be temporarily stored and transported along a string or array of MOS capacitors. The basic storage unit is called a pixel and is made up of several MOS capacitors. In almost all CCDs, charges are stored either directly at the oxide–semiconductor interface (surface-channel device), or deeper within an epitaxial layer (buried-channel device). Theoretically, a surface-channel CCD has a larger charge capacity, but it also is prone to noise arising from interface states at the boundary.

The CCD is operated by varying voltages applied to the surface electrodes. Typically, the CCD is kept for a long period in an integration state, where photon-induced electrons accumulate in the potential wells under the CCD pixels. After the integration finishes, the voltages are changed to transport the charge from one capacitor to the next. This sequence of moving charge packets by potential clocking is sometimes called "bucket-brigade" charge transfer.

The storage unit of the CCD is the MOS capacitor and it is possible to deplete, invert, or form a surface accumulation layer in the MOS capacitor by simply changing the surface potential. CCDs operate in the so-called "deep depletion" mode when the surface layers are fully depleted. For a buried-channel CCD with an n-type buried layer within a p-type epilayer, this would require the application of a positive bias to the surface electrodes. The equations governing the one-dimensional calculations of the potential distribution in the MOS capacitor are [1]:

$$\frac{d^2\psi}{d_x^2} = 0 \qquad -d < x < 0$$

$$\frac{d^2\psi}{d_x^2} = -q \cdot \frac{N_D}{\varepsilon_s} \qquad 0 < x < t$$

$$\frac{d^2\psi}{d_x^2} = q \cdot \frac{N_A}{\varepsilon_s} \qquad t < x < x_p$$

$$(56.20)$$

where ψ is the potential, N_A is the substrate acceptor density, N_D is the donor density in the epilayer, x_p is the depletion edge in the p-type substrate, and ε_s is the semiconductor permittivity. Note also that d is the oxide thickness and the origin ($x = 0$) is located at the oxide–semiconductor boundary.

The boundary conditions will be:

$$\psi(x = -d) = V_G \qquad \frac{d\psi}{d_x}\bigg|_{x=t^-} = \frac{d\psi}{d_x}\bigg|_{x=t^+}$$

$$\varepsilon_{ox} \cdot \frac{d\psi}{d_x}\bigg|_{0^+} = \varepsilon_s \cdot \frac{d\psi}{d_x}\bigg|_{0^+} \qquad \psi(x = t^-) = \psi(x = t^+)$$

$$\psi(x = 0^-) = \psi(x = 0^+) \qquad \psi(x = t + x_p) = 0$$

$$(56.21)$$

and the solutions are:

$$\psi = V_G - E_{ox} \bullet (x + d) \qquad\qquad -d < x < 0$$

$$\psi = \psi_{max} - q \bullet N_D \bullet (x - x_n)^2 / (2\varepsilon_s) \qquad 0 < x < t$$

$$\psi = q \bullet N_A \bullet (x - t - x_p)^2 / (2\varepsilon_s) \qquad t < x < t + x_p \tag{56.22}$$

where E_{ox} is the electric field in the oxide layer, V_G is the gate bias voltage, and x_n is the position of the potential maximum.

Figure 56.13 shows a typical potential profile across the MOS capacitor in a buried-channel CCD. Note the presence of a potential maximum, ψ_{max}, where the electrons will reside. It is given by:

$$\psi_{max} = \psi_J \bullet (1 + N_A/N_D) \tag{56.23}$$

where $\psi_J = q \bullet N_A \bullet x_p^2 / (2\varepsilon_s)$. Since x_p increases linearly with N_D, the potential maximum ψ_{max} and hence the charge storage capability also increases for a heavily doped epilayer. The derivation of the potential profile in a surface-channel CCD will be similar, but with d set to zero.

Charge Transport

Charge transport in a CCD refers to the transfer of charges along the MOS capacitors (i.e., from one pixel to the next). There are several clocking schemes used, generally divided according to how many external voltage regions are applied per pixel, ranging from one (uni-phase) to four-phase.

Computation of charge transport in a CCD requires solving the two-dimensional Poisson equation with appropriate boundary conditions to obtain the potential distribution within the pixel for each separate phase within the clocking cycle. Numerical techniques using finite-difference or finite-element methods have been used. For charge transfer in an n-type epilayer, the electron flux, $F(x)$, is given by:

$$F(x) = n_s(x, t) \bullet \upsilon(E(x)) - D(E(x)) \bullet (n_s(x + \Delta x, t) - n_s(x, t)) = \Delta x \tag{56.24}$$

where $n_s(x, t)$ is the electron density, $E(x)$ is the electric field in the x-direction, $\upsilon(E(x))$ is the field-dependent velocity, and $D(E(x))$ is the electron diffusivity. Figure 56.14 shows the time evolution of charges along a CCD. Some "smoothing" of the output charge profile is often observed and, for high-

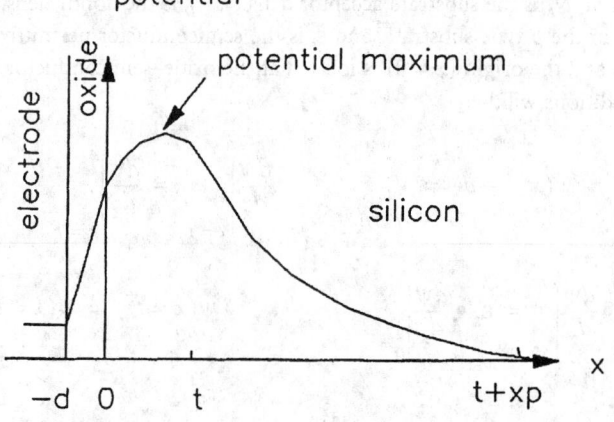

FIGURE 56.13 Potential profile across MOS capacitor.

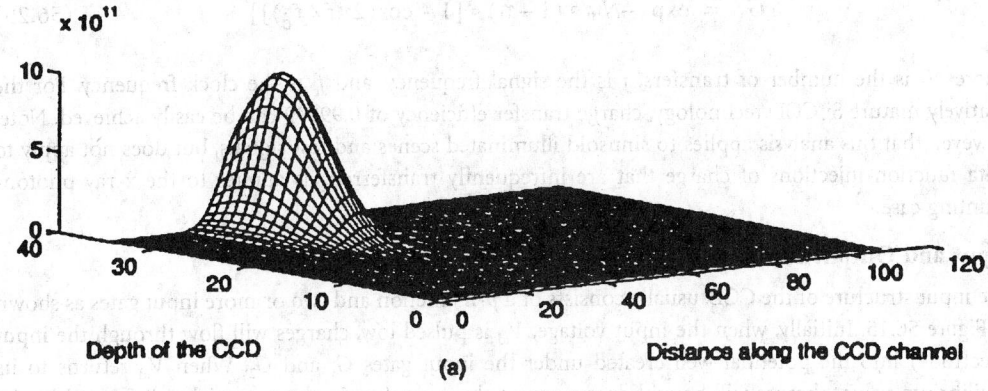

Depth of the CCD   (a)   Distance along the CCD channel

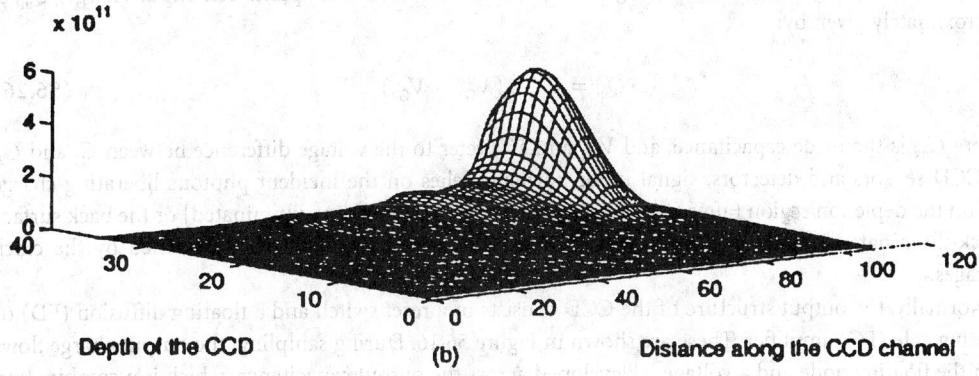

Depth of the CCD   (b)   Distance along the CCD channel

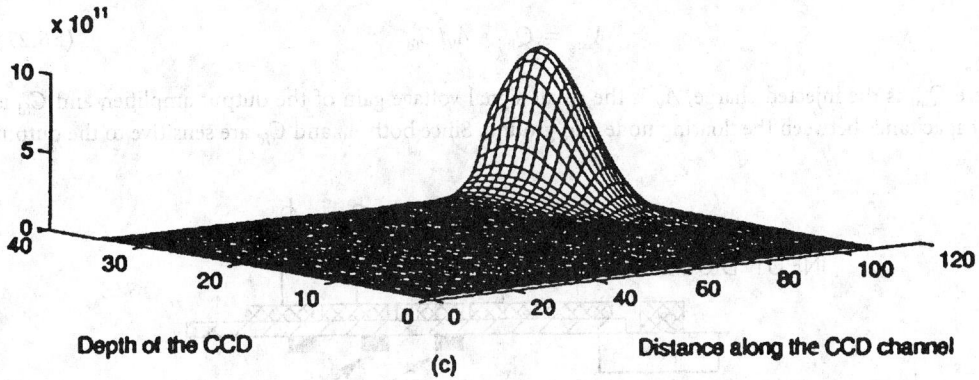

Depth of the CCD   (c)   Distance along the CCD channel

FIGURE 56.14 Time evolution of charges along a CCD: (a) $t = 200$ ps; (b) $t = 400$ ps; (c) $t = 600$ ps.

speed devices, velocity saturation will be important. Computations of charge transfer are sometimes carried out using a equivalent circuit model for the CCD in a SPICE-type simulator. It has been shown to offer both faster computation time and the ability to include external support circuits into the model.

Charge transfer efficiency, η, is the key figure-of-merit in a CCD, at least comparable to quantum efficiency. A common technique to determine η is to measure the amplitude response of the CCD using a network analyzer. η is then related to the amplitude response of the output, G_V through the following relationship:

$$G_V = \exp\{-N_T \bullet (1 - \eta) \bullet [1 - \cos(2\pi f / f_c)]\} \qquad (56.25)$$

where N_T is the number of transfers, f is the signal frequency, and f_c is the clock frequency. For the relatively mature Si CCD technology, charge transfer efficiency of 0.99999 can be easily achieved. Note, however, that this analysis applies to sinusoid illuminated scenes and test signals, but does not apply to delta function injections of charge that are infrequently transferred, as applies to the x-ray photon-counting case.

Input and Output Structures

The input structure of the CCD usually consists of a *p-n* junction and two or more input gates as shown in Figure 56.15. Initially, when the input voltage, V_{id} is pulsed low, charges will flow through the input (electrode) into the potential well created under the input gates G_1 and G_2. When V_{id} returns to its equilibrium value, charges will be withdrawn except those residing in the potential well formed by the potential difference between G_1 and G_2. Charge injection is now complete and a drop in the potential Φ_1 at the transfer gate will allow charges to enter into the first CCD pixel. The input charge, Q_{in}, is approximately given by:

$$Q_{in} = C_{ox} \bullet (V_{G1} - V_{G2}) \qquad (56.26)$$

where C_{ox} is the oxide capacitance, and V_{G1} and V_{G2} refer to the voltage difference between G_1 and G_2. In CCD sensors and detectors, signal input normally relies on the incident photons liberating charge within the depletion region (after traversing either the front gates [front-illuminated] or the back surface [back-illuminated]). Charges so created will be collected in the potential wells formed by the clock voltages.

Normally, the output structure of the CCD consists of a reset switch and a floating diffusion (FD) or floating gate (FG) amplifier. These are shown in Figure 56.16. During sampling, the output charge flows into the floating node and a voltage is developed across the output capacitance, which is a combination of the depletion capacitance of the floating node and the input gate capacitance of the source follower. In general, the voltage output, V_{sig}, can be expressed as:

$$V_{sig} = Q_{inj} \bullet A_V / C_{fd} \qquad (56.27)$$

where Q_{inj} is the injected charge, A_V is the small-signal voltage gain of the output amplifier, and C_{fd} is the capacitance between the floating node and ground. Since both A_V and C_{fd} are sensitive to the output

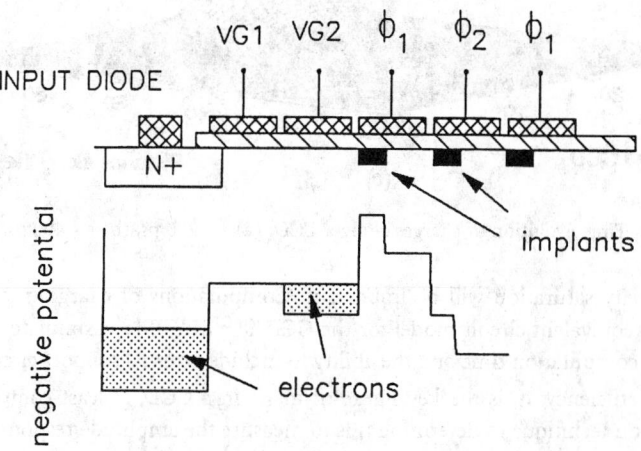

FIGURE 56.15 Input structure of a CCD.

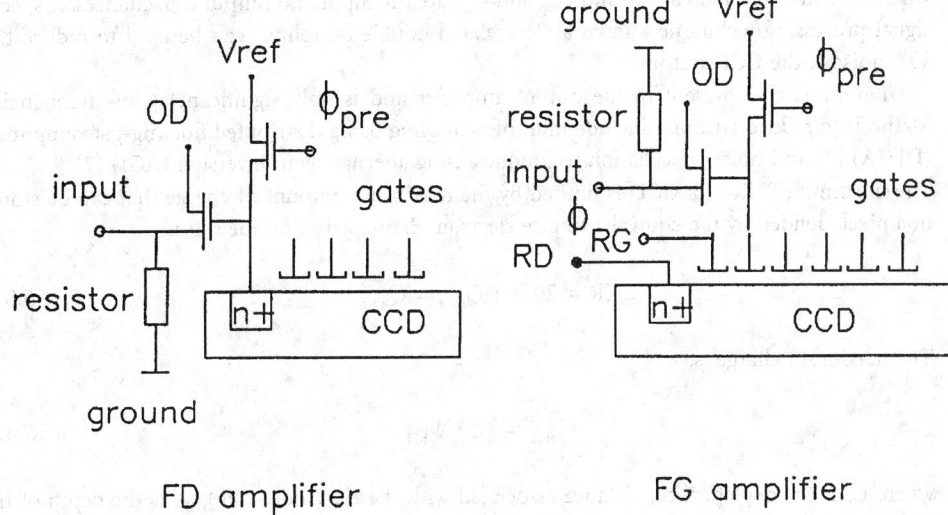

FIGURE 56.16 Output structures of a CCD.

charge and temperature, this type of output structure is not generally linear. Improvements in the output structure can be achieved using either a floating gate amplifier or Miller feedback at the output to reduce parasitic capacitances.

Noise in CCDs

CCD noise will degrade the SNR. The following are important noise sources in CCDs:

1. Thermal noise, or dark current, is due to thermally excited electrons that accumulate in the pixels during the integration. In applications where the clocking time is short compared to the integration time, the equivalent noise election, n_{th}, is given by:

$$n_{th} = \sqrt{J_d \bullet A_E \bullet t_{int}} \qquad (56.28)$$

where J_d is the leakage current density, A_E is the area of the transport electrode, and t_{int} is the integration time. A lowering of the operating temperature will normally reduce the leakage current density and hence the equivalent noise electron.

2. Bulk traps are the dominant noise sources for buried-channel CCDs and the equivalent noise electron from this source is:

$$n_{trap} = \sqrt{N_t \bullet V_{sig} \bullet Nt \bullet e^{-T_t/\tau_e}[1 - e^{-T_t/\tau_e}]} \qquad (56.29)$$

where V_{sig} is the volume of the charge packet under the transfer electrode, N_t is the density of the bulk trap states, T_t is the transfer time, and τ_e is the electron emission time constant. It can be observed that the equivalent noise electron increases as the size of the charge packet increases, an indication that bulk traps do not generally affect the SNR in CCDs. In x-ray photon counting applications some effects of bulk trapping noise have been seen.

3. Both the input and output of a CCD involve charge storage across capacitors that will be affected by fluctuations in the circuit. The equivalent noise electron for this process is:

$$n_{inp} = \sqrt{kT \bullet C_{inp}} \quad n_{out} = \sqrt{kT \bullet C_{out}} \qquad (56.30)$$

where kT is the thermal voltage and C_{inp} and C_{out} are the input and output capacitances. A special signal processing technique known as "correlated double sampling" can be used to reduce this kTC noise at the CCD output.

1/f noise is also present in the output amplifier and is only significant at low frequencies. Methods to reduce 1/f noise include multiple sampling using distributed floating-gate amplifiers (DFGA) [2] and noise cancellation techniques using alternate gain inversion (AGI) [7].

4. The maximum SNR of a CCD is limited by the maximum amount of charge that can be stored in a pixel, divided by the equivalent noise electron. Expressed in decibels, it is:

$$\text{SNR} = 20 \log(Q_{max} = Q_{noise}) \tag{56.31}$$

The maximum charge is:

$$Q_{max} = C_e \cdot \psi_{BH} \tag{56.32}$$

where C_e is the depletion capacitance associated with the electrode, and ψ_{BH} is the depth of the potential well.

CCD Power Dissipation

CCD power dissipation is primarily linked to the charging and discharging of the gate capacitances. This power is mainly consumed by the clock driver and is given by:

$$P_{clock} = C_c \cdot V^2 \cdot f_c \tag{56.33}$$

where C_c is the clock line capacitance associated with the electrodes, V is the clock voltage, and f_c is the clock frequency. For $f_c = 100$ MHz and $V = 5$ V, typical values of P_{clock} are 50 µW per pixel.

In addition to P_{clock}, the on-chip power dissipation per pixel is approximately given by:

$$P_{CCD} \approx nq(f_c L)^2/\mu_e + 2nqf_c(\psi_{BH} - nq/(2C_e)) \tag{56.34}$$

where n is the number of electrons in a pixel, L is the pixel length, and μ_e is the electron mobility. The overall power dissipation in a CCD is therefore the sum of P_{clock} and P_{CCD}.

CCDs Applications to Light Sensing

Optical Imaging and Spectroscopy

The many fine properties of CCDs have made them the detector of choice for the recording and readout of images at both high and low light levels. The applications include (1) reconnaissance, both civilian and defensive, (2) scene and personnel monitoring, (3) robotics, and (4) astronomy. In astronomy, CCD cameras have found their way into almost all observatories in the world, both for direct readout of images and as the readout cameras for spectrographs.

Perhaps the most exacting application of CCDs has been to astronomy, a discipline in which the characteristics of the CCD need to be pushed to their limits in order to gain the optimum performance. As well as the attempt to optimize this performance in order to gain as much science as possible from the images, it has also been crucial to calibrate the devices as precisely as possible and to fully understand their behavior. In this sense, astronomy has been the "driver" for the scientific development of CCDs. Large-scale commercial applications, such as video recorders, monitoring cameras, etc., have driven the need for improved manufacturing yield, blemish-free operation, uniformity in performance, and reliability.

In order to illustrate the importance of CCDs to exacting scientific requirements, this section will focus largely on the applications to astronomy.

The early demonstration of the performance of a commercial CCD at the focus of a telescope quickly led to the realization of their potential for both space and ground-based astronomical cameras and instruments. Fortunately, this development in the early 1970s was just in time for the first major optical observatory in Earth orbit, the Hubble space telescope (HST). Perhaps the most spectacular application of CCDs to visible light imaging has been within the cameras on board this telescope, i.e., the wide field and planetary cameras built at the Jet Propulsion Laboratory. The CCDs in these cameras consisted of 800×800 pixel arrays, with each pixel of size 25 µm. The first version of this camera contained eight thinned, back-illuminated devices that were more or less fully depleted. An accumulation layer at the back surface was achieved by flooding the devices with UV light from a lamp internal to the instrument. This accumulation layer allowed the collection of charge resulting from the photoelectric absorption of blue light, and accelerated the charge to the frontside potential wells. Furthermore, the devices were coated in a down-converting coronene phosphor that converted UV light into the yellow-green wavelength band, thus allowing a 20% efficiency in the UV. These devices thus had relatively high quantum efficiency from the UV through to about 1 µm, vitally important for an instrument collecting photons from galaxies billions of light years away in the universe. As well as their high quantum efficiency, these devices had high charge transfer efficiency and relatively low readout noise at the time of their development. All of these characteristics have since been superseded—at first by an improved version of the camera installed in the HST in late 1993, and then in the Advanced Camera for Surveys, to be installed in 1999. These later instruments had larger format, higher quantum efficiency, and lower readout noise. During the 1980s, devices were developed with these large formats and with readout noise levels approaching one electron.

Starting with modest arrays of size on the order of a few hundred pixels in the 1970s, the devices that became available in the 1990s were as large as 4096 pixels square. As well as in space, these devices have been employed at the prime foci of the world's largest telescopes to give images covering more than 10 arc minutes on a side while still sampling the atmospheric-limited resolution (arcsecond or subarcsecond) adequately. Such devices, when exposed to the sky for up to an hour through a broadband filter, can produce images with tens of thousands of objects for statistical studies in astronomy.

Typical pixel readout rates for astronomical cameras are of the order of 50 kHz, taking tens of seconds for a full readout. Integration times for astronomical CCD cameras can vary from fractions of a second to tens of minutes or even times in excess of an hour, usually limited by the background noise induced by cosmic ray particles and their secondaries. These cosmic ray "events" can be removed by taking at least two frames of data and cross-comparing them. This is important even on the ground, where the radioactive background and muon-induced events total about one per cm^2 per minute; but the cosmic-ray background in space is about one per cm^2 per second, so that camera exposures are rarely longer than 10 to 20 min. Sophisticated computer algorithms have been developed in order to remove the effects of this background radiation.

Of special importance to applications in space research is the packaging of the devices. Contamination of the cold CCD surface by as much as a monolayer of a heavy molecule will effectively render the CCD useless in the vacuum-UV, for example. Careful attention must therefore be paid to the local instrument environment. The CCDs on the HST were in hermetically sealed packages, and the contamination problem was transferred to the lenses covering the CCD packages. In order to achieve optimum performance, the overall camera has to be designed to satisfy the exacting requirements of the CCD, from the point of view of thermal control and stability, absence of electronic interference, and rigorous attention to the elimination of water or heavy molecule contamination. For space application, it may also be advantageous to surround the CCD package with a cosmic-ray shield, such as the 1 cm (0.4 inch) of tantalum used in the CCD cameras on the HST and the Galileo mission to Jupiter. Such radiation shields have to be designed with caution, lest they introduce more secondary particles and induced radioactivity than the primary protons that they stop.

Finally, much effort is put into the processing of astronomical CCD data [24]. Special techniques have been developed to calibrate the pixel-to-pixel nonuniformity in quantum efficiency (a function of wavelength), both on large and small scales. With some effort, this nonuniformity can be calibrated to levels below 1%. Although it may be easy to expose the CCD to a diffuse source such that there are at least 10,000 electrons per pixel, the corresponding flatfield calibration accuracy may be compromised at low light levels. For astronomical applications, especially in space, most of the pixels may receive only tens of electrons or fewer during an exposure. Low-level traps (including those caused by cosmic radiation) will then manifest themselves in the form of charge-transfer inefficiencies over localized areas or columns. It may be important to calibrate the CCD at the same exposure levels as those typically encountered during the science observation. One way of effecting this is to use hundreds or thousands of frames of data taken in different parts of the sky, and rejecting the astronomical objects in them. These "sky frames" can then be used to produce a "super sky-flat," which is the average of the individual frames of data.

CCD X-ray Imaging Spectroscopy

Scientific applications such as astronomy have been responsible for driving improvements in CCD technology. In the x-ray domain, for example, CCDs are starting to be employed for medical radiography, where the digital imaging capability and high sensitivity allows for lower patient doses and online image processing. As high-resolution readout detectors of dispersive x-ray spectrometers, CCDs are becoming widely used in the new generation of high brilliance synchrotron beamlines. As spectrometers, they are also being considered in some applications as replacement for Si(Li) diodes, where the higher operating temperatures and improved resolution and efficiency down to x-ray energies below 1 keV leads to applications of interest to biological science and technology.

In contrast with optical imaging applications with many photons per charge packet, to measure x-ray spectral information directly with a CCD single photons per pixel per image frame are required. To use the CCD as a nondispersive x-ray spectrometer requires that the correspondence between the magnitude of the electron charge packet generated initially by the x-ray photon, and the signal measured at the CCD output, must be maintained. This places a very stringent requirement on the efficiency of charge transfer. However, even before the process of charge transfer is initiated, the physics of the charge collection may also degrade the energy measurement process.

The absorption process for a photon of energy E begins with the ejection of a photoelectron, of energy $E - E_B$, where E_B is the binding energy of the appropriate silicon atom electron shell. The range of the photoelectron may be only a fraction of 1 μm. In a few percent of cases, a silicon K shell fluorescence photon may be emitted, and this has a range of about 10 μm, so that there is a finite probability of the event energy splitting into more than one pixel.

Eventually, a proportion of the energy is converted to free electrons, the rest into phonons. The average energy to create an electron-hole pair is roughly constant at about 3.6 eV of incident photon energy per pair produced in silicon. If this free electron charge packet is created in the depletion layer of the CCD, it is promptly drifted under the influence of the electric field to the buried-channel collection site. During this drift time (t_d), the charge can laterally diffuse within a radius ~ $\sqrt{2Dt_d}$, where D is the diffusion constant. Except for the charge clouds originating deep in the depletion layer, this radius is small compared with the typical pixel size. If the charge is generated in a field-free layer *outside* the depleted volume, it will radially diffuse until a fraction recombines or reaches the depletion layer. The latter fraction then starts to drift with the same additional lateral drift as the depletion layer charge packets.

The pixel boundaries are loosely defined by the asymmetric fields created by electrode biasing schemes and surface channel stop implants, and not hard physical boundaries. Hence, any lateral spreading processes may allow some splitting of the initial charge packet between pixels. Furthermore, charge *loss* may be experienced either by the partial recombination of events when liberated deep below the depletion layer, or if the fraction of split charge is too low to be recognized against the device noise level.

In astronomical applications, in space-borne observatories, there is a continuous low-level background of charged particles. Rather than liberating point-like charge clouds, they liberate a population of signal electrons along their tracks throughout the silicon. They may be easily discriminated against x-rays if these tracks are highly skewed with respect to the silicon surface and cross many pixels. Also, if the track is long enough, the magnitude of total charge cloud generated may be large compared with the typical x-ray charge packet.

These features impose a requirement to perform event recognition and analysis. To perform this on-ground would require transmission of all pixel data, including empty pixels, which for megapixel CCD frames generated on second time scales is quite infeasible. Performing this recognition on-board first requires a comparison against some lower threshold. Selection of this level is critical—too high and some split charge may be neglected with a consequent degradation of energy resolution. Too low a threshold, and many spurious events will be counted. If there is a drift or change in the zero-energy signal upon which this threshold is applied, the energy scale may be misregistered, and/or the relative fraction of events selected at different x-ray energies may be unknowingly altered. Simulations show that for the potential energy resolution of CCDs, spectral analysis of cosmic plasmas will demand a calibration of relative detection efficiency versus energy to ~3%. If the threshold value is set at about 4σ times the readout noise, this calibration may be degraded by as little as a single digital bit of a commonly used 12-bit ADC in the CCD readout electronics. Thus, the realtime digital processing of events is required to be fast and complex.

A further complication of the event splitting is that to avoid pile-up in CCD frames, the probability of having multiple events per pixel is made more stringent by requiring surrounding pixels to have no signal, and allow this recognition process. Typically, an event rate of 1 photon per 200 pixels per CCD frame is therefore imposed, but this places a rather low limit on attainable count rate performance. Especially with high-resolution focusing optics of observatories such as AXAF[5], this can be much lower than for many previous experiments. If the core of the point spread function of the mirror is not to be degraded by this effect, then special readout formats that reduce the imaged area, in order to accelerate frame readout times, must be employed.

Future Improvements to CCDs

Backside CCDs

The useful spectral range of conventional CCD detectors, although quite broad, is limited in part by the presence of the polysilicon gates on the front surface. As is illustrated in Figure 56.17, photons with

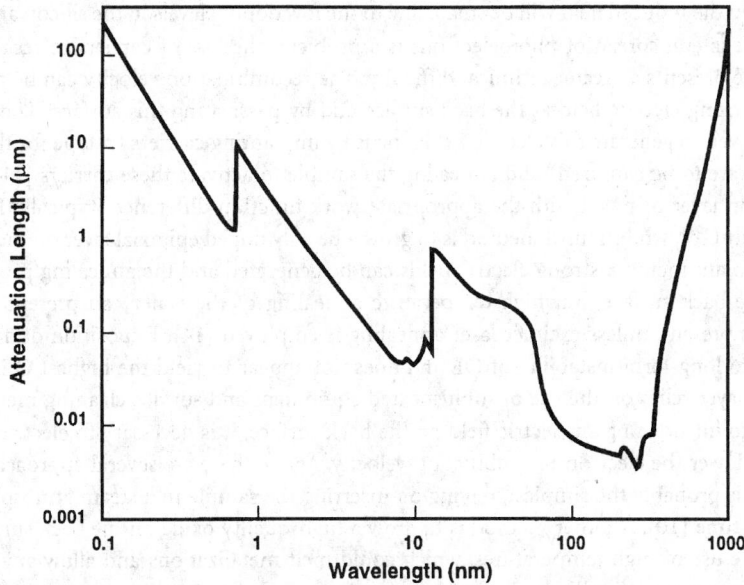

FIGURE 56.17 Attenuation length in silicon in the soft x-ray through near-infrared spectral range. The polysilicon gates of front-illuminated devices, which are typically ~0.4 μm thick, are strongly absorbing at wavelengths for which the attenuation length is less than the gate thickness, viz. wavelengths (~3 mn < λ < ~400 nm). Thin-gate and back-illuminated CCDs, with much thinner dead layers, offer improved detection efficiency in the soft x-ray and UV portions of the spectrum. (From References 8 and 9.)

wavelengths in the range ~3 nm < λ < ~400 nm have attenuation lengths in silicon less than the typical thickness of the gate structure (~0.4 μm). Photons in this spectral range are therefore absorbed before they can enter the photosensitive volume of the detector. For conventional CCDs, the detection efficiency is no more than a few percent in this band. If the gates can be made thinner by an order of magnitude, or avoided altogether, then the detection efficiency can be improved dramatically in both the very soft x-ray and the UV. Efforts following the former approach are described below. The latter strategy, which requires illumination of the back surface of the CCD (that is, the surface opposite the gates) is discussed here.

There are a number of techniques that can be used to produce back-illuminated devices [10–12], but they must all deal with two constraints: lateral diffusion of the photoelectrons before they are captured in the buried channel of a specific pixel and the tendency for photoelectrons to recombine at the back surface of the Si.

The lateral diffusion of electrons can lead to a loss of resolution of the CCD since they may cross the boundary between one pixel and another, leading to an erroneous assignment of the origin of the photon that created the electron. This problem is minimized by thinning the *back-illuminated CCD* to thicknesses that are on the order of the depletion depth of the Si, so the least drift occurs before the electron is captured in the potential well of the buried channel. This thinning is accomplished by a combination of methods, including mechanical grinding and polishing and wet etching. Handling of back-illuminated CCDs after thinning, however, is a problem since the remaining Si is between 10 and 100 μm thick, and is not strong enough to support rough handling, or even the intrinsic stresses arising from the initial fabrication of the CCD. To circumvent this problem, rim thinning or frame thinning can be used; in the first case, a rim several millimeters wide around the circumference of the wafer is protected during the thinning process, while in the latter, a region 1 or 2 mm wide around each device is protected.

Lateral diffusion can also be reduced by imposing an electric field through the thickness of the thinned membrane; but if this is done by imposing an electrode on the back surface, then the device will have a reduced QE (quantum efficiency) due to absorption in this layer. The presence of a depletion region in the CCD will also give rise to an electric field, which reduces the lateral diffusion of electrons; but if the desired depletion region is on the order of tens of micrometers, in order to image high-energy x-rays or IR photons, then the induced field will be small, due to the low doping levels in the silicon and Gauss' law.

Although the lateral spread of photoelectrons is a problem, the loss of carriers to recombination at the back surface presents a greater technical difficulty. The recombination velocity can be minimized by introducing a strong electric field at the back surface and by passivating this surface. There are several methods employed to generate this electric field: one is by implanting carriers (*p*-type for the case where photoelectrons are to be captured) and annealing the sample to activate these carriers [11], a second is to deposit a thin layer of metal with the appropriate work function difference (typically Pt or another transition metal) [10], while a third method is to grow a heavily doped epitaxial layer on the back surface [13]. In the implant method, strong electric fields can be generated and the annealing process can help to passivate the back surface, but high-temperature annealing of the wafer can present a problem if metallization is present, unless excimer laser annealing is employed [14]. Deposition of a thin metallic layer can lead to long-term instability of QE and does not appear to yield the highest QEs. Deposition of an epitaxial layer relies on the use of sophisticated equipment and surface cleaning methods.

In addition to introducing an electric field on the back surface, it is necessary to electrically passivate that surface to lower the electron recombination velocity. Again, there are several approaches. The flash gate technique is probably the simplest, relying on inserting the sample in a steam atmosphere, but it is not stable over time [10]. Another method is to grow a high-quality oxide on the back surface [11], but this requires the use of high temperatures, which could melt metalizations and allow any dopant layer to diffuse too deeply into the silicon.

Although the aim of the backside treatment process is always to maximize the fraction of charge collected from the vicinity of the back surface, there are subtle differences in requirements that depend on the application. In the UV (100 nm < λ < 400 nm), the extremely small attenuation lengths (see Figure 56.17) place a premium on nonnegligible (>~50%) charge collection efficiency very close to the back surface. Therefore, in UV sensors, the field in the immediate vicinity of the back surface is extremely

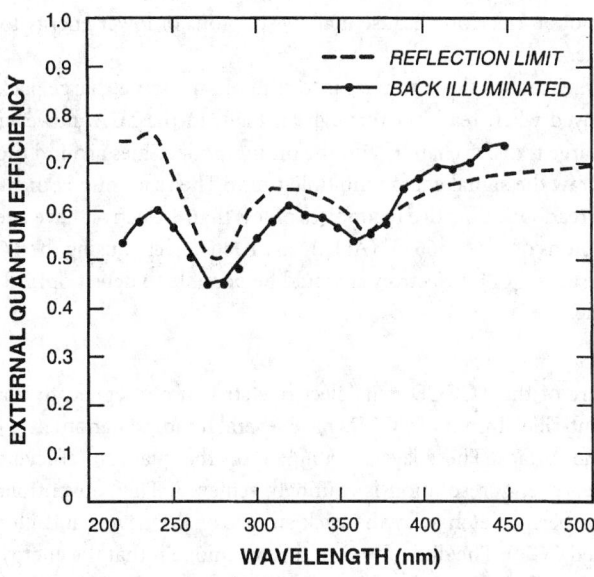

FIGURE 56.18 External quantum efficiency of back-illuminated CCD using an implanted and annealed back side. For comparison purposes, the external quantum efficiency of an ideal detector is shown, after modeling for the antireflection coating present on the back-illuminated detector.

important. Moreover, the real part of the index of refraction of silicon is quite high in the UV and optical bands, and Fresnel reflections limit the external detection efficiency that can be achieved. Thus, significant improvements in the UV response of back-illuminated CCDs have been obtained via the application of antireflection coatings as a part of the backside treatment process [12].

In the very soft x-ray range (2 nm $< \lambda <$ 100 nm), the attenuation length can be larger by a factor of 10 than in the UV. Moreover, the range of the secondary photoelectrons is large enough to be comparable to the attenuation length [15], so the physics of the secondary ionization process becomes important. Thus, in the very soft x-ray regime, it is the average properties of the device over scales of a few hundred nanometers that determine performance. However, for spectroscopic applications, the theoretical limit of device performance cannot be reached unless the charge collection efficiency substantially exceeds ~90%. Thus, in the soft x-ray, one requires very high collection efficiency over a relatively large volume in the vicinity of the back surface.

Any backside fabrication techniques will probably require compromises to be made, but it is possible to achieve very high external QE in the UV, as shown in the accompanying Figure 56.18, and in the soft x-ray region where 77% QE has been achieved at 277 eV [16]. Further improvements are necessary to achieve good results in the vacuum UV and theoretical limits for energy resolution in the soft x-ray. These improvements are possible, but will probably depend on the continued development of sophisticated processing tools.

Thinned Gate CCDs

Recently, a new type of CCD was developed that was optimized for spectroscopy below 1 keV. This detector, called a thin gate CCD (or TGCCD), contained two novel features: floating gate output amplifiers (FGAs or "skippers") and thin gates (electrodes) [2]. The goal was to develop a detector that had Fano-limited energy resolution with nonnegligible QE over the 200 eV to 1000 eV bandpass. This section briefly describes this detector and compares it with more conventional CCDs.

Most conventional CCDs use a floating diffusion amplifier (FDA) to read out the signal charge. The readnoise of the FDA has improved from several hundred electrons (rms) in the 1970s to less than 3 electrons in the last few years [3]. The energy resolution of CCDs over the entire soft x-ray bandpass is now limited by factors other than readnoise. It is desirable to push the readnoise down from a few electrons to a fraction of an electron because it allows photons of lower energy to be detected in single-photon counting mode.

One method to further reduce the readnoise is to read out the same charge packet multiple times. The signal charge is destroyed when read out through an FDA. In the FGA, however, an insulating gate is placed between the charge transfer channel and the output node. Gates around the output node are used to read out and withdraw the signal charge multiple times. The readnoise is then reduced by the square root of the number of readouts. Effective readnoises of less than 1 electron have been demonstrated using the FGA [2,4], and photons of energy 66 eV (Al L_α) have been detected using the TGCCD. If the readnoise could be pushed to a fraction of an electron, it would be possible to detect optical photons in the single-photon counting mode.

Lower Readnoise

The other novel feature of the TGCCD is its electrode structure designed to maximize soft x-ray QE. Most conventional front-side illuminated CCDs have several hundred nanometers of SiO_2/Si_3N_4 insulator and Si electrode on the surface. These layers strongly limit the quantum efficiency of the device below 1 keV, but Fano-limited energy resolution is routinely achieved. The conventional solution to increase the low-energy QE has been to etch away the back surface of the device and illuminate it from behind (a back-side illuminated CCD). The difficulty with this technique is that the energy resolution is seriously degraded due to charge diffusion and poor charge collection.

The thin gate CCD is a front-side illuminated device with a unique gate structure that gives reasonable soft x-ray QE (0.20 at 277 eV), while retaining the good energy resolution of front-side illumination. The detector is a three-phase CCD. The first two phases (comprising only 1/3 of the active area of the device) are covered with 400 nm of polysilicon (or just poly) in the usual way. A third layer of poly is deposited on top of the first two and acts as a bus for a fourth (40-nm thick) layer of poly that defines the third clock phase. Two-thirds of the active area of the detector is therefore covered with only the insulator (100 nm of SiO_2 and Si_3N_4) and this fourth layer of poly. The measured energy resolution of this device is 34 eV at C K_α (277 eV). This is approximately a factor of 3 better than the best energy resolution obtained with the ACIS back-side illuminated CCDs [5]. Simulations show that an astrophysical spectrum obtained with TGCCD contains as much or more information than one obtained with a back-side illuminated CCD [6].

Radiation Damage

CCDs are often used in environments in which they are subject to ionizing radiation, and this radiation can degrade their performance. In particular, radiation damage plays an important role in the performance of most space-borne detectors. The Van Allen belts contain protons and electrons with energies of tens of MeV. These particles can produce damage by several mechanisms. Protons can generate vacancy-interstitial pairs in the Si lattice, which diffuse freely, even at cryogenic temperatures, and it is possible for the vacancy to form a complex with P atoms in the *n*-Si buried channel that traps electrons. The trap reduces CTE by preventing signal electrons from leaving the pixel during the clocking cycle. Electrons can also produce vacancies (although with much lower cross-section than the protons), but they can also create dangling bonds at the SiO_2–Si interface above the buried channel, leading to a shift in operating voltage in the CCD and contributing to noise.

The P–V complexes remove electrons from a charge packet in the conduction band of silicon. In the worst-case limit, when the illumination levels are so low that each charge packet transferred encounters each trap in its empty trap state, the degradation of CTE is approximately 5×10^{-7} rad^{-1} (radiation doses are expressed here in total ionizing rads in Si, although it should be noted that the fraction of proton energy that goes into production of vacancy-interstitial pairs, the nonionizing energy loss, is about 1/2000 of the total ionizing dose [17].) This degradation is significant: if the device is subjected to 1000 rad yr^{-1} and there are 1000 transfers necessary to clock out the charge, then 40% of the charge packet would be lost due to proton-induced defects. At higher illumination levels, the traps greatly complicate the device response, particularly in single-photon counting applications. In this case, for example, the spectral response of the CCD becomes a function of the spatial distribution of the incident radiation.

This loss of CTE can be circumvented in several ways, including the use of increased shielding, a fat zero, annealing out the damage, implanting a narrow trough along the direction of transfer in the buried channel, or operating the device at a different (generally lower) temperature.

Relatively thick shielding, up to the equivalent 4 to 5 cm (1.6 to 2 inch) of aluminum, can provide significant attenuation (factors of several) of proton dose encountered in the radiation belts [18]. A fat zero is a uniform charge added to all the pixels of the CCD, usually induced by flooding the CCD with a low light level. Although a fat zero can reduce the CTE loss by filling the traps so they cannot be occupied by electrons from the charge packet of interest, they also add to the noise of the system. The P–V complex will dissociate at temperatures around 135° C [19], but the annealing is incomplete and may interfere with other requirements or constraints of the satellite mission. Using photolithography, a narrow trough or notch can be included in the buried channel parallel to the direction of transfer during fabrication, so charge packets are kept constrained and are less likely to interact with traps generated randomly across the buried channel [20]. It is not difficult to introduce a channel that is approximately 2 μm wide and can handle a charge packet of tens of thousands of electrons; this can decrease the CTE loss to 7×10^{-8} rad^{-1} or less. Reducing the operating temperature slows the kinetics of trap emptying [21], so a single electron can be lost from the charge packet and fill the trap for the integration time of the frame. The effectiveness of this process depends on the clock speed and integration time, as well as the ability to operate the CCD at temperatures that approach −150° C, but CTE loss can be as low as 10^{-8} rad^{-1} if troughs are included with low-temperature operation. Cosmic rays and electrons can also cause damage by displacing Si atoms from the lattice, but the flux of cosmic rays is much lower than protons in low Earth orbits and the electrons are much less effective in creating a vacancy. The approaches to hardening against proton damage will also be effective with these latter two particles.

The major effect of electrons on CCDs is to shift the flatband voltage of the CCDs and MOSFETs, but for doses around 1000 rads, this change is on the order of 10 mV. If further hardening of the gate oxide is desired, it can be accomplished using established methods [22]. The passage of electrons and other energetic charged particles through a device also creates free carriers that constitute an interfering signal. One electron-hole pair is created for each 3.65 eV absorbed in the Si, and charged particles can deposit charge packets of up to a full-well level and more, depending on the species, energy, and the active depth of the device. In astronomy, such events are an annoyance, even in terrestrial observatories, and are removed by comparing images of the same scene. In some applications, such as x-ray spectroscopy, these events are more serious because they can masquerade as desired signals unless careful analysis of the data is performed to exclude them. On the other hand, this sensitivity to charged particles can actually be used to advantage. The known conversion between particle energy deposition and liberated charge means that CCDs can be used as spectroscopic detectors, and such an application has been proposed for inertial-confinement fusion diagnostics [23].

Radiation hardening of CCDs has made rapid progress in the last few years, allowing devices to have much longer lives for scientific applications. Fabrication and operational changes have increased the hardness to displacement damage and the process requirements of high-quality ICs have led to a gate dielectric that is quite robust to ionizing radiation.

References

1. C.K. Kim, The physics of charge-coupled devices, in M.J. Howes and D.V. Morgan, (Eds.), *Charge-Coupled Devices and Systems,* Wiley, New York (1979).
2. R.P. Kraft, D.N. Burrows, G.P. Garmire, J.A. Nousek, J.R. Janesick, and P.N. Vu, Soft x-ray spectroscopy with sub-electron readnoise charge-coupled devices. *Nucl. Instr. & Meth.,* A361, 372–383, 1995.
3. D.N. Burrows, G.D. Berthiaume, M.A. Catalano, G.P. Garmire, C. Larkin, F. Marks, J.A. Nousek, and G.M. Weaver, Penn State imaging x-ray spectrometer, EUV, X-ray, and gamma-ray instrumentation for astronomy and atomic physics, eds. C.J. Hailey and O.H.W. Siegmund, *Proc. SPIE,* 1159, 92–104, 1989.

4. C.E. Chandler, R.A. Bredthauer, J.R. Janesick, J.A. Westphal, and J.E. Gunn, Sub-electron noise charge coupled devices. *SPIE Symp. on Electronic Imaging, Proc. SPIE*, 1242, 238–251, 1990.

5. R.P. Kraft, D.N. Burrows, G.P. Garmire, and J.A. Nousek. Thin-gate front side-illuminated versus back side-illuminated charge-coupled devices for X-ray astronomy. *Astrophysical J. Lett.*, 466, L51–L54, 1996.

6. M.C. Weisskopf, S.L. O'Dell, R.F. Elsner, and L.P. Van Speybroeck, Advanced X-ray Astrophysics Facility (AXAF): an overview. *Proc. SPIE*, 2515, 312–329, 1995.

7. Y. Matsunaga and S. Ohsawa, Analysis of low signal level characteristics for high-sensitivity CCD charge detector. *Trans. Electron Devices*, 39, 1465–1468, 1992.

8. D.F. Edwards, *Handbook of Optical Constants of Solids*, E.P. Palik, Ed., New York: Academic Press, 1985, 547.

9. B.L. Henke, P. Lee, T.J. Tanaka, R.L. Shimabukuro, and B.K. Fujikawa, Low energy X-ray diagnostics. D.T. Attwood, B.L. Henke, Eds., *AIP Conf. Proc. No. 75*, New York: American Institute of Physics, 1981, p. 340.

10. J.R. Janesick, D. Campbell, T. Elliott, and T. Daud, Flash technology for charge-coupled-device imaging in the ultraviolet. *Opt. Eng.*, 26: 852–863, 1987.

11. B.E. Burke, J.A. Gregory, R.W. Mountain, J.C.M. Huang, M.J. Cooper, M.J., and V.S. Dolat, High-performance visible/UV CCD imagers for space-based applications. *Proc. SPIE* 1693, 86–100, 1992.

12. M.P. Lesser, Improving CCD quantum efficiency, *Inst. in Astronomy VIII, Proc. SPIE*, 2198, 782–791, 1994.

13. M.E. Hoenk, P.J. Grunthaner, F.J. Grunthaner, R.W. Terhune, M. Fattahi, and H.-F. Tseng, Growth of a delta-doped silicon layer by molecular beam epitaxy on a charge-coupled device for reflection limited ultraviolet quantum efficiency. *Appl. Phys. Letts.* 61, 1084–1086, 1992.

14. C.M. Huang, B.E. Burke, B.B. Kosicki, R.W. Mountain, P.J. Daniels, D.C. Harrison, G.A. Lincoln, N. Usiak, M. A. Kaplan, and A.R. Forte, A new process for thinned, back-illuminated CCD imager devices. *Proc. Intl. Symp. VLSI Technol., Syst., Appl.*, New York: IEEE, 1989, 98–101.

15. F. Scholze and G. Ulm, Characterization of a windowless Si(Li) detector in the photon energy range 0.1–5 keV. *Nucl. Instr. Meth.*, A339, 49–54, 1994.

16. G. Prigozhin, M. Bautz, S. Kissel, G. Ricker, S. Kraft, F. Scholze, R. Thornagel, and G. Ulm, Absolute measurement of oxygen edge structure in the quantum efficiency of X-ray CCDs, *IEEE Trans. Nucl. Sci.*, 44, 970–975, 1997.

17. M.J. Cantella, B.E. Burke, J.A. Gregory, D.C. Harrison, E.D. Savoye, and B.-Y. Tsaur, Large silicon staring-array sensors, *Infrared Focal Plane Arrays IIIA*, J. A. Jamieson (Ed.), Washington: Ballistic Missile Defense Organization, 1994, 10-1–10-76.

18. K.C. Gendreau, M.W. Bautz, and G.R. Ricker, Proton damage in X-ray CCDs for space applications: ground evaluation techniques and effects on flight performance. *Nucl. Inst. Meth.* A335, 318–327, 1993.

19. M. Hirata, H. Saito, and J. Crawford, Effect of impurities on the annealing behavior of irradiated silicon. *J. Appl. Phys.* 38, 2433–2438, 1967.

20. T.S. Villani, W.F. Kosonocky, F.S. Shallcross, J.V. Groppe, G.M. Merais, J.T. O'Neill III, and B.J. Esposito, Construction and performance of a 320 × 244-element IR-CCD imager with PtSi Schottky-barrier detectors. *Proc. SPIE*, 1107, 9–21, 1989.

21. K.C. Gendreau, G.Y. Prigozhin, R.K. Huang, and M.W. Bautz, A technique to measure trap characteristics in CCDs using X-rays. *IEEE Trans. Electron Dev.*, 42, 1912–1917, 1995.

22. P.V. Dressendorfer, Effects of radiation on microelectronics and techniques for hardening. In T. P. Ma and P. V. Dressendorfer (Eds.), *Ionizing Radiation Effects in MOS Devices and Circuits*, New York: John Wiley & Sons, 1989, 333.

23. B.E. Burke, R.D. Petrasso, C.-K. Li, and T.C. Hotaling, Use of charge-coupled device imagers for charge-particle spectroscopy, *Rev. Sci. Inst.*, 68: 599–602, 1997.

24. G.H. Jacoby (Ed.), *CCDs in Astronomy*, San Francisco: Astronomical Society of the Pacific, 1990.

57

Densitometry Measurement

Joseph H. Altman
Pittsford, New York

57.1 Introduction

As the name indicates, densitometry is the measurement of optical density (O.D.). Optical density, in turn, can be broadly defined as a measure of the attenuation of radiant flux by some sort of optical element that can be transmitting, reflecting, or both. Densitometry is most widely applied in photographic science, and measurements in this field are covered by a four-part ANSI/ISO standard [1]. The treatment here follows that document. The measurement of transmission and reflection from the standpoint of optics has been discussed by Palmer [2]. This chapter uses the single word density, without the adjective, and uses the words "light" and "flux" interchangeably (although density can be measured in the UV and IR also, of course).

It turns out that, in practice, the measured density of a given sample can be affected significantly both by the design of the equipment and the nature of the sample. Therefore, the first rule of practical densitometry is that the sample must be measured in a way that is meaningful with respect to its intended use.

57.2 Monochrome Transmission Density

It is convenient to start by discussing transmitting samples, without the color aspect. Such samples can be either black-and-white photographic films or filter layers of various sorts. If light falls on such an element, a fraction of the flux is reflected from the first surface. The remainder penetrates the surface, and a part of this flux is absorbed. What is neither reflected nor absorbed exits the layer, as shown in Figure 57.1. Let these fractions be denoted, r, a, and t, where lower-case t indicates internal transmittance. If the element is in air, reflection can occur at the rear surface also, but this can be neglected. By the conservation of energy, one obtains

$$r + a + t = 1 \qquad (57.1)$$

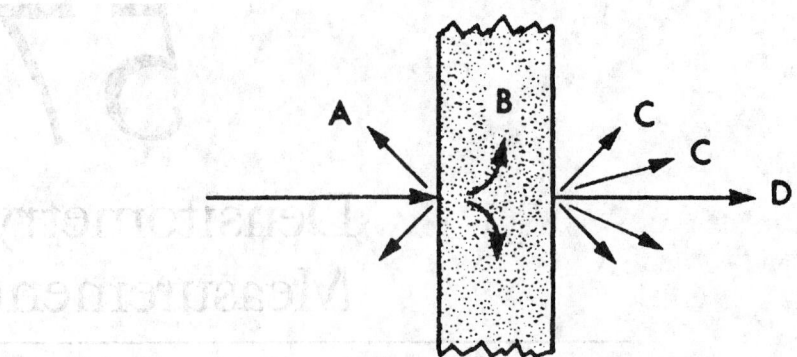

FIGURE 57.1 Cross-section of emulsion layer showing (A) reflection, (B) absorption, (C), diffuse transmission, and (D) specular transmission.

The fractions r and t can be measured independently by spectrophotographic techniques, so that a is readily determined. This fraction is termed the absorptance of the layer, and its measurement is also discussed by Palmer [2].

Some applications require the measurement of absorptance, but in most practical work, what is important is the fraction of the incident light that exits the element, regardless of where attenuation takes place. Therefore, transmittance is defined by:

$$T = \frac{\Phi_e}{\Phi_i} \qquad (57.2)$$

where e and i refer to emergent and incident flux, respectively. Note that all losses are lumped together. In silver halide photographic layers, the first-surface reflection is usually small; in thin-film filters, a significant proportion of the total loss may occur by reflection. Following Hurter and Driffield [3], the English pioneers of densitometry, one can define the transmission density D

$$D = -\log_{10} T = \log_{10}\frac{1}{T} \qquad (57.3)$$

Density is therefore a nonlinear dimensionless quantity ≥ 0. McCamy has published an extensive review of densitometry and sensitometry [4].

The operation of a practical densitometer can be discussed with reference to Figure 57.2. The system consists of a light source, an aperture to define the area being measured, a sensor, logging circuitry, and a suitable readout display. In many instruments, the output is fed to a computer.

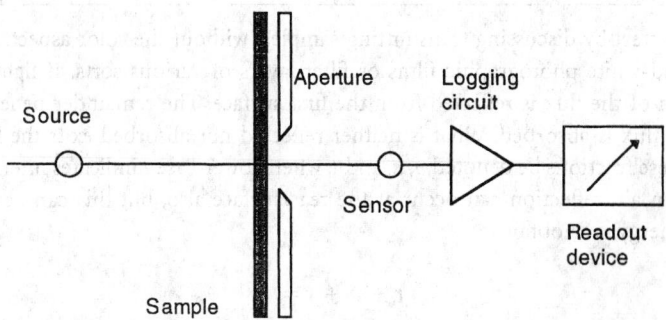

FIGURE 57.2 Schematic of transmission densitometer.

To measure transmittance/density, a reading is first taken with no sample over the aperture. This reading is a measure of the incoming light Φ_i. A second reading with the sample in place then determines Φ_e. It should be clearly understood that what the system actually does is to sense the flux passing through the aperture in the two cases, and then displays the negative logarithm of the ratio. This fact explains some of its operating characteristics. Starting from this basis, for example, McCamy [5] has treated measurement errors for the case when the sample is not uniform ("wedged") over the aperture.

The basic procedure also means that any source having enough power to operate the sensor can be used, and filters and apertures can readily be interchanged; it is only necessary to "zero" the instrument before making a measurement. The measuring aperture is usually of the order of 1 mm in diameter. If the area of this aperture is reduced to, say, 0.1 mm², the device becomes a microdensitometer. Such instruments present special problems not discussed here but treated in the literature [6].

Now consider the complications introduced by the nature of samples, especially photographic samples. Developed black-and-white photographic layers consist of discrete grains of silver metal dispersed in a thin layer of gelatin. Such grains act as scatterers, which means that some of the photons exiting the layer will not be traveling in their original directions. Such photons are identified by "C" in Figure 57.1. Clearly, the measured transmittance/density will depend on how much of this scattered light reaches the sensor. The effect of scattering can be quite large, and therefore in order to obtain values that are both reproducible and meaningful it is necessary to control what is termed the "geometry" of the system, i.e., the angular subtenses of the influx and efflux beams at the sample.

In Figure 57.3, Θ_i and Θ_e are these half-angles, respectively, referred to the normal to the element surface. The current standard [1b] prescribes two configurations, the first of which is as follows: $\Theta_i = 5°$; $\Theta_e = 90°$. The reverse of this configuration, i.e. $\Theta_i = 90°$; $\Theta_e \leq 5°$, yields the same density readings and is also permitted by the standard. This case is termed "diffuse density" and is the most important practical case because most commercial densitometers conform to this mode. Essentially, all the light exiting the sample, whatever its direction, reaches the sensor and is evaluated. This configuration simulates the case of contact printing a photographic negative, or of viewing a transparency on a diffuse illuminator. The practical construction of such an instrument is described below.

The second case occurs when the sample is illuminated, and the emergent flux collected, by lenses of finite aperture. In principle, in this case, the angles Θ_i and Θ_e can vary between, say, 5° and 90° and may differ from one another, but the standard specifies two sets of optics with matched apertures of $f/1.6$ and $f/4.5$. These relative apertures are representative of motion picture projectors and microfilm readers, respectively. The corresponding half-angles are 18.2° and 6.4°. (For a lens in air, $f$no. $= 1/(2\sin\Theta)$, where Θ is the half-angle shown in Figure 57.3.) This case is termed "projection" density.

It is useful to define a ratio

$$Q' = \text{Projection density/diffuse density} \tag{57.4}$$

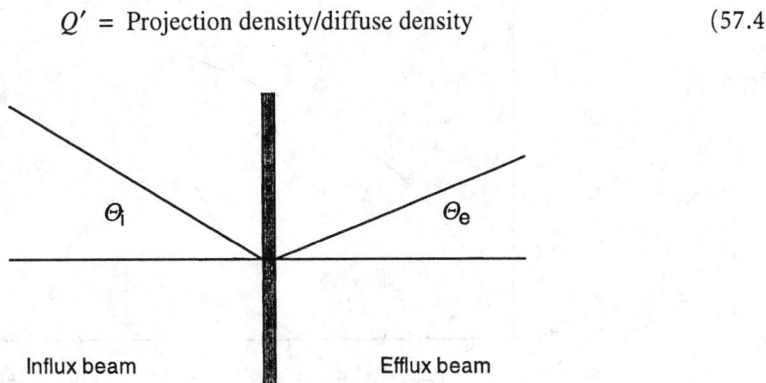

FIGURE 57.3 Angular subtenses of influx and efflux beams at sample.

which has been termed the "effective Callier coefficient." As would be expected, projection density is greater than diffuse density for a given sample so that $Q' = 1$. The exact ratio between the two densities depends on the scattering characteristics of the sample and the *f*-numbers involved. Only limited information on Q' has appeared in the literature [7] but, as Figure 57.4 shows, this parameter behaves as might be expected.

The dependence of the effective density on aperture shown in Figure 57.4 may be important in practice. Amateur photographers are aware that a B&W negative usually produces a more contrasty print in a condenser enlarger than it does in a diffuser enlarger, and this behavior can easily be deduced from Eq. (57.4). It may also be important in microdensitometry, since these instruments measure a projection density and a wide range of apertures may be encountered in various instruments. In such cases it is usually recommended that the data be reduced to diffuse density. This is easily done by measuring a sufficiently large sample in both projection and diffuse densitometers. It might be remarked that the scattering characteristics of individual films can vary significantly, and Q' should be measured for the samples at hand. Except for very low densities, Q' is relatively unchanged as the diffuse density varies.

For completeness, it is noted that earlier editions of density standards listed two other forms of density. The first of these was the case of angle Θ_i = angle $\Theta_e \leq 5°$. This case was termed "specular density" and was intended to simulate the use of the attenuating element in a collimated beam. The second case was the case of angle Θ_i = angle $\Theta_e = 90°$. This type of density was termed "doubly diffuse," and was intended to simulate the use of a negative in certain contact printers. Neither of these densities is important in modern practice; discussions of them can be found in the literature. [8]

In a color film, the "grains" are actually tiny volumes of dyed gelatin. Since the index of refraction of such a "grain" is only slightly different from the surround, it does not act as a scatterer. Thus, the density differences found on color films as a function of densitometer geometry are usually negligible and often ignored in practice.

In recent years, the photographic industry has introduced a new type of grain, the "tabular" grain, whose thickness is much smaller than the dimensions in the plane parallel to the coating surface. It has been demonstrated that, in the undeveloped state, these grains scatter light significantly less than the older "3-D" types of grains. No studies of developed T-grains have appeared in the literature, but it seems possible that developed B&W layers consisting of T-grains may scatter less than the traditional types of layers, thus reducing the sensitivity of density measurement to instrument geometry. (The data of Figure 57.4 were obtained with "3-D" grains.)

One can now turn to the optical configuration of a practical diffuse densitometer. From an optical standpoint, a good way to collect all the flux exiting the sample would be to use an integrating sphere,

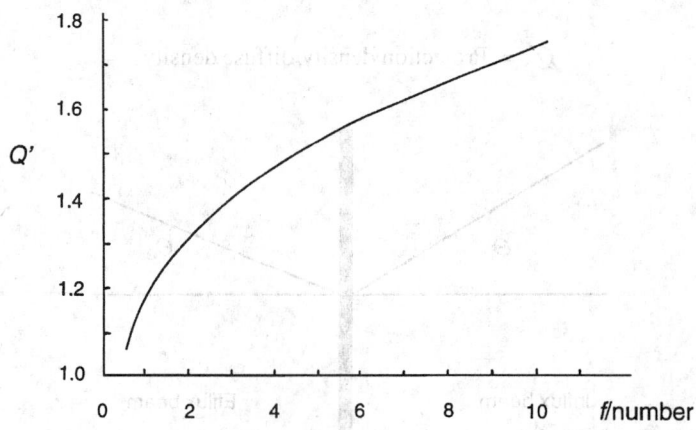

FIGURE 57.4 Variation of the effective Callier Q'-factor with the *f*-numbers of the influx and efflux beams, which were matched.

and instruments have been built using such spheres. However, in an actual contact-printing case there will be some light reflected back to the negative from the paper surface. Any retroreflection from an integrating sphere will of course be quite different from that coming from a paper surface. The better to simulate the contact-printing case, the present ANSI/ISO standard specifies that, in diffuse densitometers, the emulsion side of the sample shall be in contact with opal glass, which both acts as the diffuser and provides the desired back reflection. The optics of a diffuse densitometer are shown in Figure 57.5.

In the system shown, the sample is illuminated by diffuse light, which is the reverse of what is shown in Figure 57.3; however, as noted, diffuse density optics are reversible. Weaver studied the differences between opal-glass and integrating-sphere densities and found that because of the interreflections, the values were slightly lower in the opal-glass case [9]. The difference was ~0.04 density unit for very low densities, decreasing to 0.01 to 0.02 for samples whose densities were between 2.0 and 3.0. The opal-glass construction was also adopted in the standard because such instruments are easier to build and maintain. Care must be taken, however, to ensure that the diffuser meets the standard's specifications.

In actual operations, it is too time-consuming to zero the instrument before each reading. Likewise, the fraction Φ_e/Φ_i is rarely calculated specifically; the instrument simply provides a reading based on the amount of flux reaching the sensor. Because of the possibility of electronic drift however, it is advisable to zero the instrument periodically, unless it is known to be dependably stable. Likewise, the readings produced for some accurately known calibration sample of high density should be checked, a procedure known as "sloping" the instrument. Standard samples are available for this purpose; see the Appendix.

57.3 Monochrome Reflection Density

By definition, the reflectance factor

$$R = \frac{\Phi_r}{\Phi_o} \tag{57.5}$$

where Φ_r is the flux returned to the sensor from the sample, and Φ_o is the flux returned by a "perfectly reflecting, perfectly diffusing material located in place of the specimen" [1d]. Reflection density is then

$$D_R = -\log_{10} R = \log_{10}\frac{1}{R} \tag{57.6}$$

The optical configuration specified in the standard is shown in Figure 57.6. With reference to the normal to the specimen surface, one beam subtends an angle of $0 \pm 5°$ and the other an angle of $45° \pm 5°$. Also, the 45° beam is annular. As in the case of transmission optics, the influx and efflux beams are interchangeable. When the influx beam is at 45°, the system is termed "45/0"; when the efflux beam is at 45°, the designation is "0/45." The angle between the two beams was selected to avoid the specular reflection from the surface. The 45° beam is made annular to minimize the effects of any texture pattern that might be embossed on the sample surface. The standard also specifies that when measured, samples

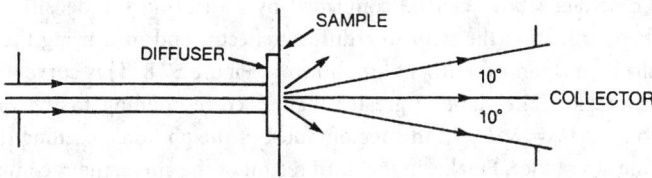

FIGURE 57.5 Optical system for measuring ISO diffuse density.

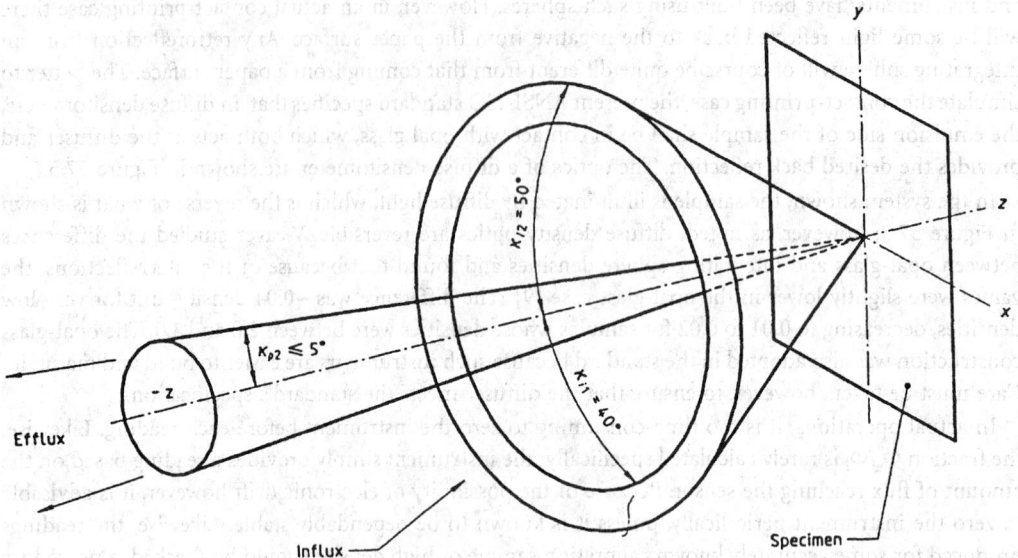

FIGURE 57.6 Annular geometry for measurement of reflection density.

shall be backed by a black diffusing material having a reflection density not less than 1.5. This procedure is specified to improve reproducibility, especially for thin samples.

It will be obvious that a reflection densitometer cannot be "zeroed" by taking a reading with no sample in place. Instead, as prescribed by the Standard [1d], calibration samples that have been measured on other primary instruments are used both to zero and slope the densitometer. Formerly, coatings of MgO_2 or $BaSO_4$ were used as reference "whites," but these layers are difficult to use. An ASTM standard [10] discusses the preparation of reference whites from pressed powders. For routine work, "plaques" consisting of stove enamel of various gray levels on a metal substrate are widely used. These plaques are very durable, but their physical form is quite different from actual paper samples. Photographic paper strips that have been calibrated in primary instruments can also be used. Physically, these strips are exactly like the samples to be read, but they are very easily damaged and must be used with care. Any scratch in the surface can cause a specular reflection to reach the sensor, which will produce a false reading.

Actually, the reflection of light from a photographic paper is more complicated than might appear at first glance. A typical B&W paper is shown in cross section in Figure 57.7. As with film samples, the emulsion layer consists of Ag grains suspended in gelatin. The support is a diffuse reflector. It turns out that reflection to the sensor occurs in three ways, as sketched. (The specular reflection is omitted in Figure 57.7.) Note that the component r_2 traverses the emulsion layer twice and r_3 at least four times, and that r_1 does not traverse the layer. For low densities, all three components contribute to the reflection reading. As the population of grains in the emulsion layer grows, the multiply-reflected beam r_3 rapidly becomes negligible. With continued increases in the grain population, the component r_2 also reduces to insignificance, and only the r_1 component remains. This component, being a surface reflection, remains essentially constant at about $0.005I_0$.

The mechanism described above can be confirmed by measuring the densities of a test strip by transmission and then cementing the strip to a diffuse reflector and measuring the same densities by reflection. The results of such an experiment are shown in Figure 57.8. This curve has three regions, as expected. In the first region, the slope is greater than 2, corresponding to the rapid loss of the r_3 component. In region 2, $\Delta D_R = 2\Delta D_T$; in this region, most of the photons reaching the sensor are those that have traversed the layer twice. Finally, in the third region of the curve, the r_2 component has become negligible, and only the surface component r_1 remains significant. Since, as noted, this component $\sim 0.005I_0$, the maximum reflection density attainable in a photographic paper should be about 2.0 to 2.3,

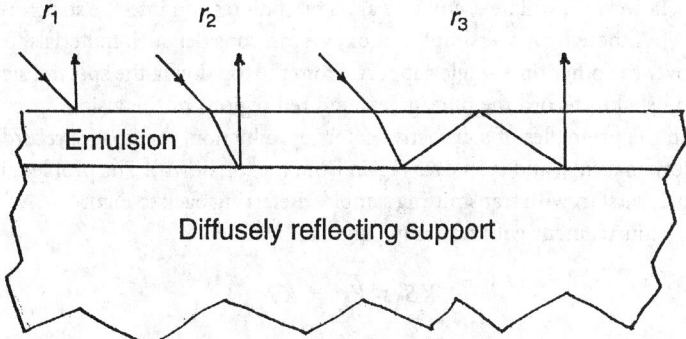

FIGURE 57.7 Cross-section of photographic paper showing different reflection paths for influx beam: r_1, surface reflection; r_2, direct reflection from support to sensor; and r_3, one or more internal reflections before reaching sensor.

and this is found to be the case for actual papers. (We have assumed a glossy paper. If the paper surface is textured, as it is in some products, specular reflections may be directed to the sensor from microareas and the maximum density attainable on the sample will drop, in some cases significantly.) The curve of Figure 57.8 also becomes very important in color reflection work, as is discussed below.

57.4 Color Transmission Densitometry

So far, we have assumed either that the measuring beam was monochromatic or that the sample was neutral, i.e., that transmittance or reflectance was constant across the spectrum of interest. Actually, the assumption of neutrality is reasonable for many B&W materials. However, of course, the densitometry of color images is extremely important, and this aspect is discussed below.

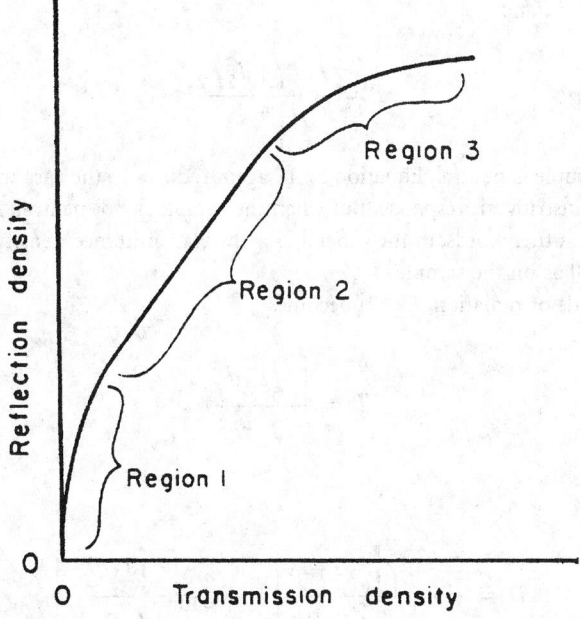

FIGURE 57.8 Densities measured on a film sample, both by transmission and after cementing the emulsion side to a diffusing reflector.

Before doing so, however, it will be useful to make a brief digression into the structure of photographic color materials [11]. Although an oversimplification, we can consider such materials to be three separate films coated one over the other on a single support. From the top down, the spectral sensitivities of these three layers are adjusted to record the blue, green, and red regions of the visible spectrum, respectively. In accordance with the principles of subtractive color reproduction, the images recorded in these layers are formed of yellow, magenta, and cyan dyes (again from the top down). The problem is to densitometer these dyes. As before, we start with transmitting samples. Referring back to Figure 57.2, for any wavelength λ, the reading of the instrument with no sample in place will be

$$KS_\lambda s_\lambda F_\lambda = KJ_\lambda \qquad (57.7)$$

where K = proportionality constant, S_λ = spectral power of the influx beam at λ, s_λ = relative spectral sensitivity of the receiver at λ, and F_λ, = transmittance of any filters in the beam at λ. (The transmission of the instrument optics may be significant, but this can be included with S_λ, since it is a fixed characteristic of the device.) The product J is termed the "response" or the "spectral product" of the instrument.

If a sample whose transmission at wavelength λ is T_λ is now placed over the aperture, the reading will be

$$KS_\lambda s_\lambda F_\lambda T_\lambda = KJ_\lambda T_\lambda \qquad (57.8)$$

and the measured transmittance reduces to T_λ, as it should. Thus, in the monochromatic case, the measured value is independent of the system response. When the influx beam contains two discrete wavelengths, the no-sample reading becomes

$$K(J_1 + J_2) \qquad (57.9)$$

and with the sample in place it will be

$$K(J_1 T_1 + J_2 T_2) \qquad (57.10)$$

The measured transmittance

$$T = \frac{J_1 T_1 + J_2 T_2}{J_1 + J_2} \qquad (57.11)$$

Note that when the sample is neutral, Equation 57.11 again reduces to the case where the transmittance is independent of the instrument response. But when the sample is not neutral, Equation 57.11 can no longer be simplified. In other words, in the general case the transmittance/density values depend on the system response as well as on the sample.

For continuous bands of radiation, Eq.(9) becomes

$$T = \frac{\int J_\lambda T_\lambda d\lambda}{\int J_\lambda d\lambda} \qquad (57.12)$$

and

$$D = -\log_{10}\left[\frac{\int J_\lambda T_\lambda d\lambda}{\int J_\lambda d\lambda}\right] = \log_{10}\left[\frac{\int J_\lambda d\lambda}{\int J_\lambda T_\lambda d\lambda}\right] \qquad (57.13)$$

The integration limits are set by the distributions.

From the above analysis, it is clear that in order to obtain meaningful density values, the spectral characteristics of the densitometer must be made equal to those of the receiver that will "view" the sample in actual use. Furthermore, of course, the actual receiver will vary from application to application. The densitometer characteristics are set by inserting appropriate filters in the beam. A number of spectral distributions are specified in Reference 1c as follows:

1. For reflection densitometry, the influx spectrum shall be ICI Illuminant A, which is essentially 2856 K.

2. For transmission densitometry the influx spectrum shall be Illuminant A modified by an infrared absorbing filter, as tabulated in the standard.

3. For samples to be viewed by an observer, the spectral characteristic of the system shall match the relative luminosity function $V(\lambda)$. This case is termed "visual" density.

4. For color negative films, a set of three distributions in the red, green, and blue regions of the spectrum is specified. This set is identified Status M, and approximates the spectral sensitivities of the three layers of color print materials.

5. For measuring red, blue, and green densities of color materials other than color negatives, a set of three responses identified Status A is provided.

6. To evaluate color images to be used in graphic arts processes, two sets of distributions are specified and are identified as Status T and Status E.

7. In the microfilm industry, prints are often made onto diazo or vesicular films. A narrow distribution centered at 400 nm is specified for this application, and the resulting densities are termed ISO printing densities Type 1.

8. A spectral distribution covering the range $\lambda \approx 360$ to 540 nm is provided for measuring samples to be printed onto B&W photographic papers. Such densities are designated ISO printing densities Type 2.

9. Status I response consists of three passbands centered at 420, 535, and 625 nm ±5 nm. This set is used in evaluating graphic arts materials such as process inks on paper.

10. A narrowband response centered at 800 nm ± 20 nm is provided for measuring effective densities to S-1 photosurfaces, such as those used in optical sound systems. Densities measured with this response are identified "ISO type 3."

These responses are shown in Figures 57.9a–e, which are reproduced from the current standard.

The next problem in color densitometry stems from the fact that the three layers are superimposed and from the nature of the dyes. The spectrophotometer curves for a typical dye set are shown in Figure 57.10. In this figure, the lower three curves refer to the dyes measured individually, while the upper curve shows the result of superimposing them. The values for the individual dyes are termed "spectral analytical densities," and those for the tri-pack are termed "spectral integral densities." Since superimposed densities may be considered to add, at any wavelength the spectral integral value is the sum of the three analytical values. Note that all three of the individual dyes absorb light outside the spectral regions in which they are supposed to work.

This unwanted absorption can cause problems in color reproduction and, more to the point for this chapter, in densitometry, because straightforward measurements of a real film at any wavelengths yield integral values. In many cases, these integral values are required. In the manufacture and processing of films, however, it may become necessary to determine the densities in an individual layer. It is not sufficient to coat and measure a layer by itself, since layers coated in a tri-pack may respond differently from the same layers operated singly.

This problem is solved by taking advantage of two rules called the "additivity rule" and the "proportionality rule." The first of these is merely the rule that densities add. The proportionality rule is an extension of Beer's law. Consider an element, such as a glass cell or a layer of a color film, containing a dye whose concentration can be varied. Beer's law states that

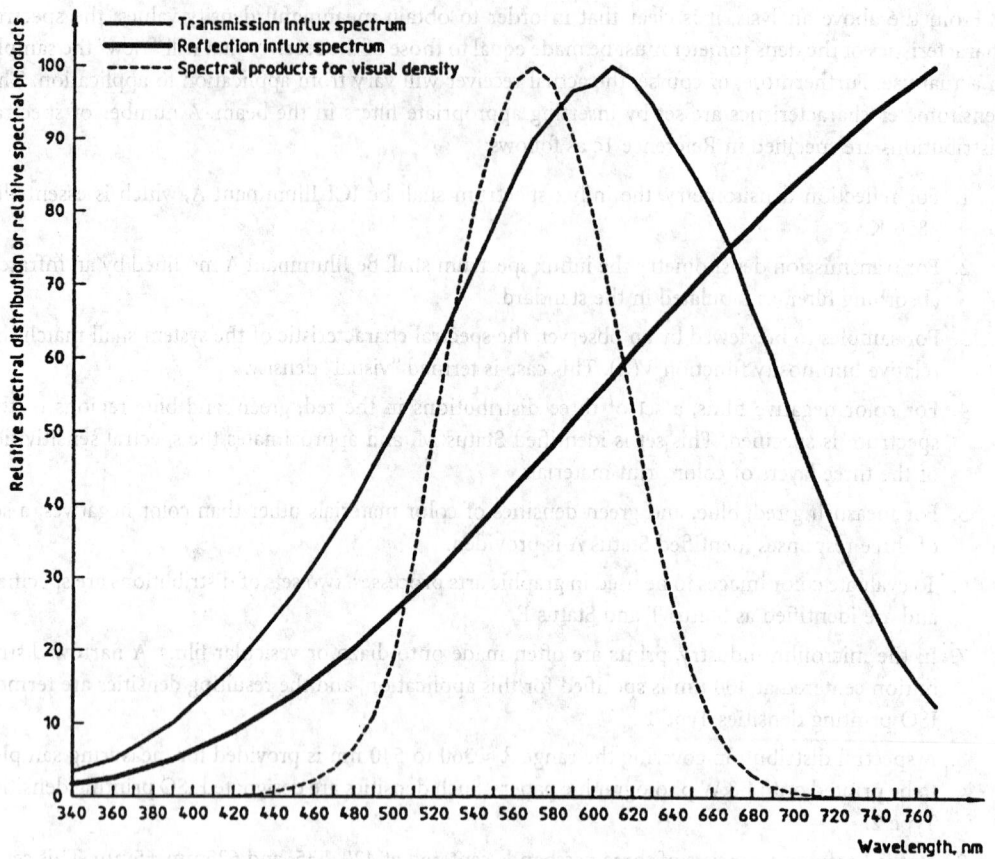

FIGURE 57.9a Spectral products for ISO density and relative spectral power distributions for influxes. *(continues)*

$$T_\lambda = e^{-\beta_\lambda c} \tag{57.14}$$

where c is the dye concentration and β_λ is the extinction coefficient at wavelength λ. The coefficient β varies with wavelength, of course, but will be constant at a given wavelength (unless changes in dye concentration produce chemical reactions.) Strictly speaking, Beer's law applies to the internal transmittance, but in the case of photographic layers, the surface reflectance is negligible. From Equation 57.14, it follows that

$$\ln T_\lambda = -\beta_\lambda c \tag{57.15}$$

and therefore,

$$D_\lambda = 0.434 \beta_\lambda c \tag{57.16}$$

Consider now the density at two wavelengths for any given dye sample. Clearly,

$$\frac{D_{\lambda 1}}{D_{\lambda 2}} = \frac{\beta_{\lambda 1}}{\beta_{\lambda 2}} \tag{57.17}$$

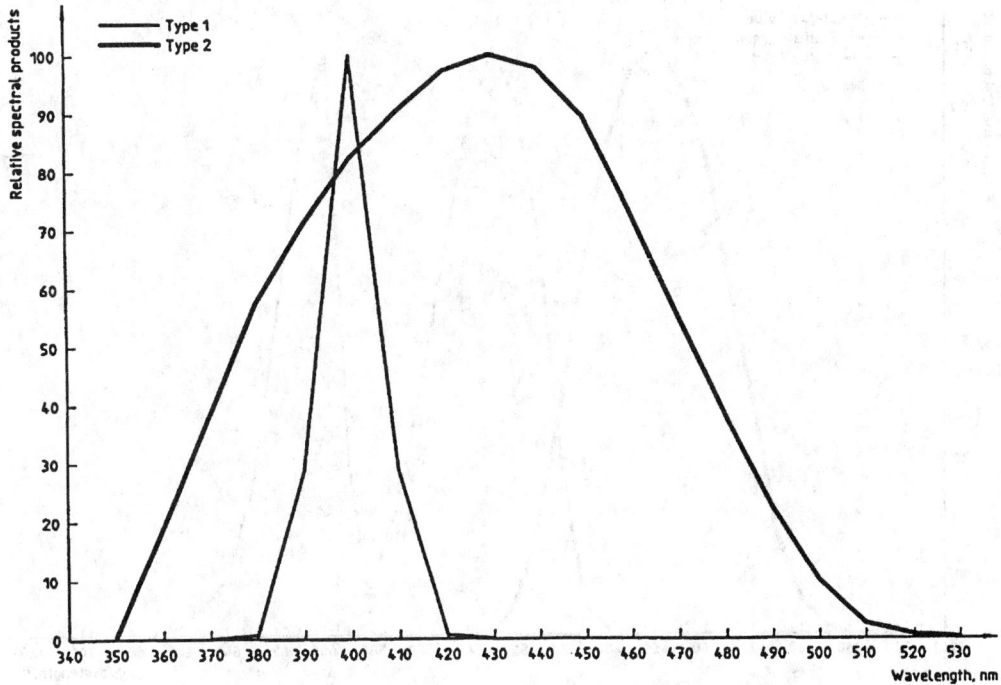

FIGURE 57.9b Spectral products for type 1 and 2 densities. *(continues)*

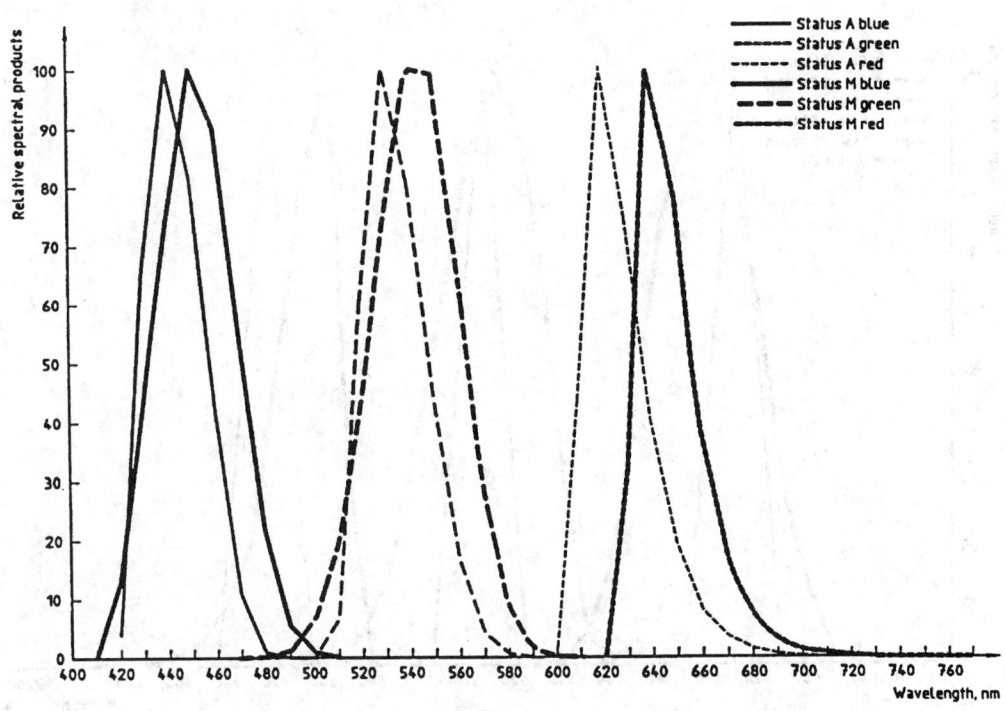

FIGURE 57.9c Spectral products for Status A and M densities. *(continues)*

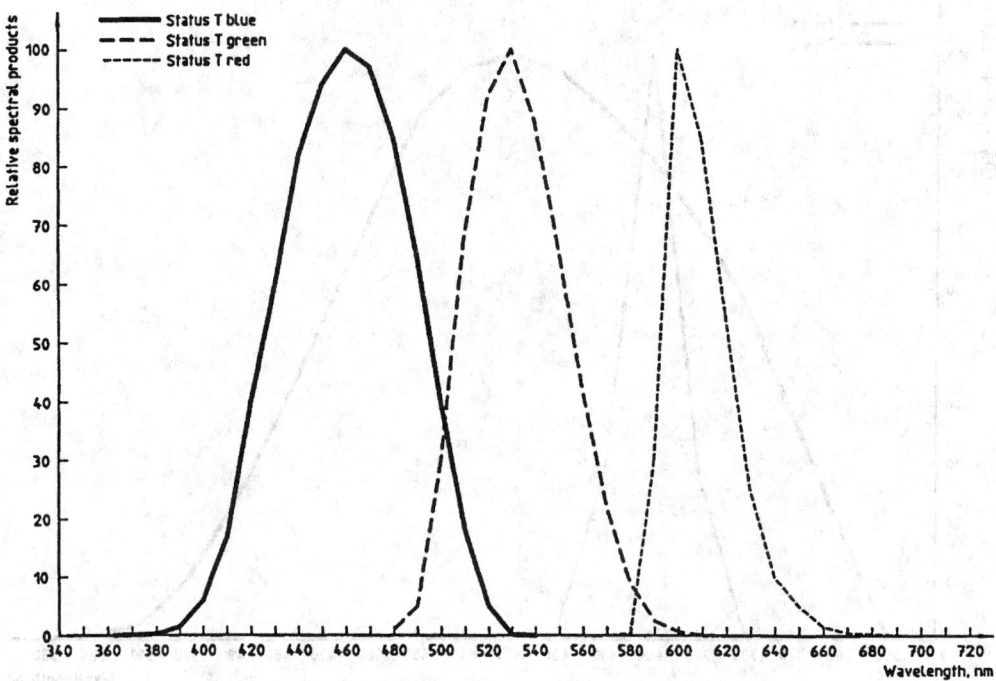

FIGURE 57.9d Spectral products for Status T densities. *(continues)*

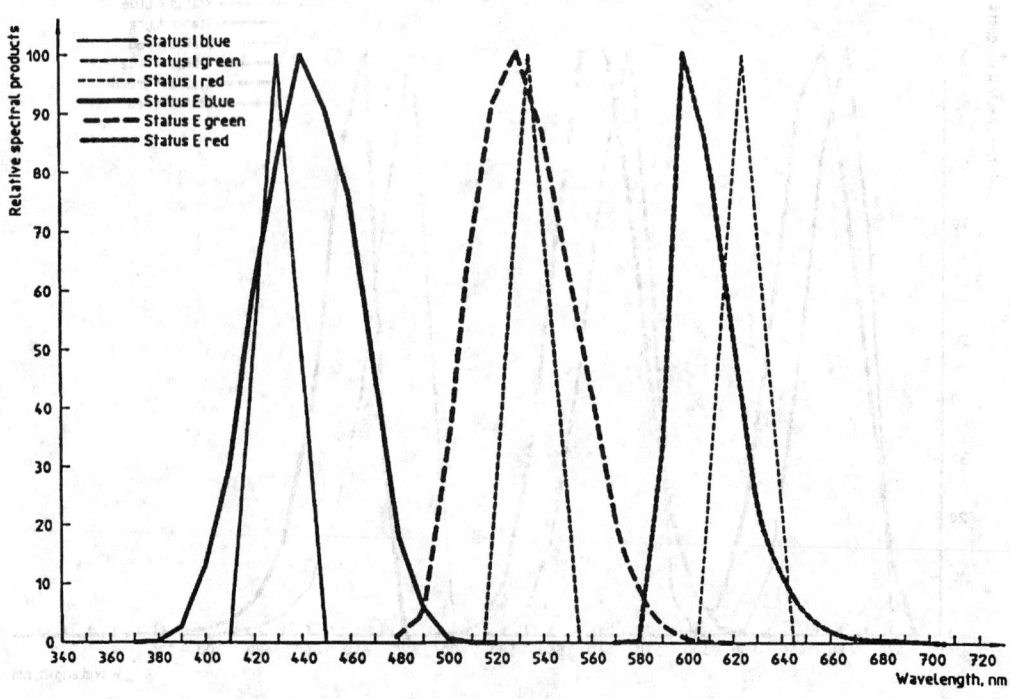

FIGURE 57.9e Spectral products for Status E and I densities.

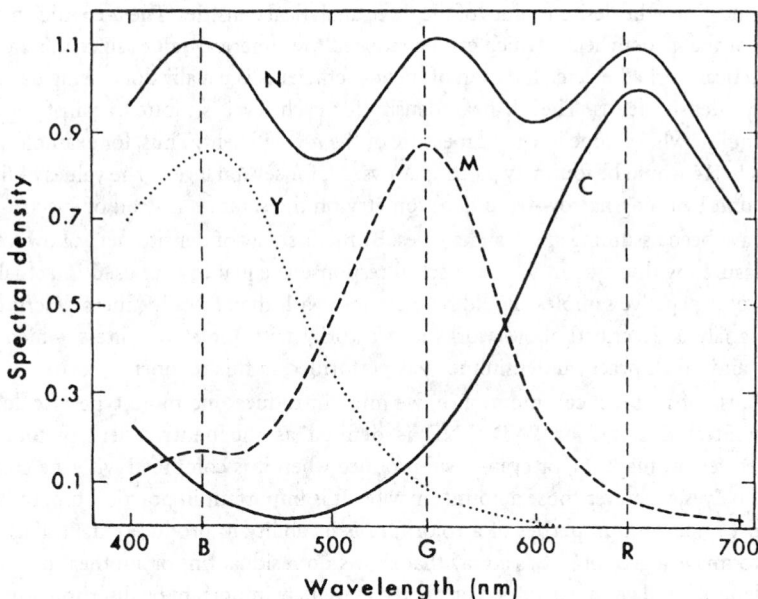

FIGURE 57.10 Spectral density distributions of the C. M, and Y components as functions of wavelength for a hypothetical color film, and of the composite absorber (N).

Thus, for a given dye sample, the ratio of the densities at any two wavelengths is fixed regardless of dye concentration (i.e., assuming Beer's law holds), and this is a statement of the proportionality rule.

As noted, straightforward density measurements on a real sample yield integral densities. The proportionality rule can be used to derive analytical densities in the following manner [12]. Assume spectral densities for simplicity. With reference to Figure 57.10, we select three wavelengths at or near the peak absorptances of the three dyes. Consider first the blue wavelength. By the additivity rule, the blue integral density is

$$D_b = Y_b + M_b + C_b \tag{57.18}$$

where Y_b, M_b, and C_b are the blue analytical densities of the yellow, magenta, and cyan dyes, respectively. But by the proportionality rule, M_b/M_g = constant, where M_g is the analytical density of the magenta dye measured at the peak wavelength in the green. The value of this ratio is readily determined from a spectrophotometric curve of the dye. Denoting such ratios a_n, for measurements at the three peak wavelengths we have

$$D_b = Y_b + a_1 M_g + a_2 C_r$$

$$D_g = a_3 Y_b + M_g + a_4 C_r$$

$$D_r = a_5 Y_b + a_6 M_g + C_r \tag{57.19}$$

This constitutes a set of three equations with three unknowns. The solution has the form

$$Y_b = b_1 D_b + b_2 D_g + b_3 D_r$$

$$M_g = b_4 D_b + b_5 D_g + b_6 D_r$$

$$C_r = b_7 D_b + b_8 D_g + b_9 D_r \tag{57.20}$$

where Y_b, M_g, and C_r are the desired values of the three analytical densities. The b-coefficients are algebraic combinations of the a-coefficients, which can be reduced to numerical values since the a-coefficients are known. In practical work, the determination of the a-coefficients is usually done using data for a number of different dye density levels. The off-peak density for each level is plotted against the peak density, giving a straight line whose slope is a good measure of the a-coefficient. Thus, for example, the coefficient a_1 in Equation 57.14 would be found by plotting M_b vs. M_g for several levels. The values of the coefficients can also be found by doing narrow-band densitometry on the samples instead of spectrophotometry.

So far, we have been assuming spectral densities. By the first law of densitometry, however, any sample should be measured with a system whose spectral response is equal to that used in actual practice—in other words, color negative samples should be measured with the Status M filters described before, etc. However, the analysis described above works satisfactorily with the status filters, which are relatively narrow-band, and much practical densitometry is performed in this manner.

Finally, because of its practical importance, we must introduce one more type of color density: the "equivalent neutral density," or END. END is defined as the neutral density that a given dye deposit—either yellow, magenta, or cyan—will produce when it is combined with the correct amounts of the other two dyes, whatever those amounts may be. It is important in practical film building, because one of the most important properties of a color film is its ability to produce a neutral as a neutral—in other words, to image a gray object as a gray that shows no residual tint of another color. Such a film is said to be "balanced," and good balance is considered extremely important by discriminating users. When the ENDs of the three layers of a color film are equal, the image will indeed be balanced, and color film sensitometry may be done in terms of ENDs. The END is a form of analytical density because it is a property of an individual layer. Originally, ENDs were measured in a special densitometer, but they are now calculated from the normal analytical densities by an equation of the form

$$\text{END} = m[\text{analytical spectral density}] + K \qquad (57.21)$$

where the constants m and K are determined by comparing the visual densities of satisfactory neutral images with the corresponding analytical spectral densities. In some cases, a second-order equation has been found to yield better results [12b].

57.5 Color Reflection Densitometry

Color reflection densitometry follows the general principles set forth above for the B&W reflection case. Since color prints are for the most part intended to be viewed by a human observer, in many cases a visual integral density provides the needed information. The film builder or process controller, however, may need information about the analytical densities of his layers, and here the approach used for transmitting samples does not work. The reason it fails is that, for such samples, Beer's law fails, as is shown by the curve of Figure 57.8. If the law held, the reflection density would be proportional to the dye concentration (see Equation 57.11).

Pinney and Vogelsong [13] have described a method of obtaining reflection analytical densities in such cases. The method involves the empirical determination of a calibration curve similar to Figure 57.8 for the material at hand, relating the reflection densities of a dye deposit to the transmission densities. The reflection integral densities of the sample are then measured and converted to transmission densities. The analytical values for these transmission densities are calculated using the method described above for such samples. Finally, the derived transmission values are converted back to reflection analyticals by going through the calibration curve in the reverse direction.

One last comment on the densitometry of color print papers should be added. In the commercial production of color prints, it may be extremely important to measure chemical stain in what should be the white areas of the picture. The passbands of the Status A filters are not well placed to monitor such stain, and it may be necessary to make additional measurements.

57.6 Densitometry of Halftone Patterns

Halftone patterns are used in the printing industry to reproduce continuous-tone images using two levels of ink: either ink or no ink. Various gray levels are produced by printing a pattern of repetitive "dots" too small to be resolved by the unaided eye. Essentially, the size of the dots is varied to produce a gray scale. Halftone patterns to be measured may be transmitting or reflecting; monochrome or color. A number of standards for densitometry in the graphic arts have been approved or are under development [14]. These standards provide much practical guidance, and are consistent with ISO-5. Densitometry in this field is an example of the basic mechanism by which the instrument operates—comparing the flux reaching the sensor with and without the sample in place. When the sample is a halftone, the flux reaching the sensor depends on the fractions of the area that are dense and clear. Adopting the notation of Ref. 13b and assuming a transmitting sample, the measured density will be

$$D_t = -\log_{10}\overline{T} = -\log_{10}[fT_s + (1-f)T_b] \qquad (57.22)$$

where D_t = measured density of pattern, $\overline{T}$ = average pattern transmittance, f = fraction of pattern area that is dense, T_s = transmittance of "solid" areas (essentially equal to dot density), and T_b = transmittance of clear areas.

It is interesting to calculate the measured density of a halftone pattern half of which is perfectly dense ($D = \infty$; $T = 0$) and half of which is perfectly clear ($D = 0$; $T = 1$). The calculated value for $D \approx 0.3$; and this value will indeed be found if such a pattern is densitometered.

If D_t, T_s, and T_b are known, Equation 57.22 provides an easy way to determine the fractional area covered by the pattern dots with an ordinary densitometer, and this is often done in graphic arts work. In practical work, T_s and T_b are measured in terms of density also, and this can be done satisfactorily on large areas. Since by definition $T = 10^{-D}$, the solution of Equation 57.22 for f can be written

$$f = \frac{1 - 10^{-(D_t - D_b)}}{1 - 10^{-(D_s - D_b)}} \qquad (57.23)$$

In this form, the equation is known as the Murray-Davies equation.

57.7 Summary

A densitometer measures the flux reaching the sensor with a sample in place, and calculates the ratio of this value to that obtained either with no sample in place (in the transmission case) or with a reference white in place (in the reflection case). The instrument then displays the negative log of this value.

The observed reading may depend significantly on the characteristics of the instrument, i.e., on the

1. angular substance of the influx beam,
2. angular substance of the efflux beam,
3. spectral power distribution of the influx beam, and
4. spectral sensitivity of the receiver.

For a complete specification of the density in a given case, these four parameters should be indicated. ISO 5.1. (Ref.1a) provides a standardized notation.

References

1. International Standard, *Photography Density Measurements,* International Organization for Standardization, Case Postale 56-CH-1211 Geneve 20, Switzerland.

   a. Part 1: Terms, Symbols and Notations ISO5/1-1984.
   b. Part 2: Geometric Conditions for Transmission Density ISO5-2, 3rd ed., 1991.
   c. Part 3: Spectral Conditions. ISO5-3, 2nd ed., 1995.
   d. Part 4: Geometric Conditions for Reflection Density. ISO5-4, 2nd ed., 1995.
2. J. L. Palmer, *The Handbook of Optics, 2nd ed.*, New York: McGraw-Hill, 1995, Chapter 25. This chapter contains an extensive bibliography.
3. F. C. Hurter and V. C. Driffield, *J. Soc. Chem. Ind. London*, 9, 455, 1890.
4. C. S. McCamy, History of sensitometry and densitometry, In E. Ostroff (ed.), *Pioneers of Photography*, Springfield VA: The Society for Imaging Science and Technology, 1987, Chap. 17, pp. 169–188.
5. C. S. McCamy, *J. Opt. Soc. Am.*, 66, 350, 1976.
6. J. C. Dainty and R. Shaw, *Image Science*. New York: Academic Press, 1974, Chap. 9.
7. H. C. Schmitt and J. H. Altman, *Appl. Opt.*, 9, 871, 1970.
8. J. H. Altman, in T. H. James (ed.), *The Theory of the Photographic Process, 4th ed.*, New York: MacMillan, 1977, Chap. 17.
9. K. S. Weaver, *J. Opt. Soc. Am.*, 40, 534, 1950.
10. *Standard Practice for Preparation of Pressed Powder White Reflection Transfer Standards*, ASTM E-259-93, West Conshohocken PA: American Society for Testing Materials.
11. For a discussion of color photography, see R. W. G. Hunt, *The Reproduction of Colour, 5th ed.*, England: Fountain Press, 1995.
12. a. R. M. Evans, W. T. Hanson, and W. L. Brewer, *Principles of Color Photography*, New York: John Wiley& Sons, 1953.
   b. P. Kowaliski, in T. H. James (ed.), *The Theory of the Photographic Process, 4th ed.*, New York: MacMillan, 1977, Chap. 18.
13. J. E. Pinney and W. F. Vogelsong, *Phot. Sci. Eng.*, 6, 367, 1962.
14. American National Standards, New York: American National Standards Institute. A number of additional standards were under development by the Committee for Graphic Arts Technology Standards (CGATS) as of late 1997.
   a. CGATS.4-1993: *Graphic Technology, Graphic arts reflection densitometry measurements, Terminology, Equations, Image Elements and Procedures.*
   b. CGATS. 9-1994: *Graphic Technology, Graphic Arts Transmission Densitometry Measurements, Terminology, Equations, Image Elements and Procedures.*

Appendix

A. Sources of Densitometers

Gretag–Macbeth
617 Little Britain Road
New Windsor, NY 12553-6148

X-Rite, Inc.
3100 44th Street SW
Grandville, MI 49418

Camag
Sonnenmattstr 11
CH-4132 Muttenz
Switzerland

Note: Makers of densitometers are listed in the *Photonics Buyers Guide*, published annually by Laurin Publishing Co., Inc., Berkshire Common, P.O. Box 4949, Pittsfield, MA 01202-4949.

B. Aids to densitometry

1. Status and Other Filters
   Gretag–Macbeth

   X-Rite, Inc.

   Eastman Kodak Co.
   Scientific Imaging Products
   343 State Street
   Rochester NY 14650

2. Standard Reference Materials, Calibration Samples
   Gretag–Macbeth

   X-Rite, Inc.

   Graphic Communications Association
   100 Dangerfield Road
   Alexandria VA 22314

   PSI Associates
   3000 Mount Read Boulevard
   Rochester, NY 14616

   Lucent Technologies
   235 Middle Road
   Henrietta NY 14467

   National Institute of Standards and Technology
   Gaithersburg MD 20899

   Eastman Kodak Co. Provides densities on film calibrated to ±5% or ±0.02 density unit, whichever is greater.

58

Colorimetry

Robert T. Marcus
Datacolor International

58.1 Introduction

Imagine how dull a world without color would be. Until the 1960s, many products were available in only a limited variety of colors. Consumers demand a variety of colors, and the materials available today allow manufacturers to meet those demands.

For many centuries, color was controlled by master color matchers adjusting the color of their products visually in natural daylight. It was about the 1950s when color measuring instruments began to make a significant impact in the manufacturing process, and by the 1970s they were commonly used in most industries. Color measuring instruments are used mostly for quality control but also for computerized color matching systems, process control, evaluation of raw materials, and as an aid to solving color problems.

While it is common to speak of a red car or a red light, color is a perception—not an intrinsic property of the object or the light. Color perception is influenced by the light source, the reflectance or transmittance properties of the object, the eye, and the brain. Color is one aspect of appearance. Gloss and texture are other aspects.

Color perception is three dimensional, i.e., three terms are needed to describe a color. Hue, lightness (sometimes called value), and chroma are one set of terms often used to describe color. Hue distinguishes blue from green from yellow, etc. Lightness distinguishes light colors from dark colors—for example, a light blue fabric from a dark blue fabric. Chroma, the most difficult of the three terms to understand, describes how different a color is from gray—for example, distinguishing a pastel green from a bright green. If the two greens are of the same hue (one is neither yellower nor bluer than the other) and have the same lightness, then the pastel green would be described as having a low chroma and the bright green as having a high chroma. Figure 58.1 is a diagram illustrating the three dimensions of color.

Daylight, fluorescent lamps, and incandescent lamps are widely used light sources for color evaluation. The perceived color of an object changes as the light changes. The human visual systems attempt to compensate for the change in light source and hold the color constant. Light booths provide standardized and controlled light sources for the visual evaluation of color. Most light booths contain a simulated daylight lamp, a cool white fluorescent lamp, an incandescent lamp, and an ultraviolet lamp for detecting fluorescence. By use of a switch, a light booth can be used to examine how the perception of a colored material changes with different light sources.

Two objects may appear the same when viewed under one light source, but different when viewed under another light source. This effect, called metamerism, is one of the major industrial problems for color matching. Metamerism usually occurs when attempts are made to produce objects that are the same color but made out of different materials, such as trying to match the dyed textile interior of a car with paint on the exterior and the plastic on the dashboard.

Colorimetry, the measurement of color, attempts to quantify the perception of color. The Commission Internationale de l'Éclairage (International Commission on Illumination, or CIE) is a voluntary organization of scientists and engineers from all over the world who are interested in light and color. The recommendations constituting modern colorimetry were first published by the CIE in 1931 and have been regularly updated since then [1].

Electromagnetic radiation (x-rays, gamma rays, light, and radio waves) irradiates the Earth constantly. Visible light is the name given to the electromagnetic radiation that the human eye perceives. The wavelength of visible light ranges from about 380 nm to about 780 nm. Sunlight is a mixture of all the wavelengths of light. Water vapor can spatially separate the light into its various wavelengths—the rainbow. Prisms and diffraction gratings can also spatially separate light into its component wavelengths.

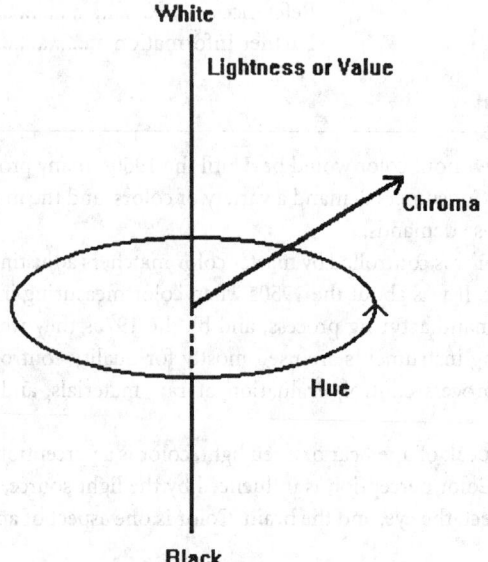

FIGURE 58.1 The three dimensions of color.

Light sources can be described numerically by their spectral power distribution, the relative amount of power the source emits at each wavelength of interest. A light source may emit power at wavelengths below (ultraviolet "light") or above (infrared "light") those of visible light. Ultraviolet radiation is important to colorimetry because it can cause fluorescence. Infrared radiation is the basis of "heat lamps" but is not important in colorimetry. Spectroradiometers measure the spectral power distribution of light sources.

Illuminants and sources are sometimes confused. Sources are actual physical entities that produce visible radiation, whereas an illuminant may only be a numerical table of values of a spectral power distribution. Initially, the CIE recommended three light sources for colorimetry in 1931. Source A, which is still in use, is an incandescent, tungsten filament light. An illuminant is the spectral power distribution of a light source. Thus, Illuminant A is the spectral power distribution of Source A. An illuminant may be defined, even when a source for that illuminant does not exist. Examples of illuminants without sources are the D series of illuminants recommended by the CIE. The D illuminants represent various phases of daylight. Illuminant D65 represents average daylight and is the most common illuminant used in colorimetry. No sources were recommended for the D series of illuminants.

Fluorescent lights have great commercial importance. Cool white fluorescent lamps are the most common light sources in offices in the United States. There are a variety of other types of fluorescent lamps used in stores and offices. The CIE also recommended a series of illuminants to represent fluorescent lamps. F2 represents cool white fluorescent lamps. Figure 58.2 shows the spectral power distributions of Illuminants A, D65, and F2.

Color vision and perception is complex and has been extensively studied. Ninety-two percent of men and 99.5 percent of women have "normal" color vision. The eye's lens focuses images on the light-sensitive retina. Rod cells make up the majority of the retina and are sensitive to low levels of illumination (night vision). Cone cells provide color vision and are located in a small area of the retina called the *foveal pit*. There are three types of cone cells. One type of cone cell has peak sensitivity to blue light, one type to green light, and one type to red light. Signals from the cone cells are transmitted to the brain where they are processed into color perceptions.

Color perception is an extremely complex phenomenon. For example, the background on which a material is viewed can have a major effect on the perceived color of that material. The ambient light to which the eye becomes adapted also influences the color of materials. Basic colorimetry, the topic of this chapter, provides the rather simple concept of dealing with the measurement of single independent colors. Most of industrial color control is adequately described using basic colorimetry. Advanced colorimetry attempts to use physical measurements to describe the perceived color of a material when viewed in a complex scene.

All colors are perceived by stimulating combinations of the three cones. Computer monitors and color television tubes produce colors by lighting combinations of red, green and blue phosphors. In 1931 when

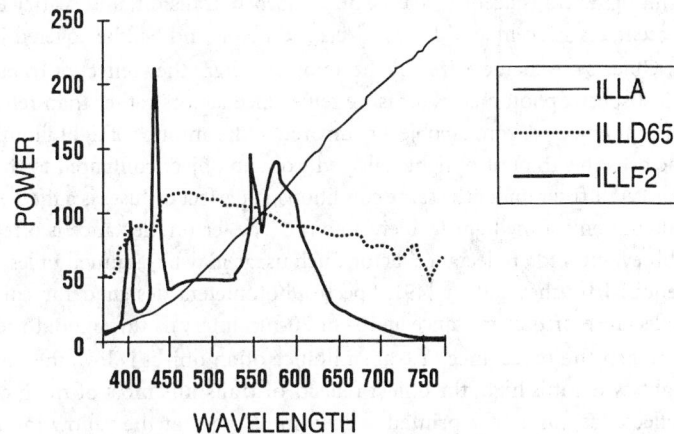

FIGURE 58.2 The spectral power distribution of CIE Standard Illuminants A, D65, and F2.

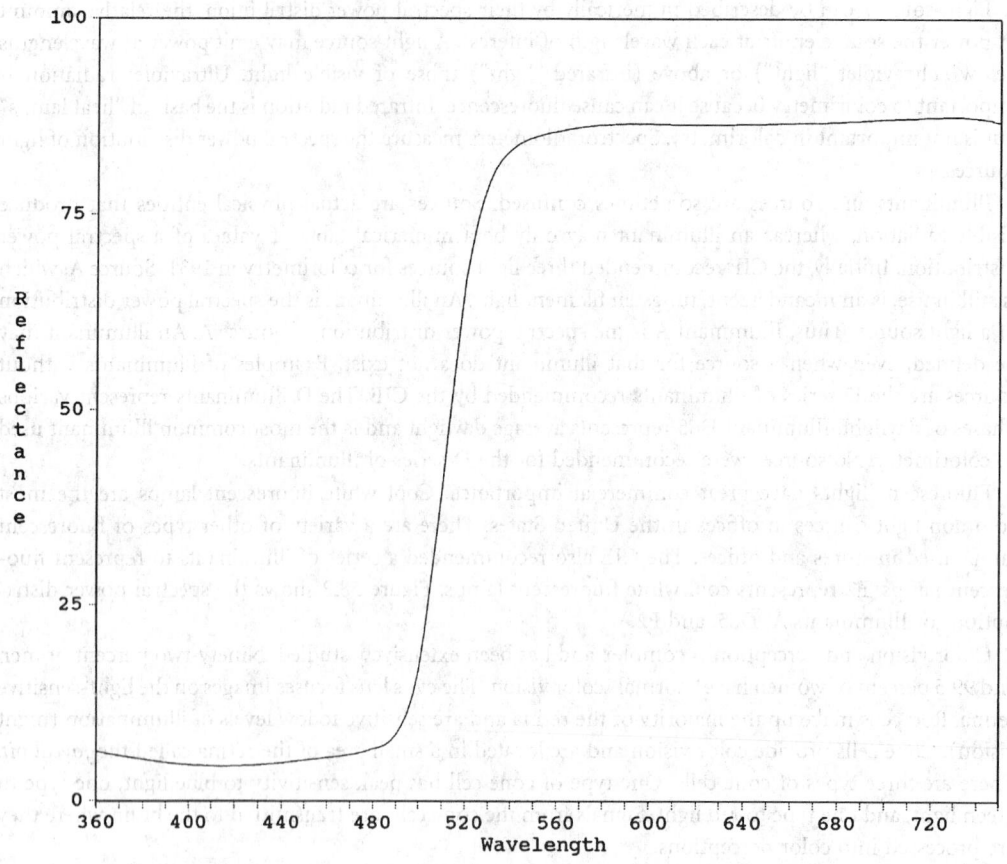

FIGURE 58.3 The spectral reflectance curve of a printed yellow ink.

the CIE was developing the system for modern colorimetry, they transformed the data from real experiments that had human observers match each wavelength of visible light with combinations of red, green and blue lights, to three mathematical imaginary "lights" labeled X, Y, and Z. All colors can be matched by varying amounts of X, Y, and Z. The amounts of each X, Y, and Z imaginary light that must be mixed together to match a color are called the tristimulus values (see "The CIE Standard Observers," page 58-6).

Most objects absorb, transmit, or reflect (scatter) light. Transparent objects absorb and transmit light. Opaque objects absorb and reflect light. Light sources emit light. Fluorescent objects absorb, reflect, and emit light. Translucent or hazy objects absorb, transmit, and scatter light. Measuring the color of fluorescent, translucent, and hazy objects is difficult and will be covered in later sections.

Objects are characterized by the amount of light they reflect or transmit at each wavelength of interest. Most spectrophotometers measure reflectance factors rather than reflectance. Reflectance is the amount of light reflected from an object compared to the amount of light illuminating that object. The reflectance factor is the amount of light reflected from an object compared to the amount of light reflected from a perfect diffuser under the same conditions. A perfect diffuser is a theoretical material that diffusely reflects 100 percent of the light incident upon it. The term reflectance is often used in a general sense, or as an abbreviation for, reflectance factor. Such usage may be assumed unless the term reflectance is specifically required by the context [49]. Spectrophotometers designed for color measurement usually measure reflectance or transmittance at 10- or 20-nm intervals throughout the visible spectrum.

When the reflectance or transmittance of an object is low, the object absorbs most of the incident light; when it is high, the object reflects or transmits most of the incident light. Figure 58.3 shows the reflectance curve of a printed yellow ink. Note that the yellow ink absorbs light in the blue portion

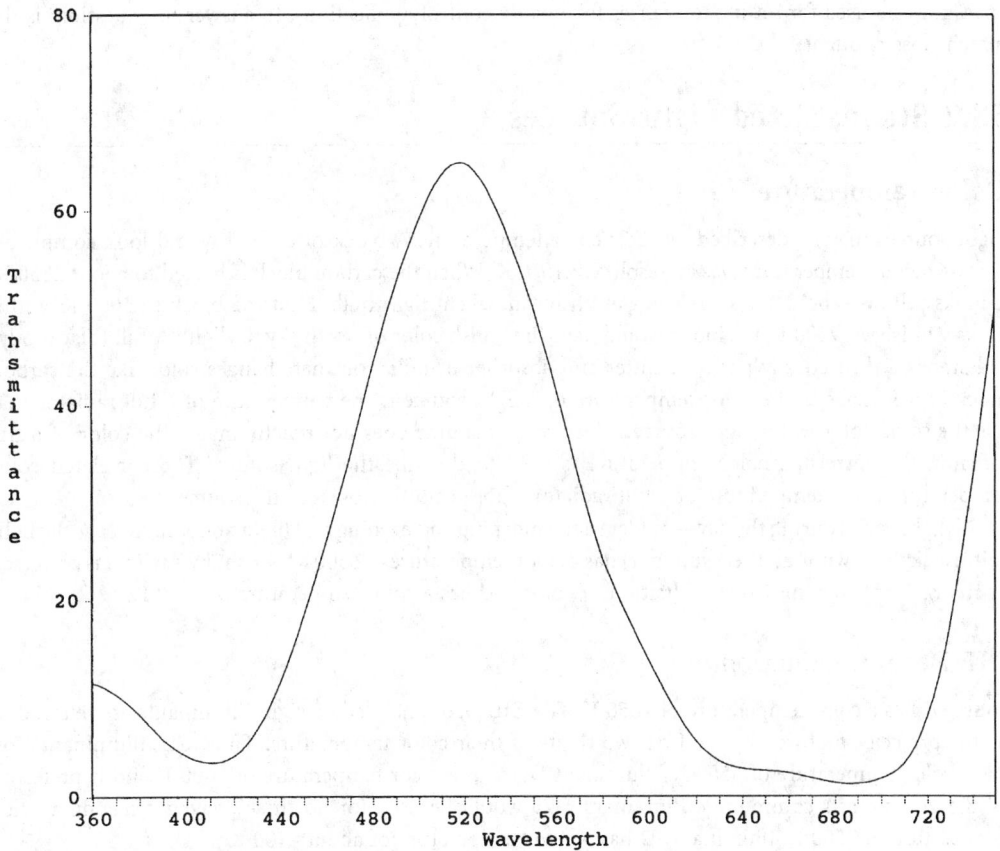

FIGURE 58.4 The spectral reflectance curve of a green transparent plastic.

of the spectrum and reflects light in the yellow and red portions of the spectrum. A green plastic (Figure 58.4) absorbs light in the blue and red portions of the spectrum and transmits light in the green portion.

The tristimulus values of an object can be calculated by combining the reflectance or transmittance of the object with the spectral power distribution of an illuminant and the color matching functions of a Standard Observer. The yellow ink's tristimulus values for Illuminant D65 and the 10 Degree Standard Observer are X = 66.62, Y = 69.72, and Z = 7.03. Those for the green plastic are X = 21.03, Y = 39.40, and Z = 24.06. An object's tristimulus values will change with the illuminant. For example, the tristimulus values for the yellow object for Illuminant A and the 10 Degree Standard Observer are X = 91.15, Y = 77.83, and Z = 2.80.

Pairs of objects are said to match when their tristimulus values are the same. However, since the calculation of the tristimulus values included the source and observer as well as the object, when one of these changes, the objects may no longer match, i.e., they may have different tristimulus values. Metameric colored objects have the same tristimulus values for the illuminant under which they match but different tristimulus values for illuminants where they do not match.

Color measurements are most often made to determine quantitatively whether or not the colors of two objects or batches of material are the same. But what happens when they do not match? Color difference equations were developed to quantify the difference. Starting with the objects' tristimulus values, a color difference equation will calculate the total color difference, ΔE or DE, and its component parts—the differences in lightness (ΔL or DL), chroma (ΔC or DC) and hue (ΔH or DH) or the differences in lightness, yellowness-blueness (Δb or Db) and redness-greenness (Δa or Da). Numerical color differences may be used for setting tolerances for quality control applications, to answer the question: is the match close enough?

58.2 Standardized Light Sources

Color Temperature

Light sources may be described by their color temperature. A block of carbon would look completely black when its temperature was at absolute zero, 0 K. When the carbon block is heated to about 2850 K, it looks yellow—about the same color as an incandescent light bulb. Heat the block to 5000 K, and it looks whitish. At 7500 K, the block would have the bluish color of north sky daylight. A full (blackbody) radiator is a theoretically perfect emitter and absorber of radiation that changes color like the carbon block just described. The color temperature of a light source is the temperature of a full radiator that has the same color as the light source. When a light source does not exactly match the color of a full radiator, the correlated color temperature is used to describe the light source. The correlated color temperature is the temperature of a full radiator whose color is closest to the source.

Daylight varies during the day—redder in the morning and evening and bluer at noon. Average daylight (diffuse skylight without direct sunlight) has a color temperature of 6500 K. North sky daylight is preferred by many people for the visual evaluation of color and has a color temperature of 7500 K.

CIE Recommendations

Source A has a color temperature of 2856 K. The CIE recommended daylight illuminants are referred to by the prefix D, followed by the first two digits of their color temperature. Thus, CIE Illuminant D65 has a color temperature of 6500 K. Illuminant D50 has a color temperature of 5000 K and is preferred by the graphic arts community. Illuminant D75 would be used for north sky daylight having a color temperature of 7500 K. Illuminant F2 has a color temperature of about 4100 K.

58.3 The CIE Standard Observers

The CIE adopted two Standard Observers based on color matching experiments. The CIE 2 Degree Standard Observer was recommended in 1931, and the CIE 10 Degree Standard Observer was recommended in 1964.

In the CIE experiments, observers having normal color vision matched spectrum colors with combinations of red, green, and blue light. Figure 58.5 illustrates the experimental setup used in the Standard

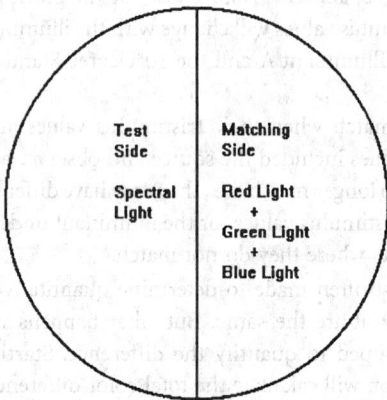

FIGURE 58.5 Experimental field of view for determining the CIE Standard Observers.

Observer experiments. One-half of a circular field was illuminated with the spectrum color, while the other half was illuminated with a mixture of red, green, and blue lights. The observer adjusted the amounts of red, green, and blue until the mixture matched the spectrum color. Unfortunately, not all of the spectrum colors could be matched with combinations of the red, green, and blue lights used in the experiment. In those cases, one of the lights had to be moved so that it illuminated the test field. By "diluting" the spectrum color with one of the lights, the resultant color could be matched with the remaining two lights. The amount of light used to dilute the spectrum color was considered to be a negative amount. To avoid having color matching functions with negative amounts of light, three "imaginary" lights (X, Y, and Z) were created by performing a mathematical transformation. Color matching functions are the amounts of X, Y, and Z required to match the colors of the spectrum and are used in the calculation of tristimulus values. Figure 58.6 shows the color matching functions, ($\bar{x}$, $\bar{y}$, and $\bar{z}$) for the 10 Degree Standard Observer.

In the first experiments, the circular field viewed was projected on the foveal pit and subtended a solid angle of 2°. This is about the size of a dime held at arm's length. The 2 Degree Standard Observer developed from these experiments is very useful when viewing small fields, such as the signal lights of ships or small colored chips. Industrial color matchers view larger fields, such as two 5 × 12-inch panels. The 10 Degree Standard Observer should be used when viewing larger fields. Observers for those experiments viewed a 10° visual field, which is about the size of a fist held at arm's length.

The standard observers represent combinations of the color matching functions of a number of observers. Standard observer color matching experiments are tedious and difficult to do. Few people can do them with reproducibility. An individual's color matching functions are likely to vary from that of a CIE Standard Observer. Although these differences do not generally present a problem, they can affect the evaluation of metameric samples.

58.4 Calculating Tristimulus Values

Tristimulus values for a reflecting samples are calculated from the following equations:

$$X = k \int_{380}^{780} R_\lambda S_\lambda \bar{x}_\lambda d\lambda \cong k \sum_{380}^{780} R_\lambda S_\lambda \bar{x}_\lambda \Delta\lambda \qquad (58.1)$$

$$Y = k \int_{380}^{780} R_\lambda S_\lambda \bar{y}_\lambda d\lambda \cong k \sum_{380}^{780} R_\lambda S_\lambda \bar{y}_\lambda \Delta\lambda \qquad (58.2)$$

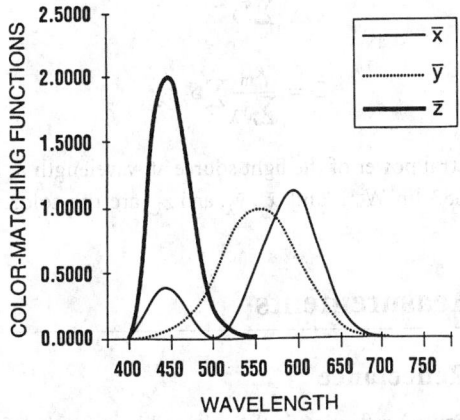

FIGURE 58.6 The color matching functions for the CIE 10 Degree Standard Observer.

$$Z = k \int_{380}^{780} R_\lambda S_\lambda \bar{z}_\lambda d\lambda \cong k \sum_{380}^{780} R_\lambda S_\lambda \bar{z}_\lambda \Delta\lambda \tag{58.3}$$

$$k = \frac{100}{\int_{380}^{780} S_\lambda \bar{y} d\lambda} \cong \frac{100}{\sum_{380}^{780} S_\lambda \bar{y} \Delta\lambda} \tag{58.4}$$

in which S_λ is the relative spectral power distribution of the illuminant at wavelength λ, R_λ is the reflectance factor of the sample at wavelength λ, and $\bar{x}_y$, $\bar{y}_\lambda$, and $\bar{z}_\lambda$ are the color matching functions of the observer at wavelength λ. For transmitting objects, substitute the transmission factor, T_λ, of the sample at wavelength λ for the reflectance factor. The factor k normalizes Y so that it will equal 100.00 for a perfect reflector or transmitter, i.e., R_λ or T_λ is 100.00 at all wavelengths of interest. The summations in the above equations are only valid if the reflectance or transmittance of the sample is measured at wavelength intervals of 1 nm or 5 nm from 380 to 780 nm.

Many commercial spectrophotometers measure wavelength intervals of 10 or 20 nm. To accurately calculate tristimulus values of samples measured with those instruments, the following equations should be used:

$$X = \sum W_{x\lambda} R_\lambda \tag{58.5}$$

$$Y = \sum W_{y\lambda} R_\lambda \tag{58.6}$$

$$Z = \sum W_{z\lambda} R_\lambda \tag{58.7}$$

in which $W_{x\lambda}$, $W_{y\lambda}$, and $W_{z\lambda}$ are weighting factors designed for 10 and 20 nm wavelength intervals. T_λ can be substituted for R_λ for transmitting samples. Although a number of weighting factors have been developed over the years [2–7], those recommended by ASTM in E 308 Standard Practice for Computing the Colors of Objects by Using the CIE System [7] are recommended.

Tristimulus values for a light source can be calculated easily from measurements taken at 1 or 5 nm intervals from the following equations:

$$X = \frac{k_m}{\sum \bar{y}_\lambda} \sum S_\lambda \bar{x}_\lambda \tag{58.8}$$

$$Y = \frac{k_m}{\sum \bar{y}_\lambda} \sum S_\lambda \bar{y}_\lambda \tag{58.9}$$

$$Z = \frac{k_m}{\sum \bar{y}_\lambda} \sum S_\lambda \bar{z}_\lambda \tag{58.10}$$

in which S_λ is the relative spectral power of the light source at wavelength λ, k_m is the maximum spectral luminous efficacy function (683 lm W^{-1}), and $\bar{x}_y$, $\bar{y}_\lambda$, and $\bar{z}_\lambda$ are the color matching functions for the standard observer.

58.5 Reflectance Measurements

Specular and Diffuse Reflectance

Specular (sometimes called *regular*) reflection is the mirror-like reflection from an object. If you shine a beam of light on a mirror, it will be entirely reflected at the same angle on the opposite side of a normal plane to the mirror's surface. However, if you shine a beam of light on a pellet of compressed barium sulfate ($BaSO_4$) powder, it will enter the surface, be scattered a number of times, and exit the pellet in many directions. This is called diffuse reflection. Glossy and semiglossy materials contain a combination of specular and diffuse reflection as shown in Fig. 58.7. Observers usually discount the specular reflection when visually evaluating the color of a material.

Illuminating and Viewing Geometries for Reflectance

Instruments designed for measuring the color of reflecting objects consist of an illuminator, a sample holder, and a receiver. The CIE has recommended four illuminating and viewing geometries for making reflectance measurements.

Bidirectional geometries illuminate the sample with a narrow beam of light and view the sample with a receiver having a narrow entrance field. In the most commonly used bidirectional geometry, the sample is illuminated at a 45° angle (±2°) from the sample's normal and viewed along the sample's normal (±10°). The other recommended geometry illuminates the sample along its normal and views the sample at 45° from its normal. Bidirectional geometries measure only diffuse reflection and can be sensitive to the surface texture of the sample. The two bidirectional geometries produce equivalent results [8, 9].

In the most common diffuse geometry, the illuminator includes an integrating sphere. An integrating sphere is a hollow metal sphere coated with a highly reflecting white coating with openings for the light source, the sample, and the receiver. The instrument's light source projects a beam of light onto the integrating sphere's wall. The light is reflected many times by the sphere's wall, and the sample is illuminated from all angles. When measuring in the diffuse/normal mode, d/0, the receiver views the sample along its normal. When the receiver is positioned in this manner, the specular reflection is directed back towards the light source and is not measured.

For some applications, it is useful to remove the specular reflection off of the sample's surface from the reflection of the light that is reflected back from the interior of the sample. This can be done by moving the receiver 6 or 8° from the sample's normal as shown in Fig. 58.8. A specular inclusion port is placed at the equal angle on the opposite side of the sample's normal. By placing a white plug having the same reflectance as the sphere wall in the port, the specular reflection can be included in the measurement. By using a light trap at the specular inclusion port, some or all of the specular reflection can be excluded from the measurement. The narrower the specular reflection peak, i.e., the glossier the material, the more specular reflection is excluded. Specular included measurements are normally abbreviated SCI, while specular excluded measurements are abbreviated SCE. This geometry is referred to as diffuse/near-normal but still abbreviated d/0. When the specular reflection is included, this geometry is sometimes referred to as total/normal, t/0.

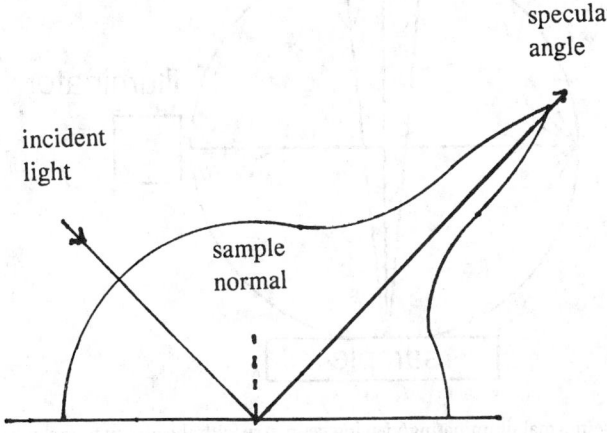

FIGURE 58.7 A semiglossy material showing a combination of specular and diffuse reflection.

The last of the CIE recommended illuminating and viewing geometries is the normal/diffuse or near-normal/diffuse, 0/d and 0/t. This is the reverse of the d/0 geometry. The illuminator illuminates the sample along its normal, or slightly off of its normal. For this geometry, the integrating sphere is part of the receiver. Light reflected from the sample is captured by the integrating sphere and the remaining optics of the receiver views the sphere wall. The two diffuse geometries produce equivalent results [8, 9].

Monochromatic and Polychromatic Illumination

An instrument's illuminator may illuminate the sample with either a narrow band of wavelengths, 1 to 10 nm wide, called monochromatic illumination, or a wide band of wavelengths, usually simulating a daylight illuminant and called polychromatic illumination. For nonfluorescent samples, either illumination method can be used with equivalent results. However, for fluorescent samples, only polychromatic illumination can produce valid results.

Sample Texture and Bidirectional Geometries

Bidirectional illuminating and viewing geometries can be very sensitive to surface texture and any polarization of reflected light. Keeping the illuminator and receiver in the same plane maximizes this sensitivity. To reduce or eliminate this sensitivity, circumferential or annular illumination (or receiving) can be used. When an illuminator provides light (or the receiver possesses responsivity) at many points distributed uniformly around a 45° cone centered at the sample's normal, we refer to circumferential illumination (or viewing). When the illuminator provides light continuously and uniformly around the cone, we refer to annular illumination (or viewing).

Which Illuminating and Viewing Geometry Is Best?

"Which illuminating and viewing geometry is best for color measurement?" is a difficult question to answer [9–12]. For matte samples, all of the geometries produce equivalent results. For high-gloss samples, the diffuse geometries with the specular reflection excluded provide measurements very close to the bidirectional geometries. For semigloss samples, the problem becomes more complex. The two bidirectional geometries are similar to the way a person evaluates color visually and are often thought

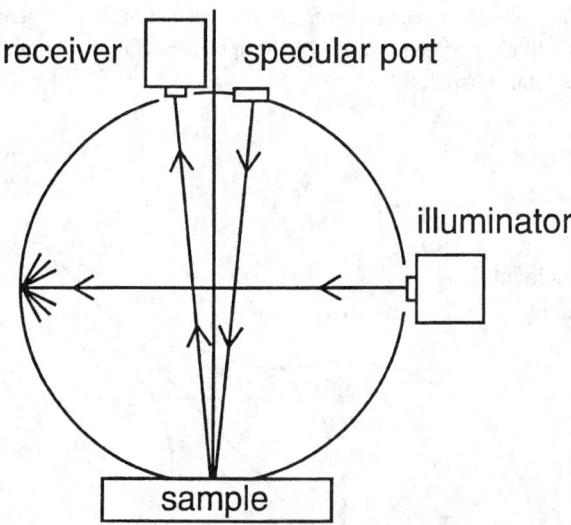

FIGURE 58.8 The diffuse/normal illuminating/viewing geometry with the ability to include or exclude the specular surface reflection.

to agree better with visual evaluation. On the other hand, the diffuse geometries measured with the specular reflection included minimize the effect of the sample's texture and gloss and are quite useful for computerized color matching. Rather than purchasing multiple instruments, most users select the geometry most suited to their major needs and compromise on other measurements.

Spectrophotometers

Spectrophotometers are used to measure an object's reflection characteristics throughout the visible spectrum. Double-beam spectrophotometers monitor a reference standard to compensate dynamically for fluctuations in source output, detector response, and atmospheric absorption to increase the instrument's stability [13]. Improvements in electronics and optics have allowed single-beam spectrophotometers to achieve the stability of double-beam instruments [14]. Single-beam instruments with an integrating sphere require a correction for the reduction of sphere efficiency caused by sample absorption [14].

Reflectance measurements are referred to as if they take place at a single wavelength. In actuality, a spectrophotometer has a finite spectral bandwidth or bandpass centered about that wavelength. Some instruments will have a bandpass as narrow as 1 nm, while others may exceed 20 nm. Bandpass is important because it influences the color measurement. The extent of this influence depends on the sample being measured [15, 16]. The spectral bandpass should be equal to the wavelength increment used in the calculation of tristimulus values to obtain the best color measurement results [17].

Spectrophotometers must be standardized before making reflectance factor measurements. The high point of the measurement scale is set by measuring a white standard of known reflectance factor [18]. The zero point is set by measuring a light trap or a black calibration standard. Single-beam, integrating sphere spectrophotometers may also require a sphere wall calibration, which is often done with a gray standard.

Some spectrophotometers allow the user to vary the size of the area measured. For most applications, the largest area possible should be measured. A number of documentary standards exist for making and reporting reflectance measurements [9, 17, 19-25].

Colorimeters and Spectrocolorimeters

Colorimeters were developed in the 1920s to 1930s as a less expensive alternative to spectrophotometers for quality control and color difference applications. They are simple to use and directly measure a sample's tristimulus values or related color coordinates. Three or four filters modify the light source and attempt to duplicate a Standard Illuminant/Standard Observer combination. Because of the difficulty in matching the CIE Illuminant and Standard Observer functions, they are less accurate than spectrophotometers for determining a sample's tristimulus values. Colorimeters determine the color difference between two samples more accurately than they determine tristimulus values, and they are often called color-difference meters. Since only one Standard Illuminant/Standard Observer combination is usually possible, colorimeters cannot be used to determine metamerism. Standards also exist for making colorimeter measurements [24–26].

A new class of instruments, spectrocolorimeters, began to appear in the 1980s. Spectrocolorimeters are spectrophotometers that only output tristimulus values or related color coordinates. They are less expensive and often have fewer options (such as variable area of view) than fully functional spectrophotometers. However, because they are spectrophotometers, they are capable of measuring metamerism.

Sample Preparation

Accurate color measurement is often dependent on sample preparation. Ideally, a sample for reflectance measurement is flat, has a uniform gloss and texture, is opaque, and is nondirectional. Always strive for sample preparation techniques that are reproducible. Consult standard test methods [24], standard practices [19, 27–29], books and articles [30–33] for help and advice on sample preparation.

58.6 Transmittance Measurement

Regular and Diffuse Transmittance

When a beam of light passes through a "transparent" material along its normal, the intensity of the beam will be decreased by absorption, but the direction of the beam will be unchanged. This is called regular transmittance. When a beam of light passes through a hazy or translucent material along its normal, the material scatters light and spreads the beam. This is called diffuse transmittance.

Illuminating and Viewing Geometries for Transmittance

The CIE has recommended three illuminating and viewing geometries for transmittance measurements. [1] Most transmittance measurements are made with spectrophotometers designed for analytical chemistry applications that use a normal/normal, 0/0, geometry. The illuminator directs the incident beam along the sample's normal, and the receiver views the sample along its normal. Only regular transmittance can be measured accurately using this geometry.

Regular and diffuse transmittance can be measured using the normal/diffuse, 0/d, geometries (or the equivalent diffuse/normal, d/0), which the CIE also recommended for reflectance measurements. Regular transmittance is measured by keeping the sample as far away as possible from the integrating sphere, whereas diffuse transmittance is measured by placing the sample in contact with the sphere. Instruments designed specifically for transmittance measurements would have only a sample or illuminator port and a receiver port. Instruments designed for reflectance measurements can be used for making transmittance measurements by placing a white material in the sample port and using the specular included mode of measurement. The illuminator or receiver port would then serve as the sample port.

The third geometry recommended by the CIE, diffuse/diffuse, d/d, is not often used for industrial color measurement. One integrating sphere is used to illuminate the sample, and a second integrating sphere is used to view the sample.

The two bidirectional reflectance geometries (45/0 and 0/45) have also been used for making regular transmittance measurements.

Monochromatic and Polychromatic Illumination

Instruments for measuring transmittance may have either monochromatic or polychromatic illumination. The transmittance of nonfluorescent samples can be measured using either type of illumination, but fluorescent samples can only be measured using polychromatic illumination.

Standardizing Instruments for Transmittance Measurements

Three techniques exist for setting the high end of the measurement scale, a transmittance factor of 1.0. Each technique produces different results, so it is important to document the method used.

Setting the instrument to read a transmittance factor of 1.0 with no sample in the sample compartment is the easiest technique. The transmittance measurements will then be relative to air. This technique is often used when solid samples are being measured.

Transmittance measurements of solid materials may also be made relative to a clear blank of similar material. To make these measurements, the clear blank is placed in the sample compartment before the instrument is standardized.

When liquids are to be measured, a holder containing solvent or nonabsorbing liquid of the same refractive index as the liquid to be measured should be placed in the sample compartment before standardizing the instrument. This eliminates any effects of the holder's transmittance.

Sample Preparation

Ideal samples will be flat with parallel sides. Liquid sample holders should be made from optical glass. Because transmittance will change with sample thickness and the concentration of colorant, it is important that sample holders and blanks used for standardization have the same path length or thickness as the sample. When two different samples must be compared, they should be prepared using the same technique and have the same thickness.

It is extremely difficult to measure the transmittance of curved materials because the curvature of the object may act as a lens and deflect the incident light away from the receiver. If the curvature is not too great, it may be possible to make a diffuse transmittance measurement.

58.7 Color Difference Calculations

A number of equations have been developed over the years for calculating the color difference between two objects [1, 10, 24, 30, 34, 35, 36]. One of the most common equations is the CIELAB recommended by the CIE in 1976 [1].

First the CIELAB coordinates, L^* (lightness), a^*, b^*, C^*_{ab} (chroma) and h_{ab} (hue angle) are calculated with Equations 58.11 through 58.15.

$$L^* = 116\, f(Y/Y_n) - 16 \tag{58.11}$$

$$a^* = 500\, [f(X/X_n) - f(Y/Y_n)] \tag{58.12}$$

$$b^* = 200\, [f(Y/Y_n) - f(Z/Z_n)] \tag{58.13}$$

$$C^*_{ab} = [a^{*2} + b^{*2}]^{1/2} \tag{58.14}$$

$$h_{ab} = \arctan[b^*/a^*] \tag{58.15}$$

in which X, Y, and Z are tristimulus values and the subscript n refers to the tristimulus values of the perfect diffuser for the given illuminant and standard observer; $f(X/X_n) = (X/X_n)^{1/3}$ for values of (X/X_n) greater than 0.008856 and $f(X/X_n) = 7.787(X/X_n) + 16/116$ for values of (X/X_n) equal to or less than 0.008856; and the same with Y and Z replacing X in turn. The hue angle is 0° along the $+a^*$ axis, 90° along the $+b^*$ axis, 180° along the $-a^*$ axis, and 270° along the $-b^*$ axis.

The total color difference, ΔE^*_{ab}, and its components—the lightness difference, ΔL^*, the chroma difference, ΔC^*_{ab}, and the hue difference, ΔH^*_{ab}—are calculated using Equations 58.16 through 58.21.

$$\Delta L^* = L^*_{trial} - L^*_{standard} \tag{58.16}$$

$$\Delta a^* = a^*_{trial} - a^*_{standard} \tag{58.17}$$

$$\Delta b^* = b^*_{trial} - b^*_{standard} \tag{58.18}$$

$$\Delta E^*_{ab} = [(\Delta L^*)^2 + (\Delta a^*)^2 + (\Delta b^*)^2]^{1/2} \tag{58.19}$$

$$\Delta C^*_{ab} = C^*_{ab\ trial} - C^*_{ab\ standard} \tag{58.20}$$

$$\Delta H^*_{ab} = [(\Delta E^*_{ab})^2 - (\Delta L^*)^2 - (\Delta C^*_{ab})^2]^{1/2} \tag{58.21}$$

A negative value of ΔL^* means the trial is darker than the standard, and a negative value of ΔC^*_{ab} means the trial has a lower chroma than the standard. When the hue angle, h_{ab} of the trial is greater than that of the standard, the sign of ΔH^*_{ab} is positive, and vice-versa. The total color difference and its components can then be used in setting color tolerances [19, 37].

Researchers have been making modifications to the CIELAB color difference equation in an attempt to have the color difference more closely correlate with visually observed color differences. The CMC(l:c) color difference equation [50–52] has gained great acceptance in the textile industry and is being tested in other areas. Starting with the CIELAB color differences, the CMC(l:c) color differences are calculated with Equations 58.22 through 58.28.

$$\Delta E = \left[\left(\frac{\Delta L^*}{lS_L}\right)^2 + \left(\frac{\Delta C^*_{ab}}{cS_C}\right)^2 + \left(\frac{\Delta H^*_{ab}}{S_H}\right)^2\right]^{1/2} \tag{58.22}$$

$$S_L = \frac{0.040975L^*}{1 + 0.01765L^*} \tag{58.23}$$

unless $L^* < 16$, in which case $S_L = 0.511$,

$$S_C = \frac{0.0638C^*_{ab}}{1 + 0.0131C^*_{ab}} + 0.638 \tag{58.24}$$

$$S_H = (FT + 1 - F)S_C \tag{58.25}$$

$$F = \left(\frac{(C^*_{ab})^4}{(C^*_{ab})^4 + 1900}\right)^{1/2} \tag{58.26}$$

$$T = 0.36 + \text{abs}[0.4\cos(35 + h_{ab})] \tag{58.27}$$

unless h is between 164° and 345° when

$$T = 0.56 + \text{abs}[0.2\cos(168 + h_{ab})] \tag{58.28}$$

in which the notation "abs" indicates the absolute (i.e., positive,) value of the term inside the square brackets. When $l = c = 1$, the formula quantifies the perceptibility of color differences. Optimum values of l and c for quantifying the acceptability of a color match must be determined for the material being measured. The textile industry has found the optimum values to be $l = 2$ and $c = 1$ [51].

In 1994, the CIE recommended a new color difference equation, CIE94 [36], which is similar to the CMC equation. CIE94 color differences are calculated using Equations 58.29 through 58.33. The perceived color-difference, ΔV, is related to the measured color difference through an overall sensitivity factor, k_E.

$$\Delta V = \frac{1}{k_E}\Delta E^*_{94} \tag{58.29}$$

$$\Delta E^*_{94} = \left[\left(\frac{\Delta L^*}{k_L S_L}\right)^2 + \left(\frac{\Delta C^*_{ab}}{k_C S_C}\right)^2 + \left(\frac{\Delta H^*_{ab}}{k_H S_H}\right)^2\right]^{1/2} \tag{58.30}$$

$$S_L = 1 \tag{58.31}$$

$$S_C = 1 + 0.045C^*_{ab} \tag{58.32}$$

$$S_H = 1 + 0.015C^*_{ab} \tag{58.33}$$

The overall sensitivity factor, k_E, is used to account for variation in the illuminating and viewing conditions. A person in the textile industry who is using CMC(2:1) and would like to compare the results with CIE94 would set $k_L = 2$ and $k_C = k_H = 1$, i.e., CIE94(2:1:1).

The improvement in correlating with visual assessments of color difference could result in either the CMC(l:c) or the CIE94 replacing the CIELAB color difference equation. Hunter and Harold [30] detail many of the older color difference equations, many of which are still in use. A history of the development of color metrics was written by Richter [53].

58.8 Special Cases

Fluorescent Samples

Fluorescent materials not only reflect light but also emit light. Light absorbed at some wavelengths is emitted at longer wavelengths. The amount of light emitted depends on the intensity and the spectral power distribution of the source. Because of the emission of light, spectrophotometers that illuminate a fluorescent material with monochromatic light cannot be used, because the light emitted at the longer wavelengths will be measured as if it had been reflected from the material. Reflectance measurements of the material illuminated by polychromatic light will include the emitted light at the proper wavelengths. If the instrument's light source is a good representation of the illuminant, the measurement will be indicative of the observed color. Techniques have been recommended by the CIE for assessing the quality of daylight simulators [55]. Measured reflectance factors at the wavelengths of emittance may be greater than 1.0. Special fluorescent calibration standards must be used to accurately measure these materials [38]. The 45/0 or 0/45 bidirectional geometries are recommended for measuring fluorescent materials [39, 40].

The complete analysis of a fluorescent material requires that the emittance be separated from the reflectance. Spectrofluorimeters were designed to analyze fluorescent samples. These instruments illuminate the sample with monochromatic light. Reflected and emitted rays pass through a second monochromator to isolate the receiver wavelengths. By viewing the sample at the same wavelength as it is illuminated, the true reflectance of the sample can be determined. Either the excitance spectra or the emittance spectra can be studied by setting each monochromator at different wavelengths. Techniques have also been developed to correct spectrophotometric measurements of fluorescent samples [54, 56].

Metallic and Pearlescent Samples

Materials that contain metallic and/or pearlescent pigments are goniochromatic, i.e., they change color with the illuminating and viewing geometry. Goniospectrophotometers are needed to measure these materials. A goniospectrophotometer illuminates (or views) the sample at a fixed angle, usually 45°, and views (or illuminates) the sample at three or more angles. The position of the receivers (or illuminators) is described by the aspecular angle, the viewing angle measured from the direction of the specular reflectance, which is equal and opposite the angle of illumination. In a goniospectrophotometer that illuminates the sample at 45° and views at three angles, one of the viewing angles would be near the specular reflection at about 25°, the second at the CIE recommended bidirectional angle of 0° and the third would be far away from the specular reflectance at about 70°.

Goniospectrophotometry is still in its infancy and the CIE recommendations and ASTM standards for making these measurements are still under development.

Retroreflecting Samples

Highway signs and high visibility clothing are examples of retroreflectors. Light shining on a retroreflector is returned in directions very close to the illumination angle. Most retroreflectance measurements are made in connection with highway safety. There are only a few specially built instruments for measuring retroreflection. A projector is usually used to illuminate the sample, and a teleradiometer or a radiometric telecolorimeter is used to view the sample. Tristimulus values can be calculated from the retroreflectance factors. Standard practices and test methods exist for specifying the illuminating and viewing geometries and making retroreflectance measurements [41–45].

Lamps, Light Sources, and Displays

Spectral radiometers and radiometric colorimeters were designed to measure lamps, light sources, and displays. These instruments are similar to spectrophotometers and colorimeters but do not need an illuminator, because the sample being measured emits light.

Lamps and light sources can either be measured directly or by measuring the reflectance of a stable white reflecting surface being illuminated by the lamp or light source. To measure televisions, computer monitors, and other similar devices, the emitted light must usually be imaged directly on the optics of the spectral radiometer or radiometric colorimeter [46, 47].

Hazy and Translucent Materials

Hazy and translucent materials are measured by placing them in contact with an integrating sphere and measuring their diffuse transmittance. A haze index can be calculated from four diffuse transmittance measurements on a reflecting spectrophotometer [48]. Two of the measurements are made with a white material in the reflectance sample port, and two measurements are made with a light trap in the port. Special instruments called *hazemeters* were designed to make haze measurements.

58.9 Instrument Manufacturers

Costs of color measuring instruments vary significantly. The more accurate instruments with the best repeatability and reproducibility can be expected to cost more. Table 58.1 lists the price ranges for various types of color measuring instruments. There is a classification in the table called spectral analyzers. Although these instruments may provide information that makes them appear to be colorimeters or spectrophotometers, they use measurement techniques not traditionally associated with those instruments. For example, several use LEDs instead of more traditional light sources. Since the instruments vary so widely in capability, the purchaser must ensure that a particular instrument has the capability to make the measurements desired with sufficient accuracy, repeatability, and reproducibility.

TABLE 58.1 Color Measuring Instruments and Their Costs

Instrument type	Price range in US $
Colorimeters	4,000–15,000
Goniospectrophotometers	12,500–30,000
Hazemeters	13,000–15,000
Radiometric Colorimeters	2,000–10,000
Retroreflectometers	3,500–15,000
Spectral Analyzers	1,500–2,500
Spectrofluorimeters	30,000–40,000
Spectrophotometers	2,500–20,000
Spectroradiometers	2,500–40,000

Table 58.2 provides contact information for a number of manufacturers of color measuring instruments. The manufacturers listed in this table design instruments specifically for color measurement. Thus, manufacturers of analytical spectrophotometers are excluded even if their instruments have color measuring capabilities. The table is also limited to manufacturers with a major presence in the U.S.

TABLE 58.2 Instrument Manufacturers

BYK-Gardner USA	**Color Savvy, Ltd.**
Rivers Park II, 9104 Guilford Road	305 S. Main Street
Columbia, MD 21046-2729	Springboro, OH 45066
(301) 483-6500	(513) 748-9160
Spectrophotometers, colorimeters, hazemeters	Colorimeters, spectrophotometers
ColorTec	**Datacolor International**
28 Center Street	5 Princess Road
Clinton, NJ 08809	Lawrenceville, NJ 08648
(908) 735-2248	(609) 924-2189
Spectral analyzers	Spectrophotometers, goniospectrophotometers
Datacolor International	**Gamma Scientific Co.**
5 Princess Road	8581 Aero Drive
Lawrenceville, NJ 08648	San Diego, CA 92123
(609) 924-2189	(619) 279-8034
Spectrophotometers, goniospectrophotometers	Spectroradiometers, retroreflectometers
GretagMacbeth	**Hunter Associates Laboratory, Inc.**
617 Little Britain Road	11491 Sunset Hills Road
New Windsor, NY 12553-6148	Reston, VA 20190
(914) 565-7660	(703) 471-6870
Spectrophotometers, goniospectrophotometers	Spectrophotometers, colorimeters
Labsphere, Inc.	**Light Source, Inc.**
P.O. Box 70, Shaker Street	4th floor, 4040 Civic Center Drive
North Sutton, NH 03260-0070	San Rafael, CA 94903
(603) 927-4266	(415) 446-4200
Spectroradiometers, radiometric colorimeters, spectrofluorimeters	Spectrophotometers, spectroradiometers
Minolta Corporation	**Photo Research Inc.**
101 Williams Drive	9330 DeSoto Avenue
Ramsey, NJ 07446	Chatsworth, CA 91311
(201) 825-4000	(818) 341-5151
Spectrophotometers, spectroradiometers, colorimeters, radiometric colorimeters, spectrofluorimeters	Spectroradiometers
X-Rite, Inc.	
3100 44th St. SW	
Grandville, MI 49418	
(616) 534-7663	
Spectrophotometers, spectrocolorimeters, goniospectrophotometers	

58.10 Defining Terms

For a more extensive collection of terms relating to color and appearance, the reader should refer to ASTM E 284 Standard Terminology of Appearance [49].

Chroma: Attribute of color used to indicate the degree of departure of the color from a gray of the same lightness.

CIE: The abbreviation for the French title of the International Commission on Illumination, Commission Internationale de l'Éclairage.

CIE standard observers: The ideal colorimetric observer data adopted by the CIE to represent the response of the average human eye, when light-adapted, to an equal-energy spectrum. The standard observer adopted in 1931 was developed from data obtained with a 2° field of vision and is commonly called the "2° standard observer." The standard observer adopted in 1964 was developed from data obtained with a 10° annular field of vision and is commonly called the "10° standard observer."

CIE tristimulus values: Amounts of the three mathematical lights necessary in a three-color additive mixture required for matching a color in the CIE System. They are designated X, Y, and Z. The illuminant and standard observer color matching functions must be designated.

colorimetry: The science of color measurement.

hue: The attribute of color perception by means of which a color is judged to be red, orange, yellow, green, blue, purple, or intermediate between adjacent pairs of these, considered in a close ring (red and purple being an adjacent pair.) White, black and grays possess no hue.

illuminant: A mathematical description of the relative power emitted by a real or imaginary light source at each wavelength in its emission spectrum.

lightness: (1) The attribute of color perception by which a nonself-luminous body is judged to reflect more or less light. (2) The attribute by which a perceived color is judged to be equivalent to one of a series of grays ranging from black to white.

metamerism: Property of two specimens that match under a specified illuminator and to a specified observer and whose spectral reflectances or transmittances differ in the visible wavelengths.

perfect reflecting diffuser: Ideal reflecting surface that neither absorbs nor transmits light, but reflects diffusely, with the radiance of the reflecting surface being the same for all reflecting angles, regardless of the angular distribution of the incident light.

reflectance: Ratio of the reflected radiant or luminous flux to the incident flux in the given conditions. The term reflectance is often used in a general sense or as an abbreviation for reflectance factor. Such usage may be assumed unless the above definition is specifically required by the context.

reflectance factor: Ratio of the flux reflected from the specimen to the flux reflected from the perfect reflecting diffuser under the same geometric and spectral conditions of measurement.

References

1. CIE, *CIE Publication 15.2, Colorimetry, 2nd ed.*, Commission International de l'Éclairage (CIE), Central Bureau of the CIE, Vienna, 1986. Available from USNC/CIE Publications, c/o TLA–Lighting Consultants, Inc., 7 Pond Street, Salem, MA 01970.

2. W. H. Foster, Jr., R. Gans, E. I. Stearns and R.E. Stearns, Weights for calculation of tristimulus values from sixteen reflectance values. *Color Engineering*, **8**(3), 25–47, 1970.

3. E. I. Stearns, The determination of weights for use in calculating tristimulus values. *Color Research and Application*, **6**, 210–212, 1981.

4. E. I. Stearns, Calculation of tristimulus values and weights with the revised CIE recommendations. *Textile Chemist and Colorist*, **17**(8), 162/53–168/59, 1985.

5. H. S. Fairman, The calculation of weight factors for tristimulus integration. *Color Research and Application*, **10**, 1199-1203, 1985.

6. F. W. Billmeyer, Jr. and H. S. Fairman, CIE method for calculating tristimulus values. *Color Research and Application*, **12**, 27–36, 1987.

7. ASTM, ASTM E 308 Standard Practice for Computing the Colors of Objects by Using the CIE System. *Annual Book of ASTM Standards*, American Society for Testing and Materials, 100 Barr Harbor Drive, West Conshohocken, PA 19428-2959.

8. F. W. Billmeyer, Jr. and R. T. Marcus, Effect of illuminating and viewing geometry on the color coordinates of samples with various surface textures. *Applied Optics*, **8**, 1763–1768, 1969.

9. ASTM, ASTM E 179 Standard Guide for Selection of Geometric Conditions for Measurement of Reflection and Transmission Properties of Materials. *Annual Book of ASTM Standards*, American Society for Testing and Materials, 100 Barr Harbor Drive, West Conshohocken, PA 19428-2959.

10. R. W. G. Hunt, *Measuring Colour, 2nd ed.*, Chichester, West Sussex, England: Ellis Horwood Limited, 1991.

11. D. C. Rich, The effect of measuring geometry on computer color matching. *Color Research and Application*, **13**, 113–118, 1988.

12. T. J. Mabon, Color measurement of plastics: which geometry is best. Presented at the Regional Technical Conference of the Society of Plastics Engineers, Inc., *Color Tolerances: Measuring Up to Today's Standards*, Cherry Hill, NJ, September 14–16, 1992.

13. J. C. Zwinkels, Errors in colorimetry caused by the measuring instrument. *Textile Chemist and Colorist*, **21**(2), 23–29, 1989.

14. R. H. Stanziola, H. Hemmendinger, and B. Momiroff, The Spectro Sensor: a new generation spectrophotometer. *Color Research and Application*, **4**, 157–163, 1979.

15. D. Strocka, Are intervals of 20 nm sufficient for industrial colour measurement?. Presented at the 2nd AIC Congress "Colour 73", York, England, July, 1973. Long abstract: R. W. G. Hunt, ed., *Colour 73*, New York: Halsted Press, Div. John Wiley & Sons, 1973.

16. H. Schmelzer, Influence of the design of instruments on the accuracy of color-difference and color-matching calculations. *J. of Coatings Technology*, **58**(739), 53–59, 1986.

17. ASTM, ASTM E 1164 Standard Practice for Obtaining Spectrophotometric Data for Object-Color Evaluation. *Annual Book of ASTM Standards*, American Society for Testing and Materials, 100 Barr Harbor Drive, West Conshohocken, PA 19428-2959.

18. E. C. Carter, F. W. Billmeyer, Jr., and D. C. Rich, *Guide to Material Standards and Their Use in Color Measurement*, ISCC Technical Report 89-1. Available from The Inter-Society Color Council, Suite 301, 11491 Sunset Hills Road, Reston, VA 22090.

19. SAE, SAE J1545 Recommended Practice for Instrumental Color Difference Measurement for Exterior Finishes, Textiles, and Colored Trim. Society of Automotive Engineers, 3001 W. Big Beaver, Troy, MI 48084.

20. ASTM, ASTM E 429 Standard Test Method for Measurement and Calculation of Reflecting Characteristics of Metallic Surfaces Using Integrating Sphere Instruments. *Annual Book of ASTM Standards*, American Society for Testing and Materials, 100 Barr Harbor Drive, West Conshohocken, PA 19428-2959.

21. ASTM, ASTM E 805 Standard Practice for Identification of Instrumental Methods of Color or Color-Difference Measurement of Materials. *Annual Book of ASTM Standards*, American Society for Testing and Materials, 100 Barr Harbor Drive, West Conshohocken, PA 19428-2959.

22. ASTM, ASTM E 1331 Standard Test Method for Reflectance Factor and Color by Spectrophotometry Using Hemispherical Geometry. *Annual Book of ASTM Standards*, American Society for Testing and Materials, 100 Barr Harbor Drive, West Conshohocken, PA 19428-2959.

23. ASTM, ASTM E 1349 Standard Test Method for Reflectance Factor and Color by Spectrophotometry Using Bidirectional Geometry. *Annual Book of ASTM Standards*, American Society for Testing and Materials, 100 Barr Harbor Drive, West Conshohocken, PA 19428-2959.

24. AATCC, AATCC Test Method 153 Color Measurement of Textiles: Instrumental. *AATCC Technical Manual/1993*, American Association of Textile Chemists and Colorists, P.O. Box 12215, Research Triangle Park, NC 27709-2215.

25. NPES, American National Standard CGATS.5 Graphic Technology—Spectral Measurement and Colorimetric Computation for Graphic Arts Images, NPES The Association for Suppliers of Printing and Publishing Technologies, 1899 Preston White Drive, Reston, VA 22091-4367.

26. ASTM, ASTM E 1347 Standard Test Method for Color and Color-Difference Measurement by Tristimulus (Filter) Colorimetry. *Annual Book of ASTM Standards*, American Society for Testing and Materials, 100 Barr Harbor Drive, West Conshohocken, PA 19428-2959.

27. ASTM, ASTM D 3925 Standard Practice for Sampling Liquid Paints and Related Pigmented Coatings. *Annual Book of ASTM Standards*, American Society for Testing and Materials, 100 Barr Harbor Drive, West Conshohocken, PA 19428-2959.

28. ASTM, ASTM D 3964 Standard Practice for Selection of Coating Specimens for Appearance Measurements. *Annual Book of ASTM Standards*, American Society for Testing and Materials, 100 Barr Harbor Drive, West Conshohocken, PA 19428-2959.

29. ASTM, ASTM D 823 Standard Practices for Producing Films of Uniform Thickness of Paint, Varnish, and Related Products on Test Panels. *Annual Book of ASTM Standards*, American Society for Testing and Materials, 100 Barr Harbor Drive, West Conshohocken, PA 19428-2959.

30. R. S. Hunter and R. W. Harold, *The Measurement of Appearance, 2nd edition*, New York: John Wiley & Sons, 1987.

31. R. L. Connelly, Sr., Preparation and mounting textile samples for color measurement, in *Color Technology in the Textile Industry,* edited by G. Celikiz and R. G. Kuehni, American Association of Textile Chemists and Colorists, P.O. Box 12215, Research Triangle Park, NC 27709, 1983.

32. C. Wilson and E. I. Stearns, The spectrophotometric reflectance measurement of small samples, in *Color Technology in the Textile Industry* edited by G. Celikiz and R. G. Kuehni, American Association of Textile Chemists and Colorists, P.O. Box 12215, Research Triangle Park, NC 27709, 1983.

33. E. I. Stearns and W. B. Prescott, Measurement of translucent cloth samples, in *Color Technology in the Textile Industry* edited by G. Celikiz and R. G. Kuehni, American Association of Textile Chemists and Colorists, P.O. Box 12215, Research Triangle Park, NC 27709, 1983.

34. G. Wyszecki and W. S. Stiles, *Color Science, Concepts and Methods, Quantitative Data and Formulae, 2nd ed.*, New York: John Wiley & Sons, 1992.

35. ASTM, ASTM D 2244 Standard Test Method for Calculation of Color Differences from Instrumentally Measured Color Coordinates. *Annual Book of ASTM Standards*, American Society for Testing and Materials, 100 Barr Harbor Drive, West Conshohocken, PA 19428-2959.

36. CIE, *CIE Technical Report 116, Industrial Colour-Difference Evaluation*, Commission International de l'Éclairage (CIE), 1995. Available from USNC/CIE Publications, c/o TLA–Lighting Consultants, Inc., 7 Pond Street, Salem, MA 01970.

37. ASTM, ASTM D 3134 Standard Practice for Establishing Color and Gloss Tolerances. *Annual Book of ASTM Standards*, American Society for Testing and Materials, 100 Barr Harbor Drive, West Conshohocken, PA 19428-2959.

38. D. Gundlach and H. Terstiege, Problems in measurement of fluorescent materials. *Color Research and Application*, **19**, 427–436, 1994.

39. ASTM, ASTM E 991 Standard Practice for Color Measurement of Fluorescent Specimens. *Annual Book of ASTM Standards*, American Society for Testing and Materials, 100 Barr Harbor Drive, West Conshohocken, PA 19428-2959.

40. R. A. McKinnon, Methods of measuring the colour of opaque fluorescent materials. *Rev. Prog. Coloration*, **17**, 56–60, 1987.

41. ASTM, ASTM E 808 Standard Practice for Describing Retroreflection. *Annual Book of ASTM Standards*, American Society for Testing and Materials, 100 Barr Harbor Drive, West Conshohocken, PA 19428-2959.

42. ASTM, ASTM E 809 Standard Practice for Measuring Photometric Characteristics of Retroreflectors. *Annual Book of ASTM Standards*, American Society for Testing and Materials, 100 Barr Harbor Drive, West Conshohocken, PA 19428-2959.

43. ASTM, ASTM E 810 Standard Test Method for Coefficient of Retroreflection of Retroreflective Sheeting. *Annual Book of ASTM Standards*, American Society for Testing and Materials, 100 Barr Harbor Drive, West Conshohocken, PA 19428-2959.

44. ASTM, ASTM D 4061 Standard Test Method for Retroreflection of Horizontal Coatings. *Annual Book of ASTM Standards*, American Society for Testing and Materials, 100 Barr Harbor Drive, West Conshohocken, PA 19428-2959.

45. ASTM, ASTM E 811 Standard Practice for Measuring Colorimetric Characteristics of Retroreflectors Under Nighttime Conditions. *Annual Book of ASTM Standards*, American Society for Testing and Materials, 100 Barr Harbor Drive, West Conshohocken, PA 19428-2959.

46. ASTM, ASTM E 1336 Standard Test Method for Obtaining Colorimetric Data from a Video Display Unit by Spectroradiometry. *Annual Book of ASTM Standards*, American Society for Testing and Materials, 100 Barr Harbor Drive, West Conshohocken, PA 19428-2959.

47. ASTM, ASTM E 1455 Standard Practice for Obtaining Colorimetric Data from a Video Display Unit Using Tristimulus Colorimeters. *Annual Book of ASTM Standards*, American Society for Testing and Materials, 100 Barr Harbor Drive, West Conshohocken, PA 19428-2959.
48. ASTM, ASTM D 1003 Standard Test Method for Haze and Luminous Transmittance of Transparent Plastics. *Annual Book of ASTM Standards*, American Society for Testing and Materials, 100 Barr Harbor Drive, West Conshohocken, PA 19428-2959.
49. ASTM, ASTM E 284 Standard Terminology of Appearance. *Annual Book of ASTM Standards*, American Society for Testing and Materials, 100 Barr Harbor Drive, West Conshohocken, PA 19428-2959.
50. F. J. J. Clarke, R. McDonald, and B. Rigg, Modification to the JPC79 colour-difference formula. *J. Soc. Dyers and Colourists*, **100**, 128–132 and 281–282, 1984.
51. British Standards Institution, BS6923:1988 British Standard Method for Calculation of small colour differences. Available from British Standards Institution, 2 Park Street, London W1A 2BS, England.
52. AATCC, AATCC Test Method 173 CMC: Calculation of Small Color Differences for Acceptability. *AATCC Technical Manual/1993*, American Association of Textile Chemists and Colorists, P.O. Box 12215, Research Triangle Park, NC 27709-2215.
53. M. Richter, The development of color metrics. *Color Research and Application*, **9**, 69–83, 1984.
54. R. A. McKinnon, Methods of measuring the colour of opaque fluorescent materials. *Rev. Progress Coloration*, **17**, 56–60, 1987.
55. CIE, *CIE Publication 51, A Method for Assessing the Quality of Daylight Simulators for Colorimetry*, Commission International de l'Éclairage (CIE), Central Bureau of the CIE, Vienna, 1982. Available from USNC/CIE Publications, c/o TLA–Lighting Consultants, Inc., 7 Pond Street, Salem, MA 01970.
56. F. W. Billmeyer, Jr., Metrology, documentary standards, and color specifications for fluorescent materials. *Color Research and Application*, **19**, 413–425, 1994.

Further Information

American Society for Testing and Materials, *ASTM Standards on Color and Appearance Measurement, Sponsored by ASTM Committee E-12 on Appearance of Materials, 5th ed.*, Philadelphia: American Society for Testing and Materials, 1996.

F. W. Billmeyer, Jr., and M. Saltzman, *Principles of Color Technology, 2nd edition*, New York: John Wiley & Sons, 1981.

The Inter-Society Color Council, Suite 301, Sunset Hills Road, Reston, VA 22090; phone (703) 318-0263; fax (703) 318-0514.

Color Research and Application, published by John Wiley & Sons, 605 Third Avenue, New York, NY 10158

59

Optical Loss

Halit Eren
Curtin University of Technology

59.1 Basic Concepts

Different properties of optics are used in many instrumentation and measurement devices. Applications vary from photographic imaging to high-speed data transmission via fibers. Once the light is generated and propagated from a source, it can be expanded, condensed, collimated, reflected, polarized, filtered, diffused, absorbed, refracted, scattered, etc. for manipulation and processing. Some of these manipulations are done to serve a particular purpose, and some results from physical characteristics of the light and optical properties of the media. In many applications, the intensity of light at the beginning will not be the same at the end due propagation characteristics and losses. It is important to point out that the word "light" is usually taken as radiation in the range of from about 380 nm to about 800 nm. This understanding limits important applications to those in the near-infrared and ultraviolet regions of the spectrum. In here, when the term "light" is used, the IR and UV portions of the spectrum are not left out.

The interpretation and treatment of "optical loss" is different for each individual instrumentation and measurement activity. It entirely depends on the purpose and method of measurements. For example, scattering of the light (radiation) from clouds is a useful property for determining atmospheric characteristics, whereas scattering of light in optical fibers may not be desirable due to resulting losses in power and decrease in efficiency. While many instrumentation systems (e.g., imaging) make use of scattering, absorption, refraction, and reflection as useful properties, in others these are considered to be mere losses that cause undesirable attenuations. In recent years, fiber optics has attracted considerable attention, and most of the following discussion on optical loss is centered on this subject. Likewise, this chapter concentrates on optical losses in fiber optics to clearly demonstrate the fundamental principles. Nevertheless, it is important to recognize that the same as losses in fiber optics may not be regarded as losses in other applications. It will be shown here that losses in fiber optics have useful properties in determining the optical fiber characteristics.

Theoretically, light propagation in fibers can be treated in a number of ways. For example, in treating propagation by modes, the fibers are viewed as optical waveguides, whereas treatment by rays is an approximate description of fibers with diameters much greater than the wavelength. In this chapter, both approaches will be used as appropriate.

An optical fiber is a circular dielectric glass (some polymers are also used) waveguide that can efficiently transport optical energy, and the information in the energy, usually by using the principle of total internal reflection. In some cases, it consists of a central glass core surrounded by a concentric cladding material with a slightly lower ($\approx 1\%$) refractive index. Since the core has higher index of refraction than the cladding, light is confined to the core if the angular condition of total internal reflection is met. Attenuation of the light begins the moment light enters the fiber. The acceptance and transmission of light depends greatly on the angle at which the light enters the fiber. The angle must be less than the *critical acceptance angle* of the particular fiber being used, as shown in Figure 59.1.

There are three basic types of fibers: single-mode step-index, multimode graded index, and multimode step index, as shown in Figures 2a, 2b, and 2c, respectively. The characteristics of optical losses in these three types of fibers vary slightly due to differences in construction and the nature of propagation of light. For example, in the case of multimode step-index fibers, light striking the core–cladding junction at an angle greater than the angle of internal reflection passes through and becomes absorbed by the opaque jacket. This represents a significant source of attenuation, limiting the injection efficiency at the transmitting end.

Single-mode fibers, Figure 59.2a, are used in transmitting broadband signals over large distances. Their attenuation is generally very small, and their transmission band is large. Owing to material properties, low attenuations can be expected for wavelengths around 1.3 to 1.6 μm. Additional attenuation arises from splices and fiber bending.

Multimode graded-index fibers have medium size cores (50 to 100 μm) and refractive indices that decrease radially outward. The two optical materials with different refractive indices are mixed together in such a way that the index of refraction decreases smoothly with distance from the fiber axis. The graded index causes the light to gradually bend back and forth across the axis in sinusoidal manner when very small injection angles are used. This greatly reduces the light losses from the fiber and results in relatively better bandwidth and efficiency. The graded-index fibers also allow the use of simpler splicing techniques.

Good-quality fibers are typically made of pure silica with index modifying dopants such as GeO_2. While bandwidth is the primary consideration in the use of fiber optics in communication applications, light attenuation characteristics are equally important. The overall quality of a fiber optic light guide is determined by its light transmissivity, defined as the ratio of the output light to that put into the fiber.

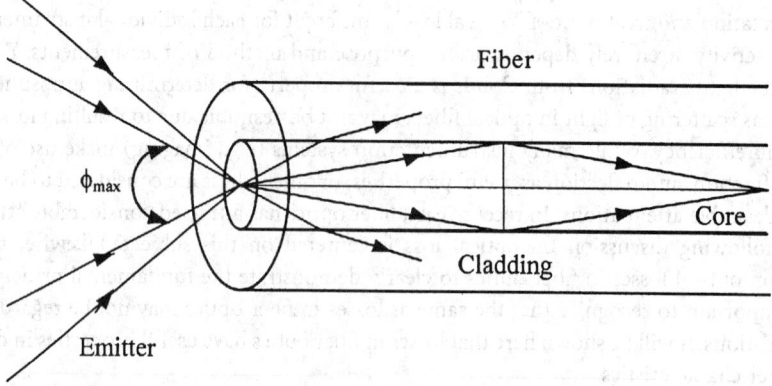

FIGURE 59.1 An optical fiber is a circular dielectric glass waveguide that can efficiently transport optical energy. It consist of a central glass core surrounded by a concentric cladding material. An important requirement for the connection of a light source is that a sufficient amount of useful light must be coupled into the fiber. The core has a higher index of refraction than the cladding; therefore, light is confined to the core only if the angular condition of total internal reflection is met. The acceptance and transmission of light depend greatly on the angle at which the light rays enter the fiber, and it must be less than the *critical acceptance angle* of a particular fiber.

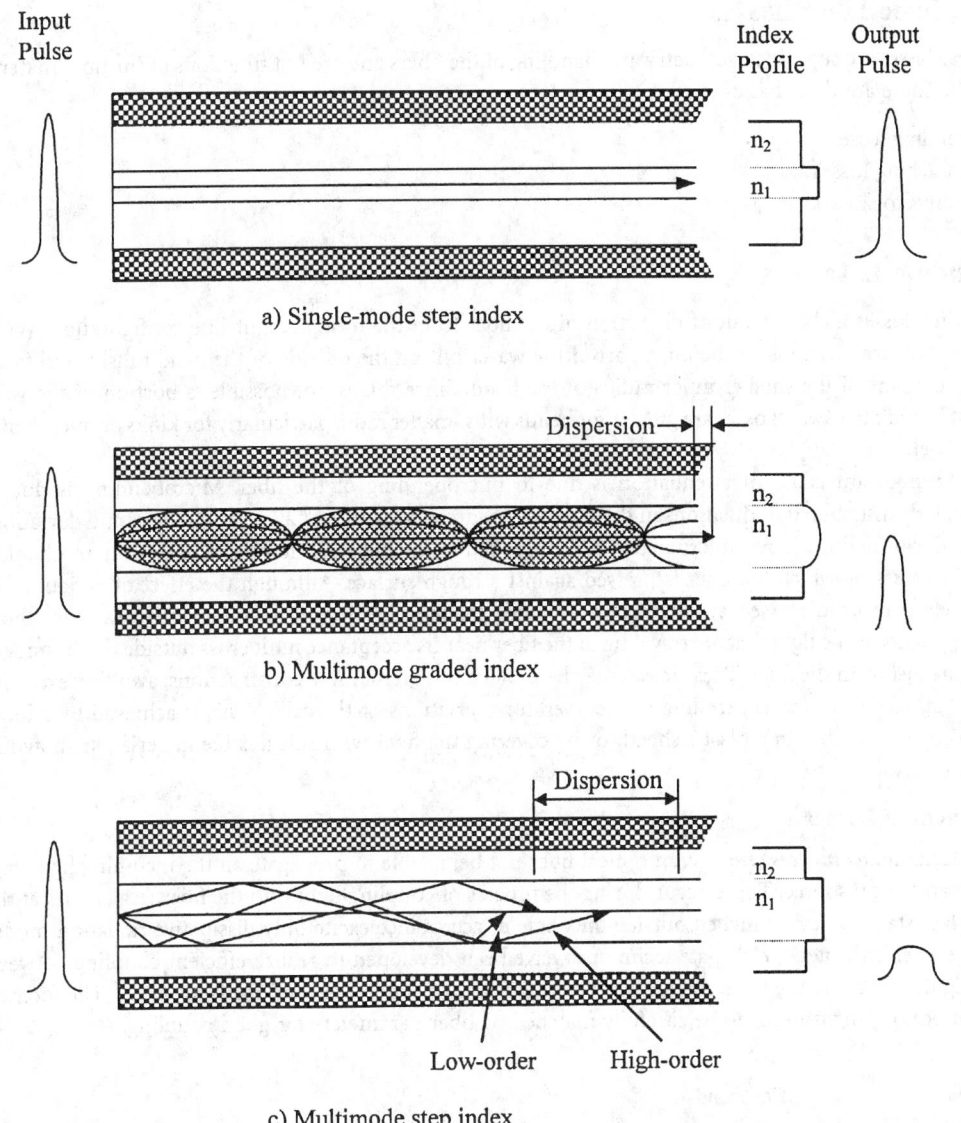

Input
Pulse

Index
Profile

Output
Pulse

n_2
n_1

a) Single-mode step index

Dispersion→ ←

n_2
n_1

b) Multimode graded index

Dispersion

n_2
n_1

Low-order   High-order

c) Multimode step index

FIGURE 59.2 Three basic types of fibers. (a) Single-mode step-index fibers, used in transmitting broadband signals over large distances, (b) multimode graded index, obtained by mixing two optical materials with different refractive indices together in such a way that the index of refraction decreases smoothly with distance from the fiber axis, and (c) multimode step index fibers, in which the refractive index changes between fiber and cladding, and rays striking the core-cladding junction reflect back in the glass.

LEDs are the primary light source used in fiber-optic links for bit rates up to 200 Mb/s and distances up to 2 km. For higher bit rates and longer distances, diode lasers are preferred. LEDs used in fiber-optic applications operate at three narrowly defined wavelength bands 650, 820 to 870, and 1300 nm. The choice of wavelength is also determined by the transmission characteristics of optical fibers.

59.2 Optical Loss Mechanisms in Optical Fibers

There are two basic categories of sources of light loss in the fiber optic systems: extrinsic and intrinsic losses.

Extrinsic Fiber Losses

These losses are specific to geometry and handling of the fibers and are not functions of the fiber material itself. There are three basic types:

bending losses
launching losses
connector losses

Bending Losses

Bending losses are the result of distortion of the fiber from the ideal straight-line configuration. While the light is traveling inside the fiber, part of the wavefront on the outside of the bend must travel faster than the part of the smaller inner radius of the bend. Since this is not possible, a portion of the wave must be radiated away. Losses are greater for bends with smaller radii, particularly for kinks or microbends in a fiber.

An important cause of attenuation is due to microbending of the fiber. Microbending is due to irregularly distributed undulations in the fiber with radii of curvature of a few millimeters and deviations from the mean line of a few micrometers, as exemplified in Figure 59.3. Microbends arise from mechanical tensile forces by which the fiber is pressed against a rough surface. Although the effect of variations in diameter can be discussed at length by waveguide theory, here it will be sufficient to say that those components of the light that are traveling in the fiber near its acceptance limit cross outside this boundary and are lost from the fiber. These losses may be avoided by careful cable constructions, avoiding excessive mechanical forces, and controlling the temperature variations of the cable. This is achieved by a loose encasing of the fiber in a plastic sheath or by covering the fiber with soft flexible material, as shown in Figure 59.4.

Launching Losses

The term *launching loss* refers to an optical fiber not being able to propagate all the incoming light rays from an optical source. These occur during the process of coupling light into the fiber (e.g., losses at the interface stages). Rays launched outside the angle of acceptance excite only dissipative radiation modes in the fiber. In general, elaborate techniques have been developed to realize efficient coupling between the light source and the fiber, mainly achieved by means of condensers and focusing lenses. The focused input beam of light needs to be carefully matched by fiber parameters for good coupling.

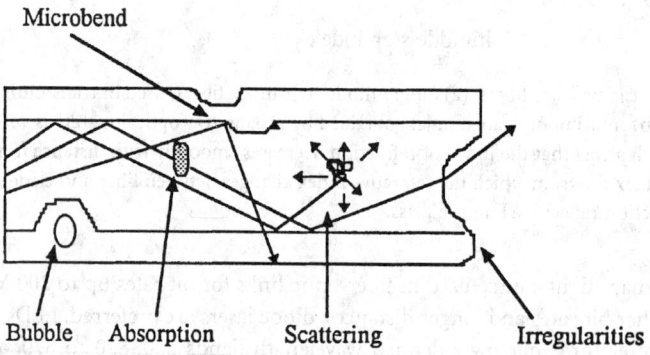

FIGURE 59.3 An important cause of attenuation is due to microbending of the fiber. This is due to irregularly distributed undulations in the fiber and from mechanical tensile forces when the fiber is pressed against a rough surface. Absorption losses are caused by the presence of impurities such as traces of metal ions (e.g., Cu^{2+}, Fe^{3+}) and hydroxyl (OH^-) ions. Despite careful manufacturing techniques, fibers can be inhomogeneous, having disordered, amorphous structures. Power losses due to scattering are caused by such imperfections in the core material and irregularities between the junction and cladding.

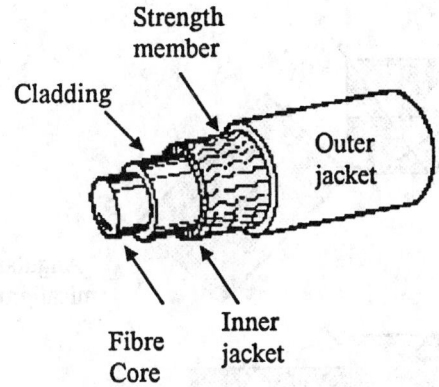

FIGURE 59.4 Some fiber losses can be avoided by careful cable construction, eliminating excessive mechanical forces, and controlling the temperature variations of the cable assembly. This is achieved by a loose encasing of the fiber in a plastic sheath or by covering the fiber with soft flexible materials. Most optical fibers constructed for communication purposes have inner and outer jackets for protection and strength.

Equally, once the light is transmitted through the fiber, output fiber characteristics must also match the output target characteristics to be able to couple the largest proportion of the transmitted light. This can be done by a suitable focusing lens arrangements in the output end.

There are also initial face (Fresnel) losses due to reflections at the entrance aperture. The Fresnel losses are greater if the fiber/source is air coupled. Hence, most optical couplings to a fiber utilize index matching materials, thus reducing coupling loss substantially.

Connector Losses

Connector losses are associated with the coupling of the output of one fiber with the input of another fiber, or couplings with detectors or other components. The significant losses may arise in fiber connectors and splices of the cores of the joined fibers having unequal diameters or misaligned centers, or if their axes are tilted. Mismatching of fiber diameters causes losses that can be approximated by $-10 \log(d/D)$. There are other connection losses such as offsets or tilts or air gaps between fibers, and poor surface finishes. Some of these are illustrated in Figure 59.5.

To take full advantage of fiber characteristics in transmission systems of very low intrinsic attenuation, the contribution of losses from other sources must also remain very small. The attenuation $a_s(d)$ due to coupling efficiency may be expressed as:

$$a_s(d) = -10 \text{ dB } \log\eta(d) \tag{59.1}$$

where $\eta(d)$ is the coupling coefficient.

In general, the positions and shapes of the fiber cores are controlled to tight manufacturing tolerances. Fibers having attenuations greater that 1 dB/km are rarely used in communication networks. Nevertheless, the attenuation of badly matched fibers may exceed 1 dB/km per connector or splice if they are badly handled during installation stages. A good coupling efficiency requires precise positioning of the fibers to center the cores. The simplest way to avoid connector losses is by splicing the two ends of the fibers permanently, either by gluing of by fusing at high temperatures.

Losses in gaps can be viewed as a type of Fresnel loss because existing air space introduces two media interfaces and their associated Fresnel reflection losses. In this case, there are two major losses to be considered. The first loss takes place in the inner surface of the transmitting fiber, and the second loss occurs due to reflections from the surface of the second fiber. One way of eliminating these losses is by introducing a coupler that matches the optical impedances of the two materials. This arrangement results in matched reflection coefficients, which is analogous to matching of impedances.

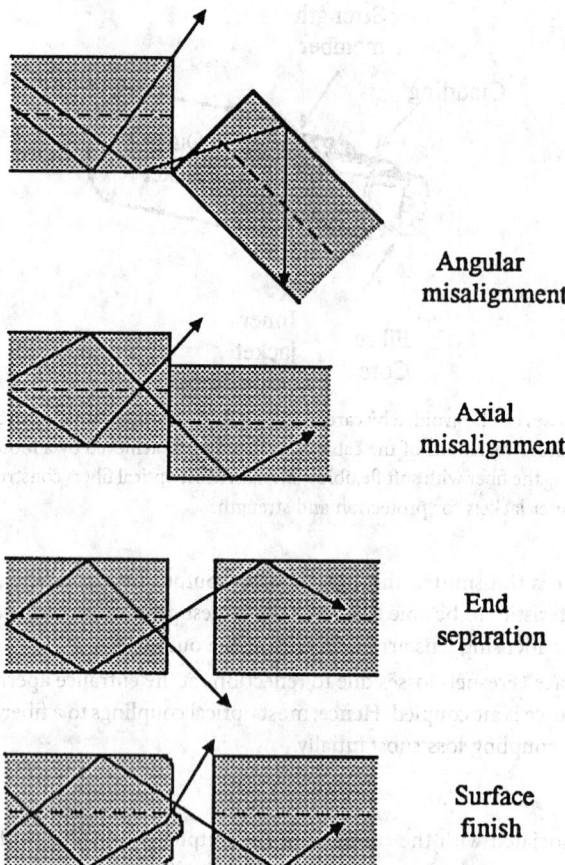

Angular misalignment

Axial misalignment

End separation

Surface finish

FIGURE 59.5 Connector losses are associated with the coupling of the output of one fiber with the input of another fiber or other components. Significant losses may arise in fiber connectors and splices of the cores of the joined fibers having unequal diameters, misaligned centers, tilted axes, and air gaps between fibers. In practical applications, fibers are permanently spliced by gluing or fusing at high temperatures.

Intrinsic Fiber Losses

Intrinsic fiber losses are those associated with the fiber optic material itself, and the total loss is proportional to length L. Once inside the fiber, light is attenuated primarily because of absorption and scattering; therefore, these are the primary causes of the losses.

Absorption Losses

As in the case of most transmissive systems, light loss through absorption in an optical fiber tends to be an exponential function of length. Absorption loss is caused by the presence of impurities such as traces of metal ions (e.g., Cu^{2+}, Fe^{3+}) and hydroxyl (OH^-) ions. Optical power is absorbed in the excitation of molecular vibrations of such impurities in the glass, as illustrated in Figure 59.3. One characteristic of absorption is that it occurs only in the vicinity of definite wavelengths corresponding to the natural oscillation frequencies or their harmonics of the particular material. In modern fibers, absorption losses are almost entirely due to OH^{-1} ions. The fundamental vibration mode of these ions corresponds to $\lambda = 2.73$ μm and the harmonics at 1.37 and 0.95 μm. It is possible to employ dehydration techniques during manufacturing to reduce presence of OH^{-1} ions.

Unlike scattering losses, which are relatively wideband effects, absorption losses due to each type of impurity act like a band-suppression filter, showing peak absorption at well defined wavelengths.

Scattering Losses

Despite the careful manufacturing techniques, most fibers are inhomogeneous that have disordered, amorphous structures. Power losses due to scattering are caused by such imperfections in the core material and irregularities between the junction and cladding as shown in Figure 59.3.

Inhomogeneities can be either structural or compositional in nature. In structural inhomogeneities, the basic molecular structure has random components, whereas, in compositional inhomogeneity, the chemical composition of the material varies. The net effect from either inhomogeneity is a fluctuation in the refractive index. As a rule of thumb, if the scale of these fluctuations is on the order of $\lambda/10$ or less, each irregularity acts as a scattering center. This is a form of Rayleigh scattering and is characterized by an effective absorption coefficient that is proportional to λ^{-4}. Rayleigh scattering can be caused by the existence of tiny dielectric inconsistencies in the glass. Because these perturbations are small with respect to the waves being propagated, light striking a Rayleigh imperfection scatters in all directions. Scattering losses are less at longer wavelengths, where the majority of the transmission losses are due to absorption from impurities such as ions. Rayleigh scattering losses are not localized, and they follow a distribution law throughout the fiber. However, they can be minimized by having low thermodynamic density fluctuations.

A small part of the scattered light may scatter backward, propagating in the opposite direction. This backscattering has important characteristics and may be used for measuring fiber properties. Usually, the inhomogeneities in the glass are smaller than the wavelength λ of the light. The scattering losses in glass fibers approximately follow the Raleigh scattering law; that is, they are very high for small wavelengths and decrease with increasing wavelength.

In general, optical losses in the glass cause the optical power in a fiber to fall off exponentially with the length L of the fiber,

$$P(L) = P(0) \ 10^{\ (-\alpha L/10 dB)} \tag{59.2}$$

where $P(0)$ = optical power that couples to the fiber, $P(L)$ = power remaining after length L, and α is the attenuation coefficient indicating the rate of loss of optical power in dB/km.

The product αL is called the *attenuation of the fiber*. An attenuation of 10 dB means that the optical power $P(L)$ at the end of the fiber is only 10% of the initial power $P(0)$. A 3-dB attenuation gives 50%, and 1 dB is about 80%.

A typical attenuation coefficient α against wavelength λ is shown in Figure 59.6 for common low-loss fused silica fiber. The optical losses for wavelengths below 1 µm are mainly due to Rayleigh scattering. At larger wavelengths absorption losses are important, notably at 1.4 µm through absorption by OH^{-1} ions. Above 1.6 µm, absorption due to impurities becomes dominant. Because of attenuations, only limited wavelength ranges are appropriate for optical data transmission.

Although intrinsic fiber losses can be associated with the core index n_f, the core index has an important role in determining the propagation time delay of optical signals. The propagation time delay t_p may be expressed by

$$t_p = n_f \ L/c \tag{59.3}$$

where c = velocity of light in the fiber, and L = fiber length.

Another type of loss in optical fibers occurs due to the propagation of light at different angles. Light propagating at shallow angles is called *low-order mode,* and light propagating at larger angles is called *high-order mode.* For a given length of fiber, the high-order modes reflect more often and cover longer distances than the low-order modes. Therefore, high-order modes suffer more losses, thus causing modal dispersions. The modal dispersion is one of the primary cause of rise time degradation for increasing fiber wavelengths. In addition, propagation time varies with index of refraction, so different wavelength components of the source spectrum have different travel times, thus causing *chromatic dispersion.*

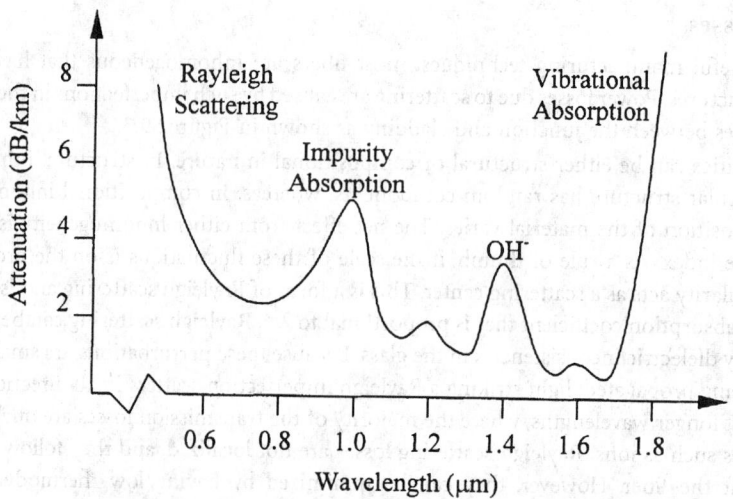

FIGURE 59.6 Attenuation characteristics of a typical optical fiber. The attenuation coefficient α varies with wavelength for all low-loss fused silica fibers due to Rayleigh scattering and impurity absorptions. The optical losses for wavelengths below 1 μm are mainly due to scattering. At greater wavelengths, absorption losses are important, notably at 1.4 μm through absorption by OH^{-1} ions. Above 1.6 μm absorption due to impurities becomes dominant.

59.3 Optical Time Domain Reflectometry Method

Optical time domain reflectometers (OTDRs) are used mainly for link testing. In this instrument, optical power is launched into the fiber, and the reflected power associated with Rayleigh scattering and other backscattering mechanisms are measured in the sending end, as shown in Figure 59.7. Manufacturers usually provide the instrument with customized analysis software and optical modules to be integrated into a computer. OTDR complements attenuation measurements by measuring the backscattering. This permits not only the attenuation of the complete fiber to be measured, but also different attenuation coefficients of fiber segments, as well as optical losses in connectors and splices. Furthermore, it indicates the location of such optical losses as well as the length of the fiber.

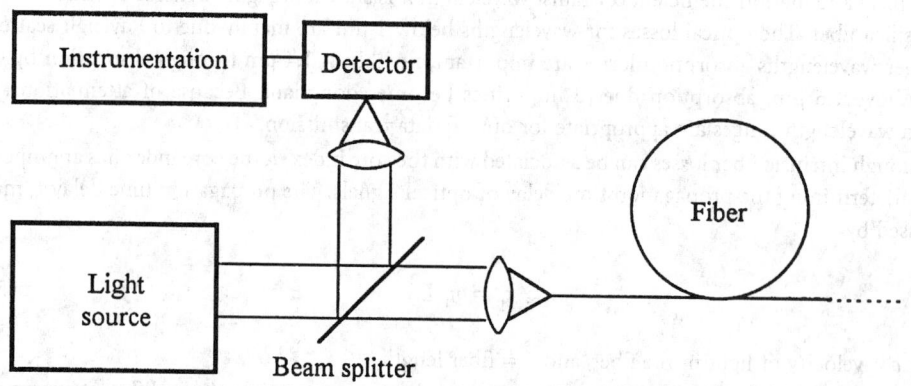

FIGURE 59.7 Optical time domain reflectometers (OTDRs) are used in link tests. The optical power is launched into the fiber, and the reflected power associated with Rayleigh scattering is measured from the same sending end. OTDRs usually are offered with customized analysis software and optical modules to be integrated into a computer. OTDR complements attenuation measurements by measuring the backscattering. This permits not only the attenuation of the complete fiber to be measured but also different attenuation coefficients of fiber segments, as well as optical losses in connectors and splices.

In OTDRs, as Figure 59.8 illustrates in block diagram form, short, intense laser pulses with duration $\Delta t = 10$ to 100 ns at peak power $M_{ax}(0)$ are coupled to the fiber. Backscattering echoes are detected from different regions of the fiber. In traversing the fiber, the power of these impulses decreases exponentially with the length of the fiber, predominantly due to scattering. A small portion of the scattered light reverses its direction and returns to the transmitter. The returned signal is uncoupled from the fiber by means of a beam splitter, to be processed further. The time history of the returned signals can be expressed as:

$$P_R(t) = K_R \, M_{ax}(0) \, \Delta t \, 10^{(-2\alpha L/10 \text{ dB})} \tag{59.4}$$

where K_R = the backscattering factor of the fiber.

The backscattering factor depends on scattering and numerical aperture. It is small, such that $P_R(t)$ is reduced as compared to $M_{ax}(0)$, typically by a factor of 50 to 60 dB. Despite this reduction, it is possible to measure the characteristics of fibers that are several kilometers long. In many cases, light pulses with high energy contents are used, along with sensitive receivers based on avalanche photodiodes.

The evolution with time of the backscattering signal $P_R(t)$ is given in Figure 59.9. If the attenuation coefficient and backscattering factors were constant throughout the length of the fiber, a curve that decreases exponentially from left to right would result. However, some power is reflected back from the end of the fiber because of some discontinuity. This appears as a sharp pulse at the right-hand side of the curve. In practical fibers, local optical losses as well as continuous losses occur due to imperfect connectors and splices. From the location and height of the steps, the position and magnitude of the local losses can easily be identified. The length of the fiber can also be obtained from timing of this pulse.

Other disturbances in the propagation of the light are also revealed in the backscattering signal. For example, variations in the attenuation coefficient of spliced fiber segments can be seen as slope changes in the $P_R(t)$ curve. In such cases, all information of importance can be drawn from the backscattering signal to enable the calculation of the attenuation and the local attenuation coefficients. A practical advantage of this method is that in measurements only one end of the fiber needs to be accessible; therefore, measurements can be done on optical cables that have already been laid.

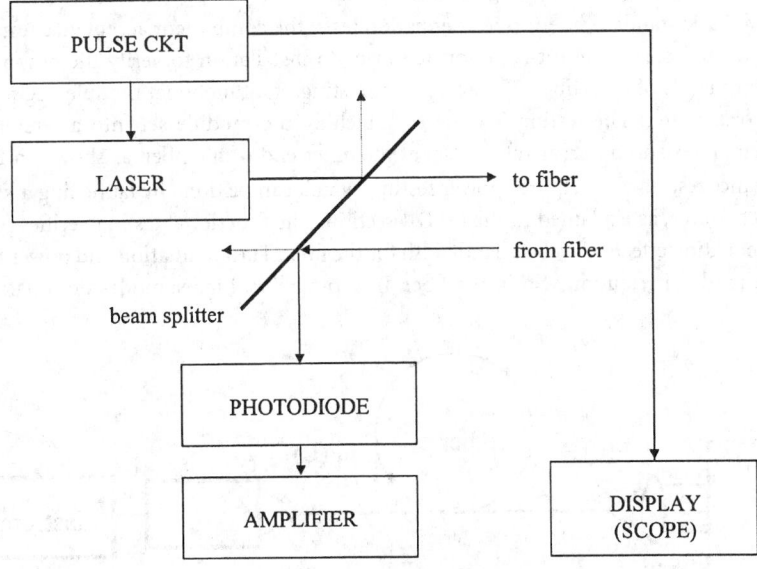

FIGURE 59.8 OTDRs usually come with a laser source, with modules that allow wavelength selection. The short, intense laser pulses with high peak power are coupled to the fiber. Backscattering echoes are detected from different regions of the fiber. The returned signal is uncoupled from the fiber by means of a beam splitter, to be processed further for analysis.

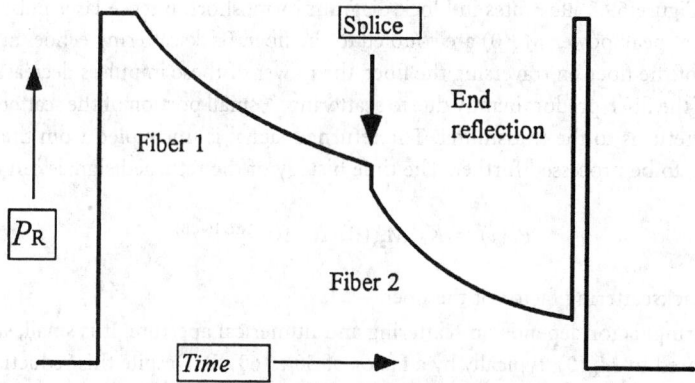

FIGURE 59.9 A typical example of the evolution of a backscattering signal in time. An exponential curve that decreases from left to right is obtained due to attenuation coefficients and backscattering factors. Some power reflects back from the end of the fiber and other discontinuities such as glass-air boundaries, appearing as sharp pulses. From the location and height of these returned sharp pulses, the positions and magnitudes of the local losses can easily be identified. The change of slope of curve also indicates the change in attenuation coefficient if different fibers are used.

59.4 Standard Field Fiber Optic Attenuation Test

In practical applications of optical fibers, it is necessary to have quantitative knowledge on the whole range of properties. Most important properties are length, attenuation, and bandwidth of the fibers, along with external diameter, core diameter, numerical aperture, and refractive index profile. The actual performance of installed links may be different from the desired performance. Therefore, it is important that both individual components and the entire assembled system undergo testing to verify compliance with the required operations. Additional testing may be conducted over the life of the system to ensure continued functional operation over long periods of time.

There are two basic standard fiber optic attenuation tests: the component acceptance test and the link test. The component acceptance test is performed prior to installation to verify the power and performance acceptance levels of each fiber. The acceptance testing of a functional module begins with power testing of the transmitter. The testing is done by attaching the module set into a reference link and verifying the data rates and bit error rates taken at the other end of the fiber, as shown in Figure 59.10.

Fiber acceptance testing also requires power testing, which can be done by launching a known power from a reference source, as explained in the OTDR section. The functional testing verifies power budget of coupling attenuation effects as well as bandwidth for the fiber. The attenuation and power transmission depend on the mode distribution within the fiber. In short fibers, higher modes dominate, whereas in

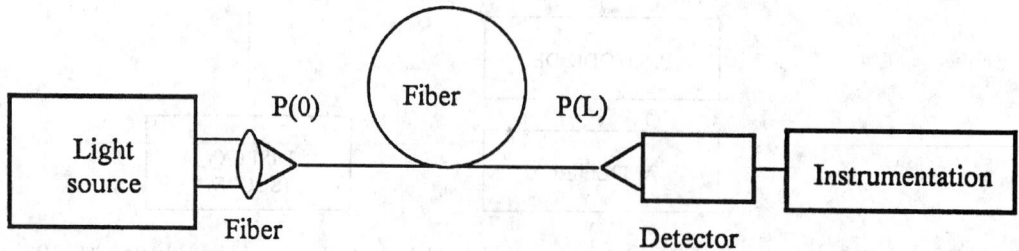

FIGURE 59.10 The measurements of transmission properties of laid optical fibers are obtained by end-to-end tests. The testing is done by attaching the module set into a reference link and verifying the data rates and bit error rates taken at the other end of the fiber.

long fibers, power is more concentrated in lower order modes. Mode stripping is often used by introducing small bends into the fiber to correlate the performance of tested fiber to actual in-service performance.

In addition to acceptance tests, fibers undergo other tests such as pull strength, breaking, humidity resistance, prolonged tension, and bend tests. Connectors, splitters, combiners, and fiber amplifiers need to be tested frequently to minimize power transmission losses.

In almost all fiber optics, low noise and very sensitive photodetectors are used in measurements. Semiconductor photodiodes made from silicon are suitable for measurements with wavelengths below 1 μm. For larger wavelengths, other detectors are used, such as those made of germanium.

Semiconductor photodiodes work via the internal photoelectric effect; that is, they absorb photons of energy $h\nu$ containing the incoming light beam power P and emit a number of electrons proportional to the number of photons, creating a current:

$$i_p = P\eta e/h\nu \tag{59.5}$$

where h = Planck's constant, ν = frequency of the absorbed light ($\nu = c/\lambda$), η = a constant of proportionality or quantum efficiency ($\eta < 1$), and e = the charge of an electron.

Photodetectors are used to measure the attenuation in fibers by measuring the optical power in the input $P(0)$ and power $P(L)$ at the end of the of the fiber. From Equation 59.1, the attenuation of the fiber can be calculated as:

$$a_0 = \alpha L = -10 \log P(L)/P(0) \tag{59.6}$$

When photodetectors are used Equation 59.4 can be written as

$$a_0 = -10 \log i_p(L)/i_p(0) \tag{59.7}$$

It is worth noting that optical power is proportional to current i_p, not to i_p^2.

Light-emitting diodes (LEDs) or incandescent halogen lamps are suitable for use as light sources for attenuation measurements. If LEDs are used, several interchangeable ones are needed, suitable to different wavelengths to make measurements. On the other hand, halogen lamps yield a wide spectrum of radiation, from which light of various wavelengths can be filtered with a monochromator that uses optical filters, prisms, or diffraction gratings. With an arrangement of this kind, the attenuation of the fiber can be measured as a function of the wavelength of the light.

To measure small optical powers precisely for the purpose of determining attenuation, the photodetector is usually connected to a selective amplifier, and the light source is modulated at low frequency (e.g., 400 Hz) using a rotating disk called a *chopper*. The selective amplifier amplifies only the similarly modulated components of the photocurrent, so that detector noise and the influence of background light are suppressed. For this purpose, frequency-selective level meters or lock-in amplifiers are often used. In both cases, the frequency and phase of the modulation are fed back in the form of reference voltage.

59.5 Out-of-Plane Scattering and Polarization Methods

The art of scatter measurements has long been evolved to established forms within the optics industry. Scatter methods provide extremely sensitive measurements in many diverse applications. There are two basic instruments developed for this purpose; the scatterometer and the polarimeter.

The basic setup for a scatterometer is given in Figure 59.11. Measured scatter is a good indicator of surface quality as well as discrete defects. However, the scattered signals are generally small compared to the specular beams, and they can vary by several orders of magnitude in just a few degrees. Therefore, scatter measurements need sophisticated instrumentation and the signal processing is more complex than many other optical techniques. Usually, for successful applications, the system specifications and measurements must be given in terms of accepted, well defined quantities. Scatter methods are used

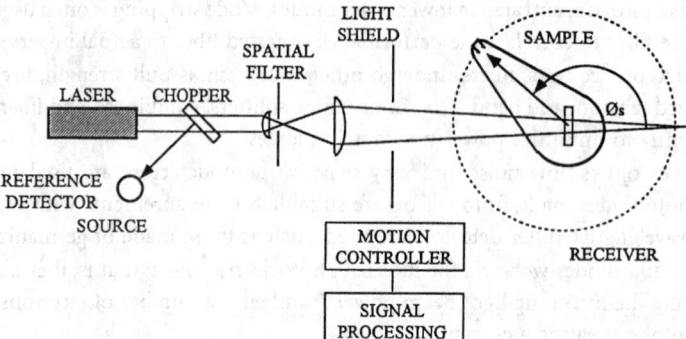

FIGURE 59.11 A typical scatterometer. In this particular type, the source is fixed in position, and the sample is rotated to the required incident angle. The receiver is rotated about the sample in the plane of incidence. In other types, the source and receiver may be fixed, and the sample is rotated so that the scatter pattern moves past the receiver. Scatterometers are offered with full supporting software.

routinely as a quality check on optical components in fiber-optic applications. Conversion of surface scatter data to other required formats, such as surface roughness, is common practice. Out-of-plane measurements and polarization-sensitive measurements are areas currently experiencing rapid advancements.

Another instrument in common use is the polarimeter, which senses the polarization of scattered light. The scattering characteristics of a sample are generally described by its bidirectional reflectance distribution function. The reflectance distribution function is the ratio of scattered flux in a particular direction to the flux of an incident beam. The scattered light is often a sensitive indicator of surface conditions. A small amount of surface roughness may reduce the specular power by less than 1% while increasing the scattered power by orders of magnitude. The retardance, attenuation, and depolarization of scattered light similarly provide sensitive indicators of conditions, such as uniformity of refractive index, orientation of surface defects, texture, strain, subsurface damage, coating microstructure, and the degree of multiscattering. Among many other methods, the use of prisms helps polarization and depolarization of the scattered light.

Instrument manufacturers are listed in Table 59.1, along with contact information.

TABLE 59.1 List of Manufacturers

AXSYS Communications	**Chiu Technical Corp.**
P.O. Box 571	252 Indian Head Rd.
Danielson, CT 06239-0571	Kingspark, NY 11754
Tel: (203) 774-4102	Tel: (516) 544-0606
Fax: (203) 774-4783	Fax: (516) 544-0809
Cuda Products Corp.	**Fiberoptic Technology Inc.**
6000-T Power Ave.	13 Fiber Rd.
Jacksonville, FL 32217-2279	Pomfret, CT 06258
Tel: (904) 737-7611	Tel: (800) 433-5248, (203) 928-0443
Fax: (904) 733-4832	Fax: (800) 543-2558, (203) 928-7664
Fiber Options	**INCOM Inc.**
80-T Orville Dr.	P.O. Drawer G.
Bohemia, NY 11716-2533	Southbridge, MA 01550-0528
Tel: (800) 739-9105, (516) 567-8320	Tel: (508) 765-9151
Fax: (516) 567-8322	Fax: (508) 765-0041
PHILTEC Inc.	**VICON Fiber Optics Corp.**
P.O. Box 359	90 Secor Lane
Arnold, MD 21012	Pelham Manor, NY 10803
Tel: (410) 757-4404	Tel: (800) 828-2071, (914) 738-5006
Fax: (410) 757-8138	Fax: (914) 738-6920

Additional Information

M. Bass, *Handbook of Optics-Fundamentals, Techniques and Design*, 2d ed., Vols. I and II, 1995, New York: McGraw-Hill.

Optical Fiber Communication Conference Proceedings, OFC/IOOC'93, Washington, DC: Optical Society of America.

L. D. Green, *Fiber Optic Communication*, 1993, Boca Raton, FL: CRC Press.

M. W. Burke, *Image Acquisition-Handbook of Machine Vision Engineering*, Vol. 1, 1996, Oxford, U.K.: Chapman & Hall.

Additional Information

N. Bass, ed., *Handbook of Optics: Fundamentals, Techniques, and Design*, 2d ed., Vols. I and II, 1995, New York, McGraw-Hill.

Optical Fiber Communication Conference, OFC/IOOC'9..., Washington, DC, Optical Society of America.

J. Gowar, *Fiber Optic Communications*, 1993, Boca Raton, Fla., CRC Press.

M. Bass, P. Enoch, eds., *Handbook of Medicine Vision Engineering*, Vol. 1, 1998, Oxford, U.K., Chapman & Hall.

60

Polarization Measurement

Soe-Mie F. Nee

U.S. Naval Air Warfare Center

60.1 Basic Concepts of Polarization

Polarization of *light* is a property of electromagnetic (EM) waves, which include heat, microwaves, radio waves, and x-rays. An EM wave has orthogonal electric and magnetic fields associated with it which vibrate in directions perpendicular to the direction of propagation. The *electric field* of a sinusoidal EM wave, in particular, can always be decomposed into two orthogonal components; each component has an *amplitude* and a *phase*. The amplitude is the maximum value of the field component, and the light *intensity* is proportional to the square of the amplitude. The phase, referred to a fixed position or time, tells what part of the cycle the electric field is vibrating in. G. G. Stokes pointed out in 1852 that these two orthogonal components do not interfere in amplitude but are additive according to vector algebra [1]. When the two orthogonal components are in phase, the EM wave is *linearly polarized*. When the two orthogonal components have the same amplitude and a relative phase of 90°, the EM wave is *circularly polarized*. In general, an EM wave has arbitrary amplitudes and phases for the two orthogonal fields and is elliptically polarized. The concept of polarization ellipse and the descriptions for the polarization of an EM wave in terms of Jones vectors and Stokes vectors are given in the subsection, "Polarization of an EM Wave," and also in References 1 through 9.

    Light is composed of an ensemble of EM waves. A group of EM waves traveling in the same direction can have some linearly polarized waves, some circularly polarized waves, and some elliptically polarized

0-8493-8347-1/99/$0.00+$.50
© 1999 by CRC Press LLC

waves. When they are combined, resulting light can be unpolarized, partially linearly polarized, or partially elliptically polarized. Unpolarized light occurs when there are no fixed directions of the electric field and also no fixed phase relations between the two orthogonal field components. In general, light is partially polarized and can be decomposed into unpolarized light and elliptically polarized light. These concepts are described in terms of Stokes vector in the subsection, "Polarization of Light," and also in References 1 through 6.

Polarized light can be produced by passing light through a polarizer. An ideal polarizer transmits only light whose electric field is parallel to the transmission axis of the polarizer and rejects light with the orthogonal field. Polarization of light can be observed by stacking two polarizers together and turning one with respect to the other. The transmitted light intensity through these two polarizers will vary, and at some particular positions it will vanish. In this case, light is linearly polarized after passing through the first polarizer. When the second polarizer is turned until its axis is perpendicular to the axis of the first polarizer, light cannot pass through the second polarizer. The transmitted intensity varies according to the square of the cosine of the angle between the two polarizers [5, 9–12]. Real polarizers are not perfect and transmit light with minimum intensity I_{min} when the polarizer axis is perpendicular to the polarization of purely linearly polarized incident light. This is caused by the small depolarization of a polarizer [12]. Depolarization is a mechanism that turns polarized light into unpolarized light and is the opposite effect of polarization. The maximum transmitted intensity I_{max} occurs when the polarizer axis is parallel to the incident polarization direction. The extinction ratio of a polarizer is defined as I_{min}/I_{max}. Other relations for polarizers can be found in References 9 through 12.

Besides the polarizer, another basic element in polarization measurements is the phase retarder or wave plate. A phase retarder changes the relative phase between the two orthogonal fields of an EM wave [4–11]. The change of relative phase between the two orthogonal components is called the *phase retardation* or *retardance*. The retardance of a quarter-wave retarder is 90°, and that of a half-wave retarder is 180°. Circularly polarized light can be generated by passing linearly polarized light through a quarter-wave plate whose axis is at 45° with respect to the incident linear polarization direction. A half-wave plate may change the polarization direction of linearly polarized light. In general, a phase retarder changes linearly polarized light into elliptically polarized light. Representations of the optical response of polarizers, retarders, and other materials in terms of the Müeller matrix and Jones matrix are given in the subsection "Polarization by the Response of a Medium" and also in References 4 through 8 and 12 through 24.

Polarization is generated by the anisotropic response of materials and/or anisotropic geometry of systems. The mechanisms for producing polarization include preferential absorption in a dichroic material, reflection and transmission at oblique incidence, double refraction in a birefringent material, diffraction by grating or wires, and scattering by particles [2–11, 13]. These properties can be utilized to make polarizers and phase retarders. For example, a wire-grid polarizer is made of parallel fine conducting wires. When light is incident on a wire-grid polarizer with the grid period smaller than the wavelength, the electric field parallel to the wires is shorted and absorbed so that only the electric field perpendicular to the wires passes through the polarizer. In a dichroic polarizer, anisotropic molecules are aligned in a preferential direction so that absorption is very different for the two orthogonal directions referred to the alignment direction. The nonpreferential field is absorbed by the molecules in the medium, while the preferential field passes through the medium [5, 8]. In a prism polarizer, the two orthogonal fields are separated by double refraction in a birefringent crystal, and the unwanted polarization is deflected away by the special geometry of a prism. A material is birefringent when it has different refractive indices for different field directions. When a light beam passes through a birefringent slab, a phase retardation is generated. Birefringent slabs can be used to make phase retarders or wave plates. Different kinds of polarizers and retarders are described in detail in References 5 and 9 through 11.

When special arrangements of polarizers and phase retarders are combined with a light source and a detector, polarized light can be generated and analyzed. Such an optical system is called a *polarimeter* or an *ellipsometer*. The subsection "Principles of Polarimetry" discusses the generation and analysis of polarized light and the operational principles for Polarizer-Sample-Analyzer (PSA) ellipsometry and

Polarizer-Compensator-Sample-Analyzer (PCSA) ellipsometry using the intensity approach associated with Stokes vectors and Müeller matrices [4, 17–23, 25, 26]. A phase retarder is also called a *compensator* because it was introduced into a polarimeter to compensate the phase change by a sample. The intensity approach was chosen because intensity, but not electric field, is measured in most experiments, and also because the electric field approach cannot treat depolarization. However, the electric field approach is convenient to use for highly polarized light when depolarization does not cause appreciable errors in the measurement. Discussion of ellipsometry using the electric field approach can be found in References 4, 15, 16, and 27 through 33].

Polarization effects are widely applied in modern optical technologies. The electro-optic modulator and shutter are based on tunable birefringence by applying a high voltage across a birefringent crystal to modulate the phase of transmitted light and hence to achieve intensity modulation [4, 7, 9, 34, 35]. Liquid crystal displays use similar principles [36]. Birefringence can also be modulated by the photo-elastic effect [37, 38]. The magneto-optical readout for laser disks utilizes the magneto-optical Kerr effect that generates phase retardation upon reflection from magnetic materials [34, 39]. Other applications of polarization are fiber optics, nonlinear optics, material characterization, medical optics, and many other fields. All of these applications utilize the anisotropic nature of materials or the anisotropic geometry of systems. This chapter is concerned with the application of polarization on material characterization. In this application, a polarimeter or ellipsometer is used to measure optical properties and surface properties of materials and thin films [40–48, see also Chapter 61, "Refractive Index"]. In the subsection "Polarization Instrumentation and Experiments," different components of polarimeters are discussed using an example of an automated reflection null ellipsometer, and two sample experiments are described to measure birefringence of a birefringent slab and the optical constants of a material.

60.2 Polarization of an Electromagnetic Wave

The electric field $E(z, t)$ of a monochromatic EM wave propagating along the z-direction with a frequency ω and an angular wave-number k can be decomposed into two orthogonal components E_x and E_y, and represented by

$$\mathbf{E}(z, t) = \hat{x}E_x(z, t) + \hat{y}E_y(z, t)$$
$$E_x(z, t) = a_x e^{i(\omega t - kz + \delta_x)}$$
$$E_y(z, t) = a_y e^{i(\omega t - kz + \delta_y)}$$

(60.1)

where a_x and δ_x are the amplitude and phase, respectively, for E_x, and a_y and δ_y are for E_y [1–4]. k is related to wavelength λ by $k = 2\pi/\lambda$. In vacuum, $k = \omega/c$. Let $\delta = \delta_y - \delta_x$ be the relative phase between E_y and E_x. Then Equation 60.1 can be simplified to

$$\mathbf{E}(z, t) = (\hat{x}\, a_x + \hat{y}\, a_y\, e^{i\delta})e^{i(\omega t - kz + \delta_x)}$$

(60.2)

Polarization Ellipse

It is often convenient to express **E** in terms of a complex variable. The observed field is actually the real part of **E**. The projection of $\text{Re}[\mathbf{E}(z, t)]$ with $\delta_x = 0$ onto the xy-plane at $z = 0$ is given by

$$\mathbf{E}(0, t) = \hat{x}\, a_x \cos\omega t + \hat{y}\, a_y \cos(\omega t + \delta)$$

(60.3)

The loci of $\mathbf{E}(0, t)$ with $a_x = 3$, $a_y = 2$ and different values of δ are shown in Figure 60.1. For $\delta = 0$, the locus of **E** is a line with a slope of a_y/a_x. The EM wave is linearly polarized when E_x and E_y are in phase with each other. In other cases, the loci are ellipses, which are called *polarization ellipses*, and the EM wave is elliptically polarized. The instantaneous electric field can be visualized by drawing an arrow from the origin to a point on an ellipse. The electric field direction rotates in the clockwise direction for $0 < \delta < 180°$, while in the counter clockwise direction for $-180 < \delta < 0°$. When $\delta = \pm90°$, the axes of the ellipse correspond to the x and y coordinate axes. If $a_x = a_y$, this ellipse then becomes a circle, and the EM wave is circularly polarized. The convention in ellipsometry defines the right-handed circularly polarized wave as the one whose field rotates in the clockwise direction with $\delta = 90°$ [4, 5]. A left-handed circularly polarized wave thus corresponds to $\delta = -90°$ for a counterclockwise rotating electric field.

In Figure 60.1, a polarization ellipse is specified by a set of three parameters: a_x, a_y, and δ. The ellipse can also be specified by the other set of three parameters: the major axis a, the minor axis b, and the orientation angle ϕ of the major axis measured from the x-axis. Figure 60.2 shows the geometry of an ellipse with these parameters. Parameters a and b can also be expressed in terms of the ellipticity e and ellipticity angle ε, defined by $e = b/a = \tan\varepsilon$. For linear polarization, $\delta = 0$, $b = 0 = e = \varepsilon$, and $\tan\phi = a_y/a_x$. For $\phi = 0$, the major and minor axes of the ellipse always correspond to the coordinate axes. For circular polarization, $a = b = a_x = a_y$ and $\delta = \pm90°$. Right-handed circularly polarized light has a positive ellipticity with $e = 1$ and $\varepsilon = 45°$, and left-handed circularly polarized light has a negative ellipticity with $e = -1$ and $\varepsilon = -45°$. In general, ϕ represents the orientation of the ellipse, and ε indicates the shape of the ellipse and the direction of field rotation. References 2 through 7 give more details about this subject.

Jones Vector and Stokes Vector

The electric field expressed in vector form in Equation 60.1 can also be expressed as a column matrix. A polarization ellipse depends on a_x, a_y, δ_x, and δ_y, but not on k and ω. By neglecting the common factor of $e^{i(\omega t - kz)}$ in both E_x and E_y, the Jones vector is defined as [4, 7, 8]

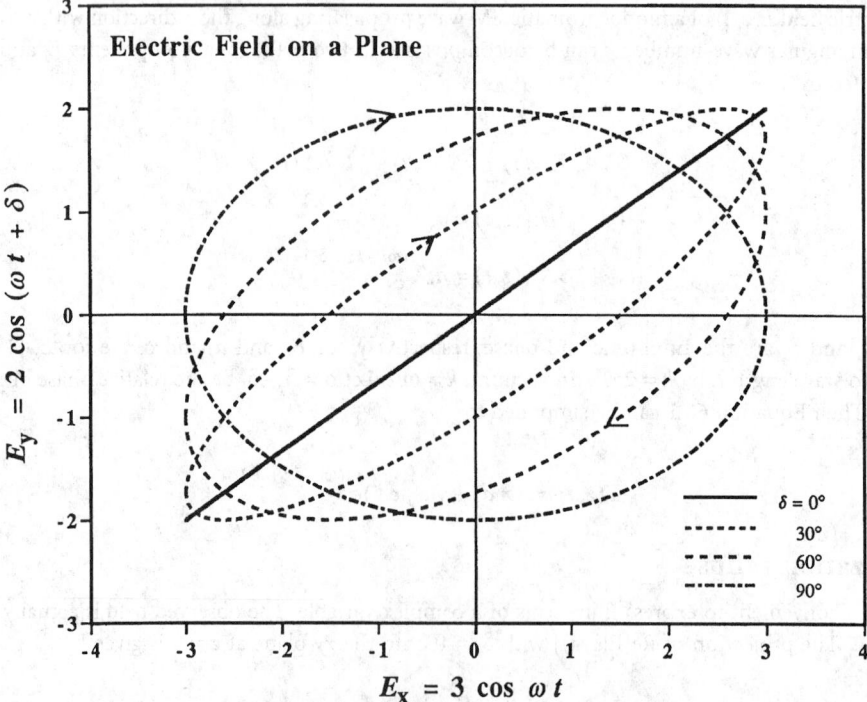

FIGURE 60.1 Projection of the electric field of an EM wave with amplitudes $a_x = 3$, $a_y = 2$ and different values of phase retardation δ onto the $z = 0$ plane. Most of these loci are ellipses and reduce to lines or circles in special cases. The electric field changes in the clockwise direction for δ between 0 and 180°.

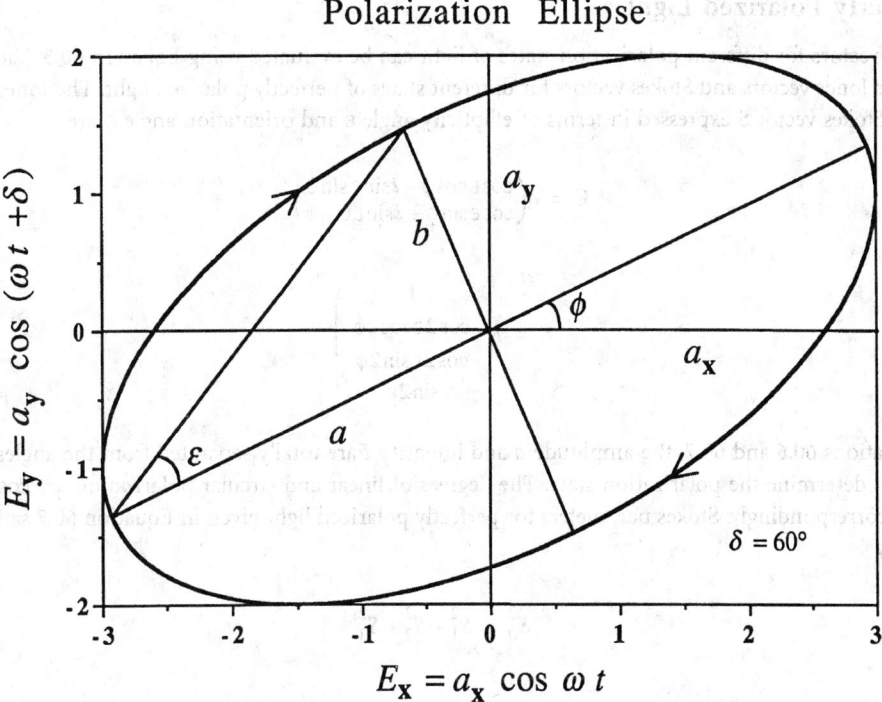

FIGURE 60.2 Characteristic parameters for a polarization ellipse. a_x and a_y are the field amplitudes in the x- and y-directions, and δ is the phase retardance; a and b are the major and minor axes of the ellipse, ϕ is the orientation of the major axis with respect to the x-axis, and ε is the ellipticity angle which is equal to $\tan^{-1} (b/a)$. A polarization ellipse can be characterized by (a_x, a_y, δ), (a, b, ϕ) or (I, ϕ, ε), where I is the intensity of the EM wave.

Both elements of a Jones vector are complex numbers. Jones algebra is convenient for describing perfectly polarized light. Since a light sensor measures only intensity but not electric field in most cases, the Stokes vector is more convenient to use in metrology. The Stokes parameters are four intensity-based parameters used to describe the polarization state of light, represented by S_0, S_1, S_2, S_3, or by I, Q, U, V [1–6, 14]. The Stokes vector is the set of these Stokes parameters, defined as [4, 14]

$$\mathbf{S} = \begin{pmatrix} S_0 \\ S_1 \\ S_2 \\ S_3 \end{pmatrix} \equiv \begin{pmatrix} \langle E_x E_x^* \rangle + \langle E_y E_y^* \rangle \\ \langle E_x E_x^* \rangle - \langle E_y E_y^* \rangle \\ \langle E_x E_y^* \rangle + \langle E_y E_x^* \rangle \\ i \langle E_x E_y^* \rangle + i \langle E_y E_x^* \rangle \end{pmatrix} \tag{60.5}$$

For an EM wave, the average bracket in Equation 60.5 represents the time average. $\langle E_x E_x^* \rangle = I_x$ is the intensity of the component of light linearly polarized in the x-direction. Similarly, $\langle E_y E_y^* \rangle = I_y$. All of the Stokes parameters are real numbers and are measurable. For an ensemble of EM waves, the average brackets represent both time and ensemble averages.

Perfectly Polarized Light

Stokes vectors for different polarization states of light can be evaluated using Equation 60.5. Table 60.1 lists the Jones vectors and Stokes vectors for different states of perfectly polarized light. The Jones vector **E** and Stokes vector **S** expressed in terms of ellipticity angle ε and orientation angle ϕ are

$$\mathbf{E} = a \begin{pmatrix} \cos\varepsilon\cos\phi - i\sin\varepsilon\sin\phi \\ \cos\varepsilon\sin\phi + i\sin\varepsilon\cos\phi \end{pmatrix} \tag{60.6}$$

$$\mathbf{S} = I \begin{pmatrix} 1 \\ \cos 2\varepsilon\cos 2\phi \\ \cos 2\varepsilon\sin 2\phi \\ \sin 2\varepsilon \end{pmatrix} \tag{60.7}$$

In Equations 60.6 and 60.7, the amplitude a and intensity I are totally separated from the angles ϕ and ε which determine the polarization state. The degrees of linear and circular polarization are $\cos 2\varepsilon$ and $\sin 2\varepsilon$, correspondingly. Stokes parameters for perfectly polarized light given in Equation 60.7 satisfy the identity

$$S_0^2 = S_1^2 + S_2^2 + S_3^2 \tag{60.8}$$

60.3 Polarization of Light

Light is composed of an ensemble of EM waves. A single EM wave has a certain electric field direction and phase. Unpolarized light can be visualized as an ensemble of EM waves with random field directions and phases. The field direction and phase for unpolarized light can not be defined then. The description of light in terms of Jones vector is therefore inadequate to describe the polarization of unpolarized light. For an ensemble of many EM waves, the electric field components in the Stokes vector given by Equation 60.5 is the sum of the corresponding components for all waves. In particular, for an ensemble of incoherent EM waves, the Stokes vectors for individual waves are additive:

$$\mathbf{S} = \sum_{i=1}^{N} \mathbf{S_i} \tag{60.9}$$

TABLE 60.1 Jones Vectors and Stokes Vectors for Different Polarization States for Perfectly Polarized Light

Polarization	Linear	Linear	Linear	Circular	Elliptical
Direction	0°	90°	±45°	right/left	
Phase	0°	0°	0°	±90°	δ
Jones vector	$\begin{pmatrix} 1 \\ 0 \end{pmatrix}$	$\begin{pmatrix} 0 \\ 1 \end{pmatrix}$	$\frac{1}{\sqrt{2}}\begin{pmatrix} 1 \\ \pm 1 \end{pmatrix}$	$\frac{1}{\sqrt{2}}\begin{pmatrix} 1 \\ \pm i \end{pmatrix}$	$\begin{pmatrix} a_x \\ a_y e^{i\delta} \end{pmatrix}$
Stokes vector	$\begin{pmatrix} 1 \\ 1 \\ 0 \\ 0 \end{pmatrix}$	$\begin{pmatrix} 1 \\ -1 \\ 0 \\ 0 \end{pmatrix}$	$\begin{pmatrix} 1 \\ 0 \\ \pm 1 \\ 0 \end{pmatrix}$	$\begin{pmatrix} 1 \\ 0 \\ 0 \\ \pm 1 \end{pmatrix}$	$\begin{pmatrix} a_x^2 + a_y^2 \\ a_x^2 - a_y^2 \\ 2a_x a_y \cos\delta \\ 2a_x a_y \sin\delta \end{pmatrix}$

For an ensemble of EM waves with identical $\phi_i = \phi$ and $\varepsilon_i = \varepsilon$, the resultant polarization is still the same as the individual wave, as indicated by Equations 60.6 and 60.7, regardless of whether these waves are coherent.

Unpolarized and Partially Polarized Light

If an ensemble consists of randomly oriented linearly polarized waves, all $\varepsilon_i = 0$ and ϕ_i are random, then $S_1 = S_2 = S_3 = 0$, according to Equations 60.7 and 60.9. Light is thus unpolarized, and $S = I(1, 0, 0, 0)$. If an ensemble consists of elliptically polarized waves with the same orientation $\phi_i = \phi$ or $\phi + \pi$ and perfectly random ellipticity angle ε_i, then the Stokes vector is $I(1, 0, 0, 0)$, and light is also unpolarized. Thus, the Stokes vector S in Equation 60.9 already implies the sense of the ensemble average of polarization. The average brackets in Eq. (5) can be considered as both the time average and ensemble average for incoherent waves.

For unpolarized light with $S_1 = S_2 = S_3 = 0$, Equation 60.8 does not hold. In general, light is partially polarized, i.e., part of it is perfectly polarized and the rest is unpolarized [1–6]. Stokes parameters for arbitrary polarizations satisfy

$$S_0^2 \geq S_1^2 + S_2^2 + S_3^2 \tag{60.10}$$

The degree of polarization is given by

$$\mathcal{P} = \frac{\sqrt{S_1^2 + S_2^2 + S_3^2}}{S_0} \tag{60.11}$$

Perfectly polarized light has $\mathcal{P} = 1$, and unpolarized light has $\mathcal{P} = 0$. Partially polarized light has $0 < \mathcal{P} < 1$. The intensity of the polarized part is $\mathcal{P}I$, and the intensity of the unpolarized part is $I(1 - \mathcal{P})$. By the superposition concept, the Stokes vector for partially polarized light can be obtained from Equation 60.7 as

$$S = I \begin{pmatrix} 1 \\ \mathcal{P}\cos 2\varepsilon \cos 2\phi \\ \mathcal{P}\cos 2\varepsilon \sin 2\phi \\ \mathcal{P}\sin 2\varepsilon \end{pmatrix} \tag{60.12}$$

The degree of linear polarization $\mathcal{P}_L$ and circular polarization $\mathcal{P}_C$ are

$$\begin{cases} \mathcal{P}_L = \mathcal{P}\cos 2\varepsilon = \dfrac{\sqrt{S_1^2 + S_2^2}}{S_0} \\[3mm] \mathcal{P}_C = \mathcal{P}\sin 2\varepsilon = \dfrac{S_3}{S_0} \end{cases} \tag{60.13}$$

Polarization by the Response of a Medium

Jones Matrix

To measure polarization, light must interact with a medium to give a response. The response of a polarizer is to pass one polarization and reject the orthogonal one. The response of a phase retarder is to change the relative phase between the two polarizations. A medium can be any optical component, a test sample or any object under investigation. The response of a medium relates the state of output light to the state of incident light. The polarization state of light can be described by a complex vector EM field, a Jones vector, or a Stokes vector [4, 5, 7, 8]. Let an incident EM wave be specified by a complex field or Jones vector (E_x, E_y), and the output field be (E_x', E_y'), the general relations between these two fields are

$$\mathbf{E}' = \begin{pmatrix} E_x' \\ E_y' \end{pmatrix} = \begin{pmatrix} r_{xx} & r_{xy} \\ r_{yx} & r_{yy} \end{pmatrix} \begin{pmatrix} E_x \\ E_y \end{pmatrix} = \mathbf{J}\,\mathbf{E} \tag{60.14}$$

$\mathbf{J}$ in Equation 60.14 is a 2×2 matrix that relates the input Jones vector $\mathbf{E}$ to the output Jones vector $\mathbf{E}'$ and is called the Jones matrix. The response of a medium is characterized by the elements of the Jones matrix, r_{xx}, r_{xy}, r_{yx}, and r_{yy}, which are all complex numbers.

Principal Coordinate System

Since the directions of an electric field are different in different rotated coordinate systems, $\{r_{ij}\}$ are not unique. For many symmetric media, there exists a coordinate system in which r_{xy} and r_{yx} are zero, and r_{xx} and r_{yy} are called the eigenvalues for $\{r_{ij}\}$. This is the principal coordinate system or principal frame whose x- and y-axes are the two principal axes. Finding the principal frame is an eigenvalue problem. If the incident polarization is along one of the principal-axis $\hat{x}$, then the output polarization is still along $\hat{x}$. In the principal frame, $\{r_{ij}\}$ are called the coefficients of response. For example, $\{r_{ij}\}$ may represent the reflection coefficients for a reflection response, or the scattering coefficients for a scattering response, etc. In the principal coordinate system, Equation 60.14 is simplified to

$$\mathbf{E}' = \begin{pmatrix} E_x' \\ E_y' \end{pmatrix} = \begin{pmatrix} r_{xx} & 0 \\ 0 & r_{yy} \end{pmatrix} \begin{pmatrix} E_x \\ E_y \end{pmatrix} = \mathbf{J}(0)\mathbf{E} \tag{60.15}$$

$\mathbf{J}(0)$ is the diagonalized Jones matrix in the principal frame. For a polarizer, the principal axes are the transmission and extinction axes. The former is assigned to the x-axis. For a phase retarder, the principal axes are the fast- and slow-axes. The phase change for the EM wave with its field along the fast-axis is larger than the slow-axis. The fast-axis is usually assigned to the x-axis. The principal Jones matrices $\mathbf{p}$ for a perfect polarizer, and $\mathbf{c}$ for a perfect wave plate with a retardance τ are [4, 5, 7, 8]

$$\mathbf{p} = \begin{pmatrix} 1 & 0 \\ 0 & 0 \end{pmatrix} \qquad \mathbf{c} = \begin{pmatrix} e^{i\tau/2} & 0 \\ 0 & e^{-i\tau/2} \end{pmatrix} \tag{60.16}$$

Ellipsometric Parameters

For reflection from a surface, the two principal axes are the s- and p-polarizations. The s-polarization field is along the y-axis which is chosen to be perpendicular to the plane of incidence. The p-polarization field is along the x-axis which is in the plane of incidence. The complex reflection coefficients for these two polarizations are designated as $r_p = r_{xx}$ and $r_s = r_{yy}$. The ellipsometric parameters ψ and Δ are defined by [4, 15, 16]

$$\frac{r_p}{r_s} = \frac{r_{xx}}{r_{yy}} = \tan\psi \exp(i\Delta) \tag{60.17}$$

Δ is the phase change between reflected and incident light. If the electric field direction of incident light is given by ϕ_0, then the field direction ϕ of reflected light can be obtained from $\tan\phi = \tan\phi_0/\tan\psi$. At the Brewster angle, where $r_p = 0$ and $\psi = 0$, then $\phi = 90°$, and reflected light is vertically polarized. Equation 60.17 can also be applied to transmissive systems. Since a vacuum does not change the polarization of light, the ellipsometric parameters are $\psi = 45°$ and $\Delta = 0°$. A perfect polarizer with the polarization along the x-axis has $\psi = 90°$ and $\Delta = 0°$, a perfect quarter-wave plate has $\psi = 45°$ and $\Delta = 90°$, and a perfect half-wave plate has $\psi = 45°$ and $\Delta = 180°$.

Müeller Matrix

The Jones calculus is convenient for perfectly polarized light and a nondepolarizing response [8]. If unpolarized light is incident on a sample, the Jones vector can not describe the field direction and phase for unpolarized light. The Stokes vector and Müeller matrix are more convenient to use in treating polarization for general cases. A relation between the output Stokes vector **S′** and the input Stokes vector **S** is

$$\mathbf{S'} = \mathbf{M} \mathbf{S} \tag{60.18}$$

The matrix **M** that relates the input and output Stokes vectors is called a Müeller matrix. **M** is a 4×4 matrix of real numbers.

For the general transformation of electric field given by Equation 60.14, the components M_{ij} of **M** can be derived from Equations 60.5, 60.14, and 60.18. The expressions for M_{ij} have been obtained by van de Hulst [13] and are also given as Equation (2.243) of Reference 4. In a measurement, the EM waves of output light may come from many different area or volume elements of a medium, so that statistical averages must be considered in the evaluation of M_{ij}. The ensemble average of **M** can still be expressed by the same expressions, with the ensemble average bracket applying to all M_{ij}. To make the Müeller matrix meaningful, the new subscripts of Equation 60.14 are reassigned as 1: xx, 2: yy, 3: xy, 4: yx. Subscripts 1 and 2 correspond to the copolarized response, and subscripts 3 and 4 correspond to the cross-polarized response. The ensemble average of any two of the response coefficients is called a *correlation function* for these coefficients. Let us define the self-correlation functions to be $2F_j$ and the cross-correlation functions to be $G_{jm} + i\, g_{jm}$ as follows:

$$\begin{cases} 2F_j \equiv \langle r_j r_j^* \rangle \\ G_{jm} + i\, g_{jm} \equiv \langle r_j r_m^* \rangle \end{cases} \quad j,m = 1,2,3,4; j \neq m \tag{60.19}$$

The cross-correlation functions have the properties of $G_{jm} = G_{mj} = \mathrm{Re}\,\langle r_j r_m^* \rangle$, and $g_{jm} = -g_{mj} = \mathrm{Im}\,\langle r_j r_m^* \rangle$. All F_j, G_{jm}, and g_{jm} are real numbers. **M** is then

$$\mathbf{M} = \begin{pmatrix} F_1 + F_2 + F_3 + F_4 & F_1 - F_2 - F_3 + F_4 & G_{13} + G_{24} & g_{13} - g_{24} \\ F_1 - F_2 + F_3 - F_4 & F_1 + F_2 - F_3 - F_4 & G_{13} - G_{24} & g_{13} + g_{24} \\ G_{14} + G_{23} & G_{14} - G_{23} & G_{12} + G_{34} & g_{12} - g_{34} \\ -g_{14} + g_{23} & -g_{14} - g_{23} & -g_{12} - g_{34} & G_{12} - G_{34} \end{pmatrix} \tag{60.20}$$

The upper left quadrant of **M** corresponds to the self-correlation terms. The lower right quadrant corresponds to the cross-correlations between the two co-polarized responses and between the two cross-polarized responses. The upper right and lower left quadrants correspond to the cross-correlations between the co-polarized and cross-polarized responses.

Principal Müeller Matrix

For the Jones matrix in the principal frame given by Equation 60.15, the cross-polarized responses are zero, so that the M_{jm} in the upper right and lower left quadrants of Equation 60.20 are zero. **M** can be expressed in terms of ψ and Δ using Equations 60.17 and 60.20 as [4, 17–19]

$$\mathbf{M} = R\begin{pmatrix} 1 & -\cos 2\psi & 0 & 0 \\ -\cos 2\psi & 1 & 0 & 0 \\ 0 & 0 & \sin 2\psi \cos\delta & \sin 2\psi \sin\Delta \\ 0 & 0 & -\sin 2\psi \sin\Delta & \sin 2\psi \cos\Delta \end{pmatrix} \tag{60.21}$$

where $R = (r_{xx}\,r_{xx}{}^* + r_{yy}\,r_{yy}{}^*)/2$. For reflection, R is the average reflectance, and for transmission, R is the average transmittance. For a vacuum, $\psi = 45°$ and $\Delta = 0°$, **M** is a unit matrix. Using Equation 60.21 or Equations 60.16, 60.19, and 60.20 directly, matrix **P** for a perfect polarizer ($\psi = 90°$, $\Delta = 0°$) and matrix **C** for a perfect wave plate ($\psi = 45°$, $\Delta = \tau$) in the principal frame are obtained as [4, 5, 20]

$$\mathbf{P} = \frac{1}{2}\begin{pmatrix} 1 & 1 & 0 & 0 \\ 1 & 1 & 0 & 0 \\ 0 & 0 & 0 & 0 \\ 0 & 0 & 0 & 0 \end{pmatrix} \qquad \mathbf{C} = \begin{pmatrix} 1 & 0 & 0 & 0 \\ 0 & 1 & 0 & 0 \\ 0 & 0 & \cos\tau & \sin\tau \\ 0 & 0 & -\sin\tau & \cos\tau \end{pmatrix} \tag{60.22}$$

Depolarization

A very interesting example is the perfectly random response. Analogous to the conditions used for incoherent scattering [17–19], the random response coefficients δr_j have the properties that

$$\begin{cases} \langle \delta r_j \rangle = 0 \\ \langle \delta r_j \delta r_m^* \rangle = \langle |\delta r_j|^2 \rangle \delta_{j,m} \qquad j,m = 1,2,3,4 \\ \langle |\delta r_j|^2 \rangle = \text{constant} \end{cases} \tag{60.23}$$

The first line states that all δr_j are each averaged to zero, so that they would not appear in the average of Equation 60.14. By substitution of r_j of Equation 60.19 by δr_j and using the conditions for δr_j given by Equation set 60.23, the correlation functions F_j and $G_{jm} + i\, g_{jm}$ can be evaluated. The second line of Equation set 60.23 states that all δr_j are uncorrelated with one another, so that all G_{ij} and g_{ij} are zero. The third line states that δr_j are isotropic, so that all F_j are the same. Eventually, all $M_{jm} = 0$ except $M_{00} = 4$ F_1. The depolarization matrix **D** for a perfectly random response is then

$$\mathbf{D} = \begin{pmatrix} 1 & 0 & 0 & 0 \\ 0 & 0 & 0 & 0 \\ 0 & 0 & 0 & 0 \\ 0 & 0 & 0 & 0 \end{pmatrix} \tag{60.24}$$

D is an ideal depolarizer as defined in Reference 20. The general Müeller matrix of Equation 60.20 satisfies the physical condition [14, 24]

$$\sum_{i,j=0}^{3} M_{ij}^2 \le 4M_{00}^2 \tag{60.25}$$

The sum of all the squares of the elements of **D** is M_{00}^2. **D** satisfies the inequality of criterion in Equation 60.25. The matrix of Equation 60.21 is nondepolarizing such that output light is still perfectly polarized if incident light is perfectly polarized. The equality in Equation 60.25 holds for **M** of Equation 60.21. This section discusses optical components and samples that are nondepolarizing. References 12 and 17 through 19 give more details about the Müeller matrices for samples that exhibit both polarization and depolarization properties.

Coordinate Transformation

In polarimetric measurements, polarizers and retarders are frequently rotated to desired positions. When a component is rotated, the incident field is not changed, but the representations of this field in the principal and laboratory coordinate systems are different. Transformations of the electric fields, Stokes vectors, Jones and Müeller matrices between these two coordinate systems or frames are basic exercises in polarimetry. Let the laboratory frame axes be x- and y-axes, and the principal frame axes be x'- and y'-axes, and the principal frame is rotated to an angle α with respect to the laboratory frame, as shown in Figure 60.3. The Jones vector (E_x', E_y') in the principal frame is related to (E_x, E_y) in the laboratory frame by

$$\mathbf{E}' = \begin{pmatrix} E_x' \\ E_y' \end{pmatrix} = \begin{pmatrix} \cos\alpha & \sin\alpha \\ -\sin\alpha & \cos\alpha \end{pmatrix} \begin{pmatrix} E_x \\ E_y \end{pmatrix} = \mathbf{r}(\alpha)\mathbf{E} \tag{60.26}$$

The rotation matrix $\mathbf{r}(\alpha)$ is the 2×2 matrix in Equation 60.26 for transformation of Jones vectors. The inverse transform is given by $\mathbf{E} = \mathbf{r}^{\mathrm{T}}(\alpha)\,\mathbf{E}'$, where the superscript T denotes the transpose of a matrix. In Figure 60.3, $\mathbf{E}'$ appears to be turned by an angle of $-\alpha$, since the coordinate system is rotated by an angle α. The Faraday rotation matrix that rotates **E** by an angle of α is equivalent to $\mathbf{r}(\alpha)$.

One can substitute $r_1 = r_2 = \cos\alpha$ and $r_3 = -r_4 = \sin\alpha$ into Equations 60.19 and 60.20 to construct the rotation matrix $\mathbf{R}(\alpha)$ for transformation of a Stokes vector **S** to a coordinate system oriented at an angle α.

$$\mathbf{R}(\alpha) = \begin{pmatrix} 1 & 0 & 0 & 0 \\ 0 & \cos 2\alpha & \sin 2\alpha & 0 \\ 0 & -\sin 2\alpha & \cos 2\alpha & 0 \\ 0 & 0 & 0 & 1 \end{pmatrix} \tag{60.27}$$

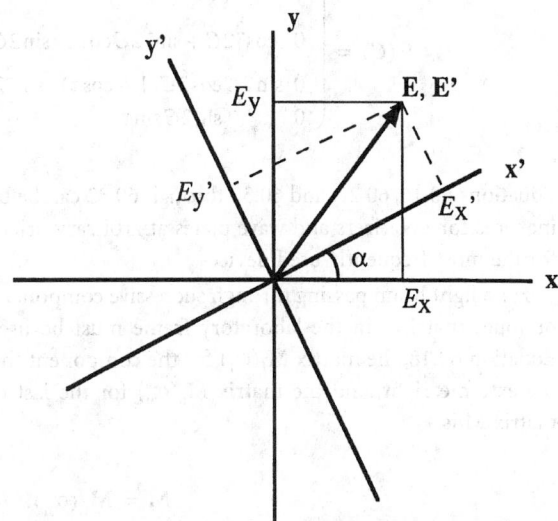

FIGURE 60.3 Coordinate transformation for the electric field components (E_x, E_y) in the laboratory system (x, y) and the components (E_x', E_y') in the principal coordinate system (x', y'). The principal frame is oriented at an angle a with respect to the laboratory frame.

The transformations between Stokes vector **S′** in the principal frame and **S** in the laboratory frame are

$$\begin{cases} \mathbf{S'} = \mathbf{R\,S} \\ \mathbf{S} = \mathbf{R^T\,S'} \end{cases} \tag{60.28}$$

The transformation of a Müeller matrix **M**(0) in the principal frame to **M**(α) in the laboratory frame can be obtained by a similarity transformation:

$$\mathbf{M}(\alpha) = \mathbf{R^T}(\alpha)\mathbf{M}(0)\mathbf{R}(\alpha) \tag{60.29}$$

Equation 60.29 can also be used for the transformation of Jones matrix **J**(0) in the principal coordinate frame to **J**(α) in the laboratory frame, provided that **M** is replaced by **J** and **R**(α) by **r**(α) in Equation 60.29.

The Jones matrix for a polarimetric component orientated at an angle α is

$$\mathbf{J}(\alpha) = \begin{pmatrix} r_{xx}\cos^2\alpha + r_{yy}\sin^2\alpha & \sin\alpha\cos\alpha(r_{xx} - r_{yy}) \\ \sin\alpha\cos\alpha(r_{xx} - r_{yy}) & r_{xx}\sin^2\alpha + r_{yy}\cos^2\alpha \end{pmatrix} \tag{60.30}$$

The Müeller matrix **P**(*P*) for a perfect polarizer oriented at an angle *P* and **C**(*C*) for a perfect compensator with a retardance τ at an angle *C* are [4, 6, 21]

$$\mathbf{P}(P) = \frac{1}{2}\begin{pmatrix} 1 & \cos 2P & \sin 2P & 0 \\ \cos 2P & \cos^2 2P & \sin 2P\cos 2P & 0 \\ \sin 2P & \sin 2P\cos 2P & \sin^2 2P & 0 \\ 0 & 0 & 0 & 0 \end{pmatrix} \tag{60.31}$$

$$\mathbf{C}(C) = \begin{pmatrix} 1 & 0 & 0 & 0 \\ 0 & \cos^2 2C + \sin^2 2C\cos\tau & \sin 2C\cos 2C(1 - \cos\tau) & -\sin 2C\sin\tau \\ 0 & \sin 2C\cos 2C(1 - \cos\tau) & \sin^2 2C + \cos^2 2C\cos\tau & \cos 2C\sin\tau \\ 0 & \sin 2C\sin\tau & -\cos 2C\sin\tau & \cos\tau \end{pmatrix} \tag{60.32}$$

Equations 60.16, 60.22, and 60.30 through 60.32 can be used to calculate the Jones matrices and Müeller matrices for polarizers and wave plates at arbitrary orientations. Table 60.2 lists some of these matrices for the most frequently used devices.

For a light beam passing through successive components oriented at different angles, Müeller matrices or Jones matrices in the laboratory frame must be used for successive multiplications. According to Equation 60.18, the matrix $\mathbf{M}_1(\alpha_1)$ for the component that light first passes through should be placed at the extreme right, and the matrix $\mathbf{M}_n(\alpha_n)$ for the last component at the extreme left. The combined matrix **M** is

$$\mathbf{M} = \mathbf{M_n}(\alpha_n)...\mathbf{M_1}(\alpha_1) \tag{60.33}$$

TABLE 60.2 The Jones Matrices and Müeller Matrices for Perfect Polarizer and Wave Plates at Different Orientation Angles

Device	Angle	Jones matrix	Müeller matrix
Polarizer	0°	$\begin{pmatrix}1&0\\0&0\end{pmatrix}$	$\dfrac{1}{2}\begin{pmatrix}1&1&0&0\\1&1&0&0\\0&0&0&0\\0&0&0&0\end{pmatrix}$
Polarizer	90°	$\begin{pmatrix}0&0\\0&1\end{pmatrix}$	$\dfrac{1}{2}\begin{pmatrix}1&-1&0&0\\-1&1&0&0\\0&0&0&0\\0&0&0&0\end{pmatrix}$
Polarizer	±45°	$\dfrac{1}{2}\begin{pmatrix}1&\pm1\\\pm1&1\end{pmatrix}$	$\dfrac{1}{2}\begin{pmatrix}1&0&\pm1&0\\0&0&0&0\\\pm1&0&1&0\\0&0&0&0\end{pmatrix}$
λ/4-plate	0°	$\begin{pmatrix}e^{i\pi/4}&0\\0&e^{-i\pi/4}\end{pmatrix}$	$\begin{pmatrix}1&0&0&0\\0&1&0&0\\0&0&0&1\\0&0&-1&0\end{pmatrix}$
λ/4-plate	±45°	$\dfrac{1}{\sqrt{2}}\begin{pmatrix}1&\pm i\\\pm i&1\end{pmatrix}$	$\begin{pmatrix}1&0&0&0\\0&0&0&\pm(-1)\\0&0&1&0\\0&\pm1&0&0\end{pmatrix}$
λ/2-plate	0°	$\begin{pmatrix}1&0\\0&-1\end{pmatrix}$	$\begin{pmatrix}1&0&0&0\\0&1&0&0\\0&0&-1&0\\0&0&0&-1\end{pmatrix}$
λ/2-plate	±45°	$\begin{pmatrix}0&\pm1\\\pm1&0\end{pmatrix}$	$\begin{pmatrix}1&0&0&0\\0&-1&0&0\\0&0&1&0\\0&0&0&-1\end{pmatrix}$

60.4 Principles of Polarimetry

Polarimetry is a method for measuring the polarization of light and the polarization response of materials. An optical system used for such purposes is called a polarimeter or an ellipsometer. To measure the polarization response of a sample, polarized light is generated and incident on the sample. By examining the polarization states of both incident and reflected or transmitted light, the characteristics of a sample can be determined. A schematic diagram of a polarimeter used to measure the polarization response of a sample is shown in Figure 60.4. The light source and polarizer are used to generate polarized light, and the analyzer and detector are used to analyze the polarization of light [4, 20]. An analyzer is a polarizer used to analyze polarized light.

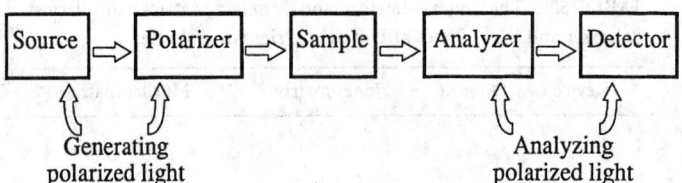

FIGURE 60.4 Schematic diagram of a polarimeter to measure polarization response. The light source and polarizer are used to generate polarized light, and the analyzer and detector are used to analyze the state of polarization of the light.

Analysis of Polarized Light

Measurement of polarization of light is essential in polarimetry, since polarized light to be examined is not limited to that generated in a laboratory. The instrument to measure the four Stokes parameters is called a photo-polarimeter or a Stokesmeter. To measure linear polarization, pass the light beam through a linear analyzer oriented at angle $A = 0°$, $90°$, and $\pm45°$, and measure the corresponding intensities I_x, I_y, I_+ and I_-. To measure circular polarization, first pass the light beam through a quarter-wave retarder with $C = 0°$, then through an analyzer oriented at $A = \pm45°$, and measure the intensities I_R and I_L. The pair of quarter-wave retarder and analyzer constitutes a circular analyzer. A detector measuring intensity corresponds to an operation given by a row vector $\mathbf{I} = (1, 0, 0, 0)$. The combined operation of a detector following an analyzer is $\mathbf{IA} = 0.5 (1, \cos 2A, \sin 2A, 0)$. The operations for the linear and circular analyzers on a Stokes vector S and the intensities obtained are given in Table 60.3. The Stokes parameters can be obtained from the difference and sum of the intensities for these pair operations, and are given by

$$
\mathbf{S} = \begin{pmatrix} S_o \\ S_1 \\ S_2 \\ S_3 \end{pmatrix} = \begin{pmatrix} I_x + I_y \\ I_x - I_y \\ I_+ - I_- \\ I_R - I_L \end{pmatrix} \tag{60.34}
$$

Equation 60.34 is a general expression that is good for any polarization states and is also the operational principle for most Stokesmeters. The four-detector Stokesmeter designed by Azzam is an exception that contains no moving components and can measure the four Stokes parameters in real time [26].

Generation of Polarized Light

Characterization of polarization response of a sample requires incident polarized light whose polarization state is controllable. A convenient source is a laser, which may be constructed to emit polarized light directly without the help of extra devices. A half-wave plate may be used to rotate the laser polarization

TABLE 60.3 Intensities I for a Light Beam with Stokes Parameters S_0, S_1, S_2, and S_3 Analyzed by Linear and Circular Analyzers. A circular analyzer consists of a quarter-wave plate oriented at $C = 0°$, followed by an analyzer oriented at $A = \pm45°$.

Analyzer	Linear	Linear	Circular
C (°)	NA	NA	0
A (°)	0, 90	45, −45	45, −45
Operation	0.5(1, ±1, 0, 0)	0.5(1, 0, ±1, 0)	0.5(1, 0, 0, ±1)
Intensity	I_x, I_y	I_+, I_-	I_R, I_L
$I =$	$(S_0 \pm S_1)/2$	$(S_0 \pm S_2)/2$	$(S_0 \pm S_3)/2$

to a desired direction by placing the fast-axis bisecting the new and old directions, as shown in Figure 60.5a. Light generated from a lamp and a monochromator is usually partially polarized [12]. To generate linearly polarized light at an angle *P*, a linear polarizer oriented at an angle *P* is placed behind the monochromator as shown in Figure 60.5b. To generate circularly polarized light, first generate linearly polarized light, and then put a quarter-wave plate behind with the fast-axis oriented at an angle of 45° or −45° with respect to the linear polarization as shown in Figure 60.5c. Such a combination of polarizer and quarter-wave plate is called a circular polarizer. When a phase retarder has an arbitrary retardance or is placed at an arbitrary angle relative to the polarizer, elliptically polarized light is then generated. Given an incident Stokes vector of (S_0, S_1, S_2, S_3), the Stokes vector S' for polarized light generated by the polarizers mentioned above can be obtained from Equations 60.18 and 60.31 through 60.33. The obtained S' are listed in Table 60.4. Note that S' is not directly proportional to S_0 unless incident light is unpolarized. Care must be taken in generating polarized light in an ellipsometer because incident light is rarely completely unpolarized.

Polarizer–Sample–Analyzer Ellipsometry

An ellipsometer is an instrument to measure the ellipsometric parameters ψ and Δ of a sample. It can be used for both reflection and transmission. An ellipsometer is usually referred to as the reflection system, and a polarimeter as the transmissive system [4]. Different ellipsometers are designed to measure different responses for different kinds of samples. It is important to know about the sample when designing an experiment. The simplest ellipsometer is a polarizer–sample–analyzer (PSA) ellipsometer. A more general one is a polarizer–compensator–sample–analyzer (PCSA) ellipsometer. Figure 60.6 shows

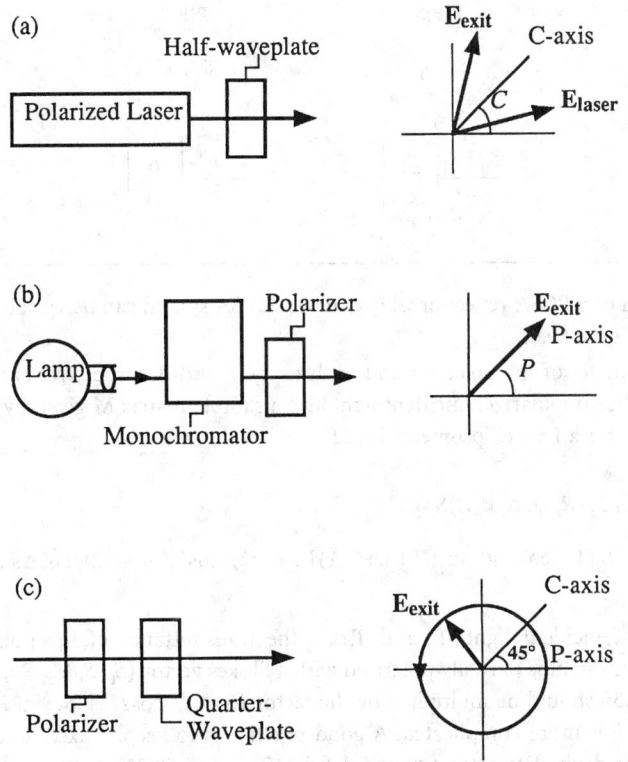

FIGURE 60.5 Generation of light linearly polarized at a desired direction using (a) a laser source and a half-wave plate and (b) a lamp, monochromator, and a polarizer, plus (c) generation of circularly polarized light using a polarizer and a quarter-wave plate.

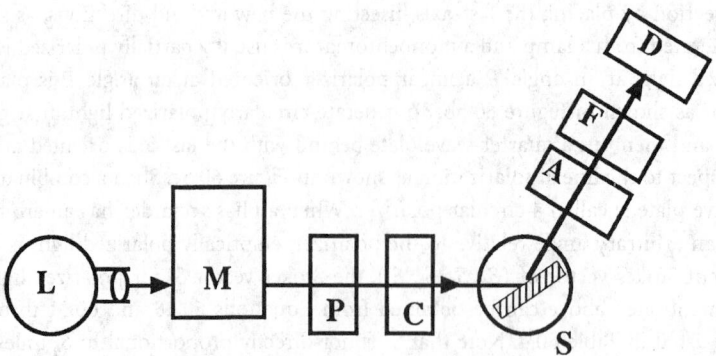

FIGURE 60.6 Schematic diagram of a PCSA ellipsometer.

TABLE 60.4 Stokes Vectors S′ for Linearly and Circularly Polarized Light Generated by Specific Combinations of a Polarizer Oriented at an Angle P and a Quarter-Wave Retarder at an Angle C. The incident Stokes vector is $S = (S_0, S_1, S_2, S_3)$.

Polarizer	Linear	Linear	Circular
$P(°)$	0, 90	45, −45	−45, 45
$C(°)$	NA	NA	0
Operation	P(P)S	P(P)S	Q(C) P(P) S
Stokes Vector	$S_x′, S_y′$	$S_+′, S_-′$	$S_R′, S_L′$
S′	$\dfrac{S_0 \pm S_1}{2}\begin{pmatrix}1\\ \pm 1\\ 0\\ 0\end{pmatrix}$	$\dfrac{S_0 \pm S_2}{2}\begin{pmatrix}1\\ 0\\ \pm 1\\ 0\end{pmatrix}$	$\dfrac{S_0 \pm S_3}{2}\begin{pmatrix}1\\ 0\\ 0\\ \pm 1\end{pmatrix}$

a schematic diagram of a PCSA reflection ellipsometer. A PSA system can be visualized by removing the compensator C in Figure 60.6.

Let the oriented angles of the polarizer and analyzer be P and A, respectively, as measured from the plane of incidence. For unpolarized incident light and a sample matrix $\mathbf{M}$ given by Equation 60.21, the measured intensity I for a PSA ellipsometer is [22]

$$
\begin{aligned}
I(P, A) &= \mathbf{I\,A}(A)\,\mathbf{M}(R, \psi, \Delta)\mathbf{P}(P)\mathbf{S} \\
&= T_p T_a R I_0 [1 - \cos 2\psi(\cos 2P + \cos 2A)] + \cos 2P \cos 2A + \sin 2\psi \cos \Delta \sin 2P \sin 2A
\end{aligned}
\tag{60.35}
$$

I_0 is the intensity of incident light, T_p and T_a are the transmittance of the polarizer and analyzer, respectively. If incident light is partially polarized with a Stokes vector (S_0, S_1, S_2, S_3), then the right-hand side of Equation 60.35 should be multiplied by the factor $(S_0 + S_1 \cos 2P + S_2 \sin 2P)/S_0$. In such a case, the dependence on P is more complicated. A good practice is to keep P fixed and vary only A. Many different ways can be devised to extract ψ and Δ from Equation 60.35, such as the Stokes polarimeter, null polarimeter, and rotating-analyzer ellipsometer.

For a Stokes polarimeter, P is set at 45° or -45°, and A at 0°, 90°, and ±45°. The ellipsometric parameters ψ and Δ can be solved from the four equations evaluated at these P and A positions via the relations

$$\left\{ \begin{aligned} \cos 2\psi &= \frac{I_y - I_x}{I_y + I_x} \\ \sin 2\psi \cos\Delta &= \frac{I_+ - I_-}{I_+ + I_-} \end{aligned} \right. \tag{60.36}$$

For a null polarimeter, set $P = \pm 45°$, and vary A to find the null positions. This method is excellent for reflection from transparent materials with Δ equal to 0 or π. The value of ψ is related to the null position $A_\pm$ as [22]

$$\tan 2A_\pm = \mp \tan 2\psi \cos\Delta \qquad \text{for } P = \pm 45° \tag{60.37}$$

Average of $A_\pm$ at $P = \pm 45°$ can eliminate errors from the misalignment of analyzer and polarizer.

In a rotating-analyzer ellipsometer (RAE), the analyzer is rotated at an angular frequency ω_r. Set $P = 45°$ or $-45°$, and $A = \omega_r t$. The measured intensity is

$$I = T_p R T_a I_0 [1 - \cos 2\psi \cos 2\omega_r t + \sin 2\psi \cos\Delta \sin 2\omega_r t] \tag{60.38}$$

The Fourier coefficients, being equal to $-\cos 2\psi$ and $\sin 2\psi \cos\Delta$, can be recovered from the demodulated signals or from a fast Fourier transform (FFT) technique [28, 29]. However, the measured Δ cannot be distinguished from $-\Delta$. This system can be fully automated for realtime operation. With a white light source and monochromator, an RAE can serve as a spectroscopic ellipsometer [15, 16, 28, 29].

Polarizer–Compensator–Sample–Analyzer Ellipsometry

In a PCSA or a PSCA ellipsometer, a compensator of retardance τ is inserted in front of or following the sample. Figure 60.6 shows a schematic diagram of a PCSA ellipsometer. For unpolarized incident light in a PCSA system, the measured intensity for general conditions of P, C, and A is given by [21]

$$I(P, C, A) = \mathbf{I}\,\mathbf{A}(A)\,\mathbf{M}(R, \psi, \Delta)\,\mathbf{C}(C)\,\mathbf{P}(P)\,\mathbf{S}$$

$$= T_p T_c R T_a I_0 [Y_0 + Y_1 \cos 2A + Y_2 \sin 2A] \tag{60.39}$$

where

$$\left\{ \begin{aligned} Y_0 &= \{1 - \cos 2\psi [\cos 2C \cos 2(P - C) - \cos\tau \sin 2C \sin 2(P - C)]\} \\ Y_1 &= \{-\cos 2\psi + [\cos 2C \cos 2(P - C) - \cos\tau \sin 2C \sin 2(P - C)]\} \\ Y_2 &= \sin 2\psi \{\cos\Delta [\sin 2C \cos 2(P - C) + \cos\tau \cos 2C \sin 2(P - C)]\} - \sin\tau \sin\Delta \sin 2(P - C) \end{aligned} \right. \tag{60.40}$$

If incident light is partially polarized, then I_0 in Equation 60.39 should be replaced by $(S_0 + S_1 \cos 2P + S_2 \sin 2P)$. For a PSCA system, interchange P and A in Equations 60.39 and 60.40. These formulas can be used to design different kinds of PCSA ellipsometers by choosing different conditions and different types of modulation. For certain special conditions, Equations 60.40 can be greatly simplified.

Null Ellipsometry

Consider a PCSA null ellipsometer (NE) in which the compensator is a perfect quarter-wave retarder ($\tau = 90°$) which is set at $C_\pm = \pm 45°$. The measured intensity is

$$I(P, C_{\pm}, A) = T_p T_c RT_a I_0 [1 - \cos 2\psi \cos 2A \pm \sin 2\psi \sin 2A \cos \Delta \pm (2P - 90°)] \qquad (60.41)$$

An NE is an instrument to find the null positions in order to determine ψ and Δ. The four zones that will null the intensity in Equation 60.41 and the null positions are listed in Table 60.5. The null positions of A give ψ directly, and the null positions of P give Δ directly. Although ψ and Δ can be determined from measurements in only one zone, systematic errors caused by imperfect components, misalignment, and partially polarized incident light can be nonnegligible. By taking the average of four zones, many of these linear systematic errors can be cancelled [4, 21, 30, 31]. To look for the null positions manually is a slow process; automation of the nulling process can speed up the measurements. Different methods can be used to automate the NE: (1) both polarizer and analyzer are controlled by servo-motors with the feedback from the detector to find the null intensity, (2) the intensity is digitized and fed into a computer which is used to find the null positions, (3) Faraday rotators are used to rotate and modulate the polarization directions of light incident on and reflected from the sample to get the slopes of intensity versus angle until the slopes are zero at the null positions [4, 27]. The advantage of NE is its simplicity in obtaining ψ and Δ. Also, its accuracy is unbeatable by other kinds of ellipsometry.

TABLE 60.5 The Null Positions of Polarizer Angle P and Analyzer Angle A for the Four Different Zones at $C = \pm 45°$ in a Null Ellipsometer.

Zone	$C(°)$	P	A
1	−45	−45° + Δ/2	ψ
2	45	−45° − Δ/2	ψ
3	−45	45° + Δ/2	−ψ
4	45	45° − Δ/2	−ψ
Average	$\psi = (A_1 + A_2 - A_3 - A_4)/4$,	$\Delta = (P_1 - P_2 + P_3 - P_4)/2$	

Phase-Modulated Ellipsometry

A phase-modulated ellipsometer (PME) uses a phase retarder whose retardance is modulative. For a PME, a good choice of P, C, and A in Equations 60.39 and 60.40 is $P = 0°$, $C = 45°$, and $A = 45°$. The intensity is then

$$I(\tau) = T_p T_c RT_a I_0 [1 - \cos 2\psi \cos \tau + \sin 2\psi \sin \Delta \sin \tau] \qquad (60.42)$$

The retardance τ is modulated, and the modulated intensity is detected. If τ is modulated according to $\tau = \tau_o \cos \Omega_t$, then $I(\tau)$ can be expanded in a Fourier series with the Fourier coefficients, depending on $\cos 2\psi$, $\sin 2\psi \cos \Delta$ and the Bessel functions of τ_o. These coefficients can be recovered from the demodulated signals, and the values of ψ and Δ can then be solved. τ can be modulated by electro-optic modulation using the Pockels effect, or piezoelectric modulation using the photoelastic effect [32, 33]. These modulations are fast, and demodulation using a lock-in amplifier is convenient. The advantage of a PME is that it contains no moving components and offers real-time measurement.

60.5 Polarization Instrumentation and Experiments

Figure 60.7 shows a schematic diagram of a PCSA reflection null ellipsometer. The instrument is composed of five systems: the source system, polarimetric system, sample system, detection system, and computer system for automatic control, data acquisition, and processing.

The simplest source is a polarized laser. The monochromatic source system in Figure 60.7 is usually used in visible, ultraviolet and infrared spectrometers. Light from a lamp source L is focused by a lens system or a spherical mirror system onto the entrance slit of a grating monochromator M. Light leaving

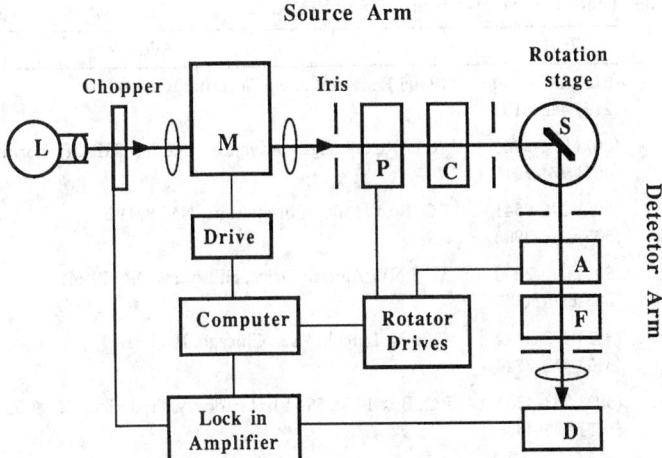

FIGURE 60.7 Schematic diagram of a PCSA reflection null ellipsometer, which is composed of five systems: the source system, polarimetric system, sample system, detection system, and computer system. The symbols are L = light source, M = monochromator, F = wavelength filter, and D detector. The symbols P, C, S, and A have their customary meanings.

the exit slit of M is collimated by another lens system and a set of iris apertures. The long-pass filter F in the detector arm is used with a grating monochromator to remove undesired short-wavelength radiation. Choices of monochromators include a grating monochromator, prism monochromator or Fourier transform spectrometer. A spectrometer is a necessary component for a spectroscopic ellipsometer. Synchrotron radiation is also a continuum source and is used to replace the lamp in the vacuum ultraviolet region [25]. The synchrotron radiation beam is intense and polarized. Grazing incidence reflection optics are usually used to avoid absorption in the components.

The polarimetric system shown in Figure 60.7 is a PCSA ellipsometer. For a PSA ellipsometer, remove the compensator C. For a PCSCA system, add another compensator in the detector arm [20]. It is better to mount the polarizers and phase retarders on automatic rotators so that their orientations can be easily aligned and read. For simple experiments, manual rotators can also do the job. References 10 and 11 give detailed descriptions and references for different kinds of polarizers and phase retarders. Many of the well known optical companies sell polarizers and wave plates (retarders) for use in the visible, ultraviolet, and near infrared spectral regions. The Buyers Guide of Laser Focus World [49] and the Photonics Buyers' Guide [50] list companies that manufacture and sell polarizers, phase retarders and polarimeters. Tables 60.6 and 60.7 list some companies that make these products. Commonly used polarizers in the visible are Glan prisms and dichroic sheets or plates. The best polarizer in the visible is the calcite Glan-Thompson prism, which has a very small extinction ratio and a large acceptance angle. It is more difficult to find a good broadband polarizer in the mid-infrared spectral region (λ: 3 to 5 µm). Wire-grid polarizers are good in the long-wave infrared region ($\lambda > 8$ µm). For a broadband phase retarder, a Babinet-Soleil compensator is convenient, since it can be set to any retardance value using a micrometer adjustment. Inexpensive wave plates are good for laser sources. Phase modulation can be achieved by modulation of the birefringence of electro-optic or photo-elastic retarders.

The sample system depends on the type of experiment to be performed. Components on the detector side of Figure 60.7 are normally mounted on a rail that is rotatable about the axis of the rotation stage. The sample holder S on the rotation stage should have enough degrees of freedom for easy alignment. In a reflection ellipsometer, the *x*-axes for *P, C,* and *A* should be well aligned to lie in the plane of incidence. A transmission polarimeter is much simpler to align.

The detection system consists of a detector and a noise suppression system. Diode detectors for use in the visible and near-infrared spectral regions are inexpensive. Photomultipliers for the visible and

TABLE 60.6 Companies That Make Polarizers and Phase Retarders

Company	Tel/Fax	Address
Cleveland Crystals	(216) 486-6100 (216) 486-6103	19306 Redwood Ave., Cleveland, OH 44110
Corning, Inc.	(607) 974-7966 (607) 974-7210	POLARCOR Team, Advance Materials, HP-CB, Corning, NY 14831
CVI Laser	(505) 296-9541 (505) 298-9908	P.O. Box 11308, Albuquerque, NM 87192
Hinds Instruments	(503) 690-2000 (503-690-3000	3175 NW Aloclek Drive, Hillsboro, OR 97124
Karl Lambrecht	(312) 472-5442 (312) 472-2724	4204 N. Lincoln Ave., Chicago, IL 60618
Meadowlark	(303) 833-4333 (303) 833-4335	P.O. Box 1000, 5964 Iris Parkway, Frederick, CO 80530
Molectron, Inc.	(503) 620-9069 (503) 620-8964	7470 SW Bridgeport Rd., Portland, OR 97224
New Focus, Inc.	(408) 980-8088 (408) 980-8883	2630 Walsh Ave., Santa Clara, CA 95051-0905
Rocky Mountain Instrument	(303) 651-2211 (303) 651-2648	1501 S. Sunset St., Longmont, Colorado 80501
Special Optics	(201) 785-4015 (201) 785-0166	P.O. Box 163, Little Falls, NJ 07424
Tower Opt. Corp.	(201) 305-9626 (201) 305-1175	130 Ryerson Ave., Wayne, NJ 07470
II-VI, Inc.	(412) 352-1504 (412) 352-4980	375 Saxonburg Blvd., Saxonburg, PA 16056

TABLE 60.7 Companies That Make Ellipsometers

Company	Tel/Fax	Address
Gaertner Scientific	847-673-5006 847-673-5009	8228 McCormick Blvd., Skokie, IL 60076
Instrument SA, Inc.	908-494-8660 908-494-8796	6 Olsen Ave., Edison, NJ 08820
J. A. Woolam Co., Inc.	402-477-7501 402-477-8214	650 J. Street, Suite 39, Lincoln, NE 68508
Leonard Research	937-426-1222 937-426-3642	2792 Indian Riffle Rd., Beavercreek, OH 45440 P.O. Box 607, Beavercreek, OH 45434-0607
Rudolph Research	201-691-1300 201-691-5480	1 Rudolph Rd., P.O. Box 1000, Flanders, NJ 07836
SOPRA	(1) 47 81 09 49 (1) 42 42 29 34	26, rue Pierre Joigneaux, 92270 Bois-Colombes, FRANCE
Tencor Instrument	415-969-6767 415-969-6731	2400 Charleston Rd., Mountain View, CA 94043-9958

near-infrared regions have high sensitivity, and cooled semiconductor detectors give good performance in the mid-infrared to far-infrared regions. In Figure 60.7, the noise suppression system includes a chopper and a lock-in amplifier. The chopper modulates incident light intensity, and the intensity detected by detector D is demodulated by a lock-in amplifier. This combination eliminates most broadband noise and greatly improves the signal-to-noise ratio. RAE and PME have their own modulations, and do not need a chopper.

A computer system provides the automatic functions to control the polarizers, retarder, and mono-chromator, and to acquire and process data. A computer system is essential for making accurate and rapid measurements. In Figure 60.7, the computer records the polarizer angle P, the analyzer angle A, the intensity I from the detector, and the wavelength λ of the monochromator. The data of $I(P)$ and $I(A)$ can be used to find the null positions of P and A. Then the computer controls the drivers to move P or A to the null positions. In a spectroscopic RAE, data of $I(t)$ are recorded as the analyzer is rotating. The computer uses the FFT program to find the Fourier coefficients, solves for ψ and Δ, records the results, and then drives the monochromator to a new wavelength and repeats the process. A spectroscopic PME uses similar computer process as RAE, besides the different modulation and demodulation.

Measurement of Birefringence

Retardance can be measured using a transmission PSA ellipsometer. A wave plate is a good sample for this experiment. The retardance δ for a birefringent slab of thickness d and principal refractive indices ne and no is

$$\delta = \frac{2\pi d(n_e - n_o)}{\lambda} \tag{60.43}$$

in the absence of multiple reflections. For a wave plate, the value of ψ is close to 45°, and the Δ value is equal to δ. Use a lamp source with a monochromator to scan the wavelength. Put a polarizer at 45° and an analyzer at –45° with respect to the fast-axis of the wave plate [40]. When λ is scanned, the transmitted intensity I through the PSA ellipsometer will vary, in proportion to $(1 + \cos\delta)$ according to Equation 60.35. I is a maximum when δ is 0°, and is a minimum when $\delta = 180°$. From the wavelengths at which a maximum or a minimum intensity occurs, the birefringence $n_e - n_o$ can be determined.

Measurement of Optical Constants

A reflection ellipsometer can be used to measure optical constants n and k of materials. Light is incident obliquely on a sample at an angle of θ. The refractive index n and the extinction coefficient k can be calculated from the measured ψ and Δ values using the following formula [2, 4, 16, 22]

$$(n + ik)^2 = \sin^2\theta \left\{ 1 + \tan^2\theta \left[\frac{1 - \tan\psi\exp(i\Delta)}{1 + \tan\psi\exp(i\Delta)} \right]^2 \right\} \tag{60.44}$$

If an automated system is not available, try an NE to obtain ψ and Δ manually. Automated systems such as an RAE or a PME can take data much faster. For transparent materials, Δ is either 0 or π, the simple null polarimeter with a PSA system offers satisfactory results [22]. The method is effective near the Brewster angle region. At the Brewster angle θ_B, $\psi = 0$, then $n = \tan\theta_B$ according to Equation 60.44. For metals whose ψ is large when $\Delta = -90°$, the principal angle ellipsometry (PAE) can be used [2]. At the principal angle θ_P, $\cos\Delta = 0$, Equation 60.44 can be simplified to

$$(n + ik)^2 = \sin^2\theta_P[1 + \tan^2\theta_P(\cos4\psi + i\,\sin4\psi)] \tag{60.45}$$

In PAE, the principal angle θ_P is searched and then ψ is measured. The Stokes polarimeter with a PSA system is suitable to search θ_P and measure ψ for PAE.

Determination of optical constants using reflection ellipsometry is subject to errors caused by surface roughness, natural oxides and surface contamination [15, 16, 41–48]. These effects can be corrected by assuming that there is an effective layer on the surface and then using a least-square regression method to fit the ellipsometric data to the appropriate model [16, 43–48]. Ellipsometry is also used to measure refractive index and thickness of a thin film on a bulk substrate whose optical constants are known [4,

15, 16, 27]. Other applications of polarimetry and ellipsometry can be found in recent proceedings about polarization [51–53].

Acknowledgment

The author would like to thank Dr. J. M. Bennett for her review of the original manuscript and her suggestions especially for the introduction. This work was supported partially by the "Polarizer Standard and Metrology Program" funded by the U.S. Naval Warfare Assessment Division.

References

1. G. G. Stokes, "On the composition and resolution of streams of polarized light from different sources," *Trans. Cambridge Phil. Soc.*, 9, 399–416, 1852, also as paper 10 of ref. [8].
2. M. Born and E. Wolf, *Principles of Optics, 5th ed.*, Oxford: Pergamon Press, 1975, Chapters 1, 2, 13, 14.
3. J. D. Jackson, *Classical Electrodynamics, 2nd ed.*, New York: John Wiley & Sons, 1975, Sections 7.1–7.5.
4. R. M. A. Azzam and N. M. Bashara, *Ellipsometry and Polarized Light, 2nd ed.*, Amsterdam: North Holland, 1987.
5. W. A. Shurcliff and S. S. Ballard, *Polarized Light*, Princeton: van Nostrand, 1964.
6. K. Serkowski, "Polarization of starlight," in Z. Kopal (ed.), *Advances in Astronomy and Astrophysics*, Vol. 1, New York: Academic Press, 1962, 289–352.
7. G. R. Fowles, *Introduction to Modern Optics, 2nd ed.*, New York: Holt, Reinhart and Winston, 1975, Chapter 6.
8. W. Swindell (ed.), *Polarized Light*, Benchmark Papers in Optics, Vol. 1, Stroudburg, PA: Dowden, Hutchinson & Ross, 1975.
9. F. A. Jenkins and H. E. White, *Fundamentals of Optics, 4th ed.*, New York: McGraw-Hill, 1976, Chapters 24–28.
10. J. M. Bennett, "Polarization," in M. Bass, E. S. Van Stryland, D. R. Williams, and W. L. Wolfe (eds.), *Handbook of Optics*, Vol. I, New York: McGraw-Hill, 1995, Chapter 5; and "Polarizers," ibid., Vol. II, Chapter 3.
11. H. E. Bennett and J. M. Bennett, "Polarization," in W. G. Driscoll and W. Vaughan (eds.), *Handbook of Optics*, New York: McGraw-Hill, 1978, Chapter 10.
12. S. F. Nee, Chan Yoo, Teresa Cole and Dennis Burge, "Characterization of Imperfect Polarizers Under Imperfect Conditions," *Appl. Opt.* 37, 54–64, 1998.
13. H. C. van de Hulst, *Light Scattering by Small Particles*, New York: Wiley, 1957, Chapter 5.
14. K. Kim, L. Mandel, and E. Wolf, "Relationship between Jones and Müeller matrices for random media," *J. Opt. Soc. Am. A* 4, 433–437, 1987.
15. D. E. Aspnes, "Spectroscopic ellipsometry of solids," in B. O. Seraphin (ed.), *Optical Properties of Solids: New Developments*, Amsterdam: North Holland, 1976, Chapter 15.
16. D. E. Aspnes, "The accurate determination of optical properties by ellipsometry," in E. D. Palik (ed.), *Handbook of Optical Constants of Solids I*, Orlando: Academic Press, 1985, Chapter 5.
17. S. F. Nee, "Polarization of specular reflection and near-specular scattering by a rough surface," *Appl. Opt.*, 35, 3570–3582, 1996.
18. S. F. Nee, "The effects of incoherent scattering on ellipsometry," in D. H. Goldstein and R. A. Chipman (eds.), *Polarization Analysis and Measurement, Proc. SPIE* 1746, 119–127, 1992.
19. S. F. Nee, "Effects of near-specular scattering on polarimetry," in D. H. Goldstein and D. B. Chenault (eds.), *Polarization Analysis and Measurement II, Proc. SPIE* 2265, 304–313, 1994.
20. R. A. Chipman, "Polarimetry," in M. Bass, E. S. Van Stryland, D. R. Williams, and W. L. Wolfe (eds.), *Handbook of Optics*, Vol. II, New York: McGraw-Hill, 1995, Chapter 22.
21. S. F. Nee, "Error reductions for a serious compensator imperfection for null ellipsometry," *J. Opt. Soc. Am. A*, 8, 314–321, 1991.

22. S. F. Nee and H. E. Bennett, "Accurate null polarimetry for measuring the refractive index of transparent materials," *J. Opt. Soc. Am.* A, 10, 2076–2083, 1993.

23. R. M. A. Azzam and N. M. Bashara, "Ellipsometry with imperfect components including incoherent effects," *J. Opt. Soc. Am.*, 61, 1380–1391, 1971.

24. E. S. Fry and G. W. Kattawar, "Relationships between elements of the Stokes matrix," *Appl. Opt.*, 20, 2811–2814, 1981.

25. J. Barth, R. L. Johnson, and M. Cardona, "Spectroscopic ellipsometry in the 6–35 eV region," in E. D. Palik (ed.), *Handbook of Optical Constants of Solids II*, Boston: Academic Press, Chapter 10, 1991.

26. R. M. A. Azzam, "Arrangement of four photodetectors for measuring the state of polarization of light," *Opt. Lett.* 10, 309–311, 1985; U.S. Patent 4,681,450 1987.

27. R. M. A. Azzam, "Ellipsometry," in M. Bass, E. S. Van Stryland, D. R. Williams, and W. L. Wolfe (eds.), *Handbook of Optics*, Vol. II, New York: McGraw-Hill, 1995, Chapter 27.

28. D. E. Aspnes, "Fourier transform detection system for rotating-analyzer ellipsometers," *Opt. Commun.*, 8, 222–225, 1973.

29. D. E. Aspnes, "High precision scanning ellipsometer," *Appl. Opt.*, 14, 220–228, 1975.

30. R. M. A. Azzam and N. M. Bashara, "Unified analysis of ellipsometry errors due to imperfect components, cell-window birefringence, and incorrect azimuth angles," *J. Opt. Soc. Am.*, 61, 600–607, 1971.

31. D. E. Aspnes, "Measurement and correction of first-order errors in ellipsometry," *J. Opt. Soc. Am.*, 61, 1077–1085, 1971.

32. S. N. Jasperson and S. E. Schnatterly, "An improved method for high reflectivity ellipsometry based on a new polarization modulation technique," *Rev. Sci. Inst.*, 40, 761–767, 1969.

33. J. C. Kemp, "Piezo-optical birefringence modulators: New use for a long-known effect," *J. Opt. Soc. Am.*, 59, 950–954, 1969.

34. J. Wilson and J. F. B. Hawkes, *Optoelectronics: An Introduction*, Englewood Cliffs: Prentice-Hall, 1983, Chapters 1, 3.

35. T. A. Maldonado, "Electro-optic modulators," in M. Bass, E. S. Van Stryland, D. R. Williams, and W. L. Wolfe (eds.), *Handbook of Optics*, Vol. II, New York: McGraw-Hill, 1995, Chapter 13.

36. S. T. Wu, "Liquid crystals," in M. Bass, E. S. Van Stryland, D. R. Williams, and W. L. Wolfe (eds.), *Handbook of Optics*, Vol. II, New York: McGraw-Hill, 1995, Chapter 14.

37. G. L. Cloud, *Optical Methods of Engineering Analysis*, Cambridge, England: Cambridge University Press, 1995, Parts I, II.

38. I. C. Chang, "Acousto-optic devices and applications," in *Handbook of Optics*, Vol. II, in M. Bass, E. S. Van Stryland, D. R. Williams, and W. L. Wolfe (eds.), New York: McGraw-Hill, 1995, Chapter 12.

39. M. Mansuripur, *The Physical Principles of Magneto-Optical Recording*, Cambridge, England: Cambridge University Press, 1995, Chapter 6.

40. J. H. Shields and J. W. Ellis, "Dispersion of birefringence of quartz in the near infrared," *J. Opt. Soc. Am.*, 46, 263–265, 1956.

41. D. K. Burge and H. E. Bennett, "Effect of a thin surface film on the ellipsometric determination of optical constants," *J. Opt. Soc. Am.*, 54, 1428–1433, 1964.

42. C. F. Fenstermaker and F. L. McCrackin, "Errors arising from surface roughness in ellipsometric measurement of the refractive index of a surface," *Surf. Sci.*, 16, 85–96, 1969.

43. D. E. Aspnes, "Studies of surface, thin film and interface properties by automatic spectroscopic ellipsometry," *J. Vac. Sci. Technol.*, 18, 289–295, 1981.

44. M. E. Pedinoff and O. M. Stafsudd, "Multiple angle ellipsometric analysis of surface layers and surface layer contaminants," *Appl. Opt.*, 21, 518–521, 1982.

45. J. P. Marton and E. C. Chang, "Surface roughness interpretation of ellipsometer measurements using the Maxwell-Garnett theory," *J. Appl. Phys.*, 45, 5008–5014, 1974.

46. D. E. Aspnes, J. B. Theeten, and F. Hottier, "Investigation of effective medium models of microscopic surface roughness by spectroscopic ellipsometry," *Phys. Rev. B*, 20, 3292–3302, 1979.

47. D. E. Aspnes, "Optical properties of thin films," *Thin Solid Films*, 89, 249–262, 1982.

48. S. F. Nee, "Ellipsometric analysis for surface roughness and texture," *Appl. Opt.*, 27, 2819–2831, 1988.

49. *Laser Focus World Buyers Guide, 1996*, PennWell Publishing Co., 1421 South Seridan, Tulsa, OK 74101.

50. *The Photonics Buyers' Guide, 1996, Book II*, Laurin Publishing Co., Berkshire Common, P.O. Box 4949, Pittsfield, MA 01202–4949.

51. R. A. Chipman (ed.), *Polarization Considerations for Optical Systems II*, Proc. SPIE, 1166, San Diego, 1989.

52. D. H. Goldstein and R. A. Chipman (eds.), *Polarization Analysis and Measurement*, Proc. SPIE, 1746, San Diego, 1992.

53. D. H. Goldstein and D. B. Chenault (eds.), *Polarization Analysis and Measurement II*, Proc. SPIE, 2265, San Diego, 1994.

61

Refractive Index Measurement

G. H. Meeteen
Schlumberger Cambridge Research

61.1 Introduction

Light is electromagnetic radiation of wavelength about 450 to 700 nm, in which region the human eye is sensitive and refractive index measurements are commonly made. The refractive index or indices of a substance describe an important part of its interaction with electromagnetic radiation. Refractive index is a dimensionless quantity, real for transparent materials and complex if there is absorption. It generally depends on the direction of light relative to axes of the material; such substances, e.g. many crystals, are anisotropic and possess more than one refractive index. In general, the index is a tensor (3×3 matrix) with up to 9 components. Many substances are isotropic, e.g., liquids, glasses, and other noncrystalline materials, and one refractive index is sufficient. Some substances are optically inhomogeneous and possess refractive index fluctuations over distances comparable in size with a wavelength. They scatter light and appear milky or turbid. Many foodstuffs and drinks are optically heterogeneous. Refractometry is, fortunately, a robust technique in the face of material complexities. It has widespread application to many industries and materials. This chapter treats the refractive index as the principal measurand; the complications of real materials may require extra attention. Excluded are strongly absorbing materials such as metals in which absorption may be more important than refraction. The appropriate measurement technique is then ellipsometry, in which the amplitude and phase of the reflected light are measured, as the polarization and the angle of incidence are varied. Some refractometric methods are essentially ellipsometric, but for transparent samples transmission methods are also important. Refractive index measurements of high precision, e.g. 10^{-4} to 10^{-5}, are possible with relatively simple techniques. Although ocular instruments remain popular, semiconductor technology has been applied to refractive index measurement in recent years, improving the measurement speed and precision. One of the limits to improvement in precision is the sample's absorption or heterogeneity, a factor considered herein. Several online instruments exist, and the application of refractive index measurements to complex industrial fluids has been recently described [1]. Fiber-optical methods do not yet possess the precision of conventional methods but offer the possibility of remote sensing.

Refractive index is measured for many reasons. It is clearly important to know the refractive index of materials used for their clarity, such as glasses and solid plastics. In complex fluids such as drinks or foods, the refractive index is a measure of dissolved or submicronic material. The Brix scale relates refractive index to sugar concentration. Common industrial applications are to microemulsions to measure their oil/water ratio, to antifreeze to check the glycol/water ratio, and to inaccessible liquids such as the electrolyte of rechargeable cells [2]. The clinical applications of light have stimulated interest in biotissue refractometry [3], and refractometry is useful for the analysis of small samples of biofluids.

61.2 Physical Principles

The electromagnetic nature of light and the atomic origin of refraction are described herein only in enough detail to be useful to refractometry; detailed descriptions of them are readily found in optics textbooks [4].

Monochromatic light can be described as an electromagnetic radiation of frequency v (Hz) and free-space wavelength λ_0. In free space (vacuum), the phase velocity c_0 is given by $c_0 = v\lambda_0$. In a medium or material substance, the phase velocity changes to $c = v\lambda$, the refractive index n of that substance then being defined as $n = c_0/c$. As the frequency is unchanged by entering the medium, $n = \lambda_0/\lambda$. In an electromagnetic wave, the vibration directions of the electric and the magnetic fields are 90° apart, and both fields vibrate at 90° to the wave's direction of propagation. In transparent and nonmagnetic materials, refractive index is determined primarily by the interaction of the electric field of the wave with the permittivity ε of the medium. This quantity is written as $\varepsilon = \varepsilon_r \varepsilon_0$, where ε_r is the relative permittivity of the medium, and ε_0 is the permittivity of free space being $10^7/(4\pi c_0^2)\text{F m}^{-1}$, or about 8.854 pF m^{-1}. In the wave theory of light, ε_r is determined by the strength of elastic interaction between the wave's electric field and the bound charges (electrons and protons) in the atoms and molecules of the medium. For nonmagnetic transparent materials, $n = \sqrt{\varepsilon_r}$ can be shown if both quantities are measured at the same frequency, and this relation is also closely true if the absorption is weak. Absorption can be ascribed to the inelasticity of the interaction of the wave's electric field with the bound charges, some of the wave's energy being turned to heat during propagation.

The time (t) and distance dependence of the electric field E of a wave propagating in the z direction can be written

$$E = E_0 \exp i(\omega t - k_0 z) \tag{61.1}$$

where $\omega = 2\pi v$ = the pulsatance or angular frequency, $k_0 = 2\pi/\lambda_0$ = the free-space wave number, and $i = \sqrt{-1}$. The field E in Equation 61.1 is complex, and the physical field is understood to be the real part of the complex one, i.e. $E = E_0 \cos(\omega t - k_0 z)$. The wave amplitude E_0 is the maximum excursion of E, both quantities being coparallel and normal to the direction of propagation z. The optical power per unit area (radiance) conveyed by the wave is proportional to $|E|^2$, or also $|E_0|^2$. The phase velocity c is given by ω/k, which the above relations show to be c_0/n. Waves in absorbing media are attenuated as z increases. This is described by a complex refractive index in Equation 61.1, i.e., writing $n = n' - n''$ so that the field oscillates in space and time as before, but decays exponentially with propagation distance z as $(-k_0 n'' z)$. The cause of the index n'' may be absorption, scattering, or both. It will be clear from context whether n refers to the real part of the complex quantity, or to the complex quantity itself. For the most part, this chapter deals with transparent or only slightly absorbing materials, where n'' is very small.

Equation 61.1 shows that n determines the phase of a wave, and this is the basis of the interferometric methods of refractometry, most used for either highly precise measurements on liquids or solids, or for gases where the refractive index is close to 1. Other common refractometric methods rely on the reflection or transmission of light at the interface between the sample and a material of known refractive index. Reflection and refraction of light at optically smooth interfaces are treated (Fresnel relations) in optics textbooks. A description follows that should suffice for the experimental refractometrist.

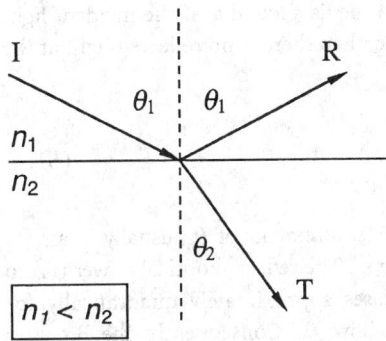

FIGURE 61.1 Light beams reflected (R) and transmitted or refracted (T) at the interface between two refractive indices, n_1 and n_2, the second being greater than the first.

Simple laws apply to the incident, reflected, and transmitted beams at a planar interface between two transparent media. For brevity, we consider only isotropic media. A parallel-collimated incident light beam produces a parallel-collimated reflected and transmitted beam. All beams are coplanar and define a plane of incidence, which is orthogonal to the planar interface. The interface normal also lies in the plane of incidence. Figure 61.1 shows the incident, reflected, and transmitted beams in the plane of incidence. In the first medium, the angle of incidence and reflection made to the interface normal are equal; both θ_1. The angle of transmission θ_2 into the second medium is given by Snell's law

$$n_1 \sin\theta_1 = n_2 \sin\theta_2 \qquad (61.2)$$

where the subscripts refer to the first and second medium. If n_1 is known and the angles are measured, n_2 can be obtained. Equation 61.2 is the basis of deviation refractometry, *q. v.*

If $n_1 > n_2$, Equation 61.2 shows that real angles of transmission do not exist for $\theta_1 > \arcsin(n_2/n_1)$; i.e., θ_2 is imaginary. The incident and reflected beams are then of equal power, transmitted light is absent, and the reflection is said to be total. However, although in the second medium no wave propagates, there is an evanescent wave close to the interface, which is nonpropagating and therefore transmits no power. The critical angle is given by

$$\theta_c = \arcsin\!\left(\frac{n_1}{n_2}\right) \qquad (61.3)$$

(See Figure 61.2.) Reversing the light direction, Figure 61.3 shows that a beam traveling parallel to the interface in the least refractive index medium enters the second medium at a transmission angle also given by θ_c. A full analysis (Fresnel's equations) shows that, in the region of the critical angle, the reflected or transmitted optical power changes very rapidly with angle of incidence. If there is no absorption, there is actually a discontinuity of reflected or transmitted radiance at θ_c. This makes θ_c easy to measure accurately and gives the critical angle or Abbe method of refractometry used by most commercial refractometers.

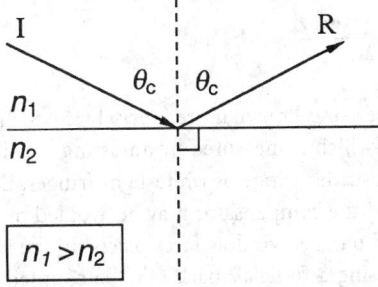

FIGURE 61.2 Total reflection at the critical angle of the incident beam (I) in the first medium of greater refractive index than the second. No light is transmitted into the second medium for angles exceeding the critical angle.

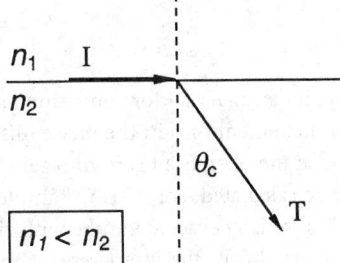

FIGURE 61.3 Total refraction of incident beam (I) transmitted as beam T into the medium of lesser refractive index at the critical angle. No light exists for angles in the second medium bigger than the critical angle.

The Fresnel laws of reflection at a plane interface between two media show that if the incident light is linearly polarized with its electric field in the plane of incidence, then there is no reflected light at the Brewster angle θ_B given by

$$\theta_B = \arctan\left(\frac{n_2}{n_1}\right) \tag{61.4}$$

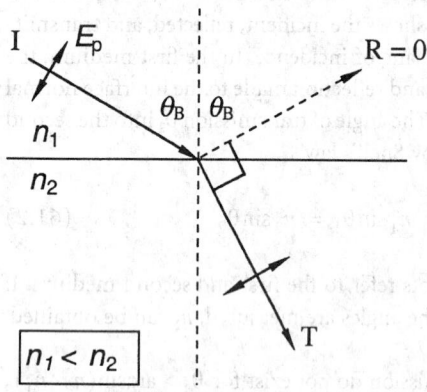

FIGURE 61.4 Light incident in the first medium at the Brewster angle, and polarized with its electric vector parallel to the plane of incidence, is transmitted (T) wholly into the second medium with zero reflection R.

(See Figure 61.4.) Measurements of θ_B, usually in air (n_1 close to 1), will give n_2. The reflected optical power is zero at θ_B and it increases approximately quadratically for angles below and above θ_B. Consequently, the Brewster angle method is regarded to be of lower precision than the critical angle method, and so is relatively less used. However, unlike the critical angle method, it does not require a reference medium of higher refractive index than the sample.

If the sample is a finely divided particulate solid or has an irregular interface, none of the preceding methods is useful. Fresnel's equations show the reflectance of any shape of interface to be zero if the refractive index difference across the interface is also zero. In the method of refractive index match, the turbidity or scattering from a cloud of particles suspended in a liquid is measured, and the refractive index of the liquid is found, which gives minimum turbidity or scattering. The liquid should not swell or interact with the interior of the solid.

61.3 Techniques

Interferometry

The sample, with parallel input and output faces, is put in the sample beam of a two-beam interferometer (e.g. Rayleigh, Michelson, or Jamin). From Equation 61.1, a sample of length z and refractive index n causes a phase lag of $2\pi n z / \lambda_0$, compared with $2\pi n_{air} z / \lambda_0$ for the reference beam in air. The phase difference between the two beams is thus

$$\delta = \frac{2\pi z(n - n_{air})}{\lambda_0} \tag{61.5}$$

The refractive index of air is 1 for approximate purposes, or is known more precisely [5]. Sample insertion causes a step increment of δ in the phase difference, which is measured by adjusting a calibrated phase compensator in the reference beam to regain the zero-order fringe of white-light fringes. Equation 61.5 enables n to be calculated. For gaseous samples, use of the compensator may be avoided by allowing the gas to enter an initially evacuated tube with rigid and parallel window faces placed in the sample beam. As the gas enters slowly, the interference fringes passing a fudicial mark may be counted. Each whole fringe corresponds to a change in δ of π radian. The refractive index of a gas depends, through its density, on pressure and temperature. The number of fringes between a vacuum ($n = 1$) and the gas at a given pressure and temperature will then give the gas refractive index.

In interferometry, the effect of optical attenuation (via absorption or scattering) by a sample of complex refractive index decreases the fringe brightness but does not affect the phase. Thus, only the real part the complex refractive index is measured.

About 0.1 of a fringe can be judged by eye; electro-optic methods can measure 10^{-6} of a fringe. The method's potential precision is very high. However, an interferometer is delicate and expensive, and good sample quality (homogeneity, face parallelism) is necessary. A difference in dispersion (wavelength dependence) of the sample and compensator refractive index can be a problem. For a sufficiently large number of fringes, the dispersion difference between the sample and compensator can make the white-light fringe pattern and zero-order fringe invisible [6]. One solution is to use a compensator more closely matched to the sample dispersion. A cheaper option is to reduce the sample length, with some loss of precision. A major use of the interference method is for gases, where $n - 1$ is about 1000 times less than $n - 1$ of solids and liquids. It is also used in solution differential refractometry, e.g. to measure the concentration in the eluted solvent relative to the pure solvent. A stable folded Jamin interferometer for refractive index measurements is described by Moosmüller and Arnot [7]. Liquid refractive index measurements using a Michelson interferometer are described by Richerzhagen [8].

Deviation Methods

Lateral and angular deviation methods make use of Snell's law (Equation 61.2). The incident beam is usually in air, of refractive index n_{air}.

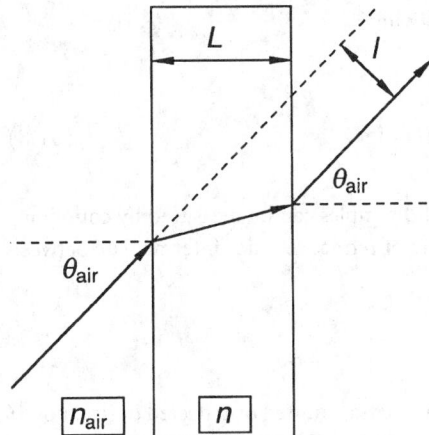

FIGURE 61.5 Lateral deviation l of a light beam on passage through a parallel-sided refractile sample slab of thickness L.

Lateral deviation l occurs for a beam of light transmitted through a parallel-sided sample of thickness L (see Figure 61.5). If the incident angle is θ_{air}, Snell's law (Equation 61.2) gives

$$\frac{n}{n_{air}} = \left[1 + \left(\frac{\cos\theta_{air}}{\sin\theta_{air} - \dfrac{l}{L}}\right)^2\right]^{\frac{1}{2}} \sin\theta_{air} \qquad (61.6)$$

If θ_{air} and L can be measured the most accurately, the precision of the method depends on the accuracy of measuring l. Modern electro-optic beam-displacement detection devices permit a precision of a few micrometers in l, giving a typical precision in n of about 0.001. The lateral deviation method is suitable for samples in sufficiently thick sheet form.

In the angular deviation method, parallel-collimated light is incident on one face of the sample of refractive index n in triangular prism form with vertex angle A, the vertex line of the prism being normal to the plane of incidence (see Figure 61.6). The exit beam deviates angularly from the direction of the incident beam. At the angle of minimum deviation D, the analysis is simplest Equation 61.1, showing that

$$\frac{n}{n_{air}} = \frac{\sin\left(\dfrac{A + D}{2}\right)}{\sin\left(\dfrac{A}{2}\right)} \qquad (61.7)$$

For $A = 60°$, a 1 minute (1') uncertainty in D gives 0.0003 for the corresponding precision in n. Angular deviation requires the sample to be a less convenient shape than the parallel-sided sheet of the lateral deviation method. However, measurements of D and A can be made simply using an ocular goniometer. Lateral or angular deviation methods can be applied to liquid samples in a suitable cell with walls having parallel sides. Such cell walls will, however, give some lateral deviation but not cause angular deviation if used to make a hollow prism to contain a liquid sample. In the Hilger-Chance angular deviation

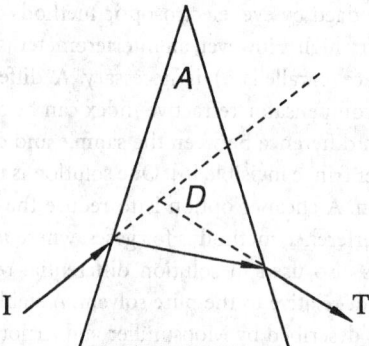

FIGURE 61.6 Angular deviation D of a light beam on passage through a prism of vertex angle A. If the interior beam forms an isosceles triangle with A at the vertex, then D is the angle of minimum deviation.

refractometer [9], the liquid sample is contained in a glass 90° V-block of refractive index n_{glass}, with parallel outer faces. A parallel-collimated light beam incident normally on one outer face of the block suffers angular deviation D (not minimum), Equation 61.2, giving

$$n^2 = n_{glass}^2 + (n_{glass}^2 - \sin^2 D)^{\frac{1}{2}} \sin D \tag{61.8}$$

For a 1′ uncertainty in D, the precision in n is about 10^{-5}. Solid samples can be measured by coupling a prism of the sample into the V-block using a thin film of liquid of refractive index intermediate between that of the sample and the prism.

Critical Angle Method

In the region of the critical angle, Fresnel's equations show that the transmitted or reflected optical power varies discontinuously with angle. Thus, θ_c is accurately measurable and a refractive index precision of about 10^{-4} to 10^{-5} is obtained. Liquid samples are placed directly onto one face of the refractometer prism of known refractive index. Solid samples need only one face to be flat and optically polished, which is coupled to face of the refractometer prism using a liquid film of intermediate refractive index. For a transparent sample, either the transmission or the reflection mode can be used, as θ_c is the same for both (see Figures 61.2 and 61.3). In the transmission mode, the sample–prism interface is illuminated diffusely through the sample, the critical angle then being the largest angle of transmission in the prism. In the reflection mode, useful for highly scattering or absorbing samples, the prism–sample interface is illuminated diffusely from within the prism. The critical angle is then where the reflectance changes discontinuously with angle of reflection within the prism, being total for $\theta \geq \theta_c$. In the transmission mode, the transmittance from the sample into the prism is zero for $\theta \geq \theta_c$ and finite for $\theta \leq \theta_c$ whereas, in the reflection mode, typically a few percent of the light is reflected for $\theta \leq \theta_c$, and 100% is reflected for $\theta \geq \theta_c$. If the interface is illuminated with diffuse light, the transmission mode gives a dark–bright contrast at the critical angle, with a dim–bright contrast for the reflection mode. For ocular observation, this difference can make the transmission method preferable. Automatic instruments invariably use the reflection mode, this being necessary for samples that attenuate by absorption or scattering. For such samples, however, the discontinuity of reflectance is replaced by a continuous function with a maximum gradient at an angle θ_{max} that differs increasingly from θ_c as the sample becomes more absorbing [10–12]. A separate measurement of the absorption index is required before θ_c can be obtained from θ_{max}. Attempts to improve the precision of critical angle refractometry of nontransparent samples are expected to become increasingly limited by this effect.

The critical angle method, however, is robust and simple to use. Despite the problems associated with optically attenuating samples, it is likely to persist as the preferred method in commercial refractometers.

Brewster Angle Method

Here, the angle of reflection θ_B is measured for which there is no linearly polarized light reflected from the sample surface. Equation 61.4 then gives the sample refractive index, where n_1 is the refractive index of the medium containing the incident and reflected light, usually air. The light is polarized such that its electric field vector vibrates parallel to the plane of incidence (see Figure 61.4). A linearly polarized laser is a suitable source of monochromatic light. Nonpolarized sources require polarizing with a Glan-Thomson prism or a quality dichroic polarizer. With a 1 mW HeNe laser and ocular judgement of the minimum reflectance, a precision in θ_B of about 0.1° can be obtained, giving a precision in the sample refractive index of about 0.005. Precision improvement by factors of 10 to 100 can be expected if photometric analysis replaces the eye. The Brewster angle method is then comparable in precision to the critical angle method. Unlike the critical angle method, however, it does not require a reference material of higher refractive index than the sample. If the sample is absorbing, the reflectance zero at θ_B is replaced by a reflectance minimum at an angle θ_{min}. It can be shown that a given absorption produces less error in the Brewster angle method than in the critical angle method [6, 12]. Despite these good features of the Brewster angle method, a commercial Brewster angle refractometer is not known to this author.

Index Match Method

The refractive index match method does not require the sample to be in any special shape or form. Thus, it might be finely divided, i.e., a powder. Assuming the particles to be isotropic and homogeneous, then the reflected radiance, the angular or lateral deviation, and the relative phase shift of transmitted light all become zero at refractive index match. If the sample is absorbing, the refractive index of the surrounding liquid which gives the minimum visibility still gives closely the real part of the sample's refractive index. Similarly, if the sample is heterogeneous or anisotropic, the method gives the average refractive index of the sample. The liquid should not dissolve, permeate, or react in any way with the solid sample.

For macroscopic samples a precision of about 0.005 in n has been estimated [6] from the index-match method, where a simple ocular estimation of match is employed. For a matching liquid of refractive index n_{liq}, the reflectance from an interface varies approximately as $(n - n_{liq})^2$. This quadratic dependence makes the sensitivity of the method decrease as the match is approached, and makes it difficult in practice to judge whether n_{liq} should be increased or decreased to improve the match.

Microscopy and Scattering Methods

The refractive index of individual particles large enough to image in a microscope can also be measured by the index-match method. Central or oblique illumination modes can be used corresponding to the Becke, and Schroeder-Van der Kolk methods, respectively, which utilize null deviation at refractive index match. A precision of 0.0002 in n is quoted [13] for the Becke line method. Whether n_{liq} is lesser or greater than the sample's refractive index n is shown by the appearance of the image.

The phase difference between light transmitted through and around the particle is proportional to $n - n_{liq}$, so that the phase changes sign each side of match. This is used in measurement of refractive index of particles by phase-contrast or interference microscopy, in which the image is brighter or darker than the background according to whether $n > n_{liq}$ or $n < n_{liq}$. These methods can measure the refractive index of particles below the conventional resolving limit, as the details of an object need not be fully resolved in order to discern match or otherwise. Users of microscopy in refractometry should refer to texts on phase-contrast and interference microscopy. Newer forms of optical microscopy (e.g., scanning confocal) have greater resolving power and can be used for refractive index matching of finer particles.

For a suspension of particles smaller than about 1 μm, the optical turbidity and light scattering are zero at refractive index match, leading to turbidimetric or nephelometric methods of measuring the particle refractive index. For particles of colloidal size, it is important to avoid aggregation or flocculation, and suspension destabilization. Thus, chemical restrictions can preclude the attainment of match, particularly if the particle's refractive index is large compared with n_{liq}. In such cases, the turbidity E is measured for changes in n_{liq} which do not destabilize the suspension, and the value of n_{liq} is found by extrapolation to $E = 0$. Extrapolation to obtain this value of n_{liq} is difficult unless the experimental parameters can be arranged to have a linear relation. Theory and experiments described by Griffin and Griffin [14] showed that linearity is obtained if n_{liq}/E is plotted versus n_{liq}.

Anisotropic and Heterogeneous Materials

If a plane face is made from a general optically anisotropic sample, and this face is then offered to the prism of a critical angle refractometer, two critical angles can be simultaneously observed. One can be enhanced over the other through the use of a linear polarizer rotated in the incident or refracted light path. The interpretation of these critical angles, in the general case, is not simple, and the reader is referred to analyses [15–17]. Similar complexity is visited upon the Brewster angle method if the sample is anisotropic, and again the reader is referred to Reference 6 for works on ellipsometry.

Highly turbid samples, such as polymer latexes and other concentrated colloidal fluids, have been investigated using the critical angle method [1, 18–20]. It appears that if the particle size is less than about half a wavelength, the critical angle method remains successful for volume fractions up to at least 0.5. For high concentrations of larger particles, there is a rapid degradation of the reflectance or transmittance discontinuity at the critical angle. This angle is replaced by a region where the reflectance or transmittance may vary in an unexpected way as the particle size and concentration increase.

The effect of heterogeneity on the reflection of polarized light, and on the Brewster angle, is also affected by sample heterogeneity [21, 22]. Whereas for a homogeneous sample the parallel-polarized reflectance at the Brewster angle is zero, the effect of sample heterogeneity on the wavelength scale is to cause finite reflectance. Comparison of the Brewster angle and the critical angle methods for heterogeneous materials awaits investigation.

Gate and coworkers have measured the refractive index of emulsion paint films [23] and coated paper [24] by a non-Brewster angle reflection method. Both materials were strongly heterogeneous on the wavelength scale and so (in air) had a strong nonspecular or diffuse reflectance. Some specular reflection remained, however, and the method derives the refractive index of the surface of the film from the ratio of the specular reflectances at a non-Brewster angle for polarizations parallel and normal to the plane of incidence.

The application of refractometry to heterogeneous and absorbing samples is clearly a developing art and, although it is clear that refractometry offers valuable information, the refractometrist should use existing instruments critically.

61.4 Review of Refractometers

The author's initial review of refractometer manufacturers revealed about 30 companies, and the true number may be more than 50. A very large number of general- and special-purpose instruments are offered by each company, e.g., 31 different models by the Atago Co., Ltd. It is clearly not possible to review, compare, and advise on individual models. The aspiring refractometrist should do this according to needs and budget. The generalities that follow should assist.

Most instruments use the critical angle (or Abbe) method. They thus have a prism of known refractive index with a face to which the sample is offered. The larger manufacturers offer three types of critical angle instrument:

1. The portable hand-held models, which require no power or battery and are customized and calibrated for Brix (sugar), urine, antifreeze, cutting oil, etc. Their refractive index precision is typically between 10^{-3} and 10^{-4}, but their range is small owing to their particular application. The user looks into an eyepiece and reads a scale on which is superposed the dark–light boundary

caused by transmission or reflection in the region of angle each side of the critical angle; often termed the critical edge. The price varies greatly between models and manufacturers, typically U.S. $200 to $400.

　　Some manufacturers offer a digital readout portable models, of similar weight (less than 1 kg) and precision, and battery powered. These cost roughly U.S. $1000 to $2000.

2. The laboratory (Abbe) refractometer, which has a refractive index precision of typically 10^{-4} to 10^{-5} and covers a wide range of refractive index, typically 1.3 to 1.75 in a low refractive index instrument, and 1.4 to 1.85 in a high refractive index model. Such instruments are calibrated so that the temperature can be varied using an external liquid bath, and the wavelength can be varied using external light sources. Very roughly, their weight, cost, and precision are about 10 times that of the portable instruments in (1) above. They require the user to estimate the position of the critical edge on a graduated scale viewed through an eyepiece.

3. The digital refractometer. Compared with those in (2) above, these have similar precision and general facilities, a cost typically a few times greater, but with the advantage of a digital or electronic readout of the data. The upper refractive index offered by several manufacturers appears to be restricted to about 1.55.

4. The process refractometer is a version of those described in (3) above, but where the prism is mounted to contact a fluid in a reactor or a pipe, and viewed via a fiber-optic cable. Thus, high pressures and temperatures can be accommodated. The prism is usually of sapphire to avoid wear of the optical face, and the precision generally tend to be somewhat lower than those described in (2) and (3) above.

References

1. M. Mohammadi, Colloidal refractometry: meaning and measurement of refractive index for dispersions; the science that time forgot, *Adv. Colloid Interface Sci.*, 62, 17–29, 1995.
2. J. E. Geake and C. Smalley, A simple linear hand-held refractometer, *J. Phys. E: Sci. Instrum.*, 16, 608–610, 1983.
3. H. Li and S. Xie, Measurement method of the refractive index of biotissue by total internal reflection, *Applied Optics*, 35, 1793–1795, 1996.
4. M. Born and E. Wolf, *Principles of Optics*, 6th ed., Oxford: Pergamon, 1980.
5. P. E. Ciddor, Refractive index of air: new equations for the visible and near infrared, *Applied Optics*, 35, 1566–1573, 1996.
6. G. H. Meeten, Refraction and extinction of polymers, in *Optical properties of polymers*, Ed. G. H. Meeten, London: Elsevier Applied Science, 1986.
7. H. Moosmüller and W. P. Arnott, Folded Jamin interferometer: a stable instrument for refractive index measurements, *Optics Lett.*, 21, 438–440, 1996.
8. B. Richerzhagen, Interferometer for measuring the absolute refractive index of liquid water as a function of temperature at 1.064 μm, *Applied Optics*, 35, 1650–1653, 1996.
9. R. S. Longhurst, *General and Physical Optics*, 3rd ed., London: Longmans, 1973.
10. G. H. Meeten, A. N. North, and F. M. Willmouth, Errors in critical-angle measurement of refractive index of optically absorbing materials, *J. Phys. E: Sci. Instrum.*, 17, 642–643, 1984.
11. P. R. Jarvis and G. H. Meeten, Critical angle measurement of refractive index of absorbing materials, *J. Phys. E: Sci. Instrum.*, 19, 296–298, 1986.
12. G. H. Meeten, Refractive index errors in the critical-angle and Brewster-angle methods applied to absorbing and heterogeneous materials, *Meas. Sci. Technol.*, 8, 728–733, 1997.
13. C. P. Saylor, *Advances in Optical and Electron Microscopy*, Vol. 1, Ed. R. Barer and V. E. Cosslett, London: Academic Press, 1966.
14. M. C. A. Griffin and W. G. Griffin, A simple turbidimetric method for the determination of the refractive index of large colloidal particles applied to casein micelles, *J. Colloid Interface Sci.*, 104, 409–415, 1985.

15. F. Yang, G. W. Bradberry, and J. R. Sambles, A method for the optical characterization of thin uniaxial samples, *J. Modern Optics*, 42, 763–774, 1995.

16. F. Yang, G. W. Bradberry, and J. R. Sambles, Critical edge characterization of the optical tensor of a uniaxial crystal, *J. Modern Optics*, 42, 1241–1252, 1995.

17. F. Yang, J. R. Sambles, and G. W. Bradberry, A simple procedure for characterizing uniaxial media, *J. Modern Optics*, 42, 1477–1458, 1995.

18. G. H. Meeten and A. N. North, Refractive index measurement of turbid colloidal fluids by transmission near the critical angle, *J. Meas. Sci. Technol.*, 2, 441–447, 1991.

19. J. E. Geake, C. S. Mill, and M. S. Mohammadi, A linear differentiating refractometer, *J. Meas. Sci. Technol.*, 5, 531–539, 1994.

20. G. H. Meeten and A. N. North, Refractive index measurement of absorbing and turbid fluids by reflection near the critical angle, *J. Meas. Sci. Technol.*, 6, 214–221, 1995.

21. E. K. Mann, E. A. van der Zeeuw, Ger J. M. Koper, P. Schaaf, and D. Bedeaux, Optical properties of surfaces covered with latex particles: comparison with theory, *J. Phys. Chem.*, 99, 790–797, 1995.

22. E. A. van der Zeeuw, L. M. C. Sagis, and Ger J. M. Koper, Direct observation of swelling of non-cross-linked latex particles by scanning angle reflectometry, *Macromolecules*, 29, 801–803, 1996.

23. L. F. Gate and J. S. Preston, The specular reflection and surface structure of emulsion paint films, *Surface Coatings International*, 8, 312–330, 1995.

24. L. F. Gate and D. J. Parsons, The specular reflection of polarised light from coated paper, *Products of Papermaking: Transactions of the Tenth Fundamental Research Symposium held at Oxford: September 1993*, Vol 1, 263–283, Leatherhead: Pira International 1994.

62

Turbidity Measurement

Daniel Harrison
John Carroll University

Michael Fisch
John Carroll University

62.1 Introduction

For nearly 50 years, turbidity measurements have been used to perform a wide variety of physical measurements. These include the determination of particle concentration per unit volume when the scattering cross-section per particle is known, the determination of particle size (scattering cross section) when the concentration of particles is known [1–3], and the determination of some of the critical exponents associated with second-order and nearly second order phase transitions [4–7].

The basic ideas necessary to understand turbidity measurements are fairly simple. In the absence of reflection losses, when a weak beam of light passes through a dielectric sample, the two processes most responsible for reducing the intensity of the transmitted beam are absorption and scattering. The reduction in transmitted light intensity due to scattering is called the sample's turbidity. Extinction includes the effects of both absorption and scattering. The Beer-Lambert or Lambert law describes the effects of both absorption and turbidity on the transmitted intensity. This law is written as

$$I_T = I_0 \exp{-(\alpha + \tau)l} \tag{62.1}$$

where I_T = intensity of the light transmitted through the sample, I_0 = intensity of the light incident on the sample, α = absorption coefficient per unit length, τ = turbidity per unit length, and l = length of the light path in the sample. As discussed below, it is more correct to use powers rather than intensities, thus this equation should be written as:

$$P_T = P_0 \exp{-(\alpha + \tau)l} \tag{62.2}$$

The most general situation is to have an absorbing medium with absorbing and scattering particles embedded within it; however, the focus here is on the simpler case of nonabsorbing medium and consider the two most common cases: nonabsorbing particles in medium and scattering caused by fluctuations in the medium itself.

0-8493-8347-1/99/$0.00+$.50
© 1999 by CRC Press LLC

Instruments generally fall into two categories: commercial and laboratory constructed. The commercial units are of two general types: (a) attachments to spectrophotometers, and (b) white-light turbidity meters, which operate under ambient conditions. Anyone interested in the former should consult the catalog of accessories for the instrument in question. Those interested in the latter should check under "nephelometers" in scientific supply catalogs. Units of this type perform their designed function admirably and probably are sufficient for routine work. However, they may need modification for the more specialized measurements performed in many research laboratories. For this reason, it is common to construct special laboratory instruments. These come in at least two types, most commonly either single-beam or dual-beam instruments. The single-beam instrument usually has an intensity-stabilized light source, whereas the dual-beam instrument corrects for drift in the light source and reflection losses by either electronically or mathematically taking the ratio of the transmitted light power and a reference beam power.

The balance of this chapter is divided into three parts. The first part briefly discusses the physical basis of turbidity measurements and demonstrates how such measurements may be used to infer the scattering cross section of the particles in the solution or the concentration of the scatters in the solution. The second section discusses turbidity of pure fluids and shows how certain critical exponents may be determined from such measurements. The last section discusses laboratory instruments and the relative trade-offs involved in such instruments.

62.2 Extinction and Turbidity: Particles in a Nonabsorbing Medium

Suppose that electromagnetic radiation is incident upon a slab of medium consisting of randomly positioned particles and that the transmitted radiation is detected as shown in Figure 62.1. For the present discussion, assume that the source and detector are in the medium. The radiation that is received by the detector will be less than that incident on the slab because of the presence of the particles in the medium—the particles have caused extinction (or attenuation) of the beam. The extinction of the radiation depends on two physical processes, scattering and absorption, whereas the turbidity depends only on scattering. In scattering, there is no change in the total energy of the radiation; rather, some of

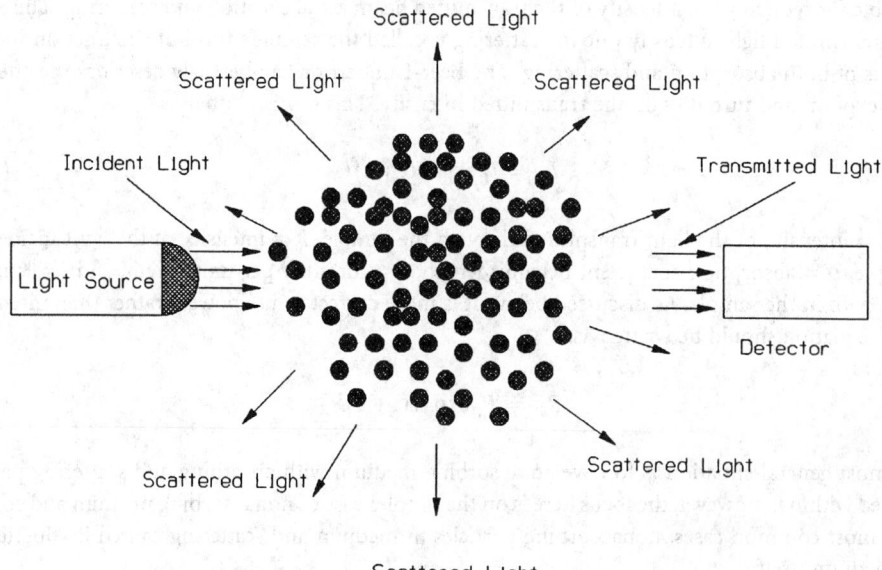

FIGURE 62.1 Idealized experiment indicating the physical basis of turbidity.

the incident radiation is redistributed away from the incident direction. In absorption, some of the energy of the incident beam is transformed into other energy forms. This extinction of the beam is rather complicated; it will, in general, depend on the chemical composition, size, shape, number, and orientation of the particles; the chemical composition of the medium; and the frequency and polarization of the incident radiation [2,3].

In discussions of scattering one often focuses on the *cross sections*, C_i, of the particles. On the basis of conservation of energy one must write:

$$C_{ext} = C_{scatt} + C_{abs} \qquad (62.3)$$

where C_{ext}, C_{scatt}, and C_{abs} = cross sections for extinction, scattering, and absorption, respectively. These cross sections all have the dimension of area, and in all cases a larger cross section (area) indicates a larger effect, and a smaller cross section a smaller effect. In general the cross section will depend on all the intensive quantities that describe the extinction. Finally, note that for nonabsorbing media, $C_{abs} = 0$.

To see how this relates to the measured turbidity, assume now that the medium is nonabsorbing, that all the particles are nonabsorbing and have the same scattering cross section (C), and that there are n of these particles per unit volume. Then, one obtains the following expression for the transmitted light power:

$$P_T = P_0 \exp{-(nCl)} \qquad (62.4)$$

where P_T = transmitted power, and P_0 = incident power. We have switched from intensity to power because power is the quantity detected, and power is independent of the details of the spatial distribution of electromagnetic radiation. Upon noting this change and comparing this to Eq. 62.2, when $\alpha = 0$, it is clear that the turbidity can be expressed as

$$\tau = nC \qquad (62.5)$$

Thus, a measured turbidity can be related to C_{scatt} if n is known, or n may be determined if C_{scatt} is known. The difficulty is then in determining C_{scatt}. There are a number of approaches that can be used. For well characterized particles, theoretical expressions may be determined [2]. In the absence of such theory, measurements may be made as a function of concentration and then, since $\tau = nC$, C may be determined. The generalizations to account for nonmonochromatic beams and extinction by a collection of noninteracting particles of the same type but different sizes are straightforward. The simplest generalization to include different sizes assumes that there is one parameter, ζ, that describes the distribution of particle size and the corresponding C_{scatt} [1]. That is, let $C_{scatt}(\zeta)$ be the scattering cross section for a particle characterized by the parameter (for instance, radius) ζ, and let the number per unit volume with parameter between ζ and $\zeta + d\zeta$ be $N(\zeta)d\zeta$, such that

$$n = \int_0^\infty N(\zeta)d\zeta$$

Then

$$\tau = \int_0^\infty C_{scatt}(\zeta)N(\zeta)d\zeta \qquad (62.6)$$

A similar integral exists when the incident radiation is not monochromatic, but, in this case, the average is over the wavelength variation of the incident radiation.

62.3 Turbidity Due to Density Fluctuations in Pure Fluids

It is well known that even in very pure and well filtered liquids, in which there are essentially no particles, there is still some reduction in the detected power due to scattering [3]. The explanation is that there are always thermal fluctuations in the dielectric constant of the media, ε, and it is these fluctuations from the mean dielectric constant that lead to the scattering of light. Through application of thermodynamic and statistical mechanical principles [3], the following expression for the turbidity results:

$$\tau = \frac{8\pi^3}{2\lambda^4}\left[\rho\left(\frac{\partial \varepsilon}{\partial \rho}\right)_T\right]^2 kT\beta_T \tag{62.7}$$

where ρ = density, ε = dielectric constant, k = Boltzmann's constant, T = absolute temperature, λ = wavelength of the incident light, and β_T = isothermal compressibility. Near a second order phase transition the fluctuations become correlated over larger distances, and this expression must be generalized to include the effects of this increase in the *correlation length*, ξ. In this case the turbidity is given by [5]

$$\tau = A\pi\beta_T\left[\frac{2\alpha^2 + 2\alpha + 1}{\alpha^3}\ln(1 + 2\alpha) - \frac{2(1 + \alpha)}{\alpha^2}\right] \tag{62.8}$$

where

$$A = \frac{\pi^2}{\lambda^4}\left[\rho\left(\frac{\partial \varepsilon}{\partial \rho}\right)_T\right]^2 kT \qquad \text{and} \qquad \alpha \equiv 2\left(\frac{2\pi n}{\lambda}\xi\right)^2$$

where n is the (mean) index of refraction of the fluid. Both the isothermal compressibility and the correlation length exhibit approximate power-law behavior of the form

$$\beta_T, \xi \sim \left(\frac{|T_c - T|}{T_c}\right)^{-x_i} \tag{62.9}$$

near these phase transitions. Thus, a measurement of $\tau(T)$ allows the determination of the critical exponents, x_i, and the values of β_T and ξ far from the phase transition. This technique has been generalized to binary liquid mixtures, liquid crystals, and other systems. In these cases, the isothermal compressibility is replaced by the appropriate generalized susceptibility. There are also, in some cases, small corrections to the above expressions. These can also be found in References 8–10.

62.4 Design of Laboratory Instruments

When more exact measurements are required than can be obtained using commercial instruments, and other enhancements such as mK temperature and/or pressure controlled sample chamber are required, it is common to construct a turbidity instrument. There are basically two types of instruments (single-beam and dual-beam). With both types of instruments, both ac and dc detection schemes are possible. Generally, the long-term time stability of synchronous ac detection techniques is superior to dc techniques. This section discusses all of these techniques.

Single-Beam Instruments: Optics

The basic optical design of a single-beam instrument is shown in Figure 62.2. The light source is a low power (1 to 5 mW) vertically polarized He-Ne laser. It is important to use a polarized laser. The

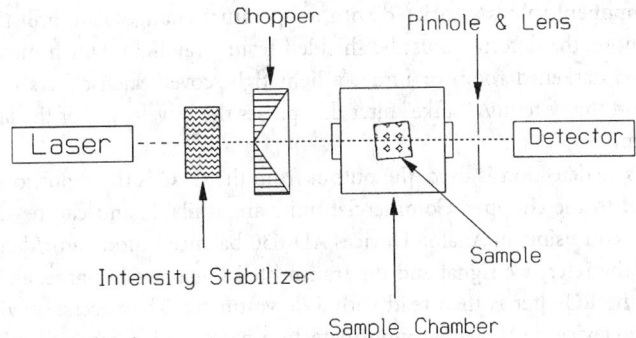

FIGURE 62.2 Block diagram of the optics for a single-beam turbidity apparatus.

polarization direction of an unpolarized laser wanders, and this can lead to time-dependent reflection coefficients from the various surfaces in this instrument, and hence reduced performance. The beam passes through a laser intensity stabilizer. This device reduces long-term drift in the laser power to ≤ 0.1 percent, thus allowing single-beam operation. Table 62.1 shows several commercial intensity stabilizers. To reduce any possible heating of the sample due to absorption, the beam is then passed through a neutral density filter, which attenuates the beam further to a level of 50 to 100 μW. The next item in the beam line depends on the detection scheme. For ac operation a light chopper is inserted into the beam line, while for dc detection nothing else is required before the sample. An output from the chopper is used for synchronous detection. The sample chamber then follows. This chamber requires an input and an output port and a design that precludes interference of any beam with other reflected beams. Other design details are determined by the materials to be studied and any required external factors that are to be controlled. After the beam has been transmitted through the sample, it passes through either a pinhole or a lens-pinhole combination [11, 12]. This results in a small angle of acceptance for the detector which, in turn, reduces the scattered light at small angles from the transmitted beam that reaches the detector. A detector then completes the optical train. For most applications, a *pin* photodiode is an excellent, inexpensive detector; however, in some applications, other types of detectors may be superior.

TABLE 62.1 Commercial Laser Intensity Stabilizers

Manufacturer/Supplier	Model	Approximate Price	Comment
Thor Labs 435 Route 206 P.O. Box 366 Newton, NJ 07860-0366 www.thorlabs.com	CR200	$1200	Good low-cost, special-purpose unit
Cambridge Research and Instrumentation 21 Erie St. Cambridge, MA 02139 (617) 491-2627	LPC-VIS	$5000	Outstanding for general use

Single-Beam Systems: Electronics

For both synchronous and dc detection, the output of the photodiode, which is a current device, must be transformed into a voltage using a current-to-voltage converter (CVC). It is important not to use just a resistor for this function. The simplest workable current-to-voltage converter uses an FET-input operation amplifier such as an LM-11 and one resistor. The circuit, also known as a transimpedance amplifier, is standard and can be found in many sources [13].

When dc detection is used, the output voltage from the CVC is read on a voltmeter. The relationship between this voltage and the turbidity will be discussed below. The stability of dc operation is greatly improved by maintaining the photodiode and the electronics at a stable temperature. The drifts in detector efficiency and in component values associated with temperature changes can limit the accuracy of the instrument. Furthermore, the detector must be shielded from stray light, which means the instrument must be operated in a darkened room or under a light tight cover. Such covers may cause excessive heating of the laser and the system. A "spike" filter that passes the wavelength of the laser may also assist in solving this problem.

When synchronous ac detection is used, the output from the CVC is the input to a lock-in amplifier which is synchronized to the chopper. Commercial units are available and can be used. One can also construct a simple lock-in using an Analog Devices AD-630 balanced modulator/demodulator, a phase shifter to ensure that the reference signal and the transmitted signal are in phase, and an RC filter. The output voltage from the RC filter is then read with a dc voltmeter. The necessary wiring diagrams are available from Analog Devices [14]. Synchronous detection has the advantage of having lower $1/f$ noise than dc circuits. Also, synchronous detection can discriminate against signals not synchronized with the reference, and hence offers much greater immunity to room lights and other stray light sources. The advantages of such detection are discussed elsewhere [13, 15].

Single-beam techniques are most useful for measuring changes in turbidity. To see this, note that the voltage output from the current-to-voltage converter or the lock-in, V_t, is given by $V_t = AP_T$, where A is an instrument constant that includes factors such as current-to-voltage gain, light power level to current conversion in the photodiode, reflection losses from the various interfaces, incident light power, gain of the lock-in, and so on. Now, P_T is given by Equation 62.2, hence V_{dc}, the measured dc voltage may be written as:

$$V_{dc} = B\exp{-(\alpha + \tau)} \tag{62.10}$$

where $B = AP_0$ is another constant, and we have included absorption. B may be determined by putting a solution of known extinction in the instrument and solving Equation 62.10 for B. Then, knowing l, α, B, and V_{dc}, the turbidity may be found by solving Equation 62.10 for τ, that is

$$\tau = \frac{1}{l}\ln\left(\frac{V_{dc}}{B}\right) - \alpha \tag{62.11}$$

This technique is rather limited if a large number of different samples must be compared. For a given sample, the change in turbidity is easily measured as a function of externally controlled parameters. However, putting samples into and taking them out of the instrument usually produces small random errors that slightly change the instrumental constant A and degrade performance. In some cases, the difference in turbidity between two samples may be small and result in a situation in which one must deal with a small difference of large numbers. This situation may be rectified using dual-beam instruments.

Dual-Beam Instruments: Optics

The optical design of a dual-beam, two-detector instrument is shown in Figure 62.3. The first several parts of the optical train are the same as for the single-beam instrument, and the same comments apply in this case. The first real difference is that there need not be an intensity stabilizer in the dual-beam system. However, inclusion of the intensity stabilizer may improve system performance. The second difference is the inclusion of a beam splitter which splits the laser beam into two parts. We have found that a microscope slide works well as a beam splitter [16]. One of these beams (usually the reflected beam) becomes a reference beam that passes through a reference sample. The other passes through the sample as in the single-beam instrument. The beams from these two paths go through lens-pinhole combinations and fall on detectors. All of the comments that apply to the single-beam instrument also apply here. The associated electronics are also identical. Once more, both dc and ac detection may be used. Experience in our laboratory shows that the best results are obtained when the sample signal is

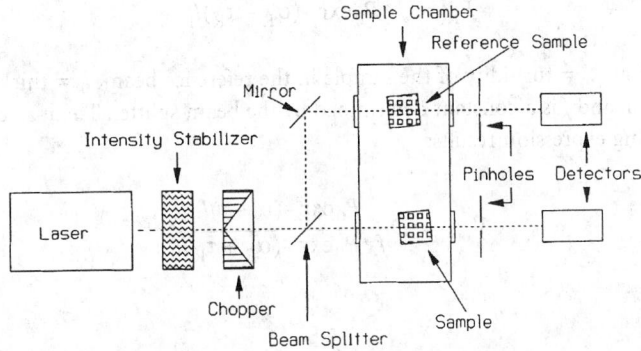

FIGURE 62.3 Block diagram of the optics for a dual-beam, two-detector turbidity apparatus.

simultaneously divided by the reference signal using an integrated circuit divider. A modification that uses two beams and one detector is shown in Figure 62.4. This modified design eliminates the need for absolute stability of two separate detectors but requires somewhat more complex optics.

The following analysis assumes that the measured output is the ratio of the sample signal voltage to the reference signal voltage. Let the voltage output from the detector's CVC be V_{sig}, and the voltage output of the reference signal detector's CVC be V_{ref}. Then

$$V_{sig} = AP_T \tag{62.12}$$

and

$$V_{ref} = CP_R \tag{62.13}$$

where A and C = instrumental constants totally analogous to the instrumental constant A above, and P_R = the power of the reference beam at the detector. It is important to realize that A and C are constant, but they need not be identical. The output voltage ratio, V_{out} is given by:

$$V_{out} = \frac{V_{sig}}{V_{ref}} = \frac{AP_T}{CP_R} = D\frac{P_T}{P_R} \tag{62.14}$$

where $D = A/C$ is a constant. Note that, in this arrangement, common multiplicative noise on the two voltages will tend to cancel out. Now P_T can be replaced by Equation 62.1, and P_R is given as

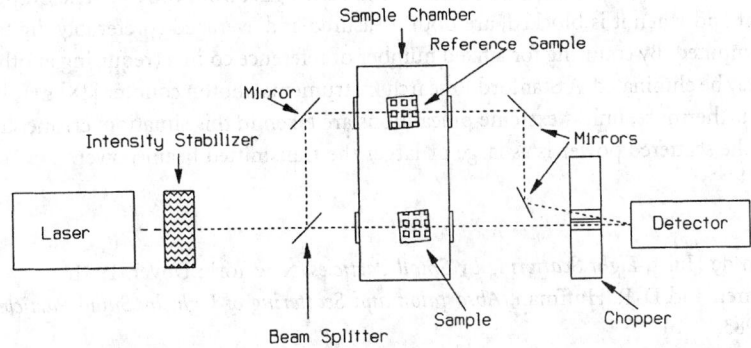

FIGURE 62.4 Block diagram of the optics for a dual-beam, one-detector turbidity apparatus.

$$P_R = f * P_0 \exp{-(\alpha_R + \tau_R) l_R} \tag{62.15}$$

where α_R = absorption, τ_R = turbidity of the sample in the reference beam, l_R = thickness of the sample in the reference beam, and f is a constant determined by the beam splitter. Then, assuming that l_R is the same as l the following expression results:

$$V_{out} = D \frac{P_0 \exp{-(\alpha + \tau) l}}{f * P_0 \exp{-(\alpha_R + \tau_R) l}} \tag{62.16}$$

and thus:

$$\tau = -\frac{1}{l} \ln\left(\frac{V_{out}}{D'}\right) + \tau_R - \alpha + \alpha_R \tag{62.17}$$

where $D' = D/f$. Often, this can be considerably simplified when the reference sample is a pure liquid because then τ_R is of order 10^{-5}, and α_R is totally negligible. Normally, the reference sample is chosen so that its optical properties approach those of the sample in the case where the concentration of particles is low or the system is far from a phase transition. The constant D' may be determined by using the same sample in both the reference and the signal arms of the instrument—in this case $V_{out} = D'$.

Dual-Beam Instruments: Electronics

The electronics for dc detection consists of two CVC converters, one for each detector. The voltage outputs from these may be read directly. However, experiments indicate that greater stability is achieved by simultaneously dividing these two signal outputs as discussed above. In our laboratory, this is accomplished using an Analog Devices AD-532 internally trimmed integrated circuit divider. These can be purchased with different accuracies. In the present application, they typically perform slightly better than the factory specifications indicate. The circuit details are available from Analog Devices [17].

The electronics for ac detection generally consist of two synchronous detectors followed by division of the sample signal by the reference signal. Once more, this may be accomplished using two Analog Devices AD-630 balanced modulator/demodulators or two commercial lock-in amplifiers. Division can be accomplished in the same way as in the dc technique section.

62.5 Limitations

Highly turbid samples, which transmit very little light, may be difficult to measure using a photodiode. In such cases, the photodiode in the sample arm may be replaced with a photomultiplier (PM). In this case, a chopper can be used to reduce the effects of the dark signal from the PM. The outputs, when the signal is present and when it is blocked, are both measured and averaged (preferably digitally), and the difference is computed. By counting for a fixed number of reference counts (requiring another PM), drift in the source may be eliminated. A Stanford Research Instruments photon counter [18] may be configured in this mode. Furthermore, unless extreme precautions are taken in this situation, erroneous results may occur because the scattered power is as large or larger the transmitted beam power.

References

1. H. C. van de Hulst, *Light Scattering by Small Particles*, New York: Dover, 1981.
2. C. F. Bohren and D. R. Huffman, *Absorption and Scattering of Light by Small Particles*, New York: Wiley, 1983.
3. M. Kerker, *The Scattering of Light and other Electromagnetic Radiation*, New York: Academic Press, 1969.

4. P. Calmettes, I. Laguës, and C. Laj, *Phys. Rev. Lett.*, 28: 478–480, 1972.

5. V. G. Puglielli and N. C. Ford, *Phys. Rev. Lett.*, 25: 143–147, 1970.

6. D. Beysens, A. Bourgou, and P. Clamettes, *Phys. Rev. A*, 26: 3589–3609, 1982.

7. J. Rouch, A. Safouane, P. Tartaglia, and S. H. Chen, *Phys. Rev. A*, 37: 4995–4997, 1988.

8. R. F. Chang, H. Burstyn, and J. V. Sengers, *Phys. Rev. A*, 19: 866–882, 1979.

9. R. Pecora (ed.), *Dynamic Light Scattering*, New York: Plenum, 1985.

10. D. L. Sidebottom and C. M. Sorensen, *J. Chem. Phys.*, 89: 1608–1615, 1988.

11. P. Walstra, *Br. J. Appl. Phys.*, 16: 1187–1192, 1965.

12. R. O. Gumprecht and C. M. Sliepcevich, *J. Phys. Chem*, 57: 90–95, 1953.

13. P. Horowitz and W. Hill, *The Art of Electronics, 2nd ed.*, New York: Cambridge University Press, 1989.

14. AD630 balanced modulator/demodulator, in *Analog Devices 1992 Special Linear Reference Manual*, 2-35 to 2-42. Analog Devices has an automated literature delivery system. The fax number for this service is (800) 446–6212.

15. R. A. Dunlap, *Experimental Physics Modern Methods*, New York: Oxford University Press, 1988.

16. R. J. Nash and M. R. Fisch, *Rev. Sci. Instr.*, 60: 3051–3054, 1989.

17. AD532 internally trimmed integrated circuit multiplier, In *Analog Devices 1992 Special Linear Reference Manual*, 2-9 to 2-14.

18. SR400 Dual Channel Photon Counter, *Stanford Research Systems*, (408) 744-9040, [online]. Available World Wide Web: http://www/srsys.com/srsys/.

63

Laser Output Measurement

Haiyin Sun
Coherent Auburn Group

63.1 Introduction

A laser is a device that emits an optical beam. A laser beam carries a certain amount of optical power. Lasers emit beams in two different ways: continuous wave (CW) or pulsed. A CW laser emits a steady power as shown in Figure 63.1. CW laser power is measured in terms of watts. The power of commonly used CW lasers ranges from a fraction of milliwatts for a small helium neon (He-Ne) laser to tens of kilowatts for a carbon dioxide (CO_2) laser. A pulsed laser emits a pulsed power as shown in Figure 63.2. The pulse can be repeated. The pulse duration and repetition rate vary for lasers of different types and are adjustable for some types of lasers. Pulsed laser power is more easily measured in terms of energy per pulse such as joules per pulse. The term "per pulse" is usually omitted. The energy of commonly used pulsed lasers ranges from a few picojoules for a semiconductor laser to tens of megajoules for a semiconductor laser array. The pulse repetition rate ranges from a single pulse for an excimer laser or

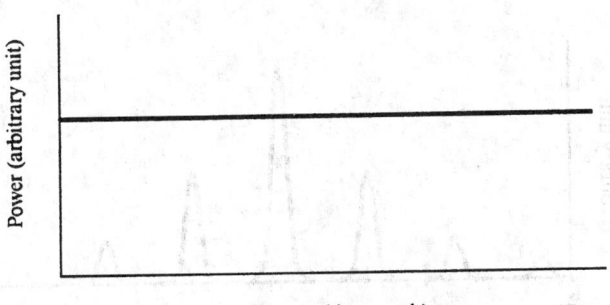

FIGURE 63.1 A CW laser emits a steady power.

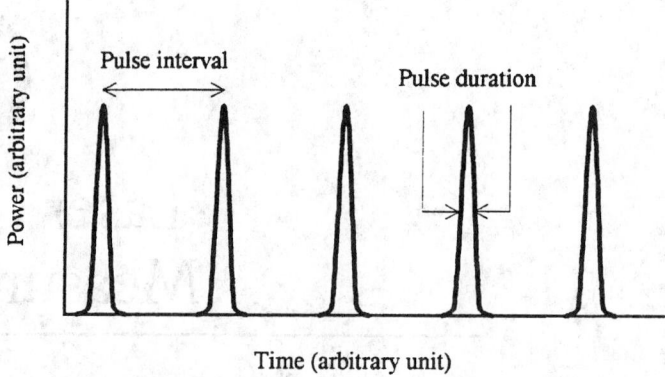

FIGURE 63.2 A pulsed laser emits a pulsed power.

an X-ray laser to hundreds of megahertz for a neodymium:YLF laser. The pulse duration ranges from tens of femtoseconds for a Ti:Sapphire laser up to continuous wave for a CO_2 or He-Ne laser.

A laser beam may contain more than one wavelength (monochromatic) component. One wavelength component may have an optical intensity different from that of another wavelength component. Optical spectral intensity is defined as power per unit wavelength. The optical intensity-wavelength profile of a laser beam is known as the "spectrum." Spectrum measurement usually means measuring the spectral profile. The absolute value of the optical spectral intensity is in fact not important. Figure 63.3 shows a typical spectrum of a multimode semiconductor laser. The wavelength components are inside several bands. Each band may be a "longitudinal mode," and the width of a band is known as the "mode linewidth." Figure 63.4 shows a typical spectrum of a single mode He-Ne laser. There is only one longitudinal mode with a very narrow linewidth. The power of a laser beam can be calculated by integrating over the laser spectrum.

The wavelength of a laser is referred to as the *central wavelength* of the spectrum. But the term "central" is usually omitted. For example, the wavelength is 780 nm for the semiconductor laser shown in Figure 63.3 and is about 632.8 nm for the He-Ne laser shown in Figure 63.4. The wavelength of commonly used lasers ranges from 270 nm in ultraviolet for a neodymium:YAG laser, or even lower for an X-ray laser, to 10.6 μm in infrared for a CO_2 laser. Many lasers have a beam with single longitudinal mode and very narrow linewidth like that shown in Figure 63.4. The spectrum measurement for these lasers then virtually reduces to wavelength measurement.

Power, spectrum, and wavelength are the three most important parameters describing a laser. Laser manufacturers should provide information about these three parameters of their lasers. Some lasers have adjustable power, spectrum, or wavelength. Users often need to measure these parameters to ensure

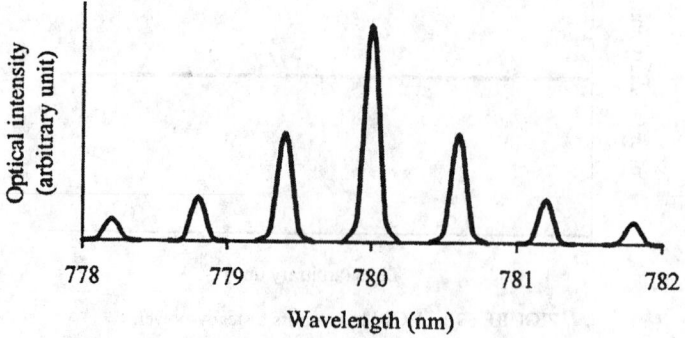

FIGURE 63.3 Spectrum of a multimode semiconductor laser.

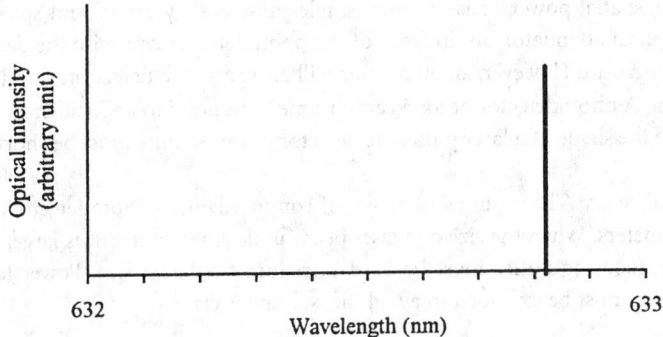

FIGURE 63.4 Spectrum of a single-mode helium neon laser.

appropriate use of these lasers. This chapter discusses the principles and techniques involved in the measurement of laser power, spectrum, and wavelength.

Readers interested in knowing more about laser working principles and characteristics can read laser textbooks such as Reference 1, available in many libraries.

63.2 Measurement of Laser Power

A laser power meter can measure laser power or energy. Figure 63.5 shows the scheme of a laser power meter. The three basic components of a laser power meter are a photodetector, an electronic conditioner, and a display device. The photodetector detects the laser beam under measurement and outputs an electrical response proportional to the laser power. The electronic conditioner processes the response and provides a digital or analog signal for displaying. The display device displays the measurement result in terms of watts or joules. To a large extent, the characteristics of the photodetector used determines the performance of a laser power meter. When selecting or using a laser power meter, the following issues are of primary concern:

1. *Spectral response range.* A photodetector has a certain spectral response range that limits the spectral range of a laser power meter. Some photodetectors have a wavelength-dependent spectral response. Calibration is necessary when using a laser power meter with such a photodetector to measure the power at different wavelengths.
2. *Power range.* The detection threshold and damage threshold of a photodetector usually determine the power range of a laser power meter. Incident laser power lower than the detection threshold can cause measurement error, while incident power higher than the damage threshold can cause permanent damage to the photodetector. Specifically, power range includes CW power range, peak

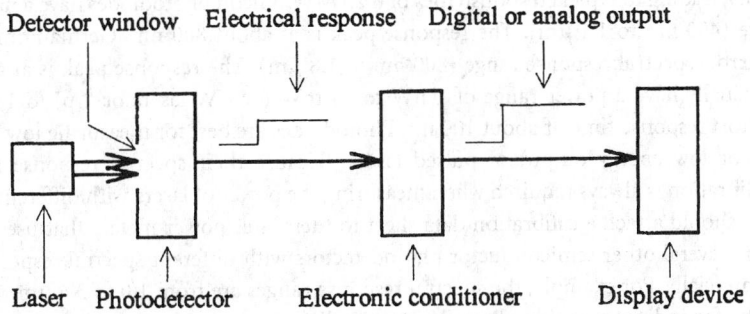

FIGURE 63.5 Scheme of a laser power meter.

power range, spatial power density range, single-pulse energy range, and spatial energy density range. An optical attenuator put in front of the photodetector can raise the damage threshold to as high as megawatt. However, an attenuator will also raise the detection threshold.

3. *Response time.* A photodetector needs a certain time to respond to an incident laser beam. In order to determine the shape of a laser pulse, the detector response time must be shorter than the width of the pulse.

4. *Detector window size.* The input window size of commonly used photodetectors are from a few to tens of millimeters. When the size of a laser beam under measurement is larger than the detector window size, focusing optics must be used to reduce the beam size. Power loss caused by the focusing optics must be excluded to avoid measurement error.

Photodetectors

Three photodetectors are commonly used in laser power meters. They are thermopiles, photodiodes, and pyroelectric probes.

Thermopiles

A thermopile is usually a light absorber disk onto which a ring of thermocouples has been deposited. The absorber converts the laser power incident on it into heat and generates between the absorber and a heat sink a temperature difference proportional to the incident laser power. The thermocouples generate and output an electrical response proportional to the temperature difference. Thermopiles usually have a long response time of a few seconds, a broad and flat spectral response range from 200 nm to 20.0 μm, and a power range of 1 mW to 5 kW for CW lasers or 0.01 J to 300 J for pulsed lasers. Because of the flat spectral response, calibration is independent of wavelength. Thermopiles are primarily used to measure moderate to high power output of CW lasers, moderate to high energy output of single-shot pulsed lasers, and the energy output of pulsed lasers with a repetition rate higher than 10 Hz.

Photodiodes

Silicon and germanium photodiodes are widely used as photodetectors. A photodiode absorbs the photons (laser beam) incident on it and utilizes the photon energy to create free carrier pairs (electrons and holes). These free carriers form a response current in an external circuit. The responsivity of a photodiode in amps/watt is given by

$$R\left(\frac{A}{W}\right) = \eta_D \frac{e}{h\nu} \tag{63.1}$$

where R = current produced by the photodiode for per watt of incident power, η_D = detection efficiency, e = electron charge (1.6×10^{-19} C), and $h\nu$ = photon energy. Since a typical photon energy is 2 eV ≈ 3 $\times 10^{-19}$ J, $e/h\nu$ is of the order of 0.5 A/W. Normal detection efficiencies η_D exceed 0.5 at optical wavelengths (500 to 800 nm), leading to typical responsivities of 0.25 A/W. Silicon photodiodes have a narrow spectral response range (400 nm to 1.1 μm). The response peak is at about 800 nm. Germanium photodiodes also have a narrow spectral response range (800 nm to 1.8 μm). The response peak is at about 1.5 μm. Photodiodes usually have a power range of 1 nW to 50 mW for CW lasers or 1 pJ to 1 μJ for pulsed lasers, and a short response time of about 100 ms. Photodiodes are best for measuring low power output of CW lasers or low energy output of pulsed lasers. Because their spectral response is wavelength dependent, calibration is always required when measuring the power of lasers with different wavelengths. Manufacturers should attach a calibration data sheet to their laser power meters that use a photodiode as the detector. Several other semiconductor photodetectors with different spectral response ranges are available commercially. For example, the spectral response ranges are from 1.0 to 3.6 μm, 1.0 to 5.5 μm, and 2 to 22 μm for indium arsenide photodetectors, indium antimonide photodetectors, and mercury cadmium telluride photodetectors, respectively.

Pyroelectric Probes

A pyroelectric probe uses a ferroelectric material that is electrically polarized at a certain temperature. The material is placed between two electrodes. Any change in temperature of the material caused by the absorption of laser power produces a response electric current in the external circuit. Pyroelectric probes are primarily used to measure the energy of pulsed lasers because they only respond to the rate of temperature change. Pyroelectric probes usually have a spectral response range from 100 nm to 100 μm, a response time as short as a few picoseconds, and a pulsed energy range from 10 nJ to 20 J.

Integration Spheres

Integration spheres are designed to collect the power of highly divergent laser beams, such as semiconductor laser beams, since these beams can overfill the input window of a photodetector and cause considerable measurement error. Figure 63.6 shows the scheme of an integration sphere. The hollow spherical cavity has a diffusive internal wall and at least two windows. The reflectivity of the internal wall is high and slightly wavelength dependent. The highly divergent laser beam under measurement is incident into the sphere from one window. The photodetector of a laser power meter is mounted on another window. A baffle is used to prevent the photodetector being directly hit by the incident beam. The sphere can collect all the incident laser power and convert the power into a diffusive radiation proportional to the power. The laser power meter measures the radiation and displays the laser power under measurement based on the calibration data of the integration sphere.

Readers interested in knowing more about laser power meters can read References 2 through 4 or contact manufacturers. Section 8.1 of this handbook provides more information about photodetectors.

63.3 Measurement of Laser Spectrum

An instrument that can measure a laser spectrum is known as an optical spectrum analyzer. An optical spectrum analyzer consists of two basic components: an optical device and a laser power meter. The optical device can select and output a certain wavelength band of a polychromatic laser beam incident on the device. The central wavelength of the output wavelength band can be scanned over a certain range by the scan of the optical device. The laser power meter measures the power of the optical device output. The width of the output wavelength band usually does not change. Therefore, the power measured by the power meter is proportional to the optical spectral intensity. Since the absolute value of the optical spectral intensity is not important in the measurement of spectrum, the laser power meter sometimes can be as simple as a photodiode combined with a voltmeter. As the optical device is scanned, the power meter outputs the power (optical spectral intensity) as a function of the wavelength, and thereby measures the spectrum. Diffraction gratings and scanning Fabry-Perot interferometers (SFPI) are two widely used optical devices for spectrum analysis.

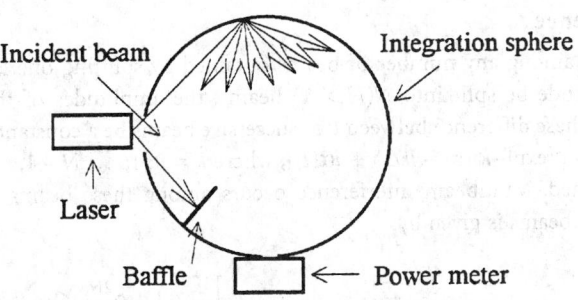

FIGURE 63.6 Scheme of an integration sphere.

Light Interference

Most optical spectrum analyzers make measurements utilizing light interference. The wave theory of light can explain interference. A single mode narrow linewidth laser beam can be described by its electric field $E(z)$:

$$E(z) = A\exp[-i\phi(z) + i\alpha(t)] \tag{63.2}$$

where A = amplitude, $\phi(z) = 2\pi z/\lambda$ = phase, λ = laser wavelength, z = coordinate in the direction of beam propagation, and $\alpha(t)$ = a phase factor that varies fast with time t. The intensity of the laser beam is given by:

$$I = |E(z)|^2 = A^2 \tag{63.3}$$

Two-Beam Interference

A beam splitter such as a partially transparent plate can split a laser beam into two described by:

$$E_1(z) = A_1\exp[-i\phi(z) + i\alpha(t)] \tag{63.4}$$

$$E_2(z) = A_2\exp[-i\phi(z) + i\alpha(t)] \tag{63.5}$$

where A_1 and A_2 are the amplitudes. Let the two beams propagate through two different distances, z_1 and z_2, respectively, and then be recombined into one. The intensity of the recombined beam is:

$$I = |E_1(z_1) + E_2(z_2)|^2 = A_1^2 + A_2^2 + 2A_1A_2\cos\frac{2\pi(z_2 - z_1)}{\lambda} \tag{63.6}$$

Equation 63.6 shows that I varies sinusoidally as $|z_2 - z_1|$ is varied. Such a phenomenon is known as *two-beam interference*. When $|z_2 - z_1| = m\lambda$ (m is any integer), I takes the maximum value of $(A_1 + A_2)^2$ and the situation is called "constructive interference." When $|z_2 - z_1| = (m + 1/2)\lambda$, I takes the minimum value of $(A_1 - A_2)^2$, and the situation is called *destructive interference*. Combining two beams with the same wavelengths but from two different lasers results in:

$$I = A_1^2 + A_2^2 + 2A_1A_2\cos\left[\frac{2\pi(z_2 - z_1)}{\lambda} + \alpha_2(t) - \alpha_1(t)\right] \tag{63.7}$$

where $\alpha_1(t)$ and $\alpha_2(t)$ = two different phase factors of the two lasers, respectively. For broadband lasers, $\alpha_1(t)$ and $\alpha_2(t)$ are uncorrelated, and they vary rapidly and randomly. The last term at the right-hand side of Equation 63.7 can contribute to I only in a time-averaged way, and the time average of this term is zero. For narrowband lasers, a beat is obtained at the differential frequency of the two lasers, and I varies sinusoidally in time for any fixed $z_2 - z_1$. The time average of the last term in Equation 63.7 is also zero. Thus, I always equals a constant of $A_1^2 + A_2^2$, no matter how $|z_2 - z_1|$ is varied.

Multibeam Interference

Interference can occur among any number of beams obtained by splitting one laser beam. Let a laser beam with unit amplitude be split into $N(N > 1)$ beams, the amplitudes of these N beams fall off progressively, and the phase difference between two successive beams be a constant $\Delta\phi$. These beams can be described by $E_k(z) = \rho^k\exp[-i\phi(z) - ik\Delta\phi + i\alpha(t)]$, where $k = 0; 1;...; N - 1$, and $\rho < 1$. Then, these N beams are recombined. Multibeam interference occurs among these beams. The intensity of the recombination of these beams is given by:

$$I = \left|\exp[-i\phi(z) + i\alpha(t)]\sum_{k=0}^{N-1}\rho^k\exp(-ik\Delta\phi)\right|^2 = \frac{1 + \rho^{2N} - 2\rho^N\cos(N\Delta\phi)}{1 + \rho^2 - 2\rho\cos(\Delta\phi)} \tag{63.8}$$

Equation 63.8 shows that I is a function of ρ and N, and is a periodic function of $\Delta\phi$. More information about the wave theory of light and light interference can be found in many advanced optics textbooks such as Reference 5.

Diffraction Gratings

A diffraction grating can spatially disperse a polychromatic laser beam into its monochromatic components and is the most widely used optical device for analyzing optical spectrum covering a relatively wide range. Several different types of diffraction gratings are available. A planar reflective diffraction grating is a collection of many small, identical and slit-shaped grooves ruled on a planar high reflective surface. Figure 63.7 shows two grooves of such a grating, an incident laser beam with three wavelength components $\lambda_1 < \lambda_2 < \lambda_3$ and three diffraction beams marked by $m = 0$, 1, and 2, respectively. The beams are in the plane defined by the grating normal and the groove normal. The grooves illuminated by the incident beam generate diffraction beams with the same amplitude since the grooves are identical. Multibeam interference occurs among the diffraction beams. The phase difference $\Delta\phi$ between two adjacent diffraction beams of a grating is given by:

$$\Delta\phi = \frac{2\pi}{\lambda}d[\sin(\theta_1) + \sin(\theta_2)] \qquad (63.9)$$

where d = groove period, θ_1 = angle between the incident beam and the grating normal, θ_2 = angle between the diffraction beams and the grating normal (θ_2 is not marked in Figure 63.7.), and $d[\sin(\theta_1) + \sin(\theta_2)]$ = path difference between two adjacent diffraction beams. Constructive interference among the diffraction beams occurs at

$$\Delta\phi = 2m\pi \qquad (63.10)$$

where m = an integer known as the *diffraction order*. Combining Equations 63.9 and 63.10 results in

$$d[\sin(\theta_1) + \sin(\theta_2)] = m\lambda \qquad (63.11)$$

Equation 63.11 is known as the *grating equation* and shows that, for a given θ_1 and m, θ_2 is a function of λ. θ_2 gives the diffraction beams a propagation direction in which I has the maximum value. The

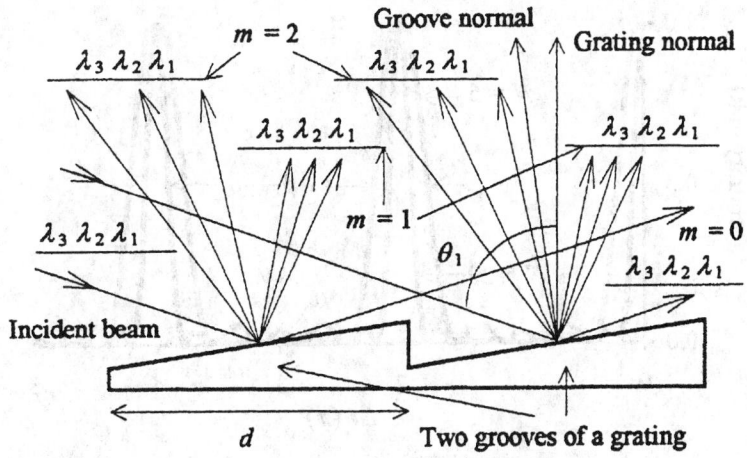

FIGURE 63.7 Two grooves of a planar reflective diffraction grating. The grating disperses an incident beam containing three wavelengths, $\lambda_1 < \lambda_2 < \lambda_3$. There are three diffraction beams with diffraction order $m = 0$, 1, 2.

angular dispersion resolution of a grating is defined by $d\theta_2/d\lambda$ and can be obtained by differentiating equation (63.11). The result is:

$$\frac{d\theta_2}{d\lambda} = \frac{m}{d\cos(\theta_2)} \tag{63.12}$$

where θ_1 is assumed to be a constant. Large $d\theta_2/d\lambda$ is often desired and can be obtained by the use of a small d, a large θ_2 close to 90°, and a large m. Usually, θ_2 must be smaller than 80° to maintain the proper functioning of the grating and $d \approx \lambda$; Equation 63.11 shows that the largest possible value of m is 2. Equations 63.11 and 63.12 also show that $m = 0$ leads to $\theta_2 = -\theta_1$ and $d\theta_2/d\lambda = 0$. That means that, in the zero diffraction order, the grating does not disperse the incident beam, and all the diffraction beams propagate in the same direction, $\theta_2 = -\theta_1$. Commonly used diffraction gratings have a groove density from 300/mm ($d \approx 3.33$ μm) to 2400/mm ($d \approx 420$ nm). The corresponding dispersion resolution for $m = 1$ is from $d\theta_2/d\lambda = 4 \times 10^{-4}$ rad/nm to $d\theta_2/d\lambda = 3.4 \times 10^{-3}$ rad/nm. The intensity of the diffraction beams of a grating can also be described by Equation 63.8 with $\rho = 1$. That is:

$$I = \frac{\sin^2\left(\frac{N\Delta\phi}{2}\right)}{\sin^2\left(\frac{\Delta\phi}{2}\right)} \tag{63.13}$$

Equation 63.13 is plotted in Figure 63.8 for $N = 5$ and 20, respectively. Equation 63.13 shows that I is a periodic function of $\Delta\phi$. I reaches maximum of N^2 at $\Delta\phi = 2m\pi$ and zero at $\Delta\phi = 2k\pi/N$ (k is any integer, but $k \neq mN$). The fringe width $\Delta\phi_w$ of I is given by

$$\Delta\phi_w = 2\left[2m\pi - \frac{2(mN-1)\pi}{N}\right] = \frac{4\pi}{N} \tag{63.14}$$

Equation 63.14 shows that the resolution of a grating is proportional to N, since the resolution can be defined as $1/\Delta\phi_w$.

Most advanced optics textbooks (e.g., Reference 5) study gratings, and manufacturers' catalogs (e.g., Reference 6) give a product-oriented description of gratings.

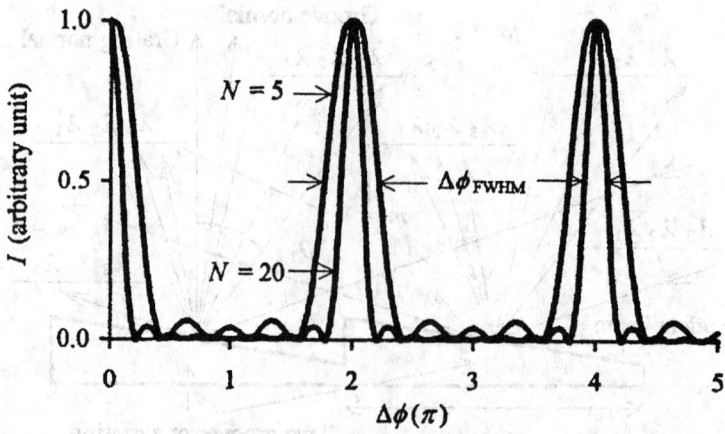

FIGURE 63.8 Multibeam interference intensity I is a periodic function of the phase difference $\Delta\phi$ between two adjacent beams. I is plotted for beam number $N = 5$ and 20, respectively. The full fringe width at half maximum power $\Delta\phi_{FWHM}$ of I decreases as N is increased.

Monochromators

Monochromators are instruments widely used for optical spectrum analysis. Various types of monochromators have been developed. Figure 63.9 shows the scheme of a simple monochromator. The laser beam under measurement is incident through the entrance slit, two concave spherical mirrors collimate the beam, the grating disperses the beam, then one concave spherical mirror focuses the beam on the exit slit, and a laser power meter measures the power of the beam passing through the exit slit. The grating is mounted on a rotator. The position of the two slits are fixed. For any given grating orientation, θ_1 and θ_2 are known. The wavelength λ of the diffraction beams passing through the exit slit can be calculated using Equation 63.11. By rotating the grating and recording the power measured as a function of the corresponding λ, we can measure the spectrum. The widths of the two slits are Δs_1 and Δs_2, respectively, and are adjustable. The three mirrors image the entrance slit on the exit plane. The image width $\Delta s_1'$ of the entrance slit is proportional to Δs_1. Equation 63.15 gives the measurement resolution of a monochromator:

$$\Delta s \frac{d\lambda}{ds} = \frac{\Delta s\, d\lambda}{f\, d\theta_2} \tag{63.15}$$

where $d\lambda/ds$ is the linear dispersion resolution, Δs equals the larger of $\Delta s_1'$ and Δs_2. $d\lambda/d\theta_2$ is the inverse of the angular resolution of the grating used, and f is the focal length of the focusing mirror. Equation 63.15 shows that reducing the widths of the two slits can increase the measurement resolution. However, the slits must be wide enough to allow enough laser power passing through for measurement. The grating rotation angular resolution also affects the measurement resolution of a monochromator. The resolution of commonly used monochromators is from 0.1 to 1 nm. Monochromators usually have a moderate price and measurement resolution. A product-oriented description of monochromators can be found in manufacturers' catalogs such as Reference 7.

Scanning Fabry-Perot Interferometer

Another widely used optical spectrum analyzer is a scanning Fabry-Perot interferometer (SFPI). An SFPI consists of two slightly wedged transparent plates with flat surfaces as shown in Figure 63.10. The two inner surfaces of the plates are set parallel to each other and are high-reflecting coated. The distance D between the two inner surfaces can be adjusted by a piezoelectric device. The two outer surfaces have a

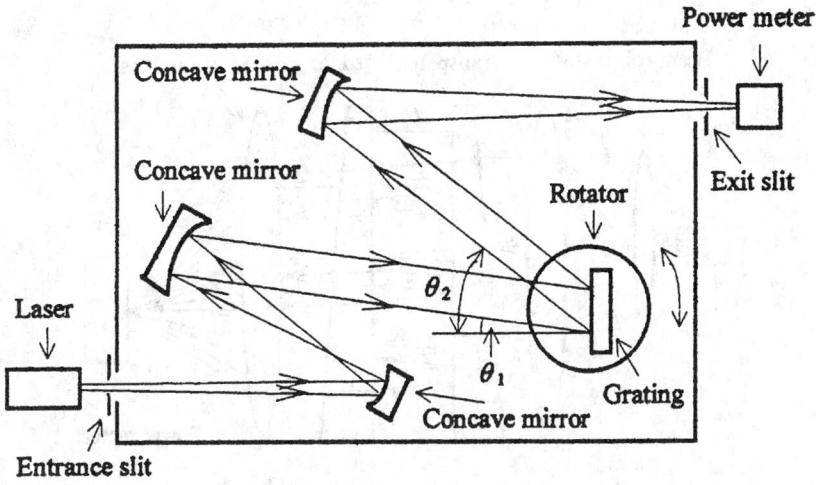

FIGURE 63.9 Scheme of a simple monochromator.

small angle between them, so that reflections of the two outer surfaces can not interfere with the reflections of the two inner surfaces. The medium between the two plates is usually air with a unit refractive index. The laser beam under measurement is collimated by a lens and incident on the SFPI. The beam transmitted through the SFPI is focused by another lens onto the photodetector of a laser power meter. When a beam with unit amplitude is incident at angle θ on the inner surface of an SFPI, multiple reflections take place at the two inner surfaces and produce a series of transmitted beams whose amplitudes fall off progressively. In Figure 63.10, only three incident rays with a few reflections are plotted. The phase difference between two successive transmitted beams is:

$$\Delta\phi = \frac{4\pi D}{\lambda\cos(\theta)} \tag{63.16}$$

Interference occurs among the amplitude of the transmitted beams. The intensity of the combination of the transmitted beam is given by:

$$I = \left| T \sum_{k=0}^{\infty} R^k \exp(-ik\Delta\phi) \right|^2 \tag{63.17}$$

where R and $T = 1 - R$ = the power reflectivity and transmission coefficient of the inner surfaces of the SFPI, respectively, and the reflectivity of the outer surfaces of the SFPI is neglected. The laser power meter measures the power of the transmitted beams. Note that Equation 63.8 can be reduced to Equation 63.17 by letting the amplitude $\rightarrow T$, $\rho \rightarrow R < 1$, and $N \rightarrow \infty$. Therefore, the I obtained in Equation 63.17 can also be described by the curve shown in Figure 63.8. The full width at half the maximum (FWHM) power of I can be found by solving Equation 63.17. The result is:

$$\Delta\phi_{\text{FWHM}} = 4\sin^{-1}\left(\frac{1-R}{2R^{1/2}}\right) \tag{63.18}$$

When R is close to 1, the right-hand side of Equation 63.18 is close to zero. We have:

$$\Delta\phi_{\text{FWHM}} = \frac{2(1-R)}{2R^{1/2}} \tag{63.19}$$

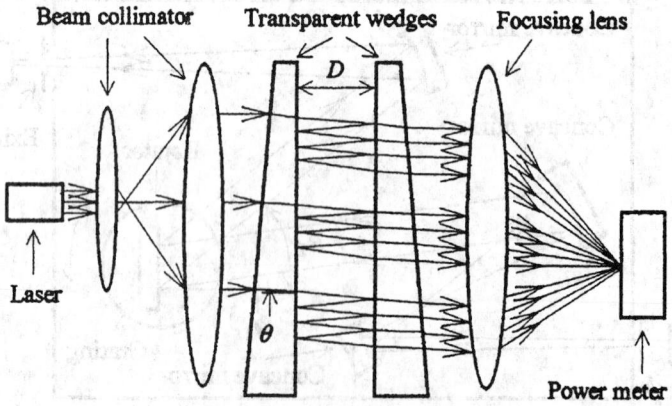

FIGURE 63.10 Scheme of a scanning Fabry-Perot interferometer. Only three incident rays and a few reflections are plotted.

Equation 63.19 shows that the fringe width of I reduces as R is increased, because larger R results in more reflections between the two inner surfaces of the SFPI. In an SFPI, the beam is usually arranged to incident at normal on the inner surface. θ becomes zero. Combining Equations 63.10 and 63.16 results in:

$$\lambda = \frac{2D}{m} \tag{63.20}$$

where λ is the wavelength at which the peak of I appears. Figure 63.8 shows that an SFPI functions like a multibandpass filter. The central wavelengths of the bands (fringes) are given by Equation 63.20 and can be adjusted by adjusting D. The FWHM of the bands is given by Equation 63.19. The spacing between two adjacent bands in terms of wavelength can be obtained by differentiating Equation 63.20 with respect to m, eliminating m and letting $\Delta m = 1$. The result is:

$$\Delta\lambda = \frac{\lambda^2}{2D} \tag{63.21}$$

$\Delta\lambda$ is known as the free spectral range (FSR) of an SFPI. The laser beam under measurement must have a wavelength bandwidth smaller than the FSR to ensure that the beam can transmit only through one band. When D is scanned, the transmitted wavelength is scanned, and the laser power meter measures and outputs the beam spectrum. The measurement resolution of an SFPI is limited by $\Delta\phi_{FWHM}$. Most SFPIs have an $R > 0.95$ to reduce the $\Delta\phi_{FWHM}$. The characteristics of an SFPI can be described by the fringe finesse F, defined as:

$$F = \frac{2\pi}{\Delta\phi_{FWHM}} = \frac{\pi R^{1/2}}{1-R} \tag{63.22}$$

where 2π is the period of the fringes given by Equation 63.10 in terms of phase. An SFPI usually has a resolution of FSR/100. For $\lambda = 500$ nm and $D = 5$ mm, the FSR is 0.025 nm, and the resolution is 0.00025 nm. Compared with monochromators, SFPIs are more expensive, and they have much higher resolution and a much smaller wavelength range.

More information about SFPIs can be found in many advanced optics textbooks such as Reference 5 and Chapter 6.5 of this handbook. A product orientated description of Fabry-Perot interferometers can be found in manufacturers' catalogs such as Reference 8.

63.4 Measurement of Laser Wavelength

For single-mode and narrow-linewidth lasers, the spectrum measurement reduces to wavelength measurement. Laser wavelength can be measured to a higher accuracy by the use of techniques that are not much different from those used to measure the spectrum. A laser wavemeter is an instrument designed to measure the wavelength without knowing the details of the spectrum.

Michelson CW Laser Wavemeter

Figure 63.11 shows the scheme of a widely used Michelson CW laser wavemeter that consists of a Michelson interferometer (MI), a laser power meter, and a computer system for data processing. The MI has two optical arms formed by a beam splitter and two mirrors, respectively. Mirror M_1 can be moved by a stepper motor driving system, and thereby the arm length z_1 can be changed. The position of mirror M_2 is fixed, and the arm length z_2 cannot be changed. Two lenses collimate the laser beam under measurement. The beam splitter splits the collimated beam into two. The two beams propagate in the

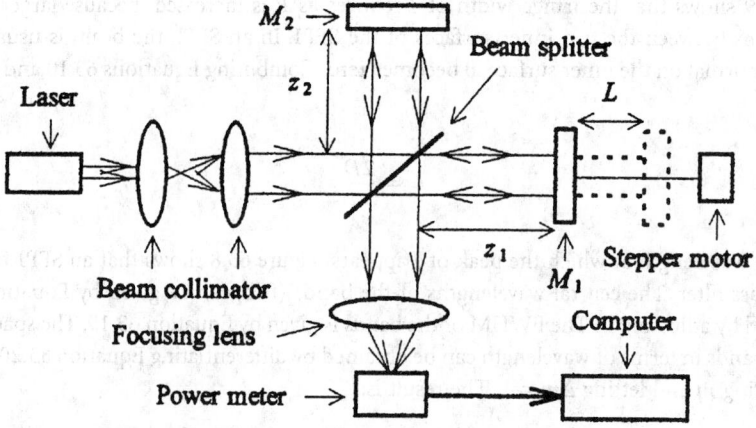

FIGURE 63.11 Scheme of a Michelson CW laser wavemeter.

two arms and are reflected by the two mirrors back to the beam splitter, respectively. Then the beam splitter recombines the two beams. Interference occurs between these two beams. Another lens focuses the combined beams onto the photodetector of a laser power meter. The computer system processes the output data of the laser power meter. The intensity I of the two combined beams can be described by Equation 63.6. As M_1 is moved, z_1 is changed, the path difference $|z_2 - z_1|$ between the two optical arms varies, and I varies periodically. The computer counts the number of the varying period of I, known as *counting the fringes*. Equation 63.23 relates the wavelength λ under measurement, M_1 moving distance L, and the counted fringe number $m + \Delta m$ by:

$$(m + \Delta m)\lambda = 2L \tag{63.23}$$

where m = an integer, Δm = a fraction, and the factor of 2 is introduced because the beam round trip distance is considered. A He-Ne laser with accurately known wavelength λ_H is used as a calibration source. For the He-Ne laser, there is the relation:

$$(m_H + \Delta m_H) = 2L \tag{63.24}$$

where m_H is another integer and Δm_H is another fraction. Combining Equations 63.23 and 63.24 results in:

$$\lambda = \frac{m_H + \Delta m_H}{m + \Delta m}\lambda_H \tag{63.25}$$

For $L = 500$ mm and $\lambda \approx 500$ nm, Equations 63.23 and 63.24 give $m \approx 2 \times 10^6$ and $m_H \approx 2 \times 10^6$. If Δm and Δm_H can be counted to an accuracy of 0.1, Equation 63.25 can provide six significant digits. Thus, λ can be calculated to a relative accuracy of about 10^{-6}. The commonly used commercial Michelson CW laser wavemeters have a relative measurement accuracy from 10^{-4} to 10^{-7}. The measurement range is usually from 400 nm to 1.1 μm, limited by the spectral range of the silicon photodetector used.

Steadily moving M_1 over a distance of 500 mm or so can take several seconds, and the computer is counting fringes during the entire moving period of M_1. Several seconds is much longer than the pulse duration of most pulsed lasers. Thus, the computer will miss fringes in the period of time between two successive pulses when measuring the wavelength of a pulsed laser, and the measurement result will be erroneous. To measure the wavelength of pulsed lasers, the laser wavemeter must not have any moving parts and must be capable of taking data instantaneously.

More information about MIs can be found in many advanced optics textbooks such as Reference 5 and Chapter 6.5 of this handbook. Reference 9 presents a comprehensive study about the design and development of a MI CW laser wavemeter that achieves an accuracy of a few parts in 10^9.

Pulsed Laser Wavemeter

Figure 63.12 shows the scheme of a Fizeau pulsed laser wavemeter, which consists of a Fizeau wedge, a charge coupled device (CCD) linear sensing array with 1024 pixels, a computer system for data processing, and an optical collimation system. The Fizeau wedge is made of a transparent material such as glass. The wedge has two flat surfaces. The angle α between the two surfaces is a few milliradians. The optical thickness $l(x)$ of the wedge is about 2 mm and varies linearly along the wedge. The optical system collimates the laser beam under measurement. The collimated beam is incident on the wedge. Both surfaces of the wedge reflect the incident beam. The two reflected beams form tens of spatial interference fringes on the CCD array. The CCD array detects the interference fringes and sends the data to the computer for processing. A Fizeau wavemeter does not have any moving parts and can measure the wavelength of pulsed lasers. It can be shown that the fringe period p is proportional to the wavelength λ under measurement by:

$$p = F_1(\alpha)\lambda \tag{63.26}$$

where $F_1(\alpha)$ is a function of α. A He-Ne laser with accurately known wavelength λ_H is used separately as a calibration source. The He-Ne laser also forms spatial interference fringes on the CCD array with a fringe period p_H proportional to λ_H by:

$$p_H = F_1(\alpha)\lambda_H \tag{63.27}$$

The data processing algorithm used in Fizeau wavemeters has two steps. λ is first calculated by combining Equations 63.26 and 63.27:

$$\lambda_1 = p\frac{\lambda_H}{p_H} \tag{63.28}$$

In Equation 63.28, the symbol λ_1 denotes the wavelength calculated in the first step because λ_1 is still not accurate enough. The data processing algorithm can calculate the fringe period p and p_H to an accuracy of higher than $10^{-4}p$ and $10^{-4}p_H$ utilizing the 1024 sensing data provided by the CCD array. Thus, Equation 63.28 can provide λ_1 to an accuracy higher than $10^{-4}\lambda_1$, and λ must fall somewhere inside the range:

$$\lambda_1 - 10^{-4}\lambda_1 < \lambda < \lambda_1 + 10^{-4}\lambda_1 \tag{63.29}$$

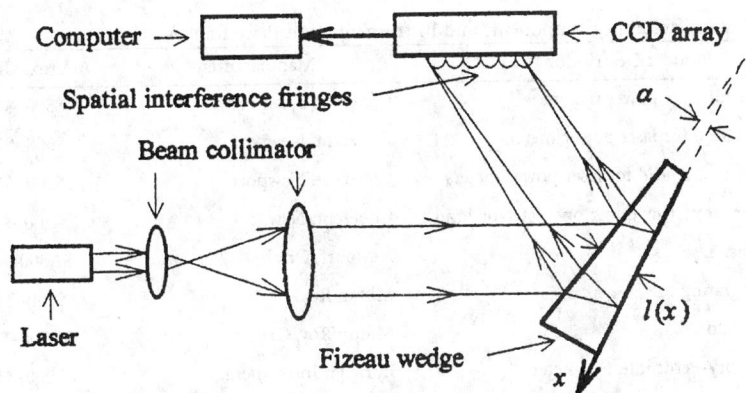

FIGURE 63.12 Scheme of a Fizeau pulsed laser wavemeter.

Any point on the CCD array corresponds to a point on the wedge with a certain optical thickness $l(x)$. It can be shown that at a given point on the CCD array, there is such a relation for λ and λ_H that:

$$(m + \Delta m)\lambda = F_2[l(x)] \tag{63.30}$$

$$(m_H + \Delta m_H)\lambda_H = F_2[l(x)] \tag{63.31}$$

where m and m_H are two integers and are unknown, Δm and Δm_H are two fraction orders at this point on the CCD array and can be measured, $F_2[l(x)]$ is a function of $l(x)$, and $l(x)$ is the thickness of the wedge at the point corresponding to the point on the CCD array. In the second step of the data processing, λ is calculated by combining Equations 63.30 and 63.31. The result is:

$$\lambda_2 = \frac{m_H + \Delta m_H}{m + \Delta m}\lambda_H \tag{63.32}$$

In Equation 63.32, we use the symbol λ_2 to denote the wavelength calculated, because λ_2 is still not necessarily equal to λ. A number of test m and m_H values are inserted into Equation 63.32 and result in a number of different λ_2 values. It can be shown that only one combination of m and m_H can result in a λ_2 that falls inside the range of Equation 63.29. This λ_2 is accepted as λ. m and m_H are of the order of 10^5, and Δm and Δm_H can be measured to an accuracy of 0.01. Therefore, Equation 63.32 can provide λ to an accuracy of $10^{-6}\lambda$. The commercial Fizeau pulsed wavemeters have a measurement accuracy of $10^{-5}\lambda$. The measurement spectral range is from 400 nm to 1.1 μm, limited by the CCD array, which is made of silicon material. More information about Fizeau wedges can be found in many advanced optics textbooks such as Reference 5 and Chapter 6.5 of this handbook. Readers interested in learning more about the design and development of Fizeau pulsed wavemeters will find Reference 10 to be a good starting point.

63.5 Instrumentation and Components

Table 63.1 lists a few companies manufacturing photodetectors, laser power meters, diffraction gratings, monochromators, scanning Fabry-Perot interferometers, Michelson CW laser wavemeters, and Fizeau pulsed laser wavemeters. The price ranges of these instruments and components are also listed in Table 63.1. Table 63.2 lists the address of these companies. These two tables are far from exhaustive. Interested readers could consult two excellent books, *Laser Focus World Buyers Guide* [11] and *Photonics Buyers' Guide* [12]. These two books are published annually and contain up-to-date information about most optical and laser manufacturers in U.S.A and their products.

TABLE 63.1 Manufacturers, Components and Instruments, and Price Ranges

Product Description	Manufacturer	Approx. Unit Price
Thermopile for laser power meters	Coherent, Newport	$500–$2,000
Photodiode head for laser power meters	Coherent, Newport	$500–$1,000
Pyroelectric probe head for laser power meters	Coherent, Newport	$900–$1,200
Laser power meter (including one detector head)	Coherent, Newport	$1,00–$5,000
Integration sphere	Newport, Oriel	$1,300–$2,500
Diffracting grating	Milton Roy	$100–$4,000
Monochromator	Milton Roy, Oriel	$1,000–$7,000
Scanning Fabry-Perot interferometer	Burleigh Instruments	$7,000–$16,000
Michelson CW laser wavemeter	Burleigh Instruments	$7,000–$25,000
Fizeau pulsed laser wavemeter	New Focus	$1,200

TABLE 63.2 Addresses of Companies Listed in Table 63.1

Burleigh Instruments, Inc.	Coherent, Inc.	Milton Roy Instruments
Burleigh	Photonics Division	820 Linden Ave.
Fishers, NY 14453	2303 Lindbergh St.	Rochester, NY 14625
Tel: (716) 924-9355	Auburn, CA 95602	Tel: (716) 248-4000
	Tel: (530) 889-5365	
New Focus, Inc.	Newport Corp.	Oriel Corp.
2630 Walsh Ave.	1791 Deere Ave.	P.O. Box 872, 250 Long Beach Blvd.
Santa Clara, CA 95051	Irvine, CA 92714	Stratford, CT 06497
Tel: (408) 980-8088	Tel: (800) 222-6440	Tel: (203) 377-8282

63.6 Defining Terms

Continuous wave (CW) laser power: Laser power that does not vary with time.
Pulsed laser power: Laser power that lasts only a short period of time.
Laser spectrum: Laser power-wavelength profile.
Laser wavelength: The spatial period of a laser electric field.

References

1. J. Hecht, *The Laser Guidebook, 2nd ed.*, New York, McGraw-Hill, Inc., 1992
2. R. DeMeis, Choose the right detector and meter to test beam strength, *Laser Focus World*, 31 (6): 105–113, 1995.
3. W. DeCosta, Power and energy meters meet measurement needs, *Laser Focus World*, 31 (5): 199–203, 1995.
4. G. Shelmire, How to make accurate laser output measurements, *Laser Focus World*, 29 (4): 241–248, 1993.
5. M. Born and E. Wolf, *Principles of Optics, 6th ed.*, Oxford, Pergamon Press, 1980.
6. C. Palmer (ed.), *Diffraction grating handbook, 2nd ed.*, Rochester, Milton Roy Instruments, 1993.
7. Product catalog, Oriel Corp., Stratford, CT.
8. Product catalog, Burleigh Instruments, Inc., Fishers, NY.
9. J. Monchalin, M. Kelly, J. Thomas, N. Kurnit, A. Szoeke, F. Zernike, P. Lee, and A. Javan, Accurate laser wavelength measurement with a precision two-beam scanning Michelson interferometer, *Appl. Opt.*, 20, 736–757, 1981.
10. B. Faust and L. Klynning, Low-cost wavemeter with a solid Fizeau interferometer and fiber-optic input, *Appl. Opt.*, 30, 5254–5259, 1991.
11. *Laser Focus World Buyers Guide*, Nashua, NH, PennWell Publishing Co., 1996.
12. *Photonics Buyers' Guide*, Pittsfield: Laurin Publishing Co., 1996.

64

Vision and Image Sensors

Stanley S. Ipson
University of Bradford

Chima Okereke
University of Bradford

Vision is the act of seeing, a human capability derived from the combination of the image forming optical system of the eye, the array of light sensitive receptors in the retina of the eye, and the information processing capacity of the retina and human brain. Applications for instruments with a similar ability to sense a scene include broadcast television, monitoring industrial processes, quality control during manufacture, viewing inaccessible or hazardous places, aiding medical diagnosis, and remote sensing from satellites and space probes, to name but a few. Each application often has particular requirements more critical than others, and the problem is to find the most economical solution satisfying these most closely.

The human eye is a familiar imaging system, highly optimized to aid survival under the naturally occurring range of illumination conditions. It is useful to describe its basic characteristics [1] to provide a benchmark against which machine vision can be compared. Light intensity can be measured using radiometric quantities such as radiant power, which is the radiant energy transferred per second in watts; irradiance, which is the radiant power falling onto a surface of unit area in W/m^2; and radiance, which is the radiant power leaving unit area of a surface per unit solid angle in $W/st/m^2$. It can also be measured using photometric units that take into account the variation in response of a standard human eye with wavelength, the CIE (Commission Internationale de l'Éclairage) response curve. The photometric equivalent of radiant power is luminous flux, measured in lumens (lm); the photometric equivalent of irradiance is illuminance, measured in lm/m^2 or lux; and the photometric equivalent of radiance is luminance measured in $lm/st/m^2$ or cd/m^2. At the wavelength of peak sensitivity (555 nm) in the CIE sensitivity curve, the conversion factor from radiometric to photometric units is 680 lm/W.

The eye has a response (observed brightness) to incident light intensity which is roughly logarithmic and is capable of adapting, given time, to an enormous range of different illuminance levels, from full sunlight (100,000 lux) to starlight (0.001 lux). The illumination generally recommended for surfaces in an office environment (about 100 lux) represents a comfortable working level. It takes about 30 min for the eye to become fully dark adapted and a further 3 min to adapt again to increased lighting levels. Despite this adaption capability, when viewing any one point in a scene under bright light conditions, the eye is capable of discerning only about 25 different intensity levels. The eye achieves its enormous range of brightness adaptation partly through the action of a variable size iris (about 2 to 8 mm in diameter), but mainly through the presence of two types of light sensitive receptor in the retina of the eye.

At high illumination levels, corresponding to photopic vision, the eye perceives color due to the excitation of three different types of cone receptors. There is considerable overlap in the spectral sensitivity curves of these red, green, and blue receptors, which have peak responses near 600, 550, and 500 nm, respectively. The overall photopic response is greatest at a wavelength near 555 nm in the yellow green region of the spectrum and falls to zero toward the red and the blue ends of the spectrum at about 750 and 380 nm, respectively. At low levels of illumination, corresponding to scotopic vision, only the rod receptors are excited with a peak luminous efficiency of about 1700 lm/W near 510 nm in the green region of the spectrum. The sensitivity to contrast of the dark-adapted eye is much poorer than that of the light-adapted eye. The transition between photopic and scotopic vision is gradual, with both excited over a luminance range from about 0.001 to 0.1 cd/m². The photopic response extends higher to the glare limit, about five orders of magnitude brighter, while the scotopic response extends lower to the threshold of vision, about three orders of magnitude lower. The photopic (CIE) and scotopic spectral responses of the eye are shown in Figure 64.1 with the response of a typical silicon-based sensor for comparison. The latter has a spectral response that is quite different from that of the human eye, extending well beyond the red end of the visible spectrum toward 1100 nm, with maximum sensitivity around 800 nm. Assuming the average spectral response of silicon, an irradiation of 1 µW/cm² corresponds to an illumination of about 0.22 lux.

The distribution of the cones and rods over the inner surface of the eyeball is nonuniform, with the cones most densely packed in the region of the fovea, a circular region about 1.5 mm in diameter situated toward the rear of the eyeball. To achieve sharpest vision, the eye muscles automatically rotate the eyeball so that the object of interest in the scene is imaged onto the fovea. The separation of the approximately 400,000 cones in the fovea is such that a normal young eye is able to distinguish alternating black and white bands, each 1 mm wide at a distance of about 5 m, corresponding to an angular resolution of about 0.2 mrad. The resolution of the dark-adapted eye is very much poorer because, although the maximum number of rods per square millimeter is similar to that of cones, several rods are connected to a single nerve end, whereas only a single cone is connected to a nerve end.

The overall performance of the eye is difficult to equal. However, a machine vision system can surpass it in individual respects such as sensitivity to infrared and ultraviolet radiations invisible to the eye, speed of response (which can be as short as nanoseconds), and sensitivity to low light levels corresponding to the detection of individual photons of radiation over periods of several hours. Machine vision charac-

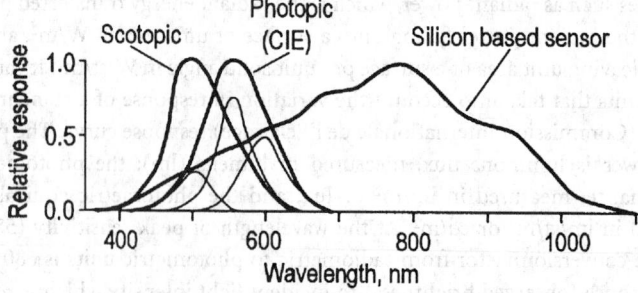

FIGURE 64.1 The linear responses, normalized to unity, of a typical bright-adapted human eye (photopic), a typical dark-adapted eye (scotopic) and a typical charge-coupled-device sensor to different wavelengths of light. The CIE response curve is an internationally agreed response curve for the photopic vision of a standard observer which is used to convert radiometric measurements to photometric. The individual responses of the red, green, and blue cones are also shown, drawn using thinner lines, beneath the photopic curve.

teristics depend on the type of image sensor employed and the modifying effects of additional components such as image intensifiers or optical fiber scopes. The machine vision equivalent of the eye includes a lens to project a 2-D image of the 3-D object of interest onto a sensor that transforms the light energy into an electrical signal. In this form, it may then be transmitted to a remote location, subjected to computer analysis, or displayed on a television screen. A rectangular 2-D image can be sensed using three different approaches. A single, small sensor can be moved in a zig-zag or raster fashion to sense the light intensity on a grid of points covering the whole image. A second approach is to use a line sensor composed of many individual sensors. If it is equal in length to one side of the image, it need only be moved in steps equal to the width of the individual sensors, a distance equal to the other side to cover the whole image. The third approach is to use an area sensor comprising a 2-D array of individual sensors. Whatever method is used, each individually sensed small region of the image is called a *picture element* or *pixel* and, in general, the resolution of the imaging system improves with increasing number of pixels. Although the first two methods require relative movement between sensor and scene, they are not limited by the sensor in the amount of resolution that can be achieved along the direction of motion. In addition, the inevitable variation in response between different sensors is easier to correct than with comparable area sensors, because there are fewer sensors. Two common applications of line sensors are monitoring objects on moving conveyer belts and scanning documents. In many cases, however, it is not practical to arrange relative movement of image and sensor, and the majority of image sensors in use are monochrome or color area sensors designed for use in television cameras. Monochrome and color cameras both use the same basic sensors, which have an inherently broad spectral response. Color sensitivity is achieved with the aid of color filters. A common approach is to deposit color filters directly on the sensor surface in a mosaic or stripe pattern. This is a cheap and compact solution but results in a lower spatial resolution in each color, compared with the resolution of the equivalent unfiltered sensor, and loses much of the incident light through absorption in the filters. Better sensitivity and resolution are obtained using three-sensor cameras that incorporate an optical arrangement (based on dichroic mirrors) that separates the incident light into three components—say red, green, and blue—which are each directed at a different sensor. However, this approach is both bulky and expensive, because it requires three sensors rather than one, and precision optics to align the images correctly on the individual sensors.

Most area sensors currently sold are designed to produce electrical signals compatible with either the 525-line American television standards (RS-170 for monochrome and NTSC for color) or the 625-line European television standards (CCIR for monochrome and PAL for color) [2]. These are interlaced television standards in which each complete image or frame is made up of two fields, each containing either odd or even numbered lines from the frame. Half the remaining lines, making up the odd total number in the frame, appear in each field, ensuring that the two fields interlace properly on the television display. Interlacing is used to avoid picture flicker without having to double the rate of information transmitted. According to the Ferry-Porter law, the critical frequency below which flicker is observed depends on the logarithm of the luminance of the picture highlights. At the brightness levels corresponding to normal viewing, flicker would be observable in pictures interrupted at 25 Hz or 30 Hz, but it is reduced to an acceptable level at the repetition rates of the fields (60 Hz for NTSC and 50 Hz for PAL). A consequence of the Ferry-Porter law is that an NTSC picture can be about six times brighter than a PAL picture and still be acceptable in terms of flicker. Many companies supply frame grabber computer boards that sample the TV-compatible voltage signals (about 1 V range) at a frequency of at least 10 MHz (for at least 512 samples per line), quantize the analog voltage to 256 or more discrete levels (8 bits), and store the resulting digital data for computer analysis. Interlaced operation is not ideal for some applications and, at higher cost, cameras are available that provide higher frames rates, higher resolutions, or progressive scan (no interlacing). These usually produce a standard digital output signal for easy input into a computer. A very wide range of sensors and cameras are currently commercially available, based on several different types of technology. Although each technology has particular advantages, the characteristics of different devices based on a specific technology can vary significantly. For example, the spectral responses of photodiode-based sensors generally extend to shorter wavelengths than those of CCD (charge-coupled device)-based sensors, but some CCD devices are available with extended blue-end responses. Improved and/or cheaper sensors and cameras are appearing on the market all the time and, when considering a new imaging system, it is wise to review the characteristics of the devices currently available in comparison with the requirements of the application in order to select the most suitable. One of the first steps in the creation of a machine vision system for a new application is the design or selection of an appropriate optical system, and this is the subject of the next section. Later sections discuss the various sensor technologies that are available and other related devices that can be used to improve sensitivity or to allow image sensing in restricted spaces such as body or machine cavities.

64.1 Image Formation

The machine vision optical system has the tasks of matching the required field of view to the dimensions of the sensor and gathering sufficient light to achieve a sensor signal that has adequate signal-to-noise ratio (SNR) while maintaining the image sharpness required by the application. Resolutions of area sensors are sometimes specified by manufacturers as the maximum number of horizontal TV lines that can be distinguished. This value may be obtained from measurements of the visibility of a test chart (EIA test pattern), or it may simply be the number of horizontal pixels divided by 1.33, in the case of a monochrome sensor. In special circumstances, measurements to subpixel accuracy can be achieved by interpolation between pixel values, and Reference 3 describes the measurement of a knife edge to about 0.1 pixel accuracy. As a rule of thumb, however, for size measurement applications, the sensor should have a number of pixels at least equal to twice the ratio of the largest to smallest object sizes of interest [4], and a lens is then selected to provide the required magnification and working distance. A lens operating in a single medium such as air is characterized [5] by two focal points and two principal planes as indicated in Figure 64.2. The principal planes coincide with the lens center in the case of an ideal thin lens but may be separated by +20 to −10 mm (the negative sign indicating reversed order), depending on the design, in multielement lenses. They can be determined by first locating the two focal points and then measuring along the axis from these, distances equal to the focal length. A lens of focal length f produces an image in best focus at a distance l_i when the object is at distance l_o where

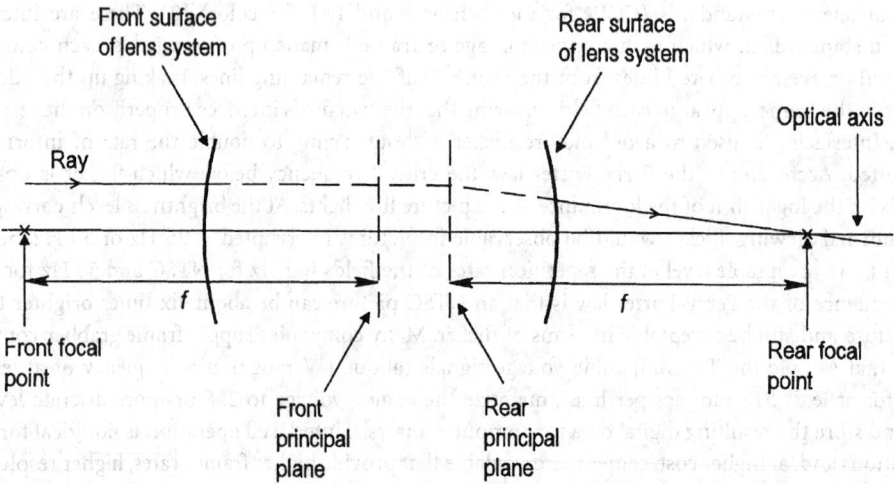

FIGURE 64.2 The trace of a ray from a distant object through a typical multielement lens system showing typical positions of focal points and principal planes. The distance from the back focal length to the lens flange mount is 17.3 mm for C-mount lenses, 12.5 mm for CS-mount lenses, and 46.5 mm for 35 mm photographic format Nikon F (bayonet) mount lenses. Adapters are available to accommodate the differences in distances and fittings of the different mounting systems.

$$\frac{1}{l_o} + \frac{1}{l_i} = \frac{1}{f} \tag{64.1}$$

and the distances are measured to the corresponding principal planes. The image magnification m, defined as the ratio of image to object size, is equal to the ratio of the image to object distances and is related to the total distance between object and image D_T by

$$D_T = \frac{F(m+1)^2}{m} + D_N \tag{64.2}$$

where D_N is the separation between the principal planes. Lenses generally have a focusing adjustment of 5 to 10% of the focal length, and extension rings between the lens and sensor are required for image magnifications greater than about 0.05. The lens extension required is simply the product of the magnification and the focal length. The depth of field F_o of an imaging system [6] is the displacement of the object along the optic axis that produces no significant degradation in the sharpness of the image. The corresponding movement of the image is the depth of focus. In the case of a lens with f/number $f_\#$ (equal to the ratio of focal length to lens iris diameter) and an image sensor with pixel size P, it is given by

$$F_o = 2f_\# \frac{P(m+1)}{m^2} \tag{64.3}$$

Television cameras sensors are manufactured to standard sizes such as 1, 2/3, 1/2, or 1/3 in. The size is defined to be twice the horizontal dimension of a rectangular image with 4:3 aspect ratio so that a 1-in. sensor has a width of 12.7 mm, a height of 9.5 mm, and a diagonal length of 15.9 mm. Lens sizes are similarly specified to allow easy matching of lenses to sensors. Because image distortion and sharpness worsen toward the edges of the field of view, it is permissible, for example, to use a 2/3-in. lens with a 1/2-in. sensor, but not the converse. A 35-mm camera lens, designed for a 24 by 36 mm image size, generally performs much better, at relatively low cost, than a corresponding C-mount lens supplied for a TV camera but a C-mount to Pentax, Cannon, or Nikon mount converter will then be required.

It is frequently necessary to relate lighting of a scene to camera sensitivity. Accurate calculations are difficult, and it is usually better to make a simple estimate and then make fine adjustments to the lighting or lens aperture. Manufacturers often specify camera sensitivities by quoting an illumination level in lux at the sensor faceplate. This may be the illumination required to achieve maximum signal output, some proportion of this maximum, or simply a "usable" signal level from the camera. The illumination of the sensor L_S in lux is related to the luminance of the object B in cd/m² by

$$L_S = \frac{TB\pi}{[2f_\#(m+1)]^2} \tag{64.4}$$

where losses in the lens are characterized by a transmission coefficient T. If the object is a uniform diffuse reflector (Lambertian surface) illuminated by L_O lux, then the luminance of the object is given by

$$B = L_o \frac{R}{\pi} \tag{64.5}$$

where R is the reflection coefficient of the surface. Some practical examples of radiometric calculations are given in a Dalsa application note [7].

64.2 Image Sensing

The primary functions occurring within a standard image sensor are the conversion of light photons falling onto the image plane into a corresponding spatial distribution of electric charge, the accumulation and storage of this charge at the point of generation, the transfer or readout of this charge, and the conversion of charge to a usable voltage signal. Each of these functions can be accomplished by a variety of approaches, but only the principal sensor types will be considered here. Sensors can be divided into two groups: (1) vacuum tube devices in which the charge readout is accomplished by an electron beam sweeping across the charge in a raster fashion similar to that in a television picture tube, and (2) solid-state devices based on charge-coupled devices or photodiodes. These three types of sensor will be described in the next three sections.

Television Camera Tubes

For many years, vacuum tubes [8] provided the only technology available for television applications, and they are still widely used because of the high-quality image signals they provide. Companies supplying tubes include Burle and Philips. Most modern tube cameras are based on the vidicon design whose basic components are indicated in Figure 64.3. Light from the scene is imaged by a lens onto a photoconductive target formed on the inner surface of an end window in a vacuum tube. An electron gun is placed at the opposite end of the tube to the window, and it provides a source of electrons that are focused into a beam, accelerated toward the target by a positive potential on a fine mesh placed just in front of the target, and scanned across the target by an electrostatic or magnetic deflector system. The target consists of a glass faceplate on the inner surface, upon which is placed a transparent electrically conducting coating of indium tin oxide. On top of this is deposited a thin layer of photoconductive material in a pattern of tiny squares, each insulated laterally from its neighbors. The transparent coating is connected to a positive voltage through an electrical load resistor across which the signal voltage is developed. In the absence of

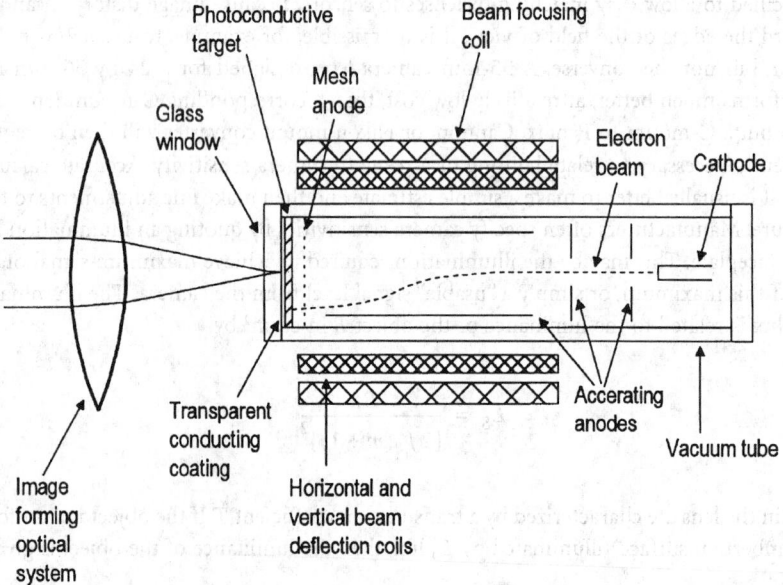

FIGURE 64.3 A schematic diagram of the major components of a vidicon camera tube. Many variants of this basic design exist utilizing different target materials to improve particular characteristics. The electron beam readout of the optically generated charge in the target makes the tube bulky and less reliable than solid-state sensors. Most tubes currently sold are for replacement in existing equipment rather than in new cameras, which almost always incorporate solid-state sensors.

light, the electron beam causes the external surface of the photoconductor to be charged to near zero potential. Light causes the resistance of the photoconductor to decrease, and its surface to acquire a more positive potential, due to the accumulation of positive charge. At each point on the surface touched by the electron beam, some of the beam current is deposited to neutralize the positive charge present due to the illumination. The rest passes through the load resistor generating an output voltage that is a function of the light intensity at that point. The precise timing of the scanning process ensures that the location of the beam is known at each instant in time.

Vidicons have some unique characteristics. They can have spectral responses extending into the ultraviolet and infrared regions of the spectrum, and they allow scan rate and adjustments in scan alignment to be made electronically by adjusting the scanning pattern. This allows very simple region-of-interest readouts. On the other hand, the electron beam readout method confers a number of disadvantages, which are not found in solid-state devices, including large size, fragility, susceptibility to shock and vibration, high power consumption and a high degree of sensitivity to external electric and magnetic fields. The principal imaging defects, which again are not present in solid-state devices, include

- lag when displaying changing images, due to slow response speed,
- image burn which is damage to the sensor surface caused by intense light,
- geometric distortion,
- drift in the apparent position of the image,
- a nonlinear relationship between signal output and light intensity which varies with target voltage.

The degree to which these effects are present depends on the type of tube and its particular construction.

The original vidicon tube design employed antimony trisulphide as the photoconductive material and a low-velocity electron beam and gave only adequate performance. Newer tubes employ the same basic design but use better electron beam readout techniques and target materials to improve characteristics such as spectral response, sensitivity, dark current, lag, and burn-in. Two tubes suitable for broadcast television use are the plumbicon developed by Philips and the saticon introduced more recently by Hitachi. The plumbicon employs a lead-oxide photoconductive target, while the saticon target comprises several layers with different combinations of selenium, arsenic, and tellurium. It can provide greater resolution than the plumbicon, but the lag performance and range of operating temperature are worse. Examples of nonbroadcast quality tubes are the newvicon, chalnicon, and the si-vidicon. The newvicon target has a double layer of cadmium and zinc tellurium, which achieves exceptional sensitivity, but lag and nonuniformity are excessive. The chalnicon employs a cadmium selenide target and provides good performance in most respects but has excessive lag. The si-vidicon target effectively consists of an array of silicon photodiodes that are reverse biased and store charge released by the incident light. The resulting tube has very high sensitivity and little lag, and it is virtually burn proof. However, it exhibits excessive blooming and nonuniformity. The silicon-intensified camera, or SIT camera, is a modification of the si-vidicon camera that employes a photocathode that generates electrons when struck by light photons. These electrons are accelerated by a voltage of several hundred volts and focused on the target of the si-vidicon sensor that produces an output in the normal way. The high-speed electrons landing on the target produce large numbers of electron-hole pairs and a corresponding larger response. The resulting SIT sensor is used in very low light level applications such as tracking satellite debris and can produce useful video images at illumination levels down to 0.001 lux. By comparison, a solid-state CCD sensor (without an image intensifier) typically requires about 1 lux to produce a useful video signal.

Charge-Coupled Devices

A CCD is fabricated on a single crystal wafer of p-type silicon and consists of a one- or two-dimensional array of charge storage cells on centers typically about 10 μm apart. Each cell has several closely spaced electrodes (gates) on top, separated from the silicon by an insulating layer of silicon dioxide. The charge is stored under one of the electrodes, and its location within the cell is defined by the pattern of positive voltages applied to the electrodes. By applying a coordinated sequence of clock pulses to all the electrodes in the array, packets of stored charge (of between 10 and 10^6 electrons) are transferred from one cell to the next until they finally reach a sensing amplifier (floating gate diffusion) which generates a voltage signal proportional to charge, usually assumed to be 1 μV per electron. The result is a device that has an inherently linear variation of output voltage with light from the minimum useful level set by noise, to the maximum useful level set by saturation of the output amplifier, or the limited capacity of the charge storage cells. The dynamic range of the device is defined as the ratio of the maximum output signal to the output resulting from noise. Manufacturers published figures may use peak-to-peak or root-mean-square noise values (typically five times smaller) in this calculation, but the former is more relevant for imaging applications.

Any light penetrating into the underlying silicon generates electron-hole pairs. The holes are swept away to the substrate electrode while the electrons accumulate beneath the nearest electrode in a potential well created by the applied voltage. The sensitivity of silicon is of the order of 1 μA of generated charge per microwatt of incident light. The cells intended to function as light-sensitive photosites have electrodes made of semitransparent polysilicon so that the light can penetrate into the storage region, whereas those intended to function only as part of a shift register for charge transfer are covered by an opaque surface layer. Due to manufacturing imperfections, the photosites do not have perfectly uniform characteristics, and a photoresponse nonuniformity (PRNU) of about 5 to 10% is fairly typical. This is easy to measure using uniform sensor illumination and its effects can be removed if necessary by calibration. The basic spectral response of the silicon extends from 200 to 1100 nm, but the first figure is typically reduced to 450 nm by absorption in the surface layers of the CCD photosites. Infrared radiation penetrates deeper into the silicon than shorter wavelengths, and charge created by an infrared photon may be collected by a different cell to the one entered by the photon. This reduces the resolution of the device, and if infrared operation is not required but the illumination contains infrared (from a tungsten lamp for example), then an infrared reflecting filter (a hot-mirror filter) is often used. CCD cells also accumulate charge linearly with time due to thermally generated electrons produced within the cells and at electrode interfaces. Like the photoresponse, this dark signal varies from cell to cell and can be compensated for by calibration. These thermally generated contributions are most significant for low light level applications and can be reduced by cooling the sensor using either a thermoelectric cooler, a Joule Thomson cooler, or a liquid nitrogen dewar. The dark signal reduces by 50% for every 7°C reduction in temperature, and at −60°C, produced by a Peltier cooler, the dark signal is typically reduced to about one electron per pixel per second. Another important temperature dependent characteristic of the CCD sensor which improves with cooling is the noise floor of the output amplifier which is proportional to $T^{1/2}$ and typically equivalent to about 300 electrons at room temperature. A particular CCD device used in astronomy, for example, and operated at about −110°C has a readout noise of about 10 electrons, a dark current less than 0.3 electrons per minute, and a quantum efficiency for converting visible photons into electrons of between 70 and 80%. Light can be integrated for periods of hours, compared with the approximately 1/8 to 1/4 s integration period of the dark-adapted eye. Compared with photographic film previously used for low-light level imaging in astronomy, CCDs are from 10 to 100 times more sensitive, linear in response rather than nonlinear, and have a much greater dynamic range so that both faint and bright objects can be recorded in the same exposure.

The short-exposure, high-frequency performance of CCD devices is limited by another effect. The process of transporting charge in a CCD sensor is not 100% efficient and, in practice, a small amount of charge is left behind at each transfer to the next cell, contributing noise and degrading resolution. This effect limits the maximum clock frequency and the maximum number of transfers in a CCD sensor. Manufacturers' data sheets for commonly available CCD sensors quote values ranging from 0.9995 to 0.99999 for the charge-transfer efficiency or CTE of a single transfer. There are many variations in CCD technology. For example, virtual-phase CCDs [9] have some of the electrodes replaced by ion-implanted regions, resulting in improved blue response and higher sensitivity because of the removal of some of the blocking surface gates, and simpler drive circuitry because of the reduction in number of gates per cell. A manufacturing technique known as *pinning* can be used to passivate the interface states, which are the biggest contribution to the dark signal, producing an order of magnitude improvement in dark signal as well as improved quantum efficiency and CTE. A signal-processing technique called *correlated double sampling* can also be applied to the output signal from the sense amplifier to improve the readout noise performance.

Linear Charge-Coupled Devices

The basic structure of a linear CCD sensor is shown in Figure 64.4. It consists of a line of up to several thousand photosites and a parallel CCD shift register terminated by a sensing amplifier. Each photosite is separated from a shift register cell by a transfer gate. During operation, a voltage is applied to each photosite gate to create empty storage wells which then accumulate amounts of charge proportional to the integral of the light intensity over time. At the end of the desired integration period, the application of a transfer pulse causes the accumulated charge packets to be transferred simultaneously to the shift register cells through the transfer gates. The charges are clocked through the shift register to the sensing amplifier at a rate of up to 20 MHz, producing a sequence of voltage pulses with amplitudes proportional to the integrated light falling on the photosites. In practice, it is common for shift registers to be placed on both sides of the photosites, with alternate photosites connected by transfer gates to the right and left registers. This halves the number of cells in each register and the time required to clock out all the data. Another 2× reduction of transfer time is achieved if each shift register is split in two with an output amplifier at each end. There is a limit, typically between 10^5 to 10^6 depending on photosite size and dimensions, to the number of electrons that can be stored in a particular cell, beyond which electrons start to spill over into adjacent cells. The saturation charge in electrons is roughly 1000 to 2000 times the area of the photosite in square micrometers. This spread of charge or blooming is a problem with images containing intense highlights. It is reduced by about a factor of 100 by adding antiblooming gates between adjacent photosites and transfer gates and channel stops between adjacent photosites. The voltage on the antiblooming gates is set at a value that allows surplus charge to drain away instead of entering the transfer gates and shift register. By clocking this voltage, variable integration times that are less than the frame pulse to frame pulse exposure time can also be attained.

Area Charge-Coupled Devices

Three basic architectures are used in area CCDs and are illustrated in Figure 64.5. The simplest is the full-frame CCD, consisting of an imaging area separated from a horizontal CCD shift register by a transfer

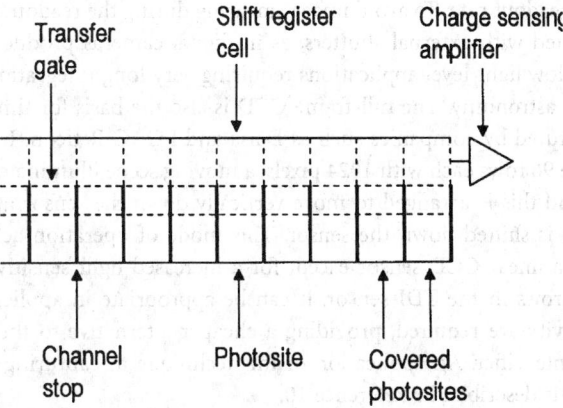

FIGURE 64.4 The architecture of a basic linear CCD showing the arrangement of photosites, transfer gates, shift register cells, and sensing amplifier. The latter produces a sequence of voltage pulses each proportional to the charge accumulated in one of the photosites. Although these pulses are often displayed on an oscilloscope during setting up of the optics and illumination, in normal use they are digitized to 8 or 12 bits, and the resulting values are stored in memory. In practice, most CCDs have two or more covered cells at each end of a line of photosites to allow the dark current to be monitored and subtracted during signal processing. Applying uniform illumination to the CCD enables the relative response of the individual photosites to be measured and compensated for.

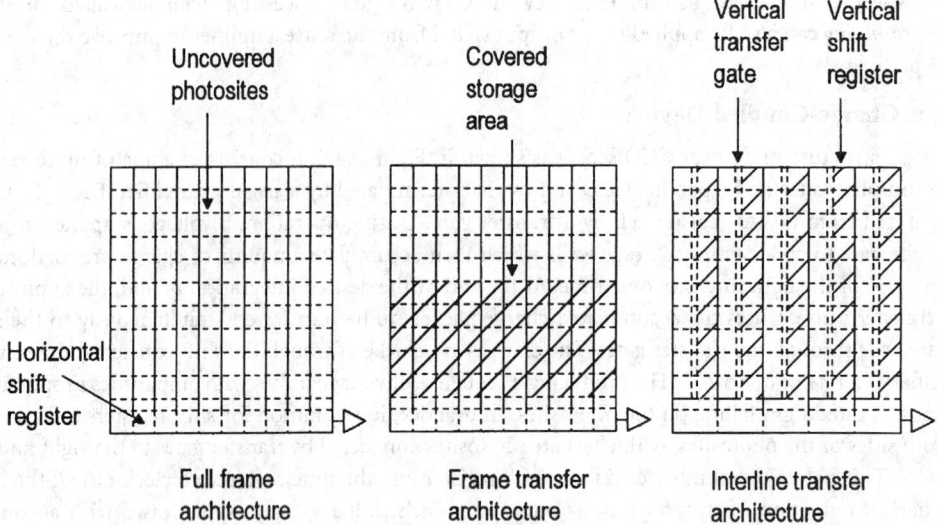

Full frame
architecture

Frame transfer
architecture

Interline transfer
architecture

FIGURE 64.5 The basic architectures used in area CCDs. FF devices fully utilize the available area for image detection and therefore achieve the largest number of pixels, currently about 5000 × 5000, but need an external shutter. FT devices are normally used at standard video rates without an external shutter. The ILT architecture allows exposure times down to less than 0.1 ms. The lower fill factor of ILT devices compared with FF and FT devices results in higher values of modulation-transfer-function at the Nyquist limit. This increases the visibility of aliasing artefacts in images containing high spatial frequencies. A low-pass optical filter must be used if aliasing is a problem.

gate. In the imaging area, each photosite is one stage of a vertical shift register separated from neighboring shift registers by channel stops and antiblooming structures. During the light integration period, the vertical clocks are stopped, creating potential wells which collect photoelectrons. At the end of this period, the charge is clocked out vertically, one row at a time, into the horizontal shift register. The charge in the horizontal shift register is then very rapidly shifted toward the output amplifier by the application of a horizontal clock signal. An example full-frame CCD is the RA1001J from EG&G Reticon, which has pixels arranged in a 1024 × 1024 configuration and dual horizontal shift registers and outputs to achieve a 30 frame per second readout rate. To avoid image smearing during the readout period, full-frame CCD sensors must be operated with external shutters, as in digital cameras produced by Kodak and other companies, or used in low light level applications requiring very long integration times compared with the readout time, as in astronomy. The full-frame CCD is also the basis for time delay and integration (TDI) sensor manufactured by companies such as Dalsa and EG&G Reticon. In the application of this device, which may have 96 rows each with 1024 pixels, a moving scene illuminates a region on the sensor only one pixel high, and this is arranged to move vertically down the sensor at the same rate that the generated photocharge is shifted down the sensor. This mode of operation achieves an output signal identical to that from a linear CCD sensor except for a increased light sensitivity proportional to the number of horizontal rows in the TDI sensor. It can be appropriate in applications where both high speed and high sensitivity are required, providing a cheaper alternative to the combination of a line sensor and an image intensifier. An application of this technique to capturing images of web systems moving at up to 5 m/s is described in Reference 10.

The requirement for an external shutter is greatly reduced in the frame-transfer CCD sensor by the provision of a light-shielded storage section into which the entire two-dimensional image charge is shifted at a much higher rate (limited primarily by CTE considerations) than is possible in a full-frame CCD. The charge can then be read from the storage region during the next integration period without any further image smearing. Some sensors, such as the EG&G Reticon RA1102 device, have the storage area divided into halves placed on opposite sides of the imaging area. This improves performance by halving the maximum number of transfers required to reach the nearest storage region. The same reduction occurs automatically in sensors designed for interlaced video applications where each integration period corresponds to one video field, and only half the number of rows in the frame is required at any one time. For example, to produce an interlaced video frame containing 576 image lines (CCIR standard), a frame transfer sensor with only 288 rows of storage is required. By changing the clock signals, the odd field can be displaced vertically by half a line width relative to the even field. This ensures that the odd and even lines contain different information and reduces aliasing, because the cell width is twice the separation between the lines in the frame. Some of the companies that produce frame-transfer CCD sensors and cameras include Cohu, Dalsa, EG&G Reticon, EEV, Kodak, Philips, and Thomson-CSF.

Image smear is virtually eliminated by the interline-transfer (ILT) architecture in which each column of photosites has an adjacent light-shielded vertical CCD shift register into which the charge is transferred by a transfer pulse. The contents of all the vertical shift registers are then shifted simultaneously one pixel at a time into a horizontal shift register where they are then rapidly shifted to an output amplifier. This approach makes it easy to implement exposure control and achieve short integration times, but it introduces *dead space* between the active pixels, reducing the sensitivity of the image sensing area and increasing aliasing effects compared with frame-transfer sensors. For the latter, the fill factor, which is the percentage of the imaging area which is light sensitive, can be close to 100%, whereas it is usually less than 50% for interline-transfer devices. Localized bright objects tend to produce vertical streaks in an ILT device because strong light can leak under the narrow light shield covering the vertical shift registers, causing image smearing similar to that in a full-frame device. This reduces the usefulness of ILT sensors for scenes containing pronounced highlights. For interlaced operation, two adjacent pixels, for example 1 and 2, 3 and 4, etc., are transferred to a single shift register cell on one field, and in the next field pixels 2 and 3, 4 and 5, etc., are transferred together. This is rather similar to the interlaced operation of a frame transfer CCD. The primary advantages of the ILT sensor are low noise and good exposure control, providing true *stop-motion* control on every field, because all photosites integrate light over the same period of time. Many companies manufacture ILT CCD sensors and cameras, including Hitachi, NEC, Panasonic, Pulnix, and Sony.

The frame transfer and interline transfer approaches both have performance advantages and disadvantages, and the hybrid frame-interline transfer (FIT) approach [11] combines some of the advantages of both. This architecture includes a light-shielded field storage area between an interline imaging area and the horizontal output shift register. With this arrangement, the charge associated with the whole image is first transferred horizontally into the interline storage area, which facilitates exposure control. The charge is then transferred at maximum speed (as in the FT sensor) into the field storage area, which minimizes the occurrence of vertical streaking. For example, the NEC microPD 3541 array clocks the vertical registers at 100 times the normal rate for an ILT sensor. This gives a 20-dB improvement in streaking threshold and an overall threshold of 80 dB, making streaking effects less than other optical effects such as lens flare. On the other hand, the FIT approach does not improve the fill factor, and aliasing artefacts associated with ILT sensors and the noise levels are somewhat higher, and the CTE reduced, because of the higher clock frequencies. Manufacturers of FIT sensors and cameras include JVC, Panasonic, and Sony.

Photodiode Arrays

Because photodiode arrays generally have less extensive electrode structures over each sensing element compared with CCD arrays, the spectral response is smoother and extends further at the blue end of the spectrum. The peak quantum efficiency is also higher ranging from 60 to 80% compared with 10 to 60% for photogates, leading to almost twice the electric output power for a given light input power. Photodiode arrays would therefore appear to be attractive alternatives to CCD array sensors but, in practice, CCD sensors have lower noise levels because they do not have reverse-bias leakage current. A photodiode consists of a thin surface region of P-type silicon formed in an N-type silicon substrate. A negative voltage applied to a surface electrode reverse-biases the P-N junction and creates a depletion region in the N-silicon containing only immobile positive charge. When the electrode is isolated, the P-N junction is left charged and is effectively a charged capacitor. Light penetrating into the depletion region creates electron-hole pairs which, with dark current, discharge the capacitor linearly with time. The penetration depth in silicon increases with wavelength and the depletion region should be wide enough to absorb all wavelengths of interest. Dark current and most of the noise sources operating in the photodiode increase with reverse bias but can be reduced by employing a thicker P-type region which allows a wide depletion region to be achieved with low bias voltage. However, this also degrades the blue-end response of the photodiode. To achieve good blue and UV response along with low bias voltage operation, a three-layer structure comprising thin P-type, intrinsic, and N-type substrates is employed. The intrinsic layer is so pure that the depletion region extends halfway across it at zero bias and extends right across it at a small reverse bias voltage. This structure provides photodiodes with excellent linearity, noise performance, and speed of response at low operating voltages. At the end of an integration period, the states of charge of the photodiodes are measured, and image sensors based on two different types of readout approach are commercially available. These are the serially switched (sometimes called self-scanned) photodiode (SSPD) arrays and the charge-coupled photodiode (CCPD) arrays illustrated in Figure 64.6.

Serially-Switched Photodiode Arrays

Associated with each photodiode in the sensor array is an MOS (metal-oxide semiconductor) switch which, when turned on by a digital voltage level applied to its control line, connects the photodiode to a readout amplifier. Each control line is connected to one of the outputs of a digital shift register and, shifting a bit through the register, sequentially reads the charge on each photodiode in the array. This type of readout is very flexible, and random readout is achieved by replacing the digital shift register by an address decoder connected to an address bus as in the SR series linear arrays from EG&G Reticon. 2-D arrays of photodiodes are connected in a configuration similar to a cross-point switching matrix with a switch and diode at each cross point and separate vertical and horizontal shift registers. To scan the array, the vertical shift register turns on a complete row of switches, which causes the photodiodes in the corresponding row to dump their charge into vertical signal lines. These are in turn connected to the output amplifier by a set of horizontal switches controlled by the horizontal shift register. Switched arrays of photodiodes are made using processes similar to those employed in the manufacture of dynamic RAM, and this along with their simple structure and small size, yields higher densities and lower manufacturing costs. However, the SSPD approach has one shortcoming, namely the large capacitances of the readout buses. This reduces the speed of the device, increases the noise, and reduces the dynamic range, which are all generally significantly worse than those of CCD sensors. The dynamic range for

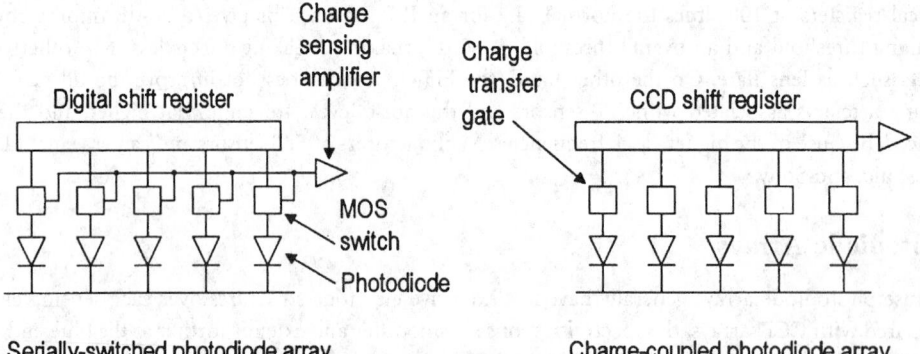

FIGURE 64.6 Readout architectures associated with photodiode array sensors. In the SSPD approach shown, each photodiode is connected in turn via an MOS switch to the sensing amplifier. This eliminates smearing, because the charge does not move between photosites. The device noise and speed characteristics are poorer than CCD sensors because of nonuniformity in the MOS switches and the large capacitance of the sensing bus. The CCPD approach combines the excellent blue end response of the photodiode with the low noise transfer characteristics of the CCD shift register to produce very high-quality linear devices.

example is typically 250 to 300 compared to more than 2500 for commercially available CCD sensors. SSPD arrays also suffer from fixed-pattern noise due to nonuniformities in the switches, which can be as high as 20% of the saturated output. The SSPD approach does have unique advantages:

- The readout method is extremely flexible and can be programmable.
- It is relatively free from blooming effects, because the sensors can be optimally isolated from each other.
- The blue response is good, which is particularly important for spectroscopic applications and for studio quality color television cameras.

Manufacturers supplying SSPD-based sensors include EG&G Reticon, Hamamatsu and VLSI Vision.

Charge-Coupled Photodiode Arrays

Although individual photodiode elements are superior to individual CCD photogate elements in terms of dynamic range, the dynamic range of the SSPD is limited by fixed pattern noise, switching noise and the difficulty of transferring charge packets over the high capacitance output bus. Replacing the horizontal digital shift registers by a low noise CCD shift register significantly improves the dynamic range of the device and the resulting hybrid design is the charge-coupled photodiode array. During operation, the charges on an entire row of photodiodes are simultaneously transferred via MOS switches into the analog CCD shift register and then shifted out in sequence to the output amplifier. By providing a reset pulse (from a second vertical digital shift register) to eliminate all pre-stored charge in one row a fixed interval before the charge is transferred to the CCD shift register, it is relatively easy to implement an electronic equivalent of a focal plane shutter [12] and control exposures in increments of one horizontal scan period. However, images of moving objects will be subject to distortions similar to those found using photographic cameras with focal-plane shutters. In the case of 2-D arrays there are inefficiencies in the movement of small charge packets over the vertical transport lines which get worse the smaller the packet. The solution adopted is to turn a small charge into a large one by pre-charging the lines with a priming charge equal to 10 to 15% of the maximum charge capacity before transferring the photo-charge to the CCD in a process known as charge-primed transfer. This requires several MOS switches forming a sequence of transfer gates and storage capacitors between each vertical transport line and CCD cell and depends critically on correct phasing of a large number of timing signals for proper operation. The CCPD approach appears to be most successful for linear arrays which do not require a vertical transport bus. Manufacturers supplying linear CCPD based sensors include Dalsa and EG&G Reticon.

64.3 Image Intensifiers

When the available solid-state image sensors do not have enough sensitivity for the scene illumination, it is necessary either to use a more sensitive tube camera or to amplify the available light. Companies supplying intensified cameras include Cohu, Kodak, Pulnix, and Philips. Intensified cameras are standard video cameras fitted with image intensifiers that increase the number of available photons falling onto the image sensor. In addition to allowing operation over a wide range of lighting of the order of 0.0001 lux from overcast starlight to 100,000 lux of full sunlight, intensified cameras can be used to stop motion in nanosecond frame times, count photons, and observe images with very high intrascene dynamic ranges. However, this increased flexibility is achieved at a considerable increase in cost, usually some loss in resolution, reduced lifetime, and increases in geometrical distortion, lag, and smear. Image intensifier tubes (IITs) are usually classified as generation I, II, and III devices.

Generation I Tubes

In a generation I intensifiers, illustrated in Figure 64.7, the incoming photons strike a multialkali photocathode within a vacuum tube, and the resulting photoelectrons emitted into the vacuum are accelerated and focused onto a phosphor coating on the rear wall of the tube. The impact of the high-energy

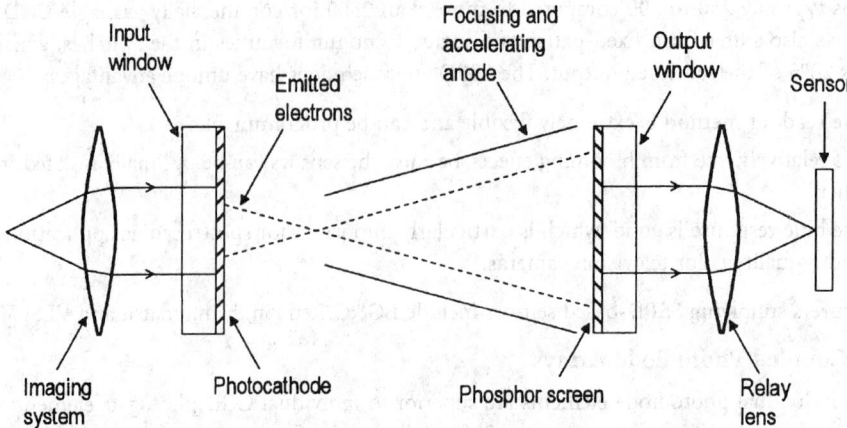

FIGURE 64.7 A schematic diagram showing the major components of a generation I image intensifier tube coupled to a sensor. Combined with a CCD sensor, the sensitivity of a SIT tube camera is achieved but with much higher reliability and greatly reduced lag, geometrical distortion, and power consumption. An example is the Cohu 5512 generation I optical-fiber coupled intensified CCD camera, sensitive down to 0.00015 lux and with 460 TV lines horizontal resolution, currently costing about $12,000.

electrons on the phosphor creates many electron-hole pairs that in turn generate light at a higher level than that striking the photocathode. The electron gain depends strongly on the accelerating voltage and also on the photocathode and phosphor materials, and it is typically around 200 with a 12 kV accelerating voltage. These tubes tend to be bulkier than later-generation tubes and have relatively low gains, although tubes can be cascaded to achieve higher gains. They are also relatively intolerant of bright sources and rapid changes of scene illumination, with increased tendency toward highlight blooming. The light output from the intensifier tube is transferred to the solid-state sensor by either a relay lens or a fiber-optic coupler. The latter is shorter and much more efficient (typically 60% compared with 6%), but requires a sensor whose protective glass or quartz window has been replaced by the fiber-optic coupler resting against the sensor surface. The narrow gaps between couplers and between the coupler and sensor are filled by an index-matching optical grease. The spectral response of the photocathode should be chosen to match the scene requirements, while the narrower phosphor response should be optimally matched to the image sensor for greatest sensitivity [13]. In general, for largest output, a phosphor with the maximum persistence that is acceptable in terms of image lag should be selected. But, if the scene includes object motion, a shorter-persistence phosphor may be necessary to avoid image smear. For night vision, P20 and P39 phosphors are frequently used. The former emits 550 nm light with very high luminous efficiency (65%) and relatively short persistence (0.2 ms), while the latter has a persistence of about 80 ms, which reduces high-frequency jitters. For high-speed imaging, the yellowish-green P46 phosphor has a persistence of only 0.16 µs, while the purplish-blue P47 phosphor has a persistence of only 0.08 µs. However, both these phosphors have luminous efficiencies of only 3%. The resolution of an intensified camera is generally quoted in line pairs per millimeter (lp/mm) and is the harmonic mean of the individual resolutions of the intensifier (typically 20 to 30 lp/mm), the coupler (typically 80 lp/mm), and the image sensor (typically 20 to 50 lp/mm).

Generation II and III Tubes

Generation II intensifiers are similar to generation I devices except that gain is achieved using a microchannel plate (MCP) instead of an accelerating potential difference in vacuum (see Fig. 64.8). Generation III intensifiers are similar to generation II except that the multialkali photocathode is replaced by a gallium arsenic solid-state structure. The resulting device has double the gain of a type II device and improved SNR and resolution. The heart of both types is an MCP disk about 25 mm dia. and 0.5 mm thick, consisting of an array of millions of glass tubes (called *microchannels*) with holes about 10 µm diameter

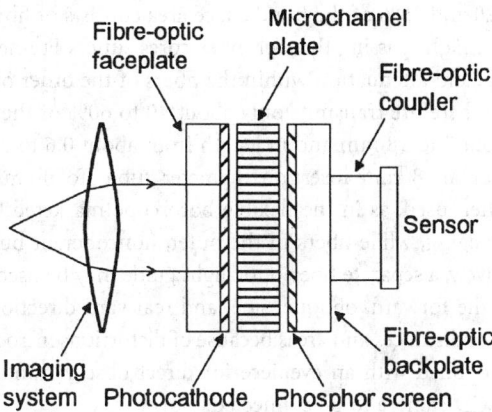

FIGURE 64.8 The basic components of a generation II or III intensified-image sensor. These are typically two and four times more sensitive than generation I intensifiers. Narrow vacuum gaps between the photocathode, microchannel plate, and phosphor screen achieve proximity focusing. An example is the Philips I800 low-light video camera, which couples an XX1410 generation image intensifier with an FT800 1/2-in., 760 × 576 resolution frame-transfer CCD at a cost of around $8000.

coated with a secondary electron emitting substance. The two faces of the MCP are coated with conducting layers, allowing a voltage (typically 50 to 900 V) to be applied between the entrance and exit of each microchannel. An energetic electron from the photocathode enters a microchannel and produces secondary electrons on striking the wall. These electrons are accelerated by the axial electric field and, in turn, dislodge more secondary electrons. This process is repeated many times down the channel. The number of electrons continues to increase until either the end of the channel is reached or current saturation occurs because the electron cloud is so dense that it inhibits further secondary emission. The electrons exit the MCP and may be accelerated further before striking a phosphor screen. The light generated is then transported to the image sensor by fiber-optic couplers. A thick phosphor layer increases light gain but allows electrons and light scattering in the phosphor, which causes some loss of resolution. Improved performance is achieved by integrating the phosphor into the fiber-optic bundle in a so-called *intagliated construction*. Each glass core in the fiber-optic bundle is etched out to a predetermined depth, leaving a small cup of cladding glass at the entrance to each fiber, which is filled with phosphor. Electrons striking the phosphor produce light that is confined to one fiber, minimizing crosstalk and improving resolution. The overall light amplification achieved is of the order of 10^4, and higher gains can be achieved by stacking multiple MCPs and by increasing the accelerating voltages. The MCP has a nonlinear response, because the current density increases at an exponential rate until limited by saturation. The IIT can be gated in only a few nanoseconds by pulsing the cathode potential from the cutoff level (a few volts positive) to the normal operating voltage. This provides a high-speed shutter capability for stopping motion as well as a means of exposure control. Generation II devices have high gain and dynamic range and are easy to gate on and off, but they have relatively low output levels compared with generation I devices. The Kodak EktaPro 1000 high-speed video camera uses a generation II intensifier as a high-gain first stage coupled to a generation I intensifier to provide the required level of light output.

64.4 Fiber Optic Scopes

It is sometimes necessary to acquire images within confined and inaccessible spaces. In some cases, this can be achieved with the aid of mirrors, but when the image must be conducted around bends, a fiberscope may be needed. This consists of a semirigid flexible sheathing around a bundle of optical fibers, each about 10 μm diameter aligned coherently so that fibers have the same relative location at entrance and exit of the bundle. An image formed on one end of the bundle therefore appears undistorted but dimmer at the other end. Between 20 and 35% of the bundle face area consists of fiber cladding and epoxy filler and does not transmit light. Light passing through fiber cores suffers Fresnel reflection losses of about 4% on entering and leaving, and attenuation within the fibers of the order of 10 to 15% for each meter of length. A 1 m fiberscope therefore transmits only about 40 to 60% of the incident light. Fiberscopes range in diameter from about 2 to 18 mm and in length from about 0.6 to 3 m, with a minimum bend radius of about 75 mm for an 8 mm insertion diameter tube. To illuminate the object, a coaxial arrangement of fibers is often used, as in the flexible boroscope marketed by Edmund Scientific. The inner fibers are used for imaging, while fibers in the outer, noncoherent bundle are used to transport light to the object. Alternatively, a separate fiber-optic light guide may be used. Fiberscopes may be fitted with optics for viewing in the forward, oblique, side, and rearward directions but may not allow easy quantitative measurements of distances and areas because of distortions introduced by wide-angle optics. Fiberscopes are generally supplied with an eyepiece for direct observation, but C-mount couplers are often available to allow a CCD camera to be connected.

64.5 Components and Trends

A wide range of image sensors and cameras based on tube and solid-state technologies is now available, and a few examples are listed in Table 64.1 to give an indication of some of the currently available devices and their approximate prices, if purchased in the U.K. It is advisable to contact local suppliers to determine actual prices and availability. Contact information for some companies is listed in Table 64.2. Manufacturers are continuing the incremental development of sensors with greater resolutions, readout rates, and signal-to-noise performances, and new application areas such as high-definition television (HDTV) and digital photography should cause prices to come down as the markets for these products expand. Some examples of this type of development are as follows.

- EG&G Reticon is currently developing a 1000 frames per second, 1024 × 1024 interline-transfer CCD sensor for high-speed imaging.
- Thomson Components (France) has developed the THX-31163 CCD sensor specifically for use in HDTV. This chip is designed to output 1152 × 1260 pixel frames in 40 ms at a transfer frequency of 47 MHz.
- Manufacturers such as Kanimage, Kodak, and Leaf Systems supply digital cameras based on very high-resolution area array sensors. The Kodak professional range of color and monochrome digital cameras, for example, incorporate CCD full-frame transfer sensors with resolutions ranging from 1024 × 1536 to 2036 × 3060 into standard 35 mm photographic cameras.

A separate trend is the development of smart cameras and smart sensors [14] for industrial and scientific applications. There are two basic motivations. One is to correct for imperfect behavior of sensors (such as pixel to pixel nonuniformity associated with MOS devices), and the other is to reduce the communications or processing bottlenecks associated with handling very large amounts of image data. A smart camera is a single-image acquisition subsystem incorporating image sensor, analog-to-digital conversion, and microprocessor that provides processed image data in a standard format such as SCSI or RS-423. Pulnix already supply a series of smart linescan cameras that implement application functions such as object detection, size measurement, and go, no-go comparison. Another example is the imputer 3 [15] from VLSI Vision incorporating a CMOS (complementary MOS) area sensor with 512 × 512 digital resolution and an i960 32-bit reduced instruction set processor. Using VLSI technology, but not CCD technology, it is possible to integrate both image acquisition and low-level processing onto a single chip to achieve a kind of electronic retina. One way of constructing such chips is to integrate photodetector and processing element at the photosite, but this produces very poor fill factors. This can be alleviated to some extent by using microlenses to focus the light onto the sensitive region. An alternative approach is to implement the photodetection and processing as separate arrays on the chip. This allows a large fill

TABLE 64.1 Cameras, Sensors, and Other Components

Manufacturer	Designation	Function	Specification	Approximate price
Cohu	4712	Monochrome TV camera	1/2-in. FT CCD, 754 × 484 pixels, 0.04 lux minimum, RS-170 output	$2,000
Cohu	4110	Digital output camera	1/2-in. FT CCD, 755 × 484 pixels, 0.01 lux minimum, 8-bit per pixel RS-422 outputs	$4,000
Cohu	4912	Monochrome TV camera	1/2-in. ILT CCD, 768 × 494 pixels, 0.02 lux minimum, RS-170 output	$1,000
Cohu	2252	Color TV camera	1/2-in. microlens ILT CCD, CMYG mosaic filter, 0.3 lux minimum, NTSC/Y-C/RGB outputs	$1,400
Edmund Scientific	A52999	Fiberscope	8 mm diameter, bend radius 75 mm, 60° field, 10 to 100 mm focus	$1,200
EG&G Reticon	D Series	Line sensors	256, 512, 1024, or 2048 pixel CCPD arrays	$80 to $2,400
EG&G Reticon	LC1911	Line scan camera	10 MHz data rate, D series sensors, RS-422 outputs	$2400
EG&G Reticon	RA1001	Area sensor	FF CCD, 1024 × 1024 pixels, dual outputs, 30 frames per second	$2,500 to $16,000, depending on grade
EG&G Reticon	TD Series	TDI CCD array	1024 × 96 pixels, single output	$400
Hamamatsu	S5464-1024Q	Spectroscopic line sensor	CMOS SSPD 1024 array, 2.5 mm by 25 μm pixels, quartz window	$1,600
Kodak	DCS 410	Color digital camera	FF CCD, 1524 × 1012 pixels, mosaic filter, Nikon N90S camera	$8,800
Kodak	DCS 460	Color digital camera	FF CCD, 3060 × 2036 pixels, mosaic filter, Nikon N90S camera	$4,0000
Panasonic	GP-US502	Remote head color camera	3 1/2-in. ILT CCDs, 768 × 494 pixels, 3 lux minimum, NTSC/Y-C/RGB outputs	$7,500
Philips	XX1410	Intensifier	18 mm generation II	$1,600
Philips	XQ3427	Tube	2/3-in. plumbicon, 400 TV lines at 55% modulation	$3,200
Philips	XQ1270	Tube	2/3-in. vidicon, resolution 500 TV lines	$120
VLSI Vision	Imputer 3	Intelligent camera	2/3-in. CMOS sensor, 512 × 512 × 8-bit, 32-bit RISC CPU	$4,000
VLSI Vision	IDS	Development system	Software and hardware to create Imputer 3 applications	$7,200

TABLE 64.2 Some Manufacturers of Image Sensors and Cameras

Burle Industries Inc., Tube Products Division 1000 New Holland Avenue Lancaster, PA 17601-5688 Tel: (800) 366-2875 Fax: (717) 295-6096	Cohu Inc., Electronics Division 5755 Kearny Villa Rd. San Diego, CA 92123 Tel: (619) 277-6700 Fax: (619) 277-0221
Dalsa Inc. 605 McMurray Road Waterloo, Ontario, Canada N2V 2E9 Tel: (519) 886-6000 Fax: (519) 886-8023	Eastman Kodak Company 343 State Street Rochester, NY 14650 Tel: (800) 235-6325 Internet: http://www.kodak.com
Edmunds Scientific Company, International Dept. Meg DiMinno Barrington, NJ 08007 Tel: (609) 573-6263 Fax: (609) 573-6882	EEV Ltd. Waterhouse Lane Chelmsford, Essex CM12QU, U.K. Tel: 01245 453652 Fax: 01245 492492
EG&G Reticon, Western Regional Sales 345 Potrero Avenue Sunnyvale, CA 94086 4197 Tel: (408) 738-4266 Fax: (408) 738-6979	Hamamatsu Photonix (U.K.) Ltd. 2 Gladbeck Way Windmill Hill, Enfield, Middx. EN2 7JA Tel: 0181 367 3560 Fax: 0181 367 6384
Hitachi Denshi (U.K.) Ltd. 14 Garrick Ind. Ctr. Irving Way, Hendon London NW9 6AQ Tel: 0181 2024311 Fax: 0181 2022451	NEC Electronics (U.K.) Ltd. Sunrise Parkway Linford Wd, Milton Keynes, Bucks. MK14 6NP Tel: 01908 691133 Fax: 01908 670290
Panasonic Industrial Europe U.K. Willoughby Road Bracknell, Berks RG12 4FP Tel: 01344 853087 Fax: 01344 853706	Philips Components Ltd. Mullard House Torrington Place London WC1E 7HD Tel: 071 5806633 Fax: 071 4362196
Pulnix America Inc. 1330 Orleans Drive Sunnyvale, CA 94089 Tel: (408) 747-0300 Fax: (408) 747-0660	Sony Broadcast & Professional Europe Image Sensing Products Schipolweg 275 1171 PK Badhoevedorp Amsterdam, The Netherlands Tel: 020 658 1171 Tel: (U.K.) 01932 816300
Thomson-CSF Unit 4, Cartel Business Centre Stroudley Road Basingstoke, Hants RG4 0UG, U.K. Tel: 01256 843323 Fax: 01256 23172	VLSI Vision Ltd., Aviation House 31 Pinkhill Edinburgh EH12 7BF, U.K. Tel: 0131 5397111 Fax: 0131 5397141

factor but requires high-frequency data paths between the two arrays. The development of smart image sensors is still in its infancy, but nevertheless the creation of an artificial eye is now on the horizon.

References

1. T. N. Cornsweet, *Visual Perception*, New York: Academic Press, 1970.
2. CCIR, *Characteristics of Monochrome and Colour Television Systems*, Recommendations and Reports of the CCIR, Vol. XI, Part 1: Broadcasting Service (Television), Section IIA, 1982.

3. W. Booth, S. S. Ipson and Y. Li, The application of machine vision and image simulation to the fabrication of knives used for cutting leather, *Proceedings of ACCV'95, Second Asian Conference on Computer Vision*, 5–8 December 1995, Singapore, II574–II578.

4. R. K. Hopwood, Design considerations for a solid-state image sensing system, Reprint from *Proceedings of SPIE*, 230, 72–82, 1980 In *EG&G Reticon 1995/6 Image Sensing and Solid State Products*.

5. S. F. Ray, *Applied Photographic Optics*, 2nd. ed., Oxford, U.K.: Focal Press, 1994

6. Depth of field characteristics using Reticon's image sensing arrays and cameras, Application note 127 In *EG&G Reticon 1995/6 Image Sensing and Solid State Products*.

7. Practical radiometry, Application note in *Dalsa CCD Image Capture Technology, 1996–1997 Databook*.

8. S. J. Lent, Pickup tubes and solid-state cameras, in K. G. Jackson and G. B. Townsend (eds.), *TV & Video Engineers Reference Book*, Oxford, U.K.: Butterworth-Heinemann, 1991.

9. L. Sheu and N. Kadekodi, Linear CCDs, Advances in linear solid-state sensors, *Electronic Imaging*, August 1984, 72–78.

10. S. C. Chamberlain and P. T. Jenkins, Capturing images at 1000 feet per minute with TDI, *Photonics Spectra*, 24(1), 155–160, 1990.

11. D. A. Rutherford, A new generation of cameras tackles tomorrow's challenges, *Photonics Spectra*, 23(9), 119–122, 1989.

12. A. Asano, MOS sensors continue to improve their image, *Advanced Imaging*, 42–44f, 1989.

13. C. L. Rintz, Designing with image tubes, *Photonics Spectra*, 23(12), 141–143, 1989.

14. J. E. Brignell, Smart sensors, in *W. Gopel, J. Hesse and J. N. Zemel (eds.), Sensors: A Comprehensive Review, Volume I: Fundamentals*, Weinheim, WCH Publications, 1989

15. O. Vellacott, VLSI Vision Review note, CMOS in camera, *IEE Review*, 40(3), 111–114, 1994.

Further Information

M. W. Burke, *Image Acquisition: Handbook of Machine Vision Engineering, Vol. I,* London: Chapman and Hall, 1996. Provides comprehensive coverage of lighting, imaging optics and image sensors.

Dalsa CCD Image Capture Technology, 1996–1997. Databook includes useful application notes and technical papers on image sensors and TDI.

Radiation Measurement

65

Radioactivity Measurement

Bert M. Coursey

Ionizing Radiation Division, Physics Laboratory, NIST

65.1 Radioactivity

Radioactivity is the phenomenon of emissions of neutral or charged particles, or electromagnetic radiations from unstable atomic nuclei. The more common types of radiations are listed in Table 65.1. Several naturally occurring (primordial) radionuclides and some produced by cosmic rays in the earth's atmosphere are given in Table 65.2. Methods of producing other unstable nuclei are discussed below. Radioactive isotopes from man-made sources and from naturally occurring nuclides are widely used in health sciences, industry, and academic research.

TABLE 65.1 Characteristics of Nuclear Radiations [8]

Type	Origin	Process	Charge	Mass [MeV]	Spectrum (energy)
α-particles	Nucleus	Nuclear decay or reaction	+2	3727.33	Discrete [MeV]
β⁻-rays (negatrons)	Nucleus	Nuclear decay	−1	0.511	Continuous [keV-MeV]
β⁺-rays (positrons)	Nuclear	Nuclear decay	+1	0.511	Continuous [keV-MeV]
γ-rays	Nucleus	Nuclear deexcitation	0	0	Discrete [keV-MeV]
X-rays	Orbital electrons	Atomic deexcitation	0	0	Discrete [eV-keV]
Internal conversion electrons	Orbital electrons	Nuclear deexcitation	−1	0.511	Discrete [high keV]
Auger electrons	Orbital electrons	Atomic deexcitation	−1	0.511	Discrete [eV-keV]
Neutrons	Nucleus	Nuclear reaction	0	939.57	Continuous or discrete [keV-MeV]
Fission fragments	Nucleus	Fission	≅20	80–160	Continuous (bimodal) 30–150 MeV

Radioactivity was first discovered by Henri Becquerel in 1896. His student Marie Curie and her husband Pierre Curie were the first to chemically separate the radioactive elements polonium and radium. Becquerel and the Curies received the Nobel Prize in physics in 1903 for their pioneering work in nuclear and radiochemistry. The historical development of the field of radioactivity is covered in a number of

TABLE 65.2 Some Cosmogenic and Naturally Occurring Radionuclides

Radionuclide	Half Life
^{3}H	12.35 years
^{7}Be	53.3 days
^{14}C	5730 years
^{40}K	1.28×10^9 years
^{232}Th	1.405×10^{10} years
^{235}U	7.04×10^8 years
^{238}U	4.47×10^9 years

standard texts, including *The Atomic Nucleus* [1] and *Radioactivity and Its Measurement* [2], and in *Radioactivity Measurements: Principles and Practice* [3]. Other excellent texts are available for students in nuclear medicine [4], in radiochemistry and nuclear chemistry [5, 6], in nuclear and particle physics [7, 8], and a popular general textbook *Radiation Detection and Measurement* by Knoll [9] for nuclear engineering and health physics.

Radioactivity is defined by the International Commission on Radiation Units and Measurements [10] in terms of the *activity*, A, of an amount of a radionuclide in a particular energy state at a given time. Mathematically, it is defined as the quotient of dN by dt, where dN is the number of spontaneous nuclear transformations from that energy state in the time interval dt, thus

$$A = \frac{dN}{dt} \tag{65.1}$$

The unit of activity in the International System (SI) of units is the becquerel (Bq), which is equal to the unit reciprocal second (s^{-1}). In many fields the older unit, the curie (Ci), is still in use, where 1 Ci = 3.7 $\times 10^{10}$ Bq (exactly).

The activity of an amount of radionuclide is given by the product of the decay constant, λ, and the number of nuclei present at time t, thus

$$A = \lambda N \tag{65.2}$$

The reciprocal of λ is the mean life of the nuclide. More commonly, one computes the time necessary for one half of the nuclei to decay, and since this is an exponential process, the half-life is given by

$$T_{1/2} = \frac{\ln 2}{\lambda} = \frac{0.69315}{\lambda} \tag{65.3}$$

The activity at any time t can be computed using the initial activity A_0 and the decay time t according to

$$A = A_0 \exp\left(-0.69315 \frac{t}{T_{1/2}}\right) \tag{65.4}$$

The term *specific activity* is in use in several applied fields and one should be careful to look at the units in any particular application. It is normally defined in terms of the amount of activity per unit mass of the element. For example, the nominal activity of ^{14}C in modern carbon (that found in the biosphere) is 0.3 Bq g^{-1}. In nuclear medicine, however, specific activity is widely used to express the amount of activity per unit quantity of the radioactively labeled compound. For example, one might produce ^{11}C in a cyclotron and oxidize it to carbon monoxide CO with a specific activity of 1 MBq ^{11}C μmol^{-1} (CO) (megabecquerel of activity per micromole of substance).

Stable nuclei are those nuclei in their ground state that have a proper number of neutrons and protons to balance the nuclear forces between constituents and thus do not decay spontaneously. Radioactive nuclei may be divided into three categories:

- those that have an excess of neutrons over protons,
- those that are neutron deficient, and
- those that are in excited nuclear states.

A schematic showing the most prominent simple transformations between elements of atomic number Z and neutron number N are given in Figure 65.1. The atomic number Z is the number of protons, and the isotopic mass number A is the total number of nucleons (protons plus neutrons). The notation for an isotope—for example, one of krypton—is $^{A}_{Z}Kr$. In this chapter, isotopes will be identified only by the mass number and the element symbol, e.g., ^{241}Am.

Particle Emission Radioactivity

Alpha particles are ^{4}He nuclei (2 protons plus 2 neutrons) and are usually emitted by heavy nuclei ($A > 150$). Alpha particles are monoenergetic and usually have energies between 4 and 6 MeV. Table 65.3 lists several alpha emitters, including some at lower and higher energies that can be used for the energy calibration of instruments. Alpha particles carry a +2 charge, and their detection with a variety of instruments is described in detail in Chapter 67, on charged-particle detectors. Alpha-particle pulse-height distributions for several radionuclides of interest in the nuclear fuel cycle are shown in Figure 65.2. The alpha-particle pulse-height distribution for an ^{241}Am source obtained with a high-resolution detector is given in Chapter 66 (Figure 66.16).

TABLE 65.3 Alpha-Particle-Emitting Radionuclides

Radionuclide	Half-Life	Energies (MeV)
^{148}Gd	90 years	3.183
^{241}Am	432.2 years	5.486, 5.443
^{210}Po	138 days	5.305
^{242}Cm	163 days	6.113, 6.070

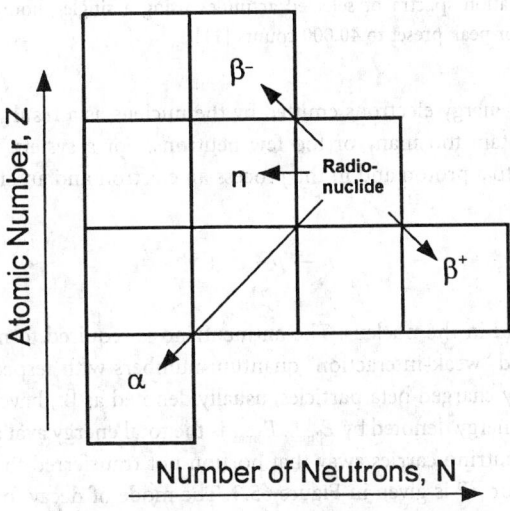

FIGURE 65.1 Characteristics of nuclear radiations [8].

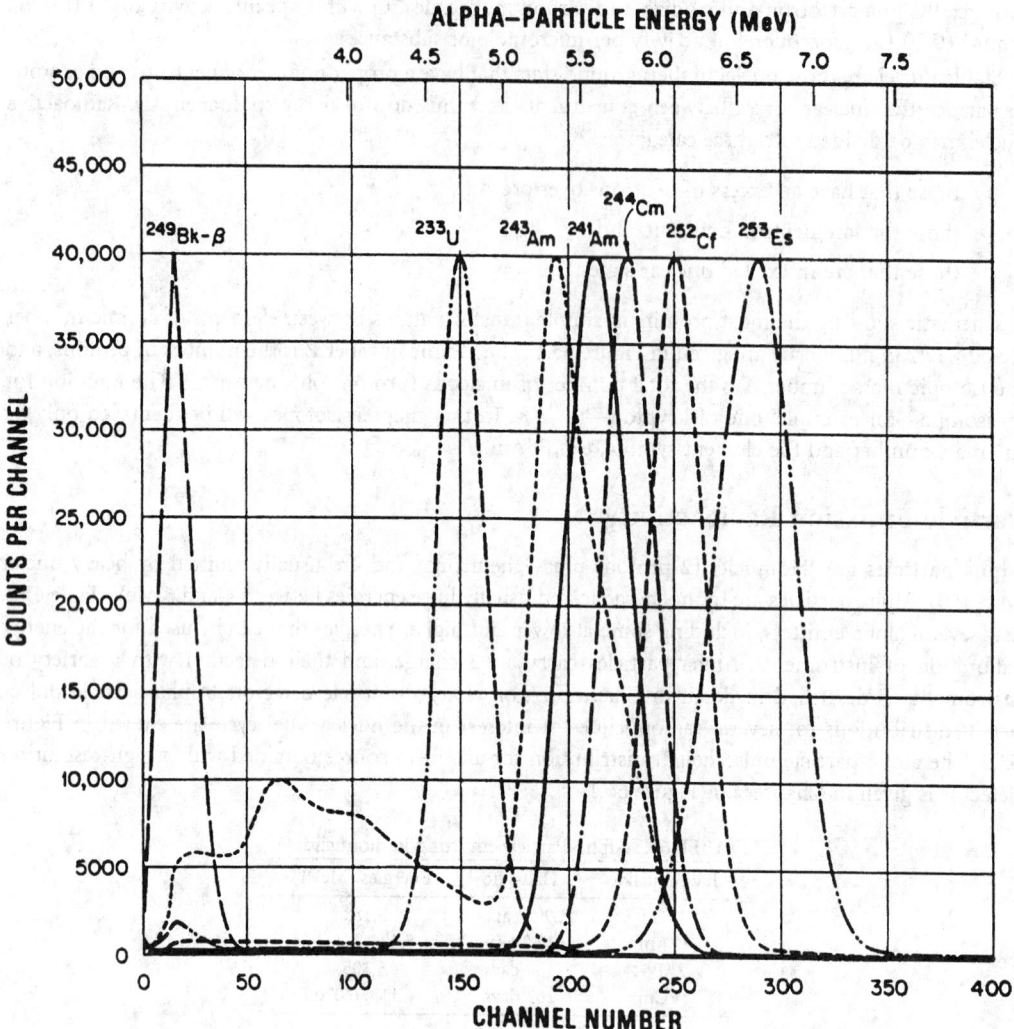

FIGURE 65.2 Liquid-scintillation spectra of selected actinides using a single-phototube detector with the peak channel counts for each major peak preset to 40,000 counts [11].

Beta particles are high-energy electrons emitted by the nucleus as a result of the "weak force" interactions in nuclei that contain too many or too few neutrons. For a system with excess neutrons, the neutron is transformed into a proton and in the process an electron and an antineutrino $\bar{\nu}$ are emitted

$$n \rightarrow p + e^- + \bar{\nu} \tag{65.5}$$

The proton remains behind in the nucleus. The antineutrino is required to maintain the conservation of momentum, energy, and "weak-interaction" quantum numbers with respect to the emitted electron [5]. The emitted negatively charged beta particles, usually denoted as β^-, have a continuum of energies from zero to a maximum energy denoted by $E_{\beta max}$. $E_{\beta max}$ is the total energy available for the kinetic energy of the electron. The antineutrino carries away that portion not transferred to the electron. An example of an emission spectrum for ^{32}P is given in Figure 65.3. The mode of decay for neutron deficient nuclei is quite similar, except that the proton is converted to a neutron with the emission of a positive electron or *positron* (β^+) and a neutrino, which may be written as

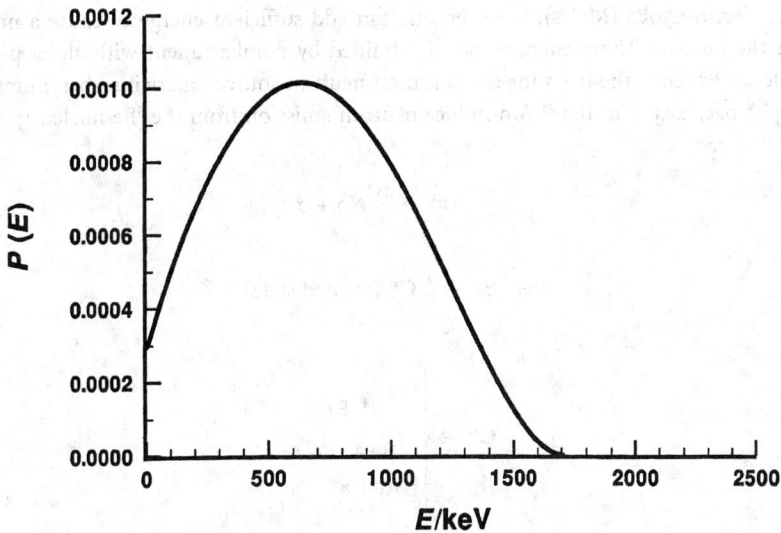

FIGURE 65.3 Theoretical distribution of beta particles as a function of energy for ^{32}P [12].

$$p \to n + e^+ + \nu \tag{65.6}$$

Although it is unusual, some radioisotopes decay by both negatron β^- and positron β^+ emission. For example, in the decay of the 13-day half-life ^{126}I one has

$$^{126}\text{I} \to {}^{126}\text{Te} + e^+ + \nu \tag{65.7}$$

$$^{126}\text{I} \to {}^{126}\text{Xe} + e^- + \overline{\nu} \tag{65.8}$$

Table 65.4 gives a short list of pure beta-particle emitters.

TABLE 65.4 Pure Beta-Particle-Emitting Radionuclides

Radionuclide	Half-Life	Maximum Energy (keV)	Average Energy (keV)
		Negatron (β^-) Emitters	
^{3}H	12.35 years	18.6	5.69
^{14}C	5730.0 years	156.5	49.5
^{32}P	14.29 days	1710	695
^{33}P	25.34 days	249	77
^{35}S	87.44 days	167	49
^{89}Sr	50.5 days	1492	583
^{90}Sr	28.5 years	546	196
^{90}Y	64.0 hours	2283	934
		Positron (β^+) Emitters[a]	
^{11}C	20.38 min.	960	386
^{18}F	109.71 min.	633	250
^{22}Na	2.602 years	545	216

a. These nuclides exhibit a small amount of electron-capture (EC) decay.

Neutrons are constituents of the nucleus that have a mass number of unity and carry no charge. In normal radioactive decay processes, neutrons are not emitted from the nucleus. With energies exceeding a few million electron volts (MeVs), however, one can add sufficient energy to cause a neutron to be ejected from the nucleus. These energies can be attained by bombardment with alpha particles from radioactive decay, which is the basis for the common neutron source americium-beryllium (AmBe) in which the alpha particles from the ^{241}Am induce neutron emission from the ^{9}Be nucleus:

$$^{241}\text{Am} \rightarrow {}^{237}\text{Np} + \alpha \qquad (65.9)$$

$$\alpha + {}^{9}\text{Be} \rightarrow {}^{13}\text{C*} \text{ (excited state)} \qquad (65.10)$$

$$^{13}\text{C*} \rightarrow \begin{cases} {}^{12}\text{C*} + n \\ {}^{8}\text{Be} + \alpha + n \\ 3\alpha + n \end{cases} \qquad (65.11)$$

In this case, the ^{13}C nucleus is formed in a highly unstable excited state and rapidly disintegrates by several processes, all of which lead to neutron emission. The neutron spectrum from a ^{241}AmBe source is shown in Figure 65.4. It is very difficult to measure neutron spectra directly, because it is necessary to deconvolve a complicated detector response. Figure 65.4 is the "standard" spectral shape recommended in recent national and international standards [13, 14]. The units on the y-axis are explained in Reference 13.

Neutrons are also readily produced in high-energy particle accelerators. Various combinations of accelerated particles and target materials are used to produce neutron beams with selected energies (e.g., 2 and 14 MeV). Neutrons from isotopic sources, accelerators and nuclear reactors can then be used in such applications as boron neutron capture therapy to treat cancer, or in metal fatigue analysis. The neutrons can also be used to produce radionuclides for other applications.

Protons with unit positive charge and unit mass number are simply ^{1}H nuclei stripped of the single orbital electron. Spontaneous emission of protons by nuclei is an exotic decay mode that has been

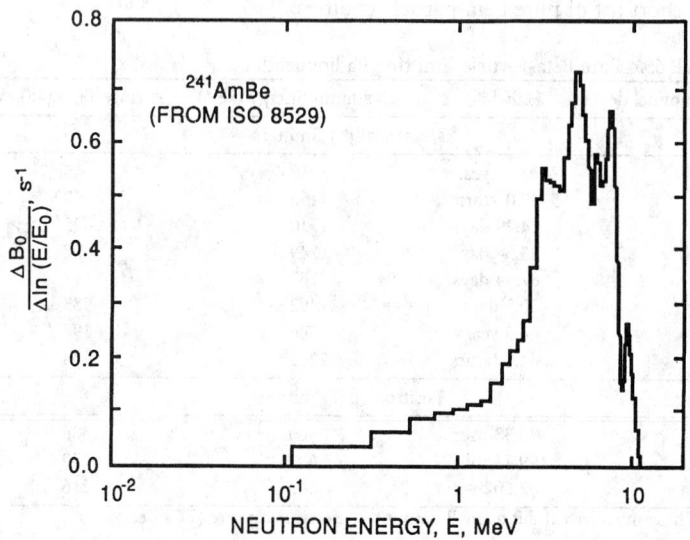

FIGURE 65.4 Neutron spectrum of an ^{241}AmBe radioisotopic source as given in the American National Standard [13] and the International Standard [14]. The units used on the y-axis are explained in Reference 13.

observed in only a few cases involving extremely neutron-deficient nuclei [15]. However, beams of protons can be made in positive-ion, charged-particle accelerators by injecting high-purity hydrogen gas, stripping the electrons, and then accelerating the protons to high energies. When these protons are incident on stable target nuclei, a variety of nuclear reactions are possible. The product nuclides are normally radioactive and neutron deficient.

Fission fragments are the large-mass debris formed when a high-Z nuclide such as ^{235}U spontaneously (or by absorption of a slow neutron) breaks up into two or more smaller nuclides in a process known as nuclear fission. The best known fissionable nuclei ^{235}U and ^{239}Pu fission as a result of capturing a neutron. For a given nucleus, the fission fragments are distributed according to a mass distribution which has two maxima. For ^{235}U fission, the maxima occur at A = 90 and A = 130, resulting in fission-product radionuclides such as ^{90}Sr, ^{95}Zr, ^{131}I and ^{137}Cs. A small number of high-Z radionuclides decay by spontaneous fission (e.g. ^{248}Cm). This decay process usually competes with alpha-particle decay and is mostly limited to high-Z nuclides.

Electromagnetic Emission Radioactivity

Gamma rays are photons emitted during nuclear deexcitation processes. These gamma-ray transitions may be from a metastable excited state, or between levels in a daughter nucleus. Two examples are shown in Figure 65.5 for the 6-hour half-life ^{99m}Tc and the 2-day ^{111}In. The large majority of gamma rays from fission products and man-made radionuclides have energies between 20 keV and 2 MeV. A list of gamma-ray emitters used for instrument calibrations are given in Table 65.5. The probability of gamma-ray emission in a particular radionuclide decay (P_γ) is also given in the table. The rate of gamma-ray emission is given by the product $P_\gamma A_0$.

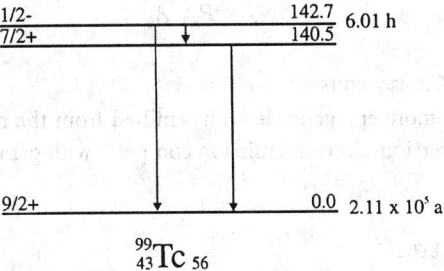

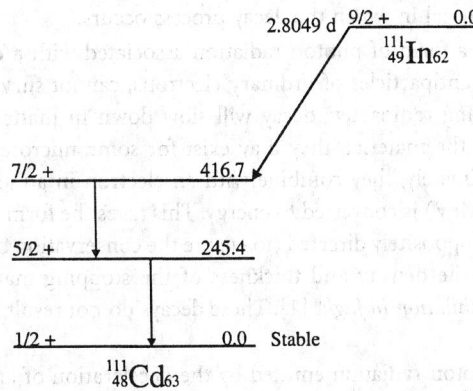

FIGURE 65.5 Simplified nuclear decay schemes for the radionuclides ^{99m}Tc and ^{111}In. The vertical arrows indicate transitions between nuclear excited states. Gamma-ray emission competes with internal conversion processes for each transition [8].

TABLE 65.5 Gamma-Ray-Emitting Radionuclides Used for Detector Efficiency Calibrations

Radionuclide	Half-Life (days)	Energy (keV)	Gamma Rays per Decay, P_γ
^{109}Cd	462.6	88	0.0363 ± 0.0002
^{57}Co	271.79	122	0.8593 ± 0.0027
^{139}Ce	137.64	166	0.7987 ± 0.0006
^{203}Hg	46.595	279	0.8148 ± 0.0008
^{113}Sn	115.09	392	0.6489 ± 0.0013
^{85}Sr	64.849	514	0.984 ± 0.004
^{137}Cs	1.102×10^4	662	0.851 ± 0.002
^{88}Y	106.63	898	0.940 ± 0.003
^{60}Co	1925.5	1173	0.9990 ± 0.0002
^{60}Co	1925.5	1332	0.99983 ± 0.00001
^{88}Y	106.63	1836	0.9936 ± 0.0003

Characteristic **X-rays** are photons emitted during atomic relaxation processes. X-rays are emitted from radionuclides when electrons are involved in the high-energy processes in the nucleus. In the *electron capture* process, for example, the nucleus captures an electron (usually a K-shell electron, since it is closest to the nucleus), and a proton and electron form a neutron. This process leaves a K-shell vacancy, and a characteristic X-ray from the daughter nucleus can be emitted as orbital electrons from higher shells fill the vacancy. An example is the decay of the 2.6-year half life ^{55}Fe

$$^{55}\text{Fe} + e^- \rightarrow {}^{55}\text{Mn} + X \tag{65.12}$$

The rate of Mn K X-ray emission from an ^{55}Fe source of activity A_0 is

$$N_K = P_{KX} A_0 \tag{65.13}$$

where P_{KX} = probability of K X-ray emission.

Conversion electrons are monoenergetic electrons emitted from the nucleus in isomeric transitions between nuclear levels. Conversion-electron emission competes with gamma-ray emission as a mode of nuclear deexcitation.

Interactions with Matter

Before examining in detail the manner in which ionizing radiations can interact, consider two sources of photon radiations, annihilation radiation and bremsstrahlung, which are associated with radioactive decay but depend on the material in which the decay process occurs.

Annihilation radiation is a form of photon radiation associated with a class of radioactive decays. Positrons, since they are the antiparticles of ordinary electrons, cannot survive long in normal matter. Thus, positrons emitted during radioactive decay will slow down in matter until they reach thermal equilibrium. (Depending on the material, they may exist for some microseconds in a state of matter known as positronium.) Ultimately, they combine with an electron in an annihilation event in which their combined mass (1.022 MeV) is converted to energy. This takes the form of two annihilation quanta of 0.511 MeV each which are oppositely directed (to ensure the conservation of momentum). Depending on the positron energy and the density and thickness of the stopping material, there is a slight but calculable probability of *annihilation in flight* [1]. These decays do not result in characteristic 0.511-keV quanta.

Bremsstrahlung is the photon radiation emitted by the deceleration of an electron in the Coulomb field of an atom. Thus, bremsstrahlung radiation is present during all beta decay processes as the emitted β particles (both negatrons and positrons) slow down in matter; it has a continuum of photon energies extending up to the maximum beta-particle energy. The shape of this continuum depends on the nature of the stopping material, as is described in the following section. Bremsstrahlung radiation represents a particular safety hazard for high-energy pure-beta particle emitters such as ^{32}P (14.4 day, 1.710 MeV β^- max). At higher energies, bremsstrahlung production is almost linear in electron energy and proportional to Z^2 of the stopping material. Thus, to minimize bremsstrahlung, one uses low-Z shielding materials such as plastic.

Now, consider the more general ways in which ionizing radiations from radioactive decay interact with matter. They interact by collisions with the electrons and nuclei along their path. The nature of these interactions is characterized by the charge, mass, and energy of the incident ionizing photon or particle, as well as the charge and mass of the traversed matter. In the energy region of interest, electromagnetic radiation (gamma-ray and X-ray photons) interact with matter principally by three processes: photoelectric absorption, Compton scattering, and pair production. In the photoelectric process, which dominates at lower energies, the photon transfers its energy to an atomic electron, which is then ejected from the atom with an energy equal to that of the incident photon minus the electron's binding energy. In Compton scattering (incoherent scattering), the photon loses a fraction of its energy to an atomic electron, and the scattered photon emerges generally in a direction different from that of the incident photon. Higher-energy photons will undergo multiple Compton scatter events until the process is finally terminated by a photoelectric absorption. For photons with energies exceeding 1.022 MeV, the process of pair production can occur whereby an electron-positron pair is formed. The positron produced will ultimately annihilate with the production of two 0.511-MeV photons (or in flight). The three interaction processes compete as a function of photon energy, electron density, and nuclear charge of the stopping material. Above 2.044 MeV, triplet production can occur [16], but this is a low-probability process, and few radionuclides emit gamma rays above 2 MeV.

Quantitative measures of the photon interactions in matter are attenuation coefficients based on cross sections for specific interactions [8,9]. The total narrow-beam attenuation coefficient μ is given by the sum

$$\mu = \mu_{\text{photoelectric}} + \mu_{\text{Compton}} + \mu_{\text{pair production}}$$

The attenuation process for a beam of photons traversing a slab of matter is an exponential function of the form

$$I = I_0 \exp(-\mu l) \tag{65.14}$$

where I and I_0 = intensities of the transmitted beam and the incident beam, respectively, l = distance traveled in matter, and μ = the linear attenuation coefficient. A useful procedure is to express distances in terms of the mass thickness—the product of density ρ and thickness l. The beam transmission equation can then be rewritten as

$$I = I_0 \exp\left[\frac{-\mu}{\rho}\right](\rho l) \tag{65.15}$$

where μ/ρ is the *mass attenuation coefficient*. The mass attenuation coefficient, in contrast to the linear attenuation coefficient, does not depend on the density of the absorber, but only on its composition. The mass attenuation coefficient as a function of energy for water and lead are shown in Figures 65.6 and 65.7. This discussion of photon interactions has been limited to narrow-beam (or ideal) geometries. In practice, most radioactive sources emit a broad beam of radiation such that the detector registers not only events from the incident beam but also those scattered through large angles, and these effects must be included in computing detector response [9].

Charged particles interact with matter in many different ways. These include:

1. inelastic collisions involving ionization and excitation of atomic electrons of the material
2. inelastic collisions involving bremsstrahlung production in the field of the atom
3. elastic scattering in the field of the atom

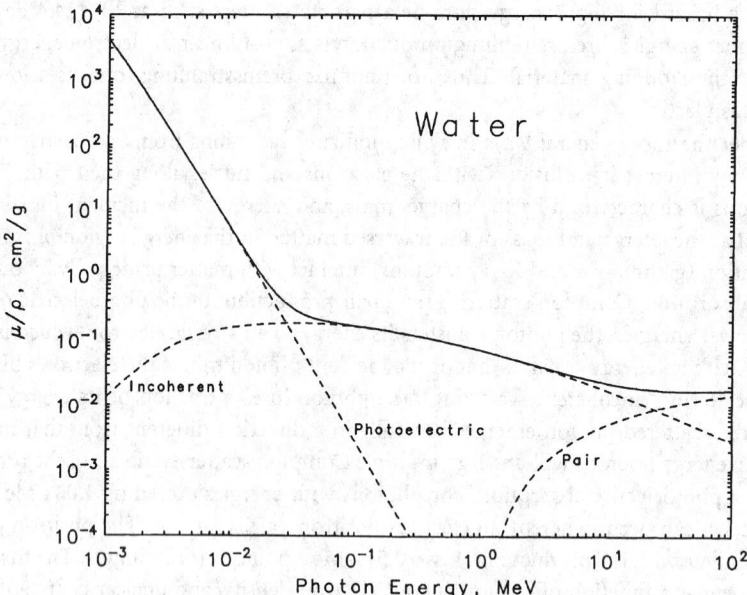

FIGURE 65.6 Narrow-beam mass attenuation coefficients for water as a function of photon energy [17].

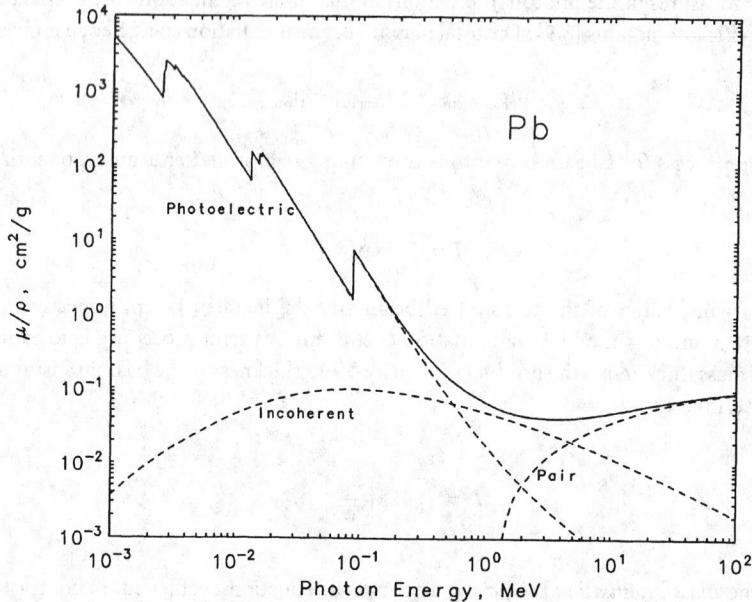

FIGURE 65.7 Narrow-beam mass attenuation coefficients for lead as a function of photon energy [17].

4. nuclear reactions
5. emission of Cerenkov radiation (photons in the visible and ultraviolet regions)

Electrons and positrons, due to their much smaller mass, behave differently from protons, alpha particles, and low-Z charged particles. Two quantitative parameters used to describe inelastic interactions of charged particles as they slow down in matter are *stopping power* and *range*. The total mass stopping power, S/ρ, is defined by the ICRU [10] as the quotient of dE by ρdl, where dE is the energy lost by a charged particle in traversing a distance dl in the material of density ρ. For energies for which the nuclear interactions can be neglected, the total mass stopping power can be represented as the sum of an electronic (ionization and excitation) term and a radiative (bremsstrahlung) term as follows:

$$\frac{S}{\rho} = \frac{1}{\rho}\left(\frac{dE}{dl}\right)_{elec} + \frac{1}{\rho}\left(\frac{dE}{dl}\right)_{rad} \tag{65.16}$$

Figure 65.8 illustrates the contributions of the electronic term and the radiative term to the total mass stopping power for electrons in lead and water over the energy range from 10^{-2} MeV to 10^3 MeV. For alpha particles, the radiative term can be neglected, but the nuclear interactions cannot. The nuclear interactions are pronounced at lower energies, while at higher energies the inelastic collision process dominates. This is illustrated in Figure 65.9 for alpha-particle interactions in lead and water.

The range of a charged particle of a given energy in a material is an important parameter in designing detectors and shielding materials. In general, a charged particle does not travel in a straight line as it slows down. However, the total rectified path length traveled by the charged particle, called the mean range, is given by

$$r_0 = \int_0^E \frac{\rho}{S}dE \tag{65.17}$$

where E = incident electron energy. The range as given in Equation 65.17 has units of g cm^{-2}. Range in water and lead as a function of incident particle energy for electrons and for alpha particles is given in Figures 65.10 and 65.11, respectively.

Neutron Interactions

Since neutrons are uncharged, they can travel some distance in matter before they interact. This interaction is normally a "nuclear reaction" in which the neutron is absorbed, scattered, or produces a nuclear reaction. The probability of each of these interactions occurring is described by the capture, scatter, and nuclear reaction cross sections. Secondary particles produced by neutrons are usually high-energy charged

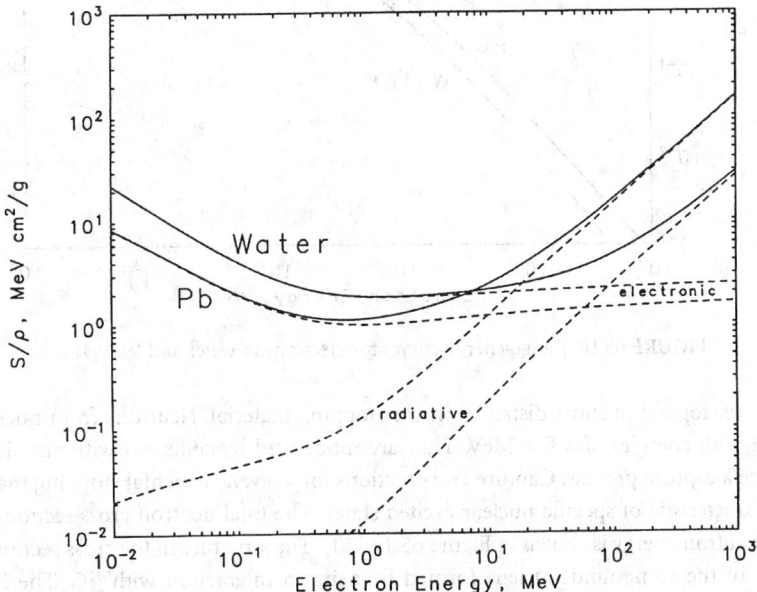

FIGURE 65.8 Stopping power for electrons in water and lead as a function of electron energy [18].

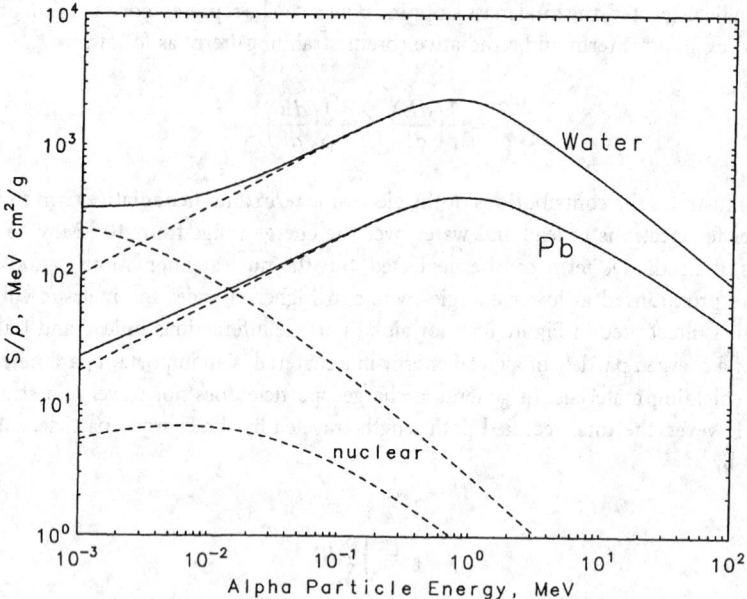

FIGURE 65.9 Stopping power for alpha particles in water and lead as a function of alpha-particle energy [19]

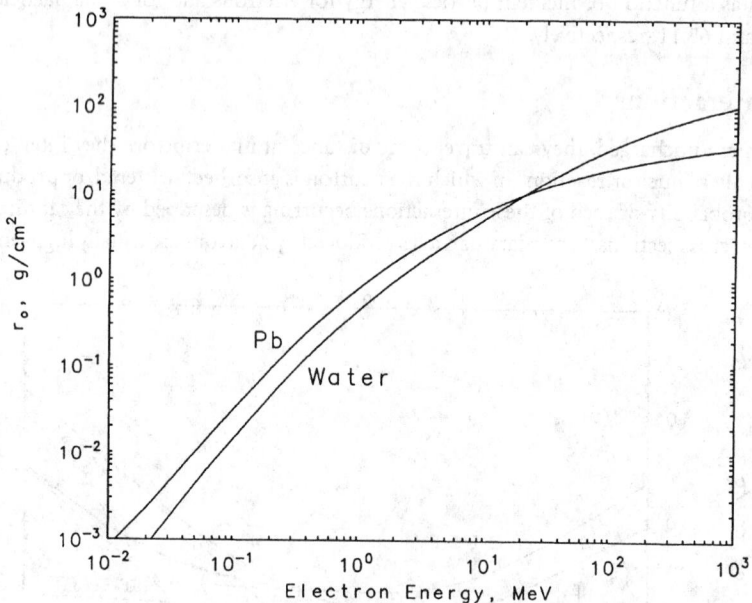

FIGURE 65.10 Range-energy curves for electrons in water and lead [18].

particles that are stopped in short distances in the stopping material. Neutrons from nuclear reactions normally start with energies of a few MeV. They are moderated by collisions with atomic nuclei until they experience a capture process. Capture cross sections for a given elemental stopping material exhibit resonances characteristic of specific nuclear excited states. The total neutron cross section of carbon as a function of neutron energy is shown in Figure 65.12 [20]. The structure in the cross section is the result of resonances in the compound nucleus formed by neutron interaction with ^{12}C. The intensity of a neutron beam in an absorber will decrease exponentially (analogous to a photon beam) with absorber thickness. The attenuation coefficient will include the scatter and capture components.

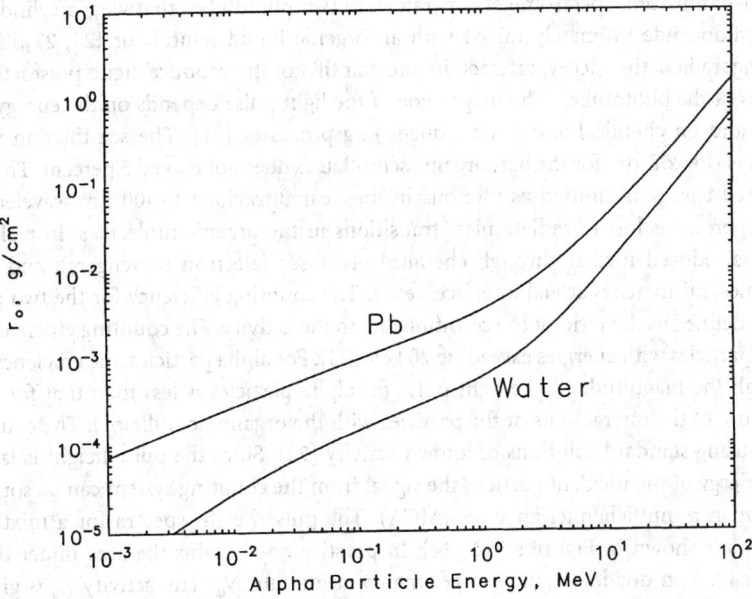

FIGURE 65.11 Range-energy curve for alpha particles in water and lead [19].

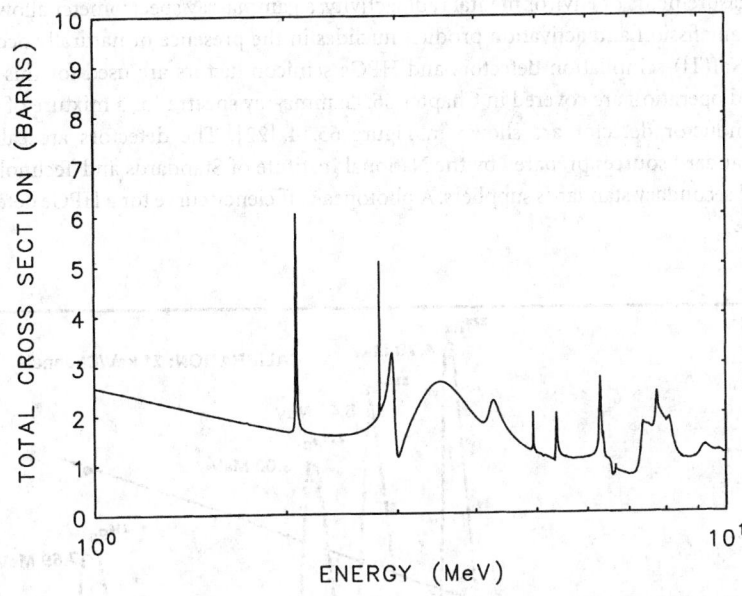

FIGURE 65.12 Total neutron cross section for carbon as a function of neutron energy [20]. The *barn* is the traditional unit of area used to describe neutron cross sections (1 *barn* = 10^{-28} m^2).

Radioactivity Measurements

Measurements of electromagnetic radiations and charged particles are discussed in Chapters 66 through 69. However, there are certain classes of instruments that are widely used for the measurement of radioactivity. Three examples will be dealt with here: alpha- and beta-particle liquid-scintillation spectrometers, high-purity germanium (HPGe) gamma-ray spectrometers, and gamma-ray dose calibrators.

Most liquid-scintillation spectrometers operate with two phototubes that view a cylindrical vial containing the radionuclide intimately mixed with an organic liquid scintillator [21, 22]. The radiations emitted during radioactive decay interact in the scintillator to produce light pulses that strike the photocathodes of the phototubes. The magnitude of the light pulse depends on the energy deposited by the radiation and on chemical and optical quenching processes [21]. The scintillation yield for high energy electrons (low dE/dx) for the best organic scintillators does not exceed 5 percent. That is, 5 percent of the deposited energy is emitted as photons in the near ultraviolet (~400 nm wavelength) and the remaining 95 percent is lost to radiationless transitions in the organic molecules. In real samples, this photon yield is reduced further through chemical processes (electron scavengers), and optical losses (colored samples, mismatches at vial interfaces, etc.). The counting efficiency for the two phototubes in coincidence is defined as the ratio of the counting rate to the activity. The counting efficiency approaches unity for beta particles with energies exceeding 20 keV [23]. For alpha particles, the efficiency is essentially unity (although the magnitude of the light pulse for alpha particles is less than that for beta particles due to the nature of the interactions of the particles with the organic scintillator). The counting systems are calibrated using standard solutions of known activity [22]. Since the pulse height is largely proportional to the energy of the incident particle, the signal from the counting system can be sorted according to pulse height in a multichannel analyzer (MCA). The pulse height spectra for a mixture of alpha-particle emitters is shown in Figure 65.13 [24]. In practice one obtains the area under the peak for a given radionuclide and divides by time to get the counting rate N_α. The activity A_0 is given by N_α/ε_c, where ε_c is the coincidence counting efficiency.

The activity of mixtures of gamma-ray emitters is needed at many stages in the nuclear fuel cycle. For example, in measurements of environmental radioactivity, a gamma-ray spectrometer allows rapid determination of many fission and activation product nuclides in the presence of naturally occurring radioactivity. Both NaI(Tl) scintillation detectors and HPGe semiconductors are used for this purpose, and their design and operation are covered in Chapter 66. Gamma-ray spectra for a mixture of radionuclides with a semiconductor detector are shown in Figure 65.14 [22]. The detectors are calibrated using radioactivity standard sources prepared by the National Institute of Standards and Technology or one of the commercial secondary standards suppliers. A photopeak efficiency curve for a HPGe detector is shown in Figure 65.15.

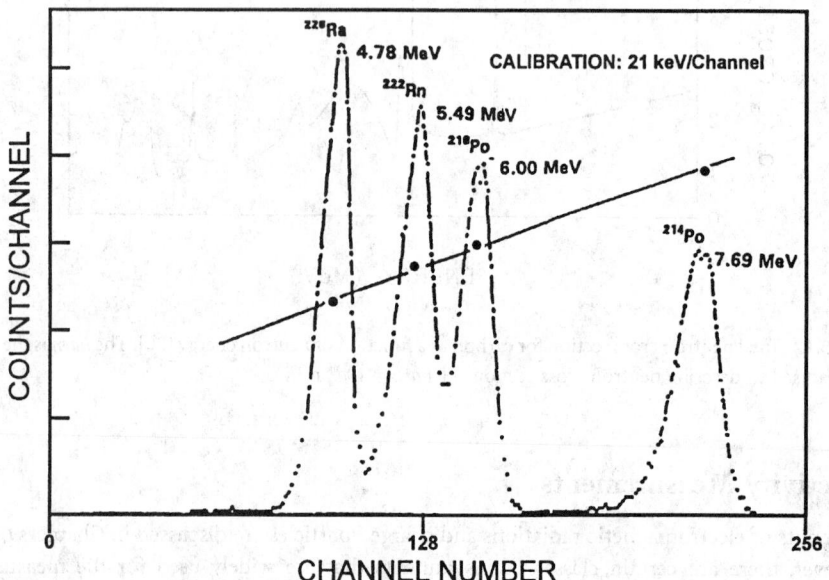

FIGURE 65.13 Liquid scintillation spectrum of mixture of alpha-particle-emitting radionuclides [24].

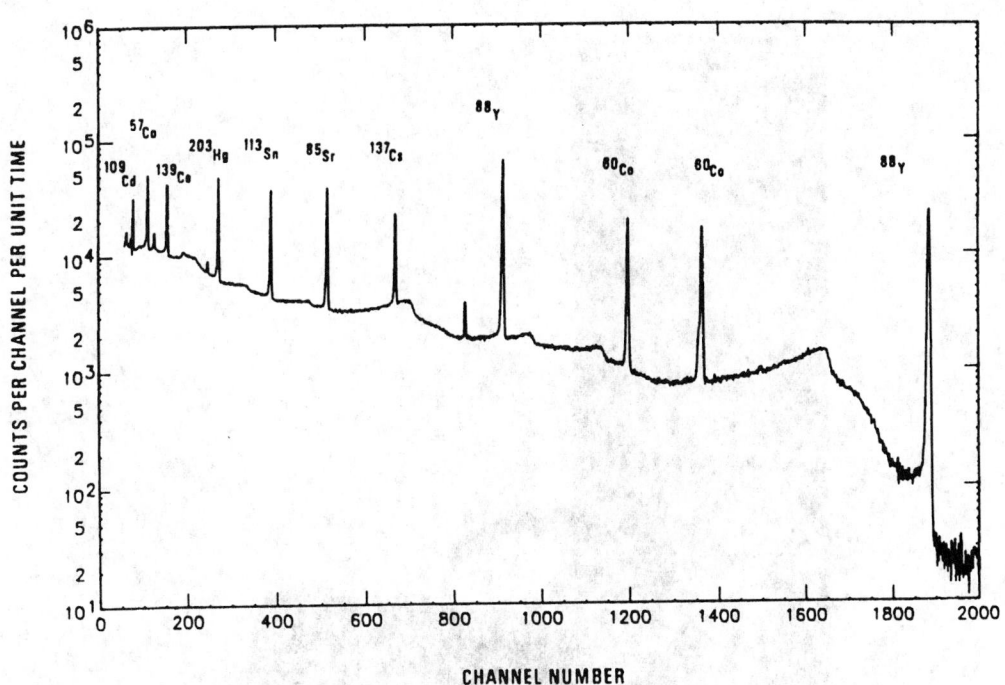

FIGURE 65.14 Gamma-ray spectrum for a solution standard of mixed radionuclides obtained with a Ge(Li) semiconductor detector [22].

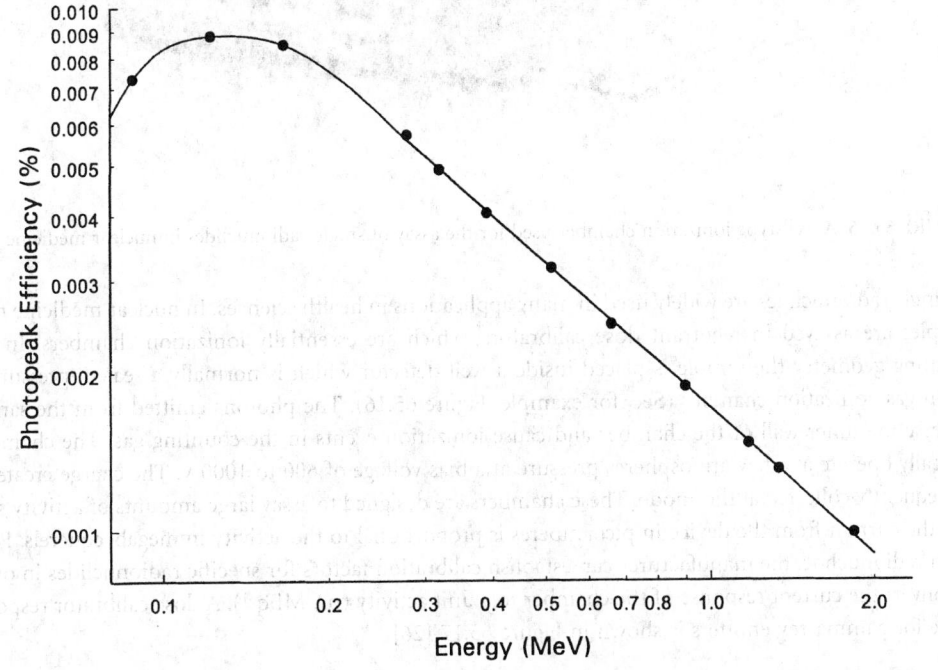

FIGURE 65.15 Photopeak efficiency as a function of energy for a 90 cm³ HPGe detector for a mixed radionuclide gamma-ray standard of the type listed in Table 65.5. A 5 mL solution source was positioned at 10 cm from the side of a p-type semiconductor [25].

FIGURE 65.16 A well-type ionization chamber used for the assay of single radionuclides in nuclear medicine [26].

Single radionuclides are widely used in many applications in health sciences. In nuclear medicine most samples are assayed in reentrant dose calibrators, which are essentially ionization chambers. In this counting geometry the sample is placed inside a well detector which is normally a sealed, pressurized argon gas, ionization chamber (See, for example, Figure 65.16). The photons emitted from the sample traverse the inner wall of the chamber and cause ionization events in the counting gas. The chambers typically operate at a few atmospheres pressure at a bias voltage of 600 to 1000 V. The charge created is subsequently collected at the anode. These chambers are designed to assay large amounts of activity such that the current from the device in picoamperes is proportional to the activity in megabecquerels. For a given radionuclide, the manufacturer can establish calibration factors for specific radionuclides in order to convert the current response of the chamber to a unit activity (pA MBq^{-1}). A dose calibrator response curve for gamma-ray emitters is shown in Figure 65.17 [26].

References

1. R.D. Evans, *The Atomic Nucleus,* New York: McGraw-Hill, 1955.

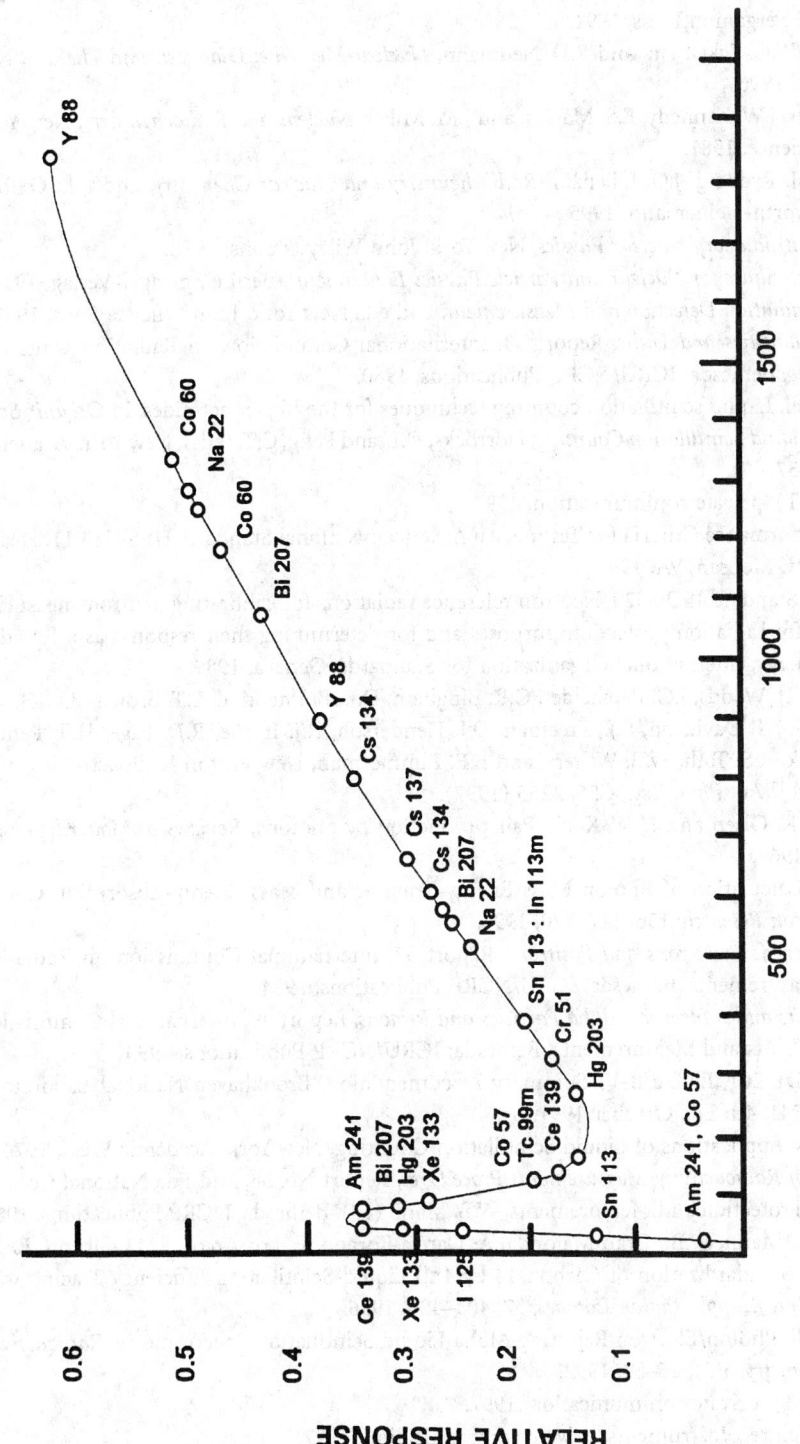

FIGURE 65.17 Photon response as a function of energy for a pressurized reentrant ionization chamber [26].

2. W.B. Mann, R.L. Ayres, and S.B. Garfinkel, *Radioactivity and Its Measurement*, Oxford, U.K.: Pergamon Press, 1980.
3. W.B. Mann, W.B., A. Rytz, and A. Spernol, *Radioactivity Measurements: Principles and Practice*, Oxford, U.K.: Pergamon Press, 1991.
4. J.C. Harbert, W.C. Eckelman, and R.D. Neumann, *Nuclear Medicine: Diagnosis and Therapy*, New York: Thieme, 1996.
5. G. Friedlander, J.W. Kennedy, E.S. Macias, and J.M. Miller, *Nuclear and Radiochemistry*, New York: Wiley-Interscience, 1981.
6. G. Choppin, J. Rydberg, J.O. Liljenzin, *Radiochemistry and Nuclear Chemistry*, 2nd Ed., Oxford, U.K.: Butterworth-Heinemann, 1995.
7. K.S. Krane, *Introductory Nuclear Physics*, New York: John Wiley & Sons, 1988.
8. W.R. Leo, *Techniques for Nuclear and Particle Physics Experiments*, Berlin: Springer-Verlag, 1992.
9. G.F. Knoll, *Radiation Detection and Measurement*, 2nd ed., New York: John Wiley & Sons, 1989.
10. *Radiation Quantities and Units*, Report 33, International Commission on Radiation Units and Measurements, Bethesda: ICRU/NCRP Publications, 1980.
11. W.J. McDowell, Liquid scintillation counting techniques for the higher actinides, in *Organic Scintillators and Liquid Scintillation Counting*, Horrocks, D.L. and Peng, C.T. (eds), New York: Academic Press, 1971, 937.
12. R. Collé (NIST), private communication, 1997
13. Personnel Performance Criteria for Testing, An American National Standard, HPS N13.11, Health Physics Society, McLean, VA, 1993.
14. International Standard ISO 8529, Neutron reference radiations for calibrating neutron-measuring devices used for radiation protection purposes and for determining their response as a function of neutron energy, International Organization for Standards, Geneva, 1989.
15. C.N. Davids, P.J. Woods, J.C. Batchelder, C.R. Bingham, D.J., Blumenthal, L.T. Brown, B.C. Busse, L.F. Conticchio, T. Davinson, S.J., Freeman, D.J. Henderson, R.J. Irvine, R.D. Page, H.T. Pentill, D., Seweryniak, K.S. Toth, W.B. Walters, and B.E. Zimmerman, New Proton Radioactivities, [165, 166, 167]Ir and [171]Au, *Phys. Rev.*, C55, 2255 (1997).
16. J.W. Motz, H.K. Olsen and H.W. Koch, Pair production by photons, *Reviews of Modern Physics*, 41, 581–639, 1969.
17. S.M. Seltzer, Calculation of Photon Mass Energy-Transfer and Mass Energy-Absorption Coefficients, *Radiation Research*, 136, 147–170, 1993.
18. *Stopping Powers for Electrons and Positrons*, Report 37, International Commission on Radiation Units and Measurements, Bethesda: ICRU/NCRP Publications, 1984.
19. *Stopping Powers and Ranges for Alpha Particles and Protons*, Report 49, International Commission on Radiation Units and Measurements, Bethesda: ICRU/NCRP Publications, 1993.
20. P.F. Rose, ENDF-201, ENDF/B-VI Summary Documentation, Brookhaven National Laboratory, BNL-NCS-17541, 4th Ed., October 1991.
21. D.L. Horrocks, Applications of Liquid Scintillation Counting, New York: Academic Press, 1974.
22. *A Handbook of Radioactivity Measurements Procedures*, Report No. 58, 2nd Ed., National Council on Radiation Protection and Measurements, W.B. Mann (ed), Bethesda, NCRP Publications, 1985.
23. Coursey, B.M., Mann, W.B., Grau Malonda, A. Garcia-Torano, E. Los Arcos, J.M., Gibson, J.A.B. and Reher, D. Standardization of Carbon-14 by $4\pi\beta$ Liquid-Scintillation Efficiency Tracing with Hydrogen-3. *Int. J. Appl. Radiat. Isotopes*, 37, 403–408, 1986.
24. W.J. McDowell, Photon/Electron Rejecting Alpha Liquid Scintillation Spectrometry, *Radioactivity and Radiochemistry*, 3(2), 26–54, 1992.
25. J. Cessna (NIST), private communication, 1997.
26. M.A. Dell, Capintec, Instruments, private communication, 1997.

66

Radioactivity Measurement

Larry A. Franks
Sandia National Laboratories

Ralph B. James
Sandia National Laboratories

Larry S. Darken
Oxford Instruments, Inc.

Radioactivity was first discovered by Henry Becquerel in 1896 when he noticed that photographic plates became fogged after exposure to uranium atoms. In addition to uranium nuclei, many other naturally occurring and man-made isotopes are known to be radioactive and decay by emitting uncharged particles (gamma rays and neutrons) and charged particles (alpha or beta particles) from their nuclei. Over the past few decades, there has been a growing need for monitoring, locating, and imaging radioactive sources in a wide variety of medical, environmental, industrial, space, scientific, and national security applications. Many of the applications rely on the use of commercially available radiation detectors, whereas others require development of new detectors to meet system requirements related to sensitivity, power, size, portability, ruggedness, maintenance, radiation hardness, and energy resolution.

The signals generated by a radiation-sensing system depend on (a) the mechanisms in which the incident radiation interacts with the detector material and (b) the scheme used to readout the interaction. In general, there are three types of radiation sensors in use today: gas-filled detectors, scintillation devices, and semiconductor detectors. The types of radiation sensors can also be divided into two groups according to whether they can measure the energy of the emitted gamma-rays, X-rays, or charged particles. Whenever an energy-resolving capability is desired, a large number of information carriers must be generated for each absorbed pulse to reduce the statistical limit on energy resolution. This is best accomplished by use of semiconductor detectors, although some spectral information can be achieved through the use of scintillators and gas-filled detectors.

The typical unit of measure of radioactivity is the becquerel (Bq), which is defined as the rate of one disintegration per second of a decaying isotope. Another common measure of radioactivity is the curie, which can be obtained by multiplying the becquerel by 3.7×10^{10}.

0-8493-8347-1/99/$0.00+$.50
© 1999 by CRC Press LLC

This chapter is divided into sections on gaseous, scintillation, and semiconductor detectors. Separate sections describing semiconductor detectors that must be cooled to cryogenic temperatures for operation and those capable of operation at ambient temperatures are presented. There are several texts devoted to the subject of radiation detection [1–7], and the reader is directed to these books for more detailed information on the principles of detector operation, device performance, problems limiting detector performance, areas of current research and development, and applications.

66.1 Gaseous Detectors

Ionization in a gas resulting from the interaction of X and gamma radiation is the basis for a wide variety of radiation detectors. The versatility of gas phase detection arises from the great flexibility in detector sizes and shapes that can be produced, their relatively low production cost, and particularly, the ability to perform a multitude of measurement tasks. Gas detectors can be separated into three distinct types: ionization chambers, Geiger-Müller counters (tubes), and proportional counters with numerous variations of each type. All utilize the ions and electrons created by the interaction of an incident photon with the detector gas, directly or indirectly, to produce an output signal. They differ in the characteristics of the electric field and nature of the output signal.

Ionization chambers can be operated in either a current mode or pulsed mode, although the current mode is more common. They operate by collecting (with an applied electric field) all the charge generated by the original ionizing event and differ in this way from both proportional and Geiger-Müller counters. The latter, normally operated in a pulsed mode, have output signals that result from amplification of the original ion pairs by gas multiplication processes.

General Operating Principles

Fundamental to the operation of gas detectors is the generation of electron-ion pairs and their movement through the surrounding gas under the influence of an applied electric field. It is convenient to summarize basic features of these processes before turning to specific detector designs.

W Values

The energy (W) required to produce an electron-ion pair in a gas depends on the gas, the type of radiation (and its energy). W values for fast electrons in common filling gases range from 26.4 eV per ion pair in argon to 41.3 eV per ion pair in helium. The presence of nonionizing energy loss processes accounts for the W values greatly in excess of the ionization energy. Fluctuations in the number of pairs produced from photons of the same energy are of significance in pulse-mode operation. The variance is generally less than expected based on Poisson statistics and is accounted for by an empirical constant, the Fano factor (typically 0.1 to 0.2).

Charge Transport

The motion of free electrons and ions in the gas under the influence of the electric field (E) is quite different. The drift velocity (v) of the more massive ions is a linear function of E/p where p is the gas pressure. It can be expressed as

$$v = \frac{\mu E}{p} \tag{66.1}$$

The proportionally constant μ is the mobility, which depends on the type of gas and the charge of the ion. Values are typically in the region of 1000 cm^2/V-s/mm Hg. The electron mobility is normally about 1000 times ion values; the electron drift velocity is not linear in E/p [8].

Electric Field Effects

The amplitude of a pulse resulting from the interaction of a photon with the wall or fill gas depends strongly on the voltage applied to the detector and serves to distinguish the three detector types. A plot of pulse amplitude as a function of applied voltage is shown in Figure 66.1. The plateau following the initial steep segment is the region of ion chamber operation. It is a region where the electric field is sufficient to reduce recombination of the original pairs to an acceptable value and further voltage increases yield no more charge (as it has all been collected) and ion saturation is established. Assuming complete charge collection, the output current accurately represents the rate ion pairs are being produced. This is the basis for ionization chamber operation.

The rapidly rising portion following the plateau marks the onset of gas multiplication (the initial electrons can acquire enough energy between collisions to generate further ionization). In the initial segment, the multiplication process is linear; that is, the charge collected is proportional to the number of original ion pairs. This is the defining characteristic of proportional counter operation. The proportionality eventually is ended at higher voltages by space charge effects caused by positive ions.

At yet higher voltages, the space charge becomes sufficient to reduce the electric field below the multiplication threshold and no further multiplication takes place. Thus, a condition is reached where the same number of positive ions are produced for all initial ion-pair populations, and the pulse amplitude is independent of the initial conditions. This characterizes Geiger-Müller counter operation.

Further details on the operation of these devices may be found in texts by Knoll [1], Price [9], Attix and Roesch [10], and Tait [11]. The monograph by Rossi and Staub [2] contains a lengthy discussion of ionization chambers together with data on the physics of the transport process.

Ionization Chambers

Ionization chambers have been designed for numerous X- and gamma-ray measurement applications. They are frequently employed in radiation survey instruments. They are found in parallel plate, cylin-

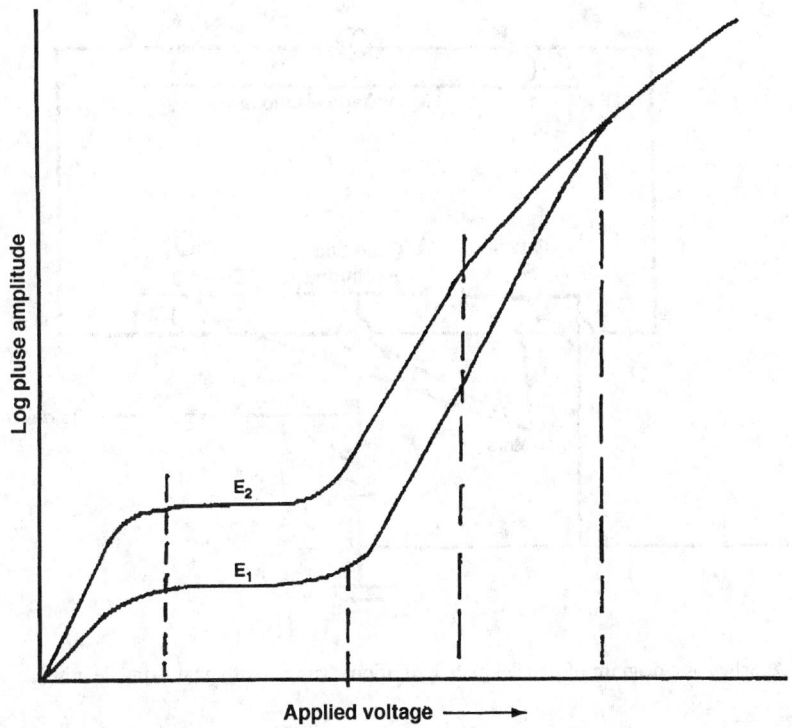

FIGURE 66.1 Distinct operating regions of gas-filled detectors. E_1 and E_2 depict pulses from photons of two energies.

drical, and spherical geometries. Essential features of their design can be found in the parallel plate chamber shown schematically in Figure 66.2. The design shown contains an optional guard ring that helps to define the active volume of the chamber. The ring is normally maintained near the collector electrode potential. (Guard rings are also employed in very low current designs to reduce leakage current.) Because the current for the ionization chamber is low, typically 1 nA or less, considerable care must taken with the insulators to minimize leakage currents.[1]

Because of the low output current, special care must be taken with the readout system. Both dc and dynamic-capacitor types are used. Only electrometers of the highest quality should be used. Dynamic-capacitor (or vibrating-reed) varieties provide more stable operation. They are normally the choice for very low current applications.

Several special-purpose ion chambers are notable. The "free air" chamber, a parallel-plate variant, is valued for accurate gamma ray exposure measurements, particularly when absolute measurements are required. This is achieved by collimation of the incoming gamma flux and by an internal design that ensures compensation for ionization produced outside the sensitive volume by secondary electrons. These chambers are limited to energies below about 100 keV, however. Cavity ionization chambers are widely used for dosimetry purpose. To function in this manner, it is necessary that they be designed to meet the requirement of the Bragg-Gray principle [12]. This states that the absorbed dose in a medium can be determined from the ionization produced in a small gas-filled cavity in the medium. The cavity dimension must be small compared to the range of the ionizing particles generated in the material so that the particle flux is essentially unperturbed. Of particular interest for health physics applications are

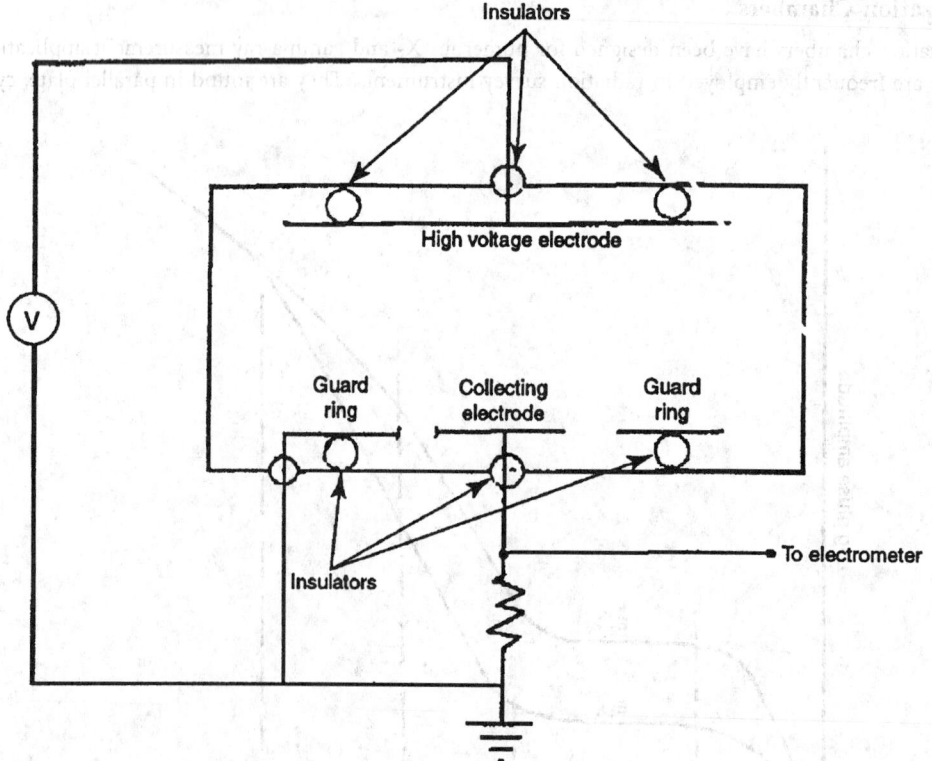

FIGURE 66.2 Schematic diagram of parallel-plate ionization chamber with guard ring defining active volume.

---

[1]Note also that ionizing radiation can cause both permanent and transient resistivity losses in insulating materials and should be considered in high-flux or high-fluence applications.

tissue equivalent chambers where the wall material is a plastic that closely simulates the absorption properties of tissue.

Proportional Counters

Proportional counters utilize gas multiplication to amplify the charge of the initial ion pair population and have the important characteristic that the charge associated with the amplified pulse is proportional to the ion pairs produced initially. They are normally operated in a pulse mode. They can be operated sealed or in a gas flow mode. The sealed style is most common for X- and gamma-ray applications. While they are found in a number of geometries, cylindrical is the most common shape. A typical cylindrical design is shown in Figure 66.3. This relatively simple design features a central wire that is maintained at high voltage and a surrounding metal container that serves as the cathode. The electric field in this geometry is given by

$$E(r) = \frac{V}{r}\ln\left(\frac{b}{a}\right) \tag{66.2}$$

where V = anode-cathode voltage, a = anode wire radius, and b = inner radius of the cathode.

It should be noted that $E(r)$ depends on the diameter of the anode wire, allowing the threshold electric field needed for multiplication to be obtained at relatively modest voltages. The multiplication process is confined to a small volume surrounding the anode where the field exceeds the multiplication threshold. The characteristics of several common counting gases are given in Table 66.1.

TABLE 66.1 Characteristics of Common Proportional Counter Gases

Gas	W (eV per ion pair)	Fano Factor	Resolution (%) @ 5.9 keV	
			Calculated	Measured
Ne + 0.5% Ar	25.3	0.05	10.1	11.6
Ar + 0.5% C_2H_2	20.3	0.075	9.8	12.2
Ar+ 10% CH_4	26.0		12.8	13.2

Source: adapted from Knoll [1].

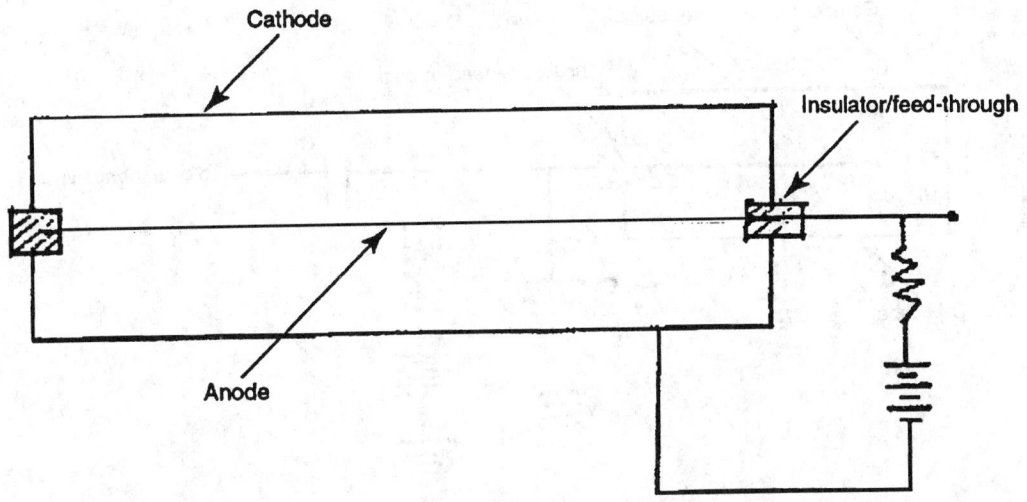

FIGURE 66.3 Schematic diagram of cylindrical proportional counter.

Proportional counters can be used for both photon and particle measurements. In photon applications they are particularly valued for spectroscopy in the low energy X-ray region. Their energy resolution and detection efficiency is generally inferior to semiconductor detectors such as lithium-drifted silicon or mercuric iodide but they offer large surface areas, reliable operation, and low cost. The energy resolution ΔE for given gas mixture for a specific photon energy can be estimated from the statistical limit given by

$$\Delta E = 2.35\left[W\frac{(F+\beta)}{E}\right]^{1/2} \tag{66.3}$$

where F = Fano Factor, β = variance factor (typically about 0.5) [1], E = energy in eV, and W = energy in eV required to produce an ion pair.

Energy resolution for a proportional counter at 5.9 keV can be expected to be on the order of 10 percent (Table 66.1).

Geiger-Müller Counters

Geiger-Müller counters (more commonly known as G-M counters or tubes) remain among the most widely used means of detecting X and gamma radiation. They have high sensitivity, are rugged, and offer low cost. Also of importance is the large amplitude of the output pulse (several volts, typically), which greatly simplifies the readout. Because of these factors, they are the detectors of choice for a variety of commercial gamma survey instruments. Like proportional counters, they utilize gas amplification to generate an output pulse, although here the output is independent of the initial number of ion pairs. They are counters only and not applicable to spectroscopy. A G-M counter with typical readout is shown schematically in Figure 66.4.

G-M counters typically use a noble gas filling, most frequently argon or helium. The gas pressure is normally in the region of a few tenths of an atmosphere. At this pressure, a typical counter would require about 1000 V.

The process that terminates multiplication in a G-M counter has important counting implications. The positive ions space charge and resultant subthreshold electric field (for gas multiplication) persist for some time after the discharge is terminated. As a result, there is an interval following each pulse when pulses from subsequent input gamma rays are not produced or have less than full amplitude. This is

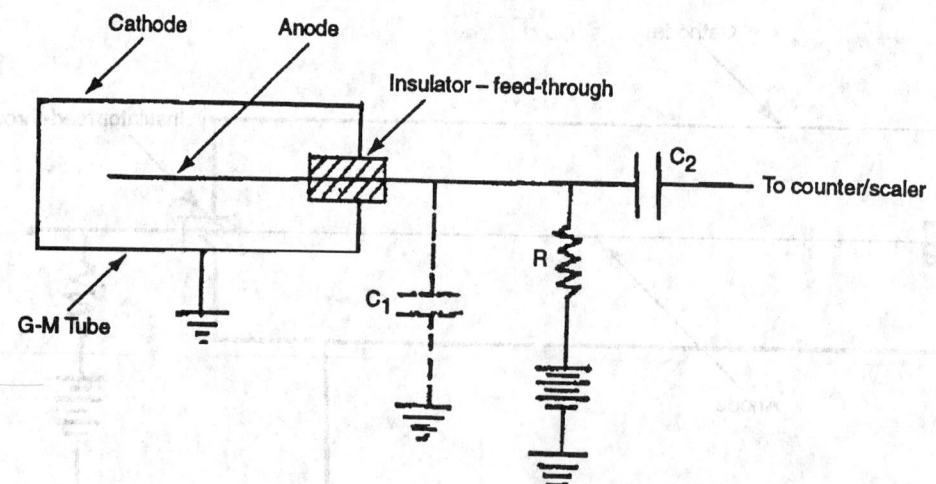

FIGURE 66.4 Typical counting circuit using a G-M tube. C_1 represents the combined tube and wiring capacitance, C_2 is a blocking capacitor that isolates the high voltage from the counter/scaler. The time constant of the circuit is RC_1.

illustrated in Figure 66.5. The period between a full-amplitude pulse and the next discharge of any size is the dead time of the G-M counter (T_d). The time from the initial full amplitude pulse until another full amplitude pulse can be produced is the recovery time. Dead times in G-M counters are on the order of 100 μs, with recovery time several times that. If the dead time is known and independent of count rate, the true count rate R_t may be determined in most cases from the observed rate R_o by the expression

$$R_t = \frac{R_o}{1 - R_o R_d} \qquad (66.4)$$

Availability

Ionization chambers, Proportional counters, and Geiger-Müller counters are standard commercial products. They are available as stand-alone detectors requiring the user to supply ancillary electronics or as part of an instrument. Custom design devices are also available.

Scintillation Detectors

Scintillators, one of the oldest means of detecting gamma radiation, remain the method of choice for a multitude of counting and spectroscopy applications, particularly when counting efficiency rather than high-energy resolution is the primary objective. They are also widely used to record transient radiation events and in timing applications. While available in solid, liquid, and gas phases, solids and liquids are preferred for gamma-ray applications. Scintillators have the common property of converting energy absorbed from the incident gamma ray into visible or near visible light. The scintillation detector thus consists of a scintillator element, in which gamma ray energy is converted to optical photons, and some form of a photocell, normally a photomultiplier tube, to convert the optical photons into an electrical signal for processing by ancillary electronics.

Desired properties of a scintillator include high transparency to its own optical emission, efficient conversion of the absorbed gamma energy into optical output, short duration of the output, and output

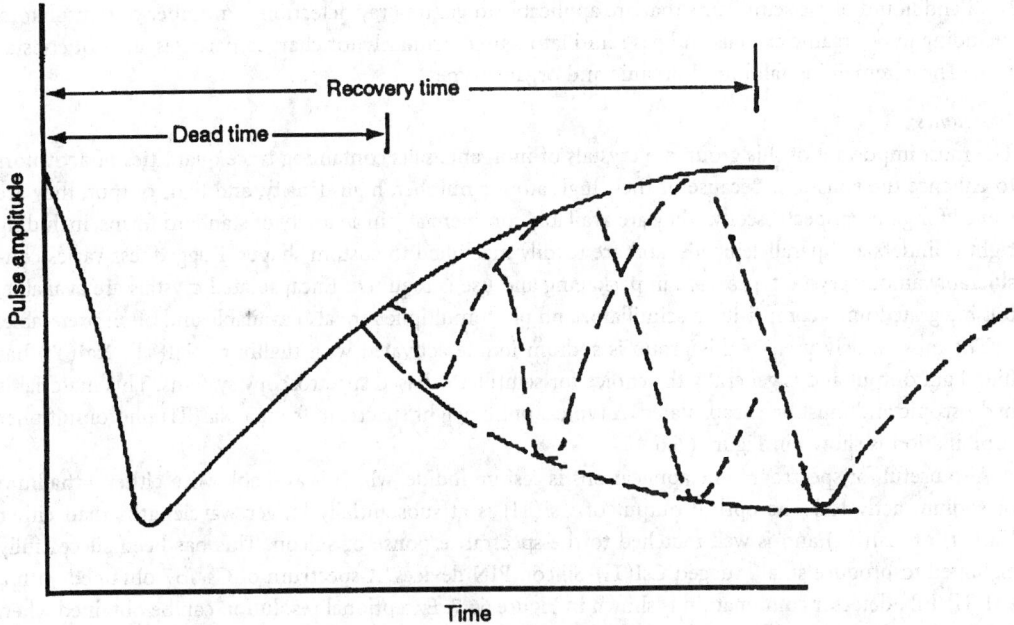

FIGURE 66.5 Illustration of dead time (Td) in a typical G-M counter.

that is proportional to the amount of energy absorbed. Other useful properties include the ability to be produced in large sizes, ability to be machined, stable output over a wide range of environmental conditions, and efficient gamma-ray absorption.

Scintillation Process

The excess energy contained by the scintillator as a result of the gamma-ray absorption is dissipated largely through nonradiative processes and appears as heat. A small fraction decays radiatively and appears as visible and near visible light. That fraction of the absorbed energy converted to optical photons is referred to as the scintillation efficiency, which varies from <1 to ≈10 percent for the more efficient scintillators.

The optical emission originates from electronically excited atomic and molecular states. The emission is broadband, frequently exceeding 50 nm at the half-intensity points. Temporally, it is characterized by a very fast rise, followed by exponential decay. For most purposes, the light pulse can be adequately represented by

$$I(t) \propto N \left[e^{-\frac{t}{\tau}} - e^{-\frac{t}{\tau_R}} \right] \tag{66.5}$$

where $I(t)$ is intensity at time t, and N is related to the total number of scintillation photons. τ_R is the time constant associated with the pulse rise time, and τ is the time constant for decay. The time required for the intensity to fall to $1/e$ of the maximum value is the pulse decay time. Optical pulses from organic scintillators are often described by the full width of the pulse at the half maximum of intensity. Further details of the scintillation process can be found in the text by Knoll [1] and, in particular, the treatise by Birks [13].

Scintillator Types

As previously noted, the scintillation process is observed in solid, liquid, and gas phases, but gamma applications are limited mainly to solids and liquids. This brief discussion is confined to the most common solid and liquid phase scintillants that are applicable to gamma ray detection. A number of scintillators, including pure organic crystals and gas scintillators used primarily for charged particles, are not considered. Those remaining fall into inorganic and organic types.

Inorganics

The most important of this group are crystals of inorganic salts containing trace quantities of activators to enhance the emission. Because of their high atomic number, high density, and light output, they are valued for gamma spectroscopy. They are available commercially in a variety of standard forms, including right cylinders and parallelepipeds, and are readily machined to custom shapes. Ruggedness varies considerably among crystal types; care in packaging and use is required. Encapsulated crystals are available, but integrated units comprising a scintillator and photomultiplier are also available and often preferable.

The most widely used of this group is sodium iodide activated with thallium, NaI(Tl). NaI(Tl) has high light output and is generally the choice for scintillator-based spectroscopy systems. This material is hydroscopic and must be encapsulated. A typical pulse height spectrum from a NaI(Tl) photomultiplier combination is shown in Figure 66.6.

Also useful for spectrographic applications is cesium iodide, which is available with either a thallium or sodium activator. The optical output of CsI(Tl) is at substantially longer wavelengths than either NaI(Tl) or CsI(Na) and is well matched to the spectral response of silicon. This has been successfully exploited to produce small, rugged CsI(Tl)-Silicon PIN devices. A spectrum of Cs-137 obtained with a CsI(Tl)/PIN detector combination is shown in Figure 66.7. Exceptional resolution can be obtained when CsI(Tl) is used with a mercuric iodide photocell. With this combination resolution at 662 keV of better than 5 percent has been reported [14].

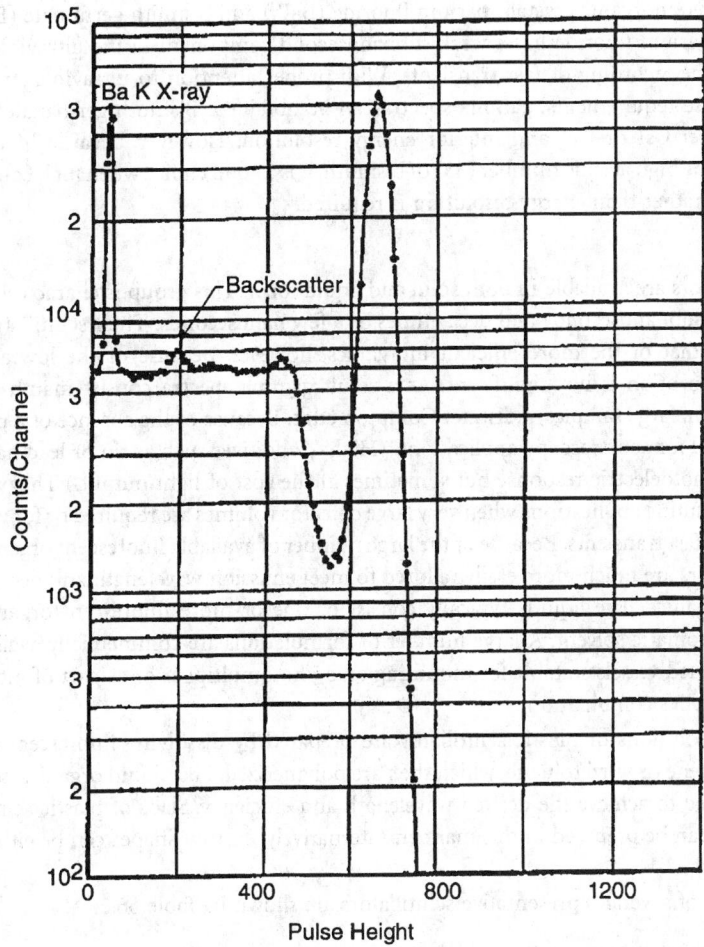

FIGURE 66.6 Typical energy spectrum obtained with a NaI(T$_1$) scintillator/photomultiplier combination in response to a Cs-137 source (662 keV).

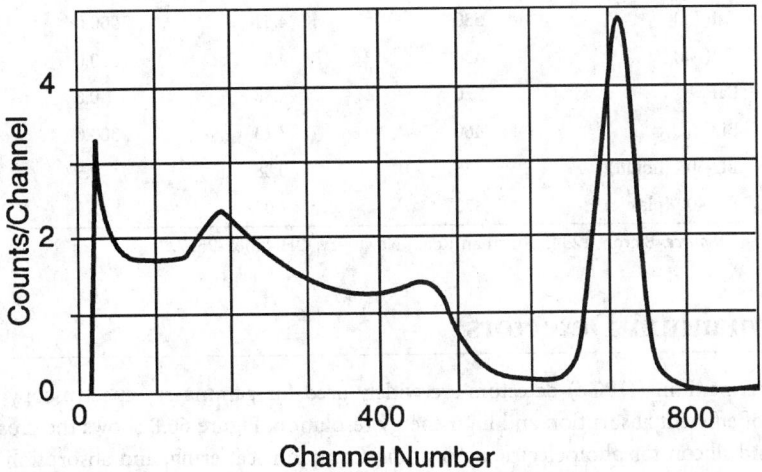

FIGURE 66.7 Energy spectrum of Cs-137 (662 keV) obtained with a CsI(Tl)/PIN combination. Specifications CsI(Tl): 1 cm × 2 cm × 2 cm; PIN: 1 cm × 2 cm.

Two unactivated inorganic crystals, barium fluoride (BaF_2) and bismuth germinate ($Bi_4Ge_3O_{12}$), warrant comment. Barium fluoride has a weak subnanosecond component in the ultraviolet that is useful for timing and the recording of fast transients. With proper attention to ultraviolet transmission and photocell response requirements, gamma spectra can be obtained. Bismuth germanate has lower light output than either CsI or NaI and inferior energy resolution. However, because of its high density (7.1 g/cm^3) and the high atomic number (83) of bismuth, it is a good choice when high counting efficiency per unit volume rather than energy resolution is required.

Organics

Organic scintillators are available in both solid and liquid form. This group is characterized by very fast rise times (often subnanosecond) and decay times of a few nanoseconds. Their scintillation efficiency is about one-third that of the more efficient inorganics, however. Because of the low atomic number, photoelectric absorption is not significant. As a result gamma spectra consist mainly of a Compton continuum with, in large samples, a distinct Compton edge. Because of the absence of a photopeak, they are not well suited for spectroscopy applications. (High-Z additives such as tin or lead have been shown to enhance the photoelectric response but sometimes at the cost of light output.) They are particularly well suited for counting applications when very large detector volumes are required or for wide bandwidth measurements of fast transients. Because of the large number of available fluorescent organic compounds, organic scintillators are much more easily tailored to meet emission wavelength and decay time requirements than inorganics. The liquids typically consist of one or more fluorescent organic compounds dissolved in an aromatic solvent. A large number of formulations are commercially available. They can be obtained in sealed vessels suitable for mounting on a photomultiplier or as part of integrated scintillator–photomultiplier combination.

Solid organic solutions or plastic scintillator are prepared by dissolving fluorescent compounds in monomers of styrene or vinyl toluene, which then are polymerized. Like liquid organics, several different solutes can be used to achieve the desired wavelength and efficiency. Slabs of plastic scintillator several meters in length can be prepared in this manner. Alternatively, custom shapes can be either cast directly or machined.

Characteristics of several representative scintillators are shown in Table 66.2.

TABLE 66.2 Properties of Common Scintillators

Scintillator	Wavelength of Emission Maximum (nm)	Density (g/cm^3)	Principle Decay Constant (ns)
NaI(T1)	415	3.67	230.0
CsI(T1)	530	4.51	1000.0
CsI(Na)	430	4.51	630.0
BaF$_2$	220	4.88	0.6
Bi$_4$Ge$_3$O$_{12}$	460	7.13	300.0
BC-505 (liquid)	425	0.88	2.5
BC-400 (plastic)	423	1.032	2.4

Source: Bicron, 12345 Kinsman Road, Newberry, OH 44065-9577.

66.2 Germanium Detectors

High-purity germanium (HPGe) detectors are widely used for gamma-ray spectroscopy due to their combination of efficient absorption and high energy resolution. Figure 66.8 shows the cross sections of germanium and silicon for photoelectric absorption, Compton scattering, and absorption by electron-positron pair production in several materials used for solid-state nuclear ration detectors. Attenuation is significantly stronger in germanium than in silicon. Over much of the gamma spectrum, the dominant

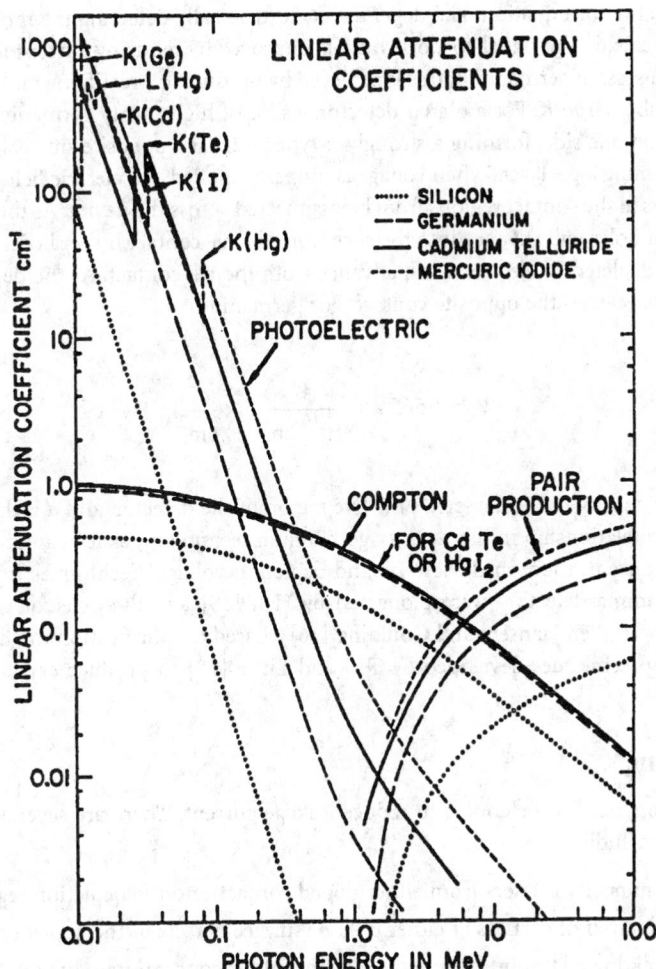

FIGURE 66.8 Attenuation coefficients vs. energy in common semiconductor materials [4].

interaction is Compton scattering. However, it is principally the stronger photoelectric absorption in germanium that makes it more suitable than silicon for gamma-ray spectroscopy. In the typical size germanium detector, a gamma ray may be scattered several times before it is photoelectrically absorbed. Thus, the energy of the gamma ray is primarily transmitted directly to a small number of electrons. These energetic electrons in turn interact with electrons in the valence bands to create mobile pairs of electrons and holes. In a detector with sufficiently large volume, the average number of electron-hole pairs N produced by an absorbed gamma-ray of energy E becomes independent of the details of the initial reaction path and varies linearly with E as follows:

$$N = \frac{E}{\varepsilon} \tag{66.6}$$

This relationship is more broadly valid and is the foundation of energy spectroscopy of gamma rays using semiconductors, gases, and cryogenic liquids (ε depending on the material). While ε is independent of the gamma-ray energy (and is also virtually the same for energy deposited by charged particles), ε in germanium does increase slightly with decreasing temperature, as does the energy gap. At 77 K, ε is 2.96 eV, and the energy gap is 0.72 eV.

Practical exploitation of Equation 66.6 depends on electronically detecting the motion of the ionized charge in an electric field. The signal-to-noise ratio is improved by reducing current flow in the detector from other mechanisms. In germanium, this is achieved by producing a rectifying and a blocking contact and by cooling to about 100 K. For a planar detector, a slice of high-purity germanium is diffused with the donor lithium on one side, forming a strongly *n*-type layer. The opposite side is implanted with the acceptor boron, forming a p+ layer. When voltage is properly applied, the electric field direction prevents the majority carriers in the contact regions from being injected across the device. As the voltage is applied, a region depleted of holes will advance into the slice from the n+ contact if the slice is *p*-type. If the slice is *n*-type, a region depleted of electrons will advance from the p+ contact. At the depletion voltage V_d, the depletion region reaches the opposite contact. For germanium,

$$V_d = 565 V \times \frac{N_e}{10^{10} \text{ cm}^{-3}} \times \frac{d^2}{\text{cm}^2} \tag{66.7}$$

N_e is the net charge density in the depleted or active region of the detector, and d is the thickness of this region. This is a key relationship in high-purity germanium technology, as it quantifies the effect of the residual impurity concentration on device size and depletion voltage. Techniques to grow germanium pure enough for gamma detectors were pioneered by Hall [15] and the detector group at Lawrence Berkeley Laboratory (Haller, Hansen, and Goulding [16]), based on purification methods of Pfann [17] and crystal-growing techniques developed by Teal and Little [18] to produce crystals for germanium transistors.

Leakage Current

Germanium detectors need to be cooled to reduce leakage current. There are several potential sources of leakage current, including

- Diffusion of minority carriers from either doped contact into the depletion region
- Thermal generation of carriers at either bulk or surface defects in the depletion region
- Electrical breakdown at points where the electric field is concentrated due to irregularities in the contact geometry, large-scale inhomogeneities in the bulk, or surface states.

Current will also be generated if the detector is not shielded from room-temperature infrared radiation. Background nuclear radiation from materials near the detector and cosmic radiation also generate leakage current.

Germanium detectors are typically liquid-nitrogen cooled and operated between 85 K and 100 K. In this temperature range, leakage current is typically less than 40 pA in "good" detectors and is not a significant contributor to system noise (400 to 900 eV). Leakage current increases with temperature and eventually becomes the predominate noise component. Pehl, Haller, and Cordi [19] reported a leakage current driven system noise of 2 keV at 150 K and 7 keV at 170 K for an 8 cm³ planar detector. These authors also reported that, above about 120 K, the leakage current had an activation energy of approximately one-half the bandgap and attributed this to generation at mid-gap surface states. Below 120 K, the temperature dependence was milder.

A typical detector/cryostat configuration is shown in Figure 66.9. The detector resides in an evacuated cryostat and is cooled by means of a copper rod inserted into a liquid-nitrogen dewar. The first stage of amplification is an FET, also cooled, positioned nearby the detector. Mechanical fixturing is designed to stabilize the detector and the mechanisms for contacting it, to provide a cooling path from the detector to liquid nitrogen, and to electrically insulate the high-voltage contact.

A variety of detector geometries are shown in Figure 66.10. These different electrode configurations allow efficiency and energy resolution to be optimized for different gamma-ray energies and applications. For example, the detector in Figure 66.10(c) minimizes noise by the lower capacitance of its electrode

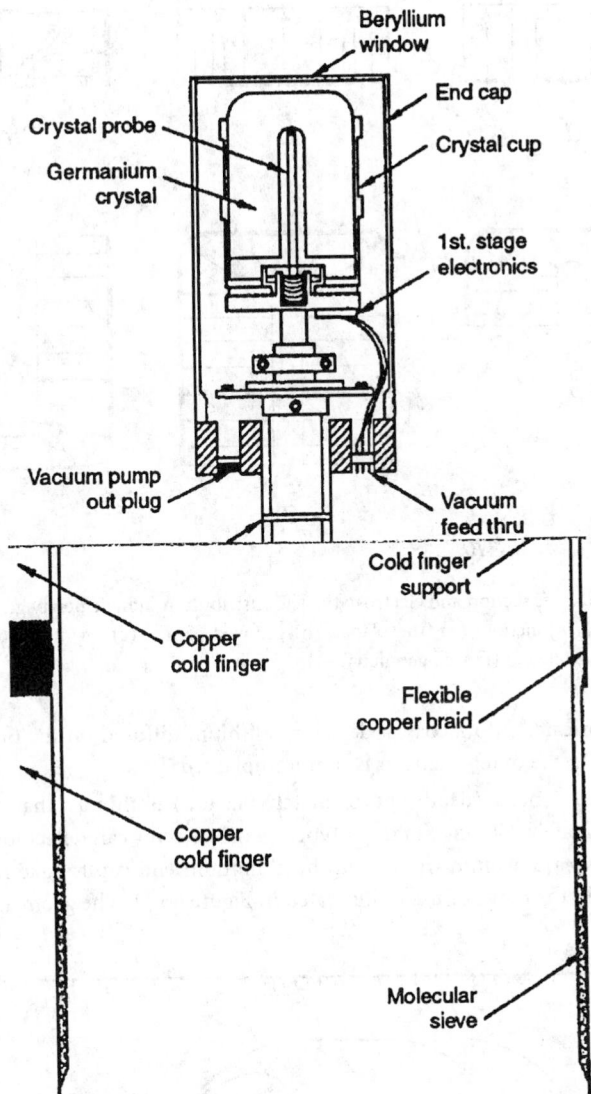

FIGURE 66.9 Schematic cross section of a dipstick cryostat. (Darken and Cost, 1993, reprinted with permission of Oxford Instruments, Inc.

configuration at the expense of the reduced stopping power. Thus, this detector would be more suitable for lower-energy gamma rays.

Coaxial Detectors

The detector type shown in Figures 66.9 and 66.10(e) has a closed-end coaxial geometry. Nearly all of the largest-volume (active volumes of 100 cm³ to 800 cm³) HPGe detectors are of this type. This electrode geometry reduces both capacitance and depletion voltage with respect to a planar detector of the same volume. This latter benefit relaxes the constraint on material purity. In addition, charge collection distances are shortened, and the uncontacted surface area, frequently troublesome in processing, is reduced. Also, the HPGe is grown by the Czochralski technique and is therefore nearly cylindrical, even before machining. It is important, however, to note that the reduction in depletion voltage is realized only when the device is contacted so that it depletes from the outer contact to the inner contact. Thus,

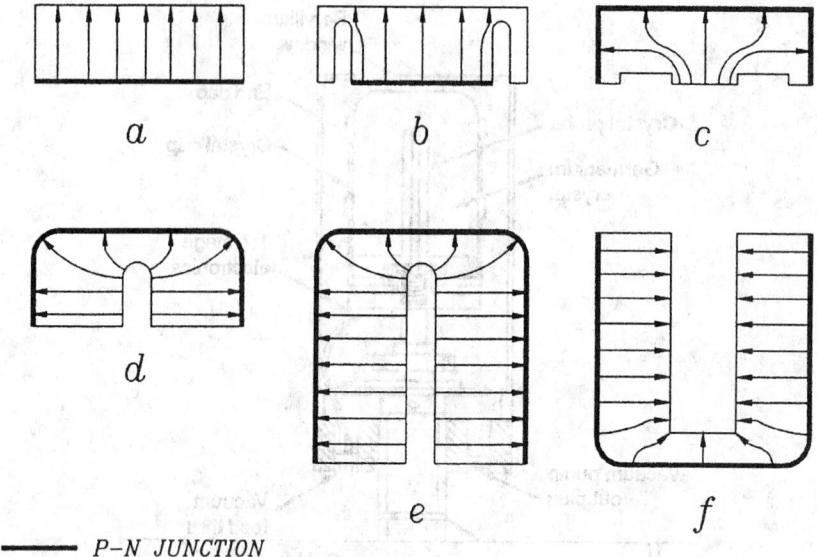

—— *P–N JUNCTION*

FIGURE 66.10 Schematic cross section and electrostatic field distribution in high-purity germanium detectors. The dark line represents the *p-n* junction: (a) true planar, (b) grooved planar, (c) low capacity planar, (d) truncated coaxial, (e) closed-end coaxial, and (f) well geometry.

p-type HPGe to be fabricated into a coaxial detector is lithium diffused on the outer diameter and, in the case of *n*-type HPGe, the outer diameter is boron implanted.

The boron-implanted contact (depth approximately 0.2 μm) is thinner than the lithium-diffused contact (depth approximately 750 μm), so the *n*-type coaxial detector can detect lower-energy radiation and is usually built with a beryllium window in the aluminum end cap to take full advantage of this feature. The difference in the range of use is illustrated in Figure 66.11. The geometric asymmetry of the

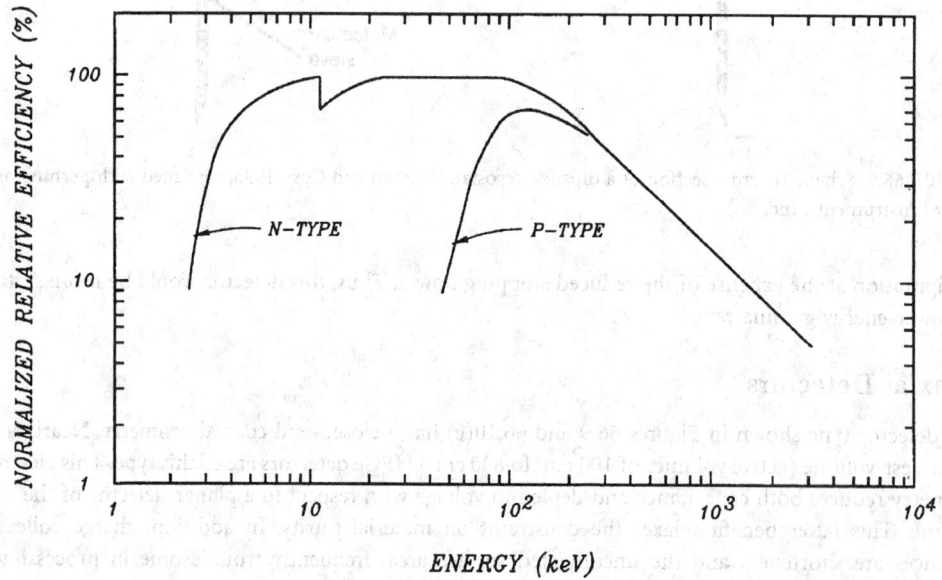

FIGURE 66.11 Relative absorption efficiencies for typical *n*- and *p*-type detectors. (Darken and Cox, 1993, reprinted with permission of Oxford Instruments, Inc.)

contacting electrodes in the coaxial detector makes charge collection more dependent on the carriers (electrons or holes) traversing to the inner contact. As more gamma rays are absorbed near the outer contact, the carriers traversing to the inner contact must travel on average a longer distance. Also, charge traversal near the inner contact is particularly effective in inducing current in the external circuit [20]. Thus, the p-type coaxial detector with positive bias on the outer electrode is more sensitive to hole collection, and the n-type coaxial detector with negative bias on the outer electrode is more sensitive to electron collection. This is a crucial consideration in applications where the hole collection is going to be degraded during use by exposure to fast neutrons or other damaging radiation. The superior neutron damage resistance of the electrode biasing polarity on n-type coaxial detectors was demonstrated by Pehl et al [21].

A typical gamma-ray spectrum of a Co^{60} source taken with a coaxial HPGe detector is shown in Figure 66.12. The salient features are the full-energy peaks at 1.17 MeV and at 1.33 MeV, and the lower energy plateaus due to incomplete energy absorption of Compton-scattered gamma rays. The peak-to-Compton ratio [22] is generally 40 to 100, depending on the size and quality of the detector. The 1.33 MeV peak is shown separately in Figure 66.13. The energy resolution measured as the full width at half the peak maximum (FWHM) for typical coaxial germanium detectors is between 1.6 and 2.1 keV for 1.33 MeV gamma rays, again depending on the size and quality of the detector. The variance in the peak L^2 (FWHM $= 2.35 \times L$, L being the standard deviation for a Gaussian distribution) can be divided into three additive components: the electronic noise component L_N^2, a component reflecting the variance in the number of electron-hole pairs created L_F^2, and a component due to incomplete charge collection L_T^2,

$$L^2 = L_N^2 + L_F^2 + L_T^2 \qquad (66.8)$$

$$L_F^2 = \varepsilon EF \qquad (66.9)$$

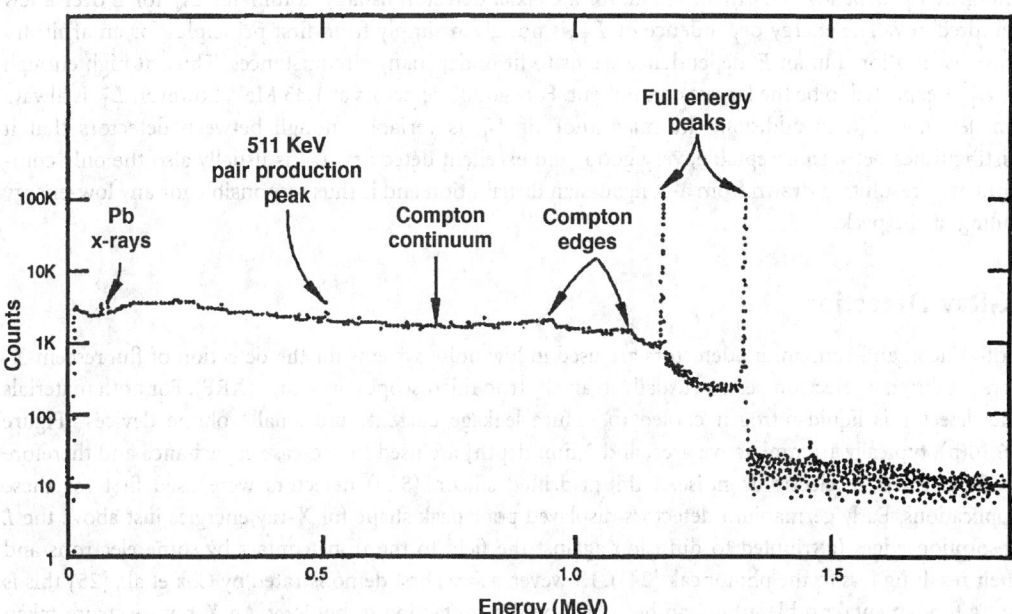

FIGURE 66.12 A ^{60}Co spectrum collected with a 15% p-type detector showing typical features of germanium detector spectrum.

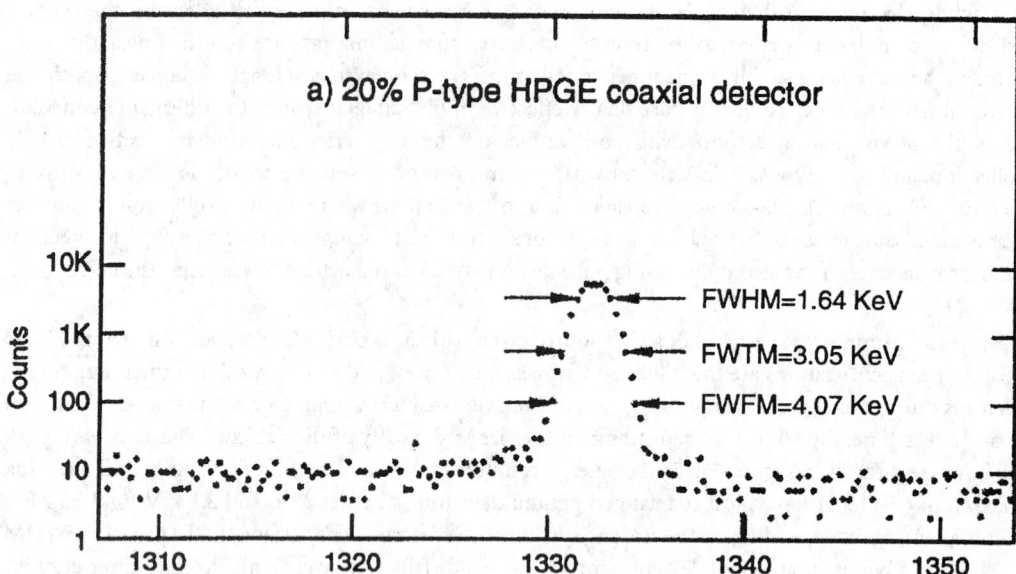

FIGURE 66.13 A ^{60}Co spectrum collected with a 22% relative efficiency *p*-type detector. (Darken and Cox, 1993, reprinted with permission of Oxford Instruments, Inc.)

F is called the Fano factor and has been experimentally determined to be no greater than 0.08 for germanium [23]. $F < 1$ implies that electron-hole pair creation events are not uncorrelated. L_T^2 is usually dominated by the trapping of electrons and holes at defect sites. However, shorter electronic shaping times, lower electric fields, and larger detectors accentuate ballistic deficit (loss of collected charge in the external electronics due to the finite traversal time of the electrons and holes across the detector). L_N^2 is independent of gamma ray E and is the dominant resolution limiting factor at low energies. L_F^2 depends linearly on E and, for a coaxial detector, usually dominates L_N^2 for E over a few hundred keV. The energy dependence of L_T^2 is not given simply from first principles for an arbitrary trap distribution, but an E^2 dependence seems to fit under many circumstances. Thus, at high enough E, L_T^2 is expected to be the largest component. For "good" detectors at 1.33 MeV, however, L_T^2 is always smaller than L_F^2. In addition, the magnitude of L_T^2 is variable enough between detectors that it distinguishes between acceptable, very good, and excellent detectors. L_T^2 is usually also the only component of resolution drawn from a nongaussian distribution and is thus responsible for any low-energy tailing of the peak.

X-Ray Detection

Both silicon and germanium detectors are used in low noise systems for the detection of fluorescent X-rays produced by electron beams (usually in an electron microscope) or X-rays (XRF). For both materials the detector is liquid-nitrogen cooled to reduce leakage current, and small volume devices [Figure 66.10(b), typically 10 mm² active area, and 3 mm depth] are used to decrease capacitance and therefore to further reduce electronic noise. Lithium-drifted silicon (SiLi) detectors were used first for these applications. Early germanium detectors displayed poor peak shape for X-ray energies just above the *L* absorption edges (attributed to diffusion against the field to the front contact by some electrons and their resulting loss to the photopeak [24]). However, as was first demonstrated by Cox et al., [25] this is not a fundamental problem but can be solved by the contacting technology. An X-ray spectrum taken with a HPGe detector is shown in Figure 66.14. Germanium has the advantages with respect to silicon of a smaller ε (2.96 eV per pair versus 3.96 eV per pair at 77 K) for better energy resolution and a higher *Z* (32 versus 14) for better photoelectric absorption of higher-energy X-rays.

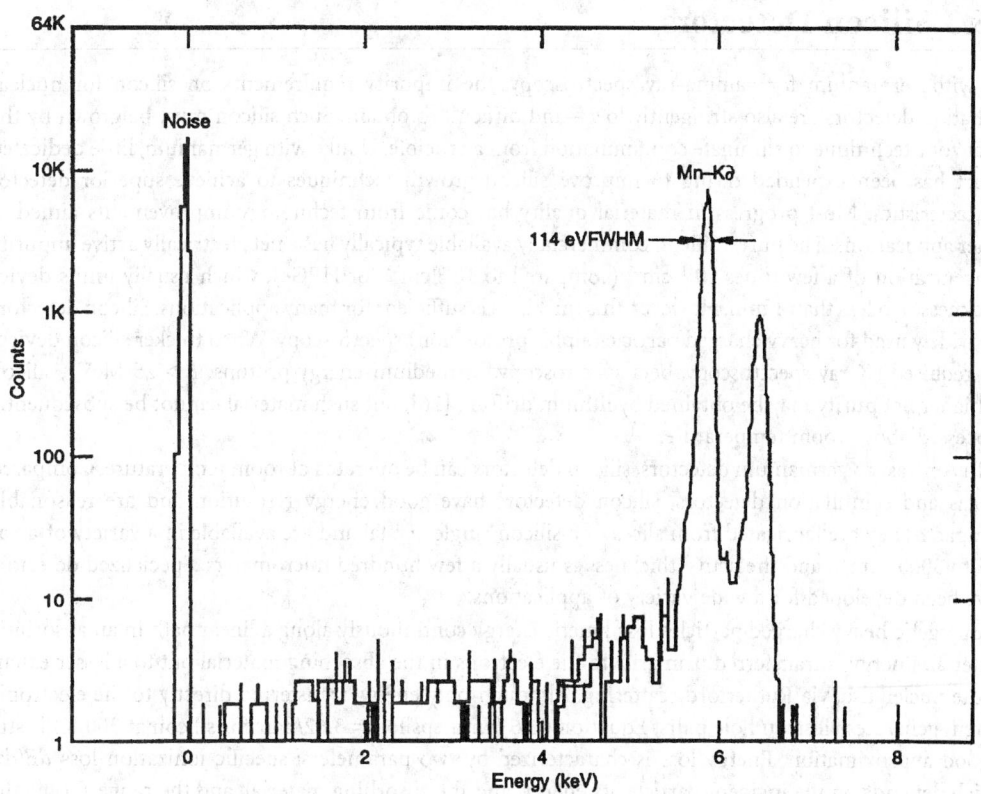

FIGURE 66.14 Manganese X-ray spectrum from ^{55}Fe source collected with an HPGe detector. (Darken and Cox, 1993, reprinted with permission of Oxford Instruments, Inc.)

Current Status of HPGe Detector Technology

High-purity germanium detectors are a mature commercial technology. Process development in crystal growing and diode fabrication have been conducted in private industry where significant advances are proprietary. However, the results of technological advances in these areas are quite evident in the continual improvement in the size, performance, and availability of HPGe detectors. Maximum photopeak efficiency for HPGe gamma-ray detectors is doubling every 6 to 8 years. Concurrently, energy resolutions are moving toward the theoretical limits of Equation 66.8 as the concentrations of trapping centers are reduced.

The reliability as well as the performance of germanium gamma-ray detectors has also continued to improve, although this is harder to quantify. Cryostats have been redesigned to reduce virtual and direct leaks, reduce microphonics, implement modular design, and to improve ruggedness. Detector makers are also making more serious attempts to offer models with reduced backgrounds by judicious design changes and careful selection of materials.

New applications for gamma-ray spectroscopy have emerged. The HPGe detector industry has recently supplied over 100 detectors each to two different experimental facilities (GAMMASPHERE in the United States, and EUROBALL in Europe), where they were arranged spherically in a modular fashion around the target of an ion accelerator to study the decay of nuclei from excited states of high angular momentum.

For users of single-detector systems, developments in the pulse processing electronics necessary for data acquisition and in the hardware and software for data analysis have resulted in both more compact and more flexible systems. Plug-in cards for a personal computer are available that not only contain the functions of the ADC and multichannel analyzer, but the high-voltage power supply and amplifier as well. Software developments also allow for control of many pulse-processing parameters that were previously set manually.

66.3 Silicon Detectors

As with germanium for gamma-ray spectroscopy, the impurity requirements on silicon for nuclear radiation detectors are also stringently low—and difficult to obtain. Such silicon must be grown by the float zone technique to eliminate contamination from a crucible. Unlike with germanium, little dedicated effort has been expended trying to improve silicon growth techniques to achieve superior detector characteristics. Most progress in material quality has come from technology improvements aimed at other applications. The purest silicon commercially available typically has a net electrically active impurity concentration of a few times 10^{11} cm^{-3} (compared to 10^{10} cm^{-3} for HPGe), which usually limits device thicknesses to less than 1 mm. However, this thickness is sufficient for many applications. Silicon detectors are widely used for heavy charged particle (alpha, proton, ion) spectroscopy. When thicker silicon devices are required (X-ray spectroscopy, beta spectroscopy, or medium-energy protons: $E > 25$ MeV), silicon of higher net purity may be obtained by lithium drifting [26], but such material cannot be subsequently processed above room temperature.

In contrast to germanium detectors, silicon detectors can be operated at room temperature. Compared to gas and scintillation detectors, silicon detectors have good energy resolution and are reasonably compact. They are fabricated from slices of a silicon single crystal and are available in a variety of areas (25 to 3000 mm^2), and the active thickness is usually a few hundred micrometers. Specialized detectors have been developed for a wide variety of applications.

Energetic heavy charged particles lose kinetic energy continuously along a linear path in an absorbing material. Energy is transferred primarily to the electrons in the absorbing material but to a lesser extent to the nuclei also, via Rutherford scattering. Although only energy transferred directly to the electronic system generates electron hole pairs, Equation 66.6 (with epsilon = 3.62/pair for silicon at 300 K) is still a good approximation. Energy loss is characterized by two parameters: specific ionization loss dE/dx, which depends on the incident particle, its energy, and the absorbing material, and the range R (i.e., the penetration depth of the particle), which determines the detector thickness required for complete energy absorption. The continuous nature of energy loss leads to substantial window effects.

Diffused Junction Detector

Silicon detectors can be generically categorized by the type of rectifying contact employed. The diffused junction detector is fabricated by diffusing phosphorus from the gas phase into p-type silicon. This is a high-temperature (900 to 1200° C) operation that is prone to introducing faster diffusing metals into the bulk that can act either as generation centers increasing leakage current, or as trapping centers degrading charge collection. The thickness of the diffused region, from 0.1 to 2.0 μm, also presents a dead layer to incident particles that is reduced in alternative technologies. Nonetheless, these detectors find use due to their ruggedness and economy.

Surface Barrier Detector

Surface barrier junctions are fabricated by either evaporating gold onto n-type silicon or aluminum onto p-type silicon. A typical entrance window is equivalent to 80 nm of silicon. The rectification properties depend on the charge density of surface states of the silicon and of the thin oxide layer over the silicon, as well as on the evaporated metal. The wafer is epoxied in an insulating ring before metallization. The finished detector is encapsulated in a can that has a front window for particle entry and a single contact in the back for the combined function of applying bias and for extracting the signal pulse. Devices can be operated either in the partially depleted or totally depleted mode. As fabrication is entirely at room temperature, there is no opportunity for metal contamination by diffusion. Generally, surface barrier detectors have lower leakage current, and less system noise than a diffused junction detector of comparable area and depth. However, detectors currently fabricated by ion implantation have still lower leakage current and electronic noise, together with a thinner and more rugged front contact. On the other hand, implanted detectors are not available in the same range of active thicknesses as surface barrier detectors. Below 100 μm and above 500 μm, only surface barrier detectors are currently available. Surface barrier detectors can be made in small quantities with rather simple equipment.

Ion Implanted Detectors

A simplified representation of ion implanted detector fabrication is shown in Figure 66.15. The first successful implementation of silicon planar processing to silicon detectors was reported by Kemmer [27]. The procedure starts with the thermal growth of an oxide film on a high-purity, *n*-type silicon wafer. Windows are then opened in the oxide by photolithographic techniques. The front contact area is implanted with boron to form the rectifying contact, and arsenic is implanted into the backside. The wafer is then annealed to activate the implant, and aluminum is evaporated on both sides to reduce sheet resistivity. Typical entrance windows are 50 nm silicon equivalent. Electrical connections are made by wire bonding to the aluminum layers. Finished detectors are canned in a manner similar to surface barrier detectors. More than one detector can be fabricated on the same wafer using the appropriate masks during photolithography. In fact, quite elaborate detector geometries can be achieved via photolithography. The detector in Figure 66.15 is actually a strip type.

This ion implantation planar process technology is well suited for mass production of wafer sizes compatible with the rest of the silicon industry. Minimum wafer diameters are now 4 or 5 in. At this diameter, breakage during fabrication is an issue for thicknesses less than 150 μm. For thicknesses greater than 500 μm, the availability of enough sufficiently pure material to justify the cost of photolithographic masks is an issue. Ion implanted detectors can be baked at 200° C to reduce outgassing. This is a significant improvement over surface barrier detectors, which irreversibly degrade by device processing above room temperature. This is a useful feature, as most heavy charged particle spectroscopy is done in a vacuum.

Leakage currents, at room temperature, are typically 1 to 10 nA per cm² active area and per 100 μm depletion depth. These values represent an order of magnitude reduction in leakage current with respect to surface barrier detectors. Two factors are relevant. Passivation of silicon surfaces by thermal oxidation

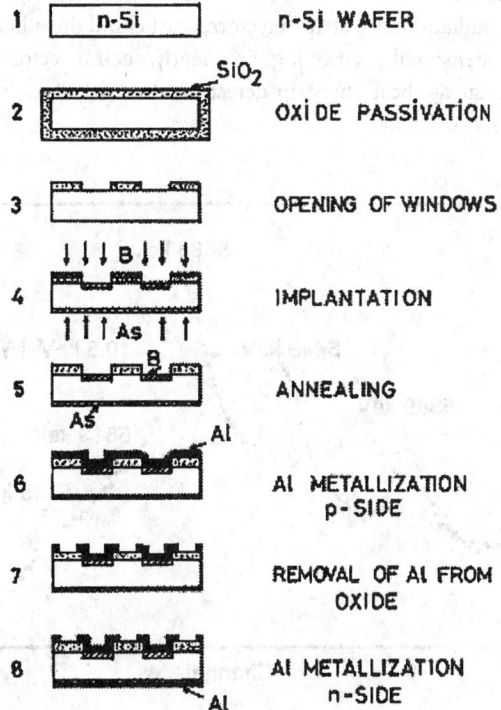

FIGURE 66.15 Steps in the fabrication of passivated planar silicon diode detectors. (From Ref. 30.)

is extremely effective in reducing leakage current around the rectifying contact. Also, the bulk generation current is reduced by the gettering of metal impurities during the high-temperature oxidation. Float zone silicon for radiation detectors usually has a minority carrier lifetime longer than 1 ms and this can be increased an order of magnitude during detector fabrication [28]. Thus, not only is leakage current reduced, but potential charge collection problems are also eliminated.

Energy Resolution

A typical spectrum of an Am-241 alpha particle source taken with an ion implanted detector is shown in Figure 66.16. While the factors considered in Equation 66.8 for germanium gamma-ray spectrometers are still valid, additional considerations also apply. In particular, if the source is moved closer to the detector to improve collection efficiency, larger differences in the angle of incidence will produce peak broadening due to larger variation in effective window thickness. Even when the source is sufficiently distanced from the detector, there will still be spatial variations in window thickness, as well as some variation in energy lost escaping from the source and traversing to the detector.

Another source of peak broadening is the variation in the small amount of particle energy lost during Rutherford scattering. This energy is transmitted directly to the scattering nuclei and does not generate electron-hole pairs, and a small pulse deficit results. These events are relatively few but large and therefore contribute disproportionately to peak variance. The FWHM contribution of this effect on a 6-MeV alpha particle peak has been estimated to be 3.5 keV [29].

Spatial Resolution

The uninterrupted progress of the semiconductor silicon industry in achieving both larger wafers and smaller device features has allowed the development of larger and more complex silicon detectors that can provide position information in addition to (or instead of) energy information. Spatial detection can be obtained by fabricating detectors as pixels (two-dimensional) or strips (one-dimensional) on the same wafer. For penetrating radiation, two strip detectors, one behind the other but with the strip pattern rotated 90°, provide two-dimensional positioning. Frequently, such detectors are individually designed and fabricated for a particular application. Strip detectors, drift detectors, and CCD (charge-coupled device) detectors will be discussed here.

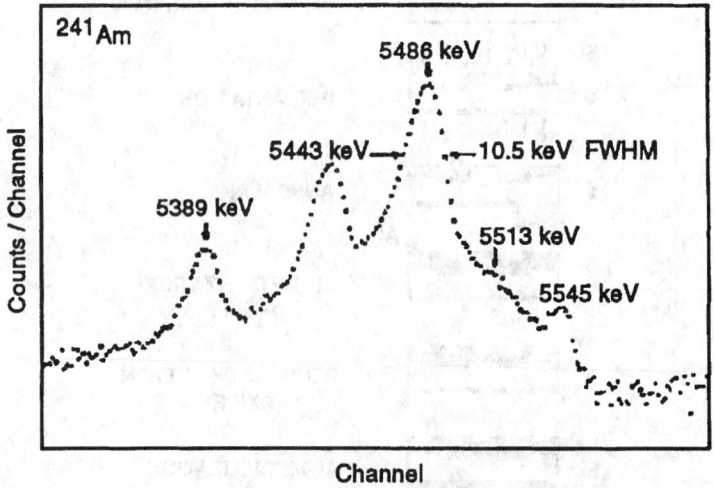

FIGURE 66.16 Spectrum of a ^{241}Am alpha-particle source (log scale) measured with an IP detector (25 mm^2 area, 300 μm thick) at room temperature. Resolution at 5.486 MeV is 10.6 keV (FWHM). (From Ref. 30).

Strip Detectors

Silicon strip detectors are currently fabricated on silicon wafers (typically approximately 300 μm thick) by using photolithographic masking to implant the rectifying contact in strips [30]. The strips usually have a pitch on the order of 100 μm and a width less than half of this size to minimize strip-to-strip capacitance and hence electronic noise [31]. The device is biased past depletion, and the back blocking contact is continuous. Each strip requires, in principle, its own signal processing electronics; however, charge division readout (capacitive or resistive) can reduce the number of amplifiers by a factor of 10. Detectors are fabricated in rectangular segments from a single wafer and can be ganged together if a larger area is needed.

Strip detectors are well established in high-energy physics experiments for reconstruction on the micron scale of the tracks of ionizing particles. The particles being tracked result from the collision of accelerated particles with a target and are highly energetic ($>10^{10}$ eV). Frequently, experimental interest is focused on short-lived particles created in the collision but which decay before they can be directly detected. Spatial resolution of the decay vertex from the original collision is necessary to detect such a particle and to determine its lifetime.

The requirements of new high-energy experiments and advances in silicon technology have produced much evolution and innovation in the strip detector concept. For example, a double-sided microstrip detector with an oxide-nitride-oxide capacitor dielectric film has been reported [32]. The use of intermediate strips to improve spatial resolution has become common [33], and the biasing network has been integrated onto the detector [34].

Drift Detectors

Silicon drift detectors were first proposed by Gatti and Rehak [35] as an alternative to silicon strip detectors in high-energy physics experiments. The primary motivation was to significantly reduce the number of readout channels. Drift detectors have subsequently been adapted for X-ray spectroscopy. These detectors are usually fabricated on n-type silicon wafers with holes collected to either a $p+$ contact on the back side of the detector, or to concentric annular $p+$ contacts on the front side. The detector is depleted from both sides. The reverse bias applied to the $p+$ annular rings is varied in such a way that electrons are collected radially in a potential energy trough to an $n+$ anode at the center of the detector on the front side.

A cross section through a circular drift detector is shown in Figure 66.17. The electron collecting anode ring surrounds the integrated FET used for the first stage of signal amplification. Enough negative bias

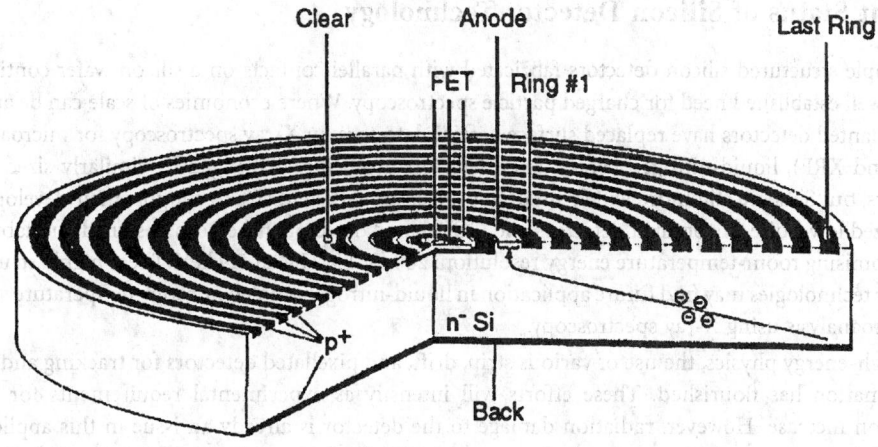

FIGURE 66.17 Cross section of a cylindrical silicon drift detector with integrated n-channel JFET. The gate of the transistor is connected to the collecting mode. The radiation entrance window for the ionizing radiation is the non-structured backside of the device. (Lechner et al., 1982).

is applied to the back contact (actually the entrance window) to deplete the wafer to the anode, which is near ground potential. At the same time, negative bias, progressively increasing in magnitude, is applied from the ring next to the anode (near ground potential) to the outermost ring, which is maintained at about two times the bias of the back contact. These applied biases deplete the detector in such a way that there is an electrostatic potential minimum for electrons that varies in depth across the detector from right under the front surface at the anode to near the back contact at the last ring. Ionized electrons will drift first to this minimum, then drift radially to the anode as shown in Figure 66.17. A feature of this contacting arrangement is that the anode capacitance, and hence amplifier series noise is low and nearly independent of the active area of the detector.

Silicon drift detectors have been designed in several different topologies of various sizes for different experimental needs. Spatial resolution for tracking and vertexing of high-energy particles is obtained by segmenting the cathode (for angular position) and analysis of signal rise time (for radial position [36]). Drift detectors with integrated electronics have been demonstrated for high-resolution room-temperature X-ray spectroscopy [37].

CCD Detectors

The design of CCD (charge coupled device) detectors has similarities to the silicon drift detector [35]. The CCD detector is normally fabricated on an n-type silicon wafer depleted both from the backside with a continuous p+ contact on the back, and from p+ CCD registers on the front. Reverse bias voltages are such that the wafer is totally depleted and the electron potential minimum is about 10 μm below the CCD registers. After an ionizing event, holes are collected to the p+ contacts, and electrons are trapped under a nearby register, then transported down a channel of registers by properly clocked voltage pulses to the registers. Each channel has its own readout anode, which can be made small to minimize capacitance, a prerequisite for minimizing noise. The first stage of amplification is frequently integrated onto the same wafer. Spatial resolution is limited to the register (pixel) size. Brauniger et al. [38] described initial results on a 6 × 6 cm CCD array of 150 × 150 μm pixels intended for satellite X-ray imaging. The system also had an energy resolution of 200 eV FWHM for 5.9 keV X-rays at room temperature.

Silicon pixel detectors have also been designed using other highly integrated device structures to optimize particular performance aspects such as timing resolution. Pixel detectors using MOS transistors [39] and using reverse-biased diodes with individual readout circuitry [40] have been described.

Present Status of Silicon Detector Technology

The simple structured silicon detectors fabricated with parallel contacts on a silicon wafer continue to serve a well established need for charged particle spectroscopy. Where economies of scale can be applied, ion implanted detectors have replaced surface barrier detectors. In X-ray spectroscopy for microanalysis (SEM and XRF), liquid-nitrogen-cooled Si:Li detectors are being challenged by similarly sized HPGe detectors, but Si:Li are still more widely used. In projects of sufficient size to support their development, specialized low-noise silicon drift detectors and CCD-based detectors have been designed and fabricated with promising room-temperature energy resolution: 200 eV FWHM at 5.9 keV. These highly structured detector technologies may find future application in liquid-nitrogen-cooled or room-temperature systems for microanalysis using X-ray spectroscopy.

In high-energy physics, the use of various strip, drift, and pixellated detectors for tracking and vertex determination has flourished. These efforts will intensify as experimental requirements for spatial resolution increase. However, radiation damage to the detector is already an issue in this application, and higher luminosity beams will only increase the problems. Nevertheless, it appears that the continuing need of the high-energy physics community for a higher number and density of signal paths forecasts continued reliance on the ever-improving integration technology of the semiconductor silicon industry.

66.4 Room-Temperature Semiconductors

Applications arise that require energy resolution beyond the capability of scintillator systems and where cryogenically cooled semiconductors are not suited. Examples include detector probes for monitoring restricted areas, monitoring at remote sites where replenishing the coolant is impractical, spectral imaging, and many portable instrument applications. There is available a class of semiconductor detectors that satisfy many such needs by providing energy resolution substantially better than the best scintillators (although inferior to cooled semiconductors) while operating at ambient temperature. In addition to spectroscopy, these devices are also useful for counting applications where high detection efficiency per unit volume is required. In these applications, the devices are operated in pulse mode wherein the charge associated with single-photon absorption events is recorded. They also can be operated in a current mode in the manner of a solid-state ion chamber. In their current stage of development, room-temperature detectors are limited in size and best suited for the energy region below 1 MeV.

The room-temperature detectors are distinguished from cryogenic semiconductors by the magnitude of the energy gap that separates the normally vacant conduction band from the highest filled band. If this energy gap is small, as is in the case of silicon (1.14 eV) and germanium (0.67 eV), electrons can be thermally stimulated across the bandgap at room temperature. The resultant current competes with the gamma-ray-generated signal precluding room-temperature operation of germanium and high-resolution applications of silicon. Thermally stimulated current is reduced to acceptable levels at bandgaps energies of about 1.4 eV and above. This phenomenon has been successfully exploited in the development of room-temperature detector materials including cadmium zinc telluride (acronym, CZT), cadmium telluride (CdTe), and mercuric iodide (HgI_2).

Theory of Operation

Operating principles of room-temperature detectors are similar to those governing the more familiar cryogenic semiconductor devices. Gamma radiation is absorbed in the material and generates electron-hole pairs that move under the influence of an applied electric field to contacts and external electronics for processing and production of the familiar pulse-height spectrum. The process is shown schematically in Figure 66.18. Fundamental to the charge transfer process is the carrier mobility (μ) and the carrier life time (τ). The product $\mu\tau E$ defines a drift length (λ) that should be long compared to the intercontact dimensions. Owing to the substantially higher average atomic number of the room-temperature detector materials in the gamma absorption cross sections, the probability of gamma ray absorption is much

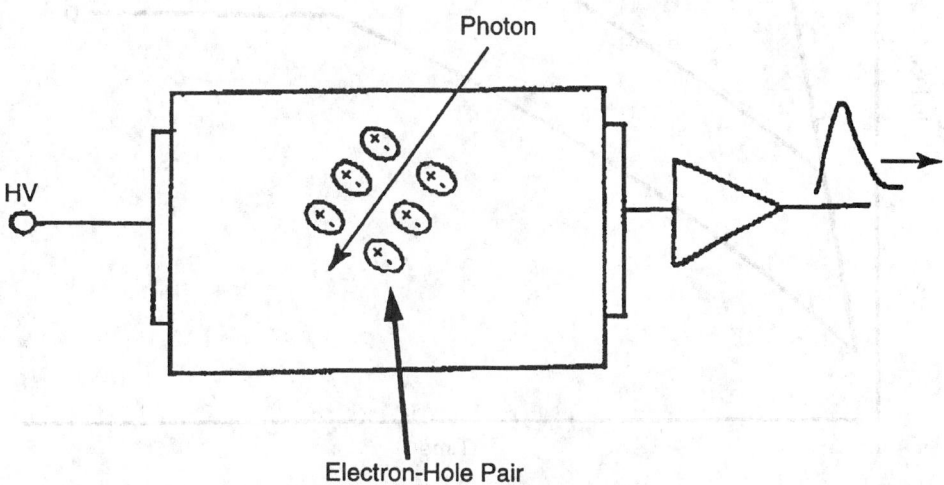

FIGURE 66.18 Schematic illustration of charge generation in a planar detector.

higher than in silicon or germanium (Figure 66.8). As a result, room-temperature detectors provide greater detection efficiency per unit thickness.

The energy required to produce an electron-hole pair (ε) is typically a few times the energy bandgap of the material. In silicon where the bandgap is 1.14 eV, the energy to produce an electron-hole pair (ε) is about 3.5 eV. The absorption of a 1 MeV photon in silicon thus produces about 285,000 pairs. Values of ε for room-temperature materials are in the region 4.2 to 5.0 eV per e-h pair (see Table 66.3) and, consequently, fewer electron-hole pairs are generated per unit of absorbed energy. Complete collection of the charge is desired, although charge trapping, which may not affect the two carrier types equally, prevents this in most cases. The drift length for holes (λ_h) in these materials is often less than the intercontact dimensions and creates a condition where the collection efficiency depends on the photon interaction depth. This is phenomenon is illustrated in Figure 66.19, where induced charge from single gamma absorption events originating at various depths in the material is plotted as a function of time. The initial fast-rising segment is due to the more mobile electrons; the slower component is due to holes. In this example, hole trapping is assumed and is manifest in the curvature of the hole segment. The charge collection efficiency (η) can be derived from the Hecht relation [41]. For a photon absorbed at a distance x from the cathode of a planar detector of thickness L operated with a uniform electric, the relationship becomes

$$\eta = \frac{\lambda_e}{L}\left[1 - \exp\frac{(L-x)}{\lambda_e}\right] + \frac{\lambda_h}{L}\left[1 - \exp\frac{-x}{\lambda_h}\right] \qquad (66.10)$$

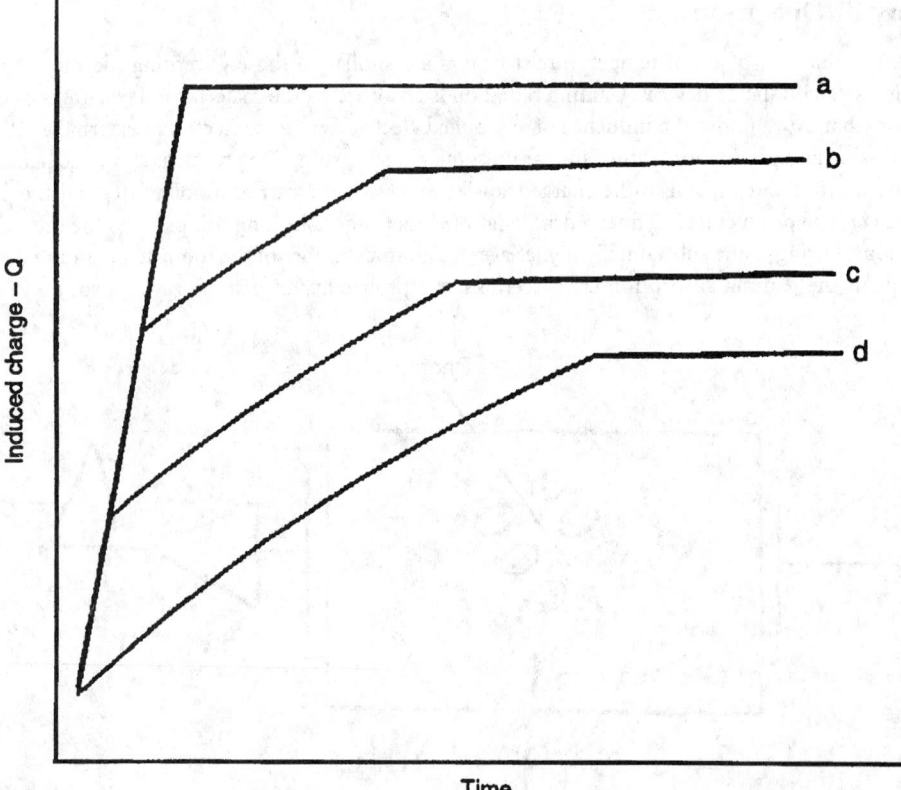

FIGURE 66.19 Charge collection in planar detector for single-photon interaction in a planar detector. Curves a through d depict the charge from photon interactions at increasing depths below the cathode.

TABLE 66.3 Physical Parameters of Common Room-Temperature Semiconductor Materials

Material	E_g (eV)	Z	ε (eV)	ρ (Ω)	$(\mu\tau)_e$ (cm²/V)	$(\mu\tau)_h$ (cm²/V)
Cadmium zinc telluride	1.65	48	5.0	10^{11}	1×10^{-3}	6×10^{-6}
Cadmium telluride	1.5	50	4.4	10^{9}	3.5×10^{-3}	2.3×10^{-4}
Mercuric iodide	2.13	62	4.2	10^{13}	1×10^{-4}	4×10^{-5}

Note: E_g = bandgap energy, Z = average atomic number, ε = energy to create an electron-hole pair, and ρ = resistivity.

Source: Semiconductors for Room-Temperature Radiation Detector Applications, R.B. James, T.E. Schlesinger, P. Siffert, and L.A. Franks (eds.), Materials Research Society, Vol. 32, Pittsburgh, PA, 1993.

The dependence of the collection efficiency on interaction depth reduces energy resolution and without mitigation would limit high resolution to thin devices. Fortunately, methods have been developed that permit high-energy resolution to be achieved in relatively thick samples. As with cooled semiconductor detectors, the energy resolution of the combined detector-electronics system is normally specified by the full width of a monoenergetic spectral peak at its half amplitude points (ΔE). The FWHM is in turn related to the variance in the peak L^2 (see Equation 66.8). It is useful to note that the energy resolution is related to the reciprocal of the product $\mu\tau$.

Operational Considerations

Important physical parameters for the leading room-temperature detectors are summarized in Table 66.3. Detectors are available with surface area of a few square centimeters and thicknesses up to about 1 cm. The performance of detectors based on the different materials varies considerably, as can the performance for detectors of the same material. The choice of specific detector material is normally dictated by the application. The exceptionally high resistivity and high photoelectric cross section in mercuric iodide permit good resolution and high efficiency in the X-ray region, particularly below 10 keV. For example, ΔE of 4 percent has been reported [42] with typical values in the region of 10 percent. For applications in the region of 0.5 MeV, trade-offs between efficient gamma absorption and resolution may be required. If energy resolution is the primary concern, thinner devices that minimize charge trapping are generally required. Considerable progress is being made in achieving both high efficiency and resolution in the region of 0.5 MeV, particularly with CZT. For example, resolution of better than 3 percent has been achieved in 1-cm thick detector at 511 keV, and about 5 percent at 662 keV in a 2.5 cm thick device [4]. Improvements in material quality can be expected further improve the performance of thick detectors, as the $\mu\tau$ values are further increased.

Procedures have been developed to overcome many of the thickness and surface area limitations of currently available devices. For instance, electronic circuits have been developed that permit the operation of planar arrays that provide spectral resolution approaching that of single units while providing substantially greater area [43]. Similarly, high gamma absorption efficiency with useful spectral resolution has been obtained with stacks of thin spectrometer-quality detectors [44]. Additionally, pulse processing and single charge collection procedures have been demonstrated that enhance spectrometer performance. This development can be expected to substantially improve the availability and price of spectrometer grade devices. Further details concerning the performance of these devices, as well as electronic processing and design details, are available in the literature [2, 3, 45, 46].

Detectors based on the materials in Table 66.3 are available commercially. Due to the evolving nature of this technology, it is recommended that buyers' guides be consulted for suppliers and current availability.

66.5 Prices and Availability

The detectors described in this chapter are available commercially. Their prices vary widely, depending on type, size, and performance. Gaseous detectors are normally in the range of a few hundred dollars for standard designs. Scintillator-photomultiplier combinations range from about a thousand to several thousand dollars, depending on size and resolution. Room-temperature semiconductor detectors range from less than one hundred dollars for small, low-resolution devices to over a thousand dollars for large (1 cm^3), high-resolution devices. Pricing of coaxial HPGe detectors is based largely on their gamma ray efficiency, which is specified relative to a 3 × 3-in. sodium iodide scintillator at 1.33 MeV. Coaxial detectors are available with relative efficiencies up to about 150 percent with cost in the area of several hundred dollars per percent efficiency. Planar HPGe detectors are normally less expensive than coaxial designs. In either case, the price includes cryostat, dewar, and preamplifier. Cryogenic silicon detectors are available in area up to several tens of square millimeters. Cost ranges to >$10,000, depending on size, performance, and complexity of design.

References

1. G. F. Knoll, *Radiation Detection and Measurement,* New York: John Wiley and Sons, 1979.
2. R. B. Rossi and H. H. Staub, *Ionization Chambers and Counters,* New York: McGraw-Hill, 1949.
3. M. J. Weber, P. Lecoq, R. C. Ruchti, C. Woody, W. M. Yen, and R. Y. Zhu (eds.), *Scintillator and Phosphor Materials,* Vol. 348, Materials Research Society, Pittsburgh, PA, 1994.
4. T. E. Schlesinger and R. B. James (eds.), *Semiconductors for Room-Temperature Nuclear Detector Applications,* Vol. 43, Semiconductors and Semimetals, San Diego, CA: Academic Press, 1995.
5. R. B. James, T. E. Schlesinger, P. Siffert, and L. Franks (eds.), *Semiconductors for Room-Temperature Radiation Detector Applications,* Vol. 302, Materials Research Society, Pittsburgh, PA, 1993.
6. J. Fraden, *AIP Handbook of Modern Sensors,* American Institute of Physics, New York, 1993.
7. M. Cuzin, R. B. James, P. F. Manfredi, and P. Siffert (eds.), *Proceedings of the 9th International Workshop on Room Temperature Semiconductor X- and Gamma-Ray Detectors,* Grenoble, France: Sept. 18–22, 1995, *Nucl. Instru. and Meth.,* 380, 1996.
8. T. E. Bortner, G. S. Hurst, and W. G. Stone, Drift velocities of electrons in some commonly used counting gases, *Rev. Sci. Instru.,* 28(2), 103, 1957.
9. W. J. Price, *Nuclear Radiation Detection,* New York: McGraw-Hill, 1958
10. F. H. Attix and W. C. Roesch, *Radiation Dosimetry,* 2nd Ed., New York: Academic Press, 1966.
11. W. H. Tait, *Radiation Detection,* Boston: Butterworth, 1980.
12. G. N. Whyte, *Principles of Radiation Dosimetry,* New York: John Wiley & Sons, 1959.
13. J. B. Birks, *Theory and Practice of Scintillation Counting,* New York: MacMillan, 1964.
14. J. Markakis, High resolution scintillation spectroscopy with HgI$_2$ as the photodetector, *Nucl. Instru. Meth.,* A263, 499, 1988.
15. R. N. Hall, Chemical impurities and lattice defects in high-purity germanium, *IEEE Trans. Nucl. Sci.,* 260, NS-21, 260–272, 1974.
16. E. E. Haller, W. L. Hansen, and F. S. Goulding, Physics of ultra-pure germanium, *Adv. In Physics* 30(1), 93–138, 1981.
17. W. G. Pfann, *Zone Melting,* New York: John Wiley & Sons, 1966.
18. G. K. Teal, and J. B. Little, Growth of germanium single crystals, *Phys. Rev.,* 78, 647, 1950.
19. R. H. Pehl, E. E. Haller, and R. C. Cordi, Operational characteristics of germanium detectors at higher temperatures, *IEEE Trans. Nucl. Sci.,* NS-20, 494, 1973.
20. L. S. Darken and C. E. Cox, High-purity germanium detector, p. 23, in *Semiconductors for Room-Temperature Nuclear Detector Applications,* New York, Academic Press, 1995.
21. R. H. Pehl, N. W. Madden, J. H. Elliott, T. W. Raudorf, R. C. Trammell, and L. S. Darken, Jr., Radiation damage resistance of reverse electrode GE coaxial detectors, *IEEE Trans. Nucl. Sci.,* NS-26, 321, 1979.

22. *IEEE Test Procedures for Germanium Detectors for Ionizing Radiation,* ANSI/IEEE Standard 325–1989.

23. R. H. Pehl and F. S. Goulding, Recent observations on the fano factor in germanium, *Nucl. Instru. Meth.,* 81, 329–330, 1970.

24. J. Lacer, E. E. Haller, and R. C. Cordi, Entrance windows in germanium low-energy X-ray detectors, *IEEE Trans. Nucl. Sci.,* NS-24, 53, 1977.

25. C. E. Cox, B. G. Lowe, and R. Sareen, Small area high purity germanium detectors for use in the energy range 100 eV to 100 keV, *IEEE Trans. Nucl. Sci.,* 35, 28, 1988.

26. E. M. Pell, Ion drift in an *n-p* junction, *J. Appl. Phys.,* 31, 291, 1960.

27. J. Kemmer, Fabrication of low noise silicon radiation detectors by the planar process, *Nucl. Instru. Meth.,* 169, 499–502, 1980.

28. J. Kemmer and G. Lutz, New detector concepts, *Nucl. Instru. Meth.,* A235, 365–377, 1987.

29. G. D. Alkhazov, A. P. Komar, and A. Vorob'ev, Ionizing fluctuations and resolution of ionization chambers and semiconductor detectors, *Nucl. Instru. Meth.,* 48, 1–12, 1967.

30. J. Kemmer, P. Burger, R. Henck, and E. Heijne, Performance and applications of passivated ion-implanted silicon detectors, *IEEE Trans. Nucl. Sci.,* NS-29, 733, 1982.

31. T. Dubbs, S. Kashigin, M. Kratzer, W. Kroeger, T. Pulliam, H. F.-W. Sadrozinski, E. Spencer, R. Wichmann, M. Wilder, W. Bialas, W. Daabrowski, Y. Unno, and T. Oshugi, Noise determination in silicon micro strips, *IEEE Trans. Nucl. Sci.,* 42, 1119, 1996.

32. Y. Saitoh, T. Akamine, M. Inoue, J. Yamanaka, K. Kadoi, R. Takano, Y. Kojima, S. Miyahara, M. Kamiaya, H. Ikeda, T. Matsuda, T. Tsuboyama, H. Ozaki, M. Tanaka, H. Iwasaki, J. Haba, Y. Higashi, Y. Yamada, S. Okuno, S. Avrillon, T. Nemota, I. Fulunishi, and Y. Asano, Fabrication of a double-sided silicon microstrip detector with an ONO capacitor dielectric film, *IEEE Trans. Nucl. Sci.,* 43, 1123, 1996.

33. P. Chochula, V. Cindro, R. Jeraj, S. Macek, D. Zontar, M. Krammer, H. Pernegger, M. Pernicka, and C. Mariotti, Readout of a Si strip detector with 200 μm pitch, *Nucl. Instru. Meth.,* A377, 409–411, 1996.

34. T. I. Westgaard, B. S. Avset, N. N. Ahmed, and L. Eversen, Radiation hardness of punch-through and FET biased silicon microstrip detectors, *Nucl. Instru. Meth.,* A377, 429–434, 1996.

35. E. Gatti and P. Rehak, Semiconductor drift chamber—an application of a novel charge transport scheme, *Nucl. Instru. Meth.,* 225, 608–614, 1984.

36. P. Rehak, J. Walton, E. Gatti, A. Longoni, M. Sampietro, J. Kemmer, H. Dietl, P. Holl, R. Klanner, G. Lutz, A. Wylie, and H. Becker, Progress in semiconductor drift detectors, *Nucl. Instru. Meth.,* A248, 367–378, 1986.

37. P. Lechner, S. Eckbauer, R. Hartman, S. Krisch, D. Hauff, R. Richter, H. Soltau, L. Struder, C. Fiorini, E. Gatti, A. Longoni, and M. Sampietro, Silicon drift detectors for high resolution room temperature X-ray spectroscopy, *Nucl. Instru. Meth.,* A377, 346–351, 1996.

38. H. Brauninger, R. Danner, D. Hauff, P. Lechner, G. Lutz, N. Meidinger, E. Pinotti, C. Reppin, L. Struder, and J. Trumper, First results with the *pn*-CCD detector system for the XMM satellite mission, *Nucl. Instru. Meth.,* 129, 1993.

39. K. Misiakos and S. Kavadias, A pixel segmented silicon strip detector for ultra fast shaping at low noise and low power consumption, *IEEE Trans. Nucl. Sci.,* 43, 1102, 1996.

40. E. Beauville, C. Cork, T. Earnest, W. Mar, J. Millaud, D. Nygen, H. Padmore, B. Turko, G. Zizka, P. Datte, and N.H. Xuong, A 2D smart pixel detector for time resolved protein crystallography, *IEEE Trans. Nucl. Sci.,* 43(3), 1243, 1996.

41. H. K. Hecht, Zum mechanisms des lichtelektrischen primarstromes in isolierenden kristallen, *Z Phys.,* 77, 235–245, 1932.

42. J. S. Iwanczyk, B. E. Patt, Y. J. Wang, and A. K. Khusainov, Comparison of HgI_2, CdTe, and Si (*p-i-n*) X-ray detectors, *Nucl. Instru. Meth.,* A380, 186–192, 1996.

43. V. Gerrish, private communication, 1995.

44. R. Olsen, R. B. James, A. Antolak, C. Wang, *Proceedings of the 1994 International Nuclear Materials Management Conference*, 23, 589, 1994.

45. B. E. Patt, J. S. Iwanczyk, G. Vikelis, and Y. J. Wang, New gamma-ray detector structures for electron only charge carrier collection utilizing high-Z compound semiconductors, *Nucl. Instr. Meth.*, A380, 276–281, 1996.

46. P. N. Luke, Electrode configuration and energy resolution in gamma-ray detectors, *Nucl. Instr. Meth.*, A380, 232–237, 1996.

67

Charged Particle Measurement

John C. Armitage
Ottawa–Carleton Institute for Physics, Carleton University

Madhu S. Dixit
Centre for Research in Particle Physics, Carleton University

Jacques Dubeau
Centre for Research in Particle Physics, Carleton University

Hans Mes
Centre for Research in Particle Physics, Carleton University

F. Gerald Oakham
Centre for Research in Particle Physics, Carleton University

67.1 Introduction

Interaction of Charged Particles with Matter

There are a number of subatomic charged particles that can be detected by their interaction with matter. These include protons (hydrogen nuclei), β particles (fast electrons), α particles (helium nuclei), light ions, heavy ions, and fission fragments. Some of these particles are emitted by natural radioactivity on Earth or originate from space and reach the Earth in the form of cosmic rays; others are due to human activities (nuclear industry, accelerators). The kinetic energy of charged particles is given in units of electron-volts (eV), where 1 eV = 1.6×10^{-19} J. Naturally occurring α and β particles from radioactive decay have energies up to 10 MeV. Ions, such as Cl^+, are accelerated to 35 MeV and higher in beam analysis procedures. Energetic particles that originate from space can have energies up to 10^{12} GeV. Charged particles traversing matter lose kinetic energy until they eventually come to a halt. In the case of electrons, protons, α particles, or light ions, most of this energy is lost through interactions with the electrons of the target. Heavier ions also lose a significant amount of kinetic energy through direct collisions with the nuclei of the constituent atoms. However, for all types of charged particles of concern here, ionization of the target atoms is the dominant mode of excitation of the detection medium. This process is also the main source of radiation damage in biological tissue.

Energy Loss of Protons and Ions

The quantity of energy deposited by an incident energetic charged particle in a detection medium is a function of the atomic mass, A_{med}, the atomic number, Z_{med}, and the density of target atoms that it meets

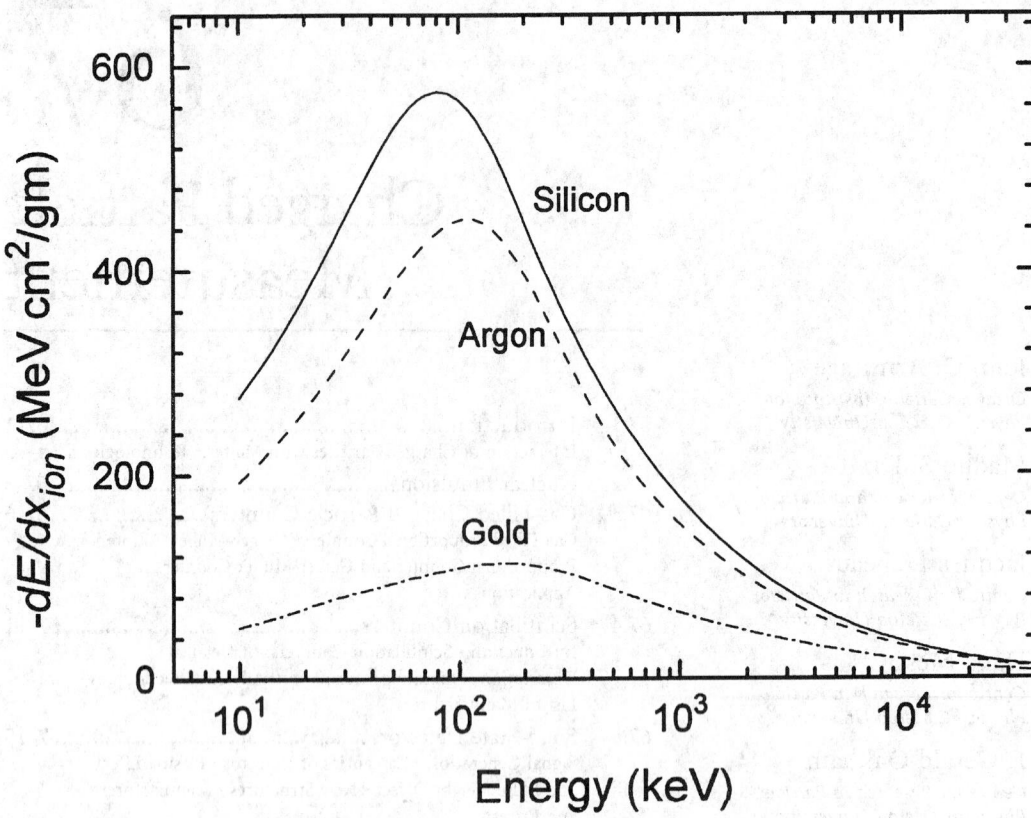

FIGURE 67.1 The electronic energy loss curves for protons in three different absorber elements are shown as a function of kinetic energy. The target element with the lower Z_{med} displays the highest stopping power due to its low average ionizing potential. These curves were obtained using the SRIM computer program. (Reference 2.)

along its path. Hence, the unit of dimension used in this field is the mass thickness, and it is measured most often in units of g cm^{-2}. This is equal to the geometric thickness or path length x, multiplied by the density of the detector volume, ρ. It follows that the rate of energy loss of a particle in the detection volume is $dE/(\rho\,dx)$ and it is given in units of keV (g cm^{-2})$^{-1}$ or MeV (g cm^{-2})$^{-1}$. The energy loss due to ionization is characterized by the Bethe-Bloch equation:

$$-\frac{1}{\rho}\frac{dE}{dx_{ion}} = D\frac{z^2 Z_{med}}{A_{med}}\frac{c^2}{v^2}\left[\ln\frac{2m_0 v^2}{I_{med}} - \ln\left(1 - \frac{v^2}{c^2}\right) - \frac{v^2}{c^2}\right] \; \text{MeV}\Big/\left(\text{g cm}^{-2}\right)^{-1} \quad (67.1)$$

with $D = 0.307$ MeV/(g cm^{-2}), c is the velocity of light (3.0×10^8 m s^{-1}). The incident ion is described by z and v, its charge and velocity, respectively. The atoms of the target are further defined by I_{med} ($\sim 16\,Z_{med}^{0.9}$ eV), their mean ionization energy. The energy loss is characterized by a peak at low energy and a minimum at higher energies where a particle of $z = 1$ is referred to as a minimum ionizing particle (MIP). The energy loss curves of protons in Si, Ar, and Au are shown on Figure 67.1. The energy loss expressed in MeV (g cm^{-2})$^{-1}$ is lower for higher Z_{med} materials, where the ratio Z_{med}/A_{med} becomes less than 0.5 and I_{med} is large. However, when multiplied by the material's density, the energy loss in MeV cm^{-1} shows a reverse trend; it is greater for high Z_{med} materials because they have a high density. The rate of energy loss varies to the second power of the charge z of the incident ion and as the inverse square of its kinetic energy ($1/v^2$ term). Equation 67.1 holds for protons and α particles of energy above 500 keV and 1 MeV, respectively, in Si where the target is described by an average ionization energy. It is not applicable for ions for which $z > 2$, where stripping must be considered. Ziegler and co-workers [1] give a very useful description of scaling rules that allow the proton energy loss to be scaled to any incident ion and target atoms. They have included these rules in a software package called SRIM [2]. Their empirical relations allows for ion stripping and gives a more exact representation of the energy loss than Equation 67.1 in the case of heavy ions and for low-velocity light ions. Alternatively, one can use published $dE/(\rho\, dx)_{ion}$ plots [3]. Interaction of fast charged particles with the nuclei of the constituent atoms of a target produces an additional contribution to the energy loss, $dE/(\rho\, dx)_{nucl}$, the nuclear stopping power. For protons and α particles, the electronic stopping power dominates by a factor of 1000 over the nuclear stopping power for most energies of interest. For ions, the nuclear stopping power may dominate at low energies (<1 MeV), with its contribution decreasing at higher energies and becoming negligible. The range of a particle, or its path length, is also a very important parameter in the selection of a detector. It is given by:

$$R(E_0) = \int_0^{E_0} \frac{dE}{dE/dx} \tag{67.2}$$

where dE/dx in this case is the total stopping power $dE/dx_{ion} + dE/dx_{nucl}$. Range–energy curves for various ions in silicon are shown on Figure 67.2.

Energy Loss of Electrons

The case of electron propagation through matter is more complicated. The electron collides with bodies having a mass equal to it (other electrons) or much heavier (nuclei). The collisions cause it to undergo many changes in direction. This "random walk" type of behavior is superimposed on the forward motion of the electron. Along its random path an electron loses energy through ionization in a way similar to ions. At increasing kinetic energy, electrons also lose energy through the emission of X-rays, called bremsstrahlung radiation. In a target constituted of atoms of charge Z_{med} and below a critical energy $E_c \sim 817/Z_{med}$ MeV, electrons lose a greater fraction of their energy through ionization. Above E_c, the energy loss due to bremsstrahlung X-ray emission dominates and is given by:

$$-\frac{1}{\rho}\frac{dE}{dx}\bigg|_{rad} = B\frac{Z_{med}^2\, E}{A}\left(\ln\left(\frac{183}{Z_{med}^{1/3}}\right) + \frac{1}{18}\right) \text{MeV}\Big/\left(g\,cm^{-2}\right)^{-1} \tag{67.3}$$

where $B = 1.40 \times 10^{-3}$ for E in MeV, and for the molar weight of the target A in grams. Equation 67.3 applies for specified energy ranges, and the interested reader should consult the available literature [4].

The electron range is defined as the distance separating the point of entry in the detection medium to the point where the electron stops. It is much shorter than its actual pathlength and can be approximated by $R(E)$ (Equation 67.4) for E ranging from 0.01 Mev to 3 Mev [5]. Ultimately, the problem of electron propagation through materials has to be addressed using Monte Carlo programs (see References 6 and 7), which follow the individual electron histories to a specified energy or geometrical cutoff.

$$R(E) = 0.2115 Z_{med}^{0.26}\, E^n \left(g\,cm^{-2}\right) \tag{67.4}$$

$n = 1.265 - 0.0954 \ln(E)$

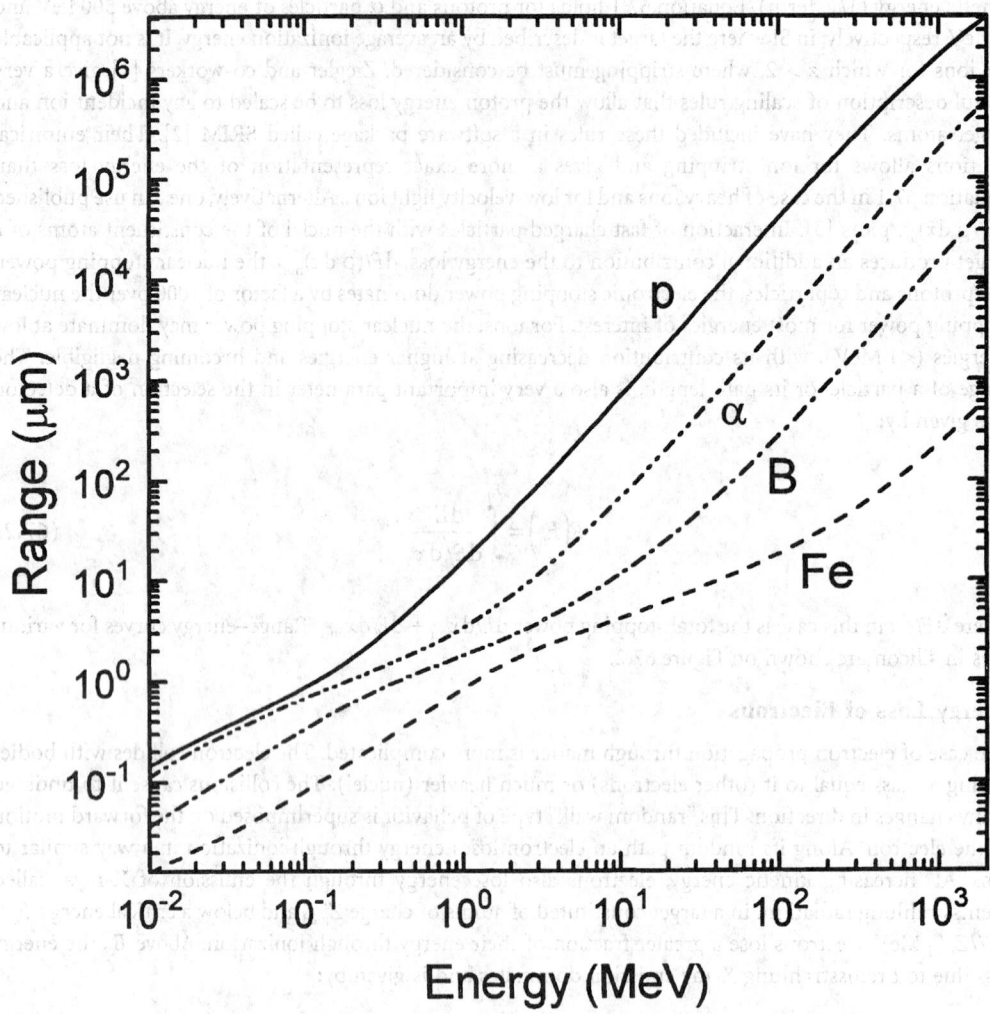

FIGURE 67.2 Ions of widely different masses have widely different ranges in a target element such as silicon. These curves for protons (p), α particles, boron ions (B), and iron (Fe) were obtained using the SRIM computer program. (Reference 2.)

Technologies

The most appropriate detection technique selected depends on the particle under study, its energy, and the type of information sought, such as source activity, position or energy measurement, or particle identification. For all cases reviewed here, except nuclear emulsions, the detection process is active and accompanied by electronic signal processing. Nuclear emulsions permanently record the ionization tracks left by traversing particles and can only be examined after their development. The active technologies include: gas-filled detectors (ionization chambers, proportional counters, Geiger-Müeller counters), scintillators, and solid-state detectors. The active volume consists of a radiation-sensitive material that is raised to an excited state through ionization by the passage of charged particles. For active detectors, this ionization is sensed by the external electronics as an induced charge in the case of gas-based and solid-state detectors, or as visible or near-visible light in the case of scintillators. The sensors described here are the front-ends of complex detection systems.

Nuclear emulsions are a special preparation of photographic emulsion designed to record "tracks" left by ionizing particles as they traverse the emulsion. The emulsion is sensitive up to the time it is processed, and the image is latent for many months or years if the emulsion is cooled. Once processed, the emulsion is a permanent record of the ionizing radiation that traversed it. In a situation where a prompt signal is not required, the nuclear emulsion may be a suitable detector. Its primary advantage is that it is compact, is self-contained, and has high spatial resolution. Since no associated equipment is needed during exposure, it is well suited for remote applications, such as studying cosmic rays and for recording the cumulative exposure to ionizing radiation in space.

Gas-filled detectors were among the first instruments designed for radiation detection. A charged particle passing through a gas-filled container leaves a trail of ionization electrons and ions. With a suitably configured external electric field, the ionization can be used to produce an electrical signal for charged particle detection. Depending on the magnitude of the applied electric field, a variety of ionization phenomena can arise in gases which have been used over the years to develop different types of gaseous charged particle detectors; for example, ionization chambers, proportional counters, Geiger-Müeller counters, spark chambers, etc. Spark chambers are rarely used these days because of electrical noise and limited counting rate ability. The other three devices are used primarily as instruments for radiation protection and monitoring. Their advantage is that they are inexpensive, easy to operate, and are available commercially. In addition, many variants of gas-filled proportional counters have been developed for use in experimental nuclear physics and elementary particle physics. These, however, need to be specifically designed and fabricated for particular applications.

Scintillation counters consist of a scintillation material, a photodetector, and a light guide to link the two. Charged particles crossing the scintillator material deposit energy in the material, which responds by emitting light isotropically. A large fraction of the light can be trapped in the material by internal reflection and directed toward a photodetector, such as a photomultiplier tube. This kind of detector can be used at very high counting rates and can achieve efficiencies approaching 100%. Components for scintillation counters are readily available and it is possible to purchase assembled counters.

Solid-state detectors are large semiconductor diodes made of very high resistivity material and are operated under reverse-bias. The small bandgap in these material ensures that the kinetic energy of the incident charged particle is efficiently converted into an electric signal through the creation of electron-hole pairs. A signal is induced on the metal contacts as the charge carriers drift under the applied electric field present in the depletion region. Their advantages include a fast response, good energy resolution, and compact size.

Table 67.1 presents some radiation sensing applications, the particles under study, and the suggested sensor–system combination. The following subsections present each sensing technology in more detail.

TABLE 67.1 Detection Applications and Suggested Instruments

Detector Type	Typical Applications	Advantages	Shortcomings
Nuclear emulsion	Cosmic ray studies, dosimetry, low statistic (single track) studies	Passive detector, high spatial resolution, unattended operation, particle identification, and energy measurement	No time information, manual scanning, chemical development
Gas-filled counter	Radiation monitoring, dosimetry	Low density, large area at low cost, nuclear and particle physics experiments, realtime particle identification	Needs ultra-pure flammable gases, high voltages required, signals need amplification
Scintillation counter	Energy measurement, particle counting, triggering	Large volumes, flexible shapes, fastest timing, high efficiency	Poor spatial resolution, poor energy resolution
Solid-state detector	Energy measurement, particle identification, position measurement, radiation monitoring, beam analysis techniques	High-energy resolution, compact size	High cost per area, signal needs amplification

67.2   Nuclear Emulsions

Nuclear emulsions are a suspension of silver halide grains, primarily AgBr, in a gelatin medium. The silver halide grains are small, on the order of 0.2 μm to 0.5 μm and constitute about half the volume and 80% of the total weight. The "pellicles" of nuclear emulsion range from 25 μm to 600 μm in thickness and can be as large as 70 cm by 30 cm. The pellicles may be supported on a glass or plastic plate during exposure, or stacked to increase the thickness of the emulsion detector, in which case the pellicles will be placed on a supporting plate as part of processing. If pellicles are stacked for an exposure, then a reference grid is usually printed on the bottom surface of each pellicle to assist in following tracks from one pellicle to the next.

The ionizing radiation crossing a grain of silver halide leaves the grain in an excited state. During development, the excited grains are reduced to elemental silver. The undeveloped grains are washed out during the fixing process. Due to their thickness and to ensure uniform response, a special development process, involving presoaking the emulsions at a reduced temperature, must be used. After development, the nuclear emulsion will have shrunk to about 40% of its original thickness. The thickness is also sensitive to the ambient humidity, and the plates are normally kept in humidity controlled storage (~50% relative humidity). The shrinkage needs to be taken into account for most measurements.

The track image in the emulsion is viewed with a microscope with magnification in the range of 10×10 to 10×100. With the aid of a special microscope, having a calibrated micrometer vertical focus adjustment and a special high-precision stage with calibrated micrometer x- and y-movements, the characteristics of the tracks can be determined and used to identify most particles and to measure their momentum. The dE/dx is measured by determining the grain density along the track. The energy or momentum is determined from the range of the particle or from multiple scattering. Additional information can also be obtained from the density of associated delta rays. These measurements can usually be combined to uniquely determine the charge (but not the sign) and momentum of the particle and the mass in the case of a singly charged particle. The interested reader may find additional information in References 8 through 10.

Nuclear emulsions are currently produced by Ilford Photographic (Ilford Research Laboratory, Ilford Ltd., London, England) and Fuji Photo Film Co. Ltd. (Tokyo, Japan). Several types of emulsion are available with different sensitivities to ionizing radiation.

67.3   Gas-Filled Charged Particle Counters

A gas-filled counter detects the ionization electrons and positive ions produced by a charged particle passing through a gas volume. An electrostatic field established between a pair of electrodes in a gas-filled container causes the electrons to move toward the anode and the ions toward the cathode. The charge movement produces an electrical pulse that can be detected using suitable electronics. A gas-filled counter can operate as an ionization counter, a proportional counter, or a Geiger–Müller counter. Figure 67.3 shows the three operating regions with characteristics determined by the electric field region in which they operate.

Ionization Counter/Chamber: At low fields, the recombination of electron-ion pairs reduces the detector signal, making it electric field dependent. With increasing field, nearly all the charges can be collected before they recombine. In an ionization counter, the signal-to-noise ratio is quite unfavorable, except when used for the detection of heavily ionizing particles (e.g., alpha particles). For radiation monitoring, the device is generally used in integrating mode as an ionization chamber. The current signal in an ionization chamber is proportional to the rate of ionization produced by charged particles interacting in the detector gas volume.

Proportional Counter: At still higher fields, the electrons acquire enough energy between collisions with gas molecules to produce secondary ionization in the gas. The secondary electrons in turn can produce further ionization along their path, leading to an electron avalanche that stops only when all the electrons are collected at the anode. This large increase in the number of primary electrons leads to

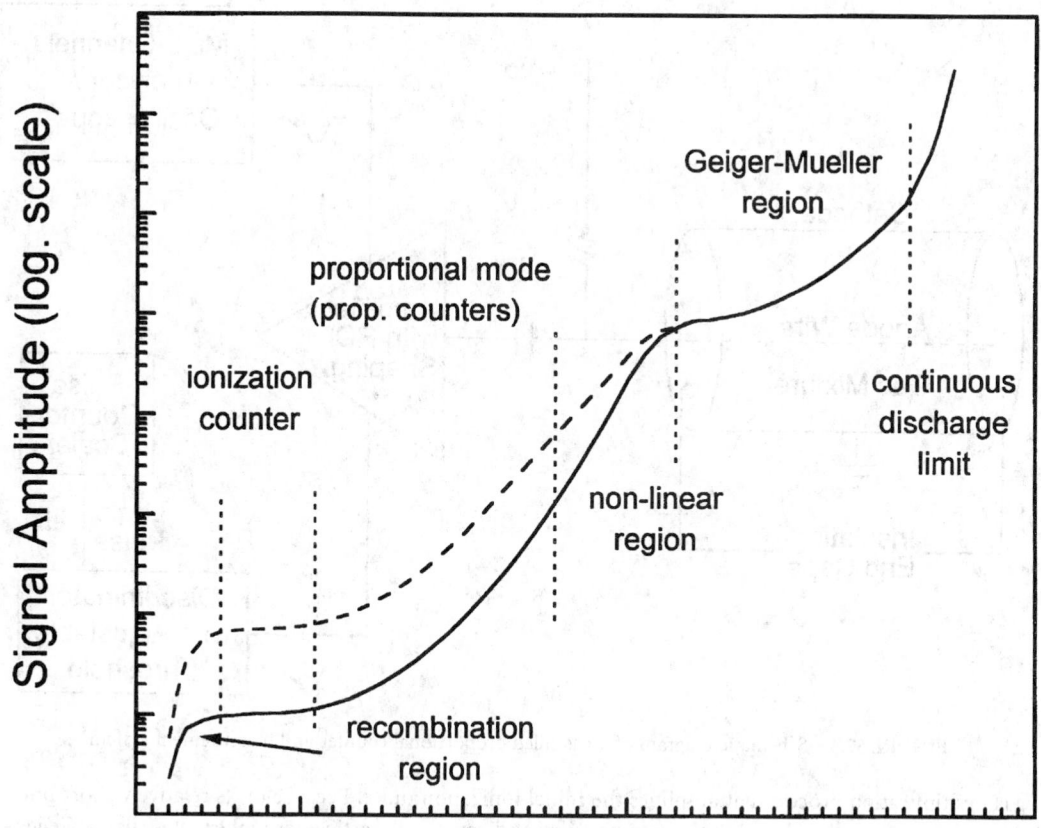

FIGURE 67.3 Typical behavior at atmospheric pressure for a gas-filled counter. Three principal operating regions are identified: ionization region, proportional region, and Geiger-Müller region. The lower curve is for minimum ionizing charged particles and the upper one for heavily ionizing ones. At low operating voltages, the ions and electrons recombine. Continuous discharge occurs at high voltages above the Geiger-Müller region of operation.

a detector signal substantially larger than that produced by a pulse ionization counter. The signal amplitude in a proportional counter is directly proportional to the ionization energy deposited in the gas by the ionizing particle. The proportional counter can detect charged particles without loss of gain at incident particle rates approachnig 10^4 s^{-1} for each millimeter of anode wire length.

Geiger–Müller Counter: Increasing the electric field still further results in increasing loss of proportionality until the counter enters the Geiger-Müller region of operation. In the Geiger-Müller region of operation, a single ionizing event induces an electrical discharge along the entire length of the anode wire. The output pulse amplitude is large and independent of the energy deposited by the ionizing particle in the gas. The Geiger-Müller counter has a relatively long dead time of a few hundred microseconds, and the detector saturates at counting rates above a few thousand particles per second.

Proportional counters have many advantages over ionization chambers and Geiger-Müller counters, both of which are now used mainly as radiation monitors in the laboratory. A gas flow proportional counter is sensitive to the ionization produced in a gas volume by a single charged particle. It can be used for counting charged particles, to measure their energy loss in the gas, and to measure their track coordinates. The proportional counter is more sensitive than an ionization counter because of a built-in

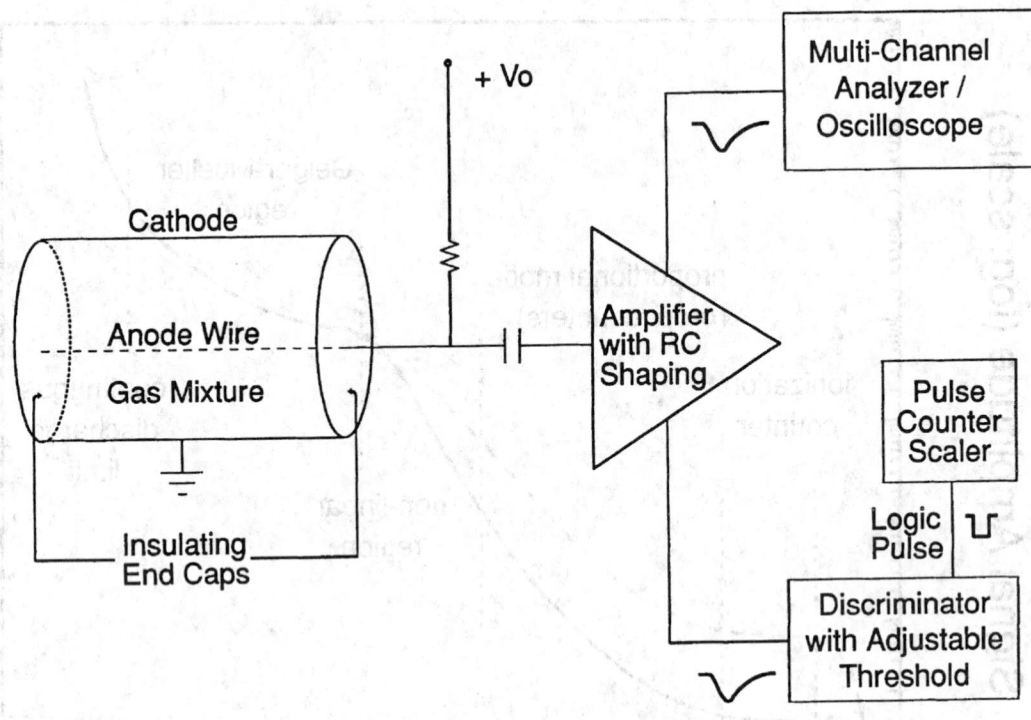

FIGURE 67.4 Schematic diagram of a gas-filled proportional counter and measurement system.

gas multiplication process that amplifies the initial ionization pulse. Because of its relatively short pulse duration, a proportional counter recovers quickly and can count particles at higher rates than a Geiger-Müller counter with its long dead time.

Gas-Filled Proportional Counters

Basic Design and Operating Principle

Figure 67.4 shows the schematic diagram of a gas-filled proportional counter and associated readout electronics. An electrically conducting gas-filled tube serves as the proportional counter cathode. A fine wire 20 to 25 μm in diameter (usually made of gold-plated tungsten) at the center of the tube, supported by insulators on both ends, forms the anode. For reasons of mechanical and electric stability, the anode wire is strung under a tension of several tens of grams, depending on the wire length and operating voltage of the detector. Mylar windows thin enough to allow passage of α and β particles close the gas volume. The cathode is usually kept at ground potential and a positive voltage V_0 is applied to the anode. The electrostatic field is radial with magnitude given by:

$$E = V_0 / \left(r \ln(b/a) \right) \tag{67.5}$$

where r is the distance from the central axis, a is the radius of the anode wire, and b is the radius of the tube. The $1/r$ dependence leads to the existence of an intense electric field near the anode wire; for example, at $r = 20$ μm, the electric field is 25 MV m^{-1} for a 20-μm diameter anode wire held at 1.5 kV within a cathode tube of radius 1 cm. A charged particle passing through the detector ionizes the gas molecules along its path. Under the influence of electric field, the primary ionization electrons start drifting toward the anode. In the strong anode field region a few wire radii from the anode surface, the electrons produce additional ionization along their paths, which leads to the production of an electron avalanche. The large increase in number of primary electrons is called gas multiplication or gas gain, which becomes significant in the high-field region near the anode. The gas gain is related to the mean free path of the electrons in the gas for secondary ionizing collisions. The electron mean free path is a function of the electric field and the gas pressure. Gas gains as high as 10^8 can be achieved by proper choice of gas mixture and mechanical construction before reaching the limit imposed by electric break-down in the gas. Most commonly used proportional counter systems have gas gains in the range of 10^4 to 10^6 [4,11,12,].

The voltage pulse observed on the anode is the result of a change in the electrostatic potential energy of the system. Most of the potential energy change is due to the movement of positive ions away from the high anode field region. The contribution of electrons to the total signal is quite small since most of the avalanche electrons traverse only a short distance before being collected by the anode. The fast movement of the ions away from the intense electric field region near the anode produces an initial rapid rise of the signal. About 50% of the pulse height is reached in roughly 0.1% of the total time, followed by a slow logarithmic rise lasting a few hundred microseconds. The signal is proportional to the total amount of ionization produced in the gas; that is, the product of the gas gain and the number of primary electron-ion pairs. The proportional counter energy resolution is determined by statistical fluctuations in the number of ions and by the statistical nature of the gas gain process itself. Deviations from proportionality occur when the space charge density of the positive ion sheath around the anode wire becomes large enough to locally reduce the electric field. The space charge effects become appreciable, depending on the gas gain at counting rates above 10^4 s^{-1} mm^{-1} of anode wire length.

Fill Gases for Proportional Counters

The choice of proportional counter fill gases is dictated by practical considerations. If low gas gain is acceptable, such as for the detection of α particles or heavy ions, both of which produce extremely high specific ionization (initial number of electron-ion pairs produced per cm), almost any gas (even air) can be used. For other particles, one requires a gas that has high specific ionization, high gas gain, low operating voltage, good proportionality, high single particle counting rate ability, and long operating life time. Noble gases such as argon meet the criteria of reasonable gas gain at low operating voltage and high specific ionization. However, the ultraviolet photons produced by excited argon atoms can cause secondary electron emission from cathodes, leading to electrical breakdown at higher operating voltages. The problem can be cured by adding a quenching polyatomic gas with rotational and vibrational levels that can readily absorb the ultraviolet photons. The addition of a quenching gas permits stable high gain operation of proportional chambers. A mixture of 90% argon and 10% methane (CH_4) called P10 is a commonly used proportional counter gas. Several different gas mixtures exist that are capable of achieving proportional gas gains in the neighborhood of 10^6.

The specific ionization for a gas can be calculated from a knowledge of dE/dx_{ion}, the rate of charged particle energy loss in the gas and W the mean energy needed to create a single electron-ion pair. The properties of some common proportional counter gases are shown in Table 67.2. The values of dE/dx_{ion}

TABLE 67.2 Characteristics of Gases Commonly Used in Proportional Counter Gas Mixtures

Gas	W (eV)	dE/dx_{ion} (keV/cm)	Specific Ionization (ion pairs/cm)
Ar	26	2.44	94
Xe	22	6.76	307
CO_2	33	3.01	91
CH_4	28	1.48	53
C_4H_{10}	23	4.50	195

Note: dE/dx_{ion} and specific ionization values are for minimum ionizing charged particles in gas at atmospheric pressure.

and the specific ionization shown in the table are for minimum ionizing charged particle tracks. The specific ionization for a gas mixture is the average specific ionization calculated with partial pressures of the component gases as weights.

The use of organic quenching gases as quenchers in proportional chamber gas mixtures can lead to the formation and deposition of polymers on the electrodes that reduce the operating life of a proportional chamber. To improve chamber lifetime and reduce aging, nonpolymerizing agents such as methylal are frequently added in small quantities to the gas mixture. For proportional counters with sealed gas volumes, the detector performance will degrade due to the contamination of gas in a sealed volume caused by component outgassing. Cleanliness and careful choice of components during fabrication can minimize these effects. The performance of a sealed proportional counter will also degrade with the degradation of the quenching gas component, which can happen when the device is exposed to large fluxes of radiation or after prolonged use. To minimize these problems, proportional counters are often used with external gas supply systems capable of refreshing the gas continuously.

Proportional Counter Operation and Readout

It takes a few hundred microseconds to develop the full voltage pulse in a proportional counter after a leading edge risetime of a few hundred nanoseconds. Even with a gas gain of 10^4 to 10^5, the pulse amplitude is only a few millivolts. Further amplification and shaping is needed to obtain signals suitable for measurements. Often, low-noise amplifiers with built-in RC differentiation to shorten the output pulses to 1 µs or less are used (see Figure 67.4). These improve the ability of proportional counters to measure particles at high counting rates without pulse pileup distortions caused by closely spaced pulses overlapping in time. The energy loss of a charged particle in a gas can be measured by measuring the pulse height of the anode signal with an oscilloscope or a multichannel analyzer. For charged particle counting, a discriminator can be used to produce logic pulses that can be counted with a scaler. For coincidence applications, timing resolution of a few nanoseconds is typical for a proportional counter. Good RF shielding and careful electric grounding are essential to maximize the sensitivity of most proportional counters.

The performance of a proportional counter is determined by operating voltage. At lower operating voltages, the best linearity is achieved at the expense of lower gas gain and, consequently, a low signal-to-noise ratio and low efficiency for particle detection. Increasing the voltage leads to larger gain, better signal-to-noise ratio, reduced linearity, faster aging, and susceptibility to electric sparking and damage. In general, the lowest possible anode voltage consistent with achieving good particle detection efficiency is preferred for the operation of proportional counters. This can be accomplished by choosing the operating voltage roughly in the middle of the high-voltage (HV) plateau for the device. On the HV plateau, the counting efficiency of the device remains constant for any variations of anode high voltage [12]. Typical proportional counter operating voltage plateaus are reached in the voltage range of 1.5 kV to 2.5 kV.

Proportional Counters: Advanced Techniques

Many variants of the simple, single wire design have been developed for nuclear physics and particle physics experiments [4,11,12]. Multiwire proportional counters were invented by George Charpak in 1968. In a multiwire proportional chamber, a parallel equidistant row of anode wires is enclosed between two cathode planes. Each anode acts as an independent proportional counter which is read out using suitable electronics. Segments of charged particle tracks can be localized in a multiwire proportional counter in the anode wire plane in a direction perpendicular to the anodes with accuracies on the order of 0.5 mm. A second coordinate along the length of the anode can also be obtained by a measurement of induced signals on a cathode that is segmented in strips in a direction normal to the anode wires. Much more precise particle track coordinate measurements with accuracies of tens of micrometers are possible in another variant called the drift chamber. This is achieved from a knowledge of drift velocities of electrons in the chamber gas as a function of electric field and by measuring the drift time of the ionization electrons to the anode. Complete three-dimensional determination of particle tracks is possible in a device known as the time projection chamber, which is a combination of a multiwire proportional chamber and a drift chamber. Gas microstrip detectors [13] improve on the high count rate abilities and spatial resolution of a multiwire proportional chamber using closely spaced anode strips that are printed photolithographically on a rigid substrate. Gas microstrip detectors with anode spacings of 200 μm are able to achieve spatial resolution of under 40 μm and count rate abilities in excess of $10^6/(s\ mm^2)$.

Geiger-Müller Counters

Basic Design and Operation

A Geiger-Müller counter [14] is similar in construction to a proportional counter. Operating at electric fields above the proportional region, it is a controlled discharge counter that produces a large output pulse of a few volts for every ionizing event in the gas independent of the amount of energy deposited in the gas. The ultraviolet photons from the electron avalanche in a Geiger-Müller counter induce an electrical discharge along the entire length of the anode wire. Gas mixtures used in Geiger-Müller counters generally contain a noble gas mixed with ethyl alcohol or halogen containing a vapor such as ethyl bromide, which acts as a quencher and eventually stops the discharge. After the discharge, the Geiger-Müller counter has a large dead time on the order of a few hundred microseconds, while the excess ionization is swept out of the chamber, during which it cannot register another pulse. If ethyl alcohol is the quenching agent, the gradual decomposition of molecules causes the counter characteristics to degrade with usage and the counter eventually ceases to function. Counters with a halogen compound used as a quenching agent do not suffer from this defect; however, because halogens are electronegative and capture electrons, such counters suffer from the problem of variable output signal delays. Since the Geiger-Müller counter pulses are several volts high, they require no further amplification for measurement purposes.

Proportional Counter and Geiger-Müller Counter Applications

Proportional counters and Geiger-Müller counters are often used in instruments for radiation protection and monitoring. Depending on the construction, they can be used for both the detection of α and β particles. The simple Geiger-Müller counter is perhaps more commonly used as a radiation monitor because it is easier to use and it has high efficiency, close to 100%. However, because of its relatively long dead time, a Geiger-Müller counter becomes less useful as a monitor in high radiation environments where particle rates exceed a few thousand counts per second. Also, a Geiger-Müller counter provides no energy information. For high counting rate applications or when energy discrimination is needed, it is better to use a proportional counter. Unfortunately, they can be cumbersome to use if an external gas supply is required. Geiger-Müller counters and proportional counter systems can be purchased from many different vendors (see Table 67.3). Prices vary from $100 to $500 for Geiger-Müller and proportional tubes, and up to $2000 for complete systems and accessories (e.g., survey meters, scalers, ratemeters, analyzers, etc).

67.4 Scintillation Counters

Introduction

The use of scintillation light is one of the original methods used to detect charged particles. In 1911, Rutherford used an activated zinc sulfide screen to detect α particles. The flashes of light produced in the screen were counted by a dark-adapted observer. Today, scintillation counters are usually made of either plastic, liquid, or crystal materials, and the resulting signal is detected electronically. The principle of operation of these devices is that the energy deposited by charged particles moving through the scintillator material causes some of the molecules to be raised to an excited state. On returning to the ground state, the molecules emit light. The decay time of the light emitted depends on the fluorescence

TABLE 67.3 List of Manufacturers for Gas-Filled Proportional
Counters and Geiger-Müller Tubes

Aptec Engineering Ltd.	Oxford Instruments Inc.
East 50 B, Kaldari Road	601 Oak Ridge Turnpike
Concord, Ontario L4K 4N8, Canada	P.O. Box 2560
Phone: +1 (905) 660-5373	Oak Ridge TN 37831-2560
Fax: +1 (905) 660-9693	Phone: +1 (615) 483-8405
	Fax: +1 (615) 483-5891
Canberra Industries, Inc.	
800 Research Parkway	Panasonic
Meriden, CT 06540	Two Panasonic Way
Phone: +1 (203) 238-2351	Cesaucus, NJ 07094
Fax: +1 (203) 235-1347	Phone: +1 (201) 392-6044
	Fax: +1 (201) 392-4315
E G & G Berthold	
P.O. Box 100163, D-75312	TGM Detectors Inc.
Bad Wildbad, Germany	160 Bear Hill Road
Phone: +49 708 1177-140	Waltham, MA 02154-1075
Fax: +49 708 1177-100	Phone: +1 (617) 890-2090
	Fax: +1 (617) 890-4711
E G & G Ortec	
100 Midland Road	Victoreen
Oak Ridge, TN 37831	6000 Cochran Road
Phone: +1 (615) 482-4411	Cleveland, OH 44139
Fax: +1 (615) 483-0396	Phone: +1 (216) 248-9300
	Fax: +1 (216) 248-9301
Ludlam Measurements Inc.	
P.O. Box 810	
501 Oak Street	
Sweetwater, TX 79556	
Phone: +1 (915) 235-5494	
Fax: +1 (915) 235-4672	

time of the scintillator. This can be very short, producing signal pulses as short as a few nanoseconds and thus allowing this type of detector to be used at very high counting rates. The light produced by the scintillator is usually detected using a photomultiplier. An appropriate choice of scintillation medium and detection method can produce a highly effective charged particle counter with an efficiency of essentially 100%.

Scintillation Counters: A Detailed Description

A scintillation counter consists of the scintillation medium, a light detector, and a light guide to transfer the light from one to the other (see Figure 67.5). Information on each of the components of a scintillation detector follows [4].

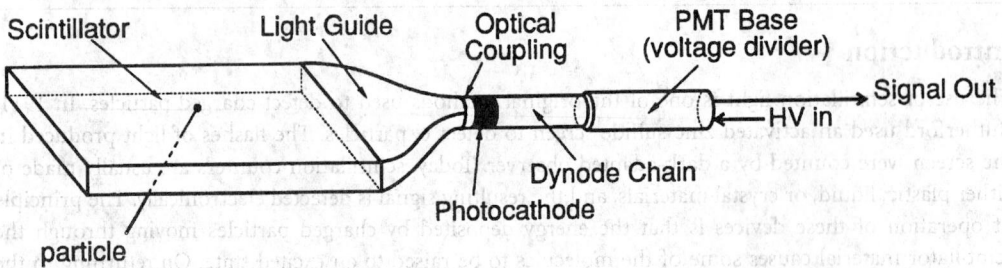

FIGURE 67.5 A typical scintillator system consisting of a scintillation medium, light detector, and light guide.

Scintillation Medium

Plastic or liquid scintillators are usually used for the detection of charged particles. Crystal scintillation media such as sodium iodide or BGO (bismuth germanate) are normally used for photon detection or applications in which energy resolution is important. Plastic detectors are inexpensive and have proved to be a reliable way of detecting charged particles. Large areas can be covered using adjacent sheets of plastic scintillator. Such an array of detectors is known as a hodoscope. Primary light output from a plastic scintillator is at short wavelengths, below 400 nm. Light at these wavelengths has a short attenuation length in the plastic, so wave-shifting dyes are employed to convert the light to longer wavelengths. Thus, a plastic scintillator contains two or more additives such as polystyrene doped with *p*-terphenyl and tetraphenyl-butadiene. Proprietary scintillators are available in a large variety of shapes and sizes from companies such as Bicron Corporation. Plastic scintillators will darken upon exposure to radiation in excess of about 10^3 grays, which will lead to inefficiencies. Liquid scintillation materials, such as toluene, are useful for constructing large volume detectors. Such devices are used in high radiation areas as liquid scintillators are inherently more radiation resistant. It is also possible to replace the liquid if radiation damage occurs. The main problem with this kind of device has been maintaining a leak-proof container and the resulting hazard associated with the toxicity of the liquid.

Light Coupling to Photon Detectors

The light produced in the scintillator is emitted isotropically. A portion of this light is trapped in the scintillator by total internal reflection. For a small detector, it is possible to couple the end of the scintillator directly to a phototube. For larger detectors (exceeding the width of the photodetector), a light guide made of a UV-transparent acrylic plastic is recommended. Acrylic plastic is widely available; Plexiglas by Rohm is one example. Light guides can be made of one piece of shaped plastic but maximum efficiency can be realized using adiabatic (constant area) light guides that minimize light losses. These light guides are quite artistic (see Figure 67.6) and are made from acrylic strips heated and then bent to the appropriate shape. The cut edges are flame polished to produce a smooth surface for good internal reflection. To

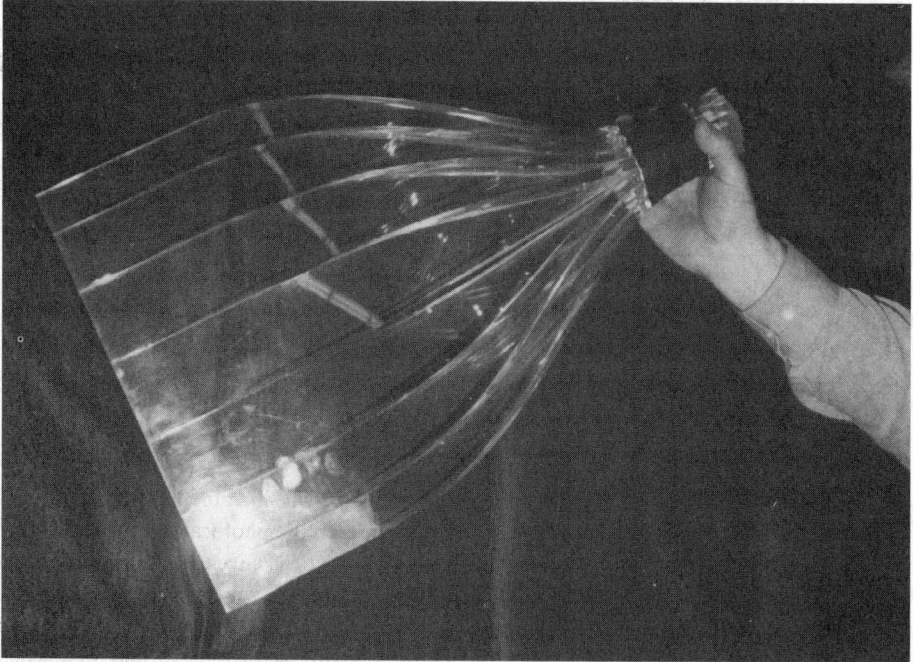

FIGURE 67.6 Photograph of adiabatic Plexiglas light guide. (Courtesy of Science and Technology Centre, Carleton University, Ottawa.)

match large scintillators to a small area detecting device, a light guide known as a "Winston cone" [15] is used on the light detector. The shape of the Winston cone minimizes light losses.

Light Detection and Readout

The standard method of detecting light from a scintillator is the use of photomultiplier (PM) tubes. These consist of a light-sensitive photocathode and an amplification stage. The photocathode is a thin layer deposited on the inside of a vacuum tube that has a good probability (>10%) of emitting an electron when struck by a photon. These electrons are electrostatically focused on a secondary electrode (dynode), which is the first element of the amplification structure. The voltage between the photocathode and this first dynode is sufficient to give the electrons enough energy to eject two to four electrons from the dynode. This process is continued through a series of dynodes resulting in a large gain, up to 10^8 in a tube with 14 dynodes. The output pulse is read out from a connection to the anode or the last dynode. The photomultiplier has connection pins corresponding to each dynode. Voltages are supplied to these via a resistive divider network contained in a "base" or "divider." The base requires one HV input and provides one or two signal outputs. Details of divider networks are available from PM tube manufacturers. Further information on the operation of phototubes can be found in Reference [16].

In operation, PM tubes are sensitive to magnetic fields and it is usually necessary to provide magnetic shielding. A tube of a single layer of mu-metal is often sufficient to shield the PM tube from the Earth's magnetic field. For applications where there are higher ambient fields, a shield of layered tubes can be used consisting of two mu-metal shields and up to two soft iron shields. They should extend approximately one tube diameter beyond the PM tube. In applications where an axial field is unavoidable, a bucking coil can be inserted between the two iron layers. A suitable choice of coil current can counteract the axial field.

Scintillator Detector Construction

Assembly of scintillators, light guides, and PM tubes is usually done with an optical cement. It is important in the preparation of the adhesive to remove air bubbles which would scatter the light. It is also possible to make a demountable joint using optical grease or vacuum grease and a mechanical fixture. Such a joint can be used to simplify the replacement of phototubes. Light levels produced in the scintillator are extremely low; therefore, detector assemblies must be enclosed in light-tight envelopes. For plastic scintillators, this is most easily arranged by wrapping with aluminum foil and protecting the foil with black plastic sheets held in place with black adhesive tape. In large systems of counters, it is useful to install LEDs in each scintillator to monitor the operation of the counter. A system for powering the LED with a short duration pulse allows a test of the counter and readout system.

Components for scintillator systems can be purchased from many manufacturers, some of which are listed at the end of this section. Prices vary widely, depending on both the details of the detector and the quantity of components ordered. Typical prices would be $600 for a piece of scintillator 0.5 cm × 45 cm × 60 cm, $200 for a voltage divider and $500 to $1200 for a 12-stage, 5-cm diameter phototube. Liquid scintillators are available for $145 per liter. Prices are in U.S. dollars. The cost of custom light guides is dominated by the available labor rate. There are some suppliers (such as Bicron) that will provide complete custom detectors consisting of scintillator, light guide, PM tube, and divider.

Operating Information

A minimum ionizing particle crossing a 1-cm thick scintillator will liberate about 10^4 photons. With typical scintillators and light guides, about 10% of these will reach the photocathode. The efficiencies of photocathodes to produce photo-electrons is usually greater than 10%. Given PM tube gains of 10^8, this leads to 10^{10} electrons at the anode, allowing analog pulses of about a volt to be produced. The pulses typically have a rise time of about 1 ns and a fall time of 5 ns. The trailing edge of the pulses can be shortened to approximately 2 ns by attaching a short grounded line to the output (clip line). The anode outputs are normally connected to a discriminator that produces a standard digital pulse for all analog pulses exceeding a fixed threshold voltage. More sophisticated circuits, such as constant fraction discriminators, can be used for best timing resolution. Using these devices along with a scintillator geometry that provides good light ouput, a timing resolution of the order of 300 ps (1σ) can be achieved. The excellent signal characteristics of scintillation counters allow them to count efficiently at rates exceeding 10 MHz.

Background counts in a scintillator are caused by cosmic rays and noise in the tube. An average rate from cosmic rays is 1 min^{-1} cm^{-2} sr^{-1}. Tube noise tends to have a lower amplitude than pulses resulting from the scintillator and will have a rate dependent on the choice of threshold. A counter should be checked to see if its count rate is consistent with cosmic ray backgrounds. High count rates are indicative of a light leak, which can be detected by shining a flashlight on the counter in a darkened room. Diligent use of aluminum foil and black tape will fix the leaks. In applications measuring particles that are energetic enough to pass through a counter, system noise can be reduced by using two layers of scintillator and putting their signals in coincidence. Noise can also be minimized by optimal setting of the tube HV and discriminator threshold. If the discriminator threshold is too low, then a large number of tube noise pulses are picked up.

Advanced Techniques

A plastic scintillation counter hodoscope can also be used to obtain position information. The spatial resolution of such a device is limited by the size of the individual detectors. For example, thin finger-shaped counters can be used for measuring beam profiles. Larger "paddles" 10 or 20 cm wide can be used to cover large areas. It is also possible to measure impact position along a counter using phototubes at both ends of a long scintillator and measuring the difference in arrival time of the pulses at the two tubes. Higher precision spatial resolution is possible using scintillating fibers with diameters of around 1 mm. Such a device has a large number of channels and requires segmented cathode phototubes or solid-state readout devices. One such device is the VLPC (visible light photon counter), which is a produced by Rockwell International and is a development from a solid-state photomultiplier. To keep noise low, VLPCs are used at cryogenic temperatures.

List of Manufacturers

Table 67.4 shows a selection of potential suppliers of components for scintillator systems.

67.5 Solid-State Detectors

Solid-state detectors have proved to be very versatile in the detection of charged particles. They have been particularly useful for energy measurements of α and β particles, light and heavy ions, and fission fragments in applications that include nuclear spectroscopy, radionuclide identification both in the environment and laboratory, dosimetry, radiation level monitoring, and non-destructive testing using particle beam techniques. In high-energy physics, solid-state detectors are used mainly as high-precision tracking detectors in a particle counting mode. Most of the devices used in all these applications are based on silicon technology except for energy measurements of β particles (>250 keV), where germanium-based detectors are also used.

Signal Generation

A charged particle moving through a semiconductor crystal loses energy according to the stopping processes described previously (Equation 67.1). The energy lost is spent promoting electrons from the valence band to the conduction band of the crystal. This leaves free electrons in the conduction band and free holes in the valence band. Under the action of an electric field, the free electrons and holes drift in opposite directions and give rise to an electrical current that can be sensed by external electronics.

TABLE 67.4 List of Manufacturers for Scintillator System Components

	Scintillator	Bases/Dividers	Phototubes	Systems
Bicron	x	x		x
DEP scientific	x			
Hamamatsu		x	x	
Philips		x	x	
pol. hi. tech.	x			
Burle			x	
Thorn E.M.I.		x	x	x

Bicron
12345 Kinsman Road
Newbury, OH, 44065-9677
Phone: +1 (216) 564-2251
Fax: +1 (216) 564-8047

Burle Industries
100 New Holland Avenue
Lancaster, PA 17601-5688
Phone: +1 (717)-295-6000
Fax: +1 (717)-295-6096

DEP Scientific
P.O. Box 60
9300 AB Roden
The Netherlands
Phone: 3150-5018808
Fax: 3150-5013510

Hamamatsu Photonics
325-6, Sunayama-cho
Hamamatsu City, 430, Japan
Phone: +81 53-452-2141
Fax: +81 53-456-7889

Philips Photonics
100 Providence Pike
Slatersville, RI 02876
Phone: +1 (401)-762-3800
Fax: +1 (401) 767-4493

pol. hi. tech
S. p. Turananse
Km 44,000
67061 Carsolo (AQ) – Italy
Phone: (0863) 997798
Fax: (0863) 995868

Thorn EMI
Bury Street
Ruislip, Middlesex, HA4 7TA, U.K.
Phone: +44 1895 630771
Fax: +44 1895 635953

The amount of charge produced is very small and the semiconductor must be depleted of majority carriers in order for the signal to be measurable. This is accomplished using a diode structure operated with a reverse-bias voltage. A profile of a "typical" diode detector is given in Figure 67.7. The size of the active region of the detector, the depletion region, is determined by the reverse bias applied and the residual fixed charges. In the case where the material has a very high resistivity (~500 Ωcm or more), the built-in field of the junction can be sufficient to generate a useful depletion layer without any reverse bias.

For a quantity of energy E_0 deposited through electronic interactions, the number of free electron-hole pairs created is $N_p = (E_0)/\varepsilon$, where ε is the average energy to produce one pair, a characteristic of the semiconductor, similar to the W value for gas detectors. Typical values of ε are small, 3.62 eV for Si at 300 K and 2.96 eV for Ge at 77 K. The variance of the number of pairs created is FN_p, where F is the Fano factor [17], which is of the order of 0.1 for Ge and Si. Hence, semiconductor materials are very efficient in converting radiation energy to an electric signal and the pulse-to-pulse variation in the signal size is smaller than predicted by Poisson statistics, where $F = 1$. When all the charge carriers are collected (i.e., the electrons reach the anode side while the holes reach the cathode side), a total of N_p electrons are available to the front end-electronics for amplification and shaping.

Categories of Detector

Device Details

Commercial solid-state detectors (i.e., silicon or germanium devices) are available from a number of manufacturers (see Table 67.5). The following silicon devices are available: silicon surface barrier detectors (SSBD), ion-implanted silicon detectors, diffused junction silicon detectors, *pin* photodiodes, and lithium drifted silicon detectors (Si(Li)). Some of the manufacturers rename the different types using their own

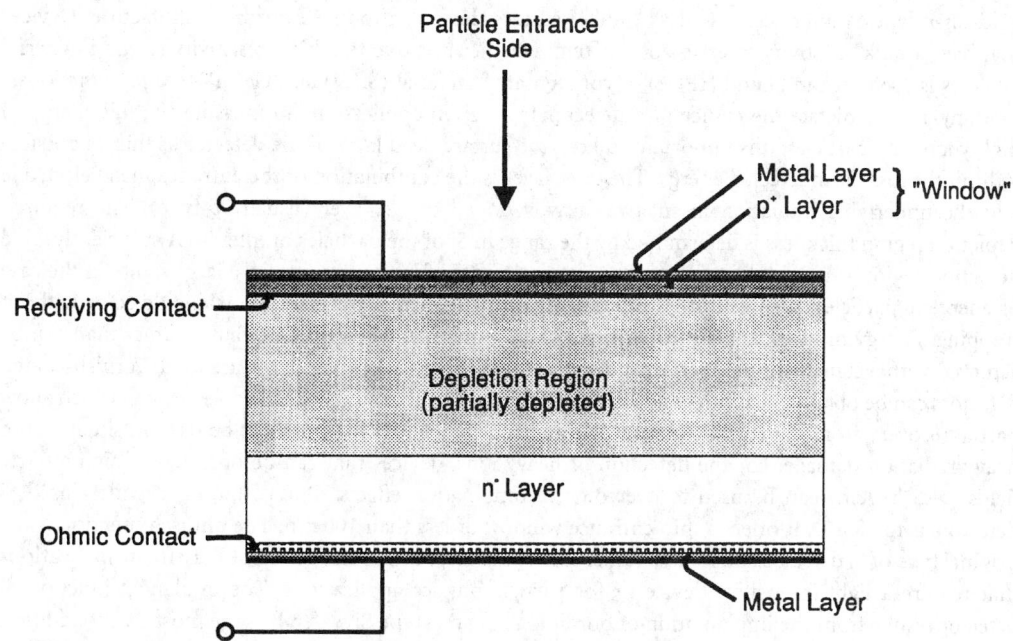

FIGURE 67.7 A solid-state diode detector consists of a thin current blocking layer, a sensitive volume of high resistivity material, and a back ohmic contact. A near uniform field may be established in the sensitive volume.

TABLE 67.5 Device Details

Type	Area (mm²)	Sensitive thickness (μm)	Dead Layers* (nm)	Particles detected	Prices 1996 US$	Manufacturers
SSBD	10–1000	10–5000	80 225 (back)	p, α, β light and heavy ions	400 to 6000	a b c d
Si, ion implanted	Up to 3000	100–1000	50	p, α, β light and heavy ions	300 to 3000	a b c d
Si pin photodiode	1–900	150–400	100	p, α, β light and heavy ions	10 to 300	e f g h
Si(Li)	25–200	to 5000	1000 Be +200 Si	β, 0.3–3 MeV	~7500	a b c d
HPGe	75–800	to 10000	2500 Be +300 Si	β, 0.3–3 MeV	~9000	a b c d
Silicon strip	See manufacturer for details					c d g i

Note: * Values in nm Si unless stated otherwise.

a. EG&G Ortec (U.S.A.), b. Oxford Instruments Inc. (U.S.A.), c. Eurysis Mesures (France), d. Canberra (Australia), e. EG&G Optoelectronics (Canada), f. Silonex Inc. (Canada), g. Hamamatsu Photonics K.K. (Japan), h. Centronic Inc. (U.S.A.), i. VTT Electronics (Finland).

brand names. The first three types actually lend themselves to nearly the same applications: ion detection. Si(Li) detectors are mostly used for soft X-ray detection, but can also be used for full energy measurement of β particles from 250 keV to 3 MeV. Germanium devices are available as lithium-drifted germanium detectors (Ge(Li)) and high-purity germanium detectors (HPGe). As with the Si(Li) detector, the primary function of germanium detectors is for photon detection but they can also be used to detect β particles. Solid-state detectors are either *pn* or *pin* diodes. Silicon devices fabricated from a starting material of about 500 Ωcm (i.e., ion-implanted Si and diffused junction Si) have the *pn* structure. The SSBD, also made from Si of a few 100's Ωcm, is in fact a hybrid of Schottky junction and *pn* junction due to its gold-on-*n* type Si interface as well as "*p*-like" surface defects introduced during manufacture. Devices that have a bulk resistivity in excess of 1 kΩcm, whether it is due to a high-resistivity starting material (such as is used for pin Si and HPGe) or compensated material (Si(Li) and Ge(Li)), are *pin* junctions.

Many aspects dictate the choice of a detector for a given application. To measure the full energy of highly ionizing particles, it is important to keep the surface dead layer of the detector as thin as possible to limit the loss of undetected energy. The dead layer is the combination of the entrance metal electrode and the underlying, radiation insensitive, heavily doped top Si layer (usually *p*-type). The required depletion region thickness is determined by the range in Si of the particles of interest. A sufficiently wide depletion region is required for full energy measurements of ions, or to obtain a large signal in the case of energetic particles, such as minimum ionizing particles, which traverses the whole detector without stopping. Energy measurements of β particles often require an active volume that is larger than can be achieved with a standard *p-n* junction and so Ge(Li), Si(Li) or HPGe detectors are used. Varnish-coated detectors can be obtained to withstand exposures to rugged environments. Bakeable devices, which allow partial recovery from damage, or inexpensive and replaceable photodiodes can be used in situations of heavy radiation damage. For the detection of heavy ions, devices that can be operated at high electric fields (>15 kV/cm) can be used to overcome recombination effects. One of the most widely used Si detectors is the SSBD. It offers a thin entrance window of less than 100 nm. The diffused junction is the original type of Si detector and it has a rather thick dead layer of 250 nm Si. It is still useful in applications that require a light-insensitive device or for particle physics applications. The implanted junction Si detector results from the implantation of boron ions in an *n*-type Si wafer. It is the most recent addition to commercial Si detectors and it offers a thinner entrance window (less than 50 nm) and a larger surface area than SSBD (see Table 67.5). All these devices can be operated in either partially or fully depleted mode. The thickness of the back dead layer is only relevant to fully depleted, transmission type detectors for energy loss (ΔE) measurements. Recently, high-quality, inexpensive Si *pin* diodes for visible and UV photon detection have been found to be suitable replacements for SSBD in many applications. The thickness of the dead layer is 100 nm, twice that of ion implanted Si detectors, and the resisitivity is as high as 2 to 8 kΩcm for the more expensive versions.

Detector Systems

Three practical systems of increasing sophistication are considered next. The first setup, depicted in Figure 67.8, is used to measure the energy spectrum and the rate of incoming charged particles. One

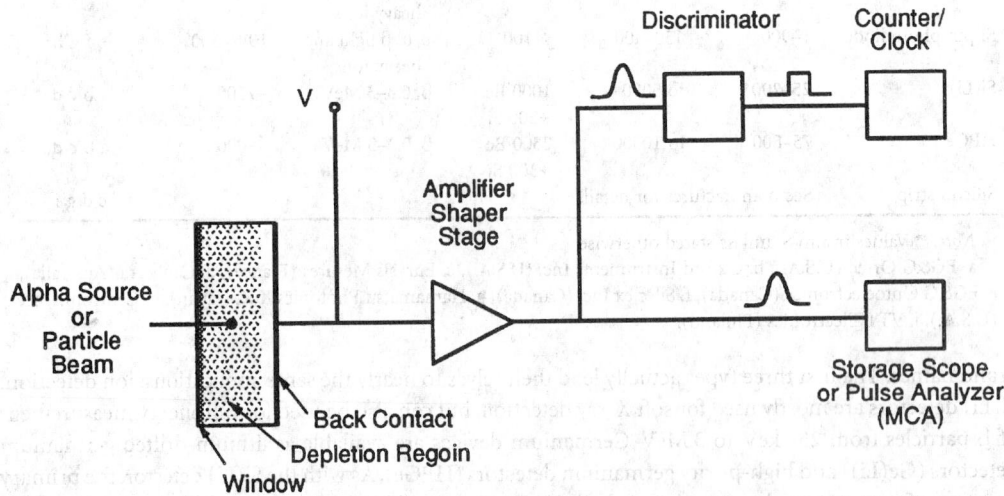

FIGURE 67.8 Using a pulse counter, this simple setup is used to measure the activity of a radioactive sample or the intensity of an accelerator beam. With the addition of a storage oscilloscope or an MCA (multichannel analyzer), the energy deposited in the detector by each incoming charged particle may be measured.

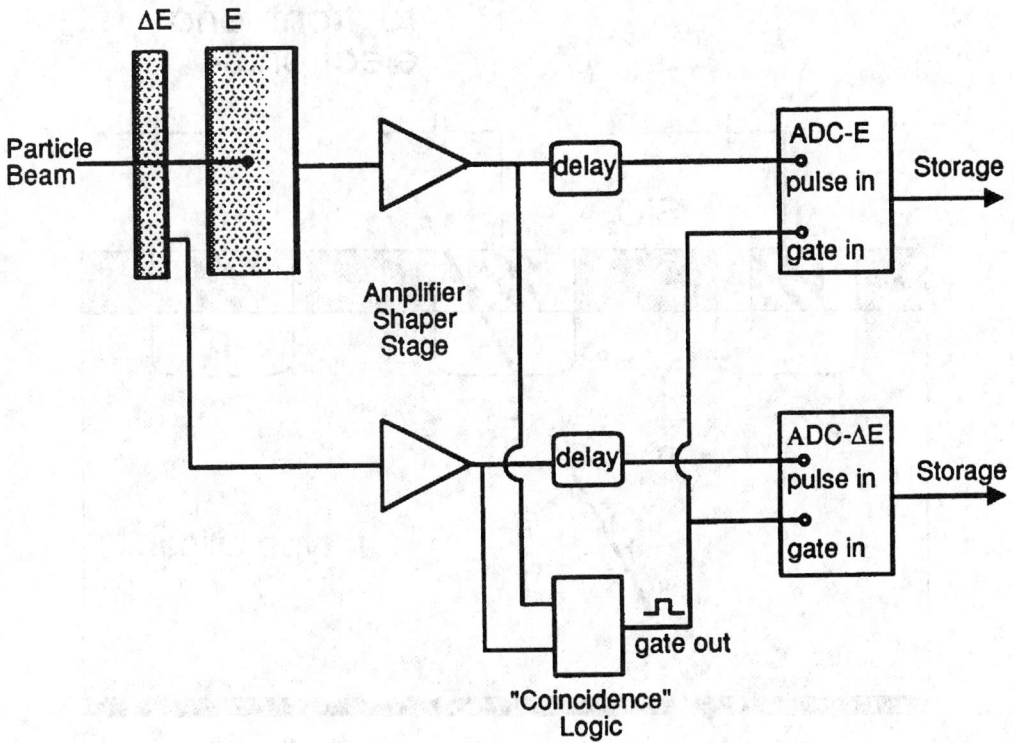

FIGURE 67.9 Knowledge of the amount of energy deposited in a transmission-type detector (ΔE) positioned in line with a full energy measuring detector (E) allows the identification of a particle due to the unique nature of dE/dx and E combination.

possible application is the measurement of the α activity from a natural source or from a wipe test sample. This simple count rate measurement requires a pn silicon detector or photodiode, a discriminator, and a pulse counter. By adding a pulse analyzer, called a multichannel analyzer (MCA in Figure 67.8), the same system can be used to record the energy of the α particles or the full energy of light ions, such as those encountered at an accelerator site where beam analysis of materials is performed. In this case, the suggested detector is a good quality SSBD or an ion-implanted detector with a thin window and operated at a voltage sufficient to generate a depletion region at least as wide as the range of the ions of interest. The second setup (Figure 67.9) shows a thin transmission type (ΔE) detector (open front and back), in front of a thick, fully stopping, (E) detector. This configuration is called a ΔE-E telescope and it supplies a simultaneous measure of the energy loss and the full energy of particles going through both devices. This information then uniquely identifies the particle. Finally, Figure 67.10 presents a position-sensitive silicon strip detector for which the localization of a particle hit to better than 5 μm in one dimension is possible [18]. Such a detector is used in particle physics experiments. In many applications, combinations with other detector types, such as scintillators, are also possible.

Detectors with sensitive thicknesses of 100 μm to 300 μm are used in many nuclear spectroscopy and high-energy physics applications. For typical values of applied voltages, full electron collection occurs in less than 10 ns, while hole collection takes less than 20 ns. The amount of charge in a pulse is very small, of the order of 1.5 (10^6) e for the full energy of an α particle and 100 e per μm of path in silicon of a minimum ionizing particle. Depending on the application, the signal processing electronics can be optimized for low noise or fast response. In spectroscopy applications, where precise measurement of the total charge is important, one uses low noise, charge integrating preamplifiers, which offer output gains of the order of 1 mV fC^{-1}. The preamplifiers are optimized for specific ranges of detector capacitance. For detectors of 5 pF and smaller, noise values of less than 1 keV (equivalent energy loss signal in Si)

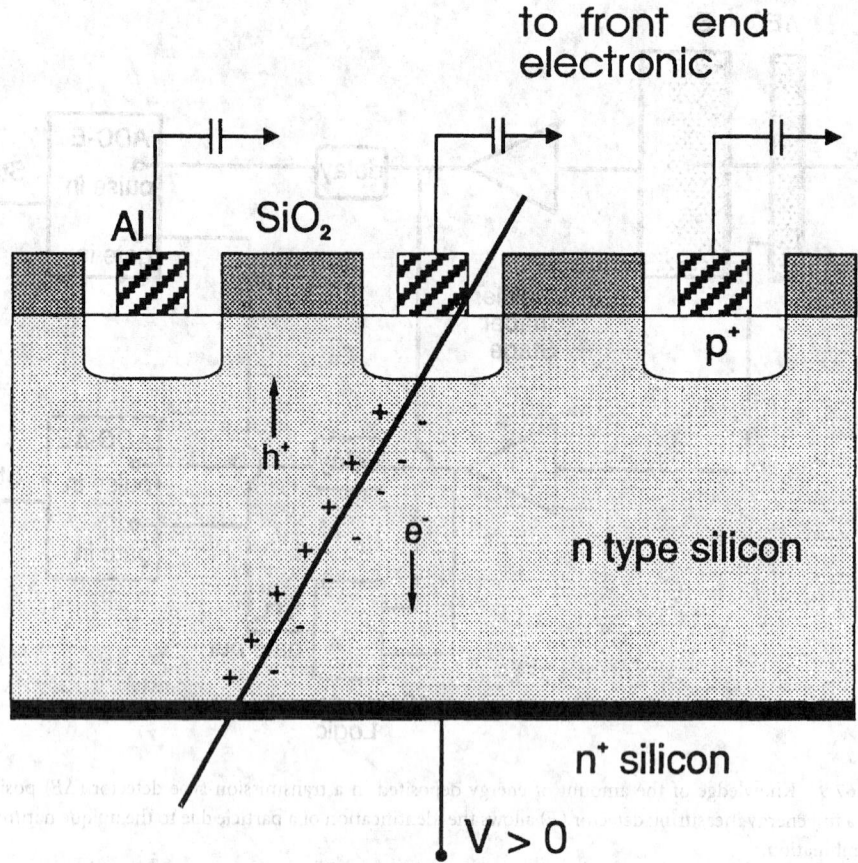

FIGURE 67.10   A silicon strip detector, such as the one depicted here, can be used to localize the hit position of a minimum ionizing particle.

can be achieved. In the case of preamplifiers optimized for larger detector capacitances (200 to 1000 pF), the noise increases with a slope of ~0.02 keV pF^{-1}. Preamplifiers produce a "step" pulse with an amplitude proportional to the input signal charge. The rise time of the pulse varies with the detector capacitance, from 5 ns to 10 ns for <10 pF to 100 ns for 1000 pF. The pulse decay time is determined by internal circuit components and usually falls in the range of 50 μs to 300 μs. The signal is further amplified and shaped using an integrating-differentiating amplifier to provide a pulse of gaussian shape whose width is of the order of 1 μs. This pulse duration limits the counting rate to much less than 10^5 cps before an unreasonably high counting deadtime is encountered. In high-energy physics experiments, the signal processing electronics coupled to position-sensitive detectors are optimized for a high counting rate capability and good timing resolution. Silicon integrated circuits containing a large number of amplifiers are specially designed for such applications. The individual amplifiers provide large output gain (15 mV fC^{-1}) and output pulses whose width can be as small as 20 ns (baseline to baseline), at the cost of an increase in noise (6 keV Si-equivalent). Count rates near 10^6 cps for every detection channel can be achieved with such systems.

Plasma Effects and Pulse Height Defect

Under some conditions, one may observe a loss in proportionality between the energy lost by the particle crossing the detector and the signal. This is called the pulse height deficit (PHD) and it is the sum of three components: E_W, E_N, and E_R, where E_W is the energy lost in the window or dead layer of the detector and E_N is the energy lost to nuclear scattering (nonionizing). The third term, E_R, is harder to define and is due to a deficit in the charge collection. Ions heavier than the proton create a plasma with a charge density as high as 10^{19} e^-/cm^3 which shields the charge from the collecting field and causes charge loss through electron-hole recombination in the plasma to occur. The PHD increases roughly with the particle mass and energy, and ranges from 1% to 3% for ions such as oxygen at 20 MeV.

New Structures

Modern silicon processing technologies allow the design of intricate structures which find applications in charged particle detection, particularly in high-energy physics experiments. These include position sensitive detectors, with strip electrodes or pixel segmentation, and large-volume silicon drift chambers. Front-end electronics can also be fabricated on the same wafer as the detecting structure itself. Close coupling provides noise reduction. The reader should be aware that new materials, such as GaAs, CVD diamond, natural diamond, cadmium telluride, HgI_2, and hydrogenated amorphous silicon, are being used in prototype detectors. These are beyond the scope of this chapter.

Manufacturers and Prices

A list of typical devices and pricing for each category of solid-state detector is given in Table 67.5. The names of some manufacturers are given but the interested reader is also encouraged to consider other sources.

References

1. J.F. Ziegler, J.P. Biersack, and U. Littmark, *The Stopping and Range of Ions in Solids*, New York: Pergamon Press, 1985.
2. J.F. Ziegler, *SRIM: The Stopping and Range of Ions in Matter*, IBM-Research Yorktown U.S.A., 1996 (computer software).
3. U. Littmark and J.F. Ziegler, *Handbook of Energetic Ions in All Elements*, Volume 6 of The stopping and ranges of ions in matter, J.F. Ziegler (ed.), New York: Pergamon Press, 1980.
4. R. Fernow, *Introduction to Experimental Particle Physics*, Cambridge, U.K.: Cambridge University Press, 1986.
5. J.E. Bateman, M.W. Waters, and R.E. Jones, Spatial resolution in xenon filled MWPC X-ray imaging detector — a computing physics approach, *Nuclear Instruments and Methods*, A135, 235-239, 1976.
6. W.R. Nelson, H. Hirayama, and D.W.O. Rodgers, *The EGS4 Code System*, SLAC-Report-265, 1985.
7. J.A. Halbleib, R.P. Kensek, T.A. Melhorn, and G.D. Valdez, *ITS Version 3.0- The Integrated Tiger Series of Coupled Electron-Photon Transport Codes*, Sandia National Laboratory, Albuquerque, NM: 1992.
8. C.F. Powell, P.H. Fowler, and D.H. Perkins, *The Study of Elementary Particles by the Photographic Method: An Account of the Principal Techniques and Discoveries, Illustrated by an Atlas of Photomicrographs*, New York: Pergamon Press, 1959.
9. Walter H. Barkas, *Nuclear Research Emulsions*, New York: Academic Press, 1963.
10. C. Plante and J. Hebert, The p/mc dependence of the grain density of tracks in the Fuji ET-7B nuclear emulsion, *Nuclear Instruments and Methods*, B108, 99-113, 1996.
11. W. R. Leo, *Techniques for Nuclear and Particle Physics Experiments*, 2nd ed., New York: Springer-Verlag, 1994.
12. N. Tsoulfanidis, *Measurement and Detection of Radiation*, New York: McGraw Hill, 1981.
13. A. Oed, Position-sensitive detector with microstrip anode for electron multiplication with gases, *Nuclear Instruments and Methods*, A263, 351, 1988.

14. R. W. Williams, Chapter 1.3 – Gas Filled Counters, *Methods of Experimental Physics, Nuclear Physics A,* 5, 1961.

15. H. Hinterberger and R. Winston, Efficient light coupler for threshold cerenkov counters, *Review of Scientific Instruments,* 37, 1094, 1966.

16. *Photomultiplier Tubes, Principle and Applications,* available from Philips Photonics, International Marketing, BP 520, F-19106, France.

17. G.F. Knoll, *Radiation Detection and Measurement,* New York: John Wiley & Sons, 1979.

18. A. Peisert, Silicon microstrip detectors, *Instrumentation on High Energy Physics,* Vol. 9, editor F. Sauli, Singapore: World Scientific, 1992.

68

Neutron Measurement

Steven M. Grimes
Ohio University

Neutrons have characteristics that make them of particular importance in technology and research. Since they are uncharged, they are able to enter the nucleus at very low energy. Furthermore, the lack of energy losses through ionization permits deep penetration into materials.

This latter characteristic makes detection of neutrons more complicated than detecting protons or alpha particles. The energy of these charged particles can easily be determined by detecting them in ionization chambers, proportional counters, or scintillators, and ultraprecise measurements can also utilize the steering effect of magnetic fields. Detection efficiencies approach 100%. Neutrons, on the other hand, are not directly detected by these detectors and some means of converting the neutron through a nuclear reaction to a charged particle or gamma ray must be used.

Nuclear reactions occur when neutrons interact with nuclei (including protons). When a neutron interacts with a nucleus, the following processes are possible:

1. Elastic scattering occurs if the neutron simply changes its direction without giving the nucleus any intrinsic excitation energy. The transfer of energy is analogous to that in a collision of two billiard balls and is governed by the laws of mechanics.
2. Capture reactions are those in which the neutron is absorbed by the nucleus to form a heavier nucleus. In this case, the energy released comes in the form of one or more gamma rays.
3. Fission occurs in many cases if a heavy nucleus (mass $> A = 230$) is struck by a neutron. The nucleus divides into two smaller nuclei with a huge energy release ($E > 100$ MeV, where 1 MeV = 10^6 eV = 1.6×10^{-13} J).
4. Inelastic scattering occurs if a neutron transfers energy to the target nucleus and leaves it in an excited state. This energy is usually emitted in the form of a gamma ray, although in some cases positron-electron emission occurs, and in many cases, a number of low-energy gamma rays are emitted rather than one higher-energy gamma ray.
5. (n,z) reactions can also occur, where the neutron is absorbed by the nucleus and a charged particle (usually a proton or alpha particle) is emitted.

Each of these processes could, in principle, be used to detect neutrons. Unfortunately, in many cases, one cannot ensure that only one type of reaction occurs and even when this condition is met, the resulting pulse spectrum does not always allow determination of the energy of the neutron.

68.1   Detector Types

Detectors Based on Elastic Scattering

Detectors based on elastic scattering are attractive because the elastic scattering cross-section is large (usually about half of the total interaction cross-section). A fundamental problem with elastic scattering is that the recoil energy as a fraction of the incident energy ranges from 0 (at 0° scattering angle) to $4A/(A + 1)^2$ (for 180° scattering angle), where A is the mass number of the target. For $A = 1$ (hydrogen), this recoil energy ranges from 0% to 100%, but for even as light a target as carbon ($A = 12$), the maximum energy of recoil is less than 30%. The fact that a continuous range of energies is produced even from a monoenergetic neutron beam makes the efficiency difficult to calculate, since it depends on accurate knowledge of the cutoff threshold of the electronics. The same feature complicates determining the energy distribution of neutrons when a range of energies is present.

Hydrogen-containing counters are nonetheless frequently used in neutron physics. Below about 500 keV, the pulse height is usually too small to be useful, but above this energy this type of counter is effective. Proportional counters or ion chambers filled with methane can be used in situations where fast timing is not required. Hydrogen-containing scintillators have poorer energy resolution but better timing information and are available in large volumes. Scintillators are subject to a threshold but can be made highly efficient ($\epsilon > 50\%$) if the volume is sufficiently large. For either proportional counters or scintillators, gamma rays can produce background pulses through the Compton effect, but some liquid scintillators have the capability of identifying the two types of pulses produced by neutrons and gamma rays, respectively, allowing discrimination against gamma rays. The excellent timing characteristics of scintillators have resulted in extensive use of these detectors for time-of-flight spectrometers. In these spectrometers, the energy of the neutron is deduced by timing the neutron over a measured flight path. For typical flight paths ($L < 10$ m), a precision of a few nanoseconds is required.

The fundamental limitation of counters of this type is that the pulse height in the detector is not uniquely related to the neutron energy. If the detector is not used in a time-of-flight spectrometer, the energy of the neutron cannot be deduced. A related problem is that the efficiency is determined by an electronic cutoff, which is difficult to determine accurately. Finally, detectors that include the entire angular distribution do not have directional sensitivity.

All of these drawbacks can be removed if only a limited portion of the angular range is sampled. This is the case in a recoil counter telescope. The spectrometer consists of a polyethylene radiator at one end of an evacuated cylinder. Near the other end are either one or, more typically, two detectors. Neutrons enter the telescope from the end holding the radiator. A small fraction of them produce recoil protons at small angles to the beam. A two-detector system is arranged so that the particles traverse both. Requiring a coincidence in pulses reduces backgrounds. These detectors register protons scattered by the neutron beam at angles near 0° to the original beam. The recoil energy of such protons is $E_n \cos^2 \theta$, where E_n is the neutron energy and θ is the angle of recoil. This implies that for a telescope which subtends an angle θ around 0°, the energy range of recoil protons will be E_n to $E_n \cos^2 \theta$. It can be shown that the average energy is simply $(\frac{1}{2}) E_n (1 + \cos^2 \theta)$. Obviously, as θ becomes small, the proton energy reflects the neutron energy. The smaller θ becomes, however, the lower the efficiency of the telescope. Typical telescopes of this type have efficiencies of 10^{-5} to 10^{-6}. It is obvious that only neutrons arriving along the axis will give protons of the right energy. Neutrons coming from the backward hemisphere will not be able to produce protons that traverse the spectrometer. The drawback of a low efficiency is partly compensated by the fact that the efficiency can be accurately calculated and the fact that the energy spectrum can be deduced.

Detectors Based on Capture Reactions

Capture reactions have small cross-sections at energies above 1 MeV. At lower energies, the cross-section typically varies as $1/\sqrt{E}$, so it can be very large at low energies. Further, the capture process is exoergic, typically yielding about 7 MeV. This energy is usually divided among two to four gamma rays. The efficiency of gamma-ray detectors is usually much less than 100% and detecting the gamma rays in coincidence requires large solid angles. It is essentially impossible to use the capture process to determine the energy of individual neutrons in a flux of neutrons of varying energies.

An alternative use of the capture process is in a passive detector. A foil of the appropriate material can be placed at the location where the flux is to be determined. The target must be one that produces a radioactive residual nucleus when a neutron is captured. Thus, ^{28}Si would not be appropriate, since ^{29}Si is stable, but ^{27}Al would be a possibility, since ^{28}Al is radioactive. After the measurement is complete, the foil is removed and placed in a counting area. A detector then counts the beta particles or gamma rays emitted. Of particular importance is the choice of target to ensure an appropriate halflife. A very short halflife would allow the decay of the radioactive nuclei before they could be counted, while a very long halflife (>1 month) might require an excessive length of time to be counted.

The complicated energy dependence of the capture cross-section makes use of the capture process mostly useful for relative measurements. If two measurements are made of spectra with the same relative distribution of neutrons with energy, a measurement of the ratio of activities will give the ratio of neutron intensities. An important advantage to this technique is that small foils can be used, allowing the determination of the flux in a small volume. Other radiation, such as gamma rays, usually does not interfere in such a measurement. An important disadvantage is that the flux is not obtained until after the counting is completed; this might be as long as a month after the measurement. An additional disadvantage has already been mentioned. If the relative energy dependence of the two spectra is not the same, the relative ratio of neutrons will not be correct.

Detectors Based On Fission

The fission process is relatively unique, since an energy release of over 100 MeV occurs even when the incoming neutron has an energy below 1 eV. Fission cross-sections for gamma rays are small, so most of the pulses will be due to neutrons. There are two types of fission targets. Those with even proton number (Z) and even neutron number (N) have an energy threshold for fission. The cross-section for fission is normally negligible below 2 to 3 MeV. Other targets for which either N or Z or both are odd have no threshold energy. These isotopes (e.g., ^{235}U) normally have a low energy cross-section dependence of $1/\sqrt{E}$, giving a very large efficiency for neutrons below 100 keV. Fission detectors are designed either for $E_n > 2$ MeV (in which case ^{238}U would be an appropriate choice) or for the entire energy range, in which case the efficiency is highest at low energies.

Fission foils must be fairly thin to allow the escape of the fission fragments into the counting volume. This typically results in low efficiencies ($<10^{-6}$). Fission chambers also require the license to possess radioactive isotopes. The energy release in the fission process is so large that the pulse height in the counter cannot be used to infer the neutron energy. Reasonably good timing information ($\Delta t < 5$ ns) allows the use of fission chambers in time-of-flight spectrometers.

Counters Based on Inelastic Scattering

Inelastic scattering results in a transfer of energy from the neutron to the target nucleus through excitation of an excited energy state. This will then result in the production of one or more gamma rays. Gamma rays can be detected in germanium detectors with good energy resolution. Unfortunately, the efficiency of a spectrometer utilizing a germanium detector and inelastic scattering is quite small. This technique is being examined as a possible approach to identifying contraband materials in luggage, since both the neutrons and the gamma rays they produce are highly penetrating.

Although most gamma ray decays occur rapidly ($<10^{-6}$ s), some gamma-ray decays are inhibited by angular momentum coupling and have halflives longer than 1 s. These could be used in a detector based on inelastic scattering just as the capture reaction is used. Two nuclei with such states are ^{89}Y and ^{103}Rh. Unlike the capture reaction, such isomeric transitions will not be effective in detecting low-energy neutrons, since there is a threshold for populating such states (approximately equal to the energy of excitation, ignoring recoil effects).

Detectors Based on (n,p) or (n,α) Reactions

Detectors utilizing (n,p) or (n,α) reactions can be passive, in which a final product nucleus is radioactive, or active, in which a proton or alpha is detected. For a discussion of the passive type, see Section 68.2. Direct detection of the charged particle can occur in a proportional counter or a scintillator. As in the case of fission, many (n,p) and (n,α) reactions have an energy threshold, which prevents their occurring until this energy is exceeded. Others have no threshold (they are exoergic) and have low energy cross-sections that vary as $1/\sqrt{E}$. These targets and the reactions that occur are:

1. $^3He(n,p)^3H$
2. $^6Li(n,\alpha)^3H$
3. $^{10}B(n,\alpha)^7Li$

Reaction (1) is often utilized in a proportional counter filled with ^3He. The second reaction can be used in a scintillator made of glass which has a loading of up to 9% ^6Li. For reaction (3), the gas BF_3 can be used in a proportional counter.

Each of these reactions is particularly effective in detecting neutrons below 1 MeV and especially thermal neutrons ($E \sim 0.025$ eV). Few charged particles are produced by neutrons at these energies from the counter housing or glass constituents, so the spectra are characterized by minimal interference from other materials. A common problem is that the low-energy neutrons that one seeks to measure are accompanied by a flux of thermal neutrons, which one does not wish to measure. In this case, the counter efficiency will be so high for the thermal neutrons that they may swamp the events one wishes to detect. An obvious solution to this problem is to use a thin layer of cadmium or boron to shield the detector, since these materials have very high thermal absorption cross-sections. Each of these reactions is much less effective at energies above 10 MeV, since backgrounds due to charged particles in the housing or in the glass can be comparably large. Although counters utilizing these reactions can be approximately 100% efficient at thermal energies, they are often less than 5% efficient at energies above 3 MeV.

There is an interest in having a counter that can detect neutrons with virtually constant efficiency independent of energy. A counter that approximately meets this characteristic is the long counter. This counter consists of a BF_3-filled proportional counter embedded in a polyethylene moderating cylinder. Those neutrons directly incident on the BF_3 chamber with very low energies will be detected with approximately 100% efficiency, while neutrons with $E_n > 1$ MeV will be detected with much smaller efficiency (<10%). By surrounding the BF_3 counter with polyethylene, the higher energy neutrons can be slowed down and the net efficiency increased, although some may be captured in the polyethylene. Very low energy neutrons will be more likely to be captured in the polyethylene, so the detection efficiency for these neutron energies is reduced. Study of the detailed energy dependence of the efficiency of such counters has resulted in a design that has certain cavities cut in the shielding on the side from which the neutrons enter. This allows some of the lower energy neutrons a shorter path to the counter. The fact that neutrons are detected after moderation means that the counter cannot give precise timing information, nor can it be used to determine neutron energy spectra. It also cannot determine the neutron flux over a small cross-sectional area but gives an average over 40 to 100 cm^2. It also will have a slightly different efficiency for neutrons coming from different directions. A sample design is shown in Figure 68.1. Note that since the counting element needs only to have an efficiency that rises sharply at low energy, a long counter can be designed using a ^3He ion chamber instead of a BF_3 tube.

Other (n,z) reactions do not have positive Q values (are not exoergic). These have thresholds and, in fact, usually have small cross-sections for a few MeV beyond the threshold. An example is $^{27}Al(n,\alpha)^{24}Na$. This reaction cannot occur for neutrons of less than 3 MeV and has a very small cross-section below 4.5 MeV. It does lead to a radioactive final nucleus and can be used as a passive detector. Unlike the detectors based on (n,γ) reaction, it will not detect low energy neutrons.

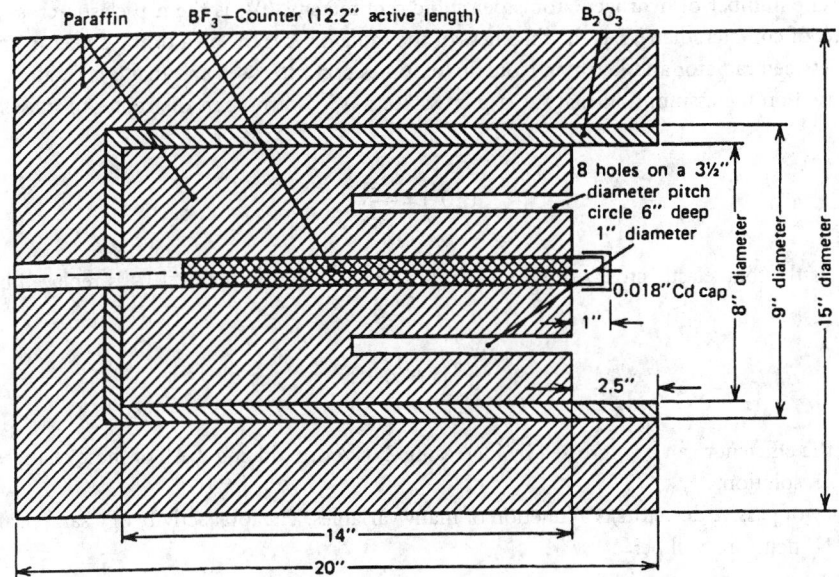

FIGURE 68.1 Cross-section of a typical long counter (described in M.H. McTaggart, AWRE NR/A1/59 (1958)). Figure is taken from K.H. Beckurts and K. Wirtz, *Neutron Physics*, Springer Verlag (1964). (Used with permission.) Neutrons incident from the right are scattered by the paraffin, causing them to lose energy. The BF$_3$ counter is efficient primarily for neutrons below 1 keV. The holes and the B$_2$O sections balance the efficiency as a function of energy.

An alternative reaction that can be used in a passive detector is the (n,2n) reaction. This typically has a threshold of about 8 MeV but has a large cross-section once the threshold is exceeded. Note that just as is found for (n,γ) or (n,z) reactions, not all nuclei reached in (n,2n) reactions are unstable.

68.2 Efficiency Calculations

Detectors based on np scattering have an efficiency that depends on the n-p elastic cross-section. For a proportional counter or scintillator, an approximate expression for the efficiency is

$$\epsilon = \left(1 - e^{-n\sigma_H L}\right)\left(1 - \frac{B}{E_n}\right) \tag{68.1}$$

This expression involves the parameters n, the number of hydrogen atoms per cm^3; σ_H, the n-p elastic cross-section; B, the electronic cutoff (in energy equivalent units); and E_n, the neutron energy. This expression ignores multiple scattering, so it would not include pulses produced by two scattering events in succession. It also does not include carbon interactions. For energies below 4.4 MeV, carbon interactions are almost entirely elastic and do not result in absorption or pulses large enough to detect. Particularly at energies above 10 MeV, carbon interactions not only remove neutrons from the beam but also, through (n,α) or (n,3α) events, can produce pulses that can be detected. Calculating the efficiency in this energy range can be done with Monte Carlo codes, but the efficiency will depend on knowledge of the ^{12}C (n,α) and ^{12}C(n,3α) cross-sections as well as the relation between alpha pulse heights and proton pulse heights as a function of energy.

Recoil telescopes have an efficiency which has the form:

$$\epsilon = \frac{n\sigma\left(0°\right)L\,A}{d^2} \tag{68.2}$$

Here, n is the number of hydrogen atoms per cubic centimeter, $\sigma(0°)$ is the n-p elastic-cross section at $0°$ (in units of cm^2/steradian), L is the thickness of the radiator, A is the area of the detector, and d is the distance between radiator and detector. Note this assumes that $n\sigma L \ll 1$ and that θ is small. A corrected version based on the assumption that the n-p cross-section is isotropic in the center of mass has:

$$\sigma\!\left(0°\right) = \frac{\sigma_{np}}{\pi} \tag{68.3}$$

where σ_{np} is the total elastic cross section and, if θ is not sufficiently small, $\sigma(0°)$ is replaced by:

$$\frac{\sigma_{np}}{2\pi}\!\left(1+\cos\theta\right) \tag{68.4}$$

Note that the efficiency can be increased by increasing L or A or reducing d. All of these changes degrade the energy resolution.

Efficiency of passive detectors is a function of many variables. The total activity of a sample after being struck by N_n neutrons will be:

$$n\sigma L\, N_n \frac{\tau}{T}\!\left[1-e^{-\frac{T}{\tau}}\right] \tag{68.5}$$

where n is the number of atoms per cm^3, σ is the appropriate cross-section (e.g., for n, γ), L is the thickness of the foil, T is the time over which the irradiation took place, and τ is the average lifetime of the radioactive nuclei produced. This expression assumes that $n\sigma L \ll 1$ and that the source strength was uniform over the time T.

If the counter has an area A, is at a distance d from the activated foil, has an efficiency ϵ_r, and measures the activity from a time t_1 after the bombardment to a time t_2, the efficiency ϵ_n will be

$$\epsilon_n = \left(n\sigma L\right)\frac{\tau}{T}\!\left(1-e^{-\frac{T}{\tau}}\right)\!\left(\frac{A\,\epsilon_r}{4\pi d^2}\right)\!\left(e^{-\frac{t_1}{\tau}} - e^{-\frac{t_2}{\tau}}\right) \tag{68.6}$$

Each of the factors in parenthesis is less than one, so the typical efficiency of such a monitor is 10^{-4} to 10^{-6}.
Fission chambers have an efficiency of:

$$\epsilon_n = n\sigma_f L \tag{68.7}$$

where n is the number of atoms per cm^3 in the foil, σ_f is the fission cross-section, and L is the thickness of the foil. At low energies, σ_f may be large enough so that the above expression is greater than 0.1. In that case,

$$\epsilon_n = \left(1-e^{-n\sigma L}\right)\frac{\sigma_f}{\sigma} \tag{68.8}$$

where σ is the total cross-section for neutrons.
The efficiency for a lithium- or boron-containing scintillator is:

$$\epsilon_n = \left(1 - e^{-\left(n_1\sigma_1 + n_2\sigma_2\right)L}\right)\left(\frac{n_1\sigma_1}{n_1\sigma_1 + n_2\sigma_2}\right) \tag{68.9}$$

n_1 is the density of lithium or boron atoms per cm^3, σ_1 is the (n,α) cross-section, n_2 is the density of atoms other than lithium or boron in the scintillator, and σ_2 is the average absorption cross-section for these atoms. If $(n_1\sigma_1 + n_2\sigma_2)\, L \ll 1$, this becomes:

$$\epsilon_n = n_1\sigma_1 L \tag{68.10}$$

Note that in each case where a term of the form $(1 - e^{-n\sigma L})$ appears, the average of this quantity is not $(1 - e^{-n<\sigma>L})$ if σ has energy fluctuations over an energy bin that are substantial. Here, $<\sigma>$ is the average cross-section over the energy bin. This limit is particularly important at low energies, where isolated resonances cause large fluctuations in σ and where the limit of small $n\sigma L$ may not be appropriate. In this case, the expression for the efficiency should be evaluated separately for individual energies in the bin and the resultant efficiencies averaged.

68.3 Summary

A summary of the features of various counters is presented in Table 68.1. More exhaustive treatment of the subject of neutron detectors can be found in References 1 and 2. Table 68.2 lists some suppliers of neutron detectors.

TABLE 68.1 Characteristics of Various Neutron Detectors

Detector Type	Gives Energy Spectrum	Small Size	Directional Information	Insensitive to Gamma Rays	Detects Neutrons Below 500 keV	Timing Information	High Efficiency
Hydrogen-containing counters							
Proportional counter	No	Yes	No	Somewhat	No	No	Yes
Scintillator	No[a]	Yes	No	Liquid Scintillator Can Reject Gamma Rays	No	Yes	Yes
Proton recoil telescope	Yes	Yes	Yes	Yes	No	Yes	No
Passive (n,γ) detector	No	Yes (smaller than others)	No	Yes	Yes	No	No
Fission detector	No[a]	Yes	No	Yes	Yes	Yes	No
Passive inelastic-scattering detector	No	Yes	No	No	No	No	No
Detectors based on (n,z) reactions							
Passive detector	No	Yes	No	Yes	Yes	No	No
Scintillator (lithium glass)	No[a]	Yes	No	No	Yes	Yes	Yes
Long counter	No	No	No	Yes	Yes	No	Yes

[a] Can give energy information when used in time-of-flight spectrometer.
[b] Timing information is considered to be available if the time is determined to within less than 10^{-6} s.
[c] High efficiency means $\geq 1\%$.

TABLE 68.2 Commercial Suppliers of Neutron Detectors

Supplier	Product
BICRON 12345 Kinsman Road Newbury, OH 44065 Tel: (216) 564-2251 Fax: (216) 564-8847	Plastic scintillators Liquid scintillators with pulse shape discrimination Assemblies, phototubes
EG&G ORTEC 100 Midland Road Oak Ridge, TN 37831-0895 Tel: (423) 482-4411 Fax: (423) 483-0396	He-3 detector and electronics Li-6 detector and electronics
NE Technology, Inc. Princeton Corporate Plaza 1 Dee Park Drive, Suite L Mamouth Junction, NJ 08852 Tel: (908) 329-1177 Fax: (908) 329-2221 Email: netec@delphi.com	Plastic scintillators Liquid scintillators with pulse shape discrimination Assemblies, phototubes
Reuter Stokes 8499 Darrow Road Twinsburg, OH 44087 Tel: (216) 425-3755 Fax: (216) 425-4045	He-3 proportional counters Fission chambers
REXON 24500 Highland Point Beachwood, OH 44122 Tel: (216) 292-7373 Fax: (216) 292-7714	Plastic scintillators Liquid scintillators with pulse shape discrimination Assemblies, phototubes
N. Wood Counter Laboratory, Inc. P.O. Box 509 Chesterton, IN 46304 Tel: (219) 926-3571 Fax: (219) 926-3571	Boron triflouride gas tubes

Defining Terms

Ionization chamber: A detector that collects the charge produced by ionization as a charged particle passes through the chamber.

Proportional counter: A detector that collects the charge produced by ionization with linear amplification as a charged particle passes through the chamber.

Scintillator: A detector that converts ionization energy to a short light pulse.

Compton effect: A process by which a gamma ray scatters from an electron on an atom. As a result, the electron is given a recoil velocity, producing ionization and yielding a pulse in an ionization chamber or proportional counter.

Isomeric transaction: A gamma ray decay of a nucleus in which the rate of decay is greatly slowed by restriction imposed by angular momentum coupling.

Time-of-flight spectrometer: A detection device for neutrons in which a neutron detector is placed a known distance from the neutron source and the energy of the neutron is deduced from the time the neutron takes to traverse this path.

Q value: A parameter expressing whether a nuclear reaction leads to a final state of higher final mass $(Q < 0)$ or of lower final mass $(Q > 0)$ than the initial state. Elastic scattering reactions have $Q = 0$.

Efficiency: The efficiency of a neutron detector is its probability of detecting a neutron incident on the detector. Efficiency is obviously a number between 0 and 1 and generally depends on neutron energy.

References

1. J.B. Marion and J.L. Fowler (eds.), *Fast Neutron Physics, Part I*, Interscience Publishers (New York, 1960).
2. G.F. Knoll, *Radiation Detection and Measurement*, John Wiley & Sons (New York, 1979).

a Q value. A parameter representing whether a nuclear reaction leads to a bound state of higher final mass ($Q < 0$) or allows that mass ($Q > 0$) than the initial state that the scattering reactions have $x = 0$.

Fission σ. The efficiency neutron kernel is its probability of detecting a neutron incident on the detector. Efficiency is obviously a number between 0 and 1 and generally depends on neutron energy.

References

1. F.B. Martin and L.J. Forcier, *Nuclear Reactor Power*, A.V. Interscience Publishers, New York, 1985.

2. C.P.T. et al., *Theory of Neutron and Matter Interaction*, John Wiley & Sons, New York, 1975.

69

Dosimetry Measurement

Brian L. Justus
Naval Research Laboratory

Mark A. Miller
Naval Research Laboratory

Alan L. Huston
Naval Research Laboratory

69.1 Radiation Dosimetry Quantities and Units 69-1
69.2 Thermoluminescence Dosimetry 69-3
69.3 Ionization Chamber Dosimeters 69-7
69.4 Film Dosimetry 69-7
69.5 Track-Etch Dosimetry 69-9
Alpha Track Dosimetry • Fast Neutron Track Dosimetry
69.6 Bubble Dosimetry 69-12
69.7 Electronic Personal Dosimeters 69-13

Radiation dosimetry is a field of radiation detection devoted to the quantitative measurement of the physical changes that occur in matter upon exposure to ionizing radiation. Radiation dosimetry is performed on a routine basis to ensure the occupational safety of workers when there exists a risk of radiation exposure. Such routine personal dosimetry is used to monitor and limit the long-term occupational exposure to workers and to assist with the assessment of the dose received in the event of an accidental exposure. Some occupations that have an associated risk of radiation exposure include medical personnel performing clinical X-ray diagnostic and radiotherapy procedures, and military and civilian personnel involved in power plant maintenance and operation. Radiation dosimetry is also an important tool in a wide range of environmental monitoring and industrial processing applications. The scope of this chapter is limited to discussion of the most widely used and commercially available technologies for dosimetry of ionizing radiation for personal protection; however, it should be recognized that many of the same technologies are used with equal effectiveness for other dosimetry applications.

69.1 Radiation Dosimetry Quantities and Units

The goal of radiation dosimetry is to quantify the amount of energy that is deposited in matter upon interaction with ionizing radiation. Ionizing radiations fall into three categories: charged particles such as beta and alpha particles, neutral particles such as neutrons, and electromagnetic radiation such as gamma rays and X-rays. The mechanism and the efficiency of the energy deposition depends on the type and the energy of the radiation as well as on the composition of the absorbing material. In particular, the biological consequences to living tissue following radiation exposure are quite dependent on the type of radiation. In order to provide a useful measure of the biological damage that might be expected to occur upon energy deposition in tissue, a system of units and standards has been developed that takes into account the differing biological effectiveness of different types of radiation. Over the years, these units and standards have undergone extensive revision and this process is still evolving. However, the units and standards in current use have been defined in detail in recent reports [1,2] published by the

0-8493-8347-1/99/$0.00+$.50
© 1999 by CRC Press LLC

TABLE 69.1 Radiation Weighting Factors for Various Types of Radiation

Type of Radiation	Quality Factor
Photons, all energies	1
Electrons and muons, all energies	1
Neutrons, <10 keV	5
Neutrons, 10 keV to 100 keV	10
Neutrons, >100 keV to 2 MeV	20
Neutrons, >2 MeV to 20 MeV	10
Neutrons, >20 MeV	5
Protons, other than recoil protons, >2 MeV	5
Alpha particles, fission fragments and heavy nuclei	20

From Reference 1.

International Commission on Radiation Units and Measurements (ICRU) and the International Commission on Radiological Protection (ICRP) and are the subject of a recent review article [3]. A brief summary of the fundamental units is provided below.

The quantity of primary interest in radiation dosimetry is the amount of energy that is absorbed per unit mass. The amount of energy imparted to a volume of matter is the difference between the sum of the energies of all the charged and uncharged particles entering the volume and the sum of the energies of all of the charged and uncharged particles leaving the volume, plus any change in the rest mass of the matter. The mean energy imparted per unit mass is defined as the absorbed dose, D, expressed in (SI) units as J kg^{-1}. The special name for the unit of absorbed dose is the gray (Gy). If the mass exposed to radiation is at a point in tissue, knowledge of the amount of absorbed dose alone is not sufficient to judge the biological effectiveness of the charged particles producing the absorbed dose. A quality factor, Q, is used to weight the absorbed dose to provide an estimate of the relative hazard of different energies and types of ionizing radiation. Q is a dimensionless factor that converts absorbed dose to dose equivalent, H, at a point in tissue. The dose equivalent has the same (SI) units, J kg^{-1}, as the absorbed dose, but the special name for the unit of dose equivalent is the sievert (Sv). If the amount of absorbed dose received on average over a specific tissue or organ is of interest, the quality factor is called the radiation weighting factor, w_R [1]. Table 69.1 lists values of w_R for various types and energies of radiation. The equivalent dose, H_T, in a tissue or organ is defined as the average absorbed dose in the given organ due to a particular type of radiation and is scaled by w_R. Thus, the total equivalent dose in a specific organ, T, due to exposure by each radiation type, R, is given by:

$$H_T = \sum_R w_R D_{T,R}$$

(69.1)

The equivalent dose also uses the special unit of sievert. For example, if an organ is exposed to neutrons of energy less than 10 keV, and receives an absorbed dose of 10 Gy (10 J kg^{-1}), the equivalent dose is determined, using Table 69.1 and Equation 69.1, to be 50 Sv (50 J kg^{-1}). Since the biological effectiveness of the exposure also depends on the specific organs targeted, dimensionless tissue weighting factors, w_T, have been defined that reflect the probability of damage resulting from a given equivalent dose. The effective dose, E, for an exposed individual is then the sum of the weighted equivalent doses for all tissues and is given by the expression:

$$E = \sum_T w_T H_T$$

(69.2)

The effective dose also uses the special unit of sievert. It should be pointed out that despite the adoption of nomenclature and definitions for equivalent dose and effective dose, for all practical purposes the quantities measured and reported for dosimetry applications are the dose equivalents defined in terms of the quality factor, Q. A further quantity for reporting purposes is the personal dose equivalent, $H_p(d)$, the dose equivalent in soft tissue, at an appropriate depth, d, below a specified point in the body. For weakly penetrating radiation, such as beta particles, a depth of 0.07 mm, or $H_p(0.07)$ is used. Eye doses are reported for 3 mm depth ($H_p(3)$) and deep doses are reported for 10 mm depth ($H_p(10)$). The personal dose equivalent is typically measured with a dosimeter that is covered with the appropriate thickness of tissue equivalent material. The dosimeter is worn on the surface of the body and is calibrated in a phantom. A phantom is a medium, such as poly(methyl methacrylate), that mimics the attenuating characteristics of the human body.

Another unit for absorbed dose, found in older publications and in current use in the United States, is the rad. One rad equals 100 erg g^{-1}. Similarly, another unit for dose equivalent is the rem. To convert from this system to the international standard (SI) units, the following conversions are used: 1 Gy = 100 rad and 1 Sv = 100 rem.

When an ionization chamber is exposed to X-ray or gamma ray photons, the amount of ionization produced in the air inside the chamber is called the exposure, X. The (SI) unit of exposure is C kg^{-1} and the special unit is called the roentgen, R. Originally, R was defined as the amount of X-ray or gamma radiation that produces 1 esu of charge of either sign in 0.001293 g of air (the mass of 1 cm^3 of air at STP). The conversion from R to C kg^{-1} is 1 R = 2.58×10^{-4} C kg^{-1}. When air is exposed to radiation, the average energy required to form an ion pair is 33.97 J C^{-1}. For an exposure of 1 R, the absorbed dose in air is 8.77×10^{-3} J kg^{-1}.

69.2 Thermoluminescence Dosimetry

Thermoluminescence dosimetry is perhaps the most widely used and cost-effective technique for radiation dosimetry. It is for many organizations the technique of choice for routine monitoring of occupational radiation exposure. Thermoluminescence dosimetry is also widely used in medicine to determine patient exposure as a result of X-ray diagnostic procedures and cancer radiotherapy treatments. The dose ranges of interest for these applications can be roughly defined as 0.1 to 1 mGy for personal dosimetry applications, 1 to 100 mGy for clinical X-ray diagnostics, and 1 to 10 Gy for medical radiotherapy applications. Thermoluminescence dosimeter (TLD) phosphors are commercially available that exhibit a linear dose response and are capable of accurately measuring the absorbed dose for all the applications and doses mentioned above. The radiation-sensitive element of a TLD is a small quantity, typically less than 100 mg, of an inorganic crystal doped with metal impurities, called activators. The activators provide the crystal with the energy storage capacity as well as the luminescent properties that are required for the crystal to function as a thermoluminescent phosphor upon exposure to ionizing radiation. The details of the energy storage and thermoluminescence mechanisms, despite intense research over many decades, are not well understood. Complex, long-range, many-body interactions between activator sites and their immediate environment may be involved. Despite this mechanistic uncertainty, the thermoluminescence phenomenon has been used reliably by following well-established heuristic procedures and by referring to a simple and intuitive model. This model assumes that the activators provide point defects, known as traps and luminescence centers, in the crystal lattice. Upon exposure to ionizing radiation, electrons and holes are captured in metastable states at the trap centers by local potential minima until such time that the electrons and holes are thermally stimulated to overcome the electric potential. The electrons and holes can then recombine, with the emission of photons, at the luminescence centers. It is apparent from this model that thermoluminescence dosimetry is a passive, integrating technique. Further, the technique shows no dose rate dependence and for most applications the TLD phosphors can be reused many times.

A comprehensive review of the properties and applications of a wide range of TLD phosphors has been published recently [4]. The characteristics of the phosphors that are of most interest for dosimetry applications include the sensitivity and the dynamic range of the TL response, the response to different types of radiation over a broad energy spectrum, the fading characteristics, reusability, and the reproducibility of the response. The TL response of the phosphor to ionizing radiation is indicated by measurement

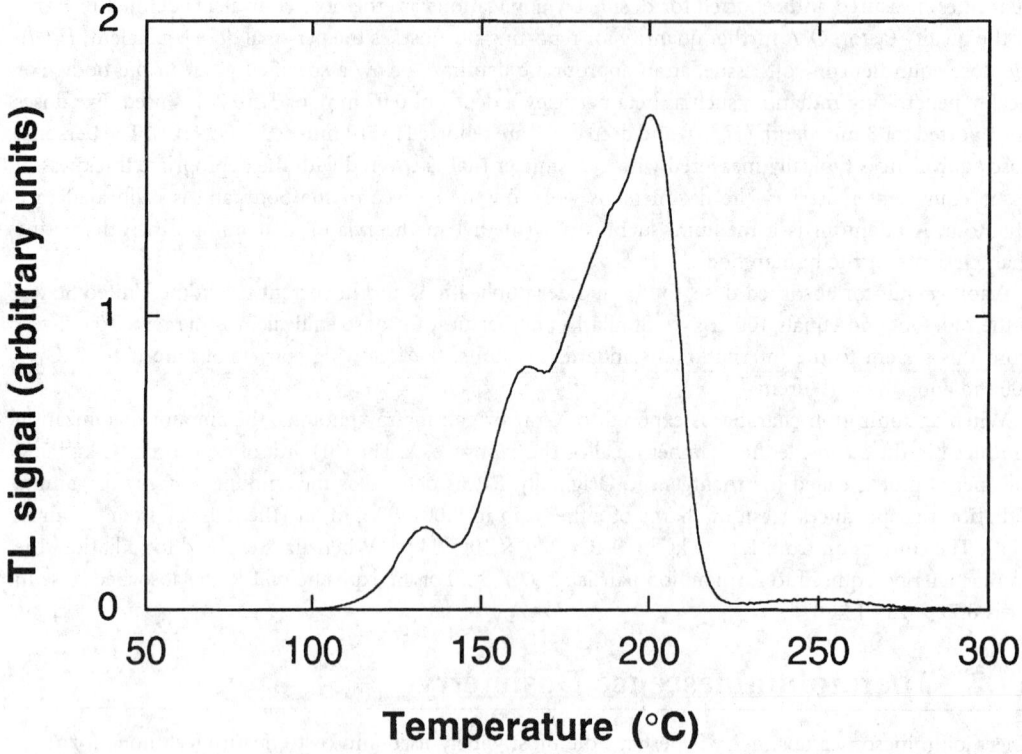

FIGURE 69.1   Thermoluminescence glow curve of LiF phosphor activated with trace amounts of Mg^{2+} and Ti^{4+} (TLD-100). The phosphor was exposed to a 2-Gy dose of ^{60}Co gamma rays and the glow curve was read using a Harshaw 2000 A/B TLD reader.

of the glow curve. A glow curve is a plot of the TL intensity vs. temperature, measured as the phosphor is heated. A representative glow curve of a commercially available LiF TLD is shown in Figure 69.1. The glow curve usually exhibits one or more glow peaks that correspond to the release and recombination of electron and hole traps of differing trap depth and stability. Thermoluminescence dosimetry is best accomplished using phosphors having glow peaks in the range 200°C to 250°C. Peaks in this temperature range are necessary to provide room-temperature trap storage times of several months. Although all traps fade with time, rapid fading of the stored information occurs from traps having glow peak temperatures below 200°C, effectively precluding practical dosimetry. Glow peaks having temperatures greater than 250°C require quite high temperatures (>300°C) to release all the filled traps. In this case, blackbody radiation from the hot phosphor and other ancillary hot matter within the sample readout chamber (sample planchet, hot gas) interferes with and can easily overwhelm the TL signal due to actual trap recombination. The response of TLD phosphors as a function of absorbed dose must be carefully calibrated for all doses of interest. This is necessary because the response deviates from linear above a threshold dose, different for each phosphor, and, ultimately, the TL signal saturates and the phosphors suffer irreversible damage at very high doses. The TL response per absorbed dose as a function of energy, for low-LET electromagnetic radiation (gamma ray and X-ray), is generally fairly uniform. The efficiency for higher LET radiation can vary and should be calibrated as a function of energy. Since personal dosimetry applications require the measurement of doses absorbed by human tissue, tissue equivalent TLDs (effective atomic number closely matches that of tissue) are attractive because the energy dependence of the TL response closely matches that of tissue.

The most commonly used TLD phosphors are lithium fluoride (LiF), calcium fluoride (CaF_2), lithium borate ($Li_2B_4O_7$), calcium sulfate ($CaSO_4$), and aluminum oxide (Al_2O_3) activated with trace quantities

FIGURE 69.2 Thermoluminescence dosimeter body badge. A standard TLD badge contains three LiF phosphor chips. A fourth chip and a fast neutron track-etch foil can be added if required. (Courtesy of Landauer, Inc.)

of transition metal or rare earth metal ions. TLD phosphors are available in a variety of forms, including powders, compressed chips, Teflon-impregnated disks, single crystals, extruded rods and thin films. TLD phosphors have been developed that respond primarily to gamma- and X-radiation, while others have been developed that, in addition, respond strongly to thermal neutrons. Unfortunately, there are no TLD phosphors having neutron sensitivity without gamma sensitivity. Therefore, in a mixed field of neutrons and gamma rays, TLDs are used in pairs, with one TLD having primarily gamma-ray sensitivity and the other having gamma-ray and thermal neutron sensitivity. The difference in the TL signal between the two TLDs of such a pair is assumed to represent the thermal neutron dose. Neutron sensitivity is enhanced, for example, in LiF phosphors by incorporating a high concentration of ^{6}Li since ^{6}Li has a much higher thermal neutron cross-section than ^{7}Li (TLD-600 contains 96% ^{6}LiF while TLD-700 has essentially no ^{6}Li). The manufacturers of TLDs exert considerable effort to maintain the uniformity of their products from batch to batch, a task of some difficulty since the characteristics of phosphors are a sensitive function of not only the composition but also the details of the manufacturing process. Thus, the properties of TLD phosphors of similar composition can vary considerably, and both the manufacturers of phosphors as well as the dosimetry service industries typically test and sort chips into groups based on their measured parameters. A representative commercially available TLD badge is shown in Figure 69.2. This standard badge contains three LiF TLDs and is capable of measuring gamma- and X-ray doses from 0.1 mGy to 10 Gy and energetic beta radiation doses from 0.4 mGy to 10 Gy. These badges are typically returned to the manufacturer on a monthly or quarterly basis for readout.

The actual practice of thermoluminescence dosimetry requires some care to achieve accurate and reproducible results. For example, TLD-100 traditionally has been a very widely used phosphor, due to its tissue equivalence, that can yield accurate and fairly sensitive dose information. However, the energetics of the trap population in TLD-100, as well as the trap dynamics and intertrap communication with time, are extremely complex phenomena. Reproducible dosimetry requires that the sensitivity remain unchanged after each use. This is accomplished by careful adherence to a detailed protocol that may include a pre-irradiation anneal (for example, 400°C for 1 h followed by 80°C for 10 to 24 h), a post-irradiation anneal to 125°C, and finally heating to readout the TL signal. It is common for individual TLD users to develop their own unique protocols for annealing and reading TLD chips. Excellent dose reproducibility (<5%) can be expected with care by following a well-defined protocol and using a set of individually calibrated TLD chips on a single, well-maintained TLD reader.

TLD instrumentation is commercially available from a number of suppliers. The choice of instrumentation depends largely on the form of the dosimeter phosphor chosen (compressed chip, Teflon disk, etc.), the desired sensitivity, and the degree of automation desired. The key element in any TLD reader is the heating method. The heating method should provide for controlled, reproducible heating and, if speed of processing is important, rapid cooling. Contact heating of the TLD on a planchet through which an electric current flows is a widely used heating technique. Because the surface of the TLD is not perfectly smooth, more uniform heating can be performed using noncontact methods, such as heating with a stream of hot gas or heating with infrared light. Laser heating is another noncontact method that permits rapid heating ($>1000°C\,s^{-1}$) using small quantities of TLD phosphor, yet permits very high signal-to-noise ratios [5]. Regardless of the heating method, blackbody radiation from the metal planchet and/or the TLD chip must be attenuated by colored glass and dielectric coated optical filters. The signal-to-noise ratio is also improved by the elimination of spurious TL signals that arise due to static charges or contamination on the surface of the TLD. Spurious TL signals are reduced by purging the sample compartment with a low flow of inert gas, such as dry nitrogen. Light detection is typically accomplished with a photomultiplier tube, using either a DC or photon counting mode of operation. The glow curve is collected using any of the heating methods described, and the TL signal can be reported as either a glow peak height or as the integrated area of a glow peak.

The conventional TLD technology described above requires bulk heating of the TLD to high temperatures (400°C to 600°C) in order to read the stored dose information. Once the TLD has been read, the dose information is permanently erased. If any malfunction in the equipment occurs during readout, the data cannot be recovered. Alternative TLD technologies have been developed that do not completely erase the dose information and permit additional, successive readouts of the absorbed dose. This characteristic can eliminate accidental loss of dose information and can also provide, if desired, an archival record of exposure. Two such dosimeters, utilizing laser readout methods, are commercially available. These dosimeters function based on optical phenomena known as phototransferred thermoluminescence (PTTL) [6] and radiophotoluminescence (RPL) [7]. A commercial dosimeter (Landauer, Inc.) based on the phenomenon of PTTL is also known as a cooled optically stimulated luminescence (COSL) dosimeter. In preparation for readout of the dose information, the dosimeter is first cooled to liquid nitrogen temperature. It is then irradiated with ultraviolet laser light to phototransfer electrons from deep traps to shallow traps. The thermoluminescence that results from the release of electrons from these very shallow traps, as the dosimeter is allowed to warm to room temperature, provides a very sensitive measure of the absorbed dose. Since it is not necessary to heat the dosimeter to high temperatures, the trapped charges are not fully annealed and the process can be repeated several times. A commercial RPL dosimeter, also known as the flat glass dosimeter (Toshiba Glass Co., Ltd.) uses a silver-activated phosphate glass that is read out by a pulsed ultraviolet laser. The manufacturer employs a special luminescence analysis protocol since the glass exhibits a prompt luminescence even in the absence of ionizing radiation exposure. RPL dosimeters are nevertheless capable of performing sensitive and reproducible dose measurements. New optical dosimetry technologies, based on the phenomenon of room-temperature optically stimulated luminescence (OSL), are being developed [8,9] and promise to provide both superior sensitivity and multiple readout capability.

Companies that provide TLD and other personal dosimetry products and services are listed in Table 69.2. This list is not intended to be comprehensive, nor is it intended to represent an endorsement by the U.S. government.

69.3 Ionization Chamber Dosimeters

Ionization chamber-based radiation dosimeters are among the most widely used and the most accurate instruments available for determining radiation exposure and absorbed dose. The operational principle of the ionization chamber is based on the formation and the collection of ion pairs that result from the interaction of energetic charged particles that pass through gases contained in a well-defined volume and electric field. Several types of ionization chambers are available, including real-time radiation field monitors, and integrating devices that accumulate dose information for extended periods of time. Real-time devices monitor the radiation-induced currents while integrating devices record changes in a static electric field. A wide array of ionization chambers, intended for applications in medical physics, equipment calibration, and personal dosimetry, are available. Small volume (<1 mL), precision ionization chambers are used by medical physicists for radiotherapy applications. These chambers are generally tissue or air equivalent and while very accurate, have low sensitivity. Larger volume (<1000 mL) ionization chambers are used for applications such as field standardization and diagnostic X-ray machine calibration that require high sensitivity in addition to excellent accuracy. For personal dosimetry applications, compact integrating ionization chambers, about the size of a ballpoint pen, termed "pocket chambers" are used. Some pocket chamber dosimeters provide a method for direct visual readout based on the electrostatic deflection of a charged quartz fiber relative to a calibrated scale. Initially, the dosimeter is fully charged, representing the zero dose setting. Exposure to a radiation source causes a loss of charge on the fiber and spring tension causes the fiber to partially return to its uncharged position. A lens system incorporated into the dosimeter in conjunction with a scale can be used to determine the absorbed dose. A commercially available direct-reading dosimeter is illustrated schematically in Figure 69.3. These instruments can measure gamma and X-ray doses from a fraction of a milligray up to several gray. Greater accuracy is possible using indirect readout, pocket chambers that utilize precision electrometers to measure the change in the potential between the cathode and anode following exposure to radiation. These dosimeters are particularly useful for measuring low-level (<mGy) exposures. Another type of ionization chamber, used primarily for radon detection, utilizes the voltage drop across an electret element to determine radiation dose (Rad-Elec, Inc.). An electret is an electrically poled insulator material, such as Teflon, that retains surface charges for long periods of time. Ion pairs produced by radiation exposure are separated and collected, resulting in a net decrease in the electret charge. Changes in the electret voltage can then be calibrated in terms of radiation exposure.

69.4 Film Dosimetry

Photographic films have been used for many years for radiography and personal dosimetry applications. However, due to the greater convenience, superior sensitivity, faster turnaround, and reusability of TLD personal dosimeters, film emulsion personal dosimeters have in many cases been displaced by TLD dosimeters. A notable example of this trend is the dosimetry program of the United States Navy, which has turned exclusively to TLD dosimeters. Despite this, film remains useful for specific applications that require analysis of an image or a charged particle track and they remain the most convenient dosimetry method available if a permanent record of the exposure is desired. Radiographic films are sensitive to a variety of radiation sources, including photons (X-ray and gamma ray), charged particles (electrons, protons, and alpha particles), and neutrons (slow and fast). A major problem associated with the use of radiographic films for dosimetry purposes is that the energy response is not at all flat. A number of approaches have been used to flatten the response, including, for example, the incorporation of a

TABLE 69.2   Vendors of Personal Dosimetry Products and Services

Vendor	Product or Service
Bicron NE 6801 Cochran Road Solon, OH 44139 Tel: (216) 248-7400	TLD phosphors, TLD readers
Panasonic Industrial Comp. Applied Technologies Group 2 Panasonic Way/7E-4 Secaucus, NJ 07094 Tel: (201) 348-5339	TLD phosphors, optically heated TLD readers, electronic dosimeters
Landauer, Inc. 2 Science Road Glenwood, IL 60425-1568 Tel: (708) 755-7000	TLD, film, track-etch services, COSL dosimeter
ICN Dosimetry Service P.O. Box 19536 Irvine, CA 92713 Tel: (714) 545-0100	TLD, film services
Eberline Dosimetry Service 5635 Jefferson St, NE Albuquerque, NM 87109-3412 Tel: (505) 345-9931	TLD readers, TLD service
Dosimeter Corporation of America 5 Eastmans Road Parsippany, NJ 07054 Tel: (800) 322-8258	Pocket dosimeters, electronic dosimeters
Toshiba Glass Co., Ltd. 3583-5 Kawashira Yoshida-Cho Haibara-Gun Shizuoka-Ken Japan Tel: 0548-32-1217	RPL glass dosimeter
Siemens Environmental Systems Limited Sopers Lane, Poole, Dorset, BH17 7ER United Kingdom Tel: 44 1202 782779	Electronic dosimeter
SAIC Commercial Products 4161 Campus Point Court San Diego, CA 92121 Tel: (619) 458-3846	Electronic dosimeters
RADOS Technology, Inc. 6460 Dobbin Road Columbia, MD 21045 Tel: (410) 740-1440	TLD dosimeters, electronic dosimeters
SE International, Inc. 436 Farm Road Summertown, TN 38483-0039 Tel: (615) 964-3561	Pocket dosimeters

TABLE 69.2 (continued) Vendors of Personal Dosimetry Products and Services	
Vendor	Product or Service
Bubble Technology Industries Hwy. 17, Chalk River Ontario, Canada, K0J 1J0 Tel: (613) 589-2456	Bubble dosimeters
Apfel Enterprises 25 Science Park New Haven, CT 06511 Tel: (203) 786-5599	Bubble dosimeters

scintillator dye such as *p*-terphenyl in the emulsion [10]. This approach has the effect of flattening the response curve dramatically between 0.1 MeV and 1.0 MeV, as well as increasing the sensitivity of the film. More commonly, in a manner similar to that used in multiple chip TLD badges, a series of metal filters, applied to different portions of the film, is used to provide energy discrimination. The film is then analyzed according to the sensitivity of each film segment for a particular radiation energy range. Figure 69.4 is a photograph of a commercially available film dosimeter badge that incorporates six absorbing filters to discriminate beta-, gamma-, and X-radiation. The dose range that can be measured is 0.1 mGy to 5 Gy for gamma- and X-radiation and 0.4 mGy to 10 Gy for energetic beta particles. Radiographic films consist of a thin layer of gelatin, approximately 10 μm to 20 μm thick, containing silver halide microcrystals, coated onto a polymer or glass substrate. Exposure to a source of radiation sensitizes the microcrystallites, creating a latent image that darkens visibly following development of the film. The radiation dose is determined by measuring the extent of the darkening with a densitometer. The densitometer measures the optical density (OD) of the film by comparing the transmission, T, of light through the exposed film with the transmission, T_0, of an identical film that has not been exposed to radiation. The optical density is a logarithmic function, $OD = \log(T_0/T)$, and can be measured accurately for values up to approximately 3. Developed film typically has a background OD of about 0.1 due to scattered light. Undeveloped film is quite sensitive to environmental factors such as light, temperature and, in particular, moisture and must be carefully protected while in use. Film dosimeters are available that can also perform thermal and fast neutron dosimetry. The response of film to thermal neutrons is enhanced by the use of a film-converter such as a 0.5 to 1 mm thick Cd foil. The foil absorbs thermal neutrons, yielding gamma rays that interact with the emulsion. Nuclear track emulsions are films that are used for dosimetry of fast neutrons in a manner quite similar to the track-etch detectors described in the next section. Nuclear track emulsions are generally much thicker than radiographic film emulsions and contain higher concentrations of silver halide microcrystals. Recoil protons transfer energy to the microcrystallites in the emulsion, creating latent images of the tracks of the protons. The range of charged particles in the emulsion is generally quite small, compared to photons, so that practically all of the energy is deposited in the film along the tracks. Nuclear track emulsions are no longer widely used due largely to the extremely high threshold, ~0.7 MeV, required for track visualization. Fast neutron dosimetry with film badges is most often accomplished by using a track-etch detector in conjunction with the radiographic film emulsion.

69.5 Track-Etch Dosimetry

Solid-state track detectors exploit the damage that occurs in dielectric materials upon exposure to ionizing charged particles. Energy is deposited in the material along a track defined by the trajectory of the ionizing particle. The damage is manifested as pits that develop on the surface of the material upon etching by chemical and electrochemical techniques. The pits are clearly evident using a simple optical microscope and the track density can be estimated using either manual or automated pit counting methods. The

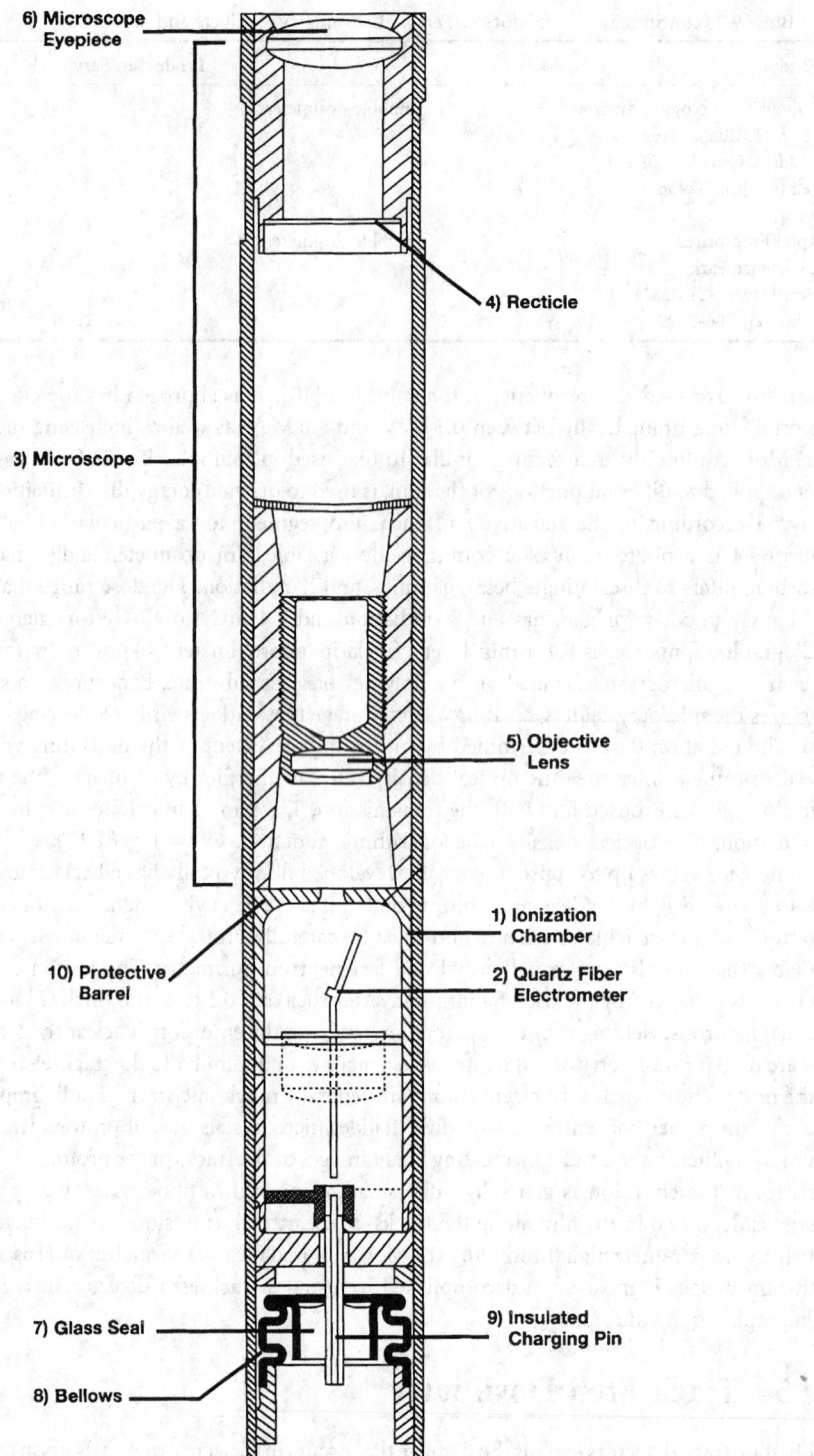

FIGURE 69.3 Schematic of direct reading pocket ionization chamber. The radiation dose is determined by measuring the deflection of a movable quartz fiber electrode. The deflection of the fiber is determined optically by projecting the image of the fiber onto a reticle using a microscope objective. (Courtesy of Dosimeter Corp. of America.)

FIGURE 69.4 Film badge for personal dosimetry. (Courtesy of Landauer, Inc.)

track density is a direct measure of the radiation dose (for many particles, there is one track formed per particle), and track-etch detectors are very effective and sensitive dosimeters. At present the most important track-etch dosimetry applications utilize organic polymer films for alpha track detection (radon monitoring) and fast neutron personal dosimetry. The principal attractive features of track-etch detectors are their extreme simplicity and low cost, good sensitivity, near tissue equivalence, small size, passive operation, and insensitivity to fast electrons and gamma rays. Track-etch detection is a favorite dosimetry technique of graduate students worldwide because of its simplicity; however, commercially available track-etch detectors are manufactured with extreme care to provide users with uniform polymer films and reproducible dose response.

Alpha Track Dosimetry

Commonly used solid-state track detectors that possess sensitivity suitable for alpha particle dosimetry include plastics such as cellulose nitrate, polycarbonate, and polyallyl diglycol carbonate [11]. Since track-etch detection is a surface phenomenon, care must be exercised to avoid scratching the surface of the polymer film, as any surface imperfections can result in false positive signals. The surfaces of the polymer film should also be protected from soiling since dust, dirt, grease, water, or any other foreign material can attenuate the alpha particles and prevent track registration in the detector. Polymer track-etch detectors do not fade significantly at room temperature; however, if the temperature is elevated such that the polymer softens, then the track damage can be annealed and etching will not yield a pit. Typically, radon track dosimeters are allowed to accumulate the alpha dose due to radon and its daughters for extended periods (up to a year) before being returned to the manufacturer for determination of the track density and the corresponding alpha particle activity. Dosimetry of other radioisotopes that emit alpha particles can also be performed; however, the response of the detector is dependent on the isotope detected. For example, alpha track detectors have been used for monitoring plutonium and americium contamination in soil and provide a convenient and cost-effective radiation survey method for decontamination and decommissioning activities [12].

Fast Neutron Track Dosimetry

The performance of polymer track-etch detectors used for personal dosimetry of fast neutrons is typically enhanced by the use of adjacent layers of a hydrogen-rich polymer, known as a proton radiator [13]. Fast neutron dosimetry is made possible by the neutron-proton elastic scattering process that occurs in the track-etch detector and the radiator. The recoil protons produced by this process are responsible for generating the damage tracks in the polymer film and thus provide an indirect measurement of the neutron flux. Although the polymers used for track-etch detection of neutrons, such as CR-39, contain hydrogen and provide for neutron-proton elastic scattering, the use of a proton radiator, such as polyethylene (having a much higher concentration of hydrogen), significantly improves the performance of the dosimeter [14]. The energy spectrum of the protons that leave the radiator and reach the detector depends on both the neutron energy and the thickness of the radiator. In addition, the total number of protons emitted depends on the neutron energy and the radiator thickness. When the thickness of the radiator equals the range of the most energetic proton emitted, then the condition of protonic equilibrium is met. Under this condition, the greatest number of protons are emitted and the detector has the greatest sensitivity. However, at protonic equilibrium, the sensitivity also displays the greatest variation in the energy response. A flatter energy response can be obtained using thinner radiators that avoid the condition of protonic equilibrium. This is obtained, however, at the expense of the sensitivity. Track-etch detection can be used for dosimetry of fast neutrons at nuclear power plants, particle accelerators and due to exposure from unmoderated isotope sources such as californium-252 and americium-241 beryllium.

69.6 Bubble Dosimetry

Another method for fast neutron personal dosimetry is bubble detection. Bubble dosimeters are manufactured using a clear, elastic polymer gel as a host matrix, supporting a dispersion of nanometer-scale superheated droplets of a liquid, such as Freon. Neutrons incident on the polymer gel generate recoil protons in an elastic scattering process, followed by deposition of the proton energy along a track, in a process similar to that described in the discussion of track-etch detectors. However, in the track-etch detectors, the tracks must be visualized by a separate etching procedure. The charged particles in a bubble dosimeter deposit their energy directly into the superheated droplets, causing the droplets to effectively explode into much larger gas bubbles. This process, which can be readily heard, is immediate, generating bubbles that can easily be seen by the naked eye, provided the bubble diameter exceeds a critical diameter required in order for the bubble to persist stably in the gel. The fast neutron dose is determined by counting the total number of bubbles generated. Bubble counting can be performed manually, by visual inspection of the dosimeter, or by automated means. Commercial bubble dosimeters are available that use either optical or acoustic automated counting techniques. Dosimeters have been developed using droplet materials that have differing energy thresholds, thereby providing the dosimeters with limited spectroscopic capabilities. Bubble dosimeters have been developed that are compensated to provide a temperature-independent response. Fast neutron bubble dosimeters have several advantages in common with track-etch detectors, including good sensitivity, near tissue equivalence, passive operation, small size, and insensitivity to thermal neutrons and gamma radiation. Additional advantages include immediate bubble visualization (no need for a separate track developing step), limited reusability, sensitivity adjustability, a flat dose response over a wide energy range, utilization of the entire volume of the dosimeter, and angular response superior to that of track-etch detectors. Bubble dosimeters are more

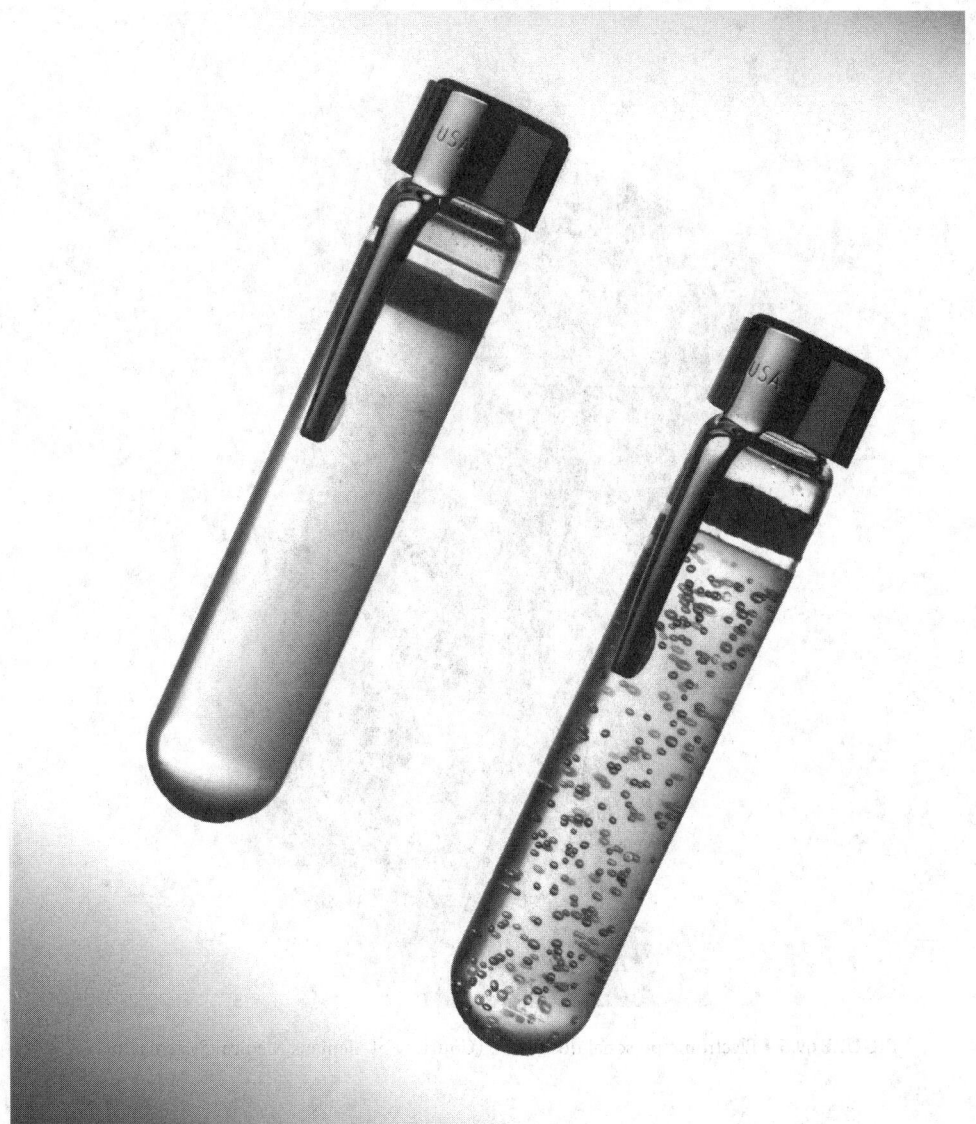

FIGURE 69.5 Bubble dosimeters for personal dosimetry of fast neutrons. The dosimeter on the left is shown before exposure to fast neutrons. For comparison, a dosimeter exposed to fast neutrons is shown on the right. Bubbles are clearly evident throughout the volume of the dosimeter. (Courtesy of Siemens Medical Systems, Inc.)

expensive than track-etch dosimeters and are best utilized when the energy of the neutron field is known. Figure 69.5 is a photograph of commercially available bubble dosimeters, before and after exposure to neutrons. A comprehensive review of bubble dosimeters, including new developments, has been published recently [15].

69.7 Electronic Personal Dosimeters

All of the personal dosimeters discussed to this point are passive devices that require no external power source. The radiation dose is determined upon completion of the sampling period by subjecting the dosimeter to a separate processing step. The nature of the processing depends on the dosimeter, but examples include heating of a TLD, chemical etching of a foil, development of an emulsion, or simply

FIGURE 69.6   Electronic personal dosimeter. (Courtesy of Siemens Medical Systems, Inc.)

counting bubbles. The electronic personal dosimeter (EPD), by contrast, is an active device that uses silicon diode detector technology to provide real-time measurements of radiation dose. The EPD has gained popularity in recent years and has mounted a serious challenge to thermoluminescence dosimetry as the preferred official dosimeter of record. EPDs can provide extremely sensitive real-time measurements of both the total dose and the dose rate. In addition, the EPD can be designed to provide an audible alarm if the total dose or dose rate exceed user defined settings. While the quality and characteristics of EPDs from different manufacturers can vary significantly, some EPDs offer accurate dose measurements over a wide range of doses, dose rates and energies, and perform on a par with TLD dosimeters. A commercially available EPD is shown in Figure 69.6. This unit uses three silicon diodes to provide deep and shallow dose information for a wide range of gamma, beta, and X-radiation energies. EPDs typically have a display for manual readout of dose data, but can be used with an automated reader for convenient archiving of data and for rezeroing the dosimeter for repeat use. Personal dosimeters are envisioned that would combine active and passive technologies, providing the accuracy and convenience of real time dose monitoring in addition to the reliability of passive, cumulative dose measurements.

Defining Terms

Radiation dosimetry: The quantitative measurement of the physical changes that occur in matter upon exposure to ionizing radiation.

Absorbed dose (D): The mean energy absorbed per unit mass.

Dose equivalent (H): The mean energy absorbed per unit mass, scaled with a quality factor, Q, that provides an estimate of the relative biological hazard of the radiation.

Equivalent dose (H_T): The mean absorbed dose in an organ, T, scaled by a radiation weighting factor, w_R.

Effective dose (E): The total mean absorbed dose for an exposed individual. Obtained by summing the equivalent doses for all tissues, each scaled by a tissue weighting factor, w_T.

Personal dose equivalent $H_p(d)$: The dose equivalent in soft tissue at a depth, d, below a specified point in the body. Commonly reported as shallow dose ($d = 3$ mm) and deep dose ($d = 10$ mm).

References

1. ICRP Publication 60, Ann. ICRP 21(1-3), 1990 *Recommendations of the International Commission on Radiological Protection*, Oxford: Pergamon Press; International Commission on Radiological Protection, 1991.
2. ICRU Report 51, *Quantities and Units in Radiation Protection Dosimetry*, Washington, D.C.: International Commission on Radiation Units and Measurements, 1993.
3. W. K. Sinclair, The Present System of Quantities and Units for Radiation Protection, *Health Physics*, 70, 781-786, 1996.
4. S. W. S. McKeever, M. Moscovitch, and P. D. Townsend, *Thermoluminescence Dosimetry Materials: Properties and Uses*. Ashford, England: Nuclear Technology Publ., 1995.
5. P. Braunlich, Present State and Future of TLD Laser Heating, *Radiation Protection Dosimetry*, 34, 345-351, 1990.
6. S. D. Miller and P. A. Eschbach, Optimized Readout System for Cooled Optically Stimulated Luminescence, *Radiation Effects and Defects in Solids*, 119-121, 15-20, 1991.
7. E. Piesch, B. Burgkhardt, and M. Vilgis, Progress in Phosphate Glass Dosimetry: Experiences and Routine Monitoring with a Modern Dosimetry System, *Radiation Protection Dosimetry*, 47, 409-414, 1993.
8. S. W. S. McKeever, M. S. Akselrod, and B. G. Markey, Pulsed Optically Stimulated Luminescence Dosimetry using α-Al$_2$O$_3$:C, *Radiation Protection Dosimetry*, 65, 267-272, 1996.
9. B. L. Justus, S. Rychnovsky, M. A. Miller, K. J. Pawlovich, and A. L. Huston, Optically Stimulated Luminescence Radiation Dosimetry using Doped Silica Glass, *Radiation Protection Dosimetry*, 74, 151-154, 1997.
10. R. A. Dudley, Dosimetry with Photographic Emulsions, in Frank H. Attix and William C. Roesch (eds.) *Radiation Dosimetry, Vol. II*. New York: Academic Press, 1966.
11. G. Espinosa, R. B. Gammage, and K. E. Meyer, Response of Different Polyallyl Di-Glycol Carbonate Materials to Alpha Particles, *Radiation Meas.*, 26, 173-177, 1996.
12. K. E. Meyer, R. B. Gammage, C. S. Dudney, S. Reed-Walker, P. Kotrappa, R. V. Wheeler, and M. Salasky, Field Measurements of Plutonium and Americium Contamination in Soils at the Nevada Test Site using Passive Alpha Detectors, *Radioactivity and Radiochemistry*, 5, 26-41, 1994.
13. J. L. Decossas, J. C. Vareille, J. P. Moliton, and J. L. Teyssier, Theoretical Study and Calculation of the Response of a Fast Neutron Dosemeter Based on Track Detection, *Radiation Protection Dosimetry*, 5, 163-170, 1984.
14. S. Sadaka, L. Makovicka, J. C. Vareille, J. L. Decossas, and J. L. Teyssier, Study of a Polyethylene and CR 39 Fast Neutron Dosemeter II: Dosimetric Efficiency of the Device, *Radiation Protection Dosimetry*, 16, 281-287, 1986.
15. H. Ing, R. A. Noulty, and T. D. McLean, Bubble Detectors — A Maturing Technology, *Radiation Measurements*, 27, 1-11, 1997.

Further Infomation

F. H. Attix, *Introduction to Radiological Physics and Radiation Dosimetry.* New York: John Wiley & Sons, 1986.

K. Becker, Dosimetric Applications of Track Etching, in Frank H. Attix (ed.) *Topics in Radiation Dosimetry, Supplement* 1. New York: Academic Press, 1972.

J. W. Boag, Ionization Chambers, in Frank H. Attix and William C. Roesch (eds.) *Radiation Dosimetry, Vol. II.* New York: Academic Press, 1966.

G. Shani, *Radiation Dosimetry Instrumentation and Methods.* Boca Raton, FL: CRC Press, 1991.

K. R. Kase, B. E. Bjarngard, and F. H. Attix (eds.) *The Dosimetry of Ionizing Radiation, Vol. III.* New York: Academic Press, 1990.

G. F. Knoll, *Radiation Detection and Measurement.* New York: John Wiley & Sons, 1989.

Chemical
Variables
Measurement

70
Composition Measurement

Michael J. Schöning
Institut für Schicht-und Ionentechnik
Forschungszentrum Julich

Olaf Glück
Institut für Schicht-und Ionentechnik
Forschungszentrum Julich

Marion Thust
Institut für Schicht-und Ionentechnik
Forschungszentrum Julich

Mushtaq Ali
The National Grid Company plc

Behrooz Pahlavanpour
The National Grid Company plc

Maria Eklund
Nynas Naphthenics AB

E.E. Uzgiris
General Electric Research and
Development Center

J.Y. Gui
General Electric Research and
Development Center

Mushtaq Ali
The National Grid Company plc

C.K. Laird
School of Applied Chemistry,
Kingston University

70.1 Electrochemical Composition Measurement

Michael J. Schöning, Olaf Glück, and Marion Thust

Electrochemical analysis in liquid solutions is concerned with the measurement of electrical quantities, such as potential, current, and charge, to gain information about the composition of the solution and the reaction kinetics of its components. The main techniques are based on the quantitative determination of reagents needed to complete a reaction or the reaction products themselves. Four traditional methods of electrochemistry are described here (Figure 70.1): potentiometry, voltammetry, coulometry, and conductometry. Potentiometry implies the measurement of an electrode potential in a system in which the electrode and the solution are in electrochemical equilibrium. Voltammetry is a technique in which the potential is controlled according to some prescribed function while the current is measured. Coulometry involves the measurement of charge needed to completely convert an analyte, and conductometry determines the electrical conductivity of the investigated test solution. The practical applications of these measurement techniques for analytical purposes range from industrial process control and environmental monitoring to food analysis and biomedical diagnostics. Both the analytical methods and their instrumentation as well as recent trends, such as electrochemical sensors are discussed.

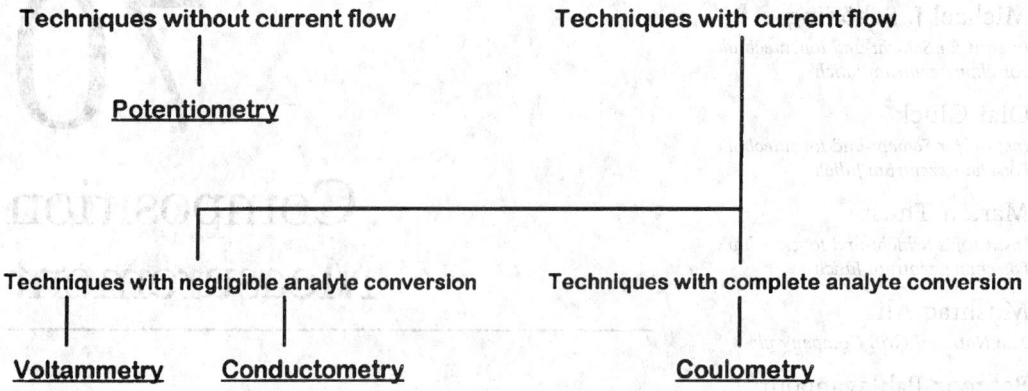

FIGURE 70.1   Electrochemical methods described in this section.

Basic Concepts and Definitions

Electrodes and the Electrical Double Layer

In electrochemistry, *electrodes* are devices for the detection of charge transfer and charge separation at phase boundaries or for the generation and variation of the charge transfer and separation with an impressed current across the phase boundary. One important feature of electrodes is a potential difference across the electrode/electrolyte phase boundary. At this interface, the conduction mechanism changes since electrode materials conduct the current via electrons whereas electrolytes conduct via ions. To understand the processes that lead to the formation of the potential difference, it is helpful to consider first an atomistic model, which was given by Helmholtz. It leads to the idea of an *electrical double layer*.

If an electrode is immersed in an electrolyte solution, the bulk regions of the two homogeneous phases — the electrode material and the electrolyte — are in equilibrium. This means that far away from the phase boundary (>1 µm), the sum of the forces on the particles is zero and charges are distributed homogeneously. Since the cohesion forces that bind the individual particles together in the bulk are significantly reduced at the surface of the electrode, particles in this region will have less neighbors or neighbors from the other phase. Thus, close to the phase boundary, the equilibrium conditions are drastically different from the equilibrium conditions in the bulk of the electrolyte. This change in the equilibrium of forces on particles at the interface can lead to an *interfacial tension*. In addition, the surface of a condensed phase usually has different electrical properties than the bulk phase, for example, due to the accumulation of free charge on the surface of an electrically charged solid. Besides, the orientation of dipoles in the surface region and adsorption of ions and dipoles from the electrolyte can lead to a change in the electrical properties. This excess charge from ions, electrons, and dipoles produces an electrical field that is accompanied by a potential difference across the phase boundary. The region in which these charges are present is termed the *electrical double layer*. The formation of an electrical double layer at interfaces is a general phenomenon but only the electrode/electrolyte interface will be considered here in more detail.

According to the hypothesis of Helmholtz, the electrical double layer has the character of a plate capacitor, whose plates consist of a homogeneously distributed charge in the metal electrode and ions of opposite charge lying in a parallel plane in the solution at a minimal distance from the surface of the electrode [1]. Modern conceptions are based on the assumption that the electron cloud in the metal extends to a certain degree into a thin layer of solvent molecules in the immediate vicinity of the electrode surface. In this layer, the dipoles of the solvent molecules (e.g., H_2O) are oriented to various degrees toward the electrode surface. Ions can accumulate in it due to electrostatic forces or be adsorbed specifically on the electrode through van der Waals and chemical forces. These substances are called *surface-active substances* or *surfactants*. The sum of oriented solvent molecules and surfactants in the immediate

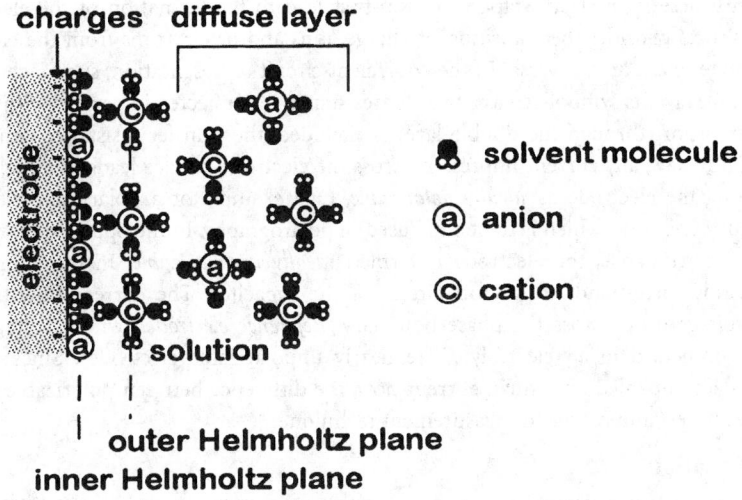

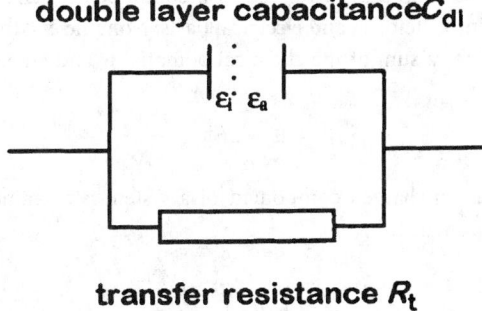

FIGURE 70.2 Electrical double layer according to the Helmholtz model and equivalent circuit representation.

vicinity of the electrode is considered as one layer. The plane through the centers of these molecules and ions parallel to the electrode surface is termed *inner Helmholtz plane* (Figure 70.2). If only electrostatic attraction is taken into account, ions from the solution can approach the surface to a distance given by their primary solvation sheaths. This means that at least a monomolecular solvent layer remains between the electrode and the solvated ion. The plane through the centers of these ions is called *outer Helmholtz plane*, and the solution region between the electrode surface and this outer Helmholtz plane is called *Helmholtz* or *compact layer*. In reality, electrostatic forces cannot retain ions at a minimal distance from the electrode surface. Due to thermal motion, the excess charge is smeared out in the direction of the electrolyte bulk to form a *diffuse layer*, also termed the *Gouy-Chapman layer*. It describes the region between the outer Helmholtz plane and the bulk of the solution. In concentrated electrolyte solutions (approx. 1 mol/L), the diffuse layer is as thin as the inner Helmholtz plane and may be considered as rigid. In highly dilute solutions, its thickness can be as large as 100 nm. As in the early model of Helmholtz, the double layer acts as a capacitor [2]. Here, two different dielectric layers with permittivities ε_i and ε_o represent the region between the electrode surface and the inner Helmholtz plane and the region between the inner and the outer Helmholtz plane, respectively (Figure 70.2).

In addition to these ideal electrostatic processes that lead to the formation of the electrical double layer, one has also to consider the transition of charge, ions and/or electrons from the electrode phase into the electrolyte phase or vice versa. In the equivalent circuit representation, such a charge transport through the double layer is symbolized as a transfer resistance R_t connected in parallel with the capacitor. If any charge transport through the double layer is excluded, the transfer resistance is nearly infinite. According to Ohm's law, any current impressed across the electrode surface leads to a high polarization voltage determing the electrode as *ideally polarizable*. One example of a polarizable electrode is the dropping mercury electrode, which is frequently used in polarography. In the opposite case with a nearly vanishing transfer resistance, the electrode is termed *ideally unpolarizable*. In the equivalent circuit representation, this corresponds to a short-circuit of the capacitor. The current flow then does not influence the voltage drop across the phase boundary. *Reference electrodes*, whose voltage have to be constant when immersed in an electrolyte, are nearly unpolarizable electrodes. Since every voltage measurement is accompanied by a small current flow, the difference between polarizable and unpolarizable electrodes is very important in measurement technique.

The Nernst Equation

If the electrode phase and the electrolyte phase contain a common ion, the potential difference across the phase boundary is determined by the effective concentration (activity) of this ion in the solution. This fact is described quantitatively by the *Nernst equation* and will be derived in the following. If one mole of ions of a species i has to be transferred from a given reference state outside into the bulk of an electrically charged phase work must be expended to overcome the chemical bonding forces and the electrical forces. This work is given by the electrochemical potential $\tilde{\mu}_i$. Since the chemical interactions of a species with its environment always possess electric components, generally the electrochemical potential cannot be separated into chemical and electrical parts. Nonetheless, the electrochemical potential is frequently given formally as a sum of the chemical potential μ_i and an electrostatic work $zF\phi$:

$$\tilde{\mu}_i = \mu_i + zF\phi \tag{70.1}$$

The chemical potential μ_i of an uncharged component of a system is the amount of Gibbs energy G inherent in 1 mol of that component [3]:

$$\mu_i = \left(\frac{\partial G}{\partial n_i}\right)_{p,T} \tag{70.2}$$

Here, n_i is the number of moles of the given component. In the case of a dilute solution, the chemical potential of a component i is

$$\mu_i = \mu_i^0 + RT \ln c_i \tag{70.3}$$

μ_i^0 denotes the standard chemical potential and c_i the concentration of the species i, R is the gas constant, and T is the absolute temperature. The values of standard chemical potentials can be found in standard textbooks of thermodynamics and in tables of physicochemical constants under the name standard molar Gibbs energies. μ_i^0 is independent of the concentration c_i. In concentrated electrolytes, the concentration c_i has to be replaced by the respective activity a_i. The activity a_i is given by the relationship $a_i = \gamma c_i$, where γ is the activity coefficient which is a correction factor for non-ideal behavior. In the second term of Equation 70.1, z denotes the charge number of the ion i, F is the Faraday constant, and ϕ is the *inner electric potential*, which is, in general, the electric work necessary for the transfer of a unit charge: for example, 1 coulomb, from infinity to a given site.

The inner electric potential may consist of two components, an *outer electric potential* ψ and a *surface electrical potential* χ. Whereas the outer electrical potential of a phase is produced by excess electric charge supplied from outside, the surface electric potential is an effect of electric forces at the interface which leads to the electrical double layer introduced above. The difference of the outer potentials of the electrode (e) and the solution (s):

$$\psi_e - \psi_s = \Delta\psi \tag{70.4}$$

is termed *Volta potential difference* and is the only measurable quantity. Neither the difference of the surface potentials of the appropriate phases $\Delta\chi$ nor the difference of the inner electric potentials:

$$\Delta\phi = \Delta\psi + \Delta\chi \tag{70.5}$$

defined as *Galvani potential difference*, can be measured directly. Strictly speaking, even the Volta potential difference between the solution and the electrode is a not measurable quantity since only the Volta potential difference between two electrodes can be measured. To determine the potential of the solution phase, one has to dip an electrode in the solution. This, however, creates a new electrode/solution interphase and, consequently, one measures the sum of two potential differences. This is the reason for the lack of absolute potentials in electrochemistry. Therefore, one uses a reference electrode that has a known potential relative to a standard electrode.

In thermodynamic equilibrium, the electrochemical potentials of the considered species are equal in both phases. For a charged particle i that may cross the phase boundary solution/electrode, this means

$$\mu_{i,s}^0 + RT\ln a_{i,s} + z_i F\phi_s = \mu_{i,e}^0 + RT\ln a_{i,e} + z_i F\phi_e \tag{70.6}$$

and therefore in equilibrium the Galvani potential difference is given by:

$$\Delta\phi = \phi_e - \phi_s = \frac{\mu_{i,s}^0 - \mu_{i,e}^0}{z_i F} + \frac{RT}{z_i F}\ln\frac{a_{i,s}}{a_{i,e}} \tag{70.7}$$

Since the chemical standard potential of the respective phases are constants, the first term in Equation 70.7 can be expressed as a standard Galvani potential difference $\Delta\phi^0$:

$$\Delta\phi = \Delta\phi^0 + \frac{RT}{z_i F}\ln\frac{a_{i,s}}{a_{i,e}} \tag{70.8}$$

For metal electrodes, the activity of the metal atoms M and that of the electrons in the electrode phase equal unity per definition. Thus, for an electrode reaction of type:

$$M_s^{z+} + ze^- \leftrightarrow M_e \tag{70.9}$$

Equation 70.8 becomes the *Nernst equation*

$$\Delta\phi = \Delta\phi^0 + \frac{RT}{zF}\ln a_s \tag{70.10}$$

which gives the relation between the activity of the potential determining ion a_s and the Galvani potential difference $\Delta\phi$. Using base 10 logarithm, the Nernst equation is given as:

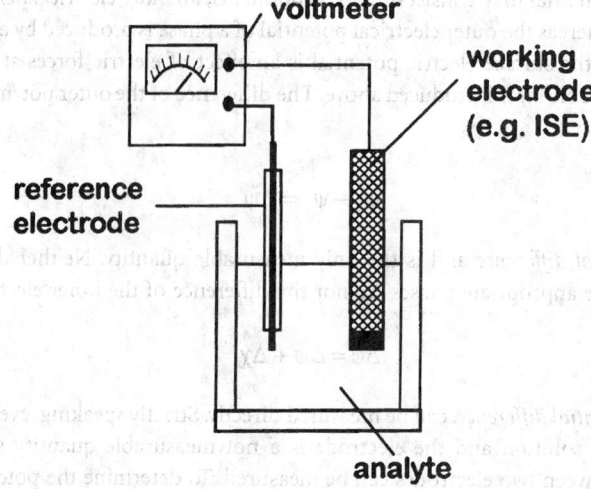

FIGURE 70.3 Schematic of an electrochemical cell with a working electrode and a reference electrode immersed in the test solution (electrolyte).

$$\Delta\phi = \Delta\phi^0 + \frac{RT \cdot 2.3}{zF}\log a_s = \Delta\phi^0 + k \cdot \log a_s \qquad (70.11)$$

where k is called *Nernst constant*.

The classical form of the Nernst equation (Equation 70.10) can be formulated more generally for a redox reaction. If a_{ox} and a_{red} are the activities of the oxidized and reduced form of the considered ion, the Galvani potential difference is given as:

$$\Delta\phi = \Delta\phi^0 + \frac{RT}{zF}\ln\frac{a_{ox}}{a_{red}} \qquad (70.12)$$

In *potentiometry*, the activity of a certain ion can be determined directly by the measurement of the equilibrium Galvani potential difference of a suitable electrode (*direct potentiometry*). On the other hand, changes of the activity of the detected ion and equivalence points can be detected in titration reactions (*potentiometric endpoint titration*).

After this rather theoretical definition of the Galvani potential difference, the question arises how to *measure* this potential difference between the bulk of the electrode and the solution. Since a potential difference cannot be measured with only one electrode, a second one must be immersed in the solution. Both are connected to a voltmeter, to complete the *electrochemical cell* (Figure 70.3). An electrochemical cell generally consists of two (or more) electrodes immersed in an analyte. Thus, in some of the old literature, a single electrode is often referred to as a *half-cell* and its potential is called *half-cell potential*. In modern electrochemistry, usually the term *electrode potential* is used. An electrochemical cell is in a current-free state during potentiometric measurements (e.g., with an *ion-selective electrode* (ISE)), but may also supply electric energy (a galvanic cell) or accept electric energy from an external source (an electrolytic cell). Since a second electrode potential arises at the phase boundary second electrode/electrolyte, only the sum of at least two Galvani potential differences can be measured. A separation into the two individual parts is impossible. Hence, the function of the second electrode, named reference electrode, is to act as an electrode of constant potential against which variations in the potential of the measuring electrode in various samples can be measured. In the Nernst equation, the Galvani potential ϕ is then replaced by E, the symbol for measurable voltages.

Classification of Electrodes

Electrodes are termed *reversible electrodes* if they transfer electrons and ions with negligible impedance. Therefore, under current, the electrochemical potential of electrons, ions, and neutral species do not change across the different interfaces that may exist in an electrode. Otherwise, the electrode is not suitable to measure thermodynamic (equilibrium) quantities such as ion activity. Since a distribution equilibrium of charged species is considered here, the electrode and the solution phase must have at least one charged species in common. Depending on the number of equilibria being involved in the forming of the electrode potential, reversible electrodes can be divided into different groups:

1. *Electrodes of the first kind.* These may be cationic or anionic electrodes at which equilibrium is established between the atoms or molecules in the electrode material and the respective cations or anions in the solution. According to the Nernst equation, the equilibrium Galvani potential difference is here determined by the activity of the considered ion in the solution. Examples for electrodes of the first kind are ISEs including metal and amalgam electrodes and the hydrogen electrode.
2. *Electrodes of the second kind.* These electrodes consist of three phases. A metal wire is covered by a layer of its sparingly soluble salt which usually has the character of a solid electrolyte (e.g., Ag and AgCl). This wire is immersed in a solution containing a soluble salt of the anions of this solid electrolyte (e.g., KCl). Here, the equilibrium between the Ag atoms in the metal and the anions in the solution is established through two equilibria: The first one is given between the metal and the cation in its sparingly soluble salt; for example,

$$Ag \leftrightarrow Ag^+ + e^- \tag{70.13}$$

   and the second one between the anion in the sparingly soluble salt and the anion in the solution; for example,

$$AgCl \leftrightarrow Ag^+ + Cl^- \tag{70.14}$$

   The electrode potential of electrodes of the second kind is rather insensitive to small current flows. Thus, they are often used as reference electrodes.
3. *Electrodes of the third kind.* In this electrode, the sparingly soluble salt contains a second cation that also forms a sparingly soluble compound with the common anion but with a higher solubility product than the electrode metal compound (e.g., Ag_2S and PbS). Here, the electrode potential depends on the activity of this cation in the solution.
4. *Oxidation-reduction (redox) electrodes.* They consist of an inert metal such as Pt, Au, or Hg that is immersed in a solution of two soluble oxidation forms of a single substance (e.g., Fe^{3+} and Fe^{2+}). Thus, for the electrode reaction:

$$Fe^{3+} + e^- \leftrightarrow Fe^{2+} \tag{70.15}$$

   the Nernst equation is:

$$E_{Fe^{3+}/Fe^{2+}} = E^0_{Fe^{3+}/Fe^{2+}} + \frac{RT}{F} \ln \frac{a_{Fe^{3+}}}{a_{Fe^{2+}}} \tag{70.16}$$

according to Equation 70.10. Here, E is termed the *electrode potential* and E^0 is designated the *standard electrode* (or *redox*) *potential* of the electrode reaction if it is measured versus the *standard hydrogen electrode* (SHE). The subscripts of E and E^0 denote the redox couple of the considered

TABLE 70.1 Some Standard Electrode
Potentials and Redox Potentials

Electrode or Half-Cell Reaction	E^0 (V)
$Li^+ + e^- \leftrightarrow Li$	−3.0403
$K^+ + e^- \leftrightarrow K$	−2.931
$Ca^{2+} + 2e^- \leftrightarrow Ca$	−2.868
$Mg^{2+} + 2e^- \leftrightarrow Mg$	−2.372
$Al^{3+} + 3e^- \leftrightarrow Al$	−1.662
$Zn^{2+} \ 2e^- \leftrightarrow Zn$	−0.762
$Fe^{2+} + 2e^- \leftrightarrow Fe$	−0.447
$Pb^{2+} + 2e^- \leftrightarrow Pb$	−0.1264
$AgCl + e^- \leftrightarrow Ag + Cl^-$	0.22216
$Hg_2Cl_2 + 2e^- \leftrightarrow 2Hg + 2Cl^-$	0.26791
$Cu^{2+} + 2e^- \leftrightarrow Cu$	0.3417
$I_2 + 2e^- \leftrightarrow 2I^-$	0.5353
$Fe^{3+} + e^- \leftrightarrow Fe^{2+}$	0.771
$Ag^+ + e^- \leftrightarrow Ag$	0.7994
$Tl^{3+} + 2e^- \leftrightarrow Tl^+$	1.2152
$2Cl^- \leftrightarrow Cl_2 + 2e^-$	1.35793
$Ce^{4+} + e^- \leftrightarrow Ce^{3+}$	1.610

electrode reaction. The standard redox potential is a measure of the reducing or oxidizing ability of a substance. If one considers, for example, two systems 1 and 2 with their respective standard redox potentials E_1^0 and E_2^0, system 1 is a stronger oxidant than system 2 if $E_1^0 > E_2^0$. This means that in a mixture of the solutions of these two systems where originally the activities of the reduced forms equal that of the oxidized forms ($a_{red}^1 = a_{ox}^1$ and $a_{red}^2 = a_{ox}^2$), an equilibrium will be established with $a_{ox}^2 > a_{red}^2$ and $a_{red}^1 > a_{ox}^1$. The experimentally determined standard potentials of well-known redox systems are listed in Reference 4. Table 70.1 gives some examples. In redox electrodes, the metal acts as a medium for the electron transfer between the two forms. In contrast to electrodes of the first kind, the solution should not contain ions of the electrode metal in order to avoid an additional Galvani potential difference at the electrode determined by the activity of the electrode metal ions in the solution. This disturbing ion activity is negligible if the standard potential of the electrode metal is a few 100 mV higher than the redox potential to be measured. Thus, mainly platinum electrodes ($E_{Pt^{2+}/Pt}^0 = 1.20$ V) and gold electrodes ($E_{Au^+/Au}^0 = 1.42$ V) are used as redox electrodes.

Reference Electrodes

The potential of an ion-selective electrode (ISE) is always measured with respect to a reference electrode. Ideally, the reference electrode should not cause chemical changes in the sample solution, or vice versa. It should maintain a constant potential relative to the sample solution, regardless of its composition. In practice, any changes of its potential with composition should be at least as small as possible and reproducible. Reference electrodes with liquid junctions, strictly speaking reference electrode *assemblies*, consist of a reference element immersed in a filling solution (often called bridge solution) contained within the electrode. This reference element should possess a fixed activity of the ion defining the potential of the element with respect to the filling solution. The electric contact between the electrode and the sample solution is made by the liquid junction consisting of a porous plug or a flow restriction which permits the filling solution to flow very slowly into the sample.

At the junction between the two electrolyte solutions, ions from both solutions diffuse into each other. Since different ions have different mobilities, they will diffuse at different rates. Thus, a charge separation will occur related in size to the difference in mobilities of the anions and cations in the two solutions. This charge separation produces a potential difference across the junction called the *liquid junction potential* [5]. In reference electrodes, usually the bridge solution is given a slightly higher pressure than the sample so that the solution, often concentrated potassium chloride, flows out relatively rapidly into the sample, and diffusion of the sample back into the salt bridge is impeded. If the bridge solution is concentrated enough, it is assumed that variations in the liquid junction potential due to the varying composition of the sample are suppressed. This is the basis on which the reference electrode assembly is used. Since the potential of the whole assembly E_{ref} is the sum of the potential of the reference element E_r in the bridge solution and the liquid junction potential E_j:

$$E_{ref} = E_r + E_j \qquad (70.17)$$

any change in the liquid junction potential appears as a change in the potential of the assembly. An extra liquid junction potential must be included if a *double junction* reference electrode is considered. When an analysis using a cell with an ion-selective electrode is carried out, standard solutions are used to calibrate the ISE. A change in the liquid junction potential that occurs when the standard solutions are replaced by the sample is termed the *residual liquid junction potential* and constitutes an error in the analytical measurement. The needed constancy of the potential can be approached by a suitable choice of standards and/or sample pretreatment, and by the use of a proper bridge solution and the best physical form of the liquid junction.

Several types of liquid junctions exist from which the best ones with regard to stability and reproducibility are complicated to realize in practice, and the worst ones are easy to use but much less stable and reproducible. Most of the commercial reference electrodes with adequate properties possess *restrained diffusion junctions* where the most common junctions available are the ceramic plug, the asbestos wick or fiber, two types of ground sleeve junction, and the palladium annulus junction (Figure 70.4). For a very large majority of applications with ion-selective electrodes, a ceramic plug will perform adequately. The flow rate of the bridge solution into the sample solution is sometimes called leak rate and is given

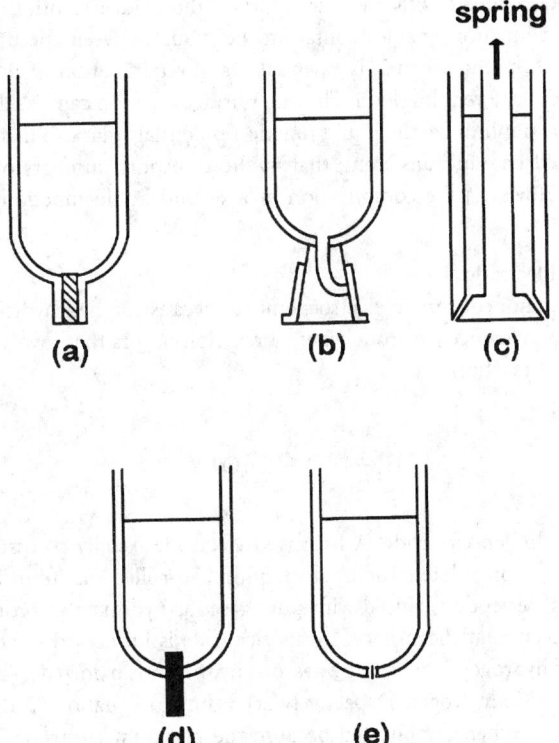

FIGURE 70.4 Different types of liquid junctions: (a) ceramic plug, (b) ground glass sleeve (type 1), (c) ground glass sleeve (type 2), (d) asbestos wick, (e) palladium annulus.

in mL per 5 cm head of bridge solution per day. The head of bridge solution is measured as the height of the surface of the bridge solution above the surface of the sample. In order to work satisfactorily, the surface of the bridge solution of all these restricted junction devices has to be at least 1 cm above the sample solution. Otherwise, if the bridge solution falls too low, the junction and the bridge will become contaminated by species diffusing from the sample. The bridge solution has then to be replaced. For the same reason, reference electrodes should be stored, when not in use, with the junction immersed in bridge solution.

Whereas the ceramic plug and the asbestos wick and fiber (Figure 70.4(a) and (d)) have relatively slow flow rates of about 0.01 to 0.1 mL per 5 cm head of bridge solution per day, ground sleeve junctions of type (*b*) have a flow rate of 1 to 2 mL. On the other hand, the flow rates of different asbestos wick junctions may vary by a factor up to 100 and the liquid junction potential may have a day-to-day (in)stability of ±2 mV under the favorable conditions of a junction between strong potassium chloride solution and an intermediate pH buffer. Under the same conditions, ground glass sleeve junctions of type (*b*) and the little-used palladium annulus junction show stabilities of ±0.06 mV and ±0.2 mV, respectively. It is worth mentioning that palladium annulus junctions may partly respond as a redox electrode in strong oxidants (e.g., 0.2 *M* $KMnO_4$ in 0.05 *M* H_2SO_4) and mild or strong reductants (e.g., 0.5 M $SnCl_2$ in 1 *M* HCl). In such samples, reference electrodes with palladium or platinum annulus junctions should not be used. Although ground glass sleeve junctions have inconveniently high flow rates and the bridge solution needs to be replenished frequently, these junction types have found particular use in applications where the junction has the tendency to clog, such as measurements in protein solutions. However, the stability of the liquid junction potential appears to be relatively poor in fast-flowing sample solutions and may be very sensitive to sample flow rate. Asbestos wick junctions are particularly liable to blockage and should consequently be used in clear solutions only.

In *double-junction reference electrodes*, the filling solution in which the reference element is immersed (reference solution) makes contact with another solution, the bridge solution, by means of a liquid junction. A second liquid junction enables contact to be made between the bridge solution and the sample. Such electrodes are useful when it is essential that contamination of the sample by the inner filling solution must be kept at a very low level. The outer bridge solution can be selected to be compatible with the sample. In order to minimize the liquid junction potentials that can drift and cause instability, the bridge solution should be equitransferent; that is, the transport numbers of its anion and cation should be nearly equal. However, the complication of a second liquid junction in the cell should be avoided if possible.

The Standard Hydrogen Electrode.
Aqueous solutions are of major concern in electrochemistry because of their hydrogen ion content. Thus, it is advantageous to use a reference electrode where a reaction occurs that involves the participation of hydrogen ions. One of this reactions is:

$$\frac{1}{2}H_2 + H_2O \leftrightarrow H_3O^+ + e^- \tag{70.18}$$

Figure 70.5 shows a hydrogen electrode. A hydrogen electrode usually consists of a platinum sheet covered by a thin layer of sponge-like structured platinum, so-called platinum black, that has a high specific surface area. This electrode is rinsed with pure gaseous hydrogen in order to form a complete layer of adsorbed H_2 molecules at the surface. If this electrode is immersed in an electrolyte, it acts as an electrode consisting of hydrogen at which the gaseous hydrogen is oxidized to hydronium ions or the hydronium ions are reduced to hydrogen, respectively, according to Equation 70.18. The real mechanism of this electrode process is rather complicated because the platinum electrode is in contact with the hydronium ions in the solution as well as with the gaseous hydrogen that is bubbled through it. Thus, the final equilibrium between the gaseous hydrogen, the dissolved hydronium ions, and the electrode phase consists of several successive equilibrium steps which can be found in Reference 6. To calculate the

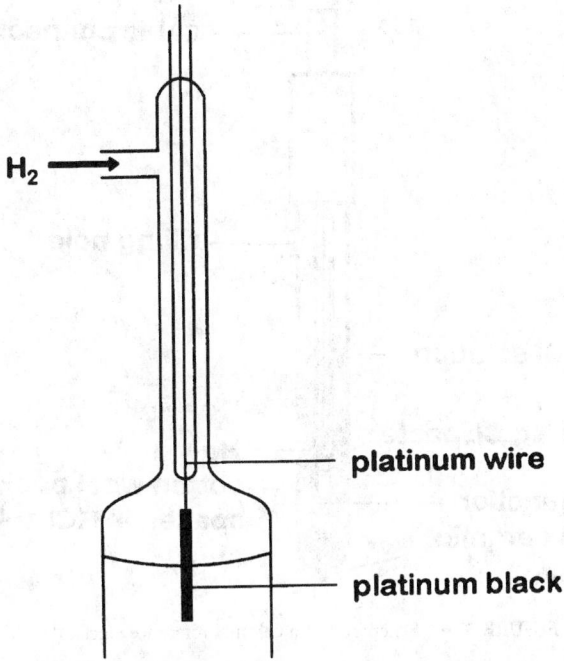

H₂ →

platinum wire

platinum black

FIGURE 70.5 Schematic of a hydrogen electrode.

potential of a hydrogen electrode, which is strictly speaking the difference between the potential of the electrode and that of the solution, one has to consider the electrochemical potentials of the respective phases. The chemical potential of gases is usually expressed in terms of the pressure p instead of the molar concentration c. Due to the elementary relationship $pV = nRT$ for ideal gases, where V is the volume of the gas and n is the amount of moles the pressure, p is proportional to the molar concentration $c = n/V$. Thus, according to Equation 70.3:

$$\mu_{H_2} = \mu_{H_2}^0 + RT \ln \frac{p_{H_2}}{p_{H_2}^0} \tag{70.19}$$

where μ_{H_2} and $\mu_{H_2}^0$ are the pressure and standard pressure of hydrogen, respectively. In the case of moderate ion concentrations, the chemical potential of the solvent water is equal to its standard chemical potential. Hence, the potential difference between the electrode and the solution is, according to Equation 70.7:

$$\Delta\phi = \frac{\mu_{H_3O^+}^0 - \frac{1}{2}\mu_{H_2}^0 - \mu_{H_2O}^0}{F} + \frac{RT}{F}\ln c_{H_3O^+} - \frac{RT}{2F}\ln \frac{p_{H_2}}{p_{H_2}^0} \tag{70.20}$$

This equation is generally valid for hydrogen electrodes. The electrode is called *standard hydrogen electrode* (SHE) if the molar concentration is such that the activity of the hydronium ions is unity ($a_{H_3O} = 1$) and the pressure of hydrogen is equal to its standard pressure. Hence, for an SHE, the second and third term in Equation 70.20 vanish. The combination of standard chemical potentials in the first term of Equation 70.20 is defined as zero. Consequently, the total potential difference across the interface SHE/electrolyte is equal to zero *by definition* at any temperature. Since standard hydrogen electrodes are very difficult to prepare, they are not used as reference electrodes in practice. However, electrode potentials are usually standardized with respect to the SHE and their values are thus called "on the hydrogen scale."

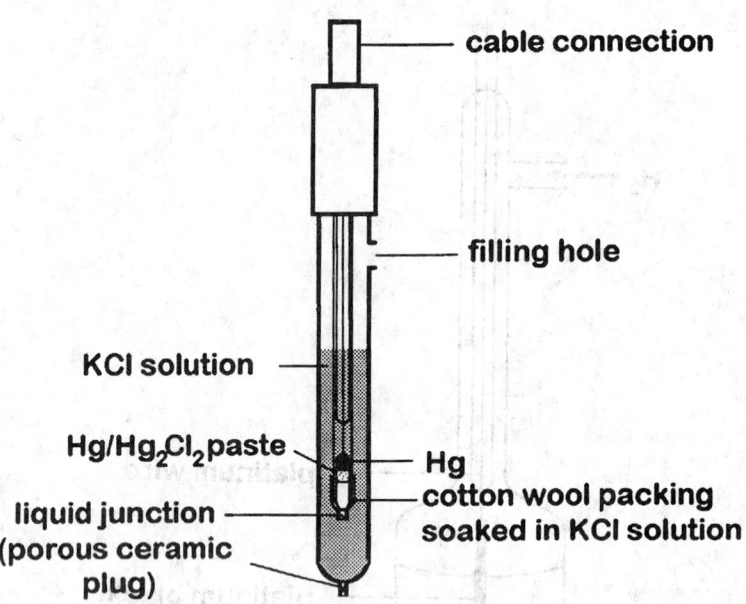

FIGURE 70.6   Schematic of a calomel reference electrode.

The Calomel Electrode.

The calomel electrode is the most common of all reference electrodes. It consists of a pool of mercury that is covered by a layer of mercurous chloride (calomel, Hg_2Cl_2). The calomel is in contact with a reference solution that is nearly always a solution of potassium chloride, saturated with mercurous chloride. Thus, the calomel electrode is a typical electrode of the second kind. Figure 70.6 shows a typical arrangement of a commercial calomel electrode assembly where the electrode is inverted, with the mercury uppermost, and packed into a narrow tube. Depending on the strength of the potassium chloride solution used, the electrode is called saturated calomel electrode (SCE), 3.8 M or 3.5 M calomel electrode, respectively. Potassium chloride is used as reference solution because it gives rise to a small liquid junction potential at the outer liquid junction of the electrode, i.e., the liquid junction with the sample. Hence, potassium chloride is a suitable reference solution as well as a good bridge solution. Furthermore, mercurous chloride has a very low solubility in potassium chloride solutions, regardless of concentration. The electrode reaction of a calomel electrode is:

$$Hg + Cl^- \leftrightarrow \frac{1}{2}Hg_2Cl_2 + e^- \qquad (70.21)$$

Its standard potential, including the liquid junction, is 0.2444 V vs. SHE at 25°C for the SCE, and 0.2501 V for the 3.5 M calomel electrode according to Reference 7. Further data are given, for example, in Reference 8.

The components of a calomel electrode are chemically stable except for the mercurous chloride, which significantly disproportionates at temperatures above 70°C according to the equation:

$$Hg_2Cl_2 \leftrightarrow Hg + HgCl_2 \qquad (70.22)$$

Hence, potential drift occurs and life time decreases with increasing working temperature. On the other hand, calomel electrodes can be used at temperatures down to –30°C if 50% glycerol is added to the potassium chloride solution.

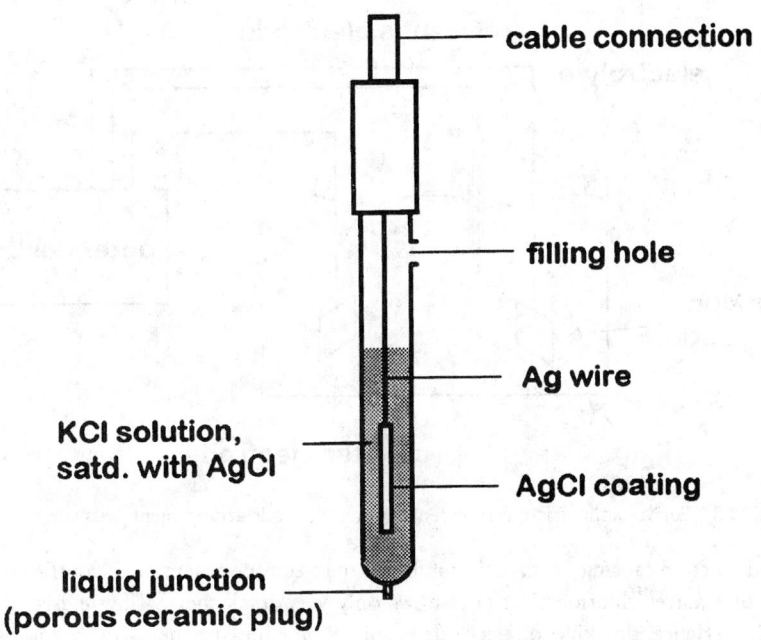

cable connection

filling hole

Ag wire

KCl solution,
satd. with AgCl

AgCl coating

liquid junction
(porous ceramic plug)

FIGURE 70.7   Schematic of a Ag/AgCl reference electrode.

Impurities in the potassium chloride solution, such as bromide and sulfide ions as well as redox agents and complexants, cause a small shift in the electrode potential. Nevertheless, the measurement of potential differences is not affected. However, the most unsatisfactory feature of the performance of the calomel electrode is its thermal hysteresis that occurs if the electrode filling material is not in thermal equilibrium or if the electrode and the sample have different temperatures. Thus, temperature stability during the storage and measurements is very important. In any cases where the temperature of the reference electrode or the sample has to be varied, it is thus usually better to use a silver/silver chloride electrode instead of a calomel electrode.

The Silver/Silver Chloride Electrode.
The silver/silver chloride electrode consists of a silver wire or plate that is coated with silver chloride. For the same reasons as with the calomel electrode, this phase is in contact with a strong potassium chloride solution, here saturated with silver chloride. Figure 70.7 shows the diagram of a typical Ag/AgCl reference electrode. Since this kind of reference electrode is the simplest and for many applications the most satisfactory one, it is commonly used as internal reference electrode of pH electrodes and other ion-selective electrodes. Besides, Ag/AgCl electrodes can be easily prepared in the laboratory. In contrast to mercury-based electrodes, the Ag/AgCl electrode does not contain toxic chemicals and is therefore recommendable for measurements in food.

The major problem with the Ag/AgCl electrode is the considerably high solubility of AgCl in concentrated potassium chloride solution. Thus, especially for the use at high temperatures, a sufficient excess of solid silver chloride must be present in the reference solution. This can be achieved, for example, through the addition of a few drops of diluted silver nitrate solution. Otherwise, silver chloride will dissolve off the electrode until saturation is reached. As a consequence, the electrode potential will drift and the lifetime of the electrode will be shortened. However, in contrast to the calomel electrode, the Ag/AgCl electrode can be used successfully up to 125°C. Its electrode potential is very stable in the long term in pure potassium chloride solutions, but is affected by impurities like redox reagents and species that react with the silver chloride, as with the calomel electrode. Unlike the calomel electrode, in the Ag/AgCl electrode, the concentration of the electrode coating in the bridge solution is rather high. Thus, a greater amount of reaction products (e.g., solid silver sulfide) may arise in the reference solution and

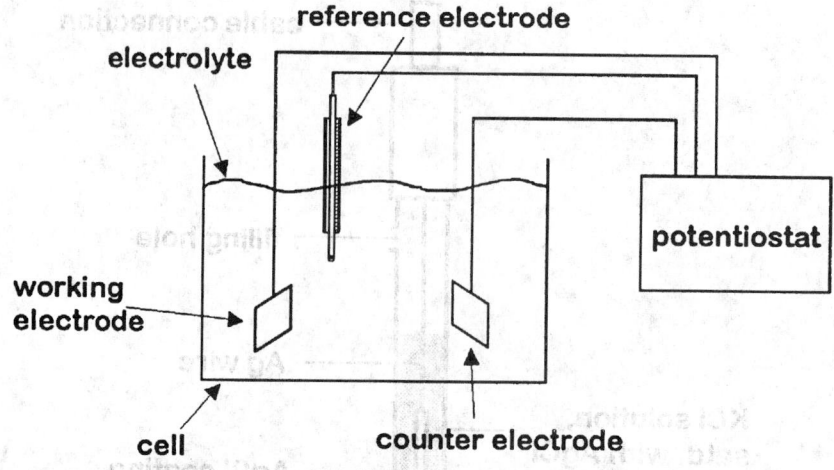

FIGURE 70.8   For voltammetric measurements a three electrode arrangement is usually employed.

block the liquid junction causing drift and instability of the electrode potential. In contrast to the calomel electrode, the silver/silver chloride electrode shows only very small thermal hysteresis effects that are usually negligible. Hence, this kind of electrode is suitable for measurements in samples with varying temperatures. Ag/AgCl electrodes are relatively insensitive to polarization. The standard potentials, including the liquid junction potentials of saturated and 3.5 M silver/silver chloride electrodes, at 25°C are 0.1989 V and 0.2046 V, respectively according to Reference 9. As with the calomel electrode, the nomenclature of the electrodes is derived from the potassium chloride concentration of the respective reference solution.

Voltammetry

The basic concept of *voltammetry* is the measurement of the current i at a redox electrode as a function of the electrode potential E. During the experiment, the electrode is immersed in a solution that contains an electroactive species; that is, a species that can undergo an electrode reaction (standard redox potential E^0). The electrode potential is changed from a value $E_1 < E^0$ to a value $E_2 > E^0$ or vice versa in a manner that is predetermined by the operator. Thus, during the measurement, the electrochemical equilibrium shifts from the oxidized (reduced) form of the analyte to the reduced (oxidized) form. The resulting charge transfer across the interface electrode/solution can be observed as a current flow, which is termed *faradaic.*

Instrumentation

Voltammetric measurements are usually performed with a cell arrangement of three electrodes (Figure 70.8). The redox electrode at which the electrode processes occurs is called *working electrode*. Its potential is measured against a suitable reference electrode, often Ag/AgCl or calomel. To adjust the potential difference between the working and the reference electrode to a certain value, a current is forced through the working electrode. Because the current and the electrode potential are related functionally, this current is unique. However, the current through the reference electrode must be kept as small as possible. Therefore, a third electrode called *auxiliary electrode* or *counter electrode* is usually employed to close the current circuit. It should be emphasized that there are two circuits: one in which the current flows and which contains the working and the auxiliary electrode and another, and a current-free one in which the potential difference between the working and the reference electrode is measured. Since almost no current flows through the reference electrode, its potential can be regarded as constant and the measured change in potential equals the potential change of the working electrode. The current through the working electrode, and thus its potential, can be adjusted by controlling the voltage between the working and the auxiliary electrode. This task is performed by an instrument called a *potentiostat*, which basically consists of a voltage source and a high-impedance feedback loop. With a function generator, that may be integrated into the potentiostat, the potential–time course can be predetermined. Modern potentiostats are controlled by a PC and offer the possibility to program many different potential–time courses. Thus, they allow the performance of several voltammetric techniques, as are discussed below. The measured current can be displayed as a function of the electrode potential or of time using a strip-chart or *xy*-recorder or a PC.

There are two possibilities to operate an electrochemical cell. In so-called *batch cells*, the electrolyte solution rests stationary during the measurement, whereas in *flow-through cells*, it flows across the electrode. Between two measurements with different solutions the cell must be cleaned in order to remove residues of the preceding measurement's solution that could disturb the new measurement. The electrochemical cell is usually built of glass or teflon because of these materials' chemical inertness.

The chemical inertness is also important for the choice of the working electrode because the electrode must not change during the measurement. Common materials are gold, platinum, and mercury. Several kinds of carbon electrodes (e.g., glassy carbon) are also used but are often covered with gold or mercury. An advantage of the solid-state electrodes is their easy handling. They can be employed as planar or as wire electrodes. Further, with the noble metal electrodes, substances having a more positive redox potential than mercury can be investigated [10]. However, the use of mercury electrodes offers distinct advantages and the voltammetric techniques using mercury electrodes are extremely well developed. These techniques play a major role in electroanalytical methods and are summarized under the term *polarography.*

In polarography, mercury is either used as a *thin mercury film electrode* (TMFE) or as a *hanging mercury drop electrode* (HMDE). The HMDE can be a *stationary mercury drop electrode* (SMDE) or a *dropping mercury electrode* (DME). The drop is produced from a thin capillary with an inner diameter that can range from several ten to a few hundred micrometers. The size of an SMDE is held constant during the measurement, whereas a DME constantly grows during its lifetime until it falls from the capillary due to its weight.

The main advantages of mercury drop electrodes are their good reproducibility and their high *overpotential* for the hydrogen evolution; that is, the fact that hydrogen evolution is inhibited and thus occurs at much higher potentials than would be expected from the standard potential. The good reproducibility is achieved because a new drop can easily and rapidly be produced from the capillary for each measurement. Hence, the contamination of the electrode with substances from a preceding measurement and from impurities in the solution is near zero. However, a drawback of HMDEs is their relative mechanical instability, which can be a problem in flow-through cells, in field measurements, and if the solution is stirred.

Stirring of the solution is often applied during the measurement if the supply of reactive species at the electrode should be enhanced. However, this forced convection affects the electrode current. Moreover, the electrolyte is often stirred and bubbled with an inert gas like nitrogen or argon before voltammetric measurements are carried out to remove dissolved oxygen. This is usually necessary to reduce background currents from oxygen reduction and to prevent undesirable oxidation or precipitation of solution components. Because the electrode currents, especially in trace and ultra-trace analysis, can be quite small, it is common to place the cell in a Faradaic cage to shield it from electromagnetic stray fields. Coaxial cables are then used for the electric connections from the cell to the instruments.

Principles of Voltammetry

Actually, the electrode current measured in voltammetry is a sum of two currents that arise due to different processes. Besides the faradaic current i_f, a capacitive current i_c results from changes in the *double layer charging*. Although the faradaic current is a direct measure for the rate of the electrode reaction, several effects usually occur that have to be considered.

Diffusion Limitation of the Faradaic Current.
The decrease of the analyte concentration at the electrode surface due to an electrode reaction must be balanced by the diffusion of species from the bulk solution. In most measurements, the consumption of reactive species is faster than the supply by diffusion and the effect of *diffusion limitation* of the faradaic electrode current is observed. To understand this important point, the time-dependent concentration profile of the analyte has to be calculated using *Fick's laws*. The electrode current can then be derived as a function of time. According to Fick's first law, the flux j of the analyte at the point r and at the time t is proportional to the gradient of the analyte concentration c:

$$j(r,t) = -D\nabla c(r,t) \tag{70.23}$$

The proportionality factor D is called the *diffusion coefficient*. At the electrode surface, the flux must be equal to the number of moles N converted per unit of time and surface area by the electrode reaction:

$$j(0,t) = dN/dt \tag{70.24}$$

The faradaic current i_f is related to dN/dt according to:

$$i_f = nFA\,dN/dt \tag{70.25}$$

where n is the number of electrons involved in the reaction of a single analyte particle, F is the Faraday constant, and A is the surface area of the working electrode. The *Nernst diffusion layer* model assumes that within a layer of thickness δ, the analyte concentration depends linearly on the distance from the electrode surface until it reaches the bulk concentration c_0. For simplicity, the diffusion problem is often considered to be one-dimensional as it is the case for a planar working electrode in a cylindrical cell. The combination of Equations 70.23 to 70.25 then gives:

$$i_f = nFAD\big((c_0 - c_e)/\delta\big) \tag{70.26}$$

where c_e is the concentration at the electrode surface. For a sufficiently large difference between the applied potential and the standard potential of the analyte's redox couple, all species reaching the electrode surface by diffusion are immediately converted and the faradaic current reaches a maximum. In this case, the analyte concentration c_e at the electrode surface can be regarded as zero.

The diffusion profile and thus the dependence of δ from time can be obtained by solving the differential equation that is known as Fick's second law:

$$\frac{\partial c(r,t)}{\partial t} = D\nabla^2 c(r,t) \tag{70.27}$$

where ∇^2 is the Laplacian operator. For *linear diffusion* — that is, one-dimensional diffusion as it was considered in Equation 70.26 — the solution of Equation 70.27 with the appropriate boundary conditions $(c_e(t=0) = c_0;\ c_e(t>0) = 0;\ c(x > \delta = c_0)$ yields:

$$\delta = \mathrm{sqrt}(\pi D t) \tag{70.28}$$

Combination with Equation 70.26 leads to the *Cottrell equation*:

$$i_f(t) = nFAc_0\,\mathrm{sqrt}(D/\pi t) \tag{70.29}$$

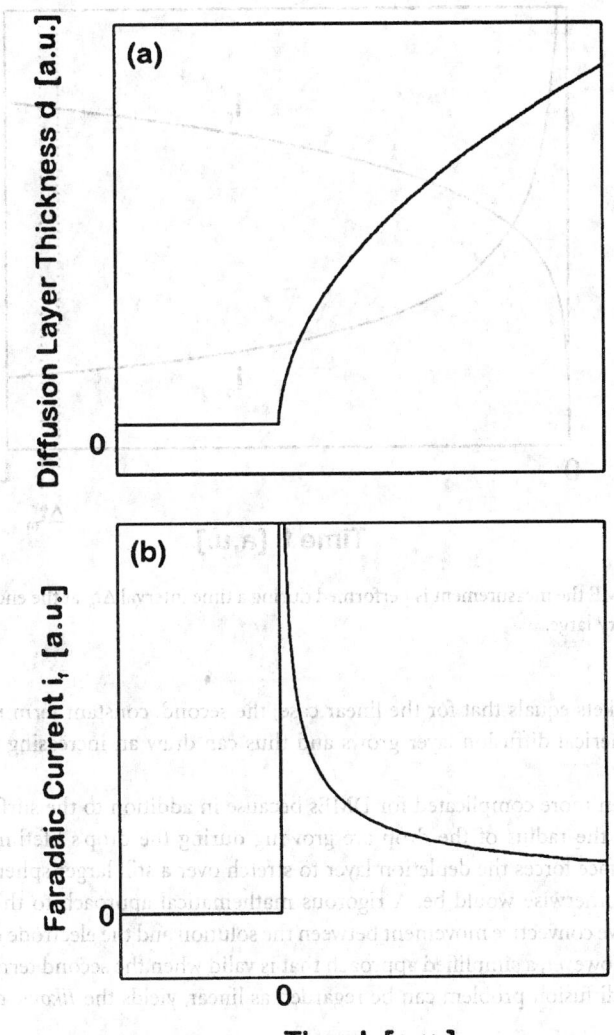

FIGURE 70.9 At a planar electrode, the diffusion-layer thickness increases with $t^{1/2}$ (a) whereas the diffusion-limited current decreases with $t^{-1/2}$ (b).

After reaching a maximum value, the current decreases with $t^{-1/2}$ and is proportional to c_0, whereas the diffusion layer thickness increases with $t^{1/2}$ (Figure 70.9).

For a spherical electrode of radius r_0, as it is the case for HMDEs, one has to change to spherical coordinates and Fick's second law becomes:

$$dc(r,t)/dt = D\left[d^2c(r,t)/dr^2 + 2/r\,dc(r,t)/dr\right] \qquad (70.30)$$

where $r > r_0$ is the radial distance from the electrode center. The solution of Equation 70.30 with the appropriate boundary conditions $c(r,0) = c_0$, $\lim(r \rightarrow \infty)\,c(r,t) = c_0$, $c(r_0, t > 0) = 0$ yields the current-time relation

$$i_f(t) = nFADc_0\left[1/(\pi Dt)^{1/2} + 1/r_0\right] \qquad (70.31)$$

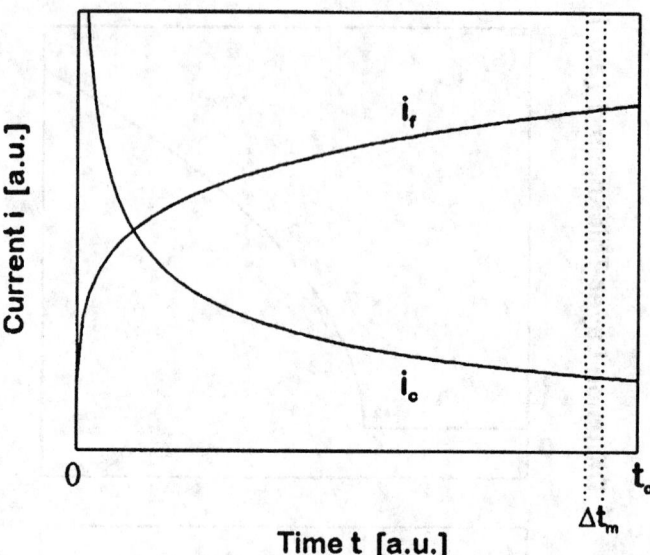

FIGURE 70.10 At a DME the measurement is performed during a time interval Δt_m at the end of the drop's lifetime when the ratio i_f/i_c is very large.

The first term in brackets equals that for the linear case; the second, constant term reflects the fact that the surface of the spherical diffusion layer grows and thus can draw an increasing number of reactive species.

The situation is even more complicated for DMEs because in addition to the surface of the diffusion layer, the surface and the radius of the drop are growing during the drop's lifetime. At any time, the growing electrode surface forces the depletion layer to stretch over a still larger sphere, which makes the layer thinner than it otherwise would be. A rigorous mathematical approach to this is rather difficult [11] because the relative convective movement between the solution and the electrode during drop growth must be considered. However, a simplified approach that is valid when the second term in Equation 70.31 is negligible, and the diffusion problem can be regarded as linear, yields the *Ilkovic equation*

$$i_f\left(t\right) = 708 n D c m^{2/3} t^{1/6} \tag{70.32}$$

where m is the mercury flow rate (mass/time) from the capillary. Consequently, the current increases during the lifetime t_d of the drop (*drop time*), whereas it decreases with time in the other arrangements that have been described. Figure 70.10 depicts this current–time relation of a DME with the characteristic current plateau at the end of the drop's lifetime.

In the considerations that have been made above, analyte transport by convection and migration in the electric field have been neglected. Convection can be regarded as absent if the solution is unstirred and if the working electrode rests motionless. However, in longer-lasting measurements, convective mass transport can play a role due to arising inhomogeneities in the density of the solution. Furthermore, if a DME is employed, the growth of the drop may cause a considerable convection of the solution. When the drop falls off, it stirs the surrounding solution and the depletion effect almost vanishes. Consequently, every drop is born in an almost homogeneous environment. The migration of electrically charged analyte particles due to the electric field in the solution can easily be suppressed using an inert supporting electrolyte with a concentration that is much larger than the analyte concentration. Since all charged species contribute to the migration current, the migration of the analyte species can then be neglected.

Double-Layer Charging Current.

A process that affects all kinds of voltammetric measurements is the flow of *capacitive current*. The accumulation of charge on one side of the electrode/solution interface causes the necessity of a mirror charge on the other side. Hence, a change of the electrode potential (i.e., in the electrode charging) causes a corresponding flux of charged particles between the double layer and the bulk solution. Therefore, the interface has a certain capacitance that is called the double-layer capacitance. The resulting *double-layer charging current* i_c is superimposed on the faradaic current and often perturbs its measurement. In analytical techniques, one is often concerned with the reduction of the capacitive/faradaic current ratio. However, the actual measurement of the double-layer capacitance is demanding and requires the technique of *impedance spectroscopy*, as described, for example, in References 12 and 13.

Irreversible Electrode Processes.

Another assumption that has been made implicitly is that the rate of the electrode reaction is very fast in comparison to the supply of analyte by diffusion (*reversible electrode process*). Under this condition, all analyte species reaching the electrode are immediately converted. However, if the reaction rate is too slow, the consumption of reactive species is compensated by the diffusion of the analyte (*irreversible electrode process*) and thus, the concentration at the electrode surface never drops to zero. The electrode current is then determined by the reaction rate and the calculations above do not hold. In practice, the situation is sometimes complicated if so-called *quasi-reversible* electrode processes with intermediate reaction rates occur. Although this concept of electrochemical *reversibility* is a simplification, it is a suitable working basis and can be summarized in the following statement: In a given electrochemical experiment, an electrode process that follows the *Nernst equation* at *any time* is called *reversible*.

Influence of Adsorption, Catalysts, and Chemical Reactions.

Besides the diffusion and reaction rate, some other processes can influence the electrode current. *Adsorption* of the analyte or its reaction product on the electrode changes the double-layer capacitance or can passivate the electrode surface and thus lower the current. Moreover, if a species serves as a *catalyst*, it may shift the equilibrium potential. In the case that the catalyst returns the product of the electrode reaction back into the initial form of the analyte, the analyte concentration at the electrode surface will always be large and thus increases the limiting current and shifts the equilibrium. All these *catalytic currents* are subject to analytical studies. Besides adsorption and catalysis, complicated scenarios occur if the electrode reaction is followed by a chemical reaction whose product itself undergoes an electrode reaction within the observed potential range.

Techniques

The several voltammetric (i.e., *potential-controlled*) techniques differ just in the manner in which the electrode potential is varied with time. The potential can be changed in distinct steps, in a continuous sweep, or it can be pulsed or superimposed with an ac signal. In addition, the rate of potential change can be varied. The characteristics, advantages, and drawbacks of the most important techniques will be discussed in the following sections. Special attention will be given to polarography due to its practical importance in electroanalysis. Besides, the emphasis will be on reversible electrode processes because only they allow the realization of *analytical* investigations, on which this chapter is focused.

Amperometry.

If in a *potential step* experiment the working electrode potential is abruptly changed from a constant value E_1 where faradaic processes do not occur to another constant value E_2 where the electrochemical equilibrium is on the side of the oxidized or reduced form of the analyte, then a faradaic current begins to flow (Figure 70.11(a)). In the case that the difference between the applied potential and the standard potential E^0 of the analyte's redox couple is sufficiently large, the effect of *diffusion limitation* sets in and a further increase of the potential difference yields no increase in the electrode current. The current is then called *limiting current*. The current–time relationship follows the Cottrell equation (Equation 70.29), with the current decreasing while the diffusion layer thickness increases.

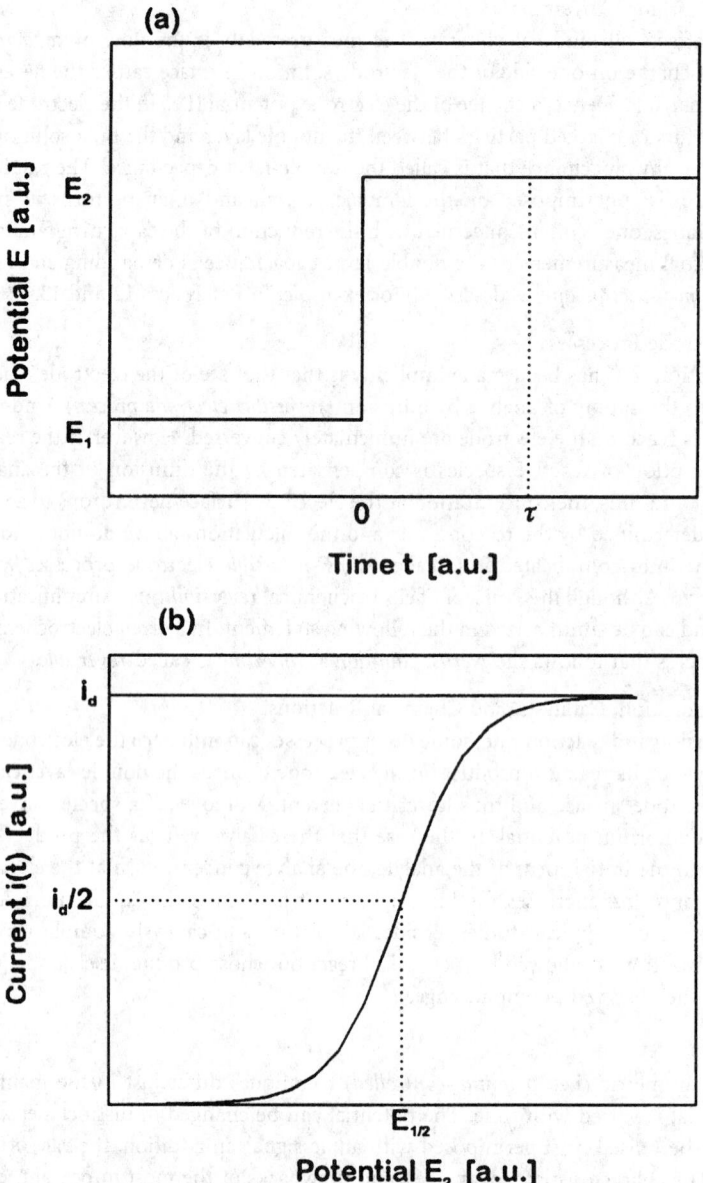

FIGURE 70.11 In potential step techniques, the current is measured a fixed time τ after the potential step (a). The measurement of $i(\tau)$ for different potential steps ΔE yields a wave-shaped current–potential relation with a half-wave-potential $E_{1/2} \approx E^0$ (b). The maximum current is proportional to the analyte's bulk concentration c_0.

If the diffusion layer thickness could be held constant, then from Equation 70.26, it follows that the current would not decrease with time but remain at a constant value. This can be accomplished if the solution is stirred or flows across the electrode in a proper way. According to Equation 70.26 (with $c_e = 0$), the current then is proportional to the analyte concentration in the solution.

The described method corresponds to the electroanalytical technique called *amperometry* [14], with the exception that in this the potential step is omitted and the electrode current is measured at a fixed potential E at which the analyte undergoes an electrode reaction and the faradaic current is in the limited region. The solution usually crosses the electrode in a laminar flow, keeping the diffusion layer thickness constant.

Because the electrode current is proportional to the concentration of the analyte, only two measurements are needed for calibration. The base current is measured in an analyte-free solution and a second measurement is performed at a known analyte concentration. It should be mentioned that amperometry cannot only be used to determine liquid and ionic components of a solution but also to measure the amount of dissolved gas in a liquid. Moreover, with modified electrochemical cells, even gas analysis can be accomplished.

The main disadvantage of amperometry is its poor selectivity. Given a certain analyte and operating at a higher potential than the corresponding standard potential E^0, all components of the solution with a standard potential smaller than E also contribute to the faradaic current. Operating at a potential $E < E^0$, the same problem occurs if substances with a standard potential larger than E are present in the solution. For this reason, amperometry is preferably carried out in solutions containing only one electroactive substance or, if possible, at a potential at which only one substance is involved in an electrode reaction. If this is impossible, the selectivity can often be enhanced by covering the working electrode with a membrane which, in comparison to the diffusion rate of the analyte through the membrane, is virtually impermeable for the interfering substances.

In addition to analytical purposes, amperometric methods can also be used to investigate reaction constants of chemical reactions. In *reversed potential step techniques*, the first potential step is followed by a second one in the opposite direction, often back to the initial value. The reaction product B of the first step is then reconverted into the original analyte A. However, if the first electrode reaction is followed by an additional chemical reaction, a certain part of B is converted into a product C before the reversed step is applied. Therefore, the current during the reversed step is reduced. The ratio of the electrode currents during the forward and reversed steps depends on the reaction constant of the chemical reaction. Because the reconversion of B into A is required, batch arrangements without convection of the electrolyte are used for reversed step methods. Otherwise, a large part of B would be flushed away from the electrode surface and could not be reconverted.

Amperometric Titration.
In *amperometric titration techniques* [15], a titrant that reacts with the analyte is added to the analyte solution. During the titration, the limiting current is measured as a function of the volume of titrant added. The titrant has to be chosen such that the reaction product is not reducible or oxidizable at the applied potential and, hence, does not contribute to the current.

If the analyte as well as the titrant are electroactive at the applied potential, then the current flow will be large at the beginning of the measurement and decreases linearly with the volume of the titrant added, because both, the analyte and the titrant, are consumed by the reaction. The concentration of electroactive species then diminishes until the analyte is totally consumed. Further addition of titrant leads to a linearly increasing current because the titrant is no longer consumed. In the plot of the current versus the volume of titrant added, the point where the slope changes is called *endpoint* of the titration. From the corresponding amount of titrant added and the stoichiometry of the reaction the original volume of analyte can be computed. If only the analyte is electroactive, then from the endpoint the current will not increase but remain zero. If only the titrant undergoes an electrode reaction, the current will be zero until all analyte is consumed and then will linearly increase from the endpoint. In practical operation, the slope of the current does not change abruptly due to background currents and the endpoint has to be determined by extrapolation of the two linear regions.

In contrast to the majority of other electrochemical techniques, amperometric titration offers the advantage that even analytes which are not reducible or oxidizable can be determined using the oxidation-reduction characteristics of the titrant. Moreover, it is possible to analyze systems that have no measurable standard potential but can be electrolyzed.

Sampled-Current Voltammetry.
Consider a *potential step* experiment like the one in the section next to previous one. If the potential difference between E_2 and E^0 is too small, the electrode reaction is not so efficient that the analyte concentration at the electrode surface becomes zero (i.e., $c_e > 0$ in Equation 70.26). Within this region,

the current depends on the applied potential. However, even in this situation, a depletion effect occurs so that the current always decreases with time. Recording the current i for different values of E_2 at a fixed time τ after switching the potential (*sampled-current voltammetry*), a sigmoidal (wave-shaped) curve is obtained (Figure 70.11(b)).

The shape of this curve can also be calculated by exactly solving the diffusion problem. A wave rising from a baseline to the diffusion-limited current i_d is obtained. In the common case, the diffusion coefficients of the analyte and its redox partner are nearly equal the *half-wave potential* $E_{1/2}$, where $i = i_d/2$ is almost identical with the standard potential E^0. Therefore, $E_{1/2}$ is often used in qualitative analysis to determine the analyte. Quantitative information about the analyte concentration is obtained from the maximum current (Cottrell current), which according to Equation 70.29 is proportional to c_0.

The influence of the double-layer charging current has been neglected thus far, but is worth considering. It obeys the equation:

$$i_c = \Delta E / R_s \exp\left(-t/\left(R_s C_{dl}\right)\right) \tag{70.33}$$

where ΔE is the potential step width, R_s the solution resistance, and C_{dl} is the double-layer capacitance. Although the measurement of R_s and C_{dl} is not trivial, one can obtain qualitative information from this formula. Comparison of Equations 70.33 and 70.29 yields that the capacitive current decreases exponentially while the faradaic current decreases according to $t^{-1/2}$. Consequently, the electrode current is measured a sufficiently long time after the potential step when the capacitive current has largely decayed, whereas the faradaic current is still significant. In polarography with DMEs, the growth of the electrode surface alters the temporal decrease of the double-layer charging current according to:

$$i_c \sim m^{2/3} t^{-1/3} \tag{70.34}$$

whereas the faradaic current increases according to $t^{1/6}$ (Equation 70.32). The current is measured shortly before the drop falls off (Figure 70.10).

The lower detection limit amounts to 10^{-5} to 10^{-6} mol/L for the determination of organic and inorganic analytes. The half-wave potential of different substances should be at least 100 mV apart for a simultaneous determination.

Linear Sweep and Cyclic Voltammetry.

In *linear sweep voltammetry* (LSV), the electrode potential is changed *continuously* from an initial to a final value at a constant rate $v = dE/dt$, such that $E(t) = E_1 \pm vt$. Starting at a potential E_1 where no faradaic process occurs, a current begins to flow when the electrode potential comes into the vicinity of E^0. The current rises to a maximum and then decreases due to the depletion effect (Figure 70.12). The solution of the diffusion equations, which yields the shape of the $i–E$ wave, can only be found numerically. For the electrode process to always follow the Nernst equation and thus be reversible, the sweep rate must not be too high (e.g., $v < 100$ mV s^{-1}). The peak potential E_p can then be calculated to be:

$$E_p = E_{1/2} \pm 1.1 \cdot \left(RT/nF\right) = E_{1/2} \pm \left(28.0/n\right) \text{mV} \left(\text{at } 25°C\right) \tag{70.35}$$

The positive sign in Equation 70.35 applies to an anodic sweep (from negative to positive potential with $v > 0$) and the negative sign to a cathodic one (from positive to negative potentials with $v < 0$). The peak current is given by:

$$i_p = 0.446 nFA \left(nF/RT\right)^{1/2} D^{1/2} c_0 v^{1/2} \tag{70.36}$$

Thus, the peak current is proportional to the bulk concentration c_0 of the analyte and depends on the sweep rate according to $v^{1/2}$.

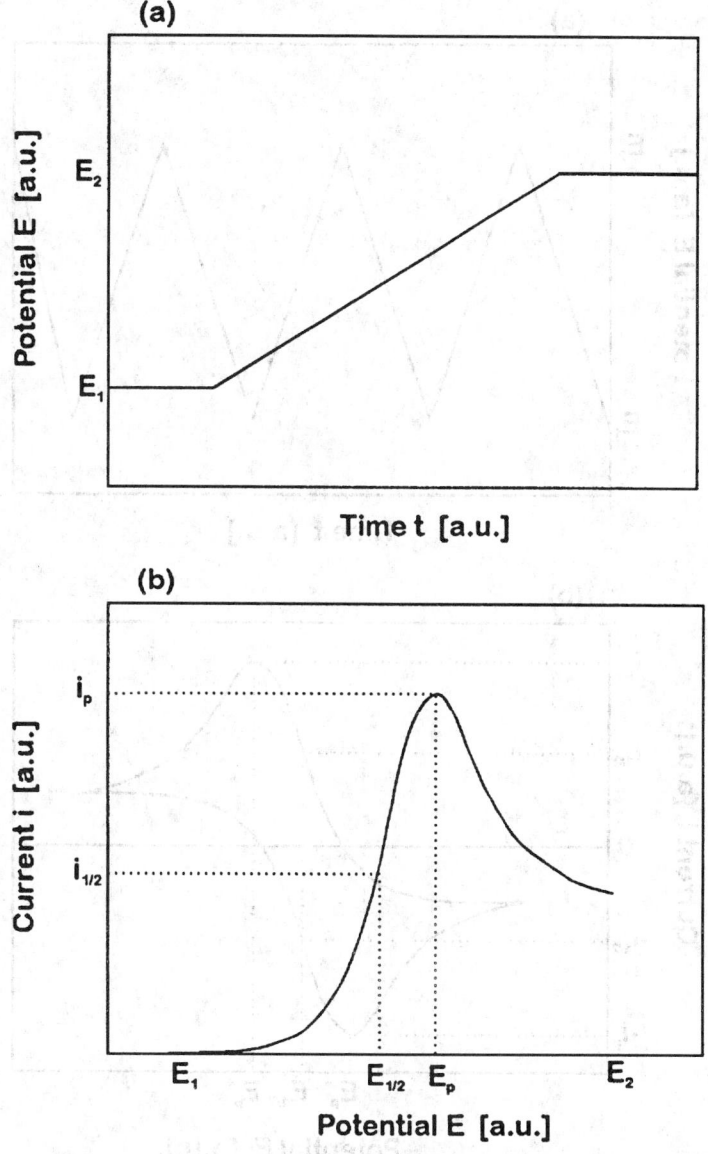

FIGURE 70.12 In LSV the potential varies linearly with time (a). The current–potential relation yields a peak-shaped curve with a half-wave potential $E_{1/2} \approx E^0$ (b). The peak current is proportional to the analyte's bulk concentration c_0.

Another contribution to the measured current is the capacitive double-layer-charging current i_c, which always flows in LSV due to the continuos change of potential. It can be calculated using the equation:

$$i_c = C(dE/dt) = Cv \qquad (70.37)$$

which yields a proportionality to v while the faradaic peak current is proportional to $v^{1/2}$. Thus, for the faradaic current to dominate the measurement, the sweep rate should not be chosen too large. A sweep rate of 100 mV s^{-1} can be regarded as an upper limit. Moreover, the surface area of the working electrode must be taken into consideration. Rough electrodes have a much larger active than geometric surface area and thus a very large capacitance. Therefore, small, very smooth electrodes should be chosen.

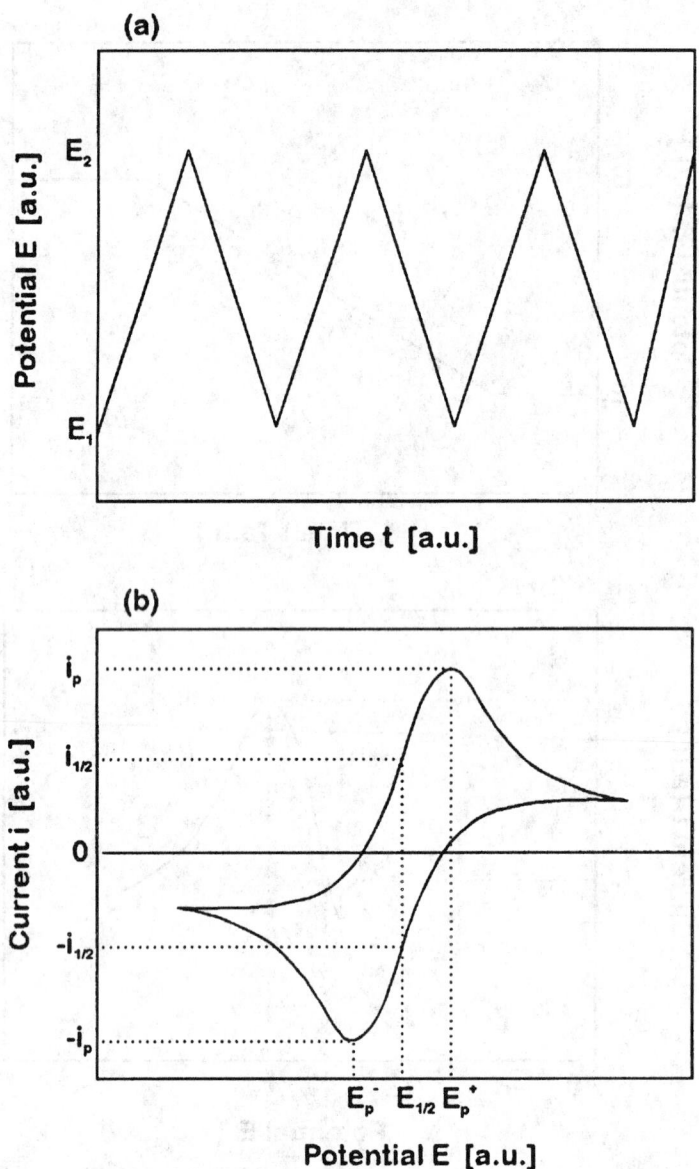

FIGURE 70.13 In CV, the potential is swept forth and back between two fixed values (a). The current–potential relation yields a peak-shaped curve with a half-wave potential $E_{1/2} \approx E^0$ (b). The peak current is proportional to the analyte's bulk concentration c_0. For totally reversible systems, the peak currents of the forward and the backward sweep are equal in magnitude but of opposite sign.

A variation of LSV is a technique called *cyclic voltammetry* (CV). Here, the electrode potential is swept forth and back between two potentials E_1 and E_2 (Figure 70.13(a)). Although the bulk concentration of the reaction product is essentially zero, its concentration at the electrode surface after the first sweep is quite large. In the backward sweep, the reaction product of the analyte is converted into the analyte again. The current flows in the opposite direction and using an *xy*-recorder, an *i–E* curve is obtained (Figure 70.13(b)). From Equation 70.35, it follows that for reversible processes, the peak potentials of the forward and backward sweep have a distance of $(56/n)$ mV at room temperature. Therefore, cyclic voltammetry is a favorable method for the investigation of the reversibility of a system. If the electrode current totally decays in the forward sweep, the analyte concentration has dropped to zero and the product concentration at the electrode surface is about c_0. Ideally, the peak current during the reverse scan should be equal (with reversed sign) to the peak current of the forward sweep.

Although the theory of LSV and CV measurements is very promising, the methods have several practical limitations. One is the frequently insufficient stability of the i–E characteristic during the first cycles in CV. However, after 5 to 10 cycles, it tends to become highly reproducible. Yet, one must be careful deriving quantitative information from these later cycles because the initial and boundary conditions of the diffusion problem have changed and convective mass transport may already play a role. Thus, the equations developed for LSV cannot be used. Another problem that concerns both LSV and CV is the potential drop that occurs in the solution between the working and the reference electrode and which leads to a distortion of the shape of the i–E wave. This error increases with increasing current flow. Thus, the rate of change v of the electrode potential is not really constant, as has been assumed in the boundary conditions for solving the diffusion equations. Furthermore, the quantitative information is usually obtained from the position E_p and the height i_p of the current peak where the error is maximum. Finally, the determination of the peak height itself is sometimes problematic due to difficulties in the extrapolation of the baseline. For all these reasons, it may be advisable to verify the results of quantitative analysis with additional methods. Nevertheless, on easy terms, the lower detection limit of LSV and CV in quantitative analysis can amount to 10^{-7} mol/L with a resolution of about 50 mV.

Besides the analysis of faradaic processes, LSV and CV are favorable techniques for the investigation of the adsorption of species on the electrode surface [16]. In such adsorption processes, the current is called *pseudocapacitive current*. Although it is a charge transfer across the interface, it exhibits many of the properties of a pure capacitive current. The current–potential wave has a very similar shape as for faradaic processes. If Θ denotes the coverage ($0 \leq \Theta \leq 1$) and q_1 the charge that is required to form a monolayer of a species, the pseudocapacitive current i_a can be expressed as:

$$i_a = q_1 \left(d\Theta/dt \right) = q_1 \left(d\Theta/dE \right) \left(dE/dt \right) = C_a v \qquad (70.38)$$

where C_a is called the *adsorption pseudocapacitance*. The calculation of C_a yields that the pseudocapacitance does not depend on v. Therefore, at any potential the current is proportional to the sweep rate ($i \sim v$). The peak potential gives information about the adsorption kinetics. In contrast to faradaic CV, it has the same value for the forward and the backward sweep.

Pulse Techniques.

Voltammetric pulse techniques are derived from potential step experiments to suppress the capacitive currents during the measurement. A potential step that can vary in amplitude and sign is periodically repeated and superimposed with a potential ramp. The current is measured at the end of the step when the double-layer charging current has largely decayed.

Normal Pulse Voltammetry. In *normal pulse techniques*, periodic voltage pulses with an increasing amplitude from pulse to pulse are superimposed on a constant potential. A typical pulse duration is about 50 ms and the current is measured during a time interval Δt_m of about 10 to 15 ms at the end of each pulse. Between two pulses there is a waiting period of a few seconds (Figure 70.14). In polarography with a DME, each drop is dislodged directly after the pulse and thus used for just one measurement.

Because normal pulse voltammetry equals a series of potential-step measurements with increasing step widths, the current obeys to Equation 70.29 and the evaluation of the measured current values can be carried out using the sampled-current method. In comparison with the step technique, the lower detection limit is enhanced for one to two orders of magnitude up to 10^{-6} and 10^{-7} mol/L [17]. The peak resolution is about 100 mV.

Square-Wave Voltammetry. In *square-wave* techniques, a periodic rectangular voltage is superimposed on a linearly rising potential ramp. The measuring interval lies at the end of a pulse when the capacitive current can be neglected (Figure 70.15(a)). Typical pulses have frequencies between 200 Hz and 250 Hz and an amplitude of $\Delta E_p = 5$ to 30 mV [17]. The capacitive current is suppressed even more effectively if the pulse is tilted to decrease during the pulse period. No pulse tilt is required if the potential ramp is

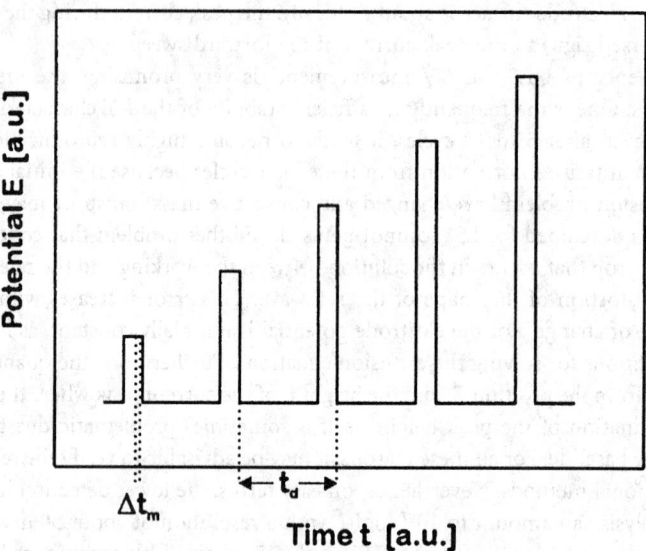

FIGURE 70.14 Normal pulse voltammetry equals a series of potential-step measurements with increasing step widths. The current is measured during a time interval t_m near the end of the pulse. In polarography with a DME, the drop is dislodged after each measurement. The drop's lifetime is denoted by t_d.

stepped (staircase ramp) instead of a linear ramp. The voltage pulse is then applied on the plateau of the stepped ramp (Figure 70.15(b)).

After rectification of the measured current values, one obtains peak-shaped i–E-curves. The peak potential corresponds to the half-wave potential of LSV and, thus, to the standard potential of the analyte's redox couple. The peak current i_p depends on the frequency and amplitude ΔE of the voltage pulses and obeys:

$$i_p \sim n^2 D \Delta E c_0 \qquad (70.39)$$

where the frequency dependence is included in the proportionality constant. The lower detection limit is in the range of 10^{-8} mol/L and the peak resolution amounts to 40 to 50 mV.

Shorter analysis times are achieved if very short and relatively large rectangular pulses with a duration $t_p = 5$ to 10 ms and an amplitude of $\Delta E = 50$ mV are superimposed on a stepped potential ramp with the same duration but smaller potential steps of about 10 mV. The potential can then be scanned at extremely high rates of up to 1200 mV s^{-1}. However, the sensitivity decreases because the ratio of faradaic to capacitive currents is lowered by the short pulse times.

Differential Pulse Voltammetry. *Differential pulse methods* are the most important ones in analytical voltammetry. Periodically repeated rectangular voltage pulses with a constant amplitude ΔE of several 10 mV are superimposed on a stepped potential ramp (Figure 70.16). The pulse duration Δt_p is about 5 to 100 ms [17]. Between two pulses, the potential is held constant for a few seconds. The current is measured in a short time interval ($\Delta t_m \approx 1$ to 20 ms) directly before a pulse is applied and for the same duration near the pulse end. If a DME is used, the drop is knocked off mechanically between two pulses and each drop serves for just one measurement.

For the evaluation, the difference between the two measured current values Δi that corresponds to one pulse is recorded as a function of the base potential. A peak-shaped curve is obtained with a maximum very close to the half-wave potential $E_{1/2}$. The peak height is proportional to the analyte concentration in the bulk:

$$\Delta i_p \sim nFA \left(D / \pi t_p \right)^{1/2} c_0 \qquad (70.40)$$

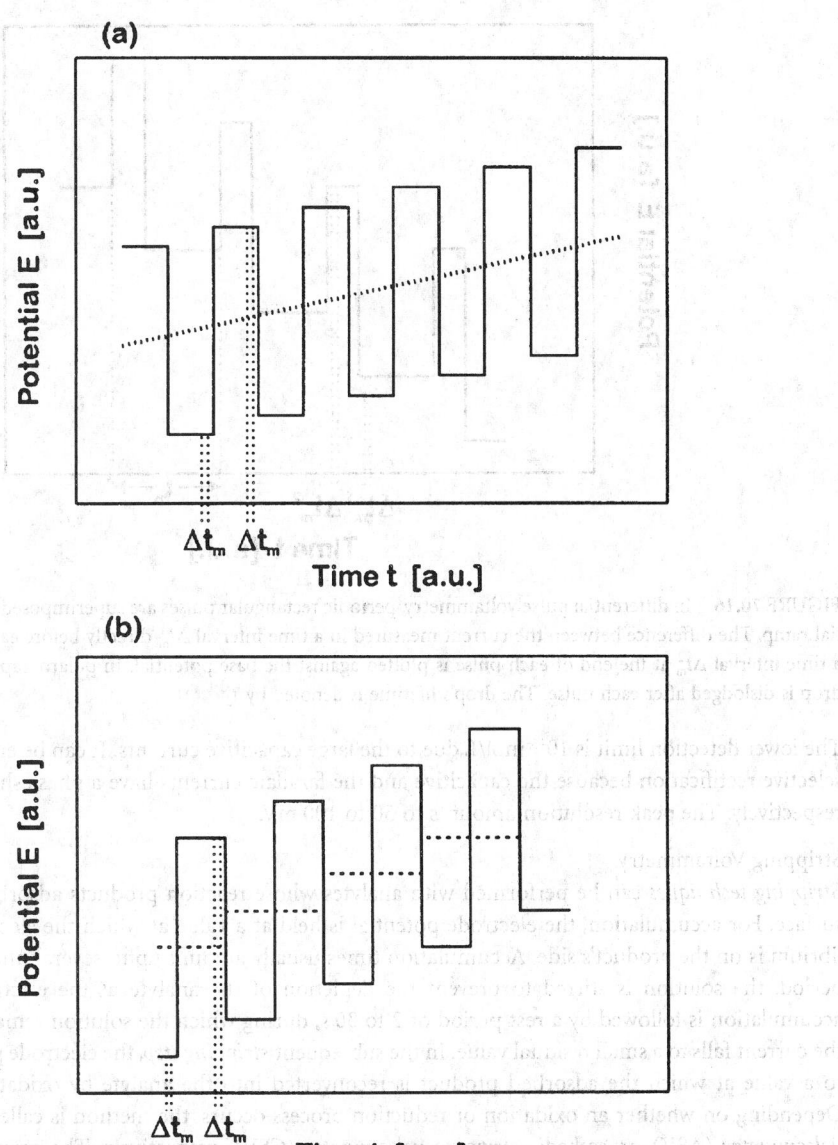

FIGURE 70.15 In square-wave voltammetry, a periodic rectangular voltage pulse is superimposed on (a) a linearly changing potential ramp (dotted line) or (b) on a stepped ramp (dashed curve). The current is measured during a time interval Δt_m at the end of each pulse.

With differential pulse measurements, a lower detection limit of 10^{-8} mol/L and a resolution of 50 to 100 mV can be achieved.

Alternating Current Voltammetry.

Alternating current techniques are similar to differential pulse methods. A linear potential ramp is modulated with a low frequency ($f \sim 50$ Hz) sinusoidal alternating voltage of small amplitude ($\Delta E \sim 50$ mV) [17]. The amplitude of the resulting alternating current is plotted against the base potential. A peak-shaped curve is obtained with a maximum that is proportional to the bulk concentration of the analyte:

$$i_p \sim \Delta E f^{1/2} c_0 \qquad (70.41)$$

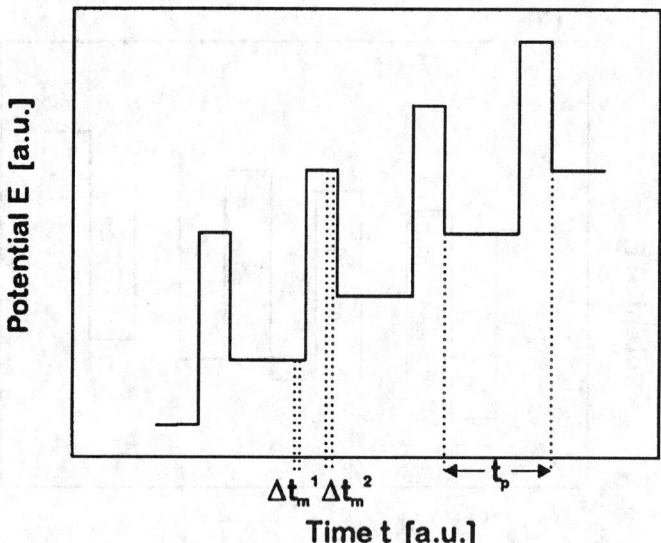

FIGURE 70.16 In differential pulse voltammetry, periodic rectangular pulses are superimposed on a stepped potential ramp. The difference between the current measured in a time interval Δt_m^1 directly before each pulse and during a time interval Δt_m^2 at the end of each pulse is plotted against the base potential. In polarography with a DME, the drop is dislodged after each pulse. The drop's lifetime is denoted by t_d.

The lower detection limit is 10^{-5} mol/L due to the large capacitive currents. It can be enhanced by phase-selective rectification because the capacitive and the faradaic currents have a phase shift of 90° and 45°, respectively. The peak resolution amounts to 50 to 100 mV.

Stripping Voltammetry.

Stripping techniques can be performed with analytes whose reaction products adsorb on the electrode surface. For accumulation, the electrode potential is held at a value at which the electrochemical equilibrium is on the product's side. Accumulation times usually amount up to several minutes. During this period, the solution is stirred to prevent the depletion of the analyte at the electrode surface. The accumulation is followed by a rest period of 2 to 30 s, during which the solution remains unstirred and the current falls to a small residual value. In the subsequent *stripping step*, the electrode potential is shifted to a value at which the adsorbed product is reconverted into the analyte by oxidation or reduction. Depending on whether an oxidation or reduction process occurs, the method is called *anodic stripping voltammetry* (ASV) or *cathodic stripping voltammetry* (CSV), respectively. The stripping step can be performed in various manners [18] of which the linear sweep method shall be exemplarily discussed here. It yields a peak-shaped *i–E* curve with a maximum at:

$$E_p = E_{1/2} - 1.1\,RT\big/nF \tag{70.42}$$

and a peak height that is proportional to the bulk analyte concentration according to:

$$i_p \sim n^{3/2} v^{1/2} c_0 \tag{70.43}$$

Different substances can be determined in successive experiments with an adequate choice of the accumulation potentials. For the first measurement, the accumulation potential is chosen to allow adsorption of only one species; in the next experiment, the first and one further analyte adsorb, and so on. For the simultaneous determination of two or more substances their peak potentials should be at least 150 mV apart.

A special case of the stripping techniques is *adsorptive stripping voltammetry* (AdSV). Here, the analyte is deposited in the form of metal chelates or organic molecules. For the formation of metal chelates, a complexing agent is added to the electrolyte or the surface of a solid-state electrode is modified with it. The stripping current is then due to the oxidation or reduction of the central atom or the ligand of the metal chelate complex. With this method, organic and organometallic compounds can be determined in the ultratrace range.

A crucial point in stripping analysis is the reproducibility. All experimental parameters have to be selected very carefully. In particular, the electrode surface must not be changed significantly by the adsorption and dissolution processes. Therefore, HMDEs are frequently employed for stripping analysis. A new drop is produced for each measurement. Another advantage of mercury electrodes is the fact that not only their surface but rather the hole bulk is used for the accumulation of analyte species. Consequently, more material can be collected. This leads to an enhanced lower determination limit which can be below 10^{-8} mol/L. Comprehensive monographs about stripping techniques are given in References 18 and 19.

Applications

Analytical applications of voltammetry concern the determination of (heavy) metal cations, typical anions (halides, pseudohalides), organometallic, and organic compounds in the 10^{-4} to 10^{-9} mol/L concentration range. Therefore, they are established in several fields like environmental, medical, food, and water analysis. A disadvantage is the usually labor-intensive sample preparation necessary, for example, to disintegrate ions from complexes, to adjust the pH of the solution, or to remove interfering species like oxygen and organic molecules. Principally, the preparation of the electrode (surface) is also crucial. However, commercially available equipment is well developed not only to enhance determination limits, sensitivity, selectivity, and reproducibility, but also to reduce the expense for electrode and cell preparation. Moreover, sample and electrode preparation can be automated to a certain degree by devices which pump different solutions for cleaning, conditioning and analysis through the cell setup. A further improvement of the instrumentation is the use of *microelectrodes* with dimensions of 1 to 100 μm. Because their dimensions are small in comparison with the diffusion length of the analyte, even for planar microelectrodes the diffusion is rather hemispherical than linear. Therefore, the depletion effect is less strong and the faradaic current is increased. Moreover, planar microelectrodes can be rotated (*rotating disk electrode*, RDE) to intensify convection and the solution can be stirred with ultrasound. Another advantage of microelectrodes is the possibility to realize several electrodes in a close neighborhood, so-called *electrode arrays*. They serve as *one* electrode if they are held at *one* potential and exhibit an improved signal-to-noise ratio due to the better diffusion conditions. In contrast, if different potentials are applied at different electrodes, the simultaneous determination of different species is possible. These techniques have just become commercially available as electrochemical detectors, for example, for high performance liquid chromatography (HPLC). In this arrangement, the different species in the solution are separated by the HPLC and flow through the detector cell one after the other. Thus, interference between different analytes is minimized. The selectivity can often be further improved by the use of membrane-covered microelectrodes. The well-known *Clark oxygen sensor* and different biochemical sensors represent promising examples of this application in amperometry. Moreover, it opens up new possibilities for the creation of microelectrode arrays.

Due to the high analytical potential and the relatively low costs of voltammetric methods in comparison with spectroscopic techniques, all aspects of voltammetry are still subject of intense research. Current efforts concern the *miniaturization* of the whole cell, including microchannels, microvalves, micropumps, and microelectrodes by means of precision mechanics and *micromachining techniques* [20]. They employ fabrication methods of silicon planar technology and LIGA technique (Lithographie, Galvanoformung, Abformung). Thin-film techniques like physical and chemical vapor deposition (PVD, CVD) allow the fabrication of electrodes with a thickness in the submicrometer range and with lateral dimensions from the micrometer to the nanometer range. One goal is the realization of a *microsystem* with the sensitive components (i.e., the electrodes) and microelectronics integrated on a single chip. The electronics could serve as a first stage of signal amplification and information processing. Although first demonstrator devices have already been presented [21] it is still quite a long way to commercially available systems.

Potentiometry

Potentiometry implies the measurement of an electrode potential in a system in which the electrode and the solution are in electrochemical equilibrium. Thus, the potential becomes the dependent variable, for example, as a function of time. In potentiometry, the current is attempted to be kept as small as possible; ideally, it should be zero. Potentiometry implies known fluxes (i.e., concentration gradients at the electrode surface) and thus information on the composition of the sample. In this section, potentiometry is related to the measurement of potentials, where the voltage source is a form of a galvanic cell, consisting of a measuring electrode and a reference electrode (in general, electrodes of the second kind). The principles of *direct potentiometric measurements* as well as *potentiometric titrations* will be described.

Ion-Selective Electrodes

The equipment required for potentiometric analysis includes a measuring electrode, also called an *ion-selective electrode* (ISE) or *indicator electrode*, and a reference electrode. In addition to the sensitivity, the most important characteristic of the ISE is given by its *selectivity*. Depending on the type of membrane, ISEs can be classified into four different groups: *glass electrodes*, *solid-state electrodes*, *liquid-membrane electrodes*, and miscellaneous *combined electrodes*. For all ISEs, the validity of the Nernst equation could be proved.

Glass Electrodes.
The most common *glass electrode* is the *pH electrode*, widely used for hydrogen ion determination. The pH glass electrode consists of a thin, pH-sensitive *glass membrane* sealed to the bottom of an ordinary glass tube. The tube is filled with a solution of hydrochloric acid (e.g., 0.1 M HCl) that is saturated with silver chloride. A silver wire, connected to an external potential-measuring device, is immersed in this solution. Note that the internal HCl concentration is constant and, thus, the internal potential (inner surface of glass membrane) of the pH electrode is fixed. Only the potential that occurs between the outer surface of the glass bulb and the test solution responds to pH changes. To measure the hydrogen ion concentration of the test solution, the glass electrode (indicator electrode) must be combined with an external reference electrode, which is required for all kinds of ISE determination. Often, pH glass electrodes are available as a combination of the indicator electrode and an internal reference electrode (e.g., Ag/AgCl in saturated KCl solution) as schematically shown in Figure 70.17.

The composition of the glass membrane clearly influences the sensitivity of the pH electrode. Usually, three-component systems of, for example, $SiO_2/Na_2O/CaO$ are employed [22]. The pH dependence can be expressed by the Nernst equation (Equation 70.11). At room temperature (T = 25°C), Equation 70.11 can be simplified by:

$$E = E^0 + 59.1 \text{ mV pH} \tag{70.44}$$

where E^0 is the standard Galvani potential with respect to the SHE. Thus, the measured potential is a linear function of pH within an extremely wide range (10 to 14 decades). The selective pH response of the pH ISE is due to the ion exchange process, in particular, due to the replacement of sodium ions in the glass membrane (m) by protons in the solution (s), and vice versa:

$$H^+_{(s)} + Na^+_{(m)} \Leftrightarrow H^+_{(m)} + Na^+_{(s)} \tag{70.45}$$

The sodium ion exchange is also responsible for the *alkaline error* of pH electrodes in solution with pH greater than 10. In spite of the high resistance of the glass membrane against chemical attack, one

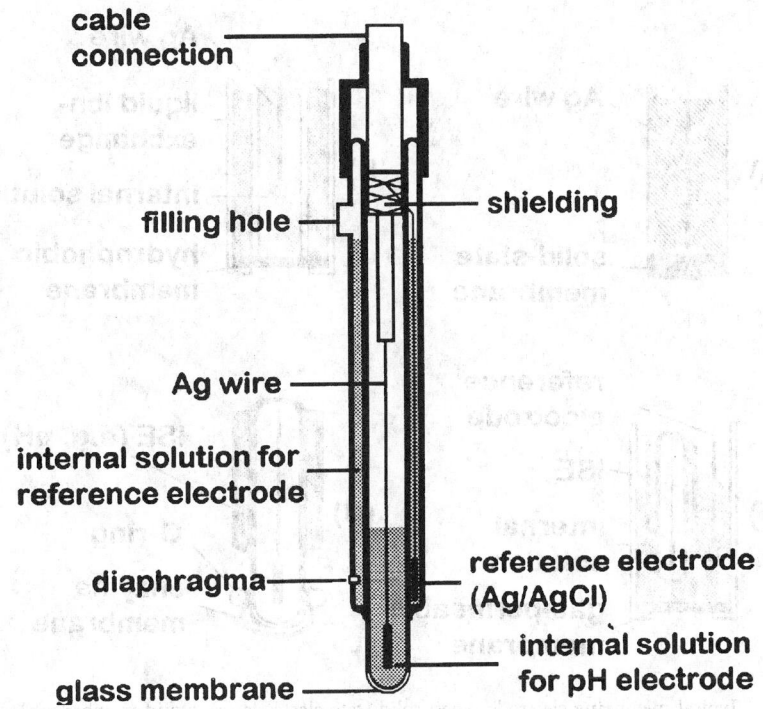

FIGURE 70.17 Combination pH glass electrode with an integrated Ag/AgCl reference electrode.

has to deal with deviations (alkaline error) from the linear pH dependence. This error (i.e., the sensitivity toward alkali-metal ions) can be greatly reduced if Na_2O is replaced by LiO_2. Because pH glass electrodes can be used in the presence of substances that interfere with other electrodes (e.g., proteins, oxidants, reductants, and viscous media), they have a wide range of applications. Typical fields are the clinical and food analysis, environmental monitoring (e.g., industrial waste, acidity of rain), and process control (e.g., fermentation, boiler water, galvanization and precipitation).

The employment of glass membranes prepared with different glass compositions allows an electrode response sensitive to cations. For example, sodium-, potassium-, and ammonium-selective glasses consist of a mixture of Na_2O, Al_2O_3, and SiO_2 in various proportions (aluminosilicate glasses). Using specific compositions and mixtures of chalcogenides, ion-selective *chalcogenide glass electrodes* with sensitivities toward monovalent ions (e.g., Ag^+, Tl^+, F^-, Cl^-, Br^-, I^-) and double-charged species (e.g., Cu^{2+}, Pb^{2+}, Cd^{2+}, Hg^{2+}, S^{2-}) can be prepared [23]. However, in all cases, some sensitivity to charged species (e.g., H^+ ions) remains. The electrode potential under these conditions is described by the *Nikolsky-Eisenmann* equation

$$E = E^0 \pm \frac{RT}{zF}\ln\left(a_i + K_{ij}a_j^{z_i/z_j}\right) \tag{70.46}$$

where z_i, z_j, and a_i, a_j are the ionic charge and activity of the primary or determined (i) and the interfering (j) ion. K_{ij} is the *selectivity coefficient*. It is a measure of the ISE ability to discriminate against the interfering ion. A small value of K_{ij} indicates an ISE with a poor selectivity.

Solid-State Electrodes.
The glass membrane of an ISE can be replaced by a single or a mixed crystal, or a polycrystalline (pressed) pellet (Figure 70.18(a)). With respect to their membrane composition, *solid-state electrodes* are divided into *homogeneous* and *heterogeneous membrane electrodes*.

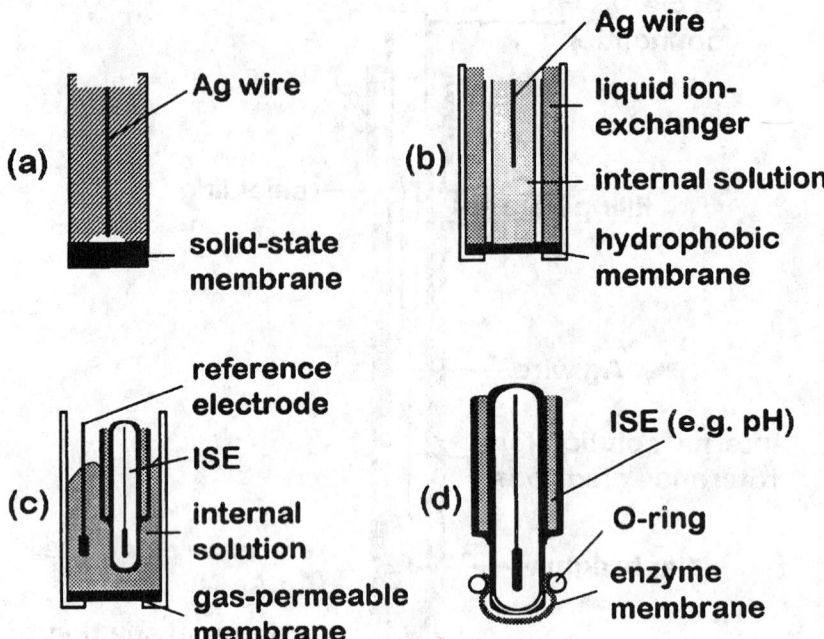

FIGURE 70.18 Typical membrane electrode types: solid-state electrode (a), liquid-membrane electrode (b), gas-sensing electrode (c), and enzyme-based electrode (d).

A typical single-crystal electrode (homogeneous membrane electrode) is the fluoride-sensitive ISE, which contains a LaF_3 crystal doped with Eu^{2+}. The crystal with a thickness of about 2 mm is sealed into the bottom of a plastic tube. The internal solution (0.1 M of NaF and NaCl) controls the potential at the crystal inner side by means of an Ag/AgCl wire as reference electrode. In contact with the test solution at the crystal outer side, an electrochemical equilibrium is established, proportional to the fluoride ion activity. This is due to an ion exchange process at the phase boundary membrane/electrolyte. In particular, fluoride ions from the membrane are replaced by fluoride ions from the solution and vice versa, where the fluoride ions can migrate from one lattice defect to another inside the crystalline membrane. Further homogeneous membrane electrodes are silver halide electrodes, where the respective silver halide (AgCl, AgI, AgBr, Ag_2S) is pressed into a pellet, placed in a tube, and contacted via a silver wire. In these substances, silver ions are accordingly able to migrate. Such electrodes have been successfully used for the selective determination of chloride, bromide, iodide, silver, and sulfide ions. Likewise, if the pellets contain Ag_2S together with the silver halides or mixtures of PbS, CdS, and CuS, solid-state electrodes sensitive toward Pb^{2+}, Cd^{2+}, Cu^{2+}, and SCN^- can be realized. Moreover, the general problem of light sensitivity and high membrane resistance can be reduced by the additional use of Ag_2S.

Instead of the pressed pellets, the ion-selective material can be incorporated into an organic polymer matrix, like silicon rubber, carbon paste, or paraffin. In heterogeneous membrane electrode preparation, a mixture of the precipitate (e.g., AgI/Ag_2S) and polysiloxane is homogenized, and the polymerization is carried out. The resulting disks are fixed on the end of a tube and the internal solution (e.g., 0.1 M KI) is contacted via a Ag/AgCl wire. *Coated-wire electrodes* represent another possibility. They can be manufactured by coating an appropriate polymeric membrane onto a conducting wire. Often, the conductor (Pt, Ag, Cu, or graphite) is dipped in a solution of polymer (e.g., polyvinylbenzylchloride (PVC) or polyacrylic acid) and the active substance. These electrodes allow the determination of K^+, Na^+, amino acids, and some drugs (e.g., cocaine). In addition to their simple miniaturization, the preparation is easy and inexpensive. However, further work is necessary to improve their analytical performance with regard to reproducibility and long-term stability.

Liquid-Membrane Electrodes.

Liquid-membrane electrodes base on two different membrane-active components, *solid ion-exchanger* and *complex-forming neutral-charged carriers*. They permit the determination of several polyvalent cations as well as certain anions. The sensor membrane (10 to 100 μm thickness) is usually prepared of a plasticized PVC containing the organic sensor-active component that is insoluble in water. A Ag/AgCl wire is immersed into the internal reference solution. The liquid-membrane electrode differs from the glass electrode only in that the test solution is separated from the solution with the known target ion activity by a hydrophobic membrane, instead of the glass layer (Figure 70.18(b)). As membrane materials besides PVC, teflon, sintered glass, filtering textile, or disks can be employed to hold the organic layer.

Liquid-membrane electrodes with ion-exchangers have been realized for the determination of, for example, Ca^{2+}, K^+, BF_4^-, ClO_4^-, IO_4^-, SCN^-, I^-, Br^-, Cl^-, HCO_3^-, $H_2PO_4^-$, and NO_3^-. On the other hand, the synthesis of compounds containing individual cavities of molecule-sized dimensions results in complex-forming neutral-charged carriers. These *ionophores* (e.g., crown ethers like cyclic polyether, depsipeptides like valinomycin, and macrotetrolides like nonactin and monactin) are capable of enveloping various target ions reversibly in their pockets. For example, valinomycin membranes show a high K^+ selectivity. Many cyclic and monocyclic carriers with remarkable ion selectivities have been successfully developed for the determination of Li^+, Cs^{2+}, Ca^{2+}, Na^+, NH_3^+, Mg^{2+}, Ag^+, Hg^{2+}, SCN^-, and $H_2PO_4^-$ [24]. For all kind of membranes, a high molecular weight (i.e., a slight overpressure) prevents the quick intrusion of the test solution inside. Hence, the electrode's lifetime is limited as a consequence of diffusion of the sensor-active component into the analyte (*leaching out*).

Combined Electrodes.

Two different types of *combined electrodes* will be presented here: *gas-sensing electrodes* and *enzyme-based electrodes*. Gas-sensing electrodes can be used to determine solutions of gases. They consist of an inner sensing element, normally a suitable ISE with an electrolyte solution (0.1 M), surrounded by a gas-permeable membrane (Figure 70.18(c)). On immersion of this ISE, the gas-permeable membrane contacts the liquid of the gas which diffuses through it, and the resultant internal solution will be examined with the ISE. The partial pressure of the gas attains an equilibrium between the test solution/membrane and the membrane/ISE phase boundary. For example, the determination of carbon dioxide, which diffuses through the semipermeable membrane, lowers the pH values of the inner solution:

$$CO_2 + H_2O \Leftrightarrow HCO_3^- + H^+ \tag{70.47}$$

Such pH changes are detected by the ISE, in this case by a pH-sensitive glass electrode. Semipermeable membrane materials are polytetrafluorethylene, polypropylene, or silicone rubber. The internal solution contains sodium chloride and an electrolyte with the corresponding ion that is determined. Gas-sensing electrodes have been realized for gases dissolved in solution, such as NH_3, NH_4Cl, CO_2, H_2CO_3, $NaHCO_3$, NO_2, $NaNO_2$, SO_2, H_2SO_3, $K_2S_2O_5$, CN, SCN, Cl_2, Br_2, I_2, and H_2S.

Enzyme electrodes are based on the coupling of an enzymatic membrane with any type of appropriate ISE. The enzyme converts (*catalyzes*) the analyte (*substrate*) to be determined extremely selective into an ionic product. The latter can be detected by the known ISE (Figure 70.18(d)). The coupling of the enzyme can be carried out by several *immobilization procedures*, such as entrapping in a gauze or gel, adsorptive or covalent binding, and cross-linking. A typical example for the operation of an enzyme electrode is given by the urea electrode. The enzyme urease hydrolyzes urea in order to liberate ammonium ions:

$$CO(NH_2)_2 + H_2O + 2H^+ \xrightarrow{\text{urease}} 2NH_4^+ + CO_2 \tag{70.48}$$

Either the alteration of the pH by a pH ISE or the variation of the NH_4^+ concentration by an ammonium-sensitive gas electrode can be detected. Likewise, penicillin, glucose, lactate, phenol, creatinine, cholesterol, salicylate, or ethanol will be catalyzed by means of the respective enzyme. Using different biological

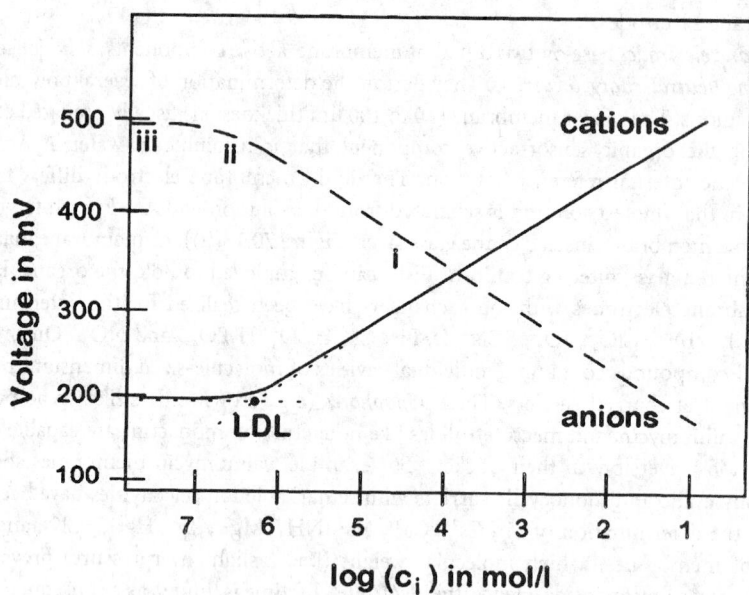

FIGURE 70.19   Schematic calibration curves for anions and cations (LDL: lower detection limit).

components (enzymes, cells, tissues, antibodies, receptors, or nucleic acids), a wide variety of analytically important substances for clinical, environmental, and food analysis can be determined. However, disadvantages of this type of electrode are its slow response time (several minutes) and the insufficient stability in the long term.

Instrumentation and Measurement.

For potentiometric measurements, one uses an indicator electrode (ISE) versus a reference electrode and a *potentiometer*, also called *pH meter* or *ion meter*. Owing to the high resistance of the ISE membranes (e.g., 5 to 500 MΩ for the glass membrane), a potentiometer with a high input resistance is required. Modern potentiometers consist of an electronic digital voltmeter with a suitable operational amplifier, scaled directly to pH units or mV, with a resolution of better than ±0.002 pH and ±0.1 mV. They may range from simple hand-held instruments for field applications to more convenient laboratory models. Frequently, potentiometers include a bias control that can be adjusted to correspond to the temperature of the test solution (*automatic temperature compensation*).

Direct Potentiometry. *Direct potentiometric* measurements can be performed for the determination of ionic species for which an appropriate ISE is available. A schematic measuring set up for direct potentiometry is shown in Figure 70.3. The measuring technique is quite simple: comparing the potential of the ISE in the test solution with its potential in a known *standard solution*. That means, before the determination, the ISE must be calibrated in solutions of known concentration of the chosen ionic species. Thus, for the ion determination to be made, at least two to three reference solutions are necessary which differ by two to five concentration decades. Typical resulting *calibration curves* for anions and cations are plotted in Figure 70.19. The curves can be separated into three distinct regions: (1) the straight part corresponding to the Nernstian slope (i.e., the sensitivity of the ISE), (2) the curve portion, and (3) the horizontal part below the lower detection limit, where almost no sensitivity exists. The *lower detection limit* (LDL) of the ISE is defined as the concentration at which the extrapolated horizontal portion of the graph intersects the extrapolated Nernstian portion of the graph.

For practical applications, there are two aspects to be dealt with: often a *total ionic strength adjuster buffer* (TISAB) is added to both the standard solutions and the test solution (same temperature) to achieve comparable ionic strengths. Then, the potential difference can be assigned to the equivalent concentration of the calibration curve. Various methods for calibration calculations are described by, for

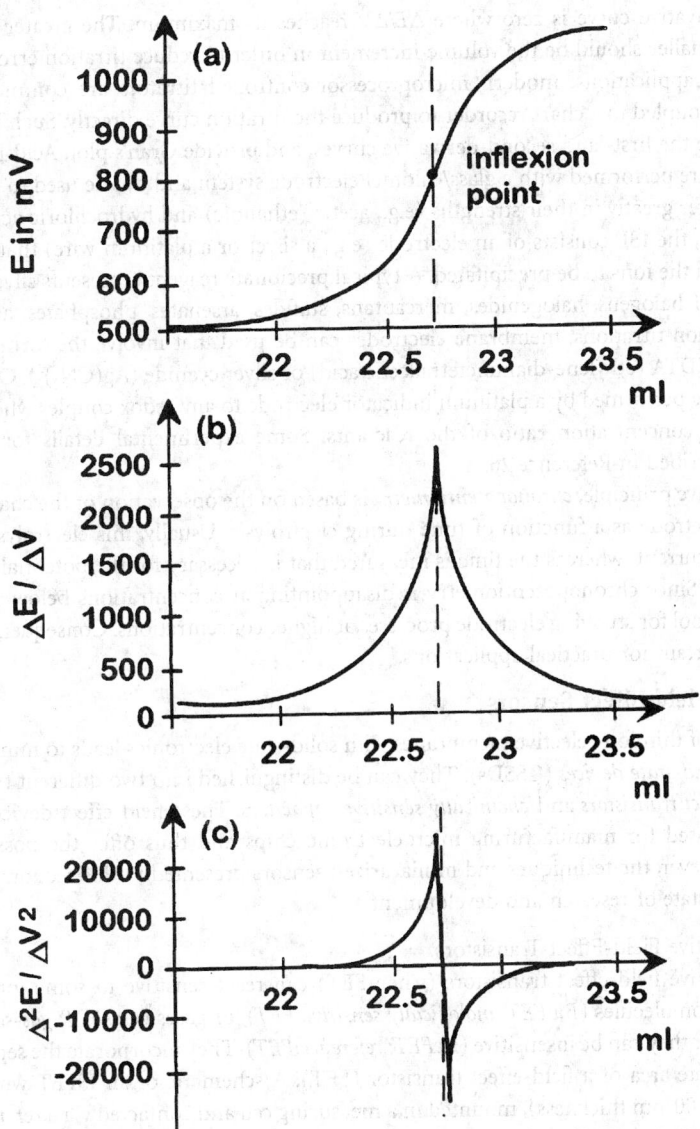

FIGURE 70.20 Characteristic potentiometric titration curve (a), first-derivative curve (b), and second-derivative curve (c).

example, Gran's plot or the standard addition method [25]. Because all measurements take place in dilute solutions ($\leq 0.1\ M$), ion concentrations can be used in the Nernst equation instead of ion activities.

Potentiometric Titrations. *Potentiometric titrations* can be applied in the fields of *acid-base, precipitation, complex-formation*, and *redox reactions*. Therefore, the ISE is used in combination with a reference electrode in order to establish the *equivalence point* in a titration curve. A typical S-shaped potentiometric titration curve, where the electrode potential is plotted versus the reagent volume (*titrant*) is given in Figure 70.20(a). The titrant is added to the initial solution which is stirred, and the ISE records the potential value at equilibrium. The equivalence point (*endpoint*) of the reaction is reached when a sudden change in the potential of the ISE occurs. The midpoint in the curve (i.e., the steeply rising portion) is termed *endpoint* or *inflexion point*. It can be evaluated by analytical methods, namely the first- and second-derivative curve (Figure 70.20(b) and (c)). The first-derivative curve gives the potential change per unit change in volume of reagent and depicts the endpoint at the maximum of the inflexion point.

The second-derivative curve is zero where $\Delta E/\Delta V$ reaches its maximum. The greater the slope at the endpoint, the smaller should be the volume increment in order to reduce titration errors.

For practical applications, modern microprocessor-controlled titrators are commercially available (*auto-titrator*), coupled to a chart recorder to produce the titration curve directly. Such instruments also allow to evaluate the first- and second-derivative curves, and provide Gran's plot. Acid-base (neutralization) titrations are performed with a glass/calomel electrode system and can be used to titrate a mixture of acids that differ greatly in their strengths (e.g., acetic (ethanoic) and hydrochloric acids). For precipitation titrations, the ISE consists of an electrode (e.g., a silver or a platinum wire) that quickly reaches equilibrium with the ions to be precipitated. A typical precipitate reagent represents silver nitrate for the determination of halogens, halogenides, mercaptans, sulfides, arsenates, phosphates, and oxalates. For complex formation titrations, membrane electrodes can be used that involve the formation of soluble complexes, like EDTA (ethylene-diaminetetraacetic acid) or silver cyanide ($Ag(CN_2)^-$). Oxidation-reduction titrations are performed by a platinum indicator electrode to any redox couples where the potential depends on the concentration ratio of the reactants. Some experimental details for potentiometric titration are described in Reference 26.

As an alternative principle, *chronopotentiometry* is based on the observation of the change in potential of a working electrode as a function of time during electrolysis. Usually, this electrolysis is performed with a constant current, whereas the time is measured that is necessary for the potential to go from one level to another. Since chronopotentiometry is disappointing at concentrations below 10^{-4} mol/L, it is only a powerful tool for studying electrode processes at higher concentrations. Consequently, this method is not very important for practical applications.

Ion-Sensitive Field-Effect Sensors

The integration of thin ion-selective membranes with solid-state electronics leads to miniaturized *chemically sensitive solid-state devices* (CSSDs). They can be distinguished into two different types: *chemically sensitive field-effect transistors* and *chemically sensitive capacitors*. These field-effect devices are based on the technology used for manufacturing microelectronic chips and thus offer the possibility of mass production. However, the techniques and miniaturized sensors presented in this section are in the most cases still in the state of research and development.

Chemically Sensitive Field-Effect Transistors.

Chemically sensitive field-effect transistors (ChemFET) can react sensitive to some ions (*ISFET: ion-sensitive FET*), biomolecules (*BioFET: biologically sensitive FET*), or gases (*GasFET: gas-sensitive FET*) in aqueous media, or they can be insensitive (*ReFET: reference FET*). They incorporate the sensor membrane directly on the gate area of a field-effect transistor (FET). A schematic of an ISFET with an SiO_2 gate insulator (about 100-nm thickness), mounted in a measuring cell and contacted via a reference electrode, is given in Figure 70.21(a). When the sensor membrane is placed into contact with the test solution of the ion to be detected, a potential shift (ΔV) occurs. The charge density at the interface solution/membrane changes because of the chemical interaction with the ions, and this potential affects the drain current (I_D) flowing between source (S) and drain (D) of the transistor. After calibration of the ISFET with standard solutions of known ion activity, the variation of I_D can be used to determine the ion concentration in the test solution (Figure 70.21(b)). Often, the ISFET is operated in a feedback loop (e.g., the *constant charge mode*, Figure 70.21(c)) and the voltage V_M needed to maintain I_D at a fixed value represents the sensor response. The sensor response can be described by the same Nernst and Nikolsky equations that characterize conventional ion-selective electrodes.

The operation principle of ChemFETs can be derived from the essential electronic behavior of *MOSFET* (*metal-oxide-semiconductor FET*) devices [27], where the drain current I_D is expressed by:

$$I_D = K_d\left(\left(V_G - V_T\right)V_D - \frac{V_D^2}{2}\right) \tag{70.49}$$

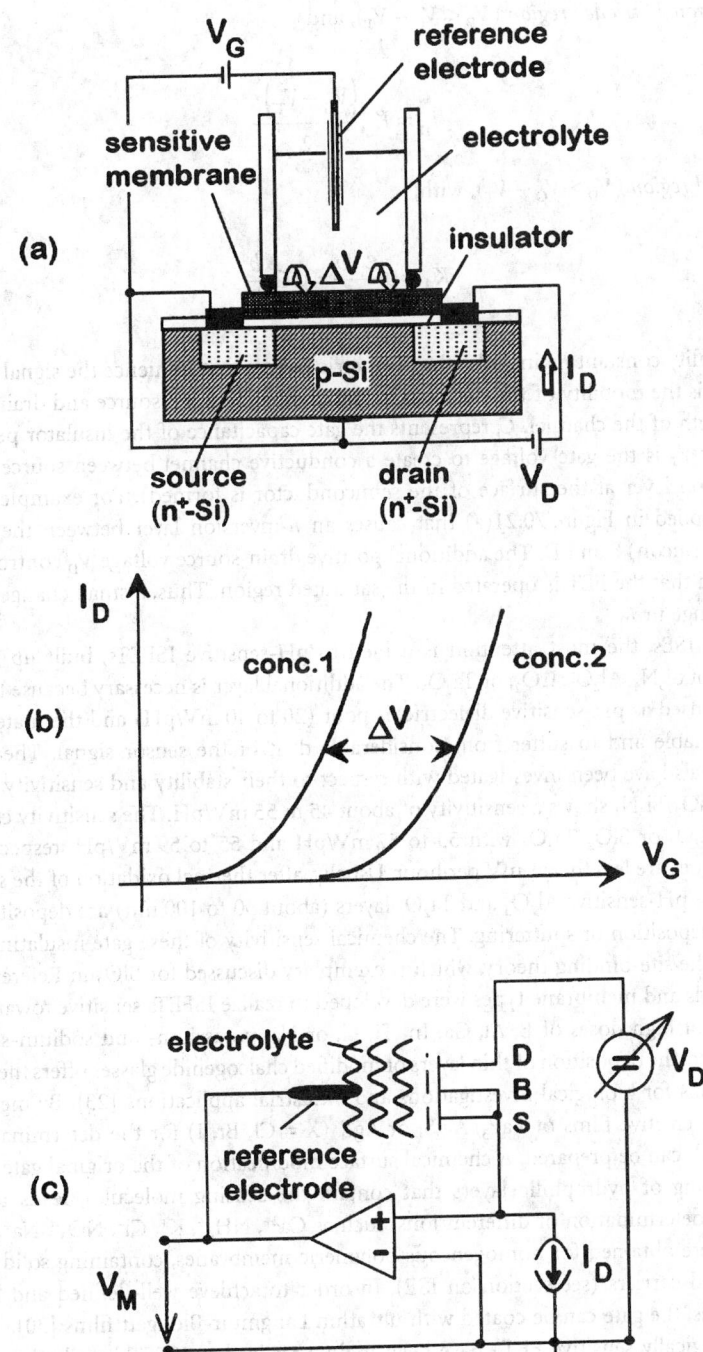

FIGURE 70.21 ISFET configuration (a), input characteristic (b), and schematic circuit of CCM (c). The metallic gate from a MOSFET (metal-oxide-semiconductor FET) is replaced by the arrangement sensitive membrane/test solution/reference electrode (V_G: gate-source voltage, V_D: drain-source voltage).

for the *nonsaturated "triode" region* ($V_D < V_G - V_T$), and

$$I_D = K_d \frac{\left(V_G - V_T\right)^2}{2}$$

(70.50)

for the *saturated region* ($V_D > V_G - V_T$), with:

$$K_d = \mu \frac{\varepsilon_i}{d_i} \frac{b}{L} = \mu C_i \frac{b}{L}$$

(70.51)

The proportionality constant K_d includes the geometric factors that influence the signal characteristic of the MOSFET. μ is the mobility of the electrons in the channel between source and drain, b is the width, and L is the length of the channel. C_i represents the gate capacitance of the insulator per unit area. The *threshold voltage* V_T is the gate voltage to create a conductive channel between source and drain (i.e., when an inversion layer at the surface of the semiconductor is formed). For example, a positive gate voltage V_G is applied in Figure 70.21(a) that causes an n-inversion layer between the two n^+-regions (highly n-doped silicon) S and D. The additional positive drain-source voltage V_D controls the measured current in a kind that the FET is operated in the saturated region. Thus, a small change in V_G results in a significant change in I_D.

Like with the ISEs, the most attention is gained to pH-sensitive ISFETs, built up of SiO_2 and an additional layer of Si_3N_4, Al_2O_3, IrO_2, or Ta_2O_5. The additional layer is necessary because the pH response of SiO_2, initially used as pH-sensitive dielectric, is poor (20 to 40 mV/pH) and the material was indeed found to be unstable and to suffer from considerable drift of the sensor signal. Therefore, different insulating materials have been investigated with respect to their stability and sensitivity. For example, a double layer of SiO_2/Si_3N_4 shows a sensitivity of about 45 to 55 mV/pH. The sensitivity can be improved by using SiO_2/Al_2O_3 or SiO_2/Ta_2O_5 with 53 to 57 mV/pH and 55 to 59 mV/pH, respectively. Also, the reported drift values are less than 1 mV per hour. Usually, after thermal oxidation of the silicon to realize the SiO_2 layer, the pH-sensitive Al_2O_3 and Ta_2O_5 layers (about 30 to 100 nm) are deposited by means of chemical vapor deposition or sputtering. The chemical sensitivity of these gate insulating materials can be explained by the site-binding theory, which is exemplary discussed for SiO_2 in Reference 28.

Several methods and membrane types were developed to realize ISFETs sensitive toward various ions. By implantation of high doses of B, Al, Ga, In, Ti, Li, or Na, potassium- and sodium-sensitive ISFETs were achieved. Also, the deposition of thin layers of modified chalcogenide glasses offers the determination of heavy metal ions for biological investigations and industrial applications [23]. By means of vacuum evaporation, ion-sensitive films of LaF_3, Ag_2S, or AgX (X = Cl, Br, I) for the determination of F^-, Cl^-, Br^-, I^-, Ag^+, and S^{2-} can be prepared. A chemical surface modification of the original gate insulator (e.g., the covalent linking of hydrophilic layers that contain the sensing molecule), leads to organic gate materials for the determination of different ions, such as Ca^{2+}, NH_4^+, K^+, Cl^-, NO_3^-, Na^+, Ag^+, etc. [29]. Similar results were obtained for homogeneous polymeric membranes, containing solid ion-exchanger or neutral-charged carriers (see section on ISE). In order to achieve well-defined and highly ordered sensor membranes, the gate can be coated with ultrathin Langmuir-Blodgett films [30].

BioFETs (biologically sensitive FET) have been mainly realized as ISFET-based enzyme sensors, so-called *EnFETs* (*enzyme FET*). The EnFET directly corresponds to the enzyme ISE and detects the potentiometric response to either the concentration change in one of the products or reactants catalyzed by the enzyme. Frequently, EnFETs consist of a pH ISFET with the individual enzyme layer for the determination of, for example, glucose, penicillin, urea, creatinine, adenosin, acetylcholine, etc. Dual pH-sensitive FETs on the same chip can be formed as an ISFET and an EnFET, where the latter one is loaded with the active enzyme. The ISFET serves as reference and the differential output signal is insensitive to pH changes. The demand of compatibility with integrated circuit technology provides enzymatic membranes that can be photolithographically patterned (e.g., photocrosslinkable materials). A review of

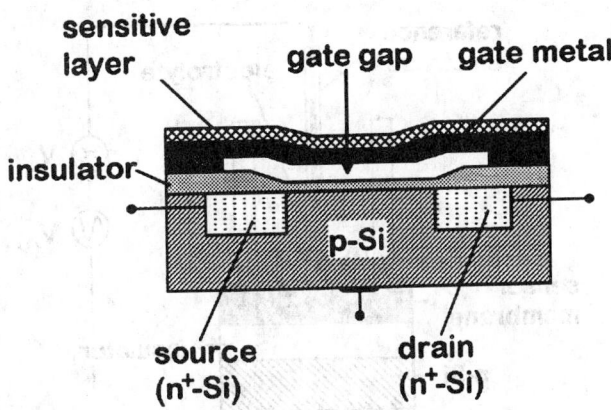

FIGURE 70.22 Schematic of a suspended gate FET (SGFET).

different categories of sensitive films and coatings and basic concepts of chemically sensitive field-effect transistors are given in Reference 31.

ReFETs (reference FETs) consist of a sensor surface that is as insensitive as possible to all kinds of substances in the test solution. Thus, a differential pair of an ISFET and a ReFET eliminates perturbations, like temperature and potential of the analyte. Appropriate materials to cover the ISFET surface with an insensitive layer are blocking materials, such as teflon or different polymers (e.g., parylene, polyacrylate, PVC). However, not well-defined potential processes as well as some ion exchange will result in nonideal behavior. Alternative concepts use nonblocking polymer membranes with a fixed membrane potential or quasi-ReFETs with a delayed pH response. The most promising approaches are the application of an inert metallic layer or wire in a differential ISFET setup as a *quasi-reference electrode*, and the miniaturization of conventional reference electrodes. For example, by means of physical vapor deposition methods, Ag/AgCl electrodes were miniaturized on silicon chips inside anisotropically etched cavities [31].

The basic mechanism of gas-sensitive FETs (GasFETs) is due to the chemical modification of the *electron work function* of a metal-insulator-semiconductor field-effect structure, for example, of a *suspended gate FET* (*SGFET*) as schematically shown in Figure 70.22. The SGFET contains an additional insulator, the "*gap*" within the gate structure, which consists of a vacuum, a gas or a nonconducting liquid. As gate metal, usually a platinum layer or mesh is used. The chemically sensitive layer on top of this structure, for example palladium, exhibits sensitivity toward hydrogen. The hydrogen molecules adsorb and dissociate atoms (H_a) on the metal surface (Pd), depending on their partial pressure, as well as desorb from the metal surface by recombination into H_2 and reacting with oxygen to form water:

$$H_2 \Leftrightarrow 2H_a \quad \text{and} \quad 4H_a + O_2 \Leftrightarrow 2H_2O \tag{70.52}$$

The adsorbed atoms diffuse rapidly to the inner surface gap/insulator where they become polarized and form an interface dipole layer, resulting in a potential drop. For example, SGFETs with Pd, operated at 100 to 140°C, are sensitive to H_2, CO, and H_2S in the ppm range, whereas an increased operating temperature up to 240°C allows the detection of alcohols (methanol, ethanol, propanol, butanol). To achieve selectivity, the surface of the suspended gate can be modified by inorganic or organic layers. Ammonia sensitivity can be achieved by catalytic metals such as Pt, Ir, Ru, or Rh. By the deposition of organic layers like polypyrole, sensitivities to alcohols and aromatic hydrocarbons are achieved. Several related devices based on SGFETs are explained in Reference 32.

Chemically Sensitive Capacitors.
Sensors on the basis of capacitive field-effect structures are much simpler to fabricate than chemically sensitive FETs, and consequently they are favorable for laboratory use. Such *EIS* (*electrolyte-insulator-semiconductor*) *structures* correspond to *MIS* (*metal-insulator-semiconductor*) *capacitors* and their operation

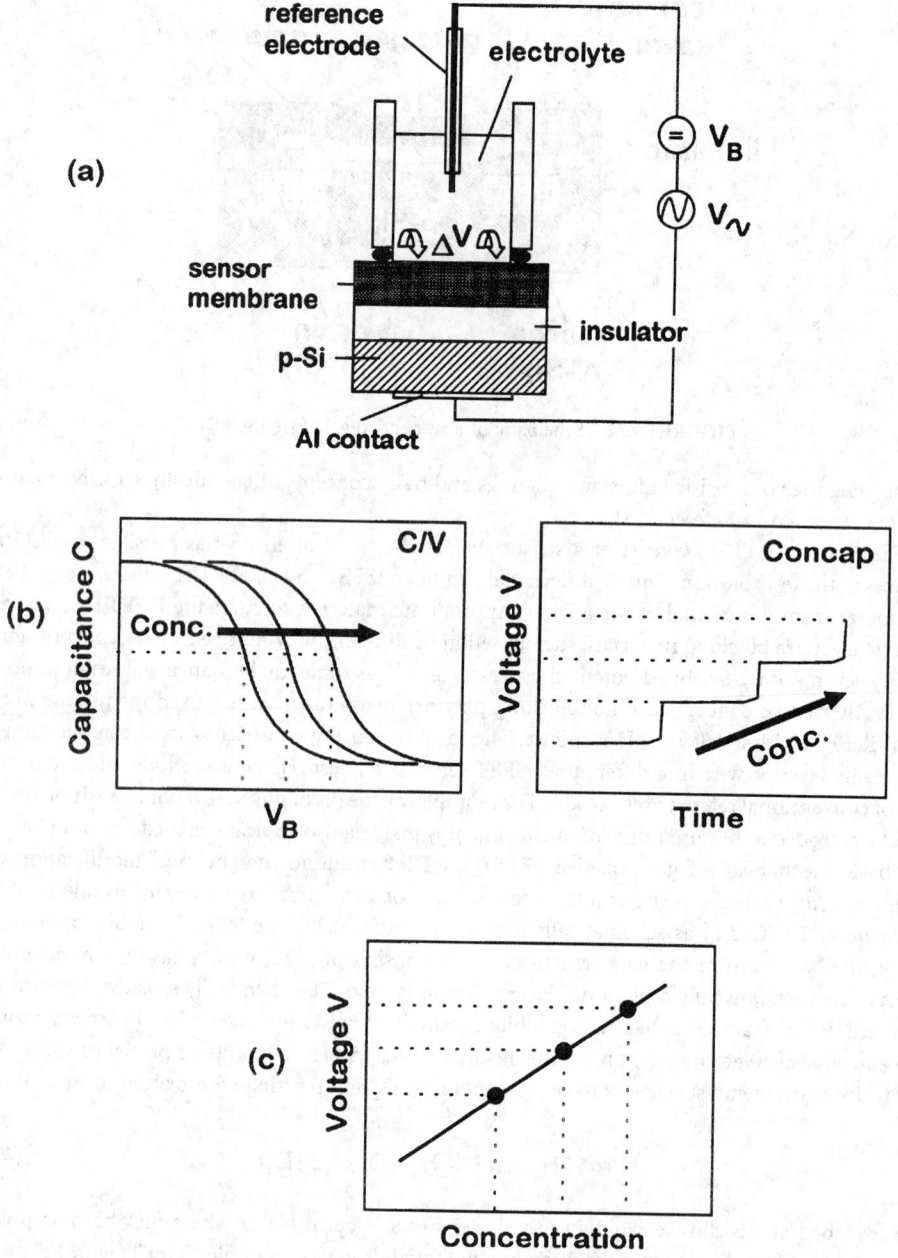

FIGURE 70.23 Schematic of an EIS (electrolyte-insulator-semiconductor) structure (a). Measurement of the EIS sensor in the C/V (capacitance/voltage) and the Concap (constant capacitance) mode (b), and resulting calibration curve (c).

principle can be derived from the fundamental MIS devices [33]. A schematic build-up of an EIS structure and the measuring principle is given in Figure 70.23(a). The sensor consists of a *p*- or *n*-type semiconductor (silicon) covered by a thermally grown SiO_2 insulating layer (<100 nm) and the sensor membrane that is directly immersed into the test solution. Usually, the sensor is contacted via a reference electrode.

Its physical properties can be explained by the charge carrier distribution at the insulator/semiconductor interface, which is controlled by both an external dc voltage (V_B) and an electrochemical interaction between the test solution and the sensor membrane (ΔV). For a *p*-Si substrate, a negative V_B ($V_B < 0$)

on the reference electrode accumulates mobile charge carriers (i.e., positive holes) at the Si/SiO$_2$ interface (*accumulation*). When V_B becomes positive ($V_B > 0$), the holes are displaced from the interface, forming a space charge region (*depletion*) at the semiconductor surface. If the potential gets more positive ($V_B \gg 0$), an inversion layer of accumulated electrons at the interface is created (*inversion*). The electrical behavior is given by the small-signal capacitance of the EIS structure. Depending on the applied V_B and a superimposed ac voltage (e.g., 1 kHz, 20 mV), a characteristic C/V (*capacitance/voltage*) curve results (Figure 70.23(b), left). The integral capacitance C, corresponding to V_B, is given by:

$$\frac{1}{C} = \frac{1}{C_M} + \frac{1}{C_I} + \frac{1}{C_S} \tag{70.53}$$

where C_M, C_I, and C_S are the capacitance values of the sensor membrane, the insulator, and the space charge region, respectively, with:

$$C = \frac{\varepsilon_0 \varepsilon_r}{d} A \tag{70.54}$$

where A is the area, d the thickness, ε_r the dielectric permittivity, and ε_0 the dielectric constant. Due to the electrochemical interaction (ΔV), a horizontal shift of the C/V curve is provided, depending on the change of the ion concentration in the test solution. As resulting measuring signal (calibration curve), the shift can be evaluated at a fixed capacitance value within the linear region of the C/V curves (e.g., 60% of the maximum capacitance, Figure 70.23(c)). Using a feedback circuit, the measured capacitance can be adjusted at a fixed value in the *Concap* (*constant capacitance*) mode (Figure 70.23(b), right). Thus, potential shifts can be recorded directly.

Chemical and biological sensing EIS structures with different organic and inorganic sensor membranes have been developed within the last few years. They consist of nearly identical sensor membrane materials and compositions as ISFETs, ranging from inorganic pH-sensitive layers (e.g., Si$_3$N$_4$, Al$_2$O$_3$, Ta$_2$O$_5$) or crystalline films (e.g., LaF$_3$, silver halides) over organic Langmuir-Blodgett films to enzymatic layers (e.g., urease, penicillinase). Much effort has been done in order to improve the limiting long-term stability that is often disclosed by FET devices in permanent contact with the analyte. Novel approaches pursue a further optimization with regard to the preparation (e.g., due to specific immobilization procedures) or the deposition of the sensor membrane in order to raise the sensor performance. For example, an extremely long-term stable pH sensor was developed by the suggestion of the pulsed laser deposition (PLD) process as the thin-film preparation method. The EIS structure consists of a layer sequence of Al/p-Si/SiO$_2$/Al$_2$O$_3$, where no degradation of the pH sensitivity during a measurement period of 2 years was found [34].

Like GasFETs, MIS (metal-insulator-silicon) capacitors and *MIS Schottky diodes* are also available as gas-sensitive devices. For the MIS capacitor, a concentration-dependent dipole layer is detected as a shift of the C/V curve. To reduce the drift of these devices, additional insulating layers, such as Al$_2$O$_3$, Si$_3$N$_4$, or Ta$_2$O$_5$ can be deposited between the metal layer and the SiO$_2$ insulator. Experimental results of Pd/Al$_2$O$_3$/SiO$_2$/Si structures show sensitivities of 25 mV ppm^{-1} around 1 ppm [35]. Schottky barrier diodes consist of a thin insulating layer (e.g., 2 nm SiO$_2$) between the metallic gate (e.g., Pd) and the semiconductor, in order to allow the current to pass through it. By variation of the metallic gate films, different sensitivities can be achieved, comparable to those of the SGFETs.

Practical Applications and Limitations.

ChemFETs possess significant advantages over classical ISEs, such as a high-input impedance that consequently eliminates the need of shielding wires and the need for voltmeters. The small sensor area includes the possibility of multiple sensor applications (*sensor arrays*) on a single chip. Moreover, temperature compensation is possible. However, most of these sensors are exposed to a chemically very reactive environment and therefore, a highly long-term stable protection (encapsulation) of the electronics from the analyte is required. The instability of the materials used induces sensor drifts of several millivolts per day. In some cases, attachment and fixation of the sensor membranes must be improved. To take the advantage of miniaturized FET devices, there is also the necessity of a small reference electrode. For ChemFETs, there exist two approaches for successful commercialization: dealing with small sample volumes for biomedical use (e.g., intracellular measurements) and the high-volume fabrication for a low-price market (e.g., environmental and process monitoring, agriculture and food analysis, leak detectors). The employment of capacitive EIS and MIS sensors offers besides the more easier manufacturing technique distinct advantages concerning the improved mechanical and electrochemical stability and sensor lifetime.

Conductometry

In addition to potentiometry, *conductometric analysis* represents the most important nonfaradaic method. *Conductometry* is based on the measurement of the electrical conductance of an electrolyte solution, which directly depends on the number of positively and negatively charged species in the solution. This analysis method is limited due to its nonselective nature, because all ions in the solution will contribute to the total conductance. Nevertheless, *direct conductance measurements* play an important role in the analysis of binary water/electrolyte mixtures, for example, in chemical water monitoring. The technique can also be applied to ascertain the endpoint detection in *conductometric titrations* for the determination of numerous substances.

Measurement of Conductance and Instrumentation

The *conductance* G of a solution is the reciprocal of the electrical resistance R and has the units of siemens (S) that correspond to ohm^{-1} (Ω^{-1}). The conductance of an uniform sample with the length l and cross-sectional area A is given by:

$$G = \kappa \frac{A}{l} \tag{70.55}$$

where the proportionality constant $\kappa = 1/\rho$ (ρ: resistivity) describes the *conductivity* (*specific conductance*) of the solution, expressed in units of S cm^{-1}. The *equivalent conductivity* Λ (*molar conductivity*) of a solution is defined as the conductivity due to one mole, measured between two electrodes which are spaced 1 cm apart, and is:

$$\Lambda = \frac{1000 \, \kappa}{c} \tag{70.56}$$

where c corresponds to the concentration of the solution in mol L^{-1}. The units of Λ are S cm^{-1}mol^{-1}. Equation 70.55 permits the calculation of the molar conductivity for a solution of known concentration by considering the experimental values of κ. The molar conductivity Λ, i.e., the mobility of ions in solution, is mainly influenced by interionic effects for strong electrolytes and the degree of dissociation for weak solutions. For strong electrolytes, the molar conductivity increases as the dilution is increased. By linear graphical extrapolation for diluted solutions of strong electrolytes, a limiting value is defined as *molar conductivity at infinite dilution* Λ_0. At infinite dilution, the interionic attraction is nil, the ions are independent of each other, and the total conductivity is:

$$\Lambda_0 = \lambda^0_+ + \lambda^0_- \tag{70.57}$$

where λ^0_+ and λ^0_- are the ionic molar conductivities of the cations and anions, respectively, at infinite dilution. For weak electrolytes, due to the nonlinear relationship between Λ and c, a graphical extrapolation cannot be made. Typical values for the limiting molar conductivities for various species in water are listed in Table 70.2.

TABLE 70.2 Molar Conductivity at Infinite Dilution Λ_0 (Ω^{-1} cm^2 mol^{-1})

Cations λ_+^0		Anions λ_-^0	
H$^+$	349.8	OH$^-$	198.3
Na$^+$	50.1	F$^-$	55.4
K$^+$	73.5	Cl$^-$	76.3
Li$^+$	38.7	Br$^-$	78.1
NH$_4^+$	73.4	I$^-$	76.8
Ag$^+$	61.9	NO$_3^-$	71.5
N(CH$_3$)$_4^+$	44.9	ClO$_4^-$	67.3
Ca^{2+}	119.0	C$_2$H$_3$O$_2^-$	40.9
Mg^{2+}	106.2	HCO$_3^-$	44.5
Cu^{2+}	107.2	AcO$^-$	40.9
Zn^{2+}	105.6	SO$_4^{2-}$	160.0
Ba^{2+}	127.7	CO$_3^{2-}$	138.6
Pb^{2+}	139.0	C$_2$O$_4^{2-}$	148.4
Fe^{3+}	204.0	PO$_4^{3-}$	240.0
La^{3+}	208.8	Fe(CN)$_6^{4-}$	442.0

The equipment needed for measuring the conductivity includes an electric power source, a cell containing the solution, and a suitable measuring bridge. The electric power source consists of an alternating current source that produces signals of about 1 kHz in order to eliminate effects of faradaic current. The measurement is performed by a Wheatstone bridge arrangement. Modern *conductivity meters* supply the alternating current and allow the measurement in a wide range of conductivities (0,001 μS cm^{-1} to 1300 mS cm^{-1}). Additional electronics eliminate disturbing capacitance effects and offer automatic range switching. An integrated temperature sensor corrects automatically conductivities to their value at 25°C. The conductivity cell consists of a pair of electrodes placed in a defined geometry to each other. Usually, the electrodes are platinized to increase their effective surface (high capacitance). Thus, disturbing faradaic currents are minimized. For accurate conductivity determination, the precise area of the electrodes A and their distance apart d, the *cell constant K*, must known exactly. Therefore, the cell constant ($K = A/d$) must be evaluated by calibration with a solution of accurately known conductivity (e.g., a standard KCl solution). Details of calibration standards and concepts of conductivity cells are given in Reference 36.

Applications of Conductometry

Direct Conductometric Measurement.
In spite of the insufficient selectivity of *direct conductometric measurements*, the high sensitivity of this procedure makes it an important analytical tool for certain applications. The specific conductivity of pure water (distilled or deionized) is about 5×10^{-8} S cm^{-1}, and the smallest trace of ionic impurity leads to a large increase in conductivity by an order of magnitude and more. Therefore, conductometric monitoring is employed where a high purity of water is required (e.g., laboratories, semiconductor processing, steam-generating power plants, ion exchanger). Conductometric measurements are widely used to control pollution of rivers and lakes, and in oceanography to control the salinity of sea water.

Conductometric Titrations.
In *conductometric titrations*, the reaction is followed by means of conductometry and is used for locating endpoints (i.e., the equivalence point (EP) in acid-base titrations (neutralization titration)). To define the titration curve, at least three or four measurements before and after the EP are required. The obtained data of the conductivity are plotted as a function of the titrant volume, and the EP is given as the intersection of the two linear extrapolated fractions. A characteristic titration curve of a strong acid (hydrochloric acid) with a strong base (sodium hydroxide) is depicted in Figure 70.24. The solid line represents the resulting titration curve, whereas the broken lines indicate the contribution of the individual species. By adding NaOH to the solution, the hydrogen ions are replaced by the equivalent number

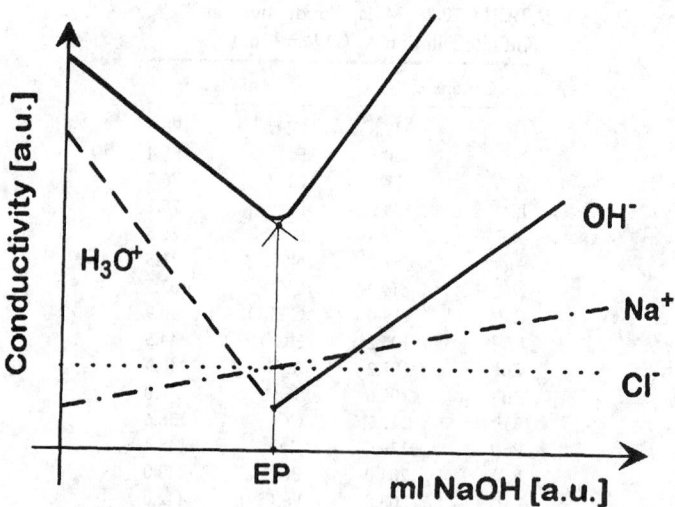

FIGURE 70.24 Conductometric titration of a strong acid (HCl) with a strong base (NaOH). The equivalence point is represented by EP.

of less mobile sodium ions (and $H^+ + OH^- \rightarrow H_2O$). As a result, the conductivity decreases to lower values. The solution exhibits its lowest conductivity at the equivalence point, where the concentrations of hydrogen and hydroxide ions are at the minimum. Further addition of NaOH reverses the slope of the titration curve, since both the sodium ion concentration and hydroxide ion concentration increase.

Due to the high linearity between the conductance and the volume of the added species, this method possesses a high accuracy and can be employed in dilute as well as in more concentrated solutions. In contrast to potentiometric titration methods, the immediate equivalence point region has no strong significance. Thus, very weak acids, such as basic acid and phenol can be titrated. Moreover, mixtures of hydrochloric acid or another strong acid and acetic (ethanoic) acid or any other weak acid can be titrated with a weak base (e.g., aqueous ammonia, acetate) or with a strong base (e.g., sodium hydroxide). Moreover, precipitation and complex-formation titrations of, for example, sodium chloride with silver nitrate are possible. For practical applications, the volume of the solution should not change appreciably during the titration. Therefore, the titrating reagent may be 20 to 100 times more concentrated than the solution being titrated, whereas the latter should be as diluted as practicable. For additional examples of analytical procedures and results of conductometric titrations, see Reference 37.

Oscillometry.

In order to investigate electrolyte solutions with high resistivities and dielectric constants, *high-frequency titration* (*oscillometry*) can be performed at 10^5 Hz to 10^7 Hz. For that, a specific measuring cell is required, where the metal electrodes encircle the outside of a glass container. In this arrangement, the electrodes are not in contact with the test solution, which is advantageous for dealing with corrosive materials. Oscillometric measurements can be employed for the determination of binary mixtures of nonionic species, where the dielectric behavior predominates (e.g., ethanol/nitrobenzene, benzene/chlorobenzene, and alcohol/water). Further practical examples are EDTA titrations and the determination of thorium (Th^{4+}) with sodium carbonate, beryllium (Be^{2+}) with sodium hydroxide, and hydrocarbons (e.g., benzene). However, the instrumentation as well as the interrelations are more complicated than for the classical conductivity method. Thus, oscillometry gets only significance for specific applications, where the presence of the electrodes interferes.

Conductometric Sensors.

Depending on the demanded size and geometry, miniaturized cells with two or more electrodes (e.g., a four-electrode conductivity meter) as well as contactless cells are commercially available as *conductometric sensors.* The contactless methods use *capacitive* and *inductive conductivity cells,* which are advantageous to circumvent electrochemically caused electrode reactions. Conductivity cells can be coupled as detectors to ion chromatographic systems for measuring ionic concentration in the eluate. For this, special *micro-conductivity cells* with a volume of about 1.5 μL have been developed.

Within the last 10 years, two aspects of conductometric applications became important: *conductometric gas sensors* and the use of *conductometric chemiresistors* as sensors. In the former, a phase change that transfers the gaseous component into a solution is necessary (e.g., by a bubbler nebulizer). All methods deal with acidic gases, such as HCl, SO_2, or CO_3, or with alkaline gases like NH_3. Also, organic halogens can be detected after their conversion into HCl or HF. By means of integrated circuit technology, thin metal films can be photolithographically patterned as interdigital electrodes onto semiconductor substrates with insulating dielectric layers of SiO_2 or Si_3N_4. Both the thin metal films and additionally deposited organic layers on top of the metallic films can lead to a change of the total resistance by variation of the ionic composition of the reacting solution. For chemiresistors, the organic layer usually consists of an ion-selective polymer layer or a Langmuir-Blodgett membrane, for biosensors enzymatic layers are used (see the ISE section). Such sensors allow the determination of different gaseous components, such as CO, NO_2, H_2S, SO_2, or NH_3, as well as the detection of biologically relevant species like urea, glucose, penicillin, and choline chlorides. Although several companies offer such gas analyzer systems, conductometric sensors and chemiresistors are still in the state of research and development.

Coulometry

Coulometry represents an electroanalytical method, where the analyte is specifically and completely converted due to direct or indirect electrolysis. The quantity of electricity (in coulombs) consumed by this reaction, the charge, is measured. A fundamental requirement of coulometry is that the species in the solution interact with 100% current efficiency; that is, the reaction corresponds to Faraday's law. According to this condition, there exist two alternatives: the analyte participates in the electrode reaction (*primary* or *direct coulometric analysis*), and the analyte reacts with a reagent, generated by an electrode reaction (*secondary* or *indirect coulometric analysis*). Two general techniques — *controlled-potential coulometry* and *coulometric titration* (*controlled-current coulometry*) — are used for coulometric analysis.

Controlled-Potential Coulometry

In this method, the potential of the working electrode is held at a constant value compared to a reference electrode. The resulting current is adjusted continuously to maintain the desired potential. The substance being determined reacts without involvement of other components in the sample. The reaction is completed when the current has practically decreased to zero. To measure the charge, a potentiostat, an instrument for measuring the time-dependent current, and a current-time integrating device are used. Modern potentiostats have a built-in electronic coulometer and allow extremely accurate determinations. Otherwise, one can use free-standing coulometers.

Controlled-potential coulometry has been widely employed for the determination of various metal ions, such as Cu, Bi, Cd, Zn, Ni, Co, Pu, and U. To apply this method, current–voltage diagrams must be available for the oxidation reduction system to be measured as well as for any reaction system at the working electrode. Current–voltage diagrams can be obtained by plotting the measured current versus the cathode-reference electrode potential. To fulfill the requirement of the 100% current efficiency in generation, it is necessary to control the potential of the working electrode. With regard to their determination, the metals are deposited at controlled potentials with a mercury cathode as working electrode and a silver wire or a platinum cylinder as anode. Typical applications are the electrolytic determination and synthesis of organic compounds like acetic acid and picric acid. Further, controlled-potential coulometry is frequently used for monitoring the concentration of constituents in gas or liquid streams, typically small oxygen contents. Here, the reduction of oxygen takes place within the pores of a porous silver cathode:

$$O_2(g) + 2H_2O + 4e^- \Leftrightarrow 4OH^- \qquad (70.58)$$

Using a cadmium sheet (m) as anode, the electrode reaction in solution (s) is:

$$Cd(m) + 2OH^- \Leftrightarrow Cd(OH)_2(s) + 2e^- \tag{70.59}$$

The quantity of the electricity (current) is passed through a standard resistor and converted to a voltage signal. Hence, the oxygen concentration is proportional to the recorded potential drop. Controlled-potential coulometry needs relatively long electrolysis times, although it proceeds virtually unattended with automatic coulometers. With a multimeter, changes in the range from 1 ppm to 1% can be dissoluted. Thus, controlled-potential coulometry permits analysis with an accuracy of a few tenths of a percent.

Coulometric Titration (Controlled-Current Coulometry)

Controlled-current coulometry maintains a constant current throughout the reaction period. Here, an excess of a redox buffer substance must be added in such a way that the potential does not cause any undesirable reaction. That means the product of the electrolysis of the redox buffer must react quantitatively with the unknown substance to be determined. *Coulometric titrations* need an electrolytically generated titrant that reacts stoichiometrically with the analyte to be determined. As in controlled-potential coulometry, 100% current efficiency is required. The current is accurately fixed at a constant value and the quantity of electricity can be calculated by the product of the current (in amperes) and the time (in seconds) using endpoint detection. In principle, any endpoint detection system that fits chemically can be used; for example, chemical indicators (color change), and potentiometric, amperometric or conductometric procedures. For coulometric titrations the instrumentation consists of a titrator (constant-current source, integrator) and a cell. As the constant-current source, an electronically controlled amperostat is preferably used. The integrator measures the product of current and time (i.e., the number of coulombs). The electrolysis cell, filled with the solution from which the titrant will be generated electrolytically and the solution to be titrated, is schematically shown in Figure 70.25. The generator electrode, at which the reagent is formed, possesses a large surface area (e.g., a rectangular strip of platinum). The auxiliary electrode (e.g., a platinum wire) is in contact with an appropriate electrolyte of higher concentration than the solution to be titrated. It is isolated from the analyte by a sintered disk or some other porous media. This is required to avoid the interference of additional products generated at the second electrode. To circumvent these limitations of internal generation, an external generator cell is often used.

Typical applications of coulometric titrations are neutralization titrations, precipitation and complex-formation titrations, and oxidation-reduction titrations. Neutralization titrations can be employed for both weak and strong acids and bases. The former can be performed with hydroxide ions generated at a platinum anode by the reaction:

$$2H_2O + 2e^- \Leftrightarrow 2OH^- + H_2(g) \tag{70.60}$$

the latter one with hydrogen ions by the reaction:

$$H_2O \Leftrightarrow \tfrac{1}{2}O_2(g) + 2H^+ + 2e^- \tag{70.61}$$

A working (*generator*) electrode of silver as anode offers the determination of Cl^-, Br^-, I^-, and mercaptans in solution(s). For bromide, the reaction becomes:

$$Ag + Br^-(s) \Leftrightarrow AgBr(s) + e^- \tag{70.62}$$

Similar precipitation and complex-formation titrations as well as oxidation-reduction titrations are described in Reference 38.

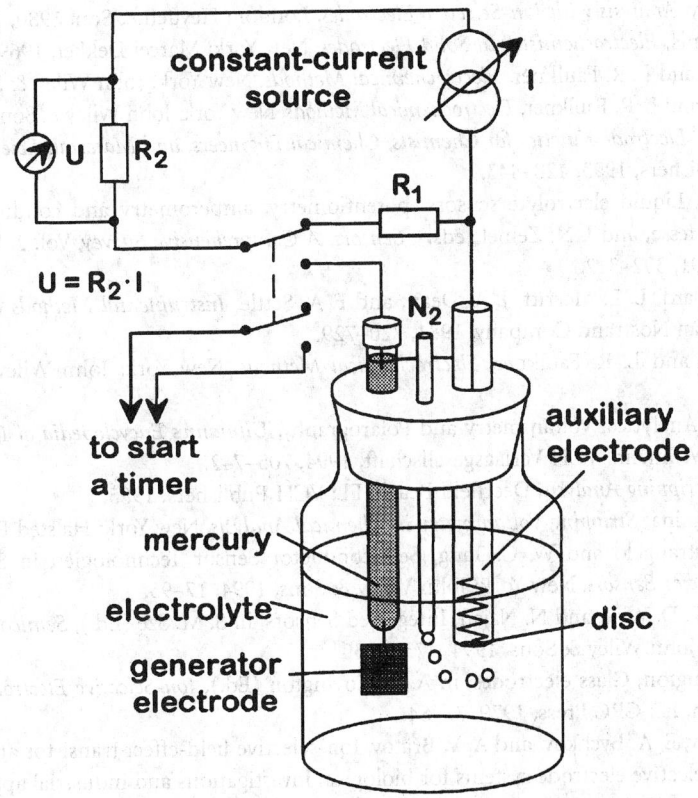

FIGURE 70.25 Coulometric titration cell with working electrode and auxiliary electrode, and equivalent circuit diagram (schematically).

Coulometric titrations possess some practical advantages: no standard solutions are required and unstable reagents can be generated or consumed immediately, small amounts of titrants can be electrically quantified with high accuracy, pretitration is possible, and the method can be readily adapted to automatic remote control. Thus, with respect to controlled-potential coulometry a wider field of practical applications exists. Often, automatic titrators for multipurpose and single analysis employ potentiometric endpoint detection. Examples are sulfur dioxide monitors and water titrators (*Karl Fischer*). For more detailed information concerning applications of coulometry, see Reference 39.

References

1. P. H. Rieger, *Electrochemistry*, Englewood Cliffs, NJ: Prentice-Hall, 1987, 70–80.
2. J. Wang, *Analytical Electrochemistry*, New York: VCH Publishers, 1994, 17–20.
3. J. Koryta and K. Šatulík, *Ion-Selective Electrodes*, Cambridge, UK: Cambridge University Press, 1983, 12–14.
4. R. C. Weast and M. J. Astle (eds.), *CRC Handbook of Chemistry and Physics*, Boca Raton, FL: CRC Press, 1982.
5. W. E. Morf, *The Principles of Ion-Selective Electrodes and Membrane Transport*, Amsterdam: Elsevier, 1981, 64–73.
6. J. Koryta, *Ions, Electrodes and Membranes*, Chichester, UK: John Wiley & Sons, 1991, 82–86.
7. P. L. Bailey, *Analysis with Ion-Selective Electrodes*, London: Heyden & Son, 1980, 15–18.
8. J. Koryta, J. Dvorák, and L. Kavan, *Principles of Electrochemistry*, Chichester: John Wiley & Sons, 1993, 176–177.

9. P. L. Bailey, *Analysis with Ion-Selective Electrodes*, London: Heyden & Son, 1980, 18–22.
10. R. N. Adams, *Electrochemistry at Solid Electrodes*, New York: Marcel Dekker, 1969, 19–29.
11. A. J. Bard and L. R. Faulkner, *Electrochemical Methods*, New York: John Wiley & Sons, 1980, 147.
12. A. J. Bard and L. R. Faulkner, *Electrochemical Methods*, New York: John Wiley & Sons, 1980, 316–366.
13. E. Gileadi, *Electrode Kinetics for Chemists, Chemical Engineers, and Materials Scientists*, New York: VCH Publishers, 1993, 428–443.
14. F. Oehme, Liquid electrolyte sensors: potentiometry, amperometry and conductometry, in W. Göpel, J. Hesse, and J. N. Zemel (eds.), *Sensors: A Comprehensive Survey*, Vol. 2, Weinheim: VCH Verlag, 1991, 302–312.
15. H. H. Willard, L. L. Merritt, J. A. Dean, and F. A. Settle, *Instrumental Methods of Analysis*, New York: D. Van Nostrand Company, 1981, 720–729.
16. A. J. Bard and L. R. Faulkner, *Electrochemical Methods*, New York: John Wiley & Sons, 1980, 186–190.
17. G. Henze, Analytical Voltammetry and Polarography, *Ullmann's Encyclopedia of Industrial Chemistry*, B5, Weinheim: VCH Verlagsgesellschaft, 1994, 705–742.
18. J. Wang, *Stripping Analysis*, Deerfield Beach, FL: VCH Publishers, 1985.
19. Kh. Z. Brainina, *Stripping Voltammetry in Chemical Analysis*, New York: Halsted Press, 1974.
20. C. H. Mastrangelo and W. C. Tang, Semiconductor Sensor Technologies, in S. M. Sze (ed.), *Semiconductor Sensors*, New York: John Wiley & Sons, 1994, 17–95.
21. K. Najafi, K. D. Wise, and N. Najafi, Integrated Sensors, in S. M. Sze (Ed.), *Semiconductor Sensors*, New York: John Wiley & Sons, 1994, 473 – 530.
22. A. K. Covington, Glass electrodes, in A. K. Covington (Ed.), *Ion-Selective Electrode Methodology*, Boca Raton, FL: CRC Press, 1979, 77–84.
23. Y. G. Vlasov, E. A. Bychkov, and A. V. Bratov, Ion-selective field-effect transistor and chalcogenide glass ion-selective electrode systems for biological investigations and industrial applications, *Analyst*, 119, 449–454, 1994.
24. A. K. Covington and P. Davison, Liquid ion exchange types, In A. K. Covington (Ed.), *Ion-Selective Electrode Methodology*, Boca Raton, FL: CRC Press, 1979, 85–110.
25. G. H. Jeffrey, J. Bassett, J. Mendham, and R. C. Denney, *Vogel's Textbook of Quantitative Chemical Analysis*, London: Longman Scientific & Technical, 1989, 572, 604.
26. G. H. Jeffrey, J. Bassett, J. Mendham, and R. C. Denney, *Vogel's Textbook of Quantitative Chemical Analysis*, London: Longman Scientific & Technical, 1989, 580–590.
27. M. J. Madou and S. R. Morrison, *Chemical Sensing with Solid State Devices*, San Diego, CA: Academic Press, 1989, 325–332.
28. M. J. Madou and S. R. Morrison, *Chemical Sensing with Solid State Devices*, San Diego, CA: Academic Press, 1989, 332–358.
29. D. N. Rheinhoudt, Application of supramolecular chemistry in the development of ion-selective ChemFETs, *Sensors and Actuators*, B6, 179–185, 1992.
30. M. J. Schöning, M. Sauke, A. Steffen, M. Marso, P. Kordoš, H. Lüth, F. Kauffmann, R. Erbach, and B. Hoffmann, Ion-sensitive field-effect transistors with ultrathin Langmuir-Blodgett membranes, *Sensors and Actuators*, B26-27, 325–328, 1995.
31. M. J. Madou and S. R. Morrison, *Chemical Sensing with Solid State Devices*, San Diego, CA: Academic Press, 1989, 361–366.
32. M. Josowicz and J. Janata, Suspended gate field-effect transistor, in T. Seiyama (ed.), *Chemical Sensor Technology*, Vol. 1, Amsterdam: Elsevier, 1988, 167–175.
33. S. M. Sze, *Physics of Semiconductor Devices*, New York: John Wiley & Sons, 1981, 332–379.
34. M. J. Schöning, D. Tsarouchas, A. Schaub, L. Beckers, W. Zander, J. Schubert, P. Kordoš, and H. Lüth, A highly long-term stable silicon-based pH sensor fabricated by pulsed laser deposition technique, *Sensors and Actuators*, B35, 228–233, 1996.
35. M. Armgarth and C. I. Nylander, Field-effect gas sensors, in W. Göpel, J. Hesse, and J. N. Zemel (eds.), *Sensors: A Comprehensive Survey*, Vol. 2, Weinheim: VCH Verlag, 1991, 509–512.

36. F. Oehme, Liquid electrolyte sensors: potentiometry, amperometry and conductometry, in W. Göpel, J. Hesse, and J. N. Zemel (eds.), *Sensors: A Comprehensive Survey*, Vol. 2, Weinheim: VCH Verlag, 1991, 317–328.
37. D. A. Skoog and D. M. West, *Principles of Instrumental Analysis*, Philadelphia, PA: Saunders College, 1980, 648–651.
38. D. A. Skoog and D. M. West, *Principles of Instrumental Analysis*, Philadelphia, PA: Saunders College, 1980, 598–599.
39. E. A. M. F. Dahmen, *Electroanalysis*, Amsterdam: Elsevier 1986, 218–224.

Further Information

P. T. Kissinger and W. R. Heinemann, *Laboratory Techniques in Electroanalytical Chemistry*, 2nd ed., New York: Marcel Dekker 1996.

P. J. Gellings and H. J. M. Bouwmeester, *The CRC Handbook of Solid State Electrochemistry*, Boca Raton, FL: CRC Press, 1997.

J. Bockris and S. Khan, *Surface Electrochemistry: A Molecular Level Approach*, New York: Plenum Press, 1993.

C. Brett and A. Brett, *Electrochemistry: Principles, Methods and Applications*, New York: Oxford University Press, 1993.

70.2 Thermal Composition Measurement

Mushtaq Ali, Behrooz Pahlavanpour, and Maria Eklund

Thermal analysis is the measurement of a physical parameter as a function of temperature. The area comprises several techniques where thermogravimetry, thermometric titrimetry, thermomechanical analysis, differential thermal analysis, differential scanning calorimetry, and some specialized techniques are among those described in this section. Applications are found in the characterization of organic materials, both solids and liquids. Materials of interest can be polymers, mineral and synthetic oils, lubricants, greases, paper (cellulose), and pharmaceuticals The material to be analyzed (typically 10 mg) must be isolated and subjected to thermal treatment, hence the technique is destructive. The obvious advantage is that the thermal profile, or structure of a large specimen can be investigated. The disadvantage is that the small sample size can give rise to excessive statistical errors. However, recent advances in microcalorimetry techniques to look at slow degradation of pharmaceuticals allow thermal analysis to be performed at room temperature on samples up to a few grams without destruction of the sample. A general schematic of thermal analysis apparatus is shown in Figure 70.26.

The history of the development of thermal analysis methods from the sixteenth century is the subject of a number of excellent papers by Mackenzie [1–3], Wendlandt [4], and Keattch [5]. Lavoiser and Laplace [6] were pioneers in the development of thermal analysis by their practical approach.

The International Confederation for Thermal Analysis and Calorimetry (ICTAC) has produced definitive guidelines regarding nomenclature and calibration [7-10].

Factors Affecting Results

The five factors affecting thermal analysis can be remembered by the acronym S.C.R.A.M. [11]. This refers to the **S**ample, **C**rucible, **R**ate of heating, **A**tmosphere, and **M**ass. See Table 70.3.

Thermogravimetry

Thermogravimetry (TG) or TGA (thermogravimetric analysis) [8] is a technique in which the mass of the sample is monitored against time or temperature while the temperature of the sample, in a specified atmosphere, is programmed.

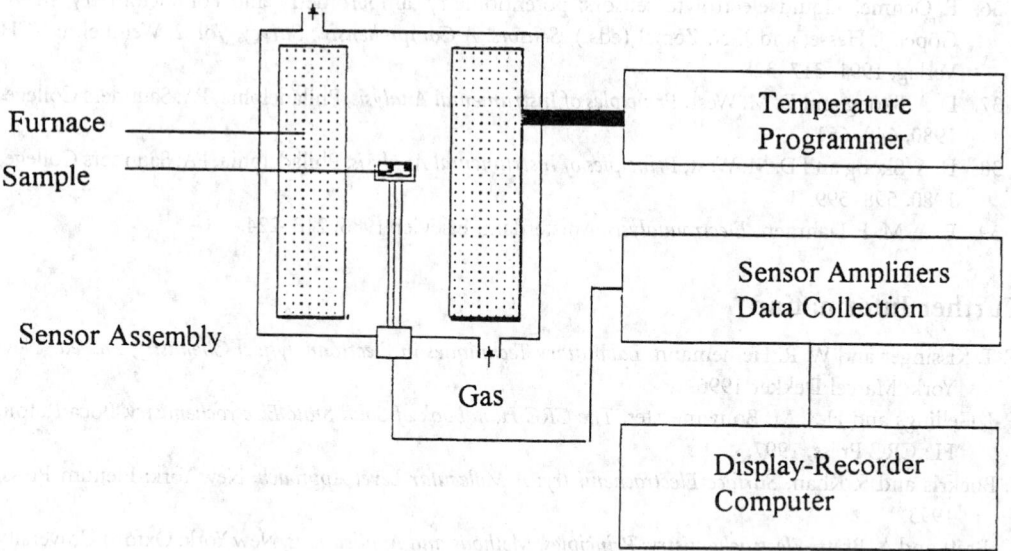

FIGURE 70.26 Schematic thermal analysis apparatus. The figure shows the essential components of a generalized thermal analysis apparatus.

TABLE 70.3 Factors Affecting Thermal-Analysis

Factor	Details
Sample	History of sample and preparative technique used can affect the curve and the presence of trace impurities (in some cases) may catalyze decompositions. Particle size can alter shape of curve (e.g., by surface reaction).
Crucible	The crucible (or sample holder) material should be such that it does not react with the sample or catalyze a reaction. The geometry of the sample holder may affect the results [12].
Rate of heating	Thermal lag: rate of heat transfer between furnace and all parts of the sample are not instantaneous. Therefore, care should be taken when working at different heating. Corrections can be applied [12].
Atmosphere	Various effects [12], including dissociation of sample.
Mass of sample	Size and packing density.

Note: Details the five main areas which would affect analysis of a sample via thermal experiments. The effects can be marked and would certainly affect repeatability also.

Derivative Thermogravimetry (DTG) shows the change in mass per unit time as a function of temperature.

Apparatus

The apparatus is referred to as a thermobalance or thermogravimetric analyzer. There are a number of configurations: horizontal, simultaneous (TGA-DTA), and vertical. The vertical design provides better sensitivity and weight capacity. The thermobalance consists of five essential components: furnace, temperature regulator, weighing mechanism, atmosphere controller, and recording system.

Calibration

Small furnaces can be calibrated by a method [8] using Curie points of a range of metals and alloys. The Curie point is the temperature at which a ferromagnetic material loses its ferromagnetism. At the Curie point, the magnetic force is reduced to zero and an apparent mass change is observed.

The study of the reactions can be divided into the stages: (1) intermediates and (2) products of reaction, (3) energetics of reaction, and (4) the reaction kinetics. Stages (1) and (2) can be readily studied by TG and DSC.

For example, the decomposition of calcium oxalate monohydrate shows three distinct steps. The first around 200°C with a loss of 12.4% corresponds to dehydration, while those at 500°C and 800°C match with a loss of CO and CO_2. These are confirmed by analysis of residues.

Kinetics of Reaction Including Measurement of α and $d\alpha/dt$.
The use of thermogravimetry as a means for the elucidation of the reaction kinetics is attractive. The nature of solid-solid interactions is quite complex [14] and will not be discussed in this section.

Consider an endothermic solid-state reaction:

$$A\left(\text{solid}\right) \rightarrow B\left(\text{solid}\right) + C\left(\text{gas}\right) \tag{70.63}$$

During the course of the reaction, there is a mass loss, combined with the loss of gas. Heat absorption also occurs. This process can be modeled. However, it should be noted that the equation (although generally applicable) are not valid for all cases. Methods and mathematical treatment of results are given in the papers by Šatava, Šesták, and Škvára [15-18].

Static (isothermal) and dynamic methods can used in a kinetic study of the weight change. The former is based on the determination of the degree of transformation at constant temperature as a function of time. The latter is the determination of the degree of transformation as a function of time during a linear increase of temperature. The static method is probably better suited for obtaining information about the slowest process, the reaction order, and reaction mechanism. The dynamic method is better if data on the kinetics of the reaction from a single curve for the whole temperature range is required. Comparisons between both methods have shown comparable results with respect to precision [19].

The extent of a reaction ξ may be defined [20] by Equation 70.64.

$$n_B = n_{B,0} + \nu_B \xi \tag{70.64}$$

where n_B = Amount of substance B
$n_{B,0}$ = Amount of substance B at $t = 0$
ν_B = Stoichiometric number of B (positive number if B is a product and negative if B is a reactant)

For solid-state reactions, the changes in the portion reacted α are followed with respect to time. Therefore, the rate of reaction can be defined by Equation 70.65.

$$\text{Rate} = d\alpha/dt \tag{70.65}$$

For solution reactions (referring to Equation 70.63), the change in concentration C_B of B is followed. Since the rate of reaction varies with time (even at constant temperature) at a value of α Equation 70.66 is derived.

$$\text{Rate} = d\alpha/dt = k_T, f\left(\alpha\right) \tag{70.66}$$

where k_T = the rate constant at temperature T
$f(\alpha)$ = mathematical expression in α

It should be noted that the form of $f(\alpha)$ sometimes alters part way through a reaction.

If hyphenated group-specific techniques are employed to study a reaction simultaneously (e.g., as is the case in TGA and FTIR), the IR-active species may not contribute the greatest mass loss and therefore the values of α will not be the same.

There are many equations relating the rate of solid-state reactions to α and they have been summarized by Sestak and Berggren [21].

A general *integrated* kinetic equation is given in Equation 70.67.

$$g(\alpha) = k_T t \tag{70.67}$$

where $g(\alpha) = \int d\alpha / f(\alpha)$.

The rate constant k_T can be calculated from the Arrhenius equation given in Equation 70.68.

$$k_T = A \exp\left(-E_A/RT\right) \tag{70.68}$$

where E_A = Activation energy (J mol^{-1})
  A = Pre-exponential factor
  R = Molar gas constant, 8.314 J (K mol)$^{-1}$

Measurement of α and $d\alpha/dt$

Consider a thermogravimetric curve consisting of one step. α at a particular time can be found using:

$$\alpha = m_i / \left(m_i - m_f\right) - m_t / \left(m_i - m_f\right) \tag{70.69}$$

The differential is hence:

$$d\alpha/dt = -\left[d\,m_t/dt/\left(m_i - m_f\right)\right] \tag{70.70}$$

This states that the rate of reaction can be measured from the slope of the mass–time curve. Since dm_t/dt is already measured by the DTG curve, $d\alpha/dt$ can be found directly from the curve.

Combination of a number of the equations discussed [13] gives:

$$\ln d\alpha/dt - \ln\left(f(\alpha)\right) = \ln\left(A/\beta\right) - E_A/RT \tag{70.71}$$

where $\beta = dT/dt$.

Thermometric Titrimetry

Thermometric titration is the measurement of the temperature change in a system as a function of time or volume of titrant. The technique consists of the measurement of the change in temperature as the titrant is added to it, under near adiabatic or more commonly referred to as isoperibol conditions. The experiments are typically carried out in a small dewar flask submerged in a well-controlled constant-temperature bath. The method can be used to study oxidation-reduction, complexation, precipitation, and neutralization reactions in aqueous solvents. Publications by Zenchelsky [22] and Jordan [23] review the technique in detail.

The basic principle is that a free energy change occurs in the system [24], and is based on the measurement of the free energy-dependent term:

$$\Delta G^\ominus = -RT \ln K \tag{70.72}$$

where $\Delta G^\ominus$ = Change in free energy under standard conditions
  R = Molar gas constant
  T = Temperature in kelvin
  K = Equilibrium constant for the system at the temperature T

A calorimetric method (entropy titration) for the determination of ΔG, ΔH, and ΔS from one thermometric titration has been described by Christensen et al. [24].

Thermomechanical Analysis

Thermomechanical analysis relates to techniques where deformation is measured as a change in either volume or length. The deformation is plotted against temperature when a sample is heated under a controlled temperature program. Thermodilatometry measures the dimensional changes as a function of time under negligible loads. Thermomechanical analysis (TMA) is similar to thermodilatometry, but also provides information regarding penetration, extension, and flexure using various types of loads on the test specimen. In dynamic mechanical analysis (DMA), the test specimen is subjected to a sinusoidally modulated stress under specified temperature. The viscoelastic response of a material is then monitored under tensile, compressive, shear, or torsional load [25].

Apparatus

A typical instrument for thermal mechanical analysis is called a dilatometer and is equipped with a linear variable differential transformer (LVDT). The displacement of the sample is transferred to the LVDT via a rod (probe) that is unaffected by heat and dimensional changes. A zero weight is accomplished for thermodilatometry by a float system so that a minimum of a load is subjected to the sample. The sample is placed on a sample holder in an oven. A force is applied through the probe in TMA and DMA. The sample cylinder and the probe are independently connected to the measuring device. The top of the probe is also connected to a balance arm. Probe movement and sample length changes are detected. The recorded signals are time, temperature, dimensional changes, and load. Various probes are available, depending on the analysis needs. Expansion, compression, and penetration probes are standard. Tension, three-point bending, and cubical expansion probes are available. Measuring temperature from $-150°C$ to $600°C$ or even up to $1500°C$ is possible, depending on the instrument.

Calibration of Probe

The temperature is usually the measured quantity in thermomechanical analysis. Therefore, calibrating the temperature axis is important. Thermomechanical analyzers can be temperature calibrated according to ASTM standard test method E 1363 [26]. An equation is developed for a linear correlation of the experimentally obtained program temperature and the actual melting temperature for known melting standards (i.e., mercury, water, tin, benzoic acid). A penetration probe is used to obtain the onset temperatures for two melting standards. The two-point calibration assumes the relationship Equation 70.73 between the actual specimen temperature (T_t) and the observed extrapolated onset temperature (T_0). S and I are the slopes and intercept, respectively, in the TMA thermal curve (Figure 70.27).

$$T_t = \left(T_0 \times S\right) + I \tag{70.73}$$

Thermomechanical methods are generally applied on solid, shaped samples like polymeric products. Special clamps are used for testing of soft samples made of rubbers, adhesives, fats, etc. Films and fibers can be tested using clamps. Liquid polymers are tested on support. DMA is used for detecting α, β, and γ transitions in cured epoxy systems [25]. Thermomechanical analysis, TMA, is used for measuring the volume change of bitumen. Scratching and crack propagation at low temperatures is simulated. This is useful when investigating asphalt paving materials [27]. The thermal expansion coefficient of linear expansion is calculated from the slope of the expansion-temperature curve. This is obtained under zero load in thermodilatometry mode. Thermodilatometry can also provide information on phase changes, sintering, and chemical reactions. Softening temperatures are measured using small-diameter tips on the probe under a load (TMA). This sensitive technique is also used for the measurement of heat distortion temperatures and glass transition temperatures of polymers [28].

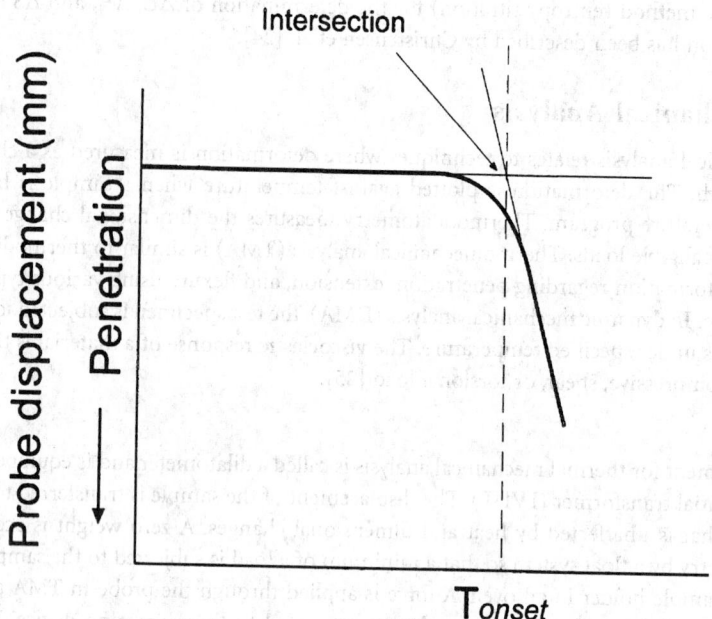

FIGURE 70.27 The calibration of a TMA instrument is a two-point method. There is an assumption that there is a relationship $T_t = (T_o \times S) + I$ between the actual specimen temperature and the onset temperature (Equation 70.73).

Differential Thermal Analysis and Differential Scanning Calorimetry

DTA is the detection of the temperature difference between the sample holder and the reference holder using the electromotive force of the thermocouples, which are attached to the holders. The sample and reference are subjected to a controlled temperature program. The differential is output as the DTA signal. DSC is similar to a DTA in construction, but the DSC measures the difference in heat flow rate to the sample and the reference. Consequently, more information is received on the thermodynamic behavior of the material using DSC. Quantitative DTA is also addressed as a DSC. This definition results in that the major part of all differential thermal analyses performed today uses DSC. An application is found for combined TGA/DTA analysis for kinetic evaluation of petroleum products [29].

Apparatus

The DTA apparatus has a sample and a reference cell subjected to the same temperature program. The measuring device consisting of a thermocouple or any temperature measurement device placed in each cell, measuring the difference in temperature. Operating temperature range is ambient to 1000°C or higher depending on the construction of the instrument and sample pan material. Differential scanning calorimetry is originally defined as individually heated cells. Equal temperature is maintained in the cells, giving an electrical signal proportional to the power needed. The DSC curves represent the rate of energy absorption. Today, most DTA units that can be calibrated to give calorimetric response are called DSC [25]. For qualitative applications, both classical DTA and DSC are equally good. In quantitative work, the DSC is claimed to be better at low heating rates [30].

Calibration and Reference Materials

The dynamic nature of thermal analysis requires a calibration and standard compound to be able to relate results obtained by different instruments. Temperature calibrations can be done using a range of selected materials. Different materials are chosen depending on the temperature range. Common standards are 1,2-dichloroethane, indium, silver sulfate, and quartz. Other organic compounds, metals, inorganic nitrates, sulfates, or chromates are also used [25].

Theory of DTA and DSC

The measured quantity is ΔT, the difference between the temperature of the sample and the reference material. In Equation 70.74, T_S is the temperature of the sample and T_R is the temperature of the reference material.

$$T_S - T_R \equiv \Delta T \tag{70.74}$$

The result is presented as a plot of ΔT against T under a stated temperature program, the differential thermal curve. An endothermic process is then shown as a negative signal. A quantity of material decomposition or the enthalpy of the process is obtained from the area of the peak. It is then, in fact, a calorimetric analysis and the technique is referred to as differential scanning calorimetry.

In DTA, heat transfer to a sample and reference causes a difference in temperature ΔT, which can be related to the energy of any transition of the sample.

$$\Delta H = K \left(\text{peak area} \right) \tag{70.75}$$

For heat flux DSC, a similar process occurs, whereas in power compensated DSC, electric heating is supplied to the sample and reference to keep their temperatures as close as possible. For best calorimetric accuracy, the constant K should vary little with T.

Specialized Techniques

Thermoelectrometry

Electrical properties such as resistance/conductance and capacitance can be measured as a function of temperature. A variation that can measure the generated EMF is called thermovoltaic detection [8].

Modulated DSC

In MDSC, the heating rate is modulated. This is performed using a small alternating power supply in combination with the standard programmed heating. The heating program is given by the equation:

$$T = T_0 + \beta t + B \sin\left(\omega t\right) \tag{70.76}$$

and the heat flow is given by:

$$dq/dt = C_p \left[\beta \omega \cos\left(\omega t\right) + f\left(t, T\right) + C \sin\left(\omega t\right) \right] \tag{70.77}$$

where T_0 = Initiation temperature
B = Amplitude of temperature modulation
ω = Angular frequency = $(2\pi f)$
C_p = Heat capacity
$f(t,T)$ = Kinetic response (average)
C = Amplitude of response to sine-wave modulation

Simultaneous Techniques

Each of the techniques discussed above provides information about the sample. However, a synergistic effect exists, in that, the total amount of information obtained (by using techniques simultaneously) regarding the sample is greater than the sum of the information from the individual techniques.

Evolved Gas Analysis.

This allows the identification of gases evolved during thermal analysis and is performed by replacing the detector with a mass spectrometer or FTIR. An alternative technique is to precede the detector by passing gases evolved during the thermal analysis through a gas chromatograph.

Thermomicroscopy.

This can be incorporated under thermoptometry (a family of techniques that measure changes of an optical property with temperature change). Thermomicroscopy uses observations under a microscope.

Applications (Including the Analysis of Electrical Insulating Materials)

Oxidative Stability of Oils and Greases and Polymers.

Oxidative degradation of oils upon heating can be monitored using a DSC apparatus. The detected onset *time* or *temperature* of the exotherm can be taken as a measure of the thermal/oxidative stability of the oil. The detected onsets are a strong function of the sample size, instrument sensitivity, kinetics, and scan rate. This enables DSC to be used in an oxidation test. Isothermal high-pressure DSC (PDSC) has been used to characterize the oxidative stability and the oxidation mechanisms of lubricants [34,35]. A PDSC works at pressures up to 3.5 MPa of a selected gas, using a wide temperature range. The technique is useful in the development of new lubricants with improved thermal and oxidative properties. The influence of metal catalysis on oil oxidation can be determined using PDSC. The volatile degradation products have been determined using combined PDSC–GC/MS (gas chromatography–mass spectrometry) [36]. PDSC gives information about relative oxidation stability used for comparing the lifetime of oils [37,38]. It has been a good technique for evaluating the thermal and oxidative stability of lubricating oils [34,39]. PDSC has also been used for evaluating deposit-forming tendencies of liquid lubricants [39].

Volatilization occurs when a low-boiling oil is heated, especially at high temperatures. This leads to uncontrolled changes in composition. It also affects the size and shape of the DSC exotherm, causing imprecise determination of the oxidation onset [34]. Use of high pressure in the DSC cell reduces volatility and evaporation interference with it. Added to this, the onset value is shifted to lower temperatures [34]. The onset becomes better defined and the peak size increases [40].

DSC is a fast technique for oxidation stability testing. This is a great advantage in the quality control of electrical insulating oils. Experimental evaluations of transformer insulating oils have shown ranking to be possible. The remaining lifetime of inhibited oils may correlate to the oxidation induction time [41]. Important parameters in PDSC are sample weight, pressure, and temperature program and have to be carefully considered before applying the technique. The sample pans must not be overlooked. Results are significantly influenced by variations in metallurgy, due to the catalytic and inhibiting effects of various metals. Oxidation induction time of lubricating greases can be determined by ASTM method D5483-93.

Predicting the Lifetime of a Product.

Estimating the lifetime of a product typically uses some form of accelerated testing. TGA decomposition kinetics can be used to arrive at aging stability information and lifetime predictions in relatively short timescales (hours compared months in conventional oven aging). The sample (e.g., insulating paper, etc.) can be heated through its decomposition at several heating rates and the weight loss as a function of temperature recorded. The activation energy is calculated from a plot of log heating rate versus the reciprocal of the temperature for a constant decomposition level. The activation energy is subsequently used to calculate kinetic parameters such as specific rate constant (k) or halflife times, as well as to estimate the lifetime of the material at a given temperature. DTA and DSC have been employed in the electric industry to study polymeric insulation and for the determination of dielectric stability and lifetime prediction.

Thermal Analysis and Stability of Materials.
TGA is widely employed in the determination of thermal stability of materials and analysis of their composition. The thermal history of electric cable insulation has been determined using DSC [42]. Thermal analysis techniques have greatly improved the quality control and inspection of electric cables. Hyphenated techniques have been employed in the analysis of trace components in electrical insulation [42-44]. ASTM method D3386-84 standardizes measurement of coefficient of linear thermal expansion of electrical insulating materials, while D3850-84 refers to the rapid determination of thermal degradation of solid electrical insulating materials by thermogravimetric methods.

Mechanical Stress Determinations.
Longitudinal mechanical stresses, frozen into electric cable insulation during the fabrication process, can produce "shrink back." This causes the insulation to shrink away from freshly cut cable ends, to varying degrees. TMA can be used to determine these stresses and has been found more versatile than the traditional BS6469 shrinkage measurement [45,46].

Evolved Gas Detection and Evolved Gas Analysis.
The main use of EGD is to distinguish between phase transitions and endothermic decompositions (e.g., coordination chemistry). It has been used for the analysis of effluents [47]. Thermogravimetric analysis coupled with FTIR has been used to establish the failure mechanisms of electrical insulating materials [48]. EGA is also used for assessing the thermal endurance of polymeric materials and is of particular value in thermosetting polymers used in the electric industry [49].

Investigation of Polymeric Systems.
Thermogravimetry can be applied to the study of polymer processes (pyrolysis, oxidative degradation, volatilization, absorption, adsorption, and polymerization) in which a change in weight occurs. The degree of crystallinity provides information regarding the thermal history of a polymer and can be measured by DSC. Physical and mechanical properties of polymers are related to the degree of crystallinity [50]. Thermophysical property measurements and analysis of additives in polymers can also be performed using thermal analysis techniques [51]. ASTM method D4000-89 can be used for the identification of plastic materials.

Pharmaceutical Applications.
Calorimetric purity determinations are used in the pharmaceutical industry. The concentration of the impurity is regarded as inversely proportional to its melting point. Therefore, an increase in the sample's impurity content decreases the melting point and broadens the melting range. DTA can also be used but DSC is preferred since it also gives the ΔH_f (heat of fusion) of the melt [52]. The DSC method is based on the van't Hoff equation. A compound may exist in various crystal forms DSC and TG are used to characterize polymorphs and assess the stability of the compounds. DSC has been used for investigating the effect of inhibitors with model membranes [53]. Drug incompatibility is defined as "an interaction between two or more components to produce changes in the chemical, physical, microbiological, or therapeutic properties of the preparation" [54]. DTA and DSC are used to record reactions as a function of temperature and investigate drug compatibility [55]. Recent advances in microcalorimetry have allowed nondestructive analysis at room temperature [56,57]. The technique is gaining popularity in the pharmceutical industry and also in the study of ballistics.

Characterization of Greases and Lubricants.
Greases and lubricants are, in application, exposed to high temperatures in both inert and in oxidizing atmosphere. Material losses due to evaporation and loss or alteration due to thermal cracking or oxidation of the molecular structure are possible. The various aging reactions are usually inhibited by additives. Thermogravimetry, differential thermoanalysis, and differential scanning calorimetry are used as test

instruments, but the overall difficulty is to find methods that correlate with real thermal aging of the greases and lubricants.

The peak onset and peak maximum temperatures from DTG, DTA, and DSC curves are used or the peak onset from TGA curve. The evaporation behavior of greases is the most used parameter, but wax content, glass temperature, and cloud point are other characteristics of greases that are studied using thermoanalytical techniques [27].

In the Noack test of evaporative loss (DIN 51 581), the sample is held at 250°C for 60 min in a air flow. The sample is weighed before and after treatment. The cause of the weight loss is not clear, whether it is evaporation of parts of the original sample or evaporation of oxidative degradation products. The question has arisen whether isothermal thermogravimetry could replace the Noack test. This would provide continuous loss information during the thermal exposure. It has been shown that there is a higher weight loss in the thermobalance than in the Noack test at equal test conditions. The deviation is caused by the difference in surface:volume ratio between the two methods [27].

Oxidation studies of low boiling lubricant or lubricating oils do not give representative results. This due to the evaporation or the oil and low boiling oxidative degradation products. A TGA curve of a lubricant produced in an air atmosphere does not always represent the oxidation reaction. The use of an elevated oxygen or air pressure in DSC has been shown to reduce sample evaporation due to a increased evaporation temperature and increase the rate of the oxidation. Several papers deal with this technique, which has found application in the characterization of lubricants [27,34-41,58].

Insulation Paper/Cellulose.

The rate of weight loss on pyrolysis of cellulosic materials has applications to engineering problems in many industries. On heating, cellulose undergoes a number of linked physical and chemical changes [59]. Properties such as weight, strength, crystallinity, and enthalpy are affected.

Thermogravimetric analysis can be used to perform a collective measurement of the weight loss due to the production of H_2O, CO, and CO_2 during degradation. Of course, the measurement will include evaporation of other pyrolysis products. The enthalpy changes can also be measured by DSC. These methods are very useful in determining the temperature range at which physical and chemical processes occur. The rate of these processes can be determined by using DTG (derivative thermogravimetry).

Acknowledgments

The authors would like to thank P. Haines, Chapman and Hall, and TA Instruments for their advice and permission.

Defining Terms

Thermal analysis: A group of techniques in which a property of the sample is monitored against time or programmed temperature (in a specified atmosphere).
Derivative: Techniques where a measurement or calculation of the first derivative is performed.
Differential: Techniques where a difference in a property is measured.

References

1. R.C. Mackenzie, A history of thermal analysis. *Thermochim. Acta*, 73, 249, 1984.
2. R.C. Mackenzie, Origin of thermal analysis. *Israel J. Chem.*, 22, 203-205, 1982.
3. R.C. Mackenzie, Early thermometry and differential thermometry. *Thermochim. Acta*, 148, 57-62, 1989.
4. W.W. Wendlandt, The development of thermal analysis instrumentation 1955–1985. *Thermochim. Acta*, 100, 1-22,1986.
5. C.J. Keattch and D. Dollimore, *Introduction to Thermogravimetry*, Heyden, London, 1975.
6. A.L. Lavoisier and P.S. de Laplace, *Mem. R. Acad. Sci.*, Paris, 355, 1784.

7. R.C. Mackenzie, C.J. Keattch, D. Dollimore, J.A. Forrester, A. Hodgson, and J.P. Redfern, Nomenclature in thermal analysis II. *Talanta,*19,1079-1081,1972.

8. J.O. Hill, *For Better Thermal Analysis and Calorimetry III.* ICTA, 1991.

9. R.C. Mackenzie, Nomenclature in thermal analysis III. *J. Thermal Anal.,* 8(1), 197-199, 1975; *Thermochim. Acta,* 28, 197, 1975.

10. R.C. Mackenzie, Nomenclature in thermal analysis, in *Treatise on Analytical Chemistry,* P.J. Elving, (Ed.), Part 1, Vol. 12, John Wiley & Sons, New York, 1983, 1-16.

11. P.J. Haines, *Thermal Methods of Analysis,* Blackie/Chapman and Hall, London, 1995.

12. E.L. Charsley, J.P. Davies, E. Gloeggler, N. Hawkins, G.W.H. Hoehne, T. Lever, K. Peters, M.J. Richardson, I. Rothemund, and A. Stegmayer, *J. Thermal Anal.,* 40, 1405-1414, 1993.

13. G.M. Lukaszewski, Accuracy in thermogravimetric analysis. *Nature,* 194, 959, 1962.

14. W.E. Garner, *The Chemistry of the Solid State,* Butterworths, London, 1955.

15. V. Šatava, *Silikáty,* Pouziti Terografickych Metod ke Studiu Reakcni Kinetiky, 5(1), 68, 1961. (The Thermographic merod [sic] of Determination of Kinetic Data).

16. V. Šatava and J. Šesták, Kinetika Analysa Termogravimetrickych Da, *Silikáty,* 8(2), 134, 1964. (Kinetic Analysis of Thermogravimetric Measurements), Source: un-numbered English-language contents page preceding p.93 in *Silikaty,* 8(2), 1964.

17. V. Šatava and F. Škvára, Mechanism and kinetics of the decomposition of solids by a thermogravimetric method. *J. Amer. Ceram. Soc.,* 52, 591-595, 1969.

18. J. Šesták, A review of methods for the mathematical evaluation kinetic data from nonisothermal and isothermal thermogravimetric measurements. *Silikáty,* 11, 153-190, 1967.

19. J. Šesták, Errors of kinetic data obtained from thermogravimetric curves at increasing temperature. *Talanta,* 13, 567, 1966.

20. M.L. McGashan, *Physico-Chemical Quantities and Units,* RSC, London, 1968, 39.

21. J. Šesták and G. Berggren, Kinetics of the mechanism of solid state reactions at increasing temperatures, *Thermochim. Acta,* 3, 1-12, 1971.

22. S.T. Zenchelsky, Thermometric Titration, *Anal. Chem.,* 32, 289R, 1960.

23. J. Jordan, *Handbook of Analytical Chemistry,* L Meites, Ed., McGraw-Hill, New York, 1963, Sec. 8-3.

24. L.S. Bark and S.M. Bark, *Thermometric Titrimetry,* International series of monographs in Analytical Chemistry, Vol. 33, Pergamon Press, 1969; J.J. Christensen, R.M. Izatt, L.D. Hansen, and J.A. Partridge, Entropy titration. A calorimetric method for the determination of ΔG, ΔH, ΔS from a single thermometric titration. *J. Phys. Chem.,* 70(6), 1966 and 40(1), 1968.

25. D. Dollimore, Thermoanalytical instrumentation, in G.W. Ewing (Ed.), *Analytical Instrumentation Handbook,* Marcel Dekker, New York, 1990.

26. ASTM Designation: E 1363-90, Standard test method for Temperature Calibration of Thermomechanical Analyzers. 1990.

27. H. Kopsch, *Thermal Methods in Petroleum Analysis,* VCH,Verlagsgesellschaft mbH, Weinheim, 1995.

28. H.H. Willard, L.L. Merritt Jr., J.A. Dean, and F.A. Settle Jr., *Instrumental Methods of Analysis,* 6th ed., Wadsworth Publishing Company, Belmont CA, 1981.

29. ASTM Designation E 698-79, Standard Test Method for Arrhenius Kinetic Constants for Thermally Unstable Materials. 1979.

30. M.I. Pope and M.D. Judd, *Differential Thermal Analysis,* Heyden & Sons Ltd., London, 1980.

31. W.W. Wendlandt, Thermoelectrometry — a review of recent thermal analysis applications. *Thermochim. Acta,* 73, 89-100, 1984.

32. P.D. Garn and G.D. Anthony, Repetitive gas chromatographic analysis of thermal decomposition products. *Anal. Chem.,* 39,1445-1448, 1967.

33. M.R. Holdiness and R. Mack, Evolved gas analysis by mass spectrometry: a review, 1 *Thermochim. Acta,* 75, 361-399, 1984.

34. S.M. Hsu, A.L. Cummings, and D.B. Clark, Society of Automotive Engineers SAE, Technical Paper 821252, 1982.

35. R. Schumacher, Practical thermoanalysis in tribology, *Tribology International,* 25(4), 259-270, 1992.

36. A. Zeman, DSC cell — a versatile tool to study thermooxidation of aviation lubricants. *Journal of Synthetic Lubricants*, 5, 133-148, 1988.

37. E. Gimzewski, A multi-sample high pressure DTA for measuring oxidation induction times. *Thermochim. Acta*, 170, 97-105, 1990.

38. R.E. Kauffman and W.E. Rhine, *Journal of the Society of Tribologists and Lubrication Engineers*, 44, 154-161, 1988.

39. Y. Zhang, P. Pei, J.M. Perez, and S.M. Hsu, *Journal of the Society of Tribologists and Lubrication Engineers*, 48, 189-195, 1992.

40. R.L. Blaine, Thermal-analytical characterization of oils and lubricants, *American Laboratory*, Reprint, January, 22, 18-20, 1974. Vol. 6; No. 1.

41. M. Eklund, Literature review of DSC oxidation tests on petroleum products. Report to International Electrotechnical Commision Technical Committee, April 10, 1996.

42. J.W. Billing, Thermal history of cable insulation revealed by DSC examination. *IEEE DMMA Conference*, 289, 309-312, June 1988.

43. R.L. Hutchinson, Thermal analysis to spectroscopy, an overview of analytical instrumentation for electrical insulating materials. *Proceeding of the 17th Electrical/Electronics Insulation Conference*, Boston, MA, 1985.

44. M.T. Baker, S. O'Connor, and J.F. Johnson, Hyphenated analysis for trace components in electrical insulations. *Proceeding of the 17th Electrical/Electronics Insulation Conference*, Boston, MA, 1985.

45. J.W. Billing and D.J. Groves, Treeing in mechanically strained h.v. cable polymers using conducting polymer electrodes. *Proc. Institution Elec. Eng.*, 121, 1451-1456, 1974.

46. BS6469 1992 Insulating and sheathing materials of electric cable Part 1 section 1.3 (equivalent to IEC 811-1-3. 1985 + Al:1990).

47. W. Lodding (Ed.), *Gas Effluent Analysis*, Edward Arnold, London, 1967.

48. M. Ali, J.M. Cooper, S.J. Fitton, and S.P. McCann, The Development of techniques for the analysis of materials, *The 7th INSUCON 1994, BEAMA International Electrical Insulation Conference*, 131-135.

49. M. Ali, J.M. Cooper, S.G. Swingler, and S.P. Waters, Simultaneous thermal and infrared analysis of insulating resins, *IEE 6th International Conference on Dielectric Materials measurements and Applications*, Manchester, 363, 77-80, 1992.

50. E.L. Charsley and S.B. Warrington (Eds.), *Thermal Analysis — Techniques and Applications*, RSC Special Publication No. 117, 1992.

51. E.A. Turi (Ed.), *Thermal Characterization of Polymeric Materials*, Academic Press, New York, 1981.

52. L. Kofler and A. Kofler, *Thermomikromethoden zur Kennzeidung Organisher Stoffe und Stoffgemische*, Verlag Chemie, Weinheim, 1954.

53. D.R. Reid, L.K. MacLachlan, R.C. Mitchell, M.J. Graham, M.J. Raw, and P.A. Smith, Spectroscopic and physicochemical studies on the interactions of reversible hydrogen ion. *Biochim. Biophys. Acta*, 1029, 24-32, 1990.

54. A. Wade (Ed.), *Pharmaceutical Handbook, 19th edition*, Pharmaceutical Press, London, 1980, 28.

55. J.L. Ford and P. Timmins, *Pharmaceutical Thermal Analysis*, Academic Press, New York, 1989.

56. R.J. Willson, A.E. Beezer, J.C. Mitchell, and W. Loh, Determination of thermodynaic and kinetic parameters from isothermal heat conduction microcalorimetry: applications to long-term-reaction studies. *J. Phys. Chem.*, 99, 7108-7113, 1995.

57. D.L. Hansen, Instrument selection for calorimetric drug stability studies, *Pharm. Technol.*, 20(4), 64-65, 68, 70, 72, 74, 1996.

58. ASTM Designation D 5483-93, Standard Test Method for the Oxidation Induction Time of Lubricating Greases by Pressure Differential Scanning Calorimetry.

59. D. Dollimore and J.M. Hoath, The preparation and examination of partially combusted cellulose chars, *Theromchim. Acta*, 45, 103-113, 1981.

70.3 Kinetic Methods

E.E.Uzgiris and J.Y.Gui

Kinetic methods involve the measurement of chemical reactions or processes in a time-dependent manner. Rates of dynamic processes are measured rather than the properties of a system at equilibrium. Of course, this approach is a central one for the study of chemical reactions and reaction mechanisms; however, it has much value in analytic chemistry; that is, in the determination of the composition of materials. This fact has been recognized for some time, but in recent years there has been a resurgence in interest in the use of kinetic methods in analytic chemistry. There have been several world congresses on this subject, numerous monographs [1,2], and the number of papers on kinetic methods has dramatically increased in the last decade [3].

Why is there such an interest? After all, there are many analytic procedures that are quite general and sensitive. As a specific example, consider the analysis for various metals in environmental samples. Metal ions can be detected by numerous means such as by ion selective electrodes, atomic flame spectroscopy, or ion coupled plasma spectroscopy, yet there is abundant literature on metal detection by catalyzed reactions in a kinetic manner [1-3]. In this case, the method of choice is dictated by cost of analysis, speed, sensitivity, and convenience. Furthermore, certain molecular species may be difficult to discriminate from others in conventional analysis. In this case, with a proper reaction, the kinetic approach is a powerful tool in detecting such constituents. Finally, the kinetic approach is the only method capable of elucidating the nature of binding sites in molecular binding because the determination of an equilibrium association constant alone is insufficient to elucidate mixed binding sites [4]. It is also the principal means of identifying short-lived intermediate species in a reaction [5].

Thus, kinetic methods comprise an important group of methods available for the analysis of substances. In some cases, kinetic methods offer unique advantages as in the study of mixed binding sites, in the delineation of competing species, and in the determination of short-lived intermediates. In other cases, kinetic methods offer speed and convenience, and low cost, as for example in such applications as clinical analysis and environmental field analysis.

In a broader sense, time-dependent changes in chemical, physical, and biological processes are universal. Because equilibrium may not be achieved in certain processes, time-dependent effects must be considered and accounted for in a satisfactory manner for analytic determinations to be accurate and reproducible. In some instances, for reasons of speed of analysis, kinetic rates are measured rather than equilibrium values. The range of time dependencies can range from picoseconds, studied with mode-locked lasers, to seconds or minutes, studied with batch mixing procedures. Kinetic methods encompass a broad range of processes and time domains. The methods of simple chemical reactions can often be applied to complicated biological processes. This is possible because often one reaction in a group of coupled reactions controls the overall rate of the process.

Kinetic methods have been classified according to different criteria. The most common classification is based on whether the method involves a catalyst. This is so because reactions are frequently quite slow. In such cases, a catalyst must be added to speed up the reactions and make rate determinations practical. In other instances, the catalyst is the analyte itself. There are two major groups of catalysts: enzymatic and nonenzymatic. Another common classification of kinetic methods is based on whether the reaction proceeds in a homogeneous or heterogeneous system. Most of the discussion will be focused on homogeneous liquid and heterogeneous liquid–solid systems because these comprise the majority of kinetic analytical methods that have been developed. Presented in Table 70.4 are classifications based on the above criteria along with example reactions.

Theoretical Aspects

A reaction involving species A and B proceeds to a product with a rate constant, k', such that the rate of change of species A is given by

TABLE 70.4 Classification of Kinetic Methods Based on System and Catalyst

System	Catalyst	Reaction examples
Homogeneous	Enzymatic	Hydrolysis
		Electron transfer
	Nonenzymatic	Redox
		Complexation
		Chemiluminescence
	No catalyst	Redox
		Chemiluminescence
Heterogeneous	Enzymatic	Immunoezymatic
		Electrode reactions
		Electrocatalysis
		Fluorescence
	Nonenzymatic	Electrode reactions
		Electrocatalysis
		Fluorescence
	No catalyst	Fluorescence
		Radioimmunoassay

$$-dA/dt = k'\left[A\right]\left[B\right] \tag{70.78}$$

where the brackets denote concentration. If the species A is of interest, then the reactant B can be in excess, in which case changes in [B] can be ignored. A pseudo-first-order reaction can be written:

$$-d\left[A\right]/dt = k\left[A\right] \tag{70.79}$$

where $k = k'$ [B] and the time evolution of [A] is just:

$$\left[A\right] = \left[A\right]_0 e^{-kt} \tag{70.80}$$

The product, P, which is the species that is usually detected, evolves as:

$$\left[P\right] = \left[A\right]_0\left(1 - e^{-kt}\right) \tag{70.81}$$

The species A can be expressed in terms of product by:

$$\left[A\right] = \left[P\right]_\infty - \left[P\right] \tag{70.82}$$

where $\left[P\right]_\infty = \left[A\right]_0$. By measuring [P] as a function of time, the initial concentration of A can be deduced from a plot of:

$$\ln\left[A\right] = \ln\left[A\right]_0 - kt \tag{70.83}$$

In this way, a calibration curve can be generated against which an unknown sample can be measured for the content of species A.

In case the reaction is of a different order, the time-dependent plots for determining $[A]_0$ take on a different form. For example, in a second-order reaction, the rate of change of $[A]$ is given by:

$$-d[A]/dt = k[A][A] \qquad (70.84)$$

where, as before, the reactant $[B]$ is considered in excess and its time dependence can be assumed to be negligible. The calibration curve is now:

$$1/[A] = 1/[A]_0 + kt \qquad (70.85)$$

Clearly, the order of the reaction under study must be known for a correct analysis. There are straightforward ways to determine the order by varying the initial concentration of $[A]$ and noting the initial velocity of the reaction. A plot of the initial velocities versus initial $[A]$ will reveal the order of the reaction [1].

One of the strong points of kinetic methods is that closely related species that may be difficult to resolve by other means can be resolved by kinetic measurements. This is particularly true when enzyme reactions are employed. Enzymatic reactions are extremely sensitive to molecular structure and closely related structural analogs may have significantly different kinetics. For example, consider species A and B going through a reaction to a product but each having a different rate constant, k_a and k_b. The detected product is given as a sum of the two components by:

$$[P]_\infty - [P] = [A]_0 \exp(-k_a t) + [B]_0 \exp(-k_b t) \qquad (70.86)$$

Then, by computer fitting or graphical analysis of a semilog plot of

$$\ln\{[P]_\infty - [P]\} = \ln[A]_0 - k_a t + \ln[B]_0 - k_b t \qquad (70.87)$$

one can extrapolate to $t = 0$ and determine, $[A]_0$ and $[B]_0$.

The important case of catalyzed reactions must be considered separately as there are important differences from the case of uncatalyzed reactions considered above. The catalyst is usually the species to be determined as it often is a metal ion or nonorganic ion of interest. Usually the catalyst combines with the reactant species $[B]$ in a very fast reaction with a given equilibrium constant to give:

$$[C] + [B] \rightleftarrows [CB] + [Y] \qquad (70.88)$$

Here, CB, the reactant B bound to C, reacts with A with a much faster rate than if B is unbound. This develops because of the reduction of the activation energy provided by the catalyst C in combination with B as discussed below. Thus,

$$[CB] + [A] \rightleftarrows [P] + [Y] \qquad (70.89)$$

This more complex kinetics simplifies to pseudo-first-order if one considers only the initial rates of the reaction. The initial velocity of the indicator product, P, takes the simple form:

$$V_0 = d[P]/dt = K'[C]_0 + K'' \qquad (70.90)$$

where K' and K'' are constants. A calibration curve for C can thereby be generated through initial velocity measurements.

Enzyme Reactions

Enzymes are a class of proteins that catalyze reactions with exquisite specificity. The activity of certain enzymes is in itself of great importance in clinical diagnosis, but enzymes can be useful in determining substrate concentration — also very important for clinical applications and for environmental analysis. The rates of enzyme reactions are directly proportional to enzyme concentration; however, there is a saturation of reaction rates with increasing substrate concentration. This saturation effect must be considered when analyzing such reactions. The essential feature of enzyme reactions involves the enzyme, the substrate, the enzyme–substrate complex, and the product. It is the formation of the enzyme–substrate complex that leads to the saturation kinetics [5]. The reaction can be represented as follows:

$$\left[E\right]+\left[S\right]\underset{k_1}{\overset{k_2}{\rightleftharpoons}}\left[ES\right]\underset{k_3}{\rightleftharpoons}\left[P\right]+\left[E\right] \tag{70.91}$$

where the reaction to form the enzyme–substrate complex is reversible as indicated by the arrows and k_1 is the forward rate and k_2 is the backward, dissociation rate, and there is no reversion of product to substrate in the initial stages of reaction.

With the condition that initially P ~ 0, and setting d[ES]/dt = 0, it is easy to show that:

$$\left[ES\right]/\left[E\right]\left[S\right]=1/K_M \tag{70.92}$$

where K_M is the Michaelis–Menton constant. Now, since the velocity of the reaction (and here one considers the initial velocity only) is given by:

$$V_i = k_3\left[ES\right] \tag{70.93}$$

and one can define a maximum velocity such that:

$$V_{imax} = k_3\left[E\right]_{tot} \tag{70.94}$$

The velocity is maximum when all of the enzyme binding sites are filled with substrate. Solving for [ES] and using [E] = [E]$_{tot}$ – [ES], one obtains:

$$V_i = k_3\left[E\right]_0\left[S\right]/\left(K_M+\left[S\right]\right) \tag{70.95}$$

This is the functional form that expresses saturation kinetics with respect to substrate concentration. Generally, for determination of activities, enzyme reactions are performed in a fully saturated regime (i.e., [S] $\gg K_M$); otherwise, the kinetic rates need corrections and the Michaelis–Menton constant must be known or needs to be determined. For determination of substrate concentration, the analysis must account for the nonlinearity of V_i with respect to [S].

Enzyme activity is defined in terms of units, rate of formation of product under given conditions, since the protein content in the enzyme preparation can be misleading — not all of the enzymes in a preparation need be active. Because enzymes are proteins, and in some cases rather delicate ones, great care must be exercised in handling and storing. The activities of enzymes are very sensitive to pH, salinity, and temperature. All of these factors must be precisely controlled for reliable kinetic determinations.

Temperature Dependence

Rate constants obey the Arrhenius relation:

$$k = A \exp\left\{-E_a/RT\right\} \tag{70.96}$$

where E_a is the activation energy, R is the gas constant, T is absolute temperature, and A is a prefactor term. Knowledge of the activation energy allows for the extrapolation of a kinetic rate to any temperature. It is the lowering of this activation energy that is at the heart of catalysis and enzymatic reactions. Because of the exponential dependence, a reduction of the activation energy can lead to a rate constant increase of many orders of magnitude.

The prefactor A is determined by some collision frequency. However, in general, reactions proceed slower than the collision theory would predict. This is because, in addition to collisional frequency, there are also configurational and entropic terms that play a role in determining A. Nevertheless, it is useful to consider the concept of diffusion-controlled reactions. Here, it is the collisional frequency that dominates the reaction. In that situation, it is possible to utilize the diffusion theory of random motion in a medium to derive A such that:

$$A_{\text{diff}} = 4\pi\left(r_{ij}\right)\left(D_i + D_j\right) N_0/1000 \tag{70.97}$$

where N_0 is Avogadro's number, D_i, and D_j, are the diffusion constants for species **i** and species **j**, and r_{ij} is the encounter distance. For D of the order of 1.5×10^{-5} cm^2 s^{-1}, which is a value appropriate for small molecules, A_{diff} is 10^{10} M^{-1} s^{-1}. Reactions involving protonation or the OH$^-$ ion proceed at this rate, but only a few enzyme-substrate complex formation reactions approach the diffusion limited rate [6].

Experimental

The kinetic methods can be further classified according to experimental approaches as presented in Table 70.5.

TABLE 70.5 Classification of Kinetic Methods Based on Mixing Technique or Equilibrium Perturbation

Technique	Methods
Batch mixing (for slow reactions)	Stirring in cuvette or flask
Flow mixing (fast reactions)	Continuous flow
	Accelerated flow
	Pulsed flow
	Stopped flow
Thermodynamic jump	Temperature jump
	Pressure jump
	Electric current jump
	Concentration jump
Periodic relaxation	Cyclic voltammetry
	Dielectric relaxation
Pulse relaxation	Time resolved fluorescence
	Time resolved phosphoresence
	Flash photolysis
	Pulse NMR
	Pulse EPR

The principal instrumental elements of a kinetic apparatus are the mixing chamber, timing device and control of data acquisition, and detector. Automation and computer controls have allowed kinetic measurements to be done routinely and with great accuracy for even very fast reactions. We consider those aspects of instrumentation unique to the problem of mixing and proper fast sampling — the essential issues of the experimental method. The other components of instrumentation are beyond the scope of this chapter; the readers may refer to the monographs for more details on those topics [1,2].

Although the nature of kinetic measurements does not require absolute quantitation of a product, it does require care in accurate timing and fast mixing of reactants. For slow reactions, the mixing chambers can be closed systems without any need for elaborate devices or techniques to initiate the reaction of interest. So called "batch mixing" can be done in ordinary optical cuvettes with a suitable magnetic stirring rod or mixing plunger. These straightforward experimental techniques are not discussed here; rather, the time domain for which kinetic methods require specialized equipment will be considered. This domain is in the region of 1 ms to 1 s, for example. Reactions with time constants in this domain are very common in current applications of kinetic methods.

Mixing Methods

In the so-called open systems, there are three approaches to initiating and monitoring reactions: (1) continuous flow, (2) pulse and accelerated flow; and (3) stopped-flow.

In method (1), the reactants are brought together into a capillary under fast flow conditions and the product is monitored (by a photodiode for example) along the length of the capillary, thus tracing out the kinetics in so far as the time dependence of the reaction is transformed into distance along the capillary by:

$$t = d/v \qquad (70.98)$$

where v is the flow velocity, and d is the distance along the capillary after the junction in which the reactants are introduced. A high flow rate ensures a high Reynolds number condition and the achievement of turbulent flow and good mixing in the capillary. This method has the disadvantage of requiring rather high molar extinction coefficient for the product to achieve sensitivity and the high consumption of sample and reactant. In addition, multiple measurements along the tube are required to trace out the kinetics.

Method (2), pulsed and accelerated flow, was devised to address these deficiencies. By accelerating the flow, it is possible to do a single point measurement: the kinetics can be deconvoluted from the known change of flow as a function of time. In addition, integrated detection can be used in which the light path of the detector and source look down the flow tube, thus affording much greater sensitivity by virtue of a long absorption path length. Rather small quantities of analyte and reactant are consumed by this method because the flow is not continuous and a single point measurement is sufficient for the measurement of the kinetic parameters.

Method (3), the stopped-flow method has all the advantages of method (2), is simpler analytically, and can measure even faster kinetics. In this method, reactant and analyte are combined from two syringes driven simultaneously by a push block as shown schematically in Figure 70.28. As the stop syringe plunger hits a precalibrated stop position, the flow is halted. Data are accumulated after the flow is stopped, free from effects of flow turbulence and other time-dependent interferences. Dead times (i.e., the time between inception of mixing and start of measurements) can be as short as 0.5 ms. The steps involved in the measurements can be automated for multi-sample, high throughput applications. A particularly simple stopped-flow system has been described by Harvey [7]. The drive syringes are standard 10-mL syringes that are manually pushed by a plunger. The mixing chamber is at the bottom of a 3-cm² observation cell. As the mixed solution enters the cell, it pushes up a float past the level of the light beam by which the reaction is monitored. After the measurement, the spent solution is displaced by pushing down on the float. This very simple approach is adequate for reactions slower than some 100 ms or so.

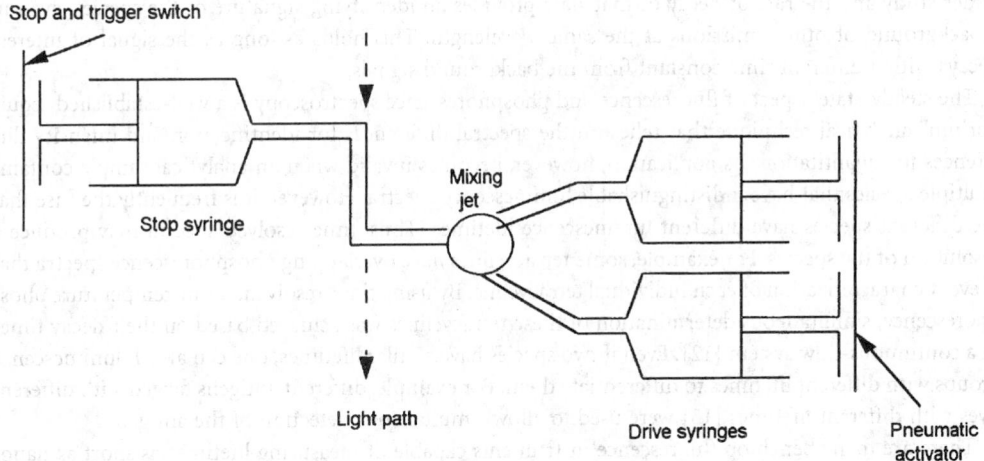

FIGURE 70.28 Schematic diagram of stopped-flow system. The reactant solutions are taken up into the two drive syringes as shown. The charging of the syringes is accomplished by valves and reservoirs not shown in the schematic for simplicity. As the activator plunger pushes the solutions through the mixing chamber, which is designed for efficient and fast mixing through tangential injection and turbulent flow (as in the Dionex Corp. system, for example), the old spent solution in the observation chamber is forced out into the stop syringe. The stop syringe plunger hits a stop, which causes immediate cessation of flow and activates the data acquisition system, which may be an oscilloscope, strip chart, or computer. The observation chamber shown here is oriented parallel to the light beam path for maximum pathlength and maximum sensitivity to absorption changes. The deadtime of such a system can be as low as 0.5 ms and the quantity of solutions required can be as low as 100 to 500 µL. At the end of a measurement, the stop syringe is purged and the drive syringes are recharged for another measurement cycle.

Reactions as fast as 1 ms can be measured by the stopped-flow technique. Fully automated sampling and data acquisition systems have been implemented [8]. Computer-controlled, three-wave valves are used to charge up the drive syringes and to flush them clean between measurements. Very fast reactions such as those involved in the folding of proteins have been studied in this way [9].

Relaxation Methods

An entirely different approach to kinetics is to probe the reactions of two reacting species that are in equilibrium by perturbing the equilibrium by a sudden change of temperature or pressure. These methods, known as relaxation methods, have as their virtue the ability to resolve kinetics in the very fast time regime much shorter than 1 ms [10]. If the equilibrium is disturbed, the relaxation to the new equilibrium state will proceed with a time constant τ given by:

$$1/\tau = k_1 + k_{-1} \qquad (70.99)$$

where k_1 and k_{-1} are the forward and back reaction rate constants between the two species, respectively. The magnitude of the response depends on the enthalpy change with temperature or volume change with pressure of the particular reaction under study.

In the temperature jump method, a pulse of energy is supplied to the sample, either by a current pulse if the solution is conducting or by a light pulse if the solution is absorptive at a suitable wavelength. Light pulses can be made extremely short with a suitable laser source — nanoseconds to picoseconds — and this approach lends itself to the examination of the very fast molecular processes such as the intermediate states in photoreception [11].

These methods are well suited for the study of fast reactions but less useful for compositional analysis. However, a type of relaxation that is well known (i.e., fluorescence and phosphorescence) have become very valuable analytical tools. In such methods, a light pulse populates and excited state of molecules under study and the rate of decay of that state provides an identifying signature of the species. even in a background of other emissions at the same wavelength. This holds as long as the signal of interest decays with a different time constant from the background signals.

The steady-state aspect of fluorescence and phosphorescence spectroscopy is a well-established "equilibrium" analytical technique that relies on the spectral differences for identification and intensity differences for quantitation. Its application, however, becomes invalid when an analytical sample contains multiple species that have indistinguishable luminescence spectra. However, it is frequently the case that the different species have different luminescence lifetimes. Thus, time-resolved spectra may produce a resolution of the species. For example, some tetracyclines have overlapping phosphorescence spectra that prevent characterization of each individual tetracycline. By using time-resolved, room-temperature phosphorescence, simultaneous determination of these tetracyclines was achieved based on their decay times in a continuous-flow system [12]. Even if two species have similar lifetimes, one can attach luminescence groups with different lifetimes to differentiate them. For example, different antigens tagged with different dyes with different lifetimes [13] were used to allow simultaneous detection of the antigens.

There are many benchtop fluorescence instruments capable of measuring lifetimes as short as nanoseconds. However, most of them are capable of monitoring only one specific wavelength as a function of time. Recently, fast optical spectrometers have been developed that have nanosecond time resolution over the entire visible spectrum [3].

Catalytic Reactions

Catalytic methods are based on the kinetic determination of catalyzed reactions. Such reactions can be extremely sensitive when the catalyst is the analyte. For example, chemiluminescence reactions of the oxidation of luminol by hydrogen peroxide catalyzed by metal ions provides extremely low detection limits for Co(II), Cu(II), Ni(II), Cr(III), and Mn(II). It should be pointed out that the term "catalyst" is loosely defined here as a substance that modifies the rate of a reaction without altering its equilibrium. Thus, the term "catalyst" includes the notion of promotion, inhibition, and, of course, true catalysis in which the catalyst remains chemically unchanged at the end of the reaction. Catalysts are usually categorized into two groups: enzymatic and nonenzymatic. Discussed below are overviews of catalytic-based kinetic methods applied in both homogeneous and heterogeneous systems.

Homogeneous Systems

Most applications of homogeneous kinetic methods are based on rate determination of catalyzed indicator (or substrate) reactions. Most frequently, the catalyst is the analyte to be determined, although, occasionally, it may serve simply as a reagent. Enzymes are one special type of catalyst. They are proteins possessing a very high degree of specificity. For example, certain enzymes can only exert catalytic actions on particular chemical bonds or steric isomers. Homogeneous enzymatic methods are widely used in clinic diagnoses to determine enzyme activity as well as enzyme substrate concentrations. The theoretical aspects of enzyme kinetics have been discussed in the previous section. Analytical applications for both enzyme activity determination and enzyme substrate detection can be found in the literature [3,14].

Homogeneous nonenzymatic catalytic methods are mainly applied for detection of metal ions, and other simple inorganic and organic species [3,15]. There are three major types of indicator reactions: redox, chemiluminescence, and complexation. One popular redox indicator reaction is the reduction of hydrogen peroxide by iodide catalyzed by metal ions (Fe, Mo, W, and Zr) that are also the analytes. The most common chemiluminescence indicator reaction is the decomposition of luminol (5-amino-2,3-dihydrophthalazine-1,4-dione) accompanied by the generation of luminescence at 425 nm. This decomposition is achieved through the oxidation of the doubly charged anion by the oxidant in this reaction (metal ions in most cases.) Although the oxidant in this case is consumed during the reaction, it is often termed a "catalyst" in the literature because its consumption is negligible in the time frame of the initial rate measurement, principally because of the ultrasensitivity of the chemiluminescence measurement. For complexation reactions, there are two main groups: ligand-exchange and complex-formation reactions. They are less studied compared with the above two indicator reactions but have promising application in the determination of non-transition metals. For example, a reaction involving ligand exchange can be used to detect 0.4 ppm Ca. The most widely used detection technique for the complexation indicator reaction is UV/VIS absorbance.

Heterogeneous Systems

Many kinetic methods depend on the application of different heterogeneous catalysis processes where the catalytic reaction takes place at the interface between two immiscible phases, usually between the liquid–solid phases. The discussion here focuses on two main areas of heterogeneous catalysis that are important in chemical analysis. The first encompasses immobilized enzymes in which the labeled enzymes are either physically or chemically attached onto a solid surface. The measurement of surface enzyme activity is then related to the analyte concentration. The second is the area of electrocatalysis, in which chemical reactions occur at the interface of an electrode and an electrolyte solution. The catalyst in this case is the charged electrode surface in either the intrinsic state or in a chemically modified state. The analyte concentration in the solution is determined by the electrode dynamic current.

The most widely used format for immobilized immunoenzymatic techniques is known as ELISA (enzyme-linked immunosorbent assay). This type of assay combines the great selectivity provided by specific antibody–antigen recognition, the high sensitivity provided by enzymatic amplification, and general applicability provided by the use of common detection methods. It has proven to be a very powerful technique for simple, rapid, and cost-effective trace analysis and is widely used today in clinical diagnosis [16], drug screening [17], food safety inspection [18], and environmental analysis [19,20].

ELISA can be operated in several different modes, depending on the nature of analyte, sample environment, and requirements on speed, cost, and detection limits. Different assays are usually classified according to their operating procedure (competitive or noncompetitive), to the signal detection technique used (calorimetric, luminescent, electrochemical, or radioactive), or to the physical arrangement of the antibody–antigen binding structure (single layer or sandwich layers). For a more detailed description, the reader is referred to several references [21,22].

A typical immunoassay procedure involves three steps: (1) immobilization of antibodies onto a solid surface, (2) competitive binding of analytes and enzyme-tagged conjugates to the antibody sites, and (3) rate measurement of a substrate reaction catalyzed by the enzyme. In most cases, only the latter two steps operate in a kinetic mode. Illustrated in Figure 70.29 is the chemiluminescence ELISA developed for rapid field analysis for PCBs (polychlorobiphenyls) in which the kinetic response of the enzymatic reaction enables the quantitative determination of PCB concentration [19]. First, a solid support of specified material and format is chosen based on the analysis requirement. The support surface is then treated with protein-A, a procedure to allow for the immobilization of antibodies in the proper orientation as shown in Figure 70.29. The third step is to immobilize the antibodies onto the protein-A coated surface. Then an enzyme–antigen conjugate (specifically the bromobiphenyl–alkaline phosphatase) is introduced to the well so as to saturate all of the antibody binding sites. After thorough rinse with pH 7 buffer solution, these conjugate treated well-plates are ready for use in analysis of samples. The analysis of PCB-containing samples proceeds simply by adding the PCB-containing solution into the well for a fixed time to allow the PCBs to displace the previously bound enzyme conjugates. The higher the PCB concentration in solution, the higher will be the displacement of the enzyme conjugates in a given amount of time. After a fixed time, the well is then thoroughly rinsed and a chemiluminescence substrate is added. Under the catalysis of alkaline phosphatase, the substrate is transformed into a luminescent species that is then detected. The initial luminescence generation rate or the total intensity within a fixed time is proportional to the surface alkaline phosphatase, and thus inversely related to the PCB concentration, as shown by the results in Figure 70.30.

Electrocatalytic reactions have been widely used for measuring chemical variables for electroactive species. However, not all electroactive species can be measured by electrochemical methods because for some species the electrode reaction kinetics may be very slow. A simple example is the reduction of molecular oxygen (O_2) at bare Pt electrodes in an aqueous solution. Oxygen cannot be reduced at the thermodynamic potential of the electrode. In this case, one can apply a large overpotential to drive the O_2 reduction. Unfortunately,

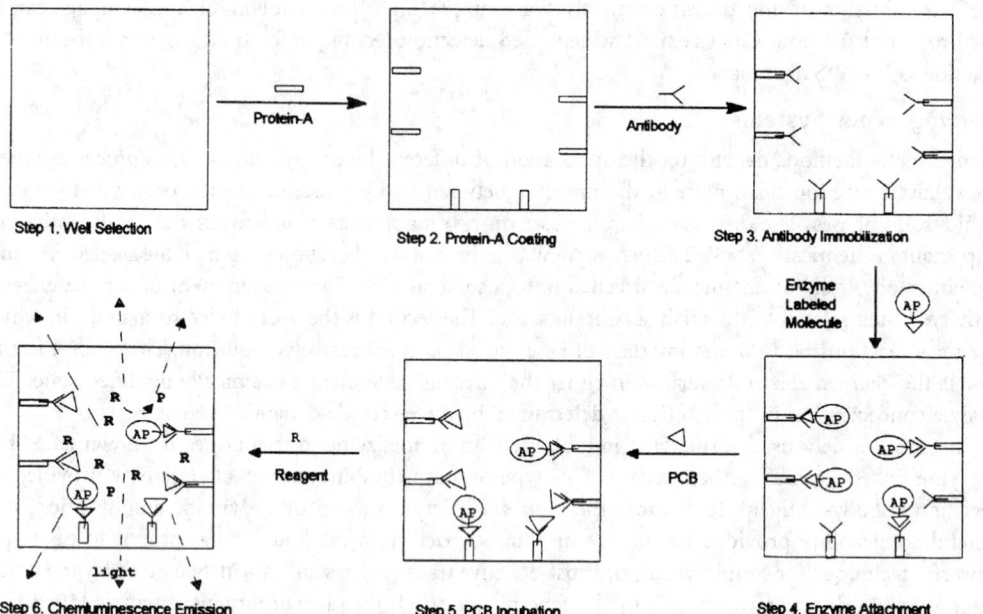

FIGURE 70.29 Pictorial presentation of chemiluminescence immunoassay.

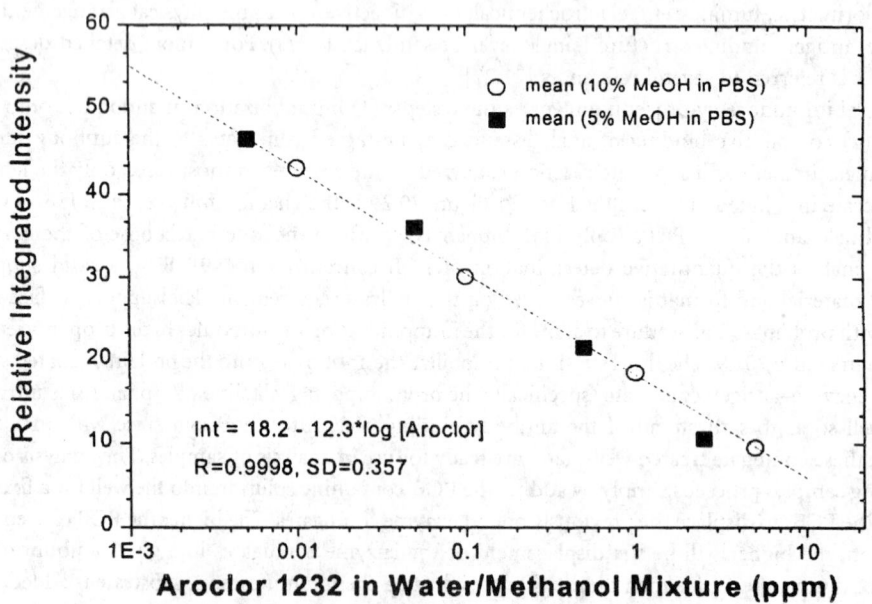

FIGURE 70.30 Dependence of chemiluminescence relative intensity on PCB Aroclor concentration. Plotted are chemiluminescence signals integrated during the first minute of enzymatic reaction (adamantyl dioxetane decomposition catalyzed by alkaline phosphatase in pH 10 buffer.) Samples contain various amount of Aroclor 1232 in pH 7 PBS buffer solution containing 5% (open circles) or 10% (solid squares) methanol.

in many cases, a large overpotential cannot be used because of the limited available potential window or because of the interference from other electroactive species. Thus, to overcome this problem, electrochemists have chemically modified electrode surfaces in order to accelerate electron transfer rates at the electrode–solution interface [23].

Chemical modification is produced by coating a monolayer of atoms, molecules, or thin layers of polymers onto the electrode surface. These surface-attached molecules may or may not be electrochemically active, but they can accelerate electrode kinetics for the target analyte. When the surface species is electrochemically active, it is termed a mediator; when inactive, it is called a promoter. For example, cytochrome-c, like many other large macromolecules, has a large electron transfer rate in a homogenous solution phase, but it exhibits extremely slow electron transfer kinetics at many metal electrode surfaces. Eddowes and Hill, as well as Gui and Kuwana [24], have successfully demonstrated that by adsorbing a monolayer of heteroaromatic molecules such as 4,4′-bipyridyl and *trans*-1,2-bis(4-pyridine)ethylene onto Au or Pt electrode surfaces, electron transfer kinetics of cytochrome-c is significantly promoted. Recently, Dong, Cotton, and co-workers [25] have used a halide-modified Au electrode to study cytochrome-c electrode kinetics. They adsorb different halides onto the Au electrode and find that they all can accelerate the electron transfer rate for cytochrome-c and the promoting effort is of the order of $F^- < Cl^- < Br^- < I^-$. Various theories for the above phenomenon have been proposed. One possible explanation is that cytochrome-c and related electron transfer molecules can adsorb onto bare electrode surfaces in undesired orientations. Besides the above atomic and molecular modified electrodes, lipid modified electrodes [26] have also shown some promoting effect for cytochrome-c electron transfer. Direct immobilization of cytochrome oxidase in a lipid bilayer at an Au electrode has resulted in electrochemical reactivity of cytochrome-c in solution [27].

Electrochemical methods can also be applied to analyze electrochemically inactive species. The electrochemical immunoassay is a typical example. It combines the great selectivity provided by specific antibody–antigen recognition, the sensitivity provided by catalytic amplification, and the simplicity of electrochemical detection. It has proven to be a useful technique for measuring chemical variables for biological, clinical, and environmental samples. There are many forms of electrochemical assays: homogeneous vs. heterogeneous, competitive vs. noncompetitive, enzymatic vs. nonenzymatic, simple vs. sandwich. Details can be found in References 21, 28, and 29.

The great advantage of the electrochemical immunoassay compared with enzyme modified electrode methods is that it is a universal method and can be configured to analyze wide range of analytes, regardless of their electrochemical reactivity. For example, Heineman and co-workers have used this technique to detect dioxin, with a detection limit of one attomole using alkaline phosphatase as enzyme to convert 4-aminophenyl phosphate to the electroactive species 4-aminophenol [28]. They also used multiple metal labels rather than enzyme labels for simultaneous detection of multiple analytes [29].

Noncatalytic Reactions

As stated earlier, most kinetic-based analytical methods are catalytic systems. Noncatalytic systems have more limited applications because equilibrium methods are usually adequate in providing the necessary accuracy and sensitivity, and the noncatalytic kinetic methods do not provide any advantages of sensitivity. However, kinetic methods have been found to be more valuable or even the only choice in some special cases, as illustrated by the following examples:

1. When a sample contains hard-to-separate interference species that demand laborious and time-consuming separation before final measurement with a classic equilibrium method. A kinetic method may provide a simpler and faster determination by not requiring a prior separation.
2. When the equilibrium method is based on a very slow reaction or a reaction cannot proceed to completion due to side reactions. In this case, an initial rate measurement is much preferred and may be the only method of analysis.
3. When the species of interest has an extremely short lifetime, such as in the case of a reaction intermediate.

Table 70.6 lists some vendors of appropriate apparatus and assay kits for performing kinetic determinations.

TABLE 70.6   Companies Providing Kinetic Instrumentation or Kinetic Assay Materials

Instrument	Company
ELISA apparatus	Dynatech Laboratories
	14340 Sullyfield Circle
	Chantilly, VA 22021
Immunoenzymatic assays	Becton Dickinson Microbiology Systems
	P.O. Box 243
	Cockeysville, MD 21030
Enzyme assay kits, ELISA kits	Pierce Chemical Co.
	3747 N Meridian Rd.
	P.O. Box 117
	Rockford, IL 61105
Electroanalytic instruments	EG&G Princeton Applied Research
	P.O. Box 2565
	Princeton, NJ 08543
Stopped-flow apparatus	Dionex Corp
	1228 Titian Way
	Sunnyvale, CA 44088
Spectrometer with stopped-flow attachment	On-Line Instruments, Inc.
	130 Conway Drive
	Bogart, GA 30622
Time-resolved spectrometers	Perkin-Elmer Corp.
	761 Main Ave.
	Norwalk, CT 06859

Note: As examples, one company is listed for each category. For more complete listings, the reader is referred to the latest buyer's guide of *Analytic Chemistry*.

Defining Terms

Antibody: One of a class of immunoglobins produced by an animal's immune response to antigens (i.e., substances foreign to the body). Antibodies bind to molecular determinants of the antigen with great specificity.

Catalyst: A substance that accelerates a chemical reaction but is not itself consumed by the reaction.

Enzyme: A protein molecule that catalyzes reactions with great specificity.

Substrate: That which is being transformed by an enzyme-mediated reaction.

References

1. D. Perez-Bendito and M. Silva, *Kinetic Methods in Analytic Chemistry*, England: Ellis Horwood, 1988.
2. A. Mottola, *Kinetic Aspects of Analytical Chemistry*, New York: John Wiley & Sons, 1988.
3. (a) H. A. Mottola and D. Perez-Bendito, Kinetic determinations and some kinetic aspects of analytical chemistry, *Anal. Chem.*, 66, 131R-162R, 1994; (b) Kinetic determinations and some kinetic aspects of analytical chemistry, *ibid.*, 68, 257R-289R, 1996.
4. S. F. Feldman, E. E. Uzgiris, C. M. Penny, J. Y. Gui, E. Y. Shu, and E. B. Stokes, Evanescent wave immunoprobe with high bivalent antibody activity, *Biosensor & Bioelectronics*, 10, 423-434, 1995.
5. R. J. H. Clark and R. E. Hester, *Time Resolved Spectroscopy*, New York: John Wiley & Sons, 1989.
6. I. Tinoco, Jr., K. Sauer, and J. C. Wang, *Physical Chemistry, Principles and Applications in Biological Sciences*, 2nd edition, Englewood Cliffs, NJ: Prentice-Hall, 1985.
7. R. A. Harvey, A simple stopped-flow photometer, *Anal. Biochem.*, 29, 58, 1969.
8. S. R. Crouch, F. J. Holler, P. K. Notz, and P. M. Beckwith, Automated stopped-flow systems for fast reaction-rate methods, *Appl. Spectrosc. Rev.*, 1, 165, 1977.

9. M. S. Briggs and H. Roder, Early hydrogen-binding events in folding reaction of ubiquitin, *Proc. Natl. Acad. Sci. USA*, 89, 2017-2021, 1992.

10. P. Fasella and G. G. Hammes, A temperature jump study of aspartate aminotransferase, *Biochemistry*, 6, 1798-1804, 1967.

11. H. Shichi (Ed.), *Biochemistry of Vision*, New York: Academic Press, 1983.

12. F. Alava-Moreno, Y.-M. Liu, M. E. Diaz-Garcia, and A. Sanz-Medel, Kalman filtering-aided time-resolved solid-surface room temperature phosphorimetry for simultaneous determination of tetracyclines in solution, *Mikrochim. Acta*, 112, 47-54, 1993.

13. J. Choo, E. Cortez, J. Laane, R. Majors, R. Verastegui, and J. R. Villarreal, Far-infrared spectra and ring-puckering potential energy functions of two oxygen-containing ring molecules with unusual bonding interactions, *Proc. SPIE-int. Soc. Opt. Eng.*, 2089, 538-539, 1993.

14. Chapter 3 of Reference 2.

15. (a) Chapter 2 of Reference 1; (b) G. G. Guilbault, in *Treatise on Analytical Chemistry*, I. M. Kolthoff and P. Elving (Eds.), 2nd ed., Part I, Vol. 1, Chapter 11, New York: John Wiley & Sons, 1978.

16. D. S. Hage, Immunoassays, *Anal. Chem.*, 65, 420R-422R, 1993.

17. T. A. Brettell and R. Saferstein, Forensic science, *Anal. Chem.*, 65, 293R-310R, 1993.

18. S. K. C. Chang, P. Rayas-Duarte, E. Holm, and C. McDonald, Food, *Anal. Chem.*, 65, 334R-363R, 1993.

19. (a) J. Y. Gui, S. F. Feldman, E. Y. Shu, D. R. Berdahl, and E. B. Stokes, Chemiluminescence immunoassay for rapid PCB analysis, *Real-Time Analysis*, 1, 45-55, 1995; (b) J. Y. Gui, D. R. Berdahl, E. Y. Shu, J. J. Salvo, S. F. Feldman, and E. B. Stokes, *Chemiluminescence Immunoassay for PCB Detection.* U.S. Patent No. 5,580,741, Dec. 3, 1996.

20. J. M. Van Emon and R. O. Mumma (Eds.), *Immunochemical Methods for Environmental Analysis: 198th National Meeting of the American Chemical Society*, ACS Symposium Series, Miami Beach, FL, Sept. 10-15, 1989.

21. (a) C. P. Price and D. J. Newman, *Principles and Practice of Immunoassay*, New York: Stockton Press, 1991; (b) T. T. Ngo (Ed.) *Electrochemical Sensors in Immunological Analysis*, New York: Plenum Press, 1987.

22. (a) E. Harlow and D. Lane, *Antibodies — A Laboratory Manual*, New York: Cold Spring Harbor Laboratory, 1988; (b) A. L. Ghindilis, P. Atanasov, and E. Wilkins, Enzyme-catalyzed direct electron transfer: fundamentals and analytical applications, *Electroanalysis*, 9, 661-674, 1997; (c) B. Liedberg, C. Nylander, and I. Lundstrom, Biosensing with surface plasmon resonance — how it all started, *Biosensors and Bioeletronics*, 10, i-ix, 1995.

23. M. D. Ryan, E. F. Bowden, and J. Q. Chambers, Dynamic electrochemistry: methodology and application, *Anal. Chem.*, 66, 360R-427R, 1994.

24. (a) M. J. Eddowes and H. A. O. Hill, Novel method for the investigation of the electrochemistry of metalloproteins: cytochrome c, *J. Chem. Soc. Chem. Commun.*, 771, 1977; (b) Y. Gui and T. K. Kuwana, Electrochemistry and spectroelectrochemistry of cytochrome c at platinum, *J. Electroanal. Chem.*, 226, 199-209, 1987.

25. (a) T. Lu, X. Yu, S. Dong, C. Zhou, S. Ye, and T. M. Cotton, Direct electrochemical reactions of cytochrome c at iodine-modified electrodes, *J. Electroanal. Chem.*, 369, 79-86, 1994; (b) X. Qu, J. Chou, T. Lu, S. Dong, and C. Zhou, T. M. Cotton, Promoter effect of halogen anions on the direct electrochemical reaction of cytochrome c at gold electrodes, *J. Electroanal. Chem.*, 381, 81-85, 1995.

26. (a) Z. Salamon and G. Tollin, Chlorophyll-photosensitized electron transfer between cytochrome c and a lipid-modified transparent indium oxide electrode, *Photochem. Photobiol.*, 58, 730-736, 1993; (b) P. Bianco, and J. Haladjian, Control of the electron transfer reactions between c-type cytochromes and lipid-modified electrodes, *J. Electrochim. Acta*, 39, 911-916, 1994.

27. J. K Cullison, F. M. Hawkridge, N. Nakashima, and S. Yoshikawa, A study of cytochrome c oxidase in lipid bilayer membranes on electrode surfaces, *Langmuir*, 10, 877-882, 1994.

28. N. Kaneki, Y. Xu, A. Kumari, H. B. Halsall, W. R. Heineman, and P. T. Kissinger, Electrochemical enzyme immunoassay using sequential saturation technique in a 20 ml capillary: dioxin as a model analyte, *Anal. Chim. Acta,* 287, 253-258, 1994.

29. (a) W. R. Heineman, H. B. Halsall, K. R. Wehmeyer, M. J. Doyle, and D. S. Wright, Immunoassay with electrochemical detection in methods of biochemical analysis, *Methods of Biochemical Analysis,* 32, 345-393, 1987; (b) M. J. Doyle, H. B. Halsall, and W. R. Heineman, Heterogeneous immunoassay for serum proteins by differential pulse anodic stripping voltammetry, *Anal. Chem.,* 54, 2318-2322, 1982.

70.4 Chromatography Composition Measurement

Behrooz Pahlavanpour, Mushtaq Ali, and C. K. Laird

During the early development of modern analytical chemistry, the study of natural materials was a primary concern of organic chemists and biologists. A major problem facing these scientists was the formulation of methods to separate and analyze the complex mixtures encountered in biological research.

Chromatography (literally "color-writing") is a physical or physicochemical technique for separation of mixtures into their components on the basis of their molecular distribution between two immiscible phases. One phase is stationary and is in a finely divided state to provide a large surface area relative to volume. The second phase is mobile and is caused to move in a fixed direction relative to the stationary phase. The mixture is transported in the mobile phase, but interaction with the stationary phase causes the components to move at different rates.

Origination of the technique, early in the 20th century, is generally attributed to Tswett, who separated plant chlorophylls by allowing solutions in petroleum ether to percolate through a vertical glass tube or column, packed with calcium carbonate. The separated components formed colored bands that were later isolated. Chromatography was adapted for qualitative or quantitative analysis of mixtures by inclusion of a suitable detector at the downstream end of the column and allowing the separated components to pass completely (elute) through the column and detector.

To analyze a sample, a suitable volume is injected into the stream of mobile phase or onto the upstream end of the column and the output of the detector is continuously monitored. The composition of the stream (eluent) passing through the detector then alternates between the pure mobile phase and mixtures with each of the components of the sample. The output record of the detector (chromatogram), plotted as a graph of response vs. time, shows a series of deflections or peaks, spaced in time and each related to a component of the mixture. For a given column, mobile phase, and set of operating conditions, the time for a component to pass through the column (retention time) is characteristic and can be used to identify the component. The peak area is proportional to the concentration of the component in the mobile phase.

In modern instrumental applications of chromatography, the stationary phase is either a solid or a liquid, and the mobile phase either a liquid or a gas. The various types of chromatography are classified according to the particular mobile and stationary phases employed. The solid stationary phase may be a granular solid packed in a tube (column), or coated as a thin layer on a suitable supporting plate (thin layer chromatography, TLC). Liquid stationary phases may be coated onto granular solids or bonded as a thin film to the inner wall of a capillary tube. In gas chromatography, the mobile phase is a gas (carrier gas), and the stationary phase is either a high-boiling liquid (gas-liquid chromatography, GLC) or a solid (gas-solid chromatography, GSC). In liquid chromatography (LC), the mobile phase is a liquid and the stationary phase is either a solid (liquid-solid chromatography) or a second liquid, immiscible with the mobile phase, coated on a granular solid (liquid-liquid chromatography).

Principles

Chromatographic theory is given in general textbooks [1,2], and also in specialized texts on different types of chromatography [3,4]. Chromatographic separation involves continuous interchange of solute molecules between the mobile and stationary phases. Four principal processes are involved: adsorption, liquid-liquid partition, ion exchange, and size exclusion. In gas chromatography, the predominant processes are adsorption, while liquid chromatography may involve all four processes. Where liquid-liquid partition is the predominant separation mechanism, the sample components are eluted in order of increasing boiling points.

Since the analyte is transported in the mobile phase, chromatography is limited to solutes that are distributed between the two phases. In practice, this means that gas chromatography is limited to substances that are thermally stable in the vapor phase and are volatile at temperatures up to the maximum operating temperature of the GC column (about 350 to 400°C for most columns and packings, although some metal columns can be operated at higher temperatures). Liquid chromatography can be used for analysis of thermally labile and high molecular weight materials such as polymeric materials and proteins, at temperatures below their boiling point and that of the eluent.

The separating power of a chromatographic column is described by analogy with distillation separation processes. It is given as the number of theoretical separation plate (either per meter of column length or in total for the column), and depends on its length, internal diameter, and on the stationary phase employed. The height equivalent to a theoretical plate (HETP) value may also be quoted. Separating power is enhanced by use of long, narrow-bore columns with packings of the finest possible mesh size to allow intimate contact between the mobile and stationary phases, but these columns require higher operating pressures to overcome the column resistance, and analysis times increase with column length.

The chromatographic separation process is highly dependent on the temperature of the column, and temperature effects can be related to the temperature dependence of the distribution or adsorption equilibria of the solute between the stationary and mobile phases. However, in practice, the choice of column operating temperature involves a compromise between resolution and speed of analysis. Liquid chromatography is commonly carried out with the column at ambient temperature, although applications requiring operation at temperatures up to 100°C are becoming more common. In gas chromatography, the column temperature has a major influence on the speed of elution and the separation of sample components. Gas chromatography can be carried out at a single controlled temperature (isothermal) or the oven temperature can be increased during the analysis in one or more linear ramps (temperature programming). Temperature programming speeds the elution of later components relative to early ones and enables mixtures containing a range of components to be separated more quickly than would be possible with isothermal operation. The broadening of later peaks due to diffusion in passing through the column is also minimized by temperature programming.

Gas Chromatography

A block diagram of a gas chromatograph is shown in Figure 70.31. The essential components are the column or columns; the carrier gas supply and flow and pressure controllers to enable carrier gas to be delivered to the column at a constant, controlled, and known rate; and the detector or detectors and associated electronics and data recording and processing system. An injector or facility for introducing suitable volumes of sample must be provided at the upstream end of the column. The column must be contained in an environment whose temperature can be held at a constant known value or heated or cooled at known rates. Temperature control in the range 25°C to 400°C (–0.1°C) and heating and cooling rates of 0.1°C to 40°C min^{-1} are typically required. Subambient operation at temperatures down to about –30°C may be required for separation of some volatile materials or for certain specialized eluents such as liquid carbon dioxide. Both injectors and detectors must be temperature controlled to allow rapid volatilization of the sample in the injector and to prevent condensation in the detector.

Columns

In gas chromatography, the processes involved in separation are predominantly adsorption and liquid-liquid partition when the eluent is liquid CO_2. Separation is almost entirely dependent on the nature of the stationary phase, with the gas phase acting mainly as an inert carrier. Separations of permanent gases

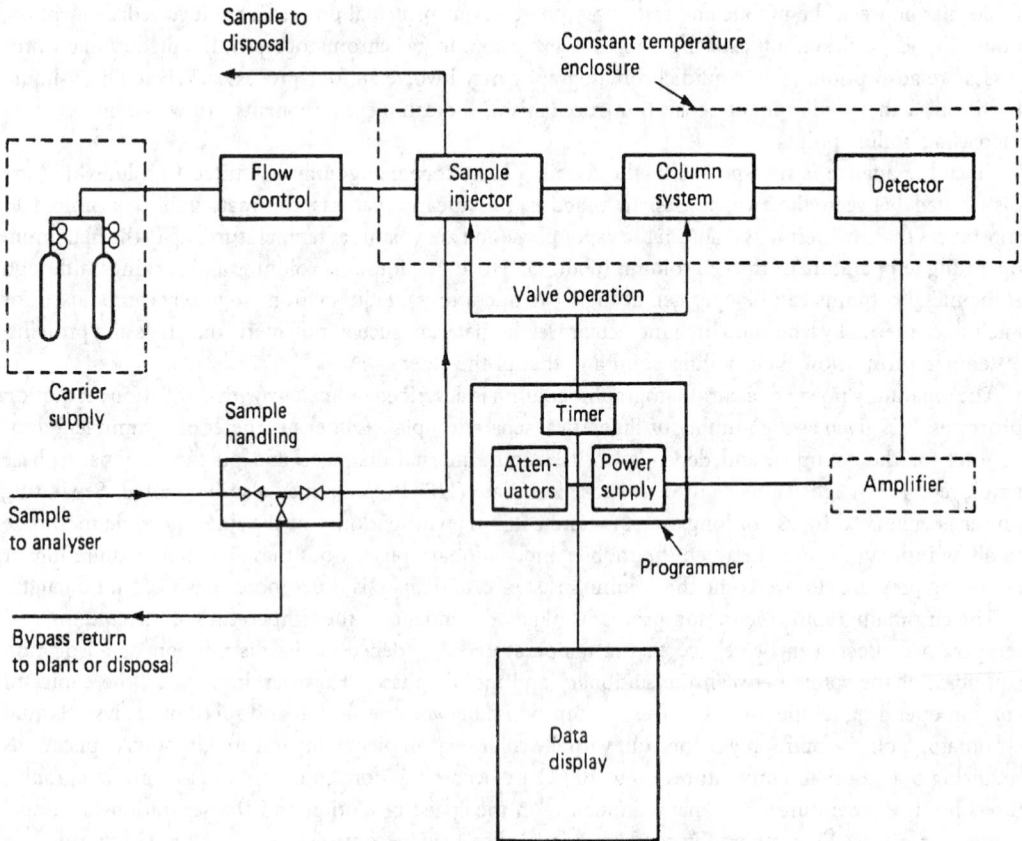

FIGURE 70.31 Functional diagram of process gas chromatography. The system consists of gas flow control, sample injection, separation of the components through the column, and a detector. Carrier gas at high pressure is used to move the sample through the column.

are carried out by gas-solid chromatography (GSC) using adsorbents such as silica gel, alumina, or synthetic zeolites (molecular sieves, particularly MS 5A and MS 13X) as the stationary phase. Proprietary porous polymers, such as Porapak (styrene-divinylbenzene copolymer), Chromosorb, and Tenax (polymer of 2,6-diphenyl-*p*-phenylene oxide), and various "carbon molecular sieves" are also used, and some of these materials are also used in separation of liquid samples.

In gas-liquid chromatography (GLC), the stationary phase is a high-boiling liquid, coated to a few percent by weight on an inert granular support such as silica, firebrick, diatomaceous earth, or Teflon. A wide range of liquids, gums, and waxes have been employed that provide stationary phases which are usable over different temperature ranges and with different polarities. Examples include silicone oils and gums, hydrocarbons, polyphenyl ethers, high molecular weight polymeric alcohols, etc.

Originally, in both GLC and GSC, the stationary phase or coated support was packed into a column, typically a glass or stainless steel tube, 1 to 3 m long, and coiled to fit the chromatograph oven. Developments in column technology have led to the gradual, but not yet complete, replacement of the packed GC columns by capillary columns. These columns are usually formed from drawn silica tubing, typically 10 to 100 m long, 0.2 to 0.5 mm o.d. with an outer protective coating of polyimide or, for operation above about 350°C, aluminium. The stationary phase is often a silicone oil — for example, polydimethylsiloxane, which instead of being coated on a granular support is present as a film, 0.1 to 5 μm thick, chemically bonded to the inner wall of the column (wall-coated, open tubular or WCOT column). Such open tubular columns operate at lower carrier gas pressures [typically 34 to 69 kPa (5 to

10 psig) instead of 138 to 340 kPa (20 to 50 psig) and carrier gas flow rates (1 mL min^{-1} or less instead of 20 to 30 mL min^{-1})] than packed columns. Capillary GC columns can typically have 3000 theoretical plates per meter (50,000 plates per column), compared with 1000 plates per meter or 2000 plates per column for a typical packed column. The capacity of the column (i.e., the size of sample that can be separated) depends on the thickness of the film of stationary phase, but is of course smaller than for a packed column and is typically in the microgram to nanogram range. Columns with thicker films have higher capacity, but lower resolving power, than thin film columns. Chemical bonding reduces the loss of stationary phase ("column bleed"), especially during temperature programming, and bonded columns are almost essential for critical applications such as coupled gas chromatography–mass spectrometry. Although low- or medium-polarity general-purpose capillary columns have high performance and can be used with a variety of samples, customized columns are available with stationary phases developed and optimized for particular analyses.

As an alternative to coating the inner wall of the capillary column with liquid stationary phase, the column wall can be coated with finely divided support, which is itself coated with stationary phase (support-coated open tubular or SCOT column). SCOT columns are one category of the more general group of porous-layer, open-tubular (PLOT) capillary columns where the inner wall of the capillary is coated or bonded with the stationary phase. PLOT columns are available with a range of solid adsorbents, including Porapak, molecular sieve, carbon molecular sieve, and alumina suitable for separation of mixtures of permanent gases and gaseous hydrocarbons, and bring the separating power of capillary columns to gas-solid chromatography. However, the difficulty of reproducibly injecting gas samples into these columns has meant that packed columns are often still favored for separation of gaseous samples.

Carrier Gas

The theory of the influence of carrier gas on the separation process was given by van Deemter, Zuidwerg, and Klinkenberg [5]. The van Deemter equation combines rate theory and plate theory and gives the relation between carrier gas velocity, u, and HETP, H, for a given carrier gas and column. The equation has the form

$$H = A + B/u + Cu \qquad (70.100)$$

where A, B, and C are constants. A depends on the particle diameter and irregularity of column packing; B depends on the tortuosity of the channels and the first power of the diffusion coefficient of solute molecules in the gas phase; and C depends on the distribution coefficient of the solute, the ratio of stationary-phase and gas-phase volumes, the effective film thickness, and inversely on the diffusion coefficient of the solute in the gas phase.

Plots of H versus u show a minimum value for H corresponding to an optimum carrier gas flow rate, where:

$$u_{opt} = \left(B/C\right)^{1/2} \qquad (70.101)$$

The van Deemter equation shows that, for a given column, a carrier gas of higher molecular weight can give a more efficient separation (lower value of H_{min}) than one of lower molecular weight, and H_{min} occurs at higher gas velocities for carrier gases of lower molecular weight. However, the equation refers to a single solute, and since chromatography involves separation of several solutes, the optimum carrier gas velocity is necessarily a compromise. The van Deemter equation also shows that for low molecular weight carrier gases, particularly hydrogen and helium, the minimum is less pronounced; that is, that carrier gas flow is less critical to column performance, and these two gases are the preferred choice, especially for capillary chromatography. Where hydrogen or helium are not suitable for the detector in use, a separate "make-up" gas supply is provided at the downstream end of the column; for example, nitrogen make-up is necessary for capillary operation of an electron capture detector.

The H vs. u curve is not symmetrical, carrier gas flow rates being more critical at values below u_{opt} than above. Thus, analyses may be speeded by increasing carrier gas flow rates above the optimum without much deterioration in column performance, but operating at flow rates that are too low leads to a relatively rapid loss of separating power.

Detectors

Detectors for gas chromatography should ideally have high sensitivity, rapid and reproducible response, and a wide range of linear response to concentration. Early detectors had universal or near universal response to all solute molecules; more recently, the emphasis has been on development of detectors with some selectivity to particular groups of compounds.

Thermal Conductivity Detector (TCD).
The thermal conductivity detector or katharometer was one of the earliest GC detectors and utilizes the change in thermal conductivity of a gas mixture with composition. The detector consists of either two or four electrically heated filaments, or for highest sensitivity especially at low temperatures, thermistors. The filaments or thermistors are connected in a Wheatstone bridge circuit, with external resistors to complete the bridge if there are only two sensing elements. The filaments or thermistors are mounted in a metal block to provide thermal stability, and provided with channels to allow the effluent from the GC column, and a separate, controlled "reference" flow of pure carrier gas to pass over the sensors or pairs of sensors. The loss of heat from the filaments depends on the filament temperature and on the conductivity of the surrounding gas. The katharometer can be operated under constant current or constant voltage conditions, or feedback circuitry can be used to maintain the filament resistance constant; but in each case, changes in gas composition lead to an out-of-balance voltage in the Wheatstone bridge circuit.

The TCD is a universal detector; however, it is less sensitive than other detectors such as the FID, and is principally used for detection and measurement of permanent gases such as oxygen, argon, nitrogen, carbon monoxide, and carbon dioxide, which either cannot be measured by the FID or require special pretreatment of the effluent gas (see below). It can be shown that the sensitivity is greatest when the filaments are operated at the maximum possible current, and when the difference in conductivity between the carrier gas and sample components is greatest. Both these conditions are fulfilled by use of helium or, better, hydrogen as the carrier gas, as these two gases have higher thermal conductivities than other common gases.

Flame Ionization Detector (FID).
The flame ionization detector (FID) is one of a group of gas detectors in which changes in the ionization current inside a chamber are measured. The ionization process occurs when a particle of high energy collides with a target particle that is thus ionized. The collision produces positive ions and secondary electrons that can be moved toward electrodes by application of an electric field, giving a measurable current, known as the ionization current, in the external circuit.

The FID utilizes the fact that, while a hydrogen–oxygen flame contains relatively few ions, it does contain highly energetic atoms. When trace amounts of organic components are added to the flame, the number of ions increases and a measurable ionization current is produced. The effluent from the GC column is fed into a hydrogen–air flame. The flame jet serves as one electrode, and a second collector electrode is placed above the flame. A potential is applied between the electrodes. When sample molecules enter the flame, ionization occurs — yielding a current that can be amplified and recorded.

The main reaction in the flame is:

$$> CH + \tfrac{1}{2} O_2 \rightarrow -CHO + e^-$$
(70.102)

However, the FID also gives a small response to substances not containing hydrogen, such as CCl_4 and CS_2. It is probable that the reaction above is preceded by hydrogenation to form CH_4 or CH_3 in the reducing part of the flame. In addition to the ionization reactions, recombination also occurs, and the response of the FID is determined by the net overall ionization process.

The FID is a mass-sensitive detector; that is, the response is proportional to the amount of organic material entering the detector per unit time. For many substances, the response is effectively proportional to the number of carbon atoms present in the flame, and the detector sensitivity can be expressed as the mass of carbon per second required to give a detectable signal. A typical figure is 10^{-11} g C s^{-1}.

The FID responds to practically all organic molecules. It is robust, has high sensitivity, good stability, and wide range of linear response, and is widely used.

The FID is insensitive to inorganic molecules and water. However, it can be used for measurement of carbon oxides (CO and CO_2) by mixing the effluent from the GC column with a controlled stream of hydrogen. The mixed gas is passed over a heated catalyst to convert the CO or CO_2 to methane (methanation), followed by FID measurement of the methane generated. This allows GC determination of these gases at lower concentrations than can be detected by the thermal conductivity detector.

Photoionization Detector (PID).
The photoionization detector has some similarities to the FID, and like the FID, it responds to a wide range of organic and also to some inorganic molecules. An interchangeable sealed lamp produces monochromatic radiation in the UV region. Molecules having ionization potentials less than the energy of the radiation can be ionized on passing through the beam. In practice, molecules with ionization potentials just above the photon energy may also be ionized, due to a proportion being in excited vibrational states. The ions formed are driven to a collector electrode by an electric field, and the ion current is measured by an electrometer amplifier.

The flame in the FID is a high-energy ionization source and produces highly fragmented ions from the molecules detected. The UV lamp in the PID is of lower energy, leading to the predominant formation of molecular ions. The response of the PID is therefore determined mainly by the ionization potential of the molecule, rather than the number of carbon atoms it contains. In addition, the ionization energy in the PID can be selected by choice of the wavelength of the UV source, and the detector can be made selective in its response. Commonly available UV lamps for the PID have energies of 11.7, 10.2, and 9.5 eV. The ionization potentials of N_2, He, CH_3CN, CO, and CO_2 are above the energy of all the lamps, and the PID does not respond to these gases. The 10.2-eV lamp is particularly useful as it allows ionization, and thus detection of alkynes and alkenes (except ethene), but not alkanes.

The PID is highly sensitive, typically to picogram levels or about 1 order of magnitude more sensitive than an FID, and has a wide linear range. Any of the commonly used carrier gases is suitable, although some gases (e.g., CO_2) absorb UV radiation and their presence may reduce the sensitivity of the detector. The main disadvantage of the detector is the fragility of the UV lamp, the need for periodic cleaning of the UV window, and the difficulty in cleaning the window if the detector becomes heavily contaminated.

Electron Capture Detector (ECD).
The electron capture detector is an ionization chamber in which molecules of electronegative species are allowed to attach to or "capture" electrons that have been slowed to thermal velocities by collision with inert gas molecules. The detector consists of a cell containing an emitting radioactive source (usually ^{63}Ni) and purged with inert gas. Electrons emitted from the source are slowed to thermal velocities (thermalized) by collision with the gas molecules and are eventually collected by a suitable electrode, giving rise to a standing current in the cell. If molecules with greater electron affinity are introduced into the cell, some of the electrons are "captured," forming negative ions that are more massive and less mobile than the free electrons, and the current in the cell is reduced. This effect is the basis of the electron capture detector.

Originally, the ECD was operated under dc conditions, potentials up to 5 V being used; but under some conditions, space charge effects produced anomalous results. Modern detectors operate under constant current conditions and use a pulsed supply, typically 25 V to 50 V. The pulse width and/or frequency are varied by feedback circuitry to maintain the ionization current in the cell at a constant level. This extends the linear range of the detector response and ensures optimum response for a range of molecules. The ECD must be used with a suitable inert gas, usually either an argon–methane mixture or nitrogen, either as the chromatograph carrier gas or (more usually) as an auxiliary gas supply.

The ECD is extremely sensitive to electronegative species, particularly halogenated molecules. It is widely used in the analysis of pesticides and some trace atmospheric components such as halocarbons, halogenated solvents, and nitrous oxide. The selectivity and extreme sensitivity is valuable, but the use of a radioactive source is a disadvantage. In certain cases, the detector response is highly sensitive to the cell temperature. The cell may be contaminated by "dirty" samples, and cleaning can be difficult or impossible.

Flame Photometric Detector (FPD).
In the flame photometric detector (FPD), the column effluent is passed through a fuel-rich hydrogen–air or hydrogen–oxygen flame, where the sample molecules are broken into simple molecular species and excited to higher electronic states. Under these conditions, most organic and other volatile compounds containing sulfur or phosphorus produce chemiluminescent species. The excited species return to their ground state, emitting characteristic molecular band spectra. The emission is monitored by a photomultiplier through a suitable filter, thus making the detector selective to either sulfur or phosphorus.

The FPD is most commonly used as a detector for sulfur-containing species. In this application, the response is due to the formation of excited S_2 molecules, S_2^*, and their subsequent chemiluminescent emission. The original sulfur-containing molecules are decomposed in the hot inner zone of the flame, and sulfur atoms are formed, which combine to form S_2^* in the cooler outer zone. As the S_2^* revert to their ground state, they emit light in a series of bands in the range 300 to 450 nm, with the most intense bands at 384.0 nm and 394.1 nm. The 394-nm band is monitored.

The FPD is highly sensitive (typically 10^{-11} g S s^{-1} or 10^{-12} g P s^{-1}), selective, and relatively simple. However, the response is nonlinear, given by:

$$I_S = I_0 \big[S \big]^n \qquad (70.103)$$

where I_S is the observed emission intensity (photomultiplier tube output), $[S]$ is the mass-flow rate of sulfur atoms, and n is a constant (value of 1.5 to 2, depending on flame conditions). Some systems incorporate circuitry to produce a linear output over 2 or 3 orders of magnitude concentration range.

Other Detectors.
A number of other GC detectors are available, but are less frequently used than those listed above. Examples include the nitrogen phosphorus detector, which is selectively sensitive to analytes containing those elements, and is used in pesticide analysis; and the helium ionization detector, in which the ionization process is due to highly energetic metastable helium atoms. The helium ionization detector is the only GC detector that permits analysis of the permanent gases at ppb levels. However, it is difficult to use, as the response is highly susceptible to the presence of trace impurities in the helium carrier gas. Additionally, gas chromatographic separation can be combined with detector systems specifically developed to measure a particular analyte. For example, trace levels of carbon monoxide in air can be measured by passing the effluent of the GC column through a heated bed of mercuric oxide. Carbon monoxide reduces the mercuric oxide, liberating mercury vapor, which is detected with high sensitivity by UV atomic absorption spectrometry.

Sample Injection

The purpose of the injection system is introduce defined and reproducible aliquots of sample into the chromatograph column. To minimize loss of resolution by diffusion, the sample must be injected as quickly as possible, and as a sharply defined slug. For analysis of liquids using packed columns, the injector is usually a zone heated to a temperature to ensure rapid volatilization of the sample, swept by carrier gas and fitted with a silicone rubber septum cap. Liquid injections are made by microsyringe through the septum cap. The syringe can be manual or controlled by an autosampler. Gas injections can also be made by syringe, but this is unsatisfactory for quantitative work as the compressive effect of the column head pressure in the injector makes syringe injections of gas unreproducible, and valve injection is preferable.

The injection valve is a six-port changeover valve that allows a fixed volume of gas, defined by a length of tubing (the sample loop), to be connected in either one of two gas streams with only momentary interruption of either stream. The valve is connected in the carrier gas stream just upstream of the column. The sample loop is filled with sample gas and the valve is operated to connect the loop into the carrier gas stream.

Injection into capillary columns is more difficult than into packed columns due to their smaller sample capacity. An internal standard (i.e., a compound similar to the analyte but separated from it by the chromatograph) is often added to the sample to enable correction to be made for random differences in injection volume. The most widely used injection technique is the so-called split–splitless injector. In the "split" mode, an aliquot of liquid sample is injected into a heated zone at the head of the column. The injector is swept by carrier gas at constant pressure. A valve allows a variable but known proportion of the carrier gas to flow to waste. Carrier gas flow rates through the column are typically 1 to 2 mL min^{-1} and the "split" flow is typically 30 to 50 mL min^{-1}. Thus, only a few percent of the aliquot of sample injected actually passes through the column. In the "splitless" mode, the valve controlling the waste stream of carrier gas is momentarily closed for a fixed time after injection, thus increasing the amount of sample transferred to the column for increased sensitivity.

Spit/splitless injectors are relatively simple and can be used with conventional syringes. However, the injector must be carefully designed to obtain reproducible results, and there may be some discrimination (particularly loss of high boiling components) in samples containing components with a wide range of boiling points. Techniques where the sample is injected directly onto the column, without splitting, can give less discrimination and better sensitivity and reproducibility, but are more difficult to use. Examples include cold on-column or programmed temperature injectors. In this technique, a length of column at the head is cooled to trap the sample. The injector is subsequently flash-heated to a high temperature to release the sample.

Alternative injection techniques include headspace analysis, where the vapor in equilibrium with a volatile liquid is sampled, and purge-and-trap techniques, where volatile components are purged from a liquid sample by a stream of inert gas. The purged components are trapped in a cooled zone at the head of the GC column and subsequently released by rapid heating for chromatographic separation and determination.

Liquid Chromatography

In early applications of liquid chromatography, the adsorbent was contained in a vertical column through which the liquid phase was passed under gravity. The performance of such systems is limited by the tortuosity of the passage of the liquid phase through the column and by the efficiency and speed of solute exchange between the mobile and stationary phases. Column performance is enhanced by use of microparticulate packings; but to achieve reasonably rapid separations in such systems, the mobile phase must be pumped through the column under pressure. The technique is known as high-performance (or sometimes high-pressure) liquid chromatography (HPLC).

High-Performance Liquid Chromatography (HPLC)

Figure 70.32 provides an illustration of process high-performance chromatography.

Columns.
HPLC columns are formed from precision-bore stainless steel tubing, 30 cm to 3 cm long, with 25 cm and 12.5 cm being the most common lengths. Standard columns have bores in the range 3.0 mm to 4.6 mm. Narrow-bore (2 mm) or microbore (1 mm) columns give increased sensitivity and higher chromatographic performance, together with reduced consumption of mobile phase, but require higher operating pressures and specially designed detector systems. The columns are packed under controlled

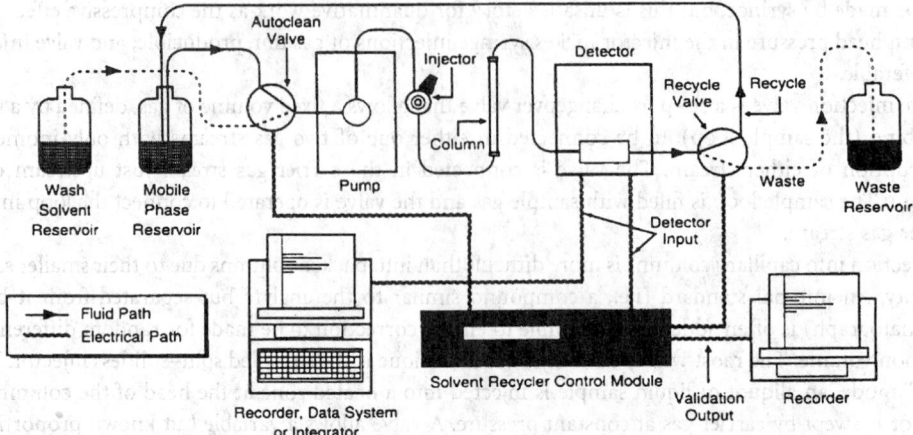

FIGURE 70.32 Functional diagram of process high-performance chromatography. The system consists of pump, sample injector, separation column, and detector. Mobile phase is liquid and wash solvent is used for cleanup. The used solvent can be recovered.

conditions with granular packing material of controlled size range, particle shape, and porosity. The "standard" particle diameter for packing materials is 5 μm, as it allows good column perfomance at moderate operating pressures, but columns packed with 10 μm or 3 to 4.5 μm particles can also be used.

A wide range of general-purpose and specialist packing materials have been developed for HPLC. The most widely used base material is silica, as it has high efficiency and physical rigidity, good solvent compatibility, and can be bonded with organosilanes for reversed-phase chromatography (see below). Other base materials include ceramics such as alumina, polymers, and graphitic carbon. Alumina has better pH stability than silica, but cannot be bonded with organosilanes. Polymers have limited organic solvent compatibility but tolerate strong alkali; they are less robust than silica and cannot be used at high operating pressures.

In liquid–solid chromatography, the samples are retained by adsorption on the support surface. The support may be coated with a liquid phase (liquid–liquid chromatography). In bonded phase HPLC, the support (usually silica, although alumina has also been used) is derivatized with a functional group covalently attached to the surface. In both coated and bonded phases, the separation is predominantly by partition; but like their counterparts in gas chromatography, bonded phases are inherently more stable than coated phases and can be used with a range of solvent and buffer systems.

In "normal" or traditional liquid chromatography, the stationary phase is polar and the mobile phase relatively nonpolar. The polar surface may be silica, or may be modified by chemical bonding of a suitable functional group. In reversed-phase chromatography, the stationary phase is nonpolar and the mobile phase is relatively polar. Reversed-phase chromatography is almost always carried out on modified silica columns, with octadecylsilyl (ODS or C18) or C8 groups being most commonly used, and the development of these systems has been largely responsible for the present popularity of HPLC.

Mobile Phase.

In contrast to gas chromatography, the mobile phase in HPLC plays a vital role in the separation process, with the rate and order of elution being determined by the relative polarities of the mobile and stationary phases, and by the nature of the sample components. In normal phase chromatography, the eluting power of the mobile phase increases with polarity; while in reversed-phase chromatography, the reverse situation applies, and eluting power decreases with increasing solvent polarity. Most of the common solvents can be used as mobile phase, but *n*-hexane is a common nonpolar solvent, while polar solvents in reversed-phase chromatography are often mixtures of water and methanol or acetonitrile. When a single solvent or mixture of fixed composition is used throughout a separation, the process is referred to as isocratic. In gradient elution procedures, the composition of the solvent is changed continously during the separation process. The change in solvent composition can be either linear with time or according to a predetermined profile. Gradient elution has some similarities in its effects to temperature programming in gas chromatography.

In HPLC, the mobile phase is pumped through the column. Flow rates for mobile phase are 0.5 to 10 mL min⁻¹ for conventional columns and 50 µl to 5 mL min⁻¹ for narrow-bore columns at pressures up to 48263 kPa (7000 psi). The flow rate must be precisely controlled and the outlet stream must be pulse-free. Metered mixing of up to four solvent streams may be required for gradient elution and column washing. HPLC pumps must be capable of delivering a pulse-free stream of mobile phase, at pressures up to 48263 kPa (7000 psi). Variations in mobile phase flow rate lead to irreproducibility in chromatographic retention times, and pulses in the flow give noise in the detector baseline.

To ensure optimum pump performance and to remove dissolved oxygen, the mobile phase must be degassed. Degassing can be by vacuum, but periodic purging with helium is more usual. Dissolved oxygen causes noise and drift in UV detectors, quenching in fluorescence detectors, and high background currents and noise in electrochemical detectors.

Detectors.

Performance requirements for HPLC detectors are similar to those in GC, and the detector system should be sensitive, have a rapid and reproducible response, and have a wide linear range of response. The following types are most commonly used:

1. Refractive index: Detectors based on measurement of the refractive index of the mobile phase are applicable to a wide range of solutes. However, they are temperature sensitive, are generally less sensitive than UV or fluorescence detectors, and cannot easily be used with gradient elution systems.
2. UV-VIS: UV or visible detection is the most common detection technique. Photometers or spectrophotometers measure the absorption of UV or visible radiation in the range 190 nm to 700 nm by the solute molecules. Detectors can be either fixed or variable wavelength. Fixed-wavelength detection at 254 nm is suitable for many solutes. The response is linear according to the Beer Lambert law and detection limits are subnanogram in favorable cases. "HPLC grade" solvents that have been specially purified to remove UV-absorbing impurities may be required.
3. Diode array: The diode array, or photodiode array (PDA), detector allows the UV or UV–VIS spectrum of the mobile phase to be repeatedly scanned during the elution. The resulting spectra are stored by the data system and are useful for checking or monitoring peak purity during elution and may provide some information on the identity of the analyte. Diode array detectors are typically two or three times less sensitive than conventional, single-wavelength UV or visible detectors.
4. Fluorescence: Detection of suitable molecules by fluorescence is typically one or two orders of magnitude more sensitive than UV–VIS detection, allowing detection down to low picogram levels in favorable cases. The fluorescence cell volume can be made as small as 5 µL, making fluorescence detection suitable for use with narrow-bore columns.
5. Electrochemical: Electroactive (oxidizable or reducible) substances can be detected by electrochemical techniques. The flow-cell volume may be as low as 1 µL, making the detectors suitable for narrow-bore columns. By choice of working conditions, the detector can be made specific to particular compounds or groups of compounds. Examples of suitable compounds for electrochemical detection include aromatic amines, phenols, and chlorinated phenols.

Injectors.

The injection process is critical in obtaining good performance in HPLC. Injection may be made directly by syringe; but for optimum results, valve injection is essential and is now almost universal. The injection valve is a six-port changeover valve, which allows a defined volume of liquid sample, contained in a sample loop, to be injected into the mobile phase. The sample loop typically has a volume in the range of 10 to 100 µL for conventional HPLC columns, and is loaded by a syringe.

The Measurement, Instrumentation, and Sensors Handbook

Ion Chromatography (IC)

Ion chromatography (IC) is a variant of HPLC in which inorganic and some organic cations and anions are separated on columns packed with high-efficiency pellicular ion-exchange resins. Anion separator columns are resin-based with positively charged fixed ionic sites, usually quaternary amines. The eluent is commonly dilute aqueous sodium hydroxide or sodium carbonate/sodium bicarbonate mixture. Cation separator columns have negatively charged fixed ionic sites, usually sulphonic acid groups, with methane sulphonic acid as eluent. Although conductimetric, amperometric (polarographic), UV–VIS photometric, and fluorescence detectors can all be used in ion chromatography, conductimetric detection is the most commonly used technique. When conductimetric detection is used, the sensitivity of the measurement is increased by reduction of the conductivity of the mobile phase (suppression) before detection. The conductivity is suppressed by conversion of the eluent to the corresponding acid or base form in a suppressor column, or membrane suppressor, downstream of the separation column. Thus, for anion analysis, the suppressor is a cation exchanger that replaces sodium ions in the eluent with hydrogen ions. Ion chromatography allows detection and measurement of anions and cations in solution, typically down to ppb ($\mu g\ L^{-1}$) levels. The common inorganic anions can be determined in a single sample aliquot. The technique is particularly useful for routine analysis of water and environmental samples.

Gel Permeation and Size Exclusion Chromatography

Gel permeation chromatography (GPC) and size exclusion chromatography (SEC) separate sample molecules on the basis of their effective molecular size in solution in aqueous or nonaqueous media. Column packings are porous materials with pores in controlled size ranges and can be either resin (e.g., sytrene-divinylbenzene copolymers) or silica based. The solute molecules interact with the column packing. Small molecules are "trapped" in the pores and pass through the column more slowly than larger molecules that do not interact so strongly. GPC was developed for polymer chemistry and is useful for the determination of the molecular weight distribution of polymers, but the techniques are also finding other applications; for example, size separation of small organics and petrochemicals and in sequential analysis and sample cleanup of environmental samples.

Hyphenated Techniques

Chromatographic methods give powerful techniques for separation of mixtures. However, the commonly used detectors give little information about the identity of the separated components, and identification from retention times may be incomplete or ambiguous. The so-called "hyphenated techniques," in which chromatographic separation is combined with another, usually spectrometric, technique, have been developed to combine chromatographic separating power with spectrometric identification.

Gas Chromatography–Mass Spectrometry (GC–MS)

Developments in column, mass spectrometer, and computer technology have made GC–MS the most widely used of the hyphenated techniques. Modern systems commonly incorporate the following features:

1. Capillary column gas chromatograph, with the column effluent fed directly to the ion source of the mass spectrometer through a heated transfer line
2. Miniaturized quadrupole or quadrupole-type mass spectrometer, optimized for use as a GC detector, and limited to this application. The mass spectrometer may have a limited mass range (maximum mass 650 or 700), corresponding to the maximum molecular weight of compounds that can commonly be analyzed by gas chromatography.
3. Control of the mass spectrometer and gas chromatograph, data recording, and mass spectral library searching by dedicated computer.

Such so-called "benchtop" GC–MS systems typically have sensitivities in the nanogram to picogram range, and unit mass spectral resolution. They give a valuable means of identification and quantitative determination of a variety of analytes, particularly in complex matrices such as biological or environmental samples. However, for applications requiring the highest sensitivity or mass spectrometric resolution, gas chromatographs coupled to high-resolution magnetic sector mass spectrometers must still be used. If it is necessary to use a packed chromatograph column, the carrier gas flow rate is likely to be too high for direct coupling to the mass spectrometer, and an interface, which allows selective removal of carrier gas, will be required.

Gas Chromatography–Infrared Spectrometry (GC–IR)

Coupling a gas chromatograph with an infrared spectrometer is an alternative to GC–MS. The effluent from the GC column is fed to a miniaturized flow-through absorption cell (light pipe) where the infrared spectra are measured. Fourier transform infrared (FTIR) spectrometry is used to achieve the combination of high spectral scan rates and resolution required to measure spectra from peaks eluting from capillary columns. Computer control and data processing are used, together with computer matching with stored libraries of IR spectra. The IR spectra are vapor-phase, and differ in some respects from liquid- or solid-phase spectra. GC–IR is to some extent complementary to GC–MS in that some molecular properties, particularly those relating to overall structure or shape, which may be lost in the fragmentation process in the mass spectrometer, may be identified in the IR spectra. However, the technique is some three or more orders of magnitude less sensitive than GC–MS, and has received less attention. GC-IR-MS can be used for particularly complex mixtures.

Liquid Chromatography–Mass Spectrometry (LC–MS)

Coupling a mass spectrometer to a liquid chromatograph, in principle, offers the advantages of GC–MS to the greater range of materials that can be analyzed by liquid, compared to gas, chromatography. However, interfacing a liquid chromatograph to a mass spectrometer is more difficult than a gas chromatograph; and while GC–MS is a well-established technique, LC–MS is only now becoming a useful routine method.

The principal difficulty in interfacing LC and MS lies in the very different physical conditions required for operation of the two techniques. Liquid chromatography uses relatively large quantities of liquid mobile phase, which may include inorganic buffers, while the mass spectrometer operates under vacuum. An interface must be used to selectively remove mobile phase before sample can be introduced to the mass spectrometer. Direct liquid injection and moving belt interfaces have been used but are unreliable. Three main types of interface are currently in use:

1. Thermospray. The LC effluent is passed through a probe, heated to 350°C to 400°C, in an evacuated region just outside the source of the mass spectrometer. The mobile phase, which often includes a volatile buffer such as ammonium ethanoate, is vaporized, and the sample molecules are ionized by a chemical ionization (CI) process. A series of lenses focuses a proportion of the ions into the mass spectrometer, while the solvent is pumped away. Under these conditions, the ionization is soft, i.e., there is little fragmentation in the mass spectrometer and the principal peak in the mass spectra is the molecular ion, M^+, MH^+, or MNH_4^+. This can be useful for molecular weight determination, but fragmentation may be increased; and in some cases, sensitivity enhanced, by including a filament in the ion source to give electron impact (EI) mass spectra. The thermospray interface is suitable for ionic and nonvolatile compounds. However, it requires the use of volatile buffers and the spectra are dependent on the solvent matrix.
2. Particle beam interface. The particle beam interface utilizes the principle of momentum to separate the solvent from the heavier solute molecules. The column effluent, mixed with helium in some designs, is passed through a series of chambers under pressure, exiting through a nozzle. As the effluent emerges, the solvent is vented while the solute molecules continue on their original trajectory and pass into the ion source of the mass spectrometer. The interface allows the use of a standard ion source in either EI or CI modes. It can therefore be used to produce standard library-searchable mass spectra, and is the only commonly used interface where this facility is routinely available. The interface requires some sample volatility, and some compounds (such as complex sugars) do not give satisfactory spectra.

3. Electrospray. In the electrospray interface, the column effluent is mixed with a nebulizing gas and passed through a jet nebulizer into a high-voltage electric field. Drop formation and ionization, by a chemical ionization process, occur. The ions enter the mass spectrometer through a capillary tube charged to a different voltage from the remainder of the interface, while the solvent is pumped away. Electrospray interfaces have, thus far, mainly been used for very high molecular weight analytes, although systems for lower molecular weights are being developed. Atmospheric pressure chemical ionization (APCI) is somewhat similar to electrospray except that the ionization process takes place at atmospheric pressure. Both systems can be used for polar compounds, molecular weights up to 100,000 Daltons, and are highly sensitive. However, ionization and separation involve a complex series of mechanisms, and setup and operation of the system, and interpretation of the spectra produced, may be difficult.

Applications in the Electricity Industry

Dissolved Gas Analysis (DGA)

Gas chromatography analysis of gases dissolved in transformer oil has been used for condition monitoring since the early 1970s [6–8]. The large volumes of gas often generated during a transformer fault have been used to trip mechanical relay for some 60 years [9]. It was later realized that if gases are evolved from the oil in sufficient quantities to operate a Buchholz relay, then slowly developing faults would also produce decomposition gases that would be dissolved in the oil. They only appear in the Buchholz at the end of a complicated system of interchange between the gases contained in bubbles rising to the surface and the less soluble atmospheric gases dissolved in the oil. It should therefore be possible to detect any incipient faults which may be present in the transformer early by analysis of the gases dissolved in the oil, using a gas chromatograph. Thermal and electric faults in a transformer produce various characteristic gases that are, to some extent, soluble in the oil. Extraction and GC analysis of dissolved gases can be used for monitoring of transformer condition. Dissolved gas analysis has been accepted as an important and vital condition monitoring technique for power transformers [10–12].

Oil samples can be collected from the equipment using syringes, bottles, or other sampling techniques, as described in IEC 567 [13]. The analysis requires extraction of the dissolved gases from the oil and then injection into a GC. The details of extraction of the gases from the oil are given in IEC 567. Generally, gases are extracted under vacuum using a mercury Toepler pump and the total volume of the extracted gas is measured by bringing to atmospheric pressure. The gases are then separated and determined by gas chromatography. An automated mercury-free instrument can also be used for extraction of dissolved gases [14] and other techniques have been used for extraction of the gases from the oil [15]. A static headspace sampling technique has been combined with capillary gas chromatography to allow dissolved gases and furan-related compounds to be determined in power transformer oils in a single GC run [16].

Regardless of the technique used for the extraction of the gases from the oil, gas chromatography is used for analysis of the gases. Alternative techniques such as mass spectrometry, although very sensitive, have not been used for routine analysis. Infrared spectrometry has been used for detection of gaseous hydrocarbons and carbon oxides [17]. The technique is rapid and accurate, and the detector is very stable. However, it cannot detect hydrogen and atmospheric gases. Hydrogen is a very important incipient gas for transformer condition monitoring. The ratio of oxygen to nitrogen dissolved in the oil is also used as an indication of oil or paper degradation in the transformer (the oxygen is used by reaction with the cellulose).

The gas chromatograph used for analysis of the gases is usually dual channel with FID and TCD detectors. A Porapak column is used for separation of hydrocarbons, and a methanizer is used for converting carbon oxides to methane followed by FID detector, while hydrogen, oxygen and nitrogen are separated on a molecular sieve column and measured by TCD. Other arrangements such as column switching, back flushing should be used if a single detector is going to be used. In such cases, a TCD is usually used as a detector. Infrared detection of hydrocarbons follow by TCD is an alternative arrangement.

A combination of Porapak and molecular sieve column is used for separation of hydrogen, oxygen, and nitrogen, followed by TCD detection. This arrangement requires a flash backflush system to prevent carbon dioxide from entering the molecular sieve column — where it would be so strongly adsorbed that it would require prolonged heating at a high temperature to remove it. The system is capable of detecting 1 ppm hydrocarbons and carbon oxides and 5 ppm hydrogen in the oil. Oxygen and nitrogen in the oil are usually present at high concentration and therefore their detection does not present any problem. High concentrations of acetylene gas in the oil may present some problems, such as poisoning the methanizer catalyst and it may stay in the column for a long time. In such cases, a longer isothermal time and higher oven temperature for cleaning of the column is the recommended technique.

Water has been determined in transformer oils with an accuracy better than 3%, precision better than 4% at the 10-ppm level and detection limit of 0.3 ppm, by headspace sampling and capillary chromatography with TCD detection. The technique could be automated using the headspace GC system proposed for dissolved gas analysis [18].

Furfuraldehyde Analysis (FFA)

Under normal operating conditions, the insulation system of transformers gradually deteriorates and produces various degradation byproducts. Thermal degradation, or aging of the paper insulation, is one of the most important factors in limiting the lifetime of a transformer. Detection and analysis of the degradation byproducts have been widely used to evaluate and monitor the degradation state of the insulation. The aging process of the paper is accompanied by the production of several byproducts — mainly carbon monoxide, carbon dioxide, and furfurals. Carbon oxides can be monitored by DGA, but their production is not specific to paper degradation. The measurement of furfurals could provide an early indication of paper degradation and their analysis by HPLC has been used as a tool for the monitoring of transformer performance [19-22]. A spectrophotometric method has also been used for analysis of furfuraldehyde [23]. The method is only capable of measuring furfuraldehyde and not other furanic compounds present in the oil.

The oil is dissolved in cyclohexane for HPLC analysis and passed through a solid-phase silica cartridge, where the furfurals, phenol, and m-cresol (which are products of degradation of phenol-formaldehyde resins in the transformer) are retained. The remaining oil in the cartridge is removed by washing with the solvent. The furfurals, phenol, and m-cresol are extracted from the cartridge with water:acetonitrile. The collected extracts are analyzed by HPLC. For separation of furfuraldehyde and other compounds, a C18 column is recommended. A UV detector at 276 nm is usually employed for detecting these compounds. The use of a UV photodiode array detector gives improved discrimination of products in what is often a complex chromatogram at the expense of some slight loss of sensitivity.

Analysis of Antioxidants in Oil

The presence of antioxidants is a key factor in controlling the oxidation of an insulating oil. Their use results in substantial savings by prolonging the oil service life and slowing down the transformer aging process. A large number of antioxidant additives are used, and HPLC is a useful technique for quantitative determination of their concentrations in insulating oil [24,25]. Thin-layer chromatography (TLC) has also been used for quantitative determination of antioxidant in insulating oil [29].

HPLC is also used for evaluation of the quality of mineral insulating oil [26,27]. The presence of certain characteristic chemical compounds is of considerable importance in the electric industry. This technique is also used to identify the presence of some byproducts of oil under electric stress, such as x-wax and fluoresence materials. Polar compounds, such as acids and aldehydes, are also products of oxidation of oil and they can also be determined by HPLC. Early detection of such products is important in transformer condition monitoring to provide prior warning of a developing fault and to enable appropriate corrective action to be taken.

HPLC is also used for health and safety monitoring of the polyaromatic hydrocarbon (PAH) content of transformer oils [28]. The toxic and carcinogenic nature of PAH compounds is well established.

Molecular Weight Distribution of Insulating Paper

Gel permeation chromatography has been used to measure the change in molecular weight distribution of insulating paper during aging [30]. This technique has been used for measurement of the degree of polymerization of the paper, but requires a sampling of the paper in the transformer, which is not generally practical. Measurement of paper degradation by analysis of products dissolved in the oil, such as FFA, although much easier, is indirect and therefore dependent on a knowledge of the history of the transformer and its components. GPC can provide direct information of the state of the paper and the average molecular weight of the cellulosic chains, and currently is mostly used as a forensic analysis tool for the investigation of failures.

Two techniques are available for getting the paper into solution for analysis. Direct dissolution into dimethylacetamide/8% lithium chloride is possible for most forms of cellulose, followed by dilution to 1% lithium chloride for analysis. However, high levels of lignin (3–4%) in the paper interfere with the dissolution process. The alternative is to derivatize the cellulose to the tricarbanilate in pyridine. The product can then be analyzed in solution in tetrahydrofuran, but there is evidence in the literature that the derivatization process itself degrades the paper.

Polychlorinated Biphenyls

Polychlorinated biphenyls (PCBs) are synthetic materials with dielectric and chemical properties that in the past made them attractive alternatives to petroleum-based products. However, they have since been identified as environmental pollutants and possible health hazards. PCBs are not used today, but their use in the past has led to widespread contamination of mineral oil with PCBs. Current legislation requires that oil in service shall have a PCB content of less than 50 ppm. Therefore, the PCB content of insulating oil in the transformer must be measured quantitatively and capillary GC is the technique currently used [31].

The sample is diluted in hexane and deoxygenated. A small volume of the resulting solution is injected into a narrow-bore capillary gas chromatographic column. The capillary column separates the PCBs into individual or small groups of overlapping congeners. Their presence in the effluent is measured by an electron capture detector (ECD).

Feedwater and Boiler Water Analysis

Ion chromatography (IC) is well suited to the analysis of highly pure water such as boiler feedwater, and often gives better precision, sensitivity, and speed of analysis than established techniques. For example, IC has been used for the determination of carbonic acid in steam-condensate cycles [32].

Morpholine is added to the thermal cycle of some CANDU reactors, and has been determined, together with its amine breakdown products, by reversed-phase HPLC on a C18 column with visible detection at 456 nm [33].

Other Applications

The size and scope of the electricity supply industry imply that it has a major effect on the environment. Chromatographic techniques are widely applied in environmental monitoring and research, and the electricity industry is both a major user of standard techniques and sponsor of research.

Defining Terms

Adsorption: The noncovalent attachment of one substance to the surface of another.
Analyte: The substance that is being analyzed.
Baseline resolution: Separation of components at the peak base (no overlap of any peak area).
Chromatography: The physicochemical technique for separation of mixtures into their components.
Column: A steel, glass, or plastic tube containing the stationary phase.
Detector: A device for monitoring the separated compounds from the chromatography by sensing chemical or physical properties of the sample.

Eluent: The moving solvent in a chromatograhic column.

Elute: To travel through and emerge from the column.

Gel permeation chromatography (GPC): A mode of LC in which samples are separated according to molecular size.

HPLC: High-performance liquid chromatography.

IC: Ion chromatography.

Mobile phase: The following solvent.

Stationary phase: The material that is contained in the column and does not move during the chromatographic process.

Acknowledgments

The authors would like to thank Perkin-Elmer and Phenomenex for permission to reproduce diagrams.

References

1. C.F. Poole and S.A. Schuette, *Contemporary Practice of Chromatography,* Elsevier Science BV, Amsterdam, 1984.
2. R.L. Grob (ed.), *Modern Practice of Gas Chromatography, 3rd ed.,* New York; John Wiley & Sons, 1995.
3. G. Guiochon and C.L. Guillemin, Quantitative gas chromatography for laboratory analyses and on-line process control, *J. Chromatogr. Library,* Vol. 42, Elsevier Science, Amsterdam, 1988.
4. J. Weiss, *Ion Chromatography, 2nd ed.,* VCH Publishers, Cambridge, U.K., 1994.
5. J.J. van Deemter, F.J. Zuiderweg, and A. Klinkenberg, Longitudinal diffusion and resistance to mass transfer as causes of non-ideality in chromatography, *Chem. Eng. Sci.,* 5, 271, 1956.
6. P.S. Pugh and H.H. Wagner, Detection of incipient faults by gas analysis, *TAIEE,* 80, 189-195, 1961.
7. L.C. Aicher and J.P. Vora, Gas analysis — a transformer diagnostic tool, *Allis Chalmers Electrical Review,* 28, 22-24, 1963.
8. E. Dornenburg and W. Strittmatter, Monitoring oil-cooled transformers by gas analysis, *Brown Boveri Review,* 61(5), 238, 1970.
9. M. Buchholz, The Buchholz protection system and its application in practice, *E.T.Z.,* 49, 1257-1260, 1928.
10. R. Muller, K. Potthoff, and K. Soldner, The analysis of gases dissolved in the oil as a means of monitoring transformers and detecting incipient faults, *CIGRE,* paper 12-02, 1970.
11. E. Dornenburg and K. Schober, Determination and analysis of gases formed in transformers. Evaluation of gas analysis results, *CIGRE,* paper 15-01, 1972
12. J.R. Booker, Experience with incipient fault detection utilizing gas chromatographic analysis, *Doble Client Fortieth Annual International Conference,* 10-5.1, 1973.
13. IEC publication 567, *Guide for the sampling of gases and oil from oil-filled electrical equipment and for the analysis of free and dissolved gases,* 1992.
14. B. Pahlavanpour, Mercury-free extraction of dissolved gases in transformer oil, *AVO International Technical Conference,* Dallas, 1996.
15. T.J. McGarvey, An automated system for analysis of gases dissolved in electrical insulating oil by headspace gas chromatography, *Doble Client Fifty-Seventh Annual International Conference,* 10-8.1, 1990.
16. Y. Leblanc, R. Gilbert, J. Jalbert, M. Duval, and J. Hubert, Determination of dissolved gases and furan-related compounds in a single chromatographic run by headspace/capillary gas chromatography, *J. Chromatogr. A,* 657, 111-118, 1993.
17. J. Jalbert, S. Charbonneau, and R. Gilbert, A new analytical method for the determination of moisture in transformer oil samples, *Sixty Third Annu. Conf. Doble Clients,* Boston, MA, March 25-29, 1996.

18. B. Pahlavanpour, S. McCann, and M. Ali, Preliminary investigation into the use of Fourier transform infrared analysis of continuous monitoring of dissolved gases in transformer oil, *Natl. Grid Tech. Rep.*, TR(T)38, 1993.

19. P.J. Burton, J. Graham, A.C. Hall, J.A. Laver, and A.J. Oliver, Recent developments by CEGB to improve the prediction and monitoring of transformer performance, *CIGRE*, paper 12-09, 1984.

20. X. Chendong, Monitoring paper insulation aging by measuring furfural contents in oil, *Proc. 5th Int. Symp. High Voltage Engineering*, 139, 1991.

21. A. DePablo, R. Andersson, H.J. Knab, B. Pahlavanpour, M. Randoux, E. Serena, and W. Tumiatti, Furanic compounds analysis as a tool for diagnostic and maintenance of oil paper insulation system, *CIGRE*, 110-09, 1993.

22. J. Unsworth and F. Mitchell, Degradation of electrical insulating paper monitoring with high performance liquid chromatography, *IEEE Trans. Elec. Insul.*, 25, 737, 1990.

23. B. Pahlavanpour and G. Duffy, Development of a rapid spectrophotometriy method for analysis of furfuraldehyde in transformer oil as an indication of paper aging, *CEIDP Annual Report*, 493, 1993.

24. C. Lamarre and A. Gendron, Analyse quantitative par chromatographie liquide haute performance du di-*tert*-butyl-2,6-*para*-cresol dans les huiles de transformateurs, *J. Chromatogr.*, 464, 448-452, 1989.

25. M. Duval, S. Lamotte, D. Gauchon, C. Lamarre, and Y. Giguette, Determination of low-improved additives in new and aged insulating oil by gel permeation chromatography, *J. Chromatogr.*, 244, 169-173, 1982.

26. M. Duval and C. Lamarre, The characterization of electrical insulating oils by high-performance liquid chromatography, *IEEE Trans. Electr. Insul.*, E12, 340-348, 1977.

27. M. Duval, C. Lamarre, and Y. Giguere, Reversed-phase high-performance liquid chromatographic analysis of polar oxidation products in transformer oils, *J. Chromatogr.*, 284, 237-280, 1984.

28. A.N. Gachanja, Analysis of polycyclic aromatic hydrocarbons by liquid chromatography, *Chromatography and Analysis*, February–March, 5-7, 1993.

29. IEC Publication 666, *Detection and determination of specified anti-oxidant additives in insulating oil*, 1979

30. D.J. Hill, T.T. Le, M. Darveniza, and T. Saha, A study of degradation of cellulosic insulation materials in a power transformers. Part 1. Molecular weight study of cellulose insulation paper, *Polymer Degradation and Stability*, 48, 79-87, 1995.

31. B. Pahlavanpour and G. Duffy, Analysis polychlorinated biphenyls in transformer oils by gas chromatography, *Natl. Grid Publication*, RDC/0045/R91, 1991

32. S. Charbonneau, R. Gilbert, and L. Lepine, Determination of carbonic acid in steam-condensate cycle samples using non-suppressed ion chromatography, *Anal. Chem.*, 67, 1204-1209, 1995.

33. C. Lamarre, R. Gilbert, and A. Gendron, Liquid chromatographic determination of morpholine and its thermal breakdown products in steam-water cycles at nuclear power plants, *J. Chromatogr.*, 467, 249-258, 1989.

71

pH Measurement

Norman F. Sheppard, Jr.
Gamera Bioscience Corporation

Anthony Guiseppi–Elie
*Virginia Commonwealth University
and Abtech Scientific, Inc.*

The measurement of pH is arguably the most widely performed test in the chemical laboratory, reflecting the importance of water as a ubiquitous solvent and reactant. In the 90 years since the first use of an electrode to determine hydrogen ion concentration, the glass electrode and its variants have matured into routine tools of analytical and process chemists. Yet, there continue to be developments that promise to broaden the scope and reach of these measurements. Among recent developments are miniature pH-sensitive field-effect transistors (pHFETS) being incorporated into pocket-sized pH "pens," metal/metal oxide pH sensors for measurements at high temperatures and pressures, and flexible fiber-optic pH sensors for measuring pH within the body. This chapter discusses electrochemical and optical methods for pH measurement, and is by necessity limited in scope; readers interested in a more comprehensive treatment may wish to refer to the recent text by Galster [1].

71.1 Definition of pH

In its most common interpretation, pH is used to specify the degree of acidity or basicity of an aqueous solution. Historically, pH was first defined as the negative logarithm of the hydrogen ion concentration (*pondus Hydrogenii*, literally hydrogen exponent), to simplify the handling of the very small concentrations (on the order of 10^{-7} moles liter^{-1}) encountered most commonly in nature. This definition of pH is expressed as Equation 71.1, where $[H^+]$ is the molar concentration of solvated protons in units of moles per liter.

$$pH = -\log\left[H^+\right] \tag{71.1}$$

As a greater understanding of the behavior of ionic solutes in solution developed, chemists recognized that the measurement techniques used to determine hydrogen ion concentration were in fact measuring the hydrogen ion activity, often referred to as the "effective concentration." This led to the adoption of the more rigorous definition of pH as the negative logarithm of the hydrogen ion activity in solution,

$$pH = -\log a_{H^+} = -\log \gamma\left[H^+\right] \tag{71.2}$$

TABLE 71.1   pH of Common Substances

Substance	pH
Soft drinks	2.0–4.0
Lemon juice	2.3
Vinegar	2.4–3.4
Wine	2.8–3.8
Beer	4.0–5.0
Cow's milk	6.3–6.6
Pure water	7.0
Blood	7.3–7.5
Sea water	8.3

Note: D. R. Lide (Ed.), *CRC Handbook of Chemistry and Physics, 74th ed.,* Boca Raton, FL: CRC Press, 1993.

where a_{H^+} is the hydrogen ion activity and γ is the activity coefficient. The two definitions expressed in Equations 71.1 and 71.2 are equivalent in dilute solution where concentration approximates activity, a_{H^+}. In practice, the routine measurement of pH is not accomplished by the direct determination of the hydrogen ion activity. Rather, pH is determined relative to one or more standard solutions of known pH.

The hydrogen ion activity of common substances ranges over many orders of magnitude, as evidenced from the tabulation presented in Table 71.1. Water, in the absence of other chemicals that might alter its equilibrium, contains 10^{-7} molar hydronium ions [H^+] and 10^{-7} molar hydroxonium ions [OH^-] at 25°C. Under these idealized conditions, water has a pH of 7.0, and is said to be neutral.

$$2H_2O \xleftrightarrow{K_W} H_3O^+ + OH^- \qquad K_W = \left[H_3O^+\right]\left[OH^-\right] = 10^{-14} \qquad (71.3)$$

The addition of acids or bases to pure water increases or decreases the hydrogen ion activity, respectively. The resulting pH depends on a number of factors, such as the concentration of the added acid or base and the strength as quantified by its dissociation constant [2].

71.2 Electrochemical Methods of pH Measurement

Electrochemical measurement of pH utilizes devices that transduce the chemical activity of the hydrogen ion into an electronic signal, such as an electrical potential difference or a change in electrical conductance. The following sections review electrochemical pH measurement, with emphasis on the glass membrane electrode. Also discussed is the operation of hydrogen ion-selective pHFETs, metal/metal oxide electrodes, and other approaches used in specialized applications. Table 71.2 provides a listing of sources, features, and key properties of selected, commercially available pH electrodes.

The Glass Membrane Indicator Electrode

The most widely used method for measuring pH is the glass membrane electrode. As illustrated schematically in Figure 71.1(a), a pH meter measures the electrical potential difference (voltage) that develops between a glass membrane pH *indicator* electrode and a *reference* electrode immersed in the sample to be tested. The indicator and reference electrodes are commonly combined into a single, functionally equivalent, probe, referred to as a combination electrode. The glass membrane of the indicator electrode develops a pH-dependent potential, as a result of ion-exchange between hydrogen ions in solution and univalent cations in the glass membrane. The sensitivity of the glass electrode membrane potential to changes in pH is small, so a suitably designed reference electrode and a high input impedance meter are required in order for the potential to be precisely measured.

The construction of a typical pH indicator electrode is illustrated in Figure 71.1(b). The glass membrane at the tip of the electrode, which acts as the transducer of pH, is approximately 0.1 mm thick. One

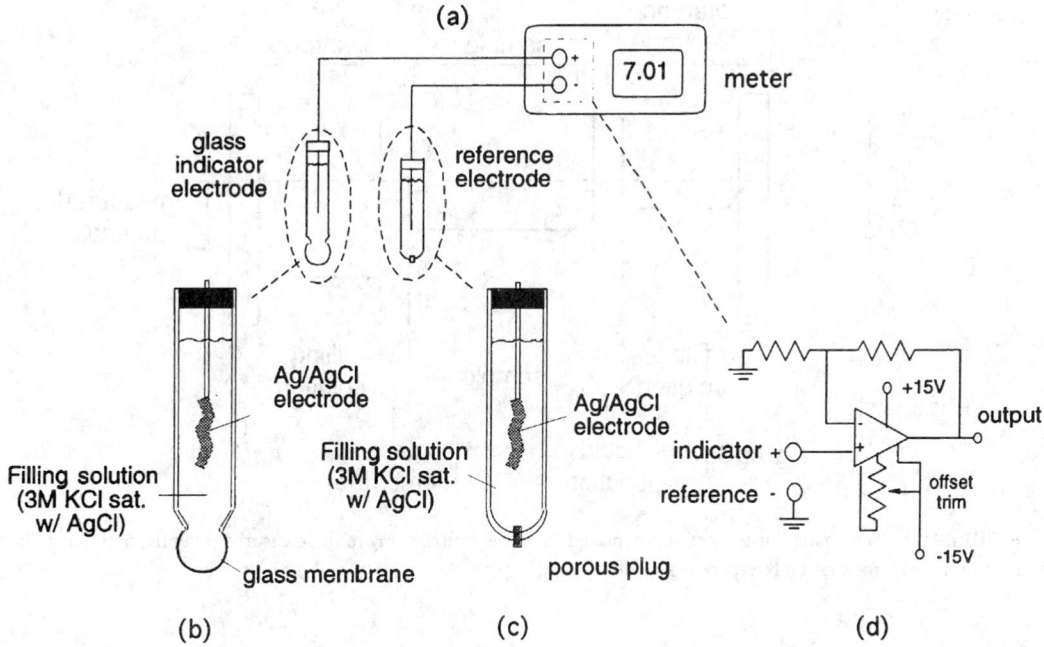

FIGURE 71.1 pH measurement using a glass membrane electrode: (a) measurement system comprising a pH meter, indicator, and reference electrodes; (b) indicator electrode construction; (c) reference electrode construction; and (d) amplifier circuit.

side of the membrane contacts the sample while the other contacts the electrode filling solution, an electrolyte of defined composition and pH. A reference element (e.g., a silver chloride coated silver wire) immersed in the filling solution makes a stable electrical contact between the potential measuring circuitry of the pH meter and the electrolyte in contact with the inner side of the glass membrane.

The reference electrode illustrated in Figure 71.1(c) provides a means of making electrical contact between the meter and the sample that is in contact with the external side of the indicator electrode's pH-sensitive glass membrane. Like the indicator electrode, the body of the reference electrode is filled with an electrolyte into which a reference element is immersed. The electrode also has a means, such as a porous ceramic frit, by which the reference electrode filling solution can make physical and electrical contact to the sample. The importance of this "liquid junction" cannot be understated, as it is a potential source of sample contamination, measurement errors, and reliability problems.

The electrical potential difference between the pH indicator electrode and the reference electrode provides a measure of pH. Figure 71.2 provides an illustration of the potential differences, the origin of which will be discussed below, that contribute to the measured potential. The main function of the pH meter of Figure 71.1(a) is to measure this difference, with a precision of 0.1 mV or better. The amplifier circuit of Figure 71.1(d) illustrates how this may be done. Due to the high electrical resistance of the indicator electrode's glass membrane, the meter must have a correspondingly high input impedance. Most pH meters currently sold contain built-in microprocessors that simplify pH measurement by performing and storing calibrations, doing diagnostics, and implementing temperature compensation.

Glass Membrane Indicator Electrode Construction

The hydrogen-ion selective glasses used to construct pH indicator electrodes are formed by fusing silica, alkali metal oxides, and alkali earth oxides. The silica component, SiO_2, makes up approximately 70% of the glass. The alkali metal oxide contributes mobile ions such as sodium or lithium, which act as electrical charge carriers, or are exchanged for protons in the hydrated glass layer. Components such as calcium oxide (CaO) are added to incorporate multivalent ions into the glass; these act to modify the network structure of the glass, imparting characteristics such as processability and chemical resistance.

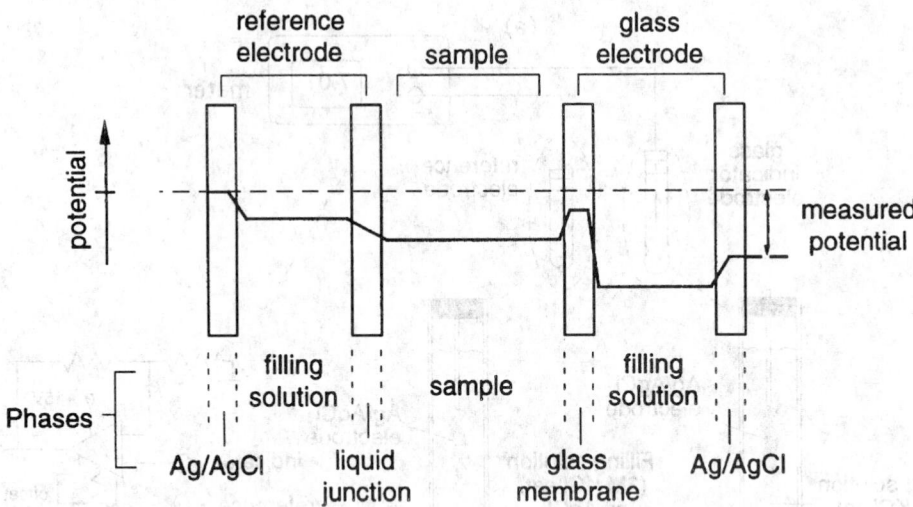

FIGURE 71.2   Schematic illustration of potential differences contributing to the measured potential between a glass membrane electrode and a reference electrode.

Glass Membrane Indicator Electrode Response to pH

When the membrane glass is immersed in aqueous solution, a gel-like hydrated layer on the order of 100 nm thick forms at the surface of the glass. Within this hydrated layer, the exchange of sodium ions in the glass for hydrogen ions in the solution produces a potential difference between the hydrated layer and the solution which depends on the hydrogen ion activity. The net potential developed across the entire glass membrane, ϕ_{net}, is the sum of boundary potentials, ϕ_b, at the inner (reference element side) and outer (sample side) extents of the membrane.

$$\phi_{net} = \phi_{b,inner} - \phi_{b,outer} \tag{71.4}$$

Under conditions where, at the outer surface, the exchange of sodium ions for hydrogen ions is complete (typically for pH less than 10), the hydrogen ion activity at the surface of the gel layer will be constant. The boundary potentials will then depend in a logarithmic manner on the hydrogen ion activity of the solution contacting the membrane, a function known as the Nernst equation.

$$\phi_b = constant' + \frac{RT}{F} \ln a_{H^+} \tag{71.5}$$

The net potential developed across the membrane is then proportional to the log of the ratio of the hydrogen ion activities of the sample and the electrode filling solution. The hydrogen ion activity of the filling solution is constant, so the membrane potential will depend linearly on pH.

$$\phi_{net} = constant'' + \frac{RT}{F} \ln \frac{a_{H^+,sample}}{a_{H^+,inner}} = constant''' - 2.3 \frac{RT}{F} pH \tag{71.6}$$

The dependence of membrane potential on pH is proportional to absolute temperature and, at 25°C, the factor 2.3 RT/F is equal to 59 mV per pH unit. Many pH meters that perform two point calibrations calculate a slope and display it as a percentage of this theoretical value. For an electrode in good condition, this will be in the range of 90% to 105%; values considerably different from this indicate a problem with the electrode.

While the membrane potential responds principally to changes in hydrogen ion activity, it also responds to sodium and other monovalent cations. These effects are most noticeable at high pH where hydrogen ion activity is small, and is referred to as alkaline error. A more detailed analysis of the membrane response, which considers the effects of interfering ions, can be found in reference texts [3].

Reference Electrodes

A reference electrode, such as that pictured in Figure 71.1(c), is used to make a stable, low resistance electrical contact between the external measuring circuit and the sample, establishing a reference potential against which the indicator electrode can be referred. In the metallic conductors of the pH meter circuitry, current is carried by electrons, while in the sample electrolyte current is carried by ions. The difference between reference electrodes is primarily the oxidation or reduction reactions that effect charge transfer across the reference element/electrolyte interface. The species participating in these reactions must be present in the electrolyte contained within the body of the reference electrode. As a result, the composition of the reference electrode filling solution is, in general, different from that of the sample, and the physical contact between the two dissimilar electrolytes forms what is known as a liquid junction. The two phase boundaries (reference element/filling solution and filling solution/sample) necessary to make electrical contact to the sample each introduce an additional potential, which adds to that of the indicator electrode, but proper design of the reference electrode ensures that these potentials remain constant and can be calibrated out.

Reference Elements

The most widely used reference electrode for pH measurement is the silver/silver chloride electrode. The electrode is constructed from a silver wire that has been coated with silver chloride and immersed in an electrolyte saturated with silver chloride. Current is readily passed across the electrode/electrolyte interface by the reduction of silver chloride to form silver metal and chloride ion.

$$AgCl + e^- \leftrightarrow Ag^0 + Cl^- \tag{71.7}$$

At equilibrium, the phase boundary potential developed at the interface depends in a Nernstian manner on the chloride ion activity of the filling solution. Maintaining a constant chloride ion activity in the filling solution ensures that this potential remains constant. This type of electrode is easily and reproducibly constructed, and the phase-boundary potential has a smaller temperature coefficient than the calomel reference electrode. The principal disadvantages of the Ag/AgCl electrode include the possibility that samples, biological samples in particular, may be contaminated with silver ion, and that precipitates (silver sulfide and nitrate) may form that can clog the liquid junction.

Another commonly used reference electrode is the calomel electrode, based on the reduction of mercuric chloride to mercury.

$$HgCl_2 + 2e^- \leftrightarrow Hg^0 + 2Cl^- \tag{71.8}$$

Like the silver/silver chloride electrode, the potential of the calomel electrode depends on the chloride ion activity of the filling solution. It is more stable than the silver/silver chloride electrode, due to favorable reaction kinetics, and is therefore preferred for high-precision electrochemical determinations. However, the operating temperature of a calomel reference electrode is limited to 70°C, compared to greater than 100°C for the silver/silver chloride electrode.

A number of reference electrodes use a platinum reference element. Ross described a reference electrode based on the reduction of iodine to iodide [4]. The nature of the redox reaction differs from the Ag/AgCl and Hg/HgCl$_2$ electrodes, in that both the iodine and iodide are soluble in the filling solution, yielding an electrode potential that is relatively insensitive to temperature. The hydrogen reference electrode consists of a platinum electrode immersed in aqueous solution, over which hydrogen gas is bubbled. The platinum catalyzes the reduction of hydrogen ion to hydrogen gas. Historically, the standard hydrogen electrode has served as the primary reference electrode for precise electrochemical determinations. The role played by the hydrogen electrode in pH measurement, however, is primarily in its use as an indicator electrode for the determination of the hydrogen ion activity of primary pH standard solutions.

Liquid Junctions

The necessary contact between the filling electrolyte of the reference electrode and the sample forms a liquid junction, and results in the development of a junction potential that depends on the compositions of the two electrolytes. The junction potential can be minimized by filling the reference electrode with an electrolyte having high concentrations of an anion and cation of comparable diffusivities. For Ag/AgCl and calomel reference electrodes, potassium chloride at concentrations ranging from $3M$ to saturation is most commonly used as the filling electrolyte, as the diffusivities of the two ions are approximately equal. The stability of the liquid junction potential depends on the constancy of the interface between the filling electrolyte and the sample. A number of approaches exist to allow the filling electrolyte to controllably leak into the sample for this purpose. These include cracked bead, porous ceramic, and annular sleeve junctions. Leak rates range from less than $1 \, \mu L \, h^{-1}$ (cracked bead) to greater than 100 (annular sleeve) $\mu L \, h^{-1}$.

Selection, Use, and Care of Reference Electrodes

Reference electrode selection is dictated by the sample to be studied and the measurement conditions, such as temperature. Silver/silver chloride electrodes are preferred over calomel for general-purpose use, and at high temperatures (70°C or greater). Calomel electrodes are preferred for high-precision determinations, and where silver contamination of the sample presents a problem. An Ag/AgCl reference electrode incorporating a double junction is also another means of preventing silver contamination of the sample. The type of liquid junction and its flow rate is important; a high flow rate may contaminate the sample and deplete the electrode, while a low flow rate may lead to clogging. Viscous, semisolid samples or low ionic strength samples require a high flow rate junction. The reference electrolyte fill port should remain open during use of the electrode to insure adequate flow through the junction. Leaving the fill port open at all times will prevent any contaminants from entering the body of the reference electrode and reduce the likelihood of junction clogging, at the expense of more frequent refilling. "Low maintenance" reference electrodes use a polymer gel saturated with electrolyte within the body of the electrode, and will perform properly if stored in filling solution.

Instrumentation

The pH meter pictured in Figure 71.1(a) measures the potential developed between the pH indicator and reference electrodes, from which the pH of the sample is determined using a previously established calibration and possibly the sample temperature. The measurement of the potential, which may range in magnitude up to a few hundred millivolts, is complicated by the large electrical resistance presented by the glass membrane. This may range from 100 MΩ to greater than 1000 MΩ, and necessitates the use of high input impedance amplifiers with FET input stages if the glass membrane potential is to be accurately measured. While the relatively low cost and high performance of commercially available meters obviates the need for amplifier construction, Figure 71.1(d) presents a simple amplifier suitable for use with a glass electrode. The amplifier circuitry of commercial pH meters incorporates additional functions to improve the accuracy and stability of the measurement, such as a driven shield to reduce noise pickup and chopper stabilization to reduce drift.

Calibration

The glass indicator electrode is calibrated using standard buffer solutions of known pH. A two point calibration procedure is illustrated schematically in Figure 71.3. A pair of standards is chosen to bracket

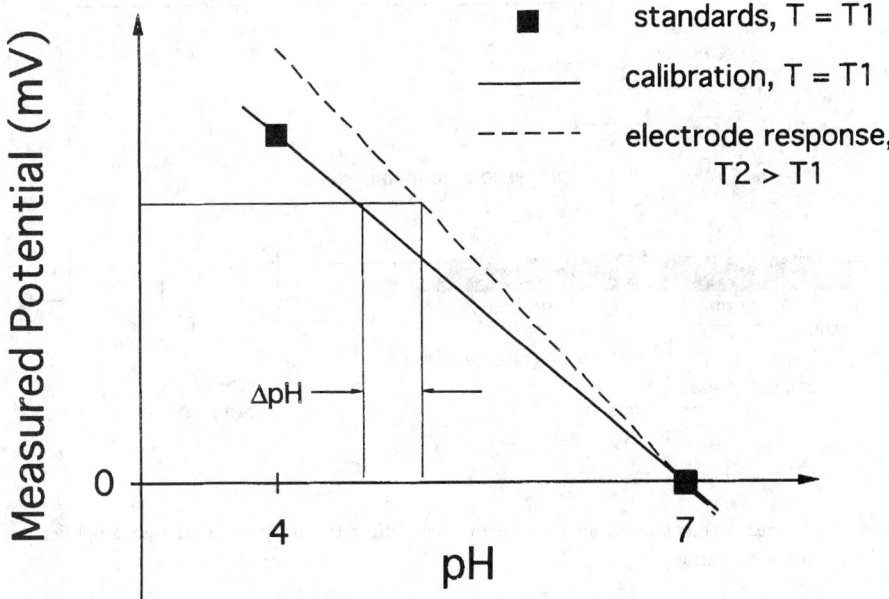

FIGURE 71.3 Calibration and temperature compensation of a glass pH electrode.

the pH range of interest. The response of the electrode is measured in each, and a calibration function is determined by linear interpolation. The slope can be compared to that expected from the Nernst equation to provide an indication of the performance of the electrode. The newest microprocessor-controlled pH meters simplify calibration by automatically recognizing the pH of standards, and constructing calibration curves from as many as five buffers.

Temperature Compensation

The potential developed across the pH indicator glass electrode membrane is temperature dependent, with a temperature coefficient of approximately 0.3% per °C, as follows from Equation 71.6. The effect of temperature on the electrode calibration is illustrated in Figure 71.3. Most pH meters have provision for temperature compensation, which corrects the slope of the measured potential versus pH calibration. Manual compensation permits the user to dial in the temperature at which the measurement is going to be made. Meters equipped with automatic temperature compensation (ATC) use a platinum resistance thermometer to directly measure the temperature of the sample. The instrumentation within the meter then corrects the calibration function such that the millivolt reading is correctly interpreted as the pH of the sample at the measurement temperature.

pHFETs

A relatively recent development in pH measurement is the introduction of systems based on the use of ion-selective field-effect transistors (ISFETs) as the sensing element. ISFETs, of which the hydrogen ion-sensitive pHFET is one variant, are derived from the metal-oxide-semiconductor FET (MOSFET), the basic building block of integrated circuits. These silicon "chips" combine a pH-responsive membrane much like that of the glass electrode with the amplification of a field-effect transistor. The integral amplification and small size have led to the development of inexpensive, battery-powered, pocket-sized pH measurement systems. These devices have found unique and expanding niches, including the food industry where the measurement of pH using breakable glass electrodes presents an unacceptable safety hazard, the measurement of the pH of gels, pastes, and slurries, and for the measurement of strongly alkaline solutions where conventional glass bulbs respond to the sodium ions and give an erroneously low reading. The following

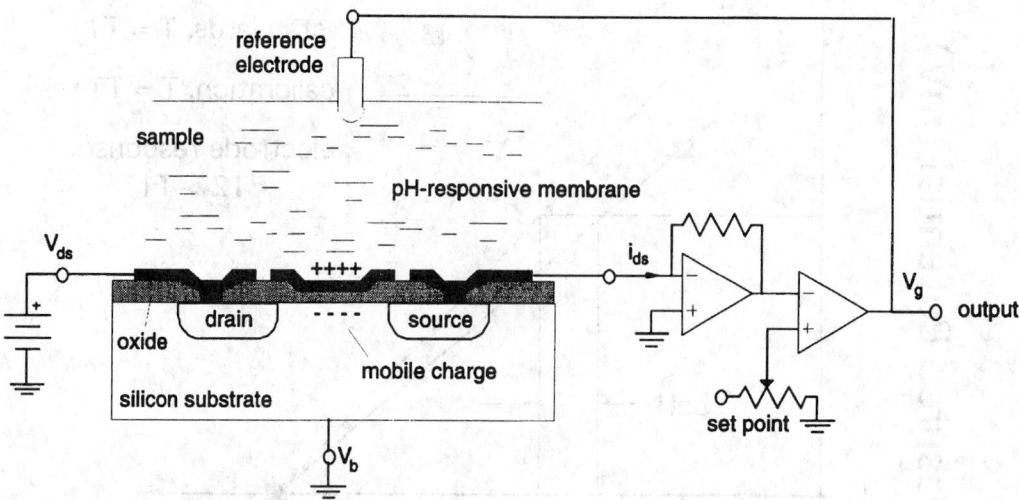

FIGURE 71.4 pH measurement using an ion-selective field effect transistor, including an amplifier circuit for constant drain current operation.

sections cover the basic operation of these devices, instrumentation, and applications. Additional details can be found in reference texts [5].

Construction and Operation

A schematic cross-section of a pH-sensitive ISFET is presented in Figure 71.4. The pHFET differs from a MOSFET in that the metal gate of the MOSFET is replaced by a pH-responsive membrane material such as silicon nitride, aluminum oxide, or tantalum oxide, which contacts the sample solution directly. As with the glass electrode, electrical contact is made to the sample through a reversible reference electrode. A suitable voltage applied to the reference electrode (relative to the silicon substrate) will charge the capacitor formed by the solution, insulating layers, and silicon substrate, and create mobile charge in the channel region. A potential simultaneously applied between the drain and source electrodes will result in current flow. Using a "charge imaging" model [6], this drain current, i_{ds}, can be described by:

$$i_{ds} = A\, V_{ds}\, Q_c = A\, V_{ds}\, C_2 \left(V_g - V_T - \frac{V_{ds}}{2} \right)$$ (71.9)

where the constant A includes geometrical factors, V_{ds} is the voltage applied to the drain, and Q_c is the mobile channel charge. The mobile charge is a function of V_g, the voltage applied to the reference electrode; V_T, the "threshold voltage" needed to produce mobile charge, and C_2, the capacitance of the gate region. The threshold voltage includes a number of terms, one of which is the phase-boundary potential at the interface between the sample and the pH-responsive insulating layer. Adsorption of protons at the surface of this layer leads to a Nernstian dependence of this potential on hydrogen ion activity with the result being that changes in pH modulate the drain current of the device.

$$i_{ds} = A\, V_{ds}\, C_2 \left(V_g - V_T + 2.3\, \frac{RT}{F}\, \mathrm{pH} - \frac{V_{ds}}{2} \right)$$ (71.10)

Instrumentation

Equation 71.10 illustrates the dependence of ISFET current on pH. The preferred method of operation of the pHFET is to operate at a constant drain current. A circuit for doing this is illustrated schematically in Figure 71.4. A transconductance amplifier converts the drain current to a voltage, which is referenced against a setpoint. The output of the comparator drives the reference electrode, V_g. Since the mobile channel charge Q_c is constant, as are V_{ds} and $V_{T'}$, changes in the comparator output directly reflect changes in the hydrogen ion activity. That is, a 59-mV change in output corresponds to a change in pH of one unit. Interface circuitry similar to that of Figure 71.4 allows the device to be connected directly to a glass electrode pH meter. As with the glass electrode, temperature compensation is required if the device is to be used at temperatures different from that of calibration.

Other Electrochemical pH Electrodes

There are a number of other approaches for electrochemical measurement of pH [1]. Two that deserve mention, as they are commercially available, either directly or incorporated within another product, are metal/metal oxide and liquid membrane electrodes.

Metal/metal Oxide pH Sensors

A pH indicator electrode with application to measurement at high temperature and pressure can be constructed from metals coated with an oxide. These metal electrodes may take the form of a wire, polished disk, or sputtered thin film, on which an oxide has been formed through thermal oxidation, chemical vapor deposition, or electrochemical oxidation. Electrodes have been constructed from systems including W/W_2O_3, Sb/Sb_2O_3, Pt/PtO_2, Ru/RuO_2, Pb/PbO_2, and Ir/IrO_2. In aqueous solution, the metal oxide can be reduced to a lower oxidation state with the consumption of a proton in solution. Using an iridium/iridium oxide (Ir/IrO_2) electrode as an example, this redox couple is believed to be represented by an Ir(III)/Ir(IV) half-cell reaction of the form:

$$2\ IrO_2 + 2\ H^+ + 2\ e^- \leftrightarrow Ir_2O_3 + H_2O \qquad (71.11)$$

The Ir/IrO_2 electrode shows a near Nernstian response of -59 mV per pH unit [7]. The chemical resistance, high temperature and pressure performance, the nonglass construction, and the considerable potential for miniaturization suggests that these pH sensors will find application in areas not typically employed by glass membrane electrodes or pHFETS.

Liquid Membrane Electrodes

Another type of electrochemical pH electrode is based on polymeric "liquid membranes," most commonly used to construct ion-selective electrodes for ions such as potassium [8]. The membranes consist of a plasticized polymer film into which an ionphore has been incorporated. The ionophore is a molecule that selectively binds and transports a given ion across the membrane, making the membrane selectively permeable to the ion of interest. The ionophore N-tridodecylamine binds hydrogen ions and has been used in the construction of a pH sensor incorporated into a disposable cartridge device for measuring blood electrolytes [9].

71.3 Optical Methods of pH Measurement

For some applications, optical methods offer advantages over the use of the glass electrode or other electrochemical devices for the measurement of pH. Organic dye molecules with pH-dependent spectral properties have been routinely used for decades in acid-base titrations and in pH indicator papers. These dyes, many of natural origin, have more recently been put to use as indicators to measure localized pH within living cells and in the development of fiber-optic probes for measuring pH within the body. The following sections cover the basic principles and major applications of optical pH measurement; more detailed treatments may be found in Reference 10.

TABLE 71.2   Sources, Features, and Key Properties of Selected pH Electrodes

Manufacturer	Reference	Range	Accuracy pH Units	Cost ($U.S., 1997)	Key Features
		Glass combination electrodes			
ABTECH Scientific, Inc.[a] Miniature Combination pH electrode, CPE 905-X	Ag/AgCl	0–14	±0.1	$285	Handheld Refillable bulb Min. vol. = 10 μL
ATI-Orion, Inc.[a] Combination Electrode, 8102BN	Proprietary	0–14	±0.03	$245	0–100°C
Brinkman[a] FUTURA Plus Combination Electrode, 39539	Ag/AgCl	0–14	±0.05	$132	Benchtop Handheld Refillable bulb
Corning[a] High Performance Comb. Electrode, 476146	Ag/AgCl "scavenger"	0–14	±0.05	$225	Increased sensitivity TRIS compatible Rapid response
Hanna[b] pHep 3 Pocket pH tester	Ag/AgCl fabric junction	0–14	±0.1	$52	Handheld, waterproof Auto 2 pt. cal. Auto temp. comp.
Hanna[c] Combination Electrode Research Grade HI1270	Ag/AgCl	0–12	±0.1	$115	Handheld Field portable Sealed rugged epoxy body
Hach[c] Hach One Electrode with Temperature Sensor, 48600-00	Ag/AgCl	0–14	±0.05	$185	Benchtop, handheld Field portable, Free-flowing reference junction 0–100°C
Microelectrodes, Inc.[a] Micro-Combination pH Probe MI-410	Ag/AgCl	0–14	±0.1	$185	Handheld Min. vol. = 5 μL
Radiometer[a] PHC2401	Ag/AgCl	0–12	±0.1	$185	General purpose
Sensorex[a] Combination Electrode 450CUS	Ag/AgCl	0–14	±0.05	$65	Epoxy body 0–100 psig Flat tip for use on moist surfaces 0–100°C
		ISFET electrodes			
ATI-Orion[a,c] pHuture Sure-flow electrode, 616500	Ag wire	0–14	±0.02	$326[d] $393[e]	Built-in temperature probe
Bioanalytical Systems, Inc.[b] pH Boy, MF8960	Ag/AgCl	2–10	±0.1	$180	Handheld, stores dry Min. vol. = 50 μL
Corning[c] ISFET electrode 476395	Ag/AgCl	0–14	±0.02	$320	Handheld, stores dry Min. vol. = 20 μL Auto temp. comp. 0–60°C
IQ Scientific Instruments[a] IQ 200	Ag/AgCl KCl gel	0–14	±0.01	$229[d] $695[e]	Handheld, sterilizable Auto. temp. comp. −5–+105°C Automatic two point calibration Min. vol. = 50 μL
Sentron[b] S1001	Ag/AgCl fabric junction	0–14	±0.01	$259[d] $695[e]	Handheld, stores dry Man. 2 or 3 pt. cal Replaceable sensor tip, 0–60°C Min. vol. = 50 μL (1 drop)
		Metal/metal oxide electrodes			
Cypress Systems, Inc.[c] METOXY	Ag/AgCl	0–14	±0.01	$1299[b]	Auto. temp. comp. 0–100°C Automatic two point calibration Min. vol. = 50 μL

TABLE 71.2 (continued)   Sources, Features, and Key Properties of Selected pH Electrodes

[a] General-purpose meter.
[b] Integrated sensor and meter.
[c] Dedicated meter (specific to manufacturer).
[d] Replacement electrode.
[e] Electrode and converter for use with general-purpose meter.

Indicator Dyes

Optical measurement of pH is based on the use of organic dye molecules that are weak acids or bases. The loss or gain of a proton changes the electronic structure of the molecule, producing a measurable change in the manner in which the molecule interacts with light. This interaction can be the absorption of light at a particular wavelength, or fluorescence by one form of the molecule. An equilibrium exists between the acid and base forms of the dye, whether free in solution or chemically attached to a supporting membrane, and can be described by the equilibrium:

$$Dye^z + H^+ \xleftrightarrow{K_a} Hdye^{z+1} \qquad K_a = \frac{a_{Dye^z} \, a_{H^+} \, a}{a_{Dye^{z+1}}} \qquad (71.12)$$

where z represents the valence of the molecule. The interrogation of the solution or membrane with light will produce a response that is weighted by the relative proportions of the acid and base forms of the molecule, which in turn depends on pH. Inspection of Equation 71.12 shows that the ratio of the two forms, and hence color, varies continuously with hydrogen ion activity. The sensitivity of the measurement is greatest when the acid and base forms of the dye are present in approximately equal concentrations. This will occur when the pH of the solution is close to the pK_a ($-\log [K_a]$) of the dye. Limitations of the human eye restrict detectable changes in color to a tenfold excess of one species over the other. This corresponds to a change of ± 1 pH unit. Thus, an indicator with a pK_a of 1×10^{-5} will display a color change if the solution in which it is dissolved changed from 4 to 6 pH units. The pH of interest therefore dictates selection of the particular dye.

   An understanding of possible interactions between the indicator dye and the sample is important in the effective use of optical indicator dyes for pH measurement [11]. Factors such as temperature, electrolyte concentration, and the presence of organic solvents may cause a shift in pK_a of the indicator dye; the shift could be as much as one or more pH units. As weak acids and bases, the addition of indicator dyes to a sample can change the pH, particularly in weakly buffered samples. This effect is important when using pH papers or optical sensors to measure small samples, where the potential exists for the buffering capacity of the sample to be exceeded by the amount of dye immobilized in the paper or on the sensor. As a general rule, potential errors can be minimized by calibrating in solutions similar in composition to the sample to be measured.

Absorption Indicator Dyes

The conjugate acid and base forms of absorption indicator dyes differ in the characteristic wavelengths at which they absorb light energy. A number of common absorption indicator dyes are listed in Table 71.3, and their pK_as span the range of pH from approximately 1 to 13. The absorption spectra of a solution of the dye phenol red at a number of different pHs is shown in the inset of Figure 71.5. The acid form of the dye has an absorption maxima at $\lambda_{max} = 435$ nm, while the base form absorbs maximally at $\lambda_{max} = 565$ nm. As the pH of a phenol red solution is increased from 6 toward the pK_a of 7.9, the equilibrium of Equation 71.12 shifts to the basic form of the dye. The relative heights of the absorption peaks change, reflecting a shift in concentrations of the two forms. For quantitative pH measurement, the absorption of light by the indicator is measured by a spectrometer or reflectometer at a specific wavelength or narrow set of wavelengths corresponding to the wavelength of maximum absorption, λ_{max} for the acidic or basic

TABLE 71.3 Acid-Base Indicator Dyes

Indicator	pK_a	λ_{max} (nm) Acid Form	Base Form
Thymol blue	1.7	544	430
Methyl orange	3.4	522	464
Bromphenol blue	3.9	436	592
Bromcresol green	4.7	444	617
Methyl red	5.0	530	427
Chlorophenol red	6.0	—	573
Bromcresol purple	6.3	433	591
Bromthymol blue	7.1	433	617
Phenol red	7.9	433	558
Cresol red	8.2	434	572
Phenolphthalein	9.4	—	553
Thymolphthalein	10.0	—	598

Note: J. A. Dean and N. A. Lange (Eds.), *Lange's Handbook of Chemistry,* 14th ed., New York: McGraw-Hill, 1992.

form of the chromophoric dye molecule. Beer-Lambert's law describes the proportionality between absorbance, A, and the concentration of the indicator, c_{dye}, in moles per liter.

$$A = \log\left(\frac{I}{I_0}\right) = \varepsilon b c_{dye} \tag{71.13}$$

The quantity I/I_0 represents the ratio of the intensity of light transmitted through the sample to that incident on the sample, ε the extinction coefficient of the dye molecule, and b the path length. The

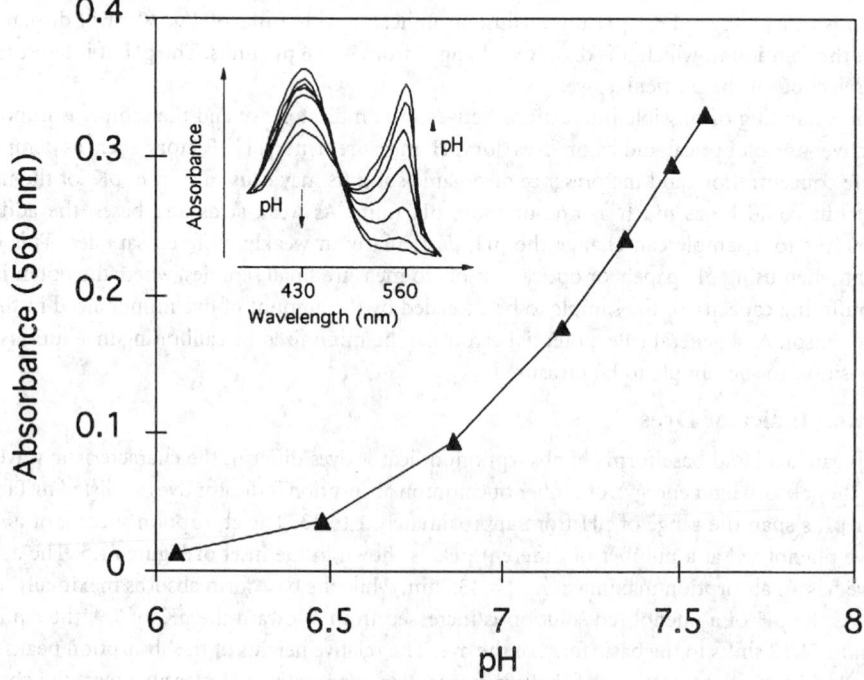

FIGURE 71.5 Magnitude of the absorption of phenol red at 565 nm as a function of pH. Inset: absorption spectra at different pHs.

TABLE 71.4 Fluorescent pH Indicator Dyes (for physiological applications)

Indicator	pK_a	Excitation Maximum (nm)	Emission Maximum (nm)
BCECF[a]	7.0	482	520
Fluorescein	6.4	490	515
HCC[b]	6.9–7.0	410	455
HPTS[c]	6.8–7.3	465	520

[a] 2′,7′-bis-(2-carboxyethyl)-5-(and-6)-carboxyfluorescein.
[b] (7-hydroxycoumarin-3-carboxylic acid).
[c] (8-hydroxy-1,3,6-pyrenetrisulfonate).

magnitude of the absorption peak heights at 565 nm, plotted in Figure 71.5 as a function of pH, reflects the changing concentration of the basic form of the dye as the equilibrium of Equation 71.12 shifts.

Fluorescent Indicator Dyes

Fluorescent indicator dyes absorb light of a particular color (or wavelength) and re-emit some of the absorbed energy as light of a different color. Absorption of light by the indicator promotes the molecule from the ground state energy, E_g, to a higher energy state, E_1. Subsequent processes such as molecular collisions lead to a transition to a lower energy excited state, E_2. The molecule can then emit a photon of energy $E_2 - E_g$, resulting in the return of the molecule to the ground state from this intermediate state. The emitted light is at a longer wavelength than the exciting wavelength, and the difference is known as the Stokes shift. In principle, a fluorescence measurement is more sensitive than an absorption measurement because the only light measured by the detector originates from fluorescing molecules. Table 71.4 lists a number of common fluorescent pH indicators used for measurement of pH in the physiological range. Fluorescein is widely used because the absorption maximum at 490 nm of the fluorescent dianion is readily excited by the 488-nm emission of argon ion lasers.

Indicator Papers

Indicator papers are a simple, rapid, and inexpensive means of measuring pH when the precision of an instrumental measurement is not necessary. These are constructed from a strip of paper or plastic that has been impregnated with one or more absorption indicator dyes chosen to span the pH range of interest. The dyes are generally covalently attached to the strip, to prevent contamination of the sample by leaching of the dyes. A strip for indicating pH in the range of 5.5 to 9.0 can be constructed using the dyes bromocresol purple, bromothymol blue, and phenol red [1]. The pH of a sample is determined to a precision of typically 0.5 units by comparing the color of the strip to a color calibration chart provided by the manufacturer.

Fiber-Optic pH Probes

Optical pH sensors, often referred to as optrodes, represent some of the most sophisticated pH sensors, finding use for remote sensing in the body or industrial plants due to their small size and lack of electrical connections. They are typically constructed by immobilizing an indicator dye at the tip of a light guide formed from one or more optical fibers, which are used to couple light between the indicator and the measurement instrumentation, as illustrated schematically in Figure 71.6(a). The resulting probes can be made very small, and with high-quality optical fibers, pH can be measured over considerable distances in electrically noisy environments that would interfere with potentiometric-type electrodes.

Construction and Instrumentation

The pH indicator dye must be immobilized in close proximity to the fiber tip. The method of immobilization impacts the sensor's response, response time, long-term stability, and mechanical integrity. The dye can be immobilized to a solid support, such as a membrane or porous glass bead, which is then

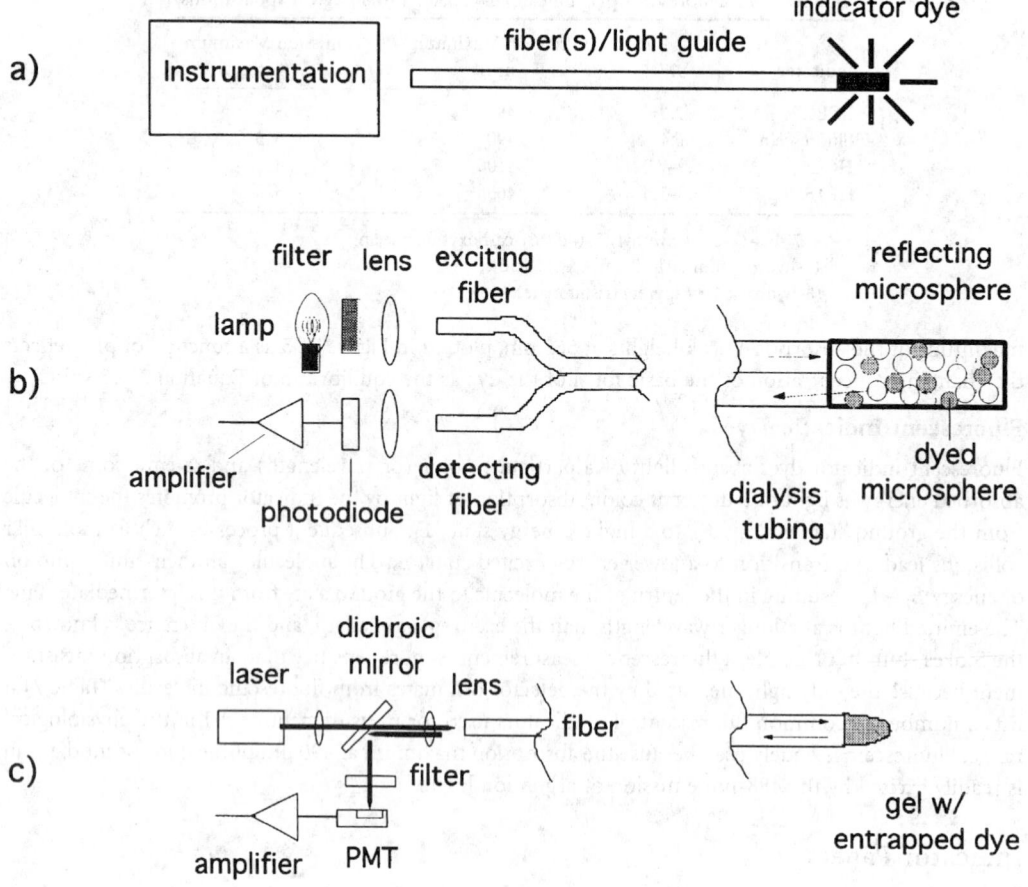

FIGURE 71.6 Optical fiber sensors: (a) generic; (b) based on absorption indicator dyes [13]; and (c) based on fluorescent indicator dyes [14].

attached to the end of the fiber. Indicator dyes have been covalently linked to the tip of glass fibers, or entrapped within polymer films formed at the tip using light energy from the fiber to initiate polymerization. The construction of probe tips and the instrumentation used for a fiber-optic pH probe depend on the type of indicator being used. Unlike the glass electrode, which has remained largely unchanged for more than 75 years, there is considerable variety in the design of fiber-optic pH sensors. While a complete review of the literature is beyond the scope of this work (interested readers may wish to consult Wolfbeis [10] and a recent review of chemical sensors by Janata [12]), the following examples illustrate the approaches to construct optical fiber pH sensors using absorption and fluorescent dyes.

Absorption optrodes measure the change in intensity of the light returned from the fiber tip/sensing region relative to the interrogating light of specific (or a narrow band) wavelength. These require a separate collection fiber or fibers and a means to reflect the light, such as a reflector or the use of scatterers such as polystyrene beads. In the example of Figure 71.6(b) [13], polyacrylamide beads containing the indicator dye phenol red were mixed with white polystyrene latex microspheres that served as scatterers. The beads were contained within dialysis tubing attached to the end of a pair of plastic optical fibers, one used for excitation, the other to collect scattered light. Light from a lamp was filtered to alternately select the λ_{max} of the base form of the dye (565 nm), and a wavelength (600 nm) where the absorbance is pH independent as a reference. (The isosbestic point at 480 nm could also have served as a reference.) The intensity of light scattered back into the second detector fiber was quantified by a photodiode. The ratio of the scattered intensities at the two wavelengths was then related to the pH of the sample.

Fluorescent indicator optrodes measure the Stokes-shifted fluorescence emission by the indicator and may use a single fiber to both interrogate and collect signal-carrying light. The amount of fluorescent pH indicator at the fiber tip must be maximized since fluorescence is emitted omnidirectionally and only a small fraction of the emitted fluorescence will be captured by the fiber. In the example of Figure 71.6(c) [14], the indicator dye was entrapped within a polymer gel formed at the tip of a glass fiber. The source of excitation light is typically a high-intensity lamp or a laser. If a lamp, its output is filtered to select a narrow band of wavelengths at the absorption maximum of the specific form of the indicator, that is then introduced into the fiber to excite the fluorophore at the tip. Emitted light collected by the fiber exits the fiber and is incident on a dichroic mirror, which reflects the long-wavelength light and passes shorter wavelengths. A filter then selects a narrow band of emission wavelengths, and finally, the light is detected. Due to the relatively small light intensities, the detector is typically a photomultiplier tube rather than a photodiode.

71.4 Frontiers of pH Measurements

Because of its wide-ranging importance, there is likely to be continued innovation in both approaches and opportunities for pH measurements. Some important developments, such as pHFETs and metal/metal oxide pH sensors, have already occurred and are now achieving noteworthy commercialization. Investigators continue to seek new materials for constructing transducers. Examples include electroconductive (electroactive and inherently conductive) polymers such as polyaniline [15] and polyelectrolyte hydrogels [16], which form pH-responsive membranes suitable for the construction of microsensors.

A noteworthy innovation is the application of pH measurement devices as transducers that are integrated into more complex analytical systems directed at other analytes. A pH indicator dye forms the basis of the pCO_2 measurement in an indwelling fiber-optic catheter [17]. Because protons are generated and/or consumed in many biological reactions involving enzymes, pH sensors have found use as transducers in biosensor devices.

An expanding area for pH measurement is likely to be the study of biological systems. The light-addressable potentiometric sensor is a device that measures the pH change resulting from the metabolic activity of cells, with application to examining cellular response to toxins and infectious agents [18]. The measurement of intracellular or subcellular pH using pH-sensitive fluorescent indicators [19] can provide insights into the physiology of the cell.

Acknowledgments

AGE thanks Allage Associates, Inc. and ABTECH Scientific, Inc. for financial support.

References

1. H. Galster, *pH Measurement: Fundamentals, Methods, Applications, Instrumentation.* New York: VCH Publishers, 1991.
2. D.A. Skoog, D.M. West, and F.J. Holler, *Fundamentals of Analytical Chemistry,* 7th ed., Philadelphia, PA: Saunders College Publishing, 1996.
3. A.J. Bard and L.R. Faulkner, *Electrochemical Methods: Fundamentals and Applications,* New York: John Wiley & Sons, 1980.
4. J.W. Ross, *Temperature Insensitive Potentiometric Electrode System.* U.S. Patent No. 4,495,050, 1982.
5. J. Janata, *Principles of Chemical Sensors,* New York: Plenum Press, 1989.
6. S.D. Senturia, The role of the MOS structure in integrated sensors, *Sensors and Actuators,* 4, 507-526, 1983.
7. M. F. Yuen, I. Lauks, and W.C. Dautremont-Smith, pH dependent voltammetry of iridium oxide films, *Solid State Ionics,* 11(1), 19-29, 1983.

8. D. Ammann, *Ion-selective Microelectrodes: Principles, Design and Applications* New York: Springer Verlag, 1986.

9. S.N. Cozzette, G. Davis, I.R. Lauks, R.M. Mier, S. Piznik, N. Smit, P. Van Der Werf, and H.J. Wieck, *Process for the Manufacture of Wholly Microfabricated Biosensors*, U.S. Patent No. 5,466,575, 1995.

10. M.J.P. Leiner and O. Wolfbeis, Fiber optic pH sensors, in O. Wolfbeis (Ed.) *Fiber Optic Chemical Sensors and Biosensors.* Vol. 1, Boca Raton, FL: CRC Press, 1991.

11. J. Janata, Do optical sensors really measure pH?, *Anal. Chem.*, 59, 1351-1356, 1987.

12. J. Janata, M. Josowicz and D.M. DeVaney, Chemical sensors, *Anal. Chem.*, 66, 207R-228R, 1994.

13. J.I. Peterson, S.R. Goldstein, R.V. Fitzgerald, and D.K. Buchwald, Fiber optic pH probe for physiological use, *Anal. Chem.*, 52, 864-869, 1980.

14. C. Munkholm, D.R. Walt, F.P. Milanovich, and S.M. Klainer, Polymer modification of fiber optic chemical sensors as a method of enhancing fluorescence signal for pH measurement, *Anal. Chem.*, 58, 1427-1430, 1986.

15. A. Guiseppi–Elie, G.G. Wallace, and T. Matsue, Chemical and biological sensors based on electrically conducting polymers, in T. Skotheim, R. Elsenbaumer, and J.R. Reynolds (Eds.) *Handbook of Conductive Polymers, 2nd edition,* Chap. 34, p. 963, New York: Marcel Dekker, 1996.

16. N.F. Sheppard Jr., M.J. Lesho, P. McNally, and A.S. Francomacaro, Microfabricated conductimetric pH sensor. *Sens. and Act. B*, 28, 95-102, 1995.

17. J.B. Yim, G.E. Khalil, R.J. Pihl, B.D. Huss and G.G. Vurek, *Apparatus for Continuously Monitoring a Plurality of Chemical Analytes through a Single Optical Fiber and Method of Making*, U.S. Patent No. 5,098,659, 1992.

18. J.W. Parce, J.C. Owicki, K.M. Kercso, G.B. Sigal, H.G. Wada, V.C. Muir, L.J. Bousse, K.L. Ross, B.I. Sikic, and H.M. McConnell, Detection of cell-affecting agents with a silicon biosensor, *Science*, 246, 243-247, 1989.

19. R. Haugland (Ed.), *Handbook of Fluorescent Probes and Research Chemicals, 6th ed.,* Chap. 23, Eugene, OR: Molecular Probes, Inc., 1996.

72

Humidity and Moisture Measurement

Gert J.W. Visscher

*Institute of Agricultural and
Environmental Engineering*

Whether one likes it or not, water and water vapor can be found everywhere. Because of the asymmetrical distribution of their electric charge, water molecules are easily adsorbed on almost any surface, where they are present as a mono- or multimolecular layer of molecules. Water vapor in the air or any other gas is generally called *humidity*; in liquids and solids, it is usually designated as *moisture*. The determination of humidity and moisture, as in prediction of floods, fog, conditions for the appearance of plant diseases, etc., is of great economic importance. Stored foodstuffs or raw materials may dry up at low humidity or get moldy at high humidity. In many industrial processes, the measurement of moisture and humidity is important for the maintenance of the optimum conditions in manufacturing. Humidity and moisture content can be expressed in a number of ways, and the number of methods for measuring them is even greater. An engineer whose main concern is to avoid condensation no matter where in his system will be interested in the *dewpoint* of the gas flow. A chemist may be interested in the mere *quantity* of water vapor, whereas in a printing-office or a storage-room, the *relative humidity* is of more importance.

Water vapor is one of the constituent gases of the Earth's atmosphere, the total pressure P of which is, according to Dalton's law, the sum of the partial pressures. This means that:

$$P = P_{N_2} + P_{O_2} + P_{H_2O} + P_{\text{other gases}} \qquad (72.1)$$

Like other gases, water vapor can be considered to behave as an ideal gas, except near saturation. In average environmental conditions, water can also be present in the liquid and solid phase, the reason to speak of water vapor rather than of water gas. An empty space in equilibrium with a flat water (or ice) surface can, at a given temperature, hold a well-defined maximum quantity of water vapor [1,2,3]. When this *saturation vapor pressure* is reached, any further addition of water vapor results in condensation. In the presence of air molecules at atmospheric pressure, the saturation vapor pressure is about 0.4% higher (expressed by the so-called enhancement factor [4]). The saturation vapor pressure for water is about 611 Pa at 0°C, 2339 Pa at 20°C, and 7383 Pa at 40°C. So, one can say that the average water vapor pressure in the Earth's atmosphere around us ranges from about a half to a few percent of the barometric pressure.

There are several ways to express humidity:

1. The *vapor pressure* is that part of the total pressure contributed by the water vapor.
2. The *absolute humidity* (or vapor concentration or vapor density) is the mass of water vapor per unit of volume. Effects of temperature and pressure are, except near saturation, according to gas laws.
3. The *relative humidity* is the ratio of the actual vapor pressure and the saturation vapor pressure at the prevailing temperature. It is usually expressed as a percentage. Since the maximum water vapor pressure depends on temperature, the relative humidity (r.h.) also depends, at a given water content, on temperature. At constant temperature and a given water content, the r.h. is, according to the equation for P mentioned above, dependent on total pressure.
4. The *dewpoint temperature* is the temperature to which a gas must be cooled, at constant pressure, to achieve saturation. When the condensate is ice, it is called *frost-point*. It is, with unchanging composition of the gas, independent of temperature. It changes with pressure since P_{H_2O} is proportional to P. Of course, condensation will occur if saturation vapor pressure is reached.
5. The *mixing ratio* is the mass of water vapor per unit mass of dry gas, usually expressed in grams per kilogram. If the ratio is related to a unit mass of humid air, it is called the *specific humidity*.
6. The *mole fraction* is the ratio of the number of moles of a component to the total number of moles present.

Conversions between the different parameters used to be cumbersome. They are now becoming standard since the introduction of the microprocessor.

Concentrations of water in a *liquid* or a *solid* are normally given in kg/kg. Except in soil physics, volumetric units are rarely used. The expression *equilibrium relative humidity* (e.r.h.) refers to a condition where there is no net exchange of water vapor between a moisture-containing material (paper, medicines, foodstuffs, tobacco, seeds, etc.) and its environment. It is the equivalent for *water activity*, a_w, used in the fields of biology or food technology, generally expressed as a ratio rather than a percentage (i.e., 0.6 instead of 60%).

It is probably difficult to find a material that is inert to water molecules and with which it would be impossible, with some physical method, to measure the presence of water. Water molecules change the length of organic materials, the conductivity and weight of hygroscopic materials and chemical absorbents, and in general the impedance of almost any material. Water absorbs infrared as well as ultraviolet radiation. It changes the color of chemicals, the refractive index of air and liquids, the velocity of sound in air or electromagnetic radiation in solids, and the thermal conductivity of gases as well as that of liquids and solids. More fundamentally, the water content can be measured by removing the water (vapor) from the sample and measuring the change of weight (or the change of pressure in a gas). Other fundamental principles are the evaporation from a water surface into the stream of sample gas (*psychrometer*) and the cooling of the gas sample until condensation is detected. Microwave absorbance, the measurement of capacitance and nuclear magnetic resonance have found application in the measurement of moisture in liquids and solids.

After an engineer has decided which parameter has to be measured, he has to realize a few things [5].

First, what is the minimum range of operation required? Over-specification can be expensive. Besides, instruments suitable for drying processes at high temperatures and, at the same time, trace detection of water vapor in dry gases do not exist.

TABLE 72.1 Methods for Measuring Humidity and Moisture

Method	g, l, s[a]	Range	Manufacturer[b]	Approx. Price $
Mechanical (hair)	g	0–100% rh	Lambrecht, Thies, Haenni, Jumo, Sato, Casella, Pékly et Richard	0.3–1 k
Condens.dewpoint	g	–80/+100°C dp	General Eastern, Michell Instr., EG&G, E+H[3], MBW, Protimeter, Panametrics	2–20 k
Dry and wet bulb	g	10–100% rh	Lambrecht, Thies, Haenni, ASL, Jenway, Casella, Ultrakust, IMAG-DLO	0.5–3 k
Lithium chloride	g	–45/+95°C dp	Honeywell, Jumo, Lee Engineering, Siemens, Philips, Weiss	0.7–2 k
Polymer (capac.)	g, s	0–100% rh	Vaisala, Rotronic, Testo, Hycal, Panametrics, Novasina, EE Elektronik, Chino, Lee Integer	0.5–2.5 k
Electrical (others)	g, s	0–100% rh	PCRC, General Eastern, Rotronic, Chino, Elmwood, Shinyei Kaisha	0.5–2.5 k
Thermal conductivity	g	0–130 g/m^3	Shibaura Electronics Co. Ltd	0.1–1 k
Al_2O_3/silicon	g, l	–80/+20°C dp	E+H, Gen. Eastern, Panametrics, Michell Instr., MCM, Shaw	1.5–3 k
Phosphorous pentoxide	g	0.5–10000 ppm	Anacon, Beckman, Dupont	2–6 k
Crystal oscillator	g	0.02–1000 ppm	Dupont	25–30 k
Infrared absorbance	g, l	0-50 ppm up to 65 °C dp	Siemens, H&B, ADC, Anacon, Kent, Horiba, Sieger, Beckman, Li-Cor	5–15 k
Infrared reflectance	s	0.02–100%	Anacon, Infrared Engin., Moisture Systems Corp., Pier Electronic, Zeltex, Bran & Luebbe, Bühler	5–15 k
NMR	l, s	0.05–100%	Oxford Anal. Instr., Bruker	10–30 k
Neutron moderation	s	>0.5%	Kay Ray, Berthold, Nuclear Ent.	8–25 k
Microwave attenuation	s	0–85%	Mütec, Scanpro, Kay Ray Inc., BFMRA	12–35 k
TDR	s	0–100%	IMKO GmbH, Campbell Sci. Inc., Soil Moisture, Tektronix	5–20 k
FD	s	0–100%	ΔT Devices, IMAG-DLO, VITEL, Troxler	0.1–10 k

[a] g, l, s: gas/liquid/solid.

[b] The table is inevitably incomplete; a manufacturer not mentioned may deliver a high-quality instrument (see also Reference 9).

[c] Abbreviations used: E+H: Endress+Hauser; MBW: MBW Elektronik AG; ASL: Automatic Systems Laboratories Ltd; PCRC: Physical and Chemical Research Corporation; MCM: Moisture Control and Measurement Ltd; H&B: Elsag Bailey Hartmann & Braun; ADC: Analytical Development Company; IMAG-DLO: Institute for Agricultural and Environmental Engineering.

Second, unlike a temperature sensor, a humidity sensor can, at least in air, essentially not be shielded from its direct environment. The question of how to measure humidity cannot be separated from the measuring problem in question (contamination, condensation, etc.). A question might be: is there a danger of condensation before or after the period of real measurement?

Third, the accuracy that may be expected in the field of hygrometry is considerably lower than that in other fields of measurement. One should be careful not to ask or to expect accuracies better than 2% or 3% r.h. or 0.5°C in dewpoint. Before deciding to use a certain method or to buy a particular instrument, it is very useful to go once more through things like humidity as well as temperature and pressure range, possible contaminants in the process, and accuracy and response time really needed. After this a consideration of frequency of service and calibration, cost of sensor replacement, etc. has to be made. At the end, it is sensible to find a supplier or manufacturer that is willing to think along with the customer. Table 72.1 provides a selective listing of the methods, ranges, and manufacturers for measuring humidity and moisture. The methods are discussed below.

72.1 Gases

There are several methods of measuring humidity, the most important of which are described here. The scope of the present survey is limited. More can be found in the literature [6,7,8,9,10]. Reference 9 covers measurement in liquids and solids as well.

The Gravimetric Method

The gravimetric method is the most fundamental way of accessing the amount of water vapor in a moist gas. In a gravimetric hygrometer, the water vapor is frozen out by a cold trap or absorbed by a chemical desiccant and weighed, while the volume or the mass of dry gas is measured directly. Since the result of a measurement gives the average value over an extended time, the instrument is used in combination with a humidity generator, capable of producing a gas of constant humidity. The method is used for primary standards of, among others, NIST in the U.S., NPL in the U.K., and NRLM in Japan. Achievable accuracies are approximately 0.1% to 0.2% in mixing ratio, or 0.04°C in the range of −35°C to +50°C dewpoint, increasing to 0.08°C at +80°C and 0.15°C at −75°C. The operation of such a standard requires high skill and sophisticated hardware.

Precision Humidity Generators

For less elaborate calibration work, a precision humidity generator is preferred. The best ones are reported to have an accuracy comparable to that of a gravimetric standard [11]. Therefore, such a generator may be considered as a primary standard as well. Three practical methods to produce an atmosphere of known humidity are described by Hasegawa [12]: the two-flow, the two-temperature, and the two-pressure method. Briefly, in the first method, a test chamber is fed by two streams of air, one being dry, the second one saturated with water at a known temperature. The resulting humidity can be calculated from the two flow rates. The two-temperature method uses air that has been saturated with water vapor at a well-known temperature, after which the air is heated to a higher temperature. In the two-pressure method, air is saturated with water vapor at an elevated pressure, and expanded isothermally to a lower, normally atmospheric pressure. Both temperature and pressure of the saturator and the test chamber are measured accurately. In general, precision generators are not transportable, so intercomparisons have to be made with a transfer standard of high accuracy. A good, if not the only, choice is a standard mirror dewpoint meter.

The Condensation Dewpoint Hygrometer

The saturation vapor pressure in air increases with temperature (Figure 72.1). This means that the air under test can be cooled to a temperature where it is just saturated with water vapor. If this is done at constant pressure and specific humidity, the true dewpoint temperature is obtained. In practice, a sample of the gas is usually drawn over a thermoelectrically cooled metal mirror. The mirror is cooled until dew or frost is detected, by optical means. In some cases, where the mirror is replaced by an inert substrate, the formation of dew is detected by electrical means or the use of surface acoustic waves. The temperature is maintained such that the thickness of the deposit is neither increasing nor decreasing. The highest accuracy of a transfer standard can be expected to be 0.03°C to 0.05°C in the range of −20°C to +40°C dewpoint. Industrial optical condensation dewpoint hygrometers claim accuracies up to 0.2°C, which may be true in case of a clean mirror; in practice 0.5°C is often more realistic. Advantages of this principle are its fundamental nature and the wide range: dewpoints to 90°C under ambient temperature can be measured. One of the disadvantages is the susceptibility of the mirror to contaminants, especially soluble salts. The sensor may measure the dewpoint of another condensable vapor if its dewpoint is above that of the water. A good control of the mirror temperature requires a temperature difference between dewpoint and ambient temperature of the sensing head. This means that at high relative humidities, the gas must be heated and measured outside the proces stream. At temperatures below 0°C, there may be

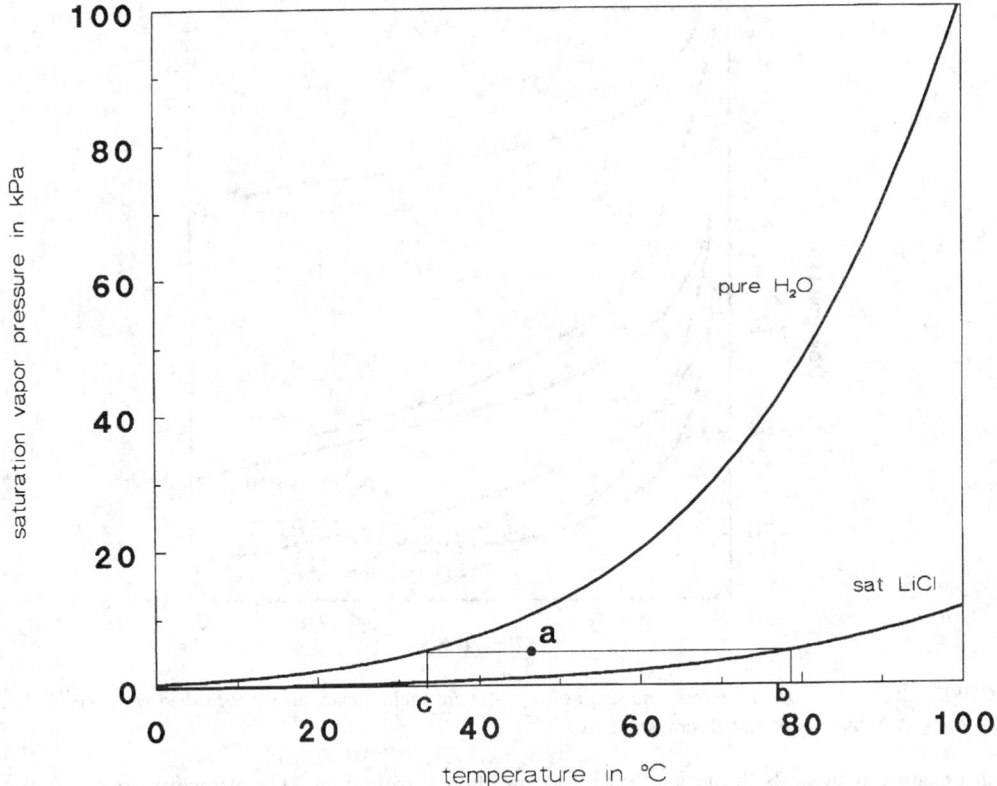

FIGURE 72.1 Saturation water vapor pressure above pure water and above a saturated solution of lithium chloride. In order to be in equilibrium with a gas sample at condition **a**, a saturated solution of LiCl has to be heated to temperature **b**. A free water surface must be cooled to a temperature **b**, as is the case in a mirror dewpoint hygrometer.

supercooled water instead of ice on the mirror. Observations have shown that below –25°C, the deposit will usually be ice. As the saturation vapor pressures over water and ice differ, one has to know the nature of the condensed layer; thus, for accurate measurements, a microscope is required. A rule of thumb is that the difference between water and ice on the mirror means a difference in dewpoint of one tenth of the temperature in degrees Celsius below zero, the dewpoint above water being the lower. Where low water content has to be measured, special attention should be given to the material and cleanliness of the pipes used. Stainless steel, polished at the internal surface, is to be preferred. The lower the moisture content, the more significant the effects are (Figure 72.2).

The Psychrometer

Two thermometers are ventilated with the gas of unknown humidity. One sensor, the dry bulb, measures the gas temperature t. The other sensor, the wet bulb, is surrounded by a wick saturated with pure water. The energy required to evaporate water into the air stream cools the wet bulb to a temperature t_w. The vapor pressure e in the sampled gas is calculated with the psychrometer equation:

$$e = e_w - A \cdot P \cdot \left(t - t_w \right) \qquad (72.2)$$

where e_w is the saturated vapor pressure at temperature t_w, P is the total atmospheric pressure, and A is the psychrometer coefficient. A depends on ventilation speed, dimensions of the wet bulb, and radiative

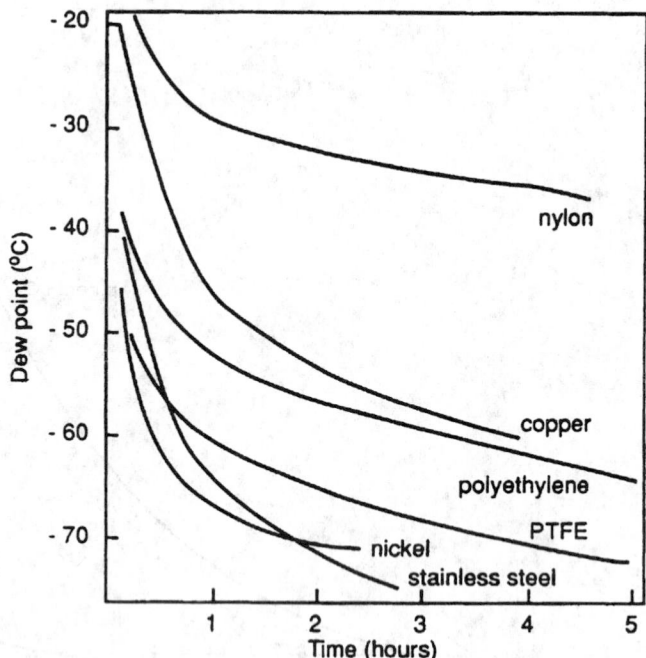

FIGURE 72.2   Illustration of the moisture given off by different tubing materials when flushed with very dry gas after being at ambient humidity. (From Reference 8.)

heat exchange between both dry and wet bulbs and their surroundings. The attractiveness of the psychrometric method lies in the fact that it is a direct and relatively simple method, with a theoretically strong basis. The accuracy of the method is determined by the accuracy of both dry and wet bulb sensors, the maintenance of a minimum ventilation speed, and a clean wick. With a ventilation speed of more than 3 m s^{-1} and a bulb diameter of 3 mm to 5 mm, A can, according to recent investigations, be assumed to be $(6.35 \pm 0.15) \times 10^{-4}$°C^{-1}, for Assmann-type psychrometers with a polished internal screen as well as sensors in a transverse air stream in a "black" radiation environment [13,14]. The radiation environment for axially ventilated psychrometers is of crucial importance. Especially when $t\text{-}t_w$ is measured directly the psychrometer is the preeminent instrument to measure near or at 100% r.h. The principle can be used up to a wet bulb temperature of 100°C at atmospheric pressure; dry bulb temperature may exceed 100°C (up to 165°C, Ultrakust, Germany). Depending on the dry bulb temperature, the wick may have problems with water supply at strong evaporation. Frozen wet bulbs, already possible at ambient temperatures below 9°C, can also lead to problems. Generally, this method adds water vapor to the atmosphere, which might be undesirable in specific applications.

Mechanical Hygrometers

Although mechanical hygrometers are losing ground, they are still widely used, mainly in room conditions. The principle relies on the elongation with r.h. of mainly human hair, textiles, or plastic fibers, the effect of which can be amplified mechanically to move a pen on a recorder. The best accuracy is 2% to 3% r.h. (for hair, in the range between 35% and 95% r.h., if regenerated at regular intervals); in general, it is wise not to expect better than 5% r.h. In the case of hair, one must be aware of the fact that the hair may be in a state different from that during calibration. Hair exhibits a dry and a wet curve; the transition takes place below 35% r.h. Once at the dry curve, the instrument may read to 20% r.h. too high if it was calibrated at the wet curve as is usually done. If the instrument is kept overnight under a wet cloth, the wet curve calibration will be re-established. The response time strongly depends on temperature: it ranges from a few minutes at 20°C to 20 or 30 min at −10°C. Temperature limits are −60°C to +90°C.

TABLE 72.2 Equilibrium Relative Humidities (%) Over Some Saturated Salt Solutions

Salt	Temperature (°C)			
	10	20	30	40
	Relative Humidity (%)			
LiCl	11.3	11.3	11.3	11.2
$MgCl_2$	33.5	33.1	32.4	31.6
$Mg(NO_3)_2$	57.4	54.4	51.4	48.4
NaCl	75.7	75.5	75.1	74.7
KCl	86.8	85.1	83.6	82.3
K_2SO_4	98.2	97.6	97.0	96.4

Although the initial costs of a mechanical hygrometer are low, the long range costs of calibration and maintenance are considerable.

The Lithium Chloride Dewpoint Meter

Addition of a hygroscopic soluble salt to pure water decreases the equilibrium saturation vapor pressure above the solution (*Raoult's law*). In Figure 72.1, this change is illustrated for a saturated lithium chloride solution, which has an equilibrium relative humidity of about 11% (see Table 72.2). A gas sample at condition *a* is in equilibrium with a saturated LiCl solution of a (higher) temperature *b*. (It is also in equilibrium with pure water at a lower temperature *c*; note the similarity with a mirror dewpoint meter.) The application of this principle led to a simple and effective sensor. A fabric sleeve over a bobbin with a bifilar winding of inert electrodes is coated with a dilute solution of lithium chloride. As the bobbin is heated by an alternating current, its resistance increases sharply at the point where the surface begins to dry out, the heating stops, the sensor begins to cool, attracts water vapor, etc., until an equilibrium temperature has been reached. This is measured by, for example, a resistance thermometer sensor in the bobbin. The sensor is simple, rugged, relatively cheap and, after some calamity, it can be reactivated by recoating it. The dewpoint range goes from –40 to +90°C, with a claimed accuracy of about 0.5°C. This estimate is often too optimistic because the influence of ambient temperature cannot be neglected. Moreover, LiCl has hydrates with one, two, three, and five molecules of water per molecule of salt. In the dewpoint regions between 34°C and 41°C, the ambiguity leads to a possible error in that area of up to 1.5°C in dewpoint. Below –12°C, the error can rise to 3.5°C. Low flow rates can cause stratification around the sensing surface, high rates can cool it too much, resulting in too high and too low readings, respectively. Flow rates somewhere between 0.05 and 1 m s⁻¹ are generally recommended. Response times of commercially available sensors are on the order of minutes. Disadvantages include: the lower limit lies at 11% relative humidity, the sensor is washed out by accidental immersion, and the power supply should not be turned off accidentally. From an ionic standpoint, the sensor can be considered precontaminated and therefore, according to the manufacturer [5], relatively insensitive to contamination. Another source [9] reports a number of gases, like sulfur vapors, ammonia, high concentrations of carbon dioxide, chlorine, hydrogen sulfide, and condensable hydrocarbons, that could attack the lithium chloride sensor.

Electric Relative Humidity Sensors

Talking about numbers, the category of electric r.h. sensors is certainly by far the greatest part. The sensors are generally small, fast responding, do not dissipate heat, and can be used in confined spaces. Until 1975, the rights in this field were almost exclusively claimed by two resistance types: the Dunmore and the Pope sensor. The Dunmore sensor uses a dilute lithium chloride solution in a polyvinylacetate binder on an insulating substrate, with the danger of washout at saturation. The resistance of the sensor, measured between a bifilar grid, is a function of the r.h. of the surrounding air. This also applies to the Pope sensor, where a polystyrene substrate itself is the sensitive part of the sensor, after treatment with sulfuric acid. This sensor is less sensitive to washout, and has a wider range (15% to 99% r.h.) than the

Dunmore type. Modern bulk sensors, based on the measurement of the change in resistance, are manufactured by, for example, Shinyei Kaisha and Elmwood Sensors. Since 1970 a lot of work has been done in the development of capacitive sensors, where the stability, the influence of temperature, the temperature range, and the susceptibility for condensation conditions could be greatly improved. Since the 1980s, this sensor type has surpassed the mentioned resistance types. Basically, this type of sensor consists of a thin polymer layer between two electrodes of various materials. The types differ mainly in polymer type and electrode material, resulting in sensors with different characteristics to withstand pollution and temperatures up to 190°C. A drawback that still seems not completely under control is the fact that a polymer swells at high humidities, causing an undesirable shift in capacitance of the sensor.

Aluminium Oxide Hygrometers

Briefly, a sheet or wire of aluminium is anodized, producing a thin layer of water-sensitive pores. Subsequently, a conductive, water-permeable gold film is deposited over it. The radius of the pores is such that the sensor is specific for water molecules, and the amount absorbed is directly related to the dewpoint of the gas. The dewpoint range goes from −110 to 20°C, measured as a change in capacitance between the aluminium base and the gold electrode. The sensor works up to high pressures (30 MPa). The response time depends on the dewpoint, ranging from seconds at 10°C to minutes below −40°C. Stated accuracies are ±1°C at higher dewpoints to ±3°C at −100°C. The sensors show slow drift, and recalibration at least twice a year is generally recommended. Where chemical attack of the aluminium can be expected, silicon capacitive type sensors may be an alternative. A commercially available type of MCM (Moisture Controls & Measurement Ltd., U.K.) is temperature controlled, and has a very short response time: less than 15 s to a level of 1 ppm. Both aluminium and silicon sensors can withstand immersion in water.

Coulometric Method

In an electrolytic hygrometer, water in a sample stream is quantatively absorbed by a phosporous pentoxide layer. At the moment water molecules are present, the probe becomes electrically conducting. With a dc voltage over the sensor, water is electrolyzed and, according to Faraday's law, a current with a well-defined magnitude occurs (1 mA $\triangle$ 0.0935 µg H_2O/s). For the measurement of the mixing ratio, the flow through the cell must be measured accurately. Further calibration is, in principle, not necessary. At higher flow rates and moisture content (3000 ppm to 10,000 ppm by volume), the water vapor may not be absorbed quantitively, so this method is especially suited for low water contents, beginning at 1 ppm. The response time is about 1 min. Typical measurement uncertainties are 10% at 1 ppm, decreasing to 5% at higher values.

Crystal Oscillator

Even lower concentrations (down to 0.02 ppm by volume) can be detected with a hygroscopic coating on the surface of a quartz crystal. The resonant frequency of the crystal is a function of the mass of the coating — in other words, the moisture content of the gas. The crystal is alternately exposed to dry and humid gas, and the shift in frequency is measured. The response time is 1 min. Typical measurement uncertainty is 1 ppm or 5%. The instrument is relatively expensive (Du Pont).

Infrared Methods

Like any heteroatomic gas, water vapor absorbs radiation in the infrared region. So, if the gas is led through the optical path between an infrared source and a detector, there is a reduction of the transmitted radiation. The source can be dispersive, that is, generated by a monochromator, or nondispersive, wideband radiation, generated by a heated tungsten or nichrome wire. In the monochromatic mode, the transmittance ratio at two different wavelengths is measured. In the nondispersive (NDIR) method, usually another path through a reference gas is taken. The detector can be gas-filled, solid-state, or pyroelectric, with adequate optical or gas filters in the right wave bands. A popular detector was and still is the Luft type. A Luft detector has two radiation-absorbing chambers containing the specific gas of interest, separated by a thin membrane. Since the detector is a dynamic device, the radiation paths are chopped, allowing the measurement of the change in capacitance. This method allows measurement over a wide range, from ppm level up to saturation. It can be used in corrosive gases, the concentration of which can be measured simultaneously by the use of another detector. The instrument must be calibrated at regular intervals; typical measurement uncertainty is 1%. Response times to less than 1 s are possible, depending on the attainable refreshment time in the measuring chamber. Since the absorption depends on the number of atoms, transmittance depends on pressure. The method used to be relatively expensive, even compared with a mirror dewpoint meter, but developments in the field of inexpensive semiconductor detectors and better optical filters have reduced prices. Further developments are going on in the direction of fiber-optic probes.

Miscellaneous Methods in Air

- The zirconia cell, acting as a battery on the presence of oxygen ions, is generally used in a mixture of air and steam. The sensor can be mounted directly in a hot gas stream (600/1700°C). Acceptable accuracies are reached at dew points of 70°C or higher.
- The frequency of a signal, generated by an acoustic source, depends on the mixing ratio. Measurements in gases up to 250°C, to dewpoints of about 70°C (0 to 30 vol.%, in special cases to 100 vol.%, Mahlo).
- The Lyman alpha hygrometer uses the 121-nm emission line of hydrogen, which is strongly absorbed by water vapor. Extremely small response times (milliseconds) can be obtained. The instrument is relatively expensive.
- The difference in thermal conductivity of dry and wet air allows the measurement of absolute humidity with two sensitive temperature sensors in a Wheatstone bridge (Shibaura).

72.2 Liquids and Solids

Gravimetric

Drying of a material at a controlled temperature and taking the difference in weight before and after drying is the most fundamental method, greatly improved by microprocessor-based instruments. The assumption is that the loss in weight is caused by water only and that no other volatile components have been removed. Another problem can arise if water present by surface adsorption or crystal water is removed.

Karl Fischer Method

A chemical method for the determination of water in solids and organic solvents is the Karl Fischer method. The Karl Fischer reagent is composed of iodine, sulfur dioxide, pyridine, and methanol. Addition of this reagent to water causes a chemical reaction in which, in excess of the other components, one mole of iodine is used for each mole of water. The reagent is added in a controlled way to a mixture of reagent and sample, while a current between to electrodes is measured. If all the water has been used, a sudden change in the current through the mixture is observed.

Infrared Techniques

The infrared technique is applied to liquids and solids as well, in the wavelength bands of 1.45, 1.94, and 2.95 μm. For liquids, the transmission mode described is used, leading to particular problems because of the small optical path needed. Even smaller transmittance paths would be necessary for solids, one of the reasons to choose reflectance from the surface as a measure for water content. The surface has to be representative for the material in question. The system needs to be calibrated for each material. Concentrations as low as 0.02% to 100% can be measured. In case of specular reflection, the method cannot be used. A relatively new development is the attenuated total reflectance crystal (ATR) method. The crystal is inserted in the liquid, which may be opaque or semiopaque, or a slurry. A beam with two wavelengths, for reference and measurement entering the crystal is reflected on the internal surface, penetrating the solution one-half wavelength at each reflection point. The amplitude of the reflected signal decreases at each reflection. The number of reflections is determined by the length of the crystal and the angle of incidence, so the sensitivity can be changed. The measuring range is claimed 0% to 100% water in solutions; measuring water in emulsions requires some precautions. The response time is negligible.

Microwave Absorbance

Microwave absorbance is generally used in materials with a more or less constant composition, apart from the moisture content. Microwave radiation from a low-power, solid-state generator is absorbed by the sample and detected by a solid-state detector. The commonly used frequencies where water strongly absorbs are 1 to 2 GHz and 9 to 10 GHz, the latter being less dependent on the composition of the material. Operating ranges from 1% to 70% of water are mentioned, with achievable accuracies of ±0.5% of water. The attenuation is influenced by bulk density, bulk material, and temperature. The path between source and detector should not contain any metallic material.

NMR

Hydrogen atoms in the field of a permanent magnet are allowed, according to quantum mechanics, to have some defined orientations in that field. To shift an atom from one orientation to another requires a defined amount of energy, dependent on the strength of the magnetic field. If electromagnetic radiation at the right frequency is applied, resonance of the hydrogen atoms occurs, and a loss in frequency power can be detected. It is specific for all hydrogen atoms, so interference with liquids other than water in the sample can be expected. Temperature and flow must be controlled. Magnetic materials must be avoided. Measuring ranges of 0.05% to 100% have been reported [9].

Neutron Moderation

This method is, like NMR, specific to hydrogen atoms. Neutrons of high energy are slowed by nuclei of hydrogen atoms. The main components are a detector of slow neutrons, next to a source of fast neutrons. The measuring range goes from 0.5%. The sensor can be made very rugged. The measured volume is a sphere of up to tens of centimeters in diameter. The method is dependent on the bulk density of the material, but largely independent of the properties of the material being analyzed [9]. The method is not suitable for foodstuffs. It suffers from necessarily severe government regulations and low acceptance by the public and the operator.

TDR

Time domain reflectometry (TDR) measures the propagation velocity of electrical pulses, mainly between 1 MHz and 1 GHz. This method is well established, especially for the measurement of water content in soil. With certain limitations, it is proven suitable for automatic installations. The main disadvantage is the complexity of the data analysis [15].

FD

The general features of the FD (frequency domain) technique are comparable to those of TDR; however, there are some important differences. The dielectric properties of the material (soil, concrete, grain, oil, etc.) can be measured at a single frequency. The ability to choose a whole range of frequencies makes the frequency domain (FD) sensor suitable for spectroscopic measurements. A sensor of this type has recently been developed by IMAG-DLO, Wageningen (The Netherlands). It is simple and inexpensive, and available for many purposes. The sensor uses a single application-specific integrated circuit (ASIC) and is suitable for the measurement of other properties of materials as well [15].

Measurement of Thermal Conductivity

The thermal conductivity of a material is related to the amount of water it contains. Heat pulses are supplied by a needle probe, and the cooling of the needle after ending the pulse is measured. It is a simple and inexpensive method, that needs calibration for the material in which it is going to be used.

Water Activity or Equilibrium Relative Humidity

A material enclosed in a measuring chamber is, after some time, in equilibrium with its environment. The moisture content of the material can be derived from the so-called adsorption isotherms for that specific material, which must be determined experimentally. The method is used for many materials, such as foodstuffs, chemicals, grains, seeds, etc., with electric humidity sensors described earlier. In cases where the water potential of living material like potatoes and leaves must be measured, thus at very high relative humidities, the junction of a small thermocouple is cooled by an electric current and, after the current has been turned off, used as an unventilated wet bulb psychrometer. It is an excellent method that requires skill.

72.3 Formulae

A relatively simple equation for the calculation of the saturation vapor pressure $e_w(t)$ in the pure phase with respect to water is the *Magnus formula*:

$$\ln e_w\left(t\right) = \ln 611.2 + 17.62t/\left(243.12 + t\right) \tag{72.3}$$

where t is the temperature in °C (on ITS-90) and $e_w(t)$ is in pascals. Equation (72.3) covers the range between −45°C and +60°C, with a maximum standard deviation of 0.3%. Over ice Equation (72.3) changes to:

$$\ln e_i\left(t\right) = \ln 611.2 + 22.46t/\left(272.62 + t\right) \tag{72.4}$$

covering the range between −65°C and 0°C with a standard deviation of less than 0.5%.

Equation 72.3 can easily be converted for the calculation of the dewpoint t_d:

$$t_d = \left[243.12 \ln\left(e/6.112\right)\right]/\left[17.62 - \ln\left(e/6.112\right)\right] \tag{72.5}$$

where e is the saturation vapor pressure e_w at dewpoint temperature t_d. The standard deviation in the range mentioned is less than 0.02K. The *frost point* t_f can be calculated from Equation 72.6:

$$t_f = \left[272.62 \ln\left(e/6.112\right)\right]/\left[22.46 - \ln\left(e/6.112\right)\right] \tag{72.6}$$

where e is the saturation vapor pressure e_i at frostpoint temperature t_f, in the range mentioned within a standard deviation of 4 mK.

More accurate formulae, given by Hyland and Wexler, updated by Sonntag can be found in Reference 7. The uncertainties mentioned above, given by Sonntag [7], are valid for the pure water system. In the presence of air, the uncertainties can only be maintained if the water vapor enhancement factor is taken into account.

72.4 Calibration

A few words about calibration, a most important and too often forgotten subject in humidity and moisture measurement. The gravimetric method has already been described, as well as some principles of humidity generators, which could be used as standards. Another method is the humidity chamber, with both temperature and humidity control, where an instrument is compared with a calibrated standard instrument like a mirror dewpoint meter or a psychrometer. Saturated (or diluted) salt solutions can be used to create a constant relative humidity in a confined space. Some values are listed in Table 72.2 (from Reference 16). The user should have knowledge of some critical factors like temperature, etc. for the application of this method. Permeation tubes are used where a repeatable, accurate, and low concentration flow is required. Unlike with temperature or pressure sensors, it is, except at the highest national level, not realistic to expect a reference or standard for humidity with an accuracy a factor 5 or 10 times better than the required accuracy for a measuring instrument. A good calibration guarantees traceability to a higher standard. This means, in general, that the calibration should be performed by a national or accredited laboratory. A rule of thumb for the intervals of calibration of humidity meters is as follows: condensation dewpoint meters and psychrometers require calibration once every 1 to 2 years; electric relative humidity sensors need calibration every 6 to 12 months; and less stable types like aluminium oxide sensors must be calibrated every 6 months, or sooner if desired [8]. A comprehensive treatment of calibration of hygrometers and attainable accuracies has been given by Wiederhold [10].

72.5 Developments

- Fiber optics are beginning to find their way in hygrometry, resulting in explosion-proof models. Changes in refractive index by the use of micropores are used in an instrument recently developed by Ultrakust (Germany). Fiber lengths go to 1000 m. The measuring range is 0% to 20% r.h. in a dewpoint range of −90°C to 20°C.
- A rapidly developing field is that of the monolithic integrated circuit sensors. Hycal (U.S.) manufactured a capacitive relative humidity sensor, while a recent commercial development in Germany reports the use of a monolithic integrated sensor instead of a mirror in a miniature dewpoint meter, with an accuracy of better than 0.5°C. Since the use of porous silicon allows the development of smart sensors, developments in that area will undoubtedly continue [17].
- The progress in the theoretical background of many water-related processes, made possible by an advanced FD sensor mentioned earlier, is of great importance for future development of simple, cheap, stable, and reliable sensors that can be used for many purposes.

References

1. R. W. Hyland and A. Wexler, Formulations for the thermodynamic properties of the saturated phases of H_2O from 173.15K to 473.15K. *ASHRAE Trans.* 89(2A), 500-519, 1983a.
2. A. Wexler, Vapor pressure formulation for ice. *J. Res. National Bureau of Standards*, 81A, 5-20, 1977.
3. *1993 ASHRAE Handbook Fundamentals SI Edition*, American Society of Heating, Refrigerating and Air Conditioning Engineers, Inc., Atlanta, GA, 1993.
4. L. Greenspan, Functional equations for the enhancement factors for CO_2-free moist air. *J. National Bureau of Standards*, 80A, 41-44, 1976.

5. Anonymous, HANDBOOK, Selecting humidity sensors for industrial processes. General Eastern Instr. Corp., 1982.

6. Moisture and humidity measurement. *Proc. 1985 Int. Symp. Moisture and Humidity*, Washington, D.C., April 15-18, 1985.

7. D. Sonntag, Advancements in the field of hygrometry. (Review Article) *Meteorologische Zeitschrift*, N.F., 3, 51-66, 1994.

8. *A Guide to the Measurement of Humidity*, The Institute of Measurement and Control, London, 1996.

9. K. Carr-Brion, *Moisture Sensors in Process Control*, Elsevier Applied Science Publishers, London and New York, 1986.

10. P. R. Wiederhold, *Water Vapor Measurement*, Marcel Dekker, New York, 1997. (Software included).

11. M. Stevens and S. A. Bell, The NPL standard humidity generator: an analysis of uncertainty by validation of individual component performance. *Meas. Sci. Technol.*, 3, 943-952, 1992.

12. S. Hasegawa, National basis of accuracy in humidity measurements. *Proc. 1985 Int. Symp. Moisture and Humidity*, Washington, D.C., 15-19, April 15-18, 1985.

13. R. G. Wylie and T. Lalas, Detailed determination of the psychrometer coefficient for the wet cylinder in a transverse airstream and an analysis of its accuracy. *Technical Paper No. 7* (CSIRO Division of Applied Physics, Sidney, Australia), 1981.

14. G. J. W. Visscher, Standard psychrometers: a matter of (p)references. *Meas. Sci. Technol.*, 5, 1451-1461, 1995.

15. M. A. Hilhorst and C. Dirksen, Dielectric water content sensors: time domain versus frequency domain. *Proc. Symp. TDR Environmental, Infrastructure and Mining Applications*, 143-153, March 1994.

16. L. Greenspan, Humidity fixed points of binary saturated aqueous solutions. *J. National Bureau of Standards*, 81A(1), 89-96, 1977.

17. G. M. O'Halloran, M. Kuhl, P. J. Trimp, and P. J. French, A humidity sensitive capacitor based on a porous silicon dielectric. *Proc. 1996 National Sensor Conf.*, Delft, The Netherlands, March 20-21, 1996.

73

Environmental Measurement

John D. Garrison
San Diego State University

Stephen B. W. Roeder
San Diego State University

Michael Bennett
Willison Associates

Kathleen M. Leonard
The University of Alabama in Huntsville

Jacqueline Le Moigne
NASA/GSFC - Code 935

Robert F. Cromp
NASA/GSFC - Code 930

73.1   Meteorological Measurement

John D. Garrison and Stephen B. W. Roeder

Meteorological measurements are measurements of the physical properties of the atmosphere. These measurements are made at all elevations in the troposphere and the stratosphere. For measurements made at elevations above ground or tower level, instruments can be carried aloft by balloons, rockets, or airplanes. Ground radar is used to detect the presence of water in the form of droplets or ice crystals at all elevations and the winds associated with them. Lidar (optical radar) of selected wavelengths is used to detect the presence and amount of aerosols and other constituents of the atmosphere and to determine cloud height. Instruments on satellites measure properties of the atmosphere at all elevations.

Quantities measured are: temperature, pressure, humidity, wind speed, wind direction, visibility, the presence and amount of precipitation, cloud amount, cloud opacity, cloud type, cloud height, broadband solar (or shortwave) radiation, longwave radiation, ultraviolet radiation, and net radiation, sunshine duration, turbidity, and the amounts of trace gases such as NO, NO_2, SO_2, and O_3. Some of the methods and instruments used to measure a number of these variables are discussed in this handbook in sections on pressure, temperature, humidity and moisture content, and air pollution monitoring. Additional information of interest is in sections on resistive sensors, inductive sensors, capacitive sensors, satellite navigation and radiolocation, and the sections under signal processing.

Meteorological measurements are made at individual sites, at several or many sites forming a local network, or at much larger networks. Much of the emphasis now is on global networks covering the entire northern and southern hemispheres. Individuals and groups can make measurements for their own purposes or they can use data provided by the various weather services. Weather service data are

0-8493-8347-1/99/$0.00+$.50
© 1999 by CRC Press LLC

stored in archives that can cover many years of measurements. The U.S. National Climatic Data Center, Asheville, NC, has archived data produced from measurements at U.S. weather stations and other National Oceanic and Atmospheric Administration (NOAA) measurement sources, including satellites. It also has non-U.S. data. These data can be purchased. The Web site where information concerning NOAA data can be obtained is: <http://www.ncdc.noaa.gov/>. The U.S. National Renewable Energy Laboratory has information on solar radiation and wind at Web site: <http://www.rredc.nrel.gov />. Data at Web sites can often be retrieved by anonymous File Transfer Protocol (FTP).

Fabrication of meteorological instruments is usually done by companies specializing in these instruments. It is generally not economically feasible for individuals to fabricate their own instruments, unless their particular application cannot use commercially available instruments. The problem usually reduces to the determination of which commercial instruments to purchase. This determination depends on cost, durability, accuracy, maintenance requirements, ease of use and the form of the output signal. A number of instrument manufacturers and distributors market complete weather stations.

The instruments used for meteorological measurements are fabricated for use in the special environment of the atmosphere. This environment varies with the latitude and longitude of the site and the elevation above the ground and above sea level. A common requirement is that the instruments be protected from adverse conditions that can cause errors in the measurement of the meteorological variables. For proper operation, some instruments (e.g., spectroradiometers) need to be in a temperature-controlled environment. The sun's heating can cause errors. Solar radiation can cause weathering of the instruments and shorten their useful life. Moisture from precipitation or dew can affect measurements adversely and also cause weathering and corrosion of the instruments. Blowing dust or sand can cause weathering of the instruments and affect the operation of mechanical parts. Insects, birds, and ice can also affect instruments adversely. Some gaseous constituents of the atmosphere can be corrosive. Packaging of the sensors and housing of the instruments in enclosures can protect them, but this protection must not interfere with the measurement of the meteorological variable. Packaging a sensor and putting the sensor in protective housing generally increases the response time of the sensor to changes in the meteorological variable it is measuring. Solar heating can be reduced by covering the sensor, its protective packaging, or housing with a white coating. Generally, housing or enclosures used to protect the instruments should be well ventilated. Sometimes a fan is used to draw air through the housing or enclosure to reduce solar heating and make the air more representative of the outside air. Loss of measurement caused by loss of electric power can be avoided by having backup power from batteries or motor-generators.

Another common requirement of meteorological measuring instruments is that they be calibrated before installation in the field. This is usually done by the manufacturer. For some applications, it may be important that the calibrations are traceable to NIST (National Institute of Standards and Technology) standards. The calibration should be checked routinely. Some instruments are constructed to be self-calibrating.

Generally, there is no reason to measure many of the meteorological variables to high precision. The temperature, humidity, and wind, for example, can vary in relatively short distances and sometimes in relatively short times by amounts that are large compared to the accuracy of the measuring instruments. This is especially true near ground level. The exact value of a meteorological variable at a particular site has little meaning unless the density of measuring sites is very high. A mesoscale network with high density of measuring sites over a relatively small area might be of interest for ecological studies. Larger-scale or global networks have widely separated sites. For these networks it is important to install all of the instruments in a standard manner so as to reduce the effect of local fluctuations on the variation of the meteorological variables from site to site. Usually, this is done by placing them on a level, open area away from surrounding buildings and other obstructions and at a fixed distance above the ground. The ground should be drained ground and not easily heated to high temperature by the sun. In comparing temperatures, one must be aware of the "heat island effect" of large cities; furthermore, an increase in the degree of urbanization of a given site over time may affect the interpretation of the temperature trends observed.

The current trend is toward automatic measurement of meteorological variables at unattended sites with automatic data retrieval and computer processing and analysis of the data. Large-scale networks consisting of many stations covering a large area (e.g., the Northern Hemisphere) are used for regional, national, and global weather forecasting.

Measurement of the Atmospheric Variables

Temperature

The mean temperature of the atmosphere for each hour of the day at a particular site has a fairly regular annual and a diurnal variation when this mean temperature is an average over many years for each hour and day of the year. The temperature at a given site and time is a superposition of the mean temperature and the fluctuations from this mean temperature caused by current cloud and wind conditions, the past history of the air mass passing over the site, and interannual variations that are not yet well understood.

More detailed sources of information on temperature measurement include the earlier section in this handbook on temperature measurement and References 1 through 5. The common methods of measuring atmospheric temperature include the following:

Electric Resistance Thermometer (RTD).
The variation of the resistance of a metal with temperature is used to cause a variation in the current passing through the resistance or the voltage across it. The electric circuit used for the measurement of temperature can utilize a constant current source, and temperature is determined from the voltage across the resistance after the circuit is calibrated. Alternatively, a constant voltage source can be used with the current through the resistance determining the temperature. Once the instrument is calibrated, it is generally expected to keep its calibration as long as electric power is supplied to the instrument. Commonly, platinum RTD thermometers are made of a fixed length of fine platinum wire or a thin platinum film on an insulating substrate. The variation of the resistance as a function of temperature is approximately linear over the range of temperature found in meteorological measurements. The quadratic correction term is quite small. The accuracy and reproducibility of the measurements and the ease of using an electric signal for transmission of data from remote unmanned sites makes electric resistance thermometers desirable for meteorological applications. Platinum is the best metal to use. With careful calibration and good circuit design, platinum resistance thermometers can measure temperature to a small fraction of a degree, much better than the accuracy needed for meteorological measurements.

Thermistors.
Thermistors usually consist of an inexpensive mixture of oxides of the transition metals. The log of their resistance varies inversely with temperature. The change in resistance with temperature can be 10^3 to 10^6 times that of a platinum resistance thermometer. Their change in resistance with temperature is used to determine temperature in the same manner as metal resistance thermometers. They are somewhat lower in cost and are somewhat less stable than platinum resistance thermometers.

Bimetallic Strip.
Bimetallic strips are discussed elsewhere in an earlier section. They are usually used for casual monitoring of inside and outside temperatures at dwellings and office buildings and for heating and cooling controls. The accuracy is generally about ±1°C. They are low in cost.

Liquid in Glass Thermometer.
These are a well-known method of measuring temperature. They are usually used for casual monitoring of inside and outside temperatures at dwellings or office buildings. These thermometers are more difficult to read than meter or dial readings of temperature, do not lend themselves to electric transmission of their readings, and are easily broken. Their cost can be low.

Pressure

One standard atmosphere of pressure corresponds to 1.01325×10^5 pascals (N m^{-2}) (14.6960 pounds per square inch, 1.01325 bars, 1013.25 mbars, 760.00 mm Hg, or 29.920 in. Hg). This is approximately the mean atmospheric pressure at sea level. Atmospheric pressure at sea level usually does not deviate more than ±5% from one standard atmosphere. Atmospheric pressure decreases with altitude. Altitude measurements in airplanes are based on air pressure measuring instruments called altimeters. At about 5500 m (18,000 ft), the atmospheric pressure is half its sea level value. The following instruments are used to measure atmospheric pressure.

Mercury Manometer

Originally, barometric pressure was measured with a mercury manometer. This is a tube, about 1 m in length, filled with mercury and inverted into an open dish of mercury. The height of the column of mercury that the external pressure maintains in the tube is a measure of the external air pressure. Hence, one standard atmosphere is 760 mm Hg. While accurate, this device is awkward and has been replaced for general use.

Aneroid Barometer.

It consists of a partially evacuated chamber that can expand or contract in response to changing external pressure. The evacuated chamber is often a series of bellows, so that the expansion and contraction occurs in one dimension. Basic aneroid barometers, which are still in use, have a mechanical linkage to a pointer giving a reading on a dial calibrated to read air pressure. High-quality mechanical barometers can achieve an accuracy of 0.1% of full scale. Aneroid barometers can also give electronic readout and eliminate the mechanical linkage; this is more the standard for serious meteorological measurements. In one method, a magnet attached to the free end of the bellows is in proximity to a Hall effect probe. The Hall probe output is proportional to the distance between the magnet and the Hall probe.

Barometric pressure is also measured with an aneroid type of device that consists of a rigid cylindrical chamber with a flexible diaphragm at its end. A capacitor is created by mounting one fixed plate close to the diaphragm and a second plate mounted on the diaphragm. As the diaphragm expands or contracts, the capacitance changes. Calibration determines the pressure associated with each value of capacitance. A range of 800 to 1060 millibars with an accuracy of ±0.3 millibars for ground-based measurements is typical. Setra Corporation produces this type of instrument for the U.S. National Weather Service ASOS network, the latter produced by AAI Systems Management Incorporated. The ASOS network is discussed below. Measurement of pressure is also discussed elsewhere in this handbook.

Humidity

Instruments that determine the density or pressure of water in vapor form in the atmosphere, generally either measure relative humidity or they measure dewpoint temperature. The pressure of water vapor just above a liquid water surface when the vapor is in equilibrium with the liquid water is the saturated vapor pressure of the water. This saturated vapor pressure increases with the temperature and equals atmospheric pressure at the boiling temperature of water. Relative humidity is the ratio of the vapor pressure in air to the saturated vapor pressure at the temperature of the air. Relative humidity is usually expressed in percent, which is this ratio times 100. The dewpoint temperature is the temperature to which the air must be lowered, so the vapor pressure in the air is the saturated vapor pressure with the relative humidity at 100%. Knowledge of water vapor density is used in weather prediction and in global climate modeling. It also affects light transmission through the atmosphere. Relative humidity is an important meterological variable. The temperature–dewpoint difference is an indicator of the likelihood of fog formation and can be used to estimate the height of clouds. More detailed sources of information on humidity measurement include the section on humidity in this handbook and References 6 to 9. Three common methods of measuring the vapor density in the atmosphere are given below.

The Chilled Mirror Method.

Chilled mirror instruments for measuring the dewpoint temperature are not sold by most instrument companies. A chilled mirror instrument developed by Technical Services Laboratory is used in the U.S. National Weather Service ASOS network discussed below. It has a mirror cooled by a solid-state thermoelectric cooler (using the Peltier effect) until water vapor in the air just starts condensing on the mirror. This condensation is detected using a laser beam reflecting from the mirror. When the reflected beam is first affected by the condensed water vapor, the temperature of the mirror is the dewpoint temperature. The mirror temperature is controlled to remain at the dewpoint temperature by an optic bridge feedback loop. The mirror is a nickel chromium surface plated on a copper block. The temperature of the block is measured to ±0.02% tolerance by a platinum resistance thermometer imbedded in the block. An identical platinum resistance thermometer measures ambient air temperature. Outside air is drawn through the protective enclosure surrounding the instrument by a fan, so that the effect of solar heating on the measured values of the dewpoint temperature and ambient temperature is negligible and so that outside air is tested. The dewpoint temperature and ambient temperature are measured between −60 and +60°C to an accuracy of 0.5°C rms. Dewpoint errors are somewhat larger below 0°C. To avoid errors that might arise from deterioration of the reflective properties of the mirror, the mirror should be inspected periodically, particularly in dirty or salty environments. This method is of higher cost than other methods of measuring the amount of water vapor in the atmosphere.

Thin Film Polymer Capacitance Method.
The capacitance is formed with a thin polymer film as dielectric placed between two vapor-permeable electrodes. Water vapor from the air diffuses into the polymer, changing the dielectric constant of the dielectric and thus the capacitance. The capacitance can be measured electrically by comparison to fixed capacitance reference standards. The measured value of the capacitance is related to the relative humidity by calibration. Instruments using these capacitive sensors can measure relative humidity between 0 and 100% at temperatures between about −40 and +60°C to about ±2% of relative humidity. These sensors can be made very small for incorporation into integrated circuits on silicon chips [9]. They are low in cost. Usually, instruments measuring humidity also measure temperature separately. The circuits used to measure relative humidity using a thin polymer capacitance yield an electric output signal (often 0 V to 5 V), which lends itself to remote transmission of the relative humidity.

Psychrometric Method.
This method is discussed in the earlier section of this handbook on humidity. Errors are introduced if the water is contaminated, if the water level in the reservoir supplying water to the wick becomes low, or the reservoir runs dry. In extremely dry environments, it can be difficult to keep the wick wet, while salty environments can change the wet bulb reading. Accuracy is affected by air speed past the wet bulb. Because of these disadvantages, psychrometers have generally been replaced by more convenient methods of measuring humidity.

Wind Speed, Wind Direction, and Wind Shear

Anemometer.
Weather stations commonly employ a 3-cup anemometer. This consists of a vertical axis rotating collar with three vanes in the form of cups. The rotation speed is directly proportional to wind speed. Figure 73.1 shows an instrument of this type. An alternative to the cup anemometer is a propeller anemometer in which the wind causes a propeller to rotate. There are several ways to obtain an electrical signal indicating the speed: a magnet attached to the rotating shaft can induce a sinusoidal electrical impulse in a pickup coil; a Hall effect sensor can be used; or the rotating shaft can interrupt a light beam, generating an electric pulse in a photodetector. Rotating anemometers can measure wind velocities from close to 0 up to 70 m s^{-1} (150 mph).

Ultrasonic Wind Sensor.
This sensor has no moving parts. Wind speed determination is as follows. An ultrasonic pulse emitted by a transducer is received by a nearby detector and the transit time calculated. Next, the transit time is measured for the return path. In the absence of wind, the transit times are equal; but in the presence of wind, the wind component along the direction between the transmitter and receiver affects the transit time. Three such pairs, mounted 120° apart, enable calculation of both the wind speed and direction. Heaters in the transducer heads minimize problems with ice and snow buildup. The absence of moving parts eliminates the need for periodic maintenance.

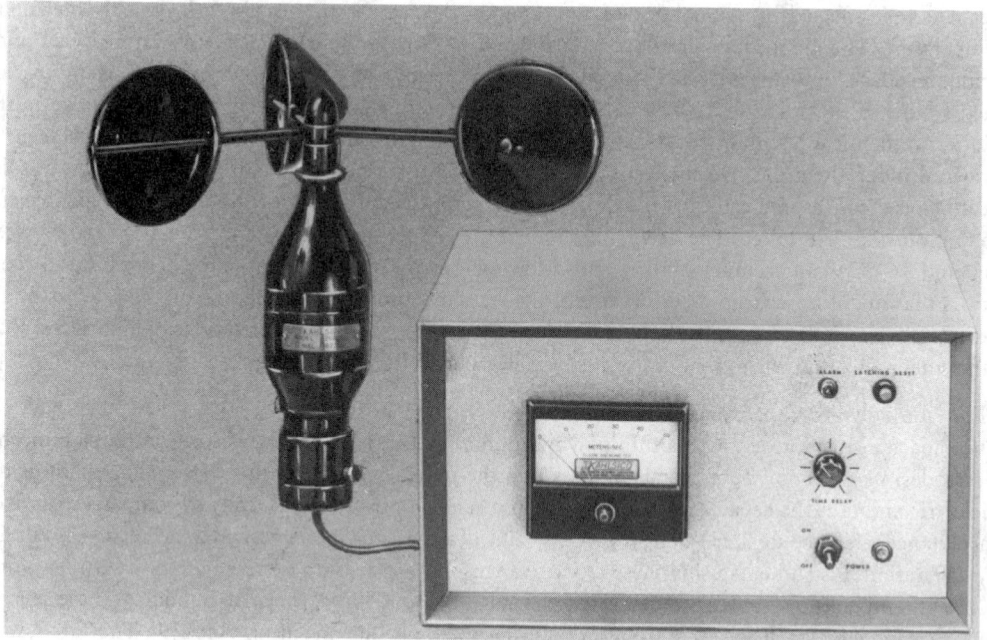

FIGURE 73.1   A cup type anemometer for measuring wind speed. (Courtesy of Kahl Scientific Instrument Corp.)

Wind Direction.

Wind direction sensors are generally some variant of the familiar weather vane. Sensitivity is maintained by constructing the weather vane to rotate on bearings with minimal resistance. Electronic readout can be achieved using a potentiometer (a "wiper" contact connected to the vane slides over a wire-wound resistor). The resistance between the contact and one end of the wire resistor indicates the position of the vane. Alternative methods of readout include optical and magnetic position sensors. Positional accuracy is ±5%.

Combination Wind Speed and Direction Sensor.

A combination wind speed and direction sensor can be made in which a propeller anemometer is mounted on a weather vane. The vane keeps the propeller device pointed into the wind. Alternatively, two propeller anemometers, rigidly mounted in a mutually perpendicular arrangement can be used to determine direction and magnitude of the horizontal wind simultaneously. Rotating anemometers and weather vanes are susceptible to ice and snow buildup and can be purchased with heaters. They need periodic maintenance.

Wind Shear.

Wind shear occurs when wind direction and/or strength change significantly over a short distance. This can occur in horizontal or vertical directions or sometimes in both. Measurement of wind shear conditions is particularly important at airports. Wind shear is determined by comparing readings made at the center of the airfield with measurements made at the periphery. An automated system to perform this function, entitled Low Level Wind Shear Alert System (LLWAS), is found at some airports. Wind shear can also be detected by doppler radar. Doppler radar used by the U.S. National Weather Service is discussed below.

Precipitation

Precipitation measuring instrumentation includes devices that measure the presence of precipitation (precipitation sensors), those that determine the quantity of precipitation, those that measure rate of precipitation, and those that measure both quantity and rate.

Precipitation Presence Sensors.

These sensors usually consist of two electric contacts in close proximity. Moisture causes electric conduction that is detected by a circuit monitoring conductance. A typical application consists of a circuit board consisting of a grid of two arrays of strips separated by small gaps. If the surface of the detector is heated, then only current precipitation will be detected and dew will not form to affect the measurement.

Rain Gages.

These instruments measure amount of rainfall. A simple rain gage can consist of a cylinder, a funnel, and an inner collection tube of much smaller diameter than the funnel for amplification of the height of rain accumulation. The height of the water column in the inner tube is converted to total rainfall. Typical graduations on the tube enable determining rain accumulation to 0.025 cm (0.01 in.) of rain.

A tipping-bucket rain gage enables the measurement of both volume and rate of rainfall. A large funnel concentrates the precipitation, which is directed into one of two small buckets. When that bucket fills, it tips out of the way and empties, closing a switch to record the event, and another empty bucket moves into its place. Typical tipping-bucket gages respond to each 0.025 cm (0.01 in.) of rain. In conditions of snow and freezing rain, tipping-bucket rain gages can be equipped with heaters on the funnel to reduce snow and ice to water. The internal components are also heated to prevent refreezing. Reported accuracy for tipping-bucket rain gages is ±0.5% at 1.2 cm h^{-1} (0.5 in h^{-1}). Frise Engineering Company produces a tipping-bucket rain gage used in the ASOS network discussed below.

Highest accuracy rain gages that collect and concentrate precipitation should have their collection surfaces made of a plastic with a low surface tension for water. This minimizes losses from surface wetting.

Rain gages exist that do not rely on collection methods. Optical rain gages utilize an infrared beam. Drops falling through this beam induce irregularities in the beam that can be interpreted in terms of precipitation rate. This type of sensor is used for the precipitation identification or present weather sensor used in the ASOS network discussed below.

Solar Radiation

The mean annual intensity of solar radiation above the atmosphere (extraterrestrial solar radiation) continues to be measured to obtain a more precise value and to look for variations in the sun's energy output. It is called the solar constant. The solar constant is close to 1367 W m^{-2}. The intensity of solar radiation above the atmosphere varies approximately sinusoidally over the year with an amplitude of close to 3.3% of the solar constant and a maximum near the first of January. This variation arises from the variation of the distance of the Earth from the sun. The sun has a spectral distribution that is roughly that of a blackbody at 5777 K with a peak of the spectrum at a wavelength of about 500 nm. Solar radiation is attenuated by scattering and absorption in the atmosphere. Attenuation is greater at wavelengths corresponding to absorption bands of certain gases in the atmosphere. On a clear day near noon, the solar intensity at the Earth's surface can be as high a 1000 W m^{-2}. Figure 73.2 shows a typical sea level solar spectrum with the sun about 48° away from the vertical (see, for example, References 10 and 11). This is called an air-mass 1.5 spectrum, because the distance the radiation travels through the atmosphere is 1.5 times the distance when the sun is vertical. About 99% of the spectrum is in the range of wavelengths shown by the figure. Measurements of the solar constant and the solar spectrum are scientific measurements made with specialized instruments. Additional information can be found in Iqbal [11] and Coulson [12].

Solar radiation instruments for general use in the field measure direct radiation from the sun, total or global radiation coming from the sky hemisphere, and diffuse or sky radiation (global radiation with the direct radiation removed). Solar radiation measuring instruments can be broadband instruments that measure the combined solar intensity (irradiance) at all wavelengths, or they can be spectrally selective instruments that measure the intensity at different wavelengths or in different wavelength bands. Only the much more common broadband instruments are discussed here. The instruments used for everyday measurement are field instruments. Field instruments are first class if they are of higher quality and provide greater accuracy and reliability (at higher cost).

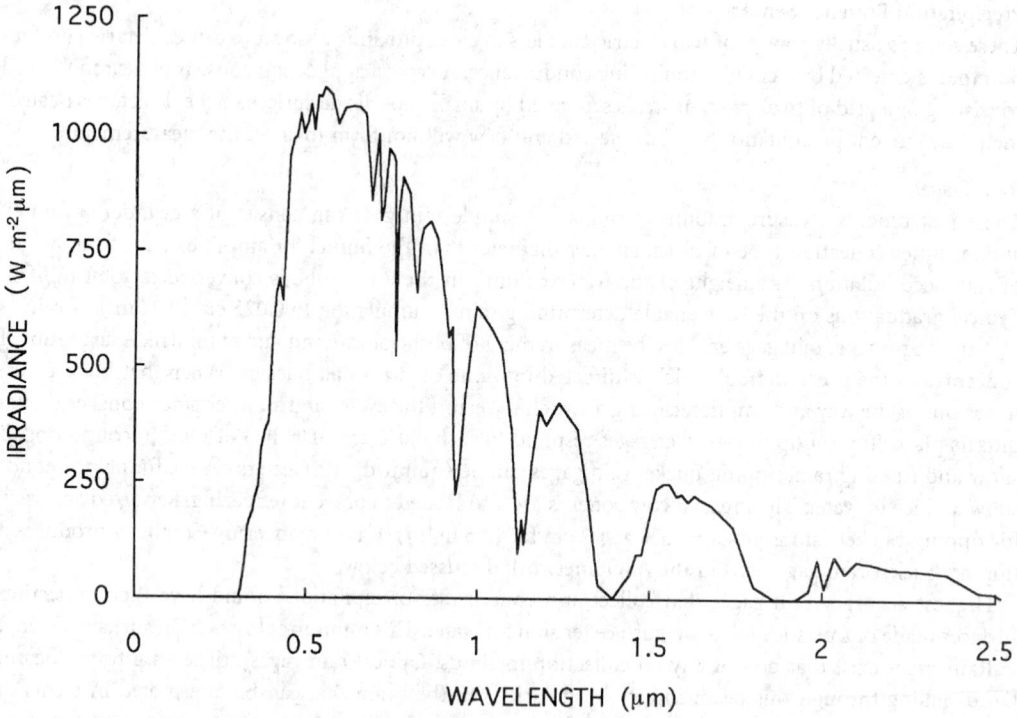

FIGURE 73.2 The intensity of solar radiation as a function of wavelength for a pathlength through the atmosphere of 1.5 times the vertical path length. The dips in the spectrum are molecular absorption bands.

Direct radiation is radiation coming directly from the sun without scattering in the atmosphere. Instruments measuring direct solar radiation usually include radiation coming from the sky out to an angular distance of about 3° away from the center of the sun's disk. They are called pyrheliometers. The radiation coming from clear sky near the sun, rather than from the solar disk, is the circumsolar radiation. This radiation can be subtracted from the pyrheliometer measurement for a more precise determination of direct radiation. This correction is often not made for routine measurements. The clear sky correction is calculated using the angular distribution of the intensity of the circumsolar radiation [13,14].

Instruments measuring global radiation are installed in a level position with a plane sensor facing up toward the sky. These instruments measure solar radiation coming from the whole sky hemisphere. They are called pyranometers. Global radiation measuring instruments should be sited in a level elevated area with no obstructions obscuring the sky hemisphere.

Diffuse radiation is the radiation coming from the sky hemisphere with the direct radiation subtracted. Pyranometers are used for measuring diffuse solar radiation and should be mounted in the same manner as pyranometers used for measurement of global radiation. They have an occulting (shade) disk or shadow band to prevent direct solar radiation from reaching the radiation sensor. The measurement of diffuse radiation involves correcting the pyranometer measurement for the part of the sky radiation shielded from the sensor by the occulting disk or shadow band. For clear skies, the occulting disk correction is calculated using the angular distribution of intensity of the circumsolar radiation. Corrections for partially cloudy and cloudy skies depend on the particular cloud conditions. Corrections for the shadow band are often determined by temporarily replacing the shadow band with an occulting disk when the sky is clear and when it is overcast. Corrections for measurements under other sky conditions can be determined by interpolation. The shadow band correction is discussed by LeBaron et al. [15]. The occulting disk must have a tracking system to make the disk follow the sun over the sky. The shadow band removes solar radiation received from a narrow swath of the sky along the path the sun follows during the day. The shadow band must be adjusted regularly during the year as the path of the sun changes over the seasons. The more common solar radiation measuring instruments include the pyranometer and pyrheliometer.

The Pyranometer.

The sensor is usually a thermopile, consisting of a number of thermocouples in series, with alternate junctions heated by the sun. The unheated junctions are near ambient temperature. This is sometimes arranged by putting the unheated junctions in thermal contact with a white surface. Heating by the sun is accomplished by placing the junctions in contact with a matte black surface of high heat conductivity or by a black coating on the junctions. The blackened surface has a constant high solar absorptance (usually ~99%) over the solar spectrum. A constant high solar absorptance over the solar spectrum is important. The solar spectrum at the surface of the Earth varies with the time of day and year and the amount of clouds, because of the spectrally dependent scattering and absorption of solar radiation by the atmosphere. An absorbing surface whose absorption of solar radiation varies with wavelength will cause the sensor to have a different sensitivity for different wavelengths of the solar spectrum. Some less expensive pyranometers use a silicon photovoltaic sensor (solar cell) to measure solar radiation. These sensors have zero sensitivity above about 1.2 μm and the spectral response below 1.2 μm is not constant. This limits the accuracy of measurements of solar intensity with photovoltaic sensors. Instruments with a thermopile sensor can use a combination of thermopiles and resistors to compensate for the variation of the output of a single thermopile with temperature. The hemispherical windows of pyranometers are usually made of a special glass which transmits solar radiation of wavelengths between about 0.3 and 2.8 μm. This includes ~99% of the solar intensity. The absorbing surface must have a cosine response as a function of angle away from the normal to the surface (Lambert law response), and a flat response as a function of azimuth around the normal to the absorbing surface, for the global radiation to be measured correctly. The degree to which the pyranometer response is linear follows the cosine law, and is temperature, spectrum, and azimuthally independent, determines whether the instrument is a first class instrument. Figure 73.3 shows a Kipp and Zonen (Netherlands) first-class pyranometer.

FIGURE 73.3   A class 1 pyranometer used for the measurement of global and diffuse solar radiation. (Courtesy of Kipp & Zonen Division of Enraf-Nonius Co.)

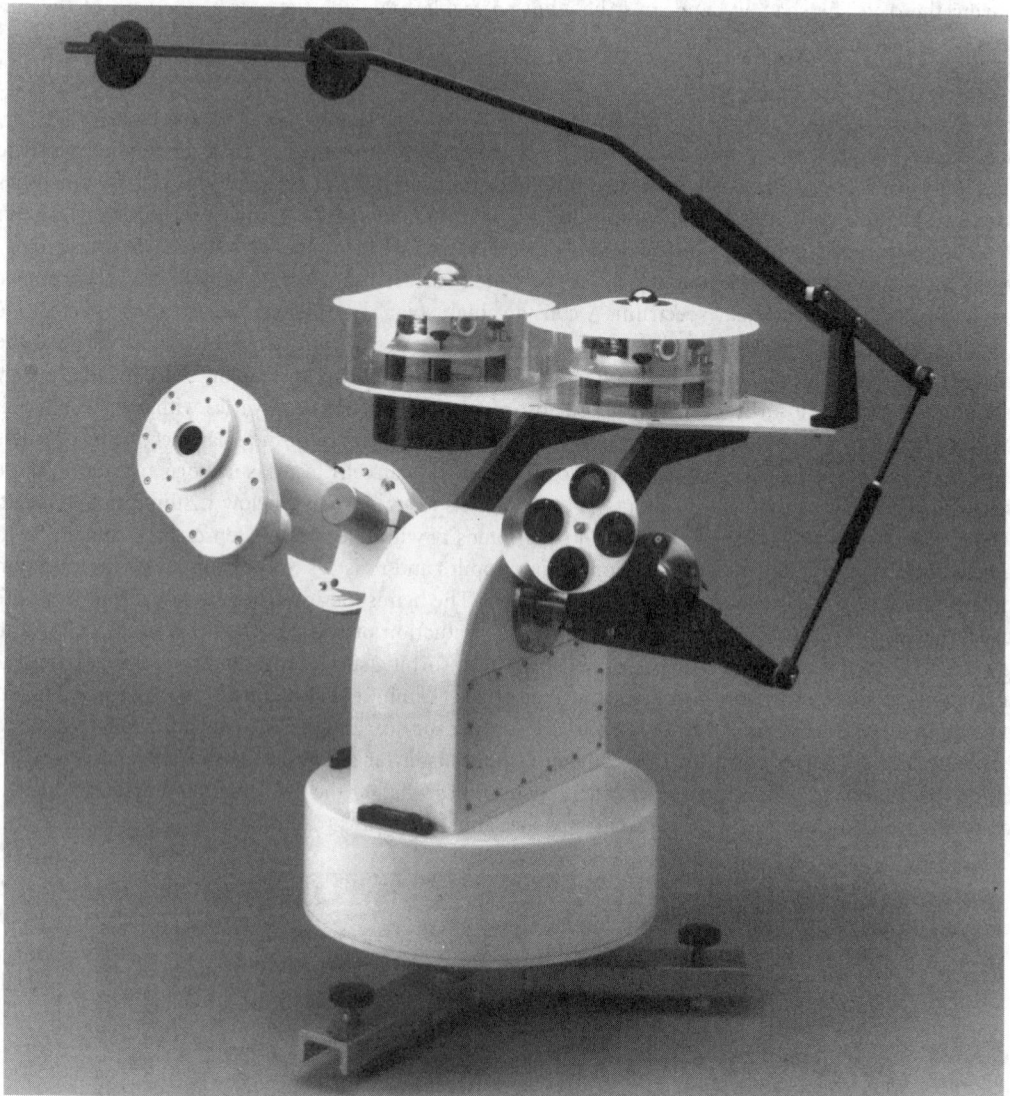

FIGURE 73.4   An automatic solar tracker with normal incidence pyrheliometer and cavity radiometer mounted on it. The tracker is coupled to two shade disks that shield direct sunlight from a first class pyranometer and precision infrared radiometer facing the sky hemisphere. (Courtesy of Eppley Laboratories.)

The Pyrheliometer.

The field pyrheliometers usually have temperature-compensated thermopile sensors with flat spectral response and linear output. The sensor is placed at the bottom of an internally blackened, diaphragmed, collimator tube that limits the angular acceptance for solar radiation to be in the range of about 5° or 6° total acceptance angle. The pyrheliometer is mounted on an equatorial mount tracker that keeps the direct radiation from the sun parallel to the axis of the collimator tube. Figure 73.4 shows an Eppley Laboratory (U.S.) normal incidence pyrheliometer (on right) and a cavity radiometer (on left) mounted on a solar tracker. Also shown are a first-class pyranometer and precision infrared radiometer shielded from direct radiation from the sun by shade disks coupled to the tracker.

FIGURE 73.5 The xenon flash instrument that determines atmospheric visibility at an ASOS system station. This system is produced by AAI Systems Management Incorp.

Visibility

Visibility in meteorology is a measure of how far one can see through the atmosphere.

Visibility Sensors.
Measurement of forward scattering of light by aerosols in a given sample volume is used to determine visibility. A pulsed beam of near-infrared light (typically 850 to 880 nm wavelength) generated by an infrared-emitting diode is projected into the ambient air. A detector (usually a silicon photodiode), placed between 1 and 2 m away and oriented between 30° and 40° off the axis of the pulsed beam, samples the scattered light. The intensity of scattering is proportional to the atmospheric extinction coefficient. The extinction coefficient (k) is related to visual range (VR) by VR = 5/k (km). Visual ranges measurable by such a system lie between about 0.3 km. and 60 km. Figure 73.5 shows the visibility sensor used in the ASOS network. A xenon flash lamp source is used with this system.

Transmissometers.
These measure the attenuation of a light beam to determine visibility.

Lidar Measurements

Lidar is optical radar in which a laser emits pulses of electromagnetic radiation. Scattered radiation that returns to an optimally tuned detector at the laser site for observation are called echos. The time from emission of the pulse to return of the echo determines the distance of the scatterer. The echos are sampled at all times between pulses, and thus all distances up to the maximum range. This provides information about the atmosphere along the pulse path. The strength of the echo is connected with the density, size, and shape of the scattering particles. Laser pulses are linearly polarized. Spherical scatterers return echos that have the same linear polarization. Nonspherical scatterers depolarize the echos. Thus, polarization of the echo also provides information concerning the scatterers. Using high spectral resolution lidar

(HSRL), the scattering by air molecules can be separated from the scattering by aerosols. Differential absorption lidar (DIAL) measures the concentration of gaseous species in the atmosphere. Stephens [16] discusses the HSRL and DIAL methods. Zhao et al. [17] present an application of the DIAL method to the determination of O_3 concentration in the atmosphere. Because of their high power output, ruby, dye, neodymium, yttrium-aluminum-garnet (YAG), and CO_2 lasers are commonly used for lidar. The use of lidars in the observation of aerosols injected into the stratosphere by volcanic eruption, and their subsequent decay has become quite common (see Post et al. [18] and Jager et al. [19] and other articles in the same journal issue).

Clouds

Observer Estimation.

For many years, observers at weather stations have estimated the amount, type, and sometimes the opacity of low clouds, middle clouds, and high clouds. Estimation is usually every hour or similar period with the estimate in tenths or eighths (octa) of the sky covered. Satellite, radar, and ceilometer data are replacing observer estimation. Observer estimates have been archived.

Ceilometers.

Ceilometers measure the altitude of clouds. Most ceilometers are based on lidar technology using an infrared pulsed laser diode, usually GaAs. Laser pulses are reflected back (echos) to a receiver, usually a Si diode detector. The round-trip time is converted into cloud height. Common ceilometers can measure cloud base altitudes to 4000 m (12,000 ft). Recent instruments developed by Vaisala can measure cloud bases up to 23,000 m (75,000 ft) with an error of ±1%. The Vaisala ceilometer is used in the ASOS network discussed below. Maintenance includes cleaning of the window. Heaters and blowers respond to automatic sensing units to clear precipitation from the window and control instrument temperature.

Companies producing and marketing meteorological instruments are presented in Table 73.1. For information on additional companies marketing meteorological instruments in the U.S., see *Thomas Register of American manufacturer's* and *Thomas Register Catalog File* found in most major libraries. For manufacturers in other countries, search library catalog files under "manufacturers."

United States Weather Service Facilities

The U.S. National Weather Service, the U.S. Federal Aviation Administration, and the U.S. Department of Defense have joined in an upgrade of weather instrumentation and analysis for U.S. weather stations. Other national weather services are upgrading their facilities. The new U.S. National Weather Service system has four components to obtain weather data throughout the United States and its possessions:

1. The Automated Surface Observing Systems (ASOS): These systems, produced by AAI Systems Management Incorporated, are installed at over 850 sites. Each system usually measures visibility, surface temperature, dewpoint temperature, pressure, wind speed and direction, visibility, cloud height, precipitation identification, freezing rain, and precipitation accumulation. Instruments used for each ASOS site are discussed above with other meteorological measuring instruments. Figure 73.6 shows the arrangement of instruments at an ASOS site.
2. Doppler weather surveillance radar (NEXRAD — WSR-88D): Doppler radar are produced by Lockheed-Martin Corporation and installed at approximately 160 sites in the U.S. Doppler radar surveys the weather out to about 300 km from the radar. The standard radar scans angles above the horizon up to about 6° in six ~1° intervals with a 360° azimuth. Rain scans go up to about 20°. Radar echos are reflection of the pulsed radar microwave signal from water drops in the form of rain or cloud. Ice crystals and snow reflect radar pulses with lower echo strength. Doppler radar can detect short-lived, possibly catastrophic events such as tornados, downbursts, and flash floods in real-time. The doppler shift of the echo determines the radial component of the wind velocity at the height and distance of the source of the reflected echo. The strength of the echo provides information on the precipitation rate. Powerful computers and sophisticated computer programs

TABLE 73.1   Companies Marketing Meteorological Instrumentation

AAI Systems Management Incorporated
  (Automated Surface Observing System)
P.O. Box 238
Hunt Valley, MD, 21030-0238
Phone: (410) 785-0282

Atmospheric Instrumentation Research, Inc.,
  (T, P, RH, RADS)
8401 Baseline Road, Dept. T
Boulder, CO 80303
Phone: (303) 499-1701, Ext. 300

Belfort Instrument Co., (VIS)
727 South Wolfe Street
Baltimore, MD 21231
Phone: (800) 937-2353

Davis Instrument Mfg. Co., (T, P, RH, W, PR)[a]
4701 Mount Hope Drive
Baltimore, MD 21215
Phone: (800) 548-9409

Electric Speed Indicator, (T, P, W)
12234 Triskett Rd., Dept. T-16
Cleveland, OH 44111-2519
Phone: (216) 251-2540

Eppley Laboratory, Inc.
12 Sheffield Ave., (SOL)
P.O. Box 419
Newport, RI 02840
Phone: (402) 847-1020, Fax: (401) 847-1031

Frise Engineering Co. (PR)
2800 Sisson Street
Baltimore, MD 21211
Phone: (410) 235-8524

Handar, Inc., (T, P, RH, W, PR, SOL, VIS, CEIL)[a]
1288 Reamwood Ave.
Building T, Sunnyvale, CA 94089
Phone: (800) 955-7367

Hycal Division, Honeywell Corp., (T, RH)
9650 Telstar Avenue
El Monte, CA 91731
Phone: ((818) 444-4000

Kahl Scientific Instrument Corp.,
  (T, P, RH, W, PR, SOL, VIS)
P.O. Box 1166
737 South Main Street
El Cajon, CA, 92022
Phone: (619) 444-2158

Kipp & Zonen Division, (SOL)
Enraf-Nonius Co., P.O. Box 507
L 2600 AM Delft, Roentgenweg 1
NL 2624 BD Delft, Netherlands
Phone: 011 31 15 269 8500
U.S.A.: 390 Central Ave., Bohemia, NY, 11716
Phone: (800) 229-5477

Lockheed-Martin Corp.
  (Weather Surveillance Doppler Radar)
365 Lakeville Road
Greatneck, NY 11020
Phone: (516) 574-1404

Qualimetrics, Inc.,
  (T, P, RH, W, PR, SOL, VIS, CEIL)[a]
1165 National Drive
Sacramento, CA 95134
Phone: (800) 806-6690

Rainwise, Inc., (T, P, RH, W, PR, SOL)[a]
P.O. Box 443
Bar Harbor, ME 04609
Phone: (800) 762-5723

Scientific Technology, (W, PR, VIS)
265 Perry Parkway, Suite 14
Gaithersburg, MD 20877-2141
Phone: (301) 948-6070

Setra Corp., (P)
45 Nagog Park
Acton, MA 01720
Phone: (800) 25-SETRA

Vaisala Oy, PL 26 FIN-00421,
  (T, P, RH, W, PR, SOL, VIS, CEIL, RADS)[a]
Helsinki, Finland
U.S.A.: 100 Commerce Way
Woburn, MA,01801-1068
Phone: (617) 933-4500

Viz Manufacturing Co., (RADS)
335 East Price St.
Philadephia, PA 19144
Phone: (215) 844-2626

R. M. Young, Co., (T, P, W, RH, PR)
2801 Aero Park Dr.
Traverse City, MI 49686
Phone: (616) 946-3980

Note: Symbols: T = temperature, P = pressure, RH = relative humidity, W = wind, PR = precipitation, SOL = solar radiation, VIS = visibility, LID = lidar, CEIL = ceilometer, RADS = radiosonde.
[a] Complete weather station available.

FIGURE 73.6   A view of an ASOS system sited at an airport near San Diego. The instrument of Figure 73.5 is seen in the foreground. A stand for the freezing rain sensor is not used. The next instrument is the ceilometer. Behind that is a metal box containing connections to the sensors and instruments to transmit the data to a remote collection and processing station. Along the same line of instruments beyond this is the precipitation identification or present weather sensor. This is followed along the same line by the sensor measuring ambient and dewpoint temperatures and a tipping bucket rain gage. To the north of the line of instruments is a pole with a cup anemometer, weather vane, flashing red lights, and lightning rod at the top.

are used to analyze the radar signals. Additional discussion of the use of radar for weather observations can be found in Stephens [16]. NEXRAD is discussed in References 20 to 23, for example.

3. Radiosonde upper air sounding: The northern hemisphere has a network of about 700 radiosonde sites, including the U.S. network, where soundings are generally made twice daily. The northern hemisphere is largely decoupled from the southern hemisphere for weather development. Radiosondes provide information on wind velocity, temperature, and humidity at all elevations up to 30 to 40 km. A temperature sensor, humidity sensor, and pressure sensor are carried aloft by

balloon at 0000 and 1200 UTC (±1 h). The wind velocity at different altitudes is measured by determining the radiosonde position at all times during the flight. The horizontal position of the radiosonde can be obtained by OMEGA, LORAN, or VLF global positioning systems. Additional information on global positioning is found in the section of this handbook on satellite navigation and radiolocation. Pressure can indicate the altitude of the radiosonde. Radar or GPS can measure the radiosonde position in three dimensions. Radiotheodolites can track the radiosonde to receive the data transmitted by radio transmitter. The radiosonde manufactured by Viz Manufacturing Co. called Bsonde is used by the U.S. National Weather Service. The instruments are mounted on a lightweight white styrofoam container. The temperature sensor is a thin, 50-mm long rod thermistor painted white to minimize heating by solar radiation. It is mounted outside the container. It measures temperature to ±0.5°C. The container has an air flow duct for ventilation of the relative humidity sensor mounted inside. The relative humidity sensor consists of an insulating strip coated with a film of carbon that measures relative humidity from 0 to 100% to ±5%. The pressure sensor is a nickel, C-span aneroid pressure sensor coupled mechanically to a 180 contact baroswitch providing the output pressure signal. Pressure is measured from 1060 mb to 5 mb with an accuracy of ±0.5 mb. Sensor signals are sent to the receiving station by an amplitude modulated transmitter operating from 1660 to 1700 MHz. The Vaisala Inc. radiosonde sensors used by the U.S. National Weather Service differ from Bsonde by use of a capacitance temperature sensor, a capacitance with polymer dielectric humidity sensor, and a capacitive aneroid pressure sensor.

4. Satellite remote sensing: Meteorological variables are deduced from measurements of the electromagnetic radiation coming from the atmosphere and the surface of the Earth using satellite based sensors. Usually, three or more different detector wavelength bands are used. The interpretation is difficult because the electromagnetic radiation can come from many levels of the atmosphere or the Earth's surface, all at different temperatures, and may have undergone multiple scattering in the atmosphere and Earth reflection. The energy coming from the atmosphere and the Earth's surface is mostly reflected and scattered solar radiation for wavelengths up to about 4 μm. Longer wavelength energy is mostly associated with thermal emission from the Earth and its atmosphere. Thermal emission from the atmosphere and the surface of the Earth consists of overlapping blackbody spectra of different temperatures. Absorption spectra characteristic of various gases in the atmosphere and their temperatures are superimposed on the overlapping blackbody spectra. Visible wavelengths from about 0.4 μm to 0.7 μm are useful for determining cloud coverage and cloud type. Infrared wavelengths from about 6 μm to 7 μm can be used to determine water vapor density. Infrared in the range from about 10 μm to 12 μm can be used to look at high clouds. It is also useful for night observation. Satellite measurements show the motion and development of the clouds and storm systems. More detailed discussion is available [16,24-27].

References

1. McGee, T. D., *Principles and Methods of Temperature Measurement*, New York, John Wiley & Sons, 1988.
2. Nicholas, J. V., *Traceable Temperatures: An Introduction to Temperature Measurement and Calibration*, New York, John Wiley & Sons, 1994.
3. Quinn, T. J., *Temperature*, New York, Academic Press, 1983.
4. Schooley, J. F., *Thermometry*, Boca Raton, FL, CRC Press, 1986.
5. Wilson, R. E., Temperature, In Ross, S. D. (Ed.), *Handbook of Applied Instrumentation*, Malabar, FL, Robert E. Krieger Publishing, 1982.
6. Hickes, W. F., Humidity and Dew Point, In Ross, S. D. (Ed.), *Handbook of Applied Instrumentation*, Malabar, FL, Robert E. Krieger Publishing, 1982.
7. Moisture and Humidity Measurement and Control in Science and Industry, *Proceedings of the 1985 International Symposium on Moisture and Humidity, Washington, D.C., April 15-18, 1985*, Research Triangle Park, NC, Instrument Society of America.

8. Silverthorne, S. V., Watson, C. W., and Baxter, R. D., Characterization of a Humidity Sensor that Incorporates a CMOS Capacitance Measuring Circuit, *Sensors and Actuators*, 19, 371-383, 1989.

9. Marvin, C. F., *Psychrometric Tables for Obtaining the Vapor Pressure, Relative Humidity, and Temperature of the Dew Point from Readings of the Wet- and Dry-Bulb Thermometers*, Washington, D.C., U.S. Government Printing Office, 1941.

10. Hulstrom, R., Bird, R., and Riordan, C., Spectral Solar Irradiance Data Sets for Selected Terrestrial Conditions, *Solar Cells*, 15, 365-391, 1985.

11. Iqbal, M., *An Introduction to Solar Radiation*, New York, Academic Press, 1983.

12. Coulson, K. L., *Solar and Terrestrial Radiation: Methods and Measurements*, New York, Academic Press, 1975.

13. Zerlaut, G. A., Solar Radiation Measurements: Calibration and Standardization Efforts, In Boer, K. W. and Duffie, J. A. (Eds.), *Advances in Solar Energy*, Vol. 1, Boulder, CO, American Solar Energy Society, 1983.

14. Major, G., *Circumsolar Correction for Pyrheliometers and Diffusometers*, WMKO/TD-NO. 635, Geneva, Switzerland, World Meteorological Organization, 1994.

15. LeBaron, B. A., Michalsky, J. J., and Perez, R., A Simple Procedure for Correcting Shadowband Data for All Sky Conditions, *Solar Energy*, 44, 249-256, 1990.

16. Stephens, G. L., Remote Sensing of the Lower Atmosphere, Oxford, U.K., Oxford University Press, 1994.

17. Zhao, Y., Howell, J. N., and Hardesty, R. M., Transportable Lidar for the Measurement of Ozone Concentration and Aerosol Profiles in the Lower Troposphere, Atlanta, GA, *SPIE International Symposium on Optical Sensing for Environmental Monitoring, SPIE Proceedings*, 2112, 310-320, October 11-14, 1993.

18. Post, M., Grund, C., Langford, A., and Proffitt, M., Observations of Pinatubo Ejecta Over Boulder, Colorado by Lidars of Three Different Wavelengths, *Geophys. Res. Lett.*, 19, 195-198, 1992.

19. Jager, H., The Pinatubo Eruption Cloud Observed by Lidar at Garmisch-Partenkirchen, *Geophys. Res. Lett.*, 19, 191-194, 1992.

20. *Next Generation Weather Radar: Results of Spring 1983 Demonstration of Prototype NEXRAD Products in an Operational Environment*, NEXRAD Joint System Program Office, Government Publication R400-N49, September 1984.

21. *Next Generation Weather Radar Product Description Document*, NEXRAD Joint System Program Office, Government Publication R400-PD-202, December 1986.

22. *A Guide for Interpreting Doppler Velocity Patterns*, NEXRAD Joint System Program Office, Government Publication R400-DV-101, October 1987.

23. Heiss, W. H., McGrew, D. L., and Sirmans, D., Nexrad: Next Generation Weather Radar (WSR-88D), *Microwave J.*, 33, 79-80, 1990.

24. Burroughs, W. J., *Watching the World's Weather*, Cambridge, U.K., Cambridge University Press, 1991.

25. Carleton, A. M., *Satellite Remote Sensing in Climatology*, Boca Raton, FL, CRC Press, 1991.

26. Houghton, J. T., Taylor, F. W., and Rodgers, C. D., *Remote Sounding of Atmospheres*, Cambridge, U.K., Cambridge University Press, 1984.

27. Scorer, R. S., *Cloud Investigation by Satellite*, New York, Halsted Press, 1986.

73.2 Air Pollution Measurement

Michael Bennett

There is a huge range of atmospheric pollutants and for a given pollutant there may be a several commercially available detection systems. In a document of this size, it is clearly not feasible to discuss in detail the theory of operation of every possible system. After some general remarks about air pollution monitoring, a broad introduction to the physics of molecular spectroscopy (the most widely used detection principle) will be presented. The detection of individual pollutants will then be discussed. Most of the practical examples given will refer to British experience, since this reflects the author's background. The principles involved, however, are universally applicable. The EPA's Web site (Table 73.5) provides a useful gateway to American legislation and practice.

Before spending money on a detection system, it is essential that the user understands the purpose of the measurement. This will affect both the choice of instruments and the location of monitoring sites. In general, active monitors with short response times will be more expensive than passive, slow-response instruments. A survey intended to determine spatial variations of an ambient pollutant might therefore be best to employ very many cheap, slow-response instruments. A classic example would be current national surveys to measure indoor radon pollution. In the U.K., the National Radiological Protection Board (NRPB) will supply householders with two passive detectors, each consisting of a strip of plastic inside a protective container. These are left in a living room and a bedroom for 3 months, at the end of which they are mailed back to the NRPB for analysis. (The number of α-particle trails in the plastic is counted). The householder is then reassured (or not) as to the safety of the dwelling; the authorities, meanwhile, can build up a national map of Rn concentration in relation to geology, building type, etc. The essence of the system is that the individual detectors are so cheap that the authorities can afford many thousands, such a number being necessary to provide realistic spatial coverage.

Identification of a particular source in the presence of background emissions, however, requires a sensor with a response time of minutes, or 1 h at most, supplemented by meteorological measurements. This need arises because, at temperate latitudes, the wind direction changes by 15° in 1 h and 90° in 24 h. Unless a source dominates local pollution, its impact is unlikely to be demonstrable through daily sampling.

Some applications may require response times of seconds or less. Odors are a very common cause of complaint. In this case, the complainant perceives fluctuations in concentration over periods of a few seconds, while he may become inured to steady concentrations over periods of minutes. It is also the case that many toxic gases are more damaging in high doses over short periods than at a steady concentration over some hours [1]. Flammability, of course, also depends on peak rather than mean concentrations.

Central to obtaining reliable information from such measurements is the protocol for the siting of instruments. This naturally depends on the objectives of the survey. The U.K. national survey for smoke and SO_2 [2], for example, was highly successful in demonstrating the effects on background pollutant concentrations of the 1956 Clean Air Act and (from the late 1960s) the availability of natural gas (cf. Figure 73.7). Sites were chosen so as not to be dominated by individual local sources, being typically in backyards or on the roofs of municipal buildings. By the early 1980s, however, urban air pollution had become increasingly dominated by road traffic emissions and the earlier network of sites no longer seemed appropriate [3,4]. Many roadside measurements have now also been made in addition to more general urban or rural surveys [5]. It should be noted that a *typical* site is not necessarily a *representative* one [6]. The former is what we experience about our daily lives. The latter is chosen through a protocol so that measurements from one site can be compared with those from another. Depending on the application, representative sites may be more or less typical.

A protocol is important not only for comparing one place with another, but also for comparisons between different times. One should not be dismissive of the old technology of the national SO_2 survey; the conservatism of the design allows one to rely on the time series of concentrations over the last 60 years. Clearly, there are advantages in keeping up with the latest technology. But, if interested in long-term trends, one must be very sure of the relative responses of "improved" instruments and methods.

The deployment of fast-response instruments around individual sources gives rise to somewhat different problems. Given the high cost of such instruments, one must be sure that statistically useful measurements of environmental impact will be made over the course of the survey. Concentrations should ideally be measured simultaneously upwind and downwind of the source. For an elevated buoyant source, it is desirable to measure downwind concentrations over a range of distances; atmospheric dispersion models are not yet so foolproof that the calculated distance from the stack to the peak ground-level

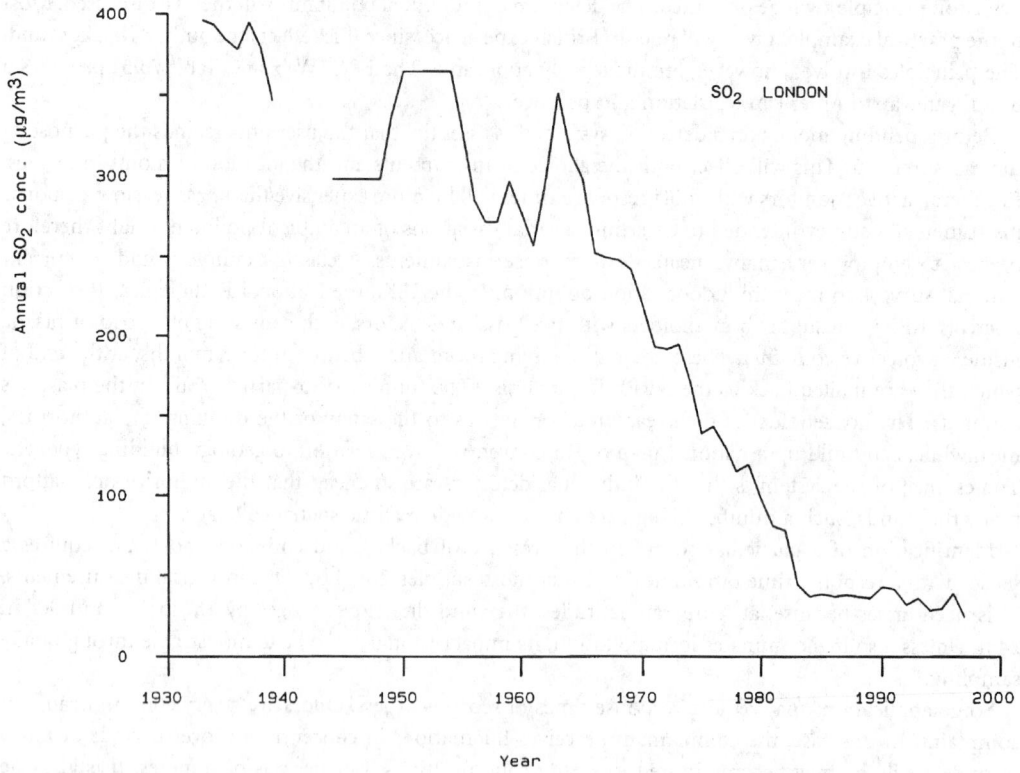

FIGURE 73.7   Annual average SO_2 concentrations in Central London, 1933–1997. The curve shows the average from two sites.

concentration can be relied on. At a given point on the ground, pollution from such a source can typically be detectable on only a few percent of hours. Mobile laboratories are therefore commonly used to enhance the capture of data, but it is then essential that some statistical and climatological analysis be applied to extrapolate the measurements obtained to long-term means at fixed target points on the ground. The protocol and analysis should be planned before the instruments are purchased.

Spectroscopic Principles

The most common principle employed for the detection of contaminants in air involves the interaction of the trace species with light. The advantages are obvious: with careful choice of system, the interaction can be made specific to the chosen molecule, and the photon will then transport itself from the point of interaction to the detector. Reference 7 provides a general undergraduate-level introduction to molecular spectroscopy, while Reference 8 describes specialized applications to atmospheric measurement.

The specificity of the interaction of light with a molecule arises from quantum theory. Light is absorbed or emitted as photons, of energy hc/λ, while individual atoms or molecules can only exist at discrete levels of energy or angular momentum. Light is thus only absorbed or emitted by gaseous species at discrete values of wavenumber, $1/\lambda$ (usually quoted in cm^{-1}). Generally, different molecular energy levels arise through three processes:

- Rotational. A molecule can spin with an angular momentum equal to a discrete value of $h/2\pi$. Transitions between angular momentum states are typically equivalent to photon energies of 1 to 100 cm^{-1}, i.e., microwave wavelengths.
- Vibrational. The atoms in a molecule can vibrate relative to one another. Transition energies between modes are typically on the order of 10^3 to 10^4 cm^{-1}, i.e., infrared wavelengths.

- Electronic. The electrons can occupy different orbitals within an atom. Transitions between these would typically take place at UV wavelengths.

Photons can initiate transitions between combinations of the above levels; observed molecular spectra thus tend to be extremely complicated, having structure over all scales of wavenumber. Atomic spectra, for example, from a monatomic gas like Hg, are simpler, since there are no vibrational or rotational modes. Nevertheless, even for molecular spectra, "selection rules" permit only a limited number of transitions to occur; a single photon can only effect a transition if the initial and final dipole moments of the molecule are different. (The direction of the change of dipole depends on the polarization of the associated photon.) Conservation of angular momentum also requires that a molecule's spin can only change by one unit at a time.

Photon energies are broadened by temperature (which causes a Doppler shift in frequency) and by pressure (which shortens the lifetime of the excited state). In practice, of course, any commercial measurement system also has a finite resolution. Parts of the spectrum may thus contain so many possible transitions that it becomes impractical to distinguish individual lines. The specificity of molecule–photon interactions promised by quantum theory is thus seen to be rather limited in practice. Spectroscopic identification of a particular species relies on the existence of resolvable structure in an accessible part of the spectrum where there is no serious interference from other common gases. In general, this search must be solved individually for each analyte — with no guarantee of the existence of a satisfactory solution.

Absorption Techniques

Detection of a specific molecule may be through either its absorption or emission of light at a particular wavelength. Consider monochromatic light of flux I passing through a gas of concentration χ. In so far as each interaction between a photon and a molecule is an independent event, the rate of interaction is separately proportional to the number of photons and the number of gas molecules. Thus,

$$\frac{\partial I\left(x,\lambda\right)}{\partial x} = -\sigma\left(\lambda\right) I\left(x,\lambda\right) \chi\left(x\right) \tag{73.1}$$

where $\sigma(\lambda)$ is the absorption cross-section. For constant gas concentration, this can be integrated as a function of path length, x, to give the Beer-Lambert law:

$$I\left(x,\lambda\right) = I\left(0,\lambda\right) e^{-\sigma\left(\lambda\right)\chi x} \tag{73.2}$$

The argument of the negative exponential is known as the optical density at this wavelength. The gas burden, χx, along the optical path is thus given by:

$$\chi x = \frac{\log\left(I\left(0,\lambda\right)/I\left(x,\lambda\right)\right)}{\sigma\left(\lambda\right)} \tag{73.3}$$

and, since $\sigma(\lambda)$ can be measured in the laboratory, one now has a measurement of the gas concentration.

The advantage of absorption techniques is that they require minimal disruption to the observed system. This is of benefit in ambient monitoring where, for example, a long-path measurement can be set up over many hundreds of meters without problems of safety or power. Equally, in emissions monitoring, it is advantageous to be able to measure gas concentrations *in situ* in the flue. Long-path techniques can also be applied in point samplers through the use of a multipass cell (e.g., a White cell [8]). The response time would then be limited by the volume of the cell in relation to the sample throughput.

Because of instrumental offsets and interference from other species, measurement of transmission at a single frequency is unlikely to give an accurate estimate of the gas concentration. More commonly, absorption is measured at several frequencies. Standard techniques include:

- Differential Optical Absorption Spectroscopy (DOAS). A broadband source is used in conjunction with a high-resolution spectroscope tuned around distinctive absorption features of the target gas. The difference between online and offline absorption then gives the gas concentration. If the measured spectrum is digitized, it can be fitted to the target spectrum, allowing some correction for interferent species. In a related analog technique, "Correlation Spectrometry," several online and offline signals are taken simultaneously from a spectrometer. Regressing the signal against a known absorption spectrum using an spectral mask gives the target gas burden.
- Fourier Transform Infrared (FTIR). The output from the gas cell is put into an interferometer. The Fourier transform of the output signal as a function of phase lag is then the absorption spectrum. Fitting programs can be applied to estimate the mix of pollutants responsible for this spectrum.
- Non-Dispersive Infrared (NDIR). Optical filters are applied to limit the input spectrum to a window where interference effects from other gases are small. No further spectral separation ("dispersion") is then applied. Gross absorption by the target gas is measured and can be calibrated by switching gas cells into the light path.
- TLDAS (Tunable Laser Diode Absorption Spectroscopy). This uses a very narrow-band source (viz. a tunable diode laser) and a broad-band receptor. The frequency of the source will typically be oscillated close to an absorption line of the target gas; the first harmonic of the signal is then proportional to the integrated concentration along the light path.

Emission Techniques

The converse of looking for the absorption of light of a given frequency by the target gas is to excite molecules of the gas and then examine the light emitted as they return to their ground state. The signal is passed through a narrow-band filter and measured with a photomultiplier tube. There are several standard techniques, including:

- Flame photometry. A flame (typically of H_2) is burned in the sample gas. The heat breaks up and ionizes the target molecules, which then relax to their ground state. Since one is now looking at fragments of the molecule rather than the molecule itself, the technique is not completely specific. Its advantage is that it tends to be very fast, being ultimately limited by the timescale for the ions to pass through the flame. A related technique, FID (Flame Ionization Detection), although not strictly spectroscopic, measures the conductivity of the flame that arises from such ionization.
- Chemiluminescence. A reactive gas is added to the sample gas. Light from the excited products of the reaction is detected.
- UV fluorescence. The sample gas is excited with UV light and the subsequent fluorescence measured. A related technique, PID (Photoionization Detection), is to measure the ionization current arising from such UV irradiation. This then detects all species whose first ionization potential is less than the photon energy of the lamp. The technique is extremely fast [9].

Particulate Sampling

Sampling of particulate matter in air gives rise to a different set of considerations from the sampling of trace gases. Note that:

- Spectroscopic methods are unlikely to be effective. The interaction of light with small particles is caused by Rayleigh or Mie scattering [10] and tends not to show very specific behavior as a function of particle composition.
- The aerodynamic behavior of a particle is a strong function of its size. Particles of diameter less than 10 μm (PM10) tend to travel with the air flow. Since, moreover, their Brownian diffusivity

is very small, they exhibit very slow deposition [11]. At less than 5 μm, such particles may be inhaled deeply into the lungs; as such, they are known as "respirable aerosol."
- Particles of diameter greater than 10 μm exhibit significant inertial and gravitational effects. They can thus deposit through impaction as the air flows around small obstacles. For particles larger than 50 μm, sedimentation becomes dominant.

The sampling method must thus be tuned to the range of particle sizes that one wants to monitor. For a general review, see the book by Vincent [12].

Slow Ambient Monitoring

The initial discussion focuses on systems with sampling times of 1 day and upward. These would typically be used for background monitoring.

Diffusion Tubes

A diffusion tube is the classic inexpensive, slow-response instrument that can be deployed in large numbers to quantify spatial variations of ambient pollutants. Typically, it consists of a sealed perspex tube with a removable cap and an active substrate at the closed end. It is deployed in the field with the open end downward and, after the prescribed exposure time (typically 7 days) it is returned to the laboratory for analysis. The system was originally developed as a personal monitor of NO_2 exposure [13] but has since been widely used for ambient urban surveys.

Assuming that the substrate is a perfect trap for the target gas, the diffusive flux of gas into the tube is given by $DA\chi/L$, where D is the diffusivity of the target species, A the internal cross-sectional area of the tube, and L the length of the tube. So long as the diffusivity remains constant over the period of measurement, the total deposition to the substrate is therefore a measure of the mean concentration over the period of exposure. Theoretically, D should vary with temperature and there may also be some circulation within the tube arising from wind across its mouth. A typical commercially available model has $L = 71$ mm and an i.d. of 12 mm, so such circulation is suppressed. Studies have shown acceptable correlation between samples from diffusion tubes and long-term means from adjacent point samplers [14].

The system employed for NO_2 is a substrate of triethanolamine deposited on a stainless steel mesh. The sample is dissolved in orthophosphoric acid and the nitrite detected colorimetrically using the Greiss/Saltzman technique. Systems are also available for SO_2, benzene, xylene, toluene, fluoride, chloride, bromide, cyanide, and nitrate. A list of participating laboratories in the U.K. can be obtained from NETCen.

Drechsel Bottles

The classic bubbler (a Drechsel bottle) consists of a Pyrex® bottle containing a solution through which filtered sample gas is bubbled. Capture of the sample gas can be enhanced by passing it through a sintered plug, so that bubble size is reduced. For the detection of SO_2, a solution of hydrogen peroxide is used so that:

$$H_2O_2 + SO_2 \rightarrow 2H^+ + SO_4^{2-} \qquad (73.4)$$

In polluted areas, the H^+ ions are then detected by titration, it being assumed that all the acidity comes from atmospheric SO_2. In cleaner, rural areas, this is insufficiently sensitive (NH_3 emissions from livestock can give apparently negative SO_2 concentrations!) and the sulfate ions are measured using ion chromatography. Sampling times are typically 24 h for a sensitivity of a few ppb.

As implemented in the U.K. national surveys of smoke and SO_2, eight such bottles are mounted in a case with a separate pump and flowmeter. At some preset time each day, the airflow is switched from one bottle to the next; at the end of the week, the operator then visits, replaces all the bottles, and returns the old ones to the laboratory for analysis. Depending on the time of the visit relative to the switching time, the sample on the day of the visit can be spread over two bottles. In this system, the air is first drawn through a separate filter (Whatman Grade 1) for each day of measurement. This captures particulate in the range 5 to 25 μm diameter; the blackness of the stain is measured photometrically to provide a 24-h sample of "black smoke."

The system is simple, inexpensive (current list prices are less than $3000), and requires minimal operator training. It was therefore possible to install many hundreds of such sites and run them for a considerable period. In central London, for example, such systems were first installed in 1933. Annual mean SO_2 concentrations were then found to be around 400 μg m^{-3}; 60 years later, this has dropped to less than 40 μg m^{-3}. Clearly, there has been considerable improvement. Such measurements have now been considerably scaled down, with 252 sites currently active in the U.K. [5].

In practice, the real cost of such a system is the manpower and the site. These are small if the system is sited in a municipal building and operated by existing staff. At a remote site, however, a secure cabin with power must be provided and it must be visited weekly. The practical difficulties of identifying, securing, and maintaining a site should never be underestimated when designing a monitoring network.

Deposition Gages [15]

Traditionally in the U.K., measurements of atmospheric dust have been made using the passive British Standard dust deposition gage, while measurements of "smoke" have been made using the active filter system described in the previous section. The standard dust deposition gage consists essentially of a bowl of diameter 300 mm and depth 225 mm mounted 1.2 m above the ground; the sample is washed into a bottle beneath the gage, from where it is filtered, dried, and weighed. After having been in use for the better part of a century, the collection efficiency of the standard gage was tested and found to be, in fact, rather poor. In very light winds, 50% of particulate at 100 μm and 80% at 200 μm might be captured, but these efficiencies fall off very rapidly with increasing wind speed.

More recent work has shown that a gage shaped like an upside-down Frisbee® has a far superior capture performance. With the addition of a foam substrate, collection efficiencies in excess of 80% in light winds, even for particle diameters as small as 50 μm, can be achieved and these efficiencies remain good at moderate wind speeds. Such gages are now commercially available.

Active Aerosol Sampling

Measurement of particulate matter of aerodynamic diameter less than 100 μm in air requires an active sampling system.

A conventional high-volume sampler for total suspended particulate (TSP) consists of a filter and a powerful blower. The system is mounted in a housing with a pitched roof, the air being drawn in under the eaves (i.e., at a height of about 1.1 m). Air sampling rates are in excess of 1 m^3 min^{-1}. The filter can be changed and weighed, or otherwise analyzed, after 24 h.

Size discrimination is introduced into such samplers by requiring the air to follow a tortuous path. Small particles then follow the streamlines, while large particles are deposited. As a first stage, the high-volume sampler may have an inlet impaction chamber, which only allows the PM10 fraction to be to captured in the filter.

In a more sophisticated system, the filter can be replaced with a cascade impactor. This consists of a stack of plates with successively smaller perforations. The holes are staggered so that air passing through one hole impacts on the following plate. As the perforation size diminishes, the air speed increases and the size of particle that can escape diminishes. By covering each plate with a removable membrane, the particulate captured in each size fraction can be analyzed and, in principle, weighed. See Figure 73.8.

Low-flow systems (1 m^3 h^{-1}) are also available, which permit measurement of PM10 or PM 2.5. Inlet heads, certified by the EPA, are available for various cutoff diameters. Reference 16 quotes capture efficiencies for such devices as a function of particle size.

Active or passive aerosol sampling systems are marketed by, among others, Charles Austin Pumps, Graseby-Andersen, and Casella.

FIGURE 73.8 Cascade impactor with a 10 μm inlet head operating at a rural site in eastern England.

Fast Ambient Monitoring

The alternative to a mechanically based, slow-response sampler that must be visited regularly is an automatic, fast-response system, self-calibrating if possible, which is logged on-site and may be interrogated remotely. Routine visits are then minimized, while visits can be made urgently if there is an alarm. Instruments currently available for various common gases are listed below. Most of these systems are designed to be bench- or rack-mountable, weighing typically in the range of 15 to 30 kg. Auto-calibration versions are available for most of them. Most manufacturers also supply appropriate logging systems with user-friendly software. The quoted measurement range is from the lower detectable limit (LDL, typically 2σ) to the maximum concentration. (The full range is not necessarily achievable on a single range setting.) The response time is typically the rise time to 95% of final value; with some instruments, there may also be a significant lag time. Prices were quoted in the U.K. in August 1996; prices of individual models are not quoted since any useful comparison would depend on detailed specifications in relation to the needs of the individual user. Quoted prices of nominally similar instruments show a surprisingly wide spread: it is worth shopping around.

TABLE 73.2   Selection of Manufacturers of Point Samplers with Their Appropriate Model Numbers

Manufacturer	SO_2	NO_x	O_3	CO	CO_2
Dasibi	4108	2100	1008	3008[a]	
Horiba	APSA-360	APNA-360	APOA-360	APMA-360[b]	
Monitor Labs	ML9850	ML9841A	ML9810	ML9830[a]	ML9820[a]
Signal-Ambitech	Ambirak	Ambirak	Ambirak	Ambirak[c]	
Thermo-Unicam	43C	42C	49C	48C[a]	41[a]

   [a] Rotating wheel containing CO (or CO_2) and N_2 gas cells is used to modulate the signal.
   [b] Modulation is achieved by alternating between sample and reference gas in the absorption chamber.
   [c] Photoacoustic detection. Modulated change in pressure due to IR absorption in sample relative to reference cell is measured.

SO_2

Most ambient monitors for SO_2 now use UV fluorescence; when irradiated, SO_2 molecules re-emit light in the range 220 to 240 nm. The measured light intensity is then proportional to the SO_2 concentration. Currently available systems have response times of order 1 to 4 min, LDL of order 0.5 to 1 ppb, and maximum ranges of 1 to 100 ppm. Prices start at around $9000. Most manufacturers also supply a converter that allows H_2S to be measured as a separate channel.

NO_x

Most ambient monitors for NO_x now employ chemiluminescence. Ozone generated within the instrument is mixed with the sample air. NO in the sample then reacts very rapidly to form NO_2, with the emission of IR radiation (peaking at 1200 nm). This is a first-order reaction. IR emission is therefore proportional to the NO concentration. Total NO_x can be measured by first passing the sample air over a catalytic converter to reduce any NO_2 present to NO. By alternating this conversion, or through the use of a dual channel system, NO_2 concentrations may be found by difference. Inevitably, the measurement of NO_2 will be noisier than that of NO or NO_x separately. Currently available systems have response times of order 1 to 4 min, LDL of order 0.5 to 1 ppb, and maximum ranges of 10 to 100 ppm. Prices start at around $9000. Most manufacturers also supply a converter that allows NH_3 to be measured as a separate channel.

O_3

Most ambient monitors for O_3 now use modulated UV absorption. O_3 has a broad absorption spectrum in the UV, with a peak around 254 nm. Practical instruments alternate the gas in the absorption cell between sample air and a de-ozonized reference gas; from the Beer-Lambert law, the log ratio of the signal must then be proportional to the O_3 concentration. Currently available systems have response times of order 20 s to 2 min, LDL of order 0.5 to 2 ppm, and maximum ranges of 1 to 200 ppm. Prices start at around $7000.

CO and CO_2

The standard method of measuring ambient CO or CO_2 is to use modulated IR absorption (NDIR); modulation of the signal allows detector offsets to be removed. Currently available systems have response times of order 15 s to 2 min, LDL of order 0.05 to 0.1 ppm for CO and <2 ppm for CO_2, and maximum ranges of 10^2 to 10^4 ppm. Prices start at around $9000.

Table 73.2 provides a selective listing of manufacturers of point samplers.

Hydrocarbons

The instruments available for the detection of ambient hydrocarbons tend to be somewhat more diverse than for the gases listed above. The most widely used technique is flame ionization detection (FID). A hydrogen flame is burned in the sample gas between two electrodes and the flame conductivity measured.

This tends to be proportional to the concentration of ·CH ions in the flame, although the calibration varies between hydrocarbon species. Most commercial instruments allow CH_4 to be measured separately by first passing the sample gas across an oxidation catalyst. Non-methane hydrocarbons (NMHC) are thereby removed from the sample; they can be estimated by difference from the total hydrocarbon (THC) channel. In principle, a FID is very fast, with some commercial instruments having response times as short as 1 s. There is, however, a tradeoff between sensitivity and response time, a sampling time of several minutes being required to give an LDL of 20 ppb.

Most of the manufacturers listed in Table 73.2 supply FIDs for ambient monitoring. Dasibi offers an instrument (Model 302) that combines an FID with a gas chromatograph, thereby giving good speciation. Photovac manufactures a hand-held FID (MicroFID), which is designed to be intrinsically safe and intended for, for example, workplace monitoring or leak detection. Prices for the standard instruments start at about $12500.

An alternative technique for hydrocarbons is photoionization detection (PID). This is nonspecific, with the detector also responding to inorganic gases of low ionization potential (e.g., NH_3). It is, however, very fast (with response times from 0.2 s for an LDL of 10 ppb) and the instruments can be made to be economical, lightweight, and intrinsically safe. A typical application might be as a hand-held leak detector in a chemical plant, or as a personal monitor of toxic solvents. Instruments are marketed by Casella, Photovac, and RAE Systems.

Wet Chemiluminescence

In addition to the above dry systems, wet chemiluminescence systems for NO_x and O_3 are available from Unisearch Associates. In these instruments, sample air is drawn over a wick containing a proprietary solution that reacts with the target gas. Light emitted in the reaction is monitored. The system is fast (response time ≤ 0.5 s), sensitive (LDL ≤ 0.1 ppb), and lightweight. It is, however, somewhat more expensive than the dry systems.

Aerosols

Measuring atmospheric aerosol concentrations at ambient levels in real time is more demanding than measuring trace gases because the pollutant is concentrated into a relatively small number of tiny particles, inaccessible to spectroscopic examination. Methods based on the scattering of laser light are available (e.g., Grimm, Model No. 1.104), although they must be calibrated to the specific aerosol being sampled. Direct gravimetric methods are also employed, based either on beta-absorption or on an oscillating filter. The sensitivity required for such a measurement can be appreciated if one considers a sampler drawing 3 L min^{-1} that is intended to measure aerosol concentrations over a sampling period of 6 min. A detection limit of 5 µg m^{-3} then requires that the sample mass be measured to a precision of 90 ng.

In the beta-absorption technique (e.g., Horiba, Model APDA-360), the sample air is drawn through a filter exposed to beta-radiation. As atmospheric aerosol is deposited on the filter, its transparency to the radiation gradually diminishes. Differentiating the transmitted signal with respect to time gives an aerosol concentration at close to real time. Commercial instruments have an automatic system for regularly changing filters, which are then available for subsequent chemical analysis.

In the TEOM (Tapered Element Oscillating Microbalance; Rupprecht and Patashnick, Model No. 1400a) system, the filter is balanced on a hollow tube through which the sample air is drawn. The tube is clamped at its base and induced to vibrate at a frequency that depends on the mass of the filter. Again, the differential of the filter mass with respect to time provides the aerosol concentration.

Whichever system is employed, care must be taken in the choice of inlet port so that the desired fraction of ambient particulate is sampled. Typically, a PM10 sampling head would be employed.

Remote Monitoring

While point samples of air pollution can be very valuable, they are limited both in their position, which is normally near ground level, and in representing a single point rather than a broader spatial sample. A variety of optical techniques are available that permit more general measurement of pollution in the atmosphere.

Long-Path Measurements

With the use of a broadband source and a retroreflector, spectroscopic analysis of the returned light can provide spatially averaged measurements of a wide range of pollutants. Two commonly used systems are Hawk (from Siemens Environmental Systems) and OPSIS.

For ambient monitoring, OPSIS applies DOAS to an absorption spectrum ranging between the UV and the near-IR. The spectrum is digitized by scanning the output from the spectrometer and sophisticated fitting routines are then applied to detect target species in the presence of other absorbers and scatterers. Standard systems are available for monitoring three (NO_2, SO_2, O_3) or five (+ toluene, benzene) species, with the three-species system costing from about \$80,000. Modules can be added for a range of trace gases. The system can also be used for emissions monitoring. An absorption band in the IR can then be used and the calibration must be optimized to the optical density of the particular flue.

Hawk employs a broadband source and an oscillating interference filter. Since the spacing of the lines of the filter normal to the light path varies with the cosine of the angle of incidence, the positions of the absorption lines must also oscillate. The system is tuned to lie close to an absorption line of a target gas; the differential signal then measures the gas concentration. A given instrument is thus limited to one, or at most a few closely related analytes; the system is particularly good for alkanes. Instrument costs start at about \$16000, but there is no fixed price since costs depend on the complexities of the spectrum of the desired target species.

Overhead Burden

If one is interested in concentrations well away from the ground, a retroreflector may be impractical. In this case, spectral measurements of sunlight scattered from the sky can provide us with an estimate of the total overhead burden of a pollutant. "Cospecs" (Correlation Spectrometers) have been available since the early 1970s for the measurement of SO_2 or NO_2 burden. These are most valuable in conjunction with mobile surveys of elevated emissions, since they allow the operator to know when he is beneath the plume. With very careful calibration, and a knowledge of the wind speed, they may also be used to estimate the pollutant flux. The instrument has no absolute zero level and so has difficulty distinguishing very broad plumes. They are still made to order by Barringer Research.

The classic measurement of overhead pollutant burden is that of stratospheric O_3 using the Dobson spectrophotometer. This measures the difference in irradiance at two wavelengths 20 nm apart close to the peak of UV absorption by O_3; other things being equal, this difference is proportional to the overhead burden of O_3. The system was used to detect the ozone hole over Antarctica [17]. The detection method has remained unchanged since the original instruments of 60 years ago, although the optics have been refined and the valve electronics have now been replaced with solid-state. The continuity of method was important in giving confidence in the surprising observation of the ozone hole in the mid-1980s. Instruments are available from Ealing Electro-Optics.

Lidar

Lidar (Light Detection And Ranging) is the optical equivalent of radar [8,18]. A pulse of laser light is directed into the atmosphere. Backscattered light is collected by a telescope and directed to a photomultiplier tube. The strength of the return signal corresponding to a given range $R = ct/2$ is given by the Lidar equation:

$$V(R) = \frac{CW}{R^2}\, \sigma_b n(r) \exp\left\{-2\int_0^R \left(\sigma(\lambda)\chi(r) - \sigma_e n(r)\right) dr\right\} \tag{73.5}$$

where C is a system constant, W the pulse energy, $n(R)$ the density of scatterers, σ_b and σ_e, respectively, the backscatter and extinction cross-sections for scattering, and $\sigma(\lambda)$ the absorption cross-section from a tracer gas of concentration χ. If the extinction is small, scanning the laser beam through an elevated plume allows a cross-section of the scatterer density to be built up. The range resolution of this cross-section depends on the frequency at which the signal is digitized; a 60-MHz digitizer gives a range resolution of 2.5 m. The time resolution of the measurement depends on the pulse repetition rate of the laser. At 30 Hz, a 2-s scan would normally give adequate lateral spatial resolution. *A priori* calibration for particulate density requires knowledge of σ_b, which in turn depends on the particle size and refractive index. In practice, an independent gravimetric measurement must be made at some point in the scanned area.

Alternatively, shots can be alternated between two nearby wavelengths having very different values of $\sigma(\lambda)$. Neglecting variations in the other terms, the Lidar equation can be rearranged to give:

$$\chi(R) = \frac{1}{2\left(\sigma(\lambda_1) - \sigma(\lambda_2)\right)} \frac{\partial}{\partial R} \log\left(\frac{V_2(R)}{V_1(R)}\right) \qquad (73.6)$$

Since the differential absorption $\sigma(\lambda_1) - \sigma(\lambda_2)$ can be measured in the laboratory, one now has a remote, range-resolved measurement of the tracer gas. This technique is known as DIAL (Differential Absorption Lidar) [8] and has been applied to a wide range of atmospheric pollutants (SO_2, NO_2, O_3, CH_4, hydrocarbons, Hg, etc.). Its particular strength is in identifying and quantifying fugitive emissions, for example, CH_4 from cattle or leaks in refinery plant [19]. Because of the greater subtlety of the measurement, it does not in general have the spatial or temporal resolution of a simple backscatter Lidar. For a given DIAL system at a given range, the product of spatial, temporal, and concentration resolution is approximately fixed; optimal performance in one is at the cost of worse performance in the others. If the signal is rather weak, there is a temptation to average measurements over a period greater than the timescale for a significant change in the target gas concentration. Great care should be taken in this case since the nonlinearity of Equation 73.6 can then lead to very serious distortions [20].

DIAL services are available from, for example, the National Physical Laboratory (NPL) and Siemens Environmental Systems in the U.K. Elight in Berlin supplies complete DIAL systems.

Emissions Monitoring

Industrial plants are, in general, authorized to release effluent to the atmosphere on the condition that such emissions are monitored reliably enough to satisfy the regulator that authorized limits have not been exceeded. This has increasingly led to the continuous monitoring of prescribed substances, with measurements being automatically displayed and archived. Aside from satisfying regulatory conditions, such monitoring can also be of value to the operator in controlling his plant. Note that any measured value of pollutant concentration must be converted to standard conditions of temperature, pressure, humidity, and excess O_2 before comparison with a regulatory standard.

Isokinetic Sampling of Aerosol

The gold standard for the measurement of aerosol emissions is to take a sample of the effluent and weigh the particulate therein. This is done by introducing a sample line into the flue, pumping the gas through a filter, and measuring the mass gained per metered volume of flue gas. It is essential that the gas speed at the sample inlet is the same as the flow velocity in the flue; if it is too fast (or slow), large particulate will be under (or over) sampled. To ensure a representative sample, such measurements must be made at several diameters downstream of any bends or confluences in the flue. It is also necessary to take several samples (4 or 8) over the cross-section of the flue. Even taking every precaution, the precision of the method may be as poor as 20% of the mean value. In practice, the nominal calibration against continuous monitors can be found to be *very much* worse than this from trial to trial.

Since each sample can take 30 min, this procedure clearly does not give a real-time measurement of particulate load. In poorly designed plants, there may also be severe practical difficulties in gaining access to the flue with heavy, powered sampling equipment at a sufficient height for the flow to be uniform. Nevertheless, this is the only direct gravimetric method of measuring particulate concentration in a flue. For this reason, regular (at least annual) such measurements would be required for the calibration of any continuous monitors.

Aerosol Sampling: *In situ* Methods

In situ methods are particularly attractive for emissions measurements because the pollutant is measured directly without any possibility of loss or transformation in the path from flue to instrument.

The most obvious way of continuously measuring the dust load in a flue is to pass a beam of light across the flue and measure the obscuration. It can be assumed that the transmission obeys the Beer-Lambert law and that the optical density is proportional to the particulate load. Practical instruments, of course, require a degree of sophistication to allow for instrumental offsets. It will not generally be possible to turn off the plant at regular intervals to check the instrument zero! Typically, the light source will be modulated and a modulated signal measured. Purge air is normally blown over transmitter and receptor windows (and the retroreflector if used) in order to keep the light path clear. The sensitivity of the system will nevertheless drift with time (and will change with the composition of particulate), so regular gravimetric calibration is essential.

Such cross-duct monitors are simple and robust, and they provide a measurement across the full width of the flue. They are limited to optical densities in the approximate range 0.01 to 2.5. Systems are available from Codel, Land Combustion, Grimm, United Sciences, and others.

For very narrow ducts, or for very low particle densities, a more sensitive technique must be chosen. As with ambient particulate monitoring, it is possible to use sideways scattering from a light beam. The transmission and reception optics are now set up so that there is no direct path between the two. Light scattered from particles in the flue can, however, be detected. This gives a true zero level in the absence of particulate, and a signal proportional to particle density. Care must be taken to ensure that the light path is representative of the particle distribution in the flue. By switching the light path, it might be possible to route it through a filter of known optical density, thereby permitting online calibration. Such systems are available from Erwin-Sick and from United Sciences.

An alternative technique is the triboelectric monitor. In this device, a metal rod is inserted across the flow. Particles striking the rod transfer static charge to it. The leakage current (or, better, the rms component thereof) is thus proportional to the particle flux along the duct. This method is fast, simple, and robust; it is little affected by the accumulation of dirt on the probe and is insensitive to the presence of steam (which would, of course, preclude optical measurements of particle density). In very wet flows, there are possible problems with leakage across the insulated base of the probe; these can be overcome with the use of purge air or extended insulation. It should be noted that, with this device, calibration is required not only for gravimetric particle density but also for the flow speed in the duct. If this speed varies significantly, the implicit particle density will be in error. Systems are available from PCME.

Gas Sampling: Spectroscopic Methods

The cross-beam technique can be applied to *in situ* sampling for pollutant gases. Practical devices are usually based on nondispersive absorption of IR or UV. DOAS can also be used. Sophisticated instrumental design is necessary to ensure:

- Minimal interference from other species present in the effluent. In combustion gases, for example, high concentrations of H_2O, CO_2, and acid gases may be present. Measurement of CO in the IR involves the use of an absorption window at a wavelength of about 4.7 µm between the absorption bands of CO_2 and H_2O.
- Minimal interference from IR emissions from the hot flue gases. Typically, a modulated system would be used.
- Insensitivity to instrumental drifts (dirty windows, misaligned optics, aging of source and detector). This can be achieved with dual light-path systems, where only one path is sensitive to the target gas. The differential signal then gives a robust measurement of concentration in the flue. This can be calibrated online by switching in cells containing a known concentration of the target gas.

Table 73.3 provides a selected listing of manufacturers of cross-duct emission systems.

TABLE 73.3 Selection of Manufacturers of Cross-duct Emission Systems

Supplier	Technique	Gases
Codel	NDIR	Multigas, CO, SO_2, NO/NO_x, HCl, H_2O, hydrocarbons, NO_2, CO_2, NH_3, VOCs
Erwin-Sick[a]	DOAS (UV)	SO_2, NO, NO_2, NH_3
Land Combustion	NDIR	CO
Monitor Labs[a]	NDIR	CO, CO_2, H_2O
	Modulated UV	NO, SO_2
OPSIS	DOAS	Many

[a] In these systems, the beam is enclosed within an optical probe inserted into the duct. Sample gas diffuses in through a ceramic filter. Besides protecting the optics, this permits calibration by injecting reference gases into the measurement cavity.

Gas Sampling: Extractive Techniques

If flue gases are to be sampled by an extractive technique, great care must be taken in delivering the sample from the flue to the monitor. Most of the techniques used for extractive emissions monitoring are the same as those for ambient gas monitoring; in many cases, even the same instruments might be used. Since process gases tend to be hotter, wetter, dustier, and more acidic than ambient air, there will often be a need to change the condition of the sample so that it is acceptable to the monitor. The aim of sampling technology is to deliver an acceptable sample to the instrument at a target gas concentration that bears a known relation to the gas concentration in the flue [21].

In general, the sample can be conditioned by dilution, by filtering, by cooling, or by drying. The techniques chosen will depend on the gas being measured and on the instrument being used. At its simplest, conditioning may amount to the sample gas being diluted with several hundred times its volume of clean air. Normal ambient monitors could then be used, but separate measurements would have to be made of O_2 and H_2O in order to express the measured concentration in standard form.

To avoid clogging of the sample line, it is usual to filter the sample as soon as practically possible. To avoid condensation, the normal practice is to keep the line heated. If, however, the instrument cannot tolerate flue gas temperatures, cooling will be necessary at some stage. If this is done without dilution, water will condense out and subsequent measurements will be on a dry gas basis. In itself, this is advantageous, but any acid gases will remain with the water and be lost. Alternatively, the sample can be dried by passing it through a permeation drier. Such driers cannot tolerate acid gases. SO_2, for example, must therefore be measured on a wet gas basis, with temperature and condensation being controlled by appropriate dilution.

Considerable skill is involved in designing and implementing continuous extractive emissions monitoring systems. Instruments are available for hydrocarbons and for the conventional range of stack gases from Bernath Atomic (THC), Horiba, Land Combustion, Monitor Labs, Servomex, Signal-Ambitech, and Thermo-Unicam.

Table 73.4 provides the addresses of the various manufacturers of air pollution measurement instrumentation, and Table 73.5 provides Web site addresses for Table 73.4.

Defining Terms

Aerosol: Particulate matter so finely divided in a gas that it remains in suspension. The aerodynamic diameter of an aerosol particle is conventionally defined as the diameter of a sphere of density 1 g cm^{-3} which would have the same sedimentation velocity.

Burden: The integrated concentration of an analyte along a defined path.

Concentration: The mass or volume of an analyte in a unit sample volume. The volume may need to be defined in terms of temperature, pressure, humidity etc.

Deposition: The transfer of material from the atmosphere to a surface.

Monitoring: A routine series of measurements made for control purposes.

Protocol: A written list of instructions for obtaining a reliable measurement.

TABLE 73.4   Addresses for Manufacturers of Air Pollution Measurement Instrumentation

Barringer Research 1730 Aimco Blvd. Mississauga, Ontario L4W 1Vl, Canada Tel: (905) 238-8837 Fax: (905) 238-3018	Bernath Atomic Gottlieb-Daimler Str., 11-15 D-3015 Wennigsen, Germany Tel: 49 (5103) 7093 Fax: 49 (5103) 709298	Casella (London) Regent House, Wolseley Road Kempston, Bedford MK42 7JY, U.K. Tel: 1 (1234) 841441 Fax: 1 (1234) 841490
Charles Austin Pumps Royston Road West Byfleet KT14 7NY, U.K. Tel: 44 (1932) 355277 Fax: 44 (1932) 351285	Codel International Station Road Bakewell Derbyshire DE45 1GE, U.K. Tel: 44 (1629) 814351 Fax: 44 (1629) 814619	Dasibi Environmental 506 Paula Ave. Glendale, CA 91201 Tel: (818) 247-7601 Fax: (818) 247-7614
Ealing Electro-Optics Greycaine Road Watford WD2 4PW, U.K. Tel: (1923) 242261 Fax: (1923) 34220	ELIGHT Warthe Str. 21 D-14513 Teltow-Berlin, Germany Tel: 49 (3328) 39500 Fax: 49 (3328) 395099	Environmental Protection Agency 401 M St., SW Washington, D.C. 20460
Erwin-Sick Optik-Elektronic Nimburger Str.11 D-79276 Reute Germany Tel: 49 (7641) 4690 Fax: 49 (7641) 469149	Graseby-Andersen 4801 Fulton Ind. Boulevard Atlanta, GA 30336 Tel: (404) 691 1910 Fax: (404) 691 6315	Grimm Labortechnik Dorfstraße 9 D-83404 Ainring, Germany Tel: 49 (8654) 5780 Fax: 49 (8654) 57810
Horiba 1080 E. Duane Ave. Ste. A Sunnyvale, CA 94086 Tel: (408) 730-4722 Fax: (408) 730-8675	Land Combustion Dronfield Sheffield S18 6DJ, U.K. Tel: 44 (1246) 417691 Fax: 44 (1246) 290274	Monitor Labs 74 Inverness Drive East Englewood, CO 80112-5189 Tel: (303) 792-3300 Fax: (303) 799-4853
NETCen Culham Abingdon OX14 3DB, U.K. Tel: 44 (1235) 463133 Fax: 44 (1235) 463011	National Physical Laboratory Queen's Road Teddington, TW11 0LW, U.K. Tel: 44 (181) 943 7095 Fax: 44 (181) 943 6755	NRPB (Radon Survey) Chilton Didcot OX11 0RQ, U.K. Tel: 44 (1235) 831600 Fax: 44 (1235) 833891
OPSIS AB Box 244 S-244 02 Furulund Sweden Tel: 46 (46) 738510 Fax: 46 (46) 738370	PCME Ltd. Stonehill, Huntingdon, PE18 6EL, U.K. Tel: 44 (1480) 455611 Fax: 44 (1480) 413500	Photovac Europe Sondervang 19 DK4100 Ringsted Denmark Tel: 45 (5767) 5008 Fax: 45 (5767) 5018
RAE Systems 680 West Maude Ave. #1 Sunnyvale, CA 94086 Tel: (408) 481-4999 Fax: (408) 481-4998	Rupprecht and Patashnick 25 Corporate Cir. Albany, NY 12203 Tel: (518) 452-0065 Fax: (518) 452-0067	Servomex Co.Inc. 90 Kerry Place Norwood, MA 02062 Tel: (781) 769-7710 Fax: (781) 769-2834
Siemens Environ. Systems Sopers Lane Poole BH17 7ER, U.K. Tel: 44 (1202) 782553 fax. 44 (1202) 782335	Signal-Ambitech Regal Way Faringdon SN7 7BX, U.K. Tel: 44 (1367) 242660 Fax: 44 (1367) 242700	Spring Innovations 216 Moss Lane Bramhall, Stockport SK7 1BD, U.K. Tel: 44 (161) 440 0082 Fax: 44 (161) 440 9127

TABLE 73.4 (continued)   Addresses for Manufacturers of Air Pollution Measurement Instrumentation

Thermo-Unicam	United Sciences	Unisearch Associates
P.O. Box 208	5310 North Pioneer Rd.	222 Snidercroft Rd.
York Street	Gibsonia, PA 15044	Concord, Ontario
Cambridge	Tel: (412) 443-8610	L4K 1B5, Canada
CB1 2SR U.K.	Fax: (412) 443-7180	Tel: (905) 669-3547
Tel: 44 (1223) 374234		Fax: (905) 669-8652
Fax: 44 (1223) 374338		

Note: For reasons of space, only a primary address has been given for each manufacturer. Many of these companies will have subsidiaries or agents in your own country: a fax should elicit the local address.

TABLE 73.5   Useful Website Addresses for Organizations Listed in Table 73.4

http://www.barringer.com/
http://www.bernath-atomic.com/
http://www.casella.co.uk/
http://www.pentol.com/codel.html
http://www.dasibi.com/
http://www.ealing.com/
http://www.epa.gov/
http://www.elight.de/
http://www.sick.de/
http://www.graseby.com/
http://www.horiba.com/
http://www.landinst.com/
http://www.monitorlabs.com/
http://www.aeat.co.uk/netcen/
http://www.npl.co.uk/
http://www.pcme.co.uk/
http://www.perkin-elmer.com/photo/   (Photovac)
http://www.raesystems.com/
http://www.rpco.com//index.htm   (Rupprecht & Patashnick)
http://www.servomex.com/
http://www.siemens.co.uk/
http://www.spring-innovations.co.uk/
http://www.unicam.co.uk/

References

1. R.F. Griffiths, The effect of uncertainties in human toxic response on hazard range estimation for ammonia and chlorine, *Atmos. Environ.,* 18, 1195-1206, 1984.
2. Warren Spring Laboratory, *National Survey of Air Pollution 1961-1971, Vol. 1,* ISBN 11 410150 7, London: HMSO, 1972.
3. D.J. Ball and R. Hume, The relative importance of vehicular and domestic emissions of dark smoke in Greater London in the mid-1970s, the significance of smoke shade measurements and an explantation of the relationship of smoke shade to gravimetric measurements of particulate, *Atmos. Environ.,* 11, 1065-1073, 1977.
4. M. Bennett, C. Rogers, and S. Sutton, Mobile measurements of winter SO_2 levels in London 1983-84, *Atmos. Environ.,* 20, 461-470, 1986.
5. V. Bertorelli and R.G. Derwent, Air quality A to Z: a directory of air quality data for the United Kingdom in the 1990s, *Meteorological Office,* ISBN 0 86180 317 5, 1995.
6. R.E. Munn, *The Design of Air Quality Monitoring Networks,* London: Macmillan, 1981.
7. C.N. Banwell and E.M. McCash, *Fundamentals of Molecular Spectroscopy,* London: McGraw-Hill, 4th ed., 1994.

8. M.W. Sigrist (Ed.), *Air Monitoring by Spectroscopic Techniques*, New York: Wiley-Interscience, 1994.
9. R. F. Griffiths, I, Mavroidis, and C.D. Jones, The development of a fast-response portable photo-ionization detector: a model of the instrument's response and validation tests in air, *J. Measurement Sci. Technol.*, 8, 1369–1379, 1998.
10. H.C. van de Hulst, *Light Scattering by Small Particles*, New York: Wiley, 1957.
11. J.H. Vincent, *Aerosol Sampling. Science and Practice*, New York: John Wiley & Sons, 1989.
12. G.A. Sehmel, Particle and gas dry deposition: a review, *Atmos. Environ.*, 14, 983-1011, 1980.
13. E.D. Palmes, A.F. Gunniston, J. DiMattio, and C. Tomczyk, Personal sampler for nitrogen dioxide, *J. Amer. Ind. Hyg. Assoc.*, 37, 570-577, 1976.
14. G.W. Campbell, J.R. Stedman, and K. Stevenson, A survey of nitrogen dioxide concentrations in the United Kingdom using diffusion tubes, July-December 1991, *Atmos. Environ.*, 28, 477-486, 1994.
15. D.J. Hall, S.L. Upton, and G.W. Marsland, Designs for a deposition gage and a flux gage for monitoring ambient dust, *Atmos. Environ.*, 28, 2963-2979, 1994.
16. D. Mark and D.J. Hall, Recent developments in airborne dust monitoring, *Clean Air*, 23, 193-217.
17. J.C. Farman, B.G. Gardiner, and J.D. Shanklin, Large losses of total ozone reveal seasonal ClO_x/NO_x interactions, *Nature*, 315, 207-210, 1985.
18. D.J. Carruthers, H. Edmunds, M. Bennett, P.T. Woods, M.J.T. Milton, R. Robinson, B.Y. Underwood, and C.J. Franklin, Validation of the UK-ADMS dispersion model and assessment of its performance relative to R-91 and ISC using archived Lidar data, *Dept. of the Environment*, Research Report No. DOE/HMIP/RR/95/022, March 1996.
19. R.H. Partridge, P.T. Woods, M.J.T. Milton, and A.J. Davenport, Gas standards and monitoring techniques for measurements of vehicle and industrial emissions, occupational exposure and air quality, *Monitor '93*, 23-35, Manchester: Spring Innovations Ltd., 1993.
20. M. Bennett, The effect of plume intermittency upon differential absorption Lidar measurements, *Atmos. Environ.*, 32, 2423-2427, 1998.
21. K. Honner, Continuous extractive gaseous emissions monitoring, *Monitor '93*, 53-56, Manchester: Spring Innovations Ltd., 1993.

73.3   Water Quality Measurement

Kathleen M. Leonard

Maintaining and verifying water quality is important in many environmental applications. The most obvious is in drinking water applications; but industrial and municipal wastewater, natural surface/groundwater, industrial process waters, and closed-loop control systems all require a certain range or maximum concentration of species to operate properly. Therefore, the development of instrumentation for monitoring and detecting these contaminants is a highly competitive area. Of course, the chemical species of interest depend on the ultimate use of the water, but this chapter section deals with some of the water quality sensors and instrumentation being currently used.

Table 73.6 provides a list of the most commonly monitored chemical species in the area of water quality. The third column categorizes the parameters in the general categories of drinking water, wastewater (municipal), industrial, stormwater, and ambient water quality. The "ultimate reference" source for accepted methods of determining chemical concentrations in water is known as the *Standard Method for the Examination of Water and Wastewater* [1]. This book has been updated regularly since 1905 and contains the current techniques recommended by Water Environment Federation (WEF) and the U.S. Environmental Protection Agency (EPA).

The next chapter subsection introduces the theory behind the some of the major types of water quality sensors, including electrical, optical, and chemical separation. Since each technique has its own particular water quality applications, a brief discussion of advantages and disadvantages is included. This section is limited to commercially available instruments commonly used both inside and outside the laboratory,

TABLE 73.6 Common Chemical Species Monitored in Water Quality

Chemical Parameter	Chemical Symbol	Primary Use[a]
Ammonium	NH_4	DW, WW
Arsenic	As	DW
Barium	Ba	DW
Bicarbonate	HCO_3^-	DW, WW
Cadmium	Cd	DW, WW
Calcium	Ca^{2+}	DW
Carbonate	CO_3^{2-}	DW
Chlorinated Hydrocarbons	Various herbicides and pesticides	DW, SW
Chloride	Cl^-	DW, SW
Fluoride	F^-	DW
Chromium	Cr	DW
Dissolved Oxygen	O_2	WQ
Hydrogen (pH)	H^+	DW, WW, WQ
Iron	Fe	DW
Lead	Pb	DW
Magnesium	Mg^{2+}	DW
Mercury	Hg	DW
Nitrate	NO_3^-	DW, WW, SW
Nitrogen	N	DW, WW
Petrochemicals	Various	
Phosphates	PO_4^{3-}	WW, SW, WQ
Potassium	K^+	WW
Selenium	Se	DW
Silver	Ag	DW
Sodium	Na^+	DW
Sulfate	SO_4^{2-}	DW, WW
Gross Measures		
Alkalinity	As $CaCO_3$	DW
Biochemical Oxygen Demand	BOD	WW, WQ
Conductivity		
Chemical Oxygen Demand	COD	WW, WQ
Hardness	As $CaCO_3$	DW
Particle counts		WW, WQ
pH	H^+	DW, WW, WQ
Total Organic Carbon	TOC	SW, WW, WQ
Other Parameters		
Color		
Dissolved Solids	In ppm	SW, DW
Turbidity	In NTU	DW, WQ
Microbiological contaminants	Counts/100 mL	DW, WW, WQ
Radiological contaminants	In curies	DW

[a] DW: Drinking water, WW: Waste water, SW: Storm water, WQ: Water Quality.

although emphasis is placed on field (i.e., portable) instruments. Future trends will be addressed in the conclusion section.

Theory

Electrical Methods of Analysis

There are various electric mechanisms employed for determining water quality concentrations. For example, in the area of electrochemical sensors, there are potentiometric, amperometric, and conductometric devices [2]. The pH sensor for determining hydrogen ion concentration is probably the most widely used instrument in this category.

Potentiometric Sensors.

This type of sensor is based on the relationship of the electrochemical cell and the chemical activity of a sample, based on the general form of the Nernst equation:

$$E_{cell} = E^0 - \left(RT/nF\right) \ln \left\{ \left(red\right)/\left(ox\right)\right\}$$ (73.7)

for the general reaction: ox + $ne^- \leftrightarrow$ red. ox and red indicate the oxidized and reduced species [3]. The term RT/F has a value of 0.059. E^0 is the standard cell potential and E_{cell} is the adjusted cell potential. In short, the activities of the oxidized and reduced species determine the potential of the electrode. If the relationship of the reacting species is known, it is used to measure concentration of that species in another solution. In most electrochemical instruments, it is important to maintain a reference electrode whose potential remains constant for all cell conditions or one that can be easily calibrated for other species. These sensors are commonly known as ion-selective electrodes (ISE) in environmental applications (also known as ion-sensitive field-effect transistors or ISFET). They are all based on the activity level of a specific ion within a solution. The key to the ISE is the use of an ion-specific membrane with channel size proportional to the concentration of the ion. It is necessary to maintain a reference potential to ensure that the "ion concentration will be directly related to the substrate potential [4]." Specific electrodes can be based on a gas electrode, metal electrode, oxidation–reduction electrode, membrane electrode, glass, liquid membrane, crystalline membrane, or electrode with metal contacting a slightly soluble salt [5]. Examples of this type of sensor and instrumentation for gross measurements include pH sensors, dissolved oxygen, hardness, and dissolved solids.

Commercially available ion-selective electrodes include sensors for ammonia, chlorides, cyanide, iodide, fluoride, nitrates, potassium, and sodium. The instruments labeled as multiparameter water quality monitors, such as that available from Solomat, are usually composed of one meter with separate probes for each measurement parameter. However, some of these multiparameter monitors have been upgraded with a multichannel probe that contains the separate probes within a single housing.

A membrane probe in popular usage is the dissolved oxygen probe based on the principle of polarography. It consists of a gas-permeable membrane over a silver anode and a platinum cathode within a cavity full of electrolyte (usually KCl). Figure 73.9 shows the placement of the anode and electrode in a typical polarographic membrane probe. Since the oxygen will be reduced at the cathode, a current is induced. The amount of oxygen can then be correlated to the current when a potential is applied to the anode and cathode.

The pH instrument is one of the most commonly used potentiometric sensors in water quality applications. It is based on a glass electrode that develops a potential related to hydrogen ion activity of the solution. Since this probe was initially developed in the early part of the 1900s, there are numerous probes and meters on the market. Small, battery-operated, hand-held units are excellent choices for fieldwork. However, for more precise applications, advanced (i.e., expensive) laboratory-grade instruments are also available. Most pH probes require standardization and may have other upkeep requirements after prolonged usage.

Amperometric Sensors.

The amperometric classification of electrical sensors is based on the measurement of current through a working electrode. An empirical relationship can be used to enhance the performance of these electrodes. An example of such a probe is an electrode for chlorine, consisting of a silver anode and platinum cathode within an electrolyte reservoir. A chemically specific membrane allows only the ions of interest within the probe cavity near the cathode. Reduction occurs at the cathode and then the silver anode reacts with the electrolyte and, through oxidation, positively charged ions produce a current.

Conductometric Sensors.

The conductometric sensors are based on measurements of concentration of a sample due to modulation in conductivity. For example, conductivity is usually measured with a direct reading of electrical conductivity in water, which is then correlated with the number of ions present. The probe usually

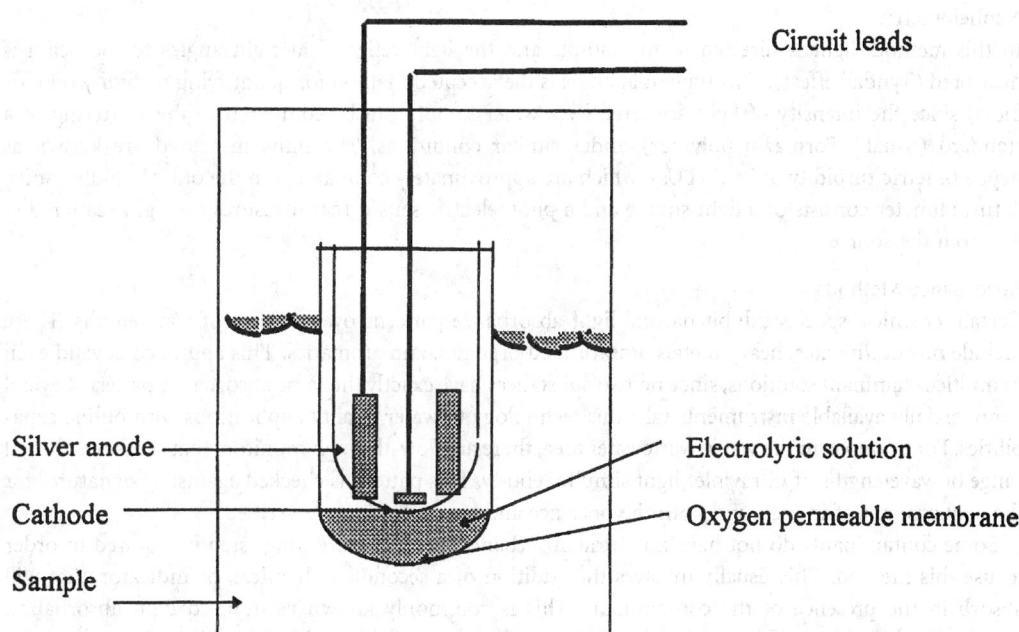

FIGURE 73.9 Simplified diagram of dissolved oxygen sensor based on polarography. The oxygen-permeable membrane surrounds the anode and cathode to deter other chemical species from entering the electrolytic solution.

consists of two voltage electrodes (+ and −) and two current electrodes (also + and −) supported by some type of housing. The newer instruments have a temperature compensation option and data storage. Most will also provide a reading of total dissolved solids. Conductometric sensors are also available for resistivity, salinity, and temperature measurements.

The most commonly available dissolved oxygen (DO) sensors are galvanic electrodes with replaceable membranes. The current is proportional to the DO concentration under steady-state conditions [1]. One of the drawbacks of this type of probe has been the need for periodic calibration, either in air or a saturated medium. Some of the newer products are self-calibrating, thus eliminating this error source. There are many companies that sell these products and the prices range from modest ($300) to high, depending on extras. If the application will be messy, (e.g., wastewater), the added expense for the deluxe model, such as a self-cleaning probe may be realized very quickly. A company that has been producing DO instruments for many years is YSI. It has a "deluxe" model with microprocessor control and RS232 interface for continuous reading.

Optical Sensors

This category involves any type of measurement system that relies on an optical property to measure a chemical concentration, including reflection, colorimetry, fluorescence, or absorption of light. The general equation relating wavelength of light (λ) to the energy of a photon (E) is given by the Planck equation:

$$E = hc/\lambda \tag{73.8}$$

where h is Planck's constant (6.626×10^{-34} J s) and c is the velocity of light (3.0×10^{8} m s^{-1}). Some of these mechanisms are included in the category of "gross optical measurements." This category includes any method that uses reflection/refraction of light beams, colorimetry, or absorption of light. For example, in order to determine the efficiency of a water treatment system due to deteriorating filter capacity or to optimize coagulant dosages, a particle counting method is used. This type of measurement can give not only quantity, but also size ranges. This method is helpful in guarding against the intrusion of cryptosporidium cysts and guardia in a drinking water facility.

Nephelometry.

In this method, light is directed to the sample and the light reflected at right angles to the beam is measured (Tyndall effect). This type of analysis is the accepted method for quantifying turbidity (cloudiness) since the intensity of light scattered by a water sample can be compared to the scattering of a standard (usually Formazin polymer) under similar conditions. The units measured are known as nephelometric turbidity units (NTUs), which are approximately comparable to the older "candle" units. A turbidimeter consists of a light source and a photoelectric sensor that measures the light scattered at 90° from the source.

Absorbance Methods.

Certain chemical species exhibit natural light absorbance patterns over a range of wavelengths. These include nitrates/nitrites, heavy metals, unsaturated organics, and aromatics. This approach is valid even in multicontaminant solutions, since no two substances have exactly the same absorbance pattern. Several commercially available instruments take this technology to water quality applications with online capabilities. For example, in the water/wastewater area, there are flow-through monitors that measure a broad range of wavelengths of ultraviolet light simultaneously. This pattern is checked against a "signature" for the contaminant of interest and both the presence and concentration are verified.

Some contaminants do not have a natural absorbance, so a "conditioning" step is required in order to use this method. This usually involves the addition of a secondary chemical, or indicator, that will absorb in the presence of the contaminant. This is commonly known as induced light absorbance. Ammonia, phosphates, and chlorine absorbance probes are available with this conditioning step.

There are several advantages to using absorbance, including low maintenance, especially when compared to ISEs, high reliability, and automatic compensation for turbidity (which can cause havoc in other systems). One of the major drawbacks of the available models is the need to buy a separate unit for each contaminant, at a comparatively higher cost than ISE.

Colorimetry.

This simple method works by correlating the color of a solution with the concentration of a specific chemical. The theory is that light absorption (A), which is the amount of light intensity (I/I_0) absorbed, as is related to concentration based on Beer's Law [5]:

$$A = \log\left(I/I_0\right) = k^n C \tag{73.9}$$

where k^n is the constant for a particular solution and C is the concentration of the solution. Beer's law is valid for most water quality ranges of concentration, but it can be verified using dilutions of a specific contaminant. The first and still simplest colorimetric methods involve the use of color-comparitor tubes (or Nessler tubes) for comparing a range of standards to water samples. However, this method is susceptible to large discrepancies due to human error in selecting variations of tints of a color. An instrument that improves upon the accuracy of colorimetry is a photoelectric colorimeter. A light source is directed through a filter (making it monochromatic), then passed through the sample cell and on to the photoelectric detector. The obvious advantage over the comparitor is the elimination of the human factor for identifying variations in color. Another adaptation is the spectrophotometer, which employs a diffraction grating to select the wavelength for a wide variety of uses. It is a very versatile instrument since both the incoming and outgoing wavelengths are selectable. Newer models incorporate computer chips for automatic programming and data storage.

Colorimetric tests include single analyte meters for chlorine and DO, and direct readout multianalyte photometers for many parameters. For example, the DPD colorometric method is commonly used to measure total chlorine concentrations in water and wastewater. However, it must be used with caution, since turbidity, other organic contaminants, monochloromine, etc. will cause interference.

Fluorimetry.
This method utilizes the fluorescence, either natural or induced, of a compound. Fluorescent chemicals absorb radiation of a specific wavelength and emit at another. Fluorescent tracers are commonly used in water quality studies to determine direction of flow for both surface and groundwater systems. Fluorescent tracers include Rhodamine B, Fluorescein, and Pontacyl Pink B. The detection instrument is known as a fluorimeter, which range from moderate cost for a stationery (single chemical) wavelength model (<$4000) to high cost for a unit with variable wavelength and detection capabilities.

Remote Fiber-Optic Spectroscopy.
In the area of *in situ* chemical analysis, remote fiber spectroscopy (RFS) shows much promise for a variety of monitoring applications. RFS is a fiber-optic application of existing spectroscopic techniques that can be applied to the monitoring of chemical species in remote locations. In this method, light of an appropriate frequency is launched into a single optical fiber or fiber bundle. The light is guided to the region of interest through the fiber by the mechanism of total internal reflection. An optical sensor (optrode) designed for a specific chemical or chemical group is positioned at either the distal end of the fiber, or a length along the fiber. The altered signal is returned through the same fiber to the spectrometer for signal decoupling and analysis. The major advantages to fiber-optic chemical sensors (FOCS) include their small diameter, *in situ* nature, resistance to harsh environments, and imperviousness to electric interference. These properties make FOCS ideal for many environmental applications.

The distinguishing feature of most RFS systems is the type of sensor employed. There are basically two approaches to the sensor, either intrinsic or extrinsic. The intrinsic method utilizes the fiber itself as the active element, for example, using change in index of refractions from the fiber to the media as the parameter being measured. The extrinsic method uses the fiber optic only as light pipe, and usually involves incorporating a chemical onto the fiber as the active element. Most of the related literature for fiber-optic chemical sensors (FOCS) deals with design and advantages of various types of optrodes, that exploit either absorbance, surface enhanced Raman scattering, or fluorescence. Although FOCS have been mentioned in literature for the past decade, there has been little success in developing a multicontaminant sensor for practical applications.

The direct fluorescence approach (measuring the natural fluorescence of the media) is an attractive technique for applications where aromatic chemicals are to be monitored. However, since many chemicals that act as contaminants are nonfluorophores, a fluorescent intermediate can be incorporated in the FOCS. Ideally, the intermediate should be sensitive to either a group of chemicals or a specific compound, and exhibit a spectral response that could be exploited to identify the compound.

As mentioned previously, the standard pH instrumentation is based on a glass pH electrode; however, optical methods have obvious advantages in corrosive conditions and fluctuating temperatures [6]. Optical-based pH sensors are generally based on the changes in the absorbance or fluorescence of an indicator dye that has been immobilized on a fiber optic. Recently, pH sensors based on the immobilization of flurorescein dye in a sol-gel matrix have been developed. Sol-gel chemistry is a method of producing a porous glass matrix at a low temperature using a hydroloysis of an alkoxide (usually TEO). An example of such a probe is a pH probe based on fluorescein dye in a sol-gel matrix [7].

Optical sensors for O_2 and dissolved oxygen, which are based on fluorescence or phosphorescence quenching, have the advantage of size, nonconsumption of oxygen, and resistance to interference due to flow rates or sample stirring over traditional dissolved oxygen electrodes. The major limitations is finding an indicator dye with a strong sensitivity to oxygen quenching. There are many such studies being performed; however, most are still in the research stage. For example, polycyclic aromatic hydrocarbons such as pyrene and long-wave absorbing dyes have been investigated due to their ability to be quenched by oxygen [6]. A fiber optics-based sensor that is on the market detects VOCs, such as carbon tetrachloride, TCE, and BTX in water systems (FiberChem). A similar application is a fluorescence-based system for measuring petroleum products in the groundwater.

Other Sensors

Infrared Spectrophotometry.

Most organic chemical compounds exhibit absorption in the infrared. This attribute can be used to determine atomic groupings based on quantum mechanics [5]. An example is a unit that measures total organic content (TOC) using UV-promoted persulfate oxidation and nondispersive infrared detection. This methodology allows for online TOC monitoring. Correlations can be made to convert values of TOC to biochemical oxygen demand (BOD) and chemical oxygen demand (COD) for particular samples. Drawbacks to this method include pretreatment to remove inorganic carbon and the high cost of instrumentation.

Respirometry.

The standard method to calculate BOD_5 is to measure oxygen consumption (based on DO concentration) over a 5-day period. However, there are other respirometers based on extremely small differences in DO or CO_2 concentrations using external gas sensors. This method is great for respirometry rates of animals, biodegradable contaminants, and sediment oxygen demand studies.

Chemical Separation Techniques

Separation techniques are based on the phase partitioning of molecules within a mixture. Gas chromatography refers to a mobile phase of a vapor, while liquid chromatography refers, of course, to a liquid mobile phase. There are several commonly used types such as gas-liquid chromatography, high-performance liquid chromatography (HPLC), and ion chromatographs. An excellent reference book for these methods is *Chemistry for Environmental Engineering* [5]. The advantages of these instruments are the ability to detect components in complex mixtures and high sensitivity. Chromatography instruments have long been considered standard in environmental laboratories, but the size of the instruments has been a drawback for field applications. However, size constraints are being challenged with the availability of equipment with size reductions of over 90%.

Mass Spectrometry.

In environmental applications, these are usually used in conjunction with gas chromatography and are referred to as GC/MS. The attraction of these instruments is the ability to detect a variety of compounds and also determine the mass fragmentation patterns of a complex organic structure [1]. The MS essentially ionizes the substances with an electron beam. The ions are then accelerated through a series of lenses according to their mass-to-charge ratios. The charged fragments are detected by a electron multiplier, which results in a mass spectra for the particular compound. In addition to the old-fashioned lab-scale GC/MS, there are portable units now commercially available that can be taken into the field for on-site environmental characterizations. These are very useful for ambient air toxics analysis, process diagnostics, and "real-time" groundwater plume movement. However, a newer technique of capillary electrophoresis (CE) has been gaining in popularity over capillary GC since it requires less solvent and can be extremely sensitive.

Atomic Absorption Spectrometry.

Since metals have their own characteristic absorption wavelengths, they are usually measured by the atomic absorption method where a light beam is directed into a sample being aspirated onto a flame. The resulting light is then sent to a monochromator with a detector that measures the amount of light absorbed by the flame. This method exhibits high sensitivity and is applicable to all of the commonly found metals in water. The drawback of this type of instrument is that it is not portable and is more of a laboratory method.

Instrumentation and Applications

Table 73.7 lists some of the instruments that are commercially available for specific water applications. This is not an exhaustive list of all types and manufacturers, but is indicative of the chemical parameters

TABLE 73.7 Instruments for Remote Water Quality Analysis

Product Name	Water Quality Parameters	Company
Biological Analyzers: portable		
HMB-IV-S	Bacteria and fungi	H & S Enterprises
Ion-Selective Electrodes: Single Paramter and Combination Types		
DO 201	Dissolved oxygen with automatic cleaning and calibration	HF Scientific, Inc
D63/5440D	Dissolved oxygen	Great Lakes Instruments
96 Series ion*plus*	(Fl, Cl, Br, Ca, Cu, I, Pb, Au, SO_2)	Orion
97 Series ion*plus*	(Ca, NO_2, NO_3, P)	Orion
SCAMP	DO, temperature	Precision Measurement Engineering
D63/5440D	Dissolved Oxygen	Royce Instrument Corporation
520c	w/ammonia, nitrate, lead, chloride, electrodes	Solomat
FPA 200 Series	NH_3, Fl, Cy, NO_3, Na	Tytronics Inc.
FPA 300/400 Series	Acids, alk, hardness, Fe, sulfide	Tytronics Inc.
YSI 5000 series	DO	YSI Inc.
Nephelometers: Turbidity		
1110-TUX	Solomat	
Model 2600	Mindata	
Conductivity Sensors		
DataSonde3 Multiprobe	Conductivity, TDS, salinity, resistivity	Hydrolab
OS 200	Ocean Sensors	
Fast Conductivity sensor	Precision Measurement Engineering	
EC200	Conductivity	Greenspan Technology Pty Ltd
YSI 5000 series	DO	YSI Inc.
Optical		
PPC 200	Particle count	HF Scientific, Inc.
DRT- 200E	Turbidity	HF Scientific, Inc.
711	Turbidity and suspended solids	Royce Instrument Corporation
pH probes: with data storage and computer compatability		
		Royce Instrument Corporation
PerpHecT Line		Orion
Respirometry		
Micro-Oxymax	Oxygen, CO_2, NH_3, CO	Columbus Instruments International Corp.
Spectrometer: In-line types only		
Laser Diode Spectrometer 3000		ALToptronic
FiberChem Sensor	TCE, BTEX, VOCs	FCI
FPA 1000 Series	Chlorine, oils in water, H_2S	Tytronics
UV Absorbance Instruments		
ChemScan	Nitrate, ammonia, iron, turbidity	Applied Spectrometry
Multiparameter Systems		
AQUALAB	Phosphate, ammonia, nitrate, pH, conductivity	Greenspan Technology
DataSonde	Temperature, conductance, TDS, resistivity	Hydrolab
Model 1260	pH, ISE, mV, ORP, cond., D.O., temp., BOD	Orion
WP 4007	(4 parameters at once, interchangeable)	Solomat
WP803	(32 channels)	Solomat
YSI 5000	BOD	YSI, Incorporated

TABLE 73.8 Companies That Make Water Quality Sensors

Applied Spectrometry Associates, Inc. W226 N55G Eastmound Drive Waukesha, WI 53186 Tel: (414) 650-2280	Ocean Sensors 9883 Pacific Heights Blvd., Suite E San Diego, CA 92121 Tel: (619) 450-4640
FiberChem Incorporated 509-376-5074	Precision Measurement Engineering 1827 Hawk View Drive Encinitas, CA 92024 Tel: (619) 942-5860
Great Lakes Instruments 9020 West Dean Road P.O. Box 23056 Milwaukee, WI 53224	Solomat 26 Pearl Street Norwalk, CT 06850 Tel: (203) 849-3111
Greenspan Technology Pty Ltd. 24 Palnerin Street Warwick, Queensland, 4370 Australia Tel: 61-76-61-7699	Royce Instrument Corporation 13555 Gentilly Road New Orleans, LA 70129 Tel: (800) 347-3505
HF Scientific 3170 Metro Parkway Fort Meyers, FL 33916-7597 Tel: (941) 337-2116	Tytronics Inc. 25 Wiggins Avenue Bedford, MA 01739-2323
H & S Enterprises 148 South Dowlen #120 Beaumont, TX 77707	YSI, Incorporated 1725 Brannum Lane Yellow Springs, OH 45387 Tel: (800) 765-4974
Hydrolab Corporation P.O. Box 50116 Austin, TX 78763	

and types of instruments that are useful for field use. For example, the chemical separations category is very sparse in the table; although there are numerous laboratory instruments, very few are useful in the field.

Many of the manufacturers provide helpful technical information on theory and applications for their products. For example, Hach Company has many excellent publications for a variety of water/wastewater applications. They sell and develop instruments that vary from inexpensive kits (e.g., for colorimetry) to spectrophotometers with computer compatibility. Additionally, due to the broad range of diversity in the water quality area, there are many small companies that produce instruments for specific applications. Table 73.8 provides contact information for each of the companies listed in Table 73.7. Another excellent source of information is the annual listing of industrial manufacturers by Water Environment Federation.

Data Evaluation

In the past, the better instruments may have had a self-storing data system, usually in the form of paper graph or a retrievable magnetic tape. Although this allowed for continuous sampling, it was not without problems. The ease of obtaining data, *in situ* data gathering, and digital data acquisition are all attributes that are attractive in environmental applications. New technology has made these wishes a reality. For example, radio telemetry is a wireless technology to link remote sensors to dedicated computers or Supervisory Control and Data Acquisition (SCADA) systems. This capability allows for truly automated data acquisition and real-time uplinks. Other systems can be hardwired to computers (usually via RS 232 boards) or through modems to a remote acquisition site.

Trends in Water Quality Measurements

Analytical Chemistry publishes a biennial review dealing with recent literature in the area of water analysis [8]. These review articles, while not in-depth on a single topic, are quite extensive in topic matter and really illuminate trends in environmental sensing. For example, in a recent review, the area of *in situ* analysis contains a significant amount of coverage. Since many water quality applications would benefit from *in situ*, real-time data, it will be an area of research and development growth for the next 5 years.

Defining Terms

A light absorption
c velocity of light (3.0×10^8 m s^{-1})
C concentration of the solution
E energy of a photon
E_{cell} cell potential adjusted
E^0 standard cell potential
F Faraday's constant (96,500 C/equivalent)
I light intensity
I_0 initial light intensity
k'' constant for a particular solution
h Planck's constant (6.626×10^{-34} J s)
n number of moles
ox   oxidized species
R ideal gas law constant (8.31 J mol^{-1} K^{-1})
red   reduced species
T absolute temperature
λ wavelength of light

References

1. *Standard Methods for the Examination of Water and Wastewater*, 17th edition, American Public Health Association, American Public Health Association, American Water Works Association and Water Environment Federation, Washington, D.C., 1990.
2. J. Janata, M. Josowicz, and D. DeVaney, Chemical sensors, *Anal. Chem.*, 66, 207R-228R, 1994.
3. D. Snoeyik, V. Jenkins, J. Ferguson, and J. Leckie, *Water Chemistry*, Third Edition, Wiley, New York, 1980.
4. D. Banks, Microsystems, Microsensors & Microactuators: An Introduction, www.ee.surrey.a...al/ D.Banks/usys_i.html, 1996.
5. C. Sawyer, P. McCarty, and G. Parkin, *Chemistry for Environmental Engineering*, McGraw-Hill, New York, 1994.
6. O. Wolfbeis, Ed., *Fiber Optic Chemical Sensors and BioSensors*, CRC Press, Boca Raton, FL, Vol. 1, 1991, 359.
7. P. Wallace, Y. Yang, and M. Campbell, Towards a distributed optical fiber chemical sensor, *Chemical, Biochemical and Environmental Fiber Sensors VII, SPIE*, Vol. 2508, 1996.
8. P. McCarthy, R. Klusman, S. Cowling, and J. Rice, Water Analysis, *Anal. Chem.*, 67-12, 525R-562R, 1995.

Further Information

HACH, *Water Analysis Handbook*, Second edition, Hach Company, Loveland, CO, 1992.
Water Environment Federation, Alexandria, VA, Annual Review.
USA BlueBook, Utility Supply of America, Northbrook, IL, 1-800-548-1234.

73.4   Satellite Imaging and Sensing

Jacqueline Le Moigne and Robert F. Cromp

What Can Be Seen from Satellite Imagery

Satellite imaging and sensing is the process by which the electromagnetic energy reflected or emitted from the Earth (or any other planetary) surface is captured by a sensor located on a spaceborne platform. The Sun as well as all terrestrial objects can be sources of energy. Visible light, radio waves, heat, ultraviolet and X-rays are all examples of electromagnetic energy. Since electromagnetic energy travels in a sinusoidal fashion, it follows the principles of wave theory, and electromagnetic waves are categorized by their wavelength within the electromagnetic spectrum. Although it is continuous, different portions of the electromagnetic spectrum are usually identified and referred to as (from shorter to longer wavelengths): cosmic rays, γ-rays, X-rays, ultraviolet, visible ([0.4 μm, 0.7 μm]), near-infrared (near-IR), mid-infrared (mid-IR), thermal infrared (above 3 μm), microwave ([1 mm, 1 m]), and television/radio wavelengths (above 1 m). Figure 73.10 shows the electromagnetic spectrum and these subdivisions.

General Sensor Principles

Sensors are often categorized as "passive" or "active." All energy observed by "passive" satellite sensors originates either from the Sun or from planetary surface features, while "active" sensors, such as radar systems, utilize their own source of energy to capture or image specific targets.

Passive and Active Sensors.
All objects give off radiation at all wavelengths, but the emitted energy varies with the wavelength and with the temperature of the object. A "blackbody" is an ideal object that absorbs and reemits all incident energy, without reflecting any. If one assumes that the Sun and the Earth behave like blackbodies, then according to the Stefan-Boltzmann law, their total radiant exitance is proportional to the fourth power of their temperature. The maximum of this exitance, called dominant wavelength, can be computed by Wien's Displacement law (see References 1 to 4 for more details on these two laws). These dominant wavelengths are 9.7 μm for the Earth (in the infrared portion of the spectrum) and 0.5 μm for the Sun (in the green visible portion of the spectrum). It implies that the energy emitted by the Earth is best observed by sensors that operate in the thermal infrared and microwave portions of the electromagnetic spectrum, while Sun energy that has been reflected by the Earth predominates in the visible, near-IR, and mid-IR portions of the spectrum. Most passive satellite sensing systems operate in the visible, infrared, or microwave portions of the spectrum. Since electromagnetic energy follows the rules of particle theory, it can be shown that the longer the wavelength, the lower the energy content of the radiation. Thus, if a given sensing system is trying to capture long wavelength energy (such as microwave), it must view large areas of the Earth to obtain detectable signals. This obviously is easier to achieve at very high altitudes — thus the utility of spaceborne remote sensing systems.

   The most common active satellite sensor is radar (acronym for "radio detection and ranging"), which operates in the microwave portion of the electromagnetic spectrum. The radar system transmits pulses of microwave energy in given directions, and then records the reflected signal received by its antenna. Radar systems were initially employed by the military as a reconnaissance system because their main advantage was to operate day or night and in almost any weather condition. They are very important in satellite

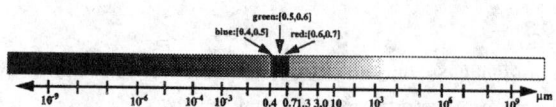

FIGURE 73.10   Electromagnetic spectrum.

remote sensing because microwave radiations are hardly affected by atmospheric "screens" such as light rain, clouds, and smoke. The time it takes for the radar signal to return to the satellite is also measured by instruments such as altimeters which are very useful in determining surface height measurements.

Polar Orbiting and Geostationary Earth Sensing Satellites.
Satellite remote sensing systems are also characterized by the different Earth orbiting trajectories of a given spacecraft. These two modes are usually referred as "polar orbiting" and "geostationary" (or "geosynchronous") satellites. A polar orbit passes near the Earth's North and South poles. Landsat, SPOT, and NOAA are near-polar satellites; their orbits are almost polar, passing above the two poles and crossing the equator at a small angle from normal (e.g., 8.2° for Landsat-4 and -5). If the orbital period of a polar orbiting satellite keeps pace with the Sun's westward progression compared to the Earth rotation, these satellites are also called "sun-synchronous." This implies that a sun-synchronous satellite always crosses the equator at the same local sun time. This time is usually very carefully chosen, depending on the application of the sensing system and the type of features that will be observed with such a system. It is often a tradeoff between several Earth science disciplines such as atmospheric and land science. Atmospheric scientists prefer observations later in the morning to allow for cloud formation, whereas the researchers performing land studies prefer earlier morning observations to minimize cloud cover.

A geostationary satellite has the same angular velocity as the Earth so its relative position is fixed with respect to the Earth. Examples of geostationary satellites are the GOES ("Geostationary Operational Environmental Satellite") series of satellites that orbit at a constant relative position above the equator.

Sensor Characteristics

Spectral Response Patterns.
The design of new satellite instruments is based on the principle that targets of interest can be identified based on their spectral characteristics. For example, different Earth surface features, such as vegetation or water, present very distinctive reflectance or emittance curves that are a function of the energy wavelength. These curves are often called the "spectral signatures" of the objects being observed. Although these curves are very representative of each feature and can help identify them, they do not correspond to unique and absolute responses. Because of different reasons, such as atmospheric interactions, temporal or location variations, the response curves of a given object observed under different conditions might vary. For this reason, these curves are often called "spectral response patterns" instead of "spectral signatures." Figure 73.11 shows an example of such reflectance patterns for several features: fir tree, clear lake water, barley, and granite.

Atmospheric Interactions.
Earth satellite sensors are designed to take into consideration the fact that all observed radiation must pass at least once through the atmosphere; therefore, the energy interactions of the atmosphere must be considered during the design phase. The distance through which the radiation passes through the atmosphere is called "path length." The effect of the atmosphere depends on the extent of the path length and on the magnitude of the energy signal. The two main atmospheric effects are known as "scattering" and "absorption."

Scattering is the unpredictable redirection of radiation by particles suspended in the atmosphere. The type and the amount of scattering mainly depend on the size of the particles but also on the wavelength of the radiation and the atmospheric path length. If these particles are smaller than the radiation wavelength, this effect is known as "Rayleigh scatter." This scattering especially affects the shorter visible wavelengths of the sunlight (i.e., blue visible wavelength) and it explains why the sky appears blue to the human eye. In the evening, when the path length is longer, the effect of the Rayleigh scatter is only visible on the longer red wavelengths of the sunlight and the sky appears red or orange. If the particles are about the size of the radiation wavelength, the scatter is known as "Mie scatter"; this scattering effect is often due to water vapor and dust. When the atmospheric particles are larger than the radiation wavelengths, a "nonselective scatter" occurs; all visible wavelengths radiations are scattered equally and this type of scattering explains why clouds appear white.

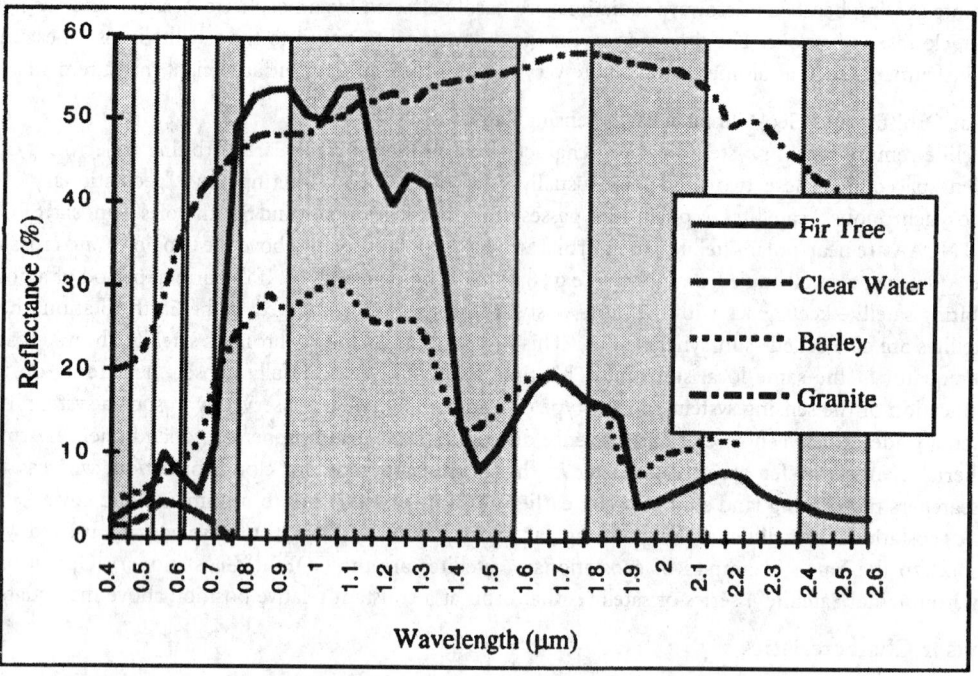

FIGURE 73.11 Examples of spectral response patterns for four different types of features, fir tree, clear water, barley (example of crop), and granite (example of rock). White areas show the portions of the spectrum corresponding to the 7 channels of Landsat-Thematic Mapper (TM-4&5).

Atmospheric absorption occurs in specific wavelengths at which gases such as water vapor, carbon dioxide, and ozone absorb the energy of solar radiations instead of transmitting it. "Atmospheric windows" are defined as the intervals of the electromagnetic spectrum outside these wavelengths, and Earth remote sensors usually concentrate their observations within the atmospheric windows. As an example, white areas of Figure 73.11 show the portions of the spectrum (i.e., the "channels" or "bands") from visible to mid-IR used by the Landsat-Thematic Mapper (TM).

Spectral, Radiometric, Spatial, and Temporal Resolutions.
Although the spectral response patterns are not absolute, they play an important role in the design of new sensors. When a new sensor is being designed, the type of features to observe and the accuracy with which they will be mapped define which wavelengths are of interest, the widths of the wavelength intervals to be used, what is the accuracy to be achieved in these bandwidths, and what is the "smallest" or "faintest" feature that might be detected by the sensor. Following the examples of Figure 73.11, the best wavelength interval to distinguish between vegetation and granite will be the [1.55 μm, 1.75 μm] wavelength interval. The above sensor requirements correspond to the "resolutions" of the sensor by which it is usually identified — spectral, radiometric, spatial, and temporal resolutions. The term "resolution" is usually employed to define the smallest unit of measurement or granularity that can be recorded in the observed data. The spectral resolution of a sensor is defined by the bandwidths utilized in the electromagnetic spectrum. The radiometric resolution defines the number of "bits" that are used to record a given energy corresponding to a given wavelength. The spatial resolution corresponds to the area covered on the Earth's surface to compute one measurement (or one picture element, "pixel") of the sensor. The temporal resolution (or frequency of observation), defined by the orbit of the satellite and the scanning of the sensor, describes how often a given Earth location is covered by the sensor.

Signal-to-Noise Ratio.
Sensors are also characterized by their signal-to-noise ratio (SNR) (i.e., the noise level relative to the strength of the signal). In this case, the "noise" usually refers to variations of intensity that are detected by the sensor and that are not caused by actual variations in feature brightness. If the noise level is very high compared to the signal level, the data will not provide an accurate representation of the observed features. At a given wavelength λ, SNR is a function of the detector quality, the spatial resolution of the sensor, as well as its spectral resolution (see Reference 1 for a detailed formula). To maintain or improve the signal-to-noise ratio and therefore improve the radiometric resolution of the sensor, a tradeoff must be made between spatial and spectral resolutions; in particular, improving spatial resolution will decrease the spectral resolution. Of course, other factors such as atmospheric interactions will also affect the SNR.

Multispectral and Hyperspectral Sensors.
The remote sensing industry is experiencing a rapid increase in the number of spectral bands of each sensor. The first Landsat sensors (Landsat-1 and 2) were designed with four bands in the visible and near-IR portions of the spectrum. Landsat-4 and 5 were refined with seven bands from visible to thermal-IR. Then, Landsat-6 and 7 were planned with an additional panchromatic band, which is highly sensitive over the visible part of the spectrum. In general, most Earth remote sensors are *multispectral*; that is, they utilize several bands to capture the energy emitted or reflected from Earth features. The addition of panchromatic imagery, which usually has a much better spatial resolution than multispectral imagery in the visible part of the spectrum, provides higher quality detail information. Multispectral and panchromatic data, usually acquired simultaneously, are co-registered and can be easily merged to obtain high spatial and spectral resolution. Co-registered multispectral-panchromatic imagery is available from sensors such as the Indian satellite sensor, IRS-1, and the French sensor, SPOT.

Ideally, if a sensor had an infinite number of spectral channels (or bands), each observed area on the ground (or pixel) could be represented by a continuous spectrum and then identified from a database of known spectral response patterns. Adding more bands and making each of them narrower is the first step toward this ideal sensor. But, as previously explained in the previous section, due to technology limitations, it was very difficult until recently to increase the number of bands without decreasing the SNR. Due to recent advances in solid state detector technology, it is now possible to increase significantly the number of bands without decreasing the SNR, thus seeing the rise of new types of sensors, called *hyperspectral*. Although the boundary between multispectral and hyperspectral sensors is sometimes defined as low as 10 bands, hyperspectral imaging usually refers to the simultaneous detection in hundreds to thousands of spectral channels. The aim of hyperspectral sensors is to provide unique identification (or "spectral fingerprints") capabilities for resolvable spectral objects. Potential applications include agricultural yield monitoring, urban planning, land use mapping, mining and mineral deposits, disaster relief/assessment, tactical military operations, and forest fire protection management. The NASA Airborne Visible InfraRed Imaging Spectrometer (AVIRIS) simultaneously collects spectral information from visible to infrared ranges (from 0.4 µm to 2.5 µm) in 224 contiguous spectral bands. Each band has an approximate bandwidth of 10 nm (or 0.01 µm), with a spatial resolution of about 20 m. The instrument flies aboard a NASA ER-2 airplane at approximately 20 km above sea level. The science objectives of the AVIRIS project are mainly directed toward understanding processes related to the global environment and climate change.

Super-Resolution.
When data is collected from a satellite sensor, each sample of information represents an area on the ground that might correspond to several features with different spectral responses and the final information represents a "mixture" of several disparate information. When a sensor does periodic imaging over the same area, the direction of observation slightly changes, and even when successive data are correctly registered (i.e., in a perfect correspondence), two respective samples might represent two slightly different areas on the ground. Super-resolution is an area of research that aims at combining such

information from different directions but taken under similar lighting conditions in the goal of improving spatial and spectral resolutions. As an example, some recent work [5] utilizes a Bayesian method for generating sub-pixel resolution composites from multiple images with different alignments. Software products have also been proposed by some commercial systems, such as ERDAS-Imagine and ENVI, to help in the unmixing of spectral bands (see Reference 6 for more detail).

Direct Readout Data

Examples of different spectral and spatial resolutions are given in Table 73.9 with a summary of current Earth remote sensing systems that operate from visible to thermal-IR wavelengths. Table 73.9 provides information about their bandwidths and their spatial resolutions. The table also indicates if the data from these sensors can be acquired by direct readout. Sensors can transmit data in two modes. The first mode, called "direct readout," transmits data as soon as it "sees" it, and any receiving station within that satellite footprint can receive this data. In the second mode, the sensor records whatever it sees for playback at a later time, and specialized receiving stations (or "ground stations") are required to receive this data since it is transmitted at higher rates than direct readout. The data received by these ground stations can cover a larger extent. Historically, the cost of acquiring and processing Earth remotely sensed data has limited satellite data collection to a small number of expensive ground stations around the world, operated by the owners of the satellites. Direct readout sensors were mostly confined to meteorological applications that required timely data such as weather forecasting, severe weather identification and tracking, and disaster prediction and assessment. But this situation is changing; due to new technology, costs have been greatly reduced, and direct readout data that was initially expensive and beyond the reach of the nontraditional user, is now generating a growing interest. A small industry has evolved to design, install, and upgrade ground stations around the world that acquire direct readout data.

Typical Attributes Measured from Space

This section does not intend to be an exhaustive up-to-date description of all Earth and space applications of satellite imaging and sensing, but rather it presents a few representative applications and their associated satellite sensors. The References section as well as the World Wide Web (WWW) offer more extensive references to other applications and sensors.

Earth Science Applications.
Over the past few decades, a number of international global change research programs have been initiated whose goals are to understand the relation between human activities and the global Earth systems processes and trends. Mission To Planet Earth (MTPE [7]) is a multi-agency program, whose goal is to achieve this kind of understanding, especially through improved satellite observations. As part of NASA's Mission to Planet Earth program, the Earth Observing System (EOS [8]) will launch, over the next 2 decades, several platforms of sensors aimed at ecology, oceanography, geology, snow, ice, hydrology, cloud, and atmospheric studies. Each platform will carry one or several instruments, thus globally covering a wide range of spectral, spatial, and temporal resolutions. Europe and Japan have similar programs, such as the ADEOS (ADvanced Earth Observation Satellite [9]), developed by the Japanese space agency, NASDA, in collaboration with France and the U.S. ADEOS, launched in 1996, flew for 9 months before it failed, and included remote sensing instruments for observing the Earth's atmosphere, land surfaces, and oceans. Other examples of such programs are the ERS and ENVISAT satellites from ESA (European Space Agency) and the Indian satellites, IRS. ERS-1 and 2 were launched in 1991 and 1995, respectively, and ENVISAT will be launched in 1999. All these satellites carry on the same platform different instruments making simultaneous observations. The first satellite in the IRS series was launched in 1988, and in 9 years, India has designed and launched six remote sensing satellites. For more information on the IRS series, see Reference 10.

In all the above programs, studies concentrate on global processes occurring in the atmosphere (especially lower parts of the atmosphere), on the Earth surface (terrestrial studies), and in the oceans (hydrospheric studies). Results of all these studies will contribute to international programs such as the

TABLE 73.9 Summary of the Main Current Earth Science Satellite Data Operating from UV to Thermal IR ("D": Direct Broadcast)

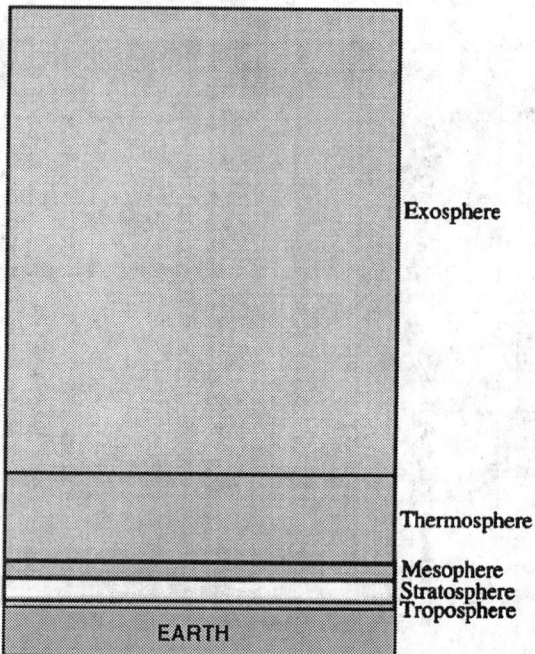

FIGURE 73.12   Simplified diagram of the different layers of the atmosphere.

World Climate Research Programme, the International Geosphere-Biosphere Programme, and the International Human Dimensions of Global Environmental Change Programme.

Several of these instruments show application promise for regional and local community interests, in helping farmers to monitor and control their agricultural productivity (weather, disease control), in early warning and in rescue efforts in case of severe storms (e.g., hurricanes), and in predicting the spread of diseases based on vegetation data combined with socio-economic data.

Examples of Atmospheric Studies.

One of the key issues in climate research is to understand global atmospheric changes and how human activity affects the composition and chemistry of the Earth's atmopshere. To create accurate models, a large number of multiyear global studies must be conducted.

The atmosphere is divided into several layers. From the Earth's surface up to interplanetary space, which starts at about 1000 km, these layers are called troposphere, stratosphere, mesosphere, thermosphere, and exosphere. Figure 73.12 shows a simplified diagram of the different atmospheric layers. Each of these layers is characterized by differences in chemical composition that produce variations in temperature. The two lower layers, troposphere (up to 10 km) and stratosphere (10 to 50 km above the Earth), are particularly important since 99% of the water vapor in the atmosphere is included in the troposphere, and 90% of the ozone of the atmosphere is included in the stratosphere. All weather phenomena occur within the troposphere, with some turbulence sometimes extending to the lower stratosphere. The concentration of ozone, which should stay mainly concentrated in the stratosphere, is being studied in both the troposphere and stratosphere.

Ozone is a relatively unstable molecule made up of three oxygen atoms. Depending on the altitude where it is found, ozone is referred to as "good" or "bad" ozone. The largest concentration of ozone is located in the stratosphere, at an altitude between 20 km and 30 km, and plays a major role in the evolution and the protection of life on Earth. Since it absorbs most of the harmful ultraviolet radiation from the Sun, stratospheric ozone protects life on Earth. When found closer to the Earth's surface, ozone may be harmful to lung tissue and plants. Recent studies have shown that the proportions of ozone in the air are increasing compared to decreasing amounts of protective ozone. However, studies still need to determine if these changes are due to human activity or if they are part of regular natural cycles.

The mission of the Total Ozone Mapping Spectrometer (TOMS) is to provide global measurements of total column ozone as well as of sulfur dioxide on a daily basis. The TOMS instrument measures the reflectivity of the atmosphere in six near-UV wavelengths (see Table 73.9) and provides differential UV absorption and surface reflectivity data. From these measurements, total ozone is computed by searching precomputed albedo tables, which depend on solar zenith angle, view angle, latitude, surface reflectance, and surface pressure; a lower amount of radiation measured by TOMS corresponds to higher concentrations of ozone. Maps of volcanic eruptions are a byproduct of TOMS sulfur dioxide measurements. The first TOMS instrument was flown on Nimbus 7 in 1978; successive ones were launched on a Russian Meteor spacecraft in 1991, on an Earth Probe satellite in 1994, and on the Japanese ADEOS satellite in 1996. See References 11 to 16 for more information on TOMS and ozone measurements.

The TOMS measurements are also being compared to the ozone measurements provided by the NOAA (National Oceanic and Atmospheric Administration) series of the Television Infrared Observing Satellite (TIROS) Operational Vertical Sounder (TOVS) data. The TOVS sounding unit consists of three instruments, including the High-Resolution Infrared Sounder-2 (HIRS-2) whose channels are shown in Table 73.9. TOVS-type instruments have been flying since 1978. These instruments provide information about the structure of the atmosphere, vertical temperature and moisture profiles, as well as cloud amounts and heights. Through analysis of this data, the TOVS Pathfinder data set is created and contains 74 layers of measurements on attributes such as temperature, water vapor, ozone level, precipitation, cloud coverage, etc. taken at various atmospheric pressure levels (e.g., 1000 mb, 850 mb, etc.). A full global coverage of TOVS data is produced twice daily, and a 16-year global data set for climate studies is being gathered [17].

Another satellite that obtains atmospheric data is the Upper Atmosphere Research Satellite (UARS). UARS was launched in 1991 and performs a comprehensive study of the stratosphere and furnishes important new data on the mesosphere and the thermosphere. UARS operates 585 km above the Earth in a near circular orbit inclined 57° to the equator. This orbit permits UARS sensors to provide global coverage of the stratosphere and mesosphere and measurements are made approximately every 36 days. The ten UARS chemistry and dynamics sensors are making measurements of temperature, pressure, wind velocity, and gas species concentrations. All these simultaneous measurements will help define the role of the upper atmosphere in our climate and its variability.

The Tropical Rainfall Measuring Mission (TRMM, [18]) is a joint project between the United States and Japan. The goal of this project is to measure precipitation at tropical latitudes and to provide accurate mapping of tropical rainfall. The mission consists of three instruments: a precipitation radar, a multichannel microwave radiometer, and a visible-infrared scanner. The data provided by TRMM will be very important to verify and develop climate models.

The French space agency, CNES, has also developed the POLDER (POLarization and Directionality of the Earth's Reflectances) instrument, which flew on ADEOS. This is the first French/Japanese cooperative project in the area of Earth observation. A second, identical instrument is to be flown on ADEOS-2, successor to ADEOS, in 1999. POLDER is a wide field-of-view imaging radiometer that will provide global, systematic measurements of spectral, directional, and polarized characteristics of the solar radiation reflected by the Earth/atmosphere system, as well as aerosols, land and sea surfaces, and water vapor measurements.

NOAA's AVHRR (Advanced Very High Resolution Radiometer [19]) is very useful to study biomass burning in the tropics, and the interactions of smoke particles with clouds. More generally, information from the five AVHRR channels (see Table 73.9) is integrated into clouds and climate models.

Weather images are an everyday occurance televised all over the world. Several weather satellites are operated by several countries. In the U.S., NASA and NOAA are operating the GOES series of geostationary satellites, which provide global weather data every 30 min since 1974, positioned at 36,000 km above the Earth. GOES image and sounder data are also used for climate studies. In Europe, the Meteosat weather satellites are developed and launched by ESA, and financed and owned by Eumetsat, an international organization of 17 European weather services. Meteosat-1 was launched in 1977, followed by five others in 1981, 1989, 1991, and 1993. Three of them are currently in service, each equipped with an imaging radiometer. Table 73.9 shows the spectral ranges of operations of these two series of geostationary satellites. Several channels in the visible, water vapor, and thermal-IR spectral bands provide important information about cloud coverage, storm formation and evolution, as well as Earth radiation.

Examples of Terrestrial Studies.

Land Cover Applications. There are two basic types of data considered most important for global change research [20]; the data for documenting and monitoring global change, and the data for discovering the dynamical interplay among the various elements that define our environment. Previous studies show that global studies of land transformations require extrapolation among several scales (spatial, radiometric, and temporal). This extrapolation is especially important to control the minimum detectable change, whether spatial, spectral, or temporal. This accuracy in change detection, which is based on the properties of the sensing systems [21], can be especially essential in distinguishing between nature- and human-induced changes.

Getting accurate quantitative information about the distribution and the areal extent of the Earth's vegetation formations is a basic requirement in understanding the dynamics of the major ecosystems. Among all land transformations most critical to study for global change research, the assessment of tropical forests is one of the most important [22-25]. The tropical forest biome forms 7% of the Earth land surface, and its extensive loss could have a major impact on the future of the Earth (habitat fragmentation, species extinction, soil degradation, global climatic modifications, etc.). Previous studies have shown that in the last 2 decades, 50% of the areal extent of tropical forests might have been lost to deforestation [23]. At present, there is a wide range of estimates of the areal extent of tropical forests and of their rates of deforestation. Therefore, there is a great need to produce accurate and up-to-date measurements concerning the Tropical Forest worldwide. A range of different sensors must be utilized for such applications.

Other examples of land cover applications include agriculture and crop forecasting, water urban planning, rangeland monitoring, mineral and oil exploration, cartography, flood monitoring, disease control, real estate tax monitoring, detection of illegal crops, etc. In many of these applications, the combination of remote sensing data and Geographic Information Systems (GISs; see References 26 and 27) show great promise in helping the decision-making process.

Most instruments utilized to observe land features are on-board low Earth orbit satellites and are multispectral sensors with two or three bands in the visible part of the spectrum and at least one band in the infrared [6].

The Landsat series of satellites is the oldest land monitoring satellite system. Initiated in 1967, the Earth Resource Technology Satellites (ERTS) program was planning a series of six satellites to perform a broad-scale, repetitive survey of the Earth's land areas. After the launch in 1972 of the first ERTS-1, the ERTS program was renamed "Landsat." As of 1997, five Landsat satellites have been launched, each one carrying two instruments. The payload of Landsat-1 and -2 included a Return Beam Vidicon (RBV) camera and a Multispectral Scanner (MSS), while Landsat-4 and -5 still use the MSS and the Thematic Mapper (TM). The first RBV system consisted of three television-like cameras with a ground resolution of 80 m, each looking respectively at the green, red, and near-infrared portions of the spectrum. On Landsat-3, the RBV was 30 m panchromatic. MSS quickly became of primary interest due mainly to its capability of producing multispectral data in a digital format. The four MSS spectral bands are shown in Table 73.9, and the spatial resolution of MSS data is about 80 m. Very early on, the utility of MSS data was recognized for such applications as agriculture, mapping, forest monitoring, geology, as well as water resource analysis. The same MSS system was kept on Landsat-4 and -5, but the RBV system was replaced by the TM system. Like MSS, TM is a multispectral scanner, but includes spatial, spectral, and radiometric improvements over MSS. With seven bands instead of four (see Table 73.9), TM covers a larger portion of the visible wavelengths, and includes two mid-IR and one thermal-IR bands. Data are quantized over 256 levels (8 bits) instead of the 64 levels for MSS, and the spatial resolution of a TM pixel is about 30 m. TM data is usually chosen to perform classification of land-cover features, man-made or natural. In vegetation and change detection applications, leaf segmentation is studied with TM visible channel data, while cell structure can be seen in near-IR, and leaf water content is found in the mid-IR channel data. The two mid-IR bands (5 and 7) are also useful for geologic applications. All the Landsat satellites are placed in low Earth orbit (at an altitude of about 900 km for Landsat-1 to -3 and 705 km for Landsat-4 and -5) and in a near-polar, sun-synchronous orbit. Landsat-4 and -5 cross the equator at 9:45 a.m. to hopefully take advantage of cloud-free imagery. Landsat-4 and -5 have a 16-day repeat cycle and their orbits are 8 days out of phase. Landsat-6 failed to achieve orbit in 1993; Landsat-7 is planned to be launched in 1998 and includes an improved TM instrument, the Enhanced Thematic Mapper (ETM), which will also include a panchromatic band at a spatial resolution of 15 m. For a more in-depth description of Landsat systems, see References 1, 29, and 30; for more applications and analysis of Landsat data, see Reference 30.

As previously mentioned, NOAA's AVHRR is primarily used for atmospheric applications but is also utilized for land surface applications. Having a near-polar, sun-synchronous orbit (at 833 km above the Earth's surface), the AVHRR instrument provides global data with a 1.1-km spatial resolution at nadir, and includes five bands, with daily or twice-daily (for thermal-IR) coverage. Since 1978, AVHRR data are available at full resolution (called Local Area Coverage (LAC)) or subsampled to 4 km (known as Global Area Coverage (GAC)). Because of its high temporal resolution, the AVHRR instrument is very useful in applications such as flood, storm, or fire monitoring, as well as volcanic eruption. Because of its global area coverage, AVHRR is also often utilized for studying geologic or physiographic features, vegetation conditions and trends at a global, continental or regional level, snow cover mapping, soil moisture analysis, and sand storms and volcanic eruptions worldwide. A popular parameter extracted from AVHRR data is the Normalized Difference Vegetation Index (NDVI), computed from GAC data as: NDVI = (Channel 2 – Channel 1)/(Channel 1 + Channel 2). GAC data is processed daily and composited on a weekly basis to produce a global map showing vegetation vigor. An example of NDVI applications is the monitoring of the Sahara desert extent. AVHRR data are also used for sea surface temperature.

The first Système Pour l'Observation de la Terre (SPOT), designed by CNES, was launched in 1986. SPOT-2 and SPOT-3 were launched, respectively, in 1990 and 1993. The SPOT satellites fly in a near-polar, sun-synchronous low Earth orbit at an altitude of 832 km and cross the equator at 10:30 a.m. SPOT's repeat cycle is 26 days, but due to its off-nadir viewing capability (viewing angle up to 27°), SPOT has a "revisiting" capability with which the same area can be seen up to five times every 26 days. This off-nadir viewing capability also enables some stereo imaging possibilities. The SPOT payload includes two identical high-resolution-visible (HRV) imaging instruments that can be employed in panchromatic or multispectral modes, with pointing capabilities. Spectral coverage of these two modes is given in Table 73.9. The panchromatic mode is 10 m spatial resolution, while multispectral data have a 20 m spatial resolution. Whereas Landsat is a scanning mirror-type instrument, SPOT employs a push-broom system with a linear array of detectors simultaneously acquiring all data pixels in one image line, which minimizes geometric errors. SPOT data are very useful for applications involving small features. Due to its increased spatial resolution, revisit and pointing capabilities, simultaneous panchromatic and multi-spectral data, and stereo data capabilities, SPOT opens a new range of applications, such as topographic mapping, studies of earthflows (e.g., land, rock, and mudslides), urban management, and military applications. SPOT-4 is planned for launch in 1998, and SPOT-5 in 2002. Among the planned improvements, a mid-IR channel will be added to SPOT-4, which will also carry a new AVHRR-type instrument, the European Vegetation instrument.

Since 1988, India has launched a series of five satellites, the IRS series. These satellites were designed in support of India's agriculture and exploration businesses, and they seem to be successful in this challenge of bringing remote sensing to the users (see Reference 10). For land applications, IRS-1A, -1B, and -1C all carry the LISS instrument, which is a multispectral scanner very similar to Landsat-TM. LISS-2 acquires imagery in four bands similar to bands 1 to 4 of Landsat-TM (from visible to near-IR) at the spatial resolution of 36.5m (see Table 73.9 for wavelengths description). LISS-3, carried on IRS-1C, also acquires imagery in four bands, but the visible blue band has been suppressed and replaced by a mid-IR band similar to TM/band 5. Due to their similarity to Landsat data, IRS/LISS-2 data could be used as complements or replacements to Landsat data if needed until Landsat-7 is launched. IRS-1C also carries a 5-m panchromatic imstrument whose data are co-registered with LISS-2 data. For more details on IRS data, see References 10 or 27.

Other instruments are also available. JERS-1, designed by Japan, was launched in 1992, and its payload includes both an SAR instrument and an optical-imaging system; see Table 73.9 for its spectral channels from visible to mid-IR wavelengths, with spatial resolutions of 18 m and 24 m. MOMS, the German Modular Optoelectronic Multispectral Scanner, has been flying as a research instrument on U.S. Space Shuttle missions, and has a spatial resolution ranging from 4.5 m to 13.5 m; see Reference 27 for more details on these different instruments.

Among the first EOS intruments to be launched is the Moderate Resolution Imaging Spectrometer (MODIS). MODIS is being developed to provide global monitoring of the atmosphere, terrestrial eco-system, and oceans, and to detect climate change. MODIS will cover the visible to infrared portions of the spectrum with 36 channels at spatial resolutions of 250 m to 1 km. Many interesting land studies will be performed by fusing together AVHRR, Landsat, and MODIS data.

The fusion of several of these types of data is becoming a very important issue [26]. Already, sensors such as SPOT or LISS-3 present the advantage of acquiring co-registered panchromatic and multispectral data. It would be of great interest to combine data from sensors with different spectral and spatial resolutions, as well as different viewpoints. The combination of coarse-resolution viewing satellites for large area surveys and finer resolution sensors for more detailed studies would offer the multilevel information necessary to assess accurately the areal extent of features of interest (e.g., tropical forests). The fusion of multispectral data with SAR data would provide information on ground cover reflectance with the shape, roughness, and moisture content information from SAR. Of course, multidata fusion requires very accurate registration of the data, as will be described in Section 2.4.

Geologic Studies. Other examples of terrestrial studies are the mapping of geologic features, such as geologic faults and earthquake sites, or volcanic eruptions. Although many geologic features lie beneath the surface of the Earth, remote sensing (aerial or satellite) provides a valuable tool to perform geologic mapping, landforms and structures analysis, as well as mineral exploration. This is due to the fact that topography and soil properties provide clues to underlying rocks and structural deformations. Landsat and SPOT gather data about the effects of subsurface geologic phenomena on the surface. These data are especially useful to recognize some specific landforms (such as volcanoes), to depict topographic features, discriminate some geologic facies and rock unit distribution patterns and more generally provide regional overviews of surface geology. In mineral exploration, rock or soil alteration can be detected by spaceborne sensors and may indicate the presence of mineral deposits or oil reservoirs. Other types of sensors that are very useful for geologic applications are radar sensors, such as the two radar systems, SIR-C and X-SAR, carried on the Space Shuttle Endevour in 1994. These sensors captured in real-time the eruption of a volcano in Russia and an earthquake in Japan [31]. For more information on geologic applications, see References 2 and 32.

Geophysics Studies. Other satellites, such as the LAEGOS-1 and -2, have proved very useful in geophysics for the study of the Earth's gravity field, tectonic plate motion, polar motion, and tides. LAGEOS sensors are reflector orbs covered with laser beams. For more information on these studies, see References 33 and 34.

Examples of Ocean Studies.

Oceans cover 75% of the Earth's surface and contain most of the energy of the planet. Although their role in climate evolution is very important, it is still poorly understood. By understanding chemical, physical, and biological processes in oceans, scientists will be able to model the interactions between oceans and the atmosphere and determine how these interactions affect Earth temperature, weather, and climate.

An example of interaction between oceans and the atmosphere is illustrated by the phenomenon known as El Niño/Southern Oscillation, which occurs in the tropical Pacific Ocean, usually around Christmas time. El Niño is due to a mass of warm water, usually located off Australia which moves eastward toward equatorial South America. El Niño develops every few years (observed on average every 4 years to a maximum of 7 years), and alters the weather in Australia, Africa, South Asia, and the tropical parts of the Americas. By understanding how winds and waves move in the tropical Pacific, scientists have been able to predict the El Niño phenomenon up to 1 year in advance. Similar phenomena are being studied in the Atlantic Ocean, where patterns seem to move much more slowly.

Besides being used to create global models, and in storm and weather forecasting, ocean data are also very important for day-to-day applications such as ship routing, oil production, and ocean fishing.

Ocean Color. Ocean color data are critical for the study of global biogeochemistry and to determine the ocean's role in the global carbon cycle and the exchange of other critical elements and gases between the atmosphere and the ocean [35,36]. It is thought that marine plants remove carbon from the atmosphere at a rate equivalent to terrestrial plants, but knowledge of interannual variability is very poor. For most oceans, the color observed by satellite in the visible part of the spectrum varies with the concentration of chlorophyll and other plant pigments present in the water. Subtle changes in ocean color usually indicate that various types and quantities of microscopic marine plants (i.e., phytoplankton are present in the water); the more phytoplankton present, the greater the concentration of plant pigments and the greener the water.

The recently launched (October 1997) Sea-viewing Wide Field-of-view Sensor (SeaWiFS,), which is a part of MTPE, provides quantitative data on global ocean bio-optical properties to the earth science community. SeaWiFS is a follow-on sensor to the Coastal Zone Color Scanner (CZCS), which ceased operations in 1986. See Table 73.9 for a channel description of these two sensors; notice that all channels are concentrated in the [0.4,0.7] interval of the electromagnetic spectrum.

Other sensors for ocean color are the imaging spectrometer for ocean color applications MOS-IRS, launched on the Indian Remote Sensing Satellite IRS-P3 in March 1996, and the imaging spectrometer MOS-PRIRODA, launched aboard the Russian multisensor remote sensing module PRIRODA and docked to space station MIR in April 1996.

Ocean Dynamics. By studying ocean circulation and sea levels trends, scientists will be able to create global maps of ocean currents and of sea surface topography. Since sea surface height and sea level variations are related to sea surface temperatures, the monitoring of mean sea levels enables the gathering of evidence that can measure global warming or El Niño-type events. For example, conditions related to El Niño may result in a change in sea surface height of 18 cm or greater [37].

TOPEX/Poseidon (T/P) is an important collaboration between U.S./NASA and France/CNES. T/P uses radar altimetry to provide 10-day maps of the height of most of the ice-free oceans' surface. Circling the world every 112 min, the satellite gathers data for 3 to 5 years, and could be operational for 10 years. The T/P satellite was launched in August 1992 on an Ariane rocket. TOPEX measures the height of the ocean surface, as well as changes in global mean sea level. From these altimetry data, global maps of ocean topography are created, from which speed and direction of ocean currents are computed worldwide. Changes in mean sea level are monitored and currently are viewed mostly as related to natural ocean variability and not climate change. Climate change must be studied over a much longer time series of altimeter data. T/P also enables study of tides, waves geophysics, and ocean surface winds.

Sea winds are also being studied with scatterometers such as the NASA Scatterometer (NSCAT) and the soon to be launched EOS Scatterometer, SearWinds. These high-frequency radar instruments measure the reflected signals from the ocean surface to detect wind speed and direction.

ERS-1 is another satellite utilized to measure ocean dynamics. ERS-1 was launched in 1991 on a sun-synchronous, near-polar low-Earth orbit at an altitude of 780 km. ERS-1 orbits the Earth in 100 min and covers the entire planet in 3 days. Its payload consists of two specialized radars and one infrared sensor. The Active Microwave instrument, consisting of a synthetic aperture radar and wind scatterometer, produces extremely detailed images of 100 km swath of the Earth's surface, with a spatial resolution of

20 m. The radar altimeter provides accurate range to sea surface and wave heights, and the along-track scanning radiometer constructs detailed pictures of the thermal structure of the seas and oceans from surface temperature measurements at an accuracy of less than 0.5°C. ERS-1 images are also utilized for land applications where the instruments need to "look through" the cloud cover.

The study of sea ice with passive and active microwave sensors is also very important and additional reading in this topic can be found in References 14 and 38.

A Few Examples of Space Science Applications.

Astronomical satellites have been developed to observe far distant objects that are usually beyond the range of ground-based instruments. They explore phenomena in the solar system, and beyond. Satellite observation of astronomical objects is also less sensitive to atmospheric interactions and can achieve higher accuracy than ground-based measurements. This section will give a brief description of the most important space science satellites.

The first astronomical satellite to be put into synchronous orbit was the International Ultraviolet Explorer (IUE) laboratory. IUE was launched in 1978 under a joint program involving NASA, ESA, and the United Kingdom. In more than 15 years of service, IUE gathered observations on more than 10,000 celestial objects. A program for coordinating its observations with those of the ROSAT satellite has been carried out under the title RIASS (Rosat-IUE All-Sky Survey). ROSAT, the Roentgen Satellite, is a joint collaboration between Germany, the U.S., and the U.K., and was launched in 1990. It is an X-ray observatory that carries two instruments, the X-ray telescope and the wide field camera.

The Infrared Astronomical Satellite, IRAS, is a joint project of the U.S., the U.K., and the Netherlands. The IRAS mission was intended to provide a survey of infrared point sources (from 12 to 100 μm), but has also produced very high-quality image data. MSX (the Mid-Course Space Experiment), ISO (the Infrared Space Observatory), and SIRTF (the Space InfraRed Telescope Facility) are other examples of recently or soon-to-be launched sensors that provide an even finer resolution.

Hipparcos (High Precision Parallax Collecting Satellite) is an astronomy satellite launched in August 1989, with the purpose of determining the astrometric parameters of stars with unprecedented precision. After a life of 4 years, Hipparcos has produced two catalogs. The Hipparcos Catalogue provides position, parallax, and proper motion measurements with accuracy of 2 milliarcsec at 9 mag for over 120,000 stars. The Tycho Catalogue is the result of somewhat less precise astrometric measurements for some 1 million stars.

COBE, the Cosmic Origin Background Explorer developed by NASA, was launched in 1989. Designed to measure the diffuse infrared and microwave radiation from the early universe, it carried three instruments: a Far Infrared Absolute Spectrophotometer (FIRAS), a Differential Microwave Radiometer (DMR), and a Diffuse Infrared Background Experiment (DIRBE). The first full-sky coverage was completed in 1990.

The Hubble Space Telescope (HST) is one of the most well-known astronomical satellites. It was built as a joint NASA/ESA project, and was launched in 1990 as a long-term space-based observatory. The heart of the system is a large reflector telescope 2.4 m in diameter. All the instruments on-board the HST use the light gathered by the reflector telescope. Current HST instruments are the Wide/Field Planetary Camera 2 (WFPC2), the Space Telescope Imaging Spectrograph (STIS), the Near-Infrared and Imaging Spectrograph (NICMOS), and the Faint Object Camera, FOC, provided by ESA. These different instruments can observe astronomical objects from UV to IR wavelengths. In 1993, the HST was serviced to correct a preliminary fault affecting the mirror with a corrective optical apparatus named COSTAR. Despite the preliminary mirror fault, and even more after correction, the HST has achieved much better results than those from observatories on Earth. Since it is located above the Earth's atmosphere (at 600 km), the HST produces highly detailed images of the stars and can detect objects beyond the range of ground-based instruments. Observations with the HST are scheduled as a space-based observatory according to worldwide astronomers' proposals.

ASCA, the Advanced Satellite for Cosmology and Astrophysics, is the product of a Japan/U.S. collaboration. Launched in 1993, this X-ray astronomy mission was still operational in 1997, and carries four large-area X-ray telescopes with arc minute resolution. ASCA data are being archived and can be searched and retrieved online at the High Energy Astrophysics Science Archive Research Center, HEASARC. GRO (Gamma Ray Observatory), and AXAF (the Advanced X-ray Astrophysics Facility) are other examples of space sensors which operate in this spectrum range.

Management and Interpretation of Satellite Data

Satellite sensors gather the electromagnetic energy reflected or emitted from Earth (or any other planetary) surface features. This energy is then converted into a digital representation that is visualized by a user and interpreted either visually or with a computer. This section summarizes some preliminary ideas on how the digital representation is formed and the basic types of data processing necessary before any further interpretation of the data. For more details on the processing of remote sensing data, see References 39 to 42.

Fundamental Data Levels

After transmission from the satellites, raw data are usually processed, calibrated, archived, and distributed by a ground-based data system. Most of NASA satellite data products are classified in the following data levels [7]:

- *Level 0 data* are the reconstructed raw instrument data at full resolution.
- *Level 1A data* are reconstructed, time-reference raw data, with ancillary information including radiometric and geometric coefficients.
- *Level 1B data* are corrected Level 1A data (in sensor units).
- *Level 2 data* are derived geophysical products from Level 1 data, at the same resolution and location; for example, atmospheric temperature profiles, gas concentrations, or winds variables.
- *Level 3 data* correspond to the same geophysical information as Level 2, but mapped onto a uniform space-time grid.
- *Level 4 data* are model output or results from analysis of lower-level data.

Image Restoration

Ideally, the scene as viewed and recorded by a sensor would be an exact rendering of the features within the sensor's viewing extent, represented as a spectral curve indicating the amount of energy reflecting/radiating for each point in a scene for a range of given wavelengths. From an engineering standpoint, this is impossible, however, because each image is discretized into a finite number of pixels. Variability defines nature, so each pixel will map into a region of the scene that contains a number of features, each producing its own unique spectral curve. The spectral signature recorded for a pixel is a function of these features and their relative sizes within the region covered by the pixel. The spectral response itself is also discretized into a finite number of bandwidths, where each bandwidth covers a small continuous band of the spectrum. The sensor records for each pixel the amount of energy observed for each band. This number itself, referred to as a Digital Number (DN), must be represented in a finite amount of computer memory, such as 8 bits, meaning that each band records activity as a whole number ranging from 0 to 255.

In practice, a number of events outside human control affect the quality of the observation, such as atmospheric scattering, variations in sun angle, high albedo, and instrument errors. Depending on the application, it may be desirable to correct for the presence of thin clouds within an image. The process of image restoration attempts to control and correct for these conditions [42].

Electromechanical effects due to the instrument itself can be discovered due to their periodic nature (such as caused by the repeated motion of a push-broom, or the revolving of a mirror, or the physical process of gathering calibration points). A Fourier transform applied to an image from a sensor undergoing periodic interference exhibits strong noise spikes. A filter can then be used to remove the offending data. Unfortunately, this also removes any good data that happens to fall at the same frequency, although normally this is but a small portion of the data. Data outages and instrument recorder failures appear as streaks in the image parallel with the scanline, and can be discovered by comparing the respective readings of the pixels in the surrounding scanlines of the image.

To account for the atmospheric effects of Rayleigh and aerosol scattering, an estimate of the portion of the signal that is due to the atmosphere is computed and subtracted from the recorded value. The reflectance of water in the near-infrared region of the spectrum should be effectively zero, so the value to subtract for the near-IR band corresponds to the reading of the sensor observed over clear open water. To compute values to be subtracted for each of the other spectral components, a histogram should be formed for each band of a number of sample readings over clear open water. The lowest reading in each band is then used as an estimate of the value to subtract from each pixel to account for the atmospheric effect. In addition, information derived from TOVS, balloon readings, or the atmospheric correction software 5S can be useful in dealing with atmospheric effects.

Data Compression

Data compression is one of the most important tools to overcome the problems of data transmission, storage, and dissemination [43]. Data compression methods are usually classified as either lossless or lossy. With a lossless data compression scheme, the original data can be reconstructed exactly without any loss; in a lossy compression scheme, original data are reconstructed with a degree of error. For transmission from the satellite to the ground station, a lossless data compression must be utilized. For browsing purposes, lossy compression enables quick searches through large amounts of data. A compression scheme is also characterized by its compression ratio, that is, the factor by which the amount of information which represents the data is reduced through compression. For earth science data, lossless compression schemes provide compression ratios up to 2 or 3, while lossy techniques can reduce the amount of information by a factor of 20 or more without degrading the visual quality of the data.

Among the lossless compression methods, the Joint Photographic Experts Group, JPEG, developed a lossless compression method that is based on a predictor, an entropy encoder for prediction error, and an entropy code specifier. Another lossless compression scheme is the Rice algorithm, which can adapt to data of any entropy range. It is based on a preprocessor that spatially decorrelates the data, followed by a variable length encoder. This algorithm gives some of the best compression ratios among all lossless methods, and has been implemented on VLSI chips at NASA.

JPEG has also developed a lossy method based on the Discrete Cosine Transform (DCT). Other methods such as vector quantization or wavelet compression provide either lossless or lossy compressions. In a vector quantization technique, a dictionary of representative vectors, also called a codebook, and all data are encoded relative to the codebook. In this method, the one-time encoding step is computationally expensive but the decoding step at the user end is fast and efficient. Vector quantization is also utilized in a progressive scheme for "quick look"/browsing purposes. In a subband/wavelet compression method, signals are decomposed using quadrature mirror or wavelet filters [44]. Most energy is contained in the low-frequency subbands and high compression ratios can be obtained by compressing the high-frequency information.

For more information or references on data compression techniques, see Reference 43.

Image Registration

In studying how the global environment is changing, programs such as Mission to Planet Earth [7] or the New Millennium program [45] involve the comparison, fusion, and integration of multiple types of remotely sensed data at various temporal, radiometric, and spatial resolutions. Results of this integration can be utilized for global change analysis, as well as for the validation of new instruments or of new data analysis. The first step in this integration of multiple data is registration, either relative image-to-image registration or absolute geo-registration, to a map or a fixed coordinate system. Another case of image registration is co-registration of multiple bands of one sensor. When the detectors of each spectral band have different spatial locations on the satellite's focal plane, there could be misregistration between each band's raw image [46,47].

Currently, the most common approach to image registration is to extract a few outstanding characteristics of the data, which are called *control points* (CPs), *tie-points,* or *reference points.* The CPs in both images (or image and map) are matched by pair and used to compute the parameters of a geometric transformation. Most available systems follow this registration approach; and because automated procedures do not always offer the needed reliability and accuracy, current systems assume some interactive choice of the CPs. But such a point selection represents a repetitive, labor- and time-intensive task that becomes prohibitive for large amounts of data. Also, since the interactive choice of control points in satellite images is sometimes difficult, too few points, inaccurate points, or ill-distributed points might be chosen, thus leading to large registration errors. A previous study [48] showed that even a small error in registration can have a large impact on the accuracy of global change measurements. For example, when looking at simulated 250 m spatial resolution MODIS (Moderate Resolution Imaging Spectrometer) data, a 1-pixel misregistration can produce 50% error in the computation of the Normalized Difference Vegetation Index (NDVI). So, for reasons of speed and accuracy, automatic registration is an important requirement to ease the workload, speed up the processing, and improve the accuracy in locating a sufficient number of well-distributed accurate tie-points.

Automatic image registration methods can be classified into two types: those that follow a human approach, by first extracting control points, and those that take a more global approach. Among the first methods, the most common features utilized as control points are the centers of gravity of regions — with or without region attributes such as areas, perimeters, ellipticity criteria, affine-invariant moments, and inter-regions distances. More recently, features extracted from a wavelet decomposition have also been utilized, such as maxima and minima of wavelet coefficients, high-interest points, or local curvature discontinuities. A few methods utilize Delaunay triangulation methods to progressively increase the number of accurate control points. For the methods that do not match individual pairs of control points, the transformation is either found by correlation or by optimization, in the spatial or in the frequency domain. When in the spatial domain, correlation or optimization is performed either in the original data or on edge gradient data. Other methods propose a global image matching of edge segments or vectors linking feature points. Some recent research has also focused on the use of wavelets for global image registration. More complete surveys of image registration methods can be found in References 47, 49, and 50.

Dimension Reduction

The first step in analyzing multichannel data is to reduce the dimension of the data space. It is particularly important when the analysis method requires a training step, for example, supervised classification (see next section). The main issue in this case has often been referred as "the Curse of Dimensionality" [52]. If the original data has a large number of bands (e.g., for hyperspectral data), theoretical studies have shown that a very large training set should be utilized; but using a large training set deteriorates the estimation of the kernel density. To solve this problem, various dimension reduction schemes enable to perform classification in a smaller-dimensional subspace. Since the information contained in multiple channels is often redundant, it is possible to decorrelate spectrally the channels and reduce the number of channels to be analyzed without losing any information. Principal Component Analysis (PCA) and Projection Pursuit are the most common techniques for dimensionality reduction. For more information on these methods, refer to References 39 through 42.

Data Mining

One objective of the NASA-initiated Mission to Planet Earth is to gather sufficient data to enable scientists to study the Earth as a dynamic system, resulting in a better understanding of the interactions between humans, the atmosphere, and the biosphere [8]. The episodic nature of most interesting events would cause them to be missed if the data were not being gathered continuously. Comprehensive data sets allow scientists to construct and evaluate complex models of many Earth-related processes. But currently, due to computation and time constraints, only a small percentage of the data gathered by remote sensing is actually viewed by an individual user. Data-gathering missions tend to be multidisciplinary, so different aspects of the data sets are pertinent to different researchers.

Data mining can be defined as the process by which data content is automatically extracted from satellite data, enabling a scientist to query the data holdings based on high-level features present within an image [51]. Given the projected large volumes of data, it is not feasible to rely solely on conventional data management paradigms. Standard methods of segmenting images are inadequate as standalone techniques for image recognition, regardless of the speeds of processing, because there are no general methods for automatically assigning meaningful semantics to any homogeneous regions that are isolated. Metadata derived directly from the image header are not rich enough to enable robust querying of a database in most instances, but limit a user to retrieving all images at a given latitude/longitude during some time period, for example, regardless of the image quality or the unique features existing due to some unexpected set of circumstances. New approaches based on techniques such as image classification (described in the next section) are now feasible, due to the phenomenal increases in computing speed, the availability of massively parallel architectures, and the breakthroughs in signal processing.

An example of data mining is the browsing of 15 years of TOVS data with two complete coverages per day, which would require looking through 10,958 scenes per attribute. Of the several products generated, the scientists are primarily interested in browsing those with some given resolution. After locating a browse product that seems to indicate an interesting structure or phenomenon, a scientist might then retrieve this data temporally, or any supporting data set for further analysis. Scientists using the TOVS data sets desire a more intelligent form of querying so they can quickly and easily find relevant data sets that are pertinent to their research. Certain TOVS observations are more "interesting" than others, and the definition of "interesting" is a combination of objective fact and subjective opinion. Data mining is applicable here to aid in evolving a retrieval heuristic based on an individual scientist's definition of "interestingness." In one approach, the scientist could prepare a representative set of images that are labeled as positive or negative instances of "interesting," and a machine learning system (e.g., neural network, genetic algorithm) could perhaps be trained to classify the remaining images in the TOVS data set according to this definition. In a second approach, the scientists could be asked to identify explicitly structural features within the images that make them interesting, and image processing routines could then be applied to detect images with these features. Over time, a scientist could provide feedback to the heuristic classifier to improve its performance. Both approaches require that the underlying representation language (structures, bin size, spatial and temporal relationships) be robust and flexible enough to permit an appropriate level of expression.

Classification

Image classification is the task of developing a statistical model that labels every potential point in some multidimensional space. A *parametric classifier* assumes that data are described by some underlying parameterized probability density function (PDF). A training set of representative data from the domain is then used to supply appropriate values. For example, if a Gaussian or normal distribution is assumed, then the means, standard deviations, and joint-covariance matrix can be computed from the training data. A *nonparametric* or *unsupervised classifier* is typically used when there is insufficient knowledge about the type of underlying PDF for the domain. Self-organizing classifier models, such as certain kinds of neural networks, are also considered nonparametric classifiers when they make no *a priori* assumptions about any PDF.

In a *statistical* or *supervised classifier*, knowledge about the distribution of the data is utilized to assign a label to an unclassified pixel. Using "ground-reference data," a training set of known points is created. A prototype vector can then be calculated as the mean of all samples for each of the classes. Assuming a Gaussian distribution in each of the channel readings for a given class, the standard deviation for each class is computed based on the sample. Then, the lowest distance from the given feature prototypes to an unclassified point determines the class of this incoming point. As simple and elegant as this approach might appear, in actuality its utility is limited. Features are not so discernible from a random labeling of an image. Thus, although this algorithm is inaccurate, it is consistent in its mislabelings. The algorithm's deterministic nature and underlying use of a continuous function combine to produce predictable behavior. In general, this algorithm labels all points similarly if they fall within the same neighborhood in the feature space.

Other parametric classifiers follow the *Maximum Likelihood Decision Rule*, which allows the construction of discriminant functions for the purposes of pattern classification. For more details on this technique, refer to Reference 53.

The classifiers discussed above are, by definition, required to assign an unclassified pixel to the one nearest class. No measurement of the distance to that class or proximity to other classes is recorded, and no information on the confidence of the labeling is provided. *Fuzzy classifiers*, on the other hand, are not obligated to pigeonhole a pixel into a single class. Instead, the pixel is assigned a degree of membership for each possible class. Intuitively, and indeed for mathematical tractability, the pixel's memberships must sum to one, and the degree of membership for a given class must be between 0 and 1, inclusively. Two examples of fuzzy classifiers are given below. The *Fuzzy Nearest Neighbor* nonparametric classifier places an unclassified vector in the dominant class of its k-closest training vectors. If no class has an outright majority, then distances to the nearby vectors for each class which tied are summed and the unclassified vector is placed in the class with the minimum sum. The *Fuzzy Decision Tree Classifier* utilizes a decision tree as the data structure that encapsulates knowledge of what to do given a set of conditions. See Reference 51 for more information on this method. Although this algorithm is conceptually simple, it is only recently that it has become computationally feasible due to the need to search the tree for each unclassified pixel to locate the nearest path. The search algorithm can also be sped up by running the algorithm on a parallel architecture such as a Single Instruction Multiple Data (SIMD) machine.

Many researchers have investigated the use of *neural networks* for automatically extracting metadata from images [54]. Many different neural network models have been considered, but with respect to performance accuracy, the backpropagation training technique has shown to be the best classifier [55]. The *backpropagation algorithm* is the most common method for training a neural network, and is the backbone of much of the current resurgence of research into neural nets [56]. With respect to pattern recognition, backpropagation can be considered to be a nonparametric technique for estimation of a posteriori probabilities.

Accuracy Assessment

A measurement derived solely from satellite imagery is of questionable use unless the technique employed for computing that measurement on those data has been validated. A technique that appears to work accurately on satellite imagery over some given location at some given time may perform abysmally on data from the same sensor at another location, or for the same location at another time. The reasons for this are many: through the course of a year, the sun angle changes causing different lighting conditions; from pass to pass, the viewing angle of the instrument can be different; with seasonal changes, surface reflectance varies due to weather conditions and the alteration of land cover as crops appear in different stages; atmospheric conditions fluctuate; and the sensor and spacecraft themselves age and possibly perform differently.

The key factor in any accuracy assessment of remote sensing data is the method and source used for dertermining what the satellite sensor is actually viewing. This ground reference data is gathered independent of the remote sensing data itself. There are several sources that can be construed as ground reference data, and each source has its own degree of accuracy. The most obvious is an actual site visit to the area of interest. What is observed, also known as "ground truth," is recorded and compared to the digital rendition of the same spatial extent. This approach usually has a high degree of accuracy, but it is often prohibitively expensive. Depending on the time between the on-site ground reference gathering and the imaging of the area, the validity of the ground reference data may be lessened due to anthropomorphic or natural influences. The shorter the life of the feature being measured, the more difficult it is to find or gather meaningful time-critical ground reference data. If ground reference data is not available, it may be possible to perform photointerpretation with some degree of success. This itself depends on the knowledge of the photointerpreter, and the availability and suitability of a display device for viewing the image data and recording the photointerpreter's assessment. Another approach is to compare the digital image with other sources of ground reference data such as air photos or appropriate reference maps, provided the feature of interest is detectable using those sources. The degree of correspondence between the ground reference data and the measurement derived from the sensor data can then be compared for accuracy. In the worst case, the lack of adequate/accurate ground reference data requires using an unsupervised clustering approach that is usually less accurate but much cheaper to produce.

The Future in Satellite Imaging and Sensing

Success of future earth and space science missions depends on increasing the availability of data to the scientific community who will be interpreting space-based observations, and on favoring interdisciplinary research for the analysis and the use of this data. One of the main challenges in the future of satellite imaging and sensing will be to handle, archive, and store all of these data in a way that can be easily accessible and retrieved by anyone who needs to use them. Systems such as the EOS Data and Information System (EOSDIS) [57,58], will require that over 1 terabyte per day be collected and processed into several levels of science data products within several hours after observation. After 15 years, the estimated amount of collected, processed, analyzed, and stored data will equal about 11,000 terabytes. Also at NASA, efforts are underway to design an advanced information system, based on an object-oriented database, with the express purpose of developing, incorporating, and evaluating state-of-the-art techniques for handling EOS-era scientific data challenges [59].

Another challenge will be to analyze this tremendous amount of data, and to find out new ways to fuse, integrate, and visualize this data. In particular, research in fast computational capabilities, such as field programmable gate arrays (FPGAs), will be of great importance.

On the other hand, the wide distribution of satellite data to the general public will be facilitated by regional distribution systems such as the Regional Application Centers, RACs [60,61], whose goal is to provide local users, such as industry, agriculturalists, urban planners, regional communities, with local and "on-time" information about regional applications.

Satellite imaging and sensing is a field with a history of more than 2 decades, but is still in full expansion. The future in satellite imaging and sensing will see developments in several areas. The next millennium will see an explosion of commercial satellite systems and the profusion of satellite data, which will have economic and sociopolitical implications. As of this writing, over 30 commercial Earth sensing satellites are either being planned or being built. MTPE and EOS will generate unprecedented amounts of diverse resolution data. The future will also see the development of locally directed satellite systems, in answer to specific applications for specific areas of the planet. Telecommunications will also be a large part of the space market. In space, after the large success of the Mars Pathfinder mission, exploration of distant planets will see a flourishing of distant satellite systems providing unprecedented amounts of data to analyze regarding other planets' surface features, atmospheric, and magnetic properties. The understanding of other planets will also enable scientists to learn more about the Earth comparatively to other planets such as Mars, and to build a comprehensive data set to aid in planning future missions. The Mars Global Surveyor is an example of such as a mission; it will map the entire planet Mars by taking high-resolution pictures of the surface. The future might see a 10-year NASA program that will send pairs of Surveyor-like orbiters and Pathfinder-like landers to Mars every 26 months. In order to gather novel and interesting data, this type of mission will need an increasing amount of on-board processing that will perform mission planning, image processing and understanding, as well as data compression and fusion. The design of systems including on-board processing will require new computational capabilities, such as reconfigurable hardware and parallel processors, as well as new developments in intelligent systems. In the near future, satellite imaging and sensing is a field that will produce unprecedented information about the Earth, its environment, and our solar system.

Acknowledgments

The authors would like to thank William J. Campbell for his support and for his useful comments upon reviewing our paper, Bob Mahoney for providing the spectral libraries used to generate Figure 73.11, and all the anonymous or nonanonymous authors of Web pages that we consulted during the research part of this endeavor. In particular, the online remote sensing tutorial by N.M. Short, edited by J. Robinson at the URL http://code935.gsfc.nasa.gov/Tutorial/TofC/Coverpage.html, and the list of selected links on remote sensing compiled by U. Malmberg and found at the URL http://www.ssc.se/rst/rss/index.html were very useful.

References

1. T.M. Lillesand and R.W. Kiefer, *Remote Sensing and Image Interpretation*, Second edition. John Wiley & Sons, New York, 1987.
2. J.B. Campbell, *Introduction to Remote Sensing*, Second edition. The Guilford Press, 1996.
3. A.P. Cracknell and L.W.B. Hayes, *Introduction to Remote Sensing*, Taylor & Francis, London, New York, 1991.
4. R. Greeves, A.Anson, and D. Landen, *Manual of Remote Sensing*, American Society of Photogrammetry, Falls Church, VA, 1975.
5. P. Cheeseman, B. Kanefsky, R. Kraft, J. Stutz, and R. Hanson, *Super-Resolved Surface Reconstruction from Multiple Images*, Technical Report FIA-94-12, NASA/Ames Research Center, Artificial Intelligence Branch, Oct. 1994.
6. T.E. Bell, Harvesting Remote-Sensing Data, *IEEE Spectrum*, 32(3), 24-31, 1995.
7. *1995 MTPE EOS Reference Handbook*, Editors G. Asrar and R. Greenstone, EOS Project Science Office, Code 900, NASA/Goddard Space Flight Center.
8. *1993 EOS Reference Handbook*, G. Asrar and D. Dokken, Eds., available from the Earth Science Support Office, Document Resource Facility, 300 D Street, SW, Suite 840, Washington, D.C. 20024; Telephone: (202) 479-0360.
9. *SCOPE: SCenario for Observation of Planet Earth*, Publication of the National Space Development Agency of Japan, NASDA, 1995.
10. K.P. Corbley, Multispectral Imagery: Identifying More than Meets the Eye, *Geo Info Systems*, 38-43, June 1997.
11. A.J. Krueger, The Global Distribution of Total Ozone: TOMS Satellite Measurements, *Planetary and Space Sciences*, 37(12), 1555-1565, 1989.
12. S. Muller and A. J. Krueger, Analysis and Comparison of Ozone Maps Obtained by TOMS and TOVS During the Map/Globus 1983 Campaign, *Planetary and Space Sciences*, 35(5), 539-545, 1987.
13. A.J. Krueger, Nimbus-7 Total Ozone Mapping spectrometer (TOMS) Data During the GAP, France, Ozone Intercomparisons of June 1981, *Planetary and Space Sciences*, 31(7), 773-777, 1983.
14. C.L. Parkinson, *Earth from Above. Using Color-Coded Satellite Images to Examine the Global Environment*, University Science Books, Sausalito, CA, 1997.
15. A.M. Thompson, The Oxyding Capacity of the Earth's Atmosphere: Probable Past and Future Changes, *Science*, 256, 1157-1165, 1992.
16. M. Schoeberl, J. Pfaendtner, R. Rood, A. Thompson, and B. Wielicki, *Atmospheres Panel Report to the Payload Panel, Palaeogeography, Palaeoclimatology, Palaeoecology* (Global and Planetary Change Section) 98, 9-21, Elsevier Science Publishers B.V., Amsterdam, 1992.
17. J. Susskind, J. Rosenfield, and D. Reuter, Remote Sensing of Weather and Climate Parameters from HIRS2/MSU on TIROS-N, *Journal of Geophysical Research*, 89(D3), 4677-4697, 1984.
18. J. Simpson (Ed.), *TRMM: The Satellite Mission to Measure Tropical Rainfall: Report of the Science Steering Group*, NASA Publication, August 1988.
19. K.B. Kidwell, *NOAA Polar Orbiter Data Users Guide*, National Oceanic and Atmospheric Administration, December 1991.
20. J. R. G. Townshend (Ed.), *Improved Global Data for Land Applications. A Proposal for a New High Resolution Data Set*, Global Change Report No. 20, Report of the Land Cover Working Group of IGBP-DIS, 1992.

21. J.R.G. Townshend and C.O. Justice, Selecting the Spatial Resolution of Satellite Sensors Required for Global Monitoring of Land Transformations, *Int. J. Remote Sensing*, 9, 187-236, 1988.

22. *TREES, Tropical Ecosystem Environment Observations by Satellites.* Strategy Proposal 1991-1993, Commission of the European Communities, Joint Research Centre, Institute for Remote Sensing Applications.

23. D. Skole and C.J. Tucker, Tropical Deforestation and Habitat Fragmentation in the Brazilian Amazon: Satellite Data from 1978 to 1988, *Science*, 260, 1905-1910, 1993.

24. C.J. Tucker, B.N. Holben, and T.E. Goff, Intensive Forest Clearing in Rondonia, Brazil, as Detected by Satellite Remote Sensing, *Remote Sensing of Environment*, 15, 255-261, 1984.

25. J.P. Malingreau, C.J. Tucker, and N. Laporte, AVHRR for Monitoring Global Tropical Deforestation, *Int. J. Remote Sensing*, 10(4&5), 855-867, 1989.

26. M. Ehlers, Integrating Remote Sensing and GIS for Environmental Monitoring and Modeling: Where Are We?, *Geo Info Systems*, 36-43, July 1995.

27. T. Cary, A World of Possibilities: Remote Sensing Data for Your GIS, *Geo Info Systems*, 38-42, September 1994.

28. N.M. Short, *The Landsat Tutorial Workbook: Basics of Satellite Remote Sensing*, Scientific and Technical Information Branch, National Aeronautics and Space Administration, Washington, D.C., 1982.

29. D.L. Williams and A. Jenetos, *Landsat-7 Science Working Group Report*, July 1993.

30. N.M. Short, P.D. Lowman, Jr., S.C. Freden, and W.A. Finch, Jr., *Mission to Earth: Landsat Views of the World*, Scientific and Technical Information Office, National Aeronautics and Space Administration, Washington, D.C., 1976.

31. D.L. Evans, E.R. Stofan, T.D. Jones, and L.M. Godwin, Earth from Sky, *Scientific American*, 271(6), 70-75, December 1994.

32. N.M. Short, *Geomorphology from Space: A Global Overview of Regional Landforms*, Scientific and Technical Information Branch, National Aeronautics and Space Administration, Washington, D.C., 1986.

33. D.E. Smith, R. Kolenkiewicz, P.J. Dunn, S.M. Klosko, J.W. Ronnins, M.H. Torrece R.G. Williamson, E.C. Pavlis, N.B. Douglas, and S.K. Fricke, *Lageos Geodetic Analysis*, SL7.1, NASA Thnical Meorandum 104549, September 1991.

34. S.C. Cohen and D.E. Smith, LAGEOS Scientific Results, *Journal of Geophysical Research*, 90, 9217-9220, 1985.

35. S.B. Hooker, W.E. Esaias, G.C. Feldman, W.W. Gregg, and C.R. McClain, Volume 1, An Overview of SeaWiFS and Ocean Color, in *SeaWiFS Technical Report Series*, S.B. Hooker (Ed.), NASA Technical Memorandum 104566, Vol. 1, July 1992.

36. EOS- Ocean Color: Availability of the Global Data Set, *Transactions of the Geophysical Union*, 70(23), June 1989.

37. *TOPEX/Poseidon: Decoding the Ocean*, French Space Agency/CNES Report, December 1993, available from Centre National d'Etudes Spatiales, 2 Place Maurice Quentin, 75039 Paris Cedex 01, France.

38. D.K. Hall and J. Martinec, *Remote Sensing of Ice and Snow*, Chapman and Hall, London, 1985.

39. P.H. Swain and S.M. Davis, *Remote Sensing: The Quantitative Approach*, McGraw-Hill, New York, 1978.

40. P.M. Mather, *Computer Processing of Remotely Sensed Images*, paperback edition, John Wiley & Sons, Chichester, 1989.

41. B. Jähne, *Digital Image Processing. Concepts, Algorithms and Scientific Applications*, Springer Verlag, New York, 1991.

42. J.G. Moik, *Digital Processing of Remotely Sensed Images*, NASA Publication SP-431, 1979.

43. J.C. Tilton and M. Manohar, Earth Science Data Compression Issues and Activities, *Remote Sensing Reviews*, 9, 271-298, 1994.

44. S. Mallat, A Theory for Multiresolution Signal Decomposition, *IEEE Pattern Analysis and Machine Intelligence*, PAMI-11(7), 674-693, 1989.

45. E.K. Casani, *The New Millennium Program: Positioning NASA for the Ambitious Space and Earth Science Missions of the 21st Century*, Albuquerque, NM, JPL Technical Report, October 1995.

46. J. Le Moigne, N. El-Saleous, and E. Vermote, Iterative Edge- and Wavelet-Based Image Registration of AVHRR and GOES Satellite Imagery, *Image Registration Workshop*, IRW97, NASA/GSFC, Greenbelt, Nov. 20-21, 1997.

47. J. Le Moigne, W.J. Campbell, and R.F. Cromp, An Automated Parallel Image Registration Technique of Multiple Source Remote Sensing Data, submitted to the *IEEE Transactions on Geoscience and Remote Sensing*, June 1996.

48. J. Townshend, C.O. Justice, C. Gurney, and J. McManus, The Impact of Misregistration on Change Detection, *IEEE Transactions on Geoscience and Remote Sensing*, 30, 1504-1060, 1992.

49. L. Brown, A Survey of Image Registration Techniques, *ACM Computer Survey*, 24(4), 1992.

50. L.M.G. Fonseca and B.S. Manjunath, Registration Techniques for Multisensor Remotely Sensed Imagery, *Journal of Photogrammetry Engineering and Remote Sensing*, 62, 1049-1056, 1996.

51. R.F. Cromp and W.J. Campbell, Data Mining of Multidimensional Remotely Sensed Images, invited paper in *Proceedings 2nd Int. Conf. on Information and Knowledge Management*, Washington, D.C., 471-480, November 1993.

52. D.W. Scott, The Curse of Dimensionality and Dimension Reduction, in *Multivariate Density Estimation: Theory, Practice, and Visualization*, John Wiley & Sons, New York, Chapter 7, 195-217, 1992.

53. H.C. Andrews, *Introduction to Mathematical Techniques in Pattern Recognition*, Wiley-Interscience, New York, 1972.

54. W.J. Campbell, S.E. Hill, and R.F. Cromp, Automatic Labeling and Characterization of Objects Using Artificial Neural Networks, *Telematics and Informatics*, 6(3-4), 259-271, 1989.

55. S.R. Chettri, R.F. Cromp, and M. Birmingham, Design of Neural Networks for Classification of Remotely Sensed Imagery, *Telematics and Informatics*, 9(3/4), 145-156, 1992.

56. J. Hertz, A. Krogh, and R. Palmer, *Introduction to the Theory of Neural Computation*, Addison-Wesley, Redwood City, CA, 1991.

57. *EOS Data and Information System (EOSDIS)*, NASA, Washington, D.C., available from the Earth Science Support Office, Document Resource Facility, 300 D Street, SW, Suite 840, Washington, D.C. 20024; Telephone: (202) 479-0360, May 1992.

58. *DAAC/DADS Internal Design Documents*, edited by Goddard DAAC, Code 902.2, NASA Goddard Space Flight Center.

59. R.F. Cromp, W.J. Campbell, and N.M. Short, Jr., An Intelligent Information Fusion System for Handling the Archiving and Querying of Terabyte-Sized Spatial Databases, *International Space Year Conference on Earth and Space Science Information Systems*, American Institute of Physics, 1992.

60. W.J. Campbell, N.M. Short, Jr., P. Coronado, and R.F. Cromp, Distributed Earth Science Validation Centers for Mission to Planet Earth, *8-th International Symposium*, ISMIS'94, Charlotte, NC, October 1994.

61. W.J. Campbell, P. Clemens, J. Garegnani, R.F. Cromp, and P. Coronado, Applying Information Technologies to Facilitate Information Access and Regional Development, *Proceedings of the Technology 2007 Workshop*, Boston, MA, September 1997.

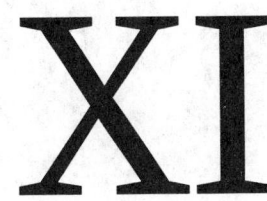

Biomedical Variables Measurement

XI

Biomedical Variables Measurement

74

Biopotentials and Electrophysiology Measurement

Nitish V. Thakor
Johns Hopkins School of Medicine

74.1 Introduction

This chapter reviews the origins, principles, and designs of instrumentation used in biopotential measurements, in particular for the electrocardiogram (ECG), the electroencephalogram (EEG), the electromyogram (EMG), and the electrooculogram (EOG). These biopotentials represent the activity of the respective organs: the heart, brain, muscle, and eyes. The biopotentials are acquired with the help of specialized electrodes that interface to the organ or the body and transduce low-noise, artifact-free signals. The basic design of a biopotential amplifier consists of an instrumentation amplifier. The amplifier should possess several characteristics, including high amplification, input impedance, and the ability to reject electrical interference, all of which are needed for the measurement of these biopotentials. Ancillary useful circuits are filters for attenuating electric interference, electrical isolation, and defibrillation shock protection. Practical considerations in biopotential measurement involve electrode placement and skin preparation, shielding from interference, and other good measurement practices.

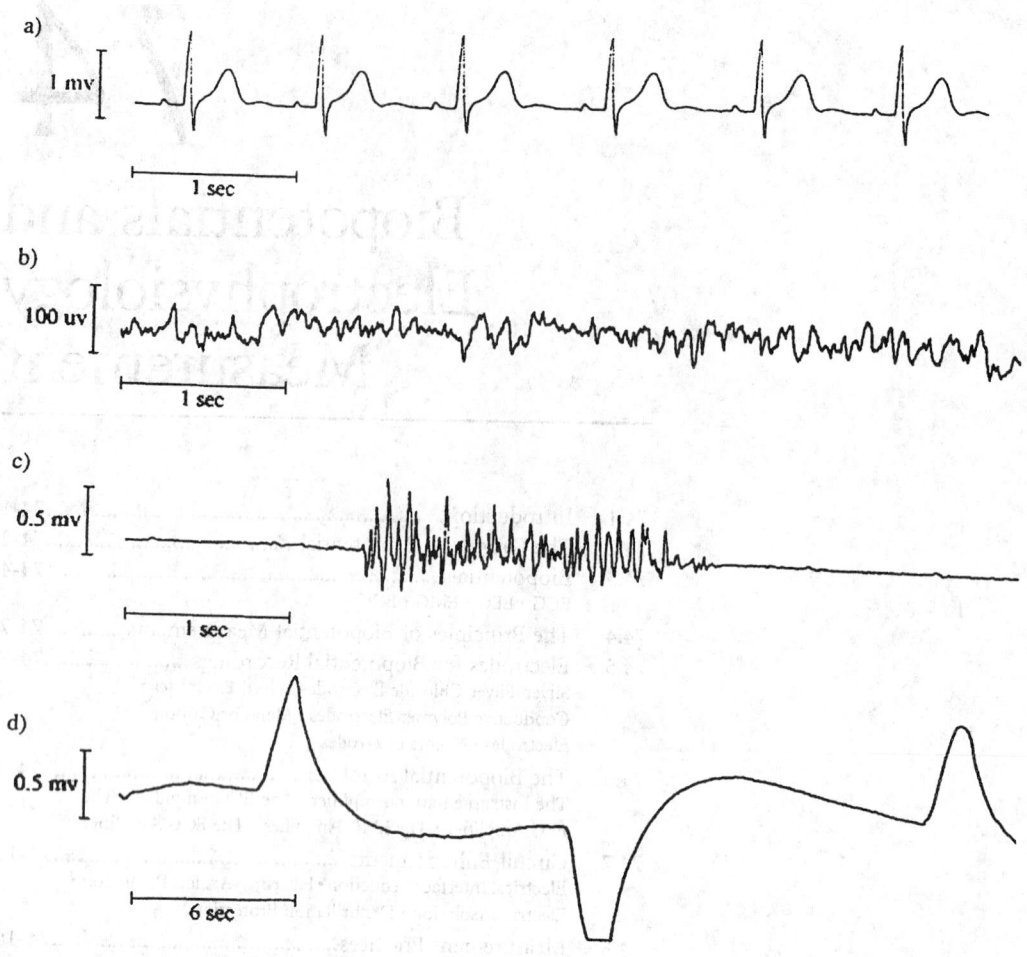

FIGURE 74.1   Sample waveforms: (a) ECG, normal sinus rhythm; (b) EEG, normal patient with open eyes; (c) EMG, flexion of biceps muscles; (d) EOG, movement of eyes from left to right.

74.2   The Origins of Biopotentials

Many organs in the human body, such as the heart, brain, muscles, and eyes, manifest their function through electric activity [1]. The heart, for example, produces a signal called the electrocardiogram or ECG (Figure 74.1a). The brain produces a signal called an electroencephalogram or EEG (Figure 74.1b). The activity of muscles, such as contraction and relaxation, produces an electromyogram or EMG (Figure 74.1c). Eye movement results in a signal called an electrooculogram or EOG (Figure 74.1d), and the retina within the eyes produces the electroretinogram or ERG. Measurements of these and other electric signals from the body can provide vital clues as to normal or pathological functions of the organs. For example, abnormal heart beats or arrhythmias can be readily diagnosed from an ECG. Neurologists interpret EEG signals to identify epileptic seizure events. EMG signals can be helpful in assessing muscle function as well as neuromuscular disorders. EOG signals are used in the diagnosis of disorders of eye movement and balance disorders.

The origins of these biopotentials can be traced to the electric activity at the cellular level [2]. The electric potential across a cell membrane is the result of different ionic concentrations that exist inside and outside the cell. The electrochemical concentration gradient across a semipermeable membrane results in the Nernst potential. The cell membrane separates high concentrations of potassium ion and low concentrations of sodium ions (along with other ions such as calcium in less significant proportions)

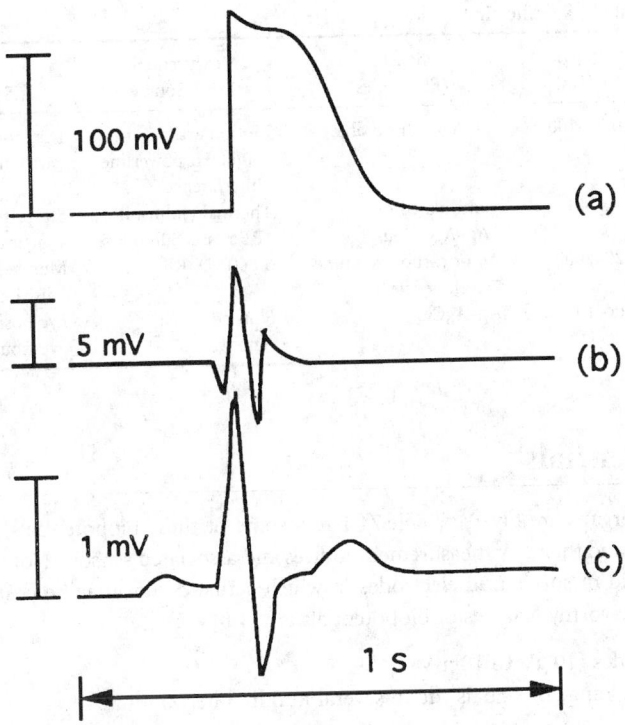

FIGURE 74.2 Schematic showing origins of biopotentials: (a) an action potential from a heart cell (recorded using a microelectrode); (b) the electrogram from the heart surface (recorded using an endocardial catheter); and (c) the ECG signal at the chest (recorded using surface electrodes).

inside a cell and just the opposite outside a cell. This difference in ionic concentration across the cell membrane produces the resting potential [3]. Some of the cells in the body are excitable and produce what is called an action potential, which results from a rapid flux of ions across the cell membrane in response to an electric stimulation or transient change in the electric gradient of the cell [4]. The electric excitation of cells generates currents in the surrounding volume conductor manifesting itself as potentials on the body.

Figure 74.2 illustrates the continuum of electrophysiological signals from the (a) heart cells, (b) myocardium (the heart muscle), and (c) the body surface. Each cell in the heart produces a characteristic action potential [4]. The activity of cells in the sinoatrial node of the heart produces an excitation that propagates from the atria to the ventricles through well-defined pathways and eventually throughout the heart; this electric excitation produces a synchronous contraction of the heart muscle [5]. The associated biopotential is the ECG. Electric excitation of a neuron produces an action potential that travels down its dendrites and axon [4]; activity of a massive number of neurons and their interactions within the cortical mantle results in the EEG signal [6]. Excitation of neurons transmitted via a nerve to a neuromuscular junction produces stimulation of muscle fibers. Constitutive elements of muscle fibers are the single motor units, and their electric activity is called a single motor unit potential [7]. The electric activity of large numbers of single motor unit potentials from groups of muscle fibers manifests on the body surface as the EMG. Contraction and relaxation of muscles is accompanied by proportionate EMG signals. The retina of the eye is a multilayered and rather regularly structured organ containing cells called rods and cones, cells that sense light and color. Motion of the eyeballs inside the conductive contents of the skull alters the electric potentials. Placing the electrode in the vicinity of the eyes (on either side of the eyes on the temples or above and below the eyes) picks up the potentials associated with eye movements called EOGs. Thus, it is clear that biopotentials at the cellular level play an integral role in the function of various vital organs.

TABLE 74.1 Biopotentials, Specifications, and Applications

Source	Amplitude (mV)	Bandwidth (Hz)	Sensor (Electrodes)	Measurement Error Source	Selected Applications
ECG	1–5	0.05–100	Ag–AgCl disposable	Motion artifact, 50/60 Hz powerline interference	Diagnosis of ischemia, arrhythmia, conduction defects
EEG	0.001–0.01	0.5–40	Gold-plated or Ag–AgCl reusable	Thermal (Johnson) RF noise, 50/60 Hz	Sleep studies, seizure detection, cortical mapping
EMG	1–10	20–2000	Ag or carbon, stainless steel, needle	50/60 Hz, RF	Muscle function, neuromuscular disease, prosthesis
EOG	0.01–0.1	dc–10	Ag–AgCl	Skin potential motion	Eye position, sleep state, vestibulo-ocular reflex

74.3 Biopotentials

Biopotentials from organs are diverse. Table 74.1 lists some of these biopotentials, their representative clinical applications, and their key measurement indices and associated sensors. Note that all acquisitions are made with the aid of specialized electrodes in which actual design may be customized for specific needs. The most noteworthy features of biopotentials are [1,8]

- Small amplitudes (10 μV to 10 mV),
- Low frequency range of signals (dc to several hundred hertz)

The most noteworthy problems of such acquisitions are

- Presence of biological interference (from skin, electrodes, motion, etc.),
- Noise from environmental sources (power line, radio frequency, electromagnetic, etc.).

These signal acquisition challenges and problems for each of the biopotentials are considered in greater detail below.

ECG

ECG signals are acquired by placing electrodes directly on the torso, arms, and legs (Figure 74.3a). The activity on the body surface is known to reflect the activity of the heart muscle underneath and in its proximity. A clinically accepted lead system has been devised and is called the 12-lead system [9, 10]. It comprises a combination of electrodes taking measurements from different regions designated limb leads, the precordial leads, and the chest leads. Limb leads derive signals from electrodes on the limbs, and are designated as leads I, II, and III. Precordial leads are designated aVR, aVL, and aVF, and are derived by combining signals from the limb leads. The remaining six leads, V1, V2, ...V6, are chest leads. Together, ECGs from these various leads help define the nature of the activity on a specific part of the heart muscle: for example, ischemia (impaired oxygen supply to the muscle) or infarction (damage to the muscle) on the left side of the chest may be noticeable in lead III.

The ECG signals at the surface of the body are small in amplitude, which make the measurements susceptible to artifacts [11], generated by the relative motion of the electrode and the skin as well as by the activity of the nearby muscles. An important consideration in good ECG signal acquisition is the use of high-quality electrodes [12]. Electrodes made out of silver coated with silver chloride or of sintered Ag–AgCl material, are recommended. An electrolytic gel is used to enhance conduction between the skin and the electrode metal. Artifacts at the electrode–skin contact as well as electromagnetic interference from all sources must be minimized [13]. Since ECG instruments are often used in critical-care environments, they must be electrically isolated for safety [14] and protected from the high voltages generated by defibrillators [15].

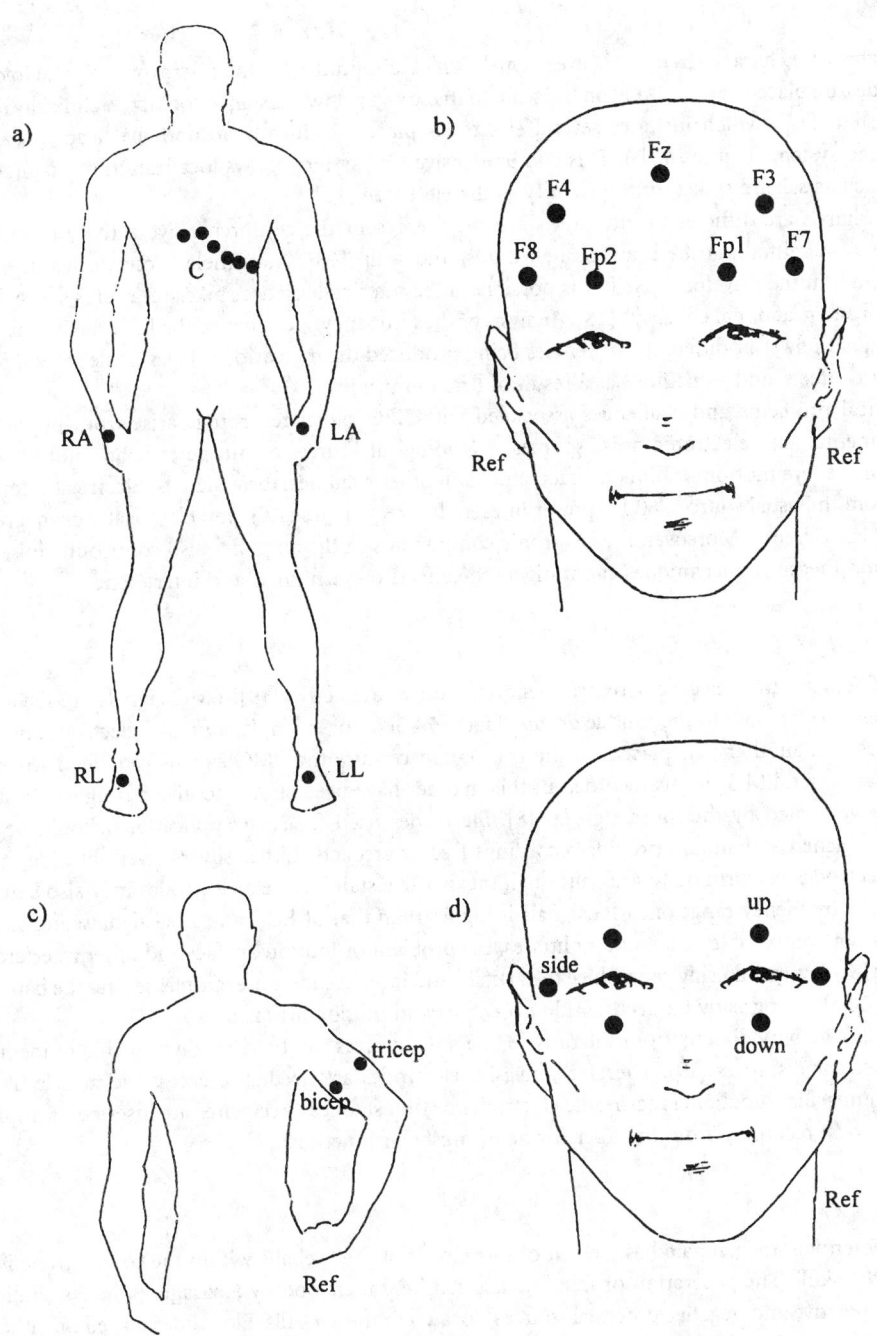

FIGURE 74.3 Schematics showing how biopotential signals are recorded from the human body. (a) ECG: 12-lead ECG is recorded using right arm (RA), left arm (LA), left leg (LL), right leg reference (RL), and six chest (C) electrodes. (b) EEG: selected electrode locations from the standard 10-20 EEG lead system with ears used as reference. (c) EMG: recording electrodes on the biceps and triceps with an independent reference. (d) EOG: electrodes above or below (up–down) and the sides of the eyes along with an independent reference.

ECG biopotential amplifiers find use in many monitoring instruments, pacemakers, and defibrillators [16]. ECG signal acquisition is also useful in many clinical applications including diagnosis of arrhythmias, ischemia, or heart failure.

EEG

EEG signals are characterized by their extremely small amplitudes (in the microvolt range). Gold-plated electrodes are placed very securely on the scalp to make a very low resistance contact. A clinically accepted lead system [17], which includes several electrodes placed uniformly around the head, is called the 10-20 lead system (Figure 74.3b). This comprehensive lead system allows localization of diagnostic features, such as seizure spikes, in the vicinity of the electrode [18].

EEG signals are difficult to interpret since they represent the comprehensive activity of billions of neurons transmitted via the brain tissues, fluids, and scalp [18]. Nevertheless, certain features can be interpreted. In the waveform itself, it is possible to see interictal seizure spikes or a full seizure (such as petit mal and grand mal epilepsy) [18]. Analysis of the frequency spectrum of the EEG can reveal changes in the signal power at different frequencies being produced during various stages of sleep, as a result of anesthetic effects, and sometimes as a result of brain injury [17].

Practical problems and challenges associated with EEG signal recordings arise from physiological, environmental, and electronic noise sources. Physiological sources of interference are motion artifact, muscle noise, eye motion or blink artifact, and sometimes even heartbeat signals. Electrical interference arises from the usual sources: 60 Hz power lines, radio frequencies (RF), and electrically or magnetically induced interference. Moreover, the electronic components in the amplifier also contribute noise. Good design and measuring techniques can mitigate the effects of such noise and interference.

EMG

Muscle fibers generate electric activity whenever muscles are active [19]. EMG signals are recorded by placing electrodes close to the muscle group (Figure 74.3c). For example, a pair of electrodes placed on the biceps and another pair placed on the triceps can capture the EMG signals generated when these muscles contract. EMG signals recorded in this manner have been shown to give a rough indication of the force generated by the muscle group [8]. Electrodes used for such applications should be small, securely attached and should provide recordings free of artifacts. Either silver–silver chloride or gold-plated electrodes perform quite well, although inexpensive stainless steel electrodes may also suffice.

Since the frequency range of EMG signals is higher than that of ECG and EEG signals, and since the signals are of comparable or larger amplitudes, the problem of motion artifact and other interference is relatively less severe. Filtering can reduce the artifact and interference: for example, setting the bandwidth to above 20 Hz can greatly reduce the skin potentials and motion artifacts.

Recording activity directly from the muscle fibers themselves can be clinically valuable in identifying neuromuscular disorders [19]. Therefore, invasive electrodes are needed to access the muscle fibers or the neuromuscular junction. Fine-needle electrodes or thin stainless steel wires are inserted or implanted to obtain local recording from the fibers or neuromuscular junctions [7].

EOG

Electric potentials are generated as a result of movement of the eyeballs within the conductive environment of the skull. The generation of EOG signals can be understood by envisaging dipoles (indicating separated positive and negative potential sources) located in the eyeballs. Electrodes placed on either side of the eyes or above and below them pick up the potentials generated by the motion of the eyeball (Figure 74.3d). This potential varies approximately in proportion to the movement of the eyeballs, and hence EOG is sometimes used to study eye positions or disorders of eye movement and balance (a reflex called vestibulo-ocular reflex affects the nystagmus of the eye). Similarly, saccades inherent in eye motion as well as blinking of the eyelids can produce changes in the EOG signal.

This signal is small (10 to 100 µV) and has low frequencies (dc to 10 Hz) [8]. Hence, an amplifier with a high gain and good low frequency response and dc stability is desirable. Additionally, the electrode–gel combination should be such that it produces low levels of junction potential, motion artifacts, and drift in the dc signal [20]. Practical problems associated with dc drift, motion artifacts, and securing electrodes in the vicinity of the eyes make their long-term use problematic. Nevertheless, EOG signals can be useful clinically in acute studies of human disorders, and therefore careful acquisition of the signal followed by appropriate analysis is used to interpret the EOG potentials.

Other biopotential recording techniques follow similar principles of measurements. The electrode design should be specifically adapted to the source of the signal. A thorough effort is required to minimize the noise and interference by improving electrode design and placement and optimizing the amplifier circuit. Good electrode attachment along with selective filtering at the amplifier can help obtain relatively noise-free recording. The design principles and practical considerations are described below.

74.4 The Principles of Biopotential Measurements

The unifying principles of biopotential recordings involve

- Electrode design and its attachment suited to the application;
- Amplifier circuit design for suitable amplification of the signal and rejection of noise and interference;
- Good measurement practices to mitigate artifacts, noise, and interference.

74.5 Electrodes for Biopotential Recordings

Electrodes for biopotential recordings are designed to obtain the signal of interest selectively while reducing the potential to pick up artifact. The design should be pragmatic to reduce cost and allow for good manufacturing and reliable long-term use. These practical considerations determine whether high-quality but reusable electrodes made of silver or gold or cheaper disposable electrodes are used [20].

Silver–Silver Chloride Electrodes

The classic, high-quality electrode design consists of a highly conductive metal, silver, interfaced to its salt, silver chloride, and connected via an electrolytic gel to the human body [21]. Silver–silver chloride–based electrode design is known to produce the lowest and most stable junction potentials [1, 20]. Junction potentials are the result of the dissimilar electrolytic interfaces, and are a serious source of electrode-based motion artifacts. Therefore, additionally, an electrolytic gel typically based on sodium or potassium chloride is applied to the electrode. A gel concentration in the order of 0.1 M (molar concentration) results in a good conductivity and low junction potential without causing skin irritation.

Reusable silver–silver chloride electrodes (Figure 74.4a) are made of silver disks coated electrolytically by silver chloride [1], or, alternatively, particles of silver and silver chloride are sintered together to form the metallic structure of the electrode. The gel is typically soaked into a foam pad or is applied directly in a pocket produced by the electrode housing. The electrode is secured to the skin by means of nonallergenic adhesive tape. The electrode is connected to the external instrumentation typically via a snap-on connector. Such electrodes are well suited for acute studies or basic research investigations.

Disposable electrodes are made similarly, although the use of silver may be minimized (for example, the snap-on button itself may be silver coated and chlorided). To allow for a secure attachment, a large foam pad attaches the electrode body with adhesive coating on one side (Figure 74.4b). Such electrodes are particularly suited for ambulatory or long term use.

Gold Electrodes

Gold-plated electrodes (Figure 74.4c), which have the advantages of high conductivity and inertness desirable in reusable electrodes, are commonly used in EEG recordings [1]. Small reusable electrodes are designed so that they can be securely attached to the scalp. The electrode body is also shaped to make a recessed space for electrolytic gel, which can be applied through a hole in the electrode body [18]. The electrodes are attached in hair-free areas by use of a strong adhesive such as colloidon or securely attached with elastic bandages or wire mesh. Similar electrodes may also be used for recording EMG, especially

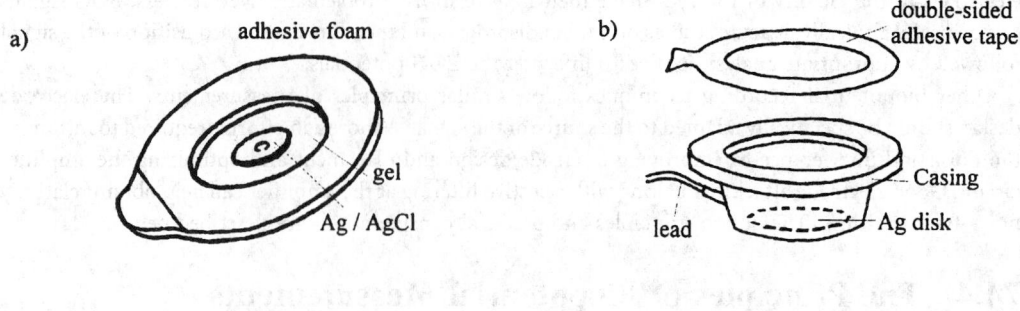

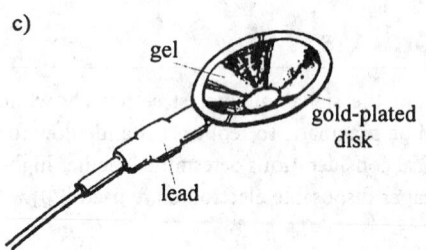

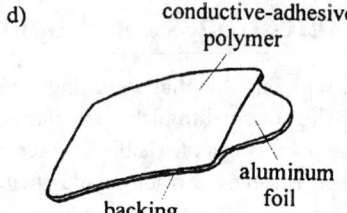

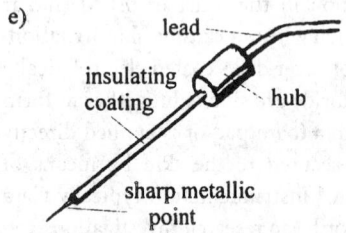

FIGURE 74.4 Examples of electrodes used in biopotential recordings: (a) disposable Ag–AgCl electrode, (b) reusable Ag–AgCl disk electrode, (c) gold disk electrode, (d) disposable conductive polymer electrode, and (e) needle electrode.

when a great deal of motion is expected. Disadvantages of using gold electrodes over silver–silver chloride electrodes include greater expense, higher junction potentials, and greater susceptibility to motion arti-facts [20]. On the other hand, gold electrodes maintain low impedance, are inert and reusable, and are good for short-term recordings as long as a highly conductive gel is applied and they are attached securely.

Conductive Polymer Electrodes

It is often convenient to construct an electrode out of a material that is simultaneously conductive and adhesive [20]. Certain polymeric materials have adhesive properties and by attaching monovalent metal ions can be made conductive. The polymer is attached to a metallic backing made of silver or aluminum foil, which allows electric contact to external instrumentation (Figure 74.4d). This electrode does not need additional adhesive or electrolytic gel and hence can be immediately and conveniently used. The conductive polymeric electrode performs adequately as long as its relatively higher resistivity (over metallic electrodes) and greater likelihood of generating artifacts are acceptable. The higher resistivity of the polymer makes these electrodes unsuitable for low-noise measurement. The polymer does not attach as effectively to the skin as does the conventional adhesive on disposable ECG electrodes built with a foam base and, furthermore, the potentials generated at the electrode–skin interface are more readily disturbed by motion. Nevertheless, when the signal level is high and when restricting the subject movement minimizes artifact, the polymeric electrode offers a relatively inexpensive solution to biopotential recording.

Metal or Carbon Electrodes

Although other metals such as stainless steel or brass electrodes [21] are used rather infrequently now because high-quality noble metal electrodes or low-cost carbon or polymeric electrodes are so readily available, historically these metallic electrodes were used in laboratory or clinical settings because of their sturdy construction and reusability. Electrode gel is applied to the metal electrode which is fastened to the body by means of a rubber band. These electrodes have the potential for producing very high levels of artifact and are bulky and awkward to use, but do offer the advantage of being reusable and tend to be inexpensive. Carbon or carbon-impregnated polymer electrodes are also used occasionally (although they are mainly used as electrical stimulation electrodes) [20]. These electrodes have a much higher resistivity and are noisier and more susceptible to artifacts, but they are inexpensive, flexible, and reusable and are thus chosen for applications such as electric stimulation or impedance plethysmography. For these applications, gel is usually not applied and the electrodes are used in "dry" form for easy attachment and removal.

Needle Electrodes

Needle electrodes (Figure 74.4e) comprise a small class of invasive electrodes, used when it is absolutely essential to record from the organ itself. The most common application is in recording from muscles or muscle fibers [8]. A metallic, typically steel, wire is delivered via a needle inserted at the site of the muscle fiber. The wire is hooked and hence fastens to the muscle fiber, even as the needle is removed. Small signals such as motor unit potentials can be recorded in this manner [7]. For research applications, similar needle or wire electrodes are sometimes connected directly to the heart muscle. Since such electrodes are noninvasive, their use is limited to only highly specialized and supervised clinical or research applications.

74.6 The Biopotential Amplifier

Biopotentials exhibit small amplitudes and low frequencies [22]. Moreover, biopotential measurements are corrupted by environmental and biological sources of interference. Therefore, the essential, although not exhaustive, design considerations include proper amplification and bandwidth, high input impedance, low noise, and stability against temperature and voltage fluctuations. The key design component of all biopotential amplifiers is the instrumentation amplifier [21]. However, each biopotential acquisition instrument has a somewhat differing set of characteristics, necessitating some specialization in the design of the instrumentation amplifier. Table 74.2 summarizes the circuit specialization needed in various biopotential amplifiers, with the ECG amplifier used as the basic design.

The Instrumentation Amplifier

The instrumentation amplifier is a circuit configuration that potentially combines the best features desirable for biopotential measurements [8], namely, high differential gain, low common mode gain,

TABLE 74.2   Distinguishing Features and Design Consideration for Biopotentials

Biopotential	Distinguishing Feature	Exclusive Amplifier Design Consideration	Additional Features Desired
ECG[a]	1 mV signal, 0.05–100 Hz BW[b]	Moderate gain, BW, noise, CMRR, input R	Electrical safety, isolation, defibrillation protection
EEG	Very small signal (microvolts)	High gain, very low noise, filtering	Safety, isolation, low electrode–skin resistance
EMG	Higher BW	Gain and BW of op amps	Postacquisition data processing
EOG	Lower frequencies, small signal	dc and low drift	Electrode–skin junction potential, artifact reduction

[a] The ECG signal acquisition is considered as the standard against which the other acquisitions are compared.
[b] BW = bandwidth.

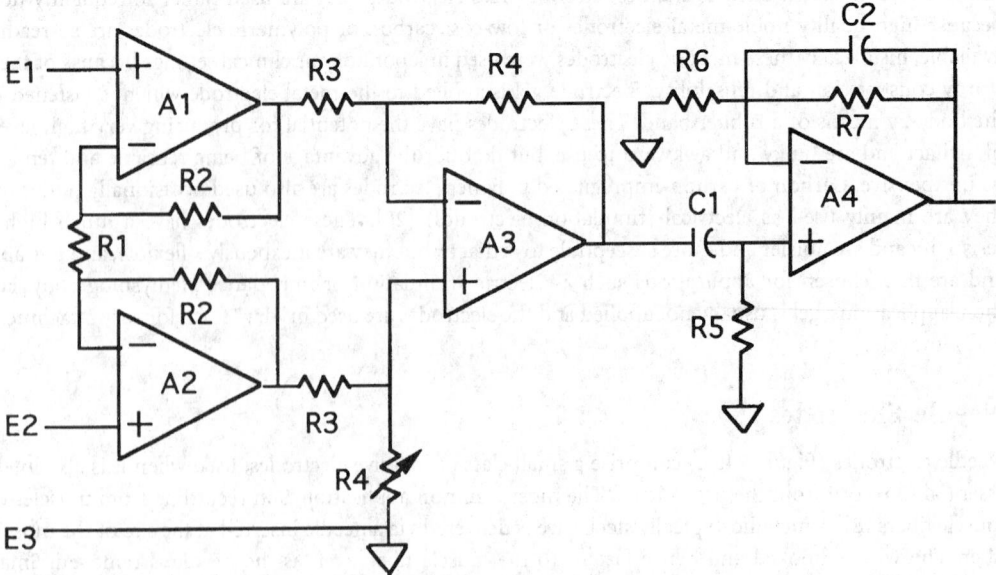

FIGURE 74.5   The instrumentation amplifier. This amplifier has a very high input impedance, high CMRR, and a differential gain set by the resistors in the two amplifier stages. The gain of the first stage (amplifiers A1 and A2) is $1 + 2R2/R1$, the second stage (amplifier A3) is $R4/R3$, and the third stage (amplifier A4) is $1 + R7/R6$. The lower corner frequency is $1/(2\pi R5 C1)$ and the upper corner frequency is $1/(2\pi R7 C2)$. The variable resistor R is adjusted to maximize the CMRR. Electrodes E1 and E2 are the recording electrodes while E3 is the reference or the ground electrode.

high common mode rejection ratio (CMRR), and high input resistance [23]. Figure 74.5 shows the design of the basic instrumentation amplifier. The basic circuit design principles have been described elsewhere [23,25]. The instrumentation amplifier is constructed from operational amplifiers, or op amps, which have many of the desirable features listed above [24]. The front end of the amplifier has two op amps, which consists of two noninverting amplifiers that have been coupled together by a common resistor $R1$. The gain of the first stage is $(1 + 2R2/R1)$. The second stage is a conventional differential amplifier with gain of $-(R4/R3)$. This design results in the desired differential gain distributed over two stages of the amplifier. It also achieves a very high input resistance as a result of the noninverting amplifier front end. It exhibits a very high CMRR as a result of the differential first stage followed by a second-stage differential amplifier. The CMRR is enhanced by adjusting one of the matching resistors and by selecting high CMRR op amps. This instrumentation amplifier is a key design component universal to many biosensor interfaces and almost all biopotential instruments [22].

The ECG Amplifier

The ECG amplifier can readily be designed using the instrumentation amplifier as the principal building block. Active filters with a lower corner frequency of 0.05 Hz and an upper corner frequency of 100 Hz are also typically added [8].

ECG amplifiers are needed in many applications, such as monitoring in cardiac intensive-care units, where safety and protection are of paramount importance. Because the possibility of a direct or low-resistance access to the heart via catheters or intravenous lines exists in such settings, very small electric leakage currents can be fatal. Consequently, leakage from the amplifier is required to be below the safety standard limit of 10 μA [14]. Additionally, safety of the patient is achieved by providing electrical isolation from the power line and the earth ground, which prevents passage of leakage current from the instrument to the patient under normal conditions or under reasonable failure conditions. Electrical isolation is achieved by using transformer or optical coupling components [9], although it is important to remember that any such design should preserve the bandwidth and linearity of the amplifier. ECG amplifiers are also likely to be operated in circumstances where defibrillators might be used; thus, the amplifier circuit must be protected against the high defibrillation voltages and must be augmented by circuit components such as current-limiting resistors, voltage-limiting diodes, and spark gaps [15].

The EEG Amplifier

The distinguishing feature of an EEG amplifier is that it must amplify very small signals [8]. The amplifier gain must be suitably enhanced to deal with microvolt or lower levels of signals. Furthermore, all components of the amplifier must have a very low thermal noise and in particular low electronic (voltage and current) noise at the front end of the amplifier. EEG amplifiers used in clinical applications again must be electrically isolated and protected against high defibrillation voltages, similar to the ECG amplifier.

The EMG Amplifier

EMG amplifiers are often used in the investigation of muscle performance, neuromuscular diseases, and in building certain powered or smart prostheses. In such applications, slightly enhanced amplifier bandwidth suffices. In addition, postprocessing circuits are almost always needed. For example, a rectified and integrated EMG signal has been shown to give a rough indication of the muscle activity, approximately related to the force being generated at the location of the EMG electrode [8].

The EOG Amplifier

The EOG signal is small in amplitude and consists of very low frequencies. Therefore, an EOG amplifier must not only have a high gain, but also a very good low frequency, or even dc, response. This frequency response also makes the amplifier potentially susceptible to shifts in the junction potential at the skin–electrode interface and to drift in the electronic circuit characteristics. In addition to using good electrodes (Ag–AgCl) and gel (high conductivity), some type of active dc or drift cancellation or correction circuit design may be necessary.

74.7 Circuit Enhancements

The basic biopotential amplifier described above, along with the specific design considerations for each biopotential, can yield a signal acquisition of acceptable quality in most laboratory settings. In practice, however, further enhancements are always necessary to achieve acceptable clinical performance in novel applications. These enhancements include circuits for reducing electric interference, filtering noise, reduction of artifacts, electrical isolation of the amplifier, and electrical protection of the circuit against defibrillation shocks [9].

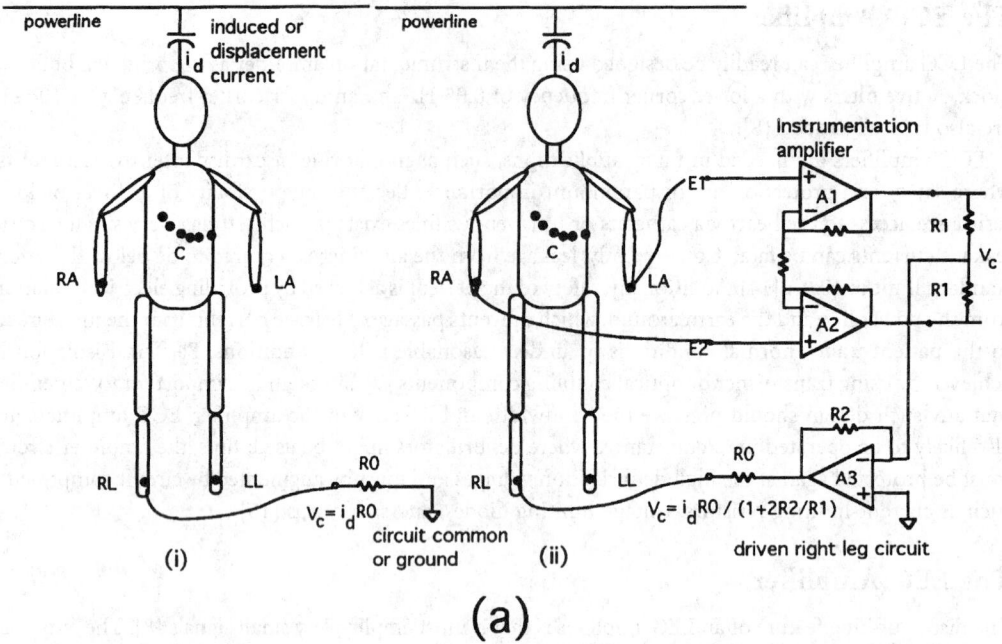

FIGURE 74.6 Circuit enhancements for biopotential measurements. (a) The schematic on the left shows electric interference induced by the displacement current i_d from the power line. This current flows into the ground electrode lead generating common-mode voltage V_c. The driven right leg circuit on the right uses negative feedback into the right leg electrode to reduce the effective common-mode voltage. (b) Amplifier front end filters — T1: RF choke; $R0$ and $C0$: RF filter; $R1$ and $C1$: high-pass filter; $R2$ and $C2$: low-pass filter. (c) Notch filter for power line interference (50 or 60 Hz): twin T notch filter in which notch frequency is governed by $R1$, $R2$, $R3$, $C1$, $C2$, and $C3$, and notch tuning by $R4$. (d) Baseline restoration circuit: the high-pass filter capacitor $C1$ is discharged by field effect transistor F when activated manually or automatically by a baseline restoration pulse. (e) Electrical isolation: transformer coupled using the transformer T (top) or optical using the diode D and the photodetector P (bottom). Note that the isolator separates circuit common on the amplifier side from the Earth ground on the output side. (f) Electrical protection circuit: resistance R limits the current, reverse-biased diodes D limit the input voltage, and the spark gap S protects against defibrillation pulse-related breakdown of the isolation transformer T.

Electrical Interference Reduction

Environmental electric interference is always present, especially in urban hospital environments. It is desirable to eliminate interference before it enters the amplifier, for example, by proper shielding of the subject, leads, and the instrument and by grounding the subject and the instrument. Sources of interference include induced signals from power lines and electric wiring; RF from transmitters, electric motors, and other appliances; magnetically induced currents in lead wires; and so on [13]. Interference induced on the body common to the biopotential sensing electrodes is called the common mode interference (as distinguished from the biopotential that is differential to the sensing electrodes). If the induced current is i_d and the resistance to ground is $R0$, then the common mode interference potential is $V_c = i_d R0$. The common mode interference is principally rejected by a differential or instrumentation amplifier with a high CMRR. Further improvement is possible by use of the "driven right leg circuit." The right leg lead, by standard convention, is used as the ground or the circuit reference. The driven right leg circuit employs the clever idea of negative feedback of the common mode signal into this lead. The common mode signal is sensed from the first stage of the instrumentation amplifier, amplified and

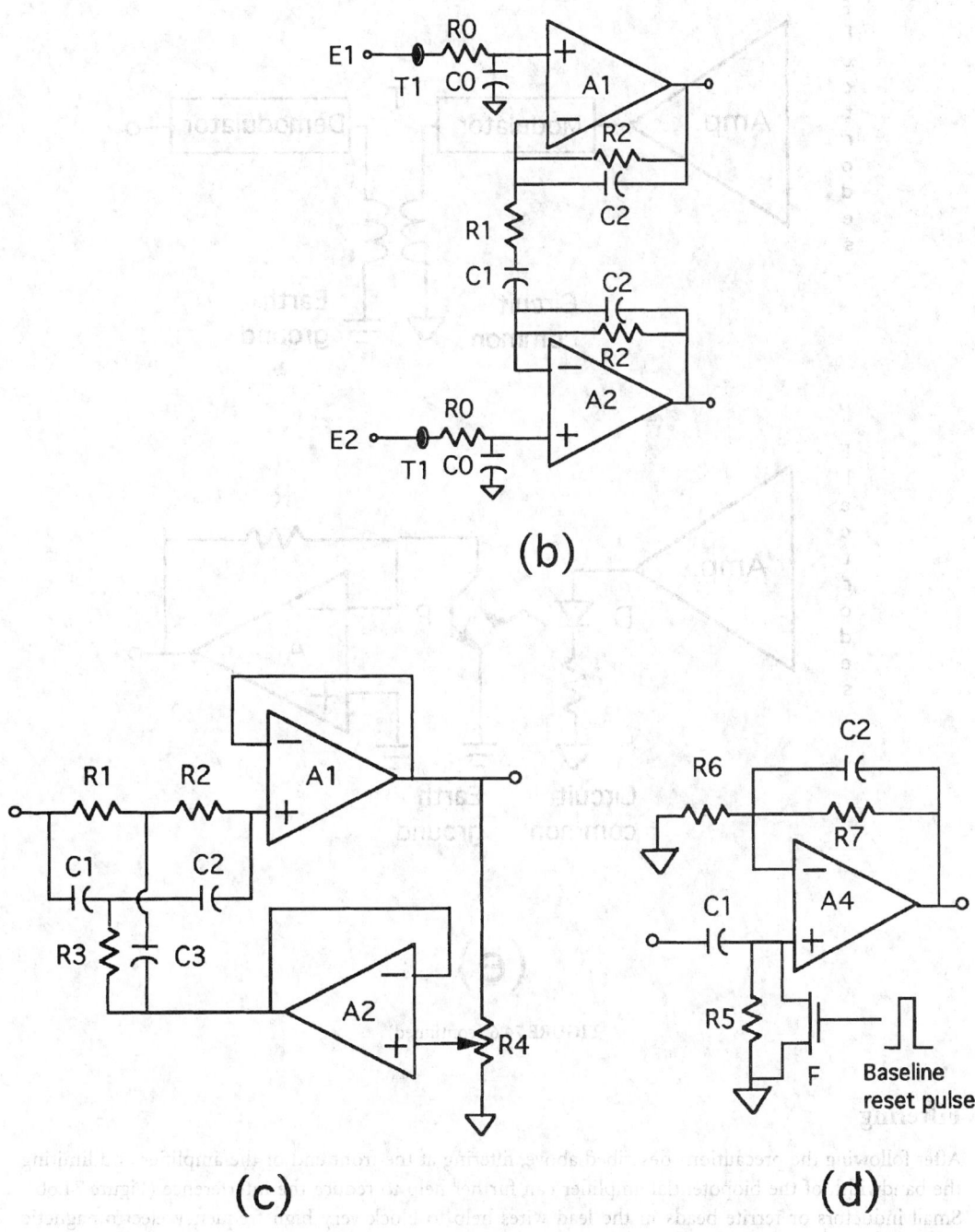

(b)

(c)

(d)

FIGURE 74.6 (continued)

inverted, and fed back into the right leg lead (Figure 74.6a). At this stage the common mode signal is reduced to $(i_d R0)/(1 + 2R2/R1)$. Thus, the common mode interference is greatly reduced at its source. The driven right leg circuit along with a high CMRR of the amplifier and filtering permit very high quality biopotential measurements.

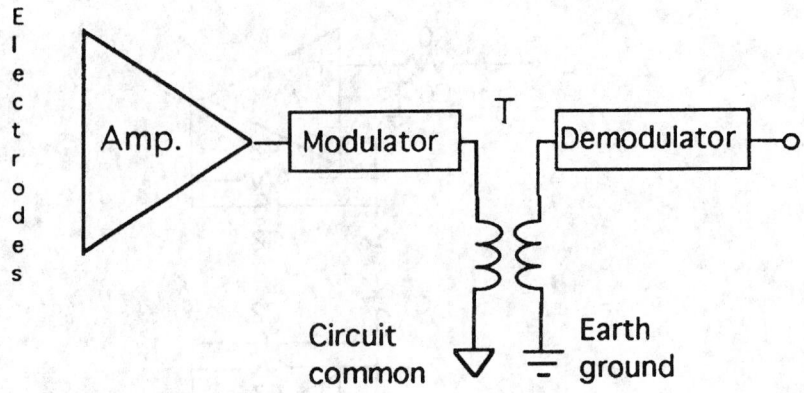

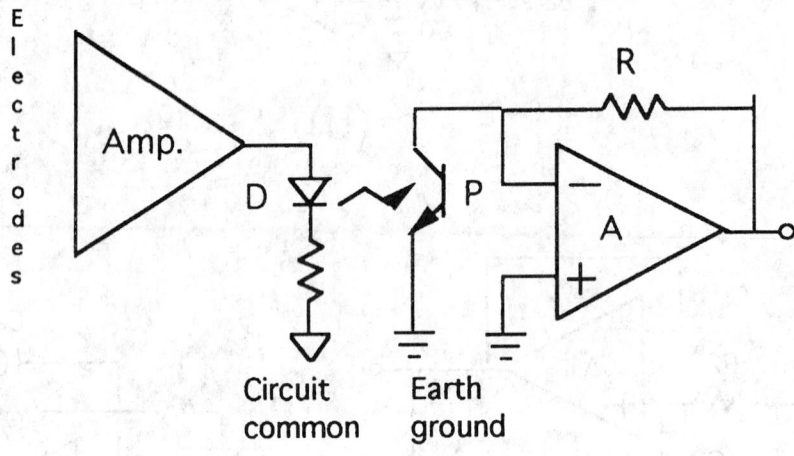

(e)

FIGURE 74.6 (continued)

Filtering

After following the precautions described above, filtering at the front end of the amplifier and limiting the bandwidth of the biopotential amplifier can further help to reduce the interference (Figure 74.6b). Small inductors or ferrite beads in the lead wires help to block very high frequency electromagnetic interference. Small capacitors between each electrode lead and ground filter the RF interference. Bandwidth limitation can be imposed at each stage of the amplifier. Because dc potentials arising at the electrode–skin interface must be blocked well before the biopotential is amplified greatly (otherwise, the amplifier could saturate), use of high-pass filtering in the early stages of amplification is recommended. Low-pass filtering at several stages of amplification is recommended to attenuate residual RF interference as well as muscle signal interference. Power line interference at 50 or 60 Hz and their harmonics clearly poses the biggest problem in biopotential measurement [11,13]. Sometimes it may be desirable to provide

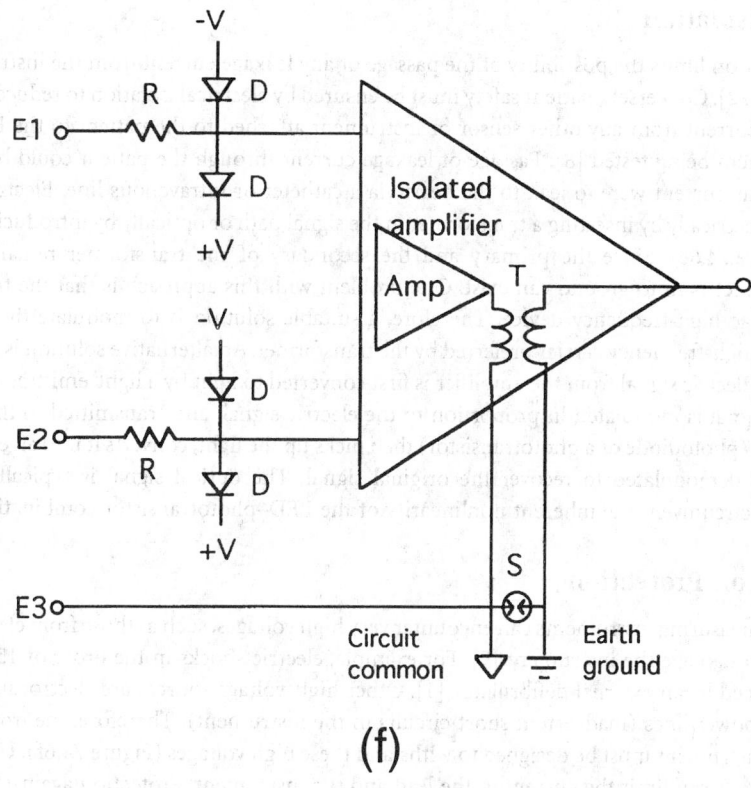

(f)

FIGURE 74.6 (continued)

a 50 or 60 Hz notch filter to remove the power line interference (Figure 74.6c), an option that is often available with low-level signal (EEG, EOG) measuring instruments. The risk of a distorted biopotential signal arises when a notch filter is used and this may affect diagnosis. Filtering should, therefore, be used selectively.

Artifact Reduction

One principal source of artifact is the potential arising at the electrode–skin interface [11]. Slow changes in the baseline can arise due to changes in the junction potential at this interface and, in some instances, can cause a temporary saturation of the amplifier [9]. This event is detected manually or automatically (by quickly discharging the high-pass capacitor in the amplifier to restore the baseline; Figure 74.6d). Movement of the subject or disturbance of the electrode can produce motion artifacts [11], which can be reduced by filtering the signal, but as suggested above, such filtering, typically high pass, can severely distort the biopotential being measured. Alternatively, computerized processing may be necessary to identify an artifact and delete it from display and processing. Of note, a biopotential source could be the desired one in one case, but an unwanted artifact in another case. For example, EOG signal resulting from blinking of eyes can produce a rather significant artifact in EEG recordings. Similarly, EMG signals become unwanted artifacts in all other non-EMG biopotential measurements. ECG monitoring must especially account for EMG artifact for high-fidelity recording. Another example is the pacemaker pulse. Since a pacemaker pulse can be detected and amplified as a short (about 2 ms) pulse preceding a QRS complex, it can be mistakenly interpreted as a heartbeat by some circuits for automatically determining heart rate. Special circuits must be designed to identify and delete this artifact [9].

Electrical Isolation

Electrical isolation limits the possibility of the passage of any leakage current from the instrument in use to the patient [22]. Conversely, patient safety must be ensured by electrical isolation to reduce the prospect of leakage of current from any other sensor or instrument attached to the patient to the Earth ground of the instrument being tested [8]. Passage of leakage current through the patient could be harmful or even fatal if this current were to leak to the heart via a catheter or intravenous line. Electrical isolation can be done electrically by inserting a transformer in the signal path or optically by introducing an optical coupler (Figure 74.6e). Since the primary and the secondary of the transformer remain electrically isolated, no direct path to ground can exist. One problem with this approach is that the transformer is inherently an ac high-frequency device. Therefore, a suitable solution is to modulate the biopotential signal using a high-frequency carrier preferred by the transformer. An alternative solution is to use optical isolation. The electric signal from the amplifier is first converted to light by a light-emitting diode (LED). This optical signal is modulated in proportion to the electric signal, and transmitted to the detector. A photodetector (photodiode or a phototransistor) then picks up the light, converts it into an electric signal, which is then demodulated to recover the original signal. The optical signal is typically pulse code modulated to circumvent the inherent nonlinearity of the LED–phototransistor combination.

Defibrillation Protection

Biopotential-measuring instruments can encounter very high voltages, such as those from electric defibrillators, that can damage the instrument [9]. For example, electric shocks in the order of 1500 to 5000 V may be produced by an external defibrillator [1]. Other high-voltage sources are electrocautery (used in surgery) and power lines (inadvertent short circuits in the instrument). Therefore, the front end of the biopotential instrument must be designed to withstand these high voltages (Figure 74.6f). Use of resistors in the input leads can limit the current in the lead and the instrument. Protection against high voltages is achieved by the use of diodes or Zener diodes. These components conduct at 0.7 V (diode conduction voltage) or 10 to 15 V (depending on the Zener diode breakdown voltage), thus protecting the sensitive amplifier components. Since it is more likely that protection against higher voltages will be needed, low-pressure gas discharge tubes such as neon lamps are also used. They break down at voltages on the order of 100 V, providing an alternative path to ground for the high voltages. As a final line of protection, the isolation components (optical isolator or transformer) must be protected by a spark gap that activates at several thousand volts. The spark gap ensures that the defibrillation pulse does not breach the isolation.

74.8 Measurement Practices

Biopotential measurements are made feasible, first of all, by good amplifier designs. High-quality biopotential measurements require use of good electrodes and their proper application on the patient, along with good laboratory or clinical practices. These practices are summarized below.

Electrode Use

Various electrodes best suited for each biopotential measurement were described earlier. First, different electrodes by virtue of their design offer distinguishing features: more secure (use of strong but less-irritant adhesives), more conductive (use of noble metals such as silver and gold), less prone to artifact (use of low-junction-potential materials such as Ag–AgCl). Electrode gel can be of considerable importance in maintaining a high-quality interface between the electrode metal and the skin. High conductivity gels, in general, help reduce the junction potentials along with the resistance (they tend, however, to be allergenic or irritating and hence a practical compromise in terms of electrolyte concentration must be found) [20]. Movement of the electrode with respect to the electrode gel and the skin is a potential source of artifact (Figure 74.7a). Such movements can change the electrode junction to skin potentials, producing

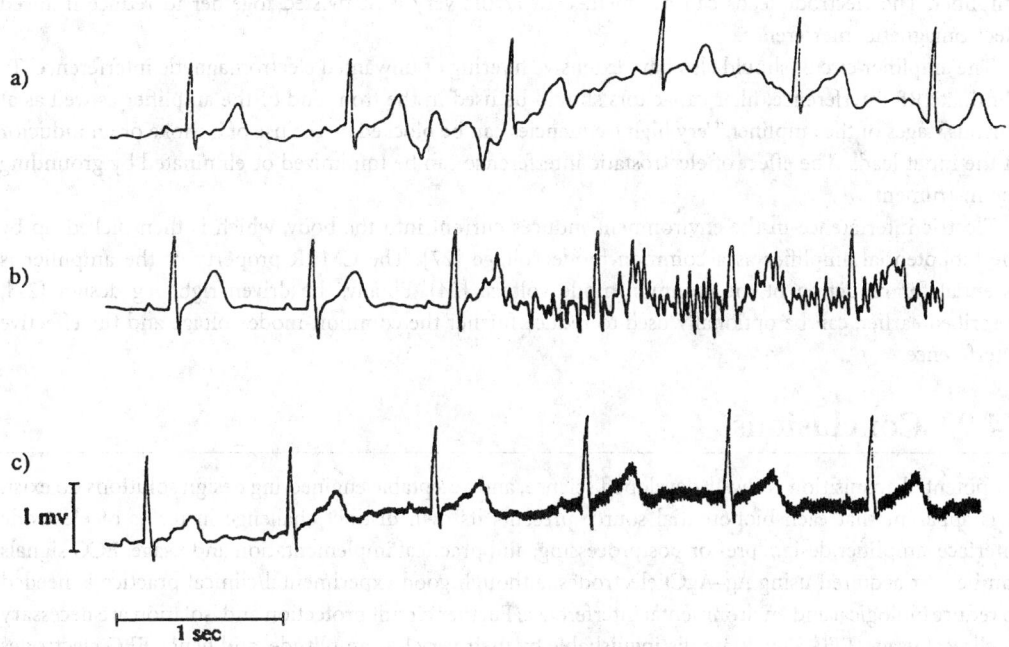

FIGURE 74.7 Examples of electric interference in biopotential recordings: (a) ECG signal with baseline changes and motion artifacts, (b) muscle signal interference, (c) electromagnetic interference (60 Hz power line and RF).

motion artifacts [21]. Placement above bony structures where there is less muscle mass can reduce unwanted motion artifact and EMG interference (Figure 74.7b). Electrodes must be securely attached, for example, with stress loops secured away from the electrode site, so that motion artifact can be reduced. In certain instances, the electrodes may be essentially glued to skin, as in the case of EEG measurements.

Skin Preparation

The potentials existing at the skin surface, attributable to potentials at the membranes of cells in the epidermal layers of the skin, can result in a large dc potential (which can be a significant problem in EOG measurements). Any disturbance of the skin by motion, touching, or deformation can cause this potential to change and result in motion artifacts (Figure 74.7a). Sweat glands in the epidermis can also contribute varying extents of skin resistance and skin potential. Such potentials and artifacts can be reduced by abrading the epidermal skin. A mild abrasion by sandpaper or its equivalent can significantly reduce skin resistance and skin potential and thereby reduce artifact [26]. A less traumatic, but somewhat less effective approach, is to use an alcohol swab or similar skin-cleansing solution to wet and clean the skin surface to remove debris, oils, and damaged or dead epidermal cells. Sometimes, as with EEG measurements where very low signals are recorded and very low noise is permitted, skin resistance must be significantly lowered, perhaps to below 2 kΩ [18]. Obviously, reduced motion or muscle activity while measurement is carried out also helps.

Reduction of Environmental Interference

Electromagnetic interference radiated from the power lines, RF interference from machines, induced magnetic field in the leads, and electric currents induced on to the body are all potential sources of environmental interference (Figure 74.7c). Shielding of the amplifier along with the electrode and the lead, and in certain extreme conditions, shielding of the subject (for example, when taking magnetic field measurements from the body) can greatly help reduce the signals picked up by or induced into the amplifier. The electrode leads can be shielded or at the very least twisted together to reduce induced electromagnetic interference.

The amplifier circuit should also have extensive filtering of unwanted electromagnetic interference. To eliminate RF interference, filter capacitors should be used in the front end of the amplifier as well as at various stages of the amplifier. Very high frequencies can be blocked by the use of a choke or an inductor at the input leads. The effect of electrostatic interference can be minimized or eliminated by grounding the instrument.

Electric interference in the environment induces current into the body, which is then picked up by the biopotential amplifier as a common-mode voltage [27]. The CMRR property of the amplifier is essential for reduction of the common-mode voltage [24]. Finally, the driven right leg design [27], described earlier, can be optionally used to reduce further the common-mode voltage and the effective interference.

74.9 Conclusions

Biopotential acquisition is a well-developed science, and acceptable engineering design solutions do exist. It is apparent that each biopotential source presents its own distinct challenge in terms of electrode interface, amplifier design, pre- or postprocessing, and practical implementation and usage. ECG signals can be best acquired using Ag–AgCl electrodes, although good experimental/clinical practice is needed to reduce biological and environmental interference. Further circuit protection and isolation are necessary in clinical usage. EEG signals are distinguishable by their very low amplitude, and hence EEG electrodes must be securely attached via a very small electrode–skin resistance and the amplifier must exhibit exceptionally low noise. For EMG acquisition, electrodes are needed that can be attached for long periods of time to the muscle groups under study. The EMG signal inevitably needs postprocessing, such as integration, to derive a measure of muscle activity. EOG signals have small amplitudes and are characterized by dc or low frequencies. Skin–electrode potentials and dc drift of the amplifier are, therefore, important considerations.

These biopotential measurement principles are applicable to a variety of conventional as well as emerging applications. For example, although ECG acquisition is used mainly in cardiac monitors, it is also of interest and importance in implantable pacemakers and defibrillators. EEG acquisition is useful in the detection of seizure spikes and study of sleep patterns and it may also be used to identify cortical dysfunction after trauma or stroke. EMG acquisition is used in diagnosing neuromuscular diseases. Interesting attempts have been made to use EMG for controlling prostheses. EOG has been helpful in diagnosing vestibulo-oclular disorders and also has been studied as a way of operating communication devices (pointing) used by quadriplegics. The measurement and instrumentation principles described in this chapter would be applicable, with some modifications, to these emergent applications.

References

1. L. A. Geddes and L. E. Baker, *Principles of Applied Biomedical Instrumentation*, 3rd ed., New York: Wiley, 1989.
2. R. Plonsey, *Bioelectric Phenomena*, New York: McGraw-Hill, 1969.
3. R. Plonsey and R. C. Barr, *Bioelectricity*, New York: Plenum, 1988.
4. R. C. Barr, "Basic electrophysiology," in *The Biomedical Engineering Handbook*, Bronzino J., Ed., Boca Raton, FL: CRC Press, pp. 101–118, 1995.
5. D. Durrer et al., "Total excitation of the isolated human heart," *Circulation*, 41, 899–912, 1970.
6. P. L. Nunez, *Electric Fields of the Brain*, New York: Oxford University Press, pp. 484, 1981.
7. K.-A. Henneberg, "Principles of electromyography," in *The Biomedical Engineering Handbook*, Bronzino J. D., Ed., Boca Raton, FL: CRC Press, pp. 191–200, 1995.
8. J. G. Webster, Ed., *Medical Instrumentation: Application and Design*, 3rd ed., New York: Wiley, 1998.

9. N. V. Thakor, "Electrocardiographic monitors," in *Encyclopedia of Medical Devices and Instrumentation*, Webster J. G., Ed., New York: Wiley, pp. 1002–1017, 1988.

10. H. V. Pipberger et al., "Recommendations for standardization of leads and specifications for instruments in electrocardiography and vector cardiography," *Circulation*, 52, 11–31, 1975.

11. J. G. Webster, "Reducing motion artifacts and interference in biopotential recording," *IEEE Trans. Biomed. Eng.*, 31, 823–826, 1984.

12. N. V. Thakor and J. G. Webster, "Electrode studies for the long-term ambulatory ECG," *Med. Biol. Eng. Comput.*, 23, 116–121,1985.

13. J. C. Huhta and J. G. Webster, "60-Hz interference in electrocardiography," *IEEE Trans. Biomed. Eng.*, 20, 91–101, 1973.

14. Anonymous, "American National Standard Safe Current Limits for Electromedical Apparatus," *ANSI/AAMI*, vol. SCL 12/78, 1978.

15. Anonymous, "American National Standard for Diagnostic Electrocardiographic Devices," *ANSI/AAMI*, vol. EC11-1982, 1984.

16. N. V. Thakor, "From Holter monitors to automatic defibrillators: developments in ambulatory arrhythmia monitoring," *IEEE Trans. Biomed. Eng.*, 31, 770–778, 1984.

17. A. S. Gevins and M. J. Aminoff, "Electroencephalography: brain electrical activity," in *Encyclopedia of Medical Devices and Instrumentation*, Webster, J. G., Ed., New York: Wiley, pp. 1084–1107, 1988.

18. E. Niedermeyer and F. Lopes da Silva, *Electroencephalography*, Baltimore: Urban, Schwarzenberg, 1987.

19. C. J. De Luca, "Electromyography," in *Encyclopedia of Medical Devices and Instrumentation*, Webster J. G., Ed., New York: Wiley, pp. 1111–1120, 1988.

20. H. Carim, "Bioelectrodes," in *Encyclopedia of Medical Devices and Instrumentation*, Webster J. G., Ed., New York: Wiley, pp. 195–226, 1988.

21. M. R. Neuman, "Biopotential electrodes," in *Medical Instrumentation: Application and Design*, Webster J. G., Ed., 3rd ed., New York: Wiley, 1988.

22. J. H. Nagle, "Biopotential amplifiers," in *The Biomedical Engineering Handbook*, Bronzino J. D., Ed., Boca Raton, FL: CRC Press, pp. 1185–1195, 1995.

23. S. Franco, *Design with Operational Amplifiers*, New York: McGraw-Hill, 1988.

24. P. Horowitz and W. Hill, *The Art of Electronics*, 2nd ed., Cambridge, England: Cambridge University Press, 1989.

25. W. J. Jung, *IC Op Amp Cookbook*, 3rd ed., Indianapolis, IN: Howard W. Sams, 1986.

26. H. W. Tam and J. G. Webster, "Minimizing electrode motion artifact by skin abrasion," *IEEE Trans. Biomed. Eng.*, 24, 134–139, 1977.

27. M. R. Neuman, "Biopotential amplifiers," in *Medical Instrumentation: Application and Design*, Webster J. G., Ed., 3rd ed., New York: Wiley, 1998.

75

Blood Pressure Measurement

Shyam Rithalia
University of Salford

Mark Sun
NeoPath, Inc.

Roger Jones
Primary Children's Medical Center

75.1 Introduction

Blood pressure measurements have been part of the basic clinical examination since the earliest days of modern medicine. The origin of blood pressure is the pumping action of the heart, and its value depends on the relationship between cardiac output and peripheral resistance. Therefore, blood pressure is considered as one of the most important physiological variables with which to assess cardiovascular hemodynamics. Venous blood pressure is determined by vascular tone, blood volume, cardiac output, and the force of contraction of the chambers of the right side of the heart. Since venous blood pressure must be obtained invasively, the term *blood pressure* most commonly refers to arterial blood pressure, which is the pressure exerted on the arterial walls when blood flows through the arteries. The highest value of pressure, which occurs when the heart contracts and ejects blood to the arteries, is called the systolic pressure (SP). The diastolic pressure (DP) represents the lowest value occurring between the ejections of blood from the heart. Pulse pressure (PP) is the difference between SP and DP, i.e., PP = SP − DP. The period from the end of one heart contraction to the end of the next is called the cardiac cycle. Mean pressure (MP) is the average pressure during a cardiac cycle.

Mathematically, MP can be decided by integrating the blood pressure over time. When only SP and DP are available, MP is often estimated by an empirical formula:

$$MP \approx DP + PP/3 \qquad (75.1)$$

Note that this formula can be very inaccurate in some extreme situations. Although SP and DP are most often measured in the clinical setting, MP has particular importance in some situations, because it is the driving force of peripheral perfusion. SP and DP can vary significantly throughout the arterial system whereas MP is almost uniform in normal situations.

The values of blood pressure vary significantly during the course of 24 h according to an individual's activity [1]. Basically, three factors, namely, the diameter of the arteries, the cardiac output, and the state or quantity of blood, are mainly responsible for the blood pressure level. When the tone increases in the muscular arterial walls so that they narrow or become less compliant, the pressure becomes higher than normal. Unfortunately, increased blood pressure does not ensure proper tissue perfusion, and in some instances, such as certain types of shock, blood pressure may seem appropriate when peripheral tissue perfusion has all but stopped. Nevertheless, observation or monitoring of blood pressures affords dynamic tracking of pathology and physiology affecting the cardiovascular system. This system in turn has profound effects on the other organs of the body.

75.2 Measurement Techniques

The basis of any physiological measurement is the biological signal, which is first sensed and transduced or converted from one form of energy to another. The signal is then conditioned, processed, and amplified. Subsequently, it is displayed, recorded, or transmitted (in some ambulatory monitoring situations). Blood pressure sensors often detect mechanical signals, such as blood pressure waves, to convert them into electric signals for further processing or transmission. They work on a variety of principles, for example, resistance, inductance, and capacitance. For accurate and reliable measurements a sensor should have good sensitivity, linearity, and stability [2].

75.3 Indirect Blood Pressure Measurement

Indirect measurement is often called noninvasive measurement because the body is not entered in the process. The upper arm, containing the brachial artery, is the most common site for indirect measurement because of its closeness to the heart and convenience of measurement, although many other sites may have been used, such as forearm or radial artery, finger, etc. Distal sites such as the wrist, although convenient to use, may give much higher systolic pressure than brachial or central sites as a result of the phenomena of impedance mismatch and reflective waves [3]. An occlusive cuff is normally placed over the upper arm and is inflated to a pressure greater than the systolic blood pressure. The cuff is then gradually deflated, while a detector system simultaneously employed determines the point at which the blood flow is restored to the limb. The detector system does not need to be a sophisticated electronic device. It may be as simple as manual palpation of the radial pulse. The most commonly used indirect methods are auscultation and oscillometry, each is described below.

Auscultatory Method

The auscultatory method most commonly employs a mercury column, an occlusive cuff, and a stethoscope. The stethoscope is placed over the blood vessel for auscultation of the Korotkoff sounds, which defines both SP and DP. The Korotkoff sounds are mainly generated by the pulse wave propagating through the brachial artery [4]. The Korotkoff sounds consist of five distinct phases. The onset of Phase I Korotkoff sounds (first appearance of clear, repetitive, tapping sounds) signifies SP and the onset of Phase V Korotkoff sounds (sounds disappear completely) often defines DP [5].

Observers may differ greatly in their interpretation of the Korotkoff sounds. Simple mechanical error can occur in the form of air leaks or obstruction in the cuff, coupling tubing, or Bourdon gage. Mercury can leak from a column gage system. In spite of the errors inherent in such simple systems, more mechanically complex systems have come into use. The impetus for the development of more elaborate detectors has come from the advantage of reproducibility from observer to observer and the convenience of automated operation. Examples of this improved instrumentation include sensors using plethysmographic principles, pulse-wave velocity sensors, and audible as well as ultrasonic microphones [6].

The readings by auscultation do not always correspond to those of intra-arterial pressure. [5]. The differences are more pronounced in certain special occasions such as obesity, pregnancy, arteriosclerosis,

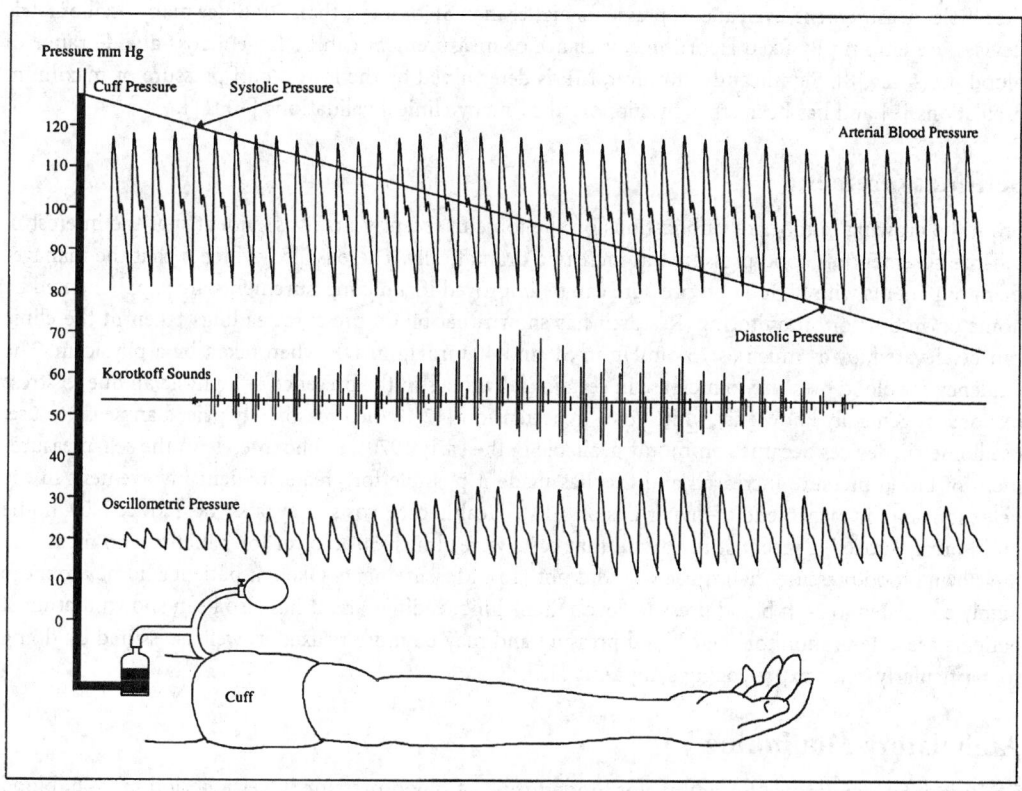

FIGURE 75.1 Indirect blood pressure measurements: oscillometric measurement and auscultatory measurement. (Adapted from Current technologies and advancement in blood pressure measurements — review of accuracy and reliability, *Biomed. Instrum. Technol.*, AAMI, Arlington, VA (publication pending). With permission.)

shock, etc. Experience with the auscultation method has also shown that determination of DP is often more difficult and less reliable than SP. However, the situation is different for the oscillometric method where oscillations caused by the pressure pulse amplitude are interpreted for SP and DP according to empirical rules [7].

Oscillometric Method

In recent years, electronic pressure and pulse monitors based on oscillometry have become popular for their simplicity of use and reliability. The principle of blood pressure measurement using the oscillometric technique is dependent on the transmission of intra-arterial pulsation to the occluding cuff surrounding the limb. An approach using this technique could start with a cuff placed around the upper arm and rapidly inflated to about 30 mmHg above the systolic blood pressure, occluding blood flow in the brachial artery. The pressure in the cuff is measured by a sensor. The pressure is then gradually decreased, often in steps, such as 5 to 8 mmHg. The oscillometric signal is detected and processed at each step of pressure. The cuff pressure can also be deflated linearly in a similar fashion as the conventional auscultatory method.

Figure 75.1 illustrates the principle of oscillometric measurement along with auscultatory measurement. Arterial pressure oscillations are superimposed on the cuff pressure when the blood vessel is no longer fully occluded. Separation of the superimposed oscillations from the cuff pressure is accomplished by filters that extract the corresponding signals. Signal sampling is carried out at a rate determined by the pulse or heart rate [7]. The oscillation amplitudes are most often used with an empirical algorithm to estimate SP and DP. Unlike the Korotkoff sounds, the pressure oscillations are detectable throughout the whole measurement, even at cuff pressures higher than SP or lower than DP. Since many oscillometric devices use empirically fixed algorithms, variance of measurement can be large across a wide range of blood pressures [8]. Significantly, however, MP is determined by the lowest cuff pressure of maximum oscillations [9] and has been strongly supported by many clinical validations [10, 11].

Self-Measurement

From the growing number of publications on the topic in recent years, it is evident that the interest in self-measurement of blood pressure has increased dramatically. There is also evidence that the management of patients with high blood pressure can be improved if clinic measurements are supplemented by home or ambulatory monitoring. Research has shown that blood pressure readings taken in the clinic can be elevated, by as much as 75 mmHg in SP and 40 mmHg in DP, when taken by a physician. The tendency for blood pressure to increase in certain individuals in the presence of a physician due to stress response is generally known as "white-coat" hypertension [12]. When reasonably priced and easy to use, oscillometric devices became commonly available in the early 1970s, public interest in the self-measurement of blood pressure increased and this has made it possible for greater patient involvement in the detection and management of hypertension [13]. Health care costs may also be reduced by home monitoring. Indeed, a recent study found that costs were almost 30% lower for patients who measured their own blood pressure than those who did not [14]. Measurements taken at patient's home are more highly correlated to 24-h blood pressure levels than clinic readings are. It has also been shown that most patients are able to monitor their blood pressure and may be more relaxed as well as assured by doing so, particularly when experiencing symptoms [15].

Ambulatory Monitoring

There is great significance for ambulatory monitoring of blood pressure. Over a period of 24 h, blood pressure is subject to numerous situational and periodic fluctuations [1]. The pressure readings have a pronounced diurnal rhythm in an individual, with a decrease of 10 to 20 mmHg during sleep and a prompt increase on getting up and walking in the morning. Readings tend to be higher during working hours and lower at home and they depend on the pattern of activity. After a bout of vigorous exercise or strenuous work, blood pressure may be reduced for several hours. The readings may be raised if the patient is talking during the measurement period. Smoking a cigarette and drinking coffee, especially if they are combined, may both raise the pressure [16]. When assessing the efficacy of antihypertensive drugs, ambulatory blood pressure monitoring can provide considerable information and validation of the drug treatment [17].

Although the technique of noninvasive ambulatory blood pressure monitoring was first described more than three decades ago, it has only recently become accepted as a clinically useful procedure for evaluation of patients with abnormal regulation of blood pressure. It gives the best evaluation for patients who have white-coat hypertension. Technical advances in microelectronics and computer technology have led to the introduction of ambulatory monitors with improved accuracy and reliability, small size, quiet operation, and reasonable low price. They can take and store several hundred readings over a period of 24 h while patients may not be compromised with their normal activities, thus becoming usable for purposes of clinical diagnosis [18]. Theoretically, ambulatory monitoring can provide information about the level and variability covering the full range of blood pressure experienced during day-to-day activities. It is now recognized to be a very useful procedure in clinical practice since blood pressure varies significantly during the course of 24 h, especially useful in detecting white-coat hypertension. However, many studies have found that accuracy of monitoring using current ambulatory monitors is acceptable only when patients are at rest but not during physical activity [19] or under truly ambulatory conditions. Report of error codes during operation in the latter situations is much higher [20].

TABLE 75.1 AHA Acceptable Bladder Dimensions for Arm of Different Sizes[a]

Cuff	Bladder Width (cm)	Bladder Length (cm)	Arm Circumference Range at Midpoint (cm)
Newborn	3	6	≤6
Infant	5	15	6–15[b]
Child	8	21	16–21[b]
Small adult	10	24	22–26
Adult	13	30	27–34
Large adult	16	38	35–44
Adult thigh	20	42	45–52

[a] There is some overlapping of the recommended range for arm circumferences in order to limit the number of cuffs; it is recommended that the larger cuff be used when available.

[b] To approximate the bladder width:arm circumference ratio of 0.40 more closely in infants and children, additional cuffs are available.

Adapted from the *Recommendations for Human Blood Pressure Determination by Sphygmomanometers*, Dallas: American Heart Association, 1993. With permission.

Cuff Size

Both the length and width of an occluding cuff are important for accurate and reliable measurement of blood pressure by indirect methods. A too-short or too-narrow cuff results in false high blood pressure readings. Several studies have shown that a cuff of inappropriate size in relation to the patient's arm circumference can cause considerable error in blood pressure measurement [21]. The cuff should also fit around the arm firmly and comfortably. Some manufacturers have designed cuffs with a fastener spaced so that a cuff of appropriate width only fits an arm of appropriate diameter. With this design, the cuff will not stay on the arm during inflation unless it fits accordingly.

According to American Heart Association (AHA) recommendations [5], the width of the cuff should be 40% of the midcircumference of the limb and the length should be twice the recommended width. Table 75.1 presents the AHA cuff sizes covering from neonates to adults.

Recommendations, Standards, and Validation Requirements

The AHA has published six editions of the AHA recommendations for indirect measurement of arterial blood pressure. The most recent edition [5] included the recommendations of the joint national committee on the diagnosis, evaluation, and treatment of hypertension for classifying and defining blood pressure levels for adults (age 18 years and older) [22], as shown in Table 75.2. The "Report of the Second Task Force on Blood Pressure Control in Children" [23] offered classification of hypertension in young age groups from newborns to adolescents, as shown in Table 75.3.

The AHA recommendations provide a systemic step-by-step procedure for measuring blood pressure, including equipment, observer, subject, and technique. It extends considerations of blood pressure recording in special populations such as infants and children, elderly, pregnant and obese subjects, etc. It also provides recommendations of self-measurement or home measurement of blood pressure, as well as ambulatory blood pressure measurement.

The Association for the Advancement of Medical Instrumentation (AAMI) and American National Standard Institute (ANSI) published and revised a national standard [24, 25] for evaluating electronic or automated sphygmomanometers. This standard established labeling requirements, safety and performance requirements, and referee test methods for electronic or automated sphygmomanometers used in indirect measurement of blood pressure. Specific requirements for ambulatory blood pressure monitors were also included. Recently, AAMI/ANSI amended this SP10 standard to include neonatal devices as well [26]. Some of the specific requirements, procedures, and limits were modified to fit neonatal

TABLE 75.2 Recommendations of the Joint National Committee on the Diagnosis, Evaluation, and Treatment of Hypertension for Classifying and Defining Blood Pressure Levels for Adults (age 18 years and older)[a]

Category	Systolic Pressure (mm Hg)	Diastolic Pressure (mm Hg)
Normal[b]	<130	<85
High normal	130–139	85–89
Hypertension[c]		
Stage 1 (mild)	140–159	90–99
Stage 2 (moderate)	160–179	100–109
Stage 3 (severe)	180–209	110–119
Stage 4 (very severe)	≥210	120

[a] Not taking antihypertensive drugs and not acutely ill. When systolic and diastolic pressures fall into different categories, the higher category should be selected to classify the individual's blood pressure status. For instance, 160/92 mmHg should be classified as stage 2, and 180/120 mmHg should be classified as stage 4. Isolated systolic hypertension is defined as a systolic blood pressure of 140 mmHg or more and a diastolic blood pressure of less than 90 mmHg and staged appropriately (e.g., 170/85 mmHg is defined as stage 2 isolated systolic hypertension). In addition to classifying stages of hypertension on the basis of average blood pressure levels, the clinician should specify presence or absence of target-organ disease and additional risk factors. For example, a patient with diabetes and a blood pressure of 142/94 mmHg plus left ventricular hypertrophy should be classified as having "stage 1 hypertension with target-organ disease (left ventricular hypertrophy) and with another major risk factor (diabetes)." This specificity is important for risk classification and management.

[b] Optimal blood pressure with respect to cardiovascular risk is less than 120 mmHg systolic and less than 80 mmHg diastolic. However, unusually low readings should be evaluated for clinical significance.

[c] Based on the average of two or more readings taken at each of two or more visits after an initial screening.

Adapted from *The fifth report of the Joint National Committee on Detection, Evaluation, and Treatment of High Blood Pressure (JNCW), Arch. Intern. Med.,* 153, 154–183, 1993.

applications, such as the maximum cuff pressure, ranges of age and weight, reference standards for validation, minimum sample size of data, etc. The overall system efficacy for both neonatal and adult devices requires that for systolic and diastolic pressures treated separately, the mean difference between the paired measurements of the test system and the reference standard shall be ±5 mmHg or less, with a standard deviation of 8 mmHg or less.

For manual or nonautomated indirect blood pressure measuring devices, ANSI/AAMI SP9 standard [27] applies.

The British Hypertension Society (BHS) also published and revised a protocol for assessing accuracy and reliability of blood pressure measurement using automatic and semiautomatic devices [28, 29]. Many automatic and semiautomatic devices, including ambulatory devices, have been evaluated according to the BHS protocol. Such evaluation provided a quality-control mechanism for manufacturers and an objective comparison for customers. However, there are many more devices available on the market, which have not been accordingly evaluated. Different from the AAMI SP10 standard in which either indirect or direct blood pressure may be used as a reference standard, the BHS protocol relies exclusively on references of sphygmomanometric blood pressure measurement, and does not recommend comparison with intra-arterial blood pressure values [30]. This could make accurate validation of ambulatory devices difficult because sphygmomanometric measurements during exercise and under ambulatory conditions are not accurate [31].

Significantly, the BHS protocol emphasized the need on special-group validation, such as children, pregnancy, and the elderly for the intended use. It also emphasized the need for validation under special circumstances, such as exercise and posture. The accuracy criteria use a grading system based on the percentages of test instrument measurements differing from the sphygmomanometric measurements by ≤5, ≤10, and ≤15 mmHg for systolic and diastolic blood pressure, respectively, as shown in Table 75.4.

TABLE 75.3 Classification of Hypertension in the Young by Age Group[a]

Age Group	High Normal (90–94th percentile) mmHg	Significant Hypertension (95–99th percentile) mmHg	Severe Hypertension (>99th percentile) mmHg
Newborns (SBP)			
7 d	—	96–105	≥106
8–30 d	—	104–109	≥110
Infants (≥2 y)			
SBP	104–111	112–117	≥118
DBP	70–73	74–81	82
Children			
3–5 y			
SBP	108–115	116–123	≥124
DBP	70–75	76–83	≥84
6–9 y			
SBP	114–121	122–129	≥130
DBP	74–77	78–85	≥86
10–12 y			
SBP	122–125	126–133	≥134
DBP	78–81	82–89	≥90
13–15 y			
SBP	130–135	136–143	≥144
DBP	80–85	86–91	≥92
Adolescents (16–18 y)			
SBP	136–141	142–149	≥150
DBP	84–91	92–97	≥98

[a] SBP indicates systolic blood pressure; DBP, diastolic blood pressure.

Adapted from the Report of the Second Task Force on Blood Pressure Control in Children — 1987, *Pediatrics*, 79, 1–25, 1987. With permission.

TABLE 75.4 Grading Criteria of the 1993 British Hypertension Society Protocol[a,b]

Grade	Absolute Difference between Standard and Test Device (mmHg) ≤5	≤10	≤15
	Cumulative Percentage of Readings		
A	60	85	95
B	50	75	90
C	40	65	85
D		Worse than C	

[a] Grades are derived from percentages of readings within 5, 10, and 15 mmHg. To achieve a grade all three percentages must be equal to or greater than the tabulated values.

[b] Grading percentages changed from the 1990 British Hypertension Society protocols due to changes in sequential assessment of blood pressure references. See original publications for details (1990, 1993).

Adapted from The British Hypertension Society protocol for the evaluation of blood pressure measuring devices, *J. Hypertension*, 11 (Suppl. 2), S43–S62, 1993. With permission.

Manufacturer, Product, Price, Efficacy, and Technology

The annual publication of the *Medical Device Register* is a comprehensive reference work that provides a wealth of detailed information on U.S. and international medical devices, medical device companies, OEM suppliers, and the key personnel in the industry. Blood pressure devices are listed in the sphygmomanometer directory. Price information of specific models for some providers is also published. For example, A&D Engineering, Inc. listed price from $51.95 (model UA701) to $179.95 (model UA-751) for a whole line of sphygmomanometers in the 1997 *Medical Device Register* [32]. Since technology and market can change rapidly, models, features, specifications, and prices may change accordingly. More specific and updated information may be available by contacting the manufacturers or distributors directly.

Table 75.5 lists only a limited number of indirect blood pressure devices from a literature review. Many of the listed blood pressure devices have multiple evaluation studies and only a few study results are presented here. In view of reference standards for comparison, although direct and indirect methods yield similar measurements, they are rarely identical because the direct method measures pressure and the indirect method is more indicative of flow [5]. Egmond et al. [33] evaluated the accuracy and reproducibility of 30 home blood pressure devices in comparison with a direct brachial arterial standard. They found average offsets of all tested devices amounted to –11.7 mmHg for systolic and 1.6 mmHg for diastolic blood pressure, which were close to those of the mercury sphygmomanometer (–14.2 mmHg for systolic and –0.1 mmHg for diastolic pressure), indicating a significant difference between the two assessment standards. When selecting a blood pressure device for a specific application, the evaluation using the reference that is of a common practice in the intended population or environment may be practically more informative, since that reference has been the common basis for decision making in blood pressure diagnosis and treatment.

Different evaluation results for the same brand product can also be due to different versions of a model used for validation, where a later version may have performance improvement over the earlier one [34]. Another source of discrepancies can come from utilizing different study protocols or only partially following the same protocols. It is recommended that the original clinical evaluation report be carefully examined in determining the desired efficacy that may meet the users' requirements. If the devices were FAD approved for marketing in the U.S., one may request a copy of their clinical validation report directly from the manufacturer.

In addition to the fundamental categories such as intended use, efficacy, and acquisition technology, listed in Table 75.5, many other categories are also very important in evaluating, selecting, purchasing, using, and maintaining blood pressure devices. These include but are not limited to the following items: measurement range of each pressure (systolic, diastolic, and mean) for each mode of intended use (i.e., neonates, children, adults); maximum pressure that can be applied by the monitor and cuff for each mode of intended use; cuff size range for the target population of the intended use; cost; measurement and record failure rate; noise and artifact rejection capability; mode of operation (manual, automatic, semiautomatic); data display; recording, charting, reporting, and interfacing; physical size and weight; power consumption; operation manual; service manual; labels and warnings; etc.

Advancement of Indirect Blood Pressure Measurement

Since the introduction of Dinamap™, an automated blood pressure monitor based on the oscillometric principle [9], many variants of oscillometric algorithms were developed. However, the fixed or variable fractions of the maximum oscillations are still the fundamental algorithms of the oscillometry [10, 53, 54]. Typically, mean blood pressure was determined by the lowest cuff pressure with greatest average oscillation [11]. Systolic and diastolic blood pressure were determined by the cuff pressure with the amplitude of oscillation being a fixed fraction of the maximum. Performance of the algorithms may be improved by introducing a greater level of complexity or variables into considerations. The Dinamap™ 1846SX (Critikon, Tampa, FL) oscillometric device offered two measurement modes. The normal mode uses two matching pulse amplitudes at each cuff pressure step to establish an oscillometric envelope or curve. Therefore, measurement time is heart rate dependent. The second mode, which the manufacturer refers to as "stat mode," is capable of faster determination by disabling the dual pulse-matching algorithm which was designed for artifact rejection. The stat mode does not appear to compromise accuracy in anesthetized patients [55], in which rapid measurement of blood pressure is often more desirable, particularly during induction and management of anesthesia.

TABLE 75.5 Survey of Indirect Blood Pressure Device Manufacturer, Product, Intended Use, and Efficacy

Manufacturer	Model	Technology	Intended Use	Reference Standard MC/AC[a]	BHS Protocol Grading SBP	BHS Protocol Grading DBP	AAMI SP10 Comparison (Device − Reference) mmHg SBP	AAMI SP10 Comparison (Device − Reference) mmHg DSP	Other Validations (Device − Reference) mmHg SBP	Other Validations (Device − Reference) mmHg DBP	Other Validations (Device − Reference) mmHg MBP	Ref.
A&D, Tokyo, Japan	TM-2420/TM-2020	Korotkoff	Health Care: Ambulatory	MC	D	D	−4 ± 11	−2 ± 11				35
	TM-2420 Version 7	Korotkoff	Health Care: Ambulatory	MC	B	B	−1.8 ± 5.0	−3.5 ± 6.8				36
Colin Medical Instruments, Plainfield, NJ	ABPM 630	Korotkoff (primary mode) Oscillometry (backup mode)	Health Care: Ambulatory	AC MC AC MC					1.4 ± 7.1 −0.4 ± 4.6 4.5 ± 6.6 1.9 ± 4.0	−0.1 ± 5.6 −6.0 ± 5.9 −1.2 ± 6.3 −6.9 ± 5.1		37
Del Mar Avionics, Irvine, CA	Pressurometer IV	Korotkoff (ECG R-wave gating)	Health Care: Ambulatory	MC MC	C	D	−2 ± 11	−3 ± 11	1.2 ± 7.3	−2.2 ± 5.7		38 39
Disetronic Medical Systems AG Burgdorf, Switzerland	CH-Druck/Pressure Scan ERKA	Korotkoff	Health Care: Ambulatory	MC	A	A	−3 ± 4	−2 ± 4				40
Novacor, France	DIASYS 200	Korotkoff	Health Care: Ambulatory	MC	C	C	−1 ± 8	0 ± 8				41
Oxford Medical, Abingdon, Oxford, U.K.	Medilog ABP	Korotkoff	Health Care: Ambulatory	AC MC			−8 ± 8 −4 ± 6	6 ± 6 −2 ± 8				42
Disetronic Medical Systems AG, Burgdorf, Switzerland	Profilomat	Korotkoff	Health Care: Ambulatory	MC	B	A	−3 ± 5	−1 ± 5				43
SpaceLabs Medical, Redmond, WA	90207	Oscillometry	Health Care: Ambulatory	MC	B	B	−1 ± 7	−3 ± 6				44
Suntech Medical Instruments, Raleigh, NC	Accutracker II (v30/23)	Korotkoff (ECG R-wave gating)	Health Care: Ambulatory	MC	A	C	−1.3 ± 6.5	−4.5 ± 7.3				45[b]
Tycos-Welch-Allyn, Arden, NC	QuietTrack	Korotkoff	Health Care: Ambulatory				0.3 ± 5.0	−1.5 ± 7.5				46[c]
Colin Medical Instruments, San Antonio, TX	BP8800MS	Oscillometry Oscillometry	Health Care: Children Health Care: Adults	MC MC			3.2 ± 6.0 2.8 ± 5.4	−0.8 + 5.2 0.0 ± 4.9				47

TABLE 75.5 (continued)   Survey of Indirect Blood Pressure Device Manufacturer, Product, Intended Use, and Efficacy

Manufacturer	Model	Technology	Intended Use	Reference Standard MC/AC[a]	BHS Protocol Grading SBP	BHS Protocol Grading DBP	AAMI SP10 Comparison (Device – Reference) mmHg SBP	AAMI SP10 Comparison (Device – Reference) mmHg DSP	Other Validations (Device –Reference) mmHg SBP	Other Validations (Device –Reference) mmHg DBP	Other Validations (Device –Reference) mmHg MBP	Ref.
Critikon, Tampa, FL	Dinamap 1846SX	Oscillometry	Health Care: Neonates, Children, Adults	AC					−8.8 ± 11.2	1.6 ± 8.9	−1.8 ± 9.7	48
	Dinamap portable monitor	Oscillometry	Health Care: Neonates, Children, Adults	MC	B	D	−1 ± 7	−6 ± 7				49[d]
SpaceLabs Medical, Redmond, WA	Oscillometric Blood Pressure Monitor	Oscillometry	Health Care: Neonates,	AC	B	B	0.1 ± 4.3	2.7 ± 4.8				50[e]
			Health Care: Children, Adults	MC	B	B	−0.6 ± 5.9	0.9 ± 6.4				
Ohmeda, Denver, CO	Finapres 3700	Volume-clamping	Health Care: Continuous Monitoring	AC					−8.4 ± 8.6	−1.1 ± 7.0	−6.8 ± 6.7	48
Terumo, Tokyo, Japan	ES-H51[f]	Korotkoff (primary mode)	Health Care: Routine Clinical	MC	A	A	0.7 ± 2.9	0.3 ± 2.6				51[g]
		Oscillometry (backup mode)		MC	B	A	−0.3 ± 5.7	−0.3 ± 4.3				

Matsushita, Osaka, Japan	Denko EW 160	Oscillometry	Self Care: Home Measurement	MC	1.8 ± 5.2	−1.7 ± 5.5	52
Nissei, Tokyo, Japan	DS 91[f]	Korotkoff	Self Care: Home Measurement	MC	−2.5 ± 7.4	2.8 ± 10.8	52
Omron, Tokyo, Japan	HEM 439[f]	Korotkoff	Self Care: Home Measurement	MC	−0.2 ± 5.3	6.2 ± 9.9	52
	HEM 719K	Korotkoff	Self Care: Home Measurement	MC	−2.3 ± 5.6	2.4 ± 4.7	52
	401C[c]	Oscillometry	Self Care: Home Measurement	MC	−1.6 ± 7.7	2.4 ± 6.1	52
Sharp, Osaka, Japan	MB 305H[f]	Korotkoff	Self Care: Home Measurement	MC	0.5 ± 4.5	9.6 ± 14.3	52
	MB 500A	Oscillometry	Self Care: Home Measurement	MC	−1.8 ± 6.7	0.7 ± 6.3	52
A&D, Tokyo, Japan	Takeda UA 751	Oscillometry	Self Care: Home Measurement	MC	−4.1 ± 5.6	0.4 ± 7.8	52

[a] MC: mercury column; AC: arterial catheter.

[b] Data quoted for the standing position; grading was the same as for pooled data of three positions (supine, seated, and standing).

[c] Data quoted for the three positions of supine, seated, and standing.

[d] Efficacy quoted was determined in adult population.

[e] Efficacy quoted was determined in neonate and adult populations, respectively.

[f] Semiautomatic; all other listed are automatic.

[g] Only partially followed AAMI and BHS protocol and only validated one size (median) of three cuffs (small, median, large).

Adapted from Current technologies and advancement in blood pressure measurements-review of accuracy and reliability, *Biomed. Instrum. Technol.*, AAMI, Arlington, VA (publication pending). With permission.

Another variant of the oscillometric algorithm was developed by Protocol Systems [56]. In addition to using pulse amplitude for primary artifact rejection, it further calculated impulse value, a principal area of pulse waveform, in constructing an oscillometric curve. This curve is smoothed by employing a Kalman filter that also provides an expected mean and acceptable upper and lower bounds of prediction for the principal area of subsequent pulse waveform. Smoothing of the oscillometric curve is accomplished by using the difference between the predicted and calculated area data of pulse waveform for each cuff pressure step. Blood pressures are derived from the final smoothed oscillometric curve.

In more recent study, oscillometric algorithms using an artificial neural network have been reported to produce better estimates of reference blood pressures than the standard oscillometric algorithm [57]. By using neural network training and processing, subtle features and nonlinear relationships of the oscillometric envelope have been modeled. Empiricism of the oscillometric fixed fraction criteria is overcome and variances of measurements are greatly reduced.

Because of its low risk and cost, noninvasive continuous blood pressure monitoring represents another need in critical-care monitoring to supplement invasive arterial catheterization. A significant development in this field is the arterial counterpulsation principle, proposed by Penaz [58], and further developed by two major groups of people [59, 60]. Finapres™, a continuous finger arterial blood pressure monitor was engineered and developed by Ohmeda, Denver, CO. Many clinical evaluation reports of these devices have been published since then.

Recently, a number of other continuous blood pressure monitors have been made commercially available. Examples of these are Cortronic APM770 [61], which monitors pulsation of the brachial artery with a slightly pressurized arm cuff and calibrates it to a continuous pressure waveform; Sentinel ARTRAC 7000 [62], which monitors pulse transit time and correlates that to pressure change; Colin CBM-3000 and JENTOW (Colin Electronics, Komaki, Japan) [63, 64] and Nellcor NCAT N-500 (Nellcor, Hayward, CA) [65], which are tonometric devices monitoring the radial artery pulse waveform by a matrix pressure sensor. All of these monitors require a frequent calibration reference. Except for a few favorable reports with the tonometric method and devices, many reports so far are unfavorable. Nevertheless, noninvasive continuous monitoring represents an important and growing field of biomedical sensor and instrumentation research and development. Continuous monitors, which maintain cuff pressure, must periodically relieve pressure to prevent the risk of venous congestion, edema, swelling, and tissue damage.

75.4 Direct Blood Pressure Measurement

Direct measurement is also called invasive measurement because bodily entry is made. For direct arterial blood pressure measurement an artery is cannulated. The equipment and procedure require proper setup, calibration, operation, and maintenance [66]. Such a system yields blood pressures dependent upon the location of the catheter tip in the vascular system. It is particularly useful for continuous determination of pressure changes at any instant in dynamic circumstances. When massive blood loss is anticipated, powerful cardiovascular medications are suddenly administered, or a patient is induced to general anesthesia, continuous monitoring of blood pressures becomes vital.

Most commonly used sites to make continuous observations are the brachial and radial arteries. The femoral or other sites may be used as points of entry to sample pressures at different locations inside the arterial tree, or even the left ventricle of the heart. Entry through the venous side of the circulation allows checks of pressures in the central veins close to the heart, the right atrium, the right ventricle, and the pulmonary artery. A catheter with a balloon tip carried by blood flow into smaller branches of the pulmonary artery can occlude flow in the artery from the right ventricle so that the tip of the catheter reads the pressure of the left atrium, just downstream. These procedures are very complex and there is always concern of risk of hazard as opposed to benefit [67].

Invasive access to a systemic artery involves considerable handling of a patient. The longer a catheter stays in a vessel, the more likely an associated thrombus will form. The Allen's test can be performed by pressing on one of the two main arteries at the wrist when the fist is clenched, then opening the hand to see if blanching indicates inadequate perfusion by the other artery. However, it has proved an equivocal predictor of possible ischemia [68]. In the newborn, when the arterial catheter is inserted through an umbilical artery, there is a particular hazard of infection and thrombosis, since thrombosis from the catheter tip in the aorta can occlude the arterial supply to vital abdominal organs. Some of the recognized contraindications and complications include poor collateral flow, severe hemorrhage diathesis, occlusive arterial disease, arterial spasm, and hematoma formation [69].

In spite of well-studied potential problems, direct blood pressure measurement is generally accepted as the gold standard of arterial pressure recording and presents the only satisfactory alternative when conventional cuff techniques are not successful. This also confers the benefit of continuous access to the artery for monitoring gas tension and blood sampling for biochemical tests. It also has the advantage of assessing cyclic variations and beat-to-beat changes of pressure continuously, and permits assessment of short-term variations [70, 71].

Catheter–Tubing–Sensor System

A large variety of vascular catheters exist. Catheter materials have undergone testing to ensure that they have a minimal tendency to form blood clots on their surface. The catheter chosen may be inserted percutaneously over a hollow stylet into the blood vessel. Guide wires can be useful to facilitate longer or larger-diameter catheters into vessels, after the guide wires have been placed through a smaller catheter or needle. Less often, entry to a vessel requires a "cutdown," a direct exposure of the vessel after a skin incision. Ultrasonic devices may assist locating the vessels not readily apparent at the skin surface.

Although pressure sensors can be located at the catheter tip, this presents a problem for calibration if left in place and a clot forms near the tip of the catheter, damping the pressure signal. Instead, most catheters connect to an external pressure sensor via fluid-filled low-compliance tubing. The signal from the sensor then undergoes transformation for display or recording. The sensor may take one of several forms, from a variable resistance diaphragm to a silicon microchip. A basic system can consist of an intravascular catheter connected to a rigid fluid-filled catheter and tubing which communicates the pressure to an elastic diaphragm, the deflection of which is detected electrically. There is a direct relationship between the deflection of the diaphragm and the voltage. The higher the voltage, the greater the pressure. Continuous low-rate infusion of heparinized saline is carried out to keep the catheter patent or free from coagulation. The advent of disposable sensor kits have greatly simplified the clinical use of intravascular monitoring [72]. With the cost continually being lowered with the development of semiconductor industry, disposable sensors become more and more cost-effective.

Although direct recording is considered the most accurate method, its accuracy may be limited by variations in the kinetic energy of the fluid in the catheter or dynamic frequency response of the measurement system. The hydraulic link between the patient and the sensor is the major source of potential errors and hazard for the monitoring. Damping and degrading the system's natural frequency, caused by trapped air bubbles, small catheters, various narrow connections, compliant and too long tubing, and too many components connected, are the two characteristic problems with a pressure sensor system. Extreme care should be exercised to eliminate all air bubbles from the fluid to provide adequate dynamic response. The sensor should be zeroed at the level of the heart to eliminate hydrostatic error [73]. A fast flush testing is easy to use for inspection of the dynamic response of the whole system of catheter–tubing–sensor. It can also help direct adjustments for the system to minimize dynamic artifacts [74, 75].

75.5 Reproducibility, Accuracy, and Reliability Issues and Recommendations for Corrective Measures

For each blood pressure assay technique, there is an issue of reproducibility of measurements given approximately similar conditions. Reproducibility quantifies the internal uncertainty of an individual method and instrument, whereas accuracy quantifies the external uncertainty when compared with a reference. Table 75.6 presents estimated uncertainties of reproducibility for three blood pressure–measuring techniques: auscultation, oscillometry, and umbilical arterial catheter [50]. When dealing with

TABLE 75.6 Estimated Uncertainties of Reproducibility for Blood Pressure Measuring Techniques of Auscultation, Oscillometry, and Umbilical Arterial Catheter[a]

	Auscultation	Oscillometry[b]	Umbilical Arterial Catheter
Neonate			
Systolic pressure (mmHg)	N/A	3.3	2.2
Diastolic pressure (mmHg)	N/A	3.4	1.8
Adult			
Systolic pressure (mmHg)	2.8	3.2	N/A
Diastolic pressure (mmHg)	2.2	3.5	N/A

[a] From Reference 50.
[b] Evaluated from SpaceLabs Medical Oscillometric monitor [50].

blood pressure measurement, it is important to bear in mind that even for standard methods, there is a certain amount of nonrepeatable random error. Consequently, taking the average of repeated measurements or multiple readings is always advised before any serious recommendation or management is made.

Table 75.7 presents a review of common problems associated with accuracy and reliability in both indirect and direct blood pressure measurements. Consequences of these problems are analyzed and recommendation of preventive action or alternative solutions are provided. Hazard or safety analyses and review are also very important.

75.6 Blood Pressure Impact, Challenge, and Future

Hypertension is one of the most common and important risk factors of health in industrialized countries [82]. It is the leading cause of death in the U.S. It is treatable by a variety of effective medications. It can cause serious damage to the heart and arteries leading to cardiac infarct, stroke, or renal failure. Significant sudden changes in blood pressure may also precede a major physiological catastrophe such as cardiac arrest. There is now almost universal acceptance that basic physiological parameters such as blood pressure should always be monitored in the clinical setting.

There has been increasing interest in automatic blood pressure monitoring devices in recent years and some clinicians are now advising patients to record their blood pressure at home over a period of up to 3 months before starting antihypertensive medication [83]. Self-monitoring of blood pressure has become very common with the development of microchip technology and oscillometric monitors. The patients no longer have to learn how to listen for Korotkoff sounds. This has also removed bias and observer errors, allowing more accurate measurement than by conventional techniques using a stethoscope and a mercury sphygmomanometer [84].

Special populations have unique blood pressure assessment requirements. Newborns require miniaturized equipment. The act of taking a blood pressure in a newborn may stimulate a series of movements causing motion artifact. The very obese may be hard to fit properly with a cuff at the upper arm, if the upper arm is too conical rather than cylindrical. In pregnancy, auscultatory and oscillometric methods, although useful to follow trends, may correspond poorly with central pressures [85] and even the proper Korotkoff sound (IV or V) to designate as diastolic pressure is uncertain [86].

Observing blood pressures has limitations. It may suggest what is happening with blood volume, but sometimes does not reveal that blood volume has become inadequate until circulatory collapse has occurred. Venous and left atrial pressures are often used in an attempt to clarify blood volume problems but with uncertain results [87]. Similarly, a satisfactory blood pressure does not always indicate adequate tissue perfusion. Some medications that increase blood pressure can do so at the expense of general perfusion. Since blood pressure is measured at specific sites in the arterial tree, if circulation has become nonhomogeneous, such as can happen in arteriosclerosis, the region distal to the arteriosclerosis can be compromised without warning from blood pressure readings sampled at another site. Even mean blood pressure, so useful otherwise, can fail in these circumstances.

TABLE 75.7 Common Issues of Accuracy and Reliability in Blood Pressure Measurement and Recommendations of Preventive Action or Alternative Solution in Both Indirect and Direct Measurements

Source	Problem	Result	Recommendation
		Indirect Measurement	
Subject	Obesity, peripheral edema, peripheral vascular disease	Weak Korotkoff sounds and diminished sound transmission may reduce the accuracy and reliability of auscultatory measurement; oscillometric measurement may also be affected	Verify with a second indirect method such as oscillometry; direct blood pressure measurement may be elected to use in severe conditions that indirect measurement does not warrant sufficient accurate and reliable measurement
	Shock, severe peripheral vasoconstriction, diminished peripheral circulation resulting from shunting of blood to central organs; Korotkoff sounds and pulses may be absent even in presence of normal pressure [76]	Any of the indirect methods, including auscultatory, oscillometric and Doppler techniques may not provide accurate and reliable reading; indirect measurement may be impossible or may give misleading results	Direct measurement should be considered
	Arrhythmias, respiratory effect	Pronounced variation in beat-to-beat blood pressure and waveform	Take multiple measurements and average
	Subject shivering, pain, anxiety, discomfort, motion artifact	Shivering and motion artifact may cause either false high or false low reading, whereas pain, anxiety, and discomfort may cause false high reading	Minimize pain, anxiety, and discomfort; reduce shivering and movement
	Physical activity within 5 min of measurement; talking, moving, arm unsupported, back unsupported, legs dangling, and any other isometric activities	False high reading that does not reflect subject's resting blood pressure	Subject should rest at least 5 min in the same position that blood pressure is going to be taken; subject should not talk and involve any isometric activities during measurement; arm should be supported at heart level
	Arm supported at above heart level	Hydrostatic pressure causes false low reading by 0.78 mmHg for each centimeter of offset [77]	Support the arm with midpoint of upper arm at heart level
	Arm supported at below heart level	Hydrostatic pressure causes false high reading by 0.78 mmHg for each centimeter of offset	Support the arm with midpoint of upper arm at heart level
	White-coat hypertension during clinical measurement	Psychological or stress response causes blood pressure temporarily elevated and unrepresentative of subject's true condition	Take multiple self-measurements at home or ambulatory monitoring as desired and provide record to care providers
	"Pseudo-hypertension" with calcified or stiffened arteries	Reduced arterial compliance, often occurring in the elderly, causes cuff blood pressure falsely too high or unable to be measured accurately	Use Osler maneuver for screening; direct method is recommended for those who test positive [78,79]
Operator	Hose kinked	Will cause reading error or operation failure	Rearrange hose to avoid kink
	Cuff used too narrow for arm	Will cause false high reading	Select appropriate cuff size that its width encircles 40% of arm circumference
	Cuff used too wide for arm	May cause false low reading; may not fit on arm	Select appropriate cuff size that its width encircles 40% of arm circumference

TABLE 75.7 (continued)   Common Issues of Accuracy and Reliability in Blood Pressure Measurement and Recommendations of Preventive Action or Alternative Solution in Both Indirect and Direct Measurements

Source	Problem	Result	Recommendation
	Cuff wrapped too loosely	Will cause false high reading; may introduce artifact of inter cuff–arm abrasion if placed for long-term monitoring	Cuff should be snugly applied; one should not be able to insert two fingers between the cuff and arm for adult
	Cuff wrapped too tightly	May cause false low reading; will restrict and impair limb circulation if placed for long-term monitoring	Cuff should be snugly but not restrictively applied; one should be able to insert one finger between the cuff and arm for adults
	Cuff pressure inflated too high	Patient discomfort; may induce increase in systolic blood pressure during inflation period, so called "cuff-inflation hypertension" [80]	Inflate cuff pressure to 30 mmHg above palpatory blood pressure
	Cuff pressure inflated too low	Will either miss or have false low systolic pressure reading	Inflate cuff pressure to 30 mmHg above palpatory blood pressure
	Cuff pressure deflated too fast	May degrade the accuracy of the reading	Deflate cuff pressure at 2–4 mmHg per heart beat or 3 mmHg/s
	Cuff pressure deflated too slow	May cause discomfort or forearm congestion	Deflate cuff pressure at 2–4 mmHg per heart beat or 3 mmHg/s
	Repeated cuff pressure measurement too frequently	May cause discomfort and forearm congestion	A sufficient time should elapse (at least 60 s) before the next reading to allow the return of normal circulation
	Miss identifying auscultatory gap between systolic and diastolic pressure	Will cause false low systolic or false high diastolic pressure	Listen to Korotkoff sounds carefully for a wide pressure deflation range or use oscillometric method
	Stethoscope head or sensor not over the brachial artery	Will not hear clear sounds or detect sufficient signal for blood pressure determination	Place the stethoscope head or sensor over the brachial artery at least 1.5 cm above the antecubital fossa
	Noise and artifact created by accidentally touching or bumping the cuff, hose, stethoscope, or sensor	May cause inaccurate reading or failure of reading	Avoid incidence of extraneous noise and artifact
Equipment	Leaky hose, bladder/cuff, or pneumatic components	Will cause inaccurate reading or failure in operation	Require service or replace equipment
	Faulty valves	Will cause inaccurate reading or failure in operation	Require service or replace equipment
	Limited selection for different size of cuffs	Will cause false low or false high reading if cuff is too large or too small, respectively	Manufacturer should provide appropriate label/labeling for the intended use and arm size; blood pressure measurement beyond the intended use of the device should be warned against and prohibited
	Device zero-shifted, out of calibration	Will create systematic bias or uncertainty in blood pressure reading	Require routine calibration and maintenance
		Direct Measurement	
Subject	Subject position change (e.g., body position change, bed lowered or elevated, etc.) in relation to pressure sensor	Subject heart level change in relation to pressure sensor will introduce bias of hydrostatic pressure in blood pressure recording	Move the sensor zero port to the heart level and zero the sensor/monitor
	Catheter whip in pulmonary artery, catheter impact in aorta or ventricle	Catheter whip can result in superimposed waves of ±10 mmHg; catheter impact can cause high-frequency transients to occur in waveform [81]	Catheter whip and catheter impact are difficult to prevent; evaluation of pressure waveform and reading should consider the effect of these events

TABLE 75.7 (continued) Common Issues of Accuracy and Reliability in Blood Pressure Measurement and Recommendations of Preventive Action or Alternative Solution in Both Indirect and Direct Measurements

Source	Problem	Result	Recommendation
	Subject severe shivering, pain, anxiety, discomfort, moving	Severe shivering and moving may cause artifact on blood pressure waveform whereas pain, anxiety, and discomfort may elevate blood pressure	Minimize pain, anxiety, and discomfort, reduce shivering and moving
Operator	Tubing kinked	Will change dynamic response of tubing system and distort pressure waveform	Use short and low compliant tubing, and place tubing appropriately to avoid kink
	Sensor zero port higher than heart level when zeroing	Hydrostatic pressure causes false low pressure measurement by 0.78 mmHg for each centimeter of offset	Move the sensor zero port to heart level and zero the sensor/monitor
	Sensor zero port lower than heart level when zeroing	Hydrostatic pressure causes false high pressure measurement by 0.78 mmHg for each centimeter of offset	Move the sensor zero port to heart level and zero the sensor/monitor
	Air bubbles entrapped in the tubing system	Air bubbles will decrease natural frequency and increase damping coefficient; therefore they damp and distort the waveform, causing high-frequency components to loss in pressure waveform	Eliminate air in both tubing system and flush solution bag; light tapping while fluid is filling the tubing system is an effective method for removing air
	Tubing too long, too thin, and with too many components	All of these will degrade the system dynamic response and result in distorted waveform and erroneous reading	Use tubing of large inner diameter, short length, and reduce the number of components as much as possible
	Connectors not tightly connected	Will decrease natural frequency of tubing system and cause pressure waveform to be distorted	Check loose luer-lock connection and cracked connection; replace cracked components and secure tight connection of all components
	Failure to flush the arterial line adequately after blood draw	May cause the catheter tip partially clotted by the blood and pressure waveform over damped and distorted	Flush the arterial line adequately; may need to replace with a new catheter if dynamic response cannot be improved to meet the minimum requirement
	Failure to zero the sensor/monitor after subject position change in relation to pressure sensor	Subject heart level change in relation to pressure sensor will introduce bias of hydrostatic pressure in blood pressure recording	Move the sensor zero port to the heart level and zero the sensor/monitor
	Failure to provide constant infusion of anticoagulation/saline solution	May cause catheter tip partially clotted by the blood and pressure waveform overdamped and distorted	Check the constant infusion device to have sufficient flow rate; flush the arterial line adequately; may need to replace with a new catheter if dynamic response cannot be improved to meet the minimum requirement
	Failure to test dynamic response at least once a shift and anytime after blood draw or component change	This leaves the system dynamic performance unknown, which may affect the accuracy of systolic pressure the most, diastolic pressure the second; mean pressure is hardly affected	Routinely perform the fast flush test to evaluate the dynamic response visually according to Gardner's chart of natural frequency vs. damping coefficient [73]
Equipment	Not equipped with an appropriate flush device	May not be able to generate quality test waveform to evaluate the adequacy of dynamic response of the catheter–tubing–sensor system	Select appropriate flush device that permits fast flush test for the system dynamic response
	Tubing or component not transparent	Unable to see entrapped air bubbles	Use transparent tubing and components

TABLE 75.7 (continued)　　Common Issues of Accuracy and Reliability in Blood Pressure Measurement and Recommendations of Preventive Action or Alternative Solution in Both Indirect and Direct Measurements

Source	Problem	Result	Recommendation
	Tubing, sensor, or constant flush device too compliant	Will decrease natural frequency of the system and cause pressure waveform to be distorted	Use only high-quality and low-compliance tubing, sensor, and constant flush device
	Stopcocks not tightly sealed	Will decrease natural frequency of the system and cause pressure waveform to be distorted	Replace with tightly sealed, high-quality stopcocks
	Monitor failure to zero the sensor electronically, sensor zero drift, or pressure amplifier zero drift	Will introduce unknown offset or bias in pressure measurement	Require service or replacement of the equipment
	Natural frequency and damping coefficient of the catheter–tubing–sensor system failure to meet minimum dynamic response requirement	Fidelity of pressure waveform recording suffers and accuracy of systolic and diastolic pressure measurement degrades	Need to optimize the catheter–tubing–sensor system by replacing part or all of the components; use low-compliance pressure sensor, tubing, and all other components; use short and large tubing and reduce the number of components as much as possible
	Blood pressure monitor failure to identify special events such as sensor zeroing, fast flush testing, blood drawing, as well as artifacts	Blood pressure monitor displays false digital reading of blood pressure without warning sign or error message	Health care provider needs to exercise care in viewing the digital results with waveform display; quality control or screening process is needed in dealing with monitoring database

Adapted from Current technologies and advancement in blood pressure measurements-review of accuracy and reliability, Biomed. Instrum. Technol., *AAMI, Arlington, VA (publication pending). With permission.*

In spite of inherent problems, observation of blood pressure through both old and new technologies retains more than enough usefulness to have remained an essential aspect of patient care. The promise of improved technology to solve problems such as those of motion artifacts, noninvasive continuous monitoring, long-distance telemetry, rapid analysis of accumulated or concurrent data, and assessment of new inaccessible regions of blood flow represent continued challenges for future biomedical research and development.

Recently, exciting research has revealed that comparing pressures taken at the arm and the ankle results in a simple but extremely useful index for assessment of lower extremity vascular disease, with implications for general cardiovascular risk factors [88]. The possibility of obtaining noninvasive blood pressures from arteries in the forehead by stick-on oscillometric patches has also been proved. At least in anesthetized patients, the forehead noninvasive blood pressure corresponded reasonably well with central arterial pressures [89]. Finger blood pressure monitors have found some applications in continuous ambulatory and sleep blood pressure assessments [90]. A technology that is capable of continuously monitoring brachial or even central blood pressure continues to be a clinical demand and future challenge.

References

1. T. G. Pickering, G. A. Harshfield, H. D. Klienert, S. Blank, and J. H. Laragh, Blood pressure during normal daily activities, sleep, and exercise, *J. Am. Med. Assoc.,* 247, 992–996, 1982.

2. L. A. Geddes, *The Direct and Indirect Measurement of Blood Pressure,* Chicago: Year Book Medical Publishers, 1970.

3. Y. Saul, F. Aristidou, D. Klaus, A. Wiemeyer, and B. Losse, Comparison of invasive blood pressure measurement in the aorta with indirect oscillometric blood pressure measurement at the wrist and forearm, *Z. Kardiol.*, 84(9), 675–685, 1995.

4. W. Dock, Occasional notes — Korotkoff sounds, *N. Engl. J. Med.*, 302, 1264–1267, 1980.

5. *Recommendations for Human Blood Pressure Determination by Sphygmomanometers*, Dallas: American Heart Association, 1993.

6. S. J. Meldrum, Indirect blood pressure measurement, *Br. J. Clin. Equip.*, 1, 257–265, 1976.

7. K. Yamakoshi, Non-invasive techniques for ambulatory blood pressure monitoring and simultaneous cardiovascular measurement, *J. Ambulat. Monit.*, 4, 123–143, 1991.

8. P. G. Loubser, Comparison of intra-arterial and automated oscillometric blood pressure measurement methods in postoperative hypertensive patients, *Med. Instrum.*, 20, 255–259, 1986.

9. M. Ramsey, Non-invasive automatic determination of mean arterial pressure, *Med. Biol. Eng. Comput.*, 17, 11–18, 1979.

10. L. A. Geddes, M. Voelz, C. Combs, D. Reiner, and C. F. Babbs, Characterization of the oscillometric method for measuring indirect blood pressure, *Ann. Biomed. Eng.*, 10, 271–280, 1982.

11. M. Ramsey, Blood pressure monitoring: automated oscillometric devices, *J. Clin. Monit.*, 7, 56–67. 1991.

12. S. D. Pierdomenico, A. Mezzetti, D. Lapenna, M. D. Guglielmi, L. Mancini, L. Salvatore, T. Antidormi, F. Costantini, and F. Cuccurullo, "White-coat" hypertension in patients with newly diagnosed hypertension: evaluation of prevalence by ambulatory monitoring and impact on cost of heath care. *Eur. Heart J.*, 16, 692–697, 1995.

13. P. R. Wilkinson and E. B. Raftery, Patients' attitudes to measuring their own blood pressure, *Br. Med. J.*, 1, 824, 1978.

14. T. G. Pickering, Utility of 24 h ambulatory blood pressure monitoring in clinical practice, *Can. J. Cardiol.*, 11 (Suppl H), 43H–48H, 1995.

15. P. E. Nielsen and J. Badskjaer, Assessment of blood pressure in hypertensive subjects using home readings, *Dan. Med. Bull.*, 28, 197–200, 1981.

16. S. Mann, R. I. Jones, M. W. Millar-Craig, C. Wood, B. A. Gould, and E. B. Raftery, The safety of ambulatory intraarterial pressure monitoring: a clinical audit of 1000 studies, *Int. J. Cardiol.*, 5, 585–597, 1984.

17. J. M. Grin, E. J. McCabe, and W. B. White, Management of hypertension after ambulatory blood pressure monitoring, *Ann. Intern. Med.*, 118, 833–837, 1993.

18. M. Bass, Ambulatory blood pressure monitoring and the primary care physician, *Clin. Invest. Med.*, 14, 256–259, 1991.

19. W. B. White, P. Lund-Johansen, and P. Omvik, Assessment of four ambulatory blood pressure monitors and measurements by clinicians versus intraarterial blood pressure at rest and during exercise, *Am. J. Cardiol.*, 65, 60–66, 1989.

20. J. A. Staessen, R. Fagard, L. Thijs, and A. Amery, and participants in the fourth international consensus conference on 24-hour ambulatory blood pressure monitoring, *Hypertension*, 26(1), 912–918, 1995.

21. H. Alexander, M. L. Cohen, and L. Steinfeld, Criteria in the choice of an occluding cuff for the indirect measurement of blood pressure, *Med. Biol. Eng. Comput.*, 15, 2–10, 1977.

22. The fifth report of the Joint National Committee on Detection, Evaluation, and Treatment of High Blood Pressure (JNCW), *Arch. Intern. Med.*, 153, 154–183, 1993.

23. Task Force on Blood Pressure Control in Children, Report of the Second Task Force on Blood Pressure Control in Children — 1987, *Pediatrics*, 79, 1–25, 1987.

24. *American National Standard for Electronic or Automated Sphygmomanometers ANSI/AAMI SP10 — 1987*, Arlington, VA: Association for the Advancement of Medical Instrumentation, 1987.

25. *American National Standard for Electronic or Automated Sphygmomanometers ANSI/AAMI SP10 — 1992*, Arlington, VA: Association for the Advancement of Medical Instrumentation, 1992.

26. *Amendment to ANSI/AAMI SP10 — 1992: American National Standard for Electronic or Automated Sphygmomanometers, ANSI/AAMI SP10A — 1996,* Arlington, VA: Association for the Advancement of Medical Instrumentation, 1996.

27. *American National Standard for Non-Automated Sphygmomanometers ANSI/AAMI SP9 — 1986,* Arlington, VA: Association for the Advancement of Medical Instrumentation, 1986.

28. E. O'Brien, J. Petrie, W. Littler, M. Sweit, P. L. Padfield, K. O'Malley, M. Jamieson, D. Altman, M. Bland, and N. Atkins, The British Hypertension Society protocol for the evaluation of automated and semi-automated blood pressure measuring devices with special reference to ambulatory systems, *J. Hypertens.,* 8, 607–619, 1990.

29. E. O'Brien, J. Petrie, W. Littler et al., The British Hypertension Society protocol for the evaluation of blood pressure measuring devices, *J. Hypertens.,* 11 (Suppl. 2), S43–S62, 1993.

30. G. Mancia and G. Parati, Commentary on the revised British Hypertension Society protocol for the evaluation of blood pressure measuring devices: a critique of aspects related to 24-hour ambulatory blood pressure measurement [commentary], *J. Hypertens.,* 11, 595–597, 1993.

31. J. Conway, Home blood pressure recording, *Clin. Exp. Hypertens.,* 8, 1247–1274, 1986.

32. *Medical Device Register,* Montvale, NJ: Medical Economics Company, 1, III-997, 1995.

33. J. Egmond, J. Lenders, E. Weernink, and T. Thien. Accuracy and reproducibility of 30 devices for self-measurement of arterial blood pressure, *Am. J. Hypertens.,* 6, 873–879, 1993.

34. Y. Imai, S. Sasaki, N. Minami et al., The accuracy and performance of the A&D TM 2421, a new ambulatory blood pressure monitoring device based on the cuff-oscillometric method and the Korotkoff sound technique, *Am. J. Hypertens.,* 5, 719–726, 1992.

35. E. O'Brien, F. Mee, N. Atkins, and K. O'Malley, Accuracy of the Takeda TM-2420/TM-2020 determined by the British Hypertension Society Protocol, *J. Hypertens.,* 9, 571–572. 1991.

36. P. Palatini, M. Penzo, C. Canali, and C. Pessina, Validation of the A&D TM-2420 Model 7 for ambulatory blood pressure monitoring and effect of microphone replacement on its performance, *J. Ambulatory Monit.,* 4, 281–288, 1991.

37. W. White, P. Lund-Johansen, and J. McCabe, Clinical evaluation of the Colin ABPM 630 at rest and during exercise: an ambulatory blood pressure monitor with gas-powered cuff inflation, *J. Hypertens.,* 7, 477–483, 1989.

38. E. O'Brien, F. Mee, N. Atkins, and K. O'Malley, Accuracy of the Del Mar Avionics Pressurometer IV determined by the British Hypertension Society Protocol, *J. Hypertens.,* 9, 567–568, 1991.

39. S. Santucci, E. Cates, G. James, Y. Schussel, D. Steiner, and T. Pickering, A comparison of two ambulatory blood pressure monitors, the Del Mar Avionics Pressurometer IV and the SpaceLabs 90202, *Am. J. Hypertens.,* 2, 797–799, 1989.

40. E. O'Brien, F. Mee, N. Atkins, and K. O'Malley, Short report: accuracy of the CH-Druck/Pressure Scan ERKA ambulatory blood pressure measuring system determined by the British Hypertension Society Protocol, *J. Hypertens.,* 10, 1283–1284, 1992.

41. E. O'Brien, F. Mee, N. Atkins, and K. O'Malley, Accuracy of the Novacor DIASYS 200 determined by the British Hypertension Society Protocol, *J. Hypertens.,* 9, 569–570, 1991.

42. G. Manning, S. Vijan, and M. Millar-Craig, Technical and clinical evaluation of the Medilog ABP non-invasive blood pressure monitor, *J. Ambulatory. Monit.,* 7, 255–264, 1994.

43. E. O'Brien, F. Mee, N. Atkins, and K. O'Malley, Accuracy of the Profilomat determined by the British Hypertension Society Protocol [short report], *J. Hypertens.,* 10, 1285–1286, 1992.

44. E. O'Brien, F. Mee, N. Atkins, and K. O'Malley, Accuracy of the SpaceLabs 90207 determined by the British Hypertension Society Protocol [short report], *J. Hypertens.,* 9, 573–574, 1991.

45. R. Taylor, K. Chidley, J. Goodwin et al., Accutracker II (version 30/23) ambulatory blood pressure monitor: clinical validation using the British Hypertension Society and Association for the Advancement of Medical Instrumentation standards, *J. Hypertens.,* 11, 1275–1282, 1993.

46. W. White, W. Susser, G. James et al. Multicenter Assessment of the QuietTrak Ambulatory Blood Pressure Recorder According to the 1992 AAMI Guidelines, *Am. J. Hypertens.,* 7, 509–514, 1994.

47. J. Ling, Y. Ohara, Y. Orime et al., Clinical evaluation of the oscillometric blood pressure monitor in adults and children based on the 1992 AAMI SP-10 standards, *J. Clin. Monit.*, 11, 123–130, 1995.

48. M. Gorback, T. Quill, and M. Lavine, The relative accuracies of two automated noninvasive arterial pressure measurement devices, *J. Clin. Monit.*, 7, 13–22, 1991.

49. E. O'Brien, F. Mee, N. Atkins, and K. O'Malley, Short report: accuracy of the Dinamap portable monitor, model 8100 determined by the British Hypertension Society Protocol, *J. Hypertens.*, 11, 761–763, 1993.

50. M. Sun, J. Tien, R. Jones, and R. Ward, A new approach to reproducibility assessment: clinical evaluation of SpaceLabs Medical oscillometric blood pressure monitor, *Biomed. Instrum. Technol.*, 30, 439–448, 1996.

51. Y. Imai, J. Hashimoto, N. Minami et al., Accuracy and performance of the Terumo ES-H51, a new portable blood pressure monitor, *Am. J. Hypertens.*, 7, 255–260, 1994.

52. Y. Imai, K. Abe, S. Sasaki et al., Clinical evaluation of semiautomatic and automatic devices for home blood pressure measurement: comparison between cuff-oscillometric and microphone methods, *J. Hypertens.*, 7, 983–990, 1989.

53. P. H. Fabre, Determination de la pression arterielle maxima par la methode oscillometrique, *C. R. Soc. Biol.* (Paris), 951–952, 1922.

54. H. Benson and J. A. Herd, Oscillometric measurement of arterial blood pressure, *Circulation*, Suppl. 3, 39–40, 1969.

55. M. Gorback, T. Quill, and D. Graubert, The accuracy of rapid oscillometric blood pressure determination, *Biomed. Instrum. Technol.*, 24, 371–374, 1990.

56. C. Nelson, T. Dorsett, and C. Davis, *Method for Noninvasive Blood Pressure Measurement by Evaluation of Waveform-Specific Area Data*, United States Patent 4,889,133, Dec. 26, 1989.

57. S. Narus, T. Egbert, T. K. Lee, J. Lu, and D. Westenskow, Noninvasive blood pressure monitoring from the supraorbital artery using an arterial neural network oscillometric algorithm, *J. Clin. Monit.*, 11, 289–297, 1995.

58. J. Penaz, Photoelectric measurement of blood pressure, volume and flow in the finger, *Digest of 10th Internat. Conf. Med. Biol. Eng.*, p. 104, Dresden, 1973.

59. K. H. Wesseling, B. de Wit, J. J. Settels, and W. H. Klawer, On the indirect registration of finger blood pressure after Penaz, *Funkt. Biol. Med.*, 1, 245–250, 1982.

60. K. Yamakoshi and A. Kamiya, Noninvasive automatic monitoring of instantaneous arterial blood pressure using the vascular unloading technique, *Med. Biol. Eng. Comput.*, 21, 557–565, 1983.

61. J. R. de Jong, R. Tepaske, G. J. Scheffer, H. H. Ros, P. P. Sipkema, and J. J. de Lange, Noninvasive continuous blood pressure measurement: a clinical evaluation of the Cortronic APM 770, *J. Clin. Monit.*, 1, 18–24, 1993.

62. C. Young, J. Mark, W. White et al., Clinical evaluation of continuous noninvasive blood pressure monitoring: accuracy and tracking capabilities, *J. Clin. Monit.*, 11, 245–252, 1995.

63. O. Kemmotsu, M. Ueda, H. Otsuka, T. Yamamura, D. C. Winter, and J. S. Eckerle, Arterial tonometry for noninvasive, continuous blood pressure monitoring during anesthesia, *Anesthesiology*, 75, 333–340, 1991.

64. O. Kemmotsu, M. Ueda, H. Otsuka, T. Yamamura, A. Okamura, T. Ishikawa, D. C. Winter, and J. S. Eckerle, Blood pressure measurement by arterial tonometry in controlled hypertension, *Anesth. Analg.*, 73, 54–58, 1991.

65. N. R. Searle, J. Perrault, H. Ste-Marie, and C. Dupont, Assessment of the arterial tonometer (N-CAT) for the continuous blood pressure measurement in rapid arterial fibrillation, *Can. J. Anaesth.*, 40, 388–93, 1993.

66. S. V. S. Rithalia, Measurement of arterial blood pressure and pulse, *J. Tissue Viability*, 4, 44–47, 1994.

67. J. E. Dalen and R. C. Bone, Is it time to pull the pulmonary artery catheter? *J. Am. Med. Assoc.*, 276, 916–918, 1996.

68. E. V. Allen, Methods of diagnosis of chronic occlusive arterial lesions distal to the wrist with illustrative cases, *Am. J. Med. Sci.*, 178, 237–244, 1929.

69. F. M. Ducharme, M. Gauthier, J. Lacroix, and L. Lafleur, Incidence of infection related to arterial catheterization in children: a prospective study, *Crit. Care Med.*, 16, 272–276, 1988.

70. S. S. Moorthy, R. K. Stoelting, and R. D. King, Delayed cyclic variations (oscillations) in pressure in a critically ill patient, *Crit. Care Med.*, 11, 476–477, 1983.

71. S. V. S. Rithalia, *Non-Invasive Measurement of Blood Gases in Critically Ill Adults*, Ph.D. thesis, University of London, 1982.

72. S. Cunningham and N. McIntosh, Blood pressure monitoring in intensive care neonates, *Br. J. Intens. Care*, 2, 381–388, 1992.

73. R. M. Gardner, Equivalence of fast flush and square wave testing of blood pressure monitoring systems, Direct Blood Pressure Measurement–Dynamic Response Requirements, *Anesthesiology*, 54, 227–236, 1981.

74. B. Kleinman, S. Powell, P. Kumar, and R. M. Gardner, The fast flush does measure the dynamic response of the entire blood pressure monitoring system, *Anesthesiology*, 77, 1215–1220, 1992.

75. R. M. Gardner and K. W. Hollingsworth, Optimizing ECG and pressure monitoring, *Crit. Care Med.*, 14, 651–658, 1986.

76. J. N. Cohn, Blood pressure measurement in shock: mechanism of inaccuracy in auscultatory and palpatory methods, *J. Am. Med. Assoc.*, 199, 118–122, 1967.

77. M. Sun and R. Jones, A hydrostatic method assessing accuracy and reliability while revealing asymmetry in blood pressure measurements, *Biomed. Instrum. Technol.*, 29, 331–342, 1995.

78. F. H. Messerli, H. O. Ventura, and C. Amodeo, Osler's maneuver and pseudohypertension, *N. Engl. J. Med.*, 312, 1548–1551, 1985.

79. F. H. Messerli, The age factor in hypertension, *Hosp. Prac.*, 15, 103–112, 1986.

80. J. Kugler, N. Schmitz, H. Seelbach, J. Rollnik, and G. Kruskemper, Rise in systolic pressure during sphygmomanometry depends on the maximum inflation pressure of the arm cuff, *J. Hypertens.*, 12, 825–829, 1994.

81. A. R. Nara, M. P. Burns, and W. G. Downs, *Biophysical Measurement Series: Blood Pressure*, Redmond, WA: SpaceLabs Medical Inc., 1993.

82. J. A. Blumenthal, W. C. Siegel, and M. Appelbaum, Failure of exercise to reduce blood pressure in patients with mild hypertension. *J. Am. Med. Assoc.*, 266, 2098–2104, 1991.

83. N. M. Kaplan, Misdiagnosis of systemic hypertension and recommendations for improvement, *Am. J. Cardiol.*, 60, 1383–1385, 1987.

84. D. W. McKay, N. R. C. Campbell, A. Chockalingam, L. Ku, C. Small, and F. Washi, Self-measurement of blood pressure: assessment of equipment. *Can. J. Cardiol.*, 11, 29H–34H, 1995.

85. M. A. Brown, M. L. Buddle, M. Bennett, B. Smith, R. Morris, and J. A. Whitworth, Ambulatory blood pressure in pregnancy: comparison of the Spacelabs 90207 and Accutracker II monitors with intra-arterial recordings, *Am. J. Obstet. Gynecol.*, 173(1), 218–223, 1995.

86. M. A. Brown, L. Reiter, B. Smith, M. L. Buddle, R. Morris, and J. A. Whitworth, Measuring blood pressure in pregnant women; a comparison of direct and indirect methods, *Am. J. Obstet. Gynecol.*, 171(3), 661–667, 1994.

87. A. Hoeft, B. Schorn, A. Weyland et al., Beside assessment of intravascular volume status in patient undergoing coronary bypass surgery, *Anesthesiology*, 81, 76–86, 1994.

88. A. B. Newman, K. Sutton-Tyrrell, and L. H. Kuller, Lower extremity arterial disease in older hypertensive adults, *Arteriosclerosis Thrombosis*, 13(4), 555–562, 1993.

89. T. K. Lee, T. P. Egbert, and D. R. Westenskow, Supraorbital artery as an alternative site for oscillometric blood pressure measurement, *J. Clin. Monit.*, 12(4), 293–297, 1996.

90. B. P. Imholz, G. J. Langenwouters, G. A. van Montrans, G. Parati, J. van Goudoever, K. H. Wesseling, W. Wieling, and G. Mancia, Feasibility of ambulatory, continous 24-hour finger arterial pressure recording, *Hypertension*, 21(1), 65–73, 1993.

76

Blood Flow
Measurements

Per Ask
Linkoping University

P. Åke Öberg
Linkoping University

76.1 Doppler Measurements

Ultrasound Doppler

The ultrasound Doppler measurements [1, 2] are based on a principle discovered by the Austrian physicist Christian Doppler 1842 [3], who theoretically predicted that a wave, backscattered from a moving object, will be shifted in frequency. The principle was verified by two hornblowers, one aboard a moving train and the other standing still, with the ability to assess the pitch of the sound.

To obtain a Doppler signal from a fluid, the fluid must contain scattering particles, which in the case of blood are the blood cells. The size of a red blood cell is about 2×7 μm, which means that the scatterers are much smaller than the wavelength of the ultrasound. Hence, a diffuse scattering of the ultrasound will occur (Rayleigh scattering). The scattering from tissues surrounding the heart and vessels usually gives a much larger signal (20 to 40 dB) than that from blood in motion. The velocity of tissue motion is usually much lower than that of blood. This contribution can therefore be suppressed by high-pass filtering. In recent years ultrasound contrast agents (consisting of gas-filled shells) have been introduced to increase the blood flow signal. Figure 76.1 illustrates the ultrasound Doppler principle. An ultrasound beam is sent toward a moving object. The beam hits the object and returns to the receiver with a Doppler-shifted frequency carrying information about the velocity of the object.

The Doppler shift f_d of an ultrasound signal with the nominal frequency f_c is given by

$$f_d = 2f_c \cdot \frac{v}{c} \tag{76.1}$$

where v is the velocity component in the direction of the ultrasound beam and c is the speed of sound in the medium which is in the range of 1500 to 1600 m s^{-1} in soft tissue and usually set to 1540 m s^{-1}. The frequency f_c is in the range of 2 to 10 MHz, which gives a wavelength between 0.15 and 0.77 mm.

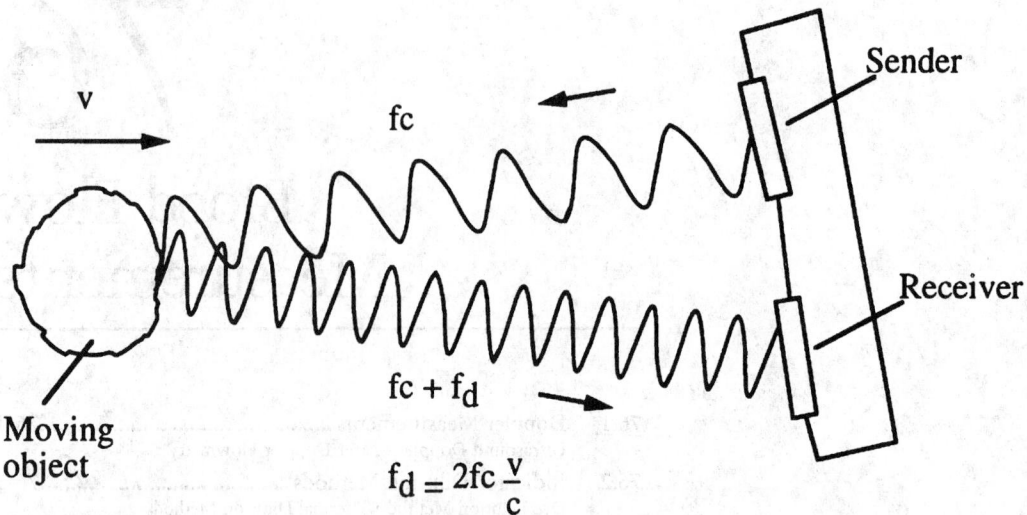

FIGURE 76.1   For Doppler ultrasound, the moving object shifts the received frequency.

The peak systolic flow velocity in the heart and larger vessels is normally 0.5 to 1 m s^{-1}, resulting in a Doppler shift in the range of 1.3 to 13 kHz (depending on the ultrasound frequency f_c). The Doppler sound is therefore audible, which is helpful for the investigator in identifying vessels and phenomena of interest.

The most straightforward ultrasound investigation is to use a continuous wave system. The ultrasound beam is focused by a suitable ultrasound transducer geometry and by a lens. By this arrangement a narrow beam can be arranged and the backscattered information can come from any section along the beam.

In order to know from where along the beam the blood flow data are collected, a pulsed Doppler system has to be used. The transducer sends four to eight cycles of the ultrasound signal; a specified time later a gate is opened and the transducer will act as a reviewer. By using a preset time delay between the sending and the reviewing signal only flow information from a certain depth will be collected. The depth is obviously determined by the delay time and the propagation velocity in the tissue. Flow velocity information is obtained by spectral estimation of the Doppler signal.

By multirange gating, blood flow at various points along the ultrasound beam can be measured. By scanning with an ultrasound beam within a sector, a two-dimensional velocity field can be presented. By color coding the information a color Doppler flow image can be obtained. By scanning in one additional orthogonal plane, a three-dimensional flow image can be created.

An important medical application of the ultrasound Doppler is in the study of heart valve flow when one might suspect stenosis or leaking flow in the valves. The pumping ability of the heart can be assessed from the general flow patterns. Regions with arteriosclerotic obstructions can be localized in the peripheral vessels.

Laser Doppler Flowmetry

The Doppler principle is also utilized in blood flow measurements in the microcirculatory bed in which laser Doppler flowmetry measures the blood perfusion. Laser light from a gas or semiconductor laser is launched into an optical fiber which leads the light to the tissue. (Figure 76.2). Photons are reflected, scattered, and absorbed in the tissue matrix and those that hit moving red blood cells become Doppler shifted, whereas those that are reflected in stationary structures are refracted without any change in frequency. A part of the photons returning to the fiber-optic system will be conducted by the pickup fiber to the detector. Shifted and nonshifted photons are mixed at the surface of a square-law photodetector. According to elementary wave mechanics this type of mixing (coherent detection) results in the

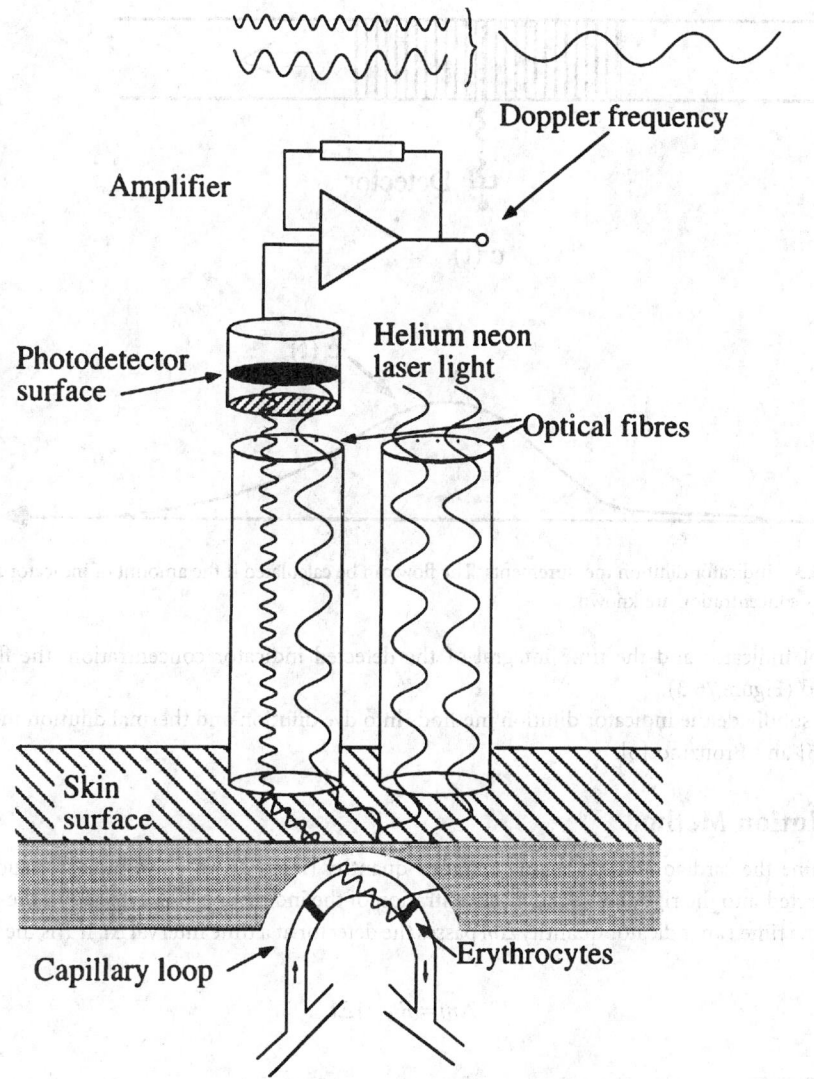

FIGURE 76.2 For laser Doppler flowmetry, the moving blood cells shift the received frequency.

sum and difference frequencies. The frequency of the "difference wave" is proportional to the average velocity of the red blood cells. The amplitude of the same signal is proportional to the number of moving scatterers in the tissue volume. A velocity distribution will result in a Doppler spectrum, usually in the range of 30 to 12,000 Hz.

Variants of the method utilizing fiber optics, airborne beams, microscope-based setups, and color-coded imaging scanners are described in the literature. For a review, see Shepherd and Öberg [4].

76.2 Indicator Dilution Methods

Cardiac output measurement is one of the most essential heart performance measures. The rest and exercise flows carry important diagnostic information. The monitoring of cardiac output is very important for the critically ill patient.

The principle is that an indicator is injected upstream in the circulation. Mixing with the circulating blood volume occurs and the indicator concentration is detected downstream. By knowing the added

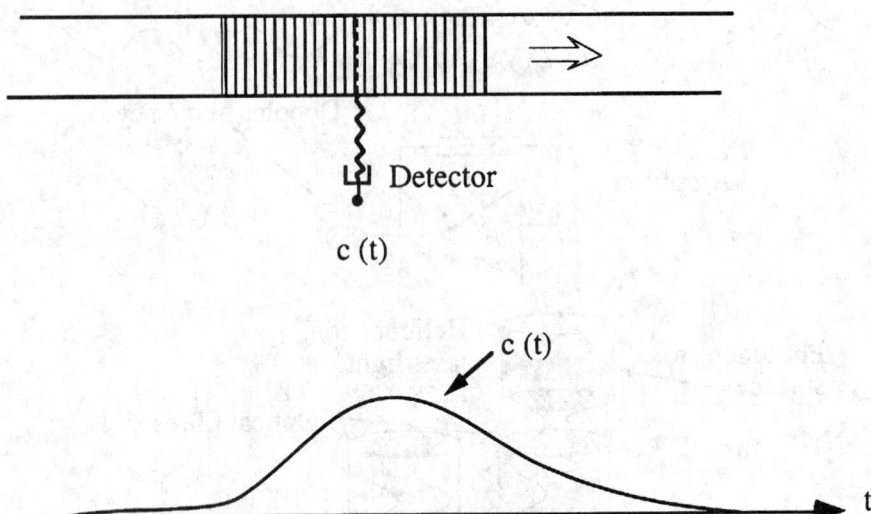

FIGURE 76.3   Indicator dilution measurements. The flow can be calculated if the amount of indicator and the time course of its concentration are known.

quantity of indicator and the time integral of the detected indicator concentration, the flow can be determined (Figure 76.3).

We can subdivide the indicator dilution methods into dye dilution and thermal dilution methods. See Webster [5] and Bronzino [6].

Dye Dilution Method

To determine the cardiac output (l/min) a known quantity (mass m) of a dye indicator such as Evans Blue is injected into the right heart and a concentration of the indicator $c(t)$ is detected in the pulmonary artery. At the time t an indicator quantity Δm passes the detector at a time interval Δt. If F is the blood flow,

$$\Delta m = F\, c\!\left(t\right)\Delta t \tag{76.2}$$

or by integration

$$F = \frac{m}{\int c\!\left(t\right)dt} \tag{76.3}$$

Thus, only the amount of indicator added and the time integral of the downstream concentration, assuming good mixing, has to be known to be able to calculate the flow, i.e., the cardiac output.

Thermal Dilution Method

The thermal dilution method is a variant of the indicator dilution method family. A thermal dilution catheter is placed with the injection outlet in the right atrium of the heart and with a temperature sensor in the pulmonary artery (Bronzino [6] and Weissel et al. [7]).

A chilled solution of dextrose in water or saline solution is used for the injection and causes transient decrease in the pulmonary artery temperature $T(t)$. The blood flow (cardiac output) can be calculated from

$$F = \int T(t)\,dt\,\rho_b c_b = V\left(T_b - T_i\right)\rho_i c_i K \tag{76.4}$$

or

$$F = \frac{V\left(T_b - T_i\right)}{\int T(t)\,dt}\left(\rho_i c_i / \rho_b c_b\right) K \tag{76.5}$$

where V is the injected volume, T_b the undisturbed temperature in the pulmonary artery, T_i the temperature of the indicator, r_i and r_b the density of the indicator and the blood, c_i and c_b the specific heat of the indicator and the blood and K is a correction factor that takes heat transfer along the catheter into account.

The thermodilution technique is the standard technique for the monitoring of cardiac output in critically ill patients.

76.3 Plethysmography

The word *plethysmography* means methods for recording volume changes of an organ or a body part. Depending on the technique used, strain gage, impedance, and optical techniques can be used for the volume determination. See Webster [5] and Figure 76.4.

Strain Gage Plethysmography

The classical strain gage plethysmography is used to study circulation in the lower extremities from changes in the circumference of the legs.

Small-diameter silicone rubber tubes, filled with mercury or other types of conductive liquids, are placed around the circumference of the leg. Changes in the latter can be directly related to electrical impedance changes of the silicone rubber tube. If a cylindrical cross section of the leg is assumed, volume changes should be proportional to the total circumference, times the change in impedance.

The strain gage tubes (Figure 76.4) are positioned around the lower part of the leg, an inflatable cuff is placed around the upper part (above the knee) and inflated to 40 to 50 mmHg, i.e., above the venous pressure that will cause the outflow of blood to cease. The increased volume of the leg is therefore proportional to the arterial inflow. The latter is determined from the initial slope. Flow is volume increase per time unit.

When the cuff pressure is released, blood will flow out from the leg via the veins. The time course of the volume change will be related to venous function. If there is venous thrombosis, the decline in the volume curve will be slower.

Impedance Plethysmography

Bioelectric impedance measurements have a history dating back to the 1940s. The reason impedance is useful for detection of volume changes is that different tissues in the body have different resistivity. Blood is one of the best conductors among the tissues of the body.

Impedance plethysmography has its most established applications in respiratory monitoring in newborn infants and for detection of venous thrombosis. Less-established applications are cardiac output measurements, peripheral blood flow studies, and body composition assessments.

A constant current with a frequency of 50 to 100 kHz and an amplitude of 0.5 to 4 mA rms is applied via skin electrodes (Figure 76.4). Influences from skin impedance are eliminated by the use of the four electrode technique.

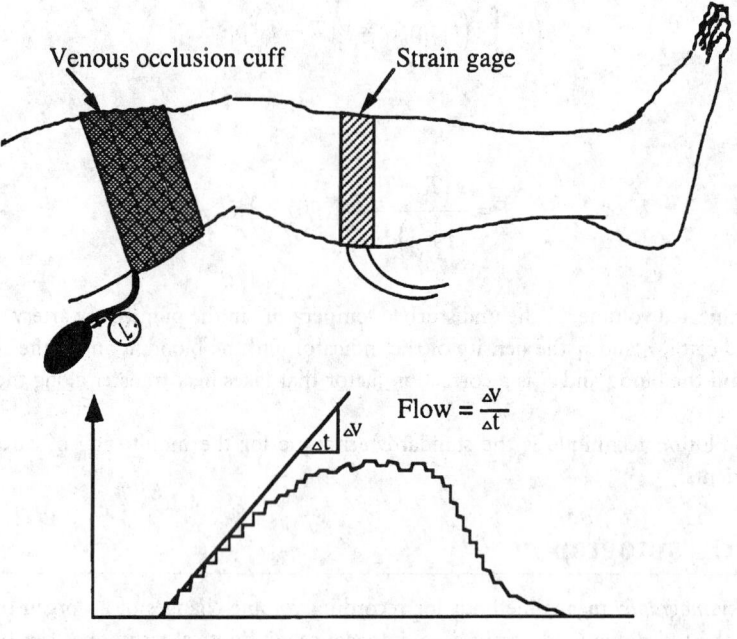

$$\text{Impedance} \sim \frac{\text{Voltage}}{\text{Current}}$$

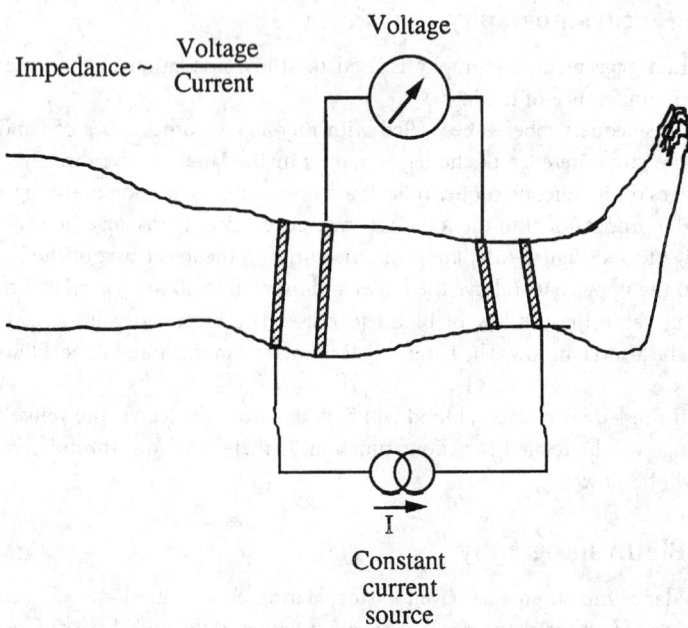

FIGURE 76.4   For occlusion plethysmography, increased volume stretches the strain gage (top) and decreases the impedance (bottom).

The measurement object can be described with a conduction object with the constant impedance Z_0 in parallel with a time-varying impedance ΔZ. The impedance ΔZ is represented with a column of a conducting media with resistivity ρ and the length L. If the changes in ΔZ are small in comparison to those of Z_0, the volume changes can be obtained as:

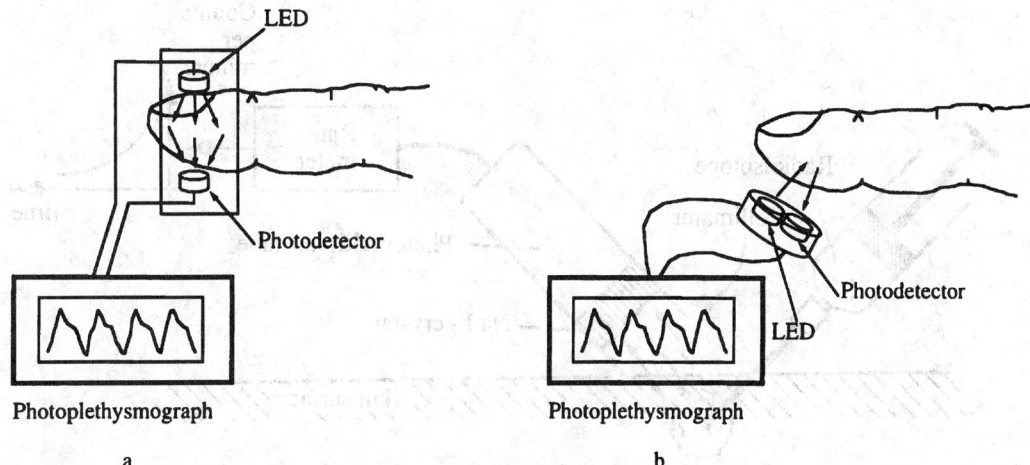

FIGURE 76.5 For photoplethysmography, increased blood decreases received light in (a) transmission mode and (b) reflection mode.

$$\Delta V \approx \rho \frac{L^2}{Z_0^2} \Delta Z \qquad (76.6)$$

This technique can be used in the same way as the strain gage plethysmoghraphic method to study circulation in the leg.

For the measurement of cardiac output, Equation 76.6 or a modified formula can be used [8]. The method seems to work rather well in normal persons for relative change, but for patients with cardiac disease the cardiac output estimation might be poor.

Photoelectric Plethysmography

Hertzman and Spealman [9] and Hertzman [10] were the first to use the descriptive term *photoelectric plethysmography* (PPG). The first reports on the successful use of the principle were published in the middle of 1930 by Molitor and Kniazuk [11]. The principle on which PPG is based is simple, although the underlying detailed optical mechanisms remain unknown. A beam of light is directed toward the part of the tissue in which blood flow (or volume) is going to be measured. (Figure 76.5). Reflected, transmitted, and scattered light leaving this volume is collected and focused on a photodetector. A signal modulated by the attenuation or scattering of light in the blood volume can be recorded. Two different components can be derived from the detector. One is pulsatile and synchronous with the heartbeat (the ac component), the other is a constant voltage (the dc component). The physiological significance of the two signals is still under debate, but they reflected the blood volume and the orientation of erythrocytes during the cardiac cycle.

PPG has been used mainly for monitoring blood perfusion in skin, venous reflux conditions, and skin flaps during plastic surgery. Challoner [12], Roberts [13], and Bernstein [14] have reviewed the methodology and applications of PPG.

76.4 Radioisotopes

Kety [15] introduced the principle of tissue clearance of rapidly diffusing inert isotopes for blood flow measurements. An extensive theoretical treatment is given by Zierler [16]. In most applications, lipid-soluble gases like ^{133}Xe and ^{85}Kr have been used. These isotopes rapidly diffuse from blood to tissue and

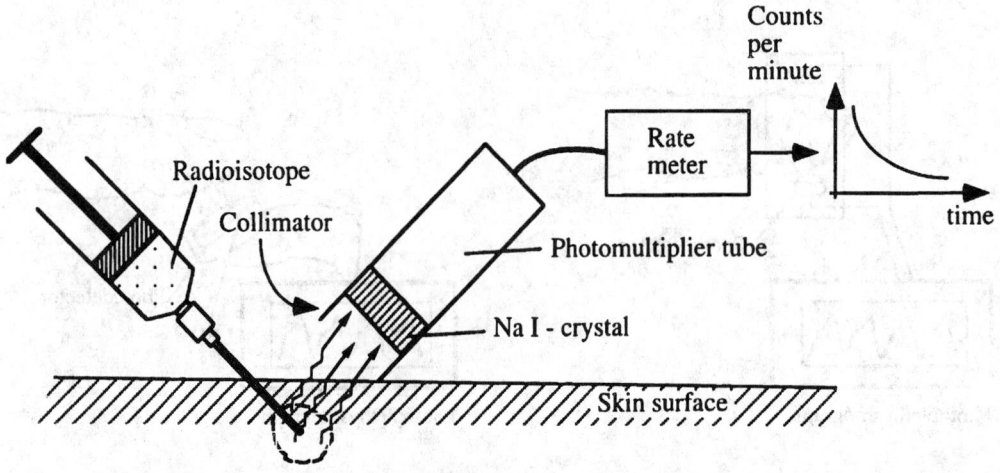

FIGURE 76.6   For isotope measurement of microcirculatory blood flow, radioactivity decreases with washout.

a rapid equilibration of the isotope concentration within a tissue volume takes place. Figure 76.6 illustrates the measurement principle. Their elimination from a microcirculatory bed is related to the blood flow rate. If the tissue is uniformly and constantly perfused, the activity of the isotope decays monoexponentially with time. The elimination of the isotope can be described by the equation

$$C(t) = C_0 \exp(-kt) \tag{76.7}$$

where $C(t)$ and C_0 are the tissue concentrations at times t and at the onset of the injection C_0. k is the clearance constant related to the local blood flow by the relation

$$k = \ln 2 / t_{1/2} \tag{76.8}$$

in which $t_{1/2}$ is half the time of decay. Blood flow Q (in $k \cdot s \cdot 100$ ml $\cdot$ min^{-1} $\cdot$ 100 g^{-1}) can then be derived from the formula

$$Q = k \cdot s \cdot 100 \tag{76.9}$$

where s denotes the tissue–blood partition coefficient. The indicator is administered via an injection into the tissue volume, or through passive diffusion after deposition on the surface of the tissue volume under study.

   The advantage of clearance methods is that they can be applied to the study of all kinds of tissue blood flow problems. One of the disadvantages is that the method does not give a continuous measurement of flow. In addition, the clearance curves are sometimes difficult to interpret. The trauma caused by injection of the isotope into the tissue seriously disturbs the flow, as shown by Holloway [17] and Sejrsen [18]. In spite of these shortcomings, the isotope clearance method has been applied extensively to the study of skin and tissue blood flow in experimental as well as clinical problem areas.

76.5   Thermal Convection Probes

Thermal convection probes have been developed specifically for skin blood flow measurements. Gibbs [19] pioneered the field by describing a probe in the form of a needle. Hensel and Bender [20] and van

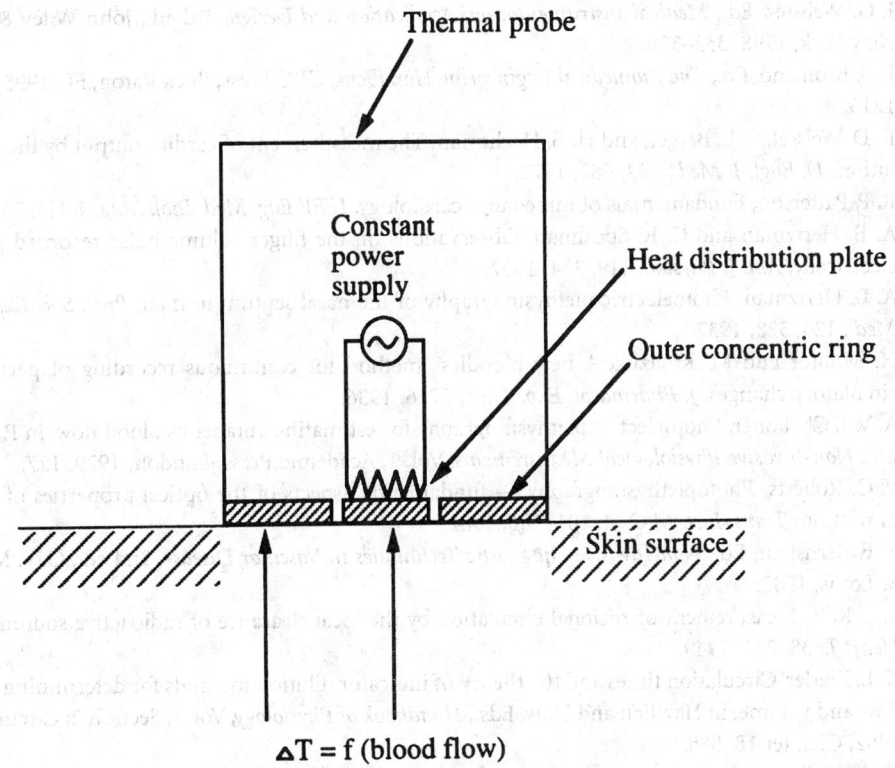

FIGURE 76.7   For thermal convection measurement of blood flow, heat dissipation increases with blood flow.

de Staak et al. [21] developed noninvasive variants by designing probes that can be positioned at the surface of the tissue.

All methods measure the rate of removal of heat from the tissue volume under the probe. A relation exists between the blood flow rate and the rate by which heat dissipates from the tissue volume under study. The sensing unit is usually designed around a central metal disk and a concentric outer ring between which a temperature difference is established. (Figure 76.7). The two rings are thermally and electrically isolated from each other and both are in contact with the tissue. The temperature difference between the two rings is a measure of the blood flow under the probe.

A temperature difference of 2 to 3°C is usually established between the inner disk and the outer annulus. The central disk is heated with an electric current and kept at a constant temperature that only by 1 or 2°C exceeds the resting temperature of the tissue under study. Thermal probes have not been extensively used because of their extreme nonlinear properties and the difficulties in their practical use, i.e., the contact pressure sensitivity. Another difficulty is the highly variable thermal characteristics of the skin.

References

1. L. Hatle and B. Angelsen, Eds., *Doppler Ultrasound in Cardiology. Physical Principles and Clinical Applications*, Lea & Febiger, Philadelphia, 1985.
2. S. Webb, Ed., *The Physics of Medical Imaging*, Adam Hilger, Philadelphia, 1988.
3. C. Doppler, quoted by D. N. White, Johann Christian Doppler and his effects — a brief history, *Ultrasound Med. Biol.*, 8, 583-591, 1982.
4. A. P. Shepherd and P. Å. Öberg, Eds., *Laser Doppler Blood Flowmetry*. Kluwer Academic Publishers, Boston, 1990.

5. J. G. Webster, Ed., *Medical Instrumentation. Application and Design*, 3rd ed., John Wiley & Sons, New York, 1998, 332–338.

6. J. D. Bronzino, Ed., *The Biomedical Engineering Handbook*, CRC Press, Boca Raton, FL, 1995, 1212-1216.

7. R. D. Weissel, R. L. Berger, and H. B. Hechtman, The measurement of cardiac output by thermodilution, *N. Engl. J. Med.*, 292, 682, 1975.

8. R. P. Patterson, Fundamentals of impedance cardiology, *IEEE Eng. Med. Biol. Mag.*, 8 (1), 35, 1989.

9. A. B. Hertzman and C. R. Spealman, Observations on the finger volume pulse recorded photoelectrically, *Am. J. Physiol.*, 119, 334, 1937.

10. A. B. Hertzman, Photoelectric plethysmography of the nasal septum in man, *Proc. Soc. Exp. Biol. Med.*, 124, 328, 1937.

11. H. Molitor and M. Kniazuk, A new bloodless method for continuous recording of peripheral circulatory changes, *J. Pharmacol. Exp. Ther.*, 57, 6, 1936.

12. A. V. J. Challoner, Photoelectric plethysmography for estimating cutaneous blood flow, in P. Rolfe, Ed., *Non-Invasive Physiological Measurements*, Vol. 1, Academic Press, London, 1979, 127.

13. V. C. Roberts, Photoplethysmography — fundamental aspects of the optical properties of blood in motion, *Trans. Inst. M.C.*, 4, 101–106, 1982.

14. E. F. Bernstein, Ed., *Non-Invasive Diagnostic Techniques in Vascular Disease*, 2nd ed., C. V. Mosby, St Louis, 1982.

15. S. S. Kety, Measurement of regional circulation by the local clearance of radioactive sodium, *Am. Heart J.*, 38, 321, 1949.

16. K. L. Zierler, Circulation times and the theory of indicator dilution methods for determining blood flow and volume, in Havillan and Dow, Eds., *Handbook of Physiology*, Vol. 1, Section 2: Circulation, 1962, Chapter 18, 585.

17. G. A. Holloway, Jr., Laser Doppler measurements of cutaneous blood flow, in P. Rolfe, Ed., *Non-Invasive Physiological Measurements*, Vol. 2, Academic Press, London, 1983.

18. P. Sejrsen, Measurement of cutaneous blood flow by freely diffusible radioactive isotopes. Methodological studies on the washout of Krypton-85 and Xenon-133 from the cutaneous tissue in man, *Danish Med. Bull.*, 18 (Suppl. 3), 9, 1971.

19. F. A. Gibbs, A thermoelectric blood flow recorder in the form of a needle, *Proc. Soc. Exp. Biol.*, N.Y., 31, 141, 1933.

20. H. Hensel and F. Bender, Fortlaufende Bestimmung der Hautdurchblutung am Menschen mit einem elektrischen Wärmeleitmesser, *Pflugers Arch.*, 263, 603, 1956.

21. W. J. B. M van de Staak, A. J. M. Brakkee, and H. E. de Rijke-Herweijer, Measurement of the thermal conductivity of the skin as an indication of skin blood flow, *J. Invest. Dermatol.*, 51, 149–154, 1968.

77

Ventilation Measurement

L. Basano
Università di Genova

P. Ottonello
Università di Genova

77.1 Ventilation

This section deals mainly with **spirometry**, i.e., with the measurement of volumes and flows associated with respiration. Spirometric tests (which embody useful information about parameters related to pulmonary function) are often used for diagnostic purposes in conjunction with other measurements; Figure 77.1 and Table 77.1 show spirometric quantities of clinical interest [1].

The classification of spirometric instruments may be based on different criteria. For example, one may consider *open-circuit* devices (the subject takes a full inspiration or expiration before connecting to the meter) and *closed-circuit* devices (the subject remains connected to the apparatus during one or several respiratory cycles). Another distinction concerns whether the instrument is a *portable* one and is mainly used for monitoring purposes or is a *diagnostic* one whose purpose is to provide an accurate value that may be compared with a reference value. As for the nature of the physical functions more directly investigated, a subdivision into *volume measurements* and *flow measurements* is important, even though, using some care, it is sometimes possible to shift from one class to the other by time-differentiation or time-integration procedures. Another technique deals with the direct evaluation (as a function of frequency) of the complex impedance of the respiratory system viewed as a suitable network of resistors, capacitors, and inductors. This method has been the subject of considerable investigation in the last two decades, thanks to the advent of computers and to their use in the spectral analysis of signals. This topic will be discussed in the final section.

As a general comment about measurements associated with ventilation, it may be said that there are few other fields in applied science where a correct *operation procedure* is virtually as important as the reliability of the measuring instruments themselves. See, for example a study conducted on nearly 6000 patients affected by some airflow obstruction (Reference 2 and references therein, especially n. 26).

Volume Measurements

In this case, gas volumes associated with the respiratory process are the main target of investigation; the principal instruments that have been used so far in the clinical routine and in research activity are [3]

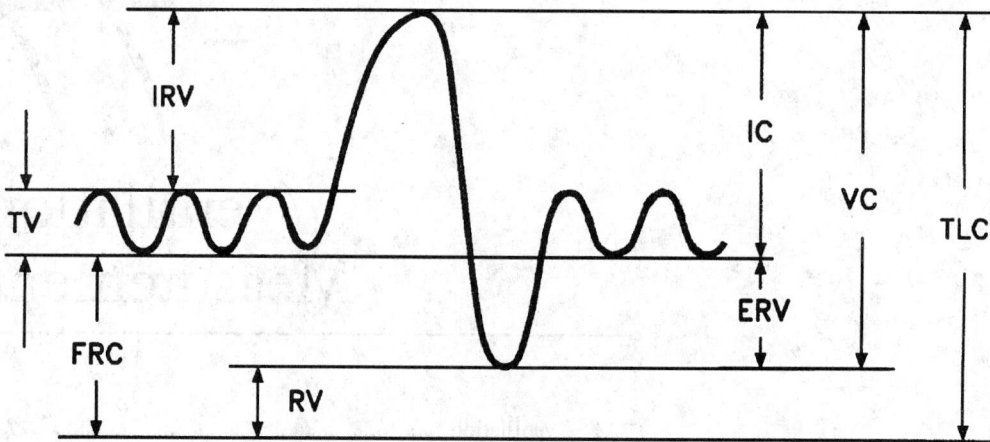

FIGURE 77.1 Spirometric trace illustrating the definitions of some significant quantities commonly evaluated in spirometry (see also Table 77.1).

TABLE 77.1 Common Spirometric Quantities

Vital capacity	VC
Inspiratory capacity	IC
Inspiratory reserve volume	IRV
Expiratory reserve volume	ERV
Tidal volume	TV
Functional residual capacity	FRC
Residual volume	RV
Total lung capacity	TLC

1. *Spirometer:* An expandable chamber whose volume is monitored during inspiration or expiration. The subject is instructed to blow into a conduit communicating with the chamber; the latter may consist of a bell, a piston, or more often a bellows (as in the once ubiquitous Vitalometer®).

2. *Turbine meter:* Based on the principle that air blown through the inlet produces the rotation of a turbine connected to a revolution counter.

3. *Impedance plethysmograph:* Based on the measurement of resistance (*strain gage plethysmograph*) or of inductance (*inductance plethysmograph*); in both cases, the impedances of an elastic coil wrapped around the subject's chest and one wrapped around the subject's abdomen are monitored during respiration.

4. *Total body plethysmograph:* A kind of sealed telephone booth inside which the subject sits; the pressure inside the box is sensed and converted to volume values.

Flow Measurements (Pneumotachometry)

In this case the airflow through the upper airways is directly measured; in principle, the flow could be evaluated by time differentiating the spirometer records, but this method would suffer from limitations due to nonlinearity effects and to errors related both to poor frequency response and to some hysteresis of the volume meter. The reverse procedure is in fact more common: since they are normally connected to an electronic integrating device, flowmeters can be employed as volume-measuring instruments as well. Calibration is normally performed by discharging a syringe (of known capacity) through the flowmeter; the response of the latter is time integrated to check whether it corresponds to the calibration volume. It is convenient to repeat the calibration procedure a few times, with the syringe discharged each time at a different speed, in order to detect possible effects due to deviations from linearity of the flow sensor.

A useful summary of standard specifications governing the performance of several types of pneumotachometers (PTM) together with a concise description of the more common PTM types can be found in Reference 4. Conventional devices for measuring respiratory flow are

1. *Linear Resistance* PTM (LRPTM): Evaluates the pressure difference generated by the (laminar) airflow across a fixed hydrodynamic resistance. This procedure is based on the Poiseuille equation and is analogous to evaluating current by measuring voltage across a known resistor and using Ohm's law. The resistive element may consist either of a bundle of tubelets (Fleisch-type) or of a wire mesh screen (Lilly-type).
2. *Hot Wire* PTM (HWPTM): A very thin wire heated by an electric current, cooled by the flowing gas; the rate at which heat is conveyed away from the wire depends on the fluid flow; a variant of this instrument uses a heated film in lieu of the heated wire.
3. *Ultrasonic* PTM (UPTM): Based on the principle that the speed of a beam of ultrasounds exchanged between a pair of transducers is increased or decreased as it propagates through the moving air; the variations of the time of flight of the ultrasonic beam is related to the average speed of the flowing air.
4. *Vortex-Shedding* PTM (VSPTM): Basically measures the frequency at which vortices are generated in the wake of a suitably shaped obstacle (the "bluff body") exposed to the flow.

Respiratory Impedance Measurements

In recent years, considerable interest has been devoted to the impedance of the respiratory system, a parameter that contains information about the morphology of our breathing apparatus and plays a role analogous to the impedance of an electrical circuit. **Respiratory impedance** may be obtained by connecting the patient's upper airways to an alternating pressure generator and by evaluating the ratio of the measured pressure to the measured flow; this can be done at several excitation frequencies, using a method that is referred to as the forced oscillation technique.

77.2 Instrumentation: Principles and Description

Measurements of Volume

Spirometer

A widely used spirometer is the bell-type (see Figure 77.2): the volume of the lungs is monitored by the position of a light cylindrical bell (possibly equipped with counterweights) connected to the patient's mouth. The bell spirometer is rather cumbersome but, thanks to its great reliability, is an ideal tool for calibration purposes as well as for comparing data recorded at different centers. In its turn, the bell spirometer needs periodic calibration at least every 3 months; this may be performed by means of a 3-L syringe equipped with electronic volume readout [5]. The syringe must be accurate to within 25 mL and the spirometer should be able to measure volumes of at least 8 L with an accuracy of at least ±3% or 0.050 L, whichever is greater, with flows betweeen zero and 14 L/s [6].

Turbine Meter

Its operating principle is quite straightforward. The revolutions of the turbine wheel, whose speed is proportional to the flow, are counted by electronic sensors (optoelectronics or Hall effect based) and processed to give volume or flow values. The main limitation of the turbine meter is that low flow values are underestimated since a greater fraction of the air slips past the wheel as the flow rate decreases; it also displays poor frequency response and can be used only in unidirectional flows. Although these features make the device unsuitable for accurate laboratory measurements, it is used extensively to monitor the ventilation of patients in intensive care and in portable instruments, thanks to its overall (sensor and pulse processing) simplicity and low cost.

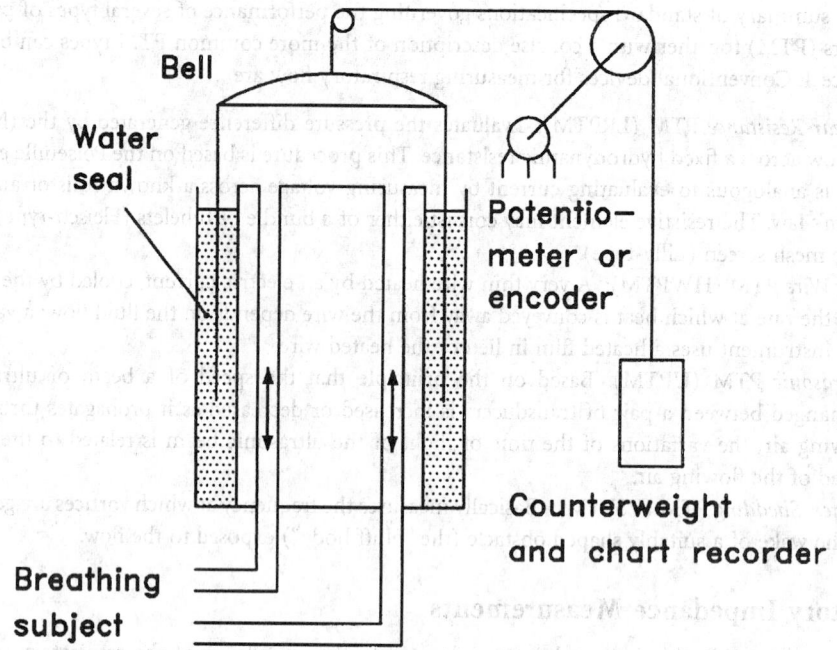

FIGURE 77.2   Schematic view of the classical bell spirometer.

Inductance Plethysmograph

Like the spirometer, the pneumotachometer, and the total-body plethysmograph, this technique allows for noninvasive monitoring of human ventilation but it does not require a connection to the airways. Two elastic belts with serpentine inductors encircle the torso, the first around the rib cage, the other around the abdomen. Each inductor is part of the resonant tank circuit of a free-running oscillator (resonant frequency from 0.2 to 1 MHz) whose output is optically coupled to the demodulator circuit (for the purpose of electrical isolation of the subject from the equipment, usually operated by the ac power line). Ventilation causes the inductor cross section areas to vary changing their inductances and thus varying the frequency of the two oscillators (FM modulation).

Strain Gage Plethysmograph

Its principle of operation is similar to that of the inductance plethysmograph; in this case, the sensing element consists of a strain gage whose resistance is varied as the gage is stretched and released during respiration.

Total-Body Plethysmograph (Constant-Volume Box)

This is a sealed chamber that allows the measurement of thoracic gas volume (TGV) by exploiting the pressure–volume relation of a fixed quantity of ideal gas. A schematic illustration of the method for a constant-volume plethysmograph is shown in Figure 77.3. The balloon-shaped object represents the subject's respiratory system, the volume of which varies during breathing. In the course of the measurement operation (which lasts only a few seconds), the shutter appearing in Figure 77.3 must remain closed in order that the alveolar pressure may be identified with the pressure measured at the mouth. In the following, the subscript A (for alveolar) denotes the gas in the lungs and the subscript C denotes the gas in the chamber.

For moderate respiratory acts, the air contained in the lungs at a given time can be treated as an approximately isothermal gas (temperature 37°C, water vapor saturated) while the air in the chamber is more appropriately treated [1] as an adiabatic gas. By differentiating the ideal gas equation for the isothermal air in the lungs,

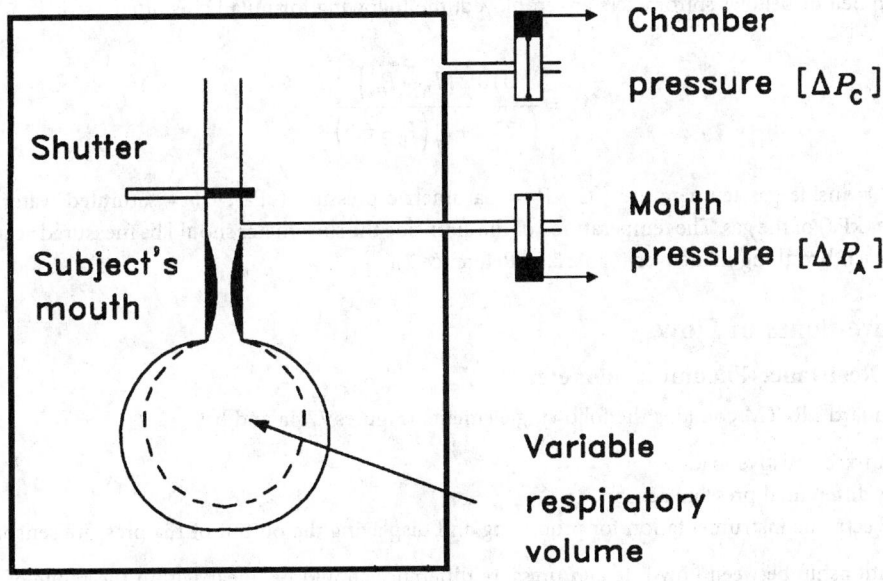

FIGURE 77.3 The respiratory system of a patient sitting inside an airtight chamber is schematized as an expandable balloon. The quantities of air inside and outside the balloon are separately constant: only their volumes and pressures change as the subject attempts to breathe against the closed shutter.

$$V_A = -P_A \frac{\Delta V_A}{\Delta P_A}$$

while for the adiabatic air in the chamber,

$$\Delta V_C = -V_C \frac{\Delta P_C}{\gamma P_C}$$

where γ is the usual ratio between molar heat capacities: $\gamma = c_p/c_V$.

Since the system is globally closed ($\Delta V_A = -\Delta V_C$), the lung volume V_A is related to the alveolar pressure P_A (which is taken to be the barometric pressure minus the water vapor pressure at 37°C) and to the differential pressures ΔP_C and ΔP_A (respectively equal to the chamber pressure fluctuation and to the alveolar pressure fluctuation) by the formula:

$$V_A = kP_A \frac{\Delta P_C}{\Delta P_A}$$

The coefficient k can be obtained through a calibration procedure that creates sinusoidal volume variations by means of a reciprocating pump. The pressure changes appearing on the right-hand side of the last equation may be measured by sending the outputs of the two sensors of Figure 77.3 to a data acquisition module installed in a host computer, typically a personal computer.

Correcting to Standard Conditions

Measurements of gas volumes make reference to lung BTPS (body temperature and pressure, saturated with water vapor) conditions. Volume measurements (V_A) made at ATP (ambient temperature and pressure) conditions should be corrected to BTPS conditions (V_B).

When bell or bellows spirometers are employed, the following formula [1] is often used:

$$V_{B} = \frac{310.2\left(P_{B} - P_{W}\right)}{\left(273.2 + t\right)\left(P_{B} - 6.3\right)} \cdot V_{A}$$ (77.1)

where t = inside gas temperature (°C), P_{B} = barometric pressure (kPa); P_{W} = saturated water vapor pressure (kPa) of the gas. The temperature t of the air inside the spirometer should be measured accurately during each breathing maneuver (Table 3 in Reference 7).

Measurements of Flow

Linear Resistance Pneumotachometer

The standard LRPTM contains the following elements (Figures 77.4a and b):

1. A fixed resistive load;
2. A differential pressure sensor;
3. Electronic instrumentation for processing and displaying the output of the pressure sensors.

The relationship between flow rate and pressure difference should be linear within the range of useful flow rates; the maximum flow value, determined by the onset of turbulence, and the linearity range are given by the manufacturer's specifications.

It is good practice to calibrate the PTM periodically, by connecting it in series with a bell spirometer, and then use it without changing the geometry of the immediately adjacent tubing. In fact, accurate flow rate measurements are possible only if information about the flow–pressure characteristics over the useful flow range is available; furthermore, this information should have been obtained in the same conditions in which the flowmeter will be used [8].

Figure 77.4(a) depicts the Fleisch PTM, whose resistive element consists of a bundle of capillary tubes. The function of the heater is to avoid the condensation of water vapor inside the tubes; the two chambers of the differential pressure sensor are connected at the two ends of the resistive element, whose resistance is lower than 0.1 kPa s/L (1 cm H_2O s/L). Since typical spirometric flows, for quiet breathing and slow maneuvers, are of the order of 1 L/s, they generate pressure differentials of about 1 cm H_2O. Considerably stronger flows, exceeding 10 L/s, may be produced by adults in some maneuvers; in these cases, Fleisch PTMs of larger dimensions should be employed in order to keep the spirometric flow within the linearity regime.

Figure 77.4(b) illustrates the Lilly PTM, whose resistive element consists of a wire mesh and is of the order of 0.05 kPa s/L (0.5 cm H_2O s/L) to which pressure differentials of the same order as the Fleisch are associated. The Lilly PTM is much less exposed than the Fleisch to the risk of nonlinearity at higher flow rates.

The linear resistance PTMs, in conjunction with an appropriate differential pressure transducer, an amplifier, and an analog or, more often, digital integrator, form the most widely used instrument for the combined measurement of flow and volume. In fact, recent recommendations on the standardization of lung function tests are limited to Fleisch and Lilly tachometers only [1].

Hot Wire Pneumotachometer

The HWPTM works on the principle that the cooling rate of a heated wire depends on the speed at which the surrounding fluid is flowing [9-11]. The rate at which thermal energy is lost by the wire may be written as

$$Q = hS\left(T_{w} - T\right)$$ (77.2)

where Q is the heat transfer rate; S is the surface area of the wire; T_{w} is the wire mean temperature; T is the temperature of the fluid surrounding the wire; h is the transfer coefficient between sensor and fluid and is defined by the equation above.

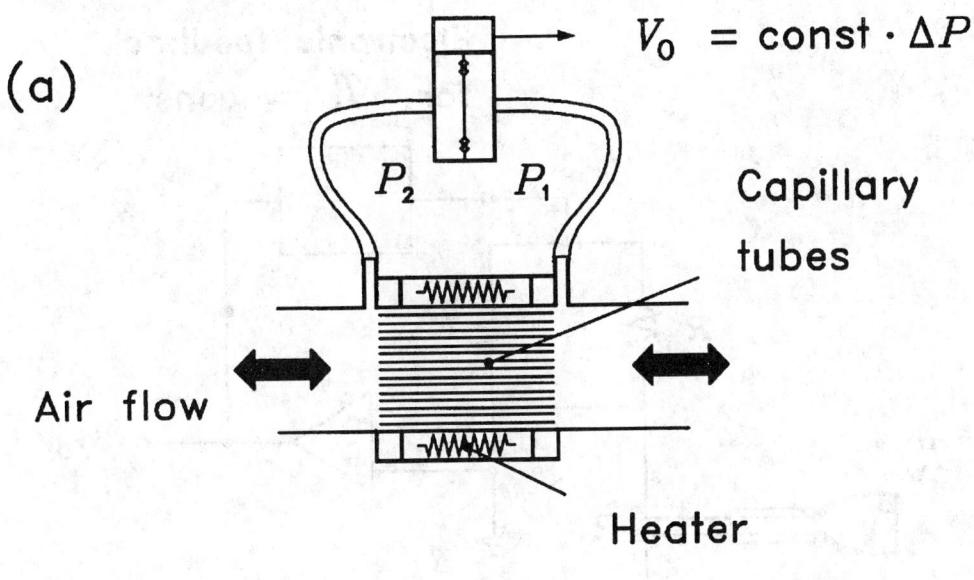

(a) $V_0 = \text{const} \cdot \Delta P$

P_2 P_1

Capillary tubes

Air flow

Heater

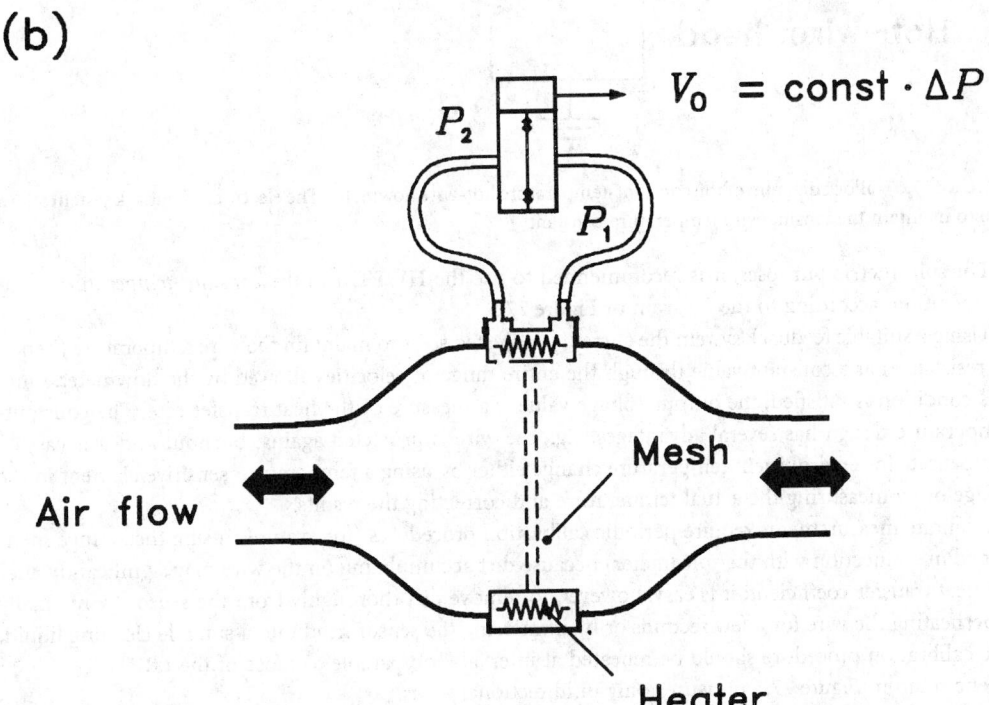

(b) $V_0 = \text{const} \cdot \Delta P$

P_2

P_1

Mesh

Air flow

Heater

FIGURE 77.4 (a) The Fleisch (capillary) pneumotachograph. (b) The Lilly (wire-mesh screen) pneumotachograph. In both sensors, the task of the heater is to reduce water vapor condensation.

The speed v of the fluid is linked to h by an important empirical relation whose analytical form has withstood the test of time; however, owing to the complexity of the formula and to the number of physical quantities involved in it, it is convenient to make direct reference to Reference 11 where all necessary details can be found.

It is wise to remember, in fact, that from the operational point of view the last word on these instruments is always entrusted to a sound calibration procedure.

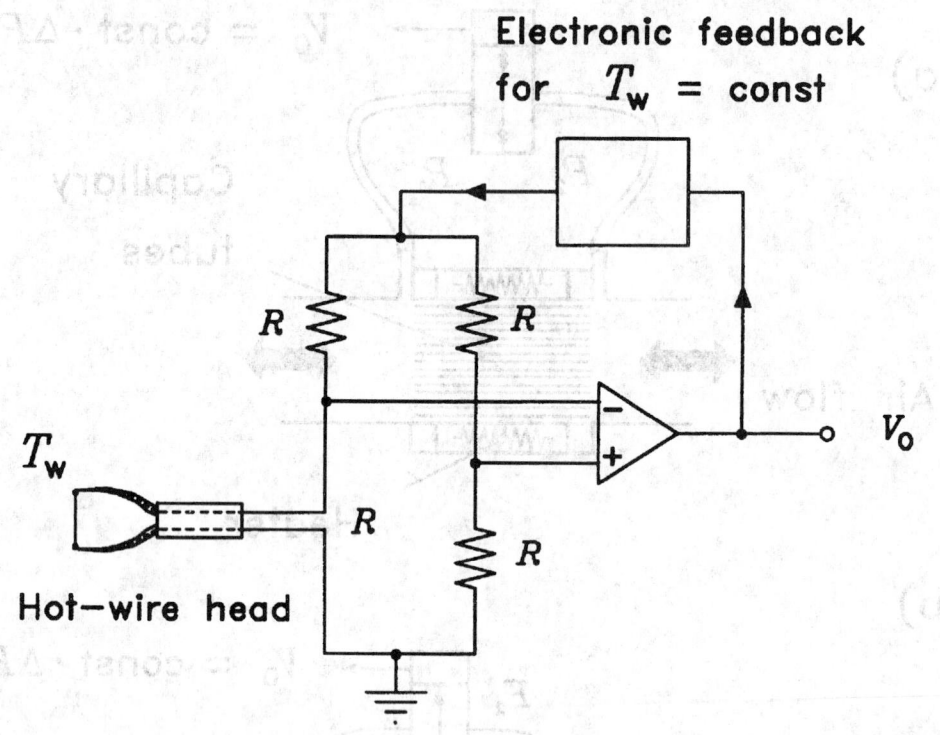

FIGURE 77.5 Block diagram of the constant-temperature hot-wire flowmeter. The electronic feedback system allows one to maintain the sensing wire temperature constant.

For spirometric purposes, it is recommended to use the HWPTM in the *constant temperature* mode of operation, according to the diagram of Figure 77.5.

Using a suitable feedback system the current is adjusted so as to maintain the wire temperature (hence, its resistance) at a constant value through the entire range of velocities allowed by the flowmeter. Once this condition is satisfied, the output voltage value is a measure of the heat transfer rate. The constant-temperature design has several advantages, e.g., the wire is protected against burnout and it is easy to compensate for environment temperature change either by using a temperature-sensitive element in the bridge or by measuring the actual temperature and correcting the results.

Accurate measurements require periodic calibration procedures (for example, using the syringe mentioned in connection with the spirometer) because dirt accumulating on the wire may significantly alter the heat transfer coefficient; it is easy, however, to remove dirt thoroughly from the sensor by manually superheating the wire for a few seconds or by immersing the sensor head into a suitable cleaning liquid. The calibration procedure should be repeated at intervals comparable to those of the LRPTM.

The setup in Figure 77.5 measures only unidirectional flows.

Hot-Film Pneumotachometer

More recently, hot-film PTMs have been developed which can be used in place of the hot wire, the rest of the PTM remaining unchanged. The sensing element is deposited onto a nonconducting support (quartz) by vacuum sputtering, which ensures uniform thickness (about 0.1 μm) of the sensing element (platinum or nickel). In these sensors, heat can be conducted through the substrate and lost by convection to the ambient gas. The length-to-diameter ratio of the sensor is smaller; consequently, the temperature distribution along the sensor is less uniform. These drawbacks, which are more theoretical than real since calibration is always necessary, are largely compensated for by a lower fragility and sensitivity to particulate contamination.

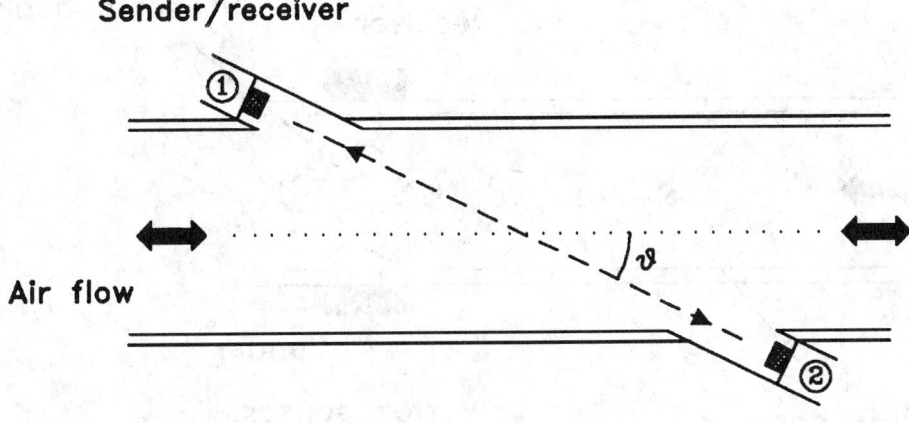

FIGURE 77.6 Transit time ultrasonic pneumotachograph. Owing to the large Q factor of the resonators, the central resonant frequencies of the two piezoelectric crystals should closely match.

Ultrasonic Pneumotachometer

The basic working principles of UPTMs [10] are (1) sound is sped up or slowed down as it propagates through a moving medium, (2) the back and forth transit time of a sound signal is related to flow velocity and turns out to be largely independent of the acoustic velocity of the medium. UPTMs can be classified according to whether they use *pulsed* or *continuous* ultrasound signals.

In the first case, the transmitter is driven by a short pulse of sine waves; the round-trip transit time of individual pulses or sequences of pulses is measured.

In the second case, a continuous ultrasonic signal is transmitted along a closed path and either the phase shift or the frequency shift is measured. As a specific example, a pulsed transit time UPMT (Figure 77.6) is described.

Two piezoelectric crystal transducers are recessed into the wall of a conduit at an angle ϑ to the flow axis. Provisions are made in order to reduce the level of acoustical and electromagnetic external interferences. Moreover, moderate heating (at about 40°C) prevents water from condensing on their surfaces. The two devices are made to function alternately as transmitter and receiver of a short burst of 50 to 200 kHz sound wave.

With no gas flowing through the conduit, the time required for sound transmission in either direction is the same. When gas is flowing, the two times are, respectively,

$$t_{12} = L/\left(c + v \cdot \cos\vartheta\right) \qquad t_{21} = L/\left(c - v \cdot \cos\vartheta\right)$$

where L is the transmitter–receiver distance and c is the speed of sound.

The gas velocity v turns out to be independent of the actual value of c (which in turn depends on the gas type and on the working conditions and can range from around 200 m s^{-1} to nearly 1000 m s^{-1}); an expression for v is easily obtained from the last equations by assuming $c^2 \gg v^2$

$$v \approx \frac{L}{2\cos\vartheta} \cdot \frac{\Delta t}{t_{\mathrm{a}}^2} \tag{77.3}$$

where $\Delta t = t_{21} - t_{12}$ and $t_{\mathrm{a}} = (t_{21} + t_{12})/2$.

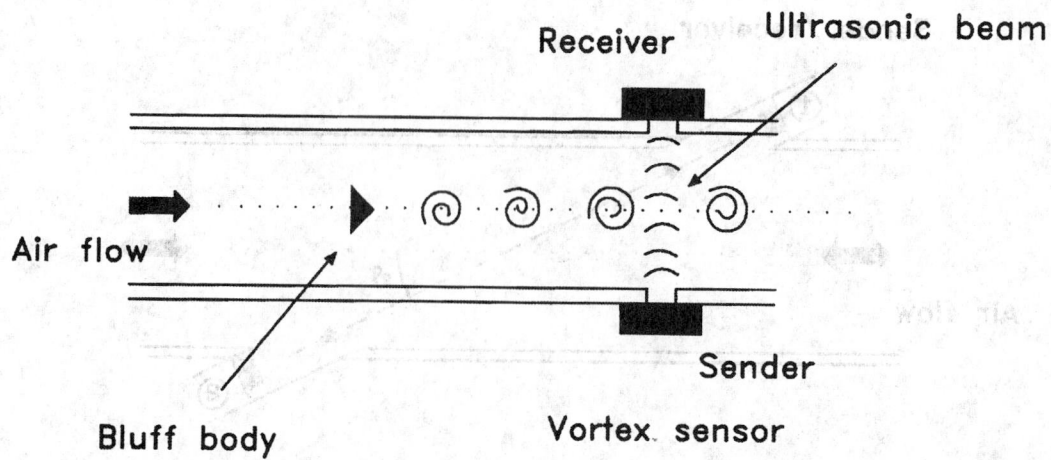

FIGURE 77.7 A vortex shedding flowmeter equipped with an ultrasound vortex sensor. Vortices generated by the bluff body are sensed by the ultrasonic beam located downstream. In the optical model, the functions of vortex generation and counting are both taken on by an optical fiber.

The flowmeter measures the average flow velocity of the gas along the path of the ultrasound beam.

In using the UPTM it should be remembered that the working frequency is very close to the resonant frequencies of the two crystals which should be closely matched: small differences would cause a strong reduction of the instrument sensitivity owing to the sharpness of the resonance curves.

Vortex-Shedding Pneumotachometer

When the Reynolds number is sufficiently high, vortices of regular periodicity are born downstream of a bluff body located in the fluid flow (see Figure 77.7).

The frequency at which these eddies are generated is a linear function of flow velocity and depends weakly on the dimensionless Strouhal number [12,13]. Different types of VSPTMs use different methods for measuring the vortex frequency; to this end, ultrasonic beams and optical fibers are widely employed.

Ultrasonic: An acoustic beam (Figure 77.7) is sent across the flow by an ultrasound transmitter located on one side of the tube (downstream of the bluff body); an ultrasound receiver is located on the opposite side of the tube. In this case, eddies generated by the bluff body act as modulators of the sound wave intensity. By doubling the setup [13] the ultrasonic VSPTM can also be used for bidirectional flows.

Optical fiber: In this model, an optical fiber is stretched across the tube and acts both as the bluff body and the vortex frequency sensor [14]. The light from an LED enters the fiber at one end, the other end being coupled to a photodiode. The intensity of the light propagating through the fiber is modulated by the mechanical vibrations of the fiber produced by the eddies moving with the fluid and fluctuates in step with the vortex generation frequency.

Both in the ultrasonic and the optical fiber VSPTM, the fluctuating intensity of the beam (respectively, sound and light) is processed to output a flow indication.

The main limitation of all types of VSPTMs is a "blind zone" at low fluid flows corresponding to the absence of vortices at low Reynolds numbers. In compensation, these devices are quite insensitive to the thermodynamic conditions of the gas and to the presence of particulate matter. Similarly to the case of HWPTMs, the last word on VSPTMs is entrusted to a sound calibration procedure.

Measurement of Volume Using a PTM Plus an Integrator

Another kind of plethysmograph can be built by combining a PTM and a digital integrator that calculates volume as follows:

$$V(\tau) = \int_0^\tau \dot{v}(t)\,dt \qquad (77.4)$$

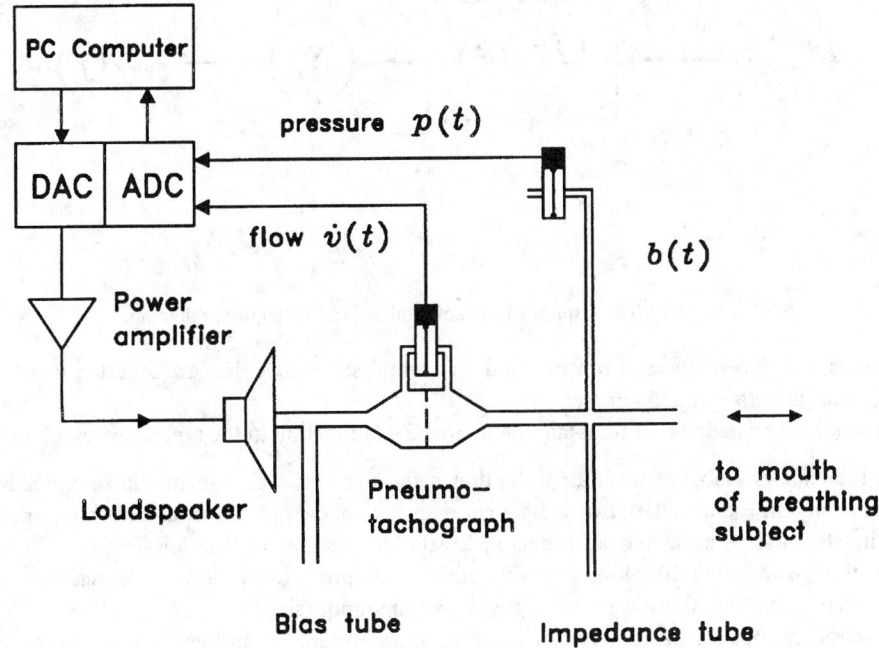

FIGURE 77.8 Experimental setup for the computer-based measurement of respiratory input impedance.

where $\dot{v}(t)$ is the instantaneous flow. The problems exhibited by this kind of instrument are the same as those of the PTM around which it is built, since the process of integration is invariably digital; moreover, correction to BTPS requires considerably more care [1]. Finally, when a heated PTM is employed, correction to BTPS is often done by assuming instantaneous thermalization of the gas within the tachometer (this means that the gas passing through the instrument assumes completely and without delay the same temperature of the tachometer [1]); this assumption may be quite unrealistic, especially in the case of the wire mesh screen type. The correction should thus be different for different types of PTMs.

The Forced Oscillation Technique

The forced oscillation technique (FOT) is a modern tool [15] used to gain information on the structural and mechanical properties of the respiratory system by measuring how the latter responds to an externally imposed excitation. When linearity is assumed, well-known correlation techniques (borrowed from conventional signal analysis) can be exploited to measure the system complex impedance $Z_R(f)$, which is defined as the ratio of complex input pressure difference to complex output flow (i.e., the ratio of amplitudes and the difference of phases) and is equal to the inverse of the system transfer function $H_R(f)$.

Depending on the part of the respiratory system actually investigated, i.e., on the positions where the excitation pressure difference is applied and the flow is measured, different values for $Z_R(f)$ can be obtained:

- *Input impedance:* Both the excitation pressure (measured relatively to the pressure on the body) and flow are measured at the mouth;
- *Transfer impedance:* The excitation pressure is measured at the mouth, whereas flow is measured at the thorax (or vice versa) by means of a body plethysmograph.

An experimental setup for the measurement of respiratory input impedance is shown in Figure 77.8. A suitable software provides

- Generation of a pseudorandom sequence (typically obtained by summing the first 25 harmonics of 2 Hz with random phases) which, after digital-to-analog conversion and proper power amplification, drives a large (20 to 30 cm diameter) loudspeaker;

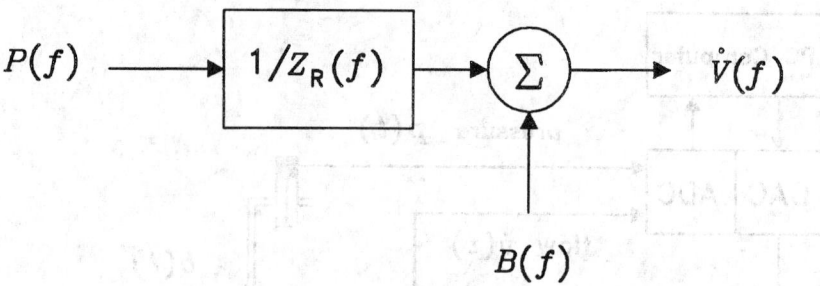

FIGURE 77.9   The respiratory system modeled in the frequency domain.

- Acquisition of two blocks of pressure and flow samples: 2×4096 data are collected in a 16-s run at a sampling rate of 256 sample/s;
- Processing of data in order to obtain the desired $Z_R(f)$ function in the typical interval 2 to 48 Hz.

The bias tube allows a flow of fresh air to be drawn through the system to minimize rebreathing of expired air. The impedance tube (2 m long, 2 cm diameter) allows spontaneous breathing; moreover, because its electrical equivalent is an inductor, it acts like a shunt for the low-frequency breathing components, thus reducing their level of disturbance on the pressure and flow measurements (noise), without affecting in a significant way the imposed pressure input (signal).

The respiratory apparatus is modeled as a system whose input is a random noise pressure $p(t)$ and whose output is a flow signal $\dot{v}(t)$, both corrupted by a breathing disturbance $b(t)$ uncorrelated to $p(t)$. Any text on random signal analysis [16] may provide useful information on the mathematical quantities employed in this section.

Using the following Fourier transform (FT) quantities:

- $P(f)$ = FT of the random noise pressure input $p(t)$
- $B(f)$ = FT of the subject's breathing noise $b(t)$
- $\dot{V}(f)$ = FT of the linear system output $\dot{v}(t)$

we have (Figure 77.9)

$$\dot{V}(f) = P(f) \cdot 1/Z_R(f) + B(f) \tag{77.5}$$

A valid estimate of $Z_R(f)$ cannot be obtained from this relation because of the presence of the large breathing disturbance. According to the theory of signal processing, a nonbiased estimate of $Z_R(f)$ can be obtained by using autospectra and cross spectra of input $[p(t)]$ and output $[\dot{v}(t)]$ signals; if $G_{PP}(f)$ and $G_{\dot{V}\dot{V}}(f)$ the autopower spectrum of pressure and flow, and $G_{\dot{V}P}(f)$ their cross power spectrum, the estimated impedance $Z_M(f)$ becomes

$$Z_M(f) = \frac{G_{PP}(f)}{G_{\dot{V}P}(f)} = \frac{G_{PP}(f)}{\dfrac{1}{Z_R(f)} \cdot G_{PP}(f) + G_{BP}(f)} \tag{77.6}$$

which becomes almost coincident with the true value $Z_R(f)$ when a long time average allows sensible reduction of the contribution of the $G_{BP}(f)$ term.

Moreover, since the systematic bias depends mainly on the presence of the breathing noise on both pressure and flow signals, all efforts should be made to reduce this disturbance as early as possible, i.e., at the sensor level. This is achieved by using the impedance tube and by keeping the value of the pneumotachograph impedance as low as possible. Further reduction is usually obtained by high-pass analog filtering (2 Hz corner frequency) before the ADC and/or by digital filtering before FFT calculations. It is worth noting that the choice of a pseudorandom excitation makes possible the use of a digital comb filter (a multiple narrow bandpass, i.e., a bandpass for each 2 Hz harmonic) [17].

When the FOT is used, it is necessary to ascertain that the coherence [16] between input and output, expressed by the γ^2 function defined below, is sufficiently large:

$$\gamma^2(f) = \frac{\left|G_{\dot{V}P}(f)\right|^2}{G_{\dot{V}\dot{V}}(f) \cdot G_{PP}(f)} \tag{77.7}$$

As a rule of the thumb, the calculated $Z_R(f_0)$ (the impedance for a particular value of frequency) is accepted only when the corresponding $\gamma(f_0)$ is larger than 0.9 to 0.95.

Actually, the value of $\gamma(f)$ is the only form of control used; at any rate, some care should be taken to interpret the coherence function as an index for the reliability of respiratory impedance data [18, 19]. Obviously, higher values of the coherence function can be achieved by increasing the pressure input amplitude, but this may lead to an increase of nonlinearity effects [20].

In conventional setups, flow is evaluated by measuring the pressure drop ΔP that develops across the pneumotachograph, whose impedance is known and usually small with respect to the subject's input impedance Z_R. Since these two impedances are in series, the differential pressure ΔP measured by the pressure sensor is small compared with the pressure P applied to both sides of the transducer. Accurate measurements then require the use of highly symmetric sensors [21], where symmetry is expressed by the value of the common-mode rejection ratio, CMRR = 20 log $P/\Delta P$. In addition to this, however, other sources of error (such as the finite impedance of pressure sensors [22]) are usually present which render the correction procedure ineffective.

All these problems can be globally overcome by the following dynamic calibration procedure [23] that requires two simple preliminary measurements and the availability of a reference impedance Z_{REF}.

If Z_M^∞ denotes the impedance measured when the measuring system is occluded ($Z_R = \infty$), Z_M^{REF} denotes the impedance measured when a known reference impedance is used ($Z_R = Z_{REF}$) and Z_M denotes the impedance measured when the subject is connected to the measuring device, then the subject's corrected impedance is given by

$$Z_R = Z_{REF} \frac{1/Z_M^{REF} - 1/Z_M^\infty}{1/Z_M - 1/Z_M^\infty} \tag{77.8}$$

Z_M^∞ and Z_M^{REF} need not be evaluated each time a new subject is connected to the device, but only when some physical change (e.g., length variations of connecting tubes, replacements of sensors, etc.) has occurred since the last time the instrument was used.

Reference Impedance

The availability of a reference impedance is recommended [24]:
1. To correct the measured impedance using the procedure described above;
2. To compare measurements obtained by different devices, different techniques, and/or different groups.

For these purposes, a compact calibrator [25] has been proposed which displays the following features:

- It is simply reproducible;
- Its impedance value is not too far from typical human values;
- It is tractable from the mathematical point of view;
- Its lowest resonances fall outside the usual working frequency range.

TABLE 77.2   Complex Values of Calibration Impedance

Frequency (Hz)	Impedance (10^5 Pa s m^{-3})		
	Real	Imaginary	Modulus
0	3.09	0.00	3.09
5	3.11	1.11	3.30
10	3.14	2.17	3.82
15	3.20	3.22	4.54
20	3.28	4.26	5.37
25	3.37	5.28	6.26
30	3.47	6.29	7.18
35	3.58	7.28	8.12
40	3.70	8.26	9.05
45	3.82	9.23	9.98
50	3.94	10.17	10.91

The calibrator consists of a bundle of 30 tubelets (each 20.0 cm long and of 2.00 mm inside diameter) and its complex impedance is given in Table 77.2.

77.3 Future Perspectives

Measurements in ventilation strongly depend on the availability of reliable sensors and detectors (pressure, temperature, optical and ultrasonic beams, etc.). The use of these components has been rising steadily in the last decade and this trend is expected to continue in the near future. The foreseeable consequences of this trend on the development of dedicated instrumentation may be

1. The increasing use of transducers of various kinds in all spirometric measurements.
2. The universal use of personal computers (and specialized software) to process and display data.
3. The growing importance of calibration procedures, because the accurate response of a particular transducer is usually reproducible but not always theoretically predictable; on the other hand, this is fundamental to all measurement systems.

Defining Terms

Plethysmography: Method employed to evaluate volume variations, with particular reference to human organs.

Pneumotachography: Methodology for measuring gas flows.

Respiratory impedance: Complex ratio between input pressure difference and induced flow at a particular frequency, when the respiratory apparatus is viewed as a complex network of linear components (resistive, inductive, capacitive).

Spirometry: The measurement of volume changes of the ventilatory system (lungs and chest wall) usually inferred from the movement of gas to and from the system.

Ventilation: Physical interaction of a body of air with the respiratory system, either through spontaneous or mechanically assisted breathing.

References

1. P.H. Quanjer, G.J. Tammeling, J.E. Cotes, O.F. Pedersen, R. Peslin, and J.C. Yernault, Lung volumes and forced ventilatory flows, *Eur. Resp. J.*, 6: 5–40, 1993.
2. R.A. Wise, J. Connett, K. Kurnow, J. Grill, L. Johnson, R. Kanner, P. Enright, and the Lung Health Study Group, Selection of spirometric measurements in a clinical trial, the Lung Health Study, *Am. J. Respir. Crit. Care Med.*, 151: 675–681, 1995.

3. T.F. McAinsh, Ed., *Physics in Medicine & Biology Encyclopedia*, Oxford, U.K.: Pergamon Press, 1986.
4. D.I. Plaut and J.G. Webster, Ultrasonic measurement of respiratory flow, *IEEE Trans. Biomed. Eng.*, BME-27: 549–558, 1980.
5. W.S. Linn, J.C. Solomon, H. Gong, Jr., E.L. Avol, W.C. Navidi, and J.M. Peters, Standardization of multiple spirometers at widely separated times and places, *Am. J. Respir. Crit. Care Med.*, 153: 1309–1313, 1996.
6. American Thoracic Society, Standardization of Spirometry 1994 Update, *Am. J. Respir. Crit. Care Med.*, 152: 1107–1136, 1995.
7. L.R. Johnson, P.L. Enright, H.T. Voelker, and D.P. Tashkin, Volume spirometers need automated internal temperature sensors, *Am. J. Respir. Crit. Care Med.*, 150: 1575–1580, 1994.
8. K.E. Finucane, B.A. Egan, and S.V. Dawson, Linearity and frequency response of pneumotachometers, *J. Appl. Physiol.*, 32: 121–126, 1972.
9. N.H. Cook and E. Rabinowicz, *Physical Measurement and Analysis*, Reading, MA: Addison-Wesley, 1963.
10. L. Marton and C. Marton, Eds., *Methods of Experimental Physics. Fluid Dynamics.* Vol. 18, New York: Academic Press, 1981.
11. L.M. Fingerson, Thermal anemometry, current state, and future directions, *Rev. Sci. Instrum.*, 65: 285–300, 1994.
12. D.F. White, A.E. Rodely, and C.L. McMurtrie, The vortex shedding flowmeter, in R.B. Dowdell, Ed., *Flow— its Measurement and Control in Science and Industry*, Pittsburgh, PA: Instrument Society of America, 967–974, 1974.
13. Ch. Buess and W. Guggenbuhl, Ultrasonic airflow meters in medical application, *Proc. Annu. Conf. IEEE Eng. Med. Biol. Soc.*, 13–16, 1987.
14. S. Webster, R. McBride, J.S. Barton, and J.D.C. Jones, Air flow measurement by vortex shedding from multimode and monomode optical fibers, *Meas. Sci. Technol.*, 3: 210–216, 1992.
15. E.D. Michaelson, E.D. Grassman, and W.R. Peters, Pulmonary mechanics by spectral analysis of forced random noise, *J. Clin. Invest.*, 56: 1210–30, 1975.
16. J.S. Bendat and A.G. Piersol, *Random Data: Analysis and Measurement Procedures*, New York: Wiley & Sons, 1986.
17. R. Farré and M. Rotger, Filtering the noises due to breathing in respiratory impedance measurements, *Eur. Respir. Rev.*, 1: 196–201, 1991.
18. E. Oostveen and A. Zwart, Reliability of the coherence function for rejecting respiratory impedance data, *Eur. Respir. Rev.*, 1: 218–221, 1991.
19. H. Franken, J. Clément, and K.P. van de Woestijne, Systematic and random errors in the determination of respiratory impedance by means of the forced oscillation technique: a theoretical study, *IEEE Trans. Biomed. Eng.*, BME-30: 642–651, 1983.
20. M. Rotger, R. Peslin, R. Farré, and C. Duvivier, Influence of amplitude, phases and frequency content of pseudorandom pressure input on impedance data and their variability, *Eur. Respir. Rev.*, 1: 178–182, 1991.
21. R. Peslin, P. Jardin, C. Duvivier, and P. Begin, In-phase rejection requirements for measuring respiratory input impedance, *J. Appl. Physiol: Respir. Environ. Exercise Physiol.*, 56: 804–809, 1984.
22. R. Farré, R. Peslin, D. Navajas, C. Gallina, and B. Suki, Analysis of the dynamic characteristics of pressure transducers for studying respiratory mechanics at high frequencies, *Med. Biol. Eng. Comput.*, 27: 531–537, 1989.
23. R. Farré, D. Navajas, R. Peslin, M. Rotger, and C. Duvivier, A correction procedure for the asymmetry of differential pressure transducers in respiratory impedance measurements, *IEEE Trans. Biomed. Eng.*, BME-36: 1137–1140, 1989.
24. M. Cauberghs and K.P. van de Woestijne, Calibration procedure of the forced oscillation technique, *Eur. Resp. Rev.*, 1: 158–162, 1991.
25. L. Basano and P. Ottonello, A calibrator for measurements of respiratory impedance, *Meas. Sci. Technol.*, 6: 982–987, 1995.

78

Blood Chemistry Measurement

Terry L. Rusch
Marshfield Medical Research Foundation

Ravi Sankar
University of South Florida

78.1 Introduction

The study of blood and its effects on sustaining life date back to the start of understanding human anatomy. There is a diverse background associated with blood chemistry, and knowledge of the makeup and purpose of blood and its components is growing by leaps and bounds. The typical discussion and history of blood chemistry relate to oxygen transport and its effects. This chapter will deal mainly with the measurement of oxygen related to health and critical-care applications, with limited descriptions of noncritical health status measurements. The beginning of blood analysis and the measurement of a person's health status were not practically and conveniently performed until technology advanced. It has been within the last 40 years that routine monitoring began in critical care. Critical-care health monitoring is the focus of this discussion because it has been one of the main motivations for developing analytical measurements. However, most critical-care monitoring has led to diagnostic tools and functional assessment of blood and its components.

Current health care professionals use a variety of common analytical measurements to assess and maintain a person's health. The measurements of interest include hemoglobin, hematocrit, blood gases (O_2 and CO_2), pH, glucose, and concentrations of electrolytes. Blood gas analysis is divided into a wide variety of specific tests. The tests include arterial, mixed venus, and **transcutaneous** oxygen tension; carbon dioxide tension; oxygen and carbon dioxide concentration; oxygen saturation. Each test has a specific purpose in diagnostics. Continuous real-time monitoring of oxygen saturation by pulse oximetry measurements is routinely used in anesthesia and critical-care settings. However, to determine heart and lung efficiency or true tissue oxygenation, other measurements are required. Different measurements are required because of the complex nature of the human body, and no one measurement gives a complete picture of overall status. For example, arterial oxygen saturation does not give a true cardiac output efficiency nor does it give a true tissue oxygenation state. Measurements can also be very misleading. During normal conditions, the measured values correspond very well. For example, under a given temperature and pH, oxygen saturation corresponds to oxygen tension. A fixed relationship occurs under these conditions. This does not mean the same relationship applies to all circumstances. A change in temperature or pH will shift the oxygen tension to a saturation relationship.

78.2 Background

Most blood chemistry measurements have a long history. In the late 1800s the hydrogen (H^+) electrode was used to measure pH and, in 1925, John Peters at Yale showed a relationship between pH and CO_2 content. The measurement of blood gas started in the early 1920s with *in vitro* analysis of oxygen saturation. It was not until much later, however, that common patient care measurements were accepted. For example, it was not until 1952 in Copenhagen that pH monitoring was performed on artificially ventilated patients. The monitoring of pH was the start of the clinical use of blood gas monitoring. Until that time, physiological measurements were performed only in the laboratory setting. Electrochemical measurements were more common, and performed with a Clark electrode or similar electrode arrangement. Leland Clark made a working model of an oxygen electrode around 1950, which introduced a practical method of measuring oxygen tension. The probe was originally developed to aid in heart bypass instruments. Not long after the Clark electrode was developed, optical transmission analysis tools were developed. The measurement of oxygen saturation, called oximetry, was one of the early optical measurements. Oximetry has a long history dating back more than 50 years to the 1930s. Matthes and Millikan recorded the earliest noninvasive reading around 1935, and Glen Millikan introduced the name **oximetry** in 1942. A real research effort was started during World War II as a part of military aviation. It took almost 60 years for pulse oximetry to become a standard piece of equipment in operating rooms, critical-care units, and emergency health care. On January 1, 1990, the American Society of Anesthesiologist (ASA) made intraoperative monitoring with pulse oximetry a standard [1]. Despite the acceptance of pulse oximetry and its benefits, there still has been little change in pulse oximetry techniques. A good historic perspective on most blood gas analysis is written in a series of articles in the *Journal of Clinical Monitoring* and in *International Anesthesiology Clinics* [2-7].

Oxygen Delivery

As Severinghaus and Astrup [6] note in the history of blood gas analysis, nothing is more important than oxygen supply for life. Therefore, it can be argued that oxygen content and consumption should be one of the most important things to monitor in critical and unconscious patients. It is important to look at the mechanisms for oxygen transport. All human tissue needs oxygen to function, and it is therefore critical to understand how oxygen is delivered. **Hemoglobin** is the binding agent for oxygen in blood, but it is necessary to oxygenate the blood and move it to the tissue that uses the oxygen. A simple explanation is as follows. Oxygen is diffused into blood via the lungs. Blood is then circulated by a system of **arteries**, **veins**, and **capillaries** where the oxygen is diffused to tissue. The tissue also diffuses carbon dioxide to blood in the capillaries, which transports back to the lungs, where the carbon dioxide is diffused to the lungs for expiration. Figure 78.1 shows a diagram of the general makeup of the components of interest. This is a very simplistic view of blood and oxygen transport. There are many other regulatory factors, such as sodium, potassium, calcium, pH, glucose, and so forth, but this simplistic view is a good starting point to understand the measurements that follow. It is also important to note that several measurement options and locations exist for gathering the same or similar data. This makes understanding the interaction and importance of different measurements difficult. For example, arterial and venus blood oxygen tensions are interdependent, but have different readings. Different physical states and health problems cause varying interdependence. A simple requirement to identify whether or not cell tissue is being properly oxygenated becomes very complex. Oxygen delivery to tissue is determined by the differential in partial pressure between cell tissue and capillaries. The partial pressure difference determines the rate and efficiency of diffusion. However, the rate is also dependent on blood flow, concentration of hemoglobin, and the saturation of oxygen. No individual parameter or measurement gives a complete picture of cell oxygen consumption.

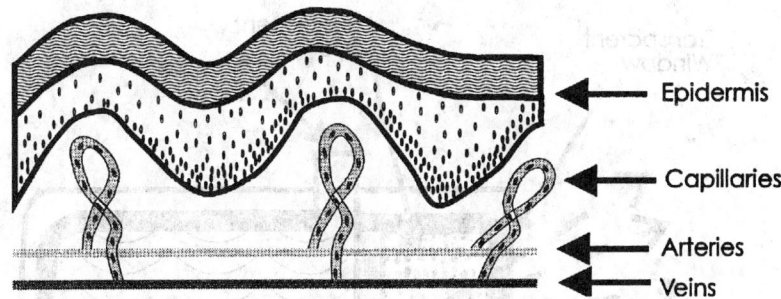

FIGURE 78.1 Components involved in oxygen transport. Oxygen is delivered to the epidermis by arteries and exchanged through the capillaries. Veins return the by-product.

Blood Hemoglobin

Blood contains many components, but hemoglobin is the main transporter of oxygen (O_2) to tissue. In addition, blood is responsible for removing the by-product of oxygen expenditure, carbon dioxide (CO_2). Hemoglobin is a good binding agent for oxygen, producing oxyhemoglobin (O_2Hb), also called oxygenated hemoglobin. Hemoglobin is a combination of four peptide chains, each containing several hundred amino acids. Hemoglobin is composed of two elements, heme and globin. Heme is the iron pigment of the electrolytes, and globin is a simple protein. Hemoglobin contains about 6% heme and 94% globin. Hemoglobin is the principal component of red blood cells. Unbound hemoglobin is called deoxyhemoglobin (RHb), or reduced hemoglobin. Hemoglobin also binds to produce additional components, such as, carboxyhemoglobin (COHb), and methemoglobin (MetHb). Methemoglobin is a result of oxidation of ferrous iron in hemoglobin [8]. There are other compounds that are found in blood, but they are usually in very low concentrations and are of interest in more specialized cases.

Measurement Methods

Most blood chemistry analysis measurements are derived from one of two methods: electrochemical and optical. There are other specific measurement techniques, such as gas chromatography, but they are more of a specialized measurement and will not be addressed. Electrochemical measurements are based on the Clark electrode with the blood component or electrolyte of interest using a different ion-specific electrode. The Clark electrode consists of an electrode in a medium. The first electrode for oxygen tension measurements was a platinum electrode and a silver anode. The current or voltage generated at the electrode is measured and is proportional to the amount of that component. The relationship is usually a calculated concentration derived from empirical data. For optical measurements, there are many approaches, but all are based on absorption, reflection, scattering, and fluorescence techniques. One optical probe has been called an optode. The optode is a cross measurement of optical fluorescence and electrochemical measurement. Figure 78.2 shows a diagram of one configuration for an optode arrangement. In an optode, an ion-selective membrane is used to diffuse an ion or compound into the fluorescent dye measurement chamber. The ion is associated with a fluorescent dye, excited with a light source, and the emitted fluorescent light is measured. Many measurements can be obtained, such as oxygen tension, carbon dioxide tension, pH, potassium concentration, calcium concentration, and chloride concentration.

Measurements are listed in one of four categories, (1) invasive or (2) noninvasive measurements and (3) continuous or (4) periodic measurements. **Invasive** measurements can be continuous or periodic, and can be performed *in vivo* or *in vitro*. Invasive **catheter** sensors can record real-time data continuously with no loss of blood sample. **Intra-arterial** measurements can be drawn periodically for minimally invasive measurements with no sample loss, or samples can be withdrawn and discarded. Periodic samples can also be drawn for analysis and measured in a physically different location at a later time. Noninvasive measurements can also be continuous or periodic, but are generally on patient, real-time measurements.

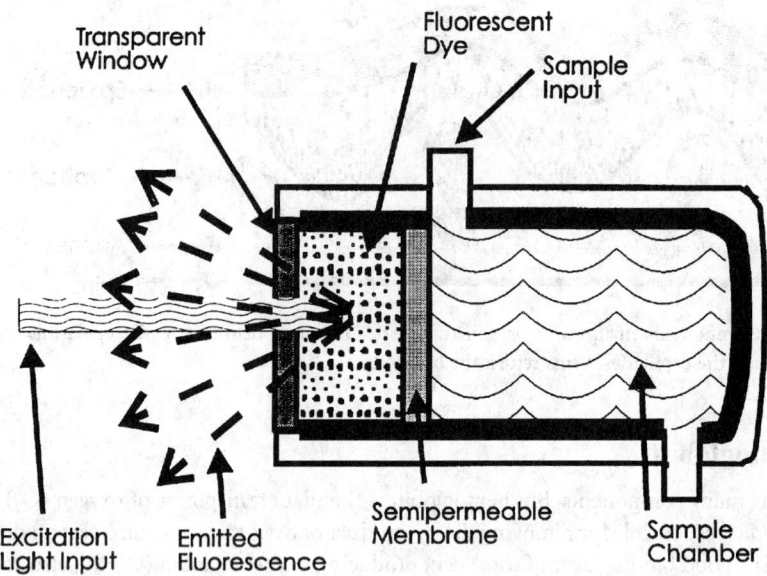

FIGURE 78.2   Chemical fluorescent optode for oxygen tension measurements. Light input through a window excites encased dye. The sample interacts with the dye through a semipermeable membrane. The sample is introduced to the dye via a flow-through chamber.

Noninvasive measurements are always preferred over invasive measurements, because invasive measurements increase risk of infection and usually mean some delay between the time of acquisition and the time results are available. However, noninvasive measurements are not always as accurate as invasive measurements.

78.3 Measurements and Techniques

There are typically several measurement techniques that yield the same or similar data. Each technique will give information about different physical states, such as cardiac efficiency or tissue oxygenation state. For example, oxygen saturation and partial pressure of oxygen are related. However, the exact relationship is dependent on pH and temperature. A text by Kenneth McClatchey on *Clinical Laboratory Medicine*, the International Federation of Clinical Chemistry (IFCC), and instrument operator's manuals are good references for laboratory protocols relating to invasively drawn samples [9,10]. Sample preparation and handling protocols, as well as standard measurement levels and physical states relating to excess or deficient readings are detailed. The range, price, and suppliers for instrumentation are extensive. A standard midrange laboratory blood gas analyzer can cost from $5,000 to $50,000. The more common measurements are described below.

Hemoglobin and Hematocrit Concentration

The total concentration of hemoglobin (CtHb) or **hematocrit** (Hct) indicates the oxygen-carrying potential of blood. The combination of amount of hemoglobin, partial pressure of oxygen or percent oxygen saturation, and rate of flow of hemoglobin determines the amount and efficiency of tissue oxygenation. Invasively drawn samples are typically used to measure hemoglobin and hematocrit values.

Hemoglobin

The total hemoglobin concentration is not the same as the red blood cell count, because red blood cells have different amounts of hemoglobin. The total hemoglobin concentration is measured optically by absorptive intensity at the isosbestic point. The Beers–Lambert law, also referred to as Beer's law, shown in Equation 78.1 regulates the absorptive property of a substance.

$$I_t = I_0\, e^{-DC\alpha_e} \tag{78.1}$$

I_t is the transmitted intensity, I_0 is the incident intensity on the sample, D is the distance light travels through the substance, C is the concentration of the solution, and α_e is the extinction coefficient at a specified wavelength. The isosbestic point is the crossover point in extinction curves for oxyhemoglobin and reduced hemoglobin. The wavelength of 805 nm is an isosbestic point, and at this wavelength the absorption is independent of hemoglobin type. If blood is assumed to be composed of only O_2Hb and RHb, then the absorbance at the isosbestic point determines the total concentration of Hb. The most common error in assuming only O_2Hb and RHb is during elevated COHb. COHb, however, generally is optically indistinguishable from O_2Hb for absorption. An alternative approach to isosbestic measurement is to calculate CtHb. Assuming only O_2Hb and RHb, the total Hb concentration is simply the sum of the two concentrations. A more accurate sum is shown in Equation 78.2.

$$CtHb = O_2Hb + RHb + COHb + MetHb \tag{78.2}$$

Here the four most common Hb derivatives are used to calculate a more accurate CtHb. The individual Hb derivatives are measured as outlined below.

Hematocrit

Hematocrit is the volumetric fraction occupied by red blood cells and is generally measured by conductivity. Hematocrit is also referred to as the packed cell volume (PVC). Hematocrit can be determined by conductivity based on the plasma ion content. Hematocrit does not contribute to the conductivity and therefore is inversely proportional to the conductivity. Hematocrit can also be determined optically in various ways. One way is through optical density measurements, where the total optical density is the sum of optical absorbance and optical scattering density. The total optical density is linearly proportional to Hct, and, at clinically relevant Hct levels of 20 to 40%, scattering is dominant over absorption. One optimization study showed that an optimum wavelength of 624 nm, at a measured angle of 90° from the incident light, gave an inverse linear intensity to the Hct level [11].

Oxygen Tension

Oxygen tension or the partial pressure of oxygen (PO_2) is a common measure of oxygenation states. The partial pressure of oxygen in hemoglobin determines how well oxygen is delivered to the cell tissue of the body. The more common partial pressure reference is arterial partial pressure of oxygen (PaO_2). If the partial pressure of oxygen is higher than the surrounding tissue, oxygen is diffused to the tissue. If the partial pressure is lower than the tissue partial pressure, no oxygen is diffused to the tissue, and tissue damage can start to occur. Equation 78.3 shows Henry's law.

$$C = \alpha_s * PO_2 \tag{78.3}$$

C is the concentration of oxygen (O_2), α_s is the solubility coefficient, and PO_2 is the partial pressure of oxygen. In diffusion, a partial pressure between the tissue and blood supply is trying to maintain equilibrium. Diffusion will occur until the two partial pressures are equal. Equation 78.4 shows diffusion equilibrium.

$$PO_2 = C_1/\alpha_{s1} = C_2/\alpha_{s2} \tag{78.4}$$

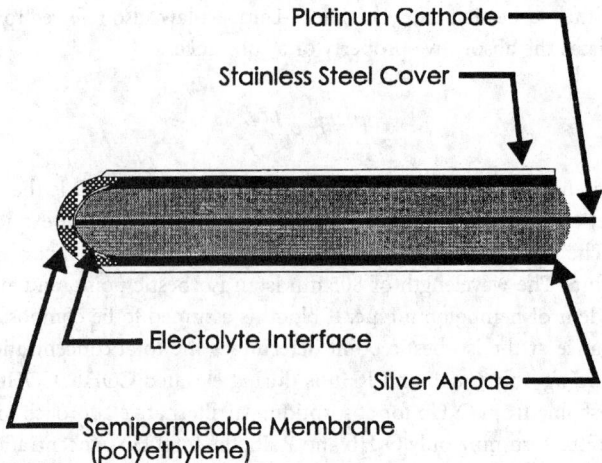

Platinum Cathode

Stainless Steel Cover

Electolyte Interface

Silver Anode

Semipermeable Membrane
(polyethylene)

FIGURE 78.3 Clark-type needle electrode for PO_2 measurements. Oxygen ion interaction through the membrane tip allows the ion concentration to be measured by dissimilar metals.

The rate of oxygen diffusion is dependent on the difference in partial pressure. The larger the difference in partial pressure, the faster the diffusion rate.

The partial pressure difference is partially determined by the oxygen delivery (D_O) rate. Oxygen delivery can be identified as the concentration of arterial oxygen minus the concentration of venus oxygen, times the rate of blood flow (R), as shown in Equation 78.5.

$$D_O = \left(C_a O_2 - C_v O_2 \right) * R \tag{78.5}$$

$C_a O_2$ is the arterial oxygen concentration, and $C_v O_2$ is the venus oxygen concentration. The oxygen delivered, however, is not the oxygen delivered to the tissue. Some oxygen is transpired through the skin. The transpired oxygen allows an alternative approximation measurement of arterial oxygen tension. The more common oxygen tension measurement is arterial oxygen tension. Oxygen tension can be measured electrochemically, transcutaneously, or optically. The techniques are listed below.

Electrochemical PO$_2$ Measurements

Electrochemical measurements of PO_2 are obtained using a basic Clark electrode with a platinum electrode and a silver/silver chloride reference electrode. Figure 78.3 shows a diagram of one type of Clark electrode. The electrode provides a path for the reduction in Equation 78.6.

$$O_2 + 4e^- \rightarrow 2O^- \tag{78.6}$$

The platinum electrode has an affinity with oxygen. If a reference PO_2 at the electrode is known, preferably zero, then the current depends only on the oxygen tension variations of the sample. Early electrodes were consumption measurements where the oxygen is removed and the sample is altered. Oxygen was attracted to the electrode and a current proportional to the oxygen content was observed. Improved electrodes use a semipermeable, constant diffusion membrane. An electrolyte is used with the membrane to improve response, longevity, and stability. An exposed electrode will become coated in whole blood and the sensitivity will degrade. The membrane and electrolyte allow the diffusion of oxygen without the sample directly contacting the electrode. In addition, a properly designed membrane reduces the stirring effect noted by Clark and others using an exposed electrode. The size and location of electrodes have been extensively studied. The exact setup for the electrode can be varied for optimum performance in different

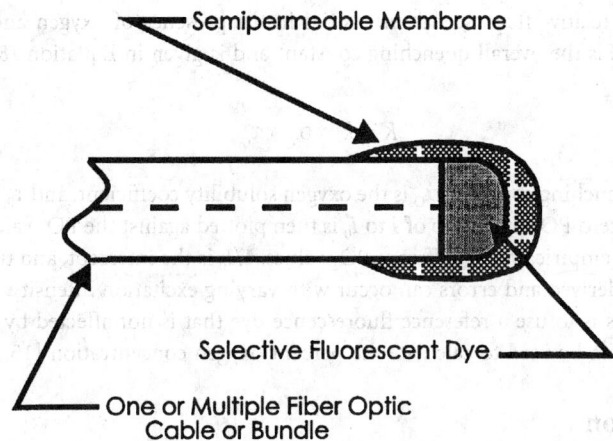

FIGURE 78.4 Chemical fluorescent fiber-optic probe. Dye is encased on a fiber tip by a membrane. Light enters the dye through the fiber and the emitted fluorescence is returned via the fiber. The amount of fluorescence is an indication of the interaction between the dye and the substance in which the fiber tip is placed.

areas. For example, a less permeable diffusion layer reduces the stirring effect and makes the probe more stable, but the result is a slower response time [3,6,12,13].

Transcutaneous Partial Pressure of Oxygen

Partial pressure and oxygen saturation are not always true indicators of actual tissue oxygen consumption. It has been known since 1851 that oxygen is respired from living tissue. A combination of heat and optical means can be used to determine the amount of oxygen expired and correlated to arterial partial pressure under controlled conditions. This method of measuring arterial oxygen partial pressure is called trans-cutaneous oxygen partial pressure ($P_{tc}O_2$). There are, however, many complications in the measurement of $P_{tc}O_2$. Many studies have tried to correlate arterial oxygen saturation with the measured expired oxygen. The most accurate measurements are made by maintaining a constant temperature as high as 45°C, to assure perfusion. This causes the complication of burns to skin, and sensors need to be moved on a regular basis.

Measurements can be performed polargraphically using Clark electrodes or alternatively by mass spectroscopy. Because of the nature of $P_{tc}O_2$, adult measurements are not common, but premature infant hemoglobin and skin are more responsive to $P_{tc}O_2$ measurements, which have had a place in monitoring neonatal oxygenation [7,14].

Optical-Based PO_2

The are many variations on optical-based blood gas measurements. The basic optical measurement for PO_2 is a **fluorescent** measurement. A fiber-optic cable is used to excite a fluorescent dye remotely. The excited dye emits a higher-wavelength signal, known as Stokes shift. The emitted signal is measured and is correlated to the PO_2 value. Figure 78.4 shows a configuration for a fluorescent sensor. Oxygen is used as a fluorescent quencher. Oxygen has a fluorescence-quenching property and attenuates the fluorescent intensity. A zero state can be measured by filling the sample chamber with a zero oxygen concentration mixture. The measured intensity is compared with the incident light source and is used to calculate the concentration of the sample. For oxygen tension, the measurement is done by a fluorescent electrochromic dye, such as pyrenebutyric acid, with an ion-selective membrane such as silicone rubber. The oxygen tension is than calculated using the Stern–Volmer quenching formula, shown in Equation 78.7, and the PO_2 value is calculated empirically.

$$I_0/I = 1 + K*PO_2 \quad \text{or} \quad I\left(PO_2\right) = I_0/\left(1 + K*PO_2\right) \tag{78.7}$$

$I(PO_2)$ and I_0 are the relative fluorescence intensities in the presence of oxygen and in the absence of oxygen, respectively. K is the overall quenching constant and is given in Equation 78.8.

$$K = k^+ * \alpha_O * \tau_0 \qquad (78.8)$$

k^+ is the collisional quenching constant, α_O is the oxygen solubility coefficient, and τ_0 is the mean lifetime of the excited state at zero PO_2. The ratio of I to I_0 is then plotted against the PO_2 value. This plot is used to calibrate the sensor empirically for different PO_2 values. $1/I_0$ is the intercept, and the slope is K/I_0. The system is empirically derived and errors can occur with varying excitation intensity. One way to correct for intensity variations is to use a reference fluorescence dye that is not affected by oxygen quenching. The same procedure can be used to calculate the percent oxygen concentration [15,16].

Oxygen Saturation

There are several readings that are interrelated regarding oxygen saturation. The most common measurements are arterial blood saturation (S_aO_2), mixed venus blood saturation (S_vO_2), and **photoplethysmogram** arterial blood saturation (S_pO_2). S_pO_2 measurements, also called pulse oximetry measurements, are arterial measurements, but are only related to true arterial saturation. Care must be taken in knowing which measurements are actually taken, because of inaccuracies in various measurements. Saturation measurements can be performed invasively using reflectance or fluorescence oximetry, or noninvasively by transmission or reflectance photoplethysmogram readings.

There is an important relationship between oxygen saturation and partial pressure. A plot shown in Figure 78.5 of arterial oxygen saturation vs. arterial partial pressure is called the oxygen hemoglobin

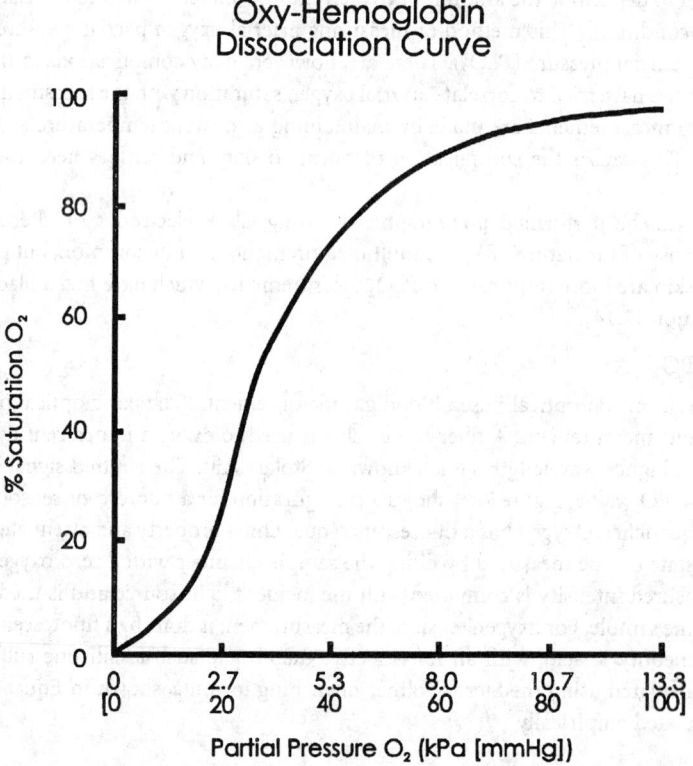

FIGURE 78.5 Oxygen saturation vs. PO_2 oxygen hemoglobin dissociation curve. The partial pressure shows a sigmoidal relationship to the percent oxygen saturation.

dissociation curve. Partial pressure falls at a linear rate and is a good indicator of changing oxygen delivery to cell tissue. The percent oxygen saturation is nonlinear with respect to partial pressure in a sigmoidal relationship. A partial pressure of 13.3 kPa (100 mmHg) is effectively 100% saturation. An increase in partial pressure above 13.3 kPa (100 mmHg) indicates free oxygen not bound to hemoglobin and in that state does not contribute much to tissue oxygenation. This means the percent saturation gives a delayed notice of desaturation. In a patient with healthy lungs, partial pressure values of greater than 24.0 kPa (180 mmHg) can be measured. A 50% decrease in partial pressure could go undetected using the percent oxygen saturation. The oxygen dissociation curve, as mentioned previously, is dependent on pH and temperature. Under normal conditions, the relationship between partial pressure and saturation is maintained, but under altered conditions, such as low perfusion, the relationship can deviate.

Roughton and Severinghaus [18] describe a computation to approximate the dissociation curve based on the Hill equation. The Hill equation is given in Equation 78.9.

$$Y = \left[\left(PO_2 / P_{50} \right)^n \right] / \left[1 + \left(PO_2 / P_{50} \right)^n \right] \tag{78.9}$$

Y is the oxygen saturation, PO_2 is the oxygen partial pressure, and P_{50} is the partial pressure at 6.7 kPa (50 mmHg). The P_{50} point is determined by the temperature and pH of the sample. Temperature and pH correction equations are also presented in References 5, 7, and 17 through 19.

Transmission Oximetry

Continuous S_aO_2 and S_vO_2 measurements are used to help evaluate whether oxygen is being adequately delivered to tissue. Oxygen is either consumed or expired. If oxygen is delivered to tissue and not consumed, then tissue oxygenation is not adequate. Early oximeters used invasively drawn samples and light transmission to measure oxygenation. The basic concept in oximetry is to transmit light through a blood sample, and blood absorbs a determined amount of light according to Beer's law, shown in Equation 78.1. For oximetry applications, hemoglobin is assumed to be composed of only two substances, oxygenated hemoglobin or oxyhemoglobin (O_2Hb) and deoxygenated or reduced hemoglobin (RHb). This is a very simplistic approach, but it is the basis for most oximetry measurements. For *in vivo* measurements, the path length for the light is constant and known. Both the O_2Hb and RHb can be measured simultaneously by using two separate wavelengths. Flash lamps and filter wheels were originally used to illuminate blood samples, and the transmitted signals were measured. Current technology uses light emitting diodes (LEDs) and alternate on and off cycles.

If two light wavelengths are used and the two substances have different extinction coefficients (α_e), or equivalently attenuation coefficients (σ), then the percentage of each substance can be calculated. The extinction coefficients of hemoglobin are well documented. Figure 78.6 is a plot of the extinction coefficients for oxyhemoglobin, reduced hemoglobin, carboxyhemoglobin, and methemoglobin. The two wavelengths of light yielding good results for oximetry are the red (660 nm) and infrared (940 nm) wavelengths. Red has the largest difference between the two extinction curves, and infrared has maximal difference after the isosbestic point where the two extinction curves cross. The functional arterial oxygen saturation (S_aO_2) is calculated using Equation 78.10 given the concentration of O_2Hb and RHb.

$$\text{Functional } S_a O_2 = C_o / \left(C_o + C_r \right) * 100 \tag{78.10}$$

The concentration of oxyhemoglobin is given by C_o and the concentration of deoxygenated hemoglobin is given by C_r. This formula must be adjusted for other contents in blood which influence measurements if present, such as carboxyhemoglobin and methemoglobin. The more accurate fractional saturation is given as the total hemoglobin concentration (tHb) in Equation 78.11.

$$\text{Fractional } S_a O_2 = C_o / \left(tHb \right) * 100 = C_o / \left(C_o + C_r + C_c + C_m \right) * 100 \tag{78.11}$$

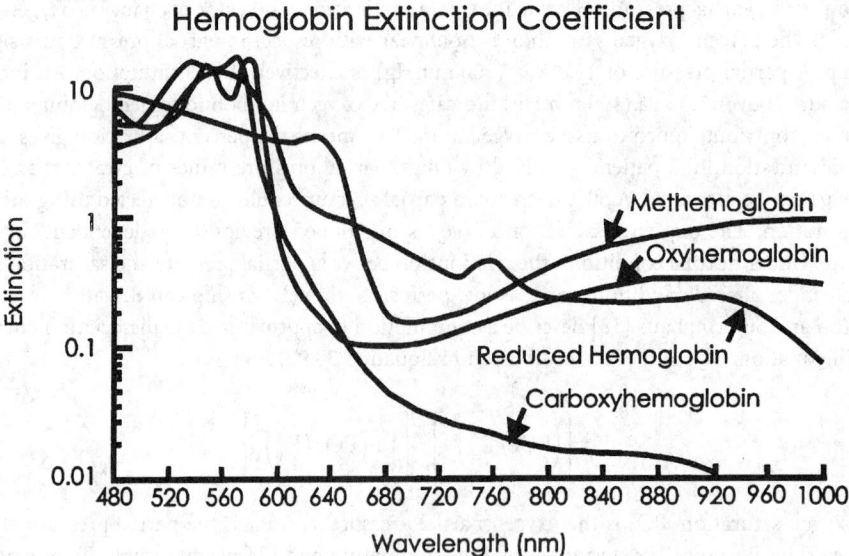

FIGURE 78.6 Hemoglobin extinction coefficients plotted by wavelength. The light extinction coefficient varies by wavelength and type of hemoglobin.

C_c is the carboxyhemoglobin (COHb) saturation and C_m is the methemoglobin (MetHb) concentration. To account for COHb and MetHb accurately, four transmission wavelengths would be required, or for accurate S_aO_2 readings the total Hb concentration needs to be known or measured [5,7,20].

Pulse Oximetry

Pulse oximetry is based on the transmission, absorption, and dispersion of light as it passes through hemoglobin. Beer's law, as stated in Equation 78.1, determines the transmission of light through a substance. For pulse oximetry, the light illuminates both arterial and venus blood and the light must traverse all tissue between light source and receiver. Figure 78.7 represents the light path, and indicates a variable (AC) path length as well as a constant (DC) path length.

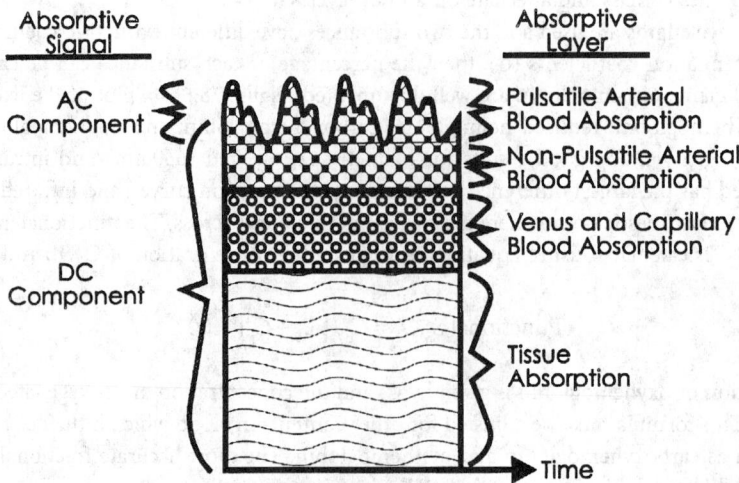

FIGURE 78.7 Absorption components encountered during transmission oximetry. The arteries, veins, capillaries, and tissue absorb light. The total light absorbed has a steady state (DC) and a varying (AC) signal component. The varying signal is due to the pulsatile arterial volume change.

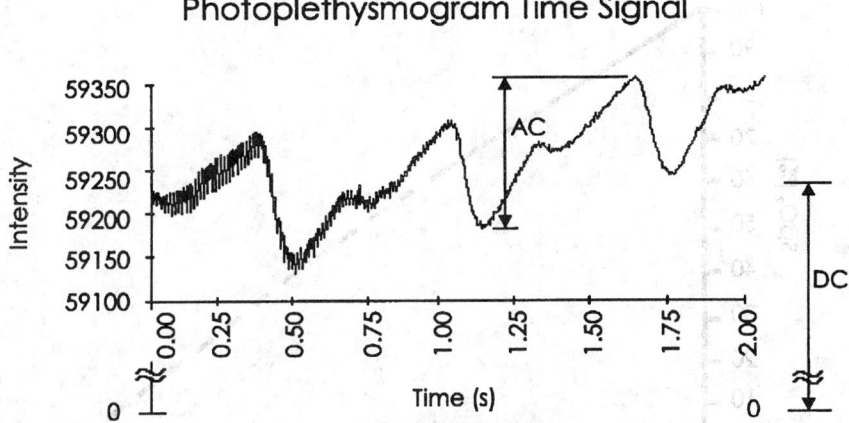

FIGURE 78.8 Photoplethysmogram with AC and DC components labeled. The light intensity transmitted through the finger varies with time. The signal contains a constant (DC) component and a varying signal (AC) component. The varying signal is due to the pulsatile arterial volume change.

To calculate the pulse oximetry digital photoplethysmogram (DPP) oxygen saturation (S_pO_2), two equations are used. An example of a DPP signal is shown in Figure 78.8 and it indicates the AC and DC DPP components. The first step is to use the red and infrared time signal to calculate an R value. The R value is the normalized ratio of the red to infrared transmitted light intensity and is shown in Equation 78.12.

$$R = \left(AC_r / DC_r \right) / \left(AC_i / DC_i \right) \tag{78.12}$$

The R value for two specific light wavelengths can be plotted against a measured S_aO_2 value, as shown in Figure 78.9. A linear approximation can then be used to calculate a S_pO_2 value. The empirical linear approximation for Figure 78.9 is listed in Equation 78.13.

$$S_pO_2 = 110 - 25R \tag{78.13}$$

The empirical approximation is used to correct for errors in the measured values. Pulse oximeters currently on the market use weighted moving average (WMA) techniques to identify the transmitted AC and DC DPP components. The DC component is the averaged signal intensity. The AC signal is computed using the WMA as a bandpass filter to single out only the AC cardiac signal.

There are many areas where pulse oximetry has limitations. One limitation was already mentioned, and that is the assumption of only two substances in hemoglobin. For most measurements, the percentage of these substances is small enough not to affect the pulse oximetry measurements. There are cases in emergency care when these substances may be present and do affect readings. At present, the user must be aware of the limitations and not use the pulse oximetry reading if the person has any of the additional substances. The solution to this problem is to use additional light sources to calculate the additional substance. The first pulse oximeters tested used a multiple-wavelength light source and could easily be implemented in new instruments.

A second limitation of pulse oximetry is background light. Because pulse oximetry currently uses transmitted light, the photodiode receiver is susceptible to ambient light. Ambient light can be from indoor lighting, sunlight, or phototherapy lights. To correct for this problem, a third light measurement can be collected with no light source, and subtracted from the transmitted intensity. Using a digital microprocessor, the subtraction can be easily performed.

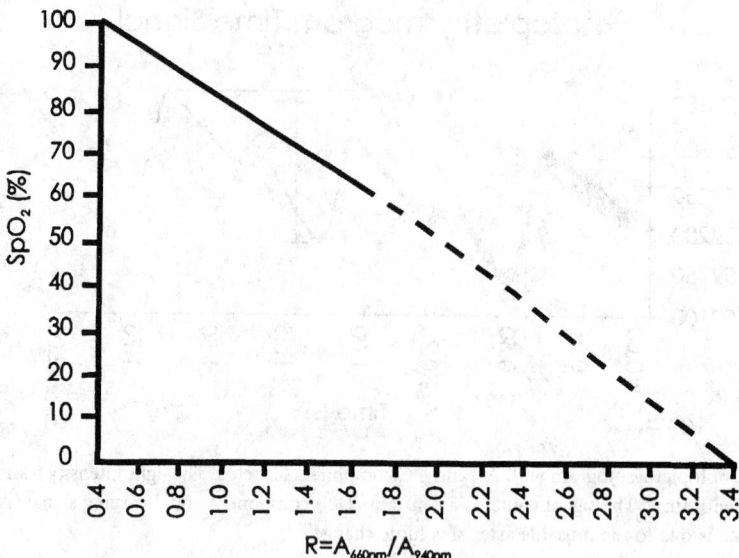

FIGURE 78.9   Measured oxygen saturation vs. *R* value. Empirically derived 660 to 940 nm ratio and oxygen saturation. The values are used to formulate a first-order equation to calculate the pulse oximetry saturation.

A third limitation is called low **perfusion**. Pulse oximetry is based on having a pulsatile signal. If the pulsatile signal is small compared with the DC signal, usually 1 count in 1000, the *R* value calculation becomes inaccurate. There are two reasons for the inaccuracy, round-off error and resolution. The round-off error can be compensated for by using more precision during the *R* value calculations.

Another limitation, which is difficult to eliminate, is motion artifact and autonomic nervous system response. Whenever there is an autonomic nervous system action, there is a transition or movement in the pulse oximetry signal. The most common technique to correct for motion artifact is averaging of consecutive measurements. Averaging works, but it slows the response time and lengthens the processing time. Another improvement would be to identify and eliminate inaccurate readings. This approach, however, is difficult to implement [7,21–23].

Reflectance Oximetry

An alternative technique to consider for DPP is reflectance probe measurements, as shown in Figure 78.10. Reflectance pulse oximetry has been addressed for *in vivo* studies by Mendelson and Ochs [24]. Because reflectance probes can be placed flat on the measurement area, they provide better shielding than a probe

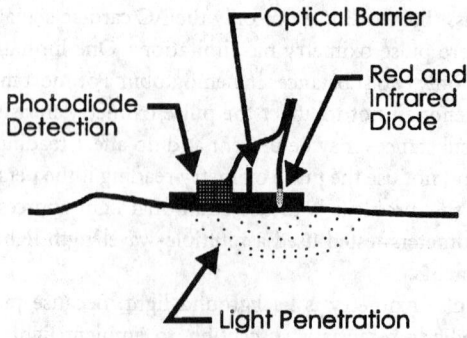

FIGURE 78.10   Reflectance oximetry probe. Red and infrared diodes illuminate the surface alternately and the reflected light is detected in close proximity by a photodiode. An optical barrier is required to eliminate direct light detection.

placed across a finger. However, reflectance measurements have less intensity than transmission measurements. Reflectance probe location and temperature effects have been analyzed. Increased temperature results in perfusion and increased signal intensity.

Schmidt et al. [25] reported an integrated circuit–based optical sensor for *in vivo* surface measurements. A set of equations was derived to estimate the amount of reflectance at a given intensity. The solution is a three-wavelength reflectance probe, with a red LED wavelength emission of 660 nm and a near and far infrared LED. The reflectance depth was calculated to be within 2 mm of the surface of the probe. The third wavelength is used to eliminate errors in measurement due to additional hemoglobin derivatives. The third wavelength was used to calculate a hematocrit value. The saturation was calculated using Equation 78.14.

$$SO_2 = A - B\left(R_{805}/R_{660}\right) \qquad (78.14)$$

A and B are constants empirically derived for a specific light source and detector under specific physiological conditions.

A current method to measure S_vO_2 is by reflectance oximetry methods. A bundled fiber-optic cable is placed through a catheter. Light enters by a fiber or fibers and the reflected light is analyzed for oxygen content using spectrophotometry. The probe is similar to the fluorescence probe shown in Figure 78.4 with no chemistry envelope. The reflectance is used to determine the oxygen concentration empirically [24–26].

pH Measurements

Blood pH is very important in sustaining life. There is a small range of values allowable to maintain life, and pH is one of the most tightly regulated parameters in the body. Typical pH measurements are take invasively either through a catheter probe or as a blood sample measurement. pH is noted as the negative decade logarithm of the molal activity of hydrogen ions. The acid and alkaline hydrogen reactions are given in Equations 70.15 and 70.16, respectively.

$$O_2 + 4H^+ + 4e \rightarrow 2H_2O \qquad (78.15)$$

$$O_2 + 2H_2O + 4e \rightarrow 4OH^- \qquad (78.16)$$

The pH level can be determined by electrochemical means. The original pH measurements were performed using hydrogen electrodes after the oxygen (O_2) was eliminated. Current electrochemical measurements use a glass electrode sensitive to hydrogen and a concentrated KCl bridge.

Optical measurements of pH are performed fluorescently. It was determined that pH corresponds to CO_2 tension. Both PCO_2 and pH can be measured using a dye sensitive to hydrogen ions and using a variant of the Henderson–Hasselbalch equation, shown in Equation 78.17.

$$I_0/I = 1 + K * H^+ \qquad (78.17)$$

I is the relative fluorescence intensity, H^+ is the hydrogen ion concentration, and I_0 and K are calibration constants.

pH fluorescence measurements can be obtained using a pH-sensitive dye, such as phenol red. The basic form of phenol red is green absorbing, and the acidic form is blue absorbing. Exciting the phenol red buffer with a green 560 nm and red 600 nm light, the ratio of intensities can be used to calculate the pH according to Equation 78.18.

$$R = k * 10^{\left[-c\big/\left(10^{-\delta}+1\right)\right]} \qquad (78.18)$$

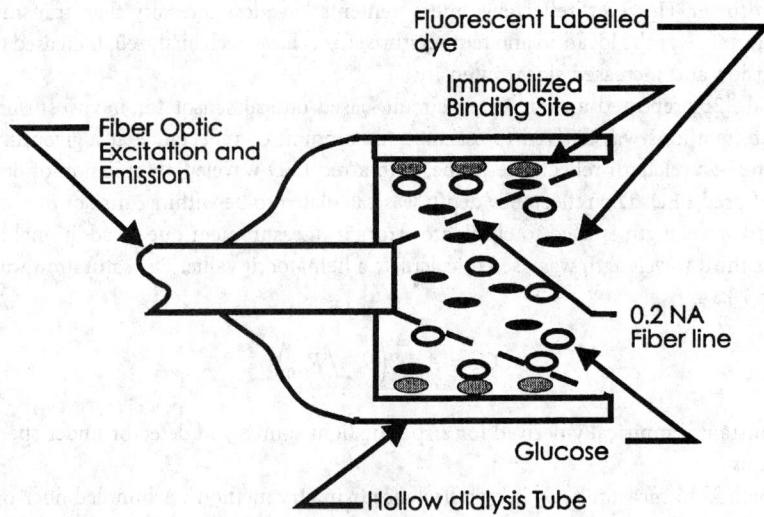

FIGURE 78.11 Optical glucose probe. A hollow tube is used to localize fluorescent dye outside the numerical aperture of the fiber. The fiber is used both to introduce excitation light and to collect emitted fluorescent levels. Only dye in the fiber view is illuminated. Dye bound to the hollow tube is not illuminated and is not detected.

k is the system optical constant, C is the green base form intensity, and δ is the difference between the pH and pK of the dye. Intensity shifts are accounted for by measuring the difference in intensity of two different excitation wavelengths. An alternative two-wavelength approach using hydroxypyrene trisulfonic acid (HTA) can be used. The basic form of HTA has a maximum excitation at a wavelength of 460 nm and an acidic maximum excitation wavelength of 410 nm. The ratio of the two fluorescent intensities at 520 nm is used to calculate the pH value [2,6,16,27].

Glucose Measurement

A common **glucose** measurement technique is by enzymatic amperometric measurements. Through selective binding, glucose and other compounds can be measured. The bound glucose changes the conductance and the glucose concentration can be determined.

Another more recent and interesting measurement technique is enzymatic optical measurements. Competitive binding between glucose and a fluorescein-labeled analog is used to determine glucose concentrations. An immobilized binding site is used to bind a fluorescent material, such as fluorescein-labeled dextran. A diagram of a glucose sensor is shown in Figure 78.11. The binding sights are fixed outside the excitation illumination. As the glucose concentration increases, the bound fluorescent material is released into the optical path. As the concentration of glucose increases, the viewed fluorescence increases and the reflected intensity increases. The reflected intensity will follow the concentration of glucose, and the actual response is measured empirically. The probe operation is reversible, allowing the probe to be reused. Reversibility is a requirement for implantable probes, which is the intended application for the optical glucose probe [27,28].

Electrolyte Concentration Measurements

In the clinical setting, it is important to monitor the concentration of various **electrolytes**. Typical measurements include sodium (Na^+), calcium (Ca^{++}), potassium (K^+), chloride (Cl^-), and magnesium (Mg^{++}). The concentrations are typically invasively drawn samples measured in a clinical laboratory. The measurement is performed using an ion-specific electrode, which is similar to the Clark electrode described previously. The clinical significance of different electrolytes is outlined in the text by

McClatchey, *Clinical Laboratory Medicine* [9]. Optical measurement techniques have been performed, but are not typically used in the clinical setting and are not covered here.

78.4 Combined Analysis Techniques

There are many combinations of instruments available. As instruments become more accurate and smaller, the trend is to simplify measurements and increase usefulness of equipment. Several selected instruments are described which have combined blood chemistry measurements into one unit.

CO-Oximetry

CO-oximetry has been noted for some time as the gold standard for oximetry readings. The more typical CO-oximeter does not measure just oxygen saturation, but several hemoglobin concentrations. For example, the Instrumentation Laboratory, Inc., IL 282 CO-oximeter uses an invasively drawn sample and measures total hemoglobin, oxyhemoglobin, carboxyhemoglobin, methemoglobin, and oxygen content in blood. The measurement technique is optical absorption. Four wavelengths at 535, 585.2, 594.5, and 626.6 nm are generated using a hollow cathode lamp. The extinction coefficients are used to calculate the concentration of RHb, O_2Hb, COHb, and MetHb. The total hemoglobin (tHb) concentration is calculated as the sum of the individual hemoglobin groups. The oxygen saturation (S_aO_2) is calculated as the concentration of oxyhemoglobin divided by the total concentration of hemoglobin. The total hemoglobin concentration, however, is usually noted as only the oxyhemoglobin concentration divided by the concentration of reduced hemoglobin and oxyhemoglobin. Therefore, care must be used in comparisons, because variations in S_aO_2 calculations can occur. The oxygen content is calculated as 1.39 times the concentration of oxyhemoglobin [29].

Intra-Arterial Probes

Intra-arterial probes have drawn much interest. They allow continuous or periodic measurements in the clinical setting in real time. These probes also offer minimally invasive measurements. The intent is to use intraarterial measurements on patients who require catheterization. The goal is a small probe that can be inserted into a catheter and not interfere with administering therapies or blood sampling. An optode-type probe is described by Shapiro et al. [30,31]. A fiber-optic probe is described by Gehrich et al. [15] for blood gas monitoring. The probe is designed to measure pH, PCO_2, and PO_2. In addition, a thermocouple is used to adjust for temperature variations, and values are normalized to standard temperatures. The sample drawing of the probe is shown in Figure 78.12. A set of chemical optical fluorescent probes is used.

78.5 Evaluation of Technology

There have been numerous measurement types and techniques used in developing an understanding of blood chemistry. The diverse background has resulted in very good indicators of how blood is used to sustain life. Despite the diverse background, many changes and advances are occurring at an impressive rate. Immediate improvements are being tested and implemented on a regular basis, and novel approaches will be seen in the near future. In addition, there is an explosion of studies being performed to understand how the body works and is put together. A brief overview of some improvements that can be expected follows.

Some basic engineering changes could be performed to enhance accuracy. Oximetry measurements could easily see engineering improvements. New techniques have been tested, but approval and acceptance are slow. There has been progress made in size, with new oximeters that can clip on the finger. The accuracy has seen only moderate improvement. Alternative probes and algorithms have been tested and even patented, but have not been marketed. One example of a patent is for Fourier analysis of pulse

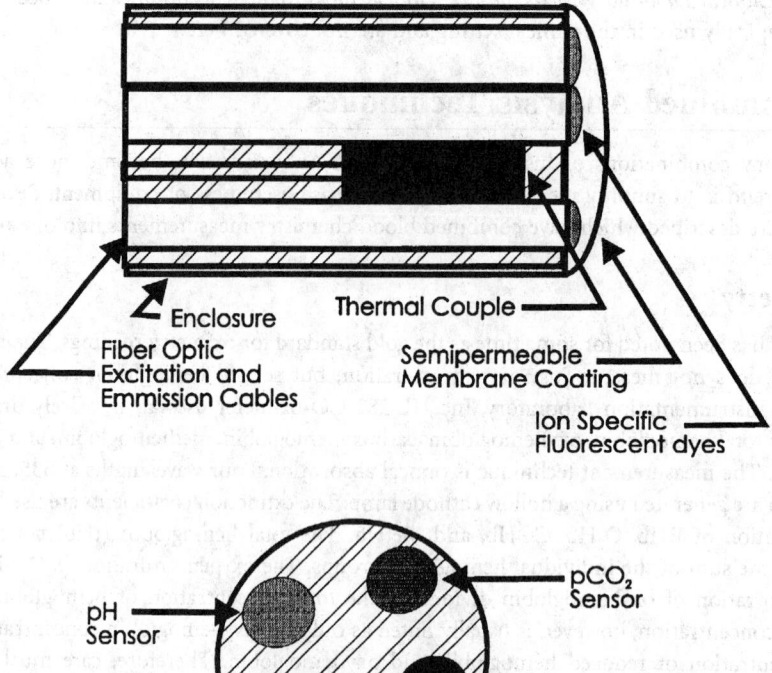

FIGURE 78.12 Combined fiber-optic sensor for pH, PCO_2, PO_2, and temperature correction measurements. Each sensor is bundled into a single enclosure and has separate detection areas.

oximetry signals for arterial saturation measurements. The spectral analysis would allow the possibility to implement some novel filter and analysis routines. This would improve stability from motion artifact. Hardware changes could also bring changes. Resolution could be improved by using more accurate photodiodes and A/D converters, or by using laser diodes for more intense and accurate light illumination. Reflectance probe advances would improve versatility.

The future of blood chemistry should see continued miniaturization. Fiber optics have become a major research direction. Fiber optics allow remote sensing and have reduced the possibility of electrical and magnetic complications for patient monitoring. Miniature fiber-optic sensors are being developed for numerous applications, and the improvement and developments can be applied to biological readings. Miniaturization not only reduces size, but can also improve stability. Small probes also allow multiple readings in one sensor. This trend has already started with blood gas probes being used for intravascular and extravascular measurements. The next trend envisioned is nanoprobes. Implantable glucose probes are already being tested. Sensor and light sources can be implemented on an integrated circuit and continue to shrink. Microsensors are already being used in biomedical application, such as blood pressure sensors and single-cell neural stimulators. Miniature chemistry and thermal-cycling laboratories are being developed into single integrated chips. The same chemical and optical arrangements can be performed for blood chemistry analysis. A complete laboratory measurement system could be at the patient location for continuous monitoring [28–30,32].

The other major anticipated change is the trend for additional and more accurate noninvasive readings. Invasive measurements can mean delays and noncontinuous monitoring, as well as increased possibility of infection. Given equal measurement accuracy, noninvasive measurements are an obvious choice over invasive readings.

It is important to note that measurement techniques will continue to improve, but the rate of growth and utilization will not grow as fast. Acceptance and utilization are dependent on cost and the effort involved in using the instruments. Cost can slow or even eliminate instrumentation from being used. However, with the increasing cost of medical care, there is interest in providing fast diagnosis and recovery for patients. This will provide incentive for improving measurement technique and accuracy.

Defining Terms

Artery: Vessel used to carry blood away from the heart.

Capillary: Semipermeable membrane use to exchange oxygen and other substances with tissue.

Catheter: Surgical instrument inserted into the body for drawing or administering fluids.

Electrolyte: A substance that dissolves into ions, becoming capable of conducting electricity.

Fluorescence: The property of emitting light when exposed to an excitation light.

Glucose (dextrose): Substance found in normal blood, which is the main source of energy in living tissue.

Hematocrit: The total concentration of hemoglobin, or packed red cells, in blood.

Hemoglobin: Oxygen-carrying pigment found in blood, which has the property of reversible oxygenation.

Intra-arterial: Within the artery.

Invasive: Pertaining to the insertion of an instrument into the body.

Oximetry: Photoelectric determination of arterial blood oxygen saturation (S_aO_2).

Perfusion: Pertaining to the passage of fluid through vessels. Low perfusion refers to reduced blood flow.

Photoplethysmogram: Chart of the volume change in an organ or limb. In oximetry it is the chart of the light passed through an ear, nose, or digit. The light intensity varies with the arterial pulse, which is related to the arterial volume.

Transcutaneous: Pertaining to entering through the skin.

Vein: Vessel used to return blood to the heart.

References

1. F.W. Cheney, The ASA closed claims study after the pulse oximeter, *ASA Newslett.* 54: 10–11, 1990.
2. J.W. Severinghaus and P.B. Astrup, History of blood gas analysis II: pH and acid-base measurements. *J. Clin. Monitoring*, 1: 259–277, 1985.
3. J.W. Severinghaus and P.B. Astrup, History of blood gas analysis IV: Leland Clark's oxygen electrode, *J. Clin. Monitoring*, 2: 125–139, 1986.
4. J.W. Severinghaus and P.B. Astrup, History of blood gas analysis V: oxygen measurement, *J. Clin. Monitoring*, 2: 175–189, 1986.
5. J.W. Severinghaus and P.B. Astrup, History of blood gas analysis VI: oximetry, *J. Clin. Monitoring*, 2: 270–288, 1986.
6. J.W. Severinghaus and P.B. Astrup, History of blood gas analysis, *Int. Anesth. Clin.*, 25 (4): 1–214, 1987.
7. K.K. Tremper and S.J. Barker, Advances in oxygen monitoring. *Int. Anesth. Clin.*, 25 (3): 1–96, 113–208, 1987.
8. *Churchhill's Illustrated Medical Dictionary*, Churchhill Livingston, New York, 1989.
9. K.D. McClatchey, *Clinical Laboratory Medicine*, Williams & Wilkins, Baltimore, MD, 1994.
10. R.W. Burnett, A.K. Covington, N. Fogh-Andersen, W.R. Kulpmann, A.H. Maas, O. Muller-Plathe, O. Siggaard-Andersen, A.L. Van Kessel, P.D. Wimberly, and W.G. Zijlstra, International Federation of Clinical Chemistry, *Eur. J. Clin. Chem. Clin. Biochem.*, 33: 247–253, 399–404, 1995.
11. J.W. Gilbert, F.P. Holladay, and H.C. Weiser, Hematocrit monitor, *Crit. Care Med.*, 17 (9): 929–966, 1989.
12. L.C. Clark, Measurement of oxygen tension, *Crit. Care Med.*, 9 (10): 690–693, 1981.
13. D.W. Lubbers, Oxygen measurement in blood and tissues and their significance: methods of measuring oxygen tension of blood and organ surfaces. *Int. Anesth. Clin.*, 4 (1): 103–122, 1966.

14. D.W. Lubbers, F. Hannebauer, and N. Opitz, Continuous transcutaneous blood gas monitoring: PCO_2-optode, fluorescence fotometric device to measure the transcutaneous PCO_2, *Birth Defects Orig. Art. Ser.*, XV (4): 123–126, 1979.

15. J.L. Gehrich, D.W. Lubbers, N. Opitz, D.R. Hansmann, W.W. Miller, J.K. Tusa, and M. Yafuso, Optical fluorescence and its application to an intravascular blood gas monitoring system, *IEEE Trans. Biomed. Eng.*, BME-33: 117–131, 1986.

16. R. Narayanaswamy, Current developments in optical biochemical sensors, *Biosensors Bioelectron.*, 6 (6): 467–475, 1991.

17. J.F. O'Riordan, T.K. Goldstick, J. Ditzel, and J.T. Ernst, Characterization of oxygen-hemoglobin equilibrium curves using nonlinear regression of the Hill equation: parameter values for normal human adults, in H.I. Bicher and D.F. Bruley, Ed., *Oxygen Transport to Tissue — IV*, Plenum Press, New York, 435–444, 1983.

18. F.J.W. Roughton and J.W. Severinghaus, Accurate determination of O_2 dissociation curve of the human blood above 98.7% saturation with data on O_2 solubility in unmodified human blood from 0° to 36°C, *J. Appl. Physiol.*, 35: 861–869, 1973.

19. J.W. Severinghaus, Simple accurate equations for human blood O_2 dissociation computations, *J. Appl. Physiol. Respir., Environ Exercise Physiol.*, 46: 599–602, 1979.

20. E.H. Wood and J.E. Geraci, Photoelectric determination of arterial oxygen saturation in man. *J. Lab. Clin. Med.*, 34: 387–401, 1949.

21. J.W. Severinghaus and J.F. Kelleher, Recent developments in pulse oximetry, *Anesthesiology* 76: 1018–1038, 1992.

22. J.A.H. Bos, W. Schelter, W. Gumbrecht, B. Montag, E.P. Eijking, S. Armbruster, W. Erdmann, and B. Lachmann, Development of a micro transmission cell for *in vivo* measurement of S_aO_2 and Hb, in *Oxygen Transport to Tissue XII*, Plenum Press, New York, 47–52, 1990.

23. Y. Mendelson and J.C. Kent, An *in vitro* tissue model for evaluating the effect of carboxyhemoglobin concentration on pulse oximetry, *IEEE Trans. Biomed. Eng.*, BME-36: 625–627, 1989.

24. Y. Mendelson and B.D. Ochs, Noninvasive pulse oximetry utilizing skin reflectance photoplethys-mography, *IEEE Trans. Biomed. Eng.*, BME-35: 798–805, 1988.

25. J.M. Schmitt, J.D. Meindl, and F.G. Mihm, An integrated circuit-based optical sensor for *in vivo* measurement of blood oxygenation, *IEEE Trans. Biomed. Eng.*, BME-33: 89–107, 1986.

26. W. Cui, L.E. Ostrander, and B.Y. Lee, *In vivo* reflectance of blood tissue as a function of light wavelength, *IEEE Trans. Biomed. Eng.*, BME-37: 623–639, 1990.

27. J.I. Peterson and G.G. Vurek, Fiber-optic sensor for biomedical applications, *Science*, 224 (4645): 123–127, 1984.

28. J.S. Schultz, S. Mansouri, and I.J. Goldstein, Affinity sensor: a new technique for developing implantable sensors for glucose and other metabolites, *Diabetes Care*, 5 (3): 245–253, 1982.

29. L.J. Brown, A new instrument for the simultaneous measurement of total hemoglobin, % oxyhe-moglobin, % carboxyhemoglobin, % methemoglobin, and oxygen content in whole blood, *IEEE Trans. Biomed. Eng.*, BME-27: 132–138, 1980.

30. B.A. Shapiro, R.D. Cane, C.M. Chomka, L.E. Bandala, and W.T. Peruzzi, Preliminary evaluation of an intra-arterial blood gas system in dogs and humans, *Crit. Care Med.*, 17 (5): 455–460, 1989.

31. B.A. Shapiro, Clinical and economic performance criteria for intra-arterial and extra-arterial blood gas monitoring with comparison with in vitro testing, *Am. J. Clin. Pathol.*, 104 (4 Suppl. 1): S100–S105, 1995.

32. C. Ajluni, Microsensors move into biomedical applications, *Electron. Design*, 44(11): 75–84, 1996.

33. J.G. Webster, Ed., *Design of Pulse Oximeters*, IOP Publishing, Bristol, U.K., 1997.

79

Medical Imaging

James T. Dobbins III
Duke University Medical Center

Sean M. Hames
Duke University Medical Center

Bruce H. Hasegawa
University of California, San Francisco

Timothy R. DeGrado
Duke University Medical Center

James A. Zagzebski
University of Wisconsin, Madison

Richard Frayne
University of Wisconsin, Madison

79.1   Introduction

Medical imaging has advanced considerably since the discovery of X-rays by Wilhelm Conrad Röntgen in 1895. Today, in addition to the continued use of X-rays for medical diagnosis, there are imaging methods that use sound (ultrasound), magnetic fields and radio waves (magnetic resonance imaging), and radionuclides (nuclear medicine). Both projection imaging and cross-sectional imaging are routinely used clinically. This chapter will describe the principles behind the various imaging modalities currently in use, and the various measurements routinely made with them.

79.2   Image Information Content

The vast majority of imaging procedures are qualitative in nature, where it is the visual presentation of anatomy that is the measurement outcome. There are also some quantitative measurements, which will be discussed in the section on nuclear medicine. However, since most imaging is concerned only with the qualitative nature of the image, a description of the salient features of image content follows.

There are three primary physical parameters of interest in image content: contrast, noise, and resolution. If these three features are known for a given image (or imaging system), then the entire physical nature of the image has been characterized. There are also psychovisual effects, such as conspicuity [1], which affect the ability of the observer to detect a particular feature, but these issues are difficult to quantitate and are outside the scope of this handbook.

The first of the physical image features, contrast, is defined as the fraction of the total image signal occupied by a particular object:

$$C = \frac{B-S}{B} \tag{79.1}$$

where S is the signal in the area of interest and B is the background signal. Contrast is determined by the properties of the object being imaged, the imaging modality, the properties of the image detector, postprocessing of the image (such as by digital processing), and the contrast of the display device.

Image noise is a measure of the stochastic nature of the image. All physical measurements, including medical images, contain a certain degree of uncertainty. In X-ray imaging, for example, the physics of X-ray production dictates that the number of X-rays incident on a unit area per unit time are random, and given by a statistical distribution known as the Poisson distribution. The greater the image noise, the less likely it is that one will observe a given object. There is a relationship between the image noise, the contrast and area of an object, and its likelihood of being observed. This is summarized in the Rose model:

$$N = \frac{k^2}{C^2 A} \tag{79.2}$$

where N is the number of quanta (such as X-rays) per unit area needed to discern an object of contrast C and area A, assuming a signal-to-noise of k. Rose found that a signal-to-noise ratio of 5 is typically required to detect a visual object reliably [2].

Resolution is the ability of an imaging system to record faithfully the range of spatial detail in an object. Recording objects with finer spatial detail requires "sharper" imaging detectors. The resolving ability of a detector is largely determined by its point-spread function. The point-spread function describes how well the imaging apparatus can record an infinitesimal point object. No detector is perfectly sharp, and some spread of the infinitesimal dot occurs — the worse the spread, the less resolving the system.

Measurement of Imaging Performance

A linear-systems approach is typically used to quantify the performance of an imaging system. The relations among contrast, noise, and resolution of an imaging system are customarily described by two functions: the modulation transfer function (MTF) and the noise power spectrum (NPS), both of which are functions of spatial frequency. The MTF is the Fourier transform of the point-spread function, and describes the inherent deterministic frequency response of the system. The NPS (also referred to as the Wiener spectrum) is proportional to the square of the Fourier transform amplitude at each frequency, and represents the variance associated with noise in the system at each particular spatial frequency. The ratio of MTF and NPS, properly normalized, is the noise equivalent quanta (NEQ), which is the square of the maximum available signal-to-noise at each spatial frequency u:

$$NEQ(u) \equiv \frac{\left(\text{large area signal}\right)^2 \cdot MTF^2(u)}{NPS(u)} \tag{79.3}$$

If the NEQ is divided by the number of incident quanta per area (e.g., the number of X-ray photons incident on the detector in X-ray imaging), the result is the detective quantum efficiency (DQE). The zero-frequency DQE is a measure of the fraction of incident quanta effectively used by the system. Alternatively, the DQE may be viewed as the efficiency with which the system utilizes the available signal-to-noise at each spatial frequency.

The actual measurement of MTF, NPS, and DQE is quite tedious, and will be only briefly summarized here. The interested reader is encouraged to consult the suggested references for the appropriate detail on these measurements. Examples of these measurements will be given for X-ray imaging.

The MTF is typically measured by imaging either a very fine slit (typically 10 to 20 µm) [3-6] or an edge [7]. The profile across the slit image is called the line-spread function (LSF). The Fourier transform of the LSF gives the MTF in the direction perpendicular to the slit. The derivative of values along the edge-response function also gives the line-spread function. Detector response typically varies with energy so it is important to specify the conditions under which MTF is measured. With X-ray imaging it is typical to use a tube voltage of 70 kV with 0.5 mm Cu filtration placed in the beam to simulate the filtering of the X-ray spectrum expected from a patient, although other measurement techniques are also found in the literature.

The NPS is measured by taking an image of a flat field, where there is no structure in the image other than noise. Contemporary methods of NPS measurement on digital systems perform a two-dimensional Fourier transform on the flat-field image [6], although when measuring the NPS of film a scanning slit is used to generate a one-dimensional NPS parallel to the direction of slit movement [8-10]. After appropriate scaling, the square of the amplitude of the two-dimensional Fourier transform is the NPS. There are many details related to measuring the NPS properly, including eliminating background trends, and the size of the region over which the Fourier transform is taken. These are all covered in detail in the references [6,8-18].

Measurement of imaging properties is easier on digital imaging systems than on film, since film must first be digitized at appropriately fine sampling intervals or else corrected for the use of one-dimensional slits [19]. However, the effects of aliasing (fictitious frequency response in a digital system due to limited sampling) makes the interpretation of MTF and NPS in digital systems more difficult than with film [11,12,20].

79.3 X-Ray Imaging

X-ray imaging requires an X-ray-generating apparatus (tube, high voltage supply, and controls) and an appropriate X-ray detector. Typical X-ray detectors include photographic film (almost always used in concert with a fluorescent screen), image intensifiers, computed radiography phosphor plates, and newer dedicated digital detectors.

The X-ray generator is basically a high-voltage step-up transformer with appropriate rectification and control circuitry. Most contemporary generators are three-phase 12-pulse, full-wave rectified to give a very low voltage ripple (3 to 10%) [21]. For procedures requiring very fast pulses of several milliseconds or less (such as coronary angiography), a tetrode-based constant-potential generator is used. The operator selects the tube kilovoltage, tube current, and exposure time appropriate for the examination of interest.

X-ray tubes contain a heated filament (which serves as the cathode) and an anode made of a tungsten/rhenium combination for conventional use or molybdenum for mammography. With the exception of dental tubes, modern clinical X-ray tubes almost always contain a rotating anode to spread the heat out over a larger area, allowing for a greater tube output without damaging the anode. Many tubes contain two filaments, a large one and a small one, depending on tube output and resolution requirements of a particular exam. Measurements on X-ray tubes and generators involve calibrations to assure that kilovoltage, tube output, and exposure time are in good agreement with the control console settings [21]. Calibration of the high voltage is done by commercially available voltage dividers, or by specially designed X-ray film cassettes with calibrating filters inside. Tube output is measured by ion chambers, and exposure time is measured either by a rotating-arm timer test tool placed over a film cassette during an X-ray exposure or by direct plotting of the exposure vs. time output of an ion chamber.

X-Ray Imaging Detectors

The most common detector for X-ray imaging is film. X-ray film is typically placed in a sandwich between two fluorescent screens (or one screen in mammography for improved visibility of small detail). Contemporary screens are made of rare earth compounds such as Gd_2O_2S, and serve to convert the X-rays to visible light which exposes the film more efficiently than X-rays alone, thus reducing patient radiation dose. The response of these screen–film combinations has good contrast at intermediate exposure ranges (as given by the film γ, or contrast ratio), but poor contrast at low or high exposures. The contrast and latitude of films are described by the characteristic curve (often referred to as the Hurter–Driffield, or HD curve). Appropriate screen–film combinations are chosen based on the anatomy to be imaged, since screen–film combinations are designed with different contrast, latitude, and exposure sensitivity characteristics [22].

A second type of X-ray detector is the image intensifier, which is used with fluoroscopy. Fluoroscopy uses a low-exposure-rate X-ray output to image a patient continuously, typically to properly position the patient for a subsequent high-exposure film image. The image intensifier (Figure 79.1) comprises a cylindrical glass enclosure, inside of which is an input screen, photocathode, focusing electrodes, accelerating anode, and output screen [21]. The X-rays are absorbed in the input screen (typically CsI), giving off light which liberates electrons from the photocathode. The photoelectrons are then accelerated to the output screen where they strike the output phosphor screen with high energy (~30 keV), giving off a bright light, which is viewed by either a video camera or motion-picture (cineradiographic) camera.

A recently developed digital X-ray detector is the photostimulable phosphor, which is referred to commonly as computed radiography [23-25]. This detector uses a special type of phosphor which stores about half of the absorbed X-ray energy in metastable states, which are read out later by laser scanning. The laser light stimulates the phosphor to emit ultraviolet light in proportion to the original X-ray exposure. The photostimulated light is then detected with a photomultiplier tube (PMT) or solid-state photodetector and digitized. The clinical apparatus (Figure 79.2) first does a prescan of the imaging plate to adjust the input range of the analog-to-digital converter based on the image histogram; the digitized signal is then logarithmically transformed and stored, displayed on a video monitor, or printed on film following optional spatial filtering and contrast adjustment.

There are also currently available or in development a variety of other digital X-ray detectors, including selenium plate detectors [26], CCD-camera detectors with fluorescent screens [27], and flat-panel arrays with amorphous silicon [28] or amorphous selenium [29] detector elements.

79.4 Computed Tomography

A diagnostic computed tomography (CT) scanner comprises an X-ray tube with collimation to provide the slice thickness, a linear array of detector elements, and a reconstruction computer. The X-ray tube and the detectors typically rotate in a gantry. The number of detectors used depends on the generation of the scanner. First-generation scanners had only one detector which was translated across the patient with the tube for each projection, and then the entire assembly was rotated to acquire the next projection view. To increase acquisition speed, second-generation scanners used several detectors in a limited fan-beam geometry. Third-generation CT scanners, which are the most common in use today, utilize a large fan array of detectors (852 elements in a current scanner), which completely encompasses the patient and allows slice acquisition times of about 1 s [21]. The X-ray tube and the detector fan array are mechanically coupled and rotate together at high speed (Figure 79.3).

The implementation of electronic slip rings, which allow continuous electric contact, has removed the physical restriction imposed by the high-voltage cables of earlier scanners. Fourth- and fifth-generation scanners have a stationary, complete ring of detectors (typically 1200 to 4800 detectors). In fourth-generation scanners the X-ray tube is rotated alone, while the fifth-generation scanner design has a focused electron beam which traverses multiple target rings. Fifth-generation scanners can acquire a slice

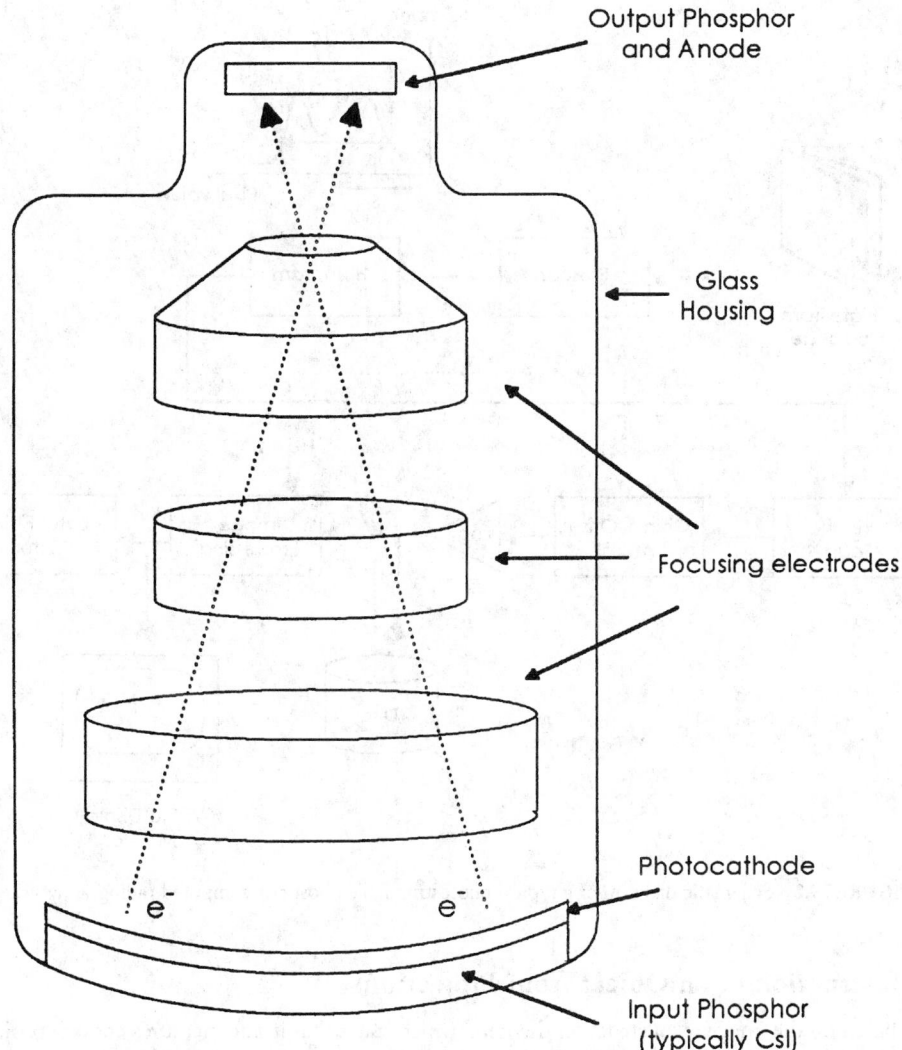

FIGURE 79.1 Schematic diagram of the major components of an image intensifier. The anode is typically at about 30 kV, and the three annular electrodes focus the beam and determine the usable area of the input surface for intensifiers having multiple formats.

fast enough (50 ms per slice and 17 slices per second) to stop cardiac motion [30,31]. These last two generations are not in common use, primarily due to high cost. In recent years, a helical-scan adaptation of third-generation scanners, allowing continuous acquisition of data over a large patient volume, has become clinically popular [32,33].

While the X-ray tubes used for CT (tube potential range 80 to 140 kV) are very similar to general radiographic tubes, the detectors are quite different from conventional radiographic detectors. Detectors used in CT are one-dimensional photon counters which must be efficient and fast. Early CT devices used scintillation detectors, which converted the X-ray energy into light photons that were counted by PMTs. Originally, single-crystal NaI was used, but it proved to be insufficient in dynamic range and had too much afterglow of scintillation light. High-pressure (25 atm) xenon gas later replaced NaI as the detector. Currently, many CT scanners use scintillating ceramics (e.g., $CdWO_4$, $(Y,Gd)_2O_3$:Eu and Gd_2O_2S:Pr,Ce) coupled to photodiodes, due to the high bulk density of the ceramics.

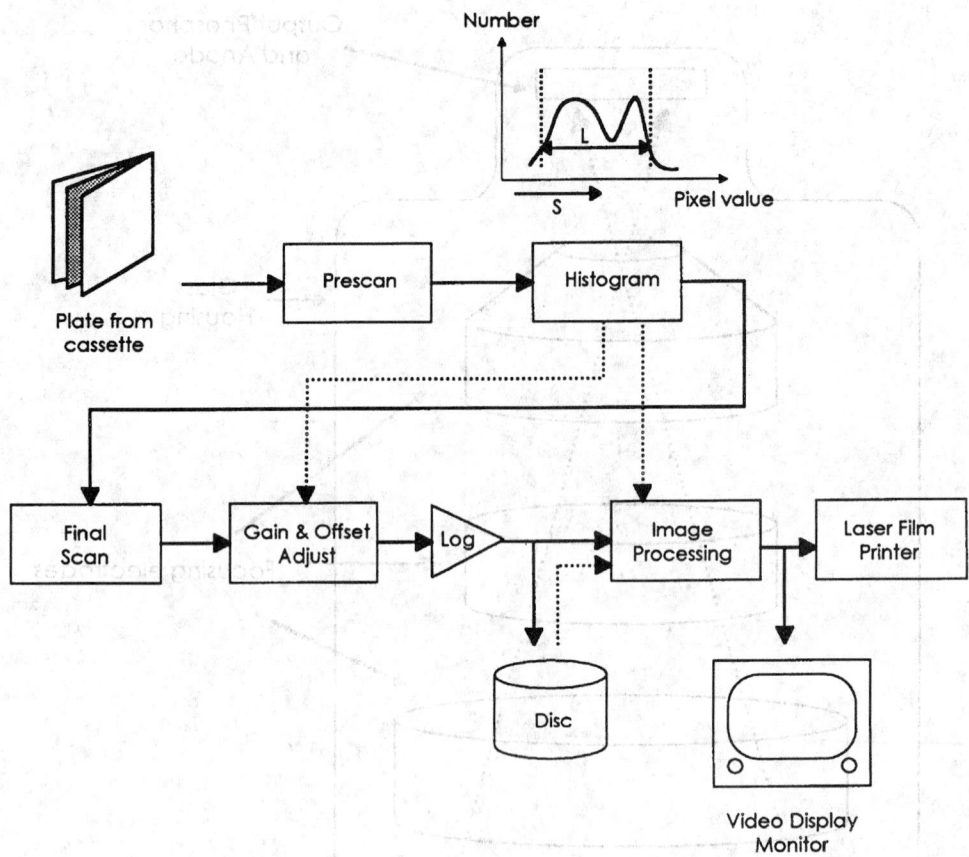

FIGURE 79.2　Schematic diagram of a typical photostimulable phosphor computed radiography system.

Reconstruction of an Object from Projections

CT is based on the image reconstruction theorem, which states that if one measures enough projections of an object, the two-dimensional distribution of that object may be reconstructed from the projection data. In CT the quantity of interest is the linear attenuation coefficient, μ, at each point in the object. The transmission of X-rays through an object of thickness x can be stated as

$$I\left(x\right)=I\left(0\right)e^{-\mu x} \tag{79.4}$$

where $I(0)$ is the incident intensity. Each ray from the focal spot of the tube to a discrete detector element is a measure of the line integral of the attenuation coefficient through the patient:

$$\lambda\left(x_r\right)=-\ln\frac{I\left(x_r\right)}{I\left(0\right)}=\int_\ell\mu\left(x_r,y_r\right)dy_r \tag{79.5}$$

where r represents the reference frame of one of the many projections through the patient.

Image reconstruction requires a method to invert Equation 79.5, in order to extract $\mu(x,y)$ of the object from the measured projection views, λ. The mathematical principles of image reconstruction from an

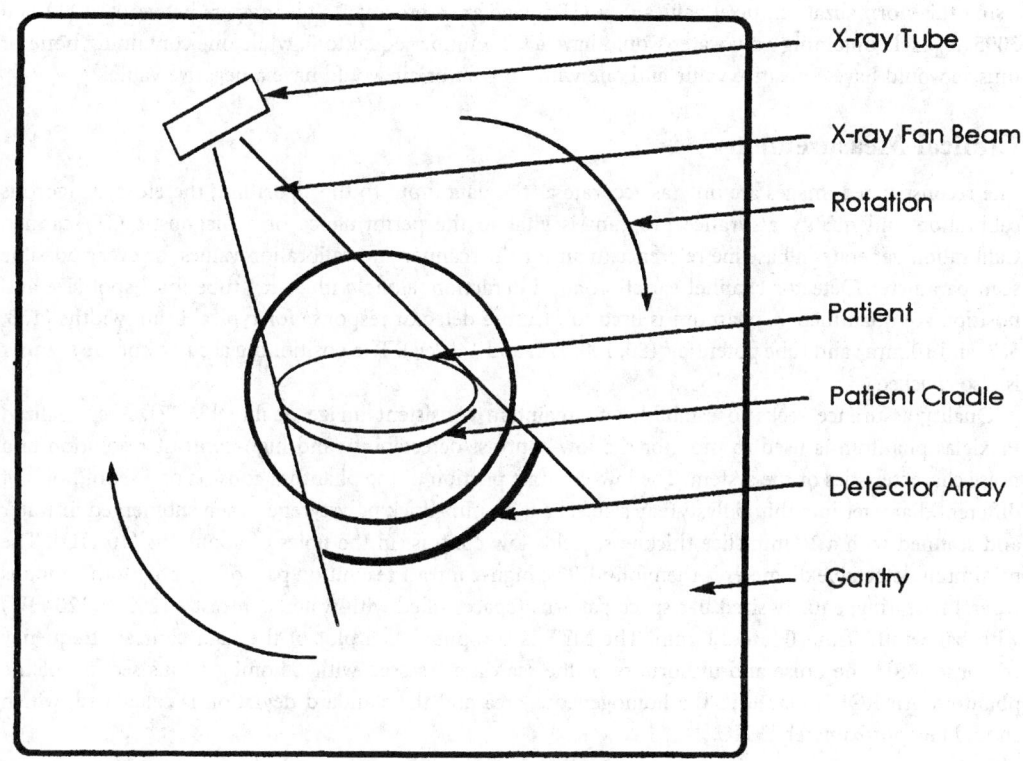

FIGURE 79.3 Orientation of components in a typical third-generation CT scanner.

infinite number of projections through an object were developed by Radon in 1917 [34]. An approximate solution to the Radon inversion, known as backprojection, was later developed because of the need for rapid computation of images in clinical CT. Backprojection involves smearing the data from each projection through the two-dimensional space of the patient, and summing over all projections. Simple backprojection yields an estimate of the patient structures, but is plagued by artifacts due to the approximate nature of the reconstruction procedure. These artifacts are successfully removed, however, by prefiltering the projection data before backprojecting. The one-dimensional prefiltering is typically performed in frequency space by multiplying by a ramp function. This technique is known as filtered backprojection, and results in more accurate reconstructions of patient anatomy [35].

There are several conditions that can reduce the quality of image reconstruction. First, an insufficient number of angular projections or incomplete sampling of the object can lead to aliasing in the reconstructed view. Second, partial volume effects occur when the object is not of homogeneous composition in a particular voxel, causing the reconstructed pixel value (CT number) to be not representative of the tissue. Third, if the acquisition is not fast enough, patient motion leads to a ghosting artifact in the reconstructed image. Last, beam hardening occurs when a high-density structure, such as the skull, significantly changes the beam energy spectrum. The result is reduced intensity of adjacent structures. Beam hardening can be reduced by slightly altering the shape of the reconstruction filter to improve the reconstruction for a particular tissue type.

To present the reconstructed data in digital format, the CT number (also known as the Hounsfield unit, HU) was developed.

$$\text{CT number} \equiv 1000 \times \frac{\mu_{pixel} - \mu_{water}}{\mu_{water}} \tag{79.6}$$

Using this normalization, pixel values in a CT image are stored as 12-bit integers between −1000 and 3095. A pixel containing only water would have a CT number equal to 0, while one containing bone or muscle would have a positive value and one with only fat or air would have a negative value.

Clinical Measurements

The reconstructed images are only as accurate as the data input to the algorithm; therefore, a rigorous calibration and quality assurance program is vital to the performance of a diagnostic CT scanner. Calibration generates a baseline reference in air for the scanner and calibration values for every possible scan parameter. Detector channel variation and interaction, along with X-ray tube focal spot size and position are quantified. A phantom is used to measure detector response for typical beam widths (1, 3, 5, 7, and 10 mm) and tube potentials (80, 100, 120, and 140 kV). The positioning accuracy of the scanner is also checked.

Quality assurance seeks to establish and maintain consistent image quality [36,37]. A specialized Plexiglas phantom is used to monitor the low-contrast detectability and high-contrast resolution and noise characteristics of the system. The low-contrast portion of the phantom consists of a set of holes of different diameter in a thin polystyrene slab. The 0.75-mm-thick polystyrene, when submerged in water and scanned with a 10 mm slice thickness, yields low contrast in the holes of about 1% (10 HU). The minimum detectable diameter is then found. The high-contrast resolution part of the phantom contains several repeating, equally sized bar/space patterns (spaces filled with water, contrast ~12% or 120 HU) with bar widths from 0.5 to 1.6 mm. The MTF is computed as a plot of the high-contrast frequency response [38]. The noise and uniformity of the scan are assessed with a homogeneous section of the phantom. An ROI is placed in the homogeneous area and the standard deviation is calculated, which should be approximately 3 HU.

79.5 Nuclear Medicine

Nuclear medicine techniques [39,40] use radiopharmaceuticals which are injected into the body to monitor or measure physiological function. Central to nuclear medicine is the role of the radiopharmaceutical as a tracer, that is, an agent with a predictable physiological action that is introduced without perturbing the function of the system. An external detector is used to record radioactivity emanating from the patient to determine the spatial distribution (and often temporal changes in concentration) of the radiopharmaceuticals in specific organs or tissues. Each radiopharmaceutical has an expected biodistribution which a radiologist evaluates to diagnose the medical status of a patient. The radiopharmaceutical can be labeled either with positron-emitting radionuclides, which produce annihilation photons, or can be labeled with "single-photon" radionuclides which emit γ-rays (or sometimes X-rays). This section considers only single-photon-emitting radionuclides, examples of which are given in Table 79.1.

The scintillation camera [41,42] is the most common device for imaging the distribution of single-photon emitting radionuclides *in vivo* (Figure 79.4). The scintillation camera incorporates a large-field (e.g., 40 by 50 cm) position-sensitive photon detector with a collimator having a large number of small parallel holes (1 to 2 mm diameter, 4 cm length) so that only photons traveling perpendicular to the detector surface are recorded. Photons emitted by the patient and passing through the collimator are absorbed by a 1-cm-thick sodium iodide scintillator coupled to an array of PMTs. The PMT signals are processed to generate signals proportional to the (x,y)-coordinates of the interaction site of the photon in the crystal. In addition, the photomultiplier tube signals are integrated to calculate the photon energy. Events falling within a specified range (typically ±7.5%) around the expected radionuclide photon energy are recorded, whereas those outside of this range are rejected as unwanted scatter or background events. An image is integrated from individual events at the calculated position and specified energy, representing detected photons emitted by the radiopharmaceutical. The camera acquires a planar projection image of the radiopharmaceutical distribution in the patient with a spatial resolution of about 1 cm. The image also can be acquired tomographically by rotating the scintillation camera around the axis of the patient.

TABLE 79.1 Examples of Tracers Used in Nuclear Medicine

Process	Tracer	Ref.
Blood Flow		
Diffusible	^{133}Xe	57
	[^{99m}Tc]-HMPAO	58
Diffusible (trapped)	[^{123}I]IMP (brain)	58
	^{201}Tl (heart),	44,59
	[^{99m}Tc]-MIBI	47
Nondiffusible (trapped)	[^{99m}Tc]macroaggregated albumin, labeled microspheres	—
Effective renal plasma flow	[^{123}I]hippuran	—
Blood Volume		
Red blood cells (RBC)	[^{99m}Tc]-RBC	45,46
Plasma	[^{125}I]-albumin	—
Transport and Metabolism		
Free fatty acids	[^{123}I]-hexadecanoic acid	44
Bile	[^{99m}Tc]-HIDA	—
Osteoblastic activity	[^{99m}Tc]-MDP	43
Glomerular filtration rate	[^{99m}Tc]-DPTA	60
Molecular Diffusion	[^{99m}TcO$_4$]	58
Receptor Systems		
Dopaminergic	[^{123}I]-IBZM	61
Cholinergic	[^{123}I]-QNB	62
Adrenergic	[^{131}I]-MIBG	63,64
Somatostatin	[^{111}In]-octreotide	65,66

Adapted from Sorenson and Phelps.[42]

This technique is called single-photon emission computed tomography (SPECT) and produces cross-sectional images representing the radiopharmaceutical concentration within the patient.

Measurement of Physiological Function

Radionuclide images can be interpreted visually or quantitatively. For example, the radiopharmaceutical [^{99m}Tc]-methylene diphosphonate (MDP) is incorporated into the bone matrix by osteoblastic activity [43]. A radiologist will inspect a nuclear medicine image for sites demonstrating focal uptake of ^{99m}Tc-MDP to determine the extent and degree of trauma, inflammation, degeneration, metastatic disease, or other skeletal disease processes. Typically, ^{99m}Tc-MDP images are interpreted visually but are not analyzed to determine the quantity of radiotracer incorporated into the skeleton.

Other nuclear medicine studies are assessed quantitatively in the sense that values extracted from the image represent the radioactivity concentration (and physiological function) in a specific organ or tissue region. Myocardial perfusion imaging with [^{99m}Tc]-hexatris-2-methoxyisobutylisonitrile (MIBI) is an example of one such "quantitative" nuclear medicine study for assessing a patient suspected of having coronary artery disease [44-46]. [^{99m}Tc]-MIBI is a lipophilic cation which accumulates in myocardial tissue roughly in proportion to blood flow [47]. Image data are acquired using SPECT to reconstruct tomograms of the myocardial concentration of [^{99m}Tc]-MIBI which are analyzed to assess regional myocardial blood flow. Although absolute measurements (μCi/g) of tissue activity are difficult (if not impossible) to obtain with SPECT, the images are interpreted "quantitatively" by extracting pixel values from the image to derive diagnostic information [48], rather than relying on "qualitative" visual interpretation of the images. Typically, ^{99m}Tc-MIBI is imaged in the "short-axis" view which presents the left ventricular myocardium in a series of annuli (or "doughnuts"). The image is analyzed using a "circumferential profile" representing the radionuclide concentration at 6° angular increments around each annular slice of the myocardium [49]. The extracted values are compared with standard values obtained from patients in whom atherosclerotic disease has been excluded by coronary angiography, thereby

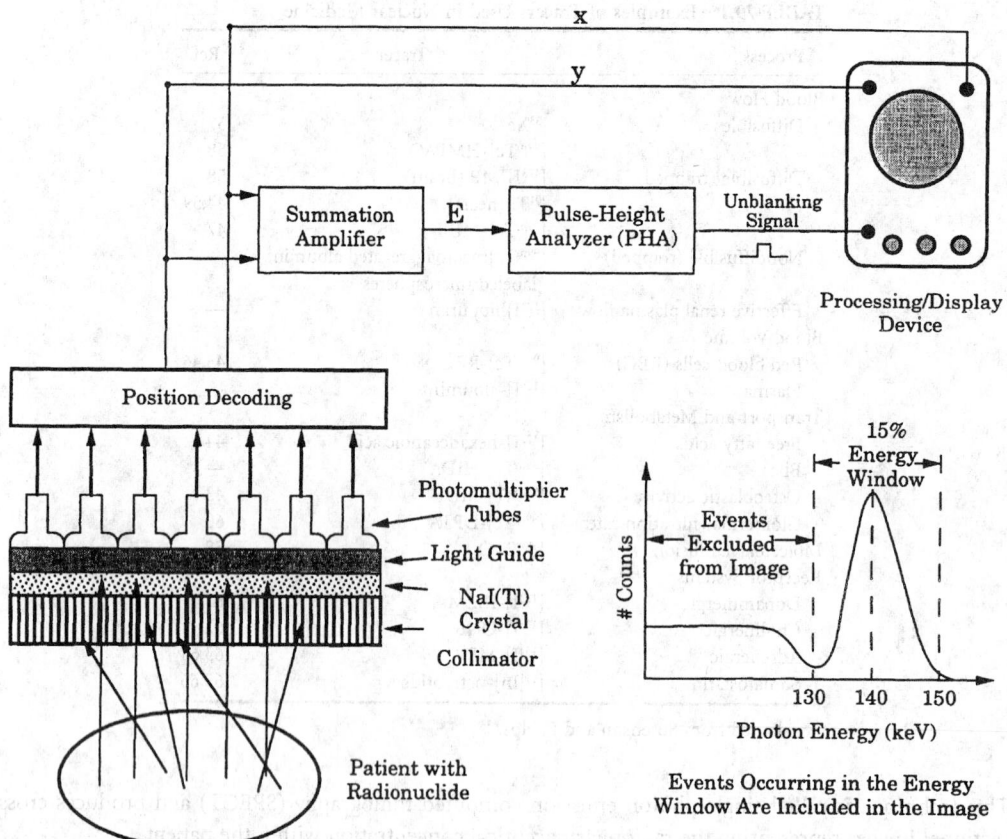

FIGURE 79.4   Scintillation camera incorporates collimator, scintillation crystal, photomultiplier tubes, and electronic circuitry to generate position (x,y) and energy (E) of photons emitted by radiopharmaceutical distribution in patient. Only events falling within a specified energy window are recorded by the processing or display device to form the nuclear medicine image.

allowing the nuclear cardiologist to assess both the presence as well as the regional extent of coronary artery disease.

Measurement of Technical Performance

Several parameters generally are measured to assess the performance of the scintillation camera [50-56]. *Spatial resolution* represents the precision with which the position of an event is localized, and can be assessed from the full-width at half-maximum (FWHM) of a profile taken across the image of a point or linear radioactive object having small dimensions in comparison to the resolution of the system. *Spatial linearity* is quantified as the accuracy with which the position of an event is localized, and represents the ability of a scintillation camera to produce a straight image of a straight object. Spatial linearity is measured as the deviation about the best-fit line in an image of a parallel line phantom or a orthogonal hole phantom, expressed as a percentage (ideally less than 1%) of the detector diameter. *Energy resolution* represents the precision with which the energy of a photon is recorded and generally is measured as the FWHM of the photopeak in an energy spectrum (number of detected photons recorded as a function of photon energy) of the radioactive source. *Flood field uniformity* assesses the ability of the camera to record a spatially uniform image when presented with a spatially uniform distribution of photons. An intrinsic measurement is performed by irradiating the uncollimated detector with the point source placed at a distance equal to at least five times the field of view of the detector. The system uniformity can be checked by irradiating the entire surface of a collimated detector with an extended source of uniform radioactivity. *Sensitivity* represents the number of photons recorded per unit of source radioactivity when the detector is operated either without (intrinsic sensitivity) or with (extrinsic sensitivity) a collimator. *Count-rate linearity* represents the ability of the camera to record a count rate proportional to the photon event rate received by the detector. At low event rates, the measured count rate increases linearly with the actual photon event rate. Because the scintillation camera acts as a paralyzable system, at higher event rates, the measured count rate is lower than that predicted from linear response. At sufficiently high event rates, the measured count rate actually can decrease with increasing photon event rate and eventually can be extinguished when imaging radioactive sources of sufficiently high activities.

79.6   Positron Emission Tomography (PET)

PET involves a physiological administration of a positron-emitting radiopharmaceutical into the human body. The principal advantage of PET over single-photon imaging is the availability of a number of physiologically relevant radiotracers that are labeled with the short-lived positron-emitting radionuclides ^{11}C ($T_{1/2}$ = 20.4 min), ^{13}N (9.96 min), ^{15}O (2.04 min), and ^{18}F (109.8 min). A typical PET center consists of a cyclotron for on-site isotope production, a radiochemistry laboratory for synthetic incorporation of the isotopes into organic molecules, and a PET scanner. PET instrumentation is described in detail in several review articles [67,68].

Principle of Coincidence Detection

The proton-rich radioisotopes used with PET imaging undergo β-decay, and emit positrons (antielectrons). A positron travels a short distance and combines with an electron from the surrounding medium. The masses of the positron and electron are converted to electromagnetic radiation in the form of two γ rays of energy 511 keV, which are emitted at nearly 180° to each other. The PET scanner utilizes multiple opposing γ detectors that surround the positron emitter, each defining a linear volume of response between the detectors. Coincidence timing circuitry enables effective localization of the decay events occurring between detector pairs, rejecting events in each detector that originate from outside the volume of response. A typical modern PET scanner employs tens of thousands of small detectors (or analogous position-coded larger detectors), yielding as many as tens of millions of such volumes of response. The coincidence principle is also utilized to measure and correct for attenuation of photons within the body, allowing the measurement of radioactivity concentration in absolute terms (i.e., Bq/mL). In this case, a separate "transmission" measurement scan is performed, using an external positron-emitting source placed adjacent to the subject yet within the volume of response. A "blank" scan is similarly acquired but without the subject in the field of view. The ratio of coincident count rates in the blank/transmission scans multiplies the corresponding coincidence counts in the emission scan to correct for attenuation along each coincidence line of response.

Detector Composition

The choice of detector material for PET scanners is influenced by a number of considerations, including scanner geometry, detection efficiency (stopping power), output signal strength (energy resolution), signal decay time (count rate capability), physical stability (i.e., hygroscopicity), availability, and cost. Inorganic scintillators are best suited for detection of the 511 keV photons. The physical properties of the two most widely used scintillators, NaI(Tl) and bismuth germanate (BGO), are shown in Table 79.2. NaI(Tl) has found application in position-sensitive detector systems that utilize a small number of large crystals observed by multiple PMTs. NaI(Tl) offers the advantages of (1) good energy resolution for effective rejection of scattered radiation, (2) good timing resolution for minimizing the coincidence-resolving time window, (3) availability of large crystals, and (4) relatively low cost. The higher stopping

TABLE 79.2   Physical Properties of Scintillators
Commonly Employed in PET Scanners

	NaI(Tl)	BGO
Density (g/cm³)	3.67	7.13
Effective atomic number	51	75
Index of refraction	1.85	2.15
Relative emission intensity	100	15
Peak wavelength (nm)	410	480
Decay constant (ns)	230	300

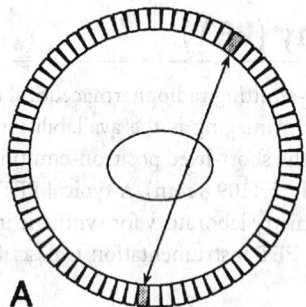
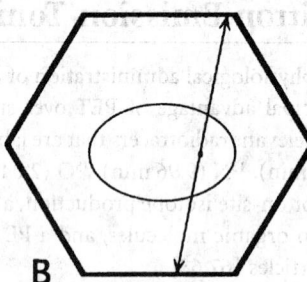

A   **B**

FIGURE 79.5   Representation of PET scanner geometries for typical scanners employing (A) multiple rings of small BGO scintillators, and (B) six large NaI(Tl) position-sensitive planar detectors. The arrows represent positron annihilation photons that are emitted 180° from each other and detected in opposing detectors. (Courtesy of Dr. T. Turkington.)

power of BGO is advantageous for detector designs that use smaller crystals with one-to-one PMTs, or position-encoded matrices of crystals [69]. The recently identified lutetium oxyorthosilicate (LSO) is a potential successor to BGO in detector block designs. LSO has a density of 7.4 g/mL, an effective atomic number of 59, a photofluorescent decay time of 40 ns, and light outputs that are ¾ that of NaI(Tl) [70].

PET Scanners

PET scanners use a number of different detector compositions and gantry configurations, each with its unique advantages and disadvantages [68]. Figure 79.5 shows two of the most common designs that are currently employed. At present, the majority of commercial designs employ a cylindrical geometry with individual BGO detector blocks arranged to form contiguous rings of detectors, each defining an image plane [71,72]. Most of these scanners have retractable lead (or tungsten) septa which are positioned between detector rings to attenuate photons that are emitted at angles not contained in the image plane. This minimizes the effect of out-of-plane scattered radiation, allowing accurate quantitation of the radioactivity distribution in each image plane by two-dimensional (tomographic) image reconstruction. With the septa retracted, all axial angles are accepted, allowing true three-dimensional volume imaging. Another scanner design uses large-area position-encoded NaI(Tl) detectors, allowing three-dimensional volume imaging [73]. In all cases, computer-assisted image reconstruction is used to produce quantitative images of radiotracer concentration in the body.

The spatial resolution of the radioactivity distributions seen in the PET image is primarily determined by the size of the detector elements. In scanners employing cylindrical detector geometry, the in-plane spatial resolution is highest in the center of the field of view (typically 4 to 5 mm FWHM of the point source response for present state-of-the-art scanners). The spatial resolution slowly degrades as the radius increases due to inadequate stopping of photons within incident detectors for nonperpendicular entrance angles. Likewise, the resolution in the axial direction is determined by the axial dimension of the detector elements.

79.7   Ultrasound Imaging

Ultrasound scanning provides a safe and noninvasive way to image the body. With this modality, brief pulses of sound are emitted by a transducer coupled to the skin surface. The sound pulse propagates through tissue at a fixed speed. Interfaces and other objects reflect portions of the acoustic energy back to the transducer, where they are detected as echoes. The ultrasound scanner forms one-dimensional, or more commonly two-dimensional, images of anatomic structures from the reflected echo patterns. In general imaging applications, ultrasound imaging uses frequencies in the 2 to 10 MHz range. Some newer ultrasound devices, for example, those used in emerging ophthalmology applications, use frequencies as high as 50 MHz [74,75].

Characteristics of Sound Waves in Tissue

The speed at which sound waves propagate through a medium depends on the density and compressibility of the medium. At 22°C, the speed of sound in air is around 300 m s^{-1}, while in fresh water it is 1480 m s^{-1}. Human soft tissues behave somewhat like water, with speeds of sound ranging from 1460 m s^{-1} for fat to 1620 m s^{-1} for muscle. The average speed of sound in tissue is taken to be 1540 m s^{-1} (1.54 mm/μs) [76,77].

Any interface, large or small, can reflect a fraction of the ultrasound energy and produce an echo. The relative amount of energy reflected depends on the change in density and compressibility at the interface; the greater the change in these properties of the materials forming the interface, the greater the amplitude of an echo. Examples of reflectors include organ boundaries, blood vessels, and small scatterers distributed more or less randomly throughout most organs. The majority of the echo data displayed on images can be attributed to this scattering process [77]. Shung [78] has reviewed experimental work on ultrasonic scattering vs. frequency in biological tissues.

As ultrasound pulses travel through tissue, they lose their strength due to attenuation. Attenuation is caused by scatter and reflection at interfaces and by absorption. For typical tissues, the amplitude of a 5-MHz beam decreases by about 50% for each centimeter traveled. The attenuation per unit distance is approximately proportional to the ultrasound frequency, so lower-frequency waves propagate greater distances through tissues than higher-frequency waves [79].

B-Mode Imagers

Figure 79.6 illustrates a typical configuration for an ultrasound imager. The operator places a handheld transducer on the skin surface of the patient. Early instruments utilized "single-element" transducers, but the majority of systems now use transducer arrays [80]. Acoustic pulses emitted by the transducer travel in well-defined beams. This beam can be "steered" in different directions, either mechanically with motors or electronically by using transducers arrays.

The same transducer detects echoes that arrive from interfaces in the body and applies them to the receiver, where they are amplified and processed for display. The instrument converts each echo signal into a dot on the display, the brightness of the dot being proportional to the echo amplitude at the transducer. The "scan converter" memory places dots in a location that corresponds to the reflector locations; information required to do this is the return time for each echo and the beam axis direction when the echo is detected.

The scanner constructs a cross-sectional image by sending out 100 to 200 such ultrasound beams, each in a slightly different direction, somewhat like a searchlight scanning the night sky. Echoes received from each beam direction are placed in the image memory using the scheme mentioned above. The entire image is updated at rates of 15 to 30 scans per second, producing a real-time image on the display monitor. This technique is referred to as B-mode imaging because echo signals simply modulate the intensity, or brightness, of the display at locations corresponding to their anatomic origin.

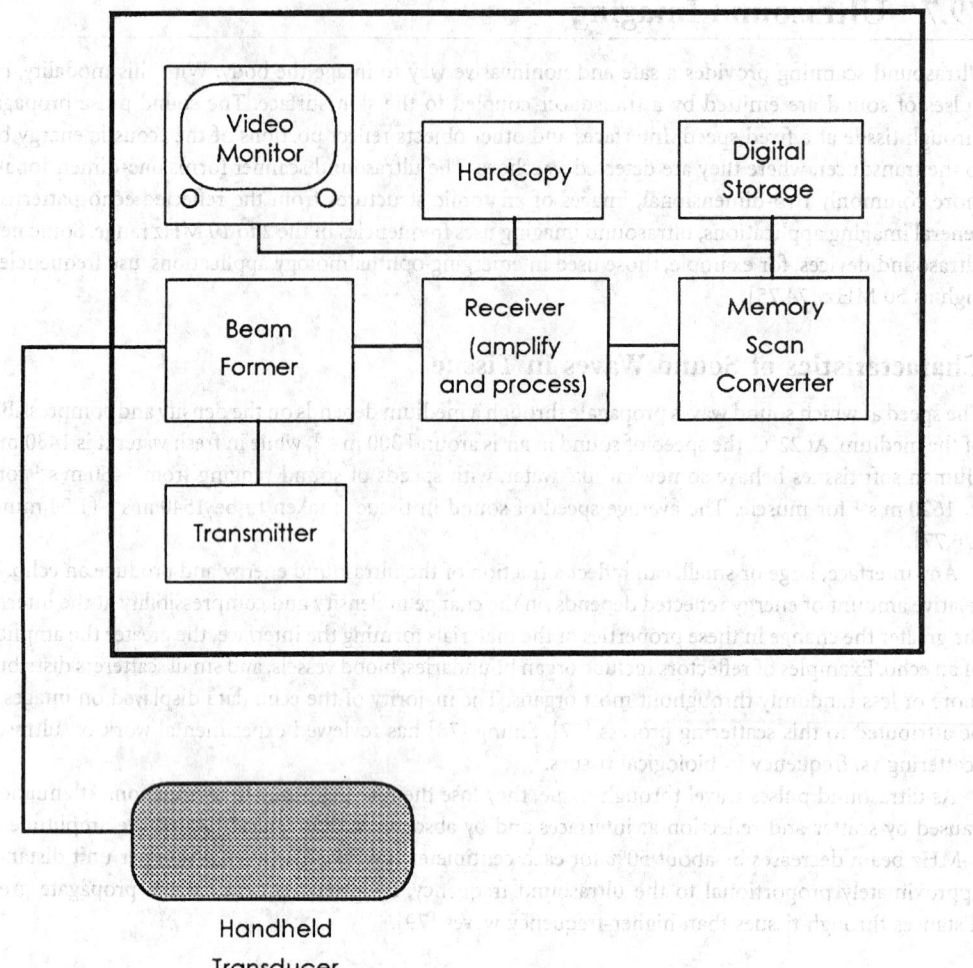

FIGURE 79.6 Components in a typical B-mode ultrasound device.

Doppler Techniques

Ultrasound instruments commonly provide Doppler records as well as B-mode images. Fundamentally, the Doppler effect is a change in the frequency of reflected waves when there is relative motion between the transducer and reflector. For motion directly toward or directly away from the transducer, the Doppler signal frequency f_d is given by

$$f_d = \frac{2 f_0 v}{c} \tag{79.7}$$

where f_0 is the frequency of the transmitted ultrasound, v is the velocity of the reflector, and c is the speed of sound. Thus, the Doppler signal frequency provides information on reflector velocity.

Continuous wave (CW) Doppler instruments consist of a transducer with separate transmitting and receiving elements, a transmitter–receiver unit, and a signal display. They extract a Doppler signal from the complex echo pattern, usually by heterodyning the echo signal with a signal that is coherent with the transmitted wave, and then low pass filtering. The most common applications are to detect and measure blood flow. With a 5-MHz ultrasound frequency and blood velocity of 50 cm/s, the Doppler signal frequency is 3.25 kHz, i.e., in the audible frequency range. A simple loudspeaker may be all that is necessary for interpreting the Doppler signal, but very often a real-time spectral analyzer is available.

Pulsed Doppler instruments are a bit more complicated, but allow the operator to define precisely the distance from the transducer from which Doppler signals are selected. In pulsed Doppler, an acoustic pulse is transmitted along a fixed beam line. Resultant echo signals are amplified and subjected to Doppler processing methods, similar to those outlined for the CW instrument. An operator-adjusted gate captures the waveform from the depth of interest, and a sample-hold device retains the value of this waveform until a subsequent pulse–echo sequence. Because the phase of the echo signal from moving reflectors changes from one pulse–echo sequence to the next, a Doppler signal can be constructed from the behavior over time of the sample-hold value.

Color Doppler Imaging

Color flow imagers may be thought of as extensions of pulsed Doppler machines. Rather than detecting Doppler signals from a single location, color flow imagers detect signals from all depths covered by the ultrasound beam, and for many beam directions. Most instruments extract and display the mean Doppler signal frequency for each location throughout the scanned field [81,82]. A color Doppler image is almost always combined with a B-mode image to provide both anatomic and flow data from the scanned plane.

Measurement of Ultrasound Instrument Performance

Defining "image quality" in ultrasound, and specifying quantifiable factors that relate to optimal B-mode imaging, is controversial to say the least. Important factors that are considered include spatial and contrast resolution, sensitivity, penetration depth, and geometric accuracy.

High-quality ultrasound imagers interrogate the scanned field using a sufficient number of individual beam lines (more than 100) such that gaps between lines can be ignored in resolution considerations. An exception may be in color flow imaging, where sparse line densities are needed for sufficient frame rates [77]. Thus, spatial resolution is dictated by the volume of the ultrasound pulse propagating through the tissue. The dimension of this pulse volume in the direction the pulse travels, i.e., the axial resolution, is determined by the duration of the pulse emitted by the transducer, while the dimension perpendicular to the beam axis, or the "lateral resolution" is determined by the beam width. Although ultrasound beam energy is concentrated near the axis, it is the nature of beams from finite-sized apertures that the intensity falls off gradually with increasing distance from the beam axis. Finally, the size of the ultrasound beam perpendicular to the image plane determines the "slice thickness," the width of the volume of tissue contributing to the echo data viewed in the image plane.

A variety of methods have been used for determining in-plane resolution. The lateral and axial dimensions of a reflector whose size is small enough that it can be considered a pointlike object are frequently used [83]. For a 3.5-MHz transducer, this "spot-size," can be as small as 0.7 mm in the axial dimension and 1 to 2 mm laterally. Smaller spot sizes are found with higher-frequency imagers, such as those using 10 MHz scan heads. Also, larger spot sizes are obtained with scanners that use fixed-focus, single-element transducers.

Slice thickness has been measured using a planar sheet of scatterers scanned with the ultrasound scanning plane intersecting the sheet at a 45° angle [83]. If the slice thickness were negligible, the image of the sheet in this projection would be a straight, horizontal line. The finite thickness of the scanned slice causes a thickening of the line; in fact, for the 45° orientation the vertical size of the image of the sheet corresponds to the slice thickness. For all ultrasound imaging systems, except annular array transducers, the slice thickness is the worst measure of spatial resolution, ranging from the 10 mm to 2 to 3 mm, depending on depth, for a 3.5-MHz transducer.

Physicians commonly use ultrasound imagers to detect cancerous tumors, for which the echoes are slightly stronger or weaker than the surrounding region. "Contrast-detail" tests [84,85] measure the smallest object that can be visualized at a fixed backscatter difference. Spherical mass detectability [86] assesses capabilities to visualize realistic focal lesions. Masses in the latter detection test are characteristic of actual tumors; furthermore, they are easily distributed throughout the scanning plane, assessing resolution at all depths.

Scanner sensitivity is an important performance feature, especially because spatial resolution can be enhanced with higher-frequency transducers. However, this is at the expense of increased ultrasound beam attenuation and poorer penetration. Although absolute measurements of sensitivity of scanners have been done [87], most centers rely upon clinically meaningful "maximum visualization distances" [83,88] for estimating and comparing sensitivity. Geometric accuracy also is important, as images frequently are used to determine structure dimensions, such as fetal head size when determining gestational age [89]. Calibration of distance measurements are done following standard protocols [83,90]; fortunately, modern scanners with digitally based image formation maintain their accuracy much better than previous systems, and many physicists maintain that tests for geometric accuracy are not crucial in routine performance assessments.

79.8 Magnetic Resonance Imaging (MRI)

Magnetic resonance (MR) imaging is a new medical imaging modality which uses magnetic fields and radiofrequency (RF) energy to produce images of the body. The technique is based on nuclear magnetic resonance (NMR) [91], which is a quantum mechanical phenomenon exhibited by atoms having either an odd number of protons or neutrons. Such atoms have a nonzero nuclear magnetic moment, μ, and will precess (or rotate) about an external magnetic field (B_0) with a frequency of $\omega_0 = \gamma B_0$, where γ is the gyromagnetic ratio which for ^{1}H is 42.57 MHz/T. A number of isotopes (including ^{1}H, ^{31}P, ^{23}Na, ^{2}H) exhibit the NMR phenomena; however, the majority of MR scanners image ^{1}H. This is because, relative to other isotopes, ^{1}H has a high inherent sensitivity and abundance in tissue. Therefore, the following discussion is limited to ^{1}H MR imaging. When placed in a B_0 field, ^{1}H nuclei align their spins either parallel or antiparallel to $\mathbf{B_0}$, with a slight excess in the lower energy parallel state. At $T = 25°C$ and $|B_0| =$ 1.5 T, an excess of ~5 in 10^6 atoms are in the parallel state (this excess increases with B_0 and T^{-1}). Because there are ~10^{23} ^{1}H per milliliter of tissue, this excess, when summed over even a small volume, results in a net magnetization, $\mathbf{M} = \Sigma \mu$.

MR Imaging Techniques

The Bloch equations [91,92] are a set of phenomenological equations that succinctly describe the evolution of the net magnetization $\mathbf{M}(\mathbf{r},t)$ during an MR imaging experiment:

$$\frac{\partial \mathbf{M}(\mathbf{r},t)}{\partial t} = \gamma \mathbf{M}(\mathbf{r},t) \times \mathbf{B}(\mathbf{r},t) - \frac{M_x(t)\hat{x} + M_y\hat{y}}{T2} + \frac{\left(M_z(t) - M_0\right)\hat{z}}{T1} \qquad (79.8)$$

where $\mathbf{M}(\mathbf{r},t) = (M_x(t), M_y(t), M_z(t))$; M_0 is the initial (or equilibrium) magnetization, $\mathbf{B}(\mathbf{r},t) = \mathbf{B_0} + \mathbf{G}(\mathbf{r},t) \cdot \mathbf{r} + \mathbf{B_1}(t)$ is the total applied magnetic field and includes terms representing the static field, $\mathbf{B_0}$, the field gradients, $\mathbf{G}(\mathbf{r},t) \cdot \mathbf{r}$; and the magnetic field component of any applied RF excitation, $\mathbf{B_1}(t)$; and T1 and T2 are the characteristic relaxation times of the tissues being imaged. The coordinate system is described in Figure 79.7. The $\mathbf{B_0}$ and $\mathbf{G}(\mathbf{r},t) \cdot \mathbf{r}$ fields are parallel to z, and the $\mathbf{B_1}$ field is orthogonal to z. Every MR imaging experiment consists of an excitation phase, in which the equilibrium magnetization is tipped away from z (the longitudinal axis) and into the transverse (x–y) plane. This is followed by a detection phase, in which the signal emitted by the excited spins is manipulated so that an echo forms. The echo-time (TE) and the repetition-time (TR) denote the time between excitation and echo formation,

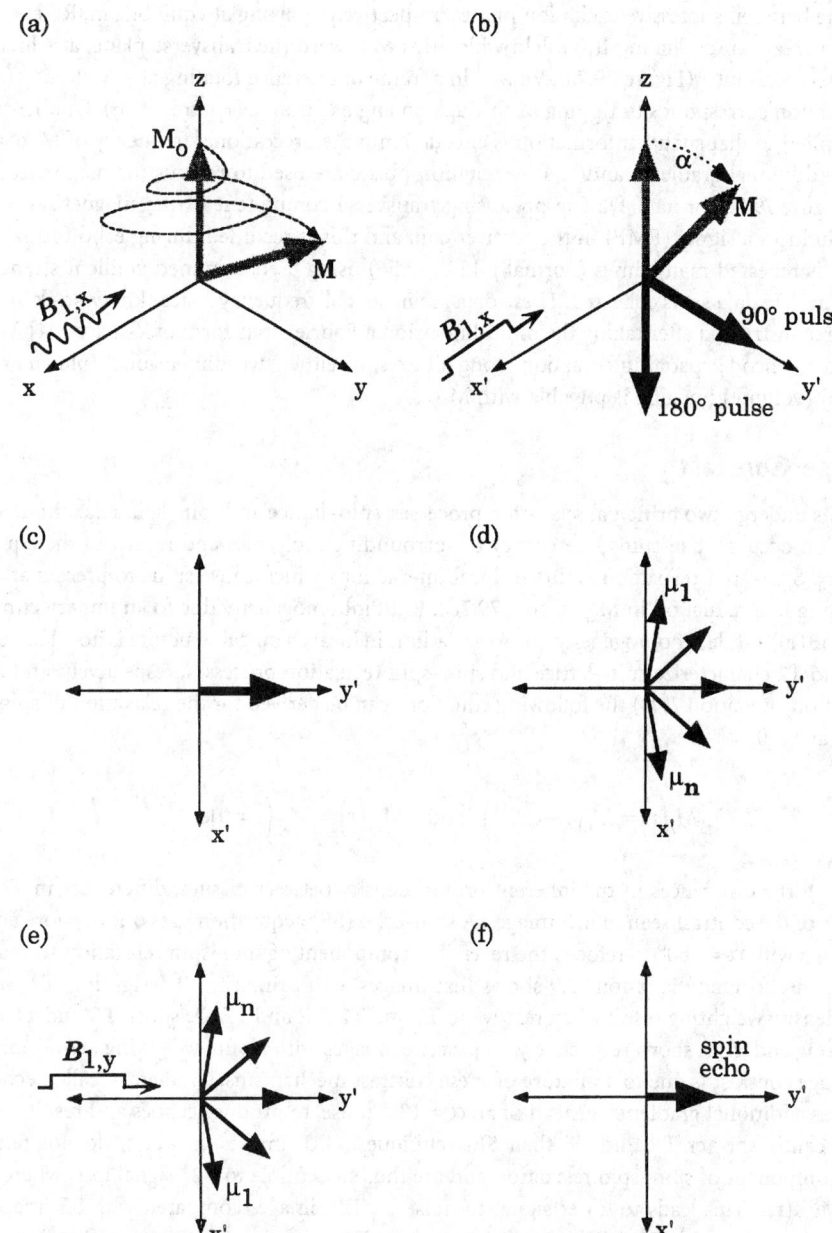

FIGURE 79.7 Graphical depiction of the MR imaging process showing the effect of the B_1-excitation pulse (a–b) and formation of a spin-echo (c–f). In (a) an RF pulse, B_1, is applied along the x axis causing the net magnetization to tip away from equilibrium, M_0, as it precesses about z. Viewed in the rotating frame of reference, this corresponds to a nutation by some angle α in the $y'-z$ plane. After application of an $\alpha = 90°$ pulse, **M** lies in the $x'-y'$ plane (c) and is subject to variations in the local magnetic field which cause the individual spins to precess at different frequencies, i.e., μ_1 and μ_n in (d). The local field variations could be due to gradients or field inhomogeneity, the latter effect leading to T2* signal loss. By applying an $\alpha = 180°$ RF pulse along y', the spins are rotated about y' (e) and are refocussed into an echo (f).

and the time between successive excitation phases, respectively. Starting at equilibrium, RF energy at ω_0 is applied to create an oscillating $\mathbf{B}_1(t)$ field which tips $\mathbf{M}$ toward the transverse plane, at which time it begins to precess about z (Figure 79.7a). Viewed in a frame of reference rotating at $-\omega_0$ about z (x', y', z), the RF excitation corresponds to tipping $\mathbf{M}$ through an angle α from z (Figure 79.7b). Gradients, $\mathbf{G}(\mathbf{r},t)$, are then applied so that spatial information is encoded into the precessional frequency of $\mathbf{M}$, $\omega = \gamma[\mathbf{B}_0 + \mathbf{G}(\mathbf{r},t) \cdot \mathbf{r}]$. Additional gradients and/or RF-excitation pulses are used to refocus the magnetization into an echo (Figure 79.7c through f). The precessing transverse component of the magnetization ($M_{xy} = M_x + jM_y$) induces a signal (EMF) in the receiver coil, and this is recorded during echo formation. The experiment is repeated many times (normally 128 to 256) using predetermined gradient strengths [93] so that a complete data set is collected. These data are in spatial-frequency space (known as $\mathbf{k}$-space) and images are reconstructed after taking the multidimensional Fourier transform of $\mathbf{k}$-space [94]. Gradients can be used to encode spatial information along all axes, so either two-dimensional (planar) or three-dimensional (volume) imaging is possible with MR.

MR Image Contrast

Excited spins undergo two principal relaxation processes, spin–lattice and spin–spin relaxation. Spin–lattice relaxation occurs when spins lose energy to surrounding molecules and return to the equilibrium position, M_0. Spin–spin relaxation is due to local interactions which cause spins to precess at different rates, resulting in a reduction in $\mathbf{M}_{xy}$ (Figure 79.7d). Field inhomogeneity due to an imperfect magnet is reversible ($\delta\mathbf{B}(\mathbf{r})$), while inhomogeneity due to variations in local chemical structure is not. The relaxation times T1 and T2 characterize spin–lattice and spin–spin relaxation processes, respectively, and from the Bloch equation (Equation 79.8) the following equations can be derived for the relaxation of spins tipped by $\alpha = 90°$ at $t = 0$:

$$M_z(t) = M_0\left(1 - e^{-t/\text{T1}}\right) \quad \text{and} \quad M_{xy}(t) = M_{xy}(t=0)e^{-t/\text{T2}} \tag{79.9}$$

Together with the differences in the inherent proton density between tissues, differences in T1 and T2 are the basis of the contrast seen in MR images. A spin-echo (SE) acquisition uses one or more additional RF excitations with $\alpha = 180°$ to refocus the reversible component of spin–spin relaxation so that one or more echoes are formed. Equation 79.9 shows that images with primarily T1 weighting, T2 weighting, or proton density weighting result when relative to T1 and T2: TR and TE are short, TR and TE are long, and TR is long and TE is short, respectively. In practice, images with a pure weighting are not obtainable because image contrast is due to a mixture of these contrast mechanisms. Gradient-recalled echo (GRE) imaging uses additional gradients, instead of an $\alpha = 180°$ pulse, to produce echoes and results in images with significantly shorter TR and TE than SE techniques. GRE images, however, do not refocus the reversible component of spin–spin relaxation and are thus susceptible to T2* signal loss, where $1/\text{T2*} = 1/\text{T2} + 2\pi\gamma|\delta\mathbf{B}(\mathbf{r})|$. This leads to lower signal-to-noise in GRE images compared with SE images. Since TR is short compared with T1, GRE images tend to have T1 weighting. If TE approaches T2*, then the T2* weighting becomes significant.

MR Instrumentation

The key components of a modern MR imaging system include a magnet, a pulse sequencer, gradient and shim coils, and an RF transmitter/receiver — the function of which are controlled by a host computer (Figure 79.8). To obtain the $\mathbf{B}_0$-field, most commercial scanners use superconducting magnets, although some special-purpose (and often lower-cost) scanners may use resistive magnets. Superconducting magnets generally have field strengths between 0.5 and 4.0 T, while resistive magnets normally have field strengths < 0.3 T. The improved signal-to-noise ratio obtained with superconducting designs is offset by the need for periodic cryogen replacement. More modern magnets, however, minimize this cost by including a cryogen reliquefier. Two sets of auxiliary gradient coils are located within the main magnet

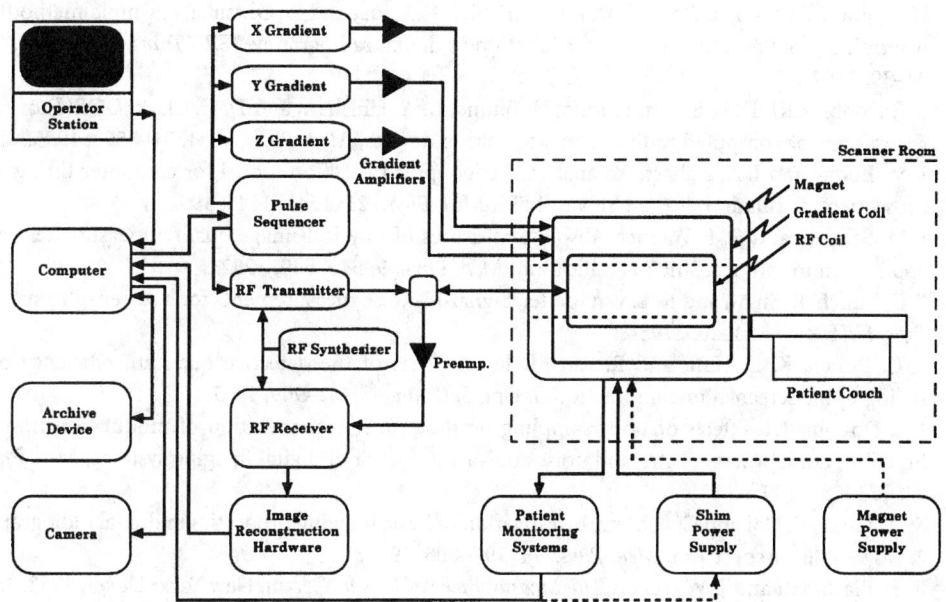

FIGURE 79.8 Overview of an MR scanner showing key components.

to provide spatially varying fields, $G(r,t)$, and to allow shimming of the B_0-field. Current gradient coil hardware can generate maximum gradients of up to 40 mT/m with rise times of 120 μs and allow fields of view from 4 to 48 cm. Shim coils improve the homogeneity of the B_0 field by decreasing $\delta B(r)$ to a few parts per million in order to minimize $T2^*$ effects and spatial distortions. Modern scanners incorporate a digital RF subsystem which excites the spins and then records the emitted signals via one or more RF coils within the magnet. An RF synthesizer is coupled to both the RF transmitter and receiver, so that synchronous detection is possible. The RF system is connected either to separate transmit and receive coil(s) or to a combined transmit/receive coil(s).

To acquire an MR image, the host computer interacts with the operator who defines the imaging parameters (such as α, TR and TE, slice location, and field of view). The parameters are then translated into instructions which are executed on a synchronous state machine known as a pulse sequencer. This device provides real-time control of the gradient and RF waveforms as well as other control functions, such as unblanking the RF receiver and enabling the ADC during an echo. Data are collected and demodulated by the receiver, and then images are reconstructed using specialized hardware built normally around a fast-array processor. The images are sent to the host computer for operator station display, archival or filming. In addition, many MR scanners incorporate devices for monitoring heart and respiration rate, and allow these signals to trigger or gate image acquisition. Future MR imaging systems will probably include higher B_0-field and gradients, and faster data processing/reconstruction hardware. In addition to current imaging apparatus, it is likely that dedicated instruments will be increasingly used to study the heart and for performing neurofunctional imaging and MR-guided interventional procedures.

References

1. G. Revesz, H. L. Kundel, and M. A. Graber, The influence of structured noise on the detection of radiologic abnormalities, *Invest. Radiol.*, 9: 479–486, 1974.
2. A. Rose, *Vision: Human and Electronic,* New York: Plenum Press, 1973.
3. K. Doi, K. Strubler, and K. Rossmann, Truncation errors in calculating the MTF of radiographic screen-film systems from the line spread function, *Phys. Med. Biol.*, 17: 241–250, 1972.
4. Modulation Transfer Function of Screen-film Systems, ICRU Report 41, 1986.

5. H. Fujita, D.-Y. Tsai, T. Itoh, K. Doi, J. Morishita, K. Ueda, and A. Ohtsuka, A simple method for determining the modulation transfer function in digital radiography, *IEEE Trans. Med. Imag.*, 11: 34–39, 1992.

6. J. T. Dobbins III, D. L. Ergun, L. Rutz, H. Blume, D. A. Hinshaw, and D. C. Clark, DQE(f) of four generations of computed radiography acquisition devices, *Med. Phys.*, 22: 1581–1593, 1995.

7. J. M. Boone and J. A. Seibert, An analytical edge spread function model for computer fitting and subsequent calculation of the LSF and MTF, *Med. Phys.*, 21: 1541–1545, 1994.

8. J. M. Sandrik and R. F. Wagner, Absolute measures of physical image quality: measurement and application to radiographic magnification, *Med. Phys.*, 9: 540–549, 1982.

9. P. C. Bunch, R. Shaw, and R. L. van Metter, Signal-to-noise measurements for a screen-film system, *Proc. SPIE*, 454: 154–163, 1984.

10. P. C. Bunch, K. E. Huff, and R. van Metter, Analysis of the detective quantum efficiency of a radiographic screen-film combination, *J. Opt. Soc. Am.*, 4: 902–909, 1987.

11. J. T. Dobbins III, Effects of undersampling on the proper interpretation of modulation transfer function, noise power spectra and noise equivalent quanta of digital imaging systems, *Med. Phys.*, 22: 171–181, 1995.

12. M. L. Giger, K. Doi and C. E. Metz, Investigation of basic imaging properties in digital radiography. 2. Noise Wiener spectrum, *Med. Phys.*, 11: 797–805, 1984.

13. R. B. Blackman and J. W. Tukey, *The Measurement of Power Spectra*, New York: Dover, 1958.

14. R. F. Wagner and J. M. Sandrik, An introduction to digital noise analysis, in A. G. Haus, Ed., *The Physics of Medical Imaging: Recording System Measurements and Techniques*, New York: American Association of Physicists in Medicine, pp. 524–545, 1979.

15. G. T. Barnes, Radiographic mottle: a comprehensive theory, *Med. Phys.*, 9: 656–667, 1982.

16. R. F. Wagner, Fast Fourier digital quantum mottle analysis with application to rare Earth intensifying screen systems, *Med. Phys.*, 4: 157–162, 1977.

17. H. H. Barrett and W. Swindell, *Radiological Imaging*. New York: Academic Press, 1981.

18. G. M. Jenkins and D. G. Watts, *Spectral Analysis and Its Applications*, San Francisco, CA: Holden-Day, 1968.

19. K. Koedooder, J. Strackee, and H. W. Venema, A new method for microdensitometer slit length correction of radiographic noise power spectra, *Med. Phys.*, 13: 469–473, 1986.

20. M. L. Giger and K. Doi, Investigation of basic imaging properties in digital radiography. 1. Modulation transfer function, *Med. Phys.*, 11: 287–295, 1984.

21. J. T. Bushberg, J. A. Seibert, E. M. Leidholdt, Jr., and J. M. Boone, *The Essential Physics of Medical Imaging*, Baltimore, MD: Williams & Wilkins, 1994.

22. M. Braun and B. C. Wilson, Comparative evaluation of several rare-Earth film-screen systems, *Radiology*, 144: 915–919, 1982.

23. M. Sonoda, M. Takano, J. Miyahara, and H. Kato, Computed radiography utilizing scanning laser stimulated luminescence, *Radiology*, 148: 833–838, 1993.

24. W. Hillen, U. Schiebel, and T. Zaengel, Imaging performance of a digital storage phosphor system, *Med. Phys.*, 14: 744–751, 1987.

25. H. Blume, Stimulable phosphor systems—technical aspects. In W. W. Peppler and A. Alter, Eds., *Proc. Chest Imaging Conf '87*, University of Wisconsin, Madison, pp. 194–201, 1987.

26. U. Neitzel, I. Maack, and S. Günther-Kohfahl, Image quality of a digital chest radiography system based on a selenium detector, *Med. Phys.*, 21: 509–516, 1994.

27. C. H. Slump, G. J. Laanstra, H. Kuipers, M. A. Boer, A. G. J. Nijmeijer, M. J. Bentum, R. Kemner, H. J. Meulenbrugge, and R. M. Snoeren, Image quality characteristic of a novel X-ray detector with multiple screen-CCD sensors for real-time diagnostic imaging. *Proc. SPIE*, 3032: 60–71, 1997.

28. L. E. Antonuk, Y. El-Mohri, J. H. Siewerdsen, J. Yorkston, W. Huang, V. E. Scarpine, and R. A. Street, Empirical investigation of the signal performance of a high-resolution, indirect detection, active matrix flat panel imager (AMFPI) for fluoroscopic and radiographic operation, *Med. Phys.*, 24: 51–70, 1997.

29. D. L. Lee, L. K. Cheung, and L. S. Jeromin, A new digital detector for projection radiography, *Proc. SPIE*, 2432: 237–249, 1995.

30. D. P Boyd, R. G. Gould, J. R. Quinn, R. Sparks, J. H. Stanley, and W. B. Herrmannsfeldt, A proposed dynamic cardiac densitometer for early detection and evaluation of heart disease, *IEEE Trans. Nucl. Sci.*, 26: 2724–2727, 1979.

31. C. H. McCollough and R. L. Morin, The technical design and performance of ultrafast computed tomography, *Radiol. Clin. North Am.*, 32: 521–536, 1994.

32. P. M. Silverman, C. J. Cooper, D. I. Weltman, and R. K. Zeman, Helical CT: practical considerations and potential pitfalls, *Radiographics*, 15(1): 25–36, 1995.

33. J. Hsieh, A general approach to the reconstruction of X-ray helical computed tomography, *Med. Phys.*, 23: 221–229, 1996.

34. J. Radon, On the determination of functions from their integrals along certain manifolds, *Ber. Verh. Sächs. Akad. Wiss. Leipzig, Math. Phys. Kl.*, 69: 262–277, 1917.

35. A. C. Kak and M. Slaney, *Principles of Computerized Tomographic Imaging*, New York: IEEE Press, 1988.

36. E. C. McCullough, J. T. Payne, H. L. Baker, Jr., R. R. Hattery, P. F. Sheedy, D. H. Stephens, and E. Gedgaudus, Performance evaluation and quality assurance of computed tomography scanner, with illustrations from the EMI, ACTA, and Delta scanners, *Radiology*, 120: 173–188, 1976.

37. E. C. McCullough and J. T. Payne, X-ray-transmission computed tomography, *Med. Phys.*, 4: 85–98, 1977.

38. R. T. Droege and R. L. Morin, A practical method to measure the MTF of CT scanners, *Med. Phys.*, 9: 758–760, 1982.

39. H. N. Wagner, Jr., Z. Szabo, and J. W. Buchanan, Eds., *Principles of Nuclear Medicine*, 2nd ed., Philadelphia, PA: Saunders, 1995.

40. M. P. Sandler, Ed., *Diagnostic Nuclear Medicine*, 3rd ed., Baltimore, MD: Williams & Wilkins, 1996.

41. R. Chandra, *Introductory Physics of Nuclear Medicine*, 4th ed., Philadelphia, PA: Lea & Febiger, 1992.

42. J. A. Sorenson and M. E. Phelps, *Physics in Nuclear Medicine*, 2nd ed., Orlando, FL: Grune & Stratton, 1987.

43. J. L. Littlefield and T. G. Rudd, [^{99m}Tc]-hydroxymethylene diphosphonate and [^{99m}Tc]-methylene diphosphonate biologic and clinical comparison: concise communication, *J. Nucl. Med.*, 24: 643, 1983.

44. G. B. Saha, W. J. MacIntyre, R. C. Brunken, R. T. Go, S. Raja, C. O. Wong, and E. Q. Chen, Present assessment of myocardial viability by nuclear imaging, *Semin. Nucl. Med.*, 26 (4): 315–35, 1996.

45. T. R. Miller and J. W. Wallis, Cardiac nuclear medicine, *Curr. Probl. Diagn. Radiol.*, 17 (5): 157–193, 1988.

46. R. J. Boudreau and M. K. Loken, Functional imaging of the heart, *Semin. Nucl. Med.*, 17(1): 28–38, 1987.

47. F. J. Wackers, D. S. Berman, J. Maddahi, D. D. Watson, G. A. Beller, H. W. Strauss, C. A. Boucher, M. Picard, B. L. Holman, and R. Fridrich, Technetium-99m hexakis 2-methoxyisobutyl isonitrile: human biodistribution, dosimetry, safety, and preliminary comparison to thallium-201 for myocardial perfusion imaging, *J. Nucl. Med.*, 30: 301–311, 1989.

48. B. M. W. Tsui, X. Zhao, E. C. Frey, and W. H. McCartney, Quantitative single-photon emission computed tomography: basics and clinical considerations, *Semin. Nucl. Med.*, 24: 38–65, 1994.

49. J. Caldwell, D. Williams, G. Harp, J. Stratton, and J. Ritchie, Quantitation of size of relative perfusion defect by single-photon emission computed tomography, *Circulation*, 70: 1048–1056, 1984.

50. Performance Measurements of Scintillation Cameras, Standards Publication No. NU1-1980, Washington D.C.: National Electrical Manufacturers Association, 1980.

51. G. Muehllehner, R. H. Wake, and R. Sano: Standards for performance measurements in scintillation cameras, *J. Nucl Med.*, 22: 72–77, 1981.

52. Scintillation Camera Acceptance Testing and Performance Evaluation, AAPM Report No. 6, New York: American Association of Physicists in Medicine, 1980.

53. Computer-Aided Scintillation Camera Testing and Performance Evaluation, AAPM Report No. 9, New York: American Association of Physicists in Medicine, 1987.

54. Performance Measurements of Scintillation Cameras, Standards Publication No. NU 1-1986, Washington D.C.: National Electrical Manufacturers Association, 1986.

55. Guide to Revised Standards for Performance Measurements of Scintillation Cameras, Washington D.C.: National Electrical Manufacturers Association, 1986.

56. L. S. Graham, Quality assurance of Auger cameras, in D. V. Rao, R. Chandra, and M. C. Graham, Eds., *Physics of Nuclear Medicine Recent Advances,* Monograph 10 of the American Association of Physicists in Medicine (AAPM); New York: American Institute of Physics, pp. 68–82, 1984.

57. L. A. O'Tuama and S. T. Treves, Brain single-photon emission computed tomography for behavior disorders in children, *Semin. Nucl. Med.,* 23 (3): 255–264, 1993.

58. G. B. Saha, W. J. MacIntyre, and R. T. Go, Radiopharmaceuticals for brain imaging, *Semin. Nucl. Med.,* 24 (4): 324–349, 1994.

59. S. Steien and J. Aaseth, Thallium-201 as an agent for myocardial imaging studies, *Analyst,* 120 (3): 779–781, 1995.

60. A. Taylor, Jr. and J. V. Nally, Clinical applications of renal scintigraphy, *Am. J. Roentgenol.,* 164: 31–41, 1995.

61. R. Schlosser and S. Schlegel, D2-receptor imaging with [123I]IBZM and single photon emission tomography in psychiatry: a survey of current status, *J. Neural Transm. Gen. Sect.,* 99 (1–3): 173–185, 1995.

62. T. Sunderland, G. Esposito, S. E. Molchan, R. Coppola, D. W. Jones, J. Gorey, J. T. Little, M. Bahro, and D. R. Weinberger, Differential cholinergic regulation in Alzheimer's patients compared to controls following chronic blockade with scopolamine: a SPECT study, *Psychopharmacology,* 121 (2): 231–241, 1995.

63. B. Shapiro, J. C. Sisson, B. L. Shulkin, M. D. Gross, and S. Zempel, The current status of meta-iodobenzylguanidine and related agents for the diagnosis of neuro-endocrine tumors, *Q. J. Nucl. Med.,* 39 (4 Suppl. 1): 3–8, 1995.

64. B. Shapiro, Imaging of catecholamine-secreting tumours: uses of MIBG in diagnosis and treatment, *Baillieres Clin. Endocrinol. Metabol.,* 7 (2): 491–507, 1993.

65. M. P. Stokkel, B. M. Kwa, and E. K. Pauwels, Imaging and staging of small-cell lung cancer: is there a future role for octreotide scintigraphy? *Br. J. Clin. Pract.,* 49: 235–238, 1995.

66. E. P. Krenning, D. J. Kwekkeboom, J. C. Reubi, P. M. Van Hagen, C. H. van Eijck, H. Y. Oei, and S. W. Lamberts, [111In]-octreotide scintigraphy in oncology, *Metab. Clin. Exp.,* 41 (9 Suppl. 2): 83–86, 1992.

67. E. J. Hoffman and M. E. Phelps, Positron emission tomography: principles and quantitation, in M. Phelps, J. Mazziotta, and H. Schelbert, Eds., *Positron Emission Tomography and Autoradiography: Principles and Applications for the Brain and Heart,* New York: Raven Press, pp. 237–286, 1986.

68. R. A. Koeppe and G. D. Hutchins, Instrumentation for positron emission tomography: tomograph and data processing and display systems, *Semin. Nucl. Med.,* 22 (3): 162–181, 1992.

69. M. E. Casey and R. Nutt, A multicrystal two dimensional BGO detector system for positron emission tomography, *IEEE Trans. Nucl. Sci.,* NS-33: 460–463, 1986.

70. C. L. Melcher and J. S. Schweitzer, Cerium-doped lutetium oxyorthosilicate: a fast, efficient new scintillator, in *Abstracts of Papers Presented at IEEE Nuclear Science Symposium,* Santa Fe, NM, November, 1991 (Abstract).

71. K. Wienhard, M. Dahlbom, L. Eriksson, C. Michel, T. Bruckbauer, U. Pietrzyk, and W. D. Heiss, The ECAT EXACT HR: performance of a new high resolution positron scanner, *J. Comput. Assist. Tomogr.,* 18: 110–118, 1994.

72. T. R. DeGrado, T. G. Turkington, J. J. Williams, C. W. Stearns, J. M. Hoffman, and R. E. Coleman, Performance characteristics of a whole body PET scanner, *J. Nucl. Med.,* 35: 1398–1406, 1994.

73. J. A. Karp, G. Muehllehner, D. A. Mankoff, C. E. Ordonez, J. M. Ollinger, M. E. Daube-Witherspoon, A. T. Haigh, and D. J. Beerbohm, Continuous-slice PENN-PET: a positron tomograph with volume imaging capability, *J. Nucl. Med.,* 31: 617–627, 1990.

74. C. J. Pavlin, K. Harasiewicz, M. D. Sherar, and F. S. Foster, Clinical use of ultrasound biomicroscopy, *Ophthalmology,* 98: 287–295, 1991.

75. D. H. Turnbull, High frequency ultrasound imaging, in L. Goldman and J. B. Fowlkes, Eds., *Medical CT and Ultrasound, Current Technology and Applications,* Madison, WI: Advanced Medical Publishing, pp. 285–297, 1995.

76. P. N. T. Wells, Propagation of ultrasound waves through tissue, in G. D. Fullerton and J. A. Zagzebski, Eds., *Medical Physics of CT and Ultrasound: Tissue Imaging and Characterization,* AAPM Monograph 6, College Park, MD American Association of Physicists in Medicine, pp. 367–387, 1980.

77. J. A. Zagzebski, *Essentials of Ultrasonic Physics,* St. Louis, MO: Mosby, 1996.

78. K. Shung, *In vitro* experimental results on ultrasonic scattering in biological tissues, in K. Shung and G. Thieme, Eds., *Ultrasonic Scattering in Biological Tissues,* Boca Raton, FL: CRC Press, pp. 291–312, 1993.

79. M. Insana, Sound attenuation in tissue, in L. Goldman and J. B. Fowlkes, Eds., *Medical CT and Ultrasound, Current Technology and Applications,* Madison, WI: Advanced Medical Publishing, pp. 15–33, 1995.

80. D. H. Turnbull, Fundamentals of Acoustic Transduction, in L. Goldman and J. B. Fowlkes, Eds., *Medical CT and Ultrasound, Current Technology and Applications,* Madison, WI: Advanced Medical Publishing, pp. 49–66, 1995.

81. K. Ferrara and G. DeAngelis, Color flow mapping, *Ultrasound Med. Biol.,* 23(3): 321–345, 1997.

82. K. J. W. Taylor, P. N. Burns, and P. N. T. Wells, *Clinical Application of Doppler Ultrasound,* 2nd ed., New York: Raven Press, pp. 19–34; 55–94, 1995.

83. *Standard Methods for Measuring the Performance of Pulse Echo Ultrasound Imaging Equipment,* Laurel, MD: American Institute of Ultrasound in Medicine, 1991.

84. H. Lopez, M. H. Loew, P. F. Butler, M. C. Hill, and R. M. Allman, A clinical evaluation of contrast-detail analysis for ultrasound images, *Med. Phys.,* 17: 48–57, 1993.

85. *Methods for Measuring Performance of Pulse-Echo Ultrasound Equipment- Part II: Digital Methods,* Laurel, MD: American Institute of Ultrasound in Medicine, 1995.

86. J. J. Rownd, E. L. Madsen, J. A. Zagzebski, G. R. Frank, and F. Dong, Phantoms and automated system for testing the resolution of ultrasound scanners, *Ultrasound Med. Biol.,* 23(2): 245–260, 1997.

87. *Standard Specification of Echoscope Sensitivity and Noise Level Including Recommended Practice For Such Measurements,* Laurel, MD: American Institute of Ultrasound in Medicine, 1991.

88. *Quality Assurance Manual for Gray Scale Ultrasound Scanners,* Laurel, MD: American Institute of Ultrasound in Medicine, 1995.

89. C. M. Rumack, S. R. Wilson, and J. W. Charboneau, *Diagnostic Ultrasound,* 2nd ed., St. Louis, MO: Mosby, 1997.

90. P. L. Carson and M. M. Goodsitt, Pulsed echo acceptance and quality control testing. In L. Goldman and J. B. Fowlkes, Eds., *Medical CT and Ultrasound, Current Technology and Applications,* Madison, WI: Advanced Medical Publishing, pp. 155–196, 1995.

91. C. P. Slichter, *Principles of Magnetic Resonance,* Berlin: Springer-Verlag, 1980.

92. F. Bloch, Nuclear induction, *Phys. Rev.,* 70: 460–485, 1946.

93. W. A. Edelstein, J. M. S. Hutchinson, G. Johnson, and T. Redpath, Spin-warp imaging and applications in whole body imaging, *Phys. Med. Biol.,* 25: 751–756, 1980.

94. W. S. Hinshaw and A. H. Lent, An introduction to NMR imaging: From Bloch equation to the imaging equation, *Proc. IEEE,* 71: 338–350, 1983.

Further Information

R. N. Bracewell, *The Fourier Transform and Its Applications,* 2nd ed., New York: McGraw-Hill, 1978.

E. O. Brigham, *The Fast Fourier Transform,* Englewood Cliffs, NJ: Prentice-Hall, 1974.

M. A. Brown and R. C. Semelka, *MRI: Basic Principles and Applications,* New York: Wiley-Liss, 1995.

J. C. Dainty and R. Shaw, *Image Science,* New York: Academic Press, 1974.

E. Fukushima and S. B. W. Roeder, *Experimental Pulse NMR. A Nuts and Bolts Approach,* Reading, MA: Addison-Wesley, 1981.

H. K. Huang, *Elements of Digital Radiology,* Englewood Cliffs, NJ: Prentice-Hall, 1987.

A. Macovski, *Medical Imaging Systems,* Englewood Cliffs, NJ: Prentice-Hall, 1983.

P. Sprawls and M. J. Bronskill, Eds., *The Physics of MRI:* 1992 *AAPM Summer School Proceedings,* Woodbury, NY: American Association of Physicists in Medicine, 1993. [An excellent overview of current MR instrumentation and clinical applications written by experts in the field.]

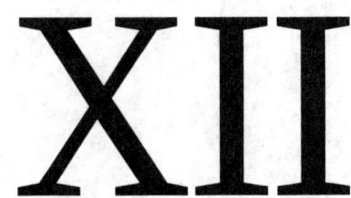

Signal Processing

80

Amplifiers and Signal Conditioners

Ramón Pallás-Areny
Universitat Politècnica de Catalunya

80.1 Introduction

Signals from sensors do not usually have suitable characteristics for display, recording, transmission, or further processing. For example, they may lack the amplitude, power, level, or bandwidth required, or they may carry superimposed interference that masks the desired information.

Signal conditioners, including amplifiers, adapt sensor signals to the requirements of the receiver (circuit or equipment) to which they are to be connected. The functions to be performed by the signal conditioner derive from the nature of both the signal and the receiver. Commonly, the receiver requires a single-ended, low-frequency (dc) voltage with low output impedance and amplitude range close to its power-supply voltage(s). A typical receiver here is an analog-to-digital converter (ADC).

Signals from sensors can be analog or digital. Digital signals come from position encoders, switches, or oscillator-based sensors connected to frequency counters. The amplitude for digital signals must be compatible with logic levels for the digital receiver, and their edges must be fast enough to prevent any false triggering. Large voltages can be attenuated by a voltage divider and slow edges can be accelerated by a Schmitt trigger.

Analog sensors are either self-generating or modulating. *Self-generating sensors* yield a voltage (thermocouples, photovoltaic, and electrochemical sensors) or current (piezo- and pyroelectric sensors) whose

0-8493-8347-1/99/$0.00+$.50
© 1999 by CRC Press LLC

bandwidth equals that of the measurand. Modulating sensors yield a variation in resistance, capacitance, self-inductance or mutual inductance, or other electrical quantities. *Modulating sensors* need to be excited or biased (semiconductor junction-based sensors) in order to provide an output voltage or current. Impedance variation-based sensors are normally placed in voltage dividers, or in Wheatstone bridges (resistive sensors) or ac bridges (resistive and reactance-variation sensors). The bandwidth for signals from modulating sensors equals that of the measured in dc-excited or biased sensors, and is twice that of the measurand in ac-excited sensors (sidebands about the carrier frequency) (see Chapter 81). Capacitive and inductive sensors require an ac excitation, whose frequency must be at least ten times higher than the maximal frequency variation of the measurand. Pallás-Areny and Webster [1] give the equivalent circuit for different sensors and analyze their interface.

Current signals can be converted into voltage signals by inserting a series resistor into the circuit. Graeme [2] analyzes current-to-voltage converters for photodiodes, applicable to other sources. Henceforth, we will refer to voltage signals to analyze transformations to be performed by signal conditioners.

80.2 Dynamic Range

The **dynamic range** for a measurand is the quotient between the measurement range and the desired resolution. Any stage for processing the signal form a sensor must have a dynamic range equal to or larger than that of the measurand. For example, to measure a temperature from 0 to 100°C with 0.1°C resolution, we need a dynamic range of at least $(100 - 0)/0.1 = 1000$ (60 dB). Hence a 10-bit ADC should be appropriate to digitize the signal because $2^{10} = 1024$. Let us assume we have a 10-bit ADC whose input range is 0 to 10 V; its resolution will be 10 V/1024 = 9.8 mV. If the sensor sensitivity is 10 mV/°C and we connect it to the ADC, the 9.8 mV resolution for the ADC will result in a 9.8 mV/(10 mV/°C) = 0.98°C resolution! In spite of having the suitable dynamic range, we do not achieve the desired resolution in temperature because the output range of our sensor (0 to 1 V) does not match the input range for the ADC (0 to 10 V).

The basic function of voltage amplifiers is to amplify the input signal so that its output extends across the input range of the subsequent stage. In the above example, an amplifier with a gain of 10 would match the sensor output range to the ADC input range. In addition, the output of the amplifier should depend only on the input signal, and the signal source should not be disturbed when connecting the amplifier. These requirements can be fulfilled by choosing the appropriate amplifier depending on the characteristics of the input signal.

80.3 Signal Classification

Signals can be classified according to their amplitude level, the relationship between their source terminals and ground, their bandwidth, and the value of their output impedance. Signals lower than around 100 mV are considered to be low level and need amplification. Larger signals may also need amplification depending on the input range of the receiver.

Single-Ended and Differential Signals

A *single-ended signal* source has one of its two output terminals at a constant voltage. For example, Figure 80.1a shows a voltage divider whose terminal L remains at the power-supply reference voltage regardless of the sensor resistance, as shown in Figure 80.1b. If terminal L is at ground potential (grounded power supply in Figure 80.1a), then the signal is single ended and grounded. If terminal L is isolated from ground (for example, if the power supply is a battery), then the signal is single ended and floating. If terminal L is at a constant voltage with respect to ground, then the signal is single ended and driven off ground. The voltage at terminal H will be the sum of the signal plus the off-ground voltage. Therefore, the off-ground voltage is common to H and L; hence, it is called the **common-mode voltage.** For example, a thermocouple bonded to a power transistor provides a signal whose amplitude depends on the temperature of the transistor case, riding on a common-mode voltage equal to the case voltage.

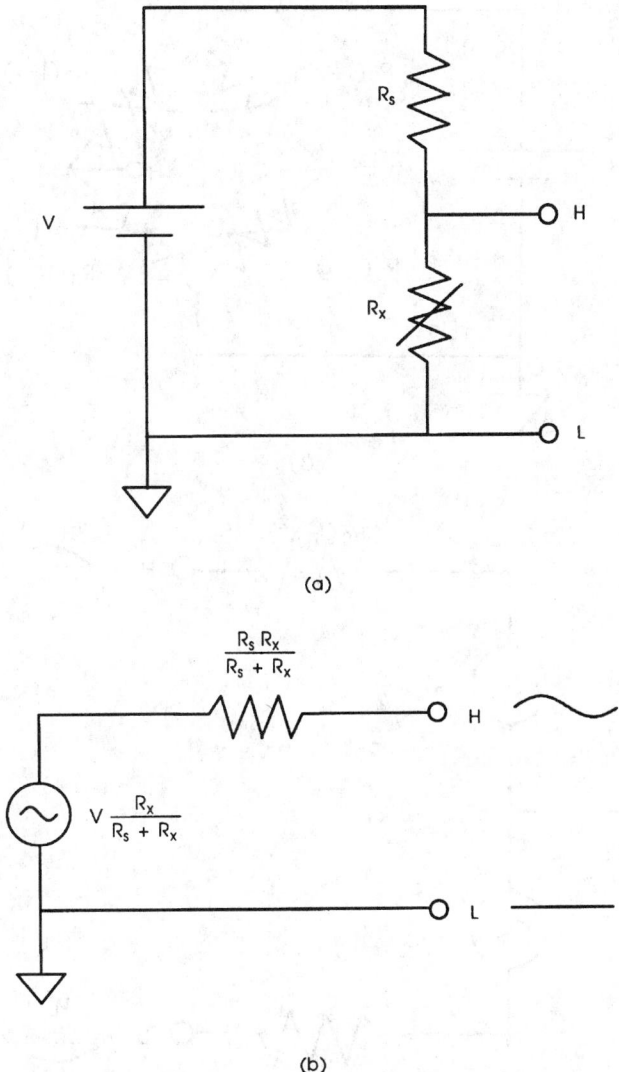

(a)

(b)

FIGURE 80.1 Classes of signals according to their source terminals. A voltage divider (a) provides a single-ended signal (b) where terminal L is at a constant voltage. A Wheatstone bridge with four sensors (c) provides a balanced differential signal which is the difference between two voltages v_H and v_L having the same amplitude but opposite signs and riding on a common-mode voltage V_c. For differential signals much smaller than the common-mode voltage, the equivalent circuit in (e) is used. If the reference point is grounded, the signal (single-ended or differential) will be grounded; if the reference point is floating, the signal will also be floating.

A *differential signal* source has two output terminals whose voltages change simultaneously by the same magnitude but in opposite directions. The Wheatstone bridge in Figure 80.1c provides a differential signal. Its equivalent circuit (Figure 80.1d) shows that there is a differential voltage ($v_d = v_H - v_L$) proportional to x and a common-mode voltage ($V_c = V/2$) that does not carry any information about x. Further, the two output impedances are balanced. We thus have a balanced differential signal with a superimposed common-mode voltage. Were the output impedances different, the signal would be unbalanced. If the bridge power supply is grounded, then the differential signal will be grounded; otherwise, it will be floating. When the differential signal is very small as compared with the common-mode voltage, in order to simplify circuit analysis it is common to use the equivalent circuit in Figure 80.1e. Some differential signals (grounded or floating) do not bear any common-mode voltage.

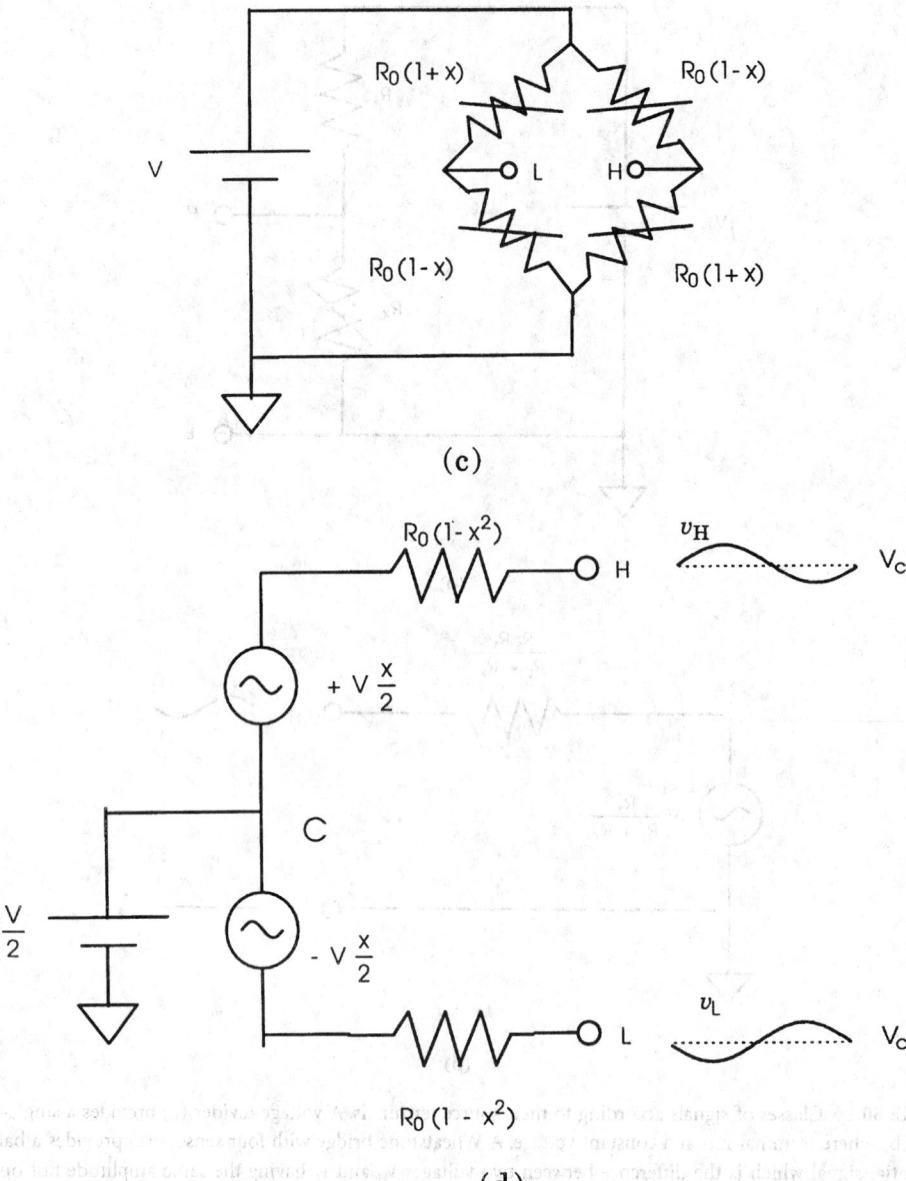

(c)

(d)

FIGURE 80.1 (continued)

Signal conditioning must ensure the compatibility between sensor signals and receivers, which will depend on the relationship between input terminals and ground. For example, a differential and grounded signal is incompatible with an amplifier having a grounded input terminal. Hence, amplifiers must also be described according to their input topology.

Narrowband and Broadband Signals

A *narrowband signal* has a very small frequency range relative to its central frequency. Narrowband signals can be dc, or static, resulting in very low frequencies, such as those from a thermocouple or a weighing

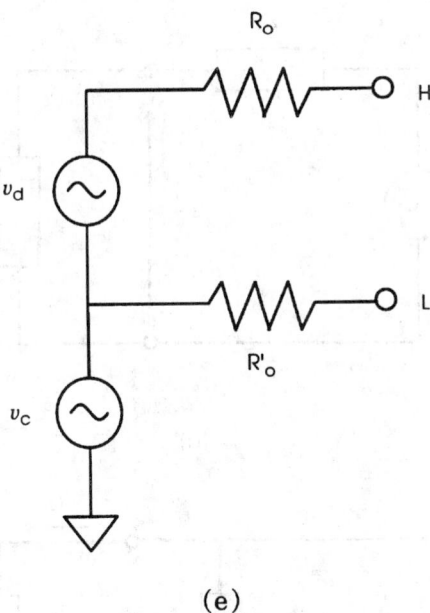

(e)

FIGURE 80.1 (continued)

scale, or ac, such as those from an ac-driven modulating sensor, in which case the exciting frequency (carrier) becomes the central frequency (see Chapter 81).

Broadband signals, such as those from sound and vibration sensors, have a large frequency range relative to their central frequency. Therefore, the value of the central frequency is crucial; a signal ranging from 1 Hz to 10 kHz is a broadband instrumentation signal, but two 10 kHz sidebands around 1 MHz are considered to be a narrowband signal. Signal conditioning of ac narrowband signals is easier because the conditioner performance only needs to be guaranteed with regard to the carrier frequency.

Low- and High-Output-Impedance Signals

The output impedance of signals determines the requirements of the input impedance of the signal conditioner. Figure 80.2a shows a voltage signal connected to a device whose input impedance is Z_d. The voltage detected will be

$$v_d = v_s \frac{Z_d}{Z_d + Z_s}$$
(80.1)

Therefore, the voltage detected will equal the signal voltage only when $Z_d \gg Z_s$; otherwise $v_d \neq v_s$ and there will be a *loading effect.* Furthermore, it may happen that a low Z_d disturbs the sensor, changing the value of v_s and rendering the measurement useless or, worse still, damaging the sensor.

At low frequencies, it is relatively easy to achieve large input impedances even for high-output-impedance signals, such as those from piezoelectric sensors. At high frequencies, however, stray input capacitances make it more difficult. For narrowband signals this is not a problem because the value for Z_s and Z_d will be almost constant and any attenuation because of a loading effect can be taken into account later. However, if the impedance seen by broadband signals is frequency dependent, then each frequency signal undergoes different attenuations which are impossible to compensate for.

Signals with very high output impedance are better modeled as current sources, Figure 80.2b. The current through the detector will be

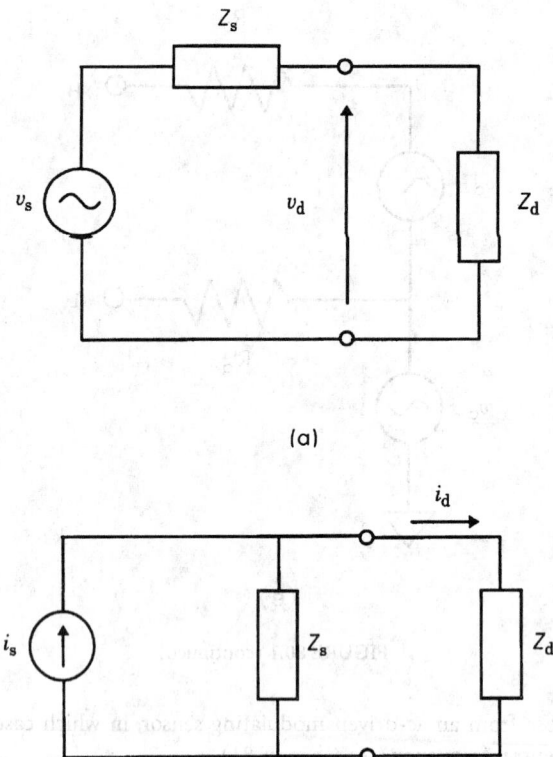

(a)

(b)

FIGURE 80.2 Equivalent circuit for a voltage signal connected to a voltage detector (a) and for a current signal connected to a current detector (b). We require $Z_d \gg Z_o$ in (a) to prevent any loading effect, and $Z_d \ll Z_s$ in (b) to prevent any shunting effect.

$$i_d = i_s \frac{Z_s}{Z_d + Z_s} \qquad (80.2)$$

In order for $i_d = i_s$, it is required that $Z_d \ll Z_s$ which is easier to achieve than $Z_d \gg Z_s$. If Z_d is not low enough, then there is a *shunting effect*.

80.4 General Amplifier Parameters

A *voltage amplifier* produces an output voltage which is a proportional reproduction of the voltage difference at its input terminals, regardless of any common-mode voltage and without loading the voltage source. Figure 80.3a shows the equivalent circuit for a general (differential) amplifier. If one input terminal is connected to one output terminal as in Figure 80.3b, the amplifier is single ended; if this common terminal is grounded, the amplifier is single ended and grounded; if the common terminal is isolated from ground, the amplifier is single ended and floating. In any case, the output power comes from the power supply, and the input signal only controls the shape of the output signal, whose amplitude is determined by the *amplifier gain*, defined as

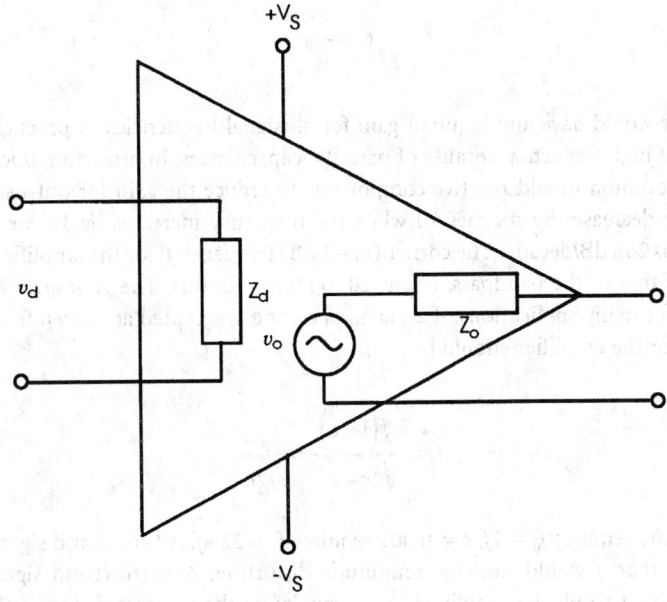

(a)

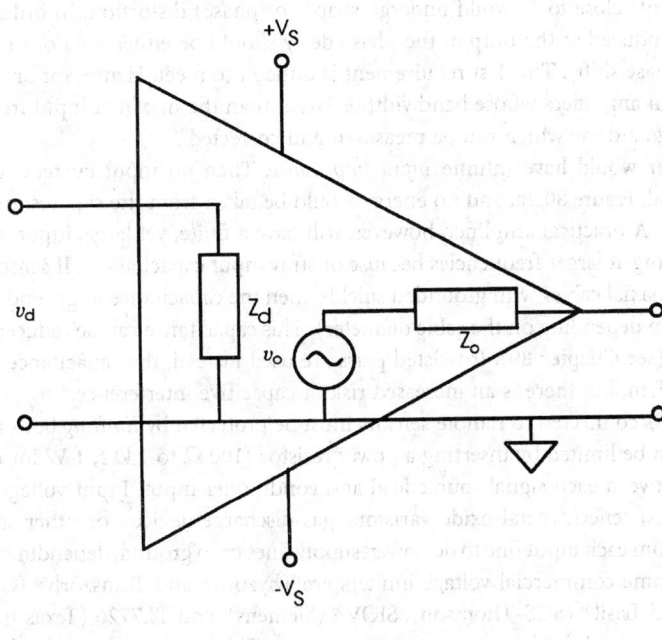

(b)

FIGURE 80.3 General amplifier, differential (a) or single ended (b). The input voltage controls the amplitude of the output voltage, whose power comes from the power supply.

$$G = \frac{v_o}{v_d} \qquad (80.3)$$

The ideal amplifier would have any required gain for all signal frequencies. A practical amplifier has a gain that rolls off at high frequency because of parasitic capacitances. In order to reduce noise and reject interference, it is common to add reactive components to reduce the gain for out-of-band frequencies further. If the gain decreases by n times 10 when the frequency increases by 10, we say that the gain (downward) slope is $20n$ dB/decade. The corner (or -3 dB) frequency f_0 for the amplifier is that for which the gain is 70% of that in the bandpass. (*Note:* 20 log 0.7 = -3 dB). The *gain error* at f_0 is then 30%, which is too large for many applications. If a maximal error ϵ is accepted at a given frequency f, then the corner frequency for the amplifier should be

$$f_0 = \frac{f(1-\epsilon)}{\sqrt{2\epsilon - \epsilon^2}} \approx \frac{f}{\sqrt{2\epsilon}} \qquad (80.4)$$

For example, $\epsilon = 0.01$ requires $f_0 = 7f$, $\epsilon = 0.001$ requires $f_0 = 22.4f$. A broadband signal with frequency components larger than f would undergo amplitude distortion. A narrowband signal centered on a frequency larger than f would be amplified by a gain lower than expected, but if the actual gain is measured, the gain error can later be corrected.

Whenever the gain decreases, the output signal is delayed with respect to the output. In the above amplifier, an input sine wave of frequency f_0 will result in an output sine wave delayed by 45° (and with relative attenuation 30% as compared with a sine wave of frequency $f \gg f_0$). Complex waveforms having frequency components close to f_0 would undergo shape (or phase) distortion. In order for a waveform to be faithfully reproduced at the output, the phase delay should be either zero or proportional to the frequency (linear phase shift). This last requirement is difficult to meet. Hence, for broadband signals it is common to design amplifiers whose bandwidth is larger than the maximal input frequency. Narrowband signals undergo a delay which can be measured and corrected.

An ideal amplifier would have infinite *input impedance*. Then no input current would flow when connecting the signal, Figure 80.2a, and no energy would be taken from the signal source, which would remain undisturbed. A practical amplifier, however, will have a finite, yet large, input impedance at low frequencies, decreasing at larger frequencies because of stray input capacitances. If sensors are connected to conditioners by coaxial cables with grounded shields, then the capacitance to ground can be very large (from 70 to 100 pF/m depending on the cable diameter). This capacitance can be reduced by using driven shields (or guards) (see Chapter 89). If twisted pairs are used instead, the capacitance between wires is only about 5 to 8 pF/m, but there is an increased risk of capacitive interference.

Signal conditioners connected to remote sensors must be protected by *limiting* both *voltage* and *input currents*. Current can be limited by inserting a power resistor (100 Ω to 1 kΩ, 1 W for example), a PTC resistor or a fuse between each signal source lead and conditioner input. Input voltages can be limited by connecting diodes, zeners, metal-oxide varistors, gas-discharge devices, or other surge-suppression nonlinear devices, from each input line to dc power-supply lines or to ground, depending on the particular protecting device. Some commercial voltage limiters are Thyzorb® and Transzorb® (General Semiconductor), Transil® and Trisil® (SGS-Thomson), SIOV® (Siemens), and TL7726 (Texas Instruments).

The ideal amplifier would also have zero *output impedance*. This would imply no loading effect because of a possible finite input impedance for the following stage, low output noise, and unlimited output power. Practical amplifiers can indeed have a low output impedance and low noise, but their output power is very limited. Common signal amplifiers provide at best about 40 mA output current and sometimes only 10 mA. The power gain, however, is quite noticeable, as input currents can be in the picoampere range (10^{-12} A) and input voltages in the millivolt range (10^{-3} V); a 10 V, 10 mA output would mean a power gain of 10^{14}! Yet the output power available is very small (100 mW). Power amplifiers are quite the opposite; they have a relatively small power gain but provide a high-power output. For both signal and power amplifiers, output power comes from the power supply, not from the input signal.

Some sensor signals do not require amplification but only *impedance transformation*, for example, to match their output impedance to that of a transmission line. Amplifiers for impedance transformation (or matching) and $G = 1$ are called buffers.

80.5 Instrumentation Amplifiers

For instrumentation signals, the so-called **instrumentation amplifier** (IA) offers performance closest to the ideal amplifier, at a moderate cost (from $1.50 up). Figure 80.4a shows the symbol for the IA and Figure 80.4b its input/output relationship; ideally this is a straight line with slope G and passing through the point (0,0), but actually it is an off-zero, seemingly straight line, whose slope is somewhat different from G. The output voltage is

$$v_{o} = v_{a} + \left(v_{os} + v_{b} + v_{r} + v_{n}\right)G + v_{ref} \tag{80.5}$$

where v_{a} depends on the input voltage v_{d}, the second term includes offset, drift, noise, and interference-rejection errors, G is the designed gain, and v_{ref} is the reference voltage, commonly 0 V (but not necessarily, thus allowing output level shifting). Equation 80.5 describes a worst-case situation where absolute values for error sources are added. In practice, some cancellation between different error sources may happen.

Figure 80.5 shows a circuit model for *error analysis* when a practical IA is connected to a signal source (assumed to be differential for completeness). Impedance from each input terminal to ground (Z_{c}) and between input terminals (Z_{d}) are all finite. Furthermore, if the input terminals are both connected to ground, v_{o} is not zero and depends on G; this is modeled by V_{os}. If the input terminals are grounded through resistors, then v_{o} also depends on the value of these resistors; this is modeled by current sources I_{B+} and I_{B-}, which represent input bias or leakage currents. These currents need a return path, and therefore a third lead connecting the signal source to the amplifier, or a common ground, is required. Neither V_{os} nor I_{B+} nor I_{B-} is constant; rather, they change with temperature and time: slow changes (<0.01 Hz) are called drift and fast changes are described as noise (hence the noise sources e_{n}, i_{n+} and i_{n-} in Figure 80.5). Common specifications for IAs are defined in Reference 3.

If a voltage v_{c} is simultaneously applied to both inputs, then v_{o} depends on v_{c} and its frequency. The *common-mode gain* is

$$G_{c}\left(f\right) = \frac{V_{o}\left(v_{d} = 0\right)}{V_{c}} \tag{80.6}$$

In order to describe the output voltage due to v_{c} as an input error voltage, we must divide the corresponding $v_{o}(v_{c})$ by G (the normal- or differential-mode gain, $G = G_{d}$). The **common-mode rejection ratio** (CMRR) is defined as

$$CMRR = \frac{G_{d}\left(f\right)}{G_{c}\left(f\right)} \tag{80.7}$$

and is usually expressed in decibels ($\{CMRR\}_{dB} = 20 \log CMRR$). The input error voltage will be

$$\frac{v_{o}\left(v_{c}\right)}{G_{d}} = \frac{G_{c}v_{c}}{G_{d}} = \frac{v_{c}}{CMRR} \tag{80.8}$$

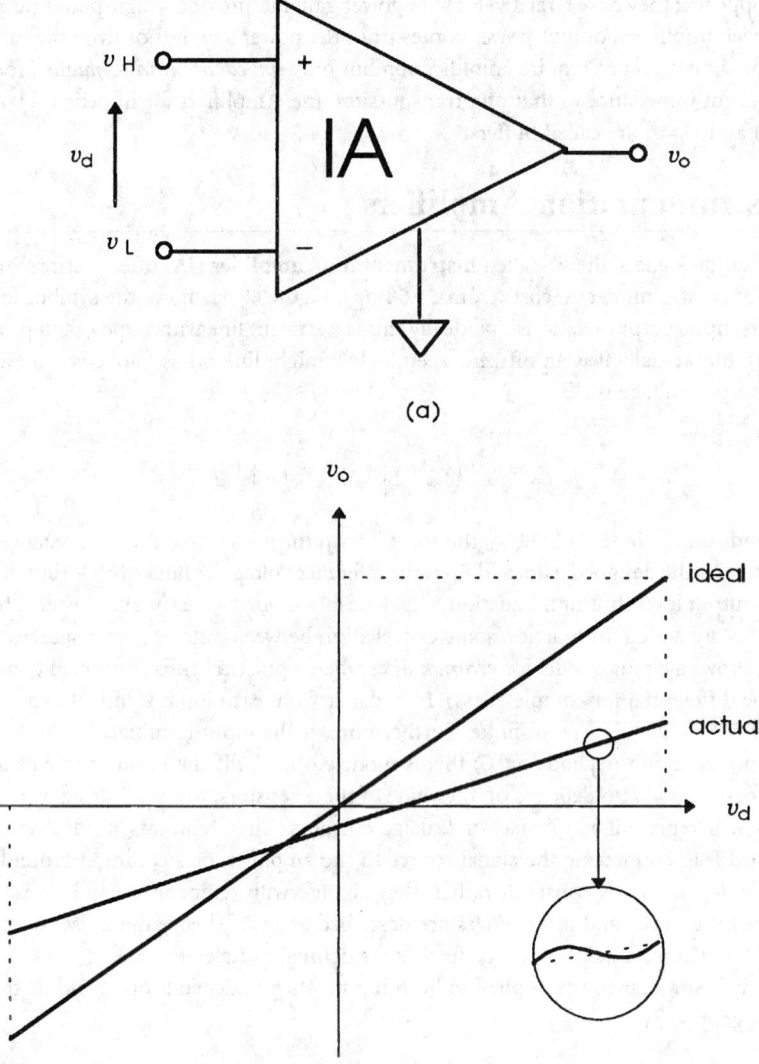

FIGURE 80.4 Instrumentation amplifier. (a) Symbol. (b) Ideal and actual input/output relationship. The ideal response is a straight line through the point (0,0) and slope G.

In the above analysis we have assumed $Z_c \ll R_o$; otherwise, if there were any unbalance (such as that for the source impedance in Figure 80.5), v_c at the voltage source would result in a differential-mode voltage at the amplifier input,

$$v_d(v_c) = v_c \left(\frac{R_o + \Delta R_o}{Z_c + R_o + \Delta R_o} - \frac{R_o}{Z_c + R_o} \right) \qquad (80.9)$$

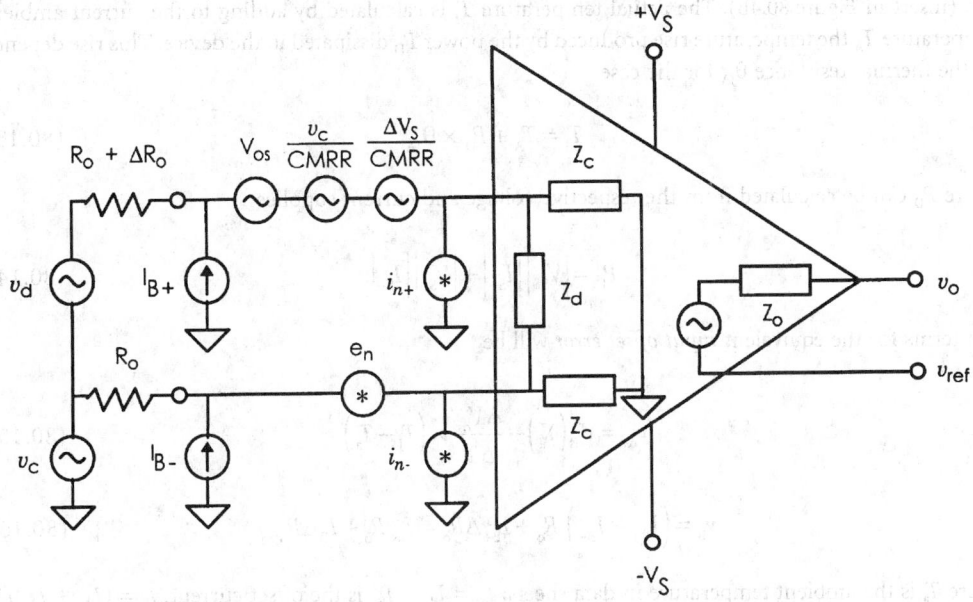

FIGURE 80.5 A model for a practical instrumentation amplifier including major error sources.

$$= v_c \frac{Z_c \Delta R_o}{\left(Z_c + R_o + \Delta R_o\right)\left(Z_c + R_o\right)} \approx v_c \frac{\Delta R_o}{Z_c}$$

which would be amplified by G_d. Then, the effective common-mode rejection ratio would be

$$\frac{1}{CMRR_e} = \frac{\Delta R_o}{Z_c} + \frac{1}{CMRR} \qquad (80.10)$$

where the CMRR is that of the IA alone, expressed as a fraction, not in decibels. Stray capacitances from input terminals to ground will decrease Z_c, therefore reducing $CMRR_e$.

The ideal amplifier is unaffected by power supply fluctuations. The practical amplifier shows output fluctuations when supply voltages change. For slow changes, the equivalent input error can be expressed as a change in input offset voltages in terms of the *power supply rejection ratio* (PSRR),

$$PSRR = \frac{\Delta V_{os}}{\Delta V_s} \qquad (80.11)$$

The terms in Equation 80.5 can be detailed as follows. Because of gain errors we have

$$v_a = v_d \left(G + e_G + \frac{\Delta G}{\Delta T} \times \Delta T + e_{NLG} \right) \qquad (80.12)$$

where G is the differential gain designed, e_G its absolute error, $\Delta G/\Delta T$ its thermal drift, ΔT the difference between the actual temperature and that at which the gain G is specified, and e_{NLG} is the nonlinearity gain error, which describes the extent to which the input/output relationship deviates from a straight line (insert in Figure 80.4b). The actual temperature T_J is calculated by adding to the current ambient temperature T_A the temperature rise produced by the power P_D dissipated in the device. This rise depends on the thermal resistance θ_{JA} for the case

$$T_J = T_A + P_D \times \theta_{JA} \tag{80.13}$$

where P_D can be calculated from the respective voltage and current supplies

$$P_D = \left|V_{S+}\right|\left|I_{S+}\right| + \left|V_{S-}\right|\left|I_{S-}\right| \tag{80.14}$$

The terms for the equivalent *input offset error* will be

$$v_{os} = V_{os}\left(T_a\right) + \frac{\Delta V_{os}}{\Delta T} \times \left(T_J - T_a\right) \tag{80.15}$$

$$v_b = \left(I_{B+} - I_{B-}\right)R_o + I_{B+}\Delta R_o = I_{os}R_o + I_B\Delta R_o \tag{80.16}$$

where T_a is the ambient temperature in data sheets, $I_{os} = I_{B+} - I_{B-}$ is the offset current, $I_B = (I_{B+} + I_{B-})/2$, and all input currents must be calculated at the actual temperature,

$$I = I\left(T_a\right) + \frac{\Delta I}{\Delta T} \times \left(T_J - T_a\right) \tag{80.17}$$

Error contributions from finite *interference rejection* are

$$v_r = \frac{v_c}{\mathrm{CMRR}_e} + \frac{\Delta V_s}{\mathrm{PSRR}} \tag{80.18}$$

where the CMRR_e must be that at the frequency for v_c, and the PSRR must be that for the frequency of the ripple ΔV_s. It is assumed that both frequencies fall inside the bandpass for the signal of interest v_d.
   The equivalent *input voltage noise* is

$$v_n = \sqrt{e_n^2 B_e + i_n^2 - R_o^2 B_{i+} + i_n^2 - R_o^2 B_{i-}} \tag{80.19}$$

where e_n^2 is the voltage noise power spectral density of the IA, i_{n+}^2 and i_{n-}^2 are the current noise power spectral densities for each input of the IA, and B_e, B_{i+}, and B_{i-} are the respective noise equivalent bandwidths of each noise source. In Figure 80.5, the transfer function for each noise source is the same as that of the signal v_d. If the signal bandwidth is determined as $f_h - f_l$ by sharp filters, then

$$B_e = f_h - f_l + f_{ce}\ln\frac{f_h}{f_l} \tag{80.20}$$

$$B_{i+} = B_{i-} = f_h - f_l + f_{ci}\ln\frac{f_h}{f_l} \tag{80.21}$$

where f_{ce} and f_{ci} are, respectively, the frequencies where the value of voltage and current noise spectral densities is twice their value at high frequency, also known as corner or 3 dB frequencies.

Another noise specification method states the peak-to-peak noise at a given low-frequency band (f_A to f_B), usually 0.1 to 10 Hz, and the noise spectral density at a frequency at which it is already constant, normally 1 or 10 kHz. In these cases, if the contribution from noise currents is negligible, the equivalent input voltage noise can be calculated from

$$v_n = \sqrt{v_{nL}^2 + v_{nH}^2} \qquad (80.22)$$

where v_{nL} and v_{nH} are, respectively, the voltage noise in the low-frequency and high-frequency bands expressed in the same units (peak-to-peak or rms voltages). To convert rms voltages into peak-to-peak values, multiply by 6.6. If the signal bandwidth is from f_l to f_h, and $f_l = f_A$ and $f_h > f_B$, then Equation 80.22 can be written

$$v_n = \sqrt{v_{nL}^2 + \left(6.6e_n\right)^2 \left(f_h - f_B\right)} \qquad (80.23)$$

where v_{nL} is the peak-to-peak value and e_n is the rms voltage noise as specified in data books. Equation 80.23 results in a peak-to-peak calculated noise that is lower than the real noise, because noise spectral density is not constant from f_B up. However, it is a simple approach providing useful results.

For signal sources with high output resistors, thermal and excess noise from resistors (see Chapter 54) must be included. For first- and second-order filters, noise bandwidth is slightly larger than signal bandwidth. Motchenbacher and Connelly [4] show how to calculate noise bandwidth, resistor noise, and noise transfer functions when different from signal transfer functions.

Low-noise design always seeks the minimal bandwidth required for the signal. When amplifying low-frequency signals, if a large capacitor C_i is connected across the input terminals in Figure 80.5, then noise and interference having a frequency larger than $f_0 = 1/2\pi(2R_o)C_i$ ($f_0 \ll f_s$) will be attenuated.

Another possible source of error for any IA, not included in Equation 80.5, is the *slew rate limit* of its output stage. Because of the limited current available, the voltage at the output terminal cannot change faster than a specified value SR. Then, if the maximal amplitude A of an output sine wave of frequency f exceeds

$$A = \frac{SR}{2\pi f} \qquad (80.24)$$

there will be a waveform distortion.

Table 80.1 lists some basic specifications for IC instrumentation amplifies whose gain G can be set by an external resistor or a single connection.

Instrumentation Amplifiers Built from Discrete Parts

Instrumentation amplifiers can be built from discrete parts by using operational amplifiers (op amps) and a few resistors. An *op amp* is basically a differential voltage amplifier whose gain A_d is very large (from 10^5 to 10^7) at dc and rolls off (20 dB/decade) from frequencies of about 1 to 100 Hz, becoming 1 at frequencies from 1 to 10 MHz for common models (Figure 80.6a), and whose input impedances are so high (up to $10^{12} \, \Omega \parallel 1$ pF) that input currents are almost negligible. Op amps can also be modeled by the circuit in Figure 80.5, and their symbol is that in Figure 80.4a, deleting IA. However, because of their large gain, op amps cannot be used directly as amplifiers; a mere 1 mV dc input voltage would saturate any op amp output. Furthermore, op amp gain changes from unit to unit, even for the same model, and for a given unit it changes with time, temperature, and supply voltages. Nevertheless, by providing external feedback, op amps are very flexible and far cheaper than IAs. But when the cost for external components and their connections, and overall reliability are also considered, the optimal solution depends on the situation.

TABLE 80.1   Basic Specifications for Some Instrumentation Amplifiers

	AD624A	AMP02F	INA110KP	LT1101AC	Units
Gain range	1–1000	1–1000	1–500	10,100	V/V
Gain error, e_G					
$G = 1$	±0.05	0.05	±0.02	n.a.	%
$G = 10$	n.s.	0.40	±0.05	±0.04	%
$G = 100$	±0.25	0.50	±0.10	±0.04	%
$G = 1000$	±1.0	0.70	n.a.	n.a.	%
Gain nonlinearity error e_{NLG}[a]					
$G = 1$	±0.005	0.006	±0.005	n.a.	%
$G = 10$	n.s.	0.006	±0.005	±0.0008	%
$G = 100$	±0.005	0.006	±0.01	±0.0008	%
$G = 1000$	±0.005	0.006	n.a.	n.a.	%
Gain drift $\Delta G/\Delta T$					
$G = 1$	5	50	±10	n.a.	μV/V/°C
$G = 10$	n.s.	50	±10	5	μV/V/°C
$G = 100$	10	50	±20	5	μV/V/°C
$G = 1000$	25	50	n.a.	n.a.	μV/V/°C
V_{os}	200 + 5/G	200	±(1000 + 5000/G)	160	μV
$\Delta v_{os}/\Delta T$	2 + 50/G	4	±(2 + 50/G)	2	μV/°C
I_B	±50	20	0.05	10	nA
$\Delta I_B/\Delta T$	±50 typ	250 typ	b	30	pA/°C
I_{os}	±35	10	0.025	0.90	nA
$\Delta I_{os}/\Delta T$	±20 typ	15 typ	n.s.	7.0	pA/°C
Z_d	1 ‖ 10 typ	10 typ	5000 ‖ 6 typ	12	GΩ
Z_c	1 ‖ 10 typ	16.5 typ	2000 ‖ 1 typ	7	GΩ
CMRR at dc					
$G = 1$	70 min	80 min	70 min	n.a.	dB
$G = 10$	n.s.	100 min	87 min	82	dB
$G = 100$	100 min	115 min	100 min	98	dB
$G = 1000$	110 min	115 min	n.a.	n.a.	dB
PSRR at dc					
$G = 1$	70 min	80 min	c	n.a.	dB
$G = 10$	n.s.	100 min	c	100	dB
$G = 100$	95 min	115 min	c	100	dB
$G = 1000$	100 min	115 min	n.a.	n.a.	dB
Bandwidth (–3 dB) (typ)					
$G = 1$	1000	1200	2500	n.a.	kHz
$G = 10$	n.s.	300	2500	37	kHz
$G = 100$	150	200	470	3.5	kHz
$G = 1000$	25	200	n.a.	n.a.	kHz
Slew rate (typ)	5.0	6	17	0.1	V/μs
Settling time to 0.01%					
$G = 1$	15 typ	10 typ	12.5	n.a.	μs
$G = 10$	15 typ	10 typ	7.5	n.a.	μs
$G = 100$	15 typ	10 typ	7.5	n.a.	μs
$G = 1000$	75 typ	10 typ	n.a.	n.a.	μs
e_n (typ)					
$G = 1$	4	120	66	n.a.	nV/$\sqrt{\text{Hz}}$
$G = 10$	4	18	12	43	nV/$\sqrt{\text{Hz}}$
$G = 100$	4	10	10	43	nV/$\sqrt{\text{Hz}}$
$G = 1000$	4	9	n.a.	n.a.	nV/$\sqrt{\text{Hz}}$
v_n 0.1 to 10 Hz (typ)					
$G = 1$	10	10	1	0.9	μVp-p
$G = 10$	n.s.	1.2	1	0.9	μVp-p
$G = 100$	0.3	0.5	1	0.9	μVp-p
$G = 1000$	0.2	0.4	1	0.9	μVp-p

TABLE 80.1 (continued)   Basic Specifications for Some Instrumentation Amplifiers

	AD624A	AMP02F	INA110KP	LT1101AC	Units
i_n 0.1 to 10 Hz (typ)	60	n.s.	n.s.	2.3	pAp-p
i_n (typ)	n.s.	400	1.8	20	fA/$\sqrt{\text{Hz}}$

Note: All parameter values are maximum, unless otherwise stated (typ = typical; min = minimum; n.a. = not applicable; n.s. = not specified). Measurement conditions are similar; consult manufacturers' data books for further detail.

 [a] For the INA110, the gain nonlinearity error is specified as percentage of the full-scale output.

 [b] Input current drift for the INA110KP approximately doubles for every 10°C increase, from 25°C (10 pA-typ) to 125°C (10 nA-typ).

 [c] The PSRR for the INA110 is specified as an input offset $\pm(10 + 180/G)$ µV/V maximum.

Figure 80.6b shows an amplifier built from an op amp with external feedback. If input currents are neglected, the current through R_2 will flow through R_1 and we have

$$v_d = v_s - v_o \frac{R_1}{R_1 + R_2} \tag{80.25}$$

$$v_o = A_d v_d \tag{80.26}$$

Therefore,

$$\frac{v_o}{v_s} = \frac{A_d\left(1+\dfrac{R_2}{R_1}\right)}{A_d + 1 + \dfrac{R_2}{R_1}} = \frac{G_i}{1 + \dfrac{G_i}{A_d}} \tag{80.27}$$

where $G_i = 1 + R_2/R_1$ is the ideal gain for the amplifier. If G_i/A_d is small enough (G_i small, A_d large), the gain does not depend on A_d but only on external components. At high frequencies, however, A_d becomes smaller and, from Equation 80.27, $v_o < G_i v_s$ so that the bandwidth for the amplifier will reduce for large gains. Franco [5] analyzes different op amp circuits useful for signal conditioning.

Figure 80.7 shows an IA built from three op amps. The input stage is fully differential and the output stage is a *difference amplifier* converting a differential voltage into a single-ended output voltage. Difference amplifiers (op amp and matched resistors) are available in IC form: AMP 03 (Analog Devices) and INA 105/6 and INA 117 (Burr-Brown). The gain equation for the complete IA is

$$G = \left(1 + 2\frac{R_2}{R_1}\right)\frac{R_4}{R_3} \tag{80.28}$$

Pallás-Areny and Webster [6] have analyzed matching conditions in order to achieve a high CMRR. Resistors R_2 do not need to be matched. Resistors R_3 and R_4 need to be closely matched. A potentiometer connected to the v_{ref} terminal makes it possible to trim the CMRR at low frequencies.

The *three-op-amp IA* has a symmetrical structure making it easy to design and test. IAs based on an IC difference amplifier do not need any user trim for high CMRR. The circuit in Figure 80.8 is an IA that lacks these advantages but uses only two op amps. Its gain equation is

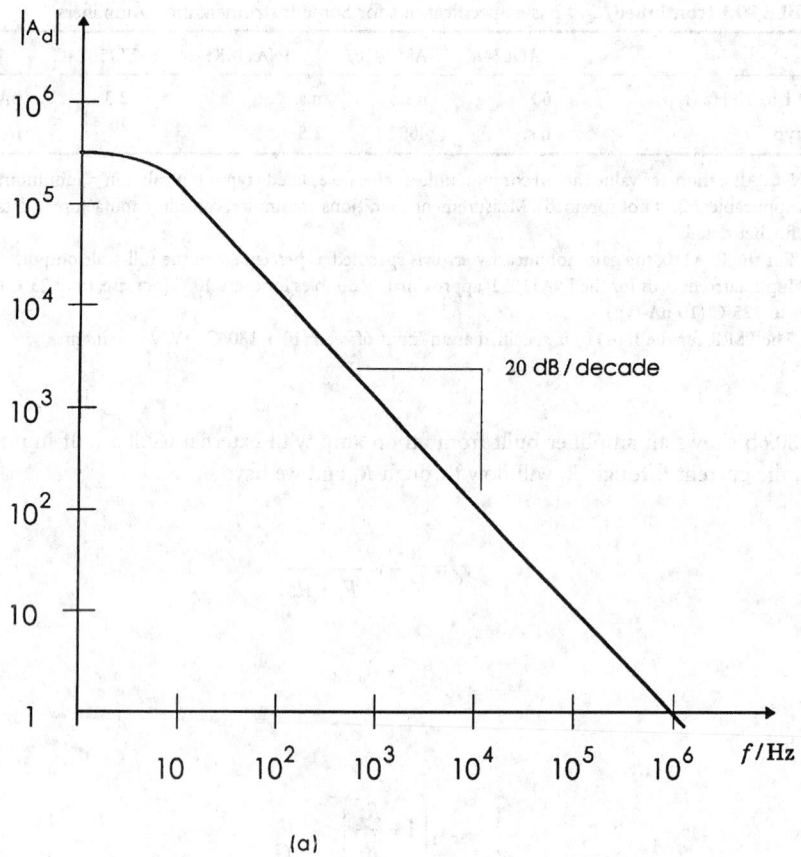

(a)

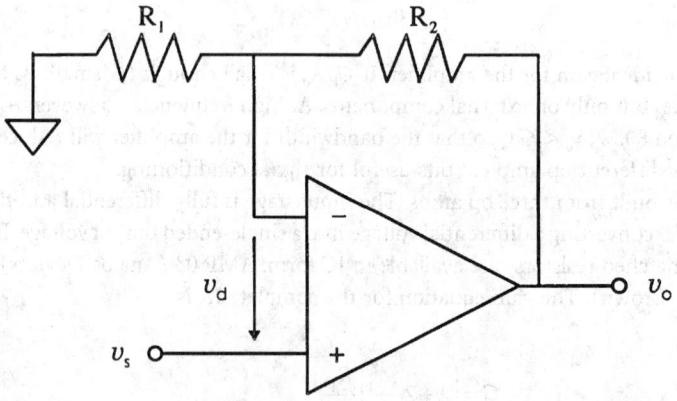

(b)

FIGURE 80.6 (a) Open loop gain for an op amp. (b) Amplifier based on an op amp with external feedback.

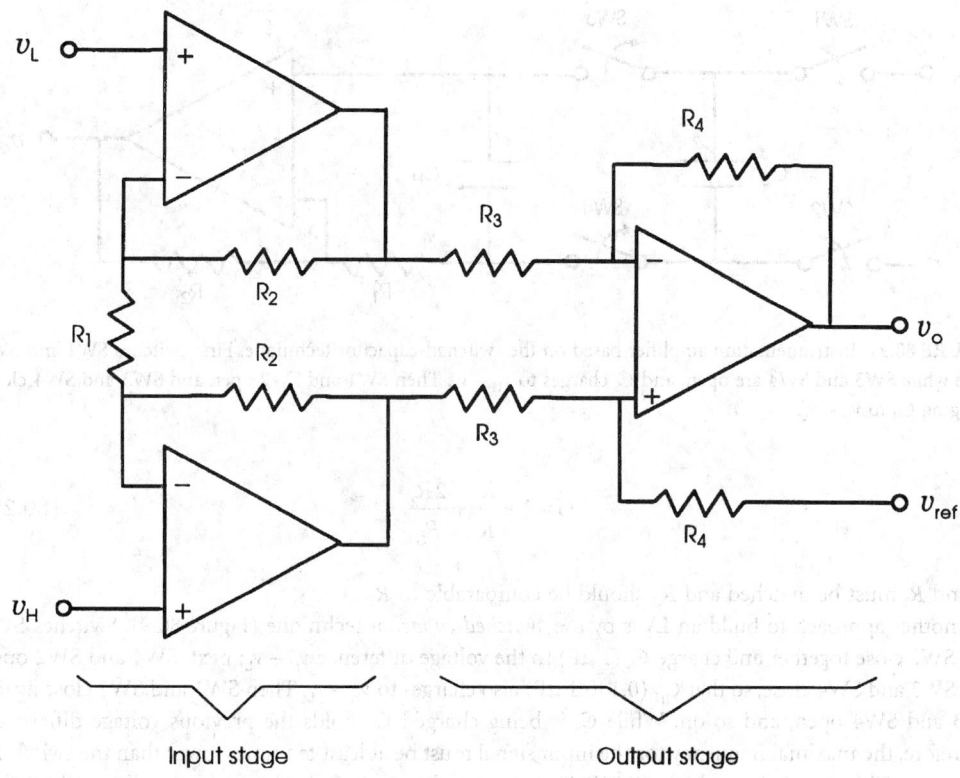

FIGURE 80.7 Instrumentation amplifier built from three op amps. R_3 and R_4 must be matched.

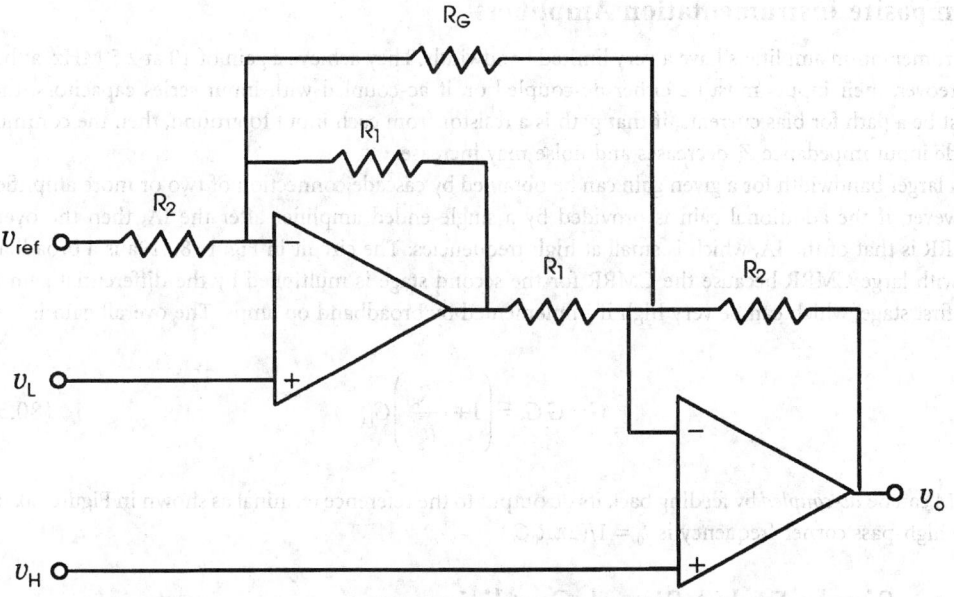

FIGURE 80.8 Instrumentation amplifier built from two op amps. R_1 and R_2 must be matched.

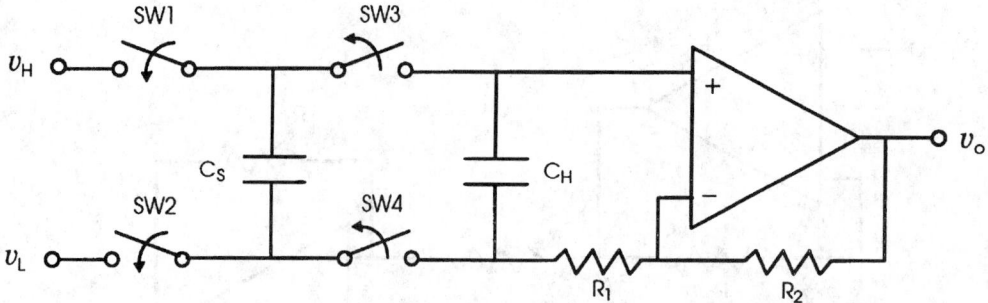

FIGURE 80.9 Instrumentation amplifier based on the switched-capacitor technique. First switches SW1 and SW2 close while SW3 and SW4 are open, and C_S charges to $v_H - v_L$. Then SW1 and SW2 open and SW3 and SW4 close, charging C_H to $v_H - v_L$.

$$G = 1 + \frac{R_2}{R_1} + \frac{2R_2}{R_G} \qquad (80.29)$$

R_1 and R_2 must be matched and R_G should be comparable to R_2.

Another approach to build an IA is by the *switched-capacitor* technique (Figure 80.9). Switches SW1 and SW2 close together and charge C_S (1 μF) to the voltage difference $v_H - v_L$; next, SW1 and SW2 open and SW3 and SW4 close, so that C_H (0.1 to 1 μF) also charges to $v_H - v_L$. Then SW1 and SW2 close again, SW3 and SW4 open, and so on. While C_S is being charged C_H holds the previous voltage difference. Therefore, the maximal frequency for the input signal must be at least ten times lower than the switching frequency. This circuit has a high CMRR because the charge at C_S is almost insensitive to the input common-mode voltage. Furthermore, it converts the differential signal to a single-ended voltage. The LTC 1043 (Linear Technology) includes two sets of four switches to implement this circuit.

Composite Instrumentation Amplifiers

Instrumentation amplifiers have a very limited bandwidth. They achieve a gain of 10 at 2.5 MHz, at best. Moreover, their inputs must be either dc-coupled or, if ac-coupled with input series capacitors, there must be a path for bias currents; if that path is a resistor from each input to ground, then the common-mode input impedance Z_c decreases and noise may increase.

A larger bandwidth for a given gain can be obtained by cascade connection of two or more amplifiers. However, if the additional gain is provided by a single-ended amplifier after the IA, then the overall CMRR is that of the IA, which is small at high frequencies. The circuit in Figure 80.10a is a broadband IA with large CMRR because the CMRR for the second stage is multiplied by the differential gain for the first stage, which can be very high if implemented by broadband op amps. The overall gain is

$$G = G_1 G_2 = \left(1 + \frac{2R_b}{R_a} \right) G_{IA} \qquad (80.30)$$

An *IA* can be *ac-coupled* by feeding back its dc output to the reference terminal as shown in Figure 80.10b. The high-pass corner frequency is $f_0 = 1/(2\pi R_0 C_0)$.

80.6 Single-Ended Signal Conditioners

Floating signals (single ended or differential) can be connected to amplifiers with single-ended grounded input. Grounded single-ended can be connected to *single-ended amplifiers*, provided the difference in

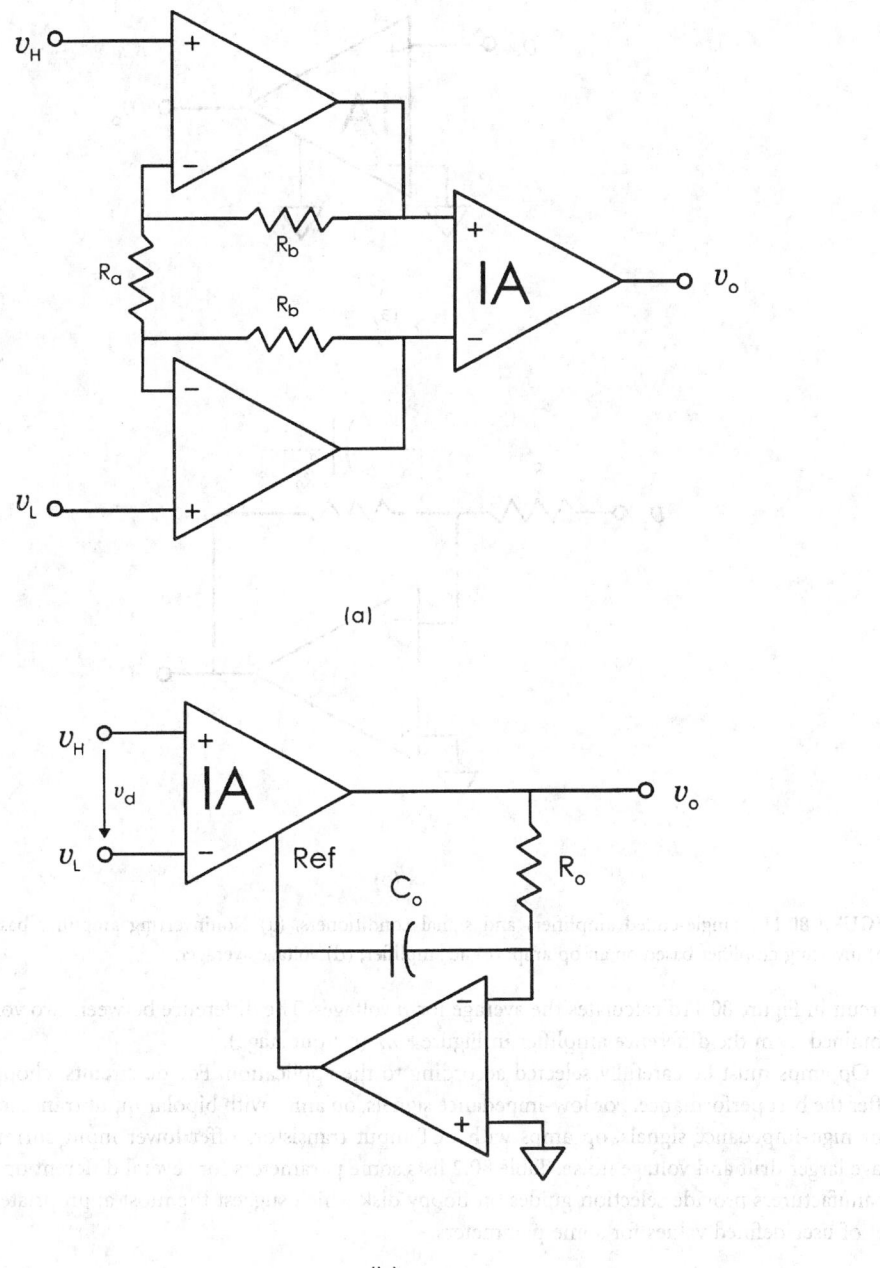

FIGURE 80.10 Composite instrumentation amplifiers. (a) Broadband IA with large CMRR; (b) ac-coupled IA.

ground potentials from signal to amplifier is not too large. Figure 80.11a shows a simple single-ended amplifier based on an IA. However, op amps are better suited than IAs for single-ended amplifiers and signal conditioners performing additional functions.

Figure 80.11b shows an *inverting amplifier* whose gain is $G = -R_2/R_1$, and whose input impedance is R_1. The capacitor on the dashed line (10 pF or larger) prevents gain peaking and oscillation. If a capacitor C replaces R_2, input signals are integrated and inverted. If C replaces R_1 instead, input signals are differentiated and inverted. The circuit in Figure 80.11c has $G = 1$ for dc and signals of low frequency relative to $f_1 = 1/(2\pi R_1 C_1)$ (offset and drift included) and $G = 1 + R_2/R_1$ for high-frequency signals. The

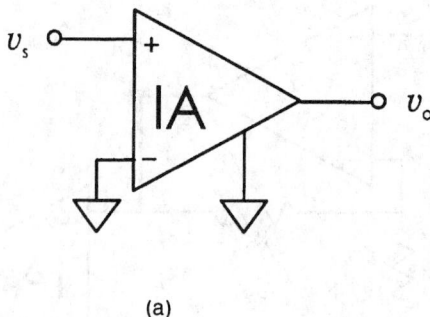

(a)

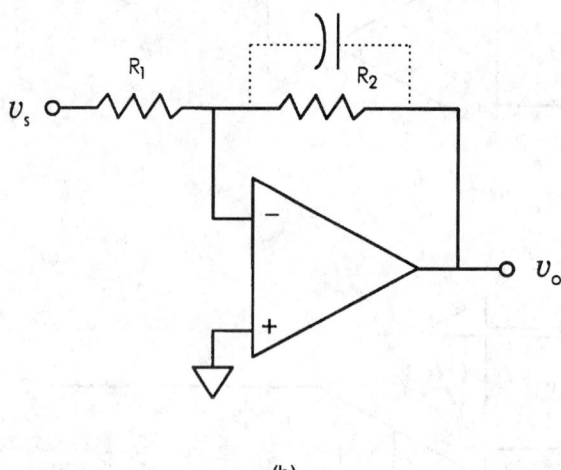

(b)

FIGURE 80.11 Single-ended amplifiers and signal conditioners. (a) Noninverting amplifier based on an IA; (b) inverting amplifier based on an op amp; (c) ac amplifier; (d) voltage averager.

circuit in Figure 80.11d calculates the average for n voltages. The difference between two voltages can be obtained from the difference amplifier in Figure 80.7 (output stage).

Op amps must be carefully selected according to the application. For dc circuits, chopper op amps offer the best performance. For low-impedance signals, op amps with bipolar input transistors are better. For high-impedance signals, op amps with FET input transistors offer lower input currents, but they have larger drift and voltage noise. Table 80.2 lists some parameters for several different op amps. Some manufacturers provide selection guides on floppy disk which suggest the most appropriate model for a set of user-defined values for some parameters.

80.7 Carrier Amplifiers

A **carrier amplifier** is a conditioner for extremely narrowband ac signals from ac-driven sensors. A carrier amplifier is made of a sine wave oscillator, to excite the sensor bridge, an ac voltage amplifier for the bridge output, a synchronous demodulator (see Chapter 84) and a low-pass filter (Figure 80.12). The NE5520/1 (Philips) are carrier amplifiers in IC form intended for (but not limited to) LVDTs driven at a frequency from 1 to 20 kHz.

Carrier amplifiers make it possible to recover the amplitude and phase of the modulating signal after amplifying the output modulated waveform from the bridge. This is useful first because ac amplifiers are not affected by offset, drift, or low-frequency noise, and therefore the bridge output can easily be

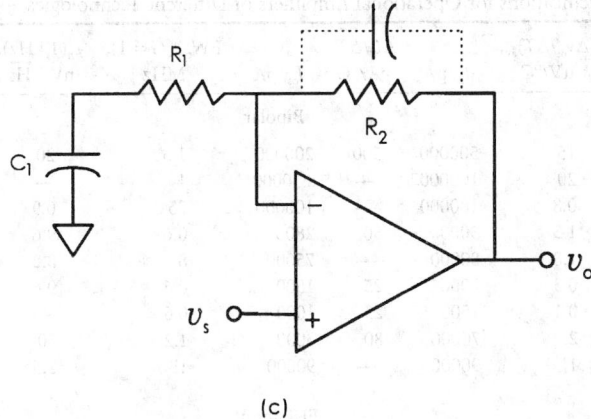

(c)

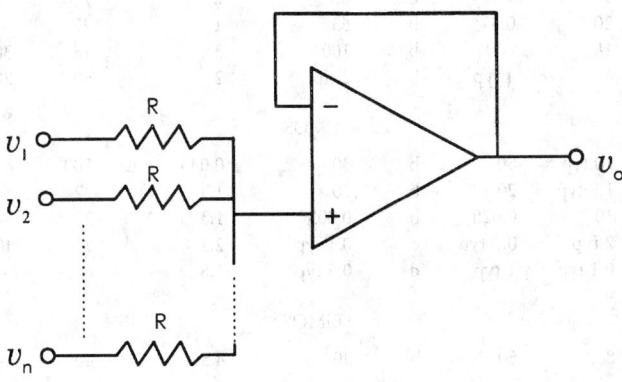

(d)

FIGURE 80.11 (continued)

amplified. Second, the *phase-sensitive demodulator* yields not only the amplitude but also the sign of the measurand. If the measurement range includes positive and negative values for the measurand, phase detection is essential.

A further advantage of carrier amplifiers is their extremely narrow frequency response, determined by the output low-pass filter. In the demodulator, the product of the modulated carrier of frequency f_c by the reference signal, also of frequency f_c, results in a baseband component and components at nf_c ($n \geq 2$). The output low-pass filter rejects components other than the baseband. If the corner frequency for this filter is f_0, then the passband for the system is $f_c \pm f_0$. Therefore, any interference of frequency f_i added to the modulated signal will be rejected if falling outside that passband. The ability to discriminate signals of interest from those added to them is described by the *series* (or *normal*) *mode rejection ratio* (SMRR), and is usually expressed in decibels. In the present case, using a first-order low-pass filter we have

$$\text{SMRR} = 20 \log \frac{v_o(f_c)}{v_o(f_i)} = 20 \log \frac{\sqrt{1 + \left(f_c - f_i\right)^2}}{f_0} \approx 20 \log \frac{\left|1 f_c - f_i\right|}{f_0} \quad (80.31)$$

TABLE 80.2 Basic Specifications for Operational Amplifiers of Different Technologies

	V_{os}, µV	$(\Delta v_{os}/\Delta T)_{av}$, µV/°C	I_B, pA	$\Delta I_B/\Delta T$, pA/°C	I_{os}, pA	$BW_{typ}(G=1)$, MHz	e_n(1 kHz), nV/√Hz	f_{ce}, Hz	$v_{n(p-p)}$, µV	i_n(1 kHz), fA/√Hz
				Bipolar						
µA741	6000	15	500000	500	200000	1.5	20	200	—	550
LM358A	3000	20	100000	—	±30000	1	—	—	—	—
LT1028	80	0.8	180000	—	100000	75	0.9	3.5	0.035	1000
OP07	75	1.3	3000	50	2800	0.6	9.6	10	0.35	170
OP27C	100	1.8	80000	—	75000	8	3.2	2.7	0.09	400
OP77A	25	0.3	2000	25	1500	0.6	9.6	10	0.35	170
OP177A	10	0.1	1500	25	1000	0.6	—	—	0.8	—
TLE2021C	600	2	70000	80	3000	1.2	30	—	0.47	90
TLE2027C	100	1	90000	—	90000	13	2.5	—	0.05	400
				FET input						
AD549K	250	5	0.1	b	0.03 typ	1	35	—	4	0.16
LF356A	2000	5	50	b	10	4.5	12	—	—	10
OPA111B	250	1	1	b	0.75	2	7	200	1.2	0.4
OPA128J	1000	20	0.3	b	65	1	27	—	4	0.22
TL071C	10000	18	200	b	100	3	18	300	4	10
TLE2061C	3000	6	4 typ	b	2 tip	2	60	20	1.2	1
				CMOS						
ICL7611A	2	10 typ	50	b	30	0.044	100	800	—	10
LMC660C	6000	1.3 typ	20	b	20	1.4	22	—	—	0.2
LMC6001A	350	10	0.025	b	0.005	1.3	22	—	—	0.13
TLC271CP	10000	2 typ	0.7 typ	c	0.1 typ	2.2	25	100	—	n.s.
TLC2201C	500	0.1 typ	1 typ	d	0.5 typ	1.8	8	—	0.7	0.6
				BiMOS						
CA3140	15000	8	50	b	30	4.5	40	—	—	—
				CMOS chopper						
LTC1052	5	0.05	30	e	30	1.2	—	—	1.5	0.6
LTC1150C	5	0.05	100	f	200	2.5	—	—	1.8	1.8
MAX430C	10	0.05	100	g	200	0.5	—	—	1.1	10
TLC2652AC	1	0.03	4 typ	d	2 typ	1.9	23	—	2.8	4
TLC2654C	20	0.3	50 typ	0.65	30 typ	1.9	13	—	1.5	4
TSC911A	15	0.15	70	—	20	1.5	—	—	11	—

Specified values are maximal unless otherwise stated and those for noise, which are typical (typ = typical, av = average; nonspecified parameters are indicated by a dash).

[a] Values estimated from graphs.
[b] I_B doubles every 10°C.
[c] I_B doubles every 7.25°C.
[d] I_B is almost constant up to 85°C.
[e] I_B is almost constant up to 75°C.
[f] I_{B+} and I_{B-} show a different behavior with temperature.
[g] I_B doubles every 10°C above about 65°C.

A power-line interference superimposed on a 10 kHz carrier will undergo an 80-dB attenuation if the output low-pass filter has $f_0 = 1$ Hz. The same interference superimposed on the baseband signal would be attenuated by only 35 dB.

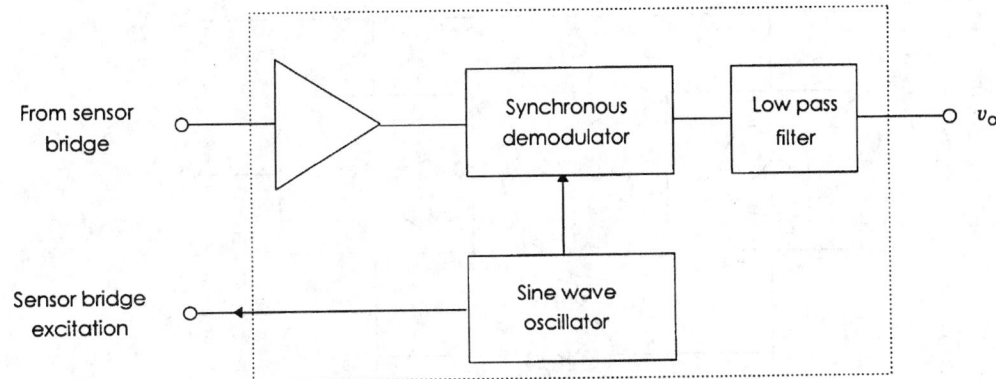

FIGURE 80.12 Elements for a carrier amplifier.

Carrier amplifiers can be built from a precision sine wave oscillator — AD2S99 (Analog Devices), 4423 (Burr-Brown), SWR300 (Thaler) — or a discrete-part oscillator, and a demodulator (plus the output filter). Some IC demodulators are based on switched amplifiers (AD630, OPA676). The floating capacitor in Figure 80.9 behaves as a synchronous demodulator if the switch clock is synchronous with the carrier, and its duty cycle is small (less than 10%), so that switches SW1 and SW2 sample the incoming modulated waveform for a very short time [7].

80.8   Lock-In Amplifiers

A *lock-in amplifier* is based on the same principle as a carrier amplifier, but instead of driving the sensor, here the carrier signal drives the experiment, so that the measurand is frequency translated. Lock-in amplifiers are manufactured as equipment intended for recovering signals immersed in high (asynchronous) noise. These amplifiers provide a range of driving frequencies and bandwidths for the output filter. Some models are vectorial because they make it possible to recover the in-phase and quadrature (90° out-of-phase) components of the incoming signal, by using two demodulators whose reference signals are delayed by 90°. Still other models use bandpass filters for the modulated signal and two demodulating stages. Meade [8] analyzes the fundamentals, specifications, and applications of some commercial lock-in amplifiers.

80.9   Isolation Amplifiers

The maximal common-mode voltage withstood by common amplifiers is smaller than their supply voltage range and seldom exceeds 10 V. Exceptions are the INA 117 (Burr-Brown) and similar difference amplifiers whose common-mode range is up to ±200 V, and the IA in Figure 80.9 when implemented by high-voltage switches (relays, optorelays). Signals with large off-ground voltages, or differences in ground potentials exceeding the input common-mode range, result in permanent amplifier damage or destruction, and a safety risk, in spite of an exceptional CMRR: a 100 V common-mode 60 Hz voltage at the input of a common IA having a 120 dB CMRR at power-line frequency does not result in a 100 V/10^6 = 100 μV output, but a burned-out IA.

Figure 80.13a shows a signal source grounded at a point far from the amplifier ground. The difference in voltage between grounds v_i not only thwarts signal measurements but can destroy the amplifier. The solution is to prevent this voltage from forcing any large current through the circuit and at the same time to provide an information link between the source and the amplifier. Figure 80.13b shows a solution: the signal source and the amplifier have separated (isolated) power supplies and the signal is coupled to the amplifier through a transformer acting as an isolation barrier for v_i. Other possible barriers are optocouplers (IL300-Siemens) and series capacitors (LTC1145-Linear Technology). Those barriers impose a large series impedance (isolation impedance, Z_i) but do not usually have a good low-frequency

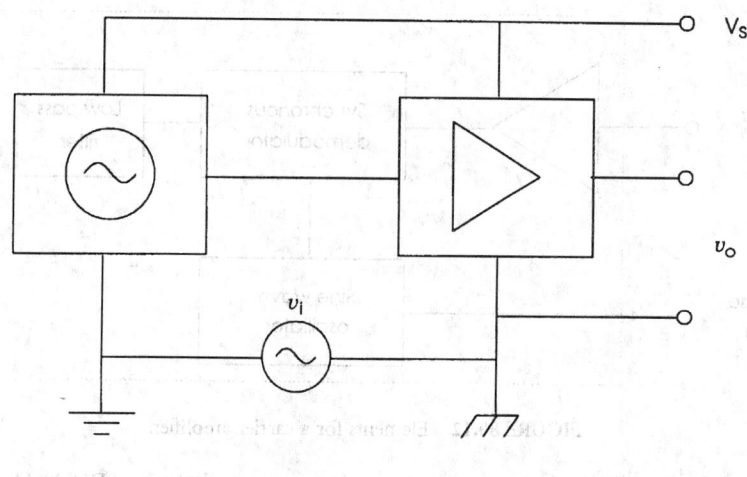

(a)

(b)

FIGURE 80.13   (a) A large difference in ground potentials damages amplifiers. (b) An isolation amplifier prevents large currents caused by this difference from flowing through the circuit.

response, hence the need to modulate and then demodulate the signal to transfer through it. The subsystem made of the modulator and demodulator, plus sometimes an input and an output amplifier and a dc–dc converter for the separate power supply, is called an **isolation amplifier**. The ability to reject the voltage difference across the barrier (isolation-mode voltage, v_i) is described by the **isolation mode rejection ratio** (IMRR), expressed in decibels,

$$IMRR = 20 \log \frac{\text{OUTPUT Voltage}}{\text{ISOLATION - MODE Voltage}} \qquad (80.32)$$

Ground isolation also protects people and equipment from contact with high voltage because Z_i limits the maximal current. Some commercial isolation amplifiers are the AD202, AD204, and AD210 (Analog Devices) and the ISOxxx series (Burr-Brown).

Table 80.3 summarizes the compatibility between signal sources and amplifiers. When grounded, amplifiers and signals are assumed to be grounded at different physical points.

80.10 Nonlinear Signal-Processing Techniques

Limiting and Clipping

Clippers or *amplitude limiters* are circuits whose output voltage has an excursion range restricted to values lower than saturation voltages. Limiting is a useful signal processing technique for signals having the information encoded in parameters other than the amplitude. For example, amplitude limiting is convenient before homodyne phase demodulators. *Limiting* can also match output signals levels to those required for TTL circuits (0 to 5 V). Limiting avoids amplifier saturation for large input signal excursions, which would result in a long recovery time before returning to linear operation.

Limiting can be achieved by op amps with diodes and zeners in a feedback loop. Figure 80.14a shows a positive voltage clipper. When v_s is positive, v_o is negative and R_2/R_1 times larger; the diode is reverse biased and the additional feedback loop does not affect the amplifier operation. When v_s is negative and large enough, v_o forward biases the diode (voltage drop V_f) and the zener clamps at V_z, the output amplitude thus being limited to $v_o = V_f + V_z$ until $|v_s| < (V_f + V_z)R_1/R_2$ (Figure 80.14b). The circuit then acts again as an inverting amplifier until v_s reaches a large negative value.

A negative voltage clipper can be designed by reversing the polarity of the diode and zener. To limit the voltage in both directions, the diode may be substituted by another zener diode. The output is then limited to $|v_o| < V_{z1} + V_{f2}$ for negative inputs to $|v_o| < V_{f1} + V_{z2}$ for positive inputs. If $V_{z1} = V_{z2}$, then the voltage limits are symmetrical. Jung [9] gives component values for several precision limiters.

Logarithmic Amplification

The dynamic range for common linear amplifiers is from 60 to 80 dB. Sensors such as photodetectors, ionizing radiation detectors, and ultrasound receivers can provide signals with an amplitude range wider than 120 dB. The only way to encompass this wide amplitude range within a narrower range is by amplitude compression. A logarithmic law compresses signals by offering equal-output amplitude changes in response to a given ratio of input amplitude increase. For example, a scaling of 1 V/decade means that the output would change by 1 V when the input changes from 10 to 100 mV, or from 100 mV to 1 V. Therefore, *logarithmic amplifiers* do not necessarily amplify (enlarge) input signals. They are rather converters providing a voltage or current proportional to the ratio of the input voltage, or current, to a reference voltage, or current.

Logarithmic conversion can be obtained by connecting a bipolar transistor as a feedback element of an op amp, Figure 80.15a. The collector current i_C and the base-emitter voltage have an exponential relationship. From the Ebers–Moll model for a transistor, if $v_{CB} = 0$, then

$$i_C = I_S \left(e^{v_{BE}/v_T} - 1 \right) \tag{80.33}$$

where $v_T = kT/q = 25$ mV at room temperature, and I_S is the saturation current for the transistor. In Figure 80.15a the input voltage is converted into an input current and the op amp forces the collector current of the transistor to equal the input current, while maintaining $v_{CB} \approx 0$ V. Hence, provided $i_C \gg I_S$, for $v_s > 0$,

$$v_o = \frac{v_T}{\log e} \log \frac{v_s}{RI_S} \tag{80.34}$$

The basic circuit in Figure 80.15a must be modified in order to provide temperature stability, phase compensation, and scale factor correction; reduce bulk resistance error; protect the base-emitter junction;

TABLE 80.3   Compatibility between Signal Sources and Conditioners

Conditioner Input / Signal source				
	Incompatible unless grounds are very close	Compatible if CMRR is large	Compatible	Compatible
	Compatible	Compatible	Compatible	Compatible
	Incompatible unless grounds are very close	Compatible if CMRR is large	Compatible for large Z_i	Compatible
	Incompatible	Compatible	Compatible for large Z_i	Compatible
	Compatible	Compatible	Compatible	Compatible
	Incompatible	Compatible if CMRR is large	Compatible for large Z_i	Compatible

Note: When grounded, signals sources and amplifiers are assumed to be grounded at different points. Isolation impedance is assumed to be very high for floating signal sources but finite (Z_i) for conditioners.

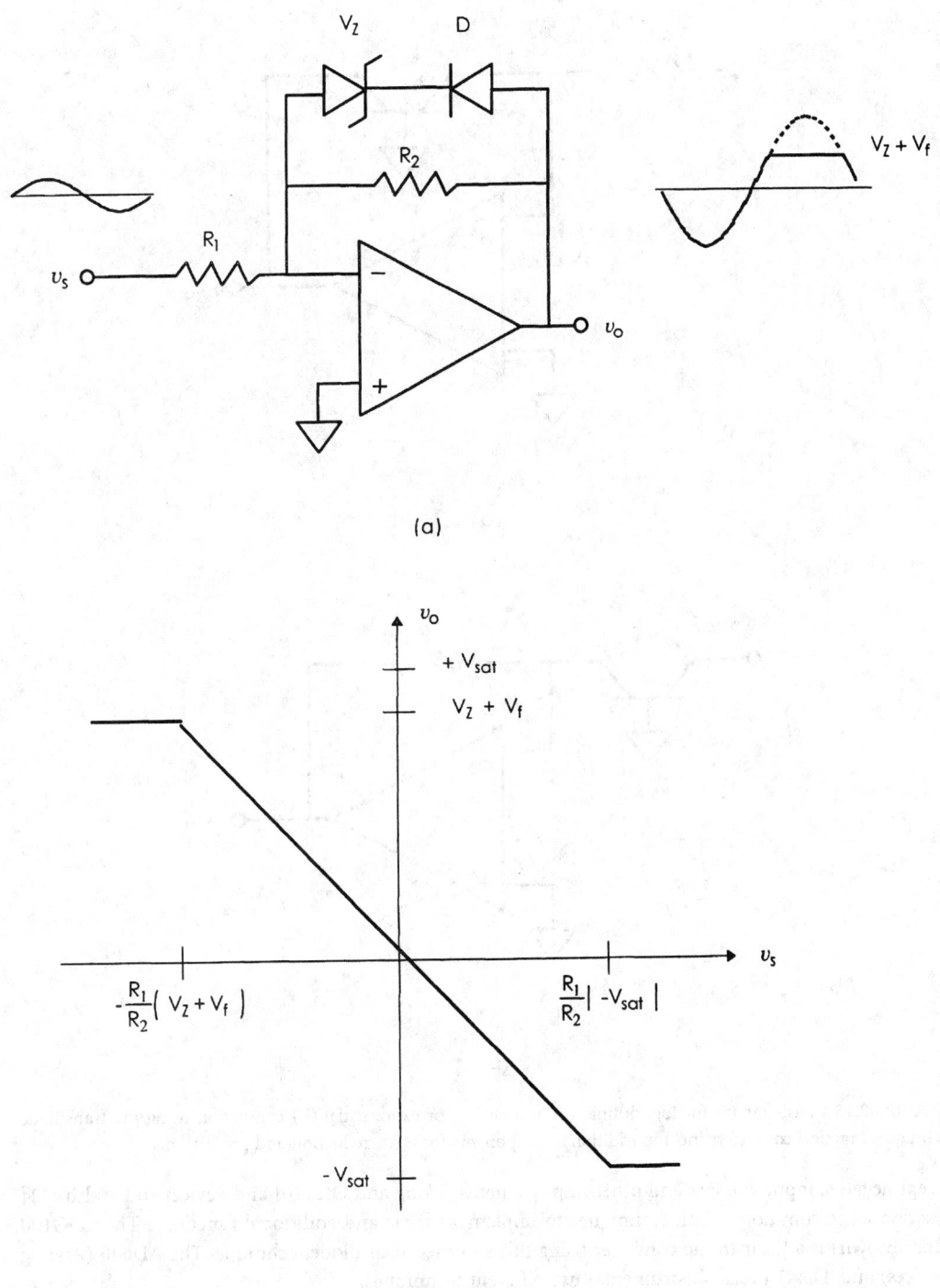

(a)

(b)

FIGURE 80.14 Voltage limiter. (a) Circuit based on op amp and diode network feedback. (b) Input/output relationship.

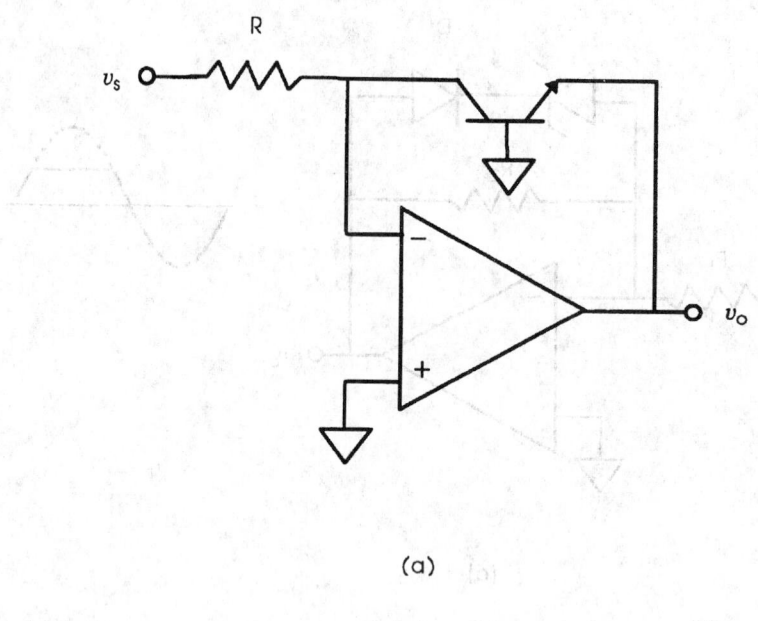

(a)

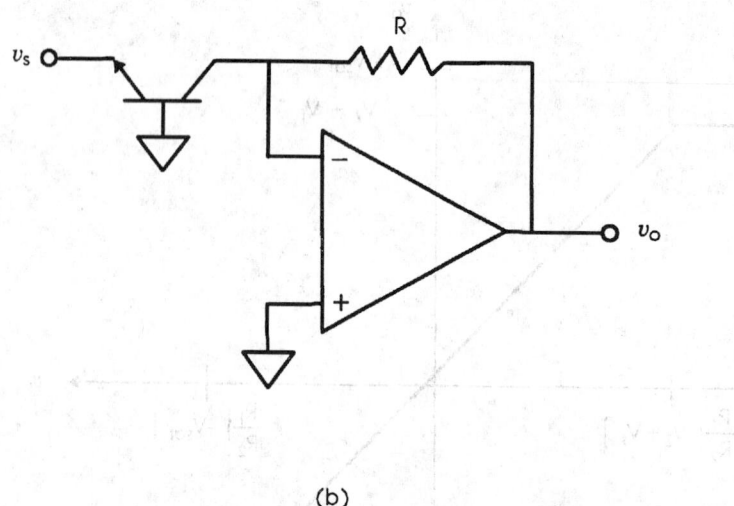

(b)

FIGURE 80.15 Basic circuit for logarithmic (a) and antilog or exponential (b) conversion using the transdiode technique. Practical converters include additional components for error reduction and protection.

accept negative input voltages and other improvements. Wong and Ott [10] and Peyton and Walsh [11] describe some common circuit techniques to implement these and additional functions. The LOG100 (Burr-Brown) is a logarithmic converter using this so-called transdiode technique. The AD640 (Analog Devices) and TL441 (Texas Instruments) use different techniques.

Figure 80.15b shows a basic *antilog* or *exponential converter* for negative input voltages. The transistor and the resistor have interchanged positions with respect to Figure 80.15b. For $v_s < 0$,

$$v_o = I_s \, R e^{v_s/v_T} \tag{80.35}$$

Positive voltages require an input *pnp* transistor instead.

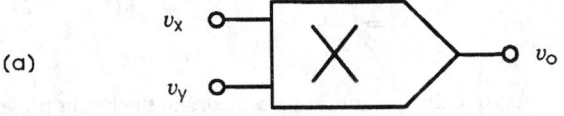

(a)

(b)

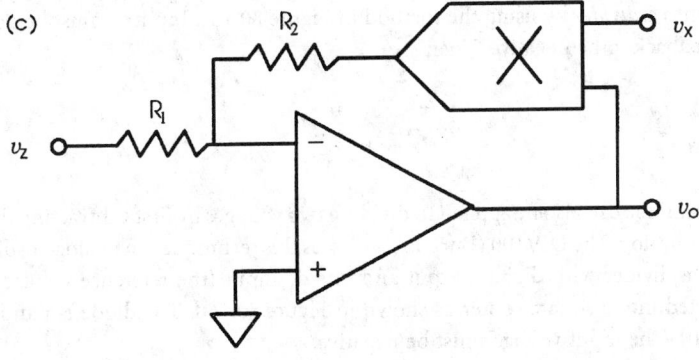

(c)

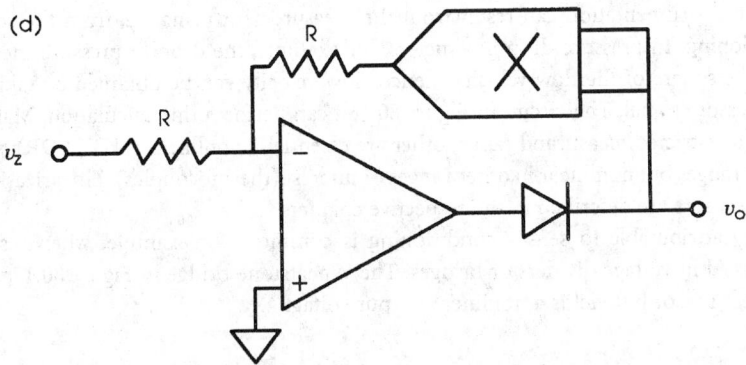

(d)

FIGURE 80.16 (a) Symbol for an analog multiplier. (b) Voltage squarer from an analog multiplier. (c) Two-quadrant analog divider from a multiplier and op amp feedback. (d) Square rooter from a multiplier and op amp feedback.

Multiplication and Division

Analog multiplication is useful not only for analog computation but also for modulation and demodulation, for voltage-controlled circuits (amplifiers, filters) and for linearization [12]. An *analog multiplier* (Figure 80.16a) has two input ports and one output port offering a voltage

$$v_o = \frac{v_x v_y}{V_m} \tag{80.36}$$

where V_m is a constant voltage. If inputs of either polarity are accepted, and their signs preserved, the device is a *four-quadrant multiplier*. If one input is restricted to have a defined polarity but the other can change sign, the device is a *two-quadrant multiplier*. If both inputs are restricted to only one polarity, the device is an *one-quadrant multiplier*. By connecting both inputs together, we obtain a voltage squarer (Figure 80.16b).

Wong and Ott[10] describe several multiplication techniques. At low frequencies, one-quadrant multipliers can be built by the log–antilog technique, based on the mathematical relationships $\log A + \log B = \log AB$ and then antilog $(\log AB) = AB$. The AD538 (Analog Devices) uses this technique. Currently, the most common multipliers use the transconductance method, which provides four-quadrant multiplication and differential ports. The AD534, AD633, AD734, AD834/5 (Analog Devices), and the MPY100 and MPY600 (Burr-Brown), are *transconductance multipliers*. A digital-to-analog converter can be considered a multiplier accepting a digital input and an analog input (the reference voltage). A multiplier can be converted into a *divider* by using the method in Figure 80.16c. Input v_x must be positive in order for the op amp feedback to be negative. Then

$$v_o = -V_m \frac{R_2}{R_1} \frac{v_z}{v_x} \tag{80.37}$$

The log–antilog technique can also be applied to dividing two voltages by first subtracting their logarithms and then taking the antilog. The DIV100 (Burr-Brown) uses this technique. An analog-to-digital converter can be considered a divider with digital output and one dc input (the reference voltage). A multiplier can also be converted into a square rooter as shown in Figure 80.16d. The diode is required to prevent circuit latch-up [10]. The input voltage must be negative.

80.11 Analog Linearization

Nonlinearity in instrumentation can result from the measurement principle, from the sensor, or from sensor conditioning. In pressure-drop flowmeters, for example, the drop in pressure measured is proportional to the square of the flow velocity; hence, flow velocity can be obtained by taking the square root of the pressure signal. The circuit in Figure 80.16d can perform this calculation. Many sensors are linear only in a restricted measurand range; other are essentially nonlinear (NTC, LDR); still others are linear in some ranges but nonlinear in other ranges of interest (thermocouples). Linearization techniques for particular sensors are described in the respective chapters.

Nonlinearity attributable to sensor conditioning is common, for example, when resistive (linear) sensors are placed in voltage dividers or bridges. The Wheatstone bridge in Figure 80.17a, for example, includes a linear sensor but yields a nonlinear output voltage,

$$v_s = V\left(\frac{1+x}{2+x} - \frac{1}{2}\right) = \frac{Vx}{2(2+x)} \tag{80.38}$$

The nonlinearity arises from the dependence of the current through the sensor on its resistance, because the bridge is supplied at a constant voltage. The circuit in Figure 80.17b provides a solution based on one op amp which forces a constant current V/R_0 through the sensor. The bridge output voltage is

$$v_s = \frac{V + v_a}{2} = V\frac{x}{2} \tag{80.39}$$

In addition, v_s is single ended. The op amp must have a good dc performance.

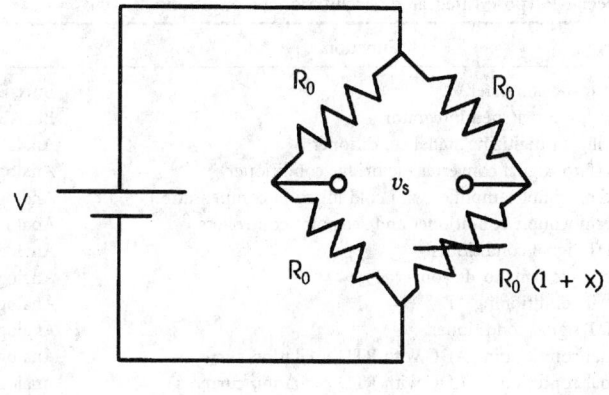

(a)

(b)

FIGURE 80.17   (a) A Wheatstone bridge supplied at a constant voltage and including a single sensor provides a nonlinear output voltage. (b) By adding an op amp which forces a constant current through the sensor, the output voltage is linearized.

TABLE 80.4   Special-Purpose Integrated Circuit Signal Conditioners

Model	Function	Manufacturer
4341	rms-to-dc converter	Burr-Brown
ACF2101	Low-noise switched integrator	Burr-Brown
AD1B60	Intelligent digitizing signal conditioner	Analog Devices
AD2S93	LVDT-to-digital converter (ac bridge conditioner)	Analog Devices
AD594	Thermocouple amplifier with cold junction compensation	Analog Devices
AD596/7	Thermocouple conditioner and set-point controllers	Analog Devices
AD598	LVDT signal conditioner	Analog Devices
AD636	rms-to-dc (rms-to-dc converter)	Analog Devices
AD670	Signal conditioning ADC	Analog Devices
AD698	LVDT signal conditioner	Analog Devices
AD7710	Signal conditioning ADC with RTD excitation currents	Analog Devices
AD7711	Signal conditioning ADC with RTD excitation currents	Analog Devices
IMP50E10	Electrically programmable analog circuit	IMP
LM903	Fluid level detector	National Semiconductor
LM1042	Fluid level detector	National Semiconductor
LM1819	Air-core meter driver	National Semiconductor
LM1830	Fluid detector	National Semiconductor
LT1025	Thermocouple cold junction compensator	Linear Technology
LT1088	Wideband rms-to-dc converter building block	Linear Technology
LTK001	Thermocouple cold junction compensator and matched amplifier	Linear Technology
TLE2425	Precision virtual ground	Texas Instruments

80.12   Special-Purpose Signal Conditioners

Table 80.4 lists some signal conditioners in IC form intended for specific sensors and describes their respective functions. The decision whether to design a signal conditioner from parts or use a model from Table 80.4 is a matter of cost, reliability, and availability. Signal conditioners are also available as subsystems (plug-in cards and modules), for example, series MB from Keithley Metrabyte, SCM from Burr-Brown, and 3B, 5B, 6B, and 7B from Analog Devices.

Defining Terms

Carrier amplifier: Voltage amplifier for narrowband ac signals, that includes in addition a sine wave oscillator, a synchronous demodulator, and a low-pass filter.

Common-mode rejection ratio (CMRR):   The gain for a differential voltage divided by the gain for a common-mode voltage in a differential amplifier. It is usually expressed in decibels.

Common-mode voltage: The average of the voltages at the input terminals of a differential amplifier.

Differential amplifier: Circuit or device that amplifies the difference in voltage between two terminals, none of which is grounded.

Dynamic range: The measurement range for a quantity divided by the desired resolution.

Instrumentation amplifier: Differential amplifier with large input impedance and low offset and gain errors.

Isolation amplifier: Voltage amplifier whose ground terminal for input voltages is independent from the ground terminal for the output voltage (i.e., there is a large impedance between both ground terminals).

Isolation Mode Rejection Ratio (IMRR):   The amplitude of the output voltage of an isolation amplifier divided by the voltage across the isolation impedance yielding that voltage.

Signal conditioner: Circuit or device that adapts a sensor signal to an ensuing circuit, such as an analog-to-digital converter.

Voltage buffer: Voltage amplifier whose gain is 1, or close to 1, and whose input impedance is very large while its output impedance is very small.

References

1. R. Pallás-Areny and J.G. Webster, *Sensors and Signal Conditioning*, New York: John Wiley & Sons, 1991.
2. J. Graeme, *Photodiode Amplifiers, Op Amp Solutions*, New York: McGraw-Hill, 1996.
3. C. Kitchin and L. Counts, *Instrumentation Amplifier Application Guide*, 2nd ed., Application Note, Norwood, MA: Analog Devices, 1992.
4. C.D. Motchenbacher and J.A. Connelly, *Low-Noise Electronic System Design*, New York: John Wiley & Sons, 1993.
5. S. Franco, *Design with Operational Amplifiers and Analog Integrated Circuits*, 2nd ed., New York: McGraw-Hill, 1998.
6. R. Pallás-Areny and J.G. Webster, Common mode rejection ratio in differential amplifiers, *IEEE Trans. Instrum. Meas.*, 40, 669–676, 1991.
7. R. Pallás-Areny and O. Casas, A novel differential synchronous demodulator for ac signals, *IEEE Trans. Instrum. Meas.*, 45, 413–416, 1996.
8. M.L. Meade, *Lock-in Amplifiers: Principles and Applications*, London: Peter Peregrinus, 1984.
9. W.G. Jung, *IC Op Amp Cookbook*, 3rd ed., Indianapolis, IN: Howard W. Sams, 1986.
10. Y.J. Wong and W.E. Ott, *Function Circuits Design and Application*, New York: McGraw-Hill, 1976.
11. A.J. Peyton and V. Walsh, *Analog Electronics with Op Amps*, Cambridge, U.K.: Cambridge University Press, 1993.
12. D.H. Sheingold, Ed., *Multiplier Application Guide*, Norwood, MA: Analog Devices, 1978.

Further Information

B.W.G. Newby, *Electronic Signal Conditioning*, Oxford, U.K.: Butterworth-Heinemann, 1994, is a book for those in the first year of an engineering degree. It covers analog and digital techniques at beginners' level, proposes simple exercises, and provides clear explanations supported by a minimum of equations.

P. Horowitz and W. Hill, *The Art of Electronics*, 2nd ed., Cambridge, U.K.: Cambridge University Press, 1989. This is a highly recommended book for anyone interested in building electronic circuits without worrying about internal details for active components.

M.N. Horenstein, *Microelectronic Circuits and Devices*, 2nd ed., Englewood Cliffs, NJ: Prentice-Hall, 1996, is an introductory electronics textbook for electrical or computer engineering students. It provides many examples and proposes many more problems, for some of which solutions are offered.

J. Dostál, *Operational Amplifiers*, 2nd ed., Oxford, U.K.: Butterworth-Heinemann, 1993, provides a good combination of theory and practical design ideas. It includes complete tables which summarizes errors and equivalent circuits for many op amp applications.

T.H. Wilmshurst, *Signal Recovery from Noise in Electronic Instrumentation*, 2nd ed., Bristol, U.K.: Adam Hilger, 1990, describes various techniques for reducing noise and interference in instrumentation. No references are provided and some demonstrations are rather short, but it provides insight into very interesting topics.

Manufacturers' data books provide a wealth of information, albeit nonuniformly. Application notes for special components should be consulted before undertaking any serious project. In addition, application notes provide handy solutions to difficult problems and often inspire good designs. Most manufacturers offer such literature free of charge. The following have shown to be particularly useful and easy to obtain: *1993 Applications Reference Manual*, Analog Devices; *1994 IC Applications Handbook*, Burr-Brown; *1990 Linear Applications Handbook* and *1993 Linear Applications Handbook* Vol. II, Linear Technology; *1994 Linear Application Handbook*, National Semiconductor; *Linear and Interface Circuit Applications*, Vols. 1, 2, and 3, Texas Instruments.

R. Pallás-Areny and J.G. Webster, *Analog Signal Processing*, New York: John Wiley & Sons, 1999, offers a design-oriented approach to processing instrumentation signals using standard analog integrated circuits, that relies on signal classification, analog domain conversions, error analysis, interference rejection and noise reduction, and highlights differential circuits.

References

1. J.S. Rose, Albert and G. Webster, *Sensors and Signal Conditioning*, New York: John Wiley & Sons, 1991.
2. Electronic Principles Amplifiers, Op Amp Schematics, New York: McGraw-Hill, 1995.
3. G. Kahn and I.O. Connor, *Instrumentation Amplifier Application Guide*, 2nd ed., Application Note, Norwood, MA: Analog Devices, 1992.
4. P.D. Hatchkinder and L.W. Couch, *Digital and Analog Communication System Design*, New York: John Wiley & Sons, 1995.
5. S.K. Franco, *Design with Operational Amplifiers and Analog Integrated Circuits*, 2nd ed., New York: McGraw-Hill, 1998.
6. R. Pallás-Areny and J.G. Webster, Common mode rejection ratio in differential amplifiers, *IEEE Trans. Instrum. Meas.*, 40, 669–676, 1991.
7. R. Pallás-Areny and J.G. Webster, A novel differential synchronous demodulator for ac signals, *IEEE Trans. Instrum. Meas.*, 45, 413–416, 1991.
8. M.J. McReady, *IC Op-Amp Cookbook*, 3rd ed., Indianapolis, IN: Howard W. Sams, 1986.
9. W.G. Jung, *IC Op Amp Cookbook*, 3rd ed., Indianapolis, IN: Howard W. Sams, 1986.
10. J.V. Wait and L.P. Huelsman, *Introduction to Operational Amplifier Theory and Applications*, New York: McGraw-Hill, 1975.
11. A.J. Peyton and V. Walsh, *Analog Electronics with Op Amps*, Cambridge, U.K.: Cambridge University Press, 1993.
12. D.H. Sheingold, *Analog-Digital Conversion Handbook*, Norwood, MA: Analog Devices, 1986.

Further Information

J. Dostál, *Operational Amplifiers*, 2nd ed., Oxford, U.K.: Butterworth-Heinemann, 1993, is a book for those in the first year of engineering degree. It covers analog and signal techniques at beginner's level, proposes simple exercises, and provides clear explanations. I have minimum of equations.

P. Horowitz and W. Hill, *The Art of Electronics*, 2nd ed., Cambridge, U.K.: Cambridge University Press, 1990. This is a highly recommended book for anyone interested in building electronic circuits without worrying about internal details of each component.

P. Horowitz and I. Robinson, *Student Manual for the Art of Electronics*, U.K.: Cambridge University Press, 1989, is an introductory guide to electronics intended mainly for students. It provides many examples and proposes many more problems. For some of all solutions are offered.

J. Dostál, *Operational Amplifiers*, 2nd ed., Oxford, U.K.: Butterworth-Heinemann, 1993, provides a comprehensive theory and practical design information including complete tables which summarizes several characteristics of circuits for most op amp applications.

T.H. Wilmshurst, *Signal Recovery from Noise in Electronic Instrumentation*, 2nd ed., Bristol, U.K.: Adam Hilger, 1990, describes various techniques for reducing noise and interference in instrumentation. No electronics are provided and some demonstrations are rather short but it provides insight into very interesting topics.

Manufacturers data books provide a wealth of information about nonuniformity. Application notes for several components should be consulted before undertaking any serious project. In addition application notes provide handy solutions to difficult problems and often imply good design. Most manufacturers offer such literature free of charge. The following have shown to be particularly useful and easy to obtain: 1992 *Applications Reference Manual, Analog Devices*, 1994; R.J. *Applications Handbook*, Burr-Brown, 1990; *Linear Applications Handbook* and 1993 *Linear Applications Handbook Vol. II*, *Linear Technology*, 1993; *Linear Applications Handbook* and *Signal Conditioning*, *National Semiconductor*; *Linear Circuits Applications Vols. 1, 2, and 3*, *Texas Instruments*.

R. Pallás-Areny and J.G. Webster, *Analog Signal Processing*, New York: John Wiley & Sons, 1999, offers a design-oriented approach to processes in instrumentation systems with standard analog integrated circuits that rely on transfer functions, and proposes equivalent circuits whose analysis increases reaction and noise reduction, and highlights different circuits.

81

Modulation

David M. Beams
University of Texas at Tyler

81.1 Introduction

It is often the case in instrumentation and communication systems that an information-bearing signal may not be in an optimal form for direct use. In such cases, the information-bearing signal may be used to alter some characteristic of a second signal more suited to the application. This process of altering one signal by means of another is known as **modulation**; the original information is called the *baseband signal*, and the signal modulated by the baseband signal is termed the *carrier* (because it "carries" the information). Recovery of the original information requires a suitable demodulation process to reverse the modulation process.

A prominent use of modulation techniques is found in radio communication. The extremely long wavelengths of electromagnetic waves at frequencies found in a typical audio signal make direct transmission impractical, because of constraints on realistic antenna size and bandwidth. Successful radio communication is made possible by using the original audio (baseband) signal to modulate a carrier signal of a much higher frequency and transmitting the modulated carrier by means of antennas of feasible size. Another example is found in the use of modems to transmit digital data by the telephone network. Digital data are not directly compatible with analog local subscriber connections, but these data may be used to modulate audible signals which may be carried over local telephone lines. Instrumentation systems use modulation techniques for telemetry (where the distances may be on the order of centimeters for implanted medical devices to hundreds of millions of kilometers for deep-space probes), for processing signals in ways for which the original signals are unsuited (such as magnetic recording of low-frequency and dc signals), and for specialized amplification purposes (carrier and lock-in amplifiers).

Techniques that modulate the amplitude of the carrier are full-carrier **amplitude modulation** (AM), reduced- or suppressed-carrier double-sideband amplitude modulation (DSB), single-sideband suppressed-carrier modulation (SSB), vestigial-sideband modulation (VSB), and on–off keying (OOK). Techniques that modulate the frequency or phase angle of the carrier include **frequency modulation**

(FM), phase modulation (PM), frequency-shift keying (FSK), and phase-shift keying (PSK). Simultaneous variation of amplitude and phase are applied in quadrature amplitude modulation (QAM). Each technique has its own particular uses. Full-carrier AM is used in radio broadcasting; VSB is used in television broadcasting. DSB appears in instrumentation systems utilizing carrier amplifiers and modulating sensors, while SSB finds use in certain high-frequency radio communications. FM is used in broadcasting (radio and television audio) and point-to-point mobile communications. OOK is commonly used to transmit digital data in optical fiber links. FSK, PSK, and QAM are found in digital communications; analog QAM carries the chrominance (color) information in color television broadcasting. The emphasis of this particular chapter will be instrumentation systems; those interested principally in communications applications could begin by consulting References 1 through 4.

81.2 Generalized Modulation

We begin by making two assumptions: (1) the highest frequency present in the baseband signal is considerably less than the carrier frequency and (2) the results derived in the following chapter pertain to sinusoidal carriers but may be extended to other periodic carrier signals (such as square waves and triangle waves). Equation 81.1 gives a general expression for a modulated sinusoidal carrier signal of radian frequency ω_c:

$$f_s(t) = A_c(t) \cos\left[\omega_c t + \phi(t)\right] \tag{81.1}$$

Information may be carried by $f_s(t)$ by modulation of its amplitude $A_c(t)$, its phase angle $\phi(t)$, or, in some cases, both (note that frequency modulation is a form of phase **angle modulation**). Equation 81.1 may be recast in an equivalent form:

$$f_s(t) = f_i(t) \cos(\omega_c t) - f_q(t) \sin(\omega_c t) \tag{81.2}$$

where $f_i(t) = A_c(t) \cos[\phi(t)]$
$\quad\quad f_q(t) = A_c(t) \sin[\phi(t)]$

Equation 81.2 gives $f_s(t)$ as the sum of a cosinusoidal carrier term with time-varying amplitude $f_i(t)$ and a sinusoidal (quadrature) carrier term with time-varying amplitude $f_q(t)$ and is thus known as the carrier-quadrature description of $f_s(t)$. The terms $f_i(t)$ and $f_q(t)$ are known, respectively, as the in-phase and quadrature components of $f_s(t)$. Carlson [1] gives the Fourier transform of a signal in carrier-quadrature form:

$$F_s(\omega) = \frac{1}{2}\left[F_i(\omega - \omega_c) + F_i(\omega + \omega_c)\right] + \frac{j}{2}\left[F_q(\omega - \omega_c) + F_q(\omega + \omega_c)\right] \tag{81.3}$$

where $F_i(\omega)$: $f_i(t)$ and $F_q(\omega)$: $f_q(t)$ are Fourier transform pairs. Notice that the spectra of both $F_i(\omega)$ and $F_q(\omega)$ are both translated by $\pm\omega_c$. Modulation of the carrier in any sense causes energy to appear at frequencies (known as *sidebands*) other than the carrier frequency. Sidebands will be symmetrically distributed relative to the carrier in all but the specialized cases of VSB and single-SSB.

81.3 Amplitude Modulation

AM that appears in instrumentation systems takes the form of double-sideband AM on which we will focus some degree of attention. VSB and SSB are encountered in communications systems but not in instrumentation systems; interested readers may refer to References 1 through 3.

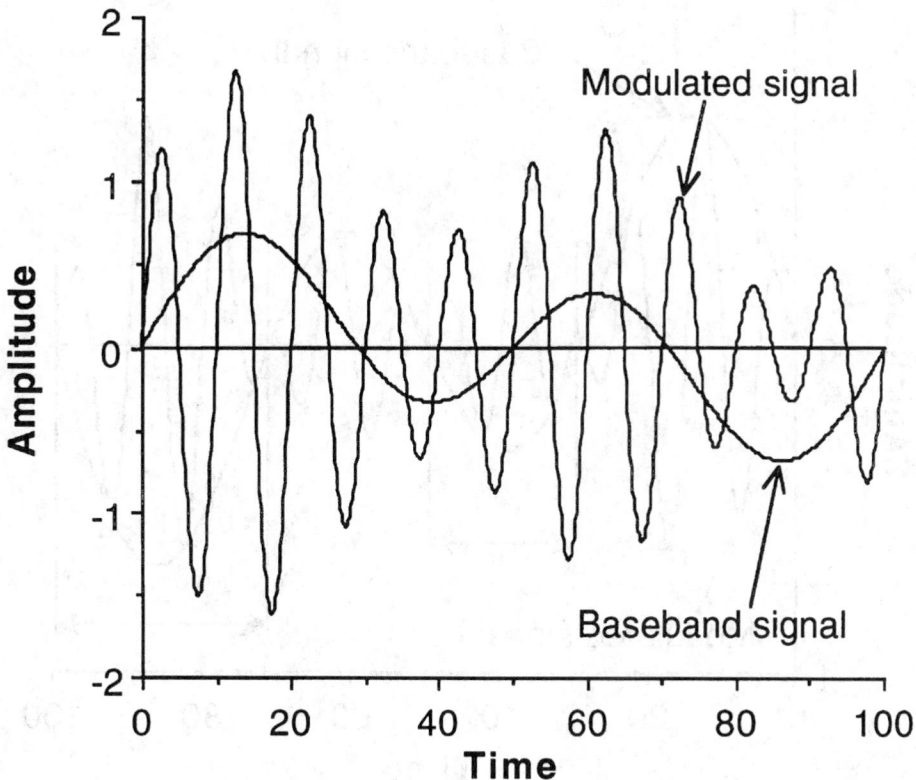

FIGURE 81.1 Time-domain representation of a baseband and the resulting full-carrier AM signal. The time scale is arbitrary.

Double-Sideband Amplitude Modulation

AM applied to a sinusoidal carrier is described by

$$f_s(t) = A_c\left[k + \mu f_m(t)\right]\cos(\omega_c t) \qquad (81.4)$$

where A_c is the amplitude of the unmodulated carrier, k is the proportion of carrier present in the modulated signal, μ is the modulation index, $f_m(t)$ is the modulating baseband signal (presumed to be a real bandpass signal), and ω_c is the carrier radian frequency. The modulation index relates the change in amplitude of the modulated signal to the amplitude of the baseband signal. The value of k ranges from 1 in full-carrier AM to 0 in suppressed-carrier double-sideband modulation. The peak value of the modulated signal is $k + \mu f_m(t)$ which may take on any positive value consistent with the dynamic range of the modulator and demodulating system; note that phase reversal of the carrier occurs if $k + \mu f_m(t)$ becomes negative. Figure 81.1 represents a full–carrier AM signal and its baseband signal. Recasting Equation 81.4 in carrier-quadrature form gives $f_i = A_c[k + \mu f_m(t)]$ and $f_q = 0$. The Fourier transform of this signal is

$$F_s(\omega) = \frac{A_c}{2}\left\{ k\delta(\omega - \omega_c) + k\delta(\omega + \omega_c) + \mu\left[F_m(\omega - \omega_c) + F_m(\omega + \omega_c)\right]\right\} \qquad (81.5)$$

where $\delta(\omega - \omega_c)$ and $\delta(\omega + \omega_c)$ are unit impulses at $+\omega_c$ and $-\omega_c$, respectively, and represent the carrier component of $F_s(\omega)$. The frequency domain representation of $F_s(\omega)$ also contains symmetric sidebands

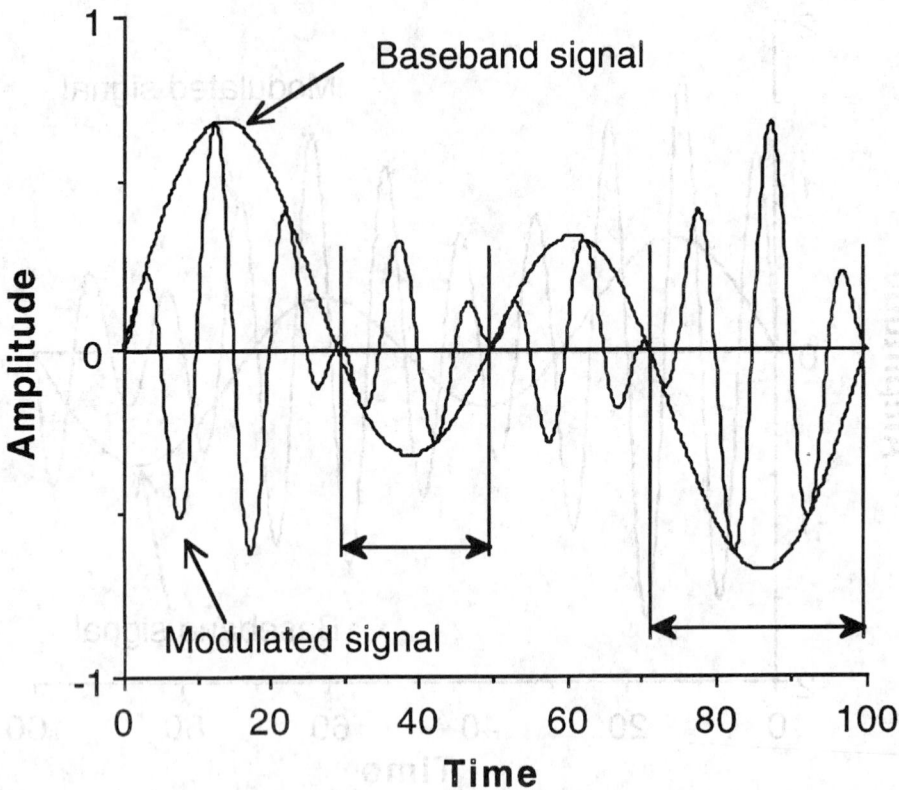

FIGURE 81.2   Time-domain representation of a DSB suppressed-carrier signal. Regions demarcated by double arrows indicate phase inversion of the modulated signal relative to the unmodulated carrier. This is in contrast to full-carrier AM in which the modulated signal is always in phase with the carrier.

about the carrier with the upper sideband arising from the positive-frequency component of $F_m(\omega)$ and the lower sideband from the negative-frequency component. A double-sideband AM signal thus has a bandwidth twice as large as that of the baseband signal.

Figure 81.2 shows a time-domain representation of a suppressed-carried DSB signal with the same baseband modulation as in Figure 81.1. The information in a full-carrier AM signal is found in the time-varying amplitude of the modulated signal, but the information carried by the suppressed-carrier DSB signal is found both in the amplitude and instantaneous phase of the modulated signal (note that the phase of the DSB signal is inverted relative to the carrier when the baseband signal is negative and in phase when the baseband signal is positive). AM is a linear modulation technique; the sum of multiple AM signals produced from a common carrier by different baseband signals is the same as one AM signal produced by the sum of the baseband signals.

Generation of Double-Sideband AM Signals

AM in radio transmitters is frequently performed by applying the modulating waveform to the supply voltage to a nonlinear radiofrequency power amplifier with a resonant-circuit load as described by Carlson [1]. Low-level AM may be achieved by direct multiplication of the carrier signal by $[k + \mu f_m(t)]$.

AM signals often arise in instrumentation systems as the result of the use of an ac drive signal to a modulating sensor. Figure 81.3 shows an example in which a balanced sensor is excited by a sinusoidal carrier. The output voltage of the differential amplifier will be zero when the sensor is balanced; a nonzero output voltage appears when the sensor is unbalanced. The magnitude of the voltage indicates the degree of imbalance in the sensor, and the phase of the output voltage relative to the carrier determines the

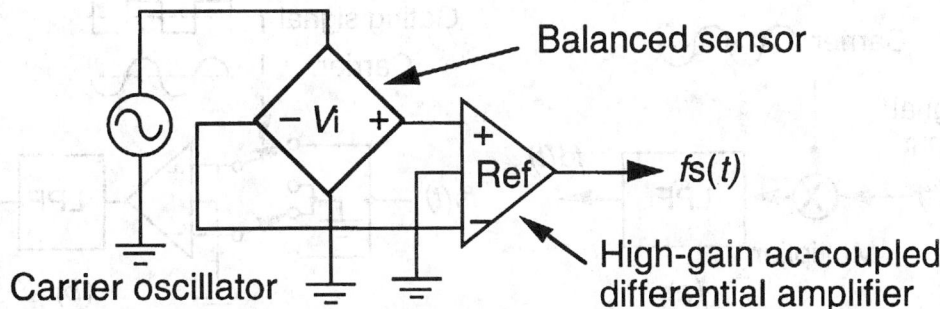

FIGURE 81.3 A balanced sensor with ac excitation and a differential-amplifier output stage. Typical sensors which might be found in this role are resistive Wheatstone bridges, differential-capacitance pressure sensors or accelerometers, or linear-variable differential transformers (LVDTs). Variation in the sensor measurand produces a DSB suppressed-carrier signal at the amplifier output.

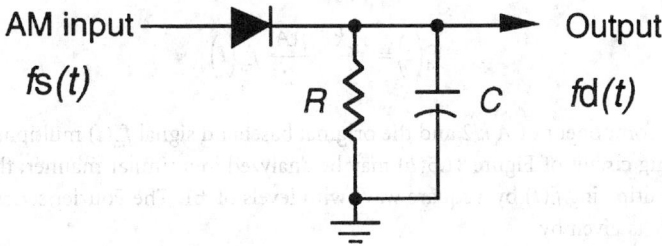

FIGURE 81.4 Envelope detector for full-carrier AM signals.

direction of imbalance. The suppressed-carrier DSB signal of Figure 81.2 would be seen at the amplifier output if we substitute the sensor measurand for the baseband signal of Figure 81.2. The technique of applying ac excitation to a balanced sensor may be required for inductive or capacitive sensors; it may also be desirable in the case of resistive sensors (such as strain-gage bridges) requiring high-gain amplifiers. In these circumstances, amplification of an ac signal minimizes both $1/f$ noise and dc offset problems associated with high-gain dc-coupled amplifiers.

Envelope Demodulation of Double-Sideband AM Signals

Full-carrier AM signals are readily demodulated by the simple envelope detector shown in Figure 81.4. The components of the RC low-pass filter are chosen such that $\omega_m \ll (1/RC) \ll \omega_c$. Envelope detection, however, cannot discriminate phase and is thus unsuitable for demodulation of signals in which phase reversal of the carrier occurs (such as reduced-carrier or suppressed-carrier signals). Synchronous demodulation is required for such signals.

Synchronous Demodulation of Double-Sideband AM Signals

Figure 81.5 shows two methods of synchronous demodulation. In Figure 81.5(*a*), the modulated signal is multiplied by $\cos(\omega_c t)$; in Figure 81.5(*b*), the modulated signal is gated by a square wave synchronous with $\cos(\omega_c t)$. Consider the multiplying circuit of Figure 81.5(*a*); the Fourier transform $F_d(\omega)$ of $f_d(t) = f_s(t)\cos(\omega_c t)$ is given by

$$F_d(\omega) = \frac{(A_c k)}{4}\left[\delta(\omega - 2\omega_c) + \delta(\omega + 2\omega_c) + 2\delta(\omega)\right] + \frac{(\mu A_c)}{4}\left[F_m(\omega - 2\omega_c) + F_m(\omega - 2\omega_c) + 2F_m(\omega)\right] \quad (81.6)$$

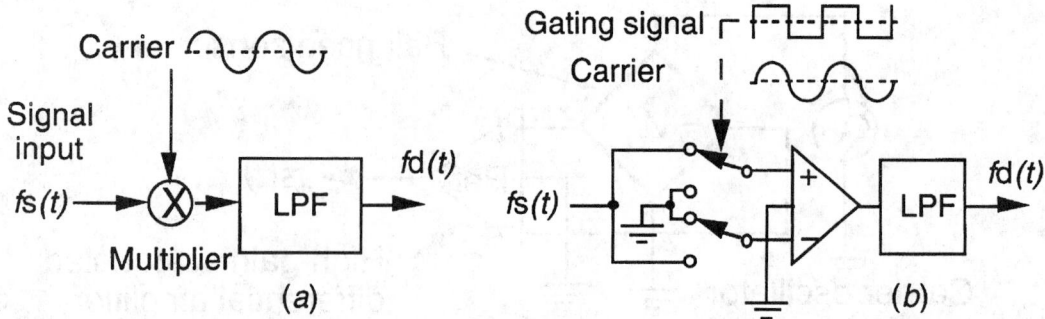

FIGURE 81.5 Multiplying (*a*) and switching (*b*) synchronous demodulators. The blocks marked **LPF** represent low-pass filters.

The spectral components translated by $\pm 2\omega_c$ may be removed by low-pass filtering; the result, translated into the time domain, is

$$f_d(t) = \frac{A_c k}{2} + \frac{\mu A_c}{2} f_m(t) \tag{81.7}$$

We thus have a dc component of $A_c k/2$ and the original baseband signal $f_m(t)$ multiplied by a scale factor of $\mu A_c/2$. The gating circuit of Figure 81.5(*b*) may be analyzed in a similar manner; the action of gating is equivalent to multiplying $f_s(t)$ by a square wave with levels of ± 1. The Fourier series representation of such a square wave is given by

$$f_g(t) = \frac{4}{\pi} \sum_{n=0}^{\infty} \frac{(-1)^n}{(2n+1)} \cos\left[(2n+1)\omega_c t\right] \tag{81.8}$$

Low-pass filtering of the product of Equations 81.4 and 81.8 gives

$$f_d(t) = \frac{2A_c k}{\pi} + \frac{2\mu A_c}{\pi} f_m(t) \tag{81.9}$$

The baseband signal is again recovered, although the scale factor is somewhat larger than that of the multiplying case. Demodulators like those of Figure 81.5(*a*) are called multiplying demodulators; circuits Figure 81.5(*b*) are known as switching demodulators. Nowicki [5] and Meade [6] discuss and compare both types of synchronous demodulators in greater detail.

We have so far made the implicit assumption that the demodulating signal is perfectly synchronized with the carrier of the modulated signal. Let us now consider the case where there exists a phase shift between these two signals. Assume that a signal expressed in carrier-quadrature form is multiplied by a demodulating signal $\cos(\omega_c t + \theta)$. The result, after suitable low-pass filtering, is

$$f_d(t) = \frac{f_i(t)}{2} \cos(\theta) + \frac{f_q(t)}{2} \sin(\theta) \tag{81.10}$$

Equation 81.10 is an important result; we see that both the level and polarity of the demodulated signal are functions of the synchronization between the demodulating signal and the modulated carrier. Synchronous demodulation is thus often called phase-sensitive demodulation. It was previously mentioned

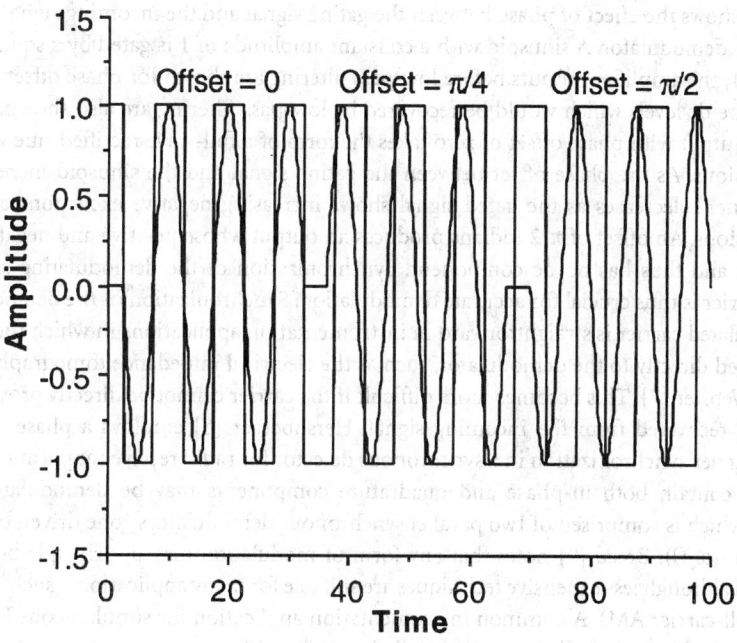

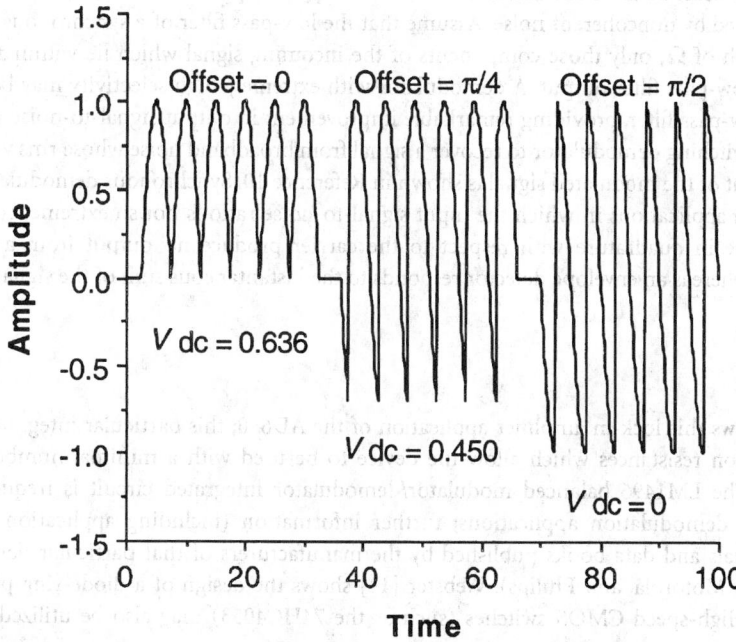

FIGURE 81.6 Gating of a constant-amplitude sinusoidal by a square wave in a switching phase-sensitive demodulator with various phase offsets between the carrier and the square wave. The upper trace shows the carrier and the square wave; the lower trace shows the signals which appear at the output of the differential amplifier in Figure 81.5(*b*). The dc output voltages of the low-pass filter are indicated for each value of phase offset.

that a DSB AM signal has no quadrature component $f_q(t)$; a phase shift of an odd integral multiple of $\pi/2$ radians between the carrier and the demodulating signal will produce a synchronous demodulator output of zero. Use of a square wave gating signal in place of the sinusoid would produce the same result as Equation 81.10 except that the recovered signals would be multiplied by $2/\pi$ instead of $1/2$.

Figure 81.6 shows the effect of phase between the gating signal and the incoming signal in a switching phase-sensitive demodulator. A sinusoid with a constant amplitude of 1 is gated by a square wave which has levels of ± 1; the amplifier outputs before low-pass filtering are shown for phase offsets of 0, $\pi/4$, and $\pi/2$ radians. The dc levels which would be recovered by low-pass filtering are also shown. Note that the demodulator output with phase offset of zero takes the form of a full-wave rectified sine which has only positive excursions. As the phase offset between the gating signal and the sinusoid increases, however, the dc component decreases as the gated signal shows increasing negative excursions and decreasing positive excursions. An offset of $\pi/2$ radians produces an output whose positive and negative excursions are symmetric and thus has no dc component. Synchronization of the demodulating signal with the modulated carrier is thus crucial for accurate demodulation. Synchronization of the demodulating signal with the modulated carrier is straightforward in instrumentation applications in which the carrier signal may be provided directly to the demodulator, such as the electrical impedance tomography applications discussed by Webster [7]. This becomes more difficult if the carrier cannot be directly provided but must be inferred or recovered from the incoming signal. Hershberger [8] employs a phase-locked loop to perform the carrier synchronization in a synchronous detector for radio receiver applications. Modulated signals which contain both in-phase and quadrature components may be demodulated by an I–Q demodulator which is comprised of two parallel synchronous demodulators (one driven by $\cos(\omega_c t)$ and the other by $\sin(\omega_c t)$); Breed [9] notes that any form of modulation may in principle be demodulated by this method although less expensive techniques are suitable for many applications (such as the envelope detector for full-carrier AM). A common instrumentation application for simultaneous I–Q demodulation is the electrical impedance bridge (often called an *LCR* bridge).

Synchronous demodulators are valuable in lock-in amplifier applications for the recovery of signals otherwise masked by noncoherent noise. Assume that the low-pass filter of a synchronous demodulator has a bandwidth of Ω; only those components of the incoming signal which lie within $\pm \Omega$ of ω_c will appear at the low-pass filter output. A demodulator with extremely high selectivity may be built by use of a narrow low-pass filter, providing remarkable improvement in output signal-to-noise ratio. The use of an AD630 switching demodulator to recover a signal from broadband noise whose rms value is 100 dB greater than that of the modulated signal is shown in Reference 10. Synchronous demodulation may also be of benefit in applications in which the input signal-to-noise ratio is not so extreme; components of noise which are in quadrature with respect to the carrier produce no output from a synchronous demodulator, whereas an envelope detector responds to the instantaneous sum of the signal and all noise components.

Examples

Figure 81.7 shows this lock-in amplifier application of the AD630; this particular integrated circuit has on-chip precision resistances which allow the device to be used with a minimal number of external components. The LM1496 balanced modulator/demodulator integrated circuit is frequently used in phase-sensitive demodulation applications; further information (including application examples) is found in manuals and data books published by the manufacturers of that particular device (National Semiconductor, Motorola, and Philips). Webster [11] shows the design of a diode-ring phase-sensitive demodulator. High-speed CMOS switches (such as the 74HC4053) may also be utilized in switching demodulators.

81.4   Angle (Frequency and Phase) Modulation

Recall from Equation 81.1 that we may modulate a carrier by varying its phase angle in accordance with the baseband signal. Consider a signal of the form:

$$f_s(t) = A_c \cos\left[\omega_c t + \Delta\phi x_m(t)\right]$$

(81.11)

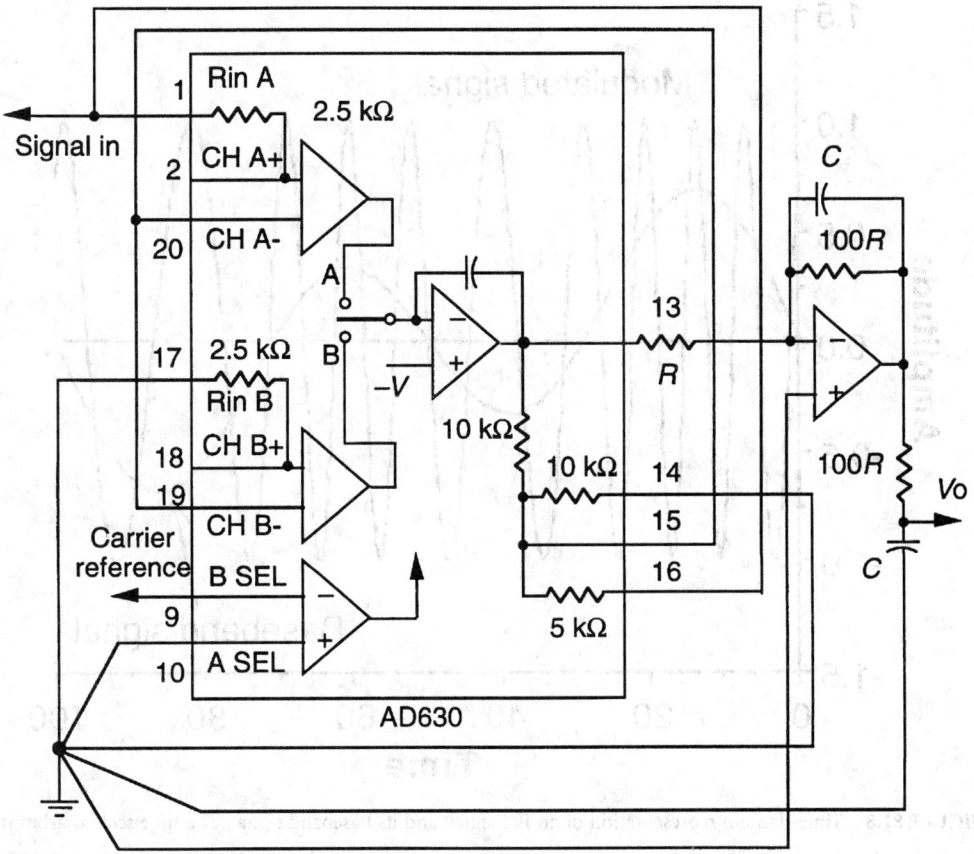

FIGURE 81.7 Lock-in amplifier circuit utilizing the Analog Devices AD630 switching balanced demodulator. This particular device is attractive for the small number of external components required and its high performance. The two-pole low-pass filter shown has a dc gain of 100 and a corner frequency of $0.00644/RC$. The principal drawback of this filter is its dc output impedance which is $100R$. This limitation could be avoided by replacing this filter by a pair of single-pole active filters or a single-amplifier two-pole filter (such as the Sallen-Key type of network). Resistances depicted within the outline of the AD630 are internal to the device itself; numbers at the periphery of the device are pin numbers. Pins labeled CH A+, CH A−, CH B+, and CH B− indicate the sense of the A and B channels of the device. Pins Rin A and Rin B are electrically connected to the CH A+ and CH B+ pins, respectively, through internal 2.5 kΩ resistances.

The instantaneous phase angle is $\omega_c t + \Delta\phi x_m(t)$; the phase relative to the unmodulated carrier is $\Delta\phi x_m(t)$. The carrier is thus phase-modulated by $x_m(t)$. The instantaneous frequency is $\omega_c + \Delta\phi[dx_m(t)/dt]$; if $x_m(t)$ is the integral of baseband signal $f_m(t)$, the instantaneous frequency becomes $\omega_c + \Delta\phi f_m(t)$. The frequency deviation $\Delta\phi f_m(t)$ is proportional to the baseband signal; the carrier is frequency modulated by $f_m(t)$. We may write the general expression for an FM signal as

$$f_s(t) = A_c \cos\left[\omega_c t + \Delta\omega \int_{-\infty}^{t} f_m(\tau)\,d\tau\right] \qquad (81.12)$$

(The change in notation from $\Delta\phi$ to $\Delta\omega$ is intended to emphasize that the signal represented by Equation 81.12 is frequency modulated.) Figure 81.8 shows a time-domain representation of an FM signal. Note the signal has constant amplitude; this is also true of a phase-modulated signal.

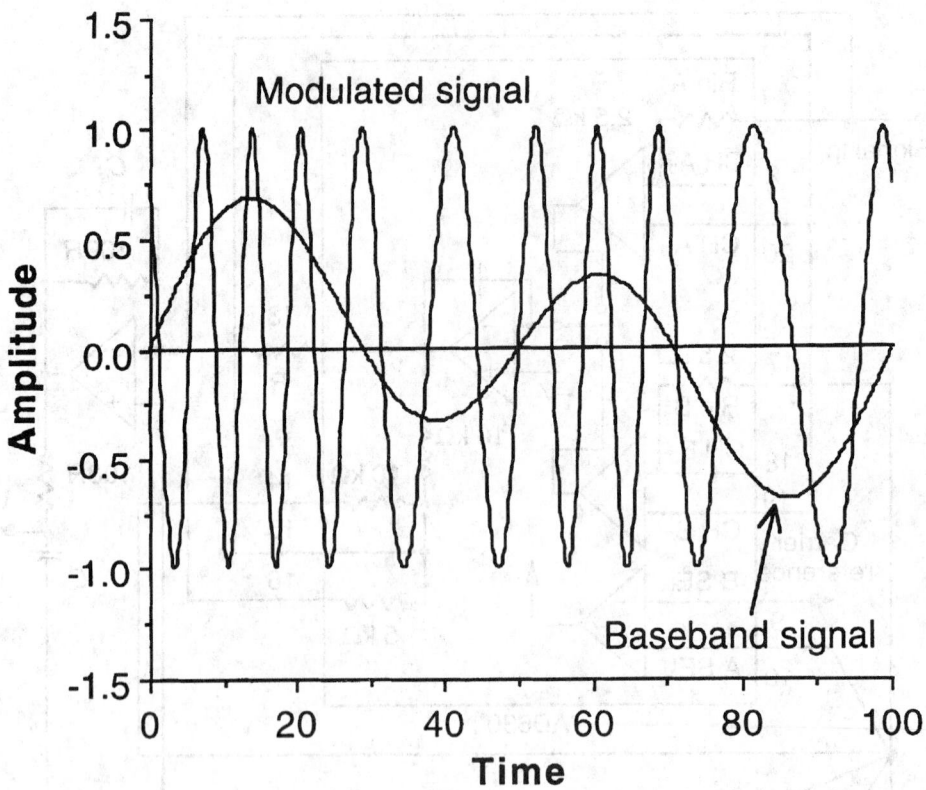

FIGURE 81.8 Time-domain representation of an FM signal and its baseband signal. The time scale is arbitrary.

Consider frequency modulation with a baseband signal $f_m = A_m \cos(\omega_m t)$; substitution into Equation 81.12 and performing the integration gives

$$f_s(t) = A_c \cos\left[\omega_c t + \beta \sin\left(\omega_m t\right)\right] \tag{81.13}$$

in which β (called the modulation index) has replaced $\Delta\omega A_m/\omega_m$. The carrier-quadrature form of Equation 81.13 is

$$f_s(t) = A_c \left\{ \cos\left[\beta \sin\left(\omega_m t\right)\right]\cos\left(\omega_c t\right) - \sin\left[\beta \sin\left(\omega_m t\right)\right]\sin\left(\omega_c t\right) \right\} \tag{81.14}$$

We note from Equation 81.14 that, unlike AM, the FM signal contains both in-phase and quadrature components and that the amplitudes of both are nonlinear functions of the modulation index. The terms $\cos[\beta\sin(\omega_m t)]$ and $\sin[\beta\sin(\omega_m t)]$ are generally expanded in terms of Bessel functions (see Schwartz [3] for detailed analysis). The Bessel function expression of an FM signal is

$$f_s(t) = A_c \sum_{n=-\infty}^{\infty} J_n(\beta)\cos\left(\omega_c + n\omega_m\right)t \tag{81.15}$$

where J_n represents a Bessel function of the first kind of order n. Beyer [12] provides tables of $J_0(\beta)$ and $J_1(\beta)$ and gives formulae for computing higher-order Bessel functions. We also note that $J_{-n}(\beta) = (-1)^n J_n(\beta)$. The approximations $J_0(\beta) = 1$ and $J_1(\beta) = (\beta/2)$ are valid for low values of the modulation index

($\beta < 0.2$); the higher-order Bessel functions are negligible under these circumstances. A carrier with sinusoidal frequency modulation with a low modulation index will show sidebands spaced at $\pm\omega_m$ about the carrier; such narrowband FM would be indistinguishable from full-carrier AM on a spectrum analyzer display (which shows the amplitudes of spectral components but not their phase relationships). As modulation index increases, however, new sideband pairs appear at $\pm2\omega_m$, $\pm 3\omega_m$, $\pm4\omega_m$, etc. as the higher-order Bessel functions become significant. The amplitude of the carrier component of $f_s(t)$ varies with $J_0(\beta)$; the carrier component disappears entirely for certain values of modulation index. These characteristics are unlike AM in which the carrier component of the modulated signal is constant and in which only one sideband pair is produced for each spectral component of the baseband signal. FM is an inherently nonlinear modulation process; the sum of multiple FM signals derived from a single carrier with individual baseband signals does not give the same result as frequency modulation of the carrier by the sum of the baseband signals. The spectrum of a phase modulated signal is similar to that of an FM signal, but the modulation index of a phase-modulated signal does not vary with ω_m. Figure 81.8 shows a time-domain representation of an FM signal and the original baseband signal.

Generation of Phase- and Frequency-Modulated Signals

FM signals may arise directly in instrumentation systems such as turbine-type flowmeters or Doppler velocity sensors. Direct FM signals may be generated by applying the baseband signal to a voltage-controlled oscillator (VCO); Sherwin and Regan [13] demonstrate this technique in a system which generates direct FM using the LM566 VCO to transmit analog information over 60-Hz power lines. In radiofrequency applications, the oscillator frequency may be varied by application of the baseband signal to a voltage-variable reactance (such as a varactor diode). Indirect FM may be generated by phase modulation of the carrier by the integrated baseband signal as in Equation 81.12; DeFrance [14] gives an example of a phase modulator circuit.

Demodulation of Phase- and Frequency-Modulated Signals

PM signals may be demodulated by the synchronous demodulator circuits previously described; they are, however, sensitive to the signal amplitude as well as phase and require a limiter circuit to produce an output proportional to phase alone. Figure 81.9 shows simple digital phase demodulators. Figure 81.10 shows three of the more common methods of FM demodulation. Figure 81.10(a) shows a quadrature detector of the type commonly used in integrated-circuit FM receivers. A limiter circuit suppresses noise-induced amplitude variations in the modulated signal; the limiter output then provides a reference signal to a synchronous (phase-sensitive) demodulator. The limiter output voltage is also coupled (via a small capacitor $C1$) to a quadrature network consisting of L, R, and $C2$. The phase of the voltage across the quadrature network relative to the limiter output is given by

$$\phi\left[V_q(\omega)\right] = \tan^{-1}\frac{-\omega L/R}{\left[1-\left(\omega/\omega_0\right)^2\right]} \tag{81.16a}$$

where

$$\omega_0 = \frac{1}{\sqrt{L\left(C_1+C_2\right)}} \tag{81.16b}$$

The variation in phase is nearly linear for frequencies close to ω_0. The phase-sensitive demodulator recovers the baseband signal from the phase shift between the quadrature-network voltage and the reference signal. The quadrature network also causes the amplitude of V_q to vary with frequency, but the variation in phase with respect to frequency predominates over amplitude variation in the vicinity of ω_0.

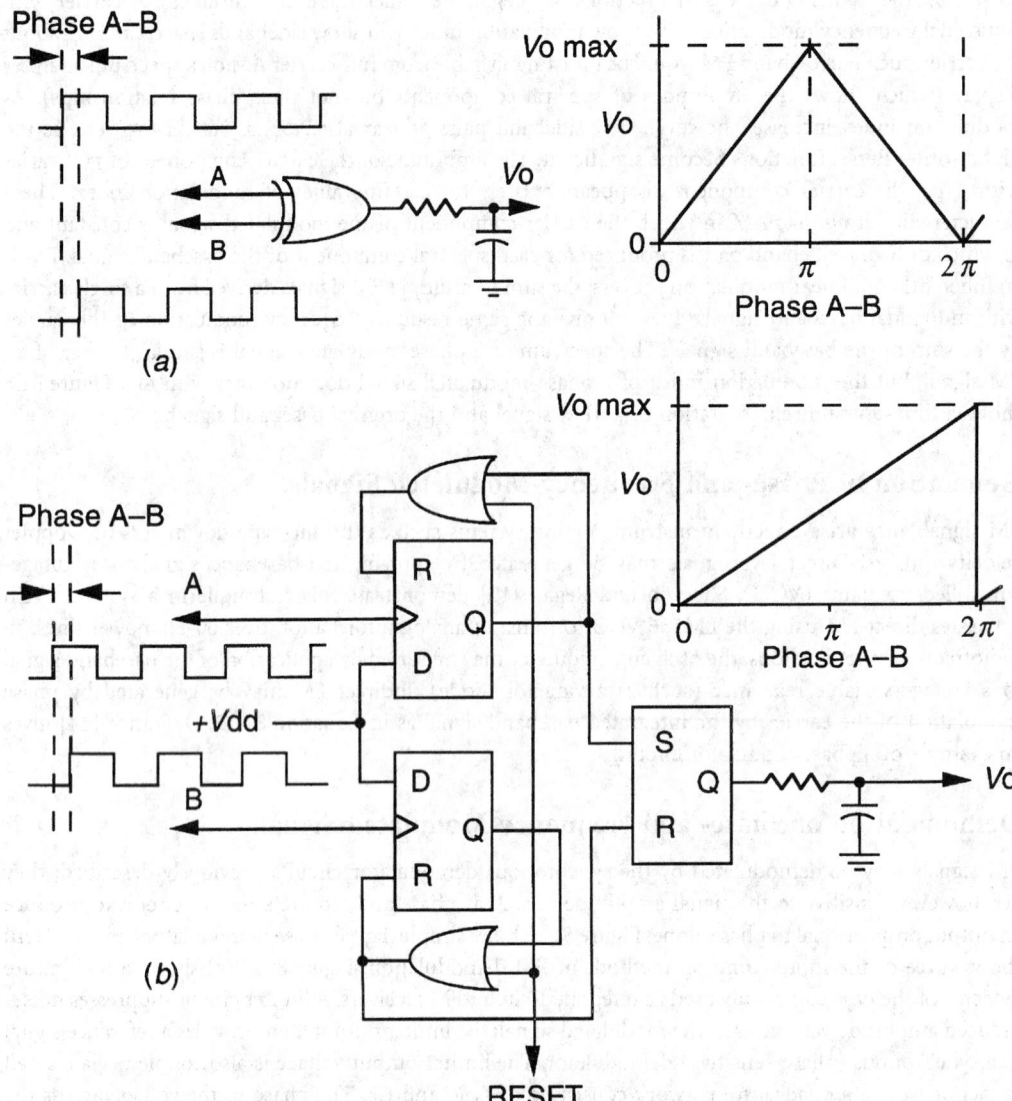

FIGURE 81.9 Examples of digital phase detectors. The exclusive-OR gate in (*a*) requires input signals with 50% duty cycle and produces an output voltage proportional to phase shift over the range of 0 to π radians (0° to 180°). Phase shifts between π and 2π radians produce an output voltage negatively proportional to phase. The edge-triggered RS flip-flop circuit in (*b*) functions with signals of arbitrary duty cycle and has a monotonic relationship between phase and output voltage over the full range of 0 to 2π radians. The circuit may be initialized by means of the RESET line. D-type flip-flops with RS capability (such as the CD4013) may be used in this circuit.

There is also generally enough voltage across the quadrature network to force the phase-sensitive demodulator into a nonlinear regime in which its output voltage is relatively insensitive to the amplitude of the voltage across the quadrature network. Figure 81.10(b) shows a frequency-to-voltage converter which consists of a monostable (one-shot) triggered on each zero crossing of the modulated signal. The pulse output of the one-shot is integrated by the low-pass filter to recreate the original baseband signal. Figure 81.10(c) shows a phase-locked loop. The phase comparator circuit produces an output voltage proportional to the phase difference between the input signal and the output of a VCO; that voltage is filtered and used to drive the VCO. Assume that $\phi_1(t)$ and $\omega_1(t)$ represent the phase and frequency of the

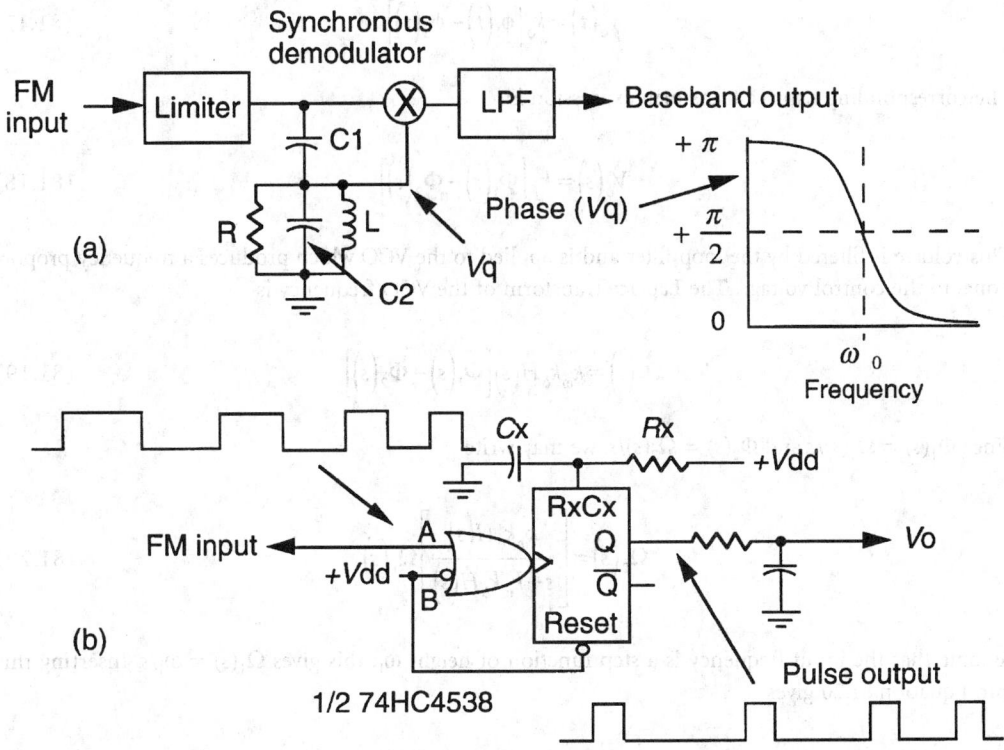

(a)

(b)

1/2 74HC4538

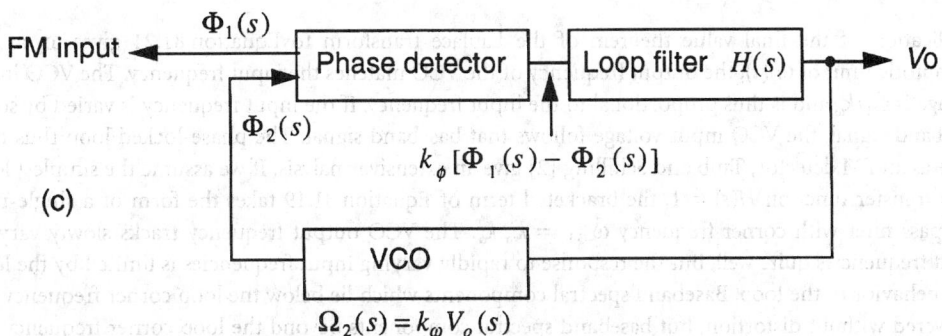

(c)

FIGURE 81.10 FM demodulators. A quadrature detector circuit is shown in (a). The phase shift of V_q (the voltage across the quadrature network consisting of L, $C2$, and R) relative to the output voltage of the limiter varies linearly with frequency for frequencies close to ω_0. The synchronous (phase-sensitive) demodulator produces an output proportional to the phase shift. A frequency-to-voltage converter is shown in (b). A monostable multivibrator (one-shot) produces a pulse of fixed width and amplitude with each cycle of the modulated signal. The average voltage of these pulses is proportional to the modulated signal frequency. The 74HC4538 contains two independent edge-triggered monostable multivibrators which may be triggered by a rising edge (as shown) or a falling edge (by applying the clock pulse to the B input and connecting the A input to V_{dd}). A phase-locked loop is shown in (c). The operation of the phase-locked loop is described in the text.

input signal as functions of time with corresponding Laplace transforms $\Phi_1(s)$ and $\Omega_1(s)$. The VCO output phase and frequency are represented by $\phi_2(t)$ and $\omega_2(t)$ with corresponding Laplace transforms $\Phi_2(s)$ and $\Omega_2(s)$. The phase detector produces an output voltage given by

$$v_\phi(t) = k_\phi \left[\phi_1(t) - \phi_2(t) \right] \tag{81.17}$$

The corresponding Laplace transform expression is

$$V_\phi(s) = k_f \left[\Phi_1(s) - \Phi_2(s) \right] \tag{81.18}$$

This voltage is filtered by the loop filter and is applied to the VCO which produces a frequency proportional to the control voltage. The Laplace transform of the VCO frequency is

$$\Omega_2(s) = k_\omega k_\phi H(s) \left[\Phi_1(s) - \Phi_2(s) \right] \tag{81.19}$$

Since $\Phi_1(s) = \Omega_1(s)/s$ and $\Phi_2(s) = \Omega_2(s)/s$, we may write

$$\Omega_2(s) = \left[\frac{k_\omega k_\phi H(s)}{s + k_\omega k_\phi H(s)} \right] \Omega_1(s) \tag{81.20}$$

Assume that the input frequency is a step function of height ω_1; this gives $\Omega_1(s) = \omega_1/s$. Inserting this into Equation 81.20 gives

$$\Omega_2(s) = \left[\frac{k_\omega k_\phi H(s)}{s + k_\omega k_\phi H(s)} \right] \frac{\omega_1}{s} \tag{81.21}$$

Application of the final value theorem of the Laplace transform to Equation 81.21 gives ω_1 as the asymptotic limit of $\omega_2(t)$; the output frequency of the VCO matches the input frequency. The VCO input voltage is ω_1/k_ω and is thus proportional to the input frequency. If the input frequency is varied by some baseband signal, the VCO input voltage follows that baseband signal. The phase-locked loop thus may serve as an FM detector; Taub and Schilling [2] give an extensive analysis. If we assume the simplest loop filter transfer function $H(s) = 1$, the bracketed term of Equation 81.19 takes the form of a single-pole low-pass filter with corner frequency $\omega_{-3dB} = k_\omega k_\phi$. The VCO output frequency tracks slowly varying input frequencies quite well, but the response to rapidly varying input frequencies is limited by the low-pass behavior of the loop. Baseband spectral components which lie below the loop corner frequency are recovered without distortion, but baseband spectral components beyond the loop corner frequency are attenuated. A single-pole RC low-pass filter is often used for the loop filter; the transfer function of the complete phase-locked loop with such a filter is:

$$\frac{\Omega_2(s)}{\Omega_1(s)} = \left[\frac{k_\omega k_\phi p_1}{s^2 + s p_1 + k_\omega k_\phi p_1} \right] \tag{81.22}$$

where $p_1 = 1/RC$. Note that the loop transfer function is of that of a second-order low-pass filter and thus may exhibit dynamics such as resonance and underdamped transient response depending on the values of the parameters k_ω, k_ϕ, and p_1.

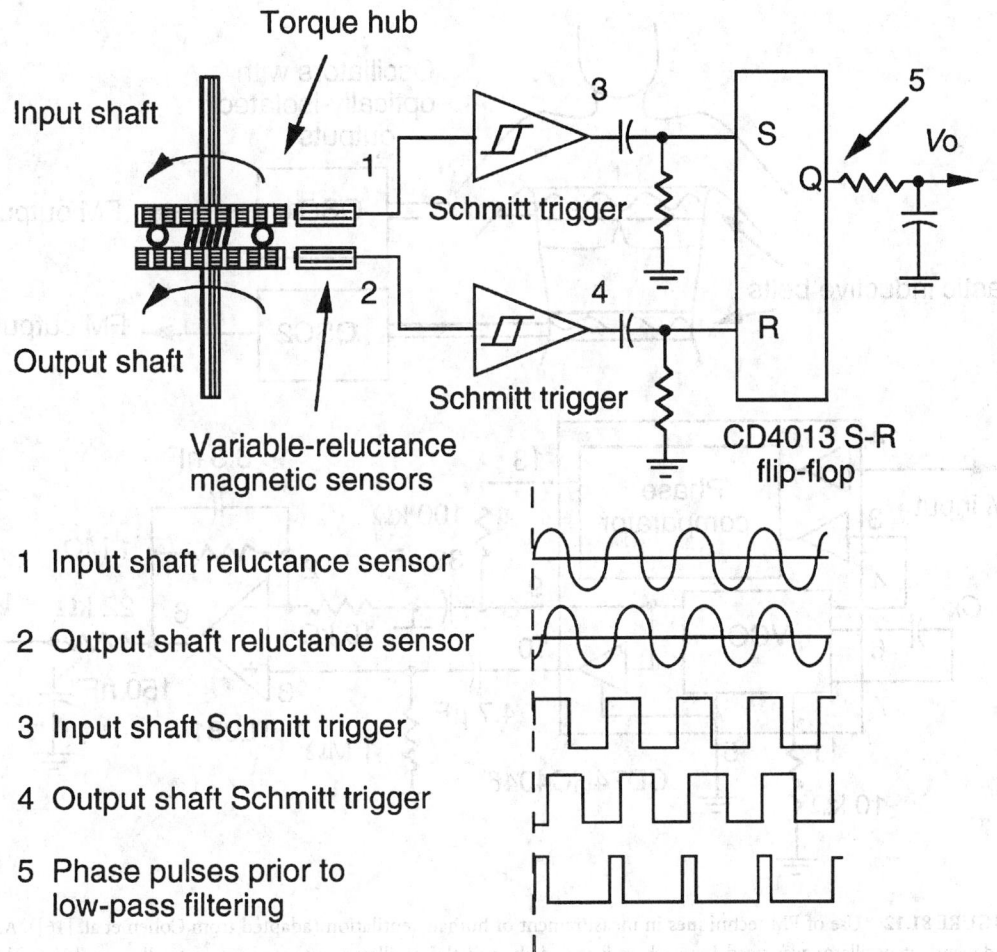

FIGURE 81.11 Use of phase-sensitive demodulation to measure torque in a rotating shaft as described by Sokol et al. [15]. Typical waveforms and the points in the circuit where they are found are indicated. The *RC* coupling to the S and R terminals of the CD4013 flip-flop provides for edge triggering.

Examples

Sokol et al. [15] describe the use of phase-sensing techniques to a shaft torque-sensing application in Figure 81.11. Two identical gears are coupled by springs in a device known as a torque hub. The teeth of the gears induce signals in variable-reluctance magnetic sensors as the shaft rotates, and the Schmitt trigger circuits transform the reluctance sensor outputs (approximately sinusoidal signals with amplitudes proportional to shaft rotational velocity) into square waves of constant amplitude. The springs compress with increasing torque, shifting the phase of the magnetic pickup signals (and hence the square waves) relative to each other. The relative phase of the square waves is translated to a dc voltage by a flip-flop phase detector and *RC* low-pass filter. The suppression of amplitude variation by the Schmitt trigger circuits causes no loss of information; the torque measurement is conveyed by phase alone.

Cohen et al. [16] in Figure 81.12 describe the use of a direct FM technique to perform noninvasive monitoring of human ventilation. Elastic belts with integral serpentine inductors encircle the chest and abdomen; these inductors are parts of resonant tank circuits of free-running radiofrequency oscillators. Ventilation causes the inductor cross-sectional areas to vary, changing their inductances and thus varying

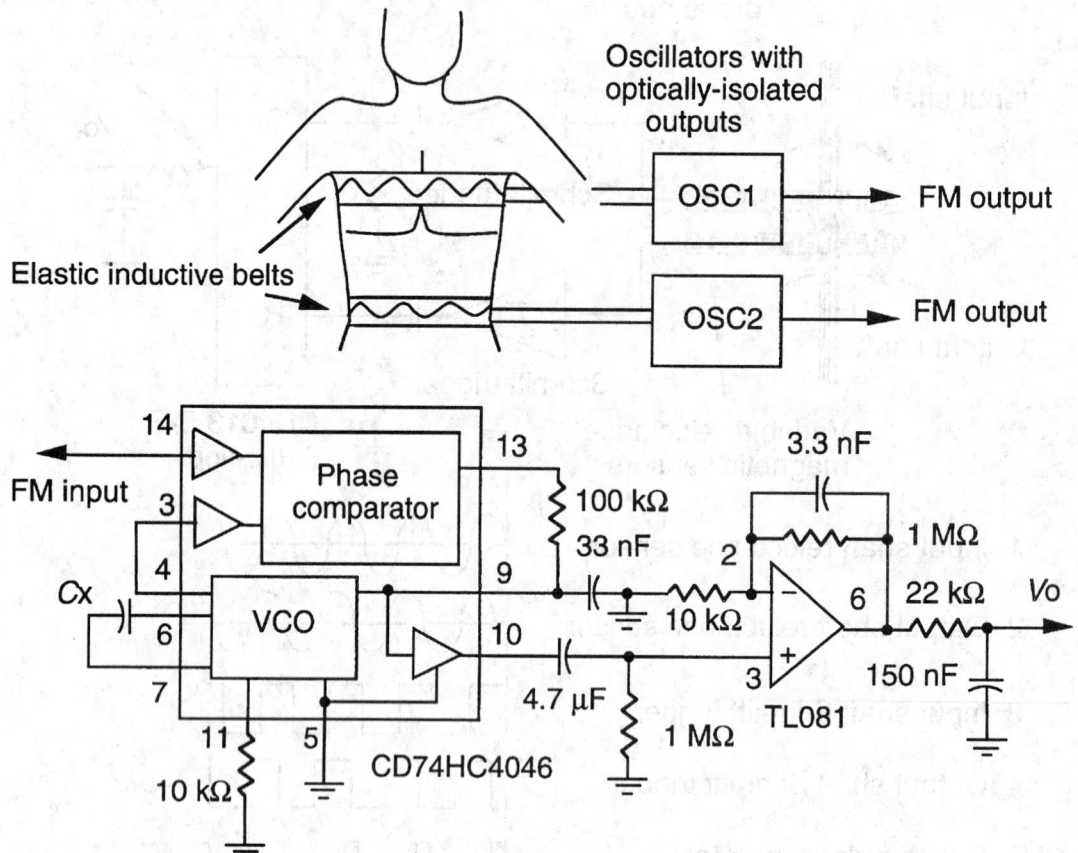

FIGURE 81.12   Use of FM techniques in measurement of human ventilation (adapted from Cohen et al. [16]). An independent oscillator was used for each inductive belt, and the oscillator outputs were optically coupled to the demodulator circuits for the purpose of electrical isolation of the subject from equipment operated from the 60 Hz ac line. One of two demodulators is shown; the value of C_x was either 470 pF for use at a nominal frequency of 850 kHz or 220 pF for use at a nominal frequency of 1.5 MHz. Numbers adjacent to the CD74HC4046 and TL081 devices are pin numbers.

the frequencies of the oscillators. Figure 81.12 also shows one of two phase-locked loop demodulator circuits which are identical except for the value of one capacitor in the VCO circuit. The phase-locked loop utilizes a single-pole RC low-pass loop filter.

Bachman [17,18] in Figure 81.13 describes a novel application of FM techniques to provide a sensing function which would otherwise be difficult to achieve (measurement of deviations in mass flow of fine seeds in an agricultural implement). A capacitive sensing cell was constructed and made part of a discriminator circuit driven at a constant frequency; the flow of seeds through the sensing capacitor shifts the resonant frequency of the discriminator and produces a change in the discriminator output voltage. An ac-coupled differential amplifier provides an output proportional to changes in the mass flow.

81.5   Instrumentation and Components

Integrated Circuits

Table 81.1 lists certain integrated circuits which may be used in application of the modulation techniques covered in this chapter. This is not an exhaustive list of all useful types nor of all manufacturers. Most

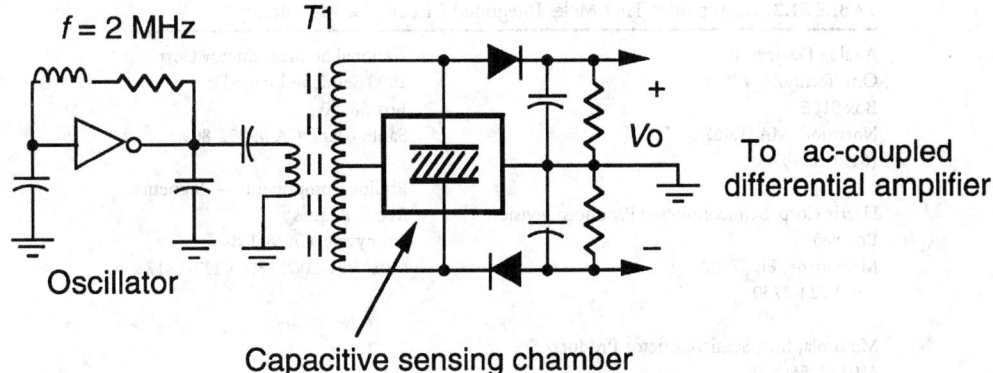

FIGURE 81.13   FM demodulator technique applied to measurement of the flow of small seeds in an agricultural machine application. The relative permittivity of the seeds causes the resonant frequency of the discriminator circuit to shift and to produce a change in the output voltage. The claimed sensitivity of this circuit is on the order of tens of femtofarads. The cross-hatched region of the sensing chamber represents the active area through which the seeds flow.

TABLE 81.1   Integrated Circuits Used in the Application of Modulation Techniques

Designation	Function	Manufacturer(s)	Approximate Price, $
AD630JN	Balanced modulator/demodulator	Analog Devices	14.78
LM1496N	Balanced modulator/demodulator	National, Motorola, Philips	1.80
AD532JH	Four-quadrant multiplier	Analog Devices	25.50
AD533JH	Four-quadrant multiplier	Analog Devices	30.88
AD633JN	Four-quadrant multiplier	Analog Devices	5.63
LM565	Phase-locked loop	National, Motorola, Philips	2.70
74HC4046	Phase-locked loop	Harris, Motorola	2.00
LM566	VCO	National, Motorola, Philips	1.65
AD650JN	Voltage-to-frequency, frequency-to-voltage converter	Analog Devices	17.00
AD652JP	Voltage–to-frequency converter	Analog Devices	17.02
AD654JN	Voltage-to-frequency converter	Analog Devices	7.26
MC3362P	FM receiver system	Motorola	3.84
MC3363P	FM receiver system	Motorola	3.84
MC4044P	Phase/frequency detector	Motorola	18.29
74HC4538	Dual retriggerable monostable (one-shot)	Harris, Motorola	2.00

integrated circuits are available in a number of packages and performance grades; the prices given are indicative of the pricing of the least expensive versions purchased in small quantities. The AD532, AD533, and AD633 four-quadrant multipliers may be used to make balanced modulator and multiplying demodulator circuits. The AD650 may be used as either a voltage-to-frequency or frequency-to-voltage converter. The Motorola MC3362P and MC3363P are single-chip narrowband FM receivers which are useful in communications or in digital telemetry via FSK. Table 81.2 provides contact information for each of these companies.

Instrumentation

Oscilloscopes and spectrum analyzers are frequently employed in analysis of modulated signals or systems utilizing modulation techniques; both types of instruments are covered elsewhere. Certain types of specialized instrumentation are available from various manufacturers. Table 81.3 shows a short list of representative types. These instruments include scalar modulation analyzers, vector signal and modulation analyzers, signal sources with analog or digital modulation capability, and instruments for analysis of television color signals (vectorscopes). Hewlett-Packard produces a line of instruments known as modulation-domain analyzers which permit the user to characterize frequency, phase, and time interval as functions of time. Table 81.4 provides contact information for each of these companies.

TABLE 81.2   Companies That Make Integrated Circuits for Modulating

Analog Devices, Inc.	National Semiconductor Corp.
One Technology Way	2900 Semiconductor Dr.
Box 9106	Box 58090
Norwood, MA 02062	Santa Clara, CA 95052-8090
(617) 329-4700	
	Philips Components — Signetics
Harris Corp. Semiconductor Products Division	811 E. Arques
Box 883	Sunnyvale, CA 94088
Melbourne, FL 37902	(408) 991-2000 (800) 227-1817
(407) 724-3730	
Motorola, Inc., Semiconductor Products Sector	
3102 N. 56th St.	
Phoenix, AZ 85018	
(602) 952-3248	

TABLE 81.3   Instruments Utilizing Modulation Techniques

Manufacturer	Model Number	Description	Price, $
Hewlett-Packard	HP 53310A	Modulation domain analyzer	10,150
	HP 5371A	Frequency and time-interval analyzer	28,550
	HP 5372A	Frequency and time-interval analyzer	31,600
	HP 5373A	Modulation domain pulse analyzer	33,650
	HP 8780A	Vector signal generator	71,400
	HP 8782B	Vector signal generator	35,700
	HP 8981B	Vector modulation analyzer	35,700
	HP 11736B	I/Q modulation tutorial software	120
I/Q Tutor	HP 89410A	Vector signal analyzer	29,050
	HP 89440A	Vector signal analyzer	52,500
	HP 89441A	Vector signal analyzer	58,150
	HP 11715A	AM/FM test source	3,015
	HP 8901A	Modulation analyzer	11,510
	HP 8901B	Modulation analyzer	16,050
Tektronix	1720	Vectorscope	3400
	1721	Vectorscope	3400
	1725	Vectorscope	3400
Rohde & Schwarz	DS 1200	TV demodulator	11,900
	SME-02	Signal generator	17,675
	SME-03	Signal generator	23,190
	SMHU-58	Signal generator	42,900
	SMT-02	Signal generator	10,085
	SMY-01	Signal generator	6335
	SMY-02	Signal generator	7625

Defining Terms

Modulation: The process of encoding the source information onto a bandpass signal with a carrier frequency f_c.

Amplitude modulation (AM):   Continuous wave modulation, where the amplitude of the carrier varies linearly with the amplitude of the modulating signal.

Angle modulation: Continuous wave modulation, where the angle of the carrier varies linearly with the amplitude of the modulating signal.

Frequency modulation (FM):   Continuous wave modulation, where the frequency of the carrier varies linearly with the amplitude of the modulating signal.

TABLE 81.4 Companies That Make Modulated Sources and Analyzers

Hewlett-Packard Co.
Test and Measurement Sector
Box 58199
Santa Clara, CA 95052-9943
(800) 452-4844

Tektronix Inc. Corporate Offices
26600 SW Parkway
Box 1000
Wilsonville, OR 97070-1000
(503) 682-3411 (800) 426-2200

Rohde & Schwarz Inc.
4425 Nicole Dr.
Lanham, MD 20706
(301) 459-8800

References

1. A. B. Carlson, *Communication Systems,* 3rd ed., New York: McGraw-Hill, 1986.
2. H. Taub and D. L. Schilling, *Principles of Communication Systems,* New York: McGraw-Hill, 1971.
3. M. Schwartz, *Information Transmission, Modulation, and Noise,* 4th ed., New York: McGraw-Hill, 1990.
4. W. Tomasi, *Advanced Electronic Communications Systems,* 2nd ed., Englewood Cliffs, NJ: Prentice-Hall, 1992.
5. D. J. Nowicki, Voltage measurement and signal demodulation, in J. G. Webster, Ed., *Electrical Impedance Tomography,* Bristol, UK: Adam Hilger, 1990.
6. M. L. Meade, *Lock-in Amplifiers: Principles and Applications,* London: Peregrinus, 1984.
7. J. G. Webster, Ed., *Electrical Impedance Tomography,* Bristol, UK: Adam Hilger, 1990.
8. D. L. Hershberger, Build a synchronous detector for AM radio, *Pop. Electron.,* 20 (4), 61, 66–71, 1982.
9. G. A. Breed, Receiver basics—part 2. Fundamental receiver architectures, *RF Design,* 17 (3), 84, 86, 88–89, 1994.
10. Anonymous, AD630 balanced modulator/demodulator, in *Analog Devices* 1992 *Special Linear Reference Manual,* pp. 2.35–2.41, 1992.
11. J. G. Webster, Ed., *Medical Instrumentation: Application and Design.,* 3rd ed., New York: John Wiley & Sons, 1998.
12. W. H. Beyer, *CRC Standard Mathematical Tables,* 28th ed., Boca Raton, FL: CRC Press, 1987.
13. J. Sherwin and T. Regan, FM remote speaker system, National Semiconductor Application Note AN-146, in *National Semiconductor Corp.* 1991 *Linear Applications Databook,* 1991.
14. J. J. DeFrance, *Communication Electronics Circuits,* 2nd ed., San Francisco, CA: Rinehart Press, 1972.
15. D. G. Sokol, R. B. Whitaker, J. J. Lord, and D. M. Beams, *Combine data center,* U.S. Patent No. 4,376,298, 1983.
16. K. P. Cohen, D. Panescu, J. H. Booske, J. G. Webster, and W. J. Tompkins, Design of an inductive plethysmograph for ventilation measurement, *Physiol. Meas.,* 15, 217–229, 1994.
17. W. J. Bachman, *Capacitive-type seed sensor for a planter monitor,* U.S. Patent No. 4,782,282, 1988.
18. W. J. Bachman, Private communication, 1995.

82

Filters

Rahman Jamal
National Instruments Germany

Robert Steer
Frequency Devices

82.1 Introduction

In its broadest sense, a filter can be defined as a signal processing system whose output signal, usually called the *response*, differs from the input signal, called the *excitation*, such that the output signal has some prescribed properties. In more practical terms an electric filter is a device designed to suppress, pass, or separate a group of signals from a mixture of signals according to the specifications in a particular application. The application areas of filtering are manifold, for example to band-limit signals before sampling to reduce aliasing, to eliminate unwanted noise in communication systems, to resolve signals into their frequency components, to convert discrete-time signals into continuous-time signals, to demodulate signals, etc. Filters are generally classified into three broad classes: *continuous-time, sampled-data*, and *discrete-time* filters depending on the type of signal being processed by the filter. Therefore, the concept of signals are fundamental in the design of filters.

A *signal* is a function of one or more independent variables such as time, space, temperature, etc. that carries information. The independent variables of a signal can either be continuous or discrete. Assuming that the signal is a function of time, in the first case the signal is called continuous-time and in the second, discrete-time. A continuous-time signal is defined at every instant of time over a given interval, whereas a discrete-time signal is defined only at a discrete-time instances. Similarly, the *values* of a signal can also be classified in either continuous or discrete.

In real-world signals, often referred to as analog signals, both amplitude and time are continuous. These types of signals cannot be processed by digital machines unless they have been converted into discrete-time signals. By contrast, a digital signal is characterized by discrete signal values, that are defined only at discrete points in time. Digital signal values are represented by a finite number of digits, which are usually binary coded. The relationship between a continuous-time signal and the corresponding discrete-time signal can be expressed in the following form:

$$x(kT) = x(t)\big|_{t=kT}, \quad k = 0,1,2,\ldots, \tag{82.1}$$

where T is called the sampling period.

Filters can be classified on the basis of the input, output, and internal operating signals. A continuous data filter is used to process continuous-time or analog signals, whereas a digital filter processes digital signals. Continuous data filters are further divided into *passive* or *active* filters, depending on the type of elements used in their implementation. Perhaps the earliest type of filters known in the engineering community are *LC* filters, which can be designed by using discrete components like inductors and capacitors, or crystal and mechanical filters that can be implemented using *LC* equivalent circuits. Since no external power is required to operate these filters, they are often referred to as *passive* filters. In contrast, *active* filters are based on active devices, primarily *RC* elements, and amplifiers. In a sampled data filter, on the other hand, the signal is sampled and processed at discrete instants of time. Depending on the type of signal processed by such a filter, one may distinguish between an *analog sampled data* filter and a *digital* filter. In an analog sampled data filter the sampled signal can principally take any value, whereas in a digital filter the sampled signal is a digital signal, the definition of which was given earlier. Examples of analog sampled data filters are switched capacitor (SC) filters and charge-transfer device (CTD) filters made of capacitors, switches, and operational amplifiers.

82.2　Filter Classification

Filters are commonly classified according to the filter function they perform. The basic functions are: low-pass, high-pass, bandpass, and bandstop. If a filter passes frequencies from zero to its cutoff frequency Ω_c and stops all frequencies higher than the cutoff frequencies, then this filter type is called an ideal **low-pass filter**. In contrast, an ideal **high-pass filter** stops all frequencies below its cutoff frequency and passes all frequencies above it. Frequencies extending from Ω_1 to Ω_2 are passed by an ideal **bandpass filter**, while all other frequencies are stopped. An ideal bandstop filter stops frequencies from Ω_1 to Ω_2 and passes all other frequencies. Figure 82.1 depicts the magnitude functions of the four basic **ideal filter** types.

So far we have discussed ideal filter characteristics having rectangular magnitude responses. These characteristics, however, are physically not realizable. As a consequence, the ideal response can only be approximated by some nonideal realizable system. Several classical approximation schemes have been developed, each of which satisfies a different criterion of optimization. This should be taken into account when comparing the performance of these filter characteristics.

82.3　The Filter Approximation Problem

Generally the input and output variables of a linear, time-invariant, causal filter can be characterized either in the time-domain through the convolution integral given by

$$y(t) = \int_0^t h_a(t-\tau) x(t) d\tau \tag{82.2}$$

or, equivalently, in the frequency-domain through the transfer function

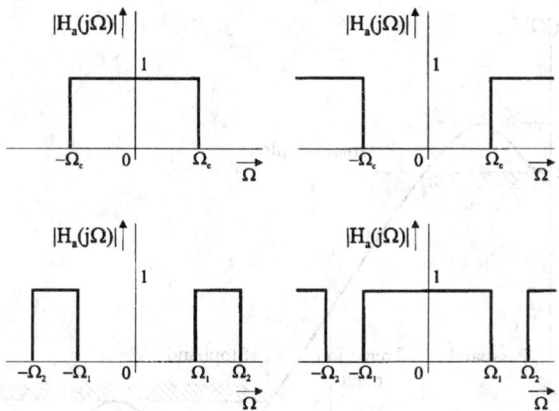

FIGURE 82.1 The magnitude function of an ideal filter is 1 in the passband and 0 in the stopband as shown for (a) low-pass, (b) high-pass, (c) bandpass, and (d) stopband filters.

$$H_a(s) = \frac{Y(s)}{X(s)} = \frac{\sum_{i=0}^{N} b_i s^i}{1 + \sum_{i=0}^{N} a_i s^i} \Leftrightarrow H_a(s) = \frac{b_N}{a_N} \prod_{i=1}^{N} \left(\frac{s - s_{0i}}{s - s_{\infty i}} \right) \qquad (82.3)$$

where $H_a(s)$ is the Laplace transform of the impulse response $h_a(t)$ and $X(s)$, $Y(s)$ are the Laplace transforms of the input signal $x(t)$ and the output or the filtered signal $y(t)$. $X(s)$ and $Y(s)$ are polynomials in $s = \sigma + j\Omega$ and the overall transfer function $H_a(s)$ is a real rational function of s with real coefficients. The zeroes of the polynomial $X(s)$ given by $s = s_{\infty i}$ are called the poles of $H_a(s)$ and are commonly referred to as the *natural frequencies* of the filter. The zeros of $Y(s)$ given by $s = s_{0i}$ which are equivalent to the zeroes of $H_a(s)$ are called the *transmission zeros* of the filter. Clearly, at these frequencies the filter output is zero for any finite input. Stability restricts the poles of $H_a(s)$ to lie in the left half of the s-plane excluding the $j\Omega$-axis, that is $\text{Re}\{s_{\infty i}\} < 0$. For a stable transfer function $H_a(s)$ reduces to $H_a(j\Omega)$ on the $j\Omega$-axis, which is the continuous-time Fourier transform of the impulse response $h_a(t)$ and can be expressed in the following form:

$$H_a(j\Omega) = |H_a(j\Omega)| d^{j\theta(\Omega)} \qquad (82.4)$$

where $|H_a(j\Omega)|$ is called the magnitude function and $\theta(\Omega) = \arg H_a(j\Omega)$ is the phase function. The gain magnitude of the filter expressed in decibels (dB) is defined by

$$\alpha(\Omega) = 20 \log |H_a(j\Omega)| = 10 \log |H_a(j\Omega)|^2 \qquad (82.5)$$

Note that a filter specification is often given in terms of its attenuation, which is the negative of the gain function also given in decibels. While the specifications for a desired filter behavior are commonly given in terms of the loss response $\alpha(\Omega)$, the solution of the filter approximation problem is always carried out with the help of the characteristic function $C(j\Omega)$ giving

$$\alpha(\Omega) = 10 \log \left[1 + |C(j\Omega)|^2 \right] \qquad (82.6)$$

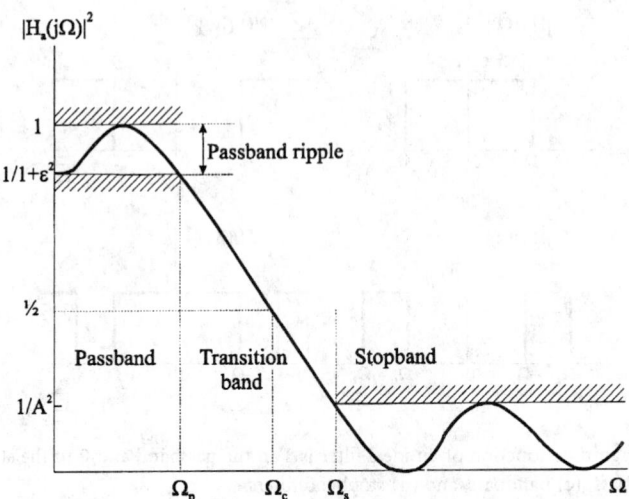

FIGURE 82.2　The squared magnitude function of an analog filter can have ripple in the passband and in the stopband.

Note that $\alpha(\Omega)$ is not a rational function, but $C(j\Omega)$ can be a polynomial or a rational function and approximation with polynomial or rational functions is relatively convenient. It can also be shown that frequency-dependent properties of $|C(j\Omega)|$ are in many ways identical to those of $\alpha(\Omega)$. The approximation problem consists of determining a desired response $|H_a(j\Omega)|$ such that the typical specifications depicted in Figure 82.2 are met. This so-called tolerance scheme is characterized by the following parameters:

Ω_p　Passband cutoff frequency (rad/s)
Ω_s　Stopband cutoff frequency (rad/s)
Ω_c　-3 dB cutoff frequency (rad/s)
ε　Permissible error in passband given by $\varepsilon = (10^{r/10} - 1)^{1/2}$, where r is the maximum acceptable attenuation in dB; note that $10 \log 1/(1 + \varepsilon^2)^{1/2} = -r$
$1/A$　Permissible maximum magnitude in the stopband, i.e., $A = 10^{\alpha/20}$, where α is the minimum acceptable attenuation in dB; note that $20 \log (1/A) = -\alpha$.

The **passband** of a low-pass filter is the region in the interval $[0,\Omega_p]$ where the desired characteristics of a given signal are preserved. In contrast, the **stopband** of a low-pass filter (the region $[\Omega_s,\infty]$) rejects signal components. The **transition** band is the **region** between $(\Omega_x - \Omega_p)$, which would be 0 for an ideal filter. Usually, the amplitudes of the permissible ripples for the magnitude response are given in decibels.

The following sections review four different classical approximations: Butterworth, Chebyshev Type I, elliptic, and Bessel.

Butterworth Filters

The frequency response of an Nth-order Butterworth low-pass filter is defined by the squared magnitude function

$$\left|H_a\left(j\Omega\right)\right|^2 = \frac{1}{1+\left(\Omega/\Omega_c\right)^{2N}} \tag{82.7}$$

It is evident from the Equation 82.7 that the Butterworth approximation has only poles, i.e., no finite zeros and yields a maximally flat response around zero and infinity. Therefore, this approximation is also called maximally flat magnitude (MFM). In addition, it exhibits a smooth response at all frequencies and a monotonic decrease from the specified cutoff frequencies.

Equation 82.7 can be extended to the complex *s*-domain, resulting in

$$H_a(s)H_a(-s) = \frac{1}{1+\left(s/j\Omega_c\right)^{2N}} \tag{82.8}$$

The poles of this function are given by the roots of the denominator

$$s_k = \Omega_c e^{j\pi\left[1/2+(2k+1)/2N\right]}, \quad k=0,1,\ldots,2N-1 \tag{82.9}$$

Note that for any *N*, these poles lie on the unit circle of radius Ω_c in the *s*-plane. To guarantee stability, the poles that lie in the left half-plane are identified with $H_a(s)$. As an example, we will determine the transfer function corresponding to a third-order Butterworth filter, i.e., $N=3$.

$$H_a(s)H_a(-s) = \frac{1}{1+\left(-s^2\right)^3} = \frac{1}{1-s^6} \tag{82.10}$$

The roots of denominator of Equation 82.10 are given by

$$s_k = \Omega_c e^{j\pi\left[1/2+(2k+1)/6\right]}, \quad k=0,1,2,3,4,5 \tag{82.11}$$

Therefore, we obtain

$$
\begin{aligned}
s_0 &= \Omega_c e^{j\pi 2/3} = -1/2 + j\sqrt{2}/2 \\
s_1 &= \Omega_c e^{j\pi} = -1 \\
s_2 &= \Omega_c e^{j\pi 4/3} = -1/2 - j\sqrt{3}/2 \\
s_3 &= \Omega_c e^{j\pi 5/3} = 1/2 - j\sqrt{3}/2 \\
s_4 &= \Omega_c e^{j2\pi} = 1 \\
s_5 &= \Omega_c e^{j\pi/3} = 1/2 + j\sqrt{3}/2
\end{aligned}
\tag{82.12}
$$

The corresponding transfer function is obtained by identifying the left half-plane poles with $H_a(s)$. Note that for the sake of simplicity we have chosen $\Omega_c = 1$.

$$H_a(s) = \frac{1}{(s+1)\left(s+1/2-j\sqrt{3}/2\right)\left(s+1/2+j\sqrt{3}/2\right)} = \frac{1}{1+2s+2s^2+s^3} \tag{82.13}$$

Table 82.1 gives the Butterworth denominator polynomials up $N=5$.

Table 82.2 gives the Butterworth poles in real and imaginary components and in frequency and Q.

TABLE 82.1 Butterworth Denominator Polynomials

Order(N)	Butterworth Denominator Polynomials of $H(s)$
1	$s + 1$
2	$s^2 + \sqrt{2}\,s + 1$
3	$s^3 + 2s^2 + 2s + 1$
4	$s^4 + 2.6131s^3 + 3.4142s^2 + 2.6131s + 1$
5	$s^5 + 3.2361s^4 + 5.2361s^3 + 5.2361s^2 + 3.2361s + 1$

TABLE 82.2 Butterworth and Bessel Poles

	Butterworth Poles				Bessel Poles (−3 dB)			
	Re	Im($\pm j$)			Re	Im($\pm j$)		
N	a	b	Ω	Q	a	b	Ω	Q
1	−1.000	0.000	1.000	—	−1.000	0.000	1.000	—
2	−0.707	0.707	1.000	0.707	−1.102	0.636	1.272	0.577
3	−1.000	0.000	1.000	—	−1.323	0.000	1.323	—
	−0.500	0.866	1.000	1.000	−1.047	0.999	1.448	0.691
4	−0.924	0.383	1.000	0.541	−1.370	0.410	1.430	0.522
	−0.383	0.924	1.000	1.307	−0.995	1.257	1.603	0.805
5	−1.000	0.000	1.000	—	−1.502	0.000	1.502	—
	−0.809	0.588	1.000	0.618	−1.381	0.718	1.556	0.564
	−0.309	0.951	1.000	1.618	−0.958	1.471	1.755	0.916
6	−0.966	0.259	1.000	0.518	−1.571	0.321	1.604	0.510
	−0.707	0.707	1.000	0.707	−1.382	0.971	1.689	0.611
	−0.259	0.966	1.000	1.932	−0.931	1.662	1.905	1.023
7	−1.000	0.000	1.000	—	−1.684	0.000	1.684	—
	−0.901	0.434	1.000	0.555	−1.612	0.589	1.716	0.532
	−0.623	0.782	1.000	0.802	−1.379	1.192	1.822	0.661
	−0.223	0.975	1.000	2.247	−0.910	1.836	2.049	1.126
8	−0.981	0.195	1.000	0.510	−1.757	0.273	1.778	0.506
	−0.831	0.556	1.000	0.601	−1.637	0.823	1.832	0.560
	−0.556	0.831	1.000	0.900	−1.374	1.388	1.953	0.711
	−0.195	0.981	1.000	2.563	−0.893	1.998	2.189	1.226

In the next example, the order N of a low-pass Butterworth filter is to be determined whose cutoff frequency (−3 dB) is $\Omega_c = 2$ kHz and stopband attenuation is greater than 40 dB at $\Omega_s = 6$ kHz. Thus the desired filter specification is

$$20 \log \left| H_a\left(j\Omega \right) \right| \leq -40, \quad \Omega \geq \Omega_s \tag{82.14}$$

or equivalently,

$$\left| H_a\left(j\Omega \right) \right| \leq 0.01, \quad \Omega \geq \Omega_s \tag{82.15}$$

It follows from Equation 82.7

$$\frac{1}{1 + \left(\Omega_s / \Omega_c \right)^{2N}} = \left(0.01 \right)^2 \tag{82.16}$$

Solving the above equation for N gives $N = 4.19$. Since N must be an integer, a fifth-order filter is required for this specification.

Chebyshev Filters or Chebyshev I Filters

The frequency response of an Nth-order Chebyshev low-pass filter is specified by the squared-magnitude frequency response function

$$\left|H_a(j\Omega)\right|^2 = \frac{1}{1 + \varepsilon^2 T_N^2(\Omega/\Omega_p)} \tag{82.17}$$

where $T_N(x)$ is the Nth-order Chebyshev polynomial and ε is a real constant less than 1 which determines the ripple of the filter. Specifically, for nonnegative integers N, the Nth-order Chebyshev polynomial is given by

$$T_N(x) = \begin{cases} \cos(N\cos^{-1} x), & |x| \le 1 \\ \cosh(N\cosh^{-1} x), & |x| \ge 1 \end{cases} \tag{82.18}$$

High-order Chebyshev polynomials can be derived from the recursion relation

$$T_{N+1}(x) = 2x T_N(x) - T_{N-1}(x) \tag{82.19}$$

where $T_0(x) = 1$ and $T_1(x) = x$.

The Chebyshev approximation gives an **equiripple** characteristic in the passband and is maximally flat near infinity in the stopband. Each of the Chebyshev polynomials has real zeros that lie within the interval $(-1,1)$ and the function values for $x \in [-1,1]$ do not exceed $+1$ and -1.

The pole locations for Chebyshev filter can be determined by generating the appropriate Chebyshev polynomials, inserting them into Equation 82.17, factoring, and then selecting only the left half plane roots. Alternatively, the pole locations P_k of an Nth-order Chebyshev filter can be computed from the relation, for $k = 1 \rightarrow N$

$$P_k = -\sin\Theta_k \sinh\beta + j\cos\Theta_k \cosh\beta \tag{82.20}$$

where $\Theta_k = (2k - 1)\pi/2N$ and $\beta = \sinh^{-1}(1/\varepsilon)$.

Note: P_{N-k+1} and P_k are complex conjugates and when N is odd there is one real pole at

$$P_{N+1} = -2\sinh\beta$$

For the Chebyshev polynomials, Ω_p is the last frequency where the amplitude response passes through the value of ripple at the edge of the passband. For odd N polynomials, where the ripple of the Chebyshev polynomial is negative going, it is the $[-1/(1 + \varepsilon^2)]^{(1/2)}$ frequency and for even N, where the ripple is positive going, is the 0 dB frequency.

The Chebyshev filter is completely specified by the three parameters ε, Ω_p, and N. In a practical design application, ε is given by the permissible passband ripple and Ω_p is specified by the desired passband cutoff frequency. The order of the filter, i.e., N, is then chosen such that the stopband specifications are satisfied.

Elliptic or Cauer Filters

The frequency response of an Nth-order elliptic low-pass filter can be expressed by

$$\left| H_a\left(j\Omega \right) \right|^2 = \frac{1}{1 + \varepsilon^2 F_N^2\left(\Omega / \Omega_p \right)} \tag{82.21}$$

where $F_N(\cdot)$ is called the Jacobian elliptic function. The elliptic approximation yields an equiripple passband and an equiripple stopband. Compared with the same-order Butterworth or Chebyshev filters, the elliptic design provides the sharpest transition between the passband and the stopband. The theory of elliptic filters, initially developed by Cauer, is involved, therefore for an extensive treatment refer to Reference 1.

Elliptic filters are completely specified by the parameters ε, α, Ω_p, Ω_s, and N

where ε = passband ripple
  a = stop band floor
  Ω_p = the frequency at the edge of the passband (for a designated passband ripple)
  Ω_s = the frequency at the edge of the stopband (for a designated stopband floor)
  N = the order of the polynomial

In a practical design exercise, the desired passband ripple, stopband floor, and Ω_s are selected and N is determined and rounded up to the nearest integer value. The appropriate Jacobian elliptic function must be selected and $H_a(j\Omega)$ must be calculated and factored to extract only the left plane poles. For some synthesis techniques, the roots must expanded into polynomial form.

This process is a formidable task. While some filter manufacturers have written their own computer programs to carry out these calculations, they are not readily available. However, the majority of applications can be accommodated by use of published tables of the pole/zero configurations of low-pass elliptic transfer functions. An extensive set of such tables for a common selection of passband ripples, stopband floors, and shape factors is available in Reference 2.

Bessel Filters

The primary objectives of the preceding three approximations were to achieve specific loss characteristics. The phase characteristics of these filters, however, are nonlinear. The Bessel filter is optimized to reduce nonlinear phase distortion, i.e., a maximally flat delay. The transfer function of a Bessel filter is given by

$$H_a\left(s \right) = \frac{B_0}{B_N\left(s \right)} = \frac{B_0}{\sum\limits_{k=0}^{N} B_k s^k}, \qquad B_k = \frac{\left(2N - k \right)!}{2^{N-k} k! \left(N - k \right)!} \qquad k = 0, 1, \ldots, N \tag{82.22}$$

where $B_N(s)$ is the Nth-order Bessel polynomial. The overall squared-magnitude frequency response function is given by

$$\left| H_a\left(j\Omega \right) \right|^2 = 1 - \frac{\Omega^2}{2N - 1} + \frac{2\left(N - 1 \right)\Omega^4}{\left(2N - 1 \right)^2 \left(2N - 3 \right)} + \cdots \tag{82.23}$$

To illustrate Equation 82.22 the Bessel transfer function for $N = 4$ is given below:

$$H_a\left(s \right) = \frac{105}{105 + 105s + 45s^2 + 10s^3 + s^4} \tag{82.24}$$

Table 82.2 lists the factored pole frequencies as real and imaginary parts and as frequency and Q for Bessel transfer functions that have been normalized to $\Omega_c = -3$ dB.

82.4 Design Examples for Passive and Active Filters

Passive *R, L, C* Filter Design

The simplest and most commonly used passive filter is the simple, first-order ($N = 1$) *R–C* filter shown in Figure 82.3. Its transfer function is that of a first-order Butterworth low-pass filter. The transfer function and –3 dB Ω_c are

$$H_a(s) = \frac{1}{RCs+1} \quad \text{where} \quad \Omega_c = \frac{1}{RC} \tag{82.25}$$

While this is the simplest possible filter implementation, both source and load impedance change the dc gain and/or corner frequency and its rolloff rate is only first order, or –6 dB/octave.

To realize higher-order transfer functions, passive filters use *R, L, C* elements usually configured in a ladder network. The design process is generally carried out in terms of a doubly terminated two-port network with source and load resistors R_1 and R_2 as shown in Figure 82.4. Its symbolic representation is given below.

The source and load resistors are normalized in regard to a reference resistance $R_B = R_1$, i.e.,

$$r_i = \frac{R_1}{R_B} = 1, \quad r_2 = \frac{R_2}{R_B} = \frac{R_2}{R_1} \tag{82.26}$$

The values of *L* and *C* are also normalized in respect to a reference frequency to simplify calculations. Their values can be easily scaled to any desired set of actual elements.

$$l_v = \frac{\Omega_B L_v}{R_B}, \quad c_v = \Omega_B C_v R_B \tag{82.27}$$

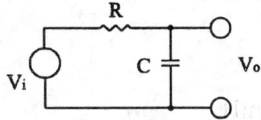

FIGURE 82.3   A passive first-order *RC* filter can serve as an antialiasing filter or to minimize high-frequency noise.

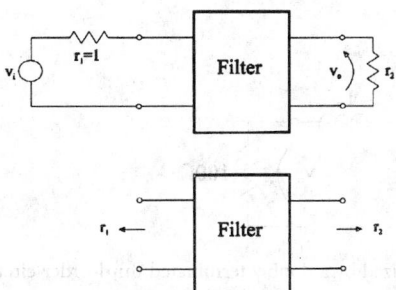

FIGURE 82.4   A passive filter can have the symbolic representation of a doubly terminated filter.

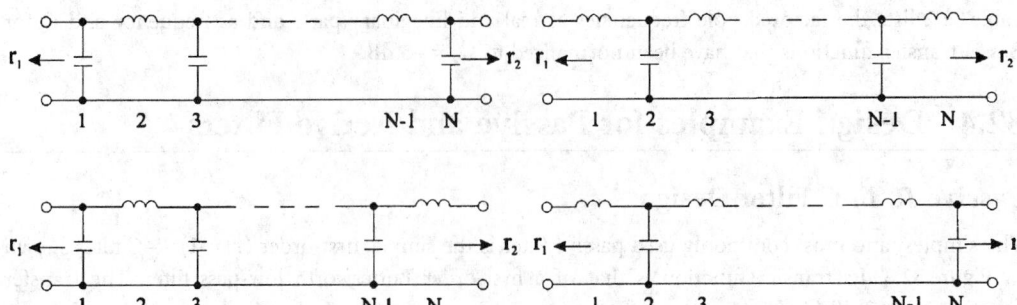

FIGURE 82.5 Even and odd N passive all-pole filter networks can be realized by several circuit configurations (N odd, above; N even, below).

Low-pass filters, whose magnitude-squared functions have no finite zero, i.e., whose characteristic functions $C(j\Omega)$ are polynomials, can be realized by lossless ladder networks consisting of inductors as the series elements and capacitors as the shunt elements. These types of approximations, also referred to as *all-pole approximations*, include the previously discussed Butterworth, Chebyshev Type I, and Bessel filters. Figure 82.5 shows four possible ladder structures for even and odd N, where N is the filter order.

In the case of doubly terminated Butterworth filters, the normalized values are precisely given by

$$a_v = 2\sin\left(\frac{(2v-1)\pi}{2N}\right), \quad v = 1, \ldots, N \qquad (82.28)$$

where a_v is the normalized L or C element value. As an example we will derive two possible circuits for a doubly terminated Butterworth low-pass of order 3 with $R_B = 100\ \Omega$ and a cutoff frequency $\Omega_c = \Omega_B = 10\ \text{kHz}$. The element values from Equation 82.28 are

$$l_1 = 2\sin\left(\frac{(2-1)\pi}{6}\right) = 1 \Rightarrow L_1 = \frac{R_B}{\Omega_c} = 1.59\ \text{mH}$$

$$c_2 = 2\sin\left(\frac{(4-1)\pi}{6}\right) = 2 \Rightarrow C_2 = \frac{2}{\Omega_c R_B} = 3.183\ \text{nF} \qquad (82.29)$$

$$l_3 = 2\sin\left(\frac{(6-1)\pi}{6}\right) = 1 \Rightarrow L_3 = \frac{R_B}{\Omega_c} = 1.59\ \text{mH}$$

A possible realization is shown in Figure 82.6.

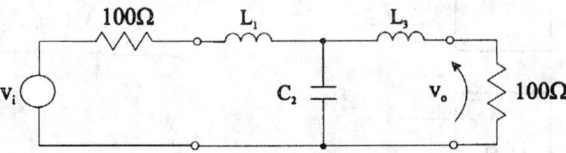

FIGURE 82.6 A third-order passive all-pole filter can be realized by a doubly terminated third-order circuit.

Table 82.3. Element Values for low-pass filter circuits

Filter Type	r_2	N = 2, Element Number		N = 3, Element Number		
		1	2	1	2	3
Butterworth	∞	1.4142	0.7071	1.5000	1.3333	0.5000
	1	1.4142	1.4142	1.0000	2.0000	1.0000
Chebyshev type I	∞	0.7159	0.4215	1.0895	1.0864	0.5158
0.1-dB ripple	1	—	—	1.0316	1.1474	1.0316
Chebyshev type I	∞	0.9403	0.7014	1.3465	1.3001	0.7981
0.5 dB ripple	1	—	—	1.5963	1.0967	1.5963
Bessel	∞	1.0000	0.3333	0.8333	0.4800	0.1667
	1	1.5774	0.4227	1.2550	0.5528	0.1922

Table 82.3 gives normalized element values for the various all-pole filter approximations discussed in the previous section up to order 3 and is based on the following normalization:

1. $r_1 = 1$;
2. All the cutoff frequencies (end of the ripple band for the Chebyshev approximation) are $\Omega_c = 1$ rad/s;
3. r_2 is either 1 or ∞, so that both singly and doubly terminated filters are included.

The element values in Table 82.3 are numbered from the source end in the same manner as in Figure 82.4. In addition, empty spaces indicate unrealizable networks. In the case of the Chebyshev filter, the amount of ripple can be specified as desired, so that in the table only a selective sample can be given. Extensive tables of prototype element values for many types of filters can be found in Reference 4.

The example given above, of a Butterworth filter of order 3, can also be verified using Table 82.3. The steps necessary to convert the normalized element values in the table into actual filter values are the same as previously illustrated.

In contrast to all-pole approximations, the characteristic function of an elliptic filter function is a rational function. The resulting filter will again be a ladder network but the series elements may be parallel combinations of capacitance and inductance and the shunt elements may be series combinations of capacitance and inductance.

Figure 82.5 illustrates the general circuits for even and odd N, respectively. As in the case of all-pole approximations, tabulations of element values for normalized low-pass filters based on elliptic approximations are also possible. Since these tables are quite involved the reader is referred to Reference 4.

Active Filter Design

Active filters are widely used and commercially available with cutoff frequencies from millihertz to megahertz. The characteristics that make them the implementation of choice for several applications are small size for low frequency filters because they do not use inductors; precision realization of theoretical transfer functions by use of precision resistors and capacitors; high input impedance that is easy to drive and for many circuit configurations the source impedance does not effect the transfer function; low output impedance that can drive loads without effecting the transfer function and can drive the transient, switched capacitive, loads of the input stages of A/D converters and low (N+THD) performance for pre-A/D antialiasing applications (as low as −100 dBc).

Active filters use R, C, A (operational amplifier) circuits to implement polynomial transfer functions. They are most often configured by cascading an appropriate number of first- and second-order sections.

The simplest first-order (N = 1) active filter is the first-order passive filter of Figure 82.3 with the addition of a unity gain follower amplifier. Its cutoff frequency (Ω_c) is the same as that qiven in Equation 82.25. Its advantage over its passive counterpart is that its operational amplifier can drive whatever load that it can tolerate without interfering with the transfer function of the filter.

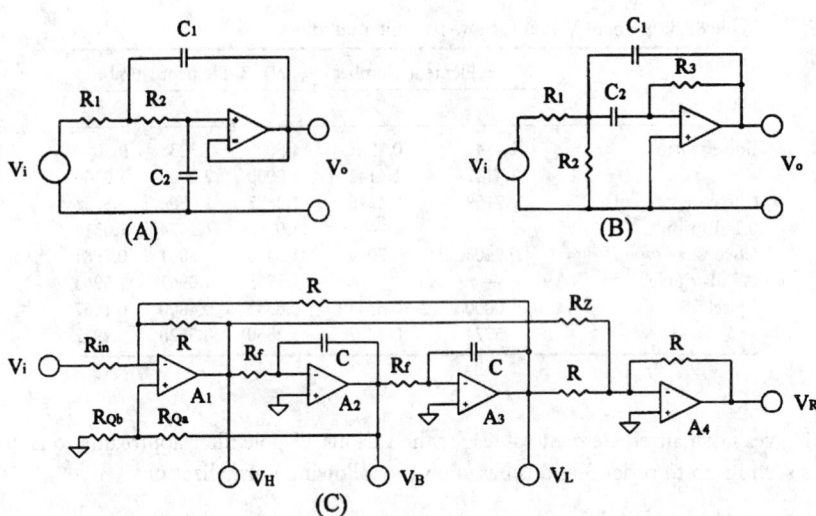

FIGURE 82.7 Second-order active filters can be realized by common filter circuits: (A) Sallen and Key low-pass, (B) multiple feedback bandpass, (C) state variable.

The vast majority of higher-order filters have poles that are not located on the negative real axis in the s-plane and therefore are in complex conjugate pairs that combine to create second-order pole pairs of the form:

$$H(s) = s^2 + \frac{\omega_p}{Q}s + \omega_p^2 \Leftrightarrow s^2 + 2as + a^2 + b^2 \qquad (82.30)$$

where $p_1, p_2 = a \pm jb$
$\quad \omega_p^2 = a^2 + b^2$

$$Q = \frac{\omega_p}{2a} = \frac{\sqrt{(a^2 + b^2)}}{2a}$$

The most commonly used two-pole active filter circuits are the *Sallen and Key* low-pass resonator, the *multiple feedback* bandpass, and the *state variable* implementation as shown in Figure 82.7a, b, and c. In the analyses that follow, the more commonly used circuits are used in their simplest form. A more comprehensive treatment of these and numerous other circuits can be found in Reference 20.

The Sallen and Key circuit of Figure 82.7a is used primarily for its simplicity. Its component count is the minimum possible for a two-pole active filter. It cannot generate stopband zeros and therefore is limited in its use to monotonic roll-off transfer functions such as Butterworth and Bessel filters. Other limitations are that the phase shift of the amplifier reduces the Q of the section and the capacitor ratio becomes large for high-Q circuits. The amplifier is used in a follower configuration and therefore is subjected to a large common mode input signal swing which is not the best condition for low distortion performance. It is recommended to use this circuit for a section $Q < 10$ and to use an amplifier whose gain bandwidth product is greater than $100 f_p$.

The transfer function and design equations for the Sallen and Key circuit of Figure 82.7a are

$$H(s) = \frac{\dfrac{1}{R_1 R_2 C_1 C_2}}{s^2 + \dfrac{1}{R_1 C_2}s + \dfrac{1}{R_1 R_2 C_1 C_2}} = \frac{\omega_p^2}{s^2 + \dfrac{\omega_p}{Q}s + \omega_p^2} \qquad (82.31)$$

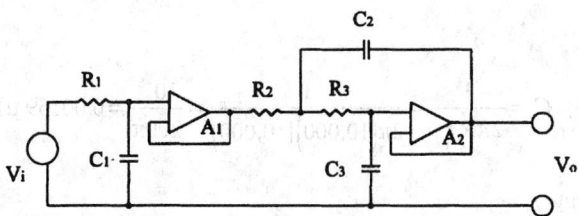

FIGURE 82.8 A three-pole Butterworth active can be configured with a buffered first-order *RC* in cascade with a two-pole Sallen and Key resonator.

from which obtains

$$\omega^2 = \frac{1}{R_1 R_2 C_1 C_2}, \quad Q = \omega_p R_1 C_2 = \sqrt{\frac{R_1 C_2}{R_2 C_1}} \qquad (82.32)$$

$$R_1, R_2 = \frac{1}{4\pi f_p Q C_2}\left[1 \pm \sqrt{1 - \frac{4Q^2 C_2}{C_1}}\right] \qquad (82.33)$$

which has valid solutions for

$$\frac{C_1}{C_2} \geq 4Q^2 \qquad (82.34)$$

In the special case where

$$R_1 = R_2 = R, \text{ then}$$
$$C = 1/2\pi R f_p, \quad C_1 = 2QC, \text{ and } C_2 = C/2Q \qquad (82.35)$$

The design sequence for Sallen and Key low-pass of Figure 82.7a is as follows:

For a required f_p and Q, select C_1, C_2 to satisfy Equation 82.34. Compute R_1, R_2 from Equation 82.33 (or Equation 82.35 if R_1 is chosen to equal R_2) and scale the values of C_1 and C_2 and R_1 and R_2 to desired impedance levels.

As an example, a three-pole low-pass active filter is shown in Figure 82.8. It is realized with a buffered single-pole *RC* low-pass filter section in cascade with a two-pole Sallen and Key section.

To construct a three-pole Butterworth filter, the pole locations are found in Table 82.2 and the element values in the sections are calculated from Equation 82.25 for the single real pole and in accordance with the Sallen and Key design sequence listed above for the complex pole pair.

From Table 82.2, the normalized pole locations are

$$f_{p1} = 1.000, \quad f_{p2} = 1.000, \text{ and } Q_{p2} = 1.000$$

For a cutoff frequency of 10 kHz and if it is desired to have an impedance level of 10 kΩ, then the capacitor values are computed as follows:

For $R_1 = 10\ k\Omega$:

from Equation 82.25, $\quad C_1 = \dfrac{1}{2\pi R_1 r_{p1}} = \dfrac{1}{2\pi (10{,}000)(10{,}000)} = \dfrac{10^{-6}}{200\pi} = 0.00159\ \mu F$

For $R_2 = R_3 = R = 10\ k\Omega$:

from Equation 82.35, $\quad C = \dfrac{1}{2\pi R f_{p2}} = \dfrac{1}{2\pi (10{,}000)(10{,}000)} = \dfrac{10^{-6}}{200\pi} = 0.00159\ \mu F$

from which

$$C_2 = 2QC = 2(0.00159)\ \mu F = 0.00318\ \mu F$$

$$C_3 = C/2Q = 0.5(0.00159)\ \mu F = 0.000795\ \mu F$$

The *multiple feedback* circuit of Figure 82.7b is a minimum component count, two-pole (or one-pole pair), bandpass filter circuit with user definable gain. It cannot generate stopband zeros and therefore is limited in its use to monotonic roll-off transfer functions. Phase shift of its amplifier reduces the Q of the section and shifts the f_p. It is recommended to use an amplifier whose open loop gain at f_p is $> 100Q^2 H_p$.

The design equations for the *multiple feedback* circuit of Figure 82.4b are

$$H(s) = -\cfrac{\dfrac{s}{R_1 C_1}}{s^2 + \dfrac{s}{R_3}\left(\dfrac{1}{C_1} + \dfrac{1}{C_2}\right) + \dfrac{(R_1 + R_2)}{R_1 R_2 R_3 C_1 C_2}} = -\cfrac{\dfrac{s\omega_p H_p}{Q}}{s^2 + \dfrac{s\omega_p}{Q} + \omega_p^2} \qquad (82.36)$$

when $s = j\omega_p$, the gain H_p is

$$H_p = \dfrac{R_3 C_2}{R_1\left(C_1 + C_2\right)} \qquad (82.37)$$

From Equation 82.36 and 82.37 for a required set of ω_p, Q, and H_p:

$$R_1 = \dfrac{Q}{C_1 H_p \omega_p}, \quad R_2 = \dfrac{Q}{\omega_p}\left(\dfrac{1}{Q^2\left(C_1 + C_2\right) - H_p C_1}\right), \quad R_3 = \dfrac{R_1 H_p\left(C_1 + C_2\right)}{C_2} \qquad (82.38)$$

For R_2 to be realizable,

$$Q^2\left(C_1 + C_2\right) \geq H_p C_1 \qquad (82.39)$$

The design sequence for a *multiple feedback* bandpass filter is as follows

Select C_1 and C_2 to satisfy Equation 82.39 for the H_p and Q required. Compute R_1, R_2, and R_3. Scale R_1, R_2, R_3, C_1, and C_2 as required to meet desired impedance levels.

Note that it is common to use $C_1 = C_2 = C$ for applications where $H_p = 1$ and $Q > 0.707$.

The *state variable* circuit of Figure 82.7c is the most widely used active filter circuit. It is the basic building block of programmable active filters and of switched capacitor designs. While it uses three or four amplifiers and numerous other circuit elements to realize a two-pole filter section, it has many desirable features. From a single input it provides low-pass (V_L), high-pass (V_H), and bandpass (V_B) outputs and by summation into an additional amplifier (A_4) (or the input stage of the next section) a band reject (V_R) or stop band zero can be created. Its two integrator resistors connect to the virtual ground of their amplifiers (A_2, A_3) and therefore have no signal swing on them. Therefore, programming resistors can be switched to these summing junctions using electronic switches. The sensitivity of the circuit to the gain and phase performance of its amplifiers is more than an order of magnitude less than single amplifier designs. The open-loop gain at f_p does not have to be multiplied by either the desired Q or the gain at dc or f_p. Second-order sections with Q up to 100 and f_p up to 1 MHz can be built with this circuit.

There are several possible variations of this circuit that improve its performance at particular outputs. The input can be brought into several places to create or eliminate phase of inversions; the damping feedback can be implemented in several ways other than the R_{Qa} and R_{Qb} that are shown in Figure 82.7c and the f_p and Q of the section can be or adjusted independently from one another. DC offset adjustment components can be added to allow the offset at any one output to be trimmed to zero.

For simplicity of presentation, Figure 82.7c makes several of the resistors equal and identifies others with subscripts that relate to their function in the circuit. Specifically, the feedback amplifier A_1, that generates the V_H output has equal feedback and input resistor from the V_L feedback signal to create unity gains from that input. Similarly, the "zero summing" amplifier, A_4 has equal resistors for its feedback and input from V_L to make the dc gain at the V_R output the same as that at V_L. More general configurations with all elements included in the equation of the transfer function are available in numerous reference texts including Reference 20.

The *state variable* circuit, as configured in Figure 82.7c, has four outputs. Their transfer functions are

$$V_L(s) = -\frac{R}{R_1(R_fC)^2}\left(\frac{1}{D(s)}\right) \tag{82.40a}$$

$$V_B(s) = \frac{R}{R_1}\left(\frac{\frac{s}{(R_fC)}}{D(s)}\right) \tag{82.40b}$$

$$V_H(s) = -\frac{R}{R_1}\left(\frac{s^2}{D(s)}\right) \tag{82.40c}$$

$$V_R(s) = \frac{R}{R_1(R_fC)^2}\left(\frac{\left(\frac{R_z}{R}\right)s^2 + 1}{D(s)}\right) \tag{82.40d}$$

where

$$D(s) = s^2 + \frac{a}{R_fC}s + \frac{1}{(R_fC)^2} = s^2 + \frac{\omega_p}{Q}s + \omega_p^2 \quad a = \frac{R_{Qb}}{(R_{Qa} + R_{Qb})}\left(2 + \frac{R}{R_1}\right) \tag{82.41}$$

Note that the dc gain at the low-pass output is

$$V_L(0) = -\frac{R}{R_i} \qquad (82.42)$$

from which obtains

$$\omega_p = \frac{1}{R_f C} \quad \text{and} \quad \frac{1}{Q} = \frac{R_{Qb}}{(R_{Qa} + R_{Qb})}\left(2 + \frac{R}{R_i}\right) \qquad (82.42)$$

The design sequence for the state variable filter of Figure 82.7c is

Select the values of R_f and C to set the frequency ω_p, the values of R_i for the desired dc gain and R_{Qa} and R_{Qb} for the desired Q and dc gain.

82.5 Discrete-Time Filters

A digital filter is a circuit or a computer program that computes a discrete output sequence from a discrete input sequence. Digital filters belong to the class of discrete-time LTI (linear time invariant) systems, which are characterized by the properties of causality, recursibility, and stability, and may be characterized in the time domain by their impulse response and in the transform domain by their transfer function. The most general case of a discrete-time LTI system with the input sequence denoted by $x(kT)$ and the resulting output sequence $y(kT)$ can be described by a set *of linear difference equations with constant coefficients.*

$$y(kT) = \sum_{\mu=0}^{N} b_\mu x(kT - \mu T) - \sum_{\mu=1}^{N} a_\mu y(kT - \mu T) \qquad (82.43)$$

where $a_0 = 1$. An equivalent relation between the input and output variables can be given through the convolution sum in terms of the impulse response sequence $h(kT)$:

$$y(kT) = \sum_{\mu=0}^{N} h(kT) x(kT - \mu T) \qquad (82.44)$$

The corresponding transfer function is given by

$$H(z) = \frac{Y(z)}{X(z)} = \frac{\displaystyle\sum_{\mu=0}^{N} b_\mu z^{-\mu}}{1 + \displaystyle\sum_{\mu=1}^{N} a_\mu z^{-\nu}} \Leftrightarrow H(z) = b_0 \prod_{\mu=1}^{N}\left(\frac{z - z_{0\mu}}{z - z_{\infty\mu}}\right) \qquad (82.45)$$

where $H(z)$ is the z-transform of the impulse response $h(kT)$ and $X(z)$, $Y(z)$ are the z-transform of the input signal $x(kT)$ and the output or the filtered signal $y(kT)$. As can be seen from Equation 82.44, if for at least one μ, $a_\mu \neq 0$, the corresponding system is recursive; its impulse response is of infinite duration — **infinite impulse response (IIR) filter**. If $a_\mu = 0$, the corresponding system is nonrecursive — **finite**

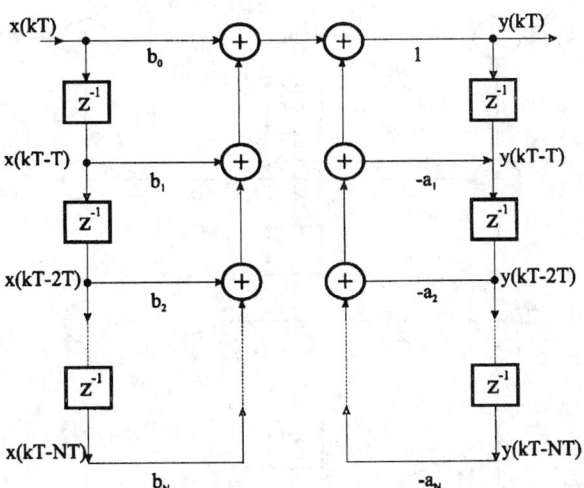

FIGURE 82.9 The difference equation of a digital filter can be realized by a direct-form I implementation that uses separate delay paths for the X and Y summations.

impulse response (FIR) filter; its impulse response is of finite duration and the transfer function $H(z)$ is a polynomial in z^{-1}. The zeros of the polynomial $X(z)$ given by $z = z_{\infty i}$ are called the poles of $H(z)$ and are commonly referred to as the *natural frequencies* of the filter. The condition for the stability of the filter is expressed by the constraint that all the poles of $H(z)$ should lie inside the unit circle, that is $|z_{\infty i}| < 1$. The zeros of $Y(z)$ given by $z = z_{0t}$ which are equivalent to the zeros of $H(z)$ are called the *transmission zeros* of the filter. Clearly, at these frequencies the output of the filter is zero for any finite input.

On the unit circle, the transfer function frequency $H(z)$ reduces to the frequency response function $H(e^{j\omega T})$, the discrete-time Fourier transform of $h(kT)$, which in general is complex and can be expressed in terms of magnitude and phase

$$H\left(e^{j\omega T}\right) = \left|H\left(e^{j\omega T}\right)\right| e^{j\theta(\omega)} \qquad (82.46)$$

The gain function of the filter is given as

$$\alpha\left(\Omega\right) = 20 \log_{10}\left|H\left(e^{j\omega T}\right)\right| \qquad (82.47)$$

It is also common practice to call the negative of the gain function the attenuation. Note that the attenuation is a positive number when the magnitude response is less than 1.

Figure 82.9 gives a block diagram realizing the difference equation of the filter, which is commonly referred to as the *direct-form I* realization. Notice that the element values for the multipliers are obtained directly from the numerator and denominator coefficients of the transfer function. By rearranging the structure in regard to the number of delays, one can obtain the canonic structure called *direct-form II* shown in Figure 82.10, which requires the minimum number of delays.

Physically, the input numbers are samples of a continuous signal and real-time digital filtering involves the computation of the iteration of Equation 82.43 for each incoming new input sample. Design of a filter consists of determining the constants a_μ and b_μ that satisfies a given filtering requirement. If the filtering is performed in real time, then the right side of Equation 82.46 must be computed in less than the sampling interval T.

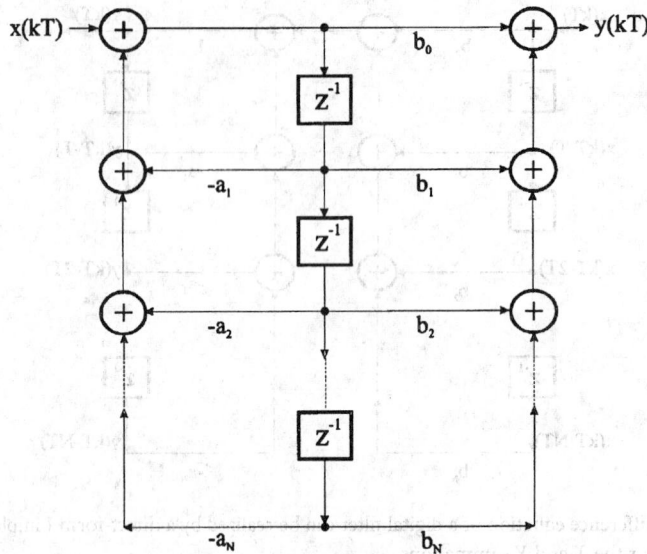

FIGURE 82.10 A direct-form II implementation of the difference equations minimizes the number of delay elements.

82.6 Digital Filter Design Process

The digital filter design procedure consists of the following basic steps:

1. Determine the desired response. The desired response is normally specified in the frequency domain in terms of the desired magnitude response and/or the desired phase response.
2. Select a class of filters (e.g., linear-phase FIR filters or IIR filters) to approximate the desired response.
3. Select the best member in the filter class.
4. Implement the best filter using a general-purpose computer, a DSP, or a custom hardware chip.
5. Analyze the filter performance to determine whether the filter satisfies all the given criteria.

82.7 FIR Filter Design

In many digital signal-processing applications, FIR filters are generally preferred over their IIR counterparts, because they offer a number of advantages compared with their IIR equivalents. Some of the good properties of FIR filters are a direct consequence of their nonrecursive structure. First, FIR filters are inherently stable and free of limit cycle oscillations under finite-word length conditions. In addition, they exhibit a very low sensitivity to variations in the filter coefficients. Second, the design of FIR filters with exactly *linear phase* (constant group delay) vs. frequency behavior can be accomplished easily. This property is useful in many application areas, such as speech processing, phase delay equalization, image processing, etc.

Finally, there exists a number of efficient algorithms for designing optimum FIR filters with arbitrary specifications. The main disadvantage of FIR filters over IIR filter is that FIR filter designs generally require, particularly in applications requiring narrow transition bands, considerably more computation to implement.

An FIR filter of order N is described by a difference equation of the form

$$y(kT) = \sum_{\mu=0}^{N} b_\mu x(kT - \mu T) \qquad (82.48)$$

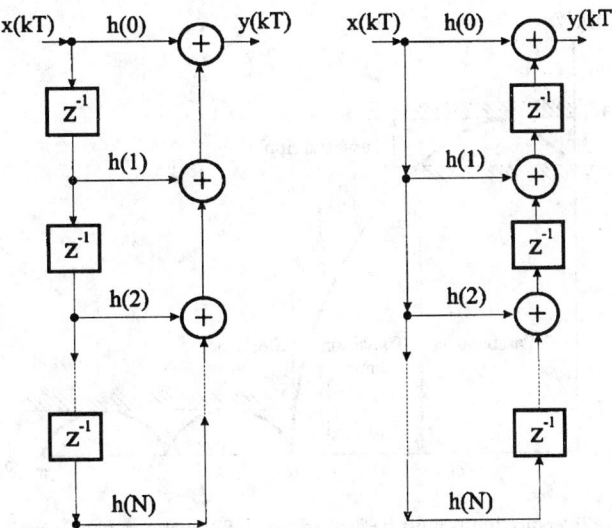

FIGURE 82.11 The sequence of the delays and summations can be varied to produce alternative direct-form implementations.

and the corresponding transfer function is

$$H(z) = \frac{Y(z)}{X(z)} = \sum_{\mu=0}^{N} b_\mu z^{-\mu} \qquad (82.49)$$

The objective of FIR filter design is to determine $N \pm 1$ coefficients given by

$$h(0), h(1), \ldots, h(N) \qquad (82.50)$$

so that the transfer function $H(e^{j\omega T})$ approximates a desired frequency characteristic. Note that because Equation 82.47 is also in the form of a convolution summation, the impulse response of an FIR filter is given by

$$h(kT) = \begin{cases} b_\mu, & k = 0, 1, \ldots, N \\ 0 & \text{otherwise} \end{cases} \qquad (82.51)$$

Two equivalent structures for FIR filters are given in Figure 82.11.
The accuracy of an FIR approximation is described by the following parameters:

δ_p passband ripple
δ_s stopband attenuation
$\Delta\omega$ transition bandwidth

These quantities are depicted in Figure 82.12 for a prototype low-pass filter. δ_p and δ_s characterize the permissible errors in the passband and in stopband, respectively. Usually, the passband ripple and stopband attenuation are given in decibels, in which case their values are related to the parameters δ_p and δ_s by

$$\text{Passband ripple (dB): } A_p = -20 \log_{10}(1 - \delta_p) \qquad (82.52)$$

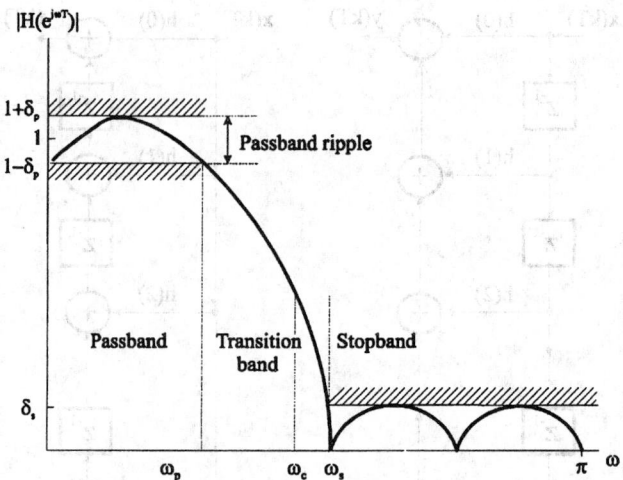

FIGURE 82.12 Tolerance limits must be defined for an FIR low-pass filter magnitude response.

$$\text{Stopband ripple } (\text{dB}): \; A_s = 020 \log_{10}(\delta_s) \tag{82.53}$$

Note that due to the symmetry and periodicity of the magnitude response of $|H(e^{j\omega T})|$, it is sufficient to give the filter specifications in the interval $0 \le \omega \le \pi$.

Windowed FIR Filters

Several design techniques can be employed to synthesize linear-phase FIR filters. The simplest implementation is based on *windowing*, which commonly begins by specifying the ideal frequency response and expanding it in a Fourier series and then truncating and smoothing the ideal impulse response by means of a window function. The truncation results in large ripples before and after the discontinuity of the ideal frequency response known as the Gibbs phenomena, which can be reduced by using a window function that tapers smoothly at both ends. Filters designed in this way possess equal passband ripple and stopband attenuation, i.e.,

$$\delta_p = \delta_s = \delta \tag{82.54}$$

To illustrate this method, let us define an ideal desired frequency response that can be expanded in a Fourier series

$$H_d\!\left(e^{jwT}\right) = \sum_{k=-\infty}^{\infty} h_d\!\left(kT\right) e^{-jk\omega T} \tag{82.55}$$

where $h_d(kT)$ is the corresponding impulse response sequence, which can be expressed in terms of $H_d(e^{j\omega T})$ as

$$h_d\!\left(kT\right) = \frac{1}{2\pi} \int_{-\pi}^{\pi} H_d\!\left(e^{jwT}\right) e^{jk\omega T} \, d\omega \tag{82.56}$$

The impulse response of the desired filter is then found by weighting this ideal impulse response with a window $w(kT)$ such that

$$h(kT) = \begin{cases} w(kT)h_d(kT), & 0 \le k \le N \\ 0, & \text{otherwise} \end{cases} \tag{82.57}$$

Note that for $w(kT)$ in the above-given interval we obtain the rectangular window. Some commonly used windows are Bartlett (triangular), Hanning, Hamming, Blackmann, etc., the definitions of which can be found in Reference 15.

As an example of this design method, consider a low-pass filter with a cutoff frequency of ω_c and a desired frequency of the form

$$H_d(e^{j\omega T}) = \begin{cases} e^{-j\omega NT/2}, & |\omega| \le \omega_c \\ 0, & \omega_c < |\omega| \le \pi, \end{cases} \tag{82.58}$$

Using Equation 82.56 we obtain the corresponding ideal impulse response

$$h_d(kT) = \frac{1}{2\pi} \int_{-\omega_c}^{\omega_c} e^{-j\omega TN/2} e^{jk\omega T} d\omega = \frac{\sin\left[\omega_c\left(kT - TN/2\right)\right]}{\pi\left(kT - TN/2\right)} \tag{82.59}$$

Choosing $N = 4$, $\omega_c = 0.6\pi$ and a Hamming window defined by

$$w(kT) = \begin{cases} 0.54 - 0.46\cos\left(2\pi kT/N\right), & 0 \le k \le N \\ 0, & \text{otherwise} \end{cases} \tag{82.60}$$

we obtain the following impulse response coefficients:

$$h(0) = -0.00748$$

$$h(1) = 0.12044$$

$$h(2) = -0.54729 \tag{82.61}$$

$$h(3) = 0.27614$$

$$h(4) = -0.03722$$

Optimum FIR Filters

As mentioned earlier, one of the principal advantages of FIR filters over their IIR counterparts is the availability of excellent design methods for optimizing arbitrary filter specifications. Generally, the design criterion for the optimum solution of an FIR filter design problem can be characterized as follows. The maximum error between the approximating response and the given desired response has to be minimized, i.e.,

$$E(e^{j\omega T}) = W_d(e^{j\omega T}) \left\| H_d(e^{j\omega T}) - |H(e^{j\omega T})| \right\| \tag{82.62}$$

where $E(e^{j\omega T})$ is the weighted error function on a close range X of $[0,\pi]$ and $W_d(e^{j\omega T})$ a weighting function, which emphasizes the approximation error parameters in the design process. If the maximum absolute value of this function is less then or equal ε on X, i.e.,

$$\varepsilon = \max_{\omega \in X} \left| E\left(e^{j\omega T}\right) \right| \tag{82.63}$$

the desired response is guaranteed to meet the given criteria. Thus, this optimization condition implies that the best approximation must have an equiripple error function. The most frequently used method for designing optimum magnitude FIR filters is the Parks–McClellan algorithm. This method essentially reduces the filter design problem into a problem in polynomial approximation in the Chebyshev approximation sense as discussed above. The maximum error between the approximation and the desired magnitude response is minimized. It offers more control over the approximation errors in different frequency bands than is possible with the window method. Using the Parks–McClellan algorithm to design FIR filters is computationally expensive. This method, however, produces optimum FIR filters by applying time-consuming iterative techniques. A FORTRAN program for the Parks–McClellan algorithm can be found in the IEEE publication Programs for DSP in Reference 12. As an example of an equiripple filter design using the Parks–McClellan algorithm, a sixth-order low-pass filter with a passband $0 \le \omega \le 0.6\pi$, a stopband $0.8\pi \le \omega \le \pi$, and equal weighting for each band was designed by means of this program.

The resulting impulse response coefficients are

$$h\left(0\right) = h\left(6\right) = -0.00596$$

$$h\left(1\right) = h\left(5\right) = -0.18459$$

$$h\left(2\right) = h\left(4\right) = 0.25596 \tag{82.64}$$

$$h\left(3\right) = 0.70055$$

Design of Narrowband FIR Filters

When using conventional techniques to design FIR filters with especially narrow bandwidths, the resulting filter lengths may be very high. FIR filters with long filter lengths often require lengthy design and implementation times, and are more susceptible to numerical inaccuracy. In some cases, conventional filter design techniques, such as the Parks–McClellan algorithm, may fail the design altogether. A very efficient algorithm called the interpolated finite impulse response (IFIR) filter design technique can be employed to design narrowband FIR filters. Using this technique produces narrowband filters that require far fewer coefficients than those filters designed by the direct application of the Parks–McClellan algorithm. For more information on IFIR filter design, see Reference 7.

82.8 IIR Filter Design

The main advantage of IIR filters over FIR filters is that IIR filters can generally approximate a filter design specification using a lower-order filter than that required by an FIR design to perform similar filtering operations. As a consequence, IIR filters execute much faster and do not require extra memory, because they execute in place. A disadvantage of IIR filters, however, is that they have a nonlinear phase response. The two most common techniques used for designing IIR filters will be discussed in this section. The first approach involves the transformation of an analog prototype filter. The second method is an optimization-based approach allowing the approximation of an arbitrary frequency response.

The transformation approach is quite popular because the approximation problem can be reduced to the design of classical analog filters, the theory of which is well established, and many closed-form design methods exist. Note that this in not true for FIR filters, for which the approximation problems are of an entire different nature. The derivation of a transfer function for a desired filter specification requires the following three basic steps:

1. Given a set of specifications for a digital filter, the first step is to map the specifications into those for an equivalent analog filter.
2. The next step involves the derivation of a corresponding analog transfer function for the analog prototype.
3. The final step is to translate the transfer function of the analog prototype into a corresponding digital filter transfer function.

Once the corresponding analog transfer function for the analog prototype is derived, it must be transformed using a transformation that maps $H_a(s)$ into $H(z)$. The simplest and most appropriate choice for s is the well-known bilinear transform of the z-variable

$$s = \frac{2\left(1 - z^{-1}\right)}{T_d\left(1 + z^{-1}\right)} \quad \Leftrightarrow \quad z = \frac{1 + \left(T_d/2\right)s}{1 - \left(T_d/2\right)s} \tag{82.65}$$

which maps a stable analog filter in the s-plane into a stable digital filter in the z-plane. Substituting s with the right-hand side of Equation 82.63 in $H_a(s)$ results in

$$H(z) = H_a\left(\frac{2\left(1 - z^{-1}\right)}{T_d\left(1 + z^{-1}\right)}\right) \quad \Rightarrow \quad H\left(e^{j\omega T}\right)\Big|_{z=e^{j\omega T}} = H_a\left(\frac{2j}{T_d}\tan\left(\frac{\omega T}{2}\right)\right) \tag{82.66}$$

As it can be seen from Equation 82.66, the analog frequency domain (imaginary axis) maps onto the digital frequency domain (unit circle) nonlinearly. This phenomena is called frequency warping and must be compensated in a practical implementation. For low frequencies Ω and ω are approximately equal. We obtain the following relation between the analog frequency Ω and the digital frequency ω

$$\Omega = \frac{2}{T_d}\tan\left(\omega T/2\right) \tag{82.67}$$

$$\omega = \frac{2}{T}\arctan\left(\Omega T_d/2\right) \tag{82.68}$$

The overall bilinear transformation procedure is as follows:

1. Convert the critical digital frequencies (e.g., ω_p and ω_s for low-pass filters) to the corresponding analog frequencies in the s-domain using the relationship given by Equation 82.67.
2. Derive the appropriate continuous prototype transfer function $H_a(s)$ that has the properties of the digital filter at the critical frequencies.
3. Apply the bilinear transform to $H_a(s)$ to obtain $H(z)$ which is the required digital filter transfer function.

To illustrate the three-step IIR design procedure using the bilinear transform, consider the design of a second-order Butterworth low-pass filter with a cutoff frequency of $\omega_c = 0.3\pi$. The sampling rate of the digital filter is to be $f_s = 10$ Hz, giving $T = 0.1$ s. First, we map the cutoff frequency to the analog frequency

$$\Omega_c = \frac{2}{0.1}\tan(0.15\pi) = 10.19 \text{ rad}/\text{s} \qquad (82.69)$$

The poles of the analog Butterworth filter transfer function $H_a(s)$ are found using Equation 82.11. As explained earlier, these poles lie equally spaced in the s-plane on a circle of radius Ω_c.

$$H_a(s) = \frac{1}{s^2 + \sqrt{2}\Omega_c s + \Omega_c^2} \qquad (82.70)$$

Application of the bilinear transformation

$$s = \frac{2(1-z^{-1})}{0.1(1+z^{-1})} \qquad (82.71)$$

gives the digital transfer function

$$H(z) = \frac{0.00002 + 0.00004z^{-1} + 0.00002z^{-2}}{1 - 1.98754z^{-1} + 0.98762z^{-2}} \qquad (82.72)$$

The above computations were carried out using Reference 9, which greatly automates the design procedure.

Design of Arbitrary IIR Filters

The IIR filter design approach discussed in the previous section is primarily suitable for frequency-selective filters based on closed-form formulas. In general, however, if a design other than standard low-pass, high-pass, bandpass, and stopband is required, or if the frequency responses of arbitrary specifications are to be matched, in such cases it is often necessary to employ algorithmic methods implemented on computers. In fact, for nonstandard response characteristics, algorithmic procedures may be the only possible design approach. Depending on the error criterion used, the algorithmic approach attempts to minimize the approximation error between the desired frequency response $H_d(e^{j\omega T})$ and $H(e^{j\omega T})$ or between the time-domain response $h_d(kT)$ and $h(kT)$. Computer software is available for conveniently implementing IIR filters approximating arbitrary frequency response functions [8,9].

Cascade-Form IIR Filter Structures

Recall that theoretically there exist an infinite number of structures to implement a digital filter. Filters realized using the structure defined by Equation 82.44 directly are referred to as direct-form IIR filters. The direct-form structure, however, is not employed in practice except when the filter order $N \leq 2$, because they are known to be sensitive to errors introduced by coefficient quantization and by finite-arithmetic conditions. Additionally, they produce large round-off noise, particularly for poles closed to the unit circle.

Two less-sensitive structures can be obtained by partial fraction expansion or by factoring the right-hand side of Equation 82.46 in terms of real rational functions of order 1 and 2. The first method leads to *parallel connections* and the second one to *cascade connections* of corresponding lower-order sections, which are used as building blocks to realize higher-order transfer functions. In practice, the cascade form is by far the preferred structure, since it gives the freedom to choose the pairing of numerators and denominators and the ordering of the resulting structure. Figure 82.13 shows a cascade-form implementation, whose overall transfer function is given by

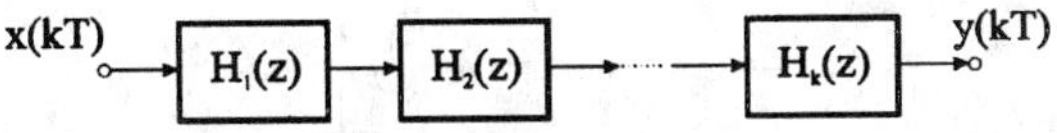

FIGURE 82.13 An IIR filter can be implemented by a cascade of individual transfer functions.

$$H\left(z\right) = \prod_{k=1}^{M} H_k\left(z\right) \tag{82.73}$$

where the transfer function of the kth building block is

$$H_k\left(z\right) = \frac{b_{0k} + b_{1k}z^{-1} + b_{2k}z^{-2}}{1 + a_{1k}z^{-2} + a_{2k}z^{-2}} \tag{82.74}$$

Note this form is achieved by factoring Equation 82.45 into second-order sections.

There are, of course, many other realization possibilities for IIR filters, such as state-space structures [9], lattice structures [10], and wave structures. The last is introduced in the next section.

82.9 Wave Digital Filters

It was shown earlier that for recursive digital filters the approximation problem can be reduced to classical design problems by making use of the bilinear transform. For wave digital filters (WDFs) this is carried one step farther in that the structures are obtained directly from classical circuits. Thus, to every WDF there corresponds an *LCR* reference filter from which it is derived. This relationship accounts for their excellent properties concerning coefficient sensitivity, dynamic range, and all aspects of stability under finite-arithmetic conditions. The synthesis of WDFs is based on the wave network characterization; therefore, the resulting structures are referred to as wave digital filters. To illustrate the basic idea behind the theory of WDFs, consider an inductor L, which is electrically described by $V(s) = sLI(s)$. In the next step we define wave variables $A_1(s)$ and $B_1(s)$ as

$$A_1\left(s\right) = V\left(s\right) + RI\left(s\right)$$
$$B_1\left(s\right) = V\left(s\right) - RI\left(s\right) \tag{82.75}$$

where R is called the port resistance. Substituting $V(s) = sLI(s)$ in the above relation and replacing s in $A_1(s)$ and $B_1(s)$ with the bilinear transform given by Equation 82.65, we obtain

$$B\left(z\right) = \frac{\left(1 - z^{-1}\right)L - \left(1 + z^{-1}\right)R}{\left(1 - z^{-1}\right)L + \left(1 + z^{-1}\right)R} A\left(z\right) \tag{82.76}$$

Letting $R = L$, the above relation reduces to

$$B\left(z\right) = -z^{-1} A\left(z\right) \tag{82.77}$$

Thus an inductor translates into a unit delay in cascade with an inverter in the digital domain. Similarly, it is easily verified that a capacitance can be simulated by a unit delay and a resistor by a digital sink. Figure 82.14 shows the digital realizations of impedances and other useful one-port circuit elements.

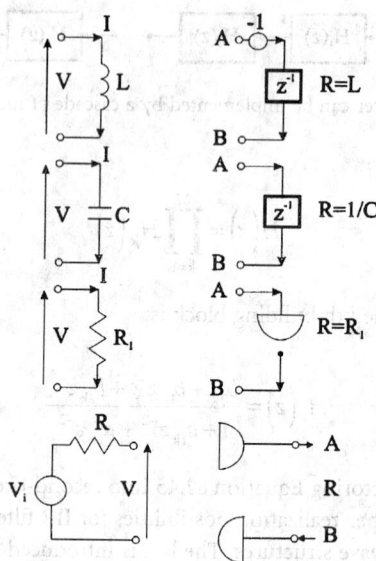

FIGURE 82.14 Digital filter implementations use functional equivalents to one port linear filter elements.

To establish an equivalence with classical circuits fully, the interconnections are also simulated by so-called wave adaptors. The most important of these interconnections are series and parallel connections, which are simulated by series and parallel adaptors, respectively. For most filters of interest, only two- and three-port adaptors are employed. For a complete design example consider Figure 82.15.

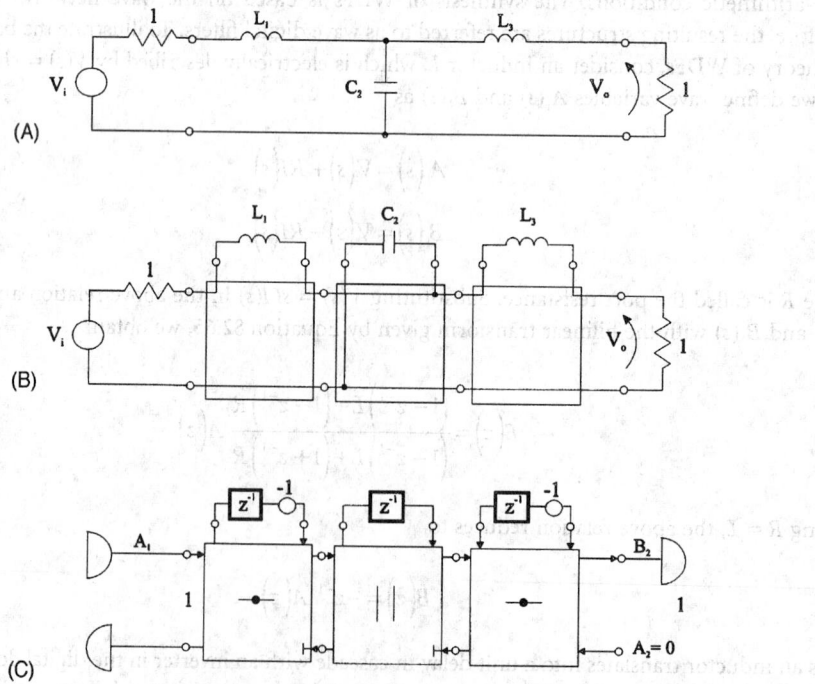

FIGURE 82.15 Digital wave filters establish equivalence with classical filter circuits by use of wave adapter substitutions: (A) *LC* reference low-pass; (B) identification of wire interconnections; (C) corresponding wave digital filter.

FIGURE 82.16 A data acquisition system with continuous time inputs and outputs uses antialias prefiltering, an A/D converter, digital signal processing, a D/A converter, and an output smoothing filter.

For a given *LC* filter, one can readily derive a corresponding WDF by using the following procedure. First, the various interconnections in the *LC* filter are identified as shown in Figure 82.15. In the next step the electrical elements in the *LC* filter are replaced by its digital realization using Figure 82.15. Finally, the interconnections are substituted using adaptors. Further discussions and numerical examples dealing with WDFs can be found in Reference 3, 13, and 14.

82.10 Anti-Aliasing and Smoothing Filters

In this section two practical application areas of filters in the analog conditioning stage of a data acquisition system are discussed. A block diagram of a typical data acquisition system is shown in Figure 82.16, consisting of an **antialiasing filter** before the analog-to-digital converter (ADC) and a smoothing filter after the digital-to-analog converter (DAC).

For a complete discrete reconstruction of a time-continuous, band-limited input signal having the spectrum $0 \leq f \leq f_{max}$, the sampling frequency must be, according to the well-known Shannon's sampling theorem, at least twice the highest frequency in the time signal. In our case, in order to be able to represent frequencies up to f_{max}, the sampling frequency $f_s = 1/T > 2f_{max}$. The necessary band limiting to $f \leq f_{max}$ of the input time-continuous signal is performed by a low-pass filter, which suppresses higher spectral components greater than f_{max}. Violation of this theorem results in alias frequencies. As a result, frequency components above $f_x/2$, the so-called Nyquist frequency, appear as frequency components below $f_x/2$. Aliasing is commonly addressed by using antialiasing filters to attenuate the frequency components at and above the Nyquist frequency to a level below the dynamic range of an ADC before the signal is digitized. Ideally, a low-pass filter with a response defined by

$$H\left(j\Omega \right) = \begin{cases} 1, & \left| \Omega \right| \leq \pi/T \\ 0, & \left| \Omega \right| \geq \pi/T \end{cases} \tag{82.78}$$

is desired to accomplish this task. In practice, a variety of techniques based on the principles of continuous-time analog low-pass filter design can be employed to approximate this "brick-wall" type of characteristic. Antialiasing filters typically exhibit attenuation slopes in the range from 45 to 120 dB/octave and stopband rejection from 75 to 100 dB. Among the types of filters more commonly used for antialias purposes are the Cauer elliptic, Bessel, and Butterworth. The optimum type of filter depends on which kinds of imperfections, e.g., gain error, phase nonlinearity, passband and stopband ripple, etc., are most likely to be tolerated in a particular application. For example, Butterworth filters exhibit very flat frequency response in the passband, while Chebyshev filters provide steeper attenuation at the expense of some passband ripple. The Bessel filter provides a linear phase response over the entire passband but less attenuation in the stopband. The Cauer elliptic filter, with its extremely sharp roll-off, is especially useful as an antialiasing filter for multichannel digitizing data acquisition systems. However, the large-phase nonlinearity makes it more appropriate for applications involving analysis of the frequency content of signals as opposed to phase content or waveform shape.

Many considerations discussed above also apply to smoothing filters. Due to the sampling process, the frequency response after the digital-to-analog conversion becomes periodic with a period equal to the sampling frequency. The quantitization steps that are created in the DAC reconstruction of the output waveform and are harmonically related to the sampling frequency must be suppressed through a low-pass filter having the frequency response of Equation 82.78 also referred to as a smoothing or reconstruction filter. While an antialiasing filter on the input avoids unwanted errors that would result from undersampling the input, a smoothing filter at the output reconstructs a continuous-time output from the discrete-time signal applied to its input.

Consideration must be given to how much antialiasing protection is needed for a given application. It is generally desirable to reduce all aliasable frequency components (at frequencies greater than half of the sampling frequency) to less than the LSB of the ADC being used. If it is possible that the aliasable input can have an amplitude as large as the full input signal range of the ADC, then it is necessary to attenuate it by the full 2^N range of the converter. Since each bit of an ADC represents a factor of 2 from the ones adjacent to it, and $20\log(2) = 6$ dB, the minimum attenuation required to reduce a full-scale input to less than a LSB is

$$\alpha < -20N\left(6\ dB\right) \tag{82.79}$$

where N is the number of bits of the ADC.

The amount of attenuation required can be reduced considerably if there is knowledge of the input frequency spectrum. For example, some sensors, for reasons of their electrical or mechanical frequency response, might not be able to produce a full-scale signal at or above the Nyquist frequency of the system and therefore "full-scale" protection is not required. In many applications, even for 16-bit converters that, in the worst case, would require 96 dB of antialias protection, 50 to 60 dB is adequate.

Additional considerations in antialias protection of the system are the noise and distortion that are introduced by the filter that is supposed to be eliminating aliasable inputs. It is possible to have a perfectly clean input signal which, when it is passed through a prefilter, gains noise and harmonic distortion components in the frequency range and of sufficient amplitude to be within a few LSBs of the ADC. The ADC cannot distinguish between an actual signal that is present in the input data and a noise or distortion component that is generated by the prefilter. It is necessary that both noise and distortion components in the output of the antialias filter must also be kept within an LBS of the ADC to ensure system accuracy.

82.11 Switched Capacitor Filters

Switched-capacitor (SC) filters, also generally referred to as analog sampled data filters, provide an alternative to conventional active-*RC* filters and are commonly used in the implementation of adjustable antialiasing filters. SC filters comprise switches, capacitors, and op amps. Essentially, an SC replaces the resistor in the more traditional analog filter designs. Because the impedance of the SC is a function of the switching frequency, one can vary the cutoff frequency of the SC filter by varying the frequency of the clock signal controlling the switching. The main advantage of SC filters is that they can be implemented in digital circuit process technology, since the equivalent of large resistors can be simulated by capacitors having small capacitance values.

When using SC filters, one must also be aware that they are in themselves a sampling device that requires antialias protection on the input and filtering on their outputs to remove clock feedthrough. However, since clock frequencies are typically 50 to 100 times f_c of the filter, a simple first or second *RC* filter on their inputs and outputs will reduce aliases and noise sufficient to permit their use with 12- to 14-bit ADCs. One need also to consider that they typically have dc offset errors that are large, vary with time, temperature, and programming or clock frequency. Interested readers may refer to References 5 and 14.

82.12 Adaptive Filters

Adaptive filtering is employed when it is necessary to realize or simulate a system whose properties vary with time. As the input characteristics of the filter change with time, the filter coefficients are varied with

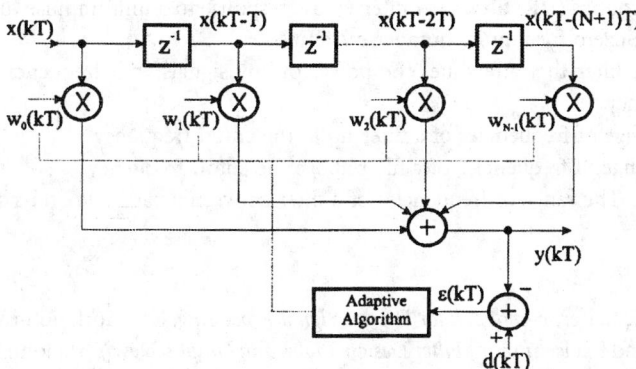

FIGURE 82.17   An adaptive filter uses an adaptive algorithm to change the performance of a digital filter in response to defined conditions.

time as a function of the filter input. Some typical applications of adaptive filtering include spectral estimation of speech, adaptive equalization, echo cancellation, and adaptive control, to name just a few. Depending on the application, the variations in the coefficients are carried out according to an optimization criterion and the adaptation is performed at a rate up to the sampling rate of the system. The self-adjustment capability of adaptive filter algorithms is very valuable when the application environment cannot be precisely described. Some of the most widely used adaptive algorithms are LMS (least-mean square), RLS (recursive least-squares), and frequency domain, also known as block algorithm. The fundamental concept of an adaptive filter is depicted in Figure 82.17.

An adaptive filter is characterized by the filter input $x(kT)$ and the desired response $d(kT)$. The error sequence $\varepsilon(kT)$ formed by

$$\varepsilon\left(kT\right) = \sum_{\mu=0}^{N-1} w_{\mu}\left(kT\right) x\left(kT - \mu T\right) \tag{82.80}$$

and $x(kT),\dots,x(kT - T(N-1))$ serve as inputs to an adaptive algorithm that recursively determines the coefficients $w_0(kT + T),\dots,w_{N-1}(kT + T)$. A number of adaptive algorithms and structures can be found in the literature that satisfy different optimization criteria in different application areas. For more detailed developments refer to References 1, 15, and 16.

Defining Terms

Antialiasing filter: Antialiasing filters remove any frequency elements above the Nyquist frequency. They are employed before the sampling operation is conducted to prevent aliasing in the sampled version of the continuous-time signal.

Bandpass filter: A filter whose passband extends from a lower cutoff frequency to an upper cutoff frequency. All frequencies outside this range are stopped.

Equiripple: Characteristic of a frequency response function whose magnitude exhibits equal maxima and minima in the passband.

Finite impulse response (FIR) filter: A filter whose response to a unit impulse function is of finite length, i.e., identically zero outside a finite interval.

High-pass filter: A filter that passes all frequencies above its cutoff frequency and stops all frequencies below it.

Ideal filter: An ideal filter passes all frequencies within its passband with no attenuation and rejects all frequencies in its stopband with infinite attenuation. There are five basic types of ideal filters: low pass, high pass, bandpass, stopband, and all pass.

Infinite impulse response (IIR) filter: A filter whose response to a unit impulse function is of infinite length, i.e., nonzero for infinite number of samples.

Low-pass filter: A filter that attenuates the power of any signals with frequencies above its defined cutoff frequency.

Passband: The range of frequencies of a filter up to the cutoff frequency.

Stopband: The range of frequencies of a filter above the cutoff frequency.

Transition region: The range of frequencies of a filter between a passband and a stopband.

References

1. S. Mitra and J. Kaiser, *Handbook for Digital Signal Processing*, New York: John Wiley & Sons, 1993.
2. E. Christian and E. Eisenmann, *Filter Design Tables and Graphs*, New York: John Wiley & Sons, 1966.
3. A. Antoniou, *Digital Filter: Analysis and Design*, New York: McGraw-Hill, 1979.
4. R. Saal, *Handbuch zum Filterentwurf [Handbook of filter design]*, Frankfurt: Allgemeine Elektricitäts-Gesellschaft AEG-Telefunken, 1979.
5. G. Temes and S. Mitra, *Modern Filter Theory and Design*, New York: John Wiley & Sons, 1973.
6. E. Cunningham, *Digital Filtering*, New York: John Wiley & Sons, 1995.
7. P. P. Vaidyanathan, *Multirate Systems and Filter Banks*, Englewood Cliffs, NJ: Prentice-Hall, 1993.
8. M. Cerna and R. Jamal, The design of digital filters using graphical programming techniques, *Proc. MessComp*, 232–238, 1995.
9. *LabVIEW Digital Filter Design Toolkit*, Austin, TX: National Instruments, 1996.
10. R. A. Roberts and C. T. Multis, *Digital Signal Processing*, Reading, MA: Addison-Wesley, 1987.
11. A. H. Gray, Jr. and J. D. Markel, Digital lattice and ladder filter synthesis, *IEEE Trans. Acoust. Speech Signal Process.*, ASSP-23: 268–277, 1975.
12. DSP Committee, IEEE ASSP, Eds., *Programs for Digital Signal Processing*, New York: IEEE Press, 1979.
13. F. Taylor, *Digital Filter Design Handbook*, New York: Marcel Dekker, 1983.
14. A. Fettweis, *Wave digital filters: theory and practice*, Proc. IEEE, 74, 270–327, 1986.
15. W. K. Chen, *The Circuits and Filters Handbook*, Boca Raton, FL: CRC Press, 1995.
16. M. L. Honig and D. Messerschmitt, *Adaptive Filters — Structures, Algorithms, and Applications*, Boston: Kluwer Academic Publishers, 1984.
17. M. Bellanger, *Digital Processing of Signals — Theory and Practice*, New York: John Wiley & Sons, 1988.
18. R. C. Dorf, Ed., *The Electrical Engineering Handbook*, Boca Raton, FL: CRC Press, 1993.
19. A. Zverev, *Handbook of Filter Synthesis*, New York: John Wiley & Sons, 1967.
20. C. Lindquist, *Active Network Design*, Long Beach, CA: Steward & Sons, 1977.

Ronney B. Panerai
University of Leicester

A. Ambrosini
Institute of Radioastronomy

C. Bortolotti
Institute of Radioastronomy

N. D'Amico
Institute of Radioastronomy

G. Grueff
Institute of Radioastronomy

S. Mariotti
Institute of Radioastronomy

S. Montebugnoli
Institute of Radioastronomy

A. Orfei
Institute of Radioastronomy

G. Tomassetti
Institute of Radioastronomy

83

Spectrum Analysis and Correlation

83.1 Spectrum Analysis and Correlation

Ronney B. Panerai

Most sensors and instruments described in previous sections of this handbook can produce continuous measurements in time or sequential measurements at fixed or variable time intervals, as represented in Figure 83.1. The temporal patterns resulting from such measurements are usually referred to as *signals*. Signals can either be *continuous* or *discrete in time* (Figure 83.1). The main objective of *spectral analysis* is to provide an estimate of the distribution of signal power at different frequencies. Spectral analysis and correlation techniques are an aid to the interpretation of signals and to the systems that generate them. These methods are now widely used for the analysis and interpretation of measurements performed in medicine, geophysics, vibration analysis, communications, and several other areas.

Although the original concept of a **signal** involves measurements as a function of time (Figure 83.1), this term has been generalized to include measurements along other dimensions, e.g., distance. In addition, signals can have multiple dimensions — the instantaneous velocity of an airplane can be regarded as a four-dimensional signal since it depends on time and three spatial coordinates.

With the growing availability of signal-processing computer packages and dedicated instruments, most readers will perform spectral analysis and correlation at the "touch of a button," visualizing results on a screen or as a computer plot. These "black-box" systems are useful for saving time and money, but users should be aware of the limitations of the fundamental techniques and circumstances in which inappropriate use can lead to misleading results. This chapter presents the basic concepts of spectral analysis and correlation based on the **fast Fourier transform (FFT)** approach. FFT algorithms allow the most efficient computer implementation of methods to perform spectral analysis and correlation and have become the most popular option. Nevertheless, other approaches, such as parametric techniques, wavelet transforms, and time-frequency analysis are also available. These will be briefly discussed and the interested reader will be directed to the pertinent literature for applications that might benefit from alternative approaches.

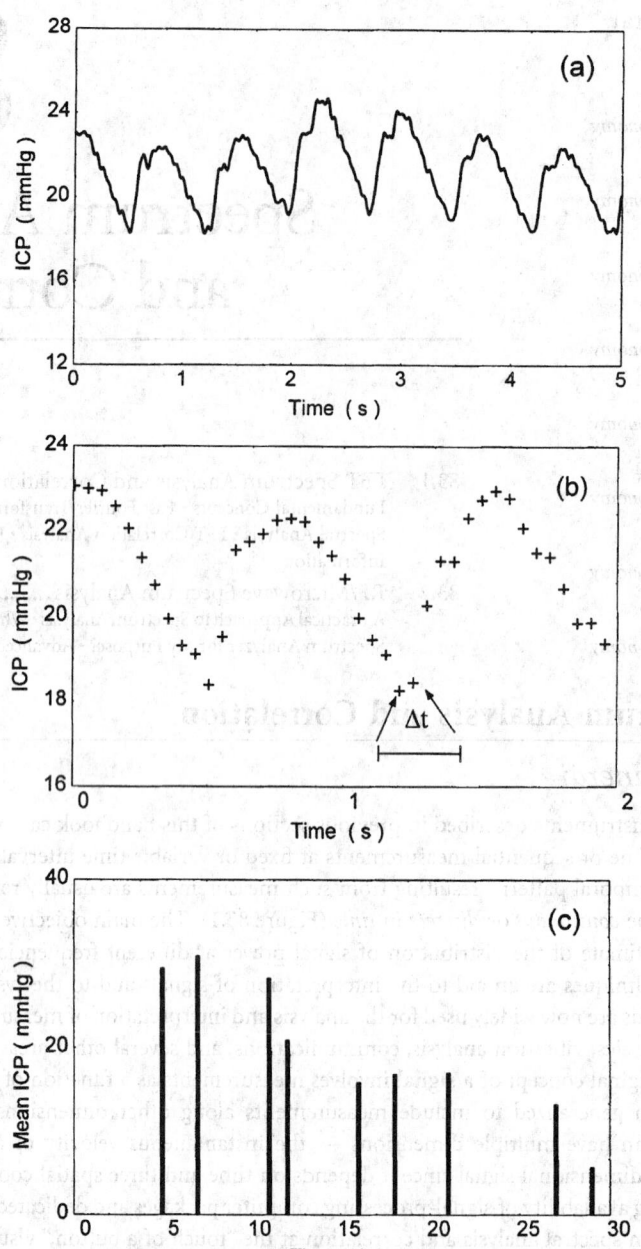

FIGURE 83.1 Examples of continuous and discrete-time signals. (a) Continuous recording of intracranial pressure in a head-injured patient. (b) Intracranial pressure measurements obtained at regular intervals of 50 ms. (c) Non-uniformly spaced measurements of mean intracranial pressure over a period of 30 h following surgery.

Fundamental Concepts

Spectral Analysis

Practical applications of spectral and correlation analysis are performed on discrete-time signals (Figure 83.1). These are obtained either from a sequence of discrete measurements or from the transformation of a continuous signal (Figure 83.1) to digital format using an **analog-to-digital converter (ADC)**.

When the latter is adopted to allow computer analysis of an originally continuous signal, two main characteristics of the ADC need to be considered. The first is the number of bits available to represent each sample, as this will determine the resolution and accuracy of the sampled signal. The second important consideration is the *sampling interval* Δt (Figure 83.1). From the *Nyquist theorem*,[1] the maximum value of Δt must be such that the *sampling frequency* $f_s = 1/\Delta t$ is at least twice the highest frequency of interest in the original signal. If this rule is not followed, spectral and correlation estimations might be considerably distorted by a phenomenon called *aliasing*.[2] Low-pass filtering before ADC is always recommended to limit the bandwidth of the continuous signal to allow the correct choice of f_s or Δt. In practice, the sampling frequency is usually much higher than the minimum required by the Nyquist theorem to provide a better visual representation of the sampled data.

Let x_n represent a discrete-time signal with samples at $n = 0, 1, 2, ..., N - 1$. The Fourier theorem[1,2] states that it is possible to decompose x_n as a sum of cosine and sine waveforms of different frequencies using an appropriate combination of amplitude coefficients. Therefore,

$$x_n = a_0 + \sum_{k=1}^{N-1} a_k \cos\left(\frac{2\pi kn}{N}\right) + \sum_{k=1}^{N-1} b_k \sin\left(\frac{2\pi kn}{N}\right) \tag{83.1}$$

where $k = 1, 2, ..., N - 1$ determines the frequency of each cosine and sine waveforms as $f_k = k/N\Delta t$. The corresponding coefficients are calculated from

$$a_0 = \frac{1}{N} \sum_{n=0}^{N-1} x_n \tag{83.2a}$$

$$a_k = \frac{1}{N} \sum_{n=0}^{N-1} x_n \cos\left(\frac{2\pi kn}{N}\right) \tag{83.2b}$$

$$b_k = \frac{1}{N} \sum_{n=0}^{N-1} x_n \sin\left(\frac{2\pi kn}{N}\right) \tag{83.2c}$$

Note that Equation 83.2a represents the mean value of x_n and that the argument $2\pi kn/N$ is the same for the *direct* (Equation 83.2) and *inverse* (Equation 83.1) **discrete Fourier transforms (DFT)**.

From Euler's formula,[3] it is possible to combine the cosine and sine terms to express the DFT in exponential form:

$$e^{j\theta} = \cos\theta + j\sin\theta \tag{83.3}$$

leading to

$$x_n = \sum_{k=0}^{N-1} c_k e^{j(2\pi kn/N)} \tag{83.4}$$

with

$$c_k = \frac{1}{N} \sum_{n=0}^{N-1} x_n e^{-j(2\pi kn/N)} \tag{83.5}$$

where c_k is now a complex value related to the original cosine and sine coefficients by

$$c_0 = a_0 \tag{83.6a}$$

$$c_k = a_k - jb_k \quad k = 1, 2, \ldots, N-1 \tag{83.6b}$$

A graphic representation of the a_k, b_k, or c_k coefficients for each value of k (or f_k) constitutes the frequency spectrum of x_n, expressing the relative contribution of different sinusoidal frequencies to the composition of x_n (Equation 83.4). Since c_k is complex (Equation 83.6b), a more meaningful physical interpretation of the spectrum is obtained with the *amplitude* and *phase* spectra, defined as

$$A_k = \left(a_k^2 + b_k^2 \right)^{1/2} = \left| c_k \right| \tag{83.7a}$$

$$\theta_k = \tan^{-1}\left(-\frac{b_k}{a_k} \right) \tag{83.7b}$$

Figure 83.2 shows the amplitude (or magnitude) and phase spectra for the signal in Figure 83.1a, sampled at intervals $\Delta t = 20$ ms. The signal was low-pass-filtered at 20 Hz before ADC. The total duration is given by $T = N\Delta t = 5$ s, corresponding to $N = 250$ samples. Before calculating the spectral coefficients, the mean value of the complete record was removed (dc term) and any linear trends were removed by fitting a straight line to the data (detrending). As will be discussed below, it is also important to apply a window to the data, to minimize the phenomenon of leakage. For $k > N/2$ both spectra present symmetrical values. This can be easily demonstrated from the fact that cosine (and sine^2) functions have even symmetry while sine has odd symmetry. From Equation 83.7 it follows that A_k and θ_k have *even* and *odd* symmetry, respectively.[4] Consequently, only half the spectral components ($k \leq N/2$) are required to give a complete description of x_n in the frequency domain.

The amplitude spectra indicates the combined amplitude of the cosine and sine terms to reconstruct x_n; the phase spectra reflects the relative phase differences (or time delays) between the sinusoidal waveforms to generate the temporal pattern of x_n. The amplitude spectra also reflects the signal power at different frequencies. For simplicity, the power spectrum can be defined as

$$P_k = A_k^2 = \left| c_k \right|^2 \tag{83.8}$$

Direct implementation of Equation 83.8, however, leads to spectral power estimates which are biased and inconsistent. More appropriate procedures for estimating the **power spectrum** (or *power density spectrum*) will be discussed later.

Parseval's theorem[5] demonstrates that the total signal energy can be computed either in time or frequency domain:

$$\frac{1}{N} \sum_{n=0}^{N-1} x_n^2 = \sum_{k=0}^{N-1} P_k \tag{83.9}$$

If x_n has zero mean, the left-hand side of Equation 83.9 is the biased estimator of signal variance.[6] Although most applications of spectral analysis concentrate on the characteristics of the amplitude or power spectra, it is important to bear in mind that the phase spectrum is also responsible for the temporal pattern of x_n. As an example, both the Dirac impulse function and white noise have a flat, constant amplitude (or

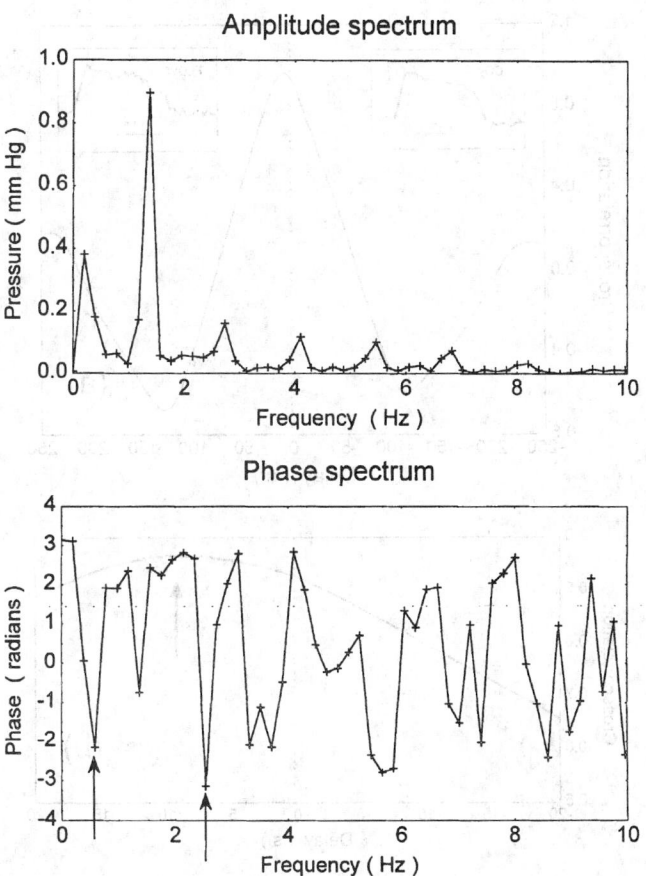

FIGURE 83.2 Amplitude and phase spectra of the intracranial pressure signal represented in Figure 83.1a after analog-to-digital conversion with a sampling interval of 20 ms. The main peak in the amplitude spectrum corresponds to the frequency of the cardiac cycle in Figure 83.1a. Wraparound of the phase spectrum is apparent in the third and 13th harmonics (arrows). Both spectra have been plotted to 10 Hz only.

power) spectra,[6] it is the difference in the phase spectra which accounts for the different morphologies in the time domain.

Interpretation of the amplitude and phase spectra of both theoretical functions and sampled data is facilitated by taking into account several properties of the DFT (Equations 83.4 and 83.5), namely, *symmetry, linearity, shifting, duality,* and *convolution.*[7] To these, a very important property of Equations 83.1 and 83.4 must be added. Since cosine and sine functions are periodic, and exist for $-\infty < t < \infty$, Equations 83.1 and 83.4 will reconstruct x_n not only in the interval of interest ($0 \leq t \leq T$) but also at all other multiple intervals $pT \leq t \leq (p+1)T$ ($p = 0, \pm 1, \pm 2, \dots$). As a consequence, spectral estimations obtained with the DFT inherently assume that x_n is *periodic* with period $T = N/\Delta t$. As discussed in the following sections, this property needs to be taken into account when performing spectral analysis with the DFT and FFT.

Correlation Analysis

The basic concept of the correlation coefficient, as a measure of the strength of linear relationship between two variables[6] can be extended to signal analysis with the definition of the **cross-correlation function** (CCF) as[5]:

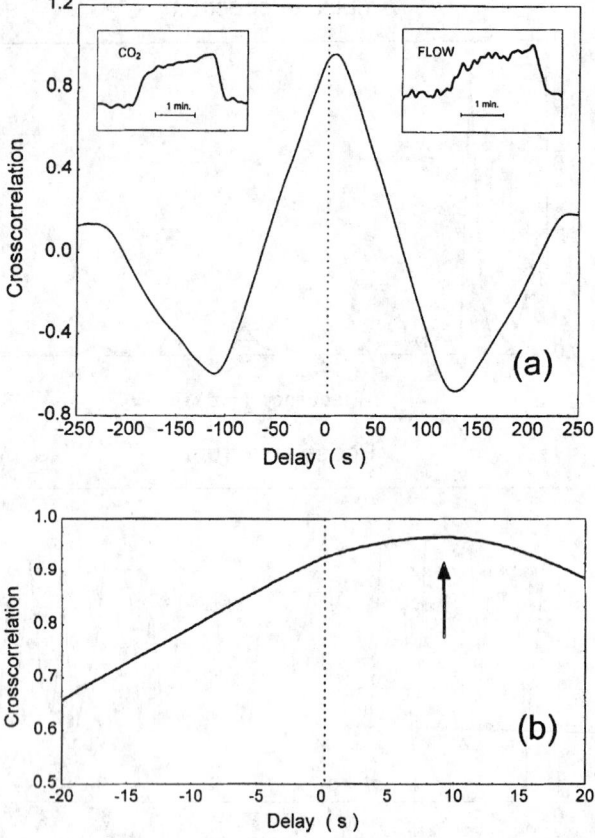

FIGURE 83.3 CCF between changes in arterial CO_2 and blood flow to the brain. Arterial CO_2 was estimated from end-tidal measurements and cerebral blood flow with Doppler ultrasound in the middle cerebral artery. (a) CCF and original signals (inserts). The cross-correlation value of approximately 1.0, observed at time delays near zero, reflects the similar temporal patterns between the two measurements. The negative cross correlations are obtained when either signal is shifted by approximately the duration of the plateau phase, which lasts 2 min. (b) Enlarging the scale around delay = 0 shows that the peak cross correlation occurs at 10 s, reflecting the time it takes for the flow to respond to the CO_2 change (Data kindly provided by Dr. Joanne Dumville, Mr. A. Ross Naylor, and Prof. David H. Evans, University of Leicester, U.K.)

$$r_{xy}(p) = \frac{1}{N} \sum_{n=0}^{N-1} x_n y_{n-p} \quad p = 0, \pm 1, \pm 2, \ldots \tag{83.10}$$

where x_n and y_n are zero-mean, discrete-time signals defined in the interval $n = 0, 1, 2, \ldots, N - 1$. For each value of p, the cross correlation is computed by shifting y_n by $p\Delta t$ and calculating the average product in Equation 83.10. If x_n and y_n are unrelated, the sum of positive and negative products will tend to zero. Conversely, is y_n tends to follow x_n, but with a time delay D, $r_{xy}(p)$ will show a peak at $p = D/\Delta t$. This property of the CCF is illustration in Figure 83.3. As noted by Bergland,[8] cross correlation can be viewed as "one signal searching to find itself in another signal."

For $y_n = x_n$, $r_{xy}(p)$ becomes the **autocorrelation function (ACF):**

$$r_{xx}(p) = \frac{1}{N} \sum_{n=0}^{N-1} x_n x_{n-p} \tag{83.11}$$

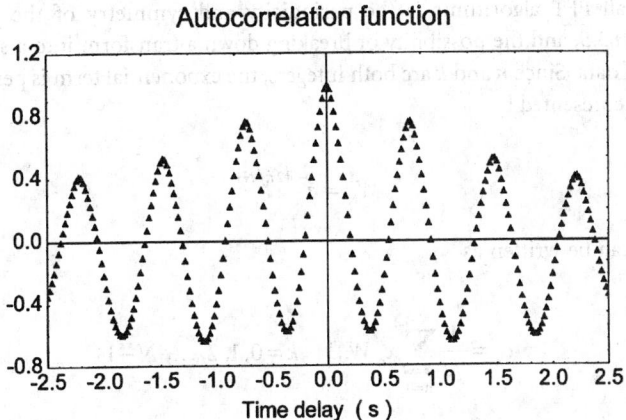

FIGURE 83.4   ACF of the discrete-time version of the signal in Figure 83.1a. The periodicity of the ACF reflects the quasi-periodic pattern of the intracranial pressure signal (Figure 83.1a).

and it is intuitive that the maximum value of $r_{xx}(p)$ occurs for $p = 0$ with

$$r_{xx}(0) = \frac{1}{N}\sum_{n=0}^{N-1} x_n^2 \qquad (83.12)$$

which represents the signal variance or total energy. Therefore, for signals with unit standard deviation, the autocorrelation peak is equal to 1.

The Wiener–Khintchine theorem[9] demonstrates that the autocorrelation function and the power spectrum constitute a Fourier transform pair, that is,

$$S_k = \sum_{p=0}^{N-1} r_{xx}(p)e^{-j(2\pi kp/N)} \qquad (83.13)$$

where S_k is usually called the autospectra of x_n.[6] Equation 83.13 indicates that it is possible to estimate the power spectra from a previous estimate of the autocorrelation function. As a transform pair, the autocorrelation function can also be derived from the autospectra by substituting S_k for c_k in Equation 83.4.

From Equation 83.11 it is clear that $r_{xx}(p)$ has even symmetry, that is, $r_{xx}(+p) = r_{xx}(-p)$. This property is apparent in Figure 83.4, which show the estimated autocorrelation function for the signal in Figure 83.1a. Another characteristic of ACF, which can be visualized in Figure 83.4, is the occurrence of secondary peaks reflecting the presence of an oscillatory component in x_n (Figure 83.1a).

Fast Fourier Transform

The FFT is not a single algorithm but rather a large family of algorithms which can increase the computational efficiency of the DFT. The main ideas behind the formulation of FFT algorithms are discussed below. A detailed description of the different algorithms that have been proposed is beyond the scope of this introduction; this can be found in References 5 through 7 and 10 through 14.

For both software and hardware implementations of Equations 83.4 and 83.5, the computational efficiency is usually expressed by the number of complex multiplications and additions required or, simply, by the *number of operations*.[10] Straight implementation of either Equation 83.4 or 83.5 leads to N^2 operations. Typically, FFT algorithms can reduce this number to $N \log_2 N$. For $N = 1024$ the FFT algorithm is 100 times faster than the direct implementation of Equation 83.4 or 83.5.

The essence of all FFT algorithms is the periodicity and symmetry of the exponential term in Equations 83.4 and 83.5, and the possibility of breaking down a transform into a sum of smaller transforms for subsets of data. Since n and k are both integers, the exponential term is periodic with period N. This is commonly represented by

$$W_N = e^{-j(2\pi/N)} \tag{83.14}$$

and Equation 83.5 can be written as

$$c_k = \frac{1}{N} \sum_{n=0}^{N-1} x_n W_N^{kn} \quad k = 0, 1, 2, \dots, N-1 \tag{83.15}$$

In many applications the terms W_N^{kn} are called **twiddle factors.** Assuming $N = 8$, calculation of the DFT with Equation 83.15 will require 64 values of W_8^{kn}. Apart from the minus sign, a simple calculation can show that there are only four different values of this coefficient, respectively: 1, j, $(1 + j)/\sqrt{2}$, and $(1 - j)/\sqrt{2}$.[4] Consequently, only these four complex factors need to be computed, representing a significant savings in number of operations.

Most FFT algorithms are based on the principle of **decimation-in-time,** involving the decomposition of the original time (or frequency) sequence into smaller subsequences. To understand how this decomposition can reduce the number of operations, assume that N is even. In this case it is possible to show that Equation 83.15 can be written as[4,5,7,11]:

$$c_k = \frac{1}{N} \sum_{r=0}^{(N/2)-1} x_r^e \cdot W_{N/2}^{kr} + \frac{1}{N} \sum_{r=0}^{(N/2)-1} x_r^o \cdot W_{N/2}^{kr} \tag{83.16}$$

where x_r^e and x_r^o represent the even- and odd-order samples of x_n, respectively. Comparing Equations 83.15 and 83.16, it is clear that the latter represents two DFTs with dimension $N/2$, involving $2(N/2)^2$ operations rather than the N^2 operations required by Equation 83.15. This process of decimation-in-time can be carried out further to improve computational performance. In the general case, N can be decomposed into q factors:

$$N = \prod_{i=1}^{q} r_i = r_1 r_2 \cdots r_q \tag{83.17}$$

The number of operations required is then[6]:

$$\text{number of operations} = N \sum_{i=1}^{q} r_i \tag{83.18}$$

In the original algorithm of Cooley and Tukey,[10] $r_i = 2$ and $N = 2^q$. In this case the theoretical number of operations required would be $2Nq = 2N \log_2 N$. As pointed out in Reference 6, further improvements in efficiency are possible because of the symmetry of the twiddle factors. The efficiency gain of most FFT algorithms using radix-2, i.e., $N = 2^q$ is

$$\text{efficiency gain} = \frac{N^2}{N \log_2 N} = \frac{N}{\log_2 N} \tag{83.19}$$

For $N = 1024$, $q = 10$ and the efficiency gain is approximately 100. Specific applications might benefit from other decompositions of the original sequence. Cases of particular interest are radix-4 and radix-8 FFTs.[14] However, as shown by Rabiner and Gold,[11] (p. 585), it is not possible to generalize the superiority of radix-8 over radix-4 algorithms.

In general, most FFT algorithms accept complex x_n sequences in Equation 83.5. By limiting x_n to the most common situation of real-valued signals, it is possible to obtain more efficient algorithms as demonstrated by Sorensen et al.[15] Uniyal[16] performed a comparison of different algorithms for real-valued sequences showing that performance is architecture dependent. For machines with a powerful floating point processor, the best results were obtained with Brunn's algorithm.[17]

The application of FFT algorithms for spectral and correlation analysis is discussed in the following sections.

FFT Spectral Analysis

For some deterministic signals, x_n can be expressed by a mathematical function and the amplitude and phase spectra can be calculated as an exact solution of Equations 83.5 and 83.7. The same is true for the power spectra (Equations 83.8 and 83.13). Examples of this exercise can be found in many textbooks.[1,2,4,5,7]

In most practical applications, there is a need to perform spectral analysis of experimental measurements, corresponding to signals which, in general, cannot be described by simple mathematical functions. In this case the spectra has to be estimated by a numerical solution of Equations 83.5 through 83.8, which can be efficiently implemented on a digital computer with an FFT algorithm. For estimation of the power spectrum, this approach is often classified as *nonparametric,* as opposed to other alternatives which are based on parametric modeling of the data such as *autoregressive* methods.[18] Considerable distortions can result from applications of the FFT unless attention is paid to the following characteristics and properties of the measured signal and the DFT/FFT.

Limited Observation of Signal in Time

Limited observation of a signal x_n in time can be seen as the multiplication of the original signal x_n^∞ by a rectangular window of duration $T = N\Delta t$ as exemplified for a single sinusoid in Figure 83.5. The DFT assumes that x_n is periodic, with period T, as mentioned previously. Instead of a single harmonic at the frequency of the original sinusoid, the power spectrum estimated with the FFT will have power at other harmonics as indicated by the spectrum in Figure 83.5c. The spectral power, which should have been concentrated on a single harmonic (Figure 83.5c, dashed line), has "leaked" to neighboring harmonics and for this reason this phenomenon is usually called *leakage*. The morphology of the distorted spectrum of Figure 83.5c can be explained by the fact that the Fourier transform of a rectangular window function (Figure 83.5b is given by a *sinc* function (sin x/x) which presents decreasing side lobes.[1,2] Multiplication in time corresponds to the convolution operation in the frequency domain.[1,2] In the general case of signals comprising several harmonics, the *sinc* functions will superimpose and the resulting spectrum is then a distorted version of the "true" spectrum. As the individual *sinc* functions superimpose to produce the complete spectrum, a *picket-fence* effect is also generated.[8] This means that spectral leakage not only adds spurious power to neighboring harmonics but also restricts the frequency resolution of the main spectral peaks. The effects of spectral leakage can be reduced by (1) increasing the period of observation and (2) multiplying the original signal x_n by a window function with a smooth transition as represented by the dashed line window in Figure 83.5b. The Fourier transform of a window function with tapered ends has smaller side lobes, thus reducing the undesirable effects leakage. A large number of tapering windows have been proposed, as reviewed by Harris.[19] As an example, the four-term Blackman–Harris window, defined as

$$ w_n = a_0 - a_1 \cos\left(\frac{2\pi n}{N}\right) + a_2 \cos\left(\frac{4\pi n}{N}\right) - a_3 \cos\left(\frac{6\pi n}{N}\right) \quad n = 0, 1, 2, \ldots N-1 \quad (83.20) $$

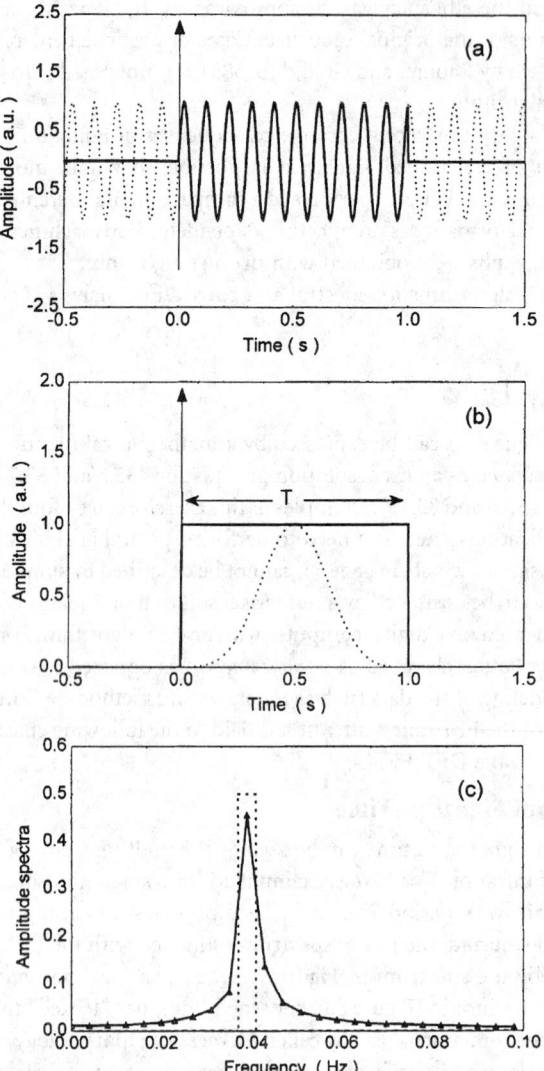

FIGURE 83.5 Effect of limited observation time T on the amplitude spectra of a sinusoidal component. (a) Observation of a single harmonic (dashed line) for a limited period of time T is equivalent to the multiplication for the rectangular function represented in (b). The Blackman–Harris window is also represented in (b) (dashed line). (c) Truncating a single harmonic produces spectral estimates smeared by *leakage* (solid line) as compared with the theoretical result (dashed line) with width equal to the frequency resolution ($f_r \approx 0.004$ Hz).

produces side lobe levels of –92 dB if the a_i coefficients are chosen as $a_0 = 0.35875$, $a_1 = 0.48829$, $a_2 = 0.14128$, $a_3 = 0.01168$.[19] Windows also play an important role in the sampling properties of power spectral estimates, as will be discussed later. Windowing attenuates the contribution of signal samples at the beginning and end of the signal and, therefore, reduces its effective signal duration. This effect is reflected by the equivalent noise bandwidth (ENBW) defined as[19]

$$\text{ENBW} = \frac{\sum\limits_{n=0}^{N-1} w_n^2}{\left[\sum\limits_{n=0}^{N-1} w_n\right]^2} \tag{83.21}$$

For a rectangular window ENBW = 1.0 and for the Blackman–Harris window (Equation 83.20) the corresponding value is 2.0. The majority of other window shapes have intermediate values of ENBW.[19]

Effects of "Zero-Padding"

Most FFT algorithms operate with $N = 2^q$ samples, the choice of q is many times critical. Since frequency resolution is inversely proportional to N, in many circumstances a value of q leading to $2^q > N$ is preferable to the option of limiting the signal to $N' = 2^{q-1}$ samples with $N' < N$. The most common and simple way of extending a signal to comply with the 2^q condition is by **zero-padding.** For signals with zero mean and with first and last values around zero, this can be accomplished by complementing the signal with Q zeros to achieve the condition $N + Q = 2^q$. For signals with end points different from zero, these values can be used for padding. If initial and final values differ significantly, a linear interpolation from the last to the first point is also a practical option. However, with the application of windowing, most signals will have similar initial and final points and these can be used for zero-padding. As discussed in the next section, zero-padding has important applications for the estimation of correlation functions via FFT. For spectral analysis, it is relatively simple to demonstrate that adding Q zeros corresponds to over sampling the N point original spectrum with a new frequency resolution which is $(N + Q)/N$ times greater than the original resolution. Consequently, although zero-padding does not introduce major distortions, it produces the false illusion of higher resolution than warranted by the available N measured signal samples.

Phase Spectrum Estimation

The use of Equation 83.7b to estimate the phase spectrum is fraught with a different kind of problem, resulting from the indetermination of the tan^{-1} function to discriminate between phase angles with absolute values greater than π. This problem is illustrated in Figure 83.2b showing that phase angles decrease continuously until reaching $-\pi$ and then "jump" to continue decreasing from the $+\pi$ value. This feature of the phase spectrum is called **wraparound.** Methods to "unwrap" the phase spectrum have been proposed,[20] but a general satisfactory solution to this problem is not available. In some cases the shifting property of the DFT[5,7] can be used to "rotate" the original signal in time, thus minimizing the slope of the phase spectrum and, consequently, the occurrence of wraparound.

Sampling Properties of Spectral Estimators

The most straightforward approach to computing the power spectrum is to use Equation 83.8. This method is known as the **periodogram.**[5] Application is limited to signals which are **stationary,** meaning stable statistical properties (such as the mean and the variance) along time. For measurements performed on *nonstationary* systems, such as speech or systems with time-varying parameters, other methods of spectral estimation are available and will be mentioned later. It is possible to demonstrate that when the period of observation T tends to infinity, Equation 83.8 gives an unbiased estimate of the power spectrum. In practice, due to finite values of T, the phenomenon of spectral leakage described above will lead to power spectral estimates which are *biased.*

The second inherent problem with the periodogram is the *variance* of the resulting spectral estimates. Assuming x_n to follow a Gaussian distribution, it follows that a_k and b_k will also be Gaussian because Equation 83.2 represents a linear transformation. Since Equations 83.7a and 83.8 involve the sum of two squared Gaussian variates, P_k will follow a χ^2 distribution with two degrees of freedom.[6] In this case the mean and the standard deviation of the power spectral estimate will be the same, *independently of the frequency considered*. As a consequence, power spectral estimates obtained from Equation 83.8, using a simple sample x_n, should be regarded as highly unreliable. In addition, the variance or standard deviation of this χ^2 distribution does not decrease with increases in sample duration N. This indicates that the periodogram (Equation 83.8) is an *inconsistent* estimator of the power spectrum.

For a χ^2 distribution with m degrees of freedom, the coefficient of variation is given by

$$CV\left[\chi_m^2\right] = \frac{\sqrt{2m}}{m} = \sqrt{\frac{2}{m}} \tag{83.22}$$

showing that it is possible to improve the reliability of power spectral estimates by increasing m. This can be achieved by replacing Equation 83.8 by[21]

$$\hat{P}_k = \frac{1}{L}\sum_{l=1}^{L} c_{k,l}^2 \quad k = 0, 1, 2, \ldots N-1 \tag{83.23}$$

with L representing a number of separate samples x_n each with length $T = N\Delta t$. If only one record of x_n can be obtained under stationary conditions, it is possible to break down this record into L segments to obtain an improved estimate of the power spectrum with variance reduced by a factor of L. However, the spectral resolution, given by $f_r = 1/T$, will be reduced by the same factor L, thus indicating an inescapable compromise between resolution and variance.

A *modified periodogram* was introduced by Welch[21] consisting of the multiplication of x_n by a triangular, or other window shape, before computing the individual spectral samples with Equation 83.5. The application of a window justifies overlapping adjacent segments of data by as much as 50%. For a signal with a total duration of N samples, the combination of overlapping with segmentation (Equation 83.23) can lead to a further reduction of the spectral variance by a factor of 11/18.

Averaging L spectral samples as indicated by Equation 83.23 represents one approach to improve spectral estimation by means of *smoothing*. A similar effect can be obtained with the **correlogram**. Equation 83.13 indicates that it is possible to estimate the power spectrum from the autocorrelation function. Limiting the number of shifts of the autocorrelation function to $p \ll N$ is equivalent to smoothing the original spectrum by convolution with the Fourier transform of a Bartlett (triangular) window.[22] As discussed in the next section, the autocorrelation function can also be computed more efficiently with the FFT and it can be shown that in this case it involves a smaller number of numerical operations than the Welch method based on the periodogram.[5]

FFT Correlation Analysis

Before considering the application of FFT algorithms to compute auto- and cross-correlation functions, it is important to discuss their sampling properties using Equations 83.10 and 83.11 as estimators. Assuming that variables x_n and y_n are not defined outside the interval $0 \leq n \leq N - 1$, it follows from Equation 83.10 that as p increases and the two functions "slide" past each other, the effective number of summed products is $N - |p|$ rather than N as implied by Equations 83.10 and 83.11. For this reason these equations are often rewritten as

$$r_{xy}(p) = \frac{1}{N-|p|}\sum_{n=0}^{N-|p|-1} x_n y_{n-1} \quad p = 0, \pm 1, \pm 2, \ldots \tag{83.24}$$

The main justification for this modification, however, is that Equations 83.10 and 83.11 lead to biased estimations of correlation functions while Equation 83.24 is *unbiased*.

Equation 83.24 normally assumes that x_n and y_n are standardized variables with zero mean and unit variance. If the mean values are different from zero, Equation 83.10 and 83.11 will produce distorted estimates with a "pyramid effect" due to the presence of the dc term. However, this effect is compensated for in Equation 83.24 and in this case the effect of the mean value is to add a constant term:

$$r_{xy}(p) = r'_{xy}(p) - m_x m_y \tag{83.25}$$

where $r'_{xy}(p)$ is the cross correlation of variables with mean values m_x and m_y, respectively.

Similarly to the DFT, Equations 83.10 and 83.11 involve N^2 operations and Equation 83.24 slightly less. Since the autocorrelation and the power spectra constitute a Fourier transform pair (Equation 83.13), the computation of correlation functions can also be sped up by means of an FFT algorithm. For the sake of generality, the **cross spectrum** of x_n and y_n can be defined as[6]

$$C_{xy}(f_k) = X(f_k) Y^*(f_k) \tag{83.26}$$

with $X(f_k)$ and $Y(f_k)$ representing the Fourier transforms of x_n and y_n, respectively. The generalized Wiener–Khintchine theorem then gives the cross-correlation function as

$$r_{xy}(p) = \sum_{k=0}^{N-1} C_{xy}(f_k) e^{-j(2\pi kp/N)} \tag{83.27}$$

Therefore, "fast" correlation functions can be computed[23] using the forward FFT to calculate $X(f_k)$ and $Y(f_k)$ and then the inverse FFT to obtain $r_{xy}(p)$ with Equation 83.27. Obviously, when autocorrelation functions are being computed with this method, only one transform is necessary to obtain the autospectra (Equation 83.13) instead of the cross spectra (Equation 83.26).

When correlation functions are computed with the FFT, it is critical to pay attention again to the periodicity of the transformed variables as an intrinsic property of the DFT. When the two functions in either Equation 83.10 and 83.11 are displaced by p samples, for periodic functions there will be nonzero products outside the range $0 \le n \le N - 1$, thus leading to significant errors in the estimated auto- or cross-correlation functions. In this case the resulting estimates are called *circular* correlations.[6] This error can be avoided by zero-padding the original signals from $n = N$ to $n = 2N - 1$ and computing the FFTs with $2N$ points. The resulting correlation functions will be noncircular and, in the range $0 \le p \le N - 1$, will agree with correlations computed with the original Equations 83.10 or 83.11. Finally, to remove bias the results of Equation 83.27 should also be multiplied by $N/(N - |p|)$ to agree with Equation 83.24.

Further Information

Software for FFT special analysis is available from multiple sources. Off-the-shelf software ranges from specialized packages for digital signal processing, such as DADiSP, to statistical packages which include FFT analysis of time series. Mathematical and engineering packages such as MATLAB also include routines for FFT spectral and correlation analysis. For a review of available options see Reference 24. For readers who want to implement their own software, FFT routines can be found in Reference 4, 13 through 15, and 18. Additional references are 25 through 27.

Hardware implementations of FFT algorithms are common in areas requiring real-time spectral analysis as in the case of blood flow velocity measurement with Doppler ultrasound. For a review of hardware implementations see References 11 and 28. Developments in this area follow the pace of change in VLSI technology.[29]

One of the limitations of the FFT is the fact that frequency resolution is the inverse of the signal observation time. Improved resolution can be obtained with *parametric* methods of spectral analysis and their application is particularly relevant when only short segments of data are available or when it is necessary to discriminate between frequency harmonics which are closely spaced in the spectrum. Broadly speaking, parametric methods assume that the data follow spectral densities with a known pole-zero structure of variable complexity, characterized by a given *model order*. All-zero models correspond to the *moving average* structure while the all-pole version represents the *autoregressive* model. The general case is the autoregressive-moving average model (ARMA). For a comprehensive review of these methods see Reference 30; further information and software implementations can be found in References 18 and 27.

Nonstationary signals present a particular problem. In cases where the signal statistical properties change relatively slowly with time, it is possible to select short segments of quasi-stationary data and to use the DFT or parametric methods to estimate the spectra as mentioned previously. However, when these changes in systems parameters or statistical moments are fast in relation to the phenomenon under observation (e.g., speech or seismic data), this approach is not feasible because of the poor frequency resolution resulting from short observation times. Methods proposed to cope with signal nonstationarity often depend on the underlying cause of nonstationary behavior.[9,31] More general methods, known as *time-frequency distributions,* are now favored by most investigators.[32] The Wigner–Ville and Choi–Williams transforms are some of the more widely used of these time-frequency distributions. In each case the signal is described by a simultaneous function of time *and* frequency and hence is graphically represented by a three-dimensional plot having time and frequency as dependent variables.

A different approach to the analysis of nonstationary data is the application of *wavelets.*[33] This alternative also has advantages in the representation of fast transients and in applications requiring data compression and pattern classification. Similarly to the sine and cosine functions, which are the basis of Fourier analysis, wavelets are orthogonal functions which can be used to decompose and reconstruct signals using a finite set of coefficients obtained by a *wavelet transform* (WT). The main difference between wavelets and sinusoids, however, is that the former are limited in time. In addition, the complete orthogonal set of wavelets can be obtained simply by expansion (or compression) and scaling of a single function, known as the *mother wavelet*. Because of their limited time duration wavelets can provide a much more synthetic decomposition of fast transients, or sharp edges in image analysis, than it is possible to obtain with the DFT. Their property of expansion/contraction of a single mother wavelet can also overcome a major limitation of the DFT, that is, to allow good frequency resolution at both low and high frequencies. For applications of the WT and commercially available software see References 34 and 35.

Defining Terms

Analog-to-digital conversion: The process of converting a continuous signal to a discrete time sequence of values usually sampled at uniform time intervals.

Autocorrelation function (ACF):   A measure of longitudinal variability of a signal which can express the statistical dependence of serial samples.

Correlogram: Numerical calculation and graphical representation of the ACF or CCF.

Cross-correlation function (CCF):   A measure of similarity between signals in the time domain which also allows the identification of time delays between transients.

Cross-spectrum: The complex product of the power spectra of two different signals.

Decimation-in-time: The process of breaking down a time series into subsequences to allow more efficient implementations of the FFT.

Discrete Fourier transform (DFT):   The usual method to obtain the Fourier series of a discrete time signal.

Fast Fourier transform (FFT):   Algorithm for the efficient computation of the DFT.

Periodogram: A family of methods to estimate the power spectrum using the DFT.

Power spectrum: The distribution of signal power as a function of frequency.

Signal: Continuous or discrete representation of a variable or measurement as a function of time or other dimension.

Stationarity: Property of signals which have statistical moments invariant with time.

Twiddle factors: Exponential term in the DFT whose periodicity allow repeated use and hence considerable savings of computation time in the FFT.

Wraparound: Overflow of phase spectral estimations above $|\pi|$ due to the uncertainty of the $\tan^{-1}$ function.

Zero-padding: Extension of a signal with zeros, constant values, or other extrapolating functions.

References

1. B.P. Lathi, *Communication Systems*, New York: John Wiley & Sons, 1968.
2. R.N. Bracewell, *The Fourier Transform and Its Applications*, Englewood Cliffs, NJ: Prentice-Hall, 1988.
3. E. Kreyszig, *Advanced Engineering Mathematics*, New York: John Wiley & Sons, 1962.
4. P.A. Lynn and W. Fuerst, *Introductory Digital Signal Processing with Computer Applications*, Chichester: John Wiley & Sons, 1989.
5. J.G. Proakis and D.G. Manolakis, *Digital Signal Processing: Principles, Algorithms, and Applications*, 2nd ed., New York: Macmillan, 1992.
6. J.S. Bendat and A.G. Piersol, *Random Data: Analysis and Measurement Procedures*, 2nd ed., New York: John Wiley & Sons, 1986.
7. A.V. Oppenheim and R.W. Schafer, *Discrete-Time Signal Processing*, Englewood Cliffs, NJ: Prentice-Hall, 1989.
8. G.D. Bergland, A guided tour of the fast Fourier transform, *IEEE Spectrum*, **6**: 41–52, 1969.
9. M.B. Priestley, *Spectral Analysis and Time Series*, London: Academic Press, 1981.
10. J.W. Cooley and J.W. Tukey, An algorithm for the machine computation of complex Fourier series, *Math. Computation*, **19**: 297–301, 1965.
11. L.R. Rabiner and B. Gold, *Theory and Application of Digital Signal Processing*, Englewood Cliffs, NJ: Prentice-Hall, 1975.
12. E.O. Brigham, *The Fast Fourier Transform and Its Applications*, 2nd ed., Englewood Cliffs, NJ: Prentice-Hall, 1988.
13. Digital Signal Processing Committee, *Programs for Digital Signal Processing*, New York: IEEE Press, 1979.
14. R.C. Singleton, An algorithm for computing the mixed radix fast Fourier transform, *IEEE Trans. Audio Electroacoust.*, **17**: 93–103, 1969.
15. H.V. Sorensen, D.L. Jones, M.T. Heideman, and C.S. Burrus, Real-valued fast Fourier transform algorithms, *IEEE Trans. Acoust. Speech Signal Proc.*, **35**: 849–863, 1987.
16. P.R. Uniyal, Transforming real-valued sequences: fast Fourier versus fast Hartley transform algorithms, *IEEE Trans. Signal Proc.*, **41**: 3249–3254, 1994.
17. G. Brunn, z-Transform DFT filters and FFT, *IEEE Trans. Acoust. Speech Signal Proc.*, **26**: 56–63, 1978.
18. S.M. Kay, *Modern Spectral Estimation*, Englewood Cliffs, NJ: Prentice-Hall, 1988.
19. F.J. Harris, On the use of windows for harmonic analysis with the discrete Fourier transform, *Proc. IEEE*, **66**: 51–83, 1978.
20. J.M. Tribolet, A new phase unwrapping algorithm, *IEEE Trans. Acoust. Speech Signal Proc.*, **25**: 170–177, 1977.
21. P.D. Welch, The use of fast Fourier transform for the estimation of power spectra: a method based on time averaging over short, modified periodograms, *IEEE Trans. Audio Electroacoust.*, **15**: 70–73, 1967.
22. R.B. Blackman and J.W. Tukey, *The Measurement of Power Spectra*, New York: Dover Publications, 1958.
23. T.G. Stockham, Jr., High-speed convolution and correlation, *1966 Spring Joint Computer Conference, AFIPS Proc.*, **28**: 229–233, 1966.
24. R. Braham, Math & visualization: new tools, new frontiers, *IEEE Spectrum*, **32**(11): 19–36, 1995.
25. W.H. Press, B.P. Flannery, S.A. Teukolsky, and W.T. Vetterling, *Numerical Recipes: The Art of Scientific Computing*, Cambridge, UK: Cambridge University Press, 1986.

26. D.M. Monro, Complex discrete fast Fourier transform, *Appl. Stat.*, **24**: 153–160, 1975.

27. S.L. Marple, Jr., *Digital Spectral Analysis with Applications*, Englewood Cliffs, NJ: Prentice-Hall, 1987.

28. G.D. Bergland, Fast Fourier transform hardware implementations — an overview, *IEEE Trans. Audio Electroacoust.*, **17**: 104–108, 1969.

29. E. Bidet, D. Castelain, C. Joanblanq, and P. Senn, A fast single-chip implementation of 8192-complex point FFT, *IEEE J. Solid-State Circuits*, **30**: 300–305, 1995.

30. S.M. Kay and S. L. Marple, Jr., Spectrum analysis — a modern perspective, *Proc. IEEE*, **69**: 1380–1419, 1981.

31. J. Leuridan, H.V. Van der Auweraer, and H. Vold, The analysis of nonstationary dynamic signals, *Sound Vibration*, **28**: 14–26, 1994.

32. L. Cohen, Time-frequency distributions — a review, *Proc. IEEE*, **77**: 941–981, 1989.

33. I. Daubechies, Orthonormal bases of compactly supported wavelets, *Commun. Pure Appl. Math.*, **41**: 909–996, 1988.

34. A. Aldroubi and M. Unser, Eds., *Wavelets in Medicine and Biology*, Boca Raton, FL: CRC Press, 1996.

35. A. Bruce, D. Donoho, and H.Y. Gao, Wavelet analysis, *IEEE Spectrum*, **33**: 26–35, 1996.

83.2 RF/Microwave Spectrum Analysis[1]

A. Ambrosini, C. Bortolotti, N. D'Amico, G. Grueff,
S. Mariotti, S. Montebugnoli, A. Orfei, and G. Tomassetti

A *signal* is usually defined by a time-varying function carrying some sort of information. Such a function most often represents a time-changing electric or magnetic field, whose propagation can be in free space or in dielectric materials constrained by conductors (waveguides, coaxial cables, etc.). A signal is said to be periodic if it repeats itself exactly after a given time T called the period. The inverse of the period T, measured in seconds, is the frequency f measured in hertz (Hz).

A periodic signal can always be represented in terms of a sum of several (possibly infinite) sinusoidal signals, with suitable amplitude and phase, and having frequencies that are integer multiples of the signal frequency. Assuming an electric signal, the square of the amplitudes of such sinusoidal signals represent the power in each sinusoid, and is said to be the power spectrum of the signal. These concepts can be generalized to a nonperiodic signal; in this case, its representation (spectrum) will include a continuous interval of frequencies, instead of a discrete distribution of integer multiples of the fundamental frequency. The representation of a signal in terms of its sinusoidal components is called Fourier analysis. The (complex) function describing the distribution of amplitudes and phases of the sinusoids composing a signal is called its Fourier transform (FT). The Fourier analysis can be readily generalized to functions of two or more variables; for instance, the FT of a function of two (spatial) variables is the starting point of many techniques of image processing. A time-dependent electrical signal can be analyzed directly as a function of time with an *oscilloscope* which is said to operate in the *time domain*. The time evolution of the signal is then displayed and evaluated on the vertical and horizontal scales of the screen.

The *spectrum analyzer* is said to operate in the *frequency domain* because it allows one to measure the harmonic content of an electric signal, that is, the power of each of its spectral components. In this case the vertical and horizontal scales read powers and frequencies. The two domains are mathematically well defined and, through the FT algorithm, it is not too difficult to switch from one response to the other. Their graphical, easily perceivable representation is shown in Figure 83.6 where the two responses are shown lying on orthogonal planes. It is trivial to say that the easiest way to make a Fourier analysis of a time-dependent signal is to have it displayed on a spectrum analyzer. Many physical processes produce

---

[1]All figures have been reproduced courtesy of Hewlett Packard, Rohde Schwarz, Hameg, Tektronix companies, and IEEE *Microwave Measurements*.

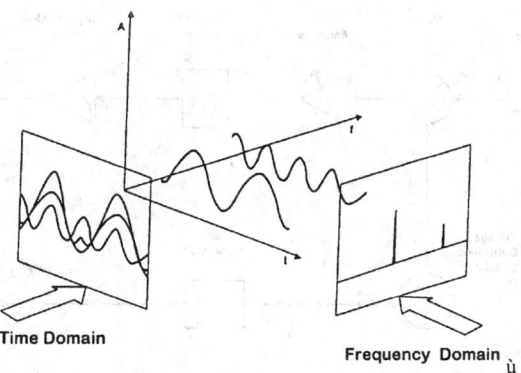

FIGURE 83.6 How the same signal can be displayed.

(electric) signals whose nature is not deterministic, but rather stochastic, or random (noise). Such signals can also be analyzed in terms of FT, although in a statistical sense only.

A time signal is said to be band-limited if its FT is nonzero only in a finite interval of frequencies, say $(F_{max} - F_{min}) = B$. Usually, this is the case and an average frequency F_0 can be defined. Although the definition is somewhat arbitrary, a (band-limited) signal is referred to as RF (radio frequency) if F_0 is in the range 100 kHz to 1 GHz and as a microwave signal in the range 1 to 1000 GHz. The distinction is not fundamental theoretically, but it has very strong practical implications in instrumentation and spectral measuring techniques. A band-limited signal can be described further as narrowband, if $B/F_0 \ll 1$, or wideband otherwise.

The first step in performing a spectral analysis of a narrowband signal is generally the so-called heterodyne downconversion: it consists in the mixing ("beating") of the signal with a pure sinusoidal signal of frequency F_L, called local oscillator (LO). In principle, mixing two signals of frequency F_0 and F_L in any nonlinear device will result in a signal output containing the original frequencies as well as difference $(F_0 - F_L)$ and the sum $(F_0 + F_L)$ frequencies, and all their harmonic (multiple) frequencies. In the pratical case, a purely quadratic mixer is used, with an LO frequency $F_L < F_0$; the output will include the frequencies $(F_0 - F_L)$, $2F_L$, $2F_0$, and $(F_0 + F_L)$, and the first term (called the intermediate frequency or IF) will be easily separated from the others, which have a much higher frequency. The bandwidth of the IF signal will be the same as the original bandwidth B; however, to preserve the original information fully in the IF signal, stringent limits must be imposed on the LO signal, because any deviation from a pure sinusoidal law will show up in the IF signal as added phase and amplitude noise, corrupting the original spectral content. The process of downconverting a (band-limited) signal is generally necessary to perform spectral analysis in the very high frequency (microwave) region, to convert the signal to a frequency range more easily handled technically.

When the heterodyne process is applied to a wideband signal (or whenever $F_L > F_{min}$) "negative" frequencies will appear in the IF signal. This process is called *double sideband* mixing, because a given IF bandwidth B (i.e., $(F_L + B/2)$ will include two separate bands of the original signal, centered at $F_L +$ IF ("upper" sideband) and $F_L -$ IF ("lower" sideband). This form of mixing is obviously undesirable in spectrum analysis, and input filters are generally necessary to split a wideband signal in several narrow-band signals before downconversion. Alternatively, special mixers can be used that can deliver the upper and lower sidebands to separate IF channels. A band-limited signal in the frequency interval $(F_{max} - F_{min}) = B$ is said to be converted to baseband when the LO is placed at $F_L = F_{min}$, so that the band is converted to the interval $(B-0)$. No further lowering of frequency is then possible, unless the signal is split into separate frequency bands by means of filters.

After downconversion, the techniques employed to perform power spectrum analysis vary considerably depending on the frequencies involved. At lower frequencies, it is possible to employ analog-to-digital converters (ADC) to get a discrete numerical representation of the analog signal, and the spectral analysis

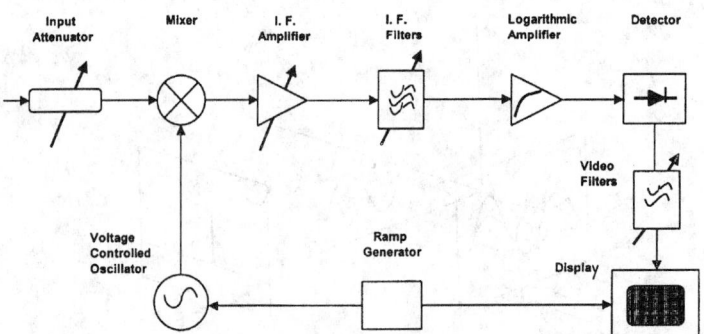

FIGURE 83.7   Block diagram of a commercial spectrum analyzer.

is then performed numerically, either by direct computation of the FT (generally via the fast Fourier transform, FFT, algorithm) or by computation of the signal autocorrelation function, which is directly related to the square modulus of the FT via the Wiener–Khinchin theorem. Considering that the ADC must sample the signal at least at the Nyquist rate (i.e., at twice the highest frequency present) and with adequate digital resolution, this process is feasible and practical only for frequencies (bandwidths) less than a few megahertz. Also, the possibility of a real-time analysis with high spectral resolution may be limited by the availability of very fast digital electronics and special-purpose computers. The digital approach is the only one that can provide extremely high spectral resolution, up to several hundred thousand channels. For high frequencies, several analog techniques are employed.

A Practical Approach to Spectrum Analysis [1]

Spectrum analysis is normally done in order to verify the harmonic content of oscillators, transmitters, frequency multipliers, etc. or the spurious components of amplifiers and mixer. Other specialized applications are possible, such as the monitoring of radio frequency interference (RFI), electromagnetic interference (EMI), and electromagnetic compatibility (EMC). These applications, as a rule, require an antenna connection and a low-noise, external amplifier. Which are then the specifications to look for in a good spectrum analyzer? We would suggest:

1. It should display selectable, very wide bands of the EM radio spectrum with power and frequency readable with good accuracy.
2. Its selectivity should range, in discrete steps, from few hertz to megahertz so that sidebands of a selected signal can be spotted and shown with the necessary details.
3. It should possess a very wide dynamic range, so that signals differing in amplitude six to eight orders of magnitude can be observed at the same time on the display.
4. Its sensitivity must be compatible with the measurements to be taken. As already mentioned, specialized applications may require external wideband, low-noise amplifiers and an antenna connection.
5. Stability and reliability are major requests but they are met most of the time.

Occasionally a battery-operated option for portable field applications may be necessary. A block diagram of a commercial spectrum analyzer is shown in Figure 83.7.

Referring to Figure 83.7 we can say that we are confronted with a radio-receiver-like superhet with a wideband input circuit. The horizontal scale of the instrument is driven by a ramp generator which is also applied to the voltage controlled LO [2].

A problem arises when dealing with a broadband mixing configuration like the one shown above, namely, avoiding receiving the image band.

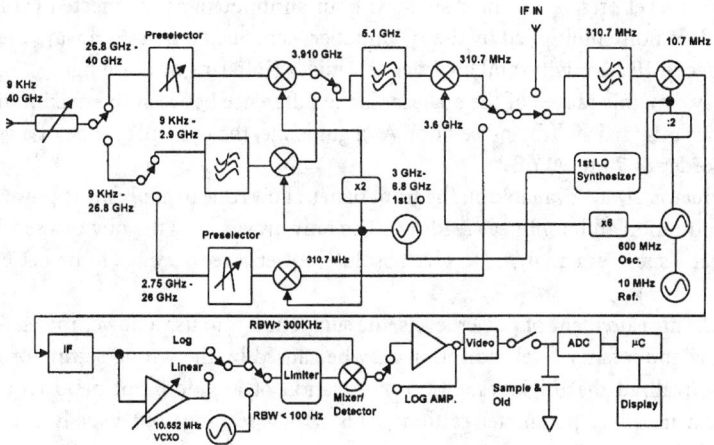

FIGURE 83.8　Standard block diagram of a modern spectrum analyzer.

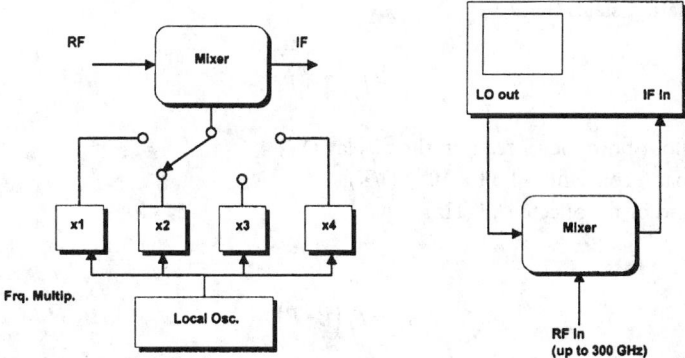

FIGURE 83.9　Encreasing the input bandwidth characteristics.

The problem is successfully tackled here by upconverting the input band to a high-valued IF. An easily designed input low-pass filter, not shown in the block diagram for simplicity, will now provide the necessary rejection of the unwanted image band.

Nowadays, with the introduction of YIG bandpass filter preselectors, tunable over very wide input bands, upconversion is not always necessary. Traces of unwanted signals may, however, show up on the display although at very low level (less than –80 dBc) on good analyzers.

A block diagram of a commercial spectrum analyzer exploiting both the mentioned principles is shown in Figure 83.8. This instrument includes a very important feature which greatly improves its performance: the LO frequency is no longer coming from a free-running source but rather from a synthesized unit referenced to a very stable quartz oscillator. The improved quality of the LO both in terms of its own noise and frequency stability, optimizes several specifications of the instrument, such as frequency determining accuracy, finer resolution on display, and reduced noise in general.

Further, a stable LO generates stable harmonics which can then be used to widen the input-selected bands up to the millimeter region. As already stated, this option requires external devices, e.g., a mixer-amplifier as shown in Figure 83.9a and b.

The power reference on the screen is the top horizontal line of the reticle. Due to the very wide dynamic range foreseen, the use of a log scale (e.g., 10 dB/square) seems appropriate. Conventionally, 1 mW is taken as the zero reference level: accordingly, dBm are used throughout.

The noise power level present on the display without an input signal connected (noise floor) is due to the input random noise multiplied by the IF amplifier gain. Such a noise is always present and varies with input frequency, IF selectivity, and analyzer sensitivity (in terms of noise figure).

The "on display dynamic range" of the analyzer is the difference between the maximum compression-free level of the input signal and the noise floor. As a guideline, the dynamic range of a good instrument could be of the order of 70 to 90 dB.

An input attenuator, always available on the front panel, allows one to apply more power to the analyzer while avoiding saturation and nonlinear readings. The only drawback is the obvious sensitivity loss. One should not expect a spectrum analyzer to give absolute power level readings to be better than a couple of dB.

For the accurate measurement of power levels, the suggestion is to use a power meter. An erratic signal pattern on display and a fancy level indication may be caused by the wrong setting of the "scan time" knob. It must be realized that high-resolution observation of a wide input band requires the proper scanning time. An incorrect parameter setting yields wrong readings but usually an optical alarm is automatically switched on to warn the operator.

The knowledge of the noise floor level allows a good valuation of the noise temperature, T_n (and therefore of the sensitivity), of the analyzer, a useful parameter on many occasions. The relations involved are as follows.

The Nyquist relation states that

$$P = k * T_n * B$$

where P = noise floor power level read on the display (W)
$\quad k$ = Boltzmann constant = 1.38×10^{-23} (J/K)
$\quad B$ = passband of the selected IF (Hz)

therefore,

$$T_n = P / \left(k * B \right)$$

Usually engineers prefer to quote the noise figure of receivers. By definition we can write

$$N = \left(T_n / T_o \right) + 1$$

where N $\qquad$ = noise factor
$\quad T_o$ $\qquad$ = 290 K
$\quad F$ (noise figure) = 10 log N

A typical F for a good spectrum analyzer is of the order of 30 dB.

It must be said, however, that the "ultimate sensitivity" of the spectrum analyzer will depend not only on its noise figure but also on the setting of other parameters like the video filter, the IF bandwidth, the insertion of averaging functions, the scan speed, the detector used, etc.

As a rough estimate a noise floor level of −130/−140 dBm is very frequently met by a good instrument.

Another criterion to select a spectrum analyzer is a good "IMD dynamic range," that is, the tendency to create spurious signals by intermodulation due to saturation.

This figure is generally quoted by the manufacturers, but it is also easily checked by the operator by injecting two equal amplitude sinusoidal signals at the input socket of the analyzer. The frequency separation between the two should be at least a couple of "resolution bandwidths," i.e., the selected IF bandwidth. As the input levels increase, spurious lines appear at the sum and difference frequencies and spacing of the input signals.

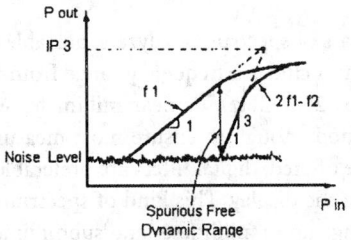 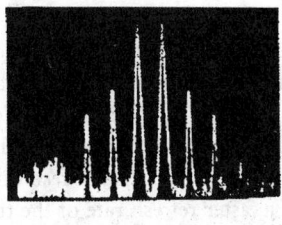

FIGURE 83.10 (a)Spurious free dynamic range. (b) Higher-order spurious.

The range in decibels between the nonoverloaded input signals on display and the barely noticeable spurious lines is known as the "spurious free dynamic range," shown graphically in Figure 83.10a, where the third-order "intercept point" is also graphically determined. If input power is increased, higher-order spurious signals appear, as shown in Figure 83.10b. The input connector of most spectrum analyzers is of the 50 Ω coaxial type. Past instruments invariably used N-type connectors because of their good mechanical and electrical behavior up to quite a few gigahertz. Today SMA or K connectors are preferred.

External millimeter wave amplifiers and converters use waveguide input terminations. As is discussed in the next section, multipurpose analyzers are available where power meter, frequency counter, tracking generator, etc. can all be housed in the same cabinet. The economic and practical convenience of these units must be weighed on a case-by-case basis.

Finally, we mention that spectrum analyzers are available equipped with AM and FM detectors to facilitate their use in the RFI monitoring applications.

What Is the Right Spectrum Analyzer for My Purpose?

Several manufacturers offer a large number of spectrum analyzer models; the choice may be made on the basis of application field (i.e., CATV, mobile telephony, service, surveillance, R&D, etc.), performance (resolution bandwidth, frequency range, accuracy, battery operation etc.), or cost.

In addition, it is important to know that most spectrum analyzers need some accessories generally not furnished as a standard: for example, a connectorized, coaxial, microwave cable is always required; a directional coupler, or power divider, or handheld sampler antenna may be very useful to pickup the signals; and a personal computer is useful to collect, store, reduce, and analyze the data.

There are four main families of RF and microwave spectrum analyzers.

Family 1

The bench instruments are top performance, but also large, heavy, and the most expensive class, intended for metrology, certification, factory reference, and for radio surveillance done by government and military institutions.

The frequency ranges span from few tens of hertz up to RF (i.e., 2.9 GHz), up to microwave region (i.e., 26.5 GHz), or up to near millimeter wavelength (i.e., 40 GHz). This class of instruments includes lower noise figures, approximately 20 dB, and may be decreased down to 10 to 15 dB with an integrated preamplifier. The synthesized local oscillator has a good phase noise (typically 10 dB better than other synthesized spectrum analyzers) for precise, accurate, and stable measurement. Also this class of instruments, by sharing the display unit, can be integrated with plug-in instruments like a power meter (for more accurate power measurements) or a tracking generator (for network analysis and mixer testing).

The interface to a computer (and a printer) such IEEE-488 or RS-232 is standard; it allows remote control and data readings; this class of spectrum analyzer often has a powerful microprocessor, RAM, and disks for storing data and performing statistical and mathematical analysis.

The best known families are the Hewlett-Packard series, 71xxxx [3] and the Rhode & Schwarz series FSxx. [4]. Indicative prices are between $50,000 and $90,000.

Family 2

Less-expensive bench instruments, the workhorse class of spectrum analyzers, portable and lightweight, are associated with a synthesized local oscillator, that includes a frequency range from few kilohertz up to RF region (i.e., 2.9 GHz), microwave region (i.e., 26.5 GHz), or near millimeter wavelengths (i.e., 40 to 50 GHz). A typical noise figure of 30 dB is good enough to ensure most measurements. A large number of filters down to few hertz of resolution are offered; digital filters are preferable to analog ones, because they give a faster refresh rate of the trace on the display. This kind of spectrum analyzer nearly always has the capability to extend the frequency range up to millimeter and submillimeter wavelengths with an external mixer. One of the most important features for a spectrum analyzer in this class is the quality of the local oscillator; it should be synthesized (PLL) to achieve stability, precision, accuracy, and low phase noise. Demodulation is also an important feature to listen to AM, FM on the loudspeaker and to display TV pictures or complex modulations onto the screen, which is often required by people working on surveillance, TV, and mobile telephone. The interface to a computer such as IEEE-488 or RS232 is standard in a large number of spectrum analyzers, and allows the remote control and data reading, storing, and manipulation.

This kind of instrument may integrate a tracking generator, a frequency counter, and other instruments that can transform the spectrum analyzer into a compact, full-featured RF and microwave laboratory.

The most popular families are the Hewlett-Packard series 856xx [3, 5], Rhode & Schwarz series FSExxx [4], Anritsu series MS26x3 [6], IFR mod. AN930 [7], and Marconi Instruments series 239x [9]. The Tektronix production should be taken in account. Prices typically span from $30,000 to $60,000.

Family 3

The entry level, a more economical class of spectrum analyzer, is intended for field use or for one specific application. If your need is mainly EMI/EMC, CATV, mobile telephone, or surveillance, perhaps you do not need the extreme stability of a synthesized local oscillator, and a frequency range up to 2 GHz may be enough; however, if you need some special functions such as "quasi peak detector" or "occupied bandwidth measurement," two functions that are a combination of a mathematical treatment with some legislative aspects, these are easily measured with a spectrum analyzer including those functions. As the normatives can change, the capability to easily upgrade the measurement software is important; some models come with a plug-in memory card, some others with 3.5" disks.

A large number of spectrum analyzer models are tailored to meet the specific needs of a customer. This is the case with the HP series 859x [3], Tektronix series 271x [10], IFR series A-xxxx [8], Anritsu MS2651 [6], and Advantest series U4x4x [4]. Costs typically are around $10, 000 to $20,000.

Family 4

The most economical class of spectrum analyzer, with prices around $2,000 to $6000, includes instruments that perform only the basic functions with a limited frequency range and filter availability and without digital capability. They are intended for service, for general-purpose measurements (i.e., IP_3, harmonic distortion) or for precertification in EMI/EMC measurements. One of the most popular series is the Hameg series HM50xx [11].

In this class are some special spectrum analyzers that come on a personal computer (PC) board. Such spectrum analyzers, generally cheap (typically $3,000 to $5,000), with frequency range up to 2 GHz, may include PLL local oscillators, tracking generators, and other advanced characteristics. The input is through a coaxial connector on the board, the output and the control is done by a virtual instrument running on the PC. One model is made by DKD Instruments [12].

Other unusual RF spectrum analyzers working in conjunction with a PC and worth noting are the instruments for EMI/EMC measurements and reduction in power lines and power cords. For this type of instrument, the core is not the hardware but the software that performs the measurement according to international standards and may guide the engineer to meet the required compatibility. An example is given by Seaward Electronic Sceptre [13].

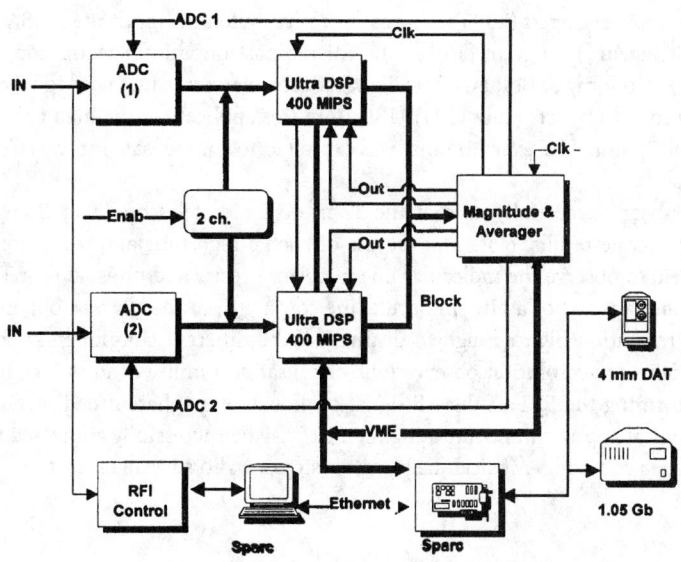

FIGURE 83.11   Block diagram of spectrum analyzer for radioastronomy.

Advanced Applications

New technological approaches and the use of spectrum analysis concepts in radioastronomy constitute some advanced spectrum analysis applications. Autocorrelators, with a typical frequency resolution of ~5/25 kHz, have been extensively used in radioastronomy. Their performance is well documented; the autocorrelation function is computed online and recorded. Later the FFT of the function is computed off line in order to get the power spectrum. Recently, the Tektronix 3054 Fourier Analyzer, based on a bank of programmable filters, was introduced as an alternative approach. The state of the art in integrated digital signal processors (DSPs) allows an alternative approach to spectrum analysis. By paralleling several of these DSPs, one is able to compute online the FFT directly on a very wide input bandwidth (several tens of megahertz).

By using this technique, high time and frequency resolution can be achieved.

A system based on the Sharp LH9124-LH9320 chip set is described in Figure 83.11. It is based on VME boards: one or two 10-bit, 40-MS/s ADCs and two boards in charge to compute the FFT of the incoming streams of data, in real time [14]. A following block computes the power and averages on board up to 64 K spectra before storing the result on disk or tape. The FFT boards are powered by one of the fastest state-of-the-art DSPs (Sharp LH9124). The overall system is controlled by an embedded FORCE 3 Sparcstation. The LH 9124 DSP works with 24+24 bits (with 6 exponent bit) in block floating point. The system architecture allows expansion of the input bandwidth and the number of channels by paralleling more DSP boards. All the computing core is housed in a VME crate and is able to produce single-sided spectra from 1024 frequency bins to 131072 bins at an input bandwidth of 12 MHz without losing data or with 56% of time efficiency at 20 MHz. Single- or double-channel operation mode is provided. In a single-channel mode the main features of the system are reported as

Input bandwidth	0.5–20 MHz
Time efficiency	100% at 12 MHz (56% at 20 MHz)
FFT size	1K, 2K, 256K (points)
Avgs out format	<256 averages → integer 24 bits
	>256 averages → float 32 bits
Windows	Hanning, Hamming, Kaiser Bessel

This spectrometer was developed (1993) as a cost-effective system for both the NASA-SETI (Search for Extraterrestrial Intelligence) program [15,16] and for radioastronomical spectroscopy [17] at the CNR Institute of Radio Astronomy of Bologna. The digital spectrometer was first used to investigate the effects of the Jupiter/SL9 comet impacts (July 1994) [18,19]. In this application, the high time resolution of the spectrometer (a 16K points FFT every 1.3 ms) was exploited to compensate for the fast planet rotational velocity Doppler shift.

The system has been successfully used at the 32-m dish radiotelescope near Bologna in many line observations with unique results. Note that the use of such a high time and resolution system in radio astronomy may help to observe the molecular line in a very precise and unusual way. The whole pattern (a couple of megahertz wide) of a NH_3 molecule line coming from the sky was obtained in flash mode with a frequency resolution high enough to distinguish the different components. The same machine can be used for high-time-resolution observations of pulsar and millisecond pulsar. In those cases, the possibility of performing the FFT of the RF signal on line allows coherent dedispersion of the pulses. This new technological approach in computing the FFT may be successfully addressed to many different fields, such as image processing, medical diagnostic systems, radio surveillance, etc.

References

1. A.E. Bailey, Ed., *Microwave Measurements*, 2nd ed., Peter Peregrins on behalf of IEEE, London, 1989.
2. Hewlett-Packard, *A.N 63* and *A.N 243*, Hewlett-Packard Company.
3. Hewlett-Packard, *Test & Measurement Catalog 1997*, pp. 224–257, Hewlett-Packard Company.
4. Rohde & Schwarz, *Catalog 96/97 Test & Measurement Products* pp. 109–145, Rohde & Schwarz Company.
5. Hewlett-Packard, *HP 8560E, HP8561E and HP8563E Portable Spectrum Analizers Technical Data*, Hewlett-Packard Company, pp. 5091–3274E.
6. Anritsu-Wiltron, *Electronic Measuring Instruments*, pp. 67–100, Anritsu-Wiltron Company.
7. IFR, *The AN 930 Microwave Spectrum Analyzer*, IFR System, Inc.
8. IFR, *The A-8000 & A-7500 Spectrum Analyzer*, IFR System, Inc.
9. Marconi Instruments, *Test and Measurement 96/97 Edition Instrument System*, pp. 4-1–4-5, Marconi Instruments Company.
10. Tektronix, *Measurement Products Catalog 1997/1998*, pp. 145–171, Tektronix Company.
11. Hameg Instruments, *Catalog 1994*, Hameg GmbH.
12. DKD Instruments, *Model 1800 Data Sheet*, DKD Instruments.
13. C. Galli, Analizzatori di spettro: rassegna commerciale, *Sel. Elettron.*, 15, 1996 [in Italian].
14. S. Montebugnoli et al, "A new 6 MHz 128000 channels spectrum analyzer," IAA-95-IAA.9.1.10. 46th International Astronautical Congress, October, Oslo, Norway, 1995.
15. S. Montebugnoli, C. Bortolotti, S. Buttaccio, A. Cattani, N. D'Amico, G. Grueff, A. Maccaferri, G. Maccaferri, A. Orfei, M. Roma, G. Tuccari, and M. Tugnoli, "Upgrade of the mini spectrum analyzer" in *5th Proceedings of the International Conference on Bioastronomy*, Capri, Vol. I, July 1996.
16. A. Orfei, S. Montebugnoli, C. Miani, J. Monari, and G. Tuccari, "The SETI facilities at the Medicina/Noto sites," (preprint) IAA-96-IAA.9.1.11, 47th International Astronautical Congress, October, Beijing, China, 1996.
17. S. Montebugnoli, C. Bortolotti, S. Buttaccio, A. Cattani, N. D'Amico, G. Grueff, A. Maccaferri, G. Maccaferri, A. Orfei, M. Roma, G. Tuccari, and M. Tugnoli, "A new high resolution digital spectrometer for radioastronomy applications," *Rev. Sci. Instrum.* 67, 365–370, 1996.
18. C. B. Cosmovici et al, "Detection of the 22 GHz line of water during and after the SL-9/Jupiter event," European SL-9/Jupiter Workshop–February, ESO Headquarters, Munchen, Germany, 1995.
19. C. B. Cosmovici, S. Montebugnoli, A. Orfei, S. Pogrebenko, and P. Colom, "First evidence of planetary water MASER emission induced by the comet/Jupiter catastrophic impact," *Planetary Space Sci.*, 44, 735–773, 1996.

84

Applied Intelligence Processing

Peter H. Sydenham
University of South Australia

Rodney Pratt
University of South Australia

84.1 Introduction

For a variety of reasons the signal that comes from the output of a sensor will usually need to be processed, as has been explained elsewhere. This may be needed for a single sensor alone, as found in the conversion from an electronic resistance temperature sensor into the equivalent digital display number.

Alternatively and increasingly so, as is now discussed, the outputs from several sensors need to be combined into a single signal to form a mapping of those sources from "many to one." An example of this would be when a set of sensors measuring many different variables is used to determine if a machine tool is correctly making a component.

There are now many ways that can be used to combine signals, each having its own features that make it the best to use in a given circumstance.

Signal-processing methods range from the well proven and mostly used method using processing based on mathematical relationships that are very precise — these are explained elsewhere. These are generally executed using digital computation and are often referred to as digital signal processing (DSP) methods.

Alternatively, it has now been shown conclusively that less quantitative methods can be also used to great effect despite their lack of complete formal mathematical formulation. These are here called the AI (applied intelligence) methods, a convention developed to distinguish man-made systems from the very broad, and oversold, use of the term *artificial intelligence.*

Many seemingly different methods exist in the latter group but, as will be shown here, they are all part of a continuum of ways that range from use of subjective to exactly objective procedures. These are not well explained as a group because they are presented in the literature as different methods used in isolation of each other. This account shows how they all tie together, thus making it easier to decide which is appropriate in a given application. They are particularly useful for handling highly nonlinear situations where algorithms cannot be realized.

While we appear to prefer the total objectivity of a mathematically formulated method of signal processing, it is now well proven that the AI methods often are better choices to use in terms of better speed of performance and often lower cost of processing. Often they are the only solution since the algorithmic approach cannot be deployed because of the lack of a suitable mathematical formulation or powerful enough processor to run the algorithm.

Signal processing in the modern instrument, therefore, will often make use of many different methods. This account is an introduction to the characteristics of the various forms and is written to assist selection. Space limitations prevent presentation of each kind in detail.

84.2 Overview of Algorithmic Methods

Traditionally the most popular method used to develop mapping models is that of mathematical modeling. The mathematical model is usually what is sought, as it provides the highest level of understanding about the subject and the most precise representation of the behavior. The major disadvantage of mathematical models is that they can quickly become so complex that implementation of these models in measurement systems is often impractical.

In this class, the single, or set of multiple, input signal(s) to the data processor is converted to the output form using tightly formulated mathematical description. This relationship is called the algorithm. Strict relationships hold; the relationship is said to be *formal*, meaning that for any given input the output will always be the same. The algorithm supports only one interpretation.

This method of signal processing is the most highly developed method and is certainly one to aim for because it is devoid of ambiguity. All will agree on how it will respond. It carries a comforting level of understanding and, thus, acceptance.

Algorithmic methods can be very accurate, traceable, and can be calibrated with relative ease and agreement. They are the basis of many instrumentation systems. The origin of their use in instrumentation goes back to the early days of computing using, first, mechanical computational machines (late 1800s to around 1930) and then analog electric devices (early 1900s to 1960s), all of which were mostly replaced by the use of digital computers commencing around 1950. All of these algorithmic methods of processing can be simplistically regarded as embodiments of a mathematical equation inside a suitable technological machine.

As the demanded complexity and performance requirements grew over time, so did the demands on the detail of the algorithm and the means to model it inside a computational machine. Mathematical description eventually reaches limits of definition as the models push the boundaries of mathematical methods and human development. Too often, this arises before adequate detail is able to be built into the model. The algorithm is then an inadequate model of the need.

As the algorithm increases in complexity, the processing power needed must be increased to maintain both fidelity and speed of processing.

Despite great advances being made in algorithm development and in computer power, the algorithmic methodology eventually encountered mathematical and technological barriers in many fields. The method is seen to not always be the best to use because of lack of an adequate algorithm or the high cost of computing.

In instrumentation, another factor also arises. Fast, detailed processing brings with it the need for increasing electrical bandwidth requirements in signal transmission. This increases implementation costs and also eventually reaches technological constraints.

84.3 Overview of Applied Intelligence Methods

Fortunately, the solutions that may overcome these limiting constraints in many circumstances were developing in other fields under the general name of artificial intelligence (now called applied intelligence in engineering), as new forms of mathematics and in other fields, such as decision theory.

Principally, a key limitation of the algorithmic method is that its unforgiving level of formalism carries with it a depth of processing exactitude that is often not warranted.

Other methods have emerged that allow vaguely subjective, as opposed to tightly objective, processing to be applied to good effect.

These AI methods have gradually gained acceptance to the degree that many are now routinely used and are supported by dedicated applications software and electronic integrated circuitry.

At first, these many alternatives were seen to be isolated methods. Gradually, the literature has shown trends to merge them in pairs. Their use in a more widely mixed form is still limited. This account seeks to give a comprehensive appreciation of the commonly met AI processing methods by placing them into relative perspective.

It is interesting to contemplate that natural world computing in animals does not appear to make much use of algorithmic methods, but does make extensive use of the methods presented here in the AI class.

The paradigm invoked here is that experience has shown that informal methods based on knowledge-based systems (KBS) can produce mappings of many inputs to one by use of less than completely formal description.

The AI methods can yield surprisingly efficient solutions to previously unsolved needs. They often can outperform algorithmic methods or carry out a similar task with far less computing power. They are all associated with multiple input processing and can be applied to forming decisions from data supplied by sensors. Each situation has to be judged on the balance between use of computing effort and effective processing.

On the downside, they lack formality and thus may be very hard to calibrate and authenticate. They, not having adequate scientific foundation and a solid formal base of operation, are not easily accepted as "sound." They are often hard to comprehend by a second party, for their description is not always adequately documented or done to any agreed convention. As their principles vary widely, they must be well understood before application is developed.

For all of these negative factors, they often are able to provide "more performance for less cost" and thus will be increasingly adopted.

Their rising level of use should not suggest the algorithmic methods will become obsolete, but more that the instrument designer now has a much larger set of processing tools available.

84.4 Mapping, in General

The high-level purpose of most signal processing is to yield knowledge of a situation so that decisions can be made.

For example, consider a health-monitoring system installed on an aircraft engine. A set of sensors of different measurand types and locations is installed at various critical points on the engine — temperatures, pressures, flow rates, metal content of the lubricating oil, and more. The data from these are collected and transmitted to a central processor using a common digital bus. The many inputs then need to be combined in some way to decide such conditions as emergency states, when to change the oil, and engine efficiency. This combination is a "mapping of many to a few."

These are not always simple mappings, for there is no adequate algorithm available to give a mathematical description for such things as degradation of oil condition. However, human intuition can be used quite effectively to obtain answers — the human mind is very capable of carrying out such mapping functions. This form of natural processing makes use of what are technically called "heuristics" — but more commonly known as "rules of thumb."

Consider the question, "How could we decide, using an automated measurement system, when loaves being made in a bakery are satisfactory to sell?" As the way to decide this almost all people asked would suggest that the weight, size, crustiness, appearance and softness inside would be the parameters that must all be satisfactory (that is, lie within a small range of values for each) to be declared suitable. Weight and size are easily measured; the others are not for they are really heuristics, as is the choice of the set of parameters.

The thought process implemented here is that the designer starts with a desire to know something about a situation. Consider how we could automatically monitor the "risk of working in a hazardous place" in order to give an alarm at set levels. In this kind of situation a study of the problem will lead to identification of key parameters. These parameters can each be assigned safety functions that express how each parameter varies with system changes. With this framework it is then possible to set up a signal-processing system that continuously calculates the risk level. This form of solution is based on ideas embodied in the wide field of decision-making theory.

The heart of application of AI methods of signal processing in instrumentation lies with appreciation of decision theory methods.

A range of multivariable mappings methods using AI ideas have emerged. Those well established in instrumentation are

- Representational measurement theory and ways sets are mapped into other sets (whose usefulness is still emerging);
- Rule and frame representation of heuristic knowledge and ways they are used to form expert systems and other KBSs;
- Binary Boolean trees as crisp logical mappings; which is the foundation of the
- Fuzzy logic method based on fuzzy set theory.

Another class of AI methodology that does not fit the same sequence, yet includes powerful methods, is those that assist optimization of the mapping setup. There are two main methods in use;

- Genetic algorithm and its use to optimize fuzzy logic and other multisensor setups, and the
- Artificial neural net, a mapping method that learns by itself, from experience, how to achieve an optimal mapping in a given situation that it has been taught to work in.

84.5   Basics of Decision Theory

Rules about a Decision-Making Process

Before entering into the detail of the AI signal-processing methods, it is necessary to develop a foundation about the ways in which decisions can be made by computers using sensed information. It is not that well appreciated, but setting up a mapping-type signal-processing situation is actually implementing a decision-making information system. General appreciation of decision making can be found in the introductory work of Kaufmann [3] .

Unlike the human brain decision maker which carries out smart thinking with ease to build a machine counterpart, an engineered object needs effective externalization of the process involved. This begins by developing appreciation of the basic rules that always apply about a decision-making situation. These have been summarized by Baker et al. [1] and condense to

1. There must be a clearly expressed criterion for making a judgment of the options available, which must be such that others will understand how the judgment was made.
2. All decisions will involve choosing alternative strategies to arrive at the best one to use. This will involve assigning score numbers to the selected parameters and deciding how to process the set of numbers.
3. A decision is made by a choice of competing alternatives in the face of given restraints. Decisions can only rarely be made in the general sense but will be made for a given set of circumstances. The complexity of a problem rises rapidly as the number of parameters rises.
4. The process used attempts to achieve some payoff as a value added or lost. It aims to rank the various mapping alternatives to advise the apparently best to use. Note that once a decision-making

mapping is built, the information about alternatives is no longer available as it will only usually embed one set of parameters as a single process.

5. A decision matrix carrying the competing parameters results. A method of combining the matrix constituents is needed, and, again, there is no singularly definitive, absolutely correct way to process the matrix.

In setting up a signal-processing mapping, these rules will need to be addressed. They will be embedded in the software of the hardware processor as its operational strategy. Considerable creativity is needed by the designer of the processor, for much of the setup of decision-making methods requires subjective human interpretation in several steps of the process. Decision making is really only needed when there is no exact and obvious answer. The devices built to mimic the human process will never be perfect. There will be much debate about which are the best methods and parameters to use. Engineers must live with this situation and make machines that will make good decisions, that are as close to perfect as possible.

Extracting Parameters

The first step in setting up a decision mapping is to understand the need. That means researching it by observation and from literature on the topic. This sets up the specific knowledge base to allow one to progress to the next step.

Then comes the need to define the key parameters of the situation. There is no organized way to develop these. They arise from inventive and innovative thought processes that seem to be based on prior learning.

To streamline this intuitive step, it is useful to apply some ordered processes that assist in externalizing appropriate parameters. Three methods are now briefly described.

Slip Writing

A group of people familiar with the problem area are read a brief statement of the problem by a person independent from the problem. An example could be "What do you think are the main parameters that a person uses to decide if a loaf of bread is fresh?"

Without much time to reflect the group is then asked to write down the key parameters that come to mind immediately as they work without talking about the problem as a group. They write down each parameter on a separate piece of paper, doing this as fast as ideas come to them. This only happens for a few minutes. The slips of paper are then collected and classified. The whole process takes around 10 min and is known as slip writing.

It will usually be found that there is common agreement about the majority of parameters with some quite unexpected ones also arising.

Slip writing a good way to find consensus. It probes the mind well and can bring out appreciation of factors that open discussion might easily inhibit. It is important in this method to decouple the thoughts of each person during the process; otherwise the real parameters may not be externalized because some people may exert influence on others of the group.

Brainstorming and Think Tanks

If participants are shown what others are thinking and are encouraged to debate issues, it is possible to gain consensus and also allow group participants the opportunity to help each other be innovative at the same time. This method works best when the participants are prepared to go into open discussion. Several similar processes are those known as brainstorming or carrying out a think-tank session.

Here a problem in need of solution is written down as a well-prepared statement by the session organizer. A team of experts, each covering the expected aspects of the problem area, are selected and sent the statement along with any supporting exhibits. Each person considers, over a few days, how he or she might contribute a solution.

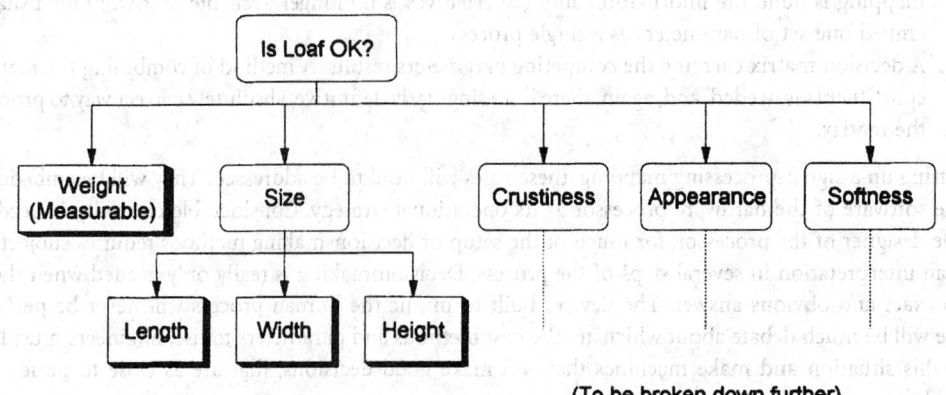

(To be broken down further)

FIGURE 84.1 Knowledge trees allow facts and their relationships to be captured in pictorial form.

The group is then assembled. The problem is first reviewed by the session leader and each person is then asked for ideas. As ideas are externalized they are recorded in very brief form — large sheets of butcher paper are suitable. These sheets must be readable by all in the group and be prepared instantly to keep up with the thoughts of the group.

It will be found that innovative ideas will arise as candidate solutions are put up and seen by others in the group. This method encourages group-driven inventiveness.

Gradually the group will settle on a few solutions that it feels have a good chance of succeeding. This list is then ordered in priority of likelihood of success, The session leader then writes up the outcomes, ready for further investigation.

Knowledge Trees

The final method to be described here for developing parameters of a decision, called knowledge trees, has the merit of ordering the relative place of parameters as well as encouraging inventiveness of solutions. It also provides a mapping structure. This procedure is based on age-old realization that we think problems through by breaking them down into ever smaller subproblems until we feel able to solve them. Overall solution is then very much a matter (but not entirely so in practice) of implementing the solution of all subproblems and combining them by a process called integration.

The need is first written down. For example, "How would we measure the quality of loaves of bread?" or in a shorter form "Is the loaf OK?"

This forms the top-level parameter of the tree given as Figure 84.1. Consideration of the situation at hand then leads to realization of the collection of parameters relevant to get this answer. These might be weight, size, crustiness, appearance, and softness. They may have been externalized by a group process, such as slip writing, or created by the individual.

Each branch on the tree is then visited to see how that parameter might be measured. Only one of these can be measured as it stands — it is reasonable to assume that weight can be measured with scales.

Size is not so easy to measure as it is expressed for there is inadequate definition. More thought will yield another level to the tree — length, width, and height — for this parameter. As linear measurements, these can also be measured with ease.

When building the branching downward, a thought process has decided how the parameters map upward and downward. Size dictates the mapping of three parameters into one, so there must also be a defined mapping model for that mapping.

Note also that to branch downward, the thought process used has actually been driven by some heuristics. Each branching has been driven by rules of some kind — but more on that in the rule-based decision-making method covered below.

Size	Crustiness	Appearance	Softness	Parameters
$\Rightarrow \dfrac{Si}{W}$	$\dfrac{C}{\Rightarrow W}$	$\dfrac{A}{\Rightarrow W}$	$\Rightarrow \dfrac{So}{W}$	Weight
	$\Rightarrow \dfrac{C}{Si}$	$\dfrac{A}{\Rightarrow Si}$	$\Rightarrow \dfrac{So}{Si}$	Size
		$\Rightarrow \dfrac{A}{C}$	$\dfrac{So}{\Rightarrow C}$	Crustiness
			$\Rightarrow \dfrac{So}{A}$	Appearance

Scores (number of times preferred)

Si	C	A	So	W
2	2	1	3	2

FIGURE 84.2 Triangle of pairs assessment is a simple way to decide which choice to make. This table gives the workings for the grading of bread as in the example of Figure 84.1

It is also easy to see why the other parameters could be measured with automated instrumentation. Softness could be assessed in terms of the squeeze factor, which is actually measurable as the compliance of the loaf at the center point of a side of the loaf. Appearance would map down the tree into color, texture, and graininess of the image. It is left to the reader to think up and draw a complete tree.

When all branch ends have been reticulated down to the point where they can be measured, the system mapping can be implemented with a human-made sensing system. The parameters are externalized and the mapping process is largely decided.

Use of tree-based thinking is a simple, yet powerful, way of keeping track of decision making for a complex situation. The recorded form also allows others to see the thought process used.

Two Examples of Decision Assistance Methods

Triangle of Pairs

Having now seen how to externalize parameters and how they might be interrelated using trees, we can move on to investigate how to set up a suitable decision-making process.

Knowing the parameters is not enough yet to design a multisensor mapping processor. The relative importance of the parameters is also a key factor. Furthermore, the tree is not the only way to combine sensor signals.

Two, of many, examples of decision assistance methods are now outlined to illustrate these points.

The first, the triangle of pairs (TOP) method, allows parameters to be ranked against others on the binary-only basis of which is preferred of each two compared. In the bread example, compare some of the parameters for their relative importance. Crustiness is preferred to weight. Softness is preferred to size and so on until all pairs have been considered. If all combinations are carried through and recorded as a matrix, a triangle of pairs results, as in Figure 84.2.

Having formed a matrix of competing parameters, the next step is to decide how the matrix can be processed. This is where much debate can occur. For the TOP method, however, the binary nature of the choices means a simple count of first preferences gives the ordered preference of parameters — at least as that person assessed it!

We will see later how the idea is extended by giving parameters a varying degree of "preference" rather that the simple binary choice allowed here.

Utility Analysis

A more fully developed method for making decisions in complex and subjective situations is one called utility analysis (UA). This is a process that can be applied to find the usefulness of a design, piece of equipment, or any similar situation where one can externalize and prioritize a set of measurable parameters. Although mostly applied to decision making as a paper study, the process is amenable to the creation of a multisensor mapping processor.

Appreciation of this process is easily obtained by working through an example. Consider the need to set up a method for grading the quality of bread made in an automated bakery.

The first step is to decide the parameters that will affect the choice finally made. We select weight, size, and appearance as the key parameters to illustrate the method. (More parameters might be needed in a real situation.) We also decide that these are important according to the relative weighting ratios of 1.0:0.2:0.8.

Next utility curves must be set up, one for each parameter. These show how the usefulness of the parameter changes as the parameter ranges.

The simplest way to obtain these graphs is to use one's own intuition, but a better way is to make use of some form of consensus-forming procedure as discussed above. Figure 84.3 shows what these three functions might look like. As a guide to their construction, if the weight of the loaf is too low, it fails to comply with legal requirements and thus has zero utility as a product below the allowed uncertainty of weight. The size factor depends on the type of bread. Here it is assumed it is for sandwich making, in which case the size can be too small for sliced meats or too big for a toaster. Appearance can only measured by mapping downward to a set of measurands; it is convenient to plot the function in more vaguely defined terms in this method.

Note that the weighting ratios have already been incorporated into the utility charts by setting their best values at the correct percentage.

It is necessary to reinforce the fact that the set of graphs is needed for the parameters, not for each case to be considered. With the charts set up, the selection process can begin.

A matrix is now created, as also shown in Figure 84.3. The actual measured weight value (960 g) for loaf 1 (under automatic inspection) is compared with the graph of weight to yield a utility of 0.4. The size is done likewise, using the size graph to get 0.2, and the appearance sensor set tells us it is between poor and good to give us 0.4. Each loaf is subjected to the same process. The other two loaves have different sets of scores.

The combined usefulness of a given loaf is now to be decided by processing the set of numbers for that loaf. Here is where some difficulty arises, because there are many ways to combine the three scores for each loaf. One way often used is simply to sum the values, as is done in the example. Note that this can be satisfactory unless a zero or other unacceptable value arises, in which case more mapping processing is needed. Assume that an acceptable combined score has been determined to be 1.8 or higher (with the best, due to the weightings, 2.0).

Having carried out the processing of the matrix, we find that the first loaf is not acceptable and the other two are equally acceptable.

What we have done is to form an automatic measuring system that can make assisted and graded decisions. There is certainly a degree of human subjectivity inherent in the method, but by recording what is taking place the task can be automated and it can also be evaluated in the light of experience, correcting choices of parameters and weightings.

Other decision-making methods exist: they use elements of the above concepts in a variety of assemblages.

The above has been a very rapid introduction to decision-making methods. These are rarely taught as the foundation of the now popularized AI signal-processing methods given next. They are, however, the conceptual basis of the AI processes.

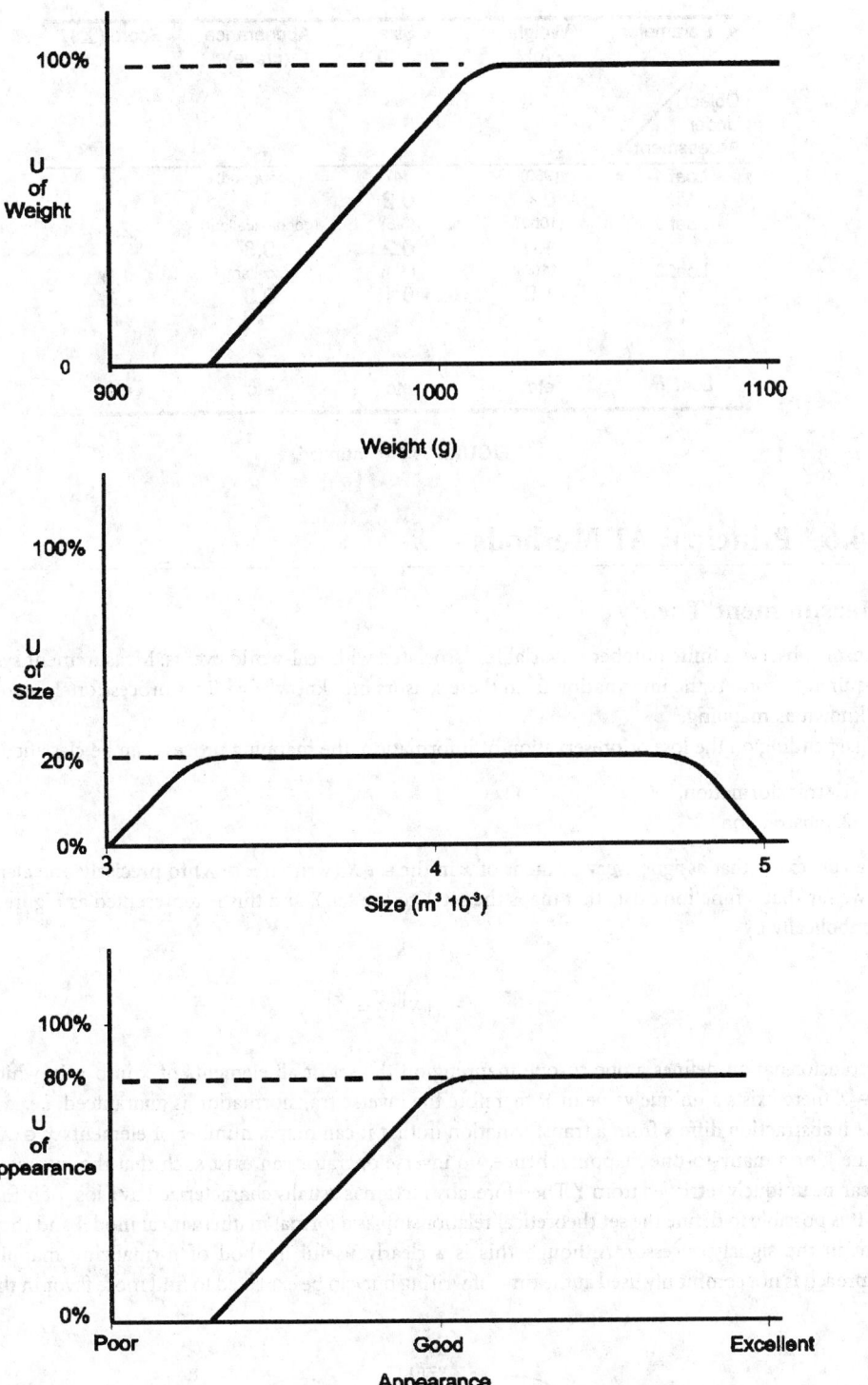

FIGURE 84.3 Utility analysis is a more-detailed way to automate loaf inspection with a set of sensors. Here are shown three of the utility functions for the example in Figure 84.1 along with the scoring matrix for loaves passing the inspection point.

Parameter Object Under Assessment	Weight (g)	Size ($m^3 \ 10^{-3}$)	Appearance (grade)	Score (Σ)
Loaf 1	(960) 0.4	(4) 0.2	(poor/good) 0.4	1.0
Loaf 2	(1000) 1.0	(3.5) 0.2	(good/excellent) 0.8	2.0
Loaf 3	(1100) 1.0	(4.9) 0.1	(excellent) 0.8	1.9
• •	•	•	•	•
Loaf 'n'	etc	etc	etc	etc

FIGURE 84.3 (continued)

84.6 Principal AI Methods

Measurement Theory

Sensors observe a finite number of variables associated with real-world events. Measurement systems are required to convert the information from these sensors into knowledge. This process often involves what is known as mapping.

Depending on the loss or preservation of information, the mapping process can be classified as either

1. transformation, or
2. abstraction.

If a rule exists that assigns every element of x in the set X (written $x \in X$) to precisely one element $y \in Y$, we say that a function exists that maps the set X to the set Y, and this is represented as Figure 84.4 and symbolically by

$$y = f\left(x\right)\left(x \in X\right)$$

A transformation defines a one-to-one mapping on the set of all elements of x into y, in which for all $x \in X$ there exists a unique value in Y; therefore the inverse transformation is guaranteed, i.e., $x = f^{-1}(y)$.

An abstraction differs from a transformation in that it can map a number of elements $x_i \in X$ into the same y, or a many-to-one mapping; hence, no inverse operator can exist such that the inverse image of X can be uniquely retrieved from Y. Therefore, abstraction is usually characterized by a loss of information.

It is possible to define the set theoretical relationship as a formal mathematical model and then embed that in the signal processor. Although this is a clearly useful method of formalizing mappings, this approach is not commonly used at the time of writing but can be expected to find more favor in the future.

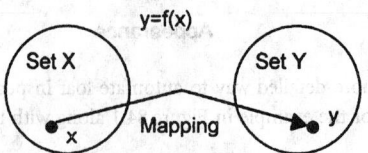

FIGURE 84.4 Pictorial representation of the set theoretical mapping process.

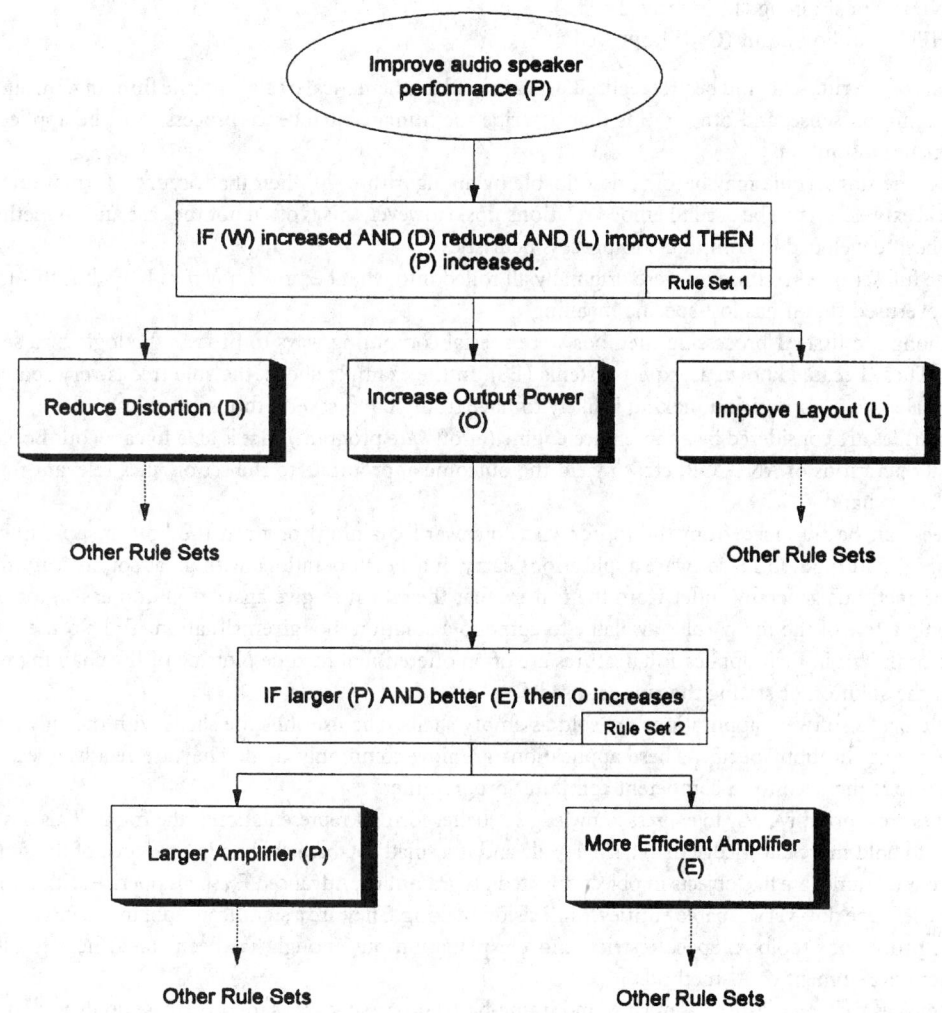

FIGURE 84.5   Part of a design knowledge tree for improving the performance of an audio speaker system.

Rule and Frame-Based Systems

In early stages of problem solving, we seem naturally to look to rules of thumb to get the solution started. Even the algorithmic methods start with this form of thinking, for one has to decide what the elements of the decision are and how they might be assembled into a strategy for implementation in the signal processor. As the rules are externalized one's thought patterns also usually build knowledge trees that show the relationship between the rules.

Consideration of the knowledge supporting the structure of a tree will reveal that the decision needed about which way to branch as one moves through a tree is actually the implementation of a rule that has relevance at that junction. Rules link parameters together.

Figure 84.5 is a tree giving some of the design options for improving the performance of an audio speaker system. This tree has been built by applying a designer's knowledge of speaker system design. The heuristic rule set for the top branch is stated as

IF   speaker output (W) increases
AND   distortion (D) is reduced

AND   positioning (L) improved

THEN   audio output (O) is improved

No part of the rule set could be first realized using formal mathematical or algorithmic thinking. Intuition, leap, common sense, and other terms that describe the human intelligence process must be applied to commence a solution.

At some stage a rule may become describable by an algorithm — when that occurs a formal mathematical expression can be used to embed relationships. However, this is often not the case and so methods have been developed in computers to process heuristics.

The full set of AI techniques were originally all rolled into what became known as KBSs but this term is so overused that it has lost specific meaning.

Among the first AI processing methods were special computing ways to process the logic of a set of rules. These became known as expert systems (ES). In the example above, the rule tree is very sparse; a practical system for decision making is likely to have from 100 to several thousand rules.

The rules are considered by an inference engine (a software program) that is able to carry out Boolean logical operations of AND, OR, etc. to yield the outcome appropriate to the set of rules relevant to the problem at hand.

Trees can be traversed from the top down (downward chaining) or from the bottom up (upward chaining) and modern ES software applications carry out these operations with great sophistication.

KBS methods generally suffer from the feature that they seem to give answers all too easily, for they use only a few of the many rules available to come to a solution in a given situation. To help users feel more confident in their application, features are often offered that include plotting of the chaining used to get the solution or stating the rule set used.

Rule-based software applications are sold as empty shells. The user fills the shells with the rule set to make the application specific. These applications are now commonly used. They are relatively easy to use without the need for a competent computer programmer.

Rules are a primitive way to express knowledge. A better form of representation is the frame. This has the ability to hold more knowledge than a single rule and is a small database about a limited area of the system of interest. Frames are like objects in object oriented programming. Advanced ES shells operate with frames.

ES shells are now very capable entities for decision making. They are a significant tool in the instrument signal processor's toolbox. Space restricts more explanation but enough has been stated here to allow further development of AI methods.

Some basic characteristics about rule- and frame-based processing are as follows (these apply variously):

- The software program is often self-explanatory as the rules can be read as (almost) normal language;
- They need considerable computer power as the rule number increases;
- They are relatively slow to yield a solution but are best used for cases where slow outcomes are applicable;
- They need special software;
- They are not that well known, so their application may be slow to find favor.

ESs soon run out of usefulness if the problem becomes complex. The computer search becomes too slow because the number of rules needed rapidly rises with problem complexity. The approach used today for large-problem-solving systems is to build an ES for each facet of the problem. These small-problem AI units are called agents. A set of agents is then combined using a conceptual software-based blackboard that calls on the agents to investigate a problem put to the system.

The chaining operation basically only carries out simple Boolean algebra operations using system parameters represented by a description called a rule. The system has no understanding of the wording of the rule. Thus, it is only processing as through the tree branching is either one way or the other. In its simplest form, it contains no concept of making that branching decision with a graded concept of which way to go.

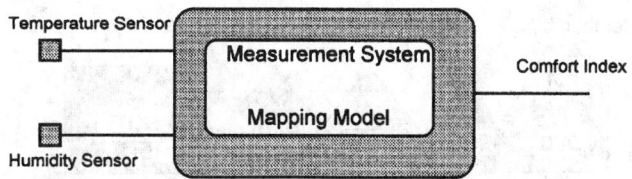

FIGURE 84.6 A simple comfort index measurement system uses temperature and humidity variables.

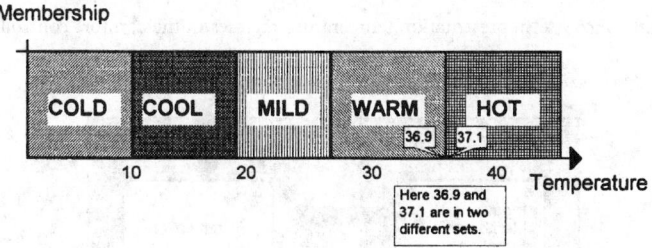

FIGURE 84.7 Conventional crisp set for the measurement system of Figure 84.6.

Real life is full of unclear logical operations. The outcome of a decision will be clearly this way or that, but just where the change arises is problematic. The changeover point is unclear because the variables of the rule are fuzzy. It is desirable to process rules with regard of their likelihood of relevance depending on the state and selection of other rules. As is explained below, the fuzzy logic method of carrying out a mapping is another way that allows the rule to have more variability than the two-state exactness of binary logic.

Fuzzy Logic

This explanation is not intended to be a rigorous mathematical examination of fuzzy sets and fuzzy logic but rather explain, through example, the application of fuzzy techniques in measurement systems, specifically in respect to mapping models. For more detail, see Mauris et al. [4].

The simple example used here is a measurement system, Figure 84.6, that maps two input sensors (temperature and humidity) into one output value (comfort index). Clearly there is no formally accepted definition of comfort index as it is a subjective assessment. One of the advantages of fuzzy sets is that they are usually intended to model people's cognitive states.

In the mid 1960s Professor Lofti Zadeh recognized the deficiencies of Boolean logic, in that its TRUE/FALSE nature did not deal well with the shades of gray that exist in real life situations.

Boolean logic uses classical set theory where an element is either viewed as entirely true or completely false $A = \{0,1\}$. These are often referred to as a crisp sets, Figure 84.7. In a crisp set the transition between sets is instantaneous, i.e., 36.9°C is considered warm whereas 37.1°C is considered hot. Hence, small changes in the input values can result in significant changes in the model output. Clearly, the real world is not like this.

Fuzzy logic uses a multivalued set where degrees of membership are represented by a number between 0 and 1 $A = [0,1]$ $\mu_A: U \to [0,1]$, where μ_A is the membership function. With fuzzy logic the transition between sets is gradual and small changes in input values result in a more graceful change in the model output, Figure 84.8.

Fuzzy Expert Systems

A fuzzy expert system, Figure 84.9, combines fuzzy membership functions and rules, in place of the often all-too-crisp Boolean logic, to reason about data. The fuzzy expert system is usually composed of three processing sections:

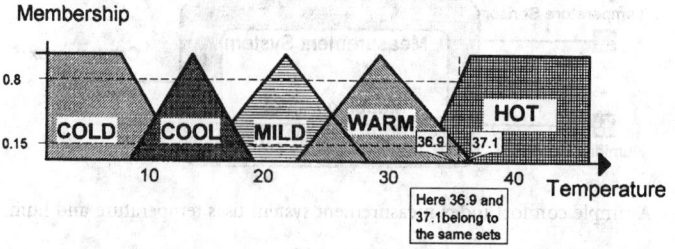

FIGURE 84.8   Fuzzy set representation temperature regimes in the comfort controller example.

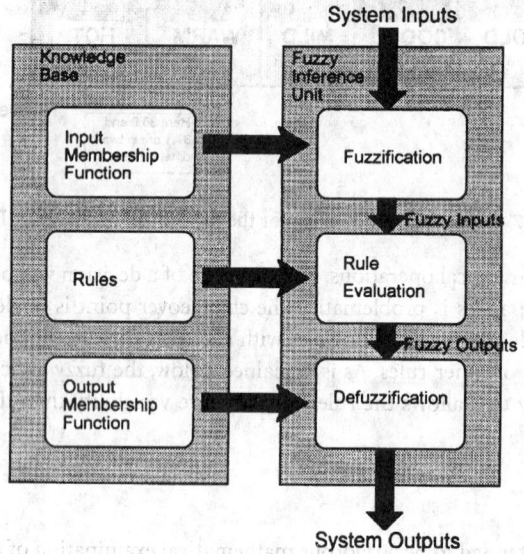

FIGURE 84.9   A fuzzy inference system combines rules in a way that allows them to be fuzzy in nature, that is, not crisp.

Step 1.   Fuzzification
Step 2.   Rule evaluation
Step 3.   Defuzzification

Step 1 — Fuzzification.

In fuzzification crisp inputs from input sensors are converted into fuzzy inputs using the membership functions in the knowledge base. A fuzzy input value is generated for each linguistic label of each input. For example, in Figure 84.8, for an input temperature of 37°C the fuzzy input is COLD(0.0), COOL(0.0), MILD(0.0), WARM(0.15), HOT(0.80). A similar set of fuzzy inputs are generated for the humidity sensor, Figure 84.10.

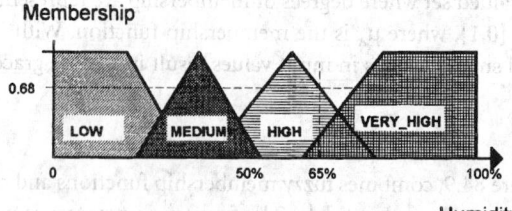

FIGURE 84.10   Humidity membership function for the controller example.

TABLE 84.1 The Fuzzy Associative Matrix Links Inputs and Outputs

Hum/Temp	Cold	Cool	Mild	Warm	Hot
Low	uncomfort.	uncomfort.	uncomfort.	uncomfort.	uncomfort.
Medium	uncomfort.	acceptable	comfortable	acceptable	uncomfort.
High	uncomfort.	acceptable	acceptable	acceptable	uncomfort.
Very_High	uncomfort.	uncomfort.	uncomfort.	uncomfort.	uncomfort.

Step 2 — Rule Evaluation.

Rules provide a link between fuzzy inputs and fuzzy outputs. Rules are usually expressed in the form of IF ... AND/OR ... THEN ... statements.

For example,

'IF' the TEMPERATURE is HOT 'AND' the HUMIDITY is HIGH 'THEN' it is UNCOMFORTABLE.
'IF' the TEMPERATURE is MILD 'AND' the HUMIDITY is MEDIUM 'THEN' it is COMFORTABLE.
'IF' the TEMPERATURE is WARM 'AND' the HUMIDITY is LOW 'THEN' it is UNCOMFORTABLE.

Rules can also be expressed in the form of a table or matrix, called the Fuzzy Associative Matrix. This matrix, Table 84.1, provides a complete description of the system performance for all combinations of inputs.

The function of the rule evaluation step is to evaluate the relative strengths or truth of each of the rules in order to determine which rules dominate. In this example the rules contain AND relationships and therefore the overall rule strength must be the minimum (MIN) value of the two strengths of the input values.

For example, at a temperature of 37°C and a humidity of 65% the rule strengths of the three example rules are

'IF' the TEMPERATURE is HOT(0.8) 'AND' HUMIDITY is HIGH(0.68) 'THEN' it is UNCOMFORT-ABLE (Rule Strength = MIN(0.8,0.68) = 0.68).
'IF' the TEMPERATURE is MILD(0.0) 'AND' the HUMIDITY is MEDIUM(0.0) 'THEN' it is COM-FORTABLE (Rule Strength = MIN(0.0,0.0) = 0.0).
'IF' the TEMPERATURE is WARM 'AND' the HUMIDITY is LOW 'THEN' it is UNCOMFORTABLE (Rule Strength = MIN(0.15,0.0) = 0.0).

All rules must be evaluated (in this case all 20 in the matrix) to determine each rule strength. If two rule strengths exist for one fuzzy output label, then the maximum (MAX) rule strength is used because this represents the rule which is most true; that is, in the previous example the fuzzy output for UNCOM-FORTABLE is the MAX(0.68,0.0) = 0.68.

Step 3 — Defuzzification.

Now that rule strengths exist for all the output fuzzy labels, a crisp output value can be determined from the fuzzy output values. The most common method used to defuzzify the fuzzy output value is the center of gravity (COG) method. The fuzzy rule strengths determined from rule evaluation are used to truncate the top of the output membership functions. Given this area curve, the COG or balance point can then be calculated. For example, in Figure 84.11, for fuzzy output values of

UNCOMFORTABLE = 0.68
ACCEPTABLE = 0.2
COMFORTABLE = 0

the COG or crisp "Comfort Index" evaluates to 25.

Fuzzy logic signal processing is now a well-developed method. It is supported by copious teaching material including those on internet and on CD ROMs provided by manufacturers of this form of special integrated circuit chip sets and setup software. Possibly its best known application has been in clothes washing machines where the wash parameters are set to suit the load. Although there are still limitations, this method can be considered to be a mature procedure.

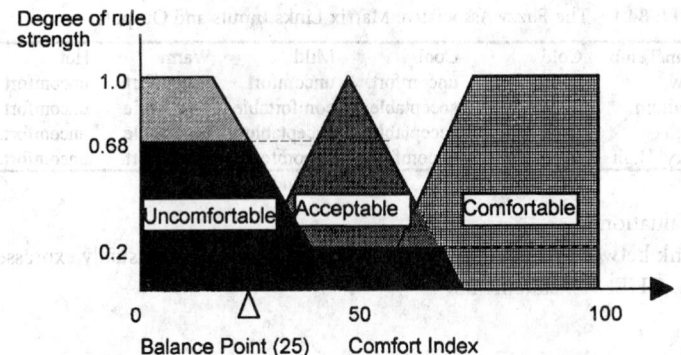

FIGURE 84.11 Output membership functions for controller. Each type of shaded area represents the three states of comfort.

Genetic Algorithms

The methods discussed thus far all have required the user to advise the system about the parameters to use. That is, they need to be taught efficient mappings.

In sharp contrast to these types of systems, there also exist other AI methods that possess the ability to learn, by themselves, what are the more optimal mapping configurations. Two main techniques are used — genetic algorithms (GAs) and artificial neural networks (ANNs). These self-learning processes both work well in certain situations and are now commonly used in signal processing. On the face of it, both seem to possess magical properties because they defy logical thought processes and a clear understanding of how they actually operate.

We begin with an overview of GAs. These make use of the basic principles by which natural selection found in living things, that is, from genetics, is able to improve gradually the fitness of species. That the genetic principles found in nature can be used in human-made applications is accredited to pioneering work of John Holland in the mid 1970s. Today, it is large field in algorithm optimization research; see Tang et al. [7]. A very simplistic description now follows to give some insight.

The concept starts with the selection of a set of features, Figure 84.12, (these can take a wide range of forms and are not just measurement parameters), that represent the essential features of a system of interest. As examples, the DNA molecule carries the code of the characteristics of living beings and a computer string can carry a coded message that represents the features of the behavior of some human-devised system.

Various types of events (crossover, mutation, inversion are commonly encountered methods) can slightly alter the code of any particular string. When this happens, the new string then represents another closely similar, but different system having new properties. Consider, next, that a number of slightly different code strings have been formed.

When a change takes place in the code of a string, it is assessed against the other strings using rules for a predecided fitness test. If improvement has occurred in the overall properties, then it is adopted as one of the full set. If not better, then it is discarded. In this way the set of strings, and thus the total system capability, gradually improves toward an optimal state.

The operational aspects of such systems are beyond description here. Suffice to say that the technique is well established — but highly specialized — and is undergoing massive international research effort in hope of alleviating certain limitations.

The first limitation is that although it is generally agreed each adopted change for the better takes the overall system capability closer to the goal being sought, there is, as yet, no theory that can be applied to show the state of maximum optimization. GAs, therefore, always have doubt associated with their solutions as to how much more improvement might be possible.

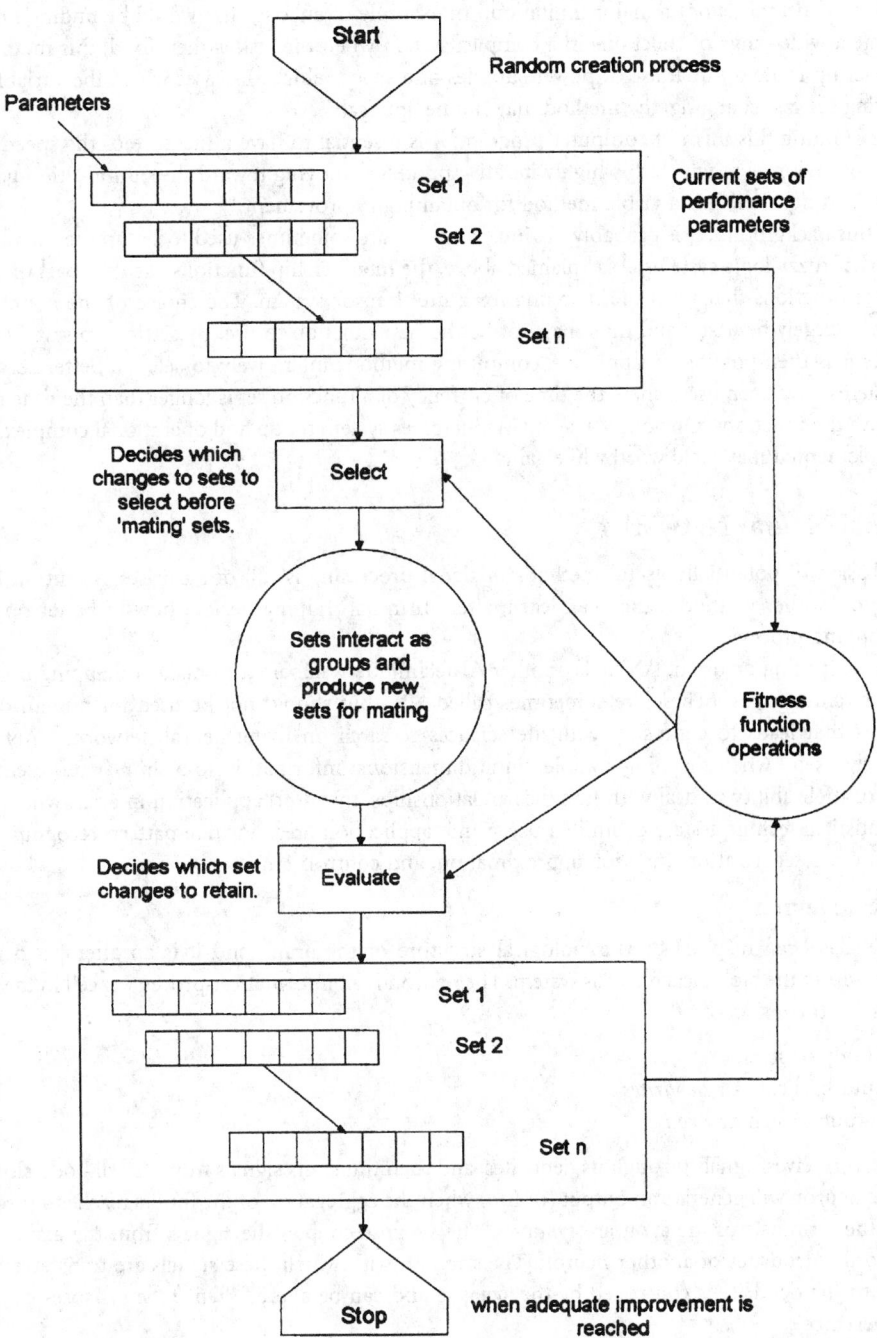

FIGURE 84.12 In the GA method sets of code strings are first modified by some form of genetic operations. They are then intercompared using fitness functions to select a better code to use in the subsequent set of strings. This iterative process is continued until some event disrupts it.

The second limitation becomes obvious when the computational demands are considered in terms of the number of comparison operations needed to be run in the improvement process. This number can be truly huge, especially as the range of options rises with increase in string length. This kind of operation usually needs very large and fast computing power. In cases where there is plenty of time to determine an improvement, this is not a major limitation. An example of effective use would be finding how best to operate a wide range of functions in a complex system when the task is then fixed; this method was used to set up more optimal use of power supplies in a space vehicle. In cases where the variables in a code string are fast changing, the method may not be applicable.

When running this form of computer program, it is essential to have a measure of the speed — the dynamic behavior — at which this highly iterative process is moving toward the optimization goal: the user can then decide if it is a viable method for obtaining improvement.

With this background, it is probably obvious why GAs are sometimes used to set up the membership functions of fuzzy logic systems. As explained above, the membership functions each are part of a set of individual functions that form the mapping for a multisensor system. The choice of the membership function is largely heuristic and thus may not be the best function to use. By setting up several sets of functions it is then possible to apply GA computing methods interatively to select a better set to use. Such systems have been used where the time of currency of a function set is longer than the time needed to improve the functions. Obviously, use of GAs increases system set up and operational complexity, but once implemented may yield worthwhile gains.

Artificial Neural Networks

The AI basket of potentially useful methods in signal processing is full of surprises. Attention is now directed to another method, which can learn, after an initial training period, how to better operate a given mapping process.

Neurocomputing or use of ANNs is another AI technique well suited to make a mapping processor in some circumstances. (These are sometimes called NNs but should not be used for human-devised systems as that leads to confusion with life sciences research on living neural networks.) ANNs are particularly useful when mapping complex multidimensional information to a simpler representation. Because of their ability to deal with nonlinear relationships, they find application in areas where traditional statistical techniques are of limited use. Some application areas include pattern recognition and classification, categorization, function approximation, and control.

Biological Neuron

The ANN has been inspired by the biological structure of the brain, and it is an attempt to mimic processes within the biological nervous system. The neuron is an information-processing cell in the brain, Figure 84.13. It consists of

1. A body or *soma*
2. Input branches or *dendrites*
3. Output branch or *axon*

The neuron receives signals through its dendrites, and then transmits signals from its cell body along the axon. The neuron will generate an output (or fire) when the aggregation of the inputs reaches a threshold level. At the terminals of the axon are *synapses*. The synapse couples the signals from the axon of one neuron to the dendrites of another neuron. The strength with which these signals are transferred from the axon to the dendrite is controlled by the synapse and can be altered; hence the synapses can learn from experience.

Desirable characteristics of neural systems include

- Massive parallelism
- Learning ability
- Ability to generalize
- Adaptability
- Fault tolerance

It is the attempt to construct machines that exhibit these characteristics that has led to the ANN methods of signal processing.

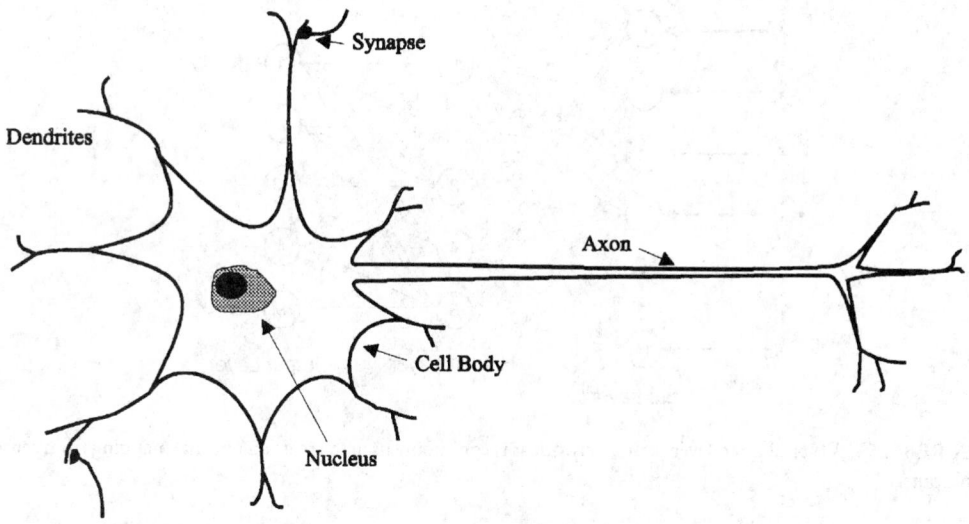

FIGURE 84.13   A biological neuron.

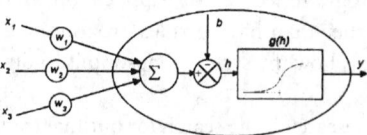

FIGURE 84.14   This neuron model is commonly used.

Artificial Neural Network

In 1943 McCulloch and Pitts proposed the first model of an artificial neuron. This has formed the basis for the generally accepted form of synthetic neuron model or processing element (PE); see Figure 84.14.
The output of the processing element y is given by

$$y = g\left[\left(\sum_i w_i \bullet x_i\right) - b\right]$$

where x_i are the PE inputs with weights (synaptic strengths) w_i, b the PE bias, and g the activation or transfer function. Many types of function have been proposed but the most popular is the sigmoid function, defined by

$$g(h) = \frac{1}{\left(1 + e^{(-\beta h)}\right)}$$

where β is the slope parameter.
By themselves the processing elements are very simple; however, when the individual elements are joined into large interconnected networks, complex relationships can be represented. Although a number of network architectures [2] exist, one of the most commonly discussed architectures found in the literature is the feed-forward multilayered perceptron. Figure 84.15 provides a simple illustration of how a multilayered ANN maps a multidimensional input vector $x_0, \dots x_{N-1}$ in an input space to a vector $y_0, \dots y_{M-1}$ in an output space.

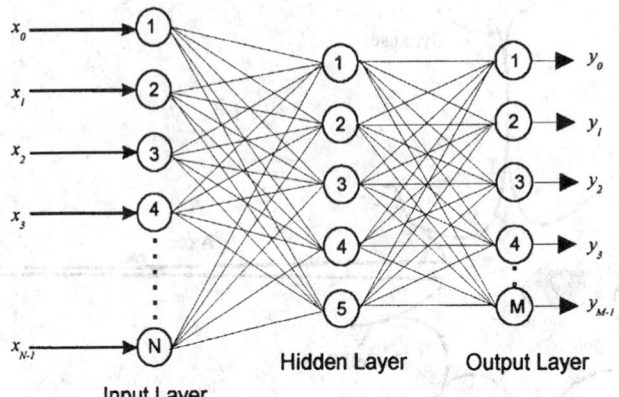

Input Layer

FIGURE 84.15 A typical three-layer neural network as is commonly used to create an optimal mapping from sensors to outputs.

Learning

A fundamental characteristic of the ANN, once it has been set up as an operational tool in software form, is that it does not need to be programmed for the application. ANNs appear to learn rules from a representative set of examples, rather than having rules programmed in by an expert. The knowledge acquired by the system that controls how the system maps input to output is held within the connection weights of the network.

The focus of extensive ongoing research is the search for optimal training techniques. These techniques tend to fall into two broad categories — supervised and unsupervised learning.

In supervised learning, a representative set of inputs is presented to the network which then modifies its internal weights in order to achieve a desired output. With unsupervised learning, the input data only are presented to the network, following which the network organizes itself through self-modification of its internal weights so that it responds differently to each input stimulus.

It is beyond the scope of this text to review all the current learning techniques for ANNs. This is well documented in the literature [8]. One popular training algorithm — backpropagation — will be discussed as an example.

Backpropagation

Backpropagation is an example of a supervised learning paradigm commonly used with multilayer perceptron network architectures. Backpropagation follows the error-correction principle and uses the error signal {d (desired output) − y (actual output)} to modify the connection weights to reduce this error. The backpropagation algorithm is implemented as follows:

1. Initialize network weights to small random values.
2. Present an input vector $x_0, \ldots x_{N-1}$ and the corresponding desired output vector $d_0, \ldots d_{M-1}$.
3. Calculate the actual output vector $y_0, \ldots y_{M-1}$ by propagating the input through the network.
4. Use a recursive algorithm starting at the output layer and adjust the weights backward by

$$w_{ij}(t+1) = w_{ij}(t) + \eta \delta_j x_i'$$

where $w_{ij}(t)$ = the weight from an input to node j at time t
 x_i' = either the output of node i or an input
 η = a gain term $(0.0 < \eta < 1.0)$
 δ_{ij} = an error term for node j

For the output layer $l = L$ the error is calculated:

$$\delta_j^L = g'\left(h_j^L\right)\left[d_j - y_j\right]$$

where h_j^L = the net input to the jth unit in the L layer

g' = the derivative of the activation function g

For hidden layers $l = (L - 1),\ldots,1$. the error is calculated:

$$\delta_j^L = g'\left(h_j^l\right)\sum_l w_{ij}^{l+1}\delta_j^{l+1}$$

where h_j^l = the net input to the jth unit in the kth layer.

g' = the derivative of the activation function g

5. Return to Step 2 and repeat for the next pattern until the error reaches a predefined minimum level.

Unfortunately, no method exists that allows the ANN to create or learn information that is not contained in the training data; that is, the ANN can only reproduce based on experience. Under certain conditions, however, the network can generalize, that is, approximate, output values for data not contained in the training set.

The neural network can be considered as a universal approximator. During supervised learning, the output eventually approximates a target value based on training data. While this is a useful function, the ability to provide output data for test cases not in the training data is more desirable. Loosely, generalization can be viewed in terms of interpolation and extrapolation based on training data. If a test case is closely surrounded by training data, then (as with interpolation) the output accuracy is generally reliable. If, however, the test case is outside of, and not sufficiently close to, training data then (as with extrapolation) the accuracy is notoriously unreliable. Therefore, if the training cases are a sufficiently large sample of the total population of possible input data so that each test case is close to a training case, then the network will adequately generalize.

While multilayer feed-forward networks are finding increasing application in a wide range of products, many design issues such as determining an optimal number of layers, units, and training set for good generalization are research topics. Current theory provides loose guidelines and many of these design issues are resolved by trial and error. Another disadvantage of ANNs is the high demand that many training algorithms can put on computing resources because of their recursive nature.

The ANN, then, provides another alternative for development of suitable mapping models for measurement systems. They can be used to describe complex nonlinear relationships using a network of very simple processing elements. The attraction of the ANN lies in its ability to learn. As long as there exists a sufficiently representative sample of input-to-output data available, the mathematical relationship of the mapping function need not be known. It is effectively taught to the network during a learning process.

Again, there exist important limitations. The worst is the time it might take to adjust the system nodes and weights to a nearly final state. This can often require considerably more time than the time-varying properties of the inputs allow. In many potential applications, they take too long to learn and are not effective. Again computational speed and power are governing factors.

Despite their shortcomings in some applications, ANNs are now a commonly used procedure to set up sensor mapping systems. Examples are banknote image detection and the increased sensitivity of the detection of aluminum in water.

84.7 Problems in Calibration of AI Processing Methods

Calibration of a measurement system is the result of using an agreed upon, often legally binding, process by which it is proven to possess a declared level of accuracy in its measurement outcome. In conventional instrument terms this implies the system can be set up and compared with a measurement method of superior performance to give its error of accuracy plus its variance from the agreed value determined with a level of uncertainty. This kind of instrument system is then accepted to have a known behavior that could be explained by the laws of physics as causal and unique in performance. The prime example, the physical standard apparatus for a parameter, can and is defined such that it will always give very closely the same outcome, even if built in different laboratories. It will have predictable behavior. At the heart of this acceptability is that it can be modeled in terms of an algorithm. All parts in it follow formal laws and have the same outcomes from implementation to implementation.

Most of this notion has to be put aside because, as explained above, AI-based instrument signal processors are built on a raft of transformations that convert subjective situations into objective ones or they carry out unexplained processes. It is, therefore, not hard to see that calibration is a major issue with this type of processor.

Processes make use of heuristics to at least start them going. The designer, when invoking any of the decision-making methods, will almost certainly not select the same rules, processes, and parameters that another designer will choose. There is a lack of consistency in AI processors. The outcomes are fuzzy, not crisp as in instrumentation that complies with a physical law. They act like humans do in that they provide a range of solutions to the same problem.

At first sight this seems to imply that we should ignore the AI possibilities for they cannot be calibrated according to long-standing metrological practices. However, their performance is often very worthy and will be workable where algorithmic methods are not. This calibration constraint must be considered in terms of human thinking, not so much in terms of physics and mathematical models.

At present, AI processing methods have been well proved in many fields, these tending to be fields in which performance does not need calibration with regard to the standards regime. Examples are the use of fuzzy logic in clothes washing machines to improve the wash by varying the wash cycle parameters, in neural network methods to aid the recognition of banknotes, and in rule-based controllers in industrial process plant controls. Genetic algorithms have been used to schedule power supply and usage in space shuttles — said to be an impossible task by any other known means. These all are typified as being needs where much experience can be devoted to ensuring they work satisfactorily.

They should normally not be the critical processing element in safety-critical situations. They can certainly be used in routine management situations where they can outperform algorithmic methods, but there they should be backed up with conventional alarms. They are often used in off-line plant control where the human and alarms are still the final arbiter. This, however, seems to be only a cautious step in our slow acceptance of new ideas.

The process of calibration here is more akin to that of conducting evaluation and validation. Does the system give the range of outcomes expected in given circumstances? Are the outcomes better than those without the processor? Could it be done as well or better by algorithmic-based processing? Is the speed it gives worth the problems it may bring? Problems in their testing, and thus calibration, are discussed by Sizemore [6]. The issues that need to be considered in the calibration of conventional instrumentation [5] are relevant to the calibration of AI-based processing but need much more care in their execution.

Such questions require consideration of the very same elements of decision theory upon which they are based to test them. They have been set up to think like humans so it is expected they will have to be calibrated and evaluated like humans — that is not at all easy.

At present, the calibration and validation of AI systems are not standardized well enough. This impedes acceptance, but standardization will improve as world knowledge of this relatively new method of processing develops to greater maturity inside instrumentation.

There will be opposition to the use of AI methods, but the performance gain they bring will ensure they are used. The forward-thinking academic measurement community is showing signs of addressing AI signal processing — but it will take time.

References

1. Baker, D., Kohler, D. C., Fleckenstein, W. O., Roden, C. E., and Sabia, R., Eds., *Physical Design of Electronic Systems*, Vol. 4, Prentice-Hall, Englewood Cliffs, NJ, 1972.
2. Jain, A. K. and Mao, J., Artificial neural networks: a tutorial, *IEEE Comput.*, pp. 31–44, 1996.
3. Kaufmann, A., *The Science of Decision Making*, Weidenfeld and Nicholson, London, 1968.
4. Mauris, G., Benoit, E., and Foulloy, L., Fuzzy sensors for the fusion of information, in *Proc. 13th IMEKO Conference*, Turin, pp. 1009–1014, 1994.
5. Nicholas, J., in *Handbook of Measurement and Control*, Vol. 3, Sydenham, P. H. and Thorn, R., Eds., John Wiley and Sons, Chichester, U.K., 1992.
6. Sizemore, N. L., Test techniques for knowledge-based systems, *ITEA J.*, XI (2), 34–43, 1990.
7. Tang, K. S., Man, K. F., Kwong, S., and He, Q., Genetic algorithms and their applications, *IEEE Signal Processing Mag.*, 13(6), 22–37, 1996.
8. Venmuri, *Artificial Neural Networks: Theoretical Concepts*, IEEE Computer Society Press, Los Angeles, CA, 1988.

85

Analog-to-Digital Converters

E. B. Loewenstein
National Instruments

85.1 Introduction

Almost every modern instrumentation system includes some form of digitizer, or *analog-to-digital converter (ADC)*. An ADC converts real-world signals (usually voltages) into digital numbers so that a computer or digital processor can (1) acquire signals automatically, (2) store and retrieve information about the signals, (3) process and analyze the information, and (4) display measurement results. A digitizing system can do these jobs with greater speed, reliability, repeatability, accuracy, and resolution than a purely analog system normally can.

The two main functions of an ADC are *sampling* and *quantization*. These two processes convert analog signals from the time and voltage continuums (respectively) into digital numbers having discrete amplitudes, at discrete times. To represent changing signals at every instant in time or at every possible voltage would take an infinite amount of storage. So for every system there is an appropriate *sampling rate* and degree of quantization (*resolution*) so that the system retains as much information as it needs about the input signals while keeping track of manageable amounts of data. Ultimately, the purpose of sampling and quantization is to reduce as much as possible the amount of information about a signal that a system must store in order to reconstruct or analyze it meaningfully.

85.2 Sampling

To prevent having to digitize an infinite amount of information, an analog signal must first be sampled. Sampling is the process of picking one value of a signal to represent the signal for some interval of time. Normally, digitizers take samples uniformly in time, e.g., every microsecond. It is not necessary to sample uniformly, but doing so has some interesting and convenient mathematical properties, which we will see later.

0-8493-8347-1/99/$0.00+$.50

Sampling is done by a circuit called a *sample-and-hold (S/H)*, which, at a sampling instant, transfers the input signal to the output and holds it steady, even though the input signal may still be changing. An S/H usually consists of a signal buffer followed by an electronic switch connected to a capacitor. At a sampling instant, the switch briefly connects the buffer to the capacitor, allowing the capacitor to charge to the input voltage. When the switch is disconnected, the capacitor retains its charge and thus keeps the sampled input voltage steady while the ADC that follows does its job. Quite often, sampling is actually done by a circuit called a *track-and-hold (T/H)*, which differs from an S/H only slightly. Whereas the S/H holds the analog signal until the next sampling instant, the T/H holds the analog signal still only until the ADC has finished its conversion cycle. After the ADC is through, the T/H reconnects the buffer to the capacitor and follows the input signal until the next sampling instant. The result is more accurate sampling, because the buffer has more time to charge the capacitor and "catch up" with (track) the input signal, which has changed since the last sampling instant. Nearly every modern ADC chip has a built-in S/H or T/H, and virtually all data acquisition systems include them.

Of course, sampling necessarily throws away some information, so the art of sampling is in choosing the right sample rate so that enough of the input signal is preserved. The major pitfall of *undersampling* (sampling too slowly) is *aliasing*, which happens whenever the input signal has energy at frequencies greater than one-half the sample rate. In Figure 85.1a, a signal (the fast sine wave) is sampled at a rate

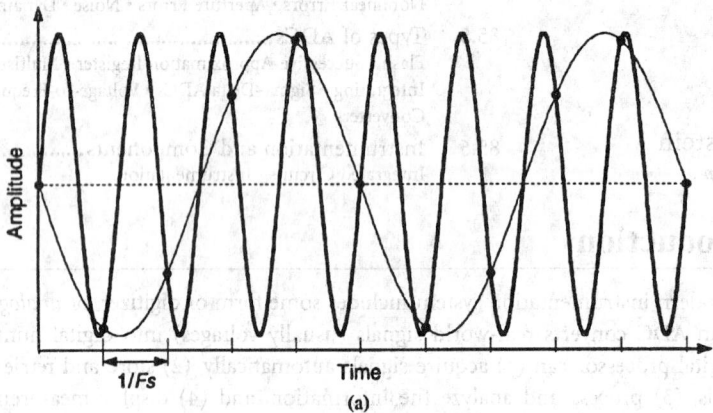

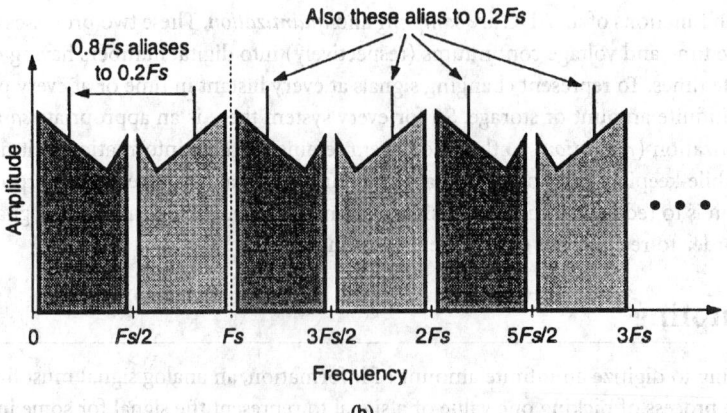

FIGURE 85.1 A demonstration of aliasing. An ADC sampling at rate *Fs* cannot distinguish between a 0.8*Fs* sine wave and a 0.2*Fs* sine wave. (a) A time-domain illustration. (b) A frequency-domain illustration. Theoretically, a sampler aliases an infinite number of 0.5*Fs*-wide frequency bands into the baseband (0 to 0.5*Fs*). Practically, finite analog bandwidth eventually limits how far out in frequency aliases can come from.

Fs, shown by the hash marks at the bottom of the graph. The sine wave has a frequency of 0.8*Fs*, which is higher than one half the sample rate (0.5*Fs*). Notice that sampling the lighter sine wave of 0.2*Fs* produces the same set of samples. The resulting sampled data is ambiguous in that we cannot tell from the data what the frequency of the incoming sine wave actually is. In fact, even though the data set appears to represent a sine wave of 0.2*Fs*, the actual signal could be any sine wave having a frequency of $(n)Fs \pm$ 0.2*Fs*, where *n* is any integer, starting with 0. So the original signal could be 0.2*Fs*, 0.8*Fs*, 1.2*Fs*, 1.8*Fs*, 2.2*Fs*, etc. (or even more than one of those). We say that 0.2*Fs* is the *alias* of a signal that may actually be at another frequency entirely. During interpretation of sampled data, it is customary to treat signals as though they occurred in the baseband (0 to 0.5*Fs*), whether or not that is the case. In general, in a system sampling at *Fs*, a signal at a frequency *F* will alias into the baseband at

$$Fa = \text{abs}\left[(n)Fs - F\right], \tag{85.1}$$

where abs denotes absolute value, $n \geq 0$, and $(n)Fs$ is the closest integer multiple of *Fs* to *F*.

Everyone has seen a demonstration of aliasing at the movies, in the form of "wagon-wheeling." As the stagecoach or wagon takes off, the wheels begin to turn, slowly at first, then faster. As the wagon speeds up, the spokes suddenly appear to be turning backward, even though the wagon is moving forward. Sometimes the spokes appear to be standing still. The reason for this is that a motion picture camera shooting film at 24 frames/s is a sampling system operating at 24 samples/s. The turning wagon wheel is a periodic signal that the camera undersamples. When the wheel begins turning just fast enough that one spoke travels at least half the distance to the next spoke in 1/24th of a second, the spokes begin to appear to move backward, and the system is aliasing. When the wheel is turning so that a spoke moves exactly the distance between two spokes in 1/24th of a second, the spokes appear to be standing still, since they all look the same to the camera.

It follows from Equation 85.1 that if we put into a sampler a signal with no energy at frequencies greater than one half the sample rate (0.5*Fs*), then aliasing will not occur. This is the essence of the Shannon sampling theorem [1], which states that, with mathematical interpolation, the complete input waveform can be recovered *exactly* from the sampled data, at all times at and in between the sampling instants, as long as the sample rate is at least twice as high as the highest frequency content in the signal. Sometimes we refer to 0.5*Fs* as the *Nyquist frequency*, because Nyquist was concerned with the maximum bandwidth of signals [2]. Similarly, twice the highest frequency content of a signal (i.e., the minimum nonaliasing sample rate) is sometimes called the *Nyquist rate*. Sample rates are specified in samples/s, or S/s, and it is also common to specify rates in kS/s, MS/s, and even GS/s.

It is not always necessary to worry about aliasing. When an instrument is measuring slow-moving dc signals or is gathering data for statistical analysis, for instance, getting frequencies right is not important. In those cases we choose the sample rate so that we can take enough data in a reasonable amount of time. On the other hand, if the instrument is a spectrum analyzer, where frequency does matter, or an oscilloscope, where fine time detail is needed, aliasing certainly is an issue. When aliased signals from beyond the frequency band of interest can interfere with measurement, an instrument needs to have an *antialias filter* before the S/H. An antialias filter is a low-pass filter with a gain of 1 throughout most of the frequency band of interest. As frequency increases, it begins to attenuate the signal; by the Nyquist frequency it must have enough attenuation to prevent higher-frequency signals from reaching the S/H with enough amplitude to disturb measurements. An efficient antialias filter must attenuate rapidly with frequency in order to make most of the baseband usable. Popular analog filters with rapid cutoff include elliptic and Chebyshev filters, which use zeros to achieve fast cutoff, and Butterworth filters (sixth order and above), which do not attenuate as aggressively, but have very flat passband response. A good book about filters is Reference 3.

Some ADCs do not need a S/H or T/H at all. If the ADC is converting a slow-moving or dc signal and precise timing isn't needed, the input may be stable enough during conversion that it is as good as

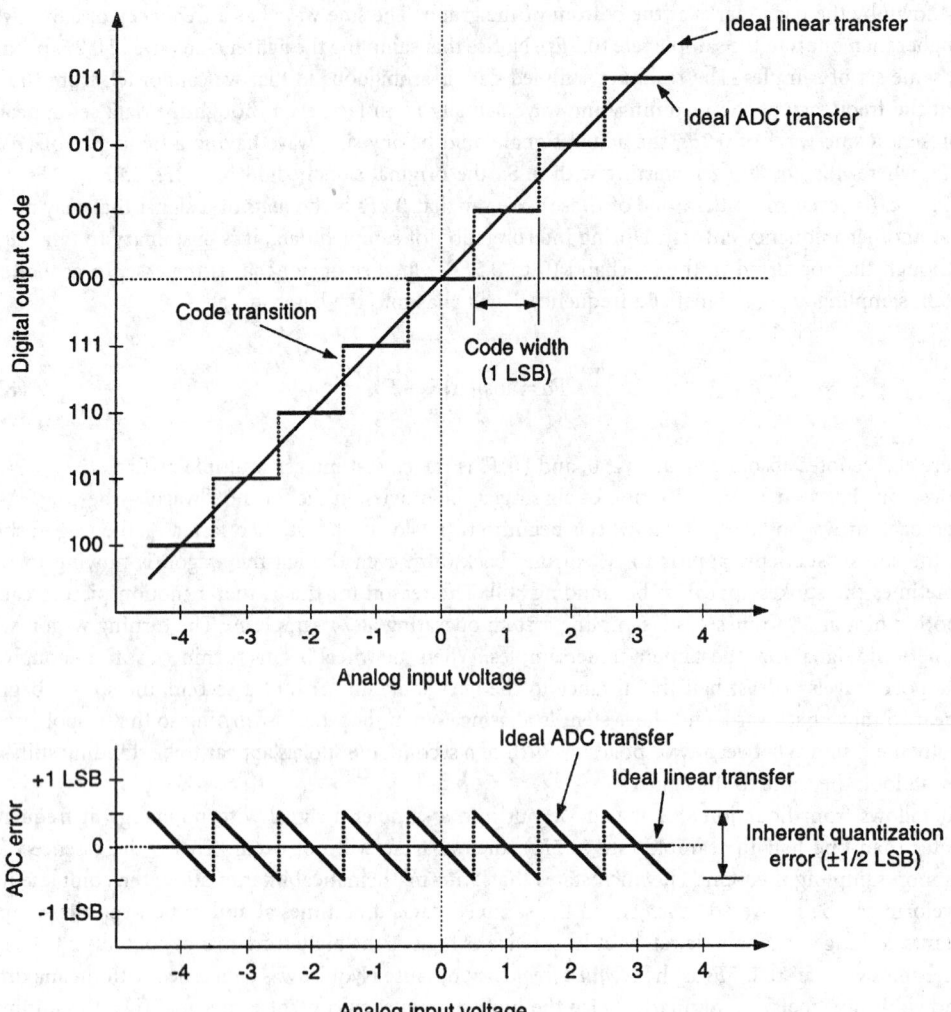

FIGURE 85.2 The ideal three-bit quantizer has eight possible digital outputs. The analog input-to-digital output transfer function is a uniform staircase with steps whose width and height are 1 LSB exactly. The bottom graph shows the ideal transfer function (a straight line) subtracted from the staircase transfer function.

sampled. There are also *integrating ADCs* (discussed later), which average the input signal over a period of time rather than sampling it. However, internally they actually sample the average.

85.3 Quantization

What sampling accomplishes in the time domain, quantization does in the amplitude domain. The process of digitization is not complete until the sampled signal, which is still in analog form, is reduced to digital information. An ADC quantizes a sampled signal by picking one integer value from a predetermined, finite list of integer values to represent each analog sample. Each integer value in the list represents a fraction of the total analog input range. Normally, an ADC chooses the value closest to the actual sample from a list of uniformly spaced values. This rule gives the *transfer function* of analog input-to-digital output a uniform "staircase" characteristic. Figure 85.2 represents a three-bit quantizer, which maps a continuum of analog input values to only eight (2^3) possible output values. Each step in the staircase has (ideally) the same width along the *x*-axis, which we call *code width* and define as 1 *LSB (least significant bit*). In this case 1 LSB is equal to 1 V. Each digital code corresponds to one of eight 1-LSB intervals making up the analog input range, which is 8 LSB (and also 8 V in this case).

Of course, we would like our measurement system to have a transfer function that is a straight line and has no steps at all. The bottom graph in Figure 85.2 is the ideal transfer function (a straight diagonal line) subtracted from the staircase function, or the *quantization error*. In an ideal ADC, the quantization error is bounded by ±½ LSB, and, over the input range, the average error is 0 LSB and the standard deviation of error is $1/\sqrt{12}$ LSB. As the bottom graph shows, the quantization error at any point is a deterministic function of the input signal.

85.4 ADC Specifications

Range and Resolution

The *input range* of an ADC is the span of voltages over which a conversion is valid. The end points at the bottom and the top of the range are called *–full-scale* and *+full-scale*, respectively. When –full-scale is 0 V the range is called *unipolar*, and when –full-scale is a negative voltage of the same magnitude as +full-scale the range is said to be *bipolar*. When the input voltage exceeds the input range, the conversion data are certain to be wrong, and most ADCs report the code at the end point of the range closest to the input voltage. This condition is called an *overrange*.

The *resolution* of an ADC is the smallest change in voltage the ADC can detect, which is inherently 1 LSB. It is customary to refer to the resolution of an ADC by the number of binary bits or decimal digits it produces; for example, "12 bits" means that the ADC can resolve one part in 2^{12} (= 4096). In the case of a digital voltmeter that reads decimal digits, we refer to the number of digits that it resolves. A "6-digit" voltmeter on a 1 V scale measures from –0.999999 V to +0.999999 V in 0.000001 V steps; it resolves one part in 2 000 000. It is also common to refer to a voltmeter that measures from –1.999999 to +1.999999 as a "6½ digit" voltmeter. Figure 85.3 compares the resolutions of common word lengths for ADCs.

Coding Conventions

There are several different formats for ADC output data. An ADC using *binary* coding produces all 0s (e.g., 000 for the three-bit converter) at –full-scale and all 1s (e.g., 111) at +full-scale. If the range is bipolar, so that –full-scale is a negative voltage, binary coding is sometimes called *offset binary*, since the code 0 does not refer to 0 V. To make digital 0 correspond to 0 V, bipolar ADCs use *two's complement* coding, which is identical to offset binary coding except that the *most significant bit* (*MSB*) is inverted, so that 100 ... 00 corresponds to –full-scale, 000 ... 00 corresponds to 0 V (*midscale*), and 011 ... 11 corresponds to +full-scale. All of the figures in this chapter depicting three-bit ADC transfer functions use two's complement coding.

Decimal-digit ADCs, such as those used in digital voltmeters, use a coding scheme call *binary-coded decimal (BCD)*. BCD data consists of a string of four-bit groups of binary digits. Each four-bit group represents a decimal digit, where 0000 is 0, 0001 is 1, and so on, up to 1001 for 9. The other six combinations (1010 through 1111) are invalid, or can be used for special information, such as the sign of the conversion.

Linear Errors

Linear errors are the largest and most common errors in an ADC and are easily corrected by simple calibrations or by additions with and multiplications by correction constants. Linear errors do not distort the transfer function; they only change somewhat the input range over which the ADC operates.

Figure 85.4 shows the transfer function of an ideal three-bit ADC with some *offset error*. The straight line joining the centers of the code transitions is raised, or offset, by 0.6 LSB, and the bottom graph shows the resulting error. Figure 85.5 shows an ideal three-bit ADC with a +25% *gain error*. The slope of the

Bits	Digits	Voltmeter "Digits"	Steps in FSR	Step size, ppm	Theoretical Dynamic Range (dB)
30	8.730		1 073 741 824	0.001	182.379
28.575	8.301	8 1/2	400 000 000	0.003	173.802
28	8.128		268 435 456	0.004	170.338
27.575	8	8	200 000 000	0.005	167.782
26	7.526		67 108 864	0.015	158.297
25.253	7.301	7 1/2	40 000 000	0.025	153.802
24.253	7	7	20 000 000	0.05	147.782
• 24	6.924		16 777 216	0.060	146.255
22	6.322		4 194 304	0.238	134.214
21.932	6.301	6 1/2	4 000 000	0.25	133.802
20.932	6	6	2 000 000	0.5	127.782
• 20	5.720		1 048 576	0.954	122.173
18.610	5.301	5 1/2	400 000	2.5	113.802
18	5.118		262 144	3.815	110.132
17.610	5	5	200 000	5	107.782
• 16	4.515		65 536	15.259	98.091
15.288	4.301	4 1/2	40 000	25	93.802
14.288	4	4	20 000	50	87.782
14	3.913		16 384	61.035	86.049
• 12	3.311		4 096	244.141	74.008
11.966	3.301	3 1/2	4 000	250	73.802
10.966	3	3	2 000	500	67.782
10	2.709		1024	976.563	61.967
8.644	2.301	2 1/2	400	2500	53.802
• 8	2.107		256	3906.25	49.926
7.644	2	2	200	5000	47.782
6	1.505		64	15625	37.885

FIGURE 85.3 Comparison of theoretical resolutions of ADCs. "Bits" refers to binary word length, and "digits" refers to decimal wordlength. • denotes popular binary word lengths. FSR is full-scale range, and theoretical dynamic range is computed from the formula $1.7609 + 6.0206n$, where n is the number of bits (see discussion of dynamic range).

line through the code transitions is 1.25 times the ideal slope of 1.00. If the slope of the line were 0.75 instead, the gain error would be –25%. The bottom graph shows the error resulting from excessive gain. Offset errors can be compensated for simply by adding a correcting voltage in the analog circuitry or by adding a constant to the digital data. Gain errors can be corrected by analog circuitry like potentiometers or voltage-controlled amplifiers or by multiplying the digital data by a correction constant.

Nonlinear Errors

Nonlinear errors are much harder to compensate for in either the digital or analog domain, and are best minimized by choosing well-designed, well-specified ADCs. Nonlinearities are characterized in two ways: differential nonlinearity and integral nonlinearity.

Differential nonlinearity (DNL) measures the irregularity in the code step widths by comparing their widths to the ideal value of 1 LSB. Figure 85.6 illustrates the three-bit ADC with some irregular code widths. Most of the codes have the proper width of 1 LSB and thus contribute no DNL, but one narrow code has a width of 0.6 LSB, producing a DNL of –0.4 LSB, and one wide code has a width of 1.8 LSB, producing a DNL of +0.8 LSB at that code. This converter would be consistent with a DNL specification of ±0.9 LSB, for example, which guarantees that all code widths are between 0.1 and 1.9 LSB.

It is possible for a code not to appear at all in the transfer function. This happens when the code has a width of 0 LSB, in which case we call it a *missing code*. Its DNL is –1 LSB. If an ADC has a single missing code, the step size at that point in the transfer function is doubled, effectively reducing the local resolution of the ADC by a factor of two. For this reason it is important for an ADC specification to declare that the ADC has *no missing codes*, guaranteeing that every code has a width greater than 0 LSB. Even if an

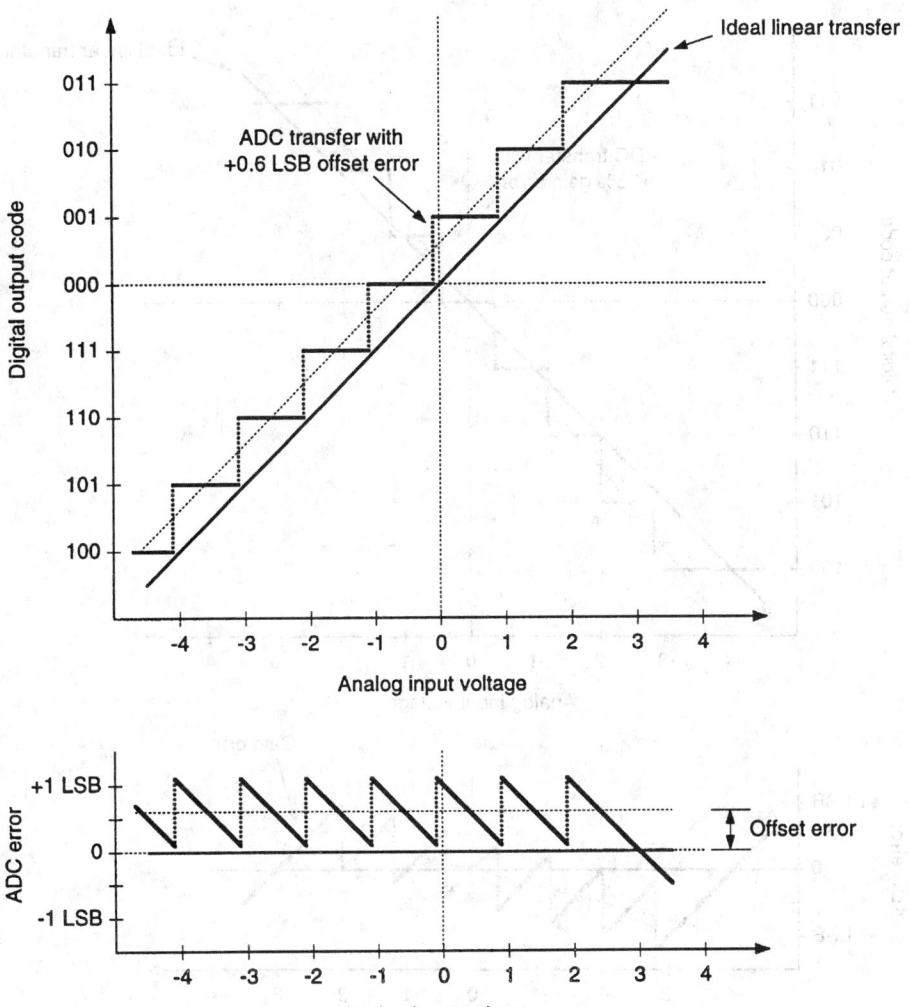

FIGURE 85.4   An ideal three-bit quantizer, only with +0.6 LSB of offset error.

ADC has missing codes, no code can have a width less than 0 LSB, so the DNL can never be worse than −1 LSB.

Integral nonlinearity (INL) measures the deviation of the code transitions from the ideal straight line, providing that the linear errors (offset and gain) have been removed. Figure 85.7 depicts an ADC with an INL error of +0.7 LSB. The offset and gain errors have been calibrated at the end points of the transfer function.

Relative accuracy (RA) is a measure of nonlinearity related to INL, but more useful. It indicates not only how far away from ideal the code transitions are, but how far any part of the transfer function, including quantization "staircase" error, deviates from ideal (assuming offset and gain errors have been calibrated at the end points). In a noiseless ADC, the worst-case RA always exceeds the worst-cast INL by ±0.5 LSB, as demonstrated in Figure 85.7. In an ADC that has a little inherent noise or has noise (called *dither*) added at the input, the RA actually improves because the addition of noise to the quantizer tends to smooth the *averaged* transfer function. Figure 85.8 shows the average of the digital output data as a function of the input voltage when 0.1 LSB rms of Gaussian random noise is intentionally added to the input. The RA improves to ±0.3 LSB from ±0.5 LSB in the noiseless case. If about 0.5 LSB rms of Gaussian noise is added, the quantization staircase becomes nearly straight. This improvement in linearity

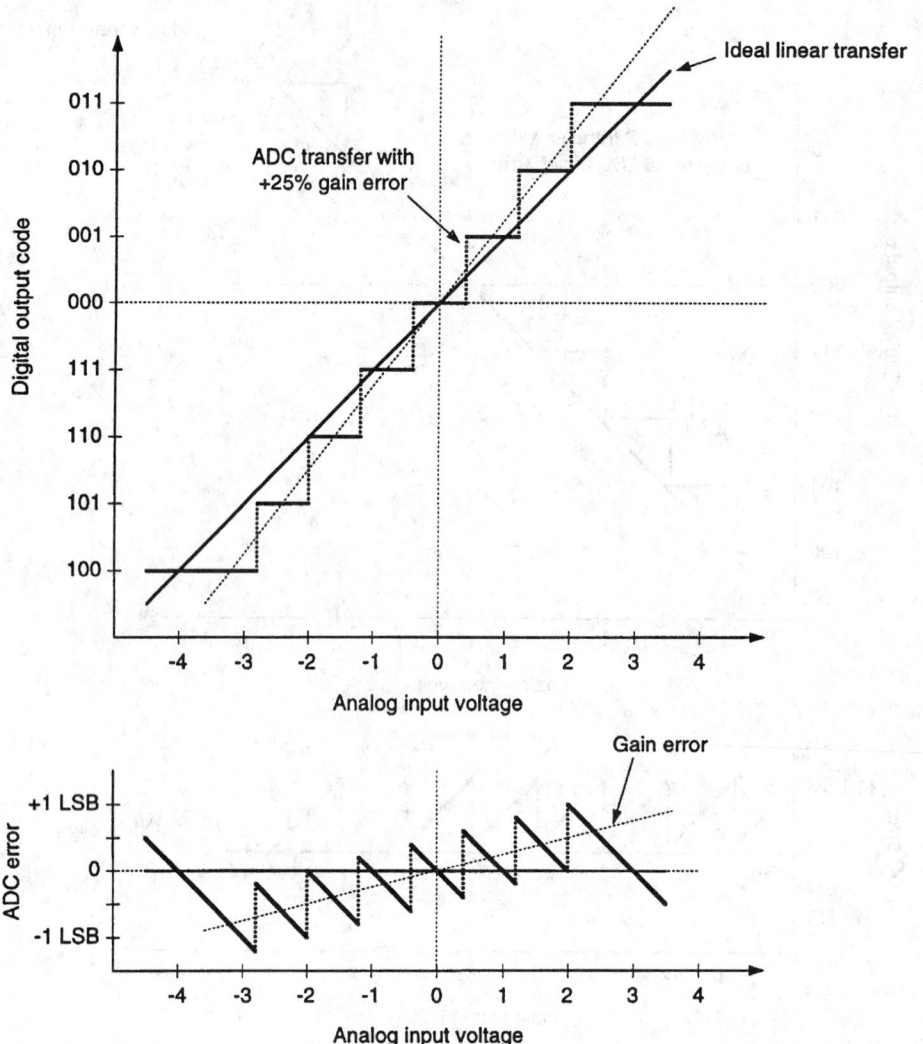

FIGURE 85.5   An ideal three-bit quantizer, only with a gain of 1.25 instead of 1.00. This represents a +25% gain error.

comes at the expense of the random error in each individual conversion caused by the noise. Adding more noise to the ADC does not improve the average quantization error much more, but it does tend to smooth out local nonlinearities in the averaged transfer function. For a good discussion of noise and dither, see Reference 4.

Aperture Errors

Aperture errors have to do with the timing of analog-to-digital conversions, particularly of the S/H. *Aperture delay* characterizes the amount of time that lapses from when an ADC (S/H) receives a convert pulse to when the sample is held as a result of the pulse. Although aperture delay (sometimes called *aperture time*) is usually specified as a few nanoseconds for an ADC or S/H by itself, this delay is usually much more than negated by the group delay in any amplifiers that precede the S/H, so that the convert pulse arrives at the S/H quite some time before the analog signal does. For instance, a typical 1 MHz bandwidth amplifier has 160 ns of delay; if the ADC or S/H it was connected to had an aperture delay of 10 ns, the effective aperture delay for the system would be −150 ns.

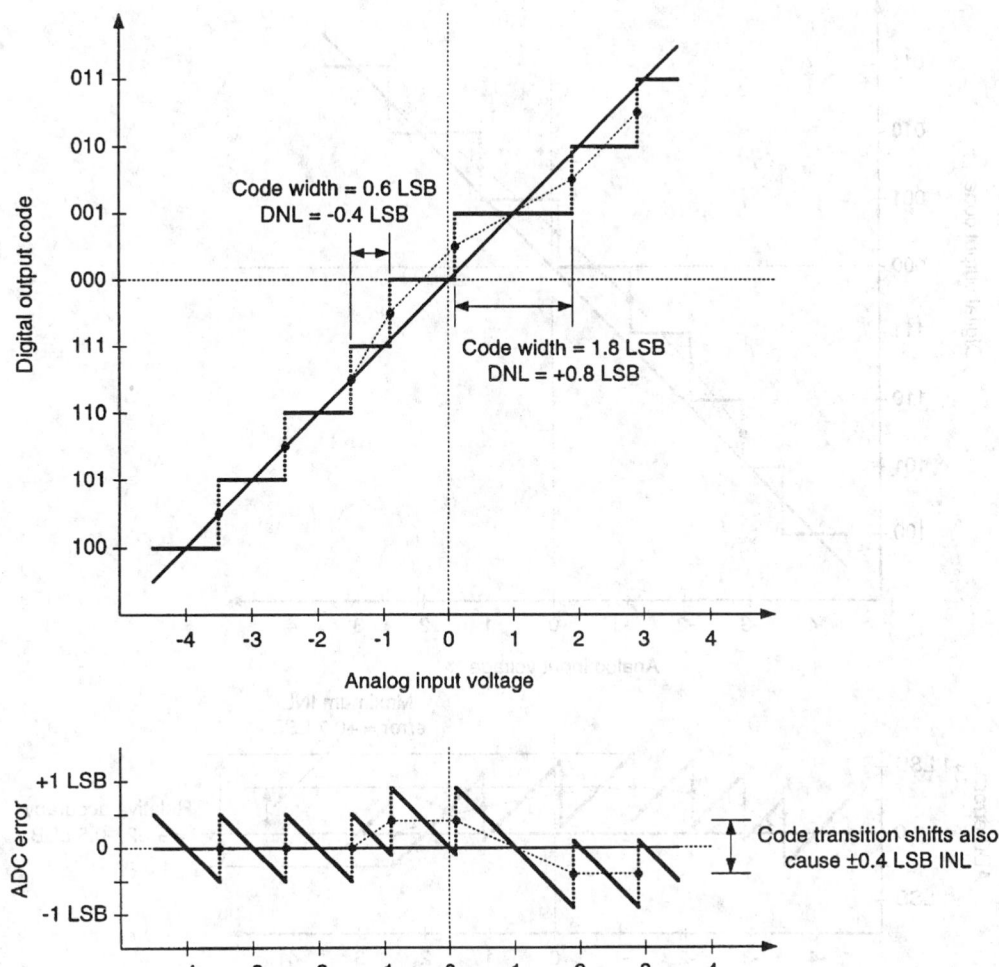

FIGURE 85.6 A three-bit quantizer with substantial DNL errors. The bottom graph illustrates the resulting INL errors.

Jitter (or *aperture jitter*) characterizes the irregularity in times at which samples are taken. If the nominal period between samples in an ADC is 1 μs, the actual time may vary from 1 μs by as much as a few hundred picoseconds or even as much as a nanosecond from cycle to cycle. Contributions to these variations can come from the crystal clock source (if included under the jitter specification), digital clock circuitry, or the S/H. Jitter is usually specified in picoseconds peak-to-peak or picoseconds rms.

Jitter interferes with measurements (particularly spectral analysis) by effectively frequency modulating the input signal by the jitter profile. A jittery ADC sampling a pure sine wave would scatter energy from the sine wave all throughout the spectrum, perhaps covering up useful spectral information. In a typical ADC, however, most of the interference from jitter tends to occur at frequencies very close to the main signal.

Noise

Noise, whether inherent in an ADC or introduced intentionally (see dither above), limits the resolution of an ADC by adding an interfering waveform to the input signal as the data is converted. Noise comes

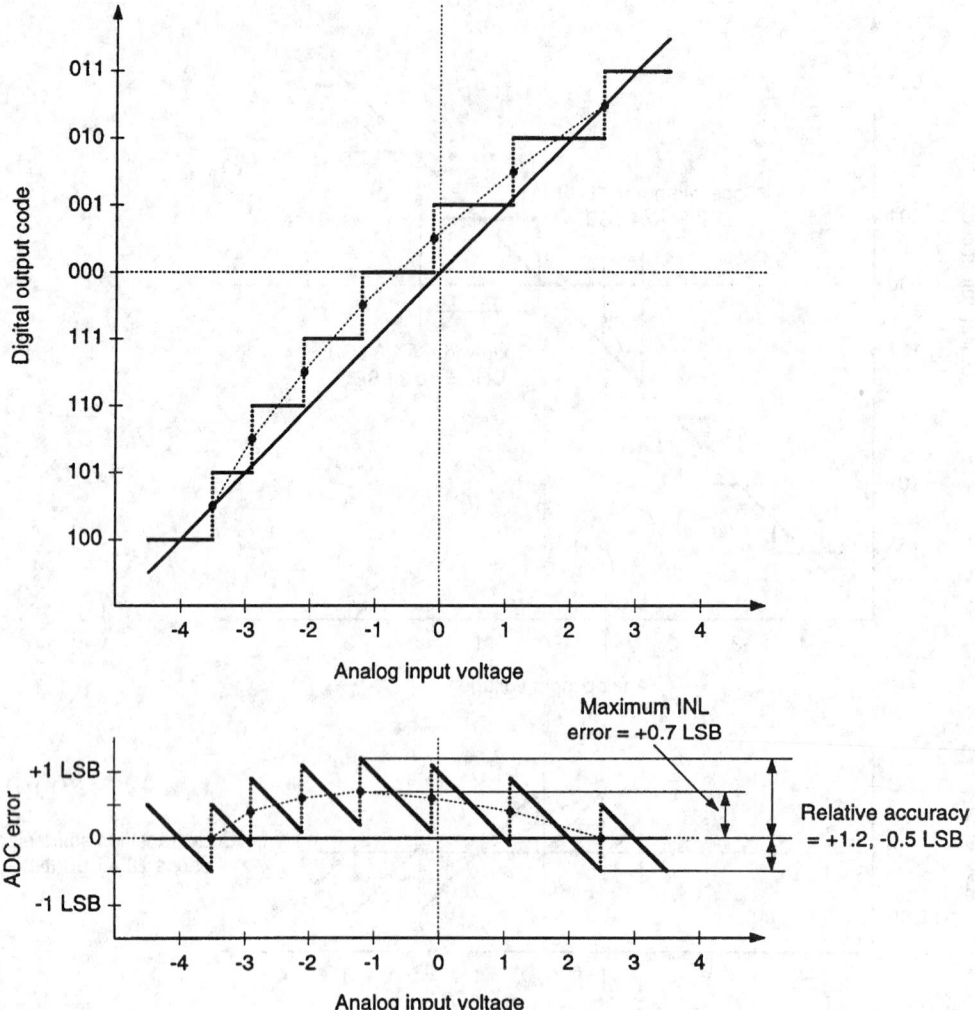

FIGURE 85.7 A three-bit quantizer with substantial INL errors. Here, the DNL error is still significant; but, for example, a 12-bit converter with 0.7 LSB of INL from a smooth error "bow" like the one above could have negligible DNL because it would have so many more steps over which to accumulate error.

from many places. The most common kind of noise is *thermal noise*, which is caused by the random nature of electric conduction in resistors and transistors. Thermal noise is worse at higher temperatures and higher resistances. Most other ADC noise is coupled electromagnetically from nearby circuitry, such as clock or logic circuits, or from routing of other input signals. Noise is usually specified in volts rms or peak-to-peak, or LSBs rms or peak-to-peak.

Quantization error (see above) can sometimes be thought of as *quantization noise*. Although quantization error is perfectly predictable with respect to the input signal, when a signal is fairly "busy" (i.e., busy enough that consecutive conversions do not tend to repeat data) the quantization error becomes chaotic, and it can be thought of it as another source of random noise, whose statistical distribution is uniform from −0.5 LSB to +0.5 LSB and whose standard deviation is $1/\sqrt{12}$ LSB. This is sometimes the dominant source of noise in spectral analysis applications.

Once noise gets into an ADC, there are ways to process out the noise if it is independent of the signal. Acquisitions of DC signals can be quieted by collecting a number of points and averaging the collection. If the noise is *white random noise*, which has equal energy density at all frequencies, averaging can reduce

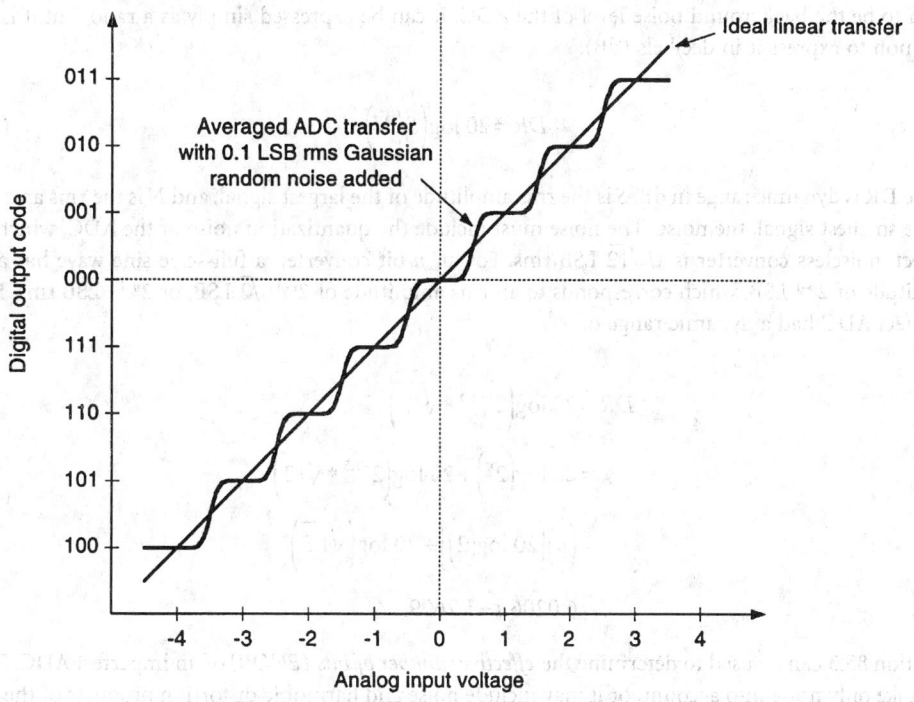

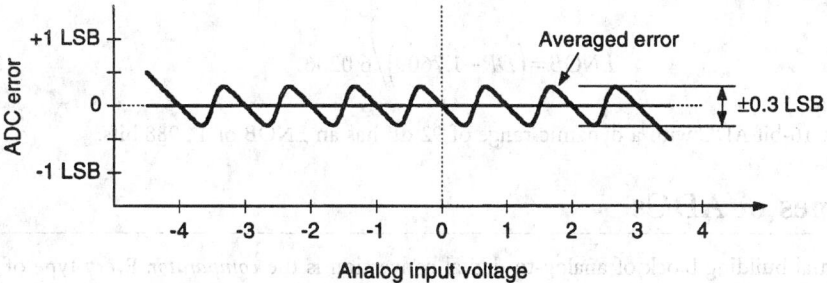

FIGURE 85.8 An ideal three-bit quantizer with 0.1 LSB rms Gaussian random noise (dither) added at the input. The relative accuracy has improved to ±0.3 LSB rms from the ±0.5 LSB expected from a noiseless quantizer. With the application of 0.5 LSB rms Gaussian noise, the transfer function becomes almost perfectly straight. Larger amounts of dither produce essentially no improvement in linearity.

the amount of noise by the square root of the number of samples averaged. The noise interfering with a repetitive waveform can be quieted by acquiring many waveforms using a level trigger and averaging the collection to produce an average waveform. Most digital oscilloscopes have waveform averaging. Quantization noise, as described above, cannot be averaged out unless other random noise is present.

The noise specifications for an ADC are for quiet, low-impedance signals at the input, such as a dead short. To preserve the noise performance of the ADCs, the user must carefully connect signals to the input with tidy cabling that keeps away from sources of electromagnetic noise. For more information on noise sources and treatment and prevention of noise, see References 5 and 6.

Dynamic Range

The *dynamic range (DR)* of an ADC is the ratio of the largest to the smallest signals the converter can represent. The largest signal is usually taken to be a full-scale sine wave, and the smallest signal is usually taken to be the background noise level of the ADC. It can be expressed simply as a ratio, but it is more common to express it in decibels (dB):

$$DR = 20 \log(S/N),$$ (85.2)

where DR is dynamic range in dB, S is the rms amplitude of the largest signal, and N is the rms amplitude of the smallest signal, the noise. The noise must include the quantization noise of the ADC, which for a perfect, noiseless converter is $1/\sqrt{12}$ LSB rms. For an n-bit converter, a full-scale sine wave has a peak amplitude of 2^{n-1} LSB, which corresponds to an rms amplitude of $2^{n-1}/\sqrt{2}$ LSB, or $2^{n-1.5}$ LSB rms. Hence a perfect ADC had a dynamic range of

$$DR = 20 \log\left(2^{n-1.5} * \sqrt{12}\right)$$

$$= 20 \log\left(2^{n}\right) + 20 \log\left(2^{-1.5} * \sqrt{12}\right)$$ (85.3)

$$= \left(n\right)\left[20 \log\left(2\right)\right] + 20 \log\left(\sqrt{1.5}\right)$$

$$\approx 6.0206n + 1.7609.$$

Equation 85.3 can be used to determine the *effective number of bits (ENOB)* of an imperfect ADC. ENOB may take only noise into account, or it may include noise and harmonic distortion products of the input signal. It is computed as

$$ENOB = \left(DR - 1.7609\right)/6.0206.$$ (85.4)

For example, a 16-bit ADC with a dynamic range of 92 dB has an ENOB of 14.988 bits.

85.5 Types of ADCs

The fundamental building block of analog-to-digital conversion is the *comparator*. Every type of ADC has at least one comparator in it, and some ADCs have many. The comparator itself is a one-bit ADC; it has two analog voltage inputs and (usually) one digital output. If the voltage at the + input is greater than the voltage at the − input, the output of the comparator is a digital 1. If the voltage at the + input is less than the voltage at the − input, the output is a digital 0 (see Figure 85.9).

Another piece that all ADCs have in common is a linearity reference. This is what a comparator in an ADC compares the input signal with in the process of conversion. It directly determines the differential and integral nonlinearities of the ADC. Examples of linearity references include capacitors (in integrating ADCs) and DACs (found in successive-approximation ADCs).

The third piece that every ADC has is a voltage reference. The reference(s) determine the full-scale input range of the ADC and are usually part of or closely associated with the linearity reference.

Flash

Flash converters are the fastest ADCs, achieving speeds near 1 GS/s and resolutions of 10 bits and below. The flash converter with n bits of resolution has $2^{n} - 1$ high-speed comparators operating in parallel (see Figure 85.10). A string of resistors between two voltage references supplies a set of uniformly spaced voltages that span the input range, one for each comparator. The input voltage is compared with all of these voltages simultaneously, and the comparator outputs are 1 for all voltages below the input voltage and 0 for all the voltages above the input voltage. The resulting collection of digital outputs from the

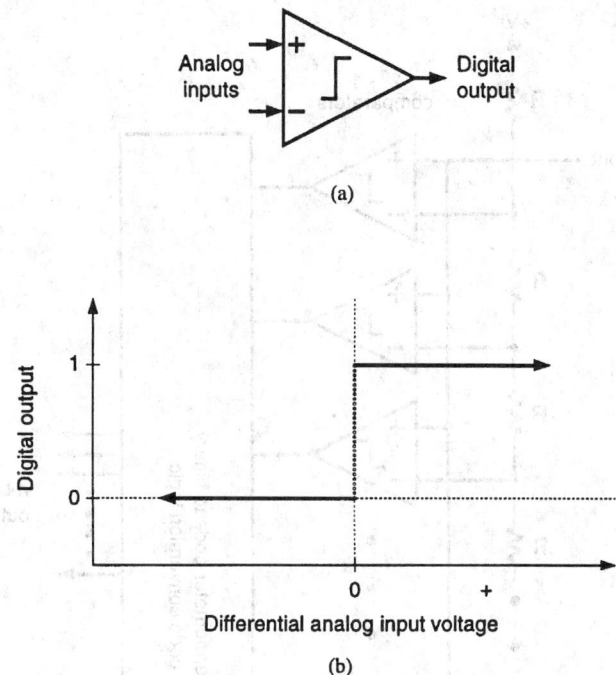

(a)

(b)

FIGURE 85.9 The comparator is the essential building block of all ADCs. (a) Comparator symbol. (b) Comparator input/output transfer function.

comparators is called a "thermometer code," because the transition between all 1s and all 0s floats up and down with the input voltage. Fast logic converts the thermometer codes to normal n-bit binary numbers.

Because of their simplicity, they are fast, but flash converters are limited to resolutions of 10 bits and below because the number of comparators and resistors goes up exponentially with resolution. Because the string resistor values typically vary only a few percent from each other in practical devices, the differential linearity of the flash ADC is quite good. But the same resistor variations can accumulate error across the input range and cause integral nonlinearity of a few LSB.

Successive-Approximation Register

Successive-approximation register (SAR) ADCs are the most common ADCs, having resolutions of 8 to 16 bits and speeds of 1 MS/s and below. They are generally low in cost, and they typically have very good integral linearity. The n-bit SAR ADC contains a high-speed n-bit DAC and comparator in a feedback loop (see Figure 85.11). The successive-approximation register sequences the DAC through a series of n "guesses," which are compared with the input voltage (Figure 85.12). As the conversion progresses, the register builds the n-bit binary conversion result out of the comparator outputs. By the end of the sequence the register has converged to the closest DAC value to the input voltage.

The speed of an SAR ADC is limited by the speed of the DAC inside the feedback loop. The DAC must settle n times to within $1/2^{-n}$ of full-scale within the conversion time of the ADC. Current SAR technology achieves 12-bit resolution at 1 MS/s and 16-bit resolution at 200 kS/s. Faster conversion at these resolutions requires multistage architectures.

Multistage

To achieve higher sample rates than SAR ADCs at resolutions of 10 to 16 bits, *multistage* ADCs (sometimes called *subranging* or *multipass* ADCs) use the iterative approach of SAR ADCs but reduce the number

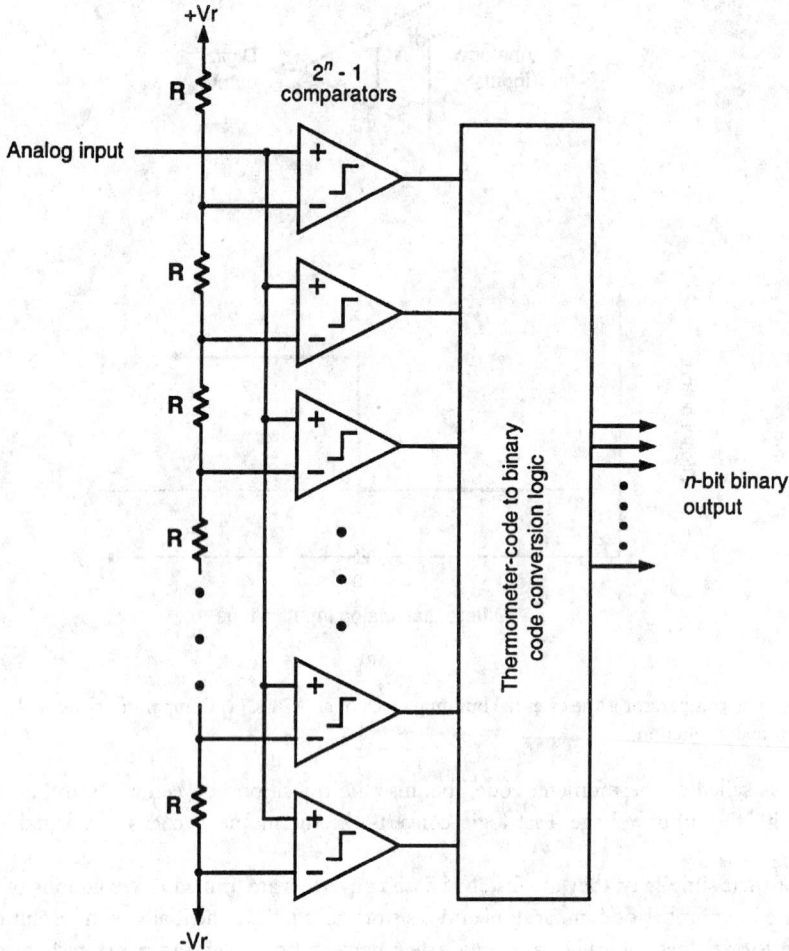

FIGURE 85.10   A flash converter has $2^n - 1$ comparators operating in parallel. It relies on the uniformity of the resistors for linearity.

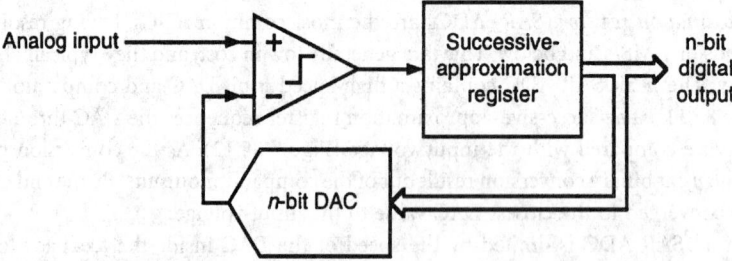

FIGURE 85.11   A successive-approximation converter has only one comparator and relies on an internal, precision DAC for linearity.

of iterations in a conversion. Instead of using just a comparator, the multistage ADC uses low-resolution flash converters (4 to 8 bits) as building blocks. Figure 85.13 illustrates an example of a 12-bit two-stage ADC built out of two flash ADCs and a fast DAC. The 6-bit flash ADC converts the residual error of the 8-bit flash ADC. The two digital outputs are combined to produce a 12-bit conversion result.

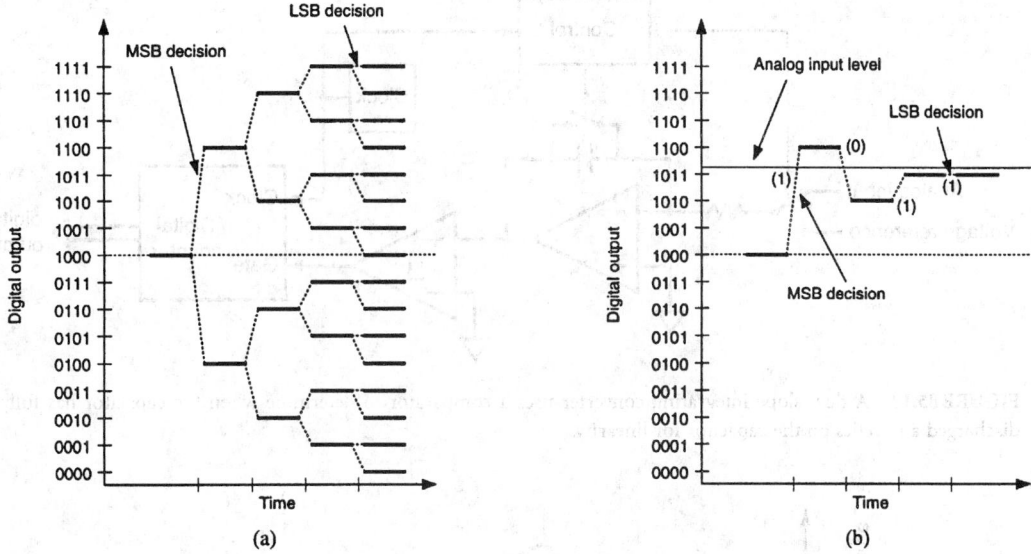

FIGURE 85.12 (a) Decision tree shows all the possible digital "guesses" of a four-bit successive-approximation converter over time. (b) Decision tree for conversion of four-bit code 1011.

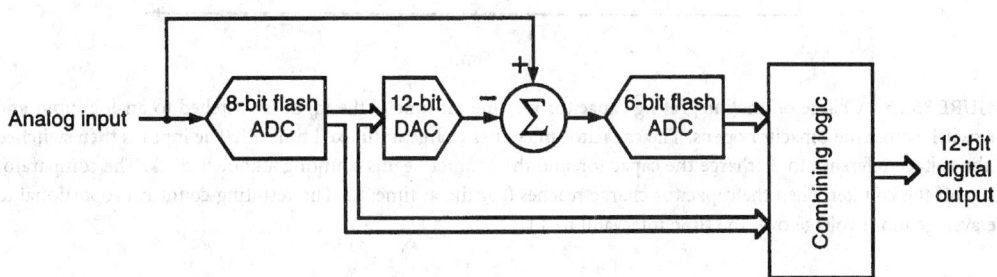

FIGURE 85.13 An example of a 12-bit multistage ADC built out of two flash ADCs and a fast DAC. The 8-bit flash ADC takes a first "guess" at the input signal and the 6-bit flash ADC converts the error in the guess, called the "residue." The 12-bit DAC actually needs to have only 8 bits, but it must be accurate to 12 bits. If the 8-bit flash ADC were perfect, the second flash ADC would only need 4 bits. But since the first flash actually may have some error, the second flash has 2 bits of "overlap."

If each flash ADC has a T/H at its input, then each stage can be converting the residual error from the previous stage while the previous stage is converting the next sample. The whole converter then can effectively operate at the sample rate of the slowest stage. Without the extra T/Hs, a new conversion cannot start until the residues have propagated through all the stages. This variation of the multistage ADC is called a *pipelined* ADC.

Integrating

Integrating converters are used for low-speed, high-resolution applications such as voltmeters. They are conceptually simple, consisting of an integrating amplifier, a comparator, a digital counter, and a very stable capacitor for accumulating charge (Figure 85.14). The most common integrating ADC in use is the dual-slope ADC, whose action is illustrated in Figure 85.15. Initially, the capacitor is discharged and so has no voltage across it. At time 0, the input to the integrator is switched to the analog input and the capacitor is allowed to charge for an amount of time, T1, which is always the same. Its rate of charging

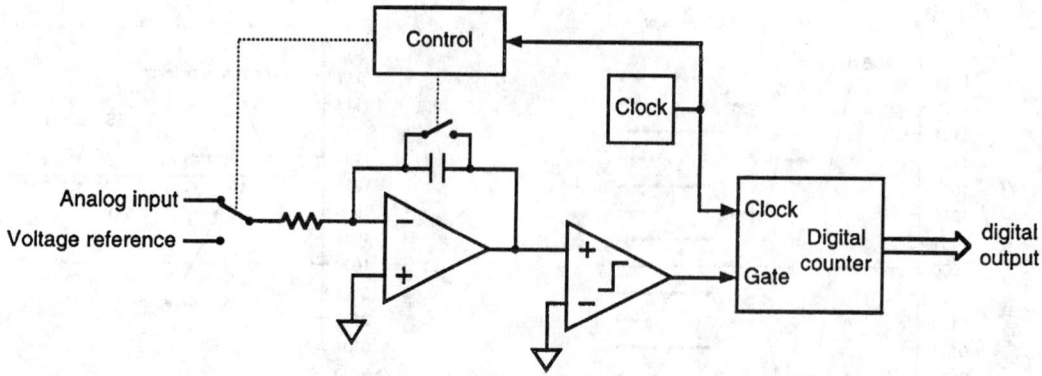

FIGURE 85.14   A dual-slope integrating converter uses a comparator to determine when the capacitor has fully discharged and relies on the capacitor for linearity.

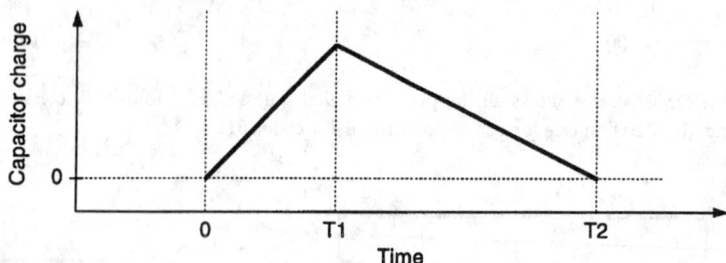

FIGURE 85.15   Charge on the integrating capacitor vs. time. At time 0, the input is switched to analog input and the switch across the capacitor opens. The capacitor integrates charge until fixed time T1. The input is then switched to the voltage reference to discharge the capacitor, and the counter begins counting a known clock. The comparator turns off the counter when the capacitor charge reaches 0 again, at time T2. The resulting count is proportional to the average input voltage over the time interval 0 to T1.

and thus its voltage at T1 are proportional to the input voltage. At time T1 the input switch flips over to the voltage reference, which has a negative value so that the capacitor will begin to discharge at a rate proportional to the reference. The counter measures how long it takes to discharge the capacitor completely. If the capacitor is of high quality, the ratio of the discharge time to the charge time is proportional to the ratio of the input voltage to the voltage reference, and so the counter output represents the analog input voltage.

An elaboration of the dual-slope ADC is the *multislope* integrating ADC. It achieves even higher resolution than the dual-slope ADC by discharging the capacitor at several progressively slower rates. At each rate, the counter is able to resolve finer increments of accumulated charge.

An important distinction between integrating converters and other ADCs is the way they sample the input voltage. Integrating converters do not sample the voltage itself; they *average* the voltage over the integration period and *then* they sample the average that is accumulated on the capacitor. This tends to reject noise that conventional sampling cannot, especially periodic noises. Most integrating ADCs operate with an integration period that is a multiple of one AC line period ($1/60$ or $1/50$ s) so that any potential interference from stray electric or magnetic fields caused by the power system is canceled.

Integrating converters are gradually being replaced in the marketplace with low-speed, high-resolution sigma–delta converters, which see. Sigma–delta converters are generally more flexible than integrating ADCs, and they are easier to use because they do not require an external charging capacitor. The resolution and speed of the two types are comparable, although integrating converters still have the highest linearity.

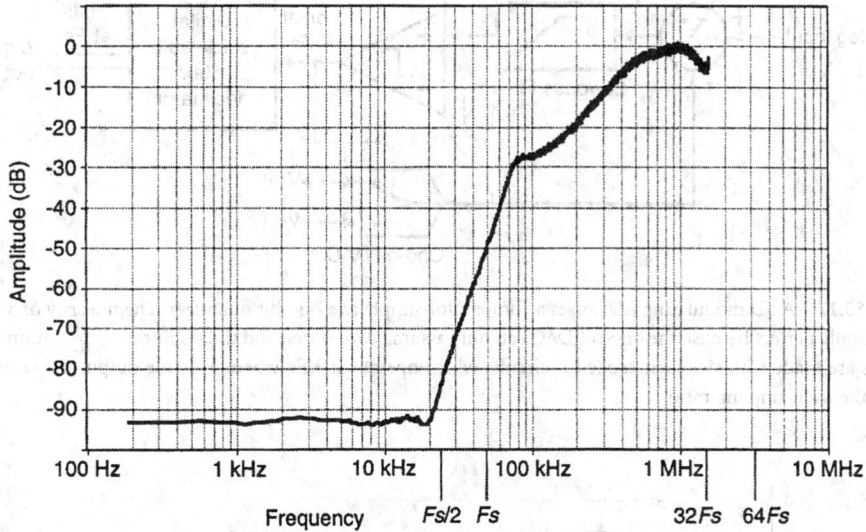

FIGURE 85.16 Spectrum of a 64-times oversampling SD ADC before the digital decimation filter. The modulator loop samples at 3.072 MS/s and the data comes out of the filter at 48 kS/s. The filter cuts off sharply at $Fs/2$, or 24 kHz, leaving only the small amount of noise left below 24 kHz.

Sigma–Delta ADCs

The *sigma–delta (SD)* ADC is quickly becoming one of the most popular types of ADC. DS ADCs typically have resolutions of 16 to 24 bits and sample rates of 100 kS/s down to 10 S/s. Because of their high resolution at 48 kS/s, they are the most common type of converters in modern digital audio equipment. DS ADCs defy intuition by quantizing initially with *very* low resolution (often one bit) at very high rates, typically 64× to 128× the eventual sample rate (called *oversampling*). The high-rate, low-resolution quantizer operates inside a feedback loop with an analog lowpass filter and a DAC to force the large amount of otherwise unavoidable quantization error (noise) to frequencies higher than the band of interest. The resulting spectral redistribution of the quantization noise is called *noise shaping*, illustrated in Figure 85.16. The low-resolution digital output of the ADC loop is fed into a digital filter that increases the resolution from the resolution of the ADC loop to the output resolution, reduces the data rate from the rate of the ADC loop to the output sample rate, and applies a low-pass digital filter, leaving only the signals in the frequency band of interest and a little inherent electronic noise.

Figure 85.17 shows how a one-bit sigma–delta ADC works. The comparator *is* the ADC, and its output is processed digitally, so that no further analog errors can accumulate. The comparator is in a feedback loop with a low-pass filter (typically third to fifth order) and a one-bit DAC. The one-bit DAC can take on only one of two values, +full-scale and –full-scale, so it is perfectly linear. The low-pass filter causes the loop gain to be high at low frequencies (the signal band of interest) and low at high frequencies. Since the error in a feedback loop is low when the gain is high and high when the gain is low, the errors dominate at high frequencies and are low in the band of interest. The result is a one-bit output whose duty cycle is proportional to the input signal. Together, the elements of the feedback loop are called a *sigma–delta modulator*.

Figure 85.18 illustrates the operation of a simple discrete-time (switched-capacitor) SD ADC. In this first-order example, the low-pass filter is just an integrator. The loop tries to force the input to the comparator back to the baseline, and the figure shows how the duty cycle of the resulting digital output reflects input signal. The digital data here have undesirable patterns which tend to repeat, called *limit cycles*. They can appear in the band of interest and interfere with the signal. Higher-order loop filters (third and above) make the bit activity so chaotic that it has no substantial limit cycles.

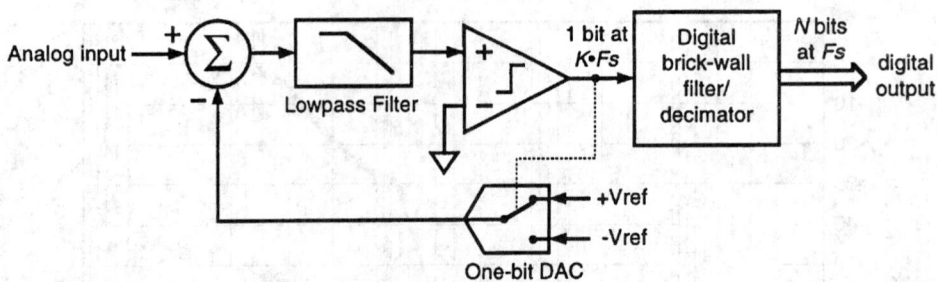

FIGURE 85.17   A SD modulating ADC uses a comparator simply as a one-bit quantizer. The linearity of a SD ADC is theoretically perfect because the one-bit DAC can only assume two values, and thus is linear by definition. Modern SD ADCs are made with switched-capacitor circuits which operate at KF_s, where F_s is the output data sample rate and K is the oversampling ratio.

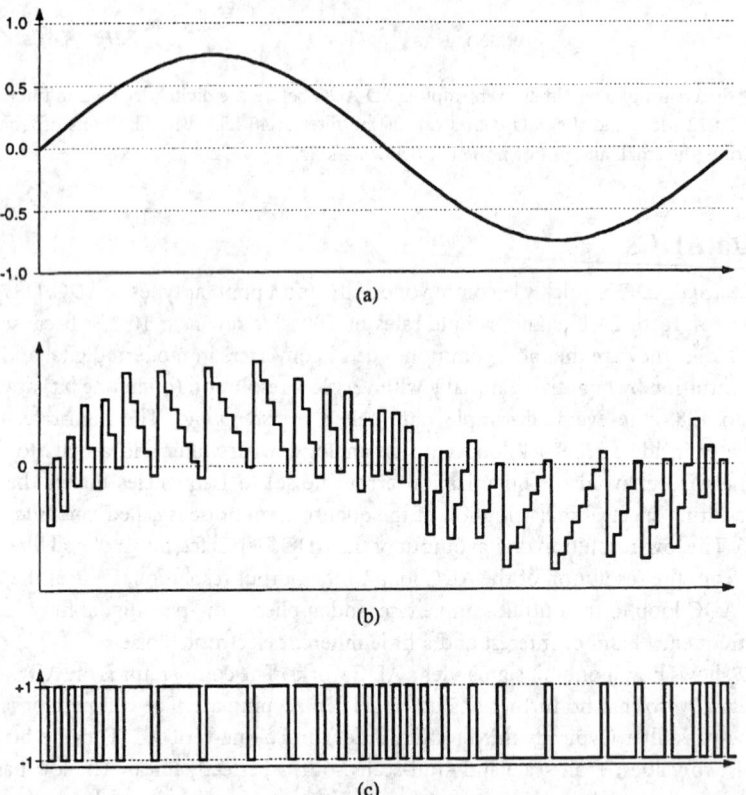

FIGURE 85.18   Behavior of a discrete-time (switched-capacitor) first-order SD modulator, where the low-pass filter is simply an integrator. In each graph, the x-axis represents time, and the y-axis represents signal level. (a) The input waveform. (b) Input to the comparator. (c) The one-bit digital comparator output. The duty cycle of this waveform corresponds to the input waveform. The digital filter and decimator recover the original waveform from this one bit.

The chief advantage of a SD converter is that it has a built-in antialias filter, and a darn good one at that. Most DS parts have a *finite-impulse response (FIR)* digital filter, which has an extremely flat frequency response in the passband and an extremely sharp cutoff, properties impossible to implement in analog filters. The ADC still needs an antialias filter to reject signals above one half the oversampling rate. But this filter is simple to build, since it has to be flat only up to one half the output sampling rate and has many octaves (all the way to near the oversampling rate) to fall off. The combination of the two filters provides watertight protection from aliases, often 96 dB of attenuation over the entire spectrum.

An important improvement of the one-bit SD is the multibit SD, wherein the comparator is replaced by a flash converter with as much as four bits of resolution. This improves the ENOB of the whole converter by several bits.

Voltage-to-Frequency Converters

Voltage-to-frequency converters (VFCs) are versatile, low-cost circuits that convert analog voltages to periodic waveforms whose frequency is proportional to the analog input voltage. A VFC is conceptually similar to an integrating converter (see above) except that the digital counter is missing and is replaced with a short-pulse generator that quickly discharges the capacitor. The voltage reference is not connected intermittently to the input; instead, it appears all the time at the minus input of the comparator instead of ground. The capacitor charges at a rate proportional to the input voltage until the voltage is equal to the voltage reference. Then the comparator trips the pulse generator, which quickly discharges the capacitor, and the cycle begins again. The periodic pulse at the comparator output can be used as the digital output.

The advantage of the VFC over conventional ADCs is that the one-bit output can be transmitted digitally, through isolation transformers, through fiber-optic cable, or through any other isolating, nonisolating, long-distance, or short-distance transmission medium. All that is needed at the receiving end to complete the analog-to-digital conversion is a digital counter, which does not need to be synchronized to the VFC itself. Sometimes, the digital conversion is not needed; a VFC can be used with an isolating transformer and a *frequency-to-voltage converter (FVC)* to create an isolation amplifier. For a good discussion of VFCs, see Reference 7.

85.6 Instrumentation and Components

Integrated Circuits

Table 85.1 lists several popular high-quality ADCs in integrated circuit form. The prices given are approximate for small quantities and for the lowest grade of part, as of mid-1996. By no means exhaustive, the list is sampling of a few of the most popular or best-performing chips of each type of ADC. Table 85.2 contains addresses, phone numbers, and internet sites for the manufacturers in Table 85.1.

Instrumentation

Plug-in data acquisition cards are becoming increasingly popular as personal computer prices come down and processor performance goes up. These cards typically contain one or more ADCs (with S/H), instrumentation amplifiers with gain and differential input, and multiplexers to switch to different inputs. Some have DACs on-board, and some have digital data and timing functions as well. Once considered low performance and hard to use, data acquisition cards have improved dramatically, equaling and in some cases exceeding capabilities of stand-alone instruments. Most come with drivers that interface to user-friendly software packages for creating easy-to-use yet custom-built computer instrumentation. Table 85.3 lists a few popular plug-in data acquisition boards and Table 85.4 lists how their manufacturers may be contacted.

TABLE 85.1   ADC Integrated Circuits

Part	Type	Sample Rate	Resolution, bits	Manufacturer	Approx. Price, $
ADC160	Integrating	1 S/s	24	Thaler	225.00
AD7714	Sigma–delta	2.62 S/s	24	Analog Devices	22.00
MAX132	Integrating	6 S/s	19	MAXIM	15.11
CS5508	Sigma–delta	20 S/s	20	Crystal	21.50
HI7190	Sigma–delta	10 S/s	24	Harris	17.85
AD1879	Sigma–delta	50 kS/s	18	Analog Devices	46.00
CS5390	Sigma–delta	50 kS/s	20	Crystal	75.30
ADS7809	SAR	100 kS/s	16	Burr-Brown	41.54
CS5101A	SAR	100 kS/s	16	Crystal	67.20
AD7893	SAR	117 kS/s	12	Analog Devices	14.00
AD976	SAR	200 kS/s	16	Analog Devices	36.50
AD7722	Sigma–delta	200 kS/s	16	Analog Devices	39.80
LTC1278	SAR	500 kS/s	12	Linear Technology	17.08
AD1385	Multistage	500 kS/s	16	Analog Devices	1053.00
ADS7819	SAR	800 kS/s	12	Burr-Brown	31.90
AD9220	Multistage	10 MS/s	12	Analog Devices	22.95
AD775	Multistage	20 MS/s	8	Analog Devices	14.00
AD9050	Multistage	40 MS/s	10	Analog Devices	39.00
AD9066	Flash	60 MS/s	6	Analog Devices	7.00
HI1276	Flash	500 MS/s	8	Harris	338.58

TABLE 85.2   Companies That Manufacture ADC Integrated Circuits

Analog Devices, Inc. One Technology Way P.O. Box 9106 Norwood, MA 02062-9106 (617) 329-4700 http://www.analog.com	Harris Corp. Semiconductor Products Division P.O. Box 883 Melbourne, FL 37902 (407) 729-4984 http://www.semi.harris.com	Maxim Integrated Products, Inc. 120 San Gabriel Drive Sunnyvale, CA 94086 (408) 737-7600 http://www.maxim-ic.com
Burr-Brown Corporation P.O. Box 11400 Tucson, AZ 85734-1400 (520) 746-1111 http://www.burr-brown.com	Linear Technology Corporation 1630 McCarthy Blvd. Milpitas, CA 95035-7417 (408) 432-1900 http://www.linear-tech.com	Thaler Corporation 2015 N. Forbes Boulevard Tucson, AZ 85745 (520) 882-4000 http://www.thaler.com
Crystal Semiconductor Corporation P.O. Box 17847 Austin, TX 78760 (512) 445-7222 http://www.cirrus.com/prodtech/crystal.html		

TABLE 85.3   Plug-In Data Acquisition Boards

Part	Type	Sample Rate	Resolution, bits	Manufacturer	Approx. Price, $
AT-A2150	Sigma–delta	51.2 kS/s	16	National Instruments	1495
AT-MIO-16XE-50	SAR	20 kS/s	16	National Instruments	995
AT-MIO-16E-10	SAR	100 kS/s	12	National Instruments	995
CIO-DAS1600/12	SAR	160 kS/s	12	ComputerBoards, Inc.	599
AT-MIO-16XE-10	SAR	100 kS/s	16	National Instruments	1995
CIO-DAS1600/16	SAR	100 kS/s	16	ComputerBoards, Inc.	699
DT-3001	SAR	330 kS/s	12	Data Translation, Inc.	995
DAS-1800AO	SAR	333 kS/s	12	Keithley Metrabyte	1299
AT-MIO-16E-1	SAR	1 MS/s	12	National Instruments	1795
FAST16-1	Multistage	1 MS/s	16	Analogic	3895

TABLE 85.4 Companies That Manufacture Plug-In Data Acquisition Boards

Analogic Corporation	Keithley Metrabyte
360 Audubon Road	440 Myles Standish Blvd.
Wakefield, MA 01880	Taunton MA 02780
(508) 977-3000	(508) 880-3000
	http://www.metrabyte.com
ComputerBoards, Inc.	
125 High Street	National Instruments Corporation
Mansfield, MA 02048	6504 Bridge Point Parkway
(508) 261-1123	Austin, TX 78730
	(512) 794-0100
Data Translation, Inc.	http://www.natinst.com
100 Locke Drive	
Marlboro, MA 01752-1192	
(508) 481-3700	
http://www.datx.com	

References

1. C. E. Shannon, Communication in the presence of noise, *Proc. IRE*, 37(1): 10–21, 1949.
2. H. Nyquist, Certain topics in telegraph transmission theory, *AIEE Transactions*, 617–644, April, 1928.
3. A. B. Williams and F. J. Taylor, *Electronic Filter Design Handbook: LC, Active, and Digital Filters*, 2nd ed., New York: McGraw-Hill, 1988.
4. S. P. Lipshitz, R. A. Wannamaker, and J. Vanderkooy, Quantization and dither: a theoretical survey, *J. Audio Eng. Soc.*, 40, 355–375, 1992.
5. H. W. Ott, *Noise Reduction Techniques in Electronic Systems*, 2nd ed., New York: John Wiley & Sons, 1988.
6. R. Morrison, *Grounding and Shielding Techniques in Instrumentation*, 3rd ed., New York: John Wiley & Sons, 1986.
7. J. Williams, Designs for High Performance Voltage-to-Frequency Converters, Application Note 14, Linear Technology Corporation, March 1986.

Further Information

M. Demler, *High-Speed Analog-to-Digital Conversion*, San Diego, CA: Academic Press, 1991.

B. M. Gordon, *The Analogic Data-Conversion Systems Digest*, Wakefield, MA: Analogic Corporation, 1981.

D. H. Sheingold, Ed., *Analog-Digital Conversion Handbook*, Englewood Cliffs, NJ: Prentice-Hall, 1986.

Crystal Semiconductor, *Delta Sigma A/D Conversion Technique Overview*, Application Note AN10, Austin, TX, 1989.

86

Computers

A. M. MacLeod
University of Abertay Dundee,
Dundee, United Kingdom

P.F. Martin
University of Abertay Dundee,
Dundee, United Kingdom

W.A. Gillespie
University of Abertay Dundee,
Dundee, United Kingdom

86.1 Introduction

Computers are an essential feature of most instrumentation systems because of their ability to supervise the collection of data and allow information to be processed, stored, and displayed. Many modern instruments are capable of providing a remote user with access to measurement information via standard computer networks.

86.2 Computer-Based Instrumentation Systems

The main features of a computer-based instrumentation system are shown in Figure 86.1. The actual implementation of such systems will depend on the application. Many commercially produced instruments such as spectrophotometers or digital storage oscilloscopes are themselves integrated computer-based measurement systems. These "standalone" instruments may be fitted with interfaces such as IEEE-488 or RS-232 to allow them to be controlled from personal computers (PCs) and also to support the transfer of data to the PC for further processing and display. Alternatively, the instrumentation system may be based around a PC/workstation or an industrial bus system such as a VME or Multibus, allowing the user the ability to customize the computer to suit the application by selecting an appropriate set of add-on cards. Recently, the introduction of laptop and notebook PCs fitted with PCMCIA interfaces with input/output capability has provided opportunities for the development of highly portable instrumentation systems.

The Single-Board Computer

The simplest form of a computer is based around the single-board computer (SBC) which contains a microprocessor, memory, and interfaces for communicating with other electronic systems. The earliest

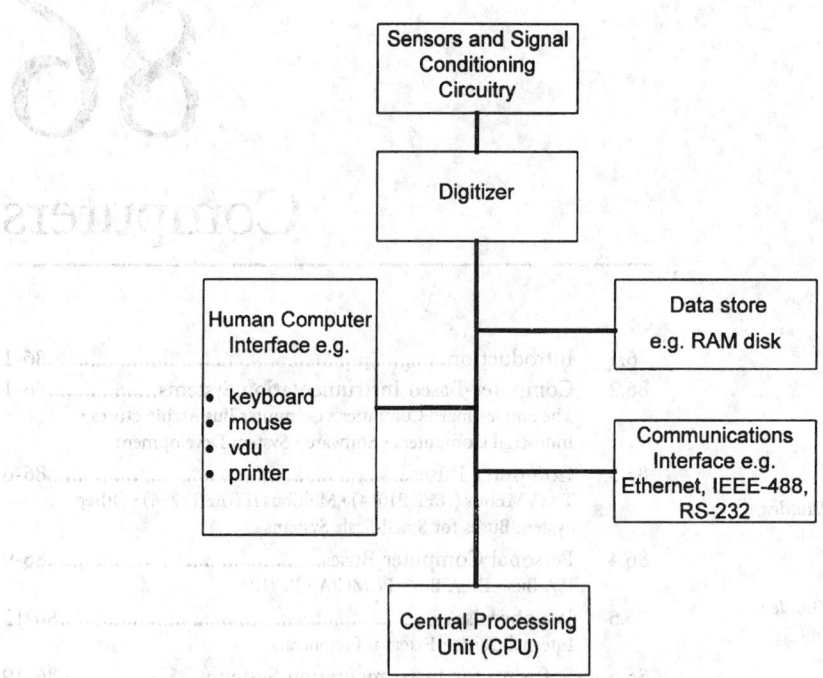

FIGURE 86.1   Elements of a computer-based instrumentation system.

form of personal computers simply comprised an SBC together with a keyboard, display, disk drives, and a power supply unit. Today, the SBC still offers a solution for the addition of limited intelligence to instrumentation systems as well as forming an element of most computer systems, e.g., as the mother-board of a PC.

An overview of a simple SBC is shown in Figure 86.2. The microprocessor, which contains the central processor unit, is responsible for executing the computer program and for controlling the flow of data around the SBC. The random access memory (RAM) acts as a temporary (i.e., volatile) storage area for both program code and data. On the motherboard of a PC the read only memory (ROM) is largely used for storing the low-level code used to access the input/output hardware, e.g., the BIOS (basic input output system). The operating system and applications software are loaded into RAM from the disk unit. In small, dedicated systems such as an oscilloscope the ROM may be used to store the program code to avoid the need for a disk drive. For small production runs or development work erasable programmable ROM (EPROM) is used as an alternative to ROM, allowing the code to be upgraded without the high cost of commissioning a new ROM chip from the manufacturers.

Data are transferred around the SBC on its data bus, which will be typically 8, 16, or 32 bits wide (corresponding to the number of bits that can be transmitted at the same time). SBCs with 8-bit data buses are less complex and consequently of lower cost than 32-bit systems and may well be the optimum solution for those process control applications which require minimal processing of 8-bit data, e.g., from temperature and position sensors. However the 32-bit bus width of most modern PCs and workstations is essential to ensure the fast operation of Windows-based applications software.

The address bus is used to identify the destination of the data on the data bus. Data transfer is usually between the microprocessor and the memory or interfaces. However, some SBCs support DMA (direct memory access) which allows data to be transferred directly between interfaces and memory without the need the information to pass through the processor. DMA is inherently faster than program-driven data transfer and is used for moving large blocks of data, e.g., loading programs from disk or the transfer of digitized video images.

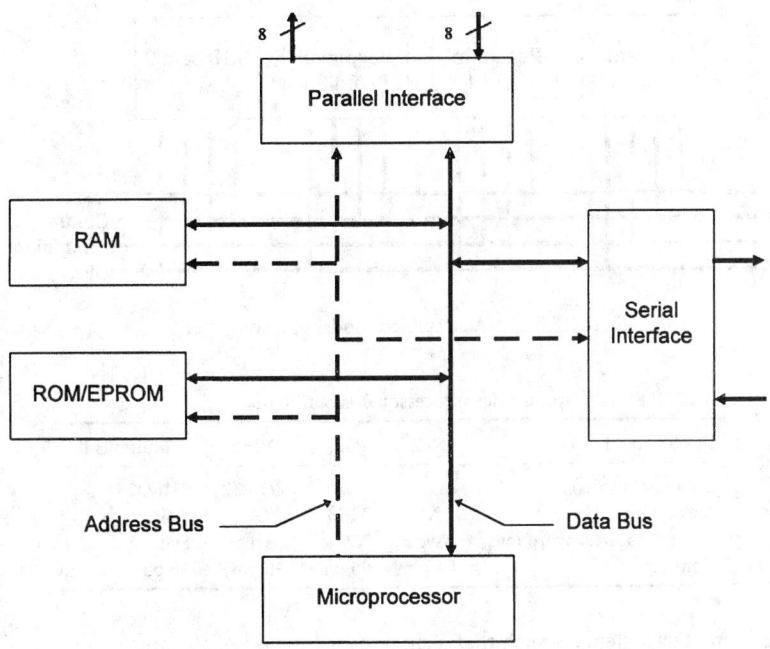

FIGURE 86.2 An overview of a single board computer.

SBCs are fitted with interfaces to allow them to communicate with other circuits. Interfaces carry out two main functions. First, they serve to ensure that the signals on the internal buses of the SBC are not affected by the connection of peripheral devices. Second, they ensure that signals can pass into and out of the computer and that appropriate voltage levels and current loading conditions are met. A well-designed interface should also provide adequate electrical protection for the SBC from transients introduced via the external connection. Parallel interfaces allow data to be transferred, usually 8 or 16 bits at a time, and contain registers that act as temporary data stores. Serial interfaces must carry out the conversion of data from the parallel format of the internal SBC data bus to and from the serial format used by the interface standard (e.g., RS-232 or RS-422). Some SBCs may contain additional interfaces to support communication with a VDU, local area network (LAN), or a disk drive. Interfaces can range in complexity from a single parallel interface chip to a LAN controller which may require a significant fraction of the SBC board area.

Computer Bus Architectures

All but the simplest computer systems contain several circuit board cards which plug into a printed circuit board backplane. The cards will include at least one SBC and a number of "add-on" cards providing functions such as interfaces to peripherals (e.g., a LAN, graphics display, disk unit) or additional memory. The actual structure of bus architectures is quite variable but the simple model shown in Figure 86.3 contains the essential features. A group of tracks will carry the data and address information with a second group of tracks being used to control the flow of data and to ensure its reliable transfer. Other tracks are reserved for the signals which provide arbitration between SBC cards to ensure that only one such card has control of the bus at any given moment. There will also be tracks which provide utility functions such as clock signals and the transmission of event or interrupt signals.

The main advantage of bus-based systems is that they help one build a complex computerized system using standard cards. By conforming to agreed standard buses (see Table 86.1 for typical examples), the system integrator can minimize problems of incompatibility and keep system costs to a minimum. The

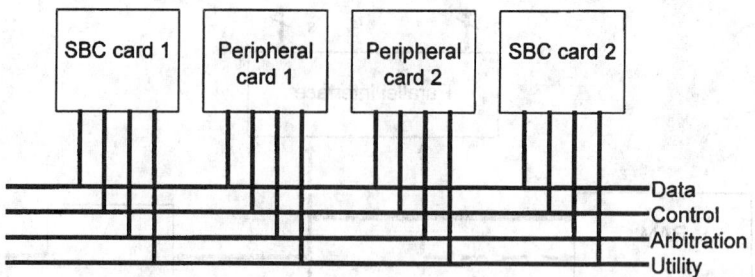

FIGURE 86.3　A simplified model of a computer bus.

TABLE 86.1　Typical Microprocessor Bus Standards

Bus Standard	STE	G96	VME	Multibus II
Data width (bits)	8	8/16	8/16/32	8/16/32
Max address (bytes)	1 M	32 M	4 G	4 G
Synchronous/asynchronous	Async	Async	Async	Sync
Connectors	64 pin	96 pin	96 pin	96 pin + 32 pin

complexity of many bus systems is such that a significant fraction of the area of each card is dedicated to providing the logic required to interface to the bus, adding appreciably to the cost. In addition, data transfer between cards, even using DMA, is relatively slow compared with transfers between chips on the same SBC. There is, therefore, a trade-off between integrating functions on a single SBC and a lack of flexibility in customising the system. An alternative to bus-based systems is to utilize processors such as transputers which are designed to be connected together directly into multiprocessor structures.

Most modern computerized systems adopt some form of distributed intelligence strategy in order to free the central processor unit to run the main application. Graphics controller cards generally employ a dedicated graphics processor able to handle the creation of the basic features such as line drawing or scaling the size of objects. LAN controller cards require intelligence to handle the network protocols and buffer the data flow to and from the network. Disk controller cards relieve the central processor of the need to handle the processes of reading and writing to the disk.

The use of a bus system with more than one SBC allows the designer to dedicate an SBC to the process of data acquisition and temporary data storage in order to ensure that the main SBC can concentrate on data processing and communication with the user. Such a strategy enables the data acquisition SBC to maintain a real-time response to making measurements while allowing the user access to processed data. In most systems one of the SBCs acts as a system controller, which has ultimate control over the use of the bus and will normally contain the bus arbitration hardware. Only one SBC at a time can obtain the authority of the controller to act as a bus master and initiate data communication. The other cards will be configured as slaves allowing them to respond to requests for information from the master. In some systems, e.g., Multibus II, bus arbitration is distributed throughout the intelligent cards.

Industrial Computers

Many instrumentation systems utilize standard PCs or workstations either fitted with add-on cards or linked to intelligent instruments via standard protocols such as IEEE-488 or RS-232. In many industrial environments there is a need to provide protection against hazards such as dust, damage by impact or vibration, and unauthorized access. The systems may have to fit more stringent electromagnetic compatibility (EMC) requirements. Industrial or ruggedized versions of desktop computers are available to meet this market. Cardframes designed to accommodate the standard industrial bus systems such as VME and Multibus can be readily customized to meet the demands of the industrial environment and ruggedized versions of PCs are also available. In designing industrial computers, care must be paid to the specification of adequate power supplies, both to provide sufficient current for the add-on cards and also to provide protection against fluctuations in the mains supply. It may also be necessary in safety-critical applications to use an uninterruptable power supply that will guarantee the operation of the system during short failures in the mains supply.

Software

All but the simplest computer systems require an operating system to support the operation of applications software. The operating system will allocate areas of memory for use by the applications programs and provide mechanisms to access system resources such as printers or displays. Some operating systems, such as MS DOS, are single-tasking systems, meaning that they will only support the operation of one program or task at a time. Instrumentation systems which simultaneously take measurements, process data, and allow the user to access information generally require a multitasking operating system, such as OS-9, UNIX, or Windows 95. In such instrumentation systems the applications software may well comprise a single program, but it may generate software processes or tasks each with their own local data and code. The multitasking operating system will allow the actual execution of these tasks to be scheduled and also provide mechanisms for communication of information between tasks and for the synchronization of tasks.

Instrumentation systems are real-time environments, i.e., their operation requires that tasks be carried out within specified time intervals; for example, the capture of data must occur at specified moments. Operating systems adopt a range of strategies for determining the scheduling of software tasks. Round-robin scheduling, for example, provides each task with execution time on a rota basis, with the amount of time being determined by the priority of the task. Operating systems which support preemptive scheduling allow high-priority tasks to become active if a specified event such as trigger signal occurs. The speed at which an operating system can switch from one task to another (i.e., context switch) is a important metric for real-time systems.

The extent to which an operating system can respond to a large number of events is extremely limited, as low-priority tasks will have a poor response time. If an instrumentation system needs to make a large number of measurements and also support other activities, such as information display, the solution is generally to use a distributed system with dedicated SBCs operating as microcontrollers to capture the data. These microcontrollers will not usually require an operating system and will each operate a single program which will be either downloaded from the master SBC or located in ROM.

Commercial packages (for examples, see Table 86.2) are readily available to support PC/workstation based instrumentation systems. A typical system will provide a Windows-type environment to control measurements and store, process, and display data and increasingly make use of the virtual-instrument concept which allows the user to configure instrumentation systems from an on-screen menu. Most packages will allow stand-alone instruments to be controlled using interface cards supporting standards such as IEEE-488 or RS-232. Some packages also support the use of add-on cards for analog-to-digital conversion or a range of control functions such as channel switching and waveform generation; however, their main restriction is the limited range of hardware configurations supported by the software supplier. Each stand-alone instrument and add-on card requires a piece of code called a device driver so that the operating system can access the hardware resources of the card and instrument. Therefore, the development of device drivers requires an intimate knowledge of both the hardware and the operating system.

System Development

The software component of any computerized instrumentation system can form a significant percentage of its total cost. This is especially true of on-off systems which require software to be developed for a specific application, where the cost of the implementation of the software and its maintenance can considerably exceed the hardware costs. In such circumstances the ability to reuse existing code and the

TABLE 86.2 Typical Data Acquisition and Display Software Packages for PC-Based Instrumentation Systems

Product	Platforms	Manufacturer
DTV	PC (Windows 3.1 or 95)	Data Translation Inc. 101 Locke Drive Marlboro, MA 01752-1192 Tel: (U.S. and Canada) (800) 525-8528 (worldwide) +1 508-481-3700
Labtech Notebook/Notebook Pro	PC (DOS, Windows 3.1 or 95)	Laboratory Technologies Corporation 400 Research Drive Wilmington, MA 01887 Tel. (U.S. and Canada) 800-8799-5228 (worldwide) +1 508-658-9972
LabVIEW	PC, Mac, Sun, Power PC	National Instruments 6504 Bridge Point Parkway Austin, TX 78730-5039 Tel: +1 512-794-0100
LabWindows	PC (DOS)	National Instruments 6504 Bridge Point Parkway Austin, TX 78730-5039 Tel: +1 512-794-0100
LabWindows/CVI	PC (Windows 3.1 or 95) Sun	National Instruments 6504 Bridge Point Parkway Austin, TX 78730-5039 Tel: +1 512-794-0100
ORIGIN	PC (Windows 3.1 or 95)	MicroCal Software Inc. One Roundhouse Plaza Northampton, MA 01060 Tel +1 413-586-2013

access to powerful development systems and debugging tools are of crucial importance. Some debug software only provides support for stepping through the operation of code written at the assembly code level. When developing device drivers, for example, access to a source-level debugger which can link the execution of code written in a high-level language to the contents of processor registers is useful.

Many SBCs that are intended for operation as stand-alone microcontrollers, instead of in bus-based systems, are often supported by a high-level language such as C or FORTH and a development environment. A common practice is to utilize a PC to develop the code which can be downloaded via a serial link to the microcontroller. Some debugging support for the execution of the code on the microcontroller is also provided. The development of powerful compact Intel 486 or Pentium microcontroller cards running operating systems such as MS DOS can greatly reduce software development time because of the ready access to PC-based software. The implementation of systems based on dedicated microcontrollers may require the use of a logic analyzer to view data on system buses.

86.3 Computer Buses

The VMEbus (IEEE P1014)

The VMEbus [1–3], utilizes a backplane fitted with two 96-pin connectors (called P1 and P2). The P1 connector provides access to all the bus signals with the exception of bits 16 to 31 of the data bus and bits 24 to 31 of the address bus, which are provided on the P2 connector. The location of cards in a VME system is important, as slot 1 must contain the system controller and the priority of the SBCs is determined by their proximity to the system controller SBC. As only 32 of the pins on the P2 connector are defined, the remaining 64 tracks on the P2 backplane may be specified by the user. VME cards may be single height (fitted with a P1 connector only) or double height (fitted with both P1 and P2 connectors). Single-height

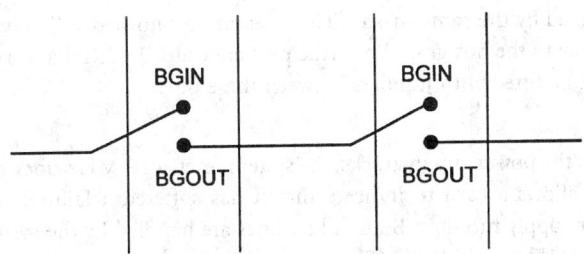

FIGURE 86.4 VMEbus arbitration daisy chain.

boards are 100 mm high and 160 mm deep and fit 3U height cardframes. Double-height boards are 233.35 mm high and 160 mm deep and fit 6U height cardframes.

VMEbus Signals

The VMEbus can be described as having four distinct groups of signals.

Data Transfer Bus (DTB).
The DTB provides 32-bit data transfer with 32-bit address information using a nonmultiplexed asynchronous approach. Transfer is supported by a simple handshake mechanism with the bus master initiating the transfer providing an address strobe (AS) signal to tell the destination card to read the address information. The master also controls two data strobe (DS1 and DS2) lines which in a write operation indicate the presence of valid data on the data bus and in a read operation that the master is ready to receive data from the slave. The slave card uses the data acknowledge signal (DTACK) in a read operation to indicate that it has placed valid data on the bus and in a write operation to confirm that it has read the data. There is clearly a danger that the bus will hang up if the slave does not respond to a request for a data transfer, so in most VME systems a watchdog timer is provided to produce a signal on the bus error (BERR) line if a DTACK signal is not detected within the time-out period.

Revision C of the VME bus specification will support 8-,16-, and 32-bit data transfers, whereas revision D supports 64-bit transfers by making use of the address bus to transfer the additional 32 bits. Data may also be transferred in blocks of up to 256 bytes without the need to retransmit the address information. Most SBCs have both master and slave interfaces allowing their memory to be accessed from other SBCs in the system. The VME bus also supports three types of addressing mode, short (16 bit), standard (24 bit), and extended (32 bit), allowing the memory decoding circuitry of cards to be minimized for small systems or for input/output cards. Six address modifier lines (AM0 to AM5) are used to indicate the addressing mode and also to give information about the mode of transfer (e.g., block transfer and whether the transfer is in privileged, supervisor, or nonprivileged mode).

Arbitration Bus.
The arbitration bus provides a mechanism for deciding which master is allowed to gain control of the bus. The arbitration bus provides four bus request lines (BR0 to BR3) and four bus grant lines. The SBC in slot 1 of the VME bus acts as a bus controller which provides arbitration using a number of scheduling algorithms. In a simple priority-based algorithm each of the bus request lines is assigned a priority, whereas in a round-robin system access to the bus cycles around the bus request lines. Each of the four bus grant lines is in fact allocated two pins on the P1 connector (see Figure 86.4). The bus grant signal therefore passes through each card (called daisy chaining); thus, in Figure 86.4 the signal enters via BG1IN and exits via BG1OUT. This scheme allows a card to intercept the bus grant signal, giving cards nearer the system controller a higher priority. A consequence of daisy chaining is that any unused card slots must be fitted with jumpers between the bus grant pins.

Priority Interrupt Bus.
The priority interrupt bus provides seven interrupt request lines (IRQ1 to IRQ7) with IRQ7 having the highest priority. The interrupt requests may be handled by several SBCs, provided that all interrupts of

the same level are handled by the same master. The interrupt acknowledge line is daisy chained though the cards in a similar way to the bus grant lines (the pins are called IACKIN and IACKOUT). Again, any unused card slots must be fitted with jumpers between these pins.

Utility Bus.
The utility bus provides the power supply tracks, the system reset, a 16-MHz clock and two system failure lines, SYSFAIL, which allows a card to indicate that it has suffered a failure, and ACFAIL, which is generated by the power supply monitor. Both failure lines are handled by the system controller SBC.

Associated with the VMEbus is VXI (*VMEbus extensions for instrumentation*). This blends the IEEE 1014 VMEbus standard with the IEEE-488 instrumentation bus using the uncommitted pins on the P2 connector to form the IEEE P1155 standard.

Multibus II (IEEE 1296)

Multibus II [3] is a synchronous bus system which is implemented on a single 96-track backplane. Cards conform to the Eurocard format of 220 × 233.65 mm; however, the bus requires only a single 96-pin DIN 41612-type connector. A second connector may be used to support a local bus or merely provide mechanical stability for a double-height card.

Unlike VME, the Multibus II standard has the following features:

1. A synchronous bus; i.e., data transfer is linked to the bus clock rather than the strobe/acknowledge control lines of the asynchronous VME system;
2. A 32-bit multiplexed address and data bus;
3. A message-passing strategy as an alternative to interrupt request lines;
4. Distributed bus arbitration rather than a central system controller (each card contains a message-passing controller, MPC, chip which participates in implementing the message passing algorithm);
5. Only intelligent cards (i.e., fitted with microprocessors) that can interface to the main bus; simple input/output (I/O) cards should make use of a local bus (e.g., the iLBX bus) using the second DIN connector.

Multibus Signals

Multibus II may be described as having five distinct groups of signals.

Central Control.
The central services module (CSM) in slot 0 is responsible for the generation of this group of signals. The CSM may be implemented on a dedicated card or incorporated onto a SBC. The CSM produces the 10-MHz bus clock as well as a range of reset signals to support both a cold and warm start as well as recovery from power supply failure.

Address/Data.
Multibus II supports a 32-bit multiplexed address/data bus (AD0 to AD31) with four parity lines (PAR0 to PAR3) providing parity information for each data byte. As in all synchronous systems, data are only sampled on an edge of the system clock, a feature that enhances noise immunity.

System Control.
There are ten system control lines (SC0 to SC9) which provide basic handshake information to assist data transfer, e.g., requester is ready or replier is not ready. In addition, these lines are used to convey information such as data width, data error indication, and whether the bus is in the request or reply phase.

Exception Signals.
Two exception signals, bus error (BUSERR) and time-out (TIMOUT), are provided. Any card detecting a data integrity problem must report this to all other cards using BUSERR. The CSM generates a TIMOUT when it detects a data communications hang up on the bus.

Arbitration Group.

The arbitration signals, which determine which card gains control of the bus, consist of a single bus request line (BREQ) and six Bus Arbitration lines (ARB0 to ARB5). To request exclusive use of the bus, a card asserts the BREQ line and provides it arbitration ID on ARB0 to ARB4. It also uses ARB5 to indicate whether the request has high priority or whether "fairness mode" is acceptable. In this latter mode the card will not make another request until after all other requesters have used the bus. Each card has the same bus arbitration logic within its MPC chip. If several cards make high-priority requests, the order of access is determined by the numerical value of the arbitration ID.

Message Passing on Multibus II

Multibus II uses message passing to implement block data transfers and interrupt requests. Each MPC chip contains a first-in first-out (FIFO) buffer, which ensures that the full bandwidth of the bus can be utilized by storing data immediately before or after transfer. In "solicited" transfers, the MPCs cooperate by warning each other that a message is to be sent. These messages may be sent as 32-byte packets in a manner that is analogous to the block transfer mechanism in VME. "Unsolicited" packets not expected by the receiving MPC are used to set up a block transfer or to act as the equivalent of interrupt request signals.

System Configuration

Multibus II employs a software configuration approach in which information such as the base memory address and arbitration ID are sent down the bus rather than by the use of jumpers or dip switches. Some information such as card type and serial number are coded on the cards themselves.

Other System Buses for Small-Scale Systems

The G64/G96 and STE standards [3] are examples of buses well suited to small industrial real-time instrumentation and control systems because of their relatively simple and hence lower-cost bus interfaces, compared with VME and Multibus. Both buses support DMA, have multiprocessor bus arbitration, and use single-height Eurocard (100 × 160 mm) cards. Prototyping cards fitted with bus interfaces are readily available and may be used to develop custom designed I/O cards. Power is supplied on the buses at +5 and ±12 V. In addition, the real-time multitasking operating system OS-9 has been ported onto SBCs which operate with these buses. Development systems that support the use of MS-DOS are also available, thus providing access to a wide range of PC-based software especially for graphics applications.

The G64 bus, which was defined by the Swiss company Gespac in 1979, specifies 64 bus lines which are mapped to rows of a DIN41612 connector. The bus has a 16-bit nonmultiplexed data bus and a 16-bit address bus; 32-bit transfers may be achieved by multiplexing the upper 16 bits of the data bus with the address bus. The G96 standard adds a further 32 bus lines by making use of the third row of a DIN41612 connector to extend the address bus width to 24 bits and provide additional interrupt request lines. The STE bus provides an unmultiplexed 8-bit data and 20-bit address bus using 64-pin DIN41612 connectors.

86.4 Personal Computer Buses

There are three main bus standards for personal computers — industry standard architecture (ISA), extended ISA (EISA), and the microchannel architecture (MCA). In addition, the Personal Computer Memory Card International Association (PCMCIA) architecture has been developed primarily for use in laptop and notebook computers.

ISA and EISA are pin compatible and are both synchronous buses with a clock rate of 8 MHz regardless of the clock rate of the main processor, whereas the MCA bus is an asynchronous bus. Slow slave add-on cards can utilize the ISA and EISA buses by using an appropriate number of wait states. The MCA architecture will not be covered in this chapter. Further details of these bus architectures are given in References 4 through 6.

ISA Bus

The original IBM PC and its clones used the standard PC bus which supported 8 data bus and 20 address bus lines and employed a 62-pin printed circuit card edge connector. When IBM introduced the PC-AT, a second 36-pin connector was added to the motherboard backplane slots to provide a 16-bit data bus and increase the number of address lines to 24. This new bus subsequently became known as the ISA bus and is capable of supporting both cards designed for the original PC bus, and cards with two connectors providing full 16-bit data access. The bus master line allows a card to take over control of the bus; however, this is generally only suited to long-term takeovers. The ISA bus supports 8-bit and 16-bit DMA transfers allowing efficient transfer of blocks of data, e.g., while loading software from disk into RAM. The ISA bus supports the I/O addressing mode of the Intel 80 × 86 range of processors with an I/O address space of 768 locations in addition to the 256 locations reserved for the motherboard.

EISA Bus

The EISA bus provides full 32-bit data and 32-bit address bus lines. ISA cards can fit into EISA bus connectors; however, cards designed for the EISA standard have bilevel edge connectors, providing double the number of contacts as an ISA card. The presence of a notch in the EISA cards allows them to be inserted farther into the motherboard socket than the ISA cards and thus mate with the additional contacts. The EISA standard increases the maximum ISA data width in DMA transfer to 32 bits and also provides a much-enhanced bus master. The following features of EISA are worthy of note.

Bus Arbitration

All EISA systems have a central arbitration control (CAC) device on the motherboard. The CAC uses a multilevel rotating priority arbitration scheme. The top-priority level rotates around three customers, the DMA controller, the dynamic memory refresh controller, and, alternately, either the main CPU or one of the bus master cards. A device that does not make a request is skipped over in the rotation process. The bus masters take it in turns to gain access to the top-priority level. Whereas ISA supported a single bus request line, the EISA standard provides a dedicated pair of request lines (MREQ0 to 14) and acknowledge lines (MAK0 to 14) for each bus master. (*Note:* Although this allows 15 bus masters to be used, in many systems the chip set which implements the CAC supports bus masters in a limited number of the EISA sockets.) The CAC supports preemption, i.e., it allows a device making a request to capture control from another device if it is next in turn. A bus master card must release the bus within 64 bus clock cycles, whereas a DMA controller has 32 clock cycles to surrender the bus.

Input/Output

The EISA bus provides 768 additional I/O locations for each slot in addition to the 768 ISA I/O locations that may be accessed from any slot. EISA cards contain nonvolatile memory to store configuration information (see below), and a minimum of 340 bytes of the I/O address space of each slot is reserved for this purpose.

System Configuration

The EISA system provides support for automatic system configuration to replace the use of jumpers and dip switches to specify parameters such as the base memory address, interrupt request line number, or DMA channel number used by each add-on card. Information on the product type and manufacturer are stored on each EISA add-on card is read by the CPU during system start-up, making it possible to identify the slots that are fitted with full EISA cards. Manufacturers of both ISA and EISA cards should supply a configuration file containing information on the programmable options that are available. When a system configuration program is run, an optimum configuration of the boards will be determined, and the configuration information written into the I/O space of each EISA card. For ISA cards, the user can be informed of the required jumper settings.

PCMCIA

The PCMCIA architecture was developed by the Personal Computer Memory Card International Association and the Japan Electronics Industry Development Association for removable add-on cards for laptop and notebook computers. Each PCMCIA socket has its own host bus adapter (HBA), which acts as an interface to the main computer bus. Cards may be plugged into PCMCIA sockets either before or after the computer has been powered up. There are three types of PC card all with the same planar dimensions (54.00 × 85.6 mm) but with differing thicknesses, namely, 3.3 mm for type I, 5.0 mm for type II, and 10.5 mm for type III. Cards and sockets are keyed to prevent them from being inserted the wrong way round. The PCMCIA standard supports cards that operate at several possible voltage levels, i.e., 5.0 V cards, 3.3 V cards, or dual-voltage 5.0 V/3.3 V cards.

Configuration

PC cards contain configuration information called the card information structure (CIS) which is stored in nonvolatile memory. Cards may be configured either on system power-up or on insertion of the card into the socket (i.e., plug and play). Configuration is carried out using a form of device driver called an enabler. The enabler may make use of two additional software services called card services and socket services. Socket services, which may be contained in ROM on the PC or loaded from disk during power-up, provide function calls to allow the HBA to be configured to cooperate with the PC card. Card services, which may be an extension to the operating system or an installable device driver, act as a server for the enabler, which performs as a client. Card services provide a range of functions such as accessing the CIS of the card, requesting system resources required by the card, and telling the enabler that a card has been inserted or removed from the socket. Enablers may be classified as dedicated to one particular card or generic, i.e., designed for a range of cards. Note that early PCMCIA cards were not designed for use with card services.

The PCMCIA Socket Interface

The PCMCIA socket comprises a 68-pin connector with 26 address lines providing access to 64 MB of memory space and a 16-bit data bus. Two V_{cc} pins and four ground pins are supplied. The maximum current that can be supplied to the card is 1.0 A with a maximum of 0.5 A from each of the two power supply pins. Release 2.x sockets apply 5.0 V to the V_{cc} pins on power-up and reduce this to 3.3 V if the card has dual-voltage capability. Cards that operate at 3.3 V only are keyed so they cannot fit into this type of socket and only into low-voltage sockets. The supply voltage provided by low-voltage sockets depends on the logic state of its voltage sense inputs. A PCMCIA socket can be configured either as a *memory only* socket or as a *memory or I/O socket*. Initially, the socket acts as a memory only socket but is converted by the enabler to a memory or I/O socket if the card is required to support I/O functions. In this mode the card can generate an interrupt request via a single IRQ pin and support both 8-bit and 16-bit I/O data transfers. DMA may be supported but not by Release 2.x systems.

PC/104

The enormous popularity of PC architecture resulted in its use in embedded systems. A need then arose for a more compact implementation of the ISA bus that could accommodate the reduced space and power requirements of embedded applications. The PC/104 specification (1992) was adopted as the base for an IEEE draft standard called the P996.1 Standard for Compact Embedded PC Modules. The key features of the PC/104 are

Size reduced to 90 × 96 mm (3.550 × 3.775 in.);
Self-stacking bus allowing modules to be "piggy-backed" and eliminating backplanes or cardframes;
Rugged 64-contact and 40-contact male and female headers replacing the PC edge connectors (64 + 40 = 104, hence PC/**104**);
Lower power consumption (<2 W per module).

TABLE 86.3   Manufacturers/Suppliers of Bus-Based Systems and Industrial PCs

System	Manufacturer/Supplier	System	Manufacturer/Supplier
VME	Wordsworth Technology Ltd 6 Enterprise Way Edenbridge, Kent TN8 6HF U.K. Tel: +44 (0) 1732 866988	STE	Arcom Control Systems Units 8-10 Clifton Road Cambridge CB1 4WH, U.K. Tel +44 (0) 1223 411200
	PEP Modular Computers, Inc. 750 Holiday Drive, Building 9 Pittsburg, PA 15220 Tel: (412) 921-3322	G64	Gespac SA 18 Chemin des Aulx 1228 Geneva, Switzerland Tel: +41 (22) 794 34 00
	BVM Ltd, Hobb Lane Hedge End Southampton, SO30 0GH, U.K. Tel: +44 (0) 1489 780144		Altek Microcomponents Ltd. Lifetrend House Heyes Lane Alderley Edge, Cheshire SK9 7LW Tel: +44 (0) 1625 584804
	Motorola, Inc. Computer Group 2900 S Diablo Way Tempe, AZ 85282 Tel: (800) 759-1107	Industrial PC	Blue Chip Technology Ltd. Chowley Oak, Tattenhall Chester, Cheshire CH3 9EX, U.K. Tel: +44 (0) 1829 772 000
VXI	National Instruments 6504 Bridge Point Parkway Austin, TX 78730-5039 Tel: (512) 794-0100		Capax Industrial PC Systems Ltd. Airport House, Purley Way Croydon, Surrey, CR0 0XZ, U.K. Tel +44 (0) 181 667 9000
Multibus	Tadpole Technology, Inc. 2001 Gateway Place, Suite 550 West, San Jose, CA 95110 Tel: (408)441-7920	PC/104	ComputerBoards, Inc. 125 High Street Mansfield, MA 02048 Tel: (508) 261-1123
	Intel Corp. 3065 Bowers Avenue Santa Clara, CA 95051		Diamond Point International (Europe) Ltd. Unit 9, North Point Business Estate, Enterprise Close Rochester, Kent ME2 4LY, U.K. Tel +44 (0)1634 718100
	Syntel Microsystems Queens Mill Road Huddersfield, HD1 3PG, U.K. Tel: +44 (0) 1484 535101		

PC/104 CPU modules range from a basic 9.6-MHz, 8088 compatible XT with one serial port and a keyboard connector to a 100-MHz, 80486DX4 with four serial ports, parallel port, IDE disk controller, display controller, Ethernet adapter, keyboard port, and up to 64 MB of on-board RAM. Pentium-based systems are also available. Systems can be customized from a wide range of modules, including data acquisition boards, solid-state disk modules, and LAN support, all in the same 3.6 by 3.8 in. stackable format.

The wide availability of software development tools for the PC, the large number of software developers familiar with the PC environment, and the ease of transferring software developed on a conventional PC to the PC/104 make this an increasingly popular format.

Table 86.3 lists some manufacturers and suppliers of bus-based systems and industrial computers.

86.5 Peripherals

Computer peripherals fall conveniently into two categories. The first category may be considered to be internal to the computer system and comprises cards plugged directly into a computer bus slot. The

TABLE 86.4 Typical Internal Peripherals

Card Type	Facilities Offered
Display adapter	Provides video and graphics facilities
Serial communications adapter	Serial communication to a similar device using the RS232, RS422, or RS485 standard
IEEE-488 Adapter	Communication with intelligent instruments using the IEEE-488 bus (GPIB)
Digital input/output (I/O)	Individual I/O lines, normally grouped as an 8-bit byte, which may be used to provide logic high and low signals (output) or sense logic high or low signals (input); inputs and outputs may be optically isolated; outputs may drive relays.
Counter/timer	Hardware counter timer allowing external pulses to be counted, digital waveforms generated or pulse widths measured; counter/timers are often available as an additional facility on digital I/O cards
Analog input	Converts an analog input voltage to an integer value which may be read by the computer; resolution typically 8, 10, 12, or 16 bits; input voltage range may be fixed or may be user selectable; conversion times vary from several seconds to tens of nanoseconds
Analog output	Produces an analog output voltage proportional to a digital input; resolution typically be 8, 10, 12, or 16 bits

second category comprises instruments external to the computer but controlled by it. These external instruments are usually themselves "intelligent," being controlled by their own CPU, and are linked to the main computer by a serial (RS-232) line or the IEEE-488 bus (GPIB). Such instruments may normally be operated in a stand-alone mode in response to their front panel controls without the requirement of an external computer.

Internal Cards

Internal cards are available to perform a wide range of functions. Table 86.4 provides a brief list of representative types. Note that both the serial interface and the IEEE-488 adapter required for the control of external "intelligent" peripherals will be fitted as internal cards. Almost all internal peripherals need to be configured before use to set up such parameters as the base address, the interrupts, and/or DMA channels used. This may involve setting jumpers or switches on the card or may be accomplished under software control using a configuration file supplied by the manufacturer. To operate the cards, bytes are written or read from appropriate addresses on the cards by the controlling SBC. The mechanism for doing this is discussed further in the section concerned with software for data acquisition. Further detail on some of the card types is given below.

Display Adapter

While graphics support is normally available as standard on a PC-based system, this is not the case with other bus-based systems such as VME or Multibus. These systems are provided with a serial port that may be connected to a terminal to provide a text-only dialogue with the operating system running on the SBC. In such circumstances the choice of display adapter will determined by the user requirements, taking into account the support for the device provided by any software packages that are to be used. If the user intends to write custom graphics software, it is essential to ensure that a graphics library is available from the vendor providing as a minimum line drawing, arc drawing, and block color fill facilities.

IEEE-488 Adapter

This device provides support for communications across the IEEE-488 bus or GPIB (general purpose-interface bus). The bus itself comprises eight data lines, five interface management lines and three handshake lines. Transfers are parallel, synchronous, and at rates up to 1.5 MB/s. The IEEE-488 bus and its applications are discussed further in the section on external instruments.

Serial Communications Adapter

This device provides support for serial communications using the RS-232, RS-422, or RS-485 standards. It is used to provide a text-based terminal for systems based on Multibus or VME and to communicate with "intelligent" instruments such as position controllers, multimeters, or storage oscilloscopes fitted with similar interfaces.

Serial adapters convert parallel data to and from a bit stream in which each data byte (5, 7, or 8 bits) is framed by a start bit, an optional parity bit and one or more stop bits. The bit stream may be sent via a circuit consisting of only two wires at rates of up to 115,200 bits per second (commonly referred to as 115200 baud). Common bit rates are 300, 600, 1200, 2400, 4800, 9600, 19200, 28800, 38400, 57600, 115200 baud. Both the transmitting and receiving adapters must be configured, normally by software, for the same baud rate, number of data bits, stop bits, and parity. Some form of flow control to prevent a receiver from being overloaded with incoming data is essential. This is accomplished either by separate handshake lines (e.g., those denoted by RTS and CTS in the standard) or by software where the receiver sends a special control byte (XOFF) back to the transmitter telling it to stop sending until it receives a second control byte (XON) to reenable it. For historical reasons the RS-232 standard does not define a bidirectional handshake procedure, and manufacturers have been forced to implement their own schemes which are not always compatible with each other.

Serial communication between devices may be

Full duplex, where either device may transmit or receive data at any time;
Half duplex, where both devices are capable of transmission or reception but only one may transmit at any instant;
Simplex, where one device is a transmitter, one is a receiver, and data can only flow in a single direction.

RS-232, developed by the Electronics Industries Association (EIA), is the oldest standard, originally developed in the early 1960s, to allow mainframe computers to communicate with terminals via modems and telephone lines. This is the origin of the names of some of the connections (e.g., RI ring indicator, DCD data carrier detect), which have no relevance in the applications considered here. A related problem is that the standard expects that the devices being connected are data terminal equipment (DTE), at one end of the link, and data communication equipment (DCE) at the other. Computers and terminals are DTE, while modems are DCE. The most commonly used revision of the standard, RS-232-C (revisions D and E also exist), was made in 1969 and is still widely used. Serial communication is made using voltage levels in the region of ±12 V, over distances up to 15 m (50 ft) at speeds up to 20000 baud.

RS-422, also developed by the EIA, is an enhancement of the RS-232 standard. Differential transmitters and receivers are employed which allow one transmitter to drive up to ten receivers, using a twisted-pair connection for each circuit, at bit rates up to 10 MBaud at distances up to 12 m (40 ft) or 100 kBaud at distances up to 1200 m (4000 ft). The RTS and CTS lines (defined in the standard) are used for flow control, while the RXD and TXD lines are used to transmit and receive data. Thus, a two-twisted-pair cable is required for duplex connection without hardware handshaking. A four-twisted-pair cable is required if hardware handshaking is used.

RS-485 is based on RS-422 and allows up to 32 driver/receiver pairs to be connected to a common data bus (two twisted pairs). Clearly, only one device can be allowed to transmit at any one time. The RTS circuit is used to disable the other transmitters connected to the bus if a device is required to transmit data. Handshaking is performed using software.

The serial interfaces on instruments are usually configured as DTE devices. We are faced with the problem of connecting one DTE device (the computer serial interface) to another (the instrument), which is not what the RS-232 standard was designed for. Furthermore, since the standard does not define a bidirectional handshake to control data flow, several incompatible handshaking schemes exist. A comprehensive survey of these is presented in Reference 7. A common solution to the DTE to DTE connection problem is the so-called *null modem*, which nothing more than a specially wired cable. Figure 86.5 shows two such connection schemes. One requires the software handshaking procedure and the other implements a bidirectional hardware handshake. The reader should refer to Reference 7 for details of other schemes and for the definitions of the mnemonics used to label the connections.

(a)   (b)

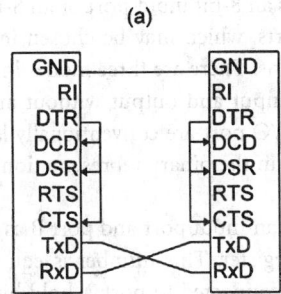
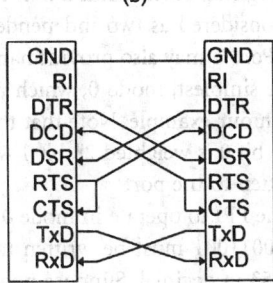

FIGURE 86.5   Two null modems for connecting DTE to DTE. In (a) all handshaking must be in software. The DTR line "fools" the serial interface that it is connected to the handshake lines of another device. Scheme (b) implements a hardware handshake. The DTR–DSR connection shows each device that the other is present. The RTS lines, connected to the DCD of the other device, which it can monitor, are used to control the flow of data in either direction.

Some common problems encountered in practice are

1. The received data are garbage. This is almost always due to the baud rate, parity, and number of stop bits not being the same at both ends of the link.
2. Data initially correct, but parts in the middle are missing. This is probably a handshake problem. The transmitting device is sending data faster than the receiver can process it.
3. No communications at all. Probably a handshake problem where the transmitter does not sense that the receiver is ready.

Digital Input and Output

These cards provide I/O lines, normally in groups of eight, which may be used to sense or generated digital signals for devices outside the computer. A group of eight input lines is referred to as an (8-bit) input port and a group of eight output lines as an (8-bit) output port. Input and output levels vary from card to card and it is best to consult the appropriate data sheet. Typically, voltages between 2.5 and 5.0 V are considered as high logic levels, whereas voltages between 0.0 and 0.5 are considered as low logic levels. These levels are sometimes referred to loosely as TTL (transistor transistor logic) levels. Note that the actual logic levels used by the various TTL families differ from these slightly. The "high" and "low" ranges may be slightly different for input and output lines. Output lines often have limited current sourcing and sinking abilities compared with TTL, and it is therefore often necessary to buffer them. It is important that voltages exceeding the maximum rated values do not appear on inputs or outputs (e.g., attempting to switch an inductive load might produce a dangerously high transient voltage at an output); otherwise, the device may be damaged. Where this is likely to be a problem, inputs and outputs should be suitably buffered or even optically isolated, which provides protection up to a few kilovolts. Outputs may also drive appropriately connected relays. I/O cards with these facilities on board are readily available.

As a minimum, an I/O card may be expected to support a control register, two I/O ports each with an associated data register, and some handshake lines. Handshake lines may sometimes be used to generate interrupts on the controlling SBC. A byte written to the control register is used to configure the I/O ports, i.e., to determine if they are to behave as input or output ports as well as to select the function, if any, of the handshake lines. It may not be possible to select the direction of optically isolated or buffered ports. Writing a byte to an output port causes a pattern of high and low voltages to appear on the lines reflecting the pattern of zeros and ones in the binary representation of the byte written. Similarly, when a byte is read from an input port, the number read is specified in binary representation by the pattern of high and low logic levels on the input lines. The following example illustrates this.

The Intel 8255 Programmable Peripheral Interface (PPI) is commonly used in digital I/O cards for the PC. Data for this device are readily available [8]. The 8255 provides three ports, denoted A, B and

C, and a control register. Ports A and B may be designated as an 8-bit input port or an 8-bit output port. Port C may be considered as two independent four-bit ports, which may be chosen independently as input or output. Port C may also provide handshake functions. There are three modes in which the chip may operate. The simplest, mode 0, which provides basic input and output without automatic hand-shaking, is used in our example. Note that the 8 bits of an I/O port are conventionally labeled bits 0 to 7. This is because bit 0 is weighted 2^0, bit 1 weighted 2^1, etc. in the binary representation of the number read from or written to the port.

To configure the PPI to operate in mode 0 with port A as an input port and port B as an output port the bit pattern 10011001 must be written to the control register. This number is equivalent to 99 in hexadecimal or 153 in decimal. Suppose now that switches connected to port A hold bits 2 and 7 high and the remaining bits low. The resulting binary pattern will be 10000100, which is equivalent to hexadecimal 84 or decimal 132. When port A is read, the number 132 (decimal) will therefore be obtained. To hold the lines connected to bits 2, 3, and 5 of port B high while leaving the remaining lines low, we see that the binary pattern 00101100 must appear at port B. 00101100 binary is equivalent to 2C hexadecimal or 44 decimal. We therefore write the decimal number 44 to port B.

Counter/Timers

These devices typically provide software programmable event counting, pulse, and frequency measurement. As output devices, they may generate a single pulse (one-shot) when a programmable number of input pulses have been counted and produce square waves of arbitrary frequency and complex duty cycles. Frequencies generated are normally based on an on-board crystal clock to provide independence from the internal clock speed of the computer. Counter/timer cards commonly support at least three independent 16-bit counters.

Common applications include:

1. Alarms. The counter is in one-shot mode and generates a single pulse on timeout. This is connected to interrupt the computer and alert the user in the middle of the currently executing task.
2. Watchdog timer. This is used to detect problems, particularly in systems which are intended to operate without operator intervention. It is similar to the Alarm described earlier except that the interrupt is used to reset the computer. In normal operation this will never occur as all software tasks executing are designed to update the counter constantly so that it never reaches its terminal count. Only if a problem develops, e.g., a software "crash," will the counter time out and the system be reset.
3. The generation of complex waveforms, e.g., for pulse width modulation. This application uses two counters in cascade, one (T1) to provide regular pulses at the carrier frequency triggering another (T2), in one-shot mode, to provide the variable duty cycle as shown in Figure 86.6.

Analog Input

An analog input card uses an analog-to-digital converter (ADC) that accepts an input voltage and supplies an integer proportional to that voltage to the computer. Many cards now are produced with on-board signal conditioning circuits that provide for variable gain either by means of switches or under program control. Cards with specialized signal conditioning circuits for common applications such as thermo-couple linearization or interfacing to strain gages and other bridge sensors, are available. Signal conditioning to protect cards destined to be used in hostile electrical environments is also available. Cheaper cards may provide fixed gain and require additional signal conditioning circuits to be provided external to the computer. Many cards also feature multiplexed inputs where one of several inputs may be selected under program control to be fed to the ADC. Some important parameters to consider in selecting a analog input board are given in Table 86.5.

At rates above a few tens of kilohertz, interrupt-driven data capture is essential to maintain speed. Faster data rates require on-board memory to avoid degrading the performance of the controlling SBC. In this case DMA may be used to transfer data to main memory and increase performance further.

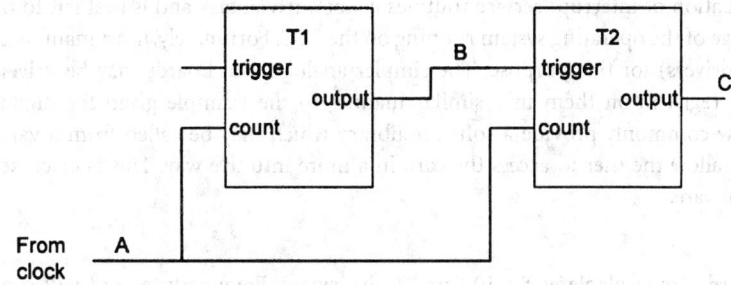

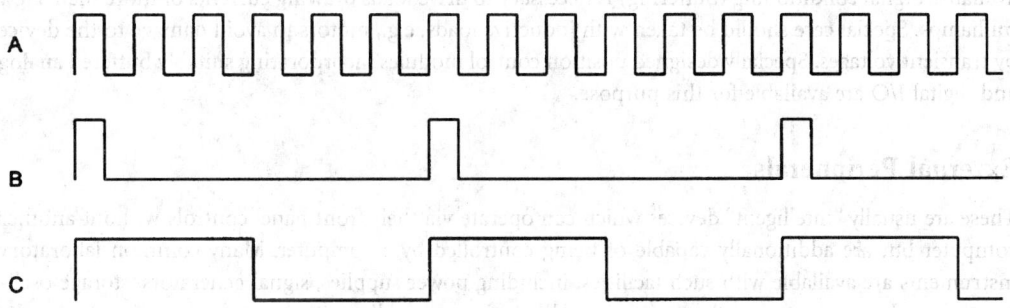

FIGURE 86.6 Using two timers to produce pulse-width modulation under software control. Both timers receive input pulses at a constant frequency from an external clock, as shown at A. Timer T1 operates in continuous mode. The trigger has no effect in this mode. The output of T1 is a single positive-going pulse when it has counted the specified (by software) number of input pulses, as shown at B. Timer T2 operates in one-shot mode. Each time it receives a trigger pulse, its output goes high for a specified (again by software) number of counts, as shown at C. In this way the frequency of the output at C is controlled by the count specified for T1 and the width of the positive-going part of C is controlled by the count specified for T2.

TABLE 86.5 Common ADC Parameters

Parameter	Description
Resolution	The smallest change in input detectable in the digital output; resolutions are commonly expressed in the number of significant bits in the digital output; hence, 8-bit resolution means 1 part in 256; 10 bit, 1 part in 1024, and 12 bit, 1 part in 4096
Linearity	The extent to which the output deviates from a linear relationship with the input; good devices will be linear to ±1 least-significant bit; i.e., the output value is guaranteed to be within ±1 of an exactly linear conversion
Range	The maximum (and minimum) input voltages; inputs may be unipolar, e.g., 0–5 V or bipolar ±5 V; voltages are specified relative to ground unless the inputs are differential, e.g., those designed for bridge sensors
Conversion speed	The time taken to convert an input voltage into a digital output, typically 1 s to 1 μs; may also be quoted as a sample rate
Linearity	The extent to which the conversion is linear, e.g., a linearity of ±1 least-significant bit means that the output value is within ±1 of the ideal linear conversion
Input impedance	The impedance between the input terminal and ground or between differential inputs.

A timer function is often incorporated to allow samples to be taken at regular intervals independently of what the controlling SBC is doing. An interrupt is generated when the conversion is complete, and an interrupt service routine is then activated to read the result of the conversion into memory. The writing and installation of interrupt service routines is not a trivial task and is best left to those with an intimate knowledge of the operating system running on the SBC. Fortunately, most manufacturers supply software (device drivers) for this purpose. The simpler analog input boards may be driven by writing values directly to registers on them in a similar manner to the example given for digital I/O cards. Manufactures now commonly provide a software library which may be called from a variety of high-level languages to allow the user to access the card in a more intuitive way. This is discussed further in the section on software.

Analog Output

Analog output cards are available as 8-, 10-, or 12-bit devices. Frequently, a card will support several channels of analog output with provision for delaying the updating of channels so that all can be updated simultaneously. Output voltages may be unipolar or bipolar and current outputs (4 to 20 mA) are also available. Signal conditioning (buffering) is necessary to drive loads drawing currents of more than a few milliamps. Special care should be taken with inductive loads, e.g., motors to avoid damage to the device by transient voltages. Specially designed position control modules incorporating suitably buffered analog and digital I/O are available for this purpose.

External Peripherals

These are usually "intelligent" devices which can operate via their front panel controls without another computer but are additionally capable of being controlled by a computer. Many common laboratory instruments are available with such facilities, including power supplies, signal generators, storage oscilloscopes, voltmeters, spectrophotometers, and position controllers. A computer can coordinate the actions of several such instruments to gather then manipulate and display data in a way which enhances the power of the instrumentation system. An almost trivial example is the use of a computer to control a signal generator and a voltmeter in order to generate the frequency response of an amplifier automatically. Such a system has an obvious role in automatic testing rigs.

Two common methods are used to control such devices: a serial link or the IEEE-488 bus. In both cases the devices are controlled by sending messages consisting of sequences of ASCII characters. Usually the sequences are chosen to have an obvious meaning, as in the example that follows, but this is not always the case — particularly with older devices where user friendliness was often sacrificed as a result of limited memory and processing power! The message sequence

"FREQ10kHz"
"SINE"
"1.0VOLTRMS"

might be used to set a signal generator to produce a 10-kHz sinusoidal signal at 1 V rms. There is little standardization in the form of device messages used although the IEEE standard 488.2 goes some way in this direction. Responses from instruments are sent in the same way, i.e., as ASCII characters so that a voltmeter might respond to a command to make a measurement with the data

"AC2.01mV"

to indicate that it was on an ac voltage range and measured 2.1 mV. Again, there is little standardization in the format of responses. Large blocks of data may be send in a binary format where possible (8-bit serial links or IEEE-488 bus) to minimize the amount of data to be transferred.

Serial Devices

Serial control of devices is accomplished using links conforming to one of the serial standards (RS-232, RS-422, or RS-485) described elsewhere in this chapter. This is a relatively simple method of control and has the advantage that much of the preliminary testing and debugging of a system can be done using a terminal or a terminal emulator program such as the public domain KERMIT available from Kermit

Distribution (Columbia University Academic Information Systems, 612 West 115th Street, New York, NY 10025, Phone: 212-854-3703). The writing of custom software that accesses the serial interface of the controlling computer is relatively easy under common operating systems including DOS, Windows, Windows 95, UNIX, and OS-9 using languages such as C, Pascal, or BASIC. It is increasingly common, particularly for DOS and Windows applications, for manufactures to provide software support for their devices.

Disadvantages of serial transfers are the relative slowness when large amounts of data are transferred, the lack of standardization in device messages, and the limited control facilities available. Advantages are the ease of testing, the simplicity of the controlling software, the relative simplicity of the interconnection scheme, and — for remote instrumentation systems — the fact that with the use of modems data can be transferred over large distances using standard telephone lines or even a radio link.

IEEE 488 Devices

The IEEE standard 488 was developed in the 1970s and rapidly became an industry standard for the interconnection and control of test equipment. This standard was modified slightly in 1987 (IEEE standard 488.1) to allow for the considerable enhancements of IEEE standard 488.2 which was introduced at the same time [9,10]. The original IEEE standard 488 specifies the electrical characteristics of the bus, the mechanical characteristics of its connectors, and a set of messages to be passed between interfaces. It does not attempt to provide any syntax or structure for communicating these messages, to specify commonly used commands, or to establish a standard for device-specific messages. These issues are addressed in IEEE standard 488.2.

The bus itself supports synchronous parallel transfers of data using three groups of lines,

A bidirectional 8-bit data bus,

Five interface management lines, and

Three handshake lines,

over distances of up to 20 m and at data transfer rates of up to 1 MB/s.

Devices on the bus are classed as *talkers*, *listeners*, or *controllers*. In general, the computer system is the bus controller which can also talk (send data) or listen (receive data). Most devices are both talkers and listeners: for example, a digital voltmeter will be a listener when receiving instructions to set the voltage range prior to making a measurement but will be a talker when returning the result of the measurement to the controller, which is itself acting as a listener. Each device on the bus must be assigned a unique address which is a number between 0 and 30. This may be done from the front panel of the device or, less conveniently, by setting switches elsewhere on the device.

It is a difficult, time-consuming, and error-prone process to write software to drive an IEEE-488 card. Purchasers of new IEEE-488 interfaces are strongly advised to obtain a device driver from the manufacturer. Such device drivers are now readily available and integrate the card into the filing system of the operating system running on the controlling computer. This allows the interface to be accessed in a natural way from high-level languages running on the controller.

86.6 Software for Instrumentation Systems

The difficulty involved in writing software for instrumentation systems depends largely on the support available from the manufacturers of the subsystems, on the operating system (if any), and on the development tools available. On one extreme, one may be working in a virtual instrument environment, such as that provided by the National Instrument LabVIEW, where software development is entirely graphical and, for small projects at least, is readily undertaken by users with little or no prior experience. On the other extreme, one is faced with the problem of developing software, which is at the very least interrupt driven and probably multitasking, for a target SBC with no resident operating system; this requires considerable expertise in software design and development together with the availability of development tools such as cross compilers and source-level debuggers.

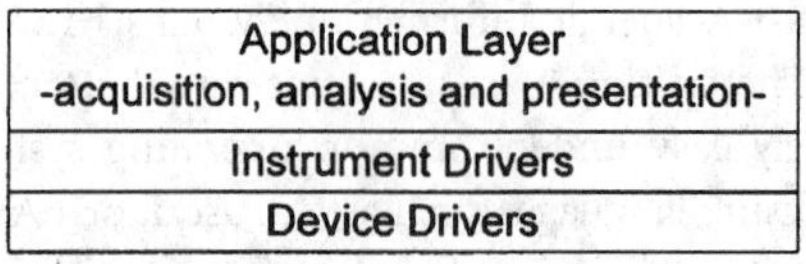

| Application Layer |
| -acquisition, analysis and presentation- |
| Instrument Drivers |
| Device Drivers |

FIGURE 86.7   A layer model for instrumentation systems.

Virtual Instruments

Figure 86.7 shows a layer model of the software for a generalized instrumentation system. The application layer handles the data acquisition, analysis, and presentation. The instrument drivers provide a mechanism for communicating with the instruments in a standard way without requiring the user to know about the often cryptic data strings which need to be sent. For example, all digital multimeters will need the facility to chose a specific input voltage range. The range coding, resolution, etc. that have to be sent to the multimeter to achieve this will vary from instrument to instrument; however, the instrument driver allows the software writer programming in the application layer to call a procedure such as

SetVoltageRange(VoltageValue)

and this procedure call is the same for all multimeters. Although some manufacturers use the term slightly differently, the instrument driver is in effect the virtual instrument. Writing *instrument drivers* is a time-consuming but not too difficult task. Instrument drivers for proprietary instrumentation software design packages are readily available from instrument manufacturers. Device drivers integrate the controlling interface (e.g., IEEE-488, RS-232, or internal card) into the operating system of the computer. Writing a *device driver* requires a detailed knowledge of the device hardware and of the computer operating system. This is a difficult task and new interfaces should be purchased with a device driver appropriate for the operating system wherever possible.

A number of development environments which are based on the virtual instrument concept are now available. These free the user from the problems of writing conventional software to control instruments and handle the data produced. Instruments appear to the software developer as "front panels" drawn on the computer screen, complete with familiar buttons, knobs, and displays. Data flow is handled by linking instruments in a block diagram using a mouse in an environment that resembles an ordinary drawing package. The software developer is working only in the application layer.

While the graphical environment allows simple systems to be developed rapidly, experienced programmers may find it restrictive. There are software development systems available that give the programmer access libraries containing instrument drivers, data analysis routines, graphics functions and data visualisations facilities in commonly used high-level languages such as C, Pascal, BASIC, and FORTRAN. Table 86.2 lists representative software packages.

Working Directly with Peripherals

It may occasionally be the case that the cost of software support for a virtual instrument development environment is not justified for a small application. Software must then be written to interface directly with the peripheral. The earlier section on digital input and output explained in principal what was necessary to program a simple interface chip. We now continue this example and show using the language C how this might be achieved.

The method of accessing the registers of peripheral cards depends on the microprocessor involved and may not even be a standard feature of the language being used. Where the peripheral forms part of the same address space as the computer memory, such as in the Motorola 680XX series, pointers can be used to read and write values in the registers. The Intel 80 × 86 series of processors often place peripherals in a separate address space which may not be accessed by pointers. In this case an extension to the language is required. Borland's Turbo C and C++ provide functions to read and write I/O mapped devices:

unsigned char inportb(int portid)
void outportb(int portid, unsigned char value)

These are used in the code fragment which implements the software for our earlier example. We assume that the 8255 PPI has base address $0 \times 1b0$ and that the program copies the value read from the input port (port A) directly to the output port (port B), until bit 0 of the input becomes zero

```
/* define the addresses of the registers for the PPI */
#define BASE 0x1b0
#define PABASE
#define PBBASE+1
#define CONTROLBASE+3

unsigned char xin;/* declare an 8-bit variable */
outportb(CONTROL, 153)/* set port A as input, port B as output */
do
{
        xin = inportb(PA);/* read the input */
        outportb(PB,xin);/* write the output */
}
while(xin & 1);      /* loop if bit 0 is still 1 */
```

This fragment also illustrates that high-level languages generally only support input and output of bytes. Masking techniques must be used to access individual bits.

The Choice of Operating System

The software development support discussed so far is typically available under DOS, Windows 3.1, Windows 95, Windows NT, and UNIX. When a system is multitasking, it has to meet stringent real-time constraints; however, none of these operating systems is particularly appropriate. DOS does not support multitasking, and the others are not optimized for real-time systems which require speedy context switching and rapid interrupt response. The most fundamental requirement of real-time applications is that ability of the system to respond to external events with very short, bounded, and predictable delays. Table 86.6 lists some important real-time operating systems and kernels.

Real-time operating systems tend not to have the mature and powerful software development support available for conventional operating systems. It is not possible simply to develop the software on a familiar operating system and then transfer the working programs to the target system. Much of the debugging, testing, and system integration will have to be done on the target itself to access the hardware. A common solution is a development system in which a conventional workstation is linked to the target system. Software is developed on the workstation using familiar tools, e.g., a Windows-based editor, support for version control, and a powerful filing system. At any time, code can be cross-compiled (i.e., compiled for the processor on the target system) and downloaded to the target. The workstation may then monitor the execution of the software running on the target processor. Features that allow the user to single-step through the source code, seen in a workstation window, while viewing the status of key variables in another are available. It is also possible to set breakpoints and allow the processes to run until one is encountered. Manufacturers of real-time operating systems are often able to provide development support of this type for their product for a variety of workstations.

TABLE 86.6 Some Important Real-Time Operating
Systems and Kernels

System	OS/Kernel	Manufacturer
OS-9	OS	Microware Systems Corporation 1900 NW 114th Street, Des Moines, IA 50325
LynxOS	OS	Lynx Real-Time Systems, Inc. 16870 Lark Avenue Los Gatos, CA 95030-2315
VxWorks	OS	Wind River Systems 1010 Atlantic Avenue Alameda, CA 94501
VRTX/OS	OS	Microtec Research 2350 Mission College Blvd. Santa Clara, CA 95054
VRTX	Kernel	Microtec Research 2350 Mission College Blvd. Santa Clara, CA 95054
iRMX	Kernel	Intel Corp. 3065 Bowers Avenue Santa Clara, CA 95051

References

1. Micrology pbt, Inc., *The VMEbus Specification Manual: Revision C1*, PRINTEX, 1985.
2. S. Heath, *VMEbus: A Practical Companion*, Boston, MA: Butterworth Heinemann, 1993.
3. J. Di Giacomo, Ed., *Digital Bus Handbook*, New York: McGraw-Hill, 1990.
4. T. Shanley and D. Anderson, *ISA System Architecture*, 3rd ed., Reading, MA: Addison-Wesley, 1995.
5. T. Shanley and D. Anderson, *EISA System Architecture*, 2nd ed., Reading, MA: Addison-Wesley, 1995.
6. T. Shanley and D. Anderson, *PCMCIA System Architecture*, 2nd ed., Reading, MA: Addison-Wesley, 1995.
7. M.D. Seyer, *RS232 Made Easy*, Englewood Cliffs, NJ: Prentice-Hall, 1984.
8. Intel Corporation, *Intel Microsystems Component Handbook*, Vol. 2.
9. IEEE Std 488.1-1987, *IEEE Standard Digital Interface for Programmable Instrumentation*.
10. IEEE Std 488.2-1987, *IEEE Standard Codes, Formats, Protocols and Common Commands*.

87

Telemetry

Albert Lozano-Nieto
Penn State University

87.1 Introduction

Telemetry is the science of gathering information at some remote location and transmitting the data to a convenient location to be examined and recorded. Telemetry can by done by different methods: optical, mechanical, hydraulic, electric, etc. The mechanical methods, either pneumatic or hydraulic have acceptable results for short distances and are used in environments that have a high level of electromagnetic interference and in those situations where, for security reasons, it is not possible to use electrical signals, for example, in explosive environments. More recently, use of optical fiber systems allows the measurement of broad bandwidth and high immunity to noise and interference. Other proposed telemetry systems are based on ultrasound, capacitive or magnetic coupling, and infrared radiation, although these methods are not routinely used. The discussion in this chapter will be limited to the most-used systems: telemetry based on electric signals. The main advantage of electric over mechanical methods is that electrically based telemetry does not have practical limits regarding the distance between the measurement and the analysis areas, and can be easily adapted and upgraded in already existing infrastructures. Electric telemetry methods are further divided depending on the transmission channel that they use as wire telemetry and wireless (or radio) telemetry. Wire telemetry is technologically the simplest solution. The limitations of wire telemetry are the low bandwidth and low transmission speed that it can support. However, it is used when the transmission wires can use the already existing infrastructure, as, for example, in most electric power lines that are also used as wire telemetry carriers. Wireless telemetry is more complex than wire telemetry, as it requires a final radio frequency (RF) stage. Despite its complexity, it is widely used because it can transmit information over longer distances; thus, it is used in those applications in which the measurement area is not normally accessible. It can also transmit at higher speeds and have enough capacity to transmit several channels of information if necessary.

Figure 87.1 displays a generic telemetry system. It consists of (not all the blocks will be always present) (1) transducers to convert physical variables to be measured into electric signals that can be easily processed; (2) conditioning circuits to amplify the low-level signal from the transducer, limit its bandwidth, and adapt impedance levels; (3) a signal-processing circuit that sometimes can be integrated in the previous circuits; (4) a subcarrier oscillator whose signal will be modulated by the output of the different transducers once processed and adapted; (5) a codifier circuit, which can be a digital encoder, an analog modulator, or digital modulator, that adapts the signal to the characteristics of the transmission channel, which is a wire or an antenna; (6) a radio transmitter, in wireless telemetry, modulated by

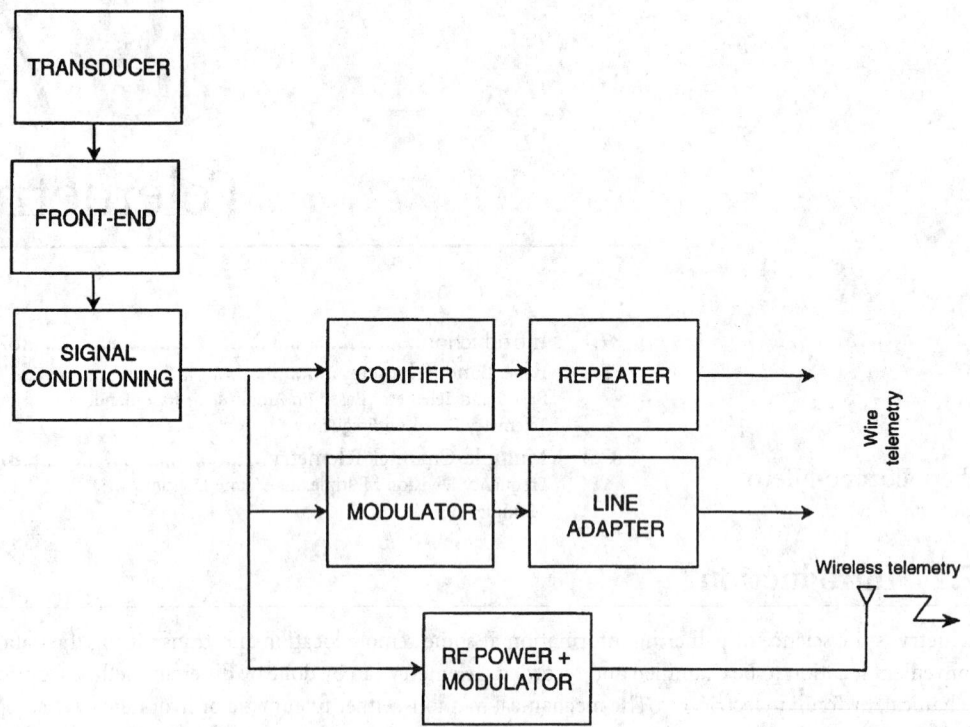

FIGURE 87.1 Block diagram for a telemetry system. Telemetry using wires can be performed in either base-band or by sending a modulated signal, while wireless telemetry uses an RF carrier and an antenna.

the composite signal; (7) an impedance line adapter, in case of wire transmission, to adapt the characteristic impedance of the line to the output impedance of the circuits connected to the adapter; and (8) for wireless communication, a transmitting antenna. The receiver end consists of similar modules. For wireless telemetry, these modules are (1) a receiving antenna designed for maximum efficiency in the RF band used; (2) a radio receiver with a demodulation scheme compatible with the modulation scheme; and (3) demodulation circuits for each of the transmitted channels. For wire telemetry, the antenna and the radio receiver are replaced by a generic front end to amplify the signal and adapt the line impedance to the input impedance of the circuits that follow. The transmission in telemetry systems, in particular wireless ones, is done by sending a signal whose analog variations in amplitude or frequency are a known function of the variations of the signals from the transducers. More recently, digital telemetry systems send data digitally as a finite set of symbols, each one representing one of the possible finite values of the composite signals at the time that it was sampled. The effective communication distance in a wireless system is limited by the power radiated by the transmitting antenna, the sensitivity fo the receiver and the bandwidth of the RF signal. As the bandwidth increases, the contribution of noise to the total signal also increases, and consequently more transmitted power is needed to maintain the same signal-to-noise ratio (SNR). This is one of the principal limitations of wireless telemetry systems. In some applications, the transmission to the receiver is done on base band, after the conditioning circuits. The advantage of base-band telemetry systems is their simplicity, although because of the base-band transmission, they are normally limited to only one channel at low speeds.

Not uncommonly, measurement system needs to acquire either different types of signals or the same type of data at different locations in the process that is being monitored. These different information signals can be transmitted using the same common carrier by multiplexing the data signals. Multiplexing allows different signals to share the same channel. Multiplexing techniques are usually considered either frequency division multiplexing (FDM) or time division multiplexing (TDM). In FDM, different subcarrier

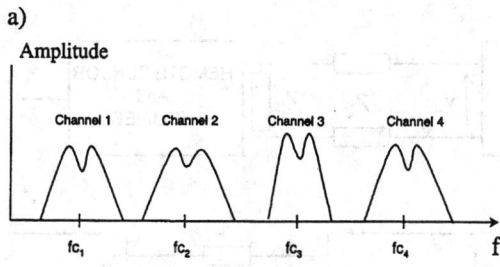

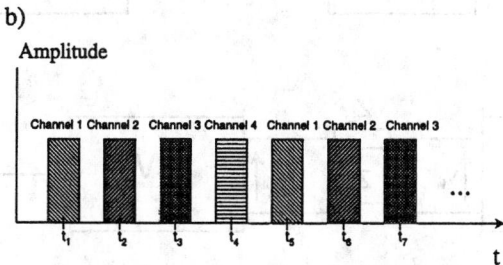

FIGURE 87.2 Basic characteristics of (a) FDM and (b) TDM signals. In FDM different channels are allocated at different subcarrier frequencies $(f_{c1}, f_{c2},\dots)$ while in TDM only one channel is transmitted at a given time. The remaining channels are transmitted sequentially.

frequencies are modulated by the different measurement channel signals, which causes the information spectrum to shift from base band to the subcarrier frequency. Then, the subcarrier frequencies modulate the RF carrier signal, which allows the transmission of all desired measurement channels simultaneously. In TDM, the whole channel is assigned entirely to each measurement channel, although only during a fraction of the time. TDM techniques use digital modulation to sample the different measurement channels at different times. Then, these samples are applied sequentially to modulate the RF carrier. Figure 87.2 illustrates these concepts by showing frequency and time graphs for FDM and TDM, respectively.

Almost all instrumentation and measurement situations are candidates for use of a telemetry link. Telemetry is widely used in space applications for either telemeasurement of a distant variable or telecommandment of actuators. In most of these types of applications, for example, in space telemetry, it is very important to design the telemetry systems to minimize the consumption of power [1]. Some land-mobile vehicles, such as trains, also use telemetry systems, either wireless or by using some of the existing power wires to transmit data to the central station and receive its commands [2]. In clinical practice, the telemetry of patients increases their quality of life and their mobility, as patients do not need to be connected to a measurement system to be monitored. Several medical applications are based on implanting a sensor in a patient and transmitting the data to be further analyzed and processed either by radio [3] or by adapted telephone lines [4] from the receiving station. Optical sensors and fiber-optic communications are used in industry to measure in environments where it is not desirable to have electric signals such as explosive atmospheres [5]. The designer of a telemetry system needs also to keep in mind the conditions in which the system will have to operate. In most of the applications, the telemetry systems must operate repeatedly without adjustment and calibration in a wide range of temperatures. Finally, as different telemetry systems are developed, the need to permit tests to be made interchangeable at all ranges increases, which require compatibility of transmitting, receiving, and signal-processing equipment at all ranges. For this reason, the Department of Defense Research and Development Squad created the Guided Missiles Committee, which formed the Working Group on Telemetry. This later became the Inter-Range Instrumentation Group (IRIG) that developed Telemetry Standards. Today, the IRIG Standard 106-96 is the primary Telemetry Standard used worldwide by both government and industry.

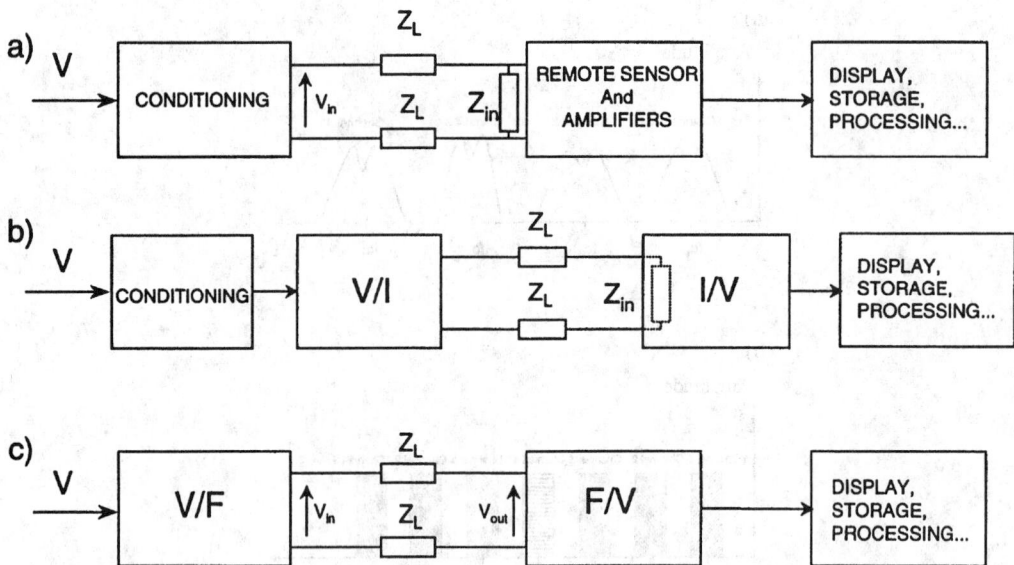

FIGURE 87.3   Different configurations for base-band telemetry. In voltage-based-base band telemetry (a) the information is transmitted as variations of a voltage signal. Current-based-base band telemetry (b) is based on sending a current signal instead of a voltage signal to neutralize the signal degradation due to the voltage divider made up by the input impedance of the receiver (Z_{in}) and the impedance of the lines (Z_L). In frequency-based base-band telemetry (c), the information is transmitted as variations of frequency which makes this system immune to noise and interference that affect the amplitude of the transmitted signal.

87.2   Base-Band Telemetry

Base-band telemetry uses a wire line to communicate the signal from the transducer after being processed and conditioned with the receiver. We will briefly describe telemetry systems based either on amplitude or frequency. More in-depth study of these base-band telemetry systems can be found in Reference 6.

Base-Band Telemetry Based on Amplitude

Voltage-Based Base-Band Telemetry

Figure 87.3a shows a simple voltage-based telemetry system. The signal from the transducer is amplified, normally to a voltage level between 1 and 15 V, and sent through a line consisting of two wires to the receiver. By making the low end of the scale 1 V, this system can detect short circuits [6]. The main problem of this configuration is the limitation on the transmission distance, which depends on the resistance of the line and the input resistance for the receiver. Also, the connecting wires form a loop that is very susceptible to interference from parasitic signals.

Current-Based Base-Band Telemetry

The limitation on transmission distance of the voltage-based system due to the impedance of the line are solved by using a current signal instead of a voltage, as is shown in Figure 87.3b. This requires an additional conversion module after the signal-processing circuits from voltage to current. At the receiver end, the signal is detected by measuring the voltage across a resistor. The most-used system in industry is the 4 to 20 mA loop. This means that 0 V is transmitted as 4 mA, while the highest voltage value is transmitted as a 20-mA current. The advantage of transmitting 4 mA for 0 V is the easy detection of an open circuit in the loop (0 mA). Other standard current values are 0 to 5, 0 to 20, 10 to 50, 1 to 5, and 2 to 10 mA. Also, voltage drops due to resistance of the wires do not affect the transmitted signal, which allows the use of thinner wires. Because this is a current mode, the parasitic voltages induced in the line

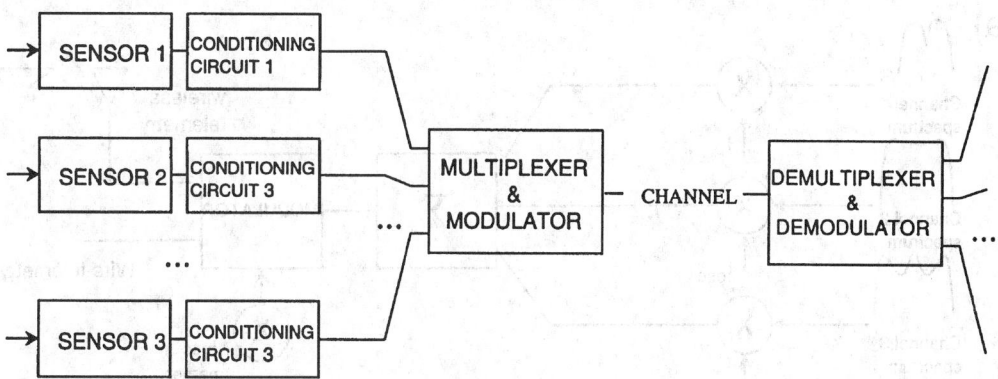

FIGURE 87.4 In multiple-channel telemetry a common transmission channel is used to transmit the measured signals from different channels using different sharing schemes.

do not affect the signal either. Current-based telemetry allows the use of grounded or floating transmitters with few modifications [6].

Base-Band Telemetry Based on Frequency

Frequency-based transmission is known to have higher immunity to noise than amplitude-based transmission. Frequency-based telemetry, shown in Figure 87.3c, is used in the presence of inductive or capacitive interference due to its immunity to noise. It also offers the possibility of isolating the receiver from the transmitter. The signal at the output of the conditioning circuit modifies the frequency of the telemetry signal, normally using a voltage-to-frequency converter. In the receiver, a frequency-to-voltage converter performs the opposite function. A special case of frequency-based telemetry is pulse telemetry, in which the modulating signal changes some characteristics of a train of pulses. Because of its importance and widespread use, pulse telemetry will be analyzed in-depth in the following sections.

87.3 Multiple-Channel Telemetry

Most of the industrial processes in which telemetry is used require the measurement of different physical variables to control the process, the measurement of only one physical variable at different locations, or normally a combination of both. In these multiple-channel measurements, base-band telemetry is not an option, as it would require building a different system for each channel. Multiple channel telemetry is achieved by sharing a common resource (transmission channel), as is shown in Figure 87.4. The sharing of the transmission channel by all the measurement channels is designated by *multiplexing*. There are two basic multiplexing techniques: FDM and TDM. In FDM, different channels are assigned to different spectral bands and the composite signal is transmitted through the communication channel. In TDM, the information for different channels is transmitted sequentially through the communication channel.

Frequency Division Multiplexing

In FDM, shown in Figure 87.5a, each measurement channel modulates a sinusoidal signal of different frequency. These sinusoidal signals are called subcarriers. Each of the modulated signals is then low-pass-filtered to ensure that the bandwidth limits are observed. After the filtering stage, all the modulated signals are fed into a summing block, producing what is known as a base-band signal. A base-band signal indicates here that the final carrier has not yet been modulated. The spectrum of the base-band signal is shown in Figure 87.5b, where it is possible to see how each measurement channel spectrum signal is allocated its own frequency. This composite signal finally modulates a carrier signal whose frequency depends on the transmission medium that is used. The signal is then fed into a transmission wire (similar

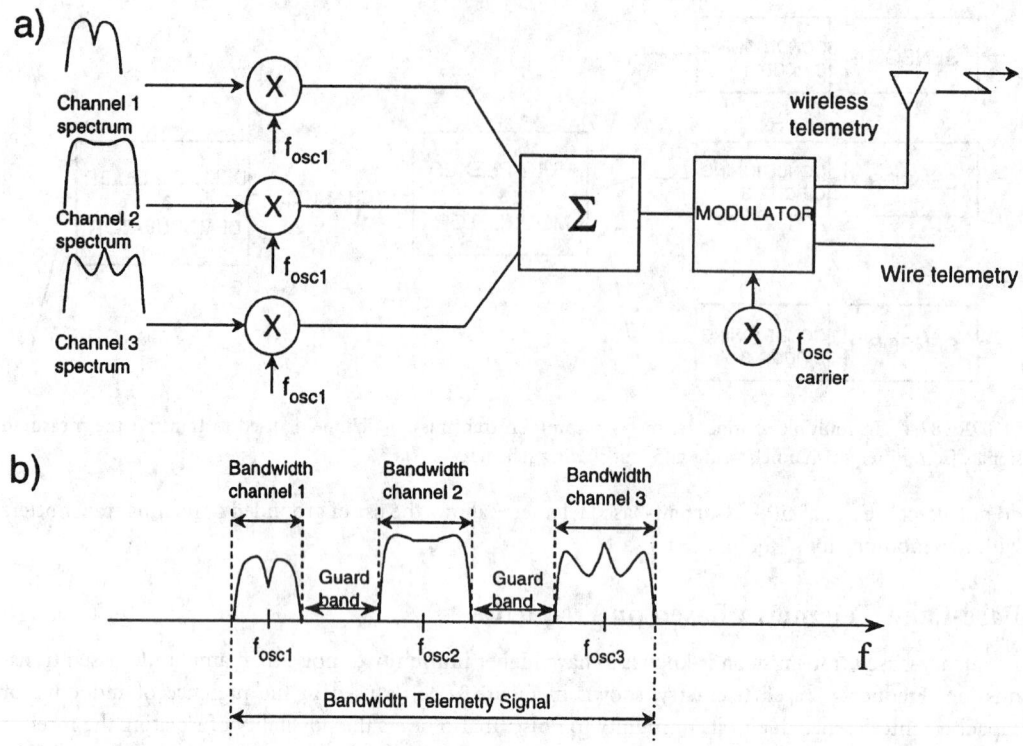

FIGURE 87.5 The different channels in an FDM system (a) are allocated at different subcarrier frequencies producing a composite signal shown in (b) that is later modulated by an RF frequency according to the transmission channel used. The guard bands limit the closeness of contiguous channels to avoid intermodulation and cross talk.

TABLE 87.1 Frequency Bands Allocated for Telemetry

Frequency band, MHz	Uses	Notes
72–76	Biotelemetry	Low power devices; restricted by Part 15 of FCC rules
88–108	Educational	Four frequencies in this band; part 90 of FCC rules
154	Industry	Band in TV channels 7–13
174–216	Biotelemetry	Low-power operations restricted to hospitals
216–222	Multiple	BW < 200 kHz
450–470	General	Telemetry as secondary basis; limited to 2 W of RF
467	Industry	Business band; limited to 2 W of RF
458–468	Biotelemetry	Band in TV channels 21–29
512–566	Biotelemetry	Low-power operations restricted to hospitals
1427–1435	Fixed	Uses in land mobile services (telemetering and telecommand)
1435–1535	Aeronautical	
2200–2290	Mobile	

to TV-broadcasting systems by cable) or, more commonly, into an antenna in the case of wireless telemetry systems. In wireless telemetry, the frequency of the carrier cannot be chosen arbitrarily, but is chosen in accordance with international agreements on the use of the electromagnetic spectrum. In the U.S., the Federal Communications Commission (FCC) is the body that regulates the allocation of frequencies for different communication services. Table 87.1 shows the most common telemetry frequency bands and their intended use. Table 87.1 is for informational purposes only, and it is not a comprehensive guide to telemetry frequencies. To find the allowed telemetry frequencies for a specific application, the maximum power allowed, and other limitations, the reader should consult the applicable FCC documents [7,8].

The allocation of bands is a process subject to change. For example, in October 1997 the FCC assigned some of the TV channel bands for patient telemetry inside hospitals, with restricted power. The FCC publishes all changes that affect frequency bands or other technical characteristics for telemetry.

At the receiver end, the carrier demodulator detects and recovers the composite base-band signal. The next step is to separate each of the subcarriers, by feeding the signal into a bank of parallel passband filters. Each channel is further demodulated, recovering the information from the transducer. The main practical problem of FDM systems is the cross talk between channels. Cross talk appears due to the nonlinearities of the electronic devices, which originates when the signal for one channel partially modulates another subcarrier in addition to the one assigned to that channel. Cross talk also originates when the spectra for two adjacent channels overlap. To avoid this effect, the subcarriers have to be chosen so that there is a separation (guard band) between the spectra of two contiguous channels. By increasing the guard band, the possibility of cross talk decreases, but the effective bandwidth also increases. The effective bandwidth equals the sum of the bandwidth of all channels, plus the sum of all the guard bands.

There are three alternative methods for each of the two modulation processes: the modulation of the measurement channel signals and the modulation of the composite signal. These methods are amplitude modulation (AM), frequency modulation (FM), and phase modulation (PM). The usual combinations are FM/FM, FM/PM, or AM/FM [6]. Here, we will analyze only on the subcarrier modulation schemes, while the modulation for the RF signal is analyzed in Chapter 81.

Subcarrier Modulation Schemes for Frequency Division Multiplexing

Subcarrier Modulation of Amplitude.

In an AM subcarrier modulation scheme, the amplitude of a particular subcarrier signal is changed according to the value of the measured channel assigned to that frequency. The resulting AM signal is given by

$$v(t) = A_c \left[1 + m(t) \right] \cos(\omega_c t)$$

where A_c is the amplitude of the carrier, $m(t)$ the modulating signal, and ω_c the frequency of the carrier.

The advantage of this type of modulation is the simplicity of the circuits that perform the modulation and the circuits required for the demodulation, in order to recover the modulating signal that carries the desired information. The percentage of modulation denotes the extent to which a carrier has been amplitude modulated. Assuming for simplicity that the modulating signal is sinusoidal of frequency ω_m, such as

$$m(t) = m \times \cos(\omega_m t)$$

the percentage of modulation (P) can be found as

$$P = m \times 100\ (\%)$$

In a more general way, the percentage of modulation (P) is expressed as

$$\frac{P}{100\%} = \frac{A_{c(max)} - A_{c(min)}}{2 A_c}$$

where $A_{c(max)}$ and $A_{c(min)}$ are the maximum and minimum values that the carrier signal achieves.

Figure 87.6 shows the spectrum of an amplitude-modulated signal, assuming that the modulating signal is a band-limited, nonperiodic signal of finite energy. Figure 87.6 shows that it consists of two

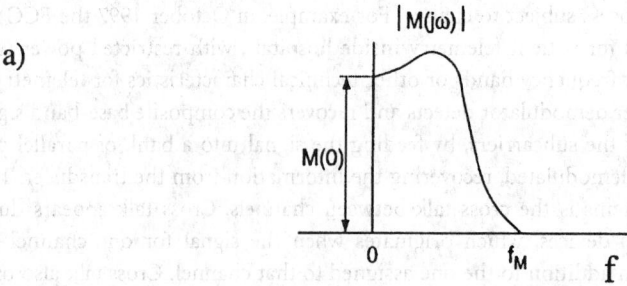

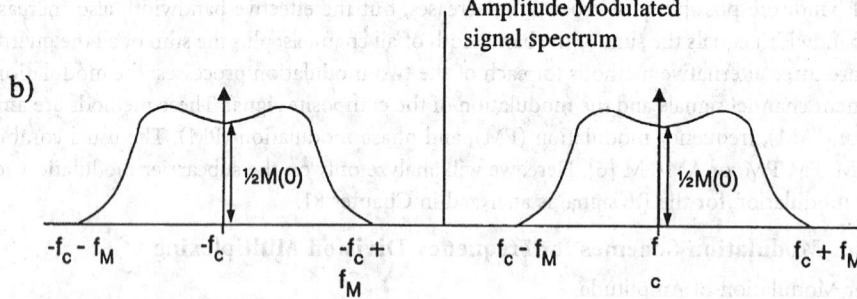

FIGURE 87.6 Resulting spectrum after amplitude modulation of a signal shown in (a). The resulting spectrum has doubled the required bandwidth, while only 0.25 of the total power is used in transmitting the desired information.

sidebands that are symmetrical in reference to the subcarrier. Figure 87.6 shows the main disadvantages of AM schemes. First, the bandwidth of the modulated channel is two times the bandwidth of the modulating signal, due to the two similar sidebands that appear. This results in an inefficient use of the spectrum. Second, the analysis of power for each of the components in Figure 87.6 shows that at least 50% of the transmitted power is used in transmitting the subcarrier, which is independent of the measured signal, as it does not contain any information. The remaining power is split between the two sidebands, which results in a maximum efficiency that it is theoretically possible to achieve of below 25%. The third main problem of AM is the possibility of overmodulation, which occurs when $m > 1$. Once a signal is overmodulated, it is not possible to recover the modulating signal with the simple circuits that are widely used for AM telemetry transmission.

The limitations of AM subcarrier modulation can be overcome using more efficient modulation techniques, such as double sideband (DSB), single sideband (SSB), and compatible single sideband (CSBB), which are also considered AM techniques. However, the complexity of these modulation systems and the cost associated with systems capable of recovering subcarrier signals modulated this way cause these not to be used in most commercial telemetry systems. Most of the available systems that use AM subcarrier techniques, use the traditional AM that has been described here, because its simplicity overcomes the possible problems of its use.

Subcarrier Modulation of Frequency.
FM (or PM) is by far the most-used subcarrier modulation scheme in FDM telemetry systems. These angle modulations are inherently nonlinear, in contrast to AM. Angle modulation can be expressed as

$$v(t) = A \cos\left[\omega_c t + \phi(t)\right]$$

where $\phi(t)$ is the modulating signal, that is, the signal from the transducers after conditioning.

It is then possible to calculate the value of the instantaneous frequency as

$$f = \frac{1}{2\pi}\frac{d}{dt}\left[\omega_c t + \phi\left(t\right)\right] = \frac{\omega_c}{2\pi} + \frac{d}{dt}\phi\left(t\right)$$

This equation shows how the signal $v(t)$ is modulated in frequency. We can analyze two parameters that can be derived from the previous equations: frequency deviation and modulation index. Frequency deviation (f_m) is the maximum departure of the instantaneous frequency from the carrier frequency. The modulation index (β) is the maximum phase deviation. The following equations show how these parameters are related. The value of the instantaneous frequency (f) is [9]

$$f = \frac{\omega_c}{2\pi} + \frac{\beta\,\omega_m}{2\pi}\cos\left(\omega_m t\right) = f_c + \beta\, f_m \cos\left(\omega_m t\right)$$

The maximum frequency deviation is Δf and is given by

$$\Delta f = \beta f_m$$

Therefore, we can write the equation for the frequency modulated signal as

$$v\left(t\right) = A\cos\left[\omega_c t + \frac{\Delta f}{f_m}\sin\left(\omega_m t\right)\right]$$

The previous equation shows that the instantaneous frequency, f, lies in the range $f_c \pm \Delta f$. However, it does not mean that all the spectral components lie in this range. The spectrum of an angle-modulated waveform cannot be written as a simple equation. In the most simple case, when the modulating signal is a sinusoidal signal, a practical rule states that the bandwidth of an FM signal is twice the sum of the maximum frequency deviation and the modulating frequency. For modulating signals commonly found in measuring systems, the bandwidth is dependent upon the modulation index; that is, as the bandwidth allocated for each channel is limited, the modulation index will also be limited.

Frequency Division Multiplexing Telemetry Standards

IRIG Standard 106-96 is the most used for military and commercial telemetry, data acquisition, and recording systems by government and industry worldwide [10]. It recognizes two types of formats for FM in FDM systems: proportional-bandwidth modulation (PBW) and constant-bandwidth modulation (CBW). It also allows the combination of PBW and CBW channels. In PBW, the bandwidth for a channel is proportional to the subcarrier frequency. The standard recognizes three classes of subcarrier deviations: 7.5, 15, and 30%. There are 25 PBW channels with a deviation frequency of 7.5%, numbered 1 to 25. The lowest channel has a central frequency of 400 Hz, which means that the lower deviation frequency is 370 Hz and the upper deviation frequency is 430 Hz. The highest channel (channel 25) has a center frequency of 560,000 Hz (deviation from 518,000 to 602,000 Hz). The center frequencies have been chosen so that the ratio between the upper deviation limit for a given channel and the lower deviation limit for the next channel is around 1.2. There are 12 PBW channels with a deviation frequency of 15%, identified as A, B, ... L. The center frequency for the lowest channel is 22,000 Hz (deviation from 18,700 Hz to 25,300 Hz), while the center frequency for the highest channel is 560,000 Hz (476,000 to 644,000 Hz), with a ratio for the center frequencies of adjacent channels being about 1.3. There are also 12 PBW channels for a deviation frequency of 30%, labeled from AA, BB, ... to LL. The center frequency for these channels is the same as that for the 15% channels.

TABLE 87.2   Characteristics of Constant Bandwidth (CBW) Channels for FDM

Channel Denomination	Frequency Deviation, kHz	Lowest Channel Center Frequency, kHz	Highest Channel Center Frequency, kHz	No. of Channels	Separation between Channels, kHz
A	±2	8	176	22	8
B	±4	16	352	22	16
C	±8	32	704	22	32
D	±16	64	1408	22	64
E	±32	128	2816	22	128
F	±64	256	3840	15	256
G	±128	512	3584	7	512
H	±256	1024	3072	4	1024

CBW channels keep the bandwidth constant and independent of its carrier frequency. There are eight possible maximum subcarrier frequency deviations labeled A (for 2 kHz deviation) to H (for 256 kHz deviation). The deviation frequency doubles from one group to the next. There are 22 A-channels, whose center frequency range from 8 to 176 kHz. The separation between adjacent channels is a constant of 8 kHz. Table 87.2 shows a summary of the characteristics of CBW channels.

IRIG Standard 106-96 gives in its appendix criteria for the use of the FDM Standards. It focuses on the limits, most of the time dependent on the hardware used, and performance trade-offs such as data accuracy for data bandwidth that may be required in the implementation of the system. The subcarrier deviation ratio determines the SNR for a channel. As a rule of thumb, the SNR varies as the three-halves power of the subcarrier deviation ratio. On the other hand, the number of subcarrier channels that can be used simultaneously to modulate an RF carrier is limited by the channel bandwidth of the RF carrier as well as considerations of SNR. Given a limited RF bandwidth, as more channels are added to the FDM system, it is necessary to reduce the deviation ratio for each channel, which reduces the SNR for each channel. It is then very important to evaluate the acceptable trade-off between the number of subcarrier channels and the acceptable SNR values. A general equation that might be used to estimate the thermal noise performance of an FM/FM channel is the following [11]:

$$\left(\frac{S}{N}\right)_d = \left(\frac{S}{N}\right)_c \left(\frac{3}{4}\right)^{1/2} \left(\frac{B_c}{F_{ud}}\right) \left(\frac{f_{dc}}{f_s}\right) \left(\frac{f_{ds}}{F_{ud}}\right)$$

where $(S/N)_d$ represents the SNR at the discriminator output, $(S/N)_c$ represents the SNR of the receiver, B_c is the intermediate-frequency bandwidth of the receiver, F_{ud} is the subcarrier discriminator output filter (at −3 dB), f_s is the subcarrier center frequency, f_{dc} is the carrier peak deviation for the subcarrier considered, and f_{ds} is the subcarrier peak deviation.

According to the Standard, the FM/FM composite FDM signal that is used to modulate an RF carrier can be of PBW format, CBW format, or a combination of both, with the only limitation that the guard bands between the channels used in the mixed format are equal or greater than the guard bands for the same channels in an unmixed format.

Time Division Multiplexing

TDM is a transmission technique that divides the time into different slots, and assigns one slot to each measurement channel. In TDM, all the transmission bandwidth is assigned entirely to each measurement channel during a fraction of the time. After the signals from the measurement channels have been low-pass filtered, they are sequentially sampled by a digital switch that samples all the measurement channels in a period of time (T) that complies with the Nyquist criteria. Figure 87.7a shows a basic block diagram for an FDM system. The output of the sampler is a train of AM pulses that contains the individual samples for the channels framed periodically, as is shown in Figure 87.7b. Finally, the composite signal modulates an RF carrier. The set of samples from each one of the input channels is called a frame. For

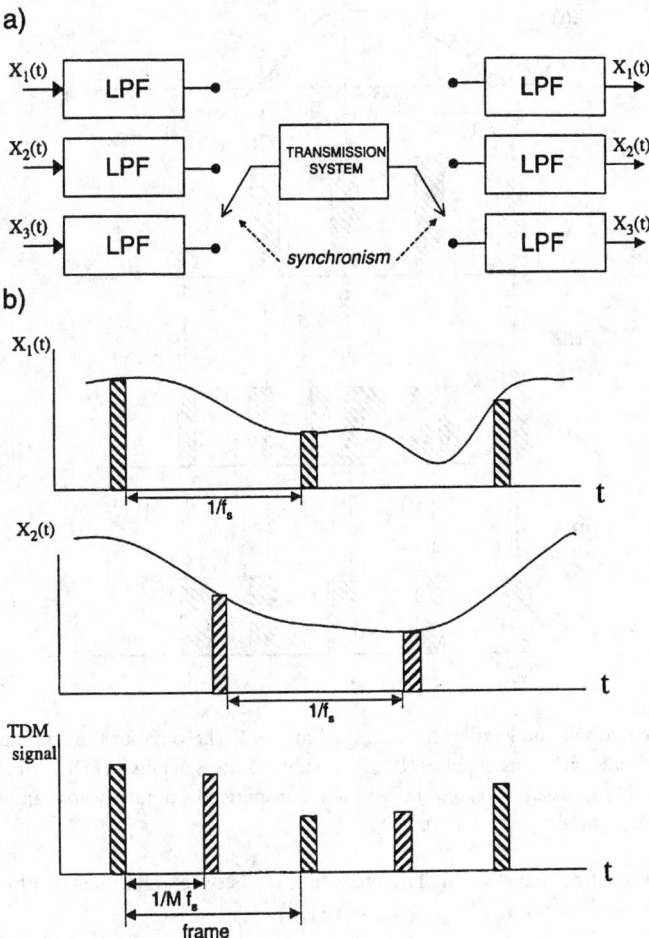

FIGURE 87.7 TDM systems (a) are based on sequentially sampling *M* different channels at a sampling frequency f_s, and sending the information for each channel sequentially (b). In TDM, the synchronism between transmitter and receiver is critical to recovery of the sampled signal. In this figure, the TDM signal is made of only two channels to increase readability. The blocks labeled LPF represent low-pass filters.

M measurement channels, the period between two consecutive pulses is $T_s/M = 1/Mf_s$, where T_s is the sampling period. The period between samples from the same channel is T_s. At the receiver end, by separating the digital signals into different channels by a synchronized demultipler and by low-pass filtering, it is possible to recover the original signal for each measurement channel.

TDM systems have advantages over FDM systems. First, FDM requires subcarrier modulators and demodulators for each channel, whereas in TDM only one multiplexer and demultiplexer are required. Second, TDM signals are resistant to the error sources that originate cross talk in FDM: nonideal filtering and cross modulation due to nonlinearities. In TDM, the separation between channels depends on the sampling system. However, because it is impossible in practice to produce perfectly square pulses, their rise and fall times are different from zero. It is then necessary to provide guard time between pulses, similar to the band guards in FDM systems. Cross talk in TDM can be easily estimated assuming that the pulse decay is exponential with a time constant (τ) approximately equal to

$$\tau = \frac{1}{2\pi B}$$

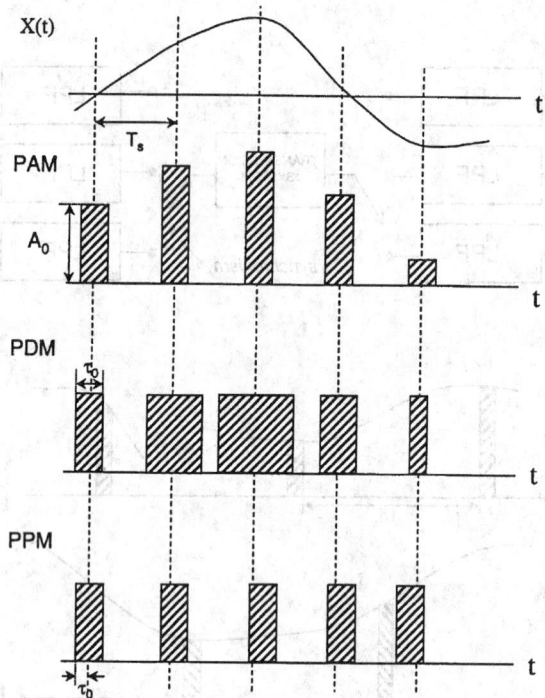

FIGURE 87.8 Different analog modulation schemes used in TDM. The variations in amplitude of the signal $x(t)$ are transmitted as amplitude variations of pulses (PAM), duration changes of pulses (PDM), or changes in the relative position of the pulses (PPM). In all the cases, the level 0 is transmitted by a pulse whose amplitude (A_0), duration (τ_0), or relative position (τ_0) is different from 0.

where B is the -3 dB channel bandwidth. The cross talk (k) between channels can be approximated as

$$k = -54.5 T_g \ (\text{dB})$$

where T_g is the minimum time separation between channels, called guard time.

A common situation in measurement systems occurs when the M signals that need to be measured have very different speeds. The channel sampling rate is determined by the fastest signal, thus needing an M-input multiplexer capable of handling signals at that sampling frequency. A convenient solution is to feed several slow signals into one multiplexer, then combine its output with the fast signal in a second multiplexer [6].

Analog Subcarrier Modulation Schemes for Time Division Multiplexing

In analog modulation for subcarriers the signal that results after the multiplexing and sampling process modulates a train of pulses. The most common methods for analog subcarrier modulation are pulse amplitude modulation (PAM), pulse duration modulation (PDM), and pulse position modulation (PPM). Figure 87.8 illustrates these three modulation schemes, where the pulses are shown square for simplicity. In analog modulation, the parameter that is modulated (amplitude, duration, or relative position) changes proportionally to the amplitude of the sampled signal. However, in PAM and PDM the values have an offset, so that when the value of the sample is zero, the pulse amplitude or the pulse width is different from zero. The reason for these offsets is to maintain the rate of the train of pulses constant, which is very important for synchronization purposes. The common characteristics of the different analog modulation schemes for pulses in TDM are (1) a modulated signal spectrum with a large low-frequency content, especially close to the sampling frequency; (2) the need to avoid overlaying

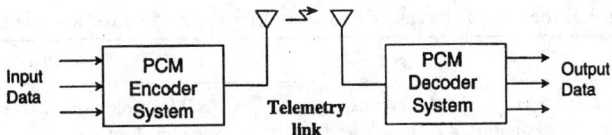

FIGURE 87.9 Block diagram showing a basic PCM link for telemetry.

between consecutive pulses in order to conserve the modulation parameters; and (3) the possibility of reconstructing the original samples from the modulated signal through low-pass filtering after demultiplexing. The reduction of noise depends on the bandwidth of the modulated signal, with this being the principal design criterion.

Pulse Amplitude Modulation.
PAM waveforms are made of unipolar, nonrectangular pulses whose amplitudes are proportional to the values of the samples. It is possible to define the modulation index using similar criteria as in analog AM. Similarly, in PAM the modulation index is limited to values less than 1.

Pulse Duration Modulation.
PDM is made of unipolar, rectangular pulses whose durations or widths depend on the values of the samples. The period between the center of two consecutive pulses is constant. The analysis of the resulting spectrum shows that it is possible to reconstruct the samples by low-pass filtering [9].

Pulse Position Modulation.
PPM is closely related to PDM, as PPM can be generated through PDM. In PPM the information resides on the time location of the pulses rather than in the pulses by themselves. It is then possible to transmit very narrow pulses to reduce the energy needed; this energy reduction is the most important advantage of PPM.

Pulse Code Modulation for Time Division Multiplexing.
All the previously analyzed subcarrier modulation schemes in telemetry systems are based on an analog signal that modulates either an analog carrier or a train of pulses. Pulse code modulation (PCM) is different: it is a digital modulation in which the measured signal is represented by a group of codified digital pulses. Two variations of PCM that are also often used are delta modulation (DM) and differential pulse code modulation (DPCM). In analog modulation schemes, the modulating signal from the transducer can take any value between the limits. If noise alters the modulating signal, it is impossible to decide its real value. Instead, if not all the values in the modulating signal are allowed, and the separation between the allowed levels is higher than the expected noise values, it is then possible to decide which were the values sent by the transmitter. This immunity against noise makes PCM systems one of the preferred alternatives for telemetry. Figure 87.9 shows the basic elements of a PCM telemetry system. A PCM encoder (or PCM commutator) converts the input data into a serial data format suitable for transmission through lines by wireless techniques. At the receiving end, a PCM decoder (or PCM decommutator) converts the serial data back into individual output data signals. PCM systems transmit data as a serial stream of digital words. The PCM encoder samples the input data and inserts the data words into a PCM frame. Words are assigned specific locations in the PCM frame, so the decoder can recover the data samples corresponding to each input signal. The simplest PCM frame consists of a frame synchronization word followed by a string of data words. The frame repeats continually to provide new data samples as the input data change. Frame synchronization enables the PCM decoder to locate the start of each frame easily.

Pulse Code Modulation Telemetry Standards.
IRIG Standard 106-96 also defines the characteristics of PCM transmission for telemetry purposes, in particular, the pulse train structure and system design characteristics. The PCM formats are divided into two classes for Standards purposes: class I and class II. The simpler types are class I, whereas the more

TABLE 87.3   Summary of the Most Relevant PCM Specifications According to IRIG 106-96

Specification	Class I	Class II
Class format support	Class I (simple formats) supported on all ranges	Class II (complex formats) requires concurrence of range involved
Primary bit representation (PCM codes)	NRZ-L, NRZ-M, NRZ-S, RNRZ-L, BiØ-L, BiØ-M, BiØ-S	Same as class II
Bit rate	10 bps to 5 Mbps	10 bps to > 5 Mbps
Bit rate accuracy and stability	0.1%	Same as class I
Bit jitter	0.1 bit	Same as class I
Bit numbering	MSB = bit number 1	Same as class I
Word length	4 to 16 bits	4 to 64 bits
Fragmented words	Not allowed	Up to 8 segments each; all segments in the same minor frame
Minor frame length	<8192 bits or <1024 words (includes synchro)	<16,384 bits (includes synchro
Major frame length	<256 minor frames	Same as class I
Minor frame numbering	First minor frame in each major frames in number 1	Same as class I
Format change	Not allowed	Frame structure is specified by frame format identification (FFI) word in every minor frame

complex types are class II. Some of the characteristics of class II systems are bit rates greater than 5 Mbit/s, word lengths in excess of 16 bits, fragmented words, unevenly spaced subcommutation, format changes, tagged data formats, asynchronous data transmission, and merger of multiple format types, among others. Table 87.3 provides a brief summary of relevant PCM specifications. Readers interested in the detailed specifications and descriptions should refer to Chapter 4 of the IRIG 106-96 Standard [10].

The following PCM codes, shown in Figure 87.10, are recognized by the IRG Standards: NRZ-L (nonreturn to zero — level), NRZ-M (nonreturn to zero — mark), NRZ-S (nonreturn to zero — space), BiØ-L (Biphase — level), BiØ-M (Bi-Phase — mark) and BiØ-S (Biphase — space). The Standard also recommends that the transmitted bit stream be continuous and contain sufficient transitions to ensure bit acquisition and continued bit synchronization. Bit rates should be at least 10 bits/s. If the bit rate is above 5 Mbit/s, the PCM system is classified as class II. In reference to the word formats, the Standard defines a fixed format as one that does not change during transmissions with regard to the frame structure, word length or location, commutation sequence, sample interval, or measurement list. Individual words may vary in length from 4 bits to not more than 16 bits in class I and not more than 64 bits in class II. Fragmented words, defined as a word divided into not more than eight segments and placed in various locations within a minor frame, are only allowed in class II. All word segments used to form a data word are constrained to the boundaries of a single minor frame. The Frame Structure allowed by the Standards for PCM telemetry specifies that data are formatted into fixed frame lengths, that contain a fixed number of equal-duration bit intervals. A minor frame is defined as the data structure in time sequence from the beginning of a minor frame synchronization pattern to the beginning of the next minor frame synchronization pattern. The minor frame length is the number of bit intervals from the beginning of the frame synchronization pattern to the beginning of the next synchronization pattern. The maximum length of a minor frame will not exceed 8192 bits nor 1024 words in class I and will not exceed 16,384 bits in class II. Minor frames consist of the synchronization pattern, data words, and subframe synchronization words if they are used. The Standard allows the use of words of different length if they are multiplexed in a single minor frame. Figure 87.11 shows a graphical representation of a PCM frame structure. Major frames contain the number of minor frames required to include one sample of every parameter in the format. Their length is defined as minor frame length multiplied by the number of minor frames contained in the major frame. The maximum number of minor frames per major frame is limited to 256.

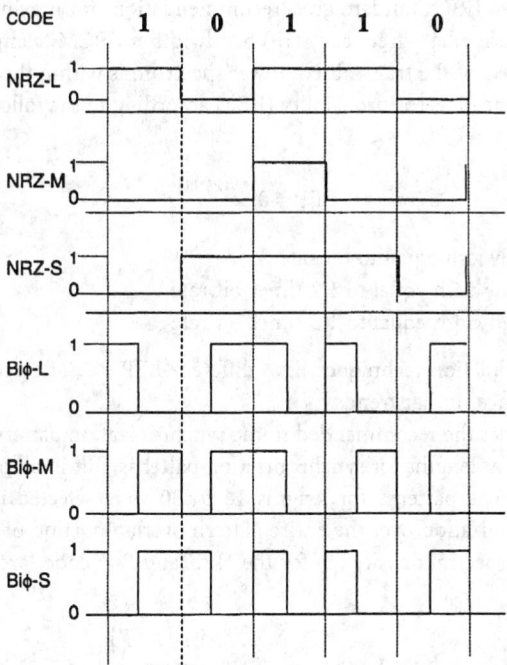

FIGURE 87.10 Different PCM codes. All lower levels in NRZ use a value different from zero. In biphase codes the information resides in the transitions rather than in the levels. In NRZ-L, a 1 is represented by the highest level, while a 0 is represented by a lower level. In NRZ-M, a 1 is represented by a change in level, while a 0 is represented by no change in level. In NRZ-S, a 1is represented by no change of level, while a 0 is represented by a change of level. In Biϕ-L, a 1 is represented by a transition to the lower level, while a 0 is represented by a transition to the higher level. In Biϕ-M, the 1 is represented by no change of level at the beginning of the bit period, while the 0 is represented by a change of level at the beginning of the bit period. In Biϕ-S, a 1 is represented by changing the level at the beginning of the bit period, while the 0 is represented by no change of level at the beginning of the bit period.

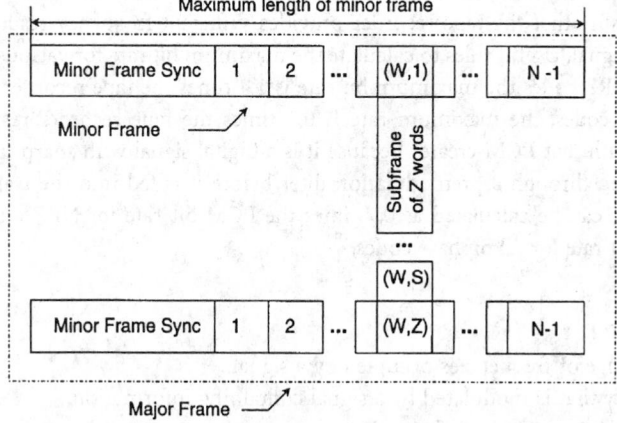

FIGURE 87.11 Structure of a PCM Frame. The maximum length of a minor frame is 8192 bits or 512 for class I and 16,284 bits for class II. A major frame contains $N \times Z$ words, where Z is the number of words in the maximum subframe, and N is the number of words in the minor frame. Regardless of its length, the minor frame synchronism is considered as one word. W is the word position in the minor frame, while S is the word position in the subframe.

Appendix C in the 106-96 IRIG Standard gives recommendations for maximal transmission efficiency in PCM telemetry. The intermediate-frequency (IF) bandwidth for PCM telemetry data receivers should be selected so that 90 to 99% of the transmitted power spectrum is within the receiver 3-dB bandwidth. The IF also has effects on the bit error probability (BEP) according to the following equation for NRZ-L PCM/FM [10]:

$$BEP = 0.5e^{(kSNR)}$$

where $k \approx -0.7$ for IF bandwidth equal to bit rate
$\qquad k \approx -0.65$ for IF bandwidth equal to 1.2 times bit rate
$\qquad k \approx -0.55$ for IF bandwidth equal to 1.5 times bit rate

Other data codes and modulation techniques have different BEP vs. SNR performance characteristics, but in any case they will have similar trends.

The Standard also specifies the recommended frame synchronization patterns for general use in PCM telemetry. There are different lengths for synchronization patterns, but in all of them the 111 is the first bit sequence transmitted. The patterns for lengths 16 to 30 were selected in order to minimize the probability of false synchronization over the entire pattern overlap portion of the ground station frame synchronization [12]. The spectral density (S) for the NRZ and BiØ codes are

$$\text{NRZ Codes} \qquad S = \frac{\sin^2\left(\pi fT\right)}{\left(\pi fT\right)^2}$$

$$\text{Biphase Codes} \qquad S = \frac{\sin^4\left(\pi fT/2\right)}{\left(\pi fT/2\right)^4}$$

The calculation of spectral densities allows the determination of the BEP for the previous type of codes assuming perfect bit synchronization. These calculations show that for the same SNR, the lowest BEP is achieved for NRZ-L and Bi codes, followed by NRZ and BiØ mark and space codes and finally for random NRZ-L codes (RNRZ-L).

Telemetry data are usually recorded onto magnetic tape for later analysis. When recording PCM data, it is important to ensure that the tape recorder provides sufficient frequency response to capture and reproduce the PCM signal. Useful rules to calculate the maximum bit rate for various PCM codes specify that for NRZ and RNRZ codes the maximum bit rate is 1.4 times the tape recorder frequency response while for all biphase codes, the maximum rate is 0.7 times the tape recorder response. To limit the transmission bandwidth that PCM creates because it is a digital signal with sharp transitions, the PCM signal is usually passed through a premodulation filter before it is fed into the transmitter input. The filter cutoff frequency can be calculated as 0.7 times the PCM bit rate for NRZ and RNRZ codes and 1.4 times the PCM bit rate for all biphase codes.

Defining Terms

Bandwidth: The range of frequencies occupied by a signal.
Carrier: A frequency that is modulated by a signal containing information.
Channel: A subcarrier that carries information.
Constant bandwidth (CBW) channel: A channel whose bandwidth is independent of its carrier frequency.
Deviation ratio: The ratio of the maximum carrier frequency deviation to the maximum data frequency deviation.
Frequency deviation: The difference between the center frequency of a carrier and its upper or lower deviation limit.

Frequency division multiplexing (FDM): A composite signal consisting of a group of subcarriers arranged so that their frequencies do not overlap or interfere with each other.

Frequency response: The highest data frequency that can be carried by the channel.

IRIG: Inter-Range Instrumentation Group of the Range Commanders Council (RCC).

Proportional bandwidth (PBW) channel: A channel whose bandwidth is proportional to its carrier frequency.

Remote switching: Telemetry consisting only of yes/no or on/off orders.

Signaling: Telemetry consisting of binary information.

Subcarrier: A carrier combined with other carriers to create a composite signal.

Subcarrier bandwidth: The difference between the upper and lower frequencies of a modulated carrier.

References

1. B.P. Dagarin, R.K. Taenaka, and E.J. Stofel, Galileo probe battery system, *IEEE Aerospace Electron. Syst. Mag.,* 11 (6): 6–13, 1996.
2. M.W. Pollack, Communications-based signaling: advanced capability for mainline railroads, *IEEE Aerospace Electron. Syst. Mag.,* 11 (11): 13–18, 1996.
3. M.C. Shults, R.K. Rhodes, S.J. Updike, B.J. Gilligan, and W.N. Reining, A telemetry-instrumentation system for monitoring multiple subcutaneously implanted glucose sensors, *IEEE Trans. Biomed. Eng.* 41: 937–942, 1994.
4. M. Rezazadeh and N.E. Evans, Multichannel physiological monitor plus simultaneous full-duplex speech channel using a dial-up telephone line, *IEEE Trans. Biomed. Eng.,* 37: 428–432, 1990.
5. Y.S. Trisno, P. Hsieh, and D. Wobschall, Optical pulses powered signal telemetry system for sensor network application, *IEEE Trans. Instrum. Meas.,* 39: 225–229, 1990.
6. R. Pallás-Areny and J.G. Webster, *Sensors and Signal Conditioning,* New York: John Wiley & Sons, 1991, 352–379.
7. Part 90: Private land mobile radio services. Subpart J: Non voice and other specialized operations (Sec. 90.238: Telemetry), *Code of Federal Regulations,* Title 47, Telecommunication. Chapter I: Federal Communications Commission.
8. Part 15: Radio frequency devices, *Code of Federal Regulations,* Title 47, Telecommunication. Chapter I: Federal Communications Commission.
9. H. Taub and D. L. Schilling, *Principles of Communication Systems,* 2nd ed., New York: McGraw-Hill, 1986.
10. IRIG, *Telemetry Standard IRIG 106-96,* Range Commander Council, U.S. Army White Sands Missile Range, 1996.
11. K. M. Uglow, Noise and bandwidth in FM/FM radio telemetry, *IRE Trans. Telemetry Remote Control,* 19–22, 1957.
12. J.L. Maury and J. Styles, Development of optimum frame synchronization codes for Goddard space flight center PCM telemetry standards, *Proc. Natl. Telemetering Conf.,* 1964.

Further Information

E. H. Higman, Pneumatic instrumentation, in B.E. Noltingk, Ed., *Instrument Technology: Instrumentation Systems,* Vol. 4, London: Butterworths, 1987.

C.H. Hoeppner, Telemetry, in R. C. Dorf, Ed., *The Electrical Engineering Handbook,* Boca Raton, FL: CRC Press, 1993.

R.S. Mackay, *Bio-Medical Telemetry,* 2nd ed., Piscataway, NJ: IEEE Press, 1993.

Telemetry Group, *Test Methods for Telemetry Systems and Subsystems. Document 118-97,* Range Commanders Council, U.S. Army White Sands Missile Range, 1997.

Telemetry Group, *Telemetry Applications Handbook Document 119-88,* Range Commanders Council, U.S. Army White Sands Missile Range, 1988.

88

Sensor Networks and Communication

Robert M. Crovella
NVIDIA Corporation

88.1 Introduction

What Is a Communication Network?

A communication network provides a system by which multiple users may share a single communication path (or medium) to exchange information. The telephone system is an example of a system containing many communication networks, which can be considered to be a single communication network as an abstract example. Communication networks are commonly used in various industries and applications to provide an economical means to allow multiple, geographically separated users to exchange information.

Ordinary Sensors vs. Networked Sensors

A definition of the function of a sensor is to map or convert one measured variable (e.g., spatial, mechanical, electromagnetic, etc.) into another — usually electric — variable or signal. This signal may then be passed to a measurement or processing system for capture and analysis, or as a direct input to some controlled process. In this case, the measured variable is represented as an electric signal. This signal must be handled individually by the measurement system, and it may also be subject to corruption from a variety of sources, such as electromagnetic interference in the case of an electric signal.

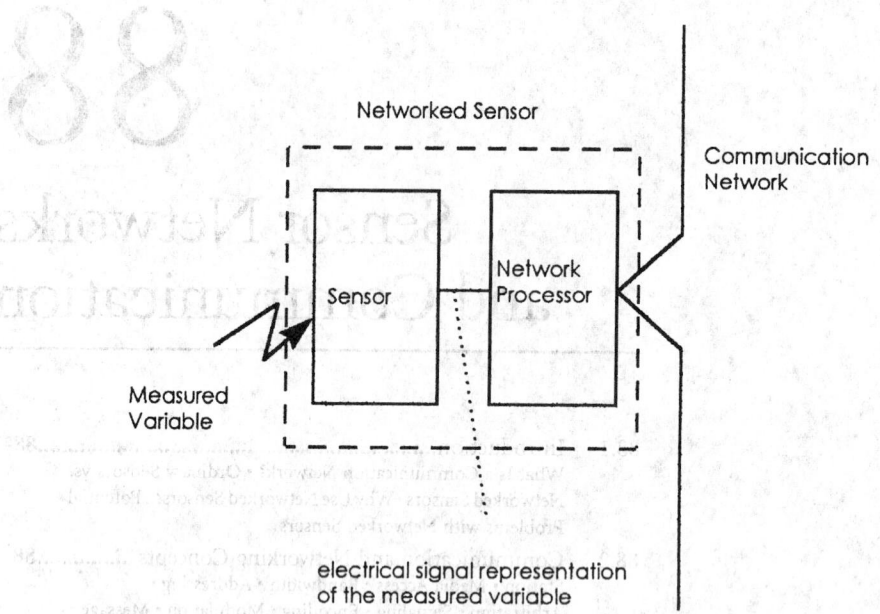

FIGURE 88.1   A networked sensor is an ordinary sensor with network communication components added.

In applications where a number of sensing devices are needed, and/or where the sensing devices are distributed geographically (or are distant from the measurement and analysis system), the application designer may wish to use a communication network to transmit sensor data from the measurement point to the measurement and analysis system. Such applications typically involve some sort of digital computing machinery at the measurement and analysis point, or as part of the control system. Figure 88.1 depicts a representative block diagram of a sensor/network system showing the relationship of the various components.

Networked sensors can be distinguished into two components: those performing the measurement function, and those components performing the communication function. In some cases, these two functions may be designed as a single unit, such that the "sensor" intrinsically includes communication capability. In other cases, an ordinary sensor may be connected to a conversion unit, which converts the output signal of the ordinary sensor into a form suitable for the network, and manages the delivery of this information on the network.

Why Use Networked Sensors?

Network communication combined with sensor technology can provide several benefits to an application, as well as to the sensor designer. The most obvious benefit of a network is the simplification of the wiring for the transmission of the signals from one place to another. For a system containing N users, the number of wires or cables T required to individually connect each user with each other user is given by Equation 88.1:

$$T = 2^{(N-1)} - 1 \qquad (88.1)$$

assuming each wire or cable can carry information in both directions between the two users connected by that cable. For more than a few users, the number of cables required (T) to provide an individual connection between each pair of users is large. Sensors are often connected to a central measurement and analysis system, and may only need to communicate with the central system. In this case, the number of individual wires or cables needed is equal to the number of sensors $(N-1)$. Even with this smaller number of cables, the wiring for a large number of sensors in some applications can be quite complex. A network may be able to reduce the total number of cables required to a much smaller number. In fact, in a sensor network, all of the sensors and the central measurement and analysis system can be connected to a single cable.

An indirect benefit of networking may be in its handling of the sensor signal. Because most modern networks are digital in nature, an analog sensor signal typically must be digitized before it can be transmitted on a network. With a networked sensor, the digitization will typically be carried out by circuitry in relatively close proximity to the sensor. As a result, the analog signal will have traveled a short distance before being converted to a digital signal. This can be a benefit in two ways. The first is that the analog signal will not suffer as much attenuation or degradation due to electric losses associated with carrying a signal over a great distance. The second is that once in digital form, the "signal" can be made relatively immune to the effects of distortion or degradation due to electromagnetic interference (EMI). Although digital transmission of signals is still subject to EMI, modern protocols and transmission systems can be designed to be very robust, using signaling that is resistant to EMI as well as using error control techniques. As a result, the effect of attenuation and disturbances can be essentially eliminated by digital transmission of the signal.

Another benefit of networking is the ability to communicate a much wider range of information — in both directions — when compared with a single cable carrying a sensor signal. With many modern networks suitable for networked sensing applications, a microprocessor is used at the sensor to manage the handling of the sensor signal and its transmission on the network. But there is generally no need to limit the microprocessor to this one function alone. The combination of the network and the microprocessor provides a platform upon which many additional functions and features can be incorporated into the networked sensor. For example, the signal of a sensor may need a certain calibration or correction function applied to it before it can be used in calculations. It may be beneficial to load into the networked sensor (through the network) a set of correction parameters or coefficients, and then have the microprocessor correct or calibrate the output of the sensor before transmitting it to the network. Sensors can be easily designed to have multiple sensing functions, such as temperature and pressure, for example. Each signal can be handled separately and transmitted separately on the network, with no need for additional connections. Sensors may be designed to store certain types of information, such as the name of the manufacturer, or certain calibration parameters determined by the manufacturer at the time of manufacture. This information can then be read out over the network and used for a variety of purposes. A sensor can even be designed to have "intelligent" functions, such as the ability to sense its environment and determine when certain parameters have been exceeded (such as operating temperature range), or report a special message containing an "alarm" when the sensor signal level exceeds a certain threshold. The combination of the network and the microprocessor leads to an endless variety of functions and features that can be added to the basic sensor technology.

Potential Problems with Networked Sensors

Networked sensors will generally require more complex circuitry than equivalent, nonnetworked sensors. A drawback of analog-to-digital (A/D) conversion and digital transmission of signals is the time and level quantization effect that A/D conversion can have on the analog signal. These effects can be mitigated with modern, high-speed A/D converters (to minimize the effect of time quantization, or the sampling effect) with the ability to convert in high resolution (i.e., using a large number of digital bits to represent the analog signal level). These drawbacks are not unique to networked sensors but rather to digitized sensor values and digital control whether or not it uses a network. Finally, the capacity of the network to carry information (the bandwidth) must be considered in any communication system. Putting a large number of sensors on a single network may overload the information-carrying capability of the network, resulting in queuing delays in the reception of sensor signals and, in some cases, lost data.

88.2 Communication and Networking Concepts

In order to be able to select an appropriate network technology, it is necessary to understand some basic terminology so that the features and capabilities of various networks and technologies can be categorized and compared.

Station

A station represents a single communicating element on a network system. Each user of the network must access the communication capability of the network via a station. Each station will typically have some implementation of the open systems interconnection (OSI) network reference model as the means of utilizing the network system.

Media Access

Media access is the method by which individual stations determine when they are permitted to transmit, or "use" the media. Media access control (MAC) is a function that is usually performed in the data link layer of the OSI reference model. Some well-known methods of media access control include carrier sense multiple access with collision detection (CSMA/CD) and token passing. CSMA/CD systems (such as Ethernet) allow all stations on a network equal access. Each station must "listen" to the network to determine periods of inactivity before transmitting. Any station wishing to use the network may begin transmitting providing the network is inactive when it checks for activity. If multiple stations attempt to transmit simultaneously, a collision occurs. This is detected by all transmitting stations, which all must immediately stop transmitting and each wait a randomly determined period of time, before attempting to use the network again. Controller area network (CAN), for example, uses a variant of CSMA/CD for media access. Token-passing systems have a logical "token" which is exchanged among stations via network messaging. The station that holds the token has permission to transmit. All other stations are only permitted to receive messages. Stations wishing to transmit but not having the token must wait until the station holding the token passes it on. Another commonly used method of media access control is master–slave. In this method, one station on the network (designated the master) is generally in charge of, and originates, all communications. Slaves only respond to the master, and only respond when the master initiates communications with them via sending a message to the slave. Profibus-FMS (see below) is an example of a protocol which uses both token passing (in some cases) and master–slave (in some cases) to control media access.

Bandwidth

Bandwidth may have several different definitions. For digital communication systems, bandwidth describes the capacity of the system to transport digital data from one place to another. This term may be applied to the raw capability of the physical and data link layers to transport message data (*raw bandwidth*, closely related to the bit rate concept) or it may be applied to the effective rate at which user-meaningful information is transported (*effective bandwidth*). The bandwidth of a given system is generally inversely proportional to the worst-case node-to-node distance. The smaller the network span, the higher its bandwidth can be.

Addressing

Addressing is a concept that assigns generally unique identifiers to each station in a network system. This identifier (the address) can then be used by the network for a variety of purposes, including identifying the origin and/or destination of messages, or arbitrating access to a shared communications medium. Another addressing or identifier concept assigns unique identifiers not to stations, but to unique pieces of data or signals that will be carried by the network. Stations then use an identifier according to what type of data they will be transmitting. Many, but not all networking methods require establishment of an explicit address for each network station.

Arbitration

Arbitration is a function closely related to MAC. Arbitration is used by some networks to define the procedure followed when multiple stations wish to use the network simultaneously.

Signaling

Signaling refers to the actual physical (e.g., electrical, optical, or other) representation of data as it is carried on the media. For example, in some networks, data elements may be represented by certain voltage levels or waveforms in the media. In other networks, data elements may be represented by the presence of certain wavelengths of light in the media. The association of all the representable data elements (e.g., 0/1 or on/off) with the corresponding signal representations in the media is the signaling scheme or method. An important signaling method where electric wires are used as the medium is differential signaling. Differential signaling represents a particular data element (1 or 0) as two different states on a pair of wires. Determining the data element requires measuring the voltage difference between the two wires, not the absolute level of the voltage on either wire. Different data elements are then represented by the (signed) voltage difference between the two wires. For example, RS-485 represents a digital 1 data element as a 5-V signal level on the first wire and a 0-V signal level on the second wire, and a digital 0 as a 0-V signal level on the first wire and 5-V signal level on the second wire. One of the principal benefits of differential signaling is that it is possible to determine the data being transmitted without knowing the ground reference potential of the transmitter. This allows the transmitter and receiver to operate reliably, even when they have different ground potentials (within limits), which is a common occurrence in communication systems.

Encoding

Encoding refers to the process of translating user-meaningful information into data elements or groups of data elements to be transported by the network system. A code book refers to the set of all relationships between user-meaningful information and data carried by the network. Encoding may occur at several levels within the OSI reference model, as user-meaningful information is transformed successively until it becomes an actual network message, produced by the data link layer. Decoding is the reverse process, whereby a network message is successively translated back into user-meaningful information.

Modulation

Modulation in a classical sense refers to a signaling technique by which data or information is used to control some combination of the frequency, phase, and/or amplitude of a carrier signal. The carrier signal carries the information to a remote receiver where it will be demodulated to retrieve the information. Modulated network systems are outside the scope of this chapter.

Message

A message is the fundamental, indivisible unit of information which is exchanged between stations. User-meaningful information will be grouped into one or more messages by the OSI network reference model.

Multiplexing

Multiplexing refers to the ability to use the media in a network to carry multiple messages or information streams "simultaneously." Multiplexed systems allow several communication channels to use the same physical wire or media. Each message or information stream may have different sources and destinations.

Multiplexing may be accomplished using a variety of means. Time division multiplexing (TDM) involves breaking access to the media into a series of time quanta. During each time quantum, the media carries a separate message or information stream. The close arrangement of time quanta allows the network media to carry multiple messages "simultaneously." Code division multiplexing (CDM) involves the separation of the code book (see Encoding) into sections. Each section of the code book provides all of the messages that will be used for a particular information stream. Therefore, a particular information stream within the network media is distinguished by all of the messages that belong to the section of the code book for that stream. Frequency division multiplexing (FDM) divides an available bandwidth of a communication channel into several frequency ranges, and assigns one information stream to each frequency range.

Protocols

A protocol is a defined method of information exchange. Protocols typically are defined at several levels within the OSI network reference model, such as at the application layer and at the data link layer. Protocols are used to define how the services provided by a particular layer are to be exercised, and how the results of these services are to be interpreted.

Service

A service represents a specific function or operation that is supported by a particular layer in the OSI network reference model. For example, an application layer service might be provided for the reading of or writing to a data element contained in another device (or station) on the network. This service might make use of a data link layer service which might be provided for supporting the exchange of a message with another device (or station) on the network.

Topology

Topology refers to the physical or geographic layout or arrangement of a network. Certain types of canonical topologies are commonly discussed in the context of networks, such as trunkline/branchline, star (or hub), ring, and daisy chain.

Bit Rate

Bit rate refers to the speed at which binary pieces of information (bits) are transmitted on a particular network. The raw bit rate of a network generally refers to the actual speed of transmission of bits on the network. The effective bit rate — or throughput — generally refers to the speed at which user information is transmitted. This number is less than or equal to the raw bit rate, depending on what percentage of the bits transmitted is used for carrying user information. The bits not carrying user information are overhead, used to carry protocol, timing, or other network information.

Duplex (Half and Full Duplex)

Half duplex refers to a communication system in which a station can either transmit information or receive information, but not both simultaneously. A full duplex network allows a station to transmit information and receive information simultaneously.

Error Control

Many network systems provide mechanisms to control errors. Error control has four aspects: prevention, detection, correction, and isolation. Error prevention may simply be shielding for the media to minimize electromagnetic disturbances, or it may be more complicated, such as signal sampling control to optimize the probability that a signal will be in the correct state when sampled. Error detection generally depends on detecting violations of protocol rules at various network levels, or violations of computed data added to a message for error control purposes. Some examples of error detection techniques are parity and cyclic redundancy check (CRC). Both methods involve the computation of additional bits of information based on the data that is contained in a message, and appending these bits to the message. For example, a particular protocol may require that the data link layer compute and append a CRC to a message prior to transmission. The receiver of the message may then also compute the CRC and compare it to the CRC which has been appended to the message. If a mismatch exists, then it is assumed an error has occurred. Error correction may take on a variety of forms. One of the simplest methods of error correction is to require that the data link layer of the transmitter retransmit a message which has been detected to have an error during transmission. This method is based on the assumption that the error was caused by a disturbance which is unlikely to occur again. Another method of error correction involves transmission of additional bits of information along with the user information in a message. These additional bits of information are computed by the transmitter to provide redundant information in the message. When fewer than a certain number of bit-level errors have occurred during the transmission of the message, the receiver is able to reconstruct the original user information accurately using the redundant information (bits) supplied within the message. Error isolation is a capability of some networks to localize the source of errors and isolate the sections of the network or the stations at which the errors have been localized. Error isolation allows the fault-free portions of the network to continue communicating even when other portions of the network have degraded to the point of generating errors.

Internetworking

There are occasions when communications between two or more points are best handled by multiple networks. This may be the case when a single network has limitations that prevent it from tying the points together (e.g., distance limits) or when multiple networks are required for other reasons (e.g., to carry different types of data). When multiple networks are used to provide communications, there may be a need to pass messages or information directly from one network to another.

A repeater may be used when the networks to be joined are logically identical, and the purpose is simply to extend the length of the network or extend its capabilities in some way. A repeater generally has no effect on messages, and simply carries all messages from one cable or port to another (i.e., a change of physical media). A repeater allows for connection of networks at the physical layer level.

A bridge is similar to a repeater, but allows for connection of networks at the data link layer level. Generally, a bridge will pass all messages from one network to another, by passing messages at the data link level.

A router usually has the function of partitioning similar networks. Two networks may be based on the same technologies and protocol, but may not be logically identical. In these cases, some, but not all, of the messages on one network may need to be carried or transported to the other network. The router has the function of determining which messages to pass back and forth based on certain rules. Functions to enable efficient, automatic routing of messages may be included in layer 3 (the network layer) of the OSI network reference model, and a router allows for connection of networks at the network layer level.

A gateway may have a function similar to a router, or it may have the function of joining dissimilar networks, i.e., networks based on dissimilar technologies and/or protocols. When functioning like a router, a gateway usually performs its discrimination at a higher protocol level than a router. When a gateway joins dissimilar networks, generally a more complex set of rules must be designed into the gateway so that message translation, mapping, and routing can occur within the gateway as it determines which messages to pass from one network to the other.

ISO/OSI Network Reference Model

The explosion in the use and types of communication networks over the last several decades has led to more precise descriptions and treatment of communication networks in general. The International

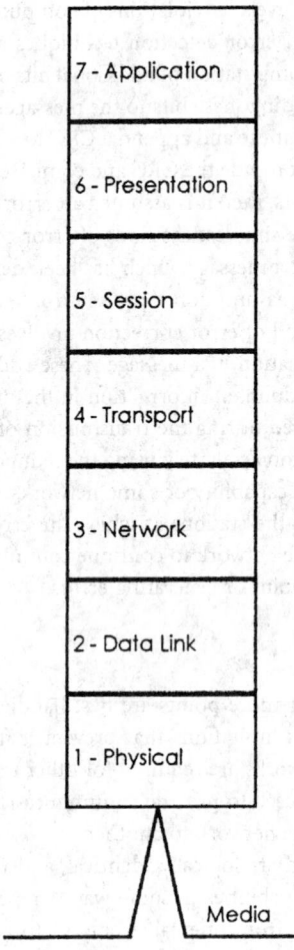

FIGURE 88.2 The ISO-OSI Seven Layer Model provides a method for segmenting communication functions

Organization for Standardization (ISO) has recognized one such method of precise description of networks, called the OSI reference model [1]. As shown in Figure 88.2, this model decomposes an arbitrary communication network into a "stack" of seven "layers." At each layer, certain types of network communication functions are described. The user of the communication system — usually another system that needs to communicate on the network — interacts with layer 7, the highest layer. The actual transmission medium (e.g., copper cable, fiber optic, free space, etc.) is connected to layer 1, the lowest layer. Most communication networks do not implement all of the layers in the reference model. In this case, formal definition, treatment, or inclusion of certain layers of the model in the actual network design are omitted. Layers 1, 2, and 7 are typically present in all networks, but the other layers may only be explicitly included or identifiable when their function is an important part of the network communications. In many sensor communication networks, the functions performed by layers 3,4,5, and 6 are "collapsed" into vestigial additions to the functions of layer 7, the application layer.

Physical Layer

The physical layer is the lowest layer of the model. This layer is responsible for converting between the symbolic or data representation of the network messages and the actual physical representation of data in the network medium. This layer specifies the behavior of the electric circuits referred to as the transmitter and the receiver. It also defines physical structures for connectors.

Data Link Layer

The data link layer, or layer 2, is responsible for several functions. This layer manages access to the network medium (MAC), structures the bits of information into well-defined groups identified as "frames" or messages, handles identification of source and destination stations on the network, and provides for error-free transmission of a message from source to destination stations, all according to the data link layer protocol. A number of standard data link layer protocols exist, which act as the basis for many of the communication networks in wide use. Ethernet, or IEEE 802.3, for example, specifies a MAC sublayer that works with the IEEE 802.2 Logical Link Control layer to form the data link layer protocol used in the majority of office information networks [2].

Network Layer

The network layer encapsulates functions related to routing of messages, both within a single network and among multiple networks. This layer typically uses addressing in a variety of forms as a key part of the functions of directing and routing messages, and the search and usage of the available communication paths.

Transport Layer

The transport layer provides any additional data transfer functions not directly provided by the data link layer for end-to-end reliable messaging. For example, some data transfer functions between stations may require the use of multiple data link layer messages to accomplish a reliable message transfer. The generation of multiple messages and the sequential disassembly, delivery, and assembly of data is accomplished by the transport layer. The transport layer also recovers from lost, duplicated and misordered messages.

Session Layer

The session layer provides for a higher level of control and management of network usage and data flow than that provided at lower layers, including opening or building up a communication channel, maintaining the channel, and closing the channel. This layer is infrequently implemented in contemporary systems.

Presentation Layer

The presentation layer provides functions to transform data from formats that are transportable by the network to the user-accessible formats that are defined in the application layer and understood in the local station.

Application Layer

The application layer, or layer 7, provides communication services directly to the user application. The usage and formatting of these services is summarized in the application layer protocol. The user interacts with the network by invoking functions and services provided by the application layer and passing data to and from the network through these services.

88.3 Network Technologies

There is a wide range of technologies in various stages of development and standardization, which address virtually all levels or layers of the ISO/OSI network reference model. One or more of the available technologies will probably suit almost any networking need. An analysis of the available technologies and their limitations will also be beneficial if it is deemed that a networking method must be designed to meet a particular application. The selection and description of technologies is by no means complete or exhaustive. The technologies presented are selected from several industries which make common use of networking to communicate sensor data. Figure 88.4 provides a comparison of selected parameters for a set of networks.

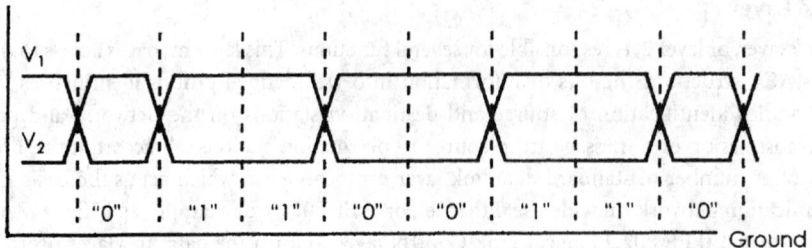

FIGURE 88.3   A sample RS-485 waveform showing voltages on differential wire pair (V_1, V_2) and superimposed bit intervals showing 0 and 1 bits. The ground reference is arbitrary within the defined signaling range.

RS-232

RS-232 (ANSI/EIA/TIA-232-E-91) is a widely used method of communication, which has been standardized in a variety of places including the Electronics Industry Association [3]. RS-232 represents elements of layer 1 of the OSI model, for communicating between two (and only two) stations. RS-232 provides a separate wire for transmission of data in each direction between the two stations, and gives the two stations different designations — data terminal equipment (DTE), and data communications equipment (DCE) — so that a method exists to distinguish which station will use which wire to transmit and receive. The signal levels for RS-232 represent a digital 1 bit as a voltage in the range of 5 to 12 V on the wire, and a digital 0 bit as a voltage of negative 5 to 12 V on the wire. RS-232 is typically implemented in a full duplex fashion, since each station can transmit to the other simultaneously using separate wires. RS-232 can be made to operate at a variety of bit rates, but typically is used at bit rates from 300 bit/s up to 115,200 bit/s.

RS-485

EIA RS-485 was made a standard in 1983, derived from the RS-422 standard. RS-485 provides for differential transmission of data on a pair of wires among 32 or more stations. Like RS-232, the standard is a layer 1 specification. RS-485 provides for half duplex communication, since a station cannot simultaneously transmit and receive independent data streams. Each station in an RS-485 system can have either a transmitter or a receiver, or both (commonly called a transceiver). Most implementations provide a transceiver. When one transceiver is transmitting, all others should be receiving (i.e., not transmitting). Which station is allowed to transmit at which time is not specified in the standard, and must be covered by a higher layer protocol (e.g., Interbus-S, Profibus-DP). Figure 88.3 shows a sample RS-485 waveform, indicating the differential nature of the signaling.

Seriplex[1]

Seriplex® is a digital, serial multiplexing system developed by Automated Process Control, Inc., in Jackson, MS. Square D Corporation purchased Automated Process Control and the rights to Seriplex in 1995, and subsequently launched Seriplex Technology Organization (STO) to manage the protocol. Seriplex is designed to be particularly efficient at handling large numbers of digital or on/off input and output points. Seriplex provides three communication wires, one for a clock signal, one for a data signal, and a ground reference. The system can be operated in two different modes (peer-to-peer and master–slave). In master–slave mode, one station is designated the master. The master synchronizes all data transmission among stations by driving a digital waveform on the clock line which all stations listen to and use for timing of transmit and receive operations. The master generates a repetitive pattern on the clock line

---

[1]Seriplex is a trademark of the Seriplex Technology Organization.

	Length	Stations	Bit Rate	Wires	Media	Topology
RS-232	30 m	2	115 kb/s	2	TP	P-P
RS-485	1200 m	32	10 Mb/s	2	TP	D-C
Seriplex	1500 m	256	200 kb/s	4	2STP	D-C,Free
AS-i	100 m	32	167 kb/s	2	UP	T-B
Interbus-S	25.6 km*	64	500 kb/s	6	3STP	Ring
CAN	450 m	64	1 Mb/s	4	2STP	T-B
4-20 mA	1000 m	2	-	2	STP	P-P
HART	1000 m	2(15)	1200 b/s	2	STP	P-P(D-C)
Profibus	9600 m	126	12 Mb/s	2	STP	D-C
Found. Fieldbus	1900 m	32	2.5 Mb/s	2	STP	D-C
LonWorks	1400 m	64	1.2 Mb/s	2	STP	D-C,Free

FIGURE 88.4 A comparison of selected parameters (maximum values) for various network technologies.** *Notes:* P-P = point to point; D-C = daisy-chain; T-B = trunkline-branchline; TP = twisted pair; STP = shielded twisted pair; UP = unshielded pair. * Maximum 400 m between stations. ** Maximum parameters for networks are not achievable simultaneously, and do not include repeaters, routers, or gateways. Maximum parameters are estimates based on available information.

which causes all stations to transmit and/or receive data on each cycle, or "scan" of the network. Each station is given an address, and uses the address along with the clock signal to determine when to drive the data line (in the case of an input point) or when to monitor the data line for valid output data (in the case of an output point). There are variations possible in implementation which allow for various clock speeds and bit rates (16, 100, and 200 kHz). Other protocol details allow for the handling of analog or multibit input and output points (by combining several bits on sequential scans together), bus fault detection, input redundancy, and communication error control using multiple scans of the network. Implementing the protocol in a sensor or other device typically requires using a Seriplex ASIC (Application Specific Integrated Circuit) which must be licensed from the STO [4].

AS-i

Actuator Sensor Interface, or AS-i, was developed by a consortium of primarily European companies interested in developing a low-cost, flexible method for connecting sensors and actuators at the lowest levels of industrial control systems. The system is managed by an independent worldwide organization [5]. The AS-i system provides a two-wire, nontwisted cable for interconnection of devices. Devices may draw current from the two wires (nominally at 24 V dc) for powering circuitry, and the data communications are modulated on top of the nominal dc level at a bit rate of 167 kHz, under the control of the master. A single parity bit per station is used for error detection. Similar to Seriplex, an AS-i device is typically implemented using a special ASIC which handles the communication.

Interbus-S

Interbus-S was developed by Phoenix Contact [6] and is controlled by the Interbus-S Club. The topology of the network is a ring, with data being sequentially shifted from point to point on the ring under the control of a network master. Each device in the ring acts as a shift register, transmitting and receiving data simultaneously at 500 kHz. The actual serial data transmission between stations conforms to RS-485. Interbus-S transmissions include a CRC for error detection. Interbus-S (Interbus-S Remote Bus) has also been extended to include a subprotocol called Interbus-Sensor Loop (or Interbus-S Local Bus). This subprotocol provides an alternate physical layer, with a single twisted pair carrying power and data on the same lines, and a reduction in the minimum size of the shift register in each station from 16 to 4 bits. Each Interbus sensor loop system can act as a single station on an Interbus-S network, or the sensor loop can be connected directly to a controller or master. Interbus-S devices are usually implemented with a special ASIC.

CAN

Controller Area Network (CAN) is a data link layer (layer 2) network technology developed by Robert Bosch Corporation [7], with an application target of onboard automotive networking. The technology is standardized in ISO 11898 [8], licensed to all major integrated circuit manufacturers, and is widely available — both as separate CAN controllers as well as CAN controllers integrated with microprocessors. As a result, CAN has been used in a variety of industries. As a data link layer technology, it is not a complete network definition. A number of physical layer options are usable with CAN (e.g., twisted pair, fiber optic, radio frequency wireless) and some have been subject to standardization (e.g., ISO 11898). Also, a number of application layer protocols have been developed for use with CAN, such as DeviceNet, Smart Distributed System (SDS), CANOpen [9], and SAE J1939 [10]. Both DeviceNet [11] and Smart Distributed System [12] have developed systems for creating networks of industrial field devices for the factory floor, including sensors and actuators.

4 to 20 mA Current Loop

The 4 to 20 mA current loop is a widely used method for transferring information from one station (the transmitter) to another station (the receiver). Therefore, this system allows for only two stations. A typical current loop system assigns a sensing range (e.g., 0 to 100°C) to the current range between 4 and 20 mA. A loop exists (i.e., two wires) between the transmitter and receiver. The transmitter can impress a certain current in the loop (using a controlled current source) so that the receiver can measure the current in the loop (e.g., by placing a small resistor in series with the loop and measuring the voltage drop across the resistor). After measuring the current, the receiver can then determine the present level of the sensed signal within the defined sensing range. This method uses current signaling, instead of voltage signaling, and therefore is relatively unaffected by potential differences between the transmitter and the receiver. This is similar to the benefit of differential (voltage) signaling, which also requires two wires. Another characteristic of this method is that it is not primarily digital in nature, as many other sensor communication systems are. The measured value can vary continuously in the range of 4 to 20 mA, and therefore can easily represent an analog sensing range, rather than a set of digital signals. Also, the signal is continuously variable and available. Another characteristic of this method is that the integrity of the loop can be verified. As long as the loop is unbroken and the transmitter is in good working order, the current in the loop should never fall below 4 mA. If the current approaches 0 mA, then the receiver can determine that a fault exists — perhaps a broken cable. These systems are widely used in various process control industries (e.g., oil refining) for connecting sensors (transmitters) with control computers. Because one station is always the transmitter and one station is always the receiver, this is a unidirectional, half duplex communication system.

HART[2]

HART® is a protocol which builds upon 4 to 20 mA communication systems. The basic idea is that additional data (beyond the basic sensor signal being carried in the current loop) can be transmitted by modulating a signal on top of the current flowing in the loop. The actual modulation method conforms closely to the Bell 202 standard for analog modem communications on telephone lines at 1200 bit/s. Because a 4 to 20 mA current loop carries a relatively slowly varying signal, it is easy to separate the 4 to 20 mA signal from the digital signal using filters. The Bell 202 standard uses continuous-phase frequency shift keying between two frequencies at up to 1200 shifts/s to modulate digital ones and zeros onto the 4 to 20 mA current loop. This method allows for bidirectional, full duplex communication between the two stations, on top of the 4 to 20 mA signal. It is also possible to configure HART communications on a network that is not carrying a 4 to 20 mA signal, in which case up to 15 devices can be connected together on the network. HART was developed by Fisher-Rosemount Corporation, and has been trans-

---

[2]HART is a trademark of the HART Communications Foundation.

ferred to an independent foundation for management [13]. Because HART is compatible with U.S. telephone systems, it can theoretically be run over the telephone line and is therefore capable of running over arbitrarily long distances.

Profibus

Profibus (PROcess FIeld BUS) is one of three networks standardized by a European standard [14]. Profibus is under the control of a global organization, PNO [15]. Profibus is an umbrella network standard which encompasses three subnetworks within the Profibus family. Profibus-DP (Distributed Periphery) is the variant which is designed specifically for communication with field devices (sensors and actuators) at the device I/O level. Profibus-PA (Process Automation) is a variant which has more capabilities designed to support the needs of device-level networking for process industries, such as oil refining. One of the capabilities of Profibus-PA is its ability to be installed in an intrinsically safe way, thus providing a higher degree of safety in environments which may be explosive or otherwise hazardous. Profibus-PA typically uses a special physical layer specification standardized under IEC 1158-2, which is used by several network systems for process automation applications. IEC 1158-2 specifies a two-wire twisted pair implementation carrying both power and data on the same two wires at 31.25 kbit/s. Profibus-FMS (Fieldbus Messaging Specification) represents the highest level implementation, which is used to link together controllers (not field or I/O devices) in a factory.

Profibus-DP systems are typically master–slave systems, where usually a single network master (the host controller) communicates with a number of slave devices (remote I/O blocks and other I/O devices). The protocol provides for cyclic exchange of I/O information as well as on-demand exchange of other types of information. Profibus-DP can be implemented on several different physical layers, including RS-485 and fiber optics, at various bit rates up to 12 Mbit/s. Profibus messages include a CRC for error detection.

Foundation Fieldbus

Foundation Fieldbus (FF) is a networking standard which has grown out of an effort within industry standards organizations, especially ISA-SP50 [16], and IEC SC65C/WG6 [17], to provide a replacement for the 4 to 20 mA analog sensor communication standard. FF provides two basic levels of networking: H1 and H2. H1 is a lower-speed system that can provide intrinsically safe (IS) operation and uses a single twisted pair to deliver both power and data communications to field devices, according to IEC 1158-2. Running at a bit rate of 31.25 kbit/s, H1 is very similar to Profibus-PA, when run on the IEC 1158-2 physical layer standard. The H1 system is designed to be able to connect hierarchically "upward" to an H2 system, which acts as the host. FF H2 can be run at either 1 or 2.5 Mbit/s on twisted pair wires, and also provides an IS option at the 1 Mbit/s rate. The H2 system can act as a network backbone in a factory environment, carrying data among various H1 systems.

WorldFIP

WorldFIP [18] is another technology of the three that were standardized in the European standard EN 50 170, running on the IEC 1158-2 physical layer. Many of the proponents of WorldFIP have embraced FF, and contributed to the development of that standard. WorldFIP is a member of the FF, and FF has incorporated many of the capabilities of WorldFIP as a result. When run on the IEC 1158-2 physical layer, WorldFIP has similar capabilities to FF.

LonWorks[3]

LonWorks® is a networking technology developed and controlled by the Echelon Corporation [19]. LonWorks is designed to be a general-purpose networking technology suitable for a variety of industries.

---

[3]LonWorks, LonTalk, and Neuron are trademarks of the Echelon Corporation.

LonWorks has been applied extensively in the building automation and control industry, as well as a variety of other industries. The core LonWorks technology for devices is contained in special integrated circuits — called Neuron® chips — which combine several microprocessors to manage the network, communications, and provide a general-purpose control environment. These chips are available from Motorola, Inc., and the Toshiba Corporation, which are licensees of the LonWorks technology. Echelon has also announced the possibility to license the LonTalk® protocol to other manufacturers for implementation in other microprocessors. LonWorks networks can be implemented on a variety of physical layers, including twisted pair at several bit rates and wireless options at 4800 bit/s, but the most common is a differential twisted pair system running at 78 kbit/s. Most of the networking details (the LonTalk protocol) are hidden from the user, and are encapsulated as functions within the general-purpose control environment. The user programs (using a language like the C programming language) the Neuron chip for each station to behave in a certain way and communicate various data items to other stations. Then, specialized tools are used to tie all of the stations together (handling addressing and other network details) to yield a functioning network. The system combines flexibility with a certain amount of ease of implementation, and can easily be applied to a variety of applications.

88.4 Applying Network Communications

Shielding

Many communication networks require shielding of the media (the cable). Shielding constitutes an electric conductor which completely encases the communication media (e.g., twisted pair) to provide protection against EMI. Shielding provides an electric conductive barrier to attenuate electromagnetic waves external to the shield, and provides a conduction path by which induced currents can be circulated and returned to the source, typically via a ground reference connection. Shields in communication systems are often grounded at only a single point. This single point of ground prevents the shield from participating in a "ground loop," which is an alternative path for current to flow between two points of potential difference connected to a common ground. Ground loops can lead to noise problems, and can be destructive if the stray currents are large enough, since a shield ground is usually not constructed to carry heavy currents.

Media

The most common media types for network systems fall into three categories: electric, optical, and electromagnetic. Electric media are based on conductors (e.g., copper wire), whereas optical media are based on optical waveguides, or fiber optics. Electromagnetic media consists of free space, or general electromagnetic wave-permeable materials, and are referred to as wireless systems. Within the category of electric media are a large variety of conductor configurations. The most common are unshielded pair, unshielded twisted pair (UTP), shielded twisted pair (STP), and coaxial (COAX). These conductor configurations have various properties which are significant for the transmission of electric signals, as well as varying degrees of immunity to EMI. As a rule of thumb, the quality of the transmission line characteristics (signal transmission and immunity to EMI) improves in the order listed. Twisted pair systems are generally easier to install, whereas coaxial and fiber-optic systems generally require more specialized tools and termination methods. Of course, wireless systems are easy to install, but attention must still be paid to the media. The characteristics of the free space such as distance and amount of EMI present must be considered for reliable operation of the network.

Bit Rate

Some networks provide only one choice of bit rate, whereas others provide user-selectable options for bit rate. Bit-rate options may be dependent on the type of media that is installed. As a rule of thumb,

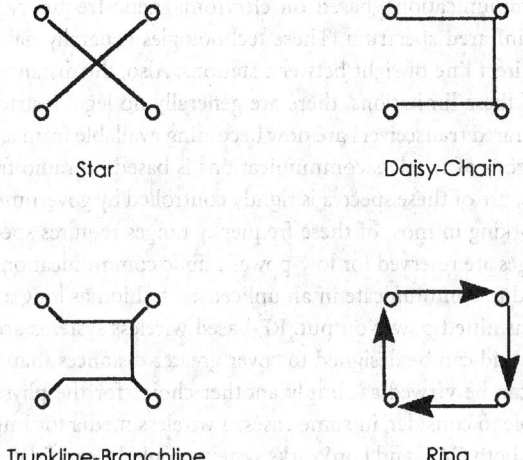

FIGURE 88.5 Some examples of the many possible network topologies using four stations.

the bit rate chosen should be the lowest possible bit rate that still supports the application requirements for speed of data transfer and overall bandwidth. This generally results in more reliable operation, and generally gives the network more immunity to minor degradations, specification violations, and EMI.

Topologies

There are a variety of network topologies that are commonly used. Topology refers to the physical arrangement and interconnection of stations by the media. Some networks can be run using several different topologies; some can only be run with a certain topology (e.g., Interbus-S requires a ring topology). The most common topologies are daisychain, trunkline–branchline, ring, and star. Variations on these exist, and networks that incorporate or can be run on highly varied topologies are sometimes called free-form, tree, or free topology networks. Figure 88.5 depicts graphically several different types of topologies. In some cases, networks require certain topologies. Deviating from these can cause degradation in network behavior (e.g., corruption of messages) or network failure.

Configuration

Most networks involve some sort of configuration. Configuration is the process of connecting stations together and assigning certain programmable parameters to each station required for proper operation of the network. The most common configurable parameter in many networks is the station address. Some networks may require other parameters to be preset, such as the communication speed, or bit rate. Some networks have the capability to autoconfigure, which means to assign parameters automatically to stations as part of the network start-up process, without explicit user intervention (e.g., Interbus-S). Many networks define various tools, which may be computer based, to assign parameters to each station in order to configure the network. In other cases, the stations may incorporate switches or other manual means to configure the necessary parameters for network operation.

88.5 Advanced Topics

Wireless Technologies

The need for networking is present even in environments where an electrical or optical cable cannot be easily distributed. This may be due to various limitations, such as difficulty in running a new cable from one building to another, or connecting to sensors in motion or on vehicles. There are two general categories of wireless communications, based on electromagnetic frequency spectra. Various wireless technologies employ the infrared spectrum. These technologies generally have transmission limited to applications that have a direct line of sight between stations. Also, the distances are generally limited to 100 m or less. Because of these limitations, there are generally no legal restrictions in employing these frequency spectra, and infrared transceivers are now becoming available from a variety of manufacturers.

The other general category of wireless communications is based on radio frequency (RF) communications. In most countries, use of these spectra is tightly controlled by governmental agencies. As a result, employing wireless networking in most of these frequency ranges requires special licensing. However, a number of frequency ranges are reserved for low-power public communications. Within these frequency ranges, devices are allowed to communicate in an unlicensed fashion as long as they transmit according to certain rules about transmitted power output. RF-based wireless systems are generally not limited to line-of-sight applications, and can be designed to cover greater distances than infrared-based systems.

Wireless technologies can be viewed as simply another choice for the physical layer media, i.e., free space. As such, it is possible to consider, in some cases, a wireless media for implementation of a variety of protocols. For example, both CAN and LonWorks systems could be candidates for wireless networking.

Fiber Optics

Another physical layer media choice is fiber-optic media. Fiber-optic media employs pulses of light delivered along a tubular waveguide (glass or plastic fiber) to transmit information from one station to another. Fiber optics enjoy some benefits over traditional copper wiring. First, attenuation of light within fiber optics is generally about an order of magnitude less than attenuation of an electric signal within a copper wire. Second, fiber-optic transmission systems can be modulated (or pulsed) at much higher frequencies, yielding greater potential bandwidths and bit rates than copper media. Finally, fiber-optic systems are generally immune to the traditional sources of EMI that can cause trouble for copper media systems. There are also limitations in present implementations of fiber-optic systems. One of the limitations is that special tools and termination techniques must be used to connect a fiber-optic cable to a sensor or field device. Second, fiber-optic "taps" are not easily created. Therefore, most fiber-optic systems are implemented in a point-to-point fashion. When multiple devices are involved in a network, each device usually acts as an optical repeater, with a fiber-optic input and a fiber-optic output port.

Network Design Considerations

Designing a network communication system from the ground up can be a lengthy undertaking, and should not be considered unless a careful review of available technologies has yielded no solutions to the particular requirements of the application. The designers must take into account a number of fundamental questions to shape the capability of the network. One topic mentioned frequently in the area of networking for control applications is the subject of determinism. This refers to the ability of the network to behave in a predictable fashion under any given set of stimuli, or external conditions. Many networks do not exhibit this characteristic. Another question to be resolved is the subject of priority, and media access. The designers must determine the conditions under which any particular station is allowed to transmit, and if multiple stations are attempting to transmit, how it will be determined which station will be given priority to transmit first. Media access methods often impact the ability of a network to behave in a deterministic fashion.

Integrating Sensors with Communications — IEEE P1451

A recent interesting development in the area of sensor networks is an effort being sponsored by the IEEE [20] out of its TC-9 committee, called IEEE P1451. This activity is working toward the development of a standard to define sensor (or transducer) interfaces to networks generically. The first part of the proposed standard, IEEE P1451.1, includes definitions for the interface between the device and the network (refer to Figure 88.1). The second part, IEEE P1451.2, includes definitions for the interface between the transducer (or sensor) and the network interface block within the device. P1451.2 includes a definition for a transducer electronic data sheet, or TEDS, which defines a summary set of information pertinent to the sensor, allowing for standardized exchange of data on the network. The proposed standard has the potential benefits to make it easier to connect a sensor to a variety of networks, and to allow similar sensors from different manufacturers to be handled in a similar fashion on the network.

References

1. *ISO/IEC 7498-1:1994 Information Technology — Open Systems Interconnection — Basic Reference Model: The Basic Model*, International Organization for Standardization (ISO), 1, rue de Varembé, Case postale 56, CH-1211 Genève 20, Switzerland, [online]. Available http://www.iso.ch/.
2. *8802-3: 1996 (ISO/IEC) [ANSI/IEEE Std 802.3, 1996 Edition] Information Technology — Telecommunications and information exchange between systems — Local and metropolitan area networks — Specific requirements — Part 3: Carrier sense multiple access with collision detection (CSMA/CD) access method and physical layer specifications*, Institute of Electrical and Electronics Engineers, 445 Hoes Lane, Piscataway, NJ 08855-1331, [online]. Available http://www.ieee.org/.
3. *ANSI/EIA/TIA-232-E-91*, Electronic Industries Association, 2500 Wilson Boulevard, Arlington, VA 22201-3834, [online]. Available http://www.eia.org/.
4. *Distributed, Intelligent I/O for Industrial Control and Data Acquisition... The SERIPLEX Control Bus, Bulletin No. 8310PD9501R4/97*, Seriplex Technical Organization, P.O. Box 27446, Raleigh, NC 27611-7446, [online]. Available http://www.seriplex.org/.
5. AS-Interface U.S.A., 5010 East Shea Blvd., Suite C-226, Scottsdale, AZ 85254, [online]. Available http://www.as-interface.com/.
6. *Interbus-S Protocol Structure, Data Sheet 0005C*, Phoenix Contact, P.O. Box 4100, Harrisburg, PA 17111, [online]. Available http://www.ibsclub.com/.
7. *CAN Specification, Version 2.0, 1991*, Robert Bosch GmbH, Postfach 50, D-7000 Stuttgart 1, Germany.
8. *ISO 11898:1993 Road vehicles—Interchange of digital information—Controller area network (CAN) for high-speed communication*, International Organization for Standardization (ISO), 1, rue de Varembé, Case postale 56, CH-1211 Genève 20, Switzerland, [online]. Available http://www.iso.ch/.
9. *CiA Draft Standard 301 (Version 3.0), CANopen Communication Profile for Industrial Systems*, CiA Headquarters, Am Weichselgarten 26, D-91058 Erlangen, Germany, [online]. Available http://www.can-cia.de/.
10. *SAE J 1939 — Recommended Practice for Serial Control and Communications Vehicle Network*, Society of Automotive Engineers, 400 Commonwealth Drive, Warrendale, PA 15096, [online]. Available http://www.sae.org/.
11. *DeviceNet Specification v1.4*, Open DeviceNet Vendors Association, 8222 Wiles Road, Suite 287, Coral Springs, FL 33067, [online]. Available http://www.odva.org/.
12. *Smart Distributed System Application Layer Protocol Specification v2.0, 1996*, Honeywell MICRO SWITCH Division, 11 West Spring Street, Freeport, IL 61032, [online]. Available http://www.sensing.honeywell.com/sds/.
13. HART Communication Foundation, 9390 Research Boulevard, Suite I-350, Austin, TX, 78759 [online]. Available http://www.ccsi.com/hart/hcfmain.html.
14. *EN 50 170 — Volume 2*, CENELEC Central Secretariat, 35, rue de Stassart, B-1050 Brussels, Belgium.
15. PROFIBUS Trade Organization U.S.A., 5010 East Shea Blvd., Suite C-226, Scottsdale, AZ 85254-4683, [online]. Available http://www.profibus.com/.
16. ISA, the International Society for Measurement & Control, P.O. Box 12277, Research Triangle Park, NC 27709, [online]. Available http://www.isa.org/.
17. *IEC 61158-2(1993-12), Fieldbus standard for use in industrial control systems — Part 2: Physical layer specification and service definition*, International Electrotechnical Commission, 3, rue de Varembé, PO Box 131, 1211 Geneva 20, Switzerland, [online]. Available http://www.iec.ch/.

18. WorldFIP Headquarters, 2, rue de Bone, 92160 Antony, France, [online]. Available http://www.world-fip.org/.
19. *005-0017-01 Rev C, LonTalk Protocol,* Echelon Corporation, 4015 Miranda Avenue, Palo Alto, CA 94304, [online]. Available http://www.echelon.com/.
20. *P1451.2, Draft Standard for a Smart Transducer Interface for Sensors and Actuators — Transducer to Microprocessor Communication Protocols and Transducer Electronic Data Sheet (TEDS) Formats,* Institute of Electrical and Electronics Engineers, 445 Hoes Lane Piscataway, NJ 08855-1331, [online]. Available http://www.ieee.org/.

Further Information

P. Z. Peebles, Jr., *Digital Communications Systems,* Englewood Cliffs, NJ: Prentice-Hall, 1987, provides a good general text on communication.

B. Svacina, *Understanding Device Level Buses,* Minneapolis, MN: TURCK Inc. (3000 Campus Dr., Minneapolis, MN 55441), 1996, is an in-depth study of the subject of communication networks for industrial field devices.

J. D. Gibson, Ed., *The Communications Handbook,* Boca Raton, FL: CRC Press, 1997, includes recent material on communication techniques.

89

Electromagnetic Compatibility

Daryl Gerke
Kimmel Gerke Associates, Ltd.

William Kimmel
Kimmel Gerke Associates, Ltd.

Jeffrey P. Mills
Illinois Institute of Technology

89.1 Grounding and Shielding

Daryl Gerke and William Kimmel

EMC (electromagnetic compatibility) is crucial to successful operation of industrial systems. Due to the increased electronic content of most industrial controls, electromagnetic interference (**EMI**) problems have increased dramatically in recent years. Two keys to EMC success are grounding and shielding. This section will briefly discuss how to implement these two crucial EMC strategies. It will also provide a general introduction to EMI problems in today's industrial electronic systems. The primary emphasis will be on practical insights and ideas gained in dealing with numerous industrial control problems.

Understanding EMI Problems

Here are three general observations on dealing with EMI problems in industrial electronics.

First, the industrial environment is harsh. The primary EMI threats are power disturbances, **RFI** (radio frequency interference), and **ESD** (electrostatic discharge). In addition, analog sensor circuits are often plagued with 50/60 Hz "ground loop" problems. Industrial electronics need more EMC care than most commercial electronics, and even more than many military systems.

Second, electronics often play a secondary role in electronics systems. Unlike a computer system, where electronics is the core technology, industrial electronics are often used to support another technology, such as chemical, mechanical, or process functions. This leads to EMC challenges when integrating the electronics to nonelectronic technologies.

Third, EMC rules and regulations are finally catching up to industrial electronics. For many years, industrial electronics were exempt from mandatory EMC rules, so unless there was an actual problem, EMC was often ignored. With the EMC directives of the European Union (**EU**) now in force, industrial electronics are no longer exempt.

Three Types of Problems

There are three aspects of the EMC problem: *emissions, susceptibility* (also known as immunity), and *self-compatibility.* **Emissions** originate within the equipment, and may upset other nearby equipment. On the other hand, external energy may upset equipment, leading to **susceptibility** (or a lack of immunity). Finally, energy internal to the system may interfere with other internal circuits, resulting in a self-compatibility problem.

Problems with both emissions and susceptibility have led to EMC regulations. Two of the best known are the **FCC** (Federal Communications Commission) regulations for emissions in the U.S., and the EU regulations for both emissions and immunity in Europe. Industrial controls have always been exempt from the FCC regulations, but they are not exempt from the EU regulations which became mandatory in January 1996.

Four Major EMC Threats

Most industrial EMC problems fall into one of four key areas: *emissions, power disturbances, radio frequency interference,* and *electrostatic discharge.* In the past, industrial systems were usually only concerned with power disturbances. Today, all four threats must be considered.

Emissions.
Emissions refer to electric energy originating within equipment that can interfere with other equipment. The prime concern of this threat is jamming nearby television receivers, which is the basis for the now mandatory EU emissions regulations. Emissions problems between industrial electronic systems, however, are rare. While it is possible to interfere with any other nearby equipment, most industrial electronics generate only minute amounts of conducted and radiated interference, well below upset thresholds for digital or analog circuits.

Emissions are best addressed at the equipment design stage. Strategies include printed circuit board design techniques, high-frequency filtering on power and signal interfaces, shielded cables, and enclosure shielding. Fixes in the field are usually limited to shielded cables or enclosures, add-on filters, and ferrite clamps on cables.

Power Disturbances.
Power disturbances can take many forms, from short transients to long sags, surges, or complete power outages. The three most serious power threats to industrial controls are transients, voltage sags, and power outages. Stray 50/60 Hz currents can cause also problems with sensitive analog circuits, particularly due to ground loops (to be discussed later). Other power disturbances, like frequency or waveform variations, often have little effect on electronic systems.

Power disturbances are very common in industrial environments. As a result, most industrial systems are pretty robust against this threat, at least at low frequencies. High-frequency threats, such as fast transients or RF on the power lines, can still cause problems. The EU tests simulate these threats with the **EFT** (electrical fast transient) and injected RF tests.

Most power disturbances are caused by nearby equipment, rather than external sources. (One critical exception is lightning, which can result in some nasty voltage and current surges). Power disturbances solutions include grounding, power filters, transient protectors, and in extreme cases, uninterruptible power systems (UPS).

Radio Frequency Interference.
RFI deals with threats in the RF range. RFI is quite common in industrial environments, and will likely get worse with the proliferation of handheld radios and cellular telephones. It is expected that that wireless LANs (local area networks) will also provide some interesting EMI challenges. There have been cases where handheld radios were banned from use due to repeated EMI problems with industrial electronics.

It turns out that the nearby handheld radio is a much bigger threat than a large commercial broadcast station several kilometers away. A key metric is electric field intensity, measured in "volts/meter." This is a function of both transmitter power, and distance from the antenna, and can be quickly predicted by the formula:

$$E(V/m) = 5.5\sqrt{PA}/d$$

where P = transmitter power in watts, A = antenna gain, and d = distance from the antenna in meters. For example, the electric field from a 1-W radio with a zero gain antenna at 1 m is about 5 V/m, while the electric field from a 10,000 W broadcast station at 1 km is about 0.5 V/m. Since unprotected equipment can fail in the 0.1 to 1 V/m range, problems can and do occur. The EU "heavy industrial" limits of 10 V/m are clearly aimed at protecting against the nearby handheld radio.

Solutions to RFI problems include high-frequency filtering on power and signal cables, shielded cables, and shielded enclosures. Analog circuits are particularly vulnerable to RFI, so they often need extra protection. Do not overlook banning radio transmitters in the immediate vicinity. Often, maintaining a 3 to 10 m distance is enough to solve the problem.

Electrostatic Discharge.
ESD refers to the sudden discharge that can occur after a gradual buildup of electric charge. ESD is most commonly associated with humans (touching controls or keyboards), but ESD can also be caused by internal arcing due to the movement of paper, plastic, etc. Internal ESD problems are increasing in industrial systems.

Although the static buildup can take a long time (seconds or even minutes), the discharge is almost instantaneous (nanoseconds or less). Furthermore, it is the sudden current, not the voltage, that is the culprit. The effect is a bit like having a dam burst — the ESD current is like water running down a mountain, destroying anything in its path. Fortunately, the current surge does not last too long, so the energy levels are not high. They are high enough, however, to damage or upset electronic devices.

The extremely fast discharge results in high frequencies well into the UHF range. At 1 ns, the transient bandwidth is over 300 MHz. As a result, it does not take a "direct ESD hit" to cause a problem. ESD upsets 5 to 10 m away are not uncommon, due to the intense electromagnetic fields associated with an ESD event. These problems are particularly insidious, since the ESD event may be occurring on a different piece of equipment.

Solutions to ESD problems include transient protection, high-frequency filtering, cable shielding, and enclosure shielding. Grounding is a very important factor in ESD protection, but it must be designed for high frequencies. Since many times ESD causes "reset" problems, extra attention to microprocessor reset circuits is beneficial.

Sources, Paths, and Receptors

A common EMI problem is gathering and organizing data. This is particularly important when trouble-shooting EMI in the field. The "source–path–receptor" model is popular. Simply stated, three elements are necessary for any EMI problem:

1. There must be a source of energy;
2. There must be a receptor that is upset by that energy;
3. There must be a coupling path between the source and receptor.

All three elements must exist at the same time, and if any one is missing, there is no EMI problem. Sometimes one can identify all three, and, other times, one can only guess. While this may seem simple, it is a useful tool to organize EMI information.

Figure 89.1 illustrates this model, giving typical sources, paths, and receptors. Several possible sources have been discussed: emissions from digital circuits, ESD, RFI from communications transmitters, and power disturbances (including lightning). Several different receptors have also been suggested: communications receivers, analog electronics, and digital electronics. Note the two types of paths: radiated and conducted. In both cases, the object is to block unwanted energy from reaching a receptor, which is done with shielding (for the radiated path) and filtering (for the conducted path).

Grounding

Grounding is probably the most important, yet least understood, aspect of EMI control. Every circuit is connected to some sort of "ground," so every circuit is affected by EMI grounding issues.

Any interference problem can be broken down into
- the SOURCE of interference
- the RECEPTOR of interference
- the PATH coupling the source to the receptor

Sources	Paths	Receptors
• Microprocessors	• Radiated	• Digital
• Video Drivers	• EM Fields	• Microprocessors
• ESD	• Crosstalk	• Reset
• Transmitters	Capacitive	• Other logic
• RF Generators	Inductive	• Low level analog
• Power Disturbances	• Conducted	• Receivers
• Lightning	• Signal	
	• Power	
	• Ground	

FIGURE 89.1 The source–path–receptor model for assessing EMI problems. All three elements must be necessary for an EMI problem to occur.

TABLE 89.1 A Ground May Work Over Wide Frequency Ranges

Type	Frequency	Typical Current Levels	Typical Duration
Power	50/60 Hz	10–1 000 A	Seconds or minutes
Lightning	300 kHz	100 000 A	Tens of milliseconds
ESD	300 MHz	10–50 A	Tens of nanoseconds
EMI	dc–Daylight	μA–A	Nanoseconds to years

What Is a Ground?

A major problem with the subject of grounding is the ambiguity of the term. Our favorite definition is one popular in the EMC community, which says that a *ground is simply a return path for current flow*. These currents can be intended, or unintended. The unintended currents are often referred to as "sneak grounds," and can cause many kinds of EMI problems. Finally, a physical connection is not even necessary at higher frequencies, where parasitic capacitance or inductance may form part of a ground path.

Different Types of Grounds

Grounds are used for many reasons, including power, safety, lightning, EMI, and ESD. Although they may share common functions, they may vary widely when it comes to frequencies and current amplitudes. Recognizing these key differences is key to understanding grounding issues.

Table 89.1 shows some frequency and amplitude requirements of several different types of grounds. Note that power and safety grounds must handle high currents, but only at low frequencies. Grounds for EMI and ESD, on the other hand, must often handle high frequencies at relatively low current levels. Lightning grounds must handle extremely high currents, but at moderate frequencies.

The frequency of transient events is calculated using the formula $f = 1/(\pi t_r)$, where f is the equivalent frequency, and t_r is the transient rise/fall time. This relationship can be derived using Fourier analysis. For example, ESD has an equivalent frequency of about 300 MHz based on a typical 1 ns rise time, and lightning has an equivalent frequency of about 300 kHz based on a 1 μs rise time.

Note that of all these types of grounds, only one actually needs an Earth connection — lightning. Other grounds may be connected to Earth by convention or for other safety reasons. For example, power neutrals are connected to Earth in many parts of the world to help provide lightning protection. On the other hand, in many other parts of the world, the power systems do not have Earth connections. When dealing with power grounding, the local safety codes will determine the proper Earth grounding methods.

TABLE 89.2 Impedance Parameters for 10-cm-Length Wires

Gage	Ω/m	μH/m	Z @ 10 kHz	Z @ 1 MHz	Z @ 100 MHz
10	0.0033	1.01	0.006	0.63	63
12	0.0052	1.05	0.007	0.66	66
14	0.0083	1.10	0.007	0.69	69
16	0.0132	1.15	0.007	0.72	72
18	0.0209	1.19	0.007	0.75	75
20	0.0333	1.24	0.008	0.78	78
22	0.0530	1.29	0.009	0.81	81
24	0.0842	1.33	0.010	0.84	84
26	0.1339	1.38	0.012	0.87	87
28	0.1688	1.40	0.019	0.88	88
30	0.2129	1.43	0.022	0.90	90

TABLE 89.3 Impedance Values for Ground Plane Impedance

Frequency	Thickness		
	0.1 mm	1 mm	10 mm
60 Hz	172 μΩ	17.2 μΩ	1.83 μΩ
1 kHz	172	17.5	11.6
10 kHz	172	33.5	36.9
100 kHz	175	116	116
1 MHz	335	369	369
10 MHz	1.16 mΩ	1.16 mΩ	1.16 mΩ
100 MHz	3.69	3.69	3.69
1000 MHz	11.6	11.6	11.6

Ground Impedances

A good ground must have a low enough impedance to minimize voltage drop in the ground system, and must provide the preferred path for current flow. The key to success is maintaining that low impedance over the entire frequency range of interest. We cannot overemphasize this point. Most EMI grounding problems are due to using the wrong approach for a given range.

The impedance of a ground conductor consists of both resistance and inductance ($Z = R + j\omega L$). For frequencies from dc through about 10 kHz, the resistance is the major factor, so heavy-gage wires are often used for low-frequency ground conductors. As the frequency increases, however, the inductance becomes the limiting factor for impedance. As a rule of thumb, the inductance for round wires is in the range of 10 nH/cm.

Table 89.2 gives the resistance, inductance, and inductive reactance for a typical wire sizes used in instrumentation power and signal circuits. Its apparent that at power and audio frequencies (dc to 10 kHz), resistance is the dominant factor in ground impedance. Thus, at low frequencies, look for ways to reduce resistance, typically by using larger wires. At frequencies above the audio range (>10 kHz), inductance becomes the dominant factor in ground impedance. Thus, at higher frequencies, look for ways to reduce the inductance of the ground path. This is accomplished by using ground planes, grids, and straps to lower the inductance.

Table 89.3 gives the impedances for solid ground planes at various frequencies. In this case, the impedances are in "ohms-per-square," which is a measure of impedance across a diagonal surface. By comparing this with Table 89.2, one can see that at high frequencies (such as 100 MHz) the ground plane impedance may be several orders of magnitude below the impedance of a wire. Furthermore, at high frequencies the thickness is not a factor, since the impedance is limited by the skin effect.

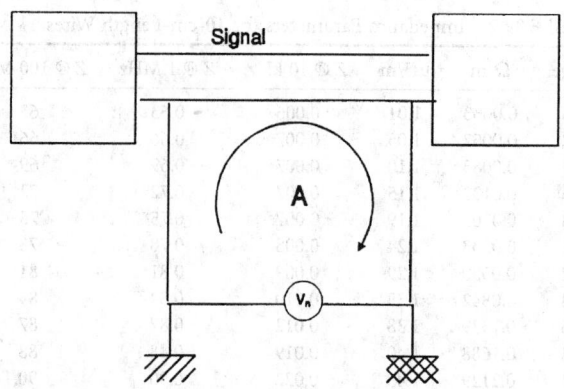

FIGURE 89.2   Typical industrial grounding situation, which also illustrates a ground loop.

Ground Topologies

Now that we have looked at ground impedance vs. frequency, we are ready to look at ground topologies vs. frequency. The impedance limitations yield two different grounding approaches, dependent on frequency. For low-frequency problems (dc to 10 kHz), single-point grounds are preferred, while at high frequencies (above 10 kHz), multipoint grounds with planes or grids become the preferred approach.

This dichotomy often causes confusion with industrial controls, but this can be minimized by determining the frequency of the EMI threat and then selecting the appropriate grounding approach. In many cases, both approaches may be necessary at the same time, leading to "hybrid" grounds, which use capacitors and inductors to alter the ground topology with frequency.

Single-Point Grounds.
At low frequencies, one can usually steer current via wires. Since the inductance is low, the limiting factor is the wire resistance itself. Furthermore, capacitive coupling from the ground wires to adjacent wires or surfaces is small, so virtually all the current follows the wiring path.

Figure 89.2 shows a typical industrial grounding scheme. Note what happens if the system is grounded at both ends. Any common noise current in the common ground path is now coupled into the circuit via the "common ground impedance." This results in the dreaded "ground loop," which will be discussed shortly. A single-point ground eliminates the ground loop, since there is no common impedance across which a common-mode voltage can be generated. Thus, single-point grounding is a very practical way to limit "ground noise" problems with the threat of low-frequency ground currents. This is very typical of 50/60 Hz currents getting into sensitive analog instrumentation circuits.

Multipoint Grounds.
Unfortunately, as the frequency increases, the inductive reactance of the wires increases. At the same time, parasitic capacitive reactance to adjacent wires or surfaces decreases, and soon it is no longer possible to maintain a true single-point ground, even if the system is wired that way. The only option left is to lower the ground path impedance, and that is accomplished with planes or grids. Furthermore, a single-point connections to a grid or plane are usually not adequate because of transmission line effects, so multipoint grounds (combined with planes/grids) become the preferred approach above 10 kHz.

Ground grids have been used for years in computer facilities, and are seeing increasing use in industrial facilities. The recommended spacing for grids is no more than $\frac{1}{20}$ of a wavelength at the highest frequency of concern. Computer room grids are often spaced about 0.7 m (about 2 ft), which meets this criteria from dc to about 25 MHz. This is very beneficial in addressing ground noise due to lightning and other power transients, which are usually in the 1 MHz range and below. But a 0.7-m grid does not help with VHF/UHF radio problems or ESD. In those cases, solid surfaces may be necessary.

Ground Loops.

Ground loops are a serious problem for sensitive analog circuits facing low-frequency threats. At high frequencies, ground loops generally do not pose serious threats if proper high-frequency precautions are taken when designing the ground system.

A ground loop exists whenever multiple ground paths exist. Unwanted currents can take unwanted paths, resulting in unwanted noise voltages at unwanted places. The problem is particularly acute with sensitive analog systems, where even a few microvolts can jam intended signals. A classic example is 60 Hz ground currents causing hum in an audio system.

Figure 89.2 shows a typical ground loop problem. Note that there must be the three conditions of any EMI problem: a source, a path, and a victim. In this case, the source can be circulating power currents, the path the common ground impedance, and the victim is often the sensitive analog circuit. With many systems problems, one cannot do anything about either the source or victim, so the solution is with the ground path. As we have already seen, single-point grounding is effective at low frequencies, and ground planes/grids are effective at higher frequencies.

If one cannot change the ground paths, one can still attack the ground loop by "breaking" it in other places. For example, transformers or optical isolators (or even fiber optics) can be used in cable connections, which will block common mode noise currents while passing intended differential mode signals. Balanced input/output (I/O) circuits can be used to "cancel" the noise through common-mode rejection. All of these are most effective at 50/60 Hz, and become less effective at higher frequencies due to parasitic capacitance.

Grounding Guidelines

By now it should be apparent that there is no magic solution for grounding. Rather, different methods and approaches are necessary for different circuits and operating conditions. Two key parameters are the threat frequency (low vs. high), and the circuit operating levels. Here are some guidelines, but keep in mind that even these may need to be modified for a particular situation.

Analog Circuits.

Since most analog circuits operate at low frequencies and are subject to low-frequency threats, single-point grounds are preferred. Typical threats are 50/60 Hz power return currents, stray switching power supply currents, and perhaps digital circuit return currents (if separate analog and digital power and grounds are not provided). Low-level analog circuits are the most vulnerable, since the signal levels are small.

Keep in mind that high-frequency threats (such as a VHF radio) to low-frequency circuits may require high-frequency grounding solutions, such as multipoint grounds. Often, this can be accomplished by using small high-frequency capacitors (1000 pF typical) which appear as a short at 100 MHz, yet still appear as a high impedance at 50/60 Hz.

Digital Circuits.

Most digital circuits today operate at relatively high frequencies, so multipoint grounds and ground planes and grids are preferred. The connections between the circuits and their grounds need to be short, fat, and direct to minimize inductance.

Digital circuits, particularly I/O circuits, are vulnerable to external high-frequency threats like RF and ESD. They are also a key source of high-frequency emissions and internal problems like cross talk. For digital circuits, multilayer boards with internal ground planes are preferred. These ground planes typically are connected to a metallic enclosure through multiple low-inductance connections.

Pay particular attention to where digital and analog circuits meet. A single-point connection is usually preferred to minimize ground loops, but installing a small resistor (1 to 10 Ω typical) or inductor (1 to 100 μH typical) at that point is often helpful in providing additional isolation. One may need to experiment with this to determine the optimum solutions.

Power Safety Grounding.

Entire books have been written about this subject, and rightly so; this is an extremely important safety issue. The key concern here is human safety and prevention of electric shock. In most parts of the world, exposed metal on line-powered equipment must be bonded to a safety grounding conductor. Furthermore, the electric wiring codes (such as the National Electrical Code in the U.S.) give very specific guidelines on how power grounding must be accomplished.

These guidelines must be followed when wiring any industrial control system, and must never be compromised by "isolated" power grounds or other similar foolishness. *Finally, if there is ever a conflict between EMI and safety grounding, the safety issues must always prevail!*

Shielding

Many systems today require at least some **shielding** for proper operation or to meet radiated emission or immunity requirements. Many engineers consider shielding purely a mechanical issue, but nothing could be farther from the truth. EMI shielding needs both an electrical and a mechanical understanding of key issues to assure success.

Two of these key issues are selecting the right material and maintaining the shielding integrity over the desired frequency range. While most people worry more about the selection, shield integrity is usually much more important. We will soon see that even very thin metallic coatings can be effective shields, yet even very small holes or penetrations can completely destroy a shield. Like grounding, shielding cannot be left to chance, and must be properly designed and implemented.

How Shielding Works

EMI shielding involves two independent mechanisms: *reflection* and *absorption.* In reflection, an electromagnetic wave bounces off the surface, just like light off a mirror. In absorption, the electromagnetic wave penetrates the material and is absorbed as it passes through, much like heat loss through an insulating wall.

Shielding effectiveness is usually expressed as follows:

$$SE\,(dB) = R(dB) + A(dB)$$

where SE is the total shielding effectiveness in dB, and R and A are the reflection and absorption losses expressed in dB. Reflection is the primary mechanism for high-frequency shielding (emissions, RFI, ESD), while absorption is the key mechanism for low-frequency magnetic field shielding. The actual formulas for calculating reflection and absorption losses are a bit complex, and beyond the scope of this chapter, but several sources are included in the further information section.

Three Types of Fields

It is customary when dealing with shielding to use three types of "fields" to explain shielding. These three fields account for differences in shielding performance due to differences in frequency and circuit impedance levels. They also explain why the same shield can behave differently for different energy sources. These are plane waves, magnetic fields, and electric fields. Figure 89.3 shows typical shielding curves for copper, with references to each type of field.

Plane Wave Fields.

If one is located greater than about $\frac{1}{6}$ wavelength from a point source, the wave impedance (ratio of electric field intensity to magnetic field intensity) is a constant 377 Ω in free space. This field is known as the "far field" or "radiation field," since real energy predominates here and propagates as a "plane wave." Since reflection losses are due to a mismatch between the wave impedance (377 Ω) and a metallic shield surface impedance (typically milliohms or less), shielding effectiveness is usually very high for plane wave sources.

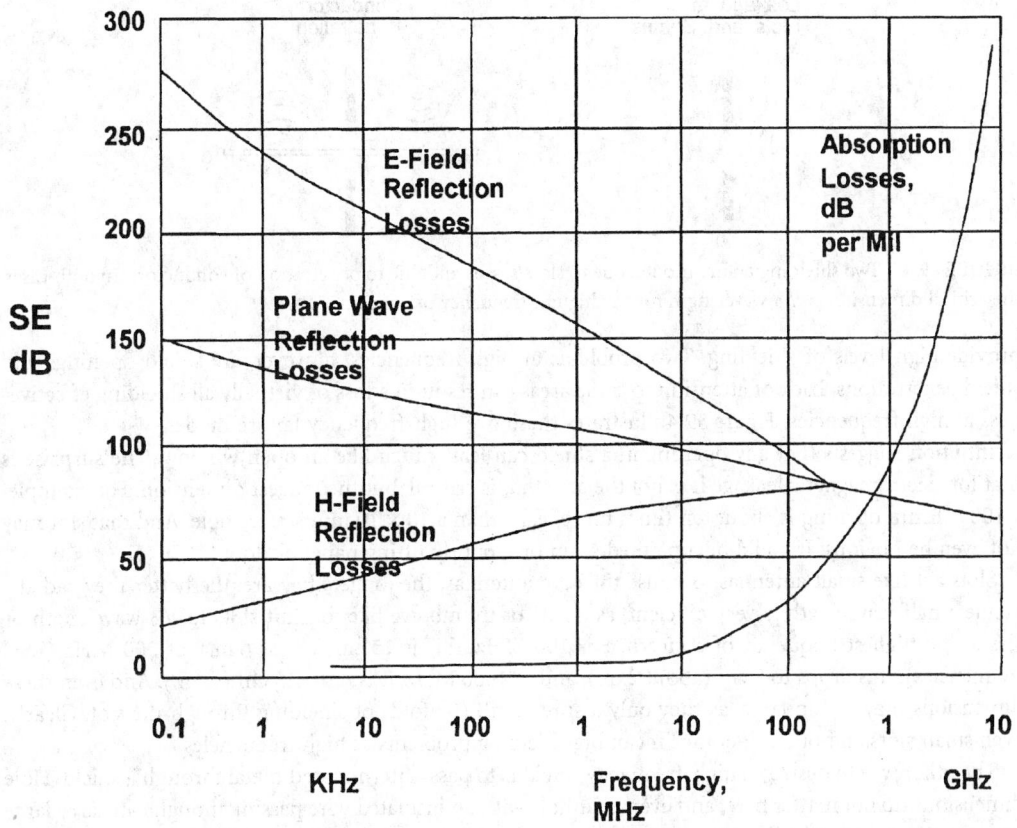

FIGURE 89.3 Typical shielding effectiveness curves for copper. Note two mechanisms (reflection and absorption) and three types of fields (electric, magnetic, and plane wave). Shielding for aluminum is almost the same as for copper.

At frequencies 30 MHz and above, once one is more than about 1 m away, one is in the plane wave region. Thus, even very thin shields work well for emissions, ESD, and RFI problems, with reflection as the prime shielding mechanism.

Electric and Magnetic Fields.
If one is located less than about ⅙ wavelength from the source, then the wave impedance is dependent on the circuit impedance. This region is known as the "near field," since reactive energy predominates here. This region is further divided into "electric" and "magnetic" fields, both dependent on source circuit impedance. For high-impedance sources (electric fields), the reflection losses are still high, but for low-impedance sources (magnetic fields), the impedance can be quite low. In the latter case, the reflection losses can become minimal.

For power line frequencies, the near field almost always predominates. As a result, materials like aluminum or copper have no reflection losses and are virtually transparent to power line magnetic fields. (As a rule, remember that aluminum foil is transparent to 60 Hz magnetic fields.) To solve this problem, permeable materials are needed to boost the electric thickness for a given physical thickness. Steel or high-permeability mu-metals are usually used to absorb (not reflect) the magnetic fields. Even so, it can still be very difficult to shield for low-frequency magnetic fields.

Why Shielding Fails

While material selection is important, other factors must also be considered. For low-frequency/low-impedance threats (power supply or power line magnetic fields), steel or other high-permeability materials are needed. For high-frequency threats, however, even very thin materials like conductive paints

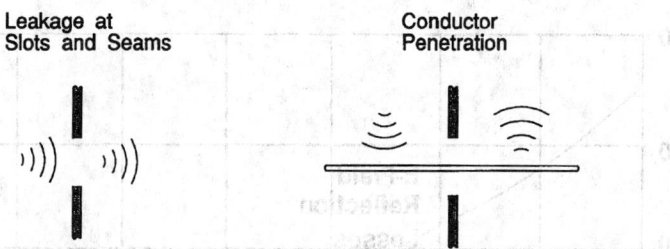

FIGURE 89.4 Two shielding failure modes, due to slots/seams and due to penetration of conductors. In both cases, the critical dimension is $1/20$ wavelength for the highest frequency of concern.

provide high levels of shielding. Two problems at high frequencies, however, are shield openings and shield penetrations. Lack of attention to these areas can result in a loss of virtually all shielding effectiveness at high frequencies. Figure 89.4 illustrates these two high-frequency failure modes.

Intuition suggests that any opening in a shield can leak, much like an open window. The surprise is that for electromagnetic leakage it is not the area that is critical, but the longest dimension. For example, a 100 × 1 mm opening is about ten times more leaky than a 10 × 10 mm square hole. And that slot may not even be obvious. It could be a painted seam or a poorly fitting panel or door.

Slots act like small antennas. Because they are antennas, the longer they are, the better they radiate. While a half wavelength is very efficient, as a rule of thumb, we like to limit slots to $1/20$ wavelength or less at the highest frequency of concern. For 100 MHz, this is 15 cm (about 6 in.); at 300 MHz (ESD frequencies), this drops to 5 cm (about 2 in.), and at 1000 MHz, it is only 1.5 cm (⅔ in.). And even these dimensions may be too large, as they only assure 20 dB (tenfold) of shielding through the slot. Clearly, even small slots and other openings mean big shielding problems at high frequencies.

The other way to destroy a high-frequency shield is to pass unterminated metal through a shield. Hole dimensions do not matter here, and even a pinhole with an insulated wire passing though can carry large amounts across the shielding barrier. The dimension that does matter is how far the penetration extends on either side of the shield. Once again, the critical distance is $1/20$ wavelength or more.

Shielding Guidelines

Now that we have looked at how shielding works (and fails), let us look at how to design good electromagnetic shields. Most of our focus will be on RF shielding in the 30 to 1000 MHz range, necessary for emissions, ESD, and RFI.

Material Selection.
We have already seen that for low-frequency magnetic interference problems, ferrous material like steel or mu-metals are necessary. Most instrumentation problems are either high impedance or high frequency in nature, so most of the time, thin conductive materials will work fine. For high frequencies, however, attention must be given to slots and penetrations.

Many enclosures today are made of plastic. For high-frequency shielding, conductive coatings also work quite well. Popular surface treatments include conductive paints, vacuum deposition, electroless plating, and even metal fabrics. Conductive plastics are also available, but they generally do not perform nearly as well as surface treatments for high frequencies.

Gasketing and Screening.
Large openings, such as ventilation ports or display areas, can be sealed with screening material. Seams or slots can be filled with conductive gaskets. In both cases, the secret is to provide complete and continuous metal-to-metal contact at all junctions. For high-frequency shielding, the connections must be almost watertight. Anything less is asking for problems.

For screening material, the smaller the openings, the higher the shielding. Window screen spacing is almost as effective as solid materials from dc to 1000 MHz, and even 5-mm (about ¼ in.) openings are often acceptable at 1000 MHz. In any case, do not exceed $1/20$ wavelength at the highest frequency of concern.

Cable Terminations and Filters.
Poor termination of shielded cables can cause big problems at high frequencies. If a shielded cable is not terminated directly at the shielding barrier, a lot of energy leaks, degrading both the cable and the enclosure shield. Pigtail connections, popular for terminating low-frequency cable shields, are particularly bothersome at high frequencies. In fact, this is a leading cause of EMI failures for RFI, emissions, and ESD. As a rule of thumb, pigtail connections should not be used on cable shields at frequencies above about 1 MHz.

Unshielded cables can also cause problems at high frequencies. In those cases, high-frequency filtering is needed directly at the interface to assure that the shield is not degraded at high frequencies. Common solutions are EMI filters on power and signal lines, or ferrite beads on the lines or cables. These must be installed as close to the shield penetration as possible. The best situation is to mount the filter directly in the shield itself, although this is not always necessary for moderate problems.

Internal Shields.
Finally, do not overlook using internal shield on critical circuits. Radio and television designers have been doing this for years, using selective shields on oscillators, power amplifiers, and the like. A classic example of this approach is the TV tuner, the most sensitive part of a television receiver. It is an inexpensive, yet highly effective shielding strategy.

Defining Terms

Conducted: Energy or interference that is propagated by a conductor, such as power, grounding, or signal interface wiring.
EFT: Electrical fast transient; a high-frequency burst of energy on power wiring.
EMC: Electromagnetic compatibility; the condition wherein electric and electronic equipment operate successfully in close proximity.
EMI: Electromagnetic interference; unwanted electric energy that may impair the function of electronic equipment.
Emissions: Electric energy emanating from an electronic source.
ESD: Electrostatic discharge; the rapid discharge that often follows a buildup of static charge.
EU: European union; formerly called the European Community.
FCC: Federal Communications Commission (U.S. government).
Ground: A return path for current.
Radiated: Energy or interference that is propagated by electromagnetic radiation through space.
RFI: Radio frequency interference; an older term for EMI, now usually used to describe interference caused by a nearby radio transmitter.
Shield: A metallic enclosure used to reduce electric or magnetic field levels.
Susceptibility: Vulnerability of electronic equipment to external sources of interference; often used interchangeably with immunity.

Further Information

D.D. Gerke and W.D. Kimmel, *EDN 's Designer's Guide to Electromagnetic Compatibility,* Newton, MA: Cahners Publishing, 1994. Basic introduction on EMC issues written by the authors. Very good for non-EMC engineers.

W.D. Kimmel and D.D. Gerke, *Electromagnetic Compatibility in Medical Equipment,* New York: IEEE Press and Interpharm Press, 1995. Detailed introduction to EMC issues in medical instrumentation, written by the authors. Applicable to most industrial control instrumentation.

W.D. Kimmel and D.D. Gerke, Internet World Wide Web Home Page [online]. Available http://www.emig-uru.com, St. Paul, MN, 1996. Wide range of useful EMC information.

R. Morrison, *Grounding and Shielding Techniques in Instrumentation,* New York: John Wiley & Sons, 1986. Detailed analog coverage, with emphasis on low-level signal issues.

H.W. Ott, *Noise Reduction Techniques in Electronic Systems,* New York: John Wiley & Sons, 1988. Good introduction to EMC issues, particularly system and analog issues.

C.R. Paul, *Introduction to Electromagnetic Compatibility,* New York: John Wiley & Sons, 1992. Written as a college text, quite analytical with very good coverage of EMC issues.

89.2 EMI and EMC Test Methods

Jeffrey P. Mills

Electric and magnetic fields must be measured for a variety of reasons. A radio or TV broadcast station is licensed to provide reliable coverage over a specified geographic area, and any properly operating receiver must pick up the signal and properly respond to it. This can be assured only if the broadcast signal is of a guaranteed minimum strength. Also, the signal must not be so strong that it interferes with a distant station sharing the same frequency. The broadcast field must be measured over its geographic area of coverage to be sure that it satisfies both criteria.

Many electric devices unintentionally radiate electromagnetic fields. Examples include

- Oscillators in superhetrodyne radio or TV receivers
- Digital logic circuits
- Switching contacts, particularly if unsuppressed
- Automotive ignition systems

Stray fields (**emissions**) from these devices can interfere with other devices, or even with the radiating device itself. This process is known as **electromagnetic interference**, commonly abbreviated EMI. Interference between two devices is known as **intersystem** EMI, whereas if a device interferes with itself it is **intrasystem EMI**. Intrasystem EMI is usually easy to spot because the device itself does not operate correctly. Intersystem EMI is usually more difficult to isolate. Its result might be a simple annoyance, such as noise on a radio and TV receiver caused by an electric vacuum cleaner or a power drill. It could, however, be much more serious; a portable radio receiver might affect aircraft navigation or critical communications.

It is also possible for a device to be **susceptible** to fields intentionally generated by a licensed transmitter such as a broadcast or mobile-radio transmitter. Examples include

- Public-address systems
- Music (high-fidelity) systems
- Telephone lines and instruments
- Digital logic circuits

Again, the result may be only an annoyance, or it could be much more serious; aircraft control surfaces have been observed to move uncontrollably due to strong electromagnetic fields. Since the fields themselves cannot be eliminated in these cases, the devices must be made immune to electromagnetic fields.

In the above cases, the interference is usually through electric and/or magnetic fields in space, so the process is known as **radiated coupling**. Another coupling path exists if two devices share the same power source. One device may generate undesired high-frequency voltages on its power leads, which then appear on the power leads of the other. The second device may then malfunction because of this high-frequency voltage. This is known as **conducted coupling**. So we must consider both radiated and conducted noise.

It is not practical to eliminate all interfering fields completely, so a compromise must be reached. A stray field will not cause EMI if it is very weak compared with the desired field, which might be the field of a broadcast signal. The permissible strength of the stray field depends on the strength of the desired field; the stronger the desired field, the more stray field can be tolerated. It also depends on the device that is being interfered with (the *victim*); some receivers can reject undesired signals better than others. Since there are many combinations of interference sources and victims, a worst-case scenario is sought that will protect most real-life situations. This occurs where the weakest legal radio or TV signal (in its licensed area of coverage) is received by the poorest available receiver.

The maximum stray field strength that causes no EMI for this worst-case scenario is incorporated into government regulations. The field actually radiated by every device must then be measured to be sure that it does not exceed this level at the nearest practical distance from it, usually 10 or 30 m. To specify and measure these field strengths accurately, the nature of electric and magnetic fields must be understood.

Unlike most electrical engineering topics, EMI control is not very precise because of the complexity of practical hardware. It is virtually impossible to predict interference more precisely than within a factor of three, and usually the margin of error is even worse. Measurements can vary significantly between two supposedly identical samples, due to slight variations in physical dimensions. If one measures the EMI resulting from two different designs, the design that exhibits less EMI is probably better, but not always. An engineer can often judge if an EMI problem exists, but one must never rely on the accuracy normally expected in other branches of electrical engineering.

Nature of Electric and Magnetic Fields

An electric field is generated by a distribution of electric charge. If the distribution changes with time, then so will the electric field. A magnetic field may be generated by a permanent magnet or by an electric current. If the permanent magnet or the current path moves, or if the current magnitude varies with time, the magnetic field will vary with time. A time-varying electric field creates a magnetic field, and conversely.

Electric fields, designated E, are normally expressed in volts per meter (V/m). Magnetic fields are designated H and expressed in amperes per meter (A/m). More often, magnetic fields are perceived as magnetic flux density, which is designated B and expressed in webers per square meter (W/m²), also known as *teslas* (T). A non-SI unit, sometimes found in older literature, is the *gauss*, equal to 10^{-4} T. Of course, any unit may be preceded by a scaling prefix such as micro or pico. In free space, B is equal to $\mu_0 H$, where, μ_0 is equal to 0.4π (approximately 1.257) µT·m/A (equivalent to µH/m).

Near a time-varying electric field source such as a charge distribution, the magnetic field is relatively weak, but it becomes stronger when observed from farther away. At a great enough distance, the ratio of E to H approaches $\sqrt{\mu/\epsilon}$, which in free space is equal to 120π (approximately 377) Ω. For a sinusoidal function of time with a frequency f, this occurs at any distance that is large compared with $\lambda/2\pi$ (approximately $\lambda/6$). Here, λ is the wavelength corresponding to f, equal to $3\cdot10^8/f$ m if f is specified in hertz. Distances much greater than $\lambda/2\pi$ are considered to be in the **far-field region**; nearer distances are in the **near-field region**. For a nonsinusoidal function of time, each Fourier frequency component must be considered separately, and the far-field region begins closer to the source for its higher frequency components.

Near a time-varying magnetic field source such as a current loop, the electric field is weak, becoming stronger when observed from a greater distance. At distances that are large compared with $\lambda/2\pi$ (the far-field region), the ratio of E to H again approaches 120π Ω.

Since $H = \sqrt{\epsilon/\mu}E$ and $B = \mu H = \sqrt{\mu\epsilon}E$ in the far-field region for either type of source, only E or B must be measured, and the other can easily be calculated from it. In free space, $\sqrt{\mu\epsilon} \approx 10^{-8}/3$ T·m/V (equivalent to s/m), so, if E is expressed in volts per meter, $B \approx 3.33E$ nT. By choice of a suitable antenna, either field can be measured. Far-field strengths are normally specified in terms of the E field, no matter whether the E or B field is measured.

Alternatively, the far-field strength may be specified in terms of **power density**, expressed in watts per square meter. This denotes the amount of radiated power passing through each square meter of a surface perpendicular to the direction away from the source. The *peak* power density P is equal to EH, and, for a sinusoidal source, the *average* power density is half this value. For a nonsinusoidal source, each frequency component must be considered separately, and the total average power is the sum of the average powers for all frequencies. Since $H = \sqrt{\epsilon/\mu}\,E$, it follows that $P = E^2/377\,\Omega$.

In regions other than the far field, the ratio of E to H varies greatly, approaching infinity for an electric field source or zero for a magnetic field source. A source may generate both electric and magnetic fields; for example, a charge moving between two electrodes causes a current to flow between them. Then the ratio of E to H may be any value at all. Therefore, at distances less than $\lambda/2\pi$ from a field source, both the E and B fields must be measured separately.

In the far-field region, both the electric and magnetic fields are perpendicular to the direction that an electromagnetic wave is propagating, and they are also perpendicular to each other. This still usually allows the fields to be oriented at many different angles with respect to the surface of the Earth. The direction of the electric field is called the **polarization** of the wave, which may be vertical, horizontal, or somewhere between. Or the wave may be **elliptically polarized**, which results from two waves that are not exactly in phase, one polarized vertically and the other horizontally. If the waves are equal in magnitude and exactly 90° out of phase, the wave is *circularly* polarized. To account for all these cases, all fields must be checked separately for vertically and horizontally polarized waves.

Measurement Antennas

Most electronic components and instruments are designed to respond to voltages or currents, not fields. To measure a field strength it is necessary to convert its effect to a voltage or a current. This is achieved by an antenna. Although many antennas are simple conductor shapes, they must be analyzed carefully if accurate quantitative measurements are desired.

A straight conductor immersed in a time-varying electric field will develop a current in it. If the conductor material is linear (the usual case), the current will be proportional to the applied electric field, so their ratio will be constant. This ratio, however, depends greatly on the geometric dimensions of the conductor and the frequency of the electric field. It must be known to calibrate the antenna.

Similarly, a closed conductive loop immersed in a time-varying magnetic field will develop a current in it. Again, if the conductor is linear, the ratio of the current to the magnetic field strength is constant but depends on the dimensions of the loop and the frequency of the magnetic field.

The easiest way to calibrate an antenna is to immerse it in a known electric or magnetic field and measure the current or voltage at the antenna terminals. The principal problem is generating the known field. To find its strength, one must use a "standard" antenna for which the current-to-field ratio can be calculated.

To calculate the required ratio, Maxwell's equations must be solved subject to the boundary conditions of the antenna conductor. For most antennas an exact closed-form solution is impossible. However, for a sinusoidally varying field encountering a straight cylindrical conductor called a **dipole antenna**, such a solution is possible, though difficult [1]. Once the solution is obtained, the required ratio becomes a simple expression if the antenna is *resonant* or *tuned*. This occurs for a precise length that is slightly less than one half the wavelength, λ, of the time-varying field. Obviously, the antenna will be resonant at only one frequency, so the ratio will be valid only for a field varying sinusoidally at that frequency. For nonsinusoidal fields, each Fourier frequency component must be measured separately, and the antenna length must be changed as different frequencies are measured. To simplify changing its length, two telescoping rods, mounted end to end, are normally used to make the dipole antenna. The measuring instrument is connected between these two rods via a transmission line.

For a given frequency, at any point on the antenna, there is a certain current I flowing in it, and there is also a certain voltage V on it with respect to ground. The ratio of these phasors, V/I, is known as the **driving-point impedance**. The precise resonant antenna length is that for which V and I are exactly in phase, i.e., for which the driving-point impedance is purely real. As mentioned above, this length is slightly less than half the wavelength, λ, and it also depends on the thickness of the telescoping rods (pp. 547–548 of Reference 1). For a rod thickness of $\lambda/400$, the resonant length is 0.476λ. The driving-point impedance of a dipole antenna of these dimensions is $64\,\Omega$. If a voltage-measuring instrument such as a radio receiver or spectrum analyzer is connected to the antenna terminals via a transmission line, and is properly matched to the $64\,\Omega$ impedance, the measured voltage V_m will be equal to $0.148\lambda E$, where E is the applied field strength and λ is the wavelength at the frequency being measured. The ratio V_m/E, equal to 0.148λ, is known as the **effective length** (l_e) of the antenna, since it relates the field strength in volts per meter to the measured terminal voltage in volts. Obviously, it is not equal to the physical antenna length but is instead approximately one third of that value. With this ratio known, the electric field strength E that causes a certain terminal voltage V_m can easily be calculated.

To simplify calculations, E is often expressed in decibels with respect to a reference field of 1 μV/m and is designated E_d. Similarly, V_m is expressed in decibels with respect to a reference voltage of 1 μV and is designated V_d. The **antenna factor** (AF) is defined as the effective length expressed in negative decibels, or AF $= -20 \log(l_e)$. Then the multiplication becomes an addition, i.e., $E_d = V_d + \mathrm{AF}$.

The above antenna factor assumes that the antenna is perfectly matched to the receiver, which implies maximum power transfer. A mismatch would change the antenna factor. Therefore, since the antenna driving-point impedance usually is not equal to the receiver input impedance, a matching circuit must be inserted between the antenna and receiver. Another essential consideration is antenna balance. Most receivers and spectrum analyzers have one input terminal grounded. If this grounded terminal is connected to one of the dipole antenna terminals, the impedances connected to the two antenna terminals will be unequal with respect to ground. This also will upset the antenna factor, since one side of the antenna will not be properly matched to the receiver. To prevent this, a balanced-to-unbalanced (**balun**) network must be inserted between the antenna and receiver. Such a circuit provides a high impedance *with respect to ground* for both input terminals, while providing the correct input impedance (such as 64 Ω) *between* its input terminals. Normally, a single network provides both the matching and balancing functions.

Unfortunately, unless the dipole antenna is precisely the correct length, its antenna factor is much more complicated. Even if the frequency being measured differs only a few percent from the antenna resonant frequency, the antenna factor becomes unpredictable and the driving-point impedance becomes complex. Thus, the electric field cannot be easily calculated from the measured terminal voltage. To achieve the simple antenna factor described above, the frequencies must be measured one at a time and the dipole antenna length properly adjusted for each frequency. It is impossible to sweep the spectrum rapidly, as when using a spectrum analyzer, unless the antenna length can somehow be varied also. This leads to mechanical difficulties and is usually impractical.

Other types of antennas, however, are less sensitive to frequency. Examples are the biconical antenna and the log-periodic antenna. A biconical antenna can perform acceptably over a range of 20 to 300 MHz, and a log-periodic antenna is useful from 300 to 1000 MHz. Their antenna factors are relatively constant, usually varying by no more than 20 dB, over their useful frequency ranges. The antenna factors are usually too difficult to calculate, but they may easily be measured simply by observing the terminal voltage resulting from a sinusoidally varying field of known strength. The known field is first measured using a tuned dipole antenna, for which the antenna factor can be calculated. The antenna factor is measured in this manner at several frequencies throughout its useful range, and the results are plotted for use with the antenna.

Unlike the tuned dipole, the biconical and log-periodic antennas do not exhibit constant driving-point impedances over their useful frequency range. Since the receiver input impedance cannot be made to follow the variation of driving-point impedance with frequency, an exact match is impossible. This affects the antenna factor just as it would for a mismatched tuned dipole. To compensate for this, the antenna factor must be measured with the antenna terminated into a known impedance, which must then be used for all measurements made with that antenna. Then the mismatch is accounted for in the antenna factor itself. The mismatch does cause the antenna to reradiate the received signal, but this effect may be minimized by performing the measurements in an open-field site, which will be discussed later.

Tuned dipole, biconical, and log-periodic antennas are *linearly polarized* antennas because they respond to only one polarization component of a propagating wave. If the antenna is oriented horizontally, only the horizontally polarized component of the wave will affect it. Similarly, only the vertically polarized component will affect a vertically oriented antenna. Thus, with two measurements, any linearly polarized antenna will detect any type of field polarization. Other types of antennas, such as the spiral antenna, are designed to detect a circularly polarized wave. They will detect vertically and horizontally polarized waves, but they could miss a wave that is circularly polarized in the reverse direction (counterclockwise instead of clockwise, for example). Consequently, circularly polarized antennas are forbidden for many types of field measurements.

All antennas discussed above respond to the electric field, *E*. As mentioned earlier, in the far-field region, the magnetic field, *B*, is simply 3.33 nT times the value of *E* expressed in volts per meter. In any other region, however, *B* is not so simply related to *E* and must be measured separately, using an antenna that responds to magnetic fields. A circular loop or coil of wire is such an antenna. The loop is cut at one point and the radio receiver or spectrum analyzer is connected between its two ends. For quantitative measurements, its antenna factor must be known. The factor can be measured by immersing the antenna in a known magnetic field and measuring its terminal voltage. To find the known magnetic field strength, the electric field is first measured, in the far-field region, using a tuned dipole antenna for which the factor is known. The magnetic field is then 3.33 nT times this value expressed in volts per meter. With the magnetic field thus determined, the antenna factor of the loop may be calculated, as required.

Measurement Environment

A major difficulty with electromagnetic field measurements is repeatability of results. Electromagnetic fields are affected by any materials in their vicinity, even by poor conductors and dielectrics. The measurement environment must therefore be carefully defined, and similar environments must be used for all comparable measurements.

The ideal environment would be one where (1) the only electromagnetic field source is the equipment under test (EUT) and (2) there is no "foreign" material at all that could affect the fields being measured. Unfortunately, the only natural location where this could be achieved is in outer space, since the Earth itself affects electromagnetic fields. Since this is impractical, attempts are made to simulate this environment on Earth.

A large outdoor open area simulates a hemisphere of free space. Such a test site is appropriately called an **open-field site**. If the conductivity, permittivity, and permeability of the Earth were constant, every open-field site would have the same effect on the electromagnetic fields radiating from the EUT. The Earth's parameters do vary, however. To compensate for this variation, a large conductive floor, or ground plane, is laid under the EUT. This causes all electromagnetic waves to be totally reflected from the ground plane, so that the Earth's properties have no effect. The ground plane must be large enough so that it appears infinite with respect to the EUT and the associated test equipment. Acceptable dimensions are $1.73d \times 2d$, where *d* is the distance between the measurement antenna and the EUT, normally 3 or 10 m. Radiated emissions must be measured in all directions from the EUT, and at various angles of inclination. This is most easily achieved by placing the EUT on a turntable, which is then rotated during the test. To allow measurement at various inclination angles, the receiving antenna height must be varied, and this is accomplished by mounting it on a halyard. A typical open-field site appears in Figure 89.5.

An open-field site provides repeatable data only if there are no nearby trees or structures that could cause undesirable reflections. Before it can be reliably used, it must be tested. This is done by generating a known electromagnetic field and measuring it. The field is normally generated by a radio frequency oscillator driving a tuned dipole antenna, for which the radiation can be calculated (pp. 237–238 of Reference 2). This radiation is then measured as though it were generated by a typical device being tested. The ratio of the voltage at the transmitting antenna terminals to that at the receiving antenna terminals is known as the **site attenuation**. If the site attenuation is within 3 dB of its calculated value, the test site is deemed acceptable.

Although an open-field site eliminates reflections, external field sources, such as licensed transmitters, still cause problems. Since electromagnetic radiation can travel thousands of miles, no open-field site will be completely free of electromagnetic fields. To eliminate the effects of these stray sources, testing must be performed inside a shielded enclosure. There, however, severe reflections occur, and measurements become inaccurate and unrepeatable.

An ideal test environment would be a shielded enclosure lined with material that does not reflect electromagnetic waves. Such an enclosure is called an *anechoic chamber*, with the understanding that the

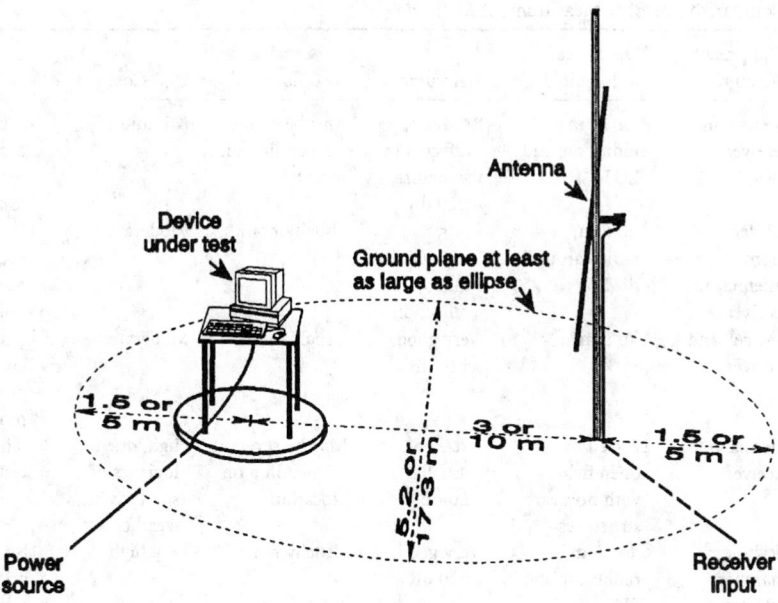

FIGURE 89.5 An open-field test site. Power and antenna cables are run under the ground plane so that they will not affect measured fields. The area outlined by the ellipse must be free of everything except the device under test, the table on which it rests, and the measuring antenna. To facilitate measuring radiation in all directions from the device, it is placed on a turntable. By rotating it during testing, and simultaneously varying the height of the receiving antenna, the direction of maximum radiation is found.

name refers to *electromagnetic* echoes. Until recently, such chambers were not practical except at very high frequencies, but improvements are constantly being made. Such an enclosure is acceptable for testing if it meets the site-attenuation requirements of a true open field. The site attenuation must be measured at several points inside the chamber, to assure that the proximity of the chamber walls has no effect. Unfortunately, such chambers are at present very expensive.

Another type of test chamber is the **transverse electromagnetic (TEM) cell**. This consists of an enlarged section of waveguide, in which the electromagnetic fields can be accurately predicted [3,4]. They are suitable only for testing small devices at relatively low frequencies. The TEM cell can be no larger than a wavelength at the frequency being tested. For example, to test at 200 MHz, the cell could be no larger than 1.5 m, or 5 ft, and the device itself must not exceed $\frac{1}{6}$ of this value, or 10 in. For small devices, however, the TEM cell is very accurate and is unaffected by stray field sources.

If a suitable anechoic chamber is not available, a device may be tested in an ordinary shielded enclosure to learn what frequencies it emits. The field strengths will be inaccurate due to the internal reflections. Then the device is tested in a true open-field site, and the suspected frequencies are measured quantitatively. Any field that exceeds the acceptable limits is then observed while the device is shut off. If it does not disappear, it is obviously not being generated by the EUT. This procedure is acceptable, although not as simple as testing inside an anechoic chamber.

Preliminary measurements may even be performed in an ordinary room. They will not be comparable with similar measurements made anywhere else, because of the effects of nearby conductors and dielectrics. Here, also, the device must be shut off to decide if any emissions are from stray external sources instead of from the EUT. This procedure provides a rough estimate of the emissions from the EUT, and it usually saves time during any later testing at a true open-field site. The various measurement methods appear in Table 89.4.

Permissible emission levels appear in the *Code of Federal Regulations* [5]. These rules assume open-field measurements, which are the most accurate possible. Even there, variations of ±6 dB are typical. Therefore, a manufacturer should allow a safety factor when performing measurements intended to assure

TABLE 89.1 Comparison of EMI Measurement Methods

Method	Equipment Required	Space Required	Accuracy	Outside Influence	Cost	Comments
Ordinary room	Antenna and receiver	3 or 10 m radius around EUT	Medium, affected by structure, ±20 dB	May be severe, depending on location	Minimum	Usually acceptable for preliminary tests
Shielded room	Shielded room, antenna, and receiver	4 to 6 m radius around EUT	Poor, ±30–40 dB due to reflections	Usually none	Moderate	Use for preliminary tests in noisy areas
TEM cell	TEM cell and receiver	1 to 3 m³	Very good, ±10 dB	Usually none	Moderate	Unusable for large EUT due to high-order modes
Open field	Antenna and receiver	17 × 20 m open field with no nearby structures	Excellent, usually ±6 dB	May be severe, depending on location	High, due to logistics of site (power, weather, etc.)	Standard test method
Shielded anechoic chamber	Anechoic chamber, antenna, and receiver	6 to 15 m radius around EUT	Very good, ±10 dB	Usually none	Very high	Use for accurate tests in noisy areas

compliance with government regulations. Otherwise, a device may pass when tested by the manufacturer but fail if later tested by the government using a supposedly identical test procedure. Since the government's measurements then prevail, the manufacturer's integrity could be questioned.

Further details on measurement techniques are available in the References 2 and 6.

Defining Terms

Antenna factor: Its effective length expressed in negative decibels.

Balun: An interface device used to isolate a dipole or other balanced antenna from the effects of a receiver having one grounded terminal.

Conducted coupling: Coupling due to voltages imposed on a shared power source.

Dipole antenna: An antenna consisting of two collinear rods with the feed line connected between them.

Driving-point impedance: The ratio of voltage to current at the driving point (normally the center) of an antenna.

Effective length: The ratio of the voltage observed at the driving-point of an antenna to the strength of its received electric field.

Electromagnetic compatibility (EMC): The capability of two or more electrical devices to operate simultaneously without mutual interference.

Electromagnetic interference (EMI): Any undesired effect of one electrical device upon another due to radiated electromagnetic fields or due to voltages imposed on a shared power source.

Elliptical polarization: Polarization of an electromagnetic wave consisting of two perpendicular electric fields of differing phase.

Emissions: Fields or conducted voltages generated by an electrical device.

Far-field region: Any location that is much farther than $\lambda/2\pi$ from an electric or magnetic field source, where λ is the wavelength at the frequency of concern.

Intersystem EMI: Electromagnetic interference between two or more systems.

Intrasystem EMI: Electromagnetic interference between two or more parts of the same system.

Near-field region: Any location that is much nearer than $\lambda/2\pi$ to an electric or magnetic field source, where λ is the wavelength at the frequency of concern.

Open-field site: A test location free of any conductors which would affect electromagnetic fields and taint the results.

Polarization: The direction of the electric field, *E*, of an electromagnetic wave.

Power density: Radiated power per unit of cross-sectional area.

Radiated coupling: Coupling due to radiated electric, magnetic, or electromagnetic fields.

Site attenuation: A measure of the degree to which electromagnetic fields at a test site are disturbed by environmental irregularities, obtained by comparing calculations with measured experimental results.

Susceptibility: The degree to which an electrical device is affected by externally generated fields or conducted voltages.

Transverse electromagnetic (TEM) cell: A relatively small test chamber in which fields can be accurately controlled by its geometric properties.

References

1. E. C. Jordan and K. G. Balmain, *Electromagnetic Waves and Radiating Systems*, 2nd ed., Englewood Cliffs, NJ: Prentice-Hall, 1968, 540–547.
2. J. P. Mills, *Electromagnetic Interference Reduction in Electronic Systems*, Englewood Cliffs, NJ: Prentice-Hall, 1993, 232–233.
3. M. L. Crawford, Generation of standard EM fields using TEM transmission cells, *IEEE Trans. Electromagn. Compat.*, EMC-16, 189–195, 1974.
4. M. L. Crawford and J. L. Workman, Predicting free-space radiated emissions from electronic equipment using TEM cell and open-field site measurements, *IEEE Int. Symp. Electromagn. Compat.*, Baltimore, MD, Oct 7–9, IEEE, Piscataway, NJ, 1980, 80–85.
5. Part 15 — Radio Frequency Devices, in *Code of Federal Regulations, Volume 47: Telecommunications*, Washington, D.C.: U.S. Government Printing Office, 1991, subpt. A-B.
6. Methods of measurement of radio-noise emissions from low-voltage electrical and electronic equipment (ANSI C63.4-1988), *American National Standards Institute (ANSI) Standards*, New York: The Institute of Electrical and Electronics Engineers, 1989.

Further Information

Guide for construction of open area test sites for performing radiated emission measurements (ANSI C63.7-1988), *American National Standards Institute (ANSI) Standards*, New York: The Institute of Electrical and Electronics Engineers, 1988.

H. W. Ott, *Noise Reduction Techniques in Electronic Systems*, 2nd ed., New York: John Wiley & Sons, 1988.

W. H. Hayt, *Engineering Electromagnetics*, 5th ed., New York: McGraw-Hill, 1989.

J. D. Kraus, *Electromagnetics*, 4th ed., New York, NY: McGraw-Hill, 1992.

S. Ramo, J. R. Whinnery, and T. Van Duzer, *Fields and Waves in Communication Electronics*, 2nd ed., New York: John Wiley & Sons, 1984.

Displays

90

Human Factors in Displays

Steven A. Murray
SPAWAR Systems Center

Barrett S. Caldwell
University of Wisconsin — Madison

90.1 Introduction

The display system is the final link between the measuring process and the user. If the display is not easy to see and easy to understand, then that process is compromised. The user's sensory capabilities and cognitive characteristics, therefore, must both be addressed in display system selection. Furthermore, display technologies and performance capabilities are easier to evaluate in the context of their intended application. Consideration of the following issues can narrow the search for candidate systems, and can prevent needless frustration during system use:

Environment. Will the display be operated in sunlight or at night?
Application. Will the display present alphanumeric data, video images, graphics, or some combination?
Task scenario. Are portability, handheld operation, or group viewing required?
System characteristics. Weight, volume, power, maintenance, cost, etc.

This chapter begins with basic treatments of light and vision. It then proceeds to discussions of visual capabilities and the display characteristics that must be matched to them.

90.2 Fundamentals of Light Measurement

The foundation metric of light is *luminous flux*, which is the rate at which light energy is emitted from a source, and is expressed in lumens (lm). *Luminous intensity* is luminous flux per unit solid angle, and its unit of measurement is the candela (cd). This is distinguished from *illuminance*, or illumination, which is simply luminous flux per unit area, expressed as lux (lx). *Luminance*, is a measure of the brightness, i.e., the amount of light, per unit area, either emitted by or reflected from a surface. Units of luminance measurement are candelas per square meter (cd/m²) or nits. Finally, *reflectance* is a unitless ratio of the amount of light striking a surface to the amount of light leaving it:

$$R = \pi \times \frac{\text{luminance}}{\text{illuminance}} \tag{90.1}$$

High reflectance can create glare, dramatically reducing visual performance.

90.3 Fundamentals of Vision

The eye functions very much like a conventional camera. Light enters the eye through a transparent *cornea* and is modulated by the *pupil*, a variable aperture opening controlled by muscles of the *iris*. The pupil grows larger in dark surroundings and smaller in bright surroundings to control the range of light intensity to the eye. Light rays are then refracted by an adjustable *lens* and brought into focus on the *retina*, where neural imaging begins. The retina contains both *cones* and *rods*, two distinctly different types of photoreceptors. Cones are concentrated near the *fovea*, or the central 2° of the visual field, and decrease rapidly with distance from this region. In contrast, rods are essentially absent in the fovea and increase in density with distance.

The eye is sensitive to three characteristics of electromagnetic radiation: (1) *brightness* (the intensity of ambient or incident light, measured in lux), (2) *hue* (the wavelength of light, measured in nm), and (3) *saturation* (relative concentration of specific hues in light, measured as a dimensionless ratio from 0 to 1). Cones are differentially sensitive to wavelength, i.e., hue, and have greater resolving power than rods because of their one-to-one mapping onto visual nerves. Cones can be further divided into three types, each maximally sensitive to a different portion of the visible light spectrum: (1) red (670 nm peak), (2) green (470 nm peak), and (3) blue (400 nm peak). Rods are more sensitive to light than cones and have many-to-one connections with the nerves that exit the eye, a feature that permits neural summation of low light signals. Human ability to discriminate differences in levels of brightness, saturation, or hue is governed by a psychophysical function known as *Weber's law*:

$$K = \frac{\Delta I}{I} \tag{90.2}$$

where ΔI is the difference, or change, in intensity, I is the initial intensity, and K is the Weber fraction. Values of K have been experimentally determined for brightness (0.079), saturation (0.019 for red), and hue (≈0.02 to 0.04, depending on the region of the visible spectrum).

Photopic vision occurs at light levels where both rods and cones are sensitive. The minimum light intensity required for photopic vision is approximately 2 lx; colors are seen in this region. As brightness decreases, a transition from photopic to scotopic vision takes place and color perception drops out gradually, a phenomenon that can impact the interpretation of color-coded information in poor light. Perception of blues and reds is lost first, then cyan and yellow-orange, and finally green, i.e., the wavelengths where the eye is most sensitive. The eye becomes most sensitive to wavelengths of about 550 nm (green) near the limit of photopic vision. *Scotopic vision* occurs at low light levels (2×10^{-7} lx to 2 lx) and primarily involves the rods; only achromatic shades of gray are seen. The transition from photopic to scotopic vision occurs slowly, requiring approximately 30 min for complete adjustment from photopic to scotopic visual sensitivity.

90.4 Visual Performance

Visual acuity is the ability to discriminate detail. The action of the lens, to change focus for objects at different distances, is called *accommodation. Minimum separable acuity*, the most common measure of discrimination, is determined by the smallest feature that the eye can detect, and is measured in terms of the reciprocal of the visual angle subtended at the eye by that feature. *Visual angle*, in minutes of arc, is calculated as

$$VA = \frac{3438H}{D} \tag{90.3}$$

where H is the height of the object and D (in the same units) is the distance from the observer. The ability to distinguish an object from its surroundings is known as *visibility*. The term is related to visual acuity, but implicitly combines factors of object size, contrast (i.e., including differences in hue and saturation), and brightness that all interact to determine true visual detection performance. On a more functional level, *readability* or *legibility* describe the ability to distinguish meaningful groups of objects (e.g., words extracted from groups of letters on a display).

Other parameters affecting visual performance include viewing angle and viewing distance. *Viewing angle* at the eye is measured from a line through the visual axis to the point being viewed, and determines where an object will register on the retina. The best image resolution occurs at the fovea, directly on the line of gaze, and visual acuity degrades with increasing angle away from this axis. *Viewing angle* at the display is the angle, in degrees, between a line normal to the display surface and the user's visual axis. The best viewing angle is, of course, on the visual axis and normal to the display surface, as luminance falls off for most displays as the angle from normal increases. Luminance reduction with viewing angle can be calculated as

$$E = E_m \cos^4 \theta \tag{90.4}$$

where E_m is the illuminance at the center of the display and θ is the viewing angle. Note that two viewing angles — at the eye and at the display — have been defined. *Viewing distance* is determined primarily by the minimum size requirements (i.e., visual angle) for objects that the user must see. A conventional reading distance is about 71 cm, although VDTs are frequently read at 46 cm. Most design criteria assume a viewing distance of between 50 and 70 cm.

Visual fatigue is an imprecise term, but one in common use, referring to the annoyance, irritation, or discomfort associated with visual tasks performed under poor conditions or for extended periods of time. A common cause of visual fatigue is *glare*, which can be due to a combination of high brightness, high reflectance, and specular (mirrorlike) reflections causing source light to reflect directly into the eye. Minimizing or eliminating glare is essential for effective display performance, and usually involves a thoughtful placement of the display, proper positioning of room lights, control of ambient light (e.g., window shades), or the use of display hoods.

90.5 Display Performance Considerations

Resolution is a measure of the smallest resolvable object on a display, and is expressed as display lines per millimeter or centimeter. Although *sharpness*, and its converse *blur*, are normally defined by the subjective reaction of the display user, sharpness has been formally measured as the ratio of the blurred border zone of letters to their stroke width [3]. Legibility is related to character quality, or readability, and depends on the sharpness of characters.

Contrast, or contrast ratio, is the measure of the luminance difference between an object and its background. While different definitions exist in the literature, luminance contrast as adopted by the International Lighting Commission (CIE) is given as

$$C_R = \frac{\text{luminance of brighter object} - \text{luminance of darker object}}{\text{luminance of brighter object}} \qquad (90.5)$$

Lower luminance displays require greater contrast to achieve the same visibility of objects. The *contrast*, or *luminance ratio* between two surfaces in the central field of vision (e.g., a display and the desk on which it rests) should be around 3:1, while the ratio between the central field and surfaces farther away (e.g., around a room) can be as high as 10:1. Ratios greater than this can induce glare. The simplest methods for contrast enhancement are the use of hooded shades or displays that can be tilted away from the offending light. Contrast-enhancing *filters*, however, can be more effective. All filters involve reducing the amount of ambient light reflected back to the user, while leaving the emitted light from the display content as unchanged as possible. Several strategies for filtering exist, including *etching* or *frosting* the display surface to break up and scatter specular reflections. *Neutral density filters* increase contrast by reducing the amount of light passing through them; ambient light must pass through twice before reaching the user's eye, while display content must only pass through once. Micromesh filters placed on the display surface limit light penetration so only rays falling perpendicular to the mesh can penetrate; this stops both specular and diffuse reflections and increases contrast. *Circular polarizers* are neutral density filters that polarize incident light, which is then prevented from returning through the filter. *Quarter-wave thin film coatings* interfere with both specular and diffuse reflections.

Gray scale refers to the number of luminance levels, or shades of gray, available in a display. The common definition is a luminance ratio of 1.4 between levels, although the eye can discriminate changes as small as 1.03 (a Weber's K value of 0.03). The number of gray shades is useful for evaluating the capability of a display to render pictorial information or the range of luminance levels that can be used for coding. The highest luminance level is determined by display capabilities, but the lowest level is determined by the luminance of the display surface when no signal is present. Bright light incident on the display can elevate this minimum level and reduce the number of usable gray shades.

Flicker is the term for detectable changes in display luminance, and occurs when the frequency of those changes is below the integrating capability of the eye. The minimum frequency at which this occurs is the *critical flicker fusion* frequency, or CFF, which depends on the luminance level of the image;, i.e., displays which do no flicker at high luminance levels may still flicker at low levels. The CFF is calculated as

$$CFF = a \log L_a + b \qquad (90.6)$$

where $a = 12.5$ for high (photopic) ambient light levels and 1.5 for low (scotopic) levels, L_a is the average luminance of the image in cd/m², or nits, and $b = 37$. This is an empirical formula, and the values for a and b are only approximate. Because the eye cannot adapt to flicker fast enough to control the light on the retina, visual irritation usually occurs where flicker is present.

Many display parameters are stated in terms of the *pixel*, or picture element. The pixel is the smallest addressable element in an electronic display, or the smallest resolvable information element seen by the user. *Refresh rate* is the frequency with which display pixels are reilluminated. Refresh rates below 50 to 80 Hz may induce perceptible flicker. The *update rate* is the frequency with which the information content of display is changed.

Linearity is the deviation of a straight line from its exact form, expressed as a percentage of total line length. Pattern distortion is the deviation of any element of a pattern (e.g., a grid) from its exact location, expressed in dimensions of the total pattern. While no specific limits are associated with these parameters,

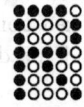

5 x 7 matrix
font

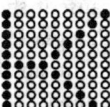

9 x 11 matrix
font

7 - segment
font

16 - segment
font

FIGURE 90.1   Examples of alphanumeric displays.

interpretation of measurement data can obviously be affected if nonlinearities are observable on the display.

Several mathematical models have been developed to quantify display *image quality* as single "figures of merit." These models, while useful, are too involved for treatment here and the reader is referred to an excellent summaries in the literature [1] for further information.

90.6   Display Content Considerations

Configurable software packages for scientific measurement (e.g., LabView™) allow great flexibility in the design of display formats. Human factors principles for display content, therefore, are as important to effective measurement as the electronic characteristics of the display itself. The following principles are an introduction to the kinds of issues that must be considered when designing display content. The interested reader is referred to Helander (1987) and Grandjean [3]) for information on human factors and design guidelines beyond those presented here.

Alphanumeric Displays

The size of a letter is its *pitch*. A general recommendation is that characters should subtend a minimum of about 12 min of arc at common reading distances. Alphanumeric displays are usually constructed of pixel arrays or segmented bars (Figure 90.1). A 5 × 7 pixel array is considered the minimum necessary to represent the ASCII character set. More pixels can increase legibility, but at higher cost for the control electronics. The seven-segment bar matrix is also common design and has good performance, but up to 16-segment arrays are available for better appearance.

Stroke width is the ratio of the thickness of the stroke to the height of the letter. Recommendations for legibility are 1:6 to 8 or 1:8 to 10 [2]. As illumination is reduced, thick letters are easier to read than thin ones. With low illumination and low contrast, letters should be boldface type with low stroke width–height ratios (e.g., 1:5).

Quantitative and Status Displays

While numeric readout displays are easy to implement in software, analog pointer displays show an advantage when it is important to observe the direction or rate of change of the values presented [4]. If the measurement application involves "more or less" or "up or down" interpretations, a straight-line or thermometer scale is preferred because it shows the measurement in relation to zero. *Moving pointers* are better able to convey minor changes in readings than fixed pointers or numeric readouts. *Scale markings* should reflect the length of the smallest unit to be read. It is desirable to have a marker for each unit, so that no interpolation is required.

Check reading indicators are used to determine whether a condition is "normal." The normal criterion point, therefore, should be clearly coded. If several indicators are grouped together, they should be arranged so that the deviant reading stands out (e.g., by indicating a different column height or dial angle, etc.).

Color is an excellent method for organizing information on a display and for locating objects rapidly. Although display users can distinguish between many different colors, they usually cannot remember more than seven to ten of them, so the number should be limited if color is going to be used as a coding dimension.

Summary

The next sections address different display technologies in light of the principles discussed here. While display guidelines are available for essentially any parameter, it is important to remember that visual perception is an integrative process. No single guideline functions alone, and display quality is usually a product of interacting needs and trade-offs.

90.7 Cathode Ray Tube Displays

The cathode ray tube (CRT, see Chapter 91) is by far the most common display technology in use today, and its widespread use in televisions and computer monitors should guarantee its continued presence throughout the foreseeable future. Advantages of CRT-based displays include (1) versatility (the same CRT can be used for alphanumerics, pictures, or graphics), (2) high-resolution capability and high luminous efficiency, extremely fast dynamic response (which can be important for rapidly changing signals), (3) extensive commercial availability (e.g., 2.5 to 64 cm, diagonally), (4) high reliability, (5) long life, and (6) relatively low cost. CRT displays can function well in high ambient illumination if filtering is used. Potential disadvantages of CRT displays are bulk (the depth of conventional CRT tubes can match or exceed their diagonal dimension, although flat CRTs are available), and vulnerability to ambient reflections and high illumination. Light falling on the smooth display surface can produce a veiling illuminance that **washes out** screen contrast and reduces the number of colors that can be perceived.

CRT Performance

Many characteristics of CRT displays depend on the type of phosphor selected for the design. Phosphor materials vary widely in their luminous efficiency, their color, and their decay time. **Decay time** interacts with display refresh rate; a phosphor with a short decay time will require a higher refresh rate to avoid observable flicker effects. Selecting a CRT with a high phosphor decay time will also result in selecting a tube with a high average luminance. Resolution depends on the **spot size** of the energized phosphor (which is effectively the thickness of the raster line). Spot size will also depend on the acceleration voltage and beam current of the cathode gun, so manufacturer's data must be noted for the voltage and current where the measurements were recorded, and compared with expected operating conditions. CRT resolution is measured in two ways. *Television lines* are the number of the most closely spaced, discernible lines that can be distinguished on an EIA (Electronic Industries Association) test chart. *Shrinking raster lines* involve a process of reducing the spacing between lines of a test pattern until no lines can be seen. This point is then expressed as the number of lines per centimeter, and is a better metric for measurement display applications. Increasing numbers of scan lines improve symbol legibility and image quality. Most conventional CRT monitors have 525 scan lines, but up to 5000-line systems are available [5]. The primary method for achieving *color CRT* displays is the use of single or multiple electron beams to energize phosphors for three primary colors — red, blue, and green. A complete range of colors is obtained by selectively energizing appropriate combinations of these three basic phosphors. Beam efficiency is reduced in this process, which means that color CRT displays are not as bright as monochrome CRT systems.

Contrast ratio is diminished by high ambient light levels, and it is often necessary to compromise between light requirements for work tasks and light levels for optimum CRT visibility. In low ambient lighting conditions, a CRT contrast ratio of 10:1 is usually attainable. A ratio of 25:1 can be achieved with contrast-enhancing filters, but at the expense of brightness.

Types of CRT Displays

In addition to the conventional, raster-scanned CRTs used for computer monitors and workstations, two variants of CRT technology should be mentioned, the direct-view storage CRT and the flat CRT. *Direct-view storage CRTs* have been designed to get around the need to refresh phosphors constantly. Direct-view systems usually add an additional, electrically charged layer — the storage element — somewhere behind the phosphor layer and an additional electron gun to maintain this charge. Displayed information is retained until it is actively erased. *Flat CRTs* have been developed to answer the need for CRT performance in a smaller physical package. The basic design technique places the electron gun at right angles to the screen and, through additional focusing circuitry or a slightly angled phosphor screen, writes the raster pattern at a high angle. Additional detail on these and other CRT designs can be found in Sherr [1].

90.8 Liquid Crystal Displays

Liquid crystal displays (LCDs, see Chapter 92) belong to the class of nonemissive technologies, i.e., displays that do not generate light of their own but control the transmission or reflection of an external light source. LCDs alter the optical path of light when an electric field is placed across the liquid crystal (LC) material.

The principal advantages of LCDs include (1) very low power consumption (important for battery-operated and portable systems such as calculators), (2) a flat display package, (3) low cost of the basic materials, and (4) excellent contrast in high ambient illumination. Some LCDs, however, have slow dynamic response (i.e., for switching display elements on and off); 100 to 500 ms rise times, for example, are visually noticeable and such systems may be unacceptable for measurement applications. Low luminance is another drawback, and can make the display difficult to read in low-light conditions without an external light source. In addition, viewing angle is limited by inherent LC characteristics, and is usually less than 45° without special designs. Many LCD features, such as switching thresholds and response times, are temperature dependent.

LCD Performance

A full range of resolution capabilities is available, from simple alphanumeric displays to systems with 63 million pixels and resolutions of 47 lines/cm [6]. LCDs are primarily used in small display applications (e.g., calculators, watches, etc.), although 53 cm diagonal, full-color video-capable displays have been developed [7].

Contrast in polarized systems is determined by the *extinction ratio* of the polarizer, i.e., the ratio of light transmitted in the parallel polarizing orientation to light transmitted in the cross-polarizing orientation. Polarizers with good extinction ratios, however, also suffer high loss of light in the transmitting orientation, so maximum brightness is traded for contrast. Contrast ratios of up to 50:1 [6] have been achieved, although 20:1 is more common.

Color displays can be achieved by placing a color mosaic over the LCD and switching the cells behind the proper combinations of mosaic holes into their transmission states. This method reduces resolution and brightness, however, as the available pixels must be assigned to each of the three primary colors. The use of thin-film transistors (TFT) as a switching technology for LCDs is the latest approach to generating large, high-resolution displays and is the subject of extensive engineering research [8].

90.9 Plasma Displays

The simplicity and durability of plasma displays (see Chapter 93) makes this technology an attractive candidate for diverse measurement needs, especially where harsh environments are expected. In addition, the switching characteristics of plasma gases have not yet been fully exploited, and this technology offers excellent potential for future engineering improvements. Plasma methods are used extensively for alpha-numeric displays in portable, laptop, and handheld computers, and for the display of video imagery. Advantages of plasma displays include enhanced memory capability, brightness, and luminous efficiency. It is also possible to retain pixels in the on state without continuous refresh signals, which means that increased brightness can be obtained for the same power and driving circuitry. This advantage also allows for excellent contrast ratios in high ambient illumination.

Plasma displays also exhibit long display life and ruggedness. It is not unusual for the display to outlast the life of the product in which it is installed, and the relatively simple panel construction can tolerate high shock and vibration environments, or extremes of temperature. Some disadvantages of plasma displays include high cost (relative to CRTs) and high power requirements (relative to LCDs). Other technologies can compete effectively with plasma devices for small alphanumeric displays in benign conditions.

Plasma Display Performance

Commercial plasma displays are available with resolutions of 40 lines/cm, and systems with almost 50 lines/cm are under development. Systems of 2 million pixels have been constructed [9]. Gray scale is achieved with dc displays by adjusting the discharge current in each cell. Displays using ac voltage can trade resolution for gray scale with spatial modulation methods, i.e., by controlling the number and location of activated pixels, rather than the level of pixel illumination. While plasma displays have good gray scale, their brightness is not yet equivalent to that of CRTs. Plasma displays show the color of the ionized gas, usually orange (i.e., where neon is used), although different gas mixtures or phosphors have been successfully used to expand the range of colors. The *hybrid ac-dc display* was designed to combine the memory capability of ac systems with the efficient matrix circuitry of dc devices [10]. The display uses both types of current to generate gas discharge; the dc component controls the pixel addressing, while the ac component controls the memory states of the cells. The *hybrid plasma–CRT display* attempts to use the gas discharge effect of the plasma panel as a matrix-addressable source of electrons for the CRT. The result is a full-color system with high brightness and good luminous efficiency.

90.10 Electroluminescent Displays

With the exception of light-emitting diodes (LEDs, see Chapter 95), electroluminescent (EL, see Chapter 94) technologies are not as prominent in the commercial arena as other types of display systems. EL materials are complex (i.e., driven and controlled by processes related to solid-state semiconductor physics) and are more difficult to work with than other display materials. Nevertheless, they offer great potential for high brightness and low cost that deserves consideration, especially as new designs become available. Matrix addressing is used for control of information display applications. EL materials are applied in two forms — powders (PEL) and thin films (TFEL) — and are controlled by both ac and dc voltages, generating four basic design approaches. Some advantages of EL displays are (1) high luminous efficiency (except ac powder designs), (2) readability in sunlight, (3) color capability, (4) compact, flat panel designs, and (5) significant potential for low-cost manufacture. A disadvantage of EL displays is that ac powder (ACPEL) systems have low luminance and contrast ratio. In addition, phosphors in powder designs scatter and reflect ambient light, reducing their contrast.

EL Display Performance

Contrast ratios of 50 to 150:1 have been demonstrated with monochrome thin-film (ACTFEL) systems, while 15 to 20:1 have been achieved with dc designs. ACTFEL designs also show excellent brightness, with demonstrated luminance levels of over 157 cd/m² (monochrome) and 26 cd/m² (color). DCPEL designs, representing newer technologies, have achieved over 100 cd/m² with monochrome designs, but may soon meet or exceed ac-based values. Resolution of EL displays is limited by the duty factor of matrix addressing; a finite amount of time is needed to energize each row, and a minimum luminance level is needed for adequate display performance, so the remaining variable becomes the number of addressable lines. A demonstrated ACTFEL display with 640 × 400 elements, with six colors and a resolution of 27 lines/cm, is typical of this technology.

90.11 Light-Emitting Diode Displays

Light-emitting diode (LED) displays (see Chapter 95) involve single-crystal phosphor materials, which distinguishes them from the polycrystal EL materials discussed in the previous section. The basic physics behind their operation is, however, quite similar. LED displays are highly versatile and well suited to a variety of measurement applications. Advantages of LED displays include high reliability and graceful degrades; individual LED elements can fail without affecting overall display performance. LEDs are rugged, for operation in harsh environments, and they are more tolerant of temperature extremes than other technologies. LEDs demonstrate better viewing angles than LCDs, and excellent brightness for visibility in sunlight. Unfortunately, LED displays also have high power consumption when packaged in large, flat panel displays, and the cost is high for the complex assembly. Optical cross talk between array elements can occur if transparent substrates are used. LEDs are the most-restricted display in terms of color range (e.g., no blue device is commercially available).

LED Display Performance

LED devices have excellent brightness, but because display brightness is also a function of the filters or magnification lenses used over the LED elements, device luminance is not, by itself, a reliable measure of overall system performance. LED displays also show very good luminance contrast. *Chrominance contrast*, however — the color difference between the LED and its background — is a factor in evaluating LED performance that is not found in other technologies. Chrominance contrast is significant because of the high saturation of most LED phosphors. It is affected by display filters, and can have significantly more influence on display performance than luminance contrast.

LED displays with resolutions of 20 to 25 lines/cm have been constructed. Flat panel displays of 38,400 discrete elements have also been demonstrated with luminance levels of around 137 cd/m² (increasing to 240 cd/m² with reduced resolution), and at least one aircraft display with 49,000 elements has been built. CRT-equivalent displays with 600 × 400 elements have also been realized with engineering development models.

Defining Terms

Decay time: The time required for the peak brightness of a phosphor to drop to a defined fraction of peak luminance; a measure of how long the phosphor remains illuminated after being energized by the electron beam.

Duty cycle: The time spent addressing each pixel during a refresh cycle; inversely proportional to the number of pixels.

Font: Refers to the form in which alphanumerics and symbols are produced.

Spot size: The size of the illuminated spot from the electron beam; limits the size of the raster line.

Transillumination: Illumination from the side of a display surface, to highlight information on the surface itself;, e.g., lighting for automobile or aircraft instruments.

Wash out: The loss of contrast (i.e., reduction in dynamic range) in an LED as the ambient light reflected off the background of the display surface approaches the light level of the active area.

References

1. S. Sherr, *Electronic Displays*, 2nd ed. New York: John Wiley & Sons, 1993.
2. M.S. Sanders and E.J. McCormick, *Human Factors in Engineering and Design*, 6th ed. New York: McGraw-Hill, 1987.
3. E. Grandjean, Design of VDT workstations, in G. Salvendy, Ed., *Handbook of Human Factors*, New York: John Wiley & Sons. 1987.
4. D.G. Payne, V.A. Lang, and J.M. Blackwell, Mixed vs. pure display format in integration and nonintegration visual display monitoring tasks, *Human Factors*, 37(3), 507–527, 1995.
5. N.H. Lehrer, The challenge of the cathode-ray tube, in L.E. Tannas, Jr., Ed., *Flat-Panel Displays and CRTs*, New York: Van Nostrand Reinhold, 1985.
6. M. Hartney, ARPA display program and the national flat panel display initiative, in *Proc. 2nd Int. Workshop on Active Matrix Liquid Crystal Displays*, IEEE, 8–15, 1995.
7. T. Morita, An overview of active matrix LCDs in business and technology, in *Proc. 2nd Int. Workshop on Active Matrix Liquid Crystal Displays*, IEEE, 1–7, 1995.
8. I-W. Wu, High-definition displays and technology trends in TFT-LCDs, *J. Soc. Inf. Display*, 2(1), 1–14, 1991.
9. L. Weber, Plasma displays, In L.E. Tannas, Jr., Ed., *Flat-Panel Displays and CRTs*, New York: Van Nostrand Reinhold, 1985.
10. Y. Amano, A new hybrid ac–dc plasma display panel, *J. Soc. Inf. Display*, 2(1), 57–58, 1994.

Further Information

E. Grandjean, *Ergonomics in Computerized Offices*, London: Taylor & Francis, 1987, is an excellent all-around treatise on the principles of effective VDT selection and use. Summarizes a wide range of research literature. If this volume is difficult to obtain, a chapter by the same author is also included in the *Handbook of Human Factors* (Reference 2).

M.G. Helander, Design of visual displays, in G. Salvendy, Ed., *Handbook of Human Factors*, New York: John Wiley & Sons, 1987, is an excellent and concise review of major human factors principles for display design and use. Includes a critical review of the foundation literature in this area.

S. Sherr, *Electronic Displays*, 2nd ed., New York: John Wiley & Sons, 1993, offers clear presentations of all important display technologies, together with a good summary of performance measurement methods for display systems. Well illustrated with a variety of commercial products.

L.E. Tannas, Jr., Ed., *Flat-Panel Displays and CRTs*, New York: Van Nostrand Reinhold, 1985, provides a thorough, yet highly readable examination of the physical principles behind essentially every major display technology. Although the technology capabilities have become dated since publication, this is well worth review.

91

Cathode Ray Tube Displays

Christopher J. Sherman

91.1 Introduction

The cathode ray tube (CRT) is unequaled in its ability to produce dynamic, quality, high-information-content imagery at high resolution. Even more impressive is that it achieves this for a lower cost per pixel than any other comparable electronic display technology. For the instrument designer requiring a high-information-content display, it offers numerous advantages. As a raw image tube, it is commonly available as an off-the-shelf item with a broad infrastructure of vendors, integrators, support, and part suppliers. Interface standards are well established and as a complete system, ready to take a standard signal input, it is available worldwide in a variety of performance ranges. To meet different application requirements, it is available in diagonal sizes from 12 mm to over 1 m with resolution from thousands of pixels to over 5 million pixels per frame. Tube characteristics improve on a yearly basis, and prices continue to decrease. Standford Resources has been tracking the CRT market for almost 20 years, and is an excellent source of information. In its latest report, it indicates that despite competition from other display technology, the CRT will remain the single largest market in the display industry. The worldwide market for CRT tubes in 1997 was 261 million units worth $26 billion U.S. This is expected to grow to 341 million units worth more than $34 billion by 2003 [1, 2]. Although there are many competing information display technologies, the CRT will be with us well into the 21st century.

0-8493-8347-1/99/$0.00+$.50
© 1999 by CRC Press LLC

91.2 History

The CRT has a rich and distinguished history. The roots of today's CRT technology extend back more than 100 years to the latter half of the 19th century. Eugene Goldstein first introduced the term *cathode rays*; John W. Hittorf, Heinrich Geissler, Julius Plücker, and others made important contributions specific to CRT technology [3]. Throughout the 19th century, researchers were interested in the nature of a luminous gas discharge and shadowy rays that occurred when a high voltage potential was applied between two electrodes in a vacuum. Sir William Crookes was an active experimentalist in this field, and early cathode ray devices came to be known as Crookes tubes. Crookes noted that a glow was generated by the surface that the rays struck; the rays themselves were not the source of light. By producing tubes with a vacuum of 1.3×10^{-4} Pa (10^{-6} torr), he eliminated the luminous gas discharge and worked directly with the cathode rays [4]. Crookes demonstrated the following: luminance depended directly upon the material properties of the surface the rays struck; a magnetic field would deflect the path of the rays; the deflection was proportional to the strength of the magnetic field; and a magnetic field could be used to focus the rays into a beam. He also suggested the rays emitted by the cathode were a stream of charged tiny particles, which were soon to be identified by Joseph John Thomson as electrons. Continuing CRT experimentation lead to the discovery of x rays by Wilhelm Konrad Röntgen in 1895.

Ferdinand Braun was the first person to envision the CRT as a tool for the display of information and is generally credited with the invention of the first device to be a direct forerunner to the modern CRT [5]. Braun in 1896 designed a CRT "indicator tube" for monitoring high frequencies in power-generating equipment. His design contained all of the same elements as today's CRTs. It utilized a cathode as an electron source, two control coils for vertical and horizontal deflection, an anode for electron acceleration and beam control, and a focusing slit. He also incorporated a phosphor screen normal to the beam for tracing its path. Although crude by later standards, this was the direct prototype for CRT oscilloscopes. In 1903 to 1905 Arthur Wehnelt added several very significant advances to Braun's design. Wehnelt developed and implemented a hot oxide–coated cathode and a beam control grid [5]. This lowered the voltage necessary to generate the electron stream and provided for much finer control of the beam current. These developments were the forerunner of the modern electron gun.

The CRT remained mostly a laboratory device until three important applications ushered it onto the center stage of information display: oscilloscopes, television, and military radar. By the early 1900s oscilloscopes were being widely used for the study of time-varying electric circuits in communication. Television was developed in the 1920s and in July of 1930 the National Broadcasting Company began the experimental broadcast of television in New York City. During World War II the CRT underwent a second wave of maturation with the advent of radar technology. By the end of the 1940s the TV-CRT was available as a consumer product. In the 1950s RCA perfected shadow mask technology and before long color became a new standard, gaining wide consumer acceptance in the 1960s. While the principal components of a CRT system have not fundamentally changed in many decades, the CRT has continued to improve.

91.3 Image Formation with a CRT

Although there are many different CRT designs, some of which are quite intricate, the process of forming an image using a CRT is straightforward. The procedure can be divided into four basic stages: beam formation, beam focusing, beam deflection, and energy conversion. These stages occur in four different regions of the CRT and follow in order from the rear of the tube (the neck) to the faceplate. As shown in Fig. 91.1, the elements at the rear of the tube are collectively called the electron gun.

Beam Generation with the Electron Gun

The cathode generates the stream of electrons used to form the image-writing beam. The traditional cathode is a metal conductor such as nickel coated with a thin layer of oxide, typically a barium strontium compound. To reduce the voltage required to generate electron emission, the cathode is heated to 700

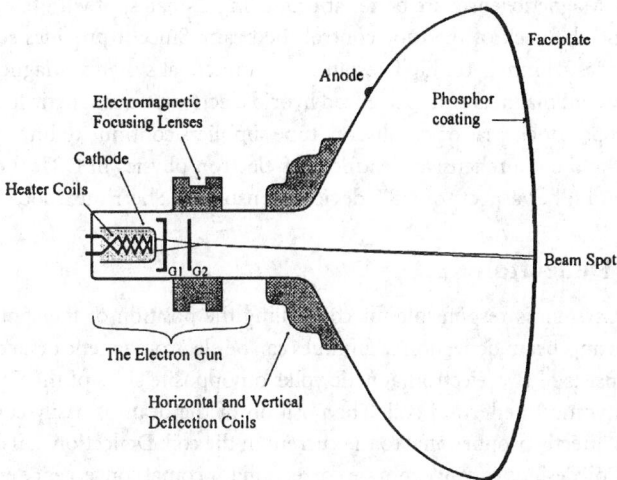

FIGURE 91.1 Example of a typical CRT with electromagnetic focus, and electromagnetic deflection. This type of tube design is usually used for demanding applications such as bright, high resolution image projection. Tubes like this are available from a number of manufacturers such as Sony, Matsushita, and Thomson.

to 1200°C. Applications that require high brightness often use more advanced and expensive dispenser cathode designs to increase the beam current while maintaining reasonable cathode life. These designs incorporate complex cathode structures and materials such as barium ceramics, molybdenum, rhenium, and tungsten. Readers interested in advanced cathode designs and additional information on CRT materials technology should consult References 6 through 10. The flow of electrons from the cathode is controlled by varying the potential between the cathode and a series of control grids commonly known as G1 (the control grid) and G2 (the acceleration grid). A voltage potential of 100 to 1000 V between G2 and the cathode creates the potential necessary to pull a stream of electrons off the cathode, forming the beam. The beam amplitude can be controlled and even completely shut off by varying the potential on G1. Thus, the voltage at G1 controls brightness because the brightness is proportional to beam current. The design of the cathode with respect to impedance and loading influences the maximum rate at which the beam can be modulated. The cathode and its associated control grids can be designed to produce a crossover flow of electrons or a laminar flow. In crossover gun designs, the emitted electrons converge to a point in front of the cathode. By using electron optics, this beam spot is imaged onto the phosphor screen. Due to inherent advantages, the crossover design is widely used. A crossover beam is narrow, making it easier to deflect than a thicker beam, and the spot can be very small, improving resolution at the screen. In theory, a laminar flow design provides for the possibility of higher beam current from a similar-sized cathode. In practice, the improvement is not usually advantageous enough to offset the added difficulty of controlling a wider beam.

Electron Beam Focusing

Beam focusing and beam current are critical in determining the final spot size and thus the resolution of the CRT. Focusing a beam of electrons is directly analogous to focusing a beam of light; the discipline is called electron optics. Concerns familiar to optical imaging such as magnification, spherical aberration, and astigmatism also confront electron optics. As CRTs become larger, and operate at higher deflection angles, spot control becomes critical. Beam focusing is achieved using either electrostatic focusing grids or electromagnetic focusing coils. Electrostatic focus is the most extensively used technique. It can be found in use in applications from television to desktop computer monitors. Electrostatic focus is achieved by applying a succession of potentials across a complex sequence of focusing grids built into the electron gun. As designers seek to improve performance further, grid designs have become intricate [11, 12]. Magnetic focus is the system of choice for all high-performance systems where resolution and brightness are design objectives. Magnetic lenses are better at producing a small spot with few aberrations. External coils in a yoke around the neck of the tube control the beam. Since it provides superior performance, electromagnetic focus is common on high-resolution commercial systems. Magnetic focus can also be achieved using permanent magnets and specialized hybrid electrostatic/magnetic focus components. Due to the tremendous impact focus has on resolution, tube suppliers continue to improve focus control [13, 14]. For an excellent and comprehensive treatment of electron physics in CRTs, beam control, detailed design discussions, and other aspects of CRT devices, consult Sol Sherr's textbook [15].

Electron Beam Deflection

The beam deflection system is responsible for controlling the position of the spot on the front face of the CRT. As with focusing, beam deflection techniques can be electromagnetic or electrostatic. A magnetic deflection system consists of two electromagnetic yolks on opposite sides of the CRT neck; a horizontal deflection coil and a vertical deflection coil. The position of the beam is easily controlled; the amount of beam deflection is directly proportional to the current in the coil. Deflection coil design also influences spot size and shape. Coil designs can incorporate correction for coma, convergence errors, and pincushion distortion [16]. Because of its low cost and efficiency, CRTs for both moderate- and high-resolution applications normally use magnetic deflection. Electrostatic deflection provides faster beam displacement but less spot size control. It is typically used in oscilloscope systems, where resolution requirements are moderate and deflection speed is paramount.

91.4 CRT Addressing: Raster Scanning vs. Stroke

Raster scanning is the most common CRT addressing technique. In raster scanning the electron beam writes each frame one line at a time. This means the horizontal deflection system requires a higher bandwidth than the vertical deflection system. High-performance raster CRTs have a horizontal bandwidth of 15 to 150 kHz, and a vertical deflection bandwidth of 30 to 180 Hz. The information is written onto the front face of the CRT from left to right and top to bottom, one line at a time. Raster scanning can be interlaced or progressive (noninterlaced). In interlace scanning, each frame of information is decomposed into two fields. The first field consists of all of the odd-numbered lines in the original frame, and the second field contains all of the even-numbered lines. Each field is scanned onto the CRT at twice the frame rate. Commercial television is interlaced. In the U.S., the frame rate is 30 Hz and the field rate is 60 Hz; this scheme is known as the NTSC standard. PAL and SECAM are two other well-known commercial formats in use today. There are numerous other standards, both analog and digital, in use worldwide, each with slightly different scan rates, resolution/addressing formats, luminance and color encoding, and timing protocols [17, 18]. Interlace scanning is an engineering compromise that conserves transmission bandwidth, electronics bandwidth, and CRT bandwidth, while maintaining acceptable performance with respect to moving imagery, resolution, and flicker. Computer monitors employ a progressive raster scan. Each frame has one field. The display is generated line by line, in order from the first pixel to the last pixel. Frame rates for the typical desktop monitor vary from 60 to 85 Hz to over 180 Hz for specialized systems.

Stroke is an alternative addressing technique that was once quite common. The name for the technique comes from the phrase "the stroke of a pen." Stroke is a point-to-point addressing system. The beam is directed to the starting point of a line, turned on, and then moved directly to the end of the line. Because there are no raster lines, stroke CRT resolution is independent of direction and is limited primarily by spot size. Stroke addressing is excellent for low-information-content screens with detailed graphic characters or simple vector graphics. It provides extremely high quality drawing, clean lines, precise detail, and high speed while conserving power and bandwidth. Stroke systems address the CRT only where there is information, not the entire screen as in raster scanning. This technique is uncommon today because it is too slow and computationally intensive for systems that have high information content, gray scale, or imagery.

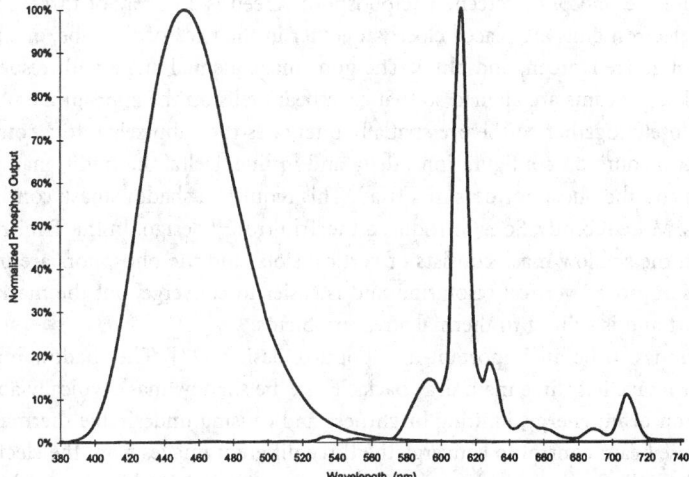

FIGURE 91.2 Phosphors can have a wide range of spectral characteristics. This graph illustrates the spectral profiles of two representative cases. The blue source is an example of a phosphor with a wide and smooth profile. The maximum output for this phosphor is in the blue at a wavelength of 452 nm. The red phosphor shows a sharper profile with several narrow peaks. The largest peak for this phosphor is at 612 nm.

91.5 The Phosphor Screen

The phosphor functions as the CRT transducer. It converts the energy of the electron beam into light. This conversion process is called cathodoluminescence. CRT phosphors are inorganic crystalline materials doped with one or more impurities called activators and coactivators. Phosphors emit light in two ways, fluorescence and phosphorescence. Fluorescence is the emission of light by the phosphor material while it is under bombardment by the electron beam. The continued emission of light after the bombardment has ceased is called phosphorescence. The length of time phosphorescence lasts is known as persistence. Persistence can vary from tens of nanoseconds to many minutes, or even hours. CRTs take advantage of both forms of cathodoluminescence. Briefly, cathodoluminescence occurs when the electron beam excites the electrons of the phosphor into higher, unstable energy states available due to the presence of the activators. When the electrons transition back to their stable states, light is emitted. The choice of phosphor depends on the requirements of the application with respect to wavelength characteristics (narrow emission spectra or broadband emission), color, brightness, resolution, and persistence. Commercial television CRTs typically make use of the following phosphor powders: $ZnS:Ag:Cl$ (blue), $Zn(Cd)S:Cu:Al$ or $ZnS:Cu:Au:Al$ (green), and $Y_2O_2S:Eu$ [19]. Television CRTs use moderately short-persistence phosphors. This ensures a new frame does not exhibit blurring due to the previous frame. Traditionally, phosphors for radar displays, where the screen is refreshed infrequently, had a mix of short and long persistence. However, with the advent of today's digital systems, the use of long-persistence phosphors has declined. The Electronics Industries Association maintains an information database of commercial phosphors for CRTs. The interested reader should consult its publication TEP116-C [20]. Figure 91.2 illustrates the spectral characteristics of several typical phosphors.

91.6 Color CRTs Using Shadow Masks

Color has been a part of the CRT world since the early 1950s. A sound understanding of human color perception is important for a detailed understanding of color in CRTs. Readers interested in an excellent comprehensive treatment of color science should consult Wyszecki and Stiles [21] or Robertson and Fisher [22] for a brief overview. The most common method of introducing color to the CRT is the three-gun shadow mask technique. The shadow mask is a metal grid of many tiny apertures held in place immediately before the phosphor screen. The phosphor screen is an array of three different phosphor subpixels. Three electron guns are placed closely together in the neck of the tube, one gun for each one of the primary colors (red, green, and blue). The guns make a small angle with respect to each other. The shadow mask and beams are aligned so that each beam falls on the appropriate phosphor dot. The dots are placed closely together so the eye spatially integrates the subpixels into a continuous blend of color. There are two common configurations, delta and in-line. Delta, the traditional method, arranges the electron guns and the phosphor dots in a triad. This requires a shadow mask consisting of a grid of circular apertures. More recently, Sony introduced the Trinitron™ design. In the Trinitron tube the guns are placed in line, the shadow mask consists of vertical slots, and the phosphors are in vertical stripes. The design offers improved vertical resolution and is easier to converge, but the mask is slightly more difficult to support and is subject to thermal stress problems.

Trinitron continues to be an important and popular design [23]. The shadow mask technique of producing a color image has three main drawbacks. First, the shadow mask typically absorbs more than 75% of the electron beam energy, limiting brightness and causing undesirable thermal effects. Second, both require precise beam control to converge the three different images from the electron guns. Third, the resolution of the display is limited by the requirements of spatial color; three subpixels are needed to make each full-color pixel.

91.7 Alternative Techniques for Realizing Color Using CRTs

There are alternative approaches to producing color with CRTs. In its own way, each seeks to improve upon one or more of the enormously successful shadow mask designs. The three most noteworthy challengers are beam index tubes, penetration phosphor tubes, and field sequential color systems. All three have been around for decades, but only field sequential color systems are commercially available.

Beam Index and Penetration Phosphor Tube Designs

A beam index tube uses a slotted mask and vertical phosphor stripes. One electron beam is focused to less than the width of a phosphor stripe. A reflected ultraviolet signal from the mask provides a feedback signal. This signal is used to index the position of the beam very accurately. Color selection is achieved by precise beam positioning and rapid beam modulation. Although beam indexing offers perfect convergence, few guns, energy savings, and high resolution, practical problems in control and manufacturing have left this approach unrealized. Penetration tubes are potentially even more advantageous since color is produced with one electron gun and no mask is required. The phosphor screen consists of a layering of several different phosphors. There are many approaches to the layering structure, but the basic principle is that each layer produces a different color. To select among the different colors, the beam energy is varied, altering its layer penetration. In theory, multiple layers of phosphors could be used to produce a full-color display. Unfortunately, this design requires high switching voltages, and the color purity is poor because of the dilution caused by leakage from the unselected phosphor layers. In the past, a few two-color systems have been produced on a limited basis with stroke writer systems (p. 147 of Reference 4).

Field Sequential Color

Field sequential color (FSC) is a different approach to realizing color using a monochrome CRT and color filters [24]. The shadow mask relies on the ability of the eye to perform a *spatial integration on a group of color elements*. Field sequential color exploits the ability of the eye to perform a *temporal integration of color fields*. Field sequential color is not a new concept; it was an early design suggestion for commercial color television. To implement field sequential color, each full-color frame is decomposed into primary color fields (red, green, and blue). Each of these color fields is displayed sequentially on a monochrome CRT while simultaneously a color filter is placed over the front of the CRT. The filters must be rapidly changed to stay in sequence with the individual fields which are displayed at three times the desired frame rate. This sequence is so quick that the eye fuses the individual color fields into a steady blend of continuous color. Field sequential color systems have all of the advantages of a monochrome CRT, namely, superior resolution and simplicity. The major historical drawbacks to field sequential color have been the difficulty in controlling a bulky, spinning, mechanical filter wheel, and two thirds of the light energy of the phosphor is thrown away in each field. In the past, some viewers have reported the ability to perceive the individual color fields intermittently; this artifact is known as "color break up." Field sequential color systems have seen a commercial rebirth due to several factors. The luminous efficiency of several phosphors has improved to the point where the light loss penalty is no longer a severe price to pay. Awkward, mechanical filter wheels have been replaced by several different types of compact, low-power, liquid crystal shutters. The use of faster frame rates and digital sampling techniques has reduced color artifacts. And finally, monochrome CRTs using field sequential color meet the growing market need for small high-resolution displays.

91.8 Image Quality and Performance

For any display technology, the ultimate test of performance and quality is how the image looks to the viewer in a real-world environment. This is highly subjective, but critical none the less. Table 91.1 highlights some important parameters associated with performance and image quality. Resolution is the most critical characteristic of any CRT. A note of caution, resolution should not be confused with addressing. A system can have a precise addressing format, but if the spot is too large the CRT resolution capability will be poor despite the addressing format. Alternatively, if the spot is precise and the addressing format is coarse, the *image* will be low resolution, but the CRT resolution will be high. Resolution is affected by almost every system parameter: spot size, focus, deflection, raster format, the mask, and beam current. For a perfectly designed system, line width and spot size will ultimately determine the resolution. Spot diameter is limited by the quality of the electron optics, the particle size and thickness of the phosphor layer, and beam current. The cross section energy distribution of the electron beam is approximately Gaussian. The higher the current, the larger the beam cross section. Scattering by individual phosphor particles also increases the spot size. Thus, high resolution is much more difficult to achieve for high-brightness applications. The champions of small spot sizes are the small 12 to 25 mm monochrome CRTs. Several commercial systems achieve ≤25 μm spot sizes, and sizes ≤15 μm have been reported [25]. Monochrome CRTs in the 53 cm class (21-in.) have spot diameters ≤150 μm. Color CRTs with shadow mask pitches ≤250 μm are available in sizes up to 21 in. at commodity prices, and this will continue to improve. Spot size does not tell the whole resolution story. In most systems something other than spot size limits performance. The scan format may be well below the performance of the CRT, the bandwidth of the support electronics is frequently less than the tube bandwidth, and, for color, the shadow mask will limit resolution more than spot size. Engineers can still design a CRT that maintains its performance even with state-of-the-art electronics. MTF (modulation transfer function) is the standard metric for resolution. MTF compares the modulation of the displayed output with the modulation of a known input signal. As the spatial frequency of the input increases, the performance of the output will show a steady roll-off in modulation depth and quality. Readers interested in an introduction to this complex topic should consult Infante's review [26]. Contrast and its related companion, gray scale, are also important to CRT image quality. Since contrast can be measured and stated in so many different terms, gray scale is probably a more practical indicator of performance from a user's standpoint. As it attempts to replace traditional film, the medical imaging community demands the most of the monochrome CRT. Medical CRTs can produce up to 256 shades of gray with 8-bit controllers. Systems are available that will do 10-bit monochrome and color. Although frequently quoted, *shades of gray* is not a true indication of CRT performance and should not be confused with true *gray scale* capability. Shades of gray is a characteristic of the image source and the capability of the electronics; gray scale reflects the capability of the CRT. Gray scale is typically defined as a series of steps in brightness in increments of

TABLE 91.1 Parameters That Influence CRT Performance and Image Quality

Resolution	Spot size and shape
	Focus accuracy
	Line width
	Addressing/scan format
	Shadow mask design
Luminance	Dynamic range
	Maximum brightness
	Gray scale in high and low ambient light
	Gray scale vs. display brightness
	Phosphor luminous efficiency
Contrast	Contrast under high ambient illumination
	Contrast under low ambient illumination
	Large area contrast
	Pixel-to-pixel contrast (small area)
Image fidelity	Frame rate
	Flicker/refresh
	Uniformity and linearity
	Aberrations (linearity, pincushion, barrel distortion, keystone, and focus)
Color	Phosphor selection
	Convergence accuracy
	Color gamut and saturation
	Color uniformity
	Color accuracy
Bandwidth	Video bandwidth (cathode design)
	Horizontal deflection design
	Vertical deflection design
	Bandwidth of the CRT supporting electronics
	Bandwidth of the signal delivery electronics

$2^{1/2}$ from black to maximum CRT output. Thus, gray scale is a realistic indicator that combines CRT dynamic range, contrast ratio, halation, and ambient reflections. More than 12 steps of gray scale is considered good performance.

   The subject of display image quality and evaluation is an important topic. In addition to resolution and gray scale, users and designers need to consider many other metrics, such as luminance, dimmability, contrast, readability, color, persistence, and convergence. Users may also be aware of visual distortions such as pincushion, barrel, and keystone. A discussion of these topics and the other issues raised by Table 91.1 is beyond the scope of a single chapter; indeed, it has been the subject of books. There are good sources of information in this area [15, 27]; in particular, Keller's [28] text is superb. It is essential to remember that no matter which metrics are used, the most single most important metric is the subjective determination of the end users. Although many factors will affect final display performance, the bottom line remains *how good does the display look* in the real-world application environment it was designed to meet.

91.9 CRT and Phosphor Lifetime

The lifetime of a CRT depends on how it is driven. The higher the beam current, the more rapidly its performance will degrade. In general, degradation is due to a failure of three parts of the CRT — the cathode, the phosphor, and the glass. Brighter displays require higher beam current, and this, in turn, requires a higher level of cathode loading. Oxide cathodes can comfortably operate with a loading range of 0.1 to 10 A/cm². This will provide a typical lifetime in the range of 10,000 h. Dispenser cathode designs can improve cathode life 2 to 3 times, but again individual cathode life depends on how much brightness the user regularly demands from the tube. Phosphor degradation is the second mechanism affecting CRT lifetime and performance. All phosphors degrade under constant electron bombardment, an effect called phosphor burning. The rate of degradation is related to beam current, anode voltage, and material

TABLE 91.2. The Strengths and Weakness of CRT Technology

Advantage	Weakness
Highest resolution technology	High power consumption
Versatile addressing formats	Weight of glass
Excellent image fidelity	Size of footprint
High speed	Emits EM radiation
Bandwidth (≥350 MHz)	Must be refreshed
Excellent contrast/gray scale	
Good selection of phosphors	
High luminous efficiency	
Bright display (up to 20,000 fL)	
Excellent color gamut	
Simplicity of interface	
Universal interface standards	
Good design flexibility	
Broad application base	
Long life and reliability	
Mature and broad knowledge base	
Worldwide sources of tubes and parts	
Inexpensive — low cost per resolution element	

parameters specific to each phosphor. Today, there is a broad range of good phosphor materials with greatly improved aging characteristics. The third mechanism is glass browning. This effect is a complicated interaction between the high energy of the electron beam and the molecular structure of the glass [10]. In recent years, glass suppliers have made great strides in providing CRT manufacturers with improved glass materials. However, phosphor aging and glass browning are still important concerns, especially in extremely bright (high beam current) applications, such as CRT projection.

91.10 CRT Strengths and Weaknesses

With a century of development behind it, the CRT has abundant advantages that make it an attractive display option for the system designer dealing with high information content or dynamic imagery. Table 91.2 lists some attributes, positive and negative, which the system designer should consider in evaluating whether of not to use CRT technology. The most notable traits of the CRT are flexibility and cost. CRTs can operate over a wide range of resolution formats from commercial television to the strict requirements of medical imaging and digital-to-film recording. One unit can easily be designed to accept a myriad of scan formats and resolutions from static text to dynamic imagery with megapixels at frame rates of hundreds of frames per second. This flexibility means the system can be upgraded later without having to redesign the display subsystem. Interfacing with the CRT is also straightforward; there are well-established standards and hardware for many scan formats from the television NTSC to the RGB standards of the personal computer industry. Finally, cost is important to emphasize; CRTs are inexpensive. The desktop CRT delivers its performance for ≤$0.00005 per color resolution element! Currently, no other high-information-content display technology can match this cost.

91.11 CRT Selection

The design requirements for (1) resolution, (2) brightness, and (3) screen size will quickly steer the system designer to a particular class of CRTs. If the design requirements are close to either television or desktop computer monitors, the choice is easy. There are literally hundreds of vendors offering integrated CRTs and electronic drivers as off-the-shelf items. Color shadow mask systems with more than 40 elements/cm are commonly available. If higher resolution is required, then medical imaging and film recorder CRTs may meet the requirements. Physically large display requirements can be supplied by projection CRTs or by tiled CRTs using video wall processors. Small to miniature CRTs provide resolutions up to 500 to 600 elements/cm and portability. If the application falls out of the mainstream, the designer will need to speak directly with CRT vendors. Table 91.3 is a partial list of CRT suppliers. Since CRTs have been designed to satisfy a broad range of specialty applications, the chances are excellent that an acceptable design will already exist. If not, the CRT is so mature that vendors working from a proven experience base can design a tube directly to customer specifications with little in the way of design problems or surprises.

91.12 Future Trends in CRT Technology

CRT technology is mature, and the industry does not foresee major surprises on the horizon. The CRT will continue to evolve, resolution will improve, screens will be flatter, deflections angles will be larger, and footprints will be reduced. The computer world continues to demand more resolution. In 1987, the 36-cm (14-in.) VGA class monitor (640 × 480) with its 300,000 pixels and 16 colors was the new standard. In 1998, the 43-cm (17-in.) desktop monitor is now capable of 1600 × 1200 performance (1.9 megapixels) with 24-bit color. The 53-cm (21-in.) monitor class, once an expensive custom device, is also widely available. The trend of increasing resolution and performance for decreasing cost shows every sign of continuing. The medical community is quickly moving to monochrome systems with 12-bit gray scale at a resolution of 2560 × 2048 (5 megapixels). All of this means the instrument designer has better resolution and fidelity for less money.

There is a quiet revolution taking place in the information display industry, which may have a significant effect on CRTs and all information display technology. There is a design trend to decouple the resolution and fidelity of the display device from its physical size. There are a number of applications motivating this change: communications, military, portable computing, simulation, and virtual reality. This is why miniature CRT technology has been more active in recent years. This is also the engine driving some of the major competitors to the miniature CRT, such as on-chip field emission displays (FED), deformable mirror devices (DMD), ferroelectric liquid crystal integrated circuit displays, and on-chip active matrix liquid crystal display (LCD) technology. If realized, these promise to be low-cost solutions because they are built on the technology foundation of the silicon chip industry. These challengers have the very real potential of eliminating the miniature CRT, and in some application areas, the LCD panel as well. However, the CRT backed by its formidable 100 years of design evolution and maturity is not standing still; its assets and market remain impressive. Although it is one of the few

TABLE 91.3 Companies That Manufacture Cathode Ray Tubes

Clinton Electronics Corporation
6701 Clinton Road
Rockford, Illinois 61111
Tel. 815-633-1444 Fax. 815-633-8712

Hitachi, Ltd.
New Marunouchi Bldg.
Marunouchi 1-chrome
Chiyoda-ku, Tyoko 100, Japan
Tel. 81-3-3212-1111 Fax. 81-3-3212-3857
U.S. address
Hitachi America, Ltd., Electron Tube Division
3850 Holcomb Bridge Road, Suite 300
Norcross, Georgia 30092-2202
Tel. 770-409-3000 Fax. 770-409-3028

Hughes Lexington, Inc.
A subsidiary of Hughes Electronics Company
1501 Newtown Pike
Lexington, Kentucky 40511
Tel. 606-243-5500 Fax. 606-243-5555

Image Systems Corporation
11595 K-tel Drive
Hopkins, Minnesota 55343
Tel. 612-935-1171 Fax 612-935-1386

Imaging & Sensing Technology Corporation
300 Westinghouse Circle
Horseheads, New York 14845-2299
Tel. 607-796-4400 Fax. 607-796-4482

ITPO Institute of Surface Engineering and Optoelectronics
Teslova 30
1000 Ljubljana, Slovenia
Tel. 386-61 1264 592/111 Fax. 386-61 1264 593

L. G. Electronics
20 Yoido-dong Youngdungpo-gu
Seoul, Korea
E-mail display2@www.goldstar.co.kr or
   monitor@www.goldstar.co.kr
U.S. address
L G Electronics, Monitors Division
1000 Silvan Avenue
Englewood Cliffs, New Jersey 07632
Tel. 201-816-2000 Fax. 201-816-2188

Matsushita Electric Industrial Ltd.
Twin 21 National Tower
1-61, Shiromi, 2-Chome, Chuo-ku,
Osaka 540, Japan
Tel. (06)908-1121
U.S. address
Panasonic Industrial Company
Computers and Communications Division
2 Panasonic Way
Secaucus, New Jersey 07094
Tel. 201-392-4502 Fax. 201-392-4441

Phillips Netherlands, BV
Phillips Components and Semiconductors
Bldg. VB Postbus 90050
5600 PB Eindhoven The Netherlands
Tel. 040-783749 Fax. 040-788399
U.S. address
Discrete Products Division
2001 West Blue Heron Blvd.
Riviera Beach, Florida 33404
Tel. 407-881-3200 or 800-447-3762 Fax. 407-881-3300

Rank Brimar, Ltd.
Greenside Way
Middleton, Manchester M24 1SN, England
Tel. 0161-681 7072 Fax. 0161-682-3818
U.S. address
25358 Avenue Stanford
Valencia, California 91355-1214
Tel. 805-295-5770 Fax. 805-295-5087

Sony Corporation
Display Systems
16550 Via Esprillo
San Diego, California 92127
Tel. 619-487-8500

Thomas Electronics
100 Riverview Drive
Wayne, New Jersey 07470
Tel. 201-696-5200 Fax. 201-696-8298

Thomson Tubes Electronics
13, avenue Morane Saulnier
Bâtiment Chavez - Vélizy Espace
BP 121/F-78148 VELIZY CEDEX France
Tel. 33-1 30 70 35 00 Fax. 33-1 30 70 35 35
U.S. address
Thompson Components and Tubes Corporation
40 G Commerce Way
Totowa, New Jersey 07511
Tel. 201-812-9000 Fax. 201-812-9050

Toshiba Corporation
Electronic Components, Cathode Ray Tube Division
1-1 Shibaura 1-Chrome, Minato-KU
Tokyo 105, Japan
Tel. 03-457-3480 Fax. 03-456-1286
U.S. address
Toshiba America, Inc.
1220 Midas Way
Sunnyvale, California 94086-4020
Tel. 408-737-9844

vacuum tubes still in use today, the traditional CRT is not doomed for obsolescence any time in the immediate future.

References

1. G. Aboud, *Cathode Ray Tubes, 1997,* 2nd ed., San Jose, CA, Stanford Resources, 1997.
2. G. Aboud, *Cathode Ray Tubes, 1997,* Internet excerpts, available http://www.stanfordresources.com/ sr/crt/crt.html, Stanford Resources, February 1998.
3. G. Shires, Ferdinand Braun and the Cathode Ray Tube, *Sci. Am.,* 230 (3): 92–101, March 1974.
4. N. H. Lehrer, The challenge of the cathode-ray tube, in L. E. Tannas, Jr., Ed., *Flat Panel Displays and CRTs,* New York: Van Nostrand Reinhold, 1985.
5. P. Keller, *The Cathode-Ray Tube, Technology, History, and Applications,* New York: Palisades Press, 1991.
6. D. C. Ketchum, CRT's: the continuing evolution, *Society for Information Display International Symposium, Conference Seminar M-3,* 1996.
7. L. R. Falce, CRT dispenser cathodes using molybdenum rhenium emitter surfaces, *Society for Information Display International Symposium Digest of Technical Papers,* 23: 331–333, 1992.
8. J. H. Lee, J. I. Jang, B. D. Ko, G. Y. Jung, W. H. Kim, K. Takechi, and H. Nakanishi, Dispenser cathodes for HDTV, *Society for Information Display International Symposium Digest of Technical Papers,* 27: 445–448, 1996.
9. T. Nakadaira, T. Kodama, Y. Hara, and M. Santoku, Temperature and cutoff stabilization of impregnated cathodes, *Society for Information Display International Symposium Digest of Technical Papers,* 27: 811–814, 1996.
10. W. Kohl, *Materials Technology for Electron Tubes,* New York, Reinhold Publishing, 1951.
11. S. Sugawara, J. Kimiya, E. Kamohara, and K. Fukuda, A new dynamic-focus electron gun for color CRTs with tri-quadrupole electron lens, *Society for Information Display International Symposium Digest of Technical Papers,* 26: 103–106, 1995.
12. J. Kimiya, S. Sugawara, T. Hasegawa, and H. Mori, A 22.5 mm neck color CRT electron gun with simplified dynamically activated quadrupole lens, *Society for Information Display International Symposium Digest of Technical Papers,* 27: 795–798, 1996.
13. D. Imabayashi, M. Santoku, and J. Karasawa, New pre-focus system structure for the trinitron gun, *Society for Information Display International Symposium Digest of Technical Papers,* 27: 807–810, 1996.
14. K. Kato, T. Sase, K. Sasaki, and M. Chiba, A high-resolution CRT monitor using built-in ultrasonic motors for focus adjustment, *Society for Information Display International Symposium Digest of Technical Papers,* 27: 63–66, 1996.
15. S. Sherr, *Electronic Displays,* 2nd ed., New York: John Wiley, 1993.
16. N. Azzi and O. Masson, Design of an NIS pin/coma-free 108° self-converging yoke for CRTs with super-flat faceplates, *Society for Information Display International Symposium Digest of Technical Papers,* 26: 183–186, 1995.
17. J. F. Fisher and R. G. Clapp, Waveforms and spectra of composite video signals, in K. Benson and J. Whitaker, *Television Engineering Handbook, Featuring HDTV Systems,* New York: McGraw-Hill Reinhold, 1992.
18. D. Pritchard, Standards and recommended practices, in K. Benson and J. Whitaker, *Television Engineering Handbook, Featuring HDTV Systems,* New York: McGraw-Hill Reinhold, 1992.
19. A. Vecht, Phosphors for color emissive displays, *Society for Information Display International Symposium Conference Seminar Notes F-2,* 1995.
20. *Optical Characteristics of Cathode Ray Tube Screens,* EIA publication TEP116-C, Feb., 1993.
21. G. Wyszecki and W. S. Stiles, *Color Science: Concepts and Methods, Quantitative Data and Formulae,* 2nd ed., New York: John Wiley & Sons, 1982.

22. A. Robertson and J. Fisher, Color vision, representation, and reproduction, in K. Benson and J. Whitaker, *Television Engineering Handbook, Featuring HDTV Systems*, New York: McGraw-Hill Reinhold, 1992.

23. M. Maeda, Trinitron technology: current status and future trends, *Society for Information Display International Symposium Digest of Technical Papers*, 27: 867–870, 1996.

24. C. Sherman, Field sequential color takes another step, *Inf. Display*, 11 (3): 12–15, March, 1995.

25. L. Ozawa, Helmet mounted 0.5 in. crt for SVGA images, *Society for Information Display International Symposium Digest of Technical Papers*, 26: 95–98, 1995.

26. C. Infante, CRT display measurements and quality, *Society for Information Display International Symposium Conference Seminar Notes M-3*, 1995.

27. J. Whitaker, *Electronic Displays, Technology, Design, and Applications*, New York: McGraw-Hill, 1994.

28. P. Keller, *Electronic Display Measurement, Concepts, Techniques, and Instrumentation*, New York: John Wiley & Sons, 1997.

Further Information

L. Ozawa, *Cathodoluminescence: Theory and Applications*, New York: Kodansha, 1990.

V. K. Zworykin and G. A. Morton, *Television: The Electronics of Image Transmission in Color and Monochrome*, New York: John Wiley & Sons, 1954.

B. Wandell, The foundations of color measurement and color perception, *Society for Information Display International Symposium, Conference Seminar M-1*, 1993. A nice brief introduction to color science (31 pages).

Electronic Industries Association (EIA), 2500 Wilson Blvd., Arlington, VA 22201 (Internet: www.eia.org). The Electronic Industries Association maintains a collection of over 1000 current engineering publications and standards. The EIA is an excellent source for information on CRT engineering, standards, phosphors, safety, market information, and electronics in general.

The Society for Information Display (SID), 1526 Brookhollow Dr., Suite 82, Santa Ana, CA 92705-5421 (Internet: www.display.org). The Society for Information Display is a good source of engineering research and development information on CRTs and information display technology in general.

Internet Resources

The following is a brief list of places to begin looking on the World Wide Web for information on CRTs and displays, standards, metrics, and current research. Also many of the manufacturers listed in Table 91.3 maintain Web sites with useful information.

The Society for Information Display	www.display.org
The Society of Motion Picture and Television Engineers	www.smpte.org
The Institute of Electrical and Electronics Engineers	www.ieee.org
The Electronic Industries Association	www.eia.org
National Information Display Laboratory	www.nta.org
The International Society for Optical Engineering	www.spie.org
The Optical Society of America	www.osa.org
Electronics & Electrical Engineering Laboratory	www.eeel.nist.gov
National Institute of Standards and Technology (NIST)	www.nist.gov
The Federal Communications Commission	www.fcc.gov

92

Liquid Crystal Displays

Kalluri R. Sarma
Honeywell, Inc.

92.1 Introduction

Liquid crystals (LC) are an important class of materials with applications ranging from display devices, optoelectronic devices, sensors, and biological and structural materials. The focus of this chapter will be on LCs for display applications. In general, most substances have a single melting point where a solid possessing a positional and orientational order changes upon melting to an isotropic liquid that has neither positional nor orientational order. However, some materials when melted from the solid state change into a cloudy liquid with orientational order at one temperature, and upon further heating change into an isotropic liquid that has no order, as shown in Fig. 92.1. Thus, an LC is a mesophase existing between the melting temperature T_m, of a crystalline phase, and clearing point T_c, of the liquid phase; i.e., below T_m, the material has a crystalline phase, above T_c, it has a liquid (isotropic) phase, and between T_m and T_c, it has a liquid crystal phase. This type of LC in which the mesophase is defined by the temperature (between T_m and T_c) is called a *thermotropic* LC. When the mesophase is defined by a solvent concentration, it is called a *lyotropic* LC. Thermotropic LCs are used for display applications. The orientational order in LC materials results in important physical properties, such as birefringence, that make these materials useful for display devices. Because LCs have the attributes of low drive voltage, low power consumption, thin form factor (flat panel displays), light weight, full-color, gray scale with a wide dynamic range, full motion video, superior image quality, and high reliability, LC displays (LCDs) are the preferred approach for battery-powered (portable) applications ranging from wristwatch displays and handheld TVs to laptop computer displays. They are also replacing cathode ray tubes (CRTs) in select applications such as avionic displays because of their high brightness and readability in sunlight.

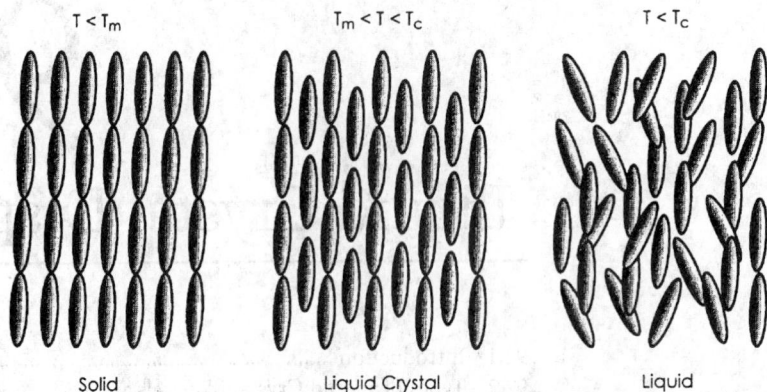

FIGURE 92.1 Illustration of a solid, an LC, and a liquid. A solid has an orientational as well as a positional order for the molecules. An LC has an orientational order only. A liquid phase is isotropic with neither positional nor orientational order.

LCs are also being used in projection display devices for head-mounted display (HMD) and for large-screen display applications. The following will discuss the various types of LC materials, their properties, LCD materials and fabrication processes, various LCD modes, and display addressing methods. There are many good general references, for example, References 1 through 7, on LCs and LCDs. At the time of this writing, LCD technology is advancing very rapidly with respect to technology development for LCD products with improved viewing angles, improved image quality, lower power consumption, and larger display sizes. The purpose of this chapter, however, is to present the basic LCD principles and technologies, as opposed to reviewing the current state of the art.

92.2 Types of Liquid Crystal Materials

Most of the LC materials are organic compounds which consist of rod-shaped or disk-shaped molecules. For display applications, LC materials with rod-shaped molecules are the most commonly used. The LC materials are broadly classified into three types (phases) — smectic, nematic, and cholesteric — according to their molecular order, as shown in Fig. 92.2. In a smectic liquid crystal, the rod-shaped molecules are arranged in layers with molecules parallel to each other. There are many different smectic phases, but smectic A and smectic C are the most common. In the smectic A LC, the molecular axis (director) is perpendicular to the layers as shown in Fig. 92.2a, and in smectic C it is tilted at an angle from the layer normal as shown in Fig. 92.2b. Also, in the nematic LC, the rod-shaped molecules are parallel to each other, but the individual molecules move relatively easily in the direction along their axis without a layer structure, as shown in Fig. 92.2c. In the cholesteric LC, the molecules are arranged in a layered fashion as in smectic LC, but the molecular axis is in the plane of each layer as shown in Fig. 92.2d. In addition, the cholesteric LC shows a helical structure in which the director n changes from layer to layer. The same LC material may have different LC phases at different temperatures. For example, the LC material may have a smectic C phase at a lower temperature, and as the temperature increases it may change to a smectic A phase and then to a nematic phase, before changing to an isotropic liquid phase at T_c.

The nematic LC is the basis for most widely used active matrix-addressed twisted nematic (TN) LCDs, and passive matrix-addressed supertwisted nematic (STN) LCDs. An example of a smectic C LC display is a passive matrix-addressed ferroelectric LCD [8]. An example of a cholesteric display is a passive matrix-addressed stabilized cholesteric texture (SCT) display with bistability [9]. A classic example of a nematic LC material is *p*-azoxyanisole (PAA) with a nematic phase in the range of 117 to 136°C.

$$CH_3 - O - \bigcirc - N = N - \bigcirc - O - CH_3$$

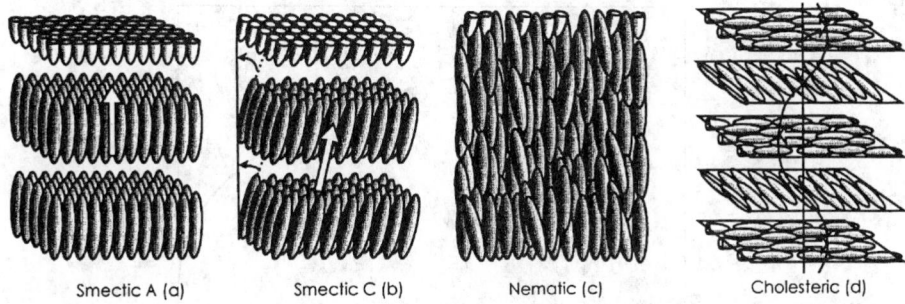

Smectic A (a)	Smectic C (b)	Nematic (c)	Cholesteric (d)

FIGURE 92.2   Molecular orientation in (a) smectic A, (b) smectic C, (c) nematic, and (d) cholesteric LC phases.

In PAA, the two benzene rings are nearly coplanar and the rigid rod is about 20 Å long and 5 Å wide. Another historical example of a nematic LC is N-(p-methoxybenzylidene-p-butylaniline) (MBBA) with a nematic phase in the range of 22 to 47°C.

$$CH_3{\diagdown}\,O-\!\langle O\rangle\!-CH=N-\!\langle O\rangle\!-CH_2\quad CH_2{\diagup}\,{}^{\diagup}CH_2{\diagdown}\,CH_3$$

MBBA has a central group that connects the two ringlike cores firmly and serves to maintain the linearity of the entire LC molecule. The terminal groups and conjugated bonds in the core are largely responsible for the dielectric, optical, and other anisotropic properties of the material. Azoxy and Schiff's base compounds are among the materials used earliest in LCDs. Because of environmental stability problems, they were replaced by biphenyl materials. More recently, phenylcyclohexane, bicyclohexane, and estercyclohexane compounds are developed to satisfy the requirements of broad temperature operation and enhanced electro-optical characteristics. The LC materials used in current LCDs are highly developed mixtures of various compounds tailored to meet the requirements of environmental stability, wide operating temperature range, proper response to the applied electric field, high electrical resistivity for matrix addressing, and fast response time.

The degree of order in the LC is an important parameter, and is defined by the *order parameter, S,* given by

$$S=\frac{1}{2}\langle 3\cos^2\theta-1\rangle \tag{92.1}$$

where θ is the angle between the molecular axis and the predominant molecular orientation $\boldsymbol{n}$. The symbol $<\ >$ represents averaging over the whole space. The predominant molecular orientation, which is known as the LC *director n* is defined as the average alignment direction of the long molecular axis. For a perfect orientational order, i.e., when all the molecules align parallel to the director, as in a perfect crystal, $\theta = 0$ and thus $S = 1$. For no orientational order, i.e., for a completely random molecular orientation, as in an isotropic liquid, $S = 0$. In a typical nematic LC, S is in the range of 0.3 to 0.7, and in a typical smectic LC it is in the range of 0.7 to 0.8, with higher values at lower temperatures. Figure 92.3 shows the temperature dependence of S for the LC material PAA [10] as an example. The order parameter decreases rapidly to a value of around 0.3, close to the clearing point T_c, and becomes zero in the isotropic state. The exact dependence of S on temperature T depends on the type of molecules considered. However, the following analytical expression derived from theory has been shown to be useful [11]:

$$S=\left(1-y\cdot T/T_c\right)^{\beta} \tag{92.2}$$

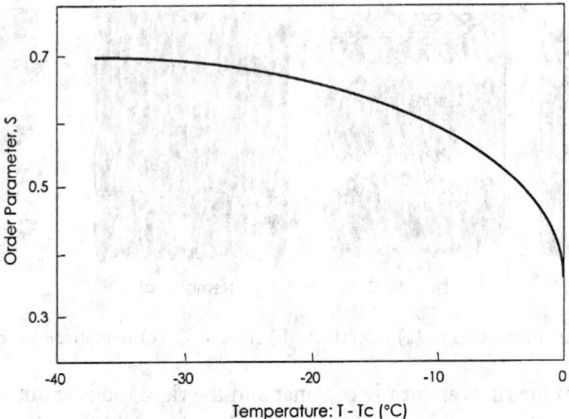

FIGURE 92.3 Temperature dependence of the order parameter S of the nematic LC P = azoxy anisole (PPA) [10]. S decreases with increasing temperature T, and rapidly approaches 0 near the clearing temperature, T_c.

where T_c corresponds to nematic–isotropic transition, y is of the order of 0.98, and β is an exponent in the range of 0.13 to 0.18 depending on the material in question.

92.3 Physical Properties of Liquid Crystals

Because of ordered structure with the molecules aligned with their long axis parallel to each other, LC molecules exhibit anisotropic properties. That is, various physical properties, such as dielectric constant ε, refractive index n, magnetic susceptibility χ, conductivity σ, and viscosity η, have different values in the direction parallel ($\parallel$) and perpendicular ($\perp$) to the molecular axis. The anisotropic physical properties, in conjunction with the ease of controlling the initial orientation (boundary condition) by surface alignment and the ease of reorienting the molecular axis by applying a voltage, is the basis for application of LCs for displays. The following will discuss the anisotropy of the dielectric constant and refractive index, elastic constants, and electro-optical characteristics which are important in the use of LCs for displays.

Dielectric Anisotropy

LC molecules exhibit dielectric anisotropy because of their permanent and induced dipoles. The dielectric anisotropy, $\Delta\varepsilon$, is expressed as

$$\Delta\varepsilon = \varepsilon_{\parallel} - \varepsilon_{\perp} \qquad (92.3)$$

where $\varepsilon_{\parallel}$ and $\varepsilon_{\perp}$ are the dielectric constants measured parallel and perpendicular to the LC director. Materials that exhibit positive dielectric anisotropy ($\Delta\varepsilon > 0$) are referred to as p-type materials and the materials that exhibit negative dielectric anisotropy ($\Delta\varepsilon < 0$) are referred to as n-type materials. The p-type LC materials tend to align themselves with their molecular axis parallel to the applied electric field, whereas n-type materials align themselves with their molecular axis perpendicular to the applied field. Generally, the dielectric constant $\varepsilon_{\parallel}$ decreases with increasing frequency [12] due to the relaxation phenomenon. However, $\varepsilon_{\perp}$ is independent of frequency over a large range of frequencies. At the crossover frequency, f_c, where $\varepsilon_{\parallel} = \varepsilon_{\perp}$, the LC material becomes isotropic. Depending on the material, this frequency, f_c, falls in the range of 100 kHz to >1 MHz. The dielectric constants $\varepsilon_{\parallel}$ and $\varepsilon_{\perp}$ also change as a function of temperature [13], as shown in Fig. 92.4 for the nematic LC cyanobiphenyl. The two dielectric constants rapidly converge as the temperature approaches T_c, where $\varepsilon_{\parallel} = \varepsilon_{\perp} = \varepsilon_{\text{isotropic}}$.

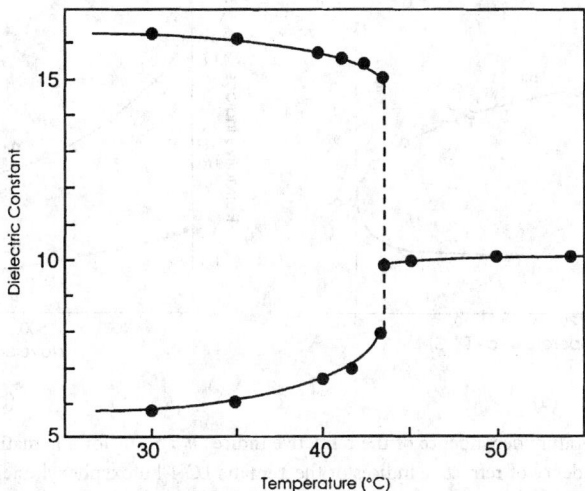

FIGURE 92.4 Temperature dependence of dielectric constant of the nematic LC cyanobiphenyl at a frequency of 100 KHz [13] exhibiting positive dielectric anisotropy ($\varepsilon_\parallel > \varepsilon_\perp$). $\varepsilon_\parallel$ decreases with temperature, whereas $\varepsilon_\perp$ increases with temperature until both are equal at the clearing temperature corresponding to the isotropic liquid.

Refractive Index Anisotropy

An LC is birefringent with anisotropic refractive indices. It has two principal refractive indices, n_o and n_e, as shown in Fig. 92.5. For the ordinary refractive index n_o, the electric field vector of the light beam oscillates perpendicular to the optic axis, and for the extraordinary refractive index n_e, the electric field vector oscillates parallel to the optic axis. In the nematic and smectic LCs, the direction of the LC director n is the optic axis of the uniaxial crystal and therefore the refractive indices for light rays with oscillations in the directions parallel and perpendicular to the director are $n_\parallel$ and $n_\perp$, respectively, i.e., $n_o = n_\perp$, $n_e = n_\parallel$, and the optical anisotropy or birefringence, Δn, is given by

$$\Delta n = n_\parallel - n_\perp = n_e - n_o \tag{92.4}$$

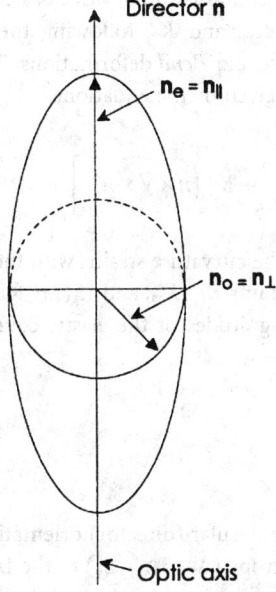

FIGURE 92.5 Refractive index anisotropy of uniaxial LC.

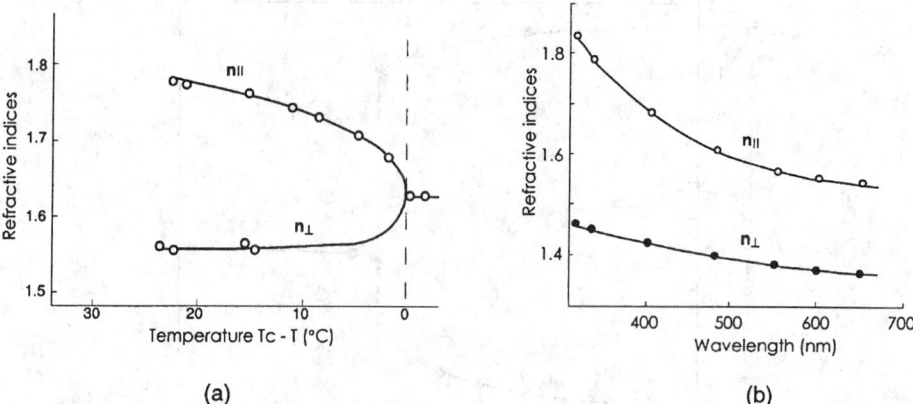

FIGURE 92.6   (a) Temperature dependence of the refractive indices $n_\parallel$ and $n_\perp$ for a nematic LC MBBA at $\lambda = 546$ nm; (b) wavelength dependence of refractive indices of the nematic LC 4-butoxyphenyl ester of 4'-hexyloxybenzoic acid at 80°C.

Fig. 92.6a and b show the temperature dependence [2] and wavelength dependence [2] of the refractive indices of typical LC materials. The dependence of refractive index on wavelength λ is generally expressed by the so-called Cauchy equation:

$$n_{o,e} = n_\infty + \alpha_{o,e}/\lambda^2$$

(92.5)

where n_∞ is the refractive index extrapolated to infinite wavelength and α is a material-specific coefficient.

Elastic Constants

In uniaxial LCs, the preferred or equilibrium orientation of the LC molecule is given by the director *n*, which may be imposed by the surface treatments at the boundary conditions or by an external field. When the LC is perturbed from an equilibrium condition by application or removal of an external field, the elastic and electrical forces determine the static deformation pattern of the LC. The transition of the director from one direction to the other induces curvature strain in the medium. Frank [14] showed that an arbitrary deformation state can be envisaged as the combination of three basic operations; *Splay, Twist,* and *Bend,* denoted by the elastic constants K_{11}, K_{22}, and K_{33} following the notation of the Oseen–Frank theory. Figure 92.7 illustrates the *Splay, Twist,* and *Bend* deformations. The elastic part of the internal free energy, *F,* of a perturbed liquid crystal is given by the equation:

$$F = \frac{1}{2}\left[K_{11}\left(\nabla \cdot n\right)^2 + K_{22}\left(n \cdot \nabla \times n\right)^2 + K_{33}\left(n \times \nabla \times n\right)^2 \right]$$

(92.6)

The free energy density is thus a quadratic function of the curvature strains with the elastic constants appearing as constants of proportionality. The elastic constants K_{11}, K_{22}, and K_{33} are temperature dependent, and decrease with increase in temperature. The magnitudes of the elastic constants, K_{ii}, can be approximated [15] by

$$K_{ii} \; \alpha S^2$$

(92.7)

Electro-Optical Characteristics

When an electric field is applied to an LC with an initial molecular (director) orientation, it will change to a new molecular orientation due to the dielectric anisotropy ($\Delta\varepsilon = \varepsilon_\parallel - \varepsilon_\perp$) of the LC. This change in

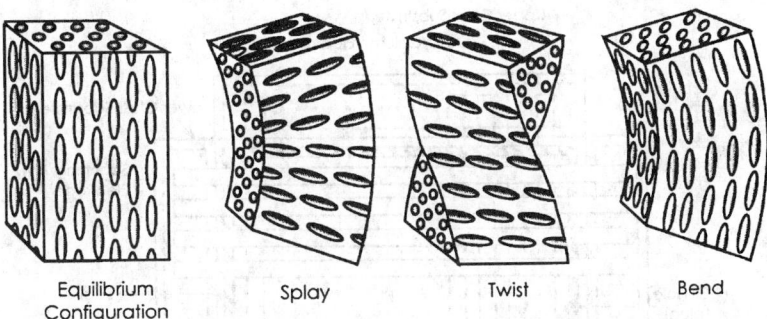

Equilibrium Configuration · **Splay** · **Twist** · **Bend**

FIGURE 92.7 Deformation of nematic LC molecules from equilibrium configuration is shown in (a). Three types of deformations — (b) splay, (c) twist, and (d) bend can describe all possible types of deformations.

molecular orientation is accompanied by a change in the optical transmission/reflection characteristics of the LC which forms the basis for LCDs. This phenomenon of an electrically driven optical modulation is known as the electro-optic effect of the LC. When an electric field E is applied to an LC, it produces an electric energy, f_e, given by

$$f_e = -\frac{1}{2}\varepsilon_\perp E^2 - \frac{1}{2}\Delta\varepsilon(n \cdot E)^2 \tag{92.8}$$

The initial molecular orientation (boundary condition achieved by surface alignment) of the LC molecules (director n) is either parallel (for $\Delta\varepsilon > 0$) or perpendicular (for $\Delta\varepsilon < 0$) to the plane of the two parallel electrodes of the display. When a field is applied across the parallel electrodes with the LC material in between, the director n orients parallel to the electric field E in $+ve$ $\Delta\varepsilon$ materials, and it orients perpendicular to the field in $-ve$ $\Delta\varepsilon$ materials. The total free energy, F, of the LC when the initial undeformed molecular orientation undergoes deformation due to the applied field is given by the sum of the electric energy f_e and the elastic energy. This transition from an undeformed state, known as *Freedericksz transition*, occurs as the field is increased to a critical field E_c. The Freedericksz transition is simply a transition from a uniform director configuration to a deformed director configuration; i.e., at any point in the LC, the order of the molecules relative to one another remains the same. The threshold electric field, E_c, is calculated by a free energy minimization technique [3], and it is given by

$$E_c = (\pi/d)\left(K_{ii}/|\Delta\varepsilon|\right)^{1/2} \tag{92.9}$$

Thus the threshold voltage V_{th} of the LC electro-optic effect is given by

$$V_{th} = E_c \cdot d = \pi\left(K_{ii}/|\Delta\varepsilon|\right)^{1/2} \tag{92.10}$$

In Equations 92.9 and 92.10, d is the thickness of the LC and k_{ii} is the appropriate elastic constant. When the field is perpendicular to the initially homogeneous orientation of the director, $K_{ii} = K_{11}$ or K_{22}. When the field is parallel to the initially homogeneous orientation, $K_{ii} = K_{33}$. In the case of a twisted orientation, $K_{ii} = K_{11} + (K_{33} - 2K_{22})/4$.

92.4 LCD Materials and Fabrication Processes

There are several types of LCDs utilizing different LC materials and LCD modes which are discussed in the next section. However, the general display assembly processes and materials are very similar for all

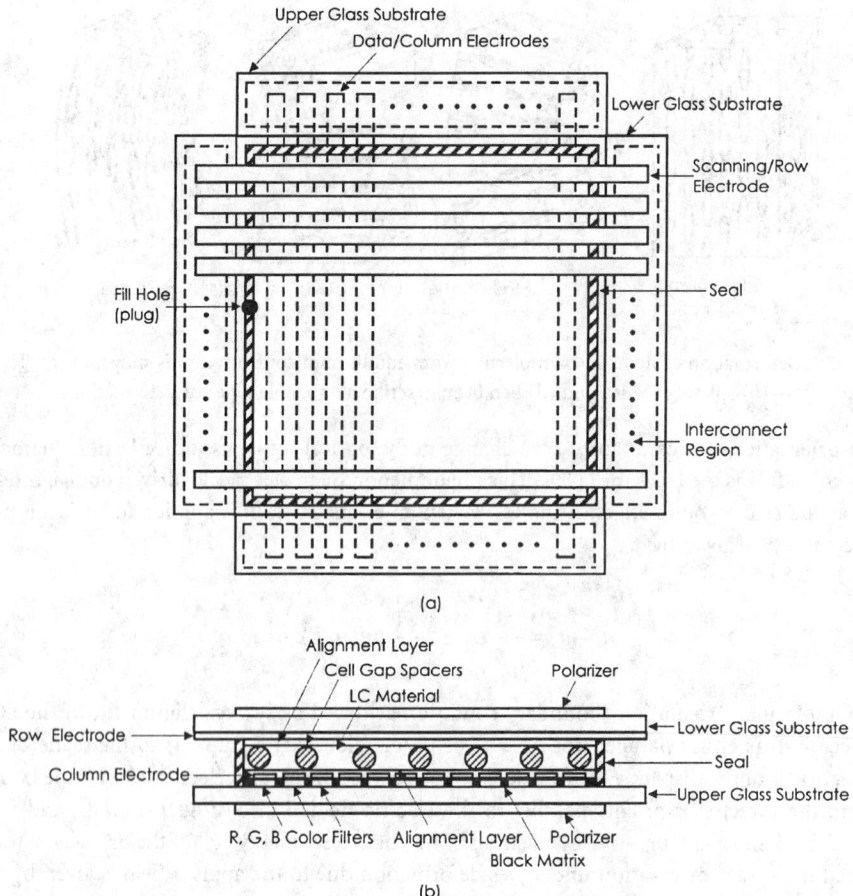

FIGURE 92.8 Plan (a) and cross-sectional view (b) of a passive matrix-addressed LCD.

these LCD modes. Figure 92.8a and b show the plan and cross-sectional view of a passive matrix-addressed LCD. The display fabrication can be broadly divided into three parts: (1) lower and upper glass fabrication processes, (2) cell assembly processes, and (3) polarizer and driver attachment and module assembly processes as illustrated in Fig. 92.9. The following will describe the various display materials and the assembly processes.

Glass Substrate

The quality of the glass substrate is important with regard to its chemical compatibility with the LC materials, surface flatness, defects, and dimensional stability under processing temperatures associated with various display fabrication steps. With a typical LCD cell gap in the range of 5 to 10 μm, the importance of the glass flatness and surface quality is clear. Glass substrate defects such as voids, scratches, streaks, and attached particles can cause electrode defects and hinder uniform LC cell spacing. Therefore, depending on the type of display, glass substrates are sometimes polished to achieve the required surface quality. Typical display glass materials include borosilicate (e.g., Corning 7059) and aluminosilicate glasses (e.g., Corning 1737), with a thickness of 0.7 or 1.1 mm.

Color Filters

In full-color LCDs, color most often is generated by use of red, green, and blue (R, G, B) color filters fabricated at each pixel, as shown in Fig. 92.8b, using a white backlight system. The color filter require-

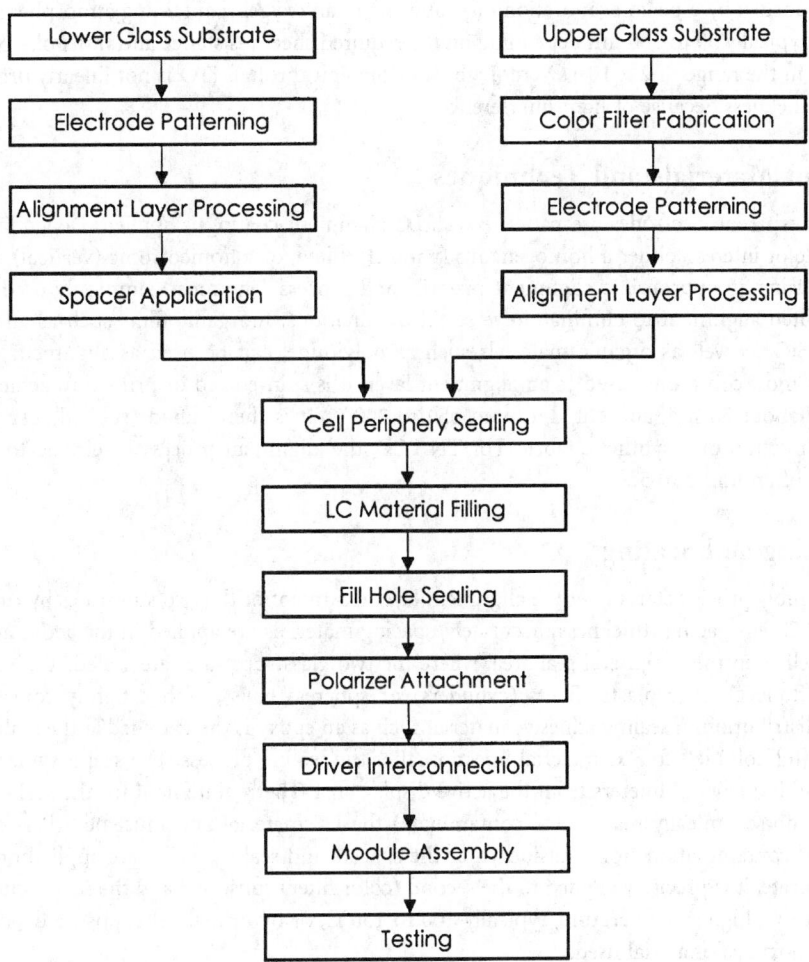

FIGURE 92.9   LCD assembly flowchart.

ments include proper spectral transmission characteristics and chemical, thermal, and dimensional stability. The display color gamut is a function of the spectral characteristics of the backlight used and the color filter transmission characteristics. By a suitable choice of these parameters, an LCD can achieve a color gamut comparable to a that of a high-quality CRT. However, trade-offs are sometimes made between color purity and transmission (brightness) characteristics of the LCD. Typical color filter thickness is about 2 μm. Color filter materials include dye and pigment dispersed polyimides and photoresists. The color filter materials are applied on the display glass by various processes, such as spin coating, printing, electrodeposition, and photolithography. First, color filter material of the first color is applied and photolithographically patterned. Then, the color filter material for the second color is processed, and then the third. A black matrix material is also applied and patterned between the color filters to block the light transmission from the interpixel regions. In some cases, a passivation layer such as low-temperature SiO_2 dielectric is deposited on the color filters to act as a barrier for impurities and to achieve a smooth surface for the subsequent transparent electrode deposition.

Transparent Electrodes

Most often, indium tin oxide (ITO) with a typical concentration of 90% In_2O_3 and 10% SnO_2 is used as the transparent conducting electrode material. The typical transmission of ITO is about 90%. It is generally deposited by e-beam evaporation or sputtering in an oxygen-containing atmosphere. The film thickness is typically 50 to 300 nm depending on the required sheet resistance and transmission. Typical resistivity is in the range of 2×10^{-4} Ω-cm. Light transmission through ITO is not linearly proportional to the film thickness because of light interference.

Alignment Materials and Techniques

After the transparent electrodes are patterned, an LC alignment coating is applied. Depending on the display mode of interest, either a homogeneous (parallel), tilted, or a homeotropic (vertical) alignment is achieved using an appropriate alignment material and process. The most commonly used TN LCD requires a tilted alignment to eliminate reverse tilt disclinations. Inorganic films such as an obliquely evaporated SiO, as well as organic materials such as polyimide, can be used as alignment materials. Polyimide is most commonly used as an alignment layer. It is spin-coated or printed to achieve a layer thickness of about 50 nm, and cured around 150 to 200°C. It is then buffed (rubbed) using a roller covered with cotton or a synthetic fabric. For TN LCD, the alignment process is selected to achieve a pretilt angle of around 2 to 5°.

Cell Spacing and Sealing

The cell assembly process starts after the alignment layer treatment of the two substrates by rubbing. To control the LC cell spacing (thickness) accurately, spacing materials are applied in the active area of the display as well as in the peripheral seal area, where the two glass substrates are sealed. Typical spacing materials include glass or plastic fibers (cylinders) or spheres (balls), with a tightly controlled size (diameter) distribution. A sealing adhesive material such as an epoxy is then applied at the seal area with an opening (fill hole) left for LC material injection after the sealing process. The seal is typically 1 to 3 mm wide and is a few millimeters from the active display area. The requirement for the seal material is that it must not chemically react with (contaminate) the LC material and must be a barrier against moisture and contamination from outside. After the spacers and seal materials are applied on the first (active) substrate, it is precisely aligned to the second (color filter) substrate, and the seal is cured either by heating it to a higher temperature, typically 100 to 150°C, or by ultraviolet exposure depending on the type of epoxy seal material used.

LC Material Filling

After the empty cell is fabricated, it is filled with the LC material. Because the cell thickness (spacing) is small ($\sim$ 5 to 10 μm), it is filled using special techniques. The most popular filling method is by evacuating the cell in a vacuum chamber, dipping the fill hole into a vessel containing the LC material, and increasing the pressure in the chamber. As the chamber pressure is raised, the cell gets filled by capillary action. After the cell is completely filled, the fill hole is capped by using an epoxy adhesive that is chemically compatible with the LC material.

External Components

The external components of an LCD include polarizers, reflectors, display drivers, and a backlight assembly. A reflective LCD uses a reflector at the back of the display, works by modulating the ambient light, and does not require backlighting. In the most commonly used transmissive mode TN LCDs, a polarizer is attached on the front as well as back surfaces of the LCD after the cell assembly is complete. Also, in the most commonly used normally white mode TN LCD, the polarizers are attached with their polarization axis crossed and along the rubbing directions of the alignment layers. The polarizer is a three-layer composite film with a stretched iodine doped polyvinyl alcohol (PVA) polarizing film in the center, and two outer films of triacetyl cellulose (TAC) for protecting the PVA film from the ambient (moisture, temperature, and harsh environment) conditions. A typical transmission range of a polarizer is 41 to 45%, with a polarization efficiency in the range of 99.9 to 99.99%.

The display row and column IC drivers are attached to the row and column bond pads of the display either by TAB (tape-automated bonding) using an ACA (anisotropic conductive adhesive), or chip on glass (COG) approaches. For backlighting transmissive LCDs, a fluorescent lamp is generally used. The R, G, B emission spectrum of the backlight and transmission spectrum of the R, G, B color filters are tuned together to achieve the desired color coordinates for the primary colors. Also, a diffuser is used to achieve uniform backlighting of the display. The backlight system may also use brightness enhancement films to tailor the light intensity distribution in the viewing cone. In addition to the above components, LCDs for specialized applications requiring enhanced performance may use a cover glass with EMI and antireflection coatings at the front, and a heater glass at the back side (between the backlight and the LCD) which facilitates low-temperature operation.

92.5 Liquid Crystal Display Modes

LC displays based on many different modes of operation have been developed. Historically, the phase change (PC) effect was discovered first in 1968 by Wysoki et al. [16]. The same year, dynamic scattering (DS) mode [17] and guest–host (GH) mode [18] were announced by Heilmeier et al. Then in 1971, the TN mode [19] was reported by Schadt and Helfrich and electrically controlled birefringence (ECB) was reported by Schiekel and Fahrenschon [20] and Hareng et al. [21]. The physical effects and the various display modes based on these effects include

Current effects:	• DS effect
Electric field effects:	• TN effect
	• STN effect
	• ECB
	• GH effect
	• Phase change effect
Thermal effects:	• Smectic effect

The DS effect is based on the anisotropy of the conductivity. Because of higher voltage operation and higher power consumption, the DS mode is not currently used. The TN and STN effects are most widely used among all the LCDs. The following will discuss various display modes.

Twisted Nematic Effect

Fig. 92.10 shows a schematic of a display based on the TN effect. It consists of nematic LC material with a positive dielectric anisotropy ($\Delta\varepsilon > 0$) with a layer thickness of about 5 μm, sandwiched between two transparent substrates with transparent electrodes. The surfaces of the transparent electrodes are coated with a polyimide alignment layer and rubbed to orient LC molecules at the substrate surfaces along the rubbing direction with a small (~3°) pretilt angle. The molecules on the two substrates are oriented 90° from each other as shown in Fig. 92.10a; i.e., the LC molecular axis rotates (twists) continuously through 90° from the first substrate to the second substrate. The TN display can be fabricated to operate in a normally black (NB) or normally white (NW) mode based on how the polarizers are attached to the outer surface of the two glass substrates. Figure 92.10a and b shows the on- and off-state of a NW mode TN LCD with crossed (orthogonal) polarizers attached with their polarization direction parallel to the LC director orientation on that substrate. Since the pitch of the twist is sufficiently large compared with the wavelength of the visible light, the direction of polarization of linearly polarized light incident normally on one surface of the display rotates through 90° by the twist of the LC molecules as it propagates through the cell and exits through the second polarizer. When a voltage is applied to the TN cell, the molecules align parallel to the direction of the field as shown in Fig. 92.10b, and the 90° optical rotatory

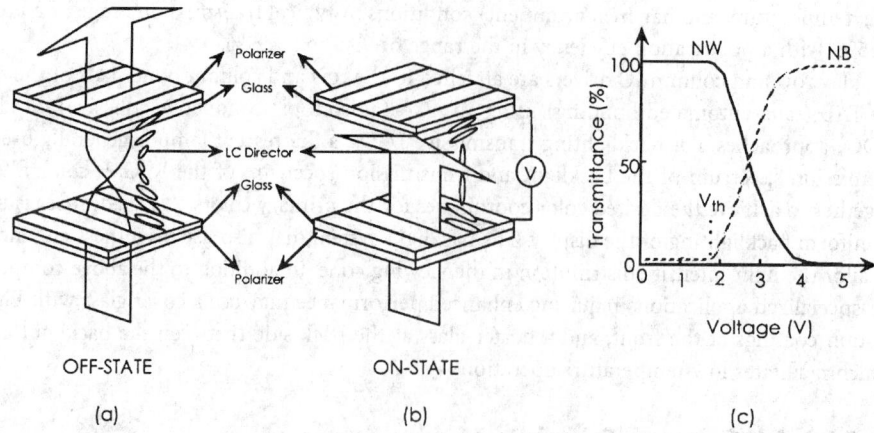

OFF-STATE   ON-STATE

(a)   (b)   (c)

FIGURE 92.10 Illustration of TN effect: (a) in the off-state, the incident polarized light is transmitted through the entrance polarizer, the 90° TN LC, and the crossed exit polarizer; (b) in the on-state; (c) the solid line shows the voltage-transmission behavior for the NW configuration, shown in (a) and (b), with crossed polarizers. The dashed line is for an NB configuration with parallel polarizers.

power is eliminated. Thus, the incident polarized light from the first polarizer is not rotated as it goes through the LC cell and gets blocked by the crossed exit polarizer.

Fig. 92.10c shows the voltage-transmission characteristics of a TN cell for NB and NW modes of operation. For an NW mode, when a sufficiently high voltage is applied, LC molecules are aligned homeotropically (parallel to the field) and there is no rotation of the electrical field vector of the polarized light. This results in complete suppression of transmission regardless of the wavelength of light. The transmission in the on-state is, however, wavelength dependent, but this does not have a significant effect on the contrast ratio, although it can influence the color balance. In the NB mode of operation, the transmission is suppressed to zero [22] for the off-state only for a monochromatic light of wavelength, $\lambda = 2d\Delta n/\sqrt{3}$. Therefore, in a practical display using a broadband backlight, there is a small amount of light leakage which lowers the display contrast ratio. In a color display, the cell gaps for the R, G, B pixels can be optimized to eliminate the light leakage and improve contrast [22].

The threshold voltage of a TN mode LCD is given by

$$V_{th} = \pi \cdot \sqrt{\left\{ \left[K_{11} + \left(K_{33} - 2K_{22} \right)/4 \right] / \varepsilon_0 \cdot \Delta\varepsilon \right\}} \tag{92.11}$$

V_{th} depends on the dielectric anisotropy and elastic constants, and is generally in the range of 2 to 3 V, with the maximum operating voltage being in the range of 5 to 8 V. This low voltage driving, coupled with low current due to high resistivity of LC materials, contributes to the very low power consumption ($\sim 1\ \mu W/cm^2$) of LCDs. The response times measured by the rise and decay times of the display τ_r, τ_d, are given by [23]:

$$\tau_d = \gamma \cdot d^2 / \Delta\varepsilon \left(V^2 - V_{th}^2 \right) \tag{92.12}$$

$$\tau_d = \gamma \cdot d^2 / \left(\Delta\varepsilon \cdot V_{th}^2 \right) \tag{92.13}$$

where γ is the rotational viscosity coefficient. The above equations show that rise time can be improved by using a thinner cell gap d, a higher $\Delta\varepsilon$, and a higher drive voltage V. Similarly, the decay time can be improved by reducing the cell gap. The turn-on time τ_r is usually shorter than the turn-off time τ_d. At room temperature, these times are of the order of 10 ms, which is adequate for many common applications such as computer and TV displays. In a TN LCD, gray scale is generated by varying the voltage using the electro-optic curve shown in Fig. 92.10c. The shallow slope of the electro-optic curve works well for the gray scale generation in active matrix-addressed displays (see next section). However, in the case of passive matrix addressing, the shallow slope greatly limits the multiplexibility (number of addressable rows) of the display, which led to the development of the STN effect.

STN Effect

In a passive matrix-addressed display, the addressability or the maximum number of addressable rows N is given by the Alt and Pleshko [24] limit:

$$V_{on}/V_{off} = \sqrt{\left(N^{1/2}+1\right)/\left(N^{1/2}-1\right)} \qquad (92.14)$$

where V_{on} and V_{off} are the rms voltages at the select and nonselect pixels. Equation 92.14 shows that as N increases V_{off} approaches V_{on} and the contrast ratio becomes 1, which makes the display not viewable. For $N = 100$, $V_{on} = 1.11V_{off}$; i.e., select voltage is only 11% higher than the nonselect voltage. This will result in a very low contrast ratio when using a TN mode with a shallow turn-on curve (Fig. 92.10c). STN displays have been developed [25, 26] to achieve a steep electro-optic curve, so that large numbers of rows can be multiplexed. The STN effect uses a twist angle of 180° to about 270° with a relatively high pretilt angle alignment. Figure 92.11 [25] illustrates the STN effect. The figure shows the voltage dependence of the midplane director tilt of a chiral nematic layer with a pretilt of 28° at both substrates. Bistability is achieved when a twist angle, ϕ, greater than 245° is used. In highly multiplexed displays twist angles in the range of 240° to 275° and tilt angles in the range of 5° to 30° are generally used. High pretilts ensure that competing distortional structure which has 180° less twist is eliminated. For a 270° left-handed twist, optimum results are achieved when the front polarizer is oriented with its polarization axis at 30° with the LC director and the rear polarizer is oriented at an angle of 60° with the projection of the director at the rear substrate. Due to interference of the optical normal modes propagating in the LC layer, the display has a yellow birefringence color in the nonselect state (yellow mode). Rotation of one of the polarizers by 90° results in a complementary image with a bright colorless state, and a blue nonselect state (blue mode). White-mode STN displays are made using retardation films. The response time of a typical STN display is on the order of 150 ms. These displays typically have a lower contrast ratio and a narrow viewing angle.

Electrically Controlled Birefringence (ECB)

This display technique is based on controlling the birefringence of the LC cell by application of an electric field. There are a number of types of this display depending on the molecular orientation of the LC cell used; examples include DAP type, homogeneous type, HAN (hybrid aligned nematic) type, and IPS (in-plane switching) type. A DAP type (homeotropic orientation) is made using an LC with a negative $\Delta\varepsilon$, sandwiched between transparent electrode substrates and placed between crossed polarizers, as shown in Fig. 92.12. In the off-state (with no electric field), the incident polarized light does not see birefringence when passing through the cell, and thus gets blocked by the crossed exit polarizer. In the on-state (when a voltage is applied), the molecular axis of the LC is inclined at an angle θ (as shown in Fig. 92.12), so the linearly polarized light becomes elliptically polarized as it passes through the cell due to birefringence. Hence a portion of the light passes through the crossed exit polarizer; the intensity I of the transmitted light through the cell is given by [20]

$$I = I_0 \sin^2 2\theta \cdot \sin^2\left(\pi \cdot d \cdot \Delta n(V)/\lambda\right) \qquad (92.15)$$

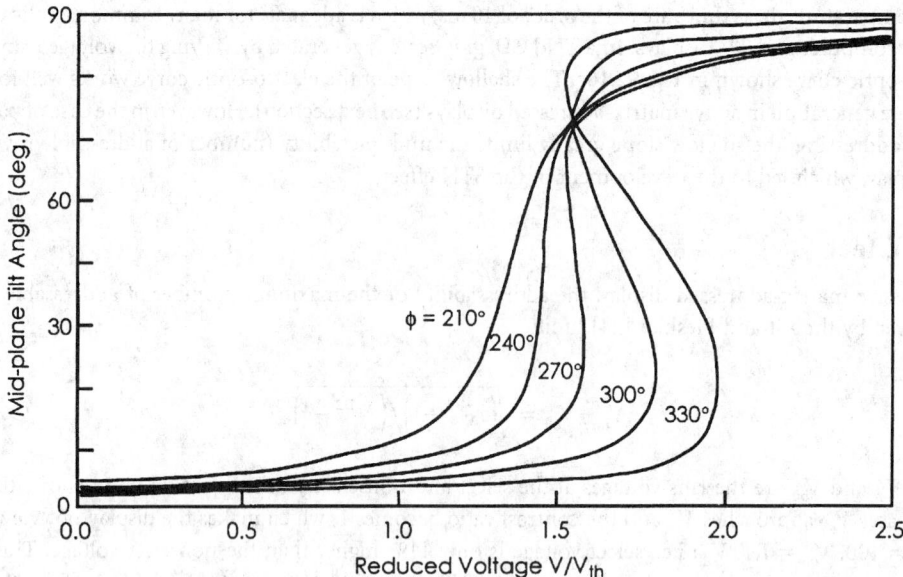

FIGURE 92.11   Calculated curves of tilt angle of local directors in the midplane of an STN cell as a function of reduced voltage V/V_{th}, where V_{th} is the Freedericksz threshold voltage of a nontwisted layer with a zero pretilt angle. The steepness of the curves increase as the twist angle, Φ, is increased, and bistability is achieved when $\Phi > 240°$.

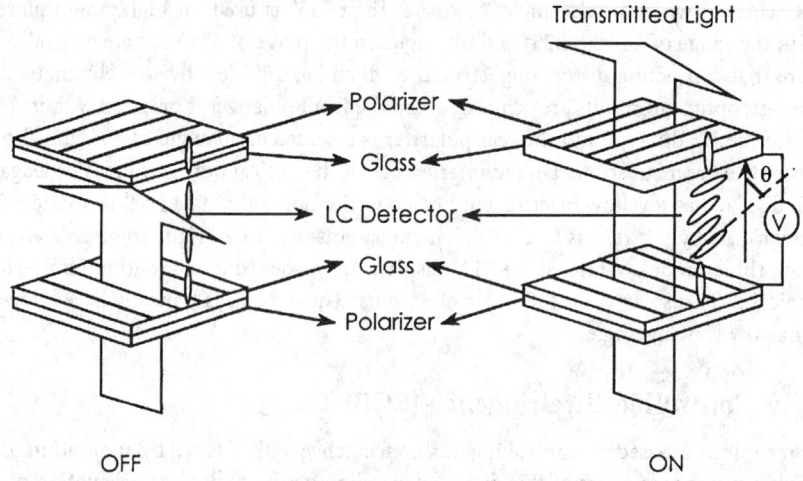

FIGURE 92.12   Illustration of an ECB display: (a) with homeotropic alignment and crossed polarizers, the off-state is black; (b) in the on-state, the output light through the LC is elliptically polarized due to the LC birefringence and the light is transmitted through the crossed polarizer.

where I_0 is the intensity of the incident light, θ is the angle between the direction of polarization of the incident light and the direction of oscillation of the ordinary light within the cell, d is the cell spacing, $\Delta n\ (V)$ is the birefringence of the cell, $d\Delta n$ is the optical phase difference, and λ is the wavelength of the incident light. The equation shows that I depends on the applied voltage and λ.

In case of the homogeneous technique, the LC cell uses positive material with a homogeneous orientation. With this method, the threshold voltage is obtained by replacing the bend elastic coefficient K_{33}, with splay elastic coefficient K_{11}. The HAN cell is characterized by a hybrid orientation cell in which the molecules are aligned perpendicular to one substrate, but parallel to the second substrate. In this mode, both positive and negative $\Delta\varepsilon$ materials can be used, and since there is no clear threshold voltage, it has the advantage of a very low drive voltage. Recently, an ECB mode based on an IPS type display has been developed to produce LCDs with extremely wide viewing angles. In the IPS mode displays, LC is homogeneously aligned and switched between on- and off-states using interdigitated electrodes fabricated on one of the display substrates [27]. The IPS mode displays use an NB mode with either positive or negative $\Delta\varepsilon$ LC materials.

Guest–Host Type

Some organic dyes show anisotropy of light absorption; i.e., they absorb more light in a specific wavelength band when the E vector of the light is parallel to the optic axis of the dye molecules, than they do when it is perpendicular. LCDs based on this principle are called guest–host (GH) displays. In these displays a small amount of a dichroic dye (guest) is mixed in the LC material (host). These dye molecules get aligned to the LC molecules; hence, their orientation can be changed (by changing the orientation of the LC molecules) by application of an electric field. When an LC material with a positive $\Delta\varepsilon$ is used, in the off-state, the E vector of the polarized light coincides with the light absorption axis of the dichroic dye; hence, light is absorbed and transmitted light is colored. When a voltage is applied for the on-state, the E vector of the polarized light is orthogonal to the absorption axis of the dye; hence, no absorption takes place, and transmitted light is not colored (white). GH LCD requires only one polarizer. Further, because the optical effect is based on absorption, the display provides a better viewing angle than a TN mode LCD.

Phase-Change Type

The phase change type of display is based on a change in molecular orientation from a helical cholesteric phase to a homeotropic nematic phase, and vice versa. For this technique, a cholesteric LC with a long helical pitch with a positive or negative $\Delta\varepsilon$ is used. No polarizers are used in this display. In the off-state of this display, the incident light passing through the cholesteric cell with a focal conic orientation is optically dispersed (scattered), and the cell looks cloudy. However, when a voltage is applied, helical structure of the cholesteric phase changes to a nematic phase with a homeotropic orientation, and the cell becomes transparent.

Thermal Effects

Thermal effect is based on a change in the electro-optical behavior due to a change in the molecular orientation of the LC when it is heated or cooled. This effect is utilized with smectic LCs with a homeotropic alignment. When this cell is heated until the isotropic phase is reached and cooled, then if the cooling is sudden, the cell becomes cloudy, whereas, if the cooling is gradual, the cell becomes transparent. These cloudy and transparent states correspond, respectively, to the focal conic and homeotropic orientations of the smectic A LC. This effect is used for large-size memory type displays, in which a laser beam is used to write the image thermally. The heating can also be accomplished by one of the transparent electrodes, while the other transparent electrode is used as a signal electrode.

92.6 Display Addressing

Display addressing (driving) techniques have a major influence on the LCD image quality. The addressing techniques can be classified in three essential types, namely, direct (static) addressing, passive matrix addressing, and active matrix addressing. In the case of low-information-content displays such as numeric displays, bar graph displays, and other fixed pattern displays, using segmented electrodes, direct addressing is used. A common example of direct-addressed displays is a numeric display using seven segmented electrodes for each digit. Each of these segmented electrodes on the front substrate and the common electrode on the back substrate are directly connected to drive signals. A voltage is selectively applied to each of the segments so that any of the digits between 0 and 9 can be displayed. For high-information-content displays, this approach becomes impractical because of the huge number of interconnects, and, hence, either passive matrix or active matrix addressing is used.

Passive Matrix Addressing

A passive matrix (PM) display comprising an LC between a matrix of transparent conducting row and column electrodes (Fig. 92.8) is the simplest and least expensive matrix-addressed LCD to manufacture. An example of a PM LCD is a color (R,G,B) VGA (video graphics array) display using the STN effect, with a pixel format of 640 ($\times$ 3 = 1920) H $\times$ 480 V, with 1920 columns and 480 rows, for a total of 2400 interconnects used for addressing a display containing 921,600 pixels. In PM addressing, the row voltages are scanned in succession with a voltage, V_r, while all the columns in a given row are driven in parallel, during the row time, with a voltage of $\pm V_c$ depending on whether the pixel is selected to be ON or OFF. As discussed above under the STN effect, the contrast ratio of PMLCDs is influenced by the Alt and Pleshko [24] addressability limitation. To enhance the operating margin for improved contrast ratio, DSTN (dual-scan STN) configuration is used in higher-information-content displays. In a DSTN, the display is separated into two halves, and the rows in each half are scanned simultaneously and synchronously, to essentially double the duty ratio of the ON pixels to increase the contrast ratio. One of the major shortcomings of the passive matrix-addressed STN display is the slow response time of the LC, which is of the order of 150 ms. This slow response time is not adequate for video applications and is barely fast enough for the graphical interface of a computer. The response time of the STN LCDs can be improved by active addressing or multiline addressing techniques [28, 29]. These techniques involve simultaneous addressing of several rows of a display to suppress the frame response problems of conventional STN LCDs.

Active Matrix Addressing

Active matrix (AM) addressing removes the multiplexing limitations [24] of the PM LCDs by incorporating a nonlinear control element in series with each pixel, and provides 100% duty ratio for the pixel using the charge stored at the pixel during the row addressing time. Figure 92.13a illustrates an active matrix array with row and column drivers and the associated display module electronics. Figure 92.13b shows a magnified view of the active matrix array in the AM LCD panel. In the figure, C_{LC} and C_S represent the pixel capacitance and the pixel storage capacitance. Typically a storage capacitor, C_S, is incorporated at each pixel to reduce the pixel voltage offset (see Equation 92.16 below) and for a broad temperature operation. Figure 92.14 shows the cross section through an AM LCD illustrating various elements of the display. Figure 92.15 shows a typical AM LCD pixel, showing the gate and data busses, thin-film transistor (TFT), ITO pixel electrode, and the storage capacitor. Fabrication of the active matrix substrate is one of the major aspects of AM LCD manufacturing. Both two-terminal devices such as back-to-back diodes, and metal-insulator-metal (MIM) diodes as well as three terminal TFTs are developed for active matrix addressing. While two-terminal devices are simple to fabricate and cost less, their limitations include lack of uniform device performance (breakdown voltage/threshold voltage) over a large display area, and lack of total isolation of the pixel when neighboring pixels are addressed. For a superior image quality AM LCDs use TFT for the active matrix device, which provides a complete isolation of the pixel from the neighboring pixels. Large-area AM LCDs use amorphous silicon (a-Si) TFTs [6], while polysilicon TFTs with integrated row and column drivers are used in small high-resolution LCDs [30].

Figure 92.16 shows the electric equivalent of a TFT-LCD pixel, display drive waveforms, and the resulting pixel voltage. As in most matrix-addressed displays with line-at-a-time addressing, the rows (gates) are scanned with a select gate pulse $V_{g,sel}$, during the frame time t_f, while all the pixels in a row are addressed simultaneously with the data voltage $\pm V_d$ during the row time t_r (= t_f/N). During the row time the select gate voltage, $V_{g,sel}$, "turns on" the TFT and charges the pixel and the storage capacitor to the data voltage V_d. After the row time, the TFT is "switched off " by application of the nonselect gate voltage, $V_{g,non-sel}$; hence, the voltage (charge) at this pixel is isolated from the rest of the matrix structure

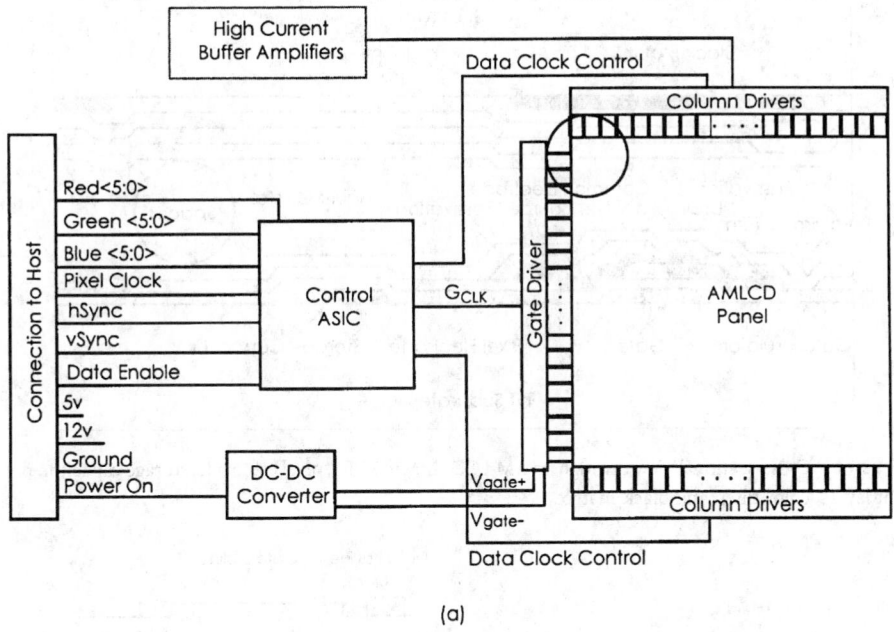

(a)

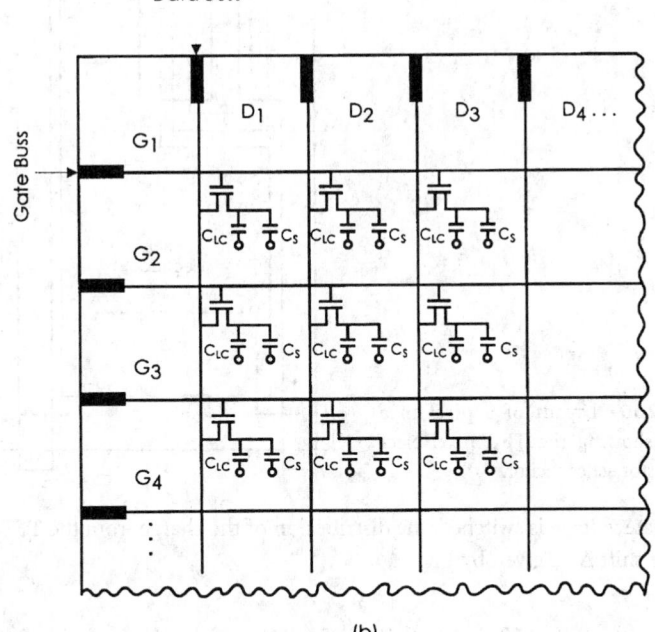

(b)

FIGURE 92.13 (a) AM LCD module electronics block diagram; (b) magnified view of the region shown in (a) illustrating active matrix TFT array.

until it is time to charge the pixel during the next frame time. Note that the LC pixel must be driven in an ac fashion with $+V_d$ and $-V_d$, during alternate frame periods, with no net dc across the pixel. A net dc voltage across the pixel results in flicker and image sticking effects [33], resulting from LC conductivity. Large and sustained dc voltages also degrade the LC material due to electrolysis. The shift in pixel voltage, ΔV_p shown in Fig. 92.16, at the end of the row time is due to the parasitic gate-to-drain capacitance, C_{gd},

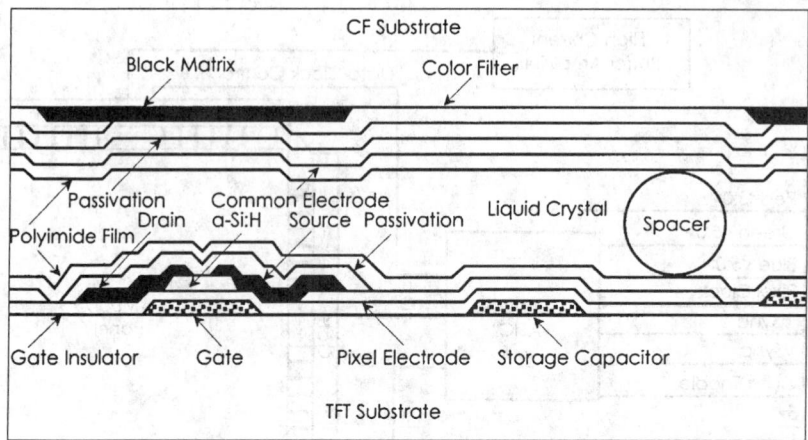

FIGURE 92.14   Cross-sectional view through an AM LCD showing TFT, pixel electrode, storage capacitor, polyimide alignment layers, color filter, and black matrix.

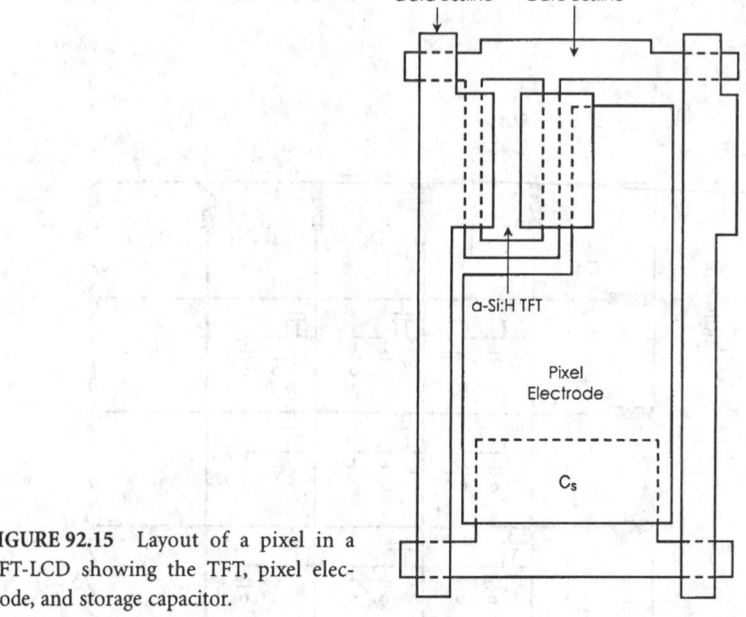

FIGURE 92.15   Layout of a pixel in a TFT-LCD showing the TFT, pixel electrode, and storage capacitor.

of the TFT. When the gate voltage is switched, the distribution of the charge from the TFT gate dielectric causes the pixel voltage shift ΔV_p, given by

$$\Delta V_p = \left(\Delta V_g\right) \cdot C_{gd} / \left(C_{gd} + C_{lc} + C_s\right) \tag{92.16}$$

For the n-channel enhancement mode a-Si TFT, this voltage shift ΔV_p is negative for both the positive and negative frames; thus, it helps pixel charging in the negative frame and hinders it in the positive frame. Further, due to increased gate bias during the negative frame, the pixel attains the data voltage much more rapidly during the addressing period. Hence, the TFT is designed for the worst-case positive frame conditions. ΔV_p is reduced by minimizing C_{gd} by decreasing the source drain overlap area of the TFT and by using a storage capacitor. Further, ΔV_p is compensated by adjusting the common electrode voltage V_{com} as shown in Fig. 92.16. Note that C_{lc} is a function of the V_p (V_{lc}) due to the dielectric

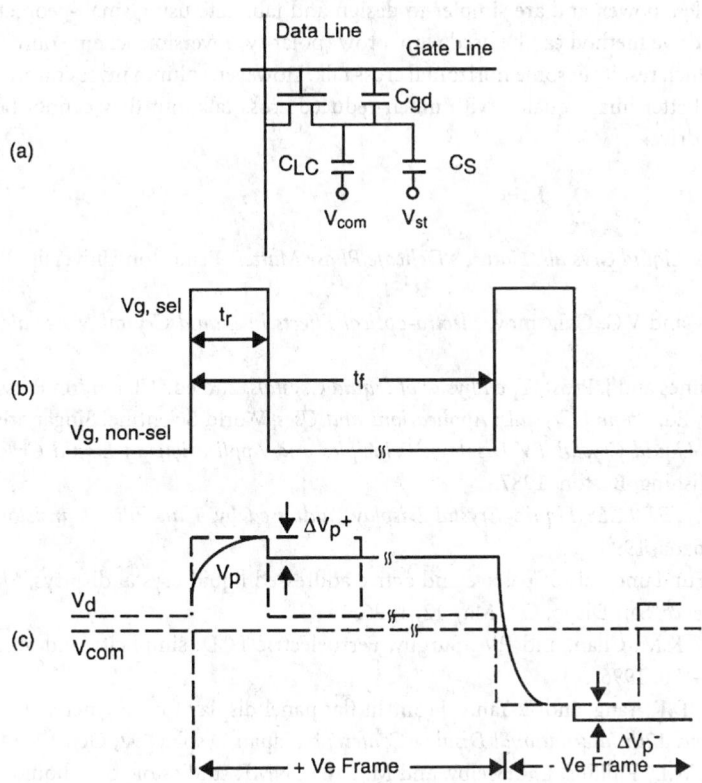

FIGURE 92.16 (a) Electric equivalent of a TFT-LCD pixel; (b) gate drive voltage waveform; (c) data voltage waveform and pixel charging behavior in the positive and negative frames.

anisotropy of the LC; and hence, adjustment to V_{com} alone does not eliminate dc for all gray levels, and modification of the gray scale voltages is required to compensate for the dielectric anisotropy of the LC.

Display Module Electronics

Figure 92.13 shows a block diagram for an AM LCD module electronics. The control block and power supply generation means are separately mounted on a PC board and connected to the row and column drivers of the LCD on one side and to the host controller on the other. The control block may include level shifters, timing generators, and analog functions in some cases; the control block takes in digital data from the host system, which is typically a graphics controller chip, and converts it into timing and signal levels required by the row and column drivers. The architecture and design of the module electronics encompassing row and column drivers have a significant impact on not only the display system cost and power consumption, but also the image quality. The LC material typically requires about 5 V to achieve optical saturation (see Fig. 92.10c). Considering the need for an ac drive, the required voltage swing across the LC material is about 10 V. To achieve this 10 V swing across the LC material, the column drivers typically use 12 V power supplies. Column driver voltage can be reduced by using a V_{com} modulation drive method. In this method, the V_{com} node (which is connected to all pixels in the display) is driven above and below a 5 V range of the column drivers. Each and every row time, the V_{com} node is alternated between a voltage above and a voltage below the 5 V output range of the column drivers. This achieves 10 V across the LC material using 5 V column drivers. This method requires additional components and consumes additional power due to the oscillation of the V_{com} node. In addition, to avoid capacitive injection problems, the row drivers usually have their negative supply modulated with the same frequency as the V_{com} node. Note, however, that compared to 10 V column drivers, 5 V column drivers consume less power, and are simpler to design and fabricate using small-geometry CMOS. The V_{com} modulation drive method can be used with a row (polarity) inversion scheme only (for elimination of pixel flicker) which results in some horizontal cross talk. However, column inversion and pixel inversion schemes provide better image quality with much-reduced cross talk, but they cannot be used with the V_{com} modulation drive.

References

1. P.J. Collings, *Liquid Crystals: Nature's Delicate Phase Matter*, Princeton University Press, Princeton, NJ, 1990.
2. L.M. Blinov and V.G. Chigrinov, *Electro-optical Effects in Liquid Crystal Materials*, Springer, New York, 1996.
3. P.G. De Gennes and J. Prost, *The Physics of Liquid Crystals*, 2nd ed., Clarendon Press, Oxford, 1993.
4. B. Bahadur, Ed., *Liquid Crystals: Applications and Uses*, World Scientific, Singapore, 1990.
5. E. Kaneko, *Liquid Crystal TV Displays: Principles and Applications of Liquid Crystal Displays*, D. Reidel Publishing, Boston, 1987.
6. T. Tsukada, *TFT/LCD: Liquid Crystal Display Addressed by Thin Film Transistors*, Gordon and Breach, Canada, 1996.
7. E. Lueder, Fundamentals of passive and active addressed liquid crystal displays, *Short Course S-1, SID Conference*, San Diego, CA, May 12, 1996.
8. P.W. Ross, L.K.M. Chan, and P.W. Surguy, Ferroelectric LCD: simplicity and versatility, *SID '94 Digest*, 147–150, 1996.
9. J.W. Doane, D.K. Yang, and Z. Yaniv, Front-lit flat panel display for polymer stabilized cholesteric textures, *Proc. 12th International Display Conference*, Japan Display '92, Oct. 12–14, 73–76, 1992.
10. J.C. Rowell, W.D. Phillips, L.R. Melby, and M. Panar, NMR studies of some liquid crystal systems, *J. Chem. Phys.*, 43, 3442–3454, 1965.
11. L. Pohl and E. Merck, Physical properties of liquid crystals, in *Liquid Crystals: Applications and Uses*, B. Bahadur, Ed., World Scientific, Singapore, 1990, 139–170.
12. P.R. Kishore, N.V.S. Rao, P.B.K. Sarma, T.F.S. Raj, M.N. Avadhanlu, and C.R.K. Murty, Field and frequency effects in nematic mixtures of negative and positive dielectric anisotropy, *Mol. Cryst. Liq. Cryst.*, 45, 3/4, 231–241, 1978.
13. D. Lippens, J.P. Parneix, and A.Chapoton, Study of 4-heptyl 4'-cyanobiphenyl using the analysis of its dielectric properties, *J. Phys.*, 38, 1465, 1977.
14. F.C. Frank, On the theory of liquid crystals, *Discuss. Faraday Soc.*, 25, 19, 1958.
15. W. H. de Jeu, *Physical Properties of Liquid Crystalline Materials*, Gordon and Breach, New York, 1980.
16. J.J. Wysocki, A. Adams, and W. Haas: Electric-field-induced phase change in cholesteric liquid crystals, *Phys. Rev. Lett.*, 20, 1024, 1968.
17. G. H. Heilmeier, L.A. Zanoni, and L.A. Barton, Dynamic scattering: a new electo-optic effect in certain classes of nematic liquid crystals, *Proc. IEEE*, 56, 1162, 1968.
18. G. H. Heilmeier and L.A. Zanoni, Guest-host interaction in nematic liquid cyrstal — a new electo-optic effect, *Appl. Lett.*, 13, 91, 1968.
19. M. Schadt and W. Helfrich, Voltage-dependent optical activity of a twisted nematic liquid crystal, *Appl. Phys. Lett.*, 18, 127, 1971.
20. M. F. Schiekel and K. Fahrenschon, Deformation of nematic liquid crystal with vertical orientation in electric fields, *Appl. Phys. Lett.*, 19, 391, 1971.
21. M. Hareng, G. Assouline, and E. Leiba, Liquid crystal matrix display by electrically controlled birefringence, *Proc. IEEE*, 60, 913, 1972.
22. C.H. Gooch and H.A. Tarry, The optical properties of twisted nematic liquid crystal structures with twist angles $\leq$ = 90°, *J. Phys. D Appl. Phys.*, 8, 1575–1584, 1975.

23. E. Jakeman and E.P. Raynes, Electro-optical response times in liquid crystals, *Phys. Lett.*, 39A, 69–70, 972.

24. P.M. Alt and P. Pleshko, Scanning limitations of liquid-crystal displays, *IEEE Trans. Electron Dev.*, ED-21, 146, 1974.

25. T.J. Scheffer and J. Nerring, A new highly multiplexable liquid crystal display, *Appl. Phys. Lett.*, 45, 1021, 1984.

26. T.J. Scheffer, *Super Twisted Nematic (STN) LCDs, SID '95 Seminar Notes*, Vol. I, M-2, 1995.

27. M. Oh-e, M. Ohta, S. Aratani, and K. Kondo, Principles and characteristics of electro-optical behavior of in-plane switching mode, in *Proc. 15th International Display Research Conf.*, Asia Display '95, 577–580, 1995.

28. T. Scheffer and B. Clifton, Active addressing method for high-contrast video-rate STN displays, *SID'92 Digest*, 228, 1992.

29. H. Muraji et al., A 9.4-in. color VGA F-STN display with fast response time and high contrast ratio by using MLS method, *SID '94 Dig.*, 61, 1994.

30. Higashi et al., A 1.8-in poly-Si TFT-LCD for HDTV projectors with 5V fully integrated driver, *SID '95 Dig.*, 81, 1995.

31. Nano et al., Characterization of sticking effects in TFT-LCD, *SID '90 Dig.*, 404, 1990.

93

Plasma-Driven Flat Panel Displays

Robert T. McGrath
The Pennsylvania State University

Ramanapathy Veerasingam
The Pennsylvania State University

William C. Moffatt
Sandia National Laboratories

Robert B. Campbell
Sandia National Laboratories

93.1 An Introduction to Plasma-Driven Flat Panel Displays

Development History and Present Status

Plasma-driven flat panel display pixels were invented by Bitzer and Slottow at the University of Illinois in 1966 [1-3]. Figure 93.1 shows one of the inventors' early designs and demonstrates its simplicity. Parallel sets of thin conducting wires are deposited on two glass substrates which are then mounted with the conductor sets perpendicular to one another as shown in the Fig. 93.1. A spacer, in this case a perforated glass dielectric, is used to maintain a gap separation of about 100 μm between the glass plates. The gap region then is filled with an inert gas, typically at a pressure of half an atmosphere. Individual pixels formed by the intersection of two conductor wires are aligned with the perforations. Pixels are illuminated by applying a voltage between two intersecting wires sufficient to initiate gas breakdown. Over the years, this basic pixel design has undergone a multitude of refinements and improvements, but the fundamental concept is still widely used.

Throughout the 1980s, plasma display products on the market were monochrome and operated with neon-based gases, directly producing within the discharge volume the red-orange (585 to 640 nm) visible photons that are characteristic of the quantum energy level structure of the neon atom. Dot matrix displays of the type shown in Fig. 93.2 were widely used [3,4]. Early work by Owens-Illinois led to improvements in glass sealing and spacer supports [5,6], and work by IBM led to improved understanding

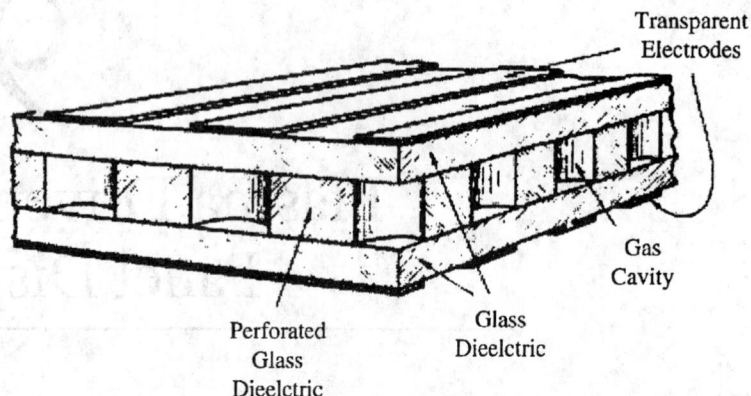

FIGURE 93.1   Structure of the ac plasma display invented at the University of Illinois. (From Bitzer, D.L. and Slottow, H.G., *AFIPS Conf. Proc.*, Vol. 29, p. 541, 1966. With permission.)

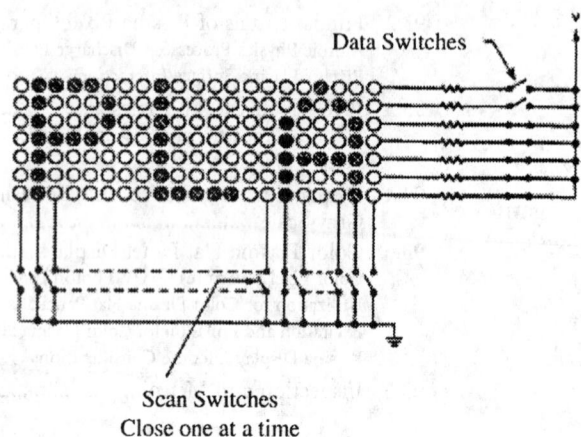

FIGURE 93.2   A simple dot matrix plasma display and data scanning switches. (From Weber, L.F., in *Flat Panel Displays and CRTs*, L.E. Tannas, Jr., Ed., Van Nostrand Reinhold, New York, 1985. With permission.)

and control of the discharge [7-15]. These advances ultimately paved the way for manufacture of large-area, high-resolution monochrome displays. The largest area plasma display panels ever manufactured were produced by Photonics Imaging. These monochrome displays had a 1-m diagonal dimension and contained over 2 million pixels with a pixel pitch of 20 pixels/cm (50 lines/in.) [4].

Advances in lithography, patterning, and phosphors have enabled continued improvement of plasma display performance and resolution. Today, many of companies offer full-color plasma flat panel displays. Table 93.1 presents a summary list compiled by the National Research Institute of display panel specifications for some of the major companies investing in plasma flat panel manufacturing [16]. According to Stanford Research, Inc., sales for plasma display panels in 1995 totaled $230 million, but projected sales for 2002 are $4.1 billion [17]. NEC projects a more aggressive market growth reaching $2.0 billion by the year 2000 and $7.0 billion by 2002 [17]. The production capacities listed in Table 93.1 represent investments committed to manufacturing as of January 1997. As these production facilities come online, color plasma flat panel display production will grow to nearly 40,000 units per month by the end of 1998, and to over 100,000 units per month by early in 2000. In 1993 Fujitsu was the first to market a high-information-content full-color plasma flat panel display, a 21-in. diagonal, ac-driven system with a 640 × 480 pixel array [17,18]. Two examples of more recent market entries are shown in Figs. 93.3

TABLE 93.1 Plasma Flat Panel Display Specifications and Manufacturer's Business Plans

Company	Product Specification			Luminecence (cd/m²)	Efficiency Specification			Factory	Capital Cost ($M)	Plan Product Ability unit/month	Target Region
	Inch	Aspect	Pixels		Contrast	lm/w	Power (W)				
Fujitsu	42	16:9	852 × 480	300	70:1	0.7	350(set) 300(panel)	Miyazaki	20	10,000	Europe (Philips), Japan
NEC	33	4:3	640 × 480	200	150:1	1.2	270(set) 190(panel)	Tamagawa, Kagoshima	5	2,000	Japan
Pioneer	40	4:3	640 × 480	400	150:1	1.2	350(set)	Kofu	5	10,000	Japan
Mitsubishi	40	4:3	640 × 480	350	200:1	0.8	350(set) 300(panel)	Kyoto	14.8	10,000	U.S.
MEC	42	16:9	852 × 480	450	150:1	10	300(panel)	Kyoto	10	5,000	Japan, U.S.
Photonics	21	5:4	1280 × 1024	100	50:1	—	300(panel)	Ohio	—	—	—
Hitachi	25	4:3	1024 × 768	150	50.1	—	250(set)	Yokohama	3	1,000	—
NHK	40	16:9	1344 × 800	93	80:1	—	—	—	—	—	—

Source: Wakabayshi, H., paper presented at *Imaging 2001: The U.S. Display Consortium Business Conference*, January 28, San Jose, CA, 1997. With permission.

FIGURE 93.3 The 40-in. diagonal dc driven plasma display from NHK. (From Mikoshiba, S., *Inf. Display*, 10(10), 21, 1994. With permission.)

and 93.4. The first example is the NHK full color, 102-cm (40-in.) diagonal, high definition television (HDTV) [19-22]. The system comprises 1,075,000 full-color pixels (1344 × 800) with a pixel pitch of 0.65 mm in both horizontal and vertical directions (15.4 pixels/cm). This pulsed, dc-driven display has a peak luminance of 93 cd/m^2, a contrast ratio of 80 to 1, and produces 256 gray levels. The display has an overall thickness of only 8 cm and weighs only 8 kg. The dimensions of the display panel itself are 87.5 × 52.0 cm with a width of only 6 mm. Shown in Fig. 93.4 is the 76-cm (30-in.) diagonal, full-color AC Plasma Display manufactured by Photonics Imaging [23-24]. The display contains an array of 1024 × 768 full-color pixels. At 16.8 pixels/cm (pixel pitch = 0.59 mm) this is the highest resolution full-color, plasma display manufactured to date. This unit has 64 gray levels per color channel and an average area (white) luminance greater than 103 cd/m^2 (30 fL).

dc and ac Plasma Pixels

As indicated above, plasma display pixels can be designed for either ac or dc operation. Figure 93.5 shows schematic diagrams for the simplest dc and ac pixel designs. In either case, sets of parallel conductor wires are deposited on glass substrates. In most cases display costs are kept low by utilizing ordinary soda-lime float glass. The two glass plates are then mounted with a separation of about 100 μm and with the conductor wire sets perpendicular to one another. The gap region between the glass plates is filled with an inert gas, which discharges and illuminates the pixels when sufficient voltage is applied across two intersecting wires.

For dc pixels, shown in Fig. 93.5a, the working gas is in direct contact with the electrodes. Electrons produced within the discharge volume flow rapidly to the anode, while ions produced flow more slowly toward the cathode. At 53.3 kPa (400 torr), a gas gap of 100 μm and an applied voltage of 200 V, the electron and ion transit times across the gap are roughly 0.2 and 20 ns, respectively. Once breakdown is initiated, the electrical resistance of the discharge is negligible. Consequently, dc operation requires that external resistors in series with each pixel be included in the circuit in order to limit the current amplitude. Often, dc pixels are operated in pulsed discharge mode with frequency modulation used to define the pixel brightness. For either ac or dc pixels, a base firing frequency of 50 kHz is typical. This frequency is too fast for the human eye to detect any *on–off* flicker, but allows sufficient flexibility for intensity and refresh control. In reviewing the literature on plasma displays, it is easy to confuse dc and ac pixels since dc pixels are often operated in pulsed mode and with electrode polarity reversal which distributes sputter damage over both electrode surfaces. The dc pixels are readily identified by conducting electrodes in direct contact with the discharge gas and the inclusion of a current-limiting resistor in the circuit for

FIGURE 93.4 The 40-in. diagonal ac driven plasma display from Photonics Imaging. (From Friedman, P.S., *Inf. Display*, 11(10), October 1995. With permission.)

each pixel. While polarity reversal is optional for dc pixel operation, it is inherently required for ac pixel operation as discussed below. Drive electronics, current limiting, gray scale, and other aspects of both dc and ac pixel operation are discussed in greater detail in subsequent sections.

Figure 93.5b shows a schematic representation of an ac plasma pixel configuration. One can see that the differences between ac and dc pixel geometry are slight; however, the resulting operational differences are significant. In the ac pixel, the conductor wires are covered with a dielectric film. Typically, lead oxide (PbO), which has a dielectric constant of about 15, is deposited at a film thickness of about 25 μm. Most ac pixels are made with a thin film (50 to 200 nm) magnesium oxide (MgO) dielectric coating covering the PbO and in contact with the working gas. This dual material dielectric film serves two principal functions, charge storage and secondary electron emission.

The exact voltage required for gas breakdown depends upon the gap width, the gas pressure, the gas composition, and MgO surface conditioning. For the pixel parameters shown in Fig. 93.5b, an externally applied voltage of about 120 to 180 V is required to initiate a discharge. In the ac pixel, once the discharge is initiated, electrons and ions flow toward the anode and cathode, respectively, as in the dc pixel. However, in the ac case, charge carriers are unable to reach the conductor wires and instead collect as a surface charge on the dielectric coating. The electric field within the gas gap is always the sum of that produced by the externally applied voltage and that produced by the surface charge. During pixel firing, if the externally applied voltage is held constant for only a few microseconds, the net electric field within the gas gap very quickly decreases (~100 to 200 ns). The gap potential

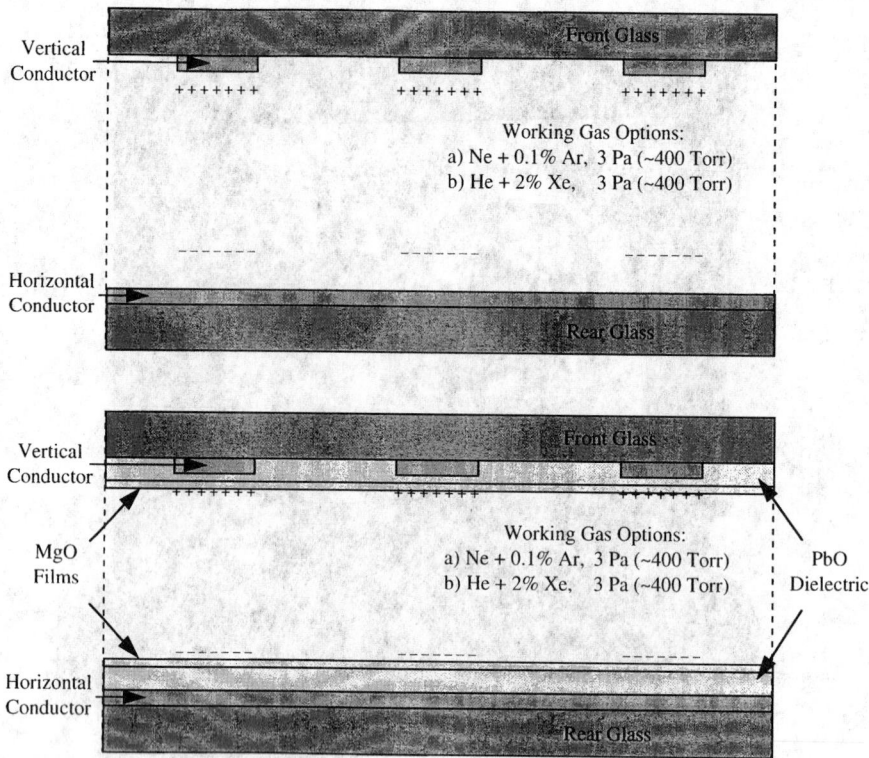

FIGURE 93.5 Schematic diagrams of (a) dc and (b) ac opposed electrode plasma pixels.

drop produced by the surface charge shields out that produced by the externally applied voltage. Eventually, the gap electric field is insufficient to sustain the discharge and the pixel turns *off*. Thus, each ac pixel is inherently self-current-limiting and, unlike the dc pixel, requires no external resistance in series with it. At the start of the next ac half cycle, the externally applied voltage is reversed. When this occurs, the voltage across the gas gap is the sum of the external voltage and the voltage produced by the surface charge established during the previous discharge. If a sufficient surface charge is present, a new discharge pulse can be initiated by application of an external voltage, which by itself would be insufficient to break down the gas. Within the new discharge, charge carriers flow quickly to reverse the polarity of the surface charge concentrations. Once again, the field within the gap is diminished and the discharge turns *off*. Storage of surface charge make ac pixels easily controllable and provides them with their inherent memory properties. The presence or absence of surface charge determines whether or not a given pixel will discharge at the onset of the next ac half cycle of the externally applied voltage. The details of how these discharge dynamics are used to write, erase, and sustain each pixel are discussed in subsequent sections, along with drive mechanisms for gray scale and for pixel array refresh.

General Attributes of Plasma Displays

Plasma-driven flat panel displays offer a number of advantages over competing display technologies. The highly nonlinear electrical behavior of each pixel, with inherent memory properties, can be used to advantage in design of the drive electronics required to refresh and to update the pixel array of the display. The simplicity of the pixel design makes large-area manufacturing problems, such as alignment and film thickness uniformity, somewhat more manageable. Relative to color active matrix liquid crystal displays

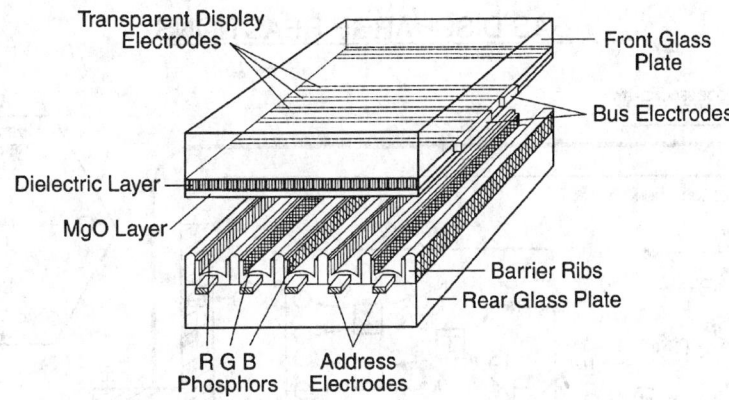

FIGURE 93.6 Structure of the ac color plasma display manufactured by Fujitsu. (From Mikoshiba, S., *Inf. Display*, 10(10), 21, 1994. With permission.)

(AMLCDs) which use a thin-film transistor (TFT) to control each pixel, less-complicated manufacturing and less-complicated drive electronics give plasma flat panel displays advantage for large-area applications. On the other hand, plasma displays require more robust drive electronics with voltages of 100 to 275 V. Plasma displays are also not well suited for portable applications since power consumption is high relative to other display technologies, but not restrictive for office or domestic use. The 76-cm (30-in.) diagonal color display manufactured by Photonics Imaging shown in Fig. 93.4 has a peak power consumption of only 300 W [23]. At high power levels, plasma-driven flat panel displays are bright enough to be readable in sunlight. The displays are also easily adjusted to a low-ambient-light condition by discharge amplitude or frequency modulation.

Plasma flat panel displays are well suited for large-area (0.5 to 5 m) applications such as videoconferencing, large meeting room displays, outdoor displays, and simulators requiring large viewing areas. Thin, high-resolution, large-area, color plasma displays are also very attractive for desktop workstation or personal computer applications requiring high-resolution graphics. Note, too, that plasma flat panel displays have very large viewing angles, greater than 160° in many designs [22-24]. For displays using metal electrodes, one often finds that the best viewing angle is slightly off normal since the front electrode wire blocks out a portion of the pixel emission. This occurs both for monochrome pixels producing visible emissions within the discharge and for color plasma displays where the viewer sees visible red, green, and blue (RGB) emissions from vacuum ultraviolet (VUV) photon-stimulated phosphors. Some manufactures have investigated use of transparent electrodes, such as indium-tin oxide (ITO), but there is a trade-off with power consumption since the conductivity of ITO is less than that of metal electrodes [18]. In contemporary designs, the metal conductor width is thin (~20 μm) and its opacity does not present a major problem.

For color pixels, the discharge gas mixture is modified to produce emissions in the VUV. In all other respects, the operational principals of the plasma discharge by the pixel are identical for color and for monochrome displays. Ideally in color plasma displays, no visible emissions are produced within the discharge itself and VUV-photostimulated luminous phosphors are used to produce the required RGB visible light. The ac color pixel design concept shown in Fig. 93.6 is that utilized by Fujitsu [18]. Long, straight barrier structures, each about 100 μm tall, are constructed parallel to and between each of the vertically oriented conductor wires on the rear glass plate. The sidewalls of these barriers are alternately coated with red, green, and blue photostimulated phosphors. Note that the Fujitsu panel employs a three-electrode, ac-driven surface discharge pixel design which is slightly more complicated than the opposed electrode ac design shown in Fig. 93.5b. This chapter will return to surface discharge configurations and other aspects of color pixel design and operation after reviewing fundamentals of the discharge physics and electrical behavior governing pixel operation.

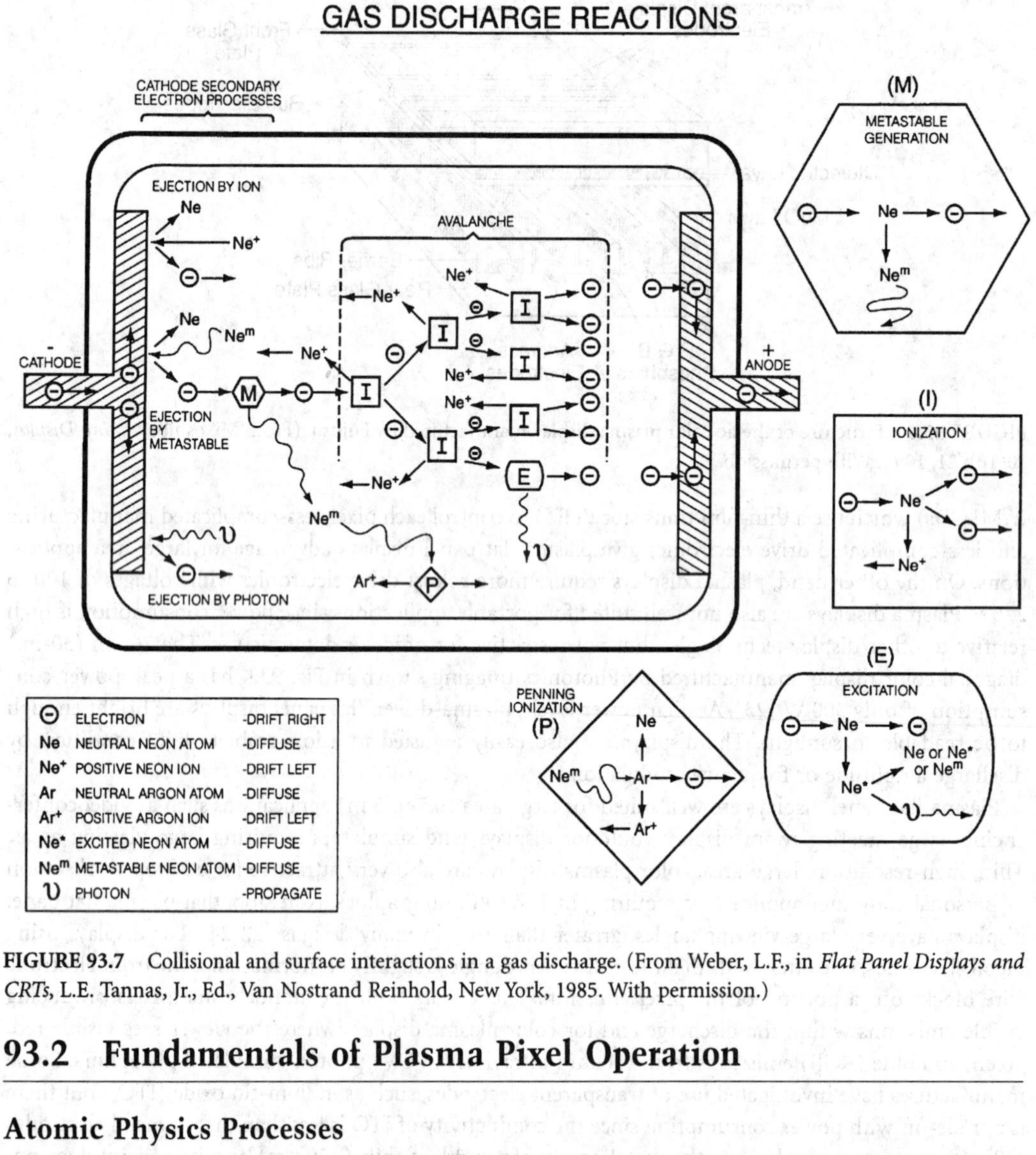

GAS DISCHARGE REACTIONS

FIGURE 93.7 Collisional and surface interactions in a gas discharge. (From Weber, L.F., in *Flat Panel Displays and CRTs*, L.E. Tannas, Jr., Ed., Van Nostrand Reinhold, New York, 1985. With permission.)

93.2 Fundamentals of Plasma Pixel Operation

Atomic Physics Processes

Although simplistic in design, the plasma display pixel is a rich environment for study of basic atomic physics, electron collisional processes, photon production and transport, and plasma–surface interactions. The coupling of these processes for a neon–argon monochrome pixel discharge was nicely summarized in the diagram from Weber which is reproduced here as Fig. 93.7 [4]. The reader interested in additional information on fundamental discharge physics is directed to one of the excellent textbooks in this field [25-27].

The physical processes governing of the behavior of the pixel discharge are closely coupled and form a closed-loop system. The discussion begins by assuming that a seed electron is resident within the gas gap and is subjected to an electric field which results from application of an externally applied voltage to the two conductors forming that pixel. Some of the gas and surface processes for production of the seed electrons will become evident as the discussion progresses. In order to ensure reliable discharge initiation, seed particles, which are either electrons or electron-producing photons or metastable atoms, are often provided by a controlled source which may be external to the pixel being fired. Some display panels include electrodes for production of seed particles at the edges of the panel outside the field of view or hidden behind opaque conductor wires. Other display panels use well-controlled temporal sequencing to ensure that nearest-neighbor pixels provide seed particles for one another [4,19,28]. Pixel addressing sequences are discussed further later in this chapter.

The transport of electrons or ions across the gas gap is a balance between field acceleration and collisional energy loss. In the example of Fig. 93.7, the gas is mostly neon (98 to 99.9%) and field-accelerated electrons will predominantly collide with Ne atoms. The quantum energy level diagram for excitation of the Ne atom is shown schematically in Fig. 93.7 [29]. Note that the lowest-lying excited state is 16.6 eV above the ground state, while the ionization energy is 21.6 eV. This means that electrons with energies less than 16.6 eV can only experience elastic collisions with the Ne atoms. When an electron is field-accelerated to an energy in excess of 16.6 eV, inelastic collisions which transfer energy from the incident electron to one of the outer-shell electrons in the Ne atom can take place. Incident electrons with kinetic energies in excess of 21.6 eV can drive ionization reactions:

$$Ne + e^- \rightarrow Ne^+ + 2e^- \tag{93.1}$$

Excitation and ionization collisions transfer energy from the electron population to the neutral atoms in the gas. At the same time, the electron population available to ionize the Ne further is increased with every ionizing event. The result is the discharge avalanche schematically shown in Fig. 93.7, which manifests itself experimentally as a rapid increase in electric current flowing in the pixel gas gap. In dc panels, an external resistor of about $R = 500$ kΩ is placed in series with each pixel. The amplitude of the externally applied voltage provided by the driving electronics, V_a, is held constant and the total voltage across the gas gap, $V_g = V_a - IR$, decreases as the circuit current, I, increases. Very quickly, a steady-state dc current in the gas gap and in the circuit is established. Brightness and gray scale are controlled by frequency modulation of the pulsed dc pixel firing using a base frequency of about 50 kHz. In ac pixel discharges, electrons and ions are driven by the applied field to the dielectric-covered anode and cathode, respectively. The buildup of charge on the dielectric surfaces shields the gap region from the field produced by the externally applied voltage. Eventually, the electric field in the gap drops below a level sufficient to sustain the discharge and the pixel turns *off*.

For electron energies greater than 16.6 eV, collisions with Ne atoms can excite outer-shell electrons in the atom to one of the numerous excited energy states shown in Fig. 93.7.

$$Ne + e^- \rightarrow Ne^{ex} + e^- \tag{93.2a}$$

$$Ne^{ex} \rightarrow Ne^* + h\nu \tag{93.2b}$$

Most of these excited states have short lifetimes ranging from fractions to tens of nanoseconds [30] and quickly decay to lower-lying atomic quantum states accompanied by the emission of a characteristic photon, indicated in Eq. 93.2 by $h\nu$, the product of Planck's constant times the photon frequency. As can be seen in Fig. 93.8, the characteristic red-orange Ne gas emissions result from electron transitions within the atom from higher-energy 2p quantum states to lower-lying 1s energy levels [30,31]. Two of the four 1s energy levels radiate to ground-emitting VUV photons with wavelengths of 74.4 and 73.6 nm. Due to quantum mechanical exclusion principles, electron decay from the other two 1s levels is more complex and depends upon fine details of the electronic wave function and upon very small perturbing interactions [31]. Consequently, decay lifetimes for these so-called metastable states are measured in seconds, which is very long relative to other dynamic physical processes governing pixel discharge behavior, such as charge or neutral particle transport. An Ne atom with an electron trapped in one of these metastable levels harbors 16.6 eV of latent energy. The metastable atom, Ne^*, is unable to dissipate its stored energy in collisions with ground-state Ne atoms, yet readily liberates its energy whenever a lower-lying energy configuration can be accessed. The principal channels in this system to lower energy configurations are Ne^* collisions with Ar or Ne^* incidence onto pixel interior surfaces.

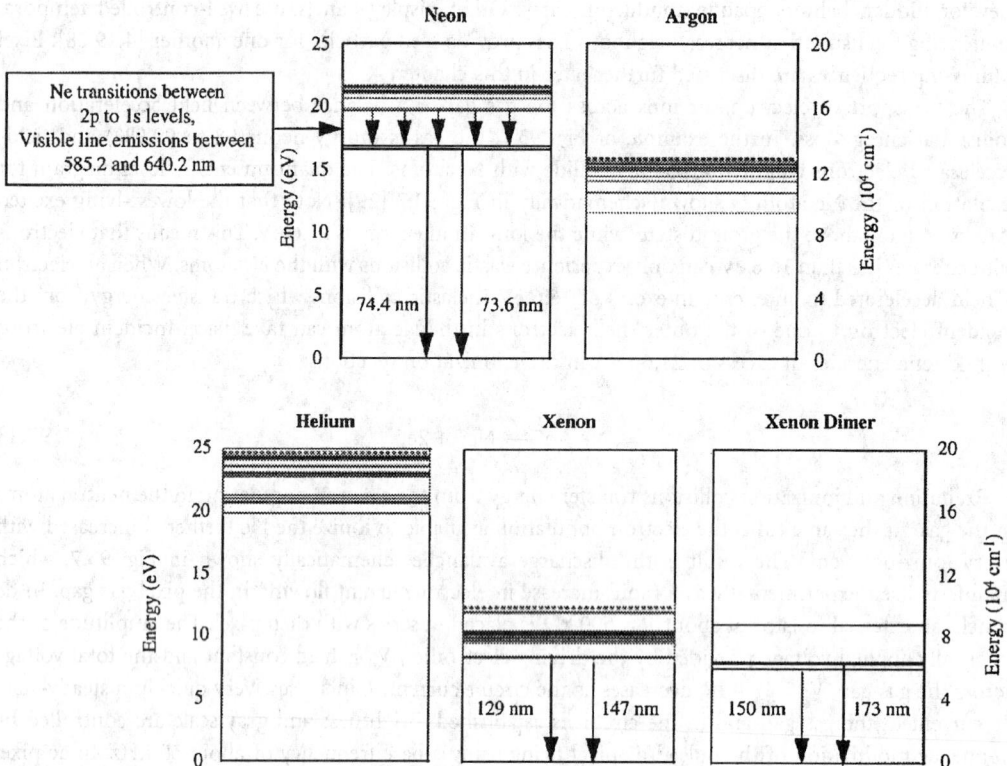

FIGURE 93.8 Quantum energy level diagrams for He, Ne, Ar, Xe, and the Xe^{2*} dimer.

Figure 93.8 shows simplified energy-level diagrams for several inert gases. The relative positioning of the allowable energy levels provides insight into the energy exchange that occurs in collisional coupling. The ionization energy of the Ar atom is 15.8 eV and lies 0.8 eV below the metastable-state Ne^*. Consequently, the Ne^* has sufficient stored energy to ionize the Ar atom:

$$Ne^* + Ar \rightarrow Ne + Ar^+ + e^- \qquad (93.3)$$

Ionizing reactions of this type are called Penning reactions, and gas mixtures that rely on metastable states of the majority gas constituent (Ne) for ionization of the minority gas constituent (Ar) are referred to as Penning gas mixtures [25,26,32]. Figure 93.9 shows the efficiency with which charge pairs are produced through ionization within Ne/Ar Penning gases containing various fractions of Ar. The curves show that for any given pressure, ion pair production per volt applied is optimal at low Ar gas fractions (0 to 10%) except for very large values of E/P, greater than 75 V/m/Pa (100 V/cm/torr), where E is the electric field strength and P is the gas pressure. Penning gas mixtures have been studied for many years. Figure 93.9 shows the original data on Ne/Ar gas breakdown published by Kruithof and Penning in 1937 [32]. An extensive volume of literature has been published on inert gas Penning processes since then, and the interested reader is referred to the excellent texts which have recently been re-released through the American Vacuum Society and MIT Press [25,26].

Plasma display pixels usually operate at pressures near 53.3 kPa (400 torr) in order to achieve sufficient photon production and brightness. Typical pixel fields are roughly 100 MV/m. Consequently, plasma pixels operate with E/P values near 18.8 V/m/Pa (25 V/cm/torr). Both charge pair production and luminous efficiency are then optimized with Ar gas fractions between 0.1 and 10%, depending upon the specifics of the pixel gas pressure, gap width, and driving voltage. For a given applied voltage, the product of the gas pressure (P) and the gas gap dimension (d) provides a measure of the balance between electron

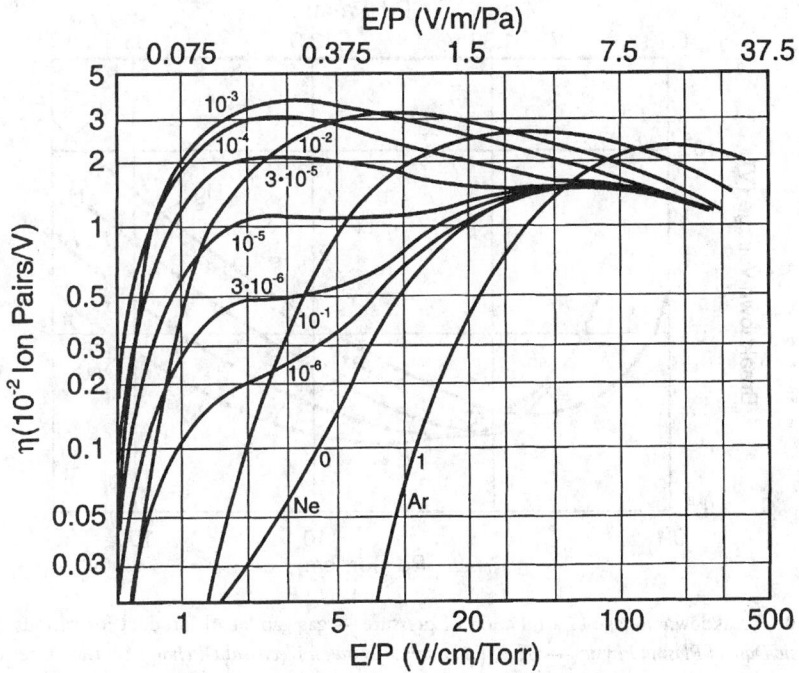

FIGURE 93.9 Ionizing collisions plotted vs. electric field strength divided by pressure. The numbers on each curve indicate the ratio of the Ar partial pressure to the total gas pressure. (From Brown, S., *Basic Data of Plasma Physics — The Fundamental Data on Electrical Discharges in Gas*, American Institute of Physics Press, New York, 1993. With permission.)

acceleration by the electric field and electron energy loss due to collisions with the background gas. Paschen curves, which plot the gas breakdown voltage vs. the Pd product, for several inert gas mixtures are shown in Fig. 93.10 [26,33,45]. In each case, minimum voltage for breakdown occurs at a value of the Pd product which is dependent upon the ionization levels, collisionality, and energy channels within the gas. For example, in Ne atomic excitation and ionization processes dominate, while in air much of the energy absorbed by the gas goes into nonionizing molecular vibration, rotation, and dissociation. For fixed pressure, the Paschen curves show that increased gap dimension lowers the electric field strength per volt applied and a large voltage is required for breakdown. On the other hand, if d is reduced for a given pressure, the electric field strength can be large, but electrons transit the gap without initiating a sufficient number of collisions to drive the type of discharge avalanche shown in Fig. 93.7. If the gas gap, d, is held fixed while pressure is varied, the shapes of the Paschen curves are again explained by electron acceleration and collisional processes. For high pressures, the mean free paths between electron collisions with the background gas atoms are short and electrons are unable to accelerate to energies sufficient to initiate ionization unless the electric field is especially strong. At low pressures, the electrons may be accelerated by the field to energies sufficient to initiate ionization, but few collisions with the background gas occur and, again, the avalanche is difficult to initiate. Penning processes are especially efficient at driving ionization. Introduction of 0.1% Ar into the neon gas lowers the minimum breakdown voltage from the value near 250 V shown in Fig. 93.10, to about 150 V. The minimum breakdown voltage occurs at a Pd product of 40 Pa-m (30 torr-cm) for this gas mixture.

Discharge Physics for Plasma Pixels

Within any discharge, electrons move very quickly, while the more massive ions move relatively slowly in comparison. In a charge-neutral plasma that is subjected to an externally applied electric field, the

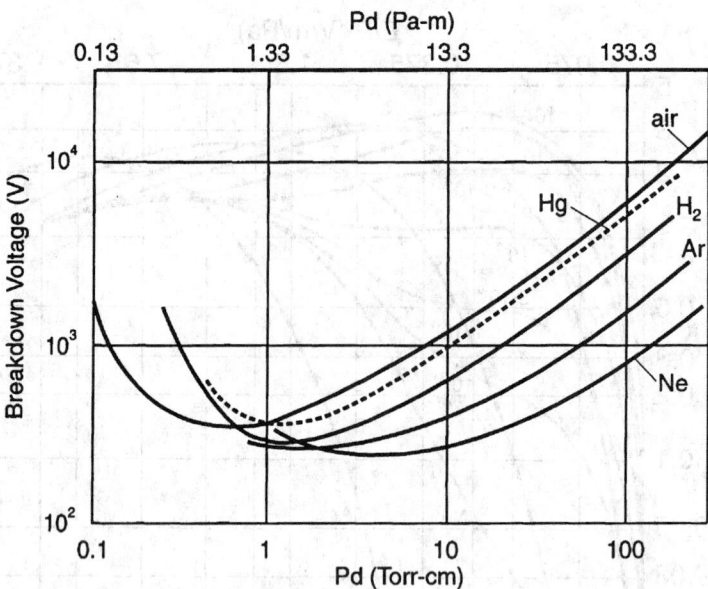

FIGURE 93.10 Breakdown voltage as a function of pressure — gas gap length product for various gases. (From Brown, S., *Basic Data of Plasma Physics — The Fundamental Data on Electrical Discharges in Gas*, American Institute of Physics Press, New York, 1993. With permission.)

mobile electrons quickly respond to the applied field and rush toward the anode. The inertia-laden ions, in a much slower fashion, begin their motion toward the cathode. Very quickly, a local charge imbalance is established as the electrons reach the anode faster than the rate of arrival of ions at the cathode. Poisson's equation

$$\nabla \cdot E(x) = 4\pi\rho(x) = 4\pi e\left(n_i(x) - n_e(x)\right) \qquad (93.4)$$

shows that a local electric field is established in response to the net positive charge density, $\rho(x)$, in the plasma region. Here, $n_i(x)$ and $n_e(x)$ are the spatial profiles of the ion and electron densities, respectively, and e is the electron charge. The field established retards the rate at which electrons flow out of any volume within the plasma column and forces them to follow the net ion motion. The ion drift motion is correspondingly accelerated, but this acceleration is smaller by a factor proportional to the mass ratio of the electron to the ion. The net ion/electron motion is called ambipolar flow and is described in detail in many basic plasma physics texts [25-27].

In steady-state dc plasma pixel discharges, the amplitude of the current flowing in the circuit and in the gas gap is defined by the value of the applied voltage and the resistor in series with the pixel. Steady-state operation dictates that charge buildup within the gap region cannot occur. The rate at which charge particle pairs arrive at the electrodes must equal their rate of production due to ionization. At the same time, the rates at which ions and electrons leave the plasma volume, arriving at the cathode and anode, respectively, must be equal. Equilibrium is sustained by establishment of the spatial potential profile within the gas gap shown in Fig. 93.11a. Due to the high electron mobility, the plasma is extremely efficient in shielding out externally applied electric fields. As a result, the potential profile is flat across the gas gap of a pixel sustaining a fully developed discharge. The entire potential drop is localized in a small zone called the sheath adjacent to each electrode. The spatial extent of the sheath is determined by the effectiveness of the electron population in shielding out the electric fields produced by the electrode potentials. The Debye length,

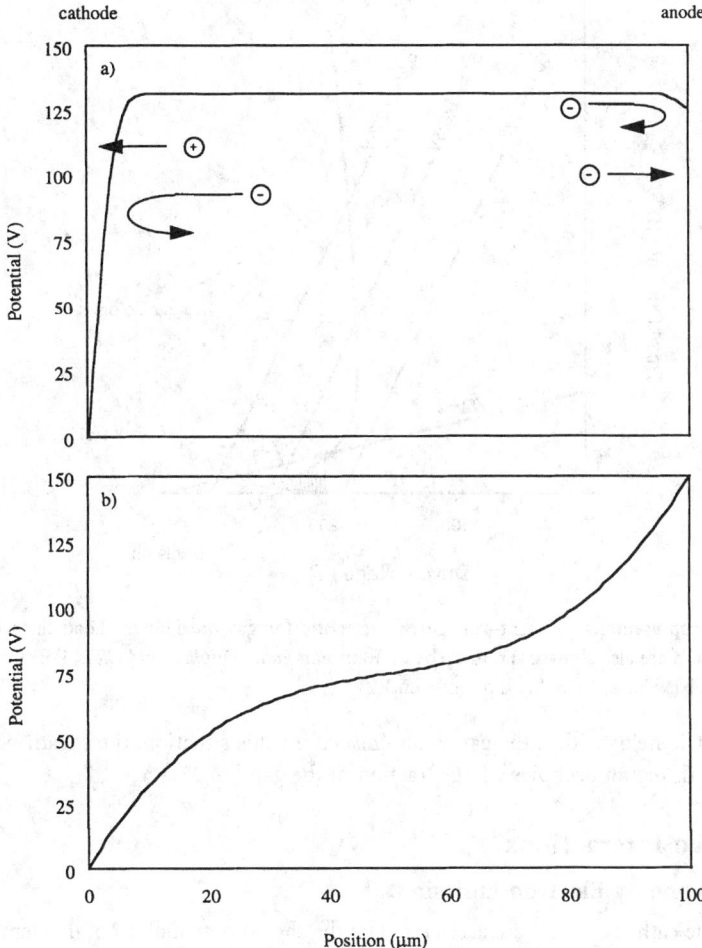

FIGURE 93.11 Potential profiles in the pixel gap region for (a) high-electron-density and (b) low-electron-density discharges.

$$\lambda_D = \sqrt{kT_e / 4\pi e^2 n_e(x)} \tag{93.5}$$

provides a measure of the shielding distance. The expression for λ_D implies that the sheath thickness increases with increasing electron temperature, T_e, and decreases as the electron density, n_e, increases. For fully developed plasma pixel discharges, the product of Boltzmann's constant and the electron temperature, kT_e, is at most a few electron volts, and n_e is of order $10^{16}/m^3$. Thus, the sheath thickness is roughly 5 μm. The potential within the plasma region adjusts, V_p, within the discharge volume rises to a value just above that of the applied voltage at the anode. Consequently, only the most energetic electrons can overcome the potential barrier at the anode which adjusts to a potential such that the rate of electron loss at the anode equals the rate of ion loss at the cathode. For ac plasma pixels, a similar potential profile is established, but changes dynamically as the pixel pulse evolves. Charge pairs incident upon the anode and cathode in ac pixels are trapped there by the dielectric film covering the conductor wires. Consequently, the potential at the discharge boundary is diminished as surface charge collects at each electrode, as shown in Fig. 93.11b. Ultimately, the discharge terminates as the electric field produced by the surface charge cancels that produced by the externally applied voltage. As the density of charge carriers is reduced near the termination of an ac pixel discharge pulse, the effectiveness of the electrons

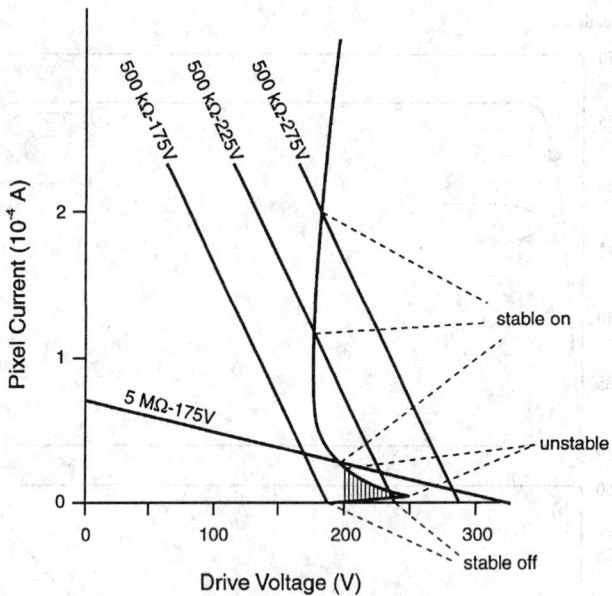

FIGURE 93.12   A representative current–voltage characteristic for gas breakdown. Load lines representative of plasma pixel operation are also shown. (From Weber, L.F., in *Flat Panel Displays and CRTs*, L.E. Tannas, Jr., Ed., Van Nostrand Reinhold, New York, 1985. With permission.)

to shield out electric fields within the gap is diminished. In this situation, the sheath potential drop is small but the sheath region occupies a large fraction of the gap [36,37].

Plasma Surface Interactions

Ion-Induced Secondary Electron Emission

Ions arriving at the cathode sheath are accelerated by the sheath potential drop. Incident ions strike the cathode with kinetic energies equal to the plasma potential, V_p, which is just over 200 V in the example shown in Fig. 93.12a. Ions incident on the cathode quickly capture an electron, additionally depositing on the cathode surface an energy equal to the recombination or ionization energy for that atom. Energy deposition on the cathode surface drives two important processes for plasma pixels — ion-induced secondary electron emission and sputtering. The first process significantly enhances the luminous efficiency of plasma pixels. The second shortens their operational lifetime as is discussed in subsequent sections.

Ion-induced secondary electron emission occurs when ion energy deposition on the surface results in electron ejection. Secondary electrons are exceptionally effective at driving discharge ionization since they gain large amounts of kinetic energy as they are accelerated across the cathode sheath and because they have ample opportunities for ionizing collisions as they traverse the entire width of the gas gap. The secondary electron emission coefficient, γ, is defined as the number of electrons ejected per incident ion [25,26]. As one would expect, γ varies with incident ion energy and with cathode material. Most ac plasma display panels take advantage of the strong secondary electron emission of MgO, which is also a good insulating material as required for surface charge storage in ac operation. Measurement of the MgO γ value is difficult, especially for low-energy ion incidence (<500 eV), and is complicated by charge buildup on the samples during the measurements [38]. Most often, relative values of secondary electron yields for different materials are deduced from discharge intensity measurements [11,12,39–42]. Chou directly measured the ion-induced secondary electron emission coefficient for MgO using a pulsed ion beam with sample surface neutralization between pulses. For ion incidence at 200 eV, he found $\gamma = 0.45$ and $\gamma = 0.05$ for Ne$^+$ and Ar$^+$, respectively [39]. Note, too, that photons and metastable atoms incident on the electrode surfaces are also capable of initiating secondary electron emission, as shown in Fig. 93.7. Since neither photons nor metastables are influenced by the electric fields within the gas gap, they propagate isotropically throughout the gas volume and are often utilized as seed particles.

Sputtering

Ions accelerated across the sheath deposit energy on the cathode surface. This often initiates sputtering, whereby an atom within the cathode material is ejected from the surface. Sputtering processes erode the cathode surface and degrade pixel performance. Contamination of the discharge by sputtered surface impurities can lead to reduction in luminous efficiency due to visible emissions from the contaminant atoms or molecules which compromise the color purity of the pixel. Unwanted surface coatings from redeposited materials can also degrade the electrical characteristics of the pixel or, in color applications, shield the phosphors from VUV photons, further degrading luminous efficiency. For argon ion, Ar^+, bombardment of MgO surfaces at 2 keV, the measured sputtering yield is slightly greater than one ejected atom per incident ion [43]. Data on sputtering yields at lower energy ion incidence are difficult to obtain. Because yields are small, large incident ion currents are required to obtain measurable signals and sample charging is once again a problem. In spite of the lack of detailed data on low-energy MgO sputtering, manufactures of ac plasma panels have been able to demonstrate display lifetimes well in excess of 10,000 h [18,23]. Shone et al. [44] have demonstrated that Rutherford backscattering of high-energy (2.8 MeV) alpha particle can be used to measure the thickness of MgO film on a PbO substrate. The film thickness accuracy obtained was ±1.5 nm. Because the technique requires a large (and expensive) particle accelerator, this technique is a very nice research tool but is ill suited for any fabrication line measurements.

93.3   Pixel Electrical Properties

Electrical Properties of dc Pixels

Figure 93.5 shows schematic diagrams and circuit models for dc and ac pixels. In the dc case, the pixel gas gap functions electrically as a variable impedance resistor. Prior to gas breakdown, the resistance is large and the pixel represents an open-circuit element. Once breakdown is initiated, the plasma is an excellent conductor and offers only modest resistance, R_p, to current flow. Since $R \gg R_p$, the circuit equation simplifies to

$$V_a = I\left(R + R_p\right) \approx IR \tag{93.6}$$

and the circuit current, I, is defined by the amplitude of the applied voltage and the size of the circuit series resistor, R. The externally applied voltage, V_a, is typically a 50-kHz square wave with a fast voltage rise time (~50 ns). The dc driving voltages range from 175 to 275 V and a typical value for the series resistor is $R = 500$ kΩ Pixel currents then range from 0.35 to 0.55 mA. Note that without a large resistance in series with the pixel, the current is limited by some physical failure such as melting of the pixel electrodes.

Figure 93.12 shows the characteristic I–V behavior of a dc pixel which has a breakdown voltage of 250 V [4]. Only a very small current due to a few stray charge carriers flows across the gas gap as the voltage increases from 0 to 250 V and the pixel remains in the *off* state. At the breakdown voltage, the situation is dynamic with the current growing rapidly and the voltage across the gas gap dropping as a result. The steady-state operating point achieved is identified by the intersection of the load line, $V_a = IR$, with the discharge I–V characteristic as shown in Fig. 93.12. For an applied voltage of 175 V the pixel is always *off*, while for $V_a = 275$ V the pixel is always *on*. For a line resistance of 500 kΩ, the bimodal operation and memory of the dc pixel at $V = 225$ V is evident in the figure. If an applied voltage of 225 V is approached from the low-voltage direction, the pixel remains *off*. If, on the other hand, a large voltage is applied and subsequently lowered to $V_a = 225$ V, then the pixel will be in an *on* state. Note that the region where the 225 V/500 kΩ load line intersects the negative resistance portion of the *I–V* characteristic is unstable. The pixel discharge will quickly transition to either the stable *on* or stable *off* operating point. As a practical matter, one should note that the negative resistance region of the *I–V* characteristic curve cannot be experimentally measured in a pixel circuit operating with a 500 kΩ series resistance. Instead, as shown in the figure, a much larger series resistor, $R = 5$ MΩ, provides a load line with slope small enough to produce stable operation in the negative resistance regime.

Electrical Properties of ac Pixels

The physical design of an opposed electrode ac pixel is shown in Fig. 93.5b. Electrically, the pixel functions as a breakdown capacitor and is described by the circuit equation:

$$V_a(t) = I(t)R + \frac{1}{C}\int_0^t I(t')dt' = I(t)R + Q(t)/C \tag{93.7}$$

where V_a is the externally applied voltage, I the circuit current, C the pixel capacitance, and Q the charge collected. For ac pixels the line resistance, R, is minimized in order to minimize power consumption and Eq. 93.7 simplifies to

$$V_a(t) = \frac{1}{C}\int_0^t I(t')dt' = Q(t)/C \tag{93.8}$$

The capacitance for each pixel is the series summation of the capacitance for each dielectric film and for the gas gap:

$$\frac{1}{C} = \frac{1}{C_{PbO}} + \frac{1}{C_{MgO}} + \frac{1}{C_{gas}} \tag{93.9}$$

In each case,

$$C_i = \frac{e_i A}{d_i} \tag{93.10}$$

where i is the material index and the surface area, A, is roughly equal to the square of the conductor wire width. As shown in Fig. 93.5b, an ac pixel is typically constructed with a PbO film of thickness $d = 25$ μm, while the thin-film MgO has thickness $d = 50$ to 200 nm. The lead oxide has a dielectric constant of roughly $\varepsilon_{PbO} = 15\varepsilon_0$, while that for MgO is $\varepsilon_{MgO} = 6\varepsilon_0$ with exact values dependent upon the film purity and microstructure [45]. Note that the MgO contribution to the total capacitance is negligible and that this material is incorporated into the design because of its excellent secondary electron emission properties. Prior to gas breakdown, the capacitance of the pixel is attributed largely to the gas gap. For 20-μm-thick conductor wires the capacitance of a pixel gas gap prior to breakdown is about 500 pF. The time derivative of Eq. 93.8 gives the circuit current:

$$I(t) = C\frac{dV(t)}{dt} \tag{93.11}$$

This charge displacement current appears as the initial large amplitude current peak in Fig. 93.13, which shows the temporal current response of a 45 × 45 ac pixel array to a single pulse within a 50-kHz square

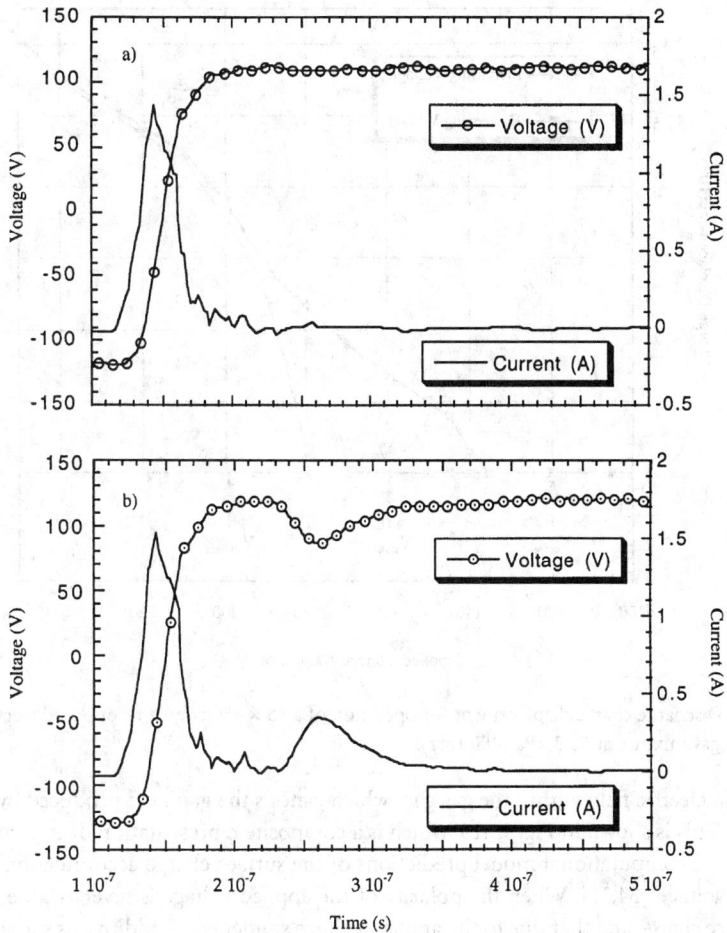

FIGURE 93.13 Voltage and current traces for a 45 × 45 array of ac plasma pixels in the (a) *on* and (b) *off* states. Drive voltage amplitudes were 117 and 127 V, respectively.

wave applied voltage pulse train. The electrical measurement shown was made using a simple induction loop probe to measure the current and a high impedance voltage probe (1 MΩ, 3 pF) to monitor the applied voltage. The signals were captured using a high-speed (300 MHz) oscilloscope.

If the applied voltage amplitude is below the gas breakdown threshold, only the capacitor charging displacement current, defined by Eq. 93.11, is observed as shown in Fig. 93.13a. If the voltage for gas breakdown is exceeded, a second current pulse due to the plasma discharge current within the gas gap is observed in the circuit, Fig. 93.13b. The plasma pulse is accompanied, of course, by strong photon emission from the gas gap region. The total charge displacement in the discharge pulse as a function of amplitude of the square wave–applied voltage is plotted in Fig. 93.14 for a helium–xenon (2%) Penning gas mixture [35]. The hysteresis or inherent memory property of the ac pixel is apparent. As the applied voltage amplitude is increased from zero to 180 V, no measurable current flows across the pixel gas gap. When no surface charge is present, below 180 V the electric field within the gap region is insufficient to drive the electron collisions into the avalanche regime. For any voltage amplitude in excess of 180 V, a gas discharge is initiated and the pixel turns *on*. If a pixel is subjected to a single voltage pulse with amplitude less than 135 V, the pixel turns *off* even if a surface charge is present.

In ac pixels, charge pairs produced during one discharge pulse collect on the surfaces of the dielectric films at the boundaries of the gas gap and are available to assist formation of the next discharge pulse in the sequence. In a fully developed ac pixel discharge, the surface charge accumulation on the dielectric

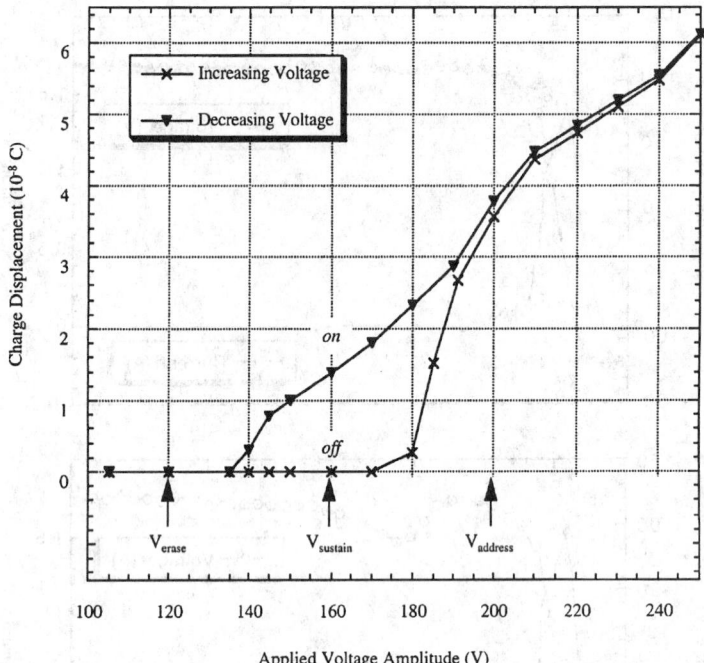

FIGURE 93.14   Discharge charge displacement for operation of a 45 × 45 array of ac opposed electrode pixels with an He – Xe (2%) gas mixture at 53.3 kPa (400 torr).

film produces an electric field within the gas gap, which cancels the gap field produced by the externally applied voltage. This is shown in Fig. 93.15, which is a composite representation of experimental current measurements and computational model predictions of the surface charge accumulation producing the surface or wall voltage [34,36]. When the polarity of the applied voltage is reversed, the potential drop due to the surface charge and that due to the applied voltage suddenly are additive as shown in the figure. The gas gap is momentarily subjected to an intense electric field which results from a potential drop roughly equal to twice the applied voltage. The presence or absence of surface charge results in the bimodal current–voltage behavior shown in Fig. 93.14.

Addressing of ac pixels is easily accomplished by taking advantage of the inherent memory of the pixel that results from this bimodal I–V behavior. For the pixel electrical properties shown in Fig. 93.14, each pixel would be continuously supplied with an ac square wave applied voltage pulse train with an amplitude of 160 V, called the sustain voltage, $V_{sustain}$. If the pixel is initially in an *off* state, it will remain so indefinitely since no surface charge is available to enhance the field produced by the sustain voltage. To turn the pixel *on*, a single high-amplitude voltage pulse, called an address (or write) pulse is delivered across the pixel electrodes. In this example, an address pulse of 200 V initiates a discharge whose charge pairs collect on the internal dielectric surfaces of the pixel. The self-limiting nature of the ac pixel is such that the surface charge concentration produced for a fully developed pixel discharge completely shields the gap region from the externally applied field. When the next sustain polarity reversal occurs, the pixel gas gap experiences a voltage equal to the sum of the sustain voltage (160 V) plus the voltage due to the surface charge produced by the previous pulse, $V_{surface}$ = 200 V in this case. The new gap voltage of 360 V is more than sufficient to initiate a second discharge and to establish a new surface charge whose polarity is opposite that of the preceding pulse. Once again, the surface charge adjusts to produce a voltage exactly canceling the field of the applied voltage. For this pulse, V_{wall} = 160 V, and the next sustain voltage polarity reversal subjects the gap to a potential difference of $V_{gap} = V_{sustain} + V_{address} = 160$ V + 160 V = 320 V, which is again sufficient to initiate a new discharge pulse. Consequently, the pixel remains in the *on* state

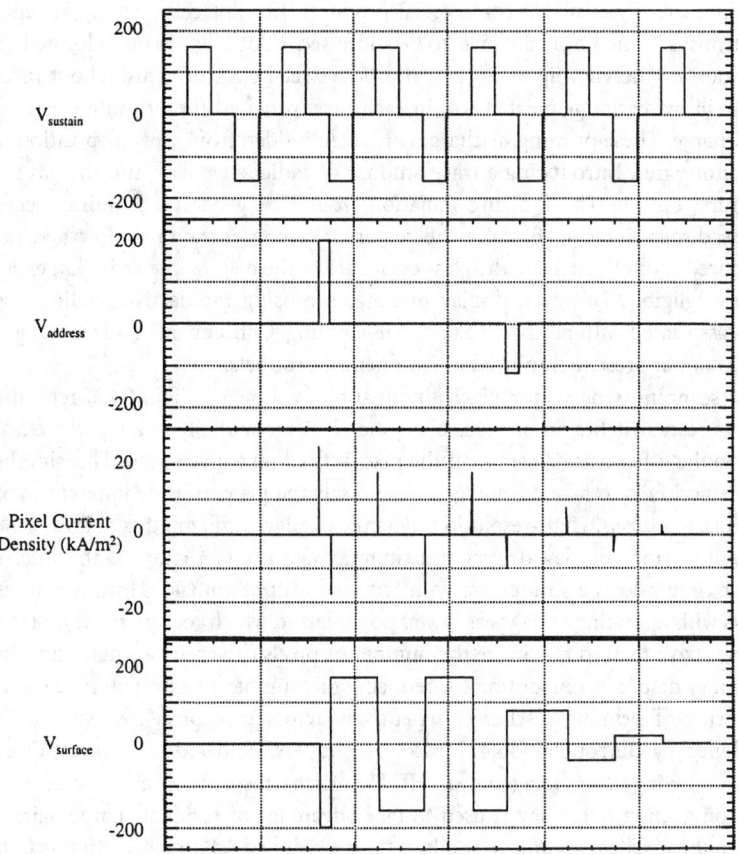

FIGURE 93.15 Sustain and address voltage waveforms for ac driven plasma pixels. The amplitude of the pixel current density and wall voltage resulting from the surface charge buildup provide a measure of the discharge intensity.

until action is taken to eliminate or diminish the surface charge buildup accompanying each discharge. This is accomplished by application of a single low-voltage pulse called an erase pulse with amplitude $V_{erase} = 120$ V for the example shown in Figs. 93.14 and 93.15. Application of the erase pulse produces a potential drop across the gas gap of $V_{gap} = V_{applied} + V_{surface} = V_{erase} + V_{surface} = 120$ V $+ 160$ V $= 280$ V. The erase pulse produces a discharge of lower intensity which is insufficient to reestablish a reversed polarity surface charge. Consequently, the erase discharge diminishes the concentration of the surface charge so that no discharge is initiated during the next pulse in the sustain applied voltage train. Ideally, the erase pulse drives the surface charge concentration identically to zero, but this rarely occurs in practice and is not essential for ac pixel operation, as can be seen in Fig. 93.15. Very low intensity discharges with negligible photon production drive the pixel to its ideal *off* state within a few ac cycles. Fortunately, these minor deviations from the ideal *off* condition have little effect on subsequent write pulses for frequency-modulated ac operation, and therefore do not affect the timing of pixel addressing and refresh which is covered in the next subsection.

93.4 Display Priming, Addressing, Refresh, and Gray Scale

Pixel priming is necessary to provide the initial source of electrons, or the priming current, required to initiate a discharge avalanche. Metastable atoms or photons can also be used as priming particles since these produce electrons via ionization of the background gas. Pilot cell priming and self-priming are two options used in currently available commercial products. In pilot cell priming, a separate cell which generates electrons is located near the pixel to be addressed. Pilot cells are often located on the periphery of the display outside the viewing area, yet can produce seed electrons throughout the display. In self-priming, an auxiliary discharge created within each pixel provides the priming electron source for the main pixel discharge. These priming discharges are often hidden from view by positioning them behind opaque conductor wires. Introducing a trace amount of radioactive Kr^{85} into the gas mixture provides a passive priming option. The ionizing radiation from Kr^{85} generates priming electrons uniformly throughout the display interior. Because the required Kr concentration is low and because the beta radiation produced cannot penetrate the glass enclosure of the display, the radiation exposure risk to the display user is negligible. However, display manufacture using radioactive seeding involves potential health hazards associated with radioactive material handling. Consequently, this seeding approach, while very effective, is not at present employed in commercial products.

A simplistic scanning scheme for pixel illumination is shown in Fig. 93.2, reproduced here from Reference 4. The scan switches on one axis open and close sequentially in a repetitive fashion, while the data switches on the other axis determine if the pixel is fired on a given scan. This simplistic refresh and data update method fails to take advantage of the discharge properties or inherent memory functions available with plasma pixels. High-resolution dynamic displays utilizing this address scheme would not be cost-competitive since display drivers constitute a significant portion of the total cost of plasma displays. Driver circuit costs also increase with required output voltage. Thus, it is desirable to design plasma displays with operating voltages as low as possible and which require the fewest number of driver chips. Designers strive then to maximize the number of pixels driven by a single chip.

For nonmemory dc pixels, one option for reducing the number of external drive switches required is to sweep the firing of priming discharges repetitively across each pixel row, such as in the self-scan circuitry developed by Burroughs [46,47]. More recently, NHK has developed a pulse memory drive scheme for its 102-cm (40-in.) diagonal dc HDTV plasma display, which is being widely used [48]. Sustain operation at high frequency is used to take advantage of residual charge pairs and metastable atoms present in the pixel gas volume as result of the preceding discharge [49]. In this fashion, each pixel is self-seeding, with seed particle populations dependent upon the time elapsed since the termination of the preceding discharge. The high-frequency operation is fast enough to take advantage of the short duration memory characteristic of the dc pixel. As the sustain voltage pulse train is applied to the electrode of a pixel, it will remain in the *on* or *off* state indefinitely until an address or erase pulse is supplied. In the NHK scheme, an auxiliary anode is used to assist in the *address* access operations. Figure 93.16a shows the block diagram of such as system, while Fig. 93.16b shows the time sequences for the scheme [48]. Note that the pulses are dc and that *on* state pulses have larger gap voltages than erase pulses. The timing sequence is critical to address a pixel selectively within the matrix. Implementation of this scheme requires (1) display anode drivers, (2) auxiliary anode drivers, (3) cathode drivers, and (4) and interfaces to decode the HDTV signals provided to the drivers.

For ac displays with memory, drivers need to provide (1) *address* (or *write*) pulses, (2) *sustain* pulses, and (3) *erase* pulses. A complex refresh scan signal is not required since a continuously supplied sustain signal, coupled with the ac pixels inherent memory, trivially maintains each pixel in either an *on* or *off* state. Pixels respond to address, sustain, and erase pulses as described in the preceding section. Similar to dc pixel dynamic control discussed above, ac pixel addressing requires well-timed application of voltage waveforms to the rows and columns of the display matrix so that the proper voltage appears across the electrodes for the pixel of interest without modifying the state of adjacent pixels. A more-detailed discussion of ac pixel addressing can be found in Reference 4 or 47.

Gray scale is achieved for dc or ac plasma displays either by modulation of the discharge current or by duty cycle modulation with fixed current. Modulating the applied voltage amplitude to vary the discharge current is not widely used because of practical limitations in effectively controlling the nonlinear response of the discharge current. However, duty cycle modulation is a viable technique both for pulse memory-driven dc displays and for ac memory displays. In either case, duty cycle modulation requires

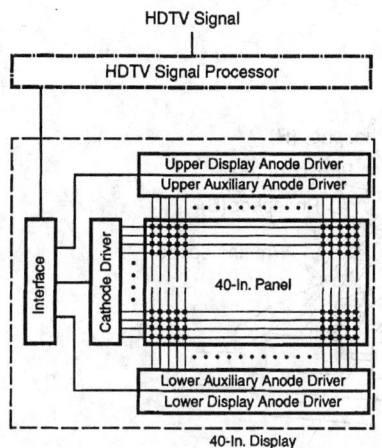
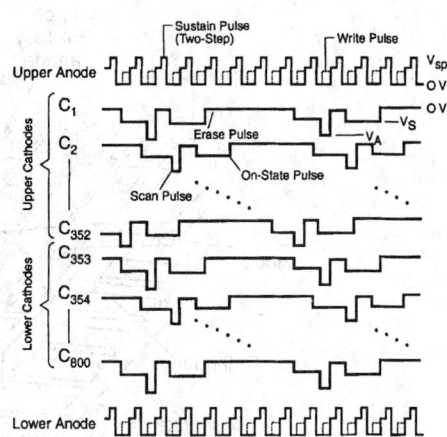

FIGURE 93.16 Pulsed memory operation of the NHK dc plasma display. (a) Block diagram of the driver system and pixel array. (b) Temporal sequences for pulsed memory operation. (From Yamamoto, T. et al., *SID' 93 Sympos. Dig.*, 165–168, 1993. With permission.)

complex circuit design for the well-timed delivery of *on* and *off* pulses. Gray scale is achieved by varying the time a pixel is *on* compared with *off* during each refresh cycle. In 50-kHz operation, a sustain half cycle is 10 μs. VUV photon emission occurs usually in less than 1 μs. For color displays the visible light emission persists much longer, with the fastest phosphors having 10% persistence times of about 5 μs. More typical phosphors have 10% persistence times in the 5 to 10 ms range [50]. If the image is updated every 40 ms, corresponding to a refresh rate of 25 images per second, then a 1/8-level brightness is achieved by having the pixel *on* for 5 ms and *off* for 35 ms during that refresh cycle. The time *on* is interspersed throughout the 40 ms refresh period by appropriate timing circuit design. For example, the NHK 102-cm (40-in.) display has a 2^8 or 256 levels of gray scale per color, providing a total of 16 million (256^3) color scale levels [48].

93.5 Color Plasma Flat Panel Displays

Color Pixel Structures

In color plasma display panels, photoluminescent phosphors provide the primary RGB optical emissions required for full-color image display. In this case, visible emissions from the discharge itself must be suppressed in order to avoid color contamination. A common approach is to utilize xenon as the minority species constituent in the Penning gas mixture of the panel. The structure and phosphor layout of the 102-cm (40-in.) diagonal color dc plasma display available from NHK is shown in Fig. 93.17, while that of the Fujitsu 53-cm (21-in.) diagonal ac color display is shown in Fig. 93.6. Each uses screen printing and hard mask or abrasive-resistant lithographic processes for conductor wire deposition, barrier structure definition, and phosphor deposition [51]. In the NHK design, the fourth section within the honeycomb color pixel structure houses a redundant green phosphor subpixel to compensate for the lower photoluminance of green phosphors relative to that of either red or blue phosphors. In a similar honeycomb dc color pixel structure, Panasonic instead incorporates a series resistor in this fourth subpixel position [20]. Printing the series resistor for each pixel on the display glass substrate complicates panel manufacturing but simplifies design requirements for the drive electronics. In the Fujitsu, the opposed electrode ac color pixel structure shown in Fig. 93.6, barrier or separation rib structures running between and parallel to each conductor wire are fabricated on the rear glass substrate. The barrier rib heights are typically 100 to 150 μm. Ac barrier rib structures and dc pixel honeycomb lattice structures are usually composed of the same PbO thick-film dielectric used to cover the conductor wires.

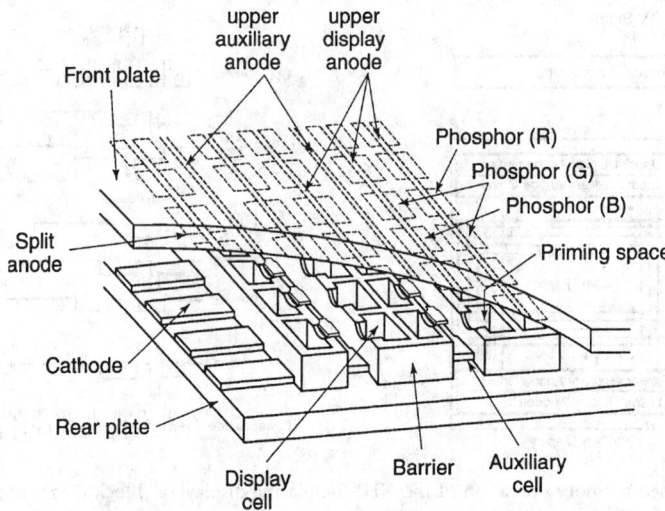

FIGURE 93.17   Structure of the 40-in. color display manufactured by NHK. (From Yamamoto, T. et al., *SID' 93 Sympos. Dig.*, 165–168, 1993. With permission.)

VUV Photon Production and Utilization for Color Plasma Flat Panel Displays

Color plasma display gas mixtures utilizing xenon as the minority species are optimized for production of VUV emissions which are used to excite RGB photoluminescent phosphors. Both neon–xenon and helium–xenon combinations are popular. The ionization energy of xenon at 12.3 eV lies below the lowest excited atomic states of either neon or helium, as shown in Fig. 93.8. Consequently, electrons accelerated by electric fields within the pixel volume preferentially impart their kinetic energy to the xenon atoms. In addition, the excited states of He or Ne produced readily transfer stored energy to the xenon atoms through ionizing Penning collision processes. Consequently, the red-orange visible emissions typical of Ne discharges are suppressed as Xe concentration is increased. Fujitsu utilizes an Ne–10% Xe working gas in its color display [18], while Photonics Imaging prefers to use an He-based background gas in its panel [23] where suppression of unwanted optical emissions from the discharge can be accomplished at somewhat lower xenon concentrations.

The tendency of xenon to fill its outermost electronic shell results in the formation of the xenon dimer molecule, Xe_2^*, whose energy states are also shown in Fig. 93.8 [52,53]. Radiative dissociation of the dimer produces photons with wavelengths near 173 and 150 nm. Figure 93.18 shows how the dimer emissions dominate the VUV spectra from He–Xe gas pixel discharges as the fraction of Xe is increased. Since VUV photons are completely absorbed by glass, the spectra shown in the figure were measured by mounting opposed electrode pixels inside of a vacuum chamber filled with the gas mix of interest. The boundaries of the panel glass were not sealed which then allowed on-edge viewing of the pixel discharges with a McPherson 0.2 m monochromator operating in conjunction with a Princeton Instrument CCD optical multichannel analyzer [34]. The background gas mix was varied to obtain the various Xe concentrations in He shown while maintaining a total gas pressure of 53.3 kPa (400 torr). At low Xe concentrations, photons from the atomic Xe $1s^4$ and $1s^5$ states dominate the emission spectra producing lines at 147 nm and, with much less intensity, at 129 nm. The Tachibana laser-induced spectroscopic measurements show the spatial and temporal evolution of the $1s^4$ Xe atomic state in He/Xe plasma display discharges [54]. Both of these atomic lines experience significant resonant absorption and reemission. Thus, the measured line intensities are strong functions of photon path length traveled and of Xe partial pressure in the background gas [55]. For the emission spectra shown in Fig. 93.18, the lithium fluoride (LiF) entrance window to the evacuated spectrometer chamber was positioned between 100 and 150 nm

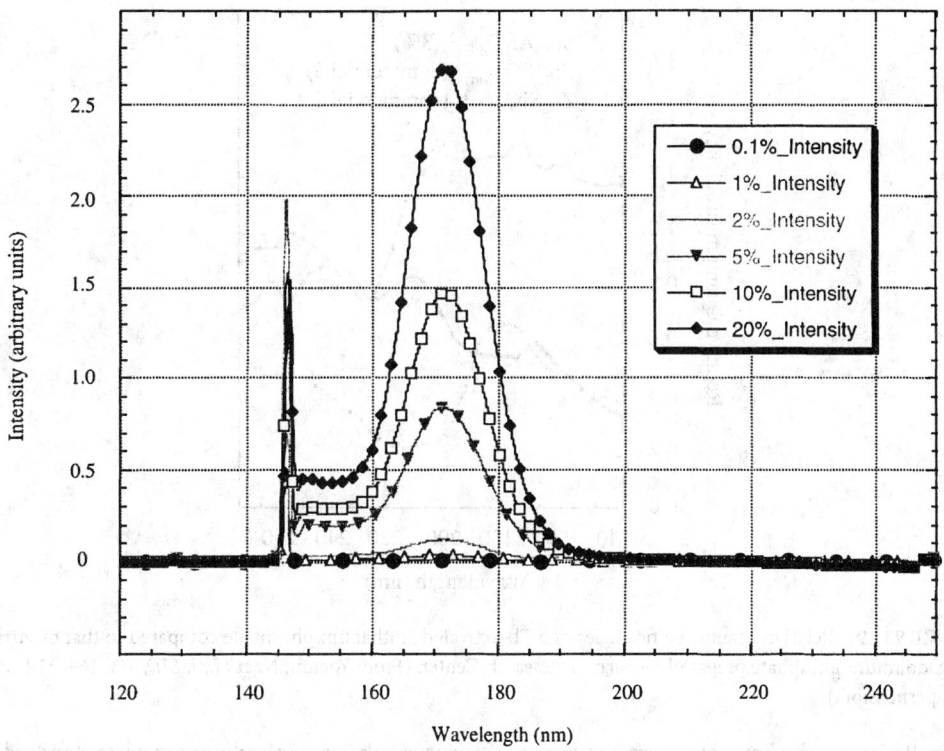

FIGURE 93.18 VUV emission spectra from opposed electrode ac plasma pixel discharges in He/Xe gas mixtures. Each spectrum was collected near the minimum sustain (or first *on*) voltage for that gas mixture, which ranged from 150 V for 0.1% Xe to 350 V for 20% Xe.

from the nearest pixel discharges, which is roughly the location of the phosphors relative to the discharge in an opposed electrode ac color display panel; for an example, see Fig. 93.5.

Recall that the optimal charge pair production per volt applied in Penning gas discharges occurs at minority species concentrations as low as 0.1%; see Fig. 93.9. However, color plasma pixels must optimize usable photon production per watt while maintaining stringent color purity requirements. Consequently, color plasma pixels typically operate with xenon concentrations ranging between 2 and 10%. Figure 93.18 shows that increased xenon concentration results in significant dimer formation and radiative emission from dimer dissociation. Since the dimer dissociation is a three-body process involving a photon and two xenon atoms, the momentum and energy conservation equations do not demand unique solutions. Consequently, emissions lines produced cover a broad spectral range spanning several tens of nanometers. Increased dimer emission is accompanied by the suppression of xenon atomic emission as energy within the atomic manifolds continues to flow toward the lowest available atomic levels; see Fig. 93.8. Note, too, that the dimer emission lines are not subject to resonant absorption. Therefore, the measured intensities shown reflect dimer emission from all pixels rows within the line of site of the spectrometer (four for the data of Fig. 93.18). In contrast, due to the strong resonance absorption of the atomic lines, more than 90% of the measured intensity of the 147-nm line is produced in the pixel row adjacent to the spectrometer window [34,55]. Care must be taken to account for these large variations in photon mean free paths when analyzing emission data.

Phosphor Excitation and Emission for Color Plasma Flat Panels

A variety of photoluminous phosphors are commercially available. Efficiencies for conversion of VUV photons to visible emissions has a complex dependence on excitation photon wavelength as can be seen

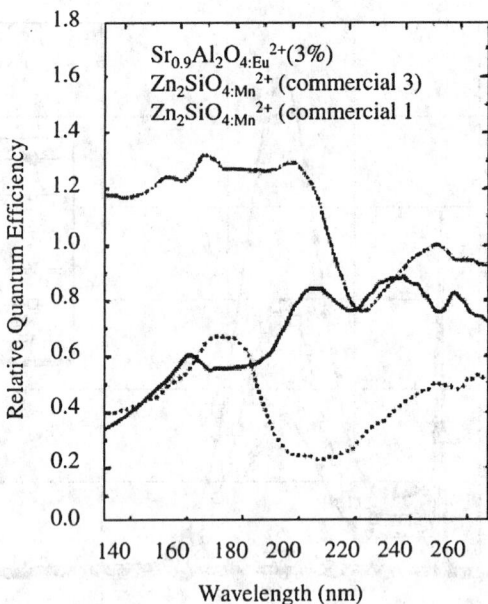

FIGURE 93.19 Relative quantum efficiencies of a Tb-activated lanthanum phosphate compared to that of yttrium and gadolinium phosphate prepared by Sarnoff Research Center. (From Yocum, N. et al., *J. SID*, 4/3, 169–172, 1996. With permission.)

in Fig. 93.19, which shows quantum conversion efficiencies relative to a sodium salicylate standard for some of the available green phosphors [50]. Conversion efficiencies for red, blue, and other green phosphors can be found in References 56 and 57. Table 93.2 provides the compositions of some selected commercially available phosphors and lists their relative quantum efficiencies for the principal emission lines of xenon discharges. Note that quantum efficiencies listed are relative values and that phosphors that convert 8.4-eV photons to visible photons near 2.3 eV have absolute efficiencies of only 27%. In principle, it is possible to produce two or more visible photons from a single high energy photon, but to date no such phosphors have been developed [58]. Table 93.2 also lists the chromaticity diagram coordinates which provide a measure of the color purity of the visible RGB emission spectra produced. The chomaticity diagram can be found in many references including Reference 59. Another consideration is the plasma display phosphor selection is persistence. Most of the phosphors listed in Table 93.2 require 5 to 13 ms for the emission intensity to decay to 10% of maximum value. For ac pixel operation at 50 kHz, each sustain voltage half cycle lasts only 10 μs while the discharge produces VUV emissions for only a small fraction of that time. Efforts are continuing for development of phosphors with faster response times. For example, Eu^{2+} green phosphors with 10% decay times of only 5 to 10 μs and with good quantum efficiencies near 173 nm have been developed [50].

Color Plasma Display Lifetime Considerations

Phosphors for plasma flat panel displays must be tolerant of the harsh environment produced by the pixel discharge. Photoluminous phosphor degradation mechanisms are at present not well understood. Contamination of the discharge by phosphor materials is a serious concern. Discharge modeling indicates that damage results principally from the accumulated fluence of photon and metastable bombardment, although fringe electric fields and prolonged surface charge accumulations could also result in ion bombardment [37]. Most ac plasma displays take advantage of MgO for enhancement of the discharge intensity by coating dielectric surfaces above the electrodes with an additional thin film of MgO. For ease of fabrication, the MgO is most often deposited using electron beam evaporation as one of the final manufacturing steps before glass seal and gas fill. If no mask is used, the MgO can also cover the

TABLE 93.2 Relative Quantum Efficiencies (QE) and Chromaticity Coordinates for Selected Phosphors [50,57]

Phosphor	Rel. QE[a] (174 nm)	Rel. QE[a] (170 nm)	Lifetime (10%) ms	x	y
NTSC green				0.21	0.71
$(La_{0.87}Tb_{0.13})PO_4$	1.1	1.4	13	0.34	0.57
$(La_{0.6}Ce_{0.27}Tb_{0.13})PO_4$	1.1	1.5	12	0.33	0.59
$(Y_{0.85}Ce_{0.1}Tb_{0.05})PO_4$	1.1	1.1	—	—	—
$(Y_{0.6}Ce_{0.27}Tb_{0.13})PO_4$	1.35	1.35	—	—	—
$(Gd_{0.87}Ce_{0.1}Tb_{0.03})PO_4$	1.0	1.1	10	0.34	0.58
$(Gd_{0.6}Ce_{0.27}Tb_{0.13})PO_4$	1.35	1.45	—	—	—
$(Ce,Tb)MgAl_{11}O_{19}$	0.9	1.4	—	—	—
$Sr_{0.9}Al_2O_4:Eu^{2+}$ (3%)	0.4	0.7	0.01	0.26	0.59
$Zn_2SiO_4:Mn$	1.2	1.3	12.5	0.21	0.72
$Zn_2SiO_4:Mn$	1.1	1.5	9.8	—	—
$Zn_2SiO_4:Mn$	0.45	0.55	5.4	—	—
$Zn_2SiO_4:Mn$	1.0	1.1	5	—	—
$Zn_3SiO_4:Mn$	1.0			0.21	0.72
$BaAl_{12}O_{19}:Mn$	1.1			0.16	0.74
$BaMgAl_{14}O_{23}:Mn$	0.92			0.15	0.73
$SrAl_{12}O_{19}:Mn$	0.62			0.16	0.75
$ZnAl_{12}O_{19}:Mn$	0.54			0.17	0.74
$CaAl_{12}O_{19}:Mn$	0.34			0.15	0.75
$YBO_3:Tb$	1.1			0.33	0.61
$LuBO_8:Tb$	1.1			0.33	0.61
$GdBO_3:Tb$	0.53			0.33	0.61
$ScBO_8:Tb$	0.36			0.35	0.60
$Sr_4 Si_8 O_8 Cl_4:Eu$	1.3			0.14	0.33
NTSC red				0.67	0.33
$Y_2O_3:Eu$	0.67			0.65	0.34
$Y_2SiO_5:Eu$	0.62			0.66	0.34
$Y_3 Al_5 O_{12}:Eu$	0.47			0.63	0.37
$Zn_8 (PO_4)_2:Mn$	0.34			0.67	0.33
$YBO_3:Eu$	1.0			0.65	0.35
$(Y,Gd)BO_8:Eu$	1.2			0.65	0.35
$GbBO_3:Eu$	0.94			0.64	0.36
$ScBO_3:Eu$	0.94			0.61	0.39
$LuBO_8:Eu$	0.74			0.63	0.37
NTSC blue				0.14	0.08
$CaWO_4:Pb$	0.74			0.17	0.17
$Y_2SiO_5:Ce$	1.1			0.16	0.09
$BaMgAl_{14}O_{23}:Mn$	1.6			0.14	0.09

[a] QEs above double rule are relative to sodium salicylate, those below relative to $Zn_2SiO_4:Mn$.

phosphors. While providing the phosphors with a protective coating, the MgO film also attenuates the intensity of the VUV photon flux available to excite the phosphors. Figure 93.20 shows VUV photon transmission as a function of wavelength through MgO films [34,60]. The measurements show that the primary atomic Xe emission lines at 129 and 147 nm are nearly 95% attenuated by an MgO film only 75 nm thick. In contrast, the dimer emission lines centered near 173 nm are much less attenuated with transmission factors of 30 to 50%, respectively, for MgO films of 200 and 75 nm. Consequently, ac plasma display designers must often achieve a balance among discharge enhancement due to MgO secondary electron emission, discharge degradation due to MgO photon absorption, and fabrication complexity associated with MgO deposition, which impacts the final cost of the display. Additionally, the designer must consider the aging or brightness degradation with time of the display, which is influenced in part by the rate of MgO sputter erosion discussed briefly below.

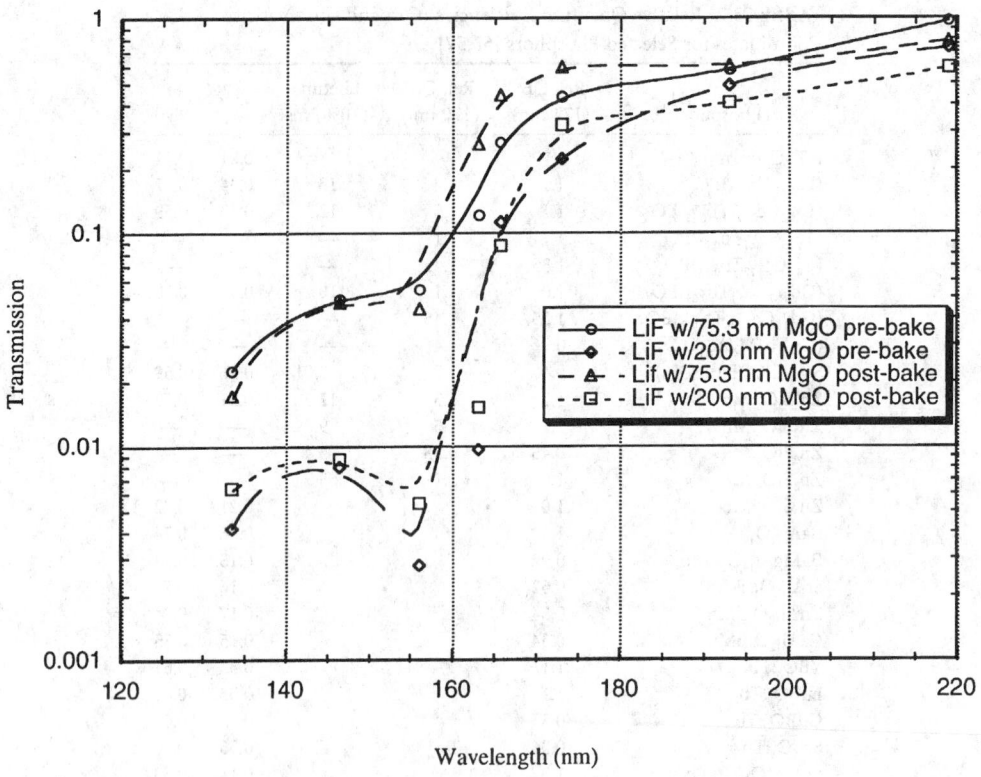

FIGURE 93.20   Photon transmission through MgO films before and after bake-out to remove water vapor.

93.6   Inspection and Metrology

Plasma display panel manufacturing uses some process steps typically found in semiconductor device manufacturing. Although the feature sizes are much larger than those required in semiconductor manufacturing, the dimensions over which feature integrity must be assured are also much larger. Confirmation that a display manufacturing process is under control or rapid quantitative characterization of the deviation from acceptable process limits is essential to producing high-quality displays with high yield. Information on a variety of length scales is required. Blanket deposition of materials onto substrates should be uniform; large deviations in thickness of the deposited material from one point on the substrate to a widely separated point should be avoided. Similarly, roughness of the deposited layers must be kept within defined limits. Finally, chemical compositions of deposited layers and their spatial homogeneity need to be measured and controlled.

The three commercial profile measuring devices, the CyberOptics, Zygo, and Leitz units, are based on laser or white light triangulation or interferometry; see Table 93.3 [61-63]. These machines are well suited for determining the profile of metallization after patterning. They do not appear to be useful for measuring the thickness of opaque layers and cannot, for example, measure the thickness of the blanket metallization before patterning. The CyberOptics unit is in wide use in the electronic packaging industry because of the low noise in its signal. Improvements in vertical resolution would increase its value for display manufacturing process control. The Zygo instrument has better vertical resolution, but has no real scanning capability. The Leitz unit is designed specifically for metrology of unfinished liquid crystal display components, and is presumably optimized for that application. Both the CyberOptics and Zygo units have a variety of heads with differing resolutions and fields of view. However, the vertical distance over which the interference can take place in the Zygo unit is very limited, so it may not be suitable for measuring features with large vertical dimensions.

TABLE 93.3 Equipment for Noncontact Profiling and Chemical Characterization of Thin Films

	CyberOptics Point Range Sensor	Zygo NewView 200	Leitz FTM400
Physical principle	Laser triangulation	Scanning white light interferometry	Laser interferometry
Maximum thickness/step height	150 μm (high resolution head)	100 μm (standard) (to 5 mm with z-drive)	70 μm
Vertical resolution	0.38 μm	100 pm	150 pm or 0.5%
Spatial resolution	>0.5 μm	0.22 μm	0.5 μm
Spot diameter/FOV	5.1 μm	140 × 110 μm	20 μm
Scan rate	10 points/s	2 or 4 μm/s	75 mm/s max.
Maximum sample size	Based on Table	10 cm × 15 cm	47 cm × 37 cm
Maximum sample weight	Based on Table	4.5 kg	4.5 kg (including holder)
Elemental detection	N.A.	N.A.	N.A.

Depending on system requirements, a useful laser metrology system can be purchased for as little as $20,000. Full-featured systems can easily cost $100,000. The bulk of the cost is, however, in the precision transport tables used for moving samples in a precisely controlled manner beneath stationary measurement heads. Since precision transport tables meeting the needs of the metrology equipment may already be part of the display production process or would require only minor upgrades to be brought in line with requirements, the cost to bring the metrology capability online may be much smaller than the figures mentioned above. Since laser head costs are low, it may also be desirable to use multiple fixed heads to speed up inspection and metrology.

Listed in Table 93.4 are the characteristics of two advanced measurement techniques for noncontact profiling and chemical analysis. To the authors' knowledge, systems of this type are not yet commercially available. They are being utilized, however, by researchers at Sandia National Laboratories and show promise for commercial usage. In beta backscattering measurements, the energy spectra of the backscattered beta particles provides an accurate measure of the elemental composition of a surface. This technique has been used to measure the thin-film thickness of trace-deposited metals on large-area surfaces (several square meters) in tokamaks [64]. With a depth resolution of about 100 pm, beta backscatter could serve as the physical basis for a high-sensitivity surface-profiling device. X-ray fluorescence measurements, routinely performed in air, can give information on film thickness and composition, and performance parameters for this technique are also listed in Table 93.4. Vacuum must be used when studying light elements like carbon, a potential undesired product of phosphor binder burnout, because the x rays produced are "soft". Although x-ray fluorescence equipment may have good lateral resolution, a larger spot size may prove useful when more averaged information such as film thickness is needed. Finally, eddy current measurements may provide a fast and reliable method for determining the thickness of the opaque, blanket metallization before patterning. This technique should be capable of providing 10% thickness measurement accuracy for conductor layers as thin as 1 μm.

TABLE 93.4 Advanced Techniques for Noncontact Profiling/Chemical Characterization of Thin Films

Physical Principle	β-Backscatter	X-Ray Microfluorescence
Maximum thickness/step height	A few mm	50 μm
Vertical resolution	About 0.1 nm	0.1 μm
Spatial resolution	Depends on pinhole size	50 μm to 2 mm
Spot diameter/FOV	Depends on pinhole size	100 μm
Scan rate	<1 min/measurement	1 spot/min
Maximum sample size	No restriction	24 cm × 21.5 cm
Maximum sample weight	No restriction	4.5 kg
Elemental detection	N.A.	<10 – 300 ppm

Many of the devices for quantitative characterization of the thickness and chemistry of the deposited layers suffer from the fact that they have a rather limited field of view. Characterization techniques are needed that will allow rapid identification of visual defects in the blanket-deposited layers and in the patterned layers produced from them. Visual inspection equipment is available for defect identification in liquid crystal displays. Manufacturers include Minato Electronics, Advantest, ADS, Photon Dynamics, and Teradyne. All use CCD devices and special algorithms for the identification of line and area defects. The defects found during high-volume plasma display manufacturing may be sufficiently similar in appearance to those found in liquid crystal display manufacturing that this equipment will prove useful with appropriate adjustments. These devices, however, are expensive.

Although researchers and equipment manufacturers believe that the equipment and techniques described above will be suitable for online process control during plasma display manufacturing, all agree that more development is required. Real parts will need to be characterized extensively on specific commercial equipment in production environments before conclusions can be drawn about the suitability of the equipment and techniques for the intended application. This is particularly true for more difficult measurements, like that of the thickness of dielectric above conductor lines. It is clear that there is no single instrument that will meet all film and feature dimensional measurement requirements and those for chemical characterization. A number of instruments will be needed to measure confidently the parameters needed for inspection, and characterize and control the manufacturing processes for plasma displays.

References

1. D.L. Bitzer and H.G. Slottow, The plasma display panel — a digitally addressable display with inherent memory, *1966 Fall Joint Computer Conf.*, Washington, D.C., *AFIPS Conf. Proc.*, 29, 541, 1966.
2. H.G. Slottow, Plasma displays, *IEEE Trans. Electron. Devices*, 23, 760–772, 1976.
3. T.N. Criscimagna and P. Pleshko, AC plasma display. Chapter 3, in *Topics in Applied Physics*, Vol. 40, *Display Devices*, Springer-Verlag, Berlin, 91–150, 1980.
4. L.F. Weber, Plasma displays, Chapter 10, in *Flat Panel Displays and CRTs*, L.E. Tannas, Jr., Ed., Van Nostrand Reinhold, New York, 1985.
5. J. Nolan, Gas discharge display panel, *1969 International Electron Devices Meeting*, Washington, D.C., 1969.
6. H.J. Hoehn and R.A. Martel, A 60 line per inch plasma display panel, *IEEE Trans. Electron Devices*, 18, 659–663, 1971.
7. O. Sahni and W.P. Jones, Spatial distribution of wall charge density in AC plasma display panels, *IEEE Trans. Electron Devices*, 25, 223–226, 1979.
8. O. Sahni and C. Lanza, Origin of the bistable voltage margin in the AC plasma display panel, *IEEE Trans. Electron. Devices*, 24, 853–859, 1977.
9. C. Lanza and O. Sahni, Numerical calculation of the characteristics of an isolated ac gas discharge display panel cell, *IBM J. Res. Dev.*, 22, 641–646, 1978.
10. O. Sahni and M.O. Aboelfotoh, The pressure dependence of the bistable voltage margin of an AC plasma panel cell, *Proc. SID*, 22, 212–218, 1981.
11. O. Sahni and C. Lanza, Importance of the secondary electron emission coefficient on E/Po for Paschen breakdown curves in AC plasma panels, *J. Appl. Phys.*, 47, 1337, 1976.
12. O. Sahni and C. Lanza, Influence of the secondary electron emission coefficient of argon on Paschen breakdown curves in AC plasma panels for neon + 0.1% argon mixture, *J. Appl. Phys.*, 47, 5107, 1976.
13. C. Lanza, Analysis of an ac gas display panel, *IBM J. Res. Dev.*, 18, 232–243, 1974.
14. O. Sahni, C. Lanza, and W.E. Howard, One-dimensional numerical simulation of ac discharges in a high-pressure mixture of Ne + 0.1% Ar confined to a narrow gap between insulated metal electrodes, *J. Appl. Phys.*, 49, 2365, 1978.

15. O. Sahni and C. Lanza, Failure of Paschen's scaling law for Ne −0.1% Ar mixtures at high pressures, *J. Appl. Phys.*, 52, 196, 1981.

16. H. Wakabayshi, Display market projections report from namora research institute, paper presented at *Imaging 2001: The U.S. Display Consortium Business Conference*, January 28, 1997, San Jose, CA.

17. K. Werner, Plasma hits the ground running, *Inf. Display*, 12(12), 30–34, 1996.

18. *Nikkei Materials & Technology*, [135], November, 1993, pp. 22–23.

19. H. Murakami and R. Toyonaga, A pulse discharge panel display for producing a color TV picture with high luminance and luminous efficiency, *IEEE Trans. Electron Devices*, 29, 988, 1982.

20. M. Ushirozawa, Y. Motoyama, T. Sakai, K. Wani, and K. Takahashi, Color dc-PDP with an improved resistor design in each cell, paper 33.2, pp. 719–722, *SID '94 Symp. Dig.*, 1994.

21. S. Mikoshiba, Color plasma displays: where are we now?, *Inf. Display*, 10(10), 21, 1994.

22. Y. Takano et al., A 40-in, DC-PDP with new pulse memory drive scheme, *SID '97 Dig.*, 1997.

23. P.S. Friedman, Are plasma display panels a low-cost technology?, *Inf. Display*, 11(10), 22–25, 1995.

24. *Inf. Display*, 10(7 & 8), 28, 1994.

25. A. von Engle, *Ionized Gases*, AIP Press, New York, 1994.

26. S. Brown, *Basic Data of Plasma Physics — The Fundamental Data on Electrical Discharges in Gas*, American Institute of Physics Press, New York, 1993.

27. F.F. Chen, *Introduction to Plasma Physics*, Plenum Press, New York, 1977.

28. D. Miller, J. Ogle, R. Cola, B. Caras, and T. Maloney, An improved performance self-scan I panel design, *Proc. SID*, 22, 159–163, 1981.

29. M.H. Miller and R.A. Roig, Transition propabilities of XeI and XeII, *Phys. Rev. A*, 8, 480, July 1973.

30. *Atomic Data and Nuclear Data Tables*, Vol. 21, No. 6, Academic Press, New York, 1978.

31. R.S. Van Dyck, C.E. Johnson, and H.A. Shugart, Lifetime lower limits for the $^3P_0$ and $^3P_2$ metastable states of neon, argon and krypton, *Phys. Rev. A*, 5, 991–993, 1972.

32. A.A. Kruithof and F.M. Penning, The Townsend ionization coefficient and some elementary processes in neon with small admixtures of argon, *Physica*, 4(6), 450, 1937.

33. F.L. Jones and W.R. Galloway, The sparking potential of mercury vapor, *Proc. Phys. Soc. Lond.*, 50, 207–212, 1938.

34. R.T. McGrath, R. Veerasingam, J.A. Hunter, P.D. Rockett, and R.B. Campbell, Measurements and simulations of VUV emissions from plasma flat panel display pixel microdischarges, *IEEE Trans. on Plasma Sci.*, 26(5), 1998.

35. P. Hines, Spectroscopic and electrical measurements on opposed electrode ac plasma display pixels, Honors engineering senior thesis, The Pennsylvania State University, University Park, 1997.

36. R. Veerasingam, R.B. Campbell, and R.T. McGrath, One-dimensional single and multipulse simulations of the ON/OFF voltages and bistable margin for He, Xe, and He/Xe filled plasma display pixels, *IEEE Trans. Plasma Sci.*, 24(6), 1399–1410, 1996.

37. R. Veerasingam, R.B. Campbell, and R.T. McGrath, Two-dimensional simulations of plasma flow and charge spreading in ac plasma displays, *IEEE Trans. Plasma Sci.*, 24, 1411–1421, 1996.

38. M. Scott, Los Alamos National Laboratory, Private Communication, 1997.

39. H. Uchike, K. Miura, N. Nakayama, T. Shinoda, and Y. Fukushima, Secondary electron emission characteristics of dielectric materials in AC-operated plasma display panels, *IEEE Trans. Electron. Devices*, 23, 1211–1217, 1976.

40. N.J. Chou and O. Sahni, Comments on "Secondary Emission Characteristics of Dielectric Materials in AC Operated Plasma Display Panels," *IEEE Trans. Electron. Devices*, 25, 60–62, 1978.

41. T. Shinoda, H. Uchike, and S. Andoh, Low-voltage operated AC plasma-display panels, *IEEE Trans. Electron. Devices*, 26, 1163–1167, 1979.

42. M. Aboelfotoh and J.A. Lorenzen, Influence of secondary-electron emission from MgO surfaces on voltage-breakdown curves in Penning mixtures for insulated-electrode discharges, *J. Appl. Phys.*, 48, 4754–4759, 1977.

43. T. Nenadovic, B. Perraillon, Z. Bogdanov, Z. Djordjevic, and M. Milic, Sputtering and surface topography of oxides, *Nucl. Instrum. Methods Phys. Res.*, B48, 538–543, 1990.

44. Schone, D. Walsh, R.T. McGrath and J.H. Burkhart, Microbeam Rutherford backscattering measurements of flat panel display thin film erosion, *Nuclear Instru. Methods Phys. Rev.-B*, 130(1–4), 543–550, 1998.

45. *CRC Handbook of Chemistry and Physics, 48th ed.*, R.C. Weast and S. M. Selby, Eds., The Chemical Rubber Co., Cleveland, OH, 1967.

46. S. Matsumoto, *Electronic Displays Devices*, John Wiley & Sons, New York, 1990.

47. G.E. Holz, The primed gas discharge cell — a cost and capability improvement for gas discharge matrix displays, *Proc. SID*, 13(2), 1972.

48. T. Yamamoto, T. Kuriyama, M. Seki, T. Katoh, H. Murakami, K. Shimada, and H. Ishiga, A 40 in. diagonal HDTV plasma display, *SID' 93 Symp. Dig.*, 165–168, 1993.

49. J. Deschamps and H. Doyeus, Plasma displays, *Phys. World*, June 1977.

50. N. Yocum, R.S. Meltzer, K.W. Jang, and M. Grimm, New green phosphors for plasma displays, *J. SID*, 4/3, 169–172, 1996.

51. H. Fujii, H. Tanabe, H. Ishiga, M. Harayama, and M. Oka, A sandblasting process for fabrication of color PDP phosphor screens, paper 37.5, 728–731, *SID '92 Symp. Dig.*, 1992.

52. M.C. Castex, Experimental determination of the lowest excited Xe_2 molecular states from VUV absorption spectra, *J. Chem. Phys.*, 74, 759–771, 1981.

53. M.R. Flannery, K.J. McCann, and N.W. Winter, Cross sections for electron impact ionization of metastable rare gas excimers (He_2^*, Kr_2^*, Xe_2^*), *J. Phys. B Mol. Phys.*, 14, 3789–3796, 1981.

54. K. Tachibana, Spatio-temperal measurement of excited $Xe(1s_4)$ atoms in a discharge cell of a plasma display panel by laser induced spectroscopic microscopy, *Appl. Phys. Lett.*, 65, 935–937, 1994.

55. T. Holstein, Imprisonment of resonance radiation in gases, *Phys. Rev.*, 72, 1212, 1947.

56. De Husk and S.E. Schnatterly, Quantum efficiency and linearity of 16 phosphors in the soft x-ray regime, *J. Opt. Soc. Am. B*, 9, 660–663, 1992.

57. J. Koike, T. Kojima, and R. Toyonaga, New tricolor phosphors for gas discharge display, *J. Electrochem. Soc. Solid-State Sci. Tech.*, 1008, June 1979.

58. P.N. Yocum, Future requirements of display phosphors from an historical perspective, *J. SID*, 4/3, 149, 1996.

59. L.E. Tannas, Jr., Ed., *Flat Panel Displays and CRTs*, Van Hostrand Reinhold, New York, 1985.

60. S.G. McLean and W.W. Duley, VUV absorption in thin MgO films, *J. Phys. Chem. Solids*, 45, 223, 1984.

61. R. Kuntz, CyberOptics, Private communication, 1-800-746-6315.

62. Zygo, Private communication, 1-860-347-8506, www.zygo.com.

63. A. Machura, Leica, Private communication, 49 (0) 6441 29-2316, Andreas.Machura@lmw.leica.com.

64. B.E. Mills, D.A. Buchenauer, A.E. Pontau, and M. Ulrickson, Characterization of deposition and erosion on the TFTR bumper limiter and wall, *J. Nucl. Mater.*, 162–164, 343–349, 1989.

94

Electroluminescent Displays

William A. Barrow
Planar Systems

94.1 Introduction

Electroluminescence (EL) is the nonthermal generation of light resulting from the application of an electric field to a substance. The light-emitting substance is generally a luminescent crystal. Most commercially available monochrome EL displays utilize ZnS:Mn as the luminescent material. EL displays have become very important in certain display markets. These include medical instrumentation, industrial control equipment, portable instrumentation, and military vehicles. The attributes of EL displays that make them attractive in these types of applications are mechanical ruggedness, relative insensitivity to ambient temperature, rapid display response, essentially unlimited viewing angle, compactness, and light weight.

There are four types of EL devices: ac thin film, ac powder, dc thin film, and dc powder. The ac thin-film display is by far the dominant device type. Some liquid crystal displays use ac powder EL for backlights. There is currently little or no commercial application of dc EL devices, either thin film or powder [1]. The focus here is on ac thin-film EL (ACTFEL) devices. While there are no widely established standard measurement techniques for the other EL device types, measurements similar to those described herein for ACTFEL devices could be applied with appropriate modifications to the other methods of device excitation.

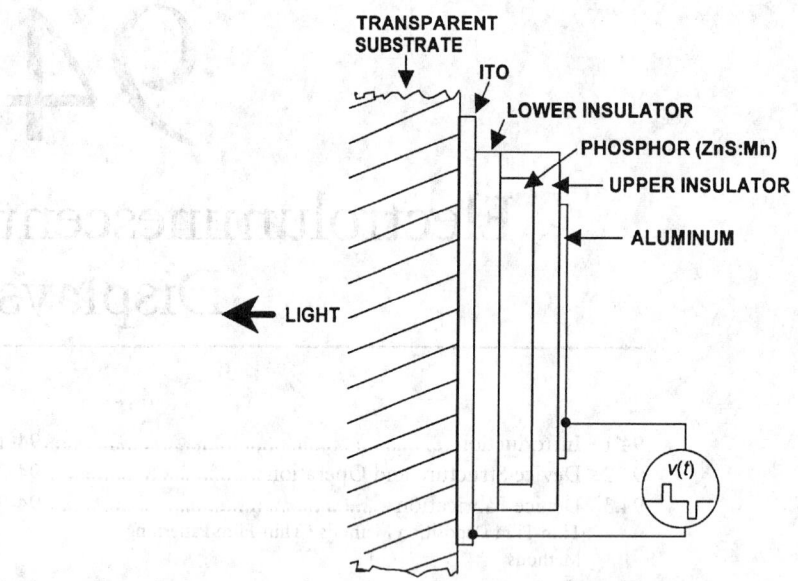

FIGURE 94.1   Schematic, cross-sectional diagram of the basic ACTFEL device structure.

94.2   Device Structure and Operation

A schematic, cross-sectional representation of the basic ACTFEL device structure [2] is shown in Fig. 94.1. The supporting substrate is usually made of very low sodium glass. If a suitable ion barrier layer is deposited between the substrate and lower electrode of the EL device, soda-lime glass can be used. The lower electrode, usually a transparent conductor such as indium tin oxide (ITO) is deposited next. The ITO is usually between 350 and 120 nm in thickness, providing sheet resistance in the range of 5 to 15 Ω/square. On top of the ITO electrode a lower insulator is deposited. This layer is typically SiON or aluminum titanium oxide (ATO). The thickness is normally around 200 nm. The phosphor layer, typically ZnS:Mn, is deposited between the lower insulator and a similar upper insulator. The phosphor thickness is typically in the range 200 to 1000 nm thickness, depending on the application. The upper electrode, typically aluminum, is deposited on top of the upper insulator. The aluminum is generally 100 to 200 nm thick.

A matrix-addressed monochrome display is created by dividing the upper and lower electrodes into orthogonal arrays of electrode stripes. The EL device is then excited locally by applying a voltage between a pair of crossing electrodes, causing an electric field to exist between them, which excites the phosphor. A color display can be created by dividing the phosphor into stripes of red-, green- and blue-emitting phosphors which are aligned with one of the sets of electrode stripes [3,4]. This is shown schematically in Fig. 94.2. Color displays can also be created by using an unpatterned phosphor which emits a broad spectrum including red, green, and blue and filtering the emission using either a patterned color filter which is aligned with the electrode stripes or a frame sequential color filter.

94.3   Device Fabrication

Thin-Film Deposition Methods

A wide variety of deposition techniques are used by various manufacturers and researchers in the fabrication of ACTFEL devices. The electrode materials ITO and aluminum are usually deposited by physical vapor deposition (PVD) techniques. ITO is almost universally deposited by dc magnetron

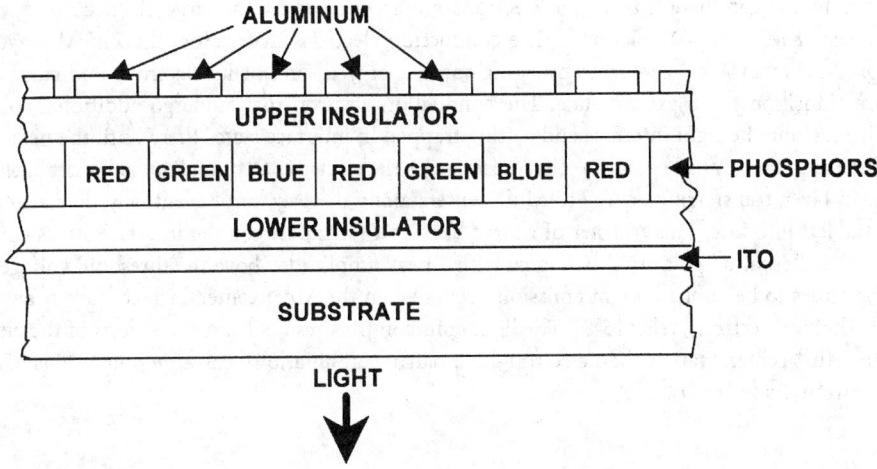

FIGURE 94.2 Schematic, cross-sectional diagram of a color ACTFEL device with the red, green, and blue primary colors produced by patterned color phosphor stripes. The color phosphor stripes are end on in this view.

sputtering from either a conductive ITO target or from a metal alloy target [5]. Optimum ITO conductivity is obtained by postdeposition annealing in a very low oxygen environment. Al is deposited either by dc magnetron sputtering from an Al metal target or by electron beam evaporation of Al metal. The insulator and phosphor layers are deposited by chemical vapor deposition (CVD) as well as PVD. Phosphor layers have been deposited by thermal as well as e-beam evaporation, sputtering, metal–organic CVD (MOCVD), and atomic layer epitaxy (ALE). Insulator layers have been deposited by e-beam evaporation, radio frequency (RF) sputtering, plasma-enhanced CVD (PECVD), and ALE.

Thin-Film Patterning Methods

Patterning of the electrodes is generally accomplished either through etching or liftoff, although some early workers patterned the upper aluminum electrodes by evaporating through a shadow mask. Aluminum is easily etched wet or dry. There are commercially available wet etches for Al. Dry etching of Al can be carried out using standard chlorine chemistries, e.g., Cl_2/BCl_3. Al can also be patterned by evaporating onto a reversed photoresist pattern and lifting off excess metal. ITO can be etched wet or dry. ITO wet etches generally consist of mixtures of HCl and HNO_3. ITO can be dry-etched in HI or HBr. Patterning of the phosphor layers is more problematic. Phosphor etches exist for most ACTFEL phosphor materials, but they are generally proprietary. Thin-film phosphors are difficult to etch because they are usually water sensitive, and some color phosphors contain heavy metals which are difficult to volatilize in a dry-etch process.

94.4 Device Operation

Luminescence Mechanisms

As the device structure shown in Fig. 94.1 indicates, ACTFEL devices are capacitively coupled. Since only displacement current can flow across the insulator layers, the drive signal must be an alternating polarity waveform. A typical alternating polarity, trapezoidal waveform is shown in Fig. 94.3. If the peak voltage, V_p, in Fig. 94.3 is larger than the **threshold voltage** of the device, V_{th}, then, when the positive pulse of the waveform is applied between the Al and ITO electrodes of the device structure shown in Fig. 94.1, the energy band diagram of the ACTFEL device will be as shown in Fig. 94.4. Electrons, which are the majority carriers in ACTFEL devices, tunnel out of the interface states on the left and into the conduction band. Once in the conduction band the electrons are accelerated by the large electric field, which is approximately 1 MV/cm = 100 kV/mm. The conduction electrons drift across the ZnS:Mn layer until they impact excite an Mn^{2+} center, transferring some energy to one of its electrons and causing it to undergo a transition to an excited state. The conduction electron may undergo additional collisions, eventually reaching the right interface and getting trapped in interface states there until the next voltage pulse, which is of the opposite polarity. This pulse causes the electrons to tunnel out and drift back across the ZnS:Mn layer, transferring energy to Mn^{2+} centers along the way, until eventually they are trapped again at the left interface. This transfer of charge back and forth between the interface states continues as long as the alternating polarity drive signal with a peak amplitude above the threshold voltage of the device continues to be applied. Light emission occurs when the Mn^{2+} centers, which have been impact excited by the hot electrons, relax [6-8]. The light emission thus results from transitions of the electrons within the Mn^{2+} centers, rather than electron–hole pair recombination near a *pn* junction as occurs in a light-emitting diode (LED).

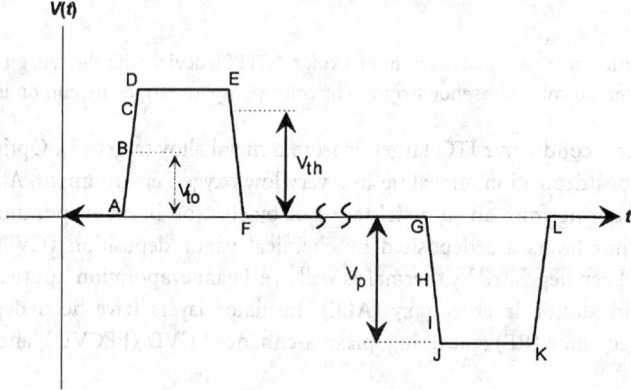

FIGURE 94.3 Typical alternating polarity, pulse drive waveform. Letters A–L mark points on the drive waveform which are referenced later in Figs. 94.5 and Fig. 94.6. The pulse width would generally be about 30 μs and the frequency would be between 60 and 500 Hz for a passive matrix-addressed display.

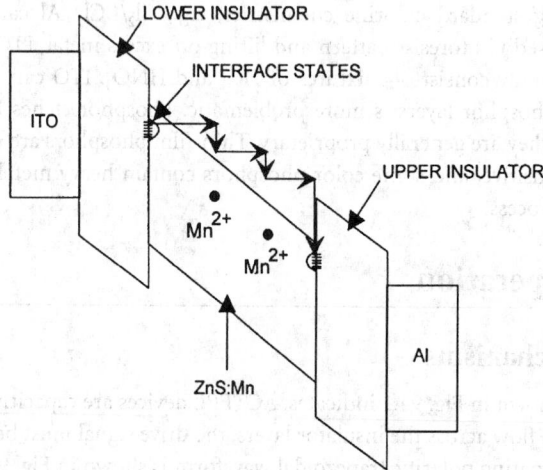

FIGURE 94.4 Energy band diagram of an ACTFEL device during the peak of the applied voltage pulse. Electrons tunnel out of insulator/phosphor interface states into the conduction band, are swept across the phosphor layer, and impact exciting emission centers as they go until they are finally thermalized and trapped at the opposite interface.

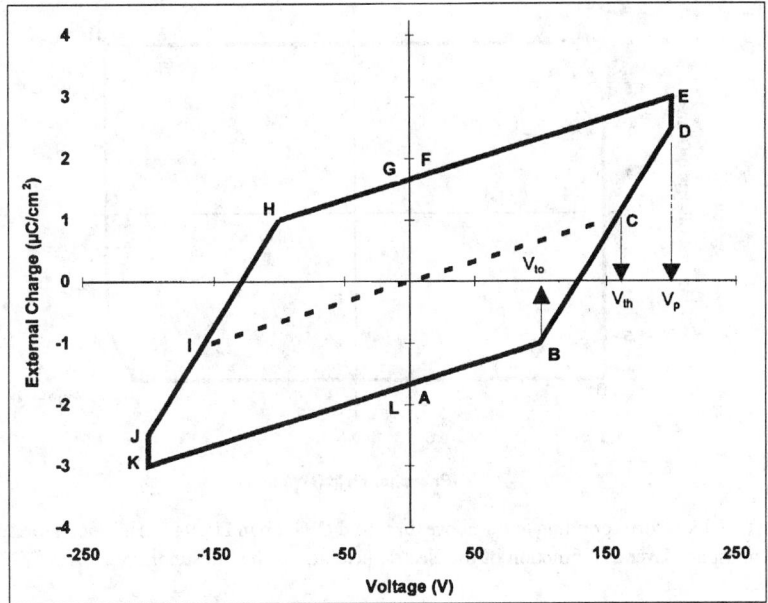

FIGURE 94.5 Idealized QV loop with no charge leakage from interface states between drive pulses. Letters A–L mark points on the QV loop which are coincident in time with the points labeled A–L on the drive waveform in Fig. 94.3. The dashed line is the QV loop for the case just below threshold. The solid line is the open loop above threshold. The area of the QV loop represents the energy dissipated by the device per cycle of the drive waveform.

Description of Charge Flow

If the external charge, Q, flowing into the ACTFEL device is plotted vs. the externally applied voltage, V, the resulting curve is called a **QV loop** [9,10]. If the amplitude of the applied voltage pulses is less than the threshold voltage of the device, the QV loop is simply a straight line with slope equal to the total capacitance of the insulator/phosphor/insulator stack. If the amplitude of the applied voltage pulses is greater than the threshold voltage of the device, the QV loop opens up. QV loops below and above threshold are shown in Fig. 94.5. Above threshold, power is dissipated in the ACTFEL device. The area encompassed by the QV loop is equal to the energy delivered to the device per period of the drive waveform.

The QV loop is measured directly. A theoretical extension of the QV loop that is sometimes used by researchers studying the physics of ACTFEL devices is a plot of the actual charge flow across the phosphor layer, Q_p, vs. the electric field across the phosphor layer, F_p. The quantities required for a plot of Q_pF_p can be calculated from the QV data if the thicknesses and dielectric constants of the insulator layers and phosphor layer are known. The actual charge flow across the phosphor is calculated by subtracting the reactive charge from the total charge. The field in the phosphor layer is calculated by adding the externally applied field to the internal **polarization field** due to the actual flow of charges across the phosphor layer. A Q_pF_p loop corresponding to the above threshold QV loop of Fig. 94.5 is shown in Fig. 94.6. The Q_pF_p loop is useful because it expresses the internal electrical characteristics of the phosphor layer during device operation.

Device Excitation

Whether the device under test is a test dot or a matrix display, the drive waveform must be ac coupled. Passively addressed matrix displays are scanned one row at a time. Data are loaded into the column drivers for a single row of pixels and the selected pixels in the row are all turned on simultaneously. The row pulse brings the voltage across each pixel in the row to a level just below threshold. The columns of selected pixels are then pulsed to bring the voltage across selected pixels to a level above threshold and the pixels are turned on. During the following frame the voltage polarities across the pixels are reversed.

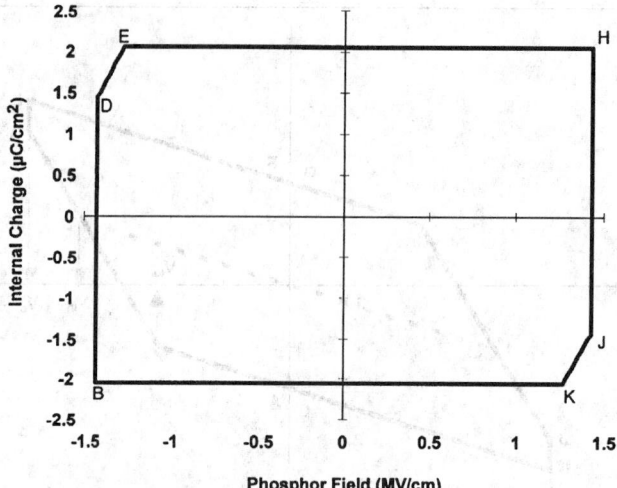

FIGURE 94.6 $Q_p F_p$ loop corresponding to the above threshold QV loop in Fig. 94.5. This loop represents the charge flow across the phosphor layer as a function of the electric field across the phosphor layer.

Each individual pixel is subjected to a drive signal similar to that shown in Fig. 94.3. To activate a pixel fully requires the voltage to be held above V_{th} for 10 to 20 μs. Since each row must be scanned in sequence, a typical display with approximately 500 rows requires at least 5 ms to scan one frame. The maximum frame rate is thus approximately 200 Hz. Displays that are addressed by an active matrix of transistors are not scanned a line at a time as passively addressed displays are, but instead are bulk driven. They have one unpatterned common electrode, usually an upper layer of ITO, to which an ac drive signal is applied. Each pixel is connected to ground through a transistor which drops a portion of the applied drive voltage when the pixel is not selected. In this type of display the drive waveform can be any ac signal of the appropriate voltage. Sine, trapezoid, and triangle waveforms have been used. An active matrix display can be driven with a frame rate higher than a passive matrix-addressed display by a factor equal to the number of rows in the display, since row-at-a-time scanning is not required. Since a light pulse is emitted for each voltage pulse, the average luminance is proportional to the frame rate, resulting in much higher luminance capability for active matrix-addressed displays.

94.5 Standard Measurements

Measurable quantities of interest include luminance, luminous efficiency, emission spectrum, latent image, and defects. The luminance of an ACTFEL display is a function of the peak voltage and frequency of the drive waveform, the intrinsic efficiency of the insulator/phosphor/insulator stack, and the operating history (aging). The efficiency is a function of the drive waveform shape and frequency as well as various device parameters. The emission spectrum is primarily determined by the phosphor host material and activators, although it is also affected by deposition and anneal conditions and in some cases by the drive voltage and frequency. **Latent image** is the burning in of a permanent image of a fixed pattern which is displayed for long periods. It can appear as a faint dark image superimposed on a bright background or as a faint bright image superimposed on a dark background. Formation of latent image is affected primarily by drive waveform symmetry and device processing parameters. Display defects include pixel, line, and mura defects.

94.6 Time-Resolved Measurements

Measurements of luminance, efficiency, and emission spectra as introduced so far involve time-averaged light emission. There are two fairly common time-resolved measurements of ACTFEL light emission:

light emission decay time and time-resolved light emission spectroscopy. The **light emission decay time**, τ, is the time it takes for the light emission from one excitation pulse to fall to $1/e$ times its initial value. The measurement of τ is started just after the trailing edge of the excitation pulse. This is necessary in order that the measured value of τ not be affected by the continuing excitation of additional emission centers, so that it represents the intrinsic relaxation time of the emission center. In devices with evaporated ZnS:Mn phosphor, τ is a strong function of the Mn concentration. It can thus be used as an analytical technique to determine the Mn concentration. Time-resolved spectroscopy is the measurement of the emission spectrum occurring during specific portions of the excitation and emission process. An example of the application of this technique is the study of the separate light pulses emitted during the leading and trailing edges of the excitation pulse with the blue-emitting phosphor SrS:Ce. Both of these techniques are frequently used to help elucidate the excitation and emission mechanisms in ACTFEL phosphors.

94.7 Test Dot Characterization

Luminance and Efficiency

A schematic representation of a measurement system for collecting luminance and efficiency vs. voltage data on a test dot is shown in Fig. 94.7. An arbitrary waveform generator provides a drive signal with the waveshape, frequency, and peak voltage determined by a control computer. The waveform generator output signal is amplified from a ± 5 V range to a ± 300 V range. A photometer, also under computer control, measures the luminance, L, of the test dot. An oscilloscope measures the voltage across the test dot and the voltage across the sense resistor. The current is calculated by dividing the voltage across the sense resistor by its resistance. The control computer can thus adjust the peak voltage up and down and collect the luminance data, as well as the waveforms representing the voltage across the test dot and the current through it. The energy dissipated, E_p, during each drive pulse is calculated as

$$E_p = \int_D I(t) \times V(t) dt \qquad (94.1)$$

where $I(t)$ and $V(t)$ are the current and voltage waveforms, respectively, and D is the duration of either the positive or negative drive pulse. This integration can be carried out by most oscilloscopes or the $I(t)$ and $V(t)$ data can be transferred to the control computer for the calculation, although this approach is generally slower. The average energy dissipated per period of the drive waveform, E, is the average of E_p for a positive pulse and a negative pulse. The average power dissipated, P, can be calculated by multiplying E by the frequency of the drive waveform. The efficiency, η, is calculated as follows:

$$\eta = \frac{\pi L A}{P} \qquad (94.2)$$

where L is the luminance in cd/m^2, A is the area of the test dot in m^2, and P is the average power in watts. Values of L and η are collected for peak voltages ranging from 10 V below threshold to 40 or 50 V above threshold. Plots of L and η vs. peak voltage for a typical device are shown in Fig. 94.8.

Charge Flow and Electric Field

The QV loop is measured using a circuit identical to that in Fig. 94.7, except that the sense resistor is replaced by a sense capacitor and the photometer is not used. The sense capacitor value is chosen to be much larger than the capacitance of the ACTFEL device so that the voltage dropped by the sense capacitor is small. Since the sense capacitor and the ACTFEL device are in series, the charges stored on them are equal. The charge, Q, on the ACTFEL device is thus

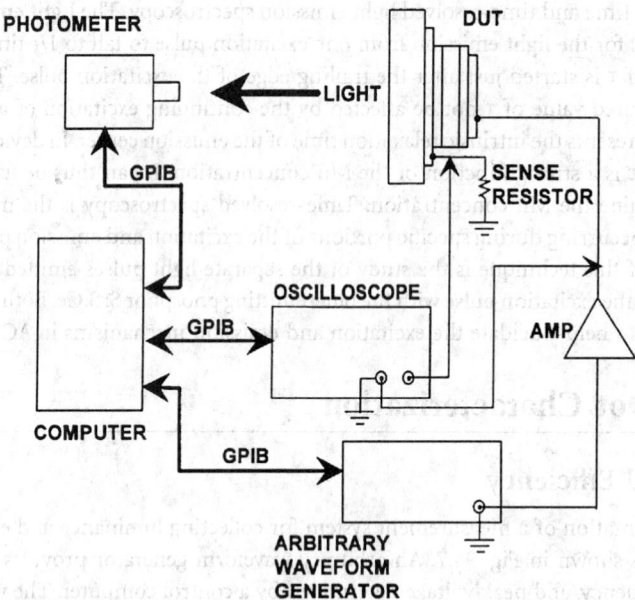

FIGURE 94.7 Schematic diagram of a system for measuring luminance and efficiency as functions of the peak drive voltage. The voltage waveform, $V(t)$, is measured by the oscilloscope at the Al electrode of the device under test. The current waveform, $I(t)$, is measured by the oscilloscope as the voltage across the sense resistor divided by its resistance. The luminance is measured by the photometer.

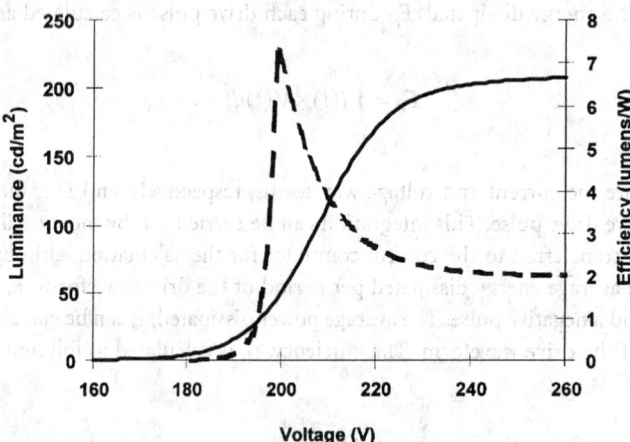

FIGURE 94.8 Plots of luminance and efficiency as functions of peak drive voltage. The solid line is the luminance and the dashed line is the efficiency.

$$Q = C_s V_s \tag{94.3}$$

where C_s is the capacitance of the sense capacitor and V_s is the voltage appearing across it. When $Q(t)$ is plotted vs. $v(t)$, the QV loop results. In the idealized QV loop shown in Fig. 94.5, V_{th} (C) is the threshold voltage, V_{to} (B) is the turn on voltage, and V_p (D,E) is the peak voltage of the drive waveform. Threshold voltage is the voltage at which the first knee in the LV curve occurs. If C_T is the total capacitance of the device below threshold, this is also the voltage at which the line $Q = C_T V$ intersects the open, above-threshold QV loop (C). In practice, it is sometimes defined as the voltage at which a certain luminance value occurs at a given frequency, e.g., the voltage at which the luminance is 1 cd/m² at 60 Hz. V_{to} is the voltage at which the slope of the QV loop changes from C_T to C_I, where C_I is the capacitance of the insulator layers.

Since a differential element of energy delivered to the ACTFEL device is $dE = V(t)dQ$, then

$$E = \int V(t)dQ \tag{94.4}$$

The energy delivered per period of the drive waveform is thus equal to the area encompassed by the QV loop. The power dissipated is just the energy per period multiplied by the frequency. In practice, the area of the QV loop is measured by numerical integration. The calculation can be carried out on the oscilloscope if it has analysis capabilities or the data can be transferred to the control computer for integration.

Generation of the Q_PF_P loop does not require any electrical measurements other than those required for the QV loop. Q_P is the charge separation across the phosphor layer and F_P is the electric field across the phosphor layer. If the thicknesses and dielectric constants of the insulator and phosphor layers are known, Q_P and F_P can be calculated from the values of $Q(t)$ and $V(t)$ on the QV loop. This is accomplished by applying the following equations [11,12]:

$$Q_p(t) = \frac{C_i + C_p}{C_i} Q(t) - C_p V(t) \tag{94.5}$$

and

$$F_p(t) = \frac{1}{d_p}\left(\frac{Q(t)}{C_i} - V(t) \right) \tag{94.6}$$

where C_i is the capacitance of the insulators, C_p is the capacitance of the phosphor layer, and d_p is the thickness of the phosphor layer.

Time-Resolved Measurements

The apparatus for measuring τ is shown in Fig. 94.9. A photomultiplier tube (PMT) is used to detect the light emission as a function of time. The drive system is set to provide relatively narrow drive pulses, typically 10 μs pulse width, at relatively low frequency, typically 60 Hz. This approach works well for phosphors with relatively long decay times, on the order of 100 μs to a few milliseconds. This is the case for many common ACTFEL phosphors such as ZnS:Mn and ZnS:Tb. Phosphors such as SrS:Ce, however, have very fast decay times and cannot be measured in this manner. In such cases the photoluminescent decay time must be measured using a pulsed laser to excite the phosphor and appropriately low RC response time of the light detection system.

The general setup for measuring time-resolved emission spectra from ACTFEL devices is shown schematically in Fig. 94.10. The oscilloscope is triggered on the drive waveform and the signal integration period is set to the region of interest. A boxcar integrator can also be used to integrate the light signal during the desired time window. The monochrometer wavelength is scanned and the emission spectrum is collected for the selected portion of the emission process.

Aging

ACTFEL devices tend to stabilize after a few tens of hours of burn-in, but can exhibit complex aging behavior during the burn-in process. The luminance vs. voltage curves for ZnS:Mn devices in which the phosphor layer is deposited by evaporation, for example, tend to shift to slightly higher voltage during burn-in [2]. Luminance vs. voltage curves for devices in which the ZnS:Mn is deposited by ALE tend to shift to slightly lower voltage [13]. Aging data is collected by measuring luminance vs. voltage at selected

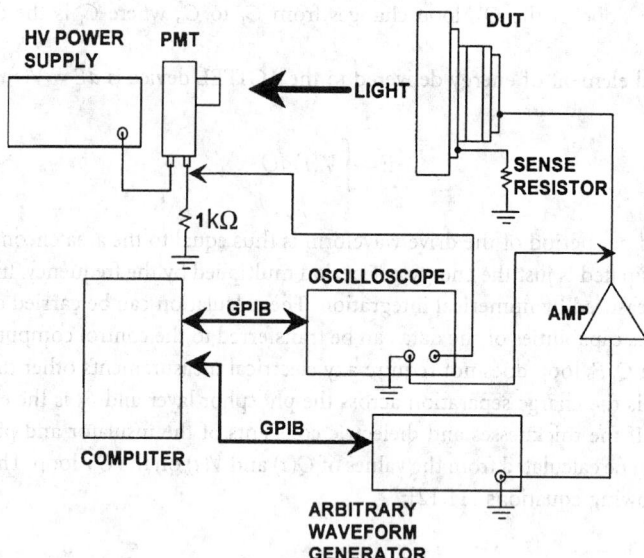

FIGURE 94.9 Apparatus for measuring the luminescent decay time of the ACTFEL phosphor. The light signal is measured by the oscilloscope across the sense resistor on the output of the PMT. The oscilloscope is triggered on the drive pulse.

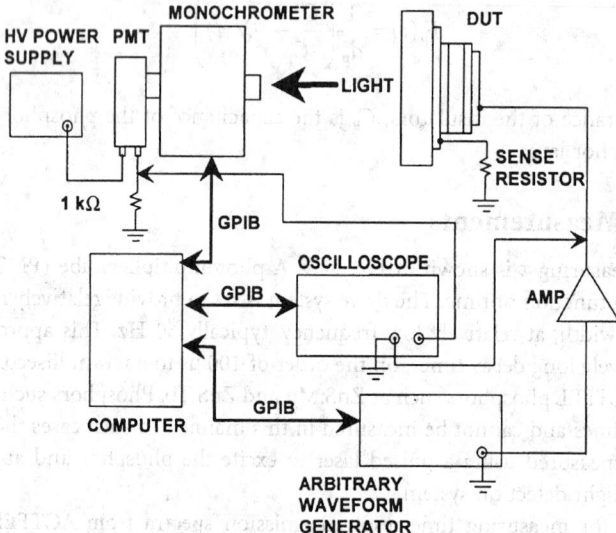

FIGURE 94.10 System for measuring the time-resolved emission spectrum of an ACTFEL device. The oscilloscope is set to integrate the light signal from the PMT during a selected time window. The monochrometer wavelength is scanned over the entire spectral range and the emission intensity data is collected from the oscilloscope.

time intervals during aging. The measurement is carried out as described earlier for luminance vs. voltage measurements. The aging is done by continuously operating the device at a fixed voltage or at a fixed voltage above threshold. The aging process can be accelerated by operating the device at higher frequency.

94.8 Characterization of Matrix-Addressed Displays

The characterization of matrix-addressed displays differs from characterization of test dots because less control of the drive waveform is readily available. The row and column drivers do not provide great flexibility, although some control can be exercised by varying the composite drive waveform and the control signals to the driver chips. These types of modifications, however, require detailed knowledge of the addressing and control electronics involved and are best left to the original display manufacturer. Measurements that are more accessible and of more general interest involve characterization of the luminance, chromaticity, uniformity, display life, latent image, and defects. For these measurements the display under test is controlled by a computer through a standard video output or a custom video display interface provided with the display. The display operating voltage and frequency are fixed. The display is placed in a dark room or enclosure. Luminance is measured with a photometer. Chromaticity is measured with a spectrophotometer or a photometer equipped with tristimulus filters. Uniformity can be measured by mounting either the display or the photometer/spectrophotometer on a translation system and collecting data at points in a regular array of locations throughout the display surface. Display life is characterized by operating the display in a full-on pattern, cycling through checkerboard patterns, etc., and making luminance measurements at exponentially increasing time intervals. Latent image is formed by displaying small blocks in various fixed locations and various fixed gray levels (if available) continuously for long periods. It is characterized by setting the entire screen to each gray level available and measuring the luminance in the locations which had blocks displayed during aging as well as unaged areas nearby. Latent image is the percent luminance difference between aged blocks and nearby unaged areas. Latent image is typically measured after aging the block pattern for 1000 h. Defects are classified into pixel, line, and mura defects. They are characterized by visual inspection with the aid of an eye loupe or by an automated flat panel inspection system.

94.9 Excitation and Measurement Equipment

Excitation of Test Dots

Test dots can be excited by any signal source that provides bipolar pulses up to a peak voltage of 300 V and sufficient current sourcing and sinking capability to charge and discharge the device capacitance. Bipolar pulse drivers with sufficient voltage and current output can be built with commercially available components. A hybrid circuit op amp, produced by Apex Microtechnology, provides sufficient voltage output and frequency response. A current boosting stage can be added to the output if the DUT capacitance is too large to drive directly. This amplifier approach is very flexible since any waveform that can be generated by the arbitrary waveform generator can be used. There are also some commercial amplifiers available that are effective for driving test dots for some measurements. They tend to have limited bandwidth and must be used with caution in situations in which the measurement is sensitive to the exact shape of the drive pulse. This would be the case, for example, in *QV* measurements and often in efficiency measurements.

Excitation of Matrix-Addressed Displays

Excitation of matrix-addressed displays is straightforward since the drive electronics are integrated with the display. Generally, only a standard ac power outlet and a computer with an appropriate video card are required. Test patterns for measurement can be created using simple computer programs.

94.10 Measurement Instruments

Measurement of Drive Voltage and Current

Digital oscilloscopes are used for all of the electrical measurements described in this chapter. Suitable instruments are available from several manufacturers, including Tektronix and Hewlett-Packard. A bandwidth of 100 MHz is more than sufficient. Waveform analysis capabilities are very helpful but not absolutely necessary. RS 232 or IEEE 488 interfaces are required for computer control and data transfer.

TABLE 94.1 Light Measurement Instruments

Instrument Type	Model	Manufacturer
Photometer	PR880	Photo Research
Photometer/spectroradiometer	PR650	Photo Research
	Pritchard 1980B	Photo Research
	GS-1280 RadOMAcam	Gamma Scientific (EG&G)
Photomultiplier tubes	—	Oriel Corp.
	—	Hamamatsu
Photodiode	PIN 10AP	UDT Sensors, Inc.
Flat panel inspection system	FIS 250	Photon Dynamics, Inc.

TABLE 94.2 Manufacturers of Light Measurement Instruments

Photo Research
9330 DeSoto Avenue, P.O. Box 2192
Chatsworth, CA 91313-2192
(818) 341-5151

Gamma Scientific (EG&G)
8581 Aero Dr.
San Diego, CA 92123-1876
(619) 279-8034

Oriel Corp.
252 Long Beach Blvd., P.O. Box 872
Stratford, CT 06497-0872
(203) 380-4200

Hamamatsu Photonics Systems Corp.
360-T Foothill Road, P.O. Box 6910
Bridgewater, NJ 08807-0910

UDT Sensors, Inc.
12525 Chadron Ave.
Hawthorne, CA 90250
(310) 978-0516

Photon Dynamics, Inc.
6325 San Ignacio Ave.
San Jose, CA 95119
(408) 226-9900

Measurement of Emitted Light

Several types of instruments are used for measuring light emission from ACTFEL displays. Photometers are used for luminance measurements. Spectrophotometers are used for measuring the emission spectrum and with suitable software can also provide luminance measurements. Time-resolved measurements are accomplished by using photomultiplier tubes or photodiode detectors. Table 94.1 lists some examples of photometers, spectroradiometers, photomultipliers, and photodiodes along with the names of the companies that manufacture them. A relatively new development for characterizing the light emission characteristics of flat panel displays, including ACTFEL displays, is the flat panel inspection system. This is a large measurement system comprising a CCD camera detector, light-tight enclosure, control computer, image processor, and specialized software. This type of system images an entire flat panel display on the CCD camera and measures luminance, chromaticity, and various defects by analyzing the image. These systems are intended for high throughput manufacturing environments and cost several hundred thousand dollars. An example of this type of system is also included in Table 94.1. Contact information for the companies listed in Table 94.1 is provided in Table 94.2.

Defining of Terms

Electroluminescence: The nonthermal generation of light resulting from the application of an electric field to a substance, usually a luminescent crystal.

Latent image: The ghost image of a previously displayed pattern which can sometimes be seen in a full field on an electronic display screen.

Light emission decay time: The time it takes for the light emission from one excitation pulse to fall to $1/e$ times its initial value.

Polarization charge: The charge trapped at the phosphor/insulator interface following the application of a drive pulse.

Polarization field: The field across the phosphor layer resulting from the polarization charge.

$Q_p F_p$ loop: The closed curve which results from plotting the internal charge flow across the phosphor layer (Q_p) vs. the electric field across the phosphor layer (F_p).

QV loop: The closed curve which results from plotting the external charge (Q) flowing into a TFEL device vs. the externally applied voltage (V).

Threshold voltage: The voltage amplitude of the drive waveform above which current flows across the phosphor layer and light is emitted from a TFEL device.

Turn-on voltage: The voltage corresponding to the first knee in the QV loop of a TFEL device. This is the voltage at which charge begins to flow across the phosphor layer. This voltage is generally less than the threshold voltage because the internal field across the phosphor layer is enhanced by the polarization field once the polarization charge has built up in the steady state.

References

1. Y.A. Ono, *Electroluminescent Displays*, Singapore: World Scientific, 1995.
2. T. Inoguchi, M. Takeda, Y. Kakihara, Y. Nakata, and M. Yoshida, "Stable high-brightness thin-film electroluminescent panels," *Dig. 1974 SID International Symposium*, 84, 1974.
3. C.N. King, R.E. Coovert, and W.A. Barrow, "Full-color 320 × 240 TFEL display panel," *Eurodisplay '87*, 14, 1987.
4. W.A. Barrow, R.E. Coovert, C.N. King, and M.J. Ziuchkovski, "Matrix-addressed full-color TFEL display," *Dig. 1988 SID International Symposium*, 284, 1988.
5. R. Tueta and M. Braguier, "Fabrication and characterization of indium tin oxide thin films for electroluminescent applications," *Thin Solid Films*, 80: 143, 1981.
6. J.D. Davidson, J.F. Wager, and I. Khormaei, "Electrical characterization and SPICE modeling of ZnS:Mn ACTFEL devices," *Dig. 1991 SID International Symposium*, 77, 1991.
7. A.A. Douglas and J.F. Wager, "ACTFEL device response to systematically varied pulse waveforms," in *Electroluminescence — Proceedings of the Sixth International Workshop on Electroluminescence*, V.P. Singh and J.C. McClure, Eds., El Paso, TX: Cinco Puntos Press, 1992.
8. D.H. Smith, "Modeling AC thin-film electroluminescent devices," *J. Lumin.*, 23: 209, 1981.
9. P.M. Alt, "Thin-film electroluminescent displays: device characteristics and performance," *Proc. SID*, 25: 123, 1984.
10. Y.A. Ono, H. Kawakami, M. Fuyama, and K. Onisawa, "Transferred charge in the active layer and EL device characteristics of TFEL cells," *Jpn. J. Appl. Phys.*, 26: 1482, 1987.
11. E. Bringuier, "Charge transfer in ZnS-type electroluminescence," *J. Appl. Phys.*, 66: 1314, 1989.
12. A. Abu-Dayah, S. Kobayashi, and J. F. Wager, "Internal charge-phosphor field characteristics of alternating-current thin-film electroluminescent devices," *Appl. Phys. Lett.*, 62: 744, 1993.
13. A. Mikami, K. Terada, K. Okibayashi, K. Tanaka, M. Yoshida, and S. Nakajima, "Aging characteristics of ZnS:Mn electroluminescent films grown by a chemical vapor deposition technique," *J. Appl. Phys.*, 72: 773, 1992.

95

Light-Emitting Diode Displays

Mohammad A. Karim
University of Tennessee, Knoxville

A light-emitting diode (LED) is a particular solid-state *p–n* junction diode that gives out light upon the application of a bias voltage. The luminescence process in this case is electroluminescence, which is associated with emission wavelengths in the visible and infrared regions of the spectrum. When a forward bias is applied to the *p–n* junction diode, carriers are injected into the depletion region in large numbers. Because of their physical proximity, the electron–hole pairs undergo a recombination that is associated with the emission of energy. Depending on the semiconductor band-gap characteristics, this emitted energy can be in the form of heat (as phonons) or light (as photons).

The solution of the Schrödinger equation for a typical crystal reveals the existence of Brillouin zones. A plot between the energy *E* of an electron in a solid and its wave vector **k** represents the allowed energy bands. It may be noted that the lattice structure affects the motion of an electron when *k* is close to $n\pi/l$ (where *n* is any integer and *l* is the crystal periodicity) and the effect of this constraint is to introduce an energy band gap between the allowed energy bands. Figure 95.1a shows portions of two *E* vs. *k* curves for neighboring energy bands within the regions $k = \pi/l$ and $k = -\pi/l$ (also known as the reduced zone).

While the upper band of Fig. 95.1 represents the energy of conduction band electrons, the curvature of the lower band can be associated with electrons having negative effective mass. The concept of negative effective mass can readily be identified with the concept of holes in the valence band. While the majority of the electrons are identified with the minima of the upper *E–k* curve, the majority of the holes are identified with the maxima of the lower *E–k* curve. The minimum value of the conduction band and the maximum value of the valence band in Fig. 95.1a both have identical *k* values. A semiconductor having such a characteristic is said to have a direct band gap, and the associated recombination in such a semiconductor is referred to as direct.

The *direct recombination* of an electron–hole pair always results in the emission of a photon. In a direct band-gap semiconductor, the emitted photon is not associated with any change in momentum (given by $hk/2\pi$) since $\Delta k = 0$. However, for some semiconducting materials, the *E* vs. *k* curve may be somewhat different, as shown in Fig. 95.1b. While the minimum conduction band energy can have a nonzero *k*, the maximum valence band energy can have $k = 0$. The electron–hole recombination in such a semiconductor is referred to as indirect.

An *indirect recombination* process involves a momentum adjustment. Most of the emission energy is thus expended in the form of heat (as phonons). Very little energy is left for the purpose of photon emission, which in most cases is a very slow process. Furthermore, since both photons and phonons are involved in this energy exchange, such transitions are less likely to occur. The interband recombination rate is basically given by

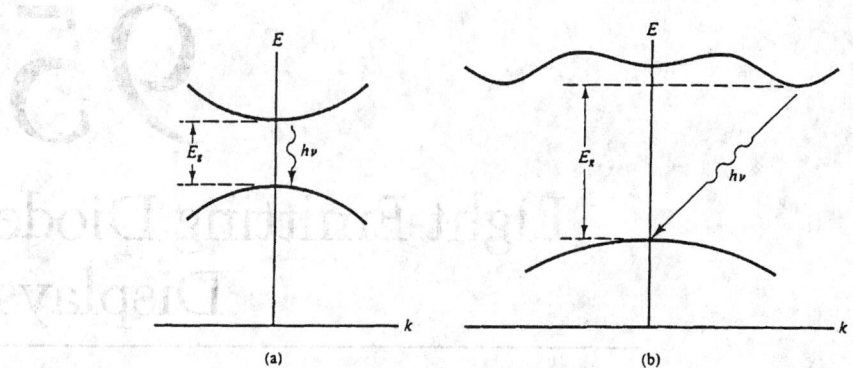

FIGURE 95.1   *E* versus *k* for semiconductors having (a) a direct band gap and (b) an indirect band gap.

$$dn/dt = B_r np \tag{95.1}$$

where B_r is a recombination-dependent constant which for a direct band-gap semiconductor is ~10^6 times larger than that for an indirect band-gap semiconductor. For direct recombination, B_r value ranges from 0.46×10^{-10} to 7.2×10^{-10} cm³/s.

All semiconductor crystal lattices are alike, being dissimilar only in terms of their band characteristics. Si and Ge both have indirect band transitions, whereas GaAs, for example, is a semiconductor that has a direct band transition. Thus, while Si and Ge are preferred for fabrication of transistors and integrated circuits, GaAs is preferred for the fabrication of LEDs.

The direct recombination (when k = constant) results in a photon emission whose wavelength (in micrometers) is given by

$$\lambda = hc/E_g = 1.24/E_g \text{(eV)} \tag{95.2}$$

where E_g is the band-gap energy. The LEDs under proper forward-biased conditions can operate in the ultraviolet, visible, and infrared regions. For the visible region, however, the spectral luminous efficiency curves of Fig. 95.2, which account for the fact that the visual response to any emission is a function of wavelength, should be of concern. It is unfortunate that there is not a single-element semiconductor suitable for fabrication of LEDs, but there are many binary and ternary compounds that can be used for fabrication of LEDs. Table 95.1 lists some of these binary semiconductor materials. The ternary semiconductors include GaAlAs, $CdGeP_2$, and $ZnGeP_2$ for infrared region operation, $CuGaS_2$ and $AgInS_2$ for visible region operation, and $CuAlS_2$ for ultraviolet region operation. Ternary semiconductors are used because their energy gaps can be tuned to a desired emission wavelength by picking appropriate composition.

Of the ternary compounds, gallium arsenide–phosphide (written as $GaAs_{1-x}P_x$) is an example that is basically a combination of two binary semiconductors, namely, GaAs and GaP. The corresponding band-gap energy of the semiconductor can be varied by changing the value of x. For example, when $x = 0$, E_g = 1.43 eV. E_g increases with increasing x until $x = 0.44$ and E_g = 1.977 eV, as shown in Fig. 95.3. However for $x \geq 0.45$, the band gap is indirect. The most common composition of $GaAs_{1-x}P_x$ used in LEDs has x = 0.4 and $E_g \approx 1.3$ eV. This band-gap energy corresponds to an emission of red light. Calculators and watches often use this particular composition of $GaAs_{1-x}P_x$.

Interestingly, the indirect band gap of $GaAs_{1-x}P_x$ (with $1 \geq x \geq 0.45$) can be used to output light ranging from yellow through green provided the semiconductor is doped with impurities such as nitrogen. The dopants introduced in the semiconductor replace phosphorus atoms which, in turn, introduce electron trap levels very near the conduction band. For example, $x = 0.5$, the doping of nitrogen increases the LED efficiency form 0.01 to 1%, as shown in Fig. 95.4. It must be noted, however, that nitrogen doping

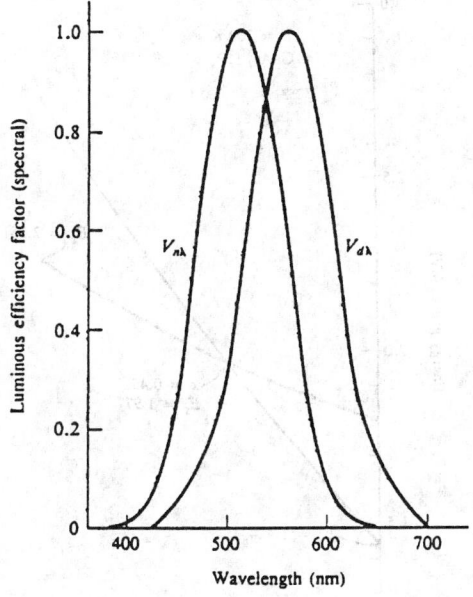

FIGURE 95.2 Spectral luminous efficiency curves. The photopic curve $V_{d\lambda}$ corresponds to the daylight-adapted case while the scotopic curve $V_{n\lambda}$ corresponds to the night-adapted case. (From Boyd, R.W., *Radiometry and the Detection of Optical Radiation*, Copyright ©1983. Reprinted by permission of John Wiley & Sons, Inc., New York.)

TABLE 95.1 Binary Semiconductors Suitable for LED Fabrication

	Material	E_g(eV)	Emission Type
III–V	GaN	3.5	UV
II–VI	ZnS	3.8	UV
II–VI	SnO₂	3.5	UV
II–VI	ZnO	3.2	UV
III–VII	CuCl	3.1	UV
II–VI	BeTe	2.8	UV
III–VII	CuBr	2.9	UV — visible
II–VI	ZnSe	2.7	Visible
III–VI	In₂O₃	2.7	Visible
II–VI	CdS	2.52	Visible
II–VI	ZnTe	2.3	Visible
III–V	GaAs	1.45	IR
II–VI	CdSe	1.75	IR — Visible
II–VI	CdTe	1.5	IR
III–VI	GaSe	2.1	Visible

shifts the peak emission wavelength toward the red. The shift is comparatively larger at and around $x = 0.05$ than $x = 1.0$. The energy emission in nitrogen-doped $GaAs_{1-x}P_x$ devices is a function of both x and the nitrogen concentration.

Nitrogen is a different type of impurity from those commonly encountered in extrinsic semiconductors. Nitrogen, like arsenic and phosphorus, has five valence electrons, but it introduces no net charge carriers in the lattice. It provides active radiative recombination centers in the indirect band-gap materials. For an electron, a recombination center is an empty state in the band gap into which an electron falls and, then, thereafter, falls into the valence band by recombining with a hole. For example, while a GaP LED emits green light (2.23 eV), a nitrogen-doped GaP LED emits yellowish green light (2.19 eV), and a heavily nitrogen-doped GaP LED emits yellow light (2.1 eV).

The injected excess carriers in a semiconductor may recombine either radiatively or nonradiatively. Whereas nonradiative recombination generates phonons, radiative recombination produces photons.

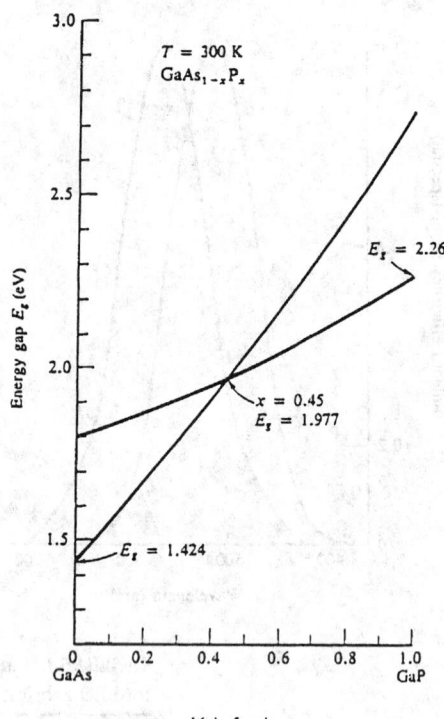

FIGURE 95.3 Band-gap energy versus x in GaAs$_{1-x}$P$_x$. (From Casey, H.J., Jr. and Parish, M.B., Eds., *Heterostructure Lasers*, Academic Press, New York, 1978. With permission.)

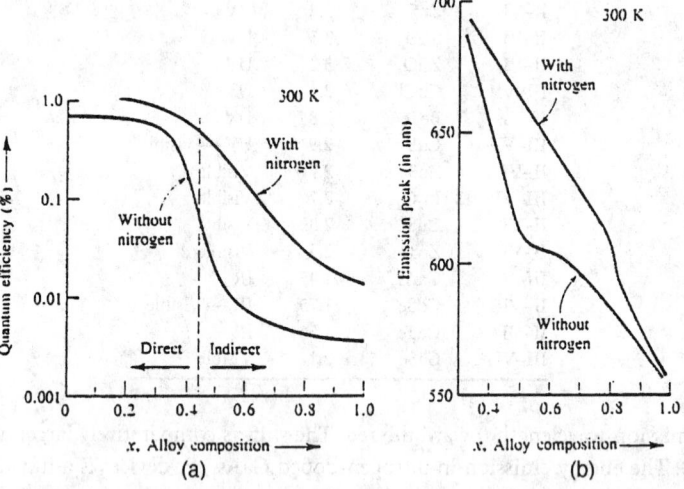

FIGURE 95.4 The effects of nitrogen doping in GaAs$_{1-x}$P$_x$: (a) quantum efficiency vs. x and (b) peak emission wavelength vs. x.

Consequently, the internal quantum efficiency η, defined as the ratio of the radiative recombination rate R_r to the total recombination rate, is given by

$$\eta = R_r / \left(R_r + R_{nr} \right) \qquad (95.3)$$

where R_{nr} is the nonradiative recombination rate. However, the injected excess carrier densities return to their value exponentially as

$$\Delta p = \Delta n = \Delta n_0 e^{-t/\tau} \tag{95.4}$$

where τ is the carrier lifetime and Δn_0 is the excess electron density at equilibrium. Since $\Delta n/R_r$ and $\Delta n/R_{nr}$ are, respectively, equivalent to the radiative recombination lifetime τ_r and the nonradiative recombination lifetime τ_{nr}, we can obtain the effective minority carrier bulk recombination time τ as

$$(1/\tau) = (1/\tau_r) + (1/\tau_{nr}) \tag{95.5}$$

such that $\eta = \tau/\tau_r$. The reason that a fast recombination time is crucial is that the longer the carrier remains in an excited state, the larger the probability that it will give out energy nonradiatively. In order for the internal quantum efficiency to be high, the radiative lifetime τ_r needs to be small. For indirect band-gap semiconductors, $\tau_r \gg \tau_{nr}$ so that very little light is generated, and for direct band-gap semiconductors, τ_r increases with temperature so that the internal quantum efficiency deteriorates with the temperature.

As long as the LEDs are used as display devices, it is not too important to have fast response characteristics. However, LEDs are also used for the purpose of optical communications, and for those applications it is appropriate to study their time response characteristics. For example, an LED can be used in conjunction with a photodetector for transmitting optical information between two points. The LED light output can be modulated to convey optical information by varying the diode current. Most often, the transmission of optical signals is facilitated by introducing an optical fiber between the LED and the photodetector.

There can be two different types of capacitances in diodes that can influence the behavior of the minority carriers. One of these is the *junction capacitance*, which is caused by the variation of majority charge in the depletion layer. While it is inversely proportional to the square root of bias voltage in the case of an abrupt junction, it is inversely proportional to the cube root of bias voltage in the case of a linearly graded junction. The second type of capacitance, known as the *diffusion capacitance*, is caused by the minority carriers.

Consider an LED that is forward biased with a dc voltage. Consider further that the bias is perturbed by a small sinusoidal signal. When the bias is withdrawn or reduced, charge begins to diffuse from the junction as a result of recombination until an equilibrium condition is achieved. Consequently, as a response to the signal voltage, the minority carrier distribution contributes to a signal current.

Consider a one-dimensional p-type semiconducting material of cross-sectional area A whose excess minority carrier density is given by

$$\delta \Delta n_p / \delta t = D_n \delta^2 \Delta n_p / \delta x^2 - \Delta n_p / \tau \tag{95.6}$$

As a direct consequence of the applied sinusoidal signal, the excess electron distribution fluctuates about its dc value. In fact, we may assume excess minority carrier density to have a time-varying component as described by

$$\Delta n_p(x,t) = \langle \Delta n_p(x) \rangle + n'_p(x) e^{j\omega t} \tag{95.7}$$

where $\langle \Delta n_p(x) \rangle$ is a time-invariant quantity. By introducing Eq. 95.7 into Eq. 95.6, we get two separate differential equations:

$$\delta^2/\delta x^2 \left(\langle \Delta n_p(x) \rangle \right) = \langle \Delta n_p(x) \rangle / (L_n)^2 \tag{95.8a}$$

and

$$\delta^2/\delta x^2\left[\Delta n_p'(x)\right]=\Delta n_p'(x)\big/\left[L_n{}^*\right]^2 \tag{95.8b}$$

where

$$L_n^*=L_n\big/\left(1+j\omega\tau\right)^{1/2} \tag{95.9a}$$

and

$$L_n=\left(D_n\tau\right)^{1/2} \tag{95.9b}$$

The dc solution of Eq. 95.8a is well known. Again, the form of Eq. 95.8b is similar to that of Eq. 95.8a and, therefore, its solution is given by

$$\Delta n_p'(x)=\Delta n_p'(0)e^{-x/L} \tag{95.10}$$

Since the frequency-dependent current $I(\omega)$ is simply a product of eAD_n and the concentration gradient, we find that

$$I(\omega)=\left|eAD_n\,dn_p'(x)\big/dx\right|_{x=0}$$
$$=I(0)\big/\left(1+\omega^2\tau^2\right)^{1/2} \tag{95.11}$$

where $I(0)$ is the intensity emitted at zero modulation frequency. We can determine the admittance next by dividing the current by the perturbing voltage. The real part of the admittance, in this case, will be equivalent to the diode conductance, whereas its imaginary part will correspond to the diffusion capacitive susceptance.

The modulation response as given by Eq. 95.11 is, however, limited by the carrier recombination time. Often an LED is characterized by its modulation bandwidth, which is defined as the frequency band over which signal power (proportional to $I^2(w)$) is half of that at $\omega=0$. Using Eq. 95.11, the 3-dB modulation bandwidth is given by

$$\Delta\omega\approx 1/\tau_r \tag{95.12}$$

where the bulk lifetime has been approximated by the radiative lifetime. Some times the 3-dB bandwidth of the LED is given by $I(\omega)=\frac{1}{2}I(0)$, but this simplification contributes to an erroneous increase in the bandwidth by a factor of 1.732.

Under conditions of thermal equilibrium, the recombination rate is proportional to the product of initial carrier concentrations, n_0 and p_0. Then, under nonequilibrium conditions, additional carriers $\Delta n=\Delta p$ are injected into the material. Consequently, the recombination rate of injected excess carrier densities is given by initial carrier concentrations and injected carrier densities as

$$R_{\Delta r}=\left[B_r\left(n_o+\Delta n\right)\left(p_o+\Delta p\right)-B_r n_o p_o\right]$$
$$=B_r\left(n_o+p_o+\Delta n\right)\Delta n \tag{95.13}$$

where B_r is the same constant introduced in Eq. 95.1. For p-type GaAs, for example, $B_r = 1.7 \times 10^{-10}$ cm³/s when $p_0 = 2.4 \times 10^{18}$ holes/cm³. Equation 95.13 is used to define the radiative carrier recombination lifetime by

$$\tau_r = \Delta n \Big/ R_{\Delta r} = \left[B_r \left(n_o + p_o + \Delta n \right) \right]^{-1} \tag{95.14}$$

In the steady-state condition, the excess carrier density can be calculated in terms of the active region width d by

$$\Delta n = J \tau_r \big/ ed \tag{95.15}$$

where J is the injection current density.

The radiative recombination lifetime is found by solving Eq. 95.14 after having eliminated Δn from it using Eq. 95.15:

$$\tau_r = \left[\left\{ \left(n_o + p_o \right)^2 + \left(4J / B_r ed \right) \right\}^{1/2} - \left(n_o + p_o \right) \right] \Big/ \left(2J / ed \right) \tag{95.16}$$

Thus, while for the low carrier injection (i.e., $n_o + p_o \gg \Delta n$), Eq. 95.16 reduces to

$$\tau_r \approx \left[B_r \left(n_o + p_o \right) \right]^{1/2} \tag{95.17a}$$

for the high carrier injection (i.e., $n_o + p_o \ll \Delta n$), it reduces to

$$\tau_r \approx \left(ed / JB_r \right)^{1/2} \tag{95.17b}$$

Equation 95.17a indicates that in highly doped semiconductors, τ_r is small. But the doping process has its own problem, since in many of the binary LED compounds higher doping may introduce nonradiative traps just below the conduction band, thus nullifying Eq. 95.12. In comparison to Eq. 95.17a, Eq. 95.17b provides a better alternative whereby τ_r can be reduced by decreasing the active region width or by increasing the current density. For the case of p-type GaAs, the radiative lifetimes vary between 2.6 and 0.35 ns, respectively, when p_0 varies between 1.0×10^{18} holes/cm³ and 1.5×10^{19} holes/cm³.

Usually, LEDs are operated at low current (≈ 10 mA) and low voltages (≈ 1.5 V), and they can be switched on and off in the order of 10 ns. In addition, because of their small sizes, they can be reasonably treated as point sources. It is, therefore, not surprising that they are highly preferred over other light sources for applications in fiber-optic data links.

Two particular LED designs are popular: *surface emitters* and *edge emitters*. They are shown in Fig. 95.5. In the former, the direction of major emission is normal to the plane of the active region, whereas in the latter the direction of major emission is in the plane of the active region. The emission pattern of the surface emitters is very much isotropic, whereas that of the edge emitters is highly directional.

As the LED light originating from a medium of refractive index n_1 goes to another medium of refractive index $n_2 (n_2 < n_1)$, only a portion of incident light is transmitted. In particular, the portion of the emitted light corresponds to only that which originates from within a cone of semiapex angle θ_c, such that

$$\theta_c = \sin^{-1} \left(n_2 / n_1 \right) \tag{95.18}$$

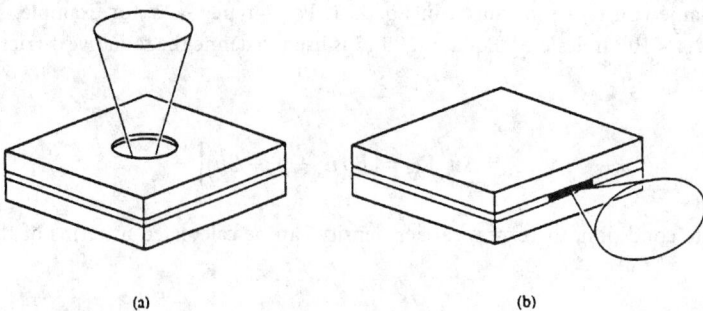

(a)   (b)

FIGURE 95.5   LED type: (a) surface emitter and (b) edge emitter.

In the case of an LED, n_1 corresponds to the refractive index of the LED medium and n_2 corresponds to that of air (or vacuum). Light originating from *beyond* angle θ_c undergoes a total internal reflection. However, the light directed from *within* the cone of the semiapex angle θ_c will be subjected to Fresnels loss. Thus, the overall transmittance T is given by

$$T = 1 - \left\{ (n_1 - n_2)/(n_1 + n_2) \right\}^2 \tag{95.19}$$

Accordingly, the total electrical-to-optical conversion efficiency in LEDs is given by

$$\eta_{\text{LED}} = T\left[(\text{solid angle within the cone})/(4\pi) \right]$$

$$= (T/2)(1 - \cos\theta_c)$$

$$= (T/4)\sin^2\theta_c \tag{95.20}$$

$$= (1/4)(n_2/n_1)^2 \left[1 - \left\{ (n_1 - n_2)/(n_1 + n_2) \right\}^2 \right]$$

Only two schemes increase the electrical-to-optical conversion efficiency in an LED. The first technique involves guaranteeing that most of the incident rays strike the glass-to-air interface at angles less than θ_c. It is accomplished by making the semiconductor–air interface hemispherical. The second method involves schemes whereby the LED is encapsulated in an almost transparent medium of high refractive index. The latter means is comparatively less expensive. If a glass of refractive index 1.5 is used for encapsulation, the LED efficiency can be increased by a factor of 3. Two of the possible encapsulation arrangements and the corresponding radiation patterns are illustrated in Fig. 95.6.

LEDs are often used in conjunction with a phototransistor to function as an optocoupler. The optocouplers are used in circumstances when it is desirable to have a transmission of signals between electrically isolated circuits. They are used to achieve noise separation by eliminating the necessity of having a common ground between the two systems. Depending on the type of coupling material, these miniature devices can provide both noise isolation as well as high voltage isolation. Figure 95.7 shows a typical case where two optocouplers are used to attain a chopper circuit. The two optocouplers chop either the positive or the negative portion of the input signals with a frequency of one half that of the control signal that is introduced at the T flip-flop. The operational amplifier provides an amplified version of the chopped output waveform. In comparison, a chopper circuit that uses simple bipolar transistors produces noise spikes in the output because of its inherent capacitive coupling.

The visible LEDs are best known for their uses in displays and indicator lamps. In applications where more than a single source of light is required, an LED array can be utilized. An LED array is a device

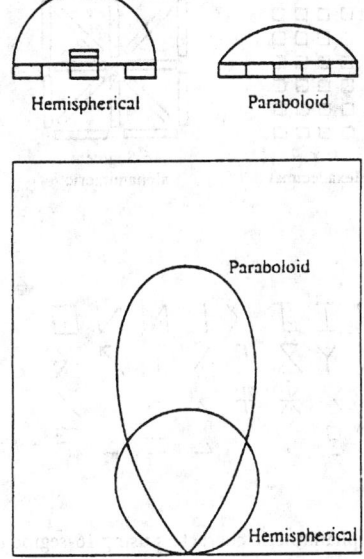

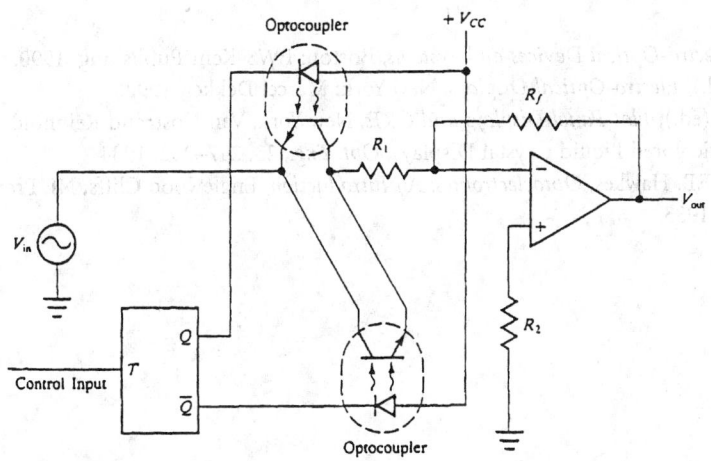

FIGURE 95.6 LED encapsulation geometries and their radiation patterns.

FIGURE 95.7 A chopping circuit with an amplifier.

consisting of a row of discrete LEDs connected together within or without a common reflector cavity. Figure 95.8a shows different LED arrangements for displaying hexadecimal numeric and alphanumeric characters, whereas Fig. 95.8b shows, for example, the possible alphanumeric characters using 16-segment displays. In digital systems, the binary codes equivalent to these characters are usually decoded and, consequently, a specific combination of LED segments are turned on to display the desired alphanumeric character.

The dot matrix display provides the most desirable display font. It gives more flexibility in shaping characters and has a lower probability of being misinterpreted in case of a display failure. However, these displays involve a large number of wires and increased circuit complexity. LED displays, in general, have an excellent viewing angle, high resonance speed ($\approx$10 ns), long life, and superior interface capability with electronics with almost no duty cycle limitation. LEDs with blue emission are not available commercially. When compared with passive displays, LED displays consume more power and involve complicated wiring with at least one wire per display element.

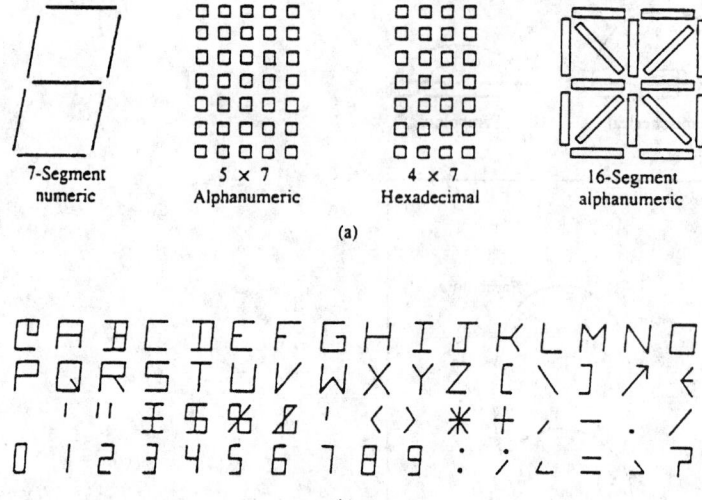

| 7-Segment numeric | 5 × 7 Alphanumeric | 4 × 7 Hexadecimal | 16-Segment alphanumeric |

(a)

(b)

FIGURE 95.8 (a) LED display formats; and (b) displayed alphanumeric characters using 16-segment displays.

References

M.A. Karim, *Electro-Optical Devices and Systems*, Boston: PWS-Kent Publishing, 1990.

M.A. Karim (ed.), *Electro-Optical Displays*, New York: Marcel Dekker, 1992.

L.E. Tannas, Jr. (ed.), *Flat-Panel Displays and CRTs*, New York: Van Nostrand Reinhold, 1985.

T. Uchida, Multicolored Liquid Crystal Displays, *Opt. Eng.*, 23, 247–252, 1984.

J. Wilson and J.F.B. Hawkes, *Optoelectronics: An Introduction*, Englewood Cliffs, NJ: Prentice-Hall International, 1985.

96

Reading/Recording Devices

Herman Vermariën
Vrije Universiteit Brussel

Edward McConnell
National Instruments

Yufeng Li
Samsung Information Systems America

96.1 Graphic Recorders

Herman Vermariën

A **graphic recorder** is essentially a measuring apparatus that is able to produce in real time a hard copy of a set of time functions with the purpose of immediate and/or later visual inspection. The curves are mostly drawn on a (long) strip of paper (from a roll or Z-fold); as such, the instrument is indicated as a strip chart recorder. The independent variable time (t) then corresponds to the strip length axis and the physical variables measured are related to the chart width. Tracings are obtained by a writing process at sites on the chart short axis (y) corresponding to the physical variables magnitudes with the strip being moved at constant velocity to generate the time axis. Graphs cannot be interpreted if essential information is absent; scales and reference levels for each physical variable recorded and for time are all a necessity. Additional information concerning the experimental conditions of the recording is also necessary and is preferably printed by the apparatus (data, investigated item, type of experiment, etc.). The capacity of the graphic recorder is thus determined by its measuring accuracy, its ability to report additional information and its graphical quality, including the sharpness of tracings, the discriminability of tracings (e.g., by different colors), and the stability of quality with respect to long-term storage. Simple chart recorders only produce tracings on calibrated paper; more-advanced graphic recorders generate tracings and calibration lines, display additional information in alphanumeric form on charts, store instrument settings and recorded data in memory (which can be reproduced on charts in diverse modes), have a built in waveform monitor screen, and can communicate with a PC via standard serial interfacing. The borderlines between these types of intelligent graphic recorders and, on the one hand, digital storage oscilloscopes equipped with a hard copy unit and, on the other hand, PC-based data acquisition systems with a laser printer or a plotter, become very unclear. The property of producing the hard copy in real time is probably the most discriminating factor between the graphic recorder and other measuring systems that produce hard copies. Graphic recorders are used for test and measurement applications in laboratory and field conditions and for industrial process monitoring and control. Graphic recorders are intensively used in biomedical measurement applications [1].

Whereas the time axis is generated by moving the chart at constant velocity, the ordinate can be marked in an analog or a digital manner. Analog recorder transducers generate a physical displacement of the writing device, e.g., a pen or a printhead. With digital transducers, moving parts are absent and the writing device is a stationary rectilinear array of equidistant writing points covering the complete width of the chart; the writing act then consists in activating the point situated at the site corresponding to the signal magnitude and putting a dot on the paper. Analog recorders thus can produce continuous lines, whereas digital recorders generate dotted lines. If ordinate and time axis resolutions are sufficient, digital recordings have excellent graphic quality regarding visual impression of continuity. Analog transducers can be used in a discontinuous mode and thus handle a set of slowly varying signals; during the scanning cycle a dot is set on the paper by the moving writing device at the sites corresponding to the magnitudes of the signal. A single digital transducer and a single analog transducer applied in the scanning mode can handle a set of signals; a single analog transducer can process only one signal in the continuous mode. For a digital transducer the number of signals recorded is essentially unlimited; it is thus programmed to draw the necessary calibration lines. With analog transducers calibrated paper is used. In this case, generally ink writing is applied and different colors ensure excellent tracing identifiability. With the digital array, dot printing can be more or less intensified for different channels or more than one adjacent points can be activated, resulting in more or less increased line blackness and thickness. However, tracing identification is usually performed by alphanumeric annotations.

In **analog graphic recorders,** the transducer can be designed in a direct mode or in a servo mode. In the direct mode the signal magnitude is directly transduced to a position of the writing device (e.g., the simple galvanometric type). In the servo mode (also called *feedback* or *compensation*) the position of the writing device is measured by a specific sensor system and the difference between the measured value and the value of the signal to be traced is applied to the transducer motor, resulting in a movement tending to zero the difference value and thus to correct the position [2,3]. In both methods the moving parts set a limit to the system frequency bandwidth. Moreover, in the feedback mode velocity and acceleration limitations may be present; thus linear system theory description of the apparatus behavior with respect to signal frequency may not be applicable. As such, the bandwidths of servo systems can be dependent on the writing width. Movement of the writing device can be generated by a rotation or by a translation. In the latter case the writing part is mechanically guided; primarily, the servo method is applied [3]. A rotation is generally obtained with a galvanometric motortype [3,4]; the galvanometer may rotate a pen, an ink jet, a light beam. The inertia of the moving part is the major parameter determining the bandwidth of the system. Translational devices allow a bandwidth of a few hertz. Higher bandwidths can be obtained with galvanometric pen types (about 100 Hz), ink jets (up to 1 kHz), and optical systems (up to 10 kHz) [1], but, being replaced by dot array recorders or data acquisition systems, these types are disappearing from the market. Major reasons are inherent errors and limitations of these analog types, maintenance needs of moving parts and ink devices, cost of photographic paper, and the lack of the possibilities of digital types.

Whereas moving parts restrict the analog recorder bandwidth, a corresponding capacity of **digital graphic recorders** is determined by the sampling frequency and the writing frequency. According to the sampling criterion the sampling frequency should be twice the highest signal frequency. This implies two samples to display a complete sine wave period, which can hardly be called a good graphic representation. Ten samples may be a minimum. The sampling frequency is a pure electronic matter; the maximal writing frequency of the dot array is the limiting factor in real time. Alternatively, if the signal spectrum exceeds the real-time bandwidth of the recorder, data can be stored at a sufficient sampling rate in memory and reproduced off-line at a slower rate which can be handled by the apparatus. Most digital recorders have this memory facility; some recorders are specifically referred to as "memory" recorders when their off-line capabilities largely exceed their online performance.

Real-time recording is primarily performed as a function of time (t–y recorders). On the other hand, x–y recording is another way of representing the data. In this case the relation between two physical variables is traced and the independent variable time is excluded (apart from the fact that, if a dashed line is used, each dash can represent a fixed time interval). In standard analog x–y recorders the chart is stationary (e.g., electrostatically fixed to the writing table); two similar analog writing transducers with identical input circuitry are assembled in the recorder. The first transducer (y) translates a long arm covering the width of the paper at which the second transducer (x) carrying the pen is moved. Evidently, recorders with memory facilities and appropriate software may produce an x–y recording in off-line mode by setting dots while the paper progresses. Recording accuracy can be formulated in similar terms as for any measuring instrument [5]. This accuracy is determined, on the one hand, by the input signal conditioning, similar for digital as well as for analog types and, on the other hand, by the accuracy of the recording transducer and its driver electronics. In the digital type, digitization bit accuracy, sampling frequency, dot array resolution, and dot writing frequency are major parameters. In the analog type, typical inconveniences of analog transducer systems can be found (such as static nonlinearity, noise and drift, dead zone and hysteresis, limited dynamic behavior); servo systems are known to be more accurate as compared with direct systems. For example, drift in analog recording can be the result of a small shift of the calibrated chart along the y-axis; the latter is excluded if the recorder draws its own calibration lines.

With respect to graphic quality, clarity and sharpness of the tracings are important (within a large range of writing velocities). Tracing quality depends on the writing velocity, i.e., the velocity of the writing device with respect to the moving paper. Evidently, the flow of writing medium (e.g., ink or heat) should be more or less adapted to this writing velocity to prevent poorly visible tracings at high velocities and thick lines at low velocities. Good identifiability of overlapping curves is essential. Sufficient dot resolution (with adequate interpolation techniques) is important in discontinuous types for easy visual inspection. A graphic recorder can be designed as a single-channel instrument or can have a multichannel input. Inputs can be standard or modular, so that the user can choose the specific signal conditioners for the application and the number of channels. A recorder can be called "general purpose" or can be assembled in a specific measuring apparatus (e.g., in biomedical applications such as electrocardiography and electroencephalography). The recorder can be portable for operation in the field or can be mounted in a laboratory rack or in a control panel. The paper can be moved vertically or horizontally on a writing table ("flat bed" recorder) and can be supplied from a roll or in Z-fold. Besides strip chart recorders and x–y recorders, circular chart recorders exist. In this case the chart rotates, one rotation corresponds to a complete measurement interval, and the chart is provided with appropriate calibration lines adapted to movement of the pens writing on it.

Apart from the low-bandwidth translational pen devices there is a decreasing interest in analog graphic recorders. They are being replaced by thermal dot array recorders or by data acquisition systems. Nevertheless, as some of them may still be manufactured and a number of apparatus may still be in use, different techniques are mentioned. A description of analog recorder principles and performances can be found in Reference 1. Galvanometric recorders apply rotational transducers. The direct as well as the servo principles are used. The direct type makes use of the d'Arsonval movement as applied in ordinary galvanometers [1, 2, 4]. Dynamically, the galvanometer acts as a mechanical resonant system and the bandwidth is thus determined by its resonant frequency, the latter being dependent on the inertia of the moving parts. Evidently, rotation gives rise to inherent errors in graphic recorders. If pen tips (perpendicular to the pen arm, thermal or ink) are used, the rotation of the pen arm fixed to the galvanometer coil occurs in a plane parallel to the chart plane, so the recording is curvilinear instead of rectilinear, introducing an error with respect to the time axis and to the ordinate axis (the ordinate value being proportional to the tangent of the rotation angle). Calibrated paper with curvilinear coordinate lines may solve this problem; nevertheless, the tracing is deformed and zero offset is critical. Rectilinear recording can be realized with pen systems, ink jets, and light beams. Rectilinear pen recording can be approximated with pen tips in case of "long-arm" pens and by mechanical rectilinearization; alternatively "knife-edge" recording is a solution [1]. In the case of ink jet and light beam recorders the rotation plane and the chart plane do not have to be parallel; writing then occurs at the intersecting line of both planes and is thus essentially rectilinear. In ink jet recording a miniature nozzle through which ink is pumped is mounted in a direction perpendicular to the axis of the galvanometer. In optical recording a sharp light beam is reflected by a small mirror connected to the galvanometric moving coil toward the photo-sensitive paper. In these methods miniaturization of the moving parts gives rise to higher resonant frequencies and thus higher bandwidths. Whereas a typical bandwidth for a galvanometric pen system is 100 Hz, the bandwidth for an ink jet system can be 1000 Hz and for an optical system 10 kHz may be reached. In the fiber-optic cathode ray tube (FO-CRT) no mechanical moving parts are present and thus there are no mechanical limits on bandwidth. The FO-CRT is essentially a one-dimensional CRT. A phosphor layer at the inside of the screen converts the cathode ray into ultraviolet (UV) light. This UV light is guided by an outer faceplate composed of glass fibers onto the photosensitive chart. As in ordinary oscilloscopes, the deflection of the spot is directly proportional to the signal applied at the input of the deflection unit. The bandwidth is determined by the driving electronics. The system can be used in scanning mode as the beam intensity is easily controlled. In the following paragraphs further details will be given on translational pen recorders and thermal dot array recorders.

Translational Pen Recorders

In **translational pen recorders** the writing device is usually a fiber-tip pen with an ink cartridge. In discontinuous applications the writing device can be a printhead with different color styli or with a colored ribbon. A manual or automatic pen lift facility is included. During recording, the writing device is translated along the y-axis as it is linked to a mechanical guidance and a closed-loop wire system. A motor and wheels system pulls the wire and thus the writing device. In some designs a motor and screw system is applied. Translational recorders are primarily designed as a servo type. The position of the pen is accurately measured, and the difference voltage between the input signal and the position magnitude (following appropriate amplification and conditioning) drives the servomotor. Servo motors can be dc or stepper types; servo electronics can be analog or digital. Position sensing can be potentiometric ("potentiometric" recorders): the pen carriage is equipped with a sliding contact on the resistor (wire wound or thick film) which covers the complete width of the paper. More recently developed methods use optical or ultrasonic principles for position sensing; with these methods contacts are absent resulting in less maintenance and longer lifetime. For example, in the ultrasonic method the pen position is sensed by a detector coil from the propagation time of an ultrasound pulse, which is imparted by a piezoelectric transducer to a magnetostrictive strip covering the chart width. Accordingly, brushless dc-motors are used in some apparatuses. In the servo system accuracy is determined for the larger part by the quality of the sensing system. A poor contact with the resistor can give rise to noise; there may be a mechanical backlash between pen tip and the sliding contact on the potentiometer. The velocity of the pen carriage is limited, about 0.5 to 2 m/s dependent on motor and mechanics design. This results in a bandwidth of the recorder depending on the amplitude of the tracing: the −3 dB frequency fits in the range from 1 to 5 Hz for a full-scale width of 200 to 250 mm. Alternatively, the pen response time to a full-scale step input is given (5 to 95% of full-scale tracing): 0.1 to 0.5 s. Overshoot of the pen step response is extremely small in accurate designs.

   In most pen recorders each tracing can cover the complete width. As such, pens must have the possibility to pass each other resulting in a small shift between adjacent pens (a few millimeters) along the time axis. In some apparatus, tracings can be synchronized with a POC-system ("pen offset compensation"); signals are digitized, stored in memory, and reproduced after a time delay correcting for the physical displacement of the pen. If immediate visual inspection is required, applying POC can be inconvenient as a consequence of this time delay. In process monitoring, slowly varying signals such as temperature, pressure, flow, etc. are followed. These signals can be handled by a single transducer in a discontinuous way; all input signals are scanned during the scanning cycle and for each signal a dot is

TABLE 96.1 Pen Recorders

Designation	Description	Manufacturer	Approximate Price (U.S.$)
LR8100	Test, meas.; 4,6,8 c. ch.; POC; printer; display; memory; analysis; alarm; interface	Yokogawa E. C.	11,500 (8 ch.)
LR102	Test, meas.; 1,2 c. ch.	Yokogawa E. C.	1,800 (2 ch.)
LR122	Test, meas.; 1,2 c. ch.; x–y; alarm; interface	Yokogawa E. C.	2,300 (2 ch.)
MC1000	Test, meas.; 4,6,8,12 c. ch.; POC; printer; waveform display; analysis; alarm; interface	Graphtec C.	20,200 (12 ch.)
BD112	Test, meas.; 2 c. ch.; POC	Kipp Z.	2,700 (2 ch.)
BD200	Test, meas.; 4,6,8 c. ch.; POC; display; x–y; alarm; interface	Kipp Z.	12,200 (8 ch.)
L250	Test, meas.; 1,2 c. ch.; POC	Linseis	2,000 (2 ch.)
L2066	Test, meas.; 1 to 6 c. ch.; POC; x–y; interface	Linseis	8,100 (6 ch.)
MCR560	Test, meas.; 2,4,6 c. ch.	W+W	3,800 (2 ch.)
DCR540	Test, meas.; 1 to 4 c. ch.; POC; display; x–y; interface	W+W	6,600 (4 ch.)
PCR500SP	Test, meas.; 2,4,6,8 c. ch.; POC; display; x–y; analysis; alarm; interface; transient option	W+W	15,900 (8 ch.)
Omega640	Test, meas.; 1,2,3 c. ch.	Omega	4,300 (3 ch.)
Omega600A	Test, meas.; 1 to 6 c. ch.; POC; printer; x–y; memory; interface	Omega	23,700 (6 ch.)
μR1000	Process mon.; 1,2,3,4 c. ch., 6 s. ch.; POC; printer; display; analysis; alarm; interface	Yokogawa E. C.	4,000 (4 c. ch.)
μR1800	Process mon.; 1,2,3,4 c. ch., 6,12,18,24 s. ch.; POC; printer; display; analysis; alarm; interface	Yokogawa E. C.	5,600 (4 c. ch.)
DR240	Process mon.; 30 s. ch.; printer; display; analysis; alarm; interface	Yokogawa E. C.	6,500 (30 s. ch.)
RL100	Process mon.; 1,2 c. ch., 6 s. ch.; printer; alarm	Honeywell	1,500 (2 c. ch.)
DPR100C/D	Process mon.; 1,2,3 c. ch., 6 s. ch.; POC; printer; display; alarm; analysis; interface	Honeywell	3,000 (6 s. ch.)
DPR3000	Process mon.; 4 to 32 s. ch.; printer; display; alarm; analysis; interface	Honeywell	7,700 (32 s. ch.)
4101	Process mon.; 1 to 4 c. ch., 6 s. ch.; POC; printer; alarm	Eurotherm	2,400 (4 c. ch.)
4180 G	Process mon.; 8,16,24,32 s. ch.; printer; waveform display; alarm; analysis; interface	Eurotherm	9,800 (32 s. ch.)
Sirec L	Process mon.; 1,2,3 c. ch.	Siemens	1,200 (3 c. ch.)

Note: c. ch. = continuous channel; s. ch. = scanned channel.

printed ("multipoint" recorder). The minimum scanning time is dependent on the moving writing device. For chart progression, dc and stepper motors are used. Calibrated paper is pulled by sprocket wheels seizing in equidistant perforations at both sides of the chart. Translational pen recorders range from simple purely analog design to intelligent microprocessor-controlled types handling a large number of channels with a broad range of control and monitor facilities (e.g., printing of a report after alarm).

Table 96.1 displays a set of translational pen recorders; some of them are equipped with a printhead. Under "Description" the major application is given: test and measurement or process monitoring. Furthermore the following are indicated: the number of continuous (c. ch.) and scanned (s. ch.) channels; the availability of POC, a printer (for additional information or for trace printing), a display for alphanumeric information (such as calibration values for each channel), or even a waveform display, data memory (allowing memory recorder functioning), x–y recording facility, alarm generation (after reaching thresholds of recorded variables), and standard serial interface options allowing communication with a PC (introduction of recorder settings, storage, and processing of recorded data, etc.). Table 96.2 gives a summary of pen recorder specifications (multipoint types also included).

Thermal Dot Array Recorders

In **thermal dot array recorders**, apart from the chart-pulling system, no moving parts are present; the writing transducer is essentially a rectilinear array of equidistal writing points which covers the total width of the paper. Although some apparatuses apply an electrostatic method [1], the thermal dot array

TABLE 96.2 Pen Recorder Specifications

Type (Test, Measurement)	Recording Width (mm)	Chart Velocity max. (mm/min)	Chart Velocity min. (mm/h)	Pen Velocity, max. (m/s)	Pen Step Response time(s)	Bandwidth (−3 dB) (Hz)	Number of Continuous Channels, max.
LR8100	250	1200	10	1.6		5	8
LR102	200	600	10	0.4	0.5	1.5	2
LR122	200	600	10	0.4	0.5	1.5	2
MC1000	250	1200	7.5	1.6			12
BD112	200	1200	6		0.2		2
BD200	250	1200	5		0.25		8
L250	250	1200	6	1	0.12	3.6	2
L2066	250	3000	1	1	0.3	2	6
MCR560	250	600	10		0.3	1.5	6
DCR540	250	600	10		0.3	1.5	4
PCR500SP	250	1200	10	2	0.15	4.5	8
Omega640	250	600	30	0.5			3
Omega600A	250	600	10	0.1			6

Type (Process Monitor)	Recording Width (mm)	Chart Velocity max. (mm/min)	Chart Velocity min. (mm/h)	Pen Step Response Time (s)	Printhead Scanning Cycle (s)	Number of Channels (max.)[a]
μR1000	100	200	5	1	10	4c, 6s
μR1800	180	200	5	1.5	10 (6s)	4c, 24s
DR240	250	25	1	—	2	30s
RL100	100	8	10	3.2	5	2c, 6s
DPR100C/D	100	100	1	1	0.6	3c, 6s
DPR3000	250	25	1	—	5	32s
4101	100	25	1	2	5	4c, 6s
4180 G	180	25	1	—	3	32s
Sirec L	100	20	1	—		3c

[a] c = continuous; s = scanned.

and thermosensitive paper are generally used. In this array the writing styli consist of miniature electrically heated resistances; thermal properties of the resistances (in close contact with the chart paper) and the electric activating pulse form determine the maximal writing frequency. The latter ranges in real-time recorders from 1 to 6.4 kHz. Heating of the thermosensitive paper results in a black dot with good long-term stability. The heating pulse is controlled in relation to the chart velocity in order to obtain sufficient blackness at high velocities. Tracing blackness or line thickness is seldom used for curve identification; alphanumeric annotation is mostly applied. With the dot array a theoretically unlimited number of waveforms can be processed; the apparatus is thus programmed to draw its own calibration lines. Different types of grid patterns can be selected by the user. Moreover, alphanumeric information can be printed for indicating experimental conditions.

Ordinate axis resolution is determined by the dot array: primarily, 8 dots/mm; exceptionally, 12 dots/mm (as in standard laser printers). The resolution along the abscissa depends on the thermal array limitations and programming. Generally, a higher resolution is used (mostly 32 dots/mm, maximally 64 dots/mm) except for the highest chart velocities (100, 200, 500 mm/s). At these high velocities and consequently short chart contact times, dots become less sharp and less black. Most of the dot array instruments are intended for high-signal-frequency applications: per channel sampling frequencies of 100, 200, and even 500 kHz are used in real time. These sampling frequencies largely exceed the writing frequencies; during the writing cycle, data are stored in memory and for each channel within each writing interval a dotted vertical line is printed between the minimal and the maximal value. For example, a sine wave with a frequency largely exceeding the writing frequency is represented as a black band with a width

TABLE 96.3 Thermal Dot Array Recorders

Designation	Description	Manufacturer	Approximate Price (U.S.$)
WR 5000	8 a. ch.; memory	Graphtec C.	17,100 (8 ch.)
WR 9000	4,8,16 a. ch.; monitor; memory; *x–y*; analysis; FFT	Graphtec C.	11,300 (4 ch.)
Mark 12	4 to 52 a. ch., 4 to 52 d. ch.; monitor; memory	W. Graphtec	30,300 (16 a. ch.)
MA 6000	2 to 16 a. ch.; monitor; memory; *x–y*; analysis; FFT	Graphtec C.	23,500 (8 ch.)
ORP 1200	4,8 a. ch., 16 d. ch.; monitor; memory; *x–y*	Yokogawa E. C.	11,600 (8 a. ch.)
ORP 1300	16 a. ch., 16 d. ch.; monitor; memory; *x–y*	Yokogawa E. C.	18,000 (16 a. ch.)
ORM 1200	4,8 a. ch., 16 d. ch.; monitor; memory; *x–y*	Yokogawa E. C.	14,100 (8 a. ch.)
ORM 1300	16 a. ch., 16 d. ch.; monitor; memory; *x–y*	Yokogawa E. C.	21,900 (16 a. ch.)
OR 1400	8 a. ch., 16 d. ch.; monitor; memory; *x–y*	Yokogawa E. C.	16,100 (8 a. ch.)
TA 240	1 to 4 a. ch.	Gould I. S.	8,500 (4 ch.)
TA 11	4,8,16 a. ch.; monitor; memory	Gould I. S.	18,900 (16 ch.)
TA 6000	8 to 64 a. ch., 8 to 32 d. ch.; monitor; memory	Gould I. S.	33,900 (16 ch.)
Windograf	2 to 4 a. ch.; monitor	Gould I. S.	10,200 (4 ch.)
Dash 10	10,20,30 a. ch.; monitor; memory	Astro-Med	22,500 (10 ch.)
MT95K2	8 to 32 a. ch., 32 d. ch.; monitor; memory; *x–y*; analysis	Astro-Med	32,600 (8 a. ch.)
8852	4 a. ch., 24 d. ch.; monitor; memory; *x–y*; analysis; FFT	Hioki E. E. C.	22,300 (4 a. ch.)
8815	4 a. ch., 32 d. ch.; memory; *x–y*	Hioki E. E. C.	4,500 (4 a. ch.)
8825	16 a. ch., 32 d. ch.; monitor; memory; *x–y*; analysis	Hioki E. E. C.	28,800 (8 a. ch.)

Note: a. ch. = analog channel; d. ch. = digital channel.

equal to the sine amplitude. In this way the graphs indicate the presence of a phenomenon with a frequency content exceeding the writing frequency. As data are stored in memory they can be reproduced at a lower rate thus revealing the actual high-frequency waveform captured. Some apparatuses use a much lower sampling rate in real time and only perform off-line: in this case the apparatus is indicated as a "memory" recorder. Digitization accuracy ranges from 8 to 16 bit, whereas the largest number of dots full scale is 4800. In this way the useful signal may be superposed on a large dc-offset: it can be written or reproduced with excellent graphic quality with the offset digitally removed and the scale adapted.

In a high-performance recorder a waveform display is extremely useful to avoid paper spoiling, in real-time and in off-line recording as well. The display is also used for apparatus settings. Signals can be calibrated and real physical values and units can be printed at the calibration lines. Via memory *x–y* plots can be obtained. Some apparatuses allow application of mathematical functions for waveform processing and analysis: original and processed waveforms can be drawn together off-line. A few types are equipped with FFT software. Computer interfacing, a large set of triggering modes (including recording at increased velocity after a specific trigger), event channels, etc. are standard facilities. Table 96.3 shows a set of thermal dot array recorders (under "Description": number of analog channels (a. ch.) and digital channels (d. ch.); waveform monitor; signal data memory, *x–y* facility, mathematical analysis, FFT) and Table 96.4 gives specifications.

Concluding Remarks

Table 96.5 gives addresses and fax and phone numbers of manufacturers of recorders mentioned in Tables 96.1 and 96.3. It should be remarked that prices mentioned in these tables hold for purchasing a complete functioning apparatus (number of channels indicated) from firms in Belgium representing the manufacturers and having provided the data sheets from which specifications were derived. With the expression "a complete functioning" apparatus a standard system is meant, thus including simple input couplers (in case of a modular design), standard RAM and analysis software, no specific options. Obviously, the list of manufacturers is incomplete. It should be mentioned that the number of manufacturers of graphic recorders is decreasing; a significant and increasing amount of applications has been taken over by

TABLE 96.4 Thermal Dot Array Recorder Specifications

Type	Recording Width (mm)	Thermal Array Resolution (dots/mm)	Chart Velocity max. (mm/s)	Chart Velocity min. (mm/h)	Maximal Writing Frequency (dots/s)	Time Axis Resolution max. (dots/mm)	Time Axis Resolution min. (dots/mm)
WR 5000	384	8	200	1	1600	64	8
WR 9000V	200	8	100	1		32	
WR 9000M	200	8	100	1		32	
Mark 12	384	8	200	1	1600	64	8
MA 6000	205	8	100	1		40	
ORP 1200	201	8	100	10	1600	32	16
ORP 1300	201	8	100	10	1600	32	16
ORM 1200	201	8	100	10	1600	32	16
ORM 1300	201	8	100	10	1600	32	16
OR 1400	201	8	250	10	6400	32	25.6
TA 240	104	8	125	36	1000	32	8
TA 11	264	8	200	36	1600	16	8
TA 6000	370	8	200	36	1600	16	8
Windograf	104	8	100	36	800	32	8
Dash 10	256	12	200	60	1200	12	6
MT95K2	400	12	500	1	2000	48	4
8852	100	8	25	10	200	16	8
8815	104	6	8	10	50	12	6
8825	256	8	20	10	200	10	10

Type	Sampling Frequency Real-Time, max. (kHz)	Bit Accuracy, max. (bits)	No., max., Channels[a]	Display Dimensions (mm)	Display Resolution (pixels)	Sampling Frequency Memory, max. (kHz)	Samples Stored/Channel, max.
WR 5000	64	14	8a	—	—	64	32 k
WR 9000V	250	12	8a	192 × 120	640 × 400	250	512 k
WR 9000M	50	14	8a	192 × 120	640 × 400	50	512 k
Mark 12	200	16	52a, 52d	97 × 77	256 × 320	200	2 M
MA 6000	500	16	16a	192 × 120	640 × 400	500	512 k
ORP 1200	100	14	8a, 16d	127[b]	320 × 240	100	32 k
ORP 1300	100	14	16a, 16d	127[b]	320 × 240	100	32 k
ORM 1200	100	14	8a, 16d	127[b]	320 × 240	100	128 k
ORM 1300	100	14	16a, 16d	127[b]	320 × 240	100	128 k
OR 1400	100	16	8a, 16d	127[b]	320 × 240	100	256 k
TA 240	5	12	4a	—	—	—	—
TA 11	250	12	16a	198 × 66	640 × 200	250	500 k
TA 6000	250	12	64a, 32d	224 × 96	640 × 200	250	500 k
Windograf	10	12	4a	178[b]	800 × 350	—	—
Dash 10	250	12	30a		256 × 64	250	512 k
MT95K2	200	12	32a, 32d			200	500 k
8852	1.6	8	4a, 24d	178[b]		100×10^3	1 M
8815	12.5	8	4a, 32d	—	—	500	30 k
8825	8	12	16a, 32d	254[b]	640 × 480	200	500 k

[a] a = analog; d = digital.
[b] Diagonal.

"paperless" recorders, i.e., data acquisition systems. Nevertheless, the possibility of generating graphs in real time remains an important feature, e.g., to provide evidence of the presence of a specific phenomenon. In recent years analog recorders have become less used and manufactured (apart from the translational pen types, especially in industrial process monitoring). Thermal array recorders have become more important: the quality and long-term stability of thermal paper have improved and cost levels are comparable with calibrated paper for ink recording. In new designs, recorders provide more capabilities

TABLE 96.5　Companies that Make Graphic Recorders

Astro-Med, Inc. Astro-Med Industrial Park, West Warwick, RI 02893, USA fax (401) 822 - 2430/phone (401) 828-4000	Kipp & Zonen, Delft BV Mercuriusweg 1, P.O. Box 507, NL-2600 AM Delft, The Netherlands fax 015-620351/phone 015-561000
Eurotherm Recorders Ltd. Dominion Way, Worthing, West Sussex BN148QL, Great Britain fax 01903-203767/phone 01903-205222	Linseis GMBH Postfach 1404, Vielitzer Strasse 43, D-8672 Selb, Germany fax 09287/70488/phone 09287/880-0
Gould Instrument Systems, Inc. 8333 Rockside Road, Valley View, OH 44125-6100, USA fax (216) 328-7400/phone (216) 328-7000	Omega Engineering, Inc. P.O. Box 4047, Stamford, CT 06907-0047, USA fax (203) 359-7700/phone (203) 359-1660
Graphtec Corporation 503-10 Shinano-cho, Totsuka-ku, Yokohama 244, Japan fax (045) 825-6396/phone (045) 825-6250	Siemens AG, Bereich Automatisierungstechnik Geschäftsgebiet Processgeräte, AUT 34, D-76181 Karlsruhe, Germany fax 0721/595-6885/phone 0721/595-2058
Hioki E. E. Corporation 81 Koizumi, Ueda, Nagano, 386-11, Japan fax 0268-28-0568/phone 0268-28-0562	Western Graphtec, Inc. 11 Vanderbilt, Irvine, CA 92718-2067, USA fax (714) 770-6010/phone (800) 854-8385
Honeywell Industrial Automation and Control 16404 North Black Canyon Hwy., Phoenix, AZ 85023, USA phone (800) 343-0228	W+W Instruments AG Frankfurt-Strasse 78, CH-4142 Münchenstein, Switzerland fax +41 (0) 6141166685/phone +41 (0) 614116477
	Yokogawa Electric Corporation Shinjuku-Nomura Bldg. 1-26-2 Nishi-Shinjuku, Shinjuku-ku, Tokyo 163-05, Japan fax 81-3-3349-1017/phone 81-3-3349-1015

and appear more intelligent, obviously leading to increased complications with respect to instrument settings and thus increased need for training and experience in the use of the instrument.

Defining Terms

Analog graphic recorder: A graphic recorder that makes use of an analog transducer system (e.g., a moving pen).

Analog recorder bandwidth: The largest frequency that can be processed by the analog recorder (−3 dB limit).

Digital graphic recorder: A graphic recorder that makes use of a digital transducer system (e.g., a fixed dot array).

Graphic recorder: A measuring apparatus that produces in real time a hard copy of a set of time-dependent variables.

Maximal sampling frequency: Maximal number of data points sampled by the digital recorder per time unit (totally or per channel).

Maximal writing frequency: Maximal number of writing (or printing) acts executed by the digital recorder per time unit.

Thermal dot array recorder: A digital recorder applying a fixed thermal dot array perpendicular to the time axis.

Translational pen recorder: An analog recorder with one or several pens being translated perpendicularly to the time axis.

References

1. H. Vermariën, Recorders, graphic, in J.G. Webster, Ed., *Encyclopedia of Medical Devices and Instrumentation*, New York: John Wiley & Sons, 1988.
2. D.A. Bell, *Electronic Instrumentation and Measurements*, 2nd ed., Englewood Cliffs, NJ: Prentice-Hall, 1994.
3. A. Miller, O.S. Talle, and C.D. Mee, Recorders, in B.M. Oliver and J.M. Cage, Eds., *Electronic Measurements and Instrumentation*, New York: McGraw-Hill, 427–479, 1975.
4. R.J. Smith, *Circuits, Devices and Systems: A First Course in Electrical Engineering*, New York: John Wiley & Sons, 1976.
5. W.H. Olson, Basic concepts in instrumentation, in J.G. Webster, Ed., *Medical Instrumentation: Application and Design*, 3rd ed., New York: John Wiley & Sons, 1998.

96.2   Data Acquisition Systems

Edward McConnell

The fundamental task of a **data acquisition system** is the measurement or generation of real-world physical signals. Before a physical signal can be measured by a computer-based system, a sensor or transducer is used to convert the physical signal into an electrical signal, such as voltage or current. Often only a plug-in data acquisition (DAQ) board is considered the data acquisition system; however, a board is only one of the components in the system. A complete DAQ system consists of sensors, signal conditioning, interface hardware, and software. Unlike stand-alone instruments, signals often cannot be directly connected to the DAQ board. The signals may need to be conditioned by some signal-conditioning accessory before they are converted to digital information by the plug-in DAQ board. Software controls the data acquisition system — acquiring the raw data, analyzing the data, and presenting the results. The components are shown in Fig. 96.1.

Signals

Signals are physical events whose magnitude or time variation contains information. DAQ systems measure various aspects of a signal in order to monitor and control the physical events. Users of DAQ systems need to know the relation of the signal to the physical event and what information is available in the signal. Generally, information is conveyed by a signal through one or more of the following signal parameters: state, rate, level, shape, or frequency content. The physical characteristics of the measured signals and the related information help determine the design of a DAQ system..

All signals are, fundamentally, analog, time-varying signals. For the purpose of discussing the methods of signal measurement using a plug-in DAQ board, a given signal should be classified as one of five signal types. Because the method of signal measurement is determined by the way the signal conveys the needed information, a classification based on this criterion is useful in understanding the fundamental building blocks of a DAQ system.

As shown in the Fig. 96.2, any signal can generally be classified as analog or digital. A digital, or binary, signal has only two possible discrete levels of interest — a high (on) level and a low (off) level. The two digital signal types are on–off signals and pulse train signals. An analog signal, on the other hand, contains information in the continuous variation of the signal with time. Analog signals are described in the time or frequency domains depending upon the information of interest. A dc type signal is a low-frequency signal, and if the phase information of a signal is presented with the frequency information, then there is no difference between the time or frequency domain representations. The category to which a signal belongs depends on the characteristic of the signal to be measured. The five types of signals can be closely paralleled with the five basic types of signal information — state, rate, level, shape, and frequency content.

Basic understanding of the signal representing the physical event being measured and controlled assists in the selection of the appropriate DAQ system.

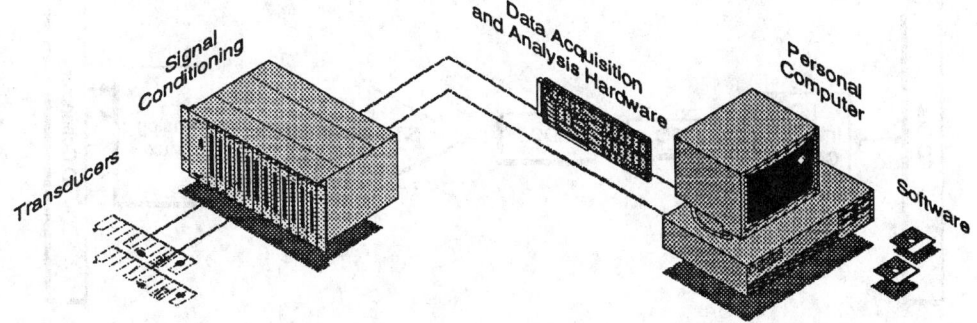

FIGURE 96.1 Components of a DAQ system.

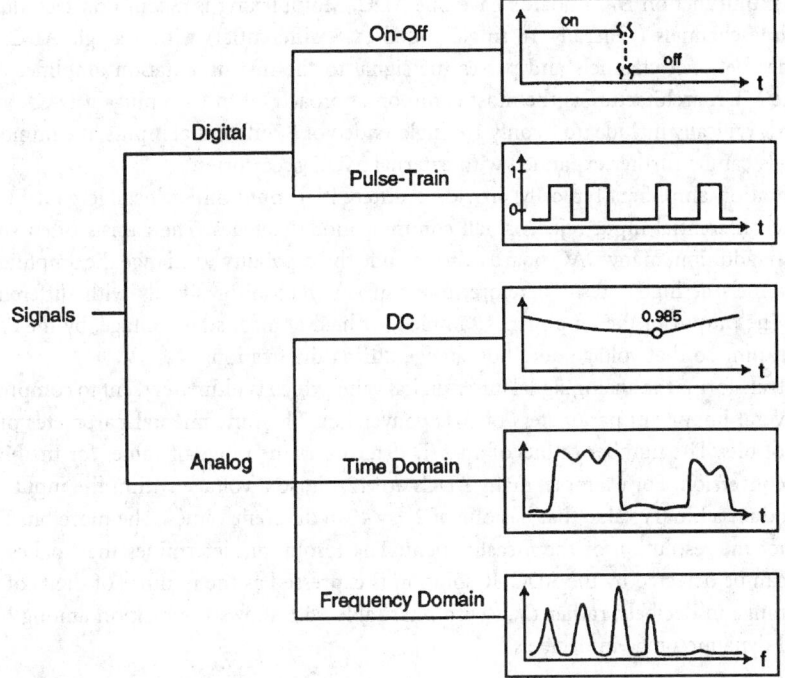

FIGURE 96.2 Classes of signals.

Plug-In DAQ Boards

The fundamental component of a DAQ system is the plug-in DAQ board. These boards plug directly into a slot in a PC and are available with analog, digital, and timing inputs and outputs (I/O). The most versatile of the plug-in DAQ boards is the multifunction I/O board. As the name implies, this board typically contains various combinations of analog-to-digital converters (ADCs), digital-to-analog converters (DACs), digital I/O lines, and counters/timers. ADCs and DACs measure and generate analog voltage signals, respectively. The digital I/O lines sense and control digital signals. Counters/timers measure pulse rates, widths, delays, and generate timing signals. These many features make the multifunction DAQ board useful for a wide range of applications.

Multifunction boards are commonly used to measure analog signals. This is done by the ADC, which converts the analog voltage level into a digital number that the computer can interpret. The analog multiplexer (MUX), the instrumentation amplifier, the sample-and-hold (S/H) circuitry, and the ADC compose the analog input section of a multifunction board (see Fig. 96.3).

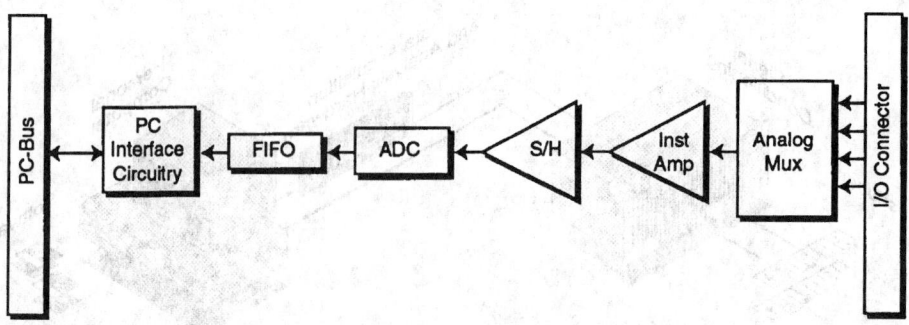

FIGURE 96.3   Analog input section of a plug-in DAQ board. *Note:* FIFO = first-in first-out buffer, S/H = sample-and-hold, Inst. Amp = instrumentation amplifier, and Mux = analog multiplexer.

Typically, multifunction DAQ boards have one ADC. Multiplexing is a common technique for measuring multiple channels (generally 16 single-ended or 8 differential) with a single ADC. The analog MUX switches between channels and passes the signal to the instrumentation amplifier and the S/H circuitry. The MUX architecture is the most common approach taken with plug-in DAQ boards. While plug-in boards typically include up to only 16 single-ended or 8 differential inputs, the number of analog input channels can be further expanded with external MUX accessories.

Instrumentation amplifiers typically provide a differential input and selectable gain by jumpers or software. The differential input rejects small common-mode voltages. The gain is often software programmable. In addition, many DAQ boards also include the capability to change the amplifier gain while scanning channels at high rates. Therefore, one can easily monitor signals with different ranges of amplitudes. The output of the amplifier is sampled, or held at a constant voltage, by the S/H device at measurement time so that voltage does not change during digitization.

The ADC transforms the analog signal into a digital value which is ultimately sent to computer memory. There are several important parameters of A/D conversion. The fundamental parameter of an ADC is the number of bits. The number of bits of an A/D determines the range of values for the binary output of the ADC conversion. For example, many ADCs are 12-bit, so a voltage within the input range of the ADC will produce a binary value that has one of $2^{12} = 4096$ different values. The more bits that an ADC has, the higher the resolution of the measurement. The resolution determines the smallest amount of change that can be detected by the ADC. Resolution is expressed as the number of digits of a voltmeter or dynamic range in decibels, rather than with bits. Table 96.6 shows the relation among bits, number of digits, and dynamic range in decibels.

TABLE 96.6   Relation Among Bits, Number of Digits, and Dynamic Range (dB)

Bits	Digits	dB
20	6.0	120
16	4.5	96
12	3.5	72
8	2.5	48

The resolution of the A/D conversion is also determined by the input range of the ADC and the gain. DAQ boards usually include an instrumentation amplifier that amplifies the analog signal by a gain factor prior to the conversion. This gain amplifies low-level signals so that more accurate measurements can be made.

Together, the input range of the ADC, the gain, and the number of bits of the board determine the minimum resolution of the measurement. For example, suppose a low-level ±30 mV signal is acquired using a 12-bit ADC that has a ±5 V input range. If the system includes an amplifier with a gain of 100, the resulting resolution of the measurement will be range/(gain * 2^{bits}) = resolution, or 10 V/(100 * 2^{12}) = 0.0244 mV.

Finally, an important parameter of digitization is the rate at which A/D conversions are made, referred to as the sampling rate. The A/D system must be able to sample the input signal fast enough to measure the important waveform attributes accurately. In order to meet this criterion, the ADC must be able to convert the analog signal to digital form quickly enough.

When scanning multiple channels with a multiplexing DAQ system, other factors can affect the throughput of the system. Specifically, the instrumentation amplifier must be able to settle to the needed accuracy before the A/D conversion occurs. With multiplexed signals, multiple signals are being switched into one instrumentation amplifier. Most amplifiers, especially when amplifying the signals with larger gains, will not be able to settle to the full accuracy of the ADC when scanning channels at high rates. To avoid this situation, consult the specified settling times of the DAQ board for the gains and sampling rates required by the application.

Types of ADCs

Different DAQ boards use different types of ADCs to digitize the signal. The most popular type of ADC on plug-in DAQ boards is the successive approximation ADC, because it offers high speed and high resolution at a modest cost.

Subranging (also called half-flash) ADCs offer very high speed conversion with sampling speeds up to several million samples per second.

The state-of-the-art technology in ADCs is sigma–delta modulating ADCs. These ADCs sample at high rates, are able to achieve high resolution, and offer the best linearity of all ADCs.

Integrating and flash ADCs are mature technologies still used on DAQ boards today. Integrating ADCs are able to digitize with high resolution but must sacrifice sampling speed to obtain it. Flash ADCs are able to achieve the highest sampling rate (gigahertz) but are available only with low resolution. The different types of ADCs are summarized in Table 96.7.

TABLE 96.7 Types of ADCs

Type of ADC	Advantages	Features
Successive approximation	High resolution	1.25 MS/s sampling rate
	High speed	12-bit resolution
	Easily multiplexed	200 kS/s sampling rate
		16-bit resolution
Subranging	Higher speed	1 MHz sampling rate
		12-bit resolution
Sigma–delta	High resolution	48 kHz sampling rate
	Excellent linearity	16-bit resolution
	Built-in antialiasing	
	State-of-the-art technology	
Integrated	High resolution	15 kHz sampling rate
	Good noise rejection	
	Mature technology	
Flash	Highest speed	125 MHz sampling rate
	Mature technology	

Analog Input Architecture

With a typical DAQ board, the multiplexer switches among analog input channels. The analog signal on the channel selected by the multiplexer then passes to the programmable gain instrumentation amplifier (PGIA), which amplifies the signal. After the signal is amplified, the sample and hold (S/H) keeps the analog signal constant so that the ADC can determine the digital representation of the analog signal. A good DAQ board will then place the digital signal in a first-in first-out (FIFO) buffer, so that no data will be lost if the sample cannot transfer immediately over the PC I/O channel to computer memory. Having a FIFO becomes especially important when the board is run under operating systems that have large interrupt latencies, such as Microsoft Windows.

Basic Analog Specifications

Almost every DAQ board data sheet specifies the number of channels, the maximum sampling rate, the resolution, and the input range and gain.

The number of channels, which is determined by the multiplexer, is usually specified in two forms — differential and single ended. Differential inputs are inputs that have different reference points for each channel, none of which is grounded by the board. Differential inputs are the best way to connect signals to the DAQ board because they provide the best noise immunity.

Single-ended inputs are inputs that are referenced to a common ground point. Because single-ended inputs are referenced to a common ground, they are not as good as differential inputs for rejecting noise. They do have a larger number of channels, however. Single-ended inputs are used when the input signals are high level (greater than 1 V), the leads from the signal source to the analog input hardware are short (less than 5 m), and all input signals share a common reference.

Some boards have pseudodifferential inputs which have all inputs referenced to the same common — like single-ended inputs — but the common is not referenced to ground. These boards have the benefit of a large number of input channels, like single-ended inputs, and the ability to remove some common-mode noise, especially if the common-mode noise is consistent across all channels. Differential inputs are still preferable to pseudodifferential, however, because differential is more immune to magnetic noise.

Sampling rate determines how fast the analog signal is converted to a digital signal. When measuring ac signals, sample at least two times faster than the highest frequency of the input signal. Even when measuring dc signals, oversample and average the data to increase the accuracy of the signal by reducing the effects of noise.

If the physical event consists of multiple dc-class signals, a DAQ board with interval scanning should be used. With interval scanning, all channels are scanned at one sample interval (usually the fastest rate of the board), with a second interval (usually slow) determining the time before repeating the scan. Interval scanning gives the effects of simultaneously sampling for slowly varying signals without requiring the additional cost of input circuitry for true simultaneous sampling.

Resolution is the number of bits that are used to represent the analog signal. The higher the resolution, the higher the number of divisions the input range is broken into, and therefore the smaller the possible detectable voltage. Unfortunately, some DAQ specifications are misleading when they specify the resolution associated with the DAQ board. Many DAQ board specifications state the resolution of the ADC without stating the linearities and noise, and therefore do not give the information needed to determine the resolution of the entire board. Resolution of the ADC, combined with the settling time, **integral nonlinearity** (INL), **differential nonlinearity** (DNL), and noise will give an understanding of the accuracy of the board.

Input range and gain determine the level of signal that should be connect to the board. Usually, the range and gain are specified separately, so the two must be combined to determine the actual signal input range as

$$\text{signal input range} = \text{range/gain}$$

For example, a board using an input range of ±10 V with a gain of 2 will have a signal input range of ±5 V. The closer the signal input range is to the range of the signal, the more accurate the readings from the DAQ board will be. If the signals have different input ranges, use a DAQ board with the feature of different gains per channel.

Data Acquisition Software

The software is often the most critical component of the DAQ system. Users of DAQ systems usually program the hardware in one of two ways — through register programming or through high-level device drivers.

Board Register-Level Programming

The first option is not to use vendor-supplied software and program the DAQ board at the hardware level. DAQ boards are typically register based; that is, they include a number of digital registers that control the operation of the board. The developer may use any standard programming language, such as C, C++, or Visual BASIC, to write series of binary codes to the DAQ board to control its operation. Although this method affords the highest level of flexibility, it is also the most difficult and time-consuming, especially for the inexperienced programmer. The programmer must know the details of programming all hardware, including the board, the PC interrupt controller, the DMA controller, and PC memory.

Driver Software

Driver software typically consists of a library of function calls usable from a standard programming language. These function calls provide a high-level interface to control the standard functions of the plug-in board. For example, a function called SCAN_OP may configure, initiate, and complete a multiple-channel scanning DAQ operation of a predetermined number of points. The function call would include parameters to indicate the channels to be scanned, the amplifier gains to be used, the sampling rate, and the total number of data points to be collected. The driver responds to this one function call by programming the plug-in board, the DMA controller, the interrupt controller, and CPU to scan the channels as requested.

What Is Digital Sampling?

Every DAQ system has the task of gathering information about analog signals. To do this, the system captures a series of instantaneous "snapshots" or samples of the signal at definite time intervals. Each sample contains information about the signal at a specific instant. Knowing the exact time of each conversion and the value of the sample, one can reconstruct, analyze, and display the digitized waveform.

Real-Time Sampling Techniques

In real-time sampling, the DAQ board digitizes consecutive samples along the signal (Fig. 96.4). According to the **Nyquist sampling theorem**, the ADC must sample at least twice the rate of the maximum frequency component in that signal to prevent aliasing. **Aliasing** is a false lower-frequency component that appears in sampled data acquired at too low a sampling rate. The frequency at one half the sampling frequency is referred to as the Nyquist frequency. Theoretically, it is possible to recover information about those signals with frequencies at or below the Nyquist frequency. Frequencies above the Nyquist frequency will alias to appear between dc and the Nyquist frequency.

For example, assume the sampling frequency, f_s, is 100 Hz. Also assume the input signal to be sampled contains the following frequencies — 25, 70, 160, and 510 Hz. Figure 96.5 shows a spectral representation of the input signal.

The mathematics of sampling theory show us that a sampled signal is shifted in the frequency domain by an amount equal to integer multiples of the sampling frequency, f_s. Figure 96.6 shows the spectral content of the input signal after sampling. Frequencies below 50 Hz, the Nyquist frequency ($f_s/2$), appear correctly. However, frequencies above the Nyquist appear as aliases below the Nyquist frequency. For example, F1 appears correctly; however, F2, F3, and F4 have aliases at 30, 40, and 10 Hz, respectively.

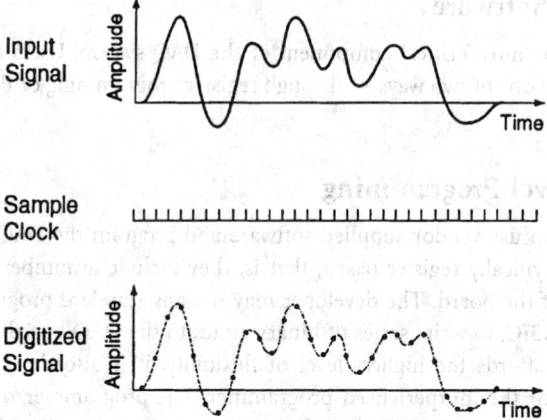

FIGURE 96.4 Consecutive discrete samples recreate the input signal.

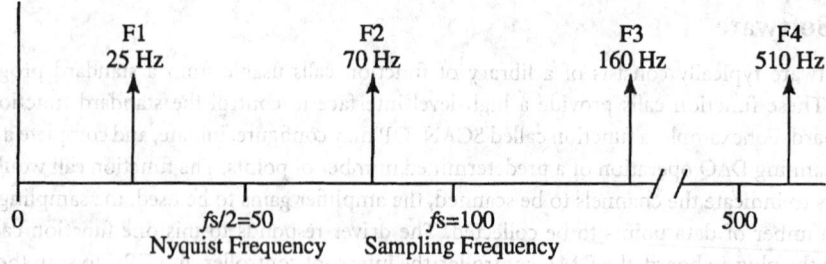

FIGURE 96.5 Spectral of signal with multiple frequencies.

The resulting frequency of aliased signals can be calculated with the following formula:

Apparent (Alias) Freq. = ABS (Closest Integer Multiple of Sampling Freq. − Input Freq.)

For the example of Figs. 96.5 and 96.6:

$$\text{Alias F2} = |100 - 70| = 30 \text{ Hz}$$

$$\text{Alias F3} = |(2)100 - 160| = 40 \text{ Hz}$$

$$\text{Alias F4} = |(5)100 - 510| = 10 \text{ Hz}$$

Preventing Aliasing

Aliasing can be prevented by using filters on the front end of the DAQ system. These antialiasing filters are set to cut off any frequencies above the Nyquist frequency (half the sampling rate). The perfect filter would reject all frequencies above the Nyquist; however, because perfect filters exist only in textbooks, one must compromise between sampling rate and selecting filters. In many applications, one- or two-pole passive filters are satisfactory. The rule of thumb is to oversample (5 to 10 times) and use these antialiasing filters when frequency information is crucial.

Alternatively, active antialiasing filters with programmable cutoff frequencies and very sharp attenuation of frequencies above the cutoff can be used. Because these filters exhibit a very steep roll-off, the DAQ system can sample at two to three times the filter cutoff frequency. Figure 96.7 shows a transfer function of a high-quality antialiasing filter.

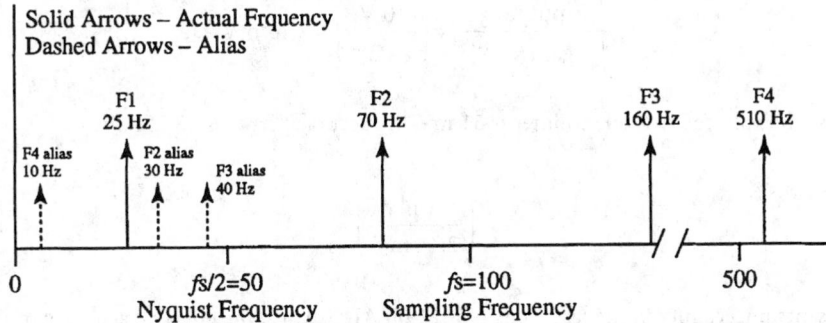

Alias F2 = |100 - 70| = 30 Hz
Alias F3 = |(2)100 - 160| = 40 Hz
Alias F4 = |(5)100 - 510| = 10 Hz

Solid Arrows – Actual Frquency
Dashed Arrows – Alias

F1
25 Hz

F2
70 Hz

F3
160 Hz

F4
510 Hz

F4 alias
10 Hz

F2 alias
30 Hz

P3 alias
40 Hz

0

fs/2=50
Nyquist Frequency

fs=100
Sampling Frequency

500

FIGURE 96.6   Spectral of signal with multiple frequencies after sampling at $f_s = 100$ Hz.

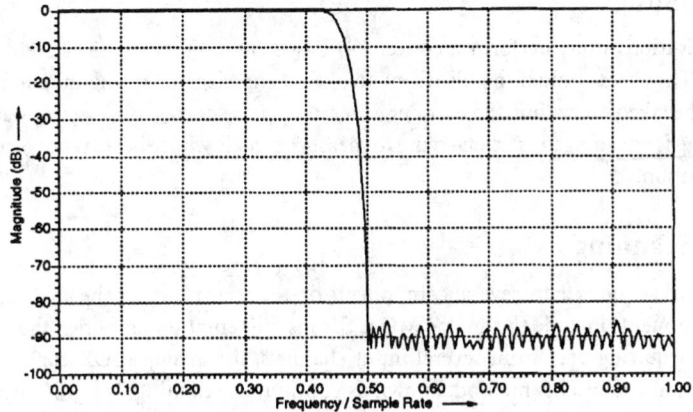

FIGURE 96.7   Magnitude portion of transfer function of an antialiasing filter.

The computer uses digital values to recreate or to analyze the waveform. Because the signal could be anything between each sample, the DAQ board may be unaware of any changes in the signal between samples. There are several sampling methods optimized for the different classes of data; they include software polling, external sampling, continuous scanning, multirate scanning, simultaneous sampling, interval scanning, and seamless changing of the sample rate.

Software Polling

A software loop polls a timing signal and starts the A/D conversion via a software command when the edge of the timing signal is detected. The timing signal may originate from the internal clock of the computer or from a clock on the DAQ board. Software polling is useful in simple, low-speed applications, such as temperature measurements.

The software loop must be fast enough to detect the timing signal and trigger a conversion. Otherwise, a window of uncertainty, also known as jitter, will exist between two successive samples. Within the window of uncertainty, the input waveform could change enough to reduce the accuracy of the ADC drastically.

Suppose a 100-Hz, 10-V full-scale sine wave is digitized (Fig. 96.8). If the polling loop takes 5 ms to detect the timing signal and to trigger a conversion, then the voltage of the input sine wave will change as much as 31 mV, [$\Delta V = 10 \sin (2\pi \times 100 \times 5 \times 10^{-6})$]. For a 12-bit ADC operating over an input range of 10 V and a gain of 1, one least significant bit (LSB) of error represents 2.44 mV:

$$\left(\frac{\text{Input range}}{\text{gain} \times 2^n}\right) = \left(\frac{10\ \text{V}}{1 \times 2^{12}}\right) = 2.44\ \text{mV}$$

But because the voltage error due to jitter is 31 mV, the accuracy error is 13 LSB.

$$\left(\frac{31\ \text{mV}}{2.44\ \text{mV}}\right)$$

This represents uncertainty in the last 4 bits of a 12-bit ADC. Thus, the effective accuracy of the system is no longer 12 bits but rather 8 bits.

External Sampling

Some DAQ applications must perform a conversion based on another physical event that triggers the data conversion. The event could be a pulse from an optical encoder measuring the rotation of a cylinder. A sample would be taken every time the encoder generates a pulse corresponding to *n* degrees of rotation. External triggering is advantageous when trying to measure signals whose occurrence is relative to another physical phenomenon.

Continuous Scanning

When a DAQ board acquires data, several components on the board convert the analog signal to a digital value. These components include the analog MUX, the instrumentation amplifier, the S/H circuitry, and the ADC. When acquiring data from several input channels, the analog MUX connects each signal to the ADC at a constant rate. This method, known as continuous scanning, is significantly less expensive than having a separate amplifier and ADC for each input channel.

Continuous scanning is advantageous because it eliminates jitter and is easy to implement. However, it is not possible to sample multiple channels simultaneously. Because the MUX switches between channels, a time skew occurs between any two successive channel samples. Continuous scanning is appropriate for applications where the time relationship between each sampled point is unimportant or where the skew is relatively negligible compared with the speed of the channel scan.

If samples from two signals are used to generate a third value, then continuous scanning can lead to significant errors if the time skew is large. In Fig. 96.9, two channels are continuously sampled and added together to produce a third value. Because the two sine waves are 90° out-of-phase, the sum of the signals should always be zero. But because of the skew time between the samples, an erroneous sawtooth signal results.

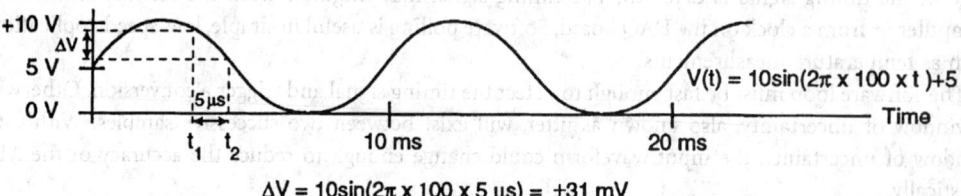

$$\Delta V = 10\sin(2\pi \times 100 \times 5\ \mu s) = \pm31\ \text{mV}$$

FIGURE 96.8   Jitter reduces the effective accuracy of the DAQ board.

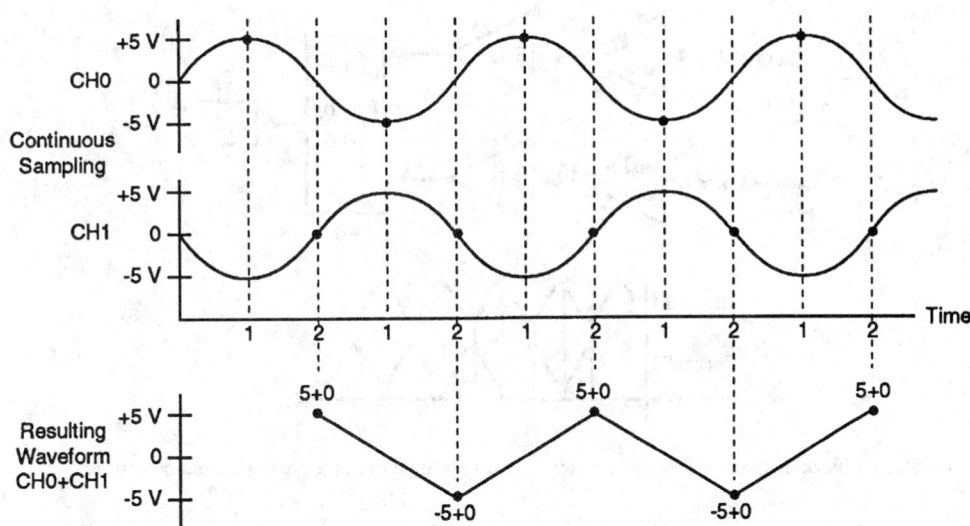

FIGURE 96.9   If the channel skew is large compared with the signal, then erroneous conclusions may result.

Multirate Scanning

Multirate scanning, a method that scans multiple channels at different scan rates, is a special case of continuous scanning. Applications that digitize multiple signals with a variety of frequencies use multirate scanning to minimize the amount of buffer space needed to store the sampled signals. Channel-independent ADCs are used to implement hardware multirate scanning; however, this method is extremely expensive. Instead of multiple ADCs, only one ADC is used. A channel/gain configuration register stores the scan rate per channel and software divides down the scan clock based on the per-channel scan rate. Software-controlled multirate scanning works by sampling each input channel at a rate that is a fraction of the specified scan rate.

Suppose the system scans channels 0 through 3 at 10 kS/s, channel 4 at 5 kS/s, and channels 5 through 7 at 1 kS/s. A base scan rate of 10 kS/s should be used. Channels 0 through 3 are acquired at the base scan rate. Software and hardware divide the base scan rate by 2 to sample channel 4 at 5 kS/s, and by 10 to sample channels 5 through 7 at 1 kS/s.

Simultaneous Sampling

For applications where the time relationship between the input signals is important, such as phase analysis of ac signals, simultaneous sampling must be used. DAQ boards capable of simultaneous sampling typically use independent instrumentation amplifiers and S/H circuitry for each input channel, along with an analog MUX, which routes the input signals to the ADC for conversion (as shown in Fig. 96.10).

To demonstrate the need for a simultaneous-sampling DAQ board, consider a system consisting of four 50-kHz input signals sampled at 200 kS/s. If the DAQ board uses continuous scanning, the skew between each channel is 5 µs (1S/200 kS/s) which represents a 270° [(15 µs/20 µs) × 360°] shift in phase between the first channel and fourth channel. Alternatively, with a simultaneous-sampling board with a maximum 5 ns interchannel time offset, the phase shift is only 0.09° [(5 µs/20 µs) × 360°]. This phenomenon is illustrated in Fig. 96.11.

Interval Scanning

For low-frequency signals, interval scanning creates the effect of simultaneous sampling, yet maintains the cost benefits of a continuous-scanning system. This method scans the input channels at one rate and

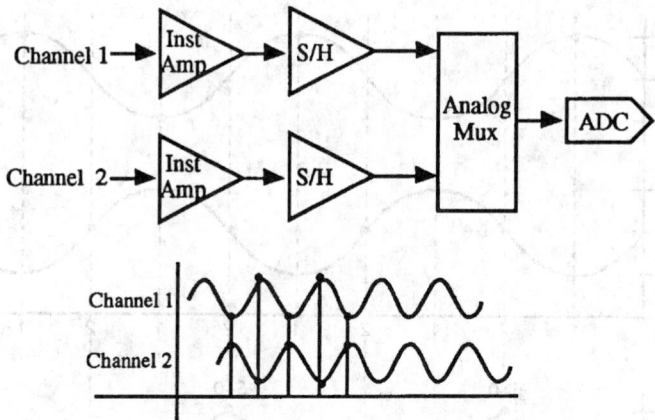

FIGURE 96.10 Block diagram of DAQ components used to sample multiple channels simultaneously.

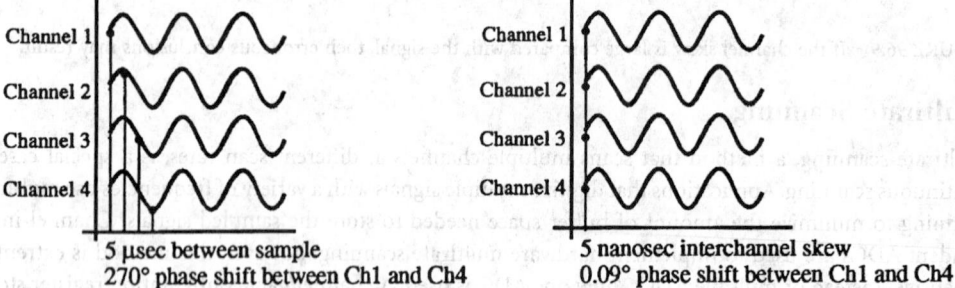

FIGURE 96.11 Comparison of continuous scanning and simultaneous sampling.

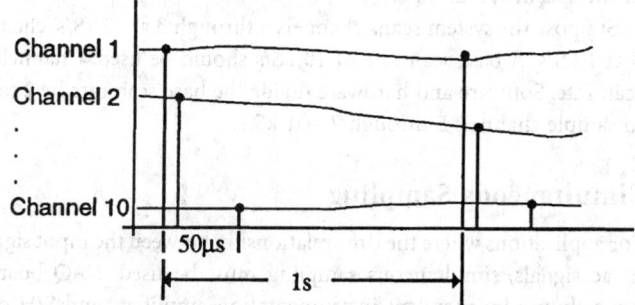

FIGURE 96.12 Interval scanning — all ten channels are scanned within 45 μs; this is insignificant relative to the overall acquisition rate of 1 S/s.

uses a second rate to control when the next scan begins. If the input channels are scanned at the fastest rate of the ADC, the effect of simultaneously sampling the channels is created. Interval scanning is appropriate for slow-moving signals, such as temperature and pressure. Interval scanning results in a jitter-free sample rate and minimal skew time between channel samples. For example, consider a DAQ system with ten temperature signals. By using interval scanning, a DAQ board can be set up to scan all channels with an interchannel delay of 5 μs, then repeat the scan every second. This method creates the effect of simultaneously sampling ten channels at 1 S/s, as shown in Fig. 96.12.

To illustrate the difference between continuous and interval scanning, consider an application that monitors the torque and RPMs of an automobile engine and computes the engine horsepower. Two signals, proportional to torque and RPM, are easily sampled by a DAQ board at a rate of 1000 S/s. The values are multiplied together to determine the horsepower as a function of time.

A continuously scanning DAQ board must sample at an aggregate rate of 2000 S/s. The time between which the torque signal is sampled and the RPM signal is sampled will always be 0.5 ms (1/2000). If either signal changes within 0.5 ms, then the calculated horsepower is incorrect. But using interval scanning at a rate of 1000 S/s, the DAQ board samples the torque signal every 1 ms, and the RPM signal is sampled as quickly as possible after the torque is sampled. If a 5-μs interchannel delay exists between the torque and RPM samples, then the time skew is reduced by 99% [(0.5 ms − 5 μs)/0.5 ms], and the chance of an incorrect calculation is reduced.

Factors Influencing the Accuracy of Measurements

How does one determine if a plug-in DAQ will deliver the required measurement results? With a sophisticated measuring device like a plug-in DAQ board, significantly different accuracies can be obtained depending on the type of board used. For example, one can purchase DAQ products on the market today with 16-bit ADCs and get less than 12 bits of useful data, or one can purchase a product with a 16-bit ADC and actually get 16 bits of useful data. This difference in accuracies causes confusion in the PC industry where everyone is used to switching out PCs, video cards, printers, and so on, and experiencing similar results between equipment.

The most important thing to do is to scrutinize more specifications than the resolution of the ADC that is used on the DAQ board. For dc-class measurements, one should at least consider the settling time of the instrumentation amplifier, DNL, **relative accuracy**, INL, and noise. If the manufacturer of the board under consideration does not supply these specifications in the data sheets, ask the vendor to provide them or run tests to determine these specifications.

Defining Terms

Alias: A false lower frequency component that appears in sampled data acquired at too low a sampling rate.

Asynchronous: (1) Hardware — A property of an event that occurs at an arbitrary time, without synchronization to a reference clock. (2) Software — A property of a function that begins an operation and returns prior to the completion or termination of the operation.

Conversion time: The time required, in an analog input or output system, from the moment a channel is interrogated (such as with a read instruction) to the moment that accurate data are available.

DAQ (data acquisition): (1) Collecting and measuring electric signals from sensors, transducers, and test probes or fixtures and inputting them to a computer for processing: (2) Collecting and measuring the same kinds of electric signals with ADC and/or DIO boards plugged into a PC, and possibly generating control signals with DAC and/or DIO boards in the same PC.

DNL (differential nonlinearity): A measure in LSB of the worst-case deviation of code widths from their ideal value of 1 LSB.

INL (integral nonlinearity): A measure in LSB of the worst-case deviation from the ideal A/D or D/A transfer characteristic of the analog I/O circuitry.

Nyquist sampling theorem: A law of sampling theory stating that if a continuous bandwidth-limited signal contains no frequency components higher than half the frequency at which it is sampled, then the original signal can be recovered without distortion.

Relative accuracy: A measure in LSB of the accuracy of an ADC. It includes all nonlinearity and quantization errors. It does not include offset and gain errors of the circuitry feeding the ADC.

Further Information

House, R., "Understanding Important DA Specifications," *Sensors,* 10(10), June 1993.

House, R., "Understanding Inaccuracies Due to Settling Time, Linearity, and Noise," *National Instruments European User Symposium Proceedings,* November 10–11, 1994, pp. 11–12.

McConnell, E., "PC-Based Data Acquisition Users Face Numerous Challenges," *ECN,* August 1994.

McConnell, E., "Choosing a Data-Acquisition Method," *Electronic Design*, 43(6), 147, 1995.

McConnell, E. and Jernigan, Dave, "Data Acquisition," in *The Electronics Handbook*, J.C. Whitaker (ed.), Boca Raton, FL: CRC Press, 1996, 1795–1822.

Potter, D. and A. Razdan, "Fundamentals of PC-Based Data Acquisition," *Sensors*, 11(2), 12–20, February 1994.

Potter, D., "Sensor to PC — Avoiding Some Common Pitfalls," *Sensors Expo Proceedings*, September 20, 1994.

Potter, D., "Signal Conditioners Expand DAQ System Capabilities," *I&CS*, 25–33, August 1995.

Johnson, G. W., *LabVIEW Graphical Programming*, New York: McGraw-Hill, 1994.

McConnell, E., "New Achievements in Counter/Timer Data Acquisition Technology," *MessComp 1994 Proceedings*, September 13–15, 1994, 492–498.

McConnell, E., "Equivalent Time Sampling Extends DA Performance," *Sensors Data Acquisition*, Special Issue, June, 13, 1995.

96.3   Magnetic and Optical Recorders

Yufeng Li

The heart of recording technology is for the process of information storage and retrieval. In addition to its obvious importance in different branches of science and engineering, it has become indispensable to our daily life. When we make a bank transaction, reserve an airplane ticket, use a credit card, watch a movie from a video tape, or listen to music from a CD, we are using the technology of recording. The general requirements for recording are information integrity, fast access, and low cost. Among the different techniques, the most popularly used ones are magnetic and optical recording.

Typical recording equipment consists of a read/write head, a medium, a coding/decoding system, a data access system, and some auxiliary mechanical and electronic components. The head and medium are for data storage and retrieval purposes, and the coding/decoding system is for data error correction. The data access system changes the relative position between the head and the medium, usually with a servo mechanism for data track following and a spinning mechanism for on-track moving. While the data access system and the auxiliary components are important to recording equipment, they are not considered essential in this chapter to the understanding of recording technology, and will not be covered. Interested readers are referred to Reference 1.

Magnetic Recording

At present, magnetic recording technology dominates the recording industry. It is used in the forms of hard disk, floppy disk, removable disk, and tape with either digital or analog mode. In its simplest form, it consists of a magnetic head and a magnetic medium, as shown in Fig. 96.13. The head is made of a piece of magnetic material in a ring shape (core), with a small gap facing the medium and a coil away from the medium. The head records (writes) and reproduces (reads) information, while the medium stores the information. The recording process is based on the phenomenon that an electric current i generates a magnetic flux ϕ as described by Ampere's law. The flux ϕ leaks out of the head core at the gap, and magnetizes the magnetic medium which moves from left to right with a velocity V under the head gap. Depending on the direction of the electric current i, the medium is magnetized with magnetization M pointing either left or right. This pattern of magnetization is retained in the memory of the medium even after the head moves away.

Two types of head may be used for reproducing. One, termed the *inductive head*, senses magnetic flux change rate, and the other, named the *magnetoresistive* (MR) *head*, senses the magnetic flux. When an inductive head is used, the reproducing process is just the reverse of the recording process. The flux coming out of the magnetized medium surface is picked up by the head core. Because the medium magnetization under the head gap changes its magnitude and direction as the medium moves, an electric

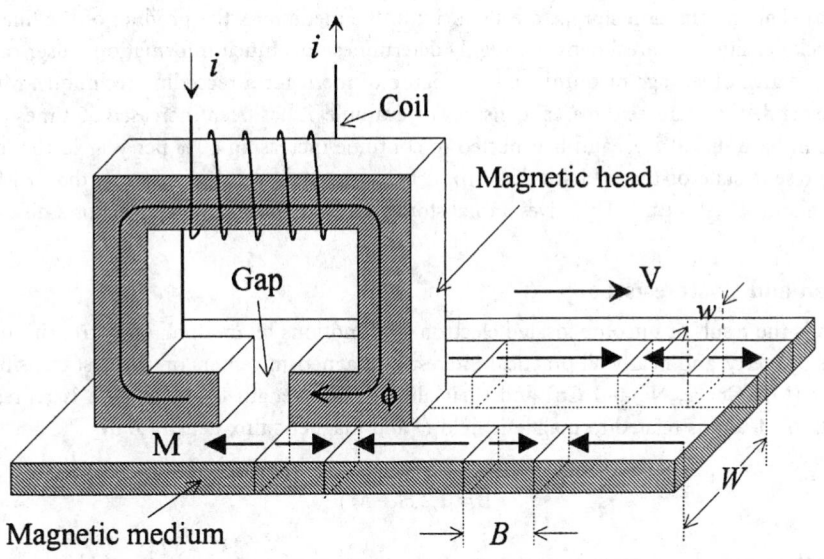

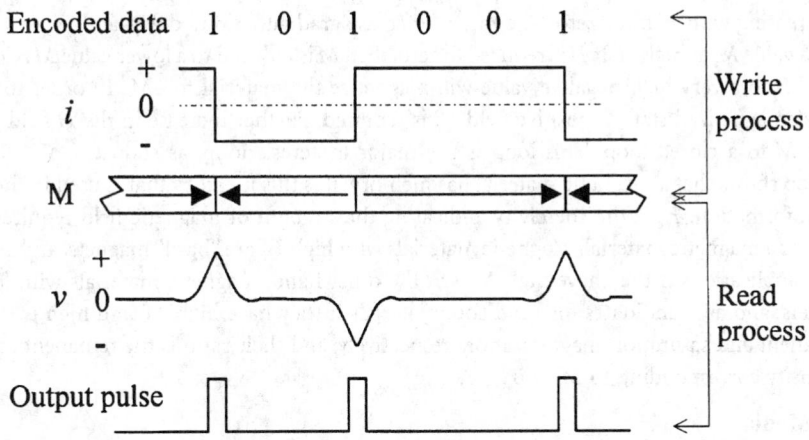

FIGURE 96.13 Conceptual diagrams illustrating the magnetic recording principle (a), and recording/reproducing process (b).

voltage is generated in the coil. This process is governed by Faraday's law. Figure 96.13b schematically shows the digital recording/reproducing process. First, all user data are encoded into a binary format — a serial of 1s and 0s. Then a write current i is sent to the coil. This current changes its direction whenever a 1 is being written. Correspondingly, a change of magnetization, termed a *transition,* is recorded in the medium for each 1 in the encoded data. During the reproducing process, the electric voltage induced in the head coil reaches a peak whenever there is a transition in the medium. A pulse detector generates a pulse for each transition. These pulses are decoded to yield the user data.

The minimum distance between two transitions in the medium is the flux change length B, and the distance between two adjacent signal tracks is the track pitch W, which is wider than the signal track width w. The flux change length can be directly converted into bit length with the proper code information. The reciprocal of the bit length is called *linear density,* and the reciprocal of the track pitch is termed *track density.* The information storage areal density in the medium is the product of the linear density and the track density. This areal density roughly determines how much information a user can store in a unit surface area of storage medium, and is a figure of merit for a recording technique. Much effort has been expended to increase the areal density. For example, it has been increased 50 times during the last decade in hard disk drives, and is expected to continue increasing 60% per year in the foreseeable future. At present, state-of-the-art hard disk products feature areal densities of more than 7 Mbits/mm² ($B < 0.1$ μm and $W < 1.5$ μm). This gives a total storage capacity of up to 6 Gbytes for a disk of 95 mm diameter.

Magnetism and Hysteresis Loop

Magnetism is the result of uncompensated electron spin motions in an atom. Only transition elements exhibit this property, and nearly all practical interest in magnetism centers on the first transition group of elements (Mn, Cr, Fe, Ni, and Co) and their alloys. The strength of magnetism is represented by magnetization M, and is related to magnetic field H and magnetic flux density B by

$$B = \mu_0 (H + M) \tag{96.1}$$

where μ_0 is the permeability of vacuum. Since M is a property of a magnetic material, it does not exist outside the magnetic material. H represents the strength acting on a magnetic material from a magnetic field which is generated either by a magnetic material or by an electric current. B is the flux density which determines the induced electric voltage in a coil. The ratio of B with and without a magnetic material is the relative permeability μ of that magnetic material.

When a magnetic field H is applied to a piece of demagnetized magnetic material, the magnetization M starts increasing with H from zero. The rate of increase gradually slows down and M asymptotically approaches a value M_s at high H. If H is reduced to zero, then M is reduced to a lower value M_r. Continuous reduction of H to a very high negative value will magnetize the material to $-M_s$. In order to bring the material to demagnetized state, a positive field H_c is required. Further increase in the H field will bring the trace of M to a closed loop. This loop is the major hysteresis loop, as shown in Fig. 96.14. The hysteresis loop shows that a magnetic material has memory. It is this memory that is used in the medium for storing information. H_c is the coercivity, indicating the strength of magnetic field required to erase the memory of a magnetic material. Magnetic materials with high H_c are "hard" magnets, and are suitable for medium applications if they have high M_r. On the other hand, magnetic materials with low H_c are "soft" magnets, and are candidates for head core materials if they have high M_s and high μ. M_r and M_s are the remanent and saturation magnetization, respectively, and their ratio is the remanent squareness. The flux density corresponding to M_s is B_s.

Magnetic Media

Magnetic media are used to store information in a magnetic recording system. In order to increase the areal density, we need to reduce flux change length B and track width w. Since B is limited by the term $M_s\delta/H_c$, where δ is the magnetic layer thickness, we can reduce B by either decreasing $M_s\delta$ or increasing H_c. However, the amplitude of the magnetic signal available for reproducing head is proportional to the term $M_s\delta w$. If we reduce track width w to increase areal density, we must increase $M_s\delta$ to avoid signal deterioration. In addition, if the magnetic layer is so thin that it causes thickness nonuniformity, more noise will appear in the reproducing process. Therefore, the major requirements for magnetic layer are high H_c, high M_r, and ease of making a uniform thin layer. Additional requirements include good magnetic and mechanical stability.

There are two groups of magnetic media. The first group is called particulate media because the magnetic materials are in the form of particles. This group includes iron oxide (γ-Fe_2O_3), cobalt-modified iron oxide (γ-Fe_2O_3+Co), chromium dioxide (CrO_2), metal particles, and barium ferrite ($BaFe_{12}O_{19}$). Some of these have been used in the magnetic recording for several decades. More recently, another group of media has been developed largely due to the ever-increasing demand for higher storage capacity

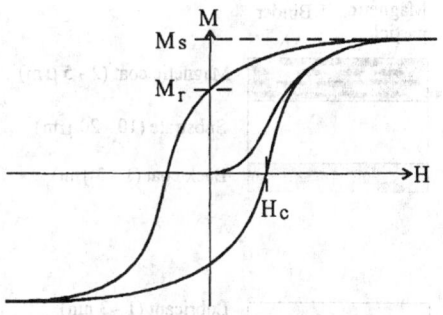

FIGURE 96.14 Hysteresis loop of a magnetic material shows the nonlinear relationship between M and H which results in magnetic memory.

TABLE 96.8 Remanence (M_r) and Coercivity (H_c) Values of Some Commonly Used Magnetic Media (some values are from Reference 5)

Group	Material	M_r (kA/m)	H_c (kA/m)	Application
Particulate	γ-Fe$_2$O$_3$	56–140	23–32	Floppy disk, audio, video, and instrumentation tapes
	γ-Fe$_2$O$_3$+Co	60–140	44–74	Floppy disk, audio, video, and instrumentation tapes
	CrO$_2$	110–140	38–58	Floppy disk, audio, video, and instrumentation tapes
	BaFe$_{12}$O$_{19}$	56	58	Floppy disk
Thin film	Co–Ni	600–1100	30–85	Hard disk
	Co–Fe	1100–1500	60–150	Hard disk
	Co–P	600–1000	36–120	Hard disk
	Co–Ni–Pt	600–1100	60–175	Hard disk
	Co–Cr–Ta	350–900	55–190	Hard disk
	Co–Cr–Pt	300–750	56–200	Hard disk

in the computer industry. This group of media is the thin-film media, where the magnetic layer can be made as a continuous thin film. Most materials in this group are cobalt-based metal alloys. Compared with particulate media, the thin-film media usually have a higher coercivity H_c, a higher remanence M_r, and can be deposited in a very thin continuous film. Table 96.8 lists H_c and M_r for some of the most popularly used particulate and thin-film media. Note that magnetic properties are affected by the fabrication process and film structure. Therefore, their values can be out of the ranges of Table 96.8 if different processes are used.

Magnetic media can be classified into three general forms of applications. Tape is the oldest form and remains an important medium today. It is central to most audio, video, and instrumentation recording, although it is also used in the computer industry for archival storage. Tape is economical and can hold a large capacity, but suffers slow access time. Hard disk is primarily used as the storage inside a computer, providing fast data access for the user, but having poor transportability. Flexible disk is designed for easy data transportation, but is limited in capacity. Besides these three general forms of applications, a hybrid of flexible and hard disk is being gradually accepted. It is a removable rigid disk capable of holding up to several gigabytes of digital data. In addition, magnetic stripes are getting wide use in different forms of cards.

The magnetic layer alone cannot be used as a medium. It needs additional components to improve its chemical and mechanical durability. Typical cross sections of a particulate magnetic tape and a thin-film hard disk are shown in Fig. 96.15. In the case of tape application, iron particles with typical size of 0.5 μm long and 0.1 μm wide are dispersed in a polymeric binder, together with solvents, lubricants, and other fillers to improve magnetic and mechanical stability. This dispersed material is then coated on an abiaxially oriented polyethylene terephthalate substrate. An optional back coat may also be applied to the other side of the substrate. The cross section of a hard disk is more complex. A high-purity aluminum–magnesium (5 wt%) substrate is diamond turned to a fine surface finish, and then electrolessly plated with a nonmagnetic nickel–phosphorus (10 at%) undercoat. This layer is used to increase the

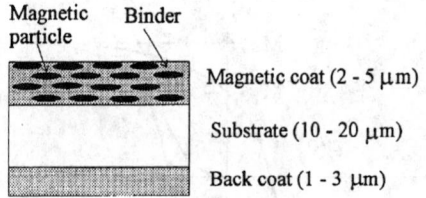

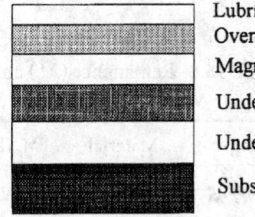

FIGURE 96.15   Cross-sectional views of a particulate magnetic tape (top) and a thin film hard disk (bottom).

TABLE 96.9   Relative Permeability (μ), Saturation Flux Density (B_s), Coercivity (H_c) and Resistivity (ρ) Values of Some Commonly Used Magnetic Head Materials at Low Frequency (some values are from Reference 5)

Material	μ	B_s (T)	H_c (A/m)	ρ ($\mu\Omega$cm)	Application
Ni–Fe–Mo	11000	0.8	2.0	100	Audio tape
Ni–Zn	300–1500	0.4–0.46	11.8–27.6	10^{11}	Floppy and hard disk drives, video and instrumentation tapes
Mn–Zn	3000–10000	0.4–0.6	11.8–15.8	10^6	Floppy and hard disk drives, video and instrumentation tapes
Fe–Si–Al	8000	1.0	2.0	85	Floppy and hard disk drives, video and instrumentation tapes
Ni–Fe	2000–4000	1.0	<10	20	Hard disk drives

hardness, reduce the defects, and improve the finish of the Al–Mg alloy, and is polished to a super surface finish. Next, an underlayer of chromium is sputtered to control the properties of the magnetic film, followed by sputtering the magnetic film. Finally, a layer of hydrogenated or nitrogenated carbon is overcoated on the magnetic film, and an ultrathin layer of perfluorinated hydrocarbon liquid lubricant is applied on top. The carbon and lubricant layers are used to improve the corrosion and mechanical resistance of the disk. For a 95-mm disk the finished product should have a surface flatness better than 10 μm and a tightly control surface roughness. In some applications, an arithmetic average roughness (R_a) of less than 0.5 nm is required.

Magnetic Heads

Magnetic heads have three functions: recording, reproducing, and erasing. Usually for stationary head applications such as tape drives, multiple heads are used to perform these functions. For moving head applications such as disk drives, a single head is employed because of the requirements of simple connections and small head mass for fast data access. Most of these heads are the inductive type, where the fundamental design is an inductive coil and a magnetic core. The general requirements for the core materials are high relative permeability μ, high saturation flux density B_s, low coercivity H_c, high electric resistivity ρ, and low magnetostriction coefficient λ. Some of the properties for the commonly used core materials are listed in Table 96.9.

The evolution of the magnetic head follows the selection of core materials, as shown in Fig. 96.16. Early heads used laminated molybdebum Permalloy (Ni-Fe-Mo, 79-17-4 wt%). These heads are inex-

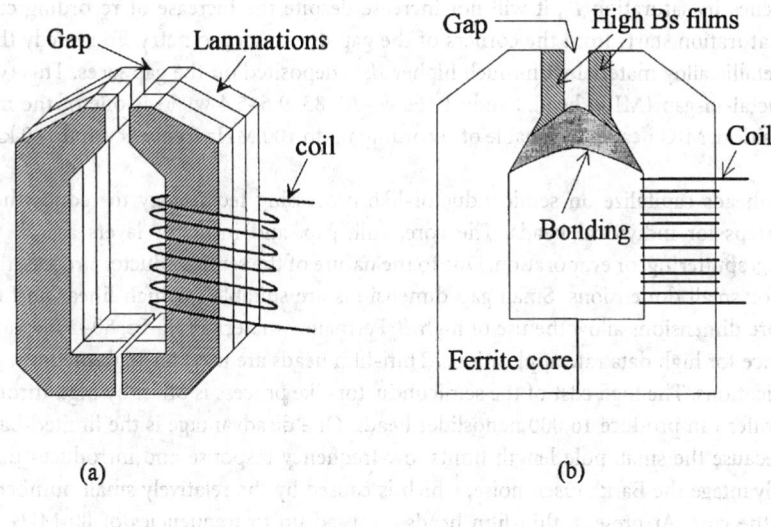

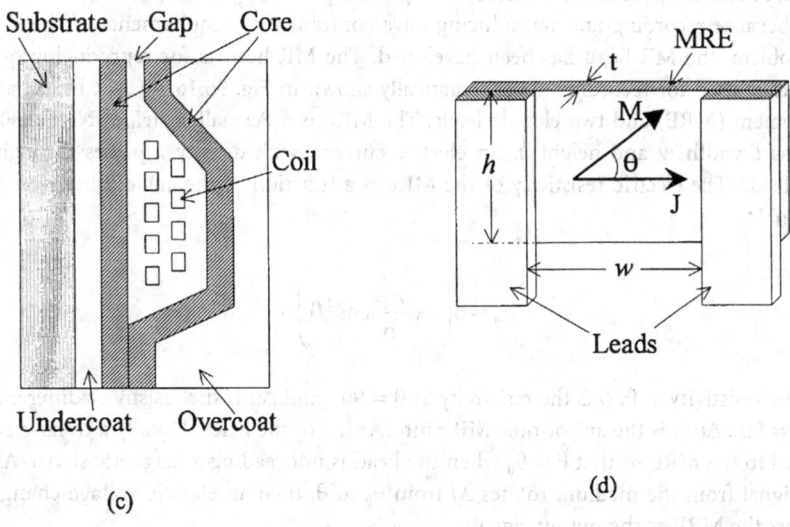

FIGURE 96.16 Schematic illustrations of (a) a laminated head, (b) cross-section of a MIG head, (c) cross-section of a thin film head, and (d) an MR sensor with leads.

pensive to make, and have low H_c and high μ and B_s. The primary drawbacks are frequency limitation, gap dimension inaccuracy, and mechanical softness. Frequency limitation is caused by the difficulty of making the lamination layer thinner than 25 μm. Eddy current loss, which is proportional to layer thickness and square root of frequency, reduces the effective permeability. As a result, laminated heads are seldom used for applications exceeding 10 MHz. Gap dimension inaccuracy is associated with the head fabrication process, and makes it unsuitable for high areal density applications. Lack of mechanical hardness reduces its usable life.

One way to reduce eddy current loss is to increase core material electric resistivity. Two types of ferrite material have high resistivity (four to nine orders higher than Permalloy) and reasonable magnetic properties: Ni–Zn and Mn–Zn. These materials are also very hard, elongating head life during head/medium contacts. The major deficiency of ferrite materials is their low B_s values. In order to record in high H_c media, high flux density B is needed in the head core. When the flux density in the core material reaches its saturation B_s, it will not increase despite the increase of recording current or coil turns. This saturation starts from the corners of the gap due to its geometry. To remedy this deficiency, a layer of metallic alloy material with much higher B_s is deposited on the gap faces. This type of head is called the metal-in-gap (MIG) head. Sendust (Fe–Si–Al, 85–9.6–5.4 wt%) is one of the materials used for the deposition. MIG heads are capable of recording up to 100 MHz frequency and 180 kA/m medium coercivity.

Thin-film heads capitalize on semiconductor-like processing technology to reduce the customized fabrication steps for individual heads. The core, coil, gap, and insulator layers are all fabricated by electroplating, sputtering, or evaporation. Due to the nature of the semiconductor process, the fabrication is accurate for small dimensions. Small gap dimensions are suitable for high linear and track density, and small core dimensions allow the use of high B_s Permalloy material (Ni–Fe, 80–20 wt%) as core with low inductance for high data rate applications. Thin-film heads are used for high medium H_c, high areal density applications. The high cost of the semiconductor-like process is offset by high throughput: a 150 × 150 mm wafer can produce 16,000 nanoslider heads. One disadvantage is the limited-band recording capability because the small pole length limits low-frequency response and introduces undershoots. A second disadvantage the Barkhausen noise, which is caused by the relatively small number of magnetic domains in the core. At present, thin-film heads are used up to frequencies of 80 MHz and medium coercivity of 200 kA/m. MIG thin-film heads are also being used for high-coercivity applications.

An inductive head is often used for both recording and reproducing. The optimal performance cannot be achieved because recording and reproducing have contradictory requirements for head design. To solve this problem, the MR head has been developed. The MR head is for reproducing only, and an inductive head is used for recording. As schematically shown in Fig. 96.16, an MR head has a magnetoresistive element (MRE) and two electric leads. The MRE is a Permalloy stripe (Ni–Fe, 80–20 wt%), with thickness t, width w, and height h. An electric current, with density J, passes through the MRE through the leads. The electric resistivity of the MRE is a function of the angle θ between J and MRE magnetization M:

$$\rho_\theta = \rho\left(1 + \frac{\Delta\rho}{\rho}\cos^2\theta\right) \tag{96.2}$$

where ρ_θ is the resistivity at θ, ρ is the resistivity at $\theta = 90°$, and $\Delta\rho$ is the resistivity difference between $\theta = 0°$ and $\theta = 90°$. $\Delta\rho/\rho$ is the anisotropic MR ratio (AMR) of the MRE. Usually a transverse magnetic field is applied to the MRE so that $\theta = \theta_0$ when the head is not reading a magnetic signal. Assume that a magnetic signal from the medium rotates M from θ_0 to θ, then an electric voltage change v will be detected across the MRE as the output signal:

$$v = Jwp\frac{\Delta\rho}{\rho}\left(\sin^2\theta_0 - \sin^2\theta\right) \tag{96.3}$$

where θ_0 is the bias angle and is set to 45° for good linearity. In practice, a longitudinal bias is also used along the MRE width direction to stabilize the magnetic domain and reduce large Barkhausen noise. To compare the output between an MR head and an inductive head, we write the inductive head output using Faraday's equation:

$$v = -nV\frac{d\phi}{dx} \tag{96.4}$$

where n is the number of the coil turns, V is the medium velocity, ϕ is the magnetic flux, and x is the coordinate axis fixed on the medium surface. Equations 96.3 and 96.4 tell us that while inductive head output is proportional to medium velocity and not suitable for low-velocity applications, the MR head can be used for either high- or low-velocity applications.

Recording Process

We can imagine the recording process in two steps. First, the magnetic flux flowing in the head core generates a fringe magnetic field around the gap. Then the magnetic field magnetizes the magnetic medium and leaves a magnetization transition in it. Partly due to the nonlinear nature of the hysteresis loop, the recording process is so complex that there has been no rigorous explanation. However, we can still obtain significant insights into the recording process by using simple models if we keep in mind the limitations.

If we set the origin of a coordinate system at the center of the gap with x axis on the head surface and y axis pointing away from the head, then the longitudinal magnetic field H_x and perpendicular magnetic field H_y of this head can be expressed by the Karlqvist approximation [2]:

$$H_x = \frac{ni}{\pi\left(g + lA_g/\mu A_c\right)} \tan^{-1}\left[\frac{yg}{x^2 + y^2 - \left(g^2/4\right)}\right] \tag{96.5}$$

$$H_y = \frac{ni}{2\pi\left(g + lA_g/\mu A_c\right)} \ln\left[\frac{\left(x - g/2\right)^2 + y^2}{\left(x + g/2\right)^2 + y^2}\right] \tag{96.6}$$

where n is the number of coil turns, i is the current in the coil, g is the gap length, l is the core length, A_g is the core cross-sectional area, μ is the relative permeability of core material, and A_c is the gap cross-sectional area. Both Eqs. 96.5 and 96.6 give accurate results for points $0.25g$ away from gap corners. Since longitudinal recording mode dominates the magnetic recording industry, we will focus on the field H_x. Equation 96.5 shows that the contours of constant H_x field are circles nesting on the two gap corners, as shown in Fig. 96.17. The greater the diameter of the circle, the weaker the magnetic field. Assume a magnetic medium, moving from left to right with a distance d above the head, has a thickness δ and a magnetization M pointing to right. At some instant the recording current turns on and generates the magnetic field H_x above the gap as depicted in Fig. 96.17. On the circumference of $H_x = H_c$, half of medium material has its magnetization reversed and half remains the same, resulting in a zero total magnetization. Since H_x has a gradient, the medium closer to the gap (inside a smaller circle) gets its magnetization reversed more completely than the medium farther away from the gap (outside a bigger circle). Therefore, magnetic transition is gradual in the medium even if the change of recording current follows a step function. Assume the original magnetization is M_r and the completely reversed magnetization is $-M_r$, this gradual change of magnetization for an isolated transition can be modeled by [3]:

$$M = \frac{2}{\pi} M_r \tan^{-1}\frac{x}{a} \tag{96.7}$$

where x is the distance from the center of transition and a is a parameter characterizing the sharpness of the transition as shown in Fig. 96.18. Assuming a thin-film medium and using the Karlqvist approximation for the head field, a is found to be [4–6]:

$$a = \frac{\left(1 - S^*\right)\left(d + \delta/2\right)}{\pi Q} + \sqrt{\left[\frac{\left(1 - S^*\right)\left(d + \delta/2\right)}{\pi Q}\right]^2 + \frac{M_r \delta\left(d + \delta/2\right)}{\pi Q H_c}} \tag{96.8}$$

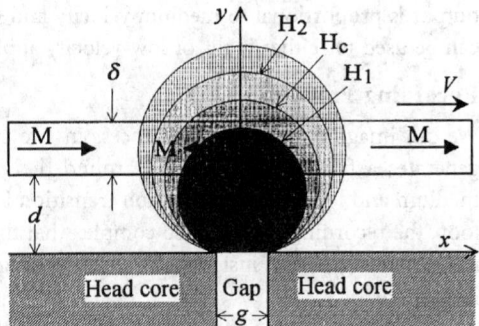

FIGURE 96.17 The constant horizontal fields of Karlqvist approximation are circles resting on the gap corners of a head, and the change of magnetization in the medium is gradual.

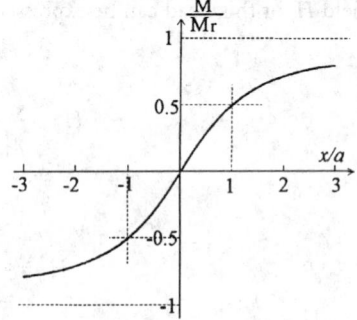

FIGURE 96.18 An isolated arctangent magnetization transition from negative M_r to positive M_r.

where S^* is the medium loop squareness and Q is the head-field gradient factor. For a reasonably well designed head, $Q \approx 0.8$. It is obvious that we want to make parameter a as small as possible so that we can record more transitions for a unit medium length. If the head gap length g and medium thickness δ are small compared with head/medium separation d, and the medium has a squareness of one, then the minimum possible value of a is [7]

$$a_m = \begin{cases} \dfrac{M_r \delta}{2\pi H_c} & \dfrac{M_r \delta}{4\pi H_c d} \geq 1 \\[3mm] \sqrt{\dfrac{M_r \delta d}{\pi H_c}} & \dfrac{M_r \delta}{4\pi H_c d} < 1 \end{cases} \tag{96.9}$$

In order to decrease the value of a and therefore increase areal density, we need to reduce medium remanence M_r, thickness δ, head/medium separation d, and to increase coercivity H_c.

Reproducing Process

In contrast to the recording process, the reproducing process is well understood. The flux density induced in the head core is on the order of a few millitesia, yielding a linear process for easier mathematical treatment. The head fringe field is the Karlqvist approximation (Eq. 96.5) and the foundation is the reciprocity theorem. For an isolated transition (Fig. 96.18) with a thin magnetic layer $\delta \ll d$, the induced electric voltage v for an inductive head is [7]:

$$v(x) = \frac{-2\mu_0 V w M_r \delta n}{\pi\left(g + lA_g/\mu A_c\right)}\left[\tan^{-1}\left(\frac{g/2 + x}{a + d}\right) + \tan^{-1}\left(\frac{g/2 - x}{a + d}\right)\right] \tag{96.10}$$

where μ_0 is the permeability of vacuum, μ is the relative permeability of the core, V is the medium velocity, w is the track width, n is the number of coil turns, g is the head gap length, d is the head/medium

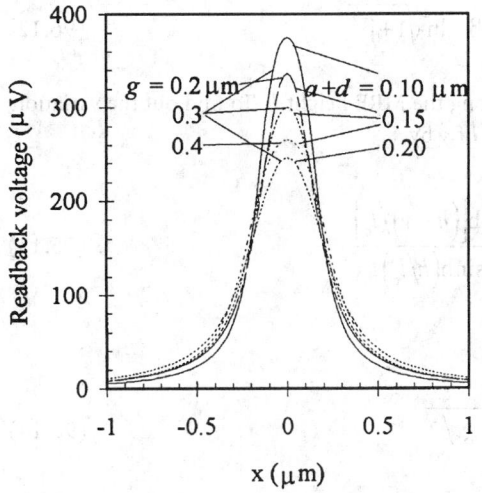

FIGURE 96.19 The reproducing voltage of an inductive head over an isolated arctangent transition shows the effects of gap length g, parameter a, and head/medium separation d.

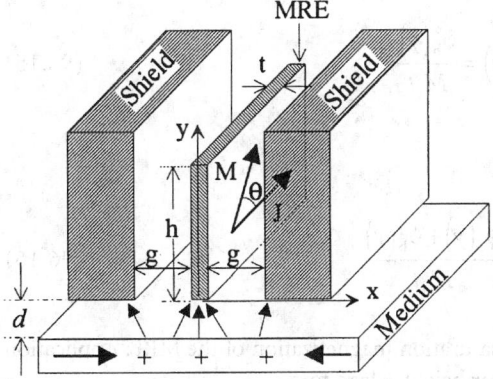

FIGURE 96.20 Schematic diagram of a shielded MR head with a shield to MRE distance g.

separation, and x is the distance between the center of the medium transition and the center of the head gap. The term $lA_g/\mu A_c$ is closely related to g for head efficiency. When a transition passes under the head, its voltage starts with a very low value, reaches a peak, then falls off again, as shown in Fig. 96.19, where the following typical values for a hard disk drive are used: $V = 20$ m/s, $w = 3.5$ μm, $M_r = 450$ kA/m, $\delta = 50$ nm, $n = 50$, $lA_g/\mu A_c = 0.1g$. The effects of g and $a + d$ are shown in Fig. 96.19. Since a greater peak voltage and a narrower spatial response are desired for the reproducing process, smaller g and $a + d$ values are helpful.

When an MR head is used for reproducing, the MRE is usually sandwiched between two magnetic shields to increase its spatial resolution to medium signals, as shown in Fig. 96.20. Since the MR head is flux sensitive, the incident flux ϕ_i on the bottom surface of the MRE should be derived as a function of the distance (x) between the center of MRE and the center of the transition [7, 8]:

$$\phi_i(x) = \frac{2\mu_0 w M_r \delta (a+d)}{\pi g} \left\{ f\left[\frac{x+(g+t)/2}{a+d}\right] - f\left[\frac{(x+t)/2}{a+d}\right] + \right.$$

$$\left. f\left[\frac{x-(g-t)/2}{a+d}\right] - f\left[\frac{(x-t)/2}{a+d}\right] \right\} \tag{96.11}$$

where g is the distance between the MRE and the shield, t is the MRE thickness, and

$$f(\beta) = \beta \tan^{-1} \beta - \ln \sqrt{1 + \beta^2} \qquad (96.12)$$

The angle between magnetization and current varies along the MRE height h. To find out the variation, we need to calculate the signal flux decay as a function of y by

$$\phi_s(y) = \phi_i \frac{\sinh\left[(h-y)/l_c\right]}{\sinh\left(h/l_c\right)} \qquad (96.13)$$

where

$$l_c = \sqrt{\mu g t / 2} \qquad (96.14)$$

Then the bias angle θ_0 and signal angle θ can be calculated by

$$\sin\theta_0(y) = \frac{\phi_b(y)}{M_s t} \qquad (96.15)$$

and

$$\sin\theta(y) = \frac{\phi_s(y) + \phi_b(y)}{M_s t} \qquad (96.16)$$

where ϕ_b is the biasing flux in the MRE and M_s is the saturation magnetization of the MRE. Application of Eqs. 96.15 and 96.16 to Eq. 96.3 and integration over height h lead to

$$v = Jw\rho \frac{\Delta\rho}{\rho} \frac{1}{h} \left[\int_0^h \sin^2\theta_0(y)\,dy - \int_0^h \sin^2\theta(y)\,dy \right] \qquad (96.17)$$

For an MR head with a 45° bias at the center and small height $h \ll l_c$, the peak voltage is [6]

$$v_p \approx \frac{9\Delta\rho JwM_r\delta(g+t)}{8\sqrt{2}\,tgM_s} \tan^{-1}\left[\frac{g}{2(a+d)} \right] \qquad (96.18)$$

The general shape of the reproducing voltage from an MR head is similar to that in Fig. 96.19.

The study of an isolated transition reveals many intrinsic features of the reproducing process. However, transitions are usually recorded closely in a magnetic medium to achieve high linear density. In this case, the magnetization variation in the medium approaches a sinusoidal wave. That is,

$$M(x) = M_r \sin\frac{2\pi}{\lambda}x \qquad (96.19)$$

where λ is the wavelength. The reproducing voltage in an inductive head becomes [9, 10]

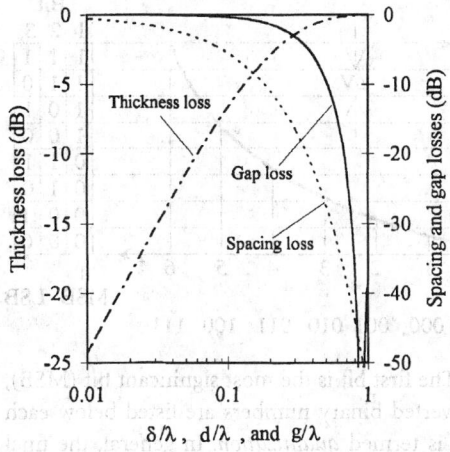

FIGURE 96.21 Spacing, thickness, and gap losses of an inductive head vs. frequency for the reproducing of a sinusoidal medium magnetization.

$$v(x) = \frac{-\mu_0 VwM_r ng}{g + lA_g/\mu A_c}\left(e^{-2\pi d/\lambda}\right)\left(1 - e^{-2\pi\delta/\lambda}\right)\left(\frac{\sin\frac{\pi g}{\lambda}}{\pi g/\lambda}\right)\cos\frac{2\pi}{\lambda}x \qquad (96.20)$$

This equation presents all the important features of the high-linear-density reproducing process. The term $\exp(-2\pi d/\lambda)$ is the spacing loss. It shows that the reproducing voltage falls exponentially with the ratio of head/medium spacing to wavelength. The second term $1 - \exp(-2\pi\delta/\lambda)$ is the thickness loss. The name of this term is misleading because its value increases with a greater medium thickness. However, the rate of increase diminishes for thicker medium. In fact, 80% of the maximum possible value is achieved by a medium thickness of 0.25λ. The last term $\sin(\pi g/\lambda)/(\pi g/\lambda)$ is the gap loss. This term is based on the Karlqvist approximation. If a more accurate head fringe field is used, this term is modified to $\sin(\pi g/\lambda)/(\pi g/\lambda)\cdot(1.25g^2 - \lambda^2)/(g^2 - \lambda^2)$ [11]. It shows a gap null at $\lambda = 1.12g$, and limits the shortest wavelength producible. These three terms are plotted in Fig. 96.21. The most significant loss comes from the spacing loss term, which is 54.6 dB for $d = \lambda$. Therefore, one of the biggest efforts spent on magnetic recording is to reduce the head/medium spacing as much as possible without causing mechanical reliability issues. For an MR head, the reproducing voltage is [11]

$$v \propto \frac{4M_r i\Delta\rho w\lambda}{ht}\left(e^{-2\pi d/\lambda}\right)\left(1 - e^{-2\pi\delta/\lambda}\right)\left(\frac{\sin\frac{\pi g}{\lambda}}{\pi g/\lambda}\right)\sin\frac{\pi(g+t)}{\lambda}\cos\frac{2\pi}{\lambda}x \qquad (96.21)$$

Digital vs. Analog Recording

Due to the nonlinearity of the hysteresis loop, magnetic recording is intrinsically suitable for digital recording, where only two states (1 and 0) are to be recognized. Many physical quantities, however, are received in analog form before they can be recorded, such as in consumer audio and instrumentation recording. In order to perform such recording, we need to either digitize the information or use the analog recording technique. In the case of digitization, we use an analog-to-digital converter to change a continuous signal into binary numbers. The process can be explained by using the example shown in Fig. 96.22. An electric signal V, normalized to the range between 0 and 1, is to be digitized into three bits. The signal is sampled at time $t = 1, 2, \ldots, 6$. At each sampling point, the first bit is assigned a 1 if the value of the continuous signal is in the top half (>0.5), otherwise assigned a 0. The second bit is assigned a 1 if the value of the continuous signal is in the top half of each half ($0.25 \leq V < 0.5$, or >0.75),

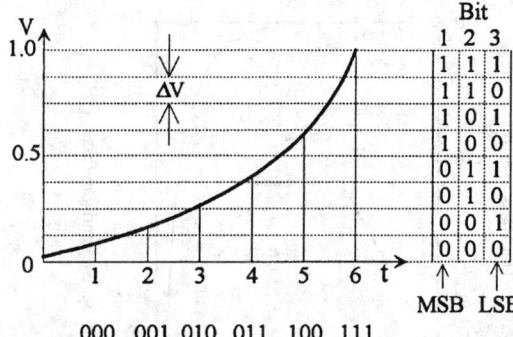

FIGURE 96.22 Schematic illustration of the quantization of a continuous signal to three bits.

otherwise assigned a 0. The third bit is assigned similarly. The first bit is the most significant bit (MSB), and the last bit is the least significant bit (LSB). The converted binary numbers are listed below each sampling point in Fig. 96.22. This process of digitization is termed *quantization*. In general, the final quantization interval is

$$\Delta V = \frac{V}{2^N} \tag{96.22}$$

Where V is the total voltage range and N is the number of bits. Because we use a finite number of bits, statistically there is a difference between the continuous signal and the quantized signal at the sampling points. This is the quantization error. It leads to a signal-to-quantization-noise ratio (SNR) [12]:

$$SNR = \frac{12 P_s}{\Delta V^2} \tag{96.23}$$

where P_s is the mean square average signal power. For a signal with uniform distribution over its full range, this yields

$$SNR = 2^{2N} \tag{96.24}$$

For a sinusoidal signal, it changes to

$$SNR = 1.5 \times 2^{2N} \tag{96.25}$$

The SNR can be improved by using more bits. This improvement, however, is limited by the SNR of the incoming continuous signal. The quantized signal is then pulse code modulated (PCM) for recording.

For analog recording, a linear relationship between the medium magnetization and the recording signal is required. This is achieved through the anhysteretic magnetization process. If we apply an alternating magnetic field and a unidirectional magnetic field to a previously demagnetized medium, and then reduce the amplitude of the alternating field to zero before we remove the unidirectional field, the remanent magnetization shows a pseudolinear relationship with the unidirectional field strength H_u up to some level. Figure 96.23 shows such an anhysteretic curve. The linearity deteriorates as H_u gets greater. In applications, the recording signal current is biased with an ac current of greater amplitude and higher frequency. Therefore, it is also termed ac-biased recording. Analog recording is easy to implement, at the price of a lowered SNR because remanent magnetization is limited to about 30% of the maximum possible M_r to achieve good linearity.

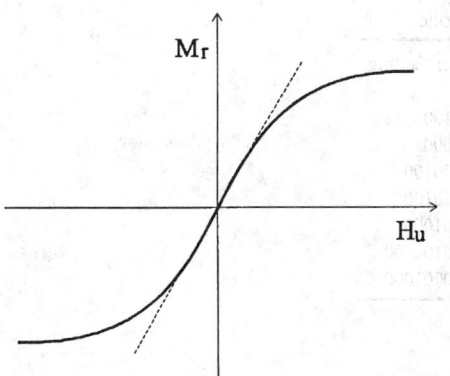

FIGURE 96.23 An anhysteretic remanent magnetization shows a pseudo-linear relationship with the applied unidirectional magnetic field to some H_u level.

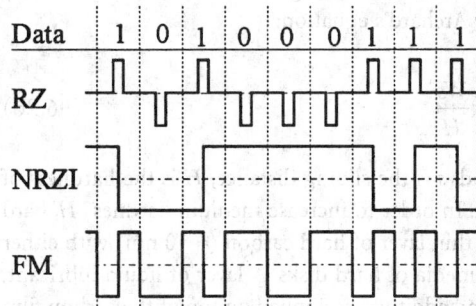

FIGURE 96.24 Comparison of some early developed codes.

Recording Codes

PCM is a scheme of modifying input binary data to make them more suitable for a recording and reproducing channel. These schemes are intended to achieve some of the following goals: (1) reducing the dc component, (2) increasing linear density, (3) providing self-clocking, (4) limiting error propagation, and (5) achieving error-free detection. There are numerous code schemes, only three of the ones developed early are shown in Fig. 96.24. The earliest and most straightforward one is the return-to-zero (RZ) code. In this scheme a positive and negative pulse is used to represent each 1 and 0, respectively, of the data. The main drawback is that direct recording over old data is not possible due to the existence of zero recording current between two data. It also generates two transitions for each bit, therefore reducing the linear density. In addition, it only uses half of the available recording current range for a transition. The non-return-to-zero-invert (NRZI) method was developed to alleviate some of these problems. It changes the recording current from one direction to the other for each 1 of the data, while making no changes for all 0s. However, it has a strong dc component and may lose reproducing synchronization if there is a long string of 0s in the input data. In addition, reproducing circuits are usually not designed for dc signal processing. In frequency modulation (FM) code there is always a transition at the bit–cell boundary which acts as a clock. There is also an additional transition at the bit–cell center for each 1 and no transition for 0s. It reduces the dc component significantly. The primary deficiency is the reduction of linear density since there are two transitions for each 1 in the data.

The most popularly used codes for magnetic recording are the run-length-limited (RLL) codes. They have the general form of $m/n(d, k)$. In these schemes, data are encoded in groups. Each input group has m bits. After encoding, each group contains n bits. In some schemes multiple groups are coded together. d and k are the minimum and maximum 0s, respectively, between two consecutive 1s in the encoded sequence. While d is used to limit the highest transition density and intersymbol interference, k is employed to ensure adequate transition frequency for reproducing clock synchronization. The encoding is carried out by using a lookup table, such as Table 96.10 for a 1/2(2,7) code [13].

TABLE 96.10 1/2(2,7) Code

Before Coding	After Coding
10	0100
11	1000
000	000100
010	100100
011	001000
0010	00100100
0011	00001000

Head/Medium Interface Tribology

As expressed in Eq. 96.20, the most effective way to increase signal amplitude, therefore areal density, is to reduce head/medium spacing d. However, wear occurs when a moving surface is in close proximity to another surface. The amount of wear is described by Archard's equation:

$$V = k\frac{Ws}{H} \tag{96.26}$$

where V is the volume worn away, W is the normal load, s is the sliding distance, H is the hardness of the surface being worn away, and k is a wear coefficient. In order to increase medium hardness H, hard Al_2O_3 particles are dispersed in particulate media and a thin layer of hard carbon ($\approx$ 10 nm) with either hydrogenation or natrogenation is coated on thin-film media of hard disks. A layer of liquid lubricant, typically perfluoropolyethers with various end groups and additives, is applied on top of the carbon film to reduce the wear coefficient k. Load is minimized to reduce wear while keeping adequate head/medium dynamic stability. For applications where the sliding distance s is modest over the lifetime of the products such as floppy disk drives and consumer tapes drives, the head physically contacts the medium during operations. In the case of hard disk application, heads are separated nominally from the media by a layer of air cushion. The head is carried on a slider, and the slider uses air-bearing surfaces (ABS) to create the air film based on hydrodynamic lubrication theory. Figure 96.25 shows two commonly used ABS. Tapers are used to help the slider take off and maintain flying stability. ABS generates higher-than-ambient pressure to lift the slider above the medium surface during operations. The tripad slider is for pseudo-contact applications while the subambient-pressure (SAP) slider is for flying (such as MR head) applications. Because the relative linear velocity between the slider and the medium changes when the head moves to different radii, a cavity region is used in the SAP slider to generate suction force to reduce flying height variation. The ABS is designed based on the modified Reynolds equation:

$$\frac{\partial}{\partial X}\left(PH^3Q\frac{\partial P}{\partial X}\right) + \frac{\partial}{\partial Y}\left(PH^3Q\frac{\partial P}{\partial Y}\right) = \Lambda_x\frac{\partial(PH)}{\partial X} + \Lambda_y\frac{\partial(PH)}{\partial Y} + \sigma\frac{\partial(PH)}{\partial T} \tag{96.27}$$

where X and Y are coordinates in the slider longitudinal and transverse directions normalized by the slider length and width, respectively, P is the hydrodynamic pressure normalized by the ambient pressure, H is the distance between the ABS and medium surface normalized by the minimum flying height, Q is the molecular slip factor, T is time normalized by the characteristic squeeze period, Λ_x and Λ_y are the bearing numbers in the x and y directions, respectively, and σ is the squeeze number. A full derivation and explanation of the Reynolds equation can be found in Reference 14. At present, high end hard disk drives feature a flying height on the order of 20 to 50 nm.

When power is turned off, the slider in the popularly used Winchester-type drives rests on the medium surface. Although the ABS and medium surface look flat and smooth, they really consist of microscopic peaks and valleys. If we model an ABS/medium contact by a flat surface and a sphere tip, the liquid lubricant on the medium surface causes a meniscus force F_m as depicted in Fig. 96.26 [15]:

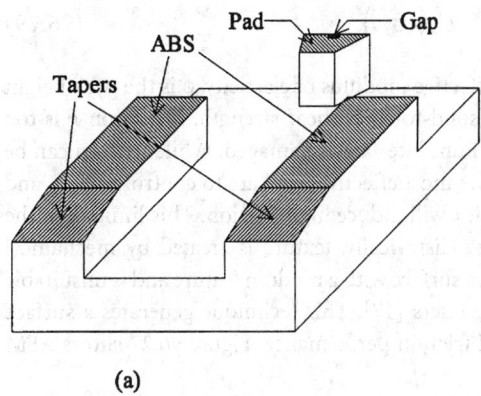

(a)

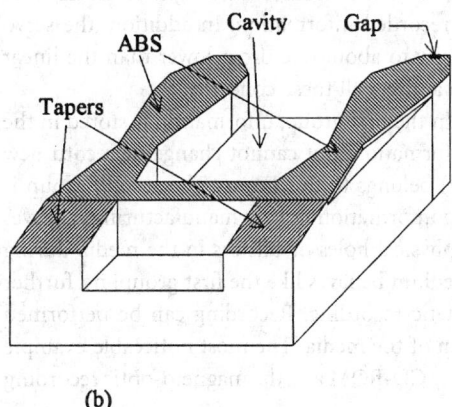

(b)

FIGURE 96.25 The ABS of (a) a tri-pad slider for pseudo-contact recording and (b) a SAP slider for conventional flying recording.

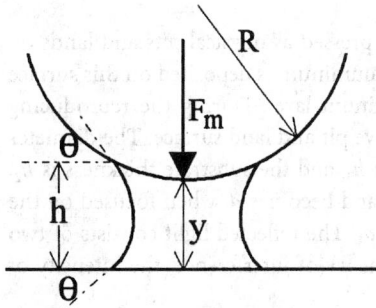

FIGURE 96.26 Formation of meniscus between a sphere tip and a flat surface.

$$F_m = \frac{4\pi R\gamma\cos\theta}{1+y/(h-y)} \qquad (96.28)$$

where R is the radius of the sphere, γ is the surface tension of the lubricant, θ is the contact angle between the lubricant and the surfaces, y is the sphere to flat surface distance, and h is the lubricant thickness. Detailed analysis [16] shows that the static friction F at a head/medium interface is a function of several parameters:

$$F = f\left(h, R, A, \eta, \gamma, \theta, E', \phi, \sigma, s\right) \tag{96.29}$$

where A is the ABS area, η is the peak density, E' is the effective modulus of elasticity, ϕ is the peak height distribution, σ is the rms peak roughness, and s is the solid-to-solid shear strength. If friction F is too large, either the drive cannot be started or the head/medium interface is damaged. While friction can be reduced practically by reducing A, γ, and increasing θ, the most effective ways are to control h, σ, η, and ϕ. Too thin a lubricant layer will cause wear, and too thick will induce high friction. This limits h to the range of 1 to 3 nm. σ is controlled by surface texture. Historically, texture is created by mechanical techniques using either free or fixed abrasives. This leaves a surface with a random feature and is unsuitable for controlling η and ϕ. Recently, people started to use lasers [17]. This technique generates a surface texture with well-defined η and ϕ to improve wear and friction performance. Figure 96.27 shows AFM images of a mechanical and a laser texture.

Optical Recording

The major obstacle to achieving higher areal density in magnetic recording is the spacing loss term. It is a great engineering challenge to keep heads and media in close proximity while maintaining the head/medium interface reliable and durable. Care must also be taken in handling magnetic media since even minute contamination or scratches can destroy the recorded information. In addition, the servo technique of using magnetic patterns limits the track density to about one order lower than the linear density. Optical recording, on the other hand, promises to address all these concerns.

Optical recording can be categorized into three groups. In the first group, information is stored in the media during manufacturing. Users can reproduce the information, but cannot change or record new information. CD-ROM (compact disk–read only memory) belongs to this group. The second group is WORM (write once read many times). Instead of recording information during manufacturing, it leaves this step to the user. This is usually achieved by creating physical holes or blisters in the media during the recording process. Once it is recorded, however, the medium behaves like the first group: no further recording is possible. The third group is similar to magnetic recording. Recording can be performed infinitely on the media by changing phase or magnetization of the media. The most noticeable example in this group is the magneto-optic (MO) technique [18]. Only CD-ROM and the magneto-optic recording are described in the following.

CD-ROM

Figure 96.28 shows the CD-ROM reproducing principle. Data are pressed as physical pits and lands on one surface of a plastic substrate, usually polycarbonate. A layer of aluminum is deposited on this surface to yield it reflective. Lacquer is then coated to protect the aluminum layer. During the reproducing process, an optical lens is used to focus a laser beam on the reflective pit and land surface. The diameter of the lens is D, the distance between the lens and the substrate is h_3, and the substrate thickness is h_2. The diameter of the laser beam is d_2 when entering the substrate, and becomes d_1 when focused on the reflective surface. The width of the pits are designed smaller than d_1. The reflected light consists of two portions: I_1 from the land and I_2 from the pit. According to the theory of interference, the intensity of the reflected light is

$$I = I_1 + I_2 + 2\sqrt{I_1 I_2}\, \cos\frac{4\pi h_1}{\lambda} \tag{96.30}$$

where λ is the wavelength of the laser and $h_1 \approx \lambda/4$ is the pit height. This leads to

$$I = \begin{cases} I_1 + I_2 - 2\sqrt{I_1 I_2} & \text{if there is a pit } (h_1 = \lambda/4) \\ I_1 + I_2 + 2\sqrt{I_1 I_2} & \text{if there is no pit } (h_1 = 0) \end{cases} \tag{96.31}$$

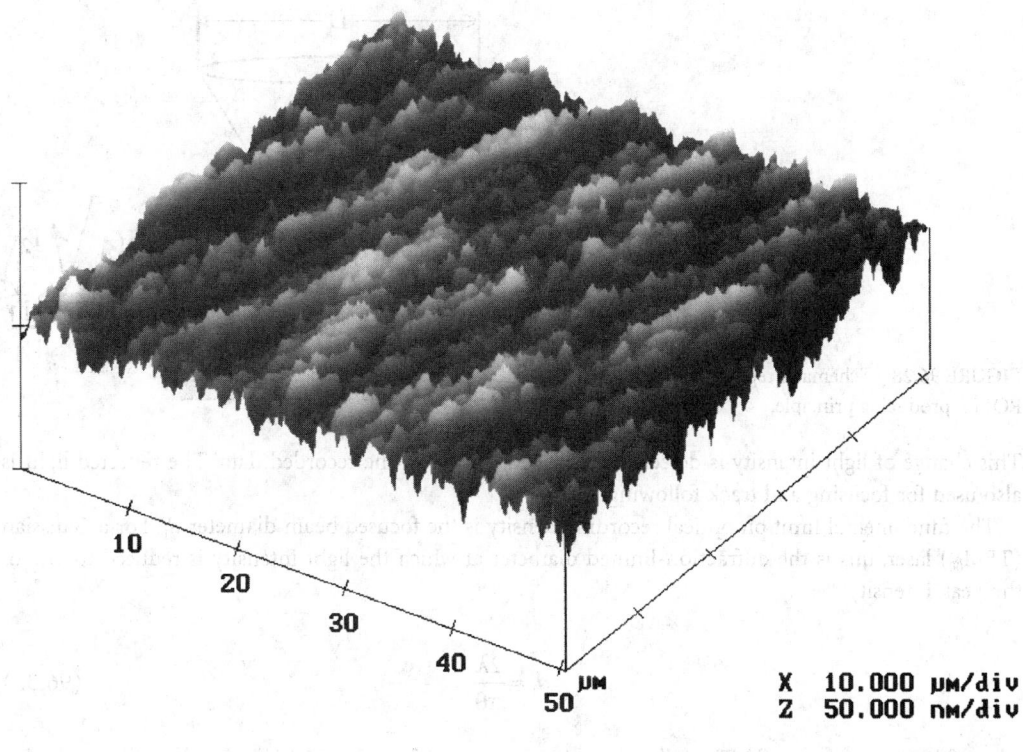

(a)

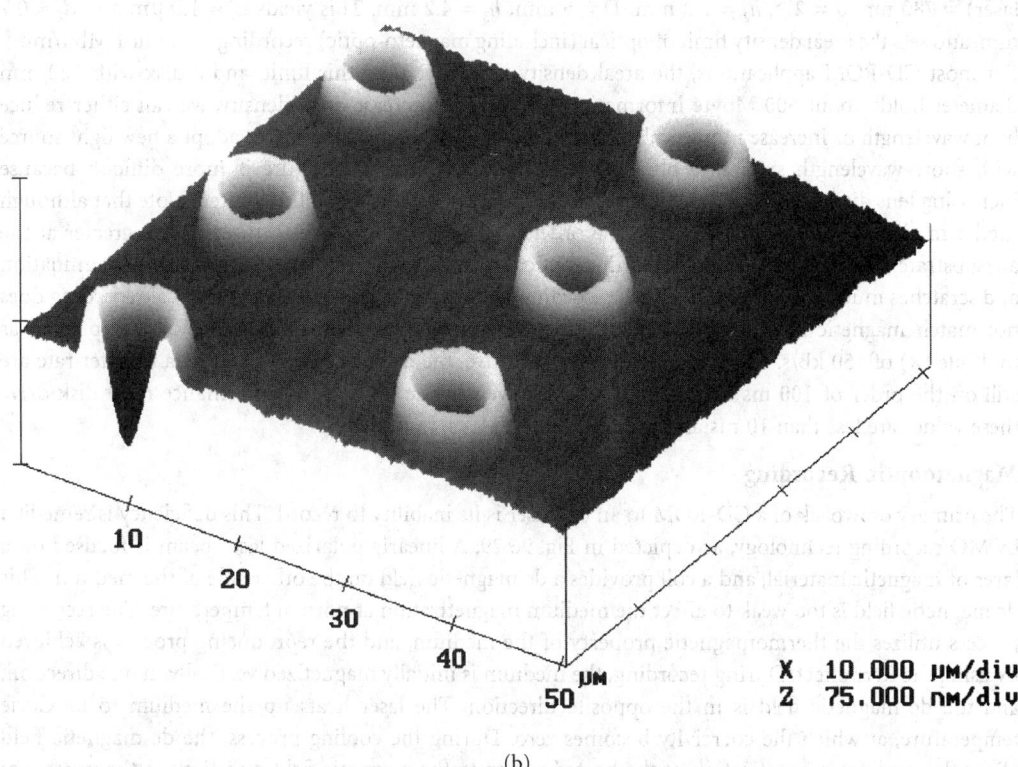

(b)

FIGURE 96.27 AFM images of (a) a mechanical texture and (b) a laser texture (Courtesy of J. Xuan).

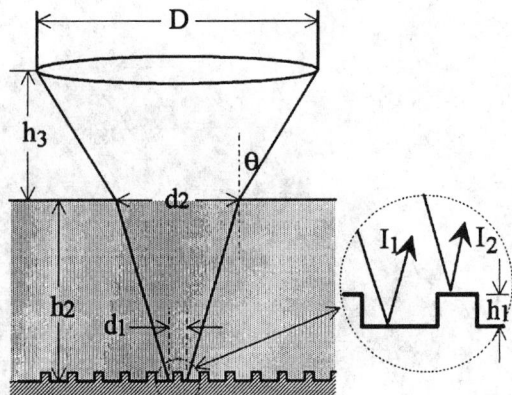

FIGURE 96.28 Schematic representation of the CD-ROM reproducing principle.

This change of light intensity is detected and decoded to yield the recorded data. The reflected light is also used for focusing and track following.

The fundamental limit on optical recording density is the focused beam diameter d_1. For a Gaussian (TEM_{00}) laser, this is the diffraction-limited diameter at which the light intensity is reduced to $1/e^2$ of the peak intensity:

$$d_1 \approx \frac{2\lambda}{\pi\theta} \tag{96.32}$$

where θ is the aperture angle. The following values are typical for a CD-ROM system: λ (gallium arsenide laser) = 780 nm, $\theta = 27°$, $h_2 = 1.2$ mm, $D = 5$ mm, $h_3 = 4.2$ mm. This yields $d_1 \approx 1.0$ µm and $d_2 \approx 0.7$ mm, and sets the areal density limit of optical (including magneto-optic) recording to about 1 Mbit/mm². For most CD-ROM applications, the areal density is smaller than this limit, and a disk with 120 mm diameter holds about 600 Mbyte information. In order to increase areal density, we can either reduce light wavelength or increase numerical aperture. Much of the effort has been to adopt a new light source with short wavelength such as a blue laser. Increasing numerical aperture is more difficult because increasing lens diameter is cost prohibitive and reducing h_2 or h_3 is reliability limited. Note that although the beam size on the focus plane is on the order of 1 µm (d_1), it is two to three orders greater at the air/substrate interface (d_2). This means that optical recording can tolerate disk surface contamination and scratches much better than magnetic recording. However, the performance of optical recording does not match magnetic recording in general. The data transfer rate of CD-ROM drives is expressed as multiple (×) of 150 kB/s. Even for a 12× CD-ROM drive, the data access time and data transfer rate are still on the order of 100 ms and 1.8 MB/s, respectively, while for a high-performance rigid disk drive these values are less than 10 ms and greater than 30 MB/s, respectively.

Magnetooptic Recording

The primary drawback of a CD-ROM to an end user is its inability to record. This deficiency is remedied by MO recording technology, as depicted in Fig. 96.29. A linearly polarized laser beam is focused on a layer of magnetic material, and a coil provides a dc magnetic field on the other side of the medium. This dc magnetic field is too weak to affect the medium magnetization at normal temperature. The recording process utilizes the thermomagnetic property of the medium, and the reproducing process is achieved by using the Kerr effect. During recording, the medium is initially magnetized vertically in one direction, and the dc magnetic field is in the opposite direction. The laser heats up the medium to its Curie temperature, at which the coercivity becomes zero. During the cooling process, the dc magnetic field aligns the medium magnetization of the heated region to the magnetic field direction. In the process of reproducing, the same laser is used with a smaller intensity. The medium is heated up to its compensation temperature, at which the coercivity becomes extremely high. Depending on the direction of the mag-

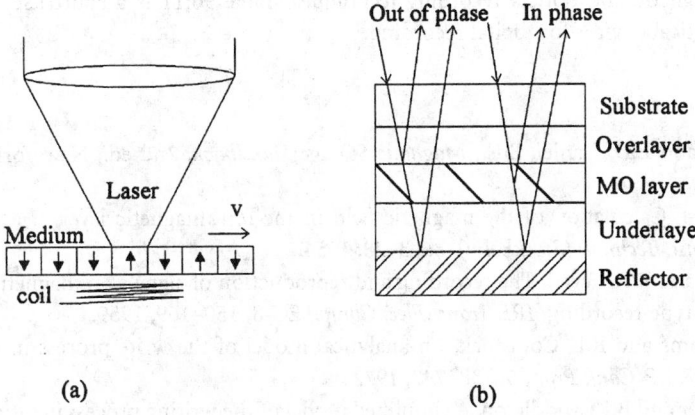

FIGURE 96.29 Schematic illustrations of (a) MO recording/reproducing and (b) quadrilayer medium cross section.

TABLE 96.11 Digital Magnetic and Optical Storage Devices

Description	Manufacturers	Approximate Price, $
Thin-film head for hard disk drive	AMC, Read-Rite, SAE	6.00–9.00
MR head for hard disk drive	AMC, Read-Rite, SAE, Seagate	8.00–12.00
Thin-film hard disk	Akashic, HMT, Komag, MCC, Stormedia	7.00–10.00
Hard disk drive	IBM, Maxtor, Quantum, Samsung, Seagate, WD	0.02–0.20/Mbytes
Floppy drive	Panasonis, Sony	20.00–40.00
Floppy disk	3M, Fuji, Memorex, Sony	0.15–0.50
Removable rigid disk drive	Iomega, Syquest	100.00–400.00
Removable rigid disk	Iomega, Maxell, Sony	5.00–20.00/100 Mbytes
Tape drive	Exabyte, HP, Seagate	100.00–400.00
Backup tape	3M, Sony, Verbatim	4.00–25.00/Gbytes
8 × CD-ROM drive	Goldstar, Panasonic	100.00–200.00
Recordable optical drive	JVC, Philips	300.00–500.00
Recordable optical disk	3M, Maxell, Memorex	3.00–15.00/650 Mbytes

netization, the polarization of the reflected light is rotated either clockwise or counterclockwise (Kerr rotation). This rotation of polarization is detected and decoded to get the data. The main disadvantage of MO recording is that a separate erasing process is needed to magnetize the medium in one direction before recording. Recently some technologies have been developed to eliminate this separate erasing process at the cost of complexity.

The medium used in MO recording must have a reasonable low Curie temperature (<300°C). The materials having this property are rare earth transition metal alloys, such as $Tb_{23}Fe_{77}$ and $Tb_{21}Co_{79}$. Unfortunately, the properties of these materials deteriorate in an oxygen and moisture environment. To protect them from air and humidity, they are sandwiched between an overlayer and a underlayer, such as SiO, AlN, SiN, and TiO_2. Another issue with the rare earth transition metal alloys is their small Kerr rotation, about 0.3°. To increase this Kerr rotation, multiple layers are used. In the so-called quadrilayer structure (Fig. 96.29b), the overlayer is about a half-wavelength thick and the underlayer is about a quarter-wavelength thick [18]. The MO layer is very thin ($\approx$3 nm). Light reflected from the reflector is out-of-phase with the light reflected from the surface of the MO layer, and is in-phase with the light reflected from the inside of the MO layer. As a result, the effective Kerr rotation is increased several times.

Compared with magnetic recording, optical recording has the intrinsic advantages of superior reliability and portability. However, its performance is inferior due to slower data access time and transfer rate. Another advantage of optical recording, higher areal density, has been disappearing or even reversing to magnetic recording. Both magnetic and optical recording will be continuously improved in the near future, probably toward different applications. Currently there are some emerging techniques that try to combine the magnetic and optical recording techniques. Table 96.11 is a short list of representative magnetic and optical devices for digital recording.

References

1. C.D. Mee and E.D. Daniel, Eds., *Magnetic Storage Handbook*, 2nd ed., New York: McGraw-Hill, 1996.
2. O. Karlqvist, Calculation of the magnetic field in the ferromagnetic layer of a magnetic drum, *Trans. R. Inst. Technol. (Stockholm)*, 86, 3–28, 1954.
3. J.J. Miyata and R.R. Tartel, The recording and reproduction of signals on magnetic medium using saturation-type recording, *IRE Trans. Elec. Comp.*, EC-8, 159–169, 1959.
4. M.L. Williams and R.L. Comstock, An analytical model of the write process in digital magnetic recording, *A.I.P. Conf. Proc.*, 5, 738–742, 1972.
5. V.A.J. Maller and B.K. Middleton, A simplified model of the writing process in saturation magnetic recording, *IERE Conf. Proc.*, 26, 137–147, 1973.
6. H.N. Bertram, *Theory of Magnetic Recording*, Cambridge, UK: Cambridge University Press, 1994.
7. C.D. Mee and E.D. Daniel, Eds., *Magnetic Recording Technology*, 2nd ed., New York: McGraw-Hill, 1996.
8. R. Potter, Digital magnetic recording theory, *IEEE Trans. Magn.*, MAG-10, 502–508, 1974.
9. W.K. Westmijze, Studies on magnetic recording, *Philips Res. Rep.*, 8, 161–183, 1953.
10. G.J. Fan, A study of the playback process of a magnetic ring head, *IBM J. Res. Dev.*, 5, 321–325, 1961.
11. J.C. Mallinson, *The Foundations of Magnetic Recording*, 2nd ed., San Diego, CA: Academic Press, 1993.
12. M.S. Roden, *Analog and Digital Communication Systems*, 4th ed., Upper Saddle River, NJ: Prentice-Hall, 1996.
13. P.A. Franaszek, *Run-length-limited variable length coding with error propagation limitation*, U.S. Patent No. 3, 689, 899, 1972.
14. W.A. Gross, L. Matsch, V. Castelli, A. Eshel, T. Vohr, and M. Wilamann, *Fluid Film Lubrication*, New York: Wiley, 1980.
15. J.N. Israelachvili, *Intermolecular and Surface Forces*, London: Academic Press, 1985, 224.
16. Y. Li and F.E. Talke, A model for the effect of humidity on stiction of the head/disk interface, *Tribology Mech. Magn. Storage Syst.*, SP-27, 79–84, 1990.
17. R. Ranjan, D.N. Lambeth, M. Tromel, P. Goglia, and Y. Li, Laser texturing for low-flying-height media, *J. Appl. Phys.*, 68, 5745–5747, 1991.
18. J. Watkinson, *The Art of Data Recording*, Oxford U.K.: Focal Press, 1994.
19. G.A.N. Connell and R. Allen, Magnetization reversal in amorphous rare-earth transition-metal alloys: TbFe, *Proc. 4th Int. Conf. on Rapidly Quenched Metals*, Sendai, Japan, 1981.

Control

97
PID Control

F. Greg Shinskey
Process Control Consultant

97.1 Introduction

Process control plays an essential role in the safe manufacture of quality products at market demand, while protecting the environment. Flow rates, pressures and temperatures within pipes and vessels, inventories of liquids and solids, and product quality are all examples of measured variables that must be controlled to meet the above objectives. While there are several means available for controlling these variables, the PID family of controllers has historically carried the major share of this responsibility and, because of their simplicity and reliability, will continue to do so in the future.

The acronym PID stands for the three principal modes of the controller, each of which bears a mathematical relationship between the controlled variable c and the **manipulated variable** m driven by the controller output. The *P*roportional mode relates changes in m to changes in c through a proportional gain. The *I*ntegral mode moves the output at a rate related to the deviation of c from its desired value, known as the set point or reference r. Finally, the *D*erivative mode moves the output in response to the time derivative or rate of change of c. Interestingly, the mathematical relationships were actually determined *after* controllers had already been created to solve process-control problems. The integral mode was initially called *automatic reset,* and the derivative mode *hyper-reset* or *pre-act* [1].

97.2 Open and Closed Loops

Figure 97.1 describes the process and controller in functional blocks connected in a loop. Inputs to the process are manipulated m and load q variables, usually rates of flow of material or energy into or out of the process. The **load** may be a single variable or an aggregate, either independent or manipulated or controlled by another controller. If independent, it is often unmeasured, with its value inferred by the level of controller output required to maintain the controlled variable at set point. **Noise** u is shown affecting the controlled variable directly, typically caused by local turbulence in flow, pressure, and liquid-level measurements, or by nonuniformity of streams whose composition is measured.

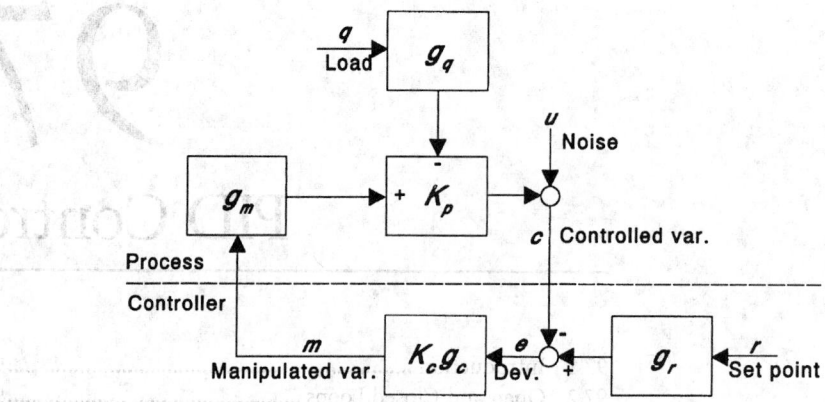

FIGURE 97.1   Process and controller connected in a loop.

Open-Loop Responses

In the absence of automatic control, the controlled variable is subject to variations in the load, and to manual adjustments to m intermittently introduced by operators. These cause variations in c following both the steady-state and dynamic functions appearing in Fig. 97.1:

$$c = K_p\left(mg_m - qg_q\right) + u \tag{97.1}$$

where K_p is the steady-state gain, g_m and g_q are the dynamic-gain vectors in the manipulated and load paths, respectively, and u is the noise level. The vectors have both magnitude and phase angle which are functions of the frequency or period of the signal passing through. In the **open loop,** variations in m are likely to be steps introduced by operators, but variations in q could take any form — steps, ramps, random variations, or cycles — depending on the source of the disturbance. Steps are easily introduced manually, and contain all the frequencies from zero to infinity; consequently, they are useful for evaluating loop response and testing the effectiveness of controllers. Noise is typified by random variations in a frequency range above the bandwidth of the control loop.

The dynamic elements typically consist of deadtime and lags. However, liquid level is the *integral* of the difference between inflow and outflow [2], in which case K_p in Fig. 97.1 is replaced by an integrator. Any difference between inflow and outflow will then be integrated, causing liquid level to continue rising or falling until the vessel limit is reached. This process has no self-regulation, and therefore cannot be left indefinitely in an open loop.

Closed-Loop Responses

In the **closed-loop,** c responds to load and set point as follows:

$$c = \frac{rg_q}{1 + 1/K_p g_m K_c g_c} - \frac{qg_q}{g_m K_c g_c + 1/K_p} + u \tag{97.2}$$

where K_c is the proportional gain of the controller, g_c is the **dynamic gain** of its integral and derivative modes, and g_r that of its set-point filter. For the typical **self-regulating** process in the open loop, c will take an exponential path following a step change in load as shown by the dashed curve in Fig. 97.2. If the loop is closed, the controller can move m to return c to r along the solid curve in Fig. 97.2, an optimum trajectory having a minimum integrated absolute error (IAE) between c and r. The leading

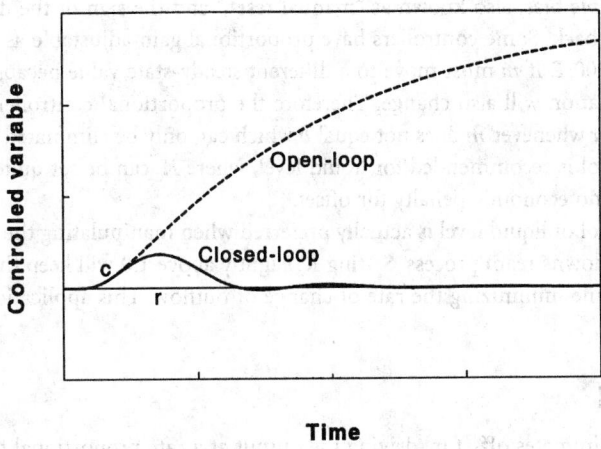

FIGURE 97.2 Closed-loop control minimizes error.

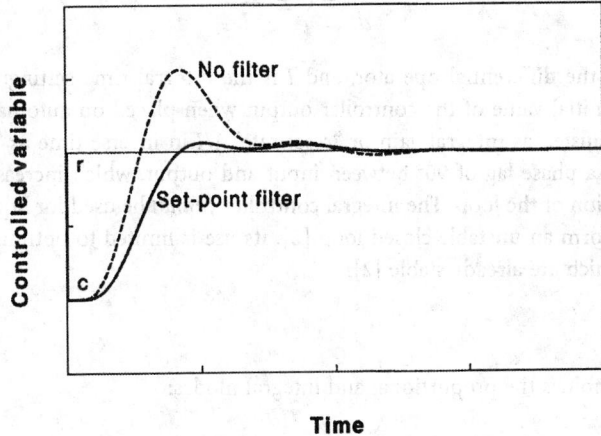

FIGURE 97.3 A lead–lag filter reduces overshoot.

edge of the curve depends on g_q and the trailing edge depends on g_m and the PID settings. If the PID settings are optimized for load response, they will usually cause c to overshoot a change in set point r, as shown by the dashed curve in Fig. 97.3. If the PID settings are then readjusted to reduce set-point overshoot, load response is extended. A preferred solution is to insert a lead-lag filter in the set-point path, which produced the solid curve in Fig. 97.3.

97.3 Mode Selection

The family of PID controllers includes P, I, PI, PD, and two PID controllers. Each has its own advantages and limitations, and therefore its preferred range of applications. Each is outlined along with an application.

Proportional Control

In a **proportional** controller, the deviation between c and r must change for the output to change:

$$m = b \pm K_c(r - c) \tag{97.3}$$

where b is an adjustable bias, also known as "manual reset," and the sign of the deviation is selected to provide negative feedback. Some controllers have proportional gain adjustable as percent **proportional band** P, where $K_c = 100/P$. If m must move to a different steady-state value because of a change in load or set point, the deviation will also change. Therefore the proportional controller allows a steady-state **offset** between c and r whenever m does not equal b which can only be eliminated by manually resetting b. Proportional control is recommended for liquid level, where K_c can be set quite high without loss of stability and there is no economic penalty for offset.

Proportional control of liquid level is actually preferred when manipulating the flow leaving a tank as the feed to a critical downstream process. Setting K_c slightly above 1.0 will keep the level in the tank for all rates of inflow, while minimizing the rate of change of outflow. This application is called **averaging level control** [2].

Integral Control

The **integral** mode eliminates offset by driving the output at a rate proportional to the deviation:

$$\frac{dm}{dt} = C \pm \frac{r-c}{I} \qquad (97.4)$$

where t is time, d is the differential operator, and I is the integral time setting; C is the constant of integration, i.e., the initial value of the controller output when placed on automatic. Some controllers have integral time adjusted as integral gain or "reset rate" $1/I$ in inverse time as "repeats per minute." Integration produces a phase lag of 90° between input and output, which increases the response time and period of oscillation of the loop. The integral controller *cannot* be used for liquid level, because two integrators in series form an unstable closed loop [2]. Its use is limited to optimizing the set points of other closed loops which are already stable [2].

PI Control

The PI controller combines the proportional and integral modes:

$$m = C \pm K_c \left[r - c + \frac{1}{I} \int (r-c)dt \right] \qquad (97.5)$$

The deviation and its integral are added vectorially, producing a phase lag falling between 0 and 90°. This combination provides stability with elimination of offset, making it the most common controller used in the fluid-processing industries. It is used almost universally, even in those applications where other controllers are better suited.

PD Control

The addition of **derivative action** to a proportional controller adds response for the rate of change of the controlled variable:

$$m = b \pm K_c \left(r - c - D\frac{dc}{dt} \right) \qquad (97.6)$$

where D is the derivative time setting. Derivative action is preferably applied only to c as indicated and not to the set point, which would only increase set-point overshoot. A pure derivative function has a dynamic gain increasing indefinitely with frequency — to avoid instability within the controller, it is usually limited to about 10 by filtering. This provides an optimum combination of responsiveness and

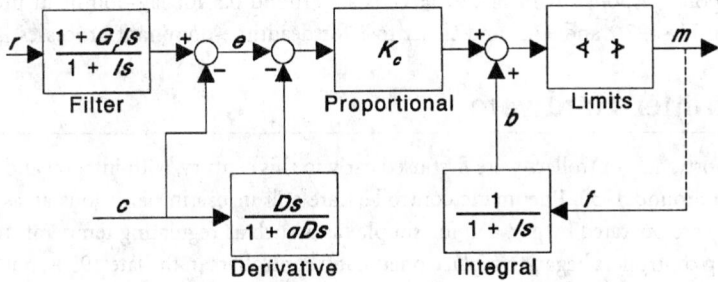

FIGURE 97.4   Commercial interacting PID controller.

noise rejection. Still, the high-frequency gain amplifies noise in flow and liquid-level measurements enough to prevent the use of the derivative mode on those loops; temperature measurements are usually noise-free, allowing it to be used to advantage there. The phase lead provided by the derivative mode reduces the period of oscillation and settling time of a loop, and improves stability.

PD controllers are recommended for batch processes, where operation begins with c away from r, and ends ideally with $c = r$ and flows at zero. The PD controller must be biased for this final output state, with D adjusted to eliminate overshoot, which for a zero-load process is permanent [2].

PID Controllers

The phase lead of derivative action more than offsets the phase lag of integral in a properly adjusted PID controller, resulting in a net phase lead and typically half the IAE of a PI controller applied to the same process. There are two types of PID controllers in common use, which differ in the way the modes are combined [3]. Early controllers combined PD and PI action in series, multiplying those functions rather than adding them. These interacting PID controllers remain in common use, even implemented digitally. Noninteracting controllers combine the integral and derivative modes in parallel, producing a more mathematically pure PID expression:

$$ m = C \pm K_c \left[r - c - \frac{1}{I} \int (r - c)dt - D\frac{dc}{dt} \right] \qquad (97.7) $$

The interacting controller can be described in the same form, but the coefficients of the individual terms are different [3]. The effective integral time of the interacting controller is $I + D$, its effective derivative time is $1/(1/I + 1/D)$, and its proportional gain is augmented by $1 + D/I$. Thus, the principal result of mode interaction is to require different PID settings for the two controllers applied to the same process.

A block diagram of a commercial interacting PID controller appears in Fig. 97.4, with transfer functions described in Laplace transforms whose operator s is equivalent to the differential operator d/dt. Integration is achieved by positive feedback of the output through a first-order lag set at the integral time. The feedback signal f is taken downstream of the high and low limits, and may be replaced with a constant or other variable to stop integration, in which case, the controller behaves as PD with remote bias. This feature is extremely valuable in preventing reset **windup,** which occurs whenever a controller with integral action remains in automatic while the loop is open, and results in a large overshoot when the loop is then closed [3]. Most noninteracting controllers lack this feature.

Derivative action combines a differentiator Ds with a lag (filter) of time constant αD, where α represents the inverse of the high-frequency gain limit of typically 10. Figure 97.4 also shows a lead-lag set-point filter having a lag time constant of integral time I and lead time of $G_r I$, which produces a gain of G_r to a step in set point. This gain is adjustable from 0 to 1, with 1 eliminating the filter and zero imposing a first-order lag. The effect of this lag is equivalent to eliminating proportional action from r in Eqs. 97.5 and 97.7, an optional feature in some PID controllers. The gain adjustment offers more flexibility in optimizing set-point response as in Fig. 97.3; G_r is set around 0.3 for lag-dominant processes such as that described in Figs. 97.2 and 97.3, and closer to 1 for deadtime-dominant processes [3].

97.4 Controller Hardware

Pneumatic proportional controllers were first used early in this century, with integral and then derivative functions added around 1930. Pneumatic controllers are still in use in hazardous areas, in remote gas fields where they are operated by gas, and for simple tasks such as regulating temperatures in buildings. Electronic analog controllers began to replace pneumatic controllers in the late 1950s, but their functions were implemented using digital microprocessors beginning around 1980. Digital control began in mainframes in the 1960s, but now is available is programmable logic controllers (PLC), personal computers (PC), and in distributed control systems (DCS).

Pneumatic Controllers

The simplest pneumatic controllers are proportional units used to control heating, ventilating, and air-conditioning (HVAC) in buildings. However, complete PID controllers with functionality similar to that shown in Fig. 97.4 are also available, used for such demanding tasks as temperature control of batch exothermic reactors and regulation of flows and pressures on offshore oil platforms. There are panel-mounted units available as well as weatherproof models used for field installation. Most have auto–manual transfer stations with bumpless transfer between the two modes of operation. Their principal limitations are a speed of response limited by transmission lines (lengths beyond 30 m are unsuitable for flow and pressure control) and lack of computing capability (required for nonlinear characterization and automatic tuning).

Electronic Controllers

The first electronic controllers mimicked their pneumatic counterparts while eliminating transmission lags. Eventually remote auto–manual transfer and remote tuning were added, and microprocessors brought these controllers calculation and logic functions, signal conditioning and characterization, and auto- and self-tuning features as well. Multiple controllers are even available in a single station for implementing cascade and feedforward systems. The bandwidth of electronic *analog* controllers extends to 10 Hz, and in some units even further. *Digital* controllers execute their calculations intermittently rather than continuously; most are limited to 10 Hz operation, which with digital filtering results in an effective deadtime of 0.1 s, reducing bandwidth to about 1.3 Hz.

Electronic controllers are used extensively in field locations and dedicated to control of individual units such as pumps, dryers, wastewater-treatment facilities — wherever only a few loops are required. Peer-to-peer communication is available in some controllers for incorporation into networks, and most can communicate with personal computers where data can be displayed on a graphic interface. Configuration (selection of scales, control modes, alarms, and other functions) can be done either via keys on the controller faceplate or through a PC.

Digital Controllers

While the stand-alone digital controller evolved from electronic analog models, centralized digital controllers originated with digital computers. Mainframe computers were first used for direct digital control (DDC) where valves were manipulated by the computer, and for supervisory control, where the computer positioned set points of analog controllers. Gradually minicomputers and microprocessors replaced the mainframe, with functions becoming *distributed* among field input–output modules, workstations, control modules, etc., in clusters and other configurations as DCS. PCs are used for some workstations, and even for direct control in some plants.

Where many loops are controlled by a single processor, the interval between samples is likely to be 0.5 s or even longer. Users tend to keep expanding the functions demanded of the processor, thereby extending the interval between samples. This reduces the bandwidth of DCS controllers to values too low for combustion and compressor controls. Some digital PID algorithms produce an *incremental* output Δm, requiring an integrator downstream to produce m. These are not available in proportional or PD action because they have no fixed bias b, only a constant of integration C which is subject to change whenever the controller is operated manually.

The PLC was originally a replacement for relay logic. Eventually PID loops were added, which generally operate much faster (e.g., 100 Hz) than other digital controllers. However, many have nonstandard algorithms and lack the functionality of other PID controllers, such as proportional and PD control, windup protection, derivative filtering, set-point filtering, nonlinear characterization, and autotuning.

97.5 Tuning Controllers

A controller is only as effective as its **tuning:** the adjustment of the PID settings relative to the process parameters to optimize load and set-point response as described in Figs. 97.2 and 97.3. Tuning is required when the controller is first commissioned on a loop, and may have to be repeated if process parameters change appreciably with time, load, set point, etc. Tuning requires the introduction of a disturbance in either the open or closed loop, and interpretation of the resulting response. *Autotuning* essentially automates manual procedures, whereas *self-tuning* can recognize and correct an unsatisfactory response without testing.

Manual Tuning

Ziegler and Nichols [1] developed the first effective tuning methods, and these are still used today. Their open-loop method steps the controller output to produce a response like the broken curve in Fig. 97.2. The apparent deadtime (delay) in the response and the steepest slope of the curve are then converted into appropriate settings for PI and PID controllers using simple formulas. The open-loop method is most accurate for lag-dominant processes, but the closed-loop method is more broadly applicable. It is based on inducing a uniform cycle by adjusting the gain of a controller with only its proportional mode effective (D is set to 0 and I at maximum). The period of the cycle and the controller gain are then used to calculate appropriate PID settings.

The open-loop method has been extended to other processes [3], but remains limited to the accuracy of step-test results. Fine tuning must be done with the loop closed: with the deviation zero, place the controller in manual, step the output, and immediately transfer to automatic — this simulates a closed-loop load change. The resulting response should appear like the solid curve in Fig. 97.2, having a first peak which is symmetrical, followed by a slight overshoot and well-damped cycle. If the overshoot is excessive, I and/or D time should be increased; in the case of undershoot, they should be decreased. (To simplify tuning of PID controllers, I and D can be moved together, keeping the I/D ratio at about 4 for noninteracting controllers and 3 for interacting controllers.) If damping is light, lower the proportional gain; if recovery is slow (producing an unsymmetrical first peak), raise the proportional gain. The set point should *not* be stepped for test purposes unless set-point response is important, as for flow controllers.

Autotuning

Some autotuning controllers use a step test in the open loop, implementing Ziegler–Nichols rules. A single pulse produces more accurate identification, and a doublet pulse is better still [3]. A closed-loop method replaces proportional cycling with relay cycling, where the output switches between high and low limits as c crosses r [3]. The period of the resulting cycle is used to set I and D, and its amplitude relative to the distance between output limits to set K_c. The principal limitation of the step and relay autotuning methods is that only two response features are used to identify a complex process. The estimated PID settings may not fit the specific process particularly well, resulting in instability in some loops and sluggishness in others. The autotuning function cannot monitor the effectiveness of its work, as self-tuning does.

Self-Tuning

A self-tuning controller need not test the process, but simply observe its closed-loop response to set-point and load changes with its current PID settings. It then performs fine-tuning as described above to bring the overshoot or damping or symmetry of the response curve closer to optimum. This may require several iterations if the PID settings are far from optimum, but can eventually converge on optimum response, and readjust as necessary whenever process parameters change. Without a test disturbance, mischaracterization is possible, especially for a sinusoidal disturbance, where detuning may result when tightening would give better control.

Defining Terms

Averaging level control: Allowing the liquid level in a tank to vary in an effort to minimize changes in its outflow.

Closed-loop response: The response of a controlled variable to changes in set point or load with the controller in the automatic mode.

Derivative action: The change in controller output responding to a rate of change in the controlled variable.

Dynamic gain: The ratio of the change in output from a function to a change in its input which varies with time or with the frequency of the input.

Integral action: The rate of change of controller output responding to a deviation between the controlled variable and set point.

Load: A variable or aggregate of variables which affects the controlled variable.

Manipulated variable: A variable changed by the controller to move the controlled variable.

Noise: A disturbance having a frequency range too high for the controller to affect.

Offset: Steady-state deviation between the controlled variable and set point.

Open-loop response: The response of a controlled variable to process inputs in the absence of control action.

Proportional action: The change in controller output responding directly to a change in the controlled variable.

Proportional band: The percentage change in the controlled variable required to drive the controller output full scale.

Self-regulation: The property of a process through which a change in the controlled variable affects either the flow into or out of a process in such a way as to reduce further changes in the controlled variable.

Tuning: Adjusting the settings of a controller to affect its performance.

Windup: Saturation of the integral model of a controller which results in the controlled variable subsequently overshooting the set point.

References

1. J. G. Ziegler and N. B. Nichols, Optimum settings for automatic controllers, *Trans. ASME,* November 1942, 759–768.
2. F. G. Shinskey, *Process Control Systems,* 4th ed., New York: McGraw-Hill, 1996, 22–25, 25–28, 192–194, 388–390.
3. F. G. Shinskey, *Feedback Controllers for the Process Industries,* New York: McGraw-Hill, 1994, 68–70, 176–178, 157–164, 148–151, 155–156.

98
Optimal Control

Halit Eren
Curtin University of Technology

98.1 Introduction

Optimal control maximizes (or minimizes) the value of a function chosen as the **performance index** or *cost function* of an operational control system. Optimal control theory, on the other hand, is the mathematics of finding parameters that cause the performance index to take an extreme value subject to system constraints.

Optimal control is applied in many disciplines, such as satellites and aerospace, aircraft and spacecraft, chemical engineering, communications engineering, robots and robotics, power systems, electric drives, computers and computer systems, etc. In many applications simple interconnections of control devices and controllers do not provide the most economic operation for which optimal control offers the solution. Hence, **optimization** is useful in obtaining the best results from a known process. If the process is operating under a steady-state condition, optimization considers the process stationary, and it is concerned only with the operating points. When the process is stationary, the resulting optimum operating point can easily be maintained by setting the predetermined set points and precalculated control parameters. Nevertheless, if the process changes from time to time, new optimum set points for the system need to be determined for each change.

The performance of a system is optimized for many reasons, such as improving the quality, increasing production, decreasing waste, obtaining greater efficiency, maximizing the safety, saving time and energy, and so on. In many optimization problems, boundary conditions are imposed by the system for safety in operations, availability of minimum and maximum power, limitations in storage capacity, capability of the operating machinery, temperature, speed, force, acceleration, indeed, for any other hard physical reason.

For the solution of optimal control problems many different methods may be used depending on the nature of the problem. For example, if the performance index and constraints can be formulated as linear algebraic functions of the controlled variables, then selection of the linear programming may be the best way to go. Simplex methods may provide a good way of solving the linear programming problem. If the equations describing the system are nonlinear, then the solution may involve nonlinear techniques or linearization of the problem in some subregions. If the problem involves determination of the largest value of a function with two or more variables, then the steepest ascent (or descent) method, sometimes termed *hill climbing* or the *gradient method* may be used.

Hill-climbing methods are still popularly used and easy to understand. They are briefly discussed here, first, to lay a good background in the sound understanding of optimal control theory.

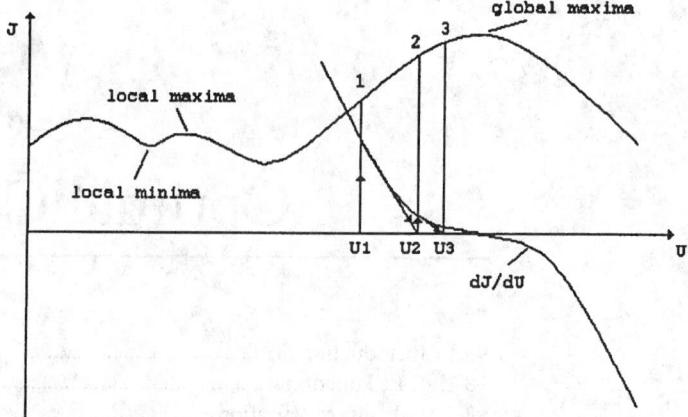

FIGURE 98.1   Hill-climbing technique illustrated in the region of absolute maxima. The sign of the first derivative of the function changes as the derivative passes through the maximum point. This figure also demonstrates the local maxima and minima that should be taken care optimal control design.

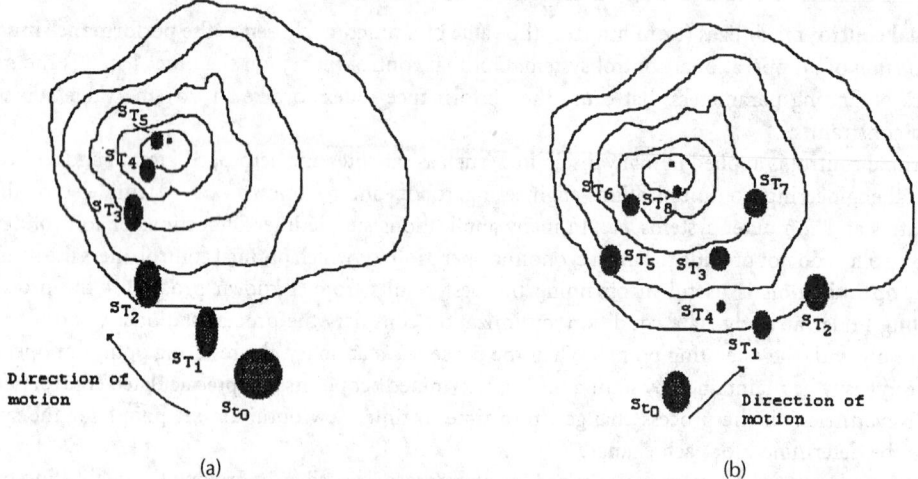

(a)   (b)

FIGURE 98.2   Examples of hill climbing method, (a) gradient, (b) random walk. All these methods use an initial estimated point where the first derivative is guessed to be zero. This point is then used, in the direction of steepest ascent or descent as the case may be, for the next point until the absolute maximum or minimum is found.

An illustration of the hill-climbing technique in the region of absolute minima is given in Fig. 98.1. This figure also illustrates a common pitfall in control theory, the local maxima and minima. As can be seen in this figure, the sign of the first derivative of the function changes as the derivative passes through the maximum point. This method commonly uses an initial estimated point where the first derivative is considered (guessed) to be zero. This point is then used, in the direction of steepest ascent or descent as the case may be, for the next point until the absolute maxima or minima is found without being trapped in local extreme points. There are many different methods to find maxima or minima by hill-climbing methods and Fig. 98.2 illustrates some examples of implementation.

Apart from hill-climbing techniques, there are many new methods available for the solution of modern optimal control problems involving hyperspaces (many variables) complex in nature and having stringent constraints. If a good mathematical model of the system is available, optimization methods may be used for finding the optimum conditions on the model, instead of seeking optimization on the actual process. This indicates that availability of the model determines the type of optimization method to be employed.

98.2 Cost Function

In designing optimal control systems, rules may be set for determining the control decisions, subject to certain constraints, such that some measure of deviations from an ideal case is minimized. That measure is usually provided by a selected performance index or cost function. Performance index is a function whose value indicates how well the actual performance of the system matches the desired performance. Appropriate selection of the performance index is important since it determines the nature and complexity of the problem. That is, if the optimal control is linear or nonlinear, stationary or time-varying will be dependent on the selected performance index. The choice of the performance index depends on system specifications, physical realizability, and the restrictions on the controls to be used. In general, the choice of an appropriate performance index involves a compromise between a meaningful evaluation of system performance and availability of feasible mathematical descriptions.

The performance index is selected by the engineer to make the system behave in a desired fashion. By definition, a system whose design minimizes the selected performance index is optimal. It is important to point out that a system optimal for one performance index may not be optimal under another performance index. In practical systems, due to possible complexities and cost of implementation, it may be better to employ approximate optimal control laws which are not rigidly tied to a single performance index.

Before starting optimization of any system, it is necessary to formulate the system by having information on system parameters and describing equations, constraints, class of allowable control vectors, and the selected performance indexes. Then the solution can proceed by determining the optimal control vector $\mathbf{u}(k)$ within the class of allowable vectors. The control vector $\mathbf{u}(k)$ will be dependent on such factors as the nature of the performance index, constraints, initial values of state, initial outputs, and the desired state as well as the desired outputs. If analytical solutions are impossible or too complicated, then alternative computational solutions may be employed.

In simple cases, errors between the desired and actual responses can be chosen as the performance index to be minimized. As an example, different descriptions of error between actual and desired responses are depicted in Fig. 98.3. The aim is to keep the errors as small as possible. The time integral of error gives the severity of the error. However, since the positive and negative errors mathematically cancel each other, absolute values must be used:

$$\text{Integral Absolute Error} = \text{IAE} = \int_0^T |e(\tau)| \, d\tau \qquad (98.1)$$

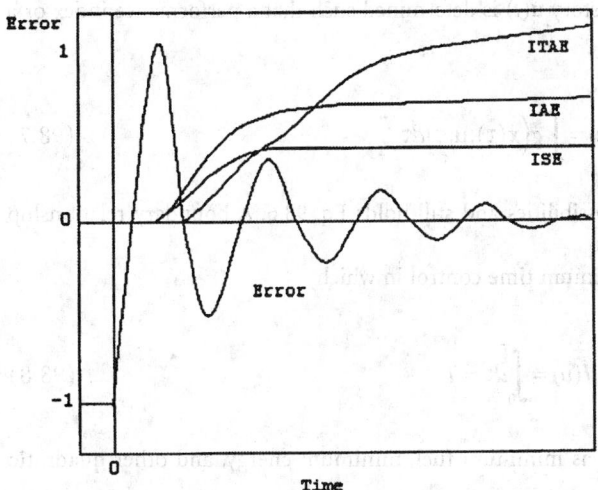

FIGURE 98.3 Description of errors between actual and desired responses. In simple optimal control applications, errors between the desired and actual responses can be chosen as the performance indexes. The aim is to keep the errors as small as possible.

In many cases it is better to use integral squared errors which take into account large errors rather than regular ones:

$$\text{Integral Squared Error} = \text{ISE} = \int_0^T e^2(\tau)d\tau \tag{98.2}$$

Time-dependent functions can be implemented to capture errors occurring late in time rather than normal transients:

$$\text{Integral Time Absolute Error} = \text{ITAE} = \int_0^T |e(\tau)|\tau d\tau \tag{98.3}$$

and

$$\text{Integral Time Squared Error} = \text{ITSE} = \int_0^T e^2(\tau)\tau d\tau \tag{98.4}$$

Mean square error, MSE, is also commonly used since slight modification leads to inclusion of statistical techniques and analysis of random noise:

$$\text{Mean Square Error} = \text{MSE} = \lim_{T \to \infty} \int_0^T 1/Te^2\tau d\tau \tag{98.5}$$

Modern optimal control theory is developed within **state-space** framework and performance indexes are more complex and comprehensive as explained below.

Suppose that the control command of a system is expressed in vectorial form as **u** and the state of the system is described by **x** (*Oxford English Dictionary* defines *state* as condition with respect to circumstances, attributes, structure, form phase or the like). Further, suppose that the rate of chance of state $\dot{\mathbf{X}}$ is a function of state **x**, control command **u**, and time t

$$\dot{\mathbf{X}} = f(\mathbf{x},\mathbf{u},t) \quad \mathbf{x}(0) = \mathbf{x}_0 \text{ known} \tag{98.6}$$

Then a control law **u(x,t)** or a control history **u**(t) is determined such that a performance index or a scalar functional

$$J(u) = \int_0^T g\big(\mathbf{x}(\tau),\mathbf{u},\tau\big)d\tau \tag{98.7}$$

takes a minimum value out of all other possibilities and still holds Eq. 98.6. A boundary relationship $\mathbf{x}(T) = \mathbf{x}_f$ must also be met as a constraint.

A most common form of $J(u)$ is the minimum time control in which

$$J(u) = \int_0^T d\tau = T \tag{98.8}$$

Many different criteria are also used, such as minimum fuel, minimum energy, and other quadratic forms

$$J(u) = \int_0^T |u(\tau)| d\tau \tag{98.9}$$

$$J(u) = \int_0^T u^2(\tau) d\tau \tag{98.10}$$

$$J(u) = \int_0^T \left(qx^2(\tau) + ru^2(\tau) \right) d\tau \tag{98.11}$$

A general term for continuous time performance index leading to optimal control is expressed as:

$$J(u(t)) = \int_0^T g(x(\tau), u(\tau)) d\tau \tag{98.12}$$

This performance index is minimized for the constraints

$$\dot{X}(t) = f(x(t), u(t), t) \quad \text{for } t \in (t_0, t_f)$$

and $x(t)$ is an admissible state, $x(t) \in X(t)$, $\forall t \in (t_0, t_f)$ are satisfied.

Slight variations of Eqs. 98.9 to 98.12 lead to mathematics of the discrete time or digital versions of optimal control.

98.3 Calculus of Variations

Calculus variations are suitable for solving linear or nonlinear optimization problems with linear or nonlinear boundary conditions. Basically, it is a collection of many different analytical methods and they are discussed differently from book to book. Here, a typical approach which leads to more general and widely used modern theories is introduced.

Consider a dynamic system operating in a time interval $t_0 < t < t_f$

$$\dot{X} = f[x, u, t] \tag{98.13}$$

where the initial state x_0 is given.

The system has n state and m control variables. The scalar function to be optimized is

$$J = k[x_f, t_f] + \int L[x, u, t] dt \tag{98.14}$$

Define a scalar, such as Hamiltonian

$$H = H[x, u, \lambda t] = L[x, u, t] + \sum \lambda_i f_i \tag{98.15}$$

λ is known as the **Lagrange multiplier**. Also, define a modified objective function:

$$J = k[x_1, t_1] + \int H dt \tag{98.16}$$

the resulting solution is optimal when

$$\lambda = -\delta H \backslash \delta x = -\delta L \backslash \delta x - \sum \lambda_i \delta f_i \backslash \delta x \qquad (98.17)$$

$$\lambda_f = \delta k \backslash \delta x \qquad (98.18)$$

$$\delta H \backslash \delta u = \delta L \backslash \delta u + \sum \lambda_i \delta f_i \backslash \delta u = 0 \quad t_0 < t < t_f \qquad (98.19)$$

The solutions of Eqs. 98.17 and 98.19 in the above form are difficult to obtain. Nevertheless, based on the above ideas more general theories can be developed such as Pontryagins maximum principle.

An advancement over the calculus variations is the Pontryagins maximum principle, which offers easier solutions and expands the range of applicability to bounded control problems. In its simplest form, this principle may be explained as follows.

Given the system

$$\mathbf{x}_i = f_i[\mathbf{x}, \mathbf{u}, t] \qquad (98.20)$$

and an objective function,

$$J = \sum k_i x_i(t_f) \qquad (98.21)$$

The maximum principle states that if the control vector u is optimum, then the Hamiltonian

$$H = \sum p_i f_i \qquad (98.21a)$$

is also maximized with respect to the control vector u over the set intervals.

$$p_i = \sum p_j \delta f_j \backslash \delta x_j \qquad (98.21b)$$

$$p_i(t_f) = k_i \qquad (98.21c)$$

where p is the adjoint or co-state vector, and

$$\delta H \backslash \delta p_i = f_i[\mathbf{x}, \mathbf{u}, t] = x_i \qquad (98.21d)$$

$$\delta H \backslash \delta p_i = \sum p_i \delta f_j \backslash \delta x_j = -p_i \qquad (98.21e)$$

The initial values of state vector x provide the remaining constants necessary to solve these equations for determining an optimum control vector. This method is also applicable to systems with complex performance indexes.

Another expansion of calculus variations is the Kalman filter. Kalman is essentially a general filtering technique that can be applied to solutions of problems such as optimal estimation, prediction, noise filtering, stochastic optimal control, and design of optimal controllers. The method has the advantage of providing estimates of variables in the presence of noise for both stationary and nonstationary processes. Most other types of noise can be treated by this method if they can be translated to Gaussian form. The technique can also be employed in systems with forcing disturbances and containing more than one noise source. Generally, systems that have either quadratic objective functions or uncorrelated Gaussian noise as the input are suitable for this technique.

A means of applying the Kalman filter to a nonlinear system is to find a good estimate of the system and then use it to define a new set of linear **state equations** which approximates the system to linear form at a normal operating point. Then the filter can be applied to the new set of linear equations. Applications of Kalman filter is endless in instrumentation and measurement systems, and other optimal control problems. It can easily be programmed on digital computers.

98.4 Riccati Equation

Some of the models in optimization problems resemble the models of traditional control theory and practice. In these models, the process and control variables are vector-valued and constrained by linear plant equations. The cost functions are in the form of quadratic costs. These type of systems are termed as **linear quadratic equations** systems or LQ systems. The theory and models of LQ systems are well developed in both deterministic and stochastic cases.

Express a quadratic optimal control problem as

$$\mathbf{x}(k+1) = \mathbf{A}\mathbf{x}(k) + \mathbf{B}\mathbf{u}(k) \quad \mathbf{x}(0) = \mathbf{x}_0 \tag{98.22}$$

where $\mathbf{x}(k)$ = state vector (n-vector)
$\mathbf{u}(k)$ = control vector (r-vector)
$\mathbf{A} = n \times n$ nonsingular vector
$\mathbf{B} = n \times r$ matrix

The aim is to find the optimal control sequence $\mathbf{u}(0), \mathbf{u}(1), \ldots, \mathbf{u}(N-1)$ that minimizes a quadratic performance index

$$J = \frac{1}{2}\mathbf{x}^*(N)\mathbf{S}\mathbf{x}(N) + \frac{1}{2}\sum \mathbf{x}^*(k)\mathbf{Q}\mathbf{x}(k) + \mathbf{u}^*(k)\mathbf{R}\mathbf{u}(k) \tag{98.23}$$

where $\mathbf{x}^*$ and $\mathbf{u}^*$ are transposes of $\mathbf{x}$ and $\mathbf{u}$ matrixes, respectively,
$\mathbf{Q} = n \times n$ positive definite or positive semidefinite symmetric matrix
$\mathbf{R} = r \times r$ positive definite or positive semidefinite symmetric matrix
$\mathbf{S} = n \times n$ positive definite or positive semidefinite symmetric matrix

Matrixes $\mathbf{Q}$, $\mathbf{S}$, and $\mathbf{R}$ are selected to weight the relative importance of the performance measures caused by the state vectors $\mathbf{x}(k)$, the final state $\mathbf{x}(N)$, and control vectors $\mathbf{u}(k)$, respectively, for $k = 0, 1, 2, \ldots, N-1$.

The initial state of the system is arbitrary but the final state $\mathbf{x}(N)$ may be fixed. If the final state is fixed, then the term $^1/_2 \mathbf{x}^*(N)\mathbf{S}\mathbf{x}(N)$ may be removed from the performance index and the terminal state $\mathbf{x}_f$ may be imposed. If the final state is not fixed, then the term $^1/_2 \mathbf{x}^*(N)\mathbf{S}\mathbf{x}(N)$ represents the weight of the performance measure to the final state.

There are many different ways of solving the above equations, one of which makes use of the concept of Lagrange multipliers. With the aid of Lagrange multipliers, the performance index may be modified as

$$L = \frac{1}{2}\mathbf{x} * (N)\mathbf{S}\mathbf{x}(N) + \frac{1}{2}\sum \mathbf{x} * (k)\mathbf{Q}\mathbf{x}(k) + \mathbf{u} * (k)\mathbf{R}\mathbf{u}(k) + \lambda * (k+1)$$

$$\left[\mathbf{A}\mathbf{x}(k) + \mathbf{B}\mathbf{u}(k) - \mathbf{x}(k+1)\right] + \left[\mathbf{A}\mathbf{x}(k) + \mathbf{B}\mathbf{u}(k) - \mathbf{x}(k+1)\right]^* \lambda * (k+1) \tag{98.24}$$

It is known that minimization of the function **L** is equivalent to minimization of performance index **J** under the same constraints.

In order to minimize, **L** needs to be differentiated with respect to vectors $\mathbf{x}(k)$, $\mathbf{u}(k)$, and $\lambda(k)$ and the results set to zero. The partial differentiation of the function **L** with respect to variables gives the following

$$\delta\mathbf{L}/\delta\mathbf{x}(k) = 0 \quad \mathbf{Q}\mathbf{x}(k) + \mathbf{A} * \lambda(k+1) - \lambda(k) = 0 \tag{98.25}$$

$$\delta\mathbf{L}/\delta\mathbf{x}(N) = 0 \quad \mathbf{S}\mathbf{x}(N) - \lambda(N) = 0 \tag{98.26}$$

$$\delta\mathbf{L}/\delta\mathbf{u}(k) = 0 \quad \mathbf{R}\mathbf{u}(k) - \mathbf{B} * \lambda(k+1) = 0 \tag{98.27}$$

and

$$\delta\mathbf{L}/\delta\lambda(k) = 0 \quad \mathbf{A}\mathbf{x}(k-1) + \mathbf{B}\mathbf{u}(k-1) - \mathbf{x}(k) = 0 \tag{98.28}$$

A close inspection of the above formulae indicates that the last equation is simply the state equation for $k = 1, 2, 3, \ldots N$. And also, the value of Lagrange multiplier can be determined by Eq. 98.26. The Lagrange multiplier is often termed the covector or the adjoint vector.

Rewriting Eqs. 98.25 and 98.27

$$\lambda(k) = \mathbf{Q}\mathbf{x}(k) + \mathbf{A} * \lambda(k+1) \tag{98.29}$$

$$\mathbf{u}(k) = -\mathbf{R}^{-1}\mathbf{B} * \lambda(k+1) \tag{98.30}$$

the state Eq. 98.22 can be expressed as

$$\mathbf{u}(k+1) = \mathbf{A}\mathbf{x}(k) - \mathbf{B}\mathbf{R}^{-1}\mathbf{B} * \lambda(k+1) \tag{98.31}$$

In order to obtain the solution to the minimization problem we need to solve Eqs. 98.29 and 98.31 simultaneously as a two-point boundary-value problem. The solutions of these two equations in this form, and the optimal values of the state vector and the Lagrange multiplier vector will lead to the values of control vector $\mathbf{u}(k)$ which will optimize the open-loop control system. However, for closed-loop control systems the Riccati transformations must be applied.

$$\mathbf{u}(k) = -\mathbf{K}(k)\,\mathbf{x}(k) \tag{98.32}$$

where $\mathbf{K}(k)$ is the $r * n$ feedback matrix. Now the Riccati equation in feedback form can be obtained by assuming that $\lambda(k)$ is written as

$$\lambda(k) = \mathbf{P}(k)\,\mathbf{x}(k) \tag{98.33}$$

where $\mathbf{P}(k)$ is an $n \times n$ matrix. Substituting Eq. 98.33 into 98.29 and 98.31 gives

$$\mathbf{P}(k)\,\mathbf{x}(k) = \mathbf{Q}\mathbf{x}(k) + \mathbf{A} * \mathbf{P}(k+1)\,\mathbf{x}(k+1) \tag{98.34}$$

$$\mathbf{x}(k+1) = \mathbf{A}\mathbf{x}(k) - \mathbf{B}\mathbf{R}^{-1}\mathbf{B} * \mathbf{P}(k+1)\,\mathbf{x}(k+1) \tag{98.35}$$

In writing Eqs. 98.34 and 98.35 the Lagrange multiplier $\lambda(k)$ has been eliminated. This is an important step in solving two-point boundary-value problems. By further manipulations it is possible to show that

$$P(k) = Q + A * P(k+1)A - A * P(k+1)B \left[R + B * P(k+1)B \right]^{-1} B * P(k+1)A \qquad (98.36)$$

Equation 98.36 is known as the **Riccati equation**.

From Eqs. 98.26 and 98.33 writing $\lambda(N) = Sx(N) = P(N) x(N)$ gives

$$P(N) = S \qquad (98.37)$$

Hence, the Riccati equation can be solved backward from $k = N$ to $k = 0$, starting from the known values of $P(N)$.

The optimal control vector $u(k)$ can now be calculated from Eqs. 98.29, 98.30, and 98.35

$$u(k) = -R^{-1}B * \lambda(k+1) = -R^{-1}B * (A*)^{-1} \left[P(k) - Q \right] x(k) = -K(k)x(k) \qquad (98.38)$$

where

$$K(k) = -R^{-1}B * (A*)^{-1} \left[P(k) - Q \right] \qquad (98.39)$$

It is worthy noting that the optimal control vector may be obtained in slightly different forms by the different manipulations of above equations, such as

$$u(k) = -R^{-1}B * \left[P^{-1}(k+1) + BR^{-1}B* \right]^{-1} Ax(k) \qquad (98.40)$$

Equations 98.38 and 98.39 indicate that optimal control law requires feedback of the state vector with time-varying gain $K(k)$. Figure 98.4 illustrates the optimal control scheme of a system based on the quadratic performance index. In practical applications, the time-varying $K(k)$ is calculated before the process begins. Once the state matrix **A**, control matrix **B**, and weighting matrices **Q, R**, and **S** are known, the gain $K(k)$ may be precomputed off-line to be used later. The control vector $u(k)$ at each stage can be determined immediately by premultiplying the state vector $x(k)$ by the known gain $K(k)$.

From the above equations, the minimum value of the performance index can also be calculated. By using the initial values

$$J_{min} = \frac{1}{2} x * (0)P(0)x(0) \qquad (98.41)$$

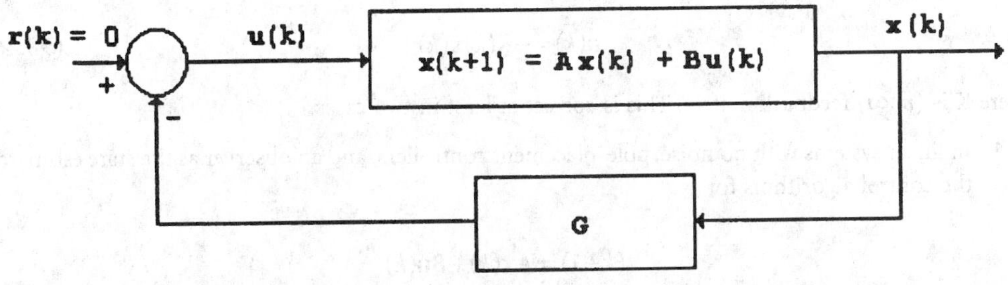

FIGURE 98.4 Optimal control scheme based on quadratic performance index. The control law requires feedback of the state vector with time varying gain $K(k)$. The gain $K(k)$ is calculated before the process begins to be used later.

The *steady-state* solutions of Riccati equations are necessary when dealing with time-invariant (steady-state) optimal controls. There are many ways of obtaining steady-state Riccati solutions; an example is given below.

$$P = Q + A * PA - A * PB(R + B * PB)^{-1} B * PA \tag{98.42}$$

The steady-state value of $\mathbf{K}$ may be found as

$$K = (R + B * PB)^{-1} B * PA \tag{98.43}$$

and the optimal control law for the steady-state operation may be expressed as

$$\mathbf{u}(k) = -(R + B * PB)^{-1} B * PA\mathbf{x}(k) \tag{98.44}$$

This section is presented for discrete-time optimal control systems rather than continuous systems, due to recent widespread use of computers and microprocessors as online and off-line control tools. Since the principles are the same, solutions can easily be extended for continuous time systems with minor modifications.

98.5 State Feedback Matrix

The state feedback method is another design technique that allows the designer to locate the poles of the system wherever they are needed. This type of approach is termed the **pole-placement method**, where the term *pole* refers to the poles of the closed-loop transfer function as in Fig. 98.5. In this method, it is assumed that **state variables** are measurable and are available for feedback. If the state is available, the system is said to be deterministic; the system is noise free and its parameters are fully known. If the state is not available, then methods such as measurement feedback laws or state estimators may be selected.

In many applications, instead of state variables, it is more convenient to use state estimates coming from an observer or Kalman filter. In this case there are three well-known methods available: the certainty equivalent, the separation, and the dual control. Certainty equivalence has considerable advantages over others since it leads to deterministic optimal control laws such Pontryagins principle or the conditional mean of the state such as Kalman filter. It leads to a practical controller, which may be built as a filter and optimal law in cascade.

State feedback controllers are relatively easy to implement. For example, in the pole-placement method the relationship of the feedback control **u** to state **x** for **linear systems** is

$$\mathbf{u}(k) = -\mathbf{K}\mathbf{x}(k) \tag{98.45}$$

In linear quadratic cases, the gain $\mathbf{K}$ is time varying:

$$\mathbf{u}(k) = -\mathbf{K}(k)\mathbf{x}(k) \tag{98.46}$$

where $\mathbf{K}$ is $(n \times r)$ feedback matrix. This is correct at least in two cases.

1. In linear systems with no noise, pole-placement controllers, and an observer as the state estimator, the control algorithms for

$$\mathbf{x}(k+1) = \mathbf{A}\mathbf{x}(k) + \mathbf{B}\mathbf{u}(k)$$

$$\mathbf{y}(k) = \mathbf{C}\mathbf{x}(k)$$

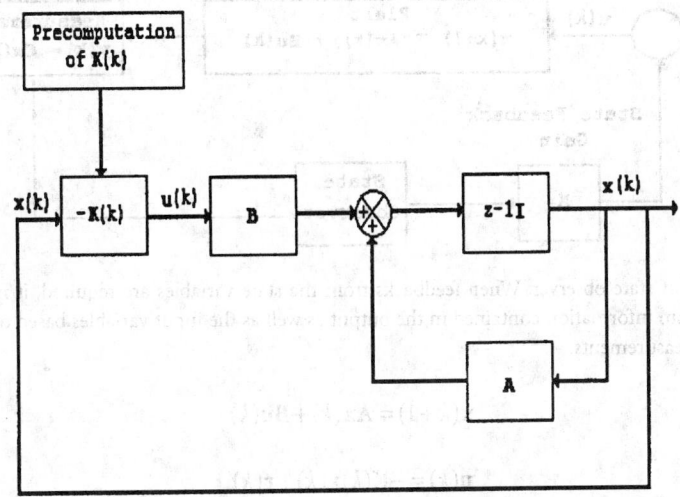

FIGURE 98.5   Pole-placement design of closed loop transfer function. This method assumes that state variables are measurable and are available for feedback.

is

$$\hat{x}(k+1) = (A - GC)\hat{x}(k) + Gy(k) + Bu\,(k) \tag{98.47}$$

$$u(k) = K\hat{x}(k) \tag{98.48}$$

where the eigenvalues of $(A - GC)$ and $(A + BK)$ are chosen to meet design specifications.

2. In linear systems, influenced by Gaussian white noise, in which the control is optimal according to a quadratic criterion, the system

$$x(k+1) = Ax(k) + Bu(k) + Gv(k) \tag{98.49}$$

$$y(k) = Cx(k) + w(k) \tag{98.50}$$

has control law of the form

$$\hat{x}\!\left(k+1/k\right) = A\hat{x}(k) + B\hat{u}(k) \tag{98.51}$$

$$\hat{x}(k+1) = A\hat{x}\!\left(k+1/k\right) + G(k+1)y(k+1) - C\hat{x}\!\left(k+1/k\right) \tag{98.52}$$

$$\hat{u}(k) = K(k)\hat{x}(k) \tag{98.53}$$

where $K(k)$ and $G(k)$ are optimal gains for the deterministic optimal controls and the Kalman filters, respectively.

In practice, not all the state variables are easily accessible, and in general only the outputs of the system are measurable. Therefore, when feedback from the state variables are required in a given design, it is necessary to observe the states from information contained in the output as well as the input variables. The subsystem that performs the observation of the state variables based on the information received from the measurements of inputs and outputs is called the state observer. Figure 98.6 shows the block diagram of such a system.

Suppose a state feedback gain **K** has been selected so that the eigenvalues of

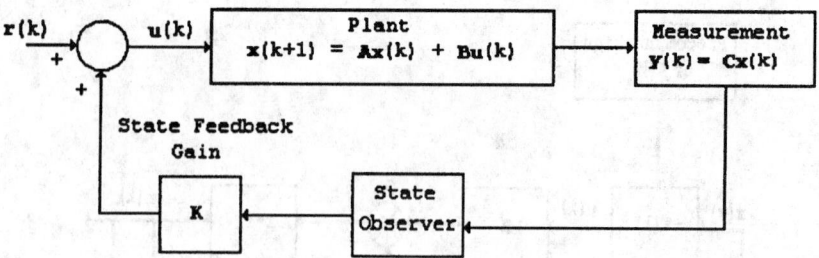

FIGURE 98.6   Use of state observer. When feedbacks from the state variables are required, it may be possible to observe the states from information contained in the output as well as the input variables based on the information received from the measurements.

$$x(k+1) = \mathbf{A}x(k) + \mathbf{B}u(k) \tag{98.54}$$

$$u(k) = -\mathbf{K}(k)x(k) + r(k) \tag{98.55}$$

are located at $\lambda_1, \lambda_2, \ldots \lambda_n$. Furthermore, assume that gain $\mathbf{G}$ in the identity observer

$$\hat{x}(k+1) = (\mathbf{A} - \mathbf{GC})\,\hat{x}(k) + \mathbf{B}u(k) + \mathbf{G}y(k) \tag{98.56}$$

is chosen such that eigenvalues of the observers are $\mu_1, \mu_2, \ldots \mu_n$.

When the observer state estimate $x(k)$ is used instead of state $x(k)$ the resulting system has a state dimension 2n, modeled by

$$\begin{bmatrix} x(k+1) \\ \hat{x}(k+1) \end{bmatrix} = \begin{bmatrix} \mathbf{A} & -\mathbf{BK} \\ \mathbf{GC} & \mathbf{A} - \mathbf{GC} - \mathbf{BK} \end{bmatrix} \begin{bmatrix} x(k) \\ \hat{x}(k) \end{bmatrix} + \begin{bmatrix} \mathbf{B} \\ \mathbf{B} \end{bmatrix} r(k) \tag{98.57}$$

A similarity transform

$$\begin{bmatrix} x(k) \\ e(k) \end{bmatrix} = \mathbf{P} \begin{bmatrix} x(k) \\ \hat{x}(k) \end{bmatrix} \tag{98.58}$$

where

$$\mathbf{P} = \begin{bmatrix} \mathbf{I} & 0 \\ \mathbf{I} & -\mathbf{I} \end{bmatrix} \tag{98.59}$$

corresponds to a change of variables to $x(k)$ and $e(k) = x(k) - x(k)$ converts this to

$$\begin{bmatrix} x(k+1) \\ e(k+1) \end{bmatrix} = \begin{bmatrix} \mathbf{A} - \mathbf{BK} & \mathbf{BK} \\ 0 & \mathbf{A} - \mathbf{GC} \end{bmatrix} \begin{bmatrix} x(k) \\ e(k) \end{bmatrix} + \begin{bmatrix} \mathbf{B} \\ 0 \end{bmatrix} r(k) \tag{98.60}$$

This system has the same eigenvalues as the original because of the nature of the similarity transforms and the eigenvalues are the solution of

$$\det\!\left(\lambda \mathbf{I} - (\mathbf{A} - \mathbf{BK})\right)\!\left(\lambda \mathbf{I} - (\mathbf{A} - \mathbf{GC})\right) = 0 \tag{98.61}$$

This requires that one of

$$\det\left(\lambda I - (A - BK)\right) = 0 \qquad\qquad (98.62)$$

$$\det\left(\lambda I - (A - GC)\right) = 0 \qquad\qquad (98.63)$$

must hold. These are also the eigenvalues of the design for the pole-placement feedback state controller $\lambda_1, \lambda_2, \ldots \lambda_n$ and of the design of the state observer $\mu 1, \mu 2, \ldots \mu n$, respectively. This indicates that the use of observer does not move the designed poles from the pole-placement algorithm.

The algorithms for pole placement observer are well known and they are part of many control design packages. Steady-state Kalman gains and optimal control gains are also commonly available in such programs. Codes for Kalman filters and LQ controllers are also easy to write.

Note: For further reading on the topic refer to the sources in references.

Defining Terms

Calculus variations: A technique which can be used for solving linear or nonlinear optimization problems with boundary conditions.

Controllability: Property of a control system such that a determined input takes every state variable from a desired initial state to desired final state.

Controller: A subsystem that assists to achieve the desired output of a plant or process.

Cost function: A function whose value indicates how well the actual performance of the system matches the desired performance.

Covector: Lagrange multiplier.

Hill climbing: A method of determining absolute maxima or minima by using an initial guess point and derivatives of the function.

Kalman Filter: A procedure which provides optimal estimates of state.

Lagrange multiplier: A mathematical expression that modifies the performance index to give an optimal solution.

Linear quadratic equation: Cost functions in quadratic form.

Linear system: A system that possesses the properties of superposition.

Optimal control: A control technique that maximizes or minimizes the cost function of a system.

Optimization: Procedure of obtaining maximum or minimum performance.

Performance index: Cost function.

Pole-placement method: A design technique that makes use of the system closed-loop properties.

Riccati equation: A set of equations that transform control equations to lead to optimal solutions.

State equation: A set of simultaneous, first-order differential equations with n variables leading to solutions of state variables.

State-space model: Mathematical expression of a system that consists of simultaneous, first-order differential equations and an output equation.

State variables: A set of linearly independent system variables so that once the values are set they can determine the value of all system variables later in time.

References

1. Whittle, P., *Optimal Control — Basics and Beyond*, John Wiley & Sons, New York, 1996.
2. Lewis, L.L and V.L. Syrmos, *Optimal Control*, 2nd ed., John Wiley & Sons, New York, 1995.
3. Ogata, K., *Discrete-Time Control Systems*, Prentice-Hall, Englewood Cliffs, NJ, 1987.
4. Kuo, B.C., *Digital Control Systems*, 2nd ed., Harcourt Brace Jovanovich, New York, 1992.

List of Manufacturers

There are no manufacturers of optimal controllers. Most mathematical packages, such as **MATLAB Optimization Toolbox, CAD/CAM** algorithms, the majority of artificial intelligence and neural network software tools, and other simulation and design packages, offer solutions to optimal control problems. Optimization tools are also part of specialized design packages addressing specific applications, such as FPGA, a language-based design, technology-specific optimization; OPTCON, optimal control algorithms for linear stochastic models; DISNEL; MIMO; and so on.

99

Electropneumatic and Electrohydraulic Instruments: Modeling of Electrohydraulic and Electrohydrostatic Actuators

M. Pachter and
C. H. Houpis
Air Force Institute of Technology

99.1 Introduction

In control systems, the output of the controller (also of the equalizer or compensator), which operates on the command (and feedback) input signals, is invariably sent to an actuator. The latter is a power element critically positioned at the plant input, and, to a large extent, determines the performance of the overall control system. Thus, the importance of the accurate modeling of the actuator element with respect to achieving and maintaining the desired control system performance over its operating range cannot be overemphasized. The impact of items such as the actuator phase lag characteristics, parameter variations and aging, sensor noise, and operating scenario (e.g., in a flight control context it is referred to as *flight conditions*) should be explicitly considered in a control system design process. To achieve a robust control system, where these items are taken into account, requires a good understanding and an accurate model of the actuation element. This chapter discusses the actuator modeling process.

99.2 Background

Power-to-weight considerations have driven actuator design to pneumatic, hydraulic, electrohydraulic, and recently, in flight control applications, electrohydrostatic configurations. In this chapter the point of view is taken that the actuator element constitutes a feedback control system in its own right. Thus, the actuator comprises of an "amplifier" element (the servo valve), a power element (the "ram"), an actuator displacement sensor, and a controller about which more will be said in the sequel. Furthermore, it is important that the actuator subsystem be considered in the context of the complete control system, e.g., a flight control system. This point of view is promulgated throughout this chapter. First, a relatively simple hydraulic actuator is considered. The focus then shifts to the modern electrohydrostatic actuator (EHA) technology which, in the flight control application, offers a high degree of combat survivability and easier maintainability since its components are collocated with the actuator. Since the EHA does not require long hydraulic lines, required maintenance time and equipment can be reduced. This also substantially reduces the profile exposed to hostile fire. In addition, actuator failures due to various causes and/or aging need to be taken into account in designing the overall control system. The design of a robust control system, utilizing an actuator, requires an accurate actuator model in order to satisfy the system performance specifications. The EHA and its flight control application are used as the "vehicle" for presenting the actuator modeling process.

Actuator Modeling Requirements

When an actuator needs to be replaced in a control system, for reasons cited previously, by one "off the shelf," the robustness of the overall control system must be unaffected by this replacement unit. Thus, the actuator design process entails

1. The derivation of an accurate mathematical model for the actuator.
2. Sensitivity analyses on the actuator control systems. This should include sensitivity to variations in load, component efficiencies, physical plant parameters, and actuator sensor noise.
3. The identification of a reasonable set of plant variations for the actuator, based on the sensitivity analysis, which is required for designing a robust actuator control system that will meet the specifications in the presence of predicted variations.

99.3 Hydraulic Actuator Modeling

The modeling of a purely hydraulic actuator employed in an irreversible flight control system is presented first. Based upon the notation shown in Fig. 99.1 the following relationships are obtained:

$$x_1 = \delta_{e_c} a \tag{99.1a}$$

$$\frac{x_v - x_f}{b + c} = \frac{x_v - x_1}{c} \tag{99.1b}$$

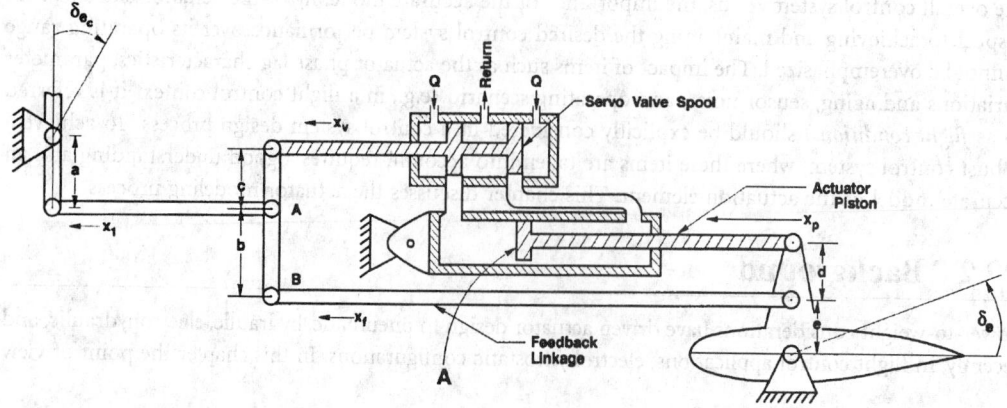

FIGURE 99.1 Irreversible control system.

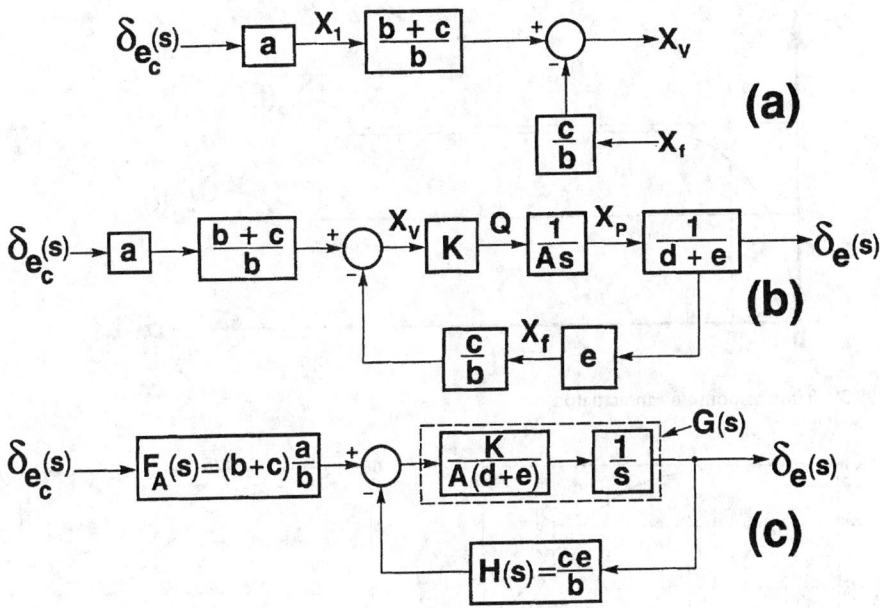

FIGURE 99.2   (a) Actuator block diagram; (b) block diagram of actuator feedback control system; (c) simplified block diagram.

which yield

$$x_v b + x_f c = x_1(b + c) \tag{99.2}$$

From these equations Fig. 99.2a is obtained. Next, the relationships

$$\frac{dx_p}{dt} = \frac{Q}{A}, \quad Q = Kx_v, \quad \delta_e = \frac{x_p}{d + e}, \quad x_f = e\delta_e \tag{99.3}$$

are utilized to yield the block diagram of a typical feedback control system (shown in Fig. 99.2b) which represents the irreversible hydraulic actuator under consideration. This block diagram is simplified to the one shown in Fig. 99.2c, which yields the first-order transfer function of the actuator:

$$G(s) = \frac{\delta_e(s)}{\delta_{e_c}(s)} = \frac{K}{\tau s + 1} \tag{99.4}$$

where the gain $K = [a(b + c)]/ce$, the time constant $\tau = [Ab(d + e)]/cKe$ of the actuator, and $0.025 < \tau < 0.1$ s. For a step input command, a typical actuator time response is shown in Fig. 99.3.

More-detailed dynamic modeling of the valve leads to higher-order hydraulic actuator models. For example, the transfer function of the currently used actuator of the F-16 aircraft is of fourth order [1] and its frequency response is compared with its first-order approximation in Fig. 99.4. It is indeed important to use the full-blown high-order actuator dynamic model in robust control system design. As Fig. 99.4 shows, the simplified first-order model fits well the high-order model in the low-frequency range. Unfortunately, a significant phase angle difference exists at the higher frequencies where the actuator will be operating, for the higher frequencies are within the bandwidth of modern robust ("high gain") control system compensators.

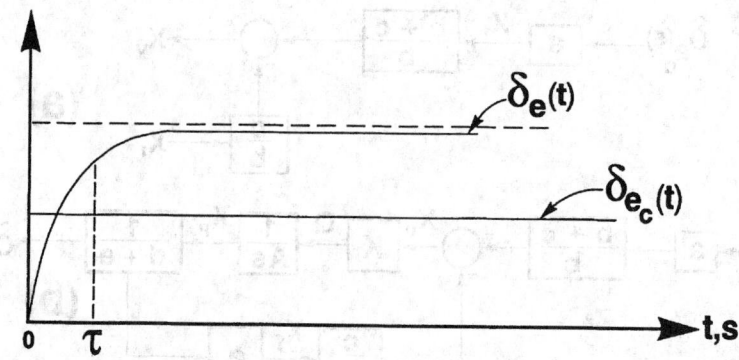

FIGURE 99.3   Time response of an actuator.

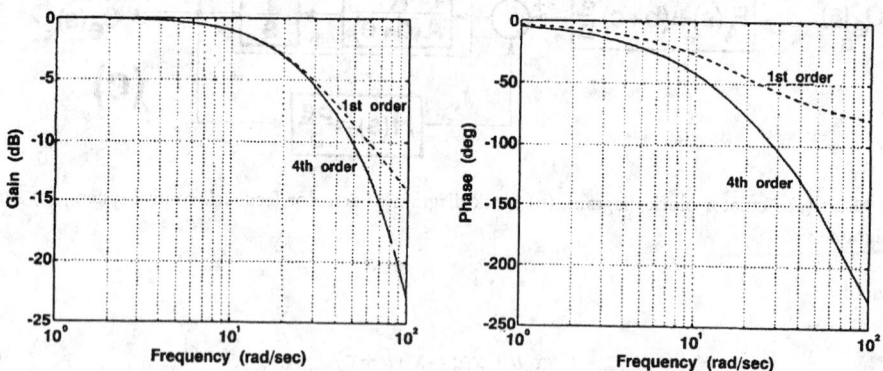

FIGURE 99.4   Comparison of first- and fourth-order actuator models.

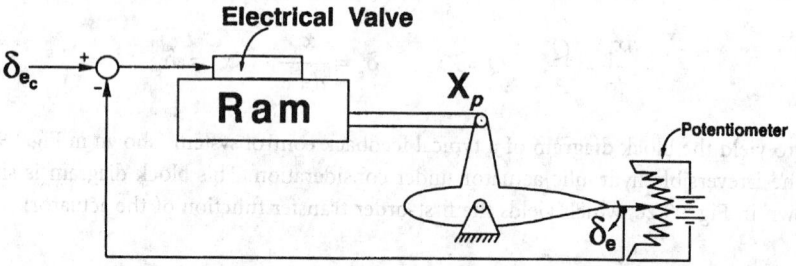

FIGURE 99.5   Electrohydraulic actuator.

The next step in the actuator state of development is then an electronic "fly-by-wire" mechanization of the feedback loop. This is shown in Fig. 99.5 where an actuator is shown that employs an electric (solenoid-controlled) valve, a potentiometer to measure the control surface deflection (actuator displacement), and where the power element of the actuator, the RAM, is an hydraulic cylinder.

99.4   EHA Actuator Modeling

The novel EHA actuator (see, e.g., Reference 2) offers a high degree of maintainability and survivability because all the actuator elements are collocated and a central hydraulic pump with the attendant high-pressure hydraulic plumbing is not needed. The modeling of the modules of the EHA is outlined first.

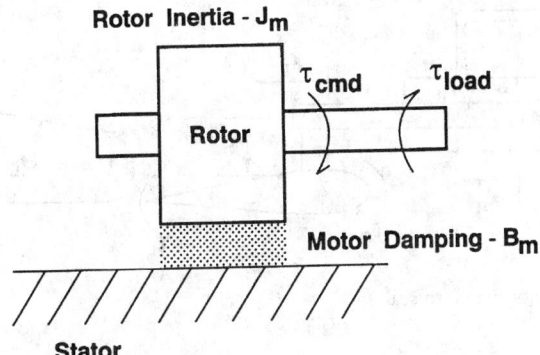

FIGURE 99.6 Motor model.

Motor

Electric motors with the rotor moment of inertia J_m and electromechanical damping B_m (see Fig. 99.6), are subject to variations in output torque and subsequent fluctuations in rotor speed ω_m. Since the torque due to the load counteracts the torque generated by the motor, perturbation may also be caused by variations of the load torque. This relationship is expressed as

$$\tau_e(s) = \tau_{cmd}(s) - \tau_{load}(s) \tag{99.5}$$

where τ_{load} is the load torque due to the differential pressure of the fluid in the pump. This results in a first-order transfer function

$$\frac{\Omega_m(s)}{\tau_e(s)} = \frac{1}{J_m s + b_m} \tag{99.6}$$

Pump and Fluid

Electric motors have a limited torque-to-mass ratio, due to the finite and limited magnetic flux density that can be generated [3]. High-pressure hydraulic systems, with the system pressure of 2000 to 5000 psi, can generate high forces resulting in more compact and higher torque-to-mass ratios than electric motors. Generally, high-pressure hydraulic systems are stiffer against the load than electric motors. Hence, the EHA utilizes a dc motor to pump high-pressure fluid into the piston chamber. The dc motor internal to the EHA converts the electric power into mechanical power. It is the pump that converts this mechanical power into hydraulic power. The hydraulic power, acting against the piston, is converted to mechanical power capable of moving the load, e.g., large flight control surfaces.

The flow rate Q_m generated by the pump is proportional to the motor speed, i.e.,

$$Q_m(s) = \frac{D_m}{2\pi} \Omega_m(s) \tag{99.7}$$

where D_m represents the pump displacement constant. The flow rate of the hydraulic fluid is primarily dependent on two factors: change in chamber volume and change in pressure due to the compressibility of the fluid effect [4]. The chamber volume changes as the piston moves through the chamber at speed dx_p/dt. The flow rate due to the changes in chamber volume is then expressed as $\pm A dx_p/dt$. Secondary fluid dynamic effects include fluid compressibility, internal leakage flow, and external leakage flow. These secondary effects are modeled by a first-order equation as follows:

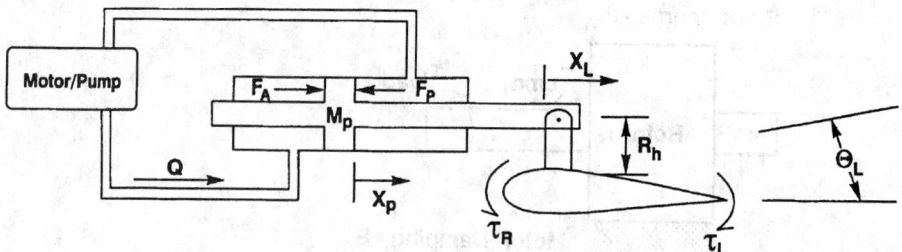

FIGURE 99.7   Simplified actuator control (not drawn to scale).

$$\delta Q(s) = Q_{\mathrm{m}}(s) - A s X_{\mathrm{p}} = \left[K_s s + C_t\right] P(s) \qquad (99.8)$$

which results in the transfer function

$$\frac{P(s)}{\delta Q(s)} = \frac{1}{K_s s + C_1} \qquad (99.9)$$

Piston and Flight Control Surface

Piston and Flight Control Surface Fundamentals

The pressure developed by the pump and fluid acts on the piston surface (see Fig. 99.7), causing the RAM to extend or retract. This force then generates a torque through a hinge to deflect the control surface. This torque has to overcome two load components: control surface inertia and aerodynamic loads. The control surface inertia is primarily due to the fact that the control surface has certain size and mass. The aerodynamic load only occurs in flight, when the air pressure over the control surface applies aerodynamic forces to it. The aerodynamic load is determined by three factors: the surface area of the flight control surface, aerodynamic loading which varies with altitude and airspeed, and the relative angle of the surface to the wind. The surface angle to the wind depends on the angle of surface deflection and on the angle of attack of the aircraft.

Piston and Flight Control Surface Dynamics

Flexible Hinge Joint Model

The magnitude of F_A acting on the piston is equal to $\wp A$, where $\wp$ is the differential pressure developed by the pump and fluid and A is the surface area of the piston. Thus, the force created by the pump and fluid can be expressed as

$$F_A = \wp A \qquad (99.10)$$

The piston dynamics, with the piston mass M_p and piston damping B_p, can be described by the second-order model:

$$F_A - F_P = \left[M_p s^2 + B_p s\right] X_P \qquad (99.11)$$

The resulting torque acting on the flight control surface due to this force imbalance can be described by

$$\tau_R = \frac{K_h}{R_h}\left(X_P - X_L\right) \qquad (99.12)$$

where K_h is the hinge stiffness constant and R_h is the hinge length. The stabilator inertia acts against the torque generated by the actuator, such that

$$\tau_R - \tau_L = \left[J_L s^2 + B_L s \right] \theta_L \qquad (99.13)$$

where τ_L is the torque created by the aerodynamic load and stabilator inertia. The variables J_L and B_L represent the mass properties of the flight control surface.

Stiff Hinge Joint Model

Equations 99.11 through 99.13 represent a rather complex model of load dynamics. The complexity of the model can be reduced if the linkage between the actuator and flight control surface is considered rigid. This is a valid assumption, since the natural frequency of the hinge for a well-designed actuation system is much greater than the actuator bandwidth of concern. Hence, the piston and load dynamics are still expressed as

$$F_A - F_P = \left[M_P s^2 + B_P s \right] X_P \qquad (99.14)$$

and

$$\tau_R - \tau_L = \left[J_L s^2 + B_L s \right] \theta_L \qquad (99.15)$$

Dividing Eq. 99.15 by R_h yields

$$F_R - \frac{\tau_L}{R_h} = \left[\frac{J_L s^2 + B_L s}{R_h} \right] \theta_L \qquad (99.16)$$

Assuming rigidity of the hinge assembly, $F_R \approx F_P$. Hence, adding Eqs. 99.14 and 99.16 results in

$$F_A - \frac{\tau_L}{R_h} = \left[M_P s^2 + B_P s \right] X_P + \left[\frac{J_L s^2 + B_L s}{R_h} \right] \theta_L \qquad (99.17)$$

where τ_L is defined as

$$\tau_L = \left(\text{Load}_{\text{aero}} \right) \theta_L \qquad (99.18)$$

Since rigidity implies that $X_P = X_L$ and $\theta_L = X_P / R_h$, Eq. 99.17 is further reduced to

$$F_e = F_A - \frac{\tau_L}{R_h} = \left[\left(M_P + \frac{J_L}{R_h^2} \right) s^2 + \left(B_P \frac{B_L}{R_h^2} \right) s \right] X_P \qquad (99.19)$$

or expressed in a transfer function form:

$$\frac{X_P(s)}{F_e(s)} = \frac{1}{s \left[\left(M_P + \frac{J_L}{R_h^2} \right) s + \left(B_P \frac{B_L}{R_h^2} \right) \right]} \qquad (99.20)$$

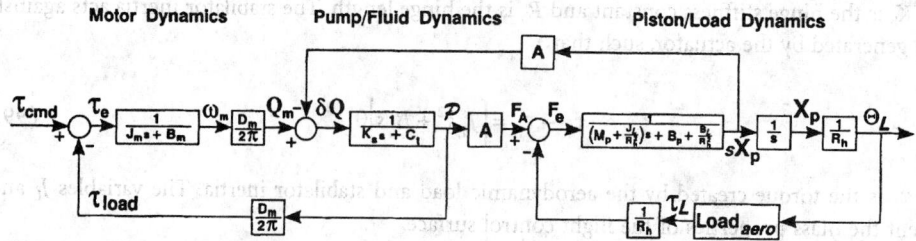

FIGURE 99.8 Bare EHA block diagram.

Equation 99.20 represents a simplified model of the load dynamics. A complete simplified design model of the bare EHA, without the controller, is shown in Fig. 99.8 using the individual component models.

Hinge Moments

The aerodynamic loads mentioned previously describe the torque opposing the piston motion. Thus, load generated by the flow of air above and below the flight control surface applies a torque to the hinge assembly, which in turn adds backpressure to the piston. The hinge moment is modeled as follows:

$$\tau_L = \bar{q} S_t R_h \left(Ch_\alpha \frac{\alpha(s)}{\theta_L(s)} + Ch_\delta \right) \theta_L \tag{99.21}$$

where S_t denotes the surface area of the control surface, $\bar{q}$ denotes the dynamic pressure, Ch_α and Ch_δ denote the hinge moment coefficients of the control surface with respect to angle of attack and surface deflection δ. The aerodynamic load dynamic pressure $\bar{q}$ is a function of airspeed U_0 and air density [2] ρ, as expressed by

$$\bar{q} = \frac{1}{2} \rho U_0^2 \tag{99.22}$$

The air density in a standard atmosphere drops exponentially with increasing altitude. The dynamic pressure $\bar{q}$ increases as the Mach number gets higher and altitude decreases. The hinge moment for modern fighters with all movable tails, where $Ch_\alpha \approx Ch_\delta$, is modeled by

$$\tau_L = \bar{q} S_t R_h Ch_\delta \left(\frac{\alpha(s)}{\theta_L(s)} + 1 \right) \theta_L \tag{99.23}$$

Aircraft Short Period Approximation

If the aircraft forward speed is assumed constant (i.e., the speed perturbations $u \approx 0$), the X force equation can be neglected since it does not significantly contribute to the short-period oscillation [5]. Thereby, the short-period approximation of the longitudinal channel of the aircraft which is relevant to the actuator dynamics is extracted and is written as

$$\begin{bmatrix} \dot{\alpha} \\ \dot{q} \end{bmatrix} = \begin{bmatrix} z_\alpha & z_q \\ m_\alpha & m_q \end{bmatrix} \begin{bmatrix} \alpha \\ q \end{bmatrix} + \begin{bmatrix} z_\delta \\ M_\delta \end{bmatrix} [\delta_e] \tag{99.24}$$

This yields a second-order minimum-phase transfer function of the form:

$$\frac{\alpha(s)}{\delta(s)} = \frac{-K(s+a)}{(s+b)(s+c)} \tag{99.25}$$

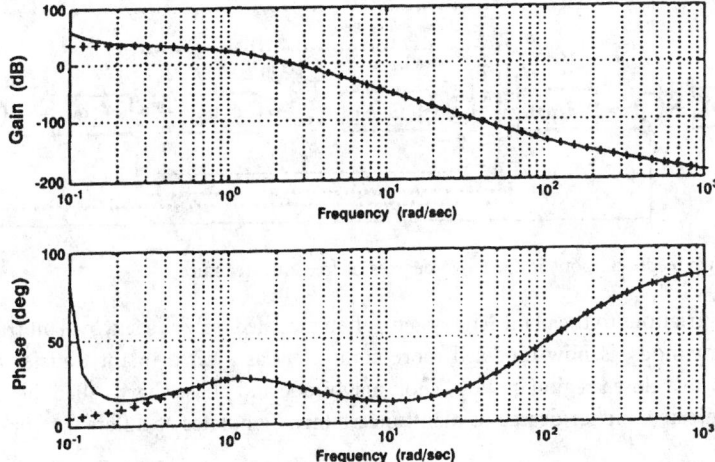

FIGURE 99.9 Comparison of full and short period approximation for longitudinal equations of motions.

For the frequency of interest, the short-period approximation closely resembles the full state model. As seen in Fig. 99.9, where the solid lines represent the full state model frequency response of the vehicle and + lines represent the short-period model frequency response of the vehicle, phase and attenuation characteristics are closely matched at high frequency. The approximation, as seen in the figure, is not valid for frequencies below 0.5 rad/s due to the effects from the slow longitudinal mode, which is referred to as the phugoid.

Quantitative Feedback Theory (QFT)

Robust QFT [3,6] compensators are ideal for obtaining the desired response from an actuator. The term *robust* in control theory implies that a system under control, in this case the actuator, remains stable throughout its operating envelope, rejects disturbances, and results in minimum degradation in the performance specifications. A QFT designed feedback system, which uses output feedback, assures that the output tracks the input values despite parametric uncertainties or disturbances. The output of a unity-gain feedback system with uncertain plant models can vary depending on the plant conditions; however, the output of a robust QFT feedback control system with uncertain plant models will not vary much. The unity-gain system, however, cannot control the tracking response, which necessitates the inclusion of a prefilter.

Plant (specifically, actuator) uncertainties may be caused by manufacturing tolerances. For example, a component A is specified to be 10 ± 0.1 mm long. In a large production run, the length of component A may be anywhere between 9.9 and 10.1 mm. This tolerance range may be substantial enough to produce noticeable variations in the plant model. Plant uncertainties may also be caused by the environment, specifically the system (actuator) operating conditions. For example, the parameters of an aircraft model display large variations depending on altitude and airspeed. Finally, the performance of various components decay over time, introducing an additional element of uncertainty.

The QFT design paradigm accounts for the plant variations in the design procedure. It is a linear design technique for designing a linear robust controller for linear or nonlinear control systems [3,6] — in this case the actuator control system. As long as all the QFT design requirements are met, the output responses are guaranteed to conform to the specification boundaries. The initial step in the design process is to establish the upper and lower performance tracking boundaries. The maximum allowable disturbance is used to determine the disturbance boundary. The phase margin angle, gain margin, or maximum peak value is used to establish the stability boundary. These combined boundaries establish the limitation of the system performance. The next step is to analyze the system to determine the parameters that will

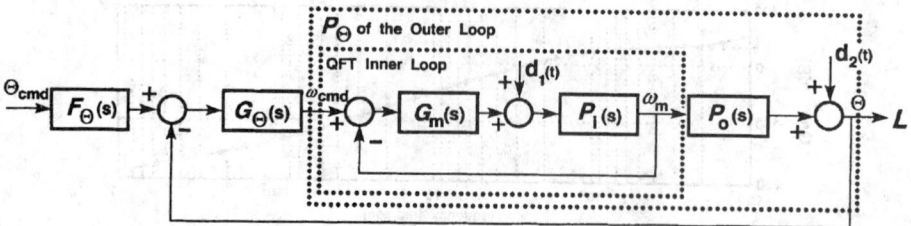

FIGURE 99.10   Two feedback loops in QFT two degrees of freedom structure.

cause noticeable variations to the plant. The system, often modeled only as a single plant transfer function in other design methods, is now modeled more completely as a *set* of plant transfer functions. The variations now form a closed region in the Nichols magnitude and phase chart, called the template; larger template sizes indicate a higher degree of plant uncertainty. Templates at different frequencies differ in size and shape.

A most important aspect of modeling is the thorough analysis of the bare actuator plant, in order to find the sources of plant variation. For an aircraft application the three largest sources of variations are the fluctuations in aerodynamic load on the flight control surfaces, motor torque, and hydraulic pump fluid pressure.

99.5 Actuator Compensator Design

QFT compensators, which are designed in the frequency domain, are ideally suited for the actuator control system. The QFT compensator processes its inputs, which are the commanded actuator displacement and the measured current actuator position, and generates a commanded signal to, e.g., the solenoid valve in a hydraulic actuator, or a current for the electric dc motor in the EHA actuator. In order to proceed with the actuator QFT compensator design, the block diagram as shown in Fig. 99.7 is transformed to the standard inner and outer QFT loop structures shown in Fig. 99.10 [2,3,6]. For this design problem the external disturbance $d_2(t)$ can be ignored, i.e., $d_2(t) = 0$, whereas the $d_1(t)$ disturbance models a possible measurement bias. The inner loop controls the angular rate of the motor while the outer loop controls the surface deflection. Since tracking in the inner loop is less important, for its function is to reduce the uncertainty level of the outer loop, the inner loop prefilter can be set to unity and the tracking performance enforcement is relegated to the prefilter F_θ of the outer loop.

99.6 Conclusion

The modeling of a hierarchy of hydraulic actuators has been presented. The modeling of modern EHAs is emphasized and dynamic models of its subcomponents are carefully developed. A systems approach to actuator modeling and design is taken and the feedback structure is highlighted. In this chapter it is stressed that, because of the complex feedback interactions at work, proper actuator modeling requires consideration of the latter in the full context of the control system in which it is to be employed. Thus, in this presentation the aerospace application is considered. At the same time the actuator is being viewed not as a component of a control system, but as a control system in its own right. Hence, it is shown that simple mechanical feedback linkages which fulfilled the role of a controlling element for the actuator can be replaced by a more flexible, high-performance, robust, full-blown electronic actuator compensator. Concerning electropneumatic instruments and sensors, these are critical elements in systems where the flow of liquids and/or gases is being controlled, e.g., in the chemical industries. In aerospace, electropneumatic sensors are used to measure airspeed, altitude, and the angle of attack and sideslip angles of the aircraft; these functions are collocated in the air data computer (ADC). Currently, modern sensors use feedback action for accuracy enhancement, and, as such, the design of modern electropneumatic instruments and sensors is conceptually similar to the design of EHAs. Detailed information on actuators and sensors may be found in References 4 and 7.

Nomenclature

A	ram cross section area
a,b,c,d,e,R_h	length of linkages
x,X	displacement
J	moment of inertia; o, ovin
K	gain
Q	flow rate
s	Laplace variable
D_m	pump displacement constant
F	force
M_p	mass of piston
B_p	piston damping
K_h	hinge stiffness constant
$\bar{q}$	dynamic pressure
S_t	aerodynamic surface area
C_h	hinge moment coefficient
U,u	airspeed
q	pitch rate
M,z,m	aerodynamic stability derivatives
P	pressure
δ	deflection angle
τ	torque
Ω,ω	rotor speed
$\mathcal{P}$	differential pressure developed by the pump
θ	deflection angle
α	angle of attack of aircraft
ρ	air density

References

1. O.R. Reynolds, M. Pachter, and C.H. Houpis, Full Envelope Flight Control System Design Using Quantitative Feedback Theory, *AIAA J. Guidance Control Dyn.*, 19, 1023–1029, 1996.
2. K.H. Kang, Electro-Hydrostatic Actuator Controller Design Using Quantitative Feedback Theory, MS thesis, Graduate School of Engineering, Air Force Institute of Technology, Wright-Patterson AFB, OH, December 1994, AFIT/GE/ENG/94D-18.
3. C.H. Houpis, M. Pachter, S. Rasmussen, and R. Sating, Quantitative Feedback Theory for the Engineer, Wright Laboratory Technical Report, WL-TR-95-3061, Wright Laboratory, Wright-Patterson AFB, OH, June 1995. (Available from the National Technical Information Service, 5285 Port Royal Road, Springfield, VA 22151, document number AD-A297574.)
4. C.W. de Silva, *Control Sensors and Actuators*, Englewood Cliffs, NJ: Prentice-Hall, 1989.
5. J.H. Blakelock, *Automatic Control of Aircraft and Missiles*, New York: John Wiley & Sons, 1991.
6. J.J. D'Azzo and C.H. Houpis, *Linear Control System Analysis and Design — Conventional and Modern*, 4th ed., New York: McGraw-Hill, 1988.
7. E.H.J. Pallet and S. Coyle, *Automatic Flight Control*, Oxford, U.K., Blackwell, 1993.

100

Explosion-Proof Instruments

Sam S. Khalilieh, P.E.

Earth Tech

100.1 Introduction

Where hazardous atmospheres can exist, electricity should be a primary concern of every engineer and system designer. Hazardous atmospheres can exist not only in the more common surroundings of industrial, chemical, and environmental facilities, but also in many less obvious environs where dust is present, where gas can accumulate, and where combustible gas-forming reactions occur. To minimize risks in such areas, it is necessary to design specific hazard-reducing electric systems. Most electric equipment is built to specific standards aimed to reduce the incidence of fires and human casualties. The majority of such incidents can be attributed to poor or defective installations, improper use of approved equipment, deteriorated equipment, and accidental applications. In combination with an explosive atmosphere, these factors can result in extremely dangerous conditions. Designing an electric system for a hazardous location requires careful planning, research, engineering, and ingenuity in using proper protection techniques to develop better applications and classifications that reduce hazards.

100.2 Fundamentals of Explosion Protection

Safety of personnel and equipment present in hazardous area should never be taken for granted. In 1913 a massive methane gas explosion in a coal mine in Glamorganshire, South Wales claimed the lives of 439 mine workers. After months of research and studies, a group of experts and scientists concluded that the explosion was caused by a small amount of electric energy stored in the circuit. This small amount of energy combined with the presence of an explosive gas and air mixture and the absence of proper protection proved to be fatal for the mine workers.

To understand the dangers associated with electric equipment in hazardous areas, one must first understand the basics. Chemically speaking, **oxidation**, combustion, and explosions are all exothermic reactions where heat is given off at different reaction speeds. For these reactions to occur, three components must be present simultaneously in certain concentrations. These components are (1) fuel (liquid, gas, or solids), (2) a sufficient amount of oxygen (air), and (3) an ignition source (electric or thermal).

Some of the ignition sources that can be potentially hazardous include (1) hot surfaces (motor windings, heat trace cable, light fixtures), (2) electric sparks and arcs (when circuits are opened and closed, short circuits), (3) mechanic sparks (friction, grinding), (4) electrostatic discharge (separation process in which at least one chargeable substance is present), and (5) radiation, compression, and shock waves.

When dealing with electric equipment in hazardous locations, it is important to understand and to be familiar with the following terms:

1. *Flash Point* — the minimum temperature at normal air pressure at which a combustible or flammable material releases sufficient vapors ignitable by an energy source. Depending on the flash point (FP), flammable liquids are divided into four classes of hazard:
   a. AI (FP < 21°C),
   b. AII (21 < FP < 55°C),
   c. AIII (55°C < FP < 100°C), and
   d. B (FP < 21°C at 15°C dissolving in water).
2. *Ignition temperature* — the minimum temperature under normal operating pressure at which a dangerous mixture ignites independently of the heating or heated element.
3. *Flammable limits* — the upper explosive limit (UEL) or the maximum concentration ratio of vapor to air mixture above which the propagation of flame does not occur when exposed to an ignition source. Here, the mixture is said to be "too rich" to burn. The lower explosive limit (LEL) is the minimum concentration ratio of vapor to air mixture below which the propagation of flame does not occur when exposed to an ignition source. Also, here, the mixture is said to be "too lean" to explode. Significant attention must be given to LEL, since it provides the minimum quantity of gas necessary to create a hazardous mixture. Generally, the flammable limits are indicated in percent by volume, which is abbreviated % vol. Note that the explosion of a mixture in the middle the UEL and the LEL is much more violent than if the mixture were closer to either limit.
4. *Maximum surface temperature* — the maximum temperature generated by a piece of electric equipment under normal or fault conditions. This temperature must be below the minimum ignition temperature of the potentially explosive surrounding atmosphere. Equipment used in hazardous locations must be clearly marked to indicate class, group, and maximum surface temperature or range referenced to 40°C (104°F) ambient temperature. Table 100.1 shows that an apparatus with a specific T class can be used in the presence of all gases having an ignition temperature higher than the T temperature class of the device. For added safety, it is recommended that the maximum surface temperature be not more than 80% of the minimum ignition temperature of the surrounding gas. The reader is cautioned not to confuse maximum working (operating) temperature with maximum surface temperature, which is measured under worst-case conditions of the electric apparatus. An electric apparatus designed to operate with a maximum ambient temperature of 70°C — even in the worst conditions of the expected temperature range — must not have a temperature rise greater than a safety margin of 10°C to be classified as T6 or 5°C for classes T3, T4, and T5 (Table 100.1).
5. *Vapor density* — the weight of a volume of pure vapor gas compared with the weight of an equal volume of dry air under the same normal atmospheric pressure and temperature. It is calculated as the ratio of molecular weight of the gas to the average molecular weight of air (28.96). Methane gas (CH_4) with molecular weight of 16 and vapor density of 0.6 tends to rise, while acetone (C_3H_6O) with molecular weight of 58 and vapor density of 2 tends to settle closer to ground levels.

In the U.S., the National Electrical Code (NEC) defines a hazardous area as "an area where a potential hazard may exist under normal or abnormal conditions because of the presence of flammable, combustible, or ignitable materials" [1]. This general description is divided into different classes, divisions, and groups to assess the extent of the hazard properly and to design and specify safe operating electric systems.

TABLE 100.1 Maximum Surface Temperature Under All Operating Conditions

Maximum Temperature		Identification Number
°C	°F	
450	842	T1
300	572	T2
280	536	T2A
260	500	T2B
230	446	T2C
215	419	T2D
200	392	T3
180	356	T3A
165	329	T2B
160	320	T3C
135	275	T4
120	248	T4A
100	212	T5
85	185	T6

Note: Surface temperature of electric apparatus during operation must not exceed limitations of the hazard present. Reprinted with permission from NFPA 70-1996, the *National Electrical Code®*, Copyright© 1995, National Fire Protection Association, Quincy, MA 02269. This reprinted material is not the complete and official position of the National Fire Protection Association, on the referenced subject which is represented only by the standard in its entirety.

The need for classification is important not only for safety, but for economic reasons as well. Proper application, good engineering, and experience can reduce the extent of the most volatile areas (Class I, Division 1) within reasonably safe distances of potential leaks and ignition sources. Under Class I, Division 1, equipment and installation costs can become an economic burden because the equipment is considerably more expensive and must pass stringent tests to ensure proper and safe operation under normal or abnormal conditions. The National Fire Protection Association (NFPA 497 A & B) [2], and the American Petroleum Institute *Recommended Practice for Classification of Locations for Electrical Installations at Petroleum Facilities* (ANSI/API RP 500) [3] are excellent resources for defining hazardous area boundaries.

100.3 Classification of Hazardous Areas

Classification of a hazardous area within a facility is usually determined by highly qualified personnel including chemical engineers, process engineers, and safety officers. Their primary objective is to determine where a potentially hazardous atmosphere exists, under what conditions it exists, and how long it exists. Careful study and design of electric installations, especially in hazardous areas, are crucial for the safe operation of electric equipment and prevention of an accidental ignition of flammable materials. The NEC, which has been adopted by many states, agencies, and companies as the basis for inspections, describes the requirements and procedures for electric installations in hazardous areas. Articles 500–504 contain the requirements of electric equipment and wiring for all voltages in locations where fire or explosion hazards may exist due to flammable gases or vapors, flammable liquids, combustible dust, or ignitable fibers or flyings.

Table 100.2 describes hazardous locations by Class, Division, and Group. The Class defines the physical form of combustible material mixed with oxygen molecules. The Division defines the probability of an

TABLE 100.2 Area Classification Based on NEC

Division 1 — Hazard Is Present under Normal Operating Conditions	
Class I	Gases and Vapor
Group A	Acetylene
Group B	Hydrogen
Group C	E.g., ethylene
Group D	E.g., methane
Class II	Combustible Dusts
Group E	Metal dust
Group F	Coal dust
Group G	Grain dust
Class III	Fibers

Division 2 — Hazard Is Present Only under Abnormal Operating Conditions

explosive fuel to air mixture being present. The Group indicates the type of vapor or dust present. The NEC gives the following definitions [1]:*

Class I, Division 1 Locations

1. Where ignitable concentrations of flammable gases or vapors can exist under normal operation conditions; may exist frequently because of repair or maintenance operations or because of leakage; and where breakdown or faulty operation of equipment or processes might release ignitable concentrations of flammable gases or vapors, and cause simultaneous failure of electric equipment.

Class I, Division 2 Locations

1. Where volatile flammable liquids or flammable gases are handled, processed, or used but where the liquids, vapors, or gases will normally be confined within closed containers or closed systems from which they can escape only in case of accidental rupture or breakdown of such containers or systems, or in case of abnormal operation of equipment;

2. Where ignitable concentrations of gases or vapors are normally prevented by positive mechanical ventilation, and where they might become hazardous through failure or abnormal operation of the ventilating equipment;

3. Adjacent to Class I, Division 1 locations, and where ignitable concentrations of gases or vapors might occasionally be communicated; unless such communication is prevented by positive-pressure ventilation from a source of clean air, and effective safeguards against ventilation failure are provided.

Class II, Division 1 Locations

1. Where combustible dust is in the air under normal operation conditions in quantities sufficient to produce explosive or ignitable mixtures;

2. Where mechanical failure or abnormal operation of machinery or equipment might cause such explosive or ignitable mixtures to be produced, and also might provide a source of ignition through simultaneous failure of electric equipment, operation of protective devices, or from other causes;

3. Where combustible dusts of an electrically conductive nature may be present in hazardous quantities.

Class II, Division 2 Locations

Where combustible dust normally is not in the air in quantities sufficient to produce explosive or ignitable mixtures, and dust accumulations normally are insufficient to interfere with the safe dissipation of heat from electric equipment, or may be ignitable by abnormal operation or failure of electric equipment.

Class III, Division 1 Locations

Where easily ignitable fibers or materials producing combustible flyings are handled, manufactured, or used.

Class III, Division 2 Locations

Where easily ignitable fibers are stored or handled. Quantities and properties of hazardous materials are the basis upon which the NEC classifies hazardous locations. Each hazardous location must be evaluated carefully to determine the appropriate classification to facilitate the design process and to help specify the correct equipment.

100.4 Enclosure Types and Requirements

Choosing the proper type of enclosure for electric equipment is important for two reasons:

1. Personnel protection against accidental contact with enclosed electric equipment.
2. Protection of internal equipment against outside harm.

Enclosures are designated by a type number indicating the degree of protection and the condition for which they are suitable. In some applications, enclosures have a dual purpose and therefore are designated by a two-part type number shown with the smaller number first (i.e., 7/9). The following enclosure types, with their enclosed equipment, have been evaluated in accordance with Underwriters Laboratories, Inc. UL 698, *Industrial Control Equipment for Use in Hazardous Locations,* and are marked to show the class and group letter designations.

Type 7 Enclosures — Type 7 enclosures are nonventilated, intended for indoor applications, and classified for Class I, Group A, B, C, D as defined in Table 100.2. The letters A to D sometimes appear as a suffix to the designation Type 7 to give the complete designation. According to UL 698, Type 7 enclosures must be designed to withstand an internal explosion pressure of specific gases and to prevent such an explosion from igniting a hazardous mixture outside the enclosure (explosion test). In addition, Type 7 enclosures fabricated from sheet steel are designed to withstand two times the internal explosion pressure for 1 min without permanent deformation and three times the explosion pressure without rupture. If constructed of cast iron, the enclosure must be capable of withstanding four times the explosion pressure without rupture or deformation. This test may be waived if calculations show a safety factor of five to one for cast metal or four to one for fabricated steel. The enclosed heat-generating devices are specifically designed to prevent external surfaces from reaching temperatures capable of igniting explosive vapor–air mixture outside the enclosure (temperature test).

Type 8 Enclosures — Type 8 enclosures are nonventilated, intended for indoor applications, and intended for Class I, Group A, B, C, D as outlined in Table 100.2. The letters A to D appear as a suffix to the designation Type 8 to give the complete designation. According to UL 698, the oil-immersed equipment must be able to operate at rated voltage and most severe current conditions in the presence of flammable gas–air mixtures without igniting these mixtures.

Type 9 Enclosures — Type 9 enclosures are nonventilated, intended for indoor applications, and classified for Class II, Group E, F, G as outlined in Table 100.2. The letters E, F, or G appear as a suffix to the designation Type 9 to give the complete designation. According to UL 698, the enclosure with its enclosed equipment is evaluated in accordance with UL 698 in effect at the time of manufacture. This evaluation includes a review of dimensional requirements for shaft opening and joints, gaskets material, and temperature rise under a blanket of dust. The device is operated at full rated load until equilibrium temperatures are reached, then allowed to cool to ambient temperature over a period of at least 30 h while continuously subjected to circulating dust of specified properties. No dust shall enter the enclosure (dust penetration test). Furthermore, Type 9 enclosures must also pass the "temperature test with dust blanket," which is similar to the temperature rise test except the circulating dust is not aimed directly at the device during testing. The dust in contact with the enclosure shall not ignite or discolor from heat, and the exterior surface temperature based on 40°C (104°F) shall not exceed specific temperatures under normal or abnormal conditions. Where gasketed enclosures are used, gaskets shall be of a noncombustible, nondeteriorating, vermin-proof material and shall be mechanically attached. Type 9 ventilated enclosures are the same as nonventilated enclosures, except that ventilation is provided by forced air from a source outside the hazardous area to produce positive pressure within the enclosure. The enclosure must also meet temperature design test.

Type 10 Enclosures — Type 10 enclosures are nonventilated and designed to meet the requirements of the U.S. Bureau of Mines which relate to atmospheres containing mixtures of methane and air, with or without coal dust present.

It is important to note that enclosures for hazardous applications are designed for specific applications and must be installed and maintained as recommended by the enclosure manufacturer, since any misapplication or alteration to the enclosure may jeopardize its integrity and may eventually cause catastrophic failure of the system. All enclosures should be solidly grounded and properly labeled with a warning sign reminding the operator of the importance of deenergizing the incoming power to the enclosure prior to its servicing.

100.5 Protection Methodologies

Choosing a protection technique that suits each application can appear complicated because safety, reliability, cost, and maintenance factors must all be considered. Over the years, few hazardous area safety protection methodologies have been used. Although methodologies differ in application and principles of operation, they all have one thing in common: to eliminate one or more components necessary for combustion. Three of the most widely used methodologies are

1. Intrinsic safety
2. Explosion-proof
3. Purging and pressurization

Intrinsic Safety

Simply stated, intrinsic safety (IS) is all about preventing explosions. IS is based on the principle of limiting the thermal and electrical energy levels in the hazardous area to levels that cannot cause an ignition of a specific hazardous mixture in its most ignitable concentration. IS pertains to the minimum ignition temperature and the minimum ignition electric energy required to cause a specific group to ignite. The energy level provided by an IS circuit is low (≈ 1 W) and is used only to power up instruments with a low energy demand. An IS circuit incorporates an intrinsically safe apparatus (field device), an associated apparatus, and an interconnecting wiring system. Designing intrinsically safe systems begins with studying the field device. This will help determine the type of associated apparatus which can be used so that the circuit functions properly under normal operating conditions, but still safe under fault conditions. Field devices can be simple, such as **resistance temperature devices (RTDs)**, thermocouples, mechanical switches, proximity switches, light emitting diodes (LEDs), or they can be nonsimple, such as transmitters, solenoid valves, and relays. A field device is considered and recognized as a "simple device" if its energy storing or generating values do not exceed 1.2 V, 0.1 A, 25 mW (or 20 µJ) in an intrinsically safe system under normal or abnormal conditions.

The simple device may be connected to an intrinsically safe circuit without further certification or approval. The fact that these devices do not have the ability to store or generate high levels of energy does not mean they can be installed in a hazardous area without modification. They must always be used with an associated apparatus to limit the amount of energy in the hazardous area, since a fault outside the hazardous area can cause sufficient high levels of energy to leak into the hazardous area. A nonsimple device (i.e., relay, transmitter) is capable of generating and storing energy levels exceeding the aforementioned values. Such devices require evaluation and approval under the entity concept (described later) to be used in conjunction with an intrinsically safe circuit. Under the entity concept, these devices have the following entity parameters: V_{max} (maximum voltage allowed), I_{max} (maximum current allowed), C_i (internal capacitance), and L_i (internal inductance). Under fault conditions, voltage and current must be kept below the V_{max} and I_{max} of the apparatus to prevent any excess heat or spark, which can be disastrous in hazardous areas. C_i and L_i indicate the ability of a device to store energy in the form of internal capacitance and internal inductance, and their value must be less than C_a and L_a of the associated apparatus (Table 100.3).

TABLE 100.3 Comparison of Entity Values of a Field Device and a Safety Barrier

Field Device		Safety Barrier
V_{max}	≥	V_{oc}
I_{max}	≥	I_{sc}
C_i	≤	C_a (maximum allowed capacitance)
L_i	≤	L_a (maximum allowed inductance)

An associated apparatus (Fig. 100.1), also known as a safety barrier, is an energy-limiting device needed to protect a field device located in a hazardous area from receiving excessive voltage or current. An associated apparatus is normally installed in a dust- and moisture-free enclosure (NEMA 4) located in a nonhazardous area, as close as possible to the hazardous area to minimize the capacitance effect of the cable. If installed in a hazardous area, the associated apparatus must be installed in an explosion-proof enclosure (i.e., NEMA 7D).

Figure 100.1 shows the three major components of a Zener safety barrier. (Note that there are other types of barriers such as isolation and repeater types.) The components are

1. The resistor, which limits the current to a specific value known as short circuit current (I_{sc}).
2. The fuse, which acts as an interrupter or protective device in case of a diode failure (fuse will blow if diode conducts).
3. The **Zener diode**, which limits the voltage to a specific value known as open circuit voltage (V_{oc}). Zener diodes are unique in their ability to conduct current under reverse bias conditions. When voltage is applied to the Zener diode in the reverse direction, a small amount of current known as leakage current is passed through. This current remains small until the bias voltage exceeds the Zener breakdown voltage. Exceeding the breakdown voltage causes the inherently high resistance of the Zener diode to drop to a very low value, thus allowing the current to increase abruptly. This sudden current increase forces the Zener diode to become a conductor, thereby diverting the excess voltage to ground. If the current continues to flow above and beyond the fuse rating, the

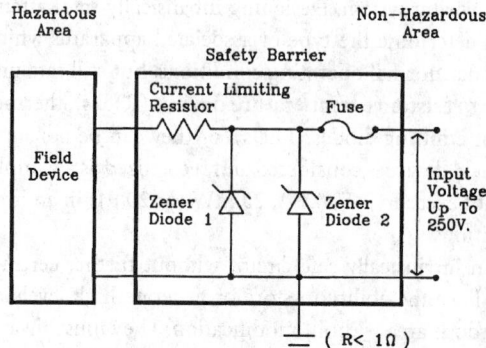

FIGURE 100.1 Major components of barrier circuit.

fuse will open and the circuit will be completely interrupted. Most safety barriers incorporate at least two diodes in parallel to provide maximum protection in case one diode fails (redundant safety).

In 1988, ANSI/UL 913 [5] allowed the use of intrinsic safety barriers with replaceable fuses as follows: "if it is accessible for replacement, and the fuse on a fuse protected shunt diode barrier shall not be replaceable by one of higher rating." The fuses are housed in tamper-proof assemblies to prevent confusion or misapplication. The diodes have specific power ratings which must not be exceeded. The Zener diodes and fuses are governed by a very specific set of parameters which allow the fuse to operate at one-third the power rating of the Zener diode and to avoid irreversible damage to the Zener diode. The power rating for the Zener diode can be determined as follows:

$$Z_w = 1.5 \times V_{oc} \times 2 \times I_f$$

where

Z_w = minimum power rating of the Zener diode
V_{oc} = maximum Zener diode open-circuit voltage
I_f = fuse current rating

Selecting the best barrier for the application depends on the field device and requires analysis to ensure proper operation of the intrinsically safe circuit under normal or abnormal conditions. Three of the more important characteristics requiring examination are (1) internal resistance, (2) rated voltage, and (3) circuit polarity. Regardless of the selected barrier, each has an internal resistance (R_i) which limits the short circuit current under fault conditions. As current passes through R_i, it creates a voltage drop across the barrier that must be accounted for ($V = IR$). The rated voltage of the safety barrier must be equal to or reasonably greater than the supply voltage. The word *reasonably* is significant because excessive supply voltage can cause the diode to conduct, rushing high current through the fuse and blowing it. The use of a regulated power supply can significantly reduce problems associated with excessive supply voltage. To complete an analysis, the circuit polarity must be established. While ac barriers can be connected with either positive or negative power supply, dc barriers can be rated to either positive or negative.

Making Field Devices Intrinsically Safe

RTDs and thermocouples can be made intrinsically safe by using isolated temperature converters (ITCs) that convert a low dc signal from the field device into a proportional 4 to 20 mA signal. These ITCs require no ground connection for the safe and proper operation of the IS circuit. Because of their ability to store energy, transmitters are considered nonsimple devices and must be approved as intrinsically safe. If they are third-party approved, their entity parameter must be carefully considered. Transmitters (4 to

20 mA) convert physical measurements in hazardous areas, such as pressure and flow, into electric signals which can be transmitted to a controller in a safe area. Depending upon the conditions, 4 to 20 mA signals can be made intrinsically safe by using a repeater barrier which duplicates the output signal to match the input signal. Repeaters can be grounded or ungrounded. Ungrounded repeater barriers are known as "transformer-isolated barriers," since the incoming voltage or signal is completely isolated from the outgoing voltage or signal via a transformer. Digital inputs, such as mechanical and proximity switches, which are simple devices, can be made intrinsically safe by using a switch amplifier. A switch amplifier is simply a relay or an optocoupler (a high-speed relay that uses optical isolation between the input and the output) that transfers a discrete signal (i.e., on/off) from the hazardous area to a safe area. Grounded safety barriers are passive devices designed specifically to prevent excessive energy in a non-hazardous area from reaching a hazardous area. These barriers can be used with most field devices. In order for such barriers to function properly, there is emphatic need for a solid, low-impedance ($<1\ \Omega$) connection to ground to prevent ground loops and induced voltages which can hinder operation of the system.

Ignition Curves

All electric circuits possess certain electric characteristics that can be classified under three categories: resistance, **inductance**, and capacitance. To some extent, all circuits possess these three characteristics. However, some of these characteristics may be so small that their effects are negligible compared with that of the others, thus the terms, resistive, inductive, and capacitive circuit. Since the concept of IS is based on the principle that a large electric current can cause an explosion in a hazardous area and the lack of it cannot, it is necessary to identify the ranges of currents that are safe and those that are dangerous.

What is a dangerous amount of electric energy? The answer lies in the ignition curves. Ignition curves are published in most IS standards, such as ANSI/UL 913 [5]. Three of the most referenced curves are shown in Fig. 100.2. The curves show the amount of energy required to ignite various hazardous atmospheric mixtures in their most easily ignitable concentration. The most easily ignitable concentration is determined by calculating the percentage of volume-to-air between the upper and lower explosive limits of a specific hazardous atmospheric mixture. In the three referenced curves, the energy levels (voltage and current) below the group curve are not sufficient to cause an ignition of the referenced group.

Since specific ignition temperature is directly related to the amount of voltage and current consumed, both V_{oc} and I_{sc} of the safety barrier must be less than V_{max} and I_{max}. When designing an intrinsically safe system, the cable resistance R (Ω/m), the inductance L (μH/m), and the capacitance C (pF/m), which are inherently distributed over the length of the cable, must be considered. The capacitance and inductance can be readily obtained from the cable manufacturer's literature. If these parameters are not available, certain default values can be used based on NFPA 493/78 (A-4-2). They are $C_c = 200$ pF/m (60 pF/ft), $L_c = 0.66\ \mu$H/m, and (0.2 μH/ft). To determine the maximum wiring distance required to ensure proper operation, the capacitance and inductance must be calculated. One common approach uses "lumped parameters," in which the voltage and current of both the intrinsically safe apparatus and the associated intrinsically safe apparatus are compared and matched according to Eqs. 100.1 and 100.2. Any deviation from either Eqs. 100.1 and 100.2 can compromise the integrity of the system and introduce hazardous conditions. The reactive parts of the system must also be considered and verified to demonstrate that C_a and L_a values of the associated apparatus are not be exceeded by the field device and the field wiring values as shown in Eqs. 100.3 and 100.4. This method, although simple and effective, tends to exaggerate the wiring capacitance and inductance effect, which can be limiting in some applications. Another method takes advantage of the relation between the cable resistance and inductance. This method can be used if the L/R ratio of the associated apparatus is higher than the calculated L/R ratio of the cable. Under these conditions, the lesser D_a (maximum allowed distance) value can be ignored and the cable length can be extended to the higher D_a value. This method is more flexible where cable length is an issue. Figure 100.3 and the following example illustrate these methods.

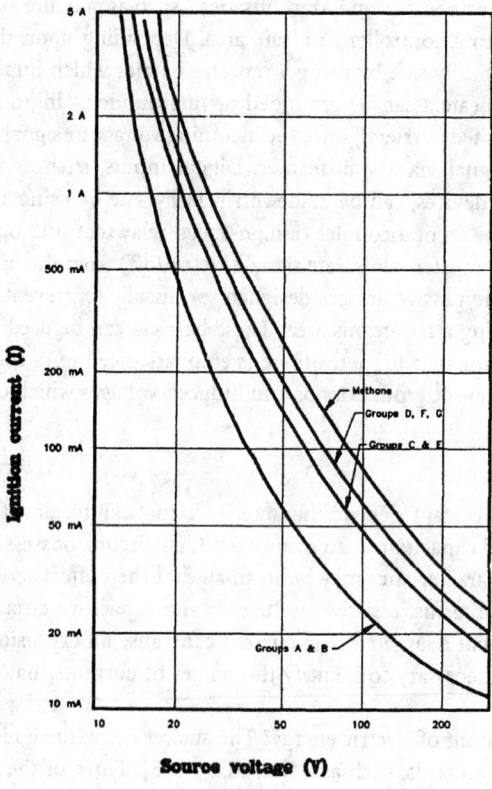

(a)

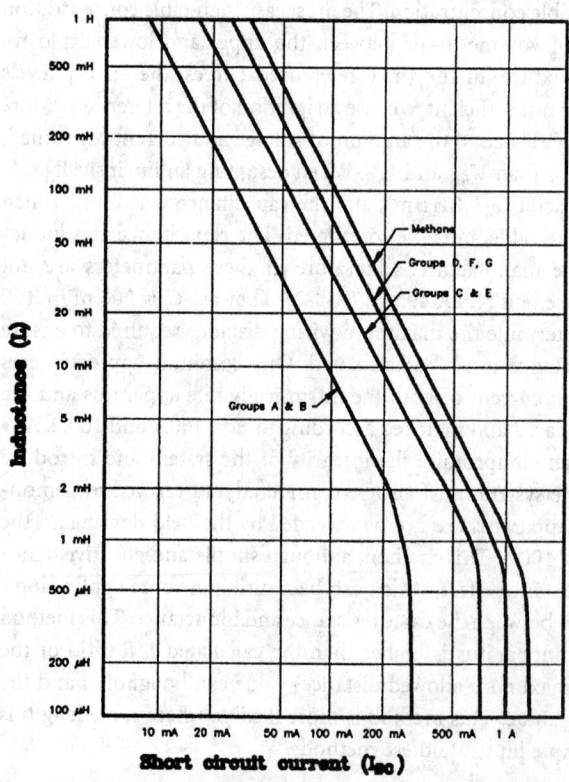

(c)

FIGURE 100.2 (a) Resistance circuit ignition curves for all circuit metals. (b) Inductance circuit ignition curves at 24 V for all circuit metals. (c) Capacitance circuit ignition curves for groups A and B for all circuit metals.

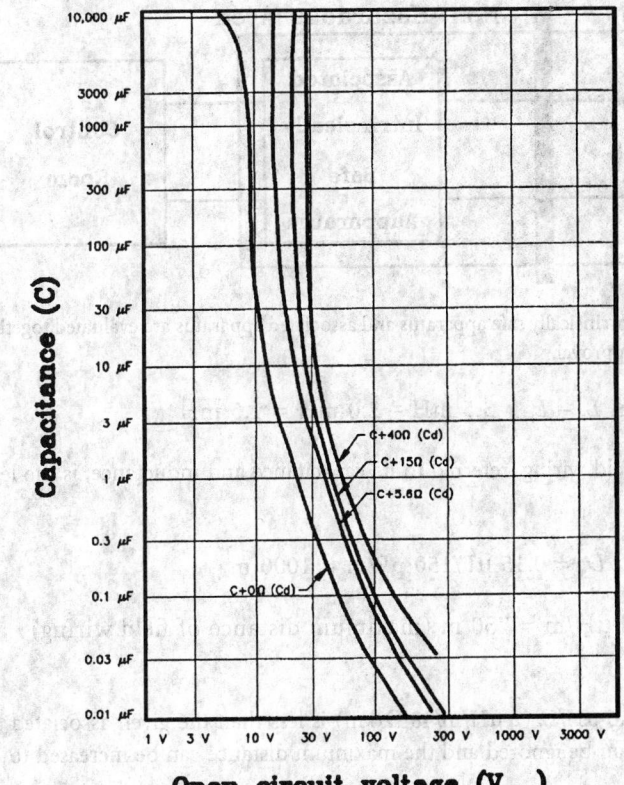

Open circuit voltage (V$_{OC}$)   FIGURE 100.2c

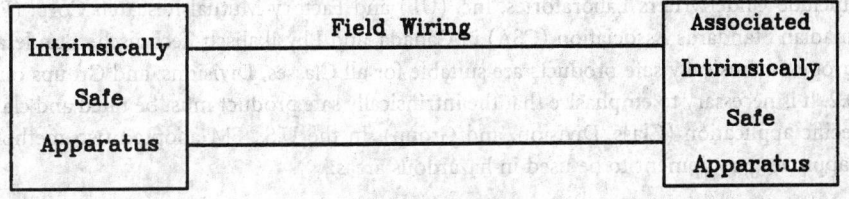

Hazardous Area		Non—Hazardous Are

V$_{min}$	=12 V	C=150 pF/m	V$_{oc}$	=24 V
V$_{max}$	=30 V	L=2.0 μH/m	I$_{sc}$	=0.3 A
I$_{max}$	=0.5 A	R=48 Ω/km	C$_a$	=0.45 μF
C$_i$	=0.3 μF		L$_a$	=3.5 mH
L$_i$	=2.0 mH		L/R	=60 μH/Ω

FIGURE 100.3 Analysis of an intrinsically safe system.

Lumped parameters method:

$$V_{oc} \le V_{max} \tag{100.1}$$

$$I_{sc} \le I_{max} \tag{100.2}$$

$$C_c \le C_a - C_i \quad 0.45\ \mu F - 0.30\ \mu F = 0.15\ \mu F \tag{100.3}$$

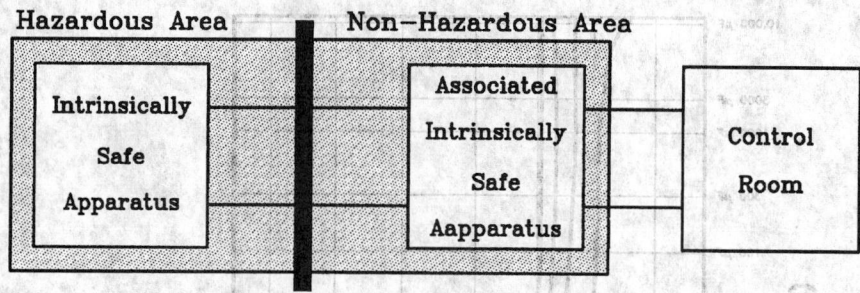

FIGURE 100.4 Loop approval. Intrinsically safe apparatus and associated apparatus are evaluated together. Shaded area indicates evaluated for loop approval.

$$L_c \leq L_a - L_i \quad 3.5 \text{ mH} - 2.0 \text{ mH} = 1.5 \text{ mH} \tag{100.4}$$

The maximum length of the field wiring, referred to its capacitance and inductance, is the lesser value of D_a.

$$D_a = 0.15 \text{ μF}/150 \text{ pF/m} = 1000 \text{ m}$$

$$D_a = 1.5 \text{ mH}/2.0 \text{ μH/m} = 750 \text{ m (maximum distance of field wiring)}$$

L/R ratio method:
Since the cable L/R ratio of 41.6 μH/Ω (2 μH/m/48 Ω/km) is less than the given associated apparatus L/R ratio, the inductive effect can be ignored and the maximum distance can be increased to 1000 m.

Certification and Approval

Although approval and certification processes help to provide safety, careful planning, designing, and engineering are still necessary. IS standards, procedures, and tests are recognized worldwide. Testing authorities include Underwriters Laboratories, Inc. (UL) and Factory Mutual Research Corp. (FM) in the U.S., Canadian Standards Association (CSA) in Canada, and Physikalisch-Technische Bundesanstalt (PTB) in Europe. Intrinsically safe products are suitable for all Classes, Divisions, and Groups outlined in Table 100.2. It is necessary to emphasize that the intrinsically safe product must be rated and classified for each specific application (Class, Division, and Group). In the U.S., FM adopted two methods for testing and approving equipment to be used in hazardous areas:

1. *Loop (System) Approval:* Where an intrinsically safe apparatus is evaluated in combination with a specific associated apparatus and is approved to be installed in this manner. Any changes to the circuit require reevaluation and certification (Fig. 100.4).
2. *Entity Approval:* Where an intrinsically safe apparatus and the associated apparatus are separately evaluated and given their own electrical entity parameters (Fig. 100.5). The correct application matches the entity parameters shown in Table 100.3. When examining the safety of a circuit, it is crucial to compare the entity values of an intrinsically safe apparatus with an associated apparatus.

Most safety barriers are entity approved for all hazardous locations. Since most field devices have the ability to store energy, they must have loop approval or entity approval for the proper construction and operation of an intrinsically safe system.

IS engineers often advocate the use of intrinsically safe equipment for the following reasons:

1. *Safety.* No explosion can occur in an intrinsically safe system under any operating condition. IS equipment operates on lower power levels and prevents shocks, excess thermal energy, and arcing. In different systems and under various scenarios, shocks, thermal energy, and arcing may cause a hazard.

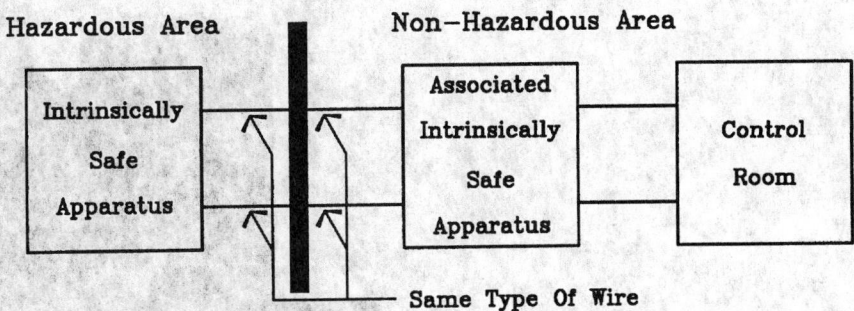

FIGURE 100.5 Entity approval. Intrinsically safe apparatus and associated apparatus are evaluated separately.

2. *Reliability.* The components and assemblies of intrinsically safe circuits are tested for reliability before they are labeled and certified. Most intrinsically safe equipment is designed with special circuitry to provide surge suppression and to prevent spikes and transients.
3. *Ease of handling and installation.* Intrinsically safe systems tend to be small and do not require expensive, bulky accessories such as enclosures, seals, and rigid metallic conduits which increase the initial investment.
4. *Economy.* In some geographic locations, facilities containing hazardous conditions must carry special liability insurance. With the proper installation of intrinsically safe circuits and equipment, the probability of an explosion is 10^{-18} [8], or nearly nonexistent. As a result, insurance rates tend to be lower.
5. *Maintenance.* Equipment may be calibrated and maintained without disconnecting power, thereby resulting in less downtime.

The wiring of intrinsically safe systems is similar to any other application, but to ensure a proper operating system, certain guidelines regarding identification and separation must be strictly followed. All intrinsically safe components including terminal blocks, conductors, and intrinsically safe apparatus must be explicitly marked and labeled. The conventional color used to identify intrinsically safe equipment is blue. In an open wiring installation, intrinsically safe conductors must be physically separated from nonintrinsically safe conductors by at least 50 mm (2 in.) so an induced voltage does not defeat the purpose of IS. Where intrinsically safe conductors occupy a raceway, the raceway should be labeled, "Intrinsically Safe Circuits." Intrinsically safe conductors should not be placed with nonintrinsically safe conductors. Where a cable tray is used, a grounded sheet metal partition may be used as an acceptable means of separation. Where intrinsically safe and nonintrinsically safe conductors occupy the same enclosure, a 50-mm (2-in.) separation must be maintained. In addition, a grounded metal partition shall be in place to prevent contact of any conductors that may come loose. Insulation deterioration of intrinsically safe conductors of different circuits occupying the same raceway or enclosure can be detrimental to the operation of the system. Intrinsically safe conductors must have an insulation grade capable of withstanding an ac test voltage of 550 V root-mean-square (rms) or twice the operating voltage of the intrinsically safe circuit. Nonintrinsically safe conductors in the same enclosure with intrinsically safe conductors must have an insulation grade capable of withstanding an ac test voltage of $2U + 1000$ V, with a minimum of 1500 V rms, where U is the sum of rms values of the voltages of the intrinsically safe conductors. A commonly used and highly recommended practice utilizes separate compartments for intrinsically safe and nonintrinsically safe conductors. In addition to physical separation of intrinsically safe conductors and nonintrinsically safe conductors, sealing of conduits and raceways housing intrinsically safe conductors is essential to prevent the passage of gases, vapors, and dusts from hazardous to nonhazardous areas. According to the NEC, seal-offs are not required to be explosion-proof. Where an associated apparatus is installed in an explosion-proof enclosure in a hazardous area, seal-offs must be explosion-proof. Although it is not required by Code, it is a good engineering practice to install explosion-proof seal-offs on conduits housing intrinsically safe conductors, as shown in Fig. 100.6.

FIGURE 100.6 Explosion-proof seal-off fitting. (Courtesy of Crouse-Hinds Division of Cooper Industries, Inc.)

Explosion-Proof Fundamentals

Explosion-proof design is a mechanical concept that relies heavily on the mechanical construction of an enclosure and the narrow tolerances between its joints, threads, and flanges to safely contain, cool, and vent any internal explosion that may occur. By definition, explosion-proof enclosures must prevent the ignition of explosive gases or vapors that may surround it (Type 7 and Type 10 enclosures only). In hazardous areas, Class I, Divisions 1 and 2, arcing devices, such as switches, contactors, and motor starters must be enclosed in an explosion-proof enclosure specifically rated for that area. Contrary to popular belief, explosion-proof enclosures are not and should not be vapor-tight. Succinctly stated, an explosion inside an enclosure must be prevented from starting a larger explosion outside the enclosure. Unlike IS, explosion-proof enclosures address the maximum internal pressure (see NEMA Type 7 enclosures). Figure 100.7 illustrates the rugged construction of a standard explosion-proof panel board.

In addition to its durability and strength, explosion-proof enclosures must also be "flame-tight." The joints or flanges must be held within narrow tolerances to allow cooling of hot gases resulting from internal explosions. In this way, if any gases are released into the outside hazardous atmosphere, they are cool enough not to cause ignition.

Explosion-proof enclosures tend to be bulky (making them easy to identify) and heavy, requiring conduit seals and careful handling. Unlike intrinsically safe equipment, explosion-proof equipment operates on normal power levels which are necessary due to the high power requirements of some circuits and equipment. With the proper equipment, installation, and maintenance, explosion-proof enclosures can safely and effectively distribute high levels of voltage and power into hazardous areas.

Where ignitable amounts of dust are present, enclosures housing electric equipment must be dust-ignition-proof. These enclosures must exclude combustible dusts from entering, while preventing arcs, sparks, or heat generated internally from igniting dust surrounding the exterior of the enclosure. These enclosures must also efficiently dissipate the heat generated internally, since many types of dust will ignite at relatively low temperatures. Unlike Class I, Division 1 explosion-proof enclosures (Type 7), Class II, Division 1 dust-ignition-proof enclosures (Type 9) are designed to prevent an explosion. Subsequently, dust-ignition-proof enclosures need not be as strong or have walls as thick as explosion-proof enclosures since there will be no internal explosion.

Purging and Pressurization

This methodology allows the safe operation of electric equipment where hazardous conditions exist and where no other methodology is applicable because of the imperative high-energy demands and actual

FIGURE 100.7 Explosion-proof lighting panelboard. (Courtesy of Crouse-Hinds Division of Cooper Industries, Inc.)

physical dimensions. This is true for large-sized motors and switchgear units where they are not commercially available for Class I, Group A and B. In addition, this methodology is used where control panels that house the instruments and electric equipment must be located in hazardous areas. Purging and pressurization is a protection method that relies on clean air or inert gas (i.e., nitrogen) to be continuously supplied to the enclosure at sufficient flow to keep the equipment adequately cooled and to provide adequate internal pressure to prevent the influx of combustible atmospheres into the enclosure. Although the enclosures are not explosion-proof, they must be relatively vapor-tight and must have adequate strength to perform safely and satisfactorily. The system consists of

1. *Clean Air (or Inert Gas) Supply:* Careful study and analysis is of crucial importance to this process since the air supplied must be reasonably free of contaminants. Finding a safe location for an air intake requires skill and ingenuity. Consulting with an HVAC specialist is recommended. Other factors such as vapor density, location, wind pattern, and surrounding environment should also be considered. Where compressors and blowers are used to supply compressed air, caution must be exercised when selecting the proper compressor or blower size and location in order to meet airflow requirements without compromising the main objective of safety and reliability.

2. *Purging:* A pressurized enclosure that has been out of service for some time tends to collect a combustible mixture. Before energizing, inert gas and positive pressure must provide a sufficient initial clean air volume to minimize the concentration of any combustible mixture that may be present. For typical applications, a flow of four times the internal volume of the enclosure is usually sufficient to minimize the concentration of combustible mixture that may exist. For unusual applications, the flow volume must be carefully calculated to ensure the success of the purging process.

3. *Pressurization:* This process uses the concept of pressure differential between the outside and the inside of the enclosure to keep flammable materials from entering. This is accomplished by maintaining a higher pressure on the inside of the enclosure. For safe operation, the protected enclosure must be constantly maintained at a positive pressure of at least 25 Pa (0.1 in. water) [2] above the surrounding atmosphere during the operation of the protected equipment.

4. *Signals and Alarms:* When positive pressure is lost, warnings and alarms are essential. Three types of pressurization and alarms can be used, depending on the nature of the controls of the enclosure and the degree of hazard outside. In addition, door interlocks are required to prevent opening of the protected enclosure while the circuit is energized.

According to NFPA 496, there are three types of pressurization. They are

Type X — Reduces the classification within the protected enclosure from Division 1 to nonclassified. This usually involves a source of ignition housed in a tight enclosure located in a potentially hazardous atmosphere. Type X requires a disconnecting means (flow or pressure switch) to deenergize power to the protected enclosure completely and automatically immediately upon failure of the protective gas supply (loss of either pressure or flow). The disconnecting means must be explosion-proof or intrinsically safe, as it is usually located in the hazardous area.

Type Y — Reduces the classification within the protected enclosure from Division 1 to Division 2. The protected enclosure houses equipment rated for Division 2 and does not provide a source of ignition. Therefore, no immediate hazard is created. Type Y requires a visual or audible alarm in case of system failure. *Word of caution:* Safeguards must be established to ensure that any malfunction in the system does not raise the external surface temperature of the enclosure to over 80% of the ignition temperature of the combustible mixture present.

Type Z — Reduces the classification within the protected enclosure from Division 2 to nonclassified. Type Z requires a visual or audible alarm to be activated if failure to maintain positive pressure or flow within the protected enclosure has been detected.

Recognizing and understanding the potential dangers associated with the use of electricity in hazardous areas is a crucial part of selecting the best protection. Techniques for most applications, where the need for different energy demands is required, will likely involve a combination of various methodologies and specialized technologies. Properly performed analysis and investigation may appear to be time-consuming and expensive, but for those who are willing to adhere to the established guidelines and solid engineering practices, the process will help ensure the highest level of compliance while yielding tremendous savings and preventing property damage and injuries.

Defining Terms

Inductance: The ability of an electric apparatus to store an electric charge (energy). An inductor will release this energy when the circuit is opened (broken).

Oxidation: The process where negatively charged ions (anions) lose electrons at the anode during electrochemical process. For anions to become neutral, they must lose electrons.

Resistance temperature device (RTD): A device that measures temperature based on change of resistance.

Zener diode: A nonlinear solid state device that does not conduct current in reverse bias mode until a critical voltage is reached. It is then able to conduct current in reverse bias without damage to the diode. Zener diodes have almost constant voltage characteristics in the reverse bias region (usual operation).

References

1. NFPA 70 (ANCI C1 — 1996), *National Electrical Code 1996*, Quincy, MA.
2. National Fire Protection Association (NFPA), Articles 493, 496, 497, Quincy, MA.

3. American Petroleum Institute 500 (RP 500), 1st ed., Washington, D.C., June 1991. *Recommended Practice for Classification of Location for Electrical Installation at Petroleum Facilities.*

4. Elcon Instruments, Inc., *Introduction to Intrinsic Safety*, 3rd printing, Norcross, 1990.

5. Underwriters Laboratories, Inc., Pub. UL 913, 4th ed., Northbrook, IL, 1988, *Standard for Intrinsically Safe Apparatus & Associated Apparatus for Use in Class I, II, III, Division 1 Hazardous Locations.*

6. Underwriters Laboratories, Inc., Pub. UL 508, *Industrial Control Equipment.*

7. Babiarz, P., Cooper Industries, Crouse-Hinds, *InTech Engineer's Notebook*, Syracuse, NY, 1994.

8. R. Stahl, Inc., RST 49, *Comprehensive Applications and Data Catalog*, Woburn, MA, 1992.

9. Oudar, J., Intrinsic safety, *J. Southern Calif. Meter Assoc.*, October, 1981.

10. Alexander, W., Intrinsically safe systems, *InTech*, April 1996.

101

Measurement and Identification of Brush, Brushless, and dc Stepping Motors

Stuart Schweid
Xerox Corporation

Robert Lofthus
Xerox Corporation

John McInroy
University of Wyoming

101.1   Introduction

There are many systems, such as robots, whose ability to move themselves or other objects is their primary purpose. There is a myriad of other systems, such as xerographic printers, where the motion is not the desired outcome but is required in the performance of the mainline objective. All of these systems require one or more "prime movers," which directly or indirectly create all motion in the system.

All of these applications have differing requirements for both the type and precision of the motion produced. There are applications, such as children's toys, that can perform well with imprecise motion requirements. Conversely, there are applications, such as printing, that require very precise motion control.

Depending on the type of motion required, a prime mover can be chosen from one of several candidates. The most ubiquitous motor types are dc brush type motors, dc brushless motors, and hybrid stepping motors. The system, which includes the prime mover, can be controlled simply as a function of time (i.e., open loop) or as a function of both time and system state (i.e., closed loop).

A closed-loop system requires some mechanism of measuring the "state" of the system. For motion systems, this is typically one or more sensors that measure position and/or velocity. In addition, for many of the control techniques, specifically, linear controllers, it is also advantageous to have a model of the system being controlled. The model can be used in conjunction with a plethora of design techniques to improve the dynamic response of the system significantly.

This chapter will describe the major modes of operations for the prime movers listed and will detail a method for measuring the open-loop response of a candidate motion system.

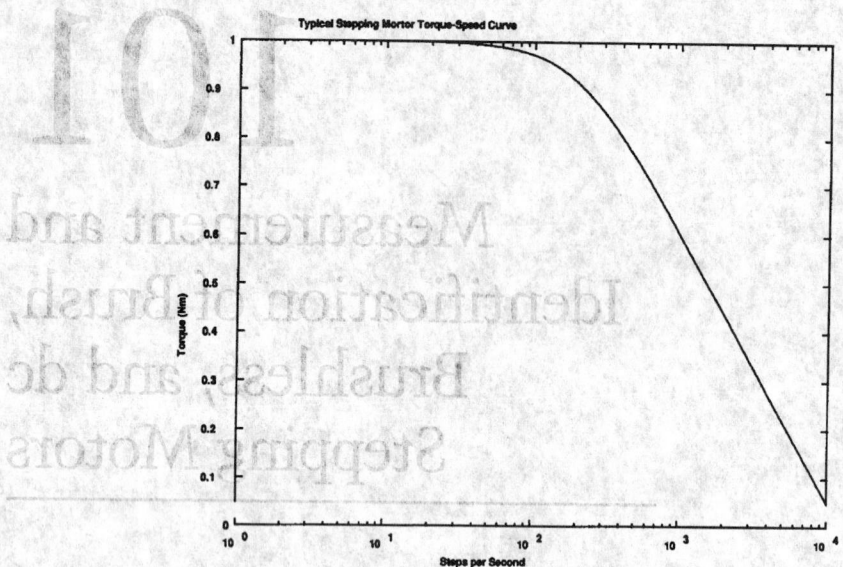

FIGURE 101.1 Typical torque–speed curve.

101.2 Hybrid Stepping Motors

Hybrid stepping motors are useful devices that provide fairly accurate positioning in the open loop: their most common mode of operation. They are successful in open-loop operation because they will remain within a commanded position increment or step as long as the motor has sufficient torque to resist any external torque applied by the system.

Since the motor operates in the open loop, it is chosen *a priori* to ensure it provides the torque necessary to remain within step. The maximum torque needed to operate the system is a function of the inertia of the system, the acceleration profile of the system, the external torque on the system, etc. Once the maximum torque required at every speed is determined, a motor can be chosen from its torque–speed curve. The torque–speed curve describes the maximum torque that can be tolerated at each speed without failing to follow a commanded step (referred to as a "losing step"). A motor whose torque–speed curve exceeds the torque requirements of the system at all speeds will be sufficient for the application. Figure 101.1 shows a typical torque–speed curve.

Hybrid stepping motors have a plurality of coils (almost always two) that, when commutated in a predefined sequence, cause the motor to turn. When the motor is operated in a mode referred to as "full stepping," the current command to each winding is constant in magnitude, but varying in polarity (positive or negative). Note that the definition of which direction of current flow is "positive" is purely arbitrary. In this mode, the motor has only four separate energizations of the winding to complete a full revolution, although clever mechanical arrangements allow the number of steps per revolution to be greatly increased: typically 200 but as large as 800.

With two distinct windings (referred to as A and B), the four possible arrangements of winding energizations are A+B+, A+B–, A–B–, A–B+. If these combinations of winding currents are commanded to the motor in the sequence listed in the previous sentence, the motor will turn.

Chopping Current Amplifiers

As previously stated, the motion of the stepper requires a current of a particular amplitude to be applied. For economic reasons, most systems have a voltage power supply available (V_{ss}) — not a current source. A "current chopper" is a lost cost closed-loop switching amplifier that regulates the current to the motor winding. It takes advantage of the fact that the motor winding has a significant inductance. The two motor windings have dynamic equations [1]

$$dI_A/dt = \left(V_A - I_A R_A + K_b \omega \sin\theta\right)/L_A$$

$$dI_B/dt = \left(V_B - I_B R_B + K_b \omega \cos\theta\right)/L_B$$

(101.1)

where I is the current winding (A), R is the current resistance (Ω), L is the current inductance (H), V is the applied voltage to the winding (V), K_b is the torque constant of the motor (Nm/A), ω is the motor velocity (rad/s), and θ is the motor position (electrical rad).

Every t_s s (where t_s is very small, i.e., 10 µs) the current chopper applies either V_{ss} or $-V_{ss}$ to the motor winding, depending on whether the desired current (I_M) is greater than or less than the measured (i.e., actual) winding current:

$$\text{if } \left(I_A > I_M\right), \quad V_A = -V_{ss} \quad \text{else} \quad V_A = V_{ss}$$

(101.2)

$$\text{if } \left(I_B > I_M\right), \quad V_B = -V_{ss} \quad \text{else} \quad V_B = V_{ss}$$

The usual requirement is that ($V_{ss} t_s/L$) is small in comparison with I_M, resulting in a the chopping range that is small when compared with the desired current. Furthermore, the mechanical system that includes the motor is typically low pass in nature, so it will have minimal response to the small-amplitude, very high frequency components present in the motor winding current.

Microstepping

Another common open-loop mode for stepper operation is microstepping. In microstepping, the commanded reference positions can be between the "full-step" positions of the motor. The commanded currents for the reference position θ_{ref} are

$$I_A = I_M \cos\left(\theta_{ref}\right)$$

$$I_B = I_M \sin\left(\theta_{ref}\right)$$

(101.3)

where θ_{ref} is the desired electrical position of the motor (in four full steps, θ_{ref} traverses one full electrical revolution). For constant velocity applications,

$$\theta_{ref} = N\omega_d t$$

(101.4)

where N is the number of electrical revolutions per mechanical revolution (one quarter the number of steps per revolution) of the motor and ω_d is the desired velocity (rad/s).

Closed-Loop Control

In addition to the open-loop uses of hybrid stepping motors, several researchers have recently invented methods of operating these motors in a closed-loop fashion. The principal difficulty arises because hybrid stepping motors are nonlinear devices, so the plethora of design techniques available in linear control theory cannot be directly applied. Hamman and Balas [1] use a fourth-order model to linearize the system about each step. This is useful in altering the undamped response of the stepper motor to a commanded change in position. Schweid [2] suggests a scheme that allows good transient behavior during constant velocity applications, but uses the complete fourth-order model. In systems that use current choppers, a second-order model is sufficient. The following control law can then be applied [3-5]:

$$I_A = I_M \cos(\theta_{ref}) - I_C \sin\theta$$
$$I_B = I_M \sin(\theta_{ref}) + I_C \sin\theta \tag{101.5}$$

The currents of Equation 101.6 linearize the model of the stepper motor, allowing common control techniques such as PID control to be applied [6]. The value I_M is used as an open-loop component to microstep the motor, while I_C is a feedback term that permits an improved dynamic response.

Position Measurement

The implementation of closed-loop control requires a method of measuring the position of the motor. Furthermore, there is the added constraint that this position measurement must be aligned with the motor construction and not have a positional offset with respect to it. This is because the control scheme of Equation 101.5 requires a command current I_C be multiplied by the sine and cosine of the motor position. One way to implement this is to have an incremental encoder with an index pulse attached to the motor. A calibration can then determine the relative position of the index pulse to the motor shaft and use this relationship in all future positional measurement.

Another scheme uses the back emf of the two coil windings to estimate the position. Some motors are constructed with additional sensing coils that allow easy measurement of these values. From Equation 101.2, the back emf of both coils can be found using

$$V_{emfA} = -K_b \omega \sin\theta = V_A - I_A R_A - L_A dI_a/dt$$
$$V_{emfB} = -K_b \omega \cos\theta = V_B - I_B R_B - L_B dI_B/dt \tag{101.6}$$

and the position could be found using

$$\theta = \operatorname{atan}(-V_{emfA}/V_{emfB}) \tag{101.7}$$

In most motors, however, V_{emfA} and V_{emfB} are not directly measurable. However, it is possible to obtain an excellent estimate of position in constant-velocity system by approximating the back emf signals through low-pass-filtered measurements of the coil voltages. One method developed at Xerox that has a U.S. patent (U.S. 5378975) is described here. Taking the Laplace transform of Equation 101.6 yields

$$V_{emfA}(s) = V_A(s) - (R_A + sL_A)I_A(s)$$
$$V_{emfB}(s) = V_B(s) - (R_B + sL_B)I_B(s) \tag{101.8}$$

Filtering both sides with low-pass analog filters matched to the dynamics of the coil produces:

$$V_{filtA}(s) = V_A(s)/(s(L_A/R_A)+1) - R_A I_A,$$
$$V_{filtB}(s) = V_B(s)/(s(L_B/R_B)+1) - R_B I_B \tag{101.9}$$

The quantities V_{filtA} and V_{filtB} are easily obtained by low-pass filtering the coil voltage with a simple RC circuit and subtracting the IR term. The subtraction is easily implemented because the current is a known input (in fact it is commanded), and the resistance is easily measured.

Assuming that the coils are matched: $L = L_A = L_B$, $R = R_A = R_B$ and the motor is moving at constant velocity, $\theta = N\omega_d t$, the back EMF voltages are sinusoidal. The low-pass-filtered versions of these voltages are given by

$$V_{filtA}(s) = \left|H\left(N\omega_d\right)\right| K_B \sin(\theta + \phi)$$

$$V_{filtB}(s) = \left|H\left(N\omega_d\right)\right| K_B \cos(\theta + \phi)$$

(101.10)

where $|H(N\omega_d)|$ is the magnitude response of the low-pass filter at the input frequency, $N\omega_d$, and $\phi = -\mathrm{atan}(N\omega_d L/R)$ is the phase shift due to the filtering. Dividing the two Equations 101.10 and taking the inverse tangent yields

$$\theta = \mathrm{atan}\left(-V_{filtA}(t)/V_{filtB}(t)\right) + \mathrm{atan}\left(N\omega_d L/R\right)$$

(101.11)

The full derivation and analysis are found in Reference 7. In addition, Reference 7 includes a Kalman filtering technique to improve the velocity measurement estimate from the position estimate. This is needed because the derivative or difference of the position measurement is highly sensitive to noise.

101.3   dc Brush and Brushless Motors

The dc servomotor is the oldest and probably the most commonly used actuator for servo systems. The name arises because a constant (or dc) voltage applied to the motor will produce motor movement. Over a linear range these motors exhibit a torque that is proportional to the current flowing in the winding. The chief disadvantage of the brush dc motor is the brush, which performs the mechanical commutation. These brushes are the first part of the motor to fail, severely limiting the reliability of dc brush motors when compared with other actuators.

Another motor that operates similarly to a dc brush motor is the dc brushless motor (a dc voltage will cause it to rotate). Its method of commutation, however, is much different. Whereas the dc motor accomplishes switching mechanically with brushes, the dc brushless motor contains Hall effect sensors that electronically control the switching. As a result, dc brushless motors have much greater life and reliability.

Pulse Width–Modulated Power Amplifier

In many systems where they operate, dc motors are required to have a varying voltage at their input in order to create the position or velocity profiles desired. Many of these systems, however, have only a single power supply V_{ss}. Rather than using analog power amplifiers that are both expensive and inefficient, a pulse width–modulated (PWM) amplifier is used. A pulse width power amplifier can deliver only V_{ss} or $-V_{ss}$ (or zero in some designs) V to the motor. Different voltages are delivered to the motor by varying the relative percentage that V_{ss} and $-V_{ss}$ are applied to the motor at a high fixed frequency. The mean or average voltage to the motor is a linear function of this relative percentage or duty cycle. The motor also receives the higher harmonics of the switched input that are integer multiples of the base frequency of the square wave. Most motors, however, have at least two poles (an electric and a mechanical) which are significantly slower than the switching frequency and thus greatly attenuate any possible effect they might have on motion. Figure 101.2 shows a PWM amplifier set to deliver 12 V to the motor.

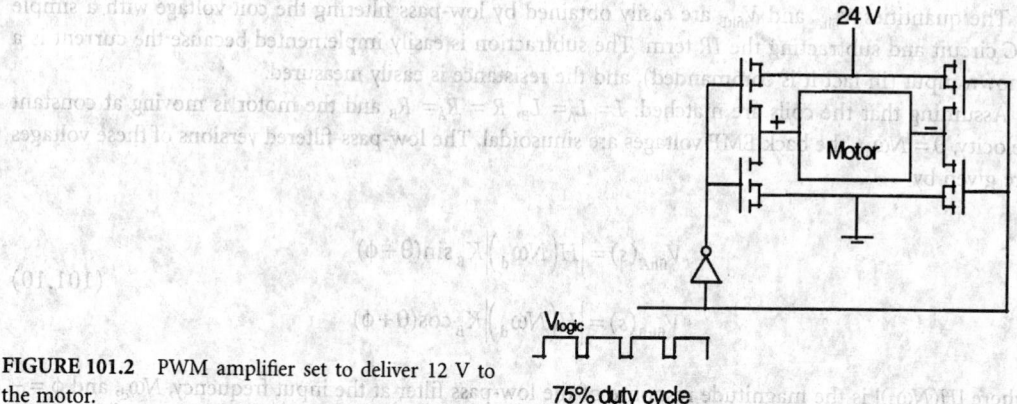

FIGURE 101.2   PWM amplifier set to deliver 12 V to the motor.   75% duty cycle

Measuring the System Dynamics

In systems with moderate or stringent motion requirements, the dc brush and brushless motors are operated in a closed-loop fashion. In order to perform in closed loop, some measure of either position or velocity is needed. A tachometer is commonly used to measure rotational speed. It outputs a voltage that is proportional to the speed of a rotating member. If a position measurement is required, an incremental encoder can be used. The incremental encoder produces a digital pulse at equal points in position. For example, a 1000 line/revolution encoder would produce a digital wave with 1000 rising (and falling) edges during one revolution.

Once the full system (including feedback sensors) is determined, a feedback control algorithm can be developed. As part of the design of many compensation algorithms, especially linear compensators such as state feedback or PID, one of the most important tasks is to have a good model of the dynamics of the system.

There are several ways to generate the model. One of the most common methods is to create the state equations of the system from the physical equations. This requires knowledge of all the parameters of the system. Sometimes these parameters are hard to measure and are not provided by the manufacturer. One example of this might be the torsional constant of a hard rubber roll in the system.

Even if all of the parameters are known, it is still desirable to have a method that measures the response of the system. This provides a way to either validate the model or to inform the designer that the model is not an accurate representation of the system.

A method is presented that permits the collection of input/output data of a system. This data can then be used to fit a frequency response curve or a transfer function model of the system. Signal analyzers, such as the HP3562A, have the ability to perform such curve fitting accurately. In addition, if the data are collected in a discrete fashion, there are commercially available software tools, such as MATLAB, that can provide model estimation in either a parametric (e.g., state space/transfer function) or nonparametric (e.g., frequency response) form. Although this procedure is described for motor systems, it is easily implemented in most linear systems.

Assume a typical closed-loop system as shown in Fig. 101.3. There are several obstacles that make it difficult to obtain the open- or closed-loop transfer function of this system. These obstacles include the inability to measure some of the outputs and the inability to control/modify certain inputs.

To acquire the frequency response of a system it is useful to perturb the system with a known input and measure the elicited response. A controllable input — one that can be externally manipulated — is required to accomplish this task. The only controllable inputs of the system in Fig. 101.3 are the torque acting upon the motor shaft and the reference command.

Changing the external torque would require mechanically attaching a calibrated brake or another motor. This is not only cumbersome, but it also can alter the system dynamics, as it introduces an inertia

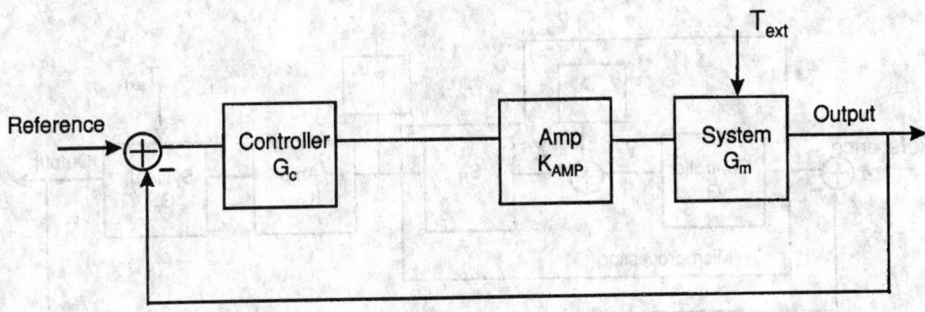

FIGURE 101.3 Typical closed-loop system.

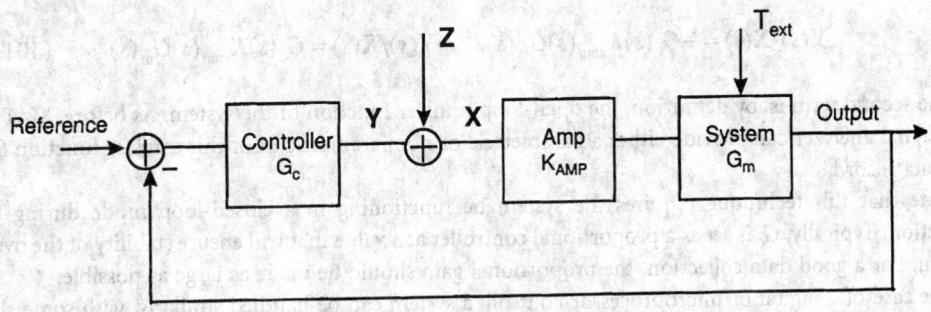

FIGURE 101.4

to the true system under measurement. Within many systems, the reference signal is a digital square wave (since the feedback element is an incremental encoder and produces a similar wave), and therefore requires a voltage-to-frequency converter if the reference is to be directly manipulated. This converter may include dynamics that can corrupt the measured data.

In addition to a controllable input, a measurable output must be available. In many servo systems, the only measurable outputs are the motor armature voltage and the position/velocity. The armature voltage may be a PWM signal that needs to be filtered to create an analog voltage. In the case of an incremental encoder, the output consists of a digital square wave generated by an incremental encoder: a frequency-to-voltage converter is required to produce an analog output. This converter may include dynamics that can corrupt the measured data.

Consider the system described in Fig. 101.4. The summing junction placed in the system has the advantage that can be added at any point that is convenient for measurement purposes — it is no longer limited to unaugmented system inputs or outputs.

In the case of analog controllers, the inputs and outputs of the summing junction are analog voltages, making them easily controlled and measured. The summing node can be easily implemented using a simple op-amp adder circuit. This introduces no appreciable dynamics into the original system and will not, therefore, skew the measured data.

When the system response is to be measured, the time-varying contributions of the other system inputs: the velocity reference and external torque, T_d, acting upon the system must be eliminated. The designer must set the velocity reference as a constant value. There is usually little control of the external torque acting on the system, and this will introduce noise into the measurement technique. To ensure a large signal-to-noise ratio (SNR), the alternative reference input (signal $z(t)$) should be made as large as possible.

The alternative reference input, $z(t)$, can stimulate the system with a swept sine input, a random input, or any other desired signal. During the stimulation, a time record is collected of the other nodes of the summing junction, $y(t)$ and $x(t)$. The transfer function between Y and X is

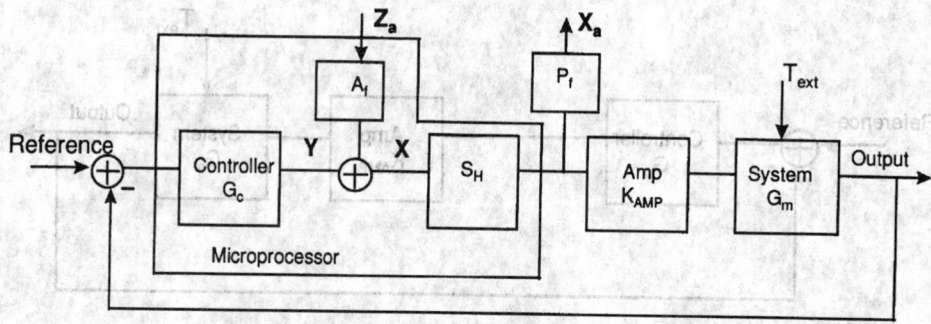

FIGURE 101.5

$$\mathbf{Y}(s)/\mathbf{X}(s) = -G_c(s)K_{amp}(s)G_m(s), \quad -\mathbf{Y}(s)/\mathbf{X}(s) = G_c(s)K_{amp}(s)G_m(s) \qquad (101.12)$$

The second term is, by definition, the open-loop transfer function of the system. As before, MATLAB or a signal analyzer can provide either a parameteric or nonparametric fit of this transfer function from the data record.

Note that this technique requires the system be functioning in a closed-loop mode during data collection. Typically, G_c is set as a proportional controller at a value that will ensure stability of the overall system. For a good data collection, the proportional gain should be made as large as possible.

The case of a digital or microprocessor-controlled system can be handled similarly, with some slight modification. Consider the system of Fig. 101.5, where a microprocessor acts as the system controller. In this instance, the summing junction can be implemented internal to the microprocessor by executing an add instruction. The problem here becomes getting the data into and out of the microprocessor. Getting the data into the microprocessor requires the addition of an analog-to-digital converter (ADC). In the block diagram, this ADC process is represented by the block A_f. A_f allows representation of any dynamics inherent in the ADC design (especially gain). The output of the microprocessor may be a digital PWM signal. This signal may need to be low-pass-filtered to create an analog voltage for the signal analyzer taking the data. The low-pass filter is represented by P_f in the Fig. 101.5. It is typically a simple *RC* circuit whose cutoff frequency is much lower than the PWM signal frequency.

Given the condition that ω_{ref} is static and external torque, T_d, is minimal, the transfer function between $\mathbf{Z}_a(s)$ and $\mathbf{X}_a(s)$ in Fig. 101.5 is

$$\mathbf{Z}_a(s)/\mathbf{X}_a(s) = \left(1 + G_c(s)S_H(s)K_{amp}(s)G_m(s)\right)\Big/\left(A_f(s)S_H(s)P_f(s)\right) \equiv \mathbf{O}(s) \qquad (101.13)$$

when $G_c = 0$, $\mathbf{Z}_a(s)/\mathbf{X}_a(s) = 1/(A_f(s)S_H(s)P_f(s)) \equiv \mathbf{C}(s)$, $(\mathbf{O}(s) - \mathbf{C}(s))/\mathbf{C}(s) = G_c(s)S_H(s)K_{amp}(s)G_m(s)$, this is the system open-loop transfer function.

A more direct approach is possible if a logic analyzer is not used but a software-based system identification package such as MATLAB is used to fit the data. The stimulus, z, can either be stored as a set of data in a lookup table in the microprocessor memory or can be generated in real time using any random number generator algorithm or other algorithm. The data Y and X at the internal summing junction can be captured via a logic analyzer or emulator if the values of $y(k)$ (the value of y at the kth sample period) and $x(k)$ are written to an external bus every sample period. Under this configuration neither a DAC nor ADC is necessary, nor is any mathematical combination of two different transfer functions required to produce the desired open-loop transfer function. The transfer function between $\mathbf{Y}(z)$ and $\mathbf{X}(z)$ is

$$\mathbf{Y(z)}/\mathbf{X(z)} = -G_c(z)S_H(z)K_{amp}(z)G_m(z)$$

$$-\mathbf{Y(z)}/\mathbf{X(z)} = G_c(z)S_H(z)K_{amp}(z)G_m(z) \qquad (101.14)$$

Again, this is the open-loop transfer function of the system, acquired using only minimal additional software and a logic analyzer. As before, the input/output set of data can be sent to any system identification algorithm to fit the transfer function.

References

1. E. Hamman, "Closed-loop control of a DC stepping motor using state space techniques," Master's thesis, Rensselaer Polytechnic Institute, Troy, NY 1983.
2. S. Schweid, "Velocity control of DC stepping motors utilizing state space theory, quasilinearization and reduced order estimation/control," Masters thesis, Rensselaer Polytechnic Institute, Troy, NY 1985.
3. J. Tal, L. Antignini, P. Gandel, and N. Veignat, "Damping a two-phase step motor with velocity coils," *Incremental Motion Control Systems and Devices Symposium*, Champaign, IL 1985, 305–309.
4. D. Reignier, "Very accurate positioning system for high speed, high acceleration motion control," *Power Conversion and Intelligent Motion*, 13, 52–62, June 1987.
5. D. Reignier, "Damping circuit and rotor encoder cut disc magnet step motor overshoot, settling time and resonances," *Power Conversion and Intelligent Motion*, 14, 64–69, April 1988.
6. S. Schweid, J. McInroy, and R. Lofthus, "Closed loop low-velocity regulation of hybrid stepping motors amidst torque disturbances," *IEEE Trans. Ind. Electron.*, 42, 316–324, 1995.
7. R. Lofthus, S. Schweid, J. McInroy, and Y. Ota, "Processing back EMF signals of hybrid step motors," *Control Eng. Practice*, 3(10), 1–10, 1995.

Appendix

Units and Conversions

B. W. Petley
National Physical Laboratory

This appendix contains several tables that list the SI base units (Table A.1), define the SI base units (Table A.2), list their derived units (Table A.3), list their prefixes (Table A.4), and list their conversion units (Table A.5).

TABLE A.1 The SI Base Units

Base quantity	Name of Base Unit	Symbol
Length	meter	m
Mass	kilogram	kg
Time	second	s
Electric current	ampere	A
Thermodynamic temperature	kelvin	K
Amount of substance	mole	mol
Luminous intensity	candela	cd

TABLE A.2 The International Definitions of the SI Base Units[a]

---

unit of length
(meter)

> The meter is the length of the path traveled by light in vacuum during a time interval of 1/299 792 458 of a second (17th CGPM,[b] 1983, Resolution 1).
>
> *Note:* The original international prototype, made of platinum-iridium, is kept at the BIPM[c] under conditions specified by the 1st CGPM in 1889.

unit of mass
(kilogram)

> The kilogram is the unit of mass: it is equal to the mass of the international prototype of the kilogram (3rd CGPM, (1901).

unit of time
(second)

> The second is the duration of 9 192 631 770 periods of the radiation corresponding to the transition between the two hyperfine levels of the ground state of the cesium-133 atom (13th CGPM, 1967, Resolution 1).

unit of electric current
(ampere)

> The ampere is that constant current which, if maintained in two straight parallel conductors of infinite length, of negligible cross section, and placed 1 meter apart in vacuum, would produce between these conductors a force equal to 2×10^{-7} newton per meter of length (CIPM, 1946, Resolution 2 approved by the 9th CGPM, 1948).
>
> *Note:* The expression "MKS unit of force" which occurs in the original text has been replaced here by "newton," a name adopted for this unit by the 9th CGPM (1948), Resolution 7.

unit of thermodynamic temperature
(kelvin)

> The kelvin, unit of thermodynamic temperature, is the fraction 1/273.16 of the thermodynamic temperature of the triple point of water (13th CGPM, 1967, Resolution 4).
>
> The 13th CGPM (1967, Resolution 3) also decided that the unit kelvin and its symbol K should be used to express an interval or a difference in temperature.
>
> *Note:* In addition to the thermodynamic temperature (symbol T), expressed in kelvin, use is also made of Celsius temperature (symbol t) defined by the equation

$$t = T - T_0$$

> where $T_0 = 273.15$ K by definition. To express Celsius temperature, the unit "degree Celsius" which is equal to the unit "kelvin" is used; in this case "degree Celsius" is a special name used in place of "kelvin." An interval or difference of Celsius temperature can, however, be expressed in kelvins as well as degrees Celsius.

unit of amount of substance
(mole)

> 1. The mole is the amount of substance of a system which contains as many elementary entities as there are atoms in 0.012 kilogram of carbon-12.
> 2. When the mole is used, the elementary entities must be specified and may be atoms, molecules, ions, electrons, other particles, or specified groups of such particles.
>
> In the definition of the mole, it is understood that unbound atoms of carbon-12, at rest, and in their ground state, are referred to.
>
> *Note:* This definition specifies at the same time the nature of the quantity whose unit is the mole.

Unit of luminous intensity
(candela)

> The candela is the luminous intensity, in a given direction, of a source that emits monochromatic radiation of frequency 540×10^{12} hertz and that has a radiant intensity in that direction of (1/683) watt per steradian (16th CGPM, 1979, resolution 3).

---

[a] The U.S. denotes the unit of length by "meter" in place of the international usage of "meter".
[b] CGPM: Conférence Général de Poids et Mesures; CIPM: Comité International des Poids et Mesures.
[c] BIPM: Bureau International des Poids et Mesures.

TABLE A.3 SI Derived Units with Special Names[a]

Derived quantity	Name	Symbol	Expressed in Terms of Other Units	Expressed in Terms of SI Base Units
Plane angle	radian	rad		$m\ m^{-1}$
Solid angle	steradian	sr		$m^2\ m^{-2}$
Frequency	hertz	Hz		s^{-1}
Force	newton	N		$m\ kg\ s^{-2}$
Pressure, stress	pascal	Pa	$N\ m^{-2}$	$m^{-1}\ kg\ s^{-2}$
Energy, work, quantity of heat	joule	J		$m^2\ kg\ s^{-2}$
Power, radiant flux	watt	W		$m^2\ kg\ s^{-3}$
Electric charge, quantity of electricity	coulomb	C		$s\ A$
Electric potential, potential difference, electromotive force	volt	V	W/A	$m^2\ kg\ s^{-3}\ A^{-1}$
Capacitance	farad	F	C/V	$m^{-2}\ kg^{-1}\ s^4\ A^2$
Electric resistance	ohm	Ω	V/A	$m^2\ kg\ s^{-3}\ A^{-2}$
Electric conductance	siemens	S	$A.V$	$m^{-2}\ kg^{-1}\ s^3\ A^2$
Magnetic flux	weber	Wb	$V\ s$	$m^2\ kg\ s^{-2}\ A^{-1}$
Magnetic flux density	tesla	T	Wb/m^2	$kg\ s^{-2}\ A^{-1}$
Inductance	henry	H	Wb/A	$m^2\ kg\ s^{-2}\ A^{-2}$
Celsius temperature	degree Celsius	°C		K
Luminous flux	lumen	lm	$cd\ sr$	$cd\ m^2\ m^{-2} = cd$
Illuminance	lux	lx	$m^{-2}\ cd\ sr$	$m^{-2}\ cd$
Activity (referred to a radio nuclide)	becquerel	Bq		s^{-1}
Absorbed dose, specific energy imparted, kerma	gray	Gy	J/kg	m^2s^{-2}
Dose equivalent, ambient dose equivalent, organ equivalent dose	sievert	Sr	J/kg	m^2s^{-2}

[a] Note that when a unit is named after a person the *symbol* takes a capital letter and the *name* takes a lowercase letter.

TABLE A.4 SI Prefixes[a]

Factor	Prefix	Symbol	Factor	Prefix	Symbol
10^{24}	yotta	Y	10^{-1}	deci	d
10^{21}	zetta	Z	10^{-2}	centi	c
10^{18}	exa	E	10^{-3}	milli	m
10^{15}	peta	P	10^{-6}	micro	μ
10^{12}	tera	T	10^{-9}	nano	n
10^{9}	giga	G	10^{-12}	pico	p
10^{6}	mega	M	10^{-15}	femto	f
10^{3}	kilo	k	10^{-18}	atto	a
10^{2}	hecto	h	10^{-21}	zepto	z
10	deca	da	10^{-24}	yocto	y

[a] The 11th CGPM (1960, Resolution 12) adopted a first series of prefixes and symbols of prefixes to form the names and symbols of the decimal multiples and submultiples of SI units. Prefixes for 10^{-15} and 10^{-18} were added by the 12th CGPM (1964, Resolution 8), those for 10^{15} and 10^{18} by the 15th CGPM (1975, Resolution 10), and those for 10^{21}, 10^{24}, 10^{-21}, and 10^{-24} were proposed by the CIPM (1990) for approval by the 19th CGPM (1991).

TABLE A.5 Conversion Factors from English Measures to SI Units [a]

Unit	Equivalent
1. Acceleration	
Acceleration of free fall, standard gravity	9.806 65 m/s^2
1 ft/s^2	0.304 8 m/s^2
1 gal	0.01 m/s^2
2. Angle	
1 second (")	4.484 81 × 10^{-6} rad
1 minute (')	2.908 9 × 10^4 rad
1 degree (°)	0.0174 532 rad
1 rad	206 264.8"
3. Area	
1 barn (b)	10^{-28} m^2
1 in.^{2}	6.451 6 × 10^{-4} m
1 ft^2	0.092 903 04 m^2
1 yd^2	0.836 127 36 m^2
1 are	100 m^2
1 acre [43560 (statute ft)^2]	4046.86 m^2
1 hectare	10 000 m^2
1 mi^2	2.590 0 × 10^6 m^2
1 square mile (based on U.S. survey foot)	2.589 998 km^2
4. Concentration, Density, Mass Density	
1 grain/gal (U.S.)	0.017 118 kg/m^3
1 lb/ft^3	16.018 46 kg/m^3
1 lb/gal (U.S.)	119.826 4 kg/m^3
1 short ton/yd^3	1186.6 kg/m^3
1 long ton/yd^3	1328.9 kg/m^3
1 oz(avdp)/in.^{3}	1730.0 kg/m^3
1 oz(avd)/gal(U.S.)	7.489 152 kg/m^3
1 lb/in.^{3}	27 680 kg/m^3
5. Energy	
1 ft lbf	1.355 818 J
1 cal$_{th}$ (thermochemical calorie)	4.184 J
1 cal$_{15}$ (15°C calorie)	4.185 5 J
1 cal$_{IT}$ [b]	4.186 8 J
1 kilocalorie (nutrition) [c]	4.186.8 J
1 watt second (W s)	1 J
1 watt hour (W h)	3600 J
1 therm (EC)	1.055 06 × 10^8 J
1 therm (U.S.)	1.054 804 × 10^8 J
1 ton TNT (equivalent)	4.184 × 10^9 J
1 BT$_{th}$	1 054.350 J
1 Btu$_{15}$	1 054.728 J
1 Btu$_{ST}$	1 055.055 852 62 J
1 quad (= 10^{15} Btu)	≈10^{18} J = 1 EJ
6. Force	
1 dyne	10^{-5} N
1 ounce-force	0.278 013 9 N
1 pound-force	4.448 222 N
1 kilogram-force	9.806 65 N
1 kip (1000 lbf)	4448.222 N
1 ton-force (2000 lbf)	8.896 443 N

Unit	Equivalent
7. Fuel consumption	
1 gallon (U.S.) per horsepower hour	$1.410\ 089 \times 10^{-9}$ m³/J
1 gallon (U.S.)/mile	2.352 15 l/km
1 gallon (U.K.)/mile	2.824 81 l/km
1 mile/gallon (U.S.), mpg	0.425 144 km/l
1 mile/gallon (U.K.)	0.354 006 km/l
1 pound per horsepower	$1.689\ 659 \times 10^{-7}$ kg/J
1 l/(100 km)	235.215/(mpg) (U.S.)
8. Length	
1 fermi	10^{-15} m = l fm
1 angstrom (Å)	10^{-10} m
1 microinch	2.54×10^{-8} m
1 mil	2.54×10^{-5} m
1 point (pt) [0.013837 in][d]	
1 pica (12 pt)	4.217 5 mm
1 inch (in.)	0.025 4 m
1 hand (4 in.)	0.101 6 m
1 foot (12 in.) (0.999998 statute ft.)	0.304 8 m
1 foot (U.S. survey)	0.304 800 6 m
1 statute foot [(1200/3937) m]	0.304 800 6 m
1 yard (yd)	0.914 4 m
1 fathom (6 ft, U.S. survey)	1.828 8 m
1 rod (16.5 statute ft)	5.029 2 m
1 chain (4 rod)	20.116 8 m
1 furlong (l0 chain)	201.168 m
1 mile (8 furlong, 5280 ft)	1609.344 m
1 statute mile (8 furlong, 5280 statute ft)	1609.347 2 m
1 nautical mile (international)[e]	1852 m
1 light year[f]	$9.640\ 73 \times 10^{15}$ m
9. Light	
1 foot-candle	10.763 91 lx
1 phot	10 000 lx
1 cd/in.²	1550.003 cd/m²
1 foot-lambert	3.426 259 cd/m²
1 lambert	3183.099 cd/m²
1 stilb	10 000 cd/m²
10. Mass	
1 pound (avdp.) (lb) (7000 gr)	0.453 592 37 kg
1 pound (troy) (5760 gr)	0.373 241 721 6 kg
1 grain (gr)	64.798 91 mg
1 scruple (20 gr)	1.296 0 g
1 pennyweight (24 gr)	1.555 174 g
1 dram (60 gr)	3.887 9 g
1 ounce (avdp) (437.5 gr)	28.349 52 g
1 ounce (troy) (480 gr)	31.103 48 g
1 carat (metric)	0.2 g
1 stone (14 lb)	6.350 29 kg
1 slug	14.593 9 kg
1 hundredweight (long)	50.802 35 kg
1 ton (short) (2000 lb)	907.184 7 kg
1 ton (long) (2240 lb)	1016.047 kg
	1.016 047 t

TABLE A.5 Conversion Factors from English Measures to SI Units
(continued)

Unit	Equivalent
Mass per Unit Length	
1 tex	10^{-6} kg/m
1 denier	$1.111\ 111 \times 10^{-7}$ kg/m
1 pound per foot	1.488 164 kg/m
1 pound per inch	17.857 97 kg/m
1 ton/mile	0.631 342 Mg/km
1 ton/1000 yd	1.111 6 kg/m
1 lb/ft	1.488 16 kg/m
Mass per Unit Area	
1 ton/mile^2	3.922 98 kg/ha
1 ton/acre	2510.71 kg/ha
1 oz/yd^2	33.905 7 g/m^2
Mass Carried × Distance (traffic factor)	
1 ton mile	1635.17 kg km
Mass carried × Distance/Volume (traffic factor)	
1 ton mile/gal (U.S.)	431.967 6 Mg km/m^3

11. Power

1 erg/s	10^{-7} W
1 ft lbf/h	$3.766\ 161 \times 10^{-4}$ W
(1 Btu$_{ST}$)	1.000 669 Btu$_{th}$
1 metric horsepower (force de cheval)	735.498 8 W
1 horsepower (550 ft lbf/s)	745.70 W
1 electric horsepower	746 W

12. Pressure, Stress

1 standard atmosphere	101 325 Pa
1 dyne/cm^2	0.1 Pa
1 torr [(101th 325/760) Pa]	133.322 4 Pa
1 N/cm^2	10 000 Pa
1 bar	100 000 Pa
1 lbf/ft^2	47.880 26 Pa
1 lbf/in^2 (psi)	6894.8 Pa
1 kgf/cm^2	98 066.5 Pa
1 cm water (4°C)	98.063 7 Pa
1 mm of mercury (0°C)	133.322 4 Pa
1 in of water (39.2°F)	249.082 Pa
1 in of mercury (60°F)	3376.85 Pa
1 ft water (39.2°F)	2988.98 Pa

13. Thermal Quantities

Fixed Points

Triple point of natural water: T_{tp}	273.16 K
Zero Celsius $(= T_0 = t_{F,0})$	273.15 K = 32°F

Temperature Conversions

Kelvin to Rankine (T_R):	$T = (5/9)T_R$
Kelvin to Celsius	$t = T - T_0$
Kelvin to Fahrenheit	$t_F = (9/5)(T - T_0) + t_{F,0}$
Celsius to Fahrenheit	$t_F = (9/5)\,t + t_{F,0}$

[Numerically: $5(\{t_F\} + 40) = 9(\{t\} + 40)$, where $\{t\}$ and $\{t_F\}$ are the numerical values of the Celsius and Fahrenheit temperatures respectively.]

Unit	Equivalent

Temperature Interval Conversions

Unit	Equivalent
1 degree centigrade	1 degree Celsius, denoted 1°C
1°C	1 K
1°F	(1/1.8) K
1°R	(1/1.8) K

Other Thermal Quantities

Unit	Equivalent
1 Btu$_{th}$/h	0.292 875 W
1 Btu$_{IT}$/h	0.293 071th 1 W
1 cal$_{IT}$/s	4.186 8 W
1 cal$_{th}$/s	4.184 W
1 cal$_{IT}$/(g °C)	4186.8 J/(kg K)
1 Btu ft/(ft^2 h °F)	1.730 735 W m^{-1}K^{-1}
1 Btu in/(ft^2 s °F)	519.220 4 W m^{-1} K^{-1}
1 clo	0.155 m^2 K/kW
1°F h ft^2/Btu	0.176 110 2 K m^2/W
1°F h ft^2/Btu·in	6.933 472 K m/W
1 Btu/lb °F ≡ 1 cal$_{ST}$/g °C	4186.8 J/kg.K

14. Torque, Moment of Force

Unit	Equivalent
1 dyne·cm	10^{-7} N m
1 kgf·m	9.806 65 N m
1 ozf·in	0.007 061 552 N m
1 lbf·in	0.112 984 8 N m
1 lbf·ft	1.355 818 N m

15. Velocity (includes speed)

Unit	Equivalent
1 foot per hour	8.466 667 × 10^{-5} m/s
1 foot per minute	0.005 08 m/s
1 knot (nautical mile per hour)	0.514 444 m/s
1 mile per hour (mi/h)	0.447 04 m/s

16. Viscosity

Unit	Equivalent
1 poise	0.1 Pa s
1 ft^2/s	0.092 903 04 m^2/s
1 lb/(ft s)	1.488 164 Pa s
1 lb/(ft h)	4.133 789 × 10^{-4} Pa s
1 lbf s/ft^2	47.880 26 Pa s
1 lbf·s/in^2	6894.757 Pa s
1 rhe	10 Pa^{-1} s^{-1}
1 slug/ft s	47.880 26 Pa s
1 stokes, St	1.0 × 10^{-4} m^2/s

17. Volume (includes capacity)

Unit	Equivalent
1 stere, st	1 m^3
1 liter^g	0.001 m^3
1 ft^3	0.028 316 8 m^3
1 in.^{3}	1.638 7 × 10^{-5} m^3
1 board foot	2.359 7 × 10^{-3} m^3
1 acre·foot	1233.48 m^3
1 dram (U.S. fluid)	3.696 7 × 10^{-6} m^3
1 gill (U.S.)	1.182 941 × 10^{-4} m^3
1 ounce (U.S. fluid)	2.957 353 × 10^{-5} m^3
1 teaspoon (tsp)^h	4.9288 922 × 10^{-6} m^3
1 tablespoon (tbsp)	1.4787 676 × 10^{-5} m^3
1 pint (U.S. fluid)	4.731 765 × 10^{-4} m^3

TABLE A.5 Conversion Factors from English Measures to SI Units
(continued)

Unit	Equivalent
1 quart (U.S. fluid)	$9.463\ 529 \times 10^{-4}\ \mathrm{m}^3$
1 gallon (U.S. liquid) [231 in.³]	$3.785\ 412 \times 10^{-3}\ \mathrm{m}^3$
1 wine barrel (bbl) [31.5 gal (U.S.)]	$0.119\ 240\ \mathrm{m}^3$
1 barrel (petroleum, 42 gal, U.S.), bbl	$0.158\ 987$
1 ounce (U.K. fluid)	$2.841\ 3 \times 10^{-5}\ \mathrm{m}^3$
1 gill (Canada & U.K.)	$1.420\ 6 \times 10^{-4}\ \mathrm{m}^3$
1 gallon (Canada & U.K.)	$4.546\ 09 \times 10^{-3}\ \mathrm{m}^3$
	$1.200\ 950$ gal (U.S.)
1 pint (U.S. dry)	$5.506\ 105 \times 10^{-4}\ \mathrm{m}^3$
1 quart (U.S. dry)	$1.101\ 221 \times 10^{-3}\ \mathrm{m}^3$
1 gallon (U.S. dry)	$4.404\ 884 \times 10^{-3}\ \mathrm{m}^3$
1 peck	$8.809\ 768 \times 10^{-3}\ \mathrm{m}^3$
1 bushel (U.S.) [2150.42 in.³]	$3.523\ 907 \times 10^{-2}\ \mathrm{m}^3$

[a]The conversion factor for a compound unit is usually not given here if it may easily be derived from simpler conversions; e.g., the conversion factors for "ft/s" to "m/s" or "ft/s²" to "m/s²" are not given, since they may be obtained from the conversion factor for "ft." Values are given to five or six significant digits except for exact values, which are usually indicated in bold type. A few former cgs measures are also included.

[b] The International Steam Table calorie of 1956.

[c] In practice the prefix kilo is usually omitted. The kilogram calorie or large calorie is an obsolete term for the kilocalorie which is used to express the energy content of foods.

[d] Typographer's definition, 1886.

[e] Originally, in 1929, the International nautical mile.

[f] Based on 1 day = 86,400 s and 1 Julian century = 36,525 days.

[g] Post 1964 value, SI symbol l or L. Between 1901 and 1964 the liter was defined as 1.000th 028 dm³.

[h] Although often given, it is doubtful whether normal usage justifies this accuracy. In Europe and elsewhere the teaspoon and tablespoon are usually exactly 5 mL and 15 mL, respectively.

References

1. CIPM, Procès-Verbaux CIPM, 49th Session, 1960, pp 71–72; *Comptes Rendues, 11th CGPM*, 1960, p. 85
2. P. Anderton and P. H. Bigg, *Changing to the Metric System*, HMSO, London, 1980.
3. *The International System of Units (SI)*, Natl. Inst. Stand. Technol., Spec. Publ. 330, 1991 ed., U.S. Government Printing Office, Washington, D.C., 1991.
4. B. N. Taylor, *Interpretation of the SI for the United States and Metric Conversion Policy for Federal Agencies*, Natl. Inst. Stand. Technol., Spec. Publ. 814, U.S. Government Printing Office, Washington, D.C., 1991.
5. E. R. Cohen, *The Physics Quick Reference Guide*, American Institute of Physics Press, New York, 1995.
6. B. N. Taylor, *Guide for the Use of the International System of Units*, 1995 ed., Natl. Inst. Stand. Technol., Spec. Publ. 811, U.S. Government Printing Office, Washington, D.C., 1995.
7. *Standard for Use of the International System of Units (SI): The Modern Metric System, IEEE/ASTM SI 10-1997*, IEEE Standards Co-ordinating Committee 14 (Revision and redesignation of ANSI/IEEE Std 268-1992 and ASTM E380), IEEE, New York: 1997.
8. *The International System of Units*, 7th ed., BIPM, France, 1998.

Index

The Measurement, Instrumentation, and Sensors Handbook